American Men & Women of Science

1995-96 • 19th Edition

The 19th edition of *AMERICAN MEN & WOMEN OF SCIENCE* was prepared by the R.R. Bowker Database Publishing Group.

Peter E. Simon, Senior Vice President, Database Publishing
Dean Hollister, Vice President, Production-Directories

Edgar Adcock, Jr., Senior Editorial Director, Directories
Karen Hallard, Managing Editor
Teresa Metzger, Senior Editor
Douglas J. Edwards, Associate Editor

Judy Redel, Research Director
Tanya Hurst, Senior Research Editor
Jeffrey Schumow, Assistant Research Editor

American Men & Women of Science

1995-96 • 19th Edition

A Biographical Directory of Today's Leaders in Physical, Biological and Related Sciences.

Volume 8 • Discipline Index

Mamye Jarrett Memorial Library
East Texas Baptist University
Marshall, Texas

R.R. BOWKER
New Providence, New Jersey

Published by R.R. Bowker, a Reed Reference Publishing Company.

Ira Siegel, President, CEO
Andrew W. Meyer, Executive Vice President
Peter E. Simon, Senior Vice President, Database Publishing
Stanley Walker, Senior Vice President, Marketing
Edward J. Roycroft, Senior Vice President, Sales

Copyright© 1994 by Reed Publishing (USA) Inc. All rights reserved. Except as permitted under the Copyright Act of 1976, no part of *American Men and Women Of Science* may be reproduced or transmitted in any form or by any means stored in any information storage and retrieval system, without prior written permission of R.R. Bowker, 121 Chanlon Road, New Providence, New Jersey, 07974.

International Standard Book Number
Set: 0-8352-3463-0
Volume I: 0-8352-3464-9
Volume II: 0-8352-3465-7
Volume III: 0-8352-3466-5
Volume IV: 0-8352-3467-3
Volume V: 0-8352-3468-1
Volume VI: 0-8352-3469-X
Volume VII: 0-8352-3470-3
Volume VIII: 0-8352-3471-1

International Standard Serial Number: 0192-8570
Library of Congress Catalog Card Number: 6-7326
Printed and bound in the United States of America.

R.R. Bowker has used its best efforts in collecting and preparing material for inclusion in this publication but does not warrant that the information herein is complete or accurate, and does not assume, and hereby disclaims, any liability to any person for any loss or damage caused by errors or omissions in *American Men and Women Of Science*, whether such errors or omissions result from negligence, accident or any other cause.

8 Volume Set

Contents

ADVISORY COMMITTEE xi

PREFACE xiii

STATISTICS xv

ABBREVIATIONS xxi

DISCIPLINE INDEX

AGRICULTURAL & FOREST SCIENCES

Agricultural Business & Management 1
Agricultural Economics 1
Agriculture, General 1
Agronomy 3
Animal Husbandry 6
Animal Science & Nutrition 7
Fish & Wildlife Sciences 10
Food Science & Technology 12
Forestry 13
Horticulture 15
Phytopathology 18
Plant Breeding & Genetics 19
Range Science & Management 20
Soils & Soil Science 20
Other Agricultural & Forest Sciences 23

BIOLOGICAL SCIENCES

Anatomy 24
Bacteriology 28
Behavior-Ethology 30
Biochemistry 33

Biology, General 55
Biomathematics 61
Biometrics-Biostatistics 62
Biophysics 64
Biotechnology 70
Botany-Phytopathology 71
Cell Biology 77
Cytology 79
Ecology 83
Embryology 91
Endocrinology 94
Entomology 95
Food Science & Technology 100
Genetics 102
Hydrobiology 111
Immunology 111
Microbiology 119
Molecular Biology 130
Neurosciences 139
Nutrition 143
Physical Anthropology 145
Physiology, Animal 146
Physiology, General 155
Physiology, Plant 162
Protein Science 166
Toxicology 166
Zoology 168
Other Biological Sciences 176

CHEMISTRY

Agricultural & Food Chemistry 178
Analytical Chemistry 180
Biochemistry 189
Chemical Dynamics 195
Chemistry, General 196
Inorganic Chemistry 203
Nuclear Chemistry 208
Organic Chemistry 210

Pharmaceutical Chemistry 227
Physical Chemistry 231
Polymer Chemistry 247
Protein Science 254
Quantum Chemistry 254
Spectroscopy 255
Structural Chemistry 256
Synthetic Inorganic & Organometallic Chemistry 257
Synthetic Organic & Natural Products Chemistry 259
Theoretical Chemistry 262
Thermodynamics & Material Properties 264
Other Chemistry 266

COMPUTER SCIENCES

Artificial Intelligence 267
Computer Sciences, General 267
Computer Engineering 274
Hardware Systems 274
Information Science & Systems 275
Intelligent Systems 276
Software Systems 278
Theory 281
Other Computer Sciences 282

ENGINEERING

Aeronautical & Astronautical Engineering 283
Agricultural Engineering 287
Artificial Intelligence 288
Bioengineering & Biomedical Engineering 288
Biological Engineering 292
Ceramics Engineering 292
Chemical Engineering 294
Civil Engineering 303
Computer Engineering 307
Electrical Engineering 308
Electronics Engineering 318
Engineering, General 323

Engineering Mechanics 326
Engineering Physics 328
Fuel Technology & Petroleum Engineering 331
Industrial & Manufacturing Engineering 332
Marine Engineering 334
Materials Science Engineering 335
Mechanical Engineering 343
Metallurgy & Physical Metallurgical Engineering 350
Mining Engineering 354
Nuclear Engineering 355
Operations Research 357
Optoelectronics 359
Polymer Engineering 360
Sanitary & Environmental Engineering 360
Systems Design & Systems Science 363
Other Engineering 366

ENVIRONMENTAL, EARTH & MARINE SCIENCES

Atmospheric Chemistry & Physics 368
Atmospheric Dynamics 368
Atmospheric Sciences, General 370
Earth Sciences, General 373
Environmental Sciences, General 374
Fuel Technology & Petroleum Engineering 380
Geochemistry 381
Geology, Applied, Economic & Engineering 383
Geology, Structural 389
Geomorphology & Glaciology 390
Geophysics 391
Hydrology & Water Resources 394
Marine Sciences, General 396
Mineralogy-Petrology 398
Oceanography 400
Paleontology 403
Physical Geography 405
Stratigraphy-Sedimentation 406
Other Environmental, Earth & Marine Sciences 408

MATHEMATICS

Algebra	409
Analysis & Functional Analysis	411
Applied Mathematics	414
Biomathematics	419
Combinatorics & Finite Mathematics	420
Geometry	421
Logic	421
Mathematical Statistics	422
Mathematics, General	426
Number Theory	434
Operations Research	435
Physical Mathematics	436
Probability	437
Topology	439
Other Mathematics	440

MEDICAL & HEALTH SCIENCES

Audiology & Speech Pathology	441
Dentistry	441
Environmental Health	442
Health Physics	445
Hospital Administration	445
Medical Devices & Medical Diagnostics	445
Medical Sciences, General	446
Medicine	450
Nursing	466
Nutrition	466
Occupational Medicine	469
Optometry	469
Parasitology	470
Pathology	471
Pharmacology	477
Pharmacy	486
Psychiatry	488
Public Health & Epidemiology	491
Surgery	494

Veterinary Medicine 497
Other Medical & Health Sciences 499

PHYSICS & ASTRONOMY

Acoustics 504
Astronomy 506
Astrophysics 509
Atomic & Molecular Physics 512
Electromagnetism 516
Elementary Particle Physics 518
Fluids 521
Health Physics 524
Mechanics 524
Medical Physics 526
Nuclear Structure 526
Optics 529
Optoelectronics 534
Physics, General 534
Plasma Physics 546
Semiconductor Devices 548
Solid State Physics 549
Spectroscopy & Spectrometry 557
Superconductors 558
Theoretical Physics 559
Thermal Physics 563
Other Physics 565

OTHER PROFESSIONAL FIELDS

Educational Administration 567
History & Philosophy of Science 570
Research Administration 572
Resource Management 578
Science Administration 580
Science Communications 582
Science Education 583
Science Policy 587
Technical Management 589

Advisory Committee

Dr. Charles Henderson Dickens
Former Executive Secretary, Federal Coordinating Council for Science, Engineering & Technology, Office of Science & Technology Policy

Mr. Alan Edward Fechter
Executive Director, Office of Scientific & Engineering Personnel, National Research Council
National Academy of Science

Dr. Oscar Nicolas Garcia
Program Director
Interactive Systems Program
National Science Foundation

Dr. Michael J. Jackson
Executive Director
Federation of American Societies for Experimental Biology

Dr. Shirley Mahaley Malcom
Head, Directorate for Education and Human Resources Programs
American Association for the Advancement of Science

Ms. Beverly Fearn Porter
Assistant to the Director for Society Relations
American Institute of Physics

Dr. William Eldon Splinter
Former Vice Chancellor for Research
University of Nebraska-Lincoln

Ms. Betty M. Vetter
Executive Director
Science Manpower Commission
Commission on Professionals in Science & Technology

Dr. Dael Lee Wolfle
Professor Emeritus
Graduate School of Public Affairs
University of Washington

Dr. Ahmed H. Zewail
Linus Pauling Professor of Chemistry & Physics
California Institute of Technology

Preface

American Men and Women of Science remains without peer as a chronicle of North American & Canadian scientific endeavor and achievement. The present work is the nineteenth edition since it was first compiled as *American Men of Science* by J. McKeen Cattell in 1906. In its eighty-nine year history *American Men and Women of Science* has profiled the careers of over 300,000 scientists and engineers. Since the first edition, the number of American scientists and the fields they pursue have grown immensely. This edition alone lists full biographies for 123,406 engineers and scientists, 6440 of which are listed for the first time. Although the book has grown, our stated purpose is the same as when Dr. Cattell first undertook the task of producing a biographical directory of active American scientists. It was his intention to record educational, personal and career data which would make "a contribution to the organization of science in America" and "make men [and women] of science acquainted with one another and with one another's work." It is our hope that this edition will fulfill these goals.

The biographies of engineers and scientists constitute seven of the eight volumes and provide birthdates, birthplaces, field of specialty, education, honorary degrees, current position, professional and concurrent experience, awards, memberships, research information and addresses for each entrant when applicable. New for the nineteenth edition is the addition of spouse's and children's names, and the option to have a Fax number or E-Mail address published. The eighth volume, the discipline index, organizes biographees by field of activity. This index, adapted from the National Science Foundation's Taxonomy of Degree and Employment Specialties, classifies entrants by 191 subject specialties listed in the table of contents of Volume 8. The index classifies scientists and engineers by state within each subject specialty, allowing the user to easily locate a scientist in a given area. Also included are statistical information and charts and recipients of the Nobel Prizes, the Craaford Prize, the Charles Stark Draper Prize, the National Medals of Science and Technology, the Fields Medal and the Alan T. Waterman Award received since the last edition.

While the scientific fields covered by *American Men and Women Of Science* are comprehensive, no attempt has been made to include all American scientists. Entrants are meant to be limited to those who have made significant contributions in their field. The names of new entrants were submitted for consideration at the editors' request by current entrants and by leaders of academic, government and private research programs and associations. Those included met the following criteria:

1. Distinguished achievement, by reason of experience, training or accomplishment, including contributions to literature, coupled with continuing activity in scientific work;

or

2. Research activity of high quality in science as evidenced by publication in reputable scientific journals; or, for those whose work cannot be published due to governmental or industrial security, research activity of high quality in science as evidenced by the judgement of the individual's peers;

or

3. Attainment of a position of substantial responsibility requiring scientific training and experience.

This edition profiles living scientists in the physical and biological fields, as well as public health scientists, engineers, mathematicians, statisticians, and computer scientists. The information is collected by means of direct communication whenever possible. All entrants receive forms for corroboration and updating. New entrants receive questionnaires and verification proofs before publication. The information submitted by entrants is included as completely as possible within the boundaries of editorial and space restrictions. If an entrant does not return the form and his or her current location can be verified in secondary sources, the full entry is repeated. References to the previous edition are given for those who do not return forms and cannot be located, but who are presumed to be still active in science or engineering. Entrants known to be deceased are noted as such and a reference to the previous edition is given. Scientists and engineers who are not citizens of the United States or Canada are included if a significant portion of their work was performed in North America.

The information in *AMWS* is also available on CD-ROM as part of *SciTech Reference Plus™*. In addition to the convenience of searching scientists and engineers, *SciTech Reference Plus™* also includes *The Directory of American Research & Technology*, sci-tech and medical related companies from the *Directory of Corporate Affiliations,* and sci-tech and medical books and serials from *Books in Print* and *Ulrich's International Periodicals Directory*. Magnetic tape leasing is also available for each of these files. For information on the CD-ROM or tape products, contact Bowker Electronic Publishing (800-323-3288). *American Men and Women Of Science* is also available for online searching through DIALOG Information Services, Inc. (3640 Hillview Ave, Palo Alto, CA 94304; 800-334-2564). Both the CD-ROM and online products allow fielded as well as key word searches of all elements of a record, including field of interest, experience, and location. Tapes and mailing lists are also available through the Cahners Direct Mail (John Panza, List Manager, Bowker Files, 245 W 17th St, New York, NY, 10011; 800-537-7930).

A project as large as publishing *American Men and Women Of Science* involves the efforts of a great many people. The editors take this opportunity to thank the nineteenth edition advisory committee for their guidance, encouragement and support. Appreciation is also expressed to the many scientific societies who provided their membership lists for the purpose of locating former entrants whose addresses had changed, and to the tens of thousands of scientists across the country who took time to provide us with biographical information. We also wish to thank Karen Strong, Sandy Cummings, Donna Colahan, Bonnie Walton, Val Harris, Debbie Wilson, Mervaine Ricks and all those whose care and devotion to accurate research and editing assured successful production of this edition.

Comments, suggestions and nominations for the twentieth edition are encouraged and should be directed to The Editors, *American Men and Women Of Science*, R.R. Bowker, 121 Chanlon Road, New Providence, New Jersey, 07974.

Karen Hallard
Managing Editor

Statistics

Statistical distribution of entrants in *American Men & Women of Science* is illustrated on the following five pages. The regional scheme for geographical analysis is diagrammed in the map below. A table enumerating the geographic distribution can be found on page xx, following the charts. The statistics are compiled by tallying all occurrences of a major index subject. Each scientist may choose to be indexed under as many as four categories; thus, the total number of subject references is greater than the number of entrants in *AMWS*.

All Disciplines

	Number	Percent
Northeast	58,716	34%
Southeast	41,472	24%
North Central	20,171	12%
South Central	12,539	7%
Mountain	11,505	7%
Pacific	26,055	15%
TOTAL	**170,458**	**100%**

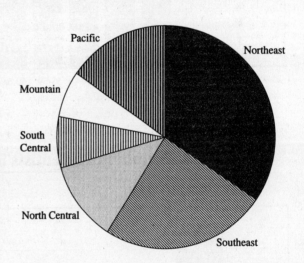

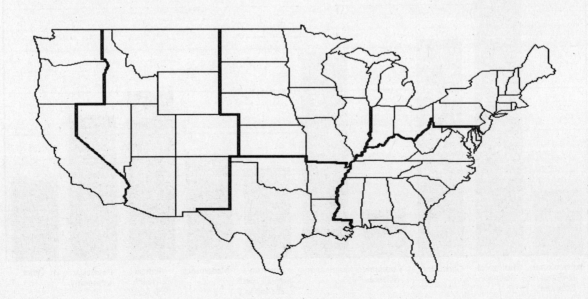

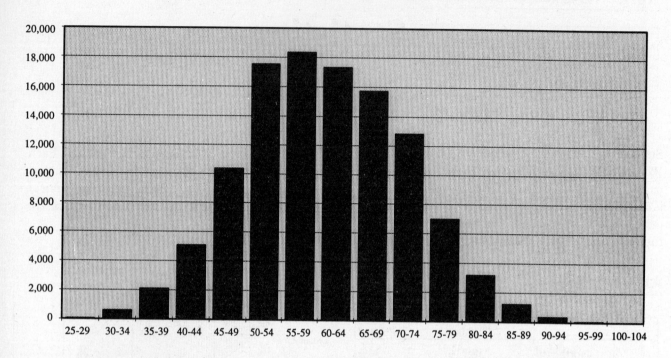

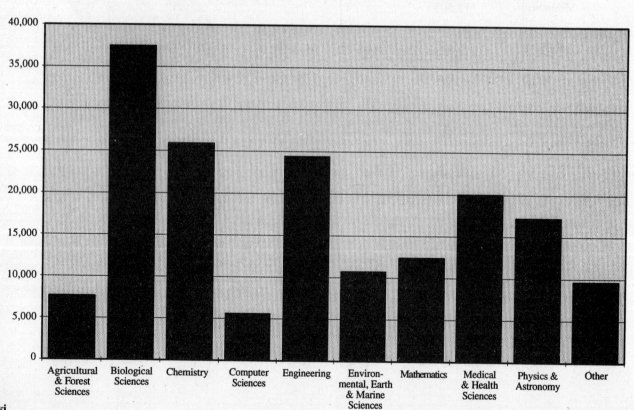

Agricultural & Forest Sciences

	Number	Percent
Northeast	1,621	21%
Southeast	2,065	27%
North Central	1,213	16%
South Central	643	8%
Mountain	744	10%
Pacific	1,314	17%
TOTAL	**7,600**	**100%**

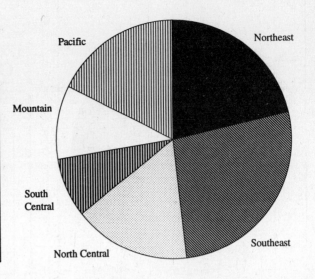

Biological Sciences

	Number	Percent
Northeast	12,357	33%
Southeast	9,318	25%
North Central	5,170	14%
South Central	2,886	8%
Mountain	2,125	6%
Pacific	5,529	15%
TOTAL	**37,385**	**100%**

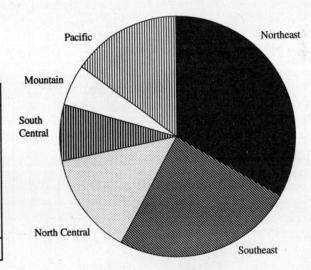

Chemistry

	Number	Percent
Northeast	10,164	39%
Southeast	6,282	24%
North Central	3,035	12%
South Central	1,785	7%
Mountain	1,350	5%
Pacific	3,217	12%
TOTAL	**25,833**	**100%**

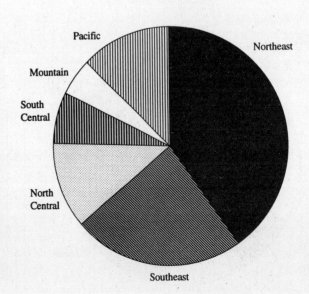

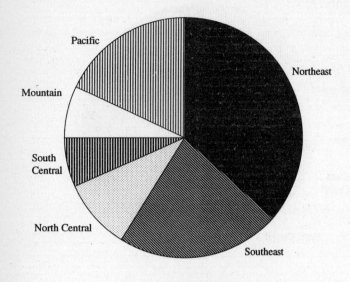

Computer Sciences

	Number	Percent
Northeast	2,001	36%
Southeast	1,257	23%
North Central	537	10%
South Central	373	7%
Mountain	384	7%
Pacific	991	18%
TOTAL	**5,543**	**100%**

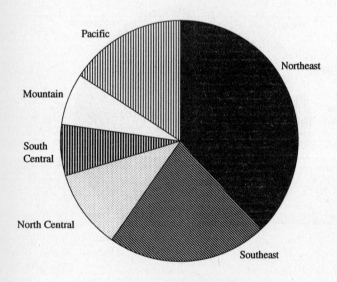

Engineering

	Number	Percent
Northeast	9,110	37%
Southeast	5,432	22%
North Central	2,514	10%
South Central	1,749	7%
Mountain	1,696	7%
Pacific	3,867	16%
TOTAL	**24,368**	**100%**

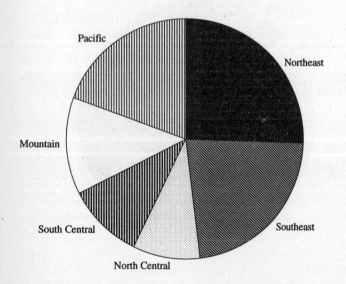

Environmental, Earth & Marine Sciences

	Number	Percent
Northeast	2,699	25%
Southeast	2,496	23%
North Central	979	9%
South Central	1,058	10%
Mountain	1,390	13%
Pacific	2,054	19%
TOTAL	**10,676**	**100%**

Mathematics

	Number	Percent
Northeast	4,375	35%
Southeast	2,781	23%
North Central	1,567	13%
South Central	938	8%
Mountain	757	6%
Pacific	1,919	16%
TOTAL	**12,337**	**100%**

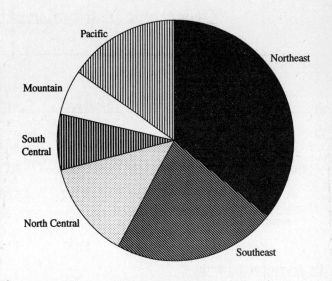

Medical & Health Sciences

	Number	Percent
Northeast	7,281	37%
Southeast	5,172	26%
North Central	2,577	13%
South Central	1,562	8%
Mountain	799	4%
Pacific	2,545	13%
TOTAL	**19,936**	**100%**

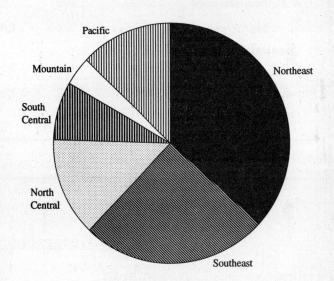

Physics & Astronomy

	Number	Percent
Northeast	5,832	34%
Southeast	3,839	22%
North Central	1,581	9%
South Central	929	5%
Mountain	1,663	10%
Pacific	3,330	19%
TOTAL	**17,174**	**100%**

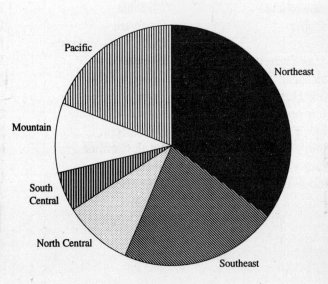

Geographic Distribution of Scientists by Discipline

	Northeast	Southeast	North Central	South Central	Mountain	Pacific	TOTAL
Agricultural & Forest Sciences	1,621	2,065	1,213	643	744	1,314	7,600
Biological Sciences	12,357	9,318	5,170	2,886	2,125	5,529	37,385
Chemistry	10,164	6,282	3,035	1,785	1,350	3,217	25,833
Computer Sciences	2,001	1,257	537	373	384	991	5,543
Engineering	9,110	5,432	2,514	1,749	1,696	3,867	24,368
Environmental, Earth & Marine Sciences	2,699	2,496	979	1,058	1,390	2,054	10,676
Mathematics	4,375	2,781	1,567	938	757	1,919	12,337
Medical & Health Sciences	7,281	5,172	2,577	1,562	799	2,545	19,936
Physics & Astronomy	5,832	3,839	1,581	929	1,663	3,330	17,174
Other Professional Fields	3,276	2,830	998	616	597	1,289	9,606
TOTAL	58,716	41,472	20,171	12,539	11,505	26,055	170,458

Geographic Definitions

Northeast
Connecticut
Indiana
Maine
Massachusetts
Michigan
New Hampshire
New Jersey
New York
Ohio
Pennsylvania
Rhode Island
Vermont

Southeast
Alabama
Delaware
District of Columbia
Florida
Georgia
Kentucky
Maryland
Mississippi
North Carolina
South Carolina
Tennessee
Virginia
West Virginia

North Central
Illinois
Iowa
Kansas
Minnesota
Missouri
Nebraska
North Dakota
South Dakota
Wisconsin

South Central
Arkansas
Louisiana
Texas
Oklahoma

Mountain
Arizona
Colorado
Idaho
Montana
Nevada
New Mexico
Utah
Wyoming

Pacific
Alaska
California
Hawaii
Oregon
Washington

Abbreviations

AAAS—American Association for the Advancement of Science
abnorm—abnormal
abstr—abstract
acad—academic, academy
acct—account, accountant, accounting
acoust—acoustic(s), acoustical
ACTH—adrenocorticotrophic hormone
actg—acting
activ—activities, activity
addn—addition(s), additional
Add—Address
adj—adjunct, adjutant
adjust—adjustment
Adm—Admiral
admin—administration, administrative
adminr—administrator(s)
admis—admission(s)
adv—adviser(s), advisory
advan—advance(d), advancement
advert—advertisement, advertising
AEC—Atomic Energy Commission
aerodyn—aerodynamic
aeronaut—aeronautic(s), aeronautical
aerophys—aerophysical, aerophysics
aesthet—aesthetic
AFB—Air Force Base
affil—affiliate(s), affiliation
agr—agricultural, agriculture
agron—agronomic, agronomical, agronomy
agrost—agrostologic, agrostological, agrostology
agt—agent
AID—Agency for International Development
Ala—Alabama
allergol—allergological, allergology
alt—alternate
Alta—Alberta
Am—America, American
AMA—American Medical Association
anal—analysis, analytic, analytical
analog—analogue
anat—anatomic, anatomical, anatomy
anesthesiol—anesthesiology
angiol—angiology
Ann—Annal(s)
ann—annual
anthrop—anthropological, anthropology
anthropom—anthropometric, anthropometrical, anthropometry
antiq—antiquary, antiquities, antiquity
antiqn—antiquarian

apicult—apicultural, apiculture
APO—Army Post Office
app—appoint, appointed
appl—applied
appln—application
approx—approximate(ly)
Apr—April
apt—apartment(s)
aquacult—aquaculture
arbit—arbitration
arch—archives
archaeol—archaeological, archaeology
archit—architectural, architecture
Arg—Argentina, Argentine
Ariz—Arizona
Ark—Arkansas
artil—artillery
asn—association
assoc(s)—associate(s), associated
asst(s)—assistant(s), assistantship(s)
assyriol—Assyriology
astrodyn—astrodynamics
astron—astronomical, astronomy
astronaut—astronautical, astronautics
astronr—astronomer
astrophys—astrophysical, astrophysics
attend—attendant, attending
atty—attorney
audiol—audiology
Aug—August
auth—author
AV—audiovisual
Ave—Avenue
avicult—avicultural, aviculture

b—born
bact—bacterial, bacteriologic, bacteriological, bacteriology
BC—British Colombia
bd—board
behav—behavior(al)
Belg—Belgian, Belgium
Bibl—Biblical
bibliog—bibliographic, bibliographical, bibliography
bibliogr—bibliographer
biochem—biochemical, biochemistry
biog—biographical, biography
biol—biological, biology
biomed—biomedical, biomedicine
biomet—biometric(s), biometrical, biometry
biophys—biophysical, biophysics

bk(s)—book(s)
bldg—building
Blvd—Boulevard
Bor—Borough
bot—botanical, botany
br—branch(es)
Brig—Brigadier
Brit—Britain, British
Bro(s)—Brother(s)
byrol—byrology
bull—Bulletin
bur—bureau
bus—business
BWI—British West Indies

c—children
Calif—California
Can—Canada, Canadian
cand—candidate
Capt—Captain
cardiol—cardiology
cardiovasc—cardiovascular
cartog—cartographic, cartographical, cartography
cartogr—cartographer
Cath—Catholic
CEngr—Corp of Engineers
cent—central
Cent Am—Central American
cert—certificate(s), certification, certified
chap—chapter
chem—chemical(s), chemistry
chemother—chemotherapy
chg—change
chmn—chairman
citricult—citriculture
class—classical
climat—climatological, climatology
clin(s)—clinic(s), clinical
cmndg—commanding
Co—County
Co—Companies, Company
co-auth—co-author
co-dir—co-director
co-ed—co-editor
co-educ—co-education, co-educational
col(s)—college(s), collegiate, colonel
collab—collaboration, collaborative
collabr—collaborator
Colo—Colorado
com—commerce, commercial
Comdr—Commander

ABBREVIATIONS

commun—communicable, communication(s)
comn(s)—commission(s), commissioned
comndg—commanding
comnr—commissioner
comp—comparitive
compos—composition
comput—computation, computer(s), computing
comt(s)—committee(s)
conchol—conchology
conf—conference
cong—congress, congressional
Conn—Connecticut
conserv—conservation, conservatory
consol—consolidated, consolidation
const—constitution, constitutional
construct—construction, constructive
consult(s)—consult, consultant(s), consultantship(s), consultation, consulting
contemp—contemporary
contrib—contribute, contributing, contribution(s)
contribr—contributor
conv—convention
coop—cooperating, cooperation, cooperative
coord—coordinate(d), coordinating, coordination
coordr—coordinator
corp—corporate, corporation(s)
corresp—correspondence, correspondent, corresponding
coun—council, counsel, counseling
counr—councilor, counselor
criminol—criminological, criminology
cryog—cryogenic(s)
crystallog—crystallographic, crystallographical, crystallography
crystallogr—crystallographer
Ct—Court
Ctr—Center
cult—cultural, culture
cur—curator
curric—curriculum
cybernet—cybernetic(s)
cytol—cytological, cytology
Czech—Czechoslovakia, Czech Republic

DC—District of Columbia
Dec—December
Del—Delaware
deleg—delegate, delegation
delinq—delinquency, delinquent
dem—democrat(s), democratic
demog—demographic, demography
demogr—demographer
demonstr—demontrator
dendrol—dendrologic, dendrological, dendrology
dent—dental, dentistry
dep—deputy
dept—department
dermat—dermatologic, dermatological, dermatology
develop—developed, developing, development, developmental
diag—diagnosis, diagnostic
dialectol—dialectological, dialectology
dict—dictionaries, dictionary
Dig—Digest

dipl—diploma, diplomate
dir(s)—director(s), directories, directory
dis—disease(s), disorders
Diss Abst—Dissertation Abstracts
dist—district
distrib—distributed, distribution, distributive
distribr—distributor(s)
div—division, divisional, divorced
DNA—deoxyribonucleic acid
doc—document(s), documentary, documentation
Dom—Dominion
Dr—Drive

E—East
ecol—ecological, ecology
econ(s)—economic(s), economical, economy
economet—econometric(s)
ECT—electroconvulsive or electroshock therapy
ed—edition(s), editor(s), editorial
ed bd—editorial board
educ—education, educational
educr—educator(s)
EEG—electroencephalogram, electroencephalographic, electroencephalography
Egyptol—Egyptology
EKG—electrocardiogram
elec—electric, electrical, electricity
electrochem—electrochemical, electrochemistry
electroph—electrophysical, electrophysics
elem—elementary
embryol—embryologic, embryological, embryology
emer—emeriti, emeritus
employ—employment
encour—encouragement
encycl—encyclopedia
endocrinol—endocrinologic, endocrinology
eng—engineering
Eng—England, English
engr(s)—engineer(s)
enol—enology
Ens—Ensign
entom—entomological, entomology
environ—environment(s), environmental
enzym—enzymology
epidemiol—epidemiologic, epidemiological, epidemiology
equip—equipment
ERDA—Energy Research & Development Administration
ESEA—Elementary & Secondary Education Act
espec—especially
estab—established, establishment(s)
ethnog—ethnographic, ethnographical, ethnography
ethnogr—ethnographer
ethnol—ethnologic, ethnological, ethnology
Europ—European
eval—evaluation
Evangel—Evangelical
eve—evening
exam—examination(s), examining
examr—examiner
except—exceptional
exec(s)—executive(s)

exeg—exegeses, exegesis, exegetic, exegetical
exhib(s)—exhibition(s), exhibit(s)
exp—experiment, experimental
exped(s)—expedition(s)
explor—exploration(s), exploratory
expos—exposition
exten—extension

fac—faculty
facil—facilities, facility
Feb—February
fed—federal
fedn—federation
fel(s)—fellow(s), fellowship(s)
fermentol—fermentology
fertil—fertility, fertilization
Fla—Florida
floricult—floricultural, floriculture
found—foundation
FPO—Fleet Post Office
Fr—French
Ft—Fort

Ga—Georgia
gastroenterol—gastroenterological, gastroenterology
gen—general
geneal—genealogical, genealogy
geod—geodesy, geodetic
geog—geographic, geographical, geography
geogr—geographer
geol—geologic, geological, geology
geom—geometric, geometrical, geometry
geomorphol—geomorphologic, geomorphology
geophys—geophysical, geophysics
Ger—German, Germanic, Germany
geriat—geriatric
geront—gerontological, gerontology
Ges—Gesellschaft
glaciol—glaciology
gov—governing, governor(s)
govt—government, governmental
grad—graduate(d)
Gt Brit—Great Britain
guid—guidance
gym—gymnasium
gynec—gynecologic, gynecological, gynecology

handbk(s)—handbook(s)
helminth—helminthology
hemat—hematologic, hematological, hematology
herpet—herpetologic, herpetological, herpetology
HEW—Department of Health, Education & Welfare
Hisp—Hispanic, Hispania
hist—historic, historical, history
histol—histological, histology
HM—Her Majesty
hochsch—hochschule
homeop—homeopathic, homeopathy
hon(s)—honor(s), honorable, honorary
hort—horticultural, horticulture
hosp(s)—hospital(s), hospitalization
hq—headquarters
HumRRO—Human Resources Research Office

ABBREVIATIONS

husb—husbandry
Hwy—Highway
hydraul—hydraulic(s)
hydrodyn—hydrodynamic(s)
hydrol—hydrologic, hydrological, hydrologics
hyg—hygiene, hygienic(s)
hypn—hypnosis

ichthyol—ichthyological, ichthyology
Ill—Illinois
illum—illuminating, illumination
illus—illustrate, illustrated, illustration
illusr—illustrator
immunol—immunologic, immunological, immunology
Imp—Imperial
improv—improvement
Inc—Incorporated
in-chg—in charge
incl—include(s), including
Ind—Indiana
indust(s)—industrial, industries, industry
Inf—Infantry
info—information
inorg—inorganic
ins—insurance
inst(s)—institute(s), institution(s)
instnl—institutional(ized)
instr(s)—instruct, instruction, instructor(s)
instrnl—instructional
int—international
intel—intellligence
introd—introduction
invert—invertebrate
invest(s)—investigation(s)
investr—investigator
irrig—irrigation
Ital—Italian

J—Journal
Jan—January
Jct—Junction
jour—journal, journalism
jr—junior
jurisp—jurisprudence
juv—juvenile

Kans—Kansas
Ky—Kentucky

La—Louisiana
lab(s)—laboratories, laboratory
lang—language(s)
laryngol—larygological, laryngology
lect—lecture(s)
lectr—lecturer(s)
legis—legislation, legislative, legislature
lett—letter(s)
lib—liberal
libr—libraries, library
librn—librarian
lic—license(d)
limnol—limnological, limnology
ling—linguistic(s), linguistical
lit—literary, literature
lithol—lithologic, lithological, lithology
Lt—Lieutenant
Ltd—Limited

m—married
mach—machine(s), machinery
mag—magazine(s)
maj—major
malacol—malacology
mammal—mammalogy
Man—Manitoba
Mar—March
Mariol—Mariology
Mass—Massechusetts
mat—material(s)
mat med—materia medica
math—mathematic(s), mathematical
Md—Maryland
mech—mechanic(s), mechanical
med—medical, medicinal, medicine
Mediter—Mediterranean
Mem—Memorial
mem—member(s), membership(s)
ment—mental(ly)
metab—metabolic, metabolism
metall—metallurgic, metallurgical, metallurgy
metallog—metallographic, metallography
metallogr—metallographer
metaphys—metaphysical, metaphysics
meteorol—meteorological, meteorology
metrol—metrological, metrology
metrop—metropolitan
Mex—Mexican, Mexico
mfg—manufacturing
mfr—manufacturer
mgr—manager
mgt—management
Mich—Michigan
microbiol—microbiological, microbiology
micros—microscopic, microscopical, microscopy
mid—middle
mil—military
mineral—mineralogical, mineralogy
Minn—Minnesota
Miss—Mississippi
mkt—market, marketing
Mo—Missouri
mod—modern
monogr—monograph
Mont—Montana
morphol—morphological, morphology
Mt—Mount
mult—multiple
munic—municipal, municipalities
mus—museum(s)
musicol—musicological, musicology
mycol—mycologic, mycology

N—North
NASA—National Aeronautics & Space Administration
nat—national, naturalized
NATO—North Atlantic Treaty Organization
navig—navigation(al)
NB—New Brunswick
NC—North Carolina
NDak—North Dakota
NDEA—National Defense Education Act
Nebr—Nebraska
nematol—nematological, nematology
nerv—nervous
Neth—Netherlands

neurol—neurological, neurology
neuropath—neuropathological, neuropathology
neuropsychiat—neuropsychiatric, neuropsychiatry
neurosurg—neurosurgical, neurosurgery
Nev—Nevada
New Eng—New England
New York—New York City
Nfld—Newfoundland
NH—New Hampshire
NIH—National Institute of Health
NIMH—National Institute of Mental Health
NJ—New Jersey
NMex—New Mexico
No—Number
nonres—nonresident
norm—normal
Norweg—Norwegian
Nov—November
NS—Nova Scotia
NSF—National Science Foundation
NSW—New South Wales
numis—numismatic(s)
nutrit—nutrition, nutritional
NY—New York State
NZ—New Zealand

observ—observatories, observatory
obstet—obstetric(s), obstetrical
occas—occasional(ly)
occup—occupation, occupational
oceanog—oceanographic, oceanographical, oceanography
oceanogr—oceanographer
Oct—October
odontol—odontology
OEEC—Organization for European Economic Cooperation
off—office, official
Okla—Oklahoma
olericult—olericulture
oncol—oncologic, oncology
Ont—Ontario
oper(s)—operation(s), operational, operative
ophthal—ophthalmologic, ophthalmological, ophthalmology
optom—optometric, optometrical, optometry
ord—ordnance
Ore—Oregon
org—organic
orgn—organization(s), organizational
orient—oriental
ornith—ornithological, ornithology
orthod—orthodontia, orthodontic(s)
orthop—orthopedic(s)
osteop—osteopathic, osteopathy
otol—otological, otology
otolaryngol—otolaryngological, otolaryngology
otorhinol—otorhinologic, otorhinology

Pa—Pennsylvania
Pac—Pacific
paleobot—paleobotanical, paleobotany
paleont—paleontology
Pan-Am—Pan-American
parasitol—parasitology
partic—participant, participating
path—pathologic, pathological, pathology

xxiii

ABBREVIATIONS

pedag—pedagogic(s), pedagogical, pedagogy
pediat—pediatric(s)
PEI—Prince Edward Islands
penol—penological, penology
periodont—periodontal, periodontic(s)
petrog—petrographic, petrographical, petrography
petrogr—petrographer
petrol—petroleum, petrologic, petrological, petrology
pharm—pharmacy
pharmaceut—pharmaceutic(s), pharmaceutical(s)
pharmacog—pharmacognosy
pharamacol—pharmacologic, pharmacological, pharmacology
phenomenol—phenomenologic(al), phenomenology
philol—philological, philology
philos—philosophic, philosophical, philosophy
photog—photographic, photography
photogeog—photogeographic, photogeography
photogr—photographer(s)
photogram—photogrammetric, photogrammetry
photom—photometric, photometrical, photometry
phycol—phycology
phys—physical
physiog—physiographic, physiographical, physiography
physiol—physiological, phsysiology
Pkwy—Parkway
Pl—Place
polit—political, politics
polytech—polytechnic(s)
pomol—pomological, pomology
pontif—pontifical
pop—population
Port—Portugal, Portuguese
Pos—Position
postgrad—postgraduate
PQ—Province of Quebec
PR—Puerto Rico
pract—practice
practr—practitioner
prehist—prehistoric, prehistory
prep—preparation, preparative, preparatory
pres—president
Presby—Presbyterian
preserv—preservation
prev—prevention, preventive
prin—principal
prob(s)—problem(s)
proc—proceedings
proctol—proctologic, proctological, proctology
prod—product(s), production, productive
prof—professional, professor, professorial
Prof Exp—Professional Experience
prog(s)—program(s), programmed, programming
proj—project(s), projection(al), projective
prom—promotion
protozool—protozoology
Prov—Province, Provincial
psychiat—psychiatric, psychiatry

psychoanal—psychoanalysis, psychoanalytic, psychoanalytical
psychol—psychological, psychology
psychomet—psychometric(s)
psychopath—psychopathologic, psychopathology
psychophys—psychophysical, psychophysics
psychophysiol—psychophysiological, psychophysiology
psychosom—psychosomatic(s)
psychother—psychoterapeutic(s), psychotherapy
Pt—Point
pub—public
publ—publication(s), publish(ed), publisher, publishing
pvt—private

Qm—Quartermaster
Qm Gen—Quartermaster General
qual—qualitative, quality
quant—quantitative
quart—quarterly
Que—Quebec

radiol—radiological, radiology
RAF—Royal Air Force
RAFVR—Royal Air Force Volunteer Reserve
RAMC—Royal Army Medical Corps
RAMCR—Royal Army Medical Corps Reserve
RAOC—Royal Army Ordnance Corps
RASC—Royal Army Service Corps
RASCR—Royal Army Service Corps Reserve
RCAF—Royal Canadian Air Force
RCAFR—Royal Canadian Air Force Reserve
RCAFVR—Royal Canadian Air Force Volunteer Reserve
RCAMC—Royal Canadian Army Medical Corps
RCAMCR—Royal Canadian Army Medical Corps Reserve
RCASC—Royal Canadian Army Service Corps
RCASCR—Royal Canadian Army Service Corps Reserve
RCEME—Royal Canadian Electrical & Mechanical Engineers
RCN—Royal Canadian Navy
RCNR—Royal Canadian Naval Reserve
RCNVR—Royal Canadian Naval Volunteer Reserve
Rd—Road
RD—Rural Delivery
rec—record(s), recording
redevelop—redevelopment
ref—reference(s)
refrig—refrigeration
regist—register(ed), registration
registr—registrar
regt—regiment(al)
rehab—rehabilitation
rel(s)—relation(s), relative
relig—religion, religious
REME—Royal Electrical & Mechanical Engineers
rep—represent, representative
Repub—Republic
req—requirements

res—research, reserve
rev—review, revised, revision
RFD—Rural Free Delivery
rhet—rhetoric, rhetorical
RI—Rhode Island
Rm—Room
RM—Royal Marines
RN—Royal Navy
RNA—ribonucleic acid
RNR—Royal Naval Reserve
RNVR—Royal Naval Volunteer Reserve
roentgenol—roentgenologic, roentgenological, roentgenology
RR—Railroad, Rural Route
Rte—Route
Russ—Russian
rwy—railway

S—South
SAfrica—South Africa
SAm—South America, South American
sanit—sanitary, sanitation
Sask—Saskatchewan
SC—South Carolina
Scand—Scandinavia(n)
sch(s)—school(s)
scholar—scholarship
sci—science(s), scientific
SDak—South Dakota
SEATO—Southeast Asia Treaty Organization
sec—secondary
sect—section
secy—secretary
seismog—seismograph, seismographic, seismography
seismogr—seismographer
seismol—seismological, seismology
sem—seminar, seminary
Sen—Senator, Senatorial
Sept—September
ser—serial, series
serol—serologic, serological, serology
serv—service(s), serving
silvicult—silvicultural, silviculture
soc(s)—societies, society
soc sci—social science
sociol—sociologic, sociological, sociology
Span—Spanish
spec—special
specif—specification(s)
spectrog—spectrograph, spectrographic, spectrography
spectrogr—spectrographer
spectrophotom—spectrophotometer, spectrophotometric, spectrophotometry
spectros—spectroscopic, spectroscopy
speleol—speleological, speleology
Sq—Square
sr—senior
St—Saint, Street(s)
sta(s)—station(s)
stand—standard(s), standardization
statist—statistical, statistics
Ste—Sainte
steril—sterility
stomatol—stomatology
stratig—stratigraphic, stratigraphy
stratigr—stratigrapher
struct—structural, structure(s)

ABBREVIATIONS

stud—student(ship)
subcomt—subcommittee
subj—subject
subsid—subsidiary
substa—substation
super—superior
suppl—supplement(s), supplemental, supplementary
supt—superintendent
supv—supervising, supervision
supvr—supervisor
supvry—supervisory
surg—surgery, surgical
surv—survey, surveying
survr—surveyor
Swed—Swedish
Switz—Switzerland
symp—symposia, symposium(s)
syphil—syphilology
syst(s)—system(s), systematic(s), systematical

taxon—taxonomic, taxonomy
tech—technical, technique(s)
technol—technologic(al), technology
tel—telegraph(y), telephone
temp—temporary
Tenn—Tennessee
Terr—Terrace
Tex—Texas
textbk(s)—textbook(s)
text ed—text edition
theol—theological, theology
theoret—theoretic(al)
ther—therapy
therapeut—therapeutic(s)
thermodyn—thermodynamic(s)
topog—topographic, topographical, topography
topogr—topographer
toxicol—toxicologic, toxicological, toxicology
trans—transaction(s)
transl—translated, translation(s)
translr—translator(s)

transp—transport, transportation
treas—treasurer, treasury
treat—treatment
trop—tropical
tuberc—tuberculosis
TV—television
Twp—Township

UAR—United Arab Republic
UK—United Kingdom
UN—United Nations
undergrad—undergraduate
unemploy—unemployment
UNESCO—United Nations Educational Scientific & Cultural Organization
UNICEF—United Nations International Childrens Fund
univ(s)—universities, university
UNRRA—United Nations Relief & Rehabilitation Administration
UNRWA—United Nations Relief & Works Agency
urol—urologic, urological, urology
US—United States
USAAF—US Army Air Force
USAAFR—US Army Air Force Reserve
USAF—US Air Force
USAFR—US Air Force Reserve
USAID—US Agency for International Development
USAR—US Army Reserve
USCG—US Coast Guard
USCGR—US Coast Guard Reserve
USDA—US Department of Agriculture
USMC—US Marine Corps
USMCR—US Marine Corps Reserve
USN—US Navy
USNAF—US Naval Air Force
USNAFR—US Naval Air Force Reserve
USNR—US Naval Reserve
USPHS—US Public Health Service
USPHSR—US Public Health Service Reserve
USSR—Union of Soviet Socialist Republics

Va—Virginia
var—various
veg—vegetable(s), vegetation
vent—ventilating, ventilation
vert—vertebrate
Vet—Veteran(s)
vet—veterinarian, veterinary
VI—Virgin Islands
vinicult—viniculture
virol—virological, virology
vis—visiting
voc—vocational
vocab—vocabulary
vol(s)—voluntary, volunteer(s), volume(s)
vpres—vice president
vs—versus
Vt—Vermont

W—West
Wash—Washington
WHO—World Health Organization
WI—West Indies
wid—widow, widowed, widower
Wis—Wisconsin
WVa—West Virginia
Wyo—Wyoming

Yearbk(s)—Yearbook(s)
YMCA—Young Men's Christian Association
YMHA—Young Men's Hebrew Association
Yr(s)—Year(s)
YT—Yukon Territory
YWCA—Young Women's Christian Association
YWHA—Young Women's Hebrew Association

zool—zoological, zoology

DISCIPLINE INDEX

AGRICULTURAL & FOREST SCIENCES

Agricultural Business & Management

ARIZONA
Gordon, Richard Seymour

ARKANSAS
Daniels, L B
Senseman, Scott Allen

CALIFORNIA
Kushner, Arthur S
Thomas, Paul Clarence

COLORADO
Gholson, Larry Estie
Lovins, Amory

DISTRICT OF COLUMBIA
Gardner, Bruce Lynn
Jennings, Vivan M

FLORIDA
Nichols, Robert Loring
Peart, Robert McDermand

IDAHO
Douglas, Dexter Richard

ILLINOIS
Prescott, Jon Michael

INDIANA
Boehlje, Michael Dean
Folkerts, Thomas Mason

IOWA
Colvin, Thomas Stuart
Harbaugh, Daniel David

KANSAS
Epp, Melvin David
Kastner, Curtis Lynn

KENTUCKY
Reed, Michael Robert

LOUISIANA
Hensley, Sess D

MARYLAND
Gadsby, Dwight Maxon
Howard, Joseph H
Luther, Lonnie W
Tallent, William Hugh

MICHIGAN
Castenson, Roger R

MISSOURI
Carlson, Wayne C
Houghton, John M
Marshall, Lucia Garcia-Iniguez
Yates-Parker, Nancy L

NEW JERSEY
Noone, Thomas Mark

NEW YORK
Gravani, Robert Bernard

NORTH CAROLINA
Johnson, Thomas

OHIO
Coleman, Marilyn A
Dembowski, Peter Vincent
Sabourin, Thomas Donald

PENNSYLVANIA
LaRossa, Robert Alan

SOUTH CAROLINA
Fischer, James Roland

TEXAS
Benedict, John Howard, Jr
Davis, Bob
Gilles, Kenneth Albert

UTAH
Butcher, John Edward

WISCONSIN
Barton, Kenneth Allen

PUERTO RICO
Rodriguez, Jorge Luis

ONTARIO
Weersink, Alfons John

QUEBEC
Deschenes, Jean-Marc

Agricultural Economics

ALABAMA
Drake, Albert Estern
Sutherland, William Neil

ARIZONA
Lord, William B

ARKANSAS
Havener, Robert D
Headley, Joseph Charles
Parsch, Lucas Dean
Thompson, Robert Lee

CALIFORNIA
Abenes, Fiorello Bigornia
Johnston, Warren E
Lin, Robert I-San
Niles, James Alfred
Norgaard, Richard Bruce
Sage, Orrin Grant, Jr
Siebert, Jerome Bernard
Squires, Dale Edward
Wallender, Wesley William

CONNECTICUT
Bentley, William Ross
Sieckhaus, John Francis

DELAWARE
Elterich, G Joachim

DISTRICT OF COLUMBIA
Beer, Charles
Ching, Chauncey T K
Darwin, Roy F
Harrington, David Holman
Jennings, Vivan M
Mellor, John Williams
Offutt, Susan Elizabeth

FLORIDA
Barnard, Donald Roy
Meltzer, Martin Isaac
Norval, Richard Andrew
Thomason, David Morton
Ward, Ronald Wayne

GEORGIA
Bhagia, Gobind Shewakram
Chiang, Tze I
Freeman, Jere Evans
Purcell, Joseph Carroll

HAWAII
Liu, Wei

IDAHO
Michalson, Edgar Lloyd

ILLINOIS
Mundlak, Yair

Seitz, Wesley Donald

INDIANA
Boehlje, Michael Dean
Doering, Otto Charles, III

IOWA
Colvin, Thomas Stuart
Johnson, Stanley R
Kliebenstein, James Bernard

KANSAS
Epp, Melvin David
Hess, Carroll V

KENTUCKY
Reed, Michael Robert

LOUISIANA
Decossas, Kenneth Miles

MAINE
Todaro, Michael P

MARYLAND
Gadsby, Dwight Maxon
Just, Richard
Weiss, Michael David

MICHIGAN
Gillingham, James Clark
Manderscheid, Lester Vincent
Niese, Jeffrey Neal

MINNESOTA
Anderson, Donald E
Munson, Robert Dean
Ruttan, Vernon W

MISSISSIPPI
Hurt, Verner C

MISSOURI
Hardin, Clifford Morris

NEBRASKA
Miller, William Lloyd

NEW JERSEY
Burns, David Jerome

NEW YORK
Herdt, Robert William
Jabbur, Ramzi Jibrail
Thompson, John C, Jr
Tomek, William Goodrich

NORTH CAROLINA
Benrud, Charles Harris
Brandt, Jon Alan

NORTH DAKOTA
Leitch, Jay A

OHIO
Adams, Dale W
Baldwin, Eldon Dean
Chern, Wen Shyong
Erven, Bernard Lee
Forster, D Lynn
Hushak, Leroy J
Larson, Donald W
Lee, Warren Ford
Meyer, Richard Lee
Rask, Norman
Shaudys, Edgar T
Vertrees, Robert Layman

OKLAHOMA
Plaxico, James Samuel

OREGON
Buccola, Steven Thomas
Miller, Stanley Frank
Schmisseur, Wilson Edward
Workman, William Glenn

PENNSYLVANIA
Alter, Theodore Roberts
Epp, Donald James
Kelly, B(ernard) Wayne
LaRossa, Robert Alan

SOUTH CAROLINA
Walker, Francis Edwin

SOUTH DAKOTA
Dobbs, Thomas Lawrence

TENNESSEE
Jumper, Sidney Roberts
Papas, Andreas Michael
Rawlins, Nolan Omri
Williamson, Handy, Jr

TEXAS
Davis, Bob
Eddleman, Bobby R
Eubank, Randall Lester
Lacewell, Ronald Dale
Nelson, A Gene
Nixon, Donald Merwin
Wendt, Charles William
Whitson, Robert Edd

UTAH
Downing, Kenton Benson
Grimshaw, Paul R
Lilieholm, Robert John
Wallentine, Max V
Workman, John Paul

VIRGINIA
Cummings, Ralph Waldo, Jr
Halvorson, Lloyd Chester
Pontius, Steven Kent
Sprague, Lucian Matthew
Woods, William Fred

WASHINGTON
Michelsen, Ari Montgomery

WEST VIRGINIA
Barr, Alfred L

WISCONSIN
Buse, Reuben Charles
Luby, Patrick Joseph
Schmidt, John Richard

WYOMING
Kearl, Willis Gordon

PUERTO RICO
Gonzalez, Gladys
Lewis, Allen Rogers

ALBERTA
Sonntag, Bernard H

BRITISH COLUMBIA
Binkley, Clark Shepard

ONTARIO
Brinkman, George Loris
Simard, Albert Joseph
Weersink, Alfons John

QUEBEC
Coffin, Harold Garth

OTHER COUNTRIES
Bywater, Anthony Colin
Malik, Mazhar Ali Khan

Agriculture, General

ALABAMA
Guthrie, Richard Lafayette
Haaland, Ronald L
Johnson, Clarence Eugene
Johnson, Loyd
Mack, Timothy Patrick

DISCIPLINE INDEX

Agriculture, General (cont)

ALASKA
Cochran, Verlan Leyerl

ARIZONA
Day, Arden Dexter
Erickson, Eric Herman, Jr
Terry, Lucy Irene

ARKANSAS
Andrews, Luther David
Bartlett, Frank David
Brown, A Hayden, Jr
Brown, Connell Jean
Clower, Dan Fredric
Horan, Francis E
Motes, Dennis Roy
Oosterhuis, Derrick M

CALIFORNIA
Abenes, Fiorello Bigornia
Ayars, James Earl
Benes, Norman Stanley
Bertoldi, Gilbert LeRoy
Bonner, James (Fredrick)
Calpouzos, Lucas
Casida, John Edward
Christensen, Allen Clare
Donnan, William W
Estilai, Ali
Garber, Richard Hammerle
Henrick, Clive Arthur
Hess, Frederick Dan
Humaydan, Hasib Shaheen
Karinen, Arthur Eli
Lovatt, Carol Jean
Lucas, William John
Myers, George Scott, Jr
Pitts, Donald James
Prend, Joseph
Raju, Namboori Bhaskara
Rammer, Irwyn Alden
Rubenstein, Howard S
Sandmeier, Ruedi Beat
Silk, Margaret Wendy Kuhn
Thomason, Ivan J
Thompson, Chester Ray
Treloar, Alan Edward
Van Bruggen, Ariena H C
Whitlock, Gaylord Purcell
Willemsen, Roger Wayne
Zalom, Frank G

COLORADO
Ball, Wilbur Perry
Dickenson, Donald Dwight
Gholson, Larry Estie
Helmerick, Robert Howard
Knutson, Kenneth Wayne
Lamm, Warren Dennis
Luebs, Ralph Edward
Thomas, William Robb

CONNECTICUT
Ahrens, John Frederick
John, Hugo Herman

DELAWARE
Green, Jerome
Smith, Constance Meta

DISTRICT OF COLUMBIA
Beer, Charles
Brown, Lester R
Jennings, Vivan M
Khan, Mohamed Shaheed
Lichens-Park, Ann Elizabeth
Mellor, John Williams
Nickle, David Allan
Parochetti, James V
Plowman, Ronald Dean
Plucknett, Donald Lovelle
Roskoski, Joann Pearl
Strommen, Norton Duane
Wilson, William Mark Dunlop

FLORIDA
De Forest, Sherwood Searle
Denmark, Harold Anderson
Kretschmer, Albert Emil, Jr
McCloud, Darell Edison
Osborne, Lance Smith
Price, Donald Ray
Rawls, Walter Cecil, Jr
Simonet, Donald Edward
Smart, Grover Cleveland, Jr
Snyder, Fred Calvin
Stoffella, Peter Joseph
Woodruff, Robert Eugene

GEORGIA
Amos, Henry Estill
Arkin, Gerald Franklin
Kanemasu, Edward Tsukasa
McMurray, Birch Lee
Minton, Norman A
Todd, James Wyatt

HAWAII
Bartosik, Alexander Michael
Gitlin, Harris Martlin
Teramura, Alan Hiroshi

IDAHO
Bondurant, James A(llison)
Callihan, Robert Harold
Douglas, Dexter Richard
McCaffrey, Joseph Peter
Michalson, Edgar Lloyd
Osgood, Charles Edgar

ILLINOIS
Bentley, Orville George
Changnon, Stanley Alcide, Jr
Clark, Jimmy Howard
Easter, Robert Arnold
Helm, Charles George
Janghorbani, Morteza
Kirby, Hilliard Walker
Knake, Ellery Louis
Olsson, Nils Ove
Parsons, Carl Michael
Patel, Mayur
Princen, Lambertus Henricus
Smiciklas, Kenneth Donald
Spahr, Sidney Louis
Troyer, Alvah Forrest
Wilson, Richard Hansel
Wolff, Robert L
Yoerger, Roger R

INDIANA
Bauman, Thomas Trost
Coalson, James A
Freeman, Verne Crawford
Gallun, Robert Louis
Gehring, Perry James
Ivaturi, Rao Venkata Krishna
Jantz, O K
Mitchell, Cary Arthur

IOWA
Black, Charles Allen
Harbaugh, Daniel David
Moore, Kenneth J
Stuckey, Richard E

KANSAS
Call, Edward Prior
Epp, Melvin David
Larson, Vernon C
Lomas, Lyle Wayne
Paulsen, Gary Melvin
Piper, Jon Kingsbury
Whitney, Wendell Keith

KENTUCKY
Householder, William Allen
Little, Charles Oran
Norfleet, Morris L
Sigafus, Roy Edward
Vogel, Willis Gene

LOUISIANA
Board, James Ellery
Cuomo, Gregory Joseph

MAINE
Barton, Barbara Ann

MARYLAND
Benbrook, Charles M
Christy, Alfred Lawrence
Cleland, Charles Frederick
Crosby, Edwin Andrew
Duke, James A
Fried, Maurice
Gadsby, Dwight Maxon
Harris, Clare I
Hopkins, Homer Thawley
Kearney, Philip C
Krizek, Donald Thomas
Matthews, Benjamin F
Menn, Julius Joel
Minnifield, Nita Michele
Nelson, Elton Glen
Pinto da Silva, Pedro Goncalves
Ross, Philip
Soto, Gerardo H
Tjio, Joe Hin

MASSACHUSETTS
Botticelli, Charles Robert
Coppinger, Raymond Parke
Deubert, Karl Heinz
Furth, David George
Kepper, Robert Edgar
Redington, Charles Bahr
Rohde, Richard Allen

MICHIGAN
Bird, George W
Henneman, Harold Albert
Hull, Jerome, Jr
Isleib, Donald Richard
Maley, Wayne A
Miller, James Ray
Thomas, John William
Uebersax, Mark Alan
Von Bernuth, Robert Dean

MINNESOTA
Marx, George Donald
Oelke, Ervin Albert
Preiss, Frederick John
Pryor, Gordon Roy
Rehm, George W

MISSISSIPPI
Hardee, Dicky Dan
Hodges, Harry Franklin
Knight, William Eric
Mutchler, Calvin Kendal
Tomlinson, James Everett

MISSOURI
Khan, Adam
Miles, Randall Jay
Pfander, William Harvey
Schumacher, Richard William

MONTANA
Morrill, Wendell Lee

NEBRASKA
Chapman, John Arthur
Dillon, Roy Dean
Francis, Charles Andrew
Grisso, Robert Dwight
Knox, Ellis Gilbert
Raun, Earle Spangler
Von Bargen, Kenneth Louis

NEVADA
Young, James Albert

NEW JERSEY
Bahr, James Theodore
Cheng, Kang
Dyer, Judith Gretchen
Krueger, Roger Warren
Markle, George Michael

NEW YORK
Bayer, George Herbert
Bellve, Anthony Rex
Broadway, Roxanne Meyer
Butler, Karl Douglas, Sr
Goodale, Douglas M
Huntington, David Hans
Jacobson, Jay Stanley
Nittler, LeRoy Walter
Raffensperger, Edgar M
Robinson, Terence Lee
Rosenberger, David A
Smalley, Ralph Ray
Thomas, Everett Dake
Topp, William Carl
Verma, Ram S

NORTH CAROLINA
Anderson, Thomas Ernest
Blake, Thomas Lewis
Carlson, William Theodore
Holm, Robert E
Lalor, William Francis
Marco, Gino Joseph
Ross, Richard Henry, Jr
Zarnstorff, Mark Edward

OHIO
Bowers, John Dalton
Cooper, Tommye
Gauntt, William Amor
Hock, Arthur George
Jalil, Mazhar
McCracken, John David
Niemczyk, Harry D
Palmer, Melville Louis
Reddy, Padala Vykuntha
Ritchie, Austin E
Scarborough, Vernon Lee
Wilson, George Rodger
Wonderling, Thomas Franklin

OKLAHOMA
Kessler, Edwin, 3rd
Klatt, Arthur Raymond
Melouk, Hassan A
Raun, Ned S
Rice, Charles Edward
Teh, Thian Hor

OREGON
Amundson, Robert Gale
Arnold, Roy Gary
Howard, William Weaver
Liston, Aaron Irving
Miller, Stanley Frank
Shock, Clinton C
Smiley, Richard Wayne
Weiser, Conrad John

PENNSYLVANIA
Aller, Harold Ernest
Bergman, Ernest L
Cerbulis, Janis
Heller, Paul R
Hull, Larry Allen
Montgomery, Ronald Eugene
Russo, Joseph Martin
Tammen, James F
Wicker, Robert Kirk

SOUTH CAROLINA
Camberato, James John
Gossett, Billy Joe
Johnson, Albert Wayne
Kittrell, Benjamin Upchurch
Manley, Donald Gene
Nolan, Clifford N

TENNESSEE
Armistead, Willis William
Caron, Richard Edward
Chiang, Thomas M
Graveel, John Gerard
Hathcock, Bobby Ray
Helweg, Otto Jennings
Young, Lawrence Dale

TEXAS
Agan, Raymond John
Bowling, Clarence C
Brokaw, Bryan Edward
Butler, Ogbourne Duke, Jr
Calub, Alfonso deGuzman
Clark, James Richard
Eddleman, Bobby R
Frederiksen, Richard Allan
Greene, Donald Miller
Hobbs, Clifford Dean
Hoffman, Robert A
Kirk, Ivan Wayne
McDonald, Lynn Dale
Palmer, William Alan
Parajulee, Megha N
Richardson, Arthur Jerold
Shotwell, Thomas Knight
Sij, John William
Stewart, Bobby Alton
Thompson, Granville Berry

UTAH
Albrechtsen, Rulon S
Clark, C Elmer
Haws, Byron Austin
Thorup, Richard M
Vandenberg, John Donald
Wallentine, Max V

VERMONT
Dritschilo, William

VIRGINIA
Drake, Charles Roy
Frahm, Richard R
Hinckley, Alden Dexter
Pontius, Steven Kent
Strauss, Michael S
Van Krey, Harry P

WASHINGTON
Butt, Billy Arthur
Peabody, Dwight Van Dorn, Jr
Smith, Samuel H
Toba, H(achiro) Harold

WEST VIRGINIA
Barr, Alfred L
Maxwell, Robert Haworth

WISCONSIN
Barnes, Robert F
Delorit, Richard John
Grunewald, Ralph
Holm, LeRoy George
Jorgensen, Neal A
Koval, Charles Francis
Stang, Elden James
Willis, Harold Lester

PUERTO RICO
Rodriguez, Jorge Luis
Rodriguez-Arias, Jorge H

ALBERTA
Bentley, C Fred
Blackshaw, Robert Earl
Hadziyev, Dimitri
Harper, Alexander Maitland
Krahn, Thomas Richard
Moyer, James Robert
Price, Mick A

BRITISH COLUMBIA
Borden, John Harvey
Mackauer, Manfred
Neilsen, Gerald Henry

MANITOBA
Grant, Cynthia Ann
Palaniswamy, Pachagounder
Smil, Vaclav

NEWFOUNDLAND
Lim, Kiok-Puan

NOVA SCOTIA
Embree, Charles Gordon

ONTARIO
Anderson, Terry Ross
Arnison, Paul Grenville
Asculai, Samuel Simon
Biswas, Asit Kumar
Catling, Paul Miles
Chong, Calvin
Coote, Denis Richard
Farnworth, Edward Robert
Hoffman, Douglas Weir
Hofstra, Gerald Gerrit
Hutchinson, Thomas C
Kevan, Peter Graham
Lister, Earl Edward
Lougheed, Everett Charles

AGRICULTURAL & FOREST SCIENCES

Singh, Rama Shankar
Svoboda, Josef
Switzer, Clayton Macfie
Turk, Fateh (Frank) M

PRINCE EDWARD ISLAND
Carter, Martin Roger
Stewart, Jeffrey Grant

QUEBEC
Barthakur, Nayana N
Belloncik, Serge
Bertrand, Forest
Coffin, Harold Garth
Deschenes, Jean-Marc
Dufour, Jacques John
Lapp, Wayne Stanley
Lemieux, Claudel
Matte, J Jacques
Willemot, Claude

SASKATCHEWAN
Storey, Gary Garfield

OTHER COUNTRIES
Edelman, Marvin
Gregory, Peter
Kaltsikes, Pantouses John
Muniappan, Rangaswamy Naicker
Rossmiller, George Eddie
Sakamoto, Clarence M
Young, Bruce Arthur

Agronomy

ALABAMA
Brown, James Melton
Coleman, Tommy Lee
Donnelly, Edward Daniel
Ensminger, Leonard Elroy
Gill, William Robert
Haaland, Ronald L
Huluka, Gobena
Palmer, Robert Gerald
Peterson, Curtis Morris
Rajanna, Bettaiya
Rogers, Howard Topping
Russel, Darrell Arden
Sutherland, William Neil
Teyker, Robert Henry
Thomas, Winfred
Trouse, Albert Charles
Truelove, Bryan
Ward, Coleman Younger

ALASKA
Mitchell, William Warren
Wooding, Frank James

ARIZONA
Ahlgren, Henry Lawrence
Allen, Stephen Gregory
Briggs, Robert Eugene
Davis, Charles Homer
Day, Arden Dexter
Dennis, Robert E
Evans, Raymond Arthur
Feaster, Carl Vance
Fink, Dwayne Harold
Fisher, Warner Douglass
Glenn, Edward Perry
Guinn, Gene
Hawk, Virgil Brown
Jackson, Ernest Baker
Kimball, Bruce Arnold
Kneebone, William Robert
Loper, Gerald Milton
McAlister, Dean Ferdinand
Metcalfe, Darrel Seymour
Minch, Edwin Wilton
Morton, Howard LeRoy
Richardson, Grant Lee
Rohweder, Dwayne A
Rubis, David Daniel
Schonhorst, Melvin Herman
Smith, Dale
Smith, Steven Ellsworth
Upchurch, Robert Phillip
Webster, Orrin John
Wierenga, Peter J

ARKANSAS
Bartlett, Frank David
Beyrouty, Craig A
Bourland, Freddie Marshall
Collins, Frederick Clinton
Collister, Earl Harold
Frans, Robert Earl
Hinkle, Dale Albert
King, John William
Lavy, Terry Lee
Motes, Dennis Roy
Musick, Gerald Joe
Oosterhuis, Derrick M
Porter, Owen Archuel
Rutger, John Neil
Scifres, Charles Joel
Sensenig, Scott Allen
Smith, Roy Jefferson, Jr
Stutte, Charles A
Talbert, Ronald Edward
Timmermann, Dan, Jr
West, Charles Patrick

Wolf, Duane Carl

CALIFORNIA
Andres, Lloyd A
Ashton, Floyd Milton
Beard, Benjamin H
Bodman, Geoffrey Baldwin
Bradford, Willis Warren
Brecht, Patrick Ernest
Breidenbach, Rowland William
Brooks, William Hamilton
Caulder, Jerry Dale
Chu, Chang-Chi
Crawford, Robert Field
Davis, Larry Alan
Domingo, Wayne Elwin
Doner, Harvey Ervin
Embleton, Tom William
Epstein, Emanuel
Estilai, Ali
Ferguson, David B
Foster, Ken Wood
Gerik, James Stephen
Grantz, David Arthur
Hagan, William Leonard
Hall, Anthony Elmitt
Hesse, Walter Herman
Hile, Mahlon Malcolm Schallig
Hills, F Jackson
Isom, William Howard
Johnson, Carl William
Lambert, Royce Leone
Laude, Horton Meyer
Lewellen, Robert Thomas
Loomis, Robert Simpson
Lorenz, Oscar Anthony
Maas, Stephen Joseph
Mikkelsen, Duane Soren
Murphy, Alfred Henry
Norris, Robert Francis
Omid, Ahmad
Oster, James Donald
Peterson, Maurice Lewellen
Phillips, Donald Arthur
Plant, Richard E
Rawal, Kanti M
Ririe, David
Ritenour, Gary Lee
Robinson, Frank Ernest
Schaller, Charles William
Schieferstein, Robert Harold
Schmitz, George William
Shih, Ching-Yuan G
Sims, William Lynn
Smith, Paul Gordon
Teuber, Larry Ross
Thomas, Richard Sanborn
Thorup, James Tat
Trumble, John Thomas
Whitehead, Marvin Delbert
Williams, William Arnold
Worker, George F, Jr
Yuen, Wing
Zary, Keith Wilfred

COLORADO
Akeson, Walter Roy
Crumpacker, David Wilson
Cuany, Robin Louis
Danielson, Robert Eldon
Dickenson, Donald Dwight
Dittberner, Phillip Lynn
Fly, Claude Lee
Haus, Thilo Enoch
Hecker, Richard Jacob
Helmerick, Robert Howard
Keim, Wayne Franklin
Klute, Arnold
Larsen, Arnold Lewis
Laughlin, Charles William
Litzenberger, Samuel Cameron
Luebs, Ralph Edward
McGinnies, William Joseph
Mickelson, Rome H
Mortvedt, John Jacob
Norstadt, Fred A
Oldemeyer, Robert King
Quick, James S
Rumburg, Charles Buddy
Sabey, Burns Roy
Schweizer, Edward E
Shaw, Robert Blaine
Shawcroft, Roy Wayne
Siemer, Eugene Glen
Simantel, Gerald M
Soltanpour, Parviz Neil
Sullivan, Edward Francis
Tsuchiya, Takumi
Willis, Wayne O
Wood, Donald Roy
Youngman, Vern E
Zimdahl, Robert Lawrence

CONNECTICUT
Washko, Walter William

DELAWARE
Boyer, John Strickland
Green, Jerome
Hill, Gideon D
Jones, Edward Raymond
Mitchell, William H
Wittenbach, Vernon Arie
Wolf, Dale E

DISTRICT OF COLUMBIA
Allen, James Ralston
Carter, Lark Poland
Eisa, Hamdy Mahmoud
Islam, Nurul
Mayes, McKinley
Schmidt, Berlie Louis
Thomas, Walter Ivan
Wiggans, Samuel Claude

FLORIDA
Aldrich, Samuel Roy
Anderson, Stanley Robert
Barnett, Ronald David
Bennette, Jerry Mac
Boote, Kenneth Jay
Bowes, George Ernest
Boyd, Frederick Tilghman
Braids, Olin Capron
Brecke, Barry John
Calvert, David Victor
Cantliffe, Daniel James
Collins, Mary Elizabeth
Dudeck, Albert Eugene
Dunavin, Leonard Sypret, Jr
Fieldhouse, Donald John
Forbes, Richard Brainard
Gale, Paula M
Gammon, Nathan, Jr
Gardner, Franklin Pierce
Gilreath, James Preston
Gorbet, Daniel Wayne
Green, Victor Eugene, Jr
Gull, Dwain D
Guzman, Victor Lionel
Hammond, Luther Carlisle
Hanlon, Edward A
Haramaki, Chiko
Hinson, Kuell
Holder, David Gordon
Horner, Earl Stewart
Johnson, Walter Lee
Kidder, Gerald
Kretschmer, Albert Emil, Jr
Lutrick, Monroe Cornelius
McCloud, Darell Edison
McKently, Alexandra H
Mislevy, Paul
Monson, Warren Glenn
Nguyen, Khuong Ba
Nguyen, Ru
Nichols, Robert Loring
Norden, Allan James
Orsenigo, Joseph Reuter
Owens, Clarence Burgess
Peacock, Hugh Anthony
Pfahler, Paul Leighton
Prine, Gordon Madison
Probst, Albert Henry
Quesenberry, Kenneth Hays
Rechcigl, John E
Rich, Jimmy Ray
Roberts, Donald Ray
Rodgers, Earl Gilbert
Ruelke, Otto Charles
Sappenfield, William Paul
Scudder, Walter Tredwell
Snyder, George Heft
Stanley, Robert Lee, Jr
Stoffella, Peter Joseph
Teare, Iwan Dale
Thompson, Buford Dale
West, Sherlie Hill
Whitty, Elmo Benjamin
Wolf, Benjamin
Woofter, Harvey Darrell
Workman, Ralph Burns

GEORGIA
Anderson, Oscar Emmett
Ashley, Doyle Allen
Bailey, George William
Boerma, H Roger
Brown, Acton Richard
Brown, Ronald Harold
Burgoa, Benali
Burton, Glenn Willard
Cummins, David Gray
Dibb, David Walter
Douglas, Charles Francis
Dowler, Clyde Cecil
Duncan, Ronny Rush
Eastin, Emory Ford
Forbes, Ian
Frere, Maurice Herbert
Gaines, Tinsley Powell
Gascho, Gary John
Hammons, Ray Otto
Hardcastle, Willis Santford
Hoogenboom, Gerrit
Hoveland, Carl Soren
Jellum, Milton Delbert
Johnson, Jerry Wayne
Kays, Stanley J
Langdale, George Wilfred
Miller, John David
Mills, Harry Arvin
Mixon, Aubrey Clifton
Morey, Darrell Dorr
Murray, Calvin Clyde
Nyczepir, Andrew Peter
Steiner, Jean Louise
Summer, Malcolm Edward
Tan, Kim H

Todd, James Wyatt
Vencill, William Keith
Weaver, James B, Jr
Widstrom, Neil Wayne
Wilkinson, Robert Eugene
Wilkinson, Stanley R
Wofford, Irvin Mirle
Wood, Bruce Wade
Worley, Ray Edward
Younts, Sanford Eugene

HAWAII
Bartholomew, Duane P
Cushing, Robert Leavitt
De la Pena, Ramon Serrano
Fox, Robert Lee
Hepton, Anthony
Nishimoto, Roy Katsuto
Osgood, Robert Vernon
Rotar, Peter P
Silva, James Anthony
Thompson, John R
Whitney, Arthur Sheldon

IDAHO
Barker, LeRoy N
Callihan, Robert Harold
Carter, John Newton
Dwelle, Robert Bruce
Eberlein, Charlotte
Haderlie, Lloyd Conn
Hansen, Leon A
Kleinkopf, Gale Eugene
Lee, Gary Albert
Lehrsch, Gary Allen
Li, Guang Chao
Murray, Glen A
Myers, James Robert
O'Keeffe, Lawrence Eugene
Osgood, Charles Edgar
Pearson, Lorentz Clarence
Sojka, Robert E
Thill, Donald Cecil
Wright, James Louis

ILLINOIS
Alexander, Denton Eugene
Bernard, Richard Lawson
Bernardo, Rex Novero
Blair, Louis Curtis
Boast, Charles Warren
Brown, Lindsay Dietrich
Buettner, Mark Roland
Burger, Ambrose William
Dudley, John Wesley
Elkins, Donald Marcum
Fink, Rodney James
Flood, Brian Robert
Fuess, Frederick William, III
Garwood, Douglas Leon
Graffis, Don Warren
Hadley, Henry Hultman
Harper, James Eugene
Hauptmann, Randal Mark
Heermann, Ruben Martin
Heichel, Gary Harold
Hoeft, Robert Gene
Holt, Donald Alexander
Howell, Robert Wayne
Huck, Morris Glen
Jackobs, Joseph Alden
Jansen, Ivan John
Johnson, Richard Ray
Kapusta, George
King, S(anford) MacCallum
Kirby, Hilliard Walker
Knake, Ellery Louis
Kurtz, Lester Touby
Leffler, Harry Rex
McGlamery, Marshal Dean
Miller, Darrell Alvin
Myers, Oval, Jr
Nicholaides, John J, III
Nickell, Cecil D
Olsen, Farrel John
Olson, Kenneth Ray
Portis, Archie Ray, Jr
Portz, Herbert Lester
Reetz, Harold Frank, Jr
Robison, Norman Glenn
Roskamp, Gordon Keith
Slife, Fred Warren
Smiciklas, Kenneth Donald
Spencer, Jack T
Sprague, George Frederick
Steele, Leon
Stickler, Fred Charles
Stoller, Edward W
Thorne, Marlowe Driggs
Troyer, Alvah Forrest
Van Riper, Gordon Everett
Wilcox, Wesley Crain

INDIANA
Alvey, David Dale
Bauman, Thomas Trost
Baumgardner, Marion F
Bucholtz, Dennis Lee
Case, Vernon Wesley
Castleberry, Ron M
Christmas, Ellsworth P
Crane, Paul Levi
Daniel, William Hugh
Foley, Michael Edward

Agronomy (cont)

Glover, David Val
Hilst, Arvin Rudolph
Hoefer, Raymond H
Housley, Thomas Lee
Johannsen, Christian Jakob
Johnston, Cliff T
Kidd, Frank Alan
Kohnke, Helmut
Lechtenberg, Victor L
Mannering, Jerry Vincent
Mengel, David Bruce
Nelson, Werner Lind
Nielsen, Niels Christian
Nyquist, Wyman Ellsworth
Parka, Stanley John
Perkins, A Thomas
Peterson, John Booth
Reiss, William Dean
Robinson, Glenn Hugh
Ross, Merrill Arthur, Jr
Sammons, David James
Schreiber, Marvin Mandel
Shockey, William Lee
Slavik, Nelson Sigman
Steinhardt, Gary Carl
Swearingin, Marvin Laverne
Tsai, Chia-Yin
Vierling, Richard Anthony
Vorst, James J
Warren, George Frederick
Williams, James Lovon, Jr
Wolt, Jeffrey Duaine
Wright, William Leland
Zimmerman, Lester J

IOWA
Addink, Sylvan
Atkins, Richard Elton
Burris, Joseph Stephen
Buxton, Dwayne Revere
Carlson, Irving Theodore
Carlson, Richard Eugene
Cavalieri, Anthony Joseph, II
Colvin, Thomas Stuart
Cruse, Richard M
Dalton, Lonnie Gene
Duvick, Donald Nelson
Fawcett, Richard Steven
Frederick, Lloyd Randall
Frey, Nicholas Martin
George, John Ronald
Green, Detroy Edward
Hall, Charles Virdus
Hatfield, Jerry Lee
Horton, Robert, Jr
Hutchcroft, Charles Dennett
Kalton, Robert Rankin
Karlen, Douglas Lawrence
Lamkey, Kendall Raye
Logsdon, Sally D
Martinson, Charlie Anton
Moore, Kenneth J
Newlin, Owen Jay
Pearce, Robert Brent
Pesek, John Thomas, Jr
Roath, William Wesley
Shibles, Richard Marwood
Skrdla, Willis Howard
Stritzel, Joseph Andrew
Svec, Leroy Vernon
Thompson, Harvey E
Tomes, Dwight Travis
Voss, Regis D
Webb, John Raymond
Whigham, David Keith
Widrlechner, Mark Peter
Woolley, Donald Grant
Wych, Robert Dale

KANSAS
Casady, Alfred Jackson
Erickson, John Robert
Feltner, Kurt C
Follett, Roy Hunter
Ham, George Eldon
Harris, Pamela Ann
Harris, Wallace Wayne
Heyne, Elmer George
Hooker, Mark L
Jacobs, Hyde Spencer
Kirkham, M B
Mock, James Joseph
Murphy, Larry S
Nilson, Erick Bogseth
Norwood, Charles Arthur
Phillips, William Maurice
Posler, Gerry Lynn
Skidmore, Edward Lyman
Sorensen, Edgar Lavell
Stahlman, Phillip Wayne
Sunderman, Herbert D
Thien, Stephen John
Vanderlip, Richard L

KENTUCKY
Bitzer, Morris Jay
Collins, Michael
Egli, Dennis B
Gentry, Claude Edwin
Hiatt, Andrew Jackson
Kasperbauer, Michael J
Lacefield, Garry Dale

Massey, Herbert Fane, Jr
Mikulcik, John D
Niffenegger, Daniel Arvid
Poneleit, Charles Gustav
Sims, John Leonidas
Smiley, Jones Hazelwood
Taylor, Norman Linn
Taylor, Timothy H
Wells, Kenneth Lincoln
Yungbluth, Thomas Alan

LOUISIANA
Ayo, Donald Joseph
Board, James Ellery
Boquet, Donald James
Borsari, Bruno
Cohn, Marc Alan
Cuomo, Gregory Joseph
Davis, Johnny Henry
Dunigan, Edward P
Faw, Wade Farris
Foret, James A
Harlan, Jack Rodney
Henderson, Merlin Theodore
Hoff, Bert John
Martin, Freddie Anthony
Miller, Russell Lee
Owings, Addison Davis
Rogers, Robert Larry
Roussel, John S
Sedberry, Joseph E, Jr
Tipton, Kenneth Warren
Waddle, Bradford Avon

MAINE
Blackmon, Clinton Ralph
Chandler, Robert Flint, Jr
Langille, Alan Ralph

MARYLAND
Aycock, Marvin Kenneth, Jr
Bandel, Vernon Allan
Beste, Charles Edward
Bruns, Herbert Arnold
Campbell, Travis Austin
Christy, Alfred Lawrence
Coffman, Charles Benjamin
Decker, Alvin Morris, Jr
Devine, Thomas Edward
Dickerson, Chester T, Jr
Elgin, James H, Jr
Fertig, Stanford Newton
Graumann, Hugo Oswalt
Hanson, Angus Alexander
Hornick, Sharon B
Klingman, Dayton L
Lamp, William Owen
Leffel, Robert Cecil
McKee, Claude Gibbons
Marten, Gordon C
Moseman, John Gustav
Mulchi, Charles Lee
Nelson, Elton Glen
Ronningen, Thomas Spooner
Schnappinger, Melvin Gerhardt, Jr
Smith, David Harrison, Jr
Snow, Philip Anthony
Soto, Gerardo H
Steffens, George Louis
Walker, Alan Kent
Weber, Deane Fay
Weil, Raymond R
Woodstock, Lowell Willard
Young, Alvin L

MASSACHUSETTS
Epstein, Eliot

MICHIGAN
Foth, Henry Donald
Inselberg, Edgar
Johnston, Taylor Jimmie
Meggitt, William Fredric
Moline, Waldemar John
Payne, Kenyon Thomas
Penner, Donald
Robertson, G Philip
Schillinger, John Andrew, Jr
Southwick, Lawrence
Tesar, Milo B
Theurer, Jessop Clair
Thomas, John William
Vitosh, Maurice Lee
Watson, Andrew John

MINNESOTA
Andow, David A
Bauer, Marvin E
Behrens, Richard
Briggs, Rodney Arthur
Burnside, Orvin C
Cardwell, Vernon Bruce
Cheng, H(wei)-H(sien)
Comstock, Verne Edward
Crookston, Robert Kent
Davis, David Warren
Elling, Laddie Joe
Gengenbach, Burle Gene
Goodding, John Alan
Hess, Delbert Coy
Hueg, William Frederick, Jr
Jones, Robert James
Lambert, Jean William
Lebsock, Kenneth L

Lofgren, James R
Lueschen, William Everett
Miller, Gerald R
Munson, Robert Dean
Otto, Harley John
Pryor, Gordon Roy
Rehm, George W
Rines, Howard Wayne
Robinson, Robert George
Stucker, Robert Evan
Sullivan, Timothy Paul
Thompson, Roy Lloyd
Walgenbach, David D
Warnes, Dennis Daniel
Wedin, Walter F
Widner, Jimmy Newton

MISSISSIPPI
Andrews, Cecil Hunter
Bagley, Clyde Pattison
Baker, Ralph Stanley
Berry, Charles Dennis
Bowman, Donald Houts
Bunch, Harry Dean
Cole, Avean Wayne
Coleman, Otto Harvey
Creech, Roy G
Davis, Richard Richardson
Delouche, James Curtis
Edwards, Ned Carmack, Jr
Heatherly, Larry G
Hodges, Harry Franklin
Kincade, Robert Tyrus
Knight, William Eric
Kurtz, Mark Edward
Lancaster, James D
Lee, Charles Richard
McWhorter, Chester Gray
Manning, Cleo Willard
Maxwell, James Donald
Peterson, Harold LeRoy
Ranney, Carleton David
Scott, Gene E
Spiers, James Monroe
Triplett, Glover Brown, Jr
Vadhwa, Om Parkash
Watson, Clarence Ellis, Jr
Watson, Vance H
Wise, Louis Neal

MISSOURI
Aldrich, Richard John
Anand, Satish Chandra
Bader, Kenneth L
Beckett, Jack Brown
Blevins, Dale Glenn
Carlson, Wayne C
Cavanah, Lloyd (Earl)
Chowdhury, Ikbalur Rashid
Davidson, Steve Edwin
Donald, William Waldie
Einhellig, Frank Arnold
Fletchall, Oscar Hale
Graham, James Carl
Gustafson, John Perry
Helsel, Zane Roger
Houghton, John M
Justus, Norman Edward
Kerr, Harold Delbert
Khan, Adam
Miles, Randall Jay
Mitchell, Roger L
Nelson, Curtis Jerome
Paul, Kamalendu Bikash
Peters, Elroy John
Poehlman, John Milton
Radke, Rodney Owen
Richards, Graydon Edward
Schumacher, Richard William
Sleper, David Allen
Tai, William
Wang, Maw Shiu
Wilson, Clyde Livingston
Woodruff, Clarence Merrill
Zech, Arthur Conrad
Zuber, Marcus Stanley

MONTANA
Aase, Jan Kristian
Blake, Tom
Brown, Jarvis Howard
Ditterline, Raymond Lee
Jackson, Grant D
McCoy, Thomas Joseph
McGuire, Charles Francis
McNeal, Francis H
Martin, John Munson

NEBRASKA
Baltensperger, David Dwight
Blad, Blaine L
Compton, William A
Doran, John Walsh
Eastin, Jerry Dean
Eastin, John A
Flowerday, Albert Dale
Francis, Charles Andrew
Frolik, Elvin Frank
Furrer, John D
Gorz, Herman Jacob
Hanway, Donald Grant
Higley, Leon George
Johnson, Virgil Allen
Knox, Ellis Gilbert

Kolade, Alabi E
Lewis, David Thomas
Louda, Svata Mary
McGill, David Park
Maranville, Jerry Wesley
Mason, Stephen Carl
Massengale, M A
Miller, Willie
Moser, Lowell E
Nelson, Darrell Wayne
Nelson, Lenis Alton
Nichols, James T
Power, James Francis
Raun, Earle Spangler
Roeth, Frederick Warren
Ross, William Max
Rumbaugh, Melvin Dale
Schmidt, John Wesley
Specht, James Eugene
Swartzendruber, Dale
Williams, James Henry, Jr

NEVADA
Gilbert, Dewayne Everett
Jensen, Edwin Harry
Leedy, Clark D

NEW JERSEY
Avissar, Roni
Gruenhagen, Richard Dale
Ilnicki, Richard Demetry
Justin, James Robert
Marrese, Richard John
Meade, John Arthur
Sprague, Milton Alan
Wilbur, Robert Daniel

NEW MEXICO
Baltensperger, Arden Albert
Banks, Philip Alan
Barnes, Carl Eldon
Chen, David J
Daugherty, LeRoy Arthur
Davis, Dick D
Finkner, Morris Dale
Finkner, Ralph Eugene
Fowler, James Lowell
Gould, Walter Leonard
Malm, Norman R
Melton, Billy Alexander, Jr
Phillips, Gregory Conrad

NEW YORK
Archimovich, Alexander S
Bergstrom, Gary Carlton
Boehle, John, Jr
Drosdoff, Matthew
Fick, Gary Warren
Goodale, Douglas M
Grunes, David Leon
Kelly, William Cary
Kennedy, Wilbert Keith
Leopold, Aldo Carl
Linscott, Dean L
Lucey, Robert Francis
Minotti, Peter Lee
Murphy, Royse Peak
Obendorf, Ralph Louis
Pardee, William Durley
Plate, Henry
Pratt, Arthur John
Reid, William Shaw
Robinson, Terence Lee
Sandsted, Roger France
Sirois, David Leon
Sorrells, Mark Earl
Szabo, Steve Stanley
Thomas, Everett Dake
Topoleski, Leonard Daniel
Vittum, Morrill Thayer
Wright, Madison Johnston

NORTH CAROLINA
Blake, Carl Thomas
Carter, Thomas Edward, Jr
Chamblee, Douglas Scales
Clapp, John Garland, Jr
Coble, Harold Dean
Collins, Henry A
Cook, Maurice Gayle
Corbin, Frederick Thomas
Cowett, Everett R
Ellis, John Fletcher
Fike, William Thomas, Jr
Flint, Elizabeth Parker
Friedrich, James Wayne
Gilbert, William Best
Gross, Harry Douglass
Hay, Russell Earl, Jr
Hebert, Teddy T
Jones, Guy Langston
Knauft, David A
Lewis, William Mason
McCollum, Robert Edmund
McLaughlin, Foil William
Miner, Gordon Stanley
Oblinger, Diana Gelene
O'Neal, Thomas Denny
Patterson, Robert Preston
Rogerson, Asa Benjamin
Thompson, Donald Loraine
Timothy, David Harry
Webb, Burleigh C
Wernsman, Earl Allen

AGRICULTURAL & FOREST SCIENCES / 5

Wilson, Lorenzo George
Worsham, Arch Douglas
Wynne, Johnny Calvin
York, Alan Clarence
Zarnstorff, Mark Edward
Zublena, Joseph Peter

NORTH DAKOTA
Bauer, Armand
Carter, Jack Franklin
Deckard, Edward Lee
Doney, Devon Lyle
Foster, Albert Earl
Frank, Albert Bernard
Halvorson, Ardell David
Halvorson, Gary Alfred
Hammond, James Jacob
Hofmann, Lenat
Lund, Hartvig Roald
Meyer, Dwain Wilber
Nalewaja, John Dennis
Seiler, Gerald Joseph
Smith, Glenn Sanborn
Weiss, Michael John
Whited, Dean Allen

OHIO
Bendixen, Leo E
Calhoun, Frank Gilbert
Cooper, Richard Lee
Eckert, Donald James
Findley, William Ray, Jr
Gauntt, William Amor
Haghiri, Faz
Henderlong, Paul Robert
Herr, Donald Edward
Hock, Arthur George
Hurto, Kirk Allen
Jackobs, John Joseph
Lafever, Howard N
Lal, Rattan
Liu, Ting-Ting Y
Long, John A
McCracken, John David
Miller, Frederick Powell
Niemczyk, Harry D
Parsons, John Lawrence
Schmidt, Walter Harold
Schwer, Joseph Francis
Smeck, Neil Edward
Streeter, John Gemmil
Stroube, Edward W
Teater, Robert Woodson
Waldron, Acie Chandler
Watson, Maurice E
Wiebold, William John
Zimmerman, Tommy Lynn

OKLAHOMA
Ahring, Robert M
Bates, Richard Pierce
Croy, Lavoy I
Edwards, Lewis Hiram
Greer, Howard A L
Huffine, Wayne Winfield
Klatt, Arthur Raymond
McMurphy, Wilfred E
Murray, Jay Clarence
Nofziger, David Lynn
Powell, Jerrel B
Reeves, Homer Eugene
Santelmann, Paul William
Stritzke, Jimmy Franklin
Tucker, Billy Bob
Verhalen, Laval M(athias)
Weibel, Dale Eldon

OREGON
Albrecht, Stephan LaRowe
Appleby, Arnold Pierce
Brun, William Alexander
Calhoun, Wheeler, Jr
Chilcote, David Owen
Crabtree, Garvin (Dudley)
Dade, Philip Eugene
Fendall, Roger K
Foote, Wilson Hoover
Frakes, Rodney Vance
Hampton, Richard Owen
Hannaway, David Bryon
Johnson, Malcolm Julius
Kronstad, Warren Ervind
Ladd, Sheldon Lane
Lee, William Orvid
Lund, Steve
Miller, Stanley Frank
Moss, Dale Nelson
Neiland, Bonita J
Rykbost, Kenneth Albert
Schweitzer, Leland Ray
Simonson, Gerald Herman
Smiley, Richard Wayne
Vomocil, James Arthur
Young, J Lowell
Yungen, John A

PENNSYLVANIA
Baker, Dale E
Baylor, John E
Berg, Clyde C
Bergman, Ernest L
Bosshart, Robert Perry
Cleveland, Richard Warren
Cole, Richard H

Duich, Joseph M
Fales, Steven Lewis
Fritton, Daniel Dale
Hall, Jon K
Harrington, Joseph Donald
Hartwig, Nathan Leroy
Hinish, Wilmer Wayne
Janke, Rhonda Rae
Johnson, Melvin Walter, Jr
Jung, Gerald Alvin
Kendall, William Anderson
Knievel, Daniel Paul
Krueger, Charles Robert
McKee, Guy William
Marriott, Lawrence Frederick
Risius, Marvin Leroy
Shenk, John Stoner
Shipp, Raymond Francis
Sprague, Howard Bennett
Waddington, Donald Van Pelt
Watschke, Thomas Lee

RHODE ISLAND
Duff, Dale Thomas
Hull, Richard James
Skogley, Conrad Richard
Wakefield, Robert Chester

SOUTH CAROLINA
Alexander, Paul Marion
Alston, Jimmy Albert
Camberato, James John
Chapman, Stephen R
Craddock, Garnet Roy
Deal, Elwyn Ernest
Franklin, Ralph E
Gossett, Billy Joe
Harvey, Lawrence Harmon
Jones, Ulysses Simpson, Jr
Jutras, Michel Wilfrid
Kittrell, Benjamin Upchurch
McClain, Eugene Fredrick
Nolan, Clifford N
Quisenberry, Virgil L
Shipe, Emerson Russell
Singh, Raghbir
Stringer, William Clayton
Wallace, Susan Ulmer

SOUTH DAKOTA
Ellsbury, Michael M
Kantack, Benjamin H
Moore, Raymond A
Reeves, Dale Leslie
Wells, Darrell Gibson

TENNESSEE
Bell, Frank F
Bond, Andrew
Boswell, Fred Carlen
Callahan, Lloyd Milton
Eskew, David Lewis
Ewing, John Arthur
Flanagan, Theodore Ross
Foutch, Harley Wayne
Fribourg, Henry August
Graveel, John Gerard
Hayes, Robert M
Josephson, Leonard Melvin
Krueger, William Arthur
Large, Richard L
Luxmoore, Robert John
Naylor, Gerald Wayne
Ramey, Harmon Hobson, Jr
Reich, Vernon Henry
Reynolds, John Horace
Skold, Laurence Nelson
Springer, Maxwell Elsworth
West, Dennis R

TEXAS
Arnold, James Darrell
Beard, James B
Bennett, William Frederick
Bovey, Rodney William
Brigham, Raymond Dale
Calub, Alfonso deGuzman
Chandler, James Michael
Collier, Jesse Wilton
Cook, Charles Garland
Craigmiles, Julian Pryor
Duble, Richard Lee
Evans, Raeford G
Gantz, Ralph Lee
Gavande, Sampat A
Golden, Dadigamuwage Chandrasiri
Hagstrom, Gerow Richard
Hauser, Victor La Vern
Ho, Clara Lin
Holt, Ethan Cleddy
Hopper, Norman Wayne
Hossner, Lloyd Richard
Jones, Ordie Reginal
Lingle, Sarah Elizabeth
McBee, George Gilbert
McDonald, Lynn Dale
McIlrath, William Oliver
Matches, Arthur Gerald
Maunder, A Bruce
Niles, George Alva
Osler, Robert Donald
Palmer, Rupert Dewitt
Reid, Donald J
Rouquette, Francis Marion, Jr

Runge, Edward C A
Sij, John William
Smith, Gerald Ray
Smith, Olin Dail
Staten, Raymond Dale
Stewart, Bobby Alton
Swakon, Doreen H D
Taylor, Richard Melvin
Trogdon, William Oren
Vietor, Donald Melvin
Wendt, Charles William
Wiegand, Craig Loren
Wiese, Allen F
Wilding, Lawrence Paul
Winter, Steven Ray
Young, Arthur Wesley

UTAH
Albrechtsen, Rulon S
Anderson, Jay LaMar
Asay, Kay Harris
Belcher, Bascom Anthony
Dewey, Wade G
Hamson, Alvin Russell
Horrocks, Rodney Dwain
Jeffery, Larry S
Openshaw, Martin David
Rasmussen, V Philip, Jr
Robison, Laren R
Thorup, Richard M
Tilton, Varien Russell

VERMONT
Clark, Neri Anthony
Matthews, David Livingston
Murphy, William Michael
Wood, Glen Meredith

VIRGINIA
Albrecht, Herbert Richard
Allen, Vivien Gore
Alley, Marcus M
Bingham, Samuel Wayne
Buckardt, Henry Lloyd
Burns, Allan Fielding
Buss, Glenn Richard
Carson, Eugene Watson, Jr
Coartney, James S
Conn, Richard Leslie
De Datta, Surajit Kumar
Dorschner, Kenneth Peter
Griffith, William Kirk
Harrison, Robert Louis
Hatzios, Kriton Kleanthis
Kroontje, Wybe
Lewis, Cornelius Crawford
O'Neill, Eileen Jane
Parrish, David Joe
Sagaral, Erasmo G (Ras)
Schmidt, Richard Edward
Smith, Townsend Jackson
Taylor, Lincoln Homer
Thiel, Thomas J
Wolf, Dale Duane

WASHINGTON
Allan, Robert Emerson
Bertramson, Bertram Rodney
Boyd, Charles Curtis
Chevalier, Peggy
Comes, Richard Durward
Dawson, Jean Howard
Ensign, Ronald D
Evans, David W
George, Donald Wayne
Henderson, Lawrence J
Hiller, Larry Keith
Johnson, Corwin McGillivray
Kleinhofs, Andris
Lauer, David Allan
Leng, Earl Reece
Maguire, James Dale
Miller, Dwane Gene
Morrison, Kenneth Jess
Morrow, Larry Alan
Ogg, Alex Grant, Jr
Papendick, Robert I
Peabody, Dwight Van Dorn, Jr
Raese, John Thomas
Rasmussen, Lowell W
Schrader, Lawrence Edwin
Schwendiman, John Leo
Swan, Dean George
Thornton, Robert Kim
Warner, Robert Lewis
Zimmermann, Charles Edward

WEST VIRGINIA
Baker, Barton Scofield
Balasko, John Allan
Singh, Rabindar Nath
Voigt, Paul Warren
Welker, William V, Jr

WISCONSIN
Albrecht, Kenneth Adrian
Andrew, Robert Harry
Barnes, Robert F
Brickbauer, Elwood Arthur
Bula, Raymond J
Cao, Weixing
Carter, Paul R
Casler, Michael Darwin
Daie, Jaleh

Delorit, Richard John
Doersch, Ronald Ernest
Doll, Jerry Dennis
Drolsom, Paul Newell
Duke, Stanley Houston
Forsberg, Robert Arnold
Goodman, Robert Merwin
Greub, Louis John
Gritton, Earl Thomas
Hagedorn, Donald James
Harpstead, Milo I
Harvey, Robert Gordon, Jr
Higgs, Roger L
Ihrke, Charles Albert
Jackson, Marion Leroy
Miller, Robert W
Oplinger, Edward Scott
Paulson, William H
Pawlisch, Paul E
Peterson, David Maurice
Strohm, Jerry Lee
Tracy, William Francis

WYOMING
Alley, Harold Pugmire
Hart, Richard Harold
Humburg, Neil Edward
Koch, David William
Kolp, Bernard J
Miller, Stephen Douglas
Schuman, Gerald E

PUERTO RICO
Rodriguez, Jorge Luis
Salas-Quintana, Salvador

ALBERTA
Andrew, William Treleaven
Bentley, C Fred
Blackshaw, Robert Earl
Briggs, Keith Glyn
Coen, Gerald Marvin
Harper, Alexander Maitland
Harris, Peter
McAndrew, David Wayne
McElgunn, James Douglas
Major, David John
Muendel, Hans-Henning
Vanden Born, William Henry
Walton, Peter Dawson
Wilson, Donald Benjamin

BRITISH COLUMBIA
Ashford, Ross
Holl, Frederick Brian
Stout, Darryl Glen

MANITOBA
Grant, Cynthia Ann
Hobbs, James Arthur
McGinnis, Robert Cameron
McVetty, Peter Barclay Edgar
Sheppard, Stephen Charles
Stobbe, Elmer Henry
Vessey, Joseph Kevin

NEWFOUNDLAND
McKenzie, David Bruce

NOVA SCOTIA
Bubar, John Stephen
Faught, John Brian
Warman, Philip Robert

ONTARIO
Alex, Jack Franklin
Buzzell, Richard Irving
Campbell, Kenneth Wilford
Chang, Fa Yan
Dhesi, Nazar Singh
Fedak, George
Ho, Keh Ming
Hope, Hugh Johnson
Horton, Roger Francis
Hume, David John
Nonnecke, Ib Libner
O'Toole, James J
Sahota, Ajit Singh
Saidak, Walter John
St Pierre, Jean Claude
Sampson, Dexter Reid
Singh, Surinder Shah
Stoskopf, N C
Switzer, Clayton Macfie
Tan, Chin Sheng
Thomas, Ronald Leslie
Warren, Francis Shirley
Zilkey, Bryan Frederick

PRINCE EDWARD ISLAND
Carter, Martin Roger
Choo, Thin-Meiw
Christie, Bertram Rodney
Kunelius, Heikki Tapani
MacLeod, John Alexander
MacLeod, LLoyd Beck
White, Ronald Paul, Sr

QUEBEC
Angers, Denis Arthur
Chung, Young Sup
Gasser, Heinz
Gervais, Paul
Klinck, Harold Rutherford

Agronomy (cont)

Lemieux, Claudel
Matte, J Jacques
Watson, Alan Kemball

SASKATCHEWAN
Austenson, Herman Milton
Cohen, Roger D H
Coupland, Robert Thomas
Hay, James Robert
Rossnagel, Brian Gordon
Scoles, Graham John
Slinkard, Alfred Eugene
Tanino, Karen Kikumi
Waddington, John
Zilke, Samuel

OTHER COUNTRIES
Britten, Edward James
Chomchalow, Narong
Couto, Walter
Hittle, Carl Nelson
Medina, Ernesto Antonio
Muchovej, James John
Rajaram, Sanjaya
Sanchez, Pedro Antonio
Simkins, Charles Abraham

Animal Husbandry

ALABAMA
Harris, Ralph Rogers
Kuhlers, Daryl Lynn
McGuire, John Albert
Moss, Buelon Rexford
Patterson, Troy B
Sewell, Raymond F
Sharma, Udhishtra Deva
Wilkening, Marvin C

ALASKA
Brundage, Arthur Lain
Husby, Fredric Martin

ARIZONA
Brown, William Hedrick
Dickinson, Frank N
Hale, William Harris
Huber, John Talmage
Murdock, Fenoi R
Rice, Richard W
Schuh, James Donald
Swingle, Roy Spencer
Taysom, Elvin David
Theurer, Clark Brent

ARKANSAS
Andrews, Luther David
Brown, A Hayden, Jr
Brunson, Clayton (Cody)
Gyles, Nicholas Roy
Harris, Grover Cleveland, Jr
Kellogg, David Wayne
McGilliard, A Dare
Noland, Paul Robert
Perry, Tilden Wayne
Stephenson, Edward Luther
Tave, Douglas

CALIFORNIA
Abenes, Fiorello Bigornia
Adams, Thomas Edwards
Anderson, Gary Bruce
Anderson, Russell K
Bauer, Robert Steven
Bradford, G Eric
Carroll, Floyd Dale
Christensen, Allen Clare
Dollar, Alexander M
Durrant, Barbara Susan
Gall, Graham A E
Garrett, William Norbert
Halloran, Hobart
Hedgecock, Dennis
Hixson, Floyd Marcus
Jacobs, John Allen
Kennington, Mack Humpherys
Laben, Robert Cochrane
Lofgreen, Glen Pehr
Prokop, Michael Joseph
Rahlmann, Donald Frederick
Siegrist, Jacob C
Sinha, Yagya Nand
Torell, Donald Theodore
Vohra, Pran Nath
Waterhouse, Howard N
Weir, William Carl

COLORADO
Johnson, Donald Eugene
Raun, Arthur Phillip
Voss, James Leo
Ward, Gerald Madison

CONNECTICUT
Brown, Lynn Ranney
Cowan, William Allen
Gaunya, William Stephen
Schake, Lowell Martin
Stake, Paul Erik

DELAWARE
Haenlein, George Friedrich Wilhelm
Hesseltine, Wilbur R
Link, Bernard Alvin

DISTRICT OF COLUMBIA
Ampy, Franklin
Cooper, George Everett
Dua, Prem Nath
Hardison, Wesley Aurel
Plowman, Ronald Dean
Wilson, William Mark Dunlop

FLORIDA
Ammerman, Clarence Bailey
Baker, Frank Sloan, Jr
Carpenter, James Woodford
Chapman, Herbert L, Jr
Combs, George Ernest
Conrad, Joseph H
Cox, Dennis Henry
Davis, George Kelso
Evans, Lee E
Gaunt, Stanley Newkirk
Hembry, Foster Glen
Koger, Marvin
Loggins, Phillip Edwards
Loosli, John Kasper
McDowell, Lee Russell
Sanford, Malcolm Thomas
Shirley, Ray Louis
Sosa, Omelio, Jr
Thomason, David Morton
Wakeman, Donald Lee

GEORGIA
Benyshek, Larry L
Cullison, Arthur Edison
Daniel, O'Dell G
Flatt, William Perry
Hill, Gary Martin
Mabry, John William
Neathery, Milton White
Newton, George Larry
Powell, George Wythe
Seerley, Robert Wayne

HAWAII
Weems, Charles William
Wyban, James A

IDAHO
Bull, Richard C

ILLINOIS
Arthur, Robert David
Carr, Tommy Russell
Clark, Jimmy Howard
Corbin, James Edward
Easter, Robert Arnold
Garrigus, Upson Stanley
Gomes, Wayne Reginald
Grossman, Michael
Harrison, Paul C
Hausler, Carl Louis
Hollis, Gilbert Ray
Hutjens, Michael Francis
Jensen, Aldon Homan
Klay, Robert Frank
Lewis, John Morgan
Lodge, James Robert
Madsen, Fred Christian
Monson, William Joye
Neagle, Lyle H
Pothoven, Marvin Arlo
Ricketts, Gary Eugene
Robb, Thomas Wilbern
Romack, Frank Eldon
Shanks, Roger D
Spahr, Sidney Louis
Vetter, Richard L
Vint, Larry Francis
Wagner, William Charles

INDIANA
Baumgardt, Billy Ray
Boston, Andrew Chester
Brown, Herbert
Forsyth, Dale Marvin
Gard, Don Irvin
Harmon, Bud Gene
Jones, Hobart Wayne
Krider, Jake Luther
Morter, Raymond Lione
Rogler, John Charles
Stewart, Terry Sanford
Stob, Martin
Thomas, Elvin Elbert
Wolfrom, Glen Wallace

IOWA
Acker, Duane Calvin
Brackelsberg, Paul O
Ewan, Richard Colin
Ewing, Solon Alexander
Freeman, Albert Eugene
Harbaugh, Daniel David
Harville, David Arthur
Hodson, Harold H, Jr
Holden, Palmer Joseph
Jurgens, Marshall Herman
Kenealy, Michael Douglas
Mente, Glen Allen
Morrical, Daniel Gene
Parrish, Frederick Charles, Jr
Rohlf, Marvin Euguene
Sell, Jerry Lee
Stevermer, Emmett J
Touchberry, Robert Walton
Warner, Donald R
Willham, Richard Lewis
Wunder, William W
Young, Jerry Wesley
Zimmerman, Dean R

KANSAS
Brethour, John Raymond
Craig, James Verne
Kropf, Donald Harris
Schalles, Robert R
Smith, Edgar Fitzhugh

KENTUCKY
Buck, Charles Frank
Hays, Virgil Wilford
Knapp, Frederick Whiton
Lane, Gary (Thomas)
Mitchell, George Ernest, Jr
Ramsay, John Martin
Stephens, Noel, Jr
Thrift, Frederick Aaron
Tucker, Ray Edwin
Varney, William York
Whiteker, McElwyn D

LOUISIANA
Cason, James Lee
Falcon, Carroll James
Franke, Donald Edward
Gassie, Edward William
Robertson, George Leven
Rusoff, Louis Leon

MAINE
Barton, Barbara Ann
Dickey, Howard Chester
Les, Edwin Paul
Musgrave, Stanley Dean

MARYLAND
Alderson, Norris Eugene
Berry, Bradford William
Engster, Henry Martin
Graber, George
Moody, Edward Grant
Norman, Howard Duane
Olson, Leonard Carl
Powell, Rex Lynn
Pursel, Vernon George
Rubin, Max
Smith, F Harrell
Stricklin, William Ray
Terrill, Clair Elman
Vandersall, John Henry
Wildt, David Edwin

MASSACHUSETTS
Borton, Anthony
Lyford, Sidney John, Jr
McColl, James Renfrew

MICHIGAN
Aulerich, Richard J
Cook, Robert Merold
Deans, Robert Jack
Emery, Roy Saltsman
Henneman, Harold Albert
Hoefer, Jacob A
Kratzer, D Dal
Lawson, David Michael
McGilliard, Lon Dee
Magee, William Thomas
Mao, Ivan Ling
Miller, Curtis C
Miller, Elwyn Ritter
Nelson, Ronald Harvey
Ritchie, Harlan
Rust, Steven Ronald
Stinson, Al Worth
Subramanian, Marappa G
Tucker, Herbert Allen

MINNESOTA
Boylan, William J
Christians, Charles J
Conlin, Bernard Joseph
Goodrich, Richard Douglas
Hansen, Leslie Bennett
Hanson, Lester Eugene
Jordan, Robert Manseau
Kromann, Rodney P
Meiske, Jay C
Miller, Kenneth Philip
Oelberg, Thomas Jonathon
Pettigrew, James Eugene, Jr
Rempel, William Ewert
Rust, Joseph William
Young, Charles Wesley

MISSISSIPPI
Bagley, Clyde Pattison
Baker, Bryan, Jr
Fuquay, John Wade
Lindley, Charles Edward
Putnam, Paul A
Thomas, Charles Hill
Tomlinson, James Everett

MISSOURI
Anderson, Ralph Robert
Day, Billy Neil
Hill, Jack Filson
Lasley, John Foster
Merilan, Charles Preston
Meyer, William Ellis
Padgitt, Dennis Darrell
Sewell, Homer B
Smith, Keith James
Smith, Nathan Elbert
Stephenson, Alfred Benjamin
Thrasher, George W
Williamson, James Lawrence

MONTANA
Burris, Martin Joe
Kress, Donnie Duane
Newman, Clarence Walter

NEBRASKA
Aberle, Elton D
Adams, Charles Henry
Ahlschwede, William T
Clanton, Donald Cather
Cundiff, Larry Verl
Dearborn, Delwyn D
Doane, Ted H
Ford, Johny Joe
Fritschen, Robert David
Gregory, Keith Edward
Guyer, Paul Quentin
Hunsley, Roger Eugene
Koch, Robert Milton
Larson, Larry Lee
Lewis, Austin James
McGaugh, John Wesley
Nielsen, Merlyn Keith
Omtvedt, Irvin T
Peo, Ernest Ramy, Jr
Van Vleck, Lloyd Dale
Ward, John K
Weichenthal, Burton Arthur
Young, Lawrence Dale

NEVADA
Cunha, Tony Joseph

NEW HAMPSHIRE
Skoglund, Winthrop Charles

NEW JERSEY
Cox, James Lee
Grover, Gary James
Horton, Glyn Michael John
Ingle, Donald Lee
Shor, Aaron Louis

NEW MEXICO
Andrew, James F
Francis, David Wesson
Holland, Lewis
Nelson, Arnold Bernard

NEW YORK
Ainslie, Harry Robert
Auclair, Walter
Bauman, Dale E
Blake, Robert Wesley
Chase, Larry Eugene
Elliot, John Murray
Foote, Robert Hutchinson
Hintz, Harold Franklin
Hogue, Douglas Emerson
Merrill, William George
Oltenacu, Elizabeth Allison Branford
Pollak, Emil John
Porter, Gilbert Harris
Poston, Hugh Arthur
Randy, Harry Anthony
Reimers, Thomas John
Rowoth, Olin Arthur
Schano, Edward Arthur
Stockbridge, Robert R
Stouffer, James Ray
Thomas, Everett Dake
Turk, Kenneth Leroy
VanSoest, Peter John
Warner, Richard G
Wellington, George Harvey

NORTH CAROLINA
Allen, Neil Keith
Burns, Joseph Charles
Clawson, Albert J
Colvard, Dean Wallace
Cornwell, John Calhoun
Glazener, Edward Walker
Gonder, Eric Charles
Hyatt, George, Jr
Johnson, George Andrew
Johnson, William Lawrence
Jones, James Robert
Knott, Fred Nelson
Leatherwood, James M
McDaniel, Benjamin Thomas
McDowell, Robert E, Jr
Porterfield, Ira Deward
Poulton, Bruce R
Vandemark, Noland Leroy
Whitlow, Lon Weidner
Young, Robert John

AGRICULTURAL & FOREST SCIENCES / 7

NORTH DAKOTA
Dinusson, William Erling
Edgerly, Charles George Morgan
Harrold, Robert Lee
Johnson, LaDon Jerome
Karn, James Frederick

OHIO
Anderson, Marlin Dean
Cline, Jack Henry
Danke, Richard John
Gauntt, William Amor
Harvey, Walter Robert
Johnson, George Robert
Kottman, Roy Milton
Ludwick, Thomas Murrell
Mahan, Donald C
Mattingly, Steele F
Murray, Finnie Ardrey, Jr
Plimpton, Rodney F, Jr
Prince, Terry Jamison
Schmidt, Glen Henry
Tyznik, William John
Wilson, George Rodger
Wilson, Richard Ferrin

OKLAHOMA
Browning, Charles Benton
Buchanan, David Shane
Gleaves, Earl William
Luce, William Glenn
Martin, Jerry, Jr
Owens, Fredric Newell
Tillman, Allen Douglas
Totusek, Robert
Turman, Elbert Jerome
Walters, Lowell Eugene
Wells, Milton Ernest
Whatley, James Arnold
Whiteman, Joe V

OREGON
Arscott, George Henry
Cheeke, Peter Robert
England, David Charles
Landers, John Herbert, Jr
Miller, Stanley Frank
Skovlin, Jon Matthew
Weswig, Paul Henry

PENNSYLVANIA
Adams, Richard Sanford
Chalupa, William Victor
Cowan, Robert Lee
Dreibelbis, John Adam
Eggert, Robert Glenn
Elkins, William L
Hagen, Daniel Russell
Hedde, Richard Duane
Mashaly, Magdi Mohamed
Morrow, David Austin, III
Muller, Lawrence Dean
Sherritt, Grant Wilson
Varga, Gabriella Anne
Wangsness, Paul Jerome
Wilson, Lowell L
Ziegler, John Henry, Jr

RHODE ISLAND
Nippo, Murn Marcus

SOUTH CAROLINA
Birrenkott, Glenn Peter, Jr
Dickey, Joseph Freeman
Edwards, Robert Lee
Lee, Daniel Dixon, Jr
Thompson, Carl Eugene

SOUTH DAKOTA
Briggs, Hilton Marshall
Britzman, Darwin Gene
Bush, Leon F
Embry, Lawrence Bryan
Emerick, Royce Jasper
Romans, John Richard
Schingoethe, David John
Wahlstrom, Richard Carl

TENNESSEE
Bletner, James Karl
Chamberlain, Charles Calvin
Griffin, Sumner Albert
Hitchcock, John Paul
Huggins, James Anthony
McFee, Alfred Frank
Miller, James Kincheloe
Scott, Mack Tommie
Shrode, Robert Ray

TEXAS
Bassett, James Wilbur
Butler, Ogbourne Duke, Jr
Byers, Floyd Michael
Calhoun, Millard Clayton
Cartwright, Thomas Campbell
Cassard, Daniel W
Cole, Noel Andy
Crenshaw, David Brooks
Dahl, Billie Eugene
Davis, Gordon Wayne
DuBose, Leo Edwin
Godbee, Richard Greene, II
Greathouse, Terrence Ray
Hanlon, Roger Thomas

Harms, Paul G
Horton, Otis Howard
Hudson, Frank Alden
Hutcheson, David Paul
Krueger, Willie Frederick
Lane, Alfred Glen
Lippke, Hagen
Mason, Tim Robert
Menzies, Carl Stephen
Preston, Rodney LeRoy
Self, Hazzle Layfette
Sherrod, Lloyd B
Sudweeks, Earl Max
Thompson, Granville Berry

UTAH
Arave, Clive W
Bennett, James Austin
Gardner, Robert Wayne
Johnston, Norman Paul
Lamb, Robert Cardon
Matthews, Doyle Jensen
Park, Robert Lynn
Richards, Clyde Rich
Thomas, Don Wylie
Wallentine, Max V
Wilcox, Clifford LaVar

VERMONT
Balch, Donald James
Dowe, Thomas Whitfield
Foss, Donald C
Slack, Nelson Hosking
Smith, Albert Matthews

VIRGINIA
Benson, Robert Haynes
Bovard, Kenly Paul
Carr, Scott Bligh
Eller, Arthur L, Jr
Etgen, William M
Evans, Joseph Liston
Fontenot, Joseph Paul
Kelly, Robert Frank
Kornegay, Ervin Thaddeus
Pearson, Ronald Earl
Webb, Kenneth Emerson, Jr
White, John Marvin
Woods, Walter Ralph

WASHINGTON
Froseth, John Allen
Hillers, Joe Karl
Kincaid, Ronald Lee

WEST VIRGINIA
Cochrane, Robert Lowe
Dunbar, Robert Standish, Jr
Horvath, Donald James
Inskeep, Emmett Keith
McClung, Marvin Richard
Thomas, Roy Orlando
Welch, James Alexander
Zinn, Dale Wendel

WISCONSIN
Dierschke, Donald Joe
Grummer, Robert Henry
Heidenreich, Charles John
Howard, W Terry
Leibbrandt, Vernon Dean
Miller, Paul Dean
Niedermeier, Robert Paul
Rutledge, Jackie Joe
Thomas, David Lee
Walton, Robert Eugene
Wittwer, Leland S

WYOMING
Botkin, Merwin P
Fisser, Herbert George
Hinds, Frank Crossman
Kercher, Conrad J
Nelms, George E

ALBERTA
Church, Robert Bertram
Lawson, John Edwin
Mears, Gerald John
Newman, John Alexander
Yeh, Francis Cho-hao

BRITISH COLUMBIA
Blair, Robert
Bowden, David Merle
Hill, Arthur Thomas
Owen, Bruce Douglas
Peterson, Raymond Glen
Stringam, Elwood Williams
Waldern, Donald E

MANITOBA
Castell, Adrian George
Elliot, James I
Grandhi, Raja Ratnam
Ingalls, Jesse Ray
Seale, Marvin Ernest

NEW BRUNSWICK
Nicholson, J W G

NOVA SCOTIA
Cock, Lorne M
MacRae, Herbert F

ONTARIO
Batra, Tilak Raj
Buchanan-Smith, Jock Gordon
Burnside, Edward Blair
Burton, John Heslop
Chambers, James Robert
Evans, Essi H
Heaney, David Paul
McAllister, Alan Jackson
Mallory, Frank Fenson
Morrison, William D
Nagai, Jiro
Orr, Donald Eugene, Jr
Rennie, James Clarence
Stone, John Bruce
Straus, Jozef

QUEBEC
Besner, Michel
Donefer, Eugene
Fahmy, Mohamed Hamed
Farmer, Chantal
Holtmann, Wilfried
Matte, J Jacques
Robert, Suzanne
Roy, Gabriel L

SASKATCHEWAN
Babiuk, Lorne A
Cohen, Roger D H
Howell, William Edwin
Knipfel, Jerry Earl

OTHER COUNTRIES
De Alba Martinez, Jorge
Lu, Christopher D
Makita, Takashi
Szebenyi, Emil

Animal Science & Nutrition

ALABAMA
Brewer, Robert Nelson
Cannon, Robert Young
Clark, Alfred James
Frobish, Lowell T
Haaland, Ronald L
Harris, Ralph Rogers
Hawkins, George Elliott, Jr
Moran, Edwin T, Jr
Moss, Buelon Rexford
Owen, Foster Gamble
Sauberlich, Howerde Edwin
Sewell, Raymond F
Voitle, Robert Allen
Wilkening, Marvin C

ALASKA
Husby, Fredric Martin
Miller, Byron F
Tomlin, Don C

ARIZONA
Anderson, Jay Oscar
Blaylock, Lynn Gail
Brown, William Hedrick
Foreman, Charles Frederick
Hale, William Harris
Huber, John Talmage
Mulder, John Bastian
Murdock, Fenoi R
Peng, Yeh-Shan
Rice, Richard W
Roberts, William Kenneth
Schmutz, Ervin Marcell
Schuh, James Donald
Sleeth, Rhule Bailey
Stull, John Warren
Swingle, Roy Spencer
Taylor, Richard G
Theurer, Clark Brent

ARKANSAS
Brown, Connell Jean
Daniels, L B
Donker, John D
Harris, Grover Cleveland, Jr
Hatfield, Efton Everett
Havener, Robert D
Keene, James H
Kellogg, David Wayne
Lee, Kwang
Lewis, Sherry M
McGilliard, A Dare
Nelson, Talmadge Seab
Noland, Paul Robert
Perry, Tilden Wayne
Stallcup, Odie Talmadge
Stephenson, Edward Luther
Sullivan, Thomas Wesley
Tollett, James Terrell
Yates, Jerome Douglas

CALIFORNIA
Abenes, Fiorello Bigornia
Anderson, Gary Bruce
Anderson, Russell K
Baldwin, Ransom Leland
Burri, Betty Jane
Cherms, Frank Llewellyn, Jr
Christensen, Allen Clare
Colvin, Harry Walter, Jr
Connor, John Michael

Dollar, Alexander M
Ducsay, Charles Andrew
Ernst, Ralph Ambrose
Garrett, William Norbert
Gordon, Malcolm Stephen
Grau, Charles Richard
Grill, Herman, Jr
Halloran, Hobart
Heitman, Hubert, Jr
Johnson, Herman Leonall
Kashyap, Tapeshwar S
Law, George Robert John
Leong, Kam Choy
Lofgreen, Glen Pehr
Lucas, David Owen
Moberg, Gary Philip
Myers, George Scott, Jr
Patton, Stuart
Prokop, Michael Joseph
Radloff, Harold David
Rahlmann, Donald Frederick
Rousek, Edwin J
Scott, Karen Christine
Sharp, Gerald Duane
Starkey, Eugene Edward
Starrett, Priscilla Holly
Trei, John Earl
Waterhouse, Howard N
Weir, William Carl
West, John Wyatt
Wilson, Barry William

COLORADO
Avens, John Stewart
Cramer, David Alan
Johnson, Donald Eugene
Kienholz, Eldon W
Lamm, Warren Dennis
Martin, Jack E
Matsushima, John K
Moreng, Robert Edward
Nett, T M
Nockels, Cheryl Ferris
Raun, Arthur Phillip
Reinbold, George W
Rollin, Bernard Elliot
Rouse, George Elverton
Seidel, George Elias, Jr
Smith, Gary Chester
Sofos, John N

CONNECTICUT
Brown, Lynn Ranney
Kinsman, Donald Markham
Pierro, Louis John
Riesen, John William
Rishell, William Arthur
Stake, Paul Erik

DELAWARE
Haenlein, George Friedrich Wilhelm
Rasmussen, Arlette Irene
Reitnour, Clarence Melvin
Salsbury, Robert Lawrence
Thoroughgood, Carolyn A

DISTRICT OF COLUMBIA
Hardison, Wesley Aurel
Lunchick, Curt
Urban, Edward Robert, Jr
Wilson, Edward Matthew
Wilson, George Donald
Zook, Bernard Charles

FLORIDA
Ammerman, Clarence Bailey
Baker, Frank Sloan, Jr
Bauernfeind, Jacob (Jack) C(hristopher)
Bertrand, Joseph E
Black, Alex
Boehme, Werner Richard
Chapman, Herbert L, Jr
Combs, George Ernest
Conrad, Joseph H
Courtney, Charles Hill
Cox, Dennis Henry
Damron, Bobby Leon
Davis, George Kelso
Douglas, Carroll Reece
Fry, Jack L
Gaunt, Stanley Newkirk
Gentry, Robert Francis
Hale, Kirk Kermit, Jr
Hammond, Andrew Charles
Hansen, Peter J
Harms, Robert Henry
Harris, Barney, Jr
Head, H Herbert
Hembry, Foster Glen
Hentges, James Franklin, Jr
Janky, Douglas Michael
Loggins, Phillip Edwards
Loosli, John Kasper
McDowell, Lee Russell
Meltzer, Martin Isaac
Miles, Richard David
Miller, Donald F
Moore, John Edward
Natzke, Roger Paul
Norval, Richard Andrew
Ott, Edgar Alton
Pate, Findlay Moye
Seals, Rupert Grant
Shirley, Ray Louis

Animal Science & Nutrition (cont)

Smith, Kenneth Leroy
Thomason, David Morton
Van Horn, Harold H, Jr
Wallace, Harold Dean
Warnick, Alvin Cropper
Wilcox, Charles Julian
Wilkowske, Howard Hugo
Wing, James Marvin
Zikakis, John Philip

GEORGIA
Amos, Henry Estill
Barb, C Richard
Britton, Walter Martin
Cherry, Jerry Arthur
Cullison, Arthur Edison
Flatt, William Perry
Fuller, Henry Lester
Hausman, Gary J
Hill, Gary Martin
Jensen, Leo Stanley
Kraeling, Robert Russell
Mabry, John William
McCullough, Marshall Edward
McGinnis, Charles Henry, Jr
Miller, William Jack
Moinuddin, Jessie Fischer
Neathery, Milton White
Neville, Walter Edward, Jr
Newton, George Larry
Powell, George Wythe
Reid, Willard Malcolm
Shutze, John V
Stuedemann, John Alfred
Sutton, William Wallace
Washburn, Kenneth W
Wyatt, Roger Dale

HAWAII
Atkinson, Shannon K C
Koshi, James H
Olbrich, Steven Emil
Ross, Ernest
Zaleski, Halina Maria

IDAHO
Bull, Richard C
Lehrer, William Peter, Jr
Montoure, John Ernest
Petersen, Charlie Frederick
Ross, Richard Henry
Schelling, Gerald Thomas

ILLINOIS
Arthur, Robert David
Bahr, Janice M
Baker, David H
Bartle, Steven Jon
Becker, Donald Eugene
Buettner, Mark Roland
Clark, Jimmy Howard
Corbin, James Edward
DiRienzo, Douglas Bruce
Easter, Robert Arnold
Fahey, George Christopher, Jr
Fernando, Rohan Luigi
Garrigus, Upson Stanley
Goodman, Billy Lee
Harrison, Paul C
Harshbarger, Kenneth E
Hetrick, John Henry
Hileman, Stanley Michael
Hollis, Gilbert Ray
Hurley, Walter L
Hutchinson, Harold David
Hutjens, Michael Francis
Jensen, Aldon Homan
Kelley, Keith Wayne
Kesler, Darrel J
Klay, Robert Frank
Knodt, Cloy Bernard
Layman, Donald Keith
Lewis, John Morgan
McKeith, Floyd Kenneth
Madsen, Fred Christian
Monson, William Joye
Moore, Clarence L
Murphy, Michael Ross
Nath, K Rajinder
Neagle, Lyle H
Parsons, Carl Michael
Pothoven, Marvin Arlo
Proctor, Jerry Franklin
Ricketts, Gary Eugene
Ridlen, Samuel Franklin
Robb, Thomas Wilbern
Robinson, James Lawrence
Scheid, Harold E
Shanks, Roger D
Thompson, David Jerome
Tuckey, Stewart Lawrence
Visek, Willard James
Waring, George Houstoun, IV
White, John Francis

INDIANA
Adams, Richard Linwood
Albright, Jack Lawrence
Anderson, David Bennett
Baumgardt, Billy Ray
Boston, Andrew Chester
Brown, Herbert
Cline, Tilford R
Coalson, James A
Elkin, Robert Glenn
Enders, George Leonhard, Jr
Forsyth, Dale Marvin
Gard, Don Irvin
Garriott, Michael Lee
Germann, Albert Frederick Ottomar, II
Gledhill, Robert Hamor
Grant, Alan Leslie
Hancock, Deana Lori
Heinicke, Herbert Raymond
Jeffers, Thomas Kirk
Jordan, Charles Edwin
Koritnik, Donald Raymond
Krider, Jake Luther
Longmire, Dennis B
Malven, Paul Vernon
Martin, Truman Glen
Matsuoka, Tats
Meyers, Michael Clinton
Outhouse, James Burton
Plumlee, Millard P, Jr
Powell, Thomas Shaw
Rogler, John Charles
Shockey, William Lee
Thomas, Elvin Elbert
Wellenreiter, Rodger Henry
Wolfrom, Glen Wallace

IOWA
Acker, Duane Calvin
Arthur, James Alan
Berger, Philip Jeffrey
Christian, Lauren L
Chung, Ronald Aloysius
Cox, David Frame
Crosby, Lon Owen
Ewan, Richard Colin
Ewing, Solon Alexander
Freeman, Albert Eugene
Harbaugh, Daniel David
Hodson, Harold H, Jr
Hoffman, Mark Peter
Holck, Gary LeRoy
Holden, Palmer Joseph
Hotchkiss, Donald K
Jacobson, Norman Leonard
Jurgens, Marshall Herman
Kenealy, Michael Douglas
Lamont, Susan Joy
Meeker, David Lynn
Mente, Glen Allen
Moore, Kenneth J
Morrical, Daniel Gene
Nelson, D Kent
Nordskog, Arne William
Robson, Richard Morris
Rohlf, Marvin Euguene
Sell, Jerry Lee
Stevermer, Emmett J
Summers, William Allen, Jr
Topel, David Glen
Young, Jerry Wesley
Zimmerman, Dean R

KANSAS
Adams, Albert Whitten
Brent, Benny Earl
Brethour, John Raymond
Harbers, Leniel H
Kastner, Curtis Lynn
Lomas, Lyle Wayne
Morrill, James Lawrence, Jr
Norton, Charles Lawrence
Sanford, Paul Everett
Stevenson, Jeffrey Smith

KENTUCKY
Baker, John P
Barnhart, Charles Elmer
Britt, Danny Gilbert
Cantor, Austin H
Chappell, Guy Lee Monty
Cromwell, Gary Leon
Haynes, Emmit Howard
Hays, Virgil Wilford
Hemken, Roger Wayne
Kemp, James Dillon
Lane, Gary (Thomas)
Little, Charles Oran
Mitchell, George Ernest, Jr
Moody, William Glenn
Olds, Durward
Robertson, John Connell
Webster, Carl David

LOUISIANA
Baham, Arnold
Borsari, Bruno
Chandler, John Edward
Cuomo, Gregory Joseph
Frye, Jennings Bryan, Jr
Humes, Paul Edwin
Johnson, William Alexander
Koonce, Kenneth Lowell
Laurent, Sebastian Marc
Nelson, Harvard G
Nickerson, Stephen Clark
Rusoff, Louis Leon
Smart, Lewis Isaac
Spanier, Arthur M
White, Thomas Wayne

MAINE
Barton, Barbara Ann
Donahoe, John Philip
Fox, Richard Romaine
Gerry, Richard Woodman
Gilbert, James Robert
Harris, Paul Chappell
Hawes, Robert Oscar
Les, Edwin Paul
Musgrave, Stanley Dean

MARYLAND
Alderson, Norris Eugene
Blosser, Timothy Hobert
Chou, Nelson Shih-Toon
Engster, Henry Martin
Estienne, Mark Joseph
Glenn, Barbara Peterson
Graber, George
Harrell, Reginal M
Heath, James Lee
Hitchner, Stephen Ballinger
Keeney, Mark
Lillehoj, Hyun Soon
Lincicome, David Richard
Luther, Lonnie W
McCurdy, John Dennis
Moody, Edward Grant
Norcross, Marvin Augustus
Olson, Leonard Carl
Oltjen, Robert Raymond
Pattee, Oliver Henry
Peterson, Irvin Leslie
Prouty, Richard Metcalf
Reiser, Sheldon
Richards, Mark P
Rubin, Max
Rumsey, Theron S
Schar, Raymond Dewitt
Sexton, Thomas John
Smith, Lewis Wilbert
Soares, Joseph Henry, Jr
Stunkard, Jim A
Thomas, Owen Pestell
Vandersall, John Henry
Wagner, David Darley
Weisbroth, Steven H
Williams, Walter Ford
Zimbelman, Robert George

MASSACHUSETTS
Lyford, Sidney John, Jr
Marcum, James Benton
Molinari, Pietro Filippo
Prior, Ronald L
Van Dongen, Cornelis Godefridus

MICHIGAN
Aulerich, Richard J
Bergen, Werner Gerhard
Carlotti, Ronald John
Chang, Timothy Scott
Coleman, Theo Houghton
Cook, Robert Merold
Cossack, Zafrallah Taha
Emery, Roy Saltsman
Flegal, Cal J
Foster, Douglas Layne
Hawkins, David Roger
Henneman, Harold Albert
Hillman, Donald
Hollingsworth, Cornelia Ann
Lawson, David Michael
McGilliard, Lon Dee
Miller, Elwyn Ritter
Purkhiser, E Dale
Rust, Steven Ronald
Saldanha, Leila Genevieve
Schwartz, Leland Dwight
Subramanian, Marappa G
Thomas, John William
Tucker, Herbert Allen
Ullrey, Duane Earl

MINNESOTA
Berg, Robert W
Caskey, Michael Conway
Conlin, Bernard Joseph
Coon, Craig Nelson
Foose, Thomas John
Goodrich, Richard Douglas
Hamre, Melvin L
Kromann, Rodney P
Marx, George Donald
Meiske, Jay C
Miller, Kenneth Philip
Oelberg, Thomas Jonathon
Parker, Ronald Bruce
Pettigrew, James Eugene, Jr
Speers, George M
Stearns, Eugene Marion, Jr
Stern, Marshall Dana
Wagenaar, Raphael Omer
Waibel, Paul Edward
Williams, Jesse Bascom

MISSISSIPPI
Bagley, Clyde Pattison
Buddington, Randal K
Cardwell, Joe Thomas
Combs, Gerald Fuson
Day, Elbert Jackson
Deaton, James Washington
Dilworth, Benjamin Conroy
Essig, Henry Werner
May, James David
Morgan, George Wallace
Pace, Henry Buford
Purchase, Harvey Graham
Putnam, Paul A
Ramsey, Dero Saunders
Tomlinson, James Everett
White, Charles Henry
Williams, Richard, Jr

MISSOURI
Baker, Joseph Willard
Buonomo, Frances Catherine
Byatt, John Christopher
Cornelius, Steven Gregory
Dub, Michael
Hammann, William Curl
Hintz, Richard Lee
Keith, Ernest Alexander
Kertz, Alois Francis
Kohlmeier, Ronald Harold
Martz, Fredric A
Morgan, Wyman
Pfander, William Harvey
Prewitt, Larry R
Savage, Jimmie Euel
Schwall, Donald V
Sewell, Homer B
Smith, Keith James
Snetsinger, David Clarence
Steinke, Frederich H
Talley, Spurgeon Morris
Thrasher, George W
Veum, Trygve Lauritz
Wilson, Edward Matthew
Winter, David F(erdinand)

MONTANA
Dynes, J Robert
MacNeil, Michael
Newman, Clarence Walter
Richards, Geoffrey Norman
Warren, Guylyn Rea

NEBRASKA
Aberle, Elton D
Ahlschwede, William T
Bennett, Gary Lee
Clanton, Donald Cather
Clemens, Edgar Thomas
Cundiff, Larry Verl
Danielson, David Murray
Dickerson, Gordon Edwin
Ford, Johny Joe
Froning, Glenn Wesley
Hunsley, Roger Eugene
Jenkins, Thomas Gordon
Koohmaraie, Mohammad
Laster, Danny Bruce
Lewis, Austin James
Pekas, Jerome Charles
Peo, Ernest Ramy, Jr
Ward, John K
Weichenthal, Burton Arthur
Yen, Jong-Tseng

NEVADA
Bailey, Curtiss Merkel

NEW HAMPSHIRE
Anderson, Donald Lindsay
Holter, James Burgess
Keener, Harry Allan
Langweiler, Marc

NEW JERSEY
Cox, James Lee
Dalrymple, Ronald Howell
Deetz, Lawrence Edward, III
De Lay, Roger Lee
Feighner, Scott Dennis
Griminger, Paul
Harner, James Philip
Horton, Glyn Michael John
Ingle, Donald Lee
Katz, Larry Steven
Leeder, Joseph Gorden
Linn, Bruce Oscar
Mehring, Arnon Lewis, Jr
Ott, Walther H
Pankavich, John Anthony
Pensack, Joseph Michael
Scanes, Colin G
Scholl, Philip Jon
Simkins, Karl LeRoy, Jr
Tudor, David Cyrus
Wells, Phillip Richard
Wilbur, Robert Daniel
Willson, John Ellis

NEW MEXICO
Galyean, Michael Lee
Holechek, Jerry Lee
Nelson, Arnold Bernard
Roberson, Robert H
Smith, Garmond Stanley

NEW YORK
Ainslie, Harry Robert
Ames, Stanley Richard
Auclair, Walter

AGRICULTURAL & FOREST SCIENCES / 9

Beermann, Donald Harold
Chase, Larry Eugene
Elliot, John Murray
Fabricant, Julius
Fletcher, Paul Wayne
Fox, Danny Gene
Hintz, Harold Franklin
Hogue, Douglas Emerson
Hutt, Frederick Bruce
Ketola, H George
Marshall, James Tilden, Jr
Merrill, William George
Morck, Timothy Anton
Mundorff-Shrestha, Sheila Ann
Oltenacu, Elizabeth Allison Branford
Ostrander, Charles Evans
Porter, Gilbert Harris
Quimby, Fred William
Randy, Harry Anthony
Rasmussen, Kathleen Maher
Reuter, Gerald Louis
Rowoth, Olin Arthur
Stillions, Merle C
Talhouk, Rabih Shakib
Topp, William Carl
Trimberger, George William
VanSoest, Peter John
Wadell, Lyle H
Warner, Richard G

NORTH CAROLINA
Allen, Neil Keith
Beam, John E
Bull, Leonard Seth
Clawson, Albert J
Cornwell, John Calhoun
Donaldson, William Emmert
Goode, Lemuel
Hamilton, Pat Brooks
Johnson, George Andrew
Johnson, William Lawrence
Knott, Fred Nelson
Leatherwood, James M
Lobaugh, Bruce
Mikat, Eileen M
Mochrie, Richard D
Olson, Howard H
Rakes, Allen Huff
Ramsey, Harold Arch
Shih, Jason Chia-Hsing
Smith, Frank Houston
Whitlow, Lon Weidner
Wise, George Herman
Young, Robert John
Zeng, Zhao-Bang

NORTH DAKOTA
Davison, Kenneth Lewis
Dinusson, William Erling
Harrold, Robert Lee
Karn, James Frederick
Nielsen, Forrest Harold
Park, Chung Sun

OHIO
Allaire, Francis Raymond
Anderson, Marlin Dean
Cline, Jack Henry
Danke, Richard John
Davis, Michael E
Dehority, Burk Allyn
Fechheimer, Nathan S
Gauntt, William Amor
Harper, Willis James
Henningfield, Mary Frances
Hines, Harold C
Latshaw, J David
Mahan, Donald C
Mikolajcik, Emil Michael
Naber, Edward Carl
Palmquist, Donald Leonard
Pate, Joy Lee
Prince, Terry Jamison
Smith, Kenneth Larry
Staubus, John Reginald
Stephens, James Fred
Tyznik, William John

OKLAHOMA
Alexander, Joseph Walker
Berry, Joe Gene
Browning, Charles Benton
Bush, Linville John
Gleaves, Earl William
Johnson, Ronald Roy
Luce, William Glenn
Martin, Jerry, Jr
Noble, Robert Lee
Owens, Fredric Newell
Raun, Ned S
Teh, Thian Hor
Thayer, Rollin Harold
Totusek, Robert
Walters, Lowell Eugene

OREGON
Adams, Frank William
Anglemier, Allen Francis
Arscott, George Henry
Bernier, Paul Emile
Church, David Calvin
Davis, Michael William
Harper, James Arthur
Holleman, Kendrick Alfred

Oldfield, James Edmund
Schmisseur, Wilson Edward
Weswig, Paul Henry

PENNSYLVANIA
Adams, Richard Sanford
Baumrucker, Craig Richard
Benedict, Robert Curtis
Buss, Edward George
Chalupa, William Victor
Cowan, Robert Lee
Farrell, Harold Maron, Jr
Flipse, Robert Joseph
Hagen, Daniel Russell
Hargrove, George Lynn
Heald, Charles William
Hedde, Richard Duane
Hershberger, Truman Verne
Keene, Owen David
Kesler, Earl Marshall
Krueger, Charles Robert
Lillie, Robert Jones
Parish, Roger Cook
Piesco, Nicholas Peter
Ramberg, Charles F, Jr
Roush, William Burdette
Shellenberger, Paul Robert
Varga, Gabriella Anne
Vasilatos-Younken, Regina
Wangsness, Paul Jerome
Wideman, Robert Frederick, Jr

RHODE ISLAND
Cosgrove, Clifford James
Donovan, Gerald Alton
Durfee, Wayne King
Meade, Thomas Leroy
Nippo, Murn Marcus
Rice, Michael Alan

SOUTH CAROLINA
Barnett, Bobby Dale
Birrenkott, Glenn Peter, Jr
Edwards, Robert Lee
Hughes, Buddy Lee
Jones, Jack Edenfield
Lee, Daniel Dixon, Jr
Skelley, George Calvin, Jr
Wise, Milton Bee

SOUTH DAKOTA
Britzman, Darwin Gene
Brown, Michael Lee
Bush, Leon F
Carlson, Charles Wendell
Costello, William James
Embry, Lawrence Bryan
Emerick, Royce Jasper
Johnson, Marvin Melrose
Males, James Robert
Parsons, John G
Schingoethe, David John
Slyter, Arthur Lowell

TENNESSEE
Barth, Karl M
Bell, Marvin Carl
Chamberlain, Charles Calvin
Chen, Thomas Tien
Demott, Bobby Joe
Fribourg, Henry August
Herting, David Clair
Hitchcock, John Paul
Kao, Race Li-Chan
Miller, James Kincheloe
Montgomery, Monty J
Papas, Andreas Michael
Richardson, Don Orland
Sachan, Dileep Singh
Wakefield, Troy, Jr
White, Alan Wayne

TEXAS
Abrams, Steven Allen
Albin, Robert Custer
Brokaw, Bryan Edward
Brown, Murray Allison
Brown, Robert Dale
Butler, Ogbourne Duke, Jr
Byers, Floyd Michael
Calhoun, Millard Clayton
Cartwright, Aubrey Lee, Jr
Cassard, Daniel W
Clark, James Richard
Cole, Noel Andy
Coleman, John Dee
Coppock, Carl Edward
Cross, Hiram Russell
Curl, Sam E
DuBose, Leo Edwin
Ellis, William C
Fineg, Jerry
Friend, Ted H
Godbee, Richard Greene, II
Greathouse, Terrence Ray
Green, George G
Helm, Raymond E
Hutcheson, David Paul
Jacob, Horace S
Johnson, Larry
Lane, Alfred Glen
Lawrence, Addison Lee
Lippke, Hagen
Mason, Tim Robert

Mellor, David Bridgwood
Mersmann, Harry John
Pond, Wilson Gideon
Preston, Rodney LeRoy
Ramsey, Clovis Boyd
Richardson, Richard Harvey
Rosborough, John Paul
Rowland, Lenton O, Jr
Ryan, Cecil Benjamin
Sherrod, Lloyd B
Standlee, William Jasper
Stewart, Gordon Arnold
Sudweeks, Earl Max
Swakon, Doreen H D
Thompson, Granville Berry
Tribble, Leland Floyd
Varner, Larry Weldon
Whiteside, Charles Hugh
Will, Paul Arthur

UTAH
Anderson, Melvin Joseph
Bohman, Verle Rudolph
Butcher, John Edward
Dobson, Donald C
Draper, Carroll Isaac
Ellis, LeGrande Clark
Ernstrom, Carl Anthon
Fonnesbeck, Paul Vance
Gardner, Robert Wayne
Gay, Charles Wilford
Hendricks, Deloy G
Lamb, Robert Cardon
Malechek, John Charles
Park, Robert Lynn
Provenza, Frederick Dan
Ralphs, Michael H
Richardson, Gary Haight
Slade, Larry Malcom
Vallentine, John Franklin
Wallentine, Max V
Warnick, Robert Eldredge

VERMONT
Carew, Lyndon Belmont, Jr
Smith, Albert Matthews
Welch, James Graham

VIRGINIA
Allen, Vivien Gore
Boyd, Earl Neal
Bragg, Denver Dayton
Carr, Scott Bligh
Denbow, Donald Michael
Eller, Arthur L, Jr
Etgen, William M
Fontenot, Joseph Paul
Frahm, Richard R
Gwazdauskas, Francis Charles
Herbein, Joseph Henry, Jr
Hohenboken, William Daniel
Holmes, Clayton Ernest
Jones, Gerald Murray
McGilliard, Michael Lon
Pearson, Ronald Earl
Polan, Carl E
Potter, Lawrence Merle
Sheppard, Alan Jonathan
Swiger, Louis Andre
Vinson, William Ellis
Webb, Kenneth Emerson, Jr
Wisman, Everett Lee
Woods, Walter Ralph

WASHINGTON
Andrews, Daniel Keller
Bearse, Gordon Everett
Beck, Thomas W
Becker, Walter Alvin
Bray, Donald James
Brekke, Clark Joseph
Carlson, James Roy
Froseth, John Allen
Hardy, Ronald W
Heinemann, Wilton Walter
Kincaid, Ronald Lee
Leung, Frederick C
Luedecke, Lloyd O
Quinn, LeBris Smith
Reeves, Jerry John
Sasser, Lyle Blaine
Thomas, John M
Towner, Richard Henry
Van Hoosier, Gerald L, Jr

WEST VIRGINIA
Anderson, Gerald Clifton
Becker, Stanley Leonard
Golway, Peter L
Hansard, Samuel L, II
Horvath, Donald James
Peterson, Ronald A
Thomas, Roy Orlando

WISCONSIN
Albrecht, Kenneth Adrian
Arrington, Louis Carroll
Bitgood, J(ohn) James
Bradley, Robert Lester, Jr
Broderick, Glen Allen
Cook, Mark Eric
Dierschke, Donald Joe
Ehle, Fred Robert
Greaser, Marion Lewis

Greger, Janet L
Hoffman, William F
Howard, W Terry
Jorgensen, Neal A
Larsen, Howard James
Leibbrandt, Vernon Dean
Maurer, Arthur James
Mertens, David Roy
Price, Susan Gail
Satter, Larry Dean
Schaefer, Daniel M
Schultz, Loris Henry
Sendelbach, Anton G
Sunde, Milton Lester
Winder, William Charles
Wittwer, Leland S

WYOMING
Hinds, Frank Crossman
Kercher, Conrad J

ALBERTA
Bowland, John Patterson
Cheng, Kuo-Joan
Clandinin, Donald Robert
Coulter, Glenn Hartman
Mathison, Gary W(ayne)
Newman, John Alexander
Price, Mick A
Robblee, Alexander (Robinson)

BRITISH COLUMBIA
Bailey, Charles Basil Mansfield
Beames, R M
Blair, Robert
Bowden, David Merle
Cheng, Kimberly Ming-Tak
Hill, Arthur Thomas
Hunt, John R
Majak, Walter
March, Beryl Elizabeth
Owen, Bruce Douglas
Quinton, Dee Arlington
Waldern, Donald E

MANITOBA
Ayles, George Burton
Castell, Adrian George
Elliot, James I
Grandhi, Raja Ratnam
Kondra, Peter Alexander
Strain, John Henry

NEW BRUNSWICK
McQueen, Ralph Edward
Nicholson, J W G

NOVA SCOTIA
Bush, Roy Sidney
Cock, Lorne M
Hamilton, Robert Milton Gregory

ONTARIO
Ainsworth, Louis
Bayley, Henry Shaw
Bray, Tammy M
Burton, John Heslop
Evans, Essi H
Fortin, Andre Francis
Gavora, Jan Samuel
Gowe, Robb Shelton
Heaney, David Paul
Irvine, Donald McLean
Ivan, Michael
Lin, Ching Y
Morrison, Spencer Horton
Morrison, William D
Orr, Donald Eugene, Jr
Osmond, Daniel Harcourt
Pang, Cho Yat
Phillips, William Ernest John
Pigden, Wallace James
Trenholm, Harold Locksley
Usborne, William Ronald
Wolynetz, Mark Stanley

PRINCE EDWARD ISLAND
Winter, Karl A

QUEBEC
Barrette, Daniel Claude
Belzile, Rene
Besner, Michel
Buckland, Roger Basil
De la Noue, Joel Jean-Louis
Donefer, Eugene
Farmer, Chantal
Gariepy, Claude
Gauthier, Gilles
Matte, J Jacques

SASKATCHEWAN
Beacom, Stanley Ernest
Bell, John Milton
Cohen, Roger D H
Crawford, Roy Douglas
Jones, Graham Alfred
Knipfel, Jerry Earl
Korsrud, Gary Olaf
Patience, John Francis
Salmon, Raymond Edward

OTHER COUNTRIES
Bywater, Anthony Colin

Animal Science & Nutrition (cont)

Fairweather-Tait, Susan Jane
Fitzhugh, Henry Allen
Jahn, Ernesto
Lu, Christopher D
Nakagawa, Shizutoshi
Santidrian, Santiago
Schmeling, Sheila Kay
Shore, Laurence Stuart
Szebenyi, Emil
Young, Bruce Arthur

Fish & Wildlife Sciences

ALABAMA
Bayne, David Roberge
Best, Troy Lee
Boschung, Herbert Theodore
Bradley, James T
Brady, Yolanda J
Causey, Miles Keith
Cowan, James Howard, Jr
Duncan, Bryan Lee
Dunham, Rex Alan
Grizzle, John Manuel
Grover, John Harris
Holler, Nicholas Robert
Huntley, Jimmy Charles
Lawrence, John Medlock
Lovell, Richard Thomas
Lovshin, Leonard Louis, Jr
Mirarchi, Ralph Edward
Moss, Donovan Dean
Phelps, Ronald P
Plumb, John Alfred
Rogers, Wilmer Alexander
Rosene, Walter, Jr
Rouse, David B
Shell, Eddie Wayne
Smitherman, Renford Oneal

ALASKA
Ahlgren, Molly O
Amend, Donald Ford
Babcock, Malin Marie
Behrend, Donald Fraser
Clark, John Harlan
Dahlberg, Michael Lee
Dean, Frederick Chamberlain
Fagen, Robert
Fay, Francis Hollis
Gard, Richard
Gharrett, Anthony John
Hanley, Thomas Andrew
Hatch, Scott Alexander
Hauser, William Joseph
Heifetz, Jonathan
Helle, John Harold (Jack)
Kessel, Brina
Klein, David Robert
LaPerriere, Jacqueline Doyle
Mathisen, Ole Alfred
Meehan, William Robert
Morrison, John Albert
Olsen, James Calvin
Paul, Augustus John, III
Pella, Jerome Jacob
Reynolds, James Blair
Rockwell, Julius, Jr
Samson, Fred Burton
Schoen, John Warren
Smoker, William Alexander
Smoker, William Williams
Snyder, George Richard
Straty, Richard Robert
Thomas, Gary Lee
Tyler, Albert Vincent

ARIZONA
Beck, John R(oland)
Brown, Bryan Turner
Dealy, John Edward
Gallizioli, Steve
Gerking, Shelby Delos
Johnson, Raymond Roy
Krausman, Paul Richard
Maughan, Owen Eugene
Minckley, Wendell Lee
Patton, David Roger
Post, George
Rinne, John Norman
Schnell, Jay Heist
Shaw, William Wesley
Smith, Andrew Thomas
Smith, Norman Sherrill
Sowls, Lyle Kenneth
Van Riper, Charles, III

ARKANSAS
Collins, Richard Arlen
Dupree, Harry K
Grizzell, Roy Ames, Jr
Jenkins, Robert M
Kilambi, Raj Varad
Tave, Douglas

CALIFORNIA
Abramson, Norman Jay
Allen, George Herbert
Anderson, Daniel William
Anderson, Lars William James
Barrett, Izadore
Baxter, John Lewis
Bayliff, William Henry
Boehlert, George Walter
Botzler, Richard George
Bray, Richard Newton
Brittan, Martin Ralph
Brownell, Robert Leo, Jr
Cech, Joseph Jerome, Jr
Chadwick, Harold King
Chamberlain, Dilworth Woolley
Chen, Lo-Chai
Chesemore, David Lee
Davis, Gary Everett
Dieterich, Robert Arthur
Ebert, Earl Ernest
Eldridge, Maxwell Bruce
Embree, James Willard, Jr
Emerson, Frederick Beauregard, Jr
Erman, Don Coutre
Eschmeyer, William Neil
Farris, David Allen
Follett, Wilbur Irving
Fridley, Robert B
Gall, Graham A E
Gerrodette, Timothy
Gilmer, David Seeley
Gordon, Malcolm Stephen
Harris, Stanley Warren
Hassler, Thomas J
Hedgecock, Dennis
Hendricks, Lawrence Joseph
Hewston, John G
Horn, Michael Hastings
Howard, Walter Egner
Iwamoto, Tomio
Jacobsen, Nadine Klecha
Jansen, Henricus Cornelis
Kask, John Laurence
Kerstetter, Theodore Harvey
Klawe, Witold L
Kope, Robert Glenn
Krejsa, Richard Joseph
Kutilek, Michael Joseph
Lenarz, William Henry
McClure, Howe Elliott
McCullough, Dale Richard
Matthews, Kathleen Ryan
Messersmith, James David
Mitchell, John Alexander
Moberg, Gary Philip
Morris, Robert Wharton
Mossman, Archie Stanton
Mullen, Ashley John
Mulroy, Thomas Wilkinson
Neal, Richard Allan
Ohlendorf, Harry Max
Orcutt, Harold George
Ratliff, Raymond Dewey
Raveling, Dennis Graff
Ridenhour, Richard Lewis
Roelofs, Terry Dean
Ryder, Oliver A
Sakagawa, Gary Toshio
Shelton, William Lee
Smith, Paul Edward
Springer, Paul Frederick
Swift, Camm Churchill
Taylor, Leighton Robert, Jr
Tribbey, Bert Allen
Vanicek, C David
Vilkitis, James Richard
Walter, Hartmut
Weintraub, Joel D
Wilcox, Bruce Alexander
Williams, Daniel Frank
Woodruff, David Scott

COLORADO
Asherin, Duane Arthur
Bailey, James Allen
Bradshaw, William Newman
Braun, Clait E
Carlson, Clarence Albert, Jr
Crouch, Glenn LeRoy
Curnow, Richard Dennis
Dittberner, Phillip Lynn
Fausch, Kurt Daniel
Flickinger, Stephen Albert
Green, Jeffrey Scott
Gregory, Richard Wallace
Hagen, Harold Kolstoe
Hein, Dale Arthur
Keith, James Oliver
Knopf, Fritz L
Lehner, Philip Nelson
Martin, Stephen George
Oldemeyer, John Lee
Parker, H Dennison
Ramsey, John Scott
Ryder, Ronald Arch
Schmidt, John Lancaster
Smith, Dwight Raymond
Stalnaker, Clair B
Sublette, James Edward
Swanson, Gustav Adolph
Thompson, Daniel Quale
Vohs, Paul Anthony, Jr

CONNECTICUT
Katz, Max
Lee, Douglas Scott
Maguder, Theodore Leo, Jr

Renfro, William Charles
Ward, Jeffrey Stuart

DELAWARE
Roth, Roland Ray
Schuler, Victor Joseph

DISTRICT OF COLUMBIA
Blockstein, David Edward
Buffington, John Douglas
Domning, Daryl Paul
Knapp, Leslie W
Lovejoy, Thomas E
Mac, Michael John
Nickum, John Gerald
Policansky, David J
Post, Boyd Wallace
Sweeney, James Michael
Thomas, Jack Ward
Tyler, James Chase
Weitzman, Stanley Howard
Wise, John P

FLORIDA
Alam, Mohammed Khurshid
Black, Kenneth Eldon
Bortone, Stephen Anthony
Branch, Lyn Clarke
Brown, Bradford E
Byrd, Isaac Burlin
Christman, Steven Philip
Davis, John Armstrong
Feddern, Henry A
Forrester, Donald J
Gilbert, Carter Rowell
Grimes, Churchill Bragaw
Gu, Binhe
Heinen, Joel Thomas
Higman, James B
Iversen, Edwin Severin
Jones, Albert Cleveland
Kale, Herbert William, II
Kitchens, Wiley M
Labisky, Ronald Frank
Leon, Kenneth Allen
Loesch, Harold Carl
McCann, James Alwyn
Marquis, David Alan
Meryman, Charles Dale
Pagan-Font, Francisco Alfredo
Porter, Richard Dee
Posner, Gerald Seymour
Powers, Joseph Edward
Prince, Eric D
Roessler, Martin A
Walker, Charles R
Wisby, Warren Jensen
Witham, P(hilip) Ross
Woods, Charles Arthur

GEORGIA
Andrews, Charles Lawrence
Briggs, John Carmon
Callaham, Mac A
Johnson, Albert Sydney, III
Marchinton, Robert Larry
Miller, George C
Payne, Jerry Allen
Schultz, Donald Paul
Scott, Donald Charles
Stober, Quentin Jerome
Winger, Parley Vernon
Zarnoch, Stanley Joseph

HAWAII
Atkinson, Shannon K C
Bardach, John E
Fast, Arlo Wade
Greenfield, David Wayne
Laurs, Robert Michael
Morin, Marie Patricia
Qin, Jianguang
Radtke, Richard Lynn
Skillman, Robert Allen
Stone, Charles Porter
Winget, Robert Newell

IDAHO
Ables, Ernest D
Congleton, James Lee
Fuller, Mark Roy
Griffith, John Spencer
Haufler, Jonathan B
Hostetter, Donald Lee
Hungerford, Kenneth Eugene
Keller, Barry Lee
Laundre, John William
MacFarland, Craig George
Nass, Roger Donald
Peek, James Merrell
Platts, William Sidney

ILLINOIS
Anderson, William Leno
Bjorklund, Richard Guy
Buck, David Homer
Burger, George Vanderkarr
Burr, Brooks Milo
Durham, Leonard
Feldhamer, George Alan
Heltne, Paul Gregory
Jahn, Lawrence A
Klimstra, Willard David
Kozicky, Edward Louis

Larimore, Richard Weldon
Larkin, Ronald Paul
Le Febvre, Eugene Allen
Lewis, William Madison
Meldrim, John Waldo
Nixon, Charles Melville
Page, Lawrence Merle
Paulson, Glenn
Prabhudesai, Mukund M
Roseberry, John L
Thomerson, Jamie E
Waring, George Houstoun, IV

INDIANA
Goetz, Frederick William, Jr
Kirkpatrick, Charles Milton
Kirkpatrick, Ralph Donald
Montague, Fredrick Howard, Jr
Schaible, Robert Hilton
Spacie, Anne
Weeks, Harmon Patrick, Jr

IOWA
Best, Louis Brown
Carlander, Kenneth Dixon
Clark, William Richard
Dinsmore, James Jay
Dowell, Virgil Eugene
Klaas, Erwin Eugene
Menzel, Bruce Willard
Moorman, Robert Bruce
Riley, Terry Zene
Summerfelt, Robert C
Wilson, Nixon Albert

KANSAS
Choate, Jerry Ronald
Eddy, Thomas A
Gipson, Philip
Halazon, George Christ
Hopkins, Theodore Louis
Kaufman, Donald Wayne
Klaassen, Harold Eugene
Robins, Charles Richard
Thompson, Max Clyde

KENTUCKY
Branson, Branley Allan
Bryant, William Stanley
Pearson, William Dean
Thompson, Marvin Pete
Timmons, Thomas Joseph
Williams, John C

LOUISIANA
Chabreck, Robert Henry
Chesney, Edward Joseph, Jr
Cordes, Carroll Lloyd
Culley, Dudley Dean, Jr
Davis, Billy J
Fingerman, Sue Whitsell
Gunning, Gerald Eugene
Hamilton, Robert Bruce
Hastings, Robert Wayne
Herke, William Herbert
Suttkus, Royal Dallas
Wakeman, John Marshall
Wright, Vernon Lee

MAINE
Coulter, Malcolm Wilford
Crawford, Hewlette Spencer, Jr
De Witt, Hugh Hamilton
Hanks, Robert William
Hunter, Malcolm Llewellyn, Jr
Mendall, Howard Lewis
Owen, Ray Bucklin, Jr
Warner, Kendall

MARYLAND
Adams, Lowell William
Albers, Peter Heinz
Barbehenn, Kyle Ray
Chou, Nelson Shih-Toon
Flyger, Vagn Folkmann
Fox, William Walter, Jr
Griswold, Bernard Lee
Heath, Robert Gardner
Hetrick, Frank M
Hill, Elwood Fayette
Hoberg, Eric Paul
Hodgdon, Harry Edward
Hoffman, David J
Houde, Edward Donald
Howe, Marshall Atherton
Leedy, Daniel Loney
Nelson, Jay Arlen
Nichols, James Dale
Pattee, Oliver Henry
Robbins, Chandler Seymour
Rothschild, Brian James
Russek-Cohen, Estelle
Sparling, Donald Wesley, Jr
Tillman, Michael Francis
Trauger, David Lee
Williams, Walter Ford

MASSACHUSETTS
Baglin, Raymond Eugene, Jr
Barske, Philip
Booke, Henry Edward
Boreman, John George, Jr
Castro, Gonzalo
Clark, Stephen Howard

AGRICULTURAL & FOREST SCIENCES / 11

Finn, John Thomas
Godfrey, Paul Jeffrey
Greeley, Frederick
Heyerdahl, Eugene Gerhardt
Larson, Joseph Stanley
Moir, Ronald Brown, Jr
Nielsen, Svend Woge
Pearce, John Bodell
Richards, F Paul
Ross, Michael Ralph
Short, Henry Laughton

MICHIGAN
Behmer, David J
Belyea, Glenn Young
Bennett, Carl Leroy
Borgeson, David P
Brown, Edward Herriot, Jr
Burton, Thomas Maxie
Cowan, Archibald B
D'Itri, Frank M
Fetterolf, Carlos de la Mesa, Jr
Field, Ronald James
Fink, William Lee
Foster, Neal Robert
Hart, John Henderson
Hartman, Wilbur Lee
Kevern, Niles Russell
Krull, John Norman
Kutkuhn, Joseph Henry
Langenau, Edward E, Jr
Latta, William Carl
Lenon, Herbert Lee
Manny, Bruce Andrew
Miller, Robert Rush
Passino, Dora R May
Patriarche, Mercer Harding
Peterson, Rolf Olin
Prince, Harold Hoopes
Robinson, William Laughlin
Schneider, James Carl
Sikarskie, James Gerard
Stanley, Jon G
Stauffer, Thomas Miel
Tack, Peter Isaac
Taylor, William Waller
Williams, Wells Eldon
Winterstein, Scott Richard

MINNESOTA
Adelman, Ira Robert
Arthur, John W
Broderius, Steven James
Buech, Richard Reed
Collins, Hollie L
Gullion, Gordon W
Hokanson, Kenneth Eric Fabian
Jannett, Frederick Joseph, Jr
Jordan, Peter Albion
Kuehn, Jerome H
Lange, Robert Echlin, Jr
McCann, Lester J
McConville, David Raymond
Mech, Lucyan David
Meyer, Fred Paul
Mount, Donald I
Rogers, Lynn Leroy
Sorensen, Peter W
Spigarelli, Steven Alan
Tanner, Ward Dean, Jr
Waters, Thomas Frank
Williams, Steven Frank

MISSISSIPPI
Arner, Dale H
Regier, Lloyd Wesley
Reinecke, Kenneth J
Robinson, Edwin Hollis
Ross, Stephen Thomas
Shields, Fletcher Douglas, Jr
Wakeley, James Stuart

MISSOURI
Baskett, Thomas Sebree
Drobney, Ronald DeLoss
Elder, William Hanna
Faaborg, John Raynor
Finger, Terry Richard
Fredrickson, Leigh H
Fritzell, Erik Kenneth
Funk, John Leon
Hanson, Willis Dale
Hunn, Joseph Bruce
Kuhns, John Farrell
Mosher, Donna Patricia
Pflieger, William Leo
Reidinger, Russell Frederick, Jr
Schoettger, Richard A
Wiggers, Ernie P

MONTANA
Ball, Irvin Joseph
Foresman, Kerry Ryan
Fritts, Steven Hugh
Gould, William Robert, III
Kendall, Katherine Clement
Lyon, Leonard Jack
Mackie, Richard John
Metzgar, Lee Hollis
Mitchell, Lawrence Gustave
O'Gara, Bart W
Rounds, Burton Ward
Schumacher, Robert E
Taber, Richard Douglas

Tennant, Donald L

NEBRASKA
Bliese, John C W
Case, Ronald Mark
Hergenrader, Gary Lee

NEVADA
Davis, Phillip Burton
Mahadeva, Madhu Narayan
Paulson, Larry Jerome
Wiemeyer, Stanley Norton

NEW HAMPSHIRE
Barry, William James
Mautz, William Ward
Smith, Roderick MacDowell
Sower, Stacia Ann
Wilson, Donald Alfred
Wu, Lin

NEW JERSEY
Applegate, James Edward
Dobson, Andrew Peter
Hu, Yaping
Landau, Matthew Paul
Pacheco, Anthony Louis

NEW MEXICO
Cooch, Frederick Graham
Davis, Charles A
Holechek, Jerry Lee
Howard, Volney Ward, Jr
Huey, William S
Johnson, James Edward
Lewis, James Chester
Rogers, John Gilbert, Jr
Schemnitz, Sanford David

NEW YORK
Ayres, Jose Marcio Correa
Brandt, Stephen Bernard
Brett, Betty Lou Hilton
Brocke, Rainer H
Brothers, Edward Bruce
Cahn, Phyllis Hofstein
Christian, John Jermyn
Conover, David Olmstead
Dickson, James Gary
Dooley, James Keith
Eckhardt, Ronald A
Elrod, Joseph Harrison
Greenhall, Arthur Merwin
Hyde, Kenneth Martin
Ketola, H George
Kiviat, Erik
Lawrence, William Mason
McHugh, John Laurence
McNeil, Richard Jerome
Madison, Dale Martin
Maestrone, Gianpaolo
Muessig, Paul Henry
Muller-Schwarze, Dietland
Oglesby, Ray Thurmond
Osterberg, Donald Mouretz
Perlmutter, Alfred
Porter, William Frank
Reuter, Gerald Louis
Ringler, Neil Harrison
Rockwell, Robert Franklin
Severinghaus, Charles William
Shumway, Sandra Elisabeth
Tillman, Robert Erwin
VanDruff, Larry Wayne
Walker, Roland
Wilkins, Bruce Tabor
Windsor, Donald Arthur

NORTH CAROLINA
Bolen, Eric George
Brown, Richard Dean
Cope, Oliver Brewern
Crowder, Larry Bryant
Dieter, Michael Phillip
Di Giulio, Richard Thomas
Doerr, Phillip David
Eggleston, David Bryan
Fairchild, Homer Eaton
Hayne, Don William
Huntsman, Gene Raymond
Lewis, Robert Minturn
Lindquist, David Gregory
Link, Garnett William, Jr
Manooch, Charles Samuel, III
Merriner, John Vennor
Nicholson, William Robert
Peters, David Stewart
Pritchard, G Ian
Rulifson, Roger Allen
Schwartz, Frank Joseph
Steinhagen, William Herrick
Stroud, Richard Hamilton
Sykes, James Enoch
Towell, William Earnest
Webster, William David
West, Jerry Lee

NORTH DAKOTA
Crawford, Richard Dwight
Krapu, Gary Lee
Lokemoen, John Theodore

OHIO
Barbour, Clyde D

Baumann, Paul C
Berra, Tim Martin
Bookhout, Theodore Arnold
Cole, Charles Franklyn
Costanzo, Jon P
Crites, John Lee
Culver, David Alan
Dabrowski, Konrad
Eastman, Joseph
Good, Ernest Eugene
Harder, John Dwight
Holeski, Paul Michael
Kinney, Edward Coyle, Jr
Lewis, Michael Anthony
Madenjian, Charles Paul
Peterle, Tony J
White, Andrew Michael

OKLAHOMA
Alexander, Joseph Walker
Chapman, Brian Richard
Hill, Loren Gilbert
Kilpatrick, Earl Buddy
Matthews, William John
Namminga, Harold Eugene
Shaw, James Harlan
Summers, Gregory Lawson
Toetz, Dale W

OREGON
Bond, Carl Eldon
Campbell, Charles J
Crawford, John Arthur
De Witt, John William, Jr
Doudoroff, Peter
Eicher, George J
Erickson, Ray Charles
Fryer, John Louis
Garton, Ronald Ray
Hall, James Dane
Hamilton, James Arthur Roy
Harville, John Patrick
Hedtke, James Lee
Hendricks, Jerry Dean
Horton, Howard Franklin
Isaacson, Dennis Lee
Krueger, William Clement
Linn, DeVon Wayne
Liss, William John
Markle, Douglas Frank
Marriage, Lowell Dean
Maser, Chris
Meslow, E Charles
Olla, Bori Liborio
Olson, Robert Eldon
Perry, Lorin Edward
Quast, Jay Charles
Rasmussen, Lois E Little
Schneider, Phillip William, Jr
Schoning, Robert Whitney
Schreck, Carl Bernhard
Sinnhuber, Russell Otto
Skovlin, Jon Matthew
Tubb, Richard Arnold
Wagner, Harry Henry
Wick, William Quentin

PENNSYLVANIA
Arnold, Dean Edward
Bursey, Charles Robert
Clark, Richard James
Feuer, Robert Charles
Mathur, Dilip
Snyder, Donald Benjamin
Yahner, Richard Howard

RHODE ISLAND
Durfee, Wayne King
Golet, Francis Charles
Gould, Walter Phillip
Husband, Thomas Paul
Laurence, Geoffrey Cameron
Lazell, James Draper
Mayer, Garry Franklin
Rice, Michael Alan
Saila, Saul Bernhard

SOUTH CAROLINA
Abercrombie, Clarence Lewis
Burrell, Victor Gregory, Jr
Dixon, Kenneth Randall
Ely, Berten E, III
Gard, Nicholas William
Harlow, Richard Fessenden
Hays, Sidney Brooks
Johnson, Robert Karl
Kennamer, James Earl
Muska, Carl Frank
Pelton, Michael Ramsay
Smith, Theodore Isaac Jogues
Sweeney, John Robert
Wood, Gene Wayne

SOUTH DAKOTA
Berry, Charles Richard, Jr
Brown, Michael Lee
Dieter, Charles David
Hamilton, Steven J
Keenlyne, Kent Douglas
Scalet, Charles George
Schmulbach, James C
Uresk, Daniel William

TENNESSEE
Benz, George William
Bulow, Frank Joseph
Cada, Glenn Francis
Chance, Charles Jackson
Dearden, Boyd L
Dimmick, Ralph W
Eddlemon, Gerald Kirk
Etnier, David Allen
Kroodsma, Roger Lee
Loar, James M
McCoy, Ralph Hines
Maier, Kurt Jay
Rose, Kenneth Alan
Sliger, Wilburn Andrew
Webb, J(ohn) Warren
Wilson, James Larry

TEXAS
Anderson, Richard Orr
Blankenship, Lytle Houston
Brown, Robert Dale
Caillouet, Charles W, Jr
Chaney, Allan Harold
Clark, Donald Ray, Jr
Darnell, Rezneat Milton
Dibble, John Thomas
Drawe, D Lynn
Fulbright, Timothy Edward
Halls, Lowell Keith
Hellier, Thomas Robert, Jr
Hubbs, Clark
Lewis, Donald Howard
Lieb, Carl Sears
Linton, Thomas LaRue
McAllister, Chris Thomas
McEachran, John D
Maki, Alan Walter
Mathews, Nancy Ellen
Neill, William Harold
Parker, Nick Charles
Schmidly, David James
Scudday, James Franklin
Stransky, John Janos
Strawn, Robert Kirk
Sullivan, Arthur Lyon
Telfair, Raymond Clark, II
Thill, Ronald E
Varner, Larry Weldon
Weller, Milton Webster
Whiteside, Bobby Gene
Zagata, Michael DeForest

UTAH
Bissonette, John Alfred
Bosakowski, Thomas
Chapman, Joseph Alan
Conover, Michael Robert
Dueser, Raymond D
Flinders, Jerran T
Hales, Donald Caleb
Helm, William Thomas
Kadlec, John A
Knowlton, Frederick Frank
Mangum, Fredrick Anthony
Maurer, Brian Alan
Rajagopal, P K
Sigler, John William
Sigler, William Franklin
Thorpe, Bert Duane
Urness, Philip Joel

VIRGINIA
Batts, Billy Stuart
Berryman, Jack Holmes
Burreson, Eugene M
Chittenden, Mark Eustace, Jr
Cordes, Donald Ormond
Cross, Gerald Howard
Dane, Charles Warren
Davis, William Spencer
Giles, Robert H, Jr
Haven, Dexter Stearns
Hosner, John Frank
Jahn, Laurence R
Kirkpatrick, Roy Lee
Loesch, Joseph G
McGinnes, Burd Sheldon
Neeper, Ralph Arnold
Neves, Richard Joseph
Reed, James Robert, Jr
Scanlon, Patrick Francis
Scattergood, Leslie Wayne
Sprague, Lucian Matthew
Streeter, Robert Glen
Sylvester, Joseph Robert
Talbot, Lee Merriam
Vaughan, Michael Ray

WASHINGTON
Allen, Julia Natalia
Anderson, James Jay
Aron, William Irwin
Barker, Morris Wayne
Bax, Nicholas John
Bell, Milo C
Bevan, Donald Edward
Burgner, Robert Louis
Catts, Elmer Paul
Clark, William Greer
Dauble, Dennis Deene
Dickhoff, Walton William
Emlen, John Merritt
Evermann, James Frederick

Fish & Wildlife Sciences (cont)

Fredin, Reynold A
Fukuhara, Francis M
Gilbert, Frederick Franklin
Ginn, Thomas Clifford
Gunderson, Donald Raymond
Hanson, Wayne Carlyle
Hard, Jeffrey John
Hardesty, Linda Howell
Hardy, Ronald W
Hayes, Murray Lawrence
Heinle, Donald Roger
Henry, Kenneth Albin
Jefferts, Keith Bartlett
Karr, James Richard
Laevastu, Taivo
Landolt, Marsha LaMerle
Low, Loh-Lee
Maltzeff, Eugene M
Manuwal, David Allen
Marion, Wayne Richard
Marshall, William Hampton
Mearns, Alan John
Nakatani, Roy E
Novotny, Anthony James
Pauley, Gilbert Buckhannan
Raymond, Howard Lawrence
Rogers, Donald Eugene
Roos, John Francis
Royce, William Francis
Salo, Ernest Olavi
Scheffer, Victor B
Schroeder, Michael Allen
Seymour, Allyn H
Shaklee, James Brooker
Silliman, Ralph Parks
Stauffer, Gary Dean
Thompson, Richard Baxter
Thorne, Richard Eugene
Utter, Fred Madison
Wedemeyer, Gary Alvin
Whitney, Richard Ralph
Woelke, Charles Edward

WEST VIRGINIA
Bullock, Graham Lambert
Henderson-Arzapalo, Anne Patricia
Michael, Edwin Daryl
Samuel, David Evan
Shalaway, Scott D
Smith, Robert Leo

WISCONSIN
Anderson, Raymond Kenneth
Baylis, Jeffrey Rowe
Copes, Frederick Albert
Dahlgren, Robert Bernard
Friend, Milton
Hickey, Joseph James
Hine, Ruth Louise
Hunt, Robert L
Keith, Lloyd Burrows
McCabe, Robert Albert
Magnuson, John Joseph
Rolley, Robert Ewell
Rongstad, Orrin James
Ruff, Robert LaVerne
Rusch, Donald Harold
Schmitz, William Robert
Stearns, Forest
Temple, Stanley A
Trainer, Daniel Olney

WYOMING
Baxter, George T
Hayden-Wing, Larry Dean
Hubert, Wayne Arthur
Stanton, Nancy Lea
Wade, Dale A

PUERTO RICO
Wilson, Marcia Hammerquist

ALBERTA
Sodhi, Navjot Singh
Speller, Stanley Wayne
Stirling, Ian G

BRITISH COLUMBIA
Albright, Lawrence John
Beacham, Terry Dale
Bunnell, Frederick Lindsley
Cheng, Kimberly Ming-Tak
Copes, Parzival
Domenici, Paolo
Donaldson, Edward Mossop
Galindo-Leal, Carlos Enrique
Hourston, Alan Stewart
Jamieson, Glen Stewart
Kennedy, William Alexander
Lane, Edwin David
Larkin, Peter Anthony
Lauzier, Raymond B
Lindsey, Casimir Charles
Margolis, Leo
Marliave, Jeffrey Burton
Newman, Murray Arthur
Perrin, Christopher John
Quinton, Dee Arlington
Reed, William J
Routledge, Richard Donovan
Wilimovsky, Norman Joseph
Zhang, Ziyang

Zwickel, Fred Charles

MANITOBA
Franzin, William Gilbert
Lawler, George Herbert
Patalas, Kazimierz
Preston, William Burton
Ward, Fredrick James

NEW BRUNSWICK
Anderson, John Murray
Erskine, Anthony J
Frantsi, Christopher
Friars, Gerald W
Hanson, John Mark
Haya, Katsuji
Kohler, Carl
Paim, Uno

NEWFOUNDLAND
Anderson, John Truman
Khan, Rasul Azim
Mercer, Malcolm Clarence
Wroblewski, Joseph S

NOVA SCOTIA
Bowen, William Donald
Dodds, Donald Gilbert
Doyle, Roger Whitney
Freeman, Harry Cleveland
Harrington, Fred Haddox
Hodder, Vincent MacKay
Kimmins, Warwick Charles
Longhurst, Alan R
Odense, Paul Holger
Rojo, Alfonso
Tremblay, Michael John
Uthe, John Frederick

ONTARIO
Ankney, C(laude) Davison
Brooks, Ronald James
Buckner, Charles Henry
Casselman, John Malcolm
Coad, Brian William
Colby, Peter J
Crossman, Edwin John
Emery, Alan Roy
Fraser, James Millan
Gibson, George
Harvey, Harold H
Ihssen, Peter Edowald
Kessler, Dan
Lavigne, David M
Law, Cecil E
McCauley, Robert William
MacCrimmon, Hugh Ross
Momot, Walter Thomas
Power, Geoffrey
Powles, Percival Mount
Regier, Henry Abraham
Reynolds, John Keith
Ryder, Richard Armitage
Savan, Milton
Smith, Donald Alan
Stasko, Aivars B
Tait, James Simpson
Tibbles, John James
Weatherley, Alan Harold

QUEBEC
Besner, Michel
Boisclair, Daniel
Gauthier, Gilles
Grant, James William Angus
Johnston, C Edward
Lapp, Wayne Stanley
Leduc, Gerard
Leggett, William C
Seguin, Louis-Roch
Shaw, Robert Fletcher
Titman, Rodger Donaldson

SASKATCHEWAN
Flood, Peter Frederick
Forsyth, Douglas John
Irvine, Donald Grant
Mitchell, George Joseph
Wobeser, Gary Arthur

OTHER COUNTRIES
Allen, Robin Leslie
Bakun, Andrew
Craig, John Frank
Crompton, David William Thomasson
Davis, William Jackson
Koslow, Julian Anthony
Ni, I-Hsun
Schmeling, Sheila Kay
Williams, Francis
Yosef, Reuven

Food Science & Technology

ALABAMA
Ammerman, Gale Richard
Dalvi, Ramesh R
Green, Nancy R
Huffman, Dale L

ARIZONA
Bessey, Paul Mack
Kline, Ralph Willard

McNamara, Donald J
Peng, Yeh-Shan

ARKANSAS
Crandall, Philip Glen
Denton, James H
Horan, Francis E

CALIFORNIA
Amerine, Maynard Andrew
Bernhard, Richard Allan
Bolin, Harold R
Boulton, Roger Brett
Chen, Ming Sheng
Chen, Tung-Shan
Clark, Walter Leighton, III
Coughlin, James Robert
Dunkley, Walter Lewis
Fleming, Sharon Elayne
Frankel, Edwin N
Gerber, Bernard Robert
Goettlich Riemann, Wilhelmina Maria Anna
Graham, Dee McDonald
Gump, Barry Hemphill
Hadley, Jeffery A
Jen, Joseph Jwu-Shan
Josephson, Ronald Victor
Kean, Chester Eugene
Lee, Chi-Hang
Lipton, Werner Jacob
Lu, Nancy Chao
Luh, Bor Shiun
Maier, V(incent) P(aul)
Meads, Philip F
Michener, H(arold) David
Montecalvo, Joseph, Jr
Nielsen, Milo Alfred
Pedersen, Leo Damborg
Prater, Arthur Nickolaus
Rabinowitz, Israel Nathan
Singleton, Vernon LeRoy
Somogyi, Laszlo P
Son, Chung Hyun
Stone, Herbert
Toma, Ramses Barsoum
Yuen, Wing

COLORADO
Avens, John Stewart
Deane, Darrell Dwight
Hess, Dexter Winfield
Long, Austin Richard
Pyler, Richard Ernst
Sofos, John N
Thomas, William Robb

CONNECTICUT
Bibeau, Thomas Clifford
Ferris, Ann M
Hall, Kenneth Noble
Kinsman, Donald Markham

DELAWARE
Hoover, Dallas Gene
Islam, Mir Nazrul
Kinney, Anthony John
Wright, Leon Wendell

DISTRICT OF COLUMBIA
Baker, Charles Wesley
Byrd, Daniel Madison, III
Cole, Margaret Elizabeth
Flora, Lewis Franklin
Miller, David Jacob
Wilson, George Donald

FLORIDA
Attaway, John Allen
Bauernfeind, Jacob (Jack) C(hristopher)
Gnaedinger, Richard H
Grimm, Charles Henry
Hardenburg, Robert Earle
Jackel, Simon Samuel
Matthews, Richard Finis
Nagy, Steven
Sabharwal, Kulbir
Sathe, Shridhar Krishna
Tamplin, Mark Lewis
Trelease, Richard Davis

GEORGIA
Brackett, Robert E
Chiang, Tze Y
Dull, Gerald G
Highland, Henry Arthur
Leffingwell, John C
Loewenstein, Morrison
Robertson, James Aldred
Schuler, George Albert
Wilson, David Merl

HAWAII
Nip, Wai Kit
Williams, David Douglas F
Wood, Betty J

IDAHO
Davidson, Philip Michael
Hibbs, Robert A

ILLINOIS
Abbott, Thomas Paul
Bookwalter, George Norman

Chen, Shepley S
Donnelly, Thomas Henry
Dunn, Dorothy Fay
Eckner, Karl Friedrich
Emken, Edward Allen
Empen, Joseph A
Flowers, Russell Sherwood, Jr
Hentges, Lynnette S W
Kuhajek, Eugene James
McKeith, Floyd Kenneth
Martin, Scott Elmore
Mateles, Richard I
Mounts, Timothy Lee
Muck, George A
Richmond, Patricia Ann
Scheid, Harold E
Sebree, Bruce Randall
Sheldon, Victor Lawrence
Singh, Laxman
Soucie, William George
Thompson, Arthur Robert
Wargel, Robert Joseph
Weingartner, Karl Ernst
Yackel, Walter Carl

INDIANA
Anderson, David Bennett
BeMiller, James Noble
Butler, Larry G
Crawford, Thomas Michael
Etzel, James Edward
Heinicke, Herbert Raymond
Ivaturi, Rao Venkata Krishna
Katz, Frances R
Lin, Ho-Mu
Nelson, Philip Edwin
Nirschl, Joseph Peter

IOWA
Dickson, James Sparrow
Johnson, Lawrence Alan
Murphy, Patricia A
Topel, David Glen

KANSAS
Bates, Lynn Shannon
Finney, Karl Frederick
Grunewald, Katharine Klevesahl
Harbers, Leniel H
Kastner, Curtis Lynn
Ponte, Joseph G, Jr

KENTUCKY
Kemp, James Dillon
Knapp, Frederick Whiton
Maisch, Weldon Frederick
Massik, Michael
Moody, William Glenn
Packett, Leonard Vasco

LOUISIANA
Broeg, Charles Burton
Constantin, Roysell Joseph
Culley, Dudley Dean, Jr
Decossas, Kenneth Miles
Dionigi, Christopher Paul
Hegsted, Maren
Johnsen, Peter Berghsey
Kadan, Ranjit Singh
Kuan, Shia Shiong
Shih, Frederick F

MAINE
Bushway, Rodney John
Todaro, Michael P

MARYLAND
Allison, Richard Gall
Bernard, Dane Thomas
Berry, Bradford William
Bluhm, Leslie
Butrum, Ritva Rauanheimo
Crosby, Edwin Andrew
Falci, Kenneth Joseph
Gilmore, Thomas Meyer
Lee, Yuen San
McBride, Gordon Williams
Metcalfe, Dean Darrel
Nelson, John Howard
Norcross, Marvin Augustus
Rechcigl, Miloslav, Jr
Salwin, Harold
Tallent, William Hugh
Webb, Byron H
Westhoff, Dennis Charles

MASSACHUSETTS
Brandler, Philip
Cardello, Armand Vincent
Gorfien, Harold
Lachica, R Victor
Learson, Robert Joseph
Salant, Abner Samuel
Taub, Irwin A(llen)
Wilkie, David Scott

MICHIGAN
Gray, Ian
Hollingsworth, Cornelia Ann
Markakis, Pericles
Muentener, Donald Arthur
Rippen, Alvin Leonard
Saldanha, Leila Genevieve
Smith, Denise Myrtle

MINNESOTA
Allen, Charles Eugene
Atwell, William Alan
Chandan, Ramesh Chandra
Fritsch, Carl Walter
Hamre, Melvin L
Lonergan, Dennis Arthur
Oelberg, Thomas Jonathon
Pour-El, Akiva
Seeley, Robert
Standing, Charles Nicholas
Sullivan, Betty J
Thenen, Shirley Warnock

MISSISSIPPI
Knight, Kathy B
McGilberry, Joe H
Marshall, Douglas Lee
Martin, James Harold
Rogers, Robert Wayne
Wang, Ping-Lieh Thomas

MISSOURI
Gordon, Dennis T
Hedrick, Harold Burdette
Heymann, Hildegarde
Lambeth, Victor Neal
Marshall, Robert T
Meyer, William Ellis
Palamand, Suryanarayana Rao
Prewitt, Larry R
Scallet, Barrett Lerner
Sidoti, Daniel Robert
Snyder, Harry E

MONTANA
Dynes, J Robert
Richards, Geoffrey Norman

NEBRASKA
Bullerman, Lloyd Bernard
Taylor, Stephen Lloyd
Taylor, Steve L
Wallen, Stanley Eugene

NEW HAMPSHIRE
Brown, Guendoline

NEW JERSEY
Adler, Irwin L
Chiang, Bin-Yea
Giddings, George Gosselin
Henrikson, Frank W
Ho, Chi-Tang
Hsu, Kenneth Hsuehchia
Hu, Ching-Yeh
Klemann, Lawrence Paul
Kleyn, Dick Henry
Lee, Tung-Ching
Smith, Robert Ewing
Tibbetts, Merrick Sawyer
Trivedi, Nayan B
Watkins, Tom R

NEW MEXICO
Del Valle, Francisco Rafael

NEW YORK
Bigelow, Sanford Walker
Bourne, Malcolm Cornelius
Cante, Charles John
Durst, Richard Allen
Efthymiou, Constantine John
Filandro, Anthony Salvatore
Gravani, Robert Bernard
Kahan, Sidney
Laster, Richard
Lewis, Bertha Ann
Moline, Sheldon Walter
Plane, Robert Allen
Reeves, Stuart Graham
Shaevel, Morton Leonard
Sherbon, John Walter
Shuler, Michael Louis
Siebert, Karl Joseph
Tzeng, Chu

NORTH CAROLINA
Beck, Charles I
Fleming, Henry Pridgen
Gold, Harvey Joseph
Humphries, Ervin G(rigg)
Johnson, George Andrew
Lineback, David R
Perfetti, Patricia F
Reed, Gerald
Webb, Neil Broyles

OHIO
Baur, Fredric John, Jr
Brooks, James Reed
Darragh, Richard T
Flautt, Thomas Joseph, Jr
Herum, Floyd L(yle)
Irmiter, Theodore Ferer
Latshaw, J David
Lee, Ken
Lee, Young-Zoon
Lindamood, John Benford
Litchfield, John Hyland
Peng, Andrew Chung Yen
Schenz, Anne Filer
Wilcox, Joseph Clifford

OKLAHOMA
Gilliland, Stanley Eugene
Seideman, Walter E
Thompson, David Russell

OREGON
Clark, James Orie, II
Holmes, Zoe Ann
Swisher, Horton Edward

PENNSYLVANIA
Benedict, Robert Curtis
Birdsall, Marion Ivens
Bowles, Bobby Linwood
Buchanan, Robert Lester
Cherry, John P
Fox, Jay B, Jr
Gadek, Frank Joseph
Juneja, Vijay Kumar
Kroger, Manfred
MacNeil, Joseph H
Mast, Morris G
Roseman, Arnold S(aul)
Rothbart, Herbert Lawrence
Rozin, Paul
Somkuti, George A
Stinson, William Sickman, Jr
Thompson, Donald B
Wagner, J Robert
Whiting, Richard Charles
Wolff, Ivan A

RHODE ISLAND
Cosgrove, Clifford James

SOUTH CAROLINA
Godfrey, W Lynn
Morr, Charles Vernon
Richards, Gary Paul
Skelley, George Calvin, Jr
Surak, John Godfrey

SOUTH DAKOTA
Schingoethe, David John

TENNESSEE
McLaughlin, Samuel Brown
Penfield, Marjorie Porter

TEXAS
Brittin, Dorothy Helen Clark
Butler, Ogbourne Duke, Jr
Chen, Anthony Hing
Engler, Cady Roy
Gardner, Frederick Albert
Greene, Donald Miller
Kubena, Karen Sidell
Phillips, Guy Frank
Ramsey, Clovis Boyd
Rooney, Lloyd William
Smith, Malcolm Crawford, Jr
Will, Paul Arthur
Yeh, Lee-Chuan Caroline

UTAH
Hendricks, Deloy G
Johnson, John Hal
Lee, Steve S

VERMONT
Bartel, Lavon L

VIRGINIA
Fellers, David Anthony
Hackney, Cameron Ray
Jones, Daniel David
LaBudde, Robert Arthur
Marcy, Joseph Edwin
Pierson, Merle Dean
Rippen, Thomas Edward
Stern, Joseph Aaron
Stewart, Kent Kallam

WASHINGTON
Ewart, Hugh Wallace, Jr
Padilla, Andrew, Jr
Pigott, George M
Powers, Joseph Robert
Pusey, P Lawrence
Rasco, Barbara A
Stansby, Maurice Earl
Widmaier, Robert George

WISCONSIN
Greger, Janet L
Ingham, Steven Charles
Kostenbader, Kenneth David, Jr
Kueper, Theodore Vincent
Lindsay, Robert Clarence
Matthews, Mary Eileen
Ogden, Robert Verl
Price, Walter Van Vranken
Schaefer, Daniel M
Scheusner, Dale Lee
Steinhart, Carol Elder
Willits, Richard Ellis

ALBERTA
Koopmans, Henry Sjoerd
Schroder, David John
Stiles, Michael Edgecombe
Whitehouse, Ronald Leslie S

BRITISH COLUMBIA
Dueck, John
Liao, Ping-huang
Li-Chan, Eunice

MANITOBA
Eskin, Neason Akiva Michael
Iverson, Stuart Leroy
Jayas, Digvir Singh
Kovacs, Miklos I P
Szekely, Joseph George

NEWFOUNDLAND
Shahidi, Fereidoon

ONTARIO
Anderson, G Harvey
Busk, Grant Curtis, Jr
Chambers, James Robert
Chu, Chun-Lung (George)
Collins, Frank William, Jr
Diosady, Levente Laszlo
Fortin, Andre Francis
Gibson, Rosalind Susan
Lips, Hilaire John
Pigden, Wallace James
Ratnayake, Walisundera Mudiyanselage Nimal
Rubin, Leon Julius
Sarwar, Ghulam
Thompson, Lilian Umale
Usborne, William Ronald
Yeung, David Lawrence

QUEBEC
Britten, Michel
Doyon, Gilles Joseph
Gariepy, Claude
Lamarche, François
Sood, Vijay Kumar

SASKATCHEWAN
Ingledew, William Michael
Knipfel, Jerry Earl
Sosulski, Frank Walter

OTHER COUNTRIES
Filadelfi-Keszi, Mary Ann Stephanie
Liu, Fu-Wen (Frank)
Lu, Christopher D
Mela, David Jason
Santidrian, Santiago

Forestry

ALABAMA
Beals, Harold Oliver
Bengtson, George Wesley
Boyer, William Davis
Davis, Terry Chaffin
DeBrunner, Louis Earl
Gilmore, Alvan Ray
Goggans, James F
Golden, Michael Stanley
Hodgkins, Earl Joseph
Holt, William Robert
Hool, James N
Huntley, Jimmy Charles
Johnson, Evert William
Larsen, Harry Stites
Livingston, Knox W
Lyle, Everett Samuel, Jr
Prakash, Channapatna Sundar
Tang, Ruen Chiu
Thompson, Emmett Frank

ALASKA
Hanley, Thomas Andrew
Juday, Glenn Patrick
Larson, Frederic Roger
Rogers, James Joseph
Slaughter, Charles Wesley
Smoker, William Alexander
Werner, Richard Allen

ARIZONA
Ahlgren, Clifford Elmer
Avery, Charles Carrington
Baker, Malchus Brooks, Jr
Boggess, William Randolph
Davis, Edwin Alden
Dealy, John Edward
Gilbertson, Robert Lee
Hull, Herbert Mitchell
Klemmedson, James Otto
Knorr, Philip Noel
Kurmes, Ernest A
Landrum, Leslie Roger
Lord, William B
McCracken, Francis Irvin
McPherson, Guy Randall
Ostrom, Carl Eric
Patton, David Roger
Reid, Charles Phillip Patrick
Robinson, William James
Ronco, Frank, Jr
Stoltenberg, Carl H
Wagle, Robert Fay
Zwolinski, Malcolm John

ARKANSAS
Baker, James Bert
Bower, David Roy Eugene

Ervin, Hollis Edward
Grizzell, Roy Ames, Jr
Ku, Timothy Tao
Shade, Elwood B
Strub, Mike Robert
Yeiser, Jimmie Lynn

CALIFORNIA
Barbour, Michael G
Beall, Frank Carroll
Becking, Rudolf (Willem)
Bensch, Klaus George
Benseler, Rolf Wilhelm
Bledsoe, Caroline Shafer
Brooks, William Hamilton
Callaham, Robert Zina
Coats, Robert N
Cobb, Fields White, Jr
Cockrell, Robert Alexander
Colwell, Robert Neil
Cooper, Charles F
Critchfield, William Burke
Davis, Lawrence S
Eriksen, Clyde Hedman
Fairfax, Sally Kirk
Fridley, Robert B
Goltz, Robert W
Helms, John Andrew
Houpis, James Louis Joseph
Jansen, Henricus Cornelis
Kelley, Allen Frederick, Jr
Kilgore, Bruce Moody
Kliejunas, John Thomas
Koonce, Andrea Lavender
Kozlowski, Theodore Thomas
Ledig, F Thomas
Levy, Barnet M
Libby, William John, (Jr)
McDonald, Philip Michael
McKillop, William L M
Murphy, Alfred Henry
Paine, Lee Alfred
Partain, Gerald Lavern
Pitelka, Louis Frank
Powers, Robert Field
Preisler, Haiganoush Krikorian
Roberts, Stephen Winston
Schniewind, Arno Peter
Storey, Theodore George
Teeguarden, Dennis Earl
Thomas, Walter Dill, Jr
Thornburgh, Dale A
Ulevitch, Richard Joel
Van Wagtendonk, Jan Willem
Vaux, Henry James
Voeks, Robert Allen
Volk, Thomas Lewis
Weatherspoon, Charles Phillip
Welch, Robin Ivor
Wilcox, W Wayne
Williams, Carroll Burns, Jr
Wilson, Carl C
Ziemer, Robert Ruhl
Zinke, Paul Joseph

COLORADO
Barney, Charles Wesley
Burns, Denver P
Dana, Robert Watson
Davidson, Ross Wallace, Jr
Fechner, Gilbert Henry
Fox, Douglas Gary
Hawksworth, Frank Goode
Hoffer, Roger M(ilton)
Kalabokidis, Kostas D
Kaufmann, Merrill R
Martinelli, Mario, Jr
Merritt, Clair
Mogren, Edwin Walfred
Pulliam, William Marshall
Ryan, Michael G
Shaw, Charles Gardner, III
Shepperd, Wayne Delbert
Shuler, Craig Edward
Striffler, William D
Tabler, Ronald Dwight
Tsuchiya, Takumi
Wangaard, Frederick Field

CONNECTICUT
Ashton, Peter Mark Shaw
Bentley, William Ross
Berlyn, Graeme Pierce
Damman, Antoni Willem Hermanus
Furnival, George Mason
Gordon, John C
Houston, David Royce
John, Hugo Herman
Smith, David Martyn
Stephens, George Robert
Ward, Jeffrey Stuart
Wargo, Philip Matthew
Winer, Herbert Isaac

DELAWARE
Kundt, John Fred

DISTRICT OF COLUMBIA
Auclair, Allan Nelson Douglas
Challinor, David
Hair, Jay Dee
Myers, Wayne Lawrence
Post, Boyd Wallace
Roskoski, Joann Pearl

Forestry (cont)

Sesco, Jerry Anthony
Smith, Richard S, Jr

FLORIDA
Blakeslee, George M
Bruns, Paul Eric
Condit, Richard
Cropper, Wendell Parker, Jr
Fatzinger, Carl Warren
Fisher, Jack Bernard
Fox, Thomas Robert
Goddard, Ray Everett
Harper, Verne Lester
Hetrick, Lawrence Andrew
Hsu, Tsong-Han
Huffman, Jacob Brainard
Janos, David Paul
Kaufman, Clemens Marcus
Long, Alan Jack
Marquis, David Alan
Outcalt, Kenneth Wayne
Smith, Wayne H
Stroh, Robert Carl
Sullivan, Edward T
Swinford, Kenneth Roberts
White, Timothy Lee

GEORGIA
Bongarten, Bruce C
Broerman, F S
Campbell, William Andrew
Dinus, Ronald John
Dwinell, Lew David
Hargreaves, Leon Abraham, Jr
Hewlett, John David
Ike, Albert Francis
Johnson, Albert Sydney, III
Kemp, Arne K
Kraus, John Franklyn
Mace, Arnett C, Jr
Nord, John C
Pepper, William Donald
Pienaar, Leon Visser
Skeen, James Norman
Sluder, Earl Ray
Sommer, Harry Edward
Steinbeck, Klaus
Ware, Kenneth Dale
Woessner, Ronald Arthur
Wood, Bruce Wade
Zarnoch, Stanley Joseph

HAWAII
Baker, Harold Lawrence
Brewbaker, James Lynn
Merriam, Robert Arnold
Skolmen, Roger Godfrey

IDAHO
Ehrenreich, John Helmuth
Fins, Lauren
Hendee, John Clare
Loewenstein, Howard
MacFarland, Craig George
Schenk, John Albright
Ulliman, Joseph James

ILLINOIS
Ashby, William Clark
Aubertin, Gerald Martin
Bhatti, Neeloo
Brown, Sandra
Dunn, Christopher Paul
Fierer, Joshua A
Guiher, John Kenneth
Holland, Israel Irving
Kirby, Hilliard Walker
Lorenz, Ralph William
Myers, Charles Christopher
Rink, George
Shen, Sinyan
Skirvin, Robert Michael
Van Sambeek, Jerome William
Walters, Charles Sebastian
Williams, David James

INDIANA
Beineke, Walter Frank
Bramble, William Clark
Byrnes, William Richard
Cassens, Daniel Lee
Chaney, William R
Gibson, Harry Gene
Holt, Harvey Allen
Hunt, Michael O'Leary
Kidd, Frank Alan
Knudson, Douglas Marvin
Moser, John William, Jr
Parker, George Ralph
Riffle, Jerry William
Senft, John Franklin

IOWA
Countryman, David Wayne
Hall, Richard Brian
Hart, Elwood Roy
Hopkins, Frederick Sherman, Jr
Mack, Michael J
McNabb, Harold Sanderson, Jr
Manwiller, Floyd George
Prestemon, Dean R
Riley, Terry Zene

Thomson, George Willis

KANSAS
Howe, Virgil K
Van Haverbeke, David F

KENTUCKY
Beatty, Oren Alexander
Coltharp, George B
Olson, James Robert
Rothwell, Frederick Mirvan

LOUISIANA
Adams, John Clyde
Allen, Arthur (Silsby)
Baldwin, Virgil Clark, Jr
Barnett, James P
Bryant, Robert L
Burns, Paul Yoder
Carpenter, Stanley Barton
Carter, Mason Carlton
Choong, Elvin T
Dell, Tommy Ray
Hough, Walter Andrew
Jewell, Frederick Forbes, Sr
Linnartz, Norwin Eugene
Lorio, Peter Leonce, Jr
Oliver, Abe D, Jr
Pezeshki, S Reza
Siegel, William Carl
Teate, James Lamar

MAINE
Ashley, Marshall Douglas
Blum, Barton Morrill
Campana, Richard John
Chacko, Rosy J
Frank, Robert McKinley, Jr
Griffin, Ralph Hawkins
Knight, Fred Barrows
Nutting, Albert Deane
Young, Harold Edle

MARYLAND
Adams, Lowell William
Brush, Grace Somers
Genys, John B
Knox, Robert Gaylord
Martin, John A
Row, Clark
Sullivan, Joseph Hearst

MASSACHUSETTS
Abbott, Herschel George
Arganbright, Donald G
Art, Henry Warren
Ashton, Peter Shaw
Bolotin, Moshe
Bond, Robert Sumner
Del Tredici, Peter James
Foster, Charles Henry Wheelwright
Larson, Joseph Stanley
MacConnell, William Preston
MacDougall, Edward Bruce
McKenzie, Malcolm Arthur
Page, Alan Cameron
Van Deusen, Paul Cook
Wilkie, David Scott
Wilson, Brayton F

MICHIGAN
Barnes, Burton Verne
Belcher, Robert Orange
Bourdo, Eric A, Jr
Carow, John
Chappelle, Daniel Eugene
Chen, Jiquan
Crowther, C Richard
Fogel, Robert Dale
Frayer, Warren Edward
Freeman, Fred W
Hanover, James W
Hart, John Henderson
Hesterberg, Gene Arthur
James, Lee Morton
Karnosky, David Frank
Kenaga, Duane Leroy
Koelling, Melvin R
Kubiske, Mark E
Lutz, Harold John
Marty, Robert Joseph
Meteer, James William
Miller, Roswell Kenfield
Niese, Jeffrey Neal
Preston, Stephen Boylan
Reed, David Doss
Ruby, John L
Simmons, Gary Adair
Sliker, Alan
Sun, Bernard Ching-Huey
White, Donald Perry
Witter, John Allen

MINNESOTA
Alm, Alvin Arthur
Bakuzis, Egolfs Voldemars
Blanchette, Robert Anthony
Boelter, Don Howard
Buech, Richard Reed
Ek, Alan Ryan
Erickson, Robert W
Frelich, Lee E
French, David W
Gertjejansen, Roland O

Gregersen, Hans Miller
Grigal, David F
Hallgren, Alvin Roland
Heinselman, Miron L
Hendricks, Lewis T
Irving, Frank Dunham
Jokela, Jalmer John
Leary, Rolfe Albert
Miller, William Eldon
Rudolf, Paul Otto
Skok, Richard Arnold
Sucoff, Edward Ira
Sullivan, Alfred Dewitt
Zasada, Zigmond Anthony

MISSISSIPPI
Amburgey, Terry L
Arner, Dale H
Doudrick, Robert Lawrence
Elam, William Warren
Foil, Robert Rodney
Friend, Alexander Lloyd
Hodges, Donald Glenn
Hodges, John Deavours
Lyon, Duane Edgar
Matthes, Ralph Kenneth, Jr
Moak, James Emanuel
Solomon, James Doyle
Switzer, George Lester
Thompson, Warren Slater
Ursic, Stanley John

MISSOURI
Brown, Merton F
Duncan, Donald Pendleton
Everson, Alan Ray
Kurtz, William Boyce
Pallardy, Stephen Gerard
Sander, Ivan Lee
Schlesinger, Richard Cary
Settergren, Carl David
Smith, Richard Chandler

MONTANA
Adams, Darius Mainard
Arno, Stephen Francis
Bay, Roger R(udolph)
Blake, George Marston
Brown, James Kerr
Carlson, Clinton E
Gatherum, Gordon Elwood
Lotan, James
Rounds, Burton Ward
Schmidt, Wyman Carl
Shearer, Raymond Charles
Steele, Robert Wilbur
Voigt, Garth Kenneth

NEBRASKA
Hergenrader, Gary Lee
Rietveld, Willis James

NEVADA
Fox, Carl Alan
Meeuwig, Richard O'Bannon
Taylor, George Evans, Jr

NEW HAMPSHIRE
Baldwin, Henry Ives
Barrett, James Passmore
Bohm, Howard A
Federer, C Anthony
Funk, David Truman
Garrett, Peter Wayne
Peart, David Ross
Safford, Lawrence Oliver
Wilson, Donald Alfred
Winch, Fred Everett, Jr

NEW JERSEY
Green, Edwin James
Hubbell, Stephen Philip
Koepp, Stephen John

NEW MEXICO
Aldon, Earl F
Gosz, James Roman
Grover, Herbert David
Heiner, Terry Charles
Hughes, Jay Melvin
Minor, Charles Oscar
Reifsnyder, William Edward

NEW YORK
Allen, Douglas Charles
Amidon, Thomas Edward
Blackmon, Bobby Glenn
Cohen, Martin William
Cote, Wilfred Arthur, Jr
Cunia, Tiberius
Dickson, James Gary
Eschner, Arthur Richard
Gladstone, William Turnbull
Greller, Andrew M
Herrington, Lee Pierce
Horn, Allen Frederick, Jr
Kohut, Robert John
Larson, Charles Conrad
Laurence, John A
Lee, Charles Northam
Leopold, Donald Joseph
Lowe, Josiah L(incoln)
McDonnell, Mark Jeffery
McNeil, Richard Jerome

Maynard, Charles Alvin
Meyer, Robert Walter
Palmer, James F
Shone, Robert Tilden
Silverborg, Savel Benhard
Stiteler, William Merle, III
Tonelli, John P, Jr
Weinstein, David Alan
Whaley, Ross Samuel
White, Edwin Henry

NORTH CAROLINA
Barden, Lawrence Samuel
Barefoot, Aldos Cortez, Jr
Booth, Donald Campbell
Brockhaus, John Albert
Brotak, Edward Allen
Bruck, Robert Ian
Bryant, Ralph Clement
Clark, James Samuel
Cooper, Arthur Wells
Cowling, Ellis Brevier
Davey, Charles Bingham
Dickerson, Willard Addison
Douglass, James Edward
Ellwood, Eric Louis
Eslyn, Wallace Eugene
Hafley, William LeRoy
Harper, James Douglas
Hart, Clarence Arthur
Haskill, John Stephen
Hastings, Felton Leo
Hellmers, Henry
Kellison, Robert Clay
Knoerr, Kenneth Richard
Kress, Lance Whitaker
Levi, Michael Phillip
Lewis, Gordon Depew
Loomis, Robert Morgan
Mott, Ralph Lionel
Peet, Robert Krug
Perry, Thomas Oliver
Richardson, Curtis John
Saylor, LeRoy C
Stambaugh, William James
Steensen, Donald H J
Stillwell, Harold Daniel
Swank, Wayne T
Swift, Lloyd Wesley, Jr
Tombaugh, Larry William
Towell, William Earnest
Zobel, Bruce John

NORTH DAKOTA
Lieberman, Diana Dale

OHIO
Boerner, Ralph E J
Brown, James Harold
Cobbe, Thomas James
Cullis, Christopher Ashley
Garland, Hereford
Gorchov, David Louis
Kriebel, Howard Burtt
Larson, Merlyn Milfred
McClaugherty, Charles Anson
Runkle, James Reade
Snyder, Gary Wayne
Sydnor, Thomas Davis
Vimmerstedt, John P
Whitmore, Frank William

OKLAHOMA
Eikenbary, Raymond Darrell
Langwig, John Edward
Lewis, David Kent
Sturgeon, Edward Earl
Walker, Nathaniel
Wittwer, Robert Frederick

OREGON
Adams, Thomas C
Amundson, Robert Gale
Bell, John Frederick
Beuter, John H
Briegleb, Philip Anthes
Brown, George Wallace
Corden, Malcolm Ernest
Demaree, Thomas L
Ethington, Robert Loren
Ferrell, William Kreiter
Hann, David William
Hansen, Everett Mathew
Heninger, Ronald Lee
Kelsey, Rick G
Krahmer, Robert Lee
McKimmy, Milford D
Maser, Chris
Mason, Richard Randolph
Merriam, Lawrence Campbell, Jr
Minore, Don
Molina, Randolph John
Myrold, David Douglas
Nelson, Earl Edward
Newton, Michael
Norris, Logan Allen
Owston, Peyton Wood
Paine, David Philip
Ripple, William John
Roth, Lewis Franklin
Rottink, Bruce Allan
Ryan, Roger Baker
Scheffer, Theodore Comstock
Skovlin, Jon Matthew

Stark, Nellie May
Stein, William Ivo
Stout, Benjamin Boreman
Suddarth, Stanley Kendrick
Sutherland, Charles F
Tappeiner, John Cummings, II
Thielges, Bart A
Trappe, James Martin
Walstad, John Daniel
Wheeler, Richard Hunting
Winjum, Jack Keith
Zaerr, Joe Benjamin

PENNSYLVANIA
Alexander, Stuart David
Beers, Thomas Wesley
Blankenhorn, Paul Richard
Bowersox, Todd Wilson
Cameron, Edward Alan
Gerhold, Henry Dietrich
Hutnik, Russell James
Jayne, Benjamin A
Jensen, Keith Frank
Klein, Edward Lawrence
Labosky, Peter, Jr
McDermott, Robert Emmet
Melton, Rex Eugene
Merrill, William
Shipman, Robert Dean
Sopper, William Edward
Steiner, Kim Carlyle
Taylor, Alan H
Towers, Barry

RHODE ISLAND
Brown, James Henry, Jr
Gould, Walter Phillip

SOUTH CAROLINA
Adams, Daniel Otis
Allen, Robert Max
Conner, William Henry
Cool, Bingham Mercur
Crosby, Emory Spear
Dunn, Benjamin Allen
Harlow, Richard Fessenden
Hook, Donal D
McGregor, William Henry Davis
Nelson, Eric Alan
Shain, William Arthur
Taras, Michael Andrew
Van Lear, David Hyde

SOUTH DAKOTA
Collins, Paul Everett
Draeger, William Charles

TENNESSEE
Barnett, Paul Edward
Cheston, Charles Edward
Core, Harold Addison
Dale, Virginia House
Kuo, Chung-Ming
Luxmoore, Robert John
Migney, Arnold Louis
Miller, Neil Austin
Norby, Richard James
Patric, James Holton
Ramke, Thomas Franklin
Schneider, Gary
Sharp, John Buckner
Taft, Kingsley Arter, Jr
Thor, Eyvind
Tillman, Larry Jaubert
Weaver, George Thomas
Wells, Garland Ray
Woods, Frank Wilson

TEXAS
Adair, Kent Thomas
Baker, Robert Donald
Bilan, M Victor
Lowry, Gerald Lafayette
McGrath, William Thomas
Merrifield, Robert G
Shreve, Loy William
Stransky, John Janos
Sullivan, Arthur Lyon
Thill, Ronald E
Van Buijtenen, Johannes Petrus
Venketeswaran, S
Walker, Laurence Colton
Watterston, Kenneth Gordon
Zagata, Michael DeForest

UTAH
Bollinger, William Hugh
DeByle, Norbert V
Downing, Kenton Benson
Fisher, Richard Forrest
Gay, Charles Wilford
Hart, George Emerson, Jr
Krebill, Richard G
Lanner, Ronald Martin
Lassen, Laurence E
Lilieholm, Robert John

VERMONT
Echelberger, Herbert Eugene
Forcier, Lawrence Kenneth
Freeman, Jeffrey VanDuyne
Gregory, Robert Aaron
Hamilton, Lawrence Stanley
Tyree, Melvin Thomas

Westing, Arthur H
Whitmore, Roy Alvin, Jr
Wilkinson, Ronald Craig

VIRGINIA
Brown, Gregory Neil
Burkhart, Harold Eugene
Burns, Russell MacBain
Callahan, James Thomas
Glasser, Wolfgang Gerhard
Hall, Otis F
Hosner, John Frank
Ifju, Geza
Kreh, Richard Edward
Krugman, Stanley Liebert
Little, Elbert L(uther), Jr
Lyon, Robert Lyndon
McElwee, Robert L
McGinnes, Burd Sheldon
Morison, Ian George
Ross, Eldon Wayne
Salom, Scott Michael
Smith, David William
Sullivan, John Dennis
Youngs, Robert Leland

WASHINGTON
Agee, James Kent
Bare, Barry Bruce
Bethel, James Samuel
Bethlahmy, Nedavia
Boyd, Charles Curtis
Brown, John Henry
Chapman, Roger Charles
Curtis, Robert Orin
Daniels, Jess Donald
DeBell, Dean Shaffer
DePuit, Edward J
Dingle, Richard William
Driver, Charles Henry
Edmonds, Robert L
Farnum, Peter
Franklin, Jerry Forest
Gaines, Edward M(cCulloch)
Gara, Robert I
Gessel, Stanley Paul
Hardesty, Linda Howell
Hatheway, William Howell
Heebner, Charles Frederick
Heilman, Paul E
Johnson, Norman Elden
Kimmey, James William
Kutscha, Norman Paul
Landsberg, Johanna D (Joan)
McCarthy, Joseph L(ePage)
Maloney, Thomas M
Marion, Wayne Richard
Megahan, Walter Franklin
Megraw, Robert Arthur
Noskowiak, Arthur Fredrick
Pierce, William R
Ritchie, Gary Alan
Rustagi, Krishna Prasad
Satterlund, Donald Robert
Schreuder, Gerard Fritz
Scott, David Robert Main
Sharpe, Grant William
Shirley, Frank Connard
Steinbrenner, Eugene Clarence
Stettler, Reinhard Friederich
Stonecypher, Roy W
Tanaka, Yasuomi
Waggener, Thomas Runyan

WEST VIRGINIA
Carvell, Kenneth Llewellyn
Crist, John Benjamin
Gillespie, William Harry
Maxwell, Robert Haworth
Peters, Penn A
Smith, Walton Ramsay
Stephenson, Steven Lee
White, David Evans
Wiant, Harry Vernon, Jr
Zinn, Gary William

WISCONSIN
Buongiorno, Joseph
Cunningham, Gordon Rowe
Einspahr, Dean William
Engelhard, Robert J
Fleischer, Herbert Oswald
Guries, Raymond Paul
Highley, Terry L
Hillier, Richard David
Johnson, Morris Alfred
Kubler, Hans
Larsen, Michael John
Larson, Philip Rodney
Lee, Chen Hui
Nelson, Neil Douglas
Patton, Robert Franklin
Quirk, John Thomas
Splitter, Gary Allen
Stearns, Forest
Vick, Charles Booker

WYOMING
Vance, George Floyd

PUERTO RICO
Wadsworth, Frank H

ALBERTA
Bella, Imre E
Dancik, Bruce Paul
Pluth, Donald John
Singh, Teja
Sodhi, Navjot Singh
Swanson, Robert Harold
Wein, Ross Wallace
Wong, Horne Richard
Yeh, Francis Cho-hao

BRITISH COLUMBIA
Alfaro, Rene Ivan
Banister, Eric Wilton
Binkley, Clark Shepard
Borden, John Harvey
Bunce, Hubert William
Chase, William Henry
Cottell, Philip L
Dobbs, Robert Curry
Drew, T John
Ekramoddoullah, Abul Kalam M
El-Kassaby, Yousry A
Etheridge, David Elliott
Gardner, Joseph Arthur Frederick
Golding, Douglas Lawrence
Haddock, Philip George
Hatton, John Victor
Holl, Frederick Brian
Hunt, Richard Stanley
Illingworth, Keith
Kennedy, Robert William
Kimmins, James Peter (Hamish)
Kozak, Antal
Lester, Donald Thomas
McLean, John Alexander
Manville, John Fieve
Marshall, Peter Lawrence
Meagher, Michael Desmond
Mitchell, Kenneth John
Namkoong, Gene
Owens, John N
Pollard, Douglas Frederick William
Prasad, Raghubir (Raj)
Preston, Caroline Margaret
Promnitz, Lawrence Charles
Reed, William J
Routledge, Richard Donovan
Shamoun, Simon Francis
Smith, Richard Barrie
Thirgood, Jack Vincent
Trussell, Paul Chandos
Von Aderkas, Patrick Jurgen Cecil
Weetman, Gordon Frederick
Worrall, John Gatland

NEW BRUNSWICK
Anderson, John Murray
Baskerville, Gordon Lawson
Bonga, Jan Max
Eidt, Douglas Conrad
Eveleigh, Eldon Spencer
Fowler, Donald Paige
Ker, John William
MacLean, David Andrew
Magasi, Laszlo P
Powell, Graham Reginald
Van Groenewoud, Herman

NEWFOUNDLAND
Bajzak, Denes
Carroll, Allan Louis
Newton, Peter Francis

NOVA SCOTIA
Bidwell, Roger Grafton Shelford
Cooper, John (Hanwell)

ONTARIO
Armson, Kenneth Avery
Basham, Jack T
Buckner, Charles Henry
Burger, Dionys
Carmean, Willard Handy
Colombo, Stephen John
Davidson, Alexander Grant
Duncan, John Robert
Farmer, Robert E, Jr
Farrar, John Laird
Fleming, Richard Arthur
Foster, Neil William
Ginns, James Herbert
Jeglum, John Karl
Johnsen, Kurt H
Jorgensen, Erik
Kayll, Albert James
Kean, Vanora Mabel
Kondo, Edward Shinichi
Lee, Hulbert Austin
Love, David Vaughan
Morrison, Ian Kenneth
Newnham, Robert Montague
Nordin, Vidar John
Redmond, Douglas Rollen
Rice, Peter (Franklin)
Setliff, Edson Carmack
Shoemaker, Robert Alan
Simard, Albert Joseph
Van Wagner, Charles Edward
Whitney, Roy Davidson
Winget, Carl Henry
Yeatman, Christopher William

QUEBEC
Bigras, Francine Jeanne
Bourchier, Robert James
Brassard, Andre
Darveau, Marcel
Daviault, Lionel
Doucet, Rene A
Fortin, Joseph Andre
Frisque, Gilles
Girouard, Ronald Maurice
Grandtner, Miroslav Marian
Hardy, Yvan J
Koran, Zoltan
Lachance, Denis
Laflamme, Gaston
Lafond, Andre
Lavallee, Andre
Majcen, Zoran
Margolis, Hank A
O'Neil, Louis C
Ouimet, Rock
Richard, Claude
Ruel, Jean-Claude

OTHER COUNTRIES
Budowski, Gerardo
Hocking, Drake
Nienstaedt, Hans
Rakowski, Krzysztof J
Resch, Helmuth

Horticulture

ALABAMA
Amling, Harry James
Ammerman, Gale Richard
Biswas, Prosanto K
Bonsi, Conrad K
Chambliss, Oyette Lavaughn
Cochis, Thomas
Gilliam, Charles Homer
Greenleaf, Walter Helmuth
Locy, Robert Donald
Norton, Joseph Daniel
Orr, Henry Porter
Perkins, Donald Young
Reynolds, Charles William
Sanderson, Kenneth Chapman
Sapra, Val T
Sayers, Earl Roger
Ward, Coleman Younger
Whatley, Booker Tillman

ALASKA
Roop, Richard Allan

ARIZONA
Alcorn, Stanley Marcus
Bessey, Paul Mack
Crosswhite, Carol D
Crosswhite, Frank Samuel
Fazio, Steve
Hitz, Chester W
Johnson, Herbert Windal
Kneebone, William Robert
Muramoto, Hiroshi
Oekber, Norman Fred
Paris, Clark Davis
Rodney, David Ross
Romney, Evan M
Schonhorst, Melvin Herman
Sharples, George Carroll
Thompson, Anson Ellis
Tucker, Thomas Curtis
Voigt, Robert Lee
Wright, Daniel Craig

ARKANSAS
Bradley, George Alexander
Einert, Alfred Erwin
Kattan, Ahmed A
Martin, Lloyd W
Moore, James Norman
Morris, Justin Roy
Motes, Dennis Roy
Reed, Hazell
Rom, Roy Curt
York, John Owen

CALIFORNIA
Amerine, Maynard Andrew
Angell, Frederick Franklyn
Baldy, Marian Wendorf
Baldy, Richard Wallace
Barham, Warren Sandusky
Bergh, Berthold Orphie
Beyer, Edgar Herman
Bitters, Willard Paul
Bowerman, Earl Harry
Brewer, Robert Franklin
Bringhurst, Royce S
Catlin, Peter Bostwick
Chambers, Robert Rood
Cook, James Arthur
Crane, Julian Coburn
Davis, Elmo Warren
Dawson, Kerry J
Dempsey, Wesley Hugh
Einset, John William
Embleton, Tom William
Epstein, Emanuel
Fridley, Robert B
Gibeault, Victor Andrew

Horticulture (cont)

Griggs, William Holland
Hagan, William Leonard
Hanson, George Peter
Harding, James A
Harrington, James Foster
Harris, Richard Wilson
Hartmann, Hudson Thomas
Hess, Charles
Houck, Laurie Gerald
Humaydan, Hasib Shaheen
Johannessen, George Andrew
Jones, Winston William
Kader, Adel Abdel
Kester, Dale Emmert
Koch, Gary Marlin
Kofranek, Anton Miles
Kohl, Harry Charles, Jr
Lewellen, Robert Thomas
Lilleland, Omund
Lipton, Werner Jacob
Little, Thomas Morton
Lorenz, Oscar Anthony
Lundeen, Glen Alfred
MacDonald, James Douglas
McNall, Lester R
Madison, John Herbert, Jr
Madore, Monica Agnes
Martin, George C
Matkin, Oris Arthur
Morris, Leonard Leslie
Munns, Donald Neville
Muth, Gilbert Jerome
Nichols, Courtland Geoffrey
Ornduff, Robert
Pitts, Donald James
Pollack, Bernard Leonard
Pollard, James Edward
Pratt, Harlan Kelley
Prend, Joseph
Puri, Yesh Paul
Rains, Donald W
Ramming, David Wilbur
Rappaport, Lawrence
Rawal, Kanti M
Reuther, Walter
Rick, Charles Madeira, Jr
Robinson, Frank Ernest
Rubatzky, Vincent E
Rubenstein, Howard S
Ryder, Edward Jonas
Ryugo, Kay
Sacher, Robert Francis
Saltveit, Mikal Endre, Jr
Simpson, John Ernest
Soost, Robert Kenneth
Stoner, Martin Franklin
Taylor, Oliver Clifton
Thomas, Paul Clarence
Thomas, Walter Dill, Jr
Thompson, David J
Thomson, John Ansel Armstrong
Tisserat, Brent Howard
Uriu, Kiyoto
Wallace, Arthur
Watson, Donald Pickett
White, Thomas Gailand
Wilkins, Harold
Wyatt, Colen Charles
Yamaguchi, Masatoshi
Zary, Keith Wilfred
Zink, Frank W

COLORADO
Alford, Donald Kay
Basham, Charles W
Brink, Kenneth Maurice
DeMooy, Cornelis Jacobus
Ells, James E
Feucht, James Roger
Hanan, Joe John
Havis, John Ralph
Holm, David Garth
Klett, James Elmer
Knutson, Kenneth Wayne
Laughlin, Charles William
Moore, Frank Devitt, III
Nabors, Murray Wayne
Stushnoff, Cecil
Townsend, Charley E
Welsh, James Ralph
Yun, Young Mok

CONNECTICUT
Ahrens, John Frederick
Ashley, Richard Allan
Carpenter, Edwin David
Jaynes, Richard Andrus
Kennard, William Crawford
Koths, Jay Sanford
Poincelot, Raymond Paul, Jr
Singha, Suman
Stephens, George Robert
Waxman, Sidney

DELAWARE
Dunham, Charles W
Green, Jerome
Riggleman, James Dale
Smith, Constance Meta

DISTRICT OF COLUMBIA
Dudley, Theodore R
Eisa, Hamdy Mahmoud
Marlowe, George Albert, Jr
Tompkins, Daniel Reuben
Wiggans, Samuel Claude

FLORIDA
Albrigo, Leo Gene
Bassett, Mark Julian
Brolmann, John Bernardus
Bryan, Herbert Harris
Bullard, Ervin Trowbridge
Burch, Derek George
Burdine, Howard William
Campbell, Carl Walter
Cantliffe, Daniel James
Carpenter, William John
Childers, Norman Franklin
Conover, Charles Albert
Crall, James Monroe
Davis, Ralph Lanier
Dean, Charles Edgar
Denmark, Harold Anderson
Drinkwater, William Otho
Dusky, Joan Agatha
Elmstrom, Gary William
Emino, Everett Raymond
Fieldhouse, Donald John
Fisher, Jack Bernard
Fitzpatrick, George
Gammon, Nathan, Jr
Gerber, John Francis
Gilreath, James Preston
Gorbet, Daniel Wayne
Gray, Dennis John
Guzman, Victor Lionel
Hall, Chesley Barker
Hanlon, Edward A
Haramaki, Chiko
Harbaugh, Brent Kalen
Hardenburg, Robert Earle
Hatton, Thurman Timbrook, Jr
Hearn, Charles Jackson
Huber, Donald John
Kender, Walter John
Koo, Robert Chung Jen
Locascio, Salvadore J
Lyrene, Paul Magnus
McConnell, Dennis Brooks
Marousky, Francis John
Martsolf, J David
Marvel, Mason E
Montelaro, James
Mortensen, John Alan
Mustard, Margaret Jean
Nell, Terril A
Overman, Amegda Jack
Ozaki, Henry Yoshio
Parsons, Lawrence Reed
Pieringer, Arthur Paul
Poole, Richard Turk, Jr
Quesenberry, Kenneth Hays
Reitz, Herman J
Rhodes, Ashby Marshall
Robitaille, Henry Arthur
Scudder, Walter Tredwell
Sheehan, Thomas John
Sherman, Wayne Bush
Showalter, Robert Kenneth
Soule, James
Stall, William Martin
Stoffella, Peter Joseph
Thompson, Buford Dale
Tucker, David Patrick Hislop
Verkade, Stephen Dunning
Wardowski, Wilfred Francis, II
Waters, Willie Estel
Wheaton, Thomas Adair
Wilfret, Gary Joe
Williams, Tom Vare
Wiltbank, William Joseph
Wolf, Benjamin
Woltz, Shreve Simpson

GEORGIA
Austin, Max E
Brown, Acton Richard
Daniell, Jeff Walter
Dekazos, Elias Demetrios
Donoho, Clive Wellington, Jr
Dull, Gerald G
Dwinell, Lew David
Forbes, Ian
Granberry, Darbie Merwin
Hendershott, Charles Henry, Jr
Jaworski, Casimir A
Jellum, Milton Delbert
Johnson, Wiley Carroll, Jr
Jones, J Benton, Jr
Kays, Stanley J
McCain, Francis Saxon
McCarter, States Marion
McLaurin, Wayne Jefferson
Payne, Jerry Allen
Phatak, Sharad Chintaman
Pokorny, Franklin Albert
Reilly, Charles Conrad
Smith, Morris Wade
Smittle, Doyle Allen
Sparks, Darrell
Thompson, James Marion
Tinga, Jacob Hinnes
Usherwood, Noble Ransom
Wiseman, Billy Ray
Wood, Bruce Wade
Worley, Ray Edward

HAWAII
Hartmann, Richard W
Heinz, Don J
Ito, Philip J
Kamemoto, Haruyuki
Murdoch, Charles Loraine
Nagai, Chifumi
Nakasone, Henry Yoshiki
Parvin, Philip Eugene
Rauch, Fred D
Sakai, William Shigeru
Tanabe, Michael John
Webb, David Thomas
Williams, David Douglas F

IDAHO
Barker, LeRoy N
Hansen, Leon A
Kleinkopf, Gale Eugene
Kolar, John Joseph
Kraus, James Ellsworth
Menser, Harry Alvin, Jr
Morris, John Leonard
Moser, Paul E
Muneta, Paul
Pavek, Joseph John
Sparks, Walter Chappel

ILLINOIS
Birkeland, Charles John
Brown, Charles Myers
Brown, Lindsay Dietrich
Culbert, John Robert
Dhaliwal, Amrik S
Garwood, Douglas Leon
George, William Leo, Jr
Green, Thomas L
Gross, Delmer Ferd
Hillyer, Irvin George
Hinchman, Ray Richard
Hymowitz, Theodore
Johnson, Glenn Richard
Jump, Lorin Keith
Kleiman, Morton
McPheeters, Kenneth Dale
Meyer, Martin Marinus, Jr
Portz, Herbert Lester
Shurtleff, Malcolm C, Jr
Simons, Roy Kenneth
Skirvin, Robert Michael
Stidd, Benton Maurice
Titus, John S
Tweedy, James Arthur
Van Sambeek, Jerome William
Williams, David James

INDIANA
Axtell, John David
Crane, Paul Levi
Emerson, Frank Henry
Erickson, Homer Theodore
Fitzgerald, Paul Jackson
Flint, Harrison Leigh
Goonewardene, Hilary Felix
Hayden, Richard Amherst
Janick, Jules
Michel, Karl Heinz
Mitchell, Cary Arthur
Moser, Bruno Carl
Patterson, Fred La Vern
Sammons, David James
Shalucha, Barbara
Tigchelaar, Edward Clarence
Wilcox, Gerald Eugene
Woodson, William Randolph
Wright, William Leland

IOWA
Addink, Sylvan
Buck, Griffith J
Denisen, Ervin Loren
Fehr, Walter R
Frey, Kenneth John
Hall, Charles Virdus
Kelley, James Durrett
Kliebenstein, James Bernard
McNeill, Michael John
Mahlstede, John Peter
Palmer, Reid G
Roath, William Wesley
Schilletter, Julian Claude
Thorne, John Carl
Widrlechner, Mark Peter

KANSAS
Albrecht, Mary Lewnes
Campbell, Ronald Wayne
Casady, Alfred Jackson
Ealy, Robert Phillip
Greig, James Kibler, Jr
Heiner, Robert E
Jennings, Paul Harry
Keen, Ray John
Thompson, Max Clyde
Wiest, Steven Craig

KENTUCKY
Barkley, Dwight G
Brown, Gerald Richard
Gray, Elmer
Hamilton-Kemp, Thomas Rogers
Kasperbauer, Michael J
Knavel, Dean Edgar
Lasheen, Aly M
Roberts, Clarence Richard
Stoltz, Leonard Paul
Williams, Albert Simpson

LOUISIANA
Ayo, Donald Joseph
Barrios, Earl P
Constantin, Roysell Joseph
Culley, Dudley Dean, Jr
Fletcher, William Ellis
Foret, James A
Hernandez, Teme P
Hogan, Le Moyne
Jones, Lloyd George
Newsom, Donald Wilson
O'Barr, Richard Dale
O'Rourke, Edmund Newton, Jr
Phills, Bobby Ray
Robbins, Marion LeRon
Wright, Johnie Algie

MAINE
Blackmon, Clinton Ralph
Chandler, Robert Flint, Jr
Eggert, Franklin Paul
Gough, Robert Edward
Hepler, Paul Raymond
McEwen, Currier

MARYLAND
Beste, Charles Edward
Bouwkamp, John C
Crosby, Edwin Andrew
Dame, Charles
Deitzer, Gerald Francis
Dickerson, Chester T, Jr
Faust, Miklos
Fogle, Harold Warman
Foy, Charles Daley
Fretz, Thomas Alvin
Fried, Maurice
Galletta, Gene John
Gross, Kenneth Charles
Handwerker, Thomas Samuel
Hirsh, Allen Gene
Hruschka, Howard Wilbur
Krizek, Donald Thomas
Link, Conrad Barnett
Long, James Delbert
Maas, John Lewis
McClurg, Charles Alan
Miller, Philip Arthur
Murphy, Charles Franklin
Odell, Lois Dorothea
Piringer, Albert Aloysius, Jr
San Antonio, James Patrick
Shanks, James Bates
Stark, Francis C, Jr
Steffens, George Louis
Sterrett, John Paul
Stoner, Allan K
Summers, Dennis Brian
Twigg, Bernard Alvin
Wang, Chien Yi
Warmbrodt, Robert Dale
Watada, Alley E
Waterworth, Howard E
Wiley, Robert Craig
Woodstock, Lowell Willard
Zimmerman, Richard Hale

MASSACHUSETTS
Bramlage, William Joseph
Del Tredici, Peter James
Devlin, Robert M
DeWolf, Gordon Parker, Jr
Irwin, Howard Samuel
Lazarte, Jaime Esteban
Lindstrom, Richard S
Lord, William John
Macnair, Richard Nelson
Mastalerz, John W
Mosher, Harold Elwood
Rosenau, William Allison
Saidha, Tekchand
Southwick, Franklin Wallburg
Tattar, Terry Alan
Tsung, Yean-Kai
Wyman, Donald

MICHIGAN
Adams, Maurice Wayne
Bowers, Robert Charles
Bukovac, Martin John
Davidson, Harold
Dennis, Frank George, Jr
Dewey, Donald Henry
Dilley, David Ross
Everson, Everett Henry
Freeman, Fred W
Harpstead, Dale D
Hogaboam, George Joseph
Honma, Shigemi
Howell, Gordon Stanley, Jr
Hull, Jerome, Jr
Kelly, John Francis
Kenworthy, Alvin Lawrence
Kessler, George Morton
Marshall, Dale Earnest
Putnam, Alan R

Ries, Stanley K
Southwick, Lawrence
Taylor, James Lee
Wittwer, Sylvan Harold

MINNESOTA
Barnes, Donald Kay
Batzer, Harold Otto
Breene, William Michael
Brenner, Mark
Choe, Hyung Tae
Davis, David Warren
Geadelmann, Jon Lee
Hackett, Wesley P
Hard, Cecil Gustav
Hertz, Leonard B
Hicks, Dale R
Lauer, Florian Isidore
Li, Pen H (Paul)
Moore, Glenn D
Nylund, Robert Einar
Pellett, Harold M
Samuelson, Charles R
Stadtherr, Richard James
White, Donald Benjamin
Widmer, Richard Ernest

MISSISSIPPI
Hanson, Kenneth Warren
Hegwood, Donald Augustine
Windham, Steve Lee

MISSOURI
Beckett, Jack Brown
Gaus, Arthur Edward
Lambeth, Victor Neal
Moore, James Frederick, Jr
Paul, Kamalendu Bikash
Rogers, Marlin Norbert
Sechler, Dale Truman
Trinklein, David Herbert

MONTANA
Evans, George Edward
Hovin, Arne William
Martin, John Munson
Skogley, Earl O

NEBRASKA
Coyne, Dermot P
Foster, John Edward
Lindsey, Marvin Frederick
Neild, Ralph E
Read, Paul Eugene

NEVADA
Gilbert, Dewayne Everett
Ruf, Robert Henry, Jr

NEW HAMPSHIRE
Eggleston, Patrick Myron
Kiang, Yun-Tzu
Loy, James Brent
Peirce, Lincoln Carret
Rogers, Owen Maurice
Rollins, Howard A, Jr
Routley, Douglas George
Wells, Otho Sylvester

NEW JERSEY
Ceponis, Michael John
Durkin, Dominic J
Funk, Cyril Reed, Jr
Hopfinger, J Anthony
Merritt, Richard Howard
Witherell, Peter Charles

NEW MEXICO
Cotter, Donald James
Finkner, Ralph Eugene
Hooks, Ronald Fred
McCuistion, Willis Lloyd
Malm, Norman R
Quinones, Ferdinand Antonio
Widmoyer, Fred Bixler

NEW YORK
Abayev, Michael
Anderson, Ronald Eugene
Barton, Donald Wilber
Bayer, George Herbert
Beil, Gary Milton
Blanpied, George David
Cummins, James Nelson
Dickson, Michael Hugh
Edgerton, Louis James
Everett, Herbert Lyman
Ewing, Elmer Ellis
Forshey, Chester Gene
Hagar, Silas Stanley
Lakso, Alan Neil
Lamb, Robert Consay
Langhans, Robert W
Law, David Martin
Leopold, Donald Joseph
Lieberman, Arthur Stuart
Loggins, Donald Anthony
Lowe, Carl Clifford
Minotti, Peter Lee
Mower, Robert G
Oberly, Gene Herman
Pardee, William Durley
Peck, Nathan Hiram
Perlmutter, Frank

Pool, Robert Morris
Powell, Loyd Earl, Jr
Price, Hugh C
Reisch, Bruce Irving
Robinson, Richard Warren
Robinson, Terence Lee
Seeley, John George
Shaulis, Nelson Jacob
Sheldrake, Raymond, Jr
Smalley, Ralph Ray
Smith, Vincent C
Sorrells, Mark Earl
Southwick, Richard Arthur
Steponkus, Peter Leo
Stiles, Warren Cryder
Straub, Richard Wayne
Sweet, Robert Dean
Topoleski, Leonard Daniel
Vittum, Morrill Thayer
Weigle, Jack LeRoy

NORTH CAROLINA
Ballinger, Walter Elmer
Blake, Thomas Lewis
Booth, Donald Campbell
Brim, Charles A
Collins, William Kerr
Cope, Will Allen
Cummings, George August
De Hertogh, August Albert
Emery, Donald Allen
Fantz, Paul Richard
Haynes, Frank Lloyd, Jr
Henderson, Warren Robert
Jackson, William Addison
Konsler, Thomas Rhinehart
Larson, Roy Axel
Miller, Carol Raymond
Mitchell, Earl Nelson
Nelson, Paul Victor
Peet, Mary Monnig
Raulston, James Chester
Romanow, Louise Rozak
Sanders, Douglas Charles
Shoemaker, Paul Beck
Skroch, Walter Arthur
Stier, Howard Livingston
Thomas, Frank Bancroft
Unrath, Claude Richard
Wilson, Lorenzo George

NORTH DAKOTA
Boe, Arthur Amos
Holland, Neal Stewart
Johansen, Robert H
Lana, Edward Peter
Scholz, Earl Walter

OHIO
Berry, Stanley Z
Buckingham, William Thomas
Cahoon, Garth Arthur
Deal, Don Robert
Ferree, David C
Foster, William J
Gould, Wilbur Alphonso
Hall, Franklin Robert
Hangarter, Roger Paul
Hartman, Fred Oscar
Hill, Robert George, Jr
Karas, James Glynn
Liu, Ting-Ting Y
Miller, Richard Lloyd
Pickering, Ed Richard
Reisch, Kenneth William
Sydnor, Thomas Davis
Tayama, Harry K

OKLAHOMA
Bates, Richard Pierce
Campbell, Raymond Earl
Taliaferro, Charles M
Wann, Elbert Van
Whitcomb, Carl Erwin

OREGON
Apple, Spencer Butler, Jr
Baggett, James Ronald
Bullock, Richard Melvin
Cameron, H Ronald
Cameron, James Wagner
Compton, Oliver Cecil
Coyier, Duane L
Crabtree, Garvin (Dudley)
Evans, Harold J
Frazier, William Allen
Fuchigami, Leslie H
Garren, Ralph, Jr
Hampton, Richard Owen
Hemphill, Delbert Dean
Jarrell, Wesley Michael
Johnson, Dale E
Lagerstedt, Harry Bert
Lawrence, Francis Joseph
Lombard, Porter Bronson
Mack, Harry John
Mielke, Eugene Albert
Mok, Machteld Cornelia
Roberts, Alfred Nathan
Rohde, Charles Raymond
Rykbost, Kenneth Albert
Schweitzer, Leland Ray
Smith, Orrin Ernest
Stevenson, Elmer Clark

Thompson, Maxine Marie
Ticknor, Robert Lewis
Toumadje, Arazdordi
Volland, Leonard Allan
Weiser, Conrad John
Westwood, Melvin (Neil)
Yungen, John A

PENNSYLVANIA
Armstrong, Robert John
Beach, Neil William
Beattie, James Monroe
Beelman, Robert B
Bergman, Ernest L
Blumenfield, David
Cole, Richard H
Craig, Richard
Daum, Donald Richard
Eck, Paul
Evensen, Kathleen Brown
Feuer, Robert Charles
Haeseler, Carl W
Heller, Paul R
Hill, Richard Ray, Jr
Hull, Larry Allen
Johnson, Littleton Wales
Marshall, Harold Gene
Russo, Joseph Martin
Shannon, Jack Corum
Shenk, John Stoner
Smith, Cyril Beverley
Stinson, Richard Floyd
Stinson, William Sickman, Jr
Tammen, James F
Tukey, Loren Davenport
Warner, Harlow Lester
Watschke, Thomas Lee
White, John W
Witham, Francis H

RHODE ISLAND
Hull, Richard James
Larmie, Walter Esmond
McGuire, John J
Shaw, Richard John

SOUTH CAROLINA
Alexander, Paul Marion
Alston, Jimmy Albert
Coston, Donald Claude
Courtney, William Henry, III
Dukes, Philip Duskin
Fery, Richard Lee
Halfacre, Robert Gordon
Nolan, Clifford N
Ogle, Wayne LeRoy
Pope, Daniel Townsend
Senn, Taze Leonard
Skelton, Thomas Eugene

SOUTH DAKOTA
Cholick, Fred Andrew
Kahler, Alex L
Lunden, Allyn Oscar
Nelson, Donald Carl
Peterson, Ronald M
Prashar, Paul D
Reeves, Dale Leslie

TENNESSEE
Collins, Jimmie Lee
Coorts, Gerald Duane
Downes, John Dixon
Halterlein, Anthony J
Hathcock, Bobby Ray
Josephson, Leonard Melvin
Sams, Carl Earnest
Smith, Rufus Albert, Jr
Southards, Carroll J
Swingle, Homer Dale

TEXAS
Bailey, Leo L
Bennett, William Frederick
Bockholt, Anton John
Calub, Alfonso deGuzman
Davies, Frederick T, Jr
Dibble, John Thomas
Fucik, John Edward
Hensz, Richard Albert
Hobbs, Clifford Dean
Holt, Ethan Cleddy
Hunter, Richard Edmund
Lineberger, Robert Daniel
Lipe, John Arthur
McDaniel, Milton Edward
McWilliams, Edward Lacaze
Miller, Julian Creighton, Jr
Nightingale, Arthur Esten
Niles, George Alva
Parsons, Jerry Montgomery
Paterson, Donald Robert
Peffley, Ellen Beth
Reed, David William
Reinert, James A
Rogers, Suzanne M Dethier
Ryall, A(lbert) Lloyd
Shreve, Loy William
Skelton, Bobby Joe
Smith, Dudley Templeton
Stansel, James Wilbert
Storey, James Benton
Tereshkovich, George
Thompson, Tommy Earl

Willingham, Francis Fries, Jr

UTAH
Anderson, Jay LaMar
Hamson, Alvin Russell
Jeffery, Larry S
Larsen, Robert Paul
Rasmussen, Harry Paul
Reimschussel, Ernest F
Seeley, Schuyler Drannan
Thomas, James H
Thorup, Richard M
Vest, Hyrum Grant, Jr
Walker, David Rudger
Walser, Ronald Herman

VERMONT
Bailey, Catherine Hayes
Klein, Deana Tarson
Matthews, David Livingston
Pellett, Norman Eugene

VIRGINIA
Ambler, John Edward
Barden, John Allan
Borchers, Edward Alan
Brubaker, Kenton Kaylor
Cocking, W Dean
Drake, Charles Roy
Horsburgh, Robert Laurie
Hortick, Harvey J
Milbocker, Daniel Clement
Smeal, Paul Lester
Starling, Thomas Madison
Ting, Sik Vung
Veilleux, Richard Ernest

WASHINGTON
Ackley, William Benton
Aichele, Murit Dean
Bienz, Darrel Rudolph
Buker, Robert Joseph
Butler, Jackie Dean
Clark, James Richard
Clore, Walter Joseph
Dean, Bill Bryan
Doughty, Charles Carter
Eighme, Lloyd Elwyn
Elfving, Donald Carl
Ewart, Hugh Wallace, Jr
Heilman, Paul E
Hiller, Larry Keith
Iritani, W M
Johnson, Charles Robert
Larsen, Fenton E
Martin, Mark Wayne
Norton, Robert Alan
Patterson, Max E
Peabody, Dwight Van Dorn, Jr
Peterson, John Carl
Proebsting, Edward Louis, Jr
Ryan, George Frisbie
Schekel, Kurt Anthony
Thornton, Robert Kim
Tukey, Harold Bradford, Jr
Williams, Max W
Withner, Carl Leslie, Jr
Wott, John Arthur

WEST VIRGINIA
Brown, Mark Wendell
Horton, Billy D
Welker, William V, Jr

WISCONSIN
Beck, Gail Edwin
Bingham, Edwin Theodore
Binning, Larry Keith
Bula, Raymond J
Cao, Weixing
Curwen, David
Daie, Jaleh
Dana, Malcolm Niven
Duewer, Raymond George
Gilbert, Franklin Andrew, Sr
Hagedorn, Donald James
Harrison, Helen Connolly
Hasselkus, Edward R
Holm, LeRoy George
Hopen, Herbert
Hougas, Robert Wayne
McCown, Brent Howard
Orth, Paul Gerhardt
Peloquin, Stanley J
Smith, Richard R
Stang, Elden James
Tibbitts, Theodore William
Tracy, William Francis

WYOMING
Bohnenblust, Kenneth E
Kolp, Bernard J

PUERTO RICO
Goyal, Megh R
Rodriguez-Arias, Jorge H
Salas-Quintana, Salvador

ALBERTA
Andrews, John Edwin
Krahn, Thomas Richard
Lynch, Dermot Roborg
McAndrew, David Wayne
Toop, Edgar Wesley

Horticulture (cont)

BRITISH COLUMBIA
Ballantyne, David John
Dabbs, Donald Henry
Daubeny, Hugh Alexander
Denby, Lyall Gordon
Dueck, John
Eaton, George Walter
Fisher, Francis John Fulton
Looney, Norman E
Neilsen, Gerald Henry
Putt, Eric Douglas
Quamme, Harvey Allen
Struble, Dean L

MANITOBA
Helgason, Sigurdur Bjorn
LaCroix, Lucien Joseph
Larter, Edward Nathan
McKenzie, Ronald Ian Hector
Staniforth, Richard John
Stefansson, Baldur Rosmund
Townley-Smith, Thomas Frederick
Wiebe, John

NEW BRUNSWICK
Coleman, Warren Kent
Young, Donald Alcoe

NOVA SCOTIA
Embree, Charles Gordon
Webster, David Henry

ONTARIO
Andersen, Emil Thorvald
Anstey, Thomas Herbert
Bishop, Charles Johnson
Childers, Walter Robert
Chu, Chun-Lung (George)
Fuleki, Tibor
Harney, Patricia Marie
Ingratta, Frank Jerry
Layne, Richard C
Liptay, Albert
Lougheed, Everett Charles
Loughton, Arthur
Marks, Charles Frank
Miles, Neil Wayne
Miller, Sherwood Robert
Olson, Arthur Olaf
O'Sullivan, John
Proctor, John Thomas Arthur
Riekels, Jerald Wayne
Svejda, Felicitas Julia
Tan, Chin Sheng

PRINCE EDWARD ISLAND
Cutcliffe, Jack Alexander
Gupta, Umesh C
Stewart, Jeffrey Grant

QUEBEC
Bigras, Francine Jeanne
Doyon, Gilles Joseph
Gasser, Heinz
Gauthier, Fernand Marcel
Girouard, Ronald Maurice
Klinck, Harold Rutherford
Lawson, Norman C

SASKATCHEWAN
Downey, Richard Keith
Goplen, Bernard Peter
Harvey, Bryan Laurence
Maginnes, Edward Alexander
Tanino, Karen Kikumi

OTHER COUNTRIES
Curtis, Byrd Collins
Filadelfi-Keszi, Mary Ann Stephanie
Holle, Miguel
Khush, Gurdev S
Liu, Fu-Wen (Frank)
Muniappan, Rangaswamy Naicker
Oertli, Johann Jakob

Phytopathology

ALABAMA
Backman, Paul Anthony
Bonsi, Conrad K
Bowen, Kira L
Chambliss, Oyette Lavaughn
Clark, Edward Maurice
Diener, Urban Lowell
Gudauskas, Robert Thomas
Haaland, Ronald L
Jacobsen, Barry James
Prakash, Channapatna Sundar
Weete, John Donald

ARIZONA
Alcorn, Stanley Marcus
Yohem, Karin Hummell

ARKANSAS
Jones, John Paul
Kim, Kyung Soo
Mason, Curtis Leonel
Reed, Hazell
Riggs, Robert D
Scott, Howard Allen

CALIFORNIA
Adaskaveg, James Elliott
Baldwin, James Gordon
Calavan, Edmond Clair
Calpouzos, Lucas
Campbell, Robert Noe
Chase, Ann Renee
Cobb, Fields White, Jr
Cromartie, Thomas Houston
Eckard, Kathleen
Epstein, Lynn
Erwin, Donald C
Gerik, James Stephen
Gilbertson, Robert Leonard
Gilmore, James Eugene
Hackney, Robert Ward
Hagan, William Leonard
Hildebrand, Donald Clair
Houck, Laurie Gerald
Jorgenson, Edsel Carpenter
Kliejunas, John Thomas
Kodira, Umesh Chengappa
Lewellen, Robert Thomas
Lucas, David Owen
MacDonald, James Douglas
Mankau, Reinhold
Nichols, Courtland Geoffrey
Noffsinger, Ella Mae
Ogawa, Joseph Minoru
Pitts, Robert Gary
Porter, Clark Alfred
Schlegel, David Edward
Starr, Mortimer Paul
Teviotdale, Beth Luise
Thomason, Ivan J
Tsao, Peter Hsing-tsuen
VanBruggen, Ariena H C
Van Bruggen, Ariena H C
Weinhold, Albert Raymond
Wilcox, W Wayne
Wilhelm, Stephen
Zary, Keith Wilfred

COLORADO
Dickenson, Donald Dwight
Hawksworth, Frank Goode
Helmerick, Robert Howard
Hill, Joseph Paul
Knutson, Kenneth Wayne
Laughlin, Charles William
Ruppel, Earl George
Wiesner, Loren Elwood
Yun, Young Mok

CONNECTICUT
Anagnostakis, Sandra Lee
Aylor, Donald Earl
Huang, Liang Hsiung
LaMondia, James A
Smith, Victoria Lynn
Walton, Gerald Steven
Wargo, Philip Matthew

DELAWARE
Carnahan, James Elliot
Delp, Charles Joseph
Howard, Richard James
Smith, Constance Meta

DISTRICT OF COLUMBIA
Auclair, Allan Nelson Douglas
Gilbert, Richard Gene
Khan, Mohamed Shaheed
Krigsvold, Dale Thomas
Smith, Richard S, Jr

FLORIDA
Agrios, George Nicholas
Blakeslee, George M
Cline, Kenneth Charles
Davis, Michael Jay
Dickson, Donald Ward
Engelhard, Arthur William
Feldmesser, Julius
Ferguson, James Scott
Fieldhouse, Donald John
Fulton, Winston Cordell
Giblin-Davis, Robin Michael
Gottwald, Timothy R
Gray, Dennis John
Howard, Charles Marion
Jones, John Paul
Miller, Thomas
Nguyen, Khuong Ba
Osborne, Lance Smith
Peacock, Hugh Anthony
Purcifull, Dan Elwood
Rich, Jimmy Ray
Ringel, Samuel Morris
Roberts, Daniel Altman
Sappenfield, William Paul
Smart, Grover Cleveland, Jr
Tsai, James Hsi-Cho

GEORGIA
Batra, Gopal Krishan
Brenneman, Timothy Branner
Chang, Chung Jan
Dinus, Ronald John
Dwinell, Lew David
Gitaitis, Ronald David
Hussey, Richard Sommers
Jackson, Curtis Rukes
Jaworski, Casimir A
Minton, Norman A
Mixon, Aubrey Clifton
Nes, William David
Nyczepir, Andrew Peter
Reilly, Charles Conrad
Stouffer, Richard Franklin
Sumner, Donald Ray
Todd, James Wyatt
Walker, Jerry Tyler
Wilson, David Merl
Wilson, Jeffrey Paul
Ziffer, Jack

HAWAII
Ko, Wen-hsiung
Ooka, Jeri Jean
Sipes, Brent Steven

IDAHO
Douglas, Dexter Richard
Halbert, Susan E
Hansen, Leon A
Harvey, Alan Eric
Roberto, Francisco Figueroa

ILLINOIS
Edwards, Dale Ivan
Ford, Richard Earl
Gessner, Robert V
Goss, George Robert
Green, Thomas L
Kirby, Hilliard Walker
Kleiman, Morton
Malek, Richard Barry
Shurtleff, Malcolm C, Jr
Sinclair, James Burton

INDIANA
Dunkle, Larry D
Huber, Don Morgan
Latin, Richard
Loesch-Fries, Loretta Sue
McNeill, Kenny Earl
Peterson, Lance George
Riffle, Jerry William
Scott, Donald Howard
Wilcox, James Raymond

IOWA
Marshall, William Emmett
Martinson, Charlie Anton
Norton, Don Carlos
Stuckey, Richard E
Yang, Xiao-Bing

KANSAS
Hetrick, Barbara Ann
Johnson, Lowell Boyden
Leslie, John Franklin
Sauer, David Bruce
Von Rumker, Rosmarie

KENTUCKY
Hendrix, James William
Pirone, Thomas Pascal
Williams, Albert Simpson

LOUISIANA
Benda, Gerd Thomas Alfred
Black, Lowell Lynn
Cotty, Peter John
Holcomb, Gordon Ernest
Jewell, Frederick Forbes, Sr
Klich, Maren Alice
Koike, Hideo
Rhoads, Robert E
Schlub, Robert Louis
Walker, Harrell Lynn

MAINE
Blackmon, Clinton Ralph
Steinhart, William Lee
Wilcox, Louis Van Inwegen, Jr

MARYLAND
Chitwood, David Joseph
Conway, William Scott
Damsteegt, Vernon Dale
Diener, Theodor Otto
Elgin, James H, Jr
French, Richard Collins
Goth, Robert W
Hadidi, Ahmed Fahmy
Kaper, Jacobus M
Krusberg, Lorin Ronald
Kulik, Martin Michael
Maas, John Lewis
Meiners, Jack Pearson
Moseman, John Gustav
Nickle, William R
Palm, Mary Egdahl
Podleckis, Edward Vidas
Raina, Ashok K
Rossman, Amy Yarnell
Schaad, Norman W
Waterworth, Howard E

MASSACHUSETTS
Holmes, Francis W(illiam)
Manning, William Joseph
Mount, Mark Samuel
Redington, Charles Bahr
Rohde, Richard Allen
Tattar, Terry Alan

Tsung, Yean-Kai

MICHIGAN
Andersen, Axel Langvad
Fulbright, Dennis Wayne
Hart, John Henderson
Morton, Harrison Leon
Theurer, Jessop Clair

MINNESOTA
Blanchette, Robert Anthony
French, David W
Leonard, Kurt John
MacDonald, David Howard
Percich, James Angelo
Rothman, Paul George
Rowell, John Bartlett
Schafer, John Francis
Widin, Katharine Douglas
Windels, Carol Elizabeth
Zeyen, Richard John

MISSISSIPPI
Abbas, Hamed K
Amburgey, Terry L
Bowman, Donald Houts
Doudrick, Robert Lawrence
Keeling, Bobbie Lee
Kurtz, Mark Edward
McLaughlin, Michael Ray
Ranney, Carleton David
Snazelle, Theodore Edward

MISSOURI
Backus, Elaine Athene
Carlson, Wayne C
Duncan, David Robert
Moore, James Frederick, Jr
Pettit, Robert Eugene
Wrather, James Allen
Wyllie, Thomas Dean

NEBRASKA
Doupnik, Ben, Jr
Lane, Leslie Carl
Miller, Willie
Steadman, James Robert
Vidaver, Anne Marie Kopecky

NEVADA
Thyr, Billy Dale

NEW JERSEY
Day, Peter Rodney
Jansson, Richard Keith
Peterson, Joseph Louis
Quinn, James Allen
Varney, Eugene Harvey

NEW MEXICO
Liddell, Craig Mason
Lindsey, Donald L
Melton, Billy Alexander, Jr

NEW YORK
Aist, James Robert
Bergstrom, Gary Carlton
Black, Lindsay MacLeod
Buckley, Edward Harland
Carroll, Robert Buck
Castello, John Donald
Goodwin, Stephen Bruce
Hagar, Silas Stanley
Jones, Edward David
Lamb, Robert Consay
Laurence, John A
Neimark, Harold Carl
Rosenberger, David A
Sinclair, Wayne A
Slack, Steven Allen
Worrall, James Joseph
Zitter, Thomas Andrew

NORTH CAROLINA
Acedo, Gregoria N
Apple, Jay Lawrence
Barnett, Ortus Webb, Jr
Bruck, Robert Ian
Cowling, Ellis Brevier
Eslyn, Wallace Eugene
Hebert, Teddy T
Huang, Jeng-Sheng
Kelman, Arthur
Klarman, William L
Kress, Lance Whitaker
Matthysse, Ann Gale
Neher, Deborah A
Romanow, Louise Rozak
Sasser, Joseph Neal
Shoemaker, Paul Beck
Sievert, Richard Carl
Sutton, Turner Bond
Therrien, Chester Dale
Triantaphyllou, Hedwig Hirschmann

NORTH DAKOTA
Lamey, Howard Arthur
Timian, Roland Gustav
Weiss, Michael John

OHIO
Bradfute, Oscar E
Curtis, Charles R
Deep, Ira Washington

Herr, Leonard Jay
Knoke, John Keith
Miller, Sally Ann
Nault, Lowell Raymond
Rowe, Randall Charles

OKLAHOMA
Barnes, George Lewis
Cassidy, Brandt George
Conway, Kenneth Edward
Hunger, Robert Marvin
Melouk, Hassan A
Sherwood, John L

OREGON
Coakley, Stella Melugin
Corden, Malcolm Ernest
Lund, Steve
Nelson, Earl Edward
Smiley, Richard Wayne
Walstad, John Daniel
Welty, Ronald Earle

PENNSYLVANIA
Ayers, John E
Carley, Harold Edwin
Christ, Barbara Jane
Fett, William Frederick
Gustine, David Lawrence
Halbrendt, John Marthon
Hickey, Kenneth Dyer
Kneebone, Leon Russell
Moorman, Gary William
Moreau, Robert Arthur
Royse, Daniel Joseph
Schipper, Arthur Louis, Jr
Snow, Jean Anthony
Wuest, Paul J
Zeiders, Kenneth Eugene

RHODE ISLAND
Gould, Mark D
Howard, Frank Leslie

SOUTH CAROLINA
Dickerson, Ottie J
Dowler, William Minor
Dukes, Philip Duskin
Epps, William Monroe
Fery, Richard Lee
Hays, Sidney Brooks
Thomas, Claude Earle
Zehr, Eldon Irvin

SOUTH DAKOTA
Ferguson, Michael William

TENNESSEE
Harrison, Robert Edwin
Southards, Carroll J
Young, Lawrence Dale

TEXAS
Bell, Alois Adrian
Browning, J(ohn) Artie
Cook, Charles Garland
Frederiksen, Richard Allan
Halliwell, Robert Stanley
Hunter, Richard Edmund
McGrath, William Thomas
Magill, Jane Mary (Oakes)
Marchetti, Marco Anthony
Robinson, Arin Forest
Standifer, Lonnie Nathaniel
Williams, Ralph Edward

UTAH
Anderson, Anne Joyce
Weber, Darrell J

VIRGINIA
Moore, Laurence Dale
Owens, Charles Wesley
Roane, Curtis Woodard

WASHINGTON
Aichele, Murit Dean
Chastagner, Gary A
Gurusiddaiah, Sarangamat
Haglund, William Arthur
Kimmey, James William
Martin, Mark Wayne
Mink, Gaylord Ira
Pusey, P Lawrence
Santo, Gerald S(unao)
Shaw, Charles Gardner
Toba, H(achiro) Harold

WEST VIRGINIA
Biggs, Alan Richard
Elliston, John E
Gallegly, Mannon Elihu
Kotcon, James Bernard

WISCONSIN
Arny, Deane Cedric
Burdsall, Harold Hugh, Jr
Chapman, R Keith
Goodman, Robert Merwin
Hagedorn, Donald James
Handelsman, Jo
Heggestad, Howard Edwin
Knous, Ted R
Owens, John Michael

Patton, Robert Franklin
Perry, James Warner

PUERTO RICO
Rodriguez, Rocio del Pilar

ALBERTA
Davidson, John G N
Harper, Frank Richard
Hiruki, Chuji
Huang, Henry Hung-Chang
Muendel, Hans-Henning

BRITISH COLUMBIA
Copeman, Robert James
De Boer, Solke Harmen
Eastwell, Kenneth Charles
Hunt, Richard Stanley
Runeckles, Victor Charles
Shamoun, Simon Francis
Utkhede, Rajeshwar Shamrao
Weintraub, Marvin

MANITOBA
Chong, James York

NEW BRUNSWICK
Bagnall, Richard Herbert
Strunz, G(eorge) M(artin)

NEWFOUNDLAND
Hampson, Michael Chisnall
Lim, Kiok-Puan

ONTARIO
Anderson, Terry Ross
Basham, Jack T
Busch, Lloyd Victor
Chiykowski, Lloyd Nicholas
Hall, Robert
Heath, Michele Christine
Ingratta, Frank Jerry
Jarvis, William Robert
McKeen, Wilbert Ezekiel
McKenzie, Allister Roy
Reyes, Andres Arenas
Seaman, William Lloyd
Setliff, Edson Carmack
Shoemaker, Robert Alan
Tu, Jui-Chang

QUEBEC
Coulombe, Louis Joseph
Girouard, Ronald Maurice
Laflamme, Gaston
Lavallee, Andre
Paulitz, Timothy Carl
Richard, Claude

SASKATCHEWAN
Duczek, Lorne
Kartha, Kutty Krishnan
Knott, Douglas Ronald
Makowski, Roberte Marie Denise
Mortensen, Knud
Tinline, Robert Davies

OTHER COUNTRIES
Chiarappa, Luigi
Hocking, Drake
Muchovej, James John
Rajaram, Sanjaya

Plant Breeding & Genetics

ALABAMA
Bonsi, Conrad K
Chambliss, Oyette Lavaughn
Prakash, Channapatna Sundar
Teyker, Robert Henry

ARIZONA
Sharples, George Carroll
Smith, Steven Ellsworth
Stith, Lee S
Thompson, Anson Ellis
Wilson, Frank Douglas

ARKANSAS
Huang, Feng Hou
Moore, James Norman

CALIFORNIA
Bergh, Berthold Orphie
Ellstrand, Norman Carl
Goldstein, Walter Elliott
Hall, Anthony Elmitt
Isom, William Howard
Jain, Subodh K
Ledig, F Thomas
Lewellen, Robert Thomas
Nichols, Courtland Geoffrey
Qualset, Calvin Odell

COLORADO
Cuany, Robin Louis
Holm, David Garth
Lapitan, Nora L
Oldemeyer, Robert King
Ruppel, Earl George

CONNECTICUT
Reed, Joseph

Zelitch, Israel

FLORIDA
Elmstrom, Gary William
Gray, Dennis John
Jones, David Alwyn
Prine, Gordon Madison
Sappenfield, William Paul
Tai, Peter Yao-Po

GEORGIA
Boerma, H Roger
Branch, William Dean
Duncan, Ronny Rush
Wessler, Susan R
Wilson, Jeffrey Paul

HAWAII
Borthakur, Dulal
Brewbaker, James Lynn
Nagai, Chifumi
Sipes, Brent Steven

IDAHO
Myers, James Robert
Wesenberg, Darrell

ILLINOIS
Bernardo, Rex Novero
George, William Leo, Jr
McPheeters, Kenneth Dale
Rice, Thomas B
Sachs, Martin M
Troyer, Alvah Forrest

INDIANA
Hoffbeck, Loren John
Ohm, Herbert Willis
Vierling, Richard Anthony
Wilcox, James Raymond

IOWA
Carlson, Irving Theodore
Duvick, Donald Nelson
Hallauer, Arnel Roy
Ivers, Drew Russell
Iwig, Mark Michael
Lamkey, Kendall Raye
Roath, William Wesley
Schnable, Patrick S
Widrlechner, Mark Peter

KANSAS
Baker, Charles H
Finney, Karl Frederick
Liang, George H
Martin, Terry Joe
Wassom, Clyde E

LOUISIANA
Caffey, Horace Rouse
Jones, Jack Earl
Owings, Addison Davis
Waddle, Bradford Avon

MAINE
Blackmon, Clinton Ralph

MARYLAND
Campbell, Travis Austin
Galletta, Gene John
Haynes, Kathleen Galante
Maas, John Lewis
Ng, Timothy J

MASSACHUSETTS
Wilkes, Hilbert Garrison, Jr

MICHIGAN
De Bruijn, Frans Johannes
Hogaboam, George Joseph
Theurer, Jessop Clair

MINNESOTA
Sentz, James Curtis

MISSISSIPPI
Berry, Charles Dennis
Creech, Roy G
Davis, Frank Marvin
Kilen, Thomas Clarence
Watson, Clarence Ellis, Jr

MISSOURI
Guilfoyle, Thomas J

MONTANA
Blake, Tom
Ditterline, Raymond Lee
McCoy, Thomas Joseph

NEBRASKA
Baltensperger, David Dwight
Francis, Charles Andrew
Gardner, Charles Olda
Hanway, Donald Grant
Haskins, Francis Arthur
Osterman, John Carl
Specht, James Eugene
Williams, James Henry, Jr

NEVADA
Thyr, Billy Dale

NEW HAMPSHIRE
Kiang, Yun-Tzu
Loy, James Brent

NEW JERSEY
Evans, David A
Krueger, Roger Warren
Noone, Thomas Mark
Quinn, James Amos

NEW YORK
Cummins, James Nelson
Earle, Elizabeth Deutsch
Goodwin, Stephen Bruce
Last, Robert L
Plaisted, Robert Leroy
Robinson, Richard Warren
Valentine, Fredrick Arthur
Ye, Guangning

NORTH CAROLINA
Goodman, Major M
Horton, Carl Frederick
Kehr, August Ernest
Oblinger, Diana Gelene

NORTH DAKOTA
Helm, James Leroy

OHIO
Cooper, Richard Lee

OKLAHOMA
Banks, Donald Jack
Edwards, Lewis Hiram
Verhalen, Laval M(athias)

OREGON
Frakes, Rodney Vance
Thielges, Bart A

PENNSYLVANIA
Ayers, John E
Berg, Clyde C
Cherry, John P
Christ, Barbara Jane
Craig, Richard
Leath, Kenneth T
Wheeler, Donald Alsop

SOUTH CAROLINA
Courtney, William Henry, III
Dille, John Emmanuel
Fery, Richard Lee
Jones, Alfred
Shipe, Emerson Russell
Thomas, Claude Earle

TENNESSEE
Allen, Freddie Lewis
Conger, Bob Vernon

TEXAS
Bird, Luther Smith
Browning, J(ohn) Artie
Burson, Byron Lynn
Cook, Charles Garland
Nguyen, Henry Thien
Peffley, Ellen Beth
Rogers, Suzanne M Dethier

UTAH
Wang, Richard Ruey-Chyi
Whitney, Elvin Dale

VIRGINIA
Buss, Glenn Richard
Sagaral, Erasmo G (Ras)
Strauss, Michael S
Veilleux, Richard Ernest

WASHINGTON
Konzak, Calvin Francis
Lee, James Moon
Peterson, Clarence James, Jr

WISCONSIN
Cramer, Jane Harris
Feirer, Russell Paul
Guries, Raymond Paul
Nelson, Neil Douglas
Simon, Philipp William

PUERTO RICO
Rodriguez, Jorge Luis
Salas-Quintana, Salvador

ALBERTA
Briggs, Keith Glyn
Woods, Donald Leslie
Yang, Rong-Cai

BRITISH COLUMBIA
El-Kassaby, Yousry A
Lanterman, William Stanley, III

MANITOBA
Aung, Taing
Kerber, Erich Rudolph
McVetty, Peter Barclay Edgar

NOVA SCOTIA
Aalders, Lewis Eldon
Chen, Lawrence Chien-Ming

20 / DISCIPLINE INDEX

Plant Breeding & Genetics (cont)

ONTARIO
Barrett, Spencer Charles Hilton
Burrows, Vernon Douglas
Buzzell, Richard Irving
Campbell, Kenneth Wilford
Johnsen, Kurt H
Kannenberg, Lyndon William
Kasha, Kenneth John
Kerr, Ernest Andrew
Reinbergs, Ernests
Sahota, Ajit Singh
Singh, Rama Shankar
Warwick, Suzanne Irene
Yeatman, Christopher William

PRINCE EDWARD ISLAND
Choo, Thin-Meiw

SASKATCHEWAN
Baker, Robert John
Knott, Douglas Ronald
Slinkard, Alfred Eugene

OTHER COUNTRIES
Hoisington, David A
Rajaram, Sanjaya

Range Science & Management

ALASKA
Hanley, Thomas Andrew

ARIZONA
Davis, Edwin Alden
Dealy, John Edward
Evans, Raymond Arthur
Johnsen, Thomas Norman, Jr
Klemmedson, James Otto
McPherson, Guy Randall
Morton, Howard LeRoy
Schmutz, Ervin Marcell
Smith, Edwin Lamar, Jr
Smith, Steven Ellsworth
Springfield, Harry Wayne

ARKANSAS
Pearson, Henry Alexander
Scifres, Charles Joel

CALIFORNIA
Connor, John Michael
Dawson, Kerry J
Green, Lisle Royal
Heady, Harold Franklin
Jansen, Henricus Cornelis
Love, Robert Merton
Murphy, Alfred Henry
Ratliff, Raymond Dewey
Williams, William Arnold

COLORADO
Asherin, Duane Arthur
Bement, Robert Earl
Branson, Farrel Allen
Cuany, Robin Louis
Dittberner, Phillip Lynn
Driscoll, Richard Stark
Hansen, Richard M
Hyder, Donald N
Lauenroth, William Karl
McGinnies, William Joseph
Oldemeyer, John Lee
Redente, Edward Francis
Rittenhouse, Larry Ronald
Shaw, Robert Blaine
Smith, Dwight Raymond
Woodmansee, Robert George

IDAHO
Callihan, Robert Harold
Clary, Warren Powell
Ehrenreich, John Helmuth
Pearson, Lorentz Clarence
Sharp, Lee Ajax
Wight, Jerald Ross

IOWA
Clark, William Richard
Morrical, Daniel Gene
Riley, Terry Zene

KANSAS
Hooker, Mark L

KENTUCKY
Baker, John P
Vogel, Willis Gene

LOUISIANA
Cuomo, Gregory Joseph
Grelen, Harold Eugene
Linnartz, Norwin Eugene

MISSISSIPPI
Bagley, Clyde Pattison

MISSOURI
Lewis, James Kelley
Mosher, Donna Patricia

Shiflet, Thomas Neal
Wiggers, Ernie P

MONTANA
Brown, James Kerr
Fisher, James Robert
Heitschmidt, Rodney Keith
MacNeil, Michael
Rounds, Burton Ward
Taylor, John Edgar

NEBRASKA
Nichols, James T
Power, James Francis
Waller, Steven Scobee

NEVADA
Davis, Phillip Burton
Eckert, Richard Edgar, Jr
Gifford, Gerald F

NEW JERSEY
Quinn, James Amos

NEW MEXICO
Aldon, Earl F
Allred, Kelly Wayne
Anderson, Dean Mauritz
Box, Thadis Wayne
Donart, Gary B
Dwyer, Don D
Fowler, James Lowell
Gibbens, Robert Parker
Herbel, Carlton Homer
Holechek, Jerry Lee
Pieper, Rex Delane

NORTH CAROLINA
Spears, Brian Merle

NORTH DAKOTA
Barker, William T
Frank, Albert Bernard
Hofmann, Lenat
Karn, James Frederick
Nyren, Paul Eric
Ries, Ronald Edward

OHIO
Risser, Paul Gillan

OKLAHOMA
Crockett, Jerry J
Sims, Phillip Leon

OREGON
Bedell, Thomas Erwin
Buckhouse, John Chapple
Crawford, John Arthur
Kelsey, Rick G
Krueger, William Clement
Maser, Chris
Poulton, Charles Edgar
Skovlin, Jon Matthew
Walters, Roland Dick

SOUTH DAKOTA
Severson, Keith Edward
Uresk, Daniel William

TEXAS
Arnold, James Darrell
Boutton, Thomas William
Bovey, Rodney William
Briske, David D
Britton, Carlton M
Dahl, Billie Eugene
Drawe, D Lynn
Fulbright, Timothy Edward
Grumbles, Jim Bob
Halls, Lowell Keith
Hauser, Victor La Vern
Kothmann, Merwyn Mortimer
Landers, Roger Q, Jr
McCully, Wayne Gunter
Owens, Michael Keith
Richardson, Richard Harvey
Schuster, Joseph L
Sosebee, Ronald Eugene
Swakon, Doreen H D
Thill, Ronald E
Varner, Larry Weldon
White, Larry Dale
Whitson, Robert Edd
Wright, Henry Albert

UTAH
Blauer, Aaron Clyde
Butcher, John Edward
Dobrowolski, James Phillip
Gay, Charles Wilford
Hull, Alvin C, Jr
Johnson, Douglas Allan
McKell, Cyrus Milo
Malechek, John Charles
Mueggler, Walter Frank
Provenza, Frederick Dan
Ralphs, Michael H
Urness, Philip Joel
Vallentine, John Franklin
Wood, Benjamin W
Workman, John Paul
Yorks, Terence Preston

VIRGINIA
Allen, Vivien Gore
Cook, Charles Wayne

WASHINGTON
DePuit, Edward J
Goebel, Carl Jerome
Hardesty, Linda Howell
Marion, Wayne Richard
Roche, Ben F, Jr
Schwendiman, John Leo

WEST VIRGINIA
Day, Thomas Arthur

WISCONSIN
Barnes, Robert F
Jorgensen, Neal A

WYOMING
Fisser, Herbert George
Hart, Richard Harold
Hayden-Wing, Larry Dean
Kearl, Willis Gordon
Laycock, William Anthony
Powell, Jeff
Schuman, Gerald E
Stanton, Nancy Lea
Sturges, David L
Tigner, James Robert

PUERTO RICO
Wilson, Marcia Hammerquist

ALBERTA
Bailey, Arthur W
Dormaar, Johan Frederik

BRITISH COLUMBIA
Quinton, Dee Arlington

SASKATCHEWAN
Cohen, Roger D H
Coupland, Robert Thomas
Waddington, John

OTHER COUNTRIES
Hocking, Drake

Soils & Soil Science

ALABAMA
Adams, Fred
Allen, Seward Ellery
Chien, Sen Hsiung
Coleman, Tommy Lee
Cope, John Thomas, Jr
Diamond, Ray Byford
Engelstad, Orvis P
Evans, Clyde Edsel
Folks, Homer Clifton
Fowler, Charles Sidney
Griffin, Robert Alfred
Guthrie, Richard Lafayette
Hajek, Benjamin F
Hill, Walter Andrew
Hiltbold, Arthur Edward, Jr
Hood, Joseph
Huluka, Gobena
Lyle, Everett Samuel, Jr
Molz, Fred John, III
Palmer, Robert Gerald
Rogers, Howard Topping
Struchtemeyer, Roland August
Sutherland, William Neil
Teyker, Robert Henry
Ward, Coleman Younger

ALASKA
Cochran, Verlan Leyerl
Drew, James
Laughlin, Winston Means
Sparrow, Elena Bautista

ARIZONA
Bohn, Hinrich Lorenz
Campbell, Ralph Edmund
Dutt, Gordon Richard
Gardner, Bryant Rogers
Heald, Walter Roland
Kimball, Bruce Arnold
Klemmedson, James Otto
Klopatek, Jeffrey Matthew
Martin, William Paxman
Morris, Gene Ray
Nakayama, Francis Shigeru
Rauschkolb, Roy Simpson
Ray, Howard Eugene
Robinson, Daniel Owen
Romney, Evan M
Shaw, Ellsworth
Stroehlein, Jack Lee
Sultan, Hassan Ahmed
Tucker, Thomas Curtis
Warrick, Arthur W
Whitt, Darnell Moses
Wierenga, Peter J
Wilson, Lorne Graham

ARKANSAS
Allen, Arthur Lee
Baker, James Bert
Bartlett, Frank David

Brown, Donald A
Lavy, Terry Lee
Legg, Joseph Ogden
Porter, Owen Archuel
Scott, Hubert Donovan
Tennille, Aubrey W
Thompson, Lyell
Walker, William M
Wells, Bobby R
Wolf, Duane Carl
Wright, Bill C

CALIFORNIA
Baker, Warren J
Bledsoe, Caroline Shafer
Bodman, Geoffrey Baldwin
Borchardt, Glenn (Arnold)
Broadbent, Francis Everett
Brown, Arthur Lloyd
Brownell, James Richard
Cannell, Glen H
Carlson, Robert Marvin
Cate, Robert Bancroft
Cliath, Mark Marshall
Coats, Robert N
Cook, James Arthur
Corwin, Dennis Lee
Crawford, Robert Field
Dalton, Francis Norbert
Doner, Harvey Ervin
Duniway, John Mason
Embleton, Tom William
Farmer, Walter Joseph
Focht, Dennis Douglass
Gerik, James Stephen
Gersper, Paul Logan
Goldberg, Sabine Ruth
Grimes, Donald Wilburn
Harmen, Raymond A
Harte, John
Hauxwell, Donald Lawrence
James, Ronald Valdemar
Jones, Milton Bennion
Jury, William Austin
Lambert, Royce Leone
Letey, John, Jr
Liebhardt, William C
Lund, Lanny Jack
Lunt, Owen Raynal
Maas, Eugene Vernon
Madison, John Herbert, Jr
Martin, James Paxman
Matson, Pamela Anne
Meredith, Farris Ray
Mikkelsen, Duane Soren
Munns, Donald Neville
Neel, James William
Nielsen, Donald R
Oster, James Donald
Paduana, Joseph A
Page, Albert Lee
Patten, Gaylord Penrod
Pitts, Donald James
Pomerening, James Albert
Powers, Robert Field
Pratt, Parker Frost
Rains, Donald W
Rateaver, Bargyla
Ratliff, Raymond Dewey
Reed, Marion Guy
Reisenauer, Hubert Michael
Rendig, Victor Vernon
Rhoades, James David
Robinson, Frank Ernest
Rolston, Dennis Eugene
Rubin, Jacob
Sage, Orrin Grant, Jr
Scow, Kate Marie
Simunek, Jiri
Smith, Robert Gordon
Spencer, William F
Sposito, Garrison
Stolzy, Lewis Hal
Thornton, John Irvin
Thorup, James Tat
Tsai, Kuei-Wu
Tullock, Robert Johns
Van Bruggen, Ariena H C
Wallender, Wesley William
Wauchope, Robert Donald
Weeks, Leslie Vernon
Whittig, Lynn D
Yates, Scott Raymond
Yen, Bing Cheng
Zicker, Eldon Louis
Zinke, Paul Joseph

COLORADO
Batchelder, Arthur Roland
Birkeland, Peter Wessel
Charlie, Wayne Alexander
Chiou, Cary T(sair)
Cole, C Vernon
Danielson, Robert Eldon
DeMooy, Cornelis Jacobus
Dickenson, Donald Dwight
Dittberner, Phillip Lynn
Doxtader, Kenneth Guy
Dubrovin, Kenneth P
Evans, Norman A(llen)
Fly, Claude Lee
Frasier, Gary Wayne
Gough, Larry Phillips
Grier, Charles Crocker

AGRICULTURAL & FOREST SCIENCES / 21

Guenzi, Wayne D
Heil, Robert Dean
Hutchinson, Gordon Lee
Johnson, Donal Dabell
Klute, Arnold
Knight, William Glenn
Lindsay, Willard Lyman
Luebs, Ralph Edward
McAuliffe, Clayton Doyle
MacCarthy, Patrick
Mickelson, Rome H
Miner, Frend John
Mortvedt, John Jacob
Mosier, Arvin Ray
Norstadt, Fred A
Peterson, Gary A
Porter, Lynn K
Pulliam, William Marshall
Redente, Edward Francis
Ryan, Michael G
Sabey, Burns Roy
Schiffman, Robert L
Schmehl, Willard Reed
Schmidt, S K
Shawcroft, Roy Wayne
Smika, Darryl Eugene
Soltanpour, Parviz Neil
Sommers, Lee Edwin
Steen-McIntyre, Virginia Carol
Viets, Frank Garfield, Jr
Westfall, Dwayne Gene
Willis, Wayne O
Woodmansee, Robert George
Young, Ralph Alden

CONNECTICUT
Ahrens, John Frederick
Damman, Antoni Willem Hermanus
Frink, Charles Richard
Guttay, Andrew John Robert
Hill, David Easton

DELAWARE
Sparks, Donald

DISTRICT OF COLUMBIA
Bronson, Roy DeBolt
Gilbert, Richard Gene
Horton, Maurice Lee
Jones, Alice J
Krigsvold, Dale Thomas
Schmidt, Berlie Louis
Soteriades, Michael C(osmas)
Swader, Fred Nicholas

FLORIDA
Albregts, Earl Eugene
Andreis, Henry Jerome
Ayers, Alvin Dearing
Blue, William Guard
Braids, Olin Capron
Calvert, David Victor
Carlisle, Victor Walter
Collins, Mary Elizabeth
Davidson, James Melvin
Denmark, Harold Anderson
Eno, Charles Franklin
Finkl, Charles William, II
Fitts, James Walter
Forbes, Richard Brainard
Fox, Thomas Robert
Gale, Paula M
Gammon, Nathan, Jr
Geraldson, Carroll Morton
Giblin-Davis, Robin Michael
Gilreath, James Preston
Graetz, Donald Alvin
Hammond, Luther Carlisle
Hanlon, Edward A
Hensel, Dale Robert
Hornsby, Arthur Grady
Hubbell, David Heuston
Hunter, Arvel Hatch
Kidder, Gerald
Lutrick, Monroe Cornelius
McNeal, Brian Lester
Mansell, Robert Shirley
Miles, Carl J
Muchovej, Rosa M C
O'Connor, George Albert
Outcalt, Kenneth Wayne
Overman, Amegda Jack
Pendleton, John Davis
Popenoe, Hugh
Prevatt, Rubert Waldemar
Pritchett, William Lawrence
Rao, Palakurthi Suresh Chandra
Rechcigl, John E
Rhoads, Frederick Milton
Sartain, Jerry Burton
Silver, Warren Seymour
Smith, Wayne H
Snyder, George Heft
Stone, Earl Lewis, Jr
Street, Jimmy Joe
Thornton, George Daniel
Wolf, Benjamin
Yuan, Tzu-Liang

GEORGIA
Anderson, Oscar Emmett
Bailey, George William
Box, James Ellis, Jr
Broerman, F S
Brown, David Smith
Bruce, Robert Russell
Brumund, William Frank
Burgoa, Benali
Burns, George Robert
Cheng, Weixin
Dibb, David Walter
Frere, Maurice Herbert
Gaines, Tinsley Powell
Gascho, Gary John
Giddens, Joel Edwin
Hendrix, Floyd Fuller, Jr
Hewlett, John David
Hook, James Edward
Ike, Albert Francis
Jaworski, Casimir A
Jones, J Benton, Jr
Kissel, David E
Langdale, George Wilfred
McLaurin, Wayne Jefferson
Parker, Albert John
Perkins, Henry Frank
Shuman, Larry Myers
Steiner, Jean Louise
Summer, Malcolm Edward
Tan, Kim H
Thomas, Adrian Wesley
Thomas, Frank Harry
Usherwood, Noble Ransom
Weaver, Robert Michael
White, Andrew Wilson, Jr
Wilkinson, Stanley R
Wilson, David Orin
Younts, Sanford Eugene

HAWAII
Ekern, Paul Chester
El-Swaify, Samir Aly
Fox, Robert Lee
Green, Richard E
Kanehiro, Yoshinori
Mahilum, Benjamin Comawas
Silva, James Anthony
Swindale, Leslie D
Tamimi, Yusuf Nimr
Tsuji, Gordon Yukio

IDAHO
Carter, David LaVere
Fosberg, Maynard Axel
Harder, Roger Wehe
Kincaid, Dennis Campbell
Lehrsch, Gary Allen
Lewis, Glenn C
Li, Guang Chao
Massee, Truman Winfield
Mayland, Henry Frederick
Michalson, Edgar Lloyd
Naylor, Denny Ve
Neibling, William Howard
Smith, Jay Hamilton
Sojka, Robert E
Thill, Donald Cecil
Trout, Thomas James
Westermann, D T
Wight, Jerald Ross

ILLINOIS
Ashby, William Clark
Aubertin, Gerald Martin
Banwart, Wayne Lee
Boast, Charles Warren
Bowles, Joseph Edward
Brown, Lindsay Dietrich
Brown, Sandra
Cook, David Robert
Fitch, Alanah
Gieseking, John Eldon
Hoeft, Robert Gene
Janghorbani, Morteza
Jansen, Ivan John
Jones, Robert L
King, S(anford) MacCallum
Klubek, Brian Paul
Leonard, Ralph Avery
Lynch, Darrel Luvene
Miller, Raymond Michael
Nash, Ralph Glen
Nicholaides, John J, III
Olson, Kenneth Ray
Pal, Dhiraj
Peck, Theodore Richard
Peters, Doyle Buren
Reetz, Harold Frank, Jr
Sheldon, Victor Lawrence
Singh, Laxman
Spomer, Louis Arthur
Stevenson, Frank Jay
Stewart, John Allan
Stucki, Joseph William
Varsa, Edward Charles
Wang, Li Chuan
Wesley, Dean E
Woolson, Edwin Albert

INDIANA
Ahlrichs, James Lloyd
Barber, Stanley Arthur
Bauman, Thomas Trost
Byrnes, William Richard
Case, Vernon Wesley
Chilgreen, Donald Ray
Christmas, Ellsworth P
Conyers, Emery Swinford
Cushman, John Howard
Fong, Shao-Ling
Franzmeier, Donald Paul
Huber, Don Morgan
Johnston, Cliff T
Lee, Linda Shahrabani
Lovell, Charles W(illiam), Jr
Low, Philip Funk
McFee, William Warren
Mausel, Paul Warner
Mengel, David Bruce
Nelson, Werner Lind
Ohlrogge, Alvin John
Phillips, Marvin W
Racke, Kenneth David
Robinson, Glenn Hugh
Roth, Charles Barron
Smucker, Silas Jonathan
Steinhardt, Gary Carl
Stivers, Russell Kennedy
Van Meter, Donald Eugene
White, Joe Lloyd
Wilcox, Gerald Eugene
Wischmeier, Walter Henry
Wolt, Jeffrey Duaine
Yahner, Joseph Edward

IOWA
Amemiya, Minoru
Black, Charles Allen
Bremner, John McColl
Cruse, Richard M
Dumenil, Lloyd C
Fenton, Thomas E
Frederick, Lloyd Randall
Hallberg, George Robert
Hanway, John Joseph
Hatfield, Jerry Lee
Horton, Robert, Jr
Karlen, Douglas Lawrence
Keeney, Dennis Raymond
Kirkham, Don
Logsdon, Sally D
Lohnes, Robert Alan
Loynachan, Thomas Eugene
Owen, Michael
Pesek, John Thomas, Jr
Schafer, John William, Jr
Scholtes, Wayne Henry
Scott, Albert Duncan
Swan, James Byron
Tabatabai, M Ali
Troeh, Frederick Roy
Voss, Regis D

KANSAS
Bidwell, Orville Willard
Follett, Roy Hunter
Hagen, Lawrence J
Ham, George Eldon
Harris, Pamela Ann
Harris, Wallace Wayne
Herron, George M
Hetrick, Barbara Ann
Kirkham, M B
Lyles, Leon
McElroy, Albert Dean
Murphy, John Joseph
Schaper, Laurence Teis
Skidmore, Edward Lyman
Smith, Floyd W
Stone, Loyd Raymond
Sunderman, Herbert D
Welch, Jerome E

KENTUCKY
Barnhisel, Richard I
Blevins, Robert L
Frye, Wilbur Wayne
Haney, Donald C
Hendrix, James William
Hourigan, William R
Johnson, Ray Edwin
Massey, Herbert Fane, Jr
Murdock, John Thomas
Peaslee, Doyle E
Phillips, Ronald Edward
Ragland, John Leonard
Rasnake, Monroe
Rothwell, Frederick Mirvan
Shugars, Jonas P
Terry, David Lee
Thomas, Grant Worthington
Wells, Kenneth Lincoln

LOUISIANA
Cain, Charles Columbus
Caldwell, Augustus George
DeLaune, Ronald D
Dunigan, Edward P
Golden, Laron E
Linnartz, Norwin Eugene
Lorio, Peter Leonce, Jr
Schlub, Robert Louis
Sedberry, Joseph E, Jr
Willis, Guye Henry
Willis, William Hillman
Wilson, John Thomas

MAINE
Cronan, Christopher Shaw
Eggert, Franklin Paul
Glenn, Rollin Copper
Tjepkema, John Dirk

MARYLAND
Axley, John Harold
Bensen, David Warren
Cady, John Gilbert
Coffman, Charles Benjamin
Fanning, Delvin Seymour
Foy, Charles Daley
Gupta, Gian Chand
Helling, Charles Siver
Hornick, Sharon B
Isensee, Allan Robert
Kaufman, Donald DeVere
Kline, Jerry Robert
Klingebiel, Albert Arnold
Kunishi, Harry Mikio
Levine, Elissa Robin
Meisinger, John Joseph
Miller, James Roland
Miller, Raymond Jarvis
Pattee, Oliver Henry
Rawlins, Stephen Last
Reuszer, Herbert William
Snow, Philip Anthony
Soto, Gerardo H
Starr, James LeRoy
Strickling, Edward
Weber, Deane Fay
Weil, Raymond R
Weismiller, Richard A
Willey, Cliff Rufus
Wise, David Haynes

MASSACHUSETTS
Barker, Allen Vaughan
Epstein, Eliot
Isgur, Benjamin
Lewis, Laurence A
Rosenau, William Allison
Sage, Joseph D
Troll, Joseph
Veneman, Peter Lourens Marinus

MICHIGAN
Brandt, Gerald H
Chaudhry, G Rasul
Christenson, Donald Robert
Dragun, James
Ellis, Boyd G
Erickson, Anton Earl
Foth, Henry Donald
Gast, Robert Gale
Gray, Donald Harford
Jones, Kenneth Lester
Kunze, Raymond J
Lucas, Robert Elmer
Meints, Vernon W
Mokma, Delbert Lewis
Mortland, Max Merle
Northup, Melvin Lee
Rieke, Paul Eugene
Ritchie, Joe T
Robertson, G Philip
Robertson, Lynn Shelby, Jr
Singh, Daulat
Tung, Fred Fu
Vitosh, Maurice Lee
White, Donald Perry
Whiteside, Eugene Perry

MINNESOTA
Adams, Russell S, Jr
Allmaras, Raymond Richard
Arneman, Harold Frederick
Boelter, Don Howard
Briggs, Rodney Arthur
Cheng, H(wei)-H(sien)
Clapp, C(harles) Edward
Clapp, Thomas Wright
Crookston, Robert Kent
Dowdy, Robert H
Farnham, Rouse Smith
Grava, Janis (John)
Grigal, David F
Gupta, Satish Chander
Krug, Edward Charles
Larson, William Earl
Latterell, Joseph J
Linden, Dennis Robert
Malzer, Gary Lee
Munson, Robert Dean
Olness, Alan
Overdahl, Curtis J
Pryor, Gordon Roy
Randall, Gyles Wade
Rehm, George W
Rust, Richard Henry
Tilman, G David

MISSISSIPPI
Balam, Baxish Singh
Bardsley, Charles Edward
Foster, George Rainey
Friend, Alexander Lloyd
George, Kalankamary Pily
Grissinger, Earl H
Heatherly, Larry G
Meyer, Lawrence Donald
Mutchler, Calvin Kendal
Nash, Victor E
Nelson, Lyle Engnar
Perry, Edward Belk
Peterson, Harold LeRoy
Pettry, David Emory
Shockley, W(oodland) G(ray)

Soils & Soil Science (cont)

Whisler, Frank Duane

MISSOURI
Blanchar, Robert W
Brown, James Richard
Chowdhury, Ikbalur Rashid
Clare, Stewart
Cox, Gene Spracher
Deming, John Miley
Hasan, Syed Eqbal
Henderson, Gray Stirling
Khan, Adam
Lambeth, Victor Neal
Miles, Randall Jay
Ponder, Felix, Jr
Radke, Rodney Owen
Richards, Graydon Edward
Schmidt, Norbert Otto
Shrader, William D
Wilson, Clyde Livingston
Woodruff, Clarence Merrill

MONTANA
Aase, Jan Kristian
Asleson, Johan Arnold
Bauder, James Warren
Brown, Paul Lawson
Ferguson, Albert Hayden
Fisher, James Robert
Jackson, Grant D
Nielsen, Gerald Alan
Olsen, Ralph A
Skogley, Earl O
Voigt, Garth Kenneth

NEBRASKA
Blad, Blaine L
Chesnin, Leon
Doran, John Walsh
Flowerday, Albert Dale
Garey, Carroll Laverne
Hoffman, Glenn Jerrald
Holzhey, Charles Steven
Hubbard, Kenneth Gene
Knox, Ellis Gilbert
Kolade, Alabi E
Leviticus, Louis I
Lewis, David Thomas
Long, Daryl Clyde
Lynn, Warren Clark
Mason, Stephen Carl
Miller, Willie
Nelson, Darrell Wayne
Nettleton, Wiley Dennis
Power, James Francis
Ross, Sam Jones, Jr
Sander, Donald Henry
Skopp, Joseph Michael
Sorensen, Robert Carl
Swartzendruber, Dale
Volk, Bob G
Wiese, Richard Anton

NEVADA
Bower, Charles Arthur
Cameron, Roy (Eugene)
Franson, Raymond Lee
Gilbert, Dewayne Everett
Leedy, Clark D
Miller, Watkins Wilford
Peterson, Frederick Forney
Thornburn, Thomas H(ampton)

NEW HAMPSHIRE
Federer, C Anthony
Friedland, Andrew J
Harter, Robert Duane
McDowell, William H
McKim, Harlan L
Safford, Lawrence Oliver

NEW JERSEY
Alderfer, Russell Brunner
Avissar, Roni
Douglas, Lowell Arthur
Esrig, Melvin I
Jumikis, Alfreds Richards
Monahan, Edward James
Padhi, Sally Bulpitt
Pramer, David
Tedrow, John Charles Fremont
Toth, Stephen John
Weissmann, Gerd Friedrich Horst

NEW MEXICO
Daugherty, LeRoy Arthur
Essington, Edward Herbert
Fowler, Eric Beaumont
Gile, Leland Henry
Grover, Herbert David
Johnson, Gordon Verle
Liddell, Craig Mason
Lindemann, William Conrad
Melton, Billy Alexander, Jr
Morin, George Cardinal Albert
Polzer, Wilfred L
Triandafilidis, George Emmanuel
Tromble, John M

NEW YORK
Abayev, Michael
Alexander, Martin

App, Alva A
Bouldin, David Ritchey
Cline, Marlin George
Drosdoff, Matthew
Francis, Arokiasamy Joseph
Gluck, Ronald Monroe
Grunes, David Leon
Harrison, William Paul
Lathwell, Douglas J
Linkins, Arthur Edward
Miller, Robert Demorest
Minotti, Peter Lee
Misiaszek, Edward T
Reid, William Shaw
Scott, Thomas Walter
Shannon, Stanton
Smalley, Ralph Ray
Stoll, Robert D
Stotzky, Guenther
Wagenet, Robert Jeffrey
Weinstein, David Alan
Welch, Ross Maynard
White, Edwin Henry
Wilks, Daniel S
Zwarun, Andrew Alexander

NORTH CAROLINA
Baird, Jack Vernon
Bohannon, Robert Arthur
Buol, Stanley Walter
Cassel, D Keith
Cook, Maurice Gayle
Cox, Frederick Russell
Cummings, George August
Daniels, Raymond Bryant
Davey, Charles Bingham
Gilliam, James Wendell
Grove, Thurman Lee
Hassan, Awatif E
Jackson, William Addison
Kamprath, Eugene John
King, Larry Dean
Knight, Clifford Burnham
McCants, Charles Bernard
McCollum, Robert Edmund
McCracken, Ralph Joseph
Miner, Gordon Stanley
Nelson, Paul Victor
Overcash, Michael Ray
Patil, Arvind Shankar
Qualls, Robert Gerald
Reed, William Edward
Robarge, Wayne
Shafer, Steven Ray
Shelton, James Edward
Spinks, Daniel Owen
Steila, Donald
Van Eck, Willem Adolph
Wahls, Harvey E(dward)
Weber, Jerome Bernard
Weed, Sterling Barg
Wollum, Arthur George, II
Zarnstorff, Mark Edward
Zublena, Joseph Peter

NORTH DAKOTA
Bauer, Armand
Doll, Eugene Carter
Enz, John Walter
Halvorson, Ardell David
Halvorson, Gary Alfred
Hofmann, Lenat
Merrill, Stephen Day
Prunty, Lyle Delmar
Vasey, Edfred H
Zubriski, Joseph Cazimer

OHIO
Basile, Robert Manlius
Calhoun, Frank Gilbert
Darrow, Robert A
Eckert, Donald James
Fausey, Norman Ray
Grube, Walter E, Jr
Haghiri, Faz
Hall, Geroge Frederick
Himes, Frank Lawrence
Hock, Arthur George
Hutchinson, Frederick Edward
Lal, Rattan
Logan, Terry James
Lyon, John Grimson
McClaugherty, Charles Anson
Miller, Frederick Powell
Munn, David Alan
Ryan, James Anthony
Sutton, Paul
Taylor, George Stanley
Vimmerstedt, John P
Watson, Maurice E
Zimmerman, Tommy Lynn

OKLAHOMA
Ahring, Robert M
Berg, William Albert
Enfield, Carl George
Gray, Fenton
Johnson, Gordon V
Jones, Randall Jefferies
Laguros, Joakim George
Lynd, Julian Quentin
Menzel, Ronald George
Morrill, Lawrence George
Nofziger, David Lynn

Puls, Robert W
Reed, Lester W
Sinclair, James Lewis
Smith, Samuel Joseph
Stone, John Floyd
Whitcomb, Carl Erwin

OREGON
Albrecht, Stephan LaRowe
Cheney, Horace Bellatti
Cochran, Patrick Holmes
Doyle, Jack David
Dyrness, Christen Theodore
Elliott, Lloyd Floren
Huddleston, James Herbert
Jarrell, Wesley Michael
Kling, Gerald Fairchild
Lu, Kuo Chin
Moore, Duane Grey
Myrold, David Douglas
Ramig, Robert E
Retallack, Gregory John
Rickman, Ronald Wayne
Rykbost, Kenneth Albert
Simonson, Gerald Herman
Volk, Veril Van
Vomocil, James Arthur
Warkentin, Benno Peter
Young, J Lowell
Youngberg, Chester Theodore
Yungen, John A

PENNSYLVANIA
Adovasio, James Michael
Baker, Dale E
Banerjee, Sushanta Kumar
Bollag, Jean-Marc
Bosshart, Robert Perry
Cheng, Cheng-Yin
Cunningham, Robert Lester
Eck, Paul
Fox, Richard Henry
Fritton, Daniel Dale
Hall, Jon K
Heddleson, Milford Raynord
Hunter, Albert Sinclair
Johnson, Leon Joseph
Koerner, Robert M
Komarneni, Sridhar
Levin, Michael H(oward)
Loughry, Frank Glade
Marriott, Lawrence Frederick
Petersen, Gary Walter
Pionke, Harry Bernhard
Senft, Joseph Philip
Waddington, Donald Van Pelt
White, John W

RHODE ISLAND
Miller, Robert Harold
Wright, William Ray

SOUTH CAROLINA
Camberato, James John
Corey, John Charles
Franklin, Ralph E
Goodroad, Lewis Leonard
Hawkins, Richard Horace
Jones, Ulysses Simpson, Jr
Kittrell, Benjamin Upchurch
Lane, Carl Leaton
Ligon, James T(eddie)
Nolan, Clifford N
Sandhu, Shingara Singh
Smith, Bill Ross
Van Lear, David Hyde

SOUTH DAKOTA
Carson, Paul LLewellyn
Ellsbury, Michael M
Fine, Lawrence Oliver
Kohl, Robert A
Moldenhauer, William Calvin
Wiersma, Daniel

TENNESSEE
Boswell, Fred Carlen
Brode, William Edward
Buntley, George Jule
Chiang, Thomas M
Foss, John E
Francis, Chester Wayne
Freitag, Dean R(ichard)
Fribourg, Henry August
Gamliel, Amir
Graveel, John Gerard
Kohland, William Francis
Large, Richard L
Lee, Suk Young
Lessman, Gary M
Lewis, Russell J
Lietzke, David Albert
Luxmoore, Robert John
Madhavan, Kunchithapatham
Mullen, Michael David
Naddy, Badie Ihrahim
Seatz, Lloyd Frank
Smalley, Glendon William
Springer, Maxwell Elsworth
Tamura, Tsuneo

TEXAS
Adams, John Edgar
Allen, Bonnie L

Anderson, Duwayne Marlo
Anderson, Warren Boyd
Armstrong, James Clyde
Bloodworth, M(orris) E(lkins)
Clementz, David Michael
Dibble, John Thomas
Dixon, Joe Boris
Dregne, Harold Ernest
Eck, Harold Victor
Elward-Berry, Julianne
Fountain, Lewis Spencer
Gavande, Sampat A
Gerard, Cleveland Joseph
Golden, Dadigamuwage Chandrasiri
Hauser, Victor La Vern
Heyman, Louis
Hipp, Billy Wayne
Ho, Clara Lin
Hoover, William L
Hossner, Lloyd Richard
Jones, Ordie Reginal
Kovar, John Alvis
Kunze, George William
Loeppert, Richard Henry, Jr
Louden, L Richard
Lowry, Gerald Lafayette
Melton, James Ray
Milford, Murray Hudson
Miyamoto, Seiichi
Neher, David Daniel
Newland, Leo Winburne
Nichols, Joe Dean
Nossaman, Norman L
Onken, Arthur Blake
Ott, Billy Joe
Perez, Francisco Luis
Rao, Shankaranarayana Ramohallinanjunda
Rubink, William Louis
Runge, Edward C A
Runkles, Jack Ralph
Stanford, Geoffrey
Stewart, Bobby Alton
Tackett, Jesse Lee
Taylor, Howard Melvin
Thompson, Louis Jean
Trogdon, William Oren
Unger, Paul Walter
Walker, Laurence Colton
Watterston, Kenneth Gordon
Weaver, Richard Wayne
Wendt, Charles William
Wiegand, Craig Loren
Wilding, Lawrence Paul
Young, Arthur Wesley
Zuberer, David Alan

UTAH
Belnap, Jayne
Blake, George Rowland
Dobrowolski, James Phillip
Fisher, Richard Forrest
Hanks, Ronald John
Hargreaves, George H(enry)
Harper, Kimball T
James, David Winston
Jeffery, Larry S
Jurinak, Jerome Joseph
Keller, Jack
Malek, Esmaiel
Miller, Raymond Woodruff
Muir, Melvin K
Nelson, Sheldon Douglas
Olsen, Edwin Carl, III
Openshaw, Martin David
Peterson, Howard Boyd
Rasmussen, V Philip, Jr
Sidle, Roy Carl
Skujins, John Janis
Smith, R L
Southard, Alvin Reid
Terry, Richard Ellis
Thorup, Richard M
Wood, Timothy E
Youd, Thomas Leslie

VERMONT
Bartlett, Richmond J
Magdoff, Frederick Robin
Murphy, William Michael

VIRGINIA
Allen, Vivien Gore
Blum, Linda Kay
Brady, Nyle C
Brown, Jerry
Burns, Allan Fielding
Carstea, Dumitru
Dawson, Murray Drayton
De Datta, Surajit Kumar
Duke, Everette Loranza
Flach, Klaus Werner
Hagedorn, Charles
Hakala, William Walter
Harlan, Phillip Walker
Herd, Darrell Gilbert
Krebs, Robert Dixon
Kreh, Richard Edward
Landa, Edward Robert
Larew, H(iram) Gordon
McClung, Andrew Colin
Malcolm, John Lowrie
Martens, David Charles
Neal, John Lloyd, Jr

O'Neill, Eileen Jane
Pasko, Thomas Joseph, Jr
Reneau, Raymond B, Jr
Rickert, David A
Rule, Joseph Houston
Thiel, Thomas J
Zelazny, Lucian Walter

WASHINGTON
Bezdicek, David Fred
Boyd, Charles Curtis
Campbell, Gaylon Sanford
Cole, Dale Warren
Cykler, John Freuler
Engibous, James Charles
Gardner, Walter Hale
Gessel, Stanley Paul
Gilkeson, Raymond Allen
Hausenbuiller, Robert Lee
Hedges, John Ivan
Heilman, Paul E
Henderson, Lawrence J
Jensen, Creighton Randall
Kittrick, James Allen
Klock, Glen Orval
Koehler, Fred Eugene
Kuo, Shiou
Landsberg, Johanna D (Joan)
Lauer, David Allan
McDole, Robert E
Megahan, Walter Franklin
Parr, James Floyd, Jr
Phillips, Steven J
Rai, Dhanpat
Reid, Preston Harding
Relyea, John Franklin
Rieger, Samuel
Routson, Ronald C
Schwendiman, John Leo
Steinbrenner, Eugene Clarence
Stevens, Todd Owen
Thornton, Robert Kim
Wildung, Raymond Earl
Wooldridge, David Dilley

WEST VIRGINIA
Keefer, Robert Faris
Kotcon, James Bernard
Ritchey, Kenneth Dale
Sencindiver, John Coe
Singh, Rabindar Nath

WISCONSIN
Barnes, Robert F
Beatty, Marvin Theodore
Bubenzer, Gary Dean
Bundy, Larry Gene
Cain, John Manford
Corey, Richard Boardman
Edil, Tuncer Berat
Gerloff, Gerald Carl
Harpstead, Milo I
Hensler, Ronald Fred
Hole, Francis Doan
Jackson, Marion Leroy
Larsen, Michael John
McIntosh, Thomas Henry
Naik, Tarun Ratilal
Norman, John Matthew
Orth, Paul Gerhardt
Paulson, William H
Peterson, Arthur Edwin
Schneider, Allan Frank
Stelly, Matthias
Walsh, Leo Marcellus
Willis, Harold Lester

WYOMING
Hough, Hugh Walter
Katta, Jayaram Reddy
Schuman, Gerald E
Smith, James Lee
Stanton, Nancy Lea
Vance, George Floyd
Williams, Stephen Earl
Zhang, Renduo

PUERTO RICO
Goyal, Megh R
Lugo-Lopez, Miguel Angel
Schroder, Eduardo C

ALBERTA
Barendregt, Rene William
Bentley, C Fred
Chang, Chi
Coen, Gerald Marvin
Cook, Fred D
Dormaar, Johan Frederik
Dudas, Marvin Joseph
Feng, Yongsheng
Foscolos, Anthony E
Janzen, Helmut Henry
McAndrew, David Wayne
McGill, William Bruce
McMillan, Neil John
Pawluk, Steve
Pluth, Donald John
Rennie, Robert John
Rice, Wendell Alfred
Robertson, James Alexander
Sommerfeldt, Theron G
Webster, Gordon Ritchie

BRITISH COLUMBIA
Davis, Roderick Leigh
Finn, William Daniel Liam
Kowalenko, Charles Grant
Lavkulich, Leslie Michael
Lowe, Lawrence E
Neilsen, Gerald Henry
Preston, Caroline Margaret
Russell, Glenn C
Schreier, Hanspeter
Slaymaker, Herbert Olav
Smith, Richard Barrie
Stevenson, David Stuart

MANITOBA
Bailey, Loraine Dolar
Grant, Cynthia Ann
Hedlin, Robert Arthur
Hobbs, James Arthur
Shaykewich, Carl Francis
Sheppard, Marsha Isabell
Sheppard, Stephen Charles
Smith, Robert Edward
Soper, Robert Joseph
Vessey, Joseph Kevin

NEW BRUNSWICK
Chow, Thien Lien
Harries, Hinrich
Van Groenewoud, Herman

NEWFOUNDLAND
Hampson, Michael Chisnall
McKenzie, David Bruce

NOVA SCOTIA
Blatt, Carl Roger
Corke, Charles Thomas
Stratton, Glenn Wayne
Warman, Philip Robert
Wright, James R

ONTARIO
Armson, Kenneth Avery
Bates, Thomas Edward
Beauchamp, Eric G
Behan-Pelletier, Valerie Mary
Beveridge, Terrance James
Bowman, Bruce T
Bunting, Brian Talbot
Csillag, Ferenc
De Kimpe, Christian Robert
Dickinson, William Trevor
Elrick, David Emerson
Foster, Neil William
Gillham, Robert Winston
Gould, William Douglas
Halstead, Ronald Lawrence
Hoffman, Douglas Weir
Jeglum, John Karl
King, Roger Hatton
Kodama, Hideomi
Lemon, Edgar Rothwell
Mack, Alexander Ross
McKeague, Justin Alexander
McKenney, Donald Joseph
Mathur, Sukhdev Prashad
Matthews, Burton Clare
Miller, Murray Henry
Morrison, Ian Kenneth
Protz, Richard
Quigley, Robert Murvin
Richards, Norval Richard
Schnitzer, Morris
Singh, Surinder Shah
Stebelsky, Ihor
Tan, Chin Sheng
Topp, Edward
Topp, G Clarke
Tu, Chin Ming
Wall, Gregory John
Williams, Peter J
Winterhalder, Keith

PRINCE EDWARD ISLAND
Carter, Martin Roger
Gupta, Umesh C
MacLeod, John Alexander
White, Ronald Paul, Sr

QUEBEC
Allen, Sandra Lee
Angers, Denis Arthur
Bonn, Ferdinand J
Bordeleau, Lucien Mario
Bourget, Sylvio-J
Broughton, Robert Stephen
Cescas, Michel Pierre
Furlan, Valentin
Laverdiere, Marc Richard
MacKenzie, Angus Finley
Ouimet, Rock
Widden, Paul Rodney
Yong, Raymond N

SASKATCHEWAN
Acton, Donald Findlay
Anderson, Darwin Wayne
Beaton, James Duncan
Campbell, Constantine Alberga
Chinn, Stanley H F
Huang, P M
Janke, Wilfred Edwin
Nelson, Louise Mary

Nuttall, Wesley Ford
Pretty, Kenneth McAlpine
Rennie, Donald Andrews
St Arnaud, Roland Joseph
Stewart, John Wray Black

OTHER COUNTRIES
Brydon, James Emerson
Couto, Walter
Dirksen, Christiaan
Muchovej, James John
Mugwira, Luke Makore
Muniappan, Rangaswamy Naicker
Oertli, Johann Jakob
Sanchez, Pedro Antonio
Shen, Chih-Kang
Simkins, Charles Abraham
Spencer, John Francis Theodore
Ugolini, Fiorenzo Cesare

Other Agricultural & Forest Sciences

ALABAMA
Campbell, Robert Terry
Clark, Edward Maurice
Coleman, Tommy Lee
Giordano, Paul M
Guthrie, Richard Lafayette
Hill, Walter Andrew
Huluka, Gobena

ALASKA
Juday, Glenn Patrick

ARIZONA
Baron, William Robert
Kauffeld, Norbert M
Knorr, Philip Noel
Waller, Gordon David

ARKANSAS
Zeide, Boris

CALIFORNIA
Adaskaveg, James Elliott
Akesson, Norman B(erndt)
Belisle, Barbara Wolfanger
Bodman, Geoffrey Baldwin
Fisher, Theodore William
Gary, Norman Erwin
Gerwick, Ben Clifford, III
Kodira, Umesh Chengappa
Mayse, Mark A
Orman, Charles
Shelly, John Richard
Taber, Stephen, III
Trevelyan, Benjamin John

COLORADO
Driscoll, Richard Stark
Hansen, Richard M
Kalabokidis, Kostas D
Massman, William Joseph
Shepperd, Wayne Delbert
Tengerdy, Robert Paul
Woodmansee, Robert George

CONNECTICUT
Anagnostakis, Sandra Lee
Connor, Lawrence John
Newton, David C
Stephens, George Robert

DELAWARE
Caron, Dewey Maurice
Cupery, Willis Eli
Kundt, John Fred
Soboczenski, Edward John

DISTRICT OF COLUMBIA
Cuatrecasas, Jose
Lunchick, Curt
Post, Boyd Wallace
Rosen, Howard Neal
Rosenberg, Norman J

FLORIDA
Dressler, Robert Louis
Gerber, John Francis
Heinen, Joel Thomas
Littell, Ramon Clarence
Lucansky, Terry Wayne

GEORGIA
Broerman, F S
Chiang, Tze I

HAWAII
Schmitt, Donald Peter

IDAHO
Callihan, Robert Harold
Michalson, Edgar Lloyd
Thill, Donald Cecil

ILLINOIS
Evans, James Forrest
Meyer, Martin Marinus, Jr
Perino, Janice Vinyard
Sinclair, James Burton
Singer, Rolf
Van Sambeek, Jerome William

Wesely, Marvin Larry

INDIANA
Bing, Richard F
Dando, William Arthur
Johnston, Cliff T
Kidd, Frank Alan

IOWA
Anderson, Todd Alan
Isely, Duane
Iwig, Mark Michael
Rohlf, Marvin Euguene
Shaw, Robert Harold

KANSAS
Baker, Charles H
Bark, Laurence Dean
Lomas, Lyle Wayne

KENTUCKY
Olson, James Robert
Vogel, Willis Gene

LOUISIANA
Baker, John Bee
Dionigi, Christopher Paul
Harbo, John Russell
Pezeshki, S Reza
Schoenly, Kenneth George

MAINE
Chandler, Robert Flint, Jr
Frank, Robert McKinley, Jr

MARYLAND
Bauer, Peter
Coffman, Charles Benjamin
Foudin, Arnold S
Heller, Stephen Richard
Rango, Albert
Uhart, Michael Scott

MASSACHUSETTS
Shawcross, William Edgerton

MICHIGAN
Bourdo, Eric A, Jr
Fulbright, Dennis Wayne
Paschke, John Donald

MINNESOTA
Abeles, Tom P
Allen, Charles Eugene
Dahlberg, Duane Arlen
Pour-El, Akiva

MISSISSIPPI
Collins, Johnnie B
Collison, Clarence H
Lloyd, Edwin Phillips

MISSOURI
Bullock, J Bruce
Hewitt, Andrew
Rash, Jay Justen

MONTANA
Behan, Mark Joseph
Caprio, Joseph Michael
Jackson, Grant D
Latham, Don Jay

NEBRASKA
Chapman, John Arthur
Knox, Ellis Gilbert

NEW HAMPSHIRE
Baldwin, Henry Ives

NEW JERSEY
Berenbaum, Morris Benjamin
Jansson, Richard Keith

NEW MEXICO
Jarmie, Nelson
Phillips, Gregory Conrad

NEW YORK
Camazine, Scott
Goodale, Douglas M
Kornfield, Jack I
Meyer, Robert Walter
Morse, Roger Alfred
Pack, Albert Boyd
Reisch, Bruce Irving
Stark, John Howard
Tonelli, John P, Jr
Wilcox, Wayne F
Wolf, Walter Alan

NORTH CAROLINA
Axtell, Richard Charles
Bobalek, Edward G(eorge)
Cubberley, Adrian H
Flint, Elizabeth Parker
Kolenbrander, Lawrence Gene
Nie, Dalin
Thomas, Richard Joseph

OHIO
Hein, Richard William
McCracken, John David
Merritt, Robert Edward

Other Agricultural & Forest Sciences (cont)

Nelson, Eric V
Vertrees, Robert Layman
Watson, Stanley Arthur

OREGON
Coop, Leonard Bryan
Denison, William Clark
Ho, Iwan
Mason, Richard Randolph
Shock, Clinton C
Young, Roy Alton

PENNSYLVANIA
Berthold, Robert, Jr
Daum, Donald Richard
Heller, Paul R
Labosky, Peter, Jr
Rogerson, Thomas Dean
Russo, Joseph Martin
Staetz, Charles Alan

RHODE ISLAND
Durfee, Wayne King

SOUTH CAROLINA
Hon, David Nyok-Sai
Quisenberry, Virgil L
Turner, John Lindsey

TENNESSEE
Baldocchi, Dennis D

TEXAS
Amador, Jose Manuel
Drees, Bastiaan M
Gates, Charles Edgar
Hillery, Herbert Vincent
Humphries, James Edward, Jr
McCully, Wayne Gunter
Smith, Dudley Templeton
Tiller, F(rank) M(onterey)

UTAH
Hansen, Lee Duane

VIRGINIA
Hamrick, Joseph Thomas
Lim, Young Woon (Peter)
Munson, Arvid W
Nichols, James Robbs
Pienkowski, Robert Louis
Youngs, Robert Leland

WASHINGTON
Campbell, Gaylon Sanford
Clement, Stephen LeRoy
Landsberg, Johanna D (Joan)
McCarthy, Joseph L(ePage)
Noel, Jan Christine
Peabody, Dwight Van Dorn, Jr
Tichy, Robert J
Tukey, Harold Bradford, Jr
Zuiches, James J

WEST VIRGINIA
Smith, Walton Ramsay

WISCONSIN
Cramer, Jane Harris
Erickson, John Ronald
Hole, Francis Doan
Holm, LeRoy George
Lillesand, Thomas Martin
Makela, Lloyd Edward
Norman, John Matthew
Springer, Edward L(ester)

WYOMING
Humburg, Neil Edward

PUERTO RICO
Lugo, Ariel E

ALBERTA
Feng, Joseph C
Goettel, Mark S
Price, Mick A
Wong, Horne Richard
Yang, Rong-Cai

BRITISH COLUMBIA
Alfaro, Rene Ivan
Gardner, Joseph Arthur Frederick
Paszner, Laszlo

MANITOBA
Amiro, Brian Douglas
Palaniswamy, Pachagounder
Pepper, Evan Harold
Singh, Harwant

NEW BRUNSWICK
Coleman, Warren Kent

ONTARIO
Baier, Wolfgang
Bright, Donald Edward
Gillespie, Terry James
McNairn, Heather Elizabeth
Plaxton, William Charles

Redhead, Scott Alan
Reeds, Lloyd George
Smith, Maurice Vernon
Summers, John David
Yan, Maxwell Menuhin

QUEBEC
Furlan, Valentin
Lemieux, Claudel
Stewart, Robin Kenny

SASKATCHEWAN
Kartha, Kutty Krishnan
Ripley, Earle Allison
Smith, Allan Edward

OTHER COUNTRIES
Hocking, Drake
Sleeter, Thomas David

BIOLOGICAL SCIENCES
Anatomy

ALABAMA
Abrahamson, Dale Raymond
Al-lami, Fadhil
Bernstein, Maurice Harry
Bhatnagar, Yogendra Mohan
Brown, Jerry William
Chibuzo, Gregory Anenonu
Chronister, Robert Blair
Dacheux, Ramon F, II
Gardner, William Albert, Jr
Gray, Bruce William
Hamel, Earl Gregory, Jr
Hand, George Samuel, Jr
Hoffman, Henry Harland
Holloway, Clarke L
Hovde, Christian Arneson
Jenkins, Ronald Lee
Kincaid, Steven Alan
Krista, Laverne Mathew
Marchase, Richard Banfield
Rodning, Charles Bernard
Shackleford, John Murphy
Speed, Edwin Maurice
Wilborn, Walter Harrison
Williams, Raymond Crawford
Wyss, James Michael

ALASKA
Ebbesson, Sven O E
Fay, Francis Hollis
Guthrie, Russell Dale

ARIZONA
Angevine, Jay Bernard, Jr
Begovac, Paul C
Bernays, Elizabeth Anna
Chiasson, Robert Breton
Hendrix, Mary J C
Krutzsch, Philip Henry
LeBouton, Albert V
McCauley, William John
McCuskey, Robert Scott
McKinley, Michael P
Markle, Ronald A
Nicolls, Ken E
Pough, Frederick Harvey
Spofford, Walter Richardson, II
Tarby, Theodore John
Witherspoon, James Donald

ARKANSAS
Burns, Edward Robert
Carter, Charleata A
Cave, Mac Donald
Chang, Louis Wai-Wah
McMillan, Harlan L
Marvin, Horace Newell
Morgans, Leland Foster
Pauly, John Edward
Powell, Ervin William
Scheving, Lawrence Einar
Sherman, Jerome Kalman
Soloff, Bernard Leroy
Tank, Patrick Wayne
Uyeda, Carl Kaoru

CALIFORNIA
Ahmad, Nazir
Akin, Gwynn Collins
Alberch, Pere
Alden, Roland Herrick
Alvarino, Angeles
Arcadi, John Albert
Armstrong, Rosa Mae
Asling, Clarence Willet
Baker, Mary Ann
Bennett, Leslie L
Bernard, George W
Beuchat, Carol Ann
Bevelander, Gerrit
Bok, P Dean
Bondareff, William
Bowers, Roger Raymond
Boyne, Philip John
Breisch, Eric Alan
Burnside, Mary Beth
Callison, George
Campbell, John Howland
Case, Norman Mondell

Chamberlain, Jack G
Chase, Robert A
Chow, Kao Liang
Clemente, Carmine Domenic
Collins, Robert C
Connelly, Thomas George
Cullen, Michael Joseph
Dalgleish, Arthur E
Dearden, Lyle Conway
Diamond, Ivan
Dixon, Andrew Derart
Dodge, Alice Hribal
Dougherty, Harry L
Drewes, Robert Clifton
Eiserling, Frederick A
Enders, Allen Coffin
Erickson, Kent L
Erpino, Michael James
Fallon, James Harry
Faulkin, Leslie J, Jr
Fierstine, Harry Lee
Fisher, Robin Scott
Fletcher, William H
Garoutte, Bill Charles
Gehlsen, Kurt Ronald
Gifford, Ernest Milton
Giolli, Roland A
Glass, Laurel Ellen
Globus, Albert
Goldstein, Abraham M B
Gray, Constance Helen
Greenleaf, Robert Dale
Hatton, Glenn Irwin
Haun, Charles Kenneth
Hayashida, Tetsuo
Hooker, William Mead
Hungerford, Gerald Fred
Hunt, Guy Marion, Jr
Hunter, Robert L
Jenkins, Floyd Albert
Jones, Edward George
Kahn, Raymond Henry
Karten, Harvey J
Keller, Raymond E
King, Barry Frederick
Kitchell, Ralph Lloyd
Klouda, Mary Ann Aberle
Ko, Chien-Ping
Kruger, Lawrence
Laham, Quentin Nadime
LaVail, Jennifer Hart
LaVail, Matthew Maurice
Lianides, Sylvia Panagos
Long, John Arthur
Lu, John Kuew-Hsiung
Lutt, Carl J
McHenry, Henry Malcolm
McMillan, Paul Junior
Marchand, E Roger
Marshall, John Foster
Maxwell, David Samuel
Menees, James H
Mensah, Patricia Lucas
Metzner, Walter
Monie, Ian Whitelaw
Mortenson, Theodore Hampton
Murad, Turhon A
Murphy, Henry D
Myrick, Albert Charles, Jr
Nelson, Gayle Herbert
Nishioka, Richard Seiji
Northcutt, Richard Glenn
Orr, Beatrice Yewer
Outzen, Henry Clair, Jr
Palade, George E
Paule, Wendelin Joseph
Plopper, Charles George
Plymale, Harry Hambleton
Ralston, Henry James, III
Reaven, Eve P
Ribak, Charles Eric
Roberts, Walter Herbert B
Rost, Thomas Lowell
Ruibal, Rodolfo
Sartoris, David John
Sawyer, Charles Henry
Scheibel, Arnold Bernard
Schmucker, Douglas Lees
Schooley, Caroline Naus
Schultz, Robert Lowell
Schweisthal, Michael Robert
Sechrist, John William
Shook, Brenda Lee
Smith, Bradley Richard
Snow, Mikel Henry
Srebnik, Herbert Harry
Stilwell, Donald Lonson
Strautz, Robert Lee
Swett, John Emery
Taylor, Anna Newman
Templeton, McCormick
Towers, Bernard
Turner, Robert Stuart
Tyler, Walter Steele
Vaughn, James E, Jr
Vijayan, Vijaya Kumari
Villablanca, Jaime Rolando
Wangler, Roger Dean
Waters, James Frederick
Wilson, Doris Burda
Wimer, Cynthia Crosby
Winer, Jeffery Allan
Wood, Richard Lyman
Young, Richard Wain

Younoszai, Rafi
Zakhary, Rizkalla
Zamenhof, Stephen
Zimmerman, Emery Gilroy

COLORADO
Billenstien, Dorothy Corinne
Charney, Michael
Davis, Robert Wilson
Dubin, Mark William
Fifkova, Eva
Finger, Thomas Emanuel
Frandson, Rowen Dale
Hanken, James
Jafek, Bruce William
Lebel, Jack Lucien
Liechty, Richard Dale
Meyer, Hermann
Moury, John David
Palmer, Michael Rule
Plakke, Ronald Keith
Rash, John Edward
Roper, Stephen David
Rudnick, Michael Dennis
Schulter-Ellis, Frances Pierce
Solomon, Gordon Charles
Whitlock, David Graham
Willson, John Tucker

CONNECTICUT
Ariyan, Stephan
Barry, Michael Anhalt
Blackburn, Daniel Glenn
Chatt, Allen Barrett
Constantine-Paton, Martha
Cooperstein, Sherwin Jerome
Crelin, Edmund Slocum
Galton, Peter Malcolm
Grasso, Joseph Anthony
Grimm-Jorgensen, Yvonne
Hand, Arthur Ralph
Matheson, Dale Whitney
Morest, Donald Kent
Mugnaini, Enrico
Roos, Henry
Schwartz, Ilsa Roslow
Wagner, Gunter Paul
Watkins, Dudley T
Yaeger, James Amos

DELAWARE
Ruben, Regina Lansing
Skeen, Leslie Carlisle

DISTRICT OF COLUMBIA
Albert, Ernest Narinder
Baldwin, Kate M
Ball, William David
Bernor, Raymond Louis
Bernstein, Jerald Jack
Bulger, Ruth Ellen
Crisp, Thomas Mitchell, Jr
De Queiroz, Kevin
Domning, Daryl Paul
Emry, Robert John
Ericksen, Mary Frances
Grand, Theodore I
Hakim, Raziel Samuel
Hayek, Lee-Ann Collins
Hayes, Raymond L, Jr
Herman, Barbara Helen
Hussain, Syed Taseer
Johnson, Thomas Nick
Kapur, Shakti Prakash
Kelly, Douglas Elliott
Koering, Marilyn Jean
Kromer, Lawrence Frederick
Murphy, James John
Norman, Wesley P
Rapisardi, Salvatore C
Slaby, Frank J
Snell, Richard Saxon
Telford, Ira Rockwood
Whitmore, Frank Clifford, Jr
Young, John Karl

FLORIDA
Allen, Ted Tipton
Bennett, Gudrun Staub
Berkley, Karen J
Berman, Irwin
Bressler, Steven L
Bunge, Richard Paul
Cameron, Don Frank
Chen, L T
Clendenin, Martha Anne
Cutts, James Harry
Demski, Leo Stanley
DeRousseau, C(arol) Jean
Dwornik, Julian Jonathan
Ericson, Grover Charles
Ewart, R Bradley
Feldherr, Carl M
Gfeller, Eduard
Goldberg, Stephen
Gordon, Kenneth Richard
Gorniak, Gerard Charles
Hinkley, Robert Edwin, Jr
Hope, George Marion
Kallenbach, Ernst Adolf Theodor
Larkin, Lynn Haydock
Leeson, Charles Roland
Leonard, Christiana Morison
Maue-Dickson, Wilma

Nolan, Michael Francis
Pearl, Gary Steven
Phelps, Christopher Prine
Reynolds, John Elliott, III
Rhodin, Johannes A G
Romrell, Lynn John
Ross, Michael H
Samuelson, Don Arthur
Saporta, Samuel
Schnitzlein, Harold Norman
Shireman, Rachel Baker
Sturtevant, Ruthann Patterson
Suzuki, Howard Kazuro
Taylor, George Thomas
Weber, James Edward
Wireman, Kenneth
Woods, Charles Arthur
Woolfenden, Glen Everett
Wyneken, Jeanette
Yakaitis-Surbis, Albina Ann

GEORGIA
Adkison, Claudia R
Bennett, Sara Neville
Benoit, Peter Wells
Binnicker, Pamela Caroline
Black, Asa C, Jr
Black, John B
Bockman, Dale Edward
Brown, Hugh Keith
Colborn, Gene Louis
Crowell-Davis, Sharon Lynn
Doetsch, Gernot Siegmar
Edwards, Betty F
English, Arthur William
Golarz-De Bourne, Maria Nelly
Gray, Stephen Wood
Gulati, Adarsh Kumar
Hawkins, Isaac Kinney
Holland, Robert Campbell
Jacobs, Virgil Leon
Kirby, Margaret Loewy
Lo, Woo-Kuen
McKenzie, John Ward
Manocha, Sohan Lall
Mulroy, Michael J
Odend'hal, Stewart
Paulsen, Douglas F
Puchtler, Holde
Richard, Christopher Alan
Scott, John Watts, Jr
Sutin, Jerome
Tigges, Johannes
Welter, Dave Allen
Williams, William Lane
Wolf, Steven L

HAWAII
Berger, Andrew John
Diamond, Milton
Kleinfeld, Ruth Grafman
Nelson, Marita Lee

IDAHO
DeSantis, Mark Edward
Eroschenko, Victor Paul
Fuller, Eugene George
Laundre, John William
Newman, Bertha L
Stephens, Trent Dee

ILLINOIS
Allin, Edgar Francis
Ariano, Marjorie Ann
Aydelotte, Margaret Beesley
Bachop, William Earl
Bakkum, Barclay W
Baron, David Alan
Bieler, Rudiger
Blaha, Gordon C
Bodley, Herbert Daniel, II
Buschmann, MaryBeth Tank
Buschmann, Robert J
Caspary, Donald M
Chiakulas, John James
Cho, Yongock
Connors, Natalie Ann
Coruccini, Robert Spencer
Costa, Raymond Lincoln, Jr
Cralley, John Clement
Daniels, Edward William
De Bruyn, Peter Paul Henry
Dinsmore, Charles Earle
DuBrul, E Lloyd
Durica, Thomas Edward
Engel, John Jay
Engel, Milton Baer
Farbman, Albert Irving
Foote, Florence Martindale
Forman, G Lawrence
Gaik, Geraldine Catherine
Geinisman, Yuri
Gibbs, Daniel
Gowgiel, Joseph Michael
Greaves, Walter Stalker
Greenough, William Tallant
Griffiths, Thomas Alan
Grimm, Arthur F
Hasegawa, Junji
Hast, Malcolm Howard
Heltne, Paul Gregory
Holmes, E(dward) Bruce
Holmes, Kenneth Robert
Kasprow, Barbara Ann

Kernis, Marten Murray
Khodadad, Jena Khadem
Kiely, Michael Lawrence
Krieg, Wendell Jordan
Lavelle, Arthur
LaVelle, Faith Wilson
Layman, Dale Pierre
Leven, Robert Maynard
McCandless, David Wayne
McNulty, John Alexander
Maibenco, Helen Craig
Mixter, Russell Lowell
Monsen, Harry
Moticka, Edward James
Nakajima, Yasuko
Naples, Virginia L
O'Morchoe, Charles C C
O'Morchoe, Patricia Jean
Orr, Mary Faith
Paparo, Anthony A
Polley, Edward Herman
Rabuck, David Glenn
Reed, Charles Allen
Rezak, Michael
Romack, Frank Eldon
Rosenberger, Alfred L
Russell, Brenda
Safanie, Alvin H
Scapino, Robert Peter
Schneider, Gary Bruce
Seale, Raymond Ulric
Siegel, Jonathan Howard
Simon, Mark Robert
Simpson, Sidney Burgess, Jr
Singer, Ronald
Sladek, John Richard, Jr
Straus, Helen Lorna Puttkammer
Sweeny, Lauren J
Thomas, Carolyn Eyster
Thurow, Gordon Ray
Towns, Clarence, Jr
Tuttle, Russell Howard
Ulinski, Philip Steven
Velardo, Joseph Thomas
Waltenbaugh, Carl
Walter, Robert John
Wezeman, Frederick H
Williams, Thomas Alan
Zaki, Abd El-Moneim Emam
Zalisko, Edward John

INDIANA
Anderson, William John
Babbs, Charles Frederick
Blevins, Charles Edward
Burr, David Bentley
Das, Gopal Dwarka
Davis, Grayson Steven
Dial, Norman Arnold
Hafner, Gary Stuart
Hinsman, Edward James
Hoversland, Roger Carl
Hullinger, Ronald Loral
Jersild, Ralph Alvin, Jr
Katzberg, Allan Alfred
Kennedy, Duncan Tilly
McKibben, John Scott
Mescher, Anthony Louis
Mizell, Sherwin
Morr, Dorothy Marie
Murphy, Robert Carl
Murray, Raymond Gorbold
Peterson, Richard George
Pietsch, Paul Andrew
Schmedtje, John Frederick
Schroeder, Dolores Margaret
Sever, David Michael
Stromberg, Melvin Willard
Stump, John Edward
Van Sickle, David C
Witzmann, Frank A

IOWA
Adams, Donald Robert
Bergman, Ronald Arly
Bhalla, Ramesh C
Carithers, Jeanine Rutherford
Christensen, George Curtis
Ciochon, Russell Lynn
Coulter, Joe Dan
Dawson, David Lynn
Dellmann, H Dieter
Ghoshal, Nani Gopal
Heidger, Paul McClay, Jr
Jacobs, Richard M
Kessel, Richard Glen
Kirkland, Willis L
Kollros, Jerry John
McInroy, Elmer Eastwood
Maynard, Jerry Allen
Meetz, Gerald David
Mennega, Aaldert
Schelper, Robert Lawrence
Searls, James Collier
Shaw, Gaylord Edward
Stromer, Marvin Henry
Thompson, Sue Ann
Tranel, Daniel T
Weingeist, Thomas Alan
Williams, Terence Heaton

KANSAS
Beary, Dexter F
Besharse, Joseph Culp

Chapman, Albert Lee
Dunn, Jon D
Enders, George Crandell
Hung, Kuen-Shan
Keller, Leland Edward
Klein, Robert Melvin
Klemm, Robert David
Mohn, Melvin P
Ollerich, Dwayne A
Redick, Mark Lankford
Sayegh, Fayez S
Vacca, Linda Lee
Walker, Richard Francis

KENTUCKY
Bhatnagar, Kunwar Prasad
Campbell, Ferrell Rulon
Cotter, William Bryan, Jr
Farrar, William Wesley
Fontaine, Julia Clare
Fuller, Peter McAfee
Gillilan, Lois Adell
Gregg, Robert Vincent
Hamon, J Hill
Herbener, George Henry
Longley, James Baird
Matulionis, Daniel H
Moody, William Glenn
Nettleton, G(ary) Stephen
Nikitovitch-Winer, Miroslava B
Rink, Richard Donald
Smith, Stephen D
Swartz, Frank Joseph
Swigart, Richard Hanawalt
Traurig, Harold H

LOUISIANA
Allen, Emory Raworth
Beal, John Anthony
Brody, Arnold R
Carpenter, Stanley Barton
Clawson, Robert Charles
Constantinides, Paris
Cook, Stephen D
Davenport, William Daniel, Jr
Gasser, Raymond Frank
Green, Jeffrey David
Hibbs, Richard Guythal
Homberger, Dominique Gabrielle
Jewell, Frederick Forbes, Sr
Kasten, Frederick H
Layman, Don Lee
Low, Frank Norman
Matthews, Murray Albert
Nickerson, Stephen Clark
Robinson, Roy Garland, Jr
Ruby, John Robert
Sarphie, Theodore G
Seltzer, Benjamin
Silverman, Harold
Specian, Robert David
Titkemeyer, Charles William
Turner, Hugh Michael
Vaupel, Martin Robert
Walker, Leon Bryan, Jr
Weber, Joseph T
Zimny, Marilyn Lucile

MAINE
Bell, Allen L
Cole, Wilbur Vose
Goodale, Fairfield
Gough, Robert Edward
Hinds, James Wadsworth
Minkoff, Eli Cooperman

MARYLAND
Anderson, Larry Douglas
Barry, Ronald Everett, Jr
Bell, Mary
Benevento, Louis Anthony
Brightman, Milton Wilfred
Broadwell, Richard Dow
Carpenter, Malcolm Breckenridge
Cotton, William Robert
Cowan, W Maxwell
De Monasterio, Francisco M
Donati, Edward Joseph
Ennist, David L
Ferrans, Victor Joaquin
Freed, Michael Abraham
Fuson, Roger Baker
Gartner, Leslie Paul
Gatti, Philip John
Gobel, Stephen
Green, Martin David
Greenhouse, Gerald Alan
Greulich, Richard Curtice
Grewe, John Mitchell
Gross, James Harrison
Harding, Fann
Harkins, Rosemary Knighton
Hein, Rosemary Ruth
Hiatt, James Lee
Hotton, Nicholas, III
Hutchins, Grover MacGregor
Kalt, Marvin Robert
Kibbey, Maura Christine
Leach, Berton Joe
Liebelt, Annabel Glockler
MacLean, Paul Donald
Margolis, Ronald Neil
Meszler, Richard M
Moreira, Jorge Eduardo

Nelson, Ralph Francis
Oberdorfer, Michael Douglas
Petrali, John Patrick
Phelps, Creighton Halstead
Platt, William Rady
Robison, Wilbur Gerald, Jr
Rose, Kenneth David
Seibel, Werner
Shear, Charles Robert
Sheridan, Michael N
Song, Jiakun
Strum, Judy May
Thorington, Richard Wainwright, Jr
Yellin, Herbert
Zirkin, Barry Ronald

MASSACHUSETTS
Ahlberg, Henry David
Albertini, David Fred
Appel, Michael Clayton
Atema, Jelle
Balogh, Karoly
Baratz, Robert Sears
Begg, David A
Belt, Warner Duane
Bieber, Frederick Robert
Bond, George Walter
Bridges, Robert Stafford
Chlapowski, Francis Joseph
Clark, Sam Lillard, Jr
Coombs, Margery Chalifoux
Damassa, David Allen
Dittmer, John Edward
Dorey, Cheryl Kathleen
Douglas, William J
Eldred, William D
Erikson, George Emil
Feldman, Martin Leonard
Frommer, Jack
George, Stephen Anthony
Gibbons, Michael Francis, Jr
Giffin, Emily Buchholtz
Gipson, Ilene Kay
Gonnella, Patricia Anne
Graybiel, Ann M
Greep, Roy Orval
Gustafson, Alvar Walter
Haidak, Gerald Lewis
Hammer, Ronald Page, Jr
Henry, Joseph L
Hoar, Richard Morgan
Hoffmann, Joan Carol
Ito, Susumu
Jacobson, Stanley
Kronman, Joseph Henry
Maran, Janice Wengerd
Marieb, Elaine Nicpon
Marks, Sandy Cole, Jr
Miller, James Albert, Jr
Murnane, Thomas William
Murthy, A S Krishna
Nandy, Kalidas
Nauta, Walle J H
Nixon, Charles William
Payne, Bertram R
Peters, Alan
Pfister, Richard Charles
Pino, Richard M
Powers, J Bradley
Rafferty, Nancy S
Raviola, Elio
Remmel, Ronald Sylvester
Rollason, Herbert Duncan
Saper, Clifford B
Sidman, Richard Leon
Smith, Dennis Matthew
Sorokin, Sergei Pitirimovitch
Stein, Otto Ludwig
Struthers, Robert Claflin
Susi, Frank Robert
Vanderburg, Charles R
Vaughan, Deborah Whittaker
West, Christopher Drane
White, Edward Lewis

MICHIGAN
Al Saadi, A Amir
Al-Saadi, Abdul A
Avery, James Knuckey
Bivins, Brack Allen
Boving, Bent Giede
Brown, Esther Marie
Brown, Roger E
Burkel, William E
Buss, Jack Theodore
Carlson, Bruce Martin
Carlson, David Sten
Castelli, Walter Andrew
Christensen, A(lbert) Kent
Coye, Robert Dudley
Coyle, Peter
Dapson, Richard W
Diaz, Fernando O
Eichler, Victor B
Falls, William McKenzie
Fischer, Theodore Vernon
Fisher, Don Lowell
Fisher, Leslie John
Floyd, Alton David
Friar, Robert Edsel
Fritts-Williams, Mary Louise Monica
Froiland, Thomas Gordon
Garg, Bhagwan D
Glover, Roy Andrew

Anatomy (cont)

Grant, Rhoda
Henry, Raymond Leo
Houts, Larry Lee
Huelke, Donald Fred
Hurst, Edith Marie Maclennan
Jampel, Robert Steven
Johnson, John Irwin, Jr
Kim, Sun-Kee
Krishan, Awtar
Lasker, Gabriel (Ward)
Lew, Gloria Maria
Lillie, John Howard
MacCallum, Donald Kenneth
McNamara, James Alyn, Jr
Meyer, David Bernard
Mizeres, Nicholas James
Moosman, Darvan Albert
Nag, Asish Chandra
Newman, Sarah Winans
Person, Steven John
Plagge, James Clarence
Pourcho, Roberta Grace
Pysh, Joseph John
Radin, Eric Leon
Rafols, Jose Antonio
Rieck, Norman Wilbur
Rupp, Ralph Russell
Seefeldt, Vern Dennis
Sippel, Theodore Otto
Strachan, Donald Stewart
Tosney, Kathryn W
Townsend, Samuel Franklin
Tweedle, Charles David
Walker, Bruce Edward
Webb, R Clinton
Wood, Pauline J
Woodburne, Russell Thomas

MINNESOTA
Alsum, Donald James
Bauer, Gustav Eric
Beitz, Alvin James
Carmichael, Stephen Webb
Czarnecki, Caroline Mary Anne
Dapkus, David Conrad
Dixit, Padmakar Kashinath
Edds, Kenneth Tiffany
Elde, Robert Philip
Erickson, James Eldred
Erlandsen, Stanley L
Fletcher, Thomas Francis
Forbes, Donna Jean
Goble, Frans Cleon
Hancock, Peter Adrian
Heggestad, Carl B
Low, Walter Cheney
Miller, Robert F
Saccoman, Frank (Michael)
Severson, Arlen Raynold
Seybold, Virginia Susan (Dick)
Sicard, Raymond E(dward)
Smithberg, Morris
Sorenson, Robert Lowell
Sundberg, Ruth Dorothy
Thompson, Edward William
Todt, William Lynn
Welter, Alphonse Nicholas
Yoss, Robert Eugene

MISSISSIPPI
Ajemian, Martin
Ashburn, Allen David
Ball, Carroll Raybourne
Haines, Duane Edwin
Lynch, James Carlyle
Martin, Billy Joe
Mihailoff, Gregory A
Roy, William Arthur
Smith, Byron Colman
Spann, Charles Henry
Walker, James Frederick

MISSOURI
Beringer, Theodore Michael
Brown, Herbert Ensign
Butterworth, Bernard Bert
Cheverud, James Michael
Cooper, Margaret Hardesty
Decker, John D
Francel, Thomas Joseph
Gavan, James Anderson
Gibbs, Finley P
Goodge, William Russell
Harvey, Joseph Eldon
Highstein, Stephen Morris
Jacobs, Allen Wayne
Julyan, Frederick John
Kort, Margaret Alexander
Krause, William John
Lake, Lorraine Frances
Loewy, Arthur D(eCosta)
Lowrance, Edward Walton
McClure, Robert Charles
McClure, Theodore Dean
McLaughlin, Carol Lynn
Menton, David Norman
Moffatt, David John
Momberg, Harold Leslie
Nyquist-Battie, Cynthia
Paull, Willis K, Jr
Peterson, Roy Reed
Price, Joseph Levering

Quay, Wilbur Brooks
Rana, Mohammed Waheeduz-Zaman
Rasmussen, David Tab
Ring, John Robert
Schreiweis, Donald Otto
Slavin, Bernard Geoffrey
Smith, Richard Jay
Spiro, Thomas
Stephens, Robert Eric
Taylor, John Joseph
Tolbert, Daniel Lee
Tumosa, Nina Jean
Woolsey, Thomas Allen
Yeager, Vernon LeRoy
Young, Paul Andrew

MONTANA
Fawcett, Don Wayne
Gajdosik, Richard Lee
Phillips, Dwight Edward

NEBRASKA
Binhammer, Robert T
Crouse, David Austin
Dalley, Arthur Frederick, II
Dossel, William Edward
Earle, Alvin Mathews
Fawcett, James Davidson
Fougeron, Myron George
Gardner, Paul Jay
Harn, Stanton Douglas
Hill, Marvin Francis
Holyoke, Edward Augustus
Leuschen, M Patricia
Littledike, Ernest Travis
Macaluso, Sr Mary Christelle
Metcalf, William Kenneth
Rosenquist, Thomas H
Sharp, John Graham
Skultety, Francis Miles
Sullivan, James Michael
Todd, Gordon Livingston
Warr, William Bruce
Yee, John Alan

NEVADA
Benes, Elinor Simson
Marlow, Ronald William
Schneider, Lawrence Kruse
Stratton, Clifford James

NEW HAMPSHIRE
Chambers, Wilbert Franklin
Meader, Ralph Gibson
Musiek, Frank Edward
Sokol, Hilda Weyl
Stahl, Barbara Jaffe

NEW JERSEY
Agnish, Narsingh Dev
Alger, Elizabeth A
Auletta, Carol Spence
Baden, Ernest
Bagnell, Carol A
Berberian, Paul Anthony
Berendsen, Peter Barney
Boccabella, Anthony Vincent
DeFouw, David O
DeProspo, Nicholas Dominick
Eastwood, Abraham Bagot
Feldman, Susan C
Ford, Daniel Morgan
Gagna, Claude Eugene
Gilani, Shamshad H
Gona, Amos G
Gona, Ophelia Delaine
Halpern, Myron Herbert
Hart, Nathan Hoult
Hess, Arthur
Hollinshead, May B
Hsu, Linda
Kleinschuster, Stephen J, III
Kmetz, John Michael
Kozam, George
Krauthamer, George Michael
Laemle, Lois K
Leung, Christopher Chung-Kit
McAuliffe, William Geoffrey
Macdonald, Gordon J
Malamed, Sasha
Mele, Frank Michael
Meyer, Rita A
Saiff, Edward Ira
Seiden, David
Tesoriero, John Vincent
Wallace, Edith Winchell
Wilson, Frank Joseph
Yu, Mang Chung

NEW MEXICO
Bourne, Earl Whitfield
Dressendorfer, Rudy
Gray, Edwin R
Karlsson, Ulf Lennart
Kelley, Robert Otis
Martin, William Clarence
Napolitano, Leonard Michael
Saland, Linda C
Thrasher, Jack D
Trotter, John Allen
Whitmore, Mary (Elizabeth) Rowe
Wood, Joseph George

NEW YORK
Aldridge, William Gordon
Alexander, A Allan
Ambron, Richard Thomas
Ames, Ira Harold
April, Ernest W
Aschner, Michael
Baker, Robert George
Becker, Norwin Howard
Beckert, William Henry
Bedford, John Michael
Benzo, Camillo Anthony
Bielat, Kenneth L
Black, Virginia H
Bothner, Richard Charles
Boucher, Louis Jack
Brandt, Philip Williams
Brody, Harold
Brophy, Mary O'Reilly
Brownscheidle, Carol Mary
Cabot, John Boit, II
Carriere, Rita Margaret
Centola, Grace Marie
Cohan, Christopher Scott
Cohn, Deirdre Arline
Colman, David Russell
Cummings, John Francis
DeLahunta, Alexander
Demeter, Steven
De Zeeuw, Carl Henri
Dibennardo, Robert
Di Stefano, Henry Saverio
Doolittle, Richard L
Dornfest, Burton S
Doty, Stephen Bruce
Drakontides, Anna Barbara
Eckert, Barry S
Edmonds, Richard H
Emmel, Victor Meyer
Evans, Howard Edward
Evans, Lance Saylor
Feagans, William Marion
Feder, Harvey Herman
Feinman, Max L
Felten, David L
Fetcho, Joseph Robert
Firriolo, Domenic
Fleagle, John G
Flood, Dorothy Garnett
Fruhman, George Joshua
Garcia, Alfredo Mariano
Gershon, Michael David
Gil, Joan
Glomski, Chester Anthony
Goodman, Donald Charles
Goodwin, Robert Earl
Greenberger, Lee M
Gresik, Edward William
Grew, John C
Grosso, Leonard
Habel, Robert Earl
Hansen, John Theodore
Hayes, Everett Russell
Henrikson, Ray Charles
Hillman, Dean Elof
Hirschman, Albert
Horn, Eugene Harold
Horst, G Roy
Jakway, Jacqueline Sinks
Kallen, Frank Clements
Kaye, Gordon I
Kaye, Nancy Weber
Kiely, Lawrence J
Kinzey, Warren Glenford
Krause, David Wilfred
Lamberg, Stanley Lawrence
Lazarow, Paul B
Lemanski, Larry Frederick
Levitan, Irwin
Lewis, Carmie Perrotta
Loy, Rebekah
Lund, Richard
Lutton, John D
McCune, Amy Reed
Maderson, Paul F A
Martin, Kathryn Helen
Mazurkiewicz, Joseph Edward
Mendel, Frank C
Miller, Richard Avery
Miller, Sue Ann
Minor, Ronald R
Mitchell, Ormond Glenn
Monheit, Alan G
Moss-Salentijn, Letty
Motzkin, Shirley M
Noback, Charles Robert
Notter, Mary Frances
Osinchak, Joseph
Padawer, Jacques
Pasik, Pedro
Pasik, Tauba
Penney, David P
Pentney, Roberta Pierson
Pesetsky, Irwin
Pfeiffer, Carroll Athey
Pham, Tuan Duc
Prutkin, Lawrence
Rasweiler, John Jacob, IV
Rhodes, Rondell H
Ringler, Neil Harrison
Riss, Walter
Robertson, Douglas Reed
Ruggiero, David A
Sack, Wolfgang Otto
Sansone, Frances Marie
Satir, Peter
Scharrer, Berta Vogel
Schecter, Arnold Joel
Schmidt, John Thomas
Schuel, Herbert
Severin, Charles Matthew
Sherman, Burton Stuart
Shriver, Joyce Elizabeth
Silverman, Ann Judith
Singh, Inder Jit
Smeriglio, Alfred John
Smith, Richard Andrew
Spence, Alexander Perkins
Springer, Alan David
Stempak, Jerome G
Stern, Jack Tuteur, Jr
Strominger, Norman Lewis
Susman, Randall Lee
Swan, Roy Craig, Jr
Szabo, Piroska Ludwig
Terzakis, John A
Tichauer, Erwin Rudolph
Tieman, Suzannah Bliss
Udin, Susan Boymel
Wahlert, John Howard
Ware, Carolyn Bogardus
Warfel, John Hiatt
Waterhouse, Joseph Stallard
Webber, Richard Harry
Wenk, Eugene J
West, William T
Whiting, Anne Margaret
Yazulla, Stephen
Zanetti, Nina Clare

NORTH CAROLINA
Ankel-Simons, Friderun Annursel
Becker, Roland Frederick
Bell, Mary Allison
Black, Betty Lynne
Bo, Walter John
Cornwell, John Calhoun
Daniel, Hal J
Duke, Kenneth Lindsay
Enlow, Donald Hugh
Everett, John Wendell
Harrison, Frederick Williams
Henson, O'Dell Williams, Jr
Johnson, Franklin M
Johnston, Malcolm Campbell
Klintworth, Gordon K
Koch, William Edward
Lauder, Jean Miles
Lawrence, Irvin E, Jr
Lay, Douglas M
Lemasters, John J
McCreight, Charles Edward
MacRae, Edith Krugelis
Miller, Inglis J, Jr
Montgomery, Royce Lee
Mortenson, Leonard Earl
O'Steen, Wendall Keith
Peach, Roy
Pollitzer, William Sprott
Putnam, Jeremiah (Jerry) L
Reedy, Michael K
Robertson, George Gordon
Robertson, James David
Sadler, Thomas William
Simons, Elwyn L
Singleton, Mary Clyde
Smallwood, James Edgar
Sulik, Kathleen Kay
White, Raymond Petrie, Jr
White, Richard Alan
Wolk, Robert George

NORTH DAKOTA
Carlson, Edward C
Hunt, Curtiss Dean
Ries, Ronald Edward
Samson, Willis Kendrick

OHIO
Ackerman, Gustave Adolph, Jr
Albernaz, Jose Geraldo
Alway, Stephen Edward
Barnes, Karen Louise
Batten, Bruce Edgar
Bloch, Edward Henry
Cardell, Robert Ridley, Jr
Chantell, Charles J
Clark, David Lee
Crafts, Roger Conant
Cruce, William L R
Crutcher, Keith A
DiDio, Liberato John Alphonse
Diesem, Charles D
Dimlich, Ruth Van Weenen
Drake, Richard Lee
Dutta, Hiran M
Eastman, Joseph
Egar, Margaret Wells
Eglitis, Irma
Ely, Daniel Lee
Fucci, Donald James
Gaughran, George Richard Lawrence
Gilloteaux, Jacques Jean-Marie A
Glenn, Loyd Lee
Goldstein, David Louis
Hall, James Lawrence
Hamlett, William Cornelius
Hayes, Thomas G

BIOLOGICAL SCIENCES / 27

Hering, Thomas M
Hikida, Robert Seiichi
Hilliard, Stephen Dale
Hines, Margaret H
Hoff, Kenneth Michael
Holtzman, David Allen
Hopfer, Ulrich
Houston, Willie Walter, Jr
Israel, Harry, III
Johnson, Thomas Raymond
Keller, Jeffrey Thomas
Kern, Michael Don
King, James S
King, John Edward
Latimer, Bruce Millikin
Liebelt, Robert Arthur
Martin, George Franklin, Jr
Massopust, Leo Carl, Jr
Meineke, Howard Albert
Moore, Fenton Daniel
Morse, Dennis Ervin
Neiman, Gary Scott
Nokes, Richard Francis
Pansky, Ben
Patton, Nancy Jane
Peterson, Ellengene Hodges
Rogers, Richard C
St Pierre, Ronald Leslie
Saksena, Vishnu P
Saul, Frank Philip
Shipley, Michael Thomas
Showers, Mary Jane C
Singer, Marcus
Stuesse, Sherry Lynn
Sucheston, Martha Elaine
Taslitz, Norman
Tornheim, Patricia Anne
Venzke, Walter George
Watanabe, Michiko
Williams, Benjamin Hayden
Wismar, Beth Louise

OKLAHOMA
Allison, John Everett
Breazile, James E
Chung, Kyung Won
Coalson, Robert Ellis
Dugan, Kimiko Hatta
Felts, William Joseph Lawrence
Howes, Robert Ingersoll, Jr
Kerley, Michael A
Knisely, William Hagerman
Lavia, Lynn Alan
Lhotka, John Francis
Martin, Loren Gene
Norvell, John Edmondson, III
Ownby, Charlotte Ledbetter
Papka, Raymond Edward
Russell, Scott D
Snow, Clyde Collins
Tomasek, James J
Wickham, M Gary

OREGON
Bartley, Murray Hill, Jr
Bell, Curtis Calvin
Carlisle, Kay Susan
Critchlow, Burtis Vaughn
Dejmal, Roger Kent
Dow, Robert Stone
Edmonds, Elaine S
Fahrenbach, Wolf Henrich
Gunberg, David Leo
Hayden, Jess, Jr
Hillman, Stanley Severin
Jenkins, Thomas William
Marshall, Frederick James
Niles, Nelson Robinson
Smith, Catherine Agnes
Weaver, Morris Eugene
Wilson, Marlene Moore
Zimmerman, Earl Abram
Zingeser, Maurice Roy

PENNSYLVANIA
AcKerman, Larry Joseph
Aker, Franklin David
Allen, Theresa O
Amenta, Peter Sebastian
Anderson, John Walberg
Aston-Jones, Gary Stephen
Beasley, Andrew Bowie
Beezhold, Donald H
Bennett, Henry Stanley
Black, Mark Morris
Boyd, Robert B
Cameron, William Edward
Caso, Louis Victor
Cauna, Nikolajs
Colony-Cokely, Pamela
Conway, John Richard
Coulter, Herbert David, Jr
Crabill, Edward Vaughn
Crouse, Gail
Crummy, Pressley Lee
De Groat, William C, Jr
De Pace, Dennis Michael
Dropp, John Jerome
Fish, Frank Eliot
Flexner, Louis Barkhouse
Gay, Carol Virginia Lovejoy
George-Weinstein, Mindy
Greene, Charlotte Helen
Greene, Robert Morris

Grove, Alvin Russell, Jr
Haroian, Alan James
Hausberger, Franz X
Hilfer, Saul Robert
Johnson, Elmer Marshall
Johnson, Robert Joseph
Kanczak, Norbert M
Kennedy, Michael Craig
Kesner, Michael H
Kochhar, Devendra M
Kriebel, Richard Marvin
Kvist, Tage Nielsen
Ladman, Aaron J(ulius)
Lambertsen, Richard H
Langdon, Herbert Lincoln
Lewis, Michael Edward
McCallister, Lawrence P
McLaughlin, Patricia J
Malewitz, Thomas Donald
Masters, Edwin M
Meyer, R Peter
Miselis, Richard Robert
Monson, Frederick Carlton
Morrison, Adrian Russel
Moskowitz, Norman
Mundell, Robert David
Munger, Bryce Leon
Nachmias, Vivianne T
Nemeth, Andrew Martin
Niewenhuis, Robert James
Niu, Mann Chiang
Ontell, Marcia
Paavola, Laurie Gail
Pelleg, Amir
Pepe, Frank Albert
Phillips, Steven Jones
Piesco, Nicholas Peter
Pratt, Neal Edwin
Pritchard, Hayden N
Raikow, Robert Jay
Ramasastry, Sai Sudarshan
Rapp, Robert
Rosenthal, Theodore Bernard
Ross, Leonard Lester
Rothman, Richard Harrison
Sachs, Howard George
Sanger, Jean M
Schuit, Kenneth Edward
Sedar, Albert William
Shea, John Raymond Michael, Jr
Short, John Albert
Siegel, Michael Ian
Smith, Allen Anderson
Sodicoff, Marvin
Sprague, James Mather
Stere, Athleen Jacobs
Sterling, Peter
Surmacz, Cynthia Ann
Swanson, Ernest Allen, Jr
Tobin, Thomas Vincent
Troyer, John Robert
Truex, Raymond Carl
Volz, John Edward
Warner, Francis James
Weisel, John Winfield
Weston, John Colby
Wideman, Robert Frederick, Jr
Wolfe, Allan Frederick
Zaccaria, Robert Anthony
Zagon, Ian Stuart

RHODE ISLAND
Calabresi, Paul
Dolyak, Frank
Ebner, Ford Francis
Goss, Richard Johnson
Hegre, Orion Donald
Janis, Christine Marie
Ripley, Robert Clarence
Strauss, Elliott William
Yevich, Paul Peter

SOUTH CAROLINA
Augustine, James Robert
Blake, Charles Albert
Debacker, Hilda Spodheim
Dougherty, William J
Fox, Richard Shirley
Hays, Ruth Lanier
Lockard, Isabel
Metcalf, Isaac Stevens Halstead
Odor, Dorothy Louise
Poteat, William Louis
Roel, Lawrence Edmund
Simson, Jo Anne V
Weymouth, Richard J
Worthington, Ward Curtis, Jr

SOUTH DAKOTA
Haertel, John David
Martin, James Edward
Moore, Josephine Carroll
Naughten, John Charles
Neufeld, Daniel Arthur
Parke, Wesley Wilkin
Rinker, George Clark
Settles, Harry Emerson

TENNESSEE
Anderson, Ted L
Atnip, Robert Lee
Aulsebrook, Lucille Hagan
Bernstorf, Earl Cranston
Bruesch, Simon Rulin

Burt, Alvin Miller, III
Corliss, Clark Edward
Dix, John Willard
Donaldson, Donald Jay
Elberger, Andrea June
Evans, James Spurgeon
Fedinec, Alexander
Gotcher, Jack Everett
Harwood, Thomas Riegel
Hasty, David Long
Hossler, Fred E
Huggins, James Anthony
King, Lloyd Elijah, Jr
Lawler, James E
Murrell, Leonard Richard
Nanney, Lillian Bradley
Peppler, Richard Douglas
Richardson, Elisha Roscoe
Rieke, Garl Kalman
Roberts, Lee Knight
Saxon, James Glenn
Schultz, Terry Wayne
Skalko, Richard G(allant)
Wilcox, Harry Hammond
Wilson, Jack Lowery
Young, Joseph Marvin

TEXAS
Adrian, Erle Keys, Jr
Ashby, Jon Kenneth
Ashworth, Robert David
Azizi, Sayed Ausim
Banks, William Joseph, Jr
Blanton, Patricia Louise
Bratton, Gerald Roy
Burton, Alexis Lucien
Callas, Gerald
Cameron, Ivan Lee
Cannon, Marvin Samuel
Carpenter, Robert James, Jr
Cavazos, Lauro Fred
Clench, Mary Heimerdinger
Coggeshall, Richard E
Cominsky, Nell Catherine
Croley, Thomas Edgar
Dafny, Nachum
Dennison, David Kee
Deter, Russell Lee, II
Dill, Russell Eugene
Dung, H C
Fife, William Paul
Finerty, John Charles
Frederick, Jeanne M
Garfield, Robert Edward
Geoghegan, William David
George, Fredrick William
German, Dwight Charles
Gibson, Kathleen Rita
Gonyea, William Joseph
Gonzalez-Lima, Francisco
Grimes, L Nichols
Hall, Charles Eric
Halpern, Salmon Reclus
Harrison, Frank
Herbert, Damon Charles
Hild, Walter J
Hill, Ronald Stewart
Houston, Marshall Lee
Hutson, James Chelton
Jacobson, Antone Gardner
Jester, James Vincent
Kendall, Michael Welt
Kennedy, Joseph Patrick
Leppi, Theodore John
Litke, Larry Lavoe
Lucas, Edgar Arthur
Marshak, David William
Matthews, James Lester
Moore, Richard Dana
Neaves, William Barlow
Norman, Reid Lynn
Payer, Andrew Francis
Pierce, Jack Robert
Pinero, Gerald Joseph
Rozier, Carolyn K
Samorajski, Thaddeus
Sampson, Herschel Wayne
Sauerland, Eberhardt Karl
Schunder, Mary Cothran
Schwartz, Colin John
Seliger, William George
Sis, Raymond Francis
Skjonsby, Harold Samuel
Steele, David Gentry
Taylor, Alan Neil
Tebo, Heyl Gremmer
Throckmorton, Gaylord Scott
Trulson, Michael E
VanderWiel, Carole Jean
Vaughan, Mary Kathleen
Warner, Marlene Ryan
Welch, Gary William
Wilczynski, Walter
Williams, Darryl Marlowe
Williams, Fred Eugene
Williams, Vick Franklin
Willis, William Darrell, Jr
Winborn, William Burt
Wordinger, Robert James
Yollick, Bernard Lawrence
Young, Margaret Claire
Zwaan, Johan Thomas

UTAH
Albertine, Kurt H
Chapman, Arthur Owen
Creel, Donnell Joseph
Jee, Webster Shew Shun
Miller, Scott Cannon
Rodin, Martha Kinscher
Schoenwolf, Gary Charles
Shultz, Leila McReynolds
Stensaas, Larry J
Stensaas, Suzanne Sperling
Stevens, Walter
Van De Graaff, Kent Marshall

VERMONT
Krupp, Patricia Powers
Stultz, Walter Alva
Wells, Joseph
Young, William Johnson, II

VIRGINIA
Allan, Frank Duane
Astruc, Juan
Atkins, David Lynn
Banker, Gary A
Brangan, Pamela J
Breil, Sandra J
Brownson, Robert Henry
Brunjes, Peter Crawford
Butler, Ann Benedict
Carson, Keith Alan
Ching, Melvin Chung Hing
Deck, James David
DeSesso, John Michael
Dessouky, Dessouky Ahmad
Eglitis, Martin Alexandris
Evans, Francis Gaynor
Fenner-Crisp, Penelope Ann
Flickinger, Charles John
Grimm, James K
Haar, Jack Luther
Harris, Thomas Mason
Heath, Everett
Humbertson, Albert O, Jr
Isaacson, Robert John
Jeffrey, Jackson Eugene
Jollie, William Pucette
Laurie, Gordon William
Leichnetz, George Robert
Muir, William Angus
Osborne, Paul James
Owers, Noel Oscar
Potts, Malcolm
Povlishock, John Theodore
Pullen, Edwin Wesley
Rubenstein, Norton Michael
Russi, Simon
Sandow, Bruce Arnold
Savitzky, Alan Howard
Scott, David Evans
Seibel, Hugo Rudolf
Sirica, Alphonse Eugene
Steeves, Harrison Ross, III
Suter, Daniel B
Talbert, George Brayton

WASHINGTON
Baskin, Denis George
Baumel, Julian Joseph
Broderson, Stevan Hardy
Churchill, Lynn
Davies, Jack
DeVito, June Logan
Frederickson, Richard Gordon
Graney, Daniel O
Halbert, Sheridan A
Harris, Roger Mason
Hendrickson, Anita Elizabeth
Herring, Susan Weller
Holbrook, Karen Ann
Kashiwa, Herbert Koro
Landau, Barbara Ruth
Luchtel, Daniel Lee
Luft, John Herman
Lundy, John Kent
Odland, George Fisher
Pietsch, Theodore Wells
Quinn, LeBris Smith
Ratzlaff, Marc Henry
Reh, Thomas Andrew
Rieke, William Oliver
Rosse, Cornelius
Sundsten, John Wallin
Tamarin, Arnold
Worthman, Robert Paul
Zamora, Cesario Siasoco

WEST VIRGINIA
Beresford, William Anthony
Bowdler, Anthony John
Brown, Paul B
Burns, John Thomas
Chang, William Wei-Lien
Culberson, James Lee
Fix, James D
Friedman, Morton Henry
Hilloowala, Rumy Ardeshir
Martin, William David
Overman, Dennis Orton
Pinkstaff, Carlin Adam
Reilly, Frank Daniel
Reyer, Randall William
Rhoten, William Blocher
Walker, Elizabeth Reed

Anatomy (cont)

Williams, Leah Ann

WISCONSIN
Anderson, Frank David
Austin, Bert Peter
Benjamin, Hiram Bernard
Bersu, Edward Thorwald
Bolender, David Leslie
Curtis, Robin Livingstone
Fallon, John Francis
Flanigan, Norbert James
Gardner, Weston Deuain
Greaser, Marion Lewis
Harrington, Sandra Serena
Jensen, Richard Harvey
Jones, Helena Speiser
Kaplan, Stanley
Kemnitz, Joseph William
Lalley, Peter Michael
Larson, Sanford J
Long, Charles Alan
Long, Sally Yates
Lough, John William, Jr
Markwald, Roger R
Oaks, John Adams
Oertel, Donata (Mrs Bill M Sugden)
Oyen, Ordean James
Pettersen, James Clark
Riley, Danny Arthur
Rouse, Thomas C
Royce, George James
Schultz, Edward
Sether, Lowell Albert
Slautterback, David Buell
Snook, Theodore
Snyder, Virginia
Sobkowicz, Hanna Maria
Thibodeau, Gary A
Weil, Michael Ray

WYOMING
Lillegraven, Jason Arthur

PUERTO RICO
Alcala, Jose Ramon
Garcia-Castro, Ivette
Kicliter, Ernest Earl, Jr
Luckett, Winter Patrick
Virkki, Niilo

ALBERTA
Cavey, Michael John
Cookson, Francis Bernard
Currie, Philip John
Famiglietti, Edward Virgil, Jr
Holland, Graham Rex
Leeson, Thomas Sydney
Levene, Cyril
Osler, Margaret J
Russell, Anthony Patrick
Sperber, Geoffrey Hilliard
Spira, Arthur William
Stell, William Kenyon
Wyse, John Patrick Henry

BRITISH COLUMBIA
Bressler, Bernard Harvey
Friedman, Constance Livingstone
Friedman, Sydney Murray
Friz, Carl T
Hollenberg, Martin James
Salt, Walter Raymond
Schwarz, Dietrich Walter Friedrich
Slonecker, Charles Edward
Todd, Mary Elizabeth
Webber, William A

MANITOBA
Bertalanffy, Felix D
Braekevelt, Charlie Roger
Klass, Alan Arnold
Nathaniel, Edward J H
Persaud, Trivedi Vidhya Nandan
Weisman, Harvey

NOVA SCOTIA
Armour, John Andrew
Cooper, John (Hanwell)
Dickson, Douglas Howard
Hansell, Margaret Mary
Hopkins, David Alan
Leslie, Ronald Allan
Rutherford, John Garvey
Saunders, Richard L de C H
Semba, Kazue
Vethamany, Victor Gladstone
Walker, Joan Marion

ONTARIO
Anderson, James Edward
Barr, Murray Llewellyn
Basmajian, John V
Begun, David Rene
Buchanan, George Dale
Buck, Robert Crawforth
Butler, Richard Gordon
Carr, David Harvey
Chapman, David MacLean
Ciriello, John
De Bold, Adolfo J
Galil, Khadry Ahmed
Gammal, Elias Bichara
Geissinger, Hans Dieter
Govind, Choonilal Keshav
Habowsky, Joseph Edmund Johannes
Ham, Arthur Worth
Hinke, Joseph Anthony Michael
Irons, Margaret Jean
Kiernan, John Alan
McMillan, Donald Burley
Melcher, Antony Henry
Metuzals, Janis
Montemurro, Donald Gilbert
Moore, Keith Leon
Romero-Sierra, Cesar Aurelio
Saunders, Shelley Rae
Scadding, Steven Richard
Singh, Roderick Pataudi
Sistek, Vladimir
Ten Cate, Arnold Richard
Winterbottom, Richard

PRINCE EDWARD ISLAND
Singh, Amreek

QUEBEC
Bendayan, Moise
Bisaillon, Andre
Blaschuk, Orest William
Bois, Pierre
Bolduc, Reginald J
Briere, Normand
Clermont, Yves Wilfred
Colonnier, Marc
Courville, Jacques
Daoust, Roger
Gagnon, Real
Garon, Olivier
Jones, Barbara Ellen
Leblond, Charles Philippe
Murphy, Richard Arthur
Nadler, Norman Jacob
Olivier, Andre
Osmond, Dennis Gordon
Parent, Andre
Paris, David Leonard
Pereira, Gerard P
Pierard, Jean Arthur
Poirier, Louis
Ramon-Moliner, Enrique
Richer, Claude-Lise
Simard, Therese Gabrielle
Stephens, Heather R
Warshawsky, Hershey

SASKATCHEWAN
Brandell, Bruce Reeves
Clayton, Hilary Mary
Fedoroff, Sergey
Flood, Peter Frederick
Munkacsi, Istvan
Newstead, James Duncan MacInnes

OTHER COUNTRIES
Armato, Ubaldo
Ben-Ze'ev, Avri
Brunser, Oscar
Donato, Rosario Francesco
Fahimi, Hossein Dariush
Ganchrow, Donald
Gertz, Samuel David
Guillery, Rainer Walter
Hashimoto, Paulo Hitonari
Helander, Herbert Dick Ferdinand
Hughes, Abbie Angharad
Ishikawa, Hiroshi
Jordan, Robert Lawrence
Komers, Petr E
Makita, Takashi
Nakane, Paul K
Oxnard, Charles Ernest
Saito, Takuma
Stallard, Richard E
Szebenyi, Emil
Wei, Stephen Hon Yin
Wenger, Byron Sylvester
Yamada, Eichi

Bacteriology

ALABAMA
Brady, Yolanda J
Hiramoto, Raymond Natsuo
Lauerman, Lloyd Herman, Jr
Montgomery, John R
Mora, Emilio Chavez
Wood, David Oliver

ARIZONA
Archer, Stanley J
Birge, Edward Asahel
Fuller, Wallace Hamilton
Gerba, Charles Peter
Reeves, Henry Courtland
Reinhard, Karl Raymond

ARKANSAS
Bates, Joseph H
Denton, James H
England, James Donald
Rank, Roger Gerald

CALIFORNIA
Ames, Giovanna Ferro-Luzzi
Andreoli, Anthony Joseph
Bayne, Henry Godwin
Biberstein, Ernst Ludwig
Bramhall, John Shepherd
Brown, Robert Lee
Brunke, Karen J
Buggs, Charles Wesley
Calderone, Julius G
Caputi, Roger William
Carlucci, Angelo Francis
Catlin, B Wesley
Chen, Benjamin P P
Chou, Tsong-Wen
Clark, Alvin John
Cooper, Robert Chauncey
Davenport, Calvin Armstrong
Delk, Ann Stevens
Delong, Edward F
Fulco, Armand J
Gaertner, Frank Herbert
Glembotski, Christopher Charles
Gold, William
Goodman, Richard E
Gunsalus, Robert Philip
Halasz, Nicholas Alexis
Heck, Joseph Gerard
Hemmingsen, Barbara Bruff
Hildebrand, Donald Clair
Hochstein, Lawrence I
Hoeprich, Paul Daniel
Ito, Keith A
Janda, John Michael
Knudson, Gregory Blair
Kustu, Sydney Govons
McClelland, Michael
Marshall, Rosemarie
Martinez, Rafael Juan
Moore, Harold Beveridge
Moyed, Harris S
Nassos, Patricia Saima
Page, Larry J
Phillips, Donald Arthur
Pickett, Morris John
Raffel, Sidney
Raj, Harkisan D
Reizer, Jonathan
Rittenberg, Sydney Charles
Romig, William Robert
Rosenberg, Steven Loren
Rosenthal, Sol Roy
Ruby, Edward George
Schiller, Neal Leander
Simons, Robert W
Smith, Kendric Charles
Speck, Reinhard Staniford
Stevens, Ronald Henry
Taylor, Barry L
Trivett, Terrence Lynn
Uyeda, Charles Tsuneo
VanBruggen, Ariena H C
Van Bruggen, Ariena H C
Vickrey, Herta Miller
Wall, Thomas Randolph
Wedberg, Stanley Edward
Weimberg, Ralph
Woodhour, Allen F
Wu, William Gay
York, Charles James
Zerez, Charles Raymond
Zlotnik, Albert

COLORADO
Boyd, William Lee
Deane, Darrell Dwight
Gabridge, Michael Gregory
Harrison, Monty DeVerl
McLean, Robert George
Morse, Helvise Glessner
Nagy, Julius G
Patterson, David
Reinbold, George W
Smith, Ralph E
Thomas, William Robb
Ulrich, John August
Vasil, Michael Lawrence

CONNECTICUT
Abeling, Edwin John
Buck, John David
Celesk, Roger A
Dingman, Douglas Wayne
Geiger, Jon Ross
Gorrell, Thomas Earl
Huang, Liang Hwang
Low, Kenneth Brooks, Jr
Lynch, John Edward
O'Neill, James F
Rupp, W Dean
Wyckoff, Delaphine Grace Rosa

DELAWARE
Eleuterio, Marianne Kingsbury

DISTRICT OF COLUMBIA
Binn, Leonard Norman
Cutchins, Ernest Charles
DeCicco, Benedict Thomas
Fanning, George Richard
Hugh, Rudolph
Kennedy, Eugene Richard
Klubes, Philip
Pittman, Margaret
Rothman, Sara Weinstein
Thiermann, Alejandro Bories
Venkatesan, Malabi M

FLORIDA
Bacus, James Nevill
Betz, John Vianney
Cain, Brian D
Clark, William Burton, IV
Dhople, Arvind Madhav
Duggan, Dennis E
Ellias, Loretta Christine
Goihman-Yahr, Mauricio
Gottwald, Timothy R
Horn, Joanne Marie
Jackson, Bettina B Carter
Martinez, Octavio Vincent
Paradise, Lois Jean
Pates, Anne Louise
Shanmugam, Keelnatham
   Thirunavukkarasu
Slack, John Madison
Smith, Paul Howard
Stuy, Johan Harrie
Wilkowske, Howard Hugo

GEORGIA
Abdelal, Ahmed T
Barbaree, James Martin
Brackett, Robert E
Bryan, Frank Leon
Caruthers, John Quincy
Chang, Chung Jan
Cherry, William Bailey
Dailey, Harry A
Dowell, Vulus Raymond, Jr
Doyle, Michael Patrick
Gerencser, Mary Ann (Aiken)
Goldsmith, Edward
Good, Robert Campbell
Hall, Charles Thomas
Hamdy, Mostafa Kamal
Hancock, Kenneth Farrell
Jones, Gilda Lynn
Kubica, George P
Morse, Stephen Allen
Pine, Leo
Reinhardt, Donald Joseph
Sawyer, Richard Trevor
Taylor, Gerald C
Tzianabos, Theodore
Van Eseltine, William Parker
Wiebe, William John
Williams, Joy Elizabeth P
Woodley, Charles Lamar
Ziegler, Harry Kirk

HAWAII
Berger, Leslie Ralph
Patterson, Harry Robert
Phillips, John Howell, Jr

IDAHO
Beck, Sidney M
Crawford, Donald Lee
Crawford, Ronald Lyle
Farrell, Larry Don
Gilmour, Campbell Morrison
Hibbs, Robert A
Hostetter, Donald Lee
Jones, Lewis William
Winston, Vern

ILLINOIS
Bahn, Arthur Nathaniel
Castignetti, Domenic
Ehrlich, Richard
Goldin, Milton
Henderson, Thomas Otis
Holzer, Timothy J
Klinger, Lawrence Edward
Klubek, Brian Paul
Laffler, Thomas G
McKinley, Vicky L
McQuistion, Thomas Evin
Madigan, Michael Thomas
Mathews, Herbert Lester
Murphy, Richard Allan
Myers, Ronald Berl
Nichols, Brian Paul
Passman, Frederick Jay
Perry, Dennis
Perry, Eugene Arthur
Sames, Richard William
Shapiro, James Alan
Shen, Linus Liang-nene
Simonson, Lloyd Grant
Tripathy, Deoki Nandan
Turek, Fred William
Weiler, William Alexander
Woese, Carl R

INDIANA
Boyer, Ernest Wendell
Claybaugh, Glenn Alan
Daily, Fay Kenoyer
Driesens, Robert James
Eisenstein, Barry I
Elliott, Robert A
Enders, George Leonhard, Jr
Hopper, Samuel Hersey
Keith, Paula Myers
Kinzel, Jerry J
Koch, Arthur Louis
Kory, Mitchell
Kulpa, Charles Frank, Jr
Levinthal, Mark
Millar, Wayne Norval

Morter, Raymond Lione
Shockey, William Lee
Van Frank, Richard Mark
Wagner, Morris
Welker, George W

IOWA
Haefele, Douglas Monroe
Hartman, Paul Arthur
Hug, Daniel Hartz
Lockhart, William Raymond
Mickelson, Milo Norval
Quinn, Loyd Yost
Rebstock, Theodore Lynn
Richardson, Robert Louis
Stanton, Thaddeus Brian
Thurston, John Robert

KANSAS
Bailie, Wayne E
Cho, Cheng T
Dykstra, Mark Allan
Gray, Andrew P
Haas, Herbert Frank
Johnson, Donovan Earl
Miller, Glendon Richard
Minocha, Harish C
Warren, Halleck Burkett, Jr

KENTUCKY
Edwards, Ogden Frazelle
Greenberg, Richard Aaron
Lillich, Thomas Tyler
Straley, Susan Calhoon
Stuart, James Glen

LOUISIANA
Bornside, George Harry
Boyd, Frank McCalla
Domingue, Gerald James, Sr
Johnson, Mary Knettles
Titkemeyer, Charles William
Willis, William Hillman
Wilson, Raphael

MAINE
Buck, Charles Elon
Gershman, Melvin
Highlands, Matthew Edward
Jerkofsky, Maryann
Pratt, Darrell Bradford
Tjepkema, John Dirk

MARYLAND
Abramson, I Jerome
Adams, James Miller
Benson, Spencer Alan
Burchard, Robert P
Chakrabarti, Siba Gopal
Chanock, Robert Merritt
Chattoraj, Dhruba Kumar
Doetsch, Raymond Nicholas
Dragunsky, Eugenia M
Ellinghausen, Herman Charles, Jr
Falkenstein, Kathy Fay
Foulds, John Douglas
Garges, Susan
Gherna, Robert Larry
Goldenberg, Martin Irwin
Gordon, Ruth Evelyn
Hanes, Darcy Elizabeth
Hanks, John Harold
Henry, Timothy James
Hill, James Carroll
Hochstein, Herbert Donald
Ivins, Bruce Edwards
Johnson-Winegar, Anna
Jones, Morris Thompson
Joseph, Sammy William
Lattuada, Charles P
McMacken, Roger
Mattick, Joseph Francis
Mohler, Irvin C, Jr
Molenda, John R
Noyes, Howard Ellis
Rosner, Judah Leon
Seto, Belinda P L
Stamper, Hugh Blair
Wagner, David Darley

MASSACHUSETTS
Amos, Harold
Canale-Parola, Ercole
Craven, Donald Edward
Crisley, Francis Daniel
Deitz, William Harris
Gibbons, Ronald J
Gilroy, James Joseph
Girard, Kenneth Francis
Gordon, Lance Kenneth
Gupta, Rajesh K
Hollocher, Thomas Clyde, Jr
Kashket, Eva Ruth
Lachica, R Victor
Litsky, Warren
Lovett, Charles McVey
MacMahon, Harold Edward
Madore, Bernadette
Malakian, Artin
Pfister, Richard Charles
Reynolds, John Theodore
Sarkar, Nilima
Sbarra, Anthony J
Stern, Ivan J

Thomas, Gail B
Watson, Stanley W

MICHIGAN
Breznak, John Allen
Carrick, Lee
Colingsworth, Donald Rudolph
Daniels, Peter John Lovell
Freter, Rolf Gustav
Jackson, Matthew Paul
Jones, Lily Ann
Juni, Elliot
Kurnit, David Martin
Loesche, Walter J
McCullough, Willard George
Martin, Joseph Patrick, Jr
Mattman, Lida Holmes
Muentener, Donald Arthur
Mulks, Martha Huard
Pecoraro, Vincent L
Portwood, Lucile Mitchell
Reddy, Chilekampalli Adinarayana
Rossmoore, Harold W
Sebek, Oldrich Karel
Shelef, Leora Aya
Wheeler, Albert Harold
Yancey, Robert John, Jr

MINNESOTA
Bauer, Henry
Beggs, William H
Berube, Robert
Haas, Larry Alfred
Johnson, Russell Clarence
Lonergan, Dennis Arthur
McKay, Larry Lee
Mizuno, William George
Needham, Gerald Morton
Porter, Frederic Edwin
Sadowsky, Michael Jay
Surdy, Ted E
Wagenaar, Raphael Omer

MISSISSIPPI
Arceneaux, Joseph Lincoln
Brown, Lewis Raymond
Sikorowski, Peter P
Wolverton, Billy Charles

MISSOURI
Barnekow, Russell George, Jr
Duncan, David Robert
Eisenstark, Abraham
Elsawi, Nehad
Elvin-Lewis, Memory P F
Finkelstein, Richard Alan
Gossling, Jennifer
Granoff, Dan Martin
Hufham, James Birk
Landick, Robert
Lozeron, Homer A
McCune, Emmett L
Miles, Donald Orval
Misfeldt, Michael Lee
Moulton, Robert Henry
Murray, Patrick Robert
Sargentini, Neil Joseph
Schmidt, Donald Arthur
Wells, Frank Edward

MONTANA
Dorward, David William
Lozano, Edgardo A
Myers, Lyle Leslie
Stoenner, Herbert George

NEBRASKA
Rodriguez-Sierra, Jorge F
Thompson, Thomas Leo
Underdahl, Norman Russell
Vidaver, Anne Marie Kopecky

NEVADA
Jay, James Monroe

NEW HAMPSHIRE
Adler, Frank Leo
Adler, Louise Tale
Jacobs, Nicholas Joseph
Kerwin, Richard Martin
Rodgers, Frank Gerald

NEW JERSEY
Beskid, George
Bruno, Charles Frank
Coriell, Lewis L
Frances, Saul
Gaffar, Abdul
Georgopapadakou, Nafsika Eleni
Green, Erika Ana
Imaeda, Tamotsu
Johnson, Layne Mark
Kestenbaum, Richard Charles
Koepp, Leila H
Koft, Bernard Waldemar
Kronenwett, Frederick Rudolph
Manjula, Belur N
Miller, A(nna) Kathrine
Morneweck, Samuel
Mukai, Cromwell Daisaku
Peterson, Arthur Carl
Regna, Peter P
Rossow, Peter William
Seneca, Harry

Shearer, Marcia Cathrine (Epple)
Solotorovsky, Morris
Tate, Robert Lee, III
Tierno, Philip M, Jr
Umbreit, Wayne William

NEW MEXICO
Hageman, James Howard
Scaletti, Joseph Victor

NEW YORK
Andersen, Kenneth J
Anderson, Kenneth Ellsworth
Beach, David H
Bopp, Lawrence Howard
Bradner, William Turnbull
Brody, Edward Norman
Bukhari, Ahmad Iqbal
Cardenas, Raúl R, Jr
Chiulli, Angelo Joseph
Corpe, William Albert
Cowell, James Leo
Delwiche, Eugene Albert
Efthymiou, Constantine John
Ellner, Paul Daniel
Enea, Vincenzo
Fabricant, Catherine G
Fendler, Janos Hugo
Goodhue, Charles Thomas
Gordon, Morris Aaron
Helmann, John Daniel
Hipp, Sally Sloan
Holmes, David Salway
Kite, Joseph Hiram, Jr
Lambert, Reginald Max
Lazaroff, Norman
McCarty, Maclyn
MacDonald, Russell Earl
Marquis, Robert E
Miller, Terry Lynn
Mohn, James Frederic
Murphy, Timothy F
Pierre, Leon L
Poindexter, Jeanne Stove
Rhee, G-Yull
Roll, David E
Sadosky, Alesia Beth
Scher, William
Shapiro, Caren Knight
Shayegani, Mehdi
Shively, Carl E
Silverman, Morris
Splittstoesser, Clara Quinnell
Splittstoesser, Don Frederick
Steinberg, Bernard Albert
Steinman, Howard Mark
Thaler, David Solomon
Trowbridge, Richard Stuart
Williams, Noreen
Wolin, Meyer Jerome
Young, William Donald, Jr
Zengel, Janice Marie

NORTH CAROLINA
Curtis, Susan Julia
Elkan, Gerald Hugh
Fulghum, Robert Schmidt
Gray, Walter C(larke)
Hopfer, Roy L
Hulcher, Frank H
Hutt, Randy
Jordan, Thomas L
King, Kendall Willard
Knuckles, Joseph Lewis
Matthysse, Ann Gale
Moody, Max Dale
Schwab, John Harris
Shepard, Maurice Charles
Silverman, Myron Simeon
Sizemore, Ronald Kelly
Sparling, Philip Frederick
Straughn, William Ringgold, Jr
Tarver, Fred Russell, Jr
Tove, Shirley Ruth
Whitmire, Carrie Ella
Wyrick, Priscilla Blakeney

NORTH DAKOTA
Berryhill, David Lee
Fillipi, Gordon Michael
McMahon, Kenneth James
Watson, David Alan

OHIO
Chapple, Paul James
Coplin, David Louis
Frea, James Irving
Frey, James R
Hunter, John Earl
Ichida, Allan A
Kagen, Herbert Paul
Kolodziej, Bruno J
Krampitz, Lester Orville
Lawrence, James Vantine
Long, Sterling K(rueger)
Modrzakowski, Malcolm Charles
Rhodes, Judith Carol
Rice, Eugene Ward
Sobota, Anthony E
Somerson, Norman L
Tabita, F Robert
Tuovinen, Olli Heikki
Walker, Richard V
Watanakunakorn, Chatrchai

OKLAHOMA
Anglin, J Hill, Jr
Booth, James Samuel
Durham, Norman Nevill
Essenberg, Richard Charles
Flournoy, Dayl Jean
Foutch, Gary Lynn
Hitzman, Donald Oliver
Sokatch, John Robert

OREGON
Albrecht, Stephan LaRowe
Anderson, Arthur W
Arp, Daniel James
Barry, Arthur Leland
Charlton, David Berry
Dooley, Douglas Charles
Elliott, Lloyd Floren
Kazerouni, Lewa
Ream, Lloyd Walter, Jr
Sandine, William Ewald

PENNSYLVANIA
Actor, Paul
Anderson, Theodore Gustave
Austrian, Robert
Bowles, Bobby Linwood
Buchanan, Robert Lester
Burdash, Nicholas Michael
Carls, Ralph A
Carpenter, Charles Patten
Castric, Peter Allen
Cinquina, Carmela Louise
Crowell, Richard Lane
Debroy, Chitrita
Duncan, Charles Lee
Fett, William Frederick
Fletcher, Ronald D
Garfinkle, Barry David
Grainger, Thomas Hutcheson, Jr
Gregory, Francis Joseph
Herbert, Michael
Juneja, Vijay Kumar
Kruse, Conrad Edward
Lee, John Cheung Han
Lehman, Ernest Dale
Levine, Elliot Myron
Lindemeyer, Rochelle G
Lindstrom, Eugene Shipman
Long, Walter Kyle, Jr
Mandel, John Herbert
Miller, Arthur James
Milligan, Wilbert Harvey, III
Millman, Irving
Osborne, William Wesley
Panos, Charles
Phillips, Allen Thurman
Poupard, James Arthur
Rest, Richard Franklin
Shirk, Richard Jay
Shockman, Gerald David
Stevens, Roy Harris
Thomulka, Kenneth William
Witlin, Bernard
Yurchenco, John Alfonso
Zimmerman, Leonard Norman

RHODE ISLAND
Jones, Kenneth Wayne
Krasner, Robert Irving
Landy, Arthur
Laux, David Charles
McConeghy, Matthew H
Tourtellotte, Mark Eton
Traxler, Richard Warwick
Wood, Norris Philip

SOUTH CAROLINA
Bryson, Thomas Allan
Graber, Charles David
Pannell, Lolita

SOUTH DAKOTA
Ferguson, Michael William
Howard, Ronald M
Kirkbride, Clyde Arnold
Prescott, Lansing M
Sword, Christopher Patrick
Weaver, Keith Eric
West, Thomas Patrick

TENNESSEE
Adler, Howard Irving
Bibb, William Robert
Bozeman, Samuel Richmond
Chiang, Thomas M
Draughon, Frances Ann
Freeman, Bob A
Gaby, William Lawrence
Gartner, T Kent
Goulding, Charles Edwin, Jr
Holtman, Darlington Frank
Quigley, Neil Benton
Rightsel, Wilton Adair
Savage, Dwayne Cecil
Shockley, Thomas E
Stevens, Stanley Edward, Jr
Woychik, Richard P

TEXAS
Bennett, Edward Owen
Blouse, Louis E, Jr
Boley, Robert B
Calkins, Harmon Eldred

Bacteriology (cont)

Collisson, Ellen Whited
Cook, Paul Fabyan
Cooper, Ronda Fern
Crawford, Gladys P
Dempsey, Walter B
Friend, Patric Lee
Goldschmidt, Millicent
Hamilton, Charleen Marie
Hejtmancik, Kelly Erwin
Ippen-Ihler, Karin Ann
Jeter, Randall Mark
Joys, Terence Michael
Kester, Andrew Stephen
Kiel, Johnathan Lloyd
LaBree, Theodore Robert
LeBlanc, Donald Joseph
Lyles, Sanders Truman
McCallum, Roderick Eugene
Matney, Thomas Stull
Payne, Shelley Marshall
Phillips, Guy Frank
Pike, Robert Merrett
Reyes, Victor E
Robertstad, Gordon Wesley
Rolfe, Rial Dewitt
Schmidt, Jerome P(aul)
Schneider, Dennis Ray
Sistrunk, William Allen
Smith, William Russell
Spellman, Craig William
Stewart, Charles Ranous
Straus, David Conrad
Sweet, Charles Edward
Walter, Ronald Bruce
Widner, William Richard
Wilson, Joe Bransford
Wood, Robert Charles
Zabransky, Ronald Joseph

UTAH
Anderson, Anne Joyce
Georgopoulos, Constantine Panos
Hayes, Sheldon P
Lighton, John R B
Nicholes, Paul Scott
Roth, John R
Spendlove, John Clifton
Torres, Anthony R
Wright, Donald N

VERMONT
Schaeffer, Warren Ira

VIRGINIA
Bauerle, Ronald H
Claus, George William
Edlich, Richard French
Hsu, Hsiu-Sheng
Kadner, Robert Joseph
Kirpekar, Abhay C
Krieg, Richard Edward, Jr
Moore, Lillian Virginia Holdeman
Nagarkatti, Prakash S
Newman, Jack Huff
O'Kane, Daniel Joseph
Raizen, Carol Eileen
Sarber, Raymond William
Srivastava, Kailash Chandra
Stokes, Gerald V
Tankersley, Robert Walker, Jr
Welshimer, Herbert Jefferson
Wilkins, Tracy Dale
Wise, Robert Irby
Yousten, Allan A

WASHINGTON
Barnes, Glover William
Drake, Charles Hadley
Grootes-Reuvecamp, Grada Alijda
Hall, Elizabeth Rose
Hazelbauer, Gerald Lee
Hurd, Robert Charles
Johnstone, Donald Lee
Luedecke, Lloyd O
Naylor, Harry Brooks
Randall, Linda Lea
Staley, James Trotter
Teresa, George Washington
Weiser, Russel Shively
Wildung, Raymond Earl

WEST VIRGINIA
Bissonnette, Gary Kent
Bullock, Graham Lambert
Chisler, John Adam
Gain, Ronald Ellsworth
Gerencser, Vincent Frederic
Woolridge, Robert Leonard

WISCONSIN
Collins, Mary Lynne Perille
Czuprynski, Charles Joseph
Deibel, Robert Howard
Hornemann, Ulfert
Kushnaryov, Vladimir Michael
McIntosh, Elaine Nelson
Myers, Charles R
Ogden, Robert Verl
Qureshi, Nilofer
Rice, Marion McBurney
Rouf, Mohammed Abdur
Schaefer, Daniel M

Snudden, Birdell Harry
Taylor, Jerry Lynn
Tufte, Marilyn Jean
Waechter-Brulla, Daryle A
Whitehead, Howard Allan
Willits, Richard Ellis

WYOMING
Maki, Leroy Robert

PUERTO RICO
Schroder, Eduardo C

ALBERTA
Cheng, Kuo-Joan
Eggert, Frank Michael
Jellard, Charles H
Langford, Edgar Verden
Rennie, Robert John
Schroder, David John
Whitehouse, Ronald Leslie S

BRITISH COLUMBIA
Albright, Lawrence John
Bismanis, Jekabs Edwards
Campbell, Jack James Ramsay
De Boer, Solke Harmen
Richards, William Reese
Syeklocha, Delfa
Trussell, Paul Chandos

MANITOBA
Lukow, Odean Michelin
Wiseman, Gordon Marcy

NEW BRUNSWICK
Robinson, John

NOVA SCOTIA
Simpson, Frederick James
Stewart, James Edward

ONTARIO
Beaulieu, J A
Blais, Burton W
Cipera, John Dominik
Clarke, Anthony John
Fitz-James, Philip Chester
Fraser, Ann Davina Elizabeth
Garcia, Manuel Mariano
Hauschild, Andreas H W
Jones, John Bryan
Jordan, David Carlyle
Kushner, Donn Jean
McKenzie, Allister Roy
Murray, Robert George Everitt
Murray, William Douglas
Rao, Salem S
Roslycky, Eugene Bohdan
Sandham, Herbert James
Speckmann, Gunter Wilhelm-Otto
Stokes, Pamela Mary
Trick, Charles Gordon
Tryphonas, Helen

QUEBEC
Bordeleau, Lucien Mario
Fournier, Michel
Frappier, Armand
Portelance, Vincent Damien
Rokeach, Luis Alberto
Yakunin, Alexander F

OTHER COUNTRIES
Dutta-Roy, Asim Kanti
Ha, Tai-You
Hosokawa, Keiichi
Modabber, Farrokh Z
Olson, John Melvin
Roy, Raman K
Wu, Albert M

Behavior-Ethology

ALABAMA
Appel, Arthur Gary
Dobson, F(rederick) Stephen
Holler, Nicholas Robert
Moyer, Kenneth Evan
Pegram, George Vernon, Jr
Richardson, Terry David
Schneider, Susan Marguerite
Sparks, David Lee
Stewart, Scott David
Taub, Edward
Wolfe, James Leonard
Wyss, James Michael

ALASKA
Ahlgren, Molly O
Fagen, Robert
Helle, John Harold (Jack)

ARIZONA
Balda, Russell Paul
Bernays, Elizabeth Anna
Burgess, Kathryn Hoy
Chapman, Reginald Frederick
Cheal, MaryLou
Figueredo, Aurelio Jose
Johnson, Robert Andrew
Klein, Sherwin Jared
Lamunyon, Craig Willis

Mulder, John Bastian
Papaj, Daniel Richard
Pepperberg, Irene Maxine
Rutowski, Ronald Lee
Schnell, Jay Heist
Slobodchikoff, Constantine Nicholas
Smith, Andrew Thomas
Smith, Robert Lloyd
Topoff, Howard Ronald

ARKANSAS
Dykman, Roscoe A
Holson, Ralph Robert
Newton, Joseph Emory O'Neal
Paule, Merle Gale
Reddy, RamaKrishna Pashuvula
Smith, Kimberly Gray

CALIFORNIA
Alberts, Allison Christine
Arnold, Arthur Palmer
Baptista, Luis Felipe
Barlow, George Webber
Benzer, Seymour
Boggs, Carol L
Bradbury, Jack W
Brown, Judith Adele
Bryson, George Gardner
Burley, Nancy
Case, Ted Joseph
Chambers, Kathleen Camille
Chapman, Loring Frederick
Chappell, Mark Allen
Cody, Martin L(eonard)
Cogswell, Howard Lyman
Coleman, Ronald Murray
Collias, Nicholas Elias
Collier, Gerald
Courchesne, Eric
Cox, Cathleen Ruth
Cummings, William Charles
Davis, William Jackson
Dingle, Richard Douglas Hugh
Dolhinow, Phyllis Carol
Efron, Robert
Endler, John Arthur
Enright, James Thomas
Farish, Donald James
Ferguson, James Mecham
Fernald, Russell Dawson
Fisler, George Frederick
Freeman, Linton Clarke
Frey, Dennis Frederick
Galin, David
Gallistel, Charles Ransom
Gambs, Roger Duane
Gary, Norman Erwin
Gazzaniga, Michael Saunders
Getz, Wayne Marcus
Haas, Richard
Hanggi, Evelyn Betty
Hardt, James Victor
Hart, Benjamin Leslie
Howard, Walter Egner
Hrdy, Sarah Blaffer
Jennrich, Ellen Coutlee
Jevning, Ron
Jones, Claris Eugene, Jr
Kavanau, Julian Lee
Kimsey, Lynn Siri
Kistner, David Harold
Knapp, Theodore Martin
Krekorian, Charles O'Neil
Kryter, Karl David
Levinson, Arthur David
Lindburg, Donald Gilson
Loeblich, Karen Elizabeth
Lovich, Jeffrey Edward
Lukens, Herbert Richard, Jr
MacMillen, Richard Edward
Martin, James Tillison
Matthews, Kathleen Ryan
Moriarty, Daniel Delmar, Jr
Morin, James Gunnar
Naitoh, Paul Yoshimasa
Nelson, Keith
Page, Robert Eugene, Jr
Powell, Jerry Alan
Price, Mary Vaughan
Randall, Janet Ann
Randle, Robert James
Riesen, Austin Herbert
Ritzmann, Ronald Fred
Rotenberry, John Thomas
Sassenrath, Ethelda Norberg
Schell, Anne McCall
Schusterman, Ronald Jay
Schwab, Ernest Roe
Scott, Michael David
Shook, Brenda Lee
Simmel, Edward Clemens
Soltysik, Szczesny Stefan
Squire, Larry Ryan
Stamps, Judy Ann
Stein, Larry
Storms, Lowell H
Sydeman, William J
Tenaza, Richard Reuben
Thompson, Paul O
Thompson, Richard Frederick
Uyeno, Edward Teiso
Vail, Patrick Virgil
Vehrencamp, Sandra Lee

Villablanca, Jaime Rolando
Waser, Nickolas Merritt
Weinberger, Norman Malcolm
Weinrich, James D
Wenzel, Bernice Martha
Williston, John Stoddard
Wimer, Cynthia Crosby
Wood, David Lee
Zaidel, Eran
Zuk, Marlene

COLORADO
Audesirk, Gerald Joseph
Audesirk, Teresa Eck
Bekoff, Anne C
Bekoff, Marc
Bernstein, Stephen
Bond, Richard Randolph
Breed, Michael Dallam
Chiszar, David Alfred
DeFries, John Clarence
Drummond, Boyce Alexander, III
Eaton, Robert Charles
Fall, Michael William
Green, Jeffrey Scott
Krear, Harry Robert
Kulkosky, Paul Joseph
Laudenslager, Mark LeRoy
Lehner, Philip Nelson
Lynch, Carol Becker
Nasci, Roger Stanley
Puck, Mary Hill
Purcell, Kenneth
Reynolds, Richard Truman
Sharpless, Seth Kinman
Tomback, Diana Francine
Troxell, Wade Oakes
Watkins, Linda Rothblum
Wiens, John Anthony
Wilson, James Russell

CONNECTICUT
Barry, Michael Anhalt
Block, Bartley C
Conklin, Harold Colyer
Craig, Catherine Lee
Davis, Michael
Goldman, Bruce Dale
Henry, Charles Stuart
Maier, Chris Thomas
Maxson, Stephen C
Miller, David Bennett
Miller, Neal Elgar
Moehlman, Patricia des Roses
Newton, David C
Novick, Alvin
Rebuffe-Scrive, Marielle Francoise
Redmond, Donald Eugene, Jr
Rettenmeyer, Carl William
Roberts, Mervin Francis
Russock, Howard Israel
Sachs, Benjamin David
Snyder, Daniel Raphael
Stevens, Joseph Charles
Tobias, Jerry Vernon
Weinstein, Curt David
Wells, Kentwood David
Wescott, Roger Williams
Wilson, William August

DELAWARE
Graham, Frances Keesler
Wood, Thomas Kenneth

DISTRICT OF COLUMBIA
Beehler, Bruce McPherson
Blockstein, David Edward
Brown, Eleanor D
Dudley, Susan D
Fox, Michael Wilson
Gladue, Brian Anthony
Gould, Edwin
Jacobs, William Wood, Jr
Jacobsen, Frederick Marius
Kleiman, Devra Gail
Malcom, Shirley Mahaley
Nickle, David Allan
Ralls, Katherine Smith
Raslear, Thomas G
Rubinoff, Roberta Wolff
Schaeff, Catherine Margaret
Wolfle, Thomas Lee
Yaniv, Simone Liliane
Young, John Karl

FLORIDA
Adams, Ralph M
Allen, Ted Tipton
Berg, William Keith
Branch, Lyn Clarke
Brockmann, Helen Jane
Brown, John Lott
Cohen, Sanford I
Condit, Richard
Dawson, William Woodson
De Lorge, John Oldham
Dewsbury, Donald Allen
Eisdorfer, Carl
Eisenberg, John Frederick
Evoy, William (Harrington)
Frederick, Peter Crawford
Gleeson, Richard Alan
Goldstein, Mark Kane
Gruber, Samuel Harvey

Gude, Richard Hunter
Heinen, Joel Thomas
Herrnkind, William Frank
Hope, George Marion
Kaufmann, John Henry
Kenshalo, Daniel Ralph
Knowlton, Nancy
Landolt, Peter John
Levey, Douglas J
Lillywhite, Harvey B
Lounibos, Leon Philip
Mackenzie, Richard Stanley
Mesterton-Gibbons, Michael Patrick
Moynihan, Martin Humphrey
Mushinsky, Henry Richard
Myrberg, Arthur August, Jr
Phillips, Michael Ian
Porter, Sanford Dee
Raizada, Mohan K
Roubik, David Ward
Rowland, Neil Edward
Salmon, Michael
Sanberg, Paul Ronald
Schneiderman, Neil
Sivinski, John Michael
Spielberger, Charles Donald
Tingle, Frederic Carley
Vander Meer, Robert Kenneth
Vierck, Charles John, Jr
Walker, Thomas Jefferson
Williams, Theodore P
Winters, Ray Wyatt
Witherington, Blair Ernest
Woolfenden, Glen Everett
Wyneken, Jeanette

GEORGIA
Bernstein, Irwin Samuel
Byrd, Larry Donald
Crowell-Davis, Sharon Lynn
Dragoin, William Bailey
Dusenbery, David Brock
Hartlage, Lawrence Clifton
Jackson, William James
McDaniel, William Franklin
Marchinton, Robert Larry
Matthews, Robert Wendell
Mulligan, Benjamin Edward
Nadler, Ronald D
Pulliam, H Ronald
Quertermus, Carl John, Jr
Saladin, Kenneth S
Smith, Euclid O'Neal
Stoneburner, Daniel Lee
Weiss, Jay M
Weissburg, Marc Joel
Wistrand, Harry Edwin
Young, Henry Edward
Zeiler, Michael Donald

HAWAII
Bitterman, Morton Edward
Blanchard, Robert Joseph
DeMartini, Edward Emile
Diamond, Milton
Freed, Leonard Alan
Kaneshiro, Kenneth Yoshimitsu
Reese, Ernst S
Vargas, Roger I

IDAHO
Fuller, Mark Roy
McCaffrey, Joseph Peter
Scott, James Michael

ILLINOIS
Abler, William Lewis
Altmann, Jeanne
Altmann, Stuart Allen
Bennett, Cecil Jackson
Berenbaum, May Roberta
Brioni, Jorge Daniel
Brown, Lauren Evans
Butler, Robert Allan
Comer, Christopher Mark
Conner, Jeffrey Keating
Donchin, Emanuel
Drickamer, Lee Charles
Elliott, Lois Lawrence
Gillette, Martha Ulbrick
Gillette, Rhanor
Goldberg, Arnold Irving
Goodrich, Michael Alan
Gramza, Anthony Francis
Grossman, Sebastian Peter
Guttman, Newman
Hausfater, Glenn
Heltne, Paul Gregory
Heybach, John Peter
Hirsch, Jerry
Irwin, Michael Edward
Jensen, Robert Alan
Kleinman, Kenneth Martin
Kruse, Kipp Colby
Lanyon, Scott Merril
Larkin, Ronald Paul
Metz, John Thomas
Moskal, Joseph Russell
Phillips, Christopher Alan
Pokorny, Joel
Pruett-Jones, Stephen Glen
Quanstrom, Walter Roy
Rabb, George Bernard
Rabinowitch, Victor

Reith, Maarten E A
Satinoff, Evelyn
Schmidt, Robert Sherwood
Smith, Douglas Calvin
Smith, Vivianne C(ameron)
Southern, William Edward
Toliver, Michael Edward
Tuttle, Russell Howard
Waring, George Houstoun, IV
Wetzel, Allan Brooke
Whelan, Christopher John
Willey, Robert Bruce
Yost, William A
Zerlin, Stanley

INDIANA
Albright, Jack Lawrence
Allen, Merrill James
Altman, Joseph
Amlaner, Charles Joseph, Jr
Brodman, Robert David
Brush, F(ranklin) Robert
Cadwallader, Joyce Vermeulen
Cooper, William Edgar, Jr
Esch, Harald Erich
Fitzhugh-Bell, Kathleen
Flanders, Robert Vern
Frommer, Gabriel Paul
Guth, S(herman) Leon
Kallman, Mary Jeanne
Lodge, David Michael
Lucas, Jeffrey Robert
McClure, Polley Ann
Rowland, David Lawrence
Rowland, William Joseph
Waser, Peter Merritt
Wasserman, Gerald Steward
Watson, Charles S

IOWA
Beachy, Christopher King
Benton, Arthur Lester
Best, Louis Brown
Fox, Stephen Sorin
Greenberg, Everett Peter
Hadow, Harlo Herbert
Johnson, Alan Kim
Kaufmann, Gerald Wayne
Moorcroft, William Herbert
Shaw, Kenneth C
Te Paske, Everett Russell

KANSAS
Cink, Calvin Lee
Clarke, Robert Francis
Craig, James Verne
Greenfield, Michael Dennis
Kaufman, Glennis Ann
Michaelis, Elias K
Michener, Charles Duncan
Reichman, Omer James
Spohn, Herbert Emil
Spradlin, Joseph E
Terman, Max R
Wolf, Thomas Michael

KENTUCKY
Cupp, Paul Vernon, Jr
Ferner, John William
Ferrell, Blaine Richard
Matheny, Adam Pence, Jr
Moore, Allen Jonathan
Sih, Andrew
Smith, Timothy Andre
Surwillo, Walter Wallace
Westneat, David French, Jr
Yeargan, Kenneth Vernon
Yokel, Robert Allen
Zolman, James F

LOUISIANA
Bauer, Raymond Thomas
Brown, Kenneth Michael
Collins, Mary Jane
Dunn, Adrian John
Homberger, Dominique Gabrielle
Jaeger, Robert Gordon
Johnsen, Peter Berghsey
Olson, Richard David
Rinderer, Thomas Earl
Riopelle, Arthur J
Waters, William F

MAINE
Butler, Ronald George
Fitch, John Henry
Francq, Edward Nathaniel Lloyd
Glanz, William Edward
Labov, Jay Brian
Marcotte, Brian Michael
Yuhas, Joseph George

MARYLAND
Aldrich, Jeffrey Richard
Anderson, David Everett
Annau, Zoltan
Baker, Frank
Ball, Gregory Francis
Barrows, Edward Myron
Barry, Ronald Everett, Jr
Batra, Suzanne Wellington Tubby
Biersner, Robert John
Blum, Harry
Borgia, Gerald

Brady, Joseph Vincent
Brauth, Steven Earle
Buchler, Edward Raymond
Buck, John Bonner
Canal-Frederick, Ghislaine R
Carter, Carol Sue
Coelho, Anthony Mendes, Jr
Costa, Paul T, Jr
Creighton, Phillip David
Desor, Jeannette Ann
Flanigan, William Francis, Jr
Forester, Donald Charles
Freed, Arthur Nelson
Galloway, William Don
Gray, David Bertsch
Groopman, John Davis
Grunberg, Neil Everett
Hanson, Frank Edwin
Heath, Martha Ellen
Hienz, Robert Douglas
Hill, James Leslie
Hodgdon, Harry Edward
Krantz, David S
Lall, Abner Bishamber
Landauer, Michael Robert
Leach, Berton Joe
Leshner, Alan Irvin
McKaye, Kenneth Robert
MacLean, Paul Donald
Marsden, Halsey M
Mishkin, Mortimer
Muul, Illar
Pare, William Paul
Popper, Arthur N
Provine, Robert Raymond
Rebach, Steve
Schein, Martin Warren
Sparling, Donald Wesley, Jr
Stricklin, William Ray
Suomi, Stephen John
Weingartner, Herbert
Whitehead, William Earl
Wirth, James Burnham
Wise, David Haynes

MASSACHUSETTS
Atema, Jelle
Bertera, James H
Bridges, Robert Stafford
Carde, Ring Richard Tomlinson
Cardello, Armand Vincent
Carey, Francis G
Coppinger, Raymond Parke
Corkin, Suzanne Hammond
Czeisler, Charles Andrew
Damassa, David Allen
Dane, Benjamin
DeBold, Joseph Francis
Dethier, Vincent Gaston
Dews, Peter Booth
Gold, Richard Michael
Griffin, Donald R(edfield)
Hall, Robert Dilwyn
Harrison, John Michael
Hauser, Marc David
Jacobs, Harry Lewis
Jearld, Ambrose, Jr
Kagan, Jerome
Kantak, Kathleen Mary
Kroodsma, Donald Eugene
Kunz, Thomas Henry
Larson, Joseph Stanley
Matthysse, Steven William
Miczek, Klaus A
Moffett, Mark William
Payne, Bertram R
Powers, J Bradley
Prescott, John Hernage
Prestwich, Kenneth Neal
Sargent, Theodore David
Scharf, Bertram
Schneider, Gerald Edward
Skavenski, Alexander Anthony
Smith, Susan May
Traniello, James Francis Anthony
Trehub, Arnold
Volkmann, Frances Cooper
Wasserman, Frederick E
West, Christopher Drane
Wilkie, David Scott
Wilson, Edward Osborne
Wolff, Peter Hartwig

MICHIGAN
Cowan, David Prime
Davis, Roger (Edward)
Douglas, Matthew M
Foster, Neal Robert
Gamboa, George John
Gillingham, James Clark
Hartmann, William Morris
Hayward, James Lloyd
Heffner, Thomas G
McCrimmon, Donald Alan, Jr
Moody, David Burritt
Moore, Thomas Edwin
Paschke, Richard Eugene
Prince, Harold Hoopes
Robinson, Terrance Earl
Schaefer, Gerald J
Silverman, Albert Jack
Stebbins, William Cooper
Stinson, Al Worth
Webb, Paul

MINNESOTA
Barnwell, Franklin Hershel
Buech, Richard Reed
Collins, Hollie L
Frydendall, Merrill J
Fukui, Hidenori Henry
Henry, Mary Gerard
Jannett, Frederick Joseph, Jr
Kornblith, Carol Lee
Mabry, Paul Davis
Meier, Manfred John
Ordway, Ellen
Overmier, J Bruce
Phillips, Richard Edward
Pusey, Anne E
Regal, Philip Joe
Roberts, Warren Wilcox
Rogers, Lynn Leroy
Smith, Thomas Jay
Sorensen, Peter W
Swift, Michael Crane
Ward, Wallace Dixon
Yellin, Absalom Moses

MISSISSIPPI
Anderson, Kenneth Verle
Dickens, Joseph Clifton
Kallman, William Michael
McKeown, James Preston
Peeler, Dudley F, Jr
Ramaswamy, Sonny B

MISSOURI
Ayyagari, L Rao
Backus, Elaine Athene
Buchanan, Bryant W
Carrel, James Elliott
Coles, Richard Warren
Dugatkin, Lee Alan
Faaborg, John Raynor
Galosy, Richard Allen
Gerhardt, H Carl, Jr
Hirsh, Ira Jean
Hunt, James Howell
Justesen, Don Robert
Losos, Jonathan B
Mathis, Sharon Alicia
Seidler, Norbert Wendelin
Sexton, Owen James
Smith, Robert Francis
Sorenson, Marion W
Sussman, Robert Wald
Tang-Martinez, Zuleyma
Taub, John Marcus
Wartzok, Douglas
Williams, Henry Warrington

MONTANA
Ball, Irvin Joseph
Fritts, Steven Hugh
Jenni, Donald Alison
Kaya, Calvin Masayuki
Lynch, Wesley Clyde
Metzgar, Lee Hollis

NEBRASKA
Joern, Anthony
Johnsgard, Paul Austin
Kamil, Alan Curtis
Rodriguez-Sierra, Jorge F
Thorson, James A

NEVADA
Hoelzer, Guy Andrew
Marlow, Ronald William
Oring, Lewis Warren
Vinyard, Gary Lee

NEW HAMPSHIRE
Barry, William James
Chabot, Christopher Cleaves
Dingman, Jane Van Zandt
McPeek, Mark Alan
Riggs, Lorrin Andrews

NEW JERSEY
Beer, Colin Gordon
Burger, Joanna
Chaiken, MarthaLeah
Chizinsky, Walter
Davis, Nancy Taggart
Edelberg, Robert
Faber, Betty Lane
Gandelman, Ronald Jay
Gochfeld, Michael
Gona, Ophelia Delaine
Gould, James L
Grant, B Rosemary
Hahn, Martin Earl
Horn, Henry Stainken
Hubbard, John W
Huber, Ivan
Jolly, Alison Bishop
Katz, Larry Steven
Kohn, Herbert Myron
Komisaruk, Barry Richard
McGuire, Terry Russell
Mihram, George Arthur
Millner, Elaine Stone
Morrison, Douglas Wildes
Murray, Bertram George, Jr
Power, Harry W, III
Saiff, Edward Ira
Stearns, Donald Edison

Behavior-Ethology (cont)

Tallal, Paula
Thompson, Robert L
White, Robert Keller
Willis, Jacalyn Giacalone

NEW MEXICO
Anderson, Dean Mauritz
Easley, Stephen Phillip
Herman, Ceil Ann
Rhodes, John Marshell
Richman, David Bruce

NEW YORK
Able, Kenneth Paul
Ader, Robert
Adkins-Regan, Elizabeth Kocher
Adler, Kraig (Kerr)
Arkin, Arthur Malcolm
Ayres, Jose Marcio Correa
Bartus, Raymond T
Beason, Robert Curtis
Berman, Carol May
Blass, Elliott Martin
Borowsky, Betty Marian
Brown, Jerram L
Brown, Robert Zanes
Camazine, Scott
Capranica, Robert R
Caraco, Thomas Benjamin
Carlson, Albert Dewayne, Jr
Chase, Ivan Dmitri
Chepko-Sade, Bonita Diane
Cohan, Christopher Scott
Cory-Slechta, Deborah Ann
DeGhett, Victor John
Diakow, Carol
Donovick, Peter Joseph
Dooley, James Keith
Durkovic, Russell George
Eickwort, George Campbell
Emlen, Stephen Thompson
Essman, Walter Bernard
Feeny, Paul Patrick
Finando, Steven J
Fox, Kevin A
French, Alan Raymond
Geary, Norcross D
Goodwin, Robert Earl
Gorlick, Dennis
Gotwald, William Harrison, Jr
Griswold, Joseph Garland
Grove, Patricia A
Hallahan, William Laskey
Halpern, Mimi
Hartung, John David
Hawkins, Robert Drake
Henderson, Donald
Hopkins, Carl Douglas
Hughes, Patrick Richard
Isaacson, Robert Lee
Jakway, Jacqueline Sinks
Johnston, Robert E
Kanzler, Walter Wilhelm
Kolmes, Steven Albert
Krauskopf, John
Krieg, David Charles
Kupfermann, Irving
Lee, Ching-Tse
Levine, Louis
Loullis, Costas Christou
Madison, Dale Martin
Mendel, Frank C
Meyer, Axel
Moller, Peter
Mozell, Maxwell Mark
Muessig, Paul Henry
Muller-Schwarze, Dietland
Mulligan, James Anthony
Nathan, Ronald Gene
Oesterreich, Roger Edward
Pasamanick, Benjamin
Pasik, Pedro
Pasik, Tauba
Peckarsky, Barbara Lynn
Pfaffmann, Carl
Phillips, Robert Rhodes
Pickering, Thomas G
Porter, William Frank
Rabe, Ausma
Reiss, Diana
Ringler, Neil Harrison
Ritter, Walter Paul
Roelofs, Wendell L
Rosenblum, Leonard Allen
Sackeim, Harold A
Salvi, Richard J
Sank, Diane
Schneider, Kathryn Claire (Johnson)
Schupf, Nicole
Seeley, Thomas Dyer
Sherman, Paul Willard
Shields, William Michael
Slaughter, John Sim
Smith, Charles James
Sobczak, Thomas Victor
Southwick, Edward E
Stewart, Margaret McBride
Sullivan, Daniel Joseph
Thomas, Garth Johnson
Thomson, James Douglas
Verrillo, Ronald Thomas
Walcott, Charles
Wasserman, Marvin
Watanabe, Myrna Edelman
Wineburg, Elliot N
Wolf, Larry Louis
Wyman, Richard L
Zitrin, Charlotte Marker

NORTH CAROLINA
Biggs, Walter Clark, Jr
Brown, Richard Dean
Case, Verna Miller
Crowder, Larry Bryant
Dixon, N(orman) Rex
Eason, Robert Gaston
Edens, Frank Wesley
Eggleston, David Bryan
Erickson, Robert Porter
Gottlieb, Gilbert
Hall, Warren G
Hall, William Charles
Herzog, Harold Albert, Jr
Howard, James Lawrence
Kamykowski, Daniel
King, Richard Austin
Lewis, Mark Henry
Logan, Cheryl Ann
Moseley, Lynn Johnson
Mueller, Helmut Charles
Myers, Robert Durant
Oppenheim, Ronald William
Powell, Roger Allen
Rulifson, Roger Allen
Soderquist, David Richard
Surwit, Richard Samuel
Vandenbergh, John Garry
Weigl, Peter Douglas
Wiley, Richard Haven, Jr
Wolk, Robert George

NORTH DAKOTA
Fivizzani, Albert John, Jr
Penland, James Granville

OHIO
Beal, Kathleen Grabaskas
Burtt, Edward Howland, Jr
Case, Denis Stephen
Corson, Samuel Abraham
DeNelsky, Garland
Downhower, Jerry F
Ely, Daniel Lee
Hangarter, Roger Paul
Holtzman, David Allen
Hostetler, Jeptha Ray
Keck, Max Johann
Knight, Walter Rea
Lacey, Beatrice Cates
McLean, Edward Bruce
Mudry, Karen Michele
Mulick, James Anton
Nault, Lowell Raymond
Orcutt, Frederic Scott, Jr
Patterson, Michael Milton
Pettijohn, Terry Frank
Radabaugh, Dennis Charles
Rovner, Jerome Sylvan
Rushforth, Norman B
Shipley, Michael Thomas
Stoffer, Richard Lawrence
Svendsen, Gerald Eugene
Taylor, Douglas Hiram
Turner, John W, Jr
Uetz, George William
Valentine, Barry Dean
Vessey, Stephen H
Vorhees, Charles V

OKLAHOMA
Baird, Troy Alan
Chapman, Brian Richard
Fox, Stanley Forrest
Lynn, Robert Thomas
Mares, Michael Allen
Mock, Douglas Wayne
Schaefer, Carl Francis
Vestal, Bedford Mather
Wells, Harrington
Young, Sharon Clairene

OREGON
Anderson, John Richard
Betts, Burr Joseph
Colvin, Dallas Verne
Cross, Stephen P
Davis, Michael William
Eaton, Gordon Gray
Fay, Warren Henry
Gallaher, Edward J
Hedrick, Ann Valerie
Hixon, Mark A
Leonard, Janet Louise
Mason, Robert Thomas
Olla, Bori Liborio
Phillips, David
Phoenix, Charles Henry
Ryker, Lee Chester

PENNSYLVANIA
Allen, Theresa O
Angstadt, Robert B
Balling, Jan Walter
Beauchamp, Gary Keith
Brenner, Frederic J
Bridger, Wagner H
Carey, Michael Dean
Carey, William Bacon
Clark, Richard James
Cole, James Edward
Cutler, Winnifred Berg
Cutt, Roger Alan
Dinges, David Francis
Doty, Richard Leroy
Eichelman, Burr S, Jr
Epstein, Alan Neil
Foxx, Richard Michael
Goldstein, E Bruce
Green, Barry George
Hoffman, Daniel Lewis
Hughes, Austin Leland
Hurvich, Leo Maurice
Keiper, Ronald R
Kendall, Philip C
Klinger, Thomas Scott
Kurland, Jeffrey Arnold
Lewis, Michael Edward
Ode, Philip E
Pion, Lawrence V
Poplawsky, Alex James
Prezant, Robert Steven
Ray, William J
Schultz, Jack C
Settle, Richard Gregg
Sidie, James Michael
Singer, Alan G
Smith, W John
Sprague, James Mather
Steele, Craig William
Stellar, Eliot
Stewart, Charles Newby
Thompson, Roger Kevin Russell
Tompkins, Laurie
Towne, William F
Vomachka, Archie Joel
Ward, Ingeborg L
Wert, Jonathan Maxwell, Jr
Williams, Timothy C
Williamson, Craig Edward
Wysocki, Charles Joseph
Yahner, Richard Howard

RHODE ISLAND
Heppner, Frank Henry
Howe, Robert Johnston
Janis, Christine Marie
Lipsitt, Lewis Paeff
Pratt, David Mariotti
Waage, Jonathan King
Wahle, Richard Andreas

SOUTH CAROLINA
Bildstein, Keith Louis
Coleman, James Roland
DeCoursey, Patricia Alice Jackson
Forsythe, Dennis Martin
Fox, Charles Wayne
Gauthreaux, Sidney Anthony
Kaiser, Charles Frederick
Lacher, Thomas Edward, Jr
Powell, Donald Ashmore
Sears, Harold Frederick

SOUTH DAKOTA
Schlenker, Evelyn Heymann

TENNESSEE
Altemeier, William Arthur, III
Ambrose, Harrison William, III
Bealer, Steven Lee
Burghardt, Gordon Martin
Cushing, Bruce S
Elberger, Andrea June
Greenberg, Neil
Kaas, Jon Howard
Lubar, Joel F
Tardif, Suzette Davis

TEXAS
Beaver, Bonnie Veryle
Belk, Gene Denton
Buskirk, Ruth Elizabeth
Collins, Anita Marguerite
Crawford, Morris Lee Jackson
Crews, David Pafford
Dalterio, Susan Linda
Formanowicz, Daniel Robert, Jr
Foster, John Robert
Friend, Ted H
Gonzalez-Lima, Francisco
Greenberg, Les Paul
Hamilton, Charles R
Hanlon, Roger Thomas
Higgins, Linden Elizabeth
Ingold, Donald Alfred
Johnson, Patricia Ann J
McCarley, Wardlow Howard
Malin, David Herbert
Marks, Gerald A
Mathews, Nancy Ellen
Moushegian, George
Peniston, Eugene Gilbert
Queller, David Charles
Richerson, Jim Vernon
Rogers, Walter Russell
Rubink, William Louis
Rylander, Michael Kent
Schreiber, Robert Alan
Sherry, Clifford Joseph
Sperling, Harry George
Stein, Jerry Michael
Strassmann, Joan Elizabeth
Suddick, Richard Phillips
Vinson, S Bradleigh
Whitsett, Johnson Mallory, II
Wilczynski, Walter
Willig, Michael Robert
Wright, Anthony Aune

UTAH
Arave, Clive W
Balph, Martha Hatch
Beck, Edward C
Bosakowski, Thomas
Cheney, Carl D
Conover, Michael Robert
Davidson, Diane West
Fleming, Donovan Ernest
Hsiao, Ting Huan
Kesner, Raymond Pierre
Messina, Frank James
Provenza, Frederick Dan
Schenkenberg, Thomas

VERMONT
Brodie, Edmund Darrell, Jr
Heinrich, Bernd
Schall, Joseph Julian

VIRGINIA
Axelson, Marta Lynne
Balster, Robert L(ouis)
Biben, Maxeen G
Block, Gene David
Bressler, Barry Lee
Brown, Luther Park
Brunjes, Peter Crawford
Carterette, Edward Calvin Hayes
Davis, Joel L
Fashing, Norman James
Fine, Michael Lawrence
Fink, Linda Susan
Gold, Paul Ernest
Graves, Hannon B
Grimm, James K
Gullotta, Frank Paul
Hornbuckle, Phyllis Ann
Jenssen, Thomas Alan
Keinath, John Allen
King, H(enry) E(ugene)
Lenhardt, Martin Louis
McCarty, Richard Charles
Marcellini, Dale Leroy
Mehner, John Frederick
Olsen, Kathie Lynn
Pienkowski, Robert Louis
Pribram, Karl Harry
Zornetzer, Steven F

WASHINGTON
Anderson, James Jay
Bowden, Douglas Mchose
Calkins, Carrol Otto
Dudley, Donald Larry
Galusha, Joseph G, Jr
Gentry, Roger Lee
Gilbert, Frederick Franklin
Gray, Robert H
Kenagy, George James
Kenney, Nancy Jane
Loughlin, Thomas Richard
Lovely, Richard Herbert
Mace, Terrence Rowley
Martin, Dennis John
Monan, Gerald E
Paulson, Dennis R
Pearson, Walter Howard
Pietsch, Theodore Wells
Schroeder, Michael Allen
Shepherd, Linda Jean
Simpson, John Barclay
Smith, Orville Auverne
Thompson, Christopher William

WEST VIRGINIA
Constantz, George Doran
Covalt-Dunning, Dorothy
Dunning, Dorothy Covalt
Marshall, Joseph Andrew

WISCONSIN
Boese, Gilbert Karyle
Boyce, Mark S
Cleeland, Charles Samuel
Curtis, Robin Livingstone
Downing, Holly Adelaide
Elfner, Lloyd F
Ficken, Millicent Sigler
Ficken, Robert W
Goldstein, Robert
Gottfried, Bradley M
Goy, Robert William
Hailman, Jack Parker
Hekmat, Hamid Moayed
Jeanne, Robert Lawrence
Kemnitz, Joseph William
Kent, Raymond D
Minock, Michael Edward
Norris, Dale Melvin, Jr
Schenk, Roy Urban
Snowdon, Charles Thomas
Strier, Karen Barbara
Takahashi, Lorey K
Wilson, Richard Howard

Yasukawa, Ken

WYOMING
Denniston, Rollin H, II
Lockwood, Jeffrey Alan
Rose, James David

PUERTO RICO
Lewis, Allen Rogers
Shapiro, Douglas York

ALBERTA
Byers, John Robert
Geist, Valerius
Koopmans, Henry Sjoerd
Michener, Gail R
Murie, Jan O
Price, Mick A
Sainsbury, Robert Stephen
Scrimgeour, Garry Joseph
Seghers, Benoni Hendrik
Sodhi, Navjot Singh
Steiner, Andre Louis
Veale, Warren Lorne

BRITISH COLUMBIA
Borden, John Harvey
Cheng, Kimberly Ming-Tak
Craig, Kenneth Denton
Dill, Lawrence Michael
Domenici, Paolo
Greenwood, Donald Dean
Healey, Michael Charles
Hill, Arthur Thomas
Jones, Richard Lamar
Miller, Edward Henry
Phillips, Anthony George
Roitberg, Bernard David
Steeves, John Douglas
Thomson, Keith A
Ward, David Mercer
Ward, Lawrence McCue

MANITOBA
Abrahams, Mark Vivian
Evans, Roger Malcolm
Hughes, Kenneth Russell
Nance, Dwight Maurice
Palaniswamy, Pachagounder
Zach, Reto

NEW BRUNSWICK
McKenzie, Joseph Addison
Seabrook, William Davidson

NEWFOUNDLAND
Carroll, Allan Louis
Green, John M

NOVA SCOTIA
Brown, Richard George Bolney
Fentress, John Carroll
Harrington, Fred Haddox
Hopkins, David Alan
Moore, Bruce Robert
Semba, Kazue
Van Houten, Ronald G

ONTARIO
Ankney, C(laude) Davison
Bailey, Edward D
Berrill, Michael
Brooks, Ronald James
Brose, David Stephen
Cade, William Henry
Caplan, Paula Joan
Collins, Nicholas Clark
Cornell, James Morris
Dagg, Anne Innis
Darling, Donald Christopher
Donald, Merlin Wilfred
Edwards, Roy Lawrence
Falls, James Bruce
Fenton, Melville Brock
Gray, David Robert
Halperin, Janet R P
Keenleyside, Miles Hugh Alston
Kimura, Doreen
Kovacs, Kit M
Leung, Lai-Wo Stan
Mallory, Frank Fenson
Mogenson, Gordon James
Montgomerie, Robert Dennis
Morris, Ralph Dennis
Noakes, David Lloyd George
Robertson, Raleigh John
Rollman, Gary Bernard
Stouffer, James L
Szabo, Tibor Imre
Vanderwolf, Cornelius Hendrik

QUEBEC
Browman, Howard Irving
Chouinard, Guy
Cloutier, Conrad Francois
Ferron, Jean H
Fitzgerald, Gerard John
Gauthier, Gilles
Grant, James William Angus
Green, David M(artin)
Hilton, Donald Frederick James
Kingsley, Michael Charles Stephen
McNeil, Jeremy Nichol
Magnan, Pierre

Milner, Brenda (Atkinson)
Prescott, Jacques
Rau, Manfred Ernst
Robert, Suzanne
Sanborne, Paul Michael
Titman, Rodger Donaldson
Vickery, William Lloyd
Vincent, C

SASKATCHEWAN
Brigham, R Mark
Doane, John Frederick
Flood, Peter Frederick
Hobson, Keith Alan
Mitchell, George Joseph
Oliphant, Lynn Wesley

OTHER COUNTRIES
Biederman-Thorson, Marguerite Ann
Chadwick, Nanette Elizabeth
Ganchrow, Donald
Glickstein, Mitchell
Hanukoglu, Israel
Hisada, Mituhiko
Holldobler, Berthold Karl
Holman, Richard Bruce
Hummel, Hans Eckhardt
Jacobson, Bertil
Komers, Petr E
Nevo, Eviatar
Oxnard, Charles Ernest
Srygley, Robert Baxter
Walther, Fritz R
Wettstein, Joseph G
Wynne-Edwards, Vero Copner
Yosef, Reuven
Zhang, Zhi-Qiang

Biochemistry

ALABAMA
Abrahamson, Dale Raymond
Alexander, Herman Davis
Aronson, Nathan Ned, Jr
Aull, John Louis
Ayling, June E
Baker, John Rowland
Baliga, B Surendra
Ball, Laurence Andrew
Baugh, Charles M
Becker, Gerald Leonard
Benos, Dale John
Bhatnagar, Yogendra Mohan
Bhown, Ajit Singh
Birkedal Hansen, Henning
Bradley, James T
Brown, Alfred Ellis
Burns, Moore J
Butler, William Thomas
Chang, Chi Hsiung
Chapatwala, Kirit D
Cherry, Joe H
Christian, Samuel Terry
Clarke, Benjamin L
Cook, William Joseph
Cornatzer, William Eugene
Daniell, Henry
Daron, Harlow Hoover
Digerness, Stanley B
Elgavish, Ada S
Elliott, Howard Clyde
Fendley, Ted Wyatt
Flodin, N(estor) W(inston)
Freeman, Bruce Alan
Goodman, Steven Richard
Guidry, Clyde R
Haggard, James Herbert
Hall, Leo McAloon
Hardman, John Kemper
Harvey, Stephen Craig
Higgins, N Patrick
Hill, Donald Lynch
Johnson, Brian John
Kaplan, Ronald S
Kochakian, Charles Daniel
Krumdieck, Carlos L
Lebowitz, Jacob
Lehman, Robert Harold
Locy, Robert Donald
Long, Calvin Lee
Longenecker, Herbert Eugene
Lorincz, Andrew Endre
McKibbin, John Mead
Magargal, Wells Wrisley, II
Marchase, Richard Banfield
Meezan, Elias
Mego, John L
Melius, Paul
Miller, Edward Joseph
Moore, Robert Blaine
Moore, William Gower Innes
Niedermeier, William
Nielsen, Brent Lynn
Otto, David A
Parish, Edward James
Paxton, Ralph
Pillion, Dennis Joseph
Prejean, Joe David
Pruitt, Kenneth M
Rivers, Douglas Bernard
Sani, Brahma Porinchu
Sauberlich, Howerde Edwin
Schaffer, Stephen Ward

Schrohenloher, Ralph Edward
Schutzbach, John Stephen
Segrest, Jere Palmer
Smith, Robert C
Smith-Somerville, Harriett Elizabeth
Strada, Samuel Joseph
Strength, Delphin Ralph
Suling, William John
Svacha, Anna Johnson
Thompson, Jerry Nelson
Thompson, Wynelle Doggett
Tomana, Milan
Wadkins, Charles L
Weete, John Donald
Wheeler, Glynn Pearce
Whikehart, David Ralph
Wilkoff, Lee Joseph
Wingo, William Jacob
Winkler, Herbert H
Winters, Alvin L
Wooten, Marie W
Yarbrough, James David

ALASKA
Behrisch, Hans Werner
Button, Don K
Duffy, Lawrence Kevin
Gharrett, Anthony John
Nakada, Henry Isao
Stekoll, Michael Steven

ARIZONA
Allen, John Rybolt
Aronson, John Noel
Barstow, Leon E
Begovac, Paul C
Bieber, Allan Leroy
Bowden, George Timothy
Brush, James S
Buck, Stephen Henderson
Caldwell, Roger Lee
Carter, Herbert Edmund
Chandler, Douglas Edwin
Cress, Anne Elizabeth
Cronin, John Read
Cusanovich, Michael A
Doubek, Dennis Lee
Eckardt, Robert E
Eskelson, Cleamond D
Fowler, Dona Jane
Fuller, Wallace Hamilton
Gendler, Sandra J
Goll, Darrel Eugene
Gordon, Richard Seymour
Guerriero, Vincent, Jr
Halpert, James Robert
Harkins, Kristi R
Harris, David Thomas
Harris, Joseph
Hazel, Jeffrey Ronald
Hendrix, Donald Louis
Hewlett, Martinez Joseph
Holmes, William Farrar
Hoober, J Kenneth
Jensen, Richard Grant
Kay, Marguerite M B
Kazal, Louis Anthony
Koldovsky, Otakar
Krahl, Maurice Edward
Larkins, Brian Allen
Law, John Harold
Lei, David Kai Yui
Liddell, Robert William, Jr
Lipke, William G
Little, John Wesley
Loper, Gerald Milton
Luchsinger, Wayne Wesley
Lukas, Ronald John
McLean, Katharine Weidman
Maddy, Kenneth Hilton
Marchalonis, John Jacob
Mardian, James K W
Mason, Merle
Milch, Lawrence Jacques
Moore, Ana M L
Moore, Thomas Andrew
Mosher, Richard Arthur
Moss, Lloyd Kent
Nardella, Francis Anthony
Peng, Yeh-Shan
Peterson, James Douglas
Potts, Albert Mintz
Powis, Garth
Price, Ralph Lorin
Reeves, Henry Courtland
Reid, Bobby Leroy
Rupley, John Allen
Seligmann, Bruce Edward
Shull, James Jay
Spizizen, John
Taylor, Richard G
Thomas, Joseph James
Tischler, Marc Eliot
Trelease, Richard Norman
Vandenberg, Edwin James
Varnell, Thomas Raymond
Vermaas, Willem F J
Vestling, Carl Swensson
Ward, Samuel
Weber, Charles Walter
Wells, Michael Arthur
Whiting, Frank M
Winters, Mary Ann
Wu, Chuanyue

Yall, Irving
Yamamura, Henry Ichiro
Yohem, Karin Hummell

ARKANSAS
Allaben, William Thomas
Badger, Thomas Mark
Benson, Ann Marie
Beranek, David T
Bhuvaneswaran, Chidambaram
Blackwell, Richard Quentin
Brewster, Marjorie Ann
Chowdhury, Parimal
Cornett, Lawrence Eugene
Davis, Virginia Eischen
De Luca, Donald Carl
Elbein, Alan D
Epstein, Joshua
Evans, Frederick Earl
Fink, Louis Maier
Forsythe, Richard Hamilton
Greenman, David Lewis
Harris, Grover Cleveland, Jr
Heflich, Robert Henry
Herring, Harold Keith
Jackson, Carlton Darnell
Jones, Robin Richard
Kadlubar, Fred F
Kaplan, Arnold
Kraemer, Louise Margaret
Lane, Forrest Eugene
Leakey, Julian Edwin Arundell
Light, Kim Edward
Millett, Francis Spencer
Morris, Manford D
Nelson, Charles A
Pynes, Gene Dale
Routh, Joseph Isaac
Sanders, Louis Lee
Sheehan, Daniel Michael
Shideler, Robert Weaver
Smith, Carroll Ward
Smith, William Grady
Sorenson, John R J
Stallcup, Odie Talmadge
Steinmeier, Robert C
Stone, Joseph
Sullivan, Thomas Wesley
Tucker, Robert Gene
Waldroup, Park William
Wennerstrom, David E
Winter, Charles Gordon
Yeh, Yunchi
York, John Lyndal

CALIFORNIA
Abbott, Mitchel Theodore
Abel, Carlos Alberto
Abrahamson, Lila
Abrash, Henry I
Ackrell, Brian A C
Aftergood, Lilla
Ahuja, Jagan N
Alexander, Caroline M
Alexander, Nicholas Michael
Alfin-Slater, Roslyn Berniece
Allen, Charles Freeman
Aloia, Roland C
Alousi, Adawia A
Altieri, Dario Carlo
Amer, Mohamed Samir
Ames, Bruce Nathan
Ames, Giovanna Ferro-Luzzi
Amy, Nancy Klein
Anand, Rajen S(ingh)
Anderson, Geoffrey Robert
Anderson, Thomas Alexander
Andreoli, Anthony Joseph
Appleman, James R
Appleman, M Michael
Arakawa, Tsutomu
Arfin, Stuart Michael
Arnon, Daniel I(srael)
Arroyave, Guillermo
Ashkenazi, Avi
Ashmore, Charles Robert
Atkinson, Daniel Edward
Ayengar, Padmasini (Mrs Frederick Aladjem)
Azhar, Salman
Babior, Bernard M
Bagdasarian, Andranik
Bailey-Serres, Julia
Baker, Nome
Balakrishnan, Krishna
Balch, William E
Baldeschwieler, John Dickson
Ballou, Clinton Edward
Balon, Thomas William
Barankiewicz, Jerzy Andrzej
Barker, Horace Albert
Barnes, Paul Richard
Barnet, Harry Nathan
Bartlett, Grant Rogers
Bartley, John C
Bartnicki-Garcia, Salomon
Bassham, James Alan
Bauer, Roger Duane
Baumgartner, Werner Andreas
Baxter, Claude Frederick
Baxter, John Darling
Beckendorf, Steven K
Becker, Joseph F
Bell, Jack Perkins

Biochemistry (cont)

Bellhorn, Margaret Burns
Beltz, Richard Edward
Benemann, John Rudiger
Benisek, William Frank
Benjamini, Eliezer
Bennett, Edward Leigh
Bennett, Raymond Dudley
Bennett, William Franklin
Benson, Andrew Alm
Benson, James R
Benson, Robert Leland
Berg, Paul
Berk, Arnold J
Berndt, Norbert
Berry, Edward Alan
Bertolami, Charles Nicholas
Bessman, Samuel Paul
Bhatnagar, Rajendra Sahai
Bhattacharya, Prabir
Bigler, William Norman
Bikle, Daniel David
Bishop, John Michael
Bjorkman, Pamela J
Black, Arthur Leo
Blackburn, Elizabeth Helen
Blankenship, James W
Blankenship, James William
Blatt, Beverly Faye
Blatt, Joel Martin
Bleil, Jeffrey D
Bokoch, Gary M
Bondar, Richard Jay Laurent
Bondy, Stephen Claude
Bonura, Thomas
Bourne, Henry R
Bowen, Charles E
Bower, Annette
Bowman, Barry J
Boyd, James Brown
Bradbury, E Morton
Bradfield, Robert B
Bradshaw, Ralph Alden
Bramhall, John Shepherd
Brandon, David Lawrence
Breidenbach, Rowland William
Brody, Stuart
Brostoff, Steven Warren
Brown, James Edward
Brown, Jeanette Snyder
Brown, Joan Heller
Brownell, Anna Gale
Brunke, Karen J
Brunton, Laurence
Brutlag, Douglas Lee
Bryant, Peter James
Buchanan, Bob Branch
Burgess, Teresa Lynn
Burton, Louis
Butler, Alison
Buttlaire, Daniel Howard
Buzin, Carolyn Hattox
Byard, James Leonard
Byers, Sanford Oscar
Cabot, Myles Clayton
Calarco, Patricia G
Camien, Merrill Nelson
Campagnoni, Anthony Thomas
Campbell, Alice del Campillo
Campbell, Judith Lynn
Cantor, Charles Robert
Cape, Ronald Elliot
Caporaso, Fredric
Carbon, John Anthony
Cardullo, Richard Anthony
Carew, Thomas Edward
Carle, Glenn Clifford
Carlisle, Edith M
Carlson, Don Marvin
Carroll, Edward James, Jr
Carson, Virginia Rosalie Gottschall
Carsten, Mary E
Castanera, Esther Goossen
Castles, James Joseph, Jr
Celniker, Susan Elizabeth
Chaffee, Rowand R J
Chamberlin, Michael John
Chan, Bock G
Chan, Sham-Yuen
Chang, Ding
Chang, Eppie Sheng
Chang, Ernest Sun-Mei
Chang, Jaw-Kang
Chang, Mei-Ping
Chapman, David J
Chaykin, Sterling
Chen, Chiadao
Chen, Ming Sheng
Chen, Stephen Shi-Hua
Chen, Tung-Shan
Cheng, Sze-Chuh
Chou, Tsong-Wen
Chow, Samson Ah-Fu
Christensen, Halvor Niels
Chuang, Hanson Yii-Kuan
Chuang, Ronald Yan-Li
Civen, Morton
Clarke, Steven Gerard
Clayton, Raymond Brazenor
Clegg, James S
Cohen, Natalie Shulman
Cohlberg, Jeffrey Allan
Cohn, Major Lloyd

Collins, James Francis
Collins, Robert C
Colvin, Harry Walter, Jr
Conn, Eric Edward
Conrad, Herbert M
Contreras, Thomas Jose
Cooper, Geoffrey Kenneth
Cooper, James Burgess
Cotman, Carl Wayne
Coty, William Allen
Cozzarelli, Nicholas Robert
Criddle, Richard S
Cromartie, Thomas Houston
Cunningham, Bruce Arthur
Cunningham, Dennis Dean
Curnutte, John Tolliver, III
Dahms, Arthur Stephen
Dahmus, Michael E
Danenberg, Peter V
Dang, Peter Hung-Chen
Daniel, Louise Jane
Danko, Stephen John
Dasgupta, Asim
David, Gary Samuel
Davies, Huw M
Davis, Frank French
Davis, Rowland Hallowell
Davis, William Ellsmore, Jr
Dekker, Charles Abram
DeLange, Robert J
Delk, Ann Stevens
Delwiche, Constant Collin
DeMet, Edward Michael
Dennis, Edward A
DeVellis, Jean
DeVenuto, Frank
Diamond, Ivan
Dietz, George William, Jr
Dietz, Thomas John
Doolittle, Russell F
Downing, Michael Richard
Dragon, Elizabeth Alice Oosterom
Draper, Roy Douglas
Dugaiczyk, Achilles
Dukes, Peter Paul
Dumus, David Paul
Dunaway, Marietta
Dunn, Michael F
Du Pont, Frances Marguerite
Durrum, Emmett Leigh
Durzan, Donald John
Duzgunes, Nejat
Dyer, Denzel Leroy
Eaks, Irving Leslie
Ecker, David John
Eckhart, Walter
Eckhert, Curtis Dale
Edelman, Gerald Maurice
Eiduson, Samuel
Eiler, John Joseph
Einset, John William
Einstein, Elizabeth Roboz
Ellis, Stanley
Ellman, George Leon
Ely, Kathryn R
Emr, Scott David
Endahl, Gerald Leroy
Eng, Lawrence F
Engvall, Eva Susanna
Eppstein, Deborah Anne
Etcheverry, Tina
Etzler, Marilynn Edith
Fahey, Robert C
Falick, Arnold M
Fan, Hsing Yun
Felton, James Steven
Fenimore, David Clarke
Feramisco, James Robert
Fernandez, Alberto Antonio
Fessler, John Hans
Fessler, Liselotte I
Fielding, Christopher J
Fife, Thomas Harley
Fink, Robert M
Finkle, Bernard Joseph
Fiorindo, Robert Philip
Fischer, Imre A
Fisher, Knute Adrian
Fishman, William Harold
Flanagan, Steven Douglas
Flashner, Michael
Fleming, James Emmitt
Fluharty, Arvan Lawrence
Forrest, Irene Stephanie
Fortes, George (Peter Alexander)
Fox, C Fred
Fox, Joan Elizabeth Bothwell
Francke, Uta
Frankel, Edwin N
Freer, Stephan T
Frey, Terrence G
Friedkin, Morris Enton
Friedlander, Martin
Friedman, Milton Joe
Fujimoto, George Iwao
Fukuda, Michiko N
Fukuda, Minoru
Fukushima, David Kenzo
Fukuyama, Thomas T
Fulco, Armand J
Gadol, Nancy
Gaertner, Frank Herbert
Gall, William Einar
Gamborg, Oluf Lind

Garcia, Eugene N
Garfin, David Edward
Garibaldi, John Attilio
Gautsch, James Willard
Geiger, Paul Jerome
Gelfand, David H
Geller, Edward
Gerhart, John C
Gibson, Thomas Richard
Giglotti, Helen Jean
Ginsberg, Theodore
Giorgio, Anthony Joseph
Giri, Shri N
Glabe, Charles G
Glasky, Alvin Jerald
Glazer, Alexander Namiot
Glembotski, Christopher Charles
Glitz, Dohn George
Gluck, Louis
Glushko, Victor
Gochman, Nathan
Gold, William
Goldbloom, David Ellis
Gonzalez, Elma
Goodman, Michael Gordon
Goodman, Myron F
Gordon, Adrienne Sue
Gordon, Harold Thomas
Gralla, Jay Douglas
Gray, Gary M
Green, Melvin Howard
Greenleaf, Robert Dale
Griffin, John Henry
Griffith, Michael James
Griffith, Owen Malcolm
Grill, Herman, Jr
Grodsky, Gerold Morton
Gruenwedel, Dieter Wolfgang
Grunbaum, Benjamin Wolf
Gunsalus, Robert Philip
Gupta, Rishab Kumar
Guthrie, Christine
Haard, Norman F
Haberland, Margaret Elizabeth
Haen, Peter John
Hafeman, Dean Gary
Hagler, Arnold T
Hainski, Martha Barrionuevo
Hajdu, Joseph
Hall, Michael Oakley
Hammerschlag, Richard
Han, Jang Hyun
Hanawalt, Philip Courtland
Hankinson, Oliver
Harary, Isaac
Harford, Joe Bryan
Harper, Elvin
Harper, Judith Jean
Hatefi, Youssef
Hatfield, G Wesley
Hathaway, Gary Michael
Hawkes, Susan Patricia
Hawkes, Wayne Christian
Hayman, Ernest Paul
Hearn, Walter Russell
Heasley, Victor Lee
Hector, Mina Fisher
Hedrick, Jerry Leo
Heeb, Mary Jo
Heftmann, Erich
Hegenauer, Jack C
Heinrich, Milton Rollin
Heinrichs, W LeRoy
Helinski, Donald Raymond
Helwig, Harold Lavern
Henderson, Gary Borgar
Henriksson, Thomas Martin
Henry, Helen L
Henry, Richard Joseph
Herber, Raymond
Herschman, Harvey R
Hershey, John William Baker
Hess, Frederick Dan
Hessinger, David Alwyn
Hill, Robert
Hines, Leonard Russell
Hjelmeland, Leonard M
Hoagland, Vincent DeForest, Jr
Hoch, Sallie O'Neil
Hochstein, Paul Eugene
Hoke, Glenn Dale
Holden, Joseph Thaddeus
Hollander, Leonore
Holley, Robert William
Holmes, David G
Holmquist, Walter Richard
Holten, Darold Duane
Holten, Virginia Zewe
Horowitz, Norman Harold
Hosoda, Junko
Hostetler, Karl Yoder
Houston, L L
Howard, Bruce David
Howard, Russell John
Hsiao, Theodore Ching-Teh
Hsieh, Philip Kwok-Young
Hsu, Robert Ying
Hsueh, Aaron Jen Wang
Hubbard, Richard W
Hubbard, William Jack
Huennekens, Frank Matthew, Jr
Huffaker, Ray C
Hugli, Tony Edward
Hullar, Theodore Lee

Hunter, Tony
Hussa, Robert Oscar
Hyman, Bradley Clark
Hyman, Richard W
Ibsen, Kenneth Howard
Insel, Paul Anthony
Itano, Harvey Akio
Jackson, Craig Merton
Jacob, Mary
Jacobson, Gail M
Jangaard, Norman Olaf
Jariwalla, Raxit Jayantilal
Jarnagin, Kurt
Jelinek, Bohdan
Jewett, Sandra Lynne
Johansson, Mats W
Johnson, Paul Hickok
Johnson, Phyllis Elaine
Johnson, Randolph Mellus
Jones, Oliver William
Jones, Theodore Harold Douglas
Jordan, Mary Ann
Joyce, Gerald F
Jukes, Thomas Hughes
Kaback, Howard Ronald
Kahlon, Talwinder Singh
Kaiser, Armin Dale
Kallman, Burton Jay
Kalman, Sumner Myron
Kalra, Vijay Kumar
Kammen, Harold Oscar
Kan, Yuet Wai
Kane, John Power
Kaneko, Jiro Jerry
Kang, Kenneth S
Kan-Mitchell, June
Karasek, Marvin A
Katz, Joseph
Kavenoff, Ruth
Kay, Robert Eugene
Keating, Eugene Kneeland
Kedes, Laurence H
Kelley, Darshan Singh
Kelley, Leon A
Kelly, Regis Baker
Kennedy, Mary Bernadette
Kenney, William Clark
Khatra, Balwant Singh
Kientz, Marvin L
Kihara, Hayato
Kim, Sung-Hou
Kinoshita, Jin Harold
Kirsch, Jack Frederick
Kirschbaum, Joel Bruce
Kishimoto, Yasuo
Klain, George J
Kliewer, Walter Mark
Klouda, Mary Ann Aberle
Koch, Bruce D
Kohler, George Oscar
Koobs, Dick Herman
Korenman, Stanley G
Korn, David
Kornberg, Arthur
Kornberg, Roger David
Koshland, Daniel Edward, Jr
Koths, Kirston Edward
Krauss, Ronald
Kun, Ernest
Kunkee, Ralph Edward
Kuo, Harng-Shen
Kurnick, Nathaniel Bertrand
Kuroki, Gary W
Kurtzman, Ralph Harold, Jr
Kushinsky, Stanley
Kuwahara, Steven Sadao
Kyte, Jack Ernst
Labavitch, John Marcus
Lad, Pramod Madhusudan
Lanchantin, Gerard Francis
Landolfi, Nicholas F
Lanyi, Janos K
Largman, Corey
Lascelles, June
Lasky, Richard David
Last, Jerold Alan
Lawrence, Paul J
Lebherz, Herbert G
Lee, Amy Shiu
Leffert, Hyam Lerner
Lehman, I Robert
Lei, Shau-Ping Laura
Lembach, Kenneth James
Lenhoff, Howard Maer
Leong, Kam Choy
Leung, Peter
Leung, Philip Min Bun
Levin, Eugene G
Levintow, Leon
Levy, Daniel
Lewis, James Clement
Lewis, Urban James
Lianides, Sylvia Panagos
Lieberman, Jack
Lin, Grace Woan-Jung
Lin, Jiann-Tsyh
Lin, Ming-Fong
Lindquist, Robert Nels
Lindsey, LeAnn L
Ling, Nicholas Chi-Kwan
Linn, Stuart Michael
Lipsick, Joseph Steven
Liu, Edwin H
Liu, Xuan

BIOLOGICAL SCIENCES / 35

Lolley, Richard Newton
Lönnerdal, Bo L
Lovatt, Carol Jean
Lovelace, C James
Lundblad, Roger Lauren
Lyon, Irving
Maack, Christopher A
McCaman, Marilyn Wales
McClelland, Michael
McClure, William Owen
McElroy, William David
McGaughey, Charles Gilbert
McIntire, Junius Merlin
McIntire, William S
McKee, Ralph Wendell
Macklin, Wendy Blair
McLaughlin, Calvin Sturgis
McMillan, Paul Junior
McNamee, Mark G
Maggio, Edward Thomas
Maier, V(incent) P(aul)
Makarem, Anis H
Makker, Sudesh Paul
Malkin, Harold Marshall
Malkin, Richard
Mann, Lewis Theodore, Jr
Manning, Jerry Edsel
Mansour, Tag Eldin
Marangos, Paul Jerome
Marcus, Frank
Markland, Francis Swaby, Jr
Marsh, James Lawrence
Martin, George Steven
Massie, Barry Michael
Matthews, Harry Roy
Matthews, Thomas Robert
Mattson, Fred Hugh
Mayron, Lewis Walter
Mazelis, Mendel
Mead, James Franklyn
Mealey, Edward H
Meares, Claude Francis
Meehan, Thomas (Dennis)
Meerdink, Denis J
Mehra, Rajesh Kumar
Melchior, Jacklyn Butler
Melis, Anastasios
Mendelson, Robert Alexander, Jr
Merchant, Sabeeha
Merriam, Esther Virginia
Mestril, Ruben
Meyerowitz, Elliot Martin
Miljanich, George Paul
Miller, Alexander
Miller, Arnold Lawrence
Miller, Jon Philip
Miller, Lloyd George
Miller, William Walter
Mitchell, Herschel Kenworthy
Miyada, Don Shuso
Mizuno, Nobuko S(himotori)
Mochizuki, Diane Yukiko
Moffitt, Robert Allan
Mohan, Chandra
Mohrenweiser, Harvey Walter
Moldave, Kivie
Montecalvo, Joseph, Jr
Montesano-Roditis, Luisa
Mooney, Larry Albert
Moore, Gerald L
Moos, Walter Hamilton
Morales, Daniel Richard
Morehouse, Margaret Gulick
Morgan, Alan Raymond
Mosteller, Raymond Dee
Moyed, Harris S
Mudd, John Brian
Mueller, Peter Klaus
Mule, Salvatore Joseph
Muralidharan, V B
Murphy, Alexander James
Murphy, Terence Martin
Nacht, Sergio
Nagel, Glenn M
Nambiar, Krishnan P
Nassos, Patricia Saima
Neidleman, Saul L
Nelson, Gary Joe
Nemere, Ilka M
Neufeld, Elizabeth Fondal
Newbrun, Ernest
Newell, Gordon Wilfred
Ngo, That Tjien
Nielsen, Milo Alfred
Nikaido, Hiroshi
Nimni, Marcel Efraim
Noller, Harry Francis, Jr
Norman, Anthony Westcott
Nothnagel, Eugene Alfred
Nussenbaum, Siegfried Fred
Nyhan, William Leo
O'Connor, Daniel Thomas
Ohms, Jack Ivan
Oldham, Susan Banks
Oliphant, Edward Eugene
Olsen, Charles Edward
Olson, Alfred C
Oppenheimer, Norman Joseph
Ortiz de Montellano, Paul Richard
Osborn, Terry Wayne
Ottke, Robert Crittenden
Ottoboni, M(inna) Alice
Overby, Lacy Rasco
Ow, David Wing
Owades, Joseph Lawrence
Oyama, Jiro
Oyama, Vance I
Packer, Lester
Paech, Christian
Palmer, Grant H
Palmer, John Warren
Papahadjopoulos, Demetrios Panayotis
Papkoff, Harold
Patterson, Paul H
Patton, John Stuart
Patton, Stuart
Paulson, James Carsten
Peckham, William Dierolf
Penhoet, Edward Etienne
Perlgut, Louis E
Perrault, Jacques
Peters, John Henry
Peterson, Charles Marquis
Peterson, Neal Alfred
Petryka, Zbyslaw Jan
Pharriss, Bruce Bailey
Phelps, Michael Edward
Philippart, Michel Paul
Phillips, David Richard
Phillips, Donald Arthur
Phleger, Charles Frederick
Pierce, John Grissim
Pierschbacher, Michael Dean
Pigiet, Vincent P
Pilgeram, Laurence Oscar
Pirkle, Hubert Chaille
Piszkiewicz, Dennis
Pitesky, Isadore
Platzer, Edward George
Poling, Stephen Michael
Popjak, George Joseph
Porter, Clark Alfred
Powanda, Michael Christopher
Powers, Dennis A
Prescott, Benjamin
Price, Paul Arms
Prusiner, Stanley Ben
Purdy, Robert H
Quail, Peter Hugh
Quistad, Gary Bennet
Rabinowitz, Israel Nathan
Rabinowitz, Jesse Charles
Rabovsky, Jean
Rabussay, Dietmar Paul
Radloff, Harold David
Rae-Venter, Barbara
Raijman, Luisa J
Rall, Stanley Carlton, Jr
Ramachandran, Janakiraman
Rao, Ananda G
Recsei, Paul Andor
Reese, Floyd Ernest
Reese, Robert Trafton
Register, Ulma Doyle
Reiness, Gary
Reizer, Jonathan
Rho, Joon H
Rice, Robert Hafling
Rich, Terrell L
Richards, John Hall
Richards, Oliver Christopher
Richmond, Jonas Edward
Rinderknecht, Heinrich
Rio, Donald C
Rittenhouse, Harry George
Roberts, Eugene
Roberts, Martin
Roberts, Sidney
Robertson, William Van Bogaert
Robinson, James McOmber
Rodgers, Richard Michael
Roepke, Raymond Rollin
Rogers, Quinton Ray
Rogoff, Martin Harold
Romani, Roger Joseph
Rome, Leonard H
Rosenberg, Abraham
Rosenberg, Steven Loren
Rosenfeld, Ron Gershon
Rosenquist, Grace Link
Rossi, John Joseph
Rothman, Stephen Sutton
Rowley, Rodney Ray
Roy-Burman, Pradip
Rubanyi, Gabor Michael
Rucker, Robert Blain
Rusay, Ronald Joseph
Russell, Percy J
Rutter, William J
Sabry, Zakaria I
Sachs, George
Sadava, David Eric
Sadee, Wolfgang
Saier, Milton H, Jr
Saltman, Paul David
Salzberg, David Aaron
Samuel, Charles Edward
Sanchez, Albert
Santi, Daniel V
Sayre, Francis Warren
Scala, James
Schachman, Howard Kapnek
Schaffer, Frederick Leland
Schaffer, Sheldon Arthur
Schapiro, Harriette Charlotte
Schekman, Randy W
Schelar, Virginia Mae
Scheve, Larry Gerard
Schick, Lloyd Alan
Schiffman, Sandra
Schimke, Robert T
Schleich, Thomas W
Schmid, Peter
Schmid, Rudi
Schmid, Sandra Louise
Schneir, Michael Lewis
Schotz, Michael C
Schraer, Rosemary
Schroeder, Duane David
Schwimmer, Sigmund
Scogin, Ron Lynn
Sears, Duane William
Seegers, Walter Henry
Seegmiller, Jarvis Edwin
Segall, Paul Edward
Segel, Irwin Harvey
Selassie, Cynthia R
Senda, Mototaka
Sensabaugh, George Frank
Serat, William Felkner
Sevall, Jack Sanders
Shah, Shantilal Nathubhai
Shannon, Leland Marion
Shapiro, Lucille
Shaw, Kenneth Noel Francis
Shen, Che-Kun James
Sherman, Linda Arlene
Shih, Jean Chen
Shively, John Ernest
Shneour, Elie Alexis
Shooter, Eric Manvers
Shore, Virgie Guinn
Short, Jay M
Shrawder, Elsie June
Shugarman, Peter Melvin
Shum, Archie Chue
Sie, Edward Hsien Choh
Siegman, Fred Stephen
Sigman, David Stephan
Simmons, Norman Stanley
Simoni, Robert Dario
Simons, Robert W
Simonsen, Donald Howard
Singer, Thomas Peter
Sinha, Yagya Nand
Sinibaldi, Ralph Michael
Sinsheimer, Robert Louis
Siperstein, Marvin David
Slater, Grant Gay
Slavkin, Harold Charles
Smith, Charles G
Smith, Emil L
Smith, Helene Sheila
Smith, Kendric Charles
Smith, Kevin Malcolm
Smith, Marion Edmonds
Smith, Martyn Thomas
Smith, Roberts Angus
Smith, Steven Sidney
Smith, Stuart
Snow, John Thomas
Snowdowne, Kenneth William
Somerville, Christopher Roland
Song, Moon K
Spanis, Curt William
Sparling, Mary Lee
Spilburg, Curtis Allen
Spray, Clive Robert
Springer, Wayne Richard
Spudich, James Anthony
Sridhar, Rajagopalan
Stallcup, Michael R
Stanczyk, Frank Zygmunt
Stanley, Wendell Meredith, Jr
Stanton, Hubert Coleman
Steinberg, Daniel
Stellwagen, Robert Harwood
Stephens, Robert James
Sterling, Rex Elliott
Stern, Robert
Sternberg, Moshe
Stevens, Ronald Henry
Stewart, Charles Jack
Still, Gerald G
Stokstad, Evan Ludvig Robert
Stone, Deborah Bennett
Strehler, Bernard Louis
Strother, Allen
Stubbs, John Dorton
Stumpf, Paul Karl
Subramani, Suresh
Sussman, Howard H
Sweetman, Lawrence
Swendseid, Marian Edna
Sy, Jose
Syvanen, Michael
Szego, Clara Marian
Szymanski, Edward Stanley
Taborsky, George
Tallman, John Gary
Tappel, Aloys Louis
Tarver, Harold
Taylor, James A
Taylor, Robert Thomas
Taylor, Susan Serota
Teresi, Joseph Dominic
Thiele, Elizabeth Henriette
Thomas, Heriberto Victor
Thomas, Richard Sanborn
Thomas, Robert E
Thompson, Chester Ray
Thomson, John Ansel Armstrong
Thornber, James Philip
Thorner, Jeremy William
Tjian, Robert Tse Nan
Tokes, Zoltan Andras
Tomlinson, Geraldine Ann
Tomlinson, Raymond Valentine
Townsend, R Reid
Tramell, Paul Richard
Traugh, Jolinda Ann
Traut, Robert Rush
Traylor, Patricia Shizuko
Treat-Clemons, Lynda George
Trifunac, Natalia Pisker
Troy, Frederic Arthur
Tsao, Constance S
Tseng, Ben Y
Tsuboi, Kenneth Kaz
Tsuji, Frederick Ichiro
Tukey, Robert H
Turgeon, Judith Lee
Tutwiler, Gene Floyd
Ulevitch, Richard Joel
Ulm, Edgar H
Unver, Ercan
Urry, Lisa Andrea
Vacquier, Victor Dimitri
Valentine, Raymond Carlyle
Vandlen, Richard Lee
Van Dop, Cornelis
Van Wart, Harold Edgar
Varki, Ajit Pothan
Varon, Silvio Salomone
Vehar, Gordon Allen
Vice, John Leonard
Vickery, Larry Edward
Victoria, Edward Jess, Jr
Vilker, Vincent Lee
Villarejo, Merna
Villarreal, Luis Perez
Vlasuk, George P
Vohra, Pran Nath
Volcani, Benjamin Elazari
Vold, Barbara Schneider
Von HUngen, Kern
Vreeland, Valerie Jane
Wade, Michael James
Wade, Richard Lincoln
Wadman, W Hugh
Walbot, Virginia Elizabeth
Walker, Howard David
Wallace, Joan M
Wallace, Robert Allan
Walter, Harry
Wang, Ching Chung
Wang, Howard Hao
Warner, Robert Collett
Warner, Thomas Garrie
Wasterlain, Claude Guy
Watson, John Alfred
Wax, Harry
Webb, David Ritchie, Jr
Weber, Arthur L
Weber, Bruce Howard
Weber, Heather R(oss) Wilson
Weber, Thomas Byrnes
Wedding, Randolph Townsend
Wegner, Marcus Immanuel
Wehr, Carl Timothy
Weimberg, Ralph
Weinberg, Barbara Lee Huberman
Weisgraber, Karl Heinrich
Weiss, Richard Louis
Wergedal, Jon E
West, Charles Allen
Whitaker, John Robert
White, Thomas James
Whitlock, Gaylord Purcell
Whitlow, Marc David
Wiktorowicz, John Edward
Wilcox, Gary Lynn
Wilcox, Ronald Bruce
Wilkinson, David Ian
Williams, Julian Carroll
Williams, Mary Ann
Wilson, Jerry Lee
Wilson, Leslie
Wilson, Lowell D
Winkelhake, Jeffrey Lee
Wolf, George
Wong, Kin-Ping
Wood, Peter Douglas
Wood, William Irwin
Wood, Willis Avery
Wu, Chuen-Shang C
Wu, Chung
Yager, Janice L Winter
Yamaguchi, Masatoshi
Yang, Heechung
Yau-Young, Annie O
Yen, Tien-Sze Benedict
Yerram, Nagender Rao
Yguerabide, Juan
Yoshida, Akira
Young, Janis Dillaha
Yu, David Tak Yan
Yu, Sharon S M
Yuwiler, Arthur
Zabin, Irving
Zaffaroni, Alejandro
Zahnley, James Curry
Zaidi, Iqbal Mehdi
Zamenhof, Stephen
Zeichner-David, Margarita
Zerez, Charles Raymond

Biochemistry (cont)

Ziboh, Vincent Azubike
Ziccardi, Robert John
Zill, Leonard Peter
Zuckerkandl, Emile

COLORADO
Abrams, Adolph
Akeson, Walter Roy
Allen, Kenneth G D
Allen, Robert H
Arend, William Phelps
Azari, Parviz
Bailey, David Tiffany
Bamburg, James Robert
Barrett, Dennis
Bechtel, Peter John
Berens, Randolph Lee
Bowden, Joe Allen
Bowles, Jean Alyce
Brennan, Patrick Joseph
Brooks Springs, Suzanne Beth
Brown, Jerry L
Bublitz, Clark
Caughey, Winslow Spaulding
Chan, Laurence Kwong-fai
Charkey, Lowell William
Charney, Michael
Church, Brooks Davis
Clarke, Steven Donald
Cooper, Dermot M F
Corcoran, John William
Deitrich, Richard Adam
Dever, John E, Jr
Dobersen, Michael J
Downing, Mancourt
Dyckes, Douglas Franz
Eley, James H
Ellinwood, William Edward
Erwin, Virgil Gene
Fahrney, David Emory
Fitzpatrick, Francis Anthony
Frerman, Frank Edward
Froede, Harry Curt
Giclas, Patricia C
Glode, Leonard Michael
Graf, George
Grainger, Robert Ball
Gramera, Robert Eugene
Grieve, Robert B
Ham, Richard George
Hamar, Dwayne Walter
Harold, Franklin Marcel
Harrill, Inez Kemble
Harrison, Merle E(dward)
Hesterberg, Thomas William
Hibler, Charles Phillip
Hinman, Norman Dean
Hogan, Christopher James
Horwitz, Kathryn Bloch
Hossner, Kim L
Jaehning, Judith A
James, Gordon Thomas
Jansen, Gustav Richard
Jones, Carol A
Kano-Sueoka, Tamiko
Kerr, Sylvia Jean
Kinsky, Stephen Charles
Kloppel, Thomas Mathew
Langan, Thomas Augustine
Lee, Virginia Ann
Lin, Leu-Fen Hou
Linden, James Carl
Lowndes, Joseph M
McCord, Joe Milton
McHenry, Charles S
Maga, Joseph Andrew
Malkinson, Alvin Maynard
Martin, Jack E
Martin, Susan Scott
Melancon, Paul R
Metzger, H Peter
Moore, Frank Archer
Mykles, Donald Lee
Panini, Sankhavaram R
Patterson, David
Pearson, John Richard
Petersen, Gene
Pizer, Lewis Ivan
Prentice, Neville
Quissell, David Olin
Reiss, Oscar Kully
Rhoads, William Denham
Roberts, Walden Kay
Ross, Cleon Walter
Saidel, Leo James
Sampson, David Ashmore
Seeds, Nicholas Warren
Seely, James Ervin
Seely, Robert J
Seibert, Michael
Sherrill, Bette Cecile Benham
Sinensky, Michael
Skelton, Marilyn Mae
Sneider, Thomas W
Solomons, Clive (Charles)
Stewart, John Morrow
Stifel, Fred B
Storey, Richard Drake
Sullivan, Sean Michael
Tabakoff, Boris
Tolbert, Bert Mills
Tomasi, Gordon Ernest

Tu, Anthony T
Virtue, Robert
Voelker, Dennis R
Weliky, Irving
Wilson, Irwin B
Winstead, Jack Alan
Woody, A-Young Moon
Woody, Robert Wayne
Yarus, Michael J
Zeiler, Kathryn Gail

CONNECTICUT
Adelberg, Edward Allen
Amacher, David E
Armitage, Ian MacLeod
Aronson, Peter S
Badoyannis, Helen Litman
Bausher, Larry Paul
Behrman, Harold R
Blue, Marie-Luise
Bondy, Philip K
Bordner, Jon D B
Bormann, Barbara-Jean Anne
Bronner, Felix
Bronsky, Albert J
Buck, Marion Gilmour
Cadman, Edwin Clarence
Canellakis, Zoe Nakos
Celesk, Roger A
Cheng, Yung-Chi
Cinti, Dominick Louis
Coleman, Joseph Emory
Crain, Richard Cullen
Crawford, Richard Bradway
Cresswell, Peter
Dannies, Priscilla Shaw
Das, Dipak K
Das, Rathindra C
Della-Fera, Mary Anne
Deutscher, Murray Paul
Dingman, Douglas Wayne
Dix, Douglas Edward
Doeg, Kenneth Albert
Dooley, Joseph Francis
Dorsky, David Isaac
Edwards, Lawrence Jay
Eisenstadt, Jerome Melvin
Engelman, Donald Max
Epstein, Paul Mark
Eustice, David Christopher
Fenton, Wayne Alexander
Fiore, Joseph Vincent
Flavell, Richard Anthony
Forbush, Bliss, III
Fordham, Joseph Raymond
Forenza, Salvatore
Forget, Bernard G
Fruton, Joseph Stewart
Galston, Arthur William
Geiger, Edwin Otto
Gerritsen, Mary Ellen
Gorrell, Thomas Earl
Grindley, Nigel David Forster
Gum, Ernest Kemp, Jr
Gunther, Jay Kenneth
Hanson, Kenneth Ralph
Havir, Evelyn A
Hebert, Daniel Normond
Heywood, Stuart Mackenzie
Hightower, Lawrence Edward
Hinman, Richard Leslie
Hobbs, Donald Clifford
Hook, Derek John
Howard, Phillenore Drummond
Huszar, Gabor
Hutchinson, Franklin
Infante, Anthony A
Janis, Ronald Allen
Jungas, Robert Leando
Keirns, James Jeffery
Kelleher, William Joseph
Kenyon, Alan J
Khairallah, Edward A
Kind, Charles Albert
Kiron, Ravi
Koe, B Kenneth
Konigsbacher, Kurt S
Konigsberg, William Henry
Krause, Leonard Anthony
Kream, Barbara Elizabeth
Kuether, Carl Albert
Lande, Saul
Leadbetter, Edward Renton
Lee, Henry C
Lee, Thomas W
Lees, Thomas Masson
Lengyel, Peter
Lentz, Thomas Lawrence
Lerner, Aaron Bunsen
Lewis, Jonathan Joseph
Loomis, Stephen Henry
Lukens, Lewis Nelson
Lustig, Bernard
McCorkle, George Maston
McGregor, Donald Neil
MacNintch, John Edwin
Marchesi, Vincent T
Marks, Paul A
Matovcik, Lisa M
Mayol, Robert Francis
Miller, Wilbur Hobart
Monro, Alastair Macleod
Moore, Peter Bartlett
Mycek, Mary J

Myles, Diana Gold
Niblack, John Franklin
Noll, Clifford Raymond, Jr
Norton, Louis Arthur
Notation, Albert David
Novoa, William Brewster
Oates, Peter Joseph
O'Looney, Patricia Anne
Osborn, Mary Jane
Ozols, Juris
Parker, Eric McFee
Paul, Jeddeo
Pawelek, John Mason
Pazoles, Christopher James
Pereira, Joseph
Peterson, Richard Burnett
Pfeiffer, Steven Eugene
Poincelot, Raymond Paul, Jr
Primakoff, Paul
Prusoff, William Herman
Pudelkiewicz, Walter Joseph
Putterman, Gerald Joseph
Rachinsky, Michael Richard
Radding, Charles Meyer
Rajendran, Vazhaikkurichi M
Rauch, Albert Lee
Ray, Verne A
Rebuffe-Scrive, Marielle Francoise
Ressler, Charlotte
Retsema, James Allan
Richards, Frank Frederick
Richards, Frederic Middlebrook
Roman, Laura M
Rosenbaum, Joel L
Rosenberg, Philip
Rossomando, Edward Frederick
Rothfield, Lawrence I
Rudnick, Gary
Rupp, W Dean
Sardinas, Joseph Louis
Schenkman, John Boris
Schmir, Gaston L
Schwartz, Pauline Mary
Schwinck, Ilse
Seligson, David
Setlow, Peter
Sigler, Paul Benjamin
Simmonds, Sofia
Smilowitz, Henry Martin
Snider, Ray Michael
Spencer, Richard Paul
Squinto, Stephen P
Steitz, Joan Argetsinger
Stowe, Bruce Bernot
Strittmatter, Philipp
Summers, William Cofield
Summers, Wilma Poos
Sun, Alexander Shihkaung
Tallman, John Francis
Tanaka, Kay
Tanzer, Marvin Lawrence
Toralballa, Gloria C
Turnipseed, Marvin Roy
Vasington, Frank D
Votaw, Robert Grimm
Walker, Frederick John
Ward, David Christian
Weissman, Sherman Morton
Wheeler, George Lawrence
Wilson, John Thomas
Wood, David Dudley
Woronick, Charles Louis
Wright, Herbert Fessenden
Xu, Zhi Xin
Yphantis, David Andrew
Zelitch, Israel

DELAWARE
Anton, David L
Billheimer, Jeffrey Thomas
Blomstrom, Dale Clifton
Bond, Elizabeth Dux
Boylen, Joyce Beatrice
Campbell, Linzy Leon
Chen, Harry Wu-Shiong
Colman, Roberta F
Crippen, Raymond Charles
Davis, Leonard George
Dennis, Don
Dhurjati, Prasad S
Diner, Bruce Aaron
Freerksen, Deborah Lynne (Chalmers)
Frey, William Adrian
Ganfield, David Judd
Gatenby, Anthony Arthur
Giles, Ralph E
Hartig, Paul Richard
Hayman, Selma
Herblin, William Fitts
Heytler, Peter George
Hodges, Charles Thomas
Holmes, Richard
Holsten, Richard David
Jackson, David Archer
Jefferies, Steven
Jenner, Edward L
Kerr, Janet Spence
Kinney, Anthony John
Koszalka, Thomas R
Lichtner, Francis Thomas, Jr
Litchfield, William John
Loomis, Gary Lee
Lorimer, George Huntly
Maloff, Bruce L(arrie)

Marrs, Barry Lee
Miles, James Lowell
Morrissey, Bruce William
Myoda, Toshio Timothy
Neubauer, Russell Howard
Pierson, Keith Bernard
Saller, Charles Frederick
Salsbury, Robert Lawrence
Salzman, Steven Kerry
Sandberg, Robert Gustave
Sands, Howard
Sheppard, David E
Singleton, Rivers, Jr
Smith, David William
Smith, Jack Louis
Snyder, Jack Austin
Stevenson, Irone Edmund, Jr
Stopkie, Roger John
Strobach, Donald Roy
Tam, Sang William
Thompson, Jeffery Scott
Thorpe, Colin
Wermus, Gerald R
White, Harold Bancroft, III
Wriston, John Clarence, Jr
Yates, Richard Alan

DISTRICT OF COLUMBIA
Alving, Carl Richard
Arcos, Joseph (Charles)
Argus, Mary Frances
Ashe, Warren (Kelly)
Attaway, David Henry
Bailey, John Martyn
Ball, William David
Beru, Nega
Blecher, Melvin
Bridges, John Robert
Brooker, Gary
Bundy, Bonita Marie
Calvert, Allen Fisher
Carroll, Alan G
Carson, Frederick Wallace
Chaput, Raymond Leo
Chen, H R
Chiang, Peter K
Chirikjian, Jack G
Cocks, Gary Thomas
Cohn, Victor Hugo
Coomes, Marguerite Wilton
Cowan, James W
Davidson, Eugene Abraham
Deutsch, Mike John
Donaldson, Robert Paul
Edwards, Cecile Hoover
Ellis, Sydney
Fanning, George Richard
Field, Ruth Bisen
Finkelstein, James David
Fishbein, William Nichols
Fleming, Patrick John
Frattali, Victor Paul
Friedberg, Felix
Gallo, Linda Lou
Gray, Irving
Hajiyani, Mehdi Hussain
Hamosh, Margit
Hayden, George A
Henderson, Ellen Jane
Hollingdale, Michael Richard
Howard, Barbara V
Irausquin, Hiltje
Jett, Marti
Jett-Tilton, Marti
Jordan, John Patrick
Kasbekar, Dinkar Kashinath
Katz, Edward
Khanna, Krishan L
Kimmel, Gary Lewis
Klubes, Philip
Kornhauser, Andrija
Krueger, Karl E
Kumar, Soma
Lai, David Ying-lun
Lakshman, M Raj
Leto, Salvatore
Ligler, Frances Smith
Malinin, George I
Matyas, Gary Ralph
Mazel, Paul
Mehler, Alan Haskell
Merritt, William D
Miller, Linda Jean
Nandedkar, Arvindkumar Narhari
Nath, Jayasree
Naylor, Paul Henry
Nishioka, David Jitsuo
Ohl, Selgo
Papadopoulos, Nicholas M
Perfetti, Randolph B
Prival, Michael Joseph
Quaife, Mary Louise
Reich, Melvin
Rennert, Owen M
Rhoads, Allen R
Richards, Roberta Lynne
Rothman, Sara Weinstein
Salzman, Lois Ann
Sarin, Prem S
Saslaw, Leonard David
Schneider, Bernard Arnold
Shank, Fred R
Shibko, Samuel Issac
Singer, Maxine Frank

Smith, Benjamin Williams
Smith, Thomas Elijah
Smulson, Mark Elliott
Southerland, William M
Speidel, Edna W
Stiehler, Robert D(aniel)
Tarantino, Laura M(ary)
Thomas, William Eric
Todhunter, John Anthony
Valassi, Kyriake V
Vanderhoek, Jack Yehudi
Venkatesan, Malabi M
Wassef, Nabila M
Whitfield, Carolyn Dickson
Wilson, Edward Matthew
Winter, William Phillips
Wolfe, Alan David
Wood, Garnett Elmer
Wright, Daniel Godwin
Wyngaarden, James Barnes
Zimmer, Elizabeth Anne

FLORIDA
Abou-Khalil, Samir
Adair, Winston Lee, Jr
Allen, Charles Marshall, Jr
Assoian, Richard Kenneth
Baker, Stephen Phillip
Barber, Michael James
Barron, Mace Gerald
Baumbach, Lisa Lorraine
Bausher, Michael George
Beckhorn, Edward John
Bell, Paul Hadley
Bellamy, Winthrop Dexter
Berry, Robert Eddy
Bieber, Theodore Immanuel
Bleiweis, Arnold Sheldon
Bliznakov, Emile George
Bock, Fred G
Bosee, Roland Andrew
Bowes, George Ernest
Brew, Keith
Briscoe, Anne M
Brown, Ross Duncan, Jr
Burchfield, Harry P
Buslig, Bela Stephen
Busse, Robert Franklyn
Cain, Brian D
Carraway, Coralie Anne Carothers
Carraway, Kermit Lee
Caswell, Anthony H
Chapman, Peter John
Cline, Kenneth Charles
Coffey, Ronald Gibson
Cohen, Pinya
Cohen, Robert Jay
Coleman, Sylvia Ethel
Connors, William Matthew
Cort, Winifred Mitchell
Coulson, Richard
Cousins, Robert John
Cowman, Richard Ammon
Crews, Fulton T
Cunningham, Glenn N
Dash, Harriman Harvey
Davenport, Thomas Lee
DeKloet, Siwo R
Dhople, Arvind Madhav
Dombro, Roy S
Downey, Kathleen Mary
Dunn, Ben Monroe
Dunn, William Arthur, Jr
Edelson, Jerome
Elliott, Paul Russell
Emerson, Geraldine Mariellen
Eoff, Kay M
Feaster, John Pipkin
Feinstein, Louis
Fisher, Waldo Reynolds
Fishman, Jack
Fontaine, Thomas Davis
Foy, Robert Bastian
Fried, Melvin
Frieden, Earl
Friedl, Frank Edward
Gander, John E
Ganguly, Rama
Gawron, Oscar
Genthner, Barbara Robyn Sharak
Giegel, Joseph Lester
Giner-Sorolla, Alfredo
Gittelman, Donald Henry
Glaser, Luis
Goldberg, Melvin Leonard
Gomatos, Peter John
Gould, Anne Bramlee
Graham, W(alter) Donald
Graven, Stanley N
Green, Harry
Greenfield, Leonard Julian
Hahn, Elliot F
Halprin, Kenneth M
Hamilton, Franklin D
Hamilton, James Guthrie
Hanks, Robert William
Hargrave, Paul Allan
Hauswirth, William Walter
Hayashi, Teru
Holt, Thomas Manning
Homann, Peter H
Howell, Ralph Rodney
Hsia, Sung Lan
Hsu, Jeng Mein

Huber, Donald John
Huijing, Frans
Humphreys, Thomas Elder
Inana, George
Isaacks, Russell Ernest
Jackson, Bettina B Carter
Jenkin, Howard M
Jensen, Roy A
Kadis, Barney Morris
Kapsalis, John George
Katchen, Bernard
Katz, Albert Barry
Kaul, Rajinder K
Kelley, George Greene
Kilberg, Michael Steven
Knapp, Francis Marion
Kopelovich, Levy
Koroly, Mary Jo
Kulwich, Roman
Lai, Patrick Kinglun
Lee, Ernest Y
Lewin, Alfred S
Light, Robley Jasper
Lim, Daniel V
Lin, Tsau-Yen
Lindsay, Raymond H
Litman, Gary William
Litosch, Irene
Luer, Carl A
Lutz, Peter Louis
McCall, John Temple
McLean, Mark Philip
McNary, Robert Reed
Makemson, John Christopher
Mallery, Charles Henry
Mans, Rusty Jay
Mayer, Marion Sidney
Mende, Thomas Julius
Menzies, Robert Allen
Merdinger, Emanuel
Miller, Kent D
Muench, Karl Hugo
Mullins, John Thomas
Nakashima, Tadayoshi
Nation, James Lamar, Sr
Neary, Joseph Thomas
Neims, Allen Howard
Ness, Gene Charles
Noble, Nancy Lee
Nonoyama, Meihan
Noonan, Kenneth Daniel
Nordby, Harold Edwin
Oberst, Fred William
O'Brien, Thomas W
Osbahr, Albert J, Jr
Ott, Edgar Alton
Pita, Julio C
Plescia, Otto John
Polson, Charles David Allen
Potter, James D
Pressman, Berton Charles
Preston, James Faulkner, III
Previc, Edward Paul
Pruzansky, Jacob Julius
Purich, Daniel Lee
Rangel Aldao, Rafael
Reid, Parlane John
Riehm, John P
Roberts, R Michael
Roberts, Thomas L
Roeder, Martin
Romeo, John Thomas
Rosenthal, Arthur Frederick
Ross, Lynne Fischer
Rossi, Anthony Michael
Rubin, Saul Howard
Ruegamer, William Raymond
Ryan, James Walter
Salas, Pedro Jose I
Samis, Harvey Voorhees, Jr
Sang, Qing-Xiang
Sarett, Herbert Paul
Sathe, Shridhar Krishna
Scarpace, Philip J
Schmidt, Robert Reinhart
Scott, Walter Alvin
Segal, Alvin
Shanmugam, Keelnatham Thirunavukkarasu
Shapiro, Jeffrey Paul
Shireman, Rachel Baker
Shirk, Paul David
Silhacek, Donald Le Roy
Silverman, David Norman
Silverstein, Herbert
Soldo, Anthony Thomas
Soliman, Magdi R I
Solomonson, Larry Paul
Soto, Aida R
Stahmann, Mark Arnold
Standaert, Mary L
Stearns, Thomas W
Stein, Abraham Morton
Stevens, Bruce Russell
Stewart, Ivan
Stoloff, Leonard
Stone, Stanley S
Storrs, Eleanor Emerett
Sullivan, Lloyd John
Taylor, Barrie Frederick
Terranova, Andrew Charles
Thiessen, Reinhardt, Jr
Tocci, Paul M
Toporek, Milton

Tsibris, John-Constantine Michael
Urbas, Branko
Vesely, David Lynn
Vickers, David Hyle
Voigt, Walter
Walker, Charles R
Wallace, Robin A
Walton, Daniel C
Wecker, Lynn
Weissbach, Arthur
Wells, Gary Neil
Whelan, William Joseph
White, Fred G
White, Roseann Spicola
Wilder, Violet Myrtle
Wilson, David Louis
Wodzinski, Rudy Joseph
Woessner, Jacob Frederick, Jr
Wood, William Otto
Woodside, Kenneth Hall
Young, David Michael
Yu, Simon Shyi-Jian
Zikakis, John Philip

GEORGIA
Abdel-Latif, Ata A
Abraham, Edathara Chacko
Adams, Robert Johnson
Agosin, Moises
Akhtak, Rashid Ahmed
Bailey, Gordon Burgess
Baumstark, Barbara Ruth
Bernstein, Robert Steven
Black, Asa C, Jr
Black, Billy C, II
Black, Clanton Candler, Jr
Boutwell, Joseph Haskell
Brackett, Benjamin Gaylord
Brewer, John Michael
Brooks, John Bill
Brubaker, Leonard Hathaway
Buccafusco, Jerry Joseph
Bustos-Valdes, Sergio Enrique
Calabrese, Ronald Lewis
Carl, G Franklin
Cavanagh, Harrison Dwight
Cherniak, Robert
Chiu, Kirts C
Claybrook, James Russell
Compans, Richard William
Cormier, Milton Joseph
Coryell, Margaret E
Cramer, Gisela Turck
Cramer, John Wesley
Curley, Winifred H
Dailey, Harry A
Danner, Dean Jay
Dembure, Philip Pito
DerVartanian, Daniel Vartan
Dirksen, Thomas Reed
Dixon, Dabney White
Doetsch, Paul William
Drummond, Margaret Crawford
Dull, Gerald G
Duncan, Robert Leon, Jr
Dure, Leon S, III
Edmondson, Dale Edward
Eriksson, Karl-Erik Lennart
Eriquez, Louis Anthony
Espelie, Karl Edward
Fales, Frank Weck
Fechheimer, Marcus
Finnerty, William Robert
Gerschenson, Mariana
Girardot, Jean Marie Denis
Glass, David Bankes
Glover, Claiborne V C, III
Green, John H
Greenberg, Jerrold
Hadd, Harry Earle
Hall, Dwight Hubert
Halper, Jaroslava
Hargrove, James Lee
Harris, Henry Earl
Heise, John J
Hicks, Heraline Elaine
Howard, John Charles
Huisman, Titus Hendrik Jan
Hunter, Frissell Roy
Iuvone, Paul Michael
Johnson, Joe
Jones, Dean Paul
Jones, George Henry
Karp, Warren S
Kiefer, Charles R(andolph)
Klee, Lucille Holljes
Kraeling, Robert Russell
Kreitzman, Stephen Neil
Krolak, John Michael
Kuck, John Frederick Read, Jr
Kuo, Jyh-Fa
Kushner, Sidney Ralph
Lambeth, J David
Leibach, Fredrick Hartmut
Lewis, Jasper Phelps
Ljungdahl, Lars Gerhard
Lyon, John Blakeslee, Jr
McCormick, Donald Bruce
McGowan, Eleanor Brookens
MacMillan, Joseph Edward
McPherson, James C, Jr
McRorie, Robert Anderson
Madden, John Joseph
Martin, Roy Joseph, Jr

Mather, Jane H
May, Sheldon William
Meagher, Richard Brian
Mendicino, Joseph Frank
Miller, Lois Kathryn
Mills, John Blakely, III
Moore, Cyril L
Morin, Leo Gregory
Moss, Claude Wayne
Myers, Dirck V
Nemeroff, Charles Barnet
Noe, Bryan Dale
Ogle, Thomas Frank
Ove, Peter
Pandey, Kailash N
Papageorge, Evangeline Thomas
Paris, Doris Fort
Peck, Harry Dowd, Jr
Peifer, James J
Pine, Leo
Powers, James Cecil
Pratt, Lee Herbert
Pressey, Russell
Puett, J David
Ragland, William Lauman, III
Reilly, Charles Conrad
Robertson, Alex F
Robinson, George Waller
Rogers, John Ernest
Sanford, Gary L
Sansing, Norman Glenn
Schadler, Daniel Leo
Schepartz, Abner Irwin
Schmidt, Gregory Wayne
Scott, David Frederick
Sgoutas, Demetrios Spiros
Shapira, Raymond
Shuster, Robert C
Sink, John Davis
Sinor, Lyle Tolbot
Smith, David Fletcher
Sophianopoulos, Alkis John
Sridaran, Rajagopala
Stevens, Ann Rebecca
Stevens, Charles David
Suddath, Fred LeRoy, (Jr)
Thedford, Roosevelt
Travis, James
Tsang, Victor Chiu Wan
Underwood, Arthur Louis, Jr
Vegotsky, Allen
Wade, Adelbert Elton
Waitzman, Morton Benjamin
Wallace, Douglas Cecil
Wampler, John E
Warren, Stephen Theodore
Whitney, J(ohn) Barry, III
Wiegel, Juergen K
Wilkinson, Robert E
Williams, Joy Elizabeth P
Williams, William Lawrence
Woodley, Charles Lamar
Woods, Wendell David
Young, Henry Edward
Ziegler, Harry Kirk

HAWAII
Ako, Harry Mu Kwong Ching
Allen, Richard Dean
Berger, Leslie Ralph
Bhagavan, Nadhipuram V
Burr, George Oswald
Chang, Franklin
Datta, Padma Rag
Fok, Agnes Kwan
Gibbons, Barbara Hollingworth
Greenwood, Frederick C
Guillory, Richard John
Howton, David Ronald
Jackson, Mel Clinton
Krupp, David Alan
Loo, Yen-Hoong
McConnell, Bruce
McKay, Robert Harvey
Mandel, Morton
Maretzki, Andrew
Matsumoto, Hiromu
Moore, Paul Harris
Morton, Bruce Eldine
Phillips, John Howell, Jr
Putman, Edison Walker
Rechnitz, Garry Arthur
Rosenthal, Eric Thomas
Scott, John Francis
Sherman, Martin
Stanley, Richard W
Stuart, William Dorsey
Sun, Samuel Sai-Ming
Tang, Chung-Shih
Van Reen, Robert
Vennesland, Birgit
Ward, Melvin A
Weems, Charles William
Yamamoto, Harry Y
Yasunobu, Kerry T

IDAHO
Armstrong, Marvin Douglas
Augustin, Jorg A L
Ayers, Arthur Raymond
Corsini, Dennis Lee
Crawford, Donald Lee
Crawford, Ronald Lyle
Dalton, Jack L

Biochemistry (cont)

Dreyfus, Pierre Marc
Dugan, Patrick R
Ellis, Robert William
Fontenelle, Lydia Julia
Hatcher, Herbert John
Keay, Leonard
LeTourneau, Duane John
McCune, Ronald William
Montoure, John Ernest
Mudumbi, Ramagopal Vijaya
Oliver, David John
Roberto, Francisco Figueroa
Robertson, Donald Claus
Rodnick, Kenneth Joseph
Tollefson, Charles Ivar
Watson, Kenneth Fredrick
Wiese, Alvin Carl
Winkel, Cleve R
Winston, Vern

ILLINOIS
Abbott, William Ashton
Agarwal, Kan L
Ahmad, Sarfraz
Aktipis, Stelios
Al-Ubaidi, Muayyad R
Amero, Sally Ann
Anderson, Byron
Anderson, David John
Anderson, Kenning M
Anderson, Louise Eleanor
Anderson, Robert Lewis
Applebury, Meredithe L
Argoudelis, Chris J
Ariano, Marjorie Ann
Armbruster, Frederick Carl
Arthur, Robert David
Aydelotte, Margaret Beesley
Baich, Annette
Baker, Harold Nordean
Baran, John Stanislaus
Barany, Michael
Baumann, Gerhard
Becker, John Henry
Becker, Michael Allen
Beecher, Christopher W W
Bekersky, Ihor
Bennett, Glenn Allen
Berman, Eleanor
Bernstein, Elaine Katz
Bernstein, Joel Edward
Berry, Robert Wayne
Betz, Robert F
Bezkorovainy, Anatoly
Bhattacharyya, Maryka Horsting
Bhatti, Neeloo
Bigelis, Ramunas
Bloomfield, Daniel Kermit
Blumenthal, Harold Jay
Bousquet, William F
Bowie, Lemuel James
Bradford, Marion McKinley
Braun, Donald Peter
Breen, Moira
Breillatt, Julian Paul, Jr
Brewer, Gregory J
Brigham, Robert B
Briskin, Donald Phillip
Brown, Peter H
Budzik, Gerald P
Buetow, Dennis Edward
Burkwall, Morris Paton, Jr
Burns, David John
Cabana, Veneracion Garganta
Calandra, Joseph Carl
Cammarata, Peter S
Casadaban, Malcolm John
Castignetti, Domenic
Ceithaml, Joseph James
Chambers, Donald A
Chandler, John W
Chassy, Bruce Matthew
Chatterton, Robert Treat, Jr
Chen, Wen Sherng
Cheung, Hou Tak
Christensen, Mary Lucas
Chuang, Tsan Iang
Chung, Jiwhey
Clark, John Magruder, Jr
Cohen, Isaac
Cole, Edmond Ray
Collins, Vernon Kirkpatrick
Conrad, Harry Edward
Constantinou, Andreas I
Coolidge, Thomas B
Cork, Douglas J
Cronan, John Emerson, Jr
Cross, John W
Cullum, Malford Eugene
Damaskus, Charles William
Dasler, Waldemar
Davis, Carl Lee
Davis, Peyton Nelson
Dawson, Glyn
Decker, Richard H
DeFilippi, Louis J
De La Huerga, Jesus
DePinto, John A
DeSombre, Eugene Robert
Devries, Arthur Leland
Dixit, Saryu N
Dodge, Patrick William

Domanik, Richard Anthony
Doughty, Clyde Carl
Drucker, Harvey
Druse-Manteuffel, Mary Jeanne
Druyan, Mary Ellen
Dubin, Alvin
Dudkiewicz, Alan Bernard
Dumas, Lawrence Bernard
Dunaway, George Alton, Jr
Ebrey, Thomas G
Eddy, Dennis Eugene
Edwards, Harold Herbert
Ehrenpreis, Seymour
Eisenhauer, Donald Alan
Emken, Edward Allen
Epstein, Wolfgang
Erve, Peter Raymond
Farnsworth, Wells Eugene
Feinstein, Robert Norman
Feldman, Fred
Fennewald, Michael Andrew
Ferrara, Louis W
Ferren, Larry Gene
Finnerty, James Lawrence
Fleisher, Lynn Dale
Flouret, George R
Foote, Carlton Dan
Ford, Susan Heim
Foster, Raymond Orrville
Fox, Jack Lawrence
Frankfater, Allen
Frenkel, Niza B
Friedman, Robert Bernard
Friedman, Yochanan
Friedmann, Herbert Claus
Fuchs, Elaine V
Gaballah, Saeed S
Gardner, Harold Wayne
Gassman, Merrill Loren
Gershbein, Leon Lee
Gettins, Peter Gregory Wolfgang
Getz, Godfrey S
Giere, Frederic Arthur
Giometti, Carol Smith
Glaser, Janet H
Glaser, Michael
Glogovsky, Robert L
Goldberg, Erwin
Goldman, Allen S
Goldstone, Alfred D
Goldwasser, Eugene
Gratton, Enrico
Graves, Charles Norman
Greene, Frank Clemson
Griffin, Martin John
Gross, Martin
Gross, Thomas Lester
Gumport, Richard I
Gunsalus, Irwin Clyde
Haas, Mark
Hac, Lucile R
Hagar, Lowell Paul
Hager, Lowell Paul
Hainline, Adrian, Jr
Halfman, Clarke Joseph
Hampel, Arnold E
Hanlon, Mary Sue
Hanly, W Carey
Harrison, William Henry
Hass, George Michael
Hathaway, Robert J
Haugen, David Allen
Hauptmann, Randal Mark
Hawkins, Richard Albert
Hawrylewicz, Ervin J
Hayashi, James Akira
Helbert, James Raymond
Held, Irene Rita
Heller, Alfred
Hendershot, William Fred
Henderson, Thomas Otis
Hiles, Richard Allen
Hiltibran, Robert Comegys
Himoe, Albert
Hjelle, Joseph Thomas
Hoerman, Kirk Conklin
Holbrook, Gabriel Peter
Holland, Louis Edward, II
Holleman, William H
Honig, George Raymond
Horwitz, Alan Fredrick
Hou, Ching-Tsang
Huffman, George Wallen
Hurley, Walter L
Ingle, Morton Blakeman
Iqbal, Zafar
Jackson, Richard W
Jacob, Samson T
Jaffe, Randal Craig
Javaid, Javaid Iqbal
Jeffay, Henry
Jiu, James
Johnson, Richard Joseph
Jonas, Ana
Jungmann, Richard A
Kagan, Jacques
Kamath, Savitri Krishna
Kanabrocki, Eugene Ladislaus
Kanofsky, Jeffrey Ronald
Kapsalis, Andreas A
Karr, Timothy Lawrence
Kass, Leon Richard
Kassner, Richard J
Kathan, Ralph Herman

Katz, Adrian I
Kaufman, Stephen J
Kemp, Robert Grant
Kempner, David H
Khan, Mohammed Abdul Quddus
Kim, Yung Dai
Kingdon, Henry Shannon
Klegerman, Melvin Earl
Klopfenstein, William Elmer
Klubek, Brian Paul
Knight, Katherine Lathrop
Koch, Elizabeth Anne
Kuczmarski, Edward R
Kuettner, Klaus E
Kumar, Sudhir
Kyncl, J Jaroslav
Lakshminarayanan, Krishnaiyer
Lamb, Robert Andrew
Lambert, Glenn Frederick
Lambert, Mary Pulliam
Lamberts, Burton Lee
Lange, Charles Ford
Lange, Yvonne
Lanyon, Scott Merril
Larson, Bruce Linder
Layman, Donald Keith
Leven, Robert Maynard
Levin, Samuel Joseph
L'Heureux, Maurice Victor
Liao, Shutsung
Loach, Paul A
Lobstein, Otto Ervin
Lockhart, Haines Boots
Lopatin, William
Lopes, John Manuel
Lorand, Laszlo
Lucher, Lynne Annette
Lushbough, Channing Harden
McCandless, David Wayne
McCorquodale, Donald James
McDonald, Hugh Joseph
Mackal, Roy Paul
Madera-Orsini, Frank
Madigan, Michael Thomas
Madsen, David Christy
Maier, George D
Mallia, Anantha Krishna
Marcotte, Patrick Allen
Margoliash, Emanuel
Markovitz, Alvin
Marr, James Joseph
Martin, Charles J
Masken, James Frederick
Matta, Michael Stanley
Mattenheimer, Hermann G W
Mehta, Rajendra G
Meredith, Stephen Charles
Miernyk, Jan Andrew
Miller, Charles G
Miller, James Edward
Miller, Robert Verne
Mistry, Sorab Pirozshah
Miyazaki, John H
Mohberg, Joyce
Molnar, Janos
Moore, Edwin Granville
Morley, Colin Godfrey Dennis
Moskal, Joseph Russell
Muck, George A
Musa, Wafaa Arafat
Myatt, Elizabeth Anne
Myers, Ronald Berl
Nakagawa, Yasushi
Nakamoto, Tokumasa
Needleman, Saul Ben
Neet, Kenneth Edward
Neuhaus, Francis Clemens
Nichols, Brian Paul
Nolan, Chris
Norris, Frank Arthur
Novak, Robert Louis
Olsen, Kenneth Wayne
Ordal, George Winford
Ort, Donald Richard
Oshiro, Yuki
Ostrow, David Henry
Padh, Harish
Pai, Sadanand V
Pal, Dhiraj
Palmer, Warren K
Papaioannou, Stamatios E
Papatheofanis, Frank John
Pasterczyk, William Robert
Pekarek, Robert Sidney
Pepperberg, David Roy
Peraino, Carl
Perlman, Robert
Peterson, Rudolph Nicholas
Plate, Charles Alfred
Plate, Janet Margaret
Portis, Archie Ray, Jr
Potempa, Lawrence Albert
Preston, Robert Leslie
Pun, Pattle Pak-Toe
Rackis, Joseph John
Radzialowski, Frederick M
Rafelson, Max Emanuel, Jr
Rao, Gopal Subba
Rao, Mrinalini Chatta
Rasenick, Mark M
Rausch, David John
Reith, Maarten E A
Ressler, Newton
Reynolds, Robert David

Richmond, Patricia Ann
Ringler, Ira
Ritter, Nadine Marie
Robbins, Kenneth Carl
Robinson, James Lawrence
Rogalski-Wilk, Adrienne Alice
Rotermund, Albert J, Jr
Rothman-Denes, Lucia B
Rowe, William Bruce
Rupprecht, Kevin Robert
Sargent, Malcolm Lee
Savidge, Jeffrey Lee
Scanu, Angelo M
Scheid, Harold E
Schenck, Jay Ruffner
Scherberg, Neal Harvey
Schlenk, Fritz
Schlueter, Robert J
Schnell, Gene Wheeler
Schultz, Richard Michael
Schumacher, Gebhard Friederich B
Schumer, William
Scott, Don
Seed, Randolph William
Sehgal, Lakshman R
Seidenfeld, Jerome
Sellers, Donald Roscoe
Shambaugh, George E, III
Shapiro, David Jordon
Shapiro, Stanley Kallick
Shaw, Paul Dale
Shen, Linus Liang-nene
Sievert, Herman William
Silver, Simon David
Simon, Selwyn
Simonson, Lloyd Grant
Sky-Peck, Howard H
Sligar, Stephen Gary
Slodki, Morey Eli
Smith, Steven Joel
Smouse, Thomas Hadley
Snyder, William Robert
Sorensen, Keld
Sorensen, Leif Boge
Spangler, Brenda Dolgin
Spitzer, Robert Harry
Splittstoesser, Walter E
Stark, Benjamin Chapman
Steck, Theodore Lyle
Stein, Herman H
Steiner, Donald Frederick
Storti, Robert V
Straus, Werner
Strauss, Bernard S
Struthers, Barbara Joan Oft
Stumpf, David Allen
Subbaiah, Papasani Venkata
Suzue, Ginzaburo
Switzer, Robert Lee
Sze, Paul Yi Ling
Tao, Mariano
Telser, Alvin Gilbert
Thompson, David Jerome
Thompson, Richard Edward
Thomson, John Ferguson
Thonar, Eugene Jean-Marie
Thorp, Frank Kedzie
Tiemeier, David Charles
Titchener, Edward Bradford
Toback, F(rederick) Gary
Tolbert, Margaret Ellen Mayo
Tomita, Joseph Tsuneki
Tookey, Harvey Llewellyn
Tripathi, Satish Chandra
Tumbleson, M(yron) E(ugene)
Turnquist, Richard Lee
Van de Kar, Louis David
Van Fossan, Donald Duane
Van Kley, Harold
Vary, James Corydon
Vary, Patricia Susan
Veis, Arthur
Vodkin, Lila Ott
Wade, David Robert
Walter, Robert John
Walter, Trevor John
Wang, Andrew H-J
Wang, Gary T
Wang, Hwa Lih
Wang, Li Chuan
Wasserman, William John
Wawszkiewicz, Edward John
Weber, Gregorio
Webster, Dale Arroy
Weinstein, Hyman Gabriel
Weiss, Samuel Bernard
Weissmann, Bernard
Welker, Neil Ernest
Wells, Warren F
Westbrook, Edwin Monroe
Westley, John Leonard
Weyhenmeyer, James Alan
Whisler, Walter William
Wiatr, Christopher Louis
Wideburg, Norman Earl
Wilbraham, Antony Charles
Wilkinson, Brian James
Williams-Ashman, Howard Guy
Wilson, Curtis Marshall
Wilson, Donald Alan
Wilson, Richard Hansel
Winkler, Martin Alan
Witmer, Heman John
Wittman, James Smythe, III

Wolf, Walter J
Womble, David Dale
Wong, Paul Wing-Kon
Wool, Ira Goodwin
Wu, Anna Fang
Wu, Tai Te
Wu-Wong, Jinshyun Ruth
Yamamoto, Hirotaka
Yu, Fu-Li
Zahalsky, Arthur C
Zak, Radovan Hynek
Zaneveld, Lourens Jan Dirk
Zaroslinski, John F

INDIANA
Ali, Rida A
Allmann, David William
Anderson, David Bennett
Ashendel, Curtis Lloyd
Asteriadis, George Thomas, Jr
Axelrod, Bernard
Baker, A Leroy
Baker, Timothy Stanley
Baldwin, William Walter
Barnhart, James William
Basu, Manju
Basu, Subhash Chandra
Bauer, Robert
Becker, Benjamin
Beeson, W Malcolm
Belagaje, Rama M
Beranek, William, Jr
Bhatti, Waqar Hamid
Biggs, Homer Gates
Blair, Paul V
Bobbitt, Jesse LeRoy
Borglum, Gerald Baltzer
Bosron, William F
Bowman, Donald Edwin
Boyer, Ernest Wendell
Brandt, Karl Garet
Bretthauer, Roger K
Bridges, C David
Brodman, Robert David
Bromer, William Wallis
Bryan, William Phelan
Bumpus, John Arthur
Butler, Larry G
Carmichael, Ralph Harry
Carrico, Robert Joseph
Castellino, Francis Joseph
Chance, Ronald E
Chandrasekaran, Rengaswami
Christner, James Edward
Clevenger, Sarah
Cole, Thomas A
Coolbaugh, Ronald Charles
Corrigan, John Joseph
Crane, Frederick Loring
Dantzig, Anne H
DeLong, Allyn F
Diamond, Steven Elliot
Diller, Erold Ray
Dilley, Richard Alan
Dillon, John Joseph, III
Donner, David Bruce
Donoho, Alvin Leroy
Duman, John Girard
Dunkle, Larry D
Dunn, Peter Edward
Eble, John Nelson
Edenberg, Howard Joseph
Edmundowicz, John Michael
Elkin, Robert Glenn
Enders, George Leonhard, Jr
Evans, Michael Allen
Fayerman, Jeffrey T
Fayle, Harlan Downing
Filmer, David Lee
Fong, Shao-Ling
Fox, Owen Forrest
Frank, Bruce Hill
Free, Alfred Henry
Frolik, Charles Alan
Fuchs, Morton S
Fuller, Ray W
Gallo, Duane Gordon
Gantzer, Mary Lou
Gardner, David Arnold
Gaur, Pramod Kumar
Gehlert, Donald Richard
Ghosh, Swapan Kumar
Gibson, David Mark
Gilham, Peter Thomas
Gledhill, Robert Hamor
Goetz, Frederick William, Jr
Goff, Charles W
Goh, Edward Hua Seng
Golab, Tomasz
Goldstein, David Joel
Gray, Peter Norman
Gunther, Gary Richard
Guthrie, George Drake
Hagen, Richard Eugene
Hancock, Deana Lori
Harper, Edwin T
Harris, Robert Allison
Hayes, Donald Charles
Hegeman, George D
Heinicke, Herbert Raymond
Heinstein, Peter
Henry, Harry James
Hermodson, Mark Allen
Herr, Earl Binkley, Jr
Herrmann, Klaus Manfred
Hershberger, Charles Lee
Hidy, Phil Harter
Ho, Nancy Wang-Yang
Ho, Peter Peck Koh
Hsiung, Hansen M
Hudock, George Anthony
Hyde, David Russell
Ingraham, Joseph Sterling
Irwin, William Elliot
Jenkins, Winborne Terry
Jilka, Robert Laurence
Johnson, Eric Richard
Kauffman, Raymond F
Kempson, Stephen Allan
Kim, Ki-Han
Kirksey, Avanelle
Kleinschmidt, Walter John
Kohlhaw, Gunter B
Koritnik, Donald Raymond
Krogmann, David William
Ladisch, Michael R
Lantero, Oreste John, Jr
Larsen, Steven H
Li, Ting Kai
Light, Albert
Lin, Renee C
Lindstrom, Terry Donald
Litov, Richard Emil
Loesch-Fries, Loretta Sue
Loudon, Gordon Marcus
Low, Philip Stewart
Lumeng, Lawrence
Lutz, Wilson Boyd
McBride, William Joseph
McDonald, James Lee, Jr
McMullen, James Robert
Mahoney, Walter C
Malacinksi, George M
Marquis, Norman Ronald
Marshall, James John
Mason, Norman Ronald
Matsumoto, Charles
Mertz, Edwin Theodore
Meyers, Michael Clinton
Millar, Wayne Norval
Moorehead, Wells Rufus
Morre, D James
Muhler, Joseph Charles
Murphy, Patrick Joseph
Natarajan, Viswanathan
Niederpruem, Donald J
Nielsen, Niels Christian
Nordschow, Carleton Deane
Nowak, Thomas
Ockerse, Ralph
Oster, Mark Otho
Ostroy, Sanford Eugene
Pace, Norman R
Parker, Herbert Edmund
Pettinga, Cornelius Wesley
Pierce, William Meredith, Jr
Probst, Gerald William
Proksch, Gary J
Puski, Gabor
Putnam, Frank William
Quackenbush, Forrest Ward
Queener, Sherry Fream
Queener, Stephen Wyatt
Ragheb, Hussein S
Rand, Phillip Gordon
Raveed, Dan
Reazin, George Harvey, Jr
Regnier, Frederick Eugene
Roach, Peter John
Rodwell, Victor William
Rogers, Kenneth Scipio
San Pietro, Anthony
Saz, Howard Jay
Schneider, Donald Louis
Schulz, Arthur R
Shaw, Walter Norman
Sherman, Louis Allen
Shields, James Edwin
Siakotos, Aristotle N
Sipe, Jerry Eugene
Skjold, Arthur Christopher
Smith, Robert William
Somerville, Ronald Lamont
Steinrauf, Larry King
Stevenson, William Campbell
Stier, Theodore James Blanchard
Stiller, Mary Louise
Stillwell, William Harry
Story, Jon Alan
Subramanian, Sethuraman
Surzycki, Stefan Jan
Swain, Richard Russell
Szuhaj, Bernard F
Taylor, Harold Leland
Termine, John David
Tischfield, Jay Arnold
Tsai, Chia-Yin
Tunnicliff, Godfrey
Van Frank, Richard Mark
Vogelhut, Paul Otto
Wagner, Eugene Stephen
Wallander, Jerome F
Wegener, Warner Smith
Weiner, Henry
Weith, Herbert Lee
White, Harold Keith
Wildfeuer, Marvin Emanuel
Williams, Gene R
Wong, David Taiwai
Wostmann, Bernard Stephan
Wright, Walter Eugene
Yan, Sau-Chi Betty
Yen, Terence Tsin Tsu
Yoder, John Menly
Young, Peter Chun Man
Zalkin, Howard
Zilz, Melvin Leonard
Zygmunt, Walter A

IOWA
Anderson, Lloyd L
Arnone, Arthur Richard
Ascoli, Mario
Benbow, Robert Michael
Bidlack, Wayne Ross
Bishop, Stephen Hurst
Bosch, Arthur James
Cain, George D
Campbell, Kevin Peter
Cavalieri, Anthony Joseph, II
Cazin, John, Jr
Celander, Evelyn Faun
Cheng, Frank Hsieh Fu
Christiansen, James Brackney
Clark, Robert A
Colilla, William
Conway, Thomas William
Cook, Robert Thomas
Crosby, Lon Owen
Cunningham, Bryce A
Dahm, Paul Adolph
Davis, Leodis
Donelson, John Everett
Dordick, Jonathan Seth
Downing, Donald Talbot
Dutton, Gary Roger
Enger, Merlin Duane
Fellows, Robert Ellis, Jr
Franklin, Robert Louis
Fromm, Herbert Jerome
Garbutt, John Thomas
Gibson, David Thomas
Goodridge, Alan G
Graves, Donald J
Greenberg, Everett Peter
Grigsby, William Redman
Gussin, Gary Nathaniel
Hammond, Earl Gullette
Horowitz, Jack
Horst, Ronald Lee
Johnson, Lawrence Alan
Kalnitsky, George
Keeney, Dennis Raymond
Kintanar, Agustin
Koerner, Theodore Alfred William, Jr
Koppelman, Ray
Lara-Braud, Carolyn Weathersbee
Lata, Gene Frederick
Lewis, Douglas Scott
Lillehoj, Eivind B
Lim, Ramon (Khe Siong)
Mahoney, Joan Munroe
Makar, Adeeb Bassili
Markovetz, Allen John
Masat, Robert James
Menninger, John Robert
Metzler, David Everett
Montgomery, Rex
Morehouse, Alpha L
Norcia, Leonard Nicholas
Oliver, Denis Richard
Olson, James Allen
Plapp, Bryce Vernon
Rebouche, Charles Joseph
Rebstock, Theodore Lynn
Rhead, William James
Richter, Erwin (William)
Robson, Richard Morris
Robyt, John F
Schmerr, Mary Jo F
Schmidt, Thomas John
Schottelius, Dorothy Dickey
Shires, Thomas Kay
Southard, Wendell Homer
Spector, Arthur Abraham
Speer, Vaughn C
Stanton, Thaddeus Brian
Stegink, Lewis D
Stellwagen, Earle C
Stewart, Mark Armstrong
Stewart, Mary E
Stinski, Mark Francis
Stromer, Marvin Henry
Swan, Patricia B
Tabatabai, Louisa Braal
Thomas, Byron Henry
Thomas, James Arthur
Tipton, Carl Lee
Wang, Wei-Yeh
Weisman, Lois Sue
Wertz, Philip Wesley
White, Bernard J
Williams, Phletus P

KANSAS
Arnold, Wilfred Niels
Bechtel, Donald Bruce
Bednekoff, Alexander G
Bode, Vernon Cecil
Bradford, Lawrence Glenn
Brown, John Clifford
Burkhard, Raymond Kenneth
Calvet, James P
Carr, Daniel Oscar
Chitnis, Parag Ram
Clegg, Robert Edward
Clevenger, Richard Lee
Cunningham, Franklin E
Dass, David
Davis, John Stewart
Davis, Lawrence Clark
Deyoe, Charles W
Dirks, Brinton Marlo
Ebner, Kurt E
ElAttar, Tawfik Mohammed Ali
Ericson, Alfred (Theodore)
Funderburgh, James Louis
Gegenheimer, Peter Albert
Grady, Harold James
Grisolia, Santiago
Grunewald, Katharine Klevesahl
Guikema, James Allen
Harris, Lewis Philip
Hedgcoth, Charlie, Jr
Helmkamp, George Merlin, Jr
Himes, Richard H
Hochman, Jerome Henry
Hopkins, Theodore Louis
Hsu, Howard Huai Ta
Hudson, Billy Gerald
Johnson, Donovan Earl
Kimmel, Joe Robert
Kinsey, William Henderson
Kitos, Paul Alan
Kramer, Karl Joseph
Lanman, Robert Charles
Leavitt, Wendell William
MacGregor, Ronal Roy
Manning, Robert Thomas
Marchin, George Leonard
Michaelis, Elias K
Michaelis, Mary Louise
Mills, Russell Clarence
Mulford, Dwight James
Murphy, John Joseph
Neufeld, Gaylen Jay
Noelken, Milton Edward
Nordin, Philip
Oppert, Brenda
Padmanabhan, Radhakrishnan
Parkinson, Andrew
Parrish, Donald Baker
Parrish, John Wesley, Jr
Ramamurti, Krishnamurti
Rawitch, Allen Barry
Reeck, Gerald Russell
Reeves, Robert Donald
Rhodes, James B
Roche, Thomas Edward
Roufa, Donald Jay
Rulifson, Willard Sloan
Sanders, Robert B
Schloss, John Vinton
Schmidt, Robert W
Seib, Paul A
Shogren, Merle Dennis
Silverstein, Richard
Smith, Larry Dean
Soodsma, James Franklin
Stetler, Dean Allen
Suzuki, Tsuneo
Takemoto, Dolores Jean
Timberlake, Joseph William
Weaver, Robert F
Wilson, George Spencer
Wong, Peter P
Yarbrough, Lynwood R
Young, George Robert
Zimniski, Stephen Joseph

KENTUCKY
Aleem, M I Hussain
Andersen, Roger Allen
Benz, Frederick W
Bhattacharya, Malaya
Bhattacharya-Chatterjee, Malaya
Bondada, Subbarao
Brown, John Wesley
Chan, Shung Kai
Cheniae, George Maurice
Cohn, David Valor
Collins, Delwood C
Dallam, Richard Duncan
Davidson, Jeffrey Neal
Dean, William L
Dickson, Robert Carl
Dunham, Valgene Loren
Farrar, William Wesley
Fell, Ronald Dean
Fonda, Margaret Lee
Futrell, Mary Feltner
Galardy, Richard Edward
Geoghegan, Thomas Edward
Goodman, Norman L
Gray, Robert Dee
Haley, Boyd Eugene
Hammond, Ray Kenneth
Hersh, Louis Barry
Hilton, Mary Anderson
Hu, Alfred Soy Lan
Humphreys, Wallace F
Jacobson, Elaine Louise
Jacobson, Myron Kenneth
Jayaram, Beby
Kargl, Thomas E
Kasarskis, Edward Joseph

Biochemistry (cont)

Kennedy, John Elmo, Jr
Knapp, Frederick Whiton
Kuc, Joseph
Lang, Calvin Allen
Lester, Robert Leonard
Levy, Robert Sigmund
Lowe, Richie Howard
McConnell, Kenneth Paul
McGeachin, Robert Lorimer
Maisch, Weldon Frederick
Mandelstam, Paul
Martin, Nancy Caroline
Mattson, Mark Paul
Noland, Jerre Lancaster
Otero, Raymond B
Packett, Leonard Vasco
Panemangalore, Myna
Pavlik, Edward John
Perlin, Michael Howard
Petering, Harold George
Prough, Russell Allen
Rao, Chalamalasetty Venkateswara
Rawls, John Marvin, Jr
Rees, Earl Douglas
Richard, John P
Rosenthal, Gerald A
Rothwell, Frederick Mirvan
Schurr, Avital
Schwert, George William
Sisken, Jesse Ernest
Slagel, Donald E
Staat, Robert Henry
Steiner, Marion Rothberg
Taylor, John Fuller
Teller, David Norton
Toman, Frank R
Vanaman, Thomas Clark
Wheeler, Thomas Jay
Williams, Arthur Lee
Winer, Alfred D
Wittliff, James Lamar

LOUISIANA

Alam, Bassima Saleh
Alam, Jawed
Alam, Syed Qamar
Bernofsky, Carl
Bobbin, Richard Peter
Bryan, Sara E
Burch, Robert Emmett
Byers, Larry Douglas
Caday, Cornelio Gacusana
Cauthen, Sally Eugenia
Chang, Simon H
Chen, Hoffman Hor-Fu
Claycomb, William Creighton
Clejan, Sandra
Clemetson, Charles Alan Blake
Cohen, William
Cohn, Marc Alan
Coulson, Roland Armstrong
Day, Donal Forest
Dessauer, Herbert Clay
Deutsch, Walter A
DiMaggio, Anthony, III
Dunn, Adrian John
Dupuy, Harold Paul
Ehrlich, Kenneth Craig
Ehrlich, Melanie
Gottlieb, A Arthur
Grimes, Sidney Ray, Jr
Griswold, Kenneth Edwin, Jr
Guthrie, John Daulton
Hamori, Eugene
Hart, Lewis Thomas
Hegsted, Maren
Henderson, Ralph Joseph, Jr
Herbert, Jack Durnin
Hill, Franklin D
Hill, James Milton
Huggins, Clyde Griffin
Hyde, Paul Martin
Jacks, Thomas Jerome
Jain, Sushil Kumar
Jazwinski, S Michal
Kastl, Peter Robert
Kennedy, Frank Scott
Kokatnur, Mohan Gundo
Kuck, James Chester
Laine, Roger Allan
Lartigue, Donald Joseph
Lee, Jordan Grey
Li, Su-Chen
Li, Yu-Teh
Lopez-Santolino, Alfredo
Lucas, Myron Cran
Mahajan, Damodar K
Marshall, Wayne Edward
Marvel, John Thomas
Miceli, Michael Vincent
Mokrasch, Lewis Carl
Moore, Thomas Stephen, Jr
Nair, C(hellappan) Rajagopalan
Newsome, David Anthony
Olmsted, Clinton Albert
Olubadewo, Joseph Olanrewaju
Ontko, Joseph Andrew
Ory, Robert Louis
Palmgren, Muriel Signe
Pryor, William Austin
Radhakrishnamurthy, Bhandaru
Reed, Brent C

Rhoads, Robert E
Rogers, Robert Larry
Rogers, Stearns Walter
Roskoski, Robert, Jr
Rudolph, Guilford George
Ruffin, Spaulding Merrick
Sarphie, Theodore G
Scher, Charles D
Schoellmann, Guenther
Settoon, Patrick Delano
Shedlarski, Joseph George, Jr
Shepherd, Hurley Sidney
Shiau, Yih-Fu
Sinha, Sudhir K
Smith, Robert Lewis
Spanier, Arthur M
Spitzer, Judy A
Srinivasan, Sathanur Ramachandran
Srinivasan, Vadake Ram
Stanfield, Manie K
Steele, Richard Harold
Stjernholm, Rune Leonard
Swenson, David Harold
Teekell, Roger Alton
Tou, Jen-sie Hsu
Tso, Patrick Po-Wing
Ullah, Abul Hasnat
Upadhyay, Jagdish M
Vedeckis, Wayne V
Vercellotti, John R
Vijayagopal, Parakat
Wang, Ting Chung
Welbourne, Tomas C
Welch, George Rickey
Wiser, Mark Frederick
Woodring, Joseph
Yeh, Kwo-Yih
Yerrapraggada, Venkat
York, David Anthony
Younathan, Ezzat Saad

MAINE

Barton, Barbara Ann
Borei, Hans Georg
Bushway, Rodney John
Cook, Richard Alfred
De Haas, Herman
Goodman, Irving
Greenwood, Paul Gene
Hill, Marquita K
Howland, John LaFollette
Kandutsch, Andrew August
Kepron, Michael Raymond
Kozak, Leslie P
McKerns, Kenneth (Wilshire)
Manyan, David Richard
Paigen, Beverly Joyce
Paigen, Kenneth
Ridgway, George Junior
Roxby, Robert
Settlemire, Carl Thomas
Sherblom, Anne P
Sidell, Bruce David
Steinhart, William Lee
Waymouth, Charity
Yonuschot, Gene R

MARYLAND

Abergel, Chantel
Ackerman, Eric J
Adams, James Miller
Adams, Junius Greene, III
Adamson, Richard H
Adelstein, Robert Simon
Ades, Ibrahim Z
Adrouny, George Adour (Kuyumjian)
Ahluwalia, Gurpreet S
Ahmed, Syed Ashrafuddin
Aksamit, Robert Rosooe
Albers, Robert Wayne
Aldridge, Mary Hennen
Allison, Richard Gall
Alsmeyer, Richard Harvey
Alvares, Alvito Peter
Amende, Lynn Meridith
Amr, Sania
Amzel, L Mario
Anderson, Larry Douglas
Anderson, Lucy Macdonald
Anderson, Norman Leigh
Anderson, Richard Allen
Anfinsen, Christian Boehmer
Anhalt, Grant James
Ansher, Sherry Singer
Argraves, W Scott
Ashwell, G Gilbert
August, Joseph Thomas
Augustine, Patricia C
Austin, Faye Carol Gould
Avigan, Joel
Bachur, Nicholas R, Sr
Baker, Carl Gwin
Baker, George Thomas, III
Balinsky, Doris
Ballentine, Robert
Barban, Stanley
Bareis, Donna Lynn
Baron, Louis Sol
Bashirelahi, Nasir
Baum, Bruce J
Beall, Robert Joseph
Beaven, Michael Anthony
Beaven, Vida Helms
Beeler, Troy James

Behar, Marjam Gojchlerner
Bellino, Francis Leonard
Benton, Allen William
Berger, Edward Alan
Berger, Robert Lewis
Berger, Shelby Louise
Berlin, Elliott
Berman, Howard Mitchell
Bessman, Maurice Jules
Bhathena, Sam Jehangirji
Bhatnagar, Gopal Mohan
Bieri, John Genther
Bigger, Cynthia Anita Hopwood
Bitman, Joel
Blackman, Marc Roy
Bleecker, Margit
Bloch, Robert Joseph
Blosser, Timothy Hobert
Bluhm, Leslie
Blum, Stanley Walter
Blumberg, Peter Mitchell
Blumenthal, Herbert
Blumenthal, Robert Paul
Boldt, Roger Earl
Bollum, Frederick James
Bonifacino, Juan S
Bono, Vincent Horace, Jr
Boone, Charles Walter
Borkovec, Alexej B
Brady, Roscoe Owen
Brand, Ludwig
Breitman, Theodore Ronald
Brewer, H Bryan, Jr
Brink, Linda Holk
Broomfield, Clarence A
Brown, Alexandra Louise
Bruns, Herbert Arnold
Bunger, Rolf
Burge, Wylie D
Bustin, Michael
Butcher, Henry Clay, IV
Butzow, James J
Cabib, Enrico
Cain, Dennis Francis
Calvert, Richard John
Carrico, Christine Kathryn
Cassman, Marvin
Catravas, George Nicholas
Catt, Kevin John
Cerveny, Thelma Jannette
Chader, Gerald Joseph
Chakrabarti, Siba Gopal
Chan, Daniel Wan-Yui
Chandra, G Ram
Chang, Lucy Ming-Shih
Chang, Michael C J
Chang, Yung-Feng
Chappelle, Emmett W
Chatterjee, Subroto
Chattoraj, Dhruba Kumar
Chaudhari, Anshumali
Chaykovsky, Michael
Chen, Chung-Ho
Chen, Hao-Chia
Chen, Winston Win-Hower
Chernick, Sidney Samuel
Chirigos, Michael Anthony
Chitwood, David Joseph
Cho-Chung, Yoon Sang
Chock, P Boon
Choi, Oksoon Hong
Chrambach, Andreas C
Chuang, De-Maw
Cimbala, Michele
Clark, Patricia
Coffey, Donald Straley
Cohen, Robert Martin
Colburn, Nancy Hall
Collins, John Henry
Colombani, Paul Michael
Colombini, Marco
Condliffe, Peter George
Constantopoulos, George
Cookson, John Thomas, Jr
Cornblath, Marvin
Costlow, Richard Dale
Cox, George Warren
Craig, Nessly Coile
Creighton, Donald John
Creveling, Cyrus Robbins, Jr
Cushman, Samuel Wright
Cutler, Gordon Butler, Jr
Cysyk, Richard L
Dacre, Jack Craven
Daly, John William
Danielpour, David
Darby, Eleanor Muriel Kapp
Das, Saroj R
Dasch, Gregory Alan
Davidian, Nancy McConnell
Davidson, Harold Michael
Davis, Julien Sinclair
Dawid, Igor Bert
Dean, Ann
Dean, Donna Joyce
De Luca, Luigi Maria
Devrcotes, Peter Nicholas
Dhyse, Frederick George
Dienel, Gerald Arthur
Dingman, Charles Wesley, II
Dintzis, Renee Zlochover
Domanski, Thaddeus John
Doukas, Harry Michael
Dudley, Peter Anthony

Dufau, Maria Luisa
Dunaway-Mariano, Debra
Dupont, Jacqueline (Louise)
Dwyer, Dennis Michael
Eaton, Barbra L
Edidin, Michael Aaron
Egan, William Michael
Eipper, Betty Anne
Eisenberg, Evan
Eisenberg, Frank, Jr
Englund, Paul Theodore
Evarts, Ritva Poukka
Fakunding, John Leonard
Falkenstein, Kathy Fay
Farrar, William L
Farrelly, James Gerard
Feldlaufer, Mark Francis
Ferguson, Earl Wilson
Ferguson, James Joseph, Jr
Fernie, Bruce Frank
Filner, Barbara
Filner, Philip
Finkelstein, David B
Finlayson, John Sylvester
Fire, Andrew Zachary
Fishman, Peter Harvey
Flanders, Kathleen Corey
Flavin, Martin
Folk, John Edward
Fornace, Albert J, Jr
Foulds, John Douglas
Fox, Barbara Saxton
Frank, Leonard Harold
Franklin, Renty Benjamin
Freire, Ernesto I
French, Richard Collins
Friedman, Leonard
Fujimura, Robert
Furfine, Charles Stuart
Gabay, Sabit
Gabriel, R Othmar
Gallo, Robert C
Gamble, James Lawder, Jr
Gantt, Ralph Raymond
Gelboin, Harry Victor
Gellert, Martin Frank
Gerard, Gary Floyd
Gerin, John Louis
Gerlt, John Alan
Gerwin, Brenda Isen
Gherna, Robert Larry
Gidez, Lewis Irwin
Ginsburg, Ann
Ginsburg, Victor
Glassman, Harold Nelson
Glazer, Robert Irwin
Goldenbaum, Paul Ernest
Goldman, David
Goldman, Stephen Shepard
Goldsmith, Paul Kenneth
Goldstein, Allan L
Golumbic, Calvin
Goor, Ronald Stephen
Gorden, Phillip
Gordon, Nathan
Graf, Lloyd Herbert
Graham, Dale Elliott
Gram, Theodore Edward
Granger, Donald Lee
Gravell, Maneth
Green, Marie Roder
Gregory, John Delafield
Gross, Kenneth Charles
Grossman, Lawrence
Gruber, Kenneth Allen
Gryder, Rosa Meyersburg
Guchhait, Ras Bihari
Guerguian, John Leo
Gupte, Sharmila Shaila
Guttman, Helene Augusta Nathan
Habig, William Henry
Hadidi, Ahmed Fahmy
Haggerty, James Francis
Hall, Richard Leland
Hallfrisch, Judith
Hamer, Dean H
Handwerker, Thomas Samuel
Hanig, Joseph Peter
Hanover, John Allan
Hansen, John Norman
Hansford, Richard Geoffrey
Harmon, Joan T
Harrison, Helen Coplan
Hayes, Dora Kruse
Hayes, Joseph Edward, Jr
Hearing, Vincent Joseph, Jr
Hegyeli, Ruth I E J
Heineman, Frederick W
Hejtmancik, James Fielding
Helmsen, Ralph John
Heming, Arthur Edward
Henderson, Louis E
Hendler, Richard Wallace
Henkart, Pierre
Herbert, Elton Warren, Jr
Herbst, Edward John
Herman, Eliot Mark
Herrett, Richard Allison
Herriott, Roger Moss
Hess, Helen Hope
Hickey, Robert Joseph
Hilmoe, Russell J(ulian)
Hobbs, Ann Snow
Hoffman, David J

Holloway, Caroline T
Holmlund, Chester Eric
Horner, William Harry
Horowitz, Jill Ann
Housewright, Riley Dee
Howe, Juliette Coupain
Huang, Charles Y
Huang, Kuo-Ping
Huang, Pien-Chien
Hubbard, Van Saxton
Huff, Sheela J
Hurd, Suzanne Sheldon
Ikeda, George J
Impraim, Chaka Cetewayo
Irwin, David
Jacobowitz, David
Jacobs, Abigail Conway
Jacobson, Kenneth Alan
Jakoby, William Bernard
Jamieson, Graham Archibald
Jandorf, Bernard Joseph
Jerina, Donald M
Johns, David Garrett
Johnson, Carl Boone
Johnson, David Freeman
Johnson, George S
Johnson-Winegar, Anna
Jones, Theodore Charles
Kao, Kung-Ying Tang
Kaper, Jacobus M
Kaplan, Emanuel
Kaplan, Stanley A
Karpel, Richard Leslie
Kaufman, Bernard Tobias
Kaufman, Elaine Elkins
Kaufman, Seymour
Kawalek, Joseph Casimir, Jr
Kearney, Philip C
Keister, Donald Lee
Keith, Jerry M
Keller, George Henry
Kelley, John Francis
Kelly, Thomas J
Ketley, Jeanne Nelson
Kibbey, Maura Christine
Kidwell, William Robert
Kim, Sooja K
Kimball, Amy Sarah
Kindt, Thomas James
Kingsbury, David Wilson
Kippenberger, Donald Justin
Kitzes, George
Klausner, Richard D
Kleinman, Hynda Karen
Kole, Hemanta Kumar
Komoriya, Akira
Korn, Edward David
Koval, Thomas Michael
Krag, Sharon S
Krakauer, Teresa
Krause, David
Kraybill, Herman Fink
Kresina, Thomas Francis
Krichevsky, Micah I
Kroll, Martin Harris
Kuff, Edward Louis
Kumaroo, Kuziyilethu Krishnan
Kundig, Fredericka Dodyk
Kunos, George
Kwiterovich, Peter O, Jr
Lai, Chun-Yen
Lakowicz, Joseph Raymond
Lambooy, John Peter
Lands, William Edward Mitchell
Landsman, David
Lane, Malcolm Daniel
Langlykke, Asger Funder
Leder, Irwin Gordon
Lee, Chi-Jen
Lee, Fang-Jen Scott
Lee, Theresa
Lee, Yuan Chuan
Lee, Yuen San
Leiter, Joseph
Leonard, Charles Brown, Jr
Leppla, Stephen Howard
Lesko, Stephen Albert
Lester, David Simon
Levander, Orville Arvid
Levenbook, Leo
Levin, Judith Goldstein
Levy, Robert
Lewis, Marc Simon
Liang, Shu-Mei
Lijinsky, William
Lin, Diane Chang
Lin, Leeuwn
Lin, Michael C
Lin, Shin
Lippel, Kenneth
Lippincott-Schwartz, Jennifer
Litman, Burton Joseph
Liu, Teh-Yung
Liverman, James Leslie
Loh, Yoke Peng
London, Edythe D
Long, Cedric William
Longfellow, David G(odwin)
Loriaux, D Lynn
Lowensohn, Howard Stanley
Luther, Lonnie W
Lutwak, Leo
Lymn, Richard Wesley
McCandliss, Russell John

McCarty, Richard Earl
McClure, Michael Edward
McCurdy, John Dennis
McDonald, Lee J
McGowan, Joan A
McGraw, Patricia Mary
Maciag, Thomas Edward
McKenney, Keith Hollis
MacKerell, Alexander Donald, Jr
McLaughlin, Alan Charles
McMacken, Roger
McNaughton, James Larry
McQuaid, Richard William
MacQuillan, Anthony M
Magnani, John Louis
Majchrowicz, Edward
Malech, Harry Lewis
Maloney, Peter Charles
Manen, Carol-Ann
Manganiello, Vincent Charles
Marcus, Carol Joyce
Margolis, Ronald Neil
Margolis, Sam Aaron
Margolis, Simeon
Margulies, Maurice
Marks, Edwin Potter
Martens, Margaret Elizabeth
Martin, Margaret Eileen
Martin, Mark Thomas
Martin, Robert G
Marwah, Joe
Mather, Ian Heywood
Mattern, Michael Ross
Matthews, Benjamin F
Mattick, Joseph Francis
Matyas, Marsha Lakes
Maurizi, Michael R
Melancon, Mark J
Melera, Peter William
Mercado, Teresa I
Merlino, Glenn T
Mertz, Walter
Metzger, Henry
Mickel, Hubert Sheldon
Middlebrook, John Leslie
Mihalyi, Elemer
Mildvan, Albert S
Miles, Edith Wilson
Minthorn, Martin Lloyd, Jr
Mitchell, Geraldine Vaughn
Mockrin, Stephen Charles
Mohla, Suresh
Montell, Craig
Moschel, Robert Carl
Moss, Bernard
Mudd, Stuart Harvey
Mueller, Helmut
Mufson, R Allan
Munn, John Irvin
Munson, Paul Lewis
Murano, Genesio
Murayama, Makio
Mushinski, J Frederic
Nair, Padmanabhan
Nakhasi, Hira Lal
Nash, Howard Allen
Nayak, Ramesh Kadbet
Nelson, John Howard
Nelson, Ralph Francis
Neufeld, Harold Alex
Newburgh, Robert Warren
Newrock, Kenneth Matthew
Newton, Sheila A
Nino, Hipolito V
Nirenberg, Marshall Warren
Noguchi, Constance Tom
Nordin, Albert Andrew
Norman, Philip Sidney
Nossal, Nancy
O'Brien, John C
O'Donnell, James Francis
Olenick, John George
Oliver, Eugene Joseph
Oppenheim, Joost J
O'Rangers, John Joseph
Ortmeyer, Heidi Karen
Osawa, Yoichi
Otani, Theodore Toshiro
Owens, Ida S
Pace, Judith G
Paddock, Jean K
Pagano, Richard Emil
Palmer, Winifred G
Papas, Takis S
Parvez, Zaheer
Passaniti, Antonino
Passonneau, Janet Vivian
Patterson, Glenn Wayne
Pedersen, Peter L
Perdue, James F
Peterkofsky, Alan
Peterson, Elbert Axel
Petrella, Vance John
Phang, James Ming
Piatigorsky, Joram Paul
Pierce, Jack Vincent
Piez, Karl Anton
Pitha-Rowe, Paula Marie
Pitlick, Frances Ann
Pogell, Burton M
Poirier, Miriam Christine Mohrhoff
Poljak, Roberto Juan
Pollard, Harvey Bruce
Pollard, Thomas Dean

Pomerantz, Seymour Herbert
Poston, John Michael
Preusch, Peter Charles
Price, Alan Roger
Prouty, Richard Metcalf
Qasba, Pradmann K
Quarles, Richard Hudson
Rabinovitz, Marco
Rakhit, Gopa
Ram, J Sri
Ramagopal, Subbanaidu
Rao, Prasad Yarlagadda
Read-Connole, Elizabeth Lee
Rechcigl, Miloslav, Jr
Reed, Warren Douglas
Rehak, Matthew Joseph
Reiser, Sheldon
Repaske, Roy
Resnik, Robert Alan
Reynolds, Kevin A
Rhee, Sue Goo
Rice, Jerry Mercer
Rice, Nancy Reed
Richert, Nancy Dembeck
Rider, Agatha Ann
Rivera, Americo, Jr
Roberts, Anita Bauer
Robinson, Cecil Howard
Robinson-White, Audrey Jean
Rochovansky, Olga Maria
Rodkey, Frederick Lee
Roeder, Lois M
Roesijadi, Guritno
Romanowski, Robert David
Rose, George David
Roseman, Saul
Rosenthal, Nathan Raymond
Roth, Thomas Frederic
Rotherham, Jean
Rubin, Martin Israel
Rubin, Robert Jay
Russell, James T
Sabol, Steven Layne
St John, Judith Brook
Salem, Norman, Jr
Samelson, Lawrence Elliot
Sampugna, Joseph
Sanslone, William Robert
Sarngadharan, Mangalasseril G
Sauer, Brian L
Saunders, James Allen
Scheibel, Leonard William
Schepartz, Saul Alexander
Scherbenske, M James
Schiaffino, Silvio Stephen
Schiffmann, Elliot
Schleif, Robert Ferber
Schnaar, Ronald Lee
Schneider, Donald Leonard
Schneider, John H
Schoenberg, Daniel Robert
Schoene, Norberta Wachter
Schrum, Mary Irene Knoller
Schwartz, Joan Poyner
Scocca, John Joseph
Seliger, Howard Harold
Seto, Belinda P L
Seydel, Frank David
Shamsuddin, Abulkalam Mohammad
Sharma, Dinesh C
Sharma, Opendra K
Shellenberger, Thomas E
Shih, Thomas Y
Siiteri, Pentti Kasper
Simic, Michael G
Simons, Samuel Stoney, Jr
Simpson, Robert Todd
Sitkovsky, Michail V
Sjoblad, Roy David
Skidmore, Wesley Dean
Slein, Milton Wilbur
Slife, Charles W
Slocum, Robert Dale
Smith, Lewis Wilbert
Smith, Sharron Williams
Smith, William Owen
Snyder, Stephen Laurie
Soares, Joseph Henry, Jr
Sobel, Mark E
Sobocinski, Philip Zygmund
Sogn, John Allen
Sokoloff, Louis
Song, Byoung-Joon
Sowers, Arthur Edward
Sporn, Michael Benjamin
Srivastava, Sudhir
Stadtman, Earl Reece
Stadtman, Thressa Campbell
Stanchfield, James Ernest
Steers, Edward, Jr
Steinert, Peter Malcolm
Stojilkovic, Stanko S
Stoolmiller, Allen Charles
Straat, Patricia Ann
Stratmeyer, Melvin Edward
Strauss, Robert R
Striker, G E
Su, Robert Tzyh-Chuan
Surrey, Kenneth
Sussman, Daniel Jesse
Svoboda, James Arvid
Sze, Heven
Tabor, Celia White
Tabor, Herbert

Talbot, Bernard
Tamminga, Carol Ann
Taniuchi, Hiroshi
Tester, Cecil Fred
Tietze, Frank
Tildon, J Tyson
Toliver, Adolphus P
Topper, Yale Jerome
Torrence, Paul Frederick
Triantaphyllopoulos, Eugenie
Tschudy, Donald P
Tso, Tien Chioh
Turner, R James
Tyree, Bernadette
Ulsamer, Andrew George, Jr
Umbreit, Thomas Hayden
Underwood, Barbara Ann
Usdin, Vera Rudin
Vance, Hugh Gordon
Vanderlinde, Raymond E
Varma, Shambhu D
Varmus, Harold Elliot
Varricchio, Frederick
Vaughan, Martha
Veech, Richard L
Venditti, John M
Venter, J Craig
Vijay, Inder Krishan
Villet, Ruxton Herrer
Vincent, Phillip G
Vonderhaar, Barbara Kay
Vydelingum, Nadarajen Ameerdanaden
Wachter, Ralph Franklin
Wagner, David Darley
Walker, Richard Ives
Walter, Donald K
Walton, Thomas Edward
Wannemacher, Robert, Jr
Waravdekar, Vaman Shivram
Warner, Huber Richard
Watkins, Clyde Andrew
Waxdal, M J
Weinbach, Eugene Clayton
Weiner, Myron
Weinstein, Constance de Courcy
Weinstein, Howard
Weir, Edward Earl, II
Weirich, Gunter Friedrich
Weiss, Joseph Francis
Westphal, Heiner J
White, Edward Austin
Whitehurst, Virgil Edwards
Whittaker, Paul
Wickerhauser, Milan
Wiech, Norbert Leonard
Wiggert, Barbara Norene
Williams, Jimmy Calvin
Willingham, Allan King
Wolf, Richard Edward, Jr
Wolff, Jan
Wolffe, Alan Paul
Wray, Granville Wayne
Wray, Virginia Lee Pollan
Wu, Henry Chi-Ping
Wu, Roy Shih-Shyong
Yamada, Kenneth Manao
Yamamoto, Richard Susumu
Yen-Koo, Helen C
Yoo, Seung Hyun
Youle, Richard James
Young, Ronald Jerome
Yu, Mei-ying Wong
Zacharius, Robert Marvin
Zielke, H Ronald
Zimmerman, Daniel Hill
Zimmerman, Steven B

MASSACHUSETTS
Abeles, Robert Heinz
Ajami, Alfred Michel
Albert, Mary Roberts Forbes (Day)
Alper, Chester Allan
Alpers, Joseph Benjamin
Alston, Theodore A
Alt, Frederick W
Anderson, Julius Horne, Jr
Antoniades, Harry Nicholas
Appel, Michael Clayton
Arnaout, M Amin
Arvan, Peter
Atkinson-Templeman, Kristine Hofgren
Auld, David Stuart
Auskaps, Aina Marija
Babayan, Vigen Khachig
Backman, Keith Cameron
Bade, Maria Leipelt
Bagshaw, Joseph Charles
Baltimore, David
Baril, Earl Francis
Barkalow, Fern J
Barngrover, Debra Anne
Bass, Norman Herbert
Beck, William Samson
Begg, David A
Bernfeld, Peter
Bikoff, Elizabeth K
Bischoff, Joyce E
Biswas, Chitra
Biswas, Debajit K
Bloch, Konrad Emil
Blout, Elkan Rogers
Bonkovsky, Herbert Lloyd
Bradley, Peter Michael
Brandts, John Frederick

Biochemistry (cont)

Brawerman, George
Breakefield, Xandra Owens
Brecher, Peter I
Brennessel, Barbara A
Brink, John Jerome
Brown, Beverly Ann
Brown, Gene Monte
Brown, Neal Curtis
Browne, Douglas Townsend
Buchanan, John Machlin
Bunn, Howard Franklin
Burkhardt, Alan Elmer
Burstein, Sumner
Cabral-Lilly, Donna
Canale-Parola, Ercole
Cantley, Lewis Clayton
Carter, Edward Albert
Caspi, Eliahu
Cheetham, Ronald D
Chen, Jane-Jane
Chen, Wenyong
Chishti, Athar Husain
Chlapowski, Francis Joseph
Chou, Iih-Nan (George)
Christopher, C William
Chu, Shu-Heh W
Cintron, Charles
Clemson, Harry C
Codington, John F
Cohen, Carl M
Cohen, Jonathan Brewer
Cohen, Nadine Dale
Cohen, Robert Elliot
Cohen, Seymour Stanley
Coleman, Peter Stephen
Collier, Robert John
Collins, Carolyn Jane
Collinson, Albert R
Comer, M Margaret
Cooper, Geoffrey Mitchell
Cornell, Neal William
Corwin, Laurence Martin
Coyne, Mary Downey
Crabtree, Donald V
Crusberg, Theodore Clifford
Cymerman, Allen
Czech, Michael Paul
D'Amore, Patricia Ann
Daniel, Peter Francis
Davidson, Betty
Davidson, Samuel James
Demain, Arnold Lester
Dennen, David W
Dennis, Patricia Ann
Desrosiers, Ronald Charles
Devlin, Robert M
DeWitt, William
Dice, J Fred
Dills, William L, Jr
Dobson, James Gordon, Jr
Doty, Paul Mead
Drysdale, James Wallace
Ducibella, Tom
Duhamel, Raymond C
Dyer, Charissa Annette
Edsall, John Tileston
Edwards, Ross B
Ehlers, Mario Ralf Werner
Eil, Charles
Ellingboe, James
Emanuel, Rodica L
Essigmann, John Martin
Fairbanks, Grant
Fasman, Gerald David
Feig, Larry Allen
Feldberg, Ross Sheldon
Field, Arthur Kirk
Fillios, Louis Charles
Fine, Richard Eliot
Fischer, George A
Flatt, Jean-Pierre
Fleming, Nigel
Foster, John McGaw
Fournier, Maurille Joseph, Jr
Fowler, Elizabeth
Fox, Irving Harvey
Fox, Thomas Oren
Francesconi, Ralph P
Franzblau, Carl
Fraser, Thomas Hunter
Frederick, Sue Ellen
Fritsch, Edward Francis
Fuller, Rufus Clinton
Furbish, Francis Scott
Furie, Bruce
Gage, L Patrick
Garg, Hari G
Garrels, James I
Gaudette, Leo Eward
Gawienowski, Anthony Michael
Gefter, Malcolm Lawrence
George, Harvey
Gergely, John
Geyer, Robert Pershing
Gill, David Michael
Gilmore, Thomas David
Goldberg, Alfred L
Goldberg, Irvin H(yman)
Goldin, Stanley Michael
Goldmacher, Victor S
Goldsby, Richard Allen
Goldstein, Lawrence S B
Goldstein, Richard Neal
Gonnella, Patricia Anne
Good, Carl M, III
Goodman, Howard Michael
Gorelick, Kenneth J
Gorfien, Harold
Green, David J
Guidotti, Guido
Gusovsky, Fabian
Hadjian, Richard Albert
Hagopian, Miasnig
Hamilton, James Arthur
Harbison, G Richard
Harris, S Richard
Harris, Wayne G
Hartman, Iclal Sirel
Hartman, Standish Chard
Haschemeyer, Audrey Elizabeth Veazie
Hastings, John Woodland
Hauser, George
Hay, Donald Ian
Hecht, Norman B
Hegsted, David Mark
Heinrich, Gerhard
Hele, Priscilla
Hepler, Peter Klock
Herrmann, Robert Lawrence
Hicks, Sonja Elaine
Hinrichs, Katrin
Hirschberg, Carlos Benjamin
Hojnacki, Jerome Louis
Holick, Michael Francis
Hollocher, Thomas Clyde, Jr
Hong, Jen Shiang
Hoskin, Francis Clifford George
Howley, Peter Maxwell
Hu, Jing-Shan
Huxley, Hugh E
Hynes, Richard Olding
Ingram, Vernon Martin
Ip, Stephen H
Irving, James Tutin
Isselbacher, Kurt Julius
Jacobson, Bruce Shell
Jameson, James Larry
Jeanloz, Roger William
Jencks, William Platt
Johal, Sarjit S
Johnson, Corinne Lessig
Jungalwala, Firoze Bamanshaw
Kadkade, Prakash Gopal
Kagan, Herbert Marcus
Kaminskas, Edvardas
Karnovsky, Manfred L
Kashket, Eva Ruth
Kashket, Shelby
Kennedy, Eugene P
Ketchum, Paul Abbott
Kidder, George Wallace, Jr
Killick, Kathleen Ann
Kilpatrick, Daniel Lee
Kim, Peter S
King, Stephen Murray
Kirschner, Marc Wallace
Kisliuk, Roy Louis
Kitz, Richard J
Kleckner, Nancy E
Knipe, David Mahan
Kobayashi, Yutaka
Kolodner, Richard David
Koul, Omanand
Krane, Stephen Martin
Kravitz, Edward Arthur
Krinsky, Norman Irving
Kubin, Rosa
Kupchik, Herbert Z
Lada, Arnold
Landis, William Joel
Leary, John Dennis
Leavis, Paul Clifton
LeBaron, Francis Newton
Ledbetter, Mary Lee Stewart
Lees, Marjorie Berman
Lees, Robert S
Lehman, William Jeffrey
Lehrer, Sherwin Samuel
Li, Jeanne B
Liberatore, Frederick Anthony
Lichtenstein, Alice Hinda
Lin, Chi-Wei
Lin, Edmund Chi Chien
Lionetti, Fabian Joseph
Lipinski, Boguslaw
Liss, Maurice
Liu, Shih-Chun
Lodish, Harvey Franklin
Loscalzo, Joseph
Lowenstein, John Martin
Lu, Renne Chen
Luna, Elizabeth J
Macchi, I Alden
McCluer, Robert Hampton
McDonagh, Jan M
MacDonald, Alex Bruce
McMaster, Paul D
McReynolds, Larry Austin
Madras, Bertha K
Mager, Milton
Maione, Theodore E
Maniscalco, Ignatius Anthony
Manning, Brenda Dale
Mariani, Henry A
Marten, James Frederick
Martz, Eric
Mautner, Henry George
Melchior, Donald L
Mohr, Scott Chalmers
Moner, John George
Moolten, Frederick London
Moreland, Robert Byrl
Morton, Mary E
Muni, Indu A
Munro, Hamish N
Murphy, John R
Neer, Eva Julia
Nelson, Donald John
Newcombe, David S
Nordin, John Hoffman
Nussbaum, Alexander Leopold
O'Brien, Richard Desmond
Offner, Gwynneth Davies
Ofner, Peter
Orkin, Stuart H
Orlando, Joseph Alexander
Orme-Johnson, Nanette Roberts
Orme-Johnson, William H
Ouellette, Andre J
Paine, Clair Maynard
Papastoitsis, Gregory
Pappenheimer, Alwin Max, Jr
Pardee, Arthur Beck
Park, Hee-Young
Park, James Theodore
Pathak, Madhukar
Paul, Benoy Bhushan
Pavlos, John
Paz, Mercedes Aurora
Pecci, Joseph
Petsko, Gregory Anthony
Pharo, Richard Levers
Pluskal, Malcolm Graham
Poccia, Dominic Louis
Porter, William L
Powers Lee, Susan Glenn
Putney, Scott David
Reif-Lehrer, Liane
Reinhold, Vernon Nye
Rendina, George A
Reyero, Cristina
Rheinwald, James George
Rice, Robert Vernon
Richardson, Charles Clifton
Richmond, Martha Ellis
Rieder, Sidney Victor
Riordan, James F
Rivera, Ezequiel Ramirez
Robbins, Phillips Wesley
Roberts, Mary Fedarko
Robinson, Trevor
Robinson, William Edward
Rogers, William Irvine
Rosemberg, Eugenia
Ross, Alonzo Harvey
Roth, Jay Sanford
Ryan, Kenneth John
Sachs, David Howard
Sacks, David B
Saide, Judith Dana
Saidha, Tekchand
Sakura, John David
Sanadi, D Rao
Sarkar, Nilima
Sarkar, Satyapriya
Sauer, Robert Thomas
Schimmel, Paul Reinhard
Schmid, Karl
Schnitzer, Jan Eugeneusz
Schwarting, Gerald Allen
Schwartz, Edith Richmond
Scordilis, Stylianos Panagiotis
Searcy, Dennis Grant
Seidel, John Charles
Seyfried, Thomas Neil
Shaeffer, Joseph Robert
Shakarjian, Michael Peter
Shashoua, Victor E
Shemin, David
Shing, Yuen Wan
Shipley, Reginald A
Short, John Lawson
Silbert, Jeremiah Eli
Sinex, Francis Marott
Sipe, Jean Dow
Sizer, Irwin Whiting
Slakey, Linda Louise
Slovacek, Rudolf Edward
Smith, Barbara D
Smith, Cassandra Lynn
Smith, Edgar Eugene
Smith, Joseph Donald
Smith, Nathan Lewis, III
Smith, Wendy Anne
Spiegelman, Bruce M
Spiro, Mary Jane
Spiro, Robert Gunter
Sreter, Frank A
Staples, Mark
Stare, Fredrick J
Stark, James Cornelius
Stein, Gary S
Stein, Janet Lee Swinehart
Stern, Arthur Irving
Stern, Ivan J
Stiles, Charles Dean
Stoeckler, Johanna D
Stollar, Bernard David
Stossel, Thomas Peter
Strauss, Phyllis R
Struhl, Kevin
Sullivan, William Daniel
Surgenor, Douglas MacNevin
Sweadner, Kathleen Joan
Szent-Gyorgyi, Andrew Gabriel
Szlyk-Modrow, Patricia Carol
Szoka, Paula Rachel
Szostak, Jack William
Talamo, Barbara Lisann
Tarlov, Alvin Richard
Tashjian, Armen H, Jr
Taunton-Rigby, Alison
Taylor, Allen
Terner, Charles
Thies, Robert Scott
Tomera, John F
Toole, Bryan Patrick
Troxler, Robert Fulton
Tucker, Katherine Louise
Tullis, James Lyman
Vallee, Bert L
Vallee, Richard Bert
Van Eikeren, Paul
Van Vunakis, Helen
Venkatesh, Yeldur Padmanabha
Vermeulen, Mary Woodall
Villee, Claude Alvin, Jr
Vitale, Joseph John
Wacker, Warren Ernest Clyde
Wald, George
Waloga, Geraldine
Walsh, Christopher Thomas
Wang, Chih-Lueh Albert
Wang, James C
Webb, Andrew Clive
Weinstein, Mark Joel
Weltman, Joel Kenneth
Westhead, Edward William, Jr
Westheimer, Frank Henry
Wharton, David Carrie
Widmaier, Eric Paul
Wildasin, Harry Lewis
Williamson, Patrick Leslie
Witman, George Bodo, III
Wohlrab, Hartmut
Woodcock, Christopher Leonard Frank
Woodman, Peter William
Wotiz, Herbert Henry
Wun, Chun Kwun
Yang, Shiow-shong
Yesair, David Wayne
Young, Delano Victor
Young, Vernon Robert
Zabrecky, James R
Zackroff, Robert V
Zieske, James David
Zimmermann, Robert Alan
Ziomek, Carol A

MICHIGAN

Abraham, Irene
Adelman, Richard Charles
Adler, John Henry
Akil, Huda
Aminoff, David
Anderson, Richard Lee
Andrewes, Arthur George
Armant, D Randall
Armstrong, Robert Lee
Bach, Michael Klaus
Ballou, David Penfield
Bandurski, Robert Stanley
Banerjee, Surath Kumar
Barald, Kate Francesca
Barbour, Stephen D
Barry, Roger Donald
Beher, William Tyers
Benjamins, Joyce Ann
Bennink, Maurice Ray
Ben-Yoseph, Yoav
Bergen, Werner Gerhard
Berk, Richard Samuel
Bernstein, Isadore A
Bertrand, Helmut
Beyer, Robert Edward
Bieber, Loran Lamoine
Bleasdale, John Edward
Block, Walter David
Bole, Giles G
Boyer, Rodney Frederick
Branson, Dean Russell
Braselton, Webb Emmett, Jr
Braughler, John Mark
Brawn, Mary Karen
Breznak, John Allen
Brooks, Samuel Carroll
Burnett, Jean Bullard
Byerrum, Richard Uglow
Campbell, Kenneth Lyle
Caraway, Wendell Thomas
Carrick, Lee
Carter-Su, Christin
Caspers, Mary Lou
Cerra, Robert Frank
Chaudhry, G Rasul
Chaudry, Irshad Hussain
Chou, Kuo Chen
Chung, Fu-Zon
Clewell, Don Bert
Cody, Wayne Livingston
Cole, George Christopher
Cole, Versa V
Conover, James H
Coon, Minor J

Cotter, Richard
Cuatrecasas, Pedro
Dabich, Danica
Datta, Prasanta
Dazzo, Frank Bryan
Deal, William Cecil, Jr
Dekker, Eugene Earl
Distler, Jack, Jr
Dixon, Jack Edward
Dolin, Morton Irwin
Doscher, Marilyn Scott
Drescher, Dennis G
Dudley, David T
Duell, Elizabeth Ann
Dusenbery, Ruth Lillian
Dyer, Richard
Eberts, Floyd S, Jr
Eble, Thomas Eugene
Einspahr, Howard Martin
Elferink, Cornelius Johan
Elhammer, Ake P
Epps, Dennis Earl
Erickson, Laurence A
Evans, David R
Fairley, James Lafayette, Jr
Ferguson-Miller, Shelagh Mary
Ferrell, William James
Fluck, Michele Marguerite
Foote, Joel Lindsley
Frank, Robert Neil
Freeman, Arthur Scott
Friar, Robert Edsel
Friedland, Melvyn
Frohman, Charles Edward
Gal, Susannah
Gee, Robert William
Gelehrter, Thomas David
Ghering, Mary Virgil
Giblin, Frank Joseph
Gilbertson, Terry Joel
Goldstein, Irwin Joseph
Goodwin, Jesse Francis
Gorman, Robert Roland
Goyings, Lloyd Samuel
Grady, Joseph Edward
Grant, Rhoda
Gray, Gary D
Greenberg, Goodwin Robert
Griffith, Thomas
Grossman, Lawrence I
Guarino, Armand John
Hajra, Amiya Kumar
Hauxwell, Ronald Earl
Heffner, Thomas G
Heinrikson, Robert L
Hinman, Jack Wiley
Hirata, Fusao
Hoch, Frederic Louis
Hoeksema, Walter David
Holland, Jack Calvin
Hollenberg, Paul Frederick
Hultquist, Donald Elliott
Hupe, Donald John
Johnson, Garland A
Johnson, Robert Michael
Joseph, Ramon R
Jourdian, George William
Kahn, A Clark
Kawooya, John Kasajja
Kent, Claudia
Kessel, David Harry
Kezdy, Ferenc J
Kimura, James Hiroshi
Kimura, Tokuji
Kindel, Paul Kurt
King, Charles Miller
Kithier, Karel
Klein, Ronald Don
Kleinsmith, Lewis Joel
Kletzien, Rolf Frederick
Klohs, Wayne D
Kominek, Leo Aloysius
Kratochvil, Clyde Harding
Krawetz, Stephen Andrew
Kupiecki, Floyd Peter
Lahti, Robert A
Lalwani, Narendra Dhanraj
Lamport, Derek Thomas Anthony
Lee, Chuan-Pu
Lee, Sungsoo C
Leng, Marguerite Lambert
Leopold, Wilbur Richard, III
Li, Li-Hsieng
Lightbody, James James
Lillevik, Hans Andreas
Lockwood, Dean H
Lomax, Margaret Irene
Louis, Lawrence Hua-Hsien
Ludwig, Martha Louise
Luecke, Richard William
Lueking, Donald Robert
McCall, Keith Bradley
McCann, Daisy S
McCauley, Roy Barnard
McConnell, David Graham
McCoy, Lowell Eugene
MacDonald, Roderick Patterson
McGroarty, Estelle Josephine
McLean, John Robert
Makinen, Kauko K
Malkin, Leonard Isadore
Marcelo, Cynthia Luz
Marks, Bernard Herman
Martin, Joseph Patrick, Jr

Massey, Vincent
Mather, Edward Chantry
Matthews, Rowena Green
Maxon, William Densmore
Medzihradsky, Fedor
Meeks, Robert G
Meisler, Miriam Horowitz
Meyer, Edward Dell
Meyerhoff, Mark Elliot
Miller, Susan Mary
Miller, Thomas Lee
Miller, William Louis
Mitchell, Robert Alexander
Mostafapour, M Kazem
Moudgil, Virinder Kumar
Nadler, Kenneth David
Nag, Asish Chandra
Nairn, Roderick
Neely, Brock Wesley
Neff, Alven William
Neidhardt, Frederick Carl
Neubig, Richard Robert
Niekamp, Carl William
Njus, David Lars
Nooden, Larry Donald
Nordby, Gordon Lee
Olson, Steven T
Orten, James M
Ovshinsky, Iris M(iroy)
Oxender, Dale LaVern
Palmer, Kenneth Charles
Parcells, Alan Jerome
Parker, Charles J, Jr
Payne, Anita H
Pecoraro, Vincent L
Petty, Howard Raymond
Petzold, Edgar
Philipsen, Judith Lynne Johnson
Podila, Gopi Krishna
Powers, Wendell Holmes
Preiss, Jack
Punch, James Darrell
Purkhiser, E Dale
Pyke, Thomas Richard
Radin, Eric Leon
Radin, Norman Samuel
Rainey, John Marion, Jr
Ram, Jeffrey L
Rapundalo, Stephen T
Raz, Avraham
Reddy, Venkat N
Reitz, Richard Henry
Renis, Harold E
Revzin, Arnold
Reynolds, Orland Bruce
Riggs, Thomas Rowland
Riley, Michael Verity
Rosen, Barry Philip
Rosenspire, Allen Jay
Rownd, Robert Harvey
Rozhin, Jurij
Rust, Steven Ronald
Sadoff, Harold Lloyd
Saltiel, Alan Robert
Saper, Mark A
Sardesai, Vishwanath M
Sarkar, Fazlul Hoque
Satoh, Paul Shigemi
Schacht, Jochen (Heinrich)
Schaefer, Gerald J
Schmaier, Alvin Harold
Scott, Ronald McLean
Sebolt-Leopold, Judith S
Sell, John Edward
Sellinger, Otto Zivko
Shappiro, David Gordon
Shastry, Barkur Srinivasa
Shichi, Hitoshi
Shipman, Charles, Jr
Shonk, Carl Ellsworth
Shore, Joseph D
Sloane, Bonnie Fiedorek
Smith, Denise Myrtle
Smith, Ellen Oliver
Smith, Eugene Joseph
Smith, Robert James
Smith, William Lee
Speck, John Clarence, Jr
Spilman, Charles Hadley
Stach, Robert William
Stanley, John Pearson
Stenesh, Jochanan
Strand, James Cameron
Stucki, William Paul
Suelter, Clarence Henry
Sun, Frank F
Swanson, Curtis James
Swaroop, Anand
Sweeley, Charles Crawford
Tannen, Richard L
Tashian, Richard Earl
Tchen, Tche Tsing
Thoene, Jess Gilbert
Tobes, Michael Charles
Tolbert, Nathan Edward
Triezenberg, Steven J
Turley, June Williams
Ueda, Tetsufumi
Van Dyke, Russell Austin
Vatsis, Kostas Petros
Venta, Patrick John
Vinogradov, Serge
Von Korff, Richard Walter
Walker, Gustav Adolphus

Walter, Richard Webb, Jr
Walz, Daniel Albert
Wang, Ching Yung
Wang, John L
Warner, Donald Theodore
Warrington, Terrell L
Watenpaugh, Keith Donald
Weinhold, Paul Allen
Weiss, Bernard
Wells, William Wood
West, Bruce David
Weymouth, Patricia Perkins
Wickrema Sinha, Asoka J
Williams, Charles Haddon, Jr
Wilson, John Edward
Winkler, Barry Steven
Winter, Harry Clark
Wolk, Coleman Peter
Wovcha, Merle G
Yamazaki, Russell Kazuo
Yanari, Sam Satomi
Ye, Qizhuang
Yocum, Charles Fredrick
Yurewicz, Edward Charles
Zannoni, Vincent G
Zemlicka, Jiri
Zile, Maija Helene

MINNESOTA
Abul-Hajj, Jusuf J
Adolph, Kenneth William
Ahmed, Khalil
Anderson, Dwight Lyman
Anderson, John Seymour
Anderson, Paul M
Arneson, Richard Michael
Bach, Marilyn Lee
Banaszak, Leonard Jerome
Barden, Roland Eugene
Baumann, Wolfgang J
Beggs, William H
Berg, Marie Hirsch
Berg, Steven Paul
Berry, James Frederick
Blomquist, Charles Howard
Boaz, David Paul
Bodine, Peter Van Nest
Bodley, James William
Breene, William Michael
Brockman, Howard Lyle, Jr
Brown, Rhoderick Edmiston, Jr
Brummond, Dewey Otto
Brunelle, Thomas E
Burton, Alice Jean
Carr, Charles William
Chandan, Ramesh Chandra
Clapper, David Lee
Cohen, Harold P
Coleman, Patrick Louis
Coon, Craig Nelson
Daravingas, George Vasilios
De Master, Eugene Glenn
Dempsey, Mary Elizabeth
Derr, Robert Frederick
Detwiler, Thomas C
Dickie, John Peter
Dousa, Thomas Patrick
Drewes, Lester Richard
Dunshee, Bryant R
Durst, Jack Rowland
Eaton, John Wallace
Edds, Kenneth Tiffany
Edstrom, Ronald Dwight
Ellefson, Ralph Donald
Ellis, Lynda Betty
Evans, Gary William
Frantz, Ivan DeRay, Jr
Frey, William Howard, II
Fuchs, James Allen
Ganguli, Mukul Chandra
Gannon, Mary Carol
Gebhard, Roger Lee
Gilboe, Daniel Pierre
Glass, Robert Louis
Goldberg, Nelson D
Gorman, Colum A
Gray, Ernest David
Greenberg, Leonard Jason
Gutmann, Helmut Rudolph
Hadler, Herbert Isaac
Hanson, Richard Steven
Hendrickson, Herman Stewart, II
Henry, Ronald Alton
Hogenkamp, Henricus Petrus C
Holman, Ralph Theodore
Homann, H Robert
Howard, James Bryant
Hurley, William Charles
Jensen, James Le Roy
Jiang, Nai-Siang
Johnson, Susan Bissen
Jones, Charles Weldon
Jones, James Donald
Joos, Richard W
Katz, Morris Howard
Kirkwood, Samuel
Koch, Robert B
Koerner, James Frederick
Kromann, Rodney P
Laakso, John William
Leung, Benjamin Shuet-Kin
Liener, Irvin Ernest
Lin, Sping
Linck, Richard Wayne

Lipscomb, John DeWald
Loh, Horace H
Longley, Robert W(illiam)
Lovrien, Rex Eugene
Lyon, Richard Hale
Mackenzie, Cosmo Glenn
Magee, Paul Terry
Malewicz, Barbara Maria
Mannering, Gilbert James
Markhart, Albert Henry, III
Mason, Harold L
Miller, Richard Lynn
Mullen, Joseph David
Neft, Nivard
Nelsestuen, Gary Lee
Nicol, Susan Elizabeth
Ober, Robert Elwood
Oegema, Theodore Richard, Jr
Orf, John W
Owen, Charles Archibald, Jr
Penniston, John Thomas
Plagemann, Peter Guenter Wilhelm
Pour-El, Akiva
Prendergast, Franklyn G
Prohaska, Joseph Robert
Ralston, Douglas Edmund
Rapp, Waldean G
Rathbun, William B
Rodgers, Nelson Earl
Rogers, Palmer, Jr
Roon, Robert Jack
Rosevear, John William
Ruschmeyer, Orlando R
Ryan, Robert J
Rychlik, Wojciech
Salisbury, Jeffrey L
Salo, Wilmar Lawrence
Salstrom, John Stuart
Sammak, Paul J
Schlenk, Hermann
Schmid, Harald Heinrich Otto
Schmidt, Jane Ann
Schwartz, Harold Leon
Sheppard, John Richard
Shier, Wayne Thomas
Smith, Quenton Terrill
Somers, Perrie Daniel
Sparnins, Velta L
Spelsberg, Thomas Coonan
Spencer, Richard L
Sperber, William H
Stankovich, Marian Theresa
Stearns, Eugene Marion, Jr
Stern, Marshall Dana
Stoesz, James Darrel
Straw, Thomas Eugene
Strehler, Emanuel Ernst
Sullivan, Betty J
Surdy, Ted E
Swaiman, Kenneth F
Szurszewski, Joseph Henry
Tautvydas, Kestutis Jonas
Thenen, Shirley Warnock
Tindall, Donald J
Tsilibary, Effie-Constantinos
Tuominen, Francis William
Ungar, Frank
Van Pilsum, John Franklin
Verma, Anil Kumar
Williams, Edmond Brady
Wilson, Michael John
Wood, John Martin
Woodward, Clare K
Yasmineh, Walid Gabriel

MISSISSIPPI
Angel, Charles
Bateman, Robert Carey, Jr
Biesiot, Patricia Marie
Brewer, Franklin Douglas
Cannon, Jerry Wayne
Dilworth, Benjamin Conroy
Eftink, Maurice R
Graves, David E
Haining, Joseph Leo
Hardy, James D
Heitz, James Robert
Ho, Ing Kang
Hoagland, Robert Edward
Kellogg, Thomas Floyd
Kennedy, Maurice Venson
Larsen, James Bouton
Long, Billy Wayne
Martin, Billy Joe
Mehrotra, Bam Deo
Morrison, John Coulter
Olson, Mark Obed Jerome
Outlaw, Henry Earl
Read, Virginia Hall
Salin, Marvin Leonard
Slobin, Lawrence I
Sulya, Louis Leon
Van Aller, Robert Thomas
Verlangieri, Anthony Joseph
Wahba, Albert J
White, Harold Birts, Jr
Wills, Gene David
Wilson, Robert Paul
Woodley, Charles Leon

MISSOURI
Ackermann, Philip Gulick
Adams, Steven Paul
Agrawal, Harish C

Biochemistry (cont)

Ahmed, Mahmoud S
Ahmed, Nahed K
Alexander, Stephen
Alkjaersig, Norma Kirstine (Mrs A P Fletcher)
Alpers, David Hershel
Apirion, David
Armbrecht, Harvey James
Arneson, Dora Williams
Ayyagari, L Rao
Bachman, Gerald Lee
Badr, Mostafa
Bajaj, S Paul
Barnes, Wayne Morris
Beinfeld, Margery Cohen
Berenbom, Max
Blake, Robert L
Blatz, Paul E
Bolla, Robert Irving
Bretton, Randolph Henry, Jr
Brot, Frederick Elliot
Brown, Barbara Illingworth
Brown, Karen Kay (Kilker)
Brown, Olen Ray
Bruns, Lester George
Burke, William Joseph
Burleigh, Bruce Daniel, Jr
Burton, Robert Main
Campbell, Benedict James
Carey, Paul L
Cenedella, Richard J
Chiappinelli, Vincent A
Chilson, Oscar P
Clare, Stewart
Cockrell, Ronald Spencer
Connolly, Daniel Thomas
Connolly, John W
Cornelius, Steven Gregory
Cornell, Creighton N
Coscia, Carmine James
Coudron, Thomas A
Crouch, Eduard Clyde
Cullen, Susan Elizabeth
Das, Manik Lal
David, John Dewood
Davis, James Wendell
De Blas, Angel Luis
Dorner, Robert Wilhelm
Ehrenthal, Irving
Eissenberg, Joel Carter
Elliott, William H
Elson, Elliot
Engle, Michael Jean
Esser, Alfred F
Evans, James William
Feather, Milton S
Feder, Joseph
Fernandez-Pol, Jose Alberto
Fisher, Harvey Franklin
Fitch, Coy Dean
Folk, William Robert
Fox, J Eugene
Franz, John Matthias
Frieden, Carl
Gale, Nord Loran
Gardner, Jerry David
Garner, George Bernard
Gay, Don Douglas
Gehrke, Charles William
Geisbuhler, Timothy Paul
Geller, David Melville
Gerhardt, Klaus Otto
Gill, James Wallace
Gledhill, William Emerson
Gold, Alvin Hirsh
Gordon, Annette Waters
Gorski, Jeffrey Paul
Gossling, Jennifer
Grant, Gregory Alan
Green, Eric Douglas
Green, Maurice
Greene, Daryle E
Grogan, Donald E
Guntaka, Ramareddy V
Hamilton, James Wilburn
Hancock, Anthony John
Harakas, N(icholas) Konstantinos
Hart, Kathleen Therese
Heard, John Thibaut, Jr
Herzig, Geoffrey Peter
Hicks, Patricia Fain
Hillman, Richard Ephraim
Hirschberg, Rona L
Ho, David Tuan-hua
Holland, Lene J
Horwitt, Max Kenneth
House, William Burtner
Howard, Susan Carol Pearcy
Huang, Jung San
Hunter, Francis Edmund, Jr
Jacob, Gary Steven
Jaworski, Ernest George
Jellinek, Max
Johnston, Marilyn Frances Meyers
Katzman, Philip Aaron
Kessler, Gerald
Khalifah, Raja Gabriel
Kim, Hyun Dju
Kim, Yee Sik
Kishore, Ganesh M
Kobayashi, George S
Koeltzow, Donald Earl
Koeppe, Owen John
Kornfeld, Rosalind Hauk
Krivi, Gwen Grabowski
Landick, Robert
Larson, Allan
Larson, Russell L
Lie, Wen-Rong
Linder, Maurine E
Longmore, William Joseph
Lowry, Oliver Howe
McCann, Peter Paul
McDaniel, Michael Lynn
McDonald, John Alexander
McMaster, Marvin Clayton, Jr
MacQuarrie, Ronald Anthony
Madl, Ronald
Marshall, Garland Ross
Martin, Arlene Patricia
Martinez-Carrion, Marino
Mathews, F(rancis) Scott
Mayo, Joseph William
Mehrle, Paul Martin, Jr
Melechen, Norman Edward
Merlie, John Paul
Miller, Herman T
Miller, Lowell D
Moore, Blake William
Morgan, William T
Morris, Roy Owen
Morris, Stephen Jon
Morton, Phillip A
Moscatelli, Ezio Anthony
Munns, Theodore Willard
Murray, Patrick Robert
Myers, Richard Lee
Nelson, Curtis Jerome
Nicholas, Harold Joseph
Nickols, G Allen
Noteboom, William Duane
O'Dell, Boyd Lee
Ortwerth, Beryl John
Padron, Jorge Louis
Peters, Ralph I
Pikaard, Craig Stuart
Pike, Linda Joy
Polakoski, Kenneth Leo
Pueppke, Steven Glenn
Quay, Wilbur Brooks
Raskas, Heschel Joshua
Rice, Charles Moen, III
Riddle, Donald Lee
Robbins, Ernest Aleck
Robins, Eli
Roodman, Stanford Trent
Rosenthal, Harold Leslie
Rovetto, Michael Julien
Sadler, J(asper) Evan
Salkoff, Lawrence Benjamin
Sanes, Joshua Richard
Schlesinger, Milton J
Schmidt, Francis J
Schulze, Irene Theresa
Seidler, Norbert Wendelin
Seltzer, Jo Louise
Sharma, Krishna
Shelton, Damon Charles
Sherman, William Reese
Shukla, Shivendra Dutt
Siehr, Donald Joseph
Silbert, David Frederick
Singh, Harbhajan
Smith, Ann
Snyder, Harry E
Staatz, William D
Starling, Jane Ann
Stern, Michele Suchard
Strauss, Arnold Wilbur
Sun, Albert Yung-Kwang
Sunde, Roger Allan
Sweet, Frederick
Teaford, Margaret Elaine
Thach, Robert Edwards
Thompson, Richard Michael
Toom, Paul Marvin
Vandepopuliere, Joseph Marcel
Warren, James C
Weber, Morton M
Weisman, Gary Andrew
Wells, Frank Edward
Welply, Joseph Kevin
Weppelman, Roger Michael
Westfall, Helen Naomi
White, Arnold Allen
Whitley, James R
Wiest, Walter Gibson
Willard, Mark Benjamin
Wixom, Robert Llewellyn
Wosilait, Walter Daniel
Yadav, Kamaleshwari Prasad
Yunger, Libby Marie
Zech, Arthur Conrad
Zenser, Terry Vernon
Zepp, Edwin Andrew

MONTANA
Blake, Tom
Dratz, Edward Alexander
El-Negoumy, Abdul Monem
Ericsson, Ronald James
Fevold, Harry Richard
Garon, Claude Francis
Hapner, Kenneth D
Hill, Walter Ensign
Jackson, Larry Lee
Jesaitis, Algirdas Joseph
Mell, Galen P
Myers, Lyle Leslie
Robbins, John Edward
Rogers, Samuel John
Roush, Allan Herbert
Stallknecht, Gilbert Franklin
Thompson, Holly Ann
Wright, Barbara Evelyn

NEBRASKA
Adrian, Thomas E
Barak, Anthony Joseph
Baumstark, John Spann
Birt, Diane Feickert
Boernke, William E
Borchers, Raymond (Lester)
Brakke, Myron Kendall
Carson, Steven Douglas
Carver, Michael Joseph
Casey, Carol A
Cavalieri, Ercole Luigi
Chollet, Raymond
Christensen, Alan Carl
Christman, Judith Kershaw
Cox, George Stanley
Curtis, Gary Lynn
Daly, Joseph Michael
Dam, Richard
Dubes, George Richard
Elthon, Thomas Eugene
Fishkin, Arthur Frederic
Gambal, David
Gardner, Charles Olda, Jr
Gessert, Carl F
Goldsmith, Dale Preston Joel
Grandjean, Carter Jules
Grotjan, Harvey Edward, Jr
Gupta, Naba K
Harman, Denham
Heidrick, Margaret Louise
Hill, Robert Mathew
Hofert, John Frederick
Holmquist, Barton
Johnson, John Arnold
Johnston, Robert Benjamin
Kimberling, William J
Klucas, Robert Vernon
Knoche, Herman William
Knoop, Floyd C
Lane, Leslie Carl
Lee, Jean Chor-Yin Wong
McClurg, James Edward
Mackenzie, Charles Westlake, III
Mackin, Robert Brian
Mahowald, Theodore Augustus
Matschiner, John Thomas
Murrin, Leonard Charles
Murthy, Veeraraghavan Krishna
Nickerson, Kenneth Warwick
O'Leary, Marion Hugh
Partridge, James Enoch
Phalen, William Edmund
Phares, Cleveland Kirk
Raha, Chitta Ranjan
Ramaley, Robert Folk
Rogan, Edward Groeniger
Rongone, Edward Laurel
Ryan, Wayne L
Scholar, Eric M
Schwartzbach, Steven Donald
Shahani, Khem Motumal
Spreitzer, Robert Joseph
Steffens, David Lee
Stevens, Sue Cassell
Swanson, Jack Lee
Taylor, Stephen Lloyd
Tuma, Dean J
Wagner, Frederick William
Watt, Dean Day
Wells, Ibert Clifton
Yee, John Alan

NEVADA
Blomquist, Gary James
Buxton, Iain Laurie Offord
Dreiling, Charles Ernest
Halarnkar, Premjit P
Harrington, Rodney E
Heisler, Charles Rankin
Hudig, Dorothy
Hylin, John Walter
Lewis, Roger Allen
Omaye, Stanley Teruo
Pardini, Ronald Shields
Peck, Ernest James, Jr
Reitz, Ronald Charles
Schooley, David Allan
Starich, Gale Hanson
Welch, William Henry, Jr
Whitfield, George Buckmaster, Jr
Winicov, Ilga
Woodin, Terry Stern

NEW HAMPSHIRE
Bresnick, Edward
Brinck-Johnsen, Truls
Chang, Ta-Yuan
Cole, Charles N
Copenhaver, John Harrison, Jr
Fanger, Michael Walter
Gray, Clarke Thomas
Gross, Robert Henry
Harbury, Henry Alexander
Hatch, Frederick Tasker
Hubbard, Colin D
Keener, Harry Allan
Kensler, Charles Joseph
Kremzner, Leon T
Kull, Fredrick J
Lienhard, Gustav E
Miller, Richard Wilson
Minocha, Subhash C
Noda, Lafayette Hachiro
Ou, Lo-Chang
Sloboda, Roger D
Smith, Edith Lucile
Smith, Samuel Cooper
Stewart, James Anthony
Teeri, Arthur Eino
Thomas, William Albert
Trumpower, Bernard Lee
Zolton, Raymond Peter

NEW JERSEY
Ahn, Ho-Sam
Alper, Carl
Anderson, David Wesley
Anderson, Ethel Irene
Arena, Joseph P
Avigad, Gad
Bahr, James Theodore
Bailin, Gary
Balazs, Endre Alexander
Barnett, Ronald E
Barton, Beverly E
Basu, Mitali
Bauman, John W, Jr
Beadling, Leslie Craig
Beck, David Paul
Beck, Jeanne Crawford
Bennun, Alfred
Benson, Charles Everett
Berberian, Paul Anthony
Berg, Richard A
Bernstein, Eugene H
Beyer-Mears, Annette
Bhagavan, Hemmige
Bieber, Mark Allan
Bird, John William Clyde
Blackburn, Gary Ray
Blood, Christine Haberern
Bolser, Donald C
Borman, Aleck
Botelho, Lynne H Parker
Boublik, Miloslav
Boyd, John Edward
Brandriss, Marjorie C
Brin, Myron
Brostrom, Charles Otto
Brostrom, Margaret Ann
Brot, Nathan
Brown, Harry Darrow
Browning, Edward T
Bruno, Charles Frank
Burg, Richard William
Burns, John J
Burrier, Robert E
Cagan, Robert H
Caldwell, Carlyle Gordon
Calvin, Harold I
Caporale, Lynn Helena
Carboni, Joan M
Carman, George Martin
Carroll, James Joseph
Carter, James Evan
Cascieri, Margaret Anne
Cash, William Davis
Cayen, Mitchell Ness
Cevallos, William Hernan
Champe, Pamela Chambers
Chao, Yu-Sheng
Chase, Theodore, Jr
Chasin, Mark
Chen, Kuang-Yu
Cheng, Kang
Chiang, Bin-Yea
Cho, Kon Ho
Choi, Sook Y
Choi, Ye-Chin
Christakos, Sylvia
Clark, Irwin
Coffey, John William
Cohen, Herman
Colby, Richard H
Collett, Edward
Comai-Fuerherm, Karen
Conney, Allan Howard
Cornell, James S
Coscarelli, Waldimero
Crane, Robert Kellogg
Cunningham, Earlene Brown
Curran, Thomas
Dalrymple, Ronald Howell
Denhardt, David Tilton
Denton, Arnold Eugene
DePamphilis, Melvin Louis
Ding, Victor Ding-Hai
Dion, Arnold Silva
Doebber, Thomas Winfield
Dougherty, Harry W
Dreyfuss, Jacques
Eades, Charles Hubert, Jr
Easterday, Richard Lee
Egan, Robert Wheeler
Elbrecht, Alex
Ellis, Daniel B(enson)
Ennis, Herbert Leo

BIOLOGICAL SCIENCES / 45

Erenrich, Evelyn Schwartz
Fagan, Paul V
Farnsworth, Patricia Nordstrom
Fawzi, Ahmad B
Feigelson, Muriel
Feigenbaum, Abraham Samuel
Feldman, Susan C
Fenichel, Richard Lee
Ferguson, Karen Anne
Fiedler-Nagy, Christa
Finkelstein, Paul
Firschein, Hilliard E
Fitzgerald, Paula Marie Dean
Fletcher, Martin J
Flint, Sarah Jane
Flora, Robert Montgomery
Foley, James E
Fong, Tung Ming
Frances, Saul
Frank, Oscar
Frantz, Beryl May
Frascella, Daniel W
Free, Charles Alfred
Frenkel, Gerald Daniel
Fresco, Jacques Robert
Freund, Thomas Steven
Fu, Shou-Cheng Joseph
Fujimore, Tohru
Gaffar, Abdul
Gaffney, Barbara Lundy
Galdes, Alphonse
Gall, Martin
Gallopo, Andrew Robert
Garcia, Maria Luisa
Gaylor, James Leroy
Gaynor, John James
Georgopapadakou, Nafsika Eleni
Ghai, Geetha R
Ghosh, Arati
Gibson, Joyce Correy
Gilfillan, Alasdair Mitchell
Gillum, Amanda McKee
Gilvarg, Charles
Ginsberg, Barry Howard
Glenn, Jeffrey K
Goldman, Emanuel
Goldsmith, Laura Tobi
Grattan, Jerome Francis
Greasham, Randolph Louis
Gyure, William Louis
Haas, Gerhard Julius
Hager, Mary Hastings
Haimovich, Beatrice
Hall, James Conrad
Hansen, Hans John
Hanson, Ronald Lee
Harris, Don Navarro
Harvey, Richard Alexander
Heaslip, Richard Joseph
Heikkila, Richard Elmer
Hill, Helene Zimmermann
Hirschberg, Erich
Hitchcock-DeGregori, Sarah Ellen
Hoffman, Allan Jordan
Huang, Dao Pei
Huang, Mou-Tuan
Huff, Jesse William
Huger, Francis P
Humayun, Mir Z
Humes, John Leroy
Humphrey, Brian
Hwang, San-Bao
Ianuzzo, C David
Inamine, Edward S(eiyu)
Ivatt, Raymond John
Jackson, Richard Lee
Jacob, Theodore August
Jakubowski, Hieronim Zbigniew
Jass, Herman Earl
Johnson, Dewey, Jr
Jones, Benjamin Lewis
Kaczorowski, Gregory John
Kadin, Harold
Kamiyama, Mikio
Kao, John Y
Katzen, Howard M
Kauffman, Frederick C
Keane, Kenneth William
Kenkare, Divaker B
Khan, Fazal R
Kirk, James Robert
Klessig, Daniel Frederick
Kongsamut, Sathapana
Koster, William Henry
Kozarich, John Warren
Kreutner, William
Krishnan, Gopal
Kroll, Emanuel
Krug, Robert M
Krystek, Stanley R, Jr
Ku, Edmond Chiu-Choon
Kumar, Suriender
Kumbaraci-Jones, Nuran Melek
Laden, Karl
Lambert, Muriel Wikswo
Lampen, J Oliver
Landau, Matthew Paul
Lane, Alexander Z
Laskin, Allen I
Laughlin, Alice
LaVoie, Edmond J
Lea, Michael Anthony
Ledeen, Robert
Lee, Catherine Coyle

Lee, Lih-Syng
Lehrer, Harris Irving
Leitz, Victoria Mary
Lenard, John
Leonards, Kenneth Stanley
Leveille, Gilbert Antonio
Levin, Wayne
Lewis, Katherine
Lin, Reng-Lang
Linemeyer, David L
Linsky, Cary Bruce
Liu, Alice Yee-Chang
Liu, Leroy Fong
Liu, Wen Chih
Logdberg, Lennart Erik
Lovell, Richard Arlington
Lu, Anthony Y H
Lunn, Charles Albert
McGuinness, Eugene T
McGuire, John L
Machlin, Lawrence Judah
Maclin, Ernest
Malathi, Parameswara
Mandel, Lewis Richard
Manne, Veeraswamy
Mannino, Raphael James
Manowitz, Paul
Margolis, Frank L
Marquez, Joseph A
Martin, Charles Everett
Marx, Joseph Vincent
Massover, William H
Mattes, William Bustin
Matthijssen, Charles
Mayhew, Eric George
Meienhofer, Johannes Arnold
Mellin, Theodore Nelson
Messing, Joachim W
Mezick, James Andrew
Miller, Douglas Kenneth
Monestier, Marc
Moore, John Baily, Jr
Morck, Roland Anton
Moreland, Suzanne
Morris, James Albert
Mosley, Stephen T
Mostillo, Ralph
Mowles, Thomas Francis
Mueller, Stephen Neil
Mullinix, Kathleen Patricia
Murthy, Vadiraja Venkatesa
Muth, Eric Anthony
Nager, Urs Felix
Naimark, George Modell
Nakhla, Atif Mounir
Nalin, Carlo M
Nash, Harold Anthony
Nemecek, Georgina Marie
Neri, Anthony
Newmark, Harold Leon
Nichols, Joseph
Nicolau, Gabriela
Niederman, Robert Aaron
Norvitch, Mary Ellen
Novick, William Joseph, Jr
O'Byrne, Elizabeth Milikin
Ofengand, Edward James
Oien, Helen Grossbeck
OSullivan, Joseph
Ott, Walther H
Ozer, Harvey Leon
Pachter, Jonathan Alan
Paisley, Nancy Sandelin
Palmere, Raymond M
Pan, Yu-Ching E
Panagides, John
Pang, David C
Papastephanou, Constantin
Paterniti, James R, Jr
Peake, Clinton J
Peets, Edwin Arnold
Pestka, Sidney
Petrack, Barbara Kepes
Pierce, George Edward
Pietruszko, Regina
Pitts, Barry James Roger
Pleasants, Elsie W
Pobiner, Bonnie Fay
Poe, Martin
Poretz, Ronald David
Prince, Roger Charles
Raetz, Christian Rudolf Hubert
Raman, Jay Ananth
Reckel, Rudolph P
Reeves, John Paul
Ren, Peter
Reuben, Roberta C
Ripley, Lynn S
Robb, Richard John
Robbins, James Clifford
Rodwell, John Dennis
Rogers, Dexter
Rosenblum, Irwin Yale
Rossow, Peter William
Rothrock, John William
Rothstein, Rodney Joel
Rowley, George Richard
Rumsey, William LeRoy
Russell, Thomas J, Jr
Ryzlak, Maria Teresa
Saldarini, Ronald John
Sanders, Marilyn Magdanz
Saunders, Robert Norman
Schaffner, Carl Paul

Schaich, Karen Marie
Scheiner, Donald M
Schlesinger, David H
Schreiner, Ceinwen Ann
Schulman, Marvin David
Schwartz, Herbert
Schwartz, Jeffrey Lee
Schwarz, Hans Jakob
Scott, William Addison, III
Seaver, Sally S
Sekula, Bernard Charles
Sessa, Grazia L
Shapiro, Bennett Michaels
Shapiro, Herman Simon
Shapiro, Irwin Louis
Shapiro, Stanley Seymour
Shatkin, Aaron Jeffrey
Sheppard, Herbert
Sheriff, Steven
Sherman, Merry Rubin
Sherman, Michael Ian
Sherr, Stanley I
Siegel, Marvin I
Silberklang, Melvin
Sills, Matthew A
Simmons, Jean Elizabeth Margaret
Simplicio, Jon
Sipos, Tibor
Skarstedt, Mark T(eofil)
Skogerson, Lawrence Eugene
Slater, Eve Elizabeth
Smith, Marian Jose
Smith, Sidney R, Jr
Snyder, Robert
Sofer, William Howard
Sohler, Arthur
Solis-Gaffar, Maria Corazon
Spano, Francis A
Spiegel, Herbert Eli
Staub, Herbert Warren
Steinberg, Malcolm Saul
Steiner, Kurt Edric
Sterbenz, Francis Joseph
Stern, William
Stiefel, Edward
Stier, Elizabeth Fleming
Stone, Charles Dean
Stone, Edward John
Sullivan, Andrew Jackson
Sutherland, Donald James
Svokos, Steve George
Swenson, Theresa Lynn
Swislocki, Norbert Ira
Szymanski, Chester Dominic
Tate, Robert Lee, III
Teipel, John William
Thomas, Kenneth Alfred, Jr
Tischio, John Patrick
Tkacz, Jan S
Tocco, Dominick Joseph
Tomeo-Dymkowski, Aline Claire
Tornqvist, Erik Gustav Markus
Triscari, Joseph
Trotta, Paul P
Tsong, Yun Yen
Udenfriend, Sidney
Umbreit, Wayne William
Vagelos, P Roy
Vander Wende, Christina
Van Inwegen, Richard Glen
Vasconcelos, Aurea C
Vellekamp, Gary John
Viebrock, Frederick William
Villafranca, Joseph John
Wagman, Gerald Howard
Wang, Bosco Shang
Wang-Iverson, Patsy
Wase, Arthur William
Wasserman, Paul Michael
Wasserman, Bruce P
Watkins, Tom R
Watrel, Warren George
Watson, Richard White, Jr
Weissbach, Herbert
Welton, Ann Frances
Williams, Jeffrey Walter
Wilson, Alan C
Wilson, Robert G
Winters, Harvey
Wolff, Donald John
Wolff, John Shearer, III
Wong, Keith Kam-Kin
Wong, Rosie Bick-Har
Wood, Alexander W
Wu, Joseph Woo-Tien
Xiong, Yimin
Yang, Chung Shu
You, Kwan-sa
Zilinskas, Barbara Ann
Zuckerman, Leo

NEW MEXICO
Anderson, William Loyd
Atencio, Alonzo C
Baker, Thomas Irving
Barton, Larry Lumir
Belinsky, Steven Alan
Brandvold, Donald Keith
Buss, William Charles
Carroll, Arthur Paul
Casillas, Edmund Rene
Coffey, John Joseph
Darnall, Dennis W
Gordon, Malcolm Wofsy

Griffin, Travis Barton
Gurd, Frank Ross Newman
Gurd, Ruth Sights
Gurley, Lawrence Ray
Hageman, James Howard
Herman, Ceil Ann
Hildebrand, Carl Edgar
Hoard, Donald Ellsworth
Jenness, Robert
Kelly, Gregory
Kemp, John Daniel
Kraemer, Paul Michael
Kuehn, Glenn Dean
Loftfield, Robert Berner
McNamara, Mary Colleen
Omdahl, John L
Pastuszyn, Andrzej
Ratliff, Robert L
Reyes, Philip
Rivett, Robert Wyman
Roberson, Robert H
Roberts, Peter Morse
Sae, Andy S W
Saponara, Arthur G
Saunders, George Cherdron
Scallen, Terence
Schoenfeld, Robert George
Seagrave, JeanClare
Smoake, James Alvin
Standefer, Jimmy Clayton
Strniste, Gary F
Thrasher, Jack D
Trotter, John Allen
Turner, Ralph B
Vander Jagt, David Lee
Vogel, Kathryn Giebler
Wild, Gaynor (Clarke)
Woodfin, Beulah Marie

NEW YORK
Aaronson, Sheldon
Aaronson, Stuart Alan
Abayev, Michael
Abell, Liese Lewis
Abood, Leo George
Abramson, Morris Barnet
Abreu, Sergio Luis
Acs, George
Aisen, Philip
Albanese, Anthony August
Albrecht, Alberta Marie
Alderfer, James Landes
Alexander, George Jay
Alexander, Renee R
Allen, Howard Joseph
Allen, Sydney Henry George
Almon, Richard Reiling
Altman, Kurt Ison
Ambron, Richard Thomas
Amiraian, Kenneth
Anchel, Marjorie Wolff
Andersen, Jon Alan
Anderson, Mary Elizabeth
Andrews, John Parray
Archibald, Reginald MacGregor
Arion, William Joseph
Armstrong, Donald
Aronson, Robert Bernard
Arvan, Dean Andrew
Asch, Harold Lawrence
Asimov, Isaac
Astill, Bernard Douglas
Atlas, Steven Alan
Auclair, Walter
Augenlicht, Leonard Harold
Augustyn, Joan Mary
Awad, Atif B
Bachvaroff, Radoslav J
Bagchi, Sakti Prasad
Baglioni, Corrado
Baird, Malcolm Barry
Balis, Moses Earl
Bambara, Robert Anthony
Banay-Schwartz, Miriam
Banerjee, Debendranath
Banerjee, Ranjit
Barany, Francis
Barber, Eugene Douglas
Bard, Enzo
Barker, Robert
Basilico, Claudio
Batt, William
Baum, George
Baum, Howard Jay
Bazinet, George Frederick
Beach, David H
Bedard, Donna Lee
Beeler, Donald A
Bellve, Anthony Rex
Belman, Sidney
Belsky, Melvin Myron
Benesch, Ruth Erica
Bensadoun, Andre
Benuck, Myron
Benzo, Camillo Anthony
Berezney, Ronald
Berl, Soll
Bernacki, Ralph J
Bernardis, Lee L
Betheil, Joseph Jay
Beyer, Carl Fredrick
Bhargava, Madhu Mittra
Bidlack, Jean Marie
Bini, Alessandra Margherita

Biochemistry (cont)

Birecka, Helena M
Birken, Steven
Bishop, Charles (William)
Bishop, David Franklin
Black, Virginia H
Blank, Martin
Bloch, Alexander
Bloch, Eric
Blosser, James Carlisle
Blume, Arthur Joel
Blumenfeld, Olga O
Blumenstein, Michael
Bonney, Robert John
Boozer, Carol Sue Neely
Bora, Sunder S
Borenfreund, Ellen
Borgese, Thomas A
Bourne, Philip Eric
Bowman, William Henry
Brachfeld, Norman
Bradlow, Herbert Leon
Braunstein, Joseph David
Bray, Bonnie Anderson
Brenner, Mortimer W
Breslow, Jan Leslie
Brewer, Curtis Fred
Briehl, Robin Walt
Brockerhoff, Hans
Brody, Edward Norman
Brody, Marcia
Brooks, Robert R
Brown, Oliver Monroe
Brownie, Alexander C
Bryan, Ashley Monroe
Bryan, John Kent
Buckley, Edward Harland
Burger, Richard Melton
Burke, G Thompson
Burke, William Thomas, Jr
Bush, Karen Jean
Bushkin, Yuri
Calame, Kathryn Lee
Calhoun, David H
Calvo, Joseph M
Campbell, Bruce (Nelson), Jr
Campbell, Thomas Colin
Cano, Francis Robert
Carter, Timothy Howard
Catapane, Edward John
Catterall, James F
Center, Sharon Anne
Cerami, Anthony
Chan, Arthur Wing Kay
Chan, Peter Sinchun
Chan, Phillip C
Chan, Samuel H P
Chang, Pei Kung (Philip)
Chang, Ta-Min
Chanley, Jacob David
Chargaff, Erwin
Charney, Martha R
Chasalow, Fred I
Chasin, Lawrence Allen
Chatterjee, Nando Kumar
Chatterjee, Sunil Kumar
Chauhan, Abha
Chauhan, Ved P S
Chen, Ching-Ling Chu
Chesley, Leon Carey
Chheda, Girish B
Chiao, Jen Wei
Chou, Ting-Chao
Christie, Nelwyn T
Christy, Nicholas Pierson
Chu, Tsann Ming
Chutkow, Jerry Grant
Clark, Virginia Lee
Clarke, Donald Dudley
Clarkson, Allen Boykin, Jr
Claus, Thomas Harrison
Clesceri, Lenore Stanke
Coderre, Jeffrey Albert
Cohen, Gerald
Colman, David Russell
Conway de Macario, Everly
Cooper, Arthur Joseph L
Corpe, William Albert
Corradino, Robert Anthony
Costa, Max
Cote, Lucien Joseph
Counts, David Francis
Cowger, Marilyn L
Cross, George Alan Martin
Cross, Richard Lester
Crystal, Ronald George
Cunningham, Richard Preston
Cunningham-Rundles, Charlotte
Daly, Marie Maynard
Damadian, Raymond
Dancis, Joseph
Danishefsky, Isidore
Darzynkiewicz, Zbigniew Dzierzykraj
Datta, Ranajit Kumar
Davenport, Lesley
Davidow, Bernard
Davidson, Mercy
Davies, Kelvin J A
De Duve, Christian Rene
De Luca, Chester
Deshmukh, Diwakar Shankar
Desnick, Robert John
Desplan, Claude

DeTitta, George Thomas
Detmers, Patricia Anne
D'Eustachio, Peter
Dickerman, Herbert W
DiMauro, Salvatore
Di Paola, Mario
Diwan, Joyce Johnson
Doering, Charles Henry
Dolan, Jo Alene
Doudney, Charles Owen
Downs, Frederick Jon
Doyle, Darrell Joseph
Drickamer, Kurt
Dubroff, Lewis Michael
Dudock, Bernard S
Duff, Ronald George
Duta, Purabi
Eberhard, Anatol
Edelman, Isidore Samuel
Ehrke, Mary Jane
Eidinoff, Maxwell Leigh
Elliott, Rosemary Waite
Ellsworth, Robert King
Elsbach, Peter
Elwood, John Clint
Elwyn, David Hunter
Elzinga, Marshall
Emmons, Scott W
Enea, Vincenzo
England, Sasha
Erlanger, Bernard Ferdinand
Erlichman, Jack
Esders, Theodore Walter
Etlinger, Joseph David
Ettinger, Murray J
Eudy, William Wayne
Evans, Mary Jo
Everhart, Donald Lee
Ewart, Mervyn H
Fasco, Michael John
Feigelson, Philip
Feinman, Richard David
Fenton, John William, II
Ferguson, John Barclay
Ferrari, Richard Alan
Ferraro, John J
Ferris, James Peter
Fewkes, Robert Charles Joseph
Fiala, Emerich Silvio
Fine, Albert Samuel
Fischbarg, Jorge
Fisher, Paul Andrew
Fishman, Myer M
Fitzgerald, Patrick James
Fleischmajer, Raul
Fletcher, Paul Wayne
Florini, James Ralph
Fondy, Thomas Paul
Fopeano, John Vincent, Jr
Forest, Charlene Lynn
Fox, Jack Jay
Fox, Thomas David
Frank, David Stanley
Frankel, Arthur Irving
Fredrick, Jerome Frederick
Free, Stephen J
Freed, Simon
Freedberg, Irwin Mark
Freedman, Aaron David
Freedman, Jeffrey Carl
Freedman, Lewis Simon
Frenkel, Krystyna
Freundlich, Martin
Fried, Victor A
Furchgott, Robert Francis
Galivan, John H
Garrick, Laura Morris
Garrick, Michael D
Garte, Seymour Jay
Gates, Robert Leroy
Gaver, Robert Calvin
Gelbard, Alan Stewart
George, Elmer, Jr
Gershengorn, Marvin Carl
Gershman, Lewis C
Gershon, Herman
Ghebrehiwet, Berhane
Giampietro, Philip Francis
Gibson, Audrey Jane
Gibson, Kenneth David
Gizis, Evangelos John
Glenn, Joseph Leonard
Goebel, Walther Frederick
Goff, Stephen Payne
Gold, Allen Morton
Goldberg, Allan Roy
Goldberger, Robert Frank
Goldman, James Eliot
Goldsmith, Lowell Alan
Goldstein, Fred Bernard
Goldstein, Gilbert
Goldstein, Jack
Goldstein, Menek
Golub, Lorne Malcolm
Golubow, Julius
Goodhue, Charles Thomas
Goodman, DeWitt Stetten
Gordon, Portia Beverly
Grauer, Amelie L
Green, Saul
Greenberg, Jacob
Greenberger, Lee M
Greengard, Olga
Greengard, Paul

Grieninger, Gerd
Griffiths, Joan Martha
Grubman, Marvin J
Grumet, Martin
Grunberger, Dezider
Gudas, Lorraine J
Gupta, Raj K
Gurel, Okan
Gurpide, Erlio
Gutcho, Sidney J
Guttenplan, Joseph B
Habicht, Jean-Pierre
Hackler, Lonnie Ross
Haines, Thomas Henry
Haldar, Dipak
Haley, Nancy Jean
Hamill, Robert W
Hanafusa, Hidesaburo
Hankes, Lawrence Valentine
Hanson, Barbara Ann
Hardy, Ralph Wilbur Frederick
Harris, Alex L
Harrison, Robert Walker, III
Harris-Warrick, Ronald Morgan
Hartenstein, Roy
Hatcher, Victor Bernard
Hausmann, Ernest
Haymovits, Asher
Haywood, Anne Mowbray
Hehre, Edward James
Heintz, Roger Lewis
Held, William Allen
Helmann, John Daniel
Hendrickson, Wayne Arthur
Henneberry, Richard Christopher
Henrikson, Katherine Pointer
Henshaw, Edgar Cummings
Heppel, Leon Alma
Herz, Fritz
Hess, George Paul
Higashiyama, Tadayoshi
Hilborn, David Alan
Hilf, Russell
Hill, Doyle Eugene
Hind, Geoffrey
Hinkle, David Currier
Hinkle, Patricia M
Hinkle, Peter Currier
Hipp, Sally Sloan
Hirschman, Albert
Hoberman, Henry Don
Hoch, George Edward
Hochstadt, Joy
Hof, Liselotte Bertha
Hoffmann, Dietrich
Hogg, James Felter
Hohmann, Philip George
Hohnadel, David Charles
Holland, Mary Jean Carey
Hollander, Joshua
Holtz, Aliza
Hommes, Frits A
Horecker, Bernard Leonard
Horowitz, Martin I
Horwitz, Susan Band
Hough, Jane Linscott
Hoyte, Robert Mikell
Hrazdina, Geza
Hsu, Ming-Ta
Huang, Cheng-Chun
Huang, Sylvia Lee
Huberman, Joel Anthony
Hudecki, Michael Stephen
Hughes, Walter Lee, Jr
Hui, Koon-Sea
Hui, Sek Wen
Iglewski, Barbara Hotham
Ikonne, Justus Uzoma
Iodice, Arthur Alfonso
Ip, Clement Cheung-Yung
Ip, Margot Morris
Isaacs, Charles Edward
Iyengar, Ravi Srinivas V
Jack, Robert Cecil Milton
Jackanicz, Theodore Michael
Jackson, Robert C
Jacobs, Laurence Stanton
Jacobs, Richard L
Jacobs, Ross D
Jacobson, Herbert (Irving)
Jainchill, Jerome
Jaken, Susan
Jamdar, Subhash C
Jayme, David Woodward
Jean-Baptiste, Emile
Jeffrey, John J
Joh, Tong Hyub
Johnson, Alan J
Jung, Chan Yong
Juo, Pei-Show
Kabat, Elvin Abraham
Kaiserman, Howard Bruce
Kaminsky, Laurence Samuel
Kaplan, Barry Hubert
Kaplan, John Ervin
Karlin, Arthur
Karpatkin, Simon
Karr, James Presby
Katsel, Pavel Leon
Katsoyannis, Panayotis G
Katz, Eugene Richard
Kazarinoff, Michael N
Keithly, Janet Sue
Kelleher, Raymond Joseph, Jr

Keller, Elizabeth Beach
Kesner, Leo
Khan, Anwar Ahmad
Khan, Nasim A
Kieras, Fred J
Kim, Giho
Kim, Young Joo
Kim, Young Tai
Kimmich, George Arthur
Kincl, Fred Allan
Kirdani, Rashad Y
Kish, Valerie Mayo
Kishore, Gollamudi Sitaram
Kiyasu, John Yutaka
Klinge, Carolyn M
Klingman, Jack Dennis
Knaak, James Bruce
Kobilinsky, Lawrence
Kochwa, Shaul
Koehn, Paul V
Kohler, Constance Anne
Koizumi, Kiyomi
Konarska, Maria Magda
Koretz, Jane Faith
Koritz, Seymour Benjamin
Kosman, Daniel Jacob
Kowalski, David Francis
Kraft, Patricia Lynn
Krakow, Joseph S
Kramer, Fred Russell
Krasna, Alvin Isaac
Krasner, Joseph
Krasnow, Frances
Kream, Jacob
Kress, Lawrence Francis
Kronman, Martin Jesse
Krulwich, Terry Ann
Kuchinskas, Edward Joseph
Kuntzman, Ronald Grover
Kuramitsu, Howard Kikuo
Lacks, Sanford
Lahita, Robert George
Lai, Fong M
Lajtha, Abel
Lamberg, Stanley Lawrence
Lamster, Ira Barry
Lang, Helga M (Sr Therese)
Lapin, Evelyn P
Largis, Elwood Eugene
LaRue, Thomas A
Last, Robert L
Lau, Joseph T Y
Lavallee, David Kenneth
Lazaroff, Norman
Le, Heng-Chun
Lee, Ching-Li
Lee, Teh Hsun
Lee, Wei-Li S
Lee-Huang, Sylvia
Leff, Judith
Lerner, Leon Maurice
Levitz, Mortimer
Levy, Hans Richard
Levy, Harvey Merrill
Lewis, Bertha Ann
Li, Heng-Chun
Li, Lu Ku
Li, Luyuan
Libby, Paul Robert
Lieberman, Seymour
Liem, Ronald Kian Hong
Lin, Mow Shiah
Lindahl, Lasse Allan
Linder, Regina
Lipke, Peter Nathan
Lis, John Thomas
Listowsky, Irving
Liu, Houng-Zung
Lloyd, Kenneth Oliver
Loeb, John Nichols
Loeb, Marilyn Rosenthal
Lonberg-Holm, Knud Karl
London, Morris
Loullis, Costas Christou
Low, Barbara Wharton
Lucas, John J
Lugay, Joaquin Castro
Luine, Victoria Nall
Lusty, Carol Jean
Lyons, Michael Joseph
McCabe, R Tyler
McClintock, David K
MacColl, Robert
McCormick, J Robert D
McCormick, Paulette Jean
McEwen, Bruce Sherman
McGuire, John Joseph
Maddaiah, Vaddanahally Thimmaiah
Madden, Robert E
Madison, James Thomas
Magid, Norman Mark
Magliulo, Anthony Rudolph
Maio, Joseph James
Maitra, Subir R
Maitra, Umadas
Maitra, Utpalendu S
Makman, Maynard Harlan
Malbon, Craig Curtis
Maley, Frank
Maley, Gladys Feldott
Malhotra, Ashwani
Malik, Mazhar N
Mandl, Ines
Manning, James Matthew

Marcu, Kenneth Brian
Marenus, Kenneth D
Marfey, Sviatopolk Peter
Margolis, Renee Kleimann
Margolis, Richard Urdangen
Margossian, Sarkis S
Marinetti, Guido V
Marks, Neville
Markus, Gabor
Marmur, Julius
Martin, Constance R
Martin, David Lee
Martin, Kumiko Oizumi
Martonosi, Anthony
Matsui, Sei-Ichi
Maturo, Joseph Martin, III
Maxfield, Frederick Rowland
Mazumder, Rajarshi
Mazur, Abraham
Meister, Alton
Meltzer, Herbert Lewis
Mendelsohn, Naomi
Merrick, Joseph M
Merrifield, Robert Bruce
Meyer, Haruko
Meyers, Marian Bennett
Miller, Herbert Kenneth
Miller, Leon Lee
Miller, Terry Lynn
Mitacek, Eugene Jaroslav
Mittag, Thomas Waldemar
Mizejewski, Gerald Jude
Model, Peter
Moline, Sheldon Walter
Monder, Carl
Mondy, Nell Irene
Monheit, Alan G
Mooney, Robert Arthur
Moorehead, Thomas J
Moos, Carl
Morck, Timothy Anton
Morin, John Edward
Morrill, Gene A
Morrison, John B
Morrison, Sidonie A
Mortlock, Robert Paul
Mosbach, Erwin Heinz
Moscatelli, David Anthony
Muller, Miklos
Muller-Eberhard, Ursula
Munk, Vladimir
Munson, Benjamin Ray
Mussell, Harry W
Nandi, Jyotirmoy
Napolitano, Raymond L
Neal, Michael William
Neeman, Moshe
Nelson, Mark Thomas
Nemerson, Yale
Nettleton, Donald Edward, Jr
Neubort, Shimon
Nisselbaum, Jerome Seymour
Norton, William Thompson
Notides, Angelo C
Obendorf, Ralph Louis
O'Connor, John Francis
Oesper, Peter
Olmsted, Joanna Belle
Olson, Robert Eugene
Ong, Eng-Bee
Oratz, Murray
Oronsky, Arnold Lewis
Papsidero, Lawrence D
Pardee, Joel David
Parham, Margaret Payne
Parker, Frank S
Pasternak, Gavril William
Patel, Mulchand Shambhubhai
Patterson, Ernest Leonard
Payne, Michael Ross
Pearlmutter, Anne Frances
Peerschke, Ellinor Irmgard Barbara
Peisach, Jack
Pellicer, Angel
Peruzzotti, George Peter
Peters, Theodore, Jr
Peverly, John Howard
Philipp, Manfred (Hans)
Pierre, Leon L
Pine, Martin J
Pinto, John
Plager, John Everett
Plummer, Thomas H, Jr
Pogo, A Oscar
Poindexter, Jeanne Stove
Polatnick, Jerome
Pollard, Jeffrey William
Popenoe, Edwin Alonzo
Portnoy, Debbi
Posner, Aaron Sidney
Praissman, Melvin
Prasthofer, Thomas W
Prestwich, Glenn Downes
Price, Frederick William
Pruslin, Fred Howard
Pullarkat, Raju Krishnan
Pullman, Maynard Edward
Pumo, Dorothy Ellen
Puszkin, Elena Getner
Pye, Orrea F
Quaroni, Andrea
Quigley, James P
Raghavan, Srinivasa
Rambaut, Paul Christopher

Rapport, Maurice M
Rathnam, Premila
Ratner, Sarah
Rattazzi, Mario Cristiano
Reddy, Bandaru Sivarama
Redman, Colvin Manuel
Reed, Roberta Gable
Reeke, George Norman, Jr
Reeves, Stuart Graham
Reichberg, Samuel Bringeissen
Reichert, Leo E, Jr
Richie, John Peter, Jr
Rieder, Conly LeRoy
Rizack, Martin A
Rizzuto, Anthony B
Roberts, James Lewis
Roberts, Jeffrey Warren
Roberts, Richard Norman
Robinson, Alix Ida
Robinson, Joseph Douglass
Roeder, Robert Gayle
Roepe, Paul David
Roffman, Steven
Rogerson, Allen Collingwood
Rosano, Thomas Gerard
Rosen, John Friesner
Rosenfeld, Louis
Rosenfeld, Martin Herbert
Rosenstein, Barry Sheldon
Rosner, William
Ross, John Brandon Alexander
Rossman, Toby Gale
Roth, Jerome Allan
Rothman, James Edward
Rothstein, Morton
Rowe, Arthur W(ilson)
Roy, Harry
Rubin, Charles Stuart
Rubin, Ronald Philip
Russell, Charlotte Sananes
Ryan, Thomas John
Sabatini, David Domingo
Sabban, Esther Louise
Sachdev, Om Prakash
Sack, Robert A
Sacks, William
Saez, Juan Carlos
Salerno, John Charles
Salvo, Joseph J
Samuels, Stanley
Sanford, Karl John
Sankar, D V Siva
Santiago, Noemi
Sarcione, Edward James
Sarma, Raghupathy
Sarma, Ramaswamy Harihara
Sassa, Shigeru
Satir, Birgit H
Sauer, Leonard A
Sauro, Marie D
Saxena, Brij B
Sayegh, Joseph Frieh
Schechter, Nisson
Scher, William
Scheuer, James
Schildkraut, Carl Louis
Schlessinger, Joseph
Schliselfeld, Louis Harold
Schneider, Allan Stanford
Schor, Joseph Martin
Schramm, Vern Lee
Schreiner, Heinz Rupert
Schubert, Edward Thomas
Schuel, Herbert
Schulman, LaDonne Heaton
Schwartz, Gerald Peter
Schwartz, Ira
Schwartz, Morton K
Schwartz, Stanley Allen
Scott, Walter Neil
Segal, Harold Lewis
Seifter, Sam
Shafer, Stephen Joel
Shafit-Zagardo, Bridget
Shalita, Alan Remi
Shalloway, David Irwin
Shapiro, Robert
Sheid, Bertrum
Shields, Dennis
Short, Sarah Harvey
Siebert, Karl Joseph
Siegelman, Harold William
Siekevitz, Philip
Siever, Larry Joseph
Silverman, Morris
Silverstein, Emanuel
Silverstein, Samuel Charles
Simon, Eric Jacob
Simon, Martha Nichols
Simon, Sanford Ralph
Simpson, Melvin Vernon
Singh, Malathy
Singh, Toolsee J
Slaunwhite, Wilson Roy, Jr
Slocum, Harry Kim
Smith, Richard Alan
Snoke, Roy Eugene
Snyderman, Selma Eleanore
Sobel, Jael Sabina
Soeiro, Ruy
Soffer, Richard Luber
Sonenberg, Martin
Sowinski, Raymond
Spaulding, Stephen Waasa

Spector, Abraham
Spector, Leonard B
Sprinson, David Benjamin
Spritz, Norton
Squires, Catherine L
Squires, Richard Felt
Srinivasan, P R
Srivastava, Bejai Inder Sahai
Stahl, William J
Stanley, Evan Richard
Stanley, Pamela Mary
Stasiw, Roman Orest
Steffen, Daniel G
Stein, Theodore Anthony
Steinman, Charles Robert
Steinman, Howard Mark
Stempien, Martin F, Jr
Sternglanz, Rolf
Stillman, Bruce William
Stohrer, Gerhard
Stracher, Alfred
Straus, David Bradley
Sturman, John Andrew
Sulkowski, Eugene
Sullivan, David Thomas
Sun, Tung-Tien
Sutherland, Betsy Middleton
Swank, Richard Tilghman
Szebenyi, Doletha M E
Szer, Wlodzimierz
Tabachnick, Milton
Tabor, John Malcolm
Tallan, Harris H
Tamir, Hadassah
Tamm, Igor
Tanenbaum, Stuart William
Tan-Wilson, Anna L(i)
Tate, Suresh S
Taub, Mary L
Teebor, George William
Teply, Lester Joseph
Testa, Raymond Thomas
Thau, Rosemarie B Zischka
Thomas, John Owen
Thompson, John Fanning
Thysen, Benjamin
Titeler, Milt
Tobkes, Nancy J
Tomashefsky, Philip
Tomasz, Alexander
Tomasz, Maria
Tometsko, Andrew M
Tooney, Nancy Marion
Toth, Eugene J
Traber, Maret G
Treble, Donald Harold
Trimble, Robert Bogue
Troll, Walter
Tropp, Burton E
Tseng, Linda
Tunis, Marvin
Turinsky, Jiri
Valinsky, Jay E
Van Buren, Jerome Paul
Van Campen, Darrell R
Verma, Devi C
Vladutiu, Georgirene Dietrich
Volkman, David J
Vratsanos, Spyros M
Wagh, Premanand Vinayak
Wainfan, Elsie
Walker, Theresa Anne
Wang, Dalton T
Wang, Jui Hsin
Wang, Peizhi
Wang, Tung Yue
Wapnir, Raul A
Warren, William A
Wasserman, Robert Harold
Watanabe, Kyoichi A
Weber, Peter B
Weinfeld, Herbert
Weinstock, Irwin Morton
Weiss, Irma Tuck
Wellner, Daniel
Wellner, Vaira Pamiljans
Wenner, Charles Earl
Westerfeld, Wilfred Wiedey
Weyter, Frederick William
Wheatley, Victor Richard
Wiberg, John Samuel
Wiesner, Rakoma
Wilk, Sherwin
Williams, Harold Henderson
Williams, William Joseph
Wilson, David Buckingham
Wilson, Karl A
Winkler, Norman Walter
Wittenberg, Beatrice A
Wittenberg, Jonathan B
Wolf, Robert Lawrence
Wolf, Walter Alan
Wolfner, Mariana Federica
Wolgemuth, Debra Joanne
Wolin, Meyer Jerome
Wong, Patrick Yui-Kwong
Woodard, Helen Quincy
Wright, Lemuel Dary
Wu, Joseph M
Wu, Ray J
Wulff, Daniel Lewis
Yang, Song-Yu
Yeagle, Philip L
Yeh, Samuel D J

Yen, Andrew
Yeung, Kwok Kam
Young, John Ding-E
Young, Roger Grierson
Yu, Shiu Yeh
Yue, Robert Hon-Sang
Yunghans, Wayne N
Zahnd, Hugo
Zakim, David
Zakrzewski, Sigmund Felix
Zanetti, Nina Clare
Zapisek, William Francis
Zengel, Janice Marie
Ziegler, Frederick Dixon
Zigman, Seymour
Zitomer, Richard Stephen
Zorzoli, Anita
Zwain, Ismail Hassan

NORTH CAROLINA
Abdullah, Munir
Agris, Paul F
Albro, Phillip William
Alphin, Reevis Stancil
Andersen, Melvin Ernest
Anderson, Marshall W
Anderson, Richmond K
Armstrong, Frank Bradley, Jr
Aurand, Leonard William
Baccanari, David Patrick
Bales, Connie Watkins
Banes, Albert Joseph
Barakat, Hisham A
Barnes, Donald Wesley
Barrett, J(ames) Carl
Barrow, Emily Mildred Stacy
Bartholow, Lester C
Bates, William K
Bell, Fred E
Bell, Robert Maurice
Berkut, Michael Kalen
Birnbaum, Linda Silber
Bloom, Kerry Steven
Bond, James Anthony
Booth, Raymond George
Bott, Kenneth F
Bray, John Thomas
Burkhart, James Gaylord
Bushey, Dean Franklin
Buttke, Thomas Martin
Cahill, Charles L
Camp Hay, Pamela Jean
Caplow, Michael
Carter, Charles Williams, Jr
Casanova, Mercedes
Caterson, Bruce
Chae, Chi-Bom
Chae, Kun
Chalovich, Joseph M
Chaney, Stephen Gifford
Chang, Kwen-Jen
Cheng, Pi-Wan
Childers, Steven Roger
Chilton, Mary-Dell Matchett
Cidlowski, John A
Clare, Debra A
Clark, Martin Ralph
Clendenon, Nancy Ruth
Coffey, James Cecil, Jr
Coke, James Logan
Coleman, Mary Sue
Cooke, Anson Richard
Cory, Joseph G
Cross, Robert Edward
Crounse, Robert Griffith
Cunningham, Carol Clem
Currie, William Deems
Curtis, Susan Julia
Daniel, Larry W
Davis, Cynthia Marie
Dawson, Jeffrey Robert
DeJong, Donald Warren
De Miranda, Paulo
Devereux, Theodora Reyling
Devlin, Robert B
Dobrogosz, Walter Jerome
Dohm, Gerald Lynis
Douglas, Michael Gilbert
Duch, David S
Duncan, Gordon Duke
Dunnick, June K
Eddy, Edward Mitchell
Edgell, Marshall Hall
Elder, Joe Allen
Eldridge, John Charles
Eling, Thomas Edward
Elion, Gertrude Belle
Ellis, Fred Wilson
Endow, Sharyn Anne
Erickson, Bruce Wayne
Erickson, Harold Paul
Errede, Beverly Jean
Failla, Mark Lawrence
Faust, Robert Gilbert
Fernandes, Daniel James
Ferone, Robert
Ferris, Robert Monsour
Findlay, John W A
Fine, Jo-David
Fjellstedt, Thorsten A
Forman, Donald T
Fridovich, Irwin
Fried, Howard Mark
Frisell, Wilhelm Richard

48 / DISCIPLINE INDEX

Biochemistry (cont)

Fyfe, James Arthur
Garlich, Jimmy Dale
Goldstein, Joyce Allene
Goz, Barry
Greene, Ronald C
Groves, William Ernest
Hall, Iris Beryl Haddon
Hamilton, Pat Brooks
Hammes, Gordon G
Hamner, Charles Edward, Jr
Harrison, John Henry, IV
Hartz, John William
Hassan, Hosni Moustafa
Hatch, Gary Ephraim
Heck, Henry d'Arcy
Hemperly, John Jacob
Hendren, Richard Wayne
Hershfield, Michael Steven
Hickey, Anthony James
Hill, Charles Horace, Jr
Hill, Robert Lee
Hitchings, George Herbert
Holbrook, David James, Jr
Holm, Robert E
Holmes, Edward Warren
Hook, Gary Edward Raumati
Horton, Carl Frederick
Horton, Horace Robert
Huang, Jeng-Sheng
Hulcher, Frank H
Ito, Takeru
Jacobson, Kenneth Allan
Jetten, Anton Marinus
Jividen, Gay Melton
Johnson, Jean Louise
Johnson, Ronald Sanders
Joklik, Wolfgang Karl
Jones, Evan Earl
Jones, Mary Ellen
Jordan, Thomas L
Kahn, Joseph Stephan
Kasperek, George James
Kaufman, Bernard
Kaufman, David Gordon
Keene, Jack Donald
Kelly, Susan Jean
King, Ann Christie
King, Jonathan Stanton
King, Kendall Willard
Kirshner, Norman
Knight, Wilson Blaine
Kodavanti, Prasada Rao S
Kodavanti, Urmila P
Kohn, Michael Charles
Korach, Kenneth Steven
Krasny, Harvey Charles
Kredich, Nicholas M
Krenitsky, Thomas Anthony
Kucera, Louis Stephen
Kwock, Lester
Lack, Leon
LaFon, Stephen Woodrow
Lee, Martin Jerome
Lee, Si Duk
Li, Steven Shoei-Lung
Lipscomb, Elizabeth Lois
Lister, Mark David
Lloyd, Norman Edward
Longmuir, Ian Stewart
Loomis, Carson Robert
Louie, Dexter Stephen
Lucier, George W
Lumb, Roger H
Lundeen, Carl Victor, Jr
McCarty, Kenneth Scott
McIlwain, David Lee
McKee, David John
McLachlan, John Alan
Marks, Richard Henry Lee
Maroni, Donna F
Marzluff, William Frank, Jr
Massaro, Edward Joseph
Means, Anthony R
Mechanic, Gerald
Meissner, Gerhard
Melnick, Ronald L
Metzgar, Richard Stanley
Miller, David S
Miller, Joseph Edwin
Miller, William Laubach
Modrich, Paul L
Moreland, Donald Edwin
Morell, Pierre
Mortenson, Leonard Earl
Mushak, Paul
Narasimhachari, Nedathur
Nayfeh, Shihadeh Nasri
Naylor, Aubrey Willard
Newton, Jack W
Nichol, Charles Adam
Niedel, James Edward
O'Brien, Deborah A
O'Callaghan, James Patrick
O'Neal, Thomas Denny
Parks, Leo Wilburn
Pattee, Harold Edward
Pearlman, William Henry
Pearlstein, Robert David
Pekala, Phillip H
Pellizzari, Edo Domenico
Penniall, Ralph
Pennington, Sammy Noel

Petes, Thomas Douglas
Pillsbury, Harold C
Pizzo, Salvatore Vincent
Poole, Doris Theodore
Posner, Herbert S
Powers, Daniel D
Prestayko, Archie William
Privalle, Christopher Thomas
Quinn, Richard Paul
Rajagopalan, K V
Ramachandran, Muthukrishnan
Ramp, Warren Kibby
Reid, Lola Cynthia McAdams
Reinhard, John Frederick, Jr
Remy, Charles Nicholas
Richardson, Stephen H
Rodbell, Martin
Roer, Robert David
Ronzio, Robert A
Ross, Jeffrey Alan
Ross, Richard Henry, Jr
Rudel, Lawrence L
Sahyoun, Naji Elias
St Clair, Richard William
Sancar, Gwendolyn Boles
Sanders, Benjamin Elbert
Sanders, Ronald L
Sassaman, Anne Phillips
Scarborough, Gene Allen
Schmidt, Stephen Paul
Schneider, E Gayle
Schneider, Howard Albert
Selkirk, James Kirkwood
Shaw, Helen Lester Anderson
Shih, Jason Chia-Hsing
Siegel, Lewis Melvin
Sizemore, Ronald Kelly
Smith, Harold Carter
Smith, Peter Blaise
Smith, Susan T
Spector, Thomas
Spicer, Leonard Dale
Spiker, Steven L
Spremulli, Gertrude H
Steege, Deborah Anderson
Stiles, Gary L
Stone, Henry Otto, Jr
Stoskopf, Michael Kerry
Strittmatter, Cornelius Frederick
Sullivan, James Bolling
Summer, George Kendrick
Sundberg, David K
Sung, Michael Tse Li
Suzuki, Kunihiko
Swaisgood, Harold Everett
Switzer, Boyd Ray
Sylvia, Avis Latham
Taggart, R(obert) Thomas
Tanford, Charles
Taylor, Martha Loeb
Telen, Marilyn Jo
Teng, Ching Sung
Theil, Elizabeth
Thompson, William Francis, III
Thurman, Ronald Glenn
Toews, Arrel Dwayne
Tourian, Ara Yervant
Tove, Samuel B
Tove, Shirley Ruth
Trainer, John Ezra, Jr
Traut, Thomas Wolfgang
Tucker, Charles Leroy, Jr
Twarog, Robert
Van Loon, Edward John
Volk, Richard James
Wachsman, Joseph T
Wagner, Robert H
Waite, Moseley
Wall, Monroe Eliot
Wallace, James William, Jr
Wallen, Cynthia Anne
Waters, Michael Dee
Watts, John Albert, Jr
Webb, Neil Broyles
Webster, Robert Edward
Weybrew, Joseph Arthur
Wheat, Robert Wayne
White, Helen Lyng
White, J Courtland
White, James Rushton
Whitehurst, Garnett Brooks
Wilson, John Eric
Wilson, William Ewing
Wise, Edmund Merriman, Jr
Wiseman, Jeffrey Stewart
Wittels, Benjamin
Wolfenden, Richard Vance
Wykle, Robert Lee
Zalani, Sunita
Zimmerman, Thomas Paul
Zucker, Robert Martin

NORTH DAKOTA
Bakke, Jerome E
Banasik, Orville James
Buckner, James Stewart
Duerre, John A
Frear, Donald Stuart
Hunt, Janet R
Jacobs, Francis Albin
Johnson, W Thomas
Klosterman, Harold J
Knull, Harvey Robert
Lambeth, David Odus

Lamoureux, Gerald Lee
Lukaski, Henry Charles
Myron, Duane R
Nelson, Dennis Raymond
Nielsen, Forrest Harold
Nordlie, Robert Conrad
Park, Chung Sun
Paulson, Gaylord D
Ray, Paul Dean
Roseland, Craig R
Rosenberg, Harry
Samson, Willis Kendrick
Schnell, Robert Craig
Sheridan, Mark Alexander
Suttle, Jeffrey Charles
Vasey, Edfred H
Waller, James R
Zimmerman, Don Charles

OHIO
Abdallah, Abdulmuniem Husein
Addanki, Somasundaram
Alben, James O
Allred, John B
Alter, Gerald M
Anderson, Robert L
Askari, Amir
Astrachan, Lazarus
Austern, Barry M
Ball, William James, Jr
Banerjee, Amiya Kumar
Banerjee, Sipra
Barber, George Alfred
Baroudy, Bahige M
Batra, Prem Parkash
Becker, Carter Miles
Behnke, William David
Behr, Stephen Richard
Behrman, Edward Joseph
Berger, Melvin
Berliner, Lawrence J
Biagini, Raymond E
Blohm, Thomas Robert
Blumenthal, Kenneth Michael
Boczar, Barbara Ann
Boggs, Robert Wayne
Boyle, Michael Dermot
Bozian, Richard C
Brattain, Michael Gene
Brecher, Arthur Seymour
Brierley, Gerald Philip
Brooks, James Reed
Brunden, Kurt Russell
Bumpus, Francis Merlin
Buzard, James Albert
Camiener, Gerald Walter
Caplan, Arnold I
Carothers, Donna June
Cassidy, Suzanne Bletterman
Caston, J Douglas
Chase, John William
Chen, I-Wen
Chengelis, Christopher P
Clark, Burr, Jr
Clark, Eloise Elizabeth
Clark, Leland Charles, Jr
Clemans, George Burtis
Cooper, Cecil
Cooper, Dale A
Coots, Robert Herman
Cornwell, David George
Couri, Daniel
Craine, Elliott Maurice
Crandall, Dana Irving
Dage, Richard Cyrus
Darrow, Robert A
Davis, Pamela Bowes
Day, Richard Allen
Dedman, John Raymond
De Fiebre, Conrad William
Denko, Charles W
Deodhar, Sharad Dinkar
Deters, Donald W
Devor, Arthur William
Dewar, Norman Ellison
Dickman, John Theodore
DiCorleto, Paul Eugene
Dollwet, Helmar Hermann Adolf
Dorer, Frederic Edmund
Dorman, Robert Vincent
Doskotch, Raymond Walter
Doyle, Richard Robert
DuBrul, Ernest
Dumaswala, Umakant J
Ehrhart, L Allen
Eichel, Herman Joseph
Ekvall, Shirley W
Evans, Helen Harrington
Evans, William R
Faber, Lee Edward
Fanger, Bradford Otto
Feller, Dennis R
Flanagan, Margaret Ann
Fleischman, Darrell Eugene
Flick, Parke Kinsey
Frajola, Walter Joseph
Francis, Marion David
Frea, James Irving
Freed, James Melvin
Fritz, Herbert Ira
Ganis, Frank Michael Gangarosa
Ganschow, Roger Elmer
Giannini, A James
Gingery, Roy Evans
Goldthwait, David Atwater

Griffin, Charles Campbell
Griffin, Richard Norman
Groce, John Wesley
Gronostajski, Richard M
Gross, Elizabeth Louise
Grossman, Charles Jerome
Gruenstein, Eric Ian
Haas, Erwin
Hagerman, Dwain Douglas
Hamilton, Thomas Alan
Handwerger, Stuart
Hanson, Richard W
Harmony, Judith A K
Hartman, Warren Emery
Hascall, Vincent Charles, Jr
Hasselgren, Per-Olof J
Heckman, Carol A
Henningfield, Mary Frances
Hering, Thomas M
Hickenbottom, John Powell
Hieber, Thomas Eugene
Highsmith, Robert F
Hiremath, Shivanand T
Hirschmann, Hans
Hitt, John Burton
Hoff, Kenneth Michael
Homan, Ruth Elizabeth
Homer, George Mohn
Hopfer, Ulrich
Horowitz, Myer George
Horrocks, Lloyd Allen
Horseman, Nelson Douglas
Hoss, Wayne Paul
Hug, George
Hunter, James Edward
Hutterer, Ferenc
Ives, David Homer
Jacobs-Lorena, Marcelo
Jacobus, William Edward
Janusz, Michael John
Jaworski, Jan Guy
Jentoft, Joyce Eileen
Joffe, Frederick M
Johnson, Carl Lynn
Johnson, J David
Johnson, Lee Frederick
Johnson, Thomas Raymond
Johnston, John O'Neal
Jones, Stephen Wallace
Jordan, Freddie L
Kaneshiro, Edna Sayomi
Kanitz, Mary Helen Hitselberger
Kao, Winston Whei-Yang
Kariya, Takashi
Kean, Edward Louis
Keller, Stephen Jay
Klein, LeRoy
Kmetec, Emil Philip
Kolenbrander, Harold Mark
Koo, Peter H
Kowal, Jerome
Krampitz, Lester Orville
Kranias, Evangelia Galani
Krueger, Robert Carl
Ku, Han San
Landau, Bernard Robert
Lane, Lois Kay
Laughlin, Ethelreda R
Lavik, Paul Sophus
Lee, Young-Zoon
Leffak, Ira Michael
Leis, Jonathan Peter
Lemon, Peter Willian Reginald
Lessard, James Louis
Liang, Tehming
Lovenberg, Walter McKay
Lukin, Marvin
McClure, Jerry Weldon
MacGee, Joseph
McIntyre, Russell Theodore
MacKenzie, Robert Douglas
McLean, Larry Raymond
McQuarrie, Irvine Gray
McQuate, John Truman
Malemud, Charles J
Mallin, Morton Lewis
Mamrack, Mark Donovan
Manning, Maurice
Mao, Simon Jen-Tan
Marzluf, George A
Means, Gary Edward
Mellgren, Ronald Lee
Merola, A John
Messineo, Luigi
Meyer, Ralph Roger
Mieyal, John Joseph
Miller, Robert Harold
Milo, George Edward
Milsted, Amy
Misono, Kunio Shiraishi
Modrzakowski, Malcolm Charles
Mohinder, Singh Kang
Morgan, Marjorie Susan
Moulton, Bruce Carl
Mukhtar, Hasan
Mukkada, Antony Joh
Mulhausen, Hedy Ann
Murray, Finnie Ardrey, Jr
Myers, Ronald Elwood
Nagy, Bela Ferenc
Neiderhiser, Dewey Harold
Newman, Howard Abraham Ira
Nuenke, Richard Harold
Nutter, William Ermal

Nygaard, Oddvar Frithjof
Ochs, Raymond S
Ogle, James D
Okerholm, Richard Arthur
Oleinick, Nancy Landy
Ottolenghi, Abramo Cesare
Pappas, Peter William
Pensky, Jack
Perry, George
Pfeiffer, Douglas Robert
Prairie, Richard Lane
Pynadath, Thomas I
Rawat, Arun Kumar
Rawson, James Rulon Young
Ray, Richard Schell
Reddy, Padala Vykuntha
Reimann, Erwin M
Rieske, John Samuel
Ritter, Edmond Jean
Ritzert, Roger William
Rodman, Harvey Meyer
Roehrig, Karla Louise
Rosenberry, Terrone Lee
Rosenfeld, Robert Samson
Rossmiller, John David
Rottman, Fritz M
Rouslin, William
Rowe, John James
Rozek, Charles Edward
Rudney, Harry
Russell, Paul Telford
Rutishauser, Urs Stephen
Rynbrandt, Donald Jay
Sable, Henry Zodoc
Saffran, Judith
Saffran, Murray
Salminen, Seppo Ossian
Samols, David R
Sawicki, Stanley George
Sayre, Richard Thomas
Scarpa, Antonio
Schaff, John Franklin
Schanbacher, Floyd Leon
Schlender, Keith K
Schroeder, Friedhelm
Schumm, Dorothy Elaine
Schwartz, Arnold
Serif, George Samuel
Servaites, Jerome Casimer
Serve, Munson Paul
Shaffer, Jacquelin Bruning
Shamberger, Raymond J
Shannon, Barry Thomas
Shertzer, Howard Grant
Silverman, Robert Hugh
Singer, Sanford Sandy
Skau, Kenneth Anthony
Skeggs, Leonard Tucker, Jr
Slonim, Arnold Robert
Smeby, Robert Rudolph
Smith, Charlotte Damron
Smith, Daniel John
Smith, Mark Anthony
Snell, Junius Fielding
Sperelakis, Nicholas
Sprecher, Howard W
Stauffer, Clyde E
Stoner, Clinton Dale
Subbiah, Ravi M T
Sundaralingam, Muttaiya
Sunkara, Sai Prasad
Swenson, Richard Paul
Swift, Terrence James
Tabita, F Robert
Tannenbaum, Carl Martin
Tejwani, Gopi Assudomal
Tennent, David Maddux
Tepperman, Katherine Gail
Thomas, Craig Eugene
Tomei, L David
Trela, John Michael
Trewyn, Ronald William
Tsai, Chun-Che
Tsai, Ming-Daw
Ugarte, Eduardo
Varnes, Marie Elizabeth
Versteegh, Larry Robert
Vester, John William
Vignos, Paul Joseph, Jr
Voss, Anne Coble
Wagner, Thomas Edwards
Walenga, Ronald W
Walker, Thomas Eugene
Wallick, Earl Taylor
Walz, Frederick George
Wang, Taitzer
Webb, Thomas Evan
Wei, Robert
Weisman, Robert A
Westerman, Philip William
Widder, James Stone
Wideman, Cyrilla Helen
Willard, James Matthew
Williams, Marshall Vance
Williams, Wallace Terry
Wilson, James Albert
Winget, Gary Douglas
Wood, Harland G
Woodworth, Mary Esther
Wright, George Joseph
Wuu, Ting-Chi
Yates, Allan James
Yiamouyiannis, John Andrew
Zealey, Marion Edward

Ziller, Stephen A, Jr
Zimmerman, Ernest Frederick
Zull, James E

OKLAHOMA
Abbott, Donald Clayton
Alaupovic, Petar
Allen, Robert Wade
Badgett, Allen A
Banschbach, Martin Wayne
Bingham, Robert J
Birckbichler, Paul Joseph
Blair, James Bryan
Bradford, Reagan Howard
Briggs, Thomas
Briscoe, William Travis
Broyles, Robert Herman
Burgus, Roger Cecil
Carpenter, Mary Pitynski
Carubelli, Raoul
Chandler, Albert Morrell
Chaturvedi, Arvind Kumar
Ciereszko, Leon Stanley, Sr
Clarke, Margaret Burnett
Coleman, Ronald Leon
Comp, Philip Cinnamon
Cox, Andrew Chadwick
Croy, Lavoy I
Cunningham, Madeleine White
Dale, George Leslie
Delaney, Robert
Dell'Orco, Robert T
Dillwith, Jack W
Duncan, Michael Robert
Dyer, David Wayne
Eddington, Carl Lee
Esmon, Charles Thomas
Essenberg, Margaret Kottke
Essenberg, Richard Charles
Ferretti, Joseph Jerome
Fish, Wayne William
Floyd, Robert A
Ford, Sharon R
Gardner, Richard Lynn
Gholson, Robert Karl
Harmon, H James
Hartsuck, Jean Ann
Hopkins, Thomas R (Tim)
Howard, Robert Eugene
Hurst, Jerry G
Johnson, B Connor
Johnson, Ronald Roy
Kampschmidt, Ralph Fred
Ketring, Darold L
Kizer, Donald Earl
Koeppe, Roger Erdman
Koh, Eunsook Tak
Leach, Franklin Rollin
Lee, Diana Mang
McCay, Paul Baker
Matsumoto, Hiroyuki
Messmer, Dennis A
Mills, John Norman
Mitchell, Earl Douglass, Jr
Morrissey, James Henry
Murphy, Marjory Beth
Niedbalski, Joseph S
Odell, George Van, Jr
Passey, Richard Boyd
Poyer, Joe Lee
Reinhart, Gregory Duncan
Reinke, Lester Allen
Roe, Bruce Allan
Sachdev, Goverdhan Pal
Sanny, Charles Gordon
Schindler, Charles Alvin
Schubert, Karel Ralph
Seeney, Charles Earl
Sherwood, John L
Short, Everett C, Jr
Silverman, Philip Michael
Smith, Eddie Carol
Tang, Jordan J N
Trachewsky, Daniel
Waller, George Rozier, Jr
Wang, Chi-Sun
Weatherby, Gerald Duncan
Wender, Simon Harold
Yu, Chang-An

OREGON
Albrecht, Stephan LaRowe
Anderson, Amy Elin
Anderson, Sonia R
Arnold, Roy Gary
Arp, Daniel James
Baisted, Derek John
Bartos, Dagmar
Bartos, Frantisek
Battaile, Julian
Beatty, Clarissa Hager
Beaudreau, George Stanley
Becker, Robert Richard
Bennett, Robert M
Bentley, J Peter
Black, John Alexander
Bonhorst, Carl W
Brandt, Howard Allen
Brodie, Ann Elizabeth
Brownell, Philip Harry
Bundy, Hallie Flowers
Carlisle, Kay Susan
Civelli, Oliver
Claycomb, Cecil Keith

Clinton, Gail M
Craig, Albert Morrison (Morrie)
Daley, Laurence Stephen
Dalton, David Andrews
Debons, Albert Frank
Dejmal, Roger Kent
DuFresne, Ann
Evans, Harold J
Freed, Virgil Haven
Gabler, Walter Louis
Gamble, Wilbert
Gatewood, Dean Charles
Gold, Michael Howard
Hare, James Frederic
Hawkinson, Stuart Winfield
Hearon, William Montgomery
Holm, Harvey William
Hoskins, Dale Douglas
Huang, Zhijian
Jones, Richard Theodore
Kabat, David
Keevil, Thomas Alan
Kilgour, Gordon Leslie
Kittinger, George William
Koop, Dennis Ray
Lis, Adam W
Lis, Elaine Walker
Lochner, Janis Elizabeth
Loomis, Walter David
MacDonald, Donald Laurie
McDuffie, Norton G(raham), Jr
Machida, Curtis A
McNulty, Wilbur Palmer
Malbica, Joseph Orazio
Malencik, Dean A
Mathews, Christopher King
Maurer, Richard Allen
Mela-Riker, Leena Marja
Miller, Lorraine Theresa
Palmes, Edward Dannelly
Rasmussen, Lois E Little
Reed, Donald James
Rigas, Demetrios A
Roselli, Charles Eugene
Russell, Peter James
Sanders-Loehr, Joann
Scannell, James Parnell
Schreiner, Roger Paul
Scott, Eion George
Shearer, Thomas Robert
Skala, James Herbert
Soderling, Thomas Richard
Somero, George Nicholls
Spencer, Elaine
Swanson, J Robert
Szalecki, Wojciech Jozef
Terwilliger, Nora Barclay
Tinsley, Ian James
Todd, Wilbert R
Toumadje, Arazdordi
Trione, Edward John
Vijayaraghavan, Srinivasan
Weiser, Conrad John
Westall, Frederick Charles
Woldegiorgis, Gebretateos
Wolf, Don Paul
Wolfe, Raymond Grover, Jr
Young, J Lowell

PENNSYLVANIA
Abrams, Richard
Abrams, William R
Adachi, Kazuhiko
Akella, Rama
Alexander, James King
Alhadeff, Jack Abraham
Allen, Arthur
Alonso-Caplen, Firelli V
Alper, Robert
Amoscato, Andrew Anthony
Angstadt, Carol Newborg
Arnott, Marilyn Sue
Asakura, Toshio
Auerbach, Victor Hugo
Avadhani, Narayan G
Axelrod, Abraham Edward
Baggott, James Patrick
Bailey, David George
Baker, Alan Paul
Baker, Wilber Winston
Baldridge, Robert Crary
Balin, Arthur Kirsner
Barranger, John Arthur
Basford, Robert Eugene
Bashey, Reza Ismail
Bates, Margaret Westbrook
Baum, Robert Harold
Baumgarten, Werner
Bayer, Margret H(elene) Janssen
Beining, Paul R
Bentley, Ronald
Bernlohr, Robert William
Bertland, Alexander U
Bertolini, Donald R
Bex, Frederick James
Bhaduri, Saumya
Bhavanandan, Veer P
Biaglow, John E
Biebuyck, Julien Francois
Bjornsti, Mary-Ann
Black, Robert Corl
Blackburn, Michael N
Blough, Herbert Allen
Boden, Guenther

Bond, Judith
Bonner, Walter Daniel, Jr
Bradley, Matthews Ogden
Brennan, Thomas Michael
Brinigar, William Seymour, Jr
Brown, William E
Buck, Clayton Arthur
Budzynski, Andrei Z
Burch, Mary Kappel
Burke, James Patrick
Burns, Richard Charles
Butler, Thomas Michael
Butler, William Barkley
Butt, Tauseef Rashid
Campbell, Iain Malcolm
Campo, Robert D
Castellano, Salvatore, Mario
Castle, John Edwards
Castric, Peter Allen
Cerbulis, Janis
Chacko, George Kutty
Chaiken, Irwin M
Chance, Britton
Cheng, Sheau-Mei
Cheng, Yih-Shyun Edmond
Chernin, Mitchell Irwin
Ch'ih, John Juwei
Chiu, Teh-Hsing
Chowrashi, Prokash K
Chun, Edward Hing Loy
Chung, Albert Edward
Clagett, Carl Owen
Clark, Charles Christopher
Cohen, Leonard Harvey
Cohn, Mildred
Cohn, Robert M
Colman, Robert W
Colony-Cokely, Pamela
Conover, Thomas Ellsworth
Cooper, Joseph E
Cooperman, Barry S
Cordes, Eugene H
Cortner, Jean A
Coss, Ronald Allen
Craven, Patricia A
Creasey, William Alfred
Cristofalo, Vincent Joseph
Crowell, Richard Lane
Cryer, Dennis Robert
Dalal, Fram Rustom
D'Alisa, Rose M
Damle, Suresh B
Daniel, James L
Das, Manjusri
Davies, Helen Jean Conrad
Davies, Robert Ernest
De La Haba, Gabriel Luis
Delluva, Adelaide Marie
Del Vecchio, Vito Gerard
Devlin, Thomas McKeown
Diamond, Leila
Di Cuollo, C John
Diven, Warren Field
DiVincenzo, George D
Dodgson, Susanna Jane
Donnelly, Thomas Edward, Jr
Dowd, Susan Ramseyer
Doyne, Thomas Harry
Draus, Frank John
Dreisbach, Joseph Herman
Dresden, Carlton F
Dressler, Hans
Duck, William N, Jr
Duker, Nahum Johanan
Dulka, Joseph John
Dumm, Mary Elizabeth
Dutton, P Leslie
Eagon, Patricia K
Edmonds, Mary P
Edwards, John R
Ehrlich, H Paul
Ellingson, John S
Ellis, Demetrius
English, Leigh Howard
Erecinska, Maria
Esfahani, Mojtaba
Evans, Audrey Elizabeth
Farrell, Harold Maron, Jr
Farren, Ann Louise
Feingold, David Sidney
Feinstein, Sheldon Israel
Fenderson, Bruce Andrew
Fenton, Marilyn Ruth
Ferrell, Robert Edward
Finn, Frances M
Fisher, Edward Allen
Flaks, Joel George
Flanagan, Thomas Leo
Fletcher, Jeffrey Edward
Fluck, Eugene Richards
Fox, Jay B, Jr
Franklin, Samuel Gregg
Franzen, James
Fraser, Nigel William
Fried, Michael Gregory
Fritz, Paul John
Furth, John J
Garfinkle, Barry David
Garwin, Jeffrey Lloyd
Gealt, Michael Alan
Ghosh, Amal Kumar
Glaid, Andrew Joseph, III
Glick, Jane Mills
Godfrey, John Carl

Biochemistry (cont)

Godfrey, Susan Sturgis
Goff, Christopher Godfrey
Gogel, Germaine E
Golder, Richard Harry
Goldfarb, Ronald H
Goldfine, Howard
Golub, Ellis Eckstein
Goodgal, Sol Howard
Goodman, David Barry Poliakoff
Gould, Robert James
Grant, Norman Howard
Grebner, Eugene Ernest
Gregory, Francis Joseph
Grindel, Joseph Michael
Grunwald, Gerald B
Gurin, Samuel
Gustine, David Lawrence
Hackney, David Daniel
Hamilton, Robert Houston
Hammel, Jay Morris
Hammerstedt, Roy H
Harding, Roy Woodrow, Jr
Hardison, Ross Cameron
Harmon, George Andrew
Hartline, Richard
Hartzler, Eva Ruth
Hass, Louis F
Hassell, John Robert
Haugaard, Niels
Hayashi, Teruo Terry
Heimer, Phyllis Marie
Heimer, Ralph
Henderson, George Richard
Henderson, Linda Shlatz
Hennessey, John P, Jr
Henry, Susan Armstrong
Herbert, Michael
Herlyn, Dorothee Maria
Hess, Eugene Lyle
Hickey, Richard James
Higgins, Michael Lee
Higgins, Terry Jay
Higman, Henry Booth
Hill, Charles Whitacre
Ho, Chien
Hoek, Joannes (Jan) Bernardus
Hoffee, Patricia Anne
Hofmann, Klaus Heinrich
Hopper, Sarah Priestly
Houck, David R
Howe, Chin Chen
Hsu, Kuo-Hom Lee
Huang, Leaf
Hudson, Alan P
Hung, Paul P
Hurst, William Jeffrey
Husic, Harold David
Idell-Wenger, Jane Arlene
Isom, Harriet C
Jackson, Ethel Noland
Jacobs, Mark
Jacobsohn, Gert Max
Jacobsohn, Myra K
Jacobson, Lewis A
Jarett, Leonard
Jen-Jacobson, Linda
Jezyk, Peter Franklin
Jim, Kam Fook
Jimenez, Sergio
Johnston, James Bennett
Jones, David Hartley
Jorns, Marilyn Schuman
Kaji, Akira
Kaji, Hideko (Katayama)
Kalf, George Frederick
Kallen, Roland Gilbert
Kane, James Francis
Kant, Jeffrey A
Karol, Meryl Helene
Kasvinsky, Peter John
Katz, Solomon H
Kayne, Marlene Steinmetz
Kefalides, Nicholas Alexander
Kennett, Roger H
Khatami, Mahin
Kim, Sangduk
Kirby, Edward Paul
Kirtley, Mary Elizabeth
Kleinman, Roberta Wilma
Knauff, Raymond Eugene
Korchak, Helen Marie
Kozinski, Andrzej
Krah, David Lee
Krause, Stephen Myron
Kritchevsky, David
Kuivaniemi, S Helena
Kwei, Gloria Y
Kwiecinski, Gary George
Lampson, George Peter
Lancaster, Jack R, Jr
La Noue, Kathryn F
Lapidus, Milton
LaRossa, Robert Alan
Layne, Porter Preston
Leboy, Phoebe Starfield
Lee, Shaw-Guang Lin
Lehman, Ernest Dale
Leibel, Wayne Stephan
Leon, Shalom A
Lerner, Leonard Joseph
Levy, Robert I
Lewbart, Marvin Louis

Lieberman, Hillel
Liebman, Paul Arno
Lien, Eric Louis
Lindstrom, Jon Martin
Little, Brian Woods
Litwack, Gerald
Logan, David Alexander
Lotlikar, Prabhakar Dattaram
Lowe-Krentz, Linda Jean
Maass, Alfred Roland
McCarl, Richard Lawrence
McCarthy, William John
McClure, William Robert
MacDonald, James Scott
Machlowitz, Roy Alan
Mack, Lawrence Lloyd
McMorris, F Arthur
Magee, Wayne Edward
Malamud, Daniel F
Manson, Lionel Arnold
Mao, James Chieh Hsia
Marcus, Abraham
Markham, George Douglas
Marks, Dawn Beatty
Marsh, Julian Bunsick
Martin, Julia Mae
Mathur, Carolyn Frances
Max, Stephen Richard
Mears, James Austin
Merkel, Joseph Robert
Metrione, Robert M
Mifflin, Theodore Edward
Miller, Elizabeth Eshelman
Miller, Gail Lorenz
Miller, James Eugene
Miller, Richard Lee
Mochan, Eugene
Monson, Frederick Carlton
Morgan, Howard E
Morris, Sidney Machen, Jr
Mortimore, Glenn Edward
Moss, Melvin Lane
Mulder, Kathleen M
Mullin, James Michael
Mumma, Ralph O
Murer, Erik Homann
Murphy, Robert Francis
Na, George Chao
Nambi, Ponnal
Nishikawa, Alfred Hirotoshi
Niu, Mann Chiang
Novak, Josef Frantisek
O'Connor, John Dennis
Oesterling, Myrna Jane
Ohnishi, Tomoko
Ohnishi, Tsuyoshi
O'Neill, John Joseph
Opas, Evan E
Opella, Stanley Joseph
Orr, Nancy Hoffner
Paik, Woon Ki
Panos, Charles
Passananti, Gaetano Thomas
Patterson, Elizabeth Knight
Paul, Harbhajan Singh
Pazur, John Howard
Peebles, Craig Lewis
Pegg, Anthony Edward
Perlis, Irwin Bernard
Persky, Harold
Pessen, Helmut
Phillips, Allen Thurman
Phillips, Marshall
Pieringer, Ronald Arthur
Platsoucas, Chris Dimitrios
Plaut, Gerhard Wolfgang Eugen
Pleasure, David
Pollack, Robert Leon
Porter, Curt Culwell
Porter, Ronald Dean
Pratt, Elizabeth Ann
Press, Linda Seghers
Prockop, Darwin J
Pruett, Patricia Onderdonk
Punnett, Thomas R
Puri, Rajinder N
Pye, Edward Kendall
Rabinowitz, Joseph Loshak
Ramachandran, Banumathi
Ramachandran, Subramania
Rao, Kalipatnapu Narasimha
Ray, Eva K
Raymond, Matthew Joseph
Reed, Ruth Elizabeth
Reinhart, Michael P
Ricciardi, Robert Paul
Rittenhouse, Susan E
Rodan, Gideon Alfred
Rose, Irwin Allan
Rose, Zelda B
Rosenblatt, Michael
Rosenbloom, Joel
Rosenfeld, Leonard M
Ross, Alta Catharine
Roth, Stephen
Rothblat, George H
Kottenberg, Hagai
Rovera, Giovanni
Royer, Garfield Paul
Rutman, Robert Jesse
Salama, Guy
Salganicoff, Leon
Sardana, Mohinder K
Savage, Carl Richard, Jr

Schepartz, Bernard
Schlegel, Robert Allen
Schleyer, Heinz
Schwartz, Elias
Scott, Dwight Baker McNair
Searls, Robert L
Seery, Virginia Lee
Seetharam, Ramnath (Ram)
Segal, Stanton
Shank, Richard Paul
Shannon, Jack Corum
Shapiro, Irving Meyer
Shaw, Leslie M J
Shiman, Ross
Shiver, John W
Shockman, Gerald David
Sidie, James Michael
Siegel, Richard C
Silver, Melvin Joel
Silverman, Robert Eliot
Singer, Alan G
Skutches, Charles L
Smith, Colleen Mary
Sorof, Sam
Soslau, Gerald
Sprince, Herbert
Stambaugh, Richard L
Stewart, Barbara Yost
Stiller, Richard L
Stinson, Edgar Erwin
Strauss, Jerome Frank, III
Suhadolnik, Robert J
Sun, James Dean
Sung, Cheng-Po
Surmacz, Cynthia Ann
Swaney, John Brewster
Sylvester, James Edward
Tatum, Charles Maris
Taylor, William Daniel
Thakur, Madhukar L
Thayer, Donald Wayne
Thayer, William
Thimann, Kenneth Vivian
Thompson, Marvin P
Tien, Ming
Tint, Howard
Tomarelli, Rudolph Michael
Tong, Winton
Touchstone, Joseph Cary
Treece, Jack Milan
Tristram-Nagle, Stephanie Ann
Tromp, Gerardus C
Tu, Chen-Pei David
Tuan, Rocky Sung-Chi
Tulenko, Thomas Norman
Tulp, Orien Lee
Tuszynski, George P
Uitto, Jouni Jorma
Ulmer, Jeffrey Blaine
Upton, K Of Virginia
Vanderkooi, Jane M
Van Rossum, George Donald Victor
Vaughan, James Roland
Vergona, Kathleen Anne Dobrosielski
Voet, Judith Greenwald
Vogel, Wolfgang Hellmut
Vogt, Molly Thomas
Wainer, Arthur
Wallace, Herbert William
Wampler, D Eugene
Warme, Paul Kenneth
Warren, Leonard
Watt, Robert M
Waxman, Lloyd H
Weber, Annemarie
Weinbaum, George
Weinberg, Eric S
Weinhouse, Sidney
Weinryb, Ira
Weiss, Benjamin
Weiss, Sidney
Westley, John William
Whitfield, Carol F(aye)
Widnell, Christopher Courtenay
Williamson, John Richard
Wilson, David F
Winkler, James David
Winsten, Seymour
Woychik, John Henry
Yellin, Tobias O
Yonetani, Takashi
Young, Franklin
Young, Peter Ronald
Yui, Katsuyuki
Yushok, Wasley Donald
Zaleski, Jan F
Zelson, Philip Richard
Zemaitis, Michael Alan
Zurawski, Vincent Richard, Jr
Zweidler, Alfred

RHODE ISLAND
Beale, Samuel I
Biggins, John
Brautigan, David L
Brown, Phyllis R
Cha, Sungman
Clough, Wendy Glasgow
Coleman, John Russell
Constantinides, Spiros Minas
Crabtree, Gerald Winston
Crowley, James Patrick
Dahlberg, Albert Edward
Dain, Joel A

Davis, Robert Paul
Dawicki, Doloretta Diane
Donovan, Gerald Alton
Fausto, Nelson
Flanagan, Thomas Raymond
Hai, Chi-Ming
Hammen, Carl Schlee
Hawrot, Edward
Hegre, Carman Stanford
Landy, Arthur
Lederberg, Seymour
Lusk, Joan Edith
Malcolm, Alexander Russell
Martin, Horace F
Meedel, Thomas Huyck
Miech, Ralph Patrick
O'Leary, Gerard Paul, Jr
Parks, Robert Emmett, Jr
Purvis, John L
Rao, Girimaji J Sathyanarayana
Rothman, Frank George
Sapolsky, Asher Isadore
Shaikh, Zahir Ahmad
Steiner, Manfred
Tourtellotte, Mark Eton
Tremblay, George Charles
Turner, Michael D
Von Riesen, Daniel Dean
Yoon, Poksyn Song
Young, Robert M

SOUTH CAROLINA
Allen, Donald Orrie
Arnaud, Philippe
Baggett, Billy
Banik, Narendra Lal
Baynes, John William
Berger, Franklin Gordon
Bishop, Muriel Boyd
Camper, Nyal Dwight
Chao, Julie
Cowgill, Robert Warren
Davis, Craig Wilson
Davis, Leroy
Dodds, Alvin Franklin
Doig, Marion Tilton, III
Dunlap, Robert Bruce
Finlay, Mary Fleming
Fowler, Stanley D
Gadsden, Richard Hamilton, Sr
Greenfield, Seymour
Hays, Ruth Lanier
Henricks, Donald Maurice
Hollis, Bruce Warren
Holstein, Arthur G
Hood, Samuel Lowry
Jonsson, Haldor Turner, Jr
Kistler, Wilson Stephen, Jr
Knapp, Daniel Roger
Knight, Anne Bradley
Kowalczyk, Jeanne Stuart
Krall, Albert Raymond
Lazarchick, John
Ledford, Barry Edward
Lin, Tu
Lopes-Virella, Maria Fernanda Leal
McCord, William Mellen
McDonald, John Kenneiy
Maheshwari, Kewal Krishnan
Miller, Ronald Lee
Mitchell, Jack Harris, Jr
Morrow, William Scot
Nelson, George Humphry
Ning, John Tse Tso
Orcutt, Donald Adelbert
Oswald, Edward Odell
Paynter, Malcolm James Benjamin
Powell, Gary Lee
Priest, David Gerard
Ritchie, Kim
Rittenbury, Max Sanford
Roel, Lawrence Edmund
Rohlfing, Duane L
Sawyer, Roger Holmes
Schmidt, Gilbert Carl
Schwabe, Christian
Shively, Jessup MacLean
Singh, Inderjit
Sodetz, James M
Spain, James Dorris, Jr
Stidworthy, George H
Stillway, Lewis William
Stratton, Lewis Palmer
Stutzenberger, Fred John
Swanson, Arnold Arthur
Taylor, Harold Allison, Jr
Turk, Donald Earle
Waldman, Alan S
Waldman, Barbara Criscuolo
Wheeler, Darrell Deane
Wilson, Gregory Bruce
Wilson, Steven Paul
Woodard, Geoffrey Dean Leroy
Wuthier, Roy Edward
Yoch, Duane Charles
Zemp, John Workman
Zimmerman, James Kenneth

SOUTH DAKOTA
Brady, Frank Owen
Brandwein, Bernard Jay
Cook, David Edgar
Dwivedi, Chandradhar
Emerick, Royce Jasper

Ferguson, Michael William
Hills, Loran C
Krueger, Keatha Kathrine
Langworthy, Thomas Allan
Lindahl, Ronald Gunnar
Neuhaus, Otto Wilhelm
Olson, Oscar Edward
Prescott, Lansing M
Small, Gary D
Thomas, John Alva
Whitehead, Eugene Irving

TENNESSEE

Airee, Shakti Kumar
Andersen, Richard Nicolaj
Anderson, Ted L
Andrews, John Stevens, Jr
Anjaneyulu, P S R
Baxter, John Edwards
Ben-Porat, Tamar
Berney, Stuart Alan
Blakley, Raymond L
Blevins, Raymond Dean
Bond, Andrew
Bradham, Laurence Stobo
Brent, Thomas Peter
Briggs, Robert Chester
Broquist, Harry Pearson
Brown, Fountaine Christine
Brown, James Walker, Jr
Bryant, Robert Emory
Bucovaz, Edsel Tony
Bunick, Gerard John
Burk, Raymond Franklin, Jr
Burr, William Wesley, Jr
Burt, Alvin Miller, III
Burtis, Carl A, Jr
Byers, Lawrence Wallace
Capdevila, H Jorge
Carlson, Gerald Michael
Carpenter, Graham Frederick
Chalkley, G Roger
Champney, William Scott
Chen, James Pai-Fun
Chen, Thomas Tien
Cheung, Wai Yiu
Chunduru, Srinivas Kishore
Chung, King-Thom
Chytil, Frank
Clauberg, Martin
Cohen, Stanley
Cohn, Waldo E
Conary, Jon Thurston
Coniglio, John Giglio
Cook, George A
Cook, Robert James
Corbin, Jack David
Costlow, Mark Enoch
Cox, Ray
Croom, Henrietta Brown
Cullen, Marion Permilla
Cunningham, Leon William
Danzo, Benjamin Joseph
Darby, William Jefferson
Das, Salil Kumar
Davenport, James Whitcomb
Desiderio, Dominic Morse, Jr
Di Pietro, David Louis
Dockter, Michael Edward
Duhl, David M
Dulaney, John Thornton
Epler, James L
Ernst-Fonberg, Marylou
Exton, John Howard
Fain, John Nicholas
Farkas, Walter Robert
Faulkner, Willard Riley
Fink, Robert David
Fleischer, Becca Catherine
Fleischer, Sidney
Francis, Sharron H
Friedman, Daniel Lester
Gates, Ronald Eugene
Geller, Arthur Michael
Gillespie, Elizabeth
Godfrey, Paul Russell
Gotterer, Gerald S
Goulding, Charles Edwin, Jr
Granner, Daryl Kitley
Guengerich, Frederick Peter
Guyer, Cheryl Ann
Hardman, Joel G
Harshman, Sidney
Hartman, Frederick Cooper
Hash, John H
Hayes, Robert M
Heimberg, Murray
Henke, Randolph Ray
Hilker, Doris M
Hill, George Carver
Hill, Robert James
Hingerty, Brian Edward
Howell, Elizabeth E
Inagami, Tadashi
Irving, Charles Clayton
Jackowski, Suzanne
Jacobson, Karl Bruce
James, Jesse
Jennings, Lisa Helen Kyle
Jones, Peter D
Joshi, Jayant Gopal
Kao, Race Li-Chan
Katze, Jon R
Kelley, Jim Lee

Kenney, Francis T
Kitabchi, Abbas E
Kono, Tetsuro
Kotb, Malak Y
Kraus, Lorraine Marquardt
Kuiken, Kenneth (Alfred)
Laboda, Henry M
Landon, Erwin Jacob
Larimer, Frank William
Lee, Kai-Lin
Lee, Ten Ching
Lerner, Joseph
Lessman, Charles Allen
Lin, Kuang-Tzu Davis
Lothstein, Leonard
Lyman, Beverly Ann
McCoy, Sue
McDonald, Ted Painter
Mani, Rama I
Mann, George Vernon
Marchok, Ann Catherine
Marnett, Lawrence Joseph
Martindale, William Earl
Maxwell, Richard Elmore
Mayberry, William Roy
Mayer, Steven Edward
Mithcell, William Marvin
Montie, Thomas C
Monty, Kenneth James
Moses, Henry A
Neal, Robert A
Nishimoto, Satoru Kenneth
Niyogi, Salil Kumar
O'Connor, Timothy Edmond
Olins, Ada Levy
Olins, Donald Edward
Ong, David Eugene
Osheroff, Neil
Pabst, Michael John
Papas, Andreas Michael
Park, Charles Rawlinson
Park, Jane Harting
Pfeffer, Lawrence Marc
Reed, Peter William
Regen, David Marvin
Rogers, Beverly Jane
Ryan, Kevin William
Sachan, Dileep Singh
Savage, Dwayne Cecil
Savage, Jane Ramsdell
Schwarz, Otto John
Senogles, Susan Elizabeth
Seyer, Jerome Michael
Shugart, Lee Raleigh
Sloane, Nathan Howard
Smith, John Thurmond
Snapper, James Robert
Snyder, Fred Leonard
Staros, James Vaughan
Stevens, Audrey L
Stevens, Stanley Edward, Jr
Stockell-Hartree, Anne
Stone, William Lawrence
Stubbs, Gerald James
Stulberg, Melvin Philip
Tam, James Pingkwan
Thakar, Jay H
Thomas, Edwin Lee
Thomason, Donald Brent
Totter, John Randolph
Touster, Oscar
Trupin, Joel Sunrise
Uziel, Mayo
Volkin, Elliot
Vroman, Hugh Egmont
Wagner, Conrad
Warnock, Laken Guinn
Washington, Arthur Clover
Waters, Larry Charles
Watterson, D Martin
Wicks, Wesley Doane
Wigler, Paul William
Wilcox, Henry G
Wimalasena, Jay
Womack, Frances C
Wood, John Lewis
Woychik, Richard P
Yang, Wen-Kuang
Yarbro, Claude Lee, Jr
Zee, Paulus

TEXAS

Abell, Creed Wills
Aggarwal, Bharat Bhushan
Allen, Julius Cadden
Anderson, John Arthur
Anderson, Robert E
Angelides, Kimon Jerry
Ansari, Guhlam Ahmad Shakeel
Arenaz, Pablo
Arlinghaus, Ralph B
Asimakis, Gregory K
Atassi, Zouhair
Atkinson, Mark Arthur Leonard
Awapara, Jorge
Awasthi, Yogesh C
Babitch, Joseph Aaron
Baldwin, Thomas Oakley
Baptist, James (Noel)
Barker, Kenneth Leroy
Barnes, Eugene Miller, Jr
Barnes, Larry D
Barnett, Don(ald) R(ay)
Barr, Charles Richard

Bartel, Allen Hawley
Beaudet, Arthur L
Beckingham, Kathleen Mary
Becvar, James Edgar
Beerstecher, Ernest, Jr
Behal, Francis Joseph
Bellion, Edward
Benedict, C R
Benedict, Chauncey
Bertrand, Helen Anne
Birnbaumer, Lutz
Boltralik, John Joseph
Bottino, Nestor Rodolfo
Bowles, William Howard
Brady, Scott T
Brand, Jerry Jay
Bryan, Joseph
Brysk, Miriam Mason
Buchanan, Christine Elizabeth
Burks, James Kenneth
Busch, Harris
Butcher, Reginald William
Butow, Ronald A
Byerley, Lauri Olson
Camp, Bennie Joe
Campbell, James Wayne
Campbell, William Jackson
Caprioli, Richard Michael
Carney, Darrell Howard
Carr, Bruce R
Carson, Daniel Douglas
Chan, Lawrence Chin Bong
Chan, Lee-Nien Lillian
Chan, Pui-Kwong
Chappell, Cynthia Lou
Chatterjee, Bandana
Chen, Shih-Fong
Chinn, Herman Isaac
Chiou, George Chung-Yih
Cimadevilla, Jose M
Cirillo, Vincent Paul
Clark, Dale Allen
Cohen, Allen Barry
Convertino, Victor Anthony
Cottam, Gene Larry
Couch, James Russell
Crass, Maurice Frederick, III
Creger, Clarence R
Criscuolo, Dominic
Dahm, Karl Heinz
Davis, Alvie Lee
Davison, Daniel Burtonn
Dawson, Earl B
Dean, David Devereaux
Deisenhofer, Johann
DeLoach, John Rooker
DeMoss, John A
Denney, Richard Max
Dennis, Joe
DeShazo, Mary Lynn Davison
Dieckert, Julius Walter
Di Ferrante, Nicola Mario
Dobson, Harold Lawrence
Doctor, Vasant Manilal
Doebbler, Gerald Francis
Douglas, Tommy Charles
Dubbs, Del Rose M
Ducis, Ilze
Dunn, Floyd Warren
Dutt, Anuradha
Dzidic, Ismet
Earhart, Charles Franklin, Jr
Edmundson, Allen B
Eichberg, Joseph
Eidels, Leon
Entman, Mark Lawrence
Eppright, Margaret
Estabrook, Ronald (Winfield)
Evans, Claudia T
Everse, Johannes
Fair, Daryl S
Fischer, Susan Marie
Fleischmann, William Robert, Jr
Foster, Donald Myers
Franzl, Robert E
Frenkel, Rene A
Friedberg, Errol Clive
Friend, Patric Lee
Furlong, Norman Burr, Jr
Gan, Jose Cajilig
Garber, Alan J
Garbers, David Lorn
Garcia, Hector D
Gatlin, Delbert Monroe, III
Geoghegan, William David
Georgiou, George
Gilbert, Brian E
Gilbert, Hiram Frazier
Gilles, Kenneth Albert
Goldschmidt, Millicent
Goldstein, Joseph Leonard
Goodman, Joel Mitchell
Gordon, Wayne Lecky
Gorelic, Lester Sylvan
Gottlieb, Paul David
Gotto, Antonio Marion, Jr
Gracy, Robert Wayne
Graff, Gustav
Griffin, James Emmett
Grimes, L Nichols
Grinnell, Frederick
Grundy, Scott Montgomery
Guentzel, M Neal
Guidry, Marion Antoine

Guirard, Beverly Marie
Gunn, John Martyn
Guo, Yan-Shi
Guynn, Robert William
Haber, Bernard
Hall, Frank Foy
Hall, Timothy Couzens
Hamilton, Charleen Marie
Hanahan, Donald James
Hardcastle, James Edward
Hardesty, Boyd A
Harding, Winfred Mood
Harper, Michael John Kennedy
Harris, Ben Gerald
Harris, Edward David
Harris, Stephen Eubank
Hedges, Dorothea Huseby
Hentges, David John
Hewett-Emmett, David
Hewitt, Roger R
Highlander-Bultema, Sarah Katherine
Hillar, Marian
Hillis, David Mark
Hodgins, Daniel Stephen
Hogan, Michael Edward
Holoubek, Viktor
Horowitz, Paul Martin
Houston, Forrest Gish
Howard, Harriette Ella Pierce
Hsueh, Andie M
Huang, Charles T L
Huang, Shyi Yi
Hurlbert, Robert Boston
Hurley, Laurence Harold
Hutcheson, Eldridge Tilmon, III
Ibanez, Manuel Luis
Ippen-Ihler, Karin Ann
Irvin, James Duard
Irwin, Louis Neal
James, Harold Lee
Jester, James Vincent
Johnson, Kenneth Maurice, Jr
Johnston, John Marshall
Jorgensen, George Norman
Joshi, Vasudev Chhotalal
Jurtshuk, Peter, Jr
Kaman, Robert Lawrence
Kantor, Harvey Sherwin
Kasschau, Margaret Ramsey
Kaufmann, Anthony J
Kennedy, Robert Alan
Kiel, Johnathan Lloyd
Kimball, Aubrey Pierce
King, Richard Joe
Kit, Saul
Kitto, George Barrie
Klebe, Robert John
Kokkinakis, Demetrius Michael
Konkel, David Anthony
Kramer, Gisela A
Krieg, Daniel R
Kubena, Leon Franklin
Kurosky, Alexander
Lacko, Andras Gyorgy
Lagowski, Jeanne Mund
Lam, Kwok-Wai
Landmann, Wendell August
Langdon, Robert Godwin
Lansford, Edwin Myers, Jr
Lawrence, Addison Lee
Lawrence, Richard Aubrey
Lee, James C
Lee, John Chung
Lever, Julia Elizabeth
Lewis, Donald Everett
Lichtenberger, Lenard Michael
Lingle, Sarah Elizabeth
Little, Gwynne E
Longenecker, John Bender
Lorenzetti, Ole John
Lospalluto, Joseph John
Lotan, Reuben
Lucid, Shannon W
Luduena, Richard Froilan
Luk, Gordon David
McCallum, Roderick Eugene
McCammon, James Andrew
McCarthy, John Lawrence, Jr
McCord, Tommy Joe
McDonald, George Gordon
Mace, Kenneth Dean
McGarry, John Denis
McKeehan, Wallace Lee
McKnight, Thomas David
MacLeod, Michael Christopher
McMillin, Jeanie
Magill, Jane Mary (Oakes)
Margolin, Solomon B
Masaracchia, Ruthann A
Mason, James Ian
Mason, Morton Freeman
Masoro, Edward Joseph
Massey, John Boyd
Masters, Bettie Sue Siler
Matthews, Kathleen Shive
Medina, Miguel Angel
Meistrich, Marvin Lawrence
Mendel, Julius Louis
Mendelson, Carole Ruth
Merrill, Gerald Alan
Mersmann, Harry John
Meyer, Franz
Michael, Lloyd Hal
Middleditch, Brian Stanley

Biochemistry (cont)

Milewich, Leon
Miller, Edward Godfrey, Jr
Miller, Joyce Mary
Miller, Sanford Arthur
Mills, Gordon Candee
Mills, William Ronald
Milner, Alice N
Miner, James Joshua
Mitra, Sankar
Mize, Charles Edward
Modak, Arvind T
Moore, Erin Colleen
Mooz, Elizabeth Dodd
Morgan, Page Wesley
Morrisett, Joel David
Mukherjee, Amal
Murgola, Emanuel J
Nakajima, Motowo
Nall, Barry T
Nelson, Thomas Evar
Newman, Robert Alwin
Nicholson, Wayne Lowell
Nishimura, Jonathan Sei
Nishioka, Kenji
Noall, Matthew Wilcox
Nordyke, Ellis Larrimore
Norton, Scott J
O'Donovan, Gerard Anthony
Olson, Merle Stratte
Oro, Juan
Overturf, Merrill L
Painter, Richard Grant
Palmer, Graham
Papaconstantinou, John
Patel, Nutankumar T
Patsch, Wolfgang
Payne, Shelley Marshall
Perkins, John Phillip
Peterson, David Oscar
Peterson, Julian Arnold
Pettit, Flora Hunter
Pirtle, Robert M
Plunkett, William Kingsbury
Poduslo, Shirley Ellen
Poenie, Martin Francis
Poffenbarger, Phillip Lynn
Poulsen, Lawrence Leroy
Powell, Bernard Lawrence
Pownall, Henry Joseph
Prager, Morton David
Prescott, John Mack
Price, Peter Michael
Quiocho, Florante A
Rainwater, David Luther
Randerath, Kurt
Rassin, David Keith
Raushel, Frank Michael
Ravel, Joanne Macow
Reagor, John Charles
Reber, Elwood Frank
Reed, Lester James
Rege, Ajay Anand
Reid, Michael Baron
Reid, Ted Warren
Reiser, Raymond
Reyes, Victor E
Richardson, Arlan Gilbert
Ridgway, Helen Jane
Riehl, Robert Michael
Riggs, Austen Fox, II
Riser, Mary Elizabeth
Ritzi, Earl Michael
Roberts, Susan Jean
Robinson, Neal Clark
Ro-Choi, Tae Suk
Roels, Oswald A
Rogers, Gary Allen
Roller, Herbert Alfred
Root, Elizabeth Jean
Rose, Kathleen Mary
Rosen, Jeffrey Mark
Ross, Doris Laune
Ross, Elliott M
Roux, Stanley Joseph
Rudolph, Frederick Byron
Sahasrabuddhe, Chintaman Gopal
Sallee, Verney Lee
Salomon, Lothar L
Sampson, Herschel Wayne
Schaffer, Barbara Noyes
Schlamowitz, Max
Schneider, Dennis Ray
Schonbrunn, Agnes
Schram, Alfred C
Schroeder, Hartmut Richard
Schroepfer, George John, Jr
Seifert, William Edgar, Jr
Serwer, Philip
Shapiro, David M
Shaw, Emil Gilbert
Shaw, Robert Wayne
Shetlar, Marvin Roy
Shive, William
Siciliano, Michael J
Siler-Khodr, Theresa M
Simpson, Evan Rutherford
Simpson, John Wayne
Sirbasku, David Andrew
Skelley, Dean Sutherland
Skinner, Charles Gordon
Skow, Loren Curtis
Smith, Edward Russell
Smith, Leland Leroy
Smith, Louis C
Smith, Michael Lew
Smith, Russell Lamar
Snell, Esmond Emerson
Snell, William J
Snyder, Gary Dean
Sordahl, Louis A
Spallholz, Julian Ernest
Sparkman, Dennis Raymond
Srere, Paul Arnold
Sridhara, S
Srivastava, Satish Kumar
Stancel, George Michael
Starcher, Barry Chapin
Stephens, Robert Lawrence
Stewart, James Ray
Stith, William Joseph
Stone, William Harold
Stoops, James King
Stouffer, John Emerson
Strinden, Sarah Taylor
Stroman, David Womack
Stroynowski, Iwona T
Sutton, Harry Eldon
Taegtmeyer, Heinrich
Taurog, Alvin
Taylor, Alan Neil
Tcholakian, Robert Kevork
Temeyer, Kevin Bruce
Thompson, Edward Ivins Bradbridge
Thompson, Ernest Aubrey, Jr
Thompson, Guy A, Jr
Tong, Alex W
Towne, Jack C
Tsai, Ming-Jer
Tsin, Andrew Tsang Cheung
Tsutsui, Ethel Ashworth
Tu, Shiao-chun
Uyeda, Kosaku
Vanderzant, Erma Schumacher
Van Dreal, Paul Arthur
Van Eys, Jan
Vela, Gerard Roland
Vogel, James John
Vonder Haar, Raymond A
Wagner, Martin James
Wakil, Salih J
Walborg, Earl Fredrick, Jr
Walker, James Benjamin
Walter, Charles Frank
Wang, Kuan
Ward, Darrell N
Webb, Bill D
Weigel, Paul H(enry)
Weinstock, George Matthew
Wells, Robert Dale
Wellso, Stanley Gordon
Wentland, Stephen Henry
Werbin, Harold
Wheeler, Michael Hugh
Whelly, Sandra Marie
Whitenberg, David Calvin
Whittle, John Antony
Wiggans, Donald Sherman
Wilkes, Stella H
Williams, Charles Herbert
Williams, Ralph Edward
Williams, William Thomas
Willms, Charles Ronald
Wilson, Golder North
Wilson, John H
Wilson, Samuel H
Wohlman, Alan
Wold, Finn
Wolinsky, Ira
Wood, Randall Dudley
Wood, Robert Charles
Worthen, Howard George
Wright, Woodring Erik
Wu, Ming-Chi
Yeoman, Lynn Chalmers
Yorio, Thomas
Young, Ryland F
Ziegler, Daniel
Ziegler, Miriam Mary

UTAH

Aird, Steven Douglas
Allred, Keith Reid
Andersen, William Ralph
Anderson, Anne Joyce
Ash, Kenneth Owen
Aust, Steven Douglas
Beck, Jay Vern
Bennett, Jesse Harland
Bradshaw, William S
Bryson, Melvin Joseph
Burnham, Bruce Franklin
Burton, Sheril Dale
Caldwell, Karin D
Casjens, Sherwood Reid
Clark, C Elmer
Czerlinski, George Heinrich
Dickman, Sherman Russell
Ellis, LeGrande Clark
Emery, Thomas Fred
Evans, Robert John
Franklin, Michael R(oger)
Franklin, Naomi C
Galster, William Allen
Gortatowski, Melvin Jerome
Gubler, Clark Johnson
Hansen, Dale J
Henderson, Lavell Merl
Heninger, Richard Wilford
Herrick, Glenn Arthur
Hill, John Mayes, Jr
Hutchings, Brian Lamar
Janatova, Jarmila
Johnson, LaVell R
Johnson, Ralph M, Jr
Klein, Sigrid Marta
Krieger, Carl Henry
Kuehl, LeRoy Robert
Lawson, Larry Dale
Linker, Alfred
McIntyre, Thomas Marty
Mangelson, Farrin Leon
Mangum, John Harvey
Miller, Gene Walker
Mohammad, Syed Fazal
Nelson, Don Harry
Nielson, Eldon Denzel
Ramachandran, Chittoor Krishna
Rasmussen, Kathleen Goertz
Rilling, Hans Christopher
Rowe, Mark J
Schweizer, Martin Paul
Scouten, William Henry
Seeley, Schuyler Drannan
Simmons, Daniel L
Simmons, John Robert
Sipe, David Michael
Stokes, Barry Owen
Straight, Richard Coleman
Sweat, Floyd Walter
Takemoto, Jon Yutaka
Torres, Anthony R
Vassel, Bruno
Velick, Sidney Frederick
Vernon, Leo Preston
Weber, Darrell J
West, Charles Donald
Williams, Roger Richard
Winder, William W
Winge, Dennis R
Wright, Donald N
Yoshikami, Doju

VERMONT

Carew, Lyndon Belmont, Jr
Chiu, Jen-Fu
Clemmons, Jackson Joshua Walter
Currier, William Wesley
Foote, Murray Wilbur
Hall, Ross Hume
Hartnett, John (Conrad)
Hayes, John William
Kelleher, Philip Conboy
Kelley, Jason
Kilpatrick, Charles William
Lamden, Merton Philip
Lechevalier, Mary P
Long, George Louis
Mann, Kenneth Gerard
Melville, Donald Burton
Meyer, William Laros
Otter, Timothy
Racusen, David
Sims, Ethan Allen Hitchcock
Sinclair, Peter Robert
Sjogren, Robert Erik
Thanassi, John Walter
Tyzbir, Robert S
Wallace, Susan Scholes
Watters, Christopher Deffner
Weed, Lawrence Leonard
Weller, David Lloyd
Woodcock-Mitchell, Janet Louise
Woodworth, Robert Cummings

VIRGINIA

Abbott, Lynn De Forrest, Jr
Alberte, Randall Sheldon
Alscher, Ruth
Anderson, Bruce Murray
Anderson, Carl William (Bill)
Appleton, Martin David
Baldwin, Robert Russel
Banks, William Louis, Jr
Barak, Eve Ida
Bender, Patrick Kevin
Benzinger, Rolf Hans
Boatman, Sandra
Bonar, Robert Addison
Bradbeer, Clive
Brandt, Richard Bernard
Brockman, Robert W
Brown, Barry Lee
Brown, Elise Ann Brandenburger
Brown, Gregory Neil
Brown, Jay Clark
Brown, Loretta Ann Port
Bruns, David Eugene
Bunce, George Edwin
Caponio, Joseph Francis
Carson, Eugene Watson, Jr
Chen, Jiann-Shin
Chin, Byong Han
Ching, Melvin Chung Hing
Chlebowski, Jan F
Clark, Mary Jane
Claus, George William
Cohen, I Kelman
Collins, James Malcolm
Coursen, Bradner Wood
Creutz, Carl Eugene
Dalton, Harry P
Deemer, Marie Nieft
De Long, Chester Wallace
DeLorenzo, Robert John
Dementi, Brian Armstead
DeVries, George H
Dewey, William Leo
Diegelmann, Robert Frederick
Dockter, Kenneth Wylie
Doherty, John Douglas
Doman, Elvira
Dunn, John Thornton
Ebel, Richard E
Elford, Howard Lee
Engel, Ruben William
Evans, Herbert John
Feher, Joseph John
Franson, Richard Carl
Fugate, Kearby Joe
Garrett, Reginald Hooker
Garrison, Norman Eugene
Gaugler, Robert Walter
Gear, Adrian R L
Gould, David Huntington
Gregory, Eugene Michael
Grider, John Raymond
Grisham, Charles Milton
Grogan, William McLean
Gruemer, Hanns-Dieter
Haber, Lynne Tracey
Hager, Chester Bradley
Hatzios, Kriton Kleanthis
Hecht, Sidney Michael
Hempfling, Walter Pahl
Hess, John Lloyd
Higgins, Edwin Stanley
Hilu, Khidir Wanni
Holloway, Peter William
Hotta, Shoichi Steven
Huang, Ching-Hsien
Huang, Laura Chi
Jennings, Allen Lee
Jones, Daniel David
Kadner, Robert Joseph
Kalimi, Mohammed Yahya
Kelly, Robert Frank
Kimbrough, Theo Daniel, Jr
King, Betty Louise
Kline, Edward Samuel
Kuemmerle, Nancy Benton Stevens
Kupke, Donald Walter
Kutchai, Howard C
Larner, Joseph
Liberti, Joseph Pollara
McGilvery, Robert Warren
McWright, Cornelius Glen
Maggio, Bruno
Mavis, Richard David
Mikkelsen, Ross Blake
Misra, Hara Prasad
Murtagh, John E
Newman, Jack Huff
Newton, William Edward
Niehaus, Walter G, Jr
Nolin, Janet M
Ogilvie, James William, Jr
Olsen, Kathie Lynn
Olson, Lee Charles
Olson, William Arthur
O'Neal, Charles Harold
Osuch, Mary Ann V
Pallansch, Michael J
Pepe, Gerald Joseph
Peters, Esther Caroline
Pinkston, Margaret Fountain
Place, Janet Dobbins
Polan, Carl E
Price, Byron Frederick
Price, Steven
Pryer, Nancy Kathryn
Raizen, Carol Eileen
Rickett, Frederic Lawrence
Roberts, Catherine Harrison
Rodricks, Joseph Victor
Roscoe, Henry George
Rosenthal, Miriam Dick
Rothberg, Simon
Rutherford, Charles
Rutter, Henry Alouis, Jr
Sando, Julianne J
Saunders, Joseph Francis
Schellenberg, Karl A
Schirch, Laverne Gene
Shelton, Keith Ray
Shukla, Kamal Kant
Sitz, Thomas O
Sobel, Robert Edward
Spangenberg, Dorothy Breslin
Spearing, Cecilia W
Stansly, Philip Gerald
Steinberg, Marcia Irene
Stepka, William
Stewart, Kent Kallam
Storrie, Brian
Stout, Ernest Ray
Swell, Leon
Swope, Fred C
Thompson, Thomas Edward
Topham, Richard Walton
Treadwell, George Edward, Jr
Trelawny, Gilbert Sterling
Van Tuyle, Glenn Charles
Vermeulen, Carl William
Villar-Palasi, Carlos

BIOLOGICAL SCIENCES / 53

White, Howard Dwaine
Wilkinson, Christopher Foster
Willett, James Delos
Williams, Luther Steward
Woo, Yin-tak
Yip, George
Young, Nelson Forsaith
Young, William W, Jr
Yu, Robert Kuan-Jen

WASHINGTON
Adman, Elinor Thomson
Ammann, Harriet Maria
Andersen, Niels Hjorth
Anderson, Larry Ernest
Aronoff, Samuel
Ash, Roy Phillip
Avner, Ellis David
Bankson, Daniel Duke
Barron, Edward J
Barrueto, Richard Benigno
Bendich, Arnold Jay
Blake, James J
Bohnert, Janice Lee
Bonner, John Franklin, Jr
Bornstein, Paul
Branca, Andrew A
Brekke, Clark Joseph
Brosemer, Ronald Webster
Brown, George Willard, Jr
Butler, Lillian Ida
Calza, Roger Ernest
Capp, Grayson L
Carpenter, Carolyn V
Carter, William G
Cataldo, Dominic Anthony
Cheevers, William Phillip
Chen, Shi-Han
Christensen, Gerald M
Churchill, Lynn
Clapshaw, Patric Arnold
Crampton, George H
Croteau, Rodney
Dailey, Frank Allen
Dale, Beverly A
Davie, Earl W
Felton, Samuel Page
Fischer, Edmond H
Fitch, Cynthia Lynn
Floss, Heinz G
Forrey, Arden W
Foster, Robert Joe
Fournier, R E Keith
Fraenkel-Conrat, Jane E
Frankart, William A
Freisheim, James Harold
Gaines, Robert D
Glomset, John A
Goheen, Steven Charles
Gordon, Milton Paul
Griswold, Michael David
Hakomori, Sen-Itiroh
Hall, Benjamin Downs
Hall, Stanton Harris
Harding, Boyd W
Harding, Joseph Warren, Jr
Harper, Alfred Edwin
Hauschka, Stephen D
Hawkins, Richard L
Hazelbauer, Gerald Lee
Herriott, Jon R
Ho, Lydia Su-yong
Horbett, Thomas Alan
Hu, Shiu-Lok
Kachmar, John Frederick
Kaplan, Phyllis Deen
Keller, Patricia J
Kelly, Jeffrey John
Ketchie, Delmer O
Klebanoff, Seymour J
Kutsky, Roman Joseph
Labbe, Robert Ferdinand
Lawrence, John McCune
Leid, R Wes
Leung, Frederick C
Lewis, Norman G
Lightfoot, Donald Richard
Loeb, Lawrence Arthur
Loewus, Frank A
Loewus, Mary W
Lygre, David Gerald
McAnally, John Sackett
McDonough, Leslie Marvin
McFadden, Bruce Alden
Magnuson, Nancy Susanne
Mahlum, Daniel Dennis
Malins, Donald Clive
Margolis, Robert Lewis
Mirkes, Philip Edmund
Monsen, Elaine R
Morris, David Robert
Narayanan, A Sampath
Neurath, Hans
Neve, Richard Anthony
Noyes, Claudia Margaret
Owen, Stanley Paul
Pall, Martin L
Parson, William Wood
Patt, Leonard Merton
Pomeranz, Yeshajahu
Powers, Joseph Robert
Prody, Gerry Ann
Randall, Linda Lea
Reeves, Raymond

Reh, Thomas Andrew
Reid, Brian Robert
Riederer-Henderson, Mary Ann
Ritter, Preston Peck Otto
Rohrschneider, Larry Ray
Rosen, Henry
Sage, Helene E
Schnell, Jerome Vincent
Schrader, Lawrence Edwin
Shepherd, Linda Jean
Shiota, Tetsuo
Short, William Arthur
Shultz, Terry D
Sieker, Larry Charles
Smerdon, Michael John
Smith, Elizabeth Knapp
Smith, Sam Corry
Spackman, Darrel H
Spence, Kemet Dean
Stahl, William Louis
Starr, Jason Leonard
Stenkamp, Ronald Eugene
Stevens, Vernon Lewis
Stevens, Vincent Leroy
Storm, Daniel Ralph
Su, Judy Ya-Hwa Lin
Thomas, John M
Thompson, Roy Charles, Jr
Thorpe, Neal Owen
Unrau, David George
Uribe, Ernest Gilbert
Utter, Fred Madison
Varanasi, Usha
Walsh, Kenneth Andrew
Walters, Ronald Arlen
Wedemeyer, Gary Alvin
White, Fredric Paul
Widmaier, Robert George
Wiley, William Rodney
Wyrick, Ronald Earl
Yancey, Paul Herbert
Yount, Ralph Granville
Yu, Ming-Ho
Zaugg, Waldo S

WEST VIRGINIA
Beattie, Diana Scott
Blaydes, David Fairchild
Bond, Guy Hugh
Brooks, James Lee
Butcher, Fred Ray
Calhoon, Donald Alan
Campbell, Clyde Del
Canady, William James
Capstack, Ernest
Coe, Elmon Lee
Diehl, John Edwin
Foster, Joyce Geraldine
Guyer, Kenneth Eugene, Jr
Hansard, Samuel L, II
Harris, Charles Lawrence
Jagannathan, Singanallur N
Kaczmarczyk, Walter J
Konat, Gregory W
Krause, Reginald Frederick
Malin, Howard Gerald
Markiw, Roman Teodor
Martin, William Gilbert
Mashburn, Louise Tull
Mashburn, Thompson Arthur, Jr
Mecca, Christyna Emma
Moffa, David Joseph
Ong, Tong-man
Quinlan, Dennis Charles
Rafter, Gale William
Reasor, Mark Jae
Reichenbecher, Vernon Edgar, Jr
Roberts, Joseph Linton
Tryfiates, George P
Van Dyke, Knox
Williams, Leah Ann
Wirtz, George H

WISCONSIN
Adler, Julius
Amasino, Richard M
Anand, Amarjit Singh
Anderson, Laurens
Ankel, Helmut K
Attie, Alan D
Balish, Edward
Bavisotto, Vincent
Becker, Wayne Marvin
Beinert, Helmut
Benevenga, Norlin Jay
Bergdoll, Merlin Scott
Bergtrom, Gerald
Bernstein, Bradley Alan
Boutwell, Roswell Knight
Broderick, Glen Allen
Brown, Raymond Russell
Brown, William Henry
Buchanan-Davidson, Dorothy Jean
Burger, Warren Clark
Burgess, Ann Baker
Burgess, Richard Ray
Burke, Thomas J
Burris, Robert Harza
Calbert, Harold Edward
Calvanico, Nickolas Joseph
Campbell, Harold Alexander
Cassens, Robert G
Chakraburtty, Kalpana
Chen, Chong Maw

Chen, Franklin M
Chen, Shao Lin
Cherayil, George Devassia
Cohen, Philip Pacy
ColÁs, Antonio E
Collins, Mary Lynne Perille
Courtright, James Ben
Cox, Michael Matthew
Dahlberg, James E(ric)
Dahms, Nancy Margaret
Deese, Dawson Charles
DeLuca, Hector Floyd
Derse, Phillip H
Dimond, Randall Lloyd
Dodson, Vernon N
Ehle, Fred Robert
Elliott, Bernard Burton
Fahien, Leonard A
Fahl, William Edwin
Feinberg, Benjamin Allen
Feirer, Russell Paul
Fischbach, Fritz Albert
Fredricks, Walter William
Frey, Perry Allen
Ganther, Howard Edward
Ghazarian, Jacob G
Gilboe, David Dougherty
Girotti, Albert William
Gleiter, Melvin Earl
Goldberger, Amy
Goodfriend, Theodore L
Gorsica, Henry Jan
Gorski, Jack
Gourse, Richard Lawrence
Greaser, Marion Lewis
Greenspan, Daniel S
Griffith, Owen W
Haas, Michael John
Hager, Steven Ralph
Harkin, John McLay
Harkness, Donald R
Heath, Timothy Douglas
Hoekstra, William George
Hokin, Lowell Edward
Hosler, Charles Frederick, Jr
Huss, Ronald John
Jacobson, Gunnard Kenneth
Joel, Cliffe David
John, Kavanakuvhiy V
Johnson, Morris Alfred
Jones, Berne Lee
Kaesberg, Paul Joseph
Kahan, Lawrence
Kaltenbach, John Paul
Karavolas, Harry J
Kasper, Charles Boyer
Keegstra, Kenneth G
Kitchell, James Frederick
Klaassen, Dwight Homer
Klebba, Phillip E
Kochan, Robert George
Kornguth, Steven E
Kramer, Elizabeth
Kushnaryov, Vladimir Michael
Lai, Ching-San
Lardy, Henry Arnold
Lemann, Jacob, Jr
Littlewood, Barbara Shaffer
Ludden, Paul W
MacDonald, Michael J
McIntosh, Elaine Nelson
Mahl, Mearl Carl
Martin, Thomas Fabian John
Mertz, Janet Elaine
Metzenberg, Robert Lee
Miller, James Alexander
Miziorko, Henry Michael
Mosesson, Michael W
Mosher, Deane Fremont, Jr
Moskowitz, Gerard Jay
Mueller, Gerald Conrad
Munroe, Stephen Horner
Myers, Charles R
Nagodawithana, Tilak Walter
Nelson, David Lee
Oaks, John Adams
Oberley, Terry De Wayne
Olson, Earl Burdette, Jr
Ordman, Alfred Bram
Pan, David
Pariza, Michael Willard
Peanasky, Robert Joseph
Perlman, Richard
Perry, Billy Wayne
Pitot, Henry C, III
Poland, Alan P
Porter, John Willard
Potter, Van Rensselaer
Pscheidt, Gordon Robert
Qureshi, Nilofer
Raines, Ronald T
Reed, George Henry
Ritter, Karla Schwensen
Roll, Paul M
Sasse, Edward Alexander
Schantz, Edward Joseph
Scheusner, Dale Lee
Schnoes, Heinrich Konstantin
Schoeninger, Margaret J
Schulz, Leslie Olmstead
Seitz, Eugene W
Selman, Bruce R
Shrago, Earl
Siegel, Frank Leonard

Siegel, Jack Morton
Sims, Peter Jay
Smith, Lloyd Michael
Smith, Milton Reynolds
Steele, Robert Darryl
Stone, William Ellis
Stratman, Frederick William
Sussman, Michael R
Suttie, John Weston
Swick, Robert Winfield
Sytsma, Kenneth Jay
Takayama, Kuni
Tewksbury, Duane Allan
Tews, Jean Kring
Tormey, Douglass Cole
Trautman, Jack Carl
Tsao, Francis Hsiang-Chian
Turkington, Roger W
Twining, Sally Shinew
Unsworth, Brian Russell
Verma, Ajit K
Wall, Joseph Sennen
Weaver, Robert Hinchman
Wege, Ann Christene
Weil, Michael Ray
Wejksnora, Peter James
Wells, Barbara Duryea
Wenzel, Frederick J
Wilken, David Richard
Willits, Richard Ellis
Witt, Patricia L
Wyse, Roger Earl
Yin, Jun-Jie

WYOMING
Asplund, Russell Owen
Bosshardt, David Kirn
Bulla, Lee Austin, Jr
Caldwell, Daniel R
George, Robert Porter
Isaak, Dale Darwin
Ji, Inhae
Ji, Tae H(wa)
Kaiser, Ivan Irvin
Lewis, Randolph Vance
McColloch, Robert James
Nunamaker, Richard Allan
Petersen, Nancy Sue
Sullivan, Brian Patrick
Villemez, Clarence Louis, Jr

PUERTO RICO
Alcala, Jose Ramon
Banerjee, Dipak Kumar
El-Khatib, Shukri M
Lluberes, Rosa P
Mandavilli, Satya N
Preston, Alan Martin
Sandza, Joseph Gerard

ALBERTA
Baker, Glen Bryan
Basu, Tapan Kumar
Bentley, Michael Martin
Bleackley, Robert Christopher
Bridger, William Aitken
Brindley, David N
Campbell, James Nicoll
Cass, Carol E
Clandinin, Michael Thomas
Collier, Herbert Bruce
Cossins, Edwin Albert
Crowe, Arlene Joyce
Davis, Norman Rodger
Dixon, Gordon H
Eggert, Frank Michael
Francis, Mike McD
Gaucher, George Maurice
Gooding, Ronald Harry
Goren, Howard Joseph
Hart, David Arthur
Hiruki, Chuji
Hollenberg, Morley Donald
Huber, Reuben Eugene
Iatrou, Kostas
Jacobson, Ada Leah
James, Michael G
Jensen, Susan Elaine
Kadis, Vincent William
Kaneda, Toshi
Kay, Cyril Max
Kuntz, Garland Parke Paul
McElhaney, Ronald Nelson
Madiyalakan, Ragupathy
Madsen, Neil Bernard
Meintzer, Roger Bruce
Nash, David
Paetkau, Verner Henry
Paranchych, William
Paterson, Alan Robb Phillips
Rattner, Jerome Bernard
Roberts, David Wilfred Alan
Rorstad, Otto Peder
Russell, James Christopher
Schroder, David John
Schultz, Gilbert Allan
Scott, Paul G
Smillie, Lawrence Bruce
Sohal, Parmjit S
Spence, Matthew W
Spencer, Mary Eileen Stapleton
Stevenson, Kenneth James
Stinson, Robert Anthony
Suresh, Mavanur Rangarajan

Biochemistry (cont)

Tamaoki, Taiki
Vance, Dennis E
Vance, Jean E
Van De Sande, Johan Hubert
Walsh, Michael Patrick
Wang, Jerry Hsueh-Ching
Weiner, Joel Hirsch
Zalik, Saul

BRITISH COLUMBIA
Aebersold, Ruedi H
Autor, Anne Pomeroy
Bragg, Philip Dell
Buckley, James Thomas
Burton, Albert Frederick
Bushnell, Gordon William
Butler, Gordon Cecil
Candido, Edward Peter Mario
Cullis, Pieter Rutter
Dennis, Patrick P
Desai, Indrajit Dayalji
Dorchester, John Edmund Carleton
Eastwell, Kenneth Charles
Emerman, Joanne Tannis
Finlay, Barton Brett
Friz, Carl T
Frohlich, Jiri J
Garg, Arun K
Godolphin, William
Green, Beverley R
Hancock, Robert Ernest William
Haunerland, Norbert Heinrich
Hochachka, Peter William
Honda, Barry Marvin
Kasinsky, Harold Edward
Kennedy, Christopher Jesse
Lee, Melvin
Li-Chan, Eunice
McBride, Barry Clarke
Mackie, George Alexander
Majak, Walter
Matheson, Alastair Taylor
Matsuo, Robert R
Mauk, Arthur Grant
Mauk, Marcia Rokus
Meheriuk, Michael
Mendoza, Celso Enriquez
Molday, Robert S
Nichols, Jack Loran
Olive, Peggy Louise
Pearce, Richard Hugh
Redfield, Rosemary Jeanne
Rennie, Paul Steven
Richards, James Frederick
Richards, William Reese
Sadowski, Ivan J
Salari, Hassan
Shrimpton, Douglas Malcolm
Singh, Raj Kumari
Skala, Josef Petr
Taylor, Iain Edgar Park
Tener, Gordon Malcolm
Thurlbeck, William Michael
Towers, George Hugh Neil
Underhill, Edward Wesley
Vincent, Steven Robert
Walker, Ian Gardner
Waygood, Ernest Roy
Weeks, Gerald
Whyte, John Nimmo Crosbie
Zbarsky, Sidney Howard

MANITOBA
Biswas, Shib D
Blanchaer, Marcel Corneille
Bodnaryk, Robert Peter
Borsa, Joseph
Burton, David Norman
Choy, Patrick C
Clayton, James Wallace
Dakshinamurti, Krishnamurti
Davie, James Ronald
Foster, Charles David Owen
Hamilton, Ian Robert
Hill, Robert D
Hougen, Frithjof W
Kanfer, Julian Norman
Kovacs, Miklos I P
Kruger, James Edward
Kunz, Bernard Alexander
Lukow, Odean Michelin
Marquardt, Ronald Ralph
Pritchard, Ernest Thackeray
Rosenmann, Edward A
Singh, Harwant
Standing, Kenneth Graham
Stevens, Frits Christiaan
Suzuki, Isamu
Wrogemann, Klaus
Yamada, Esther V
Yurkowski, Michael
Zahradka, Peter

NEW BRUNSWICK
Cashion, Peter Joseph
Chopin, Thierry Bernard Raymond
Fraser, Alan Richard
Gauthier, Didier
Haya, Katsuji
McQueen, Ralph Edward

NEWFOUNDLAND
Barnsley, Eric Arthur
Brosnan, Margaret Eileen
Burness, Alfred Thomas Henry
Davidson, William Scott
Feltham, Lewellyn Allister Woodrow
Hoekman, Theodore Bernard
Idler, David Richard
Ke, Paul Jenn
Keough, Kevin Michael William
Mookerjea, Sailen
Orr, James Cameron
Senciall, Ian Robert
Shaw, Derek Humphrey

NOVA SCOTIA
Addison, Richard Frederick
Beveridge, James MacDonald Richardson
Bidwell, Roger Grafton Shelford
Blair, Alan Huntley
Breckenridge, Carl
Brown, Robert George
Bush, Roy Sidney
Byers, David M
Chambers, Robert Warner
Cohen, William David
Cook, Harold W
Dolphin, Peter James
Gray, Michael William
Hamilton, Robert Milton Gregory
Helleiner, Christopher Walter
Holzbecher, Jiri
Kimmins, Warwick Charles
Laycock, Maurice Vivian
McFarlane, Ellen Sandra
MacRae, Herbert F
MacRae, Thomas Henry
Mezei, Catherine
Palmer, Frederick B St Clair
Ragan, Mark Adair
Wainwright, Stanley D
White, Thomas David

ONTARIO
Alexander, James Craig
Allan, Robert K
Ananthanarayanan, Vettaikkoru S
Andrews, David William
Annett, Robert Gordon
Anwar, Rashid Ahmad
Appanna, Vasu Dhananda
Arif, Basil Mumtaz
Armstrong, John Briggs
Atkinson, Burr Gervais
Bacchetti, Silvia
Baines, A D
Ballantyne, James Stuart
Barran, Leslie Rohit
Baxter, Robert MacCallum
Bayley, Stanley Thomas
Beare-Rogers, Joyce Louise
Begin-Heick, Nicole
Bennick, Anders
Benoiton, Normand Leo
Bettger, William Joseph
Bewley, John Derek
Bhavnani, Bhagu R
Birmingham, Brendan Charles
Birnboim, Hyman Chaim
Blais, Burton W
Bols, Niels Christian
Brash, John Law
Bray, Tammy M
Brown, Stewart Anglin
Brownstone, Yehoshua Shieky
Buchwald, Manuel
Burnison, Bryan Kent
Cairns, William Louis
Callahan, John William
Calleja, G(odehardo) B
Camerman, Norman
Cameron, Ross G
Campbell, James
Cann, Malcolm Calvin
Canvin, David T
Capone, John Pasquale
Carey, Paul Richard
Carroll, Kenneth Kitchener
Chaconas, George
Clarke, Anthony John
Cohen, Saul Louis
Collins, Frank William, Jr
Connell, George Edward
Connelly, Philip Walter
Cummins, W(illiam) Raymond
Cunnane, Stephen C
Dales, Samuel
Davis, Alan
Deber, Charles Michael
De Bold, Adolfo J
Dedhar, Shoukat
Deeley, Roger Graham
Dennis, David Thomas
D'Iorio, Antoine
Donisch, Valentine
Dorrell, Douglas Gordon
Dorrington, Keith John
Downer, Roger George Hamill
Dumbroff, Erwin Bernard
Ebisuzaki, Kaney
Elce, John Shackleton
Epand, Richard Mayer
Farber, Emmanuel
Ferrier, Barbara May
Fischer, Peter Wilhelm Fritz
Fish, Eleanor N
Fitt, Peter Stanley
Fitz-James, Philip Chester
Flanagan, Peter Rutledge
Flynn, Thomas Geoffrey
Forsdyke, Donald Roy
Fourney, Ronald Mitchell
Fraser, Ann Davina Elizabeth
Freeman, Karl Boruch
Fuleki, Tibor
Galsworthy, Peter Robert
Galsworthy, Sara B
Ganoza, M Celia
Gardner, David R
Gentner, Norman Elwood
Ghosh, Hara Prasad
Glick, Bernard Robert
Gold, Marvin H
Goldberg, David Myer
Gornall, Allan Godfrey
Gould, William Douglas
Greenberg, Bruce Matthew
Greenblatt, Jack Fred
Grinstein, Sergio
Gupta, Radhey Shyam
Gurd, James W
Hagen, Paul Beo
Heacock, Ronald A
Heagy, Fred Clark
Hew, Choy-Leong
Hickey, Donal Aloysius
Himms-Hagen, Jean
Hobkirk, Ronald
Hofmann, Theo
Hope, Hugh Johnson
Horner, Alan Alfred
Ingles, Charles James
Irwin, David Michael
Israelstam, Gerald Frank
Jellinek, Peter Harry
Jenkins, Kenneth James William
Jones, John Dewi
Jones, Owen Thomas
Kalab, Miloslav
Kaplan, Harvey
Kapur, Bhushan M
Keeley, Fred W
Khan, Abdul Waheed
Kisilevsky, Robert
Kluger, Ronald H
Kramer, John Karl Gerhard
Krupka, Richard M(orley)
Kuhns, William Joseph
Kuksis, Arnis
Kuroski-De Bold, Mercedes Lina
Kushner, Donn Jean
Lane, Byron George
Lau, Catherine Y
Lawford, George Ross
Layne, Donald Sainteval
Lee, David K H
Lee, Eng-Hong
Lepock, James Ronald
Letarte, Michelle
Liew, Choong-Chin
Ling, Victor
Lingwood, Clifford Alan
Little, James Alexander
Lo, Theodore Ching-Yang
Logan, David Mackenzie
McCalla, Dennis Robert
MacLennan, David Herman
McMurray, William Colin Campbell
MacRae, Andrew Richard
Magee, William Lovel
Maguire, Robert James
Mak, William Wai-Nam
Malkin, Aaron
Manolson, Morris Frank
Mathur, Sukhdev Prashad
Mellors, Alan
Metuzals, Janis
Milligan, Larry Patrick
Mishra, Ram K
Moon, Thomas William
Morand, Peter
Murray, Robert Kincaid
Murray, William Douglas
Neelin, James Michael
Nesheim, Michael Ernest
Nicholls, Doris McEwen
Nicholls, Peter
Nikiforuk, Gordon
Nozzolillo, Constance
O'Brien, Peter J
Olson, Arthur Olaf
Packham, Marian Aitchison
Painter, Robert Hilton
Perry, Malcolm Blythe
Plaxton, William Charles
Polley, John Richard
Possmayer, Fred
Proulx, Pierre R
Rabin, Elijah Zephania
Rauser, Wilfried Ernst
Reed, Juta Kuttis
Riordan, John Richard
Rixon, Raymond Harwood
Robinson, Brian Howard
Rock, Gail Ann
Rogers, Charles Graham
Rolleston, Francis Stopford
Rosa, Nestor
Roslycky, Eugene Bohdan
Ryan, Michael T
Sanwal, Bishnu Dat
Sarkar, Bibudhendra
Sarma, Dittakavi S R
Schachter, H
Schneider, Henry
Scott, Fraser Wallace
Scrimgeour, Kenneth Gray
Sen, Amar Kumar
Shah, Bhagwan G
Sikorska, Marianna
Singh, Bhagirath
Siu, Chi-Hung
Smiley, James Richard
Smith, David Burrard
Sole, Michael Joseph
Spencer, John Hedley
Sribney, Michael
Stainer, Dennis William
Stanacev, Nikola Ziva
Stavric, Stanislava
Steele, John Earle
Stephenson, Norman Robert
Stewart, Harold Brown
Storey, Kenneth Bruce
Strickland, Kenneth Percy
Suderman, Harold Julius
Szabo, Arthur Gustav
Taylor, Keith Edward
Taylor, Norman Fletcher
Tepperman, Barry Lorne
Thomas, Barry Holland
Thompson, John Eveleigh
Tinker, David Owen
Tobe, Stephen Solomon
Toews, Cornelius J
Trevithick, John Richard
Trick, Charles Gordon
Tsai, Chishiun S
Tsang, Charles Pak Wai
Tustanoff, Eugene Reno
Vafopoulo, Xanthe
Van Huystee, Robert Bernard
Vardanis, Alexander
Veliky, Ivan Alois
Viswanatha, Thammaiah
Waithe, William Irwin
Walker, Peter Roy
Warner, Alden Howard
Wasi, Safia
White, Gordon Allan
Williamson, Denis George
Wilson, David George
Wolfe, Bernard Martin
Wood, Janet Marion
Yamazaki, Hiroshi
Yang, Man-chiu
Yip, Cecil Cheung-Ching
Younglai, Edward Victor
Ziegler, Peter
Zilkey, Bryan Frederick
Zitnak, Ambrose
Zobel, Alicja Maria

PRINCE EDWARD ISLAND
Rigney, James Arthur
Suzuki, Michio

QUEBEC
Abshire, Claude James
Alavi, Misbahuddin Zafar
Anderson, William Alan
Antakly, Tony
Archibald, Frederick S
Barcelo, Raymond
Barden, Nicholas
Beaudoin, Adrien Robert
Belanger, Luc
Bellabarba, Diego
Berlinguet, Louis
Bilimoria, Minoo Hormasji
Blostein, Rhoda
Borgeat, Pierre
Boulet, Marcel
Braun, Peter Eric
Brown, Gregory Gaynor
Bukowiecki, Ludwik J
Cedergren, Robert J
Chabot, Benoit
Chafouleas, James G
Clark, Michael Wayne
Cooper, David Gordon
Cormier, Francois
Cousineau, Gilles H
Daigneault, Rejean
David, Jean
De Lamirande, Gaston
DeMedicis, E M J A
Devor, Kenneth Arthur
Dubé, François
DuBow, Michael Scott
Ducharme, Jacques R
Dupont, Claire Hammel
Dupuis, Gilles
Edward, Deirdre Waldron
Fazekas, Arpad Gyula
Ford-Hutchinson, Anthony W
Gagnon, Claude
Gariepy, Claude
Germinario, Ralph Joseph
Gianetto, Robert
Godin, Claude

Goltzman, David
Graham, Angus Frederick
Gresser, Michael Joseph
Guderley, Helga Elizabeth
Guillemette, Gaetan
Hallenbeck, Patrick Clark
Hamet, Pavel
Hancock, Ronald Lee
Heisler, Seymour
Herscovics, Annette Antoinette
Hosein, Esau Abbas
Ibrahim, Ragai Kamel
Ingram, Jordan Miles
Johnstone, Rose M
Kallai-Sanfacon, Mary-Ann
Kluepfel, Dieter
Labrie, Fernand
Labuda, Damian
Lapointe, Jacques
Leduc, Gerard
Lehoux, Jean-Guy
Lemonde, Andre
Levy, Samuel Wolfe
Lin, Sheng-Xiang
Liuzzi, Michel
MacKenzie, Robert Earl
Maclachlan, Gordon Alistair
Madiraju, S R Murthy
Manjunath, Puttaswamy
Marceau, Normand Luc
Martineau, Ronald
Mathieu, Leo Gilles
Meighen, Edward Arthur
Momparler, Richard Lewis
Morais, Rejean
Moroz, Leonard Arthur
Murgita, Robert Anthony
Murthy, Mahadi Raghavandrarao Ven
Niven, Donald Ferries
Page, Michel
Palfree, Roger Grenville Eric
Pande, Shri Vardhan
Pappius, Hanna M
Pelletier, Omer
Phan, Chon-Ton
Ponka, Premysl
Poole, Anthony Robin
Poole, Ronald John
Powell, William St John
Preiss, Benjamin
Roberge, Andree Groleau
Roberts, Kenneth David
Rochefort, Joseph Guy
Rokeach, Luis Alberto
Roughley, Peter James
Roy, Claude Charles
Roy, Robert Michael McGregor
Sairam, Malur Ramaswamy
Sandor, Thomas
Sarhan, Fathey
Schiffrin, Ernesto Luis
Schucher, Reuben
Schulman, Herbert Michael
Seidah, Nabil George
Sheinin, Rose
Shore, Gordon Charles
Shoubridge, Eric Alan
Sinclair, Ronald
Sirois, Pierre
Skup, Daniel
Slilaty, Steve N
Solomon, Samuel
Sourkes, Theodore Lionel
Srivastava, Ashok Kumar
Srivastava, Prakash Narain
Stephens, Heather R
Sygusch, Jurgen
Talbot, Pierre J
Tan, Ah-Ti Chu
Tan, Liat
Tanguay, Robert M
Taussig, Andrew
Tenenhouse, Alan M
Tenenhouse, Harriet Susie
Thirion, Jean Paul Joseph
Tonks, David Bayard
Van Gelder, Nico Michel
Van Lier, Johannes Ernestinus
Vezina, Claude
Villeneuve, Andre
Vinay, Patrick
Wang, Eugenia
Wasserman, Aaron Reuben
Willemot, Claude
Wolfe, Leonhard Scott
Yakunin, Alexander F
Yousef, Ibrahim Mohmoud
Zannis, Maria

SASKATCHEWAN
Angel, Joseph Francis
Blair, Donald George Ralph
Boulton, Alan Arthur
Craig, Burton Mackay
Davis, Bruce Allan
Dawson, Peter Stephen Shevyn
Denford, Keith Eugene
Durden, David Alan
Faber, Albert John
Gear, James Richard
Grant, Donald R
Gupta, Vidya Sagar
Harold, Stephen
Howarth, Ronald Edward

Kalra, Jawahar
Khandelwal, Ramji Lal
Korsrud, Gary Olaf
Kurz, Wolfgang Gebhard Walter
Lowden, J Alexander
McGregor, Douglas Ian
McLennan, Barry Dean
Martin, Robert O
Patience, John Francis
Richardson, J(ohn) Steven
Robertson, Hugh Elburn
Shargool, Peter Douglas
Vella, Francis
Waygood, Edward Bruce
Wood, James Douglas
Yu, Peter Hao

OTHER COUNTRIES
Abelev, Garri Izrailevich
Ahn, Tae In
Alivisatos, Spyridon Gerasimos Anastasios
Amsterdam, Abraham
Andrews, Peter Walter
Arai, Ken-Ichi
Atkinson, Paul H
Bachmann, Fedor W
Baddiley, James
Bader, Hermann
Baig, Mirza Mansoor
Baldwin, Robert William
Barel, Monique
Barouki, Robert
Bayev, Alexander A
Ben-Ze'ev, Avri
Binder, Bernd R
Boffa, Lidia C
Bogoch, Samuel
Bonting, Sjoerd Lieuwe
Bronk, John Ramsey
Brown, Barker Hastings
Bunn, Clive Leighton
Buzina, Ratko
Canellakis, Evangelo S
Cerbon-Solorzano, Jorge
Chang, Albert Yen
Chen, Jan-Kan
Chetsanga, Christopher J
Chipman, David Mayer
Choi, Yong Chun
Cooke, T Derek V
Creighton, Thomas Edwin
De Haën, Christoph
Delmer, Deborah P
De Renobales, Mertxe
Dikstein, Shabtay
Doerfler, Walter Hans
Donato, Rosario Francesco
Dreosti, Ivor Eustace
Dutta-Roy, Asim Kanti
Dworkin, Mark Bruce
Edelman, Marvin
Edelstein, Stuart J
Eggers, Hans J
Eik-Nes, Kristen Borger
Engelborghs, Yves
Fahimi, Hossein Dariush
Fawaz, George
Fritz, Irving Bamdas
Fujiwara, Keigi
Garcia-Sainz, J Adolfo
Ginis, Asterios Michael
Glick, David M
Goding, James Watson
Goldberg, Burton David
Graber, Robert Philip
Gregory, Peter
Guimaraes, Romeu Cardoso
Haegele, Klaus D
Hanukoglu, Israel
Harpur, Robert Peter
Hayashi, Masao
Heinz, Erich
Helmreich, Ernst J M
Henderson, David Andrew
Hillcoat, Brian Leslie
Hirano, Toshio
Hiyama, Tetsuo
Hooper, Nigel
Hosokawa, Keiichi
Ikehara, Yukio
Ionescu, Lavinel G
Itiaba, Kibe
Jaffe, Werner G
Jakob, Karl Michael
Jarvis, Simon Michael
Jensen, Elwood Vernon
Jost, Jean-Pierre
Ju, Jin Soon
Kang, Sungzong
Kato, Ikunoshin
Katz, Bernard
Kaye, Alvin Maurice
Kent, Stephen Brian Henry
Kobata, Akira
Koerber, Walter Ludwig
Kornberg, Sir Hans Leo
Kuhn, Klaus
Kumar, S Anand
Lee, Joseph Chuen Kwun
Lee, Men Hui
Leitzmann, Claus
Lewin, Lawrence M
Liao, Ta-Hsiu

Livett, Bruce G
Lonsdale-Eccles, John David
Loppnow, Harald
Low, Chow-Eng
Low, Teresa Lingchun Kao
Lu, Christopher D
Maass, Wolfgang Siegfried Gunther
MacCarthy, Jean Juliet
Maruta, Hiroshi
Meins, Frederick, Jr
Merlini, Giampaolo
Mezquita, Cristobal
Millar, Robert Peter
Mitchell, Peter
Mittal, Balraj
Mizuno, Shigeki
Modabber, Farrokh Z
Montgomery, Morris William
Muller-Eberhard, Hans Joachim
Munoz, Maria De Lourdes
Nagata, Kazuhiro
Olson, John Melvin
Paik, Young-Ki
Pina, Enrique
Pruss, Rebecca M
Querinjean, Pierre Joseph
Quintanilha, Alexandre Tiedtke
Rakowski, Krzysztof J
Ray, Prasanta K
Reynolds, Jacqueline Ann
Roy, Raman K
Said, Mostafa Mohamed
Schmell, Eli David
Schneider, Wolfgang Johann
Schwartz, Peter Larry
Seya, Tsukasa
Shaw, Elliott Nathan
Spierer, Pierre
Stojanovic, Borislav Jovan
Stoppani, Andres Oscar Manuel
Strous, Ger J
Subramanian, Alap Raman
Sullivan-Kessler, Ann Clare
Sung, Chen-Yu
Syze, Richard Laurence Millington
Tachibana, Takehiko
Takeuchi, Kiyoshi Hiro
Taniyama, Tadayoshi
Teleb, Zakaria Ahmed
Thiery, Jean Paul
Thorson, John Wells
Toutant, Jean-Pierre
Tsolas, Orestes
Tsou, Chen-Lu
Ueno, Hiroshi
Varga, Janos M
Vina, Juan R
Watanabe, Mamoru
Webb, Cynthia Ann Glinert
Weil, Marvin Lee
Welsh, Richard Stanley
Wenger, Byron Sylvester
Willard-Gallo, Karen Elizabeth
Wong, Jeffrey Tze-Fei
Wu, Albert M
Wu, Cheng-Wen
Wu, Felicia Ying-Hsiueh
Yamamoto, Hiroshi
Yeon-Sook, Yun
Zenk, Meinhart Hans Christian

Biology, General

ALABAMA
Appel, Arthur Gary
Beyers, Robert John
Boozer, Reuben Bryan
Carter, Charlotte
Chang, Chi Hsiung
Dacheux, Ramon F, II
Dorgan, William Joseph
Freeman, John A
Graham, Tom Maness
Hanson, Roger Wayne
Koulourides, Theodore I
McClintock, James Bruce
Michalek, Suzanne M
Pearson, Allen Mobley
Perry, Nelson Allen
Poirier, Gary Raymond
Richardson, Terry David
Sledge, Eugene Bondurant
Stock, Carl William
Tucker, Charles Eugene
Turco, Jenifer
Wilkes, James C
Williams, Louis Gressett
Williams, Michael Ledell

ALASKA
Albert, Thomas F
Proenza, Luis Mariano
Rice, Stanley Donald
Roop, Richard Allan
Schoen, John Warren
Shields, Gerald Francis
Shirley, Thomas Clifton
Willson, Mary Frances

ARIZONA
Butler, Lillian Catherine
Haase, Edward Francis
Hadley, Mac Eugene

Hildebrand, John Grant, III
Johnson, Mary Ida
Justice, Keith Evans
Kammer, Ann Emma
Keck, Konrad
Kurtz, Edwin Bernard, Jr
Lisonbee, Lorenzo Kenneth
McElhinney, Margaret M (Cocklin)
Nagle, Ray Burdell
Pommerville, Jeffrey Carl
Prior, David James
Rasmussen, David Irvin
Schaffer, William Morris
Schmidt, Justin Orvel
Slobodchikoff, Constantine Nicholas
Stahnke, Herbert Ludwig
Stini, William Arthur
Timmermann, Barbara Nawalany
Tischler, Marc Eliot
Topoff, Howard Ronald
Towill, Leslie Ruth
Van Voorhies, Wayne Alan
Verbeke, Judith Ann
Villanueva, Antonio R
Younggren, Newell A
Zegura, Stephen Luke

ARKANSAS
Alguire, Mary Slanina (Hirsch)
Buffaloe, Neal Dollison
Clower, Dan Fredric
Demarest, Jeffrey R
Evans, William L
Geren, Collis Ross
Karlin, Alvan A
McMillan, Harlan L
Scheving, Lawrence Einar
Shade, Elwood B

CALIFORNIA
Adomian, Gerald E
Aggeler, Judith
Amirkhanian, John David
Anand, Rajen S(ingh)
Anderson, Steven Clement
Antipa, Gregory Alexis
Araki, George Shoichi
Arce, Gina
Archibald, James David
Armstrong, Peter Brownell
Arnheim, Norman
Awbrey, Frank Thomas
Baker, Herbert George
Baptista, Luis Felipe
Baresi, Larry
Barker, David Lowell
Barnett, Eugene Victor
Barondes, Samuel Herbert
Bauer, Eugene Andrew
Baylink, David J
Beers, William Howard
Belisle, Barbara Wolfanger
Bell, Jack Perkins
Berg, William Eugene
Berns, Michael W
Bernstein, Gerald Sanford
Bissell, Dwight Montgomery
Bissing, Donald Ray
Bjorkman, Olle Erik
Blau, Helen Margaret
Block, Barbara Ann
Bodell, William J
Bok, P Dean
Bondy, Stephen Claude
Bournias-Vardiabasis, Nicole
Bowman, Barry J
Brokaw, Charles Jacob
Brown, Howard S
Brown, Judith Adele
Brown, Walter Creighton
Brusca, Gary J
Bryant, Susan Victoria
Buchsbaum, Ralph
Buffett, Rita Frances
Bullock, Theodore Holmes
Cain, Stanley A
Callahan, Joan Rea
Callison, George
Campbell, Bruce Carleton
Campbell, Richard Dana
Carle, Glenn Clifford
Carpenter, Frances Lynn
Carroll, Edward James, Jr
Chapman, Lloyd William
Chuong, Cheng-Ming
Church, Ronald L
Clegg, Michael Tran
Clothier, Galen Edward
Collins, Barbara Jane
Colwell, Robert Knight
Cooper, Edwin Lowell
Coppedge, James
Cosgrove, Gerald Edward
Crick, Francis Harry Compton
Cushing, John (Eldridge), Jr
Daniel, Charles Waller
Dasgupta, Asim
Dawson, John E
DePass, Linval R
DeVries, Gerald W
Diamond, Marian C
Dirksen, Ellen Roter
Doell, Ruth Gertrude
Duffey, Sean Stephen

Biology, General (cont)

Duncan, Thomas Osler
Du Pont, Frances Marguerite
Ehrlich, Paul Ralph
Emr, Scott David
Enders, Allen Coffin
Endler, John Arthur
Engelmann, Franz
Epel, David
Farish, Donald James
Fender, Derek Henry
Fernald, Russell Dawson
Fisher, Kathleen Mary Flynn
Fisher, Steven Kay
Fleming, James Emmitt
Fox, Joan Elizabeth Bothwell
Futch, David Gardner
Game, John Charles
Gascoigne, Nicholas Robert John
Gatlin, Lila L
Gaudin, Anthony J
Gautsch, James Willard
Giese, Arthur Charles
Gill, Ayesha Elenin
Gilula, Norton Bernard
Goldstein, Bernard
Goodman, Michael Gordon
Gordh, Gordon
Grey, Robert Dean
Grinnell, Alan Dale
Grobstein, Clifford
Haen, Peter John
Hahn, George
Haig, Janet
Hanes, Deanne Meredith
Hanggi, Evelyn Betty
Hanson, James Charles
Hardin, Garrett (James)
Hare, John Daniel, III
Havran, Wendy Lynn
Hayashi, Izumi
Hayflick, Leonard
Healey, Patrick Leonard
Heller, Horace Craig
Henderson, Sheri Dawn
Henry, Helen L
Hill, Rolla B, Jr
Hinegardner, Ralph
Hogue, Charles Leonard
Hungate, Robert Edward
Hurd, Paul DeHart
Iltis, Wilfred Gregor
Jain, Subodh K
Jakus, Marie A
Jeffery, William Richard
Jenkins, Kenneth Dunning
Jennrich, Ellen Coutlee
Jones, Edward George
Jones, Rebecca Anne
Kavanaugh, David Henry
Kay, Robert Eugene
Keller, Raymond E
Kelly, Amy Schick
Kendig, Joan Johnston
Kennedy, Donald
Knudsen, Eric Ingvald
Konishi, Masakazu
Kornberg, Thomas B
Lambert, Charles Calvin
Lester, Henry Allen
Lewis, Edward B
Licht, Paul
Lindstedt-Siva, K June
Lipshitz, Howard David
Logsdon, Donald Francis, Jr
Loomis, William Farnsworth, Jr
Lord, Elizabeth Mary
Lovatt, Carol Jean
Love, Robert Merton
McCarthy, Miles Duffield
McCluskey, Elwood Sturges
McElroy, William David
Mackey, James P
McPherson, Alexander
Madore, Monica Agnes
Manougian, Edward
Martin, Gail Roberta
Martin, Michael
Martinez, Joe L, Jr
Mather, Jennie Powell
Mayeri, Earl Melchior
Miles, Lindsey Anne
Millan, Jose Luis
Miller, Larry O'Dell
Miller, Marcia Madsen
Moore, John Alexander
Mortenson, Theadore Hampton
Moser, Charles R
Murdoch, William W
Murphy, Collin Grisseau
Murphy, Ted Daniel
Murphy, Terence Martin
Nafpaktitis, Basil G
Nakamura, Robert Masao
Nash, Carroll Blue
Nelson, Gregory A
Newbrun, Ernest
Nickel, Phillip Arnold
Niklowitz, Werner Johannes
Nishioka, Richard Seiji
Nuccitelli, Richard Lee
Oechel, Walter C
Oldstone, Michael Beauregard Alan
Oliphant, Edward Eugene
Oppenheimer, Steven Bernard
Ordahl, Charles Philip
Owen, Paul Howard
Pearse, John Stuart
Perrault, Jacques
Pinto, John Darwin
Pitts, Wanna Dene
Plank, Stephen J
Pollock, Edward G
Poodry, Clifton Arthur
Power, Dennis Michael
Prend, Joseph
Rae-Venter, Barbara
Readhead, Carol Winifred
Regnery, David Cook
Robertson, Richard Thomas
Rodden, Robert Morris
Roest, Aryan Ingomar
Rollins, Wade Cuthbert
Rosenberg, Marvin J
Rosenfeld, Ron Gershon
Rothman, Alvin Harvey
Rubin, Harry
Sartoris, David John
Schlapfer, Werner T
Schlinger, Evert Irving
Scott, Michael David
Scudder, Harvey Israel
Segall, Paul Edward
Shapiro, Arthur Maurice
Shepard, David C
Sheppard, Dean
Shuler, Charles F
Simon, Melvin
Singer, S(eymour) J(onathan)
Siu, Gerald
Smith, Lewis Dennis
Sperry, Roger Wolcott
Staal, Gerardus Benardus
Stanton, Toni Lynn
Starrett, Priscilla Holly
Steinhardt, Richard Antony
Stevens, Charles F
Stewart, Brent Scott
Stewart, Joan Godsil
Strehler, Bernard Louis
Strickberger, Monroe Wolf
Sutton, Dallas Albert
Swan, Lawrence Wesley
Sweeney, Beatrice Marcy
Talamantes, Frank
Thaler, M Michael
Thompson, Jeffrey Michael
Thompson, Rosemary Ann
Thorne, Robert Folger
Thornton, Robert Melvin
Timourian, Hector
Tokes, Zoltan Andras
Tokuyasu, Kiyoteru
Tordoff, Walter, III
Towner, Howard Frost
Tukey, Robert H
Tyan, Marvin L
Uyeno, Edward Teiso
Vacquier, Victor Dimitri
Vasu, Bangalore Seshachalam
Vermeij, Geerat Jacobus
Vinje, Mary M (Taylor)
Wake, David Burton
Wallace, Robert Bruce
Watt, Ward B
Wedberg, Stanley Edward
Weidner, William Jeffrey
Weston, Charles Richard
Wickler, Steven John
Wiley, Lynn M
Williams, Joseph Burton
Wills, Christopher J
Winget, Charles M
Wolfe, Stephen Landis
Wong-Staal, Flossie
Woods, Geraldine Pittman
Work, Telford Hindley
Yarnall, John Lee
Young, Joseph Hardie
Yu, Grace Wei-Chi Hu
Yund, Mary Alice
Zavortink, Thomas James
Zeichner-David, Margarita
Zomzely-Neurath, Claire Eleanore
Zucker, Irving
Zyskind, Judith W

COLORADO
Alpern, Herbert P
Anderson, Roger Arthur
Audesirk, Teresa Eck
Avens, John Stewart
Benson, Norman G
Bond, Richard Randolph
Brandom, William Franklin
Compton, Thomas Lee
Dunahay, Terri Goodman
Ellinwood, William Edward
Eschmeyer, Paul Henry
Gorell, Thomas Andrew
Grant, Michael Clarence
Ivory, Thomas Martin, III
Jones, Carol A
Katsh, Seymour
Keen, Veryl F
Kelman, Robert Bernard
Lee, Robert Edward
Lesh-Laurie, Georgia Elizabeth
Linhart, Yan Bohumil
Luttges, Marvin Wayne
Miller, Charles William
Moran, David Taylor
Moury, John David
Nornes, Howard Onsgaard
Quissell, David Olin
Ranu, Rajinder S
Schultz, Phyllis W
Shields, Lora Maneum
Snyder, Gregory Kirk
Sokol, Ronald Jay
Stendell, Rey Carl
Storey, Richard Drake
Thompson, Henry Joseph
Winchester, Albert McCombs
Winston, Paul Wolf

CONNECTICUT
Adair, Eleanor R
Aitken, Thomas Henry Gardiner
Anderson, Herbert Godwin, Jr
Ariyan, Stephan
Baldwin, John Charles
Block, Bartley C
Boell, Edgar John
Booth, Charles E
Bullock, Ward Ervin, Jr
Buss, Leo William
Carluccio, Leeds Mario
Cutler, Leslie Stuart
Dix, Douglas Edward
Doeg, Kenneth Albert
Dye, Frank J
Gay, Thomas John
Goldman-Rakic, Patricia S(hoer)
Goldsmith, Mary Helen Martin
Gorski, Leon John
Halaban, Ruth
Hanson, Earl Dorchester
Hebert, Daniel Normond
Henry, Charles Stuart
Huszar, Gabor
Jokinen, Eileen Hope
Kankel, Douglas Ray
Kiefer, Barry Irwin
Kollar, Edward James
Kosher, Robert Andrew
LaMondia, James A
Laufer, Hans
Lefrancois, Leo
Levin, Martin Allen
Loewenthal, Lois Anne
Loomis, Stephen Henry
Maguder, Theodore Leo, Jr
Matheson, Dale Whitney
Merwin, June Rae
Morowitz, Harold Joseph
Myles, Diana Gold
Pierro, Louis John
Rakic, Pasko
Roman, Laura M
Roos, Henry
Samuel, Edmund William
Satter, Ruth
Schultz, R Jack
Simpson, Tracy L
Slater, James Alexander
Somers, Michael Eugene
Stadnicki, Stanley Walter, Jr
Taigen, Theodore Lee
Trinkaus, John Philip
Vignery, Agnes M C
Wheeler, Bernice Marion
Yang, Tsu-Ju (Thomas)

DELAWARE
Cubberley, Virginia
Hoppenjans, Donald William
Kinney, Anthony John
Lotrich, Victor Arthur
Stetson, Milton H
White, Harold Bancroft, III
Williams, Mary Bearden

DISTRICT OF COLUMBIA
Brickey, Paris Manaford
Brooks, John Langdon
Burns, John McLauren
Chiang, Peter K
Chiarodo, Andrew
Coomes, Marguerite Wilton
Cox, Geraldine Anne Vang
De Queiroz, Kevin
DiMichele, William Anthony
Douple, Evan Barr
Frizzera, Glauco
Gladue, Brian Anthony
Gordon, William Ransome
Grissell, Edward Eric Fowler
Hakim, Raziel Samuel
Harshbarger, John Carl, Jr
Hoffmann, Robert Shaw
Johnson, Kurt Edward
Kleiman, Devra Gail
Kromer, Lawrence Frederick
Kumar, Ajit
McDiarmid, Roy Wallace
McEwen, Gerald Noah, Jr
Mathis, Wayne Neilsen
Melnick, Vijaya L
Norman, Wesley P
Norrbom, Allen Lee
Nowicke, Joan Weiland
Robbins, Robert Kanner
Roper, Clyde Forrest Eugene
Soffen, Gerald A(lan)
Speidel, John Joseph
Springer, Victor Gruschka
Thomas, William Eric
Wilson, Don Ellis
Wylie, Richard Michael
Zug, George R

FLORIDA
Abele, Lawrence Gordon
Abou-Khalil, Samir
Bennett, Thomas P(eter)
Branham, Joseph Morhart
Bray, Joan Lynne
Brinkley, Linda Lee
Britt, N Wilson
Castner, James Lee
Christman, Steven Philip
Clegern, Robert Wayne
Couch, John Alexander
Crump, Martha Lynn
Donahue, Roger P
Emmel, Thomas C
Feldmesser, Julius
Fernandez, Hugo Lathrop
Fitzgerald, Robert James
Fournie, John William
Fraser, Ronald Chester
Frei, Sr John Karen
Gaines, Michael Stephen
Gambrell, Lydia Jahn
Gennaro, Robert Nash
Giesel, James Theodore
Goss, Gary Jack
Gregg, James Henderson
Gupta, Virendra K
Hackman, John Clement
Hansen, Keith Leyton
Hass, Michael A
Hollinger, Thomas Garber
Iverson, Ray Mads
Jensen, Albert Christian
Kaul, Rajinder K
Keller, Laura R
Kelly, Richard Delmer
Keppler, William J
Koevenig, James L
Larkin, Lynn Haydock
Leigh, Egbert Giles, Jr
Lloyd, James Edward
Lutz, Peter Louis
McClure, Joseph A
MacDonald, Eve Lapeyrouse
Mallery, Charles Henry
Miskimen, Carmen Rivera
Morrill, John Barstow, Jr
Muller, Kenneth Joseph
Murison, Gerald Leonard
Nation, James Lamar, Sr
Nicosia, Santo Valerio
Norman, Eliane Meyer
Ocampo-Friedmann, Roseli C
Potter, Lincoln Truslow
Raizada, Mohan K
Reiskind, Jonathan
Rokach, Joshua
Romrell, Lynn John
Roux, Kenneth H
Samuelson, Don Arthur
Schwassman, Horst Otto
Selman, Kelly
Sinclair, Thomas Russell
Smart, Grover Cleveland, Jr
Smith, Albert Carl
Stechschulte, Agnes Louise
Sturtevant, Ruthann Patterson
Sumners, DeWitt L
Taylor, George Thomas
Taylor, J(ames) Herbert
Thureson-Klein, Asa Kristina
Tschinkel, Walter Rheinhardt
Waggoner, Phillip Ray
Wagoner, Dale E
Westfall, Minter Jackson, Jr
Williams, Norris Hagan
Wireman, Kenneth
Zikakis, John Philip

GEORGIA
Andrews, Lucy Gordon
Black, Jessie Kate
Boudinot, Frank Douglas
Browne, John Mwalimu
Bryans, Trabue Daley
Butts, David
Calabrese, Ronald Lewis
Carter, Eloise B
Carter, James Richard
Cavanagh, Harrison Dwight
Chin, Edward
Coward, Stuart Jess
Dashek, William Vincent
DeHaan, Robert Lawrence
Dibull, Alfred
Duncan, Robert Leon, Jr
Duncan, Wilbur Howard
Evatt, Bruce Lee
Fetner, Robert Henry
Hamada, Spencer Hiroshi
Hancock, Kenneth Farrell
Hayes, John Thompson

Haynes, John Kermit
Heise, John J
Humphreys, Walter James
Hunter, Roy, Jr
Kethley, Thomas William
Klompen, J S H
Koballa, Thomas Raymond, Jr
Krolak, John Michael
Lindsay, David Taylor
McNeil, Paul L
McPherson, Robert Merrill
Miller-Stevens, Louise Teresa
Nabrit, Samuel Milton
Parrish, Fred Kenneth
Paulsen, Douglas F
Pendergrass, Levester
Pienaar, Leon Visser
Robinson, Margaret Chisolm
Srivastava, Prakash Narain
Stouffer, Richard Franklin
Sutton, William Wallace
White, Donald Henry
Willis, Judith Horwitz
Wistrand, Harry Edwin
Wood, John Grady
Wyatt, Robert Edward
Zinsmeister, Philip Price

HAWAII
Borthakur, Dulal
Carr, Gerald Dwayne
Carson, Hampton Lawrence
Culliney, Thomas W
De Feo, Vincent Joseph
Flickinger, Reed Adams
Hapai, Marlene Nachbar
Humphreys, Tom Daniel
Kane, Robert Edward
Kaneshiro, Kenneth Yoshimitsu
Lee, Cheng-Sheng
Leung, Julia Pauline
Read, George Wesley
Siegel, Barbara Zenz
Ward, Melvin A
Yanagimachi, Ryuzo

IDAHO
Bratz, Robert Davis
Henderson, Douglass Miles
Johnson, James Blakeslee
Mincher, Bruce J
Mudumbi, Ramagopal Vijaya
Wyllie, Gilbert Alexander

ILLINOIS
Anderson, David John
Augspurger, Carol Kathleen
Bachop, William Earl
Baron, David Alan
Barr, Susan Hartline
Beck, Sidney L
Beer, Alan E
Benkendorf, Carol Ann
Berry, James Frederick
Berry, Robert Wayne
Bieler, Rudiger
Birkenholz, Dale Eugene
Bjorklund, Richard Guy
Brand, Raymond Howard
Briggs, Robert Wilbur
Carr, Virginia McMillan
Castignetti, Domenic
Chappell, Dorothy Field
Chen, Shepley S
Chung, Stephen
Constantinou, Andreas I
Cox, Thomas C
Crawford, Susan Young
Dal Canto, Mauro Carlo
Daniel, Jon Cameron
Decker, Robert Scott
Disterhoft, John Francis
Drengler, Keith Allan
Eckner, Karl Friedrich
Fast, Dale Eugene
Flinn, James Edwin
Foster, Raymond Orrville
Friedman, Yochanan
Gassman, Merrill Loren
Gerdy, James Robert
Gibbs, Daniel
Gibson, David John
Gillette, Rhanor
Gingle, Alan Raymond
Giometti, Carol Smith
Goldman, Max
Gomes, Wayne Reginald
Gonzales, Federico
Goodheart, Clyde Raymond
Grdina, David John
Howe, Henry Franklin
Huberman, Eliezer
Irwin, Michael Edward
Jasper, Donald K
Josephs, Robert
Kasprow, Barbara Ann
Kaufman, Stephen J
Koch, Alisa Erika
Lamppa, Gayle K
Layman, Dale Pierre
Lucher, Lynne Annette
McKinley, Vicky L
MacLeod, Ellis Gilmore
Maiorana, Virginia Catherine

Melin, Brian Edward
Mitchell, John Laurin Amos
Mockford, Edward Lee
Muffley, Harry Chilton
Murphy, Richard Allan
Naples, Virginia L
Natalini, John Joseph
Norstog, Knut Jonson
Nyberg, Dennis Wayne
Overton, Jane Harper
Palincsar, Edward Emil
Pappas, George Demetrios
Parker, Nancy Johanne Rentner
Peterson, Christopher Gerard
Phillips, Carleton Jaffrey
Prahlad, Kadaba V
Pruett-Jones, Stephen Glen
Rolfe, Gary Lavelle
Rosales-Sharp, Maria Consolacion
Rotermund, Albert J, Jr
Roth, Robert Mark
Scharf, Arthur Alfred
Schennum, Wayne Edward
Siegel, Ivens Aaron
Simpson, Sidney Burgess, Jr
Sorensen, Paul Davidsen
Spiroff, Boris E N
Stocum, David Leon
Straus, Helen Lorna Puttkammer
Straus, Werner
Sze, Paul Yi Ling
Thiruvathukal, Kris V
Thonar, Eugene Jean-Marie
Towle, David Walter
Tripathi, Satish Chandra
Van Valen, Leigh Maiorana
Velardo, Joseph Thomas
Wade, Michael John
Wasserman, William John
Wiatr, Christopher Louis
Yopp, John Herman
Zaki, Abd El-Moneim Emam
Zimmerman, Roger Paul

INDIANA
Baker, Edgar Gates Stanley
Bates, Charles Johnson
Borgens, Richard Ben
Broxmeyer, Hal Edward
Burkholder, Timothy Jay
Chiscon, J Alfred
Crowell, (Prince) Sears, Jr
Delfin, Eliseo Dais
Dial, Norman Arnold
DiMicco, Joseph Anthony
Dolph, Gary Edward
Gallo, Duane Gordon
Gammon, James Robert
Goff, Charles W
Grabowski, Sandra Reynolds
Hammond, Charles Thomas
Hendrix, Jon Richard
Hunt, Linda Margaret
Iten, Laurie Elaine
Jantz, O K
Jensen, Richard Jorg
Kammeraad, Adrian
Koch, Arthur Louis
Latin, Richard
Levy, Morris
McGowan, Michael James
Malacinksi, George M
Maloney, Michael Stephen
Mescher, Anthony Louis
Miller, Richard William
Mitchell, Cary Arthur
Nolan, Val, Jr
O'Connor, Brian Lee
Olson, John Bennet
Ortoleva, Peter Joseph
Ostroy, Sanford Eugene
Peterson, Laverne E
Peterson, Richard George
Raff, Robert Albert
Real, Leslie Allan
Rosenberg, Gary David
Ruesink, Albert William
Shoup, Jane Rearick
Simmons, Eric Leslie
Speece, Susan Phillips
Srivastava, Arun
Stabler, Timothy Allen
Stillwell, William Harry
Taparowsky, Elizabeth Jane
Tweedell, Kenyon Stanley
Vanable, Joseph William, Jr
Wermuth, Jerome Francis
Wilkin, Peter J
Witzmann, Frank A
Young, Frank Nelson, Jr

IOWA
Adams, Donald Robert
Allegre, Charles Frederick
Andersen, Arnold E
Anderson, Todd Alan
Bamrick, John Francis
Bonfiglio, Michael
Buettner, Garry Richard
Christiansen, Kenneth Allen
Christiansen, Paul Arthur
Cruden, Robert William
Czarnecki, David Bruce
Denburg, Jeffrey Lewis

DiSpirito, Alan Angelo
Dolphin, Warren Dean
Eagleson, Gerald Wayne
Getting, Peter Alexander
Hadow, Harlo Herbert
Hoffmann, Richard John
Horton, Robert, Jr
Isely, Duane
Jordan, Diane Kathleen
Kapler, Jospeh Edward
Kirkland, Willis L
Laffoon, John
Milkman, Roger Dawson
Rogers, Rodney Albert
Soll, David Richard
Solursh, Michael
Swanson, Harold Dueker
Te Paske, Everett Russell
Thompson, Sue Ann
Waziri, Rafiq
Williams, Norman Eugene

KANSAS
Ashe, James S
Bader, Robert Smith
Barnes, William Gartin
Beary, Dexter F
Bechtel, Donald Bruce
Bowen, Daniel Edward
Chowdhury, Mridula
Cink, Calvin Lee
Cutler, Bruce
Dahl, Nancy Ann
Dey, Sudhansu Kumar
Enna, Salvatore Joseph
Halazon, George Christ
Haufler, Christopher Hardin
Johnston, Richard Fourness
Neely, Peter Munro
Oppert, Brenda
Pierson, David W
Piper, Jon Kingsbury
Prophet, Carl Wright
Reichman, Omer James
Spooner, Brian Sandford
Weis, Jerry Samuel
Williams, Larry Gale

KENTUCKY
Bennett, Thomas Edward
Boyarsky, Lila Harriet
Calkins, John
Ettensohn, Francis Robert
Ferner, John William
Fingar, Victor H
Fontaine, Julia Clare
Harmet, Kenneth Herman
Humphries, Asa Alan, Jr
Keefe, Thomas Leeven
McLaughlin, Barbara Jean
Norfleet, Morris L
Parker, Joseph Corbin, Jr
Sabharwal, Pritam Singh
Slagel, Donald E
Timmons, Thomas Joseph
Wagner, Charles Eugene
Wolfson, Alfred M

LOUISIANA
Avault, James W, Jr
Blackwell, Meredith
Borsari, Bruno
Brody, Arnold R
Dempesy, Colby Wilson
Devall, Margaret S
Ellgaard, Erik G
Etheridge, Albert Louis
Grodner, Mary Laslie
Heins, David Carl
Heyn, Anton Nicolaas Johannes
Knesel, John Arthur
Krogstad, Donald John
Leung, Wai Yan
Levitzky, Michael Gordon
Lonergan, Thomas A
McPherson, Alvadus Bradley
Makielski, Sarah Kimball
Misra, Raghunath P
Newsome, David Anthony
Owings, Addison Davis
Reese, William Dean
Shepherd, David Preston
Specian, Robert David
Spring, Jeffrey H
Sullivan, Victoria I
Thien, Leonard B
Walters, Marian R
Webert, Henry S

MAINE
Barker, Jane Ellen
Bell, Allen L
Bennett, Miriam Frances
Blake, Richard D
Chacko, Rosy J
Champlin, Arthur Kingsley
Davisson, Muriel Trask
Herold, Richard Carl
Kornfield, Irving Leslie
Labov, Jay Brian
Lewis, Alan James
Minkoff, Eli Cooperman
Mun, Alton M
Raisbeck, Barbara

Spirito, Carl Peter
Stevens, Leroy Carlton, Jr
Tyler, Mary Stott

MARYLAND
Akiyama, Steven Ken
Albuquerque, Edson Xavier
Arnheiter, Heinz
Barker, Jeffery Lange
Bass, Eugene L
Beyer, W Nelson
Bloch, Robert Joseph
Blye, Richard Perry
Bodian, David
Britz, Steven J
Brookes, Neville
Brown, Donald D
Brown, Kenneth Stephen
Burgess, Wilson Hales
Butman, Bryan Timothy
Chappelle, Emmett W
Cheng, Sheue-yann
Chuang, De-Maw
Coleman, Richard Walter
Crosby, Gayle Marcella
Darden, Lindley
Dawid, Igor Bert
Delahunty, George
Der Simonian, Rebecca
Dixon, Dennis Michael
Donlon, Mildred A
Durum, Scott Kenneth
Elson, Hannah Friedman
Forman, Michele R
Freeman, Colette
Freiberg, Samuel Robert
Futcher, Anthony Graham
Gallin, John I
Gantt, Elisabeth
Garruto, Ralph Michael
Geis, Aelred Dean
Giacometti, Luigi
Gill, Douglas Edward
Glick, J Leslie
Grant, Philip
Gray, Paulette S
Griesbach, Robert James
Gulyas, Bela Janos
Gunn, Charles Robert
Hasselmeyer, Eileen Grace
Hathaway, Wilfred Bostock
Hausman, Steven J
Hay, Robert J
Hein, Rosemary Ruth
Heine, Ursula Ingrid
Hewitt, Arthur Tyl
Hiatt, James Lee
Highton, Richard
Holland, Mark Alexander
Horenstein, Evelyn Anne
Johns, Michael Marieb Edward
Kalberer, John Theodore, Jr
Karklins, Olgerts Longins
Kayar, Susan Rennie
Kelley, Russell Victor
Kernaghan, Roy Peter
Klein, David C
Koslow, Stephen Hugh
Kundig, Fredericka Dodyk
Lane, H Clifford
Ledney, George David
Leonard, Warren J
LeRoith, Derek
Lichtenfels, James Ralph
Lin, Shin
Lyons, Russette M
Maddox, Yvonne T
Mahoney, Francis Joseph
Maier, Robert J
Martin, George Reilly
Mattson, Margaret Ellen
Moloney, John Bromley
Moreira, Jorge Eduardo
Murphy, Douglas Blakeney
Oberdorfer, Michael Douglas
Odell, Lois Dorothea
Oka, Takami
Parakkal, Paul Fab
Peck, Carl Curtis
Piatigorsky, Joram Paul
Pienta, Roman Joseph
Pilitt, Patricia Ann
Platt, Austin Pickard
Pollard, Harvey Bruce
Popper, Arthur N
Reid, Clarice D
Rifkin, Erik
Robbins, Robert John
Rodrigues, Merlyn M
Roth, Thomas Frederic
Rotherham, Jean
Scala, John Richard
Scully, Erik Paul
Shamsuddin, Abulkalam Mohammad
Shear, Charles Robert
Sich, Jeffrey John
Sieckmann, Donna G
Slife, Charles W
Snipes, Charles Andrew
Sokolove, Phillip Gary
Sowers, Arthur Edward
Spector, Novera Herbert
Stanley, Steven Mitchell
Stojilkovic, Stanko S

Biology, General (cont)

Stratmeyer, Melvin Edward
Stringfellow, Frank
Sullivan, Joseph Hearst
Thorington, Richard Wainwright, Jr
Traystman, Richard J
Varma, Shambhu D
Vlahakis, George
Walker, Richard Ives
Wargotz, Eric S
Weimer, John Thomas
Weinreich, Daniel
Wergin, William Peter
Winslow, Raimond L
Wolf, Robert Oliver
Zelenka, Peggy Sue

MASSACHUSETTS
Ackerman, Steven J
Ahlberg, Henry David
Albertini, David Fred
Ament, Alison Stone
Arnaout, M Amin
Bade, Maria Leipelt
Bawa, Kamaljit S
Begg, David A
Berry, Vern Vincent
Block, Steven M
Bond, George Walter
Borgatti, Alfred Lawrence
Burggren, Warren William
Cheetham, Ronald D
Christensen, Thomas Gash
Cochrane, David Earle
Cohen, Alan Mathew
Cohen, Samuel H
Collier, Robert John
Cook, Robert Edward
Copeland, Donald Eugene
Copeland, Frederick Cleveland
D'Amore, Patricia Ann
D'Avanzo, Charlene
Davis, Elizabeth Allaway
Dick, Stanley
DiLiddo, Rebecca McBride
Dinsmore, Jonathan H
Dowling, John Elliott
Du Moulin, Gary C
Edgar, Robert Kent
Feder, William Adolph
Fimian, Walter Joseph, Jr
Fink, Rachel Deborah
Fitzgerald, Marie Anton
Freadman, Marvin Alan
Fulton, Chandler Montgomery
Furshpan, Edwin Jean
Garay, Leslie Andrew
Gilmore, Thomas David
Goldhor, Susan
Goldin, Stanley Michael
Golub, Samuel J(oseph)
Gonnella, Patricia Anne
Goodenough, Judith Elizabeth
Gould, Stephen Jay
Greeley, Frederick
Gross, Jerome
Haimes, Howard B
Hall, Jeffrey Connor
Hargis, Betty Jean
Hexter, William Michael
Hichar, Joseph Kenneth
Holle, Paul August
Horvitz, Howard Robert
Hubbard, Ruth
Inoue, Shinya
Intres, Richard
Kaltenbach-Townsend, Jane
Kamien, Ethel N
Kelner, Albert
Kidder, George Wallace, Jr
Kimball, John Ward
Kleckner, Nancy E
Kunkel, Joseph George
Leavis, Paul Clifton
Levi, Herbert Walter
Levin, Andrew Eliot
Levins, Richard
Lewontin, Richard Charles
Liberatore, Frederick Anthony
Liss, Maurice
Longcope, Christopher
Lyerla, Jo Ann Harding
Lyman, Charles Peirson
McPherson, John M
Margulis, Lynn
Maaland, Richard Harry
Matthews, Samuel Arthur
Maxam, Allan M
Mayr, Ernst
Millard, William James
Monette, Francis C
Muckenthaler, Florian August
Mulcare, Donald J
Mulrennan, Cecilia Agnes
Murphey, Rodney Keith
Naqvi, S Rehan Hasan
Park, Hee-Young
Pierce, Naomi Ellen
Powell, Jeanne Adele
Putney, Scott David
Raam, Shanthi
Reardon, John Joseph
Redington, Charles Bahr

Rosemberg, Eugenia
Roth, John L, Jr
Ruben, Morris P
Saidha, Tekchand
Scheltema, Amelie Hains
Schmitt, Francis Otto
Seaman, Edna
Seitz, Anna W
Shahrik, H Arto
Slavin, Ovid
Smith, Dennis Matthew
Sprague, Isabelle Baird
Steele, John H
Stoffolano, John George, Jr
Sussman, Maurice
Tauber, Alfred Imre
Taubman, Martin Arnold
Thomas, Gail B
Todd, Neil Bowman
Treistman, Steven Neal
Vaillant, Henry Winchester
Van Cleave, Victor Harold
Walker, James Willard
Webb, Andrew Clive
West, Arthur James, II
White, Alan Whitcomb
Widmaier, Eric Paul
Wilson, Edward Osborne
Wu, Jung-Tsung
Yerganian, George
Zackroff, Robert V
Zetter, Bruce Robert
Zottoli, Steven Jaynes

MICHIGAN
Allen, Edward David
Ansbacher, Rudi
Armant, D Randall
Bach, Shirley
Barald, Kate Francesca
Blinn, Lorena Virginia
Boving, Bent Giede
Bromley, Stephen C
Brown, Donald Frederick Mackenzie
Bush, Guy L
Chang, Chia-Cheng
Cohen, Sanford Ned
Conover, James H
D'Amato, Constance Joan
Fetterolf, Carlos de la Mesa, Jr
Fisher, Leslie John
Frisancho, A Roberto
Garlick, Robert L
Gay, Helen
Grudzien, Thaddeus Arthur, Jr
Hancock, James Findley, Jr
Hayward, James Lloyd
Hiscoe, Helen Brush
Hitchcock, Dorothy Jean
Houts, Larry Lee
Hudson, Roy Davage
Jackson, Matthew Paul
Koehler, Lawrence D
Laborde, Alice L
Leach, Karen Lynn
Lewis, Ralph William
Lopushinsky, Theodore
Low, Bobbi Stiers
McCauley, Roy Barnard
McIntosh, Lee
Medlin, Julie Anne Jones
Miller, Donald Elbert
Miller, James Ray
Moore, William Samuel
Morawa, Arnold Peter
Moudgil, Virinder Kumar
Murnik, Mary Rengo
Narayan, Latha
Oakley, Bruce
Olexia, Paul Dale
Oliver-Ruby, Daphna R
Otterbacher, Eric Wayne
Pavgi, Sushama
Peters, Joseph John
Rogers, Claude Marvin
Root-Bernstein, Robert Scott
Rutter, Edward Walter, Jr
San Clemente, Charles Leonard
Schmaier, Alvin Harold
Schneider, Michael J
Sloat, Barbara Furin
Snitgen, Donald Albert
Stebbins, William Cooper
Sussman, Alfred Sheppard
Taylor, John Dirk
Teeri, James Arthur
Thompson, Paul Woodard
Townsend, Samuel Franklin
Vandermeer, John H
Van Deventer, William Carlstead
Vatsis, Kostas Petros
Venta, Patrick John
Webber, Patrick John
Whitney, Marion Isabelle
Wood, Pauline V
Wynne, Michael James
Yates, Jon Arthur

MINNESOTA
Anderson, Beverly
Baker, Shirley Marie
Birney, Elmer Clea
Dewald, Gordon Wayne
Edds, Kenneth Tiffany

Ellinger, Mark Stephen
Erickson, James Eldred
Fraser, Judy Ann
Frenzel, Louis Daniel, Jr
Heidcamp, William H
Kerr, Norman Story
Kerr, Sylvia Joann
Korte, Clare A
McCulloch, Joseph Howard
McKinnell, Robert Gilmore
Milliner, Eric Killmon
Mork, David Peter Sogn
Palm, John Daniel
Pusey, Anne E
Regal, Philip Joe
Reynhout, James Kenneth
Samuelson, Charles R
Schimpf, David Jeffrey
Sehe, Charles Theodore
Shapiro, Burton Leonard
Sherer, Glenn Keith
Smith, Quenton Terrill
Spangler, George Russell
Tate, Jeffrey L
Tihon, Claude
Trachte, George J
Tyce, Gertrude Mary
Widin, Katharine Douglas
Wittry, Esperance

MISSISSIPPI
Case, Steven Thomas
Corbett, James John
Cox, Prentiss Gwendolyn
Edney, Norris Allen
Goddard, Jerome
Helms, Thomas Joseph
Kurtz, Mark Edward
McClurkin, Iola Taylor
Rockhold, Robin William
Spann, Charles Henry
Stark, William Polson
Steen, James Southworth
Sutton, William Wallace

MISSOURI
Bagby, John R, Jr
Beckett, Jack Brown
Bischoff, Eric Richard
Burke, William Joseph
De Buhr, Larry Eugene
Derby, Albert
Elliott, Dana Ray
Elsawi, Nehad
Fahim, Mostafa Safwat
Finley, David Emanuel
Francel, Thomas Joseph
Gerdes, Charles Frederick
Gossling, Jennifer
Green, Michael
Hagen, Gretchen
Hamilton, John Meacham
Hansen, Harry Louis
Hanson, Willis Dale
Hart, Richard Allen
Hayes, Alice Bourke
Ho, David Tuan-hua
Hodges, Glenn R(oss)
Huckabay, John Porter
Jones, Allan W
Kirk, David Livingstone
Kirk, Marilyn M
Krukowski, Marilyn
Larson, Allan
Lin, Hsiu-san
Lumb, Ethel Sue
Magill, Robert Earle
Nichols, Herbert Wayne
Oshima, Eugene Akio
Raven, Peter Hamilton
Roti Roti, Joseph Lee
Rowe, Vernon Dodds
Sanes, Joshua Richard
Schaal, Barbara Anna
Slatopolsky, Eduardo
Smith, James Lee
Stalker, Harrison Dailey
Stevenson, Robert Thomas
Stiehl, Richard Borg
Stombaugh, Tom Atkins
Van Essen, David Clinton
Voorhees, Frank Ray
Wang, Bin Ching
Wartzok, Douglas
Westfall, Helen Naomi
Wilkens, Lon Allan
Yonke, Thomas Richard

MONTANA
Cameron, David Glen
Carlson, Clinton E
Coe, John Emmons
DeBoer, Kenneth F
Medora, Rustem Sohrab
Mitchell, Lawrence Gustave
Pickett, James M
Pittendrigh, Colin Stephenson
Preece, Sherman Joy, Jr
Sheridan, Richard P

NEBRASKA
Ballinger, Royce Eugene
Banerjee, Mihir R
Bliese, John C W

Boernke, William E
Bolick, Margaret Ruth
Boohar, Richard Kenneth
Brooks, Merle Eugene
Casey, Carol A
Cavalieri, Ercole Luigi
Curtin, Charles Byron
Dalley, Arthur Frederick, II
Joern, Anthony
Jones, John Ackland
McCue, Robert Owen
Nair, Chandra Kunju
Platz, James Ernest
Yee, John Alan

NEVADA
Hillyard, Stanley Donald
Lowenthal, Douglas H
Mahadeva, Madhu Narayan

NEW HAMPSHIRE
Bergman, Kenneth David
Dingman, Jane Van Zandt
Dixon, Wallace Clark, Jr
Ferm, Vergil Harkness
Jackson, Herbert William
Langford, George Malcolm
Loros, Jennifer Jane
Milne, Lorus Johnson
Milne, Margery (Joan) (Greene)
Sloboda, Roger D
Stahl, Barbara Jaffe
Walker, Charles Wayne

NEW JERSEY
Abbott, Joan
Adamo, Joseph Albert
Adler, Beatriz C
Arnold, Frederic G
Auletta, Carol Spence
Bacharach, Martin Max
Belmonte, Rocco George
Berberian, Paul Anthony
Bolton, Laura Lee
Bonner, John Tyler
Borysko, Emil
Brown, Thomas Edward
Carboni, Joan M
Carlson, Richard P
Chen, James Che Wen
Chiang, Bin-Yea
Cook, Marie Mildred
Cristini, Angela
Dashman, Theodore
Davis, Bill David
Dhruv, Rohini Arvind
Durand, James Blanchard
Fangboner, Raymond Franklin
Francoeur, Robert Thomas
Garone, John Edward
Gavurin, Lillian
Gepner, Ivan Alan
Gona, Ophelia Delaine
Grant, B Rosemary
Hampton, Suzanne Harvey
Handel, Steven Neil
Hayat, M A
Hill, Helene Zimmermann
Hitchcock-DeGregori, Sarah Ellen
Hockel, Gregory Martin
Hsu, Linda
Hubbell, Stephen Philip
Jabbar, Gina Marie
Katz, Laurence Barry
Koepp, Leila H
Kuehn, Harold Herman
Ledeen, Robert
Lee, Hsin-Yi
Leone, Ida Alba
Levin, Barry Edward
Linsky, Cary Bruce
Liu, Leroy Fong
Lonski, Joseph
Loveland, Robert Edward
McCutchen, Charles Walter
Mayer, Thomas C
Meiss, Alfred Nelson
Mele, Frank Michael
Mosley, Stephen T
Muth, Eric Anthony
Nagle, James John
Neri, Anthony
Patel, Ramesh N
Perrine, John W
Petry, Robert Kendrick
Proudfoot, Bernadette Agnes
Ritchie, David Malcolm
Rosenthal, Judith Wolder
Rossow, Peter William
Sabrosky, Curtis Williams
Salerno, Alphonse
Schreckenberg, Mary Gervasia
Sherman, Michael Ian
Siegel, Allan
Sinha, Arabinda Kumar
Stein, Donald Gerald
Trotta, Paul P
Vasconcelos, Aurea C
Venable, Patricia Lengel
Vrijenhoek, Robert Charles
Wallace, Edith Winchell
Weis, Judith Shulman
Willis, Jacalyn Giacalone
Zipf, Elizabeth M(argaret)

NEW MEXICO
Allred, Kelly Wayne
Anderson, Dean Mauritz
Baca, Oswald Gilbert
Barrera, Cecilio Richard
Degenhardt, William George
Dittmer, Howard James
Groblewski, Gerald Eugene
Hillsman, Matthew Jerome
Hubby, John L
Karlsson, Ulf Lennart
Lechner, John Fred
McNamara, Mary Colleen
Sklar, Larry A

NEW YORK
Aaronson, Stuart Alan
Abbatiello, Michael James
Ablin, Richard J
Adler, Kraig (Kerr)
Agnello, Arthur Michael
Alcamo, I Edward
Allen, Howard Joseph
Alsop, David W
Angeletti, Ruth Hogue
Archimovich, Alexander S
Bablanian, Rostom
Ballantyne, Donald Lindsay
Banerjee, Partha
Banerjee, Ranjit
Bannister, Thomas Turpin
Baroni, Timothy J
Basile, Dominick V
Bellve, Anthony Rex
Berlinrood, Martin
Bini, Alessandra Margherita
Borowsky, Richard Lewis
Boyer, Barbara Conta
Boyer, John Frederick
Brenowitz, A Harry
Bretscher, Anthony P
Britten, Bryan Terrence
Brown, William Louis, Jr
Browner, Robert Herman
Bruns, Peter John
Brunson, John Taylor
Burke, William Thomas, Jr
Cardillo, Frances M
Catapane, Edward John
Chabot, Brian F
Chiao, Jen Wei
Cohen, Nicholas
Commoner, Barry
Coppock, Donald Leslie
Corpe, William Albert
Cottrell, Stephen F
Craddock, Elysse Margaret
Cramer, Eva Brown
D'Adamo, Amedeo Filiberto, Jr
Day, Ivana Podvalova
Desplan, Claude
Detmers, Patricia Anne
DiGaudio, Mary Rose
Doty, Stephen Bruce
Dranginis, Anne M
Druger, Marvin
DuFrain, Russell Jerome
Dykhuizen, Daniel Edward
Edmunds, Leland Nicholas, Jr
Ellis, John Francis
Erk, Frank Chris
Falkowski, Paul Gordon
Ferdinand, Patricia
Ferguson, John Barclay
Fischer, Richard Bernard
Fleagle, John G
Fortner, Joseph Gerald
Frair, Wayne
Friedland, Beatrice L
Gabriel, Mordecai Lionel
Garcia, Alfredo Mariano
Geisler, Grace Sr
Girsch, Stephen John
Goldsmith, Eli David
Goode, Robert P
Gorovsky, Martin A
Graziadei, William Daniel, III
Grew, John C
Griffiths, Raymond Bert
Grillo, Ramon S
Hagar, Silas Stanley
Hallahan, William Laskey
Hardy, Matthew Phillip
Hartung, John David
Hartwick, Richard Allen
Hashim, George A
Helson, Lawrence
Henderson, Thomas E
Hershey, Linda Ann
Hillman, Dean Elof
Himes, Marion
Hirsch, Helmut V B
Hirshon, Jordon Barry
Hoffman, Roger Alan
Holtz, Aliza
Hoogstraal, Harry
Hooper, George Bates
Hotchkiss, Sharon K
Hoy, Ronald Raymond
Hrushesky, William John Michael
Huang, Chau-Ting
Huang, Chester Chen-Chiu
Hudecki, Michael Stephen
Hughes, Stephen Edward
Hyde, Kenneth Martin
Ingalls, James Warren, Jr
Isenberg, George Raymond, Jr
Jacobs, Laurence Stanton
Jacques, Felix Anthony
Jahan-Parwar, Behrus
Kandel, Eric Richard
Katz, Gary Victor
Kemphues, Kenneth J
Kerr, Marilyn Sue
Kessler, Dietrich
Keston, Albert S
Kim, Untae
Klotz, Richard Lawrence
Ko, Li-wen
Koenig, Edward
Korf, Richard Paul
Kramer, Alfred William, Jr
LaFountain, James Robert, Jr
Lamberg, Stanley Lawrence
Lapin, David Marvin
Laychock, Suzanne Gale
Lein, Pamela J
Levin, Norman Lewis
Liguori, Vincent Robert
Llinas, Rodolfo
Lnenicka, Gregory Allen
Lockshin, Richard Ansel
Lossinsky, Albert S
McCann-Collier, Marjorie Dolores
McDaniel, Carl Nimitz
McDonnell, Mark Jeffery
McEwen, Bruce F
McGee-Russell, Samuel M
McKenna, Olivia Cleveland
Mandel, Irwin D
Mandl, Richard H
Marks, Neville
Mascarenhas, Joseph Peter
Mason, Larry Gordon
Meiselman, Newton
Meng, Heinz Karl
Menon, Gopinathan K(unnariath)
Mirand, Edwin Albert
Mitchell, John Taylor
Mitra, Jyotirmay
Moore, Malcolm A S
Moor-Jankowski, J
Moran, Denis Joseph
Morrill, Gene A
Motzkin, Shirley M
Muschio, Henry M, Jr
Nasrallah, Mikhail Elia
Nicholson, J(ohn) Charles (Godfrey)
North, Robert J
Novacek, Michael John
Novak, Joseph Donald
Pagala, Murali Krishna
Pattee, Howard Hunt, Jr
Pesce, Michael A
Pietraface, William John
Poindexter, Jeanne Stove
Possidente, Bernard Philip, Jr
Prestwidge, Kathleen Joyce
Privitera, Carmelo Anthony
Pumo, Dorothy Ellen
Puszkin, Saul
Quaroni, Andrea
Rabino, Isaac
Raventos-Suaraz, Carmen Elvira
Reedy, John Joseph
Reilly, Margaret Anne
Reiss, Carol S
Reuss, Ronald Merl
Rhodes, Rondell H
Rich, Abby M
Richmond, Milo Eugene
Risley, Michael Samuel
Robbins, Edith Schultz
Robinson, Beatrice Letterman
Rosenbaum, Peter Andrew
Rosenberg, Warren L
Rosendorff, Clive
Rosi, David
Ryan, Simeon P
Saez, Juan Carlos
Salpeter, Miriam Mirl
Salthe, Stanley Norman
Satir, Birgit H
Sato, Gordon Hisashi
Scalia, Frank
Scharff, Matthew Daniel
Schwartz, Norman Martin
Selsky, Melvyn Ira
Shafit-Zagardo, Bridget
Shannon, Jerry A, Jr
Sharma, Sansar C
Sharpless, George Robert
Shaw, Spencer
Shechter, Yaakov
Sherman, Paul Willard
Siegel, Charles David
Siemann, Dietmar W
Sklarew, Robert J
Sleeper, David Allanbrook
Slocum, Harry Kim
Smeriglio, Alfred John
Sokal, Robert Reuven
Spencer, Selden J
Springer, Alan David
Strand, Fleur Lillian
Styles, Twitty Junius
Sun, Tung-Tien
Talhouk, Rabih Shakib
Tamm, Igor
Thompson, Robert Poole
Tokay, Elbert
Tomashefsky, Philip
Trimble, Mary Ellen
Vasey, Carey Edward
Veith, Frank James
Vuilleumier, Francois
Walker, Philip Caleb
Wasserman, Marvin
Weiss, Klaudiusz Robert
Wells, Russell Frederick
White, James Edwin
Williams, Donald Benjamin
Williams, Ernest Herbert, Jr
Witkin, Steven S
Witkus, Eleanor Ruth
Wysocki, Annette Bernadette
Ycas, Martynas
Yen, Andrew
Yeung, Kwok Kam
Young, William Donald, Jr
Zitrin, Charlotte Marker
Zubay, Geoffrey

NORTH CAROLINA
Abramson, Jon Stuart
Adler, Kenneth B
Anderson, Charles Eugene
Bailey, Donald Etheridge
Banerjee, Umesh Chandra
Bates, William K
Bode, Arthur Palfrey
Bond, James Anthony
Brosnihan, K Bridget
Burrous, Stanley Emerson
Campbell, Robert Joseph
Camp Hay, Pamela Jean
Cheng, Pi-Wan
Edwards, Nancy C
Endow, Sharyn Anne
Feduccia, John Alan
Glasser, Richard Lee
Glassman, Edward
Goodman, Major M
Gordon, Christopher John
Gruber, Helen Elizabeth
Haning, Blanche Cournoyer
Harrison, Frederick Williams
Henson, Anna Miriam (Morgan)
Horton, James Heathman
Jenkins, Jimmy Raymond
Jenner, Charles Edwin
Jetten, Anton Marinus
Jolls, Claudia Lee
Kalmus, Gerhard Wolfgang
Kimmel, Donald Loraine, Jr
Kirk, Daniel Eddins
Lauder, Jean Miles
Laurie, John Sewall
LeBaron, Homer McKay
Lehman, Harvey Eugene
Leise, Esther M
Lemasters, John J
Lytle, Charles Franklin
McCrady, Edward, III
Martin, Virginia Lorelle
Meyer, John Richard
Moore, Martha May
Mueller, Nancy Schneider
Mukunnemkeril, George Mathew
Nicklas, Robert Bruce
O'Rand, Michael Gene
Parker, Beulah Mae
Petters, Robert Michael
Pung, Oscar J
Rabb, Robert Lamar
Randall, John Frank
Reddish, Paul Sigman
Robinson, Kent
Rustioni, Aldo
Sadler, Thomas William
Smith, Ned Allan
Spencer, Lorraine Barney
Stiven, Alan Ernest
Stuart, Ann Elizabeth
Teng, Ching Sung
Terborgh, John J
Theil, Elizabeth
Thomas, Mary Beth
Triantaphyllou, Hedwig Hirschmann
Turner, James Eldridge
Tyndall, Jesse Parker
Vogel, Steven
Watts, John Albert, Jr
White, Richard Alan
Yongue, William Henry
Zucker, Robert Martin

NORTH DAKOTA
Bleier, William Joseph
Khansari, David Nemat
Scoby, Donald Ray
Starks, Thomas Leroy

OHIO
Akeson, Richard Allan
Allenspach, Allan Leroy
Andreas, Barbara Kloha
Bank, Harvey L
Beal, Kathleen Gabaskas
Boczar, Barbara Ann
Brackenbury, Robert William
Brent, Morgan McKenzie
Caplan, Arnold I
Cerroni, Rose E
Chang, In-Kook
Chumlea, William Cameron
Clay, Mary Ellen
Costello, Walter James
Craft, Thomas Jacob, Sr
Crutcher, Keith A
Dasch, Clement Eugene
Davis, James Allen
Easterly, Nathan William
Egar, Margaret Wells
Epp, Leonard G
Fried, Joel Robert
Fry, Anne Evans
Graham, Shirley Ann
Gwatkin, Ralph Buchanan Lloyd
Hamre, Harold Thomas
Hartsough, Robert Ray
Haubrich, Robert Rice
Hitt, John Burton
Hoagstrom, Carl William
Holeski, Paul Michael
Houston, Willie Walter, Jr
Jacobs-Lorena, Marcelo
Jarroll, Edward Lee, Jr
Jegla, Dorothy Eldredge
Kaiserman-Abramof, Ita Rebeca
Karas, James Glynn
Keefe, John Richard
Kleiner, Susan Mala
Kolodziej, Bruno J
Koo, Peter H
Landis, Story Cleland
Lasek, Raymond J
Leonard, Billie Charles
Lovejoy, Owen
Luckenbill-Edds, Louise
Mao, Simon Jen-Tan
Mikula, Bernard C
Mitchell, Rodger (David)
Moore, Randy
Murthy, Raman Chittaram
Niemczyk, Harry D
Peterson, Paul Constant
Pfohl, Ronald John
Pickering, Ed Richard
Racle, Fred Arnold
Raynie, Douglas Edward
Samols, David R
Sawicki, Stanley George
Schurr, Karl M
Scott, William James, Jr
Shaffer, Charles Franklin
Sherman, Thomas Fairchild
Stambrook, Peter J
Stokes, Bradford Taylor
Tassava, Roy A
Taylor, Thomas Norwood
Thomson, Dale S
Van Ummersen, Claire Ann
Vardaris, Richard Miles
Voneida, Theodore J
Wong, Laurence Cheng Kong
Yoon, Jong Sik

OKLAHOMA
Ahshapanek, Don Colesto
Bantle, John Albert
Bastian, Joseph
Bogenschutz, Robert Parks
Fisher, John Berton
Kocan, Katherine M
Marek, Edmund Anthony
Miller, Helen Carter
Muneer, Razia Sultana
Ogilvie, Marilyn Bailey
O'Neal, Steven George
Patterson, Manford Kenneth, Jr
Radke, William John
Rainbolt, Mary Louise
Shmaefsky, Brian Robert
Thornton, John William
Young, David Allen

OREGON
Arch, Stephen William
Barmack, Neal Herbert
Beatty, Joseph John
Benedict, Ellen Maring
Blaustein, Andrew Richard
Breakey, Donald Ray
Brownell, Philip Harry
Carlisle, Kay Susan
Coakley, Stella Melugin
Dawson, Peter Sanford
Duncan, James Thayer
Gerke, John Royal
Gwilliam, Gilbert Franklin
Harris, Patricia J
Hedberg, Karen K
Henny, Charles Joseph
Kimmel, Charles Brown
King, Charles Everett
Lande, Russell Scott
McCollum, Gin
McNulty, Wilbur Palmer
Montagna, William
Orkney, G Dale
Owczarzak, Alfred
Phoenix, Charles Henry
Riffey, Meribeth M
Ruben, John Alex
Russell, Nancy Jeanne

Biology, General (cont)

Shapiro, Lynda P
Springer, Martha Edith
Walker, Kenneth Merriam
Warren, Charles Edward
Wessells, Norman Keith
Weston, James A
Willis, David Lee
Wolf, Don Paul

PENNSYLVANIA
Akruk, Samir Rizk
Baker, Kenneth Melvin
Barr, Richard Arthur
Bean, Barry
Bergstresser, Kenneth A
Bertolini, Donald R
Binkley, Sue Ann
Birchem, Regina
Botelho, Stella Yates
Bradt, Patricia Thornton
Brunkard, Kathleen Marie
Burgess, David Ray
Campbell, Graham Le Mesurier
Campo, Robert D
Chacko, Samuel K
Chepenik, Kenneth Paul
Compher, Marvin Keen, Jr
Corey, Sharon Eva
Das, Manjusri
Davies, Ronald Edgar
Debroy, Chitrita
DeMott, Howard Ephraim
Desmond, Mary Elizabeth
Diamond, Leila
DiBerardino, Marie A
Dillon, Lawrence Samuel
DiPasquale, Gene
Donawick, William Joseph
Dropp, John Jerome
Ehrlich, H Paul
Ellis, Demetrius
Erickson, Ralph O
Everhart, Leighton Phreaner, Jr
Feit, Ira (Nathan)
Floros, Joanna
Fowler, H(oratio) Seymour
Fremount, Henry Neil
George, John Lothar
Gerstein, George Leonard
Gertz, Steven Michael
Giri, Lallan
Gottlieb, Frederick Jay
Greene, Robert Morris
Grobstein, Paul
Grove, Alvin Russell, Jr
Gundy, Samuel Charles
Halbrendt, John Marthon
Hallinan, Edward Joseph
Hassell, John Robert
Hazuda, Daria Jean
Herzog, Karl A
Hillman, Ralph
Himes, Craig L
Hoffman, Albert Charles
Hoffman, Daniel Lewis
Hopper, James Ernest
Hull, Larry Allen
Hulse, Arthur Charles
Jacobsohn, Myra K
Janzen, Daniel Hunt
Jensen, Bruce David
Jim, Kam Fook
Katoh, Arthur
Kauffman, Stuart Alan
Kayhart, Marion
Kennedy, Michael Craig
King, David Beeman
Krishtalka, Leonard
LaBelle, Edward Francis
Lanni, Frederick
Lash, James (Jay) W
Lee, John Cheung Han
Leighton, Charles Cutler
Levine, Elliot Myron
Lindstrom, Eugene Shipman
Logan, David Alexander
MacDermott, Richard J, Jr
McDevitt, David Stephen
McGary, Carl T
McNamara, Pamela Dee (MacMinigal)
Maksymowych, Roman
Malsberger, Richard Griffith
Mann, Stanley Joseph
Masteller, Edwin C
Maxson, Linda Ellen R
Meek, Edward Stanley
Meriney, Stephen D
Meyer, R Peter
Miller, Richard Lee
Mintz, Beatrice
Miselis, Richard Robert
Moss, William Wayne
Murray, Marion
Nadelhaft, Irving
Neff, Stuart Edmund
Nei, Masatoshi
Ode, Philip E
Olds-Clarke, Patricia Jean
Orth, Joanne M
Paoletti, Robert Anthony
Parker, Jon Irving
Peightel, William Edgar
Phoenix, Donald R
Pitkin, Ruthanne B
Pleasure, David
Rabin, Harvey
Rappaport, Harry P
Roth, Stephen
Ruffolo, Robert Richard, Jr
Ryan, James Patrick
Sachs, Howard George
Salzberg, Brian Matthew
Sanger, Joseph William
Schuyler, Alfred Ernest
Schwartz, Jeffrey H
Settlemyer, Kenneth Theodore
Shostak, Stanley
Simmons, Daryl Michael
Smith, Allen Anderson
Smith, John Bryan
Smith, W John
Sorensen, Ralph Albrecht
Stableford, Louis Tranter
Staetz, Charles Alan
Stere, Athleen Jacobs
Stewart, Barbara Yost
Studer, Rebecca Kathryn
Suyama, Yoshitaka
Taylor, D Lansing
Telfer, William Harrison
Thompson, Roger Kevin Russell
Tobin, Thomas Vincent
Towne, William F
Travis, Robert Victor
Tristram-Nagle, Stephanie Ann
Uricchio, William Andrew
Weinreb, Eva Lurie
Weiss, Kenneth Monrad
West, Keith P
Wheeler, Donald Alsop
Woodruff, Richard Ira
Yoho, Timothy Price

RHODE ISLAND
Ellis, Richard Akers
Fausto, Nelson
Glicksman, Arvin Sigmund
Goertemiller, Clarence C, Jr
Hufnagel, Linda Ann
Rand, David McNear
Silver, Alene Freudenheim
Stern, Leo
Wilde, Charles Edward, Jr
Young, Robert M
Zarcaro, Robert Michael

SOUTH CAROLINA
Browdy, Craig Lawrence
Cromer, Jerry Haltiwanger
Dawson, Wallace Douglas, Jr
Dobbs, Harry Donald
Fitzharris, Timothy Patrick
Fox, Charles Wayne
Gee, Adrian Philip
Graef, Philip Edwin
Horn, Charles Norman
Houk, Richard Duncan
Kistler, Wilson Stephen, Jr
Krebs, Julia Elizabeth
LaBar, Martin
Manley, Donald Gene
Metcalf, Isaac Stevens Halstead
Millette, Clarke Francis
Mulvey, Margaret
Pielou, William P
Pollard, Arthur Joseph
Porcher, Richard Dwight
Rembert, David Hopkins, Jr
Revis-Wagner, Charles Kenyon
Rohlfing, Duane L
Schmidt, Roger Paul
Scott, Jaunita Simons
Shealy, Harry Everett, Jr
Singh, Inderjit
Smith, Michael Howard
Surver, William Merle, Jr
Swallow, Richard Louis
Vernberg, Winona B
Voit, Eberhard Otto
Wourms, John P

SOUTH DAKOTA
Froiland, Sven Gordon
Lindahl, Ronald Gunnar

TENNESSEE
Alsop, Frederick Joseph, III
Biggers, Charles James
Briggs, Robert Chester
Campbell, James A
Cardoso, Sergio Steiner
Chandler, Robert Walter
Chen, Chung-Hsuan
Cohn, Sidney Arthur
Coons, Lewis Bennion
Cooper, Terrance G
Cushing, Bruce S
Darden, Edgar Bascomb
Davenport, James Whitcomb
Dix, John Willard
Duhl, David M
Gallagher, Michael Terrance
Garth, Richard Edwin
Harris, John Wallace
Hasty, David Long
Hossler, Fred E
Jennings, Lisa Helen Kyle
Jeon, Kwang Wu
Kahlon, Prem Singh
Kathman, R Deedee
Kimball, Richard Fuller
McCauley, David Evan
MacFadden, Donald Lee
McGhee, Charles Robert
Mansbach, Charles M, II
Marchok, Ann Catherine
Mathis, Philip Monroe
Monaco, Paul J
Murti, Kuruganti Gopalakrishna
Neff, Robert Jack
Norden, Jeanette Jean
Norton, Virginia Marino
Nunnally, David Ambrose
Olsen, John Stuart
Orgebin-Crist, Marie-Claire
Owens, Charles Allen
Pui, Ching-Hon
Redman, Charles Edwin
Reed, Horace Beecher
Ripley, Thomas H
Roberts, Lee Knight
Rogers, Beverly Jane
Roop, Robert Dickinson
Samuels, Robert
Schilling, Edward Eugene
Shivers, Charles Alex
Smith, Arlo Irving
Yates, Harris Oliver

TEXAS
Anderson, Richard Gilpin Wood
Aufderheide, Karl John
Azizi, Sayed Ausim
Baptist, James (Noel)
Bashour, Fouad A
Belk, Gene Denton
Bhaskaran, Govindan
Bickham, John W
Buskirk, Ruth Elizabeth
Byrd, Earl William, Jr
Cameron, Guy Neil
Cate, Rodney Lee
Chan, Lee-Nien Lillian
Ciomperlik, Matthew A
Cole, Garry Thomas
Coulter, Murray W
Craik, Eva Lee
Crawford, Gladys P
Dahl, Roy Dennis
DeShaw, James Richard
Diggs, George Minor, Jr
Dronamraju, Krishna Rao
Dunbar, Bonnie Sue
Emery, William Henry Perry
Eskew, Cletis Theodore
Faulkner, Russell Conklin, Jr
Forsyth, John Wiley
Fortner, George William
Franklin, Luther Edward
Frigyesi, Tamas L
Gary, Roland Thacher
Glasser, Stanley Richard
Gottesfeld, Zehava
Grant, Verne (Edwin)
Grimes, L Nichols
Grimm, Elizabeth Ann
Guentzel, M Neal
Hacker, Carl Sidney
Handler, Shirley Wolz
Harris, Kerry Francis Patrick
Hollyfield, Joe G
Honeycutt, Rodney Lee
Hook, Magnus Ao
Howard, Harriette Ella Pierce
Hsu, Laura Hwei-Nien Ling
Hug, Verena
Irwin, Louis Neal
Johnson, Larry
Jordan, Chris Sullivan
Kalthoff, Klaus Otto
Knaff, David Barry
Kurzrock, Razelle
Larimer, James Lynn
Lee, James C
Lieb, Carl Sears
Litke, Larry Lavoe
Lockett, Clodovia
Mastromarino, Anthony John
Mecham, John Stephen
Meola, Roger Walker
Mukherjee, Amal
Muntz, Kathryn Howe
Myers, Jack Edgar
Nakajima, Motowo
Parkening, Terry Arthur
Parker, Roy Denver, Jr
Pierce, Benjamin Allen
Pinson, Ernest Alexander
Poduslo, Shirley Ellen
Pool, Thomas Burgess
Prabhakar, Bellur Subbanna
Price, Harold James
Rajaraman, Srinivasan
Redburn, Dianna Ammons
Reese, Weldon Harold
Robinson, Arin Forest
Romsdahl, Marvin Magnus
Rosenfeld, Charles Richard
Rowell, Chester Morrison, Jr
Samollow, Paul B
Sant'Ambrogio, Giuseppe
Sauer, Helmut Wilhelm
Schafer, Rollie R
Schlueter, Edgar Albert
Schmidly, David James
Schneider, Barbara G
Schwartz, Colin John
Shake, Roy Eugene
Shur, Barry David
Simpson, Beryl Brintnall
Smatresk, Neal Joseph
Smith, Cornelia Marschall
Smith, William Russell
Sohal, Rajindar Singh
Spear, Irwin
Stacy, David Lowell
Sullivan, Patricia Ann Nagengast
Sweet, Merrill Henry, II
Taylor, Albert Cecil
Taylor, D(orothy) Jane
Tcholakian, Robert Kevork
Telfair, Raymond Clark, II
Tuff, Donald Wray
Van Overbeek, Johannes
Waldbillig, Ronald Charles
Waymire, Jack Calvin
Weigel, Paul H(enry)
Wheeler, Michael Hugh
Wicksten, Mary Katherine
Wilson, Robert Eugene
Winkler, Matthew M
Wolfenberger, Virginia Ann
Worthington, Richard Dane
Wright, David Anthony
Wright, Woodring Erik
Wu, Kenneth Kun-Yu

UTAH
Bradshaw, William S
Burgess, Paul Richards
Cuellar, Orlando
Davidson, Diane West
Ellis, LeGrande Clark
Galinsky, Raymond Ethan
Hathaway, Ralph Robert
Hess, Wilford Moser (Bill)
Johnston, Norman Paul
Palmblad, Ivan G
Robbins, Robert Raymond
Schoenwolf, Gary Charles
Sharp, Joseph C(ecil)
Shearin, Nancy Louise
Straight, Richard Coleman
Stringham, Reed Millington, Jr
Vickery, Robert Kingston, Jr
Wagner, Frederic Hamilton
Warren, Reed Parley
Youssef, Nabil Naguib

VERMONT
Keller, Tony S
Krupp, Patricia Powers
Landesman, Richard
Raper, Carlene Allen
Trombulak, Stephen Christopher

VIRGINIA
Alden, Raymond W, III
Baker, Jeffrey John Wheeler
Bender, Michael E
Bentz, Gregory Dean
Bieber, Samuel
Birdsong, Ray Stuart
Bodkin, Norlyn L
Bonar, Robert Addison
Brangan, Pamela J
Britt, Douglas Lee
Brown, Luther Park
Brubaker, Kenton Kaylor
Bull, Alice Louise
Burian, Richard M
Chinnici, Joseph (Frank) Peter
Daniel, Joseph Car, Jr
Deck, James David
Decker, Robert Dean
De Haan, Henry J
Diehl, Fred A
Diwan, Bhalchandra Apparao
Eglitis, Martin Alexandris
Fink, Linda Susan
Flint, Franklin Ford
Forbes, Michael Shepard
Friedman, Ruth T
Gaines, Gregory
Garrison, Norman Eugene
Grainger, Robert Michael
Griffin, Gary J
Grimm, James K
Gross, Paul Randolph
Hamilton, Howard Laverne
Heath, Everett
Holt, Perry Cecil
Hunt, Lindsay McLaurin, Jr
Jackson, Elizabeth Burger
Johnston, Pauline Kay
Jones, Melton Rodney
King, Betty Louise
Konigsberg, Irwin R
Lutes, Charlene McClanahan
Mandell, Alan
Marshall, Harold George
Mellinger, Clair
Moncrief, Nancy D
Moyer, Wayne A

Nygard, Neal Richard
Osgood, Christopher James
Owers, Noel Oscar
Patrick, Graham Abner
Pinschmidt, Mary Warren
Price, Steven
Radice, Gary Paul
Reams, Willie Mathews, Jr
Richmond, Isabelle Louise
Scheckler, Stephen Edward
Shear, William Albert
Spears, Joseph Faulconer
Stanley, Melissa Sue Millam
Strauss, Michael S
Stroup, Cynthia Roxane
Thompson, Jesse Clay, Jr
Turner, Bruce Jay
Williams, Robert Ellis
Wingard, Christopher Jon
Wiseman, Lawrence Linden
Zornetzer, Steven F

WASHINGTON
Anderson, Leigh
Bakken, Aimee Hayes
Benson, Keith Rodney
Binder, Marc David
Boersma, P Dee
Brookbank, John Warren
Cattolico, Rose Ann
Cloney, Richard Alan
Cohen, Arthur LeRoy
Donaldson, Lauren Russell
Dube, Maurice Andrew
Eberhardt, Lester Lee
Ford, James
Fournier, R E Keith
Gaddum-Rosse, Penelope
Garber, Richard Lincoln
Goldstein, Lester
Griswold, Michael David
Guttman, Burton Samuel
Kundig, Werner
Kutsky, Roman Joseph
Leung, Frederick C
Luchtel, Daniel Lee
McMenamin, John William
Maguire, Yu Ping
Mirkes, Philip Edmund
Mittler, Robert S
Monan, Gerald E
Nameroff, Mark A
Nathanson, Neil Marc
Orians, Gordon Howell
Prothero, John W
Robinovitch, Murray R
Roos, John Francis
Rubel, Edwin W
Soltero, Raymond Arthur
Soule, Oscar Hommel
Stenchever, Morton Albert
Strand, John A, III
Su, Judy Ya-Hwa Lin
Tamarin, Arnold
Tartar, Vance
Taub, Frieda B
Webber, Herbert H
Withner, Carl Leslie, Jr
Wood, James W

WEST VIRGINIA
Bayless, Laurence Emery
Burns, John Thomas
Castranova, Vincent
Leather, Gerald Roger
Mecca, Christyna Emma
Pauley, Thomas Kyle
Pritchett, William Henry
Shalaway, Scott D
Stephenson, Steven Lee

WISCONSIN
Akins, Virginia
Albrecht, Kenneth Adrian
Armstrong, George Michael
Balsano, Joseph Silvio
Bast, Rose Ann
Bownds, M Deric
Centonze, Victoria E
Claflin, Tom O
Claude, Philippa
Cripps, Derek J
Duewer, Elizabeth Ann
Foote, Kenneth Gerald
Fossland, Robert Gerard
Gillis, John Ericsen
Greenler, Robert George
Grunewald, Ralph
Hall, Marion Trufant
Hamerstrom, Frances
Hennen, Sally
Jordan, William R, III
Karrer, Kathleen Marie
Koval, Charles Francis
Kumaran, A Krishna
Kung, Ching
Lai, Ching-San
LaMarca, Michael James
Leonard, Thomas Joseph
Light, Douglas B
Mendelson, Carl Victor
Minock, Michael Edward
Norris, Dale Melvin, Jr
Oertel, Donata (Mrs Bill M Sugden)

Perry, James Warner
Pierson, Edgar Franklin
Pruss, Thaddeus P
Rapacz, Jan
Sattler, Carol Ann
Schatten, Gerald Phillip
Seeburger, George Harold
Sonneborn, David R
Staszak, David John
Stretton, Antony Oliver Ward
Strohm, Jerry Lee
Tormey, Douglass Cole
Uno, Hideo
Verma, Ajit K
Wenzel, Frederick J
Wikum, Douglas Arnold

WYOMING
George, Robert Porter
Nunamaker, Richard Allan

PUERTO RICO
Carrasco-Serrano, Clara E
Opava-Stitzer, Susan C
Rodriguez, Rocio del Pilar

ALBERTA
Ball, George Eugene
Blackshaw, Robert Earl
Bland, Brian Herbert
Browder, Leon Wilfred
Campenot, Robert Barry
Cavey, Michael John
Langridge, William Henry Russell
Pattie, Donald L
Prepas, Ellie E
Zalik, Sara E

BRITISH COLUMBIA
Alfaro, Rene Ivan
Belton, Peter
Booth, Amanda Jane
Cavalier-Smith, Thomas
Ellis, Derek V
Finnegan, Cyril Vincent
Freeman, Hugh J
Ganders, Fred Russell
Lee, Chi-Yu Gregory
Paul, Miles Richard
Ricker, William Edwin
Shamoun, Simon Francis
Smith, Michael Joseph
Srivastava, Lalit Mohan
Sziklai, Oscar
Taylor, Iain Edgar Park

MANITOBA
Braekevelt, Charlie Roger
Hara, Toshiaki J
Huebner, Erwin
Huebner, Judith Dee
Novak, Marie Marta
Pepper, Evan Harold
Somorjai, Rajmund Lewis

NEW BRUNSWICK
Anderson, John Murray
Burt, Michael David Brunskill
Glebe, Brian Douglas
McKenzie, Joseph Addison
Pohle, Gerhard Werner
Thomas, Martin Lewis Hall

NEWFOUNDLAND
Cowan, Garry Ian McTaggart

NOVA SCOTIA
Dadswell, Michael John
Fentress, John Carroll
Ferrier, Gregory R
Hall, Brian Keith
McLaren, Ian Alexander
Vethamany, Victor Gladstone
Wassersug, Richard Joel

ONTARIO
Anderson, Terry Ross
Atkinson, Burr Gervais
Aubin, Jane E
Balon, Eugene Kornel
Bancroft, John Basil
Barron, John Robert
Battle, Helen Irene
Baum, Bernard R
Brooks, Daniel Rusk
Carlone, Robert Leo
Catling, Paul Miles
Caveney, Stanley
Chambers, Ann Franklin
Cook, David Greenfield
Cook, W H
Dedhar, Shoukat
Dodson, Edward O
Fallding, Margaret Hurlstone Hardy
Fenton, Melville Brock
Fisher, Kenneth Robert Stanley
Forer, Arthur H
Fourney, Ronald Mitchell
Harding, Paul George Richard
Harmsen, Rudolf
Hickey, Donal Aloysius
Hutchinson, Thomas C
Israelstam, Gerald Frank
Keddy, Paul Anthony

Kerbel, Robert Stephen
Lean, David Robert Samuel
Liversage, Richard Albert
MacCrimmon, Hugh Ross
McCully, Margaret E
Malkin, Aaron
Marcus, George Jacob
Masui, Yoshio
Money, Kenneth Eric
Montgomerie, Robert Dennis
Murray, William Douglas
Ouellet, Henri
Ozburn, George W
Peck, Stewart Blaine
Philogene, Bernard J R
Phipps, James Bird
Possmayer, Fred
Prevec, Ludvik Anthony
Purko, John
Raaphorst, G Peter
Rand, Richard Peter
Rapport, David Joseph
Robertson, Raleigh John
Roth, René Romain
Ruckerbauer, Gerda Margareta
Shih, Chang-Tai
Skoryna, Stanley C
Small, Ernest
Sorger, George Joseph
Sprague, John Booty
Stavric, Bozidar
Teskey, Herbert Joseph
Threlkeld, Stephen Francis H
Trevors, Jack Thomas
Tsui, Lap-Chee
Turk, Fateh (Frank) M
Ursino, Donald Joseph
Vijay, Hari Mohan
Warner, Alden Howard

QUEBEC
Alavi, Misbahuddin Zafar
Belanger, Luc
Bell, Graham Arthur Charlton
Benoit, Guy J C
Bussey, Arthur Howard
Dainiak, Nicholas
Descarries, Laurent
Digby, Peter Saki Bassett
Dubé, François
Dugre, Robert
Galbraith, David Allan
Gervais, Francine
Goltzman, David
Grad, Bernard Raymond
Guyda, Harvey John
Mailloux, Gerard
Maly, Edward J
Manjunath, Puttaswamy
Marceau, Normand Luc
Marsden, Joan Chauvin
Mulay, Shree
Pallotta, Dominick John
Roy, Robert Michael McGregor
Sattler, Rolf
Sergeant, David Ernest
Sikorska, Hanna
Sirois, Pierre
Skup, Daniel
Srivastava, Uma Shanker
Tannenbaum, Gloria Shaffer
Vincent, C
Wolsky, Alexander

SASKATCHEWAN
Basinger, James Frederick
Kartha, Kutty Krishnan
McLennan, Barry Dean
Schmutz, Josef Konrad
Scoles, Graham John
Spurr, David Tupper
Yu, Peter Hao

YUKON TERRITORY
Doermann, August Henry

OTHER COUNTRIES
Abelev, Garri Izrailevich
Anderson, Emory Dean
Andrews, Peter Walter
Bose, Subir Kumar
Braquet, Pierre G
Chan, Albert Sun Chi
Comfort, Alexander
Emmons, Lyman Randlett
Forbes, Milton L
Francke, Oscar F
Fritz, Irving Bamdas
Goldberg, Burton David
Gould-Somero, Meredith
Hageman, Steven James
Haight, John Richard
Hamilton, William Donald
Hammerman, Ira Saul
Holldobler, Berthold Karl
Jakowska, Sophie
Klein, Jan
Kobata, Akira
Lesseps, Roland Joseph
Lewis, Arnold Leroy, II
Lin, Chin-Tarng
MacLean, William Plannette, III
Marikovsky, Yehuda
Milstein, Cesar

Munoz, Maria De Lourdes
Nakagawa, Shizutoshi
Neher, Erwin
Nevo, Eviatar
Paik, Young-Ki
Rand, A Stanley
Rickenberg, Howard V
Rosso, Pedro
Saito, Takuma
Sarukhan, Jose
Schliwa, Manfred
Sleeter, Thomas David
Smokovitis, Athanassios A
Solter, Davor
Withers, Philip Carew

Biomathematics

ALABAMA
Cowan, James Howard, Jr
Hazelrig, Jane
Katholi, Charles Robinson
Mack, Timothy Patrick
Nowack, William J
Siler, William MacDowell
Turner, Malcolm Elijah

ALASKA
Fagen, Robert

ARIZONA
Arabyan, Ara
Gaud, William S
Vincent, Thomas Lange
Winfree, Arthur T

ARKANSAS
Dykman, Roscoe A

CALIFORNIA
Baldwin, Ransom Leland
Blischke, Wallace Robert
Boulton, Roger Brett
Brokaw, Charles Jacob
Case, Ted Joseph
Deriso, Richard Bruce
Desharnais, Robert Anthony
DiStefano, Joseph John, III
Dixon, Wilfrid Joseph
Edelman, Jay Barry
Evans, John W
Feldman, Marcus William
Ferguson, William E
Fernald, Russell Dawson
Fernandez, Alberto Antonio
Forsythe, Alan Barry
Freeman, Linton Clarke
Getz, Wayne Marcus
Gorman, Cornelia M
Holmquist, Walter Richard
Jenden, Donald James
Kaus, Peter Edward
Klinger, Allen
Landahl, Herbert Daniel
Lange, Kenneth L
Lein, Allen
Licko, Vojtech
Macey, Robert Irwin
Mullen, Ashley John
Nunney, Leonard Peter
Oster, George F
Powell, Thomas Mabrey
Ritz-Gold, Caroline Joyce
Rose, Michael Robertson
Rotenberg, Manuel
Waterman, Michael S
Whittemore, Alice S

COLORADO
Best, Jay Boyd
Osborn, Ronald George
Pulliam, William Marshall
Reiner, John Maximilian

CONNECTICUT
Holsinger, Kent Eugene
Lambrakis, Konstantine Christos
Palladino, Joseph L
Russu, Irina Maria
Sarfarazi, Mansoor
Wagner, Gunter Paul

DISTRICT OF COLUMBIA
Rosenberg, Edith E
Tangney, John Francis
Tucker, John Richard

FLORIDA
Barnard, Donald Roy
Bloom, Stephen Allen
Cropper, Wendell Parker, Jr
Haynes, Duncan Harold
Leigh, Egbert Giles, Jr
Meltzer, Martin Isaac
Simberloff, Daniel S
Stamper, James Harris
Stetson, Robert Franklin
Travis, Joseph

GEORGIA
Asmussen, Marjorie A
Glasser, John Weakley
Hoogenboom, Gerrit

62 / DISCIPLINE INDEX

Biomathematics (cont)
Longini, Ira Mann, Jr
Lyons, Nancy I
Pulliam, H Ronald
Throne, James Edward

ILLINOIS
Altmann, Jeanne
Barr, Lloyd
Carnes, Bruce Alfred
Charlesworth, Brian
Deysach, Lawrence George
Hausfater, Glenn
Jakobsson, Eric Gunnar, Sr
Jordan, Steven Lee
Larkin, Ronald Paul
Levine, Michael W
Lloyd, Monte
Loehle, Craig S
MacKichan, Janis Jean
Mittenthal, Jay Edward
Smeach, Stephen Charles
Trimble, John Leonard
Webber, Charles Lewis, Jr
Yokoyama, Shozo

INDIANA
Park, Richard Avery, IV
Parkhurst, David Frank
Stewart, Terry Sanford

IOWA
Dexter, Franklin, III

KANSAS
Welch, Stephen Melwood

KENTUCKY
Crowley, Philip Haney
Sih, Andrew

LOUISIANA
Schoenly, Kenneth George

MAINE
Dowse, Harold Burgess

MARYLAND
Ballentine, Robert
Blum, Harry
Corliss, John Burt
Covell, David Gene
FitzHugh, Richard
Ginevan, Michael Edward
Kroll, Martin Harris
Myers, Charles
Parker, Rodger D
Rinzel, John Matthew
Robinson, David Adair
Sastre, Antonio
Steinmetz, Michael Anthony
Ulanowicz, Robert Edward
Yaes, Robert Joel

MASSACHUSETTS
Carpenter, Gail Alexandra
Caswell, Hal
Futrelle, Robert Peel
He, Bin
Keller, Evelyn Fox
Levins, Richard
Levy, Elinor Miller
Ransil, Bernard J(erome)
Schnitzer, Jan Eugeneusz
Shipley, Reginald A
Smith, Tim Denis
Wun, Chun Kwun
Yuan, Fan

MICHIGAN
Bookstein, Fred Leon
Estabrook, George Frederick
Harrison, Michael Jay
Jacquez, John Alfred
Kootsey, Joseph Mailen
Reed, David Doss
Rosenspire, Allen Jay
Savageau, Michael Antonio

MINNESOTA
Andow, David A
Curtsinger, James Webb
Rescigno, Aldo
Tilman, G David

MISSISSIPPI
Ikenberry, Roy Dewayne

MISSOURI
Dugatkin, Lee Alan
Luecke, Richard H
Wette, Reimut
Woodruff, Clarence Merrill

MONTANA
MacNeil, Michael
Thomas, Robert Glenn

NEBRASKA
Ueda, Clarence Tad

NEW JERSEY
Cronin, Jane Smiley
Dobson, Andrew Peter
Hubbell, Stephen Philip
Levin, Simon Asher
Quinn, James Allen
Weiss, Stanley H

NEW MEXICO
Bell, George Irving
Beyer, William A
Chen, David J
Jett, James Hubert
Mewhinney, James Albert
Scott, Bobby Randolph

NEW YORK
Altshuler, Bernard
Andersen, Olaf Sparre
Blessing, John A
Blumenson, Leslie Eli
Chou, Ting-Chao
Clark, Alfred, Jr
Clark, Patricia Ann Andre
Cohen, Joel Ephraim
Durkin, Patrick Ralph
Ginzburg, Lev R
Greco, William Robert
Gurevitch, Jessica
Hastings, Harold Morris
Holland, Mary Jean Carey
Hunt, David Michael
Johnson, Lawrence Lloyd
Kilpper, Robert William
Koretz, Jane Faith
Leisman, Gerald
Millstein, Jeffrey Alan
Niklas, Karl Joseph
Okubo, Akira
Percus, Jerome K
Ringo, James Lewis
Schiff, Joel D
Schimmel, Herbert
Sellers, Peter Hoadley
Silverman, Benjamin David
Tanner, Martin Abba
Vaidhyanathan, V S

NORTH CAROLINA
Carter, Thomas Edward, Jr
Crawford-Brown, Douglas John
Crowder, Larry Bryant
Dillard, Margaret Bleick
Gold, Harvey Joseph
Kohn, Michael Charles
Overton, John H
Schaffer, Henry Elkin
Smith, Charles Eugene
Stinner, Ronald Edwin
Vaughan, Douglas Stanwood
Weiner, Daniel Lee
Weir, Bruce Spencer
Woodbury, Max Atkin

OHIO
Fraser, Alex Stewart
Gates, Michael Andrew
Geho, Walter Blair
Klapper, Michael H

OKLAHOMA
Buchanan, David Shane
Dugan, Kimiko Hatta
Wells, Harrington

OREGON
Craig, Albert Morrison (Morrie)
Udovic, Daniel

PENNSYLVANIA
Abremski, Kenneth Edward
Altschuler, Martin David
Badler, Norman Ira
Boxenbaum, Harold George
Chay, Teresa R
Chiang, Soong Tao
Cohen, Stanley
Erickson, Ralph O
Gibson, Raymond Edward
Karreman, George
Kushner, Harvey
Marks, Louis Sheppard
Neveln, Bob
Rapp, Paul Ernest
Rosen, Gerald Harris
Stong, David Bruce
Tallarida, Ronald Joseph

RHODE ISLAND
Brooks, Lisa Delevan
Costantino, Robert Francis
Hai, Chi-Ming
Marsh, Donald Jay

SOUTH CAROLINA
Bohning, Daryl Eugene
Kosinski, Robert Joseph
Voit, Eberhard Otto

TENNESSEE
Bernard, Selden Robert
Gutzke, William H N
Hallam, Thomas Guy
Rivas, Marian Lucy

Rose, Kenneth Alan

TEXAS
Brown, Barry W
Crick, Rex Edward
Davison, Daniel Burtonn
Foster, Donald Myers
Fox, George Edward
Furlong, Norman Burr, Jr
Hanis, Craig L
Hellums, Jesse David
Jansson, Birger
Ladde, Gangaram Shivlingappa
Li, Wen-Hsiung
Marks, Gerald A
Miikkulainen, Risto Pekka
Saltzberg, Bernard
Schuhmann, Robert Ewald
Srinivasan, Ramachandra Srini
Thall, Peter Francis
Thames, Howard Davis, Jr
Walter, Charles Frank
White, Robert Allen
Zimmerman, Stuart O

UTAH
Cole, Walter Eckle

VERMONT
Costanza, Michael Charles

VIRGINIA
Adams, James Milton
Chandler, Jerry LeRoy
Mikulecky, Donald C
Rubenstein, Norton Michael
Sharov, Alexei A
Tyson, John Jeanes

WASHINGTON
Anderson, James Jay
Bare, Barry Bruce
Barker, Morris Wayne
Bassingthwaighte, James B
Dailey, Frank Allen
DeMaster, Douglas Paul
Francis, Robert Colgate
Kareiva, Peter Michael
Laevastu, Taivo
Moody, Michael Eugene
Moolgavkar, Suresh Hiraji

WEST VIRGINIA
Brown, Mark Wendell

WISCONSIN
Boyce, Mark S
Carpenter, Stephen Russell
Chover, Joshua
Holden, James Edward
Jungck, John Richard
Tonellato, Peter J
Yoganandan, Narayan

ALBERTA
Beck, James S
McGann, Locksley Earl
Prepas, Ellie E
Strobeck, Curtis
Voorhees, Burton Hamilton
Whitelaw, William Albert

BRITISH COLUMBIA
Barclay, Hugh John
Hoffmann, Geoffrey William
LeBlond, Paul Henri
Peterman, Randall Martin
Welch, David Warren

MANITOBA
Gordon, Richard

NEW BRUNSWICK
Van Groenewoud, Herman

NEWFOUNDLAND
Newton, Peter Francis
Pepin, P
Wroblewski, Joseph S

NOVA SCOTIA
Moore, Bruce Robert
Rosen, Robert
Silvert, William Lawrence
Wong, Alan Yau Kuen

ONTARIO
Barr, David Wallace
Endrenyi, Laszlo
Haynes, Robert Hall
Miller, Donald Richard
Norwich, Kenneth Howard
Taylor, Peter D
Zuker, Michael

QUEBEC
Diksic, Mirko
Jolicoeur, Pierre
Legendre, Pierre
Mackey, Michael Charles
Ouimet, Rock
Outerbridge, John Stuart
Roberge, Fernand Adrien

SASKATCHEWAN
Lapp, Martin Stanley

OTHER COUNTRIES
Argos, Patrick
Bodmer, Walter Fred
Haegele, Klaus D
Kang, Kewon
May, Robert McCredie
Rio, Maria Esther
Stearns, Stephen Curtis

Biometrics-Biostatistics

ALABAMA
Best, Troy Lee
Borton, Thomas Ernest
Hazelrig, Jane
Hoff, Charles Jay
Holt, William Robert
Johnson, Evert William
McGuire, John Albert
Wallace, Dennis D
Watson, Jack Ellsworth

ALASKA
Dahlberg, Michael Lee
Fox, John Frederick
Heifetz, Jonathan
Olsen, James Calvin
Reynolds, James Blair
Snyder, George Richard

ARIZONA
Aickin, Mikel G
Allen, Stephen Gregory
Figueredo, Aurelio Jose
Gaud, William S
Jacobowitz, Ronald
Moon, Thomas Edward
Nicolls, Ken E

ARKANSAS
Brown, Connell Jean
Dunn, James Eldon
Gaylor, David William
Holson, Ralph Robert
Walker, William M
Walls, Robert Clarence

CALIFORNIA
Abramson, Norman Jay
Ahumada, Albert Jil, Jr
Anderson, Dennis Elmo
Antoniak, Charles Edward
Arthur, Susan Peterson
Bell, A(udra) Earl
Bentley, Donald Lyon
Bernstein, Leslie
Bray, Richard Newton
Brown, Byron William, Jr
Burrill, Melinda Jane
Callahan, Joan Rea
Chang, Potter Chien-Tien
Chiang, Chin Long
Clark, Virginia
Cumberland, William Glen
D'Antoni, Hector Luis
Enright, James Thomas
Farris, David Allen
Fearn, Dean Henry
Forsythe, Alan Barry
Foster, Ken Wood
Franti, Charles Elmer
Garber, Morris Joseph
Gerrodette, Timothy
Gordon, Louis
Guthrie, Donald
Hill, Annlia Paganini
Hills, F Jackson
Hubert, Helen Betty
James, Kenneth Eugene
Jerison, Harry Jacob
Klauber, Melville Roberts
Kope, Robert Glenn
Kraemer, Helena Chmura
Kuzma, Jan Waldemar
Lachenbruch, Peter Anthony
Lee, David Anson
Little, Thomas Morton
Lovich, Jeffrey Edward
Mackey, Bruce Ernest
Malone, Marvin Herbert
Marshall, Rosemarie
Massey, Frank Jones, Jr
Melnick, Michael
Merchant, Roland Samuel, Sr
Mikel, Thomas Kelly, Jr
Moore, Dan Houston, II
Moriarty, David John
Moses, Lincoln Ellsworth
Mueller, Thomas Joseph
Murad, Turhon A
Parrish, Richard Henry
Rao, Jammalamadaka S
Ray, Rose Marie
Rice, Dorothy Pechman
Rocke, David M
Rotenberry, John Thomas
Rustagi, Jagdish S
Shonick, William
Show, Ivan Tristan
Sinsheimer, Janet Suzanne

BIOLOGICAL SCIENCES / 63

Swan, Shanna Helen
Tarter, Michael E
Treloar, Alan Edward
Ury, Hans Konrad
Varady, John Carl
Wall, Francis Joseph
Ward, David Gene
Wiggins, Alvin Dennie
Williams, Stanley Clark
Wimer, Cynthia Crosby
Wyzga, Ronald Edward
Zippin, Calvin

COLORADO
Alpern, Herbert P
Angleton, George M
Bernstein, Stephen
Bowden, David Clark
Chase, Gerald Roy
Chiszar, David Alfred
Gough, Larry Phillips
Johnson, Thomas Eugene
Jones, Richard Hunn
Mielke, Paul W, Jr
Oldemeyer, John Lee
Porter, Kenneth Raymond
Sowell, John Basil

CONNECTICUT
Chan, Yick-Kwong
Holford, Theodore Richard
Holsinger, Kent Eugene
Peduzzi, Peter N
Woronick, Charles Louis

DELAWARE
Fellner, William Henry
Miller, Douglas Charles
Pell, Sidney
Schuler, Victor Joseph

DISTRICT OF COLUMBIA
Ahmed, Susan Wolofski
Ampy, Franklin
Chiazze, Leonard, Jr
Cohen, Michael Paul
Hayek, Lee-Ann Collins
Hurley, Frank Leo
Jernigan, Robert Wayne
Myers, Wayne Lawrence
Nayak, Tapan Kumar
Rao, Mamidanna S
Tyrrell, Henry Flansburg
Weiss, Ira Paul

FLORIDA
Banghart, Frank W
Barnard, Donald Roy
Brain, Carlos W
Carrier, Steven Theodore
Chew, Victor
DeRousseau, C(arol) Jean
Gaunt, Stanley Newkirk
Greenberg, Richard Alvin
Hagans, James Albert
Hagenmaier, Robert Doller
Honeyman, Merton Seymour
Jones, David Alwyn
Kastenbaum, Marvin Aaron
Littell, Ramon Clarence
Lyman, Gary Herbert
McCann, James Alwyn
McCoy, Earl Donald
Nichols, Robert Loring
Prager, Michael Haskell
Rathburn, Carlisle Baxter, Jr
Stamper, James Harris
Stroh, Robert Carl

GEORGIA
Bailey, Robert Leroy
Carter, Melvin Winsor
Cohen, Alonzo Clifford, Jr
Hogue, Carol Jane Rowland
Hutcheson, Kermit
Karon, John Marshall
Kneib, Ronald Thomas
Kutner, Michael Henry
Longini, Ira Mann, Jr
Lyons, Nancy I
Mabry, John William
McLaurin, Wayne Jefferson
Mosteller, Robert Cobb
Nichols, Michael Charles
Pepper, William Donald
Pogue, Richard Ewert
Ware, Kenneth Dale
Zarnoch, Stanley Joseph

HAWAII
Brewbaker, James Lynn
Chung, Chin Sik
Freed, Leonard Alan
Grove, John Sinclair
Krupp, David Alan
Silva, James Anthony
Simon, Christine Mae

IDAHO
Everson, Dale O
Keller, Barry Lee

ILLINOIS
Bernardo, Rex Novero

Best, William Robert
Carmer, Samuel Grant
Chen, Edwin Hung-Teh
Corruccini, Robert Spencer
Deysach, Lawrence George
Dyer, Alan Richard
Fernando, Rohan Luigi
Flay, Brian Richard
Ghent, Arthur W
Grossman, Michael
Haenszel, William Manning
Hand, Roger
Katz, Alan Jeffrey
Kruse, Kipp Colby
Lin, Lawrence I-Kuei
Loehle, Craig S
Martin, Yvonne Connolly
Mattson, Dale Edward
Myers, Charles Christopher
Norusis, Marija Jurate
Parsons, Carl Michael
Sacks, Jerome
Schaeffer, David Joseph
Seif, Robert Dale
Shanks, Roger D
Sollberger, Arne Rudolph
Thisted, Ronald Aaron
Zar, Jerrold Howard

INDIANA
Boston, Andrew Chester
Cerimele, Benito Joseph
Cooper, William Edgar, Jr
Garton, David Wendell
Hui, Siu Lui
Martin, Truman Glen
Moser, John William, Jr
Nyquist, Wyman Ellsworth
Pachut, Joseph F(rancis), Jr
Parkhurst, David Frank
Sampson, Charles Berlin
Samuels, Myra Lee
Stewart, Terry Sanford
Tamura, Roy N
Wolfrom, Glen Wallace

IOWA
Arndt, Stephan
Berger, Philip Jeffrey
Budd, Ann Freeman
Cady, Foster Bernard
Clark, William Richard
Cox, Charles Philip
Cox, David Frame
Fuller, Wayne Arthur
Hotchkiss, Donald K
Kempthorne, Oscar
Moore, Kenneth J
Remington, Richard Delleraine
Wunder, William W
Yang, Xiao-Bing

KANSAS
Fryer, Holly Claire
Kaufman, Glennis Ann
Lanman, Robert Charles
Schalles, Robert R
Slade, Norman Andrew
Stewart, William Henry
Wassom, Clyde E
Welch, Stephen Melwood

KENTUCKY
Anderson, Richard L
Hays, Virgil Wilford
Lanska, Douglas John
Moore, Allen Jonathan
O'Connor, Carol Alf
Stevenson, Robert Jan
Surwillo, Walter Wallace

LOUISIANA
Baldwin, Virgil Clark, Jr
Diem, John Edwin
Eggen, Douglas Ambrose
Koonce, Kenneth Lowell
Owings, Addison Davis
Seigel, Richard Allyn
Webber, Larry Stanford
Weinberg, Roger
Wright, Vernon Lee

MAINE
Gilbert, James Robert

MARYLAND
Abbey, Helen
Anello, Charles
Blot, William James
Brant, Larry James
Brodsky, Allen
Byar, David Peery
Chen, Tar Timothy
Chiacchierini, Richard Philip
Cutler, Jeffrey Alan
Der Simonian, Rebecca
Diamond, Earl Louis
Edwards, Brenda Kay
Elkin, William Futter
Ellenberg, Jonas Harold
Ellenberg, Susan Smith
Embody, Daniel Robert
Feinleib, Manning
Feldman, Jacob J

Fisher, Pearl Davidowitz
Freedman, Laurence Stuart
Garrison, Robert J
Ginevan, Michael Edward
Goldberg, Irving David
Greenhouse, Samuel William
Haynes, Suzanne G
Heath, Robert Gardner
Hebel, John Richard
Hill, James Leslie
Homer, Louis David
Jablon, Seymour
Jessup, Gordon L, Jr
Kimball, Allyn Winthrop
Knatterud, Genell Lavonne
Knoke, James Dean
Knox, Robert Gaylord
Kramer, Morton
Lachin, John Marion, III
Langenberg, Patricia Warrington
Lao, Chang Sheng
Meisinger, John Joseph
Mellits, E David
Milton, Roy Charles
Mosimann, James Emile
Myers, Max H
Nichols, James Dale
Osborne, J Scott, III
Pearson, Jay Dee
Pickle, Linda Williams
Rastogi, Suresh Chandra
Rider, Rowland Vance
Rosenberg, Saul H
Ross, Alan
Royall, Richard Miles
Rubinstein, Lawrence Victor
Russek-Cohen, Estelle
Schneiderman, Marvin Arthur
Shaw, Richard Franklin
Simon, Richard Macy
Smith, Lewis Wilbert
Smith, Paul John
Steinberg, Seth Michael
Suomi, Stephen John
Tarone, Robert Ernest
Tonascia, James A
Walker, Michael Dirck
Wilson, P David

MASSACHUSETTS
Ash, Arlene Sandra
Finn, John Thomas
Gelber, Richard David
Gleason, Ray Edward
Gordon, Claire Catherine
Gray, Douglas Carmon
Haight, Thomas H
Helgesen, Robert Gordon
Hoaglin, David Caster
Kamen, Gary P
Kolakowski, Donald Louis
Kunkel, Joseph George
Lagakos, Stephen William
Lavin, Philip Todd
Lemeshow, Stanley Alan
Rand, William Medden
Ransil, Bernard J(erome)
Schoenfeld, David Alan
Smith, Frederick Edward
Smith, Tim Denis
Snow, Beatrice Lee
Stanley, Kenneth Earl
Tsiatis, Anastasios A
Van Deusen, Paul Cook
Wyshak, Grace
Zelen, Marvin

MICHIGAN
Assenzo, Joseph Robert
Bennett, Carl Leroy
Bookstein, Fred Leon
Brunden, Marshall Nils
Cornell, Richard Garth
Gill, John Leslie
Grudzien, Thaddeus Arthur, Jr
Huitema, Bradley Eugene
Keen, Robert Eric
Kowalski, Charles Joseph
McCrimmon, Donald Alan, Jr
McGilliard, Lon Dee
Mao, Ivan Ling
Metzler, Carl Maust
Mohberg, Noel Ross
Moll, Russell Addison
Musch, David C(harles)
Nordby, Gordon Lee
Patterson, Richard L
Reed, David Doss
Schork, Michael Anthony
Simmons, Gary Adair
Sing, Charles F
Sokol, Robert James
Stoline, Michael Ross
Tilley, Barbara Claire
Ullman, Nelly Szabo
Winterstein, Scott Richard
Zemach, Rita

MINNESOTA
Bartsch, Glenn Emil
Boen, James Robert
Boylan, William J
Curtsinger, James Webb
Elveback, Lillian Rose

Fukui, Hidenori Henry
Gatewood, Lael Cranmer
Goldman, Anne Ipsen
Hansen, Leslie Bennett
Jacobs, David R, Jr
Johnson, Eugene A
Keenan, Kathleen Margaret
Kjelsberg, Marcus Olaf
Kvalseth, Tarald Oddvar
Le, Chap Than
Loewenson, Ruth Brandenburger
McHugh, Richard B
Martin, Margaret Pearl
O'Fallon, William M
Siniff, Donald Blair
Stucker, Robert Evan
Sullivan, Alfred Dewitt
Taylor, William F
Weckwerth, Vernon Ervin
Zelterman, Daniel

MISSISSIPPI
Gerard, Patrick Dale
Meade, James Horace, Jr
Meydrech, Edward Frank
Watson, Clarence Ellis, Jr

MISSOURI
Flora, Jairus Dale, Jr
Gaffey, William Robert
Harvey, Joseph Eldon
Heymann, Hildegarde
Hill, Jack Filson
Hintz, Richard Lee
Moulton, Robert Henry
Smith, Richard Jay
Smith, Robert Francis
Spitznagel, Edward Lawrence, Jr
Thomson, Gordon Merle
Wette, Reimut

MONTANA
Brown, James Kerr
Jurist, John Michael
Metzgar, Lee Hollis

NEBRASKA
Bennett, Gary Lee
Compton, William A
Gardner, Charles Olda
Kimberling, William J
Mumm, Robert Franklin
Schutz, Wilfred M

NEVADA
Flueck, John A
Kinnison, Robert Ray
Leedy, Clark D

NEW HAMPSHIRE
Hill, Percy Holmes
McPeek, Mark Alan
Urban, Willard Edward, Jr

NEW JERSEY
Blickstein, Stuart I
Clymer, Arthur Benjamin
Copenhaver, Thomas Wesley
Givens, Samuel Virtue
Grover, Gary James
Holland, Paul William
Hsu, Chin-Fei
Johnson, Mary Frances
London, Mark David
McGhee, George Rufus, Jr
Mather, Robert Eugene
Meagher, Thomas Robert
Mietlowski, William Leonard
Miller, Alex
Morin, Peter Jay
Ordille, Carol Maria
Ott, Walther H
Roberts-Marcus, Helen Miriam
Rodda, Bruce Edward
Sharma, Ran S
Walsh, Teresa Marie

NEW MEXICO
Finkner, Morris Dale
Johnston, Roger Glenn
Salzman, Gary Clyde
Southward, Glen Morris

NEW YORK
Allison, David B
Alvir, Jose Ma J
Blumenson, Leslie Eli
Calhoon, Robert Ellsworth
Carruthers, Raymond Ingalls
Cisne, John Luther
Dibennardo, Robert
Dudewicz, Edward John
DuFrain, Russell Jerome
Durkin, Patrick Ralph
Falk, Catherine T
Federer, Walter Theodore
Feldman, Joseph Gerald
Fleiss, Joseph L
Greco, William Robert
Hucke, Dorothy Marie
Levene, Howard
Levin, Bruce
Lininger, Lloyd Lesley
Marcus, Leslie F

Biometrics-Biostatistics (cont)

Mendell, Nancy Role
Mike, Valerie
Neugebauer, Richard
Oates, Richard Patrick
Porter, William Frank
Priore, Roger L
Robinson, Harry
Rockwell, Robert Franklin
Rohlf, F James
Schey, Harry Moritz
Schmee, Josef
Schneiweiss, Jeannette W
Shah, Bhupendra K
Shore, Roy E
Siegel, Carole Ethel
Singh, Madho
Smoller, Sylvia Wassertheil
Sokal, Robert Reuven
Sommer, Charles John
Stein, Theodore Anthony
Tanner, Martin Abba
Turnbull, Bruce William
Varma, Andre A O
Whitby, Owen
Wood, Paul Mix
Yakir, Benjamin
Yozawitz, Allan
Zielezny, Maria Anna

NORTH CAROLINA
Abernathy, James Ralph
Atchley, William Reid
Boklage, Charles Edward
Brockhaus, John Albert
Caruolo, Edward Vitangelo
Chester, Alexander Jeffrey
Coulter, Elizabeth Jackson
Dinse, Gregg Ernest
Edelman, David Anthony
George, Stephen L
Gold, Harvey Joseph
Guess, Harry Adelbert
Hartford, Winslow H
Haseman, Joseph Kyd
Hayne, Don William
Hicks, Robert Eugene
Hoel, David Gerhard
Lessler, Judith Thomasson
Marcus, Allan H
Margolin, Barry Herbert
Moll, Robert Harry
Quade, Dana Edward Anthony
Rawlings, John Oren
Reckhow, Kenneth Howland
Reice, Seth Robert
Schaaf, William Edward
Sen, Pranab Kumar
Starmer, C Frank
Stier, Howard Livingston
Swallow, William Hutchinson
Sweeny, Hale Caterson
Turnbull, Craig David
Vaughan, Douglas Stanwood
Weiner, Daniel Lee
Weir, Bruce Spencer
Wells, Henry Bradley
Williams, Ann Houston
Zeng, Zhao-Bang

NORTH DAKOTA
Lieberman, Diana Dale
Wrenn, William J

OHIO
Badger, George Franklin
Buncher, Charles Ralph
Cacioppo, John T
Caldito, Gloria C
Davis, Michael E
Deddens, James Albert
Gartside, Peter Stuart
Goel, Prem Kumar
Guo, Shumei
Heckman, Carol A
Hoy, Casey William
Khamis, Harry Joseph
Lake, Robin Benjamin
Lima, John J
Liu, Ting-Ting Y
MacLean, David Belmont
Madenjian, Charles Paul
Neff, Raymond Kenneth
Powers, Jean D
Reiches, Nancy A
Runkle, James Reade
Siervogel, Roger M
Slutzky, Gale David

OKLAHOMA
Calvert, Jon Channing
Morrison, Robert Dean
Sanderson, George Albert
Schaefer, Carl Francis
Sonleitner, Frank Joseph
Verhalen, Laval M(athias)
Weeks, David Lee

OREGON
Ary, Dennis
Beatty, Joseph John
Ferraro, Steven Peter
Frakes, Rodney Vance

Isaacson, Dennis Lee
Lu, Kuo Hwa
Matarazzo, Joseph Dominic
Petersen, Roger Gene
Phillips, Donald Lundahl
Rowe, Kenneth Eugene
Urquhart, N Scott

PENNSYLVANIA
Arrington, Wendell S
Cronk, Christine Elizabeth
Emberton, Kenneth C
Fienberg, Stephen Elliott
Gertz, Steven Michael
Gleser, Leon Jay
Gould, A Lawrence
Hendrickson, John Alfred, Jr
Iglewicz, Boris
Kesner, Michael H
Kleban, Morton H
Klinger, Thomas Scott
Krall, John Morton
Kushner, Harvey
Li, Ching Chun
Lindsay, Bruce George
Lustbader, Edward David
McLaughlin, Francis X(avier)
Maksymowych, Roman
Marazita, Mary Louise
Mason, Thomas Joseph
Menduke, Hyman
Miller, G(erson) H(arry)
Morehouse, Chauncey Anderson
Pinski, Gabriel
Rockette, Howard Earl, Jr
Roush, William Burdette
Rubin, Leonard Sidney
Rucinska, Ewa J
Schneider, Bruce E
Schor, Stanley
Shirk, Richard Jay
Singer, Arthur Chester
Spielman, Richard Saul
Stong, David Bruce
Williams, George W
Wysocki, Charles Joseph

RHODE ISLAND
Brooks, Lisa Delevan
Costantino, Robert Francis

SOUTH CAROLINA
Abercrombie, Clarence Lewis
Dixon, Kenneth Randall
Drane, John Wanzer
Gross, Alan John
Hazen, Terry Clyde
Hunt, Hurshell Harvey
Loadholt, Claude Boyd
Miller, Millage Clinton, III
Pinder, John Edgar, III
Rust, Philip Frederick
Woodard, Geoffrey Dean Leroy

TENNESSEE
Dupont, William Dudley
Federspiel, Charles Foster
Fribourg, Henry August
Gosslee, David Gilbert
Groer, Peter Gerold
Harris, Edward Frederick
Mathis, Philip Monroe
Nanney, Lillian Bradley
Somes, Grant William
Tan, Wai-Yuan
Vander Zwaag, Roger
Van Winkle, Webster, Jr
West, Dennis R
Zeighami, Elaine Ann

TEXAS
Blangero, John Charles
Bratcher, Thomas Lester
Brokaw, Bryan Edward
Carroll, Raymond James
Cooper, Sharon P
Crandall, Keith Alan
Deming, Stanley Norris
Downs, Thomas D
Ford, Charles Erwin
Formanowicz, Daniel Robert, Jr
Fucik, John Edward
Gates, Charles Edgar
Glasser, Jay Howard
Gruber, George J
Hallum, Cecil Ralph
Hawkins, C Morton
Hsi, Bartholomew P
Huffman, Ronald Dean
Johnston, Dennis Addington
Johnston, Walter Edward
Kammerer, Candace Marie
Kirk, Ivan Wayne
Kirk, Roger E
Meyer, Betty Michelson
Montagna, Paul A
Orlando, Anthony Michael
Owens, Michael Keith
Parajulee, Megha N
Ring, Dennis Randall
Stedman, James Murphey
Thall, Peter Francis
Vaughn, William King
Wehrly, Thomas Edward

Whorton, Elbert Benjamin
Wilcox, Roberta Arlene
Wilkerson, James Edward
Willig, Michael Robert

UTAH
Dueser, Raymond D
Horn, Susan Dadakis
Hunt, Steven Charles
Maurer, Brian Alan
Reading, James Cardon
Ryel, Lawrence Atwell
Sisson, Donald Victor
Turner, David L(ee)

VERMONT
Costanza, Michael Charles
Emerson, John David
Haugh, Larry Douglas
Slack, Nelson Hosking

VIRGINIA
Bailey, Robert Clifton
Carter, Walter Hansbrough, Jr
Chang, William Y B
Chase, Helen Christina (Matulic)
Choi, Sung Chil
Fashing, Norman James
Fletcher, John Caldwell
Flora, Roger E
Frahm, Richard R
Gardenier, Turkan Kumbaraci
Henry, Neil Wylie
Holtzman, Golde Ivan
Kilpatrick, S James, Jr
Leczynski, Barbara Ann
Loesch, Joseph G
McGilliard, Michael Lon
Minton, Paul Dixon
Munson, Arvid W
Reynolds, Marion Rudolph, Jr
Sharov, Alexei A
Taub, Stephan Robert
Weiss, William

WASHINGTON
Bare, Barry Bruce
Barker, Morris Wayne
Bax, Nicholas John
Benedetti, Jacqueline Kay
Breslow, Norman Edward
Chapman, Douglas George
Clark, William Greer
Conquest, Loveday Loyce
Crowley, John James
Dailey, Frank Allen
Diehr, Paula Hagedorn
Farnum, Peter
Feigl, Polly Catherine
Fisher, Lloyd D
Gilbert, Ethel Schaefer
Grizzle, James Ennis
Hard, Jeffrey John
Hatheway, William Howell
Jackson, Crawford Gardner, Jr
Kronmal, Richard Aaron
Landis, Wayne G
McCaughran, Donald Alistair
Moolgavkar, Suresh Hiraji
Perrin, Edward Burton
Richardson, Michael Lewellyn
Rustagi, Krishna Prasad
Thomas, John M
Thompson, Donovan Jerome
Thompson, Elizabeth Alison
Towner, Richard Henry
Wahl, Patricia Walker
Weiss, Noel Scott
White, Colin
Wijsman, Ellen Marie

WEST VIRGINIA
Keller, Edward Clarence, Jr
Thayne, William V
Townsend, Edwin C

WISCONSIN
Boyce, Mark S
Carpenter, Stephen Russell
Casler, Michael Darwin
He, Xi
Klein, John Peter
Klotz, Jerome Hamilton
Kowal, Robert Raymond
Long, Charles Alan
Owens, John Michael
Rolley, Robert Ewell

WYOMING
Eddy, David Maxon
Mateer, Niall John

PUERTO RICO
Bangdiwala, Ishver Surchand

ALBERTA
Bella, Imre E
Hardin, Robert Toombs
Kalantar, Alfred Husayn
Lysyk, Timothy James
Yang, Rong-Cai
Yeh, Francis Cho-hao

BRITISH COLUMBIA
Barclay, Hugh John
Campbell, Alan
El-Kassaby, Yousry A
Frazer, Bryan Douglas
Hall, John Wilfred
Lauzier, Raymond B
Mackas, David Lloyd
Mitchell, Kenneth John
Petkau, Albert John
Pielou, Evelyn C
Routledge, Richard Donovan
Sterling, Theodor David
Todd, Mary Elizabeth
Welch, David Warren

MANITOBA
Abrahams, Mark Vivian
Pip, Eva
Stephens, Newman Lloyd

NEWFOUNDLAND
Newton, Peter Francis

ONTARIO
Balakrishnan, Narayanaswamy
Bertell, Rosalie
Brown, Kenneth Stephen
Chambers, James Robert
Chapman, Judith-Anne Williams
DeBoer, Gerrit
Dunnett, Charles William
Endrenyi, Laszlo
Fleming, Richard Arthur
Forbes, William Frederick
Gent, Michael
Goldsmith, Charles Harry
Ihssen, Peter Edowald
Robson, Douglas Sherman
Sackett, David Lawrence
Schneider, Bruce Alton
Tritchler, David Lynn
Wolynetz, Mark Stanley

QUEBEC
Bailar, John Christian, III
Jolicoeur, Pierre
Kingsley, Michael Charles Stephen
Legendre, Pierre
Magnan, Pierre
Siemiatycki, Jack
Vickery, William Lloyd

SASKATCHEWAN
Spurr, David Tupper

OTHER COUNTRIES
Bernstein, David Maier
Fitzhugh, Henry Allen
Hageman, Steven James
Kaiser, Hinrich
Siu, Tsunpui Oswald

Biophysics

ALABAMA
Bartlett, James Holly
Benos, Dale John
Cheung, Herbert Chiu-Ching
Cline, George Bruce
Downey, James Merritt
Elgavish, Ada S
Friedlander, Michael J
Friedman, Michael E
Golden, Michael Stanley
Harvey, Stephen Craig
Hazelrig, Jane
Khaled, Mohammad Abu
Klip, Willem
Latimer, Paul Henry
Magargal, Wells Wrisley, II
Nordlund, Thomas Michael
Rahn, Ronald Otto
Taylor, Aubrey Elmo
Tice, Thomas Robert

ALASKA
Fink, Thomas Robert

ARIZONA
Barr, Ronald Edward
Bier, Milan
Blankenship, Robert Eugene
Brown, Michael Frederick
Butler, Byron C
Connor, William Gorden
Diehn, Bodo
Eagle, Donald Frohlichstein
Epstein, L(udwig) Ivan
Holmes, William Farrar
Idso, Sherwood B
Kessler, John Otto
Kilkson, Rein
Lindsay, Stuart
Lukas, Ronald John
Mardian, James K W
Moore, Ana M L
Moore, Thomas Andrew
Mosher, Richard Arthur
Rez, Peter
Rupley, John Allen
Scott, Alwyn C
Tollin, Gordon

BIOLOGICAL SCIENCES / 65

Vermaas, Willem F J
Wyckoff, Ralph Walter Graystone

ARKANSAS
Baker, Max Leslie
Becker, Ralph Sherman
Demarest, Jeffrey R
Everett, Wilbur Wayne
Hart, Ronald Wilson
Lindley, Barry Drew
Martin, Duncan Willis
Moss, Alfred Jefferson, Jr
Nagle, William Arthur
Steinmeier, Robert C

CALIFORNIA
Abbott, Bernard C
Amirkhanian, John David
Amis, Eric J
Arakawa, Tsutomu
Baily, Norman Arthur
Baker, Richard Freligh
Bales, Barney Leroy
Baskin, Ronald J
Bearden, Alan Joyce
Bejar, Ezra
Berardo, Peter Antonio
Berry, Christine Albachten
Berry, Edward Alan
Bhatnagar, Rajendra Sahai
Biedebach, Mark Conrad
Bjorkman, Pamela J
Blakely, Eleanor Alice
Boxer, Steven George
Brady, Allan Jordan
Breisch, Eric Alan
Bremermann, Hans J
Brokaw, Charles Jacob
Bruner, Leon James
Brunk, Clifford Franklin
Budinger, Thomas Francis
Burns, Victor Will
Buttlaire, Daniel Howard
Cardullo, Richard Anthony
Carrano, Anthony Vito
Chan, Sunney Ignatius
Chen, Yang-Jen
Chou, Chung-Kwang
Chu, William Tongil
Clayton, Roderick Keener
Codrington, Robert Smith
Connelly, Clarence Morley
Cooke, Roger
Courtney, Kenneth Randall
Crandall, Walter Ellis
Curry, Fitz-Roy Edward
Dalton, Francis Norbert
Davis, Thomas Pearse
Dea, Phoebe Kin-Kin
Deamer, David Wilson, Jr
De Gaston, Alexis Neal
Deonier, Richard Charles
Dickinson, Wade
Dill, Kenneth Austin
Dismukes, Robert Key
Dobson, R Lowry
Dubuc, Paul U
Du Pont, Frances Marguerite
Duzgunes, Nejat
Edelman, Jay Barry
Edmonds, Peter Derek
Eiler, John Joseph
Eisenberg, David
Eisenman, George
Ely, Kathryn R
Evans, Evan Cyfeiliog, III
Fain, Gordon Lee
Feher, George
Fink, Anthony Lawrence
Fletterick, Robert John
Forte, John Gaetano
Fortes, George (Peter Alexander)
Frattini, Paul L
Frey, Terrence G
Gaffey, Cornelius Thomas
Garfin, David Edward
Gill, James Edward
Glaeser, Robert M
Glick, Harold Alan
Goerke, Jon
Gofman, John William
Goldman, Marvin
Grantz, David Arthur
Griffith, Owen Malcolm
Groce, David Eiben
Hagler, Arnold T
Hall, James Ewbank
Hanawalt, Philip Courtland
Hanson, Carl Veith
Hastings, David Frank
Hathaway, Gary Michael
Hayes, Thomas L
Haymond, Herman Ralph
Hazelwood, R(obert) Nichols
Hearst, John Eugene
Heath, Robert Louis
Hegenauer, Jack C
Hemmingsen, Edvard A(lfred)
Henderson, Sheri Dawn
Hill, Terrell Leslie
Holmes, Donald Eugene
Hong, Keelung
Hooker, Thomas M, Jr
Hopfield, John Joseph

Horwitz, Joseph
Jacobs, Peter Martin
James, Thomas Larry
Jariwalla, Raxit Jayantilal
Jensen, Ronald Harry
Johnson, C Scott
Johnson, Fred M
Junge, Douglas
Kase, Kenneth Raymond
Kavenoff, Ruth
Keizer, Joel Edward
Kelly, Donald Horton
Kendig, Joan Johnston
Kenyon, George Lommel
Kiger, John Andrew, Jr
Kim, Kwang-Jin
Kim, Sung-Hou
Kitting, Christopher Lee
Klein, Melvin Phillip
Kleinfeld, A M
Kliger, David Saul
Korenbrot, Juan Igal
Kortright, James McDougall
Lad, Pramod Madhusudan
Lanyi, Janos K
Latta, Harrison
Lecar, Harold
Lees, Graham
Leif, Robert Cary
Lele, Padmakar Pratap
Lewis, Steven M
Licko, Vojtech
Lieber, Richard L
Lieberman, Robert Arthur
Lindgren, Frank Tycko
Live, David H
Loughman, William D
Lubliner, J(acob)
Lucas, Joe Nathan
Lyman, John Tompkins
Lyon, Irving
McConnell, Harden Marsden
Macey, Robert Irwin
Mackay, Ralph Stuart
McNamee, Mark G
Maestre, Marcos Francisco
Magde, Douglas
Maki, August Harold
Mandelkern, Mark Alan
Manning, JaRue Stanley
Marcus, Carol Silber
Matthews, David Allan
Matthews, Harry Roy
Meehan, Thomas (Dennis)
Mehra, Rajesh Kumar
Mel, Howard Charles
Mendelson, Robert Alexander, Jr
Milanovich, Fred Paul
Miledi, Ricardo
Miljanich, George Paul
Morales, Manuel Frank
Mortimer, Robert Keith
Mueller, Thomas Joseph
Nacht, Sergio
Nakamura, Robert Masao
Nelson, Gary Joe
Neville, James Ryan
Nichols, Alexander Vladimir
Norman, Amos
Nothnagel, Eugene Alfred
Nuccitelli, Richard Lee
O'Konski, Chester Thomas
O'Leary, Dennis Patrick
Owicki, John Callaghan
Pacela, Allan F
Paolini, Paul Joseph
Papahadjopoulos, Demetrios Panayotis
Paselk, Richard Alan
Payne, Philip Warren
Phibbs, Roderic H
Philo, John Sterner
Ray, Dan S
Reese, Robert Trafton
Rich, Terrell L
Richardson, Irvin Whaley
Ritz-Gold, Caroline Joyce
Robison, William Lewis
Ross, John
Rothman, Stephen Sutton
Ruby, Ronald Henry
Saifer, Mark Gary Pierce
Sanui, Hisashi
Sargent, Thornton William, III
Sauer, Kenneth
Schmid, Carl William
Schmid, Peter
Schmid-Schoenbein, Geert W
Schmidt, Paul Gardner
Shafer, Richard Howard
Shepanski, John Francis
Sheppard, Asher R
Shetlar, Martin David
Shopes, Bob
Siegel, Edward
Silk, Margaret Wendy Kuhn
Simmons, Norman Stanley
Singer, Jerome Ralph
Sinsheimer, Robert Louis
Smith, Emil L
Smith, Helene Sheila
Smith, Kendric Charles
Snowdowne, Kenneth William
Snyder, Robert Gene
Sokolich, William Gary

Solomon, Edward I
Sondhaus, Charles Anderson
Spitzer, Nicholas Canaday
Stannard, J Newell
Stevens, Lewis Axtell
Strickland, Erasmus Hardin
Sutherland, Robert Melvin
Thomas, Charles Allen, Jr
Thomas, Richard Sanborn
Thompson, Lawrence Hadley
Tinoco, Joan W H
Tomich, John Matthew
Toy, Arthur John, Jr
Tsien, Richard Winyu
Van Dilla, Marvin Albert
Vickery, Larry Edward
Wade, Michael James
Walter, Harry
Ward, John F
Welch, Graeme P
Wemmer, David Earl
White, Stephen Halley
Wilbur, David Wesley
Williams, Lawrence Ernest
Williams, Robley Cook
Winchell, Harry Saul
Winet, Howard
Withers, Hubert Rodney
Wong, Kin-Ping
Wyatt, Philip Joseph
Yang, Jen Tsi
Yayanos, A Aristides
Yguerabide, Juan
Zimm, Bruno Hasbrouck
Zucker, Robert Stephen

COLORADO
Barisas, Bernard George, Jr
Beam, Kurt George, Jr
Best, LaVar
Cann, John Rusweiler
Carpenter, Donald Gilbert
Cech, Carol Martinson
Clark, Benton C
Cummings, Donald Joseph
Dahl, Adrian Hilman
Dewey, Thomas Gregory
Dudek, F Edward
Elkind, Mortimer M
Fotino, Mircea
Fox, Michael Henry
Furcinitti, Paul Stephen
Gamow, Rustem Igor
Gill, Stanley Jensen
Gorson, Robert O
Johnson, James Edward
Lett, John Terence
Levine, Simon Rock
McHenry, Charles S
Mueller, Theodore Arnold
Naughton, Michael A
Pautler, Eugene L
Pettijohn, David E
Phillipson, Paul Edgar
Puck, Theodore Thomas
Sanderson, Richard James
Seibert, Michael
Solie, Thomas Norman
Sowell, John Basil
Sullivan, Sean Michael
Uhlenbeck, Olke Cornelis
Woody, Robert Wayne
York, Sheldon Stafford

CONNECTICUT
Adler, Alan David
Agulian, Samuel Kevork
Armitage, Ian MacLeod
Aylor, Donald Earl
Boulpaep, Emile L
Braswell, Emory Harold
Bronner, Felix
Brudvig, Gary Wayne
Bursuker, Isia
Caradonna, John Philip
Coleman, Joseph Emory
Crain, Richard Cullen
De Rocco, Andrew Gabriel
Engelman, Donald Max
Forbush, Bliss, III
Gagge, Adolf Pharo
Gent, Martin Paul Neville
Glasel, Jay Arthur
Golub, Efim I
Heller, John Herbert
Henderson, Edward George
Herbette, Leo G
Hoffman, Joseph Frederick
Hutchinson, Franklin
Knox, James Russell, Jr
Lambrakis, Konstantine Christos
Loew, Leslie Max
Macnab, Robert Marshall
Markowitz, David
Moore, Peter Bartlett
Morowitz, Harold Joseph
Peterson, Cynthia Wyeth
Prestegard, James Harold
Prigodich, Richard Victor
Richards, Frederic Middlebrook
Rockwell, Sara Campbell
Russu, Irina Maria
Schor, Robert
Schulz, Robert J

Schuster, Todd Mervyn
Schwartz, Ilsa Roslow
Schwartz, Tobias Louis
Sha'afi, Ramadan Issa
Shulman, Robert Gerson
Sigler, Paul Benjamin
Slayman, Clifford L
Sturtevant, Julian Munson
Terry, Thomas Milton
Waxman, Stephen George
Wolf, Elizabeth Anne
Wyckoff, Harold Winfield
Yphantis, David Andrew
Zubal, I George

DELAWARE
Bolton, Ellis Truesdale
Brenner, Stephen Louis
Brown, Barry Stephen
Diner, Bruce Aaron
Dwivedi, Anil Mohan
Ehrlich, Robert Stark
Hartig, Paul Richard
Hartzell, Charles Ross, III
Jain, Mahendra Kumar
Lichtner, Francis Thomas, Jr
Moe, Gregory Robert
Nelson, Mark James
Reuben, Jacques
Roe, David Christopher
Sharnoff, Mark

DISTRICT OF COLUMBIA
Andersen, Frank Alan
Barr, Nathaniel Frank
Chang, Eddie Li
De Levie, Robert
Douple, Evan Barr
Goodenough, David John
Gray, Irving
Hirsch, Roland Felix
Jones, Janice Lorraine
Karle, Isabella Lugoski
Kowalsky, Arthur
Ledley, Robert Steven
Malinin, George I
Misra, Prabhakar
Montrose, Charles Joseph
O'Leary, Timothy Joseph
Putzrath, Resha Mae
Rosenberg, Edith E
Salu, Yehuda
Schimmerling, Walter
Schnur, Joel Martin
Shropshire, Walter, Jr
Sinks, Lucius Frederick
Sulzman, Frank Michael
Trubatch, Sheldon L
Ward, Keith Bolen, Jr
Weiss, Ira Paul
Wood, Robert Winfield

FLORIDA
Anderson, Peter Alexander Vallance
Barber, Michael James
Beidler, Lloyd M
Biersdorf, William Richard
Block, Ronald Edward
Blum, Alvin Seymour
Boyce, Richard P
Brandon, Frank Bayard
Buss, Daryl Dean
Caswell, Anthony H
Chapman, Michael S
Cohen, Robert Jay
Colahan, Patrick Timothy
Edwards, Charles
Finlayson, Birdwell
Gilmer, Penny Jane
Gordon, Kenneth Richard
Hayashi, Teru
Haynes, Duncan Harold
Herbert, Thomas James
Himel, Chester Mora
Hornicek, Francis John
Huebner, Jay Stanley
Jacobson, Jerry I
Kasha, Michael
Kerrick, Wallace Glenn Lee
Landowne, David
Leblanc, Roger M
Lindsey, Bruce Gilbert
McConnell, Dennis Brooks
Magleby, Karl LeGrande
Makowski, Lee
Mallery, Charles Henry
Mankin, Richard Wendell
Moulton, Grace Charbonnet
Mukherjee, Pritish
Muller, Kenneth Joseph
Pepinsky, Raymond
Pressman, Berton Charles
Purich, Daniel Lee
Rhodes, William Clifford
Roberts, Sanford B(ernard)
Rose, Birgit
Salas, Pedro Jose I
Scarpace, Philip J
Schleyer, Walter Leo
Schoor, W Peter
Schulte, Alfons F
Shah, Dinesh Ochhavlal
Sharp, John Turner
Snyder, Patricia Ann

Biophysics (cont)

Stevens, Bruce Russell
Szczepaniak, Krystyna
Thorhaug, Anitra L
Williams, Theodore P
Young, David Michael
Zengel, Janet Elaine

GEORGIA
Abercrombie, Ronald Ford
Adams, Robert Johnson
De Felice, Louis John
DeHaan, Robert Lawrence
Dusenbery, David Brock
Edmondson, Dale Edward
Etzler, Frank M
Eubig, Casimir
Fechheimer, Marcus
Fox, Ronald Forrest
Godt, Robert Eugene
Green, Keith
Habte-Mariam, Yitbarek
Hall, Dwight Hubert
Joyner, Ronald Wayne
Kolbeck, Ralph Carl
Lee, John William
Little, Robert Colby
McDowall, Debra J
Moore, Cyril L
Moriarty, C Michael
Netzel, Thomas Leonard
Nosek, Thomas Michael
Pohl, Douglas George
Pooler, John Preston
Richard, Christopher Alan
Sadun, Alberto Carlo
Scott, Robert Allen
Sink, John Davis
Sophianopoulos, Alkis John
Sprawls, Perry, Jr
Stevenson, Dennis A
Wallace, Robert Henry
Wartell, Roger Martin
Wilson, William David
Yeargers, Edward Klingensmith
Yu, Nai-Teng

HAWAII
Brewbaker, James Lynn
Cooke, Ian McLean
Jameson, David M

IDAHO
D'Aoust, Brian Gilbert
Moore, Richard
Tanner, John Eyer, Jr

ILLINOIS
Aktipis, Stelios
Andriacchi, Thomas Peter
Applebury, Meredithe L
Astumian, Raymond Dean
Barany, Michael
Barr, Lloyd
Bers, Donald M
Bond, Howard Edward
Borso, Charles S
Bosmann, Harold Bruce
Briskin, Donald Phillip
Caspary, Donald M
Chan, Yun Lai
Chang, Chong-Hwan
Chignell, Derek Alan
Cohn, Gerald Edward
Cox, Thomas C
Crofts, Antony R
Dallos, Peter John
Dawson, M Joan
Debrunner, Peter Georg
DeVault, Don Charles
Dickinson, Helen Rose
Domanik, Richard Anthony
Ducoff, Howard S
Dunn, Floyd
Dunn, Robert Bruce
Dutta, Pulak
Ebrey, Thomas G
Eder, Douglas Jules
Eisenberg, Robert S
Fields, Theodore
Fleming, Graham Richard
Fox, Jack Lawrence
Fung, Leslie Wo-Mei
Geil, Phillip H
Geisler, Fred Harden
Gennis, Robert Bennett
Gettins, Peter Gregory Wolfgang
Ghiron, Camillo A
Gillette, Rhanor
Gingle, Alan Raymond
Gould, John Michael
Govindjee, M
Gratton, Enrico
Grdina, David John
Grossweiner, Leonard Irwin
Hagstrom, Ray Theodore
Helman, Sandy I
Hoffman, Brian Mark
Horwitz, Alan Fredrick
Hubbard, Lincoln Beals
Ingram, Forrest Duane
Jakobsson, Eric Gunnar, Sr
Johnson, Michael Evart

Josephs, Robert
Keiderling, Timothy Allen
Kidder, George Wallace, III
Kim, Yung Dai
Kimura, Mineo
Koushanpour, Esmail
Krafft, Grant A
Kuchnir, Franca Tabliabue
Kuczmarski, Edward R
Lange, Yvonne
Lanzl, Lawrence Herman
Lauffenburger, Douglas Alan
Lauterbur, Paul Christian
LeBreton, Pierre Robert
Levine, Michael W
Longworth, James W
Makinen, Marvin William
Miller, Donald Morton
Mittenthal, Jay Edward
Moffat, John Keith
Moretz, Roger C
Nakajima, Shigehiro
Nakajima, Yasuko
Narahashi, Toshio
Niles, Walter Dulany, II
Nilges, Mark J
O'Brien, William Daniel, Jr
Offner, Franklin Faller
Oldfield, Eric
Olsen, Kenneth Wayne
Oono, Yoshitsugu
Ovadia, Jacques
Page, Ernest
Peak, Meyrick James
Pepperberg, David Roy
Preston, Robert Leslie
Rakowski, Robert F
Reynolds, Larry Owen
Ries, Herman Elkan, Jr
Rowland, Robert Edmund
Scharf, Arthur Alfred
Scheiner, Steve
Schiffer, Marianne Tsuk
Schlenker, Robert Alison
Sherman, Warren V
Shriver, John William
Singer, Irwin
Sleator, William Warner, Jr
Sligar, Stephen Gary
Spears, Kenneth George
Stevens, Fred Jay
Taylor, Edwin William
Thurnauer, Marion Charlotte
Uretz, Robert Benjamin
Vanderkooi, Garret
Wang, Andrew H-J
Wasielewski, Michael Roman
Weber, Gregorio
Weissman, Michael Benjamin
Westbrook, Edwin Monroe
White, William
Whitmarsh, John
Wilson, Hugh Reid
Wolynes, Peter Guy
Wraight, Colin Allen
Wynveen, Robert Allen
Yeates, Donovan B

INDIANA
Allerhand, Adam
Alvager, Torsten Karl Erik
Ascarelli, Gianni
Baker, Timothy Stanley
Bakken, George Stewart
Besch, Henry Roland, Jr
Bina, Minou
Bockrath, Richard Charles, Jr
Bosron, William F
Bridges, C David
Bryan, William Phelan
Byrn, Stephen Robert
Cramer, William Anthony
Davis, Grayson Steven
Filmer, David Lee
Fong, Francis K
Frank, Bruce Hill
Friedman, Julius Jay
Haak, Richard Arlen
Hanks, Alan Ray
Havel, Henry Acken
Helrich, Carl Sanfrid, Jr
Hui, Chiu Shuen
Kelly-Fry, Elizabeth
Macchia, Donald Dean
Madden, Keith Patrick
Murray, Lionel Philip, Jr
Nowak, Thomas
Nussbaum, Elmer
Ortoleva, Peter Joseph
Pak, William Louis
Pearlstein, Robert Milton
Quay, John Ferguson
Randall, James Edwin
Schauf, Charles Lawrence
Sherman, Louis Allen
Strickholm, Alfred
Swez, John Adam
Van Frank, Richard Mark
Vogelhut, Paul Otto
Wolff, Ronald Keith

IOWA
Applequist, Jon Barr
Berg, Virginia Seymour

Campbell, Kevin Peter
Chandran, Krishnan Bala
Foss, John G(erald)
Getting, Peter Alexander
Greenberg, Everett Peter
Hatfield, Jerry Lee
Hoffman, Eric Alfred
Kawai, Masataka
Kintanar, Agustin
Mayfield, John Eric
Rougvie, Malcolm Arnold
Shen, Sheldon Shih-Ta
Six, Erich Walther
Small, Gerald J
Struve, Walter Scott
Wu, Chun-Fang
Wunder, Charles C(ooper)

KANSAS
Barton, Janice Sweeny
Colen, Alan Hugh
Friesen, Benjamin S
Gordon, Michael Andrew
Guikema, James Allen
Hersh, Robert Tweed
Kimler, Bruce Franklin
Krishnan, Engil Kolaj
Manney, Thomas Richard
Michaelis, Elias K
Valenzeno, Dennis Paul
Welling, Daniel J
Yarbrough, Lynwood R

KENTUCKY
Bowen, Thomas Earle, Jr
Butterfield, David Allan
Christensen, Ralph C(hresten)
Coffey, Charles William, II
Coohill, Thomas Patrick
Engelberg, Joseph
Getchell, Thomas Vincent
Hirsch, Henry Richard
Sayeg, Joseph A
Schilb, Theodore Paul
Syed, Ibrahim Bijli
Walker, Sheppard Matthew
Wyssbrod, Herman Robert

LOUISIANA
Azzam, Rasheed M A
Beuerman, Roger Wilmer
Clarkson, Craig William
Clejan, Sandra
Eggen, Douglas Ambrose
Hamori, Eugene
Hensley, John Coleman, II
Heyn, Anton Nicolaas Johannes
Hyman, Edward Sidney
Klyce, Stephen Downing
Levitzky, Michael Gordon
Luftig, Ronald Bernard
McAfee, Robert Dixon
Marino, A A
Miceli, Michael Vincent
Navar, Luis Gabriel
Taylor, Eric Robert
Welch, George Rickey
Wiechelman, Karen Janice

MAINE
Blake, Richard D
Hughes, William Taylor
Kirkpatrick, Francis Hubbard
Nelson, Clifford Vincent

MARYLAND
Aamodt, Roger Louis
Ackerman, Eric J
Amzel, L Mario
Andrews, Stephen Brian
Battey, James F
Bax, Ad
Bellino, Francis Leonard
Berger, Robert Lewis
Berson, Alan
Berzofsky, Jay Arthur
Bicknell-Brown, Ellen
Blaustein, Mordecai P
Blumenthal, Robert Paul
Bockstahler, Larry Earl
Brinley, Floyd John, Jr
Bronk, Burt V
Brownell, William Edward
Burt, David Reed
Bush, C Allen
Cantor, Kenneth P
Carlson, Francis Dewey
Cassman, Marvin
Charney, Elliot
Chattoraj, Dhruba Kumar
Chen, Henry Lowe
Chen, Yi-Der
Clark, Carl Cyrus
Coble, Anna Jane
Colombini, Marco
Copeland, Edmund Sargent
Coulter, Charles L
Davies, David R
Davies, Philip Winne
Davis, Julien Sinclair
Devreotes, Peter Nicholas
Dintzis, Howard Marvin
Dintzis, Renee Zlochover
Donaldson, Sue Karen

Eanes, Edward David
Eaton, William Allen
Edidin, Michael Aaron
Ehrenstein, Gerald
Eichhorn, Gunther Louis
Eipper, Betty Anne
Farrell, Richard Alfred
FitzHugh, Richard
Flower, Robert Walter
Freire, Ernesto I
Fried, Jerrold
Friedman, Fred K
Froehlich, Jeffrey Paul
Fuhr, Irvin
Gallin, Elaine K
Ganguly, Pankaj
Geduldig, Donald
Gilbert, Daniel Lee
Goldman, Lawrence
Goode, Melvyn Dennis
Gupte, Sharmila Shaila
Gutierrez, Peter Luis
Hagins, William
Hansford, Richard Geoffrey
Harris, Andrew Leonard
Hartley, Robert William, Jr
Hendler, Richard Wallace
Hirsh, Allen Gene
Ho, Henry Sokore
Hobbs, Ann Snow
Hofrichter, Harry James
Houk, Albert Edward Hennessee
Huang, Charles Y
Hybl, Albert
Iwasa, Kuni H
Jernigan, Robert Lee
Kafka, Marian Stern
Kaplan, Ann Esther
Kasianowicz, John James
Kempner, Ellis Stanley
Ketley, Jeanne Nelson
Klausner, Richard D
Lakowicz, Joseph Raymond
Larrabee, Martin Glover
Lattman, Eaton Edward
Leapman, Richard David
Lederer, William Jonathan
Lee, Byungkook
Lester, David Simon
Levinson, John Z
Levitan, Herbert
Lewis, Marc Simon
Lin, Diane Chang
Lin, Shin
Lipicky, Raymond John
Litman, Burton Joseph
Livengood, David Robert
Love, Warner Edwards
Lymangrover, John R
Lymn, Richard Wesley
Lytle, Carl David
McCally, Russell Lee
MacKerell, Alexander Donald, Jr
McLaughlin, Alan Charles
McPhie, Peter
Maddox, Yvonne T
Martin, Mark Thomas
Maslow, David E
Mazur, Jacob
Mildvan, Albert S
Miles, Edith Wilson
Mills, William Andy
Minton, Allen Paul
Monti, John Anthony
Mullins, Lorin John
Myers, Lawrence Stanley, Jr
Nelson, Ralph Francis
Noguchi, Constance Tom
Norvell, John Charles
Nossal, Ralph J
Padlan, Eduardo Agustin
Pagano, Richard Emil
Parsegian, Vozken Adrian
Pattabiraman, Nagarajan
Pedersen, Peter L
Pinto da Silva, Pedro Goncalves
Podolsky, Richard James
Pollard, Thomas Dean
Pollycove, Myron
Pratt, Arnold Warburton
Puskin, Jerome Sanford
Rakhit, Gopa
Rall, Wilfrid
Rall, William Frederick
Rifkind, Joseph Moses
Rinzel, John Matthew
Robertson, Baldwin
Robertson, James Sydnor
Rodbard, David
Rose, George David
Saba, George Peter, II
Sastre, Antonio
Schambra, Philip Ellis
Schulze, Dan Howard
Shamoo, Adil E
Shin, Yong Ae Im
Shore, Moris Lawrence
Siatkowski, Ronald E
Sinclair, Warren Keith
Sinha, Birandra Kumar
Sjodin, Raymond Andrew
Smith, Elmer Robert
Sowers, Arthur Edward
Spero, Leonard

BIOLOGICAL SCIENCES / 67

Steinert, Peter Malcolm
Steven, Alasdair C
Stevens, Walter Joseph
Straat, Patricia Ann
Swenberg, Charles Edward
Sze, Heven
Szu, Shousun Chen
Taylor, Lauriston Sale
Torchia, Dennis Anthony
Ts'o, Paul On Pong
Tung, Ming Sung
Turner, R James
Van De Merwe, Willem Pieter
Vargas, Fernando Figueroa
Vogl, Thomas Paul
Williams, Robert Jackson
Wolff, John B
Wolffe, Alan Paul
Yasbin, Ronald Eliott
Yockey, Hubert Palmer
Yoo, Seung Hyun
Yu, Leepo Cheng

MASSACHUSETTS
Adelman, William Joseph, Jr
Adelstein, Stanely James
Anderson-Olivo, Margaret
Auer, Henry Ernest
Auld, David Stuart
Bansil, Rama
Becker, David Stewart
Berg, Howard Curtis
Bethune, John Lemuel
Breckenridge, John Robert
Browne, Douglas Townsend
Brownell, Gordon Lee
Cabral-Lilly, Donna
Cameron, John Stanley
Campbell, Mary Kathryn
Cantiello, Horacio Fabio
Cantley, Lewis Clayton
Carey, Martin Conrad
Carlton, William Herbert
Castronovo, Frank Paul, Jr
Catsimpoolas, Nicholas
Champion, Paul Morris
Cohen, Carolyn
Coleman, Peter Stephen
Correia, John Arthur
Crusberg, Theodore Clifford
Curtis, Joseph C
DeRosier, David J
Doane, Marshall Gordon
Downer, Nancy Wuerth
Edwards, Ross B
Epp, Edward Rudolph
Epstein, Herman Theodore
Fairbanks, Grant
Falchuk, Kenneth H
Fasman, Gerald David
Fine, Samuel
Ford, Norman Cornell, Jr
Fossel, Eric Thor
Frishkopf, Lawrence Samuel
George, Stephen Anthony
Glimcher, Melvin Jacob
Goldman, David Eliot
Gross, David John
Hamilton, James Arthur
Hardy, William Lyle
Harrison, Stephen Coplan
Hartline, Peter Haldan
Hastings, John Woodland
He, Bin
Herzfeld, Judith
Herzlinger, George Arthur
Hoffman, Allen Herbert
Hoop, Bernard, Jr
Horowitz, Arie
Housman, David E
Inoue, Shinya
Jaffe, Lionel F
Jennings, William Harney, Jr
Jordan, Peter C H
Karasz, Frank Erwin
Kemp, Daniel Schaeffer
King, John Gordon
Klibanov, Alexander M
Koehler, Andreas Martin
Koltun, Walter Lang
Kropf, Allen
Laing, Ronald Albert
Lamola, Angelo Anthony
Landis, William Joel
Langley, Kenneth Hall
Lees, Sidney
LeFevre, Paul Green
Lehman, William Jeffrey
Lerman, Leonard Solomon
Levy, Elinor Miller
Loretz, Thomas J
Loscalzo, Joseph
McMahon, Thomas Arthur
MacNichol, Edward Ford, Jr
Marrero, Hector G
Marx, Kenneth Allan
Melchior, Donald L
Miller, Keith Wyatt
Morgan, Kathleen Greive
Neuringer, Leo J
Orme-Johnson, Nanette Roberts
Orme-Johnson, William H
Papaefthymiou, Georgia Christou
Petsko, Gregory Anthony

Phillies, George David Joseph
Platt, John Rader
Plocke, Donald J
Rich, Alexander
Rigney, David Roth
Rose, Robert M(ichael)
Rosen, Philip
Rosenblith, Walter Alter
Rothschild, Kenneth J
Sandler, Sheldon Samuel
Sauer, Robert Thomas
Scheuchenzuber, H Joseph
Schimmel, Paul Reinhard
Sekuler, Robert W
Shapiro, Jacob
Shipley, George Graham
Simons, Elizabeth Reiman
Singer, Joshua J
Small, Donald MacFarland
Smith, Peter John Shand
Solomon, Arthur Kaskel
Stanley, H(arry) Eugene
Stephens, Raymond Edward
Sternick, Edward Selby
Stossel, Thomas Peter
Tanaka, Toyoichi
Taylor, Robert E
Teeter, Martha Mary
Toner, Mehmet
Tosteson, Daniel Charles
Tripathy, Sukant K
Umans, Robert Scott
Vallee, Bert L
Verma, Surendra P
Villars, Felix Marc Hermann
Voigt, Herbert Frederick
Waloga, Geraldine
Wang, Chih-Lueh Albert
Weaver, James Cowles
Webb, Robert Howard
Webster, Edward William
Wiley, Don Craig
Woodhull-McNeal, Ann P
Yellen, Gary
Yen, Peter Kai Jen
Zaner, Ken Scott
Zimmermann, Robert Alan

MICHIGAN
Axelrod, Daniel
Babcock, Gerald Thomas
Blanchard, Fred Ayres
Carson, Paul Langford
Carter-Su, Christin
Chou, Kuo Chen
Conrad, Michael
Cowlishaw, John David
Crippen, Gordon Marvin
Dambach, George Ernest
Epps, Dennis Earl
Fenstermacher, Joseph Don
Ferguson-Miller, Shelagh Mary
Garlick, Robert L
Gates, David Murray
Goldstein, Albert
Halvorson, Herbert Russell
Hodgson, Voigt R
Horvath, William John
Hubbard, Robert Phillip
Jursinic, Paul Andrew
Kimura, Tokuji
Krimm, Samuel
Levich, Calman
Liboff, Abraham R
Lindemann, Charles Benard
McConnell, David Graham
McGrath, John Joseph
McGroarty, Estelle Josephine
Maggiora, Gerald M
Matthews, Rowena Green
Mizukami, Hiroshi
Neubig, Richard Robert
Njus, David Lars
Novak, Raymond Francis
Oncley, John Lawrence
Orton, Colin George
Ottova, Angela
Parkinson, William Charles
Peterson-Kennedy, Sydney Ellen
Petty, Howard Raymond
Phan, Sem Hin
Pollack, Gerald Leslie
Rimai, Lajos
Rogers, William Leslie
Rohrer, Douglas C
Rosenspire, Allen Jay
Sands, Richard Hamilton
Saper, Mark A
Schullery, Stephen Edmund
Sevilla, Michael Douglas
Shichi, Hitoshi
Stephenson, Robert Storer
Tepley, Norman
Tien, H Ti
Valeriote, Frederick Augustus
Webb, Paul
Wiley, John W
Wolterink, Lester Floyd
Zand, Robert

MINNESOTA
Ackerman, Eugene
Adiarte, Arthur Lardizabal
Adolph, Kenneth William

Allewell, Norma Mary
Baumann, Wolfgang J
Bloomfield, Victor Alfred
Brown, Rhoderick Edmiston, Jr
Carter, John Vernon
Evans, Douglas Fennell
Forro, Frederick, Jr
Garrity, Michael K
Goldstein, Stuart Frederick
Haaland, John Edward
Haddad, Louis Charles
Khan, Faiz Mohammad
Kluetz, Michael David
Loken, Merle Kenneth
McCullough, Edwin Charles
Meyer, Frank Henry
Miller, Robert F
O'Connor, Michael Kieran
Polnaszek, Carl Francis
Sammak, Paul J
Schmitt, Otto Herbert
Stoesz, James Darrel
Szurszewski, Joseph Henry
Taylor, Stuart Robert
Thomas, David Dale
Tsong, Tian Yow
Ugurbil, Kamil
Yapel, Anthony Francis, Jr

MISSISSIPPI
Chaires, Jonathan Bradford
Coleman, Thomas George
Dwyer, Terry M
Eftink, Maurice R
Graves, David E
Pandey, Ras Bihari

MISSOURI
Ackerman, Joseph John Henry
Ackers, Gary K
Burgess, James Harland
Cochran, Andrew Aaron
Eldredge, Donald Herbert
Esser, Alfred F
Henson, Bob Londes
Highstein, Stephen Morris
Keane, John Francis, Jr
Larson, Kenneth Blaine
Luxon, Bruce Arlie
Menz, Leo Joseph
Morris, Stephen Jon
Northrip, John Willard
Phillips, William Dale
Pickard, William Freeman
Poon, Rebecca Yuetmay
Pueppke, Steven Glenn
Roti Roti, Joseph Lee
Rowe, Elizabeth Snow
Schmitz, Kenneth Stanley
Seidler, Norbert Wendelin
Shear, David Ben
Smith, William R
Ter-Pogossian, Michel Mathew
Thomas, George Joseph, Jr
Vaughan, William Mace
Wartzok, Douglas
Wong, Tuck Chuen
Woodruff, Clarence Merrill

MONTANA
Callis, Patrik Robert
Hill, Walter Ensign
Jesaitis, Algirdas Joseph
Jurist, John Michael
Ribi, Edgar
Thomas, Robert Glenn

NEBRASKA
Chakkalakal, Dennis Abraham
Hahn, George LeRoy
Harbison, Gerard Stanislaus
Holmquist, Barton
Parkhurst, Lawrence John
Salhany, James Mitchell
Schneiderman, Martin Howard
Song, Pill-Soon

NEVADA
Barnes, George
Harrington, Rodney E

NEW HAMPSHIRE
Blanchard, Robert Osborn
Dennison, David Severin
Gross, Robert Henry
Kegeles, Gerson
Kidder, John Newell
Manasek, Francis John
Musiek, Frank Edward
Oldfield, Daniel G
Richmond, Robert Chaffee
Rosenberg, Alburt M
Stanton, Bruce Alan
Swartz, Harold M
Vournakis, John Nicholas

NEW JERSEY
Ashkin, Arthur
Becker, Joseph Whitney
Bender, Max
Bennun, Alfred
Berman, Helen Miriam
Blumberg, William Emil
Boltz, Robert Charles, Jr

Breslauer, Kenneth John
Breslow, Esther M G
Brown, Rodney Duvall, III
Brudner, Harvey Jerome
Chandler, Louis
Chen, Kuang-Yu
Connor, John Arthur
DeBari, Vincent A
Duran, Walter Nunez
Eisenberger, Peter Michael
Fitzgerald, Paula Marie Dean
Fong, Tung Ming
Fordham, William David
Fresco, Jacques Robert
Friedman, Kenneth Joseph
Gagliardi, L John
Gagna, Claude Eugene
Garcia, Maria Luisa
Hearn, Ruby Puryear
Hill, David G
Horn, Leif
Hwang, San-Bao
Isaacson, Allen
Ji, Sungchul
Jordan, Frank
Kaczorowski, Gregory John
Lee, Lih-Syng
Leonards, Kenneth Stanley
McCaslin, Darrell
Madison, Vincent Stewart
Manning, Gerald Stuart
Massover, William H
Mayhew, Eric George
Mendelsohn, Richard
Micheli-Tzanakou, Evangelia
Mims, William B
Pethica, Brian Anthony
Pilla, Arthur Anthony
Poe, Martin
Polinsky, Ronald John
Reynolds, George Thomas
Rousseau, Denis Lawrence
Roy, Herbert C
Ruknudin, Abdul Majeed
Rumsey, William LeRoy
Sheriff, Steven
Siano, Donald Bruce
Siegel, Jeffry A
Silver, Frederick Howard
Slocum, Donald Hillman
Steinberg, Malcolm Saul
Stephenson, Elizabeth Weiss
Strauss, George
Tank, David W
Teipel, John William
Tomaselli, Vincent Paul
Treu, Jesse Isaiah
Trotta, Paul P
Tzanakou, M Evangelia
Wangemann, Robert Theodore
Wassarman, Paul Michael
Witz, Gisela
Zoltan, Bart Joseph

NEW MEXICO
Bartholdi, Marty Frank
Beckel, Charles Leroy
Bell, George Irving
Burks, Christian
Coppa, Nicholas V
Cram, Leighton Scott
Crissman, Harry Allen
Diel, Joseph Henry
Fletcher, Edward Royce
Frauenfelder, Hans
Galey, William Raleigh
Goldstein, Byron Bernard
Gurd, Ruth Sights
Hanson, Kenneth Merrill
Heller, Leon
Hildebrand, Carl Edgar
Holm, Dale M
Jett, James Hubert
Kelsey, Charles Andrew
Matwiyoff, Nicholas Alexander
Omdahl, John L
Partridge, L Donald
Perelson, Alan Stuart
Raju, Mudundi Ramakrishna
Salzman, Gary Clyde
Schoenborn, Benno P
Scott, Bobby Randolph
Shera, E Brooks
Stapp, John Paul

NEW YORK
Akers, Charles Kenton
Al-Awqati, Qais
Alfano, Robert R
Almog, Rami
Anbar, Michael
Andersen, Olaf Sparre
Anderson, Lowell Leonard
Ault, Jeffrey George
Bahary, William S
Baier, Robert Edward
Baker, Robert George
Barish, Robert John
Bean, Charles Palmer
Becker, Robert O
Bell, Duncan Hadley
Bello, Jake
Benham, Craig John
Bennett, Michael Vander Laan

Biophysics (cont)

Bernhard, William Allen
Bettelheim, Frederick A
Bickmore, John Tarry
Bidlack, Jean Marie
Bigler, Rodney Errol
Bird, Richard Putnam
Birge, Robert Richards
Bisson, Mary A
Bittman, Robert
Blank, Martin
Blei, Ira
Bond, Victor Potter
Bookchin, Robert M
Bottomley, Paul Arthur
Boutjdir, Mohamed
Brand, John S
Brandt, Philip Williams
Bray, James William
Brehm, Lawrence Paul
Breneman, Edwin Jay
Brenner, Henry Clifton
Briehl, Robin Walt
Brink, Frank, Jr
Brody, Marcia
Brody, Seymour Steven
Bushinsky, David Allen
Candia, Oscar A
Carstensen, Edwin L(orenz)
Cerny, Laurence Charles
Chamberlain, Charles Craig
Chappell, Richard Lee
Cohen, Beverly Singer
Cohen, Norman
Coleman, James R
Cowburn, David
Dahl, John Robert
Damadian, Raymond
Davenport, Lesley
Day, Loren A
De Lisi, Charles
DeTitta, George Thomas
Dicello, John Francis, Jr
Diem, Max
Diwan, Joyce Johnson
Doetschman, David Charles
Dorset, Douglas Lewis
Dreizen, Paul
Duax, William Leo
Duffey, Michael Eugene
Eberle, Helen I
Eberstein, Arthur
Eickelberg, W Warren B
Eisenstadt, Maurice
Eisinger, Josef
Ely, Thomas Sharpless
Eshel, Dan
Fabry, Mary E Riepe
Fabry, Thomas Lester
Fajer, Jack
Featherstone, John Douglas Bernard
Feigenson, Gerald William
Feldman, Isaac
Fischbarg, Jorge
Fisher, Vincent J
Fishman, Jerry Haskel
Forgacs, Gabor
Foster, Kenneth William
Foster, Margaret C
Freedman, Jeffrey Carl
Friedberg, Carl E
Friedman, Helen Lowenthal
Friedman, Joel Mitchel
Gao, Jiali
Gardella, Joseph Augustus, Jr
Gardner, Daniel
Geacintov, Nicholas
Gershman, Lewis C
Gerstman, Hubert Louis
Giaever, Ivar
Gibson, Quentin Howieson
Green, Michael Enoch
Griffin, Jane Flanigen
Gross, Leo
Grumet, Martin
Gunter, Karlene Klages
Gunter, Thomas E, Jr
Gupta, Raj K
Gurpide, Erlio
Gutstein, William H
Haas, David Jean
Haines, Thomas Henry
Hamblen, David Philip
Hanson, Louise I Karle
Harper, Richard Allan
Harris, Jack Kenyon
Harris-Warrick, Ronald Morgan
Harth, Erich Martin
Haywood, Anne Mowbray
Hirschman, Shalom Zarach
Ho, John Ting-Sum
Holowka, David Allan
Hoory, Shlomo
Horowicz, Paul
Huang, Sylvia Lee
Hui, Sek Wen
Jacobsen, Chris J
Jelinski, Lynn W
Johnson, Roger A
Jung, Chan Yong
Kallenbach, Neville R
Kalogeropoulos, Theodore E
Kaplan, Ehud
Karlin, Arthur
Kass, Robert S
Keese, Charles Richard
Keng, Peter C
Khanna, Shyam Mohan
Kim, Carl Stephen
Kimmich, George Arthur
Kirchberger, Madeleine
Knight, Bruce Winton, (Jr)
Knowles, Richard James Robert
Knox, Robert Seiple
Koenig, Seymour Hillel
Koretz, Jane Faith
Koutcher, Jason Arthur
Kramer, Stephen Leonard
Kronman, Martin Jesse
Krugh, Thomas Richard
Kuki, Atsuo
Kumbar, Mahadevappa M
LaCelle, Paul (Louis)
Lange, Christopher Stephen
Laughlin, John Seth
Lawton, Richard Woodruff
Leibovic, K Nicholas
Leith, ArDean
Lerch, Irving A
Levandowsky, Michael
Levy, George Charles
Lichtman, Marshall A
Liebow, Charles
Lipson, Edward David
Lnenicka, Gregory Allen
Loew, Ellis Roger
Loretz, Christopher Alan
Loud, Alden Vickery
Low, Barbara Wharton
Lusty, Carol Jean
MacColl, Robert
McLaughlin, Stuart Graydon Arthur
Mahler, David S
Malamud, Herbert
Maniloff, Jack
Marcus, Michael Alan
Mauzerall, David Charles
Mehaffey, Leathem, III
Mehler, Ernest Louis
Meltzer, Richard Stuart
Mercer, Kermit R
Mercier, Gustavo Alberto, Jr
Metcalf, Harold
Moore, Richard Davis
Morken, Donald A
Myers, Anne Boone
Newman, Jay Edward
Newman, Stuart Alan
Noble, Robert Warren, Jr
Ockman, Nathan
Ohki, Shinpei
Ornstein, Leonard
Oster, Gerald
Paganelli, Charles Victor
Paine, Philip Lowell
Pardee, Joel David
Parker, Frank S
Parthasarathy, Rengachary
Peracchia, Camillo
Pilkis, Simon J
Pullman, Ira
Quigley, Gary Joseph
Rabbany, Sina Y
Ravitz, Leonard J, Jr
Rein, Robert
Reinstein, Lawrence Elliot
Rich, Marvin R
Rieder, Conly LeRoy
Robinson, Joseph Douglass
Roepe, Paul David
Ross, John Brandon Alexander
Rossi, Harald Herman
Rothenberg, Lawrence Neil
Rowlett, Roger Scott
Rubin, Byron Herbert
Rudy, Bernardo
Saba, Thomas Maron
Sachs, Frederick
Sachs, John Richard
Salerno, John Charles
Salvo, Joseph J
Sarma, Ramaswamy Harihara
Scheraga, Harold Abraham
Schick, Kenneth Leonard
Schiff, Joel D
Schlessinger, Joseph
Schmidt, John Thomas
Schneider, Allan Stanford
Scholes, Charles Patterson
Schwartz, Herbert Mark
Seeman, Nadrian Charles
Setlow, Jane Kellock
Setlow, Richard Burton
Shahn, Ezra
Shamos, Morris Herbert
Sharma, Minoti
Sherman, Fred
Shrager, Peter George
Siegel, Benjamin Morton
Siemann, Dietmar W
Simon, Martha Nichols
Skinner, Kathleen Mary
Smith, George David
Smith, Robert L
Snell, Fred Manget
Socolar, Sidney Joseph
Spangler, Robert Alan
Spanswick, Roger Morgan
Springer, Charles Sinclair, Jr
Starzak, Michael Edward
Stewart, Carleton C
Strekas, Thomas C
Studier, Frederick William
Subjeck, John Robert
Sullivan, W(alter) James
Sutherland, John Clark
Sweeney, William Victor
Tan, Yen T
Texter, John
Tobkes, Nancy J
Tooney, Nancy Marion
Toribara, Taft Yutaka
Turner, Douglas Hugh
Tycko, Daniel H
Uzgiris, Egidijus E
Vaidhyanathan, V S
Vroman, Leo
Wald, Alvin Stanley
Wall, Joseph S
Warden, Joseph Tallman
Webb, Watt Wetmore
Weber, David Alexander
Weeks, Charles Merritt
Weinstein, Harel
Wetmur, James Gerard
Windhager, Erich E
Winter, William Thomas
Wishnia, Arnold
Wobschall, Darold C
Xu, Zhenchun
Yalow, A(braham) Aaron
Yalow, Rosalyn Sussman
Yeagle, Philip L
Yeh, Noel Kuei-Eng
Yen, Andrew
Young, John Ding-E
Zablow, Leonard
Zadunaisky, Jose Atilio
Zaider, Marco A
Zeitz, Louis
Zelman, Allen
Zipfel, Warren Roger
Zobel, C(arl) Richard
Zwislocki, Jozef John

NORTH CAROLINA
Agris, Paul F
Allen, Nina Stromgren
Allis, John W
Anderson, Nels Carl, Jr
Bakerman, Seymour
Bear, Richard Scott
Bentzel, Carl Johan
Bereman, Robert Deane
Blackman, Carl F
Blum, Jacob Joseph
Burbeck, Christina Anderson
Burt, Charles Tyler
Carter, Charles Williams, Jr
Cleveland, Gregor George
Corless, Joseph Michael James
Coulter, Norman Arthur, Jr
Crawford-Brown, Douglas John
DeGuzman, Allan Francis
Dillard, Margaret Bleick
Edwards, Harold Henry
Elder, Joe Allen
Ellenson, James L
Erickson, Harold Paul
Faust, Robert Gilbert
Finn, Arthur Leonard
Fluke, Donald John
Guild, Walter Rufus
Hammes, Gordon G
Henkens, Robert William
Hochmuth, Robert Milo
Holzwarth, George Michael
Jacobson, Kenneth Allan
Johnson, G Allan
Juliano, Rudolph Lawrence
Kim, Ki-Hyon
Knopp, James A
Kremkau, Frederick William
Lee, Greta Marlene
Lentz, Barry R
Lieberman, Melvyn
London, Robert Elliot
Loomis, Carson Robert
McIntosh, Thomas James
McRee, Donald Ikerd
Mandel, Lazaro J
Nozaki, Yasuhiko
Padilla, George M
Pallotta, Barry S
Pearlstein, Robert David
Periasamy, Ammasi
Rabinowitz, James Robert
Roberts, Verne Louis
Roer, Robert David
Shaw, Barbara Ramsay
Shields, Howard William
Simon, Sidney Arthur
Smith, Charles Eugene
Spicer, Leonard Dale
Starmer, C Frank
Swan, Algernon Gordon
Tanford, Charles
Thompson, Nancy Lynn
Thubrikar, Mano J
Thurber, Robert Eugene
Wallen, Cynthia Anne
Wheeler, Kenneth Theodore, Jr
Williams, Myra Nicol
Witcofski, Richard Lou
Wolbarsht, Myron Lee
Wong, Fulton
Zucker, Robert Martin

NORTH DAKOTA
Ary, Thomas Edward
Moore, Vaughn Clayton

OHIO
Adragna, Norma C
Ayyangar, Komanduri M
Bahniuk, Eugene
Ball, William James, Jr
Bank, Harvey L
Becker, Carter Miles
Behnke, William David
Berliner, Lawrence J
Blackwell, John
Bobst, Albert M
Cassim, Joseph Yusuf Khan
Clark, Eloise Elizabeth
Fleischman, Darrell Eugene
Glenn, Loyd Lee
Grant, Roderick M, Jr
Gross, Elizabeth Louise
Halm, Dan Robert
Harpst, Jerry Adams
Heckman, Carol A
Hoff, Henry Frederick
Hollander, Philip B
Hopfer, Ulrich
Hopkins, Amos Lawrence
Houk, Thomas William
Ikeda-Saito, Masao
Jamieson, Alexander MacRae
Jentoft, Joyce Eileen
Johnson, Ronald Gene
Jones, Stephen Wallace
Kantor, George Joseph
Katz, J Lawrence
Kazarian, Leon Edward
Kmetec, Emil Philip
Kornacker, Karl
Kreishman, George Paul
LaManna, Joseph Charles
Lauf, Peter Kurt
Lipetz, Leo Elijah
MacIntyre, William James
McLean, Larry Raymond
Mateescu, Gheorghe D
Mattice, Wayne Lee
Messineo, Luigi
Moss, Thomas H
Nelson, Richard Carl
Paul, Richard Jerome
Pfeiffer, Douglas Robert
Rattan, Kuldip Singh
Rodgers, Michael A J
Rosenblatt, Charles Steven
Ross, Robert Talman
Rothstein, Jerome
Rudy, Yoram
Scarpa, Antonio
Schepler, Kenneth Lee
Schlosser, Herbert
Schroeder, Friedhelm
Shainoff, John Rieden
Silvidi, Anthony Alfred
Sperelakis, Nicholas
Starchman, Dale Edward
Stuehr, John Edward
Sundaralingam, Muttaiya
Swenson, Richard Paul
Tsai, Chun-Che
Von Gierke, Henning Edgar
Waterson, John R
Watson, Barry
Westerman, Philip William

OKLAHOMA
Cox, Andrew Chadwick
Harmon, H James
Hartman, Roger Duane
Hopkins, Thomas R (Tim)
Matsumoto, Hiroyuki
Meier, Charles Frederick, Jr
Reinhart, Gregory Duncan
Scott, Hugh Lawrence, Jr
Spivey, Howard Olin

OREGON
Anderson, Sonia R
Andresen, Michael Christian
Benolken, Robert Marshall
Broide, Michael Lynn
Burke, Michael John
Craig, Albert Morrison (Morrie)
Dahlquist, Frederick Willis
Ehrmantraut, Harry Charles
Evett, Jay Fredrick
Faber, Jan Job
Garlid, Keith David
Griffith, O Hayes
Haugland, Richard Paul
Hsu, Kwan
Huang, Zhijian
Johnson, Walter Curtis, Jr
Jost, Patricia Cowan
Lochner, Janis Elizabeth
Macpherson, Cullen H
MacSwan, Iain Christie

Mela-Riker, Leena Marja
Novick, Aaron
Rigas, Demetrios A
Rutten, Michael John
Schellman, John Anthony
Tobias, Cornelius Anthony
Toumadje, Arazdordi
Von Hippel, Peter Hans
Wagner, Orvin Edson
Woldegiorgis, Gebretateos

PENNSYLVANIA
Alteveer, Robert Jan George
Armstrong, Clay M
Benson, Brent W
Bering, Charles Lawrence
Blasie, J Kent
Boggs, Sallie Patton Slaughter
Bonner, Walter Daniel, Jr
Brinton, Charles Chester, Jr
Brown, Darrell Quentin
Brown, Truman Roscoe
Brunkard, Kathleen Marie
Burnett, Roger Macdonald
Byler, David Michael
Cameron, William Edward
Castor, LaRoy Northrop
Chance, Britton
Chapman, John Donald
Chay, Teresa R
Clark, T(helma) K
Cohn, Mildred
Conger, Alan Douglas
Constantinides, Panayiotis Pericleous
Cosgrove, Daniel Joseph
Deering, Reginald
Deutsch, Carol Joan
De Weer, Paul Joseph
Dodgson, Susanna Jane
Drees, John Allen
Driscoll, Dorothy H
Driska, Steven P
Englander, Sol Walter
Faber, Donald Stuart
Ferrone, Frank Anthony
Finegold, Leonard X
Fried, Michael Gregory
Fuchs, Franklin
Gogel, Germaine E
Graetzer, Reinhard
Graham, William Rendall
Harris, Lowell Dee
Hennessey, John P, Jr
Hickey, Richard James
Higgins, Joseph John
Ho, Chien
Holliday, Charles Walter
Horan, Paul Karl
Horn, Lyle William
Huang, Leaf
Jaffe, Eileen Karen
Jen-Jacobson, Linda
Jensen, Bruce David
Johnson, Ernest Walter
Johnson, Kenneth Allen
Joseph, Peter Maron
Kahn, Leo David
Karlson, Eskil Leannart
Khalil, Mohamed Thanaa
Kivlighn, Salah Dean
Kornfield, A(lfred) T(heodore)
Kozak, Wlodzimierz M
Kuo, Lawrence C
Kurland, Robert John
Lancaster, Jack R, Jr
Lankford, Edward B
LaRossa, Robert Alan
Lauffer, Max Augustus, Jr
Levine, Rhea Joy Cottler
Liebman, Paul Arno
Lindgren, Clark Allen
Llinas, Miguel
Lowenhaupt, Benjamin
McClure, William Robert
McConnell, Robert A
Matthews, Charles Robert
Mendelson, Emanuel Share
Misra, Dhirendra N
Mogus, Mary Ann
Moore, Dan Houston
Moskowitz, Gordon David
Murphy, Robert Francis
Nagle, John F
Noordergraaf, Abraham
Ochiai, Ei-Ichiro
Ohnishi, Tomoko
Ohnishi, Tsuyoshi
Owen, Charles Scott
Peachey, Lee DeBorde
Peebles, Craig Lewis
Pepe, Frank Albert
Person, Stanley R
Pessen, Helmut
Pfeffer, Philip Elliot
Pollock, John Archie
Ramachandran, N
Rapp, Paul Ernest
Rosenberg, Jerome Laib
Rosenbloom, Joel
Rottenberg, Hagai
Salama, Guy
Salganicoff, Leon
Sands, Jeffrey Alan
Scherer, Peter William

Schleyer, Heinz
Schwan, Herman Paul
Sciamanda, Robert Joseph
Selinsky, Barry Steven
Senft, Joseph Philip
Seybert, David Wayne
Sheers, William Sadler
Siegel, Richard C
Sittel, Karl
Smith, Douglas Lee
Snipes, Wallace Clayton
Stevens, Charles Le Roy
Strother, Greenville Kash
Taylor, D Lansing
Taylor, William Daniel
Tristram-Nagle, Stephanie Ann
Tu, Shu-i
Vanderkooi, Jane M
Wang, Allan Zu-Wu
Watt, Robert M
Weisel, John Winfield
Wetzel, Ronald Burnell
Wickstrom, Eric
Williamson, John Richard
Wolken, Jerome Jay
Wood, Thomas Hamil
Worthington, Charles Roy
Yamamoto, Nobuto
Yonetani, Takashi
Yushok, Wasley Donald

RHODE ISLAND
Andrews, Howard Lucius
Chapman, Kent M
Dowben, Robert Morris
Fawaz Estrup, Faiza
Fisher, Harold Wilbur
Hai, Chi-Ming
Marsh, Donald Jay
Polk, C(harles)
Stewart, Peter Arthur
Tourtellotte, Mark Eton

SOUTH CAROLINA
Larcom, Lyndon Lyle
Sanders, Samuel Marshall, Jr
Wheeler, Darrell Deane
Wise, William Curtis

SOUTH DAKOTA
Hurley, David John
Thompson, John Darrell

TENNESSEE
Adams, Paul Louis
Anjaneyulu, P S R
Auxier, John A
Bhattacharyya, Mohit Lal
Bunick, Gerard John
Chunduru, Srinivas Kishore
Churchich, Jorge E
Cunningham, Leon William
Darden, Edgar Bascomb
Dockter, Michael Edward
Dudney, Charles Sherman
Erickson, Jon Jay
Fleischer, Sidney
Frederiksen, Dixie Ward
Georghiou, Solon
Goulding, Charles Edwin, Jr
Greenbaum, Elias
Gutzke, William H N
Hingerty, Brian Edward
Incardona, Antonino L
Johnson, Carroll Kenneth
Macleod, Robert M
Morgan, Karl Ziegler
Mullaney, Paul F
Olins, Donald Edward
Randolph, Malcolm Logan
Senogles, Susan Elizabeth
Staros, James Vaughan
Steen, R Grant
Stone, William Lawrence
Stubbs, Gerald James
Thomason, Donald Brent
Tibbetts, Clark
Venable, John Heinz, Jr
Wikswo, John Peter, Jr
Williams, Robley Cook, Jr
Wondergem, Robert
Yen, Michael R T

TEXAS
Adams, Emory Temple, Jr
Alexander, William Carter
Allen, Robert Charles
Angelides, Kimon Jerry
Ansevin, Allen Thornburg
Atkinson, Mark Arthur Leonard
Baker, Robert David
Barnes, Charles M
Beall, Paula Thornton
Brady, Scott T
Brodwick, Malcolm Stephen
Brown, Arthur Morton
Butler, Bruce David
Czerwinski, Edmund William
Deisenhofer, Johann
Downey, H Fred
Eakin, Richard Timothy
Ellis, Kenneth Joseph
Espey, Lawrence Lee
Fishman, Harvey Morton

Garay, Andrew Steven
Gierasch, Lila Mary
Gilbert, Hiram Frazier
Ginsburg, Nathan
Glantz, Raymon M
Gray, Donald Melvin
Hackert, Marvin LeRoy
Hademenos, James George
Hazlewood, Carlton Frank
Hendrickson, Constance McRight
Herbert, Morley Allen
Hewitt, Roger R
Hogan, Michael Edward
Hopkins, Robert Charles
Huang, Huey Wen
Hudspeth, Albert James
Jagger, John
Jennings, Michael Leon
Jones, Carl E
Kalthoff, Klaus Otto
King, Richard Joe
Klein, David L
Krohmer, Jack Stewart
Lang, Dimitrij Adolf
Leuchtag, H Richard
Livingston, Linda
Lott, James Robert
McCammon, James Andrew
McCarter, Roger John Moore
Macfarlane, Ronald Duncan
McIntire, Larry V(ern)
McSherry, Diana Hartridge
Mahendroo, Prem P
Massey, John Boyd
Masters, Bettie Sue Siler
Meistrich, Marvin Lawrence
Meyn, Raymond Everett, Jr
Morrisett, Joel David
Murphy, Paul Henry
Nachtwey, David Stuart
Nakajima, Motowo
O'Benar, John DeMarion
Ong, Poen Sing
Parker, Cleofus Varren, Jr
Powell, Michael Robert
Reuss, Luis
Robberson, Donald Lewis
Robinson, Neal Clark
Rorschach, Harold Emil, Jr
Rupert, Claud Stanley
Scheie, Paul Olaf
Schrank, Auline Raymond
Serwer, Philip
Shalek, Robert James
Shaw, Robert Wayne
Skolnick, Malcolm Harris
Smith, Thomas Caldwell
Stull, James Travis
Taylor, Morris Chapman
Tilbury, Roy Sidney
Trkula, David
Tu, Shiao-chun
Waggener, Robert Glenn
Waldbillig, Ronald Charles
Walmsley, Judith Abrams
West, Bruce Joseph
Wills, Nancy Kay
Wilson, Samuel H
Wong, Brendan So
Wright, Ann Elizabeth
Yang, Chui-Hsu (Tracy)

UTAH
Ailion, David Charles
Billings, R Gail
Brown, Harold Mack, Jr
Burton, Frederick Glenn
Czerlinski, George Heinrich
Earley, Charles Willard
Gardner, Reed McArthur
Greenfield, Harvey Stanley
Hansen, Lee Duane
Lasater, Eric Martin
Lloyd, Ray Dix
Mohammad, Syed Fazal
Piette, Lawrence Hector
Powers, Linda Sue
Shearin, Nancy Louise
Sipe, David Michael
Spikes, John Daniel
Tuckett, Robert P
Wood, O Lew
Woodbury, John Walter
Yoshikami, Doju

VERMONT
Lipson, Richard L
Nyborg, Wesley LeMars
Otter, Timothy
Parsons, Rodney Lawrence
Reuben, John Philip
Sachs, Thomas Dudley
Spearing, Ann Marie
Tritton, Thomas Richard
Tyree, Melvin Thomas
Wallace, Susan Scholes
Webb, George Dayton

VIRGINIA
Aqualino, Alan
Averill, Bruce Alan
Biltonen, Rodney Lincoln
Bishop, Marilyn Frances
Brill, Arthur Sylvan

Bryant, Robert George
Chandross, Ronald Jay
Chlebowski, Jan F
Clarke, Alexander Mallory
Cleary, Stephen Francis
Cole, Patricia Ellen
Creutz, Carl Eugene
Cyr, W Howard
Dominey, Raymond Nelson
Eldridge, Charles A
Feher, Joseph John
Flinn, Jane Margaret
Ford, George Dudley
Fox, Kenneth Richard
Gabriel, Barbra L
Garrison, James C
Goldman, Israel David
Grogan, William McLean
Hackett, John Taylor
Ham, William Taylor, Jr
Hamilton, Thomas Charles
Hoy, Gilbert Richard
Hutchinson, Thomas Eugene
Johnson, Michael L
Keefe, William Edward
Kretsinger, Robert
Kutchai, Howard C
Liberti, Joseph Pollara
Looney, William Boyd
Maggio, Bruno
Martin, James Henry, III
Martin, Robert Bruce
Mikulecky, Donald C
Nielsen, Peter Tryon
Noble, Julian Victor
Ridgway, Ellis Branson
Ritter, Rogers C
Roper, L(eon) David
Rozzell, Thomas Clifton
Saltz, Joel Haskin
Schwab, Walter Edwin
Shukla, Kamal Kant
Somlyo, Avril Virginia
Sukow, Wayne William
Szabo, Gabor
Terner, James
Thompson, Thomas Edward
Tolles, Walter Edwin
Wickman, Herbert Hollis
Wingard, Christopher Jon
Wyeth, Newell Convers

WASHINGTON
Barlow, Clyde Howard
Bassingthwaighte, James B
Binder, Marc David
Blinks, John Rogers
Braby, Leslie Alan
Callis, James Bertram
Cellarius, Richard Andrew
Curtis, Stanley Bartlett
Detwiler, Peter Benton
Dunker, Alan Keith
Fisher, Darrell R
Glass, James Clifford
Gordon, Albert McCague
Herriott, Jon R
Hille, Bertil
Jacobson, Baruch S
Jensen, Lyle Howard
Johnson, John Richard
Kaufman, William Carl, Jr
Kehl, Theodore H
Miller, Douglas Lawrence
Nutter, Robert Leland
Palmer, Harvey Earl
Prothero, John W
Schurr, John Michael
Stern, Edward Abraham
Stirling, Charles E
Tam, Patrick Yui-Chiu
Tenforde, Thomas SeBastian
Towe, Arnold Lester
Windsor, Maurice William
Wootton, Peter
Zimbrick, John David

WEST VIRGINIA
Douglass, Kenneth Harmon
Franz, Gunter Norbert
Kartha, Mukund K
Wallace, William Edward, Jr

WISCONSIN
Anderegg, John William
Blattner, Frederick Russell
Cameron, John Roderick
DeLuca, Paul Michael, Jr
Dennis, Warren Howard
Fischbach, Fritz Albert
Greenebaum, Ben
Hendee, Wiliam R
Holden, James Edward
Hyde, James Stewart
Jackson, Meyer B
Kincaid, James Robert
Kroger, Larry A
Lai, Ching-San
Lelkes, Peter Istvan
Markley, John Lute
Martin, Thomas Fabian John
Miller, Joyce Fiddick
Nickles, Robert Jerome
Norman, John Matthew

Biophysics (cont)

Patrick, Michael Heath
Raines, Ronald T
Record, M Thomas, Jr
Reed, George Henry
Ritter, Mark Alfred
Romans-Hess, Alice Yvonne
Scheele, Robert Blain
Smith, Lloyd Michael
Wells, Barbara Duryea
Westler, William Milo
Worley, John David
Yin, Jun-Jie
Yu, Hyuk

WYOMING
Smith, William Kirby

PUERTO RICO
Copson, David Arthur
Orkand, Richard K

ALBERTA
Bazett-Jones, David Paul
Beck, James S
Challice, Cyril Eugene
Codding, Penelope Wixson
Kisman, Kenneth Edwin
Kumar, Shrawan
McElhaney, Ronald Nelson
McGann, Locksley Earl
Overfield, Robert Edward
Poznansky, Mark Joab
Roche, Rodney Sylvester
Smith, Richard Sidney
Stein, Richard Bernard
Sykes, Brian Douglas
Umezawa, Hiroomi
Van De Sande, Johan Hubert
Walsh, Michael Patrick
Weiner, Joel Hirsch

BRITISH COLUMBIA
Belton, Peter
Cushley, Robert John
Durand, Ralph Edward
Friedmann, Gerhart B
Gosline, John M
Green, Beverley R
Hoffer, J(oaquin) A(ndres)
Hoffmann, Geoffrey William
Lam, Gabriel Kit Ying
Mauk, Arthur Grant
Palaty, Vladimir
Rowlands, Stanley
Skarsgard, Lloyd Donald
Stephens-Newsham, Lloyd G
Withers, Stephen George
Wortis, Michael

MANITOBA
Barnard, John Wesley
Bihler, Ivan
Forrest, Bruce James
Froese, Gerd
Mantsch, Henry H
Polimeni, Philip Iniziato
Standing, Kenneth Graham

NEW BRUNSWICK
Edwards, Merrill Arthur
Sharp, Allan Roy

NOVA SCOTIA
Luner, Stephen Jay
McDonald, Terence Francis
O'Dor, Ronald Keith
Verpoorte, Jacob A
Wong, Alan Yau Kuen

ONTARIO
Ananthanarayanan, Vettaikkoru S
Baker, Robert G
Bayley, Stanley Thomas
Bear, Christine Eleanor
Butler, Keith Winston
Cairns, William Louis
Calleja, G(odehardo) B
Canham, Peter Bennett
Carey, Paul Richard
Cradduck, Trevor David
Cummins, W(illiam) Raymond
Cunningham, John (Robert)
Epand, Richard Mayer
Farrar, John Keith
Fenster, Aaron
Ferrier, Jack Moreland
Finlay, Joseph Bryan
Forer, Arthur H
Gardner, David R
Goldsack, Douglas Eugene
Greenblatt, Jack Fred
Greenstock, Clive Lewis
Grinstein, Sergio
Groom, Alan Clifford
Hallett, Frederick Ross
Hawton, Margaret H
Haynes, Robert Hall
Hinke, Joseph Anthony Michael
Hunt, John Wilfred
Jacobson, Stuart Lee
Jeffrey, Kenneth Robert
Johns, Harold E

Kennedy, James Cecil
Krull, Ulrich Jorg
Kruuv, Jack
Langille, Brian Lowell
Lepock, James Ronald
Leung, Philip Man Kit
Macdonald, Peter Moore
MacLennan, David Herman
Miller, Richard Graham
Millman, Barry Mackenzie
Morris, Catherine Elizabeth
Morton, Richard Alan
Nicholls, Peter
Noolandi, Jaan
Norwich, Kenneth Howard
Osborne, Richard Vincent
Ottensmeyer, Frank Peter
Paul, William
Pfalzner, Paul Michael
Raaphorst, G Peter
Racey, Thomas James
Rainbow, Andrew James
Rand, Richard Peter
Rauth, Andrew Michael
Rawlinson, John Alan
Roach, Margot Ruth
Rotenberg, A Daniel
Rothstein, Aser
Sherebrin, Marvin Harold
Siminovitch, Louis
Slater, Gary W
Smith, Ian Cormack Palmer
Song, Seh-Hoon
Stinson, Robert Henry
Szabo, Arthur Gustav
Tannock, Ian Frederick
Taylor, C P(atrick) S(tirling)
Till, James Edgar
Tong, Bok Yin
Trainor, Lynne E H

PRINCE EDWARD ISLAND
Beauregard, Marc Daniel

QUEBEC
Cohen, Montague
Dugas, Hermann
Frojmovic, Maurice Mony
Gjedde, Albert
Goldsmith, Harry L
Hallenbeck, Patrick Clark
Lamarche, François
Lehnert, Shirley Margaret
Lennox, Robert Bruce
Lin, Sheng-Xiang
Mackey, Michael Charles
Marceau, Normand Luc
Payet, Marcel Daniel
Pezolet, Michel
Podgorsak, Ervin B
Poole, Ronald John
Roberge, Fernand Adrien
Roy, Guy
Ruiz-Petrich, Elena
Schanne, Otto F
Schiller, Peter Wilhelm
Schreiner, Lawrence John
Seidah, Nabil George
Seufert, Wolf D
Sygusch, Jurgen
Van Calsteren, Marie-Rose

SASKATCHEWAN
Scheltgen, Elmer

OTHER COUNTRIES
Almers, Wolfhard
Araki, Masasuke
Azbel, Mark
Berns, Donald Sheldon
Brown, Larry Robert
Cerbon-Solorzano, Jorge
Chang, Donald Choy
Creighton, Thomas Edwin
Curtis, Adam Sebastian G
Deranleau, David A
De Troyer, Andre Jules
Engelborghs, Yves
Falk, Gertrude
Fujiwara, Keigi
Gontier, Jean Roger
Guerrero, Ariel Heriberto
Hall, Theodore (Alvin)
Hammerman, Ira Saul
Hayon, Elie M
Heinz, Erich
Hiyama, Tetsuo
Kan, Lou Sing
Kang, Sungzong
Kuehn, Lorne Allan
Kuhn, Klaus
Lewis, Aaron
Liao, Ta-Hsiu
Lieb, William Robert
Maas, Peter
Marikovsky, Yehuda
Mendelsohn, Mortimer Lester
Michel, Hartmut
Mittal, Balraj
Monos, Emil
Olson, John Melvin
Quintanilha, Alexandre Tiedtke
Rettori, Ovidio
Rinehart, Frank Palmer

Sargent, David Fisher
Scott, Mary Jean
Sellin, Lawrence C
Stoeber, Werner
Thorson, John Wells
Tipans, Igors O
Trentham, David R
Tsernoglou, Demetrius
Tsou, Chen-Lu
Van Driessche, Willy
Vartsky, David
Westerhof, Nicolaas
Whittembury, Guillermo
Wilkins, Maurice Hugh Frederick
Wu, Cheng-Wen
Wuthrich, Kurt

Biotechnology

ALABAMA
Daniell, Henry
Taylor, Kenneth Boivin

ARIZONA
Vermaas, Willem F J

ARKANSAS
Huang, Feng Hou

CALIFORNIA
Anderson, Geoffrey Robert
Ashkenazi, Avi
Balakrishnan, Krishna
Bechtol, Kathleen B
Bell, A(udra) Earl
Bennett, William Franklin
Bonner, James (Fredrick)
Brown, James Edward
Cardineau, Guy A
Cleland, Jeffrey Lynn
Danko, Stephen John
Etcheverry, Tina
Friedman, Milton Joe
Gaertner, Frank Herbert
Gilbertson, Robert Leonard
Goff, Lynda J
Hagler, Arnold T
Han, Jang Hyun
Henriksson, Thomas Martin
Hora, Maninder Singh
Kahn, Raymond Henry
Konopka, Ronald J
Lasky, Richard David
Lee, David Anson
Luh, Bor Shiun
Marcus, Frank
Martineau, Belinda M
Mitsuhashi, Masato
Nguyen, Binh Trong
Nguyen, Tue H
Ogunseitan, Oladele Abiola
Owicki, John Callaghan
Paech, Christian
Pan, Richard Y
Pitcher, Wayne Harold, Jr
Reilly, Dorothea Eleanor
Robertson, David W
Ryan, Wen
Sharma, Bhavender Paul
Shopes, Bob
Sliwkowski, Mary Burke
Sutherland, Robert Melvin
Sy, Jose
Trumble, John Thomas
Ullman, Edwin Fisher
Unver, Ercan
Whitlow, Marc David
Williams, Mark Alan
Winkler, Marjorie Everett
Zaidi, Iqbal Mehdi
Zamost, Bruce Lee

COLORADO
Lapitan, Nora L
Seely, Robert J
Todd, Paul Wilson
Vaseen, V(esper) Albert
Zeiler, Kathryn Gail

CONNECTICUT
Bayney, Richard Michael
Berg, Claire M
Brandsma, Janet Lousie
Das, Rathindra C
Fordham, Joseph Raymond
Geiger, Jon Ross
Merwin, June Rae

DELAWARE
Anton, David L
Busche, Robert M(arion)
Keeler, Calvin Lee, Jr
Myoda, Toshio Timothy
Seliskar, Denise Martha
Work, Dennis M

DISTRICT OF COLUMBIA
Boron, David J
Cole, Margaret Elizabeth
Forman, David S
Lichens-Park, Ann Elizabeth
Phillips, Terence Martyn

FLORIDA
Bolton, Wade E
Genthner, Fred J
Hiebert, Ernest
Horn, Joanne Marie
Katz, Albert Barry
Lee, Ted C K
Milne, Edward Lawrence
Muchovej, Rosa M C
Rangel Aldao, Rafael

GEORGIA
Baumstark, Barbara Ruth
Wessler, Susan R
Youngman, Philip John

HAWAII
Albert, Henrick Horst
Fitch, Maureen Meiko Masuda
Kathariou, Sophia
Nagai, Chifumi

ILLINOIS
Ausich, Rodney L
Bender, James Gregg
Bigelis, Ramunas
Burns, David John
Cote, Gregory Lee
Eisenhauer, Donald Alan
Glaser, Janet H
Gupta, Kailash Chandra
Hahn, Donald Richard
Hou, Ching-Tsang
Inglett, George Everett
Klegerman, Melvin Earl
Kurtzman, Cletus Paul
Miyazaki, John H
Padh, Harish
Prasad, Ramanujam Srinivasa
Rice, Thomas B
Schwartz, Robert David
Sternberg, Shmuel Mookie
Tiemeier, David Charles
Winkler, Martin Alan

INDIANA
Bennetzen, Jeffrey Lynn
Fayerman, Jeffrey T
Hurrell, John Gordon
Johnson, Alan L
Kulpa, Charles Frank, Jr
Lagu, Avinash L
Laskowski, Michael, Jr
Lin, Ho-Mu
McMullen, James Robert
Srivastava, Arun
Woodson, William Randolph

IOWA
Haefele, Douglas Monroe
Lamkey, Kendall Raye
Montgomery, Rex

KANSAS
Li, Yi
Liang, George H
Oppert, Brenda

KENTUCKY
Maiti, Indu B
Wang, Yi-Tin

LOUISIANA
Cotty, Peter John
Sizemore, Robert Carlen
Triplett, Barbara Ann
Ullah, Abul Hasnat

MAINE
Kepron, Michael Raymond

MARYLAND
Allnutt, F C Thomas
Balady, Michael A
Chang, Zhaohua
Dermody, William Christian
Differbach, Carl William
Hanjan, Satnam S
Harrell, Reginal M
Kasianowicz, John James
Keith, Jerry M
Ledney, George David
Lees, Andrew
Levin, Barbara Chernov
Lowke, George E
McCurley, Marian Frances
McGraw, Patricia Mary
Martin, Mark Thomas
Mischke, Barbara Suzanne
Poljak, Roberto Juan
Rao, Prasad Yarlagadda
Sagripanti, Jose-Luis
Sauer, Brian L
Schlom, Jeffrey
Shih, Thomas Y
Yang, Bingzhi

MASSACHUSETTS
Akkara, Joseph A
Bradley, Peter Michael
Cantor, Charles R
Chen, Wenyong
Fleming, Nigel
Fong, Godwin Wing-Kin

Gupta, Rajesh K
Hodge, Richard Paul
Liu, Jun S
Margolin, Alexey L
Raam, Shanthi
Read, Dorothy Louise
Rheinwald, James George
Roy, Sayon
Thomas, Aubrey Stephen, Jr
Toner, Mehmet
Torchilin, Vladimir Petrovich
Venkatesh, Yeldur Padmanabha
Weinberg, Crispin Bernard
Yang, Shiow-shong

MICHIGAN
Bhatnagar, Lakshmi
Fischer, Howard David
Hillegas, William Joseph
Janik, Borek
Niekamp, Carl William
Tsuji, Kiyoshi
Tuls, Jody Lynn Foy
Verrill, Harland Lester
Wang, Henry Y
Yang, Victor Chi-Min

MINNESOTA
Coleman, Patrick Louis
Sadowsky, Michael Jay
Sarkar, Gobinda
Skilling, Darroll Dean

MISSISSIPPI
Sadana, Ajit

MISSOURI
Ball, Timothy K
Bretton, Randolph Henry, Jr
Connolly, Daniel Thomas
Elsawi, Nehad
Gill, James Wallace
Johnson, Richard Dean
Krishnan, B Rajendra

MONTANA
Richardson, Charles

NEBRASKA
Brooks, Merle Eugene
Godfrey, Maurice
Kwang, Jimmy
Siragusa, Gregory Ross
Steffens, David Lee

NEW JERSEY
Ardelt, Wojciech Joseph
Barer, Sol Joseph
Cahn, Frederick
Coutinho, Claude Bernard
Cuffari, Gilbert Luke
Fujimore, Tohru
Glenn, Jeffrey K
Huang, Dao Pei
Ikeda, Tatsuya
Khan, Fazal R
Krueger, Roger Warren
Laskin, Allen I
Liao, Mei-June
Lomedico, Peter T
Lunn, Charles Albert
Mostillo, Ralph
Nalin, Carlo M
Noone, Thomas Mark
Patel, Ramesh N
Pierce, George Edward
Prince, Roger Charles
Raman, Jay Ananth
Silberklang, Melvin
Sipos, Tibor
Umland, Shelby Price
Worne, Howard E
Zilinskas, Barbara Ann

NEW MEXICO
Barton, Larry Lumir
Roubicek, Rudolf V

NEW YORK
Babich, Harvey
Elander, Richard Paul
Holmes, David Salway
Hui, Sek Wen
Jayme, David Woodward
Page, Gregory Vincent
Pawlowski, Philip John
Payne, Michael Ross
Prestwich, Glenn Downes
Rabino, Isaac
Reisch, Bruce Irving
Santiago, Noemi
Sulkowski, Eugene
Tsai, Gow-Jen
Warner, Garvin L
Watanabe, Myrna Edelman
Ye, Guangning

NORTH CAROLINA
Beck, Charles I
Bird, Kimon T
Boone, Lawrence Rudolph, III
Dibner, Mark Douglas
Doelling, Vivian Walter
Oliver, James David

Parikh, Indu
Shih, Jason Chia-Hsing
Sobsey, Mark David
Swaisgood, Harold Everett
Wise, Edmund Merriman, Jr

OHIO
Chapple, Paul James
Flanagan, Margaret Ann
Poggenburg, John Kenneth, Jr
Rozek, Charles Edward

OKLAHOMA
Cassidy, Brandt George
Dyer, David Wayne
Ford, Sharon R
Veltri, Robert William

OREGON
Doyle, Jack David
Kazerouni, Lewa
Seidler, Ramon John

PENNSYLVANIA
Alonso-Caplen, Firelli V
Brenchley, Jean Elnora
Brunson, Kenneth W
Burman, Sudhir
Freese, Andrew
Hennessey, John P, Jr
Herzog, David Paul
Kane, James Francis
Press, Linda Seghers
Tien, Ming
Young, Peter Ronald

SOUTH CAROLINA
Cantu, Eduardo S
Ning, John Tse Tso

TENNESSEE
Berven, Barry Allen
Brown, Gilbert Morris
Dees, Craig
Gresshoff, Peter Michael
Pfeffer, Lawrence Marc
Prokop, Ales
Stacey, Gary
Wang, Taylor Gunjin

TEXAS
Brown, Richard Malcolm, Jr
Fisher, Charles William
Guerrero, Felix David
Hall, Timothy Couzens
Hughes, Mark R
Karunakaran, Thonthi
Nelson, David Loren
Peffley, Ellen Beth
Powers, John Michael
Rege, Ajay Anand
Rogers, Suzanne M Dethier
Temeyer, Kevin Bruce

UTAH
Stokes, Barry Owen
Wang, Richard Ruey-Chyi

VIRGINIA
Haber, Lynne Tracey
Jones, Daniel David
Lubiniecki, Anthony Stanley
Srivastava, Kailash Chandra
Veilleux, Richard Ernest
Williams, Luther Steward
Wolfe, Barbara Ann

WASHINGTON
Calza, Roger Ernest
Fitch, Cynthia Lynn
Gray, Patrick William
Hall, Benjamin Downs
Hauschka, Stephen D
Lee, James Moon
Raff, Howard V
Shepherd, Linda Jean
Taylor, Dean Perron
Thompson, Elizabeth Alison

WISCONSIN
Brill, Winston J
Burgess, Richard Ray
Dahlberg, James E(ric)
Lorton, Steven Paul

PUERTO RICO
Candelas, Graciela C
Schroder, Eduardo C
Toranzos, Gary Antonio

ALBERTA
Gaucher, George Maurice
Madiyalakan, Ragupathy
Rosenberg, Susan Mary

BRITISH COLUMBIA
Eaves, Allen Charles Edward
Von Aderkas, Patrick Jurgen Cecil

MANITOBA
McVetty, Peter Barclay Edgar

ONTARIO
Appanna, Vasu Dhananda

Calleja, G(odehardo) B
Dube, Ian David
Ghosh, Hara Prasad
Grunder, Allan Angus
Haj-Ahmad, Yousef
Huber, Carol (Saunderson)
Kosaric, Naim
Legge, Raymond Louis
Patel, Girishchandra Babubhai
Sahota, Ajit Singh
Winter, Peter

PRINCE EDWARD ISLAND
Beauregard, Marc Daniel

QUEBEC
Archibald, Frederick S
Cormier, Francois
Germain, Lucie
Goldstein, Sandu M
Slilaty, Steve N

SASKATCHEWAN
Kurz, Wolfgang Gebhard Walter
Nelson, Louise Mary
Xiang, Jim

OTHER COUNTRIES
Chang, Yu-Sun
Hoisington, David A
Wuthrich, Kurt

Botany-Phytopathology

ALABAMA
Backman, Paul Anthony
Bhatnagar, Yogendra Mohan
Bowen, William R
Brewer, Jesse Wayne
Clark, Edward Maurice
Cochis, Thomas
Curl, Elroy Arvel
Darden, William H, Jr
Deason, Temd R
Diener, Urban Lowell
Estes, Edna E
Eyster, Henry Clyde
Freeman, John A
Freeman, John Daniel
Gough, Francis Jacob
Hagler, Thomas Benjamin
Haynes, Robert Ralph
Hocking, George Macdonald
Huntley, Jimmy Charles
Jacobsen, Barry James
Jones, Daniel David
Jones, Jeanette
Latham, Archie J
Lelong, Michel Georges
Lyle, James Albert
McGuire, Robert Frank
Marshall, Norton Little
Moser, Stephen Adcock
Olson, Richard Louis
Peterson, Curtis Morris
Rosene, Walter, Jr
Sharma, Govind C
Shepherd, Raymond Lee
Steward, Frederick Campion
Truelove, Bryan
Wilson, Eugene M

ALASKA
Juday, Glenn Patrick
Laursen, Gary A
Logsdon, Charles Eldon
Mitchell, William Warren
Sagers, Cynthia Louise

ARIZONA
Ahlgren, Isabel Fulton
Alcorn, Stanley Marcus
Anderson, Edward Frederick
Aronson, Jerome Melville
Baker, Gladys Elizabeth
Blinn, Dean Ward
Bloss, Homer Earl
Booth, John Austin
Caldwell, Roger Lee
Canright, James Edward
Clark, William Dennis
Crosswhite, Carol D
Crosswhite, Frank Samuel
Davis, Owen Kent
Gilbertson, Robert Lee
Gries, George Alexander
Haase, Edward Francis
Hall, Gustav Wesley
Hine, Richard Bates
Hoshaw, Robert William
Huff, Albert Keith
Hull, Herbert Mitchell
Jackson, Ernest Baker
Jackson, Stephen Thomas
Johnson, Raymond Roy
Kernkamp, Milton F
Klopatek, Jeffrey Matthew
Landrum, Leslie Roger
Leatherman, Anna D
Leathers, Chester Ray
McCleary, James A
Mason, Charles Thomas, Jr
Mogensen, Hans Lloyd

Nelson, Merritt Richard
Nigh, Edward Leroy, Jr
Pinkava, Donald John
Reeder, John Raymond
Rominger, James McDonald
Russell, Thomas Edward
Schoenwetter, James
Sommerfeld, Milton R
Stanghellini, Michael Eugene
Timmermann, Barbara Nawalany
Van Etten, Hans D
Verbeke, Judith Ann
Wagle, Robert Fay

ARKANSAS
Adams, Randall Henry
Dale, James Lowell
Evans, Raymond David
Goode, Monroe Jack
Hardin, Hilliard Frances
Harris, William M
Hutchison, James A
Jones, John Paul
Lim, Sung Man
Logan, Lowell Alvin
Meyer, Richard Lee
Riggs, Robert D
Slack, Derald Allen
Smith, Edwin Burnell
Steinkraus, Donald Curtiss
Sutherland, John Bruce, IV
Te Beest, David Orien
Templeton, George Earl
Timmermann, Dan, Jr
Tucker, Gary Edward
Walters, Hubert Jack

CALIFORNIA
Adaskaveg, James Elliott
Aly, Raza
Anderson, Dennis Elmo
Anthony, Margery Stuart
Anzalone, Louis, Jr
Armbruster, Barbara L
Ashworth, Lee Jackson, Jr
Baad, Michael Francis
Baalman, Robert J
Baker, Herbert George
Baldwin, James Gordon
Barbour, Michael G
Bega, Robert V
Benjamin, Richard Keith
Benseler, Rolf Wilhelm
Bianchi, Donald Ernest
Bizzoco, Richard Lawremce Weiss
Bowerman, Earl Harry
Brown, Richard McPike
Bruening, George
Buddenhagen, Ivan William
Burleigh, James Reynolds
Butler, Edward Eugene
Butler, John Earl
Calavan, Edmond Clair
Calpouzos, Lucas
Cande, W Zacheus
Capon, Brian
Carlquist, Sherwin
Chase, Ann Renee
Cheadle, Vernon Irvin
Christianson, Michael Lee
Collins, O'Neil Ray
Constance, Lincoln
Cooke, Ron Charles
Cox, Hiden Toy
Cunningham, Virgil Dwayne
Cutler, Hugh Carson
Danko, Stephen John
D'Antoni, Hector Luis
DeMason, Darleen Audrey
Dempster, Lauramay Tinsley
Derr, William Frederick
Desjardins, Paul Roy
DeVay, James Edson
Dimitman, Jerome Eugene
Dixon, Peter Stanley
Dodds, James Allan
Doyle, William T
Duffus, James Edward
Duncan, Thomas Osler
Duniway, John Mason
Duran, Ruben
Ebert, Wesley W
Eckard, Kathleen
Eckert, Joseph Webster
Ediger, Robert I
Einset, John William
Emboden, William Allen, Jr
Endo, Robert Minoru
English, William Harley
Epstein, Lynn
Erspamer, Jack Laverne
Esau, Katherine
Falk, Richard H
Ford, Donald Hoskins
Furumoto, Warren Akira
Game, John Charles
Garber, Richard Hammerle
Gardner, Wayne Scott
Gasser, Charles Scott
Gerik, James Stephen
Gifford, Ernest Milton
Gill, Harmohindar Singh
Goff, Lynda J
Goodman, Victor Herke

Botany-Phytopathology (cont)

Green, Norman Edward
Green, Paul Barnett
Grieve, Catherine Macy
Grillos, Steve John
Grogan, Raymond Gerald
Gumpf, David John
Hancock, Joseph Griscom, Jr
Harrington, Dalton
Harris, Hubert Andrew
Harvey, John Marshall
Heckard, Lawrence Ray
Henrickson, James Solberg
Hildebrand, Donald Clair
Hildreth, Robert Claire
Hills, F Jackson
Hirsch, Ann Mary
Hoefert, Lynn Lucretia
Howard, Dexter Herbert
Ingram, John (William), Jr
Isom, William Howard
Jones, Claris Eugene, Jr
Kado, Clarence Isao
Kantz, Paul Thomas, Jr
Kaplan, Donald Robert
Karle, Harry P
Kasapligil, Baki
Kavaljian, Lee Gregory
Keen, Noel Thomas
Keil, David John
Kendrick, James Blair, Jr
Kinloch, Bohun Baker, Jr
Kjeldsen, Chris Kelvin
Kliejunas, John Thomas
Koonce, Andrea Lavender
Kowalski, Donald T
Kurtzman, Ralph Harold, Jr
Kyhos, Donald William
Laemmlen, Franklin
Laetsch, Watson McMillan
Lang, Norma Jean
Largent, David Lee
Lathrop, Earl Wesley
Laude, Horton Meyer
Lear, Bert
Lewin, Ralph Arnold
Lewis, Frank Harlan
Lindsay, George Edmund
Long, Sharon Rugel
Lord, Elizabeth Mary
Lucas, William John
Lukens, Raymond James
McCain, Arthur Hamilton
McClintock, Elizabeth
MacDonald, James Douglas
McHale, John T
McNeal, Dale William, Jr
Madore, Monica Agnes
Mallory, Thomas E
Mathias, Mildred Esther
Menge, John Arthur
Meredith, Farris Ray
Mircetich, Srecko M
Mitchell, Norman L
Mooring, John Stuart
Moran, Reid (Venable)
Mortenson, Theodore Hampton
Moseley, Maynard Fowle
Muller, Cornelius Herman
Munnecke, Donald Edwin
Murphy, Terence Martin
Muth, Gilbert Jerome
Neher, Robert Trostle
Nelson, Klayton Edward
Nelson, Richard D(ouglas)
Neushul, Michael, Jr
Nichols, Carl William
Norris, Daniel Howard
Nyland, George
O'Neill, Thomas Brendan
Ornduff, Robert
Panopoulos, Nickolas John
Parker, Virgil Thomas
Parmeter, John Richard, Jr
Parnell, Dennis Richard
Paulus, Albert
Philbrick, Ralph
Phillips, Douglas J
Pitelka, Louis Frank
Pitts, Wanna Dene
Platt, Kathryn Ann
Pray, Thomas Richard
Prezelin, Barbara Berntsen
Price, Mary Vaughan
Purcell, Alexander Holmes, III
Quail, Peter Hugh
Quibell, Charles Fox
Raabe, Robert Donald
Rader, William Ernest
Radewald, John Dale
Raju, Namboori Bhaskara
Ramsey, Richard Harold
Rappaport, Lawrence
Rasmussen, Robert A
Reeve, Marian Enzler
Reynolds, Don Rupert
Rosen, Howard
Rosenberg, Dan Yale
Rudd, Velva Elaine
Sakai, Ann K
Saltveit, Mikal Endre, Jr
Scharpf, Robert F
Schnathorst, William Charles
Schneider, Henry
Schroth, Milton Neil
Schwegmann, Jack Carl
Scora, Rainer Walter
Shapiro, Arthur Maurice
Sherman, Robert James
Sidhu, Gurmel Singh
Silk, Margaret Wendy Kuhn
Silva, Paul Claude
Simpson, Robert Blake
Singh, Jaswant
Smith, Alan Reid
Smith, James Payne, Jr
Sparks, Steven Richard
Sparling, Shirley
Spurr, Arthur Richard
Stebbins, George Ledyard
Sterling, Clarence
Stern, Kingsley Rowland
Stoner, Martin Franklin
Sung, Zinmay Renee
Sussex, Ian Mitchell
Swatek, Frank Edward
Tavares, Isabelle Irene
Thaw, Richard Franklin
Thiers, Harry Delbert
Thomas, John Hunter
Thomas, Walter Dill, Jr
Thomson, William Walter
Tiffney, Bruce Haynes
Tisserat, Brent Howard
Travis, Robert LeRoy
Tucker, John Maurice
Uyemoto, Jerry Kazumitsu
Van Gundy, Seymour Dean
Vasek, Frank Charles
Vinters, Harry Valdis
Vinyard, William Corwin
Vogl, Richard J
Vreeland, Valerie Jane
Walch, Henry Andrew, Jr
Walker, Dan B
Walker, Dennis Kendon
Walkington, David L
Wallace, Gary Dean
Waser, Nickolas Merritt
Watterson, Jon Craig
Weathers, Lewis Glen
Webster, Grady Linder
Webster, Robert K
Weiler, John Henry, Jr
Weinhold, Albert Raymond
Weis, Arthur Edward
Wells, Kenneth
Whaley, Julian Wendell
Whitehead, Marvin Delbert
Whitney, Kenneth Dean
Wiedman, Harold W
Wilson, Kenneth Allen
Worker, George F, Jr
Yoder, David Lee
Yu, Grace Wei-Chi Hu
Zavala, Maria Elena
Zavarin, Eugene
Zedler, Paul H(ugo)
Zentmyer, George Aubrey, Jr
Zscheile, Frederick Paul, Jr

COLORADO
Altman, Jack
Anderson, Roger Arthur
Bailey, Dana Kavanagh
Baker, R Ralph
Bauer, Penelope Hanchey
Bock, Jane Haskett
Bragonier, Wendell Hughell
Davidson, Darwin Ervin
Dunahay, Terri Goodman
Ferchau, Hugo Alfred
Feucht, James Roger
Grant, Michael Clarence
Hansmann, Eugene William
Harrison, Monty DeVerl
Hartman, Emily Lou
Hess, Dexter Winfield
Hinds, Thomas Edward
Hogan, Christopher James
King, Charles C
Kreutzer, William Alexander
Larsen, Arnold Lewis
Lindsey, Julia Page
Livingston, Clark Holcomb
Longpre, Edwin Keith
McIntyre, Gary A
Martin, Susan Scott
Mitchell, John Edwards
Nisbet, Jerry J
Ranker, Tom A
Reed, John Frederick
Reeves, Fontaine Brent, Jr
Ruppel, Earl George
Shaw, Robert Blaine
Shushan, Sam
Siemer, Eugene Glen
Steinkamp, Myrna Pratt
Weber, William Alfred
Wilken, Dieter H
Wingate, Frederick Huston
Zeiler, Kathryn Gail

CONNECTICUT
Anagnostakis, Sandra Lee
Anderson, Gregory Joseph
Berlyn, Graeme Pierce
Bristol, Melvin Lee
Carluccio, Leeds Mario
Cooke, John Cooper
Duffy, Regina Maurice
Galston, Arthur William
Goodwin, Richard Hale
Gordon, Philip N
Greenblatt, Irwin M
Hickey, Leo Joseph
Horsfall, James Gordon
Huang, Liang Hsiung
Koning, Ross E
Les, Donald Henry
Lier, Frank George
McClymont, John Wilbur
Miller, Russell Bensley
Rich, Saul
Schneider, Craig William
Smith, William Hulse
Stevenson, Harlan Quinn
Taylor, Gordon Stevens
Tomlinson, Harley
Trainor, Francis Rice
Wallner, William E
Walton, Gerald Steven
Webster, Terry R
Wulff, Barry Lee

DELAWARE
Bozarth, Gene Allen
Crossan, Donald Franklin
Delp, Charles Joseph
Dill, Norman Hudson
Gallagher, John Leslie
Hadley, Bruce Alan
Hodson, Robert Cleaves
Howard, Richard James
Lichtner, Francis Thomas, Jr
Matlack, Albert Shelton
Mitchell, William H
Morehart, Allen L
Pizzolato, Thompson Demetrio
Seliskar, Denise Martha
Smith, Constance Meta

DISTRICT OF COLUMBIA
Barnes, John Maurice
Bellmer, Elizabeth Henry
DeFilippis, Robert Anthony
DiMichele, William Anthony
Donaldson, Robert Paul
Dudley, Theodore R
Elias, Thomas S
Frederick, Lafayette
Goldberg, Aaron
Hines, Pamela Jean
Horton, Maurice Lee
Krigsvold, Dale Thomas
Lellinger, David Bruce
Littler, Diane Sullivan
Littler, Mark Masterton
Meyer, Frederick Gustav
Mislivec, Philip Brian
Norris, James Newcome, IV
Robinson, Harold Ernest
Rodman, James Eric
Skog, Laurence Edgar
Smith, Richard S, Jr
Sze, Philip
Wasshausen, Dieter Carl
Wurdack, John J
Zimmer, Elizabeth Anne

FLORIDA
Agrios, George Nicholas
Aldrich, Henry Carl
Anderson, Loran C
Austin, Daniel Frank
Bartz, Jerry A
Beneke, Everett Smith
Berger, Richard Donald
Blazquez Y Servin, Carlos Humberto
Burch, Derek George
Childs, James Fielding Lewis
Cline, Kenneth Charles
Comstock, Jack Charles
Crall, James Monroe
Darby, John Feaster
Davis, Michael Jay
Dawes, Clinton John
Dean, Jack Lemuel
DeMort, Carole Lyle
Dickson, Donald Ward
Dodd, John Durrance
DuCharme, Ernest Peter
Eilers, Frederick Irving
Ellison, Marlon L
Engelhard, Arthur William
Erdos, Gregory William
Essig, Frederick Burt
Ewart, R Bradley
Fisher, Jack Bernard
Fleming, Theodore Harris
Foster, Virginia
Freeman, Thomas Edward
Friedmann, E(merich) Imre
Garnsey, Stephen Michael
Gilbert, Margaret Lois
Gilmer, Robert McCullough
Gottwald, Timothy R
Gray, Dennis John
Griffin, Dana Gove, III
Grimm, Robert Blair
Hartman, James Xavier
Hiebert, Ernest
Hopkins, Donald Lee
Janos, David Paul
Judd, Walter Stephen
Kimbrough, James W
Koch, William Julian
Koevenig, James L
Kretschmer, Albert Emil, Jr
Kucharek, Thomas Albert
Lambe, Robert Carl
Langdon, Kenneth R
Lee, David Webster
Lindberg, George Donald
Litz, Richard Earle
Luke, Herbert Hodges
McMillan, Robert Thomas, Jr
Magie, Robert Ogden
Marvel, Mason E
Meister, Charles William
Miller, Kim Irving
Miskimen, Carmen Rivera
Mullin, Robert Spencer
Mullins, John Thomas
Mustard, Margaret Jean
Nemec, Stanley
Niblett, Charles Leslie
Norman, Eliane Meyer
Norris, Dean Rayburn
Ocampo-Friedmann, Roseli C
Payne, Willard William
Phlips, Edward Julien
Pring, Daryl Roger
Purdy, Laurence Henry
Ranzoni, Francis Verne
Romeo, John Thomas
Roubik, David Ward
Rumbach, William Ervin
Schenck, Norman Carl
Schoeneweiss, Donald F
Shanor, Leland
Shepard, James F
Simons, John Norton
Smith, Alan Paul
Smith, Harlan Eugene
Sonoda, Ronald Masahiro
Stall, Robert Eugene
Stern, William Louis
Stone, Margaret Hodgman
Stone, William Jack Hanson
Sweet, Herman Royden
Taylor, Walter Kingsley
Thayer, Paul Loyd
Thompson, Neal Philip
Thorhaug, Anitra L
Tryon, Rolla Milton, Jr
Vasil, Indra Kumar
Vasil, Vimla
Volin, Raymond Bradford
Ward, Daniel Bertram
Weingartner, David Peter
Werner, Patricia Ann Snyder
Whiteside, Jack Oliver
Whittier, Henry O
Williams, Norris Hagan
Williams, Tom Vare
Zettler, Francis William

GEORGIA
Ahearn, Donald G
Andrews, Lucy Gordon
Bacon, Charles Wilson
Berry, Charles Richard
Boole, John Allen, Jr
Bozeman, John Russell
Brown, Ronald Harold
Burbanck, Madeline Palmer
Carter, Eloise B
Carter, James Richard
Christenberry, George Andrew
Darvill, Alan G
Drapalik, Donald Joseph
Faircloth, Wayne Reynolds
Fuller, Melvin Stuart
Gillaspie, Athey Graves, Jr
Gitaitis, Ronald David
Hancock, Kenneth Farrell
Hanlin, Richard Thomas
Hurd, Maggie Patricianne
Hussey, Richard Sommers
James, Charles William
Johnson, Alva William
Jones, Samuel B, Jr
Kaufman, Leo
Kochert, Gary Dean
Kuhlman, Elmer George
Kuhn, Cedric W
LeNoir, William Cannon, Jr
McCarter, States Marion
Mims, Charles Wayne
Minton, Norman A
Murdy, William Henry
Pendergrass, Levester
Porter, David
Powell, William Morton
Powers, Harry Robert, Jr
Ragsdale, Harvey Larimore
Rich, Fredrick James
Robinson, Margaret Chisolm
Romanovicz, Dwight Keith
Roncadori, Ronald Wayne
Ruehle, John Leonard
Schadler, Daniel Leo
Snyder, Hugh Donald
Sommer, Harry Edward

Stouffer, Richard Franklin
Tinga, Jacob Hinnes
Wells, Homer Douglas
Wilkinson, Robert E
Wilson, Jeffrey Paul
Wyatt, Robert Edward
Wynn, Willard Kendall, Jr
Zimmer, David E

HAWAII
Abbott, Isabella Aiona
Alvarez, Anne Maino
Apt, Walter James
Aragaki, Minoru
Carr, Gerald Dwayne
Doty, Maxwell Stanford
Hemmes, Don E
Hodgson, Lynn Morrison
Holtzmann, Oliver Vincent
Keeley, Sterling Carter
Krauss, Beatrice Hilmer
Lamoureux, Charles Harrington
Ooka, Jeri Jean
Patil, Suresh Siddheshwar
Rohrbach, Kenneth G
St John, Harold
Sakai, William Shigeru
Smith, Albert Charles
Sun, Samuel Sai-Ming
Theobald, William L
Trujillo, Eduardo E

IDAHO
Anderegg, Doyle Edward
Ayers, Arthur Raymond
Clary, Warren Powell
Davidson, Christopher
Davis, James Robert
Douglas, Dorothy Ann
Fenwick, Harry
Finley, Arthur Marion
Guthrie, James Warren
Helton, Audus Winzle
Holte, Karl E
LeTourneau, Duane John
McDonald, Geral Irving
Maloy, Otis Cleo, Jr
Morris, John Leonard
Moser, Paul E
Naskali, Richard John
Ohms, Richard Earl
Osgood, Charles Edgar
Packard, Patricia Lois
Partridge, Arthur Dean
Romanko, Richard Robert
Simpson, William Roy
Sojka, Robert E
Tylutki, Edmund Eugene
Wiese, Maurice Victor

ILLINOIS
Anderson, Roger Clark
Armstrong, Joseph Everett
Bailey, Zeno Earl
Becker, Steven Allan
Bouck, G Benjamin
Bozzola, John Joseph
Cain, Jerome Richard
Carothers, Zane Bland
Chamberlain, Donald William
Chappell, Dorothy Field
Charlesworth, Deborah
Chuang, Tsan Iang
Conner, Jeffrey Keating
Crandall-Stotler, Barbara Jean
Crane, Joseph Leland
Dziadyk, Bohdan
Edwards, Dale Ivan
Engel, John Jay
Ford, Richard Earl
Foster, Robin Bradford
Gardner, Harold Wayne
Gassman, Merrill Loren
Gessner, Robert V
Glassman, Sidney Frederick
Gray, Lewis Richard
Green, Thomas L
Grosklags, James Henry
Hadley, Elmer Burton
Hanzely, Laszlo
Hartstirn, Walter
Henry, Robert David
Hesseltine, Clifford William
Himelick, Eugene Bryson
Hinchman, Ray Richard
Hoffman, Larry Ronald
Hohn, Matthew Henry
Hooker, Arthur Lee
Huft, Michael John
Jones, Almut Gitter
Jump, John Austin
Kauffman, Harold
Keating, Richard Clark
Kessler, Kenneth J, Jr
Levin, Geoffrey Arthur
Liberta, Anthony E
Lippincott, Barbara Barnes
Lippincott, James Andrew
Lorenz, Ralph William
McPheeters, Kenneth Dale
Malek, Richard Barry
Marsh, Terrence George
Matten, Lawrence Charles
Middleton, Beth Ann
Miller, Charles Edward
Miller, Raymond Michael
Mohlenbrock, Robert H, Jr
Monoson, Herbert L
Murakishi, Harry Haruo
Murray, Mary Aileen
Myers, Ronald Berl
Nevling, Lorin Ives, Jr
O'Flaherty, Larrance Michael Arthur
Pappelis, Aristotel John
Paxton, Jack Dunmire
Perino, Janice Vinyard
Perry, Eugene Arthur
Richard, John L
Ries, Stephen Michael
Robertson, Kenneth Ray
Rouffa, Albert Stanley
Ruddat, Manfred
Schmid, Walter Egid
Scott, William Wallace
Seigler, David Stanley
Shurtleff, Malcolm C, Jr
Sinclair, James Burton
Singer, Rolf
Singh, Dilbagh
Sorensen, Paul Davidsen
Spencer, Jack T
Stein-Taylor, Janet Ruth
Stidd, Benton Maurice
Stieber, Michael Thomas
Stotler, Raymond Eugene
Sundberg, Walter James
Taylor, Roy Lewis
Thornberry, Halbert Houston
Titman, Paul Wilson
Toliver, Michael Edward
Troll, Ralph
Tuveson, Robert Williams
Ugent, Donald
Whelan, Christopher John
White, Donald Glenn
Whiteside, Wesley C
Wicklow, Donald Thomas
Wilcox, Wesley Crain
Yos, David Albert
Zimmerman, Craig Arthur

INDIANA
Abney, Thomas Scott
Adams, Preston
Bath, James Edmond
Bennetzen, Jeffrey Lynn
Bhattacharya, Pradeep Kumar
Bishop, Charles Franklin
Bracker, Charles E
Brenneman, James Alden
Brooks, Austin Edward
Bucholtz, Dennis Lee
Coolbaugh, Ronald Charles
Crouch, Martha Louise
Davies, Harold William, (Jr)
Dunkle, Larry D
Eiser, Arthur L
Emerson, Frank Henry
Fitzgerald, Paul Jackson
Foos, Kenneth Michael
Gastony, Gerald Joseph
Goonewardene, Hilary Felix
Green, Ralph J, Jr
Grove, Stanley Neal
Hammond, Charles Thomas
Heiser, Charles Bixler, Jr
Hennen, Joe Fleetwood
Huber, Don Morgan
Janutolo, Delano Blake
Jensen, Richard Jorg
Kenaga, Clare Burton
Kidd, Frank Alan
Lister, Richard Malcolm
Low, Philip Stewart
McGrath, James J
Mahlberg, Paul Gordon
Marshall, James John
Merrill, Gary Lane Smith
Nicholson, Ralph Lester
Orpurt, Philip Arvid
Pecknold, Paul Carson
Postlethwait, Samuel Noel
Roth, Jonathan Nicholas
Rothrock, Paul E
Savage, Earl John
Schoknecht, Jean Donze
Scott, Donald Howard
Shaner, Gregory Ellis
Simmons, Emory Guy
Smith, Dale Metz
Stevenson, Forrest Frederick
Susalla, Anne A
Tansey, Michael Richard
Tseng, Charles C
Watson, Maxine Amanda
Webb, Mary Alice
Whitehead, Donald Reed
Williams, Edwin Bruce
Wilson, Kathryn Jay
Wilson, Kenneth Sheridan
Winternheimer, P Louis
Yates, Willard F, Jr
Youse, Howard Ray

IOWA
Borgman, Robert P
Cazin, John, Jr
Christiansen, Paul Arthur
Czarnecki, David Bruce
Dunleavy, John M
Durkee, LaVerne H
Durkee, Lenore T
Eilers, Lawrence John
Embree, Robert William
Ford, Clark Fugier
Graham, Benjamin Franklin
Haefele, Douglas Monroe
Hodges, Clinton Frederick
Horner, Harry Theodore
Huffman, Donald Marion
Isely, Duane
Knaphus, George
Knutson, Roger M
Lersten, Nels R
Main, Stephen Paul
Martinson, Charlie Anton
Orr, Alan R
Pohl, Richard Walter
Poulter, Dolores Irma
Schnable, Patrick S
Simons, Marr Dixon
Sjölund, Richard David
Tachibana, Hideo
Tiffany, Lois Hattery
Vakili, Nader Gholi
Walker, Waldo Sylvester
Whitson, Paul David
Widrlechner, Mark Peter
Wilkinson, Daniel R

KANSAS
Albrecht, Mary Lewnes
Barkley, Theodore Mitchell
Bechtel, Donald Bruce
Browder, Lewis Eugene
Dawson, James Thomas
Edmunds, Leon K
Haufler, Christopher Hardin
Hetrick, Barbara Ann
Ikenberry, Gilford John, Jr
Keeling, Richard Paire
Kramer, Charles Lawrence
Lane, Meredith Anne
Leisman, Gilbert Arthur
Lichtwardt, Robert William
McGregor, Ronald Leighton
Martin, Terry Joe
Peterson, John Edward, Jr
Piper, Jon Kingsbury
Sauer, David Bruce
Schwenk, Fred Walter
Sperry, Theodore Melrose
Stuteville, Donald Lee
Thomasson, Joseph R
Tomanek, Gerald Wayne
Torres, Andrew M
Wells, Philip Vincent
Youngman, Arthur L

KENTUCKY
Bryant, William Stanley
Clark, Jimmy Dorral
Davis, William S
Diachun, Stephen
Dillard, Gary Eugene
Eversmeyer, Harold Edwin
Fuller, Marian Jane
Harris, Denny Olan
Herron, James Watt
Jenkins, Jeff Harlin
Keefe, Thomas Leeven
King, Joe Mack
Myers, Roy Maurice
Nicely, Kenneth Aubrey
Niffenegger, Daniel Arvid
Perlin, Michael Howard
Pirone, Thomas Pascal
Shepherd, Robert James
Shugars, Jonas P
Siegel, Malcolm Richard
Smiley, Jones Hazelwood
Thieret, John William
Thompson, Ralph Luther
Wheeler, Harry Ernest
Wiedeman, Varley Earl
Williams, Albert Simpson
Winstead, Joe Everett
Wolfson, Alfred M

LOUISIANA
Allen, Arthur (Silsby)
Berggren, Gerard Thomas, Jr
Birchfield, Wray
Black, Lowell Lynn
Bond, William Payton
Chapman, Russell Leonard
Christian, James A
Clark, Christopher Alan
Curry, Mary Grace
Damann, Kenneth Eugene, Jr
Darwin, Steven Peter
DePoe, Charles Edward
Erbe, Lawrence Wayne
Hardberger, Florian Max
Jewell, Frederick Forbes, Sr
Kalinsky, Robert George
Kenknight, Glenn
Kirk, Ben Truett
Klich, Maren Alice
Koike, Hideo
Leuck, Edwine E, II
Lieux, Meredith Hoag
Longstreth, David J
Lowy, Bernard
Lutes, Dallas D
Lynch, Steven Paul
MacKenzie, David Robert
Morris, Everett Franklin
Saigo, Roy Hirofumi
Schneidau, John Donald, Jr
Shepherd, Hurley Sidney
Sims, Asa C, Jr
Snow, Johnnie Park
Thomas, Roy Dale
Tucker, Shirley Cotter
Urbatsch, Lowell Edward
Utley, John Foster, III
Vail, Sidney Lee
Walker, Harrell Lynn
Walkinshaw, Charles Howard, Jr
Webert, Henry S
Welden, Arthur Luna
White, James Clarence
Williams, George, Jr

MAINE
Andersen, Robert Arthur
Campana, Richard John
Gelinas, Douglas Alfred
Hehre, Edward James, Jr
Lewis, Alan James
Manzer, Franklin Edward
Melching, J Stanley
Neubauer, Benedict Francis
Richards, Charles Davis
Thomas, Robert James
Wave, Herbert Edwin

MARYLAND
Altevogt, Raymond Fred
Anderson, Mauritz Gunnar
Barclay, Arthur S
Barksdale, Thomas Henry
Batra, Lekh Raj
Bean, George A
Benjamin, Chester Ray
Bereston, Eugene Sydney
Bonde, Morris Reiner
Bromfield, Kenneth Raymond
Broome, Carmen Rose
Bruns, Herbert Arnold
Collins, Ralph Porter
Corbett, M Kenneth
Damsteegt, Vernon Dale
Davis, Benjamin Harold
Davis, Robert Edward
Deahl, Kenneth Luvere
Devine, Thomas Edward
Diener, Theodor Otto
Dixon, Dennis Michael
Duke, James A
Farr, David Frederick
Faust, Maria Anna
Fravel, Deborah Rexroad
Fulkerson, John Frederick
Futcher, Anthony Graham
Gehris, Clarence Winfred
Goth, Robert W
Graham, Joseph Harry
Griesbach, Robert James
Gross, Kenneth Charles
Gunn, Charles Robert
Herman, Eliot Mark
Hiatt, Caspar Wistar, III
Hodgson, Richard Holmes
Holland, Mark Alexander
Imle, Ernest Paul
Jong, Shung-Chang
Kahn, Robert Phillip
Kantzes, James (George)
Karlander, Edward P
Kingsolver, Charles H
Kirkbride, Joseph Harold, Jr
Krauss, Robert Wallfar
Kulik, Martin Michael
Lentz, Paul Lewis
Lewis, Jack A
Lockard, J David
Lumsden, Robert Douglas
Maas, John Lewis
Menez, Ernani Guingona
Merz, William George
Miksche, Jerome Phillip
Moline, Harold Emil
Morgan, Omar Drennan, Jr
Motta, Jerome J
Ostazeski, Stanley A
Palm, Mary Egdahl
Papavizas, George Constantine
Perdue, Robert Edward, Jr
Phillips, William George
Podleckis, Edward Vidas
Price, Samuel
Reveal, James L
Rissler, Jane Francina
Rossman, Amy Yarnell
Sayre, Richard Martin
Simpson, Marion Emma
Sisler, Hugh Delane
Slocum, Robert Dale
Smith, Andrew George
Smith, William Owen
Stavely, Joseph Rennie
Terrell, Edward Everett
Uecker, Francis August
Warmbrodt, Robert Dale
Waterworth, Howard E

Botany-Phytopathology (cont)

Windler, Donald Richard
Zacharius, Robert Marvin

MASSACHUSETTS
Ahmadjian, Vernon
Ashton, Peter Shaw
Barke, Harvey Ellis
Bawa, Kamaljit S
Bertin, Robert Ian
Bigelow, Howard Elson
Boger, Edwin August, Sr
Bolotin, Moshe
Bowley, Donovan Robin
Bradley, Peter Michael
Branton, Daniel
Brennan, James Robert
Burk, Carl John
Camp, Russell R
Caruso, Frank Lawrence
Cheetham, Ronald D
Colt, LeBaron C, Jr
Creighton, Harriet Baldwin
Davis, Edward Lyon
Del Tredici, Peter James
Dent, Thomas Curtis
DeWolf, Gordon Parker, Jr
DiLiddo, Rebecca McBride
Forman, Richard T T
Frederick, Sue Ellen
Freeberg, John Arthur
Godfrey, Paul Jeffrey
Golubic, Stephan
Gruber, Peter Johannes
Haight, Thomas H
Hare, Joan Conway
Haskell, David Andrew
Hewitson, Walter Milton
Hilferty, Frank Joseph
Hoffmann, George Robert
Holmes, Francis W(illiam)
Howard, Richard Alden
Irwin, Howard Samuel
Johansen, Hans William
Jost, Dana Nelson
Kaplan, Lawrence
Kissmeyer-Nielsen, Erik
Klekowski, Edward Joseph, Jr
Lovejoy, David Arnold
Madore, Bernadette
Manning, William Joseph
Melan, Melissa A
Mish, Lawrence Bronislaw
Mount, Mark Samuel
Nickerson, Norton Hart
Nixon, Charles William
Page, Joanna R Ziegler
Palser, Barbara Frances
Pfister, Donald Henry
Primack, Richard Bart
Reese, Elwyn Thomas
Rollins, Reed Clark
Roth, John L, Jr
Sayre, Geneva
Scheirer, Daniel Charles
Schofield, Edmund Acton, Jr
Schubert, Bernice Giduz
Schultes, Richard Evans
Schuster, Rudolf Mathias
Shapiro, Seymour
Signer, Ethan Royal
Spence, Willard Lewis
Spongberg, Stephen Alan
Stein, Diana B
Stevens, Peter Francis
Thomas, Aubrey Stephen, Jr
Tomlinson, Philip Barry
Troll, Joseph
Wilce, Robert Thayer
Wilson, Brayton F
Wood, Carroll E, Jr
Woodwell, George Masters

MICHIGAN
Abou-El-Seoud, Mohamed Osman
Andersen, Axel Langvad
Anderson, William Russell
Andresen, Norman A
Barnes, Burton Verne
Beaman, John Homer
Beck, Charles Beverley
Belcher, Robert Orange
Bird, George W
Bowers, Maynard C
Bowers, Robert Charles
Burnham, Robyn Jeanette
Burton, Clyde Leaon
Crum, Howard Alvin
Dazzo, Frank Bryan
De Zoeten, Gustaaf A
Erbisch, Frederic H
Estabrook, George Frederick
Fogel, Robert Dale
Franke, Robert G
Freeman, Dwight Carl
Gilbert, William James
Glime, Janice Mildred
Halloin, John McDonell
Hampton, Raymond Earl
Hart, Lynn Patrick
Hollensen, Raymond Hans
Holt, Imy Vincent
Hooker, William James

Hurst, Elaine H
Jones, Alan Lee
Kaufman, Peter Bishop
Klos, Edward John
Knobloch, Irving William
Krueger, Robert John
Krupka, Lawrence Ronald
LaCroix, Joseph Donald
Lacy, Melvyn Leroy
Lin, Chang Kwei
Lockwood, John LeBaron
Lowry, Robert James
McMeekin, Dorothy
Murphy, Peter George
Ohlrogge, John B
Olexia, Paul Dale
Pippen, Richard Wayne
Raikhel, Natasha V
Ramsdell, Donald Charles
Reznicek, Anton Albert
Rippon, John Willard
Roeper, Richard Allen
Rogers, Alvin Lee
Rogers, Claude Marvin
Rusch, Wilbert H, Sr
Scheffer, Robert Paul
Schlichting, Harold Eugene, Jr
Schmitter, Ruth Elizabeth
Shaffer, Robert Lynn
Shontz, John Paul
Stein, Howard Jay
Steiner, Erich E
Stephens, Christine Taylor
Stephenson, Stephen Neil
Stowell, Ewell Addison
Tarapchak, Stephen J
Toczek, Donald Richard
Vande Berg, Warren James
Van Faasen, Paul
Vargas, Joseph Martin, Jr
Volz, Paul Albert
Voss, Edward Groesbeck
Wagner, Florence Signaigo
Wagner, Warren Herbert, Jr
Wells, James Ray
Wilson, Ronald Wayne
Wujek, Daniel Everett
Wynne, Michael James
Yu, Shih-An

MINNESOTA
Abbott, Rose Marie Savelkoul
Anderson, Neil Albert
Banttari, Ernest E
Bissonnette, Howard Louis
Blanchette, Robert Anthony
Burton, Daniel Frederick
Burton, Verona Devine
Bushnell, William Rodgers
Bushong, Jerold Ward
Carlson, John Bernard
Charvat, Iris
Clapp, Thomas Wright
Ezell, Wayland Lee
French, David W
Gordon, Donald
Grewe, Alfred H, Jr
Groth, James Vernon
Hill, Eddie P
Jefferson, Carol Annette
Johnson, Herbert Gordon
Kennedy, Bill Wade
Kommedahl, Thor
Krupa, Sagar
Larsen, Philip O
Lawrence, Donald Buermann
Leonard, Kurt John
Levine, Allen Stuart
Lockhart, Benham Edward
Loeffler, Robert J
McCulloch, Joseph Howard
McLaughlin, James L
Markhart, Albert Henry, III
Mason, Charles Perry
Mirocha, Chester Joseph
Monson, Paul Herman
Nyvall, Robert Frederick
O'Rourke, Richard Clair
Ownbey, Gerald Bruce
Percich, James Angelo
Raup, Hugh Miller
Rines, Howard Wayne
Roberts, Glenn Dale
Roelfs, Alan Paul
Romig, Robert William McClelland
Rowell, John Bartlett
Schafer, John Francis
Schimpf, David Jeffrey
Severin, Charles Hilarion
Silflow, Carolyn Dorothy
Silverman, William Bernard
Singer, Susan Rundell
Skilling, Darroll Dean
Skjegstad, Kenneth
Stienstra, Ward Curtis
Tolbert, Robert John
Wetmore, Clifford Major
Widin, Katharine Douglas
Wilcoxson, Roy Dell

MISSISSIPPI
Ammon, Vernon Dale
Arner, Dale H
Batson, William Edward, Jr

Cibula, William Ganley
Cliburn, Joseph William
Davis, Robert Gene
Delouche, James Curtis
Eleuterius, Lionel Numa
Filer, Theodore H, Jr
Graves, Clinton Hannibal, Jr
Hare, Mary Louise Eckles
Heatherly, Larry G
Huneycutt, Maeburn Bruce
Keeling, Bobbie Lee
McGuire, James Marcus
Pitre, Henry Nolle, Jr
Pullen, Thomas Marion
Raj, Baldev
Rosenkranz, Eugen Emil
Sciumbato, Gabriel Lon
Sherman, Harry Logan
Spencer, James Alphus
Stewart, Robert Archie, II
Trevathan, Larry Eugene
Watson, James Ray, Jr
Wooten, Jean W

MISSOURI
Althaus, Ralph Elwood
Anderson, Robert Gordon
Apirion, David
Bell, Max Ewart
Bond, Lora
Bourne, Carol Elizabeth Mulligan
Brown, Merton F
Calvert, Oscar Hugh
Castaner, David
Croat, Thomas Bernard
Crosby, Marshall Robert
Cumbie, Billy Glenn
D'Arcy, William Gerald
Davidse, Gerrit
De Buhr, Larry Eugene
Duncan, David Robert
Dunn, David Baxter
Dwyer, John Duncan
Ewan, Joseph (Andorfer)
Finley, David Emanuel
Fraley, Robert Thomas
Gentry, Alwyn Howard
Goldblatt, Peter
Goodman, Robert Norman
Gowans, Charles Shields
Graham, James Carl
Hagen, Gretchen
Hanks, David L
Hayes, Alice Bourke
Huckabay, John Porter
Kobayashi, George S
Kullberg, Russell Gordon
Lewis, Walter Hepworth
Lindhorst, Taylor Erwin
Lissant, Ellen Kern
Magill, Robert Earle
Maniotis, James
Millikan, Daniel Franklin, Jr
Moore, James Frederick, Jr
Morin, Nancy Ruth
Neely, Robert Dan
Nichols, Herbert Wayne
Pallardy, Stephen Gerard
Ponder, Felix, Jr
Raven, Peter Hamilton
Redfearn, Paul Leslie, Jr
Rhodes, Russell G
Sehgal, Om Parkash
Stevens, Warren Douglas
Tai, William
Wagenknecht, Burdette Lewis
Wallin, Jack Robb
Weber, Wallace Rudolph
Wood, Joseph M
Wyllie, Thomas Dean
Zardini, Elsa Matilde

MONTANA
Carlson, Clinton E
Carroll, Thomas William
Chessin, Meyer
Elliott, Eugene Willis
Foley, Dean Carroll
Kendall, Katherine Clement
McCoy, Thomas Joseph
Mathre, Donald Eugene
Rumely, John Hamilton
Sands, David Chandler
Sawyer, Paul Thompson
Solberg, Richard Allen
Strobel, Gary A

NEBRASKA
Becker, Donald A
Bolick, Margaret Ruth
Boosalis, Michael Gus
Brooks, Merle Eugene
Egan, Robert Shaw
Gardner, Charles Olda, Jr
Gauger, Wendell Lee
Ikenberry, Richard W
Kaplan, Sanford Sandy
Kaul, Robert Bruce
Keeler, Kathleen Howard
Kerr, Eric Donald
Langenberg, Willem G
Maier, Charles Robert
Morris, Thomas Jack
Partridge, James Enoch

Peterson, Glenn Walter
Steadman, James Robert
Sutherland, David M
Swift, Lloyd Harrison
Weihing, John Lawson
Wysong, David Serge

NEVADA
Bohmont, Dale W
Fox, Carl Alan
Gupton, Oscar Wilmot
Lupan, David Martin
Mozingo, Hugh Nelson
Niles, Wesley E
Thyr, Billy Dale

NEW HAMPSHIRE
Blanchard, Robert Osborn
Crow, Garrett Eugene
DeMaggio, Augustus Edward
Fagerberg, Wayne Robert
Fralick, Richard Allston
Jones, Richard Conrad
Mathieson, Arthur C
Minocha, Subhash C
Norton, Robert James
Rich, Avery Edmund
Rock, Barrett Nelson
Schreiber, Richard William
Shigo, Alex Lloyd
Shortle, Walter Charles

NEW JERSEY
Babcock, Philip Arnold
Brown, Thomas Edward
Cappellini, Raymond Adolph
Ceponis, Michael John
Chen, James Che Wen
Chen, Tseh-An
Davis, Spencer Harwood, Jr
Day, Peter Rodney
Deems, Robert Eugene
Di Sanzo, Carmine Pasqualino
Fairbrothers, David Earl
Froyd, James Donald
Gaynor, John James
Greenfield, Sydney Stanley
Halisky, Philip Michael
Handel, Steven Neil
Jacobs, William Paul
Kasper, Andrew E, Jr
Klessig, Daniel Frederick
Kuehn, Harold Herman
Kuhnen, Sybil Marie
Lewis, Gwynne David
Ling, Hubert
Maiello, John Michael
Markle, George Michael
Meagher, Thomas Robert
Myers, Ronald Fenner
Quinn, James Allen
Quinn, James Amos
Reid, Hay Bruce, Jr
Schwalb, Marvin N
Springer, John Kenneth
Star, Aura E
Watts, Daniel Jay
Weber, Paul Van Vranken
Younkin, Stuart G

NEW MEXICO
Allred, Kelly Wayne
Carter, Jack Lee
Dittmer, Howard James
Dunford, Max Patterson
Grover, Herbert David
Heiner, Terry Charles
Holloway, Richard George
Hooks, Ronald Fred
Hsi, David Ching Heng
Kidd, David Eugene
Lindsey, Donald L
Martin, William Clarence
Peterson, Roger Shipp
Potter, Loren David
Spellenberg, Richard (William)
Todsen, Thomas Kamp

NEW YORK
Aist, James Robert
Alves, Leo Manuel
Andrus, Richard Edward
App, Alva A
Arneson, Phil Alan
Banerjee, Partha
Banks, Harlan Parker
Baroni, Timothy J
Basile, Dominick V
Bates, David Martin
Bedford, Barbara Lynn
Beer, Steven Vincent
Bentley, Barbara Lee
Biondo, Frank X
Blasdell, Robert Ferris
Bobear, Jean B
Boothroyd, Carl William
Brandwein, Paul Franz
Braun, Alvin Joseph
Brett, Betty Lou Hilton
Buck, William R
Carroll, Robert Baker
Carroll, Robert Buck
Castello, John Donald
Chase, Sherret Spaulding

Chock, Jan Sun-Lum
Churchill, Algernon Coolidge
Cox, Donald David
Crepet, William Louis
Cronquist, Arthur John
Davis, David
De Laubenfels, David John
Dietert, Margaret Flowers
Dolan, Desmond Daniel
Dress, William John
Dudock, Bernard S
Dumont, Kent P
Earle, Elizabeth Deutsch
Erb, Kenneth
Forero, Enrique
Forest, Charlene Lynn
Fry, William Earl
Goldstein, Solomon
Goodwin, Stephen Bruce
Gordon, Morris Aaron
Green, John Irving
Greller, Andrew M
Griffin, David H
Haber, Alan Howard
Hager, Richard Arnold
Haines, John Haldor
Hammill, Terrence Michael
Hanson, Maureen Rebecca
Harman, Gary Elvan
Heidrick, Lee E
Heusser, Calvin John
Hibben, Craig Rittenhouse
Hirshon, Jordon Barry
Ho, Hon Hing
Hoham, Ronald William
Holmgren, Noel Herman
Holmgren, Patricia Kern
Howard, Harold Henry
Howell, Stephen Herbert
Hudler, George William
Humber, Richard Alan
Hunt, David Michael
Hunter, James Edward
Jones, Clive Gareth
Kimmerer, Robin Wall
Kiviat, Erik
Kohut, Robert John
Korf, Richard Paul
Koyama, Tetsuo
Larson, Donald Alfred
Latorella, A Henry
Law, David Martin
Ledbetter, Myron C
Leopold, Donald Joseph
Lorbeer, James W
Lowe, Josiah L(incoln)
Luteyn, James Leonard
McDonnell, Mark Jeffery
Mai, William Frederick
Maple, William Thomas
Marengo, Norman Payson
Marsh, Leland C
Meiselman, Newton
Metzner, Jerome
Mickel, John Thomas
Miles, Philip Giltner
Miller, Norton George
Mitchell, Richard Shepard
Mitra, Jyotirmay
Monz, Pauline
Mori, Scott Alan
Mussell, Harry W
Nee, Michael
Niklas, Karl Joseph
Nolan, James Robert
Pickett, Steward T A
Prescott, Henry Emil, Jr
Putala, Eugene Charles
Rana, Mohammad A
Randall, Eric A
Reddy, Kalluru Jayarami
Richardson, F C
Robinson, Albert Dean
Robinson, Alix Ida
Robinson, Beatrice Letterman
Rochow, William Frantz
Rogers, Scott Orland
Rosenthal, Stanley Arthur
Schumacher, George John
Seago, James Lynn, Jr
Selsky, Melvyn Ira
Settle, Wilbur Jewell
Shechter, Yaakov
Sheviak, Charles John
Silva-Hutner, Margarita
Simone, Leo Daniel
Sirois, David Leon
Slack, Steven Allen
Smith, Ora
Stalter, Richard
Stanley, Edward Alexander
Staples, Richard Cromwell
Sweeney, Robert Anderson
Tepper, Herbert Bernard
Thomson, James Douglas
Thurston, Herbert David
Titus, John Elliott
Toenniessen, Gary Herbert
Torgeson, Dewayne Clinton
Uhl, Charles Harrison
Wang, Chun-Juan Kao
Webber, Edgar Ernest
Weinstein, Leonard Harlan
Wilson, Jack Belmont

Yoder, Olen Curtis
Zabel, Robert Alger
Zaitlin, Milton
Zitter, Thomas Andrew

NORTH CAROLINA
Averre, Charles Wilson, III
Aycock, Robert
Baranski, Michael Joseph
Barker, Kenneth Reece
Bateman, Durward F
Beard, Luther Stanford
Bell, Clyde Ritchie
Benson, David Michael
Beute, Marvin Kenneth
Billings, William Dwight
Bland, Charles E
Blum, Udo
Butterfield, Earle James
Campbell, Peter Hallock
Camp Hay, Pamela Jean
Carpenter, Irvin Watson, Jr
Chopra, Baldeo K
Cowling, Ellis Brevier
Culberson, William Louis
Cutter, Lois Jotter
Dickison, William Campbell
Downs, Robert Jack
DuBay, Denis Thomas
Duncan, Harry Ernest
Echandi, Eddie
Ellis, Don Edwin
Fantz, Paul Richard
Finch, Harry C
Flagg, Raymond Osbourn
Fulcher, William Ernest
Garner, Jasper Henry Barkdoll
Gensel, Patricia Gabbey
Gooding, Guy V, Jr
Haning, Blanche Cournoyer
Hardin, James Walker
Heagle, Allen Streeter
Heatwole, Harold Franklin
Hebert, Teddy T
Henderson, Nannette Smith
Hilger, Anthony Edward
Hommersand, Max Hoyt
Horton, Carl Frederick
Horton, James Heathman
Hunt, Kenneth Whitten
Johnson, Terry Walter, Jr
Jolls, Claudia Lee
Kapraun, Donald Frederick
Khan, Sekender Ali
Kohlmeyer, Jan Justus
Krochmal, Arnold
Lapp, Neil Arden
Lewis, Paul Ollin
Lucas, George Blanchard
Lucas, Leon Thomas
McVaugh, Rogers
Main, Charles Edward
Manly, Jethro Oates
Massey, Jimmy R
Matthews, James Francis
Mayfield, John Emory
Merritt, James Francis
Miller, Carol Raymond
Mott, Ralph Lionel
Mowbray, Thomas Bruce
Neher, Deborah A
Padgett, David Emerson
Peet, Robert Krug
Philpott, Jane
Powell, Nathaniel Thomas
Radford, Albert Ernest
Reinert, Richard Allyn
Ross, John Paul
Searles, Richard Brownlee
Shafer, Steven Ray
Shoemaker, Paul Beck
Sieren, David Joseph
Sievert, Richard Carl
Spears, Brian Merle
Spencer, Lorraine Barney
Spurr, Harvey Wesley, Jr
Stone, Donald Eugene
Strider, David Lewis
Swab, Janice Coffey
Therrien, Chester Dale
Tweedy, Billy Gene
Van Dyke, Cecil Gerald
Wallace, James William, Jr
Ward, John Everett, Jr
Welch, Aaron Waddington
Wells, Charles Van
Whitford, Larry Alston
Wilbur, Robert Lynch
Winstead, Nash Nicks
Wyatt, Raymond L
Yarnell, Richard Asa
Yeats, Frederick Tinsley
Zublena, Joseph Peter

NORTH DAKOTA
Barker, William T
Esslinger, Theodore Lee
Freeman, Myron L
Freeman, Thomas Patrick
Hosford, Robert Morgan, Jr
Kiesling, Richard Lorin
LaDuke, John Carl
Lamey, Howard Arthur
Seiler, Gerald Joseph

Starks, Thomas Leroy
Suttle, Jeffrey Charles

OHIO
Alldridge, Norman Alfred
Andreas, Barbara Kloha
Benzing, David H
Boerner, Ralph E J
Byrne, John Maxwell
Cantino, Philip Douglas
Cavender, James C
Chitaley, Shyamala D
Chuey, Carl F
Collins, Gary Brent
Cooke, William Bridge
Cooperrider, Tom Smith
Coplin, David Louis
Crawford, Daniel John
Curtis, Charles R
Deal, Don Robert
De Jong, Diederik Cornelis Dignus
Delanglade, Ronald Allan
Easterly, Nathan William
Ellett, Clayton Wayne
Eshbaugh, William Hardy
Farley, James D
Fisher, T Richard
Floyd, Gary Leon
Fulford, Margaret Hannah
Garraway, Michael Oliver
Giesy, Robert
Gilbert, Gareth E
Gingery, Roy Evans
Gochenaur, Sally Elizabeth
Gorchov, David Louis
Gordon, Donald Theile
Graffius, James Herbert
Graham, Alan Keith
Graham, Shirley Ann
Hauser, Edward J P
Heimsch, Charles W
Herr, Leonard Jay
Hillis, Llewellya
Hobbs, Clinton Howard
Hoitink, Harry A J
Ichida, Allan A
Janson, Blair F
Jensen, William August
Kaufmann, Maurice John
Kolattukudy, P E
Laufersweiler, Joseph Daniel
Laushman, Roger H
Leben, Curt (Charles)
Lehmann, Paul F
Lilly, Percy Lane
Lloyd, Robert Michael
Long, Terrill Jewett
Loucks, Orie Lipton
Louie, Raymond
McGinnis, Michael Randy
Macior, Lazarus Walter
Mapes, Gene Kathleen
Mason, David Lamont
Mattox, Karl
Miller, Harvey Alfred
Moore, Randy
Mueller, Sabina Gertrude
Mulroy, Juliana Catherine
Partyka, Robert Edward
Pollack, J Dennis
Powell, Charles Carleton, Jr
Rhodes, Landon Harrison
Risser, Paul Gillan
Romans, Robert Charles
Rowe, Randall Charles
Rudolph, Emanuel David
Runkle, James Reade
Sack, Fred David
Schmitt, John Arvid, Jr
Schmitthenner, August Fredrick
Schreiber, Lawrence
Silvius, John Edward
Smith, Calvin Albert
Snider, Jerry Allen
Snyder, Gary Wayne
Stoutamire, Warren Petrie
Stuckey, Ronald Lewis
Stuessy, Tod Falor
Taylor, Edith L
Tulecke, Walt
Weidensaul, T Craig
Wickstrom, Conrad Eugene
Williams, Lansing Earl
Wilson, Kenneth Glade
Wilson, Thomas Kendrick

OKLAHOMA
Barnes, George Lewis
Boke, Norman Hill
Campbell, Thomas Hodgen
Conway, Kenneth Edward
Essenberg, Margaret Kottke
Estes, James Russell
Gregory, Garold Fay
Larsh, Howard William
Levetin Avery, Estelle
Littlefield, Larry James
Love, Harry Schroeder, Jr
Melouk, Hassan A
Nelson, John Marvin, Jr
Richardson, Paul Ernest
Russell, Scott D
Sherwood, John L
Sturgeon, Roy V, Jr

Taylor, Constance Elaine Southern
Taylor, R(aymond) John
Tyrl, Ronald Jay
Young, David Allen

OREGON
Aho, Paul E
Allen, Thomas Cort, Jr
Baker, Kenneth Frank
Bierzychudek, Paulette F
Brandt, William Henry
Brehm, Bertram George, Jr
Calvin, Clyde Lacey
Cameron, H Ronald
Carroll, George C
Chambers, Kenton Lee
Coakley, Stella Melugin
Converse, Richard Hugo
Corden, Malcolm Ernest
Daley, Laurence Stephen
Dalton, David Andrews
Denison, William Clark
Erdman, Kimball
Florance, Edwin R
Geeseman, Gordon E
Goheen, Austin Clement
Hampton, Richard Owen
Hansen, Everett Mathew
Hardison, John Robert
Higgins, Paul Daniel
Horner, Chester Ellsworth
Johnson, Dale E
Johnson, John Morris
Lang, Frank Alexander
Leach, Charles Morley
Linderman, Robert G
Lippert, Byron E
Liston, Aaron Irving
Meints, Russel H
Milbrath, Gene McCoy
Miller, Paul William
Mills, Dallice Ivan
Minore, Don
Molina, Randolph John
Moore, Larry Wallace
Neiland, Bonita J
Nelson, Earl Edward
Novak, Robert Otto
Orkney, G Dale
Powelson, Robert Loran
Quatrano, Ralph Stephen
Ream, Lloyd Walter, Jr
Retallack, Gregory John
Schmidt, Clifford LeRoy
Schreiner, Roger Paul
Sherwood, Martha Allen
Spotts, Robert Allen
Tepfer, Sanford Samuel
Trappe, James Martin
Wagner, David Henry
Wagner, Orvin Edson
Welty, Ronald Earle
Wimber, Donald Edward
Zobel, Donald Bruce

PENNSYLVANIA
Allison, William Hugh
Archibald, Patricia Ann
Banerjee, Sushanta Kumar
Barnes, William Shelley
Bartuska, Doris G
Bayer, Margret H(elene) Janssen
Biebel, Paul Joseph
Birchem, Regina
Black, Robert Corl
Bloom, James R
Bowen, Paul Ross
Bradt, Patricia Thornton
Brennan, Thomas Michael
Brunkard, Kathleen Marie
Bryner, Charles Leslie
Carley, Harold Edwin
Cavaliere, Alphonse Ralph
Christ, Barbara Jane
Cole, Herbert, Jr
Dahl, A(nthony) Orville
DeFigio, Daniel A
DeMott, Howard Ephraim
Dobbins, David Ross
Erickson, Ralph O
Fett, William Frederick
Gaither, Thomas Walter
Gustine, David Lawrence
Hickey, Kenneth Dyer
Hillson, Charles James
Hoffmaster, Donald Edeburn
Hotchkiss, Arland Tillotson
Hunter, Barry B
Jacobsen, Terry Dale
Keener, Carl Samuel
Kelley, William Russell
Kiger, Robert William
Kneebone, Leon Russell
Leath, Kenneth T
Lukezic, Felix Lee
Maksymowych, Roman
Mears, James Austin
Mikesell, Jan Erwin
Miller, Helena Agnes
Miller, Robert Ernest
Mingrone, Louis V
Montgomery, James Douglas
Moorman, Gary William
Nelson, Paul Edward

Botany-Phytopathology (cont)

Oswald, John Wieland
Parks, James C
Pell, Eva Joy
Pickering, Jerry L
Poethig, Richard Scott
Pottmeyer, Judith Ann
Pritchard, Hayden N
Pursell, Ronald A
Roia, Frank Costa, Jr
Royse, Daniel Joseph
Salch, Richard K
Salvin, Samuel Bernard
Schaeffer, Robert L, Jr
Schein, Richard David
Schipper, Arthur Louis, Jr
Schisler, Lee Charles
Schrock, Gould Frederick
Schuyler, Alfred Ernest
Settlemyer, Kenneth Theodore
Sherwood, Robert Tinsley
Sinden, James Whaples
Smith, Bruce Barton
Snow, Jean Anthony
Stephenson, Andrew George
Stewart, Elwin Lynn
Stone, Benjamin Clemens
Thimann, Kenneth Vivian
Thomas, Joab Langston
Utech, Frederick Herbert
Verhoek, Susan Elizabeth
Wheeler, Donald Alsop
Wuest, Paul J

RHODE ISLAND
Beckman, Carl Harry
Church, George Lyle
Goos, Roger Delmon
Hammen, Susan Lum
Harlin, Marilyn Miler
Hartmann, George Charles
Hauke, Richard Louis
Heywood, Peter
Howard, Frank Leslie
Hull, Richard James
Jackson, Noel
Mueller, Walter Carl
Palmatier, Elmer Arthur
Schmitt, Johanna
Swift, Dorothy Garrison
Zavada, Michael Stephan
Zuck, Robert Karl

SOUTH CAROLINA
Alexander, Paul Marion
Baxter, Luther Willis, Jr
Cowley, Gerald Taylor
Dickerson, Ottie J
Dowler, William Minor
Dukes, Philip Duskin
Fox, Charles Wayne
Guram, Malkiat Singh
Houk, Richard Duncan
Kingsland, Graydon Chapman
Marx, Donald Henry
Miller, Robert Walker, Jr
Pollard, Arthur Joseph
Porcher, Richard Dwight
Powell, Robert W, Jr
Schoulties, Calvin Lee
Sharitz, Rebecca Reyburn
Shealy, Harry Everett, Jr
Stewart, Shelton E
Strobel, James Walter
Witcher, Wesley

SOUTH DAKOTA
Brashier, Clyde Kenneth
Buchenau, George William
Hart, Charles Richard
Myers, Gerald Andy
Tatina, Robert Edward
Van Bruggen, Theodore

TENNESSEE
Artist, Russell (Charles)
Ballal, S K
Bell, Sandra Lucille
Browne, Edward Tankard, Jr
Caponetti, James Dante
Channell, Robert Bennie
Churchill, John Alvord
Cox, Edmond Rudolph, Jr
Ellis, William Haynes
Gunasekaran, Muthukumaran
Heilman, Alan Smith
Herndon, Walter Roger
Hilty, James Willard
Hughes, Karen Woodbury
Hunter, Gordon Eugene
Jennings, Lisa Helen Kyle
Johnson, Leander Floyd
Jones, Larry Hudson
Karve, Mohan Dattatreya
Kral, Robert
Murrell, James Thomas, Jr
Nall, Ray(mond) W(illett)
Olsen, John Stuart
Quarterman, Elsie
Ramseur, George Shuford
Rosing, Wayne C
Salk, Martha Scheer
Schilling, Edward Eugene
Sharp, Aaron John
Shriner, David Sylva
Stacey, Gary
Staub, Robert J
Trigiano, Robert Nicholas
Van Horn, Gene Stanley
Vredeveld, Nicholas Gene
Whittier, Dean Page
Wolf, Frederick Taylor

TEXAS
Amato, Vincent Alfred
Arnott, Howard Joseph
Arp, Gerald Kench
Averett, John E
Berry, Robert Wade
Bischoff, Harry William
Boutton, Thomas William
Bragg, Louis Hairston
Brand, Jerry Jay
Brown, Richard Malcolm, Jr
Bryant, Vaughn Motley, Jr
Burgess, Jack D
Camp, Earl D
Cole, Garry Thomas
Cox, Elenor R
Dean, David Devereaux
Diggs, George Minor, Jr
Elliot, Arthur McAuley
Ellzey, Joanne Tontz
Fearing, Olin S
Fowler, Norma Lee
Fryxell, Paul Arnold
Fucik, John Edward
Fulton, Joseph Patton
Grant, Verne (Edwin)
Griffing, Lawrence Randolph
Harris, Kerry Francis Patrick
Hatch, Stephan LaVor
Higgins, Larry Charles
Hobbs, Clifford Dean
Hoff, Victor John
Hunter, Richard Edmund
Johnston, La Verne Albert
Johnston, Marshall Conring
Judd, Frank Wayne
Keller, Harold Willard
Kennedy, Robert Alan
Kimber, Clarissa Therese
Kroschewsky, Julius Richard
La Claire, John Willard, II
Lee, Addison Earl
Loeblich, Alfred Richard, III
Lonard, Robert (Irvin)
Lyda, Stuart D
McCracken, Michael Dwayne
McFeeley, James Calvin
McGrath, William Thomas
Mahler, William Fred
Manis, Archie L
Mueller, Dale M J
Neill, Robert Lee
Northington, David K
Ortega, Jacobo
Pilcher, Benjamin Lee
Powell, Michael A
Prior, Paul Verdayne
Proctor, Vernon Willard
Roach, Archibald Wilson Kilbourne
Rosberg, David William
Schneider, Edward Lee
Shane, John Denis
Sleeth, Bailey
Smith, Don Wiley
Smith, Gerald Ray
Smith, Roberta Hawkins
Sohmer, Seymour H
Sorensen, Lazern Otto
Standifer, Lonnie Nathaniel
Stanford, Jack Wayne
Starr, Richard Cawthon
Staten, Raymond Dale
Stewart, Robert Blaylock
Sumrall, H Glenn
Suttle, Curtis
Sweet, Charles Edward
Szaniszlo, Paul Joseph
Taylor, Robert Lee
Thomas, Ruth Beatrice
Thurston, Earle Laurence
Toler, Robert William
Turner, Billie Lee
Van Alfen, Neal K
Venketeswaran, S
Waddell, Henry Thomas
Wadsworth, Dallas Fremont
Ward, Calvin Herbert
Wellso, Stanley Gordon
Williams, Kenneth Bock
Williges, George Goudie
Willingham, Francis Fries, Jr
Wilson, Hugh Daniel
Wilson, Robert Eugene
Worthington, Richard Dane
Wright, Robert Anderson

UTAH
Ball, Terry Briggs
Barkworth, Mary Elizabeth
Belnap, Jayne
Blauer, Aaron Clyde
Bollinger, William Hugh
Bozniak, Eugene George
Campbell, William Frank
Cannon, Orson Silver
Cox, Paul Alan
Dalton, Patrick Daly
Epstein, William Warren
Griffin, Gerald D
Hansen, Afton M
Harper, Kimball T
Harrison, Bertrand Fereday
Harrison, H Keith
Hess, Wilford Moser (Bill)
Holmgren, Arthur Herman
McKell, Cyrus Milo
Mumford, David Louis
Robbins, Robert Raymond
Shaw, Richard Joshua
Shultz, Leila McReynolds
Smith, Marvin Artell
Takemoto, Jon Yutaka
Tilton, Varien Russell
Treshow, Michael
Vest, Hyrum Grant, Jr
Weber, Darrell J
Whitney, Elvin Dale
Wiens, Delbert
Wood, Benjamin W
Wullstein, Leroy Hugh

VERMONT
Barrington, David Stanley
Cook, Philip W
Gregory, Robert Aaron
Hyde, Beal Baker
Jervis, Robert Alfred
Klein, Deana Tarson
Ullrich, Robert Carl
Vogelmann, Hubert Walter

VIRGINIA
Al-Doory, Yousef
Berliner, Martha D
Bodkin, Norlyn L
Bonner, Robert Dubois
Bradley, Ted Ray
Breil, David A
Chen, Jiann-Shin
Clutter, Mary Elizabeth
Couch, Houston Brown
Coursen, Bradner Wood
Decker, Robert Dean
Drake, Charles Roy
Fuller, Stephen William
Garrison, Hazel Jeanne
Gates, James Edward
Hatzios, Kriton Kleanthis
Hill, Lynn Michael
Hufford, Terry Lee
Johnson, Miles F
Jones, Johnnye M
Lacy, George Holcombe
Lawrey, James Donald
Little, Elbert L(uther), Jr
McKinsey, Richard Davis
Marshall, Harold George
Martin, William Wallace
Miller, Lawrence Ingram
Miller, Orson K, Jr
Milton, Nancy Melissa
Moody, Arnold Ralph
Moore, Laurence Dale
Owens, Vivian Ann
Parker, Bruce C
Paterson, Robert Andrew
Phipps, Patrick Michael
Place, Janet Dobbins
Porter, Daniel Morris
Porter, Duncan MacNair
Ramsey, Gwynn W
Riopel, James L
Roane, Curtis Woodard
Roane, Martha Kotila
Roshal, Jay Yehudie
Rosinski, Joanne
Runk, Benjamin Franklin Dewees
Scott, Joseph Lee
Scott, Marvin Wade
Shetler, Stanwyn Gerald
Silberhorn, Gene Michael
Skog, Judith Ellen
Tenney, Wilton R
Tolin, Sue Ann
Treadwell, George Edward, Jr
Vaughan, Michael Ray
Ware, Donna Marie Eggers
Warren, Herman Lecil
Wells, Elizabeth Fortson
Willis, Lloyd L, II
Wills, Wirt Henry
Wilson, Charles Maye
Wilson, Coyt Taylor
Winstead, Janet
Young, Donald Raymond

WASHINGTON
Ammirati, Joseph Frank, Jr
Athow, Kirk Leland
Bliss, Lawrence Carroll
Booth, Beatrice Crosby
Bristow, Peter Richard
Bruehl, George William
Carr, Robert Leroy
Chastagner, Gary A
Clark, Raymond Loyd
Cobb, William Thompson
Cook, Robert James
Covey, Ronald Perrin, Jr
Del Moral, Roger
Denton, Melinda Fay
Dietz, Sherl M
Drum, Ryan William
Edwards, Gerald Elmo
Elfving, Donald Carl
Fridlund, Paul Russell
Gabrielson, Richard Lewis
Gotelli, David M
Gould, Charles Jay
Gross, Dennis Charles
Gurusiddaiah, Sarangamat
Hadwiger, Lee A
Halperin, Walter
Haskins, Edward Frederick
Hecht, Adolph
Hendrix, John Walter
Hewitt, William Boright
Hindman, Joseph Lee
Hudson, Peggy R
Irving, Patricia Marie
Johnson, Dennis Allen
Johnson, Richard Evan
Kraft, John M
Kruckeberg, Arthur Rice
Leopold, Estella (Bergere)
Lippert, Laverne Francis
Parish, Curtis Lee
Phillips, Ronald Carl
Post, Douglas Manners
Rahn, Joan Elma
Rayburn, William Reed
Rickard, William Howard, Jr
Rogers, Jack David
Schemske, Douglas William
Sharp, Eugene Lester
Shaw, Charles Gardner
Silbernagel, Matt Joseph
Smith, Samuel H
Staley, John M
Taylor, Ronald
Thornton, Melvin LeRoy
Turner, William Junior
Waaland, Joseph Robert
Walker, Richard Battson
Ward, George Henry
Whisler, Howard Clinton
Withner, Carl Leslie, Jr

WEST VIRGINIA
Barrat, Joseph George
Bell, Carl F
Biggs, Alan Richard
Binder, Franklin Lewis
Chapman, Carl Joseph
Clarkson, Roy Burdette
Clovis, Jesse Franklin
Day, Thomas Arthur
Elkins, John Rush
Elliston, John E
Gain, Ronald Ellsworth
Gillespie, William Harry
Guthrie, Roland L
Hindal, Dale Frank
Larson, Gary Eugene
Mills, Howard Leonard
Nunley, Robert Gray
Plymale, Edward Lewis
Pore, Robert Scott
Smith, Glenn Edward
Sorenson, William George
Stephenson, Steven Lee
Studlar, Susan Moyle
Van Der Zwet, Tom

WISCONSIN
Andrews, John Herrick
Arnholt, Philip John
Bennett, James Peter
Berbee, John Gerard
Blum, John Leo
Boone, Donald Milford
Bostrack, Jack M
Bowers, Frank Dana
Crone, Lawrence John
Croxdale, Judith Gerow
DeGroot, Rodney Charles
Dibben, Martyn James
Duewer, Elizabeth Ann
Durbin, Richard Duane
Ellingboe, Albert Harlan
Evert, Ray Franklin
Fay, Marcus J
Follstad, Merle Norman
Freckmann, Robert W
Fulton, Robert Watt
Gasiorkiewicz, Eugene Constantine
Gerloff, Gerald Carl
Graham, Linda Kay Edwards
Grau, Craig Robert
Grittinger, Thomas Foster
Hagedorn, Donald James
Hanson, Earle William
Harriman, Neil Arthur
Heggestad, Howard Edwin
Hillier, Richard David
Iltis, Hugh Hellmut
Jordan, William R, III
Jowett, David
Kitzke, Eugene David
Koch, Rudy G
Larsen, Michael John
Larson, Philip Rodney

McDonough, Eugene Stowell
Maravolo, Nicholas Charles
Maxwell, Douglas Paul
Michaelson, Merle Edward
Millington, William Frank
Moore, John Duain
Nair, Gangadharan V M
Nelson, Allen Charles
Newsome, Richard Duane
Palm, Elmer Thurman
Parker, Alan Douglas
Perry, James Warner
Rice, Marion McBurney
Salamun, Peter Joseph
Sequeira, Luis
Sharkey, Thomas D
Simon, Philipp William
Smalley, Eugene Byron
Smith, Stanley Galen
Stearns, Forest
Stevenson, Walter Roe
Sussman, Michael R
Sytsma, Kenneth Jay
Tandon, Shiv R
Tews, Leonard L
Thiesfeld, Virgil Arthur
Tibbitts, Theodore William
Tiefel, Ralph Maurice
Unbehaun, Laraine Marie
Unger, James William
Wade, Earl Kenneth
Waller, Donald Macgregor
Warner, James Howard
Weber, Albert Vincent
Williams, Paul Hugh
Wimpee, Charles F
Worf, Gayle L

WYOMING
Bohnenblust, Kenneth E
Christensen, Martha

PUERTO RICO
Rodriguez, Rocio del Pilar

ALBERTA
Bakshi, Trilochan Singh
Belland, Rene Jean
Bird, Charles Durham
Campbell, John Duncan
Cormack, Robert George Hall
Cuny, Robert Michael
Degenhardt, Keith Jacob
Gorham, Paul Raymond
Harper, Frank Richard
Hickman, Michael
Hiratsuka, Yasuyuki
Laroche, Andre
Nelson, Gordon Albert
Parkinson, Dennis
Skoropad, William Peter
Suresh, Mavanur Rangarajan
Tamaoki, Taiki
Wilson, Donald Benjamin

BRITISH COLUMBIA
Bandoni, Robert Joseph
Bigelow, Margaret Elizabeth Barr
Bisalputra, Thana
Bohm, Bruce Arthur
Buchanan, Ronald James
Copeman, Robert James
De Boer, Solke Harmen
Dueck, John
Fisher, Francis John Fulton
Fushtey, Stephen George
Ganders, Fred Russell
Green, Beverley R
Hansen, Anton Juergen
Harvey, Michael John
Hebda, Richard Joseph
Hughes, Gilbert C
Kuijt, Job
Maze, Jack Reiser
Newroth, Peter Russell
Owens, John N
Paden, John Wilburn
Pepin, Herbert Spencer
Pocock, Stanley Albert John
Rahe, James Edward
Schofield, Wilfred Borden
Shaw, Michael
Stace-Smith, Richard
Towers, George Hugh Neil
Wall, Ronald Eugene
Warrington, Patrick Douglas
Weintraub, Marvin
Whitney, Harvey Stuart
Wright, Norman Samuel

MANITOBA
Atkinson, Thomas Grisedale
Bernier, Claude
Chong, James York
Dever, Donald Andrew
Dugle, Janet Mary Rogge
Harder, Donald Edward
Johnson, Karen Louise
Kerber, Erich Rudolph
Mills, John T
Olsen, Orvil Alva
Rohringer, Roland
Samborski, Daniel James

NEW BRUNSWICK
Lakshminarayana, J S S
Magasi, Laszlo P
Taylor, Andrew Ronald Argo
Whitney, Norman John

NEWFOUNDLAND
Hampson, Michael Chisnall
Scott, Peter John
Sheath, Robert Gordon

NOVA SCOTIA
Hall, Ivan Victor
McFadden, Lorne Austin
McLachlan, Jack (Lamont)
Ross, Robert Gordon
Vander Kloet, Sam Peter
Van der Meer, John Peter

ONTARIO
Alex, Jack Franklin
Allen, Wayne Robert
Anderson, Terry Ross
Argus, George William
Badenhuizen, Nicolaas Pieter
Barr, Donald John Stoddart
Barrett, Spencer Charles Hilton
Basham, Jack T
Baum, Bernard R
Benedict, Winfred Gerald
Bonn, William Gordon
Boyer, Michael George
Busch, Lloyd Victor
Catling, Paul Miles
Chaly, Nathalie
Cody, William James
Cook, Frankland Shaw
Corlett, Michael Philip
Cruise, James E
Dalpe, Yolande
Duthie, Hamish
Eckenwalder, James Emory
Fahselt, Dianne
Fedak, George
Forbes, Bruce Cameron
Gayed, Sobhy Kamel
Gerrath, Joseph Fredrick
Ginns, James Herbert
Good, Harold Marquis
Greyson, Richard Irving
Grodzinski, Bernard
Heath, Ian Brent
Heath, Michele Christine
Hellebust, Johan Arnvid
Higgins, Verna Jessie
Hughes, Stanley John
Illman, William Irwin
Jarzen, David MacArthur
Jeglum, John Karl
Jones, Roger
Kean, Vanora Mabel
Kendrick, Bryce
Kevan, Peter Graham
Krug, John Christian
Lott, John Norman Arthur
McAndrews, John Henry
McKeen, Colin Douglas
McKeen, Wilbert Ezekiel
McKenzie, Allister Roy
McNeill, John
Manocha, Manmohan Singh
Marks, Charles Frank
Martens, John William
Morton, John Kenneth
Northover, John
Nozzolillo, Constance
Olthof, Theodorus Hendrikus Antonius
Patrick, Zenon Alexander
Peterson, Robert Lawrence
Posluszny, Usher
Pringle, James Scott
Redhead, Scott Alan
Ritchie, James Cunningham
Sahota, Ajit Singh
Savile, Douglas Barton Osborne
Seaman, William Lloyd
Shoemaker, Robert Alan
Singh, Rama Shankar
Sinha, Ramesh Chandra
Soper, James Herbert
Sutton, John Clifford
Thompson, Hazen Spencer
Townshend, John Linden
Traquair, James Alvin
Tu, Jui-Chang
Van Huystee, Robert Bernard
Ward, Edmund William Beswick
Warwick, Suzanne Irene
Winterhalder, Keith
Zilkey, Bryan Frederick

PRINCE EDWARD ISLAND
Hanic, Louis A
MacQuarrie, Ian Gregor
Thompson, Leith Stanley
Willis, Carl Bertram

QUEBEC
Bessette, France Marie
Brouillet, Luc
Carlson, Lester William
Charest, Pierre M
Coulombe, Louis Joseph
Estey, Ralph Howard
Gagnon, Camilien Joseph Xavier
Gibbs, Sarah Preble
Goldstein, Melvin E
Langford, Arthur Nicol
Legault, Albert
Morisset, Pierre
Paulitz, Timothy Carl
Richard, Claude
Sackston, Waldemar E
Sattler, Rolf
Vieth, Joachim
Watson, Alan Kemball
Willemot, Claude

SASKATCHEWAN
Gruen, Hans Edmund
Harms, Vernon Lee
Haskins, Reginald Hinton
Lapp, Martin Stanley
Looman, Jan
Mortensen, Knud
Redmann, Robert Emanuel
Sawhney, Vipen Kumar
Smith, Jeffrey Drew
Steeves, Taylor Armstrong
Tinline, Robert Davies

OTHER COUNTRIES
Berg, Arthur R
Bulmer, Glenn Stuart
Chiarappa, Luigi
Cowan, Richard Sumner
De Renobales, Mertxe
Dettmann, Mary Elizabeth
Eiten, George
Erke, Keith Howard
Foard, Donald Edward
Gibbs, R Darnley
Gottlieb, Otto Richard
Gressel, Jonathan Ben
Hiyama, Tetsuo
Huisingh, Donald
Jensen, Lawrence Craig-Winston
Koch, Stephen Douglas
Kurobane, Itsuo
Lieth, Helmut Heinrich Friedrich
Lowden, Richard Max
MacCarthy, Jean Juliet
Medina, Ernesto Antonio
Meins, Frederick, Jr
Muchovej, James John
Nakayama, Takao
Pickett-Heaps, Jeremy David
Prance, Ghillean T
Rakowski, Krzysztof J
Reichstein, Tadeus
Saari, Eugene E
Seymour, Roland Lee
Smith, Douglas Roane
Steer, Martin William

Cell Biology

ALABAMA
Kayes, Stephen Geoffrey
Lincoln, Thomas M
Paxton, Ralph
Philips, Joseph Bond, III
Siegal, Gene Philip
Stephenson, Edwin Clark

ARIZONA
Begovac, Paul C
Garewal, Harinder Singh
Harris, David Thomas
Hoober, J Kenneth
Laird, Hugh Edward, II
Lamunyon, Craig Willis
Wheeler, Diana Esther
Wu, Chuanyue

ARKANSAS
Huang, Feng Hou

CALIFORNIA
Alexander, Caroline M
Almasan, Alexandru
Avila, Vernon Lee
Bacskai, Brian James
Barcellos-Hoff, Mary Helen
Bard, Frederique
Bechtol, Kathleen B
Bejar, Ezra
Bennett, C Frank
Blatt, Beverly Faye
Cardineau, Guy A
Cardullo, Richard Anthony
Caulfield, John Philip
Chang, Mei-Ping
Cherr, Gary N
Cohen, Natalie Shulman
DeLeon, Daisy Delgado
Dietz, Thomas John
Epstein, Lynn
Etcheverry, Tina
Forte, John Gaetano
Fryxell, Karl Joseph
Fukuda, Michiko N
Gehlsen, Kurt Ronald
Goff, Lynda J
Gong, Yu
Harford, Joe Bryan
Hedrick, Jerry Leo
Hoch, Sallie O'Neil
Hurkman, William James
Jadus, Martin Robert
Jordan, Mary Ann
Kane, Susan Elizabeth
King, Barry Frederick
Koch, Bruce D
Larabell, Carolyn A
Lembach, Kenneth James
Lin, Ming-Fong
Linker-Israeli, Mariana
Loughman, William D
Macklin, Wendy Blair
McMillan, Paul Junior
Martins-Green, Manuela M
Meizel, Stanley
Melnick, Michael
Miller, Arnold Lawrence
Montesano-Roditis, Luisa
Ning, Shoucheng
Nothnagel, Eugene Alfred
Owicki, John Callaghan
Patzer, Eric John
Perryman, Elizabeth Kay
Reiness, Gary
Rosenberg, Abraham
Rubas, Werner
Sanui, Hisashi
Schmid, Sandra Louise
Schwartz, Martin Alexander
Simmons, Dwayne Deangelo
Sliwkowski, Mary Burke
Sparling, Mary Lee
Springer, Wayne Richard
Stephens, Robert James
Sutherland, Robert Melvin
Thompson, Jeffrey Michael
Torres, Martine
Traynor-Kaplan, Alexis
Tsien, Roger Yonchien
Turgeon, Judith Lee
Umiel, Tehila
Vacquier, Victor Dimitri
Williams, Mark Alan
Wilson, Leslie
Woodruff, Teresa K
Yerram, Nagender Rao

COLORADO
Bamburg, James Robert
DeLa Cruz, Vidal F, Jr
Lieberman, Michael Merril
Martin, Richard Jay
Melancon, Paul R
Mykles, Donald Lee
Panini, Sankhavaram R
Pfenninger, Karl H
Poyton, Robert Oliver
Prescott, David Marshall
Seeds, Nicholas Warren
Wolf, Joe

CONNECTICUT
Blue, Marie-Luise
Bothwell, Alfred Lester Meador
Cooperstein, Sherwin Jerome
Das, Rathindra C
Dawson, Margaret Ann
Froshauer, Susan
Gross, Ian
Harwood, Harold James, Jr
Hebert, Daniel Normond
Hightower, Lawrence Edward
Jahn, Reinhard
Levin, Martin Allen
Merwin, June Rae
Tucker, Edward B

DELAWARE
Keeler, Calvin Lee, Jr

DISTRICT OF COLUMBIA
Dunkel, Virginia Catherine
Forman, David S
Henderson, Ellen Jane
Rhoads, Allen R
Vanderhoek, Jack Yehudi

FLORIDA
Assoian, Richard Kenneth
Bolton, Wade E
Colwin, Laura Hunter
Dunn, William Arthur, Jr
Hansen, Peter J
Hofer, Kurt Gabriel
Hornicek, Francis John
Keller, Thomas C S
Kopelovich, Levy
Pratt, Melanie M
Rangel Aldao, Rafael
Rhodin, Johannes A G
Sang, Qing-Xiang
Undeen, Albert Harold

GEORGIA
Brockbank, Kelvin Gordon Mashader
Compans, Richard William
Fechheimer, Marcus
Hargrove, James Lee
Kiefer, Charles R(andolph)
Lo, Woo-Kuen
Palevitz, Barry Allan
Sanford, Gary L
Sommer, Harry Edward

Cell Biology (cont)

Toney, Thomas Wesley
Wilson, Donella Joyce
Young, Henry Edward

HAWAII
Allen, Richard Dean
Fitch, Maureen Meiko Masuda
Nagai, Chifumi
Rosenthal, Eric Thomas

IDAHO
Rodnick, Kenneth Joseph

ILLINOIS
Ayres, Kathleen N
Bozzola, John Joseph
Brewer, Gregory J
Briskin, Donald Phillip
Budzik, Gerald P
Buetow, Dennis Edward
Cross, John W
Daniels, Edward William
Eisenhauer, Donald Alan
Goodheart, Clyde Raymond
Greenough, William Tallant
Gupta, Kailash Chandra
Hurley, Walter L
Johnson, Richard Joseph
Jonah, Margaret Martin
Karr, Timothy Lawrence
Kartha, Sreedharan
Katzenellenbogen, Benita Schulman
Langman, Craig Bradford
Loizzi, Robert Francis
McCorquodale, Maureen Marie
Maecker, Holden T
Nielsen, Peter James
Northup, Sharon Joan
Padh, Harish
Pappelis, Aristotel John
Radosevich, James A
Ryan, Jon Michael
Schneider, Gary Bruce
Sherwood, Robert Lawrence
Swinnen, Lode J
Ten Eick, Robert Edwin
Wasserman, William John
Weyhenmeyer, James Alan
Wezeman, Frederick H
Yamamoto, Hirotaka

INDIANA
Bumol, Thomas F
Chernoff, Ellen Ann Goldman
Pepper, Daniel Allen
Santerre, Robert Frank
Sato, Masahiko
Smith, Gary Lee
Webb, Mary Alice

IOWA
Longo, Frank Joseph
Robson, Richard Morris
Rodermel, Steven Robert
Swanson, Harold Dueker
Weisman, Lois Sue
Wiens, Darrell John

KANSAS
Ash, Ronald Joseph
Chitnis, Parag Ram
Gattone, Vincent H, II
Hochman, Jerome Henry
Johnson, Terry Charles
Upton, Steve Jay
Westfall, Jane Anne

KENTUCKY
Bondada, Subbarao
Mattson, Mark Paul
Wheeler, Thomas Jay

LOUISIANA
Alliegro, Mark C
Bautista, Abraham Parana
Clejan, Sandra
Cohn, Marc Alan
Jazwinski, S Michal
Snow, Lloyd Dale
West, David B
Wiser, Mark Frederick

MAINE
Andersen, Robert Arthur
Aronson, Frederick Rupp
Rappaport, Raymond

MARYLAND
Amende, Lynn Meridith
Blithe, Diana Lynn
Borst, Diane Elizabeth
Bowers, Mary Blair
Brock, Mary Anne
Brunswick, Mark
Buckheit, Robert Walter, Jr
Choi, Oksoon Hong
Colburn, Nancy Hall
Craig, Nessly Coile
Dermody, William Christian
Fambrough, Douglas McIntosh
Farrell, Robert Edmund, Jr
Flanders, Kathleen Corey

Gallin, Elaine K
Hanjan, Satnam S
Hla, Timothy Tun
Horowitz, Jill Ann
Kiang, Juliann G
Kibbey, Maura Christine
Klausner, Richard D
Kloetzel, John Arthur
Kole, Hemanta Kumar
Korn, Edward David
Leapman, Richard David
Lester, David Simon
Leto, Thomas L
Lo, Chu Shek
Loh, Yoke Peng
Mather, Ian Heywood
Melera, Peter William
Misra, Rohini Rita
Mosher, James Arthur
Nordin, Albert Andrew
Penno, Margaret Susan Beisel
Pinto da Silva, Pedro Goncalves
Platz, Robert Dale
Rabinovitz, Marco
Rafajko, Robert Richard
Redman, Robert Shelton
Resau, James Howard
Roberts, David Duncan
Roth, Thomas Frederic
Schulze, Dan Howard
Shamsuddin, Abulkalam Mohammad
Slife, Charles W
Small, Judy Ann
Stromberg, Kurt
Strum, Judy May
Su, Robert Tzyh-Chuan
Terasaki, Mark Ryo
Vonderhaar, Barbara Kay
Wergin, William Peter
Woolverton, Christopher Jude
Yewdell, Jonathan Wilson
Yoo, Seung Hyun
Young, Howard Alan

MASSACHUSETTS
Beggs, Alan H, III
Bernfield, Merton Ronald
Bischoff, Joyce E
Biswas, Debajit K
Bradley, Peter Michael
Branton, Daniel
Brown, Dennis
Byers, Hugh Randolph
Chen, Jane-Jane
Cochrane, David Earle
Davis, Claude Geoffrey
DeArruda, Monika Vicira
DuBose, David A
Dyer, Charissa Annette
Edwards, Ross B
Fulton, Chandler Montgomery
Gallo, Richard Louis
Gilmore, Thomas David
Goodenough, Daniel Adino
Gruber, Peter Johannes
Gullans, Steven R
Horowitz, Arie
Joyce, Nancy C
Keyes, Susan Riley
Klagsbrun, Michael
Kruskal, Benjamin A
Landis, William Joel
Larson, David Michael
Ledbetter, Mary Lee Stewart
Lian, Jane B
Little, John Bertram
Lotz, Margaret M
Miller, Andrew Leitch
Moreland, Robert Byrl
Morton, Mary E
Murrain, Michelle
Olsen, Bjorn Reino
Park, Hee-Young
Rheinwald, James George
Ruderman, Joan V
Ruprecht, Ruth Margrit
Scordilis, Stylianos Panagiotis
Smith, Peter John Shand
Sullivan, William Daniel
Thies, Robert Scott
Wheeler, Grant N
Witman, George Bodo, III
Zabrecky, James R

MICHIGAN
Anderson, Richard Lee
Armant, D Randall
Dinsdale, John Edward
Buxser, Stephen Earl
Christensen, A(lbert) Kent
Fischer, Howard David
Fluck, Michele Marguerite
Gordon, Sheldon Robert
Koehler, Lawrence D
Lalwani, Narendra Dhanraj
Ottova, Angela
Reiners, John Joseph, Jr
Schwartz, Jessica
Sebolt-Leopold, Judith S
Sloane, Bonnie Fiedorek
Sloat, Barbara Furin

MINNESOTA
Clapper, David Lee

Emeagwali, Dale Brown
Jones, Charles Weldon
Korte, Clare A
Leof, Edward B
Linck, Richard Wayne
Sicard, Raymond E(dward)
Sinha, Akhouri Achyutanand
Sorenson, Robert Lowell
Strehler, Emanuel Ernst
Tsilibary, Effie-Constantinos
West, Michael Allan

MISSOURI
Bretton, Randolph Henry, Jr
Casperson, Gerald F
Connolly, Daniel Thomas
Engel, Leslie Carroll
Guntaka, Ramareddy V
Myers, Richard Lee
Nyquist-Battie, Cynthia
Poon, Rebecca Yuetmay
Seidler, Norbert Wendelin
Staatz, William D
Thomsen, Robert Harold
Turner, John T
Weisman, Gary Andrew
Worthington, Ronald Edward

NEBRASKA
Godfrey, Maurice
Rosenquist, Thomas H
Volberg, Thomas M

NEVADA
Vig, Baldev K

NEW HAMPSHIRE
Loros, Jennifer Jane
Rochelle, Lori Gatzy
Roos, Thomas Bloom
Spiegel, Evelyn Sclufer
Spiegel, Melvin

NEW JERSEY
Axelrod, David E
Barroso, Margarida Maria R
Blood, Christine Haberern
Cahn, Frederick
Curran, Thomas
Fong, Dunne
Fujimore, Tohru
Gelfand, Israel M
Georgopapadakou, Nafsika Eleni
Guo, Jian Zhong
Kongsamut, Sathapana
Koo, Gloria C
Lambert, Muriel Wikswo
La Rocca, Paul Joseph
Liu, Alice Yee-Chang
Logdberg, Lennart Erik
Miller, Douglas Kenneth
Moral, Josefa Liduvina
Nakhla, Atif Mounir
Pachter, Jonathan Alan
Plotkin, Diane Joyce
Presky, David H
Rashidbaigi, Abbas
Silberklang, Melvin
Whiteley, Phyllis Ellen
Wille, John Jacob, Jr
Xiong, Yimin

NEW MEXICO
Bernstein, Marvin Harry
Finch, Gregory Lee
Seagrave, JeanClare

NEW YORK
Allis, C David
Asch, Bonnie Bradshaw
Badalamenti, Marie Ann
Barker, Kenneth Ray
Barnwell, John Wesley
Bisson, Mary A
Borenfreund, Ellen
Bowser, Samuel S
Bregman, Allyn A(aron)
Celada, Franco
Chadha, Kailash C
Chaudhuri, Asok
Comley, Lucy T
Davidson, Mercy
Duffey, Michael Eugene
Edmunds, Leland Nicholas, Jr
Eshel, Dan
Factor, Jan Robert
Ferdinand, Patricia
Font, Cecilio Rafael
Forest, Charlene Lynn
Fox, Thomas David
Garrick, Laura Morris
Graziadei, William Daniel, III
Gupta, Sanjeev
Hanson, Maureen Rebecca
Kaplan, John Ervin
Karl, Peter I
Keese, Charles Richard
Keng, Peter C
Li, Luyuan
Liebow, Charles
Lin, Jane Huey-Chai
Ma, Hong
Magid, Norman Mark
Maitra, Subir R

Mendelsohn, Naomi
Menon, Gopinathan K(unnariath)
Miller, Sue Ann
Mukherjee, Asit B
Najfeld, Vesna
Norin, Allen Joseph
Orlow, Seth J
Paine, Philip Lowell
Philipson, Lennart
Prasthofer, Thomas W
Repasky, Elizabeth Ann
Richter, Goetz Wilfried
Robbins, Edith Schultz
Rogers, Scott Orland
Rosenbluth, Jack
Sabatini, David Domingo
Sanders, F Kingsley
Sauro, Marie D
Spector, David Lawrence
Stillman, Bruce William
Swanson, Stephen King
Tannenbaum, Janet
Ter Kuile, Benno Herman
Thomas, John Owen
Trimble, Robert Bogue
Tulchin, Natalie
Vijayasaradhi, Setaluri
Woods, Philip Sargent
Zwain, Ismail Hassan

NORTH CAROLINA
Allen, Nina Stromgren
Andrews, Matthew Tucker
Caron, Marc G
Caruolo, Edward Vitangelo
Cosgrove, William Burnham
Davis, Cynthia Marie
Eddy, Edward Mitchell
Edwards, Nancy C
Erickson, Harold Paul
Faber, James Edward
Failla, Mark Lawrence
Gilligan, Diana Mary
Hoffman, Maureen Richardson
Lewis, Jon C
Maier, Robert Hawthorne
Maroni, Donna F
Mitchell, Ann Denman
O'Brien, Deborah A
Olden, Kenneth
Smith, Donald Eugene
Smith, Gary Keith
Thomas, Mary Beth
Tlsty, Thea Dorothy

NORTH DAKOTA
Carlson, Edward C
Reynolds, Lawrence P

OHIO
Birkenberger, Lori
Carothers, Donna June
Chapple, Paul James
Francis, Joseph William
Gruenstein, Eric Ian
Holtzman, David Allen
Kao, Winston Whei-Yang
Kmetec, Emil Philip
Liedtke, Carole M
Meyer, Ralph Roger
Michaels, John Edward
Mohinder, Singh Kang
Pate, Joy Lee
Perry, George
Rosenthal, Ken Steven
Silverman, Robert Hugh
Smith, Mark Anthony
Stalvey, John Robert Dixon
Stevenson, J(oseph) Ross
Stoner, Gary David

OKLAHOMA
Bell, Paul Burton, Jr
Conaway, Joan W
Conaway, Ronald C
Klein, John Robert
Koren, Eugen
Muneer, Razia Sultana
Williams, Judy Ann

OREGON
Anderson, Amy Elin
Quatrano, Ralph Stephen
Rutten, Michael John
Stouffer, Richard Lee

PENNSYLVANIA
Akella, Rama
Ashton, Francis T
Beezhold, Donald H
Brunson, Kenneth W
Buono, Russell Joseph
Butler, William Barkley
Camoratto, Anna Marie
Campbell, Phil G
Clawson, Gary Alan
Driska, Steven P
Floros, Joanna
Flynn, John Thomas
Fussell, Catharine Pugh
Gay, Carol Virginia Lovejoy
Goldfarb, Ronald H
Goodman, David Barry Poliakoff
Greif, Karen Faye

Hilfer, Saul Robert
Hopper, Anita Klein
Johnson, Susan E
Jonak, Zdenka L
Kassis, Shouki
Krummel, Debra A
Lazarides, Elias
Leboy, Phoebe Starfield
Linask, Kersti Katrin
McDevitt, David Stephen
McMorris, F Arthur
Mulder, Kathleen M
Murphy, Robert Francis
Peachey, Lee DeBorde
Puri, Rajinder N
Ramachandran, Banumathi
Roosa, Robert Andrew
Saporito, Michael S
Schlegel, Robert Allen
Schultz, Richard Morris
Shiman, Ross
Strayer, David S
Ulmer, Jeffrey Blaine
Vergona, Kathleen Anne Dobrosielski
Walsh, Charles Joseph
Wang, Allan Zu-Wu
Weisel, John Winfield
Weiss, Leon
Yang, Da-Ping (David)
Yoon, Kyonggeun
Yui, Katsuyuki
Zaleski, Jan F

RHODE ISLAND
Brautigan, David L
Dawicki, Doloretta Diane
Donnelly, Grace Marie
Yoon, Poksyn Song

SOUTH CAROLINA
Lopes-Virella, Maria Fernanda Leal
Pharr, Pamela Northington
Simson, Jo Anne V
Willingham, Mark C

SOUTH DAKOTA
Hurley, David John
Prescott, Lansing M

TENNESSEE
Conary, Jon Thurston
Costlow, Mark Enoch
Donaldson, Donald Jay
Hoover, Richard Lee
Joplin, Karl Henry
Lessman, Charles Allen
Lothstein, Leonard
Monty, Kenneth James
Nishimoto, Satoru Kenneth
Olins, Ada Levy
Pfeffer, Lawrence Marc
Rasch, Ellen M
Steen, R Grant
Walne, Patricia Lee
Whitt, Michael A

TEXAS
Albrecht, Thomas Blair
Aufderheide, Karl John
Barcellona, Wayne J
Beale, Elmus G
Brown, Richard Malcolm, Jr
Burghardt, Robert Casey
Carney, Darrell Howard
Dauwalder, Marianne
Deter, Russell Lee, II
Garfield, Robert Edward
Gaulden, Mary Esther
Gillard, Baiba Kurins
Gogol, Edward Peter
Griffing, Lawrence Randolph
Hamilton, Charleen Marie
Henney, Henry Russell, Jr
Huang, Shyi Yi
Hughes, Mark R
Hutson, James Chelton
Jackson, Gilchrist Lewis
Koke, Joseph R
La Claire, John Willard, II
Levine, Alan E
McKnight, Thomas David
Marshak, David William
Miller, Thomas Allen
Painter, Richard Grant
Reid, Michael Baron
Reuss, Luis
Roux, Stanley Joseph
Shain, Sydney A
Tenner, Thomas Edward, Jr

UTAH
McIntyre, Thomas Marty
White, Raymond L

VERMONT
Moehring, Thomas John

VIRGINIA
Banker, Gary A
Bloodgood, Robert Alan
Corwin, Jeffrey Todd
Deck, James David
Gorbsky, Gary James
Mikkelsen, Ross Blake

Owens, Gary
Povlishock, John Theodore
Pryer, Nancy Kathryn
Roth, Karl Sebastian
Stetka, Daniel George
Tuttle, Jeremy Ballou

WASHINGTON
Blinks, John Rogers
Hosick, Howard Lawrence
Meadows, Gary Glenn
Miller, Arthur Dusty
Pollack, Sylvia Byrne
Zakian, Virginia Araxie

WEST VIRGINIA
Sutter, Richard P

WISCONSIN
Augustine, James A
Dahms, Nancy Margaret
Dove, William Franklin
Feirer, Russell Paul
Harms, Jerome Scott
Hutz, Reinhold Josef
Lardy, Henry Arnold
Lee, Ping-Cheung
Lelkes, Peter Istvan
Lorton, Steven Paul
Potter, Van Rensselaer
Ris, Hans
Schatten, Gerald Phillip
Slautterback, David Buell
Tandon, Shiv R
Terasawa, EI

PUERTO RICO
Carrasco-Serrano, Clara E
Kuffler, Damien Paul
Santacana, Guido E

ALBERTA
Bentley, Michael Martin
Hart, David Arthur
MacIntosh, Brian Robert
Vance, Jean E

BRITISH COLUMBIA
Brandhorst, Bruce Peter
Cavalier-Smith, Thomas
Durand, Ralph Edward
Eaves, Allen Charles Edward
Setterfield, George Ambrose
Spiegelman, George Boole

MANITOBA
Kardami, Elissavet
Zahradka, Peter

NEWFOUNDLAND
Sheath, Robert Gordon

NOVA SCOTIA
MacRae, Thomas Henry

ONTARIO
Adams, Gabrielle H M
Atwal, Onkar Singh
Bear, Christine Eleanor
Buchwald, Manuel
Chambers, Ann Franklin
Chang, Patricia Lai-Yung
Fallding, Margaret Hurlstone Hardy
Foskett, J Kevin
Ghosh, Hara Prasad
Gurd, James W
Haj-Ahmad, Yousef
Heath, Ian Brent
Horton, Roger Francis
Jones, Owen Thomas
Kidder, Gerald Marshall
Lee, Eng-Hong
Legge, Raymond Louis
Mickle, Donald Alexander
O'Day, Danton Harry
Percy, Maire Ede
Scadding, Steven Richard
Siminovitch, David
Wasi, Safia
Young, Paul Gary
Zobel, Alicja Maria

QUEBEC
Blaschuk, Orest William
Chafouleas, James G
Dufresne, Louise
Germain, Lucie
Guillemette, Gaetan
Jothy, Serge
Lagace, Lisette
Morais, Rejean

OTHER COUNTRIES
Avila, Jesus
Bunn, Clive Leighton
Chang, Donald Choy
Chang, Yu-Sun
Chen, Jan-Kan
Euteneuer, Ursula Brigitte
Guimaraes, Romeu Cardoso
Hayashi, Masao
Ikehara, Yukio
Jacob, Francois
Karasaki, Shuichi

Lane, Nancy Jane
McBeath, Elena
Nagata, Kazuhiro
Sakmann, Bert
Zenk, Meinhart Hans Christian

Cytology

ALABAMA
Blalock, James Edwin
Bowen, William R
Dietz, Robert Austin
Dorgan, William Joseph
Furuto, Donald K
Gaffney, Edwin Vincent
Gray, Bruce William
Hajduk, Stephen Louis
Moore, Bobby Graham
Sapp, Walter J
Sapra, Val T
Steward, Frederick Campion
Watson, Jack Ellsworth
Wilborn, Walter Harrison
Williams, John Watkins, III
Wyss, James Michael

ARIZONA
Aposhian, Hurair Vasken
Capco, David G
Chandler, Douglas Edwin
Doane, Winifred Walsh
Endrizzi, John Edwin
Ferris, Wayne Robert
Gerner, Eugene Willard
Goll, Darrel Eugene
Grim, J(ohn)Norman
Harkins, Kristi R
Holmgren, Paul
Kischer, Clayton Ward
Pinkava, Donald John
Pommerville, Jeffrey Carl
Shimizu, Nobuyoshi
Trelease, Richard Norman

ARKANSAS
Dippell, Ruth Virginia
Evans, William L
Johnson, Bob Duell
Townsend, James Willis

CALIFORNIA
Adinolfi, Anthony M
Alfert, Max
Amoore, John Ernest
Andrus, William DeWitt, Jr
Antipa, Gregory Alexis
Arcadi, John Albert
Armstrong, Peter Brownell
Barker, Mary Elizabeth
Bartholomew, James Collins
Basbaum, Carol Beth
Beers, William Howard
Bernard, George W
Berns, Michael W
Bernstein, Emil Oscar
Bils, Robert F
Bissell, Mina Jahan
Bizzoco, Richard Lawremce Weiss
Blanks, Janet Marie
Bloom, Floyd Elliott
Bok, P Dean
Brasch, Klaus Rainer
Breisch, Eric Alan
Brodsky, Frances M(artha)
Bryant, Susan Victoria
Burns, Victor Will
Burnside, Mary Beth
Burwen, Susan Jo
Cailleau, Relda
Cande, W Zacheus
Cantor, Marvin H
Cascarano, Joseph
Clark, William R
Clothier, Galen Edward
Cohen, Morris
Connell, Carolyn Joanne
Cooper, Kenneth Willard
Cotman, Carl Wayne
Cronshaw, James
Cunningham, Dennis Dean
Daniel, Ronald Scott
Das, Nirmal Kanti
Davis, Roger Alan
De Francesco, Laura
Deitch, Arline D
Demaree, Richard Spottswood, Jr
Dirksen, Ellen Roter
Dodge, Alice Hribal
Dunnebacke-Dixon, Thelma Hudson
Dvorak, Jan
Ehrenfeld, Elvera
Eiserling, Frederick A
Estilai, Ali
Falk, Richard H
Feeney-Burns, Mary Lynette
Fisher, Knute Adrian
Fisher, Steven Kay
Flashman, Stuart Milton
Fosket, Donald Elston
Friend, Daniel S
Fristrom, Dianne
Gifford, Ernest Milton
Ginsberg, Mark Howard

Glabe, Charles G
Golder, Thomas Keith
Golub, Edward S
Gordon, Manuel Joe
Grey, Robert Dean
Griffith, Donal Louis
Hackett, Nora Reed
Haimo, Linda T
Hamkalo, Barbara Ann
Hanson, James Charles
Harris, John Wayne
Harris, Morgan
Herschman, Harvey R
Hess, Frederick Dan
Hessinger, David Alwyn
Ignarro, Louis Joseph
Jeffery, William Richard
Jenkins, Burton Charles
Jensen, Ronald Harry
Johnston, George Robert
Jones, Gary Edward
King, Eileen Brenneman
Klevecz, Robert Raymond
Kluss, Byron Curtis
Kornberg, Thomas B
Lake, James Albert
LaVail, Matthew Maurice
Lebo, Roger Van
Leif, Robert Cary
Lewiston, Norman James
Long, John Arthur
Loughman, William D
Lucas, Joe Nathan
Lyke, Edward Bonsteel
McGaughey, Charles Gilbert
McHale, John T
Mak, Linda Louise
Mayall, Brian Holden
Moore, Betty Clark
Nasatir, Maimon
Nelson-Rees, Walter Anthony
Nemere, Ilka M
Niklowitz, Werner Johannes
Ohno, Susumu
Ohnuki, Yasushi
Okada, Tadashi A
Palade, George E
Parry, Gordon
Parsons, John Arthur
Pendse, Pratapsinha C
Perkins, David Dexter
Plopper, Charles George
Pollock, Edward G
Raju, Namboori Bhaskara
Ralph, Peter
Revel, Jean Paul
Ribak, Charles Eric
Rick, Charles Madeira, Jr
Rosen, Steven David
Schechter, Joel Ernest
Schmidt, Barbara A
Schmucker, Douglas Lees
Schooley, Caroline Naus
Schraer, Rosemary
Sekhon, Sant Singh
Sercarz, Eli
Shih, Ching-Yuan G
Simpson, Larry P
Smith, Martyn Thomas
Snow, Mikel Henry
Speicher, Benjamin Robert
Stanbridge, Eric John
Stern, Herbert
Stewart, Joan Godsil
Stoeckenius, Walther
Strobel, Edward
Stryer, Lubert
Szabo, Arlene Slogoff
Szego, Clara Marian
Talbot, Prudence
Thomson, William Walter
Tokuyasu, Kiyoteru
Vacquier, Victor Dimitri
Vande Berg, Jerry Stanley
Vanderlaan, Martin
Weigle, William O
Weihing, Robert Ralph
Weinstock, Alfred
Whitney, Kenneth Dean
Wissig, Steven
Wolff, Sheldon
Wood, Richard Lyman
Zieg, Roger Grant

COLORADO
Angell, Robert Walker
Bonneville, Mary Agnes
Eisenbarth, George Stephen
Fifkova, Eva
Fotino, Mircea
Fox, Michael Henry
Gabridge, Michael Gregory
Giddings, Thomas H, Jr
Gorthy, Willis Charles
Hahn, William Eugene
Ham, Richard George
Henson, Peter Mitchell
Hogan, Christopher James
Horak, Donald L
Klymkowsky, Michael W
Lee, Robert Edward
McIntosh, John Richard
Maylie-Pfenninger, M F
Sanderson, Richard James

Cytology (cont)

Snyder, Judith Armstrong
Stack, Stephen M
Staehelin, Lucas Andrew
Stein, Gretchen Herpel
Stone, Gordon Emory
Tsuchiya, Takumi

CONNECTICUT
Anderson, Gregory Joseph
Ashley, Terry Fay
Beitch, Irwin
Blackburn, Daniel Glenn
Child, Frank Malcolm
Constantine-Paton, Martha
Cooperstein, Sherwin Jerome
Dembitzer, Herbert
Grasso, Joseph Anthony
Hand, Arthur Ralph
Hogan, James C
Jamieson, James Douglas
Kent, John Franklin
Koerting, Lola Elisabeth
Longwell, Arlene Crosby (Mazzone)
Milici, Anthony J
Moellmann, Gisela E Bielitz
Mundkur, Balaji
Myles, Diana Gold
Patton, Curtis LeVerne
Pochron, Sharon
Rakic, Pasko
Roman, Laura M
Rosenbaum, Joel L
Simpson, Tracy L
Stevenson, Harlan Quinn
Sun, Alexander Shihkaung
Trinkaus, John Philip
Vitkauskas, Grace
Wachtel, Allen W
Woods, John Whitcomb

DELAWARE
Borgaonkar, Digamber Shankarrao
Chen, Harry Wu-Shiong
Giaquinta, Robert T
Howard, Richard James
Ruben, Regina Lansing
Wagner, Roger Curtis

DISTRICT OF COLUMBIA
Albert, Ernest Narinder
Baldwin, Kate M
Ball, William David
Chapman, George Bunker
Chiarodo, Andrew
Fleming, Patrick John
Griffin, Joe Lee
Harris, Rudolph
Humphreys, Susie Hunt
Hunter, Jehu Callis
Jones, Janice Lorraine
Kapur, Shakti Prakash
Leak, Lee Virn
Mullins, James Michael
Schiff, Stefan Otto
Wrathall, Jean Rew

FLORIDA
Barkalow, Derek Talbot
Bourguignon, Lilly Y W
Braunschweiger, Paul G
Bunge, Mary Bartlett
Bunge, Richard Paul
Carraway, Coralie Anne Carothers
Chambers, Edward Lucas
Chegini, Nasser
Chen, L T
Cohen, Glenn Milton
Coleman, Sylvia Ethel
Dawes, Clinton John
Deats, Edith Potter
Ehrlich, Howard George
Gennaro, Joseph Francis
Goldberg, Walter M
Hayashi, Teru
Hinkley, Robert Edwin, Jr
Hurst, Josephine M
Klein, Richard Lester
Luykx, Peter (van Oosterzee)
MacDonald, Eve Lapeyrouse
Murison, Gerald Leonard
Nicosia, Santo Valerio
Noonan, Kenneth Daniel
Polson, Charles David Allen
Rose, Birgit
Ross, Michael H
Salas, Pedro Jose I
Schank, Stanley Cox
Scott, Walter Alvin
Selman, Kelly
Smith, David Spencer
Warmke, Harry Earl
Warren, Richard Joseph
Warren, Robert Holmes
Weber, James Edward
Wilfret, Gary Joe

GEORGIA
Adkison, Claudia R
Andrews, Lucy Gordon
Bates, Harold Brennan, Jr
Bryan, John Henry Donald
Byrd, J Rogers

Davis, Herbert L, Jr
Fritz, Michael E
Gulati, Adarsh Kumar
Hamada, Spencer Hiroshi
Hassold, Terry Jon
Haynes, John Kermit
Herman, Chester Joseph
Howard, Eugene Frank
Jones, Betty Ruth
McKinney, Ralph Vincent, Jr
Merkle, Roberta K
Ove, Peter
Parker, Curtis Lloyd
Patterson, Rosalyn Mitchell
Paulsen, Douglas F
Priest, Robert Eugene
Rasmussen, Howard
Ritter, Hope Thomas Martin, Jr
Romanovicz, Dwight Keith
Satya-Prakash, K L
Schuster, George Sheah
Sharkey, Margaret Mary
Stevens, Ann Rebecca
Welter, Dave Allen

HAWAII
Ahearn, Gregory Allen
Fok, Agnes Kwan
Hemmes, Don E
Kikudome, Gary Yoshinori
Kleinfeld, Ruth Grafman
Sagawa, Yoneo
Yang, Hong-Yi

IDAHO
D'Aoust, Brian Gilbert
Myers, James Robert
Rourke, Arthur W

ILLINOIS
Albach, Richard Allen
Albrecht-Buehler, Guenter Wilhelm
Aydelotte, Margaret Beesley
Bartles, James Richard
Becker, Robert Paul
Chang, Kwang-Poo
Chuang, Tsan Iang
Cole, Madison Brooks, Jr
Cordes, William Charles
Crang, Richard Francis Earl
Daniel, Jon Cameron
Daniels, Edward William
Decker, Robert Scott
Doyle, William Lewis
Dybas, Linda Kathryn
Farbman, Albert Irving
Garber, Edward David
Giordano, Tony
Goldman, Robert David
Grdina, David John
Griesbach, Robert Anthony
Hadley, Henry Hultman
Hanzely, Laszlo
Ingram, Forrest Duane
Kethley, John Bryan
Khodadad, Jena Khadem
Kuczmarski, Edward R
Kulfinski, Frank Benjamin
LaVelle, Faith Wilson
Lerner, Jules
Leven, Robert Maynard
Levy, Michael R
Loizzi, Robert Francis
Ma, Te Hsiu
McNulty, John Alexander
Martin, Terence Edwin
Menco, Bernard Philip Max
Millhouse, Edward W, Jr
Morgan, Juliet
Nakajima, Yasuko
Nardi, James Benjamin
Novales, Ronald Richards
O'Morchoe, Patricia Jean
Overton, Jane Harper
Pappelis, Aristotel John
Plewa, Michael Jacob
Polet, Herman
Preisler, Harvey D
Puchalski, Randall Francis
Rotermund, Albert J, Jr
Rowley, Janet D
Schleicher, Joseph Bernard
Schreiber, Hans
Seed, Thomas Michael
Spear, Brian Blackburn
Straus, Werner
Swift, Hewson Hoyt
Telser, Alvin Gilbert
Trout, Jerome Joseph
Verhage, Harold Glenn
Weber, David Frederick
Willey, Ruth Lippitt
Wolosewick, John J
Zalisko, Edward John

INDIANA
Asai, David J
Ashendel, Curtis Lloyd
BeMiller, Paraskevi Mavridis
Bracker, Charles E
Chiscon, Martha Oakley
Chrisman, Charles Larry
Glover, David Val
Goff, Charles W

Grove, Stanley Neal
Gunther, Gary Richard
Hammond, Charles Thomas
Klaunig, James E
Maloney, Michael Stephen
Moskowitz, Merwin
Palmer, Catherine Gardella
Phadke, Kalindi
Rai, Karamjit Singh
St John, Philip Alan
Schoknecht, Jean Donze
Sheffer, Richard Douglas
Sinclair, John Henry
Stuart, Jeffrey James
Tischfield, Jay Arnold
Togasaki, Robert K
Tweedell, Kenyon Stanley
Williams, Daniel Charles
Winicur, Sandra
Zilz, Melvin Leonard

IOWA
Carlson, Wayne R
Durkee, Lenore T
Kodama, Robert Makoto
Maynard, Jerry Allen
Meetz, Gerald David
Orr, Alan R
Runyan, William Scottie
Sandra, Alexander
Shaw, Gaylord Edward
Stromer, Marvin Henry
Sullivan, Charles Henry
Tomes, Dwight Travis
Welshons, William John

KANSAS
Burchill, Brower Rene
Burton, Paul Ray
Dentler, William Lee, Jr
Funderburgh, James Louis
Haufler, Christopher Hardin
Kimler, Bruce Franklin
Klein, Robert Melvin
Poisner, Alan Mark
Sarras, Michael P, Jr
Spooner, Brian Sandford
Westfall, Jane Anne

KENTUCKY
Hoffman, Eugene James
Humphries, Asa Alan, Jr
Pavlik, Edward John
Sisken, Jesse Ernest
Traurig, Harold H
Wolfson, Alfred M

LOUISIANA
Allen, Emory Raworth
Baum, Lawrence Stephen
Cowden, Ronald Reed
Di Mario, Patrick Joseph
Fuseler, John William
Gallaher, William Richard
Hurley, Maureen
Jacks, Thomas Jerome
Jeter, James Rolater, Jr
Kasten, Frederick H
Lanners, H Norbert
Lin, James C H
Lumsden, Richard
Nickerson, Stephen Clark
Pisano, Joseph Carmen
Ramsey, Paul Roger
Specian, Robert David
Wakeman, John Marshall
Weidner, Earl
Yates, Robert Doyle
Yeh, Kwo-Yih

MAINE
Bell, Allen L
Chacko, Rosy J
Cook, James Richard
Davisson, Muriel Trask
Greenwood, Paul Gene
Hunter, Susan Julia
Leiter, Edward Henry
Roberts, Franklin Lewis
Waymouth, Charity

MARYLAND
Aamodt, Roger Louis
Adelman, Mark Robert
Adler, Ruben
Attallah, A M
Barry, Sue-ning C
Bell, Mary
Berger, Edward Alan
Bodammer, Joel Edward
Bradlaw, June A
Broadwell, Richard Dow
Brown, Joshua Robert Calloway
Buck, John Bonner
Buck, Raymond Wilbur, Jr
Chang, Yao Teh
Chen, T R
Chu, Elizabeth Wann
Coe, Gerald Edwin
Cohen, Maimon Moses
Cushman, Samuel Wright
Daniels, Mathew Paul
Dwivedi, Radhey Shyam
Evans, Virginia John

Finch, Robert Allen
Frank, Martin
Gall, Joseph Grafton
Gallo, Robert C
Garfield, Sanford Allen
Gobel, Stephen
Goldsmith, Paul Kenneth
Goode, Melvyn Dennis
Greenhouse, Gerald Alan
Griesbach, Robert James
Gwynn, Edgar Percival
Hamburger, Anne W
Hanover, John Allan
Hascall, Gretchen Katharine
Haudenschild, Christian C
Hay, Robert J
Hegyeli, Ruth I E J
Heine, Ursula Ingrid
Horenstein, Evelyn Anne
Hunt, Lois Turpin
Imberski, Richard Bernard
Iype, Pullolickal Thomas
Kang, Yuan-Hsu
Kidwell, William Robert
Kirby, Paul Edward
Kloetzel, John Arthur
Krakauer, Teresa
Leach, William Matthew
Lim, David J
Lin, Diane Chang
Linden, Carol D
Lippincott-Schwartz, Jennifer
Lowensohn, Howard Stanley
McAtee, Lloyd Thomas
McGarrity, Gerard John
Malmgren, Richard Axel
Maupin, Pamela
Meszler, Richard M
Moorhead, Paul Sidney
Moreira, Jorge Eduardo
Murphy, Douglas Blakeney
Musson, Robert A
Nauman, Robert Karl
Nayak, Ramesh Kadbet
Neale, Elaine Anne
Newton, Sheila A
Pagano, Richard Emil
Phelps, Patricia C
Pitlick, Frances Ann
Pluznik, Dov Herbert
Podskalny, Judith Mary
Pollard, Thomas Dean
Price, Samuel
Read-Connole, Elizabeth Lee
Reissig, Magdalena
Resau, James Howard
Robey, Pamela Gehron
Robinson, David Mason
Robinson-White, Audrey Jean
Robison, Wilbur Gerald, Jr
Schneider, Walter Carl
Schuetz, Allen W
Shelton, Emma
Silverman, David J
Small, Eugene Beach
Smith, Gilbert Howlett
Spatz, Maria
Taylor, William George
Tjio, Joe Hin
Trump, Benjamin Franklin
Vigil, Eugene Leon
Vincent, Monroe Mortimer
Vinores, Stanley Anthony
Warmbrodt, Robert Dale
Wenk, Martin Lester
Wergin, William Peter
Wollman, Seymour Horace
Wolniak, Stephen M
Wray, Granville Wayne
Zirkin, Barry Ronald

MASSACHUSETTS
Albert, Mary Roberts Forbes (Day)
Belt, Warner Duane
Berlowitz Tarrant, Laurence
Billingham, Rupert Everett
Borek, Carmia Ganz
Bursztajn, Sherry
Chang, Amy Y
Chlapowski, Francis Joseph
Chodosh, Sanford
Cintron, Charles
Cohen, Carl M
Cohen, Samuel H
Cornman, Ivor
Curtis, Joseph C
D'Amore, Patricia Ann
Davidson, Samuel James
Deegan, Linda Ann
Dice, J Fred
Dorey, Cheryl Kathleen
Ducibella, Tom
Erikson, Raymond Leo
Fine, Richard Eliot
Fink, Rachel Deborah
Frederick, Sue Ellen
Gauthier, Geraldine Florence
Gerbi, Susan Alexandra
Gimbrone, Michael Anthony, Jr
Goldsby, Richard Allen
Haimes, Howard B
Harding, Clifford Vincent, Jr
Harrison, Bettina Hall
Hausman, Robert Edward

Hepler, Peter Klock
Herman, Ira Marc
Ito, Susumu
Kirschner, Marc Wallace
Krane, Stanley Garson
Lawrence, Jeanne Bentley
Lehman, William Jeffrey
Liss, Robert H
Margulis, Lynn
Moner, John George
Neutra, Marian R
Padykula, Helen Ann
Pardue, Mary Lou
Paul, David Louis
Pederson, Thoru Judd
Penman, Sheldon
Pino, Richard M
Poccia, Dominic Louis
Pryor, Marilyn Ann Zirk
Rafferty, Nancy S
Remillard, Stephen Philip
Reyero, Cristina
Rivera, Ezequiel Ramirez
Rollins, Reed Clark
Ryan, Una Scully
Ryser, Hugues Jean-Paul
Schneeberger, Eveline E
Schwartz, Martin Alexander
Seals, Jonathan Roger
Shahrik, H Arto
Skobe, Ziedomis
Sluder, Greenfield
Smith, Dennis Matthew
Sonnenschein, Carlos
Stephens, Raymond Edward
Stossel, Thomas Peter
Sullivan, Susan Jean
Swanson, Carl Pontius
Ting, Yu-Chen
Toole, Bryan Patrick
Trinkaus-Randall, Vickery E
Ulrich, Frank
Williams, Mary Carol
Williamson, Patrick Leslie
Woodcock, Christopher Leonard Frank
Young, Delano Victor
Zackroff, Robert V
Zeldin, Michael Hermen
Zimmerman, William Frederick

MICHIGAN
Adams, Jack Donald
Aggarwal, Surinder K
Al Saadi, A Amir
Al-Saadi, Abdul A
Andresen, Norman A
Bagchi, Mihir
Bedrossian, Carlos Wanes Menino
Bhuyan, Bijoy Kumar
Bollinger, Robert Otto
Bouma, Hessel, III
Bowers, Maynard C
Boxer, Laurence A
Choe, Byung-Kil
Cronkite, Donald Lee
Erbisch, Frederic H
Ernst, Stephen Arnold
Essner, Edward Stanley
Falls, William McKenzie
Floyd, Alton David
Gordon, Sheldon Robert
Gray, Robert Howard
Han, Seong S
Heidemann, Steven Richard
Jacobs, Charles Warren
Kim, Sun-Kee
Kleinsmith, Lewis Joel
Krishan, Awtar
Levine, Laurence
Lindemann, Charles Benard
Nag, Asish Chandra
Pence, Leland Hadley
Pourcho, Roberta Grace
Pysh, Joseph John
Reddan, John R
Rizki, Tahir Mirza
Schmitter, Ruth Elizabeth
Sicko-Goad, Linda May
Sink, Kenneth C, Jr
Tosney, Kathryn W
Unakar, Nalin J
Walker, Glenn Kenneth
Wiener, Joseph

MINNESOTA
Bauer, Gustav Eric
Beitz, Alvin James
Blanchette, Robert Anthony
Blumenfeld, Martin
Cahoon, Sr Mary Odile
Downing, Stephen Ward
Edds, Kenneth Tiffany
Elde, Robert Philip
Forbes, Donna Jean
Gooch, Van Douglas
Heidcamp, William H
Jalal, Syed M
Johnson, Ross Glenn
Korte, Clare A
Linck, Richard Wayne
Malewicz, Barbara Maria
Nelson, John Daniel
Oetting, William Starr
Palm, Sally Louise

Phillips, Ronald Lewis
Rosenberg, Murray David
Rottmann, Warren Leonard
Severson, Arlen Raynold
Sheppard, John Richard
Shier, Wayne Thomas
Silverman, William Bernard
Stadelmann, Eduard Joseph
Thompson, Edward William
Weber, Alvin Francis

MISSISSIPPI
Hare, Mary Louise Eckles
Howse, Harold Darrow
Martin, Billy Joe
St Amand, Wilbrod
Wise, Dwayne Allison

MISSOURI
Beringer, Theodore Michael
Berlin, Jerry D
Bischoff, Eric Richard
Broschat, Kay O
Brown, David Hazzard
Burdick, Allan Bernard
Cook, Nathan Howard
Goldstein, Milton Norman
Goodenough, Ursula Wiltshire
Gordon, Albert Raye
Gustafson, John Perry
Harakas, N(icholas) Konstantinos
Kimber, Gordon
Kort, Margaret Alexander
McQuade, Henry Alonzo
Melnykovych, George
Menton, David Norman
Montague, Michael James
Rifas, Leonard
Sears, Ernest Robert
Sheridan, Judson Dean
Stahl, Philip Damien
Tai, William
Taylor, John Joseph
Tolmach, L(eonard) J(oseph)

MONTANA
Bilderback, David Earl
Linner, John Gunnar
Thompson, Holly Ann

NEBRASKA
Bieber, Raymond W
Crouse, David Austin
Fawcett, James Davidson
McCue, Robert Owen
Rhode, Solon Lafayette, III
Rodriguez-Sierra, Jorge F
Rosenquist, Thomas H
Sharp, John Graham
Sheehan, John Francis
Yee, John Alan

NEVADA
Nauman, Charles Hartley
Stratton, Clifford James

NEW HAMPSHIRE
Carpenter, Stanley John
Fagerberg, Wayne Robert
Handler, Evelyn Erika
Langford, George Malcolm
Manasek, Francis John
Oldfield, Daniel G
Schreiber, Richard William
Sokol, Hilda Weyl
Stanton, Bruce Alan
Thomas, William Albert

NEW JERSEY
Bayne, Ellen Kahn
Berendsen, Peter Barney
Blackburn, Gary Ray
Boltz, Robert Charles, Jr
Ciosek, Carl Peter, Jr
Colby, Richard H
Collier, Marjorie McCann
Eastwood, Abraham Bagot
Fiedler-Nagy, Christa
Goggins, Jean A
Gray, Harry Edward
Greene, Arthur E
Gupta, Godaveri Rawat
Hansen, Hans John
Hyndman, Arnold Gene
Kelly, Robert P
Khavkin, Theodor
Kleinschuster, Stephen J, III
Kmetz, John Michael
Kung, Ted Teshih
Laemle, Lois K
Lipkin, George
Lutz, Richard Arthur
Mack, James Patrick
Malamed, Sasha
Massover, William H
Mueller, Stephen Neil
Mullins, Deborra E
Penningroth, Stephen Meader
Prensky, Wolf
Rohrs, Harold Clark
Rossow, Peter William
Ruknudin, Abdul Majeed
Schroff, Peter David
Seiden, David

Singer, Irwin I
Singer, Robert Mark
Steinberg, Malcolm Saul
Strahs, Kenneth Robert
Studzinski, George P
Tesoriero, John Vincent
Turner, Matthew X
Vena, Joseph Augustus
Wille, John Jacob, Jr
Wilson, Frank Joseph

NEW MEXICO
Bartholdi, Marty Frank
Bear, David George
Bourne, Earl Whitfield
Crissman, Harry Allen
Deaven, Larry Lee
Dunford, Max Patterson
Johnson, Neil Francis
Nelson, Mary Anne
Oliver, Janet Mary
Saland, Linda C
Stricker, Stephen Alexander
Tobey, Robert Allen
Trotter, John Allen
Vogel, Kathryn Giebler

NEW YORK
Aist, James Robert
Alsop, David W
Ambron, Richard Thomas
Anderson, O(rvil) Roger
Anversa, Piero
Baker, Raymond Milton
Barr, Charles E
Bartfeld, Harry
Bassett, Charles Andrew Loockerman
Beard, Margaret Elzada
Beckert, William Henry
Bender, Michael A
Berezney, Ronald
Bergeron, John Albert
Bleyman, Lea Kanner
Blobel, Gunter
Bogart, Bruce Ian
Brett, Betty Lou Hilton
Brooks, John S J
Brown, William J
Brownscheidle, Carol Mary
Bushkin, Yuri
Carriere, Rita Margaret
Chiquoine, A Duncan
Cline, Sylvia Good
Cox, Dudley
Cramer, Eva Brown
Darnell, James Edwin, Jr
Darzynkiewicz, Zbigniew Dzierzykraj
De Duve, Christian Rene
De Lemos, Carmen Loretta
De Luca, Chester
DeLuca, Patrick John
Deschner, Eleanor Elizabeth
De Terra, Noël
Detmers, Patricia Anne
Drakontides, Anna Barbara
Drickamer, Kurt
Dziak, Rose Mary
Earle, Elizabeth Deutsch
Eckert, Barry S
Engelhardt, Dean Lee
Etlinger, Joseph David
Faust, Irving M
Fischman, Harlow Kenneth
Fisher, Paul Andrew
Friedman, Eileen Anne
Friedman, Marc Mitchell
Gabrusewycz-Garcia, Natalia
Galinsky, Irving
Garner, James G
Gil, Joan
Godec, Ciril J
Godman, Gabriel C
Goldfischer, Sidney L
Gordon, Mildred Kobrin
Goyert, Sanna Mather
Greenberg, Jay R
Gueft, Boris
Haines, Thomas Henry
Hammill, Terrence Michael
Hanley, Kevin Joseph
Hansen, John Theodore
Hard, Robert Paul
Herman, Lawrence
Herz, Fritz
Himes, Marion
Hitzeman, Jean Walter
Hoch, Harvey C
Holtzman, Eric
Hotchkiss, Sharon K
Huang, Chester Chen-Chiu
Huberman, Joel Anthony
Humber, Richard Alan
Israel, Herbert William
Jensen, Thomas E
Johnson, Patricia R
Jones, C Robert
Kaye, Nancy Weber
Koss, Leopold George
Kreibich, Gert
Krim, Mathilde
LaFountain, James Robert, Jr
Lavett, Diane Kathryn Juricek
Ledbetter, Myron C
Liem, Ronald Kian Hong

Lin, Yue Jee
Lorch, Joan
Loud, Alden Vickery
Luck, David Jonathan Lewis
Lyser, Katherine May
McCann-Collier, Marjorie Dolores
McKeown-Longo, Paula Jean
McLean, Robert J
McMahon, Rita Mary
Magdoff-Fairchild, Beatrice
Manzo, Rene Paul
Mazurkiewicz, Joseph Edward
Merriam, Robert William
Metzner, Jerome
Millis, Albert Jason Taylor
Mitra, Jyotirmay
Moriber, Louis G
North, Robert J
Notter, Mary Frances
Novikoff, Phyllis Marie
Nur, Uzi
O'Brien, James Francis
Ohr, Eleonore A
Olmsted, Joanna Belle
Orlic, Donald
Ornstein, Leonard
Orsi, Ernest Vinicio
Padawer, Jacques
Panessa-Warren, Barbara Jean
Partin, John Calvin
Penney, David P
Phillips, David Mann
Pokorny, Kathryn Stein
Puszkin, Saul
Quigley, James P
Raventos-Suarez, Carmen Elvira
Rich, Abby M
Rieder, Conly LeRoy
Robbins, Edith Schultz
Rome, Doris Spector
Rosenberg, Warren L
Rothstein, Howard
Rothwell, Norman Vincent
Ruben, Robert Joel
Saks, Norman Martin
Schin, Kissu
Schuel, Herbert
Schuh, Joseph Edward
Shafiq, Saiyid Ahmad
Sharp, William R
Shields, Dennis
Shopsis, Charles S
Skinner, Kathleen Mary
Sklarew, Robert J
Skoultchi, Arthur
Sloboda, Adolph Edward
Slocum, Harry Kim
Soeiro, Ruy
Soifer, David
Spector, David Lawrence
Straubinger, Robert M
Sutherland, Betsy Middleton
Swinton, David Charles
Szalay, Jeanne
Sztul, Elizabeth Sabina
Tilley, Shermaine Ann
Tomasz, Alexander
Tonna, Edgar Anthony
Vanderberg, Jerome Philip
Van't Hof, Jack
Viceps-Madore, Dace I
Warner, Jonathan Robert
Wasserman, Marvin
Waxman, Samuel
Weiss, Leonard
Weissmann, Gerald
Wenk, Eugene J
Wheeless, Leon Lum, Jr
Wolgemuth, Debra Joanne
Worgul, Basil Vladimir
Yang, Shung-Jun
Younghans, Wayne N
Zaleski, Marek Bohdan
Zanetti, Nina Clare
Zeigel, Robert Francis
Zucker-Franklin, Dorothea
Zuzolo, Ralph C

NORTH CAROLINA
Allen, Charles Marvin
Banerjee, Umesh Chandra
Brown, William Roy
Browne, Carole Lynn
Chan, Po Chuen
Crouse, Helen Virginia
Faust, Robert Gilbert
Fletcher, Donald James
Gruber, Helen Elizabeth
Hackenbrock, Charles Robert
Hall, Iris Beryl Haddon
Hanker, Jacob S
Harris, Albert Kenneth, Jr
Herman, Brian
Johnston, William Webb
Juliano, Rudolph Lawrence
Kalmus, Gerhard Wolfgang
Lee, Greta Marlene
Maroni, Gustavo Primo
Matthews, James Francis
Mickey, George Henry
Misch, Donald William
Moses, Montrose James
Motwani, Nalini M
Nicklas, Robert Bruce

Cytology (cont)

Padilla, George M
Page, Rodney Lee
Phillips, Lyle Llewellyn
Preston, Robert Julian
Pukkila, Patricia Jean
Roberts, John Fredrick
Salmon, Edward Dickinson
Sheetz, Michael Patrick
Stalker, Harold Thomas
Therrien, Chester Dale
Thomas, Mary Beth
Timothy, David Harry
Ting-Beall, Hie Ping
Triantaphyllou, Anastasios Christos
Ward, Robert T
Weintraub, Robert Louis
Yongue, William Henry

NORTH DAKOTA
Freeman, Thomas Patrick
Hunt, Curtiss Dean
Jauhar, Prem P
La Chance, Leo Emery
LaDuke, John Carl
Riemann, John G

OHIO
Budd, Geoffrey Colin
Cardell, Robert Ridley, Jr
Chakraborty, Joana J(yotsna)
Cohn, Norman Stanley
Cope, Frederick Oliver
Craft, Thomas Jacob, Sr
Culp, Lloyd Anthony
DiCorleto, Paul Eugene
Dollinger, Elwood Johnson
Downey, Ronald J
Egar, Margaret Wells
Flaumenhaft, Eugene
Floyd, Gary Leon
Gao, Kuixiong
Gesinski, Raymond Marion
Gilloteaux, Jacques Jean-Marie A
Greider, Marie Helen
Gwatkin, Ralph Buchanan Lloyd
Hamlett, William Cornelius
Heckman, Carol A
Hink, Walter Fredric
Hitt, John Burton
Holderbaum, Daniel
Honda, Shigeru Irwin
Ip, Wallace
Kaetzel, Marcia Aldyth
Kaneshiro, Edna Sayomi
Keefe, John Richard
Kreutzer, Richard D
Lane, Richard Durelle
Macintyre, Malcolm Neil
Macintyre, Stephen S
Milsted, Amy
Moore, Fenton Daniel
Moore, Randy
Paddock, Elton Farnham
Parysek, Linda M
Pribor, Donald B
Przybylski, Ronald J
Robinson, John Mitchell
Samols, David R
Schwelitz, Faye Dorothy
Snider, Jerry Allen
Strauch, Arthur Roger, III
Sunkara, Sai Prasad
Tandler, Bernard
Vaughn, Jack C
Watanabe, Michiko
Yemma, John Joseph
Zambernard, Joseph
Zull, James E

OKLAHOMA
Allen, Robert Wade
Bantle, John Albert
Bell, Paul Burton, Jr
Bell, Rondal E
Birckbichler, Paul Joseph
Chung, Kyung Won
Clarke, Margaret Burnett
De Bault, Lawrence Edward
Dugan, Kimiko Hatta
Duncan, Michael Robert
Kollmorgen, G Mark
Lerner, Michael Paul
McClung, J Keith
Murphy, Marjory Beth
Papka, Raymond Edward
Ramon, Serafin
Roszel, Jeffie Fisher
Russell, Scott D
Thornton, John William

OREGON
Bajer, Andrew
Bethea, Cynthia Louise
Brenner, Robert Murray
Conte, Frank Philip
Dooley, Douglas Charles
Fahrenbach, Wolf Henrich
Florance, Edwin R
Harris, Patricia J
Haunold, Alfred
Johnson, John Morris
Montagna, William

Newman, Lester Joseph
Owczarzak, Alfred
Terwilliger, Nora Barclay
Wimber, Donald Edward

PENNSYLVANIA
Abel, John H, Jr
Alperin, Richard Junius
Anderson, Neil Owen
Atkinson, Barbara
Baker, Wilber Winston
Baumrucker, Craig Richard
Bayer, Manfred Erich
Bean, Barry
Bennett, Henry Stanley
Bibbo, Marluce
Billheimer, Foster E
Birchem, Regina
Black, Mark Morris
Brooks, John J
Burholt, Dennis Robert
Bursey, Charles Robert
Campbell, Graham Le Mesurier
Cannizzaro, Linda A
Castor, LaRoy Northrop
Cauna, Nikolajs
Colony-Cokely, Pamela
Cooke, Peter Hayman
Coss, Ronald Allen
DeCaro, Thomas F
DiBerardino, Marie A
Farber, Phillip Andrew
Flemister, Sarah C
Flesch, David C
Fluck, Richard Allen
Freed, Jerome James
Fussell, Catharine Pugh
Gabriel, Edward George
Gersh, Eileen Sutton
Gilbert, Susan Pond
Gross, Dennis Michael
Grun, Paul
Halleck, Margaret S
Hansburg, Daniel
Heald, Charles William
Hopper, James Ernest
Horan, Paul Karl
Hymer, Wesley C
Jacobsen, Terry Dale
Jarvik, Jonathan Wallace
Kennedy, Ann Randtke
Kleinzeller, Arnost
Knobler, Robert Leonard
Knudsen, Karen Ann
Koprowska, Irena
Koros, Aurelia M Carissimo
Kvist, Tage Nielsen
Ladman, Aaron J(ulius)
La Noue, Kathryn F
Levine, Elliot Myron
Levine, Rhea Joy Cottler
Lieberman, Irving
Lockwood, Arthur H
Lotze, Michael T
McDonald, Barbara Brown
Majumdar, Shyamal K
Malamud, Daniel F
Michel, Kenneth Earl
Miller, Robert Christopher
Monson, Frederick Carlton
Moorman, Gary William
Mundell, Robert David
Munger, Bryce Leon
Nichols, Warren Wesley
Nyquist, Sally Elizabeth
Oels, Helen C
Olds-Clarke, Patricia Jean
Orr, Nancy Hoffner
Paavola, Laurie Gail
Partanen, Carl Richard
Patterson, Elizabeth Knight
Peachey, Lee DeBorde
Piesco, Nicholas Peter
Porter, Keith Roberts
Pruett, Patricia Onderdonk
Quillin, Charles Robert
Roth, Stephen
Rubin, Walter
Sanger, Jean M
Sas, Daryl
Savage, Robert E
Schraer, Harald
Schuit, Kenneth Edward
Schwartz, Arthur Gerald
Sharpless, Thomas Kite
Sheffield, Joel Benson
Socher, Susan Helen
Starling, James Lyne
Storey, Bayard Thayer
Sundarraj, Nirmala
Surmacz, Cynthia Ann
Takats, Stephen Tibor
Taylor, D Lansing
Tchao, Ruy
Tilney, Lewis Gawtry
Trotter, Nancy Louisa
Vaidya, Akhil Babubhai
Vergona, Kathleen Anne Dobrosielski
Weinreb, Eva Lurie
Weisenberg, Richard Charles
Weston, John Colby
Widnell, Christopher Courtenay
Wolfe, Allan Frederick
Yip, Rick Ka Sun

RHODE ISLAND
Coleman, Annette Wilbois
Goertemiller, Clarence C, Jr
Heywood, Peter
Hufnagel, Linda Ann
Keogh, Richard Neil
Leduc, Elizabeth
Matsumoto, Lloyd H
Miller, Kenneth Raymond
Mottinger, John P
Plotz, Richard Douglas
Ripley, Robert Clarence
Rogers, Steffen Harold

SOUTH CAROLINA
Borg, Thomas K
Cheng, Thomas Clement
Dickey, Joseph Freeman
Dougherty, William J
Fitzharris, Timothy Patrick
Fowler, Stanley D
King, Elizabeth Norfleet
Odor, Dorothy Louise
Sawyer, Roger Holmes
Wourms, John P

SOUTH DAKOTA
Chen, Chen Ho
Evenson, Donald Paul
Neufeld, Daniel Arthur

TENNESSEE
Bibring, Thomas
Bogitsh, Burton Jerome
Boyles, Janet
Carlson, James Gordon
Duck, Bobby Neal
Dumont, James Nicholas
Fuhr, Joseph Ernest
Gartner, T Kent
Handel, Mary Ann
Hasty, David Long
Henson, James Wesley
Jeon, Kwang Wu
Kennedy, John Robert
Mazur, Peter
Monaco, Paul J
Mrotek, James Joseph
Mullin, Beth Conway
Olins, Donald Edward
Olsen, John Stuart
Rasch, Ellen M
Reger, James Frederick
Risby, Edward Louis
Rosing, Wayne C
Roth, Linwood Evans
Schultz, Terry Wayne
Shepherd, Virginia L
Skalko, Richard G(allant)
Swenson, Paul Arthur
Wells, Marion Robert

TEXAS
Adair, Gerald Michael
Ahearn, Michael John
Altenburg, Lewis Conrad
Anderson, Richard Gilpin Wood
Arnott, Howard Joseph
Arrighi, Frances Ellen
Bashaw, Elexis Cook
Beall, Paula Thornton
Beckingham, Kathleen Mary
Bickham, John W
Biesele, John Julius
Blystone, Robert Vernon
Brady, Scott T
Burson, Byron Lynn
Burton, Alexis Lucien
Butler, James Keith
Cannon, Marvin Samuel
Cavazos, Lauro Fred
Chang, Jeffrey Peh-I
Cockerline, Alan Wesley
Cole, Garry Thomas
Davis, Frances Maria
Davis, Walter Lewis
Elder, Fred F B
Ellzey, Joanne Tontz
Emery, William Henry Perry
Espey, Lawrence Lee
Fearing, Olin S
Goldstein, Margaret Ann
Gomer, Richard Hans
Goodman, Joel Mitchell
Gratzner, Howard G
Greenberg, Stanly Donald
Grinnell, Frederick
Hagino, Nobuyoshi
Hittelman, Walter Nathan
Hoage, Terrell Rudolph
Hoff, Victor John
Howard, Harriette Ella Pierce
Hudspeth, Albert James
Hunter, Jerry Don
Jackson, Raymond Carl
Karnaky, Karl John, Jr
Kasschau, Margaret Ramsey
Kidd, Harold J
Leffingwell, Thomas Pegg, Jr
Levinson, Charles
Litke, Larry Lavoe
McGraw, John Leon, Jr
McKeehan, Wallace Lee
Maguire, Marjorie Paquette
Mayberry, Lillian Faye
Meistrich, Marvin Lawrence
Moore, Charleen Morizot
Moyer, Mary Pat Sutter
Nations, Claude
Neaves, William Barlow
Ordonez, Nelson Gonzalo
Pathak, Sen
Peffley, Ellen Beth
Philpott, Charles William
Pinero, Gerald Joseph
Pool, Thomas Burgess
Powell, Michael A
Price, Harold James
Price, Maureen G
Rao, Potu Narasimha
Rizzo, Peter Jacob
Roberts, Susan Jean
Root, Elizabeth Jean
Sampson, Herschel Wayne
Schertz, Keith Francis
Seman, Gabriel
Shannon, Wilburn Allen, Jr
Shay, Jerry William
Sirbasku, David Andrew
Smith, James R
Smith, Willard Newell
Stevens, Clark
Stubblefield, Travis Elton
Thurston, Earle Laurence
Tomasovic, Stephen Peter
Van Dreal, Paul Arthur
Venketeswaran, S
Vonder Haar, Raymond A
Williams, Robert K
Williams, William Thomas
Wise, Gary E
Woo, Savio L C
Yin, Helen Lu

UTAH
Bollinger, William Hugh
Dewey, Douglas R
Gard, David Lynn
Kaplan, Jerry
Miles, Charles P
Miller, Scott Cannon
O'Neill, Frank John
Sites, Jack Walter, Jr
Wang, Richard Ruey-Chyi
Whitton, Leslie
Wolstenholme, David Robert
Youssef, Nabil Naguib

VERMONT
Folinas, Helen
Meyer, Diane Hutchins
Stevens, Dean Finley
Sullivan, Thomas Donald
Wells, Joseph
Woodcock-Mitchell, Janet Louise

VIRGINIA
Bahns, Mary
Barber, Albert Alcide
Bempong, Maxwell Alexander
Bender, Patrick Kevin
Breil, Sandra J
Brown, Jay Clark
Carson, Keith Alan
Collins, Jimmy Harold
Cone, Clarence Donald, Jr
DeVries, George H
Forbes, Michael Shepard
Goldstein, Lewis Charles
Gorbsky, Gary James
Grogan, William McLean
Heinemann, Richard Leslie
Herr, John Christian
Hoegerman, Stanton Fred
Jakoi, Emma Raff
Jones, Johnnye M
Laurie, Gordon William
Little, Charles Durwood, Jr
Martin, William Wallace
Merz, Timothy
Miller, Oscar Lee, Jr
Murli, Hemalatha
Owens, Vivian Ann
Rebhun, Lionel Israel
Rodewald, Richard David
Rosinski, Joanne
Saacke, Richard George
Sandow, Bruce Arnold
Schwartz, Lawrence B
Scott, Joseph Lee
Spangenberg, Dorothy Breslin
Stetler, David Albert
Teichler-Zallen, Doris
Weber, Michael Joseph

WASHINGTON
Atkinson, Lenette Rogers
Bakken, Aimee Hayes
Baskin, Denis George
Bolender, Robert P
Bornstein, Paul
Brooks, Antone L
Byers, Breck Edward
Carr, Robert Leroy
Cota-Robles, Eugene H
Drum, Ryan William
Erickson, John (Elmer)
Frederickson, Richard Gordon

Hecht, Adolph
Hendrickson, Anita Elizabeth
Horn, Diane
Humphrey, Donald Glen
Koehler, James K
Leik, Jean
McCollum, Gilbert Dewey, Jr
Margolis, Robert Lewis
Matlock, Daniel Budd
Pauw, Peter George
Prody, Gerry Ann
Quinn, LeBris Smith
Ross, Russell
Stanley, Hugh P
Stern, Irving B
Szubinska, Barbara
Waaland, Joseph Robert
Yao, Meng-Chao

WEST VIRGINIA
Nath, Joginder
Quinlan, Dennis Charles
Tolstead, William Lawrence

WISCONSIN
Allred, Lawrence Ervin (Joe)
Austin, Bert Peter
Bavister, Barry Douglas
Bitgood, J(ohn) James
Burke, Janice Marie
Estervig, David Nels
Goodman, Eugene Marvin
Graham, Linda Kay Edwards
Harb, Joseph Marshall
Karrer, Kathleen Marie
Kowal, Robert Raymond
Kushnaryov, Vladimir Michael
Lough, John William, Jr
Meisner, Lorraine Faxon
Millington, William Frank
Newcomb, Eldon Henry
Oaks, John Adams
Plaut, Walter (Sigmund)
Press, Newtol
Ruffolo, John Joseph, Jr
Schatten, Heide
Siegesmund, Kenneth A
Snyder, Virginia
Sobkowicz, Hanna Maria
Tandon, Shiv R
Thet, Lyn Aung

WYOMING
Jenkins, Robert Allan
Smith-Sonneborn, Joan

PUERTO RICO
Kozek, Wieslaw Joseph
Virkki, Niilo

ALBERTA
Cass, Carol E
Huang, Henry Hung-Chang
Kuspira, J
Larson, Ruby Ila
Longenecker, Bryan Michael
McGann, Locksley Earl
Malhotra, Sudarshan Kumar
Neuwirth, Maria
Rattner, Jerome Bernard
Shnitka, Theodor Khyam
Stevenson, Bruce R
Wagenaar, Emile B
Wolowyk, Michael Walter
Zalik, Sara E

BRITISH COLUMBIA
Aebersold, Ruedi H
Auersperg, Nelly
Berger, James Dennis
Boothroyd, Eric Roger
Brunette, Donald Maxwell
Burke, Robert D
Durand, Ralph Edward
Emerman, Joanne Tannis
Fankboner, Peter Vaughn
Molday, Robert S
Setterfield, George Ambrose
Singh, Raj Kumari
Stich, Hans F
Von Aderkas, Patrick Jurgen Cecil

MANITOBA
Aung, Taing
Chong, James York
Evans, Laurie Edward
Hamerton, John Laurence
Harder, Donald Edwald
Kerber, Erich Rudolph
Szekely, Joseph George

NEW BRUNSWICK
Bonga, Jan Max
Yoo, Bong Yul

NEWFOUNDLAND
Bal, Arya Kumar

NOVA SCOTIA
Angelopoulos, Edith W
Chen, Lawrence Chien-Ming
Dickson, Douglas Howard
Kamra, Om Perkash
Kapoor, Brij M

Wood, Eunice Marjorie

ONTARIO
Aye, Maung Tin
Beveridge, Terrance James
Britton, Donald MacPhail
Brown, David Lyle
Carr, David Harvey
Caveney, Stanley
Chaly, Nathalie
Cheng, Hazel Pei-Ling
Cummins, Joseph E
Dales, Samuel
Davidson, Douglas
Dingle, Allan Douglas
Fedak, George
Forer, Arthur H
Gorczynski, Reginald Meiczyslaw
Govind, Choonilal Keshav
Habowsky, Joseph Edmund Johannes
Hattori, Toshiaki
Heath, Ian Brent
Heath, Michele Christine
Horgen, Paul Arthur
Johnson, Byron F
Joneja, Madan Gopal
Kadanka, Zdenek Karel
Kasha, Kenneth John
Leatherland, John F
Leppard, Gary Grant
Liao, Shuen-Kuei
Lu, Benjamin Chi-Ko
Lynn, Denis Heward
McKeen, Wilbert Ezekiel
Metuzals, Janis
Moens, Peter B
Morris, Gerald Patrick
Murray, Robert George Everitt
Pelletier, R Marc
Rixon, Raymond Harwood
Rosenthal, Kenneth Lee
Rothstein, Aser
Sergovich, Frederick Raymond
Shaver, Evelyn Louise
Shivers, Richard Ray
Soudek, Dushan Edward
Stanisz, Andrzej Maciej
Sun, Anthony Mein-Fang
Thompson, John Eveleigh
Wallen, Donald George
Whitfield, James F
Zimmerman, Arthur Maurice
Zimmerman, Selma Blau
Zobel, Alicja Maria

QUEBEC
Beaudoin, Adrien Robert
Bendayan, Moise
Bennett, Gary Colin
Bergeron, John Joseph Marcel
Brailovsky, Carlos Alberto
Brawer, James Robin
Brouillet, Luc
Charest, Pierre M
Clark, Michael Wayne
Couillard, Pierre
Dhindsa, K S
Enesco, Hildegard Esper
Gibbs, Sarah Preble
Grant, William Frederick
Lafontaine, Jean-Gabriel
Meisels, Alexander
Parent, Andre
Pickett, Cecil Bruce
Reiswig, Henry Michael
Roy, Robert Michael McGregor
Schulman, Herbert Michael
Simard, Rene
Sinclair, Ronald
Stephens, Heather R
Wang, Eugenia
Wolsky, Maria de Issekutz

SASKATCHEWAN
Fowke, Lawrence Carroll
Newstead, James Duncan MacInnes
Oliphant, Lynn Wesley
Scoles, Graham John

OTHER COUNTRIES
Agathos, Spiros Nicholas
Ahn, Tae In
Amsterdam, Abraham
Araki, Masasuke
Armato, Ubaldo
Bar-Shavit, Zvi
Ben-Ze'ev, Avri
Britten, Edward James
Elsevier, Susan Maria
Fahimi, Hossein Dariush
Geiger, Benjamin
Heiniger, Hans-Jorg
Hotta, Yasuo
Ishikawa, Hiroshi
Jakob, Karl Michael
Jensen, Cynthia G
Jensen, Lawrence Craig-Winston
Jost, Jean-Pierre
Kaltsikes, Pantouses John
Kobata, Akira
Lee, Joseph Chuen Kwun
Makita, Takashi
Mittal, Balraj
Pickett-Heaps, Jeremy David

Rabinovitch, Michel Pinkus
Saito, Takuma
Schliwa, Manfred
Steer, Martin William
Strous, Ger J
Troutt, Louise Leotta
Wakayama, Yoshihiro
Yamada, Eichi

Ecology

ALABAMA
Bayne, David Roberge
Best, Troy Lee
Beyers, Robert John
Blackmon, James B
Blanchard, Richard Lee
Boyd, Claude Elson
Boyer, William Davis
DeBrunner, Louis Earl
Dobson, F(rederick) Stephen
Dunham, Rex Alan
Dusi, Julian Luigi
Golden, Michael Stanley
Guyer, Craig
Hays, Kirby Lee
Hodgkins, Earl Joseph
Holler, Nicholas Robert
Huntley, Jimmy Charles
Lehman, Robert Harold
McClintock, James Bruce
Mack, Timothy Patrick
Mettee, Maurice Ferdinand
Modlin, Richard Frank
Mullen, Gary Richard
Nelson, David Herman
Rajanna, Bettaiya
Regan, Gerald Thomas
Richardson, Terry David
Rogers, David T, Jr
Rosene, Walter, Jr
Rouse, David B
Scheiring, Joseph Frank
Stewart, Scott David
Suberkropp, Keller Francis
Tennessen, Kenneth Joseph
Wetzel, Robert George

ALASKA
Ahlgren, Molly O
Alexander, Vera
Bane, Gilbert Winfield
Behrend, Donald Fraser
Forest, Kathryn Jo
Fox, John Frederick
Gard, Richard
Goering, John James
Guthrie, Russell Dale
Hanley, Thomas Andrew
Hatch, Scott Alexander
Heifetz, Jonathan
Helle, John Harold (Jack)
Highsmith, Raymond C
Juday, Glenn Patrick
Klein, David Robert
Laursen, Gary A
MacLean, Stephen Frederick, Jr
Mathisen, Ole Alfred
Morrison, John Albert
Norton, David William
Reynolds, James Blair
Sagers, Cynthia Louise
Samson, Fred Burton
Schoen, John Warren
Shirley, Thomas Clifton
Smoker, William Williams
Snyder, George Richard
Stekoll, Michael Steven
Thomas, Gary Lee
Viereck, Leslie A
Werner, Richard Allen
West, George Curtiss
Willson, Mary Frances

ARIZONA
Ahlgren, Isabel Fulton
Beck, John R(oland)
Blinn, Dean Ward
Bradley, Michael Douglas
Brathovde, James Robert
Brown, Bryan Turner
Burgess, Kathryn Hoy
Calder, William Alexander, III
Carothers, Steven Warren
Chew, Robert Marshall
Cole, Gerald Ainsworth
Collins, James Paul
Crosswhite, Carol D
Crosswhite, Frank Samuel
Davis, Owen Kent
Dealy, John Edward
Dunlap, Donald Gene
Faeth, Stanley Herman
Fisher, Stuart Gordon
Foster, Kenneth Earl
Fritts, Harold Clark
Gaud, William S
Gerking, Shelby Delos
Haase, Edward Francis
Hammond, H David
Heed, William Battles
Hendrickson, John Roscoe
Hodges, Carl Norris

Huber, Roger Thomas
Hughes, Malcolm Kenneth
Istock, Conrad Alan
Jackson, Stephen Thomas
Johnsen, Thomas Norman, Jr
Johnson, Clarence Daniel
Johnson, Jack Donald
Johnson, Robert Andrew
Justice, Keith Evans
Kay, Fenton Ray
Klopatek, Jeffrey Matthew
Krausman, Paul Richard
Leatherman, Anna D
McGinnies, William Grovenor
McPherson, Guy Randall
Martin, Paul Schultz
Martin, Samuel Clark
Montgomery, Willson Linn
Muma, Martin Hammond
Papaj, Daniel Richard
Patten, Duncan Theunissen
Patton, David Roger
Pough, Frederick Harvey
Price, Peter Wilfrid
Reid, Charles Phillip Patrick
Rosenzweig, Michael Leo
Schaffer, William Morris
Schmutz, Ervin Marcell
Schnell, Jay Heist
Shaw, William Wesley
Smith, Andrew Thomas
Smith, Norman Sherrill
Sommerfeld, Milton R
Springfield, Harry Wayne
States, Jack Sterling
Szarek, Stanley Richard
Thomson, Donald A
Tschirley, Fred Harold
Turner, Raymond Marriner
Van Asdall, Willard
Van Riper, Charles, III
Van Voorhies, Wayne Alan
Walsberg, Glenn Eric
Washton, Nathan Seymour
Watson, Theo Franklin
Zwolinski, Malcolm John

ARKANSAS
Andrews, Theodore Francis
Bacon, Edmond James
Bednarz, James C
Burleigh, Joseph Gaynor
Dale, Edward Everett, Jr
Evans, Raymond David
Hanebrink, Earl L
Harp, George Lemaul
Logan, Lowell Alvin
Plummer, Michael V(an)
Smith, Kimberly Gray
Steinkraus, Donald Curtiss
Tave, Douglas
Thornton, Kent W
Zeide, Boris

CALIFORNIA
Addicott, Fredrick Taylor
Ainley, David George
Alberts, Allison Christine
Anderson, Daniel William
Anderson, Steven Clement
Anspaugh, Lynn Richard
Anthony, Margery Stuart
Azam, Farooq
Baalman, Robert J
Baker, Herbert George
Bakus, Gerald Joseph
Baldy, Richard Wallace
Barbour, Michael G
Beatty, Kenneth Wilson
Becking, Rudolf (Willem)
Beidleman, Richard Gooch
Bennett, Albert Farrell
Bennett, Charles Franklin
Benseler, Rolf Wilhelm
Bertsch, Hans
Beuchat, Carol Ann
Bibel, Debra Jan
Bjorkman, Olle Erik
Bledsoe, Caroline Shafer
Boggs, Carol L
Bohnsack, Kurt K
Botkin, Daniel Benjamin
Bottjer, David John
Bowers, Darl Eugene
Boyd, Milton John
Brant, Daniel (Hosmer)
Brattstrom, Bayard Holmes
Bray, Richard Newton
Brittan, Martin Ralph
Brooks, William Hamilton
Buchsbaum, Ralph
Burley, Nancy
Burnett, Bryan Reeder
Cailliet, Gregor Michel
Campbell, Bruce Carleton
Carey, James Robert
Carlucci, Angelo Francis
Carney, Heath Joseph
Carpenter, Frances Lynn
Case, Ted Joseph
Cech, Joseph Jerome, Jr
Chapin, F Stuart, III
Chappell, Mark Allen
Chen, Carl W(an-Cheng)

Ecology (cont)

Cheng, Lanna
Chesemore, David Lee
Christianson, Lee (Edward)
Coan, Eugene Victor
Coats, Robert N
Cobb, Fields White, Jr
Cody, Martin L(eonard)
Cogswell, Howard Lyman
Cohen, Anne Carolyn Constant
Coleman, Ronald Murray
Collier, Boyd David
Collier, Gerald
Colwell, Robert Knight
Connell, Joseph H
Constantine, Denny G
Cooper, Charles F
Costa, Daniel Paul
Cowles, David Lyle
Cox, Cathleen Ruth
Cox, George W
Coyer, James A
Cummings, John Patrick
D'Antoni, Hector Luis
Dasmann, Raymond Fredric
Davis, Gary Everett
Dawson, Kerry J
Dayton, Paul K
Demond, Joan
Deriso, Richard Bruce
Desharnais, Robert Anthony
Dewey, Kathryn G
Dexter, Deborah Mary
Diamond, Jared Mason
Di Girolamo, Rudolph Gerard
Dingle, Richard Douglas Hugh
Dole, Jim
Doutt, Richard Leroy
Dowell, Robert Vernon
Drewes, Robert Clifton
Duffey, Sean Stephen
Ebert, Thomas A
Edmunds, Peter James
Ehrlich, Anne Howland
Endler, John Arthur
Eriksen, Clyde Hedman
Erman, Don Coutre
Estes, James Allen
Evans, Kenneth Jack
Farris, David Allen
Firle, Tomas E(rasmus)
Fisler, George Frederick
Fletcher, Donald Warren
Ford, Richard Fiske
Fox, Laurel R
Frankie, Gordon William
Fuhrman, Jed Alan
Fusaro, Craig Allen
Gates, Gerald Otis
Gerrodette, Timothy
Getz, Wayne Marcus
Gill, Robert Wager
Gilmer, David Seeley
Given, Robert R
Goeden, Richard Dean
Gohr, Frank August
Goldman, Charles Remington
Golueke, Clarence George
Gotshall, Daniel Warren
Graves, Joseph L
Greenfield, Stanley Marshall
Greenhouse, N Anthony
Grubbs, David Edward
Hanes, Ted L
Hanson, Joe A
Hardy, Cecil Ross
Hare, John Daniel, III
Harte, John
Hassler, Thomas J
Hastings, Alan Matthew
Haverty, Michael Irving
Hazen, William Eugene
Hemingway, George Thomson
Hendler, Gordon Lee
Hendricks, Lawrence Joseph
Hespenheide, Henry August, III
Hewston, John G
Hinds, David Stewart
Hobson, Edmund Schofield
Hodges, Lance Thomas
Hodson, William Myron
Horn, Michael Hastings
Hoskins, Cortez William
Houpis, James Louis Joseph
Houston, Roy Seamands
Howard, Walter Egner
Hoyt, Donald Frank
Huffaker, Carl Barton
Hunt, George Lester, Jr
Hurlbert, Stuart Hartley
Iltis, Wilfred Gregor
Jacobsen, Nadine Klecha
Jain, Subodh K
Jennrich, Ellen Coutlee
Jones, Claris Eugene, Jr
Jones, Gilbert Fred
Jorgenson, Edsel Carpenter
Keeley, Jon E
Keil, David John
Kemp, Paul Raymond
Kilgore, Bruce Moody
Kitting, Christopher Lee
Knight, Allen Warner

Koonce, Andrea Lavender
Kuris, Armand Michael
Kushner, Arthur S
Kutilek, Michael Joseph
Lafferty, Kevin D
Lane, Robert Sidney
Lang, Kenneth Lyle
Lange, Carina Beatriz
Langenheim, Jean Harmon
Legner, E Fred
Leigh, Thomas Francis
Lewis, Anthony Wetzel
Lidicker, William Zander, Jr
Lindberg, David Robert
Lindburg, Donald Gilson
Loeblich, Helen Nina (Tappan)
Loomis, Robert Henry
Loomis, Robert Simpson
Love, Robert Merton
Lovich, Jeffrey Edward
Lowenstam, Heinz Adolf
Ludwig, John Howard
Lysyj, Ihor
McClenaghan, Leroy Ritter, Jr
McClure, Howe Elliott
McCullough, Dale Richard
McDonald, Philip Michael
Maciolek, John A
MacMillen, Richard Edward
Mahall, Bruce Elliott
Main, Robert Andrew
Major, Jack
Mattice, Jack Shafer
Maurer, Donald Leo
Melack, John Michael
Mikel, Thomas Kelly, Jr
Miller, Alan Charles
Mooney, Harold A
Moriarty, David John
Mortenson, Theodore Hampton
Mossman, Archie Stanton
Moyle, Peter Briggs
Mueller, Peter Klaus
Mullen, Ashley John
Muller, Cornelius Herman
Mulligan, Timothy James
Mulroy, Thomas Wilkinson
Murdoch, William W
Murray, Steven Nelsen
Myers, Willard Glazier, Jr
Myrick, Albert Charles, Jr
Neher, Robert Trostle
Nichols, Frederic Hone
Niesen, Thomas Marvin
Nobel, Park S
Nunney, Leonard Peter
Nybakken, James W
Oechel, Walter C
Oglesby, Larry Calmer
Ogunseitan, Oladele Abiola
Ohlendorf, Harry Max
Ornduff, Robert
Page, Robert Eugene, Jr
Parker, Virgil Thomas
Pearcy, Robert Woodwell
Phaff, Herman Jan
Phillips, Edwin Allen
Pitelka, Louis Frank
Pitts, Wanna Dene
Porcella, Donald Burke
Potts, Donald Cameron
Powell, Jerry Alan
Powell, Thomas Mabrey
Prasad, Kota S
Price, Mary Vaughan
Purcell, Alexander Holmes, III
Ralph, C(lement) John
Randall, Janet Ann
Rateaver, Bargyla
Ratliff, Raymond Dewey
Rau, Gregory Hudson
Rechnitzer, Andreas Buchwald
Reeve, Marian Enzler
Reisen, William Kenneth
Resh, Vincent Harry
Richerson, Peter James
Roberts, Stephen Winston
Robilliard, Gordon Allan
Robison, William Lewis
Rockland, Louis B
Rose, Michael Robertson
Rotenberry, John Thomas
Rothe, Karolyn Regina
Roughgarden, Jonathan David
Rundel, Philip Wilson
Sakai, Ann K
Salt, George William
Sawyer, John Orvel, Jr
Schoener, Thomas William
Schultz, Arnold Max
Schwab, Ernest Roe
Scott, Norman Jackson, Jr
Scow, Kate Marie
Shapiro, Arthur Maurice
Sherman, Robert James
Show, Ivan Tristan
Silver, Mary Wilcox
Slack, Keith Vollmer
Smith, Kenneth Lawrence, Jr
Smith, Paul Edward
Smith, Robert William
Soltz, David Lee
Soule, Dorothy (Fisher)
Sparks, Steven Richard

Stamps, Judy Ann
Sternberg, Hilgard O'Reilly
Stewart, Brent Scott
Stewart, Glenn Raymond
Stewart, Joan Godsil
Straughan, Isdale (Dale) Margaret
Strong, Donald Raymond, Jr
Sydeman, William J
Taylor, Robert Joe
Tegner, Mia Jean
Tenaza, Richard Reuben
Thornburgh, Dale A
Thum, Alan Bradley
Towner, Howard Frost
Tremor, John W
Tribbey, Bert Allen
Turner, Frederick Brown
Urry, Lisa Andrea
Vail, Patrick Virgil
Van Wagtendonk, Jan Willem
Verner, Jared
Vilkitis, James Richard
Vitousek, Peter Morrison
Voeks, Robert Allen
Vogl, Richard J
Walter, Hartmut
Warner, Robert Ronald
Waser, Nickolas Merritt
Waters, William E
Watt, Kenneth Edmund Ferguson
Weathers, Wesley Wayne
Weintraub, Joel D
Weis, Arthur Edward
Wells, Patrick Harrington
Westman, Walter Emil
Weston, Charles Richard
Weston, Henry Griggs, Jr
Wilcox, Bruce Alexander
Willemsen, Roger Wayne
Williams, Daniel Frank
Williams, Joseph Burton
Williams, Stanley Clark
Wilson, Kent Raymond
Wirtz, William Otis, II
Woodruff, David Scott
Zalom, Frank G
Zedler, Joy Buswell
Zedler, Paul H(ugo)
Zinkl, Joseph Grandjean
Zuk, Marlene

COLORADO
Anderson, David R
Apley, Martyn Linn
Bailey, James Allen
Barney, Charles Wesley
Benson, Norman G
Bock, Carl E
Bock, Jane Haskett
Bond, Richard Randolph
Bonham, Charles D
Bowers, Marion Deane
Bradshaw, William Newman
Branson, Farrel Allen
Brown, Lewis Marvin
Buckner, David Lee
Carey, Cynthia
Carlson, Clarence Albert, Jr
Cassel, J(oseph) Frank(lin)
Cohen, Ronald R H
Cole, C Vernon
Compton, Thomas Lee
Coughenour, Michael B
Crockett, Allen Bruce
Crouch, Glenn LeRoy
Cruz, Alexander
Davidson, Darwin Ervin
Driscoll, Richard Stark
Drummond, Boyce Alexander, III
Elias, Scott Armstrong
Fall, Michael William
Forester, Richard Monroe
Francy, David Bruce
Gibson, James H
Gilbert, Douglas L
Goetz, Harold
Gough, Larry Phillips
Graul, Walter Dale
Green, Jeffrey Scott
Grier, Charles Crocker
Hansmann, Eugene William
Havlick, Spenser Woodworth
Hein, Dale Arthur
Herrmann, Scott Joseph
Hoffman, Dale A
Hutchinson, Gordon Lee
Kalabokidis, Kostas D
Kaufmann, Merrill R
Keammerer, Warren Roy
Keith, James Oliver
King, Charles C
Kissling, Don Lester
Knopf, Fritz L
Krear, Harry Robert
LaBounty, James Francis, Sr
Lauenroth, William Karl
Lehner, Philip Nelson
Lindauer, Ivo Eugene
Linhart, Yan Bohumil
McClellan, John Forbes
McGinnies, William Joseph
McLean, Robert George
Mahoney, Charles Lindbergh
Martin, Stephen George

Marzolf, George Richard
Moore, Russell Thomas
Nasci, Roger Stanley
Oldemeyer, John Lee
Packard, Gary Claire
Porter, Kenneth Raymond
Pulliam, William Marshall
Redente, Edward Francis
Reed, Edward Brandt
Reynolds, Richard Truman
Ryan, Michael G
Schmidt, John Lancaster
Seilheimer, Jack Arthur
Shaw, Robert Blaine
Shepperd, Wayne Delbert
Sinensky, Michael
Smith, Dwight Raymond
Southwick, Charles Henry
Sowell, John Basil
Starr, Robert I
Stucky, Richard K(eith)
Swanson, Gustav Adolph
Thompson, Daniel Quale
Tomback, Diana Francine
Tracy, C Richard
Van Horn, Donald H
Vohs, Paul Anthony, Jr
Ward, James Vernon
Wasser, Clinton Howard
Whicker, Floyd Ward
Wiens, John Anthony
Wilson, David George
Woodmansee, Robert George
Wunder, Bruce Arnold

CONNECTICUT
Anderson, Gregory Joseph
Ashton, Peter Mark Shaw
Baillie, Priscilla Woods
Bongiorno, Salvatore F
Bormann, Frederick Herbert
Borst, Daryll C
Brewer, Robert Hyde
Buss, Leo William
Calabrese, Anthony
Carlton, James Theodore
Collins, Stephen
Craig, Catherine Lee
Damman, Antoni Willem Hermanus
DeSanto, Robert Spilka
Egler, Frank Edwin
Feng, Sung Yen
Goodwin, Richard Hale
Gorski, Leon John
Graikoski, John T
Guttay, Andrew John Robert
Haakonsen, Harry Olav
Haffner, Rudolph Eric
Henry, Charles Stuart
Holsinger, Kent Eugene
Jokinen, Eileen Hope
Katz, Max
Lee, Douglas Scott
Lier, Frank George
McClure, Mark Stephen
Maguder, Theodore Leo, Jr
Maier, Chris Thomas
Malone, Thomas Francis
Miller, Russell Bensley
Moehlman, Patricia des Roses
Niering, William Albert
Ostrander, Darl Reed
Remington, Charles Lee
Renfro, William Charles
Rettenmeyer, Carl William
Rich, Peter Hamilton
Shaw, Brenda Roberts
Siccama, Thomas G
Silander, John August, Jr
Singletary, Robert Lombard
Smith, David Martyn
Smith, Dwight Glenn
Streams, Frederick Arthur
Taigen, Theodore Lee
Vogt, Kristiina Ann
Ward, Jeffrey Stuart
Welsh, Barbara Lathrop
Wheeler, Bernice Marion
Whitlatch, Robert Bruce
Wulff, Barry Lee

DELAWARE
Cornell, Howard Vernon
Curtis, Lawrence Andrew
Dill, Norman Hudson
Eisenberg, Robert Michael
Gaffney, Patrick M
Hough-Goldstein, Judith Anne
Kalkstein, Laurence Saul
Karlson, Ronald Henry
Lotrich, Victor Arthur
Mason, Charles Eugene
Matlack, Albert Shelton
Ray, Thomas Shelby
Roth, Roland Ray
Schuler, Victor Joseph
Seliskar, Denise Martha
Sherman, John Walter
Smith, David William
Targett, Timothy Erwin

DISTRICT OF COLUMBIA
Adey, Walter Hamilton
Arcos, Joseph (Charles)

Auclair, Allan Nelson Douglas
Barber, Mary Combs
Beehler, Bruce McPherson
Blockstein, David Edward
Brooks, John Langdon
Brown, Eleanor D
Buffington, John Douglas
Burris, William Edmon
Challinor, David
Cox, Geraldine Anne Vang
Cuatrecasas, Jose
DiMichele, William Anthony
Dudley, Theodore R
Goodland, Robert James A
Gould, Edwin
Hoffmann, Robert Shaw
Littler, Diane Sullion
Littler, Mark Masterton
Lovejoy, Thomas E
McDiarmid, Roy Wallace
Malcom, Shirley Mahaley
Merchant, Henry Clifton
Nayak, Tapan Kumar
Nickum, John Gerald
Paris, Oscar Hall
Parsons, John David
Policansky, David J
Reisa, James Joseph, Jr
Rodman, James Eric
Roper, Clyde Forrest Eugene
Roskoski, Joann Pearl
Schaeff, Catherine Margaret
Silver, Francis
Sze, Philip
Terrell, Charles R
Terrell, Terry Lee Tickhill
Wilson, Don Ellis
Zucchetto, James John

FLORIDA
Abele, Lawrence Gordon
Adams, Ralph M
Alevizon, William
Alexander, Taylor Richard
Allen, Ted Tipton
Bennette, Jerry Mac
Betzer, Peter Robin
Bitton, Gabriel
Bloom, Stephen Allen
Branch, Lyn Clarke
Branham, Joseph Morhart
Brenner, Richard Joseph
Britt, N Wilson
Brower, Lincoln Pierson
Burch, Derek George
Burney, Curtis Michael
Castner, James Lee
Christman, Steven Philip
Clark, Kerry Bruce
Condit, Richard
Cooper, George Raymond
Cowell, Bruce Craig
Cropper, Wendell Parker, Jr
Crump, Martha Lynn
Dame, David Allan
Daubenmire, Rexford
Davis, John Armstrong
DeMort, Carole Lyle
Dinsmore, Bruce Heasley
Dragovich, Alexander
Feinsinger, Peter
Fitzpatrick, John Weaver
Fleming, Theodore Harris
Fourqurean, James Warren
Fox, Thomas Robert
Frank, John Howard
Frederick, Peter Crawford
Frei, Sr John Karen
Friedmann, E(merich) Imre
Giesel, James Theodore
Gilbert, Margaret Lois
Glynn, Peter W
Goldberg, Walter M
Gu, Binhe
Gupta, Virendra K
Hansen, Keith Leyton
Harris, Lawrence Dean
Harwell, Mark Alan
Hoffmann, Harrison Adolph
Hofstetter, Ronald Harold
Hollis, Mark Dexter
Hopkins, Thomas Lee
Hoy, Marjorie Ann
Jackson, Daniel Francis
James, Frances Crews
Janos, David Paul
Jenkins, Dale Wilson
Johnson, F Clifford
Jones, David Alwyn
Jones, John A(rthur)
Kale, Herbert William, II
Kaufman, Clemens Marcus
Kerr, John Polk
Kirkwood, James Benjamine
Kitchens, Wiley M
Knowlton, Nancy
Kruczynski, William Leonard
Lanciani, Carmine Andrew
Landolt, Peter John
Lawrence, Pauline Olive
Lee, David Webster
Leigh, Egbert Giles, Jr
Lessios, Harilaos Angelou
Levey, Douglas J

Lillywhite, Harvey B
Long, Alan Jack
Lounibos, Leon Philip
Lowe, Jack Ira
Lowe, Ronald Edsel
McCoy, Earl Donald
McKenney, Charles Lynn, Jr
McNab, Brian Keith
McSweeny, Edward Shearman
Mahadevan, Selvakumaran
Mariscal, Richard North
Marquis, David Alan
Maturo, Frank J S, Jr
Mayer, Foster Lee, Jr
Means, D(onald) Bruce
Miskimen, George William
Moshiri, Gerald Alexander
Murison, Gerald Leonard
Mushinsky, Henry Richard
Nelson, Walter Garnet
Nichols, Robert Loring
Nordlie, Frank Gerald
Norval, Richard Andrew
Ocampo-Friedmann, Roseli C
Odum, Howard Thomas
Ogden, John Conrad
Osborne, Lance Smith
Outcalt, Kenneth Wayne
Parker, John Hilliard
Phlips, Edward Julien
Porter, Richard Dee
Porter, Sanford Dee
Prager, Michael Haskell
Pritchard, Parmely Herbert
Putz, Francis Edward
Reid, George Kell
Reynolds, John Elliott, III
Rich, Earl Robert
Rossi, Anthony Michael
Roubik, David Ward
Schaiberger, George Elmer
Schelske, Claire L
Sheng, Yea-Yi Peter
Simberloff, Daniel S
Smith, Alan Paul
Smith, Sharon Louise
Snedaker, Samuel Curry
Steinman, Alan David
Stout, Isaac Jack
Taylor, Walter Kingsley
Telford, Sam Rountree, Jr
Tingle, Frederic Carley
Trama, Francesco Biagio
Travis, Joseph
Virnstein, Robert W
Walker, Thomas Jefferson
Walsh, John Joseph
Weber, Neal Albert
Werner, Patricia Ann Snyder
Wheeler, George Carlos
Wheeler, Jeanette Norris
Whitcomb, Willard Hall
Wilkinson, Robert Cleveland, Jr
Witherington, Blair Ernest
Woodruff, Robert Eugene
Woods, Charles Arthur
Woolfenden, Glen Everett
Young, Craig Marden

GEORGIA
Andrews, Charles Lawrence
Bozeman, John Russell
Bratton, Susan Power
Brower, John Harold
Burbanck, Madeline Palmer
Burbanck, William Dudley
Burns, Lawrence Anthony
Carter, Eloise B
Carter, James Richard
Cheng, Weixin
Coleman, David Cowan
Cotter, David James
Crossley, DeRyee Ashton, Jr
Davenport, Leslie Bryan, Jr
Davis, Robert
Dwinell, Lew David
Fey, Willard Russell
Fritz, William J
Glasser, John Weakley
Glooschenko, Walter Arthur
Golley, Frank Benjamin
Gordon, Robert Edward
Greear, Philip French-Carson
Hamrick, James Lewis
Hayes, John Thompson
Haynes, Ronnie J
Jordan, Carl Frederick
Kneib, Ronald Thomas
Lassiter, Ray Roberts
Lipps, Emma Lewis
McKeever, Sturgis
McPherson, Robert Merrill
Meentemeyer, Vernon George
Meyer, Judy Lynn
Mills, James Norman
Monk, Carl Douglas
Montgomery, Daniel Michael
Murray, Joan Baird
Newell, Steven Young
Odend'hal, Stewart
Odum, Eugene Pleasants
Parker, Albert John
Patten, Bernard Clarence
Payne, Jerry Allen

Plummer, Gayther Lynn
Porter, James W
Pulliam, H Ronald
Quertermus, Carl John, Jr
Ragsdale, Harvey Larimore
Relyea, Kenneth George
Remillard, Marguerite Madden
Scott, Donald Charles
Sharp, Homer Franklin, Jr
Shimkets, Lawrence Joseph
Shure, Donald Joseph
Skeen, James Norman
Smith, Euclid O'Neal
Snell, Terry Wayne
Stanton, George Edwin
Stober, Quentin Jerome
Throne, James Edward
Walls, Nancy Williams
Weissburg, Marc Joel
White, Donald Henry
Wiebe, William John
Wiegert, Richard G
Wood, Kenneth George
Woodall, William Robert, Jr
Wyatt, Robert Edward

HAWAII
Bardach, John E
Berg, Carl John, Jr
Bridges, Kent Wentworth
Brock, Richard Eugene
Clarke, Thomas Arthur
Coles, Stephen Lee
Culliney, Thomas W
Datta, Padma Rag
DeMartini, Edward Emile
Eddinger, Charles Robert
Freed, Leonard Alan
Hadfield, Michael Gale
Hapai, Marlene Nachbar
Howarth, Francis Gard
Hunter, Cynthia L
Landry, Michael Raymond
Liu, Yong-Biao
Medler, John Thomas
Mueller-Dombois, Dieter
Parrish, James Davis
Qin, Jianguang
Radtke, Richard Lynn
Reese, Ernst S
Roderick, George Karlsson
Rutherford, James Charles
Scheuer, Paul Josef
Sibert, John Rickard
Simon, Christine Mae
Skillman, Robert Allen
Stone, Charles Porter
Teramura, Alan Hiroshi
Tomich, Prosper Quentin
Vargas, Roger I
Winget, Robert Newell

IDAHO
Ables, Ernest D
Anderson, Jay Ennis
Bratz, Robert Davis
Clark, William Hilton
Douglas, Dorothy Ann
Fuller, Mark Roy
Goddard, Stephen
Hungerford, Kenneth Eugene
Johnson, Donald Ralph
Johnson, Frederic Duane
Johnson, James Blakeslee
Jonas, Robert James
Keller, Barry Lee
Laundre, John William
McCaffrey, Joseph Peter
MacFarland, Craig George
Minshall, Gerry Wayne
O'Keeffe, Lawrence Eugene
Pearson, Lorentz Clarence
Peek, James Merrell
Scott, James Michael
Spomer, George Guy
Trost, Charles Henry
Wyllie, Gilbert Alexander
Yensen, Arthur Eric

ILLINOIS
Altmann, Jeanne
Ames, Peter L
Anderson, Roger Clark
Ashby, William Clark
Augspurger, Carol Kathleen
Batzli, George Oliver
Berenbaum, May Roberta
Betz, Robert F
Bhatti, Neeloo
Bjorklund, Richard Guy
Bond, James Arthur, II
Brand, Raymond Howard
Brown, Lauren Evans
Brown, Sandra
Brugam, Richard Blair
Bryant, Marvin Pierce
Burger, George Vanderkarr
Burr, Brooks Milo
Carnes, Bruce Alfred
Conner, Jeffrey Keating
Corbett, Gail Rushford
Culver, David Clair
Drickamer, Lee Charles
Dunn, Christopher Paul

BIOLOGICAL SCIENCES / 85

Dunsing, Marilyn Magdalene
Durham, Leonard
Dziadyk, Bohdan
Engel, John Jay
Engstrom, Norman Ardell
Epstein, Samuel Stanley
Feder, Martin Elliott
Fell, George Brady
Flynn, Robert James
Forman, G Lawrence
Foster, Robin Bradford
Freiburg, Richard Eighme
Getz, Lowell Lee
Ghent, Arthur W
Gibson, David John
Giometti, Carol Smith
Girard, G Tanner
Greenberg, Bernard
Gustafson, Philip Felix
Hadley, Elmer Burton
Haynes, Robert C
Heaney, Lawrence R
Heltne, Paul Gregory
Herendeen, Robert Albert
Herricks, Edwin E
Hesketh, J D
Hinchman, Ray Richard
Hohn, Matthew Henry
Howe, Henry Franklin
Hunt, Lawrence Barrie
Jahn, Lawrence A
Jones, Carl Joseph
Klubek, Brian Paul
Kruse, Kipp Colby
Kulfinski, Frank Benjamin
Lamp, Herbert F
Larkin, Ronald Paul
Le Febvre, Eugene Allen
Leibold, Mathew Albert
Levenson, James B
Levin, Geoffrey Arthur
Lima, Gail M
Lloyd, Monte
Loehle, Craig S
Lorenz, Ralph William
McKinley, Vicky L
Marsh, Terrence George
Mathis, Billy John
Melin, Brian Edward
Mertz, David B
Meserve, Peter Lambert
Middleton, Beth Ann
Miller, Raymond Michael
Moll, Edward Owen
Munyer, Edward Arnold
Nelson, Harry Gladstone
Nixon, Charles Melville
Park, Thomas
Passman, Frederick Jay
Perino, Janice Vinyard
Peterson, Christopher Gerard
Phillips, Christopher Alan
Phipps, Richard L
Pruett-Jones, Stephen Glen
Robertson, Philip Alan
Rolfe, Gary Lavelle
Rosenthal, Gerson Max, IV
Rouffa, Albert Stanley
Sather, J Henry
Savitz, Jan
Schennum, Wayne Edward
Singer, Rolf
Southern, William Edward
Sprugel, George, Jr
Stahl, John Benton
Streets, David George
Stull, Elisabeth Ann
Thompson, Charles Frederick
Thompson, Vinton Newbold
Toliver, Michael Edward
Voight, Janet Ruth
Voigt, John Wilbur
Ware, George Henry
Weiler, William Alexander
Werner, William Ernest, Jr
Whelan, Christopher John
Whitacre, David Martin
Whitley, Larry Stephen
Wicklow, Donald Thomas
Willey, Ruth Lippitt
Williams, Thomas Alan
Zar, Jerrold Howard
Zimmerman, Craig Arthur
Zusy, Dennis

INDIANA
Allen, Durward Leon
Bakken, George Stewart
Berry, James William
Bjerregaard, Richard S
Brodman, Robert David
Byrnes, William Richard
Cooper, William Edgar, Jr
Daily, Fay Kenoyer
Doemel, William Naylor
Eberly, William Robert
Flanders, Robert Vern
Frey, David Grover
Gammon, James Robert
Hellenthal, Ronald Allen
Holt, Harvey Allen
Jackson, Marion T
Johnson, Michael David
Kapustka, Lawrence A

86 / DISCIPLINE INDEX

Ecology (cont)

Kirkpatrick, Charles Milton
Kirkpatrick, Ralph Donald
Lamberti, Gary Anthony
Lindsey, Alton Anthony
Lodge, David Michael
Lucas, Jeffrey Robert
McCafferty, William Patrick
McClure, Polley Ann
McIntosh, Robert Patrick
Mihaliak, Charles Andrew
Miller, Richard William
Montague, Fredrick Howard, Jr
Mueller, Wayne Paul
Mumford, Russell Eugene
Nelson, Craig Eugene
Olson, John Bennet
O'Neil, Robert James
Paladino, Frank Vincent
Park, Richard Avery, IV
Parker, George Ralph
Pelton, John Forrester
Porter, Clyde L, Jr
Randolph, James Collier
Real, Leslie Allan
Rowland, William Joseph
Russo, Raymond Joseph
Schmelz, Damian Vincent
Smith, Charles Edward, Jr
Squiers, Edwin Richard
Turner, Barbara Holman
Vorst, James J
Watson, Maxine Amanda
Weeks, Harmon Patrick, Jr
Whitaker, John O, Jr
Whitehead, Donald Reed
Willard, Daniel Edward
Williams, Eliot Churchill
Wise, Charles Davidson

IOWA

Ackerman, Ralph Austin
Bachmann, Roger Werner
Beachy, Christopher King
Berg, Virginia Seymour
Best, Louis Brown
Bovbjerg, Richard Viggo
Cavalieri, Anthony Joseph, II
Cawley, Edward T
Christiansen, Paul Arthur
Clark, William Richard
Cruden, Robert William
Czarnecki, David Bruce
Danforth, William (Frank)
Delong, Karl Thomas
Eckblad, James Wilbur
Franklin, William Lloyd
Gilbert, William Henry, III
Glenn-Lewin, David Carl
Graham, Benjamin Franklin
Hoffman, George R
Kaufmann, Gerald Wayne
Klaas, Erwin Eugene
Knutson, Roger M
Lyon, David Louis
McDonald, Donald Burt
Main, Stephen Paul
Malanson, George Patrick
Matteson, Patricia Claire
Merkley, Wayne Bingham
Poulter, Dolores Irma
Showers, William Broze, Jr
Whitson, Paul David

KANSAS

Armitage, Kenneth Barclay
Ashe, James S
Bellah, Robert Glenn
Bowen, Daniel Edward
Boyd, Roger Lee
Cink, Calvin Lee
Clark, George Richmond, II
Curless, William Toole
Durst, Harold Everett
Ealy, Robert Phillip
Fitch, Henry Sheldon
Fleharty, Eugene
Gipson, Philip
Greenfield, Michael Dennis
Hanson, Hugh
Hays, Horace Albennie
Jackson, Sharon Wesley
Johnson, John Christopher, Jr
Johnston, Richard Fourness
Kaufman, Donald Wayne
Kaufman, Glennis Ann
Kelting, Ralph Walter
Klaassen, Harold Eugene
Loudon, Catherine
Neely, Peter Munro
O'Brien, William John
Owensby, Clenton Edgar
Pierson, David W
Piper, Jon Kingsbury
Platt, Dwight Rich
Prophet, Carl Wright
Reichman, Omer James
Robel, Robert Joseph
Slade, Norman Andrew
Smith, Christopher Carlisle
Spencer, Dwight Louis
Sperry, Theodore Melrose
Terman, Max R

Wolf, Thomas Michael
Zimmerman, John Lester

KENTUCKY

Atlas, Ronald M
Barnes, Richard N
Baskin, Jerry Mack
Boehms, Charles Nelson
Bryant, William Stanley
Crowley, Philip Haney
Cupp, Paul Vernon, Jr
Dillard, Gary Eugene
Ettensohn, Francis Robert
Ferner, John William
Harmet, Kenneth Herman
Hendrix, James William
Howell, Jerry Fonce, Jr
Martin, William Haywood, III
Muller, Robert Neil
Pass, Bobby Clifton
Pearson, William Dean
Prins, Rudolph
Robinson, Thane Sparks
Sih, Andrew
Stevenson, Robert Jan
Thompson, Marvin Pete
Thompson, Ralph Luther
Thorp, James Harrison, III
Timmons, Thomas Joseph
Webster, Carl David
Westneat, David French, Jr
White, David Sanford
Wiedeman, Varley Earl
Wilder, Cleo Duke
Winstead, Joe Everett
Yambert, Paul Abt
Yeargan, Kenneth Vernon

LOUISIANA

Avent, Robert M
Bahr, Leonard M, Jr
Baldwin, Virgil Clark, Jr
Bauer, Raymond Thomas
Baumgardner, Ray K
Bounds, Harold C
Brown, Kenneth Michael
Carpenter, Stanley Barton
Chabreck, Robert Henry
Cordes, Carroll Lloyd
Cotty, Peter John
Curry, Mary Grace
Deaton, Lewis Edward
Defenbaugh, Richard Eugene
DePoe, Charles Edward
Devall, Margaret S
Dunigan, Edward P
Englande, Andrew Joseph
Fleeger, John Wayne
Gosselink, James G
Hackbarth, Winston (Philip)
Hamilton, Robert Bruce
Hastings, Robert Wayne
Heins, David Carl
Hensley, Sess D
Herke, William Herbert
Homberger, Dominique Gabrielle
Hughes, Janice S
Jaeger, Robert Gordon
Johnston, James Baker
Kalinsky, Robert George
Kinn, Donald Norman
Klich, Maren Alice
Knaus, Ronald Mallen
Leuck, Edwine E, II
Loden, Michael Simpson
Longstreth, David J
Lorio, Peter Leonce, Jr
Lynch, Steven Paul
Moore, Walter Guy
Pezeshki, S Reza
Platt, William Joshua, III
Poirrier, Michael Anthony
Ramsey, Paul Roger
Remsen, James Vanderbeek, Jr
Sammarco, Paul William
Schoenly, Kenneth George
Seigel, Richard Allyn
Smalley, Alfred Evans
Stiffey, Arthur V
Teate, James Lamar
Turner, Robert Eugene
Vidrine, Malcolm Francis

MAINE

Beals, Edward Wesley
Borei, Hans Georg
Brawley, Susan Howard
Butler, Ronald George
Coulter, Malcolm Wilford
Crawford, Hewlette Spencer, Jr
Cronan, Christopher Shaw
Davis, John Dunning
Dimond, John Barnet
Drury, William Holland
Fitch, John Henry
Francq, Edward Nathaniel Lloyd
Gilbert, James Robert
Glanz, William Edward
Greenwood, Paul Gene
Griffin, Ralph Hawkins
Haines, Terry Alan
Hidu, Herbert
Hunter, Malcolm Llewellyn, Jr
King, Gary Michael

Larsen, Peter Foster
Les, Edwin Paul
Lewis, Alan James
Mayer, Lawrence M
Mazurkiewicz, Michael
Nelson, Robert Edward
Owen, Ray Bucklin, Jr
Parker, James Willard
Schwintzer, Christa Rose
Stickney, Alden Parkhurst
Tjepkema, John Dirk
Vadas, Robert Louis
Watling, Les
Welch, Walter Raynes
Wilcox, Louis Van Inwegen, Jr

MARYLAND

Adams, Lowell William
Ballentine, Robert
Barbehenn, Kyle Ray
Barrows, Edward Myron
Barry, Ronald Everett, Jr
Batra, Suzanne Wellington Tubby
Beyer, W Nelson
Bledsoe, Lewis Jackson (Sam)
Borgia, Gerald
Boynton, Walter Raymond
Bresler, Jack Barry
Brush, Grace Somers
Bunce, James Arthur
Calhoun, John Bumpass
Cammen, Leon Matthew
Capone, Douglas George
Chen, T R
Clark, Trevor H
Coleman, Richard Walter
Cornwell, Jeffrey C
Costanza, Robert
Coulombe, Harry N
Creighton, Phillip David
Cronin, Lewis Eugene
D'Elia, Christopher Francis
Duke, James A
Ellis, David H
Elwood, Jerry William
Eny, Desire M(arc)
Faust, Maria Anna
Flittner, Glenn Arden
Flyger, Vagn Folkmann
Fowler, Bruce Andrew
Freed, Arthur Nelson
Galler, Sidney Roland
Gehris, Clarence Winfred
Gerring, Irving
Gerritsen, Jeroen
Gift, James J
Gill, Douglas Edward
Gillespie, Walter Lee
Ginevan, Michael Edward
Griswold, Bernard Lee
Harrell, Reginal M
Hassett, Charles Clifford
Hill, Jane Virginia Foster
Hines, Anson Hemingway
Hocutt, Charles H
Hodgdon, Harry Edward
Hornor, Sally Graham
Houde, Edward Donald
Howe, Marshall Atherton
Hull, James Clark
Inouye, David William
Johnson, Albert W
Kearney, Michael Sean
Kemp, William Michael
Knox, Robert Gaylord
Krywolap, George Nicholas
Kutz, Frederick Winfield
Lamp, William Owen
Leedy, Daniel Loney
McBryde, F(elix) Webster
McKaye, Kenneth Robert
McVey, James Paul
Malone, Thomas C
Martin, John A
Meredith, William G
Mihursky, Joseph Anthony
Moon, Milton Lewis
Morcock, Robert Edward
Morgan, Raymond P
Moseman, John Gustav
Mosher, James Arthur
Mountford, Kent
Murray, Laura
Muul, Illar
Nichols, James Dale
Odell, Lois Dorothea
Osman, Richard William
Paul, Robert William, Jr
Phelps, Harriette Longacre
Platt, Austin Pickard
Purcell, Jennifer Estelle
Rebach, Steve
Ritchie, Jerry Carlyle
Robertson, Andrew
Ross, Philip
Russek-Cohen, Estelle
Schneider, Eric Davis
Seliger, Howard Harold
Sellner, Kevin Gregory
Setzler-Hamilton, Eileen Marie
Sparling, Donald Wesley, Jr
Steinhauer, Allen Laurence
Stoecker, Diane Kastelowitz

Sullivan, Joseph Hearst
Todd, Robin Grenville
Trauger, David Lee
Ulanowicz, Robert Edward
Ward, Fraser Prescott
Weil, Raymond R
Weisberg, Stephen Barry
Williams, Robert Jackson
Wise, David Haynes
Yanagihara, Richard

MASSACHUSETTS

Alpert, Peter
Ament, Alison Stone
Art, Henry Warren
Baird, Ronald C
Bawa, Kamaljit S
Bazzaz, Fakhri Al
Beckwitt, Richard David
Bertin, Robert Ian
Binford, Michael W
Burk, Carl John
Burris, John Edward
Butman, Cheryl Ann
Canale-Parola, Ercole
Capuzzo, Judith M
Castro, Gonzalo
Caswell, Hal
Cheek, Dennis William
Cheetham, Ronald D
Chew, Frances Sze-Ling
Chisholm, Sallie Watson
Clark, William Cummin
Clemens, Daniel Theodore
Coler, Robert A
Colinvaux, Paul Alfred
Colt, LeBaron C, Jr
Cook, Robert Edward
Crisley, Francis Daniel
Curran, Harold Allen
D'Avanzo, Charlene
Duncan, Stewart
Edwards, Robert Lomas
Field, Hermann Haviland
Finn, John Thomas
Forman, Richard T T
Furth, David George
Gifford, Cameron Edward
Godfrey, Paul Jeffrey
Golubic, Stephan
Gregory, Constantine J
Gunner, Haim Bernard
Harbison, G Richard
Hay, Elizabeth Dexter
Healy, William Ryder
Helgesen, Robert Gordon
Hillman, Robert Edward
Hobbie, John Eyres
Hoff, James Gaven
Jahoda, John C
Jearld, Ambrose, Jr
Johanningsmeier, Arthur George
Jones, Gwilym Strong
Jost, Dana Nelson
Kazmaier, Harold Eugene
Kricher, John C
Kroodsma, Donald Eugene
Kunz, Thomas Henry
Lajtha, Kate
Larson, Joseph Stanley
Lee, Siu-Lam
Levy, Charles Kingsley
Lockwood, Linda Gail
Lovejoy, David Arnold
MacCoy, Clinton Viles
Marshall, Nelson
Mayo, Barbara Shuler
Moffett, Mark William
Moir, Ronald Brown, Jr
Moomaw, William Renken
Moore, Johnes Kittelle
Mulcahy, David Louis
Nickerson, Norton Hart
Pearce, John Bodell
Pechenik, Jan A
Perry, Alfred Eugene
Peterson, Bruce Jon
Primack, Richard Bart
Reardon, John Joseph
Reimold, Robert J
Rhoads, Donald Cave
Richards, F Paul
Rock, Paul Bernard
Ross, Michael Ralph
Ruber, Ernest
Sargent, Theodore David
Schofield, Edmund Acton, Jr
Shaver, Gaius Robert
Short, Henry Laughton
Smith, Frederick Edward
Smith, Grahame J C
Smith, Susan May
Smith, Tim Denis
Snyder, Dana Paul
Tamarin, Robert Harvey
Taylor, James Kenneth
Teal, John Moline
Telford, Sam Rountree, III
Thomas, Aubrey Stephen, Jr
Traniello, James Francis Anthony
Wasserman, Frederick E
White, Alan Whitcomb
Wigley, Roland L
Wilkie, David Scott

Williams, Ernest Edward
Woodwell, George Masters
Wright, Richard T

MICHIGAN
Allan, J David
Bajema, Carl J
Batts, Henry Lewis, Jr
Beaver, Donald Loyd
Beeton, Alfred Merle
Belyea, Glenn Young
Benninghoff, William Shiffer
Bowen, Stephen Hartman
Bowers, Maynard C
Bowker, Richard George
Branson, Dean Russell
Brewer, Richard (Dean)
Brown, Edward Herriot, Jr
Brown, Robert Thorson
Burnham, Robyn Jeanette
Burton, Thomas Maxie
Caldwell, Larry D
Cantlon, John Edward
Caswell, Herbert Hall, Jr
Catenhusen, John
Chen, Jiquan
Cowan, David Prime
Daniels, Stacy Leroy
Dapson, Richard W
Douglas, Matthew M
Driscoll, Egbert Gotzian
Engelmann, Manfred David
Estabrook, George Frederick
Evans, Francis Cope
Field, Ronald James
Fogel, Robert Dale
Foster, Neal Robert
Freeman, Dwight Carl
Gebben, Alan Irwin
Giesy, John Paul, Jr
Gillingham, James Clark
Gleason, Gale R, Jr
Glime, Janice Mildred
Golenberg, Edward Michael
Grafius, Edward John
Gross, Katherine Lynn
Hartman, Wilbur Lee
Haynes, Dean L
Hayward, James Lloyd
Hill, Richard William
Hill, Susan Douglas
Hiltunen, Jarl Kalervo
Holland-Beeton, Ruth Elizabeth
Hough, Richard Anton
Hunter, Robert Douglas
Hurst, Elaine H
Janke, Robert A
Keen, Robert Eric
Kerfoot, Wilson Charles
Kevern, Niles Russell
King, Darrell Lee
Klug, Michael J
Knudson, Vernie Anton
Kraft, Kenneth J
Kubiske, Mark E
Kurta, Allen
Kutkuhn, Joseph Henry
Lauff, George Howard
Lehman, John Theodore
Lenski, Richard E
Lopushinsky, Theodore
Low, Bobbi Stiers
McCrimmon, Donald Alan, Jr
McNabb, Clarence Duncan, Jr
MacTavish, John N
Manny, Bruce Andrew
Marsh, Frank Lewis
Martin, Michael McCulloch
Meier, Peter Gustav
Moll, Russell Addison
Murphy, Peter George
Niemi, Alfred Otto
Oldfield, Thomas Edward
Olexia, Paul Dale
Parejko, Ronald Anthony
Paul, Eldor Alvin
Peterson, Rolf Olin
Prince, Harold Hoopes
Prychodko, William Wasyl
Rathcke, Beverly Jean
Reed, David Doss
Riebesell, John F
Robertson, G Philip
Robinson, William Laughlin
Rollins-Page, Earl Arthur
Scavia, Donald
Shontz, John Paul
Siddiqui, Waheed Hasan
Simmons, Gary Adair
Smith, Bryce Everton
Solomon, Allen M
Stapp, William B
Stehr, Frederick William
Stephenson, Stephen Neil
Stoermer, Eugene F
Studier, Eugene H
Tarapchak, Stephen J
Taylor, William Waller
Teeri, James Arthur
Thompson, Paul Woodard
Tiedje, James Michael
Vanderploeg, Henry Alfred
Webber, Patrick John
Werner, Earl Edward

Winterstein, Scott Richard
Witter, John Allen
Yerkes, William D(ilworth), Jr

MINNESOTA
Abbott, Robinson S, Jr
Andow, David A
Ankley, Gerald Thomas
Baker, Robert Charles
Bakuzis, Egolfs Voldemars
Bedford, William Brian
Breckenridge, Walter John
Buech, Richard Reed
Christian, Donald Paul
Collins, Hollie L
Cooper, James Alfred
Corbin, Kendall Wallace
Cushing, Edward John
Davis, Margaret Bryan
Downing, William Lawrence
Erickson, James Eldred
Foose, Thomas John
Ford, Norman Lee
Fredrickson, Arnold G(erhard)
Frelich, Lee E
Fremling, Calvin R
Frenzel, Louis Daniel, Jr
Frydendall, Merrill J
Gorham, Eville
Grigal, David F
Heinselman, Miron L
Henry, Mary Gerard
Heuschele, Ann
Holt, Charles Steele
Jannett, Frederick Joseph, Jr
Jefferson, Carol Annette
Jordan, Peter Albion
Klemer, Andrew Robert
Krogstad, Blanchard Orlando
Lawrence, Donald Buermann
Lonnes, Perry Bert
McConville, David Raymond
McNaught, Donald Curtis
Mech, Lucyan David
Megard, Robert O
Miller, William Eldon
Ordway, Ellen
Pemble, Richard Hoppe
Perry, James Alfred
Rogers, Lynn Leroy
Rudolf, Paul Otto
Sammak, Paul J
Schimpf, David Jeffrey
Schmid, William Dale
Shapiro, Joseph
Siniff, Donald Blair
Spangler, George Russell
Spigarelli, Steven Alan
Stanley, Patricia Mary
Swain, Edward Balcom
Swift, Michael Crane
Tanner, Ward Dean, Jr
Tester, John Robert
Tilman, G David
VanAmburg, Gerald Leroy
Waters, Thomas Frank
Whiteside, Melbourne C
Windels, Carol Elizabeth
Wright, Herbert Edgar, Jr
Zischke, James Albert

MISSISSIPPI
Anderson, Gary
Arner, Dale H
Barko, John William
Cibula, William Ganley
Cooper, Charles Morris
Friend, Alexander Lloyd
Fye, Robert Eaton
Hardee, Dicky Dan
Hendricks, Donovan Edward
Hodges, John Deavours
Jackson, Jerome Alan
Keiser, Edmund Davis, Jr
Kirby, Conrad Joseph, Jr
Mylroie, John Eglinton
Parker, William Skinker
Randolph, Kenneth Norris
Reinecke, Kenneth J
Sadler, William Otho
Sherman, Harry Logan
Shields, Fletcher Douglas, Jr
Smith, James Winfred
Stewart, Robert Archie, II
Switzer, George Lester
Wakeley, James Stuart
Walley, Willis Wayne
Woodmansee, Robert Asbury

MISSOURI
Alderfer, Ronald Godshall
Aldrich, Richard John
Belshe, John Francis
Bourne, Carol Elizabeth Mulligan
Buchanan, Bryant W
Calabrese, Diane M
Carrel, James Elliott
Coles, Richard Warren
Davis, D Wayne
Dina, Stephen James
Drobney, Ronald DeLoss
Dugatkin, Lee Alan
Einhellig, Frank Arnold
Elder, William Hanna

Elliott, Dana Ray
Faaborg, John Raynor
Field, Christopher Bower
Finger, Terry Richard
Fredrickson, Leigh H
Fritzell, Erik Kenneth
Funk, John Leon
Georgiadis, Nicholas J
Gerhardt, H Carl, Jr
Gerrish, James Ramsay
Gersbacher, Willard Marion
Hanson, Willis Dale
Hawksley, Oscar
Hess, John Berger
Houghton, John M
Howell, Dillon Lee
Hunt, James Howell
Ingersol, Robert Harding
Jones, John Richard
Losos, Jonathan B
Mathis, Sharon Alicia
Mueller, Irene Marian
Myers, Richard F
Pallardy, Stephen Gerard
Reidinger, Russell Frederick, Jr
Sander, Ivan Lee
Sexton, Owen James
Shaddy, James Henry
Shiflet, Thomas Neal
Stern, Daniel Henry
Stern, Michele Suchard
Styron, Clarence Edward, Jr
Sussman, Robert Wald
Templeton, Alan Robert
Tomasi, Thomas Edward
Wartzok, Douglas
Wiggers, Ernie P
Williams, Henry Warrington
Wilson, Stephen W
Wiltshire, Charles Thomas

MONTANA
Arno, Stephen Francis
Ball, Irvin Joseph
Behan, Mark Joseph
Collins, Don Desmond
Craighead, John J
Fisher, James Robert
Fritts, Steven Hugh
Gould, William Robert, III
Habeck, James Robert
Heitschmidt, Rodney Keith
Johnson, Howard Ernest
Johnson, Oscar Walter
Kendall, Katherine Clement
Lotan, James
Lyon, Leonard Jack
McClelland, Bernard Riley
Mackie, Richard John
Manion, James J
Metzgar, Lee Hollis
Moore, Robert Emmett
Nielsen, Gerald Alan
Parsons, David Jerome
Rounds, Burton Ward
Rumely, John Hamilton
Sheldon, Andrew Lee
Smith, Jean E
Stanford, Jack Arthur
Taber, Richard Douglas
Taylor, John Edgar
Voigt, Garth Kenneth
Wright, John Clifford

NEBRASKA
Ballinger, Royce Eugene
Becker, Donald A
Bragg, Thomas Braxton
Brooks, Merle Eugene
Case, Ronald Mark
Geluso, Kenneth Nicholas
Grube, George Edward
Hergenrader, Gary Lee
Higley, Leon George
Joern, Anthony
Johnsgard, Paul Austin
Keeler, Kathleen Howard
Louda, Svata Mary
Lunt, Steele Ray
Nagel, Harold George
Peters, Edward James
Rand, Patricia June
Rowland, Neil Wilson
Schlesinger, Allen Brian
Waller, Steven Scobee
Walsh, Gary Lynn

NEVADA
Brussard, Peter Frans
Deacon, James Everett
Douglas, Charles Leigh
Fiero, George William, Jr
Fox, Carl Alan
Franson, Raymond Lee
Hoelzer, Guy Andrew
Hylton, Alvin Roy
Marlow, Ronald William
Meeuwig, Richard O'Bannon
Murvosh, Chad M
O'Farrell, Michael John
O'Farrell, Thomas Paul
Oring, Lewis Warren
Paulson, Larry Jerome
Robertson, Joseph Henry

Rust, Richard W
Starkweather, Peter Lathrop
Taylor, George Evans, Jr
Tueller, Paul T
Vinyard, Gary Lee

NEW HAMPSHIRE
Baldwin, Henry Ives
Barry, William James
Borg, Robert Munson
Croker, Robert Arthur
Eggleston, Patrick Myron
Flaccus, Edward
Friedland, Andrew J
Goder, Harold Arthur
Haney, James Filmore
Holmes, Richard Turner
Kotila, Paul Myron
McDowell, William H
McPeek, Mark Alan
Mautz, William Ward
Peart, David Ross
Safford, Lawrence Oliver
Spencer, Larry T
Wu, Lin
Wurster-Hill, Doris Hadley

NEW JERSEY
Alexander, Richard Raymond
Arnold, Frederic G
Bartha, Richard
Bruno, Stephen Francis
Burger, Joanna
Carter, Richard John
Casey, Timothy M
Cribben, Larry Dean
Cristini, Angela
Cromartie, William James, Jr
Crossner, Kenneth Alan
Crow, John H
Dauerman, Leonard
Dobi, John Steven
Dobson, Andrew Peter
Ehrenfeld, David W
Gardiner, Lion Frederick
Gochfeld, Michael
Good, Ralph Edward
Grant, B Rosemary
Grassle, John Frederick
Grassle, Judith Payne
Greenstein, Teddy
Hamby, Robert Jay
Handel, Steven Neil
Horn, Henry Stainken
Hu, Yaping
Hunter, Joseph Vincent
Isquith, Irwin R
Keating, Kathleen Irwin
Kennish, Michael Joseph
Kubersky, Edward Sidney
Lafornara, Joseph Philip
Leck, Charles Frederick
Levin, Simon Asher
Lo Pinto, Richard William
McCormick, Jon Michael
McGhee, George Rufus, Jr
McGuire, Terry Russell
Maiello, John Michael
Manganelli, Raymond M(ichael)
Martinez del Rio, Carlos
Meagher, Thomas Robert
Merrill, Leland (Gilbert), Jr
Morgan, Mark Douglas
Morin, Peter Jay
Morrison, Douglas Wildes
Murray, Bertram George, Jr
Nebel, Carl Walter
Pollock, Leland Wells
Power, Harry W, III
Pramer, David
Psuty, Norbert Phillip
Quinn, James Amos
Scherer, Robert C
Schultz, George Adam
Sebetich, Michael J
Shubeck, Paul Peter
Simpson, Robert Lee
Spiegel, Leonard Emile
Stearns, Donald Edison
Swan, Frederick Robbins, Jr
Swift, Fred Calvin
Willis, Jacalyn Giacalone
Ziegler, John Benjamin

NEW MEXICO
Anderson, Dean Mauritz
Bahr, Thomas Gordon
Bohrer, Vorsila Laurene
Brown, James Hemphill
Cole, Richard Allen
Cooch, Frederick Graham
Crawford, Clifford Smeed
Damon, Edward G(eorge)
Davis, Charles A
Degenhardt, William George
Duszynski, Donald Walter
Easley, Stephen Phillip
Findley, James Smith
Gennaro, Antonio Louis
Gosz, James Roman
Grover, Herbert David
Gutschick, Vincent Peter
Hansen, Wayne Richard
Holechek, Jerry Lee

88 / DISCIPLINE INDEX

Ecology (cont)

Jones, Kirkland Lee
Lewis, James Chester
Liddell, Craig Mason
Martin, William Clarence
Mexal, John Gregory
Molles, Manuel Carl, Jr
Pieper, Rex Delane
Rea, Kenneth Harold
Rogers, John Gilbert, Jr
Ryti, Randall Todd
Schemnitz, Sanford David
Secor, Jack Behrent
Taylor, Robert Gay
Whitford, Walter George
Zimmerman, Dale A

NEW YORK

Able, Kenneth Paul
Alexander, Maurice Myron
Allen, Douglas Charles
Allen, J Frances
Andrle, Robert Francis
Arnold, Steven Lloyd
Ayres, Jose Marcio Correa
Banerjee, Partha
Bard, Gily Epstein
Barnes, Jeffrey Karl
Beason, Robert Curtis
Bedford, Barbara Lynn
Bell, Michael Allen
Bentley, Barbara Lee
Bernard, John Milford
Bothner, Richard Charles
Boylen, Charles William
Boynton, John E
Brandt, Stephen Bernard
Breed, Helen Illick
Brocke, Rainer H
Brothers, Edward Bruce
Brown, Jerram L
Brown, Robert Zanes
Burgess, Robert Lewis
Caraco, Thomas Benjamin
Carruthers, Raymond Ingalls
Cassin, Joseph M
Cerwonka, Robert Henry
Chabot, Brian F
Christian, John Jermyn
Churchill, Algernon Coolidge
Clarke, Raymond Dennis
Coffroth, Mary Alice
Cohen, Joel Ephraim
Cole, Jonathan Jay
Confer, John L
Conover, David Olmstead
Conway, John Bell
Cox, Donald David
Crepet, William Louis
Crowell, Kenneth L
Daniels, Robert Artie
Dayton, Bruce R
De Laubenfels, David John
Dietert, Margaret Flowers
Dindal, Daniel Lee
Dooley, James Keith
Eickwort, George Campbell
Emlen, Stephen Thompson
Esser, Aristide Henri
Feeny, Paul Patrick
Fisher, Nicholas Seth
Forest, Herman Silva
Frankle, William Ernest
Fraser, Douglas Fyfe
French, Alan Raymond
Futuyma, Douglas Joel
Gallagher, Jane Chispa
George, Carl Joseph Winder
Gerard, Valrie Ann
Ginzburg, Lev R
Goodwin, Robert Earl
Green, John Irving
Greene, Kingsley L
Greenlaw, Jon Stanley
Greller, Andrew M
Gurevitch, Jessica
Habicht, Ernst Rollemann, Jr
Hairston, Nelson George, Jr
Hajek, Ann Elizabeth
Hall, Charles Addison Smith
Harman, Willard Nelson
Harrison, Richard Gerald
Haugh, John Richard
Hauser, Richard Scott
Haynes, James Mitchell
Hendrey, George Rummens
Henshaw, Robert Eugene
Herreid, Clyde F, II
Hertz, Paul Eric
Hoham, Ronald William
Holway, James Gary
Horn, Edward Gustav
Howard, Harold Henry
Howarth, Robert W
Hughes, Patrick Richard
Jarvis, Richard S
Johnson, Robert Walter
Jones, Clive Gareth
Kamran, Mervyn Arthur
Kanzler, Walter Wilhelm
Katsel, Pavel Leon
Keen, William Hubert
Kelly, John Russell

Kimmerer, Robin Wall
Kiviat, Erik
Klotz, Richard Lawrence
Kohut, Robert John
Langer, Arthur M
LaRow, Edward J
Lasker, Howard Robert
Lauer, Gerald J
Laurence, John A
Leopold, Donald Joseph
Levandowsky, Michael
Levinton, Jeffrey Sheldon
Lieberman, Arthur Stuart
Likens, Gene Elden
Loggins, Donald Anthony
Lovett, Gary Martin
Luteyn, James Leonard
McCune, Amy Reed
McDonnell, Mark Jeffery
McNaughton, Samuel J
Madison, Dale Martin
Makarewicz, Joseph Chester
Maple, William Thomas
Marra, John Frank
Mattfeld, George Francis
Maxwell, George Ralph, II
Meyer, Axel
Millstein, Jeffrey Alan
Mitchell, Myron James
Mosher, John Ivan
Muessig, Paul Henry
Muller-Schwarze, Dietland
Mytelka, Alan Ira
Naidu, Janakiram Ramaswamy
Nee, Michael
Norton, Roy Arnold
Nyrop, Jan Peter
O'Connor, Joel Sturges
Okubo, Akira
Pasby, Brian
Payne, Harrison H
Peckarsky, Barbara Lynn
Phillips, Arthur William, Jr
Phillips, Robert Rhodes
Pickett, Steward T A
Pierce, Madelene Evans
Pimentel, David
Poindexter, Jeanne Stove
Porter, William Frank
Pough, Richard Hooper
Rachlin, Joseph Wolfe
Rana, Mohammad A
Raynal, Dudley Jones
Reid, Archibald, IV
Ringler, Neil Harrison
Robinson, Myron
Rockwell, Robert Franklin
Root, Richard Bruce
Rosenbaum, Peter Andrew
Rough, Gaylord Earl
Roze, Uldis
Saks, Norman Martin
Schaffner, William Robert
Schneider, Kathryn Claire (Johnson)
Seeley, Thomas Dyer
Selleck, George Wilbur
Shields, William Michael
Siegfried, Clifford Anton
Silva, Pedro Joao Neves
Slack, Nancy G
Slobodkin, Lawrence Basil
Slusarczuk, George Marcelius Jaremias
Sohacki, Leonard Paul
Southwick, Edward E
Stalter, Richard
Steineck, Paul Lewis
Stewart, Kenton M
Stewart, Margaret McBride
Storr, John Frederick
Strayer, David Lowell
Sweeney, Robert Anderson
Thomson, James Douglas
Tietjen, John H
Tillman, Robert Erwin
Titus, John Elliott
Tobach, Ethel
Tobiessen, Peter Laws
Toenniessen, Gary Herbert
VanDruff, Larry Wayne
Vawter, Alfred Thomas
Ventura, William Paul
Vuilleumier, Francois
Wecker, Stanley C
Weinstein, David Alan
Weinstein, Leonard Harlan
White, James Edwin
Williams, Ernest Herbert, Jr
Wolf, Larry Louis
Wright, Margaret Ruth
Wurster, Charles F
Wyman, Richard L
Wynter, Carlton Elleston, Jr
Wysolmerski, Theresa
Zinder, Stephen Henry

NORTH CAROLINA

Aldridge, David William
Anderson, Roger Fabian
Anderson, Thomas Ernest
Baranski, Michael Joseph
Barden, Lawrence Samuel
Bellis, Vincent J, Jr
Billings, William Dwight
Blum, Udo

Bolen, Eric George
Boyce, Stephen Gaddy
Boyette, Joseph Greene
Brinson, Mark McClellan
Bruck, Robert Ian
Bryden, Robert Richmond
Butts, Jeffrey A
Campbell, Peter Hallock
Carson, Johnny Lee
Chester, Alexander Jeffrey
Chestnut, Alphonse F
Christensen, Norman Leroy, Jr
Clark, James Samuel
Cooper, Arthur Wells
Copeland, Billy Joe
Corliss, Bruce Hayward
Crowder, Larry Bryant
Darling, Marilyn Stagner
Davis, Graham Johnson
Derrick, Finnis Ray
Dickerson, Willard Addison
Dimock, Ronald Vilroy, Jr
Doerr, Phillip David
DuBay, Denis Thomas
Eggleston, David Bryan
Elias, Robert William
Esch, Gerald Wisler
Flint, Elizabeth Parker
Govoni, John Jeffrey
Gross, Harry Douglass
Grove, Thurman Lee
Hackney, Courtney Thomas
Hairston, Nelson George
Harper, James Douglas
Heatwole, Harold Franklin
Holland, Marjorie Miriam
Hoss, Donald Earl
Hunt, Kenneth Whitten
Jolls, Claudia Lee
Knight, Clifford Burnham
Kuenzler, Edward Julian
Lacey, Elizabeth Patterson
Lee, Si Duk
Livingstone, Daniel Archibald
Lutz, Paul E
McKinney, Frank Kenneth
McLeod, Michael John
Mason, Robert Edward
Menhinick, Edward Fulton
Miller, Robert James, II
Moore, Allen Murdoch
Mowbray, Thomas Bruce
Mozley, Samuel Clifford
Neher, Deborah A
Parnell, James Franklin
Peet, Robert Krug
Peterson, Charles Henry
Pittillo, Jack Daniel
Powell, Roger Allen
Qualls, Robert Gerald
Quay, Thomas Lavelle
Reice, Seth Robert
Richardson, Curtis John
Rittschof, Daniel
Robinson, Peter John
Romanow, Louise Rozak
Rublee, Parke Alstan
Ryals, George Lynwood, Jr
Schaaf, William Edward
Schlesinger, William Harrison
Schwartz, Frank Joseph
Searles, Richard Brownlee
Smathers, Garrett Arthur
Spears, Brian Merle
Spencer, Lorraine Barney
Stewart, Paul Alva
Stillwell, Harold Daniel
Stinner, Ronald Edwin
Stiven, Alan Ernest
Strain, Boyd Ray
Sutherland, John Patrick
Swank, Wayne T
Thayer, Gordon Wallace
Turner, Alvis Greely
Van Pelt, Arnold Francis, Jr
Ward, John Everett, Jr
Weigl, Peter Douglas
Weiss, Charles Manuel
Wentworth, Thomas Ralph
Wiley, Richard Haven, Jr
Williams, Ann Houston
Wolcott, Thomas Gordon
Wright, Stuart Joseph
Yarnell, Richard Asa
Yongue, William Henry
Zarnstorff, Mark Edward
Zeiger, Errol

NORTH DAKOTA

Clambey, Gary Kenneth
Crawford, Richard Dwight
Greenwald, Stephen Mark
Krapu, Gary Lee
Lieberman, Diana Dale
Lieberman, Milton Eugene
Lokemoen, John Theodore
Ralston, Robert D
Ries, Ronald Edward
Scoby, Donald Ray
Wrenn, William J

OHIO

Alrutz, Robert Willard
Andreas, Barbara Kloha

Barrett, Gary Wayne
Baumann, Paul C
Beatley, Janice Carson
Bieri, Robert
Boerner, Ralph E J
Brown, James Harold
Burky, Albert John
Burtt, Edward Howland, Jr
Camp, Mark Jeffrey
Case, Denis Stephen
Chesnut, Thomas Lloyd
Chuey, Carl F
Claussen, Dennis Lee
Cline, Morris George
Cobbe, Thomas James
Collins, Gary Brent
Cooke, George Dennis
Costanzo, Jon P
Crites, John Lee
Culver, David Alan
Davis, Craig Brian
Deonier, D L
Dexter, Ralph Warren
Egloff, David Allen
Elfner, Lynn Edward
Federle, Thomas Walter
Forsyth, Jane Louise
Fraleigh, Peter Charles
Francko, David Alex
Gates, Michael Andrew
Goldstein, David Louis
Gorchov, David Louis
Hamilton, Ernest Scovell
Hauser, Edward J P
Heath, Robert Thornton
Hille, Kenneth R
Hillis, Llewellya
Hoagstrom, Carl William
Hobbs, Horton Holcombe, III
Holeski, Paul Michael
Horn, David Jacobs
Hoy, Casey William
Jackson, William Bruce
Keiser, Terry Dean
Klemm, Donald J
Knoke, John Keith
Kohn, Harold William
Kowal, Norman Edward
Laing, Charles Corbett
Laufersweiler, Joseph Daniel
Laushman, Roger H
Lewis, Michael Anthony
Long, Edward B
Loucks, Orie Lipton
Lowe, Rex Loren
McCall, Peter Law
McClaugherty, Charles Anson
MacLean, David Belmont
McLean, Edward Bruce
Madenjian, Charles Paul
Martin, Elden William
Miller, Arnold I
Miller, Michael Charles
Mitsch, William Joseph
Mulroy, Juliana Catherine
Nault, Lowell Raymond
Olive, John H
Orcutt, Frederic Scott, Jr
Orr, Lowell Preston
Palmer, Brent David
Pearson, Paul Guy
Peterjohn, Glenn William
Peterle, Tony J
Phinney, George Jay
Radabaugh, Dennis Charles
Rubin, David Charles
Runkle, James Reade
Sanger, Jon Edward
Schroeder, Lauren Alfred
Shah, Kanti L
Sidman, Charles L
Snyder, Gary Wayne
Stansbery, David Honor
Stein, Carol B
Stein, Roy Allen
Stepier, Carol Ann
Stoffer, Richard Lawrence
Svendsen, Gerald Eugene
Taylor, Douglas Hiram
Thibault, Roger Edward
Thompson, John Leslie
Uetz, George William
Ungar, Irwin A
Valentine, Barry Dean
Vaughn, Charles Melvin
Vessey, Stephen H
Wali, Mohan Kishen
Wickstrom, Conrad Eugene
Williams, Patrick Kelly
Wilson, Mark Allan
Winner, Robert William
Wissing, Thomas Edward

OKLAHOMA

Baird, Troy Alan
Black, Jeffrey Howard
Buck, Paul
Burks, Sterling Leon
Chapman, Brian Richard
Crockett, Jerry J
Dorris, Troy Clyde
Fox, Stanley Forrest
Gray, Thomas Merrill
Grula, Mary Muedeking

BIOLOGICAL SCIENCES / 89

Hutchison, Victor Hobbs
Korstad, John Edward
LeGrand, Frank Edward
Love, Harry Schroeder, Jr
Lynn, Robert Thomas
McPherson, James King
Madden, Michael Preston
Mares, Michael Allen
Matthews, William John
Miller, Helen Carter
Mock, Douglas Wayne
Namminga, Harold Eugene
Nighswonger, Paul Floyd
Rice, Elroy Leon
Shaw, James Harlan
Shmaefsky, Brian Robert
Sinclair, James Lewis
Sonleitner, Frank Joseph
Sturgeon, Edward Earl
Talent, Larry Gene
Taylor, Constance Elaine Southern
Thurman, Lloy Duane
Tyler, Jack D
Vestal, Bedford Mather
Vishniac, Helen Simpson
Vitt, Laurie Joseph
Wells, Harrington
Whitcomb, Carl Erwin
Wilhm, Jerry L

OREGON
Anderson, John Richard
Anderson, Norman Herbert
Bachelet, Dominique
Beatty, Joseph John
Betts, Burr Joseph
Bierzychudek, Paulette F
Bouck, Gerald R
Bradshaw, Gay
Bradshaw, William Emmons
Buscemi, Philip Augustus
Callahan, Clarence Arthur
Carolin, Valentine Mott, Jr
Castenholz, Richard William
Chilcote, William W
Church, Marshall Robbins
Cimberg, Robert Lawrence
Clark, Mary Eleanor
Colvin, Dallas Verne
Cook, Stanton Arnold
Crawford, John Arthur
Cross, Stephen P
Davis, Michael William
Denison, William Clark
De Witt, John William, Jr
Eddleman, Lee E
Farris, Richard Austin
Ferraro, Steven Peter
Field, Katharine G
Fish, Joseph Leroy
Forbes, Richard Bryan
Frank, Peter Wolfgang
Garton, Ronald Ray
Gray, Jane
Hall, Frederick Columbus
Hansen, Everett Mathew
Hawke, Scott Dransfield
Hazel, Charles Richard
Hedrick, Ann Valerie
Henny, Charles Joseph
Hermann, Richard Karl
Higgins, Paul Daniel
Hixon, Mark A
Holzapfel, Christina Marie
Hughes, Robert Mason
Jarrell, Wesley Michael
Johnson, Michael Paul
Johnson, Samuel Edgar, II
Kogan, Marcos
Lackey, Robert T
Linn, DeVon Wayne
Liss, William John
Lotspeich, Frederick Benjamin
Lubchenco, Jane
Lyford, John H, Jr
McConnaughey, Bayard Harlow
McIntire, Charles David
Maloney, Thomas Edward
Malueg, Kenneth Wilbur
Marriage, Lowell Dean
Mason, Richard Randolph
Menge, Bruce Allan
Meslow, E Charles
Minore, Don
Moldenke, Alison Feerick
Molina, Randolph John
Murphy, Thomas A
Nebeker, Alan V
Neiland, Bonita J
Neilson, Ronald Price
Newton, Michael
Osgood, David William
Pearcy, William Gordon
Perry, David Anthony
Petersen, Richard Randolph
Peterson, Spencer Alan
Phillips, Donald Lundahl
Pizzimenti, John Joseph
Poulton, Charles Edgar
Preston, Eric Miles
Rottink, Bruce Allan
Ryan, Roger Baker
Schreiner, Roger Paul
Schrumpf, Barry James

Sherr, Barry Frederick
Sherr, Evelyn Brown
Stark, Nellie May
Stein, William Ivo
Stout, Benjamin Boreman
Tappeiner, John Cummings, II
Terraglio, Frank Peter
Thorson, Thomas Bertel
Tinnin, Robert Owen
Tubb, Richard Arnold
Udovic, Daniel
Volland, Leonard Allan
Wagner, David Henry
Waring, Richard H
Wick, William Quentin
Winjum, Jack Keith
Wood, Anne Michelle
Worrest, Robert Charles
Yamada, Sylvia Behrens
Zobel, Donald Bruce

PENNSYLVANIA
Abrahamson, Warren Gene, II
Arnold, Dean Edward
Barnett, Leland Bruce
Barone, John B
Beach, Neil William
Bosshart, Robert Perry
Bott, Thomas Lee
Bradt, Patricia Thornton
Brenner, Frederic J
Cale, William Graham, Jr
Cameron, Edward Alan
Carey, Michael Dean
Casida, Lester Earl, Jr
Clark, Richard James
Coffman, William Page
Commito, John Angelo
Cruzan, John
Cummins, Kenneth William
Cunningham, Harry N, Jr
Denoncourt, Robert Francis
Dudzinski, Diane Marie
Dunson, William Albert
Emberton, Kenneth C
Feuer, Robert Charles
Fisher, Robert L
Franz, Craig Joseph
Gilbert, Scott F
Glazier, Douglas Stewart
Good, Norma Frauendorf
Goulden, Clyde Edward
Graybill, Donald Lee
Guida, Vincent George
Haase, Bruce Lee
Hart, David Dickinson
Hartman, Richard Thomas
Heckscher, Stevens
Hendrickson, John Alfred, Jr
Hendrix, Sherman Samuel
Hickey, Richard James
Hoffman, Daniel Lewis
Hoffmaster, Donald Edeburn
Hutnik, Russell James
Illick, J(ohn) Rowland
Jacobs, George Joseph
Janzen, Daniel Hunt
Jayne, Benjamin A
Kilham, Susan Soltau
Kimmel, William Griffiths
Kirkland, Gordon Laidlaw, Jr
Kodrich, William Ralph
Koide, Roger Tai
Kuserk, Frank Thomas
Latham, Roger Earl
Lawrence, Vinnedge Moore
Legge, Thomas Nelson
Levin, Michael H(oward)
Linzey, Alicia Vogt
McCrea, Kenneth Duncan
McDiffett, Wayne Francis
McNair, Dennis M
McPheron, Bruce Alan
Manger, Martin C
Medve, Richard J
Mellinger, Michael Vance
Miller, Kenneth Melvin
Montgomery, James Douglas
Moon, Thomas Charles
Moore, John Robert
Mueller, Charles Frederick
Ostrofsky, Milton Lewis
Parker, Jon Irving
Patil, Ganapati P
Pearson, David Leander
Pottmeyer, Judith Ann
Prezant, Robert Steven
Ratzlaff, Willis
Richardson, Jonathan L
Ricklefs, Robert Eric
Rosenzweig, William David
Rymon, Larry Maring
Salamon, Kenneth J
Sanders, Robert W
Sather, Bryant Thomas
Saunders, William Bruce
Schultz, Jack C
Schuyler, Alfred Ernest
Settlemyer, Kenneth Theodore
Sheldon, Joseph Kenneth
Shipman, Robert Dean
Shontz, Charles Jack
Snyder, Donald Benjamin
Snyder, Robert LeRoy

Stauffer, Jay Richard, Jr
Stephenson, Andrew George
Taylor, Alan H
Throckmorton, Ann Elizabeth
Unz, Richard F(rederick)
Vannote, Robin L
Weiner, Jacob
Wert, Jonathan Maxwell, Jr
Wilhelm, Eugene J, Jr
Williams, Frederick McGee
Williamson, David Edward
Yahner, Richard Howard

RHODE ISLAND
Bertness, Mark David
Brooks, Lisa Delevan
Brown, James Henry, Jr
Golet, Francis Charles
Gould, Mark D
Hammen, Susan Lum
Husband, Thomas Paul
Hyland, Kerwin Ellsworth, Jr
Janis, Christine Marie
Jossi, Jack William
Lazell, James Draper
Mayer, Garry Franklin
Miller, Don Curtis
Morse, Douglass Hathaway
Nixon, Scott West
Oviatt, Candace Ann
Pearson, Philip Richardson, Jr
Pilson, Michael Edward Quinton
Rand, David McNear
Schmitt, Johanna
Shoop, C Robert
Tarzwell, Clarence Matthew
Waage, Jonathan King
Wahle, Richard Andreas
Webb, Thompson, III

SOUTH CAROLINA
Abernathy, A(twell) Ray
Banus, Mario Douglas
Biernbaum, Charles Knox
Blood, Elizabeth Reid
Brisbin, I Lehr, Jr
Brusca, Richard Charles
Congdon, Justin D
Coull, Bruce Charles
Dame, Richard Franklin
Davis, Luckett Vanderford
Dean, John Mark
DeCoursey, Patricia Alice Jackson
Dixon, Kenneth Randall
Edwards, John C
Feller, Robert Jarman
Fore, Stephanie Anne
Forsythe, Dennis Martin
Fox, Charles Wayne
Gard, Nicholas William
Gibbons, J Whitfield
Gorden, Robert Wayne
Harlow, Richard Fessenden
Hazen, Terry Clyde
Helms, Carl Wilbert
Horn, Charles Norman
Jacobs, Jacqueline E
Jagoe, Charles Henry
Kelly, Robert Withers
Kennamer, James Earl
Kosinski, Robert Joseph
Krebs, Julia Elizabeth
Lacher, Thomas Edward, Jr
Lincoln, David Erwin
McKellar, Henry Northington, Jr
McLeod, Kenneth William
Morris, James T
Mulvey, Margaret
Muska, Carl Frank
Newman, Michael Charles
Olson, John Bernard
Patton, Ernest Gibbes
Pinder, John Edgar, III
Pollard, Arthur Joseph
Revis-Wagner, Charles Kenyon
Rice, Theodore Roosevelt
Sharitz, Rebecca Reyburn
Shepard, Buford Merle
Smith, Michael Howard
Stancyk, Stephen Edward
Teska, William Reinhold
Turner, Jack Allen
Van Dolah, Robert Frederick
Weeks, Stephen Charles
Wood, Gene Wayne
Woodin, Sarah Ann
Worthen, Wade Bolton

SOUTH DAKOTA
Berry, Charles Richard, Jr
Dieter, Charles David
Diggins, Maureen Rita
Ellsbury, Michael M
Froiland, Sven Gordon
Haertel, Lois Steben
Hutcheson, Harvie Leon, Jr
Keenlyne, Kent Douglas
Kieckhefer, Robert William
Severson, Keith Edward
Tatina, Robert Edward

TENNESSEE
Abernethy, Virginia Deane
Alsop, Frederick Joseph, III

Ambrose, Harrison William, III
Amundsen, Clifford C
Auerbach, Stanley Irving
Ball, Mary Uhrich
Barnthouse, Lawrence Warner
Bradshaw, Aubrey Swift
Breeden, John Elbert
Brinkhurst, Ralph O
Brode, William Edward
Bunting, Dewey Lee, II
Cada, Glenn Francis
Campbell, James A
Chapman, Joe Alexander
Clarke, James Harold
Clebsch, Edward Ernst Cooper
Coutant, Charles Coe
Craig, Robert Bruce
Dale, Virginia House
DeAngelis, Donald Lee
Dearden, Boyd L
DeSelm, Henry Rawie
Dimmick, Ralph W
Echternacht, Arthur Charles
Eddlemon, Gerald Kirk
Elmore, James Lewis
Emanuel, William Robert
Francis, Chester Wayne
Garten, Charles Thomas, Jr
Gehrs, Carl William
Hallam, Thomas Guy
Harris, William Franklin, III
Hildebrand, Stephen George
James, Ted Ralph
Kathman, R Deedee
Kaye, Stephen Vincent
Kimmel, Bruce Lee
Kocher, David Charles
Koh, Sung-Cheol
Kroodsma, Roger Lee
Lanza, Guy Robert
Leffler, Charles William
Lessman, Gary M
Loar, James M
McBrayer, James Franklin
McCarthy, John F
McCormick, J Frank
McLaughlin, Samuel Brown
Maier, Kurt Jay
Martin, Robert Eugene
Miller, Neil Austin
Nall, Ray(mond) W(illett)
Norby, Richard James
Olson, J(erry) S
O'Neill, Robert Vincent
Parchment, John Gerald
Payne, James
Polis, Gary Allan
Quarterman, Elsie
Reed, Robert Marshall
Reichle, David Edward
Roop, Robert Dickinson
Salk, Martha Scheer
Schneider, Gary
Scott, Arthur Floyd
Sharma, Gopal Krishan
Sharp, Aaron John
Sharples, Frances Ellen
Staub, Robert J
Suter, Glenn Walter, II
Tarpley, Wallace Armell
Tolbert, Virginia Rose
Turner, Robert Spilman
Van Hook, Robert Irving, Jr
Van Winkle, Webster, Jr
Voorhees, Larry Donald
Walker, Kenneth Russell
Wallace, Gary Oren
Watson, Annetta Paule
Weaver, George Thomas
Webb, J(ohn) Warren
Willard, William Kenneth
Wiser, Cyrus Wymer
Witherspoon, John Pinkney, Jr
Woods, Frank Wilson
Yarbro, Claude Lee, Jr

TEXAS
Adams, Clark Edward
Anderson, Richard Orr
Arp, Gerald Kench
Aumann, Glenn D
Beitinger, Thomas Lee
Belk, Gene Denton
Blankenship, Lytle Houston
Boutton, Thomas William
Briske, David D
Britton, Carlton M
Browning, J(ohn) Artie
Bryant, Vaughn Motley, Jr
Buskirk, Ruth Elizabeth
Cameron, Guy Neil
Camp, Frank A, III
Chandler, James Michael
Chrzanowski, Thomas Henry
Clark, Donald Ray, Jr
Clark, William Jesse
Cotner, James Bryan
Coulson, Robert N
Darnell, Rezneat Milton
Delco, Exalton Alfonso, Jr
DeShaw, James Richard
Dickson, Kenneth Lynn
Diggs, George Minor, Jr
Dodd, Jimmie Dale

Ecology (cont)

DuBar, Jules R
Dunkle, Sidney Warren
Dyksterhuis, Edsko Jerry
Erdman, Howard E
Fitzpatrick, Lloyd Charles
Fonteyn, Paul John
Formanowicz, Daniel Robert, Jr
Foster, John Robert
Fowler, Norma Lee
Freeman, Charles Edward, Jr
Fulbright, Timothy Edward
Garner, Herschel Whitaker
Gehlbach, Frederick Renner
Gerard, Cleveland Joseph
Grumbles, Jim Bob
Guthrie, Rufus Kent
Hannan, Herbert Herrick
Harcombe, Paul Albin
Harris, Arthur Horne
Hellier, Thomas Robert, Jr
Herbst, Richard Peter
Humphries, James Edward, Jr
Inglis, Jack Morton
Ingold, Donald Alfred
Johnston, John Spencer
Jones, Clyde J
Judd, Frank Wayne
Keith, Donald Edwards
Kennedy, James H
Kennedy, Joseph Patrick
Kibler, Kenneth G
Kroh, Glenn Clinton
Kunz, Sidney Edmund
Lacy, Julia Caroline
Landers, Roger Q, Jr
Lind, Owen Thomas
Lopez, Genaro
Lynch, Daniel Matthew
McAllister, Chris Thomas
McCarley, Wardlow Howard
McCullough, Jack Dennis
McCully, Wayne Gunter
Maclean, Graeme Stanley
McMahon, Robert Francis, III
McMillan, Calvin
Maguire, Bassett, Jr
Mangan, Robert Lawrence
Mansfield, Clifton Tyler
Martin, Robert Frederick
Mathews, Nancy Ellen
Merchant, Michael Edward
Middleditch, Brian Stanley
Moldenhauer, Ralph Roy
Mueller, Dale M J
Neill, William Harold
Newman, George Allen
Nixon, Elray S
Oppenheimer, Carl Henry, Jr
Owens, Michael Keith
Parajulee, Megha N
Parker, Robert Hallett
Peacock, John Talmer
Perez, Francisco Luis
Pettit, Russell Dean
Philips, Billy Ulyses
Pianka, Eric R
Pierce, Jack Robert
Powell, Michael A
Ray, James P
Reeder, William Glase
Rennie, Thomas Howard
Richardson, Richard Harvey
Ring, Dennis Randall
Rozas, Lawrence Paul
Rylander, Michael Kent
Samollow, Paul B
Schmidly, David James
Schneider, Dennis Ray
Schroder, Gene David
Schuster, Joseph L
Scudday, James Franklin
Shake, Roy Eugene
Sissom, Stanley Lewis
Smeins, Fred E
Smith, Alan Lyle
Sterner, Robert Warner
Stewart, Kenneth Wilson
Stransky, John Janos
Strassmann, Joan Elizabeth
Telfair, Raymond Clark, II
Tomson, Mason Butler
Tuttle, Merlin Devere
Van Auken, Oscar William
Vincent, Jerry William
Walker, Laurence Colton
Waller, William T
Wendt, Theodore Mil
Whiteside, Charles Hugh
Willig, Michael Robert
Willingham, Francis Fries, Jr
Wilson, Robert Eugene
Wohlschlag, Donald Eugene
Wolfenberger, Virginia Ann
Wood, Carl Eugene
Wu, Hsin-I
Zagata, Michael DeForest

UTAH

Allred, Donald Mervin
Baer, James L
Ball, Terry Briggs
Barnes, James Ray
Belnap, Jayne
Bissonette, John Alfred
Blaisdell, James Pershing
Blauer, Aaron Clyde
Bosakowski, Thomas
Bozniak, Eugene George
Brotherson, Jack DeVon
Buchanan, Hayle
Caldwell, Martyn Mathews
Chapman, Joseph Alan
Cox, Paul Alan
Dalton, Patrick Daly
Davidson, Diane West
Dobrowolski, James Phillip
Dueser, Raymond D
Ehleringer, James Russell
Ekdale, Allan Anton
Epstein, William Warren
Fisher, Richard Forrest
Flinders, Jerran T
Gessaman, James A
Harper, Kimball T
Hayward, Charles Lynn
Helm, William Thomas
Hirth, Harold Frederick
Johnson, Douglas Allan
Kadlec, John A
Lighton, John R B
Lloyd, Ray Dix
McKell, Cyrus Milo
MacMahon, James A
Malechek, John Charles
Mangum, Fredrick Anthony
Maurer, Brian Alan
Messina, Frank James
Miller, Raymond Woodruff
Mueggler, Walter Frank
Neuhold, John Mathew
Olsen, Peter Fredric
Palmblad, Ivan G
Ralphs, Michael H
Reid, William Harper
Richardson, Jay Wilson, Jr
Rickart, Eric Allan
Rushforth, Samuel Roberts
Shultz, Leila McReynolds
Sites, Jack Walter, Jr
Skujins, John Janis
Smith, Howard Duane
Spillett, James Juan
West, Neil Elliott
White, Clayton M
Wood, Timothy E
Yorks, Terence Preston

VERMONT

Barton, James Don, Jr
Bean, Daniel Joseph
Brodie, Edmund Darrell, Jr
Dritschilo, William
Forcier, Lawrence Kenneth
Hamilton, Lawrence Stanley
Heinrich, Bernd
Henson, Earl Bennette
Jervis, Robert Alfred
Lechevalier, Mary P
McIntosh, Alan William
Potash, Milton
Schall, Joseph Julian
Spearing, Ann Marie
Westing, Arthur H
Woods, Kerry David

VIRGINIA

Andrews, Robin M
Barbaro, Ronald D
Barton, Alexander James
Bass, Michael Lawrence
Benfield, Ernest Frederick
Bishop, John Watson
Blem, Charles R
Bliss, Dorothy Crandall
Blum, Linda Kay
Bodkin, Norlyn L
Boots, Sharon G
Briggs, Jeffrey L
Britt, Douglas Lee
Britton, Maxwell Edwin
Brooks, Garnett Ryland, Jr
Brown, Luther Park
Cairns, John, Jr
Callahan, James Thomas
Chang, William Y B
Cherry, Donald Stephen
Cocking, W Dean
Courtney, Mark William
Cranford, Jack Allen
Dane, Charles Warren
Day, Frank Patterson, Jr
Ernst, Carl Henry
Evans, Gary R
Eyman, Lyle Dean
Fashing, Norman James
Fink, Linda Susan
Fisher, Elwood
Fuller, Stephen William
Gaines, Gregory
Gangstad, Edward Otis
Giles, Robert H, Jr
Gottschalk, John Simison
Gough, Stephen Bradford
Hinckley, Alden Dexter
Hufford, Terry Lee
Hurd, Lawrence Edward
Jahn, Laurence R
Jameson, Donald Albert
Jenkins, Robert Ellsworth, Jr
Jenkins, Robert Walls, Jr
Jenssen, Thomas Alan
Johnson, Philip L
Karowe, David Nathan
Keinath, John Allen
Kelly, Mahlon George, Jr
Kelso, Donald Preston
Kirkpatrick, Roy Lee
Lawrey, James Donald
Lee, James A
Levy, Gerald Frank
Little, Elbert L(uther), Jr
Logan, Jesse Alan
Luckenbach, Mark Wayne
Mellinger, Clair
Milton, Nancy Melissa
Moore, David Jay
Musick, John A
Neves, Richard Joseph
Odum, William Eugene
Osborn, Kenneth Wakeman
Patterson, Mark Robert
Peters, Esther Caroline
Pienkowski, Robert Louis
Semtner, Paul Joseph
Sharov, Alexei A
Shugart, Herman Henry, Jr
Shuster, Carl Nathaniel, Jr
Sigafoos, Robert Sumner
Simmons, George Matthew, Jr
Talbot, Lee Merriam
Terman, Charles Richard
Tiwari, Surendra Nath
Underwood, Lawrence Statton
Vaughan, Michael Ray
Ware, Stewart Alexander
Webster, Jackson Ross
Wells, Elizabeth Fortson
West, David Armstrong
Wilbur, Henry Miles
Willis, Lloyd L, II
Winston, Judith Ellen
Wood, Leonard E(ugene)
Young, Donald Raymond
Zieman, Joseph Crowe, Jr

WASHINGTON

Adamson, Lucile Frances
Agee, James Kent
Allen, Julia Natalia
Bax, Nicholas John
Becker, Clarence Dale
Berryman, Alan Andrew
Bliss, Lawrence Carroll
Boersma, P Dee
Booth, Beatrice Crosby
Braham, Howard Wallace
Bredahl, Edward Arlan
Brenchley, Gayle Anne
Calkins, Carrol Otto
Carr, Robert Leroy
Catts, Elmer Paul
Curl, Herbert (Charles), Jr
Cushing, Colbert Ellis
Dauble, Dennis Deene
DeMaster, Douglas Paul
DePuit, Edward J
Drum, Ryan William
Edmondson, W Thomas
Eickstaedt, Lawrence Lee
Emlen, John Merritt
Favorite, Felix
Fitzner, Richard Earl
Fleming, Richard Seaman
Fonda, Richard Weston
Franklin, Jerry Forest
Funk, William Henry
Gentry, Roger Lee
Gibson, Flash
Ginn, Thomas Clifford
Goebel, Carl Jerome
Gray, Robert H
Gunderson, Donald Raymond
Hansen, David Henry
Hanson, Wayne Carlyle
Hard, Jeffrey John
Harris, Grant Anderson
Hatheway, William Howell
Heinle, Donald Roger
Hicks, David L
Huey, Raymond Brunson
Irving, Patricia Marie
Johnson, Richard Evan
Kareiva, Peter Michael
Karlstrom, Ernest Leonard
Karr, James Richard
Kenagy, George James
King, James Roger
Kohn, Alan Jacobs
Landis, Wayne G
McAlister, William Bruce
McDonough, Leslie Marvin
Mace, Terrence Rowley
Mack, Richard Norton
Manuwal, David Allen
Mason, David Thomas
Mearns, Alan John
Meeuse, Bastiaan J D
Mehringer, Peter Joseph, Jr
Mills, Claudia Eileen
Morgan, Kenneth Robb
Muller-Parker, Gisele Therese
Murphy, Mary Eileen
Naiman, Robert Joseph
Napp, Jeffrey M
Nelson, Jack Raymond
Norvell, Lorelei Lehwalder
Novotny, Anthony James
Orians, Gordon Howell
Page, Thomas Lee
Paine, Robert T
Parker, Richard Alan
Paulson, Dennis R
Pearson, Walter Howard
Price, Keith Robinson
Quay, Paul Douglas
Rayburn, William Reed
Rogers, Lee Edward
Schemske, Douglas William
Schneider, David Edwin
Schroeder, Michael Allen
Schultz, Vincent
Seymour, Allyn H
Simenstad, Charles Arthur
Sluss, Robert Reginald
Soule, Oscar Hommel
Sprugel, Douglas George
Staley, James Trotter
Stevens, Todd Owen
Summers, William Clarke
Swartzman, Gordon Leni
Swedberg, Kenneth C
Tanaka, Yasuomi
Taub, Frieda B
Taylor, Peter Berkley
Templeton, William Lees
Thom, Ronald Mark
Thompson, Christopher William
Thompson, John N
Tsukada, Matsuo
Van Voris, Peter
Wiedemann, Alfred Max
Woelke, Charles Edward
Wolda, Hindrik

WEST VIRGINIA

Bissonnette, Gary Kent
Brown, Mark Wendell
Carvell, Kenneth Llewellyn
Constantz, George Doran
Day, Thomas Arthur
Keller, Edward Clarence, Jr
Kotcon, James Bernard
Lang, Gerald Edward
Leather, Gerald Roger
Ludke, James Larry
Mugaas, John Nels
Pauley, Thomas Kyle
Shalaway, Scott D
Shan, Robert Kuocheng
Smith, Robert Leo
Stephenson, Steven Lee

WISCONSIN

Adams, Michael Studebaker
Anderson, Raymond Kenneth
Bennett, James Peter
Boyce, Mark S
Carpenter, Stephen Russell
Cook, Robert Sewell
Copes, Frederick Albert
Cottam, Grant
Curtin, Charles Gilbert
Davidson, Donald William
Dukenschein, Jeanne Therese
Edgington, David Norman
Fitzgerald, George Patrick
Foote, Kenneth Gerald
Grittinger, Thomas Foster
Halgren, Lee A
Hall, Kent D
Haney, Alan William
Hardin, James William
He, Xi
Hickey, Joseph James
Hillier, Richard David
Hine, Ruth Louise
Hole, Francis Doan
Howell, Evelyn Anne
Iltis, Hugh Hellmut
Jeanne, Robert Lawrence
Johnson, Wendel J
Kitchell, James Frederick
Kitchell, Jennifer Ann
Kline, Virginia March
Krezoski, John R
Larsen, James Arthur
Larsen, Michael John
Long, Claudine Fern
McCabe, Robert Albert
McCown, Brent Howard
McDonald, Malcolm Edwin
McIntosh, Thomas Henry
Magnuson, John Joseph
Miller, David Hewitt
Minock, Michael Edward
Moermond, Timothy Creighton
Morgan, Michael Dean
Newsome, Richard Duane
Ogren, Herman August
Porter, Warren Paul
Richman, Sumner
Rolley, Robert Ewell
Rongstad, Orrin James
Rosson, Reinhardt Arthur

Ruff, Robert LaVerne
Rusch, Donald Harold
Seale, Dianne B
Smith, Stanley Galen
Stearns, Forest
Temple, Stanley A
Waller, Donald Macgregor
Warner, James Howard
Weise, Charles Martin
White, Charley Monroe
Whitford, Philip Burton
Wikum, Douglas Arnold
Yasukawa, Ken
Young, Allen Marcus
Yuill, Thomas MacKay

WYOMING
Adams, John Collins
Anderson, Stanley H
Bergman, Harold Lee
Christensen, Martha
Craighead, Frank Cooper, Jr
Diem, Kenneth Lee
Edwards, William Charles
Fisser, Herbert George
Hayden-Wing, Larry Dean
Holbrook, Frederick R
Kennington, Garth Stanford
Knight, Dennis Hal
Laycock, William Anthony
Lockwood, Jeffrey Alan
Parker, Michael
Reiners, William A
Scott, Richard Walter
Stanton, Nancy Lea
Sullivan, Brian Patrick

PUERTO RICO
Bruck, David Lewis
Lewis, Allen Rogers
Wilson, Marcia Hammerquist

ALBERTA
Addicott, John Fredrick
Agbeti, Michael Domingo
Anderson, Paul Knight
Belland, Rene Jean
Bidgood, Bryant Frederick
Blackshaw, Robert Earl
Boag, David Archibald
Byers, John Robert
Clifford, Hugh Fleming
Dancik, Bruce Paul
Davies, Ronald Wallace
Evans, William George
Freedman, Herbert I
Freeman, Milton Malcolm Roland
Fuller, William Albert
Hickman, Michael
Holmes, John Carl
Huang, Henry Hung-Chang
Lysyk, Timothy James
Macpherson, Andrew Hall
McPherson, Harold James
Michener, Gail R
Murie, Jan O
Pattie, Donald L
Schindler, David William
Scrimgeour, Garry Joseph
Seghers, Benoni Hendrik
Sodhi, Navjot Singh
Swanson, Robert Harold
Vitt, Dale Hadley
Wein, Ross Wallace
Wong, Horne Richard

BRITISH COLUMBIA
Adamson, Martin Leif
Barclay, Hugh John
Beacham, Terry Dale
Bell, Marcus Arthur Money
Buchanan, Ronald James
Bunce, Hubert William
Bunnell, Frederick Lindsley
Campbell, Alan
Chapman, Peter Michael
Chitty, Dennis Hubert
Cowan, Ian McTaggart
Dill, Lawrence Michael
Druehl, Louis D
Fisher, Francis John Fulton
Foreman, Ronald Eugene
Frazer, Bryan Douglas
Galindo-Leal, Carlos Enrique
Hartwick, Earl Brian
Healey, Michael Charles
Hebda, Richard Joseph
Jamieson, Glen Stewart
Kimmins, James Peter (Hamish)
Krajina, Vladimir Joseph
Lane, Edwin David
Lauzier, Raymond B
Levings, Colin David
Lewin, Victor
Mackas, David Lloyd
Mackauer, Manfred
McMullen, Robert David
Mathewes, Rolf Walter
Perrin, Christopher John
Perry, Richard Ian
Peterman, Randall Martin
Pielou, Evelyn C
Quinton, Dee Arlington
Reimchen, Thomas Edward

Roitberg, Bernard David
Rowe, John Stanley
Scotter, George Wilby
Stockner, John G
Strang, Robert M
Thomson, Alan John
Thomson, Keith A
Tunnicliffe, Verena Julia
Turkington, Robert (Roy) Albert
Vermeer, Kees
Walker, Ian Richard
Walters, Carl John
Wangersky, Peter John
Ward, David Mercer
Warrington, Patrick Douglas
Wellington, William George
Zhang, Ziyang
Zwickel, Fred Charles

MANITOBA
Abrahams, Mark Vivian
Amiro, Brian Douglas
Aung, Taing
Bodaly, Richard Andrew
Brunskill, Gregg John
Brust, Reinhart A
Fee, Everett John
Gee, John Henry
Guthrie, John Erskine
Huebner, Judith Dee
Iverson, Stuart Leroy
Johnson, Karen Louise
Keleher, J J
Patalas, Kazimierz
Preston, William Burton
Pruitt, William O(badiah), Jr
Rosenberg, David Michael
Sinha, Ranendra Nath
Smith, Roger Francis Cooper
Staniforth, Richard John
Turnock, William James
Ward, Fredrick James
White, Noel David George
Zach, Reto

NEW BRUNSWICK
Baskerville, Gordon Lawson
Beninger, Peter Gerard
Chopin, Thierry Bernard Raymond
Eidt, Douglas Conrad
Erskine, Anthony J
Eveleigh, Eldon Spencer
Fairchild, Wayne Lawrence
Hanson, John Mark
Harries, Hinrich
Logan, Alan
MacLean, David Andrew
Taylor, Andrew Ronald Argo
Van Groenewoud, Herman
Waiwood, Kenneth George

NEWFOUNDLAND
Anderson, John Truman
Carroll, Allan Louis
Davis, Charles (Carroll)
Haedrich, Richard L
Newton, Peter Francis
Pepin, P

NOVA SCOTIA
Bowen, William Donald
Cook, Robert Harry
Daborn, Graham Richard
Dadswell, Michael John
Doyle, Roger Whitney
Hardman, John Michael
Harrington, Fred Haddox
Longhurst, Alan R
Mann, Kenneth H
Miller, Robert Joseph
Newkirk, Gary Francis
Ogden, James Gordon, III
Silvert, William Lawrence
Specht, Harold Balfour
Stobo, Wayne Thomas
Tremblay, Michael John
Wiles, Michael

ONTARIO
Alex, Jack Franklin
Ambrose, John Daniel
Ankney, C(laude) Davison
Balon, Eugene Kornel
Banfield, Alexander William Francis
Barica, Jan M
Barrett, Spencer Charles Hilton
Baxter, Robert MacCallum
Behan-Pelletier, Valerie Mary
Beveridge, Terrance James
Birmingham, Brendan Charles
Brooks, Ronald James
Brown, Seward Ralph
Buckner, Charles Henry
Burger, Dionys
Burnison, Bryan Kent
Calder, Dale Ralph
Cameron, Duncan MacLean, Jr
Catling, Paul Miles
Cavers, Paul Brethen
Chengalath, Rama
Coad, Brian William
Collins, Nicholas Clark
Danks, Hugh Victor
Darling, Donald Christopher

Edwards, Roy Lawrence
Emery, Alan Roy
Ewins, Peter J
Falls, James Bruce
Fenton, Melville Brock
Fleming, Richard Arthur
Forbes, Bruce Cameron
Gill, Bruce Douglas
Graham, Terry Edward
Graham, William Muir
Green, Roger Harrison
Halfon, Efraim
Harcourt, Douglas George
Harvey, Harold H
Hebert, Paul David Neil
Hofstra, Gerald Gerrit
Hogan, Gary D
Hynes, Hugh Bernard Noel
Jeglum, John Karl
Jones, Philip Arthur
Jones, Roger
Kaushik, Narinder Kumar
Keddy, Paul Anthony
Kelso, John Richard Murray
Kershaw, Kenneth Andrew
Kevan, Peter Graham
Kobluk, David Ronald
Kovacs, Kit M
Krantzberg, Gail
Kudo, Akira
Lachance, Marc-Andre
Laing, John E
Lavigne, David M
Leach, Joseph Henry
Lee, David Robert
LeRoux, Edgar Joseph
Lovett-Doust, Jonathan Nicolas
McAndrews, John Henry
M'Closkey, Robert Thomas
Mackay, Rosemary Joan
McQueen, Donald James
Maycock, Paul Frederick
Merriam, Howard Gray
Middleton, Alex Lewis Aitken
Miyanishi, Kiyoko
Montgomerie, Robert Dennis
Morin, Antoine
Morris, Ralph Dennis
Morrison, Ian Kenneth
Mudroch, Alena
Munroe, Eugene Gordon
Newnham, Robert Montague
Nicholls, Kenneth Howard
Pierce, Ronald Cecil
Powles, Percival Mount
Rapport, David Joseph
Regier, Henry Abraham
Risk, Michael John
Ritchie, James Cunningham
Robertson, Raleigh John
Roberts-Pichette, Patricia Ruth
Ryder, Richard Armitage
Sale, Peter Francis
Sears, Markham Karli
Shih, Chang-Tai
Slocombe, Donald Scott
Smith, David William
Smith, Donald Alan
Smol, John Paul
Solman, Victor Edward Frick
Sprules, William Gary
Stebelsky, Ihor
Stokes, Pamela Mary
Svoboda, Josef
Taylor, William David
Tu, Chin Ming
Vollenweider, Richard A
Wallen, Donald George
Weatherley, Alan Harold
Weseloh, D V (Chip)
Williams, David Dudley
Winget, Carl Henry
Winterbottom, Richard
Winterhalder, Keith

PRINCE EDWARD ISLAND
Stewart, Jeffrey Grant

QUEBEC
Beland, Pierre
Bell, Graham Arthur Charlton
Bider, John Roger
Boisclair, Daniel
Bourget, Edwin
Brunel, Pierre
Castonguay, Martin
Cloutier, Conrad Francois
Dansereau, Pierre
Darveau, Marcel
Ferron, Jean H
Galbraith, David Allan
Gauthier, Gilles
Grandtner, Miroslav Marian
Grant, James William Angus
Hill, Stuart Baxter
Kalff, Jacob
Kingsley, Michael Charles Stephen
Lacroix, Guy
Laflamme, Gaston
Langford, Arthur Nicol
Lechowicz, Martin John
Legendre, Louis
Legendre, Pierre
Leggett, William C

Lemieux, Claudel
Levasseur, Maurice Edgar
McNeil, Jeremy Nichol
Magnan, Pierre
Majcen, Zoran
Maly, Edward J
Margolis, Hank A
Marsden, Joan Chauvin
Morisset, Pierre
Ouimet, Rock
Percy, Jonathan Arthur
Peters, Robert Henry
Rau, Manfred Ernst
Reiswig, Henry Michael
Richard, Pierre Joseph Herve
Sainte-Marie, Bernard
Sanborne, Paul Michael
Sergeant, David Ernest
Sharma, Madan Lal
Shoubridge, Eric Alan
Stewart, Robin Kenny
Titman, Rodger Donaldson
Vickery, William Lloyd
Vincent, C

SASKATCHEWAN
Bothwell, Max Lewis
Brigham, R Mark
Coupland, Robert Thomas
Evans, Marlene Sandra
Forsyth, Douglas John
Gratto-Trevor, Cheri Lynn
Hammer, Ulrich Theodore
Hobson, Keith Alan
Lehmkuhl, Dennis Merle
Maher, William J
Makowski, Roberte Marie Denise
Mitchell, George Joseph
Mukerji, Mukul Kumar
Nelson, Louise Mary
Oliphant, Lynn Wesley
Redmann, Robert Emanuel
Sarjeant, William Antony Swithin
Schmutz, Josef Konrad
Secoy, Diane Marie
Waddington, John
Walther, Alina
Zilke, Samuel

OTHER COUNTRIES
Adames, Abdiel Jose
Biederman-Thorson, Marguerite Ann
Bray, John Roger
Briand, Frederic Jean-Paul
Budowski, Gerardo
Buttemer, William Ashley
Camiz, Sergio
Caperon, John
Chadwick, Nanette Elizabeth
Chesson, Peter Leith
Cheung, Mo-Tsing Miranda
Craig, John Frank
Dahl, Arthur Lyon
Dickman, Michael David
Eiten, George
Erdtmann, Bernd Dietrich
Fowler, Scott Wellington
Gottlieb, Otto Richard
Holldobler, Berthold Karl
Jackson, Jeremy Bradford Cook
Johannes, Robert Earl
Kessell, Stephen Robert
Komarkova, Vera
Komers, Petr E
Koslow, Julian Anthony
Laurent, Pierre
Lee, Douglas Harry Kedgwin
Lieth, Helmut Heinrich Friedrich
Marsh, James Alexander, Jr
May, Robert McCredie
Medina, Ernesto Antonio
Myres, Miles Timothy
Nelson, Stephen Glen
Nevo, Eviatar
Ni, I-Hsun
O'Neill, Patricia Lynn
Paulay, Gustav
Peakall, David B
Rodriguez, Gilberto
Saether, Ole Anton
Shubert, L Elliot
Srygley, Robert Baxter
Stam, Jos
Stearns, Stephen Curtis
Tosi, Joseph Andrew, Jr
Ugolini, Fiorenzo Cesare
Wynne-Edwards, Vero Copner
Yosef, Reuven
Zhang, Zhi-Qiang

Embryology

ALABAMA
Bacon, Arthur Lorenza
Freeman, John A
Hand, George Samuel, Jr
Hood, Ronald David
Keys, Charles Everel
Lubega, Seth Gasuza
McLaughlin, Ellen Winnie
Stephenson, Edwin Clark

Embryology (cont)

ARIZONA
Bagnara, Joseph Thomas
Doane, Winifred Walsh
Hendrix, Mary J C
Kischer, Clayton Ward
Maienschein, Jane Ann
Pogany, Gilbert Claude
Terry, Lucy Irene

ARKANSAS
Johnston, Perry Max
Sheehan, Daniel Michael
Tank, Patrick Wayne

CALIFORNIA
Alberch, Pere
Anderson, Gary Bruce
Armstrong, Peter Brownell
Armstrong, Rosa Mae
Baird, John Jeffers
Bernard, George W
Blatt, Beverly Faye
Brownell, Anna Gale
Bryant, Peter James
Calarco, Patricia G
Celniker, Susan Elizabeth
Chuong, Cheng-Ming
Davidson, Eric Harris
Dawson, John E
Duester, Gregg L
Dunaway, Marietta
Dunnebacke-Dixon, Thelma Hudson
Durrant, Barbara Susan
Epel, David
Fisher, Robin Scott
Fristrom, Dianne
Fukuda, Michiko N
Glass, Laurel Ellen
Golbus, Mitchell S
Grey, Robert Dean
Hammar, Allan H
Heath, Harrison Duane
Hendrickx, Andrew George
Jeffery, William Richard
Kalland, Gene Arnold
Keller, Raymond E
King, Barry Frederick
King, Robbins Sydney
Ko, Chien-Ping
Krejsa, Richard Joseph
Laham, Quentin Nadime
Lambert, Charles Calvin
Lengyel, Judith Ann
Levine, Michael Steven
Lipshitz, Howard David
Luckock, Arlene Suzanne
Mak, Linda Louise
Martin, Gail Roberta
Martins-Green, Manuela M
Maxson, Robert E, Jr
Melnick, Michael
Menees, James H
Mertes, David H
Meyerowitz, Elliot Martin
Moore, Betty Clark
Moretti, Richard Leo
Neff, William Medina
Nuccitelli, Richard Lee
Pedersen, Roger Arnold
Plopper, Charles George
Rio, Donald C
Schweisthal, Michael Robert
Scott, Matthew P
Shook, Brenda Lee
Smith, Bradley Richard
Smith, Lewis Dennis
Sparling, Mary Lee
Spitzer, Nicholas Canaday
Stephens, Lee Bishop, Jr
Stephens, Robert James
Stockdale, Frank Edward
Sung, Zinmay Renee
Thurmond, William
Towers, Bernard
Urry, Lisa Andrea
Vaughn, James E, Jr
Vreeland, Valerie Jane
Webster, Barbara Donahue
Willhite, Calvin Campbell
Wilson, Doris Burda
Wilson, Wilfred J
Wilt, Fred H
Wood, Richard Lyman
Woods, Geraldine Pittman
Zernik, Joseph

COLORADO
Barrett, Dennis
Bekoff, Anne C
Finger, Thomas Emanuel
Ham, Richard George
Hanken, James
Lesh-Laurie, Georgia Elizabeth
Markert, Clement Lawrence
Maylie-Pfenninger, M F
Moury, John David
Runner, Meredith Noftzger
Seidel, George Elias, Jr
Torbit, Charles Allen, Jr

CONNECTICUT
Beitch, Irwin
Berry, Spencer Julian
Boell, Edgar John
Clark, Hugh
Cohen, Melvin Joseph
Constantine-Paton, Martha
Fell, Paul Erven
Fiore, Carl
Herrmann, Heinz
Infante, Anthony A
Jaffe, Laurinda A
Kent, John Franklin
Kollar, Edward James
Lee, Thomas W
Levin, Martin Allen
Morest, Donald Kent
Pierro, Louis John
Rakic, Pasko
Roman, Laura M
Rossomando, Edward Frederick
Staugaard, Burton Christian
Upholt, William Boyce

DELAWARE
Brent, Robert Leonard

DISTRICT OF COLUMBIA
Avery, Gordon B
Ball, William David
Dimond, Marie Therese
Dosier, Larry Waddell
Goeringer, Gerald Conrad
Hayes, Raymond L, Jr
Hines, Pamela Jean
Kimmel, Carole Anne
Nishioka, David Jitsuo
Van Arsdel, William Campbell, III

FLORIDA
Bray, Joan Lynne
Cameron, Don Frank
Chen, L T
Cohen, Glenn Milton
Colwin, Arthur Lentz
Colwin, Laura Hunter
Deats, Edith Potter
Gilbert-Barness, Enid F
Goldberg, Stephen
Grabowski, Casimer Thaddeus
Hinsch, Gertrude Wilma
Hopper, Arthur Frederick
Jaffee, Oscar Charles
Koevenig, James L
Muller, Kenneth Joseph
Rice, Stanley Alan
Shireman, Rachel Baker
Taylor, George Thomas
Tripp, John Rathbone
Vasil, Vimla

GEORGIA
DeHaan, Robert Lawrence
Gulati, Adarsh Kumar
Hicks, Heraline Elaine
Khan, Iqbal M
McKenzie, John Ward
Patterson, Rosalyn Mitchell
Paulsen, Douglas F
Smeltzer, Richard Homer
Sohal, Gurkirpal Singh
Whitney, J(ohn) Barry, III
Zinsmeister, Philip Price

HAWAII
Arnold, John Miller
Dalton, Howard Clark
Hadfield, Michael Gale
Haley, Samuel Randolph
Nelson, Marita Lee
Rosenthal, Eric Thomas
Wyban, James A

IDAHO
Cloud, Joseph George
Fuller, Eugene George
Stephens, Trent Dee

ILLINOIS
Aydelotte, Margaret Beesley
Bornslaeger, Elayne A
Carr, Virginia McMillan
Criley, Bruce
Dinsmore, Charles Earle
Doering, Jeffrey Louis
Dudkiewicz, Alan Bernard
Durica, Thomas Edward
Dybas, Linda Kathryn
Farbman, Albert Irving
Ferguson, Edwin Louis
Foote, Florence Martindale
Gaik, Geraldine Catherine
Goldman, Allen S
Karr, Timothy Lawrence
LaVelle, Faith Wilson
Menco, Bernard Philip Max
Mittenthal, Jay Edward
Moskal, Joseph Russell
Nardi, James Benjamin
Overton, Jane Harper
Pollack, Emanuel Davis
Rabuck, David Glenn
Ritter, Nadine Marie
Schmidt, Anthony John
Seale, Raymond Ulric
Spiroff, Boris E N
Van Alten, Pierson Jay
Williams, Thomas Alan
Zalisko, Edward John

INDIANA
BeMiller, Paraskevi Mavridis
Cadwallader, Joyce Vermeulen
Chernoff, Ellen Ann Goldman
Crouch, Martha Louise
Das, Gopal Dwarka
Davis, Grayson Steven
Denner, Melvin Walter
Hoversland, Roger Carl
Hunt, Linda Margaret
Kennedy, Duncan Tilly
Maloney, Michael Stephen
Mescher, Anthony Louis
Pietsch, Paul Andrew
Schaible, Robert Hilton
Tweedell, Kenyon Stanley

IOWA
Benbow, Robert Michael
Kieso, Robert Alfred
Kollros, Jerry John
Meetz, Gerald David
Mennega, Aaldert
Milkman, Roger Dawson
Rogers, Frances Arlene
Shen, Sheldon Shih-Ta
Solursh, Michael
Sullivan, Charles Henry
Welshons, William John
Wiens, Darrell John

KANSAS
Bode, Vernon Cecil
Conrad, Gary Warren
Dey, Sudhansu Kumar
Frost-Mason, Sally Kay
Gattone, Vincent H, II
Grebe, Janice Durr
Smalley, Katherine N
Wyttenbach, Charles Richard

KENTUCKY
Alexander, Lloyd Ephraim
Bhatnagar, Kunwar Prasad
Birge, Wesley Joe
Humphries, Asa Alan, Jr
Just, John Josef
Matulionis, Daniel H
Smith, Stephen D

LOUISIANA
Clawson, Robert Charles
Cobb, Glenn Wayne
Cowden, Ronald Reed
Gasser, Raymond Frank
Grodner, Mary Laslie
Kern, Clifford H, III
Newsome, David Anthony
Peebles, Edward McCrady
Vaupel, Martin Robert
Weber, Joseph T
Yeh, Kwo-Yih

MAINE
Bailey, Donald Wayne
Brawley, Susan Howard
Hoppe, Peter Christian
Rappaport, Raymond

MARYLAND
Ackerman, Eric J
Berger, Edward Alan
Brown, Kenneth Stephen
Chan, Wai-Yee
Daniels, Mathew Paul
Dean, Jurrien
Ebert, James David
Finch, Robert Allen
Fire, Andrew Zachary
Gall, Joseph Grafton
Goode, Melvyn Dennis
Grant, Philip
Greenhouse, Gerald Alan
Gulyas, Bela Janos
Hanover, John Allan
Hausman, Steven J
Heck, Margaret Mathilde Sophie
Hiatt, James Lee
Hoffman, David J
Imberski, Richard Bernard
Kaighn, Morris Edward
McKnight, Steven Lanier
Merlino, Glenn T
Newrock, Kenneth Matthew
Oberdorfer, Michael Douglas
Provine, Robert Raymond
Rall, William Frederick
Schnaar, Ronald Lee
Shafer, W Sue
Song, Jiakun
Stratmeyer, Melvin Edward
Strum, Judy May
White, Elizabeth Lloyd
Wolffe, Alan Paul
Wu, Roy Shih-Shyong
Zirkin, Barry Ronald

MASSACHUSETTS
Adler, Richard R
Ahlberg, Henry David
Albertini, David Fred
Begg, David A
Bernfield, Merton Ronald
Bieber, Frederick Robert
Cohen, Alan Mathew
Crain, William Rathbone, Jr
Dittmer, John Edward
Ducibella, Tom
Erickson, Alan Eric
Ernst, Susan Gwenn
Eschenberg, Kathryn (Marcella)
Fink, Rachel Deborah
Furshpan, Edwin Jean
Gibbons, Michael Francis, Jr
Gonnella, Patricia Anne
Hausman, Robert Edward
Hay, Elizabeth Dexter
Hertig, Arthur Tremain
Hinrichs, Katrin
Hoar, Richard Morgan
Horvitz, Howard Robert
Hynes, Richard Olding
Jaenisch, Rudolf
Kafatos, Fotis C
Kaltenbach-Townsend, Jane
Lazarte, Jaime Esteban
Maas, Richard Louis
Miller, Andrew Leitch
Miller, James Albert, Jr
Papaioannou, Virginia Eileen
Pardue, Mary Lou
Poccia, Dominic Louis
Rafferty, Keen Alexander, Jr
Rafferty, Nancy S
Rollason, Grace Saunders
Ruderman, Joan V
Saunders, John Warren, Jr
Sorokin, Sergei Pitirimovitch
Spiegelman, Martha
Stein, Otto Ludwig
Strauss, William Mark
Toole, Bryan Patrick
Webb, Andrew Clive
Wolf, Merrill Kenneth
Wright, Mary Lou
Wu, Jung-Tsung
Ziomek, Carol A

MICHIGAN
Abraham, Irene
Armant, D Randall
Atkinson, James William
Avery, James Knuckey
Beaudoin, Allan Roger
Boving, Bent Giede
Buss, Jack Theodore
Cather, James Newton
Duwe, Arthur Edward
Easter, Stephen Sherman, Jr
Eichler, Victor B
Fisher, Don Lowell
Fritts-Williams, Mary Louise Monica
Froiland, Thomas Gordon
Goustin, Anton Scott
Heady, Judith E
Hill, Susan Douglas
Kemp, Norman Everett
Koehler, Lawrence D
Mathews, Willis Woodrow
Sacco, Anthony G
Smith, Richard Harding
Stinson, Al Worth
Tosney, Kathryn W

MINNESOTA
Ellinger, Mark Stephen
Sherer, Glenn Keith
Shoger, Ross L
Sicard, Raymond E(dward)
Singer, Susan Rundell
Sinha, Akhouri Achyutanand
Smail, James Richard
Smithberg, Morris
Todt, William Lynn

MISSISSIPPI
Ball, Carroll Raybourne
Keiser, Edmund Davis, Jr
Lawler, Adrian Russell
Martin, Billy Joe
Roy, William Arthur

MISSOURI
Cheney, Clarissa M
Chiappinelli, Vincent A
Cryer, Philip Eugene
David, John Dewood
Decker, John D
Derby, Albert
Engel, Leslie Carroll
Ericson, Avis J
Friedman, Harvey Paul
Hamburger, Viktor
Moffatt, David John
Osdoby, Philip
Price, Joseph Levering
Sanes, Joshua Richard
Schreiweis, Donald Otto
Sharp, John Roland

BIOLOGICAL SCIENCES / 93

MONTANA
Foresman, Kerry Ryan
Thompson, Holly Ann

NEBRASKA
Dossel, William Edward
Fawcett, James Davidson
Holyoke, Edward Augustus
Lund, Douglas E
McCue, Robert Owen
Rosenquist, Thomas H
Schlesinger, Allen Brian
Smith, Paula Beth
Turpen, James Baxter

NEVADA
Tibbitts, Forrest Donald

NEW HAMPSHIRE
Ferm, Vergil Harkness
Foret, John Emil
Spiegel, Evelyn Sclufer
Spiegel, Melvin
Thomas, William Albert

NEW JERSEY
Agnish, Narsingh Dev
Arthur, Alan Thorne
Collier, Marjorie McCann
Essien, Francine B
Fangboner, Raymond Franklin
Francoeur, Robert Thomas
Frost, David
Halpern, Myron Herbert
Hampton, Suzanne Harvey
Hart, Nathan Hoult
Hollinshead, May B
Hyndman, Arnold Gene
Ivatt, Raymond John
Mitala, Joseph Jerrold
Neri, Anthony
Pai, Anna Chao
Ruknudin, Abdul Majeed
Saiff, Edward Ira
Shelden, Robert Merten
Silver, Lee Merrill
Steinberg, Malcolm Saul
Stenn, Kurt S
Szot, Robert Joseph
Tesoriero, John Vincent
Trelstad, Robert Laurence
Weis, Peddrick

NEW MEXICO
Overturf, Gary D
Stricker, Stephen Alexander

NEW YORK
Angerer, Robert Clifford
Arny, Margaret Jane
Ash, William James
Asnes, Clara F
Auclair, Walter
Bachvarova, Rosemary Faulkner
Benzo, Camillo Anthony
Black, Virginia H
Blackler, Antonie W C
Boylan, Elizabeth S
Bradbury, Michael Wayne
Brownscheidle, Carol Mary
Cairns, John Mackay
Collier, Jack Reed
Crain, Stanley M
Crotty, William Joseph
Currie, Julia Ruth
Dornfest, Burton S
Easton, Douglas P
Factor, Jan Robert
Foote, Robert Hutchinson
Fortune, Joanne Elizabeth
Fowler, James A
Frair, Wayne
Gates, Allen H(azen), Jr
Gershon, Michael David
Gordon, Jon W
Greengard, Olga
Grumet, Martin
Harris, Jack Kenyon
Holtfreter, Johannes Friedrich Karl
Hopkins, Betty Jo Henderson
Katz, Eugene Richard
Kaye, Nancy Weber
Kostreva, David Robert
Kuehnert, Charles Carroll
Lemanski, Larry Frederick
Loy, Rebekah
Lyser, Katherine May
McCann-Collier, Marjorie Dolores
McCormick, Paulette Jean
McCourt, Robert Perry
McDaniel, Carl Nimitz
Miller, Richard Kermit
Miller, Sue Ann
Mitchell, John Taylor
Munne, Santiago
Murphy, Michael Joseph
Newman, Stuart Alan
Rasweiler, John Jacob, IV
Reynolds, Wynetka Ann King
Rich, Abby M
Robbins, Edith Schultz
Roeder, Robert Gayle
Rosenbaum, Peter Andrew
Ruddell, Alanna

Schlafer, Donald H
Schuel, Herbert
Siden, Edward Joel
Sobel, Jael Sabina
Spence, Alexander Perkins
Swartz, Gordon Elmer
Szabo, Piroska Ludwig
Tabor, John Malcolm
Topp, William Carl
Udin, Susan Boymel
Waelsch, Salome Gluecksohn
Wenk, Eugene J
Wolfner, Mariana Federica
Wu, Joseph M
Zanetti, Nina Clare
Zapisek, William Francis

NORTH CAROLINA
Anderton, Laura Gaddes
Andrews, Matthew Tucker
Black, Betty Lynne
Boklage, Charles Edward
Counce, Sheila Jean
Edwards, Nancy C
Fail, Patricia A
Harris, Albert Kenneth, Jr
Johnston, Malcolm Campbell
Kalmus, Gerhard Wolfgang
Lawrence, Irvin E, Jr
Lehman, Harvey Eugene
Lewis, Susan Erskine
Lieberman, Melvyn
McLachlan, John Alan
Reid, Lola Cynthia McAdams
Sadler, Thomas William
Shafer, Thomas Howard
Sulik, Kathleen Kay
Welsch, Frank

NORTH DAKOTA
Owen, Alice Koning
Thompson, Michael Bruce

OHIO
Baker, Peter C
Beal, Kathleen Grabaskas
Birky, Carl William, Jr
Caston, J Douglas
Chakraborty, Joana J(yotsna)
Coleman, Marilyn A
Crutcher, Keith A
DuBrul, Ernest
Egar, Margaret Wells
Ferguson, Marion Lee
Gao, Kuixiong
Hamlett, William Cornelius
Herschler, Michael Saul
Hilliard, Stephen Dale
Houston, Willie Walter, Jr
Jacobs-Lorena, Marcelo
Kriebel, Howard Burtt
Lane, Roger Lee
Raghavan, Valayamghat
Rutishauser, Urs Stephen
Saksena, Vishnu P
Tepperman, Katherine Gail
Thomson, Dale S
Vorhees, Charles V
Watanabe, Michiko
Yow, Francis Wagoner
Zimmerman, Ernest Frederick

OKLAHOMA
Dugan, Kimiko Hatta
Lavia, Lynn Alan
Lhotka, John Francis
Tobin, Sara L

OREGON
Adams, Frank William
Cimberg, Robert Lawrence
Fish, Joseph Leroy
Gunberg, David Leo
Morris, John Edward
Postlethwait, John Harvey
Quatrano, Ralph Stephen
Weston, James A

PENNSYLVANIA
Aronson, John Ferguson
Brinster, Ralph L
Buono, Russell Joseph
Cohen, Leonard Harvey
DiBerardino, Marie A
Fabian, Michael William
Fenderson, Bruce Andrew
Fluck, Richard Allen
Gibley, Charles W, Jr
Grobstein, Paul
Grunwald, Gerald B
Heyner, Susan
Hilfer, Saul Robert
Hollyday, Margaret Anne
Holtzer, Howard
Idzkowsky, Henry Joseph
Jensh, Ronald Paul
Kennedy, Michael Craig
Kochhar, Devendra M
Kvist, Tage Nielsen
Lash, James (Jay) W
Leibel, Wayne Stephan
Linask, Kersti Katrin
McLaughlin, Patricia J
Meriney, Stephen D

Mezger-Freed, Liselotte
Niu, Mann Chiang
Oppenheimer, Jane Marion
Piesco, Nicholas Peter
Poethig, Richard Scott
Pollock, John Archie
Ramasastry, Sai Sudarshan
Roth, Stephen
Searls, Robert L
Short, John Albert
Telfer, William Harrison
Thomson, Keith Stewart
Tuan, Rocky Sung-Chi
Vogel, Norman William
Weston, John Colby
Zaccaria, Robert Anthony
Zagon, Ian Stuart

RHODE ISLAND
Coleman, John Russell
Fausto-Sterling, Anne
Fish, William Arthur
Meedel, Thomas Huyck
Silver, Alene Freudenheim
Weisz, Paul B

SOUTH CAROLINA
Dickey, Joseph Freeman
Finlay, Mary Fleming
Herr, John Mervin, Jr
Odor, Dorothy Louise
Sawyer, Roger Holmes
Taber, Elsie

SOUTH DAKOTA
Haertel, John David
Johnson, Leland Gilbert
Naughten, John Charles
Neufeld, Daniel Arthur
Settles, Harry Emerson

TENNESSEE
Dumont, James Nicholas
Hasty, David Long
Hoffman, Loren Harold
Lessman, Charles Allen
MacCabe, Jeffrey Allan
McFee, Alfred Frank
McGavock, Walter Donald
Mallette, John M
Popp, Raymond Arthur
Skalko, Richard G(allant)
Trigiano, Robert Nicholas
Wachtel, Stephen Shoel
Walton, Barbara Ann
Wiser, Cyrus Wymer

TEXAS
Blystone, Robert Vernon
Carson, Daniel Douglas
Collins, Russell Lewis
Dalterio, Susan Linda
Deter, Russell Lee, II
DiMichele, Leonard Vincent
Faulkner, Russell Conklin, Jr
George, Fredrick William
Gomer, Richard Hans
Grimes, L Nichols
Houston, Marshall Lee
Jacobson, Antone Gardner
Kalthoff, Klaus Otto
Litke, Larry Lavoe
Martin, Edward Williford
Pierce, Jack Robert
Roberts, Susan Jean
Sauer, Helmut Wilhelm
Schwalm, Fritz Ekkehardt
Siler-Khodr, Theresa M
Wordinger, Robert James
Zwaan, Johan Thomas

UTAH
Capecchi, Mario Renato
Jacobson, Marcus
Schoenwolf, Gary Charles
Seegmiller, Robert Earl
Wurst, Gloria Zettle

VERMONT
Evans, Hiram John
Woodcock-Mitchell, Janet Louise

VIRGINIA
Allan, Frank Duane
Black, Robert Earl Lee
Creager, Joan Guynn
Deck, James David
DeSesso, John Michael
Devine, Charles Joseph, Jr
Diehl, Fred A
Eglitis, Martin Alexandris
Harris, Thomas Mason
Jollie, William Pucette
Laurie, Gordon William
Little, Charles Durwood, Jr
Moyer, Wayne A
Palisano, John Raymond
Potts, Malcolm
Radice, Gary Paul
Reynolds, John Dick
Rubenstein, Norton Michael
Sandow, Bruce Arnold
Wiseman, Lawrence Linden

WASHINGTON
Avner, Ellis David
Bakken, Aimee Hayes
Davies, Jack
Hauschka, Stephen D
Hendrickson, Anita Elizabeth
Hille, Merrill Burr
Holbrook, Karen Ann
Hosick, Howard Lawrence
Lowe, Janet Marie
Quinn, LeBris Smith
Reh, Thomas Andrew
Sikov, Melvin Richard
Wakimoto, Barbara Toshiko
Yancey, Paul Herbert

WEST VIRGINIA
Butcher, Roy Lovell
Reyer, Randall William
Williams, Leah Ann

WISCONSIN
Auerbach, Robert
Bavister, Barry Douglas
Bersu, Edward Thorwald
Bolender, David Leslie
Fallon, John Francis
Kaplan, Stanley
Keefer, Carol Lyndon
Orsini, Margaret Ward (Giordano)
Schindler, Joel Marvin
Snook, Theodore
Snyder, Virginia
Sobkowicz, Hanna Maria

WYOMING
Lillegraven, Jason Arthur
Petersen, Nancy Sue

PUERTO RICO
Craig, Syndey Pollock, III
Luckett, Winter Patrick

ALBERTA
Campenot, Robert Barry
Cass, David D
Cavey, Michael John
Cuny, Robert Michael
Heming, Bruce Sword
Machin, Geoffrey Allan
Sperber, Geoffrey Hilliard

BRITISH COLUMBIA
Brandhorst, Bruce Peter
Burke, Robert D
Diewert, Virginia M
Finnegan, Cyril Vincent
Todd, Mary Elizabeth
Von Aderkas, Patrick Jurgen Cecil

MANITOBA
Gordon, Richard
Laale, Hans W

NEW BRUNSWICK
Whittaker, J Richard

NOVA SCOTIA
Dickson, Douglas Howard

ONTARIO
Balon, Eugene Kornel
Buchanan, George Dale
Butler, Richard Gordon
Chua, Kian Eng
Chui, David H K
Dingle, Allan Douglas
Etches, Robert J
Fallding, Margaret Hurlstone Hardy
Fisher, Kenneth Robert Stanley
Kidder, Gerald Marshall
Leibo, Stanley Paul
Liversage, Richard Albert
Rossant, Janet
Scadding, Steven Richard
Zimmerman, Selma Blau

QUEBEC
Blaschuk, Orest William
Briere, Normand
Dubé, François
Mulay, Shree
Trasler, Daphne Gay
Wolsky, Alexander

SASKATCHEWAN
Butler, Harry
Fedoroff, Sergey
Flood, Peter Frederick

OTHER COUNTRIES
Araki, Masasuke
Dworkin, Mark Bruce
Geiger, Benjamin
Helander, Herbert Dick Ferdinand
Kordan, Herbert Allen
Lesseps, Roland Joseph
Mizuno, Shigeki
Phillips, Roger Guy
Schneider, Wolfgang Johann
Shore, Laurence Stuart
Solter, Davor
Szebenyi, Emil
Tachibana, Takehiko

Embryology (cont)

Takeuchi, Kiyoshi Hiro
Thiery, Jean Paul
Toutant, Jean-Pierre
Webb, Cynthia Ann Glinert
Wenger, Byron Sylvester

Endocrinology

ALABAMA
Berecek, Kathleen Helen
Garver, David L
Paxton, Ralph
Yarbrough, James David

ARIZONA
Butler, Lillian Catherine
Hagedorn, Henry Howard
Justus, Jerry T

ARKANSAS
Allaben, William Thomas
Sheehan, Daniel Michael
Stallcup, Odie Talmadge

CALIFORNIA
Arnold, Arthur Palmer
Balon, Thomas William
Chambers, Kathleen Camille
Charles, M Arthur
Curry, Donald Lawrence
Davidson, Mayer B
DeLeon, Daisy Delgado
Engelmann, Franz
Ezrin, Calvin
Fagin, Katherine Diane
Gibson, Thomas Richard
Gordan, Gilbert Saul
Grumbach, Melvin Malcolm
Horwitz, David Larry
Hostetler, Karl Yoder
Jarnagin, Kurt
Johnson, Randolph Mellus
Kalland, Gene Arnold
Krauss, Ronald
Laird, Cleve Watrous
Lev-Ran, Arye
Licht, Paul
Lieberburg, Ivan M
Lin, Ming-Fong
Ling, Nicholas Chi-Kwan
Lu, John Kuew-Hsiung
Mellon, Synthia
Mestril, Ruben
Moberg, Gary Philip
Mohan, Chandra
Nemere, Ilka M
Nishioka, Richard Seiji
Papkoff, Harold
Parlow, Albert Francis
Perryman, Elizabeth Kay
Rosenfeld, George
Rosenfeld, Ron Gershon
Russell, Sharon May
Sawyer, Wilbur Henderson
Shelesnyak, Moses Chaim
Snow, George Edward
Szego, Clara Marian
Thomas, Eric Owen
Turgeon, Judith Lee
Van Dop, Cornelis
Vickery, Larry Edward
Walker, Ameae M
Weber, Heather R(oss) Wilson
Wechter, William Julius
Wilcox, Ronald Bruce
Williams, Hibbard E
Yund, Mary Alice
Zernik, Joseph

COLORADO
Carnahan, Gilbert William
Draznin, Boris
Hanken, James
Hofeldt, Fred Dan
Hossner, Kim L
Kano-Sueoka, Tamiko
Nett, T M
Roberts, P Elaine
Seely, James Ervin

CONNECTICUT
Goldman, Bruce Dale
Matovcik, Lisa M
Pawelek, John Mason
Rebuffe-Scrive, Marielle Francoise
Tamborlane, William Valentine

DELAWARE
Saller, Charles Frederick

DISTRICT OF COLUMBIA
Alleva, John J
Callaway, Clifford Wayne
Geelhoed, Glenn William
Hone, Jennifer
Lippman, Marc Estes
Lumpkin, Michael Dirksen
Martin, Malcolm Mencer
Naylor, Paul Henry
Vernikos, Joan
Wong, Harry Yuen Chee

FLORIDA
Chavin, Walter
Chegini, Nasser
Cooper, Denise R
Deats, Edith Potter
Gennaro, Joseph Francis
Hansen, Peter J
Howard, Guy Allen
Katovich, Michael J
McLean, Mark Philip
Mintz, Daniel Harvey
Neary, Joseph Thomas
Phelps, Christopher Prine
Rao, Krothapalli Ranga
Rosenbloom, Arlan Lee
Sang, Qing-Xiang
Scarpace, Philip J
Shapiro, Jeffrey Paul
Shirk, Paul David
Spieler, Richard Earl
Standaert, Mary L
Wallace, Robin A

GEORGIA
Barb, C Richard
Brann, Darrell Wayne
Costoff, Allen
Hoffman, William Hubert
Hollowell, Joseph Gurney, Jr
Khan, Iqbal M
Muldoon, Thomas George
Noe, Bryan Dale
Ogle, Thomas Frank
Toney, Thomas Wesley
Tyler, Jean Mary

HAWAII
Atkinson, Shannon K C
Lee, Cheng-Sheng

IDAHO
Laurence, Kenneth Allen
Mead, Rodney A
Rodnick, Kenneth Joseph

ILLINOIS
Bombeck, Charles Thomas, III
Budzik, Gerald P
Fang, Victor Shengkuen
Gaskins, H Rex
Gershbein, Leon Lee
Goldman, Allen S
Gu, Yan
Hileman, Stanley Michael
Kasprow, Barbara Ann
Katzenellenbogen, Benita Schulman
Kesler, Darrel J
Kulkarni, Bidy
Kyncl, J Jaroslav
Langman, Craig Bradford
Liao, Shutsung
Loizzi, Robert Francis
Lorenzen, Janice R
McNulty, John Alexander
Meyers, Cal Yale
Nutting, Ehard Forrest
Prinz, Richard Allen
Rabuck, David Glenn
Rubenstein, Arthur Harold
Schneider, Gary Bruce
Schwartz, Neena Betty
Shambaugh, George E, III
Singh, Sant Parkash
Steiner, Donald Frederick
Velardo, Joseph Thomas
Verhage, Harold Glenn
Walter, Robert John
Weiss, Roy E
Yelich, Michael Ralph
Zaneveld, Lourens Jan Dirk

INDIANA
Cadwallader, Joyce Vermeulen
Cherbas, Peter Thomas
Frolik, Charles Alan
Grant, Alan Leslie
Hancock, Deana Lori
Heiman, Mark Louis
Henry, David P, II
Hoversland, Roger Carl
Howard, David K
Johnson, Alan L
Moore, Ward Wilfred
Pierce, William Meredith, Jr
Turner, Charles Hall
Young, Peter Chun Man

IOWA
Ascoli, Mario
Ford, Stephen Paul
Stewart, Mary E
Trenkle, Allen H

KANSAS
Dass, David
Dey, Sudhansu Kumar
Greenwald, Gilbert Saul
MacGregor, Ronal Roy
Rawitch, Allen Barry
Smalley, Katherine N
Stevenson, Jeffrey Smith
Voogt, James Leonard
Zimniski, Stephen Joseph

KENTUCKY
Anderson, James Wingo
Buchanan, Bernard J
Just, John Josef
Miller, Ralph English
Nikitovitch-Winer, Miroslava B
Peterson, Roy Phillip
Rao, Chalamalasetty Venkateswara

LOUISIANA
Fingerman, Milton
Hansel, William
Harrison, Richard Miller
Knesel, John Arthur
Olubadewo, Joseph Olanrewaju
Saphier, David
Spring, Jeffrey H
Walters, Marian R

MAINE
Musgrave, Stanley Dean

MARYLAND
Aldrich, Jeffrey Richard
Alleva, Frederic Remo
Ball, Gregory Francis
Bhathena, Sam Jehangirji
Blithe, Diana Lynn
Chan, Wai-Yee
Cohen, Rochelle Sandra
Cutler, Gordon Butler, Jr
Dermody, William Christian
Dufau, Maria Luisa
Eastman, Richard Cyrus
Estienne, Mark Joseph
Fradkin, Judith Elaine
Freeman, Colette
Gann, Donald Stuart
Gill, John Russell, Jr
Grave, Gilman Drew
Gueriguian, John Leo
Hoeg, Jeffrey Michael
Howe, Juliette Coupain
Loriaux, D Lynn
Lutwak, Leo
Manganiello, Vincent Charles
Mills, James Louis
Nakhasi, Hira Lal
Nekola, Mary Virginia
Nisula, Bruce Carl
Ortmeyer, Heidi Karen
Raina, Ashok K
Rosen, Saul W
Roth, Jesse
Simons, Samuel Stoney, Jr
Stojilkovic, Stanko S
Stolz, Walter S
Striker, G E
Strumwasser, Felix
Vydelingum, Nadarajen Ameerdanaden
Wilder, Ronald Lynn
Wollman, Seymour Horace
Zimbelman, Robert George
Zirkin, Barry Ronald

MASSACHUSETTS
Arvan, Peter
Callard, Gloria Vincz
Chen, Wenyong
Coyne, Mary Downey
Dayal, Yogeshwar
De Cherney, Alan Hersh
Emanuel, Rodica L
Flatt, Jean-Pierre
Frisch, Rose Epstein
Gawienowski, Anthony Michael
Gustafson, Alvar Walter
Kaltenbach-Townsend, Jane
Kew, David
Kosasky, Harold Jack
McArthur, Janet W
Neer, Eva Julia
Neer, Robert M
Orme-Johnson, Nanette Roberts
Raam, Shanthi
Reppert, Steve Marion
Rosemberg, Eugenia
Rosner, Anthony Leopold
Sacks, David B
Sawin, Clark Timothy
Shamgochian, Maureen Dowd
Snyder, Benjamin Willard
Thies, Robert Scott
Thorn, George W
Villa-Komaroff, Lydia
Villee, Claude Alvin, Jr
Villee, Dorothy Balzer
Widmaier, Eric Paul
Wolfson, Adele Judith
Wright, Mary Lou

MICHIGAN
Al Saadi, A Amir
Barney, Christopher Carroll
Bleasdale, John Edward
Bone, Henry G
Carter-Su, Christin
Christensen, A(lbert) Kent
Dunbar, Joseph C
Floyd, John Claiborne, Jr
Foster, Douglas Layne
Friar, Robert Edsel
Horowitz, Samuel Boris
Hsu, Chen-Hsing
Ireland, James
Keyes, Paul Landis
Landefeld, Thomas Dale
Lew, Gloria Maria
Mccorkle, Fred Miller
Midgley, A Rees, Jr
Pavgi, Sushama
Payne, Anita H
Rivera, Evelyn Margaret
Saltiel, Alan Robert
Schwartz, Jessica
Subramanian, Marappa G
Tucker, Herbert Allen

MINNESOTA
Bauer, Gustav Eric
Bodine, Peter Van Nest
Brown, David Mitchell
Fallon, Ann Marie
Gilboe, Daniel Pierre
Goetz, Frederick Charles
Hanson, Robert C
Ryan, Robert J
Sorensen, Peter W
Sorenson, Robert Lowell
Tindall, Donald J
Towle, Howard Colgate
Wilson, Michael John

MISSISSIPPI
Hall, Margaret (Margot) Jean

MISSOURI
Armbrecht, Harvey James
Blaine, Edward Homer
Buonomo, Frances Catherine
Byatt, John Christopher
Holland, Lene J
Mooradian, Arshag Dertad
Nickols, G Allen
Rifas, Leonard
Tannenbaum, Michael Glen
Tomasi, Thomas Edward

MONTANA
Coe, John Emmons
Ericsson, Ronald James
Linner, John Gunnar

NEBRASKA
Fawcett, James Davidson
Grotjan, Harvey Edward, Jr
Mackin, Robert Brian
Veomett, George Ector

NEVADA
Potter, Gilbert David

NEW HAMPSHIRE
Sokol, Hilda Weyl
Sower, Stacia Ann

NEW JERSEY
Bagnell, Carol A
Beyer-Mears, Annette
Blumenthal, Rosalyn D
Campfield, L Arthur
Cascieri, Margaret Anne
Doebber, Thomas Winfield
Goldsmith, Laura Tobi
Gray, Harry Edward
Guo, Jian Zhong
Katz, Larry Steven
Landau, Matthew Paul
Lenard, John
Mackway-Girardi, Ann Marie
Malamed, Sasha
Meyers, Kenneth Purcell
Nakhla, Atif Mounir
Neri, Rudolph Orazio
O'Byrne, Elizabeth Milikin
Rabii, Jamshid
Royce, Paul C
Ryzlak, Maria Teresa
Scanes, Colin G
Schlesinger, David H
Seaver, Sally S
Sherman, Merry Rubin
Szot, Robert Joseph
Triscari, Joseph
Wang, Bosco Shang

NEW MEXICO
Tombes, Averett S

NEW YORK
Archibald, Reginald MacGregor
Atlas, Steven Alan
Bambara, Robert Anthony
Bardin, Clyde Wayne
Beermann, Donald Harold
Benjamin, William B
Black, Virginia H
Bloch, Eric
Bockman, Richard Steven
Bofinger, Diane P
Boozer, Carol Sue Neely
Bora, Sunder S
Brown, Lawrence S, Jr
Brown, Stephen Clawson
Campbell, Robert G
Chandran, V Ravi
Chasalow, Fred I
Christian, John Jermyn

Christy, Nicholas Pierson
Cohen, Leonard A
Dickerman, Herbert W
Dorsey, Thomas Edward
Fortune, Joanne Elizabeth
Furlanetto, Richard W
Gapp, David Alger
Gershengorn, Marvin Carl
Ginsberg-Fellner, Fredda Vita
Greif, Roger Louis
Grota, Lee J
Hardy, Matthew Phillip
Henrikson, Katherine Pointer
Hilf, Russell
Hinkle, Patricia M
Jakway, Jacqueline Sinks
Johnston, Robert E
Kim, Untae
Klinge, Carolyn M
Koide, Samuel Saburo
Kraus, Shirley Ruth
Kream, Jacob
Kristal, Mark Bennett
Laychock, Suzanne Gale
Lieberman, Seymour
Liebow, Charles
Loeb, John Nichols
McEwen, Bruce Sherman
Maddaiah, Vaddanahally Thimmaiah
Maitra, Utpalendu S
Martin, Kumiko Oizumi
Michael, Sandra Dale
Monder, Carl
Mooney, Robert Arthur
Morishima, Akira
Nisselbaum, Jerome Seymour
Orlowski, Craig Charles
Pi-Sunyer, F Xavier
Rathnam, Premila
Richman, Robert Alan
Rothstein, Howard
Sauer, Leonard A
Schneider, Bruce Solomon
Southren, A Louis
Spitz, Irving Manfred
Strain, Gladys Witt
Strand, Fleur Lillian
Sundaram, Kalyan
Thysen, Benjamin
Voorhess, Mary Louise
Zwain, Ismail Hassan

NORTH CAROLINA
Clemmons, David Roberts
Eddy, Edward Mitchell
Eldridge, John Charles
Fail, Patricia A
Hackney, Anthony C
Hedlund, Laurence William
Hodson, Charles Andrew
Korach, Kenneth Steven
Lobaugh, Bruce
Lund, Pauline Kay
Meyer, Ralph A, Jr
O'Steen, Wendall Keith
Sahyoun, Naji Elias
Smith, Donald Eugene
Surwit, Richard Samuel
Teng, Christina Wei-Tien Tu
Toverud, Svein Utheim
Tyrey, Lee
Van Wyk, Judson John

NORTH DAKOTA
Fivizzani, Albert John, Jr
Lukaski, Henry Charles
Reynolds, Lawrence P
Sheridan, Mark Alexander

OHIO
Case, Denis Stephen
Faiman, Charles
Fanger, Bradford Otto
Geho, Walter Blair
Handwerger, Stuart
Hiremath, Shivanand T
Horseman, Nelson Douglas
Jegla, Thomas Cyril
Johnston, John O'Neal
Palmer, Brent David
Pate, Joy Lee
Saffran, Murray
Stalvey, John Robert Dixon
Stevenson, J(oseph) Ross
Thornton, Janice Elaine
Turner, John W, Jr
Yocum, George David

OKLAHOMA
Koren, Eugen
Lavia, Lynn Alan
Maton, Paul Nicholas
Trachewsky, Daniel
Watson, Gary Hunter
Williams, Judy Ann

OREGON
Brenner, Robert Murray
Debons, Albert Frank
Greer, Monte Arnold
Mason, Robert Thomas
Ramaley, Judith Aitken
Roselli, Charles Eugene
Rutten, Michael John

Ruwe, William David
Stormshak, Fredrick
Stouffer, Richard Lee

PENNSYLVANIA
Baumrucker, Craig Richard
Bertolini, Donald R
Brooks, David Patrick
Butler, William Barkley
Camoratto, Anna Marie
Demers, Laurence Maurice
Finn, Frances M
Goodman, David Barry Poliakoff
Gordon, Janice Taylor
Kauffman, Gordon Lee, Jr
Kwiecinski, Gary George
Lakoski, Joan Marie
Leach, Roland Melville, Jr
Lerner, Leonard Joseph
Lien, Eric Louis
Nambi, Ponnal
Rosenblatt, Michael
Santen, Richard J
Silverman, Robert Eliot
Stewart, Charles Newby
Vasilatos-Younken, Regina
Wickersham, Edward Walker

RHODE ISLAND
Flanagan, Thomas Raymond
Jackson, Ivor Michael David
Oxenkrug, Gregory Faiva

SOUTH CAROLINA
Bates, G William
Blake, Charles Albert
Waldman, Barbara Criscuolo
Wilson, Marlene Ann

SOUTH DAKOTA
Slyter, Arthur Lowell

TENNESSEE
Anderson, Ted L
Chen, Thomas Tien
Coulson, Patricia Bunker
Danzo, Benjamin Joseph
Heimberg, Murray
Herting, David Clair
Hoffman, Loren Harold
Joplin, Karl Henry
Kitabchi, Abbas E
Lessman, Charles Allen
Sander, Linda Dian
Wimalasena, Jay
Zemel, Michael Barry

TEXAS
Barker, Kenneth Leroy
Beale, Elmus G
Bertrand, Helen Anne
Brown, Robert Dale
Bryan, George Thomas
Burghardt, Robert Casey
Byerley, Lauri Olson
Carson, Daniel Douglas
Chubb, Curtis Evans
Cooper, Cary Wayne
Dahm, Karl Heinz
Garfield, Robert Edward
George, Fredrick William
Guo, Yan-Shi
Herbert, Damon Charles
Hug, Verena
Hutson, James Chelton
Keeley, Larry Lee
King, Thomas Scott
Kitay, Julian I
Krulich, Ladislav
Kutteh, William Hanna
Norman, Reid Lynn
Ordonez, Nelson Gonzalo
Rosenfeld, Charles Richard
Roy, Arun K
Schonbrunn, Agnes
Smith, Edward Russell
Smith, Eric Morgan
Smith, Keith Davis
Stubbs, Donald William
Taurog, Alvin
Taylor, D(orothy) Jane
Thompson, Edward Ivins Bradbridge
Wilson, Jean Donald

UTAH
Rasmussen, Kathleen Goertz
Wurst, Gloria Zettle

VIRGINIA
Adler, Robert Alan
Brown, Barry Lee
Carey, Robert Munson
Edwards, Leslie Erroll
Liberti, Joseph Pollara
McCarty, Richard Charles
MacLeod, Robert Meredith
McNabb, F M Anne
Pepe, Gerald Joseph
Rohn, Reuben D
Turner, Terry Tomo

WASHINGTON
Baskin, Denis George
Beck, Thomas W

Carlberg, Karen Ann
Fujimoto, Wilfred Y
Harding, Boyd W
Hawkins, Richard L
Horn, Diane
Kenney, Nancy Jane
Porte, Daniel, Jr
Riddiford, Lynn Moorhead
Smith, Elizabeth Knapp
Steiner, Robert Alan
Thompson, Christopher William
Truman, James William
Wood, Francis C, Jr
Woods, Stephen Charles

WEST VIRGINIA
Burns, John Thomas
Butcher, Roy Lovell
Inskeep, Emmett Keith
Mugaas, John Nels
Rhoten, William Blocher

WISCONSIN
Dierschke, Donald Joe
Drucker, William D
Goldberger, Amy
Haas, Michael John
Hager, Steven Ralph
Hutz, Reinhold Josef
Lardy, Henry Arnold
Lorton, Steven Paul
Ndon, John A
Piacsek, Bela Emery
Takahashi, Lorey K
Terasawa, Ei
Warner, Eldon Dezelle

WYOMING
Ji, Inhae

PUERTO RICO
Banerjee, Dipak Kumar

ALBERTA
Lorscheider, Fritz Louis
Mears, Gerald John
Peter, Richard Ector
Rorstad, Otto Peder

BRITISH COLUMBIA
Clarke, William Craig
Donaldson, Edward Mossop
Emerman, Joanne Tannis
Hughes, Maryanne Robinson
Jones, David Robert
Kraintz, Leon
McKeown, Brian Alfred
Vincent, Steven Robert

MANITOBA
Angel, Aubie
Eales, John Geoffrey

ONTARIO
Ainsworth, Louis
Armstrong, David Thomas
Ballantyne, James Stuart
Bhavnani, Bhagu R
Bols, Niels Christian
Challis, John Richard George
Etches, Robert J
Graham, Richard Charles Burwell
Hegele, Robert A
Irwin, David Michael
Jellinck, Peter Harry
Kutcher, Stanley Paul
Leatherland, John F
Little, James Alexander
Liversage, Richard Albert
Mak, William Wai-Nam
Malkin, Aaron
Moon, Thomas William
Raeside, James Inglis
Reid, Robert Leslie
Steel, Colin Geoffrey Hendry
Tobe, Stephen Solomon
Vafopoulo, Xanthe
Vallerand, Andre L
Volpe, Robert
Weick, Richard Fred
Wiebe, John Peter

QUEBEC
Antakly, Tony
Bateman, Andrew
Benard, Bernard
Bendayan, Moise
Birmingham, Marion Krantz
Blaschuk, Orest William
Duguid, William Paris
Dussault, Jean H
Engel, Charles Robert
Farmer, Chantal
Gagnon, Claude
Garcia, Raul
Glorieux, Francis Henri
Guillemette, Gaetan
La Roche, Gilles
Manjunath, Puttaswamy
Mulay, Shree
Pearson Murphy, Beverly Elaine
Peronnet, Francois R R
Posner, Barry Innis
Roberts, Kenneth David

Sairam, Malur Ramaswamy
Solomon, Samuel
Tannenbaum, Gloria Shaffer
Viereck, Christopher

SASKATCHEWAN
Gratto-Trevor, Cheri Lynn
Khandelwal, Ramji Lal
Weisbart, Melvin

OTHER COUNTRIES
Amsterdam, Abraham
Armato, Ubaldo
Barouki, Robert
Buttemer, William Ashley
Eik-Nes, Kristen Borger
Farid, Nadir R
Garcia-Sainz, J Adolfo
Gustafsson, Jan-Ake
Hanukoglu, Israel
Henderson, David Andrew
Ichikawa, Shuichi
Ishikawa, Hiroshi
Kaye, Alvin Maurice
Kuo, Ching-Ming
Markert, Claus O
Millar, Robert Peter
Pina, Enrique
Racotta, Radu Gheorghe
Sabry, Ismail
Shimaoka, Katsutaro
Shore, Laurence Stuart
Stoppani, Andres Oscar Manuel
Vane, John Robert

Entomology

ALABAMA
Adams, Curtis H
Appel, Arthur Gary
Barnes, William Wayne
Berger, Robert S
Blake, George Henry, Jr
Brewer, Jesse Wayne
Brown, Jack Stanley
Chambliss, Oyette Lavaughn
Clark, Wayne Elden
Cross, Earle Albright, Jr
Gaylor, Michael James
Gilliland, Floyd Ray, Jr
Hays, Kirby Lee
Johnson, William E, Jr
Kouskolekas, Costas Alexander
Mack, Timothy Patrick
Mangat, Baldev Singh
Mullen, Gary Richard
Sanford, L G
Scheiring, Joseph Frank
Snoddy, Edward L
Stewart, Scott David
Tate, Lawrence Gray
Tennessen, Kenneth Joseph
Williams, Michael Ledell

ALASKA
Sagers, Cynthia Louise
Werner, Richard Allen

ARIZONA
Amin, Omar M
Bariola, Louis Anthony
Barker, Roy Jean
Bartlett, Alan C
Bernays, Elizabeth Anna
Bowers, William Sigmond
Burgess, Kathryn Hoy
Butler, George Daniel, Jr
Cazier, Mont Adelbert
Chapman, Reginald Frederick
Collins, Richard Cornelius
Crosswhite, Carol D
Cupp, Eddie Wayne
Davidson, Elizabeth West
Davis, Norman Thomas
Doane, Charles Chesley
Erickson, Eric Herman, Jr
Faeth, Stanley Herman
Figueredo, Aurelio Jose
Flint, Hollis Mitchell
Fritz, Roy Fredolin
George, Boyd Winston
Gerhardt, Paul Donald
Graham, Harry Morgan
Hadley, Neil F
Hagedorn, Henry Howard
Henneberry, Thomas James
Hewitt, George Berlyn
Hildebrand, John Grant, III
Huber, Roger Thomas
Johnson, Clarence Daniel
Johnson, Robert Andrew
Kauffeld, Norbert M
Lamunyon, Craig Willis
Langston, Dave Thomas
Law, John Harold
Loper, Gerald Milton
Minch, Edwin Wilton
Moore, Leon
Muma, Martin Hammond
Nutting, William Leroy
Papaj, Daniel Richard
Pinter, Paul James, Jr
Price, Peter Wilfrid

Entomology (cont)

Schmidt, Justin Orvel
Smith, Robert Lloyd
Spangler, Hayward Gosse
Stahnke, Herbert Ludwig
Terry, Lucy Irene
Tuttle, Donald Monroe
Ware, George Whitaker, Jr
Watson, Theo Franklin
Werner, Floyd Gerald
Wheeler, Diana Esther
Wick, James Roy
Wright, James Elbert
Zimmerman, James Roscoe

ARKANSAS
Adams, Randall Henry
Barton, Harvey Eugene
Burleigh, Joseph Gaynor
Clower, Dan Fredric
Couser, Raymond Dowell
Lancaster, Jessie Leonard, Jr
Musick, Gerald Joe
Phillips, Jacob Robinson
Price, Roger DeForrest
Steelman, Carrol Dayton
Steinkraus, Donald Curtiss
Stephen, Frederick Malcolm, Jr
Thompson, Lynne Charles
Watson, Robert Lee
Yearian, William C
Young, Seth Yarbrough, III

CALIFORNIA
Adams, Phillip A
Allen, William Westhead
Allison, William Earl
Andres, Lloyd A
Arnaud, Paul Henri, Jr
Bacon, Oscar Gray
Baker, Frederick Charles
Barnes, Martin McRae
Barr, Allan Ralph
Bell, A(udra) Earl
Bellinger, Peter F
Benson, Robert Leland
Bishop, Jack Lynn
Blanc, Francis Louis
Boggs, Carol L
Bohart, Richard Mitchell
Bohnsack, Kurt K
Bradley, Timothy Jud
Brown, Leland Ralph
Calavan, Edmond Clair
Caltagirone, Leopoldo Enrique
Campbell, Bruce Carleton
Carey, James Robert
Casida, John Edward
Chang, Ming-Houng (Albert)
Cheng, Lanna
Christensen, Howard Anthony
Cox, H C
Cromartie, Thomas Houston
Cunningham, Virgil Dwayne
Curtis, Charles Elliott
Dahlsten, Donald L
Daly, Howell Vann
DaMassa, Al John
Darst, Philip High
Dingle, Richard Douglas Hugh
Doutt, Richard Leroy
Dowell, Robert Vernon
Duffey, Sean Stephen
Dunbar, Dennis Monroe
Edwards, J Gordon
Eldridge, Bruce Frederick
Eriksen, Clyde Hedman
Falcon, Louis A
Ferguson, William E
Fisher, Theodore William
Force, Don Clement
Frankie, Gordon William
Fristrom, Dianne
Furman, Deane Philip
Garcia, Richard
Garth, John Shrader
Gary, Norman Erwin
Gaston, Lyle Kenneth
Georghiou, George Paul
Gilmore, James Eugene
Goeden, Richard Dean
Goh, Kean S
Golder, Thomas Keith
Gordh, Gordon
Grigarick, Albert Anthony, Jr
Hackwell, Glenn Alfred
Hagen, Kenneth Sverre
Hammock, Bruce Dupree
Hare, John Daniel, III
Haverty, Michael Irving
Hespenheide, Henry August, III
Hogue, Charles Leonard
Huffaker, Carl Barton
Hunter, Alice S (Baker)
Iltis, Wilfred Gregor
Jameson, Everett Williams, Jr
Jefferson, Roland Newton
Jeppson, Lee Ralph
Jorgenson, Edsel Carpenter
Judson, Charles LeRoy
Kadner, Carl George
Kaloostian, George H
Kavanaugh, David Henry
Kaya, Harry Kazuyoshi
Kimsey, Lynn Siri
Kissinger, David George
Kistner, David Harold
Koehler, Carlton Smith
Kulman, Herbert Marvin
Laidlaw, Harry Hyde, Jr
Lane, Robert Sidney
Lauck, David R
Leahy, Sr Mary Gerald
Legner, E Fred
Leigh, Thomas Francis
Lindgren, David Leonard
Linsley, Earle Gorton
Loeblich, Karen Elizabeth
Loomis, Richard Biggar
McClelland, George Anderson Hugh
McClure, Howe Elliott
McCluskey, Elwood Sturges
McMurtry, James A
March, Ralph Burton
Marrone, Pamela Gail
Maxwell, Kenneth Eugene
Mayse, Mark A
Michelbacher, Abraham E
Millar, Jocelyn Grenville
Miller, Thomas Albert
Monroe, Ronald Eugene
Morse, Joseph Grant
Mussen, Eric Carnes
Nelson, Richard D(ouglas)
Nickel, Phillip Arnold
Nowell, Wesley Raymond
Oatman, Earl R
Oldfield, George Newton
Page, Robert Eugene, Jr
Partida, Gregory John, Jr
Penny, Norman Dale
Pinto, John Darwin
Pipa, Rudolph Louis
Powell, Jerry Alan
Preisler, Haiganoush Krikorian
Purcell, Alexander Holmes, III
Rammer, Irwyn Alden
Reisen, William Kenneth
Resh, Vincent Harry
Reynolds, Harold Truman
Robertson, Jacqueline Lee
Rogoff, William Milton
Rust, Michael Keith
Ryckman, Raymond Edward
Schaefer, Charles Herbert
Schlinger, Evert Irving
Scott, Matthew P
Scudder, Harvey Israel
Sevacherian, Vahram
Shapiro, Arthur Maurice
Simkover, Harold George
Sleeper, Elbert Launee
Smith, Ray Fred
Smith, Richard Harrison
Soderstrom, Edwin Loren
Staal, Gerardus Benardus
Steffan, Wallace Allan
Stelzer, Lorin Roy
Stern, Vernon Mark
Strong, Donald Raymond, Jr
Summers, Charles Geddes
Sylvester, Edward Sanford
Tanada, Yoshinori
Thompson, Robert Kruger
Thorp, Robbin Walker
Trumble, John Thomas
Truxal, Fred Stone
Vail, Patrick Virgil
Vincent, Leonard Stuart
Wade, William Howard
Wasbauer, Marius Sheridan
Washino, Robert K
Waters, William E
Weber, Barbara C
Weis, Arthur Edward
Wells, Patrick Harrington
Williams, Carroll Burns, Jr
Wood, David Lee
Zalom, Frank G
Zavortink, Thomas James

COLORADO
Barnes, Ralph Craig
Beal, Richard Sidney, Jr
Becker, George Charles
Bowers, Marion Deane
Breed, Michael Dallam
Burns, Denver P
Drummond, Boyce Alexander, III
Elias, Scott Armstrong
Evans, Howard Ensign
Francy, David Bruce
Gholson, Larry Estie
Gregg, Robert Edmond
Gubler, Duane J
Harrison, Monty DeVerl
Harriss, Thomas T
Johnsen, Richard Emanuel
King, Charles C
Lanham, Urless Norton
Linam, Jay H
Nasci, Roger Stanley
Owen, William Bert
Polhemus, John Thomas
Roberts, P Elaine
Simpson, Robert Gene
Sublette, James Edward
Yun, Young Mok

CONNECTICUT
Aitken, Thomas Henry Gardiner
Anderson, John Fredric
Andreadis, Theodore George
Berry, Spencer Julian
Block, Bartley C
Connor, Lawrence John
DeCoursey, Russell Myles
Henry, Charles Stuart
Krinsky, William Lewis
Levine, Harvey Robert
McClure, Mark Stephen
Magnarelli, Louis Anthony
Maier, Chris Thomas
Miller, Russell Bensley
Nelson, Vernon A
Raghuvir, Nuggehalli Narayana
Remington, Charles Lee
Rettenmeyer, Carl William
Savos, Milton George
Schaefer, Carl W, II
Slater, James Alexander
Stephens, George Robert
Streams, Frederick Arthur
Wallner, William E
Weseloh, Ronald Mack
Wyman, Robert J

DELAWARE
Allen, Robert Thomas
Carnahan, James Elliot
Caron, Dewey Maurice
Cornell, Howard Vernon
Day, William H
Hough-Goldstein, Judith Anne
Koeppe, Mary Kolean
Krone, Lawrence James
Leeper, John Robert
Mason, Charles Eugene
Meade, Alston Bancroft
Miller, Douglas Charles
Soboczenski, Edward John
Weber, Richard Gerald
Wood, Thomas Kenneth

DISTRICT OF COLUMBIA
Anderson, Donald Morgan
Baerg, David Carl
Brickey, Paris Manaford
Bridges, John Robert
Burns, John McLauren
Cate, James Richard, Jr
Chapman, George Bunker
Chaput, Raymond Leo
Davis, Donald Ray
Flint, Oliver Simeon, Jr
Gagne, Raymond J
Gorham, John Richard
Grissell, Edward Eric Fowler
Hodges, Ronald William
Klein, Terry Allen
Krombein, Karl Von Vorse
Lorimer, Nancy L
Mathis, Wayne Neilsen
Nickle, David Allan
Norrbom, Allen Lee
Poole, Robert Wayne
Reed, William Doyle
Robbins, Robert Kanner
Robinson, Harold Ernest
Smith, David Rollins
Sollers-Riedel, Helen
Spangler, Paul Junior
Tompkins, George Jonathan
White, Richard Earl

FLORIDA
Agee, Herndon Royce
Allen, Jon Charles
Arnett, Ross Harold, Jr
Baranowski, Richard Matthew
Barnard, Donald Roy
Beckemeyer, Elizabeth Frances
Bick, George Herman
Bidlingmayer, William Lester
Brenner, Richard Joseph
Britt, N Wilson
Brooks, Robert Franklin
Bullock, Robert Crossley
Burditt, Arthur Kendall, Jr
Butler, Jerry Frank
Callahan, Philip Serna
Castner, James Lee
Chambers, Derrell Lynn
Clegern, Robert Wayne
Clements, Burie Webster
CmejLa, Howard Edward
Couch, Terry Lee
Cromroy, Harvey Leonard
Dame, David Allan
Davis, Dean Frederick
Denmark, Harold Anderson
Downey, John Charles
Emerson, Kary Cadmus
Fairchild, Graham Bell
Fatzinger, Carl Warren
Ferguson, James Scott
Fisk, Frank Wilbur
Flowers, Ralph Wills
Frank, John Howard
Fulton, Winston Cordell
Gerberg, Eugene Jordan
Giblin-Davis, Robin Michael
Glancey, Burnett Michael
Goss, Gary Jack
Gupta, Virendra K
Habeck, Dale Herbert
Haines, Robert Gordon
Hall, David Goodsell, IV
Hall, Donald William
Hamon, Avas Burdette
Heppner, John Bernhard
Hetrick, Lawrence Andrew
Himel, Chester Mora
Howard, Forrest William
Hoy, James Benjamin
Hribar, Lawrence Joseph
Kerr, Stratton H
Kingsolver, John Mark
Kitzmiller, James Blaine
Koehler, Philip Gene
Kuitert, Louis Cornelius
Landolt, Peter John
Lawrence, Pauline Olive
Levy, Richard
Lofgren, Clifford Swanson
Lounibos, Leon Philip
Lowe, Ronald Edsel
McCoy, Clayton William
McCoy, Earl Donald
McLaughlin, John Ross
Mankin, Richard Wendell
Mayer, Marion Sidney
Mayer, Richard Thomas
Mead, Frank Waldreth
Mendez, Eustorgio
Menzer, Robert Everett
Mitchell, Everett Royal
Mount, Gary Arthur
Mulrennan, John Andrew, Jr
Nation, James Lamar, Sr
Nayar, Jai Krishen
Nguyen, Khuong Ba
Nguyen, Ru
Nigg, Herbert Nicholas
Norval, Richard Andrew
Oberlander, Herbert
O'Meara, George Francis
Osborne, Lance Smith
Osburn, Richard Lee
Patterson, Richard Sheldon
Peters, William Lee
Pletsch, Donald James
Porter, Sanford Dee
Price, James Felix
Rathburn, Carlisle Baxter, Jr
Reiskind, Jonathan
Roberts, Donald Ray
Roberts, Richard Harris
Rossi, Anthony Michael
Roubik, David Ward
Sanford, Malcolm Thomas
Scheibner, Rudolph A
Schuster, David J
Seawright, Jack Arlyn
Shankland, Daniel Leslie
Shapiro, Jeffrey Paul
Silhacek, Donald Le Roy
Simonet, Donald Edward
Simons, John Norton
Sivinski, John Michael
Smith, Carroll N
Smittle, Burrell Joe
Sosa, Omelio, Jr
Sprenkel, Richard Keiser
Stange, Lionel Alvin
Teal, Peter E A
Terranova, Andrew Charles
Thomas, Michael Charles
Tingle, Frederic Carley
Tsai, James Hsi-Cho
Tschinkel, Walter Rheinhardt
Vickers, David Hyle
Waddill, Van Hulen
Waites, Robert Ellsworth
Walker, Thomas Jefferson
Weber, Neal Albert
Weems, Howard Vincent, Jr
Weidhaas, Donald E
West-Eberhard, Mary J
Wheeler, George Carlos
Wheeler, Jeanette Norris
Whitcomb, Willard Hall
White, Albert Cornelius
Wilkinson, Robert Cleveland, Jr
Williams, David Francis
Williams, Norris Hagan
Wilson, Louis Frederick
Woodruff, Robert Eugene
Workman, Ralph Burns
Yu, Simon Shyi-Jian
Yunker, Conrad Erhardt

GEORGIA
Allee, Marshall Craig
Amerson, Grady Malcolm
Arbogast, Richard Terrance
Atyeo, Warren Thomas
Bass, Max H(erman)
Blum, Murray Sheldon
Boush, George Mallory
Brower, John Harold
Caldwell, Sloan Daniel
Chalfant, Richard Bruce
Chamberlain, Roy William
Collins, William Erle

BIOLOGICAL SCIENCES / 97

Davis, Robert
Dutcher, James Dwight
Espelie, Karl Edward
Evans, Burton Robert
Franklin, Rudolph Thomas
French, Frank Elwood, Jr
Gilbert, Edward E
Gitaitis, Ronald David
Harris, Emmett Dewitt, Jr
Harrison, James Ostelle
Hays, Donald Brooks
Hermann, Henry Remley, Jr
Highland, Henry Arthur
Hunter, Preston Eugene
Johnson, Donald Ross
Kappus, Karl Daniel
Keirans, James Edward
Klompen, J S H
Lea, Arden Otterbein
Lewis, Wallace Joe
Lynch, Robert Earl
McPherson, Robert Merrill
Matthews, Robert Wendell
Mulaik, Stanley B
Mumtaz, Mohammad Moizuddin
Neel, William Wallace
Newhouse, Verne Frederic
Nord, John C
Nyczepir, Andrew Peter
Oliver, James Henry, Jr
Payne, Jerry Allen
Pratt, Harry Davis
Rice, Paul LaVerne
Rogers, Charlie Ellic
Sheppard, David Craig
Shure, Donald Joseph
Smith, Eric Howard
Sparks, Alton Neal
Sudia, William Daniel
Taylor, Robert Tieche
Throne, James Edward
Tietjen, William Leighton
Todd, James Wyatt
Wallace, James Bruce
Weathersby, Augustus Burns
Whitesell, James Judd
Willis, Judith Horwitz
Wiseman, Billy Ray

HAWAII
Andersen, Dean Martin
Beardsley, John Wyman, Jr
Chang, Franklin
Culliney, Thomas W
Duckworth, Walter Donald
Eschle, James Lee
Fujii, Jack K
Hardy, D Elmo
Harris, Ernest James
Howarth, Francis Gard
Little, Harold Franklin
Liu, Yong-Biao
Medler, John Thomas
Mitchell, Wallace Clark
Namba, Ryoji
Ota, Asher Kenhachiro
Roderick, George Karlsson
Sherman, Martin
Simon, Christine Mae
Tamashiro, Minoru
Vargas, Roger I

IDAHO
Anderson, Robert Curtis
Baird, Craig Riska
Barr, William Frederick
Brusven, Merlyn Ardel
Clark, William Hilton
Gittins, Arthur Richard
Halbert, Susan E
Hostetter, Donald Lee
Johnson, James Blakeslee
Klowden, Marc Jeffrey
McCaffrey, Joseph Peter
O'Keeffe, Lawrence Eugene
Osgood, Charles Edgar
Schell, Stewart Claude
Schenk, John Albright
Stark, Ronald William
Stiller, David
Stoltz, Robert Lewis
Waters, Norman Dale

ILLINOIS
Appleby, James E
Beland, Gary LaVern
Berenbaum, May Roberta
Bhatti, Rashid
Bouseman, John Keith
Chandran, Satish Raman
Chang, Kwang-Poo
Cibulsky, Robert John
Delcomyn, Fred
Estep, Charles Blackburn
Flood, Brian Robert
French, Allen Lee
Friedman, Stanley
Funk, Richard Cullen
Gibbs, Daniel
Goodrich, Michael Alan
Green, Thomas L
Greenberg, Bernard
Hamilton, Robert W
Helm, Charles George
Horsfall, William Robert
Irwin, Michael Edward
Jones, Carl Joseph
Karr, Timothy Lawrence
Kethley, John Bryan
Khan, Mohammed Abdul Quddus
Khattab, Ghazi M A
LaBerge, Wallace E
Luckmann, William Henry
MacLeod, Ellis Gilmore
McPherson, John Edwin
Maddox, Joseph Vernard
Malek, Richard Barry
Melin, Brian Edward
Metcalf, Robert Lee
Meyer, Ronald Harmon
Mockford, Edward Lee
Nappi, Anthony Joseph
Nardi, James Benjamin
Nelson, Harry Gladstone
Riegel, Garland Tavner
Robertson, Hugh Mereth
Sedman, Yale S
Selander, Richard Brent
Sternburg, James Gordon
Thayer, Margaret Kathryn
Thompson, Vinton Newbold
Toliver, Michael Edward
Unzicker, John Duane
Vandehey, Robert C
Wagner, John Alexander
Waldbauer, Gilbert Peter
Watson, David Livingston
Webb, Donald Wayne
Wenzel, Rupert Leon
Willey, Robert Bruce
Wittig, Gertraude Christa

INDIANA
Bath, James Edmond
Broersma, Delmar B
Burkholder, Timothy Jay
Chio, E(ddie) Hang
Corrigan, John Joseph
Craig, George Brownlee, Jr
Delfin, Eliseo Dais
Denner, Melvin Walter
Dunn, Peter Edward
Edwards, Charles Richard
Esch, Harald Erich
Flanders, Robert Vern
Fuchs, Morton S
Gallun, Robert Louis
Goonewardene, Hilary Felix
Gould, George Edwin
Grimstad, Paul Robert
Hellenthal, Ronald Allen
Hickey, William August
Hoefer, Raymond H
Hunt, Linda Margaret
Jantz, O K
Johnson, Michael David
Krekeler, Carl Herman
Lamberti, Gary Anthony
McCafferty, William Patrick
McGowan, Michael James
O'Neil, Robert James
Ortman, Eldon Emil
Osmun, John Vincent
Peterson, Lance George
Racke, Kenneth David
Rai, Karamjit Singh
Russo, Raymond Joseph
Schuder, Donald Lloyd
Stuart, Jeffrey James
Trosper, James Hamilton
Turpin, Frank Thomas
Ward, Gertrude Luckhardt
Wilson, Mark Curtis
Zimmack, Harold Lincoln

IOWA
Coats, Joel Robert
Dahm, Paul Adolph
Dicke, Ferdinand Frederick
Eiben, Galen J
Guthrie, Wilbur Dean
Hart, Elwood Roy
Krafsur, Elliot Scoville
Lewis, Robert Earl
Matteson, Patricia Claire
Mertins, James Walter
Mutchmor, John A
Pedigo, Larry Preston
Rowley, Wayne A
Showers, William Broze, Jr
Stay, Barbara
Stockdale, Harold James
Stoltzfus, William Bryan
Wilson, Nixon Albert
Wilson, Richard Lee
Yang, Xiao-Bing

KANSAS
Ashe, James S
Beer, Robert Edward
Blocker, Henry Derrick
Byers, George William
Cutler, Bruce
Dinkins, Reed Leon
Eddy, Thomas A
Elzinga, Richard John
Godfrey, George Lawrence
Greene, Gerald L
Greenfield, Michael Dennis
Harvey, Thomas Larkin
Hatchett, Jimmy Howell
Hopkins, Theodore Louis
Kadoum, Ahmed Mohamed
Kramer, Karl Joseph
Loudon, Catherine
Lungstrom, Leon
Marsh, Paul Malcolm
Martin, Terry Joe
Michener, Charles Duncan
Mills, Robert Barney
Owen, Bernard Lawton
Peters, Leroy Lynn
Stockhammer, Karl Adolf
Thompson, Hugh Erwin
Von Rumker, Rosmarie
Walker, Neil Allan
Whitney, Wendell Keith
Wilde, Gerald Eldon

KENTUCKY
Christensen, Christian Martin
Covell, Charles VanOrden, Jr
Dahlman, Douglas Lee
Ferrell, Blaine Richard
Freytag, Paul Harold
Gregory, Wesley Wright, Jr
Knapp, Fred William
Moore, Allen Jonathan
Nordin, Gerald LeRoy
Pass, Bobby Clifton
Rodriguez, Juan Guadalupe
Sugden, Evan A
White, David Sanford
Yeargan, Kenneth Vernon

LOUISIANA
Boudreaux, Henry Bruce
Chapin, Joan Beggs
Dakin, Matt Eitel
Fuxa, James Roderick
Graves, Jerry Brook
Hammond, Abner M, Jr
Harbo, John Russell
Hensley, Sess D
Khalaf, Kamel T
Kinn, Donald Norman
Lambremont, Edward Nelson
Long, William Henry
Miller, Albert
Oliver, Abe D, Jr
Reagan, Thomas Eugene
Rinderer, Thomas Earl
Rolston, Lawrence H
Roussel, John S
Schoenly, Kenneth George
Scott, Harold George
Spring, Jeffrey H
Stewart, T Bonner
Tucker, Kenneth Wilburn
Vidrine, Malcolm Francis
Walker, John Robert
Woodring, Joseph
Yadav, Raghunath P

MAINE
Burbutis, Paul Philip
Dimond, John Barnet
Forsythe, Howard Yost, Jr
Jennings, Daniel Thomas
Knight, Fred Barrows
McDaniel, Ivan Noel
Mazurkiewicz, Michael
Nelson, Robert Edward
Raisbeck, Barbara
Sommerman, Kathryn Martha
Storch, Richard Harry
Wave, Herbert Edwin
Yonuschot, Gene R

MARYLAND
Adams, Jean Ruth
Adler, Victor Eugene
Aldrich, Jeffrey Richard
Armstrong, Earlene
Azad, Abdu F(arhang)
Bailey, Charles Lavon
Baker, Edward William
Barnett, Douglas Eldon
Barrows, Edward Myron
Barry, Cornelius
Batra, Suzanne Wellington Tubby
Bickley, William Elbert
Borgia, Gerald
Bram, Ralph A
Burton, George Joseph
Cantelo, William Wesley
Cleveland, Merrill L
Conner, George William
Coulson, Jack Richard
Crystal, Maxwell Melvin
Dasch, Gregory Alan
Davidson, John Angus
Feldlaufer, Mark Francis
Futcher, Anthony Graham
Gonzales, Ciriaco Q
Graham, Charles Lee
Gwadz, Robert Walter
Hansford, Richard Geoffrey
Hanson, Frank Edwin
Harrison, Floyd Perry
Hassett, Charles Clifford
Hathaway, Wilfred Bostock
Herbert, Elton Warren, Jr
Inscoe, May Nilson
Joseph, Stanley Robert
Klun, Jerome Anthony
Krestensen, Elroy R
Krysan, James Louis
Kutz, Frederick Winfield
Lamp, William Owen
Marks, Edwin Potter
Menn, Julius Joel
Messersmith, Donald Howard
Miller, Douglass Ross
Murray, William Sparrow
Neal, John William, Jr
Nelson, James Harold
Nelson, Judd Owen
Nickle, William R
Parrish, Dale Wayne
Phillips, William George
Pilitt, Patricia Ann
Raina, Ashok K
Ramsay, Maynard Jack
Reddy, Gunda
Redfern, Robert Earl
Ridgway, Richard L
Riley, Robert C
Rozeboom, Lloyd Eugene
Sauer, Richard John
Schiff, Nathan Mark
Schwartz, Paul Henry, Jr
Scott, Thomas Wallace
Shapiro, Martin
Shimanuki, Hachiro
Sholdt, Lester Lance
Sonawane, Babasaheb R
Steinhauer, Allen Laurence
Stoetzel, Manya Brooke
Svoboda, James Arvid
Todd, Robin Grenville
Traub, Robert
Trpis, Milan
Vaughn, James L
Ward, Ronald A(nthony)
Weirich, Gunter Friedrich
Wise, David Haynes
Wood, Francis Eugene
Yoder, Wayne Alva

MASSACHUSETTS
Barke, Harvey Ellis
Borgatti, Alfred Lawrence
Bowdan, Elizabeth Segal
Carde, Ring Richard Tomlinson
Carpenter, Frank Morton
Cohen, Samuel H
Dalgleish, Robert Campbell
Edman, John David
Ferro, David Newton
Furth, David George
Helgesen, Robert Gordon
Kafatos, Fotis C
Kelts, Larry Jim
Kunkel, Joseph George
Lee, Siu-Lam
Leonard, David E
Levi, Herbert Walter
Moffett, Mark William
Nutting, William Brown
Peters, Thomas Michael
Smith, Grahame J C
Spielman, Andrew
Stoffolano, John George, Jr
Terwedow, Henry Albert, Jr
Traniello, James Francis Anthony
Trimmer, Barry Andrew
Wall, William James
Weaver, Nevin
Yin, Chih-Ming

MICHIGAN
Bird, George W
Bland, Roger Gladwin
Bratt, Albertus Dirk
Breznak, John Allen
Bush, Guy L
Cantrall, Irving James
Cook, David Russell
Cowan, David Prime
Douglas, Matthew M
Evans, David Arnold
Fischer, Roland Lee
Fleming, Richard Cornwell
Gamboa, George John
Gangwere, Stanley Kenneth
Grafius, Edward John
Guyer, Gordon Earl
Haynes, Dean L
Hoffman, Julius R
Hollingworth, Robert Michael
Hoopingarner, Roger A
Howe, Robert George
Howitt, Angus Joseph
Husband, Robert W
Kawooya, John Kasajja
McCall, Robert B
Meier, Peter Gustav
Merritt, Richard William
Moore, Thomas Edwin
Newson, Harold Don
Paschke, John Donald
Poston, Freddie Lee, Jr
Rizzo, Donald Charles
Roeper, Richard Allen
Ruppel, Robert Frank

Entomology (cont)

Scarborough, Charles Spurgeon
Scheel, Carl Alfred
Shappirio, David Gordon
Simmons, Gary Adair
Stehr, Frederick William
Toczek, Donald Richard
Wilmot, Thomas Ray
Witter, John Allen

MINNESOTA
Andow, David A
Baker, Griffin Jonathan
Bartel, Monroe H
Batzer, Harold Otto
Borchers, Harold Allison
Chiang, Huai C
Christian, Paul Jackson
Cook, Edwin Francis
Cutkomp, Laurence Kremer
Fallon, Ann Marie
Fremling, Calvin R
Gundersen, Ralph Wilhelm
Hamrum, Charles Lowell
Harein, Phillip Keith
Krogstad, Blanchard Orlando
Kurtti, Timothy John
Lewis, Standley Eugene
Mauston, Glenn Warren
Miller, William Eldon
Moore, Glenn D
Noetzel, David Martin
Ordway, Ellen
Peck, John Hubert
Preiss, Frederick John
Radcliffe, Edward B
Ragsdale, David Willard
Rao, Gundu Hirisave Rama
Richards, Albert Glenn
Samuelson, Charles R
Tonn, Robert James
Walgenbach, David D
Wallace, Marion Brooks

MISSISSIPPI
Bailey, Jack Clinton, II
Brewer, Franklin Douglas
Collison, Clarence H
Combs, Robert L, Jr
Davich, Theodore Bert
Davis, Frank Marvin
Dickens, Joseph Clifton
Dorough, H Wyman
Fye, Robert Eaton
Goddard, Jerome
Hardee, Dicky Dan
Heitz, James Robert
Helms, Thomas Joseph
Hendricks, Donovan Edward
Kincade, Robert Tyrus
Lago, Paul Keith
Lloyd, Edwin Phillips
McKibben, Gerald Hopkins
Mather, Bryant
Mitchell, Earl Bruce
Nebeker, Thomas Evan
Norment, Beverly Ray
Ouzts, Johnny Drew
Parrott, William Lamar
Pitre, Henry Nolle, Jr
Poe, Sidney LaMarr
Ramaswamy, Sonny B
Sikorowski, Peter P
Smith, James Winfred
Solomon, James Doyle
Srinivasan, Asoka
Stark, William Polson

MISSOURI
Backus, Elaine Athene
Braasch, Norman L
Calabrese, Diane M
Carlson, Wayne C
Carrel, James Elliott
Cohick, A Doyle, Jr
Craig, Wilfred Stuart
Davis, Jerry Collins
Elliott, Dana Ray
Enns, Wilbur Ronald
Fairchild, Mahlon Lowell
Feir, Dorothy Jean
Gerdes, Charles Frederick
Haberman, Warren Otto
Hall, David Goodsell, Sr
Hall, Robert Dickinson
Hansen, Harry Louis
Hart, Richard Allen
Huggans, James Lee
Hunt, James Howell
Ignoffo, Carlo Michael
Johannsen, Frederick Richard
Keaster, Armon Joseph
Knowles, Charles Otis
Puttler, Benjamin
Sanders, Darryl Paul
Shaddy, James Henry
Stern, Daniel Henry
Townsend, Howard Garfield, Jr
Voorhees, Frank Ray
Wilson, Stephen W
Wingo, Curtis W
Yonke, Thomas Richard

MONTANA
Dysart, Richard James
Fisher, James Robert
Gless, Elmer E
Goodwin, Ronald Hayse
Hastings, Ellsworth (Bernard)
Hendrickson, Robert Mark, Jr
Lowe, James Harry, Jr
Morrill, Wendell Lee
Onsager, Jerome Andrew
Visscher, Saralee Neumann

NEBRASKA
Ball, Harold James
Campbell, John Bryan
Foster, John Edward
Higley, Leon George
Joern, Anthony
Jones, John Ackland
Keith, David Lee
Louda, Svata Mary
Lunt, Steele Ray
Manglitz, George Rudolph
Mayo, Z B
Miller, Willie
Nagel, Harold George
Pappas, Larry George
Pruess, Kenneth Paul
Ratcliffe, Brett Charles
Raun, Earle Spangler
Staples, Robert
Witkowski, John Frederick
Young, Jerry H

NEVADA
Arnett, William Harold
Lum, Patrick Tung Moon
Murvosh, Chad M
Rust, Richard W

NEW HAMPSHIRE
Bowman, James Sheppard
Chandler, Donald Stewart
Kotila, Paul Myron
Norton, Robert James
Reeves, Roger Marcel

NEW JERSEY
Carter, Richard John
Cromartie, William James, Jr
Dyer, Judith Gretchen
Faber, Betty Lane
Forgash, Andrew John
Gupta, Ayodhya P
Hubbell, Stephen Philip
Huber, Ivan
Jansson, Richard Keith
Lashomb, James Harold
Leyson, Jose Florante Justinitane
Lovell, James Byron
Markle, George Michael
Morin, Peter Jay
O'Connor, Charles Timothy
Ostlind, Dan A
Padhi, Sally Bulpitt
Race, Stuart Rice
Sabrosky, Curtis Williams
Scholl, Philip Jon
Shubeck, Paul Peter
Swift, Fred Calvin
Williams, Roger Wright
Witherell, Peter Charles
Ziegler, John Benjamin

NEW MEXICO
Ellington, Joe J
Huddleston, Ellis Wright
Jaycox, Elbert Ralph
Kinzer, H Grant
Martignoni, Mauro Emilio
Ortiz, Melchor, Jr
Owens, John Charles
Richman, David Bruce
Tombes, Averett S
Ward, Charles Richard
Wolff, Theodore Albert

NEW YORK
Agnello, Arthur Michael
Allen, Douglas Charles
Arnold, Steven Lloyd
Balboni, Edward Raymond
Barnes, Jeffrey Karl
Benach, Jorge L
Bentley, Barbara Lee
Benton, Allen Haydon
Black, Lindsay MacLeod
Brezner, Jerome
Broadway, Roxanne Meyer
Butts, William Lester
Carruthers, Raymond Ingalls
Castello, John Donald
Conway, John Bell
Craddock, Elysse Margaret
Crowell, Robert Merrill
Davis, Alexander Cochran
Dewey, James Edwin
Eickwort, George Campbell
Feeny, Paul Patrick
Fiori, Bart J
Forbes, James
Frishman, Austin Michael
Futuyma, Douglas Joel
Garay, Gustav John
Glass, Edward Hadley
Gotwald, William Harrison, Jr
Hajek, Ann Elizabeth
Harrison, Richard Gerald
Hughes, Patrick Richard
Hughes, Stephen Edward
Humber, Richard Alan
Jamnback, Hugo Andrew, Jr
Johnson, Warren Thurston
Jones, Clive Gareth
Kamran, Mervyn Arthur
Kaplan, Martin L
Knipple, Douglas Charles
Knowlton, Carroll Babbidge, Jr
Koch, Henry George
Kolmes, Steven Albert
Kramer, John Paul
Kurczewski, Frank E
Lener, Walter
Lienk, Siegfried Eric
Lugthart, Garrit John, Jr
Main, Andrew James
Maramorosch, Karl
Miller, Howard Charles
Millstein, Jeffrey Alan
Morse, Roger Alfred
Muchmore, William Breuleux
Muka, Arthur Allen
Nakatsugawa, Tsutomu
Norton, Roy Arnold
Nyrop, Jan Peter
Pechuman, LaVerne LeRoy
Peckarsky, Barbara Lynn
Pimentel, David
Platnick, Norman I
Raffensperger, Edgar M
Redmond, Billy Lee
Renwick, J Alan A
Roberts, Donald Wilson
Roelofs, Wendell L
Rohlf, F James
Rozen, Jerome George, Jr
St Leger, Raymond John
Schaefers, George Albert
Schuh, Randall Tobias
Seeley, Thomas Dyer
Shoemaker, Christine Annette
Simeone, John Babtista
Sleeper, David Allanbrook
Soderlund, David Matthew
Southwick, Edward E
Splittstoesser, Clara Quinnell
Straub, Richard Wayne
Sullivan, Daniel Joseph
Tashiro, Haruo
Tauber, Catherine A
Tauber, Maurice Jesse
Tingey, Ward M
Trammel, Kenneth
Vasey, Carey Edward
Vawter, Alfred Thomas
Williams, Ernest Herbert, Jr

NORTH CAROLINA
Anderson, Roger Fabian
Anderson, Thomas Ernest
Atchley, William Reid
Axtell, Richard Charles
Baker, James Robert
Barnett, John William
Booth, Donald Campbell
Bradley, Julius Roscoe, Jr
Brooks, Wayne Maurice
Buxton, Jay A
Camp, Haney Bolon
Campbell, William Vernon
Clark, James Samuel
Daggy, Tom
Deitz, Lewis Levering
Dickerson, Willard Addison
Fairchild, Homer Eaton
Falter, John Max
Farrier, Maurice Hugh
George, Charles Redgenal
Gilbert, Lawrence Irwin
Glover, Sandra Jean
Guthrie, Frank Edwin
Hain, Fred Paul
Harper, James Douglas
Hastings, Felton Leo
Hill, Alfred, Jr
Hsiao, Henry Shih-Chan
Kennedy, George Grady
Knight, Clifford Burnham
Knuckles, Joseph Lewis
Koeppe, John K
Kuhr, Ronald John
Meyer, John Richard
Moore, Harry Ballard, Jr
Neunzig, Herbert Henry
Nijhout, H Frederik
Olive, Aulsey Thomas
Parker, Beulah Mae
Rabb, Robert Lamar
Redlinger, Leonard Maurice
Robertson, Clyde Henry
Robertson, Robert L
Rock, George Calvert
Rolofson, George Lawrence
Romanow, Louise Rozak
Ross, Richard Henry, Jr
Schmidt, Stephen Paul
Smith, Clyde F
Smith, Edward Holman
Sorensen, Kenneth Alan
Spears, Brian Merle
Stinner, Ronald Edwin
Thatcher, Robert Clifford
Van Geluwe, John David
Van Pelt, Arnold Francis, Jr
Weiden, Mathias Herman Joseph
Witt, Donald James
Wright, Charles Gerald
Yamamoto, Robert Takaichi
Young, David Allan

NORTH DAKOTA
Adams, Terrance Sturgis
Balsbaugh, Edward Ulmont, Jr
Callenbach, John Anton
Carlson, Robert Bruce
Charlet, Laurence Dean
La Chance, Leo Emery
Leopold, Roger Allen
McDonald, Ian Cameron Crawford
Riemann, John G
Roseland, Craig R
Schmidt, Claude Henri
Schulz, John Theodore
Tetrault, Robert Close
Weiss, Michael John
Wrenn, William J

OHIO
Arlian, Larry George
Berry, Richard Lee
Briggs, John Dorian
Burton, Ralph P
Burtt, Edward Howland, Jr
Butz, Andrew
Cannon, William Nelson, Jr
Chesnut, Thomas Lloyd
Clay, Mary Ellen
Collins, William John
Dasch, Clement Eugene
Deal, Don Robert
Denlinger, David Landis
Deonier, D L
Foote, Benjamin Archer
Goleman, Denzil Lyle
Graves, Anne Carol Finger
Graves, Robert Charles
Hall, Franklin Robert
Hink, Walter Fredric
Hintz, Howard William
Hiremath, Shivanand T
Holdsworth, Robert Powell
Holeski, Paul Michael
Horn, David Jacobs
Hoy, Casey William
Jalil, Mazhar
Klein, Michael Gardner
Klemm, Donald J
Knoke, John Keith
Kritsky, Gene Ralph
Ladd, Thyril Leone, Jr
Lindquist, Richard Kenneth
Lyon, William Francis
MacLean, Bonnie Kuseske
MacLean, David Belmont
Martinson, Candace
Miller, Richard Lloyd
Nault, Lowell Raymond
Needham, Glen Ray
Nelson, Eric V
Nielsen, David Gary
Niemczyk, Harry D
Peterson, Paul Constant
Rings, Roy Wilson
Roach, William Kenney
Romoser, William Sherburne
Rovner, Jerome Sylvan
Schurr, Karl M
Semel, Maurie
Sferra, Pasquale Richard
Shambaugh, George Franklin
Shetlar, David John
Stairs, Gordon R
Stoffer, Richard Lawrence
Tafuri, John Francis
Treece, Robert Eugene
Triplehorn, Charles A
Uetz, George William
Valentine, Barry Dean
Waldron, Acie Chandler
Weaver, Andrew Albert
Williams, Roger Neal
Wrensch, Dana Louise
Yocum, George David

OKLAHOMA
Barnes, George Lewis
Bartell, Daniel P
Brown, Harley Procter
Crowder, Larry A
Drew, William Arthur
Eikenbary, Raymond Darrell
Gray, Thomas Merrill
Hair, Jakie Alexander
Hopla, Cluff Earl
Kindler, Sharon Dean
Melouk, Hassan A
Nelson, John Marvin, Jr
Peters, Don Clayton
Price, Richard Graydon
Russell, Charles Clayton
Sauer, John Robert
Shorter, Daniel Albert

BIOLOGICAL SCIENCES / 99

Sonleitner, Frank Joseph
Starks, Kenneth James
Tobin, Sara L
Webster, James Allan
Woolever, Patricia
Wright, Russell Emery
Young, Sharon Clairene

OREGON
Anderson, John Richard
Anderson, Norman Herbert
Berry, Ralph Eugene
Bishop, Guy William
Boddy, Dennis Warren
Bradshaw, William Emmons
Carman, Glenn Elwin
Carolin, Valentine Mott, Jr
Coffey, Marvin Dale
Coop, Leonard Bryan
Crowell, Hamblin Howes
Daterman, Gary Edward
Dickason, Elvis Arnie
Ferguson, George Ray
Isaacson, Dennis Lee
Kamm, James A
Kogan, Marcos
Krantz, Gerald William
Lattin, John D
Liss, William John
Malcolm, David Robert
Mason, Richard Randolph
Moldenke, Alison Feerick
Nebeker, Alan V
Oman, Paul Wilson
Postlethwait, John Harvey
Rossignol, Philippe Albert
Ryan, Roger Baker
Stephen, William Procuronoff
Walstad, John Daniel

PENNSYLVANIA
Aller, Harold Ernest
Atwood, Mark Wyllie
Berthold, Robert, Jr
Bode, William Morris
Bray, Dale Frank
Byers, Robert Allan
Cameron, Edward Alan
Catalano, Raymond Anthony
Cherry, Edward Taylor
Coffman, William Page
Conway, John Richard
English, Leigh Howard
Frazier, James Lewis
Fredrickson, Richard William
Fusco, Robert Angelo
Hajjar, Nicolas Philippe
Hendrickson, John Alfred, Jr
Hill, Kenneth Lee
Hower, Arthur Aaron, Jr
Hull, Larry Allen
Humphreys, Jan Gordon
Joos, Barbara
Kim, Ke Chung
Lawrence, Vinnedge Moore
McNair, Dennis M
McPheron, Bruce Alan
Manley, Thomas Roy
Marks, Louis Sheppard
Masteller, Edwin C
Montgomery, Ronald Eugene
Moss, William Wayne
Mumma, Ralph O
Ode, Philip E
Pearson, David Leander
Pitts, Charles W
Presser, Bruce Douglas
Rao, Balakrishna Raghavendra
Rutschky, Charles William
Schroeder, Mark Edwin
Schultz, Jack C
Sheldon, Joseph Kenneth
Simpson, Geddes Wilson
Smilowitz, Zane
Snetsinger, Robert J
Staetz, Charles Alan
Streu, Herbert Thomas
Swisher, Ely Martin
Wheeler, Alfred George, Jr
Williamson, Craig Edward
Wolfersberger, Michael Gregg
Wooldridge, David Paul
Yendol, William G
Yoho, Timothy Price
Yurkiewicz, William J

RHODE ISLAND
Hyland, Kerwin Ellsworth, Jr
Kerr, Theodore William, Jr
LeBrun, Roger Arthur
Reichart, Charles Valerian
Waage, Jonathan King

SOUTH CAROLINA
Adkins, Theodore Roosevelt, Jr
Alverson, David Roy
Creighton, Charlie Scattergood
Davis, Luckett Vanderford
DuRant, John Alexander, III
Elsey, Kent D
Fery, Richard Lee
Forsythe, Dennis Martin
Hays, Sidney Brooks
Henson, Joseph Lawrence

Johnson, Albert Wayne
Manley, Donald Gene
Moore, Raymond F, Jr
Noblet, Raymond
Schalk, James Maximillian
Shepard, Buford Merle
Skelton, Thomas Eugene
Spooner, John D
Wimer, Larry Thomas

SOUTH DAKOTA
Ellsbury, Michael M
Kantack, Benjamin H
Kieckhefer, Robert William
McDaniel, Burruss, Jr
Sutter, Gerald Rodney
Walstrom, Robert John

TENNESSEE
Bancroft, Harold Ramsey
Brown, Carl Dee
Caron, Richard Edward
Copeland, Thompson Preston
Gerhardt, Reid Richard
Haines, Thomas Walton
Hull, George, Jr
Joplin, Karl Henry
Kring, James Burton
Lambdin, Paris Lee
Lentz, Gary Lynn
Liles, James Neil
McGhee, Charles Robert
Nelson, Charles Henry
Patrick, Charles Russell
Reed, Horace Beecher
Shipp, Oliver Elmo
Smith, Omar Ewing, Jr
Southards, Carroll J
Tarpley, Wallace Armell
Tolbert, Virginia Rose
Watson, Annetta Paule
Webb, J(ohn) Warren
White, Jane Vicknair

TEXAS
Adkisson, Perry Lee
Applegate, Richard Lee
Bailey, Leo L
Bay, Darrell Edward
Benedict, John Howard, Jr
Bhaskaran, Govindan
Bowling, Clarence C
Brook, Ted Stephens
Brooks, Derl
Bull, Don Lee
Ciomperlik, Matthew A
Cogburn, Robert Ray
Collins, Anita Marguerite
Cook, Benjamin Jacob
Coulson, Robert N
Curtin, Thomas J
Dean, Herbert A
DeLoach, Culver Jackson, Jr
Drees, Bastiaan M
Drummond, Roger Otto
Duhrkopf, Richard Edward
Dulmage, Howard Taylor
Dunkle, Sidney Warren
Durden, Christopher John
Ewert, Adam
Fincher, George Truman
Fisher, Charles William
French, Jeptha Victor
Frisbie, Raymond Edward
Gibson, William Wallace
Gillaspy, James Edward
Gilstrap, Franklin Ephriam
Gold, Roger Eugene
Goodwin, William Jennings
Greenberg, Les Paul
Hall, Clarence Coney, Jr
Hanna, Ralph Lynn
Harden, Philip Howard
Harris, Kerry Francis Patrick
Harris, Marvin Kirk
Hartberg, Warren Keith
Higgins, Linden Elizabeth
Hilliard, John Roy, Jr
Hollingsworth, Joseph Pettus
Horner, Norman V
Keeley, Larry Lee
Kennedy, James H
Klaus, Ewald Fred, Jr
Kunz, Sidney Edmund
Lopez, Genaro
Mangan, Robert Lawrence
Martin, Paul Bain
Meadows, Charles Milton
Meola, Roger Walker
Meola, Shirlee May
Merchant, Michael Edward
Morrison, Eston Odell
Nachman, Ronald James
Nettles, William Carl, Jr
Olson, Jimmy Karl
Palmer, William Alan
Parajulee, Megha N
Parker, Roy Denver, Jr
Peck, William B
Queller, David Charles
Reinert, James A
Richerson, Jim Vernon
Ring, Dennis Randall
Robinson, Arin Forest

Robinson, James Vance
Rubink, William Louis
Scanlon, John Earl
Schaffner, Joseph Clarence
Schuster, Michael Frank
Shake, Roy Eugene
Slosser, Jeffrey Eric
Smith, James Willie, Jr
Standifer, Lonnie Nathaniel
Sterling, Winfield Lincoln
Stewart, Kenneth Wilson
Stone, Jay D
Strassmann, Joan Elizabeth
Summers, Max Duane
Tuff, Donald Wray
Van Cleave, Horace William
Vinson, S Bradleigh
Walker, J Knox
Watkins, Julian F, II
Wellso, Stanley Gordon
Williams, Ralph Edward
Williams, Robert K
Wolfenbarger, Dan A

UTAH
Allred, Dorald Mervin
Amman, Gene Doyle
Anderson, Russell D
Baumann, Richard William
Booth, Gary Melvon
Cole, Walter Eckle
Davidson, Diane West
Davis, Donald Walter
Elbel, Robert E
Grundmann, Albert Wendell
Havertz, David S
Haws, Byron Austin
Hsiao, Ting Huan
Jorgensen, Clive D
Messina, Frank James
Nielsen, Lewis Thomas
Nielson, Mervin William
Richardson, Jay Wilson, Jr
Stark, Harold Emil
Tipton, Vernon John
Vandenberg, John Donald
Whitehead, Armand T
Wood, Stephen Lane

VERMONT
Freeman, Jeffrey VanDuyne
Happ, George Movius
MacCollom, George Butterick
McLean, Donald Lewis

VIRGINIA
Bressler, Barry Lee
Callahan, James Thomas
Carico, James Edwin
Cochran, Donald Gordon
Dogger, James Russell
Eaton, John LeRoy
Eckerlin, Ralph Peter
Emsley, Michael Gordon
Fashing, Norman James
Fink, Linda Susan
Fletcher, Lowell W
Foote, Richard Herbert
Gray, Faith Harriet
Grimm, James K
Haines, Kenneth A
Heikkenen, Herman John
Hinckley, Alden Dexter
Hofmaster, Richard Namon
Homsher, Paul John
Horsburgh, Robert Laurie
Jones, Tappey Hughes
Jubb, Gerald Lombard, Jr
Karowe, David Nathan
Kliewer, John Wallace
Knipling, Edward Fred
Kok, Loke-Tuck
Kosztarab, Michael
Logan, Jesse Alan
Lyon, Robert Lyndon
Mullins, Donald Eugene
Payne, Thomas Lee
Pienkowski, Robert Louis
Roberts, James Ernest, Sr
Robinson, William H
Ross, Mary Harvey
Salom, Scott Michael
Schultz, Peter Berthold
Semtner, Paul Joseph
Shaffer, Jay Charles
Sharov, Alexei A
Smith, John Cole
Smythe, Richard Vincent
Spyhalski, Edward James
Stevens, Thomas McConnell
Turner, Ernest Craig, Jr
Weidhaas, John August, Jr
Zimmerman, John Harvey

WASHINGTON
Akre, Roger David
Bay, Ernest C
Bernard, Gary Dale
Biever, Kenneth Duane
Branson, Terry Fred
Burts, Everett C
Butt, Billy Arthur
Calkins, Carrol Otto
Catts, Elmer Paul

Clement, Stephen LeRoy
Cone, Wyatt Wayne
Dolphin, Robert Earl
Eighme, Lloyd Elwyn
Gara, Robert I
Gibson, Flash
Halfhill, John Eric
Hoyt, Stanley Charles
Kareiva, Peter Michael
Klostermeyer, Edward Charles
Klostermeyer, Lyle Edward
Kraft, Gerald F
McDonough, Leslie Marvin
Meeuse, Bastiaan J D
Milne, David Hall
Moffett, David Franklin, Jr
Moffitt, Harold Roger
Paulson, Dennis R
Pike, Keith Schade
Riddiford, Lynn Moorhead
Rogers, Lee Edward
Senger, Clyde Merle
Shanks, Carl Harmon, Jr
Sluss, Robert Reginald
Smith, Stamford Dennis
Spence, Kemet Dean
Toba, H(achiro) Harold
Wolda, Hindrik

WEST VIRGINIA
Adkins, Dean Aaron
Brown, Mark Wendell
Butler, Linda
Cole, Larry King
Sheppard, Roger Floyd

WISCONSIN
Beck, Stanley Dwight
Burkholder, Wendell Eugene
Carlson, Stanley David
Chapman, R Keith
Coppel, Harry Charles
DeFoliart, Gene Ray
Dixon, John Charles
Downing, Holly Adelaide
Drecktrah, Harold Gene
Fisher, Ellsworth Henry
Ghazarian, Jacob G
Giese, Ronald Lawrence
Gojmerac, Walter Louis
Grothaus, Roger Harry
Halgren, Lee A
Hilsenhoff, William LeRoy
Jeanne, Robert Lawrence
Kennedy, M(aldon) Keith
Koval, Charles Francis
Lichtenstein, E Paul
Norris, Dale Melvin, Jr
Owens, John Michael
Ritter, Karla Schwensen
Shenefelt, Roy David
Singer, George
Willis, Harold Lester
Wittrock, Darwin Donald

WYOMING
Burkhardt, Christian Carl
Holbrook, Frederick R
Lavigne, Robert James
Lawson, Fred Avery
Lloyd, John Edward
Lockwood, Jeffrey Alan
Nunamaker, Richard Allan
Pfadt, Robert E
Stoner, Adair
Tabachnick, Walter J
Wangberg, James Keith

PUERTO RICO
Cruz, Carlos
Fox, Irving
Virkki, Niilo

ALBERTA
Byers, John Robert
Craig, Douglas Abercrombie M
Cuny, Robert Michael
Evans, William George
Goettel, Mark S
Gooding, Ronald Harry
Harper, Alexander Maitland
Heming, Bruce Sword
Hill, Bernard Dale
Iatrou, Kostas
Lysyk, Timothy James
Nelson, William Arnold
Pritchard, Gordon
Shemanchuk, Joseph Alexander
Wilkinson, Paul R
Wilson, Mark Vincent Hardman
Wong, Horne Richard

BRITISH COLUMBIA
Barclay, Hugh John
Beirne, Bryan Patrick
Belton, Peter
Borden, John Harvey
Campbell, Alan
Cram, William Thomas
Forbes, Albert Ronald
Frazer, Bryan Douglas
Haunerland, Norbert Heinrich
Jamieson, Glen Stewart
Jones, Richard Lamar

Entomology (cont)

Kobylnyk, Ronald William
Lindgren, Bo Staffan
McIver, Susan Bertha
Mackauer, Manfred
McLean, John Alexander
McMullen, Robert David
Manville, John Fieve
Mendoza, Celso Enriquez
Ring, Richard Alexander
Roitberg, Bernard David
Safranyik, Laszlo
Scudder, Geoffrey George Edgar
Struble, Dean L
Thomson, Alan John
Underhill, Edward Wesley
Wellington, William George

MANITOBA
Barker, Philip Shaw
Brust, Reinhart A
Flannagan, John Fullan
Gerber, George Hilton
Loschiavo, Samuel Ralph
Palaniswamy, Pachagounder
Preston, William Burton
Robinson, Arthur Grant
Rosenberg, David Michael
Sinha, Ranendra Nath
Turnock, William James
Westdal, Paul Harold
White, Noel David George
Wylie, Harold Glenn

NEW BRUNSWICK
Eidt, Douglas Conrad
Eveleigh, Eldon Spencer
MacGillivray, M Ellen
Seabrook, William Davidson
Smith, Harry John
Strunz, G(eorge) M(artin)
Varty, Isaac William

NEWFOUNDLAND
Bennett, Gordon Fraser
Carroll, Allan Louis
Lim, Kiok-Puan

NOVA SCOTIA
Angelopoulos, Edith W
Hardman, John Michael
Specht, Harold Balfour

ONTARIO
Angus, Thomas Anderson
Barron, John Robert
Behan-Pelletier, Valerie Mary
Bright, Donald Edward
Cade, William Henry
Cartier, Jean Jacques
Chant, Donald A
Chiykowski, Lloyd Nicholas
Collins, Nicholas Clark
Cunningham, John Castel
Danks, Hugh Victor
Darling, Donald Christopher
Davies, Douglas Mackenzie
Downer, Roger George Hamill
Downes, John Antony
Edwards, Roy Lawrence
Ellis, Clifford Roy
Fitz-James, Philip Chester
Fleming, Richard Arthur
Friend, William George
George, John Allen
Gill, Bruce Douglas
Hamilton, Kenneth Gavin Andrew
Harcourt, Douglas George
Harris, Charles Ronald
House, Howard Leslie
Howden, Henry Fuller
Jaques, Robert Paul
Jones, Philip Arthur
Judd, William Wallace
Kevan, Peter Graham
Laing, John E
Lee, Robert Maung Kyaw Win
LeRoux, Edgar Joseph
Lindquist, Evert E
Locke, Michael
McAlpine, James Francis
McEwen, Freeman Lester
Masner, Lubomir
Morgan, Alan Vivian
Morin, Antoine
Munroe, Eugene Gordon
Nesbitt, Herbert Hugh John
Oliver, Donald Raymond
Ozburn, George W
Peck, Stewart Blaine
Peterson, Bobbie Vern (Robert)
Philogene, Bernard J R
Retnakaran, Arthur
Sears, Markham Karli
Sharkey, Michael Joseph
Smetana, Ales
Smith, Jonathan Jeremy Berkeley
Steele, John Earle
Teskey, Herbert Joseph
Tobe, Stephen Solomon
Tomlin, Alan David
Watson, Wynnfield Young
Whitfield, Gary Hugh
Wiggins, Glenn Blakely
Williams, David Dudley
Wyatt, Gerard Robert
Yoshimoto, Carl Masaru

PRINCE EDWARD ISLAND
Stewart, Jeffrey Grant
Thompson, Leith Stanley

QUEBEC
Auclair, Jacques Lucien
Cloutier, Conrad Francois
Harper, Pierre (Peter) Paul
Hill, Stuart Baxter
Hilton, Donald Frederick James
Kevan, Douglas Keith McEwan
McFarlane, John Elwood
McNeil, Jeremy Nichol
Mailloux, Gerard
O'Neil, Louis C
Pilon, Jean-Guy
Quednau, Franz Wolfgang
Rau, Manfred Ernst
Sanborne, Paul Michael
Sharma, Madan Lal
Srivastava, Prakash Narain
Stewart, Robin Kenny
Vincent, C

SASKATCHEWAN
Doane, John Frederick
Ewen, Alwyn Bradley
Khachatourians, George G
Lehmkuhl, Dennis Merle
Makowski, Roberte Marie Denise
Mukerji, Mukul Kumar
Peschken, Diether Paul
Riegert, Paul William
Zacharuk, R Y

OTHER COUNTRIES
Adames, Abdiel Jose
Aiello, Annette
Bailey, Donald Leroy
Birch, Martin Christopher
Brown, Anthony William Aldridge
Daoust, Richard Alan
De Renobales, Mertxe
Gratz, Norman G
Hanson, Paul Eliot
Harbach, Ralph Edward
Holldobler, Berthold Karl
Hummel, Hans Eckhardt
Klassen, Waldemar
Knutson, Lloyd Vernon
Muniappan, Rangaswamy Naicker
Perry, Albert Solomon
Rentz, David Charles
Saether, Ole Anton
Saunders, Joseph Lloyd
Zhang, Zhi-Qiang

Food Science & Technology

ALABAMA
Chastain, Marian Faulkner
Marion, James Edsel
Navia, Juan Marcelo
Rymal, Kenneth Stuart
Singh, Bharat

ALASKA
Roop, Richard Allan

ARIZONA
Alton, Alvin John
Kaufman, C(harles) W(esley)
Kline, Ralph Willard
Nelson, Frank Eugene
Price, Ralph Lorin
Seperich, George Joseph
Sleeth, Rhule Bailey
Stiles, Philip Glenn
Urbain, Walter Mathias

ARKANSAS
Herring, Harold Keith
Kattan, Ahmed A
Lewis, Paul Kermith, Jr

CALIFORNIA
Amen, Ronald Joseph
Ashton, David Hugh
Avera, Fitzhugh Lee
Bentley, David R
Berhow, Mark Alan
Brant, Albert Wade
Butterworth, Thomas Austin
Caporaso, Fredric
Chou, Tsong-Wen
Clark, Walter Leighton, III
Collins, Edwin Bruce
Davé, Bhalchandra A
Dunkley, Walter Lewis
Feeney, Robert Earl
Finkle, Bernard Joseph
Goldstein, Walter Elliott
Holmes, David G
Hugunin, Alan Godfrey
Huskey, Glen E
Ingle, James Davis
Ito, Keith A
Jaye, Murray Joseph
Jen, Joseph Jwu-Shan
King, Alfred Douglas, Jr
Lei, Shau-Ping Laura
Levenson, Harold Samuel
Lu, Nancy Chao
Luh, Bor Shiun
Lukes, Thomas Mark
Lundeen, Glen Alfred
Mazelis, Mendel
Meggison, David Laurence
Michener, H(arold) David
Midura, Thaddeus
Moshy, Raymond Joseph
Nelson, David A
Noble, Ann Curtis
Pangborn, Rose Marie Valdes
Prater, Arthur Nickolaus
Prend, Joseph
Robertson, George Harcourt
Rockland, Louis B
Russell, Gerald Frederick
Sayre, Robert Newton
Schneeman, Barbara Olds
Schwimmer, Sigmund
Serbia, George William
Sheneman, Jack Marshall
Silberstein, Otmar Otto
Singleton, Vernon LeRoy
Spector, Sheldon Laurence
Stevenson, Kenneth Eugene
Sutton, Donald Dunsmore
Webster, John Robert
Wodicka, Virgil O

COLORADO
Church, Brooks Davis
Deane, Darrell Dwight
Harper, Judson M(orse)
Long, Austin Richard
Maga, Joseph Andrew
Pan, Huo-Ping
Schmidt, Glenn Roy
Skelton, Marilyn Mae
Smith, Gary Chester
Sofos, John N
Stone, Martha Barnes

CONNECTICUT
Agarwal, Vipin K
Bibeau, Thomas Clifford
Dobry, Reuven
Fiore, Joseph Vincent
Fordham, Joseph Raymond
Hall, Kenneth Noble
Hart, William James, Jr
Hopper, Paul Frederick
Kinsman, Donald Markham
Kotula, Anthony W
Piehl, Donald Herbert
Ranalli, Anthony William
Ross, Robert Edgar
Schoen, Herbert M
Servadio, Gildo Joseph

DELAWARE
Link, Bernard Alvin
Mitchell, Donald Gilman

DISTRICT OF COLUMBIA
Attaway, David Henry
Cole, Margaret Elizabeth
Hill, Walter Ernest
Jacobson, Michael F
Kromer, Lawrence Frederick
Kuznesof, Paul Martin
Shank, Fred R
Weinberg, Myron Simon

FLORIDA
Ahmed, Esam Mahmoud
Bacus, James Nevill
Balaban, Murat Omer
Barber, Franklin Weston
Bates, Robert P(arker)
Beckhorn, Edward John
Bernstein, Sheldon
Brown, Ross Duncan, Jr
Brown, William Lewis
Burgess, Hovey Mann
Carbonell, Robert Joseph
Cotton, Robert Henry
Dubravcic, Milan Frane
Fox, Kenneth Ian
Gill, William Joseph
Gnaedinger, Richard H
Hale, Kirk Kermit, Jr
Hankinson, Denzel J
Harris, Natholyn Dalton
Jackel, Simon Samuel
Janky, Douglas Michael
Kapsalis, John George
Koburger, John Alfred
Moore, Lynn E(lewis)
Nanz, Robert Augustus Rollins
Otwell, Walter Steven
Sabharwal, Kulbir
Showalter, Robert Kenneth
Teixeira, Arthur Alves
Trelease, Richard Davis
Vedamuthu, Ebenezer Rajkumar
Whipple, Royson Newton

GEORGIA
Ayres, John Clifton
Beuchat, Larry Ray
Bhatia, Darshan Singh
Brown, William E
Bryan, Frank Leon
Cecil, Sam Reber
Clark, Allen Varden
Dekazos, Elias Demetrios
Dennison, Daniel B
Donahoo, Pat
Doyle, Michael Patrick
Ebert, Andrew Gabriel
Eitenmiller, Ronald Ray
Green, John H
Koehler, Philip Edward
Malaspina, Alex
Nakayama, Tommy
Powers, John Joseph
Prothro, Johnnie W
Ray, Apurba Kanti
Schuler, George Albert
Smit, Christian Jacobus Bester
Swaminathan, Balasubramanian
Tybor, Philip Thomas
Worthington, Robert Earl
Young, Louis Lee
Zallen, Eugenia Malone

HAWAII
Chan, Harvey Thomas, Jr
Moser, Roy Edgar
Moy, James Hee
Nip, Wai Kit
Seifert, Josef
Waslien, Carol Irene
Yamamoto, Harry Y

IDAHO
Augustin, Jorg A L
Branen, Alfred Larry
Davidson, Philip Michael
Greene, Barbara E
Heimsch, Richard Charles
Muneta, Paul
Zaehringer, Mary Veronica

ILLINOIS
Abrams, Israel Jacob
Akin, Cavit
Auerbach, Earl
Blaschek, Hans P
Bookwalter, George Norman
Bothast, Rodney Jacob
Brigham, Robert B
Brown, Peter H
Burkwall, Morris Paton, Jr
Campbell, Michael Floyd
Chen, Wen Sherng
Cheryan, Munir
Chung, Jiwhey
Clark, John Peter, III
Cole, Morton S
Dean, Robert Waters
Dordick, Isadore
Duxbury, Dean David
Eckner, Karl Friedrich
Elliott, James Gary
Erdman, John Wilson, Jr
Flowers, Russell Sherwood, Jr
Gabis, Damien Anthony
Glenister, Paul Robson
Halaby, George Anton
Harris, Ronald David
Inglett, George Everett
Kelley, Keith Wayne
Klinger, Lawrence Edward
Kummerow, Fred August
Lebermann, Kenneth Wayne
Lechowich, Richard V
Lundquist, Burton Russell
Lushbough, Channing Harden
McKeith, Floyd Kenneth
Miller, Gregory Duane
Milner, Reid Thompson
Moeller, Theodore William
Newton, Stephen Bruington
Osri, Stanley M(aurice)
Pedraja, Rafael R
Rackis, Joseph John
Rakosky, Joseph, Jr
Reagan, James Oliver
Schanefelt, Robert Von
Scott, Don
Sehgal, Lakshman R
Sherman, Joseph E
Siedler, Arthur James
Simon, Selwyn
Smith, John
Smittle, Richard Baird
Smouse, Thomas Hadley
Soucie, William George
Steinberg, Marvin Phillip
Tompkin, Robert Bruce
Udani, Kanakkumar Harilal
Vosti, Donald Curtis
Wei, Lun-Shin
Welsh, Thomas Laurence
White, John Francis
Wistreich, Hugo Eryk
Witter, Lloyd David

INDIANA
Babel, Frederick John
Bates, Charles Johnson
Chambers, James Vernon

Cook, David Allan
Cousin, Maribeth Anne
Creinin, Howard Lee
Endres, Joseph George
Forrest, John Charles
Hagen, Richard Eugene
Henke, Mitchell C
Irwin, William Elliot
Judge, Max David
Kittaka, Robert Shinnosuke
Liska, Bernard Joseph
Litov, Richard Emil
Marks, Jay Stewart
Morris, John F
Nelson, Philip Edwin
Ogilvy, Winston Stowell
Parmelee, Carlton Edwin
Puski, Gabor
Radanovics, Charles
Stadelman, William Jacob
Tennyson, Richard Harvey
Wallander, Jerome F
Weaver, Connie Marie
Whistler, Roy Lester
Williams, Leamon Dale

IOWA
Chung, Ronald Aloysius
Hammond, Earl Gullette
Hartman, Paul Arthur
Kraft, Allen Abraham
LaGrange, William Somers
Marion, William W
Rohlf, Marvin Euguene
Sebranek, Joseph George
Walker, Homer Wayne

KANSAS
Allen, Deloran Matthew
Bowers, Jane Ann (Raymond)
Caul, Jean Frances
Cunningham, Franklin E
Dirks, Brinton Marlo
Farrell, Eugene Patrick
Finney, Karl Frederick
Fung, Daniel Yee Chak
Harbers, Carole Ann Z
Hoover, William Jay
Kropf, Donald Harris
Krum, Jack Kern
Ponte, Joseph G, Jr
Ramamurti, Krishnamurti
Setser, Carole Sue
Vetter, James Louis

KENTUCKY
Bujake, John Edward, Jr
Harmet, Kenneth Herman
Langlois, Bruce Edward
Luzzio, Frederick Anthony
Maisch, Weldon Frederick

LOUISIANA
Barkate, John Albert
Broeg, Charles Burton
Day, Donal Forest
Frank, Arlen W(alker)
Grodner, Robert Maynard
James, William Holden
Kadan, Ranjit Singh
Liuzzo, Joseph Anthony
Rao, Ramachandra M R
St Angelo, Allen Joseph
Windhauser, Marlene M
Yerrapraggada, Venkat
Younathan, Margaret Tims

MAINE
Highlands, Matthew Edward
Slabyj, Bohdan M

MARYLAND
Alsmeyer, Richard Harvey
Anderson, Mauritz Gunnar
Anderson, Sue Ann Debes
Butman, Bryan Timothy
Dame, Charles
Denny, Cleve B
Devreotes, Peter Nicholas
Gardner, Sherwin
Goodman, Sarah Anne
Graebert, Eric W
Hall, Richard Leland
Heath, James Lee
Kim, Sooja K
King, Raymond Leroy
Matthews, Ruth H
Milner, Max
Moats, William Alden
Modderman, John Philip
Molenda, John R
Porzio, Michael Anthony
Seifried, Adele Susan Corbin
Seifried, Harold Edwin
Twigg, Bernard Alvin
Vaughn, Moses William
Wabeck, Charles J
Wagner, David Darley
Westhoff, Dennis Charles
Whitehair, Leo A
Wiley, Robert Craig
Williams, Eleanor Ruth
Yeatman, John Newton

MASSACHUSETTS
Anderson, Edward Everett
Ayuso, Katharine
Babayan, Vigen Khachig
Breslau, Barry Richard
Buck, Ernest Mauro
Bustead, Ronald Lorima, Jr
Cardello, Armand Vincent
Clayton, J(oe) T(odd)
Clydesdale, Fergus Macdonald
Esselen, William B
Fagerson, Irving Seymour
Fram, Harvey
Francis, Frederick John
Goldblith, Samuel Abraham
Hultin, Herbert Oscar
Jarboe, Jerry K(ent)
Kepper, Robert Edgar
King, Frederick Jessop
Kissmeyer-Nielsen, Erik
Lampi, Rauno Andrew
Levin, Robert E
Maloney, John F
Mehrlich, Ferdinand Paul
Narayan, Krishnamurthi Ananth
Powers, Edmund Maurice
Ross, Edward William, Jr
Rowley, Durwood B
Salant, Abner Samuel
Sawyer, Frederick Miles
Short, John Lawson
Silverman, Gerald
Sinskey, Anthony J
Tan, Barrie
Tillotson, James E
Tuomy, Justin M(atthew)
Whitney, Lester F(rank)
Williams, Marion Porter
Yamins, J(acob) L(ouis)
Zapsalis, Charles

MICHIGAN
Bartholmey, Sandra Jean
Brunner, Jay Robert
Carlotti, Ronald John
Cotter, Richard
Crevasse, Gary A
Dawson, Lawrence E
Einset, Eystein
Frodey, Ray Charles
Haines, William C
Harmon, Laurence George
Hedrick, Theodore Isaac
Hollingsworth, Cornelia Ann
Jackson, John Mathews
Markakis, Pericles
Marsh, Alice Garrett
Merkel, Robert Anthony
Niekamp, Carl William
Price, James F
Purvis, George Allen
Schaller, Daryl Richard
Shelef, Leora Aya
Stine, Charles Maxwell
Uebersax, Mark Alan
Wishnetsky, Theodore
Zabik, Mary Ellen

MINNESOTA
Anderson, Ray Harold
Bauman, Howard Eugene
Behnke, James Ralph
Berube, Robert
Breene, William Michael
Busta, Francis Fredrick
Caldwell, Elwood F
Chandan, Ramesh Chandra
Christianson, George
Cronk, Ted Clifford
Daravingas, George Vasilios
Davis, Eugenia Asimakopoulos
Dunshee, Bryant R
Durst, Jack Rowland
Epley, Richard Jess
Franzen, Kay Louise
Gallaher, Daniel David
Gillett, Tedford A
Gordon, Joan
Heller, Steven Nelson
Hurley, William Charles
Jezeski, James John
Johnson, Guy Henry
Katz, Morris Howard
Labuza, Theodore Peter
Levine, Allen Stuart
McKay, Larry Lee
Meeder, Jeanne Elizabeth
Morris, Howard Arthur
Mullen, Joseph David
Noland, Wayland Evan
Opie, Joseph Wendell
Packard, Vernal Sidney, Jr
Pflug, Irving John
Rapp, Waldean G
Reineccius, Gary (Aubrey)
Reinhart, Richard D
Rodgers, Nelson Earl
Sapakie, Sidney Freidin
Sherck, Charles Keith
Slavin, Joanne Louise
Stoll, William Francis
Sullivan, Betty J
Tatini, Sita Ramayya
Thomas, Elmer Lawrence

Touba, Ali R
Vermilyea, Barry Lynn
Vickers, Zata Marie
Zottola, Edmund Anthony

MISSISSIPPI
Ford, Robert Sedgwick
Garrett, Ephraim Spencer, III
Marshall, Douglas Lee
Regier, Lloyd Wesley

MISSOURI
Bough, Wayne Arnold
Fields, Marion Lee
Heymann, Hildegarde
Hood, Larry Lee
Kluba, Richard Michael
Lo, Grace S
Marshall, Robert T
Mavis, James Osbert
Naumann, Hugh Donald
Newell, Jon Albert
Schwall, Donald V
Snyder, Harry E
Titus, Dudley Seymour
Unklesbay, Nan F
Wells, Frank Edward

MONTANA
Stanislao, Bettie Chloe Carter

NEBRASKA
Aberle, Elton D
Beach, Betty Laura
Brazis, A(dolph) Richard
Bullerman, Lloyd Bernard
Froning, Glenn Wesley
Hanna, Milford A
Hartung, Theodore Eugene
Shahani, Khem Motumal
Siragusa, Gregory Ross
Smith, Durward A
Taylor, Steve L

NEW HAMPSHIRE
Hayes, Kirby Maxwell

NEW JERSEY
Adler, Irwin L
Akerboom, Jack
Bakal, Abraham I
Bednarcyk, Norman Earle
Bedrosian, Karakian
Brogle, Richard Charles
Buck, Robert Edward
Canzonier, Walter J
Chu, Horn Dean
Dalrymple, Ronald Howell
Finley, John Westcott
Franceschini, Remo
Fulde, Roland Charles
Giddings, George Gosselin
Gilbert, Seymour George
Goodenough, Eugene Ross
Hayakawa, Kan-Ichi
Hlavacek, Robert John
Hsu, Kenneth Hsuehchia
Johnson, Bobby Ray
Jones, Benjamin Lewis
Karel, Marcus
Katz, Ira
Kirk, James Robert
Lachance, Paul Albert
Litman, Irving Ira
Lund, Daryl B
McAnelly, John Kitchel
Miller, Albert Thomas
Miller, Gary A
Montville, Thomas Joseph
Morck, Roland Anton
Otterburn, Michael Storey
Pleasants, Elsie W
Raman, Jay Ananth
Richberg, Carl George
Schaich, Karen Marie
Scharpf, Lewis George, Jr
Sekula, Bernard Charles
Smith, Robert Ewing
Solberg, Myron
Staub, Herbert Warren
Stone, Charles Dean
Supran, Michael Kenneth
Wasserman, Bruce P
Wells, Phillip Richard
Wolin, Alan George
Yu, Stephen Kin-Cheun

NEW MEXICO
O'Brien, Robert Thomas
Ray, Earl Elmer

NEW YORK
Armbruster, Gertrude D
Baker, Robert Carl
Beermann, Donald Harold
Bigelow, Sanford Walker
Borisenok, Walter A
Bourne, Malcolm Cornelius
Chang, Pei Kung (Philip)
Dibble, Marjorie Veit
Downing, Donald Leonard
Ehmann, Edward Paul
Filandro, Anthony Salvatore
Gizis, Evangelos John

Graham, Donald C W
Halpern, Bruce Peter
Hang, Yong Deng
Hickernell, Gary L
Holland, Robert Francis
Kahan, Sidney
Khan, Paul
Kosikowski, Frank Vincent
Kraft, Patricia Lynn
Kramer, Franklin
Kroenberg, Berndt
Ledford, Richard Allison
Leff, Judith
Lewis, Bertha Ann
Livingston, G E
Lugay, Joaquin Castro
March, Richard Pell
Marshall, James Tilden, Jr
Mehta, Bipin Mohanlal
Miller, Dennis Dean
Nelson, Mark Thomas
Page, Gregory Vincent
Parliment, Thomas H
Potter, Norman N
Prescott, Henry Emil, Jr
Ramstad, Paul Ellertson
Rao, Mentreddi Anandha
Regenstein, Joe Mac
Riha, William E, Jr
Schiffmann, Robert F
Shipe, W(illiam) Frank(lin)
Smith, Ora
Szczesniak, Alina Surmacka
Turkki, Pirkko Reetta
Wellington, George Harvey
White, James Carrick
Zall, Robert Rouben

NORTH CAROLINA
Beam, John E
Beck, Charles I
Dahiya, Raghunath S
DeJong, Donald Warren
DiMarco, G Robert
Fleming, Henry Pridgen
Gregory, Max Edwin
Jones, Ivan Dunlavy
Jones, Victor Alan
Lineback, David R
Lloyd, Norman Edward
May, Kenneth Nathaniel
Morse, Roy E
Neumann, Calvin Lee
Oblinger, James Leslie
Oliver, James David
Reed, Gerald
Schiffman, Susan S
Swaisgood, Harold Everett
Tarver, Fred Russell, Jr
Thomas, Frank Bancroft
Walter, William Mood, Jr
Webb, Neil Broyles
Young, Clyde Thomas

NORTH DAKOTA
Donnelly, Brendan James

OHIO
Banwart, George J
Blaisdell, John Lewis
Cahill, Vern Richard
Chism, Grady William, III
Cornelius, Billy Dean
De Fiebre, Conrad William
Gallander, James Francis
Gould, Wilbur Alphonso
Grab, Eugene Granville, Jr
Hansen, Poul M T
Hartman, Warren Emery
Holzinger, Thomas Walter
Hunt, Fern Ensminger
Hunter, James Edward
Husaini, Saeed A
Jandacek, Ronald James
Jaynes, John Alva
Joffe, Frederick M
Kristoffersen, Thorvald
Larkin, Edward P
Litchfield, John Hyland
McComis, William T(homas)
Neer, Keith Lowell
Parrett, Ned Albert
Sharma, Shri C
Skiest, Eugene Norman
Stauffer, Clyde E
Wilcox, Joseph Clifford
Youngquist, R(udolph) William
Ziller, Stephen A, Jr

OKLAHOMA
Berry, Joe Gene
Gilliland, Stanley Eugene
Guenther, John James
Henrickson, Robert Lee
Ray, Frederick Kalb

OREGON
Anglemier, Allen Francis
Arnold, Roy Gary
Crawford, David Lee
Dutson, Thayne R
Elliker, Paul R
Kazerouni, Lewa
Kennick, Walter Herbert

Food Science & Technology (cont)

**Kifer, Paul Edgar
Libbey, Leonard Morton
Mickelberry, William Charles
Morita, Toshiko N
Pearson, Albert Marchant
Scanlan, Richard Anthony
Sinnhuber, Russell Otto
Wehr, Herbert Michael
Woodburn, Margy Jeanette
Wrolstad, Ronald Earl
Yang, Hoya Y

PENNSYLVANIA
Barnes, William Shelley
Beelman, Robert B
Birdsall, Marion Ivens
Brody, Aaron Leo
Buchanan, Robert Lester
Craig, James Clifford, Jr
Dimick, Paul Slayton
Fugger, Joseph
Garwin, Jeffrey Lloyd
Hagberg, Elroy Carl
Hills, Claude Hibbard
Johnson, Littleton Wales
Johnson, Ogden Carl
Keeney, Philip G
Kroger, Manfred
MacNeil, Joseph H
Medina, Marjorie B
Miller, Arthur James
Miller, Kenneth Melvin
Mottur, George Preston
Palumbo, Samuel Anthony
Ramachandran, N
Roseman, Arnold S(aul)
Sapers, Gerald M
Segall, Stanley
Seligson, Frances Hess
Shirk, Richard Jay
Thompson, Donald B
Whiting, Richard Charles
Wohlpert, Kenneth Joseph
Ziegler, John Henry, Jr

RHODE ISLAND
Chichester, Clinton Oscar
Constantinides, Spiros Minas
Cosgrove, Clifford James
Dymsza, Henry A
Josephson, Edward Samuel
Rand, Arthur Gorham, Jr
Simpson, Kenneth L

SOUTH CAROLINA
Godfrey, W Lynn
Hollis, Bruce Warren
Lundstrom, Ronald Charles
Surak, John Godfrey

SOUTH DAKOTA
Costello, William James

TENNESSEE
Bass, Mary Anna
Collins, Jimmie Lee
Das, Salil Kumar
Draughon, Frances Ann
Jaynes, Hugh Oliver
Perry, Margaret Nutt
Prokop, Ales

TEXAS
Armstrong, George Glaucus, Jr
Arnott, Howard Joseph
Bourland, Charles Thomas
Brittin, Dorothy Helen Clark
Burns, Edward Eugene
Carpenter, Zerle Leon
Cross, Hiram Russell
Czajka-Narins, Dorice M
Dill, Charles William
Frank, Hilmer Aaron
Goodwin, Tommy Lee
Hsueh, Andie M
Hunnell, John Wesley
Johnston, Melvin Roscoe
Kendall, John Hugh
King, General Tye
LaBree, Theodore Robert
Landmann, Wendell August
Lusas, Edmund W
Matz, Samuel Adam
Mellor, David Bridgwood
Miller, Edward Godfrey, Jr
Orts, Frank A
Rooney, Lloyd William
Sistrunk, William Allen
Smith, Malcolm Crawford, Jr
Sweat, Vincent Eugene
Vanderzant, Carl
Varsel, Charles John

UTAH
Huber, Clayton Shirl
Larsen, Lloyd Don
Mendenhall, Von Thatcher
Salunkhe, Dattajeerao K
Yorks, Terence Preston

VERMONT
Mitchell, William Alexander

VIRGINIA
Axelson, Marta Lynne
Berkow, Susan E
Burnette, Mahlon Admire, III
Collins, William F
Cooler, Frederick William
Fellers, David Anthony
Flick, George Joseph, Jr
Keenan, Thomas William
Khan, Mahmood Ahmed
Lopez, Anthony
Mulvaney, Thomas Richard
Palmer, James Kenneth
Pao, Eleanor M
Phillips, Jean Allen
Rippen, Thomas Edward
Rosenfield, Daniel
Srivastava, Kailash Chandra
Stern, Joseph Aaron
Swope, Fred C
Symons, Hugh William
Wesley, Roy Lewis
Yip, George

WASHINGTON
Barbosa-Canovas, Gustavo Victor
Brekke, Clark Joseph
Drake, Stephen Ralph
Eklund, Melvin Wesley
Hard, Margaret McGregor
Hoskins, Frederick Hall
Ingalsbe, David Weeden
Lee, Donald Jack
Maguire, Yu Ping
Matches, Jack Ronald
Nagel, Charles William
Pigott, George M
Swanson, Barry Grant
Wekell, Marleen Marie

WISCONSIN
Amundson, Clyde Howard
Bard, John C
Bartholomew, Darrell Thomas
Bernstein, Bradley Alan
Bradley, Robert Lester, Jr
Buege, Dennis Richard
Chu, Fun Sun
Curwen, David
Deese, Dawson Charles
Duke, Stanley Houston
Dutton, Herbert Jasper
Fennema, Owen Richard
Freund, Peter Richard
Greaser, Marion Lewis
Grindrod, Paul (Edward)
Hartzell, Thomas H
Hefle, Susan Lynn
Hill, Charles Graham, Jr
Johnson, Dale Waldo
Kainski, Mercedes H
Kauffman, Robert Giller
Lindsay, Robert Clarence
McDivitt, Maxine Estelle
Marsh, Benjamin Bruce
Marth, Elmer Herman
Maurer, Arthur James
Moskowitz, Gerard Jay
Nagodawithana, Tilak Walter
Nelson, John Howard
Olson, Norman Fredrick
Price, Walter Van Vranken
Ryan, Dale Scott
Seitz, Eugene W
Snudden, Birdell Harry
Steinke, Paul Karl Willi
Suess, Gene Guy
Trautman, Jack Carl
Wall, Joseph Sennen
Wallenfeldt, Evert
Warner, H Jack
Weber, Frank E
Wege, Ann Christene
Weiss, Ronald
Winder, William Charles

WYOMING
Field, Ray A

PUERTO RICO
Toranzos, Gary Antonio

ALBERTA
Doornenbal, Hubert
Hadziyev, Dimitri
Hawrysh, Zenia Jean
Jackson, Harold
Jeremiah, Lester Earl
Stiles, Michael Edgecombe

BRITISH COLUMBIA
Duncan, Douglas Wallace
Earl, Allan Edwin
Kitson, John Aidan
Leichter, Joseph
Li-Chan, Eunice
Powrie, William Duncan

MANITOBA
Borsa, Joseph
Cenkowski, Stefan

Singh, Harwant
Vaisey-Genser, Florence Marion
White, Noel David George

NEWFOUNDLAND
Ke, Paul Jenn

NOVA SCOTIA
Ackman, Robert George
Tung, Marvin Arthur

ONTARIO
Blais, Burton W
Chapman, Ross Alexander
Clark, David Sedgefield
Davidson, Charles Mackenzie
Diosady, Levente Laszlo
Ferrier, Leslie Kenneth
Fuleki, Tibor
Lawford, George Ross
Lentz, Claude Peter
Loughheed, Thomas Crossley
Patel, Girishchandra Babubhai
Rayman, Mohamad Khalil
Sahasrabudhe, Madhu R
Stanley, David Warwick
Stavric, Bozidar
Thompson, Lilian Umale
Timbers, Gordon Ernest
Todd, Ewen Cameron David
Usborne, William Ronald
Van Den Berg, L
Veliky, Ivan Alois
Wood, Peter John

QUEBEC
Britten, Michel
Champagne, Claude P
David, Jean
Gagnon, Marcel
Goulet, Jacques
Idziak, Edmund Stefan
Kuhnlein, Harriet V
Moreau, Jean Raymond
Riel, Rene Rosaire
Sood, Vijay Kumar
Vezina, Claude

SASKATCHEWAN
Ingledew, William Michael
Sosulski, Frank Walter
Youngs, Clarence George

OTHER COUNTRIES
Brown, Delos D
Fairweather-Tait, Susan Jane
Filadelfi-Keszi, Mary Ann Stephanie
Ginis, Asterios Michael
Kaffezakis, John George
Knorr, Dietrich W
Montgomery, Morris William
Rio, Maria Esther
Segal, Mark
Smith, Meredith Ford
Valencia, Mauro Eduardo
Yang, Ho Seung

Genetics

ALABAMA
Barker, Peter Eugene
Briles, Connally Oran
Briles, David Elwood
Carter, Charlotte
Clement, William Madison, Jr
Cosper, Paula
Daniell, Henry
Denton, Tom Eugene
Dobson, F(rederick) Stephen
Dunham, Rex Alan
Fattig, W Donald
Finley, Sara Crews
Finley, Wayne House
Fuller, Gerald M
Garver, David L
Goggans, James F
Hardman, John Kemper
Hoff, Charles Jay
Horabin, Jamila Iddi
Hunter, Eric
Lemke, Paul Arenz
Locy, Robert Donald
Lubega, Seth Gasuza
McCombs, Candace Cragen
McDaniel, Gayner Raiford
McGuire, John Albert
Nielsen, Brent Lynn
Norton, Joseph Daniel
Olson, Richard Louis
Roozen, Kenneth James
Sayers, Earl Roger
Shepherd, Raymond Lee
Singh, Jarnail
Smith-Somerville, Harriett Elizabeth
Stephenson, Edwin Clark
Thompson, Jerry Nelson
Watson, Jack Ellsworth
Williams, John Watkins, III

ALASKA
Gharrett, Anthony John
Helle, John Harold (Jack)
Sagers, Cynthia Louise

Shields, Gerald Francis
Snyder, George Richard

ARIZONA
Allen, Stephen Gregory
Bartlett, Alan C
Bernstein, Harris
Birge, Edward Asahel
Church, Kathleen
Day, Arden Dexter
Dickinson, Frank N
Doane, Winifred Walsh
Dukepoo, Frank Charles
Endrizzi, John Edwin
English, Darrel Starr
Erickson, Robert Porter
Goldstein, Elliott Stuart
Guest, William C
Harris, Robert Martin
Heed, William Battles
Istock, Conrad Alan
Ito, Junetsu
Johnson, Herbert Windal
Kidwell, Margaret Gale
Kurth, Janice H
Little, John Wesley
McDaniel, Robert Gene
MacDonald, Marnie L
Mendelson, Neil Harland
Mount, David William Alexander
Muramoto, Hiroshi
Rasmussen, David Irvin
Rubis, David Daniel
Shimizu, Nobuyoshi
Turcotte, Edgar Lewis
Ward, Oscar Gardien
Ward, Samuel
Wilson, Frank Douglas
Woolf, Charles Martin
Zegura, Stephen Luke

ARKANSAS
Andrews, Luther David
Boulware, Ralph Frederick
Bourland, Freddie Marshall
Brown, A Hayden, Jr
Brown, Connell Jean
Clayton, Frances Elizabeth
Collins, Frederick Clinton
Collister, Earl Harold
Dippell, Ruth Virginia
Epstein, Joshua
Gyles, Nicholas Roy
Haggard, Bruce Wayne
Hansen, Deborah Kay
Reddy, RamaKrishna Pashuvula
Rutger, John Neil
Sears, Jack Wood
Smith, Edwin Burnell
Tave, Douglas
Wolff, George Louis
York, John Owen

CALIFORNIA
Abbott, Ursula K
Abelson, John Norman
Abplanalp, Hans
Adams, Jane N
Allard, Robert Wayne
Allen, Marcia Katzman
Almasan, Alexandru
Ames, Bruce Nathan
Ames, Giovanna Ferro-Luzzi
Amirkhanian, John David
Anderson, W French
Angell, Frederick Franklyn
Ayala, Francisco Jose
Bailey-Serres, Julia
Baker, Robert Frank
Baldy, Marian Wendorf
Barratt, Raymond William
Baxter, John Darling
Beard, Benjamin H
Bechtol, Kathleen B
Beckendorf, Steven K
Bell, A(udra) Earl
Belser, William Luther, Jr
Benzer, Seymour
Bergh, Berthold Orphie
Bernoco, Domenico
Bernstein, Sanford Irwin
Bertani, Giuseppe
Beyer, Edgar Herman
Blatt, Beverly Faye
Bliss, Fredrick Allen
Boggs, Carol L
Bonner, James (Fredrick)
Botstein, David
Bournias-Vardiabasis, Nicole
Bowen, Sarane Thompson
Bowling, Ann L
Bowman, Barry J
Boyd, James Brown
Boyer, Herbert Wayne
Bradford, G Eric
Brasch, Klaus Rainer
Brody, Stuart
Brown, Howard S
Brownell, Anna Gale
Bryant, Peter James
Buehring, Gertrude Case
Bullas, Leonard Raymond
Burrill, Melinda Jane
Busch, Robert Edward

Buth, Donald George
Calarco, Patricia G
Campbell, Allan McCulloch
Campbell, David Paul
Campbell, Judith Lynn
Cande, W Zacheus
Capron, Alexander Morgan
Carbon, John Anthony
Carpenter, Adelaide Trowbridge Clark
Catlin, B Wesley
Cavalli-Sforza, Luigi Luca
Celniker, Susan Elizabeth
Center, Elizabeth M
Chamberlin, Michael John
Chihara, Carol Joyce
Chow, Samson Ah-Fu
Christianson, Michael Lee
Chu, Irwin Y E
Clark, Alvin John
Clegg, Michael Tran
Cline, Thomas Warren
Close, Perry
Coffino, Philip
Cohen, Larry William
Cohen, Stanley Norman
Comings, David Edward
Cooper, Kenneth Willard
Corcoran, Mary Ritzel
Critchfield, William Burke
Daly, Kevin Richard
Davis, Brian Kent
Davis, Elmo Warren
Davis, Ronald Wayne
Davis, Rowland Hallowell
De Francesco, Laura
Dempsey, Wesley Hugh
Desharnais, Robert Anthony
Dragon, Elizabeth Alice Oosterom
Dvorak, Jan
Eastmond, David Albert
Ebert, Wesley W
Eckhart, Walter
Edgar, Robert Stuart
Ellstrand, Norman Carl
Emr, Scott David
Englesberg, Ellis
Ensign, Stewart Ellery
Epstein, Charles Joseph
Erickson, Jeanne Marie
Esposito, Michael Salvatore
Estilai, Ali
Evans, Ronald M
Falk, Darrel Ross
Farish, Donald James
Feldman, Jerry F
Fisher, Kathleen Mary Flynn
Fitch, Walter M
Flashman, Stuart Milton
Fluharty, Arvan Lawrence
Fogel, Seymour
Francke, Uta
Freeling, Michael
Friedmann, Theodore
Fristrom, Dianne
Fristrom, James W
Fryxell, Karl Joseph
Fujimoto, Atsuko Ono
Futch, David Gardner
Gall, Graham A E
Game, John Charles
Ganesan, Adayapalam T
Ganesan, Ann K
Gautsch, James Willard
Gelfand, David H
Gill, Ayesha Elenin
Gober, James William
Golbus, Mitchell S
Goodman, Corey Scott
Gordon, Manuel Joe
Gorman, Cornelia M
Graves, Joseph L
Green, Melvin Martin
Grody, Wayne William
Grunstein, Michael
Gutman, George Andre
Guze, Carol (Konrad Lydon)
Hackett, Adeline J
Hackney, Robert Ward
Han, Jang Hyun
Hanawalt, Philip Courtland
Hankinson, Oliver
Harding, James A
Harris, Mary Styles
Hastings, Alan Matthew
Hedgecock, Dennis
Helinski, Donald Raymond
Herskowitz, Ira
Herzenberg, Leonard Arthur
Hill, Annlia Paganini
Hill, Ray Allen
Hoch, James Alfred
Hogness, David Swenson
Hoopes, Laura Livingston Mays
Horowitz, Norman Harold
Hughes, Norman
Humaydan, Hasib Shaheen
Hyman, Bradley Clark
Jain, Subodh K
Jameson, David Lee
Jarvik, Lissy F
Jefferson, Margaret Correan
Jenkins, Burton Charles
Johnson, Ben Francis
Johnson, Bertil Lennart

Johnson, Carl William
Johnston, George Robert
Jones, Gary Edward
Kaiser, Armin Dale
Kane, Susan Elizabeth
Kaplan, William David
Kashyap, Tapeshwar S
Kedes, Laurence H
Kevles, Daniel Jerome
Kiger, John Andrew, Jr
King, Mary-Claire
Kinloch, Bohun Baker, Jr
Kirschbaum, Joel Bruce
Knudson, Gregory Blair
Konopka, Ronald J
Kroman, Ronald Avron
Kudo, Shinichi
Kustu, Sydney Govons
Laben, Robert Cochrane
Laidlaw, Harry Hyde, Jr
Lange, Kenneth L
Latimer, Howard Leroy
Law, George Robert John
Lebo, Roger Van
Lederberg, Esther Miriam
Ledig, F Thomas
Lefevre, George, Jr
Lengyel, Judith Ann
Levinson, Arthur David
Libby, William John, (Jr)
Lieb, Margaret
Lindsley, Dan Leslie
Lipshitz, Howard David
Long, Sharon Rugel
Loo, Melanie Wai Sue
Loughman, William D
Luckock, Arlene Suzanne
Lugo, Tracy Gross
McClelland, Michael
McClenaghan, Leroy Ritter, Jr
McEwen, Joan Elizabeth
McFarlane, John Spencer
Mack, Thomas McCulloch
McLaughlin, Calvin Sturgis
Mangan, Jerrome
Marsh, James Lawrence
Martin, David William, Jr
Martin, George Steven
Martineau, Belinda M
Martinek, George William
Matthews, Beverly Bond
Maxwell, Joyce Bennett
Melnick, Michael
Merchant, Sabeeha
Merriam, John Roger
Meyerowitz, Elliot Martin
Miller, Alexander
Mohandas, Thuluvancheri
Morse, Daniel E
Mortimer, Robert Keith
Nelson, Keith
Nelson-Rees, Walter Anthony
Neufeld, Elizabeth Fondal
Nunney, Leonard Peter
Ohno, Susumu
Ohnuki, Yasushi
Okada, Tadashi A
Ow, David Wing
Owen, Ray David
Packman, Seymour
Page, Robert Eugene, Jr
Pedersen, Roger Arnold
Pendse, Pratapsinha C
Perkins, David Dexter
Phinney, Bernard Orrin
Powers, Dennis A
Pratt, David
Puhalla, John Edward
Puri, Yesh Paul
Qualset, Calvin Odell
Quiros, Carlos F
Rachmeler, Martin
Raju, Namboori Bhaskara
Ratty, Frank John, Jr
Rawal, Kanti M
Rearden, Carole Ann
Riblet, Roy Johnson
Rick, Charles Madeira, Jr
Rimoin, David (Lawrence)
Rinehart, Robert R
Rio, Donald C
Rose, Michael Robertson
Rosen, Howard
Rosenberg, Marvin J
Rosenthal, Allan Lawrence
Ross, Ian Kenneth
Rossi, John Joseph
Rotter, Jerome Israel
Roy-Burman, Pradip
Ryder, Edward Jonas
Ryder, Oliver A
Sacher, Robert Francis
Sadanaga, Kiyoshi
Saier, Milton H, Jr
St Lawrence, Patricia
Sassaman, Clay Alan
Schekman, Randy W
Schlegel, Robert John
Schneider, Edward Lewis
Schulke, James Darrell
Scott, Matthew P
Seegmiller, Jarvis Edwin
Senda, Mototaka
Sensabaugh, George Frank

Shapiro, Larry Jay
Sharp, Gary Duane
Shultz, Fred Townsend
Sidhu, Gurmel Singh
Silagi, Selma
Silverman, Michael Robert
Simmel, Edward Clemens
Simons, Robert W
Simpson, Robert Blake
Singh, Arjun
Sinibaldi, Ralph Michael
Slatkin, Montgomery (Wilson)
Slavkin, Harold Charles
Smith, Douglas Wemp
Smith, Helene Sheila
Smith, Kendric Charles
Smith, Steven Sidney
Snow, Sidney Richard
Sokoloff, Alexander
Somerville, Christopher Roland
Soost, Robert Kenneth
Sparkes, Robert Stanley
Spence, Mary Anne
Spieler, Richard Arno
Spieth, Philip Theodore
Stallcup, Michael R
Stansfield, William D
Stormont, Clyde J
Straus, Daniel Steven
Strauss, Ellen Glowacki
Strickberger, Monroe Wolf
Strobel, Edward
Stumph, William Edward
Sung, Zinmay Renee
Syvanen, Michael
Tallman, John Gary
Taylor, Charles Ellett
Teuber, Larry Ross
Thomas, Heriberto Victor
Thompson, David J
Thompson, Lawrence Hadley
Thwaites, William Mueller
Torfs, Claudine Pierette
Valenzuela, Pablo
Van Kuyk, Robert William
Vogt, Peter Klaus
Vyas, Girish Narmadashankar
Wallace, Robert Bruce
Warren, Don Cameron
Waser, Nickolas Merritt
Wechsler, Steven Lewis
Wehr, Carl Timothy
Weinberg, Barbara Lee Huberman
Weis, Arthur Edward
Welch, James Edward
Wheelis, Mark Lewis
Whissell-Buechy, Dorothy Y E
Whitaker, Thomas Wallace
Wiebe, Michael Eugene
Wiktorowicz, John Edward
Wilcox, Bruce Alexander
Wilcox, Gary Lynn
Williams, Bobby Joe
Williams, Julian Carroll
Wills, Christopher J
Wolff, Sheldon
Woodruff, David Scott
Yager, Janice L Winter
Ying, Kuang Lin
Yoshida, Akira
Youngblom, Janey Heg-Joung
Yund, Mary Alice
Zamenhof, Patrice Joy
Zary, Keith Wilfred
Zimm, Georgianna Grevatt
Zipser, David
Zuccarelli, Anthony Joseph
Zuckerkandl, Emile

COLORADO
Balbinder, Elias
Barrett, Dennis
Bock, Jane Haskett
Brandom, William Franklin
Brinks, James S
Clark, Roger William
Crumpacker, David Wilson
DeFries, John Clarence
Deitrich, Richard Adam
Dixon, Linda Kay
Eberhart, Steve A
Fechner, Gilbert Henry
Gardiner, Katheleen Jane
Heim, Werner George
Hughes, Harrison Gilliatt
Jaehning, Judith A
Johnson, Thomas Eugene
Jones, Carol A
Kano-Sueoka, Tamiko
Kao, Fa-Ten
Keim, Wayne Franklin
Lapitan, Nora L
Linhart, Yan Bohumil
Lynch, Carol Becker
Markert, Clement Lawrence
Mitton, Jeffry Bond
Morse, Helvise Glessner
Morse, Melvin Laurance
Nash, Donald Joseph
Nora, James Jackson
Ogg, James Elvis
Parma, David Hopkins
Patterson, David
Peeples, Earle Edward

Pettijohn, David E
Prescott, David Marshall
Puck, Mary Hill
Puck, Theodore Thomas
Quick, James S
Robinson, Arthur
Schanfield, Moses Samuel
Sinensky, Michael
Stein, Gretchen Herpel
Sueoka, Noboru
Taylor, Austin Laurence
Townsend, Charley E
Tsuchiya, Takumi
Vandenberg, Steven Gerritjan
Waldren, Charles Allen
Wehner, Jeanne M
Welsh, James Ralph
Wilson, Vincent L
Wood, William Barry, III

CONNECTICUT
Adelberg, Edward Allen
Amacher, David E
Anagnostakis, Sandra Lee
Ashley, Terry Fay
Astavanis-Tsakonas, Spyridon
Bachmann, Barbara Joyce
Berg, Claire M
Berlyn, Mary K Berry
Bothwell, Alfred Lester Meador
Brandsma, Janet Lousie
Breg, William Roy
Chovnick, Arthur
Donady, J James
Eisenstadt, Jerome Melvin
Fenton, Wayne Alexander
Flavell, Richard Anthony
Froshauer, Susan
Fu, Wei-ning
Galbraith, Donald Barrett
Geiger, Jon Ross
Ginsburg, Benson Earl
Golub, Efim I
Gordon, Philip N
Greenblatt, Irwin M
Grindley, Nigel David Forster
Halaban, Ruth
Hanson, Earl Dorchester
Holsinger, Kent Eugene
Kavathas, Paula
Kidd, Kenneth Kay
Koerting, Lola Elisabeth
Kreizinger, Jean Dolloff
Les, Donald Henry
Liskay, Robert Michael
Longwell, Arlene Crosby (Mazzone)
Low, Kenneth Brooks, Jr
McCorkle, George Maston
Martinez, Robert Manuel
Matheson, Dale Whitney
Pawelek, John Mason
Pierro, Louis John
Poole, Andrew E
Potluri, Venkateswara Rao
Poulson, Donald Frederick
Powell, Jeffrey Robert
Proctor, Alan Ray
Radding, Charles Meyer
Ray, Verne A
Ricciuti, Florence Christine
Rice, Frank J
Rishell, William Arthur
Ruddle, Francis Hugh
Rupp, W Dean
Sarfarazi, Mansoor
Schacter, Bernice Zeldin
Schultz, R Jack
Schwinck, Ilse
Shapiro, Nathan
Silander, John August, Jr
Slayman, Carolyn Walch
Somes, Ralph Gilmore, Jr
Stevenson, Harlan Quinn
Summers, Wilma Poos
Truesdell, Susan Jane
Wagner, Gunter Paul
Ward, David Christian
Wheeler, Bernice Marion
Wilson, Gary August

DELAWARE
Borgaonkar, Digamber Shankarrao
Eleuterio, Marianne Kingsbury
Francis, David W
Gaffney, Patrick M
Gatenby, Anthony Arthur
Gould, Adair Brasted
Hobgood, Richard Troy, Jr
Hodson, Robert Cleaves
Hoover, Dallas Gene
Irr, Joseph David
Kundt, John Fred
Lighty, Richard William
Martin-Deleon, Patricia Anastasia
Myoda, Toshio Timothy
Pene, Jacques Jean
Sheppard, David E
White, Harold Bancroft, III

DISTRICT OF COLUMBIA
Ampy, Franklin
Bergmann, Fred Heinz
Bundy, Bonita Marie
Calvert, Allen Fisher

Genetics (cont)

Chase, Gary Andrew
Chen, H R
Dunkel, Virginia Catherine
Dutta, Sisir Kamal
Eastwood, Basil R
Gelmann, Edward P
Golin, John Euster
Haskins, Caryl Parker
Hawkins, Morris, Jr
Hill, Richard Norman
Hill, Walter Ernest
Hines, Pamela Jean
Holzman, Gerald Bruce
Kumar, Ajit
Landman, Otto Ernest
Lorimer, Nancy L
May, Sterling Randolph
Murray, Robert Fulton, Jr
Myrianthopoulos, Ntinos
Nightingale, Elena Ottolenghi
Ohi, Seigo
Poillon, William Neville
Policansky, David J
Prival, Michael Joseph
Ralls, Katherine Smith
Rothman, Sara Weinstein
Salzman, Lois Ann
Santamour, Frank Shalvey, Jr
Schaeff, Catherine Margaret
Setlow, Valerie Petit
Simopoulos, Artemis Panageotis
Thomas, Walter Ivan
Townsend, Alden Miller
Winter, William Phillips
Wrathall, Jean Rew
Zimmer, Elizabeth Anne

FLORIDA

Adams, Roger Omar
Baker, Richard H
Barnett, Ronald David
Baumbach, Lisa Lorraine
Baylis, John Robert, Jr
Binninger, David Michael
Buslig, Bela Stephen
Chase, Christine Davis
Davis, Ralph Lanier
Dean, Charles Edgar
DeBusk, A Gib
DeKloet, Siwo R
Donahue, Roger P
Duggan, Dennis E
Edwardson, John Richard
Emmel, Thomas C
Ferl, Robert Joseph
Fischler, Drake Anthony
Fresquez, Catalina Lourdes
Gaunt, Stanley Newkirk
Gilbert-Barness, Enid F
Gilman, Lauren Cundiff
Goddard, Ray Everett
Greer, Sheldon
Hall, Harlan Glenn
Hauswirth, William Walter
Hecht, Frederick
Herskowitz, Irwin Herman
Hinson, Kuell
Honeyman, Merton Seymour
Hoy, Marjorie Ann
Huijing, Frans
Ingram, Lonnie O'Neal
Johnson, F Clifford
Jones, David Alwyn
Jones, John Paul
Kaul, Rajinder K
Keppler, William J
Kitzmiller, James Blaine
Kopelovich, Levy
Kossuth, Susan
Kuhn, David Truman
Lai, Patrick Kinglun
Laipis, Philip James
Lee, Doh-Yeel
Lessios, Harilaos Angelou
Lewin, Alfred S
Linden, Duane B
Long, Alan Jack
Luykx, Peter (van Oosterzee)
McCormack, Wayne Thomas
Marcus, Nancy Helen
Martin, Franklin Wayne
Miyamoto, Michael Masao
Moffa-White, Andrea Marie
Mortensen, John Alan
Moyer, Richard W
Muench, Karl Hugo
Norden, Allan James
Osburn, Richard Lee
Peacock, Hugh Anthony
Pfahler, Paul Leighton
Polson, Charles David Allen
Reid, Parlane John
Richmond, Rollin Charles
Rife, David Cecil
Roberts, Thomas L
Roess, William B
Rumbach, William Ervin
Schank, Stanley Cox
Scott, John Warner
Seawright, Jack Arlyn
Shanmugam, Keelnatham
   Thirunavukkarasu
Shirk, Paul David
Smith, Rex L
Stine, Gerald James
Stock, David Allen
Stroh, Robert Carl
Taylor, J(ames) Herbert
Teas, Howard Jones
Tedesco, Thomas Albert
Terranova, Andrew Charles
Thomason, David Morton
Tracey, Martin Louis, Jr
Travis, Joseph
Voelly, Richard Walter
Wagoner, Dale E
Walker, Thomas Jefferson
Warren, Richard Joseph
White, Timothy Lee
Wilfret, Gary Joe
Wireman, Kenneth

GEORGIA

Adkison, Linda Russell
Anderson, Wyatt W
Andrews, Lucy Gordon
Arnold, Jonathan
Asmussen, Marjorie A
Avise, John C
Baumstark, Barbara Ruth
Bennett, Sara Neville
Bongarten, Bruce C
Brinton, Margo A
Brownell, George H
Bryan, John Henry Donald
Burk, Lawrence G
Byrd, J Rogers
Case, Mary Elizabeth
Chen, Andrew Tat-Leng
Clark, Flora Mae
Crenshaw, John Walden, Jr
Crouse, Gray F
Danner, Dean Jay
Davis, Herbert L, Jr
Dinus, Ronald John
Duncan, Ronny Rush
Dusenbery, David Brock
Elmer, William Arthur
Elsas, Louis Jacob, II
Falek, Arthur
Fleming, Attie Anderson
Fridovich-Keil, Judith Lisa
Gardner, Arthur Wendel
Giles, Norman Henry
Glover, Claiborne V C, III
Golden, Ben Roy
Hall, Dwight Hubert
Hammons, Ray Otto
Hamrick, James Lewis
Hanna, Wayne William
Hassold, Terry Jon
Hollowell, Joseph Gurney, Jr
Howe, Henry Branch, Jr
Hugus, Barbara Hendrick
Jacobs, Patricia Ann
Jinks-Robertson, Sue
Kraus, John Franklyn
Kushner, Sidney Ralph
Lee, Joshua Alexander
Lucchesi, John Charles
McCartney, Morley Gordon
Madden, John Joseph
Marks, Henry L
Mason, James Michael
Meagher, Richard Brian
Miller, John David
Miller, Paul George
Mixon, Aubrey Clifton
Neville, Walter Edward, Jr
Nickerson, John Munro
Oakley, Godfrey Porter, Jr
Oliver, James Henry, Jr
Pandey, Kailash N
Patterson, Rosalyn Mitchell
Ray, Charles, Jr
Reines, Daniel
Satya-Prakash, K L
Schierman, Louis W
Schmidt, Gregory Wayne
Shimkets, Lawrence Joseph
Sluder, Earl Ray
Snell, Terry Wayne
Summers, Anne O
Thompson, James Marion
Thompson, Peter Ervin
Wallace, Douglas Cecil
Warren, Stephen Theodore
Washburn, Kenneth W
Weaver, James B, Jr
Wessler, Susan R
Whitney, J(ohn) Barry, III
Widstrom, Neil Wayne
Wilson, Jeffrey Paul
Wistrand, Harry Edwin
Woessner, Ronald Arthur
Woodley, Charles Lamar
Youngman, Philip John

HAWAII

Albert, Henrick Horst
Ashton, Geoffrey C
Chung, Chin Sik
Dalton, Howard Clark
Goodman, Madeleine Joyce
Grove, John Sinclair
Heinz, Don J
Hunt, John A
Kathariou, Sophia
Kikudome, Gary Yoshinori
Malecha, Spencer R
Mi, Ming-Pi
Reimer, Diedrich
Roderick, George Karlsson
Rotar, Peter P
Sagawa, Yoneo
Scott, John Francis
Simon, Christine Mae
Sipes, Brent Steven
Stuart, William Dorsey

IDAHO

Christian, Ross Edgar
Dahmen, Jerome J
Farrell, Larry Don
Fins, Lauren
Forbes, Oliver Clifford
Hansen, Leon A
McCune, Mary Joan Huxley
Myers, James Robert
Pavek, Joseph John
Pearson, Lorentz Clarence
Stephens, Trent Dee
Tullis, James Earl
Winston, Vern

ILLINOIS

Alexander, Denton Eugene
Alexander, Nancy J
Allison, David C
Amarose, Anthony Philip
Aprison, Barry Steven
Ausich, Rodney L
Bailey, Zeno Earl
Baumgardner, Kandy Diane
Beck, Sidney L
Becker, Michael Allen
Bell, Clara G
Bennett, Cecil Jackson
Bernard, Richard Lawson
Bernardo, Rex Novero
Bigelis, Ramunas
Bouck, Noel
Bowman, James E
Bradford, Laura Sample
Brewen, J Grant
Briggs, Robert Wilbur
Briles, Worthie Elwood
Brockman, Herman E
Casadaban, Malcolm John
Chakrabarty, Ananda Mohan
Charlesworth, Brian
Charlesworth, Deborah
Cole, Michael Allen
Conner, Jeffrey Keating
Cummings, Michael R
Daniel, William L
Davidson, Richard Laurence
Desborough, Sharon Lee
Doering, Jeffrey Louis
Dorus, Elizabeth
Dudley, John Wesley
Englert, Du Wayne Cleveland
Esposito, Rochelle E
Farnsworth, Marjorie Whyte
Fast, Dale Eugene
Fennewald, Michael Andrew
Ferguson, Edwin Louis
Fernando, Rohan Luigi
Fleisher, Lynn Dale
Fox, James David
Francis, Bettina Magnus
Frischer, Henri
Fuchs, Elaine V
Garber, Edward David
Gardner, Jeffery Fay
Garwood, Douglas Leon
Gaskins, H Rex
Geer, Billy W
Gendel, Steven Michael
Gerdy, James Robert
Giordano, Tony
Gorsic, Joseph
Grahn, Douglas
Gravett, Howard L
Griesbach, Robert Anthony
Grossman, Michael
Grundbacher, Frederick John
Hirsch, Jerry
Holland, Louis Edward, II
Hooker, Arthur Lee
Huberman, Eliezer
Hymowitz, Theodore
Jones, Carl Joseph
Juergensmeyer, Elizabeth B
Jump, Lorin Keith
Kang, David Soosang
Katz, Alan Jeffrey
Kaufman, Thomas Charles
Keim, Barbara Howell
King, Robert Charles
Koch, Elizabeth Anne
Kurtzman, Cletus Paul
Laffler, Thomas G
Lamppa, Gayle K
Laughnan, John Raphael
Leffler, Harry Rex
Lerner, Jules
Liebman, Susan Weiss
Lopes, John Manuel
Ma, Te Hsiu
McCorquodale, Maureen Marie
Mahowald, Anthony P
Marczynska, Barbara Mary
Markovitz, Alvin
Melvold, Roger Wayne
Mets, Laurens Jan
Miller, Darrell Alvin
Miller, Robert Verne
Myers, Oval, Jr
Nadler, Henry Louis
Nadler, Steven Anthony
Nagylaki, Thomas Andrew
Nair, Shankar P
Nanney, David Ledbetter
Natalini, John Joseph
Nickell, Cecil D
Nyberg, Dennis Wayne
Panasenko, Sharon Muldoon
Patterson, Earl Byron
Phillips, Christopher Alan
Plate, Charles Alfred
Plewa, Michael Jacob
Pratt, Charles Walter
Propst, Catherine Lamb
Prout, Timothy
Rasmusen, Benjamin Arthur
Rice, Thomas B
Rink, George
Robertson, Hugh Mereth
Robison, Norman Glenn
Roderick, William Rodney
Roth, Robert Mark
Rowley, Janet D
Ruddat, Manfred
Rupprecht, Kevin Robert
Sachs, Martin M
Salisbury, Glenn Wade
Sargent, Malcolm Lee
Schreiber, Hans
Shanks, Roger D
Shapiro, James Alan
Simon, Ellen McMurtrie
Smith, Patricia Anne
Spear, Brian Blackburn
Spiess, Eliot Bruce
Spofford, Janice Brogue
Steffensen, Dale Marriott
Strauss, Bernard S
Takahashi, Joseph S
Thompson, Vinton Newbold
Tripathi, Satish Chandra
Troyer, Alvah Forrest
Tuveson, Robert Williams
Vandehey, Robert C
Vary, Patricia Susan
Vodkin, Lila Ott
Vodkin, Michael Harold
Wade, Michael John
Weber, David Frederick
Whitt, Dixie Dailey
Whitt, Gregory Sidney
Widholm, Jack Milton
Yarger, James G
Yokoyama, Shozo

INDIANA

Anderson-Mauser, Linda Marie
Axtell, John David
Bard, Martin
Beineke, Walter Frank
Bender, Harvey Alan
Bennetzen, Jeffrey Lynn
Bixler, David
Bockrath, Richard Charles, Jr
Bohren, Bernard Benjamin
Bonner, James Jose
Boston, Andrew Chester
Brush, F(ranklin) Robert
Castleberry, Ron M
Cherbas, Peter Thomas
Chiscon, J Alfred
Chrisman, Charles Larry
Christian, Joe Clark
Cole, Thomas A
Conneally, P Michael
Dantzig, Anne H
Edenberg, Howard Joseph
Eisenstein, Barry I
Engstrom, Lee Edward
Gallun, Robert Louis
Garriott, Michael Lee
Garton, David Wendell
Gastony, Gerald Joseph
Gidda, Jaswant Singh
Glover, David Val
Goff, Charles W
Goldstein, David Joel
Guthrie, Catherine Shirley Nicholson
Harvey, William Homer
Hegeman, George D
Hendrix, Jon Richard
Hershberger, Charles Lee
Hickey, William August
Hodes, Marion Edward
Hudock, George Anthony
Hyde, David Russell
Janick, Jules
Karn, Robert Cameron
Kinzel, Jerry J
Kwon, Byoung Se
Larsen, Steven H
Levinthal, Mark
McClure, Polley Ann
Martin, Truman Glen

Mertens, Thomas Robert
Nielsen, Niels Christian
Nyquist, Wyman Ellsworth
Ohm, Herbert Willis
Pak, William Louis
Palmer, Catherine Gardella
Patterson, Fred La Vern
Polley, Lowell David
Potter, Rosario H Yap
Quaid, Kimberly Andrea
Queener, Stephen Wyatt
Rai, Karamjit Singh
Reed, Terry Eugene
Rhoades, Marcus Morton
Roman, Ann
Santerre, Robert Frank
Schaible, Robert Hilton
Schwartz, Drew
Shaw, Margery Wayne
Sheffer, Richard Douglas
Simon, Edward Harvey
Skjold, Arthur Christopher
Somerville, Ronald Lamont
Spieth, John
Stewart, Terry Sanford
Stuart, Jeffrey James
Tan, James Chien-Hua
Taylor, Milton William
Tessman, Irwin
Tigchelaar, Edward Clarence
Tischfield, Jay Arnold
Tomes, Mark Louis
Tsai, Chia-Yin
Wappner, Rebecca Sue
Watson, Maxine Amanda
Weaver, David Dawson
Wilcox, Frank H
Yen, Terence Tsin Tsu
Zuckerman, Steven H

IOWA
Atherly, Alan G
Bamrick, John Francis
Benbow, Robert Michael
Berger, Philip Jeffrey
Carlson, Wayne R
Christian, Lauren L
Dalton, Lonnie Gene
Duvick, Donald Nelson
Ford, Clark Fugier
Frankel, Joseph
Grant, David Miller
Gussin, Gary Nathaniel
Hall, Richard Brian
Hegmann, Joseph Paul
Hoffmann, Richard John
Imsande, John
Jordan, Diane Kathleen
Kieso, Robert Alfred
Krafsur, Elliot Scoville
Lamkey, Kendall Raye
Lucas, Gene Allan
McNeill, Michael John
Malone, Robert Edward
Marshall, William Emmett
Meeker, David Lynn
Menninger, John Robert
Milkman, Roger Dawson
Miller, Wilmer Jay
Palmer, Reid G
Pattee, Peter A
Peterson, Peter Andrew
Rhead, William James
Robertson, Donald Sage
Rodermel, Steven Robert
Russell, Wilbert Ambrick
Schnable, Patrick S
Stieler, Carol Mae
Sunshine, Melvin Gilbert
Thorne, John Carl
Tomes, Dwight Travis
Touchberry, Robert Walton
Walker, Jean Tweedy
Wang, Wei-Yeh
Weisman, Lois Sue
Welshons, William John
Wu, Chun-Fang
Wunder, William W

KANSAS
Bassi, Sukh D
Bode, Vernon Cecil
Chitnis, Parag Ram
Clayberg, Carl Dudley
Craig, James Verne
Denell, Robin Ernest
Epp, Melvin David
Gill, Bikram Singh
Goldberg, Ivan D
Grebe, Janice Durr
Haufler, Christopher Hardin
Heiner, Robert E
Iandolo, John Joseph
Jackson, Sharon Wesley
Kaufman, Glennis Ann
Kinsey, John Aaron, Jr
Leslie, John Franklin
Liang, George H
Mock, James Joseph
Pittenger, Thad Heckle, Jr
Roufa, Donald Jay
Schalles, Robert R
Schlager, Gunther
Shankel, Delbert Merrill

Stetler, Dean Allen
Van Haverbeke, David F
Wassom, Clyde E
Weir, John Arnold
Wheat, John David
Wolf, Thomas Michael
Wolfe, Herbert Glenn

KENTUCKY
Collins, Glenn Burton
Cotter, William Bryan, Jr
Davidson, Jeffrey Neal
Dickson, Robert Carl
Goodwill, Robert
Maiti, Indu B
Martin, Nancy Caroline
Moore, Allen Jonathan
Perlin, Michael Howard
Poneleit, Charles Gustav
Rawls, John Marvin, Jr
Sheen, Shuh-Ji
Sisken, Jesse Ernest
Stuart, James Glen
Williams, Arthur Lee
Wolfson, Alfred M
Yungbluth, Thomas Alan

LOUISIANA
Adams, John Clyde
Bennett, Joan Wennstrom
Bhatnagar, Deepak
Chambers, Doyle
Christian, James A
Ellgaard, Erik G
French, Wilbur Lile
Griswold, Kenneth Edwin, Jr
Harlan, Jack Rodney
Hayes, Donald H
Humes, Paul Edwin
Jazwinski, S Michal
Karam, Jim Daniel
Kern, Clifford H, III
Koonce, Kenneth Lowell
Lee, William Roscoe
Lin, James C H
Lucas, Myron Cran
Mizell, Merle
Pelias, Mary Zengel
Ramsey, Paul Roger
Rinderer, Thomas Earl
Shepherd, Hurley Sidney
Sinha, Sudhir K
Thurmon, Theodore Francis
Tipton, Kenneth Warren
Tucker, Kenneth Wilburn
Warters, Mary

MAINE
Bailey, Donald Wayne
Bernstein, Seldon Edwin
Blake, Richard D
Chai, Chen Kang
Champlin, Arthur Kingsley
Davisson, Muriel Trask
Doolittle, Donald Preston
Dowse, Harold Burgess
Eicher, Eva Mae
Fox, Richard Romaine
Guidi, John Neil
Harris, Paul Chappell
Harrison, David Ellsworth
Hoppe, Peter Christian
Jerkofsky, Maryann
Kaliss, Nathan
LaMarche, Paul H
Les, Edwin Paul
Mobraaten, Larry Edward
Paigen, Beverly Joyce
Paigen, Kenneth
Ringo, John Moyer
Roberts, Franklin Lewis
Roderick, Thomas Huston
Russell, Elizabeth Shull
Snell, George Davis
Steinhart, William Lee

MARYLAND
Adams, Junius Greene, III
Adhya, Sankar L
Altevogt, Raymond Fred
Amato, R Stephen S
Aycock, Marvin Kenneth, Jr
Barnett, Audrey
Barnhart, Benjamin J
Baron, Louis Sol
Basta, Milan
Battey, James F
Bethke, Bruce Donald
Bias, Wilma B
Bigger, Cynthia Anita Hopwood
Bohr, Vilhelm Alfred
Borgia, Gerald
Bottino, Paul James
Boyer, Samuel H, IV
Bresler, Jack Barry
Breyere, Edward Joseph
Briggle, Leland Wilson
Brown, Kenneth Stephen
Buck, Raymond Wilbur, Jr
Camerini-Otero, Rafael Daniel
Chan, Wai-Yee
Charnas, Lawrence Richard
Chattoraj, Dhruba Kumar
Chen, T R

Childs, Barton
Coe, Gerald Edwin
Cohen, Bernice Hirschhorn
Cohen, Maimon Moses
Collins, Francis Sellers
Conner, George William
Cowan, Elliot Paul
Dean, Ann
Dean, Jurrien
Devine, Thomas Edward
DiPaolo, Joseph Amedeo
Duncan, William Raymond
Elespuru, Rosalie K
Elgin, James H, Jr
Falke, Ernest Victor
Falkenstein, Kathy Fay
Fearon, Douglas T
Felix, Jeanette S
Fire, Andrew Zachary
Gahl, William A
Galletta, Gene John
Garges, Susan
Genys, John B
Gethmann, Richard Charles
Gilden, Raymond Victor
Goldman, David
Gottesman, Michael
Graham, Dale Elliott
Gray, David Bertsch
Gray, Paulette S
Greenberg, Judith Horovitz
Griesbach, Robert James
Guss, Maurice Louis
Gwynn, Edgar Percival
Hamer, Dean H
Hansen, Carl Tams
Harrell, Reginal M
Hartman, Philip Emil
Haynes, Kathleen Galante
Hejtmancik, James Fielding
Hoeg, Jeffrey Michael
Hoffman, Harold A
Holbrook, Nikki J
Holland, Mark Alexander
Holtzman, Neil Anton
Horowitz, Jill Ann
Hoyer, Leon William
Huang, Pien-Chien
Hudson, Lynn Diane
Huppi, Konrad E
Imberski, Richard Bernard
James, Stephanie Lynn
Jones, Theodore Charles
Jordan, Elke
Kaiser-Kupfer, Muriel I
Kelly, Thomas J
Kernaghan, Roy Peter
Kety, Seymour S
Kimball, Paul Clark
Kindt, Thomas James
Korper, Samuel
Koshland, Douglas E
Kraemer, Kenneth H
Kulkarni, Ashok Balkrishna
Kwiterovich, Peter O, Jr
Kyle, Wendell H(enry)
Lacy, Ann Matthews
Landsman, David
Leffell, Mary S
Lerman, Michael Isaac
Levin, Barbara Chernov
Levine, Arthur Samuel
Lewis, Herman William
Liu, Pu
Lunney, Joan K
McBride, Orlando W
McGraw, Patricia Mary
McKenney, Keith Hollis
McKeon, Catherine
McMacken, Roger
MacQuillan, Anthony M
Margulies, David Harvey
Marsh, David George
Martin, Robert G
Mayer, Vernon William, Jr
Melera, Peter William
Merril, Carl R
Migeon, Barbara Ruben
Mischke, Barbara Suzanne
Misra, Rohini Rita
Mittal, Kamal Kant
Money, John William
Moorhead, Paul Sidney
Moseman, John Gustav
Nash, Howard Allen
Naylor, Alfred F
Newrock, Kenneth Matthew
O'Brien, Stephen James
Plato, Chris C
Potter, Michael
Price, Samuel
Ramagopal, Subbanaidu
Remondini, David Joseph
Robbins, April Ruth
Robbins, Jay Howard
Robbins, Robert John
Robison, Wilbur Gerald, Jr
Rosner, Judah Leon
Russek-Cohen, Estelle
Sack, George H(enry), Jr
Sauer, Brian L
Scheinberg, Sam Louis
Schiff, Nathan Mark
Schuellein, Robert Joseph

Seydel, Frank David
Shaw, Richard Franklin
Shearn, Allen David
Silverman, Jeffrey Alan
Small, Judy Ann
Smith, Gilbert Howlett
Smith, Hamilton Othanel
Smith-Gill, Sandra Joyce
Snope, Andrew John
Solomon, Joel Martin
Sprott, Richard Lawrence
Stonehill, Elliott H
Strathern, Jeffrey Neal
Sussman, Daniel Jesse
Swergold, Gary David
Tester, Cecil Fred
Thorgeirsson, Snorri S
Tjio, Joe Hin
Uphoff, Delta Emma
Vlahakis, George
Voll, Mary Jane
Walker, Mary Clare
Wilder, Ronald Lynn
Wolf, Richard Edward, Jr
Woods, Lewis Curry, III
Wu, Henry Chi-Ping
Yasbin, Ronald Eliott

MASSACHUSETTS
Alper, Chester Allan
Atkinson-Templeman, Kristine Hofgren
Ausubel, Frederick Michael
Bachinsky, David Robert
Backman, Keith Cameron
Banderet, Louis Eugene
Bawa, Kamaljit S
Beckwith, Jonathan Roger
Beckwitt, Richard David
Beggs, Alan H, III
Bieber, Frederick Robert
Breakefield, Xandra Owens
Brown, Beverly Ann
Cantor, Charles R
Chin, William W
Clark, Arnold M
Comer, M Margaret
Cooper, Geoffrey Mitchell
Demain, Arnold Lester
Deutsch, Curtis Kimball
Dick, Stanley
Dorf, Martin Edward
Dubey, Devendra P
Feig, Larry Allen
Fields, Bernard N
Fink, Gerald Ralph
Fowler, Elizabeth
Fox, Maurice Sanford
Fox, Thomas Oren
Fox, Thomas Walton
Fulton, Chandler Montgomery
Goetinck, Paul Firmin
Goldberg, Edward B
Goldstein, Lawrence S B
Good, Carl M, III
Haber, James Edward
Hall, Jeffrey Connor
Hartl, Daniel L
Hattis, Dale B
Hecht, Norman R
Hexter, William Michael
Hoffmann, George Robert
Holmes, Helen Bequaert
Horvitz, Howard Robert
Hotchkiss, Rollin Douglas
Housman, David E
Ives, Philip Truman
Jameson, James Larry
Kafatos, Fotis C
Kelner, Albert
Kelton, Diane Elizabeth
King, Jonathan (Alan)
Klekowski, Edward Joseph, Jr
Knipe, David Mahan
Kolakowski, Donald Louis
Kolodner, Richard David
Koul, Omanand
Kunkel, Louis M
Lawrence, Jeanne Bentley
Ledbetter, Mary Lee Stewart
Leder, Philip
Lee, Shwu-Luan
Lemontt, Jeffrey Fielding
Lerman, Leonard Solomon
Levy, Deborah Louise
Lewontin, Richard Charles
Li, Frederick P
Lovett, Charles McVey
Ludwin, Isadore
Lyerla, Jo Ann Harding
Lyerla, Timothy Arden
Ma, Nancy Shui-Fong
Maas, Richard Louis
Madhavan, Kornath
Malamy, Michael Howard
Mange, Arthur P
Marcum, James Benton
Marinus, Martin Gerard
Merritt, Robert Buell
Miller, Lynn
Mitchell, David Hillard
Moolten, Frederick London
Mulcahy, David Louis
Myers, Richard Hepworth
Nixon, Charles William

Genetics (cont)

Nogueira, Christine Pietraroia
Papaioannou, Virginia Eileen
Pardue, Mary Lou
Petri, William Hugh
Petricciani, John C
Pierce, Edward Ronald
Poteete, Anthony Robert
Read, Dorothy Louise
Reiner, Albey M
Riley, Monica
Rollins, Reed Clark
Roy, Sayon
Sager, Ruth
Schaefer, Ernst J
Seidman, Jonathan G
Seyfried, Thomas Neil
Sidman, Richard Leon
Siegel, Eli Charles
Signer, Ethan Royal
Smith, Cassandra Lynn
Snow, Beatrice Lee
Solbrig, Otto Thomas
Stein, Diana B
Strauss, William Mark
Streilein, Jacob Wayne
Struhl, Kevin
Swanson, Carl Pontius
Szoka, Paula Rachel
Szostak, Jack William
Tamarin, Robert Harvey
Tilley, Stephen George
Timberlake, William Edward
Todd, Neil Bowman
Tolan, Dean Richard
Townes, Philip Leonard
Tsuang, Ming Tso
Tsung, Yean-Kai
Vankin, George Lawrence
Warner, Carol Miller
Weinstein, Alexander
Wertz, Dorothy C
Widmayer, Dorothea Jane
Williamson, Patrick Leslie
Witman, George Bodo, III
Wolf, Merrill Kenneth
Young, Delano Victor
Zannis, Vassilis I

MICHIGAN
Aaron, Charles Sidney
Abraham, Irene
Adams, Jack Donald
Adams, Julian Philip
Allen, Sally Lyman
Al Saadi, A Amir
Aminoff, David
Arking, Robert
Bach, Shirley
Bacon, Larry Dean
Barbour, Stephen D
Baumiller, Robert C
Ben-Yoseph, Yoav
Bertrand, Helmut
Bloom, Miriam
Bouma, Hessel, III
Bowers, Maynard C
Burnett, Jean Bullard
Bush, Guy L
Butterworth, Francis M
Chang, Chia-Cheng
Chu, Ernest Hsiao-Ying
Chung, Shiau-Ta
Conover, James H
Cooper, Stephen
Corcos, Alain Francois
Cress, Charles Edwin
Crittenden, Lyman Butler
Cronkite, Donald Lee
De Bruijn, Frans Johannes
Dusenbery, Ruth Lillian
Everson, Everett Henry
Ficsor, Gyula
Fluck, Michele Marguerite
Forsthoefel, Paulinus Frederick
Freeman, Dwight Carl
Freytag, Svend O
Friedman, Thomas Baer
Froiland, Thomas Gordon
Gay, Helen
Gelehrter, Thomas David
Gentile, James Michael
Gershowitz, Henry
Golenberg, Edward Michael
Goustin, Anton Scott
Grudzien, Thaddeus Arthur, Jr
Hackel, Emanuel
Hancock, James Findley, Jr
Hanover, James W
Harpstead, Dale D
Harris, James Edward
Helling, Robert Bruce
Higgins, James Victor
Jacobs, Charles Warren
Janca, Frank Charles
Jones, Lily Ann
Karnosky, David Frank
Klein, Ronald Don
Krawetz, Stephen Andrew
Krishna, Gopala
Kurnit, David Martin
Lenski, Richard E
Levine, Myron

Lomax, Margaret Irene
McGilliard, Lon Dee
Martin, Joseph Patrick, Jr
Mayeda, Kazutoshi
Meisler, Miriam Horowitz
Miller, Curtis C
Miller, Dorothy Anne Smith
Miller, Orlando Jack
Moll, Patricia Peyser
Montgomery, Ilene Nowicki
Murnik, Mary Rengo
Neel, James Van Gundia
Neidhardt, Frederick Carl
Padgett, George Arnold
Pelzer, Charles Francis
Porter, Calvin Anthon
Rizki, Tahir Mirza
Robbins, Leonard Gilbert
Ruby, John L
Shastry, Barkur Srinivasa
Shontz, Nancy Nickerson
Siegel, Albert
Sing, Charles F
Sink, Kenneth C, Jr
Smith, David I
Smith, Ellen Oliver
Smith, P(aul) Dennis
Steiner, Erich E
Stone, Howard Anderson
Swaroop, Anand
Tashian, Richard Earl
Thoene, Jess Gilbert
Toriello, Helga Valdmanis
Triezenberg, Steven J
Trosko, James Edward
Tse, Harley Y
Venta, Patrick John
Verley, Frank A
Weiss, Bernard
Wilmot, Thomas Ray
Wolk, Coleman Peter
Wolman, Sandra R

MINNESOTA
Anderson, Victor Elving
Barnes, Donald Kay
Beck, Barbara North
Berg, Robert W
Boylan, William J
Busch, Robert Henry
Caldecott, Richard S
Cervenka, Jaroslav
Comstock, Verne Edward
Condell, Yvonne C
Conlin, Bernard Joseph
Corbin, Kendall Wallace
Curtsinger, James Webb
Dapkus, David Conrad
David, Chelladurai S
Dearden, Douglas Morey
Dewald, Gordon Wayne
Emeagwali, Dale Brown
Enfield, Franklin D
Engh, Helmer A, Jr
Fallon, Ann Marie
Fausch, Homer David
Forro, Frederick, Jr
Fukui, Hidenori Henry
Gengenbach, Burle Gene
Glass, Arthur Warren
Gordon, Hymie
Hansen, Leslie Bennett
Hanson, Daniel Ralph
Hedman, Stephen Clifford
Herforth, Robert S
Himes, John Harter
Jalal, Syed M
Jessen, Carl Roger
Johnson, Theodore Reynold
Jokela, Jalmer John
Jones, Charles Weldon
King, Richard Allen
Kline, Bruce Clayton
Korte, Clare A
Kowles, Richard Vincent
Lefebvre, Paul Alvin
Leonard, Kurt John
Lindgren, Alice Marilyn Lindell
Lofgren, James R
Lukasewycz, Omelan Alexander
Lykken, David Thoreson
McKay, Larry Lee
Magee, Paul Terry
Merrell, David John
Oetting, William Starr
Orf, John W
Palm, John Daniel
Phillips, Ronald Lewis
Rasmusson, Donald C
Reed, Elizabeth Wagner
Rehwaldt, Charles A
Reilly, Bernard Edward
Rines, Howard Wayne
Rodell, Charles Franklin
Rudolf, Paul Otto
Sarkar, Gobinda
Schacht, Lee Eastman
Shapiro, Burton Leonard
Shoffner, Robert Nurman
Silflow, Carolyn Dorothy
Singer, Susan Rundell
Snustad, Donald Peter
Spelsberg, Thomas Coonan
Spurrell, Francis Arthur

Strehler, Emanuel Ernst
Stucker, Robert Evan
Stuthman, Deon Dean
Sulerud, Ralph L
Weibust, Robert Smith
Wettstein, Peter J
White, Donald Benjamin
Witkop, Carl Jacob, Jr
Woodward, Val Waddoups

MISSISSIPPI
Berry, Charles Dennis
Case, Steven Thomas
Creech, Roy G
Doudrick, Robert Lawrence
Gupton, Creighton Lee
Jackson, John Fenwick
Jenkins, Johnie Norton
Peeler, Dudley F, Jr
St Amand, Wilbrod
Santangelo, George Michael
Thomas, Charles Hill
Watson, Clarence Ellis, Jr
Wise, Dwayne Allison
Yarbrough, Karen Marguerite

MISSOURI
Alexander, Stephen
Anderson, Robert Glenn
Apirion, David
Ayyagari, L Rao
Ball, Timothy K
Beckett, Jack Brown
Birchler, James Arthur
Bourne, Carol Elizabeth Mulligan
Brandt, E J
Brot, Frederick Elliot
Burdick, Allan Bernard
Cannon, John Francis
Casperson, Gerald F
Cheney, Clarissa M
Cheverud, James Michael
Cloninger, Claude Robert
Coe, Edward Harold, Jr
Curtiss, Roy, III
David, John Dewood
Eissenberg, Joel Carter
Engel, Leslie Carroll
Fischhoff, David Allen
Folk, William Robert
Friedman, Lawrence David
Georgiadis, Nicholas J
Goodenough, Ursula Wiltshire
Gowans, Charles Shields
Green, Philip Palmer
Gustafson, John Perry
Hansen, Ted Howard
Heard, John Thibaut, Jr
Hess, John Berger
Hill, Jack Filson
Hillman, Richard Ephraim
Hoffman, Jacqueline Louise
Horsch, Robert Bruce
Huang, Tim Hui-ming
Hufham, James Birk
Johnson, Terrell Kent
Kimber, Gordon
Klotz, John William
Krishnan, B Rajendra
Larson, Allan
Levine, Robert Paul
Lie, Wen-Rong
Lower, William Russell
Lye, Robert J
McCann, Peter Paul
McQuade, Henry Alonzo
Melechen, Norman Edward
Mohamed, Aly Hamed
Neuffer, Myron Gerald
Radford, Diane Mary
Redei, Gyorgy Pal
Riddle, Donald Lee
Salkoff, Lawrence Benjamin
Sawyer, Stanley Arthur
Schreiweis, Donald Otto
Sears, Ernest Robert
Shreffler, Donald Cecil
Sly, William S
Stalker, Harrison Dailey
Stufflebeam, Charles Edward
Tai, William
Templeton, Alan Robert
Trinklein, David Herbert
Tritz, Gerald Joseph
Wang, Richard J
Willard, Mark Benjamin

MONTANA
Allendorf, Frederick William
Blackwell, Robert Leighton
Blake, Tom
Burfening, Peter J
Burris, Martin Joe
Cameron, David Glen
Cantrell, John Leonard
Carlson, George A
Ericsson, Ronald James
Hovin, Arne William
Kress, Donnie Duane
McCoy, Thomas Joseph
McNeal, Francis H
MacNeil, Michael
Opitz, John Marius
Priest, Jean Lane Hirsch

Stimpfling, Jack Herman
Warren, Guylyn Rea

NEBRASKA
Ahlschwede, William T
Baltensperger, David Dwight
Bennett, Gary Lee
Brumbaugh, John (Albert)
Calvert, Jay Gregory
Carson, Steven Douglas
Christensen, Alan Carl
Christman, Judith Kershaw
Compton, William A
Cundiff, Larry Verl
Dickerson, Gordon Edwin
Dubes, George Richard
Eastin, John A
Godfrey, Maurice
Gorz, Herman Jacob
Gregory, Keith Edward
Harris, Dewey Lynn
Jenkins, Thomas Gordon
Kay, H David
Keeler, Kathleen Howard
Kimberling, William J
Kumar, Shrawan
Laster, Danny Bruce
Lund, Douglas E
Lunt, Steele Ray
Osterman, John Carl
Ross, William Max
Sanger, Warren Glenn
Schutz, Wilfred M
Spreitzer, Robert Joseph
Van Vleck, Lloyd Dale
Veomett, George Ector
Young, Lawrence Dale

NEVADA
Bailey, Curtiss Merkel
Brussard, Peter Frans
Vig, Baldev K
Winicov, Ilga

NEW HAMPSHIRE
Berger, Edward Michael
Bergman, Kenneth David
Cole, Charles N
Collins, Walter Marshall
Dunlap, Jay Clark
Funk, David Truman
Garrett, Peter Wayne
Green, Donald MacDonald
Gross, Robert Henry
Hatch, Frederick Tasker
Kiang, Yun-Tzu
Loros, Jennifer Jane
Minocha, Subhash C
Rogers, Owen Maurice
Soares, Eugene Robbins
Wurster-Hill, Doris Hadley

NEW JERSEY
Agnish, Narsingh Dev
Arthur, Alan Thorne
Axelrod, David E
Bacharach, Martin Max
Benson, Charles Everett
Berg, Richard A
Brandriss, Marjorie C
Byrne, Barbara Jean McManamy
Carlson, James H
Champe, Sewell Preston
Choi, Ye-Chin
Choy, Wai Nang
Cox, Edward Charles
Day, Peter Rodney
Day-Salvatore, Debra-Lynn
Dhruv, Rohini Arvind
Egger, M(aurice) David
Essien, Francine B
Fisher, Kenneth Walter
Frenkel, Gerald Daniel
Gavurin, Lillian
Gepner, Ivan Alan
Gillum, Amanda McKee
Goldman, Emanuel
Grant, B Rosemary
Grassle, Judith Payne
Hill, Helene Zimmermann
Hite, Mark
Hu, Ching-Yeh
Hu, Yaping
Huber, Ivan
Humayun, Mir Z
Jaffe, Ernst Richard
Jolly, Clifford J
Joslyn, Dennis Joseph
Kaback, David Brian
Kapp, Robert Wesley, Jr
Kirsch, Donald R
Klug, William Stephen
Krause, Eliot
Kurtz, Myra Berman
Lambert, Muriel Wikswo
Lee, Ming-Liang
Leibowitz, Michael Jonathan
Levine, Arnold J
McGuire, Terry Russell
Maloy, Joseph T
Martin, Charles Everett
Mather, Robert Eugene
Messing, Joachim W
Mezick, James Andrew

BIOLOGICAL SCIENCES / 107

Middleton, Richard B
Moyer, Samuel Edward
Nagle, James John
Newton, (William) Austin
Ozer, Harvey Leon
Pai, Anna Chao
Passmore, Howard Clinton
Plotkin, Diane Joyce
Prensky, Wolf
Ripley, Lynn S
Rosen, Dianne L
Rosenberg, Leon Emanuel
Rosenstraus, Maurice Jay
Rothstein, Rodney Joel
Rushton, Alan R
Schreiner, Ceinwen Ann
Schroff, Peter David
Scott, William Addison, III
Shapiro, Herman Simon
Shin, Seung-il
Silhavy, Thomas J
Silver, Lee Merrill
Smouse, Peter Edgar
Sofer, William Howard
Stern, Elizabeth Kay
Swenson, Theresa Lynn
Unowsky, Joel
Weisbrot, David R
Wille, John Jacob, Jr
Witkin, Evelyn Maisel

NEW MEXICO
Baker, William Kaufman
Bartholdi, Marty Frank
Burks, Christian
Chen, David J
Deaven, Larry Lee
Dillon, Richard Thomas
Dunford, Max Patterson
Fisher, James Thomas
Gurd, Ruth Sights
Holland, Lewis
Hsi, David Ching Heng
Hubby, John L
Johnson, William Wayne
Kelly, Gregory
Kraemer, Paul Michael
McCuistion, Willis Lloyd
Melton, Billy Alexander, Jr
Nelson, Mary Anne
Phillips, Gregory Conrad
Strniste, Gary F
Thomassen, David George
Wagner, Robert Philip

NEW YORK
Abdelnoor, Alexander Michael
Allis, C David
Allison, David B
Amidon, Thomas Edward
Anderson, Ronald Eugene
Ash, William James
Auerbach, Arleen D
Baglioni, Corrado
Baird, Malcolm Barry
Baker, Raymond Milton
Banerjee, Ranjit
Basilico, Claudio
Baum, Howard Jay
Bazinet, George Frederick
Beam, Carl Adams
Beardsley, Robert Eugene
Belfort, Marlene
Bell, Robin Graham
Bender, Michael A
Benjaminson, Morris Aaron
Bernheimer, Harriet P
Birshtein, Barbara K
Bishop, David Franklin
Bleyman, Lea Kanner
Bloom, Stephen Earl
Bopp, Lawrence Howard
Borowsky, Richard Lewis
Bradbury, Michael Wayne
Bregman, Allyn A(aron)
Breslow, Jan Leslie
Brody, Edward Norman
Broker, Thomas Richard
Bruns, Peter John
Bukhari, Ahmad Iqbal
Bushkin, Yuri
Buxbaum, Joel N
Calame, Kathryn Lee
Calhoon, Robert Ellsworth
Calvo, Joseph M
Campbell, Douglas Arthur
Carlson, Elof Axel
Carlson, Marian B
Catterall, James F
Chaganti, Raju Sreerama Kamalasana
Chalfie, Martin
Chang, Tien-ding
Chapman, Verne M
Chase, Sherret Spaulding
Chasin, Lawrence Allen
Chen, Ching-Ling Chu
Chepko-Sade, Bonita Diane
Christie, Nelwyn T
Clark, Virginia Lee
Cohen, Elias
Cole, Randall Knight
Craddock, Elysse Margaret
Cunningham, Richard Preston
Davidson, Mercy

Desnick, Robert John
Desplan, Claude
D'Eustachio, Peter
Dibennardo, Robert
Dietert, Rodney Reynolds
Dottin, Robert P
Doudney, Charles Owen
Dranginis, Anne M
DuFrain, Russell Jerome
Dykhuizen, Daniel Edward
Ehrman, Lee
Eldridge, Roswell
Elliott, Rosemary Waite
Emmons, Scott W
Emsley, James Alan Burns
Enea, Vincenzo
Erbe, Richard W
Erk, Frank Chris
Erlenmeyer-Kimling, L
Everett, Herbert Lyman
Falk, Catherine T
Ferguson, John Barclay
Figurski, David Henry
Fischman, Harlow Kenneth
Forest, Charlene Lynn
Fortune, Joanne Elizabeth
Fox, Thomas David
Frair, Wayne
Free, Stephen J
Friedman, Ronald Marvin
Fuscaldo, Anthony Alfred
Galinsky, Irving
Gallagher, Jane Chispa
Garner, James G
Garrick, Laura Morris
Gates, Allen H(azen), Jr
German, James Lafayette, III
Giampietro, Philip Francis
Gilbert, Fred
Ginzburg, Lev R
Gladstone, William Turnbull
Glass, Hiram Bentley
Goff, Stephen Payne
Goldberg, Allan Roy
Goldschmidt, Raul Max
Goldsmith, Lowell Alan
Goldstein, Fred Bernard
Gonzalez, Eulogio Raphael
Goodwin, Stephen Bruce
Gordon, Jon W
Grimes, Gary Wayne
Grimwood, Brian Gene
Grodzicker, Terri Irene
Grossfield, Joseph
Gurevitch, Jessica
Guthrie, Robert
Hall, Barry Gordon
Hanafusa, Hidesaburo
Hanson, Maureen Rebecca
Harford, Agnes Gayler
Harrison, Richard Gerald
Hartung, John David
Hershey, Alfred D
Hinkle, David Currier
Hirschhorn, Kurt
Hochstadt, Joy
Holmes, David Salway
Hommes, Frits A
Hoo, Joe Jie
Hotchkiss, Sharon K
Howard, Irmgard Matilda Keeler
Huberman, Joel Anthony
Hurst, Donald D
Hutt, Frederick Bruce
Jagiello, Georgiana Mary
Jainchill, Jerome
Jamnback, Hugo Andrew, Jr
Jhanwar, Suresh Chandra
Johnsen, Roger Craig
Johnson, Edward Michael
Johnson, Lawrence Lloyd
Kaelbling, Margot
Kallman, Klaus D
Kambysellis, Michael Panagiotis
Kascsak, Richard John
Katz, Eugene Richard
Kelleher, Raymond Joseph, Jr
Kelly, Sally Marie
Kemphues, Kenneth J
Kessin, Richard Harry
Khan, Nasim A
Kish, Valerie Mayo
Klinge, Carolyn M
Klinger, Harold P
Knipple, Douglas Charles
Konarska, Maria Magda
Korf, Richard Paul
Kramer, Fred Russell
Kucherlapati, Raju Suryanarayana
Lacks, Sanford
LaFountain, James Robert, Jr
Lalley, Peter Austin
Lambert, Robert John
Lange, Christopher Stephen
Last, Robert L
Latorella, A Henry
Lavett, Diane Kathryn Juricek
Lawrence, Christopher William
Lazzarini, Robert A
Leary, James Francis
Lebwohl, Mark Gabriel
Lederberg, Joshua
Leifer, Zer
Leslie, Paul Willard

Levene, Howard
Levine, Louis
Levitan, Max
Lewis, Leslie Arthur
Liao, Martha
Lilly, Frank
Lin, Yue Jee
Lindahl, Lasse Allan
Lis, John Thomas
Litwin, Stephen David
Liu, Houng-Zung
Lugthart, Garrit John, Jr
Ma, Hong
Maas, Werner Karl
McClintock, Barbara
McCormick, Paulette Jean
Marien, Daniel
Marsh, William Laurence
Maynard, Charles Alvin
Mehta, Bipin Mohanlal
Meyer, Axel
Meyers, Marian Bennett
Michels, Corinne Anthony
Miller, Morton W
Mitra, Jyotirmay
Model, Peter
Monheit, Alan G
Moore, Carol Wood
Morishima, Akira
Morse, Randall Heywood
Mount, Stephen M
Mukherjee, Asit B
Munne, Santiago
Munro, Donald W, Jr
Murphy, Donal B
Muschio, Henry M, Jr
Nasrallah, Mikhail Elia
Nitowsky, Harold Martin
Novick, Richard P
Nur, Uzi
Oltenacu, Elizabeth Allison Branford
O'Reilly, Richard
Orlow, Seth J
Ottman, Ruth
Pellicer, Angel
Pesetsky, Irwin
Pollard, Jeffrey William
Possidente, Bernard Philip, Jr
Prakash, Louise
Pumo, Dorothy Ellen
Qazi, Qutubuddin H
Rainer, John David
Rajan, Thiruchandurai Viswanathan
Ramirez, Francisco
Rattazzi, Mario Cristiano
Raveche, Elizabeth Marie
Reddy, Kalluru Jayarami
Reilly, Marguerite
Reisch, Bruce Irving
Rinchik, Eugene M
Robinson, Albert Dean
Rockwell, Robert Franklin
Rogers, Scott Orland
Rogerson, Allen Collingwood
Roll, David E
Rosenbaum, Peter Andrew
Rosenstein, Barry Sheldon
Rossman, Toby Gale
Rotheim, Minna B
Rothwell, Norman Vincent
Rowley, Peter Templeton
Ruddell, Alanna
Rudner, Rivka
Rudy, Bernardo
Russel, Marjorie Ellen
Russell, George K(eith)
Salwen, Martin J
Sank, Diane
Sechrist, Lynne Luan
Shafer, Stephen Joel
Shafit-Zagardo, Bridget
Shechter, Yaakov
Sherman, Fred
Shields, William Michael
Shows, Thomas Byron
Siegel, Irwin Michael
Silva, Pedro Joao Neves
Silverstein, Emanuel
Silverstein, Saul Jay
Singh, Madho
Sirlin, Julio Leo
Sirotnak, Francis Michael
Smith, Harold Hill
Solish, George Irving
Sorrells, Mark Earl
Spector, David Lawrence
Srb, Adrian Morris
Stanley, Pamela Mary
Steinberg, Bettie Murray
Stillman, Bruce William
Stinson, Harry Theodore, Jr
Stockert, Robert J
Studier, Frederick William
Sullivan, David Thomas
Susca, Louis Anthony
Taub, Mary L
Thaler, David Solomon
Thomas, John Owen
Thompson, Steven Risley
Tilley, Shermaine Ann
Tomashefsky, Philip
Tye, Bik-Kwoon
Valentine, Fredrick Arthur
Verma, Ram S

Vladutiu, Georgirene Dietrich
Waelsch, Salome Gluecksohn
Wallace, Donald Howard
Wang, Peizhi
Warburton, Dorothy
Wasserman, Marvin
Weitkamp, Lowell R
Weyter, Frederick William
Wiberg, John Samuel
Williamson, David Lee
Wilson, Dwight Elliott, Jr
Winchester, Robert J
Wishnick, Marcia M
Wolfner, Mariana Federica
Wolgemuth, Debra Joanne
Wulff, Daniel Lewis
Yang, Shung-Jun
Yoder, Olen Curtis
Young, Charles Stuart Hamish
Zahler, Stanley Arnold
Zaleski, Marek Bohdan
Zengel, Janice Marie
Zinder, Norton David
Zitomer, Richard Stephen

NORTH CAROLINA
Allen, Wendall E
Amos, Dennis Bernard
Anderton, Laura Gaddes
Andrews, Matthew Tucker
Atchley, William Reid
Barrett, J(ames) Carl
Barry, Edward Gail
Bates, William K
Bell, Juliette B
Bewley, Glenn Carl
Bishop, Jack Belmont
Bishop, Paul Edward
Bloom, Kerry Steven
Boklage, Charles Edward
Boone, Lawrence Rudolph, III
Bostian, Carey Hoyt
Bott, Kenneth F
Boyd, Jeffrey Allen
Boynton, John E
Burchall, James J
Burkhart, James Gaylord
Butterworth, Byron Edwin
Carter, Thomas Edward, Jr
Chae, Chi-Bom
Chaplin, James Ferris
Claxton, Larry Davis
Cockerham, Columbus Clark
Cohen, Carl
Collins, William Kerr
Cook, Robert Edward
Cope, Will Allen
Corley, Ronald Bruce
Counce, Sheila Jean
Crowl, Robert Harold
Curtis, Susan Julia
Daugherty, Patricia A
Davis, Daniel Layten
De Serres, Frederick Joseph
Dillard, Emmett Urcey
Drake, John W
Edgell, Marshall Hall
Edwards, James Wesley
Eisen, Eugene J
Endow, Sharyn Anne
Errede, Beverly Jean
Farber, Rosann Alexander
Frelinger, Jeffrey
Gerstel, Dan Ulrich
Gilbert, Lawrence Irwin
Gillham, Nicholas Wright
Glazener, Edward Walker
Gooder, Harry
Goodman, Harold Orbeck
Goodman, Major M
Grosch, Daniel Swartwood
Gross, Samson Richard
Guild, Walter Rufus
Hankins, Gerald Robert
Hanson, Warren Durward
Harris, Elizabeth Holder
Haughton, Geoffrey
Havenstein, Gerald B
Hebert, Teddy T
Heise, Eugene Royce
Hershfield, Michael Steven
Hicks, Robert Eugene
Hildreth, Philip Elwin
Holmes, Edward Warren
Horner, Theodore Wright
Hutchison, Clyde Allen, III
Issitt, Peter David
Jividen, Gay Melton
Jolls, Claudia Lee
Judd, Burke Haycock
Kellison, Robert Clay
Kerschner, Jean
Knauft, David A
Komma, Donald Jerry
Kredich, Nicholas M
Leamy, Larry Jackson
Legates, James Edward
Levings, Charles Sandford, III
Lewis, Paul Ollin
Lewis, Susan Erskine
Li, Steven Shoei-Lung
McDaniel, Benjamin Thomas
McKenzie, Wendell Herbert
Malling, Heinrich Valdemar

Genetics (cont)

Mann, Thurston (Jefferson)
Maroni, Donna F
Maroni, Gustavo Primo
Matzinger, Dale Frederick
Merritt, James Francis
Mickey, George Henry
Mitchell, Ann Denman
Modrich, Paul L
Moll, Robert Harry
Moore, Martha May
Ostrowski, Ronald Stephen
Perry, Thomas Oliver
Petes, Thomas Douglas
Petters, Robert Michael
Phelps, Allen Warner
Phillips, Lyle Llewellyn
Pollitzer, William Sprott
Preston, Robert Julian
Pukkila, Patricia Jean
Rawlings, John Oren
Robison, Odis Wayne
Sargent, Frank Dorrance
Saylor, LeRoy C
Scandalios, John George
Schaffer, Henry Elkin
Sheridan, William
Sidhu, Bhag Singh
Smith, Gary Joseph
Spencer, Lorraine Barney
Spiker, Steven L
Stone, Donald Eugene
Stuber, Charles William
Sulik, Kathleen Kay
Suzuki, Kunihiko
Swift, Michael
Taggart, R(obert) Thomas
Thompson, William Francis, III
Tice, Raymond Richard
Timothy, David Harry
Tlsty, Thea Dorothy
Tourian, Ara Yervant
Triantaphyllou, Anastasios Christos
Verghese, Margrith Wehrli
Voelker, Robert Allen
Ward, Calvin Lucian
Ward, Frances Ellen
Webster, Robert Edward
Weeks, Leo
Weir, Bruce Spencer
Welsch, Frank
Whittinghill, Maurice
Williamson, John Hybert
Wilson, James Franklin
Wong, Fulton
Wright, Clarence Paul
Wynne, Johnny Calvin
Zeng, Zhao-Bang
Zobel, Bruce John

NORTH DAKOTA

Doney, Devon Lyle
Hastings, James Michael
Hein, David William
Helm, James Leroy
Jauhar, Prem P
Joppa, Leonard Robert
La Chance, Leo Emery
McDonald, Ian Cameron Crawford
Riemann, John G
Seiler, Gerald Joseph
Sheridan, William Francis
Smith, Garry Austin
Whited, Dean Allen
Williams, Norman Dale

OHIO

Asch, David Kent
Atkins, Charles Gilmore
Banerjee, Sipra
Baroudy, Bahige M
Barr, Harry L
Beck, Doris Jean
Bhattacharjee, Jnanendra K
Birkenberger, Lori
Birky, Carl William, Jr
Burns, George W
Byard, Pamela Joy
Carver, Eugene Arthur
Cassidy, Suzanne Bletterman
Chase, John William
Clise, Ronald Leo
Cohn, Norman Stanley
Cooper, Richard Lee
Cullis, Christopher Ashley
Davis, Michael E
Dean, Donald Harry
Dickerman, Richard Curtis
Doerder, F Paul
Dollinger, Elwood Johnson
Erway, Lawrence Clifton, Jr
Fechheimer, Nathan S
Fraser, Alex Stewart
Freed, James Melvin
Fuerst, Paul Anthony
Ganschow, Roger Elmer
Gates, Michael Andrew
Gesinski, Raymond Marion
Glatzer, Louis
Goldman, Stephen L
Gregg, Thomas G
Griffing, J Bruce
Gromko, Mark Hedges
Gronostajski, Richard M
Hayes, Thomas G
Herschler, Michael Saul
Hines, Harold C
Hinton, Claude Willey
Hogg, Robert W
House, Verl Lee
Huether, Carl Albert
Jollick, Joseph Darryl
Kalter, Harold
Kao, Winston Whei-Yang
Keever, Carolyn Anne
Kontras, Stella B
Kreutzer, Richard D
Kriebel, Howard Burtt
Kurczynski, Thaddeus Walter
Lafever, Howard N
Laughner, William James, Jr
Laushman, Roger H
Loper, John C
McCune, Sylvia Ann
McDougall, Kenneth J
Macintyre, Malcolm Neil
McQuate, John Truman
Marzluf, George A
Mikula, Bernard C
Milsted, Amy
Moore, Jay Winston
Morrow, Grant, III
Nebert, Daniel Walter
Nestor, Karl Elwood
Oakley, Berl Ray
Paddock, Elton Farnham
Powelson, Elizabeth Eugenie
Powers, Jean D
Rake, Adrian Vaughan
Rayle, Richard Eugene
Rimm, Alfred A
Robinow, Meinhard
Rozek, Charles Edward
Rucknagel, Donald Louis
Schafer, Irwin Arnold
Schwer, Joseph Francis
Seiger, Marvin Barr
Shaffer, Jacquelin Bruning
Sidman, Charles L
Siervogel, Roger M
Skavaril, Russell Vincent
Stalvey, John Robert Dixon
Steinberg, Arthur Gerald
Stepien, Carol Ann
Sullivan, Robert Little
Treichel, Robin Stong
Trewyn, Ronald William
Turner, Monte Earl
Washington, Willie James
Waterson, John R
Williams, Marshall Vance
Wilson, Kenneth Glade
Woodruff, Ronny Clifford
Woodworth, Mary Esther
Wrensch, Dana Louise
Yoon, Jong Sik
Young, Sydney Sze Yih
Zartman, David Lester

OKLAHOMA

Altmiller, Dale Henry
Badgett, Allen A
Bruneau, Leslie Herbert
Buchanan, David Shane
Carpenter, Nancy Jane
Dale, George Leslie
Dyer, David Wayne
Edwards, Lewis Hiram
Elisens, Wayne John
Essenberg, Richard Charles
Hunger, Robert Marvin
LeGrand, Frank Edward
Miner, Gary David
Morrissey, James Henry
Muneer, Razia Sultana
Murray, Jay Clarence
Powell, Jerrel B
Ramon, Serafin
Schaefer, Frederick Vail
Smith, Edward Lee
Taliaferro, Charles M
Thompson, James Neal, Jr
Tobin, Sara L
Verhalen, Laval M(athias)
Wann, Elbert Van
Wells, Harrington
Young, Sharon Clairene

OREGON

Baer, Adela (Dee)
Bagby, Grover Carlton
Bernier, Paul Emile
Bradshaw, William Emmons
Buist, Neil R M
Cameron, H Ronald
Cameron, James Wagner
Civelli, Oliver
Dawson, Peter Sanford
Duffield, Deborah Ann
Field, Katharine G
Fowler, Gregory L
Gallaher, Edward J
Geeseman, Gordon E
Gold, Michael Howard
Haunold, Alfred
Hays, John Bruce
Hedrick, Ann Valerie
Ho, Iwan
Holzapfel, Christina Marie
Hoornbeek, Frank Kent
Hrubant, Henry Everett
Kabat, David
Kohler, Peter
Koler, Robert Donald
Kronstad, Warren Ervind
Ladd, Sheldon Lane
Lande, Russell Scott
Litt, Michael
Lolle, Susan Janne
Mills, Dallice Ivan
Mohler, James Dawson
Mok, Machteld Cornelia
Newell, Nanette
Newman, Lester Joseph
Novitski, Edward
Orkney, G Dale
Postlethwait, John Harvey
Prescott, Gerald H
Ream, Lloyd Walter, Jr
Roberts, Paul Alfred
Russell, Peter James
Stahl, Franklin William
Thompson, Maxine Marie
Wagner, David Henry

PENNSYLVANIA

Alperin, Richard Junius
Armstrong, Robert John
Arnott, Marilyn Sue
Avadhani, Narayan G
Barnes, William Shelley
Barranger, John Arthur
Bean, Barry
Beasley, Andrew Bowie
Berg, Clyde C
Bjornsti, Mary-Ann
Bowne, Samuel Winter, Jr
Brownstein, Barbara L
Buetow, Kenneth H
Buss, Edward George
Butt, Tauseef Rashid
Cannizzaro, Linda A
Carlton, Bruce Charles
Chernin, Mitchell Irwin
Christ, Barbara Jane
Clark, Andrew Galen
Cleveland, Richard Warren
Cooper, Jane Elizabeth
Cortner, Jean A
Corwin, Harry O
Craig, Richard
Croce, Carlo Maria
Cryer, Dennis Robert
Currier, Thomas Curtis
D'Agostino, Marie A
Debouck, Christine Marie
Deering, Reginald
Del Vecchio, Vito Gerard
Denis, Kathleen A
DiBerardino, Marie A
Dudzinski, Diane Marie
Eckhardt, Robert Barry
Eckroat, Larry Raymond
Edlind, Thomas D
Farber, Phillip Andrew
Feinstein, Sheldon Israel
Ferrell, Robert Edward
Finger, Irving
Floros, Joanna
Freed, Jerome James
Friedman, Robert David
Fuscaldo, Kathryn Elizabeth
Gabriel, Edward George
Gasser, David Lloyd
Gealt, Michael Alan
Gerhold, Henry Dietrich
Gersh, Eileen Sutton
Gilbert, Scott F
Glorioso, Joseph Charles, III
Goff, Christopher Godfrey
Gollin, Susanne Merle
Goodgal, Sol Howard
Goodwin, Kenneth
Gorin, Michael Bruce
Gots, Joseph Simon
Gottlieb, Frederick Jay
Gottlieb, Karen Ann
Grebner, Eugene Ernest
Grun, Paul
Hargrove, George Lynn
Harris, Harry
Haskins, Mark
Hedrick, Philip William
Henry, Susan Armstrong
Higgins, Terry Jay
Hill, Charles Whitacre
Hill, Richard Ray, Jr
Hoffman, Albert Charles
Hopper, Anita Klein
Hsu, Susan Hu
Hughes, Austin Leland
Infanger, Ann
Jackson, Ethel Noland
Jacobsen, Terry Dale
Jargiello, Patricia
Jarvik, Jonathan Wallace
Jenkins, John Bruner
Jervis, Herbert Hunter
Jezyk, Peter Franklin
Johnson, Melvin Walter, Jr
Johnston, James Bennett
Jones, Elizabeth W
Kamboh, Mohammad Ilyas
Kaney, Anthony Rolland
Kant, Jeffrey A
Kayhart, Marion
Kazazian, Haig H, Jr
Kennett, Roger H
King, Robert Willis
Knowles, Barbara B
Knudson, Alfred George, Jr
Kozinski, Andrzei
Kuivaniemi, S Helena
Ladda, Roger Louis
Leibel, Wayne Stephan
Li, Ching Chun
Lustbader, Edward David
McCarthy, Patrick Charles
McDonald, Daniel James
McFeely, Richard Aubrey
McKinley, Carolyn May
McNamara, Pamela Dee (MacMinigal)
McPheron, Bruce Alan
Majumdar, Shyamal K
Mann, Stanley Joseph
Marazita, Mary Louise
Marks, Louis Sheppard
Marshall, Harold Gene
Maxson, Linda Ellen R
Mezger-Freed, Liselotte
Michel, Kenneth Earl
Miller, Robert Christopher
Mode, Charles J
Morrison, William Joseph
Morrow, Terry Oran
Mullin, James Michael
Mulvihill, John Joseph
Nass, Margit M K
Nei, Masatoshi
Nichols, Warren Wesley
Ohlsson-Wilhelm, Betsy Mae
Oka, Seishi William
Olds-Clarke, Patricia Jean
Opas, Evan E
Orr, Nancy Hoffner
Partanen, Carl Richard
Patterson, Donald Floyd
Peebles, Craig Lewis
Person, Stanley R
Poethig, Richard Scott
Porter, Ronald Dean
Punnett, Hope Handler
Pursell, Mary Helen
Pyeritz, Reed E
Ricciardi, Robert Paul
Rose, Raymond Wesley, Jr
Royse, Daniel Joseph
Russell, Richard Lawson
Sanders, Mary Elizabeth
Schmickel, Roy David
Schultz, Jane Schwartz
Schwaber, Jerrold
Schwartz, Elias
Shannon, Jack Corum
Sideropoulos, Aris S
Silvers, Willys Kent
Specht, Lawrence W
Spielman, Richard Saul
Staetz, Charles Alan
Steiner, Kim Carlyle
Stephenson, Andrew George
Stolc, Viktor
Suyama, Yoshitaka
Sylvester, James Edward
Takats, Stephen Tibor
Tartof, Kenneth D
Taylor, Rhoda E
Tevethia, Mary Judith (Robinson)
Tompkins, Laurie
Tonzetich, John
Towne, William F
Tromp, Gerardus C
Tu, Chen-Pei David
Turner, J Howard
Turoczi, Lester J
Weinberg, Eric S
Weiss, Kenneth Monrad
Wenger, Sharon Louise
Wheeler, Donald Alsop
Wolfe, John Hall
Wright, James Everett, Jr
Wurst, Glen Gilbert
Wysocki, Charles Joseph
Yang, Da-Ping (David)
Yunis, Jorge J
Zink, Gilbert Leroy

RHODE ISLAND

Brooks, Lisa Delevan
Clough, Wendy Glasgow
Coleman, John Russell
Costantino, Robert Francis
Fausto-Sterling, Anne
Fischer, Glenn Albert
Gonsalves, Neil Ignatius
Hagy, George Washington
Holstein, Thomas James
Landy, Arthur
Lederberg, Seymour
Malcolm, Alexander Russell
Mottinger, John P
Quevedo, Walter Cole, Jr
Rand, David McNear
Rao, Girimaji J Sathyanarayana
Rothman, Frank George

SOUTH CAROLINA
Arnaud, Philippe
Best, Robert Glen
Byrd, Wilbert Preston
Cantu, Eduardo S
Chapman, Stephen R
Davis, Leroy
Dawson, Wallace Douglas, Jr
Dille, John Emmanuel
Ely, Berten E, III
Finlay, Mary Fleming
Fore, Stephanie Anne
Glick, Bruce
Godley, Willie Cecil
Graham, William Doyce, Jr
Jones, Alfred
Kinney, Terry B, Jr
Labanick, George Michael
Lincoln, David Erwin
McClain, Eugene Fredrick
Maheshwari, Kewal Krishnan
Mishra, Nawin C
Mulvey, Margaret
Olson, John Bernard
Pollard, Arthur Joseph
Rupert, Earlene Atchison
Rust, Philip Frederick
Sawyer, Roger Holmes
Shipe, Emerson Russell
Surver, William Merle, Jr
Taylor, Harold Allison, Jr
Waldman, Alan S
Wang, An-Chuan
Weeks, Stephen Charles
Wilson, Gregory Bruce
Worthen, Wade Bolton
Yardley, Darrell Gene

SOUTH DAKOTA
Cholick, Fred Andrew
Kahler, Alex L
Lindahl, Ronald Gunnar
Lunden, Allyn Oscar
Morgan, Walter Clifford
Weaver, Keith Eric
West, Thomas Patrick

TENNESSEE
Barnett, Paul Edward
Bell, Sandra Lucille
Biggers, Charles James
Bryant, Robert Emory
Champney, William Scott
Chi, David Shyh-Wei
Conger, Bob Vernon
Dev, Vaithilingam Gangathara
Duck, Bobby Neal
Epler, James L
Garth, Richard Edwin
Generoso, Walderico Malinawan
Gresshoff, Peter Michael
Harris, Edward Frederick
Harris, John Wallace
Henke, Randolph Ray
Hickok, Leslie George
Hochman, Benjamin
Howe, Martha Morgan
Hughes, Karen Woodbury
Jones, Larry Hudson
Kahlon, Prem Singh
Larimer, Frank William
Lin, Kuang-Tzu Davis
Lozzio, Carmen Bertucci
McCauley, David Evan
McFee, Alfred Frank
Mathis, Philip Monroe
Mosig, Gisela
Niyogi, Salil Kumar
Owens, Charles Allen
Popp, Raymond Arthur
Pui, Ching-Hon
Quigley, Neil Benton
Ramey, Harmon Hobson, Jr
Regan, James Dale
Riggsby, William Stuart
Rivas, Marian Lucy
Russell, Liane Brauch
Russell, William Lawson
Schlarbaum, Scott E
Sega, Gary Andrew
Selby, Paul Bruce
Shirley, Herschel Vincent, Jr
Sotomayor, Rene Eduardo
Summitt, Robert L
Taft, Kingsley Arter, Jr
Thor, Eyvind
Tibbetts, Clark
Wachtel, Stephen Shoel
West, Dennis R
Womack, Frances C
Wood, Henderson Kingsberry
Woychik, Richard P
Yang, Wen-Kuang
Young, Lawrence Dale

TEXAS
Adair, Gerald Michael
Alexander, Mary Louise
Allen, Archie C
Alperin, Jack Bernard
Altenburg, Lewis Conrad
Anderson, David Eugene
Arenaz, Pablo
Artzt, Karen
Aufderheide, Karl John
Barnett, Don(ald) R(ay)
Bartley, William Call
Bashaw, Elexis Cook
Bawdon, Roger Everett
Beaudet, Arthur L
Beckingham, Kathleen Mary
Benedik, Michael Joseph
Bickham, John W
Blangero, John Charles
Bowman, Barbara Hyde
Brokaw, Bryan Edward
Brown, Michael S
Burdette, Walter James
Burson, Byron Lynn
Busbee, David L
Calub, Alfonso deGuzman
Caskey, Charles Thomas
Chakraborty, Ranajit
Chalmers, John Harvey, Jr
Chan, Lawrence Chin Bong
Chan, Teh-Sheng
Christadoss, Premkumar
Collier, Jesse Wilton
Collins, Anita Marguerite
Coulter, Murray W
Cox, Rody Powell
Crandall, Keith Alan
Creel, Gordon C
Darlington, Gretchen Ann Jolly
Davison, Daniel Burtonn
Dewees, Andre Aaron
Doe, Frank Joseph
Dorn, Gordon Lee
Douglas, Tommy Charles
Dronamraju, Krishna Rao
Duhrkopf, Richard Edward
Dyke, Bennett
Earhart, Charles Franklin, Jr
Elder, Fred F B
Erdman, Howard E
Evans, Raeford G
Fanguy, Roy Charles
Fisher, Charles William
Frederiksen, Richard Allan
Fuerst, Robert
Gaulden, Mary Esther
George, Fredrick William
Gilmore, Earl C
Girvin, Eb Carl
Gold, John Rush
Goldstein, Joseph Leonard
Gottlieb, Paul David
Gratzner, Howard G
Greenbaum, Ira Fred
Greenberg, Frank
Griffin, James Emmett
Handler, Shirley Wolz
Hanis, Craig L
Hart, Gary Elwood
Hartberg, Warren Keith
Hecht, Ralph Martin
Henney, Henry Russell, Jr
Hewett-Emmett, David
Highlander-Bultema, Sarah Katherine
Hillis, David Mark
Hittelman, Walter Nathan
Hixson, James Elmer
Hoff, Victor John
Hogan, Michael Edward
Honeycutt, Rodney Lee
Hsie, Abraham Wuhsiung
Hughes, Mark R
Ippen-Ihler, Karin Ann
Jackson, Gilchrist Lewis
Jackson, Raymond Carl
Jacob, Horace S
Jeter, Randall Mark
Johnston, John Spencer
Kammerer, Candace Marie
Kaplan, Samuel
Karunakaran, Thonthi
Kidd, Harold J
Kieffer, Nat
Kimbrell, Deborah Ann
Klaus, Ewald Fred, Jr
Klebe, Robert John
Kohel, Russell James
Konkel, David Anthony
Kraig, Ellen
Kroschewsky, Julius Richard
Krueger, Willie Frederick
Kumar, Vinay
Kurosky, Alexander
Kurzrock, Razelle
LaBrie, David Andre
LeBlanc, Donald Joseph
Ledley, Fred David
Lee, James C
Lee, Shih-Shun
Lester, Larry James
Li, Wen-Hsiung
Lingle, Sarah Elizabeth
Long, Walter K
McBride, Raymond Andrew
MacCluer, Jean Walters
McCrady, William B
McDaniel, Milton Edward
McDonald, Lynn Dale
McIlrath, William Oliver
McNutt, Clarence Wallace
Magill, Clint William
Magill, Jane Mary (Oakes)
Maguire, Marjorie Paquette
Margolin, Solomon B
Mathews, Nancy Ellen
Maunder, A Bruce
Miller, Julian Creighton, Jr
Moody, Eric Edward Marshall
Moore, Charleen Morizot
Morrow, Kenneth John, Jr
Murgola, Emanuel J
Myers, Terry Lewis
Nelson, David Loren
Nguyen, Henry Thien
Nicholson, Wayne Lowell
Norwood, James S
O'Brien, William E
Parker, Nick Charles
Pathak, Sen
Perlman, Philip Stewart
Peterson, David Oscar
Pierce, Benjamin Allen
Prakash, Satya
Prasad, Naresh
Prasad, Rupi
Pratt, David R
Price, Harold James
Price, Peter Michael
Queller, David Charles
Rainwater, David Luther
Rao, Potu Narasimha
Redshaw, Peggy Ann
Riser, Mary Elizabeth
Robberson, Donald Lewis
Rose, Kathleen Mary
Rosenblum, Eugene David
Roth, Jack A
Samollow, Paul B
Sanders, Bobby Gene
Schertz, Keith Francis
Schroeter, Gilbert Loren
Schull, William Jackson
Shanley, Mark Stephen
Shay, Jerry William
Siciliano, Michael J
Siede, Wolfram
Simpson, Joe Leigh
Skow, Loren Curtis
Smith, Gerald Ray
Smith, James Douglas
Snider, Philip Joseph
Srivastava, Satish Kumar
Stansel, James Wilbert
Stewart, Charles Ranous
Stone, William Harold
Strassmann, Joan Elizabeth
Stroman, David Womack
Strong, Louise Connally
Summers, Max Duane
Sutton, Harry Eldon
Templeton, Joe Wayne
Thompson, Tommy Earl
Tierce, John Forrest
Trentin, John Joseph
Van Buijtenen, Johannes Petrus
VandeBerg, John Lee
Walker, Cheryl Lyn
Walter, Ronald Bruce
Ward, Jonathan Bishop, Jr
Weinstock, George Matthew
Wheeler, Marshall Ralph
Williams, Robert Sanders
Williams-Blangero, Sarah Ann
Willig, Michael Robert
Wilson, Carol Maggart
Wilson, Golder North
Wilson, John H
Womack, James E
Woo, Savio L C
Wright, David Anthony
Wright, Stephen E
Wright, Woodring Erik
Yamauchi, Toshio
Young, Ryland F
Zwaan, Johan Thomas

UTAH
Albrechtsen, Rulon S
Andersen, William Ralph
Arave, Clive W
Asay, Kay Harris
Belcher, Bascom Anthony
Capecchi, Mario Renato
Davern, Cedric I
Dewey, Douglas R
Dewey, Wade G
Dickinson, William Joseph
Farmer, James Lee
Franklin, Naomi C
Gardner, Eldon John
George, Sarah B(rewster)
Georgopoulos, Constantine Panos
Gesteland, Raymond Frederick
Hansen, Afton M
Hunt, Steven Charles
Jeffery, Duane Eldro
Knapp, Gayle
Koehn, Richard Karl
Lanner, Ronald Martin
Lark, Karl Gordon
Larsen, Lloyd Don
McArthur, Eldon Durant
Mullen, Richard Joseph
Okun, Lawrence M
Park, Robert Lynn
Parkinson, John Stansfield
Shumway, Lewis Kay
Simmons, Daniel L
Simmons, John Robert
Sites, Jack Walter, Jr
Skolnick, Mark Henry
Stutz, Howard Coombs
Thomas, James H
Williams, Roger Richard
Wurst, Gloria Zettle

VERMONT
Albertini, Richard Joseph
Boraker, David Kenneth
Kilpatrick, Charles William
Moehring, Thomas John
Nicklas, Janice A
Novotny, Charles
Raper, Carlene Allen
Saul, George Brandon, II
Ullrich, Robert Carl
Wilkinson, Ronald Craig
Young, William Johnson, II

VIRGINIA
Annan, Murvel Eugene
Bempong, Maxwell Alexander
Bender, Patrick Kevin
Benepal, Parshotam S
Benzinger, Rolf Hans
Berger, Beverly Jane
Bradley, Sterling Gaylen
Brosseau, George Emile, Jr
Brown, Loretta Ann Port
Brown, Russell Vedder
Burian, Richard M
Buss, Glenn Richard
Chandler, Jerry LeRoy
Chinnici, Joseph (Frank) Peter
Cifone, Maria Ann
Cyr, W Howard
DeGiovanni-Donnelly, Rosalie F
De Haan, Henry J
Esen, Asim
Falkinham, Joseph Oliver, III
Frahm, Richard R
Gaines, James Abner
Garrett, Reginald Hooker
Grant, Bruce S
Harriman, Philip Darling
Heinemann, Richard Leslie
Hill, Jim T
Hilu, Khidir Wanni
Hoegerman, Stanton Fred
Hohenboken, William Daniel
Homsher, Paul Jim
Howard-Peebles, Patricia Nell
Howes, Cecil Edgar
Huskey, Robert John
Jarvis, Floyd Eldridge, Jr
Jones, Joyce Howell
Jones, Melton Rodney
Jones, William F
Kadner, Robert Joseph
Kelly, Thaddeus Elliott
Kilpatrick, S James, Jr
Kolsrud, Gretchen Schabtach
Krugman, Stanley Liebert
Lacy, George Holcombe
Leighton, Alvah Theodore, Jr
Lutes, Charlene McClanahan
McGilliard, Michael Lon
Marlowe, Thomas Johnson
Mauer, Irving
Merz, Timothy
Miller, Oscar Lee, Jr
Muir, William Angus
Murli, Hemalatha
Nance, Walter Elmore
Nasser, DeLill
Osgood, Christopher James
Raizen, Carol Eileen
Roth, Karl Sebastian
Schulman, Joseph Daniel
Sherald, Allen Franklin
Siegel, Paul Benjamin
Stetka, Daniel George
Stevens, Robert Edward
Swiger, Louis Andre
Taub, Stephan Robert
Teichler-Zallen, Doris
Townsend, J(oel) Ives
Trout, William Edgar, III
Veilleux, Richard Ernest
Vinson, William Ellis
Wallace, Bruce
West, David Armstrong
White, John Marvin
Wolf, Barry
Wright, Theodore Robert Fairbank

WASHINGTON
Ahmed, Saiyed I
Becker, Walter Alvin
Bendich, Arnold Jay
Bigley, Robert Harry
Bogyo, Thomas P
Brooks, Antone L
Byers, Breck Edward
Carr, Robert Leroy
Cheevers, William Phillip
Chen, Shi-Han

Genetics (cont)

Cosman, David John
Daniels, Jess Donald
Erickson, John (Elmer)
Fangman, Walton L
Felsenstein, Joseph
Fialkow, Philip Jack
Fitch, Cynthia Lynn
Fournier, R E Keith
Gartler, Stanley Michael
Giblett, Eloise Rosalie
Grootes-Reuvecamp, Grada Alijda
Hall, Benjamin Downs
Hard, Jeffrey John
Hartwell, Leland Harrison
Hawthorne, Donald Clair
Hazelbauer, Gerald Lee
Hecht, Adolph
Heston, Leonard L
Hillers, Joe Karl
Hood, Leroy E
Hosick, Howard Lawrence
Humphrey, Donald Glen
Hutchison, Nancy Jean
Karp, Laurence Edward
Kleinhofs, Andris
Konzak, Calvin Francis
Kraft, Joan Creech
Kutter, Elizabeth Martin
Laird, Charles David
Lamb, Mary Rose
Landis, Wayne G
Lane, Wallace
Lasure, Linda Lee
Leng, Earl Reece
Lightfoot, Donald Richard
McClary, Cecil Fay
MacKay, Vivian Louise
Martin, Mark Wayne
Matlock, Daniel Budd
Miller, Arthur Dusty
Motulsky, Arno Gunther
Nelson, Karen Ann
Nilan, Robert Arthur
Nute, Peter Eric
Ochs, Hans D
Olson, Maynard Victor
Osborne, Richard Hazelet
Pall, Martin L
Pearson, Mark Landell
Peterson, Clarence James, Jr
Pious, Donald A
Prieur, David John
Raskind, Wendy H
Salk, Darrell John
Schellenberg, Gerard David
Schemske, Douglas William
Schroeder, Alice Louise
Shortess, David Keen
Sibley, Carol Hopkins
Stadler, David Ross
Stamatoyannopoulos, George
Stenchever, Morton Albert
Stettler, Reinhard Friederich
Stonecypher, Roy W
Taylor, Ronald
Thelen, Thomas Harvey
Thompson, Elizabeth Alison
Towner, Richard Henry
Utter, Fred Madison
Vigfusson, Norman V
Wakimoto, Barbara Toshiko
Wijsman, Ellen Marie
Yao, Meng-Chao
Young, Francis Allan
Zakian, Virginia Araxie

WEST VIRGINIA
Chisler, John Adam
Headings, Verle Emery
Kaczmarczyk, Walter J
Keller, Edward Clarence, Jr
Latterell, Richard L
Ong, Tong-man
Reichenbecher, Vernon Edgar, Jr
Thayne, William V
Ulrich, Valentin
Voigt, Paul Warren
Yelton, David Baetz

WISCONSIN
Abrahamson, Seymour
Adler, Julius
Ahlquist, Paul G
Anderson, Robert Philip
Azen, Edwin Allan
Bennett, Kenneth A
Bergtrom, Gerald
Bersu, Edward Thorwald
Bitgood, J(ohn) James
Burgess, Ann Baker
Casler, Michael Darwin
Chambliss, Glenn Hilton
Chapman, Arthur Barclay
Courtright, James Ben
Crow, James Franklin
Curtis, Robin Livingstone
Datta, Surinder P
Dimond, Randall Lloyd
Dove, William Franklin
Duewer, Elizabeth Ann
Einspahr, Dean William
Ellingboe, Albert Harlan
Forsberg, Robert Arnold
Gabelman, Warren Henry
Gourse, Richard Lawrence
Greenspan, Daniel S
Haas, Michael John
Hall, Marion Trufant
Harms, Jerome Scott
Higgs, Roger L
Hornemann, Ulfert
Hougas, Robert Wayne
Huss, Ronald John
Ihrke, Charles Albert
Jacobs, Lois Jean
Jacobson, Gunnard Kenneth
Jungck, John Richard
Kaplan, Stanley
Kermicle, Jerry Lee
Kung, Ching
Lim, Johng Ki
Littlewood, Barbara Shaffer
Long, Sally Yates
McDonough, Eugene Stowell
Meisner, Lorraine Faxon
Mertz, Janet Elaine
Miller, Paul Dean
Nagodawithana, Tilak Walter
Nelson, Oliver Evans, Jr
Pan, David
Phillips, Ruth Brosi
Porter, John Willard
Rapacz, Jan
Schindler, Joel Marvin
Seelke, Ralph Walter
Sendelbach, Anton G
Sheehy, Michael Joseph
Simon, Philipp William
Smith, Richard R
Sondel, Paul Mark
Spritz, Richard Andrew
Strohm, Jerry Lee
Susman, Millard
Sytsma, Kenneth Jay
Szybalski, Waclaw
Temin, Rayla Greenberg
Tracy, William Francis
Wejksnora, Peter James
Williams, Paul Hugh

WYOMING
Bear, Phyllis Dorothy
Petersen, Nancy Sue
Tabachnick, Walter J

PUERTO RICO
Bruck, David Lewis
Carrasco-Serrano, Clara E
Craig, Syndey Pollock, III
Squire, Richard Douglas
Virkki, Niilo

ALBERTA
Ahmed, Asad
Andrews, John Edwin
Bentley, Michael Martin
Berg, Roy Torgny
Briggs, Keith Glyn
Church, Robert Bertram
Dancik, Bruce Paul
Dixon, Gordon H
Francis, Mike McD
Fredeen, Howard T
Gooding, Ronald Harry
Hodgetts, Ross Birnie
Iatrou, Kostas
Kuspira, J
Langridge, William Henry Russell
Laroche, Andre
Larson, Ruby Ila
Machin, Geoffrey Allan
Martin, Renee Halo
Muendel, Hans-Henning
Nash, David
Roberts, David Wilfred Alan
Rosenberg, Susan Mary
Rudd, Noreen L
Sanderson, Kenneth Edwin
Scheinberg, Eliyahu
Spence, Matthew W
Strobeck, Curtis
Von Borstel, Robert Carsten
Wagenaar, Emile B
Wegmann, Thomas George
Weijer, Jan
Yang, Rong-Cai
Yeh, Francis Cho-hao
Yoon, Ji-Won

BRITISH COLUMBIA
Applegarth, Derek A
Baird, Patricia A
Beacham, Terry Dale
Cavalier-Smith, Thomas
Cooke, Fred
Druehl, Louis D
Eaves, Connie Jean
El-Kassaby, Yousry A
Friedman, Jan Marshall
Ganders, Fred Russell
Griffiths, Anthony J F
Hill, Arthur Thomas
Holl, Frederick Brian
Holm, David George
Hunt, Richard Stanley
Illingworth, Keith
Kafer, Etta (Mrs E R Boothroyd)
MacDiarmid, William Donald
Meagher, Michael Desmond
Miller, James Reginald
Monsalve, Maria Victoria
Namkoong, Gene
Redfield, Rosemary Jeanne
Sadowski, Ivan J
Salari, Hassan
Setterfield, George Ambrose
Spiegelman, George Boole
Stich, Hans F
Styles, Ernest Derek
Suzuki, David Takayoshi
Sziklai, Oscar
Tener, Gordon Malcolm
Utkhede, Rajeshwar Shamrao

MANITOBA
Aung, Taing
Ayles, George Burton
Bodaly, Richard Andrew
Campbell, Allan Barrie
Clayton, James Wallace
Dyck, Peter Leonard
Evans, Laurie Edward
Hamerton, John Laurence
Helgason, Sigurdur Bjorn
Kerber, Erich Rudolph
Kondra, Peter Alexander
Kunz, Bernard Alexander
Larter, Edward Nathan
Lewis, Marion Jean
McAlpine, Phyllis Jean
Wrogemann, Klaus

NEW BRUNSWICK
Bonga, Jan Max
Fowler, Donald Paige
Young, Donald Alcoe

NEWFOUNDLAND
Carr, Steven McEwin
Davidson, William Scott
Scott, Peter John

NOVA SCOTIA
Chiasson, Leo Patrick
Crober, Donald Curtis
Doyle, Roger Whitney
Gray, Michael William
Haley, Leslie Ernest
Kamra, Om Perkash
Newkirk, Gary Francis
Van der Meer, John Peter
Wainwright, Lillian K (Schneider)
Welch, J Philip
Zouros, Eleftherios

ONTARIO
Anstey, Thomas Herbert
Arnison, Paul Grenville
Axelrad, Arthur Aaron
Bacchetti, Silvia
Baker, Allan John
Barrett, Spencer Charles Hilton
Battle, Helen Irene
Behme, Ronald John
Bernstein, Alan
Britton, Donald MacPhail
Buchwald, Manuel
Burnside, Edward Blair
Cade, William Henry
Campbell, Kenneth Wilford
Carmody, George R
Carstens, Eric Bruce
Chaconas, George
Chaly, Nathalie
Chambers, James Robert
Chang, Patricia Lai-Yung
Childers, Walter Robert
Chui, David H K
Cinader, Bernhard
Connelly, Philip Walter
Cox, Diane Wilson
Cummins, Joseph E
Danska, Jaynes
Davidson, Ronald G
Deeley, Roger Graham
Dodson, Edward O
Dube, Ian David
Engstrom, Mark Douglas
Fedak, George
Fejer, Stephen Oscar
Fiser, Paul S(tanley)
Galsworthy, Peter Robert
Galsworthy, Sara B
Goldberg, Gary Steven
Gowe, Robb Shelton
Grunder, Allan Angus
Gupta, Radhey Shyam
Haj-Ahmad, Yousef
Harney, Patricia Marie
Haynes, Robert Hall
Heddle, John A M
Hegele, Robert A
Hickey, Donal Aloysius
Ho, Keh Ming
Hutton, Elaine Myrtle
Ihssen, Peter Edowald
Irwin, David Michael
Joneja, Madan Gopal
Kadanka, Zdenek Karel
Kalow, Werner
Kang, C Yong
Kasha, Kenneth John
Kasupski, George Joseph
Kean, Vanora Mabel
Kidder, Gerald Marshall
Krepinsky, Jiri J
Lachance, Marc-Andre
Larsen, Ellen Wynne
Lin, Ching Y
Lu, Benjamin Chi-Ko
MacLennan, David Herman
Manolson, Morris Frank
Moens, Peter B
Newcombe, Howard Borden
Percy, Maire Ede
Petras, Michael Luke
Rajhathy, Tibor
Reed, Thomas Edward
Reid, Robert Leslie
Robinson, Brian Howard
Rossant, Janet
Sampson, Dexter Reid
Saunders, Shelley Rae
Seligy, Verner Leslie
Sengar, Dharmendra Pal Singh
Sergovich, Frederick Raymond
Simpson, Nancy E
Singal, Dharam Parkash
Singh, Rama Shankar
Singh, Rama Shankar
Sinha, Raj P
Smiley, James Richard
Soltan, Hubert Constantine
Sorger, George Joseph
Soudek, Dushan Edward
Tallan, Irwin
Thompson, Margaret A Wilson
Threlkeld, Stephen Francis H
Trevors, Jack Thomas
Tsui, Lap-Chee
Uchida, Irene Ayako
Walden, David Burton
Warwick, Suzanne Irene
Waye, John Stewart
Worton, Ronald Gilbert
Yeatman, Christopher William
Young, Paul Gary

PRINCE EDWARD ISLAND
Choo, Thin-Meiw

QUEBEC
Anderson, William Alan
Brown, Douglas Fletcher
Buckland, Roger Basil
Cedergren, Robert J
Chabot, Benoit
Chafouleas, James G
Chiang, Morgan S
Chung, Young Sup
Clark, Michael Wayne
Comeau, Andre I
DuBow, Michael Scott
Fahmy, Mohamed Hamed
Fraser, Frank Clarke
Galbraith, David Allan
Gervais, Francine
Glorieux, Francis Henri
Goodfriend, Lawrence
Grant, William Frederick
Green, David M(artin)
Guttmann, Ronald David
Hallenbeck, Patrick Clark
Hamelin, Claude
Labuda, Damian
Langford, Arthur Nicol
Lapointe, Jacques
Messing, Karen
Mukerjee, Barid
Nishioka, Yutaka
Palmour, Roberta Martin
Pinsky, Leonard
Preus, Marilyn Ione
Rosenblatt, David Sidney
Roy, Robert Michael McGregor
Sarhan, Fathey
Scriver, Charles Robert
Shoubridge, Eric Alan
Sikorska, Hanna
Slilaty, Steve N
Southin, John L
Stanners, Clifford Paul
Stevenson, Mary M
Tanguay, Robert M
Thirion, Jean Paul Joseph
Trasler, Daphne Gay
Vezina, Claude
Wolsky, Alexander

SASKATCHEWAN
Baker, Robert John
Crawford, Roy Douglas
Goplen, Bernard Peter
Harvey, Bryan Laurence
Khachatourians, George G
Kraay, Gerrit Jacob
Mortensen, Knud
Rank, Gerald Henry
Rossnagel, Brian Gordon
Scheltgen, Elmer
Scoles, Graham John
Shokeir, Mohamed Hassan Kamel
Spurr, David Tupper
Vella, Francis

BIOLOGICAL SCIENCES / 111

Williams, Charles Melville

YUKON TERRITORY
Doermann, August Henry

OTHER COUNTRIES
Agarwal, Shyam S
Andrews, Peter Walter
Arai, Ken-Ichi
Arber, Werner
Bloom, Arthur David
Bodmer, Walter Fred
Bortolozzi, Jehud
Bowman, James Talton
Britten, Edward James
Chomchalow, Narong
Christodoulou, John
Collins, Vincent Peter
Curtis, Byrd Collins
Da Cunha, Antonio Brito
Doerfler, Walter Hans
Elsevier, Susan Maria
Fitzhugh, Henry Allen
Goding, James Watson
Goodman, Fred
Hoisington, David A
Hotta, Yasuo
Jakob, Karl Michael
Kahn, Albert
Kaiser, Hinrich
Kaltsikes, Pantouses John
Kang, Kewon
Kerr, Warwick Estevam
Kimoto, Masao
Klassen, Waldemar
Klein, Jan
Meins, Frederick, Jr
Mojica-a, Tobias
Morton, Newton Ennis
Nevo, Eviatar
Nienstaedt, Hans
Ou, Jonathan Tsien-hsiong
Pavan, Crodowaldo
Radford, Alan
Rajaram, Sanjaya
Salzano, Francisco Mauro
Spencer, John Francis Theodore
Spierer, Pierre
Stearns, Stephen Curtis
Stodolsky, Marvin
Subramanian, Alap Raman
Tan, Y H
Watanabe, Takeshi
Yakura, Hidetaka
Yamamoto, Hiroshi

Hydrobiology

ALABAMA
Bayne, David Roberge
Beyers, Robert John
Brown, Jack Stanley
Folkerts, George William
Richardson, Terry David
Rouse, David B

ALASKA
Alexander, Vera
Warner, Mark Clayson

ARIZONA
Blinn, Dean Ward

ARKANSAS
Jones, Joe Maxey
Oliver, Kelly Hoyet, Jr

CALIFORNIA
Conklin, Douglas Edgar
Eriksen, Clyde Hedman
Fuhrman, Jed Alan
Horne, Alexander John
Knight, Allen Warner
Landolfi, Nicholas F
McCosker, John E
Maciolek, John A
Mercer, Edward King
Slack, Keith Vollmer
Tribbey, Bert Allen

COLORADO
Cohen, Ronald R H
Fausch, Kurt Daniel
Havlick, Spenser Woodworth
LaBounty, James Francis, Sr
Ward, James Vernon
Windell, John Thomas

CONNECTICUT
Jokinen, Eileen Hope
Katz, Max
Les, Donald Henry
Whitworth, Walter Richard

DELAWARE
Curtis, Lawrence Andrew
Smith, David William

DISTRICT OF COLUMBIA
Hart, Charles Willard, Jr

FLORIDA
Belanger, Thomas V

Bowes, George Ernest
Cardeilhac, Paul T
Dinsmore, Bruce Heasley
Haller, William T
Kerr, John Polk
Lee, Ming T
McCowen, Max Creager
Nordlie, Frank Gerald
Steinman, Alan David

GEORGIA
Greeson, Phillip Edward
Meyer, Judy Lynn
Nichols, Michael Charles
Stirewalt, Harvey Lee
Stoneburner, Daniel Lee
Tietjen, William Leighton
Wallace, James Bruce
Winger, Parley Vernon

HAWAII
Fast, Arlo Wade
Qin, Jianguang

ILLINOIS
Brigham, Warren Ulrich
Brugam, Richard Blair
Haynes, Robert C
Jahn, Lawrence A
Sparks, Richard Edward
Stull, Elisabeth Ann

INDIANA
Chamberlain, William Maynard
Davies, Harold William, (Jr)
Frey, David Grover
Gammon, James Robert
Lamberti, Gary Anthony
Lodge, David Michael
Spacie, Anne

IOWA
Bovbjerg, Richard Viggo
McDonald, Donald Burt
Merkley, Wayne Bingham

KANSAS
Boles, Robert Joe

KENTUCKY
Stevenson, Robert Jan
Timmons, Thomas Joseph

LOUISIANA
Brown, Kenneth Michael
Dionigi, Christopher Paul
Heins, David Carl
Moore, Walter Guy
Poirrier, Michael Anthony
Vidrine, Malcolm Francis

MARYLAND
Elwood, Jerry William
Gerritsen, Jeroen
Pancella, John Raymond
Paul, Robert William, Jr
Picciolo, Grace Lee

MASSACHUSETTS
Binford, Michael W
Colinvaux, Paul Alfred
Colt, LeBaron C, Jr
Dacey, John W H
Guillard, Robert Russell Louis

MICHIGAN
Andresen, Norman A
Bowen, Stephen Hartman
Burton, Thomas Maxie
Douglas, Matthew M
Foster, Neal Robert
Holland-Beeton, Ruth Elizabeth
Hough, Richard Anton
Kevern, Niles Russell
Kovalak, William Paul
Scavia, Donald
Stoermer, Eugene F
Wujek, Daniel Everett

MINNESOTA
Brezonik, Patrick Lee
Davis, Margaret Bryan
Henry, Mary Gerard
Holt, Charles Steele
Klemer, Andrew Robert
McConville, David Raymond
Perry, James Alfred
Ruschmeyer, Orlando R
Straw, Thomas Eugene
Welter, Alphonse Nicholas

MISSISSIPPI
Knight, Luther Augustus, Jr

MISSOURI
Hanson, Willis Dale
Stern, Daniel Henry
Welply, Joseph Kevin

MONTANA
Kaya, Calvin Masayuki
Schumacher, Robert E

NEVADA
Deacon, James Everett

NEW HAMPSHIRE
Gilbert, John Jouett

NEW JERSEY
Dorfman, Donald
Keating, Kathleen Irwin
Morgan, Mark Douglas
Riemer, Donald Neil
Rosengren, John
Weis, Judith Shulman

NEW MEXICO
Molles, Manuel Carl, Jr

NEW YORK
Boylen, Charles William
Collins, Carol Desormeau
Forest, Herman Silva
Godshalk, Gordon Lamar
Hairston, Nelson George, Jr
Hoham, Ronald William
Johnston, Richard Boles, Jr
Klotz, Richard Lawrence
Makarewicz, Joseph Chester
Oglesby, Ray Thurmond
Osterberg, Donald Mouretz
Rhee, G-Yull
Slobodkin, Lawrence Basil
Smith, Alden Ernest
Strayer, David Lowell

NORTH CAROLINA
Brinson, Mark McClellan
Campbell, Peter Hallock
Derrick, Finnis Ray
Grove, Thurman Lee
Kamykowski, Daniel

OHIO
Cooke, George Dennis
Culver, David Alex
Federle, Thomas Walter
Francko, David Alex
Herdendorf, Charles Edward, III
Hummon, William Dale
Laushman, Roger H
Olive, John H
Staker, Robert D
Stansbery, David Honor
Stein, Carol B
Wickstrom, Conrad Eugene

OKLAHOMA
Hawxby, Keith William
Korstad, John Edward
Matthews, William John

OREGON
Buscemi, Philip Augustus
Castenholz, Richard William
Chapman, Gary Adair
Church, Marshall Robbins
Garton, Ronald Ray
Hughes, Robert Mason
Malueg, Kenneth Wilbur

PENNSYLVANIA
Arnold, Dean Edward
Bott, Thomas Lee
Bouchard, Raymond William
Coffman, William Page
Haase, Bruce Lee
Kilham, Susan Soltau
Kimmel, William Griffiths
Kleinman, Robert L P
Mickle, Ann Marie
Prezant, Robert Steven
Sanders, Robert W
Williamson, Craig Edward

SOUTH CAROLINA
Abernathy, A(twell) Ray
Spain, James Dorris, Jr
Wixson, Bobby Guinn

TENNESSEE
Brinkhurst, Ralph O
Bunting, Dewey Lee, II
Cada, Glenn Francis
Elmore, James Lewis
Maier, Kurt Jay

TEXAS
Anderson, Richard Orr
Clark, William Jesse
Dickson, Kenneth Lynn
Foster, John Robert
Hellier, Thomas Robert, Jr
Lind, Owen Thomas
Maki, Alan Walter
Neill, William Harold
Sterner, Robert Warner
Waller, William T

UTAH
Quinn, Barry George

VIRGINIA
Benfield, Ernest Frederick
Bishop, John Watson
Brehmer, Morris Leroy

Buikema, Arthur L, Jr
Chang, William Y B
Hendricks, Albert Cates
Stevens, Robert Edward
Webster, Jackson Ross
Zubkoff, Paul L(eon)

WASHINGTON
Bisson, Peter Andre
Damkaer, David Martin
Smith, Stamford Dennis
Soltero, Raymond Arthur

WEST VIRGINIA
Adkins, Dean Aaron
Bissonnette, Gary Kent
Jenkins, Charles Robert
Weaks, Thomas Elton

WISCONSIN
Brooks, Arthur S
Carpenter, Stephen Russell
Graham, Linda Kay Edwards
Magnuson, John Joseph
Mortimer, Clifford Hiley
Remsen, Charles C, III
Schmitz, William Robert
Seale, Dianne B
Verch, Richard Lee

WYOMING
Dolan, Joseph Edward

ALBERTA
Gorham, Paul Raymond
Scrimgeour, Garry Joseph
Wallis, Peter Malcolm

BRITISH COLUMBIA
Nordin, Richard Nels
Perrin, Christopher John
Walker, Ian Richard

MANITOBA
Bodaly, Richard Andrew
Flannagan, John Fullan
Hamilton, Robert Duncan
Hecky, Robert Eugene
Pip, Eva
Staniforth, Richard John

NEWFOUNDLAND
Davis, Charles (Carroll)
Sheath, Robert Gordon

NOVA SCOTIA
Chen, Lawrence Chien-Ming

ONTARIO
Barica, Jan M
Calder, Dale Ralph
Chengalath, Rama
Coad, Brian William
Colby, Peter J
Collins, Nicholas Clark
Duthie, Hamish
Fernando, Constantine Herbert
Fraser, James Millan
Harrison, Arthur Desmond
Hebert, Paul David Neil
Leach, Joseph Henry
MacCrimmon, Hugh Ross
Nicholls, Kenneth Howard
Oliver, Donald Raymond
Smol, John Paul
Taylor, William David
Vollenweider, Richard A
Wiggins, Glenn Blakely

QUEBEC
De la Noue, Joel Jean-Louis
Kalff, Jacob
Marsden, Joan Chauvin
Peters, Robert Henry

OTHER COUNTRIES
Davis, William Jackson
Dickman, Michael David
Stearns, Stephen Curtis

Immunology

ALABAMA
Abrahamson, Dale Raymond
Acton, Ronald Terry
Alford, Charles Aaron, Jr
Bennett, Joe Claude
Blalock, James Edwin
Briles, Connally Oran
Briles, David Elwood
Carrozza, John Henry
Chapatwala, Kirit D
Cooper, Max Dale
Egan, Marianne Louise
Elson, Charles O, III
Ewald, Sandra J
Furuto, Donald K
Gathings, William Edward
Gillespie, George Yancey
Hajduk, Stephen Louis
Hazlegrove, Leven S
Hiramoto, Raymond Natsuo
Hunter, Eric

Immunology (cont)

Kayes, Stephen Geoffrey
Kearney, John F
Kilpatrick, John Michael
Kiyono, Hiroshi
Klesius, Phillip Harry
Lamon, Eddie William
Lauerman, Lloyd Herman, Jr
Lausch, Robert Nagle
Lidin, Bodil Inger Maria
Lobuglio, Albert Francis
McCarthy, Dennis Joseph
McCombs, Candace Cragen
McGhee, Jerry Roger
Mestecky, Jiri
Michalek, Suzanne M
Michalski, Joseph Potter
Montgomery, John R
Niemann, Marilyn Anne
Panangala, Victor S
Patterson, Loyd Thomas
Peterson, Raymond Dale August
Pillion, Dennis Joseph
Rohrer, James William
Russell, Michael W(illiam)
Schrohenloher, Ralph Edward
Singh, Shiva Pujan
Smith, Paul Clay
Turco, Jenifer
Walia, Amrik Singh
Winters, Alvin L

ARIZONA
Aden, David Paul
Archer, Stanley J
Baier, Joseph George
Burkholder, Peter M
Cawley, Leo Patrick
DeLuca, Dominick
Halonen, Marilyn Jean
Harris, David Thomas
Hendrix, Mary J C
Hersh, Evan Manuel
Janssen, Robert (James) J
Jeter, Wayburn Stewart
Kay, Marguerite M B
Lopez, Maria del Carmen
Lukas, Ronald John
Manak, Rita C
Marchalonis, John Jacob
Meinke, Geraldine Chciuk
Nagle, Ray Burdell
Nardella, Francis Anthony
Northey, William T
Perper, Robert J
Pinnas, Jacob Louis
Roholt, Oliver A, Jr
Seligmann, Bruce Edward
Wallraff, Evelyn Bartels
Watson, Ronald Ross
Woods, Alexander Hamilton
Wu, Chuanyue

ARKANSAS
Jasin, Hugo E
Leone, Charles Abner
Milburn, Gary L
Rank, Roger Gerald
Roberts, Dean Winn, Jr
Soderberg, Lee Stephen Freeman
Thoma, John Anthony
Wennerstrom, David E

CALIFORNIA
Abel, Carlos Alberto
Aladjem, Frederick
Allison, Anthony Clifford
Allison, James Patrick
Altieri, Dario Carlo
Amkraut, Alfred A
Anthony, Ronald Lewis
Arquilla, Edward R
Ascher, Michael S
Ashkenazi, Avi
Balakrishnan, Krishna
Banovitz, Jay Bernard
Bartl, Paul
Bastian, John F
Beall, Gildon Noel
Bechtol, Kathleen B
Becker, Martin Joseph
Benjamini, Eliezer
Bennett, C Frank
Berek, Jonathan S
Bhattacharya, Prabir
Bishop, John Michael
Bishop, Nancy Horschel
Bjorkman, Pamela J
Black, Kirby Samuel
Blair, Phyllis Beebe
Blakeslee, Dennis L(auren)
Block, Jerome Bernard
Bluestein, Harry Gilbert
Bokoch, Gary M
Bonavida, Benjamin
Bond, Martha W(illis)
Bozdech, Marek Jiri
Bramhall, John Shepherd
Brandon, David Lawrence
Bremermann, Hans J
Brostoff, Steven Warren
Bruckner, David Alan
Brunell, Philip Alfred
Burnham, Thomas K
Burns, John William
Byers, Vera Steinberger
Carlson, James Reynold
Carson, Dennis A
Castles, James Joseph, Jr
Caulfield, John Philip
Cernosek, Stanley Frank, Jr
Chang, Chin-Hai
Chang, Jennie C C
Chang, Mei-Ping
Chen, Benjamin P P
Chen, Shiuan
Chenoweth, Dennis Edwin
Chesnut, Robert W
Childress, Evelyn Tutt
Chisari, Francis Vincent
Church, Joseph August
Clark, William R
Clement, Loran T
Cooper, Edwin Lowell
Cooper, Neil R
Corbeil, Lynette B
Costea, Nicolas V
Cowan, Keith Morris
Cremer, Natalie E
Curnutte, John Tolliver, III
Curtiss, Linda K
Dafforn, Geoffrey Alan
Davenport, Calvin Armstrong
David, Gary Samuel
Davies, Huw M
Davis, William Ellsmore, Jr
Delk, Ann Stevens
DeLustro, Frank Anthony
Del Villano, Bert Charles
Dennert, Gunther
Dixon, Frank James
Dixon, James Francis Peter
Dreyer, William J
Duffey, Paul Stephen
Dunnebacke-Dixon, Thelma Hudson
Dutton, Richard W
Duzgunes, Nejat
Eardley, Diane Douglas
Edgington, Thomas S
Effros, Rita B
Engvall, Eva Susanna
Ernst, David N
Etzler, Marilynn Edith
Evans, Ernest Edward, Jr
Fabrikant, Irene Berger
Fahey, John Leslie
Fathman, C Garrison
Feldman, Joseph David
Felgner, Philip Louis
Fidler, John M
Forghani-Abkenari, Bagher
Frick, Oscar L
Friou, George Jacob
Gadol, Nancy
Gale, Robert Peter
Gall, William Einar
Garfin, David Edward
Garratty, George
Garvey, Justine Spring
Gascoigne, Nicholas Robert John
Gelfand, David H
Gigli, Irma
Ginsberg, Mark Howard
Ginsberg, Theodore
Giorgi, Janis V(incent)
Gitnick, Gary L
Glembotski, Christopher Charles
Glovsky, M Michael
Golde, David William
Golub, Edward S
Golub, Sidney Harris
Gong, Yu
Good, Anne Haines
Goodman, Joel Warren
Goodman, Michael Gordon
Goodman, Richard E
Gottlieb, Michael Stuart
Green, Donald Eugene
Green, Douglas R
Greene, Elias Louis
Greenspan, John Simon
Grey, Howard M
Guenther, Donna Marie
Gupta, Rishab Kumar
Gupta, Sudhir
Gutman, George Andre
Haddad, Zack H
Haen, Peter John
Hanes, Deanne Meredith
Harris, John Wayne
Heeb, Mary Jo
Heiner, Douglas C
Henriksson, Thomas Martin
Herzenberg, Leonard Arthur
Heuer, Ann Elizabeth
Hilgard, Henry Rohrs
Holden, Howard T
Holman, Halsted Reid
Holmberg, Charles Arthur
Horwitz, Marcus Aaron
Houston, L L
Howard, Russell John
Hubbard, William Jack
Huff, Dennis Karl
Ishizaka, Kimishige
Ishizaka, Teruko
Jacob, Chaim O
Jadus, Martin Robert
Johansson, Mats W
Jones, Patricia Pearce
Jordan, Stanley Clark
Kang, Chang-Yuil
Kan-Mitchell, June
Katz, David Harvey
Kawakamit, Toshiaki
Kelley, Darshan Singh
Kermani-Arab, Vali
Kibler, Ruthann
Kiprov, Dobri D
Klaustermeyer, William Berner
Klinman, Norman Ralph
Kornfeld, Lottie
Koshland, Marian Elliott
Koths, Kirston Edward
Kubo, Ralph Teruo
Kuus-Reichel, Kristine
Lad, Pramod Madhusudan
Landolfi, Nicholas F
Landy, Maurice
Lerner, Richard Alan
Levinson, Warren E
Lewis, Susanna Maxwell
Linker-Israeli, Mariana
Linna, Timo Juhani
Linscott, William Dean
Lou, Kingdon
Lucas, David Owen
Luzzio, Anthony Joseph
McCarthy, Robert Elmer
McCormick, Robert T
McDevitt, Hugh O'Neill
MacKenzie, Malcolm R
McLaughlin-Taylor, Elizabeth
McMannis, John D
McMillan, Robert
Maino, Vernon Carter
Makinodan, Takashi
Makker, Sudesh Paul
Manclark, Charles Robert
Mann, Lewis Theodore, Jr
Marangos, Paul Jerome
Marquardt, Diana
Martin, David William, Jr
Masouredis, Serafeim Panogiotis
Mathies, Margaret Jean
Matsumura, Kenneth N
Mayron, Lewis Walter
Merchant, Bruce
Michaeli, Dov
Miller, Alexander
Miller, John Johnston, III
Mitchell, Malcolm Stuart
Mitsuhashi, Masato
Mochizuki, Diane Yukiko
Morrison, Sherie Leaver
Muller-Sieburg, Christa E
Murphy, Frederick A
Myhre, Byron Arnold
Nagy, Stephen Mears, Jr
Nakamura, Robert Motoharu
Ning, Shoucheng
Nitecki, Danute Emilija
Oh, Chan Soo
Olsen, Charles Edward
Osebold, John William
Outzen, Henry Clair, Jr
Owen, Ray David
Owen, Robert L
Parker, John William
Parkman, Robertson
Patzer, Eric John
Payne, Rose Marise
Pellegrino, Michele A
Perez, Hector Daniel
Peter, James Bernard
Peterlin, Boris Matija
Phelps, Leroy Nash
Pischel, Ken Donald
Pitesky, Isadore
Pollack, William
Prince, Harry E
Quismorio, Francisco P, Jr
Ralph, Peter
Raulet, David Henri
Rearden, Carole Ann
Reese, Robert Trafton
Reeve, Peter
Reisfeld, Ralph Alfred
Remington, Jack Samuel
Rheins, Lawrence A
Riblet, Roy Johnson
Richman, David Paul
Robbins, Dick Lamson
Rodgers, Richard Michael
Roederer, Mario
Rosenau, Werner
Rosenberg, Leon T
Rosenquist, Grace Link
Rosenthal, Sol Roy
Rounds, Donald Edwin
Rousell, Ralph Henry
Saifer, Mark Gary Pierce
Salk, Jonas Edward
Schiller, Neal Leander
Schoenholz, Walter Kurt
Schonfeld, Steven Emanuel
Sears, Duane William
Seeger, Robert Charles
Senda, Mototaka
Senyk, George
Sercarz, Eli
Sharma, Somesh Datt
Shen, Wei-Chiang
Sherman, Linda Arlene
Shively, John Ernest
Shopes, Bob
Shulman, Ira Andrew
Silverman, Paul Hyman
Singer, Paul A
Smith, Richard Scott
Smithwick, Elizabeth Mary
Snow, George Edward
Snow, John Thomas
Solomon, George Freeman
Spiegelberg, Hans L
Spitler, Lynn E
Stevens, Ronald Henry
Stiehm, E Richard
Stormont, Clyde J
Stringfellow, Dale Alan
Strober, Samuel
Suffin, Stephen Chester
Sweet, Benjamin Hersh
Szego, Clara Marian
Tachibana, Dora K
Takasugi, Mitsuo
Tamerius, John
Tan, Eng M
Taylor, Clive Roy
Tempelis, Constantine H
Terasaki, Paul Ichiro
Terres, Geronimo
Theofilopoulos, Argyrios N
Thoman, Marilyn Louise
Thueson, David Orel
Trowbridge, Ian Stuart
Ullman, Edwin Fisher
Umiel, Tehila
Unver, Ercan
Usinger, William R
Uyeda, Charles Tsuneo
Van Kuyk, Robert William
Vannier, Wilton Emile
Van Winkle, Michael George
Vaughan, John Heath
Vedros, Neylan Anthony
Vickrey, Herta Miller
Victoria, Edward Jess, Jr
Vogt, Peter Klaus
Vredevoe, Donna Lou
Vyas, Girish Narmadashankar
Wall, Thomas Randolph
Walter, Harry
Ware, Carl F
Wasserman, Stephen I
Webb, David Ritchie, Jr
Wedberg, Stanley Edward
Weigle, William O
Weliky, Norman
Wetzel, Gayle Delmonte
Whitlow, Marc David
Wiebe, Michael Eugene
Williams, Terry Wayne
Wilson, Darcy Benoit
Winkelhake, Jeffrey Lee
Witte, Owen Neil
Wofsy, Leon
Wright, Richard Kenneth
Wu, William Gay
Yamamoto, Richard
Young, Janis Dillaha
Yu, David Tak Yan
Zeleznick, Lowell D
Ziccardi, Robert John
Zighelboim, Jacob
Zlotnik, Albert
Zuniga, Martha C
Zvaifler, Nathan J

COLORADO
Arend, William Phelps
Barisas, Bernard George, Jr
Berkelhammer, Jane
Borysenko, Myrin
Brooks, Bradford O
Brooks Springs, Suzanne Beth
Bunn, Paul A, Jr
Cambier, John C
Campbell, Priscilla Ann
Chai, Hyman
Claman, Henry Neumann
Cockerell, Gary Lee
Crowle, Alfred John
De Martini, James Charles
Freed, John Howard
Gelfand, Erwin William
Giclas, Patricia C
Glode, Leonard Michael
Goren, Mayer Bear
Grieve, Robert B
Hager, Jean Carol
Harbeck, Ronald Joseph
Hayward, Anthony R
Henson, Peter Mitchell
Hudson, Bruce William
Irons, Richard Davis
Jacobs, Barbara B
Jones, Carol A
Jones, Steven Wayne
Jordan, Russell Thomas
Kirkpatrick, Charles Harvey
Larson, Kenneth Allen
Laudenslager, Mark LeRoy
Leung, Donald Yap Man
Levine, Simon Rock

Lieberman, Michael Merril
McCalmon, Robert T, Jr
Marrack, Philippa Charlotte
Martin, Richard Jay
Moorhead, John Wilbur
Repine, John E
Riches, David William Henry
Rodriguez, Eugene
Schanfield, Moses Samuel
Schooley, Robert T
Shopp, George Milton, Jr
Simske, Steven John
Smith, Ralph E
Spaulding, Harry Samuel, Jr
Talmage, David Wilson
Tengerdy, Robert Paul
Trapani, Ignatius Louis
Tyler, Kenneth Laurence
Urban, Richard William
Vasil, Michael Lawrence

CONNECTICUT
Ariyan, Stephan
Askenase, Philip William
Aune, Thomas M
Baltimore, Robert Samuel
Baumgarten, Alexander
Becker, Elmer Lewis
Blue, Marie-Luise
Booss, John
Bormann, Barbara-Jean Anne
Bothwell, Alfred Lester Meador
Brandsma, Janet Lousie
Braun, Phyllis C
Bullock, Ward Ervin, Jr
Bursuker, Isia
Cone, Robert Edward
Cornell-Bell, Ann Hall
Cresswell, Peter
Devlin, Richard Gerald, Jr
Doherty, Niall Stephen
Edberg, Stephen Charles
Epstein, Paul Mark
Faanes, Ronald
Flavell, Richard Anthony
Fong, Jack Sun-Chik
Fredericksen, Tommy L
Gomez-Cambronero, Julian
Grattan, James Alex
Horowitz, Mark Charles
Jacoby, Robert Ottinger
Janeway, Charles Alderson, Jr
Kantor, Fred Stuart
Kavathas, Paula
Kenyon, Alan J
Khan, Abdul Jamil
Kiron, Ravi
Korn, Joseph Howard
Kowolenko, Michael D
Krause, Peter James
Lacouture, Peter George
Lee, Sin Hang
Letts, Lindsay Gordon
Littman, Bruce H
Loose, Leland David
Martin, R(ufus) Russell
Mayol, Robert Francis
Newborg, Michael Foxx
Niblack, John Franklin
Otterness, Ivan George
Pfeifer, Richard Wallace
Pober, Jordan S
Polmar, Stephen Holland
Pratt, Diane McMahon
Primakoff, Paul
Puddington, Lynn
Rishell, William Arthur
Rocklin, Ross E
Ross, Martin Russell
Ruddle, Nancy Hartman
Schacter, Bernice Zeldin
Silbart, Lawrence K
Smith, Abigail LeVan
Smith, Brian Richard
Snider, Ray Michael
Spitalny, George Leonard
Walker, Frederick John
Wood, David Dudley
Woronick, Charles Louis
Yang, Tsu-Ju (Thomas)

DELAWARE
Dohms, John Edward
Fahey, James R
Frey, William Adrian
Fries, Cara Rosendale
Ganfield, David Judd
Ginnard, Charles Raymond
Gorzynski, Timothy James
Greenblatt, Hellen Chaya
Karol, Robin A
Keeler, Calvin Lee, Jr
Krell, Robert Donald
Melby, James Michael
Neubauer, Russell Howard
Ruben, Regina Lansing
Sands, Howard
Tripp, Marenes Robert

DISTRICT OF COLUMBIA
Affronti, Lewis Francis
Alving, Carl Richard
Birx, Deborah L
Bundy, Bonita Marie

Calderone, Richard Arthur
Chan, Maria M
Cocks, Gary Thomas
Gravely, Sally Margaret
Herscowitz, Herbert Bernard
Hollingdale, Michael Richard
Jett-Tilton, Marti
Johnson, Armead
Katz, Paul K
Kind, Phyllis Dawn
Krzych, Urszula
Ladisch, Stephan
Leto, Salvatore
Ligler, Frances Smith
Lumpkin, Michael Dirksen
Lyon, Jeffrey A
McCarty, Gale A
Matyas, Gary Ralph
Miller, Linda Jean
Palestine, Alan G
Pearson, Gary Richard
Phillips, Terence Martyn
Prograis, Lawrence
Richards, Roberta Lynne
Roane, Philip Ransom, Jr
Royal, George Calvin, Jr
Taylor, Diane Wallace
Tompkins, George Jonathan
Turner, Willie
Waldman, Robert H
Walters, Curla Sybil
Wassef, Nabila M
Wilson, Edward Matthew
Wright, Daniel Godwin

FLORIDA
Adams, Roger Omar
Avner, Barry P
Bernstein, Stanley H
Bliznakov, Emile George
Blomberg, Bonnie B
Bolton, Wade E
Brenner, Richard Joseph
Burke, George William, III
Chambliss, Keith Wayne
Clark, William Burton, IV
Cohen, Pinya
Crandall, Richard B
Crews, Fulton T
Day, Noorbibi Kassam
Elfenbein, Gerald Jay
Fletcher, Mary Ann
Fogel, Bernard J
Frankel, Jack William
Friedman, Herman
Gennaro, Joseph Francis
Gifford, George Edwin
Goihman-Yahr, Mauricio
Groupe, Vincent
Hadden, John Winthrop
Hoffmann, Edward Marker
Hornicek, Francis John
Jackson, Bettina B Carter
Jacobs, Diane Margaret
Johnson, Howard Marcellus
Kiefer, David John
Klein, Paul Alvin
Lai, Patrick Kinglun
Lawman, Michael John Patrick
Legler, Donald Wayne
Leon, Kenneth Allen
Lin, Tsue-Ming
Litman, Gary William
Lockey, Richard Funk
Longley, Ross E
Luer, Carl A
McArthur, William P
McCormack, Wayne Thomas
McGuigan, James E
Mahan, Suman Mulakhraj
Paradise, Lois Jean
Patarca, Roberto
Pearl, Gary Steven
Peck, Ammon Broughton
Plescia, Otto John
Pollara, Bernard
Pruzansky, Jacob Julius
Roberts, Thomas L
Roux, Kenneth H
Rowlands, David T, Jr
Schultz, Duane Robert
Scornik, Juan Carlos
Sinkovics, Joseph G
Small, Parker Adams, Jr
Smith, Dennis Eugene
Specter, Steven Carl
Stock, David Allen
Stone, Stanley S
Sweeney, Michael Joseph
Tamplin, Mark Lewis
Thorn, Richard Mark
Thuning-Roberson, Claire Ann
White, Roseann Spicola
Williams, Ralph C, Jr
Zarco, Romeo Morales

GEORGIA
Alonso, Kenneth B
Bennett, Sara Neville
Brockbank, Kelvin Gordon Mashader
Butcher, Brian T
Campbell, Gary Homer
Cavallaro, Joseph John
Cherry, William Bailey

Claflin, Alice J
Clark, Winston Craig
Colley, Daniel George
Compans, Richard William
Criswell, Bennie Sue
Damian, Raymond T
Daugharty, Harry
Duncan, Robert Leon, Jr
Evans, Donald Lee
Feeley, John Cornelius
Garver, Frederick Albert
Goldman, John Abner
Gooding, Linda R
Grayzel, Arthur I
Gulati, Adarsh Kumar
Gusdon, John Paul
Hall, Charles Thomas
Holt, Peter Stephen
Hugus, Barbara Hendrick
Hunter, Robert L
Kagan, Irving George
Kiefer, Charles R(andolph)
Knudsen, Richard Carl
Krauss, Jonathan Seth
McKinney, Roger Minor
Madden, John Joseph
Maddison, Shirley Eunice
Margolis, Harold Stephen
Martin, Linda Spencer
Mason, James Michael
Mathews, Henry Mabbett
Navalkar, Ram G
Peters, Clarence J
Pine, Leo
Pratt, Lee Herbert
Ragland, William Lauman, III
Reese, Andy Clare
Reiss, Errol
Rodey, Glenn Eugene
Roesel, Catherine Elizabeth
Roper, Maryann
Sawyer, Richard Trevor
Schierman, Louis W
Schmid, Donald Scott
Sell, Kenneth W
Sinor, Lyle Tolbot
Smith, David Fletcher
Stouffer, Richard Franklin
Stulting, Robert Doyle, Jr
Sulzer, Alexander Jackson
Tenoso, Harold John
Tsang, Victor Chiu Wan
Tunstall, Lucille Hawkins
Urso, Paul
Vogler, Larry B
Walls, Kenneth W
Ziegler, Harry Kirk

HAWAII
Benedict, Albert Alfred
Desowitz, Robert
Grothaus, Paul Gerard
Hokama, Yoshitsugi
Oishi, Noboru
Person, Donald Ames
Roberts, Robert Russell
Stuart, William Dorsey
Vann, Douglas Carroll

IDAHO
Ayers, Arthur Raymond
Congleton, James Lee
Laurence, Kenneth Allen

ILLINOIS
Andersen, Burton
Anderson, Byron
Ariano, Marjorie Ann
Arnason, Barry Gilbert Wyatt
Arroyave, Carlos Mariano
Bahn, Arthur Nathaniel
Baum, Linda Louise
Bell, Clara G
Bender, James Gregg
Bhatti, Rashid
Bluestone, Jeffrey A
Brackett, Robert Giles
Braun, Donald Peter
Brennan, Patricia Conlon
Briles, Worthie Elwood
Buchholz Shaw, Donna Marie
Cabana, Veneracion Garganta
Callaghan, Owen Hugh
Carter, George W
Chandler, John W
Chang, Chong-Hwan
Cheung, Hou Tak
Chudwin, David S
Cohen, Edward P
Conta, Barbara Saunders
Dal Canto, Mauro Carlo
Datta, Syamal K
Dau, Peter Caine
Dray, Sheldon
Dudkiewicz, Alan Bernard
Dybas, Linda Kathryn
Ekstedt, Richard Dean
Emeson, Eugene Edward
Falk, Lawrence A, Jr
Feldbush, Thomas Lee
Fenters, James Dean
Fiedler, Virginia Carol
Fitch, Frank Wesley

Friedman, Yochanan
Gaskins, H Rex
Gewurz, Henry
Goldberg, Erwin
Gotoff, Samuel P
Grundbacher, Frederick John
Halfman, Clarke Joseph
Hanly, W Carey
Hanson, Lyle Eugene
Harris, Jules Eli
Heim, Lyle Raymond
Holzer, Timothy J
Horng, Wayne J
Jelachich, Mary Lou
Johnson, Richard Joseph
Johnson, William Joseph
Jones, Carl Joseph
Jones, Marjorie Ann
Kapsalis, Andreas A
Kim, Byung Suk
Kim, Yoon Berm
Kim, Yung Dai
Klegerman, Melvin Earl
Knight, Katherine Lathrop
Koch, Alisa Erika
Kraft, Sumner Charles
Kulkarni, Anant Sadashiv
Landay, Alan Lee
Lange, Charles Ford
Lazda, Velta Abuls
Lint, Thomas F
McConnachie, Peter Ross
McLeod, Rima W
Maecker, Holden T
Marczynska, Barbara Mary
Margoliash, Emanuel
Markowitz, Abraham Sam
Mathews, Herbert Lester
Melvold, Roger Wayne
Miller, Stephen Douglas
Moskal, Joseph Russell
Moticka, Edward James
Nevalainen, David Eric
Ostrow, David Henry
Pachman, Lauren M
Plate, Janet Margaret
Potempa, Lawrence Albert
Prasad, Rameshwar
Pun, Pattle Pak-Toe
Radosevich, James A
Ratajczak, Helen Vosskuhler
Rich, Kenneth C
Robertson, Abel Alfred Lazzarini, Jr
Rogalski-Wilk, Adrienne Alice
Rowe, William Bruce
Rowley, Donald Adams
Sabet, Tawfik Younis
Sanborn, Mark Robert
Sandberg, Philip A
Schneider, Gary Bruce
Schreiber, Hans
Segre, Diego
Segre, Mariangela Bertani
Sharon, Nehama
Sherwood, Robert Lawrence
Shipchandler, Mohammed Tyebji
Shulman, Stanford Taylor
Simonson, Lloyd Grant
Sorensen, Keld
Spear, Patricia Gail
Springer, Georg F
Stevens, Fred Jay
Storb, Ursula
Tartof, David
Taswell, Howard Filmore
Thompson, Kenneth David
Tomita, Joseph Tsuneki
Turner, Donald W
Van Alten, Pierson Jay
Van Epps, Dennis Eugene
Venkataraman, M
Voss, Edward William, Jr
Waltenbaugh, Carl
Walter, Robert John
Wass, John Alfred
Weisz-Carrington, Paul
Willoughby, William Franklin
Winkler, Martin Alan
Wolf, Raoul
Wu-Wong, Jinshyun Ruth
Young, Jay Maitland
Zahalsky, Arthur C
Zeiss, Chester Raymond

INDIANA
Aldo-Benson, Marlene Ann
Ashendel, Curtis Lloyd
Basu, Manju
Bauer, Dietrich Charles
Bick, Peter Hamilton
Boguslaski, Robert Charles
Botero, J M
Brahmi, Zacharie
Diffley, Peter
Dunn, Peter Edward
Dziarski, Roman
Edwards, Joshua Leroy
Freeman, Max James
Gale, Charles
Gaur, Pramod Kumar
Ghosh, Swapan Kumar
Gregory, Richard Lee
Gunther, Gary Richard
Harvey, William Homer

Immunology (cont)

Hicks, Edward James
Ho, Peter Peck Koh
Hoversland, Roger Carl
Howard, David K
Hullinger, Ronald Loral
Hurrell, John Gordon
Ingraham, Joseph Sterling
Jenski, Laura Jean
Koppel, Gary Allen
Kreps, David Paul
Kwon, Byoung Se
Lopez, Carlos
McGowan, Michael James
McIntyre, John A
Miller (Gilbert), Carol Ann
Mocharla, Raman
Morter, Raymond Lione
Nunez, William J, III
Petersen, Bruce H
Roales, Robert R
Schmidtke, Jon Robert
Simon, Edward Harvey
Taylor, Milton William
Wagner, Morris
Wilde, Charles Edward, III
Yoder, John Menly
Zimmerman, Sarah E

IOWA
Ashman, Robert F
Ballas, Zuhari K
Boney, William Arthur, Jr
Butler, John Edward
Cheng, Frank Hsieh Fu
Clark, Robert A
Dailey, Morris Owen
Fitzgerald, Gerald Roland
Ford, Stephen Paul
Goeken, Nancy Ellen
Hartman, Paul Arthur
Hoffmann, Louis Gerhard
Hom, Ben Lin
Hsu, Shu Ying Li
Lamont, Susan Joy
Lubaroff, David Martin
Mayfield, John Eric
Meetz, Gerald David
Mengeling, William Lloyd
Miller, Wilmer Jay
Naides, Stanley J
O'Berry, Phillip Aaron
Quinn, Loyd Yost
Rebers, Paul Armand
Richerson, Hal Bates
Schmerr, Mary Jo F
Steinmuller, David
Stieler, Carol Mae
Thurston, John Robert
Weinstock, Joel Vincent
Welter, C Joseph

KANSAS
Bailie, Wayne E
Bartkoski, Michael John, Jr
Bradford, Lawrence Glenn
Brown, Alan R
Brown, John Clifford
Bryan, Christopher F
Dass, David
Draper, Laurence Rene
Gray, Andrew P
Greene, Nathan Doyle
Haas, Herbert Frank
Halsey, John Frederick
Hochman, Jerome Henry
Krishnan, Engil Kolaj
McElree, Helen
McVey, David Scott
Minocha, Harish C
Morrison, David Campbell
Parmely, Michael J
Ringle, David Allan
Suzuki, Tsuneo
Sweet, George H
Wolf, Thomas Michael
Wood, Gary Warren

KENTUCKY
Ambrose, Charles T
Bhattacharya, Malaya
Bhattacharya-Chatterjee, Malaya
Bondada, Subbarao
Espinosa, Enrique
Glassock, Richard James
Higginbotham, Robert David
Justus, David Eldon
Kaplan, Alan Marc
Kohler, Heinz
Maiti, Indu B
Ruchman, Isaac
Straley, Susan Calhoon
Thompson, John S
Wacker, Waldon Burdette
Wallace, John Howard

LOUISIANA
Bagby, Gregory John
Bautista, Abraham Parana
Cannon, Donald Charles
Choi, Yong Sung
Cohen, Geraldine H
Domer, Judith E

Domingue, Gerald James, Sr
Epps, Anna Cherrie
Espinoza, Luis Rolan
Etheredge, Edward Ezekiel
Franklin, Rudolph Michael
Fulginiti, Vincent Anthony
Gaumer, Herman Richard
Gebhardt, Bryan Matthew
Gormus, Bobby Joe
Gottlieb, A Arthur
Hastings, Robert Clyde
Hodgin, Ezra Clay
Hyslop, Newton Everett, Jr
Ivens, Mary Sue
Jerrells, Thomas Ray
Klei, Thomas Ray
Kohler, Peter Francis
Krahenbuhl, James Lee
Kuan, Shia Shiong
Lanners, H Norbert
Lehrer, Samuel Bruce
McDonald, John C
Magin, Ralph Walter
Martin, Louis Norbert
Millikan, Larry Edward
Misra, Raghunath P
Neucere, Joseph Navin
Nickerson, Stephen Clark
O'Neil, Carol Elliot
Pisano, Joseph Carmen
Salvaggio, John Edmond
Saphier, David
Shaffer, Morris Frank
Silberman, Ronald
Sizemore, Robert Carlen
Sorensen, Ricardo Uwe
Spitzer, Judy A
Strauss, Arthur Joseph Louis
Sullivan, Karen A
Thompson, James Jarrard
Wolf, Robert E

MAINE
Aronson, Frederick Rupp
Donahoe, John Philip
Eskelund, Kenneth H
Evans, Robert
Fink, Mary Alexander
Harrison, David Ellsworth
Kaliss, Nathan
Kepron, Michael Raymond
Moody, Charles Edward, Jr
Ng, Ah-Kau
Novotny, James Frank
Renn, Donald Walter
Shultz, Leonard Donald
Snell, George Davis
Weiss, John Jay
Whitaker, R Blake

MARYLAND
Abe, Ryo
Adkinson, Newton Franklin, Jr
Adler, William
Alexander, Nancy J
Amzel, L Mario
Anderson, Arthur O
Anderson, Robert Simpers
Anhalt, Grant James
Ansari, Aftab A
Aoki, Tadao
Appella, Ettore
Asofsky, Richard Marcy
Augustine, Patricia C
Austin, Faye Carol Gould
Azad, Abdu F(arhang)
Baer, Harold
Baker, Phillip John
Bala, Shukal
Balady, Michael A
Barker, Lewellys Franklin
Basta, Milan
Beaven, Michael Anthony
Beckner, Suzanne K
Bennett, William Ernest
Berkower, Ira
Berman, Howard Mitchell
Berne, Bernard H
Berzofsky, Jay Arthur
Beschorner, W E
Bhatnagar, Gopal Mohan
Biswas, Robin Michael
Blaese, R Michael
Bloom, Eda Terri
Bochner, Bruce Scott
Borsos, Tibor
Braatz, James Anthony
Breyere, Edward Joseph
Brink, Linda Holk
Brook, Itzhak
Brown, James Edward
Brunswick, Mark
Buckheit, Robert Walter, Jr
Bustin, Michael
Butman, Bryan Timothy
Carski, Theodore Robert
Caspi, Rachel R
Chen, Joseph Ke-Chou
Chen, Priscilla B(urtis)
Choi, Oksoon Hong
Chused, Thomas Morton
Cohen, Sheldon Gilbert
Cole, Gerald Alan
Coligan, John E

Collins, Frank Miles
Colombani, Paul Michael
Cowan, Elliot Paul
Cox, George Warren
Craig, Susan Walker
Creasia, Donald Anthony
Dannenberg, Arthur Milton, Jr
Das, Saroj R
Dasch, Gregory Alan
Davidson, Wendy Fay
Dean, Ann
DeVries, Yuan Lin
Dickler, Howard B
Diggs, Carter Lee
Dintzis, Renee Zlochover
Djurickovic, Draginja Branko
Dodd, Roger Yates
Donlon, Mildred A
Dragunsky, Eugenia M
Duncan, William Raymond
Durum, Scott Kenneth
Dwyer, Dennis Michael
Eckels, Kenneth Henry
Edidin, Michael Aaron
Ennist, David L
Epstein, Suzanne Louise
Esposito, Vito Michael
Evans, Charles Hawes
Falkler, William Alexander, Jr
Farrell, Robert Edmund, Jr
Fast, Patricia E
Fearon, Douglas T
Feinman, Susan (Birnbaum)
Fernie, Bruce Frank
Fex, Jorgen
Fields, Kay Louise
Finkelman, Fred Douglass
Fleisher, Thomas A
Fox, Barbara Saxton
Fraser, Blair Allen
Froehlich, Luz
Fuson, Roger Baker
Gainer, Joseph Henry
Gasbarre, Louis Charles
Gearhart, Patricia Johanna
Germain, Ronald N
Gerrard, Theresa Lee
Gery, Igal
Gilden, Raymond Victor
Glaudemans, Cornelis P
Goldman, Daniel Ware
Goldsmith, Paul Kenneth
Goldstein, Robert Arnold
Goodman, Howard Charles
Goodman, Sarah Anne
Granger, Donald Lee
Gravell, Maneth
Green, Ira
Gress, Ronald E
Griffin, Diane Edmund
Habig, William Henry
Hackett, Joseph Leo
Hale, Martha L
Hammer, Carl Helman
Hampar, Berge
Hanjan, Satnam S
Hanks, John Harold
Hanna, Edgar E(thelbert), Jr
Hanna, Michael G, Jr
Hardegree, Mary Carolyn
Hart, Mary Kate
Harvath, Liana
Haspel, Martin Victor
Hausman, Stephen J
Hearing, Vincent Joseph, Jr
Hecht, Toby T
Hellman, Alfred
Henkart, Pierre
Hess, Allan Duane
Hewetson, John Francis
Hoffeld, J Terrell
Hoffman, Thomas
Hook, William Arthur
Hoyer, Leon William
Huppi, Konrad E
Ivins, Bruce Edwards
Jaffe, Elaine Sarkin
Jakab, George Joseph
James, Stephanie Lynn
Johnson, Leslye
Johnson-Winegar, Anna
Johnston, Margaret Irene
Joseph, J Mehsen
Kagey-Sobotka, Anne
Keegan, Achsah D
Kenny, James Joseph
Kenyon, Richard H
Kickler, Thomas Steven
Krakauer, Henry
Krakauer, Teresa
Kramer, Norman Clifford
Krause, David
Krause, Richard Michael
Kresina, Thomas Francis
Kupers, Rudolf Carl
LaSalle, Bernard
Lederman, Howard Mark
Ledney, George David
Lee, Chi-Jen
Leef, James Lewis
Lees, Andrew
Leto, Thomas L
Liang, Shu-Mei
Lichtenstein, Lawrence M

Lillehoj, Hyun Soon
Lippincott-Schwartz, Jennifer
Lobel, Steven A
Longo, Dan L
Lunney, Joan K
McCarron, Richard M
McCartney-Francis, Nancy L
McClintock, Patrick Ralph
McCurdy, John Dennis
McFarland, Henry F
McFarlin, Dale Elroy
MacGlashan, Donald W, Jr
Mackler, Bruce F
MacVittie, Thomas Joseph
Mage, Michael Gordon
Mage, Rose G
Mageau, Richard Paul
Malech, Harry Lewis
Malmgren, Richard Axel
Mardiney, Michael Ralph, Jr
Maret, S Melissa
Margulies, David Harvey
Marquardt, Warren William
Marsh, David George
Martin, Mark Thomas
Matzinger, Polly
Maurer, Bruce Anthony
Meltzer, Monte Sean
Mergenhagen, Stephan Edward
Metcalfe, Dean Darrel
Metzger, Henry
Miller, Frederick William
Mittal, Kamal Kant
Mond, James Jacob
Monjan, Andrew Arthur
Morris, Joseph Anthony
Morse, Herbert Carpenter, III
Mufson, R Allan
Murrell, Kenneth Darwin
Mushinski, J Frederic
Nacy, Carol Anne
Nauman, Robert Karl
Nelson, David L
Neta, Ruth
Nordin, Albert Andrew
Norman, Philip Sidney
Notkins, Abner Louis
O'Beirne, Andrew Jon
Oppenheim, Joost J
O'Rangers, John Joseph
Order, Stanley Elias
Oroszian, Stephen
Ortaldo, John R
Ostrand-Rosenberg, Suzanne T
Ottesen, Eric Albert
Ozato, Keiko
Packard, Beverly Sue
Padarathsingh, Martin Lancelot
Parkman, Paul Douglas
Parvez, Zaheer
Paul, William Erwin
Picciolo, Grace Lee
Plaut, Marshall
Plotz, Paul Hunter
Pluznik, Dov Herbert
Pohl, Lance Rudy
Poljak, Roberto Juan
Powers, Kendall Gardner
Prendergast, Robert Anthony
Prince, Gregory Antone
Quinnan, Gerald Vincent, Jr
Ram, J Sri
Ranney, Richard Raymond
Read-Connole, Elizabeth Lee
Reitz, Marvin Savidge, Jr
Robbins, John B
Roberson, Bob Sanders
Robey, Frank A
Rose, Noel Richard
Rosenstein, Robert William
Rossio, Jeffrey L
Russell, Philip King
Samelson, Lawrence Elliot
Sandberg, Ann Linnea
Santos, George Wesley
Saxinger, W(illiam) Carl
Schleimer, Robert P
Schlom, Jeffrey
Schlueberg, Ann Elizabeth Snider
Schnittman, Steven Marc
Schricker, Robert Lee
Schulze, Dan Howard
Segal, David Miller
Seto, Belinda P L
Shaw, Stephen
Shevach, Ethan Menahem
Shih, James Waikuo
Shin, Moon L
Sich, Jeffrey John
Sieckmann, Donna G
Siegel, Jay Philip
Silverstein, Arthur M
Singer, Alfred
Siraganian, Reuben Paul
Sitkovsky, Michail V
Smith, Phillip Doyle
Smith-Gill, Sandra Joyce
Snader, Kenneth Means
Sogn, John Allen
Solomon, Joel Martin
Spero, Leonard
Stamper, Hugh Blair
Stein, Kathryn E
Stevenson, Henry C

Stobo, John David
Stone, Sanford Herbert
Strober, Warren
Stylos, William A
Suskind, Sigmund Richard
Sztein, Marcelo Benjamin
Taniuchi, Hiroshi
Taylor, Christopher E
Ting, Chou-Chik
Tomazic, Vesna J
Tosato, Giovanna
Trapani, Robert-John
Turkeltaub, Paul Charles
Urban, Joseph F
Valentine, Martin Douglas
Venter, J Craig
Vincent, Monroe Mortimer
Vogel, Stefanie N
Wahl, Sharon Knudson
Waldmann, Thomas A
Walker, Mary Clare
Walker, Richard Ives
Waxdal, M J
Weinblatt, Anita
Weislow, Owen Stuart
Weiss, Joseph Francis
Whittum-Hudson, Judith Anne
Wilder, Ronald Lynn
Williams, Jimmy Calvin
Winchurch, Richard Albert
Winkelstein, Jerry A
Wong, Dennis Mun
Wong, Donald Tai On
Woolverton, Christopher Jude
Wunderlich, John R
Yang, Bingzhi
Yarchoan, Robert
Yewdell, Jonathan Wilson
Youle, Richard James
Young, Howard Alan
Yu, Mei-ying Wong
Zatz, Marion M
Zbar, Berton
Zimmerman, Daniel Hill

MASSACHUSETTS
Abromson-Leeman, Sara R
Ackerman, Steven J
Ahmed, A Razzaque
Alderman, Edward M
Allansmith, Mathea R
Alper, Chester Allan
Anderson, Deborah Jean
Anderson, Paul Joseph
Andres, Giuseppe A
Andre-Schwartz, Janine
Argyris, Bertie
Arnaout, M Amin
Atkinson-Templeman, Kristine Hofgren
Austen, K(arl) Frank
Baltimore, David
Barlozzari, Teresa
Benacerraf, Baruj
Bikoff, Elizabeth K
Billingham, Rupert Everett
Bing, David H
Blanchard, Gordon Carlton
Bloch, Kurt Julius
Bogden, Arthur Eugene
Brennessel, Barbara A
Brown, Beverly Ann
Bursztajn, Sherry
Cantor, Harvey
Carney, Walter Patrick
Carter, Edward Albert
Carvalho, Angelina C A
Cathou, Renata Egone
Cavacini, Lisa Ann
Chang, Joseph Yoon
Chase, Arleen Ruth
Chishti, Athar Husain
Coleman, Robert Marshall
Colvin, Robert B
Crusberg, Theodore Clifford
Czop, Joyce K
Dasch, James R
David, John R
Davis, Claude Geoffrey
Dittmer, John Edward
Dorf, Martin Edward
Dubey, Devendra P
Ducibella, Tom
Du Moulin, Gary C
Dvorak, Ann Marie-Tompkins
Eisen, Herman Nathaniel
Esber, Henry Jemil
Essex, Myron
Field, Arthur Kirk
Finberg, Robert William
Fleming, Nigel
Fowler, Elizabeth
Gallo, Richard Louis
Geha, Raif S
Gerety, Robert John
Girard, Kenneth Francis
Goldmacher, Victor S
Gordon, Lance Kenneth
Greenstein, Julia L
Greiner, Dale L
Griffith, Irwin J
Gupta, Rajesh K
Haber, Edgar
Hadjian, Richard Albert
Hansen, David Elliott

Hargis, Betty Jean
Harrison, Bettina Hall
Hartner, William Christopher
Herrmann, Steven H
Hirsch, Martin Stanley
Hochman, Paula S
Huber, Brigitte T
Janicki, Bernard William
Jung, Lawrence Kwok Leung
Kaplan, Melvin Hyman
Kelley, Vicki E
Khaw, Ban-An
Kimball, John Ward
Klempner, Mark Steven
Knight, Glenn B
Kruskal, Benjamin A
Kupchik, Herbert Z
Kupiec-Weglinski, Jerzy W
Lampson, Lois Alterman
Lees, Marjorie Berman
Leskowitz, Sidney
Levine, Lawrence
Levy, Elinor Miller
Liberatore, Frederick Anthony
Litt, Mortimer
Long, Nancy Carol
MacDonald, Alex Bruce
McIntire, Kenneth Robert
Mackin, William Michael
Madoff, Morton A
Malakian, Artin
Malkiel, Saul
Margolies, Michael N
Martz, Eric
Miller-Graziano, Carol L
Mole, John Edwin
Monaco, Anthony Peter
Monette, Francis C
Mordes, John Peter
Mulder, Carel
Nagler-Anderson, Cathryn
Nisonoff, Alfred
Parker, David Charles
Pasternack, Mark Steven
Peri, Barbara Anne
Pier, Gerald Bryan
Plaut, Andrew George
Raam, Shanthi
Reif, Arnold E
Remold, Heinz G
Reisert, Patricia
Reyero, Cristina
Rodrick, Mary Lofy
Ross, Alonzo Harvey
Rule, Allyn H
Russell, Paul Snowden
Sachs, David Howard
Saravis, Calvin
Schlossman, Stuart Franklin
Schneeberger, Eveline E
Schur, Peter Henry
Schwartz, Anthony
Seidman, Jonathan G
Shakarjian, Michael Peter
Shamgochian, Maureen Dowd
Sheffer, Albert L
Sipe, Jean Dow
Smith, Daniel James
Smith, Nathan Lewis, III
Sorensen, Craig Michael
Steiner, Lisa Amelia
Stelos, Peter
Stollar, Bernard David
Stossel, Thomas Peter
Strauss, Phyllis R
Strauss, William Mark
Streilein, Jacob Wayne
Strom, Terry Barton
Strominger, Jack L
Sullivan, David Anthony
Sunshine, Geoffrey H
Tauber, Alfred Imre
Taubman, Martin Arnold
Terry, William David
Thomas, David Warren
Thomas, Gail B
Thomas, Peter
Toth, Carol Ann
Tracey, Daniel Edward
Tsung, Yean-Kai
Tulip, Thomas Hunt
Van Cleave, Victor Harold
Vanderburg, Charles R
Venkatesh, Yeldur Padmanabha
Walker, Bruce David
Warner, Carol Miller
Weetall, Howard H
Weller, Peter Fahey
Weltman, Joel Kenneth
Wieder, Kenneth J
Winn, Henry Joseph
Wolff, Sheldon Malcolm
Zabrecky, James R
Zurier, Robert B

MICHIGAN
Bach, Michael Klaus
Bacon, Larry Dean
Baragi, Vijaykumar M
Barald, Kate Francesca
Barksdale, Charles Madsen
Berger, Ann Elizabeth
Boros, Dov Lewis
Callewaert, Denis Marc

Campbell, Wilbur Harold
Carey, Thomas E
Carrick, Lee
Chang, Timothy Scott
Chensue, Stephen W
Cohen, Flossie
Cole, George Christopher
Conder, George Anthony
Distler, Jack, Jr
Dore-Duffy, Paula
Dusenbery, Ruth Lillian
Duwe, Arthur Edward
Fischer, Howard David
Fraker, Pamela Jean
Goodman, Morris
Gray, Gary D
Hackel, Emanuel
Hakim, Margaret Heath
Heppner, Gloria Hill
Hirata, Fusao
Jensen, James Burt
Johnson, Herbert Gardner
Kaplan, Joseph
Kauffman, Carol A
Kaufman, Donald Barry
Kithier, Karel
Kong, Yi-Chi Mei
Kunkel, Steven L
Lefford, Maurice J
Leon, Myron A
Lerman, Stephen Paul
Lightbody, James James
Liu, Stephen C Y
Lopatin, Dennis Edward
McCann, Daisy S
Mccorkle, Fred Miller
Miller, Fred R
Montgomery, Ilene Nowicki
Montgomery, Paul Charles
Mulks, Martha Huard
Nairn, Roderick
Narayan, Latha
Ownby, Dennis Randall
Patterson, Ronald James
Pence, Leland Hadley
Petty, Howard Raymond
Phan, Sem Hin
Poulik, Miroslav Dave
Rauch, Helene Coben
Reiners, John Joseph, Jr
Root-Bernstein, Robert Scott
Rosenspire, Allen Jay
Rowe, Nathaniel H
Sacco, Anthony G
San Clemente, Charles Leonard
Saper, Mark A
Sarkar, Fazlul Hoque
Schmaier, Alvin Harold
Sell, John Edward
Simon, Michael Richard
Singh, Vijendra Kumar
Sloat, Barbara Furin
Smith, Jerry Warren
Smith, Richard Harding
Smith, Robert James
Sundick, Roy
Swanborg, Robert Harry
Toledo-Pereyra, Luis Horacio
Truden, Judith Lucille
Tse, Harley Y
Verrill, Harland Lester
Wang, Ching Yung
Ward, Peter A
Watts, Jeffrey Lynn
Wei, Wei-Zen
Weiner, Lawrence Myron
Whalen, Joseph Wilson
Whitehouse, Frank, Jr
Wilhelm, Rudolf Ernst
Williams, Jeffrey F
Yanari, Sam Satomi
Yates, Jon Arthur
Yurewicz, Edward Charles

MINNESOTA
Abraham, Robert Thomas
Adams, Ernest Clarence
Atluru, Durgaprasadarao
Azar, Miguel M
Bach, Marilyn Lee
Beck, Barbara North
Carter, Orwin Lee
Clark, Connie
Cunningham, Julie Margaret
Dalmasso, Agustin Pascual
David, Chelladurai S
Duran, Lise Weisman
Filipovich, Alexandra H
Gleich, Gerald J
Greenberg, Leonard Jason
Hallgren, Helen M
Homburger, Henry A
Jemmerson, Ronald Renomer Weaver
Jeska, Edward Lawrence
Johnson, Arthur Gilbert
Johnson, Russell Clarence
Johnson, Theodore Reynold
Kersey, John H
Kyle, Robert Arthur
Leibson, Paul Joseph
Lennon, Vanda Alice
Lindgren, Alice Marilyn Lindell
Lukasewycz, Omelan Alexander
McKean, David Jesse

McPherson, Thomas C(oatsworth)
Markowitz, Harold
Mescher, Matthew F
Mitchell, Roger Harold
Muscoplat, Charles Craig
Nelson, Robert D
Page, Arthur R
Peterson, Phillip Keith
Ritts, Roy Ellot, Jr
Rogers, Roy Steele, III
Sharma, Jagdev Mittra
Shope, Richard Edwin, Jr
Song, Chang Won
Stromberg, Bert Edwin, Jr
Vallera, Daniel A
West, Michael Allan

MISSISSIPPI
Clem, Lester William
Cruse, Julius Major, Jr
Cuchens, Marvin A
Lewis, Robert Edwin, Jr
Lobb, Craig J
Sadana, Ajit
Sindelar, Robert D
Steen, James Southworth
Watson, Edna Sue

MISSOURI
Abdou, Nabih I
Atkinson, John Patterson
Bell, Jeffrey
Bellone, Clifford John
Birch, Ruth Ellen
Brown, Eric
Brown, Karen Kay (Kilker)
Buening, Gerald Matthew
Buonomo, Frances Catherine
Catalona, William John
Codd, John Edward
Colten, Harvey Radin
Cullen, Susan Elizabeth
De Blas, Angel Luis
Esser, Alfred F
Fernandez-Pol, Jose Alberto
Ferris, Deam Hunter
Finkelstein, Richard Alan
Friedman, Harvey Paul
Granoff, Dan Martin
Green, Theodore James
Hansen, Ted Howard
Harakas, N(icholas) Konstantinos
Herzig, Geoffrey Peter
Howard, Susan Carol Pearcy
Janski, Alvin Michael
Johnston, Marilyn Frances Meyers
Kenny, Michael Thomas
Krishnan, B Rajendra
Kulczycki, Anthony, Jr
Lafrenz, David E
Lie, Wen-Rong
Little, John Russell, Jr
Loughman, Barbara Ellen Evers
Lyle, Leon Richards
Miles, Donald Orval
Miller, Herman T
Misfeldt, Michael Lee
Mohanakumar, Thalachallour
Munns, Theodore Willard
Murray, Patrick Robert
Myers, Richard Lee
Nahm, Moon H
Olson, Gerald Allen
Parker, Charles W
Poon, Rebecca Yuetmay
Premachandra, Bhartur N
Rifas, Leonard
Roodman, Stanford Trent
Strunk, Robert Charles
Tumosa, Nina Jean
Twining, Linda Carol
Unanue, Emil R
Viamontes, George Ignacio
Wang, Richard J
Willard, Mark Benjamin

MONTANA
Braune, Maximillian O
Cantrell, John Leonard
Chesbro, Bruce W
Coe, John Emmons
Jesaitis, Algirdas Joseph
Jutila, John W
Jutila, Mark Arthur
Lodmell, Donald Louis
Munoz, John Joaquin
Myers, Lyle Leslie
Pincus, Seth Henry
Portis, John L
Reed, Norman D
Richardson, Charles
Rudbach, Jon Anthony
Speer, Clarence Arvon

NEBRASKA
Carson, Steven Douglas
Chaperon, Edward Alfred
Crouse, David Austin
Curtis, Gary Lynn
Dyer, John Kaye
Gerber, Jay Dean
Ghosh, Chitta Ranjan
Heidrick, Margaret Louise
Kay, H David

Immunology (cont)

Khan, Manzoor M
Klassen, Lynell W
Kobayashi, Roger Hideo
O'Brien, Richard Lee
Ryan, Wayne L
Sharp, John Graham
Siragusa, Gregory Ross
Smith, Paula Beth
Stratta, Robert Joseph
Turpen, James Baxter

NEVADA
Henry, Claudia
Hudig, Dorothy
Kozel, Thomas Randall

NEW HAMPSHIRE
Adler, Frank Leo
Adler, Louise Tale
Fanger, Michael Walter
Green, William Robert
Kaiser, C William
Pistole, Thomas Gordon
Thomas, William Albert
Tung, Amar S

NEW JERSEY
Anderson, David Wesley
Barton, Beverly E
Basu, Mitali
Becker, Joseph Whitney
Bendich, Adrianne
Blood, Christine Haberern
Blumenthal, Rosalyn D
Boltz, Robert Charles, Jr
Brockman, John A, Jr
Brunda, Michael J
Capetola, Robert Joseph
Caporale, Lynn Helena
Carlson, Richard P
Cavender, Druie
D'Alesandro, Philip Anthony
Das, Kiron Moy
Davies, Philip
DeBari, Vincent A
Dhruv, Rohini Arvind
Dumont, Francis T
Egan, Robert Wheeler
Ende, Norman
Fiedler-Nagy, Christa
Founds, Henry William
Gaffar, Abdul
Garcia, Maria Luisa
Gately, Maurice Kent
Gilfillan, Alasdair Mitchell
Gillette, Ronald William
Glenn, Jeffrey K
Goldenberg, David Milton
Goldstein, Gideon
Gottlieb, Alice Bendix
Green, Erika Ana
Green, Gerald
Grimes, David
Gross, Peter A
Guerrero, Jorge
Gupta, Ayodhya P
Gupta, Godaveri Rawat
Gyan, Nanik D
Heaslip, Richard Joseph
Heveran, John Edward
Hirsch, Robert L
Humes, John Leroy
Hwang, San-Bao
Kamiyama, Mikio
Kawanishi, Hidenori
Khavkin, Theodor
Kleinschuster, Stephen J, III
Koepp, Leila H
Koo, Gloria C
Krishnan, Gopal
Kumbaraci-Jones, Nuran Melek
Lehrer, Harris Irving
Letizia, Gabriel Joseph
Logdberg, Lennart Erik
McKearn, Thomas Joseph
Manjula, Belur N
Mannino, Raphael James
Mark, David Fu-Chi
Miller, Douglas Kenneth
Monestier, Marc
Moore, Vernon Leon
Mostillo, Ralph
Norvitch, Mary Ellen
Oleske, James Matthew
Panagides, John
Panush, Richard Sheldon
Park, Eunkyue
Passmore, Howard Clinton
Peterson, Arthur Carl
Ponzio, Nicholas Michael
Poretz, Ronald David
Presky, David H
Ramanarayanan, Madhava
Raychaudhuri, Anilbaran
Reckel, Rudolph P
Rivkin, Israel
Robb, Richard John
Rodwell, John Dennis
Rosenstraus, Maurice Jay
Saha, Anil
Schenkel, Robert H
Schleifer, Steven Jay
Schlesinger, R(obert) Walter
Schroff, Peter David
Seaver, Sally S
Sehgal, Surendra N
Sethi, Jitender K
Shin, Seung-il
Shulman, Sidney
Sigal, Nolan H
Sikder, Santosh K
Singer, Irwin I
Singh, Iqbal
Smith, Sidney R, Jr
Solinger, Alan M
Stark, Dennis Michael
Stolen, Joanne Siu
Tachovsky, Thomas Gregory
Taylor, Bernard Franklin
Teipel, John William
Tierno, Philip M, Jr
Tripodi, Daniel
Turner, Matthew X
Umland, Shelby Price
Wadsworth, Scott
Wang, Bosco Shang
Wang, Peng
Wang-Iverson, Patsy
Welton, Ann Frances
Whiteley, Phyllis Ellen
Williamson, Alan R
Woehler, Michael Edward
Wojnar, Robert John
Wong, Rosie Bick-Har
Yarmush, Martin Leon
Zuckerman, Leo

NEW MEXICO
Anderson, William Loyd
Banks, Keith L
Bell, George Irving
Bice, David Earl
Crago, Sylvia S
Davis, Larry Ernest
Edwards, Bruce S
Finch, Gregory Lee
Mold, Carolyn
Perelson, Alan Stuart
Rhodes, Buck Austin
Saland, Linda C
Saunders, George Cherdron
Smith, David Marshall
Sopori, Mohan L
Thrasher, Jack D
Tokuda, Sei

NEW YORK
Abdelnoor, Alexander Michael
Ablin, Richard J
Abraham, George N
Akolar, Pradip N
Al-Askari, Salah
Allen, Howard Joseph
Allen, Peter Zachary
Ames, Ira Harold
Amiraian, Kenneth
Antczak, Douglass
Arny, Margaret Jane
Ascensao, Joao L
Bachrach, Howard L
Bachvaroff, Radoslav J
Baird, Barbara A
Baker, David A
Bankert, Richard Burton
Barnwell, John Wesley
Bartal, Arie H
Bartfeld, Harry
Basch, Ross S
Baum, George
Baum, John
Bazinet, George Frederick
Bealmear, Patricia Maria
Bell, Robin Graham
Beutner, Ernst Herman
Beyer, Carl Fredrick
Bielat, Kenneth L
Birken, Steven
Birshtein, Barbara K
Bloom, Barry R
Bovbjerg, Dana H
Bowen, William H
Brandriss, Michael W
Braunstein, Joseph David
Brentjens, Jan R
Brooks, John S J
Brophy, Mary O'Reilly
Brown, Kathryn Marie
Bush, Maurice E
Bushkin, Yuri
Butler, Vincent Paul, Jr
Buxbaum, Joel N
Cabrera, Edelberto Jose
Calame, Kathryn Lee
Calvelli, Theresa A
Campbell, Samuel Gordon
Celada, Franco
Cerini, Costantino Peter
Chadha, Kailash C
Chase, Merrill Wallace
Chase, Randolph Montieth, Jr
Chauhan, Abha
Chess, Leonard
Chiao, Jen Wei
Chisholm, David R
Christian, Charles L
Chu, Tsann Ming
Cohen, Elias
Cohen, Martin William
Cohen, Nicholas
Cohn, Deirdre Arline
Coico, Richard F
Collins, Arlene Rycombel
Conway de Macario, Everly
Cooper, Norman S
Cowell, James Leo
Craig, John Philip
Cramer, Eva Brown
Crystal, Ronald George
Cunninham-Rundles, Charlotte
Curley, Dennis M
Dean, David A
DeForest, Peter Rupert
Despommier, Dickson
D'Eustachio, Peter
Dietert, Rodney Reynolds
Divgi, Chaitanya R
Duff, Ronald George
Dupont, Bo
Durkin, Helen Germaine
Durr, Friedrich (E)
Dworetzky, Murray
Efthymiou, Constantine John
Ehrke, Mary Jane
Ellner, Paul Daniel
Erlanger, Bernard Ferdinand
Evans, Mary Jo
Evans, Richard Todd
Everhart, Donald Lee
Farah, Fuad Salim
Feit, Carl
Felten, David L
Fenton, John William, II
Ferrone, Soldano
Fischel, Edward Elliot
Fisher, Clark Alan
Flanagan, Thomas Donald
Fleisher, Martin
Fleit, Howard B
Fondy, Thomas Paul
Fontana, Vincent J
Fotino, Marilena
Frair, Wayne
Frangione, Blas
Friedman, Eli A
Fritz, Katherine Elizabeth
Fuji, Hiroshi
Furmanski, Philip
Gabrielsen, Ann Emily
Genco, Robert J
Ghebrehiwet, Berhane
Gibbs, David Lee
Gibofsky, Allan
Ginsberg-Fellner, Fredda Vita
Godfrey, Henry Philip
Goldsmith, Lowell Alan
Gotschlich, Emil C
Grew, John C
Grimwood, Brian Gene
Grob, David
Gutcho, Sidney J
Habicht, Gail Sorem
Hammerling, Ulrich
Han, Tin
Hashim, George A
Hawrylko, Eugenia Anna
Henley, Walter L
Herr, Harry Wallace
Hirschhorn, Kurt
Hirschhorn, Rochelle
Ho, May-Kin
Holowka, David Allan
Holzman, Robert Stephen
Houghton, Alan N
Howard, Irmgard Matilda Keeler
Hsu, Konrad Chang
Iglewski, Barbara Hotham
Isaacs, Charles Edward
Isenberg, Henry David
Joel, Darrel Dean
Johnson, Lawrence Lloyd
Johnston, Dean
Josephson, Alan S
Kalsow, Carolyn Marie
Kaplan, Joel Howard
Kascsak, Richard John
Kim, Charles Wesley
Kim, Untae
Kim, Young Tai
Kimberly, Robert Parker
Kishore, Gollamudi Sitaram
Kite, Joseph Hiram, Jr
Kochwa, Shaul
Kratzel, Robert Jeffrey
Kream, Jacob
Krown, Susan E
Lahita, Robert George
Lamberson, Harold Vincent, Jr
Lambert, Reginald Max
Lau, Joseph T Y
Lawrence, David A
Leary, James Francis
Leddy, John Plunkett
Lee, Ching-Li
Lehrer, Gerard Michael
Leinwand, Leslie
Lepp, Cyrus Andrew
Levine, Bernard Benjamin
Lewis, Robert Miller
Lilly, Frank
Lipson, Steven Mark
Litwin, Stephen David
Lloyd, Kenneth Oliver
Lockshin, Michael Dan
Loeb, Marilyn Rosenthal
Lord, Edith M
Lossinsky, Albert S
Lutton, John D
Lyons, Michael Joseph
Macario, Alberto J L
McClintock, David K
MacGillivray, Margaret Hilda
McNamara, Thomas Francis
Macphail, Stuart
Macris, Nicholas T
Magill, Thomas Pleines
Maitra, Utpalendu S
Maleckar, James R
Manski, Wladyslaw J
Marboe, Charles Chostner
Marcu, Kenneth Brian
Marsh, William Laurence
Mayers, George Louis
Meruelo, Daniel
Michael, Sandra Dale
Miller, Frederick
Millis, Albert Jason Taylor
Mizejewski, Gerald Jude
Mohn, James Frederic
Mohos, Steven Charles
Moline, Sheldon Walter
Mongini, Patricia Katherine Ann
Moor-Jankowski, J
Morgan, Donald O'Quinn
Morse, Jane H
Morse, Stephen S(cott)
Mueller, August P
Murphy, Donal B
Muschel, Louis Henry
Nathenson, Stanley G
Neiders, Mirdza Erika
Neumann, Norbert Paul
Nisselbaum, Jerome Seymour
Noble, Bernice
Norcross, Neil Linwood
Norin, Allen Joseph
North, Robert J
Nussenzweig, Victor
Oettgen, Herbert Friedrich
Old, Lloyd John
O'Reilly, Richard
Oreskes, Irwin
Ovary, Zoltan
Packman, Charles Henry
Papsidero, Lawrence D
Perlman, Ely
Phillips-Quagliata, Julia Molyneux
Pierce, Carol S
Pitt, Jane
Pruslin, Fred Howard
Quaroni, Andrea
Quimby, Fred William
Rabbany, Sina Y
Rafla, Sameer
Raine, Cedric Stuart
Rajan, Thiruchandurai Viswanathan
Rao, Yalamanchili A K
Rapaport, Felix Theodosius
Rapp, Dorothy Glaves
Raveche, Elizabeth Marie
Raventos-Suaraz, Carmen Elvira
Reddy, Mohan Muthireval
Reimers, Thomas John
Reiss, Carol S
Repasky, Elizabeth Ann
Rinchik, Eugene M
Rosenfeld, Stephen I
Rosenfield, Richard Ernest
Rosenstreich, David Leon
Sanford, Karl John
Santiago, Noemi
Sayegh, Joseph Frieh
Scharff, Matthew Daniel
Schauf, Victoria
Schiffman, Gerald
Schoenfeld, Cy
Schwartz, Stanley Allen
Scott, David William
Sehgal, Pravinkumar B
Seiden, Philip Edward
Senitzer, David
Senterfit, Laurence Benfred
Seon, Ben K
Shapiro, Caren Knight
Shayegani, Mehdi
Siden, Edward Joel
Siegal, Frederick Paul
Silverstein, Samuel Charles
Siminoff, Paul
Singh, Toolsee J
Siskind, Gregory William
Slovin, Susan Faith
Smith, Kendall A
Socha, Wladyslaw Wojciech
Spar, Irving Leo
Spitzer, Roger Earl
Stevens, Roy White
Stewart, Carleton C
Stolfi, Robert Louis
Stoner, Richard Dean
Stutman, Osias
Sultzer, Barnet Martin
Sussdorf, Dieter Hans
Tabor, John Malcolm
Taub, Aaron M

BIOLOGICAL SCIENCES / 117

Taubman, Sheldon Bailey
Thorbecke, Geertruida Jeanette
Tilley, Shermaine Ann
Tomasi, Thomas B, Jr
Utermohlen, Virginia
Uzgiris, Egidijus E
Vaage, Jan
Valentine, Fred Townsend
Van Oss, Carel J
Vijayasaradhi, Setaluri
Vilcek, Jan Tomas
Vladutiu, Adrian O
Volkman, David J
Von Strandtmann, Maximillian
Waksman, Byron Halstead
Walcott, Benjamin
Waltzer, Wayne C
Warner, Garvin L
Weksler, Marc Edward
Wicher, Victoria
Williams, Curtis Alvin, Jr
Winchester, Robert J
Winter, Alexander J
Witkin, Steven S
Woodruff, Judith J
Wright, Samuel D
Yang, Shung-Jun
Yang, Song-Yu
Yeung, Kwok Kam
Young, John Ding-E
Zaleski, Marek Bohdan
Zauderer, Maurice
Zolla-Pazner, Susan Beth

NORTH CAROLINA
Abramson, Jon Stuart
Adams, Dolph Oliver
Amos, Dennis Bernard
Arnold, Roland R
Barnes, Donald Wesley
Bast, Robert Clinton, Jr
Bolognesi, Dani Paul
Brown, William Roy
Burleson, Gary R
Buttke, Thomas Martin
Carter, Philip Brian
Caterson, Bruce
Ciancolo, George J
Clinton, Bruce Allan
Cohen, Carl
Cohen, Harvey Jay
Cohen, Myron S
Cohen, Philip Lawrence
Collins, Jeffrey Jay
Corley, Ronald Bruce
Cronenberger, Jo Helen
Dawson, Jeffrey Robert
Day, Eugene Davis
Devlin, Robert B
Durack, David Tulloch
Edens, Frank Wesley
Eisenberg, Robert
Erickson, Bruce Wayne
Falletta, John Matthew
Fine, Jo-David
Folds, James Donald
Frank, Michael M
Frelinger, Jeffrey
Gardner, Edward, Jr
Golden, Carole Ann
Hall, Russell P
Haskill, John Stephen
Haughton, Geoffrey
Heise, Eugene Royce
Hershfield, Michael Steven
Hoffman, Donald Richard
Hoffman, Maureane Richardson
Hunt, William B, Jr
Hutt, Randy
Issitt, Peter David
James, Karen K(anke)
Kariman, Khalil
Kataria, Yash P
Keene, Jack Donald
Kirkbride, L(ouis) D(ale)
Klapper, David G
Kostyu, Donna D
Kuhn, Raymond Eugene
Lee, David Charles
Luster, Michael I
McReynolds, Richard A
Metzgar, Richard Stanley
Mitchell, Thomas Greenfield
Mizel, Steven B
Mueller, Nancy Schneider
Noga, Edward Joseph
O'Callaghan, James Patrick
Page, Rodney Lee
Palczuk, Nicholas C
Paluch, Edward Peter
Pizzo, Salvatore Vincent
Quinn, Richard Paul
Reid, Lola Cynthia McAdams
Richardson, Stephen H
Rosse, Wendell Franklyn
Sage, Harvey J
Sassaman, Anne Phillips
Schwab, John Harris
Seigler, Hilliard Foster
Silverman, Myron Simeon
Singer, Kay Hiemstra
Smith, A Mason
Smith, Gary Keith
Snyderman, Ralph

Sonnenfeld, Gerald
Strausbauch, Paul Henry
Stuhlmiller, Gary Michael
Taggart, R(obert) Thomas
Telen, Marilyn Jo
Thomas, Francis T
Thomas, Judith M
VanMiddlesworth, Frank L
Volkman, Alvin
Ward, Frances Ellen
White, Sandra L
Winfield, John Buckner
Wolberg, Gerald
Yount, William J
Zimmerman, Thomas Paul

NORTH DAKOTA
Hastings, James Michael
Khansari, David Nemat
Rosenberg, Harry
Samson, Willis Kendrick
Watson, David Alan

OHIO
Akeson, Ann L
Alexander, James Wesley
Bajpai, Praphulla K
Ball, Edward James
Ball, William James, Jr
Barriga, Omar Oscar
Barth, Rolf Frederick
Battisto, Jack Richard
Baxter, William D
Berger, Melvin
Bernstein, I Leonard
Biagini, Raymond E
Bigley, Nancy Jane
Blanton, Ronald Edward
Bonventre, Peter Frank
Boyle, Michael Dermot
Brenner, Lorry Jack
Cathcart, Martha K
Chakraborty, Joana J(yotsna)
Chang, Jae Chan
Chen, Ker-Sang
Chorpenning, Frank Winslow
Daniel, Thomas Mallon
Decker, Lucile Ellen
Durnford, Joyce M
Elmets, Craig Allan
Finke, James Harold
Francis, Joseph William
Frey, James R
Friedlander, Miriam Alice
Gao, Kuixiong
Glaser, Ronald
Greenspan, Neil Sanford
Grossman, Charles Jerome
Gupta, Manjula K
Haas, Erwin
Hamilton, Thomas Alan
Harmony, Judith A K
Henningfield, Mary Frances
Herman, Jerome Herbert
Hess, Evelyn V
Hinman, Channing L
Horton, John Edward
Houchens, David Paul
Jamasbi, Roudabeh J
Janusz, Michael John
Kapral, Frank Albert
Karp, Richard Dale
Kearns, Robert J
Keever, Carolyn Anne
Kochan, Ivan
Koo, Peter H
Kreier, Julius Peter
Krueger, Robert George
Lambert, Laurie E
Lamm, Michael Emanuel
Lane, Richard Durelle
Lang, Raymond W
Lass, Jonathan H
Lee, Young-Zoon
Lehmann, Paul F
Mahmoud, Adel A F
Malemud, Charles J
Mao, Simon Jen-Tan
Megel, Herbert
Merritt, Katharine
Metzger, Dennis W
Michael, Jacob Gabriel
Miller, Sally Ann
Modrzakowski, Malcolm Charles
Morgan, Marjorie Susan
Morris, Randal Edward
Muckerheide, Annette
Murray, Finnie Ardrey, Jr
Newman, Simon Louis
Nordlund, James John
Olsen, Richard George
Ooi, Boon Seng
Orosz, Charles George
Pate, Joy Lee
Pesce, Amadeo J
Proffitt, Max Rowland
Rheins, Melvin S
Rhodes, Judith Carol
Robinson, Michael K
Rodman, Harvey Meyer
Rosenthal, Ken Steven
Rutishauser, Urs Stephen
Saif, Linda Jean
Saif, Yehia M(ohamed)

St Pierre, Ronald Leslie
Schanbacher, Floyd Leon
Shaffer, Charles Franklin
Shannon, Barry Thomas
Sheridan, John Francis
Sidman, Charles L
Smith, Kenneth Larry
Stavitsky, Abram Benjamin
Stevenson, J(oseph) Ross
Stevenson, John Ray
Stotts, Jane
Sy, Man-Sun
Tarr, Melinda Jean
Tomei, L David
Treichel, Robin Stong
Tykocinski, Mark L
Wei, Robert
West, Clark Darwin
Whitacre, Caroline C
Widder, James Stone
Wolfe, Seth August, Jr
Wolos, Jeffrey Alan
Yen, Belinda R S
Zwilling, Bruce Stephen

OKLAHOMA
Allen, Robert Wade
Blackstock, Rebecca
Cain, William Aaron
Campbell, Gregory Alan
Comp, Philip Cinnamon
Confer, Anthony Wayne
Cunningham, Madeleine White
Epstein, Robert B
Esmon, Charles Thomas
Graves, Donald C
Hall, Leo Terry
Hall, Nancy K
Harley, John Barker
Hyde, Richard Moorehead
Kincade, Paul W
Klein, John Robert
Kollmorgen, G Mark
Lee, Lela A
Leu, Richard William
Patnode, Robert Arthur
Sherwood, John L
Teague, Perry Owen
Veltri, Robert William
Waner, Joseph Lloyd
Webb, Carol F
Wikel, Stephen Kenneth
Wolgamott, Gary

OREGON
Bagby, Grover Carlton
Bardana, Emil John, Jr
Barry, Ronald A
Bartos, Frantisek
Bayne, Christopher Jeffrey
Bennett, Robert M
Chilgren, John Douglas
Clark, David Thurmond
Koller, Loren D
Leslie, Gerrie Allen
Lochner, Janis Elizabeth
McClard, Ronald Wayne
Malley, Arthur
Newell, Nanette
Norman, Douglas James
Pirofsky, Bernard
Rittenberg, Marvin Barry
Riviere, George Robert
Ruben, Laurens Norman
Schaeffer, Morris

PENNSYLVANIA
Akella, Rama
Allen, Elizabeth Morei
Amoscato, Andrew Anthony
Atchison, Robert Wayne
Atkins, Paul C
Babu, Uma Mahesh
Badger, Alison Mary
Bagasra, Omar
Baker, Alan Paul
Ball, Edward D
Barker, Clyde Frederick
Becker, Anthony J, Jr
Becker, Robert S
Beezhold, Donald H
Berney, Steven
Bertolini, Donald R
Brooks, John J
Brown, John M
Brunson, Kenneth W
Burdash, Nicholas Michael
Caliguiri, Lawrence Anthony
Calkins, Catherine E
Callahan, Hugh James
Cancro, Michael P
Caso, Louis Victor
Cebra, John Joseph
Ceglowski, Walter Stanley
Cohen, Stanley
Creech, Hugh John
Crowell, Richard Lane
D'Alisa, Rose M
Dalton, Barbara J
Das, Manjusri
Davis, Hugh Michael
Dean, Jack Hugh
Debouck, Christine Marie
DeHoratius, Raphael Joseph

Denis, Kathleen A
Donnelly, John James, III
Douglas, Steven Daniel
Dryden, Richard Lee
Duquesnoy, Rene J
Eichberg, Jorg Wilhelm
Eisenstein, Toby K
Farrell, Jay Phillip
Fenderson, Bruce Andrew
Fenton, Marilyn Ruth
Fiedel, Barry Allen
Fireman, Philip
Fleetwood, Mildred Kaiser
Flick, John A
Gasser, David Lloyd
Gerhard, Walter Ulrich
Gilbert, Scott F
Gill, Thomas James, III
Gilman, Steven Christopher
Glorioso, Joseph Charles, III
Goldfarb, Ronald H
Greenstein, Jeffrey Ian
Gregory, Francis Joseph
Harris, Tzvee N
Harrison, Aline Margaret
Havas, Helga Francis
Hennessey, John P, Jr
Herberman, Ronald Bruce
Herlyn, Dorothee Maria
Herzog, David Paul
Higgins, Terry Jay
Hsu, Kuo-Hom Lee
Hsu, Susan Hu
Hughes, Austin Leland
Jegasothy, Brian V
Jensen, Bruce David
Jimenez, Sergio
Jonak, Zdenka L
Joseph, Jeymohan
Kamoun, Malek
Karakawa, Walter Wataru
Karol, Meryl Helene
Karush, Fred
Kennett, Roger H
Khatami, Mahin
Knobler, Robert Leonard
Knowles, Barbara B
Koffler, David
Korngold, Robert
Koros, Aurelia M Carissimo
Krah, David Lee
Kreider, John Wesley
Lalley, Edward T
Lancaster, Jack R, Jr
Lattime, Edmund Charles
Lee, John Cheung Han
Leech, Stephen H
Liberti, Paul A
Lindstrom, Jon Martin
Live, Israel
Long, Walter Kyle, Jr
Lotze, Michael T
Lubeck, Michael D
McAlack, Robert Francis
McCarthy, Patrick Charles
McKenna, William Gillies
Maguire, Henry C, Jr
Manson, Lionel Arnold
Mashaly, Magdi Mohamed
Mastrangelo, Michael Joseph
Mastro, Andrea M
Maurer, Paul Herbert
Maxson, Linda Ellen R
Merryman, Carmen F
Miller, Elizabeth Eshelman
Milligan, Wilbert Harvey, III
Millman, Irving
Misra, Dhirendra N
Mitchell, Kenneth Frank
Morahan, Page Smith
Murasko, Donna Marie
Nathanson, Neal
Natuk, Robert James
Neefe, John R
Nowotny, Alois Henry Andre
Ogburn, Clifton Alfred
Ohlsson-Wilhelm, Betsy Mae
Opas, Evan E
Owen, Charles Scott
Pearson, David D
Pelus, Louis Martin
Pepe, Frank Albert
Phillips, S Michael
Platsoucas, Chris Dimitrios
Proctor, Julian Warrilow
Pure, Ellen
Puri, Rajinder N
Raikow, Radmila Boruvka
Raymond, Matthew Joseph
Rest, Richard Franklin
Ritter, Carl A
Rockey, John Henry
Roesing, Timothy George
Romano, Paula Josephine
Rozmiarek, Harry
Rubin, Benjamin Arnold
Salerno, Ronald Anthony
Salvin, Samuel Bernard
Sanders, Martin E
Schlegel, Robert Allen
Schreiber, Alan D
Schultz, Jane Schwartz
Schwaber, Jerrold
Schwartzman, Robert M

Immunology (cont)

Shiver, John W
Siegel, Richard C
Simmons, Richard Lawrence
Six, Howard R
Smith, Jackson Bruce
Southam, Chester Milton
Steplewski, Zenon
Strayer, David S
Taichman, Norton Stanley
Tax, Anne W
Tevethia, Satvir S
Thomas, Paul Milton
Thomas, William J
Tint, Howard
Ulmer, Jeffrey Blaine
Vessey, Adele Ruth
Watt, Robert M
Weber, Wilfried T
Weiss, Leon
Wheelock, Earle Frederick
White, John R
Whiteside, Theresa L
Williams, Taffy J
Wing, Edward Joseph
Winkelstein, Alan
Yang, Da-Ping (David)
Young, Peter Ronald
Yui, Katsuyuki
Yurchenco, John Alfonso
Zarkower, Arian
Zink, Gilbert Leroy
Zmijewski, Chester Michael
Zurawski, Vincent Richard, Jr

RHODE ISLAND

Bailey, Garland Howard
Biron, Christine A
Blazar, Beverly A
Christenson, Lisa
DiLeone, Gilbert Robert
Dolyak, Frank
Knopf, Paul M
Laux, David Charles
McMaster, Philip Robert Bache
Shank, Peter R
Tourtellotte, Mark Eton
Wanebo, Harold
Yoon, Poksyn Song
Zawadzki, Zbigniew Apolinary

SOUTH CAROLINA

Arnaud, Philippe
Boackle, Robert J
Bowers, William E
Chao, Julie
Cheng, Thomas Clement
Feller, Robert Jarman
Fudenberg, H Hugh
Galbraith, Robert Michael
Gee, Adrian Philip
Ghaffar, Abdul
Glick, Bruce
Goust, Jean Michel
Graber, Charles David
Hahn, Amy Beth
Hazen, Terry Clyde
Holper, Jacob Charles
Lazarchick, John
LeRoy, Edward Carwile
Lill, Patsy Henry
Lopes-Virella, Maria Fernanda Leal
Lundstrom, Ronald Charles
Mackaness, George Bellamy
Maheshwari, Kewal Krishnan
Mathur, Subbi
Millette, Clarke Francis
Sigel, M(ola) Michael
Virella, Gabriel T
Wang, An-Chuan
White, Edward
Wilson, Gregory Bruce

SOUTH DAKOTA

Cafruny, William Alan
Hildreth, Michael B
Howard, Ronald M
Hurley, David John
Lynn, Raymond J

TENNESSEE

Altemeier, William Arthur, III
Briggs, Robert Chester
Chandler, Robert Walter
Chen, James Pai-Fun
Chi, David Shyh-Wei
Chiang, Thomas M
Chunduru, Srinivas Kishore
Cooper, Terrance G
Dees, Craig
Duhl, David M
Fishman, Marvin
Fuson, Ernest Wayne
Gallagher, Michael Terrance
Gengozian, Nazareth
Gillespie, Elizabeth
Griffin, Guy David
Gupta, Ramesh C
Harwood, Thomas Riegel
Herrod, Henry Grady
Ichiki, Albert Tatsuo
Isa, Abdallah Mohammad
Jennings, Lisa Helen Kyle

Jurand, Jerry George
Kennel, Stephen John
Kotb, Malak Y
Kuo, Chao-Ying
Meyers-Elliott, Roberta Hart
Miller, Harold Charles
Ourth, Donald Dean
Portner, Allen
Roberts, Audrey Nadine
Roberts, Lee Knight
Rouse, Barry Tyrrell
Rubin, Donald Howard
Ryan, Kevin William
South, Mary Ann
Stout, Robert Daniel
Sun, Deming
Thomas, James Ward
Townes, Alexander Sloan
Turner, Ella Victoria
Tyler, John D
Vredeveld, Nicholas Gene
Wachtel, Stephen Shoel
Walker, William Stanley
Webster, Robert G
Wust, Carl John
Yoo, Tai-June

TEXAS

Aggarwal, Bharat Bhushan
Allen, Robert Charles
Alvarez-Gonzalez, Rafael
Atassi, Zouhair
Attanasio, Roberta
Baughn, Robert Elroy
Bjorndahl, Jay Mark
Boley, Robert B
Bowen, James Milton
Capra, J Donald
Chang, Tse-Wen
Chanh, Tran C
Chappell, Cynthia Lou
Chatterjee, Bandana
Christadoss, Premkumar
Cohen, Allen Barry
Collisson, Ellen Whited
Daniels, Jerry Claude
Denney, Richard Max
Douglas, Tommy Charles
Dreesman, Gordon Ronald
Evans, John Edward
Everse, Johannes
Fahlberg, Willson Joel
Fanguy, Roy Charles
Faust, Charles Harry, Jr
Fernandes, Gabriel
Fidler, Isaiah J
Fisher, William Francis
Fleischmann, William Robert, Jr
Forman, James
Fortner, George William
Foster, Terry Lynn
Franzl, Robert E
Geoghegan, William David
Gillard, Baiba Kurins
Goldman, Armond Samuel
Goldschmidt, Millicent
Gottlieb, Paul David
Grant, John Andrew, Jr
Grimm, Elizabeth Ann
Gutterman, Jordan U
Guy, Leona Ruth
Heggers, John Paul
Hejtmancik, Kelly Erwin
Hollinger, F(rederick) Blaine
Huang, Shyi Yi
Hutson, James Chelton
Jester, James Vincent
Johnson, Alice Ruffin
Jordon, Robert Earl
Joys, Terence Michael
Kahan, Barry D
Kemp, Walter Michael
Kettman, John Rutherford, Jr
Kiel, Johnathan Lloyd
Kier, Ann B
Killion, Jerald Jay
Klimpel, Gary R
Kniker, William Theodore
Koppenheffer, Thomas Lynn
Kraig, Ellen
Kripke, Margaret Louise (Cook)
Krolick, Keith A
Kumar, Vinay
Kutteh, William Hanna
Lanford, Robert Eldon
Laughter, Arline H
Lee, John C
Lefkowitz, Doris Lynne
Loan, Raymond Wallace
Loftin, Karin Christiane
Lopez-Berestein, Gabriel
McBride, Raymond Andrew
McCallum, Roderick Eugene
McConnell, Stewart
McKenna, John Morgan
McMurray, David N
Mandy, William John
Marcus, Donald M
Marshall, Gailen D, Jr
Mavligit, Giora M
Measel, John William
Merrill, Gerald Alan
Morrow, Kenneth John, Jr
Murthy, Krishna Kesava

Nakajima, Motowo
Nash, Donald Robert
Niederkorn, Jerry Young
Nikaein, Afzal
Nishioka, Kenji
Norris, Steven James
Ogra, Pearay L
Ordonez, Nelson Gonzalo
Painter, Richard Grant
Paque, Ronald E
Patsch, Wolfgang
Paull, Barry Richard
Pellis, Neal Robert
Periman, Phillip
Pierce, Carl William
Pinckard, Robert Neal
Poduslo, Shirley Ellen
Poenie, Martin Francis
Powell, Bernard Lawrence
Prabhakar, Bellur Subbanna
Prager, Morton David
Rael, Eppie David
Rainwater, David Luther
Rajaraman, Srinivasan
Ranney, David Francis
Reyes, Victor E
Rich, Robert Regier
Rich, Susan Marie Solliday
Ritzi, Earl Michael
Robertson, Stella M
Rodkey, Leo Scott
Rossen, Roger Downey
Roth, Jack A
Rudolph, Frederick Byron
Sanders, Bobby Gene
Sanford, Barbara Ann
Schlamowitz, Max
Schmalstieg, Frank Crawford
Schmidt, Jerome P(aul)
Schneider, Sandra Lee
Sell, Stewart
Shadduck, John Allen
Shearer, William Thomas
Shillitoe, Edward John
Smiley, James Donald
Smith, Eric Morgan
Sparkman, Dennis Raymond
Spellman, Craig William
Stone, William Harold
Straus, David Conrad
Streckfuss, Joseph Larry
Stroynowski, Iwona T
Taylor, D(orothy) Jane
Templeton, Joe Wayne
Tizard, Ian Rodney
Tom, Baldwin Heng
Tong, Alex W
Trentin, John Joseph
Twomey, Jeremiah John
Ulrich, Stephen E
Valdivieso, Dario
Veit, Bruce Clinton
Villacorte, Guillermo Vilar
Vitetta, Ellen S
Wagner, Gerald Gale
Walker, David Hughes
Weir, Michael Ross
Winters, Wendell Delos
Wolinsky, Jerry Saul
Yang, Ovid Y H
Yanni, John Michael
Yeoman, Lynn Chalmers

UTAH

Araneo, Barbara Ann
Barnett, Bill B
Daynes, Raymond Austin
DeWitt, Charles Wayne, Jr
Donaldson, David Miller
Fujinami, Robert S
Hammond, Mary Elizabeth Hale
Hayes, Sheldon P
Healey, Mark Calvin
Marcus, Stanley
Mohammad, Syed Fazal
Muna, Nadeem Mitri
Nelson, Jerry Rees
North, James A
Rasmussen, Kathleen Goertz
Shelby, Nancy Jane
Sidwell, Robert William
Sipe, David Michael
Spendlove, Rex S
Straight, Richard Coleman
Thorpe, Bert Duane
Torres, Anthony R
Warren, Reed Parley
Wiley, Bill Beauford

VERMONT

Albertini, Richard Joseph
Boraker, David Kenneth
Cooper, Sheldon Mark
Fahey, John Vincent
Falk, Leslie Alan
Kelley, Jason
Nicklas, Janice A
Rappaport, Irving

VIRGINIA

Acton, Jean D
Barta, Ota
Braciale, Thomas Joseph, Jr
Braciale, Vivian Lam

Bradley, Sterling Gaylen
Brown, Barry Lee
Brown, Russell Vedder
Burger, Carol J
Chaparas, Sotiros D
DeVries, George H
Elgert, Klaus Dieter
Engelhard, Victor H
Escobar, Mario R
Esen, Asim
Eyre, Peter
Fu, Shu Man
Hager, Chester Bradley
Hard, Richard C, Jr
Herr, John Christian
Hsu, Hsiu-Sheng
Kaattari, Stephen L
Kelley, John Michael
Lubiniecki, Anthony Stanley
McWright, Cornelius Glen
Mullinax, Perry Franklin
Munson, Albert Enoch
Myrvik, Quentin N
Nagarkatti, Mitzi
Nagarkatti, Prakash S
Normansell, David E
Nygard, Neal Richard
Osuch, Mary Ann V
Place, Janet Dobbins
Sando, Julianne J
Smibert, Robert Merrall, II
Smith, Wade Kilgore
Steinberg, Alfred David
Szakal, Andras Kalman
Taylor, Ronald Paul
Tew, John Garn
Tucker, Anne Nichols
Wilkins, Tracy Dale
Wright, George Green
Wright, George Leonard, Jr

WASHINGTON

Altman, Leonard Charles
Baker, Patricia J
Barnes, Glover William
Bean, Michael Arthur
Buessow, Scott
Chen, Lieping
Clapshaw, Patric Arnold
Cosman, David John
Couser, William Griffith
Dailey, Frank Allen
Davis, William C
Eidinger, David
Ely, John Thomas Anderson
Farr, Andrew Grant
Frankart, William A
Gillis, Steven
Graves, Scott Stoll
Gray, Patrick William
Hakomori, Sen-Itiroh
Harlan, John Marshall
Hellstrom, Ingegerd Elisabet
Hellstrom, Karl Erik Lennart
Henney, Christopher Scot
Hood, Leroy E
Horn, Diane
Houck, John Candee
Hsu, Chin Shung
Hu, Shiu-Lok
Jackson, Anne Louise
Kenny, George Edward
Lang, Bruce Z
Leid, R Wes
Lynch, David H
McGuire, Travis Clinton
McIvor, Keith L
Magnuson, Nancy Susanne
Maguire, Yu Ping
Maliszewski, Charles R
Mannik, Mart
Meadows, Gary Glenn
Mittler, Robert S
Morrissey, Philip John
Nelson, Karen Ann
Ochs, Hans D
Pauley, Gilbert Buckhannan
Perlmutter, Roger M
Pious, Donald A
Pollack, Sylvia Byrne
Raff, Howard V
Rosse, Cornelius
Sheiness, Diana Kay
Shultz, Terry D
Sibley, Carol Hopkins
Urdal, David L
Wedgwood, Ralph Josiah Patrick
Weiser, Russel Shively

WEST VIRGINIA

Anderson, Douglas Poole
Barnett, John Brian
Burrell, Robert
Landreth, Kenneth S
Lewis, Daniel Moore
Mengoli, Henry Francis
Olenchock, Stephen Anthony
Reichenbecher, Vernon Edgar, Jr

WISCONSIN

Abramoff, Peter
Auerbach, Robert
Blattner, Frederick Russell
Blazkovec, Andrew A

BIOLOGICAL SCIENCES / 119

Borden, Ernest Carleton
Brown, Raymond Russell
Calvanico, Nickolas Joseph
Chusid, Michael Joseph
Click, Robert Edward
Coe, Christopher Lane
Cook, Mark Eric
Czuprynski, Charles Joseph
Datta, Surinder P
Davie, Joseph Myrten
Dimond, Randall Lloyd
Fink, Jordan Norman
Fredricks, Walter William
Fritz, Robert B
Goldberger, Amy
Grossberg, Sidney Edward
Haasch, Mary Lynn
Harms, Jerome Scott
Heath, Timothy Douglas
Hefle, Susan Lynn
Hinsdill, Ronald D
Hinshaw, Virginia Snyder
Hinze, Harry Clifford
Hirsch, Samuel Roger
Hong, Richard
Jensen, Richard Harvey
Kahan, Lawrence
Kim, Zaezeung
Klebba, Phillip E
Koethe, Susan M
Kurup, Viswanath Parameswar
Long, Claudine Fern
Manning, Dean David
Mansfield, John Michael
Marx, James John, Jr
Paulnock, Donna Marie
Rapacz, Jan
Schultz, Ronald David
Sheehy, Michael Joseph
Sidky, Younan Abdel Malik
Smith, Donald Ward
Sondel, Paul Mark
Tomar, Russell H
Truitt, Robert Lindell
Tufte, Marilyn Jean
Twining, Sally Shinew
Weidanz, William P
Witt, Patricia L
Yoshino, Timothy Phillip

WYOMING
Belden, Everett Lee
Isaak, Dale Darwin
Pier, Allan Clark

PUERTO RICO
Bolanos, Benjamin
Garcia-Castro, Ivette
Habeeb, Ahmed Fathi Sayed Ahmed
Hillyer, George Vanzandt
Kozek, Wieslaw Joseph
Lluberes, Rosa P

ALBERTA
Baron, Robert Walter
Befus, A(lbert) Dean
Bleackley, Robert Christopher
Dossetor, John Beamish
Eggert, Frank Michael
Hart, David Arthur
Jerry, L Martin
Longenecker, Bryan Michael
McPherson, Thomas Alexander
Madiyalakan, Ragupathy
Paetkau, Verner Henry
Rice, Wendell Alfred
Stemke, Gerald W
Suresh, Mavanur Rangarajan
Wegmann, Thomas George

BRITISH COLUMBIA
Aebersold, Ruedi H
Agnew, Robert Morson
Albright, Lawrence John
Chase, William Henry
Ekramoddoullah, Abul Kalam M
Ferguson, Alexander Cunningham
Hancock, Robert Ernest William
Hoffmann, Geoffrey William
Kelly, Michael Thomas
McBride, Barry Clarke
Matheson, David Stewart
Osoba, David
Salinas, Fernando A
Silver, Hulbert Keyes Belford
Teh, Hung-Sia
Tingle, Aubrey James

MANITOBA
Berczi, Istvan
Chow, Donna Arlene
Froese, Arnold
Garther, John G
Greenberg, Arnold Harvey
Kepron, Wayne
Lewis, Marion Jean
Paraskevas, Frixos
Sabbadini, Edris Rinaldo
Sehon, Alec

NEW BRUNSWICK
Bagnall, Richard Herbert

NEWFOUNDLAND
Hawkins, David Geoffrey
Michalak, Thomas Ireneusz

NOVA SCOTIA
Carr, Ronald Irving
Dolphin, Peter James
Jones, John Verrier
MacSween, Joseph Michael

ONTARIO
Baker, Michael Allen
Barber, Brian Harold
Barr, Ronald Duncan
Beaulieu, J A
Bienenstock, John
Bishop, Claude Titus
Blais, Burton W
Broder, Irvin
Campbell, James B
Chadwick, June Stephens
Chaly, Nathalie
Chechik, Boris
Cinader, Bernhard
Clark, David A
Crookston, Marie Cutbush
Dales, Samuel
Danska, Jaynes
Delovitch, Terry L
Dent, Peter Boris
Derbyshire, John Brian
De Veber, Leverett L
Dorrington, Keith John
Dosch, Hans-Michael
Dubiski, Stanislaw
Duncan, John Robert
Fish, Eleanor N
Forsdyke, Donald Roy
Freedman, Melvin Harris
Galsworthy, Sara B
Gauldie, Jack
Gorczynski, Reginald Meiczyslaw
Greenstock, Clive Lewis
Grinstein, Sergio
Gupta, Krishana Chandara
Hozumi, Nobumichi
Inman, Robert Davies
Kaushik, Azad
Kennedy, James Cecil
Kerbel, Robert Stephen
Krepinsky, Jiri J
Kuroski-De Bold, Mercedes Lina
Lang, Gerhard Herbert
Lau, Catherine Y
Lee, Eng-Hong
Letarte, Michelle
Lingwood, Clifford Alan
Logan, James Edward
McDermott, Mark Rundle
Mahdy, Mohamed Sabet
Mak, Tak Wah
Mak, William Wai-Nam
Medzon, Edward Lionel
Mellors, Alan
Metzgar, Don P
Miller, Richard Graham
Minta, Joe Oduro
Nielsen, Klaus H B
Nisbet-Brown, Eric Robert
Paul, Leendert Cornelis
Phillips, Robert Allan
Pickering, Richard Joseph
Pope, Barbara L
Pross, Hugh Frederick
Pruzanski, Waldemar
Ranadive, Narendranath Santuram
Rice, Christine E
Richter, Maxwell
Rosenthal, Kenneth Lee
Ruckerbauer, Gerda Margareta
Samagh, Bakhshish Singh
Sauder, Daniel Nathan
Sengar, Dharmendra Pal Singh
Shewen, Patricia Ellen
Sinclair, Nicholas Roderick
Singal, Dharam Parkash
Singh, Bhagirath
Singhal, Sharwan Kumar
Stanisz, Andrzej Maciej
Stemshorn, Barry William
Stewart, Thomas Henry McKenzie
Stiller, Calvin R
Strejan, Gill Henric
Szewczuk, Myron Ross
Tryphonas, Helen
Tsang, Charles Pak Wai
Underdown, Brian James
Vadas, Peter
Vijay, Hari Mohan
Waithe, William Irwin
Wilkie, Bruce Nicholson
Zimmerman, Barry
Zobel, Alicja Maria

QUEBEC
Adamkiewicz, Vincent Witold
Antel, Jack Perry
Archambault, Denis
Bernard, Nicole F
Borgeat, Pierre
Cantor, Ena D
Delespesse, Guy Joseph
Ford-Hutchinson, Anthony W
Fournier, Michel

Freedman, Samuel Orkin
Gervais, Francine
Gold, Phil
Gordon, Julius
Guttmann, Ronald David
Lamoureux, Gilles
Lapp, Wayne Stanley
Martineau, Ronald
Menezes, Jose Piedade Caetano Agnelo
Miller, Sandra Carol
Moroz, Leonard Arthur
Murgita, Robert Anthony
Osmond, Dennis Gordon
Page, Michel
Palfree, Roger Grenville Eric
Poole, Anthony Robin
Potworowski, Edouard Francois
Rokeach, Luis Alberto
Rola-Pleszczynski, Marek
Schulz, Jan Ivan
Shuster, Joseph
Siboo, Russell
Sikorska, Hanna
Sirois, Pierre
Skamene, Emil
Stevenson, Mary M
Tanner, Charles E
Thomson, David M P
Trudel, Michel D

SASKATCHEWAN
Babiuk, Lorne A
Blakley, Barry Raymond
McLennan, Barry Dean
Xiang, Jim

OTHER COUNTRIES
Aalund, Ole
Abelev, Garri Izrailevich
Acres, Robert Bruce
Adldinger, Hans Karl
Agarwal, Shyam S
Andrews, Peter Walter
Arai, Ken-Ichi
Arala-Chaves, Mario Passalaqua
Baldwin, Robert William
Barel, Monique
Barrett, James Thomas
Binder, Bernd R
Bodmer, Walter Fred
Bojalil, Luis Felipe
Bolhuis, Reinder L H
Borek, Felix
Borel, Yves
Braquet, Pierre G
Brondz, Boris Davidovich
Campbell, Virginia Wiley
Cooke, T Derek V
Dausset, Jean Baptiste Gabriel
Dianzani, Ferdinando
Dyrberg, Thomas Peter
Eggers, Hans J
English, Leonard Stanley
Farid, Nadir R
Fernandez-Cruz, Eduardo P
Fey, George
Franklin, Richard Morris
Fujimoto, Shigeyoshi
Gelzer, Justus
Gemsa, Diethard
Ghiara, Paolo
Goding, James Watson
Gorski, Andrzej
Gowans, James L
Ha, Tai-You
Hansson, Goran K
Hayry, Pekka Juha
Heiniger, Hans-Jorg
Hirano, Toshio
Hirokawa, Katsuiku
Hofmann, Bo
Hsu, Clement C S
Ide, Hiroyuki
Izui, Shozo
Jerne, Niels Kaj
Kerjaschki, Dontscho
Kessel, Rosslyn William Ian
Kimoto, Masao
Kobata, Akira
Krueger, Gerhard R F
Kruisbeek, Ada M
Kuhn, Klaus
Lamoyi, Edmundo
Lee, Men Hui
Levy, David Alfred
Lin, Chin-Tarng
Ling, Chung-Mei
Loppnow, Harald
Low, Chow-Eng
Low, Teresa Lingchun Kao
Lumb, Judith Rae H
MacDonald, Thomas Thornton
Marikovsky, Yehuda
Markert, Claus O
Maruta, Hiroshi
Maruyama, Koshi
Merlini, Giampaolo
Michaelsen, Terje E
Minowada, Jun
Morrow, William John Woodroofe
Muller-Eberhard, Hans Joachim
Munoz, Maria De Lourdes
Pan, In-Chang
Prowse, Stephen James

Quastel, Michael Reuben
Querinjean, Pierre Joseph
Ray, Prasanta K
Revillard, Jean-Pierre Remy
Romagnani, Sergio
Santoro, Ferrucio Fontes
Saravia, Nancy G
Schena, Francesco Paolo
Seya, Tsukasa
Solter, Davor
Stingl, Georg
Tachibana, Takehiko
Takai, Yasuyuki
Taniyama, Tadayoshi
Van Furth, Ralph
Van Loveren, Henk
Varga, Janos M
Vogel, Carl-Wilhelm E
Wang, Soo Ray
Watanabe, Takeshi
Webb, Cynthia Ann Glinert
Weil, Marvin Lee
Weiss, David Walter
Willard-Gallo, Karen Elizabeth
Wu, Albert M
Yakura, Hidetaka
Yamamoto, Hiroshi
Yeon-Sook, Yun
Yoshida, Takeshi

Microbiology

ALABAMA
Acton, Ronald Terry
Alford, Charles Aaron, Jr
Allen, Lois Brenda
Ammerman, Gale Richard
Backman, Paul Anthony
Ball, Laurence Andrew
Bhatnagar, Yogendra Mohan
Blalock, James Edwin
Brady, Yolanda J
Briles, David Elwood
Brown, Alfred Ellis
Carrozza, John Henry
Carter, Charlotte
Cassell, Gail Houston
Chapatwala, Kirit D
Cody, Reynolds M
Coggin, Joseph Hiram
Cook, David Wilson
Davis, Norman Duane
Dimopoullos, George Takis
Edgar, Samuel Allen
Egan, Marianne Louise
Estes, Edna E
Francis, Robert Dorl
Giambrone, Joseph James
Gillespie, George Yancey
Gottlieb, Sheldon F
Green, Margaret
Higgins, N Patrick
Hunter, Eric
Hunter, Katherine Morton
Jones, Daniel David
Jones, Jeanette
Kearney, John F
Klesius, Phillip Harry
Lauerman, Lloyd Herman, Jr
Lemke, Paul Arenz
Lidin, Bodil Inger Maria
McCaskey, Thomas Andrew
McGhee, Jerry Roger
Michalek, Suzanne M
Moser, Stephen Adcock
Nielsen, Brent Lynn
Pass, Robert Floyd
Patterson, Loyd Thomas
Plumb, John Alfred
Rivers, Douglas Bernard
Russell, Michael W(illiam)
Schulze, Karl Ludwig
Schutzbach, John Stephen
Shannon, William Michael
Siddique, Irtaza H
Singh, Shiva Pujan
Smith, Paul Clay
Smith-Somerville, Harriett Elizabeth
Stagno, Sergio Bruno
Suberkropp, Keller Francis
Suling, William John
Thomas, Julian Edward, Sr
Turco, Jenifer
Wertz, Gail T Williams
Wilkoff, Lee Joseph
Winkler, Herbert H
Winters, Alvin L
Wood, David Oliver
Worley, S D

ALASKA
Brown, Edward James
Button, Don K
Roop, Richard Allan

ARIZONA
Alcorn, Stanley Marcus
Appelgren, Walter Phon
Archer, Stanley J
Bachman, Marvin Charles
Baker, Gladys Elizabeth
Birge, Edward Asahel
Chalquest, Richard Ross

Microbiology (cont)

Froman, Seymour
Gendler, Sandra J
Gerba, Charles Peter
Hewlett, Martinez Joseph
Hoober, J Kenneth
Janssen, Robert (James) J
Jeter, Wayburn Stewart
Koffler, Henry
Ludovici, Peter Paul
Mandle, Robert Joseph
Mare, Cornelius John
Meinke, William John
Mendelson, Neil Harland
Mosher, Richard Arthur
Moulder, James William
Nelson, Frank Eugene
Pommerville, Jeffrey Carl
Reinhard, Karl Raymond
Schmidt, Jean M
Shull, James Jay
Sinclair, Norval A
Sinski, James Thomas
Spizizen, John
Sypherd, Paul Starr
Vincent, Walter Sampson
Wallraff, Evelyn Bartels
Yall, Irving

ARKANSAS

Abernathy, Robert Shields
Barron, Almen Leo
Bower, Raymond Kenneth
Bowling, Robert Edward
Champlin, William G
Denton, James H
Elbein, Alan D
Ferguson, Dale Vernon
Hardin, Hilliard Frances
Heddleston, Kenneth Luther
Heflich, Robert Henry
Hinck, Lawrence Wilson
Howell, Robert T
Kaplan, Arnold
Paulissen, Leo John
Pynes, Gene Dale
Rank, Roger Gerald
Skeeles, John Kirkpatrick
Soderberg, Lee Stephen Freeman
Steinkraus, Donald Curtiss
Sutherland, John Bruce, IV
Wennerstrom, David E
Wolf, Duane Carl
Young, Seth Yarbrough, III

CALIFORNIA

Adams, Jane N
Adaskaveg, James Elliott
Allen, Charles Freeman
Allison, Anthony Clifford
Aly, Raza
Anthony, Ronald Lewis
Ashton, David Hugh
Ayengar, Padmasini (Mrs Frederick Aladjem)
Balch, William E
Barker, Horace Albert
Bartnicki-Garcia, Salomon
Baxter, William Leroy
Beaman, Blaine Lee
Belisle, Barbara Wolfanger
Benenson, Abram Salmon
Berk, Arnold J
Berry, Edward Alan
Bertani, Giuseppe
Bertani, Lillian Elizabeth
Bibel, Debra Jan
Bishop, John Michael
Bishop, Nancy Horschel
Bissett, Marjorie Louise
Bittle, James Long
Bochner, Barry Ronald
Borchardt, Kenneth
Botzler, Richard George
Bovell, Carlton Rowland
Bowman, Barry J
Boyer, Herbert Wayne
Bradshaw, Lawrence Jack
Braemer, Allen C
Brownell, Anna Gale
Bruckner, David Alan
Brunell, Philip Alfred
Brunke, Karen J
Buchanan, Bob Branch
Buehring, Gertrude Case
Bulich, Anthony Andrew
Byatt, Pamela Hilda
Cabasso, Victor Jack
Calavan, Edmond Clair
Campbell, Judith Lynn
Canawati, Hanna N
Capers, Evelyn Lorraine
Carbon, John Anthony
Cardiff, Robert Darrell
Cardineau, Guy A
Caren, Linda Davis
Carlberg, David Marvin
Carlson, James Reynold
Catlin, B Wesley
Chamberlin, Michael John
Chan, Sham-Yuen
Chang, George Washington
Chang, Jennie C C

Chang, Robert Shihman
Chatigny, Mark A
Chen, Anthony Bartheolomew
Chen, Benjamin P P
Childress, Evelyn Tutt
Chou, Tsong-Wen
Clark, Alvin John
Cohen, Morris
Cohen, Stanley Norman
Collin, William Kent
Collins, Edwin Bruce
Connor, James D
Constantine, Denny G
Cooper, Robert Chauncey
Corbeil, Lynette B
Craig, James Morrison
Cremer, Natalie E
Crocker, Thomas Timothy
Dahms, Arthur Stephen
DaMassa, Al John
Daniels, Jeffrey Irwin
Dasgupta, Asim
Davé, Bhalchandra A
Davenport, Calvin Armstrong
Davis, Charles Edward
Davis, Mark M
Davis, Rowland Hallowell
Delong, Edward F
Del Villano, Bert Charles
Des Marais, David John
Di Girolamo, Rudolph Gerard
Dragon, Elizabeth Alice Oosterom
Duffey, Paul Stephen
Dulbecco, Renato
Duniway, John Mason
Duzgunes, Nejat
Eaton, Norman Ray
Eckhart, Walter
Ehrenfeld, Elvera
Eiserling, Frederick A
Eldredge, Kelly Husbands
Emmons, Richard William
Englesberg, Ellis
Eppstein, Deborah Anne
Epstein, Lynn
Erickson, Jeanne Marie
Evans, Ernest Edward, Jr
Fabrikant, Irene Berger
Falkow, Stanley
Farmer, Walter Joseph
Finegold, Sydney Martin
Flashman, Stuart Milton
Forghani-Abkenari, Bagher
Fox, Eugene N
Fuhrman, Jed Alan
Fukuyama, Thomas T
Fulco, Armand J
Fung, Henry C, Jr
Funk, Glenn Albert
Ganesan, Ann K
Geiduschek, Ernest Peter
Gilbertson, Robert Leonard
Giorgi, Janis V(incent)
Gitnick, Gary L
Glembotski, Christopher Charles
Gober, James William
Goehler, Brigitte Hanna
Gold, William
Gong, Yu
Goodman, Nelson
Goodman, Richard E
Gordon, Irving
Granger, Gale A
Green, Melvin Howard
Green, Richard H
Greene, Elias Louis
Grilione, Patricia Louise
Gumpf, David John
Gunsalus, Robert Philip
Gupta, Rishab Kumar
Gutman, George Andre
Hackett, Adeline J
Hadley, William Keith
Hagen, Charles Alfred
Haglund, John Richard
Haight, Roger Dean
Halde, Carlyn Jean
Hammar, Allan H
Hankinson, Oliver
Hanna, Lavelle
Hanson, Carl Veith
Hatfield, G Wesley
Haygood, Margo Genevieve
Heckly, Robert Joseph
Hedrick, Ronald Paul
Heinrich, Milton Rollin
Hemmingsen, Barbara Bruff
Henderson, Gary Borgar
Herskowitz, Ira
Heuer, Ann Elizabeth
Heuschele, Werner Paul
Hill, Douglas Wayne
Hill, Ray Allen
Hirsh, Dwight Charles, III
Hirst, George Keble
Hoeprich, Paul Daniel
Holland, John Joseph
Holmberg, Charles Arthur
Horvath, Kalman
Horwitz, David A
Howard, Russell John
Huff, Dennis Karl
Hunderfund, Richard C

Hungate, Robert Edward
Hunter, Tony
Huskey, Glen E
Hyman, Bradley Clark
Hyman, Richard W
Imagawa, David Tadashi
Ingraham, John Lyman
Jackson, Andrew Otis
Jackson, James Oliver
Janda, John Michael
Jariwalla, Raxit Jayantilal
Jawetz, Ernest
Jaye, Murray Joseph
Johansson, Karl Richard
Jones, Theodore Harold Douglas
Kallman, Burton Jay
Kang, Chang-Yuil
Kang, Kenneth S
Kang, Tae Wha
Kavenoff, Ruth
Kennedy, Elhart James
Kettering, James David
Kibler, Ruthann
Kim, Juhee
King, Alfred Douglas, Jr
Klein, Harold Paul
Klement, Vaclav
Knudson, Gregory Blair
Koshland, Daniel Edward, Jr
Kozloff, Lloyd M
Kubo, Ralph Teruo
Kuhn, Daisy Angelika
Kunkee, Ralph Edward
Kutner, Leon Jay
Lai, Michael Ming-Chiao
Lane, Robert Sidney
Lanyi, Janos K
Lascelles, June
Lechtman, Max D
Lederberg, Esther Miriam
Lester, William Lewis
Levinson, Warren E
Levintow, Leon
Lewis, William Perry
Lindberg, Lois Helen
Liu, Chi-Li
Louie, Robert Eugene
Lynch, Richard Vance, III
McCarthy, Robert Elmer
McEwen, Joan Elizabeth
Mah, Robert A
Manclark, Charles Robert
Mankau, Reinhold
Manning, JaRue Stanley
Marr, Allen Gerald
Marrone, Pamela Gail
Martin, George Steven
Masover, Gerald K
Mathies, Margaret Jean
Matthews, Thomas Robert
Merigan, Thomas Charles, Jr
Merriam, Esther Virginia
Metcalf, Robert Harker
Michener, H(arold) David
Midura, Thaddeus
Mielenz, Jonathan Richard
Miller, Martin Wesley
Miller, Sol
Miranda, Quirinus Ronnie
Mitchell, John Alexander
Morenzoni, Richard Anthony
Muller-Sieburg, Christa E
Murphy, Frederick A
Mussen, Eric Carnes
Nahhas, Fuad Michael
Nassos, Patricia Saima
Nayak, Debi Prosad
Neilands, John Brian
Nematollahi, Jay
Nguyen, Binh Trong
Nichol, Francis Richard, Jr
Nichols, Barbara Ann
Nierlich, Donald P
Nikaido, Hiroshi
Nomura, Masayasu
Obijeski, John Francis
Ogunseitan, Oladele Abiola
Oh, Jang Ok
Olson, Betty H
O'Neill, Thomas Brendan
Opfell, John Burton
Osebold, John William
Oshiro, Lyndon Satoru
Overby, Lacy Rasco
Padgett, Billie Lou
Pappagianis, Demosthenes
Penhoet, Edward Etienne
Perrault, Jacques
Peterson, George Harold
Phaff, Herman Jan
Phelps, Leroy Nash
Pitts, Robert Gary
Porter, David Dixon
Pratt, David
Prusiner, Stanley Ben
Puhvel, Sirje Madli
Purcell, Alexander Holmes, III
Quilligan, James Joseph, Jr
Rachmeler, Martin
Raj, Harkisan D
Redfield, William David
Reeve, Peter
Reilly, Dorothea Eleanor
Remington, Jack Samuel

Richman, Douglas Daniel
Riggs, John L
Rivier, Catherine L
Roantree, Robert Joseph
Rodgers, Richard Michael
Rogoff, Martin Harold
Rosenberg, Steven Loren
Ross, Ian Kenneth
Rowland, Ivan W
Ruby, Edward George
Rudnick, Albert
Russell, Ruth Lois
Sacks, Lawrence Edgar
Saier, Milton H, Jr
Samuel, Charles Edward
Saperstein, Sidney
Sapico, Francisco L
Schachter, Julius
Schaffer, Frederick Leland
Schein, Arnold Harold
Schekman, Randy W
Scherrer, Rene
Schiller, Neal Leander
Schmid, Peter
Schonfeld, Steven Emanuel
Schwerdt, Carlton Everett
Scow, Kate Marie
Semancik, Joseph Stephen
Senyk, George
Seto, Joseph Tobey
Shapiro, Lucille
Sheneman, Jack Marshall
Shum, Archie Chue
Simpson, Robert Blake
Smith, Douglas Wemp
Smith, Michael R
Soli, Giorgio
Soller, Arthur
Spahn, Gerard Joseph
Spanis, Curt William
Spotts, Charles Russell
Stallcup, Michael R
Stanbridge, Eric John
Starr, Mortimer Paul
Steenbergen, James Franklin
Stephens, William Leonard
Stevens, Jack Gerald
Stevens, Ronald Henry
Stevenson, Kenneth Eugene
Steward, John P
Stohlman, Stephen Arnold
Strauss, Ellen Glowacki
Strauss, James Henry
Stringfellow, Dale Alan
Suffin, Stephen Chester
Sutton, Donald Dunsmore
Swatek, Frank Edward
Sweet, Benjamin Hersh
Syvanen, Michael
Tachibana, Dora K
Tanada, Yoshinori
Taylor, Barry L
Taylor, John Marston
Tenenbaum, Saul
Thorner, Jeremy William
Thrupp, Lauri David
Tidwell, William Lee
Tomlinson, Geraldine Ann
Treagan, Lucy
Umanzio, Carl Beeman
Utterback, Nyle Gene
Uyeda, Charles Tsuneo
Vail, Patrick Virgil
VanBruggen, Ariena H C
Vedros, Neylan Anthony
Vice, John Leonard
Vickrey, Herta Miller
Vilker, Vincent Lee
Villarreal, Luis Perez
Vogt, Peter Klaus
Volcani, Benjamin Elazari
Vold, Barbara Schneider
Volz, Michael George
Vredevoe, Donna Lou
Vyas, Girish Narmadashankar
Wagner, Edward Knapp
Waitz, Jay Allan
Wall, Thomas Randolph
Wallace, Robert Bruce
Ward, Bess B
Wayne, Lawrence Gershon
Webb, David Ritchie, Jr
Wechsler, Steven Lewis
Wedberg, Stanley Edward
Wehrle, Paul F
Weinberg, Barbara Lee Huberman
Wheelis, Mark Lewis
White, Maurice Leopold
White, Thomas James
Whitehead, Marvin Delbert
Wiebe, Michael Eugene
Wilcox, Gary Lynn
Wilhelm, Alan Roy
Williams, Terry Wayne
Wistreich, George A
Wolf, Beverly
Wolochow, Hyman
Wood, Willis Avery
Woodhour, Allen F
Woolfolk, Clifford Allen
Wu, Jia-Hsi
Wu, William Gay
Wyatt, Philip Joseph
Yamamoto, Richard

Yayanos, A Aristides
Yen, Tien-Sze Benedict
York, Charles James
York, George Kenneth, II
Yu, David Tak Yan
Zamenhof, Stephen
Zamost, Bruce Lee
Zee, Yuan Chung
Zlotnik, Albert
Zuccarelli, Anthony Joseph

COLORADO
Andrews, Ken J
Avens, John Stewart
Barber, Thomas Lynwood
Berens, Randolph Lee
Blair, Carol Dean
Bourquin, Al Willis J
Bowles, Jean Alyce
Brennan, Patrick Joseph
Brooks, Bradford O
Bunn, Paul A, Jr
Calisher, Charles Henry
Cambier, John C
Chow, Tsu Ling
Church, Brooks Davis
Crowle, Alfred John
Danna, Kathleen Janet
De Martini, James Charles
Dobersen, Michael J
Francy, David Bruce
Frerman, Frank Edward
Gabridge, Michael Gregory
Goren, Mayer Bear
Grant, Dale Walter
Grieve, Robert B
Gubler, Duane J
Hager, Jean Carol
Harold, Franklin Marcel
Harold, Ruth L
Ivory, Thomas Martin, III
Jaehning, Judith A
Janes, Donald Wallace
Johnson, Donal Dabell
Jordan, Russell Thomas
Keller, Frederick Albert, Jr
Klein, Donald Albert
Kloppel, Thomas Mathew
Knight, William Glenn
Linden, James Carl
McClatchy, Joseph Kenneth
Mika, Leonard Aloysius
Norstadt, Fred A
Ogg, James Elvis
Pizer, Lewis Ivan
Poyton, Robert Oliver
Prentice, Neville
Ranu, Rajinder S
Rich, Marvin A
Roberts, Walden Kay
Schmidt, S K
Segal, William
Shulls, Wells Alexander
Smith, Ralph E
Sofos, John N
Talmage, David Wilson
Taylor, Austin Laurence
Tengerdy, Robert Paul
Trent, Dennis W
Tyler, Kenneth Laurence
Ulrich, John August
Updegraff, David Maule
Urban, Richard William
Vasil, Michael Lawrence
Wyss, Orville

CONNECTICUT
Aaslestad, Halvor Gunerius
Aitken, Thomas Henry Gardiner
Armstrong, Martine Yvonne Katherine
Baltimore, Robert Samuel
Berg, Claire M
Black, Francis Lee
Blogoslawski, Walter
Bonner, Daniel Patrick
Booss, John
Borgia, Julian Frank
Braun, Phyllis C
Buck, John David
Burke, Carroll N
Cameron, J A
Celesk, Roger A
Cheng, Yung-Chi
Coleman, William H
Coykendall, Alan Littlefield
Cummings, Dennis Paul
Curnen, Edward Charles, Jr
Deutscher, Murray Paul
Dingman, Douglas Wayne
Dorsky, David Isaac
Eagar, Robert Gouldman, Jr
Edberg, Stephen Charles
Eustice, David Christopher
Feldman, Kathleen Ann
Firshein, William
Forenza, Salvatore
Fried, John H
Froshauer, Susan
Geiger, Edwin Otto
Geiger, Jon Ross
Golub, Efim I
Gorrell, Thomas Earl
Hsiung, Gueh Djen
Huang, Liang Hsiung

Kemp, Gordon Arthur
Kirber, Maria Wiener
Kiron, Ravi
Kowolenko, Michael D
Krause, Peter James
LaMondia, James A
Leadbetter, Edward Renton
Lee, Sin Hang
Lentz, Thomas Lawrence
Loose, Leland David
Low, Kenneth Brooks, Jr
McGregor, Donald Neil
McKhann, Charles Fremont
Marcus, Philip Irving
Martin, R(ufus) Russell
Mazzone, Horace M
Miller, I George, Jr
Millian, Stephen Jerry
Natarajan, Kottayam Viswanathan
Nielsen, Peter Adams
Norton, Louis Arthur
Ornston, Leo Nicholas
Pazoles, Christopher James
Puddington, Lynn
Rauscher, Frank Joseph, Jr
Ray, Verne A
Repak, Arthur Jack
Retsema, James Allan
Rho, Jinnque
Romano, Antonio Harold
Ross, Martin Russell
Rothfield, Lawrence I
Routien, John Broderick
Sardinas, Joseph Louis
Sekellick, Margaret Jean
Setlow, Peter
Sieckhaus, John Francis
Slayman, Clifford L
Smith, Abigail LeVan
Terry, Thomas Milton
Treffers, Henry Peter
Truesdell, Susan Jane
Ukeles, Ravenna
Verses, Christ James
Ward, David Christian
Wernau, William Charles
Wilson, Gary August
Wyckoff, Delaphine Grace Rosa
Xu, Zhi Xin

DELAWARE
Anderson, Jeffrey John
Benton, William J
Campbell, Linzy Leon
Carberry, Judith B
De Courcy, Samuel Joseph, Jr
Dexter, Stephen C
Dohms, John Edward
Eleuterio, Marianne Kingsbury
Enquist, Lynn William
Fries, Cara Rosendale
Gray, John Edward
Herson, Diane S
Hoover, Dallas Gene
Howard, Richard James
Jackson, David Archer
Korant, Bruce David
Lockart, Royce Zeno, Jr
Maciag, William John, Jr
Marrs, Barry Lee
Myoda, Toshio Timothy
Neubauer, Russell Howard
Pene, Jacques Jean
Rosenberger, John Knox
Sariaslani, Sima
Singleton, Rivers, Jr
Smith, David William
Stopkie, Roger John
Yates, Richard Alan
Yin, Fay Hoh
Zajac, Barbara Ann
Zajac, Ihor

DISTRICT OF COLUMBIA
Affronti, Lewis Francis
Albright, Julia W
Ashe, Warren (Kelly)
Bellanti, Joseph A
Berman, Sanford
Bundy, Bonita Marie
Calderone, Richard Arthur
Chapman, George Bunker
Cocks, Gary Thomas
Cutchins, Ernest Charles
DeCicco, Benedict Thomas
Dunkel, Virginia Catherine
Frederick, Lafayette
Gemski, Peter
Gibbs, Clarence Joseph, Jr
Gravely, Sally Margaret
Halstead, Thora Waters
Hartley, Janet Wilson
Herscowitz, Herbert Bernard
Hill, Walter Ernest
Hollinshead, Ariel Cahill
Hugh, Rudolph
Katz, Edward
Kingsbury, David Thomas
Krigsvold, Dale Thomas
Kumar, Ajit
Lichens-Park, Ann Elizabeth
Madden, David Larry
Madden, Joseph Michael
Malveaux, Floyd J

Matyas, Gary Ralph
Miller, William Robert
Moscovici, Carlo
Neihof, Rex A
O'Hern, Elizabeth Moot
Okrend, Harold
Olson, James Gordon
Oyewole, Saundra Herndon
Parker, Richard H
Parrott, Robert Harold
Pearson, Gary Richard
Prival, Michael Joseph
Putzrath, Resha Mae
Rabin, Robert
Reich, Melvin
Roane, Philip Ransom, Jr
Roskoski, Joann Pearl
Sabin, Albert B(ruce)
Salzman, Lois Ann
Salzman, Norman Post
Saz, Arthur Kenneth
Sever, John Louis
Silver, Sylvia
Sreevalsan, Thazepadath
Tompkins, George Jonathan
Turner, Willie
Venkatesan, Malabi M
Walters, Curla Sybil
Wassef, Nabila M
Young, Frank E

FLORIDA
Aldrich, Henry Carl
Baer, Herman
Balkwill, David Lee
Beckhorn, Edward John
Bellamy, Winthrop Dexter
Beneke, Everett Smith
Bergs, Victor Visvaldis
Bernstein, Sheldon
Betz, John Vianney
Bitton, Gabriel
Bleiweis, Arnold Sheldon
Bliznakov, Emile George
Block, Seymour Stanton
Blomberg, Bonnie B
Brandon, Frank Bayard
Brown, Ross Duncan, Jr
Brown, Thomas Allen
Brown, William Lewis
Burney, Curtis Michael
Carvajal, Fernando
Chapman, Peter John
Chester, Brent
Clark, William Burton, IV
Cleary, Timothy Joseph
Cohen, Marc Singman
Cohen, Pinya
Coleman, Sylvia Ethel
Condit, Richard
Cort, Winifred Mitchell
Cowman, Richard Ammon
Dhople, Arvind Madhav
Dierks, Richard Ernest
Djeu, Julie Y
Durand, Donald P
Eilers, Frederick Irving
Ellis, Edwin M
Ewart, R Bradley
Farina, Joseph Peter
Farrow, Wendall Moore
Fitzgerald, Robert James
Fox, Kenneth Ian
Frankel, Jack William
Frediani, Harold Arthur
Friedman, Herman
Friedmann, E(merich) Imre
Gaskin, Jack Michael
Gennaro, Robert Nash
Genthner, Barbara Robyn Sharak
Genthner, Fred J
Gifford, George Edwin
Godzeski, Carl William
Goihman-Yahr, Mauricio
Gomatos, Peter John
Gottwald, Timothy R
Groupe, Vincent
Halbert, Seymour Putterman
Halkias, Demetrios
Hartman, James Xavier
Hauswirth, William Walter
Heinis, Julius Leo
Helmstetter, Charles E
Hiebert, Ernest
Houts, Garnette Edwin
Ingram, Lonnie O'Neal
Jenkin, Howard M
Jensen, Roy A
Kern, Jerome
Kiefer, David John
Klein, Paul Alvin
Koburger, John Alfred
Lai, Patrick Kinglun
Lim, Daniel V
Lin, Tsue-Ming
Lopez, Diana Montes De Oca
McArthur, William P
McClung, Norvel Malcolm
Makemson, John Christopher
Mans, Rusty Jay
Martinez, Octavio Vincent
Moyer, Richard W
Muchovej, Rosa M C
Nakashima, Tadayoshi

Nonoyama, Meihan
Ocampo-Friedmann, Roseli C
Paradise, Lois Jean
Patarca, Roberto
Phlips, Edward Julien
Preston, James Faulkner, III
Previc, Edward Paul
Pruzansky, Jacob Julius
Purcifull, Dan Elwood
Putnam, Hugh D
Radhakrishnan, Chittur Venkitasubhan
Ringel, Samuel Morris
Robb, James Arthur
Rubin, Harvey Louis
Sallman, Bennett
Sandoval, Howard Kenneth
Schuytema, Eunice Chambers
Scudder, Walter Tredwell
Shands, Joseph Walter, Jr
Silver, Warren Seymour
Sinkovics, Joseph G
Smith, Kenneth Leroy
Specter, Steven Carl
Spencer, John Lawrence
Stechschulte, Agnes Louise
Stock, David Allen
Streitfeld, Murray Mark
Sweet, Herman Royden
Tamplin, Mark Lewis
Taylor, Barrie Frederick
Tsai, James Hsi-Cho
Undeen, Albert Harold
Verkade, Stephen Dunning
White, Franklin Henry
White, Roseann Spicola
Widra, Abe
Wilson, William D
Wodzinski, Rudy Joseph

GEORGIA
Ahearn, Donald G
Ajello, Libero
Anderson, David Prewitt
Barbaree, James Martin
Bard, Raymond Camillo
Batra, Gopal Krishan
Beard, Charles Walter
Bennett, Sara Neville
Best, Gary Keith
Beuchat, Larry Ray
Boring, John Rutledge, III
Brinton, Margo A
Brooks, John Bill
Brown, William E
Brownell, George H
Brugh, Max, Jr
Bryans, Trabue Daley
Cassel, William Alwein
Cavallaro, Joseph John
Chang, Chung Jan
Cherry, William Bailey
Cohen, Jay O
Cole, John Rufus, Jr
Colley, Daniel George
Compans, Richard William
Cox, Nelson Anthony
Crouse, Gray F
Dailey, Harry A
David, Robert Michael
Dobrovolny, Charles George
Dowdle, Walter R
Dowell, Vulus Raymond, Jr
Doyle, Michael Patrick
Duncan, Robert Leon, Jr
Eagon, Robert Garfield
Eriksson, Karl-Erik Lennart
Eriquez, Louis Anthony
Evans, Donald Lee
Ewing, William Howell
Farmer, John James, III
Favero, Martin
Fincher, Edward Lester
Finnerty, William Robert
Garver, Frederick Albert
Gitaitis, Ronald David
Good, Robert Campbell
Gratzek, John B
Green, John H
Hall, Charles Thomas
Hatheway, Charles Louis
Hedden, Kenneth Forsythe
Herrmann, Kenneth L
Hierholzer, John Charles
Holt, Peter Stephen
Howe, Henry Branch, Jr
Hubbard, Jerry S
Ibrahim, Adloy N
Jones, Gilda Lynn
Jones, Rena Talley
Jones, Wilbur Douglas, Jr
Kaplan, William
Kellogg, Douglas Sheldon, Jr
Kerr, Thomas James
Kleven, Stanley H
Krishnamurti, Pullabhotla V
Kubica, George P
LaMotte, Louis Cossitt, Jr
Lehner, Andreas Friedrich
Ljungdahl, Lars Gerhard
Luck, John Virgil
Lukert, Phil Dean
McClure, Harold Monroe
McDade, Joseph Edward
Margolis, Harold Stephen

Microbiology (cont)

Martin, William Randolph
Mathews, Henry Mabbett
Miller, Lois Kathryn
Morse, Stephen Allen
Moss, Claude Wayne
Nahmias, Andre Joseph
Navalkar, Ram G
Newell, Steven Young
Novitsky, James Alan
Paris, Doris Fort
Payne, William Jackson
Peck, Harry Dowd, Jr
Pine, Leo
Reinhardt, Donald Joseph
Reiss, Errol
Rogers, John Ernest
Roth, Ivan Lambert
Sawyer, Richard Trevor
Schuler, George Albert
Schuster, George Sheah
Schwartz, David Alan
Shimkets, Lawrence Joseph
Shotts, Emmett Booker, Jr
Shuster, Robert C
Snyder, Hugh Donald
Spitznagel, John Keith
Stulting, Robert Doyle, Jr
Sudia, William Daniel
Suggs, Morris Talmage, Jr
Sulzer, Alexander Jackson
Summers, Anne O
Swaminathan, Balasubramanian
Taylor, Gerald C
Tenoso, Harold John
Thomason, Berenice Miller
Tornabene, Thomas Guy
Tsao, Mark Fu-Pao
Tunstall, Lucille Hawkins
Tzianabos, Theodore
Vogel, Ralph A
Volkmann, Keith Robert
Wallace, Douglas Cecil
White, Lendell Aaron
Wiebe, William John
Wiegel, Juergen K
Wilkinson, Hazel Wiley
Wolven-Garrett, Anne M
Woodley, Charles Lamar
Wyatt, Roger Dale
Youngman, Philip John
Ziegler, Harry Kirk
Ziffer, Jack

HAWAII
Alicata, Joseph Everett
Allen, Richard Dean
Berger, Leslie Ralph
Borthakur, Dulal
Fok, Agnes Kwan
Fujioka, Roger Sadao
Furusawa, Eiichi
Hall, John Bradley
Higa, Harry Hiroshi
Kathariou, Sophia
Klemmer, Howard Wesley
Loh, Philip Choo-Seng
Mandel, Morton
Marchette, Nyven John
Patterson, Gregory Matthew Leon
Person, Donald Ames
Roberts, Robert Russell
Scott, John Francis
Taussig, Steven J
Vennesland, Birgit
Ward, Melvin A

IDAHO
Beck, Sidney M
Crawford, Donald Lee
Crawford, Ronald Lyle
Davidson, Philip Michael
Dugan, Patrick R
Farrell, Larry Don
Frank, Floyd William
Hatcher, Herbert John
Heimsch, Richard Charles
Hostetter, Donald Lee
Keay, Leonard
Korus, Roger Alan
Laurence, Kenneth Allen
Lingg, Al J
McCune, Mary Joan Huxley
Roberto, Francisco Figueroa
Robertson, Donald Claus
Winston, Vern

ILLINOIS
Akin, Cavit
Albach, Richard Allen
Alexander, Nancy J
Armbruster, Frederick Carl
Bahn, Arthur Nathaniel
Bartell, Marvin H
Bender, James Gregg
Bhatti, Rashid
Bigelis, Ramunas
Bird, Thomas Joseph
Blaschek, Hans P
Blumenthal, Harold Jay
Bothast, Rodney Jacob
Braendle, Donald Harold
Braun, Donald Peter
Brennan, Patricia Conlon
Brewer, Gregory J
Brigham, Robert B
Brockman, Herman E
Burmeister, Harland Reno
Came, Paul E
Campbell, Michael Floyd
Casadaban, Malcolm John
Castignetti, Domenic
Chakrabarty, Ananda Mohan
Chang, Kwang-Poo
Chen, Shepley S
Cheng, Shu-Sing
Cheung, Hou Tak
Christensen, Mary Lucas
Chu, Daniel Tim-Wo
Chudwin, David S
Cibulsky, Robert John
Cohen, Sidney
Cooper, Morris Davidson
Corbett, Jules John
Cork, Douglas J
Crane, Anatole
Crouch, Norman Albert
DePinto, John A
Dhaliwal, Amrik S
Doering, Jeffrey Louis
Dommert, Arthur Roland
Drucker, Harvey
Duncan, James Lowell
Eachus, Alan Campbell
Ekstedt, Richard Dean
Engelbrecht, R(ichard) S(tevens)
Falk, Lawrence A, Jr
Farrand, Stephen Kendall
Fennewald, Michael Andrew
Fenters, James Dean
Flowers, Russell Sherwood, Jr
Ford, Richard Earl
Frampton, Elon Wilson
Frank, James Richard
Gabis, Damien Anthony
Galsky, Alan Gary
Gerding, Dale Nicholas
Gessner, Robert V
Girolami, Roland Louis
Goldin, Milton
Goldman, Manuel
Goodheart, Clyde Raymond
Goss, George Robert
Gould, John Michael
Gupta, Kailash Chandra
Hac, Lucile R
Hagar, Lowell Paul
Hahn, Donald Richard
Halpern, Bernard
Hanson, Lyle Eugene
Heim, Lyle Raymond
Herrmann, Ernest Carl, Jr
Holland, Louis Edward, II
Holleman, William H
Holzer, Timothy J
Hou, Ching-Tsang
Hsu, Wen-Tah
Huberman, Eliezer
Ingle, Morton Blakeman
Isaacson, Richard Evan
Jackson, George G
Jackson, Robert W
Jayaswal, Radheshyam K
Jonah, Margaret Martin
Jones, Marjorie Ann
Kaizer, Herbert
Kallio, Reino Emil
Kaye, Saul
Keudell, Kenneth Carson
Khodadad, Jena Khadem
Khoobyarian, Newton
Kim, Yoon Berm
Klegerman, Melvin Earl
Konisky, Jordan
Kurtzman, Cletus Paul
Laffler, Thomas G
Lakshminarayanan, Krishnaiyer
Lamb, Robert Andrew
Landau, William
Landay, Alan Lee
Lechowich, Richard V
Levy, Allan Henry
Lippincott, Barbara Barnes
Lippincott, James Andrew
Lopes, John Manuel
Lucher, Lynne Annette
Madigan, Michael Thomas
Markovitz, Alvin
Martin, Scott Elmore
Mateles, Richard I
Mathews, Herbert Lester
Matsumura, Philip
Melin, Brian Edward
Meyer, Richard Charles
Miller, Charles G
Miller, Raymond Michael
Miller, Robert Verne
Miller, Stephen Douglas
Moon, Robert John
Morello, Josephine A
Murphy, Richard Allan
Musa, Wafaa Arafat
Natalini, John Joseph
Nath, K Rajinder
Nath, Nrapendra
Neuhaus, Francis Clemens
Newton, Stephen Bruington
Nichols, Brian Paul
Orland, Frank J
Ostrow, David Henry
Padh, Harish
Pal, Dhiraj
Papoutsakis, Eleftherios Terry
Passman, Frederick Jay
Peak, Meyrick James
Pedraja, Rafael R
Pekarek, Robert Sidney
Perry, Dennis
Perry, Eugene Arthur
Peterson, Christopher Gerard
Peterson, David Allan
Plate, Charles Alfred
Plewa, Michael Jacob
Prakasam, Tata B S
Pratt, Charles Walter
Pumper, Robert William
Pun, Pattle Pak-Toe
Rabinovich, Sergio Rospigliosi
Rakosky, Joseph, Jr
Regunathan, Perialwar
Reilly, Christopher Aloysius, Jr
Rittmann, Bruce Edward
Roy, Dipak
Ruddat, Manfred
Rundell, Mary Kathleen
Rupprecht, Kevin Robert
Sadler, George D
Salyers, Abigail Ann
Sanborn, Mark Robert
Scherr, George Harry
Schleicher, Joseph Bernard
Schlenk, Fritz
Schnell, Gene Wheeler
Seed, Thomas Michael
Shapiro, James Alan
Shapiro, Stanley Kallick
Shechmeister, Isaac Leo
Shen, Linus Liang-nene
Sherman, Joseph E
Sherwood, Robert Lawrence
Shipkowitz, Nathan L
Silliker, John Harold
Simon, Ellen McMurtrie
Simon, Selwyn
Simonson, Lloyd Grant
Singer, Samuel
Smittle, Richard Baird
Snyder, William Robert
Spear, Patricia Gail
Starzyk, Marvin John
Stevens, Joseph Alfred
Strano, Alfonso J
Strauss, Bernard S
Sutton, Lewis McMechan
Switzer, Robert Lee
Taylor, Welton Ivan
Thompson, Kenneth David
Tomita, Joseph Tsuneki
Tompkin, Robert Bruce
Tripathy, Deoki Nandan
Tsang, Joseph Chiao-Liang
Turner, Donald W
Vary, Patricia Susan
Walter, Trevor John
Wawszkiewicz, Edward John
Weary, Marlys E
Welker, Neil Ernest
Wetegrove, Robert Lloyd
Whitt, Dixie Dailey
Wiatr, Christopher Louis
Wideburg, Norman Earl
Wilkinson, Brian James
Witmer, Heman John
Witter, Lloyd David
Wolfe, Ralph Stoner
Wolff, Robert John
Yackel, Walter Carl
Yamashiroya, Herbert Mitsugi
Yotis, William William

INDIANA
Allen, Norris Elliott
Argot, Jeanne
Aronson, Arthur Ian
Asteriadis, George Thomas, Jr
Babel, Frederick John
Baldwin, William Walter
Bauer, Carl Eugene
Bauer, Dietrich Charles
Becker, Benjamin
Begue, William John
Bockrath, Richard Charles, Jr
Boisvenue, Rudolph Joseph
Boyd, Frederick Mervin
Boyer, Ernest Wendell
Bozarth, Robert F
Bracker, Charles E
Burnstein, Theodore
Caltrider, Paul Gene
Counter, Frederick T, Jr
Cousin, Maribeth Anne
Crawford, James Gordon
Daily, Fay Kenoyer
Daily, William Allen
Day, Lawrence Eugene
Diffley, Peter
Doemel, William Naylor
Dziarski, Roman
Edenberg, Howard Joseph
Enders, George Leonhard, Jr
Fayerman, Jeffrey T
Foglesong, Mark Allen
Folkerts, Thomas Mason
Foos, Kenneth Michael
Franson, Timothy Raymond
Gale, Charles
Gavin, John Joseph
Gest, Howard
Gordee, Robert Stouffer
Gregory, Richard Lee
Gustafson, Donald Pink
Haak, Richard Arlen
Haelterman, Edward Omer
Hanson, Robert Jack
Harvey, William Homer
Hegeman, George D
Hendrickson, Donald Allen
Hershberger, Charles Lee
Hicks, Garland Fisher, Jr
Hodson, Phillip Harvey
Hoehn, Marvin Martin
Huber, Floyd Milton
Ingraham, Joseph Sterling
Kapustka, Lawrence A
Keith, Paula Myers
Kinsel, Norma Ann
Kinzel, Jerry J
Kirsch, Edwin Joseph
Kittaka, Robert Shinnosuke
Klausmeier, Robert Edward
Konetzka, Walter Anthony
Konopka, Allan Eugene
Kreps, David Paul
Kulpa, Charles Frank, Jr
Larsen, Steven H
Lees, Norman Douglas
Levine, Alvin Saul
Levinthal, Mark
Lister, Richard Malcolm
Lively, David Harryman
Loesch-Fries, Loretta Sue
Lopez, Carlos
Lovett, James Satterthwaite
McClung, Leland S(wint)
McMullen, James Robert
Mallett, Gordon Edward
Mason, Earl James
Matsuoka, Tats
Miescher, Guido
Millar, Wayne Norval
Miller (Gilbert), Carol Ann
Miller, Chris H
Minton, Sherman Anthony
Mocharla, Raman
Morse, Erskine Vance
Moskowitz, Merwin
Niederpruem, Donald J
Niss, Hamilton Frederick
Nunez, William J, III
Oster, Mark Otho
Pike, Loy Dean
Pleasants, Julian Randolph
Pollard, Morris
Queener, Stephen Wyatt
Ragheb, Hussein S
Roales, Robert R
Rogers, Howell Wade
Saz, Howard Jay
Schloemer, Robert Henry
Schoknecht, Jean Donze
Sherman, Louis Allen
Simon, Edward Harvey
Sipe, Jerry Eugene
Smith, Gary Lee
Smith, Robert William
Srivastava, Arun
Stone, Robert Louis
Summers, William Allen, Sr
Taylor, Milton William
Tessman, Irwin
Turner, Jan Ross
Wegener, Warner Smith
Weinberg, Eugene David
White, David
Whitney, John Glen
Williams, Daniel Charles
Williams, Robert Dee
Zimmerman, Sarah E
Zygmunt, Walter A

IOWA
Allison, Milton James
Anderson, Todd Alan
Bryner, John Henry
Cazin, John, Jr
Collins, Richard Francis
De Jong, Peter J
Dickson, James Sparrow
DiSpirito, Alan Angelo
Dworschack, Robert George
Fitzgerald, Gerald Roland
Frederick, Lloyd Randall
Frey, Merwin Lester
Gibson, David Thomas
Goss, Robert Charles
Greenberg, Everett Peter
Haefele, Douglas Monroe
Hanson, Austin Moe
Harp, James A
Harris, Delbert Linn
Hartman, Paul Arthur
Hausler, William John, Jr
Hughes, David Edward
Johnson, William
Kemeny, Lorant

Kieso, Robert Alfred
LaGrange, William Somers
Lambert, George
McClurkin, Arlan Wilbur
Markovetz, Allen John
Marshall, William Emmett
Martinson, Charlie Anton
Mayfield, John Eric
Mengeling, William Lloyd
Menninger, John Robert
Murphy, Mary Nadine
Naides, Stanley J
Newcomb, Harvey Russell
Nicholson, Donald Paul
O'Berry, Phillip Aaron
Packer, R(aymond) Allen
Paul, Prem Sagar
Perlman, Stanley
Pirtle, Eugene Claude
Quinn, Loyd Yost
Rodriguez, Jose Enrique
Ross, Richard Francis
Sabiston, Charles Barker, Jr
Shaw, Gaylord Edward
Smith, Claire Leroy
Songer, Joseph Richard
Stahly, Donald Paul
Stalheim, Ole H Viking
Stanton, Thaddeus Brian
Stinski, Mark Francis
Sweat, Robert Lee
Thompson, Sue Ann
Tjostem, John Leander
Van Der Maaten, Martin Junior
Van Eck, Edward Arthur
Walker, Homer Wayne
Wegner, Dennis L
Weisman, Lois Sue
Welter, C Joseph
Williams, Fred Devoe
Williams, Phletus P
Wood, Richard Lee

KANSAS
Akagi, James Masuji
Amelunxen, Remi Edward
Ash, Ronald Joseph
Bailie, Wayne E
Barnes, William Gartin
Bartkoski, Michael John, Jr
Bechtel, Donald Bruce
Behbehani, Abbas M
Bode, Vernon Cecil
Bradford, Lawrence Glenn
Buller, Clarence S
Cho, Cheng T
Coles, Embert Harvey, Jr
Consigli, Richard Albert
Decedue, Charles Joseph
Dirks, Brinton Marlo
Duncan, Bettie
Dykstra, Mark Allan
Field, Marvin Frederick
Fina, Louis R
Fung, Daniel Yee Chak
Furtado, Dolores
Goldberg, Ivan D
Guikema, James Allen
Haas, Herbert Frank
Ham, George Eldon
Harris, Pamela Ann
Hetrick, Barbara Ann
Iandolo, John Joseph
Jensen, Thorkil
Johnson, Donovan Earl
Langston, Clarence Walter
Leslie, John Franklin
Lindsey, Norma Jack
Liu, Chien
McMillen, Janis Kay
McVey, David Scott
Marchin, George Leonard
Miller, Glendon Richard
Morrison, David Campbell
Padmanabhan, Radhakrishnan
Paretsky, David
Peterson, John Edward, Jr
Ramamurti, Krishnamurti
Sarachek, Alvin
Shankel, Delbert Merrill
Stetler, Dean Allen
Takemoto, Dolores Jean
Upton, Steve Jay
Urban, James Edward
Werder, Alvar Arvid
Wong, Peter P
Wood, Gary Warren

KENTUCKY
Allen, George Perry
Andreasen, Arthur Albinus
Atlas, Ronald M
Bryans, John Thomas
Budde, Mary Laurence
Clark, Evelyn Genevieve
Crabtree, Koby Takayashi
Dalzell, Robert Clinton
Dickson, Robert Carl
Elliott, Larry P
Goodman, Norman L
Hamilton-Kemp, Thomas Rogers
Hammond, Ray Kenneth
Harris, Denny Olan
Hendrix, James William

Johnston, Paul Bruns
Keller, Kenneth F
Langlois, Bruce Edward
Lillich, Thomas Tyler
Liu, Pinghui Victor
McCollum, William Howard
Maisch, Weldon Frederick
Oetinger, David Frederick
Otero, Raymond B
Overman, Timothy Lloyd
Rodriguez, Lorraine Ditzler
Rothwell, Frederick Mirvan
Ruchman, Isaac
Squires, Robert Wright
Staat, Robert Henry
Steiner, Sheldon
Streips, Uldis Normunds
Twedt, Robert Madsen
Vanaman, Thomas Clark
Vittetoe, Marie Clare
Wacker, Waldon Burdette
Wallace, John Howard
Weber, Frederick, Jr
Williams, Arthur Lee
Wiseman, Ralph Franklin
Zimmer, Stephen George

LOUISIANA
Akers, Thomas Gilbert
Alam, Jawed
Barkate, John Albert
Bhatnagar, Deepak
Braymer, Hugh Douglas
Burns, Kenneth Franklin
Centifanto, Ysolina M
Corstvet, Richard E
Cotty, Peter John
Coward, Joe Edwin
Day, Donal Forest
Deas, Jane Ellen
Domer, Judith E
Dunigan, Edward P
Ehrlich, Kenneth Craig
Ehrlich, Melanie
Gallaher, William Richard
Gaumer, Herman Richard
Gerone, Peter John
Gormus, Bobby Joe
Gray, Raymond Francis
Greer, Donald Lee
Hart, Lewis Thomas
Hedrick, Harold Gilman
Hurley, Maureen
Huxsoll, David Leslie
Hyman, Edward Sidney
Hyslop, Newton Everett, Jr
Ivens, Mary Sue
Jamison, Richard Melvin
Johnson, Emmett John
Kadan, Ranjit Singh
Klich, Maren Alice
Koike, Hideo
Larkin, John Montague
Lembeck, William Jacobs
Leuck, Edwine E, II
Luftig, Ronald Bernard
Martin, Louis Norbert
Meyers, Samuel Philip
Monsour, Victor
O'Callaghan, Dennis John
O'Callaghan, Richard J
Pierce, William Arthur, Jr
Rhoads, Robert E
Rogers, Stearns Walter
Scher, Charles D
Shaffer, Morris Frank
Siebeling, Ronald Jon
Silberman, Ronald
Sizemore, Robert Carlen
Socolofsky, Marion David
Soike, Kenneth Fieroe
Spence, Hilda Adele
Srinivasan, Vadake Ram
Starrett, Richmond Mullins
Stevenson, L Harold
Stiffey, Arthur V
Storz, Johannes
Thompson, James Jarrard
Upadhyay, Jagdish M
Walkinshaw, Charles Howard, Jr
Weidner, Earl
Wilson, John Thomas
Wright, Maureen Smith

MAINE
Bain, William Murray
Curtis, Paul Robinson
De Siervo, August Joseph
Donahoe, John Philip
Gezon, Horace Martin
Jerkofsky, Maryann
King, Gary Michael
Lycette, R(ichard) (Milton)
Moody, Charles Edward, Jr
Nicholson, Bruce Lee
Norton, Cynthia Friend
Novotny, James Frank
Ridgway, George Junior

MARYLAND
Abramson, I Jerome
Ahmed, Syed Ashrafuddin
Albano, Marianita Madamba
Albrecht, Paul

Allen, William Peter
Allnutt, F C Thomas
Amemiya, Kei
Anderson, Arthur O
Anderson, Mauritz Gunnar
Armstrong, Earlene
Aurelian, Laure
Austin, Faye Carol Gould
Azad, Abdu F(arhang)
Baker, Phillip John
Bala, Shukal
Balady, Michael A
Barile, Michael Frederick
Baron, Louis Sol
Bassin, Robert Harris
Bausum, Howard Thomas
Beemon, Karen Louise
Benson, Spencer Alan
Berger, Edward Alan
Bernard, Dane Thomas
Berne, Bernard H
Berry, Bradford William
Biswal, Nilambar
Biswas, Robin Michael
Bloom, Eda Terri
Bluhm, Leslie
Blumberg, Peter Mitchell
Boyd, Virginia Ann Lewis
Bozeman, F Marilyn
Bradlaw, June A
Brandt, Carl David
Brandt, Walter Edmund
Brook, Itzhak
Brown, Paul Wheeler
Buckheit, Robert Walter, Jr
Burge, Wylie D
Burton, George Joseph
Butman, Bryan Timothy
Calabi, Ornella
Capone, Douglas George
Carski, Theodore Robert
Chang, Yung-Feng
Charache, Patricia
Chen, Joseph Ke-Chou
Chen, Priscilla B(urtis)
Cheng, Tu-chen
Choppin, Purnell Whittington
Clark, Trevor H
Cole, Gerald Alan
Coleman, Richard Walter
Collins, Frank Miles
Colwell, Rita R
Cook, Thomas M
Costlow, Richard Dale
Dalrymple, Joel McKeith
Daniels, William Fowler, Sr
Das, Naba Kishore
Dasch, Gregory Alan
Dashiell, Thomas Ronald
DeFrank, Joseph J
DeLamater, Edward Doane
Denny, Cleve B
Diener, Theodor Otto
Dodd, Roger Yates
Donlon, Mildred A
Eaves, George Newton
Eckels, Kenneth Henry
Edelman, Robert
Elin, Ronald John
Elwood, Jerry William
Esposito, Vito Michael
Evans, George Leonard
Faust, Maria Anna
Feinman, Susan (Birnbaum)
Fernie, Bruce Frank
Fine, Donald Lee
Finkelstein, David B
Fischinger, Peter John
Fiset, Paul
Fish, Donald C
FitzGerald, David J P
Fitzgerald, Edward Aloysius
Formal, Samuel Bernard
Foulds, John Douglas
Frasch, Carl Edward
Friedman, Robert Morris
Froehlich, Luz
Fulkerson, John Frederick
Gajdusek, Daniel Carleton
Galasso, George John
Ganaway, James Rives
Garges, Susan
Gerin, John Louis
Goldenbaum, Paul Ernest
Goldstein, Robert Arnold
Gonda, Matthew Allen
Goodman, Sarah Anne
Gordon, Milton
Gouge, Susan Cornelia Jones
Granger, Donald Lee
Gravell, Maneth
Griffin, Diane Edmund
Gruber, Jack
Grunberg, Neil Everett
Guss, Maurice Louis
Guttman, Helene Augusta Nathan
Habig, William Henry
Hackett, Joseph Leo
Hadidi, Ahmed Fahmy
Hale, Martha L
Hamilton, Bruce King
Hampar, Berge
Hanes, Darcy Elizabeth
Hanna, Edgar E(thelbert), Jr

BIOLOGICAL SCIENCES / 123

Hart, Mary Kate
Haspel, Martin Victor
Hawley, Robert John
Hearn, Henry James, Jr
Hellman, Alfred
Henry, Timothy James
Hetrick, Frank M
Hewetson, John Francis
Heydrick, Fred Painter
Hill, James Carroll
Hochstein, Herbert Donald
Hoffmann, Conrad Edmund
Hoggan, Malcolm David
Holmes, Kathryn Voelker
Holmes, Randall Kent
Hook, William Arthur
Hornor, Sally Graham
Housewright, Riley Dee
Howell, David McBrier
Hoyer, Bill Henriksen
Huebner, Robert Joseph
Hunt, Lois Turpin
Ivins, Bruce Edwards
Jakoby, William Bernard
James, Stephanie Lynn
Jaouni, Katherine Cook
Jemski, Joseph Victor
Johnson, Roger W
Johnson-Winegar, Anna
Jordan, Craig Alan
Joseph, J Mehsen
Joseph, Sammy William
Kadis, Solomon
Kaighn, Morris Edward
Kalica, Anthony R
Kaper, James Bennett
Keister, Donald Lee
Keith, Jerry M
Kelly, Thomas J
Kenyon, Richard H
Khan, Hameed A
Kimball, Paul Clark
Kingsbury, David Wilson
Kingsbury, Elizabeth W
Kole, Hemanta Kumar
Kramer, Tim R
Krause, Richard Michael
Krichevsky, Micah I
Kwon-Chung, Kyung Joo
La Montagne, John Ring
Landon, John Campbell
Langlykke, Asger Funder
LaSalle, Bernard
Lederman, Howard Mark
Ledney, George David
Lee, Chi-Jen
Lee, Fang-Jen Scott
Lee, Theresa
Leppla, Stephen Howard
Levin, Judith Goldstein
Levin, Morris A
Levy, Hilton Bertram
Lewis, Fred A
Libonati, Joseph Peter
Lobel, Steven A
Long, Cedric William
Lovett, Paul Scott
Lucas, John Paul
McCarron, Richard M
McGarrity, Gerard John
McKenney, Keith Hollis
McNicol, Lore Anne
MacQuillan, Anthony M
MacVittie, Thomas Joseph
Mageau, Richard Paul
Maloney, Peter Charles
Manaker, Robert Anthony
Margulies, Maurice
Marquardt, Warren William
Martin, Robert G
Maurer, Bruce Anthony
Mayer, Vernon William, Jr
Mergenhagen, Stephan Edward
Merz, William George
Meyers, Wayne Marvin
Minah, Glenn Ernest
Mischke, Barbara Suzanne
Moloney, John Bromley
Monath, Thomas P
Morgan, Herbert R
Morris, Joseph Anthony
Moss, Bernard
Murphy, Patrick Aidan
Murray, Gary Joseph
Narayan, Opendra
Nash, Howard Allen
Nauman, Robert Karl
Neva, Franklin Allen
Nordin, Albert Andrew
Notkins, Abner Louis
Nutter, John
O'Beirne, Andrew Jon
Olenick, John George
Oliver, Eugene Joseph
Papas, Takis S
Parkman, Paul Douglas
Pavlovskis, Olgerts Raimonds
Payton, Cecil Warren
Pearson, John William
Phillips, Grace Briggs
Pienta, Roman Joseph
Pilchard, Edwin Ivan
Pollok, Nicholas Lewis, III
Poston, John Michael

Microbiology (cont)

Powers, Kendall Gardner
Preble, Olivia Toby
Purcell, Robert Harry
Puziss, Milton
Quinnan, Gerald Vincent, Jr
Rader, Ronald Alan
Ranhand, Jon M
Ranney, Richard Raymond
Read-Connole, Elizabeth Lee
Reissig, Magdalena
Reitz, Marvin Savidge, Jr
Rhim, Johng Sik
Rice, Nancy Reed
Rizzo, Anthony Augustine
Robbins, Keith Cranston
Roberson, Bob Sanders
Rochovansky, Olga Maria
Rose, Noel Richard
Rosenstein, Robert William
Rosner, Judah Leon
Rossio, Jeffrey L
Russell, Philip King
Sack, Richard Bradley
Sagripanti, Jose-Luis
Sarma, Padman S
Saxinger, W(illiam) Carl
Schaub, Stephen Alexander
Schluederberg, Ann Elizabeth Snider
Schneider, Walter Carl
Schricker, Robert Lee
Schultz, Warren Walter
Schwab, Bernard
Scott, Thomas Wallace
Shaffer, Charles Henry, Jr
Shah, Keerti V
Shapiro, Martin
Shelokov, Alexis
Shih, James Waikuo
Sibal, Louis Richard
Sich, Jeffrey John
Sieckmann, Donna G
Silverman, David J
Simmon, Vincent Fowler
Simons, Daniel J
Sjoblad, Roy David
Slyter, Leonard L
Small, Eugene Beach
Smith, Andrew George
Smith, Dorothy Gordon
Snyder, Merrill J
Speedie, Marilyn Kay
Stadtman, Thressa Campbell
Stamper, Hugh Blair
Steinman, Harry Gordon
Steinman, Irvin David
Straus, Stephen Ezra
Strauss, Robert R
Strickland, George Thomas
Stunkard, Jim A
Su, Robert Tzyh-Chuan
Sydiskis, Robert Joseph
Sztein, Marcelo Benjamin
Taube, Sheila Efron
Taylor, David Neely
Tepper, Byron Seymour
Theodore, Theodore Spiros
Tingle, Marjorie Anne
Tonik, Ellis J
Tully, Joseph George
Turner, Thomas Bourne
Tyler, Bonnie Moreland
Umbreit, Thomas Hayden
Vadlamudi, Sri Krishna
Varmus, Harold Elliot
Vera, Harriette Dryden
Vermund, Sten Halvor
Vincent, Phillip G
Voll, Mary Jane
Wachter, Ralph Franklin
Wahl, Sharon Knudson
Walker, Mary Clare
Walker, Richard Ives
Walter, Eugene LeRoy, Jr
Walton, Thomas Edward
Weiner, Ronald Martin
Weislow, Owen Stuart
Weiss, Emilio
Westhoff, Dennis Charles
Whitehurst, Virgil Edwards
Williams, Jimmy Calvin
Winchurch, Richard Albert
Wisseman, Charles Louis, Jr
Wolf, Richard Edward, Jr
Wolff, David A
Wong, Dennis Mun
Woodman, Daniel Ralph
Woodward, Theodore Englar
Woolverton, Christopher Jude
Yanagihara, Richard
Yang, Bingzhi
Yasbin, Ronald Eliott
Young, Howard Alan
Young, Viola Mae (Horvath)
Zebovitz, Eugene
Zierdt, Charles Henry
Zimmerman, Eugene Munro

MASSACHUSETTS
Akkara, Joseph A
Allen, Mary A Mennes
Anderson, Paul Joseph
Asato, Yukio
Baker, Edgar Eugene, Jr
Barkalow, Fern J
Berry, Vern Vincent
Black, Paul H
Blacklow, Neil Richard
Blaustein, Ernest Herman
Boger, Edwin August, Sr
Borgatti, Alfred Lawrence
Brennessel, Barbara A
Broitman, Selwyn Arthur
Canale-Parola, Ercole
Carney, Walter Patrick
Chang, Te Wen
Chou, Iih-Nan (George)
Coffin, John Miller
Cohen, Joel Ralph
Cohen, Robert Elliot
Cole, Edward Anthony
Collier, Robert John
Collins, Carolyn Jane
Comer, M Margaret
Conti, Samuel F
Cooney, Charles Leland
Cooney, Joseph Jude
Cooper, Geoffrey Mitchell
Craven, Donald Edward
Crisley, Francis Daniel
Davis, Bernard David
Deitz, William Harris
Demain, Arnold Lester
Dennen, David W
DePalma, Philip Anthony
Derow, Matthew Arnold
Desrosiers, Ronald Charles
Dowell, Clifton Enders
Du Moulin, Gary C
Esber, Henry Jemil
Essex, Myron
Eubanks, Elizabeth Ruberta
Fairbanks, Grant
Fields, Bernard N
Fraenkel, Dan Gabriel
Fritsch, Edward Francis
Fuller, Rufus Clinton
Gabliks, Janis
Gerety, Robert John
Gibbons, Ronald J
Gill, David Michael
Goldberg, Edward B
Goldstein, Richard Neal
Gordon, Lance Kenneth
Gupta, Rajesh K
Halvorson, Harlyn Odell
Hanshaw, James Barry
Harlow, Edward E
Hastings, John Woodland
Hawiger, Jack Jacek
Hirsch, Martin Stanley
Hotchkiss, Rollin Douglas
Howley, Peter Maxwell
Jaenisch, Rudolf
Janicki, Bernard William
Jannasch, Holger Windekilde
Johnson, Corinne Lessig
Jordan, Harold Vernon
Jost, Dana Nelson
Kaplan, David Lee
Kelley, William S
Kelner, Albert
Kepper, Robert Edgar
Ketchum, Paul Abbott
Kibrick, Sidney
Kieff, Elliott Dan
Klempner, Mark Steven
Knipe, David Mahan
Kundsin, Ruth Blumfeld
Labbe, Ronald Gilbert
Lachica, R Victor
Lessie, Thomas Guy
Levin, Robert E
Levy, Stuart B
Lingappa, Banadakoppa Thimmappa
Lingappa, Yamuna
Litsky, Bertha Yanis
Lodish, Harvey Franklin
Mabuchi, Katsuhide
McCabe, William R
McCormick, Neil Glenn
Magasanik, Boris
Malakian, Artin
Mandels, Mary Hickox
Manning, William Joseph
Mata, Leonardo J
Mathews-Roth, Micheline Mary
Medearis, Donald N, Jr
Milani, Victor John
Miller, Judith Evelyn
Mitchell, Ralph
Mount, Mark Samuel
Mulder, Carel
Murphy, John R
Orme-Johnson, William H
Park, James Theodore
Poteete, Anthony Robert
Powers, Edmund Maurice
Previte, Joseph James
Prindle, Bryce
Putney, Scott David
Read, Dorothy Louise
Reiner, Albey M
Riley, Monica
Robinton, Elizabeth Dorothy
Rogers, Morris Ralph
Rosenberg, Fred A
Rowley, Durwood B
Ruprecht, Ruth Margrit
Rustigian, Robert
Sarkar, Nilima
Schaechter, Moselio
Schaffer, Priscilla Ann
Sharp, Phillip Allen
Shaw, Eugene
Silverman, Gerald
Sinskey, Anthony J
Smith, Cassandra Lynn
Sonenshein, Abraham Lincoln
Strauss, Phyllis R
Struhl, Kevin
Suit, Joan C
Sussman, Raquel Rotman
Szostak, Jack William
Taubman, Martin Arnold
Thayer, Philip Standish
Thorne, Curtis Blaine
Tipper, Donald John
Tyrrell, Elizabeth Ann
Van Cleave, Victor Harold
Walker, Robert W
Wallace, David H
Wilder, Martin Stuart
Wirsen, Carl O, Jr
Wun, Chun Kwun
Yang, Shiow-shong
Zetter, Bruce Robert
Zimmermann, Robert Alan

MICHIGAN
Adler, Richard
Anderson, Richard Lee
Anderson, Theodore
Bach, John Alfred
Beardmore, William Boone
Bender, Robert Algerd
Benham, Ross Stephen
Berk, Richard Samuel
Berlin, Byron Sanford
Bhatnagar, Lakshmi
Boros, Dov Lewis
Brawn, Mary Karen
Breznak, John Allen
Brockman, Ellis R
Brown, William John
Brubaker, Robert Robinson
Burgoyne, George Harvey
Buxser, Stephen Earl
Carrick, Lee
Chandra, Purna
Chang, Timothy Scott
Chaudhry, G Rasul
Choe, Byung-Kil
Clewell, Don Bert
Coburn, Joel Thomas
Cochran, Kenneth William
Cohen, Michael Alan
Cole, George Christopher
Conder, George Anthony
Cooper, Stephen
Corner, Thomas Richard
Cunningham, Charles Henry
Dardiri, Ahmed Hamed
Dazzo, Frank Bryan
De Bruijn, Frans Johannes
Denison, Frank Willis, Jr
Diglio, Clement Anthony
Domagala, John Michael
Dore-Duffy, Paula
Eisenberg, Robert C
Erlandson, Arvid Leonard
Fadly, Aly Mahmoud
Fisher, Linda E
Fitzgerald, Robert Hannon, Jr
Fogel, Robert Dale
Friedman, Stephen Burt
Fulbright, Dennis Wayne
Garvin, Donald Frank
Gerhardt, Philipp
Grady, Joseph Edward
Haines, Richard Francis
Haines, William C
Hanka, Ladislav James
Hari, V
Harmon, Laurence George
Heberlein, Gary T
Heifetz, Carl Louis
Heppner, Gloria Hill
Hoeksema, Walter David
Holt, John Gilbert
Jackson, Matthew Paul
Jacobs, Charles Warren
Jeffries, Charles Dean
Jones, Garth Wicks
Jourdian, George William
Kauffman, Carol A
Klug, Michael J
Kominek, Leo Aloysius
Kong, Yi-Chi Mei
Krabbenhoft, Kenneth Louis
Kurnit, David Martin
Lefford, Maurice J
Lenski, Richard E
Levine, Myron
Levine, Seymour
Liu, Stephen C Y
Lomax, Margaret Irene
McCauley, Roy Barnard
McGroarty, Estelle Josephine
Mack, Walter Noel
Marshall, Vincent dePaul
Mason, Donald Joseph
Meyer, Edward Dell
Miller, Brinton Marshall
Miller, Thomas Lee
Molinari, John A
Montgomery, Paul Charles
Mostafapour, M Kazem
Mulks, Martha Huard
Mychajlonka, Myron
Narayan, Latha
Nazerian, Keyvan
Neidhardt, Frederick Carl
Olsen, Ronald H
Osborn, June Elaine
Parejko, Ronald Anthony
Paschke, John Donald
Patterson, Maria Jevitz
Peabody, Frank Robert
Pence, Leland Hadley
Peterson, Ward Davis, Jr
Podila, Gopi Krishna
Preiss, Jack
Punch, James Darrell
Pyke, Thomas Richard
Rauch, Helene Coben
Reddy, Chilekampalli Adinarayana
Renis, Harold E
Reusser, Fritz
Righthand, Vera Fay
Robertson, G Philip
Robertson, John Harvey
Roeper, Richard Allen
Rogers, Alvin Lee
Rowe, Nathaniel H
Sadoff, Harold Lloyd
San Clemente, Charles Leonard
Sarkar, Fazlul Hoque
Savageau, Michael Antonio
Schuurmans, David Meinte
Sebek, Oldrich Karel
Serafini, Angela
Shelef, Leora Aya
Shipman, Charles, Jr
Skean, James Dan
Smith, Ellen Oliver
Smith, Jerry Warren
Sokolski, Walter Thomas
Steel, Robert
Stone, Howard Anderson
Trapp, Allan Laverne
Truden, Judith Lucille
Tsuji, Kiyoshi
Tung, Fred Fu
Walker, Gustav Adolphus
Walter, Richard Webb, Jr
Wang, Henry Y
Watts, Jeffrey Lynn
Wei, Wei-Zen
Weiner, Lawrence Myron
Whalen, Joseph Wilson
Whitehouse, Frank, Jr
Witz, Dennis Fredrick
Wolk, Coleman Peter
Yancey, Robert John, Jr
Yang, Gene Ching-Hua
Yokoyama, Melvin T
Zeikus, J Gregory

MINNESOTA
Anderson, Dwight Lyman
Bach, Marilyn Lee
Balfour, Henry H, Jr
Beggs, William H
Berube, Robert
Blazevic, Donna Jean
Burton, Alice Jean
Bushong, Jerold Ward
Busta, Francis Fredrick
Cleary, Paul Patrick
Dobbins, Durell Cecil
Dworkin, Martin
Emeagwali, Dale Brown
Faras, Anthony James
Fitzgerald, Thomas James
Fraser, Judy Ann
Fredrickson, Arnold G(erhard)
Gilboe, Daniel Pierre
Goldstein, Stuart Frederick
Grant, David James William
Haase, Ashley Thomson
Hanson, Richard Steven
Hapke, Bern
Henry, Ronald Alton
Hooper, Alan Bacon
Hu, Wei-Shou
Jezeski, James John
Johnson, Russell Clarence
Johnson, Theodore Reynold
Katz, Morris Howard
Kerr, Sylvia Joann
Kline, Bruce Clayton
LaPorte, David Coleman
Loken, Keith I
Lyon, Richard Hale
McDuff, Charles Robert
McKay, Larry Lee
Magee, Paul Terry
Manning, Patrick James
Miller, Richard Lynn
Nash, Peter
Nelson, Robert D
Oetting, William Starr
Oxborrow, Gordon Squires
Percich, James Angelo

Peterson, Phillip Keith
Pflug, Irving John
Plagemann, Peter Guenter Wilhelm
Pomeroy, Benjamin Sherwood
Prince, James T
Ragsdale, David Willard
Reilly, Bernard Edward
Ritts, Roy Ellot, Jr
Roberts, Glenn Dale
Robertsen, John Alan
Rodgers, Nelson Earl
Rogers, Palmer, Jr
Rohlfing, Stephen Roy
Roon, Robert Jack
Sabath, Leon David
Sadowsky, Michael Jay
Schachtele, Charles Francis
Schrank, Gordon Dabney
Schuman, Leonard Michael
Sharma, Jagdev Mittra
Shope, Richard Edwin, Jr
Silverman, William Bernard
Sperber, William H
Stanley, Patricia Mary
Straw, Thomas Eugene
Tatini, Sita Ramayya
Thenen, Shirley Warnock
Tilman, G David
Tsien, Hsienchyang
Vallera, Daniel A
Vermilyea, Barry Lynn
Wagenaar, Raphael Omer
Watson, Dennis Wallace
Widin, Katharine Douglas
Ziegler, Richard James

MISSISSIPPI
Arceneaux, Joseph Lincoln
Bardsley, Charles Edward
Brown, Lewis Raymond
Byers, Benjamin Rowe
Davis, Robert Gene
Downer, Donald Newson
Gafford, Lanelle Guyton
Garrett, Ephraim Spencer, III
Gentry, Glenn Aden
Grogan, James Bigbee
Hwang, Huey-Min
Lewis, Robert Edwin, Jr
Little, Brenda Joyce
Lobb, Craig J
Magee, Lyman Abbott
Marshall, Douglas Lee
Martin, James Harold
Mickelson, John Clair
Peterson, Harold LeRoy
Randall, Charles Chandler
Santangelo, George Michael
Sikorowski, Peter P
Snazelle, Theodore Edward
Uzodinma, John E
Wahba, Albert J
Ward, Herbert Bailey
White, Charles Henry
Williamson, John S
Yarbrough, Karen Marguerite

MISSOURI
Arens, Max Quirin
Bajpai, Rakesh Kumar
Ball, Timothy K
Bell, Jeffrey
Blenden, Donald C
Bogosian, Gregg
Branson, Dorothy Swingle
Brescia, Vincent Thomas
Brown, Karen Kay (Kilker)
Brown, Olen Ray
Burchard, Jeanette
Casperson, Gerald F
DiFate, Victor George
Eissenberg, Joel Carter
Elsawi, Nehad
Eltz, Robert Walter
Elvin-Lewis, Memory P F
Engley, Frank B, Jr
Fales, William Harold
Ferris, Deam Hunter
Fields, Marion Lee
Finkelstein, Richard Alan
Fischhoff, David Allen
Fleischman, Julian B
Folk, William Robert
Fraley, Robert Thomas
Gale, Nord Loran
Garrison, Robert Gene
Gledhill, William Emerson
Goldberg, Herbert Sam
Gossling, Jennifer
Granoff, Dan Martin
Green, Theodore James
Guntaka, Ramareddy V
Gustafson, Mark Edward
Hallas, Laurence Edward
Hamilton, Thomas Reid
Harrison, Arthur Pennoyer, Jr
Hirschberg, Rona L
Hodges, Glenn R(oss)
Jacob, Gary Steven
Kenny, Michael Thomas
Kishore, Ganesh M
Koplow, Jane
Kos, Edward Stanley
Landick, Robert

Laskowski, Leonard Francis, Jr
Lin, Hsiu-san
McCann, Peter Paul
McCune, Emmett L
McIntosh, Arthur Herbert Cranstoun
Marshall, Robert T
Medoff, Gerald
Merlie, John Paul
Miles, Donald Orval
Misfeldt, Michael Lee
Mueckler, Mike Max
Murray, Patrick Robert
Myers, Richard Lee
Naumann, Hugh Donald
Olson, Gerald Allen
Olson, Lloyd Clarence
Parisi, Joseph Thomas
Pueppke, Steven Glenn
Rice, Charles Moen, III
Riddle, Donald Lee
Riggs, Hammond Greenwald, Jr
Rogolsky, Marvin
Rosenquist, Bruce David
Sargentini, Neil Joseph
Savage, George Roland
Scherr, David DeLano
Schlesinger, Milton J
Schlessinger, David
Schmidt, Francis J
Schultz, Irwin
Schulze, Irene Theresa
Shieh, Kenneth Kuang-Zen
Solorzano, Robert Francis
Spurrier, Elmer R
Symington, Janey Studt
Tritz, Gerald Joseph
Wang, Richard J
Weber, Morton M
Weeks, Robert Joe
Wenner, Herbert Allan
Westfall, Helen Naomi
Winicov, Murray William
Yaverbaum, Sidney

MONTANA
Anacker, Robert Leroy
Bacon, Marion
Bond, Clifford Walter
Braune, Maximillian O
Combie, Joan D
Cutler, Jimmy Edward
Dorward, David William
Faust, Richard Ahlvers
Fiscus, Alvin G
Garon, Claude Francis
Jutila, John W
Jutila, Mark Arthur
McFeters, Gordon Alwyn
Munoz, John Joaquin
Nakamura, Mitsuru J
Nelson, Nels M
Newman, Franklin Scott
Parker, John C
Rudbach, Jon Anthony
Speer, Clarence Arvon
Stoenner, Herbert George
Swanson, John L
Taylor, John Jacob
Temple, Kenneth Loren
Thomas, Leo Alvon
Warren, Guylyn Rea

NEBRASKA
Baumstark, John Spann
Booth, Sheldon James
Brazis, A(dolph) Richard
Brown, Albert Loren
Bullerman, Lloyd Bernard
Calvert, Jay Gregory
Chaperon, Edward Alfred
Davis, Eldon Vernon
Doran, John Walsh
Dubes, George Richard
Dyer, John Kaye
Ghosh, Chitta Ranjan
Ikenberry, Richard W
Kelling, Clayton Lynn
Knoop, Floyd C
Kobayashi, Roger Hideo
Kolar, Joseph Robert, Jr
Kwang, Jimmy
Lane, Leslie Carl
McFadden, Harry Webber, Jr
Miller, Norman Gustav
Morris, Thomas Jack
Nickerson, Kenneth Warwick
Ramaley, Robert Folk
Rhode, Solon Lafayette, III
Sanders, Christine Culp
Sanders, W Eugene, Jr
Schwartzbach, Steven Donald
Severin, Matthew Joseph
Siragusa, Gregory Ross
Torres-Medina, Alfonso
Underdahl, Norman Russell
Van Etten, James L
Veomett, George Ector
Wallen, Stanley Eugene
Weber, Allen Thomas
White, Roberta Jean

NEVADA
DiSalvo, Arthur F
Franson, Raymond Lee

Jay, James Monroe
Kozel, Thomas Randall
Winicov, Ilga

NEW HAMPSHIRE
Adler, Louise Tale
Bent, Donald Frederick
Blakemore, Richard Peter
Blanchard, Robert Osborn
Brand, Karl Gerhard
Chesbro, William Ronald
Dunlap, Jay Clark
Gray, Clarke Thomas
Green, William Robert
Hines, Mark Edward
Jones, Galen Everts
Lubin, Martin
Pfefferkorn, Elmer Roy, Jr
Pistole, Thomas Gordon
Rodgers, Frank Gerald
Smith, Edith Lucile
Smith, Mary Bunting
Tuttle, Robert Lewis
Zsigray, Robert Michael

NEW JERSEY
Adamo, Joseph Albert
Anderson, David Wesley
Avigad, Gad
Axelrod, David E
Bardell, David
Barton, Beverly E
Baughn, Charles (Otto), Jr
Benson, Charles Everett
Bernstein, Eugene H
Bontempo, John A
Bowman, Patricia Imig
Brandriss, Marjorie C
Burg, Richard William
Byrne, Barbara Jean McManamy
Byrne, Kevin M
Carman, George Martin
Carter, James Evan
Champe, Pamela Chambers
Charney, William
Choi, Ye-Chin
Cino, Paul Michael
Cook, Elizabeth Anne
Coscarelli, Waldimero
Cuffari, Gilbert Luke
Daoust, Donald Roger
DePamphilis, Melvin Louis
Dhruv, Rohini Arvind
Dondershine, Frank Haskin
Dubin, Donald T
Eigen, Edward
Ellison, Solon Arthur
Eng, Robert H K
Ennis, Herbert Leo
Evans, Ralph H, Jr
Eveleigh, Douglas Edward
Feighner, Scott Dennis
Feldman, Lawrence A
Fernandes, Prabhavathi Bhat
Fong, Dunne
Founds, Henry William
Frances, Saul
Gaffar, Abdul
Gale, George Osborne
Gately, Maurice Kent
Genetelli, Emil J
Ghosh, Arati
Ghosh, Bijan K
Gillum, Amanda McKee
Girardi, Anthony Joseph
Goel, Mahesh C
Goldemberg, Robert Lewis
Goldman, Emanuel
Green, Erika Ana
Greenstein, Teddy
Gross, Peter A
Gullo, Vincent Philip
Haas, Gerhard Julius
Heck, James Virgil
Hendlin, David
Henry, Sydney Mark
Hirsch, Robert L
Humayun, Mir Z
Jacks, Thomas Mauro
Jakubowski, Hieronim Zbigniew
Jansons, Vilma Karina
Johnson, Layne Mark
Jones, Benjamin Lewis
Kaminski, Zigmund Charles
Keller, John Randall
Kirchoff, William F
Kirsch, Donald R
Kirsch, Nathan Carl
Klosek, Richard C
Koepp, Leila H
Kozak, Marilyn Sue
Kraskin, Kenneth Stanford
Kuchler, Robert Joseph
Kuehn, Harold Herman
Kurtz, Myra Berman
Lampen, J Oliver
Lasfargues, Etienne Yves
Laskin, Allen I
Lee, Tung-Ching
Lenard, John
Letizia, Gabriel Joseph
Leung, Albert Yuk-Sing
Levin, Joseph David
Leyson, Jose Florante Justinianne

Ling, Hubert
Louria, Donald Bruce
McAnelly, John Kitchel
McCoy, John Philip, Jr
McDaniel, Lloyd Everett
Mannino, Raphael James
Masurekar, Prakash Sharatchandra
Matthijssen, Charles
Mayernik, John Joseph
Meyers, Edward
Middleton, Richard B
Midlige, Frederick Horstmann, Jr
Miraglia, Gennaro J
Montville, Thomas Joseph
Moral, Josefa Liduvina
Neri, Anthony
Ofengand, Edward James
Osborne, Frank Harold
OSullivan, Joseph
Ozer, Harvey Leon
Padhi, Sally Bulpitt
Park, Eunkyue
Patel, Ramesh N
Perritt, Alexander M
Pestka, Sidney
Pianotti, Roland Salvatore
Pierce, George Edward
Pobiner, Bonnie Fay
Pramer, David
Prince, Herbert N
Raska, Karel Frantisek, Jr
Reilly, Hilda Christine
Ripley, Lynn S
Rosen, Marvin
Rosen, William Edward
Sall, Theodore
Schaffner, Carl Paul
Scheiner, Donald M
Schlesinger, R(obert) Walter
Schulman, Marvin David
Schwalb, Marvin N
Sehgal, Surendra N
Sekula, Bernard Charles
Shearer, Marcia Cathrine (Epple)
Shovlin, Francis Edward
Sikder, Santosh K
Simpson, Robert Wayne
Sipos, Tibor
Sohler, Arthur
Solberg, Myron
Stapley, Edward Olley
Sterbenz, Francis Joseph
Stockton, John Richard
Stollar, Victor
Stoudt, Thomas Henry
Strohl, William Allen
Swenson, Christine Erica
Tate, Robert Lee, III
Taylor, Bernard Franklin
Testa, Douglas
Tierno, Philip M, Jr
Tkacz, Jan S
Torregrossa, Robert Emile
Trivedi, Nayan B
Udem, Stephen Alexander
Unowsky, Joel
Vallese, Frank M
Voos, Jane Rhein
Walton, Robert Bruce
Wang, Bosco Shang
Wang-Iverson, Patsy
Wasserman, Bruce P
Watrel, Warren George
Watson, Richard White, Jr
Wellerson, Ralph, Jr
Werth, Jean Marie
Williams, Jeffrey Walter
Winters, Harvey
Woehler, Michael Edward
Wolff, John Shearer, III
Woodruff, Harold Boyd
Worne, Howard E
Zimmerman, Sheldon Bernard

NEW MEXICO
Baca, Oswald Gilbert
Baker, Thomas Irving
Barrera, Cecilio Richard
Barton, Larry Lumir
Botsford, James L
Cords, Carl Ernest, Jr
Davis, Larry Ernest
Gregg, Charles Thornton
Kraemer, Paul Michael
Liddell, Craig Mason
McCarthy, Charlotte Marie
McLaren, Leroy Clarence
Martignoni, Mauro Emilio
Matchett, William H
Mayeux, Jerry Vincent
Nelson, Mary Anne
O'Brien, Robert Thomas
Radloff, Roger James
Rypka, Eugene Weston
Sivinski, Jacek Stefan
Stevenson, Robert Edwin
Taylor, Robert Gay
Tokuda, Sei

NEW YORK
Abdelnoor, Alexander Michael
Ablin, Richard J
Abramowicz, Daniel Albert
Ajl, Samuel Jacob

Microbiology (cont)

Albrecht, Alberta Marie
Alcamo, I Edward
Amsterdam, Daniel
Andersen, Jon Alan
Andersen, Kenneth J
Anderson, Carl William
Anderson, Lucia Lewis
Appel, Max J
Babich, Harvey
Bablanian, Rostom
Baker, David A
Balduzzi, Piero
Banerjee, Partha
Barany, Francis
Bard, Enzo
Barksdale, Lane W
Bassett, Emmett W
Battley, Edwin Hall
Batzing, Barry Lewis
Bazinet, George Frederick
Beach, David H
Beardsley, Robert Eugene
Bedard, Donna Lee
Belly, Robert T
Benjaminson, Morris Aaron
Bergstrom, Gary Carlton
Bernheimer, Alan Weyl
Bernheimer, Harriet P
Berns, Kenneth
Bielat, Kenneth L
Biondo, Frank X
Birken, Steven
Blank, Robert H
Bopp, Lawrence Howard
Bottone, Edward Joseph
Bowen, William H
Boylen, Charles William
Breese, Sydney Salisbury, Jr
Brody, Marcia
Brown, Eric Reeder
Bruner, Dorsey William
Bukhari, Ahmad Iqbal
Bukovsan, Laura A
Burger, Richard Melton
Calhoun, David H
Calnek, Bruce Wixson
Calvo, Joseph M
Cano, Francis Robert
Carmichael, Leland E
Carp, Richard Irvin
Carter, Timothy Howard
Casals, Jordi
Castello, John Donald
Catterall, James F
Celis, Roberto T F
Cerini, Costantino Peter
Chase, Merrill Wallace
Chen, Ching-Ling Chu
Chiao, Jen Wei
Chisholm, David R
Chiulli, Angelo Joseph
Christensen, James Roger
Clark, Virginia Lee
Clesceri, Lenore Stanke
Cohen, Elias
Cole, Jonathan Jay
Collins, Arlene Rycombel
Collins, Carol Desormeau
Conway de Macario, Everly
Cooper, Norman S
Corpe, William Albert
Couch, Troy Lee
Cowell, James Leo
Cox, Dudley
Craig, John Philip
Cramer, Eva Brown
Crook, Philip George
Cunningham, Richard Preston
Danielson, Irvin Sigwald
Deibel, Rudolf
DeLuca, Patrick John
Dewhurst, Stephen
DiGaudio, Mary Rose
DiLiello, Leo Ralph
Dodge, Cleveland John
Dondero, Norman Carl
Dougherty, Robert Malvin
Dranginis, Anne M
Efthymiou, Constantine John
Ehrlich, Henry Lutz
Elander, Richard Paul
Ellner, Paul Daniel
Elsbach, Peter
Eudy, William Wayne
Evans, Mary Jo
Evans, Richard Todd
Fabricant, Catherine G
Fantini, Amedeo Alexander
Ferguson, John Barclay
Fewkes, Robert Charles Joseph
Figurski, David Henry
Finn, R(obert) K(aul)
Fisher, Clark Alan
Flanagan, Thomas Donald
Forgacs, Joseph
Fox, Sally Ingersoll
Francis, Arokiasamy Joseph
Free, Stephen J
Freedman, Michael Lewis
Frickey, Paul Henry
Friedman, Selwyn Marvin
Froelich, Ernest
Furmanski, Philip
Fuscaldo, Anthony Alfred
Garay, Gustav John
Gehrig, Robert Frank
Gelbard, Alan Stewart
George, Elmer, Jr
Gibbs, David Lee
Gibson, Audrey Jane
Gillespie, James Howard
Gilmour, Marion Nyholm H
Ginsberg, Harold Samuel
Godfrey, Henry Philip
Goff, Stephen Payne
Goldberg, Allan Roy
Goldschmidt, Raul Max
Gonsalves, Dennis
Goodhue, Charles Thomas
Gordon, Morris Aaron
Gorzynski, Eugene Arthur
Graham, Donald C W
Gravani, Robert Bernard
Graziadei, William Daniel, III
Griffin, David H
Hadley, Susan Jane
Hagar, Silas Stanley
Hajek, Ann Elizabeth
Halstead, Scott Barker
Hammill, Terrence Michael
Hare, John Donald
Harman, Gary Elvan
Hashim, George A
Hawkins, Linda Louise
Haywood, Anne Mowbray
Hechemy, Karim E
Hehre, Edward James
Held, Abraham Albert
Helmann, John Daniel
Henneberry, Richard Christopher
Hipp, Sally Sloan
Ho, Hon Hing
Hochstadt, Joy
Hoffman, Heiner
Hoffmann, Michael K
Horwitz, Marshall Sydney
Hotchin, John Elton
Hsu, Ming-Ta
Huang, Alice Shih-Hou
Humber, Richard Alan
Hurwitz, Jerard
Hutchison, Dorris Jeannette
Hutner, Seymour Herbert
Iglewski, Barbara Hotham
Iglewski, Wallace
Isaacs, Charles Edward
Isenberg, Henry David
Isseroff, Hadar
Jahiel, Rene
Jarolmen, Howard
Juo, Pei-Show
Kalsow, Carolyn Marie
Katz, Michael
Keithly, Janet Sue
Kele, Roger Alan
Keller, Dolores Elaine
Khan, Nasim A
Khan, Paul
Kilbourne, Edwin Dennis
Kim, Charles Wesley
Kim, Kwang Shin
Klein, Elena Buimovici
Klein, Richard Joseph
Knight, Paul R
Konarska, Maria Magda
Korf, Richard Paul
Kraft, William Gerald
Kratzel, Robert Jeffrey
Lacks, Sanford
Laffin, Robert James
Lahita, Robert George
Lamberson, Harold Vincent, Jr
Landsberger, Frank Robbert
Lasley, Betty Jean
Lazaroff, Norman
Ledford, Richard Allison
Lee, John Joseph
Lee, William Thomas
Leff, Judith
Lehman, John Michael
Leifer, Zer
Levandowsky, Michael
Lewis, Leslie Arthur
Liguori, Vincent Robert
Lilly, Frank
Lin, Jane Huey-Chai
Lindahl, Lasse Allan
Linder, Regina
Lindsay, Harry Lee
Linke, Harald Arthur Bruno
Linkins, Arthur Edward
Lipke, Peter Nathan
Lipson, Edward David
Lipson, Steven Mark
Loeb, Marilyn Rosenthal
Lossinsky, Albert S
Lowe, Josiah L(incoln)
Lusty, Carol Jean
Lyons, Michael Joseph
McAllister, William T
Macario, Alberto J L
McCormack, Grace
McCormick, J Robert D
McNamara, Thomas Francis
McSharry, James John
Maestrone, Gianpaolo
Magill, Thomas Pleines
Mahoney, Robert Patrick
Maio, Joseph James
Maleckar, James R
Maniloff, Jack
Manly, Kenneth Fred
Manski, Wladyslaw J
Maramorosch, Karl
Marcu, Kenneth Brian
Marquis, Robert E
Mashimo, Paul Akira
May, Paul S
Maynard, Charles Alvin
Mehta, Bipin Mohanlal
Merrick, Joseph M
Meyer, Haruko
Michaud, Ronald Normand
Milgrom, Felix
Miller, Terry Lynn
Mindich, Leonard Eugene
Misiek, Martin
Morin, John Edward
Morse, Stephen S(cott)
Mortlock, Robert Paul
Mundorff-Shrestha, Sheila Ann
Munk, Vladimir
Murphy, James Slater
Murphy, Timothy F
Muschio, Henry M, Jr
Nachbar, Martin Stephen
Napolitano, Raymond L
Neimark, Harold Carl
Nellis, Lois Fonda
Neu, Harold Conrad
Neurath, Alexander Robert
Noronha, Fernando M Oliveira
North, Robert J
Nuzzi, Robert
Oates, Richard Patrick
O'Brien, James Francis
Orsi, Ernest Vinicio
Page, Gregory Vincent
Parham, Margaret Payne
Patterson, Ernest Leonard
Peterson, Robert H F
Pfau, Charles Julius
Phillips, Arthur William, Jr
Pierce, Carol S
Pisano, Michael A
Pitt, Jane
Poiesz, Bernard Joseph
Potter, Norman N
Powell, Sharon Kay
Prince, Alfred M
Pruslin, Fred Howard
Puttlitz, Donald Herbert
Quigley, James P
Racaniello, Vincent Raimondi
Reddy, Kalluru Jayarami
Reilly, Marguerite
Rinchik, Eugene M
Rizzuto, Anthony B
Robinson, Alix Ida
Roeder, Robert Gayle
Rogerson, Allen Collingwood
Rosenfeld, Martin Herbert
Rosenstein, Barry Sheldon
Rosi, David
Rossman, Toby Gale
Rotheim, Minna B
Russel, Marjorie Ellen
Sadosky, Alesia Beth
St Leger, Raymond John
Salkin, Ira Fred
Salton, Milton Robert James
Sambrotto, Raymond Nicholas
Santoro, Thomas
Sawyer, William D
Schaeffer, James Robert
Schauf, Victoria
Scher, William
Schloer, Gertrude M
Schuster, Frederick Lee
Schwartz, Stanley Allen
Scott, Fredric Winthrop
Scranton, Mary Isabelle
Sechrist, Lynne Luan
Sehgal, Pravinkumar B
Senitzer, David
Senterfit, Laurence Benfred
Shapiro, Caren Knight
Sharp, William R
Shayegani, Mehdi
Sheffy, Ben Edward
Shuler, Michael Louis
Silva, Pedro Joao Neves
Silva-Hutner, Margarita
Silverstein, Samuel Charles
Siminoff, Paul
Simon, Robert David
Slepecky, Ralph Andrew
Smith, James Eldon
Spencer, Frank
Stamer, John Richard
Stanley, Pamela Mary
Stark, Egon
Starr, Theodore Jack
Steinberg, Bernard Albert
Steinberg, Bettie Murray
Steinkraus, Keith Hartley
Stevens, Roy White
Stotzky, Guenther
Stryker, Martin H
Su, Tah-Mun
Sudds, Richard Huyette, Jr
Sultzer, Barnet Martin
Surgalla, Michael Joseph
Swaney, Lois Mae
Swanson, Stephen King
Taber, Harry Warren
Tamm, Igor
Tanenbaum, Stuart William
Tanzer, Charles
Tendler, Moses David
Ter Kuile, Benno Herman
Testa, Raymond Thomas
Thacore, Harshad Rai
Thaler, David Solomon
Toenniessen, Gary Herbert
Trowbridge, Richard Stuart
Uzgiris, Egidijus E
Van Tassell, Morgan Howard
Vilcek, Jan Tomas
Vogel, Henry
Wang, Peizhi
Wassermann, Felix Emil
Weiner, Matei
Wetmur, James Gerard
White, James Carrick
White, James Patrick
Wiberg, John Samuel
Wicher, Konrad J
Williams, Curtis Alvin, Jr
Wimmer, Eckard
Winter, Jeanette E
Wolfson, Leonard Louis
Wolin, Meyer Jerome
Wulff, Daniel Lewis
Young, Charles Stuart Hamish
Zabel, Robert Alger
Zahler, Stanley Arnold
Zeigel, Robert Francis
Zero, Domenick Thomas
Zinder, Stephen Henry
Zolla-Pazner, Susan Beth
Zwarun, Andrew Alexander

NORTH CAROLINA

Allen, Wendall E
Arnold, Luther Bishop, Jr
Bachenheimer, Steven Larry
Banes, Albert Joseph
Barakat, Hisham A
Barker, James Cathey
Barnett, Ortus Webb, Jr
Bates, William K
Bishop, Paul Edward
Bolognesi, Dani Paul
Boone, Lawrence Rudolph, III
Bott, Kenneth F
Brown, Talmage Thurman, Jr
Burchall, James J
Buttke, Thomas Martin
Carter, Philip Brian
Chang, John Chung-Shih
Chapman, John Franklin, Jr
Clinton, Bruce Allan
Coggins, Leroy
Collins, Jeffrey Jay
Colwell, William Maxwell
Connolly, Kevin Michael
Corley, Ronald Bruce
Crawford, James Joseph L
Cromartie, William James
Dahiya, Raghunath S
Devereux, Theodora Reyling
Dobrogosz, Walter Jerome
Douros, John Drenkle
Drake, John W
Drexler, Henry
Durack, David Tulloch
Ennever, John Joseph
Estevez, Enrique Gonzalo
Failla, Mark Lawrence
Ferone, Robert
Fleming, Henry Pridgen
Folds, James Donald
Frisell, Wilhelm Richard
Fyfe, James Arthur
Gardner, Donald Eugene
Gilfillan, Robert Frederick
Golden, Carole Ann
Gonder, Eric Charles
Gooder, Harry
Hamilton, Pat Brooks
Harper, James Douglas
Hassan, Hosni Moustafa
Hill, Gale Bartholomew
Hopfer, Roy L
Huang, Eng-Shang (Clark)
Huang, Jeng-Sheng
Hutt, Randy
Jacobson, Kenneth Allan
Johnston, Robert Edward
Joklik, Wolfgang Karl
Katz, Samuel Lawrence
Keene, Jack Donald
Klapper, David G
Klein, Dolph
Knuckles, Joseph Lewis
Kohn, Frank S
Kucera, Louis Stephen
Lecce, James Giacomo
Luginbuhl, Geraldine Hobson
McNeill, John J
McReynolds, Richard A
Malling, Heinrich Valdemar
Manire, George Philip

Mitchell, Thomas Greenfield
Mizel, Steven B
Moody, Max Dale
Morrison, Ralph M
Muller, Uwe Richard
Noga, Edward Joseph
Oblinger, James Leslie
Oliver, James David
Osterhout, Suydam
Pagano, Joseph Stephen
Parks, Leo Wilburn
Perry, Jerome John
Pfaender, Frederic Karl
Phelps, Allen Warner
Phibbs, Paul Vester, Jr
Price, Kenneth Elbert
Privalle, Christopher Thomas
Pukkila, Patricia Jean
Reller, L Barth
Richardson, Stephen H
Rublee, Parke Alstan
Sandok, Paul Louis
Seed, John Richard
Shafer, Steven Ray
Shepard, Maurice Charles
Shih, Jason Chia-Hsing
Sizemore, Ronald Kelly
Sobsey, Mark David
Sonnenfeld, Gerald
Speck, Marvin Luther
Stone, Henry Otto, Jr
Tennant, Raymond Wallace
Turner, Alvis Greely
Van De Rijn, Ivo
Wachsman, Joseph T
Ward, John Everett, Jr
Webb, Neil Broyles
Webster, Robert Edward
Wheat, Robert Wayne
White, Sandra L
Willett, Hilda Pope
Williams, Ray Clayton
Wilson, Harold Albert
Wilson, James Franklin
Wise, Edmund Merriman, Jr
Wise, Ernest George
Witt, Donald James
Wolberg, Gerald
Wollum, Arthur George, II
Zeiger, Errol
Zwadyk, Peter, Jr

NORTH DAKOTA
Duerre, John A
Fillipi, Gordon Michael
Fischer, Robert George
Hastings, James Michael
Kelleher, James Joseph
Marwin, Richard Martin
Vennes, John Wesley
Waller, James R
Watson, David Alan

OHIO
Allison, David Coulter
Asch, David Kent
Astrachan, Lazarus
Atkins, Charles Gilmore
Bannan, Elmer Alexander
Banwart, George J
Baroudy, Bahige M
Battisto, Jack Richard
Berg, Gerald
Bertram, Timothy Allyn
Bigley, Nancy Jane
Blumenthal, Robert Martin
Bonventre, Peter Frank
Bowman, Bernard Ulysses, Jr
Boyer, Jere Michael
Boyle, Michael Dermot
Brady, Robert James
Brent, Morgan McKenzie
Briner, William Watson
Brueske, Charles H
Bubel, Hans Curt
Burnham, Jeffrey C
Camiener, Gerald Walter
Carmichael, Wayne William
Chen, Ker-Sang
Chorpenning, Frank Winslow
Cornelius, Billy Dean
Cox, Charles Donald
Cox, Donald Cody
Cramblett, Henry G
Croft, Charles Clayton
Daniel, Thomas Mallon
Darrow, Robert A
De Fiebre, Conrad William
Dehority, Burk Allyn
Deters, Donald W
Dewar, Norman Ellison
Docherty, John Joseph
Evans, Helen Harrington
Federle, Thomas Walter
Flaumenhaft, Eugene
Frea, James Irving
Freimer, Earl H
Geldreich, Edwin E(mery)
Glaser, Ronald
Goldstein, Gerald
Graves, Anne Carol Finger
Hageage, George John, Jr
Hamparian, Vincent
Haynes, Ralph Edwards

Hoff, John C
Hogg, Robert W
Holder, Ian Alan
Igel, Howard Joseph
Irmiter, Theodore Ferer
Jacobsen, Donald Weldon
Jalil, Mazhar
Jamasbi, Roudabeh J
Janusz, Michael John
Judge, Leo Francis, Jr
Kallenberger, Waldo Elbert
Kapral, Frank Albert
Kearns, Robert J
Kiggins, Edward M
Kirkland, Jerry J
Klein, Michael Gardner
Kohler, Erwin Miller
Kolodziej, Bruno J
Krampitz, Lester Orville
Kreier, Julius Peter
Krueger, Robert George
Kunin, Calvin Murry
Larkin, Edward P
Lass, Jonathan H
Lee, Young-Zoon
Lehmann, Paul F
Lichstein, Herman Carlton
Lingrel, Jerry B
Litchfield, John Hyland
Loper, John C
McFarland, Charles R
Maguire, Michael Ernest
Maier, Siegfried
Mallin, Morton Lewis
Martin, Scott McClung
Mayer, Gerald Douglas
Meredith, William Edward
Merola, A John
Merritt, Katharine
Meyer, Ralph Roger
Michael, William R
Mikolajcik, Emil Michael
Milo, George Edward
Modrzakowski, Malcolm Charles
Morris, Randal Edward
Morris Hooke, Anne
Muckerheide, Annette
Mukkada, Antony Job
Nankervis, George Arthur
Newman, Simon Louis
Nunn, Dorothy Mae
Oakley, Berl Ray
Ockerman, Herbert W
Oglesbee, Michael Jerl
Olsen, Richard George
Ottolenghi, Abramo Cesare
Pollack, J Dennis
Proffitt, Max Rowland
Rabin, Erwin R
Randles, Chester
Rawson, James Rulon Young
Rheins, Melvin S
Rhodes, Judith Carol
Rice, Eugene Ward
Rosen, Samuel
Rosenthal, Ken Steven
Rowe, John James
Safferman, Robert S
Saif, Linda Jean
Saif, Yehia M(ohamed)
Salminen, Seppo Ossian
Sawicki, Stanley George
Scarpino, Pasquale Valentine
Schiff, Gilbert Martin
Sheridan, John Francis
Silverman, Robert Hugh
Smith, Kenneth Larry
Smith, Roger Dean
Somerson, Norman L
Staker, Robert D
Stanberry, Lawrence Raymond
Stephens, James Fred
Stevenson, John Ray
Stotts, Jane
Stukus, Philip Eugene
Sunkara, Sai Prasad
Tabita, F Robert
Thomas, Donald Charles
Treick, Ronald Walter
Trela, John Michael
Trewyn, Ronald William
Troller, John Arthur
Tuovinen, Olli Heikki
Ugarte, Eduardo
Vestal, J Robie
Ward, Richard Leo
Washington, John A, II
Watanakunakorn, Chatrchai
Weber, George Russell
Weis, Dale Stern
Wickstrom, Conrad Eugene
Wiginton, Dan Allen
Wilcox, Joseph Clifford
Williams, Marshall Vance
Williamson, Clarence Kelly
Wise, Donald L
Wood, Harland G
Woodworth, Mary Esther
Yang, Tsanyen
Yohn, David Stewart
Zealey, Marion Edward

OKLAHOMA
Blackstock, Rebecca

Bryant, Rebecca Smith
Cain, William Aaron
Clark, James Bennett
Confer, Anthony Wayne
Conway, Kenneth Edward
Cozad, George Carmon
Cunningham, Madeleine White
Dyer, David Wayne
Ferretti, Joseph Jerome
Flournoy, Dayl Jean
Fulton, Robert Wesley
Gee, Lynn LaMarr
Gilliland, Stanley Eugene
Graves, Donald C
Hall, Leo Terry
Harmon, H James
Hatten, Betty Arlene
Hitzman, Donald Oliver
Hurley, James Edgar
Hyde, Richard Moorehead
Kocan, Katherine M
Lancaster, John
Lerner, Michael Paul
Leu, Richard William
Messmer, Dennis A
Murphy, Juneann Wadsworth
Nelson, John Marvin, Jr
Nordquist, Robert Ersel
Pfister, Robert M
Rhoades, Everett Ronald
Richardson, Lavon Preston
Rowan, Dighton Francis
Schaefer, Frederick Vail
Schindler, Charles Alvin
Schubert, Karel Ralph
Scott, Lawrence Vernon
Seideman, Walter E
Sinclair, James Lewis
Streebin, Leale E
Veltri, Robert William
Waner, Joseph Lloyd
Webb, Carol F
Wegner, Gene H
Wolgamott, Gary
Woolsey, Marion Elmer
Yu, Linda

OREGON
Barry, Arthur Leland
Barry, Ronald A
Brosbe, Edwin Allan
Castenholz, Richard William
Clark, James Orie, II
Clinton, Gail M
Doyle, Jack David
Elliker, Paul R
Elliott, Lloyd Floren
Florance, Edwin R
Fryer, John Louis
Gerke, John Royal
Hallum, Jules Verne
Holm, Harvey William
Kazerouni, Lewa
Kilbourn, Joan Priscilla Payne
Kim, Kenneth
Knittel, Martin Dean
Kuhnley, Lyle Carlton
Lee, Jong Sun
Leong, Jo-Ann Ching
McConnaughey, Bayard Harlow
Machida, Curtis A
Mattson, Donald Eugene
Millette, Robert Loomis
Miner, J Ronald
Molina, Randolph John
Morita, Richard Yukio
Morita, Toshiko N
Myrold, David Douglas
Nelson, Earl Edward
Oginsky, Evelyn Lenore
Pearson, George Denton
Rittenberg, Marvin Barry
Russell, Peter James
Sandine, William Ewald
Schaeffer, Morris
Seidler, Ramon John
Sherr, Barry Frederick
Sherr, Evelyn Brown
Siegel, Benjamin Vincent
Sistrom, William R
Starr, Patricia Rae
Taylor, Mary Lowell Branson
Taylor, Walter Herman, Jr
Van Dyke, Henry
Weaver, William Judson
Wedman, Elwood Edward
Wehr, Herbert Michael
Wood, Anne Michelle
Woodburn, Margy Jeanette

PENNSYLVANIA
Actor, Paul
Alexander, Aaron D
Alexander, James King
Atchison, Robert Wayne
Axler, David Alan
Baechler, Charles Albert
Bailey, Denis Mahlon
Baum, Robert Harold
Bayer, Manfred Erich
Bayer, Margret H(elene) Janssen
Bean, Barry
Becker, Robert S
Beilstein, Henry Richard

BIOLOGICAL SCIENCES / 127

Beining, Paul R
Benedict, Robert Curtis
Benz, Edward John
Bering, Charles Lawrence
Bernlohr, Robert William
Bernstein, Alan
Bhaduri, Saumya
Blough, Herbert Allen
Bondi, Amedeo
Bott, Thomas Lee
Boutros, Susan Noblit
Bowles, Bobby Linwood
Brenchley, Jean Elnora
Brinton, Charles Chester, Jr
Brown, William E
Buchanan, Robert Lester
Caliguiri, Lawrence Anthony
Callahan, Hugh James
Casida, Lester Earl, Jr
Castric, Peter Allen
Cheng, Yih-Shyun Edmond
Chiu, Teh-Hsing
Cohen, Gary H
Coleman, Charles Mosby
Cronholm, Lois S
Crowell, Richard Lane
Cundy, Kenneth Raymond
Daneo-Moore, Lolita
Davies, Helen Jean Conrad
Davies, Warren Lewis
Debroy, Chitrita
Deering, Reginald
Deforest, Adamadia
DeMeio, Joseph Louis
Diamond, Leila
Di Cuollo, C John
Dryden, Richard Lee
Dudzinski, Diane Marie
Eberhart, Robert J
Edlind, Thomas D
Eisenstein, Toby K
Elliott, Arthur York
Ellis, Richard John
Farber, Florence Eileen
Farber, Paul Alan
Feinstein, Sheldon Israel
Feit, Ira (Nathan)
Fletcher, Ronald D
Forbes, Martin
Fraser, Nigel William
Gadebusch, Hans Henning
Gammon, Richard Anthony
Garfinkle, Barry David
Gealt, Michael Alan
Gerhard, Walter Ulrich
Gilman, Steven Christopher
Glorioso, Joseph Charles, III
Godfrey, Susan Sturgis
Goldfine, Howard
Goldner, Herman
Gots, Joseph Simon
Gray, Alan
Gregory, Francis Joseph
Hammel, Jay Morris
Hammond, Benjamin Franklin
Hankins, William Alfred
Harmon, George Andrew
Harris, Susanna
Henderson, Earl Erwin
Hendrix, Sherman Samuel
Higgins, Michael Lee
Hilleman, Maurice Ralph
Ho, Monto
Hobby, Gladys Lounsbury
Hoffee, Patricia Anne
Hsu, Kuo-Hom Lee
Hummeler, Klaus
Hung, Paul P
Isom, Harriet C
Iwatsuki, Shunzaburo
Jackson, Ethel Noland
Jacobson, Lewis A
Jarvik, Jonathan Wallace
Johnston, James Bennett
Joseph, Jeymohan
Juneja, Vijay Kumar
Jungkind, Donald Lee
Kane, James Francis
Keenan, John Douglas
Kennedy, Harvey Edward
King, Robert Willis
Kirk, Billy Edward
Kissinger, John Calvin
Klein, Morton
Klens, Paul Frank
Kneebone, Leon Russell
Knobler, Robert Leonard
Koesterer, Martin George
Koprowski, Hilary
Krah, David Lee
Krawiec, Steven Stack
Kreider, John Wesley
Kuserk, Frank Thomas
Lancaster, Jack R, Jr
Landau, Burton Joseph
Larson, Vivian M
Lashen, Edward S
Lawrence, William Chase
Layne, Porter Preston
Lee, Chin-Chiu
Lee, John Cheung Han
Leech, Stephen H
Liegey, Francis William
Live, Israel

Microbiology (cont)

Lobo, Francis X
Logan, David Alexander
Long, Carole Ann
Long, Walter Kyle, Jr
McAlack, Robert Francis
McCarthy, Frank John
McCarthy, William John
Machlowitz, Roy Alan
McKinstry, Donald Michael
McNamara, Pamela Dee (MacMinigal)
Magee, Wayne Edward
Majumdar, Shyamal K
Malamud, Daniel F
Mandel, John Herbert
Manson, Lionel Arnold
Mast, Morris G
Mathur, Carolyn Frances
Melamed, Sidney
Miller, James Eugene
Milligan, Wilbert Harvey, III
Millman, Irving
Minsavage, Edward Joseph
Miovic, Margaret Lancefield
Mizutani, Satoshi
Moorman, Gary William
Morahan, Page Smith
Nathanson, Neal
Natuk, Robert James
Nelson, Paul Edward
Neubeck, Clifford Edward
Osborne, William Wesley
Pagano, Joseph Frank
Pakman, Leonard Marvin
Palumbo, Samuel Anthony
Patrick, Ruth (Mrs Charles Hodge IV)
Patton, William Henry
Pepper, Rollin E
Phillips, Bruce A
Plotkin, Stanley Alan
Pootjes, Christine Fredricka
Porter, Ronald Dean
Poste, George Henry
Poupard, James Arthur
Pratt, Elizabeth Ann
Provost, Philip Joseph
Rapp, Fred
Ray, Eva K
Reich, Claude Virgil
Reisner, Gerald Seymour
Resconich, Emil Carl
Rest, Richard Franklin
Ricciardi, Robert Paul
Roesing, Timothy George
Roia, Frank Costa, Jr
Romano, Paula Josephine
Roosa, Robert Andrew
Rosan, Burton
Rosenkranz, Herbert S
Rosenzweig, William David
Rozmiarek, Harry
Rubin, Benjamin Arnold
Sagik, Bernard Phillip
Salerno, Ronald Anthony
Sanders, Robert W
Santer, Melvin
Schaedler, Russell William
Schiff, Paul L, Jr
Sheffield, Joel Benson
Shockman, Gerald David
Sideropoulos, Aris S
Singh, Balwant
Six, Howard R
Skalka, Anna Marie
Slifkin, Malcolm
Slotnick, Victor Bernard
Smith, Harry Logan, Jr
Smith, James Lee
Smith, Josephine Reist
Snow, Jean Anthony
Sojka, Gary Allan
Somkuti, George A
Southam, Chester Milton
Stempen, Henry
Steplewski, Zenon
Stere, Athleen Jacobs
Stevens, Roy Harris
Suyama, Yoshitaka
Taubler, James H
Tax, Anne W
Tevethia, Satvir S
Thayer, Donald Wayne
Thimann, Kenneth Vivian
Thomas, William J
Thomulka, Kenneth William
Tint, Howard
Unz, Richard F(rederick)
Vaidya, Akhil Babubhai
Vaughan, James Roland
Vessey, Adele Ruth
Warren, George Harry
Watson, Joseph Alexander
Weinbaum, George
Whiting, Richard Charles
Wilcox, Wesley C
Wilhelm, James Maurice
Willett, Norman P
Wing, Edward Joseph
Wolf, Benjamin
Wolfe, John Hall
Wolfersberger, Michael Gregg
Yamamoto, Nobuto
Yang, Da-Ping (David)
Youngner, Julius Stuart
Yurchenco, John Alfonso
Zaika, Laura Larysa
Zimmerer, Robert P
Zubrzycki, Leonard Joseph

RHODE ISLAND

Carpenter, Charles C J
Clough, Wendy Glasgow
Crowley, James Patrick
Goos, Roger Delmon
Howard, Frank Leslie
Howe, Calderon
Jones, Kenneth Wayne
Lederberg, Seymour
Liss, Alan
McConeghy, Matthew H
Meedel, Thomas Huyck
Miller, Robert Harold
O'Leary, Gerard Paul, Jr
Prager, Jan Clement
Shank, Peter R
Sieburth, John McNeill
Swift, Dorothy Garrison
Worthen, Leonard Robert
Yates, Vance Joseph

SOUTH CAROLINA

Aelion, Claire Marjorie
Baxter, Ann Webster
Brown, Arnold
Bryan, Charles Stone
Coulston, Mary Lou
Davis, Leroy
Davis, Raymond F
Ely, Berten E, III
Galbraith, Robert Michael
Ghaffar, Abdul
Goodroad, Lewis Leonard
Gorden, Robert Wayne
Grady, Cecil Paul Leslie, Jr
Gustafson, Ralph Alan
Hayasaka, Steven S
Hazen, Terry Clyde
Henson, Joseph Lawrence
Higerd, Thomas Braden
Keyser, Peter D
Lampky, James Robert
Paynter, Malcolm James Benjamin
Reddick-Mitchum, Rhoda Anne
Richards, Gary Paul
Schmidt, Roger Paul
Shively, Jessup MacLean
Sigel, M(ola) Michael
Stutzenberger, Fred John
Vereen, Larry Edwin
Virella, Gabriel T
Wilkinson, Thomas Ross
Yoch, Duane Charles

SOUTH DAKOTA

Cafruny, William Alan
Hildreth, Michael B
Langworthy, Thomas Allan
Lynn, Raymond J
Pengra, Robert Monroe
Prescott, Lansing M
Smith, Paul Francis
Sword, Christopher Patrick
Weaver, Keith Eric
Westby, Carl A

TENNESSEE

Avis, Kenneth Edward
Bean, William J, Jr
Beck, Raymond Warren
Becker, Jeffrey Marvin
Belew-Noah, Patricia W
Ben-Porat, Tamar
Blaser, Martin Jack
Brown, Arthur
Bryant, Robert Emory
Chandler, Robert Walter
Chi, David Shyh-Wei
Chung, King-Thom
Cook, Robert James
Cooper, Terrance G
Cox, Edmond Rudolph, Jr
Cypess, Raymond Harold
Davison, Brian Henry
Dees, Craig
Draughon, Frances Ann
Farrington, Joseph Kirby
Freeman, Bob A
Hadden, Charles Thomas
Harshman, Sidney
Herting, David Clair
Hollis, Cecil George
Hougland, Arthur Eldon
Howe, Martha Morgan
Howell, Elizabeth E
Incardona, Antonino L
Isa, Abdallah Mohammad
Johnson, Charles William
Karve, Mohan Dattatreya
Karzon, David T
Katze, Jon R
Koh, Sung-Cheol
Larimer, Frank William
Mayberry, William Roy
Miller, Harold Charles
Mithcell, William Marvin
Montie, Thomas C
Mullen, Michael David
Ourth, Donald Dean
Portner, Allen
Postlethwaite, Arnold Eugene
Prokop, Ales
Quigley, Neil Benton
Roberts, Audrey Nadine
Robinson, John Price
Rouse, Barry Tyrrell
Rubin, Donald Howard
Ryan, Kevin William
Ryden, Fred Ward
Savage, Dwayne Cecil
Stacey, Gary
Stevens, Audrey L
Stevens, Stanley Edward, Jr
Todd, William McClintock
Trentham, Jimmy N
Vredeveld, Nicholas Gene
Webster, Robert G
White, David Cleaveland
Whitt, Michael A
Wiesmeyer, Herbert
Wilson, Benjamin James

TEXAS

Albrecht, Thomas Blair
Alvarez-Gonzalez, Rafael
Ascenzi, Joseph Michael
Atkinson, Mark Arthur Leonard
Baer, Richard
Baine, William Brennan
Baptist, James (Noel)
Baron, Samuel
Baseman, Joel Barry
Baughn, Robert Elroy
Bawdon, Roger Everett
Bellion, Edward
Benedik, Michael Joseph
Bettinger, George E
Black, Samuel Harold
Blouse, Louis E, Jr
Bose, Henry Robert, Jr
Bowen, James Milton
Brand, Jerry Jay
Brown, Dennis Taylor
Brown, Lee Roy, Jr
Browning, J(ohn) Artie
Brysk, Miriam Mason
Buchanan, Christine Elizabeth
Butel, Janet Susan
Castro, Gilbert Anthony
Cate, Thomas Randolph
Chan, James C
Chan, Teh-Sheng
Chanh, Tran C
Chappell, Cynthia Lou
Chatterjee, Bandana
Chen, Young Chang
Chrzanowski, Thomas Henry
Cirillo, Vincent Paul
Cole, Garry Thomas
Collisson, Ellen Whited
Cooper, Ronda Fern
Cotner, James Bryan
Crandell, Robert Allen
Crawford, Gladys P
Croley, Thomas Edgar
Davis, Charles Patrick
Davis, James Royce
Dibble, John Thomas
Donnelly, Patricia Vryling
Dorn, Gordon Lee
Dreesman, Gordon Ronald
Dubbs, Del Rose M
Dubose, Robert Trafton
Dulaney, Eugene Lambert
Dwyer, Lawrence Arthur
Earhart, Charles Franklin, Jr
East, James Lindsay
Eichenwald, Heinz Felix
Eidels, Leon
Eldridge, David Wyatt
Ellzey, Joanne Tontz
Eugster, A Konrad
Evans, John Edward
Fang, Jiasong
Fleenor, Marvin Bension
Fleischmann, William Robert, Jr
Foster, Billy Glen
Foster, Terry Lynn
Fox, George Edward
Frank, Hilmer Aaron
Freeman, Maynard Lloyd
French, Jeptha Victor
Friend, Patric Lee
Fuerst, Robert
Gardner, Earl William, Jr
Gardner, Frederick Albert
Georgiou, George
Gilbert, Brian E
Goldschmidt, Millicent
Goodman, Joel Mitchell
Grimm, Elizabeth Ann
Guentzel, M Neal
Guerrero, Felix David
Guthrie, Rufus Kent
Harris, Elizabeth Forsyth
Harris, Kerry Francis Patrick
Haye, Keith R
Heberling, Richard Leon
Heck, Fred Carl
Heggers, John Paul
Heinze, John Edward
Hejtmancik, Kelly Erwin
Henney, Henry Russell, Jr
Henry, Clay Allen
Hentges, David John
Highlander-Bultema, Sarah Katherine
Hillis, William Daniel, Sr
Holguin, Alfonso Hudson
Huang, Shyi Yi
Huber, Thomas Wayne
Humphrey, Ronald DeVere
Humphrey, Ronald Mack
Jeter, Randall Mark
Jorgensen, George Norman
Jorgensen, James H
Joys, Terence Michael
Jurtshuk, Peter, Jr
Kalter, Seymour Sanford
Kantor, Harvey Sherwin
Karunakaran, Thonthi
Kasel, Julius Albert
Kaufmann, Anthony J
Kumar, Vinay
Kurosky, Alexander
Kutteh, William Hanna
LaBrie, David Andre
Lacko, Andras Gyorgy
Lansford, Edwin Myers, Jr
LeBlanc, Donald Joseph
Lees, George Edward
Lefkowitz, Doris Lynne
Lefkowitz, Stanley S
Leibowitz, Julian Lazar
Leininger, Harold Vernon
Loan, Raymond Wallace
Loftin, Karin Christiane
McBride, Mollie Elizabeth
McCallum, Roderick Eugene
McConnell, Stewart
McCracken, Alexander Walker
McDonald, William Charles
Mace, Kenneth Dean
McMurray, David N
Mandel, Manley
Mattingly, Stephen Joseph
Mayberry, Lillian Faye
Mayor, Heather Donald
Measel, John William
Melnick, Joseph Louis
Merrill, Gerald Alan
Miget, Russell John
Moyer, Mary Pat Sutter
Moyer, Rex Carlton
Murthy, Krishna Kesava
Nash, Donald Robert
Nicholson, Wayne Lowell
Norris, Steven James
O'Donovan, Gerard Anthony
Ohlenbusch, Robert Eugene
Olson, Leroy Justin
Oujesky, Helen Matusevich
Paque, Ronald E
Payne, Shelley Marshall
Perez, John Carlos
Peterson, Johnny Wayne
Phillips, Guy Frank
Powell, Bernard Lawrence
Prabhakar, Bellur Subbanna
Prasad, Rupi
Prashad, Nagindra
Price, Peter Michael
Quarles, John Monroe
Rael, Eppie David
Reeves, James Blanchette
Rege, Ajay Anand
Reyes, Victor E
Rich, Susan Marie Solliday
Riehl, Robert Michael
Riggs, Stuart
Ritzi, Earl Michael
Rolfe, Rial Dewitt
Rolston, Kenneth Vijaykumar Issac
Romeo, Tony
Rosenblum, Eugene David
Rudolph, Frederick Byron
Safe, Stephen Harvey
Sanford, Barbara Ann
Schlech, Barry Arthur
Schmidt, Jerome P(aul)
Schneider, Dennis Ray
Schwarz, John Robert
Seman, Gabriel
Serwer, Philip
Shadduck, John Allen
Shanley, Mark Stephen
Shillitoe, Edward John
Simpson, Russell Bruce
Smith, Gerald Ray
Smith, Kendall O
Smith, Russell Lamar
Smith, William Russell
Snell, William J
Spellman, Craig William
Stenback, Wayne Albert
Stephens, Robert Lawrence
Stevens, Clark
Stewart, Charles Ranous
Stewart, James Ray
Straus, David Conrad
Streckfuss, Joseph Larry
Stroman, David Womack
Stroynowski, Iwona T
Summers, Max Duane
Sumrall, H Glenn
Suttle, Curtis
Szaniszlo, Paul Joseph

Taber, Willard Allen
Taylor, Robert Dalton
Temeyer, Kevin Bruce
Thomas, Virginia Lynn
Toler, Robert William
Tong, Alex W
Trentin, John Joseph
Trkula, David
Valdivieso, Dario
Van Alfen, Neal K
Vanderzant, Carl
Vela, Gerard Roland
Villa, Vicente Domingo
Vitetta, Ellen S
Walker, David Hughes
Walker, James Roy
Weichlein, Russell George
Weinstock, George Matthew
Welch, Gordon E
Wendt, Theodore Mil
Whitford, Howard Wayne
Williams, Robert Pierce
Wilson, Robert Eugene
Winters, Wendell Delos
Wynne, Elmer Staten
Yanni, John Michael
Young, Ryland F
Zabransky, Ronald Joseph
Zuberer, David Alan

UTAH
Barnett, Bill B
Beck, Jay Vern
Belnap, Jayne
Blauer, Aaron Clyde
Bradshaw, Willard Henry
Brierley, Corale Louise
Brierley, James Alan
Burton, Sheril Dale
Cole, Barry Charles
Crane, George Thomas
Franklin, Naomi C
Fujinami, Robert S
Georgopoulos, Constantine Panos
Harrington, Glenn William
Jensen, Marcus Martin
Johnson, F Brent
Kern, Earl R
Klein, Sigrid Marta
Lark, Cynthia Ann
Larsen, Austin Ellis
Larsen, Don Hyrum
Larsen, Lloyd Don
Lombardi, Paul Schoenfeld
Marcus, Stanley
Matsen, John Martin
Melton, Arthur Richard
Muna, Nadeem Mitri
Nelson, Jerry Rees
North, James A
O'Neill, Frank John
Post, Frederick Just
Prescott, Stephen M
Rees, Horace Benner, Jr
Roth, John R
Sagers, Richard Douglas
Sidwell, Robert William
Skujins, John Janis
Spendlove, Rex S
Stockland, Alan Eugene
Stokes, Barry Owen
Vandenberg, John Donald
Warren, Reed Parley
Welkie, George William
Wiley, Bill Beauford
Wood, Timothy E
Wullstein, Leroy Hugh

VERMONT
Craighead, John Edward
Fives-Taylor, Paula Marie
Johnstone, Donald Boyes
Kuehner, Calvin Charles
Lachapelle, Rene Charles
Lechevalier, Hubert Arthur
Lechevalier, Mary P
Ledinko, Nada
McLean, Donald Lewis
Moehring, Joan Marquart
Moehring, Thomas John
Raper, Carlene Allen
Schaeffer, Warren Ira
Sjogren, Robert Erik
Ullrich, Robert Carl
Wallace, Susan Scholes
Weed, Lawrence Leonard

VIRGINIA
Al-Doory, Yousef
Andrews, Wallace Henry
Andrykovitch, George
Ayers, William Arthur
Barak, Eve Ida
Barr, Fred S
Bates, Robert Clair
Benjamin, David Charles
Berkow, Susan E
Berliner, Martha D
Blum, Linda Kay
Boardman, Gregory Dale
Boyle, John Joseph
Braciale, Thomas Joseph, Jr
Braciale, Vivian Lam
Bradley, Sterling Gaylen

Brown, Jay Clark
Cabral, Guy Antony
Chalgren, Steve Dwayne
Chaparas, Sotiros D
Chen, Jiann-Shin
Chung, Choong Wha
Clarke, Gary Anthony
Claus, George William
Coleman, Philip Hoxie
Cordes, Donald Ormond
Coursen, Bradner Wood
Cummins, Cecil Stratford
Dalton, Harry P
Domermuth, Charles Henry, Jr
Escobar, Mario R
Evans, Herbert John
Formica, Joseph Victor
Fu, Shu Man
Fugate, Kearby Joe
Halleck, Frank Eugene
Heineken, Frederick George
Hench, Miles Ellsworth
Hoptman, Julian
Hsu, Hsiu-Sheng
Huang, H(sing) T(sung)
Johnson, James Carl
Johnson, John LeRoy
Kelley, John Michael
King, Betty Louise
Kirk, Paul Wheeler, Jr
Kirpekar, Abhay C
Krieg, Noel Roger
Kuemmerle, Nancy Benton Stevens
Lemp, John Frederick, Jr
Lewis, Cornelius Crawford
Loria, Roger Moshe
Lubiniecki, Anthony Stanley
McCombs, Robert Matthew
McCowen, Sara Moss
McCuen, Robert William
McDonald, Gerald O
McGowan, John Joseph
McVicar, John West
McWright, Cornelius Glen
Martin, William Wallace
Merchant, Donald Joseph
Moore, Walter Edward C
Munson, Albert Enoch
Myrvik, Quentin N
Nagarkatti, Mitzi
Nagarkatti, Prakash S
Nasser, DeLill
Neal, John Lloyd, Jr
Newton, William Edward
Normansell, David E
Palisano, John Raymond
Pierson, Merle Dean
Pledger, Richard Alfred
Pryer, Nancy Kathryn
Raizen, Carol Eileen
Roberts, Catherine Harrison
Roshal, Jay Yehudie
Royt, Paulette Anne
Russell, Catherine Marie
Scheld, William Michael
Scott, Marvin Wade
Shadomy, Smith
Smibert, Robert Merrall, II
Somers, Kenneth Donald
Srivastava, Kailash Chandra
Stern, Joseph Aaron
Stevens, Thomas McConnell
Stokes, Gerald V
Tankersley, Robert Walker, Jr
Tew, John Garn
Tolin, Sue Ann
Toney, Marcellus E, Jr
Trelawny, Gilbert Sterling
Tucker, Anne Nichols
Uffen, Robert L
Vermeulen, Carl William
Volk, Wesley Aaron
Wagner, Robert Roderick
Ware, Lawrence Leslie, Jr
Weber, Michael Joseph
Wilkins, Judd Rice
Wilkins, Tracy Dale
Williams, Luther Steward
Wright, George Green
Yousten, Allan A

WASHINGTON
Ahmed, Saiyed I
Baross, John Allen
Benedict, Robert Glenn
Bezdicek, David Fred
Boatman, Edwin S
Booth, Beatrice Crosby
Calza, Roger Ernest
Cheevers, William Phillip
Cho, Byung-Ryul
Cohen, Arthur LeRoy
Cosman, David John
Cota-Robles, Eugene H
Coyle, Marie Bridget
Davis, William C
Deming, Jody W
Duncan, James Byron
Eidinger, David
Eklund, Melvin Wesley
Evermann, James Frederick
Gold, Eli
Grable, Albert E
Graves, Scott Stoll

Groman, Neal Benjamin
Grootes-Reuvecamp, Grada Alijda
Gurusiddaiah, Sarangamat
Hawthorne, Donald Clair
Hu, Shiu-Lok
Irgens, Roar L
Johnstone, Donald Lee
Kenny, George Edward
Kirchheimer, Waldemar Franz
Kuo, Cho-Chou
Leid, R Wes
Lewin, Joyce Chismore
Lowe, Janet Marie
McDonald, John Stoner
McFadden, Bruce Alden
Magnuson, Nancy Susanne
Matches, Jack Ronald
Miller, Arthur Dusty
Nakata, Herbert Minoru
Nester, Eugene William
Norvell, Lorelei Lehwalder
Nutter, Robert Leland
Pacha, Robert Edward
Parish, Curtis Lee
Pierson, Beverly Kanda
Pusey, P Lawrence
Raff, Howard V
Rayburn, William Reed
Reeves, Raymond
Rohrschneider, Larry Ray
Sheiness, Diana Kay
Sherris, John C
Smith, Louis De Spain
Spence, Kemet Dean
Stevens, Todd Owen
Stokes, Jacob Leo
Stuart, Kenneth Daniel
Taylor, Dean Perron
Van Hoosier, Gerald L, Jr
Wang, San-Pin
Wekell, Marleen Marie
Whisler, Howard Clinton
Whiteley, Helen Riaboff
Wiley, William Rodney
Zakian, Virginia Araxie

WEST VIRGINIA
Albertson, John Newman, Jr
Anderson, Douglas Poole
Barnett, John Brian
Berry, Vinod K
Binder, Franklin Lewis
Bissonnette, Gary Kent
Buckelew, Albert Rhoades, Jr
Calhoon, Donald Alan
Charon, Nyles William
Cook, Harold Andrew
Hahon, Nicholas
Lewis, Daniel Moore
Lim, James Khai-Jin
Moat, Albert Groombridge
Olenchock, Stephen Anthony
Ong, Tong-man
Pore, Robert Scott
Wolf, Kenneth Edward
Yelton, David Baetz

WISCONSIN
Armstrong, George Michael
Balish, Edward
Brawner, Thomas A
Brill, Winston J
Brock, Katherine Middleton
Brock, Thomas Dale
Burris, Robert Harza
Byrne, Gerald I
Chambliss, Glenn Hilton
Clausz, John Clay
Cliver, Dean Otis
Collins, Mary Lynne Perille
Courtright, James Ben
Czuprynski, Charles Joseph
DeMars, Robert Ivan
Dick, Elliot C
Dukenschein, Jeanne Therese
Ehle, Fred Robert
Foster, Edwin Michael
Friend, Milton
Gerberich, John Barnes
Golubjatnikov, Rjurik
Gourse, Richard Lawrence
Graham, Linda Kay Edwards
Grossberg, Sidney Edward
Haasch, Mary Lynn
Hinshaw, Virginia Snyder
Hinze, Harry Clifford
Hoerl, Bryan G
Howard, Thomas Hyland
Huss, Ronald John
Jacobson, Gunnard Kenneth
Jameson, Patricia Madoline
Jeffries, Thomas William
Klebba, Phillip E
Kostenbader, Kenneth David, Jr
Kurup, Viswanath Parameswar
Kushnaryov, Vladimir Michael
Mahl, Mearl Carl
Mansfield, John Michael
Marsh, Richard Floyd
Marth, Elmer Herman
Moskowitz, Gerard Jay
Myers, Charles R
Nagodawithana, Tilak Walter
Ndon, John A

Nealson, Kenneth Henry
Nelson, Allen Charles
Nelson, Thomas Clifford
Olson, Norman Fredrick
Pariza, Michael Willard
Parker, Dorothy Lundquist
Peppler, Henry James
Remsen, Charles C, III
Rigney, Mary Margaret
Rosson, Reinhardt Arthur
Rude, Theodore Alfred
Rueckert, Roland R
Scheusner, Dale Lee
Schultz, Ronald David
Schwartz, Leander Joseph
Seelke, Ralph Walter
Seitz, Eugene W
Sieber, Fritz
Skatrud, Thomas Joseph
Smith, Clyde Konrad
Smith, Donald Ward
Sonneborn, David R
Spalatin, Josip
Steeves, Richard Allison
Steinhart, Carol Elder
Steinke, Paul Karl Willi
Szybalski, Waclaw
Takayama, Kuni
Taylor, Jerry Lynn
Temin, Howard Martin
Truitt, Robert Lindell
Tufte, Marilyn Jean
Waechter-Brulla, Daryle A
Walker, Duard Lee
Wege, Ann Christene
Weidanz, William P
Wejksnora, Peter James
Wilcox, Kent Westbrook
Williams, Anna Maria
Willits, Richard Ellis
Wimpee, Charles F
Yuill, Thomas MacKay

WYOMING
Adams, John Collins
Belden, Everett Lee
Blank, Carl Herbert
Bulla, Lee Austin, Jr
Caldwell, Daniel R
George, Robert Porter
Isaak, Dale Darwin
Pier, Allan Clark

PUERTO RICO
Bolanos, Benjamin
Colon, Julio Ismael
Copson, David Arthur
Craig, Syndey Pollock, III
Garcia-Castro, Ivette
Lluberes, Rosa P
Ramirez-Ronda, Carlos Hector
Toranzos, Gary Antonio
Torres-Blasini, Gladys

ALBERTA
Albritton, William Leonard
Athar, Mohammed Aqueel
Birdsell, Dale Carl
Campbell, James Nicoll
Cheng, Kuo-Joan
Colter, John Sparby
Costerton, J William F
Darcel, Colin Le Q
Dixon, John Michael Siddons
Fedorak, Phillip Michael
Francis, Mike McD
Fujita, Donald J
Gaucher, George Maurice
Goettel, Mark S
Greer, George Gordon
Hiruki, Chuji
Jack, Thomas Richard
Jensen, Susan Elaine
Kadis, Vincent William
Kaneda, Toshi
Langridge, William Henry Russell
Larke, R(obert) P(eter) Bryce
McCready, Ronald Glen Lang
McElhaney, Ronald Nelson
Marusyk, Raymond George
Paranchych, William
Parkinson, Dennis
Rennie, Robert John
Rice, Wendell Alfred
Rosenberg, Susan Mary
Schroder, David John
Scraba, Douglas G
Stiles, Michael Edgecombe
Wallis, Peter Malcolm
Wayman, Morris
Weiner, Joel Hirsch
Whitehouse, Ronald Leslie S
Yoon, Ji-Won

BRITISH COLUMBIA
Agnew, Robert Morson
Albright, Lawrence John
Anderson, John Donald
Blackwood, Allister Clark
Bower, Susan Mae
Chiko, Arthur Wesley
Child, Jeffrey James
Chow, Anthony W
Copeman, Robert James

Microbiology (cont)

Dempster, George
Dolman, Claude Ernest
Duncan, Douglas Wallace
Finlay, Barton Brett
Hancock, Robert Ernest William
Hawirko, Roma Zenovea
Kelly, Michael Thomas
Levy, Julia Gerwing
McBride, Barry Clarke
McCarter, John Alexander
McLean, Donald Millis
Majak, Walter
Miller, Robert Carmi, Jr
Redfield, Rosemary Jeanne
Sadowski, Ivan J
Spiegelman, George Boole
Tingle, Aubrey James
Townsley, Philip McNair
Tremaine, Jack H
Trust, Trevor John
Walden, C(ecil) Craig
Weeks, Gerald
Westlake, Donald William Speck
Yamamoto, Tatsuzo

MANITOBA
Borsa, Joseph
Burton, David Norman
Campbell, Norman E Ross
Gill, Clifford Cressey
Hamilton, Ian Robert
Hamilton, Robert Duncan
Kunz, Bernard Alexander
Lukow, Odean Michelin
Maniar, Atish Chandra
Pepper, Evan Harold
Rollo, Ian McIntosh
Ronald, Allan Ross
Shiu, Robert P C
Suzuki, Isamu
Vessey, Joseph Kevin
Wallbank, Alfred Mills
Warner, Peter
Westdal, Paul Harold

NEW BRUNSWICK
Singh, Rudra Prasad

NEWFOUNDLAND
Barnsley, Eric Arthur
Burness, Alfred Thomas Henry
Davidson, William Scott
Nolan, Richard Arthur

NOVA SCOTIA
Easterbrook, Kenneth Brian
Kind, Leon Saul
Mahony, David Edward
Ozere, Rudolph L
Rozee, Kenneth Roy
Stewart, James Edward
Stratton, Glenn Wayne
Vining, Leo Charles
White, Robert Lester
Wort, Arthur John

ONTARIO
Appanna, Vasu Dhananda
Arif, Basil Mumtaz
Barnum, Donald Alfred
Barran, Leslie Rohit
Behme, Ronald John
Beveridge, Terrance James
Bouillant, Alain Marcel
Burnison, Bryan Kent
Cairns, William Louis
Calleja, G(odehardo) B
Campbell, James B
Carstens, Eric Bruce
Chadwick, June Stephens
Chaudhary, Rabindra Kumar
Chernesky, Max Alexander
Clark, A Gavin
Clark, David Sedgefield
Clarke, Anthony John
Dales, Samuel
Dalpe, Yolande
Davidson, Charles Mackenzie
Deeley, Roger Graham
Derbyshire, John Brian
Doane, Frances Whitman
Duncan, I B R
Dutka, Bernard J
Farkas-Himsley, Hannah
Faulkner, Peter
Fish, Eleanor N
Fitt, Peter Stanley
Forsberg, Cecil Wallace
Franklin, Mervyn
Fraser, Ann Davina Elizabeth
Friesen, James Donald
Furesz, John
Galsworthy, Sara B
Gentner, Norman Elwood
Gochnauer, Thomas Alexander
Gould, William Douglas
Graham, Frank Lawson
Gregory, Kenneth Fowler
Gupta, Krishana Chandara
Gyles, C L
Hauschild, Andreas H W
Heath, Ian Brent
Hendry, Anne Teresa
Higgins, Verna Jessie
Inman, Robert Davies
Inniss, William Edgar
Iyer, Rajul V
Johnson, Byron F
Johnson-Lussenburg, Christine Margaret
Kang, C Yong
Kasupski, George Joseph
Kaushik, Azad
Kennedy, Kevin Joseph
Khan, Abdul Waheed
Lachance, Marc-Andre
Lang, Gerhard Herbert
Lee, Peter E
Liu, Dickson Lee Shen
McCurdy, Howard Douglas, Jr
Macdonald, John Barfoot
Mahdy, Mohamed Sabet
Mak, Stanley
Medzon, Edward Lionel
Metzgar, Don P
Miller, John James
Moo-Young, Murray
Murray, William Douglas
Nielsen, Klaus H B
Parker, Wayne Jeffery
Patel, Girishchandra Babubhai
Pope, Barbara L
Pross, Hugh Frederick
Rao, Salem S
Rayman, Mohamad Khalil
Rhodes, Andrew James
Richardson, Harold
Rigby, Charlotte Edith
Robinow, Carl Franz
Rogers, Charles Graham
Roslycky, Eugene Bohdan
Sabina, Leslie Robert
Sattar, Syed Abdus
Savan, Milton
Seligy, Verner Leslie
Siminovitch, Louis
Smiley, James Richard
Sorger, George Joseph
Spence, Leslie Percival
Sprott, Gordon Dennis
Stainer, Dennis William
Stavric, Stanislava
Stemshorn, Barry William
Stevens, John Bagshaw
Stevenson, Ian Lawrie
Stewart, Robert Bruce
Szewczuk, Myron Ross
Todd, Ewen Cameron David
Topp, Edward
Trevors, Jack Thomas
Truscott, Robert Bruce
Tu, Chin Ming
Tu, Jui-Chang
Vas, Stephen Istvan
Veliky, Ivan Alois
Ward, Edmund William Beswick
Wellman, Angela Myra
Winterhalder, Keith
Wood, Janet Marion
Young, James Christopher F

QUEBEC
Abshire, Claude James
Ackermann, Hans Wolfgang
Archambault, Denis
Archibald, Frederick S
Belloncik, Serge
Benoit, Guy J C
Bilimoria, Minoo Hormasji
Bordeleau, Lucien Mario
Bourgaux, Pierre
Brailovsky, Carlos Alberto
Branton, Philip Edward
Cantor, Ena D
Chabot, Benoit
Chagnon, Andre
Champagne, Claude P
Chan, Eddie Chin Sun
Chung, Young Sup
Comeau, Andre I
DeVoe, Irving Woodrow
DuBow, Michael Scott
Dugre, Robert
Evering, Winston E N D
Gervais, Francine
Gibbs, Sarah Preble
Goulet, Jacques
Graham, Angus Frederick
Hallenbeck, Patrick Clark
Hamelin, Claude
Idziak, Edmund Stefan
Joncas, Jean Harry
Jurasek, Lubomir
Kluepfel, Dieter
Knowles, Roger
Laflamme, Gaston
Leduy, Anh
Lussier, Gilles L
MacLeod, Robert Angus
Martineau, Ronald
Mathieu, Leo Gilles
Menezes, Jose Piedade Caetano Agnelo
Murgita, Robert Anthony
Niven, Donald Ferries
Paice, Michael
Portelance, Vincent Damien
Richard, Claude
Roberge, Marcien Romeo
Sheinin, Rose
Siboo, Russell
Simard, Rene
Simard, Ronald E
Skup, Daniel
Sonea, Sorin I
Stanners, Clifford Paul
Talbot, Pierre J
Taussig, Andrew
Thirion, Jean Paul Joseph
Trepanier, Pierre
Trudel, Michel D
Vezina, Claude
Weber, Joseph M

SASKATCHEWAN
Babiuk, Lorne A
Cullimore, Denis Roy
Haskins, Reginald Hinton
Ingledew, William Michael
Jones, Graham Alfred
Julien, Jean-Paul
Khachatourians, George G
Kurz, Wolfgang Gebhard Walter
Nelson, Louise Mary
Robertson, Hugh Elburn
Waygood, Edward Bruce

YUKON TERRITORY
Doermann, August Henry

OTHER COUNTRIES
Aalund, Ole
Adldinger, Hans Karl
Agathos, Spiros Nicholas
Arber, Werner
Atkinson, Paul H
Baddiley, James
Barrett, James Thomas
Bishop, David Hugh Langler
Bojalil, Luis Felipe
Bolhuis, Reinder L H
Borlaug, Norman Ernest
Cerbon-Solorzano, Jorge
Chang, Kenneth Shueh-Shen
Chang, Yu-Sun
Dianzani, Ferdinando
Doerfler, Walter Hans
Dutta-Roy, Asim Kanti
Dyrberg, Thomas Peter
Eggers, Hans J
Feldman, Jose M
Franklin, Richard Morris
Gelzer, Justus
Gjessing, Helen Witton
Graber, Robert Philip
Greenblatt, Charles Leonard
Greene, Velvl William
Gresser, Ion
Ha, Tai-You
Hiyama, Tetsuo
Hoadley, Alfred Warner
Hofmann, Bo
Hosokawa, Keiichi
Huang, Kun-Yen
Kaffezakis, John George
Kessel, Rosslyn William Ian
Kimoto, Masao
Koerber, Walter Ludwig
Kourany, Miguel
Krueger, Gerhard R F
Loppnow, Harald
Lwoff, Andre Michel
Marikovsky, Yehuda
Maruyama, Koshi
Mitruka, Brij Mohan
Mizuno, Shigeki
Mou, Duen-Gang
Nakagawa, Shizutoshi
Ou, Jonathan Tsien-hsiong
Pan, In-Chang
Qadri, Syed M Hussain
Ray, Prasanta K
Roy, Raman K
Scheid, Stephan Andreas
Shapiro, Stuart
Shubert, L Elliot
Sinha, Asru Kumar
Spencer, John Francis Theodore
Spratt, Brian Geoffrey
Stingl, Georg
Stojanovic, Borislav Jovan
Stoppani, Andres Oscar Manuel
Subramanian, Alap Raman
Sykes, Richard Brook
Taniyama, Tadayoshi
Thormar, Halldor
Weil, Marvin Lee
Weiss, David Walter
Yoshida, Takeshi

Molecular Biology

ALABAMA
Anderson, Peter Glennie
Aull, John Louis
Ball, Laurence Andrew
Bradley, James T
Chapatwala, Kirit D
Christian, Samuel Terry
Cook, William Joseph
Daniell, Henry
Dunham, Rex Alan
Egan, Marianne Louise
Fox, Sidney Walter
Furuto, Donald K
Gay, Steffen
Giambrone, Joseph James
Gillespie, George Yancey
Hajduk, Stephen Louis
Hardman, John Kemper
Harvey, Stephen Craig
Higgins, N Patrick
Horabin, Jamila Iddi
Hunter, Eric
Jenkins, Ronald Lee
Kilpatrick, John Michael
Lebowitz, Jacob
Lidin, Bodil Inger Maria
Lincoln, Thomas M
Nielsen, Brent Lynn
Oparil, Suzanne
Philips, Joseph Bond, III
Prakash, Channapatna Sundar
Roozen, Kenneth James
Sani, Brahma Porinchu
Siegal, Gene Philip
Stephenson, Edwin Clark
Umeda, Patrick Kaichi
Urry, Dan Wesley
Winters, Alvin L
Wood, David Oliver

ARIZONA
Bernstein, Carol
Bernstein, Harris
Birge, Edward Asahel
Bourque, Don Philippe
Burgess, Kathryn Hoy
Chandler, Douglas Edwin
Doane, Winifred Walsh
Gendler, Sandra J
Goldstein, Elliott Stuart
Goll, Darrel Eugene
Guerriero, Vincent, Jr
Hagedorn, Henry Howard
Hall, Jennifer Dean
Harkins, Kristi R
Harris, David Thomas
Hewlett, Martinez Joseph
Hoober, J Kenneth
Kay, Marguerite M B
Krahl, Maurice Edward
Lamunyon, Craig Willis
Little, John Wesley
Mendelson, Neil Harland
Mount, David William Alexander
Nagle, Ray Burdell
Pommerville, Jeffrey Carl
Rose, Seth David
Van Voorhies, Wayne Alan
Vermaas, Willem F J
Vincent, Walter Sampson
Ward, Samuel
Wright, Daniel Craig
Wu, Chuanyue
Yohem, Karin Hummell

ARKANSAS
Cave, Mac Donald
Cornett, Lawrence Eugene
Heflich, Robert Henry
Huang, Feng Hou
Kaplan, Arnold
Leakey, Julian Edwin Arundell
Poirier, Lionel Albert
Winter, Charles Gordon

CALIFORNIA
Abelson, John Norman
Almasan, Alexandru
Alousi, Adawia A
Ames, Giovanna Ferro-Luzzi
Anderson, W French
Anthony, Ronald Lewis
Arnheim, Norman
Arrhenius, Gustaf Olof Svante
Ashkenazi, Avi
Attardi, Giuseppe M
Babcock, Gary G
Bailey-Serres, Julia
Baker, Bruce S
Baker, Robert Frank
Balch, William E
Baluda, Marcel A
Bandman, Everett
Baxter, John Darling
Beachy, Roger Neil
Beckendorf, Steven K
Bekhor, Isaac
Bennett, C Frank
Benzer, Seymour
Berk, Arnold J
Bernstein, Sanford Irwin
Bhattacharya, Prabir
Bissell, Dwight Montgomery
Bjorkman, Pamela J
Blackburn, Elizabeth Helen
Bokoch, Gary M
Bonner, James (Fredrick)
Bower, Annette
Boyd, James Brown
Boyer, Paul Delos
Bradshaw, Ralph Alden
Bramhall, John Shepherd
Branscomb, Elbert Warren

Brasch, Klaus Rainer
Bray, Elizabeth Ann
Breidenbach, Rowland William
Briggs, Winslow Russell
Britten, Roy John
Bruening, George
Brunke, Karen J
Brutlag, Douglas Lee
Burgess, Teresa Lynn
Calendar, Richard
Campagnoni, Anthony Thomas
Campbell, Judith Lynn
Cantor, Charles Robert
Carbon, John Anthony
Cardineau, Guy A
Carrano, Anthony Vito
Celniker, Susan Elizabeth
Chamberlin, Michael John
Chan, Sham-Yuen
Chang, Jennie C C
Choudary, Prabhakara Velagapudi
Chow, Samson Ah-Fu
Clark, Alvin John
Cline, Thomas Warren
Coffino, Philip
Cohen, Edward Hirsch
Cohen, Larry William
Conner, Brenda Jean
Cooper, James Burgess
Coyer, James A
Craik, Charles S
Crisp, Carl Eugene
Cronin, Michael John
Crooke, Stanley T
Curnutte, John Tolliver, III
Curry, Donald Lawrence
Dahms, Arthur Stephen
Daniell, Ellen
Davidson, Eric Harris
Davidson, Norman Ralph
Davies, Huw M
Davis, Craig H
Davis, Ronald Wayne
Davis, Rowland Hallowell
De Francesco, Laura
DeLeon, Daisy Delgado
Delong, Edward F
Dennert, Gunther
Deonier, Richard Charles
Dickerson, Richard Earl
Dietz, George William, Jr
Dietz, Thomas John
Doi, Roy Hiroshi
Donoghue, Daniel James
Downing, Michael Richard
Dragon, Elizabeth Alice Oosterom
Dreyer, William J
Duesberg, Peter H
Duester, Gregg L
Duffey, Paul Stephen
Dugaiczyk, Achilles
Dunaway, Marietta
Dunnebacke-Dixon, Thelma Hudson
Echols, Harrison
Ecker, David John
Eckhart, Walter
Edelman, Gerald Maurice
Edelman, Jay Barry
Eiserling, Frederick A
Emr, Scott David
Eng, Lawrence F
Englesberg, Ellis
Epstein, Charles Joseph
Erickson, Jeanne Marie
Ernest, Michael Jeffrey
Esposito, Michael Salvatore
Etcheverry, Tina
Ewig, Carl Stephen
Felgner, Philip Louis
Felton, James Steven
Fessler, John Hans
Fleming, James Emmitt
Fraenkel-Conrat, Heinz Ludwig
Friedmann, Theodore
Fryxell, Karl Joseph
Fukuda, Michiko N
Fukuda, Minoru
Fulco, Armand J
Funk, Glenn Albert
Gaertner, Frank Herbert
Game, John Charles
Ganesan, Adayapalam T
Ganesan, Ann K
Gascoigne, Nicholas Robert John
Gasser, Charles Scott
Gautsch, James Willard
Gehlsen, Kurt Ronald
Geiduschek, Ernest Peter
Gelfand, David H
Gerace, Larry R
Gibson, Thomas Richard
Gilbertson, Robert Leonard
Gilula, Norton Bernard
Ginsberg, Theodore
Glaser, Donald Arthur
Glazer, Alexander Namiot
Gober, James William
Goff, Lynda J
Golde, David William
Gong, Yu
Gorman, Cornelia M
Gray, Gary M
Gray, Joe William
Grody, Wayne William

Grunbaum, Benjamin Wolf
Grunstein, Michael
Gum, James Raymond, Jr
Gunsalus, Robert Philip
Guthrie, Christine
Gutman, George Andre
Hackney, Robert Ward
Hall, Raymond G, Jr
Hamkalo, Barbara Ann
Han, Jang Hyun
Hanawalt, Philip Courtland
Haney, David N
Hankinson, Oliver
Hanson, Carl Veith
Harford, Joe Bryan
Hathaway, Gary Michael
Haygood, Margo Genevieve
Heffernan, Laurel Grace
Henriksson, Thomas Martin
Henry, Helen L
Herschman, Harvey R
Hirsch, Ann Mary
Hoch, James Alfred
Hoch, Paul Edwin
Hoch, Sallie O'Neil
Hoke, Glenn Dale
Hoopes, Laura Livingston Mays
Hosoda, Junko
Howard, Bruce David
Howard, Russell John
Huang, Anthony Hwoon Chung
Hunter, Tony
Hurd, Ralph Eugene
Hurkman, William James
Hyman, Bradley Clark
Itakura, Keiichi
Jacob, Chaim O
Jacob, Mary
Jacobson, Ralph Allen
Jadus, Martin Robert
Janda, John Michael
Jardetzky, Oleg
Jariwalla, Raxit Jayantilal
Jarnagin, Kurt
Jeffery, William Richard
Johansson, Mats W
Johnson, Paul Hickok
Jones, Kenneth Charles
Kado, Clarence Isao
Kan, Yuet Wai
Kane, Susan Elizabeth
Kang, Tae Wha
Kan-Mitchell, June
Kasamatsu, Harumi
Katz, Louis
Kedes, Laurence H
Kelley, Darshan Singh
Kelly, Regis Baker
Kennedy, Mary Bernadette
Kim, Sung-Hou
Kirschbaum, Joel Bruce
Klevecz, Robert Raymond
Knudson, Gregory Blair
Kopito, Ron Rieger
Korn, David
Koshland, Daniel Edward, Jr
Koski, Raymond Allen
Koths, Kirston Edward
Kozloff, Lloyd M
Kudo, Shinichi
Lai, Michael Ming-Chiao
Lake, James Albert
Lambert, Charles Calvin
Landolfi, Nicholas F
Landolph, Joseph Richard, Jr
Langridge, Robert
Larrick, James William
Lasky, Lawrence Alan
Last, Jerold Alan
Lebo, Roger Van
Lee, Amy Shiu
Leffert, Hyam Lerner
Lei, Shau-Ping Laura
Leighton, Terrance J
Lengyel, Judith Ann
Leung, David Wai-Hung
Levine, Michael Steven
Levinson, Arthur David
Lewis, Susanna Maxwell
Lieb, Margaret
Lindh, Allan Goddard
Lipshitz, Howard David
Lipsick, Joseph Steven
Liu, Xuan
Long, Sharon Rugel
Lowenstein, Jerold Marvin
Lundblad, Roger Lauren
Lusis, Aldons Jekabs
McClelland, Michael
McCullough, Richard Donald
McEwen, Joan Elizabeth
MacInnis, Austin J
Macklin, Wendy Blair
Mackman, Nigel
MacLeod, Carol Louise
Mangan, Jerrome
Mansour, Tag Eldin
Marshall, Charles Richard
Martin, David William, Jr
Martin, George Steven
Martineau, Belinda M
Martins-Green, Manuela M
Martinson, Harold Gerhard
Matthews, Beverly Bond

Matthews, Harry Roy
Maxson, Robert E, Jr
Meehan, Thomas (Dennis)
Mehra, Rajesh Kumar
Melis, Anastasios
Mellon, Synthia
Merchant, Sabeeha
Mestril, Ruben
Meyerowitz, Elliot Martin
Miledi, Ricardo
Miller, Arnold Lawrence
Mitsuhashi, Masato
Molinari, Robert James
Montesano-Roditis, Luisa
Morse, Daniel E
Mosteller, Raymond Dee
Nagel, Glenn M
Napolitano, Leonard Michael, Jr
Neufeld, Berney Roy
Nguyen, Binh Trong
Nierlich, Donald P
Nomura, Masayasu
Norman, Gary L
Ogunseitan, Oladele Abiola
Oh, Chan Soo
Olson, Betty H
Ordahl, Charles Philip
Orgel, Leslie E
Ow, David Wing
Pallavicini, Maria Georgina
Pan, Richard Y
Pauling, Edward Crellin
Pellegrini, Maria C
Perrault, Jacques
Peterlin, Boris Matija
Philpott, Delbert E
Phinney, Bernard Orrin
Phleger, Charles Frederick
Pilgeram, Laurence Oscar
Pischel, Ken Donald
Polit, Andres Cassard
Quail, Peter Hugh
Rabussay, Dietmar Paul
Raj, Harkisan D
Ray, Dan S
Reeve, Peter
Reitz, Richard Elmer
Reizer, Jonathan
Rettenmier, Carl Wayne
Rho, Joon H
Rice, Robert Hafling
Riggs, Arthur Dale
Rio, Donald C
Robinson, William Sidney
Rome, Leonard H
Rosen, Howard
Rosenfeld, Ron Gershon
Rosenquist, Grace Link
Rosenthal, Allan Lawrence
Roy-Burman, Pradip
Rubin, Carol Marie
Rubin, Gerald M
Ryder, Oliver A
Sadee, Wolfgang
Saifer, Mark Gary Pierce
Salser, Winston Albert
Samuel, Charles Edward
Saykally, Richard James
Schachman, Howard Kapnek
Scheffler, Immo Erich
Schekman, Randy W
Scott, Matthew P
Sedat, John William
Seegmiller, Jarvis Edwin
Senda, Mototaka
Shapiro, Lucille
Shen, Che-Kun James
Shih, Jean Chen
Shopes, Bob
Short, Jay M
Simons, Robert W
Simpson, Robert Blake
Singer, B
Sinibaldi, Ralph Michael
Sinsheimer, Robert Louis
Sjostrand, Fritiof S
Smith, Charles Allen
Smith, Douglas Wemp
Smith, Helene Sheila
Smith, Steven Sidney
Somerville, Christopher Roland
Song, Moon K
Spray, Clive Robert
Stallcup, Michael R
Stanley, Wendell Meredith, Jr
Stellwagen, Robert Harwood
Stent, Gunther Siegmund
Stephens, Robert James
Stevens, Lewis Axtell
Strauss, James Henry
Stringfellow, Dale Alan
Strobel, Edward
Stryer, Lubert
Stubbs, John Dorton
Stumph, William Edward
Subramani, Suresh
Summers, Donald F
Sung, Lanping Amy
Sung, Zinmay Renee
Sussman, Howard H
Sutherland, Robert Melvin
Sy, Jose
Syvanen, Michael
Taylor, Barry L

Thompson, Lawrence Hadley
Thorner, Jeremy William
Tisserat, Brent Howard
Tjian, Robert Tse Nan
Tobin, Allan Joshua
Traugh, Jolinda Ann
Traut, Robert Rush
Triche, Timothy J
Tseng, Ben Y
Tukey, Robert H
Urry, Lisa Andrea
Varshavsky, Alexander Jacob
Vedvick, Thomas Scott
Verheyden, Julien P H
Vickery, Larry Edward
Villarreal, Luis Perez
Vlasuk, George P
Vogt, Peter Klaus
Volcani, Benjamin Elazari
Wade, Michael James
Wall, Thomas Randolph
Wallace, Robert Bruce
Warner, Robert Collett
Warshel, Arieh
Wasterlain, Claude Guy
Waterman, Michael S
Webb, David Ritchie, Jr
Weber, Heather R(oss) Wilson
Weinberg, Barbara Lee Huberman
Weinstein, David E
Wettstein, Felix O
Wiebe, Michael Eugene
Wiktorowicz, John Edward
Wilcox, Gary Lynn
Williams, Mark Alan
Williams, Robley Cook
Willson, Clyde D
Wong, Chi-Huey
Wood, William Irwin
Woodward, Dow Owen
Wu, Sing-Yung
Yamamoto, Keith Robert
Yamamoto, Richard
Yanofsky, Charles
Yen, Tien-Sze Benedict
Yoshida, Akira
Yu, Grace Wei-Chi Hu
Yund, Mary Alice
Zabin, Irving
Zaidi, Iqbal Mehdi
Zarucki, Tanya Z
Zeichner-David, Margarita
Zernik, Joseph
Zieg, Roger Grant
Zlotnik, Albert
Zuccarelli, Anthony Joseph
Zuckerkandl, Emile
Zuker, Charles S

COLORADO
Andrews, Ken J
Balbinder, Elias
Bamburg, James Robert
Barrett, Dennis
Blair, Carol Dean
Brown, Lewis Marvin
Cech, Thomas Robert
Clark, Roger William
Danna, Kathleen Janet
DeLa Cruz, Vidal F, Jr
Dunahay, Terri Goodman
Dynan, William Shelley
Gardiner, Katheleen Jane
Gerschenson, Lazaro E
Giclas, Patricia C
Glode, Leonard Michael
Hahn, William Eugene
Hogan, Christopher James
Horwitz, Kathryn Bloch
Hossner, Kim L
Ishii, Douglas Nobuo
Jaehning, Judith A
Johnson, Gary L
Johnson, Thomas Eugene
Jones, Steven Wayne
Kano-Sueoka, Tamiko
Lapitan, Nora L
Lowndes, Joseph M
McHenry, Charles S
Maller, James Leighton
Markert, Clement Lawrence
Mattoon, James Richard
Melancon, Paul R
Morse, Helvise Glessner
Nasci, Roger Stanley
Nett, T M
Panini, Sankhavaram R
Patterson, David
Pettijohn, David E
Poyton, Robert Oliver
Prescott, David Marshall
Ranu, Rajinder S
Rash, John Edward
Roberts, P Elaine
Simon, Nancy Jane
Sinensky, Michael
Sneider, Thomas W
Stein, Gretchen Herpel
Stushnoff, Cecil
Tabakoff, Boris
Taylor, Austin Laurence
Trent, Dennis W
Urban, Richard William
Vasil, Michael Lawrence

Molecular Biology (cont)

Waldren, Charles Allen
Wood, William Barry, III
Yarus, Michael J
Zeiler, Kathryn Gail

CONNECTICUT

Aaslestad, Halvor Gunerius
Altman, Sidney
Amacher, David E
Bacopoulos, Nicholas G
Badoyannis, Helen Litman
Bayney, Richard Michael
Berg, Claire M
Bormann, Barbara-Jean Anne
Bothwell, Alfred Lester Meador
Brandsma, Janet Lousie
Braun, Phyllis C
Buck, Marion Gilmour
Cheng, Yung-Chi
Das, Rathindra C
Deutscher, Murray Paul
Dingman, Douglas Wayne
Doeg, Kenneth Albert
Dorsky, David Isaac
Eisenstadt, Jerome Melvin
Epstein, Paul Mark
Fenton, Wayne Alexander
Flavell, Richard Anthony
Forget, Bernard G
Froshauer, Susan
Fu, Wei-ning
Garen, Alan
Glasel, Jay Arthur
Golub, Efim F
Grindley, Nigel David Forster
Harewood, Ken Rupert
Heywood, Stuart Mackenzie
Hightower, Lawrence Edward
Hobart, Peter Merrill
Hutchinson, Franklin
Infante, Anthony A
Kidd, Kenneth Kay
Kiron, Ravi
Knox, James Russell, Jr
Kotick, Michael Paul
Lentz, Thomas Lawrence
Les, Donald Henry
Lewis, Jonathan Joseph
Liskay, Robert Michael
Low, Kenneth Brooks, Jr
Lucas-Lenard, Jean Marian
McCorkle, George Maston
Merwin, June Rae
Morrow, Jon S
Ortner, Mary Joanne
Osborn, Mary Jane
Parker, Eric McFee
Pfeiffer, Steven Eugene
Primakoff, Paul
Proctor, Alan Ray
Rae, Peter Murdoch MacPhail
Rosenbaum, Joel L
Rossomando, Edward Frederick
Ruddle, Francis Hugh
Rupp, W Dean
Russu, Irina Maria
Sarfarazi, Mansoor
Schepartz, Alanna
Schwartz, Ilsa Roslow
Setlow, Peter
Siegel, Norman Joseph
Smilowitz, Henry Martin
Söll, Dieter Gerhard
Squinto, Stephen P
Steitz, Joan Argetsinger
Steitz, Thomas Arthur
Summers, William Cofield
Summers, Wilma Poos
Tallman, John Francis
Upholt, William Boyce
Verses, Christ James
Vitkauskas, Grace
Walker, Frederick John
Weissman, Sherman Morton
Wilson, John Thomas
Wyman, Robert J
Xu, Zhi Xin
Zahler, Raphael
Zelitch, Israel

DELAWARE

Davis, Leonard George
Diner, Bruce Aaron
Enquist, Lynn William
Fahnestock, Stephen Richard
Freerksen, Deborah Lynne (Chalmers)
Gatenby, Anthony Arthur
Giaquinta, Robert T
Gray, John Edward
Hartzell, Charles Ross, III
Hobgood, Richard Troy, Jr
Hodson, Robert Cleaves
Holsten, Richard David
Hoover, Dallas Gene
Jackson, David Archer
Keeler, Calvin Lee, Jr
Kinney, Anthony John
Marrs, Barry Lee
Neubauer, Russell Howard

DISTRICT OF COLUMBIA

Alberts, Bruce Michael
Barker, Winona Clinton
Beru, Nega
Chapman, George Bunker
Chen, H R
Doctor, Bhupendra P
Fanning, George Richard
Garavelli, John Stephen
Gelmann, Edward P
Golin, John Euster
Hawkins, Morris, Jr
Henderson, Ellen Jane
Hill, Walter Ernest
Hollingdale, Michael Richard
Holmes, George Edward
Hone, Jennifer
Jacobson, Michael F
Jett-Tilton, Marti
Kellar, Kenneth Jon
Kornhauser, Andrija
Kumar, Ajit
Lai, David Ying-lun
Miller, Linda Jean
Ohi, Seigo
Phillips, Terence Martyn
Rothman, Sara Weinstein
Schaeff, Catherine Margaret
Setlow, Valerie Petit
Singer, Maxine Frank
Smith, David Allen
Thiermann, Alejandro Bories
Todhunter, John Anthony
Venkatesan, Malabi M
Wolfe, Alan David
Zimmer, Elizabeth Anne

FLORIDA

Allred, David R
Anderson, William McDowell
Assoian, Richard Kenneth
Balducci, Lodovico
Barber, Michael James
Baumbach, Lisa Lorraine
Binninger, David Michael
Bray, Joan Lynne
Cain, Brian D
Chase, Christine Davis
Cooper, Denise R
Cousins, Robert John
Davis, Francis Clarke
Denslow, Nancy Derrick
Dewanjee, Mrinal K
Dhople, Arvind Madhav
Eilers, Frederick Irving
Fenna, Roger Edward
Ferl, Robert Joseph
Fresquez, Catalina Lourdes
Hamilton, Franklin D
Hauswirth, William Walter
Hiebert, Ernest
Horn, Joanne Marie
Hoy, Marjorie Ann
Inana, George
Ingram, Lonnie O'Neal
Johnson, Irving Stanley
Kaul, Rajinder K
Laipis, Philip James
Lee, Ernest Y
Lee, Marietta Y W T
Lewin, Alfred S
Lim, Daniel V
McCormack, Wayne Thomas
McKently, Alexandra H
McKinney, Michael
McLean, Mark Philip
Mahan, Suman Mulakhraj
Makemson, John Christopher
Mans, Rusty Jay
Miles, Richard David
Miyamoto, Michael Masao
Moyer, Richard W
Murison, Gerald Leonard
Neary, Joseph Thomas
Nonoyama, Meihan
O'Brien, Thomas W
Patarca, Roberto
Plescia, Otto John
Polson, Charles David Allen
Previc, Edward Paul
Purich, Daniel Lee
Rangel Aldao, Rafael
Roess, William B
Shanmugam, Keelnatham Thirunavukkarasu
Shirk, Paul David
Standaert, Mary L
Stock, David Allen
Thorn, Richard Mark
Tumer, Nihal
Webster, George Calvin
Weissbach, Arthur
Wilson, David Louis
Wodzinski, Rudy Joseph

GEORGIA

Adams, Robert Johnson
Adkison, Linda Russell
Albersheim, Peter
Alonso, Kenneth B
Austin, Garth E
Baumstark, Barbara Ruth
Bowen, John Metcalf
Brann, Darrell Wayne
Brinton, Margo A
Clark, Winston Craig
Crouse, Gray F
Danner, Dean Jay
Doetsch, Paul William
Evans, Donald Lee
Evatt, Bruce Lee
Fong, Peter
Fridovich-Keil, Judith Lisa
Garver, Frederick Albert
Gerschenson, Mariana
Glover, Claiborne V C, III
Greenberg, Jerrold
Hall, Carole L
Hall, Dwight Hubert
Halper, Jaroslava
Jinks-Robertson, Sue
Jones, George Henry
Key, Joe Lynn
Khan, Iqbal M
Koerner, T J
Lehner, Andreas Friedrich
Mason, James Michael
Miller, Lois Kathryn
Noe, Bryan Dale
Pandey, Kailash N
Patel, Gordhan L
Peck, Harry Dowd, Jr
Pratt, Lee Herbert
Robinson, Margaret Chisolm
Schmidt, Gregory Wayne
Scott, June Rothman
Sinor, Lyle Tolbot
Suddath, Fred LeRoy, (Jr)
Thedford, Roosevelt
Tsang, Victor Chiu Wan
Wallace, Douglas Cecil
Warren, Stephen Theodore
Wartell, Roger Martin
Wessler, Susan R
Whitney, J(ohn) Barry, III
Willis, Judith Horwitz
Wilson, Donella Joyce
Youngman, Philip John

HAWAII

Albert, Henrick Horst
Borthakur, Dulal
Fitch, Maureen Meiko Masuda
Gibbons, Barbara Hollingworth
Gibbons, Ian Read
Hunter, Cynthia L
Kathariou, Sophia
Keeley, Sterling Carter
Kleinfeld, Ruth Grafman
Mandel, Morton
Moore, Paul Harris
Noll, Hans
Roberts, Robert Russell
Rosenthal, Eric Thomas
Sakai, William Shigeru
Scott, John Francis
Stuart, William Dorsey
Sun, Samuel Sai-Ming
Ward, Melvin A
Wood, Betty J

IDAHO

Ayers, Arthur Raymond
Elmore, John Jesse, Jr
Farrell, Larry Don
Keay, Leonard
McCune, Mary Joan Huxley
Oliver, David John
Roberto, Francisco Figueroa
Stephens, Trent Dee
Watson, Kenneth Fredrick

ILLINOIS

Ahmad, Sarfraz
Alexander, Nancy J
Amero, Sally Ann
Applebury, Meredithe L
Aprison, Barry Steven
Ausich, Rodney L
Ayres, Kathleen N
Bahn, Arthur Nathaniel
Becker, Michael Allen
Bhatti, Rashid
Blair, Louis Curtis
Blaschek, Hans P
Buetow, Dennis Edward
Burns, David John
Casadaban, Malcolm John
Chang, Chong-Hwan
Chassy, Bruce Matthew
Cheung, Hou Tak
Chiang, Kwen-Sheng
Constantinou, Andreas I
Cronan, John Emerson, Jr
Cummings, Michael R
DeFilippi, Louis J
Doering, Jeffrey Louis
Dumas, Lawrence Bernard
Engel, James Douglas
Epstein, Wolfgang
Farrand, Stephen Kendall
Ferguson, Edwin Louis
Fisher, Richard Gary
Fosslien, Egil
Fox, Jack Lawrence
Frank, James Richard
Frenkel, Niza B
Fuchs, Elaine V
Gaballah, Saeed S
Gaskins, H Rex
Gerding, Dale Nicholas
Giere, Frederic Arthur
Giordano, Tony
Glaser, Janet H
Glaser, Michael
Goldberg, Erwin
Goldman, Manuel
Gumport, Richard I
Gupta, Kailash Chandra
Hagar, Lowell Paul
Hahn, Donald Richard
Hampel, Arnold E
Haselkorn, Robert
Hauptmann, Randal Mark
Hileman, Stanley Michael
Hirsch, Lawrence Leonard
Holland, Louis Edward, II
Holzer, Timothy J
Horwitz, Alan Fredrick
Hsu, Wen-Tah
Jayaswal, Radheshyam K
Kaizer, Herbert
Kemper, Byron W
Khodadad, Jena Khadem
Kornel, Ludwig
Krafft, Grant A
Kramer, James M
Kurtzman, Cletus Paul
Laffler, Thomas G
Lamb, Robert Andrew
Lambert, Mary Pulliam
Larson, Bruce Linder
Lazda, Velta Abuls
Liao, Shutsung
Lipka, James J
Lippincott, Barbara Barnes
Lopes, John Manuel
Lucher, Lynne Annette
McCorquodale, Donald James
McCorquodale, Maureen Marie
McKinley, Vicky L
Maecker, Holden T
Margoliash, Emanuel
Markovitz, Alvin
Martin, Terence Edwin
Menco, Bernard Philip Max
Mets, Laurens Jan
Miernyk, Jan Andrew
Miller, Robert Verne
Miller, Stephen Douglas
Mitchell, John Laurin Amos
Moscona, Aron Arthur
Myers, Ronald Berl
Myers, Walter Loy
Nadler, Steven Anthony
Neet, Kenneth Edward
Nichols, Brian Paul
Oldfield, Eric
Ordal, George Winford
Panasenko, Sharon Muldoon
Papaioannou, Stamatios E
Peak, Meyrick James
Plate, Charles Alfred
Plate, Janet Margaret
Plewa, Michael Jacob
Pratt, Charles Walter
Propst, Catherine Lamb
Pun, Pattle Pak-Toe
Radosevich, James A
Rafelson, Max Emanuel, Jr
Rao, Mrinalini Chatta
Ratajczak, Helen Vosskuhler
Reichmann, Manfred Eliezer
Rice, Thomas B
Ries, Herman Elkan, Jr
Ritter, Nadine Marie
Robertson, Abel Alfred Lazzarini, Jr
Robertson, Hugh Mereth
Robinson, James Lawrence
Rogalski-Wilk, Adrienne Alice
Roizman, Bernard
Rothman-Denes, Lucia B
Rowley, Janet D
Rupprecht, Kevin Robert
Sachs, Martin M
Scherberg, Neal Harvey
Schnell, Gene Wheeler
Schultz, Richard Michael
Shen, Linus Liang-nene
Silver, Simon David
Sligar, Stephen Gary
Smith, David Waldo Edward
Spear, Brian Blackburn
Spear, Patricia Gail
Stark, Benjamin Chapman
Steiner, Donald Frederick
Stevens, Fred Jay
Storti, Robert V
Strauss, Bernard S
Strickland, James Arthur
Swinnen, Lode J
Switzer, Robert Lee
Takahashi, Joseph S
Tao, Mariano
Thompson, Kenneth David
Tiemeier, David Charles
Toback, F(rederick) Gary
Towle, David Walter
Tripathi, Satish Chandra
Tripathy, Deoki Nandan
Vary, Patricia Susan
Velardo, Joseph Thomas

Verhage, Harold Glenn
Vodkin, Lila Ott
Wang, Gary T
Wass, John Alfred
Wasserman, William John
Welker, Neil Ernest
Westbrook, Edwin Monroe
Weyhenmeyer, James Alan
Wezeman, Frederick H
Whitfield, Harvey James, Jr
Wiatr, Christopher Louis
Womble, David Dale
Wool, Ira Goodwin
Wu, Tai Te
Wu-Wong, Jinshyun Ruth
Yamamoto, Hirotaka
Yarger, James G
Yokoyama, Shozo
Yu, Fu-Li
Zimmerman, Roger Paul

INDIANA
Anderson, John Nicholas
Anderson-Mauser, Linda Marie
Ashendel, Curtis Lloyd
Axelrod, Bernard
Baird, William McKenzie
Basu, Subhash Chandra
Bauer, Carl Eugene
Belagaje, Rama M
Bennetzen, Jeffrey Lynn
Bina, Minou
Blumenthal, Thomas
Bonner, James Jose
Botero, J M
Boyer, Ernest Wendell
Cherbas, Peter Thomas
Cooper, Robin D G
Crouch, Martha Louise
Denner, Melvin Walter
Edenberg, Howard Joseph
Eisenstein, Barry I
Fayerman, Jeffrey T
Ferris, Virginia Rogers
Foley, Michael Edward
Gale, Charles
Gaur, Pramod Kumar
Gehlert, Donald Richard
Grant, Alan Leslie
Gray, Peter Norman
Guthrie, George Drake
Hancock, Deana Lori
Harvey, William Homer
Herrmann, Klaus Manfred
Hershberger, Charles Lee
Ho, Nancy Wang-Yang
Hodes, Marion Edward
Howard, David K
Hsiung, Hansen M
Huber, Paul William
Hyde, David Russell
Johnson, Alan L
Kapustka, Lawrence A
Konieczny, Stephen Francis
Larsen, Steven H
Lavender, John Francis
Lees, Norman Douglas
Levinthal, Mark
Lister, Richard Malcolm
Loesch-Fries, Loretta Sue
Mocharla, Raman
Nielsen, Niels Christian
Pak, William Louis
Pepper, Daniel Allen
Pietsch, Paul Andrew
Polley, Lowell David
Preer, John Randolph, Jr
Roman, Ann
Rossmann, Michael G
Santerre, Robert Frank
Sherman, Louis Allen
Slavik, Nelson Sigman
Smith, Gerald Floyd
Somerville, Ronald Lamont
Srivastava, Arun
Stuart, Jeffrey James
Surzycki, Stefan Jan
Termine, John David
Tessman, Irwin
Tischfield, Jay Arnold
Van Etten, Robert Lee
Van Frank, Richard Mark
Vierling, Richard Anthony
Webb, Mary Alice
Wildfeuer, Marvin Emanuel
Woodson, William Randolph
Zimmerman, Sarah E

IOWA
Ascoli, Mario
Atherly, Alan G
Benbow, Robert Michael
Conn, P Michael
Donelson, John Everett
Fitzgerald, Gerald Roland
Ford, Clark Fugier
Ford, Stephen Paul
Goodridge, Alan G
Gussin, Gary Nathaniel
Horowitz, Jack
Makar, Adeeb Bassili
Malone, Robert Edward
Mayfield, John Eric
Mengeling, William Lloyd

Menninger, John Robert
Milkman, Roger Dawson
Perlman, Stanley
Rodermel, Steven Robert
Schmerr, Mary Jo F
Schnable, Patrick S
Stanton, Thaddeus Brian
Sullivan, Charles Henry
Tomes, Dwight Travis
Trenkle, Allen H
Wendel, Jonathan F

KANSAS
Ash, Ronald Joseph
Besharse, Joseph Culp
Bradford, Lawrence Glenn
Calvet, James P
Chitnis, Parag Ram
Davis, Lawrence Clark
Decedue, Charles Joseph
Denell, Robin Ernest
Funderburgh, James Louis
Gegenheimer, Peter Albert
Goldberg, Ivan D
Grisolia, Santiago
Hedgcoth, Charlie, Jr
Iandolo, John Joseph
Johnson, Lowell Boyden
Johnson, Terry Charles
Leavitt, Wendell William
Leslie, John Franklin
Li, Yi
McVey, David Scott
Minocha, Harish C
Muthukrishnan, Subbaratnam
Padmanabhan, Radhakrishnan
Parkinson, Andrew
Rawitch, Allen Barry
Roufa, Donald Jay
Samson, Frederick Eugene, Jr
Schloss, John Vinton
Singhal, Ram P
Smith, Larry Dean
Stetler, Dean Allen
Takemoto, Dolores Jean
Weaver, Robert F
Wong, Peter P
Yarbrough, Lynwood R
Zimniski, Stephen Joseph

KENTUCKY
Bondada, Subbarao
Davidson, Jeffrey Neal
Dickson, Robert Carl
Feese, Bennie Taylor
Geoghegan, Thomas Edward
Jacobson, Elaine Louise
Lesnaw, Judith Alice
Maiti, Indu B
Martin, Nancy Caroline
Mattson, Mark Paul
Perlin, Michael Howard
Pirone, Thomas Pascal
Rao, Chalamalasetty Venkateswara
Rawls, John Marvin, Jr
Sisken, Jesse Ernest
Slagel, Donald E
Straley, Susan Calhoon
Toman, Frank R
Westneat, David French, Jr
Williams, Arthur Lee
Zimmer, Stephen George

LOUISIANA
Alam, Jawed
Bhatnagar, Deepak
Bhattacharyya, Ashim Kumar
Cardelli, James Allen
Clejan, Sandra
Cohen, J Craig
Ehrlich, Kenneth Craig
Ehrlich, Melanie
Garcia, Meredith Mason
Grimes, Sidney Ray, Jr
Haycock, John Winthrop
Ivens, Mary Sue
Jazwinski, S Michal
Leng, Wai-Choi
Moore, Thomas Stephen, Jr
O'Callaghan, Richard J
Olmsted, Clinton Albert
Rothschild, Henry
Sarphie, Theodore G
Scher, Charles D
Shepherd, Hurley Sidney
Toscano, William Agostino, Jr
Triplett, Barbara Ann
Ullah, Abul Hasnat
Vedeckis, Wayne V
Walters, Marian R
Warren, Lionel Gustave
Weidner, Earl
Wiser, Mark Frederick
Wright, Maureen Smith

MAINE
Andersen, Robert Arthur
Aronson, Frederick Rupp
McKerns, Kenneth (Wilshire)
Renn, Donald Walter
Roxby, Robert
Steinhart, William Lee

MARYLAND
Ackerman, Eric J
Adams, Junius Greene, III
Adelstein, Robert Simon
Ades, Ibrahim Z
Adhya, Sankar L
Allnutt, F C Thomas
Anderson, Norman Leigh
August, Joseph Thomas
Austin, Faye Carol Gould
Avigan, Joel
Azad, Abdu F(arhang)
Baker, George Thomas, III
Bala, Shukal
Baron, Louis Sol
Basta, Milan
Beemon, Karen Louise
Beer, Michael
Bellino, Francis Leonard
Benson, Spencer Alan
Berger, Shelby Louise
Berkower, Ira
Bethke, Bruce Donald
Bhatnagar, Gopal Mohan
Bhorjee, Jaswant S
Bigger, Cynthia Anita Hopwood
Bird, Robert Earl
Bloch, Robert Joseph
Bockstahler, Larry Earl
Bonner, Tom Ivan
Boyd, Virginia Ann Lewis
Broomfield, Clarence A
Brown, Alexandra Louise
Brown, James Edward
Buckheit, Robert Walter, Jr
Burt, David Reed
Bustin, Michael
Camerini-Otero, Rafael Daniel
Cashel, Michael
Cerveny, Thelma Jannette
Chan, Wai-Yee
Chang, Henry
Charnas, Lawrence Richard
Cheng, Tu-chen
Cho-Chung, Yoon Sang
Choppin, Purnell Whittington
Cleveland, Don W
Cohen, Gerson H
Cohen, Jack Sidney
Colburn, Nancy Hall
Coleman, William Gilmore, Jr
Colombini, Marco
Cowan, Elliot Paul
Cox, George Warren
Craig, Nessly Coile
Crouch, Robert J
Cutler, Richard Gail
Davies, David R
Dean, Ann
Dean, Donna Joyce
Dean, Jurrien
Deitzer, Gerald Francis
Devreotes, Peter Nicholas
Dintzis, Renee Zlochover
Doniger, Jay
Dufau, Maria Luisa
Dwivedi, Ram Shyam
Eckels, Kenneth Henry
Eichhorn, Gunther Louis
Eipper, Betty Anne
Eisenstadt, Edward
Elespuru, Rosalie K
Elson, Hannah Friedman
Ennist, David L
Ewing, June Swift
Fedoroff, Nina V
Feigal, Ellen G
Felsenfeld, Gary
Fields, Kay Louise
Finkelstein, David B
Fire, Andrew Zachary
Fornace, Albert J, Jr
Foulds, John Douglas
Franklin, Renty Benjamin
Fujimura, Robert
Gall, Joseph Grafton
Garges, Susan
Gartland, William Joseph
Gelboin, Harry Victor
Gellert, Martin Frank
Gerard, Gary Floyd
Goldberg, Michael Ian
Goldman, David
Goldsmith, Merrill E
Gonda, Matthew Allen
Goode, Melvyn Dennis
Graham, Dale Elliott
Gravell, Maneth
Greenhouse, Gerald Alan
Gruber, Kenneth Allen
Guss, Maurice Louis
Hadidi, Ahmed Fahmy
Hairstone, Marcus A
Hanes, Darcy Elizabeth
Hanna, Edgar E(thelbert), Jr
Hansen, John Norman
Harrington, William Fields
Hatfield, Dolph Lee
Haun, Randy S
Hearing, Vincent Joseph, Jr
Hejtmancik, James Fielding
Henry, Timothy James
Herman, Eliot Mark
Hickey, Robert Joseph

Hinton, Deborah M
Hla, Timothy Tun
Hoeg, Jeffrey Michael
Holland, Mark Alexander
Holloway, Caroline T
Horowitz, Jill Ann
Huang, Ru-Chih Chow
Hudson, Lynn Diane
Huppi, Konrad E
Imberski, Richard Bernard
Impraim, Chaka Cetewayo
Irwin, David
Ivins, Bruce Edwards
Jernigan, Robert Lee
Johnson, Alfred C
Jordan, Craig Alan
Jordan, Elke
Josephs, Steven F
Kaper, Jacobus M
Kaper, James Bennett
Karpel, Richard Leslie
Keister, Donald Lee
Keith, Jerry M
Kelly, Thomas J
Ketley, Jeanne Nelson
Kibbey, Maura Christine
Kimball, Paul Clark
Kimmel, Alan R
Klee, Werner A
Kleinman, Hynda Karen
Kraemer, Kenneth H
Krause, David
Kuff, Edward Louis
Kulkarni, Ashok Balkrishna
Kundig, Fredericka Dodyk
Kung, Shain-Dow
Kunos, George
Landsman, David
Lane, Malcolm Daniel
Larner, Andrew Charles
Lederer, William Jonathan
Lee, Chi-Jen
Lee, Fang-Jen Scott
Lee, Theresa
Leppla, Stephen Howard
Lerman, Michael Isaac
Leto, Thomas L
Levin, Barbara Chernov
Levin, Judith Goldstein
Levine, Arthur Samuel
Liang, Shu-Mei
Lillehoj, Hyun Soon
Linehan, William Marston
Lippincott-Schwartz, Jennifer
Lipsky, Robert H
Liu, Pu
Liu, Teh-Yung
Loh, Yoke Peng
Longfellow, David G(odwin)
Luborsky, Samuel William
Lunney, Joan K
McCandliss, Russell John
McCarron, Richard M
McClure, Michael Edward
McCune, Susan K
McGraw, Patricia Mary
Maciag, Thomas Edward
McKenney, Keith Hollis
McKeon, Catherine
McLachlin, Jeanne Ruth
McMacken, Roger
McNicol, Lore Anne
Malech, Harry Lewis
Maloney, Peter Charles
Manganiello, Vincent Charles
Margolis, Sam Aaron
Margulies, David Harvey
Margulies, Maurice
Marks, Edwin Potter
Martin, Malcolm Alan
Martin, Robert G
Mather, Ian Heywood
Mattern, Michael Ross
Matthews, Benjamin F
Matyas, Marsha Lakes
Max, Edward E
Melera, Peter William
Merlino, Glenn T
Merril, Carl R
Miles, Edith Wilson
Miles, Harry Todd
Miller, Frederick William
Milman, Gregory
Misra, Rohini Rita
Mockrin, Stephen Charles
Montell, Craig
Mora, Peter T
Moss, Bernard
Mouradian, M Maral
Mufson, R Allan
Mushinski, J Frederic
Nakhasi, Hira Lal
Nash, Howard Allen
Nathans, Daniel
Nayak, Ramesh Kadbet
Noguchi, Constance Tom
Noguchi, Philip D
Nossal, Nancy
Owens, Ida S
Pace, Judith G
Pastan, Ira Harry
Peterson, Jane Louise
Piatigorsky, Joram Paul
Pitha-Rowe, Paula Marie

Molecular Biology (cont)

Ramagopal, Subbanaidu
Rao, Prasad Yarlagadda
Reitz, Marvin Savidge, Jr
Repaske, Roy
Resau, James Howard
Rice, Nancy Reed
Rick, Paul David
Rose, James A
Rosenthal, Leonard Jason
Rosner, Judah Leon
Roth, Thomas Frederic
Sabol, Steven Layne
Sack, George H(enry), Jr
Safer, Brian
Sagripanti, Jose-Luis
Samuel, Albert
Sanford, Katherine Koontz
Sauer, Brian L
Saunders, James Allen
Sausville, Edward Anthony
Saxinger, W(illiam) Carl
Schaad, Norman W
Schiff, Nathan Mark
Schleif, Robert Ferber
Schneider, Walter Carl
Schoenberg, Daniel Robert
Schultz, Warren Walter
Schulze, Dan Howard
Schwartz, Joan Poyner
Scocca, John Joseph
Seifried, Adele Susan Corbin
Seto, Belinda P L
Sexton, Thomas John
Sharma, Opendra K
Shih, Thomas Y
Shin, Yong Ae Im
Simic, Michael G
Simons, Samuel Stoney, Jr
Skolnick, Phil
Slocum, Robert Dale
Small, Judy Ann
Smith, Gilbert Howlett
Sobel, Mark E
Sollner-Webb, Barbara Thea
Song, Byoung-Joon
Spero, Leonard
Srivastava, Sudhir
Stanchfield, James Ernest
Steinert, Peter Malcolm
Steven, Alasdair C
Stoecker, Diane Kastelowitz
Stonehill, Elliott H
Strathern, Jeffrey Neal
Straus, Stephen Ezra
Stromberg, Kurt
Su, Robert Tzyh-Chuan
Suskind, Sigmund Richard
Sussman, Daniel Jesse
Swergold, Gary David
Szepesi, Bela
Talbot, Bernard
Taniuchi, Hiroshi
Tester, Cecil Fred
Thorgeirsson, Snorri S
Torrence, Paul Frederick
Vahey, Maryanne T
Vandenbergh, David John
Vande Woude, George
Varmus, Harold Elliot
Varricchio, Frederick
Venkatesan, S
Villet, Ruxton Herrer
Vonderhaar, Barbara Kay
Waldmann, Thomas A
Weiner, Ronald Martin
Welsch, Federico
Westphal, Heiner J
White, Elizabeth Lloyd
Winkles, Jeffrey A
Wolf, Richard Edward, Jr
Wolffe, Alan Paul
Woodstock, Lowell Willard
Wu, Henry Chi-Ping
Yarmolinsky, Michael Bezalel
Yasbin, Ronald Eliott
Yockey, Hubert Palmer
Yoo, Seung Hyun
Young, Howard Alan
Young, Ronald Jerome
Yuan, Robert
Zimmerman, Steven B

MASSACHUSETTS

Adams, David S
Aghajanian, John Gregory
Ahmed, A Razzaque
Atala, Anthony
Auron, Philip E
Backman, Keith Cameron
Bagshaw, Joseph Charles
Baltimore, David
Banerjee, Papia T
Barkalow, Fern J
Beckwitt, Richard David
Beggs, Alan H, III
Bernhard, Jeffrey David
Bikoff, Elizabeth K
Bischoff, Joyce E
Biswas, Debajit K
Block, Steven M
Boedtker Doty, Helga
Brennessel, Barbara A
Brown, Beverly Ann
Burgeson, Robert Eugene
Bursztajn, Sherry
Carter, Edward Albert
Carvalho, Angelina C A
Cavacini, Lisa Ann
Chen, Wenyong
Chikaraishi, Dona M
Chin, William W
Chishti, Athar Husain
Coffin, John Miller
Cohen, Robert Elliot
Collins, Carolyn Jane
Collins, Tucker
Comer, M Margaret
Cooper, Geoffrey Mitchell
Coor, Thomas
Crain, William Rathbone, Jr
Davis, Claude Geoffrey
DeArruda, Monika Vicira
DeRosier, David J
Desrosiers, Ronald Charles
DeWitt, William
Dvorak, Ann Marie-Tompkins
Ehlers, Mario Ralf Werner
Eil, Charles
Eldred, William D
Emanuel, Rodica L
Ernst, Susan Gwenn
Feig, Larry Allen
Fenton, Matthew John
Fitzgerald, Marie Anton
Fournier, Maurille Joseph, Jr
Fowler, Elizabeth
Fox, Maurice Sanford
Fox, Thomas Oren
Fraser, Thomas Hunter
Fritsch, Edward Francis
Fulton, Chandler Montgomery
Furie, Bruce
Gage, L Patrick
Gallo, Richard Louis
Garrels, James I
Gefter, Malcolm Lawrence
Gilbert, Walter
Gill, David Michael
Gilmore, Thomas David
Godine, John Elliott
Goldberg, Alfred L
Goldberg, Edward B
Goldstein, Lawrence S B
Goldstein, Richard Neal
Gordon, Katherine
Griffith, Irwin J
Gullans, Steven R
Gusovsky, Fabian
Guterman, Sonia Kosow
Hecht, Norman B
Heinrich, Gerhard
Hodge, Richard Paul
Hoffmann, George Robert
Horvitz, Howard Robert
Hotchkiss, Rollin Douglas
Howley, Peter Maxwell
Hu, Jing-Shan
Hynes, Richard Olding
Ingwall, Joanne S
Isselbacher, Kurt Julius
Jameson, James Larry
Kafatos, Fotis C
Kelley, William S
Kew, David
Kilpatrick, Daniel Lee
King, Jonathan (Alan)
Kleckner, Nancy E
Knipe, David Mahan
Kolodner, Richard David
Krane, Stanley Garson
Kupiec-Weglinski, Jerzy W
Lai, Elaine Y
Lander, Arthur Douglas
Lawrence, Jeanne Bentley
Ledbetter, Mary Lee Stewart
Leder, Philip
Lee, Shwu-Luan
Lehman, William Jeffrey
Lemontt, Jeffrey Fielding
Lerman, Leonard Solomon
Levy, Stuart B
Little, John Bertram
Losick, Richard Marc
Lou, Peter Louis
Lovett, Charles McVey
Lu, Frederick Ming
Maas, Richard Louis
McNally, Elizabeth Mary
McReynolds, Larry Austin
Magasanik, Boris
Malamy, Michael Howard
Malt, Ronald A
Maniatis, Thomas Peter
Marx, Kenneth Allan
Melan, Melissa A
Meselson, Matthews
Moolten, Frederick London
Moreland, Robert Byrl
Morton, Mary E
Mount, Mark Samuel
Mulder, Carel
Nadal-Ginard, Bernardo
Neer, Eva Julia
Nogueira, Christine Pietraroia
Offner, Gwynneth Davies
Orkin, Stuart H
Palubinskas, Felix Stanley
Papaioannou, Virginia Eileen
Pardee, Arthur Beck
Pardue, Mary Lou
Park, Hee-Young
Parker, David Charles
Pederson, Thoru Judd
Pero, Janice Gay
Petri, William Hugh
Pluskal, Malcolm Graham
Poccia, Dominic Louis
Poteete, Anthony Robert
Potts, John Thomas, Jr
Ptashne, Mark
Putney, Scott David
Read, Dorothy Louise
Rheinwald, James George
Rich, Alexander
Rigney, David Roth
Riley, Monica
Roberts, Richard John
Ross, Alonzo Harvey
Roy, Sayon
Ruderman, Joan V
Ruprecht, Ruth Margrit
Sachs, David Howard
Sarkar, Nilima
Sarkar, Satyapriya
Sass, Heinz
Sauer, Robert Thomas
Schaechter, Moselio
Seidman, Jonathan G
Shaeffer, Joseph Robert
Shakarjian, Michael Peter
Sharp, Phillip Allen
Shing, Yuen Wan
Signer, Ethan Royal
Smith, Cassandra Lynn
Sonenshein, Abraham Lincoln
Sorkin, Barbara C
Springer, Timothy Alan
Stein, Diana B
Stein, Gary S
Stein, Janet Lee Swinehart
Stephenson, Mary Louise
Strauss, Phyllis R
Strauss, William Mark
Struhl, Kevin
Sussman, Maurice
Szostak, Jack William
Tashjian, Armen H, Jr
Taunton-Rigby, Alison
Thomas, Clayton Lay
Thomas, Peter
Timberlake, William Edward
Tipper, Donald John
Tolan, Dean Richard
Tonegawa, Susumu
Torriani Gorini, Annamaria
Troxler, Robert Fulton
Van Cleave, Victor Harold
Vanderburg, Charles R
Villa-Komaroff, Lydia
Wald, George
Wang, Chih-Lueh Albert
Wang, James C
Wang, Yu-Li
Webb, Andrew Clive
Weinberg, Robert A
Williamson, Patrick Leslie
Witman, George Bodo, III
Wright, Andrew
Young, Delano Victor
Zackroff, Robert V
Zannis, Vassilis I
Zimmermann, Robert Alan

MICHIGAN

Abraham, Irene
Bender, Robert Algerd
Bergen, Werner Gerhard
Bernstein, Isadore A
Bertrand, Helmut
Bhatnagar, Lakshmi
Bouma, Hessel, III
Brawn, Mary Karen
Briggs, Josephine P
Burnett, Jean Bullard
Bush, Guy L
Butterworth, Francis M
Chang, Chia-Cheng
Chaudhry, G Rasul
Chopra, Dharam Pal
Chou, Kuo Chen
Christensen, A(lbert) Kent
Datta, Prasanta
De Bruijn, Frans Johannes
Douthit, Harry Anderson, Jr
Drescher, Dennis G
Duell, Elizabeth Ann
Eaton, Leslie Charles
Einspahr, Howard Martin
Elferink, Cornelius Johan
Fernandez-Madrid, Felix
Ficsor, Gyula
Fischer, Howard David
Fisher, Linda E
Fluck, Michele Marguerite
Freeman, Arthur Scott
Fulbright, Dennis Wayne
Garvin, Jeffrey Lawrence
Gelehrter, Thomas David
Gentile, James Michael
Goldstein, Steven Alan
Goustin, Anton Scott
Grossman, Lawrence I
Hari, V
Heady, Judith E
Heberlein, Gary T
Hirata, Fusao
Hupe, Donald John
Jackson, Matthew Paul
Jones, Lily Ann
Kim, Sun-Kee
Klein, Ronald Don
Kletzien, Rolf Frederick
Krawetz, Stephen Andrew
Krishna, Gopala
Kurnit, David Martin
Lahti, Robert A
Lalwani, Narendra Dhanraj
Landefeld, Thomas Dale
Leff, Todd
Levine, Myron
Lomax, Margaret Irene
Lutter, Leonard C
McCauley, Roy Barnard
McCormick, J Justin
McGroarty, Estelle Josephine
McIntosh, Lee
Maher, Veronica Mary
Marletta, Michael Anthony
Meisler, Miriam Horowitz
Miller, Orlando Jack
Montgomery, Ilene Nowicki
Moudgil, Virinder Kumar
Mulks, Martha Huard
Neidhardt, Frederick Carl
Payne, Anita H
Pecoraro, Vincent L
Pelzer, Charles Francis
Podila, Gopi Krishna
Porter, Calvin Anthon
Preiss, Jack
Reddy, Chilekampalli Adinarayana
Reiners, John Joseph, Jr
Reusser, Fritz
Revzin, Arnold
Righthand, Vera Fay
Robbins, Leonard Gilbert
Rollins-Page, Earl Arthur
Rownd, Robert Harvey
Saltiel, Alan Robert
Sarkar, Fazlul Hoque
Savageau, Michael Antonio
Schramm, John Gilbert
Schwartz, Jessica
Sebolt-Leopold, Judith S
Shastry, Barkur Srinivasa
Siegel, Albert
Smith, David I
Smith, P(aul) Dennis
Snyder, Loren Russell
Swaroop, Anand
Tashian, Richard Earl
Theurer, Jessop Clair
Triezenberg, Steven J
Trosko, James Edward
Tse, Harley Y
Uhler, Michael David
Varani, James
Venta, Patrick John
Warner, Donald Theodore
Whalen, Joseph Wilson
Ye, Qizhuang

MINNESOTA

Adolph, Kenneth William
Anderson, Dwight Lyman
Atluru, Durgaprasadarao
Blumenfeld, Martin
Bodine, Peter Van Nest
Cleary, Paul Patrick
Drewes, Lester Richard
Ellinger, Mark Stephen
Emeagwali, Dale Brown
Fallon, Ann Marie
Fass, David N
Forro, Frederick, Jr
Francis, Gary Stuart
Gengenbach, Burle Gene
Getz, Michael John
Haase, Ashley Thomson
Hedman, Stephen Clifford
Johnson, Russell Clarence
Johnson, Theodore Reynold
Jones, Charles Weldon
Kline, Bruce Clayton
LaPorte, David Coleman
Lefebvre, Paul Alvin
Leung, Benjamin Shuet-Kin
Linck, Richard Wayne
Lindgren, Alice Marilyn Lindell
Magee, Paul Terry
Malewicz, Barbara Maria
Miller, Richard Lynn
Nelson, John Daniel
Oetting, William Starr
Reilly, Bernard Edward
Rubenstein, Irwin
Rychlik, Wojciech
Sadowsky, Michael Jay
Salisbury, Jeffrey L
Salstrom, John Stuart
Sarkar, Gobinda
Schottel, Janet L
Shoffner, Robert Nurman
Silflow, Carolyn Dorothy

BIOLOGICAL SCIENCES / 135

Strehler, Emanuel Ernst
Tautvydas, Kestutis Jonas
Tindall, Donald J
Todt, William Lynn
Tsilibary, Effie-Constantinos
Verma, Anil Kumar
Weatherbee, James A

MISSISSIPPI
Biesiot, Patricia Marie
Boyle, John Aloysius, Jr
Case, Steven Thomas
Chaires, Jonathan Bradford
Gafford, Lanelle Guyton
Graves, David E
Olson, Mark Obed Jerome
Outlaw, Henry Earl
Santangelo, George Michael
Sikorowski, Peter P
Slobin, Lawrence I
Wahba, Albert J
Wellman, Susan Elizabeth
Williamson, John S

MISSOURI
Adams, Steven Paul
Alexander, Stephen
Apirion, David
Armbrecht, Harvey James
Ayyagari, L Rao
Ball, Timothy K
Barnes, Wayne Morris
Beinfeld, Margery Cohen
Bell, Jeffrey
Bolla, Robert Irving
Bourne, Carol Elizabeth Mulligan
Bovy, Philippe R
Brandt, E J
Brown, Karen Kay (Kilker)
Cannon, John Francis
Casperson, Gerald F
David, John Dewood
De Blas, Angel Luis
DiFate, Victor George
Drahos, David Joseph
Eissenberg, Joel Carter
Elgin, Sarah Carlisle Roberts
Eliceiri, George L(ouis)
Engel, Leslie Carroll
Fernandez-Pol, Jose Alberto
Ferris, Deam Hunter
Finkelstein, Richard Alan
Fischhoff, David Allen
Folk, William Robert
Fraley, Robert Thomas
Gordon, Jeffrey I
Gorski, Jeffrey Paul
Grant, Gregory Alan
Green, Eric Douglas
Guilfoyle, Thomas J
Guntaka, Ramareddy V
Hagen, Gretchen
Hirschberg, Rona L
Holland, Lene J
Holtzer, Marilyn Emerson
Hopkins, Johns Wilson
Horsch, Robert Bruce
Huang, Tim Hui-ming
Jaworski, Ernest George
Kennell, David Epperson
Koplow, Jane
Korsmeyer, Stanley Joel
Krishnan, B Rajendra
Krivi, Gwen Grabowski
Lafrenz, David E
Landick, Robert
Lie, Wen-Rong
Lin, Hsiu-san
Lohman, Timothy Michael
Lozeron, Homer A
McCann, Peter Paul
Melechen, Norman Edward
Merlie, John Paul
Miles, Charles Donald
Mooradian, Arshag Dertad
Mortensen, Harley Eugene
Mueckler, Mike Max
Munns, Theodore Willard
Pikaard, Craig Stuart
Radford, Diane Mary
Rice, Charles Moen, III
Riddle, Donald Lee
Sadler, J(asper) Evan
Salkoff, Lawrence Benjamin
Sargentini, Neil Joseph
Schlesinger, Sondra
Schlessinger, David
Schmidt, Francis J
Scholnick, Steven Bruce
Schuck, James Michael
Sharma, Krishna
Strauss, Arnold Wilbur
Templeton, Alan Robert
Thach, Robert Edwards
Thomas, George Joseph, Jr
Wang, Richard J
Weisman, Gary Andrew
Westfall, Helen Naomi
Willard, Mark Benjamin
Wood, David Collier
Yarbro, John Williamson

MONTANA
Dorward, David William

Fevold, Harry Richard
Linner, John Gunnar
Pincus, Seth Henry
Warren, Guylyn Rea

NEBRASKA
Boernke, William E
Brumbaugh, John (Albert)
Calvert, Jay Gregory
Christensen, Alan Carl
Christman, Judith Kershaw
Cox, George Stanley
Dubes, George Richard
Elthon, Thomas Eugene
Gardner, Charles Olda, Jr
Godfrey, Maurice
Holmquist, Barton
Kay, H David
Kumar, Shrawan
Kwang, Jimmy
Leuschen, M Patricia
Mackin, Robert Brian
Morris, Thomas Jack
Partridge, James Enoch
Schneiderman, Martin Howard
Schwartzbach, Steven Donald
Sharp, John Graham
Spreitzer, Robert Joseph
Steffens, David Lee
Veomett, George Ector
Volberg, Thomas M
Weeks, Donald Paul

NEVADA
Harrington, Rodney E
Hoelzer, Guy Andrew
Hudig, Dorothy
Winicov, Ilga

NEW HAMPSHIRE
Bergman, Kenneth David
Cole, Charles N
Dunlap, Jay Clark
Gross, Robert Henry
Loros, Jennifer Jane
Rodgers, Frank Gerald
Trumpower, Bernard Lee
Vournakis, John Nicholas

NEW JERSEY
Abou-Sabe, Morad A
Adamo, Joseph Albert
Axelrod, David E
Barroso, Margarida Maria R
Barton, Beverly E
Basu, Mitali
Boublik, Miloslav
Brandriss, Marjorie C
Byrne, Barbara Jean McManamy
Byrne, Bruce Campbell
Caporale, Lynn Helena
Carboni, Joan M
Cascieri, Margaret Anne
Champe, Sewell Preston
Chen, Ling-Sing Kang
Chizzonite, Richard A
Choi, Ye-Chin
Chovan, James Peter
Cohen, Sheila M
Collett, Edward
Collier, Marjorie McCann
Cunningham, Earlene Brown
Curran, Thomas
Day, Peter Rodney
Day-Salvatore, Debra-Lynn
Denhardt, David Tilton
DePamphilis, Melvin Louis
Dubin, Donald T
Egan, Robert Wheeler
Ennis, Herbert Leo
Essien, Francine B
Feighner, Scott Dennis
Fernandes, Prabhavathi Bhat
Fitzgerald, Paula Marie Dean
Flint, Sarah Jane
Frenkel, Gerald Daniel
Fresco, Jacques Robert
Friedman, Thea Marla
Gagna, Claude Eugene
Garro, Anthony Joseph
Gaynor, John James
Ghosh, Bijan K
Gillum, Amanda McKee
Ginsberg, Barry Howard
Goldman, Emanuel
Goldsmith, Laura Tobi
Gray, Harry Edward
Guo, Jian Zhong
Hitchcock-DeGregori, Sarah Ellen
Hu, Ching-Yeh
Huang, Dao Pei
Humayun, Mir Z
Hwang, San-Bao
Inouye, Masayori
Jakubowski, Hieronim Zbigniew
Johnson, Frank Harris
Kaback, David Brian
Kamiyama, Mikio
Khan, Fazal R
Kirsch, Donald R
Klessig, Daniel Frederick
Kramer, Richard Allen
Krishnan, Gopal
Krug, Robert M

Kumar, Surinder
Kurtz, Myra Berman
Lambert, Muriel Wikswo
Laskin, Allen I
Leibowitz, Michael Jonathan
Lenard, John
Levine, Arnold J
Liebman, Jeffrey Mark
Linemeyer, David L
Liu, Alice Yee-Chang
Liu, Leroy Fong
Lunn, Charles Albert
McKearn, Thomas Joseph
Manne, Veeraswamy
Mannino, Raphael James
Margolskee, Robert F
Mark, David Fu-Chi
Martin, Charles Everett
Massover, William H
Mattes, William Bustin
Messing, Joachim W
Monestier, Marc
Mosley, Stephen T
Mostillo, Ralph
Mullinix, Kathleen Patricia
Nelson, Nathan
Nemecek, Georgina Marie
Niederman, Robert Aaron
Norvitch, Mary Ellen
Ofengand, Edward James
Ozer, Harvey Leon
Pestka, Sidney
Pieczenik, George
Plotkin, Diane Joyce
Price, C A
Rashidbaigi, Abbas
Reuben, Roberta C
Ripley, Lynn S
Robb, Richard John
Rodwell, John Dennis
Rosenblum, Irwin Yale
Rosenstraus, Maurice Jay
Rothstein, Rodney Joel
Ryzlak, Maria Teresa
Sanders, Marilyn Magdanz
Schlesinger, R(obert) Walter
Schroff, Peter David
Seaver, Sally S
Shatkin, Aaron Jeffrey
Shenk, Thomas Eugene
Sherman, Michael Ian
Shin, Seung-il
Sikder, Santosh K
Silberklang, Melvin
Silver, Lee Merrill
Sinha, Navin Kumar
Sofer, William Howard
Stenn, Kurt S
Tesoriero, John Vincent
Thomas, Kenneth Alfred, Jr
Tilghman, Shirley Marie
Trotta, Paul P
Umland, Shelby Price
Unowsky, Joel
Vasconcelos, Aurea C
Wang-Iverson, Patsy
Wassarman, Paul Michael
Wasserman, Bruce P
Weissbach, Herbert
Wennogle, Lawrence
Werth, Jean Marie
Wieschaus, Eric F
Williams, Jeffrey Walter
Xiong, Yimin
Zilinskas, Barbara Ann

NEW MEXICO
Bartholdi, Marty Frank
Belinsky, Steven Alan
Burks, Christian
Buss, William Charles
Crissman, Harry Allen
Goad, Walter Benson, Jr
Kelly, Gregory
Kuehn, Glenn Dean
Lechner, John Fred
McCarthy, Charlotte Marie
Nelson, Mary Anne
Schoenborn, Benno P
Strniste, Gary F
Thomassen, David George

NEW YORK
Aaronson, Stuart Alan
Abramowicz, Daniel Albert
Alexander, Renee R
Allfrey, Vincent George
Allis, C David
Anderson, Carl William
Angerer, Lynne Musgrave
Angerer, Robert Clifford
Asch, Bonnie Bradshaw
Asch, Harold Lawrence
Augenlicht, Leonard Harold
Axel, Richard
Bachrach, Howard L
Bachvaroff, Radoslav J
Bachvarova, Rosemary Faulkner
Badalamente, Marie Ann
Baker, David A
Bambara, Robert Anthony
Banerjee, Debendranath
Banerjee, Ranjit
Barany, Francis

Bard, Enzo
Barnwell, John Wesley
Bartus, Raymond T
Basilico, Claudio
Bauer, William R
Baum, Howard Jay
Bedard, Donna Lee
Belfort, Marlene
Bender, Michael A
Benjamin, William B
Berezney, Ronald
Birshtein, Barbara K
Bishop, David Franklin
Black, Lindsay MacLeod
Bloch, Alexander
Bofinger, Diane P
Bopp, Lawrence Howard
Boutjdir, Mohamed
Bradbury, Michael Wayne
Bregman, Allyn A(aron)
Broadway, Roxanne Meyer
Brody, Edward Norman
Broker, Thomas Richard
Brooks, John S J
Brooks, Robert R
Brunner, Michael
Bruns, Peter John
Buckley, Edward Harland
Budd, Thomas Wayne
Bukhari, Ahmad Iqbal
Burger, Richard Melton
Buxbaum, Joel N
Calame, Kathryn Lee
Calvo, Joseph M
Carlson, Marian B
Carter, Timothy Howard
Catterall, James F
Celis, Roberto T F
Chadha, Kailash C
Chan, Samuel H P
Chapman, Verne M
Chatterjee, Nando Kumar
Chaudhuri, Asok
Chen, Anne Chi
Chen, Ching-Ling Chu
Chow, Louise Tsi
Christie, Nelwyn T
Chu, Frederick Kingsome
Clark, Virginia Lee
Coleman, Paul David
Colman, David Russell
Conway de Macario, Everly
Cooper, Norman S
Coppock, Donald Leslie
Couch, Troy Lee
Counts, David Francis
Craddock, Elysse Margaret
Cross, George Alan Martin
Crowley, Thomas Edward
Crystal, Ronald George
Cunningham, Richard Preston
D'Adamo, Amedeo Filiberto, Jr
Davidson, Mercy
Davies, Kelvin J A
Day, Loren A
De Leon, Victor M
Delihas, Nicholas
Desnick, Robert John
Desplan, Claude
D'Eustachio, Peter
Dewhurst, Stephen
Draginis, Anne M
Drickamer, Kurt
Drlica, Karl
Dubroff, Lewis Michael
Dudock, Bernard S
Duff, Ronald George
Dunn, John Patrick James
Easton, Douglas P
Eberle, Helen I
Eckhardt, Ronald A
Eisinger, Josef
Emmons, Scott W
Enea, Vincenzo
Engelhardt, Dean Lee
Eshel, Dan
Evans, Irene M
Evans, Mary Jo
Figurski, David Henry
Finkelstein, Jacob Noah
Finlay, Thomas Hiram
Fischman, Harlow Kenneth
Fisher, Paul Andrew
Fishman, Jerry Haskel
Fox, Thomas David
Free, Stephen J
Freedberg, Irwin Mark
Freedman, Aaron David
Freundlich, Martin
Friedman, Selwyn Marvin
Garrick, Laura Morris
Garte, Seymour Jay
Gibson, Audrey Jane
Gibson, Kenneth David
Goff, Stephen Payne
Goldberg, Allan Roy
Goldschmidt, Raul Max
Goldstein, Mindy Sue
Gordon, Jon W
Graziadei, William Daniel, III
Greenberg, Jay R
Grodzicker, Terri Irene
Grubman, Marvin J
Grumet, Martin

Molecular Biology (cont)

Gulati, Subhash Chander
Gupta, Sanjeev
Gurel, Okan
Gutcho, Sidney J
Hajjar, David Phillip
Hall, Barry Gordon
Hanafusa, Hidesaburo
Hanson, Maureen Rebecca
Hardy, Ralph Wilbur Frederick
Harford, Agnes Gayler
Hehre, Edward James
Held, William Allen
Helmann, John Daniel
Henshaw, Edgar Cummings
Hilborn, David Alan
Hinkle, David Currier
Hochstadt, Joy
Holmes, David Salway
Horiuchi, Kensuke
Hough, Paul Van Campen
Howard, Glenn Willard, Jr
Howell, Stephen Herbert
Hsu, Ming-Ta
Huang, Alice Shih-Hou
Huberman, Joel Anthony
Iglewski, Wallace
Isseroff, Hadar
Iyengar, Ravi Srinivas V
Jacobson, Ann Beatrice
Jakes, Karen Sorkin
Janakidevi, K
Joshi, Sharad Gopal
Kaiserman, Howard Bruce
Kascsak, Richard John
Katsel, Pavel Leon
Katz, Eugene Richard
Kaye, Jerome Sidney
Kaye, Nancy Weber
Keng, Peter C
Kerwar, Suresh
Khan, Nasim A
Kish, Valerie Mayo
Kissileff, Harry R
Kline, Larry Keith
Knipple, Douglas Charles
Kobilinsky, Lawrence
Koide, Samuel Saburo
Konarska, Maria Magda
Kowalski, David Francis
Krakow, Joseph S
Kramer, Fred Russell
Krim, Mathilde
Kumar, Nirjan
Lacks, Sanford
LaFountain, James Robert, Jr
Landau, Joseph Victor
Lange, Christopher Stephen
Lanks, Karl William
Last, Robert L
Lau, Joseph T Y
Lavett, Diane Kathryn Juricek
Lazzarini, Robert A
Leary, James Francis
Lebwohl, Mark Gabriel
Ledbetter, Myron C
Lee-Huang, Sylvia
Leifer, Zer
Leleiko, Neal Simon
Lemanski, Larry Frederick
Li, Luyuan
Liao, Martha
Lin, Fu Hai
Lin, Jane Huey-Chai
Lindahl, Lasse Allan
Lipke, Peter Nathan
Lipson, Steven Mark
Lis, John Thomas
Lucas, John J
Lusty, Carol Jean
Lyman, Harvard
Lyons, Michael Joseph
Ma, Hong
Macario, Alberto J L
McCabe, R Tyler
McCann-Collier, Marjorie Dolores
McCormick, Paulette Jean
McEwen, Bruce Sherman
McFall, Elizabeth
Madison, James Thomas
Magid, Norman Mark
Maio, Joseph James
Maitra, Umadas
Maitra, Utpalendu S
Maley, Gladys Feldott
Maniloff, Jack
Manly, Kenneth Fred
Marcu, Kenneth Brian
Marenus, Kenneth D
Marians, Kenneth J
Marmur, Julius
Mascarenhas, Joseph Peter
Massie, Harold Raymond
Mazurkiewicz, Joseph Edward
Mehta, Bipin Mohanlal
Mendelsohn, Naomi
Meruelo, Daniel
Michels, Corinne Anthony
Miller, David Lee
Millis, Albert Jason Taylor
Model, Peter
Morin, John Edward
Morse, Randall Heywood

Morse, Stephen S(cott)
Mount, Stephen M
Munne, Santiago
Murphy, Timothy F
Neal, Michael William
Neimark, Harold Carl
Neubort, Shimon
Newman, Stuart Alan
Novick, Richard P
Obendorf, Ralph Louis
Olmsted, Joanna Belle
Paine, Philip Lowell
Pellicer, Angel
Pierucci, Olga
Pietraface, William John
Pogo, A Oscar
Pogo, Beatriz G T
Poiesz, Bernard Joseph
Polatnick, Jerome
Pollack, Robert Elliot
Powell, Sharon Kay
Prakash, Louise
Prasthofer, Thomas W
Price, Frederick William
Pruslin, Fred Howard
Pumo, Dorothy Ellen
Quigley, Gary Joseph
Reddy, Kalluru Jayarami
Redman, Colvin Manuel
Reichert, Leo E, Jr
Rinchik, Eugene M
Roberts, James Lewis
Roberts, Jeffrey Warren
Robinson, Alix Ida
Rodman, Toby C
Roeder, Robert Gayle
Roepe, Paul David
Rogers, Scott Orland
Rogerson, Allen Collingwood
Roll, David E
Rosenberg, Barbara Hatch
Rosenstein, Barry Sheldon
Roy, Harry
Ruddell, Alanna
Rudy, Bernardo
Russel, Marjorie Ellen
Sabatini, David Domingo
Sabban, Esther Louise
Saez, Juan Carlos
St Leger, Raymond John
Salvo, Joseph J
Sanders, F Kingsley
Sanford, Karl John
Sarma, Raghupathy
Sarma, Ramaswamy Harihara
Satir, Peter
Sauro, Marie D
Scher, William
Schlessinger, Joseph
Schloer, Gertrude M
Schneider, Bruce Solomon
Schoenfeld, Cy
Schulman, LaDonne Heaton
Seeman, Nadrian Charles
Sehgal, Pravinkumar B
Shafer, Stephen Joel
Shafit-Zagardo, Bridget
Shafritz, David Andrew
Shahn, Ezra
Shalloway, David Irwin
Sherman, Fred
Shodell, Michael J
Siden, Edward Joel
Silva, Pedro Joao Neves
Silverstein, Emanuel
Silverstein, Samuel Charles
Silverstein, Saul Jay
Simon, Martha Nichols
Sirlin, Julio Leo
Smith, Issar
Sobel, Jael Sabina
Sobell, Henry Martinique
Soifer, David
Spector, David Lawrence
Srivastava, Bejai Inder Sahai
Stanley, Pamela Mary
Steinberg, Bettie Murray
Sternglanz, Rolf
Stillman, Bruce William
Straus, David Bradley
Studier, Frederick William
Sullivan, David Thomas
Swaney, Lois Mae
Sweet, Robert Mahlon
Taber, Harry Warren
Tabor, John Malcolm
Talhouk, Rabih Shakib
Taub, Mary L
Thaler, David Solomon
Thomas, John Owen
Tilley, Shermaine Ann
Toenniessen, Gary Herbert
Tomasi, Thomas B, Jr
Tsai, Jir Shiong
Tullson, Peter C
Tye, Bik-Kwoon
Van't Hof, Jack
Vulliemoz, Yvonne
Wang, Peizhi
Wang, Tung Yue
Warner, Garvin L
Warner, Jonathan Robert
West, Norman Reed
Wiberg, John Samuel

Williams, Noreen
Wilson, David Buckingham
Wilson, Karl A
Wimmer, Eckard
Wishnia, Arnold
Wolfner, Mariana Federica
Wolgemuth, Debra Joanne
Wood, Harry Alan
Woods, Philip Sargent
Wu, Joseph M
Wulff, Daniel Lewis
Yang, Song-Yu
Ye, Guangning
Yoder, Olen Curtis
Young, Charles Stuart Hamish
Yunghans, Wayne N
Zahler, Stanley Arnold
Zaitlin, Milton
Zapisek, William Francis
Zengel, Janice Marie
Zitomer, Richard Stephen

NORTH CAROLINA
Abramson, Jon Stuart
Acedo, Gregoria N
Agris, Paul F
Anderson, Marshall W
Andrews, Matthew Tucker
Barnett, Ortus Webb, Jr
Bast, Robert Clinton, Jr
Bear, Richard Scott
Bell, Juliette B
Bell, Robert Maurice
Birnbaum, Linda Silber
Blackman, Carl F
Bloom, Kerry Steven
Boone, Lawrence Rudolph, III
Bott, Kenneth F
Boyd, Jeffrey Allen
Boynton, John E
Brown, William Roy
Burchall, James J
Burkhart, James Gaylord
Carter, Charles Williams, Jr
Chae, Chi-Bom
Chen, Michael Yu-Men
Chilton, Mary-Dell Matchett
Clinton, Bruce Allan
Coleman, Mary Sue
Corley, Ronald Bruce
Cunningham, Carol Clem
Curtis, Susan Julia
Deher, Kevin L
De Serres, Frederick Joseph
Devereux, Theodora Reyling
Devlin, Robert B
Dieter, Michael Phillip
Douglas, Michael Gilbert
Eddy, Edward Mitchell
Edgell, Marshall Hall
Endow, Sharyn Anne
Erickson, Harold Paul
Errede, Beverly Jean
Faber, James Edward
Farber, Rosann Alexander
Fernandes, Daniel James
Guild, Walter Rufus
Hankins, Gerald Robert
Harris, Elizabeth Holder
Hassan, Hosni Moustafa
Heise, Eugene Royce
Hemperly, John Jacob
Huang, Pi-Shang (Clark)
Hutchison, Clyde Allen, III
Jetten, Anton Marinus
Jividen, Gay Melton
Johnson, Ronald Sanders
Judd, Burke Haycock
Kaufmann, William Karl
Keene, Jack Donald
King, Ann Christie
Kodavanti, Urmila P
Kredich, Nicholas M
Kucera, Louis Stephen
Lapetina, Eduardo G
Lee, Greta Marlene
Lewis, Susan Erskine
Lund, Pauline Kay
Malling, Heinrich Valdemar
Maroni, Gustavo Primo
Marzluff, William Frank, Jr
Matthysse, Ann Gale
Means, Anthony R
Metzger, Richard Stanley
Mitchell, Ann Denman
Mitchell, Thomas Greenfield
Mortenson, Leonard Earl
Mott, Ralph Lionel
Negishi, Masahiko
O'Brien, Deborah A
Pagano, Joseph Stephen
Petes, Thomas Douglas
Pizzo, Salvatore Vincent
Pukkila, Patricia Jean
Ramachandran, Muthukrishnan
Reid, Lola Cynthia McAdams
Richardson, Jane S
Ross, Jeffrey Alan
Sahyoun, Naji Elias
Sancar, Aziz
Sancar, Gwendolyn Boles
Scandalios, John George
Shafer, Thomas Howard
Shaw, Barbara Ramsay

Sonntag, William Edmund
Spiker, Steven L
Stafford, Darrel Wayne
Steege, Deborah Anderson
Stiles, Gary L
Stone, Henry Otto, Jr
Suk, William Alfred
Sung, Michael Tse Li
Taggart, R(obert) Thomas
Telen, Marilyn Jo
Teng, Christina Wei-Tien Tu
Theil, Elizabeth
Thompson, William Francis, III
Tlsty, Thea Dorothy
Toews, Arrel Dwayne
Webster, Robert Edward
Wheeler, Kenneth Theodore, Jr
White, Glenn E
Wilson, Elizabeth Mary
Wise, Edmund Merriman, Jr
Witt, Donald James
Wong, Fulton
Zalani, Sunita

NORTH DAKOTA
Duerre, John A
Hastings, James Michael
Park, Chung Sun
Watson, David Alan

OHIO
Alway, Stephen Edward
Asch, David Kent
Atkins, Charles Gilmore
Banerjee, Amiya Kumar
Banerjee, Sipra
Baroudy, Bahige M
Birky, Carl William, Jr
Blanton, Ronald Edward
Blumenthal, Robert Martin
Boczar, Barbara Ann
Bonventre, Peter Frank
Brunden, Kurt Russell
Byers, Thomas Jones
Cardell, Robert Ridley, Jr
Chen, Wen Yuan
Chiang, John Y L
Cohn, Norman Stanley
Cooper, Dale A
Cope, Frederick Oliver
Coplin, David Louis
Cox, Donald Cody
Cullis, Christopher Ashley
Culp, Lloyd Anthony
D'Ambrosio, Steven M
Decker, Lucile Ellen
Dedman, John Raymond
Deters, Donald W
DiCorleto, Paul Eugene
Doerder, F Paul
Drake, Richard Lee
DuBrul, Ernest
Evans, Helen Harrington
Evans, Thomas Edward
Fenoglio, Cecilia M
Flanagan, Margaret Ann
Flick, Parke Kinsey
Floyd, Gary Leon
Francis, Joseph William
Gesinski, Raymond Marion
Goldthwait, David Atwater
Gronostajski, Richard M
Hamilton, Thomas Alan
Handwerger, Stuart
Hangarter, Roger Paul
Harpst, Jerry Adams
Hillis, Llewellya
Hiremath, Shivanand T
Horseman, Nelson Douglas
Hutton, John James, Jr
Jacobs-Lorena, Marcelo
Jegla, Thomas Cyril
Johnson, Lee Frederick
Johnson, Thomas Raymond
Kantor, George Joseph
Kao, Winston Whei-Yang
Keller, Stephen Jay
Kolattukudy, P E
Kopchick, John J
Kriebel, Howard Burtt
Lane, Richard Durelle
Laughner, William James, Jr
Lechner, Joseph H
Leis, Jonathan Peter
Luck, Dennis Noel
McQuarrie, Irvine Gray
Maguire, Michael Ernest
Mahmoud, Adel A F
Mamrack, Mark Donovan
Martin, Scott McClung
Meyer, Ralph Roger
Miller, Robert Harold
Milsted, Amy
Moffet, Robert Bruce
Nebert, Daniel Walter
Oakley, Berl Ray
Ochs, Raymond S
Oleinick, Nancy Landy
Pavelic, Zlatko Paul
Rake, Adrian Vaughan
Rhodes, Judith Carol
Rosenthal, Ken Steven
Rowe, John James
Rozek, Charles Edward

BIOLOGICAL SCIENCES / 137

Saif, Linda Jean
Samols, David R
Sawicki, Stanley George
Sayre, Richard Thomas
Scarpa, Antonio
Scott, Roy Albert, III
Sen, Ganes C
Shaffer, Jacquelin Bruning
Shannon, Barry Thomas
Shuster, Charles W
Silverman, Robert Hugh
Slonczewski, Joan Lyn
Stalvey, John Robert Dixon
Stanberry, Lawrence Raymond
Stepien, Carol Ann
Stoner, Gary David
Stukus, Philip Eugene
Sunkara, Sai Prasad
Swenson, Richard Paul
Tabita, F Robert
Tepperman, Katherine Gail
Thomas, Donald Charles
Thornton, Janice Elaine
Tomei, L David
Treichel, Robin Stong
Trewyn, Ronald William
Turner, Monte Earl
Tykocinski, Mark L
Versteegh, Larry Robert
Viola, Ronald Edward
Waterson, John R
Webb, Thomas Evan
Wiginton, Dan Allen
Williams, Marshall Vance
Wilson, Kenneth Glade
Wise, Donald L
Woodruff, Ronny Clifford
Woodworth, Mary Esther

OKLAHOMA
Allen, Robert Wade
Altmiller, Dale Henry
Bourdeau, James Edward
Broyles, Robert Herman
Cassidy, Brandt George
Chung, Kyung Won
Conaway, Joan W
Conaway, Ronald C
Delaney, Robert
Elisens, Wayne John
Epstein, Robert B
Essenberg, Richard Charles
Ford, Sharon R
Graves, Donald C
Hall, Leo Terry
Harmon, H James
Johnson, Arthur Edward
Klein, John Robert
Matsumoto, Hiroyuki
Melcher, Ulrich Karl
Morrissey, James Henry
Roe, Bruce Allan
Schaefer, Frederick Vail
Schubert, Karel Ralph
Silverman, Philip Michael
Trachewsky, Daniel
Watson, Gary Hunter
Webb, Carol F
Wolgamott, Gary

OREGON
Bagby, Grover Carlton
Bustamante, Carlos Jose
Civelli, Oliver
Clinton, Gail M
Dalton, David Andrews
Dooley, Douglas Charles
Field, Katharine G
Gold, Michael Howard
Gould, Steven James
Hays, John Bruce
Ho, Iwan
Kohler, Peter
Lolle, Susan Janne
Machida, Curtis A
Matthews, Brian Wesley
Maurer, Richard Allen
Meints, Russel H
Millette, Robert Loomis
Newell, Nanette
Novick, Aaron
Postlethwait, John Harvey
Quatrano, Ralph Stephen
Ream, Lloyd Walter, Jr
Rittenberg, Marvin Barry
Robinson, Arthur B
Russell, Peter James
Schellman, John Anthony
Soderling, Thomas Richard
Thorsett, Grant Orel

PENNSYLVANIA
Abrams, William R
Abremski, Kenneth Edward
Akella, Rama
Alonso-Caplen, Firelli V
Aynardi, Martha Whitman
Babu, Uma Mahesh
Baker, Alan Paul
Balin, Arthur Kirsner
Baserga, Renato
Behe, Michael Joseph
Bering, Charles Lawrence
Bernlohr, Robert William

Berrettini, Wade Hayhurst
Bhaduri, Saumya
Billheimer, Foster E
Billingsley, Melvin Lee
Bjornsti, Mary-Ann
Blough, Herbert Allen
Brenchley, Jean Elnora
Brinigar, William Seymour, Jr
Brownstein, Barbara L
Buono, Russell Joseph
Burnett, Roger Macdonald
Butt, Tauseef Rashid
Cannizzaro, Linda A
Carey, David J
Chang, Frank N
Cheng, Sheau-Mei
Cheng, Yih-Shyun Edmond
Chernin, Mitchell Irwin
Cherry, John P
Chu, Mon-Li
Chung, Albert Edward
Clawson, Gary Alan
Cohen, Leonard Harvey
Cohen, Stanley
Cooperman, Barry S
Currier, Thomas Curtis
Cutler, Winnifred Berg
D'Agostino, Marie A
Das, Manjusri
Debouck, Christine Marie
Debroy, Chitrita
Deering, Reginald
Denis, Kathleen A
Diorio, Alfred Frank
Diven, Warren Field
Donnelly, John James, III
Dyme Southgate, Agnes
Edlind, Thomas D
Emberton, Kenneth C
Evensen, Kathleen Brown
Farber, Florence Eileen
Feinstein, Sheldon Israel
Ferrone, Frank Anthony
Fisher, Edward Allen
Fletcher, Jeffrey Edward
Floros, Joanna
Flynn, John Thomas
Frankel, Fred Robert
Fraser, Nigel William
Freed, Jerome James
Freese, Andrew
Gealt, Michael Alan
Gilbert, Scott F
Gilman, Steven Christopher
Glorioso, Joseph Charles, III
Godfrey, Susan Sturgis
Goff, Christopher Godfrey
Goldstein, Norma Ornstein
Goodgal, Sol Howard
Gorin, Michael Bruce
Graetzer, Reinhard
Grebner, Eugene Ernest
Grunstein, Michael M
Grunwald, Gerald B
Hardison, Ross Cameron
Hendrix, Roger Walden
Henry, Susan Armstrong
Hopper, Anita Klein
Howe, Chin Chen
Hudson, Alan P
Hughes, Austin Leland
Jackson, Ethel Noland
Jacobson, Lewis A
Jarvik, Jonathan Wallace
Johnson, Kenneth Allen
Johnston, James Bennett
Jones, Barry N
Juneja, Vijay Kumar
Kamboh, Mohammad Ilyas
Kane, James Francis
Kant, Jeffrey A
Karush, Fred
Kayne, Marlene Steinmetz
Kennett, Roger H
Khatami, Mahin
King, Robert Willis
Krawiec, Steven Stack
Kuivaniemi, S Helena
Ladda, Roger Louis
Lalley, Edward T
Lawrence, William Chase
Layne, Porter Preston
Lazo, John Stephen
Leboy, Phoebe Starfield
Leibel, Wayne Stephan
Lindstrom, Jon Martin
Llinas, Miguel
Lockwood, Arthur H
Logan, David Alexander
Long, Walter Kyle, Jr
Lotze, Michael T
Lu, Ponzy
McCarthy, Patrick Charles
McClure, William Robert
McDevitt, David Stephen
McKenna, William Gillies
McLaughlin, Patricia J
McMorris, F Arthur
McNamara, Pamela Dee (MacMinigal)
McPheron, Bruce Alan
Manson, Lionel Arnold
Masker, Warren Edward
Mastro, Andrea M
Mifflin, Theodore Edward

Milcarek, Christine
Morris, Sidney Machen, Jr
Mulder, Kathleen M
Nass, Margit M K
Nemer, Martin Joseph
Niu, Mann Chiang
Oldham, James Warren
Opas, Evan E
Peebles, Craig Lewis
Perry, Robert Palese
Person, Stanley R
Phillips, Allen Thurman
Phillips, Stephen Lee
Platsoucas, Chris Dimitrios
Pollock, John Archie
Porter, Ronald Dean
Poste, George Henry
Pratt, Elizabeth Ann
Punnett, Hope Handler
Pyeritz, Reed E
Ramachandran, Banumathi
Ramachandran, N
Ricciardi, Robert Paul
Ritter, Carl A
Roesing, Timothy George
Rosenberg, Martin
Ruggieri, David Raymond
Russell, Alan James
Rutman, Robert Jesse
Salerno, Ronald Anthony
Sanger, Jean M
Sanger, Joseph William
Sazama, Kathleen
Schlegel, Robert Allen
Schwaber, Jerrold
Schwartz, Elias
Seetharam, Ramnath (Ram)
Shockman, Gerald David
Sideropoulos, Aris S
Singer, Alan G
Skalka, Anna Marie
Smith, Amos Brittain, III
Spolsky, Christina Maria
Stevens, Roy Harris
Stewart, Barbara Yost
Stolc, Viktor
Strayer, David S
Sylvester, James Edward
Taubler, James H
Tax, Anne W
Taylor, William Daniel
Tevethia, Mary Judith (Robinson)
Tien, Ming
Tromp, Gerardus C
Truneh, Alan
Tu, Chen-Pei David
Tuan, Rocky Sung-Chi
Turoczi, Lester J
Vaidya, Akhil Babubhai
Wachsberger, Phyllis Rachelle
Ware, Vassie C
Weinberg, Eric S
Weiss, Leon
Wetzel, Ronald Burnell
Whitfield, Carol F(aye)
Wickstrom, Eric
Wilhelm, James Maurice
Williams, Jacinta B
Wolfe, John Hall
Yen, Tim J
Yoon, Kyonggeun
Young, Peter Ronald
Yui, Katsuyuki
Zweidler, Alfred

RHODE ISLAND
Brautigan, David L
Clough, Wendy Glasgow
Cohen, Paul S(idney)
Coleman, John Russell
Dahlberg, Albert Edward
Fausto, Nelson
Fawaz Estrup, Faiza
Fisher, Harold Wilbur
Knopf, Paul M
Landy, Arthur
Lederberg, Seymour
Marotta, Charles Anthony
Meedel, Thomas Huyck
O'Leary, Gerard Paul, Jr
Rotman, Boris
Sapolsky, Asher Isadore
Shank, Peter R

SOUTH CAROLINA
Arnaud, Philippe
Berger, Franklin Gordon
Cantu, Eduardo S
Chao, Julie
Chao, Lee
Davis, Leroy
Ely, Berten E, III
Hahn, Amy Beth
Higerd, Thomas Braden
Kistler, Wilson Stephen, Jr
Larcom, Lyndon Lyle
Ledford, Barry Edward
Lin, Tu
Maheshwari, Kewal Krishnan
Ning, John Tse Tso
Norris, James Scott
Paddock, Gary Vincent
Richards, Gary Paul
Waldman, Alan S

Waldman, Barbara Criscuolo
Wilson, Gregory Bruce
Yardley, Darrell Gene

SOUTH DAKOTA
Brady, Frank Owen
Hildreth, Michael B
Lindahl, Ronald Gunnar
Weaver, Keith Eric
West, Thomas Patrick

TENNESSEE
Anderson, Ted L
Barnett, William Edgar
Billen, Daniel
Blaser, Martin Jack
Bryant, Robert Emory
Bucovaz, Edsel Tony
Bunick, Gerard John
Close, David Matzen
Conary, Jon Thurston
Cook, George A
Cooper, Terrance G
Duhl, David M
Ernst-Fonberg, Marylou
Fleischer, Sidney
Friedman, Daniel Lester
Granner, Daryl Kitley
Granoff, Allan
Gresshoff, Peter Michael
Griffin, Guy David
Guyer, Cheryl Ann
Hadden, Charles Thomas
Hazelton, Bonni Jane
Henke, Randolph Ray
Hingerty, Brian Edward
Howe, Martha Morgan
Howell, Elizabeth E
Incardona, Antonino L
Jarrett, Harry Wellington, III
Jeon, Kwang Wu
Jones, Larry Hudson
Joplin, Karl Henry
Katze, Jon R
Kitabchi, Abbas E
Larimer, Frank William
LeStourgeon, Wallace Meade
Lothstein, Leonard
McGavock, Walter Donald
Miller, Mark Steven
Mitchell, William Marvin
Monaco, Paul J
Mullin, Beth Conway
Nienhuis, Arthur Wesley
Nishimoto, Satoru Kenneth
Niyogi, Salil Kumar
Nowak, Thaddeus Stanley, Jr
Olins, Ada Levy
Olins, Donald Edward
Pfeffer, Lawrence Marc
Portner, Allen
Purcell, William Paul
Quigley, Neil Benton
Riggsby, William Stuart
Rubin, Donald Howard
Ryan, Kevin William
Savage, Dwayne Cecil
Skinner, Dorothy M
Stacey, Gary
Staros, James Vaughan
Stevens, Stanley Edward, Jr
Thomason, Donald Brent
Tibbetts, Clark
Touster, Oscar
Trigiano, Robert Nicholas
Trupin, Joel Sunrise
Walker, William Stanley
Washington, Arthur Clover
Watterson, D Martin
Whitt, Michael A
Wicks, Wesley Doane
Williams, Robley Cook, Jr
Woychik, Richard P
Yang, Wen-Kuang

TEXAS
Adair, Gerald Michael
Aggarwal, Bharat Bhushan
Alhossainy, Effat M
Allen, Randy Dale
Alvarez-Gonzalez, Rafael
Amoss, Max St Clair
Angelides, Kimon Jerry
Arenaz, Pablo
Arntzen, Charles Joel
Atkinson, Mark Arthur Leonard
Aufderheide, Karl John
Barker, Kenneth Leroy
Beale, Elmus G
Beaudet, Arthur L
Beckingham, Kathleen Mary
Bellion, Edward
Benedik, Michael Joseph
Bennett, George Nelson
Breen, Gail Anne Marie
Bremer, Hans
Brown, Dennis Taylor
Brysk, Miriam Mason
Buchanan, Christine Elizabeth
Byerley, Lauri Olson
Campbell, James Wayne
Carney, Darrell Howard
Carson, Daniel Douglas
Chadwick, Arthur Vorce

Molecular Biology (cont)

Chalmers, John Harvey, Jr
Chan, John Yeuk-hon
Chan, Lawrence Chin Bong
Chan, Lee-Nien Lillian
Chanh, Tran C
Chatterjee, Bandana
Collisson, Ellen Whited
Coulter, Murray W
Crandall, Keith Alan
Crawford, Isaac Lyle
Davison, Daniel Burtonn
Dempsey, Walter B
Denney, Richard Max
Donnelly, Patricia Vryling
Douglas, Tommy Charles
Dowhan, William
Earhart, Charles Franklin, Jr
Epstein, Henry F
Evans, Claudia T
Evans, John Edward
Faust, Charles Harry, Jr
Fisher, Charles William
Fleischmann, William Robert, Jr
Forrest, Hugh Sommerville
Fox, George Edward
Frederiksen, Richard Allan
Garber, Alan J
Garcia, Hector D
Garrard, William T
Georgiou, George
Gogol, Edward Peter
Gomer, Richard Hans
Goodman, Joel Mitchell
Gorelic, Lester Sylvan
Gottlieb, Paul David
Gratzner, Howard G
Griffin, James Emmett
Griffing, Lawrence Randolph
Grimm, Elizabeth Ann
Guerrero, Felix David
Hackert, Marvin LeRoy
Hagino, Nobuyoshi
Hall, Timothy Couzens
Hamilton, Charleen Marie
Hanis, Craig L
Harris, Stephen Eubank
Hecht, Ralph Martin
Heimbach, Richard Dean
Henney, Henry Russell, Jr
Hewitt, Roger R
Highlander-Bultema, Sarah Katherine
Hillar, Marian
Hillis, David Mark
Hixson, James Elmer
Ho, Begonia Y
Holoubek, Viktor
Honeycutt, Rodney Lee
Hopkins, Robert Charles
Howard, Harriette Ella Pierce
Hughes, Mark R
Ihler, Garret Martin
Ippen-Ihler, Karin Ann
Jagger, John
James, Harold Lee
Jeter, Randall Mark
Joys, Terence Michael
Kaplan, Samuel
Karunakaran, Thonthi
Kier, Ann B
Kimbrell, Deborah Ann
Kit, Saul
Konkel, David Anthony
Kraig, Ellen
Kroschewsky, Julius Richard
Kurzrock, Razelle
Kutteh, William Hanna
La Claire, John Willard, II
Lang, Dimitrij Adolf
LeBlanc, Donald Joseph
Ledley, Fred David
Lee, Chong Sung
Leuchtag, H Richard
Lever, Julia Elizabeth
Lieberman, Michael Williams
Lupski, James Richard
McBride, Raymond Andrew
McKnight, Thomas David
MacLeod, Michael Christopher
Mandel, Manley
Marsh, Robert Cecil
Masters, Bettie Sue Siler
Mayor, Heather Donald
Meltz, Martin Lowell
Miller, Thomas Allen
Mills, William Ronald
Mitra, Sankar
Murgola, Emanuel J
Nelson, David Loren
Nguyen, Henry Thien
Nicholson, Wayne Lowell
Nikaein, Afzal
Norris, Steven James
Oka, Kazuhiro
O'Malley, Bert W
Painter, Richard Grant
Patel, Nutankumar T
Payne, Shelley Marshall
Perlman, Philip Stewart
Peterson, David Oscar
Pirtle, Robert M
Podusio, Shirley Ellen
Poenie, Martin Francis

Powell, Bernard Lawrence
Price, Peter Michael
Rajaraman, Srinivasan
Rege, Ajay Anand
Ritzi, Earl Michael
Robberson, Donald Lewis
Romeo, Tony
Rose, Kathleen Mary
Rosen, Jeffrey Mark
Ross, Elliott M
Roth, Jack A
Roy, Arun K
Rupert, Claud Stanley
Sauer, Helmut Wilhelm
Saunders, Grady Franklin
Schaffer, Barbara Noyes
Schrader, William Thurber
Serwer, Philip
Shain, Sydney A
Shanley, Mark Stephen
Shay, Jerry William
Siede, Wolfram
Smith, Don Wiley
Smith, Russell Lamar
Snell, William J
Sparkman, Dennis Raymond
Stewart, Charles Ranous
Stone, William Harold
Stroman, David Womack
Stroynowski, Iwona T
Stubblefield, Travis Elton
Summers, Max Duane
Sutton, Harry Eldon
Temeyer, Kevin Bruce
Thompson, Edward Ivins Bradbridge
Thompson, Ernest Aubrey, Jr
Tomasovic, Stephen Peter
Tong, Alex W
Trkula, David
Tsai, Ming-Jer
Van Alfen, Neal K
Vinson, S Bradleigh
Walker, Cheryl Lyn
Walmsley, Judith Abrams
Walter, Ronald Bruce
Warner, Marlene Ryan
Weinstock, George Matthew
Wells, Robert Dale
Wild, James Robert
Williams, Robert Sanders
Wilson, Carol Maggart
Wilson, Samuel H
Woo, Savio L C
Wood, Robert Charles
Wright, Stephen E
Wright, Woodring Erik
Yang, Funnei
Yeh, Lee-Chuan Caroline
Yeoman, Lynn Chalmers
Yielding, K Lemone
Young, Ryland F

UTAH

Aird, Steven Douglas
Anderson, Anne Joyce
Capecchi, Mario Renato
Carroll, Dana
Farmer, James Lee
Franklin, Naomi C
Fujinami, Robert S
Georgopoulos, Constantine Panos
Gesteland, Raymond Frederick
Gurney, Elizabeth Tucker Guice
Gurney, Theodore, Jr
Herrick, Glenn Arthur
Hirning, Lane D
Huang, Wai Mun
Kemp, John Wilmer
Knapp, Gayle
Lark, Karl Gordon
Larsen, Lloyd Don
Lighton, John R B
Okun, Lawrence M
Parkinson, John Stansfield
Rasmussen, Kathleen Goertz
Roth, John R
Simmons, Daniel L
Smith, Marvin Artell
Takemoto, Jon Yutaka
Tilton, Varien Russell
Torres, Anthony R
Warters, Raymond Leon
White, Raymond L
Williams, Roger Richard
Wolstenholme, David Robert
Yoshikami, Doju

VERMONT

Albertini, Richard Joseph
Hoagland, Mahlon Bush
Kelley, Jason
Long, George Louis
Moehring, Thomas John
Nicklas, Janice A
Raper, Carlene Allen
Schaeffer, Warren Ira
Ullrich, Robert Carl
Wallace, Susan Scholes
Weller, David Lloyd
Woodcock-Mitchell, Janet Louise

VIRGINIA

Auton, David Lee
Bates, Robert Clair

Bauerle, Ronald H
Bender, Patrick Kevin
Beyer, Ann L
Blatt, Jeremiah Lion
Braciale, Vivian Lam
Brown, Jay Clark
Chen, Jiann-Shin
Chlebowski, Jan F
Cole, Patricia Ellen
Collins, James Malcolm
Courtney, Mark William
DeGiovanni-Donnelly, Rosalie F
DeLorenzo, Robert John
Eglitis, Martin Alexandris
Esen, Asim
Garrett, Reginald Hooker
Garrison, James C
Grogan, William McLean
Gross, Paul Randolph
Gwazdauskas, Francis Charles
Haber, Lynne Tracey
Hackney, Cameron Ray
Hamlin, Joyce Libby
Jakoi, Emma Raff
Kalimi, Mohammed Yahya
Kelley, John Michael
King, Betty Louise
Kretsinger, Robert
Kuemmerle, Nancy Benton Stevens
Lacy, George Holcombe
Laurie, Gordon William
Lemp, John Frederick, Jr
McGowan, John Joseph
Marron, Michael Thomas
Mikkelsen, Ross Blake
Miller, Oscar Lee, Jr
Osheim, Yvonne Nelson
Osuch, Mary Ann V
Owens, Gary
Pryer, Nancy Kathryn
Roberts, Catherine Harrison
Rutherford, Charles
Salley, John Jones, Jr
Stout, Ernest Ray
Tolin, Sue Ann
Tyson, John Jeanes
Vermeulen, Carl William
Wagner-Bartak, Claus Gunter Johann
Williams, Luther Steward

WASHINGTON

Adman, Elinor Thomson
Albers, John J
Bakken, Aimee Hayes
Bendich, Arnold Jay
Bornstein, Paul
Byers, Breck Edward
Calza, Roger Ernest
Camerman, Arthur
Carter, Barrie J
Cattolico, Rose Ann
Cheevers, William Phillip
Clapshaw, Patric Arnold
Cosman, David John
Dale, Beverly A
Deming, Jody W
Feagin, Jean Ellen
Fitch, Cynthia Lynn
Galas, David John
Gallant, Jonathan A
Garber, Richard Lincoln
Graves, Scott Stoll
Gray, Patrick William
Gurusiddaiah, Sarangamat
Hall, Benjamin Downs
Hall, Stanton Harris
Hauschka, Stephen D
Haydock, Paul Vincent
Hazelbauer, Gerald Lee
Hosick, Howard Lawrence
Hu, Shiu-Lok
King, Irena B
Kohlhepp, Sue Joanne
Kraft, Joan Creech
Kutter, Elizabeth Martin
Lamb, Mary Rose
McFadden, Bruce Alden
MacKay, Vivian Louise
Magnuson, Nancy Susanne
Margolis, Robert Lewis
Matlock, Daniel Budd
Miller, Arthur Dusty
Narayanan, A Sampath
Nathanson, Neil Marc
Norvell, Lorelei Lehwalder
Pall, Martin L
Pious, Donald A
Prody, Gerry Ann
Randall, Linda Lea
Reeder, Ronald Howard
Reeves, Jerry John
Reeves, Raymond
Riddiford, Lynn Moorhead
Rohrschneider, Larry Ray
Sage, Helene E
Salk, Darrell John
Sheiness, Diana Kay
Sibley, Carol Hopkins
Smith, Gerald Ralph
Steiner, Robert Alan
Stenkamp, Ronald Eugene
Stuart, Kenneth Daniel
Taylor, Dean Perron
Wakimoto, Barbara Toshiko

Weintraub, Harold M
Yancey, Paul Herbert
Yao, Meng-Chao
Young, Elton Theodore
Zakian, Virginia Araxie

WEST VIRGINIA

Konat, Gregory W
Reichenbecher, Vernon Edgar, Jr
Rhoten, William Blocher
Yelton, David Baetz

WISCONSIN

Allred, Lawrence Ervin (Joe)
Amasino, Richard M
Anderson, Robert Philip
Armstrong, George Michael
Attie, Alan D
Barton, Kenneth Allen
Bast, Rose Ann
Becker, Wayne Marvin
Bergen, Lawrence
Bergtrom, Gerald
Blattner, Frederick Russell
Borisy, Gary Guy
Brown, Raymond Russell
Burgess, Ann Baker
Burgess, Richard Ray
Calvanico, Nickolas Joseph
Chambliss, Glenn Hilton
Collins, Mary Lynne Perille
Courtright, James Ben
Cox, Michael Matthew
Cramer, Jane Harris
Dahlberg, James E(ric)
Dahms, Nancy Margaret
Daie, Jaleh
Dimond, Randall Lloyd
Dwyer-Hallquist, Patricia
Ellingboe, Albert Harlan
Fahl, William Edwin
Goldberger, Amy
Goodman, Robert Merwin
Gourse, Richard Lawrence
Greenspan, Daniel S
Haas, Michael John
Haasch, Mary Lynn
Harms, Jerome Scott
Hokin, Lowell Edward
Hopwood, Larry Eugene
Hornemann, Ulfert
Hutton, James Robert
Hutz, Reinhold Josef
Inman, Ross
Jacobson, Gunnard Kenneth
John, Maliyakal Eappen
Johnson, Warren Victor
Kaesberg, Paul Joseph
Kahan, Lawrence
Karrer, Kathleen Marie
Kasper, Charles Boyer
Keegstra, Kenneth G
Kermicle, Jerry Lee
Klebba, Phillip E
Kubinski, Henry A
Littlewood, Roland Kay
Markley, John Lute
Martin, Thomas Fabian John
Mertz, Janet Elaine
Moskowitz, Gerard Jay
Munroe, Stephen Horner
Nelson, David Lee
Norris, Dale Melvin, Jr
Pan, David
Patrick, Michael Heath
Rapacz, Jan
Record, M Thomas, Jr
Reznikoff, William Stanton
Ross, Jeffrey
Scheele, Robert Blain
Schindler, Joel Marvin
Seelke, Ralph Walter
Sheehy, Michael Joseph
Slautterback, David Buell
Smith, Lloyd Michael
Spritz, Richard Andrew
Sussman, Michael R
Sytsma, Kenneth Jay
Szybalski, Waclaw
Tandon, Shiv R
Verma, Ajit K
Waring, Gail L
Wejksnora, Peter James
Wells, Barbara Duryea
Wilcox, Kent Westbrook
Wimpee, Charles F
Wolter, Karl Erich

WYOMING

Ji, Inhae
Lewis, Randolph Vance
Petersen, Nancy Sue
Smith-Sonneborn, Joan

PUERTO RICO

Candelas, Graciela C
Carrasco-Serrano, Clara E
Craig, Syndey Pollock, III

ALBERTA

Bazett-Jones, David Paul
Bentley, Michael Martin
Church, Robert Bertram
Dancik, Bruce Paul

Dixon, Gordon H
Hart, David Arthur
Hiruki, Chuji
Iatrou, Kostas
Jensen, Susan Elaine
Langridge, William Henry Russell
Laroche, Andre
Lemieux, Raymond Urgel
Machin, Geoffrey Allan
Morgan, Antony Richard
Paetkau, Verner Henry
Rattner, Jerome Bernard
Rosenberg, Susan Mary
Scraba, Douglas G
Vance, Dennis E
Van De Sande, Johan Hubert
Walsh, Michael Patrick
Weiner, Joel Hirsch

BRITISH COLUMBIA
Aebersold, Ruedi H
Astell, Caroline R
Beacham, Terry Dale
Brandhorst, Bruce Peter
Burke, Robert D
Candido, Edward Peter Mario
Cavalier-Smith, Thomas
Dennis, Patrick P
Druehl, Louis D
Eastwell, Kenneth Charles
Ekramoddoullah, Abul Kalam M
Emerman, Joanne Tannis
Finlay, Barton Brett
Green, Beverley R
Griffiths, Anthony J F
Hancock, Robert Ernest William
Honda, Barry Marvin
Kafer, Etta (Mrs E R Boothroyd)
Lederis, Karolis (Karl)
Mackie, George Alexander
Matheson, Alastair Taylor
Redfield, Rosemary Jeanne
Rennie, Paul Steven
Richards, William Reese
Sadowski, Ivan J
Salari, Hassan
Shamoun, Simon Francis
Skala, Josef Petr
Smith, Michael Joseph
Spiegelman, George Boole
Tener, Gordon Malcolm
Walker, Ian Gardner
Weeks, Gerald

MANITOBA
Borsa, Joseph
Davie, James Ronald
Duckworth, Henry William (Harry)
Kunz, Bernard Alexander
Rosenmann, Edward A
Wrogemann, Klaus
Zahradka, Peter

NEW BRUNSWICK
Krause, Margarida Oliveira

NEWFOUNDLAND
Burness, Alfred Thomas Henry
Davidson, William Scott
Michalak, Thomas Ireneusz
Michalski, Chester James
Morris, Claude C

NOVA SCOTIA
Chambers, Robert Warner
Doolittle, Warren Ford, III
Gray, Michael William
MacRae, Thomas Henry
Ragan, Mark Adair
Stratton, Glenn Wayne
Vining, Leo Charles

ONTARIO
Arif, Basil Mumtaz
Asculai, Samuel Simon
Atkinson, Burr Gervais
Axelrad, Arthur Aaron
Bacchetti, Silvia
Bag, Jnanankur
Baker, Allan John
Barran, Leslie Rohit
Bayley, Stanley Thomas
Behki, Ram M
Birnbaum, George I
Birnboim, Hyman Chaim
Brown, Ian Ross
Buchwald, Manuel
Camerman, Norman
Capone, John Pasquale
Carstens, Eric Bruce
Chaconas, George
Chambers, Ann Franklin
Chui, David H K
Cinader, Bernhard
Danska, Jaynes
Davis, Alan
Dedhar, Shoukat
Deeley, Roger Graham
Dube, Ian David
Etches, Robert J
Fish, Eleanor N
Fitt, Peter Stanley
Forsberg, Cecil Wallace
Forsdyke, Donald Roy

Fourney, Ronald Mitchell
Fraser, Ann Davina Elizabeth
Freedman, Melvin Harris
Friesen, James Donald
Ganoza, M Celia
Ghosh, Hara Prasad
Glick, Bernard Robert
Goldberg, Gary Steven
Graham, Frank Lawson
Greenblatt, Jack Fred
Gupta, Radhey Shyam
Haj-Ahmad, Yousef
Haynes, Robert Hall
Hegele, Robert A
Heikkila, John J
Hickey, Donal Aloysius
Hofmann, Theo
Horgen, Paul Arthur
Hozumi, Nobumichi
Ingles, Charles James
Irwin, David Michael
Johnson-Lussenburg, Christine Margaret
Jones, Owen Thomas
Kang, C Yong
Kasha, Kenneth John
Kasupski, George Joseph
Kidder, Gerald Marshall
Krepinsky, Jiri J
Lachance, Marc-Andre
Lau, Catherine Y
Liew, Choong-Chin
Ling, Victor
Logan, David Mackenzie
Lu, Benjamin Chi-Ko
Lynn, Denis Heward
MacLennan, David Herman
Mak, Tak Wah
Mak, William Wai-Nam
Manolson, Morris Frank
Metuzals, Janis
Mickle, Donald Alexander
Mishra, Ram K
Narang, Saran A
Neelin, James Michael
Ottensmeyer, Frank Peter
Pearlman, Ronald E
Percy, Maire Ede
Przybylska, Maria
Rainbow, Andrew James
Rayman, Mohamad Khalil
Scott, Fraser Wallace
Seligy, Verner Leslie
Sells, Bruce Howard
Singal, Dharam Parkash
Singh, Bhagirath
Singh, Rama Shankar
Smiley, James Richard
Sorger, George Joseph
Spencer, John Hedley
Stavric, Stanislava
Stiller, Calvin R
Straus, Neil Alexander
Sun, Anthony Mein-Fang
Tobe, Stephen Solomon
Trick, Charles Gordon
Tu, Jui-Chang
Ulpian, Carla
Waithe, William Irwin
Walker, Peter Roy
Warner, Alden Howard
Warwick, Suzanne Irene
Waye, John Stewart
Wood, Janet Marion
Worton, Ronald Gilbert
Wu, Tai Wing
Young, Paul Gary

QUEBEC
Anderson, William Alan
Antakly, Tony
Archambault, Denis
Barden, Nicholas
Beaudoin, Adrien Robert
Belanger, Luc
Bellabarba, Diego
Brown, Gregory Gaynor
Chabot, Benoit
Chafouleas, James G
Charest, Pierre M
Chung, Young Sup
Cousineau, Gilles H
DuBow, Michael Scott
Dunn, Robert James
Fournier, Michel
Glorieux, Francis Henri
Graham, Angus Frederick
Green, David M(artin)
Jothy, Serge
Kay, Denis G
Labuda, Damian
Lagace, Lisette
Lapointe, Jacques
Lebel, Susan
Levasseur, Maurice Edgar
Madiraju, S R Murthy
Morais, Rejean
Murgita, Robert Anthony
Murthy, Mahadi Raghavandrarao Ven
Nishioka, Yutaka
Paice, Michael
Palfree, Roger Grenville Eric
Pande, Shri Vardhan
Pickett, Cecil Bruce
Ponka, Premysl

Rokeach, Luis Alberto
Sairam, Malur Ramaswamy
Sarhan, Fathey
Seufert, Wolf D
Shoubridge, Eric Alan
Simard, Rene
Skup, Daniel
Slilaty, Steve N
Stanners, Clifford Paul
Sygusch, Jurgen
Talbot, Pierre J
Tanguay, Robert M
Trepanier, Pierre
Trudel, Michel D
Wang, Eugenia

SASKATCHEWAN
Babiuk, Lorne A
Blair, Donald George Ralph
Faber, Albert John
Quail, John Wilson
Robertson, Beverly Ellis
Xiang, Jim

YUKON TERRITORY
Doermann, August Henry

OTHER COUNTRIES
Abelev, Garri Izrailevich
Adldinger, Hans Karl
Agathos, Spiros Nicholas
Ahn, Tae In
Amsterdam, Abraham
Anagnou, Nicholas P
Arai, Ken-Ichi
Arber, Werner
Argos, Patrick
Armato, Ubaldo
Arnott, Struther
Atkinson, Paul H
Avila, Jesus
Bachmann, Fedor W
Baldwin, Robert William
Barel, Monique
Barouki, Robert
Bayev, Alexander A
Ben-Ze'ev, Avri
Bishop, David Hugh Langler
Boffa, Lidia C
Bolhuis, Reinder L H
Bunn, Clive Leighton
Campbell, Virginia Wiley
Caro, Lucien
Chang, Donald Choy
Chang, Yu-Sun
Chetsanga, Christopher J
Choi, Yong Chun
Christodoulou, John
Collins, Vincent Peter
Creighton, Thomas Edwin
Curtis, Adam Sebastian G
De Haën, Christoph
Doerfler, Walter Hans
Donato, Rosario Francesco
Dworkin, Mark Bruce
Dyrberg, Thomas Peter
Edelman, Marvin
Eggers, Hans J
Elsevier, Susan Maria
Engelborghs, Yves
Fermi, Giulio
Fernandez-Cruz, Eduardo P
Fersht, Alan Roy
Franklin, Richard Morris
Ganten, Detlev
Garcia-Sainz, J Adolfo
Geiger, Benjamin
Ghiara, Paolo
Goding, James Watson
Greene, Lewis Joel
Gruss, Peter H
Guimaraes, Romeu Cardoso
Haggis, Geoffrey Harvey
Hansson, Goran K
Hanukoglu, Israel
Hashimoto, Paulo Hitonari
Hayashi, Masao
Henderson, David Andrew
Hirano, Toshio
Hofmann, Bo
Hoisington, David A
Hosokawa, Keiichi
Ikehara, Yukio
Ishikawa, Hiroshi
Jakob, Karl Michael
Kaye, Alvin Maurice
Kimoto, Masao
Klug, Aaron
Kuhn, Klaus
Kuriyama, Kinya
Lamoyi, Edmundo
Liao, Ta-Hsiu
Ling, Chung-Mei
Loppnow, Harald
Maruta, Hiroshi
Meins, Frederick, Jr
Mezquita, Cristobal
Michaels, Allan
Michaelsen, Terje E
Millar, Robert Peter
Mittal, Balraj
Mizuno, Shigeki
Nagata, Kazuhiro
Paik, Young-Ki

Richmond, Robert H
Roy, Raman K
Santoro, Ferrucio Fontes
Sargent, David Fisher
Schena, Francesco Paolo
Schmell, Eli David
Schneider, Wolfgang Johann
Sheldrick, Peter
Spencer, John Francis Theodore
Spierer, Pierre
Stark, George Robert
Stebbing, Nowell
Steer, Martin William
Stingl, Georg
Stodolsky, Marvin
Strous, Ger J
Subramanian, Alap Raman
Tan, Y H
Taniyama, Tadayoshi
Teleb, Zakaria Ahmed
Thiery, Jean Paul
Toutant, Jean-Pierre
Tsolas, Orestes
Tsou, Chen-Lu
Wakayama, Yoshihiro
Watanabe, Takeshi
Willard-Gallo, Karen Elizabeth
Wu, Felicia Ying-Hsiueh
Wuthrich, Kurt
Wyman, Jeffries

Neurosciences

ALABAMA
Berecek, Kathleen Helen
Bradley, Laurence A
Egan, Marianne Louise
Friedlander, Michael J
Garver, David L
Loop, Michael Stuart
Marchase, Richard Banfield
Nowack, William J
Oparil, Suzanne
Oyster, Clyde William
Williams, Raymond Crawford

ARIZONA
Allen, John Jeffery
Bloedel, James R
Buck, Stephen Henderson
Chapman, Reginald Frederick
Cheal, MaryLou
Davis, Thomas P
Hildebrand, John Grant, III
Hruby, Victor J
Scott, Alwyn C
Triffet, Terry

ARKANSAS
Dykman, Roscoe A
Gilmore, Shirley Ann
Holson, Ralph Robert
Komoroski, Richard Andrew
Paule, Merle Gale
Skinner, Robert Dowell
Wessinger, William David

CALIFORNIA
Ackerman, Neil
Alkana, Ronald L
Allman, John Morgan
Andrews, Russell J
Arnold, Arthur Palmer
Aserinsky, Eugene
Ashe, John Herman
Bacskai, Brian James
Beckman, Alexander Lynn
Benzer, Seymour
Berger, Ralph Jacob
Bowers, Chauncey Warner
Bradshaw, Ralph Alden
Bremermann, Hans J
Brown, Warren Shelburne, Jr
Buchwald, Jennifer S
Campagnoni, Anthony Thomas
Chambers, Kathleen Camille
Chapman, Loring Frederick
Cohen, Malcolm Martin
Cohn, Major Lloyd
Curras, Margarita C
Daunton, Nancy Gottlieb
Davis, William Jackson
Dorfman, Leslie Joseph
Edelman, Gerald Maurice
Ellman, George Leon
Eng, Lawrence F
Enoch, Jay Martin
Estrada, Norma Ruth
Fagin, Katherine Diane
Fain, Gordon Lee
Farley, Roger Dean
Fields, Howard Lincoln
Fisher, Robin Scott
Fluharty, Arvan Lawrence
Fromkin, Victoria A
Fryxell, Karl Joseph
Gall, William Einar
Gehrmann, John Edward
Getz, Wayne Marcus
Gibson, Thomas Richard
Glaser, Donald Arthur
Goldberg, Mark Arthur
Goodman, Corey Scott

Neurosciences (cont)

Greenwood, Mary Rita Cooke
Grinnell, Alan Dale
Haines, Richard Foster
Hall, Zach Winter
Hammerschlag, Richard
Hatton, Glenn Irwin
Haun, Charles Kenneth
Hefti, Franz F
Hillyard, Steven Allen
Horowitz, John M
Howd, Robert A
Huszczuk, Andrew Huszcza
Jewett, Don L
Johnson, Chris Alan
Johnson, Edwin C
Johnson, Randolph Mellus
Junge, Douglas
Katz, Barrett
Kelly, Regis Baker
Kennedy, Mary Bernadette
Kishimoto, Yasuo
Ko, Chien-Ping
Koch, Bruce D
Konopka, Ronald J
Levine, Michael S
Libet, Benjamin
Lieberburg, Ivan M
Lindsley, Donald B
Lingappa, Jaisri Rao
Litt, Lawrence
McCaman, Richard Eugene
McClure, William Owen
McCullough, Richard Donald
McGaugh, James L
McGinnis, James F
Macklin, Wendy Blair
Marg, Elwin
Marmor, Michael F
Martin, James Tillison
Matthews, Beverly Bond
Metzner, Walter
Miledi, Ricardo
Miller, Arnold Lawrence
Miller, Arthur Joseph
Mohan, Chandra
Morse, Daniel E
Mulloney, Brian
Patterson, Paul H
Peper, Erik
Perlmutter, Milton Manuel
Phelps, Michael Edward
Polit, Andres Cassard
Ralston, Henry James, III
Ramachandran, Vilayanur Subramanian
Rao, Tadimeti Seetapati
Reiness, Gary
Ribak, Charles Eric
Richman, David Paul
Ridgway, Sam H
Robertson, David W
Rome, Leonard H
Rosenberg, Abraham
Ruesch, Jurgen
Russell, Sharon May
Sadee, Wolfgang
Sawyer, Charles Henry
Schwab, Ernest Roe
Segall, Paul Edward
Sheppard, Asher R
Shih, Jean Chen
Simmons, Dwayne Deangelo
Skrzypek, Josef
Snow, George Edward
Sperling, George
Spitzer, Nicholas Canaday
Starr, Arnold
Stent, Gunther Siegmund
Stryker, Michael Paul
Tatic-Lucic, Svetlana
Terry, Robert Davis
Tewari, Sujata Lahiri
Thompson, Jeffrey Michael
Thompson, Richard Frederick
Tobin, Allan Joshua
Tomich, John Matthew
Tschirgi, Robert Donald
Tyler, Christopher William
Uyeno, Edward Teiso
Vaughn, James E, Jr
Villablanca, Jaime Rolando
Vinters, Harry Valdis
Wasterlain, Claude Guy
Westheimer, Gerald
Williston, John Stoddard
Wimer, Richard E
Yund, E William
Yuwiler, Arthur
Zucker, Robert Stephen
Zuker, Charles S

COLORADO
Bamburg, James Robert
Bekoff, Anne C
Best, LaVar
Betz, William J
Deitrich, Richard Adam
Dubin, Mark William
Eaton, Robert Charles
Finger, Thomas Emanuel
Ishii, Douglas Nobuo
Kater, Stanley B
Kulkosky, Paul Joseph

Lee, Robert Edward
Lin, Leu-Fen Hou
Lynch, George Robert
Mykles, Donald Lee
Pfenninger, Karl H
Roper, Stephen David
Seeds, Nicholas Warren
Tyler, Kenneth Laurence

CONNECTICUT
Badoyannis, Helen Litman
Barry, Michael Anhalt
Bayney, Richard Michael
Berlind, Allan
Bigland-Ritchie, Brenda
Bohannon, Richard Wallace
Charney, Dennis S
Chatt, Allen Barrett
Cohen, Lawrence Baruch
Cohen, Melvin Joseph
Davis, Michael
Denenberg, Victor Hugo
Glasel, Jay Arthur
Goldman, Bruce Dale
Heninger, George Robert
Herbette, Leo G
LaMotte, Carole Choate
Landmesser, Lynn Therese
Lentz, Thomas Lawrence
Marks, Lawrence Edward
Miller, William Henry
Morest, Donald Kent
Morrow, Jon S
Parker, Eric McFee
Potashner, Steven Jay
Ritchie, Brenda Rachel (Bigland)
Sachs, Benjamin David
Salafia, W(illiam) Ronald
Schwartz, Ilsa Roslow
Snider, Ray Michael
Squinto, Stephen P
Tallman, John Francis
Taub, Arthur
Wallace, Robert B
Waxman, Stephen George
Welch, Annemarie S

DELAWARE
Brown, Barry Stephen
Carnahan, James Elliot
Granda, Allen Manuel
Gulick, Walter Lawrence
Holyoke, Caleb William, Jr
Saller, Charles Frederick
Scott, Thomas Russell
Tam, Sang William

DISTRICT OF COLUMBIA
Alleva, John J
Chow, Ida
Cohn, Victor Hugo
Doctor, Bhupendra P
Finn, Edward J
Fishbein, William Nichols
Forman, David S
Herman, Barbara Helen
Jacobsen, Frederick Marius
Johnston, Laurance S
Joshi, Jay B
Kobylarz, Erik J
Lumpkin, Michael Dirksen
Perry, David Carter
Raslear, Thomas G
Sobotka, Thomas Joseph
Thomas, William Eric
Weinberger, Daniel R
Weiss, Ira Paul
Wylie, Richard Michael

FLORIDA
Ajmone-Marsan, Cosimo
Anderson, Peter Alexander Vallance
Barrett, Ellen Faye
Barrett, John Neil
Baumbach, Lisa Lorraine
Bennett, Gudrun Staub
Berkley, Karen J
Bunge, Mary Bartlett
Crews, Fulton T
Dombro, Roy S
Dunn, William Arthur, Jr
Evoy, William (Harrington)
Gennaro, Joseph Francis
Gleeson, Richard Alan
Goldstein, Mark Kane
Hamasaki, Duco I
Heaton, Marieta Barrow
Johnson, Irving Stanley
Knapp, Francis Marion
Lindsey, Bruce Gilbert
McKinney, Michael
Menzies, Robert Allen
Miskimen, George William
Muller, Kenneth Joseph
Munson, John Bacon
Neary, Joseph Thomas
Phelps, Christopher Prine
Potter, Lincoln Truslow
Price, David Alan
Rowland, Neil Edward
Sanberg, Paul Ronald
Tucker, Gail Susan
Tumer, Nihal
Wecker, Lynn

Wilson, David Louis
Wolgin, David L

GEORGIA
Albers, H Elliott
Barb, C Richard
Bates, Harold Brennan, Jr
Benoit, Peter Wells
Black, Asa C, Jr
Boothe, Ronald G
Bowen, John Metcalf
Brann, Darrell Wayne
Byrd, Larry Donald
Denson, Donald D
Doetsch, Gernot Siegmar
Edwards, Gaylen Lee
Fletcher, James Erving
Gibson, John Michael
Green, Robert Castleman
Holland, Robert Campbell
Isaac, Walter
Iuvone, Paul Michael
Jackson, William James
Loring, David William
McDaniel, William Franklin
Mahadik, Sahebarao P
Mahesh, Virendra B
May, Sheldon William
Mulroy, Michael J
Richard, Christopher Alan
Schweri, Margaret Mary
Sutin, Jerome
Toney, Thomas Wesley
Weissburg, Marc Joel
Wolf, Steven L
Zinsmeister, Philip Price

HAWAII
Bitterman, Morton Edward
Blanchard, Robert Joseph
Cooke, Ian McLean
Hardman, John M
Liu, Yong-Biao
Morton, Bruce Eldine

IDAHO
DeSantis, Mark Edward

ILLINOIS
Abler, William Lewis
Agarwal, Gyan C
Applebury, Meredithe L
Becker, Robert Paul
Bernstein, Joel Edward
Biller, Jose
Boshes, Louis D
Brewer, Gregory J
Brioni, Jorge Daniel
Brown, Meyer
Buschmann, MaryBeth Tank
Carr, Virginia McMillan
Comer, Christopher Mark
Dal Canto, Mauro Carlo
Daley, Darryl Lee
Dallos, Peter John
Dodge, Patrick William
Ebrey, Thomas G
Farbman, Albert Irving
Feng, Albert Shih-Hung
Gillette, Martha Ulbrick
Giordano, Tony
Glaser, Michael
Goldberg, Jay M
Gottlieb, Gerald Lane
Greenough, William Tallant
Hileman, Stanley Michael
Hoshiko, Michael S
Jensen, Robert Alan
Juraska, Janice Marie
Kleinman, Kenneth Martin
Kochman, Ronald Lawrence
Koenig, Heidi M
Krafft, Grant A
Lambert, Mary Pulliam
Lavelle, Arthur
Levine, Michael W
Lin, Chun-wel
Lit, Alfred
McKelvy, Jeffrey F
McNulty, John Alexander
Mafee, Mahmood Forootan
Manyam, Bala Venktesha
Menco, Bernard Philip Max
Muma, Nancy A
Nakajima, Shigehiro
Narahash, Toshio
Nardi, James Benjamin
Neet, Kenneth Edward
Niles, Walter Dulany, II
Offner, Franklin Faller
Pinto, Lawrence Henry
Pokorny, Joel
Pollack, Emanuel Davis
Rasenick, Mark M
Ruggero, Mario Alfredo
Satinoff, Evelyn
Schmidt, Robert Sherwood
Sladek, Celia Davis
Stumpf, David Allen
Takahashi, Joseph S
Ten Eick, Robert Edwin
Weyhenmeyer, James Alan
Wheeler, Bruce Christopher
Wilson, Hugh Reid

Wu, Chau Hsiung
Yamamoto, Hirotaka
Zimmerman, Roger Paul

INDIANA
Aprison, Morris Herman
Bayer, Shirley Ann
Chancey, Charles Clifton
Chernoff, Ellen Ann Goldman
Cohen, Marlene Lois
Conneally, P Michael
Devoe, Robert Donald
Echtenkamp, Stephen Frederick
Frommer, Gabriel Paul
Fuller, Ray W
Gehlert, Donald Richard
Gidda, Jaswant Singh
Guthrie, Catherine Shirley Nicholson
Henry, David P, II
Hyde, David Russell
Kennedy, Duncan Tilly
Leander, John David
Mason, Norman Ronald
Nichols, David Earl
Pak, William Louis
Pietsch, Paul Andrew
Rebec, George Vincent
Schulze, Gene Edward
Strickholm, Alfred
Thor, Karl Bruce
Vasko, Michael Richard
Wasserman, Gerald Steward

IOWA
Coulter, Joe Dan
Kollros, Jerry John
Myers, Glenn Alexander
Orme-Johnson, David Wear
Perlman, Stanley
Robinson, Robert George
Wu, Chun-Fang

KANSAS
Alper, Richard H
Berman, Nancy Elizabeth Johnson
Imig, Thomas Jacob
Michaelis, Mary Louise
Nelson, Stanley Reid
Norris, Patricia Ann
Pazdernik, Thomas Lowell
Samson, Frederick Eugene, Jr
Smalley, Katherine N
Voogt, James Leonard
Westfall, Jane Anne

KENTUCKY
Essig, Carl Fohl
Hirsch, Henry Richard
Legan, Sandra Jean
Mattson, Mark Paul
Olson, William Henry
Parker, Joseph Corbin, Jr
Randall, David Clark
Schurr, Avital
Slevin, John Thomas
Surwillo, Walter Wallace

LOUISIANA
Bradley, Ronald James
Caday, Cornelio Gacusana
Davis, George Diament
Dunn, Adrian John
Fingerman, Milton
Garcia, Meredith Mason
Johnsen, Peter Berghsey
Kastin, Abba J
Komiskey, Harold Louis
Kreisman, Norman Richard
Miceli, Michael Vincent
Nair, Pankajam K
Prasad, Chandan
Saphier, David
Seaman, Ronald L
Silverman, Harold
Smith, Diane Elizabeth
Strauss, Arthur Joseph Louis
York, David Anthony

MAINE
Dowse, Harold Burgess

MARYLAND
Alberts, Walter Watson
Anderson, David Everett
Andrews, Stephen Brian
Arora, Prince Kumar
Ball, Gregory Francis
Battey, James F
Bhathena, Sam Jehangirji
Bloch, Robert Joseph
Broadwell, Richard Dow
Brown, James Harvey
Brownell, William Edward
Cantor, David S
Carpenter, Malcolm Breckenridge
Charnas, Lawrence Richard
Chiueh, Chuang Chin
Cohen, Robert Martin
Cohen, Rochelle Sandra
Creveling, Cyrus Robbins, Jr
Dienel, Gerald Arthur
Dingman, Charles Wesley, II
Dubner, Ronald
Elson, Hannah Friedman

Estienne, Mark Joseph
Fechter, Laurence David
Fex, Jorgen
Fields, Kay Louise
Gainer, Harold
Garruto, Ralph Michael
Gatti, Philip John
Gilbert, Daniel Lee
Glenn, John Frazier
Gold, Philip William
Goldman, David
Goldstein, Murray
Goodwin, Frederick King
Green, Martin David
Guilarte, Tomas R
Guttman, Helene Augusta Nathan
Hallett, Mark
Hanbauer, Ingeborg
Hanley, Daniel F, Jr
Harris, Andrew Leonard
Helke, Cinda Jane
Hempel, Franklin Glenn
Henkin, Robert I
Hickey, Robert Joseph
Hill, James Leslie
Hoffman, Paul Ned
Hudson, Lynn Diane
Jacobowitz, David
Jacobson, Kenneth Alan
Johnson, Kenneth Olafur
Jordan, Craig Alan
Kandasamy, Sathasiva B
Kaufman, Elaine Elkins
Kaufmann, Peter G
Kennedy, Charles
Kety, Seymour S
King, Gregory L
Klatzo, Igor
Kuenzel, Wayne John
Kulkarni, Ashok Balkrishna
Larrabee, Martin Glover
Laterra, John J
Leach, Berton Joe
Lerman, Michael Isaac
Leshner, Alan Irvin
Lester, David Simon
Lever, John
Loh, Yoke Peng
Ludlow, Christy L
McCarron, Richard M
McCune, Susan K
McKenna, Mary Catherine
MacLean, Paul Donald
Mains, Richard E
Millar, David B
Mishkin, Mortimer
Monjan, Andrew Arthur
Montell, Craig
Moran, Jeffrey Chris
Mouradian, M Maral
Myslinski, Norbert Raymond
Nelson, Phillip Gillard
Nuite-Belleville, Jo Ann
Parsegian, Vozken Adrian
Pellmar, Terry C
Phelps, Creighton Halstead
Pickar, David
Poggio, Gian Franco
Popper, Arthur N
Pubols, Benjamin Henry, Jr
Rall, Wilfrid
Rawlings, Samuel Craig
Reese, Thomas Sargent
Rennels, Marshall L
Rogawski, Michael Andrew
Roman, Gustavo Campos
Ruchkin, Daniel S
Russell, James T
Saavedra, Juan M
Salem, Norman, Jr
Schnaar, Ronald Lee
Schramm, Lawrence Peter
Selmanoff, Michael Kidd
Sheridan, Philip Henry
Shih, Tsung-Ming Anthony
Small, Judy Ann
Song, Jiakun
Stoolmiller, Allen Charles
Strumwasser, Felix
Summers, Raymond
Sussman, Daniel Jesse
Vandenbergh, David John
Vinores, Stanley Anthony
Vogl, Thomas Paul
Webster, Henry deForest
Weight, Forrest F
Williamson, Lura C
Wilson, Wyndham Hopkins
Yu, Mei-ying Wong
Zielke, H Ronald

MASSACHUSETTS
Anderson-Olivo, Margaret
Axelson, John F
Barlow, John Sutton
Berman, Marlene Oscar
Bird, Stephanie J(ean)
Bowdan, Elizabeth Segal
Bursztajn, Sherry
Callard, Gloria Vincz
Caplan, Louis Robert
Cohen, Jonathan Brewer
Daniel, Peter Francis
DeBold, Joseph Francis

De Luca, Carlo J
Dethier, Vincent Gaston
Deutsch, Curtis Kimball
Dyer, Charissa Annette
Edwards, Ross B
Eldred, William D
Emanuel, Rodica L
Epstein, Irving Robert
Estes, William K
Fleming, Nigel
Gamzu, Elkan R
Graybiel, Ann M
Griffin, Donald R(edfield)
Gross, David John
Gusovsky, Fabian
Hall, Robert Dilwyn
Hammer, Ronald Page, Jr
Hauser, George
Hauser, Marc David
Hausman, Robert Edward
Kamen, Gary P
Kanarek, Robin Beth
Kauer, John Stuart
Kazemi, Homayoun
Keast, Craig Lewis
Kebabian, John Willis
Kornetsky, Conan
Koul, Omanand
Kravitz, Edward Arthur
Lee, Gloria
Lees, Marjorie Berman
Madras, Bertha K
Marcus, Elliot M
Marder, Eve Esther
Masek, Bruce James
Masland, Richard Harry
Merfeld, Daniel Michael
Morris, Robert
Morton, Mary E
Murphey, Rodney Keith
Murrain, Michelle
Neer, Eva Julia
Olivo, Richard Francis
O'Malley, Alice T
Palay, Sanford Louis
Papastoitsis, Gregory
Reppert, Steve Marion
Roffler-Tarlov, Suzanne K
Ross, Alonzo Harvey
Schneider, Gerald Edward
Schomer, Donald Lee
Schwarting, Gerald Allen
Schwartz, William Joseph
Shamgochian, Maureen Dowd
Sidman, Richard Leon
Sihag, Ram K
Singer, Joshua J
Smith, Peter John Shand
Solomon, Paul R
Sweadner, Kathleen Joan
Talamo, Barbara Lisann
Trimmer, Barry Andrew
Tyler, H Richard
Villa-Komaroff, Lydia
Voigt, Herbert Frederick
Weinberg, Crispin Bernard
Williams, Heather
Wolf, Merrill Kenneth
Wyse, Gordon Arthur
Yellen, Gary
Zervas, Nicholas Themistocles

MICHIGAN
Agranoff, Bernard William
Anderson, Thomas Edward
Barksdale, Charles Madsen
Barney, Christopher Carroll
Barraco, I Robin A
Bradley, Robert Martin
Casey, Kenneth L(yman)
Diaz, Fernando G
Dolan, David Francis
Dore-Duffy, Paula
Drescher, Dennis G
Dunbar, Joseph C
Easter, Stephen Sherman, Jr
Fenstermacher, Joseph Don
Fernandez, Hector R C
Foster, Douglas Layne
Freeman, Arthur Scott
Galloway, Matthew Peter
Gebber, Gerard L
Glover, Roy Andrew
Hall, Edward Dallas
Hawkins, Joseph Elmer
Heidemann, Steven Richard
Johnson, John Irwin, Jr
Krier, Jacob
Lahti, Robert A
Levine, Steven Richard
Lew, Gloria Maria
Nuttall, Alfred L
Pysh, Joseph John
Richardson, Rudy James
Schaefer, Gerald J
Schreur, Peggy Jo Korty Dobry
Stach, Robert William
Stout, John Frederick
Swaroop, Anand
Tosney, Kathryn W
Tse, Harley Y
Tweedle, Charles David
Ueda, Tetsufumi
Uhde, Thomas Whitley

Uhler, Michael David
Wiley, John W
Zamorano, Lucia Jopehina
Zand, Robert

MINNESOTA
Alsum, Donald James
Anderson, Victor Elving
Beitz, Alvin James
Berry, James Frederick
Brown, Rhoderick Edmiston, Jr
Cunningham, Julie Margaret
Drewes, Lester Richard
Dysken, Maurice William
Ghazali, Masood Raheem
Gumnit, Robert J
Iadecola, Costantino
Iaizzo, Paul Anthony
Koerner, James Frederick
Levine, Allen Stuart
Low, Walter Cheney
Messing, Rita Bailey
Miller, Robert F
Mortimer, James Arthur
Newman, Eric Allan
Overmier, J Bruce
Peterson, Phillip Keith
Rathbun, William B
Ray, Charles Dean
Sarkar, Gobinda
Seybold, Virginia Susan (Dick)
Soechting, John F
Sorensen, Peter W
Stauffer, Edward Keith
Terzuolo, Carlo A
Westmoreland, Barbara Fenn
Ziegler, Richard James

MISSISSIPPI
Alford, Geary Simmons
Dickens, Joseph Clifton
Haines, Duane Edwin
Jennings, David Phipps
Lynch, James Carlyle
Ma, Terence P
Peeler, Dudley F, Jr

MISSOURI
Agrawal, Harish C
Ahmed, Mahmoud S
Beinfeld, Margery Cohen
Cohen, Adolph Irvin
Derby, Albert
Finger, Stanley
Merlie, John Paul
Mills, Steven Harlon
Mooradian, Arshag Dertad
Natani, Kirmach
Nyquist-Battie, Cynthia
Pearlman, Alan L
Peck, Carol King
Powers, William John
Rosenthal, Harold Leslie
Sanes, Joshua Richard
Stein, Paul S G
Stephens, Robert Eric
Tannenbaum, Michael Glen
Thach, William Thomas
Tumosa, Nina Jean
Wright, Dennis Charles
Young, Paul Andrew

MONTANA
Linner, John Gunnar

NEBRASKA
Lau, Yuen-Sum
Mackin, Robert Brian
Mann, Michael David
Murrin, Leonard Charles

NEVADA
Ort, Carol

NEW HAMPSHIRE
Chabot, Christopher Cleaves
Nattie, Eugene Edward
Sower, Stacia Ann
Velez, Samuel Jose

NEW JERSEY
Allen, Richard Charles
Baker, Thomas
Banerjee, Pradeep Kumar
Cagan, Robert H
Campfield, L Arthur
Cascieri, Margaret Anne
Chaiken, MarthaLeah
Connor, John Arthur
Conway, Paul Gary
Curran, Thomas
Egger, M(aurice) David
Fangboner, Raymond Franklin
Feldman, Susan C
Fong, Tung Ming
Fryer, Rodney Ian
Gelperin, Alan
Gertner, Sheldon Bernard
Gray, Harry Edward
Gross, Charles Gordon
Hey, John Anthony
Hoebel, Bartley Gore
Hubbard, John W
Jabbar, Gina Marie

Kozikowski, Alan Paul
Laemle, Lois K
Levinson, Barry L
Malamed, Sasha
Manowitz, Paul
Margolskee, Robert F
Moyer, John Allen
Nelson, Nathan
Pachter, Jonathan Alan
Panagides, John
Rabii, Jamshid
Schlesinger, David H
Schreckenberg, Mary Gervasia
Solla, Sara A
Szewczak, Mark Russell
Tallal, Paula
Thompson, Robert L
Vargas, Hugo Martin

NEW MEXICO
Osbourn, Gordon Cecil
Savage, Daniel Dexter
Wild, Gaynor (Clarke)

NEW YORK
Abramson, Morris Barnet
Antrobus, John Simmons
Aschner, Michael
Barlow, Robert Brown
Beason, Robert Curtis
Begleiter, Henri
Bennett, Michael Vander Laan
Bidlack, Jean Marie
Bishop, Beverly Petterson
Blosser, James Carlisle
Bodnar, Richard Julius
Bolton, David C
Brody, Harold
Brown, Joel Edward
Brown, Lucy Leseur
Burchfiel, James Lee
Cabot, John Boit, II
Cammer, Wendy
Capranica, Robert R
Carpenter, David O
Chappell, Richard Lee
Chauhan, Abha
Chauhan, Ved P S
Chutkow, Jerry Grant
Cohan, Christopher Scott
Coleman, Paul David
Cooper, Arthur Joseph L
Crain, Stanley M
Davidson, Michael
Del Cerro, Manuel
Demeter, Steven
Diakow, Carol
Doty, Robert William
Drakontides, Anna Barbara
Edmonston, William Edward, Jr
Engbretson, Gustav Alan
Feldman, Samuel M
Felten, David L
Fetcho, Joseph Robert
Font, Cecilio Rafael
Gardner, Daniel
Gardner, Eliot Lawrence
Gardner, Esther Polinsky
Gershon, Michael David
Gibson, Gary Eugene
Gintautas, Jonas
Goldman, James Eliot
Gootman, Phyllis Myrna Adler
Grafstein, Bernice
Greenberg, Danielle
Greenberger, Lee M
Grob, David
Grota, Lee J
Halpern, Bruce Peter
Hankes, Lawrence Valentine
Harris-Warrick, Ronald Morgan
Hartung, John David
Henneberry, Richard Christopher
Hiller, Jacob Moses
Hof, Liselotte Bertha
Holland, Mary Jean Carey
Holloway, Ralph L, Jr
Holtzman, Eric
Hood, Donald C
Horel, James Alan
Hough, Lindsay B
Hughes, Stephen Edward
Jacquet, Yasuko F
Jakway, Jacqueline Sinks
John, E Roy
Johnston, Robert E
Karpiak, Stephen Edward
Kimelberg, Harold Keith
Knipple, Douglas Charles
Koizumi, Kiyomi
Kostreva, David Robert
Kristal, Mark Bennett
Kupfermann, Irving
Latimer, Clinton Narath
Lehrer, Gerard Michael
Leibovic, K Nicholas
Leibowitz, Sarah Fryer
Leisman, Gerald
Levinthal, Charles F
Lin, Fu Hai
Lnenicka, Gregory Allen
Low, Barbara Wharton
Luine, Victoria Nall
Lyser, Katherine May

Neurosciences (cont)

McCabe, R Tyler
McEwen, Bruce Sherman
Margolis, Richard Urdangen
Martin, David Lee
Mazurkiewicz, Joseph Edward
Mendell, Lorne Michael
Mozell, Maxwell Mark
Murphy, Randall Bertrand
Nelson, Mark Thomas
Noback, Charles Robert
Norton, William Thompson
Okamoto, Michiko
Pagala, Murali Krishna
Palmer, Eu(Gene) Charles
Pasik, Pedro
Pasik, Tauba
Pentney, Roberta Pierson
Pentyala, Srinivas Narasimha
Podleski, Thomas Roger
Possidente, Bernard Philip, Jr
Powell, Sharon Kay
Preston, James Benson
Raine, Cedric Stuart
Reeke, George Norman, Jr
Reilly, Margaret Anne
Ringo, James Lewis
Roberts, James Lewis
Rubinson, Kalman
Rudy, Bernardo
Ruggiero, David A
Sadosky, Alesia Beth
Saez, Juan Carlos
Scheinberg, Labe Charles
Scott, Sheryl Ann
Sechzer, Jeri Altneu
Seegal, Richard Field
Shapley, Robert M
Singh, Toolsee J
Smith, Gerard Peter
Soifer, David
Solomon, Seymour
Strand, Fleur Lillian
Teich, Malvin Carl
Thomas, Garth Johnson
Tieman, Suzannah Bliss
Udin, Susan Boymel
Victor, Jonathan David
Walcott, Benjamin
Wallman, Joshua
West, Norman Reed
Whitaker-Azmitia, Patricia Mack
Williams, Curtis Alvin, Jr
Williamson, Samuel Johns
Wineburg, Elliot N
Yazulla, Stephen
Yoburn, Byron Crocker
York, James Lester
Zukin, Stephen R

NORTH CAROLINA
Bell, Mary Allison
Boklage, Charles Edward
Booth, Raymond George
Campbell, Robert Joseph
Damstra, Terri
Defoggi, Ernest
Dibner, Mark Douglas
Dykstra, Linda A
Edens, Frank Wesley
Farel, Paul Bertrand
Ferris, Robert Monsour
Gerhardt, Don John
Hedlund, Laurence William
Hemperly, John Jacob
Hicks, T Philip
Hodson, Charles Andrew
Howard, James Lawrence
Kodavanti, Prasada Rao S
Leise, Esther M
Levitt, Melvin
Lewin, Anita Hana
Ludel, Jacqueline
Lund, Pauline Kay
McNamara, James O'Connell
MacPhail, Robert C
Morell, Pierre
Morgan, Kevin Thomas
O'Steen, Wendall Keith
Penry, James Kiffin
Purves, Dale
Sahyoun, Naji Elias
Smith, Ned Allan
Somjen, George G
Sonntag, William Edmund
Staddon, John Eric Rayner
Stern, Warren C
Surwit, Richard Samuel
Suzuki, Kunihiko
Tilson, Hugh Arval
Toews, Arrel Dwayne
Tyrey, Lee
Weiner, Richard D
Williams, Redford Brown, Jr
Wilson, John Eric
Wong, Fulton

NORTH DAKOTA
Penland, James Granville

OHIO
Alley, Keith Edward
Alway, Stephen Edward

Brunden, Kurt Russell
Carr, Albert A
Chovan, John David
Clark, David Lee
Cruce, William L R
Dorman, Robert Vincent
Dun, Nae Jiuum
Durand, Dominique M
Gao, Kuixiong
Gesteland, Robert Charles
Giannini, A James
Glenn, Loyd Lee
Godfrey, Donald Albert
Goodrich, Cecilie Ann
Gruenstein, Eric Ian
Heesch, Cheryl Miller
Herrup, Karl
Holtzman, David Allen
Horrocks, Lloyd Allen
Jones, Stephen Wallace
Jordan, Freddie L
Keller, Jeffrey Thomas
Kornacker, Karl
Krontiris-Litowitz, Johanna Kaye
Lacey, Beatrice Cates
Lacey, John I
LaManna, Joseph Charles
Landis, Dennis Michael Doyle
Landreth, Gary E
Lane, Richard Durelle
McQuarrie, Irvine Gray
Messer, William Sherwood, Jr
Mulick, James Anton
Myers, Ronald Elwood
Nasrallah, Henry A
Patterson, Michael Milton
Perry, George
Riker, Donald Kay
Robbins, David O
Robbins, Norman
Rosenberg, Howard C
Schwarz, Marvin
Sheridan, John Francis
Smith, Mark Anthony
Stuesse, Sherry Lynn
Thornton, Janice Elaine
Turner, John W, Jr
Waller, Hardress Jocelyn
Watanabe, Michiko
Whitacre, Caroline C
White, Robert J
Wolfe, Seth August, Jr
Zigmond, Richard Eric

OKLAHOMA
Dryhurst, Glenn
Farber, Jay Paul
Farris, Bradley Kent
Holloway, Frank A
Jones, Sharon Lynn
Norvell, John Edmondson, III
Papka, Raymond Edward
Revzin, Alvin Morton
Tobin, Sara L
Williams, Judy Ann

OREGON
Andresen, Michael Christian
Blythe, Linda L
Brown, Arthur Charles
Cereghino, James Joseph
Fahrenbach, Wolf Henrich
Gallaher, Edward J
Gwilliam, Gilbert Franklin
Leonard, Janet Louise
Lewy, Alfred James
Machida, Curtis A
Roselli, Charles Eugene
Ruwe, William David
Seil, Fredrick John
Soderling, Thomas Richard

PENNSYLVANIA
Allen, Michael Thomas
Balaban, Carey D
Barchi, Robert Lawrence
Billingsley, Melvin Lee
Bridger, Wagner H
Brooks, David Patrick
Buono, Russell Joseph
Cameron, William Edward
Clark, T(helma) K
Connor, John D
Davies, Richard Oelbaum
De Groat, William C, Jr
DeHaven-Hudkins, Diane Louise
DeKosky, Steven Trent
Dichter, Marc Allen
Eisenman, Leonard Max
Ermentrout, George Bard
Erulkar, Solomon David
Faber, Donald Stuart
Farley, Belmont Greenlee
Fernstrom, John Dickson
Fletcher, Jeffrey Edward
Freese, Andrew
Gerstein, George Leonard
Gogel, Germaine E
Goldberg, Michael Ellis
Gordon, Janice Taylor
Greenstein, Jeffrey Ian
Greif, Karen Faye
Grobstein, Paul
Grunwald, Gerald B

Hand, Peter James
Hartman, Herman Bernard
Hildebrandt, Thomas Henry
Hill, Shirley Yarde
Hollyday, Margaret Anne
Hurvich, Leo Maurice
Jameson, Dorothea
Joseph, Jeymohan
Josiassen, Richard Carlton
Kaplan, Barry B
Kauffman, Gordon Lee, Jr
Kovachich, Gyula Bertalan
Kozak, Wlodzimierz M
Kvist, Tage Nielsen
Lakoski, Joan Marie
Leavitt, Marc Laurence
Lindgren, Clark Allen
Lindstrom, Jon Martin
Lublin, Fred D
McLaughlin, Patricia J
McMorris, F Arthur
Peacock, Samuel Moore, Jr
Pollock, John Archie
Poplawsky, Alex James
Reynolds, Charles F
Ridenour, Marcella V
Robinson, Charles J
Roemer, Richard Arthur
Rubin, Leonard Sidney
Ruggieri, Michael Raymond
Russell, Richard Lawson
Salmoiraghi, Gian Carlo
Saporito, Michael S
Saunders, James Charles
Senft, Joseph Philip
Smyth, Thomas, Jr
Sodicoff, Marvin
Sprague, James Mather
Stewart, Charles Newby
Storella, Robert J, Jr
Stricker, Edward Michael
Summy-Long, Joan Yvette
Whinnery, James Elliott
Wilberger, James Eldridge
Wolstenholme, Wayne W
Wysocki, Charles Joseph
Yip, Joseph W
Zagon, Ian Stuart
Zigmond, Michael Jonathan

RHODE ISLAND
Christenson, Lisa
Flanagan, Thomas Raymond
Greenblatt, Samuel Harold
Hawrot, Edward
Hufnagel, Linda Ann
Jackson, Ivor Michael David
Knopf, Paul M
McIlwain, James Terrell
Oxenkrug, Gregory Faiva

SOUTH CAROLINA
Blake, Charles Albert
DeCoursey, Patricia Alice Jackson
Krall, Albert Raymond
Powell, Donald Ashmore
Wilson, Marlene Ann

TENNESSEE
Bealer, Steven Lee
Burt, Alvin Miller, III
Clauberg, Martin
DeSaussure, Richard Laurens, jr
Desiderio, Dominic Morse, Jr
Dettbarn, Wolf Dietrich
Elberger, Andrea June
Hoover, Donald Barry
Kirshner, Howard Stephen
Kostrzewa, Richard Michael
Krueger, James Martin
Lawler, James E
Lubar, Joel F
Miyamoto, Michael Dwight
Nowak, Thaddeus Stanley, Jr
Partridge, Lloyd Donald
Robert, Timothy Albert
Robertson, David
Sanders-Bush, Elaine
Senogles, Susan Elizabeth
Walker, William Stanley
Wikswo, John Peter, Jr

TEXAS
Achor, L(ouis) Joseph Merlin
Alkadhi, Karim A
Avery, Leon
Bachevalier, Jocelyne H
Barnes, Eugene Miller, Jr
Bjorndahl, Jay Mark
Brady, Scott T
Campbell, Carlos Boyd Godfrey
Clark, Wesley Gleason
Cooper, Cary Wayne
Dafny, Nachum
Denney, Richard Max
Dilsaver, Steven Charles
Doody, Rachelle Smith
Ducis, Ilze
Eidelberg, Eduardo
Finitzo-Hieber, Terese
Fox, Donald A
Franzl, Robert E
Gerken, George Manz
Gonzalez-Lima, Francisco

Gorry, G Anthony
Hamilton, Charles R
Ho, Begonia Y
Huffman, Ronald Dean
Irwin, Louis Neal
Jaweed, Mazher
Johnson, Kenneth Maurice, Jr
Johnston, Daniel
Keeley, Larry Lee
King, Thomas Scott
Klemm, William Robert
Kozlowski, Gerald P
Krulich, Ladislav
Leuchtag, H Richard
McAdoo, David J
Marks, Gerald A
Marshak, David William
Miikkulainen, Risto Pekka
Mikiten, Terry Michael
Norman, Reid Lynn
Oka, Kazuhiro
Pease, Paul Lorin
Peniston, Eugene Gilbert
Pirch, James Herman
Ross, Elliott M
Rylander, Michael Kent
Saltzberg, Bernard
Schaffer, Barbara Noyes
Sherry, Clifford Joseph
Skolnick, Malcolm Harris
Smith, Edward Russell
Smith, Eric Morgan
Thompson, Wesley Jay
Vinson, David Berwick
Wilczynski, Walter
Wilson, L Britt
Wong, Brendan So
Yung, W K Alfred

UTAH
Cheney, Carl D
Creel, Donnell Joseph
Fidone, Salvatore Joseph
Galster, William Allen
Gibb, James Woolley
Jacobson, Marcus
Jarcho, Leonard Wallenstein
Lasater, Eric Martin
Mullen, Richard Joseph
Okun, Lawrence M

VERMONT
Freedman, Steven Leslie
Hendley, Edith Di Pasquale
Parsons, Rodney Lawrence

VIRGINIA
Atrakchi, Aisar Hasan
Banker, Gary A
Boadle-Biber, Margaret Clare
Carson, Keith Alan
Corwin, Jeffrey Todd
De Haan, Henry J
Denbow, Donald Michael
Doherty, John Douglas
Fine, Michael Lawrence
Friesen, Wolfgang Otto
Gold, Paul Ernest
Grant, John Wallace
Gray, Faith Harriet
Grider, John Raymond
Guth, Lloyd
Katz, Alan Charles
Kurtzke, John F, Sr
Louttit, Richard Talcott
McCarty, Richard Charles
Macdonald, Timothy Lee
Maggio, Bruno
Murray, Jeanne Morris
Pellock, John Michael
Phillips, Larry H, II
Povlishock, John Theodore
Rosenblum, William I
Stein, Barry Edward
Tuttle, Jeremy Ballou
Warchol, Mark Edward
Werbos, Paul John

WASHINGTON
Artru, Alan Arthur
Baksi, Samarendra Nath
Barnes, Charles Dee
Baskin, Denis George
Beck, Thomas W
Bernard, Gary Dale
Bowden, Douglas Mchose
Calvin, William Howard
Catterall, William A
Churchill, Lynn
Clapshaw, Patric Arnold
Hawkins, Richard L
Hendrickson, Anita Elizabeth
Hille, Bertil
Kenney, Nancy Jane
Larsen, Lawrence Harold
Lovely, Richard Herbert
Mendelson, Martin
Nathanson, Neil Marc
Newmark, Jonathan
Stahl, William Louis
Steiner, Robert Alan
Truman, James William
Whatmore, George Bernard
Woods, Stephen Charles

BIOLOGICAL SCIENCES / 143

Young, Francis Allan

WEST VIRGINIA
Azzaro, Albert J
Culberson, James Lee
Gladfelter, Wilbert Eugene
Konat, Gregory W

WISCONSIN
Berman, Alvin Leonard
Bloom, Alan S
Bownds, M Deric
Boyeson, Michael George
Claude, Philippa
Curtis, Robin Livingstone
Gilboe, David Dougherty
Hamsher, Kerry de Sandoz
Hartmann, Henrik Anton
Hetzler, Bruce Edward
Hind, Joseph Edward
Hokin, Lowell Edward
Keesey, Richard E
Lelkes, Peter Istvan
Light, Douglas B
Prieto, Thomas E
Rhode, William Stanley
Sothmann, Mark Steven
Sufit, Robert Louis
Takahashi, Lorey K
Terasawa, EI
Tulunay-Keesey, Ulker
Welker, Wallace I
Wong-Riley, Margaret Tze Tung
Yoganandan, Narayan
Zuperku, Edward John

PUERTO RICO
Banerjee, Dipak Kumar
De La Sierra, Angell O
Kicliter, Ernest Earl, Jr
Kuffler, Damien Paul
Orkand, Richard K

ALBERTA
Kolb, Bryan Edward
Lukowiak, Kenneth Daniel
MacIntosh, Brian Robert
Roth, Sheldon H
Sainsbury, Robert Stephen
Spencer, Andrew Nigel
Stell, William Kenyon

BRITISH COLUMBIA
Crockett, David James
Jones, David Robert
McGeer, Edith Graef
McGeer, Patrick L
Mackie, George Owen
Pate, Brian David
Phillips, Anthony George
Schwarz, Dietrich Walter Friedrich
Sinclair, John G
Steeves, John Douglas
Vincent, Steven Robert

MANITOBA
Dakshinamurti, Krishnamurti
Glavin, Gary Bertrun

NEW BRUNSWICK
Seabrook, William Davidson
Sivasubramanian, Pakkirisamy

NOVA SCOTIA
Connolly, John Francis
Hanna, Brian Dale
Howlett, Susan Ellen

ONTARIO
Anderson, G Harvey
Barr, Murray Llewellyn
Beninger, Richard J
Bisby, Mark A
Brooks, Vernon Bernard
Brown, Ian Ross
Coscina, Donald Victor
De la Torre, Jack Carlos
Donald, Merlin Wilfred
Downer, Roger George Hamill
Govind, Choonilal Keshav
Gurd, James W
Hachinski, Vladimir
Hallett, Peter Edward
Halperin, Janet R P
Harrison, Robert Victor
Himms-Hagen, Jean
Howard, Ian Porteous
Hrdina, Pavel Dusan
Jones, Douglas L
Jones, Owen Thomas
Kalant, Harold
Kimura, Doreen
Kutcher, Stanley Paul
Leung, Lai-Wo Stan
Loeb, Gerald Eli
Mazurkiewicz-Kwilecki, Irena Maria
Pace Asciak, Cecil
Percy, Maire Ede
Pomeranz, Bruce Herbert
Regan, David M
Rollman, Gary Bernard
Schneiderman, Jacob Harry
Seguin, Jerome Joseph
Smith, Jonathan Jeremy Berkeley

Stanisz, Andrzej Maciej
Steel, Colin Geoffrey Hendry
Tasker, Ronald Reginald
Vanderwolf, Cornelius Hendrik
Weaver, Lynne C
Weick, Richard Fred
Witelson, Sandra Freedman

QUEBEC
Anctil, Michel
Antel, Jack Perry
Birmingham, Marion Krantz
Browman, Howard Irving
Capek, Radan
Chase, Ronald
Chouinard, Guy
Clark, Michael Wayne
Cohen, Monroe W
Collier, Brian
Cuello, A Claudio
Diksic, Mirko
Dykes, Robert William
Ervin, Frank (Raymond)
Gjedde, Albert
Gloor, Pierre
Jasper, Herbert Henry
Julien, Jean-Pierre
Karpati, George
MacIntosh, Frank Campbell
Murthy, Mahadi Raghavandrarao Ven
Olivier, Andre
Outerbridge, John Stuart
Palmour, Roberta Martin
Pappius, Hanna M
Paradis, Michel
Pasztor, Valerie Margaret
Robert, Suzanne
Steriade, Mircea
Tannenbaum, Gloria Shaffer
Viereck, Christopher
Young, Simon N

SASKATCHEWAN
Durden, David Alan
Irvine, Donald Grant
Johnson, Dennis Duane
Richardson, J(ohn) Steven
Wood, James Douglas

OTHER COUNTRIES
Andrej, Ladislav
Araki, Masasuke
Avila, Jesus
Cavonius, Carl Richard
Collins, Vincent Peter
Cras, Patrick
Curtis, Adam Sebastian G
De Troyer, Andre Jules
Donato, Rosario Francesco
Ganchrow, Donald
Guerrero-Munoz, Frederico
Hashimoto, Paulo Hitonari
Hirokawa, Katsuiku
Hisada, Mituhiko
Holman, Richard Bruce
Hughes, Abbie Angharad
Kawamura, Hiroshi
Kuriyama, Kinya
Ladinsky, Herbert
Levi-Montalcini, Rita
Lieb, William Robert
Livett, Bruce G
Lu, Guo-Wei
Maruta, Hiroshi
Millar, Robert Peter
Mølhave, Lars
Nabeshima, Toshitaka
Reuter, Harald
Sabry, Ismail
Sampson, Sanford Robert
Toutant, Jean-Pierre
Tung, Che-Se
Wakayama, Yoshihiro
Wenger, Byron Sylvester
Wettstein, Joseph G
Yamada, Eichi
Zapata, Patricio
Zbuzek, Vratislav

Nutrition

ALABAMA
Andrews, Frances E
Chastain, Marian Faulkner
Clark, Alfred James
Kochakian, Charles Daniel
Lovell, Richard Thomas
Menaker, Lewis
Navia, Juan Marcelo
Otto, David A
Roland, David Alfred, Sr
Sani, Brahma Porinchu
Sauberlich, Howerde Edwin
Tamura, Tsunenobu

ARIZONA
Gordon, Richard Seymour
Kaufman, C(harles) W(esley)
McCaughey, William Frank
McNamara, Donald J
Peng, Yeh-Shan
Sander, Eugene George
Stini, William Arthur

ARKANSAS
Greenman, David Lewis
Hatfield, Efton Everett
Horan, Francis E
Lewis, Sherry M
Martin, Lloyd W
Schieferstein, George Jacob
Tucker, Robert Gene

CALIFORNIA
Allerton, Samuel E
Amy, Nancy Klein
Baldwin, Ransom Leland
Barnard, R James
Betschart, Antoinette
Brandon, David Lawrence
Brown, Kenneth H
Burri, Betty Jane
Calloway, Doris Howes
Calvert, Chris
Canham, John Edward
Caporaso, Fredric
Carlisle, Edith M
Carlson, Don Marvin
Chang, George Washington
Chen, Tung-Shan
Christensen, Halvor Niels
Curry, Donald Lawrence
Dobbins, John Potter
Eckhert, Curtis Dale
Feeney, Robert Earl
Fleming, Sharon Elayne
Gietzen, Dorothy Winter
Gray, Gary M
Greenwood, Mary Rita Cooke
Grivetti, Louis Evan
Harrison, Michael R
Hawkes, Wayne Christian
Hill, Fredric William
Jacob, Mary
Johnson, Herman Leonall
Keagy, Pamela M
Keen, Carl L
Kelley, Darshan Singh
Kretsch, Mary Josephine
Lin, Grace Woan-Jung
Lönnerdal, Bo L
Lu, Nancy Chao
Mizuno, Nobuko S(himotori)
Montecalvo, Joseph, Jr
Nassos, Patricia Saima
Nelson, Gary Joe
Nielsen, Milo Alfred
Norman, Anthony Westcott
Painter, Ruth Coburn Robbins
Prater, Arthur Nickolaus
Rahlmann, Donald Frederick
Rao, Ananda G
Rodriguez, Mildred Shepherd
Rucker, Robert Blain
Russell, Gerald Frederick
Scala, James
Scott, Karen Christine
Slater, Grant Gay
Slavkin, Harold Charles
Smith, Christine H
Strother, Allen
Thomas, Heriberto Victor
Thomson, John Ansel Armstrong
Tinoco, Joan W H
Tsao, Constance S
Turnlund, Judith Rae
Yang, Meiling T
Zamenhof, Stephen

COLORADO
Allen, Kenneth G D
Ham, Richard George
Hambidge, K Michael
Panini, Sankhavaram R
Stifel, Fred B

CONNECTICUT
Bronner, Felix
Della-Fera, Mary Anne
Ferris, Ann M
Fordham, Joseph Raymond
Jensen, Robert Gordon
Khairallah, Edward A
Rebuffe-Scrive, Marielle Francoise
Ross, Donald Joseph

DELAWARE
Billheimer, Jeffrey Thomas
Kerr, Janet Spence
Rasmussen, Arlette Irene

DISTRICT OF COLUMBIA
Baker, Charles Wesley
Baltzell, Janet K
Callaway, Clifford Wayne
Canary, John J(oseph)
Frattali, Victor Paul
Harris, Suzanne Straight
Jacobson, Michael F
Pennington, Jean A T
Pla, Gwendolyn W
Prosky, Leon
Read, Merrill Stafford
Shank, Fred R
Simopoulos, Artemis Panageotis
Young, John Karl

FLORIDA
Anderson, William McDowell
Bauernfeind, Jacob (Jack) C(hristopher)
Borum, Peggy R
Cousins, Robert John
Grodberg, Marcus Gordon
Mantero-Atienza, Emilio
Miller, Donald F
Nanz, Robert Augustus Rollins
Ott, Edgar Alton
Rowland, Neil Edward
Sathe, Shridhar Krishna

GEORGIA
Berdanier, Carolyn Dawson
Carl, G Franklin
Edwards, Hardy Malcolm, Jr
Girardot, Jean Marie Denis
Hargrove, James Lee
McCormick, Donald Bruce
Mickelsen, Olaf
Mullen, Barbara J
Wang, Marian M
Warner, Harold
Woods, Wendell David

HAWAII
Ako, Harry Mu Kwong Ching
Hartung, G(eorge) Harley
Johnson, Nancy Ebersole
Waslien, Carol Irene
Wood, Betty J

IDAHO
Keim, Kathryn Sarah
Scott, James Michael

ILLINOIS
Abrams, Israel Jacob
Bhattacharyya, Maryka Horsting
Bookwalter, George Norman
Campbell, Michael Floyd
Clark, Jimmy Howard
Dasler, Waldemar
DiRienzo, Douglas Bruce
Druyan, Mary Ellen
Fahey, George Christopher, Jr
Hawrylewicz, Ervin J
Hentges, Lynnette S W
Heybach, John Peter
Inglett, George Everett
Janghorbani, Morteza
Jeffay, Henry
Jones, Michael R
Larson, Bruce Linder
Layman, Donald Keith
Lerner, Louis L(eonard)
Mallia, Anantha Krishna
Metcoff, Jack
Miller, Gregory Duane
Mobarhan, Sohrab
Reynolds, Robert David
Robinson, James Lawrence
Shambaugh, George E, III
Sutton, Lewis McMechan
Thompson, David Jerome
Velu, John G
White, Randy D
Wittman, James Smythe, III

INDIANA
Akrabawi, Salim S
Biggs, Homer Gates
Burns, Robert Alexander
Cook, David Allan
Day, Harry Gilbert
Forsyth, Dale Marvin
Frolik, Charles Alan
Leoschke, William Leroy
Montieth, Richard Voorhees
Pleasants, Julian Randolph
Schulz, Arthur R

IOWA
Barua, Arun B
Beitz, Donald Clarence
Goodridge, Alan G
Grundleger, Melvin Lewis
Lewis, Douglas Scott
Morriss, Frank Howard, Jr
Olson, James Allen
Rebouche, Charles Joseph
Roderuck, Charlotte Elizabeth
Thompson, Robert Gary
Young, Jerry Wesley

KANSAS
Harbers, Carole Ann Z
Krishnan, Engil Kolaj
Parrish, John Wesley, Jr
Ranhotra, Gurbachan Singh

KENTUCKY
Boling, James A
Chen, Linda Li-Yueh Huang
Chow, Ching Kuang
Fell, Ronald Dean
Glauert, Howard Perry
Jacobson, Elaine Louise
Kasarskis, Edward Joseph
Lee, Chung Ja
Mercer, Leonard Preston, II
Panemangalore, Myna
Petering, Harold George

Nutrition (cont)

Webster, Carl David

LOUISIANA
Alam, Bassima Saleh
Alam, Syed Qamar
Culley, Dudley Dean, Jr
Hegsted, Maren
Misra, Raghunath P
Pfeifer, Gerard David
Pryor, William Austin
Ramsey, Paul Roger
Stanfield, Manie K
Tso, Patrick Po-Wing
Udall, John Nicholas, Jr
Windhauser, Marlene M
Yerrapraggada, Venkat

MAINE
Cook, Richard Alfred

MARYLAND
Anderson, Richard Allen
Beecher, Gary Richard
Bhathena, Sam Jehangirji
Brannon, Patsy M
Caulfield, Laura E
Chakrabarti, Siba Gopal
Chang, Mei Ling (Wu)
De Luca, Luigi Maria
Edelman, Robert
Evarts, Ritva Poukka
Freedman, Laurence Stuart
Goor, Ronald Stephen
Gori, Gio Batta
Green, Martin David
Guilarte, Tomas R
Hanks, John Harold
Henkin, Robert I
Hornstein, Irwin
Howe, Juliette Coupain
Hubbard, Van Saxton
Kim, Sooja K
Knapka, Joseph J
Krebs-Smith, Susan M
Kulkarni, Ashok Balkrishna
Kwiterovich, Peter O, Jr
McGowan, Joan A
McKenna, Mary Catherine
McNaughton, James Larry
Matthews, Ruth H
Micozzi, Marc S
Nakhasi, Hira Lal
Passwater, Richard Albert
Pilch, Susan Marie
Poston, John Michael
Preusch, Peter Charles
Rechcigl, Miloslav, Jr
Salem, Norman, Jr
Sampugna, Joseph
Sanslone, William Robert
Sass, Neil Leslie
Scarbrough, Frank Edward
Smith, James Cecil
Sundaresan, Peruvemba Ramnathan
Taylor, Lauriston Sale
Taylor, Phillip R
Trout, David Linn
Vanderslice, Joseph Thomas
Varma, Shambhu D
Vydelingum, Nadarajen Ameerdanaden
Whittaker, Paul
Woods, Lewis Curry, III
Yergey, Alfred L, III

MASSACHUSETTS
Auskaps, Aina Marija
Babayan, Vigen Khachig
Barngrover, Debra Anne
Bert, Mark Henry
Clinton, Steven K
Flatt, Jean-Pierre
Geyer, Robert Pershing
Kanarek, Robin Beth
Krall, Elizabeth A
Lichtenstein, Alice Hinda
Narayan, Krishnamurthi Ananth
Nauss, Kathleen Minihan
Pober, Zalmon
Rand, William Medden
Roubenoff, Ronenn
Russell, Robert M
Sawyer, Frederick Miles
Schaefer, Ernst J
Shaw, James Headon
Stare, Fredrick J
Taylor, Allen
Zannis, Vassilis I

MICHIGAN
Cossack, Zafrallah Taha
Frisancho, A Roberto
Hunt, Charles E
Jen, Kai-Lin Catherine
Klurfeld, David Michael
Ku, Pao Kwen
McBean, Lois D
Marshall, Charles Wheeler
Saldanha, Leila Genevieve
Sardesai, Vishwanath M
Thomas, John William
Van Dyke, Russell Austin
Wishnetsky, Theodore

MINNESOTA
Barker, Norval Glen
Cleary, Margot Phoebe
Coon, Craig Nelson
Dixit, Padmakar Kashinath
Holman, Ralph Theodore
Johnson, Guy Henry
Johnson, Susan Bissen
Levine, Allen Stuart
Lukasewycz, Omelan Alexander
Meeder, Jeanne Elizabeth
Savaiano, Dennis Alan
Slavin, Joanne Louise

MISSISSIPPI
Combs, Gerald Fuson
Robinson, Edwin Hollis
Wilson, Robert Paul

MISSOURI
Chi, Myung Sun
Cornelius, Steven Gregory
Coudron, Thomas A
Czarnecki, Gail L
Hopkins, Daniel T
Shelton, Damon Charles
Sunde, Roger Allan
Vander Tuig, Jerry G
Veenhuizen, Jeffrey J
Vineyard, Billy Dale
Weisman, Gary Andrew
Williams, James E

MONTANA
Stanislao, Bettie Chloe Carter

NEBRASKA
Birt, Diane Feickert
Ferrell, Calvin L
Gessert, Carl F
Grandjean, Carter Jules
Lewis, Austin James
Yen, Jong-Tseng

NEW JERSEY
Colby, Richard H
Dalrymple, Ronald Howell
Deetz, Lawrence Edward, III
Ellenbogen, Leon
Fu, Shou-Cheng Joseph
Griminger, Paul
Hager, Mary Hastings
Keating, Kathleen Irwin
Lederman, Sally Ann
Machlin, Lawrence Judah
Martinez del Rio, Carlos
Mellies, Margot J
Mohammed, Kasheed
Morck, Roland Anton
Pleasants, Elsie W
Rosen, Robert T
Sargent, William Quirk
Staub, Herbert Warren
Stein, T Peter
Trifan, Daniel Siegfried
Triscari, Joseph
Watkins, Tom R
Weissman, Paul Morton

NEW MEXICO
Bock, Margaret Ann
Del Valle, Francisco Rafael
Dressendorfer, Rudy
Mills, David Edward
Omdahl, John L

NEW YORK
Albrecht, Alberta Marie
Ames, Stanley Richard
Awad, Atif B
Bauman, Dale E
Bernardis, Lee L
Chan, Mabel M
Chisholm, David R
Cohen, Leonard A
Cooperman, Jack M
Corradino, Robert Anthony
Cunningham-Rundles, Charlotte
De Luca, Chester
DiFrancesco, Loretta
Duta, Purabi
Finberg, Laurence
Fusco, Carmen Carrier
Garza, Cutberto
Geary, Norcross D
Goodman, DeWitt Stetten
Graham, Donald C W
Greenberg, Danielle
Haas, Jere Douglas
Haley, Nancy Jean
Henshaw, Edgar Cummings
Heymsfield, Steven B
Howard, Irmgard Matilda Keeler
Isaacs, Charles Edward
Jayme, David Woodward
Johnson, Patricia R
Kahan, Sidney
Karl, Peter I
Khan, Paul
Kishore, Gollamudi Sitaram
Klavins, Janis Viliberts
Kuftinec, Mladen M
Leleiko, Neal Simon
Lengemann, Frederick William
Lossinsky, Albert S
McGuire, John Joseph
Morrison, Mary Alice
Napoli, Joseph Leonard
Nelson, Mark Thomas
Oser, Bernard Levussove
Patterson, Ernest Leonard
Pinto, John
Rasmussen, Kathleen Maher
Repasky, Elizabeth Ann
Richie, John Peter, Jr
Rosensweig, Norton S
Sauer, Leonard A
Shayegani, Mehdi
Singh, Malathy
Sparks, Charles Edward
Steffen, Daniel G
Strain, Gladys Witt
Thysen, Benjamin
Traber, Maret G
Wapnir, Raul A
Wasserman, Robert Harold
Welch, Ross Maynard
Wu, Joseph M

NORTH CAROLINA
Bales, Connie Watkins
Bartholow, Lester C
Borgman, Robert F
Catignani, George Louis
Failla, Mark Lawrence
Gallagher, Margie Lee
Hackney, Anthony C
Hayes, Johnnie Ray
Louie, Dexter Stephen
Magee, Aden Combs, III
Morse, Roy E
Reinhard, John Frederick, Jr
Ronzio, Robert A
Schiffman, Susan S
Shaw, Helen Lester Anderson
Swaisgood, Harold Everett
Taylor, Martha Loeb

NORTH DAKOTA
Hunt, Curtiss Dean
Hunt, Janet R
Jacobs, Francis Albin
Milne, David Bayard
Nielsen, Forrest Harold
Uthus, Eric O

OHIO
Acosta, Phyllis Brown
Armstrong, Dwight W
Behr, Stephen Richard
Benton, Duane Allen
Birkhahn, Ronald H
Boggs, Robert Wayne
Burge, Jean C
Cardell, Robert Ridley, Jr
Churella, Helen R
Cooper, Dale A
Dabrowski, Konrad
Ekvall, Shirley W
Farrier, Noel John
Green, Ralph
Grooms, Thomas Albin
Hagerman, Larry M
Hecker, Art L
Hink, Walter Fredric
Jandacek, Ronald James
Lemon, Peter Willian Reginald
McCarthy, F D
Martin, Elden William
Miller, Robert Harold
Naito, Herbert K
Palmquist, Donald Leonard
Reddy, Padala Vykuntha
Richardson, Keith Erwin
Robinow, Meinhard
Smith, Kenneth Thomas
Stevenson, John Ray
Varma, Raj Narayan
Voss, Anne Coble
Webb, Thomas Evan
Weber, George Russell
Williams, Wallace Terry

OKLAHOMA
Koh, Eunsook Tak
Patterson, Manford Kenneth, Jr
Rikans, Lora Elizabeth
Seideman, Walter E
Thurman, Lloy Duane

OREGON
Hackman, Robert Mark
Sinnhuber, Russell Otto
Swank, Roy Laver
Whanger, Philip Daniel
Woldegiorgis, Gebretateos
Yearick, Elisabeth Stelle

PENNSYLVANIA
Albright, Fred Ronald
Barnes, William Shelley
Biebuyck, Julien Francois
Birdsall, Marion Ivens
Davies, Ronald Edgar
Dixit, Rakesh
Dryden, Richard Lee
Fenton, Marilyn Ruth
Green, Michael H(enry)
Jensen, Gordon L
Krummel, Debra A
Leach, Roland Melville, Jr
Mottur, George Preston
Mullen, James L
Pike, Ruth Lillian
Pollack, Robert Leon
Ramachandran, N
Rao, Kalipatnapu Narasimha
Ross, Alta Catharine
Scholz, Richard W
Seligson, Frances Hess
Skutches, Charles L
Smith, John Edgar
Sprince, Herbert
Stewart, Charles Newby
Thompson, Donald B
Tuan, Rocky Sung-Chi
Tulp, Orien Lee
Vasilatos-Younken, Regina
Wager-Page', Shirley A
Wright, Helen S
Yushok, Wasley Donald

RHODE ISLAND
Dymsza, Henry A
Nippo, Murn Marcus
Rand, Arthur Gorham, Jr

SOUTH CAROLINA
Hollis, Bruce Warren
Maurice, D V
Roel, Lawrence Edmund
Turk, Donald Earle

SOUTH DAKOTA
Grove, John Amos

TENNESSEE
Benton, Charles Herbert
Cook, Robert James
Cullen, Marion Permilla
Herting, David Clair
Jones, Peter D
Nowak, Thaddeus Stanley, Jr
Ong, David Eugene
Papas, Andreas Michael
Perry, Margaret Nutt
Robbins, Kelly Roy
Wagner, Conrad
Wakefield, Troy, Jr
Washington, Arthur Clover

TEXAS
Anthony, W Brady
Arnold, Watson Caufield
Bertrand, Helen Anne
Bielamowicz, Mary Claire Kinney
Brittin, Dorothy Helen Clark
Brown, Robert Dale
Carney, Darrell Howard
Cartwright, Aubrey Lee, Jr
Dudrick, Stanley John
Fernandes, Gabriel
Freeland-Graves, Jeanne H
Fry, Peggy Crooke
Gatlin, Delbert Monroe, III
Harris, Edward David
Haskell, Betty Echternach
Hsueh, Andie M
Huber, Gary Louis
Klein, Gordon Leslie
Kubena, Leon Franklin
Landmann, Wendell August
Lane, Helen W
Lawrence, Addison Lee
Liepa, George Uldis
Lifschitz, Meyer David
McMurray, David N
Madsen, Kenneth Olaf
Masoro, Edward Joseph
Mastromarino, Anthony John
Moyer, Mary Pat Sutter
Oberleas, Donald
Pond, Wilson Gideon
Rassin, David Keith
Reagor, John Charles
Root, Elizabeth Jean
Rudolph, Frederick Byron
Sherrod, Lloyd B
Spallholz, Julian Ernest
Wong, William Wai-Lun
Wood, Randall Dudley
Worthy, Graham Anthony James
Yeh, Lee-Chuan Caroline
Yu, Byung Pal

UTAH
Chan, Gary Mannerstedt
Graff, Darrell Jay
Hansen, Roger Gaurth
Johnston, Stephen Charles
Lawson, Larry Dale
Mahoney, Arthur W
Provenza, Frederick Dan
Ramachandran, Chittoor Krishna
Windham, Carol Thompson
Wyse, Bonita W

VERMONT
Bartel, Lavon L
Merrow-Hopp, Susan B
Tyzbir, Robert S

BIOLOGICAL SCIENCES / 145

VIRGINIA
Brandt, Richard Bernard
Burnette, Mahlon Admire, III
Christiansen, Marjorie Miner
Dannenburg, Warren Nathaniel
Feher, Joseph John
Herbein, Joseph Henry, Jr
Kirkpatrick, Roy Lee
Pao, Eleanor M
Rosenfield, Daniel
Sheppard, Alan Jonathan
Webb, Ryland Edwin

WASHINGTON
Adamson, Lucile Frances
Bankson, Daniel Duke
Childs, Marian Tolbert
Drum, Ryan William
Felton, Samuel Page
Hardy, Ronald W
Harper, Alfred Edwin
Kachmar, John Frederick
King, Irena B
Kutter, Elizabeth Martin
Lee, Donald Jack
Murphy, Mary Eileen
Pubols, Merton Harold
Sasser, Lyle Blaine
Shultz, Terry D
Swanson, Barry Grant
Yu, Ming-Ho

WEST VIRGINIA
Nomani, M Zafar
Schubert, John Rockwell

WISCONSIN
Armentano, Louis Ernst
Attie, Alan D
Baumann, Carl August
Brown, Raymond Russell
Deese, Dawson Charles
Hoekstra, William George
Jackson, Marion Leroy
Keesey, Richard E
Kemnitz, Joseph William
Kochan, Robert George
McIntosh, Elaine Nelson
Marlett, Judith Ann
Ogden, Robert Verl
Olsen, Ward Alan
Pariza, Michael Willard
Roberts, Willard Lewis
Schoeninger, Margaret J
Schulz, Leslie Olmstead
Steele, Robert Darryl
Tews, Jean Kring
Wege, Ann Christene

PUERTO RICO
El-Khatib, Shukri M
Preston, Alan Martin

ALBERTA
Basu, Tapan Kumar
Hawrysh, Zenia Jean
Mathison, Gary W(ayne)
Sohal, Parmjit S

BRITISH COLUMBIA
Freeman, Hugh J
Whyte, John Nimmo Crosbie

MANITOBA
Angel, Aubie
Ingalls, Jesse Ray
Marquardt, Ronald Ralph

NEWFOUNDLAND
Orr, Robin Denise Moore

NOVA SCOTIA
Castell, John Daniel
Hamilton, Robert Milton Gregory
Lall, Santosh Prakash
Weld, Charles Beecher

ONTARIO
Alexander, James Craig
Bayley, Henry Shaw
Beare-Rogers, Joyce Louise
Begin-Heick, Nicole
Bettger, William Joseph
Bray, Tammy M
Cheney, Margaret C
Cinader, Bernhard
Dalpe, Yolande
Fischer, Peter Wilhelm Fritz
Flanagan, Peter Rutledge
Fortin, Andre Francis
Hill, Eldon G
Holub, Bruce John
Pace Asciak, Cecil
Shah, Bhagwan G
Stavric, Bozidar
Yeung, David Lawrence

QUEBEC
Castonguay, Andre
Donefer, Eugene
Gougeon, Rejeanne
Kallai-Sanfacon, Mary-Ann
Kuhnlein, Harriet V
Pelletier, Omer

Peronnet, Francois R R
Pihl, Robert O
Srivastava, Uma Shanker
Tremblay, Angelo
Young, Simon N

SASKATCHEWAN
Patience, John Francis
Sosulski, Frank Walter

OTHER COUNTRIES
Bronk, John Ramsey
Brunser, Oscar
Crompton, David William Thomasson
Dutta-Roy, Asim Kanti
Fairweather-Tait, Susan Jane
Jaffe, Werner G
Ju, Jin Soon
Mela, David Jason
Palmer, Sushma Mahyera
Rio, Maria Esther
Said, Mostafa Mohamed
Santidrian, Santiago
Segal, Mark
Smith, Meredith Ford
Valyasevi, Aree
Vina, Juan R
Waterlow, John C

Physical Anthropology

ALABAMA
Hoff, Charles Jay
Wyss, James Michael

ALASKA
Milan, Frederick Arthur

ARIZONA
Birkby, Walter H
Bleibtreu, Hermann Karl
Merbs, Charles Francis
Olsen, Stanley John
Rittenbaugh, Cheryl K
Schoenwetter, James
Stini, William Arthur
Turner, Christy Gentry, II
Zegura, Stephen Luke

CALIFORNIA
Brodsky, Carroll M
Dolhinow, Phyllis Carol
Greenberg, Joseph H
Harrison, Gail Grigsby
Howell, Francis Clark
Johanson, Donald Carl
Jordan, Brigitte
Jurmain, Robert Douglas
Lindburg, Donald Gilson
McHenry, Henry Malcolm
McLeod, Samuel Albert
Murad, Turhon A
Napton, Lewis Kyle
Rogers, Spencer Lee
Roll, Barbara Honeyman Heath
Sarich, Vincent M
Treloar, Alan Edward
Williams, Bobby Joe
Wood, Corinne Shear
Zihlman, Adrienne Louella

COLORADO
Brues, Alice Mossie
Charney, Michael
Greene, David Lee
Hackenberg, Robert Allan
Kelso, Alec John (Jack)
Moore, Lorna Grindlay
Schanfield, Moses Samuel
Schulter-Ellis, Frances Pierce

CONNECTICUT
Heller, John Herbert
Hill, Andrew
Kidd, Kenneth Kay
Lamm, Foster Philip
Laughlin, William Sceva
Pospisil, Leopold Jaroslav

DELAWARE
Leslie, Charles Miller

DISTRICT OF COLUMBIA
Bernor, Raymond Louis
Ericksen, Mary Frances
Ortner, Donald John
Ubelaker, Douglas Henry

FLORIDA
Armelagos, George John
DeRousseau, C(arol) Jean
Lieberman, Leslie Sue
Linares, Olga F
Maples, William Ross
Wienker, Curtis Wakefield

GEORGIA
Smith, Euclid O'Neal

HAWAII
Baker, Paul Thornell
Goodman, Madeleine Joyce
Hanna, Joel Michael

Pietrusewsky, Michael, Jr

ILLINOIS
Buikstra, Jane Ellen
Corruccini, Robert Spencer
Costa, Raymond Lincoln, Jr
Giles, Eugene
Hausfater, Glenn
Klepinger, Linda Lehman
Reed, Charles Allen
Rosenberger, Alfred L
Simon, Mark Robert
Singer, Ronald
Tuttle, Russell Howard

INDIANA
Burr, David Bentley
O'Connor, Brian Lee
Todd, Harry Flynn, Jr

IOWA
Andelson, Jonathan Gary
Ciochon, Russell Lynn
Dawson, David Lynn
Staley, Robert Newton

KANSAS
Finnegan, Michael

KENTUCKY
Reid, Russell Martin
Savage, Steven Paul
Wiese, Helen Jean Coleman

MAINE
Howells, William White
Stoudt, Howard Webster

MARYLAND
Bresler, Jack Barry
Coelho, Anthony Mendes, Jr
Garruto, Ralph Michael
Micozzi, Marc S
Mischke, Barbara Suzanne
Pearson, Jay Dee
Yin, Frank Chi-Pong

MASSACHUSETTS
Gibbons, Michael Francis, Jr
Gordon, Claire Catherine
Hauser, Marc David
Hunt, Edward Eyre
Kolakowski, Donald Louis
Lamberg-Karlovsky, Clifford Charles
Moore, Sally Falk
Seltzer, Carl Coleman
Swedlund, Alan Charles
Thorp, James Wilson
Tucker, Katherine Louise
Wrangham, Richard Walter

MICHIGAN
Bajema, Carl J
Brace, C Loring
Garn, Stanley Marion
Lasker, Gabriel (Ward)
Livingstone, Frank Brown
Reynolds, Herbert McGaughey
Smith, Bennett Holly
Snow, Loudell Fromme
Weiss, Mark Lawrence
Wolpoff, Milford Howell

MINNESOTA
Himes, John Harter

MISSOURI
Beck, Lois Grant
Cheverud, James Michael
Gavan, James Anderson
Rasmussen, David Tab
Smith, Richard Jay
Sussman, Robert Wald

MONTANA
Smith, Charline Galloway

NEBRASKA
Kumar, Shrawan

NEVADA
Brooks, Sheilagh Thompson
Riley, Carroll Lavern

NEW JERSEY
Jolly, Clifford J
Tiger, Lionel

NEW MEXICO
Easley, Stephen Phillip

NEW YORK
Ascher, Robert
Berman, Carol May
Chepko-Sade, Bonita Diane
Delson, Eric
Dibennardo, Robert
Dubroff, Lewis Michael
Fleagle, John G
Geise, Marie Clabeaux
Gerber, Linda M
Gustav, Bonnie Lee
Haas, Jere Douglas
Holloway, Ralph L, Jr

Kennedy, Kenneth Adrian Raine
Kinzey, Warren Glenford
Leslie, Paul Willard
Little, Michael Alan
Mendel, Frank C
Pasamanick, Benjamin
Rightmire, George Philip
Sank, Diane
Sirianni, Joyce E
Sokal, Robert Reuven
Steegmann, Albert Theodore, Jr
Stern, Jack Tuteur, Jr
Susman, Randall Lee
Tattersall, Ian
Taylor, James Vandigriff
Townsend, John Marshall
Winter, John Henry

NORTH CAROLINA
Cartmill, Matt
Daniel, Hal J
Holcomb, George Ruhle
O'Barr, William McAlston
Pollitzer, William Sprott
Simons, Elwyn L

OHIO
Blank, John Edward
Byard, Pamela Joy
Chumlea, William Cameron
Latimer, Bruce Millikin
McConville, John Theodore
Marras, William Steven
Poirier, Frank Eugene
Saul, Frank Philip
Saul, Julie Mather
Sciulli, Paul William
Slutzky, Gale David

OKLAHOMA
Bell, Robert Eugene
Snow, Clyde Collins

OREGON
Moreno-Black, Geraldine S

PENNSYLVANIA
Blumberg, Baruch Samuel
Cronk, Christine Elizabeth
Dyson-Hudson, V Rada
Eckhardt, Robert Barry
Friedlaender, Jonathan Scott
Gottlieb, Karen Ann
Harpending, Henry Cosad
Johnston, Francis E
Katz, Solomon H
Mann, Alan Eugene
Schwartz, Jeffrey H
Siegel, Michael Ian
Tinsman, James Herbert, Jr
Weiss, Kenneth Monrad
Zimmerman, Michael Raymond

SOUTH CAROLINA
Rathbun, Ted Allan

TENNESSEE
Bass, William Marvin, III
Harris, Edward Frederick
McNutt, Charles Harrison
Tardif, Suzette Davis

TEXAS
Biggerstaff, Robert Huggins
Blangero, John Charles
Bramblett, Claud Allen
Fry, Edward Irad
Gibson, Kathleen Rita
Hixson, James Elmer
Lamb, Neven P
Malina, Robert Marion
Marshak, David William
Novak, Ladislav Peter
Race, George Justice
Steele, David Gentry
Wetherington, Ronald K
Williams-Blangero, Sarah Ann
Wohlschlag, Donald Eugene

UTAH
McCullough, John Martin

WASHINGTON
Chrisman, Noel Judson
Colby, Susan Melanie
Eck, Gerald Gilbert
Hurlich, Marshall Gerald
Lundy, John Kent
Newell-Morris, Laura
Nute, Peter Eric

WEST VIRGINIA
Fix, James D
Hilloowala, Rumy Ardeshir

WISCONSIN
Bennett, Kenneth A
Leutenegger, Walter
Oyen, Ordean James
Tappen, Neil Campbell

WYOMING
Gill, George Wilhelm

Physical Anthropology (cont)

BRITISH COLUMBIA
Booth, Amanda Jane

MANITOBA
De Pena, Joan Finkle
Kaufert, Joseph Mossman
Lewis, Marion Jean

NOVA SCOTIA
Walker, Joan Marion

ONTARIO
Anderson, James Edward
Beevis, David
Begun, David Rene
Gaherty, Geoffrey George
Helmuth, Hermann Siegfried
Hunter, W(illiam) Stuart
McFeat, Tom Farrar Scott
Saunders, Shelley Rae
Singh, Ripu Daman

QUEBEC
Kuhnlein, Harriet V

OTHER COUNTRIES
Goldstein, Marcus Solomon
Oxnard, Charles Ernest

Physiology, Animal

ALABAMA
Appel, Arthur Gary
Ardell, Jeffrey Laurence
Barker, Samuel Booth
Beasley, Philip Gene
Beckett, Sidney D
Berecek, Kathleen Helen
Bone, Leon Wilson
Boshell, Buris Raye
Bowie, Walter C
Bunch, Wilton Herbert
Carlo, Waldemar Alberto
Dacheux, Ramon F, II
Dixon, Earl, Jr
Downey, James Merritt
Elgavish, Ada S
Friedlander, Michael J
Gibbons, Ashton Frank Eleazer
Gladden, Bruce
Goldman, Ronald
Grubbs, Clinton Julian
Hageman, Gilbert Robert
Harding, Thomas Hague
Holloway, Clarke L
Jeffcoat, Marjorie K
Jenkins, Ronald Lee
Kochakian, Charles Daniel
Logic, Joseph Richard
McClintock, James Bruce
Magargal, Wells Wrisley, II
Matalon, Sadis
Modlin, Richard Frank
Neill, Jimmy Dyke
Oparil, Suzanne
Pegram, George Vernon, Jr
Philips, Joseph Bond, III
Pittman, James Allen, Jr
Pritchett, John Franklyn
Redding, Richard William
Schafer, James Arthur
Schnaper, Harold Warren
Schneyer, Charlotte A
Schoultz, Ture William
Sharma, Udhishtra Deva
Smith, Curtis R
Takahashi, Ellen Shizuko
Thomas, Julian Edward, Sr
Wit, Lawrence Carl
Yarbrough, James David

ALASKA
Babcock, Malin Marie
Feist, Dale Daniel
Miller, Lyster Keith
Paul, Augustus John, III
Proenza, Luis Mariano
Shirley, Thomas Clifton

ARIZONA
Baldwin, Ann Linda
Bloedel, James R
Blouin, Leonard Thomas
Buck, Stephen Henderson
Chandler, Douglas Edwin
Chapman, Reginald Frederick
Chen, Hsien-Jen James
Chiasson, Robert Breton
Clayton, John Wesley, Jr
Fowler, Dona Jane
Hadley, Mac Eugene
Hadley, Neil F
Hagedorn, Henry Howard
Harkins, Kristi R
Hazel, Jeffrey Ronald
Heine, Melvin Wayne
Hildebrand, John Grant, III
Johnson, Mary Ida
Kammer, Ann Emma
Kazal, Louis Anthony
Knudson, Ronald Joel

Krahl, Maurice Edward
Lei, David Kai Yui
Loy, Robert Graves
McCauley, William John
McCuskey, Robert Scott
Manciet, Lorraine Hanna
Nicolls, Ken E
Nugent, Charles Arter, Jr
Parsons, L Claire
Pickens, Peter E
Pough, Frederick Harvey
Prior, David James
Tarby, Theodore John
Tischler, Marc Eliot
Van Voorhies, Wayne Alan
Walsberg, Glenn Eric
Wegner, Thomas Norman
Wheeler, Diana Esther
Winfree, Arthur T
Witherspoon, James Donald

ARKANSAS
Allaben, William Thomas
Bridgman, John Francis
Cockerham, Lorris G(ay)
Cornett, Lawrence Eugene
Daniels, L B
Demarest, Jeffrey R
Doyle, Lee Lee
Garcia-Rill, Edgar E
Greenman, David Lewis
Harris, Grover Cleveland, Jr
Hatfield, Efton Everett
Kellogg, David Wayne
Light, Kim Edward
McGilliard, A Dare
McMillan, Harlan L
Marvin, Horace Newell
Piper, Edgar L
Rayford, Phillip Leon
Reddy, RamaKrishna Pashuvula
Soulsby, Michael Edward
Stallcup, Odie Talmadge
Tollett, James Terrell

CALIFORNIA
Adams, Thomas Edwards
Adams, William S
Ahmad, Nazir
Alberts, Allison Christine
Alexander, Natalie
Alousi, Adawia A
Amirkhanian, John David
Anand, Rajen S(ingh)
Anderson, Gary Bruce
Anderson, Geoffrey Robert
Antipa, Gregory Alexis
Arnaud, Claude Donald, Jr
Arp, Alissa Jan
Ashe, John Herman
Ashmore, Charles Robert
Austin, George M
Avila, Vernon Lee
Bacchus, Habeeb
Baker, Mary Ann
Baldwin, Ransom Leland
Balon, Thomas William
Barker, David Lowell
Barnes, Paul Richard
Barrett, Robert
Baylor, Denis Aristide
Beattie, Randall Chester
Beekman, Bruce Edward
Bennett, Albert Farrell
Bennett, Edward Leigh
Bentley, David R
Bercovitz, Arden Bryan
Berger, Ralph Jacob
Bern, Howard Alan
Bethune, John Edmund
Beuchat, Carol Ann
Bickford, Reginald G
Biglieri, Edward George
Binggeli, Richard Lee
Bloom, Floyd Elliott
Bolaffi, Janice Lerner
Book, Steven Arnold
Botstein, David
Bovell, Carlton Rowland
Bower, Annette
Brace, Robert Allen
Bradford, G Eric
Bradley, Timothy Jud
Braunstein, Glenn David
Breisch, Eric Alan
Brown, Joan Heller
Brown, Marvin Ross
Brunton, Laurence
Buchwald, Jennifer S
Buchwald, Nathaniel Avrom
Bullock, Leslie Patricia
Bullock, Theodore Holmes
Burrill, Melinda Jane
Callantine, Merritt Reece
Carew, Thomas Edward
Carlsen, Richard Chester
Carson, Virginia Rosalie Gottschall
Carstens, Earl E
Cavalieri, Ralph R
Cech, Joseph Jerome, Jr
Chalupa, Leo M
Chang, Ernest Sun-Mei
Chapman, Loring Frederick
Chappell, Mark Allen

Chopra, Inder Jit
Chow, Kao Liang
Clayton, Raymond Brazenor
Cohen, Malcolm Martin
Colvin, Harry Walter, Jr
Connell, Carolyn Joanne
Cornsweet, Tom Norman
Coty, William Allen
Covell, James Wachob
Cowles, David Lyle
Crescitelli, Frederick
Cronin, Michael John
Cronshaw, James
Cupps, Perry Thomas
Curras, Margarita C
Curry, Donald Lawrence
Dallman, Mary Fenner
Davidson, Julian M
Davis, Earle Andrew, Jr
Dawson, John E
Deller, John Joseph
Demetrescu, M
Diamond, Marian C
Doe, Richard P
Drewes, Robert Clifton
Duckles, Sue Piper
Duffey, Sean Stephen
Dunn, Arnold Samuel
Durrant, Barbara Susan
Edgren, Richard Arthur
Edmunds, Peter James
Eichenholz, Alfred
Eisenson, Jon
Eldred, Earl
Ellis, Stanley
Engel, Jerome, Jr
Engelmann, Franz
Enright, James Thomas
Ernest, Michael Jeffrey
Erpino, Michael James
Fagin, Katherine Diane
Fain, Gordon Lee
Fatt, Irving
Fernald, Russell Dawson
Fielding, Christopher J
Finch, Caleb Ellicott
Fiorindo, Robert Philip
Fisher, Robin Scott
Fisher, Steven Kay
Fleming, James Emmitt
Florsheim, Warner Hanns
Forsham, Peter Hugh
Frasier, S Douglas
Freeman, Ralph David
Freeman, Walter Jackson, III
Fronek, Kitty
Fuhrman, Frederick Alexander
Fuster, Joaquin Maria
Gabel, Joseph C
Gallistel, Charles Ransom
Ganong, William Francis
Gardin, Julius M
Garey, Walter Francis
Garoutte, Bill Charles
Garwood, Victor Paul
Geiselman, Paula J
Geyer, Mark Allen
Giri, Shri N
Gittes, Ruben Foster
Gledhill, Barton L
Glushko, Victor
Gold, Warren Maxwell
Goldsmith, Ralph Samuel
Goldstein, Bernard
Gonzalez, Ramon Rafael, Jr
Gordon, Malcolm Stephen
Gorski, Roger Anthony
Gray, Constance Helen
Gray, Sarah Delcenia
Greenwood, Mary Rita Cooke
Griffin, David William
Grindeland, Richard Edward
Grinnell, Alan Dale
Grodsky, Gerold Morton
Grover, Robert Frederic
Guillemin, Roger (Charles Louis)
Hammock, Bruce Dupree
Hance, Anthony James
Hansen, Robert John
Hatton, Glenn Irwin
Hayashida, Tetsuo
Heber, David
Hemingway, George Thomson
Hemmingsen, Edvard A(lfred)
Henzl, Milan Rastislav
Hershman, Jerome Marshall
Hoffman, Arlene Faun
Hollenberg, Milton
Holmes, William Neil
Horowitz, John M
Horvath, Kalman
Hoyt, Donald Frank
Hsueh, Aaron Jen Wang
Ingels, Neil Barton, Jr
Jacobsen, Nadine Klecha
Jaffe, Robert B
Johansson, Mats W
Jones, Edward George
Jones, James Henry
Jones, Ronald McClung
Kahler, Richard Lee
Kalland, Gene Arnold
Kaneko, Jiro Jerry

Kaplan, Selna L
Karam, John Harvey
Keil, Lanny Charles
Keller, Edward Lowell
Kelly, Amy Schick
Kelman, Bruce Jerry
Kendig, Joan Johnston
Kerstetter, Theodore Harvey
Kitzes, Leonard Martin
Kleeman, Charles Richard
Klouda, Mary Ann Aberle
Knudsen, Eric Ingvald
Ko, Chien-Ping
Konopka, Ronald J
Koopowitz, Harold
Kooyman, Gerald Lee
Koschier, Francis Joseph
Krupp, Marcus Abraham
La Ganga, Thomas S
Laiken, Nora Dawn
Lance, Valentine A
Leffert, Hyam Lerner
Lein, Allen
Leonora, John
Lester, Henry Allen
Levine, Jon David
Levine, Rachmiel
Libet, Benjamin
Lieber, Richard L
Lindsley, David Ford
Linfoot, John Ardis
Livingston, Robert Burr
Longhurst, John Charles
Lord, Elizabeth Mary
Lu, John Kuew-Hsiung
Luckock, Arlene Suzanne
McClure, William Owen
McCluskey, Elwood Sturges
McCulloch, Andrew Douglas
McCullough, Richard Donald
Macey, Robert Irwin
MacLeod, Donald Iain Archibald
McMahan, Uel Jackson, II
MacMillen, Richard Edward
Mandell, Robert Burton
Marcellino, George Raymond
Marshall, John Foster
Marshall, Louise Hanson
Martinez, Joe L, Jr
Massie, Barry Michael
Mayeri, Earl Melchior
Meerdink, Denis J
Mendel, Verne Edward
Merzenich, Michael Matthias
Metzner, Walter
Miljanich, George Paul
Miller, Arthur Joseph
Miller, Larry O'Dell
Mines, Allan Howard
Moberg, Gary Philip
Mohler, John George
Mole, Paul Angelo
Morehouse, Laurence Englemohr
Moretti, Richard Leo
Morgan, Meredith Walter
Mosier, H David, Jr
Mueller, Thomas Joseph
Nacht, Sergio
Nadel, Jay A
Nagy, Kenneth Alex
Nandi, Satyabrata
Nelson, Darren Melvin
Neville, James Ryan
Newman, Melvin Micklin
Nicoll, Charles S
Nicoll, Roger Andrew
Nishioka, Richard Seiji
Novak, Robert Eugene
Oglesby, Larry Calmer
O'Leary, Dennis Patrick
Parker, Harold R
Peckham, William Dierolf
Peitz, Betsy
Perry, Jacquelin
Peterson-Falzone, Sally Jean
Potter, Richard Lyle
Povzhitkov, Moysey Michael
Powell, Frank Ludwig, Jr
Powers, Dennis A
Pozos, Robert Steven
Rahlmann, Donald Frederick
Ramachandran, Janakiraman
Ramachandran, Vilayanur Subramanian
Ramsay, David John
Renkin, Eugene Marshall
Rich, Terrell L
Riedman, Richard M
Rivier, Catherine L
Robertson, Richard Thomas
Rosenthal, Fred
Rossi, John Joseph
Roth, Ariel A
Rothman, Stephen Sutton
Rowley, Rodney Ray
Rubanyi, Gabor Michael
Rubinstein, Lydia
Russell, Sharon May
Sanders, Brenda Marie
Sassenrath, Ethelda Norberg
Saunders, Virginia Fox
Sawyer, Charles Henry
Sawyer, Wilbur Henderson
Sayers, George
Schatzlein, Frank Charles

Scheibel, Arnold Bernard
Schlag, John
Schlapfer, Werner T
Schoenholz, Walter Kurt
Schwab, Ernest Roe
Scobey, Robert P
Segall, Paul Edward
Segundo, Jose Pedro
Selverston, Allen Israel
Sharma, Arjun D
Sharp, Gary Duane
Shelesnyak, Moses Chaim
Shneour, Elie Alexis
Shoemaker, Vaughan Hurst
Shook, Brenda Lee
Siegel, Edward T
Siggins, George Robert
Simkin, Benjamin
Sinha, Yagya Nand
Sloan, William Cooper
Smiles, Kenneth Albert
Smith, Erla Ring
Smith, Warren Drew
Snow, George Edward
Snowdowne, Kenneth William
Solomon, David Harris
Soltysik, Szczesny Stefan
Song, Moon K
Sonnenschein, Ralph Robert
Soule, Roger Gilbert
Spitzer, Nicholas Canaday
Stabenfeldt, George H
Stanczyk, Frank Zygmunt
Stanton, Hubert Coleman
Stanton, Toni Lynn
Steffey, Eugene P
Steinberg, Roy Herbert
Steinhardt, Richard Antony
Stickney, Janice Lee
Stone, Daniel Boxall
Stryker, Michael Paul
Swerdloff, Ronald S
Swett, John Emery
Thomas, Eric Owen
Thurmond, William
Timiras, Paola Silvestri
Ting, Irwin Peter
Titelbaum, Sydney
Tjioe, Djoe Tjhoo
Tormey, John McDivit
Towle, Albert
Tsien, Richard Winyu
Tullis, Richard Eugene
Tune, Bruce Malcolm
Tutwiler, Gene Floyd
VanderLaan, Willard Parker
Van Sluyters, Richard Charles
Vidal, Jacques J
Vidoli, Vivian Ann
Villablanca, Jaime Rolando
Volk, Thomas Lewis
Wagner, Jeames Arthur
Wangler, Roger Dean
Ward, David Gene
Warren, Dwight William, III
Wechter, William Julius
Weg, Ruth B(ass)
Weinberg, Daniel I
Weiner, Richard Ira
Wells, Patrick Harrington
West, John B
Wickler, Steven John
Wilson, Lowell D
Wine, Jeffrey Justus
Winer, Jeffery Allan
Winet, Howard
Wolfe, Allan Marvin
Woody, Charles Dillon
Wright, James Edward
Wright, Richard Kenneth
Yen, S C Yen
Zajac, Felix Edward, III
Zomzely-Neurath, Claire Eleanore
Zucker, Irving

COLORADO
Allen, Kenneth G D
Alpern, Herbert P
Amann, Rupert Preynoessl
Audesirk, Gerald Joseph
Banchero, Natalio
Bekoff, Anne C
Berens, Randolph Lee
Best, Jay Boyd
Betz, George
Capen, Ronald L
Cicerone, Carol Mitsuko
Conger, John Douglas
Cundiff, Milford Mel Fields
Dudek, F Edward
Eisenbarth, George Stephen
Ellinwood, William Edward
Gamow, Rustem Igor
Gersten, Jerome William
Gorell, Thomas Andrew
Hesterberg, Thomas William
Hossner, Kim L
Jacobson, Eugene Donald
Jansen, Gustav Richard
Jones, David Evan
Katz, Fred H
Korr, Irvin Morris
Krausz, Stephen
Kulkosky, Paul Joseph

Laudenslager, Mark LeRoy
Luttges, Marvin Wayne
Lynch, Carol Becker
Lynch, George Robert
McClellan, John Forbes
MacGregor, Ronald John
Martin, Alexander Robert
Moran, David Taylor
Musick, James R
Nett, T M
Neville, Margaret Cobb
Niswender, Gordon Dean
Nornes, Howard Onsgaard
Norris, David Otto
Northern, Jerry Lee
Palmer, Michael Rule
Platt, James Earl
Seidel, George Elias, Jr
Simske, Steven John
Snyder, Gregory Kirk
Sussman, Karl Edgar
Tabakoff, Boris
Tucker, Alan
Voss, James Leo
Watkins, Linda Rothblum
Wickelgren, Warren Otis

CONNECTICUT
Ballantyne, Garth H
Barone, Milo C
Barry, Michael Anhalt
Blackburn, Daniel Glenn
Blechner, Jack Norman
Block, Bartley C
Boell, Edgar John
Booth, Charles E
Boron, Walter Frank
Chapple, William Dismore
Chatt, Allen Barrett
Clark, Nancy Barnes
Cohen, Melvin Joseph
Constantine-Paton, Martha
Daw, Nigel Warwick
Della-Fera, Mary Anne
De Nuccio, David Joseph
Dillard, Morris, Jr
Doeg, Kenneth Albert
DuBois, Arthur Brooks
Emerson, Thomas Edward, Jr
Fleming, James Stuart
Forbush, Bliss, III
Gallo, Robert Vincent
Gay, Thomas John
Gerritsen, Mary Ellen
Goldman, Bruce Dale
Goldman-Rakic, Patricia S(hoer)
Goldsmith, Timothy Henshaw
Grimm-Jorgensen, Yvonne
Hitchcock, Margaret
Imbruce, Richard Peter
Janis, Ronald Allen
Kent, John Franklin
Kream, Barbara Elizabeth
Lamotte, Robert Hill
Letts, Lindsay Gordon
Loomis, Stephen Henry
Mason, John Wayne
Michael, Charles Reid
Morest, Donald Kent
Notation, Albert David
Novick, Alvin
O'Looney, Patricia Anne
Ornston, Leo Nicholas
Pilar, Guillermo Roman
Raisz, Lawrence Gideon
Rakic, Pasko
Rauch, Albert Lee
Riesen, John William
Sasaki, Clarence Takashi
Shapiro, Irving
Shepherd, Gordon Murray
Somers, Michael Eugene
Stitt, John Thomas
Swadlow, Harvey A
Taigen, Theodore Lee
Thurberg, Frederick Peter
Woody, Charles Owen, Jr
Wyman, Robert J
Zahler, Raphael

DELAWARE
Beckman, David Allen
Jain, Mahendra Kumar
Kerr, Janet Spence
Lee, Robert John
Lotrich, Victor Arthur
Melby, James Michael
Saller, Charles Frederick
Salzman, Steven Kerry
Sammelwitz, Paul H
Scott, Thomas Russell
Stephens, Gregory A
Stetson, Milton H
Taylor, Malcolm Herbert

DISTRICT OF COLUMBIA
Ahluwalia, Balwant Singh
Allen, Robert Erwin
Barker, Winona Clinton
Beck, Lucille Bluso
Becker, Kenneth Louis
Benson, Dennis Alan
Blanquet, Richard Steven
Bowling, Lloyd Spencer, Sr

Chaput, Raymond Leo
Crawford, Lester M
Crisp, Thomas Mitchell, Jr
Dimond, Marie Therese
Dudley, Susan D
Eagles, Douglas Alan
Frey, Mary Anne Bassett
Harmon, John W
Henry, Walter Lester, Jr
Holloway, James Ashley
Ison-Franklin, Eleanor Lutia
Kenney, Richard Alec
Kot, Peter Aloysius
Lavine, Robert Alan
Lumpkin, Michael Dirksen
McEwen, Gerald Noah, Jr
Malveaux, Floyd J
Murphy, James John
Packer, Randall Kent
Pearlman, Ronald C
Pennington, Jean A T
Phillips, Robert Ward
Pointer, Richard Hamilton
Rapisardi, Salvatore C
Rose, John Charles
Rosenberg, Edith E
Silva, Omega Logan
Sulzman, Frank Michael
Tidball, Charles Stanley
Tidball, M Elizabeth Peters
Van Arsdel, William Campbell, III
Wilson, Don Ellis
Wilson, Edward Matthew
Wong, Harry Yuen Chee
Wylie, Richard Michael
Wyngaarden, James Barnes
Young, John Karl

FLORIDA
Adams, Ralph M
Adams, Roger Omar
Anderson, John Francis
Baker, Carleton Harold
Barrett, Ellen Faye
Barrett, John Neil
Barron, Mace Gerald
Bassett, Arthur L
Bensen, Jack F
Besch, Emerson Louis
Biersdorf, William Richard
Blackwell, Harold Richard
Booth, Nicholas Henry
Boyd, Eleanor H
Bramante, Pietro Ottavio
Brown, Dawn LaRue
Bzoch, Kenneth R
Cassin, Sidney
Chen, Chao Ling
Clendenin, Martha Anne
Davenport, Paul W
Davidoff, Robert Alan
Dietz, John R
Dunn, William Arthur, Jr
Easton, Dexter Morgan
Ellias, Loretta Christine
Emerson, Geraldine Mariellen
Emmanuel, George
Farina, Joseph Peter
Fernandez, Hugo Lathrop
Freeman, Marc Edward
Freund, Gerhard
Friedl, Frank Edward
Galindo, Anibal H
Gaunt, Robert
Gordon, Kenneth Richard
Gorniak, Gerard Charles
Greenberg, Michael John
Gruber, Samuel Harvey
Hackman, John Clement
Hammer, Lowell Clarke
Hansen, Peter J
Harrison, Robert J
Holt, Joseph Paynter, Sr
Hope, George Marion
Jaeger, Marc Jules
Joftes, David Lion
Kalra, Satya Paul
Katovich, Michael J
Kessler, Richard Howard
Knapp, Francis Marion
Krzanowski, Joseph John, Jr
Landolt, Peter John
Lawrence, Pauline Olive
Levey, Douglas J
Levy, Norman Stuart
Lillywhite, Harvey B
Lindsey, Bruce Gilbert
Lipner, Harry Joel
Loewenstein, Werner Randolph
Lutz, Peter Louis
McCroskey, Robert Lee
McKenney, Charles Lynn, Jr
McKenzie, John Maxwell
McLean, Mark Philip
McNab, Brian Keith
Magleby, Karl LeGrande
Merritt, Alfred M, II
Michie, David Doss
Miles, Richard David
Nation, James Lamar, Sr
Nichols, Wilmer Wayne
Nordlie, Frank Gerald
Oberlander, Herbert
Otis, Arthur Brooks

Ott, Edgar Alton
Palmore, William P
Pargman, David
Penhos, Juan Carlos
Peterson, Ralph Edward
Pfeiffer, Eric A
Phelps, Christopher Prine
Pieper, Heinz Paul
Pollock, Michael L
Ramsey, James Marvin
Rao, Krothapalli Ranga
Rao, Papineni Seethapathi
Reininger, Edward Joseph
Reynolds, David George
Robinson, Gerald Garland
Rodrick, Gary Eugene
Root, Allen William
Rowland, Neil Edward
Sackner, Marvin Arthur
Schmidt-Nielsen, Bodil Mimi
Shankland, Daniel Leslie
Sharf, Donald Jack
Sharp, John Turner
Snow, Thomas Russell
Soliman, Karam Farag Attia
Spieler, Richard Earl
Stafford, Robert Oppen
Stainsby, Wendell Nicholls
Sturbaum, Barbara Ann
Sypert, George Walter
Thatcher, William Watters
Thomas, William Clark, Jr
Thompson, Ronald Halsey
Thuning-Roberson, Claire Ann
Tiffany, William James, III
Tobey, Frank Lindley, Jr
Tumer, Nihal
Vail, Edwin George
Vallowe, Henry Howard
Vander Meer, Robert Kenneth
Vesely, David Lynn
Voigt, Walter
Walker, Don Wesley
Weber, James Edward
White, Arlynn Quinton, Jr
Wilcox, Christopher Stuart
Wilde, Walter Samuel
Wright, Paul Albert
Zengel, Janet Elaine

GEORGIA
Baldwin, Bernell Elwyn
Barb, C Richard
Bhalla, Vinod Kumar
Binnicker, Pamela Caroline
Black, John B
Bond, Gary Carl
Brann, Darrell Wayne
Buccafusco, Jerry Joseph
Calabrese, Ronald Lewis
Catravas, John D
Comeau, Roger William
Costoff, Allen
Coulter, Dwight Bernard
Crim, Joe William
Dale, Edwin
Dingledine, Raymond J
Doetsch, Gernot Siegmar
Eaton, Douglas Charles
Edwards, Gaylen Lee
Ehrhart, Ina C
English, Arthur William
Franch, Robert H
Givens, James Robert
Hadd, Harry Earle
Hendrich, Chester Eugene
Hofman, Wendell Fey
Howarth, Birkett, Jr
Humphrey, Donald R
Hunt, William Daniel
Inge, Walter Herndon, Jr
Innes, David Lyn
Jones, Robert Leroy
Joyner, Ronald Wayne
Kraeling, Robert Russell
Leibach, Fredrick Hartmut
Leitch, Gordon James
Little, Robert Colby
McDaniel, William Franklin
Mahesh, Virendra B
Martin, David Edward
Michael, Richard Phillip
Miller, David Arthur
Mills, Thomas Marshall
Mokler, Corwin Morris
Mulroy, Michael J
Nadler, Ronald D
Neville, Walter Edward, Jr
Nosek, Thomas Michael
Ogle, Thomas Frank
Parks, John S
Pashley, David Henry
Phillips, Lawrence Stone
Porter, James W
Porterfield, Susan Payne
Puett, J David
Richard, Christopher Alan
Rinard, Gilbert Allen
Smith, Anderson Dodd
Smith, Jesse Graham, Jr
Snell, Terry Wayne
Sridaran, Rajagopala
Stoney, Samuel David, Jr
Thompson, Frederick Nimrod, Jr

148 / DISCIPLINE INDEX

Physiology, Animal (cont)

Wood, John Grady

HAWAII
Ahearn, Gregory Allen
Atkinson, Shannon K C
Brick, Robert Wayne
Bryant-Greenwood, Gillian Doreen
De Feo, Vincent Joseph
Gillary, Howard L
Grau, Edward Gordon
Greenwood, Frederick C
Hall, John Bradley
Hapai, Marlene Nachbar
Hartline, Daniel Keffer
Hartung, G(eorge) Harley
Krupp, David Alan
Lee, Cheng-Sheng
Nelson, Marita Lee
Pang-Ching, Glenn K
Read, George Wesley
Wayman, Oliver
Weems, Charles William
Whittow, George Causey
Yount, David Eugene
Zaleski, Halina Maria

IDAHO
Cloud, Joseph George
Fuller, Mark Roy
House, Edwin W
Kelley, Fenton Crosland
Mead, Rodney A
Mudumbi, Ramagopal Vijaya
Rodnick, Kenneth Joseph

ILLINOIS
Agin, Daniel Pierre
Ahmad, Sarfraz
Bahr, Janice M
Banerjee, Chandra Madhab
Barr, Lloyd
Barr, Susan Hartline
Bartell, Marvin H
Bartke, Andrzej
Bastian, James W
Baumann, Gerhard
Becker, Donald Eugene
Becker, John Henry
Berger, Sheldon
Berry, Robert Wayne
Bers, Donald M
Bettice, John Allen
Bhattacharyya, Maryka Horsting
Bingel, Audrey Susanna
Bombeck, Charles Thomas, III
Bruce, David Stewart
Burhop, Kenneth Eugene
Buschmann, MaryBeth Tank
Buschmann, Robert J
Caspary, Donald M
Chatterton, Robert Treat, Jr
Chesky, Jeffrey Alan
Clayton-Hopkins, Judith Ann
Cline, William H, Jr
Cohen, David Harris
Cohen, Maynard
Cox, Thomas C
Cralley, John Clement
Crystal, George Jeffrey
Delcomyn, Fred
DeSombre, Eugene Robert
Disterhoft, John Francis
Dobrin, Philip Boone
Doemling, Donald Bernard
Dunn, Robert Bruce
Dziuk, Philip J
Elble, Rodger Jacob
Enroth-Cugell, Christina
Fanslow, Don J
Fay, Richard Rozzell
Feigen, Larry Philip
Feng, Albert Shih-Hung
Ferguson, James L
Fisher, Cletus G
Flouret, George R
Ford, Lincoln Edmond
Gasdorf, Edgar Carl
Gershbein, Leon Lee
Gewertz, Bruce Labe
Gibbs, Daniel
Gibori, Geula
Giere, Frederic Arthur
Gillette, Rhanor
Gold, Jay Joseph
Goldman, Max
Goldstick, Thomas Karl
Graves, Charles Norman
Green, Orville
Griffiths, Thomas Alan
Hansen, Timothy Ray
Harrison, Paul C
Heybach, John Peter
Holmes, Kenneth Robert
Houk, James Charles
Hughes, John Russell
Hurley, Walter L
Jackson, Gary Loucks
Jaffe, Randal Craig
Janusek, Linda Witek
Jensen, Donald Reed
Jones, Stephen Bender
Kallio, Reino Emil

Katz, Adrian I
Katzenellenbogen, Benita Schulman
Kelley, Keith Wayne
Klabunde, Richard Edwin
Konisky, Jordan
Landau, Richard Louis
Langman, Craig Bradford
Layman, Dale Pierre
Layman, Donald Keith
Lee, Chung
Levitsky, Lynne Lipton
Lindheimer, Marshall D
Litteria, Marilyn
Lodge, James Robert
Malik, Asrar B
Manohar, Murli
Marczynski, Thaddeus John
Matsumura, Philip
Mayor, Gilbert Harold
Metzger, Boyd Ernest
Michael, Joel Allen
Moon, Richard C
Mulvihill, Mary Lou Jolie
Nakajima, Shigehiro
Nakajima, Yasuko
Natalini, John Joseph
Nicolette, John Anthony
Nielsen, Peter James
Novales, Ronald Richards
Nutting, Ehard Forrest
Omachi, Akira
Opgenorth, Terry John
Oscai, Lawrence B
Owen, Oliver Elon
Pappas, George Demetrios
Pederson, Vernon Clayton
Pepperberg, David Roy
Peterson, Barry Wayne
Peterson, Darryl Ronnie
Pindok, Marie Theresa
Pope, Richard M
Preston, Robert Leslie
Puchalski, Randall Francis
Radwanska, Ewa
Rakowski, Robert F
Rao, Mrinalini Chatta
Rechtschaffen, Allan
Riddle, Wayne Allen
Ritter, Nadine Marie
Romack, Frank Eldon
Rosen, Arthur Leonard
Rosenfield, Robert Lee
Rotermund, Albert J, Jr
Routtenberg, Aryeh
Rovick, Allen Asher
Roys, Chester Crosby
Ruggero, Mario Alfredo
Rymer, William Zev
Satinoff, Evelyn
Schwartz, Neena Betty
Sehgal, Lakshman R
Shearer, William McCague
Shriver, John William
Siegel, Jonathan Howard
Singer, Irwin
Skosey, John Lyle
Smith, Douglas Calvin
Stearner, Sigrid Phyllis
Steger, Richard Warren
Steiner, Donald Frederick
Sweeney, Daryl Charles
Swiatek, Kenneth Robert
Takahashi, Joseph S
Tang, Pei Chin
Thompson, Phebe Kirsten
Thonar, Eugene Jean-Marie
Tone, James N
Towle, David Walter
Tse, Warren W
Turek, Fred William
Twardock, Arthur Robert
Verhage, Harold Glenn
Visek, Willard James
Wade, David Robert
Wagner, William Charles
Wass, John Alfred
Webber, Charles Lewis, Jr
Webster, James Randolph, Jr
West, Charles Hutchison Keesor
Wetzel, Allan Brooke
White, Randy D
Wilber, Laura Ann
Wilson, Donald Alan
Winter, Robert John
Yelich, Michael Ralph
Zar, Jerrold Howard
Zemlin, Willard R

INDIANA
Amlaner, Charles Joseph, Jr
Anderson, David Bennett
Anderson, John Nicholas
Andrews, Frederick Newcomb
Babbs, Charles Frederick
Bakken, George Stewart
Begue, William John
Ben-Jonathan, Nira
Bohlen, Harold Glenn
Bottoms, Gerald Doyle
Brennan, David Michael
Bridges, C David
Brush, F(ranklin) Robert
Burkholder, Timothy Jay
Cadwallader, Joyce Vermeulen

Carter, James M
Chance, Ronald E
Clemens, James Allen
Costill, David Lee
Devoe, Robert Donald
Duff, Douglas Willard
Duman, John Girard
Dunn, Peter Edward
Dustman, John Henry
Echtenkamp, Stephen Frederick
Epstein, J Aubrey
Esch, Harald Erich
Everson, Ronald Ward
Fasola, Alfred Francis
Garriott, Michael Lee
Geddes, Leslie Alexander
Gidda, Jaswant Singh
Goetsch, Gerald D
Goetz, Frederick William, Jr
Grant, Alan Leslie
Gunther, Gary Richard
Hammel, Harold Theodore
Heiman, Mark Louis
Hill, Donald Louis
Holland, James Philip
Irwin, Glenn Ward, Jr
Johnson, Alan L
Johnston, Cyrus Conrad, Jr
Kempson, Stephen Allan
Knott, John Russell
Konetzka, Walter Anthony
Koritnik, Donald Raymond
Lesh, Thomas Allan
Macchia, Donald Dean
McClure, Polley Ann
Malven, Paul Vernon
Meyer, Frederick Richard
Neff, William Duwayne
Ochs, Sidney
Ostroy, Sanford Eugene
Outhouse, James Burton
Paladino, Frank Vincent
Paschall, Homer Donald
Peterson, Richard George
Pflanzer, Richard Gary
Powell, Richard Cinclair
Prange, Henry Davies
Reading, Rogers W
Renzi, Alfred Arthur
Roales, Robert R
Rothe, Carl Frederick
Rowland, David Lawrence
St John, Philip Alan
Stabler, Timothy Allen
Steer, Max David
Tacker, Willis Arnold, Jr
Tanner, George Albert
Thor, Karl Bruce
Turner, Charles Hall
Voelz, Michael H
Wagner, Wiltz Walker, Jr
Weinstein, Paul P
Williams, Daniel Charles
Yoder, John Menly
Zeller, Frank Jacob
Zimmack, Harold Lincoln

IOWA
Ackerman, Ralph Austin
Anderson, Lloyd L
Bhalla, Ramesh C
Campbell, Kevin Peter
Carithers, Jeanine Rutherford
Conn, P Michael
Coulter, Joe Dan
Dellmann, H Dieter
Denburg, Jeffrey Lewis
Dunham, Jewett
Eagleson, Gerald Wayne
Fellows, Robert Ellis, Jr
Fishman, Irving Yale
Ford, Stephen Paul
Getting, Peter Alexander
Heistad, Donald Dean
Hembrough, Frederick B
Henninger, Ann Louise
Hoffman, Eric Alfred
Hubel, Kenneth Andrew
Husted, Russell Forest
Kaplan, Murray Lee
Lewis, Douglas Scott
Marple, Dennis Neil
Masat, Robert James
Morrical, Daniel Gene
Morriss, Frank Howard, Jr
Randic, Mirjana
Read, Charles H
Redmond, James Ronald
Riggs, Dixon L
Schmidt, Thomas John
Shaw, Gaylord Edward
Shen, Sheldon Shih-Ta
Simpson, Robert John
Spaziani, Eugene
Stratton, Donald Brendan
Swenson, Melvin John
Thompson, Sue Ann
Waziri, Rafiq
Whipp, Shannon Carl
Williams, Dean E
Wu, Chun-Fang

KANSAS
Baranczuk, Richard John

Besharse, Joseph Culp
Bunag, Ruben David
Cheney, Paul David
Dahl, Nancy Ann
Davis, John Stewart
Dey, Sudhansu Kumar
Dunn, Jon D
Erickson, Howard Hugh
Fedde, Marion Roger
Ferraro, John Anthony
Fina, Louis R
Gattone, Vincent H, II
Goetz, Kenneth Lee
Goetzinger, Cornelius Peter
Gonzalez, Norberto Carlos
Hopkins, Theodore Louis
Johnson, Donald Charles
Keller, Leland Edward
Kramer, Karl Joseph
Leavitt, Wendell William
Longley, William Joseph
Maher, Michael John
Meek, Joseph Chester, Jr
Michaelis, Elias K
Neufeld, Gaylen Jay
Orr, James Anthony
Parrish, John Wesley, Jr
Quadri, Syed Kaleemullah
Ringle, David Allan
Rowe, Edward C
Smalley, Katherine N
Stevenson, Jeffrey Smith
Sullivan, Lawrence Paul
Tessel, Richard Earl
Vacca, Linda Lee
Valenzeno, Dennis Paul
Wilson, Fred E
Yochim, Jerome M
Zehr, John E
Zimniski, Stephen Joseph

KENTUCKY
Barber, William J
Baur, John M
Bennett, Thomas Edward
Blake, James Neal
Boehms, Charles Nelson
Boyarsky, Louis Lester
Cohn, David Valor
Creek, Robert Omer
Dutt, Ray Horn
Fell, Ronald Dean
Ferrell, Blaine Richard
Fingar, Victor H
Frazier, Donald Tha
Green, William Warden
Gross, David Ross
Harris, Patrick Donald
Hirsch, Henry Richard
Jones, Sanford L
Just, John Josef
Kargl, Thomas E
Legan, Sandra Jean
McCook, Robert Devon
McGraw, Charles Patrick
McLaughlin, Barbara Jean
Miller, Ralph English
Moody, William Glenn
Ott, Cobern Erwin
Puckett, Hugh
Rao, Chalamalasetty Venkateswara
Smothers, James Llewellyn
Spoor, William Arthur
Stamford, Bryant
Traurig, Harold H
Urbscheit, Nancy Lee
Wead, William Badertscher
Williams, Walter Michael

LOUISIANA
Anderson, Mary Bitner
Arimura, Akira
Bagby, Gregory John
Bailey, R L
Barker, Hal B
Bartell, Clelmer Kay
Battarbee, Harold Douglas
Baum, Lawrence Stephen
Bautista, Abraham Parana
Beadle, Ralph Eugene
Berlin, Charles I
Beuerman, Roger Wilmer
Caprio, John Theodore
Carter, Mary Kathleen
Christian, Frederick Ade
Coy, David Howard
Deaton, Lewis Edward
Dietz, Thomas Howard
Ely, Thomas Harrison
Fingerman, Milton
Gaar, Kermit Albert, Jr
Godke, Robert Alan
Happel, Leo Theodore, Jr
Harbo, John Russell
Harrison, Richard Miller
Isire, Clifton O, Jr
Jenkins, William L
Johnsen, Peter Berghsey
Knesel, John Arthur
Korthuis, Ronald John
Kreisman, Norman Richard
Laurent, Sebastian Marc
Levitzky, Michael Gordon
Lowe, Robert Franklin, Jr

McNamara, Dennis B
Mahajan, Damodar K
Miller, Harvey I
Nair, Pankajam K
Navar, Luis Gabriel
O'Dell Smith, Roberta Maxine
Olmsted, Clinton Albert
Olson, Richard David
Osborne, Jerry L
Peattie, Robert Addison
Porter, Johnny Ray
Rice, David A
Robertson, George Leven
St Angelo, Allen Joseph
Saphier, David
Schally, Andrew Victor
Shepherd, David Preston
Silverman, Harold
Spring, Jeffrey H
Teekell, Roger Alton
Tso, Patrick Po-Wing
Wakeman, John Marshall
Wallin, John David
Walters, Marian R
Windhauser, Marlene M
Woodring, Joseph
Yerrapraggada, Venkat
York, David Anthony

MAINE
Bain, William Murray
Bayer, Robert Clark
Bernstein, Seldon Edwin
Dowse, Harold Burgess
Gainey, Louis Franklin, Jr
Harris, Paul Chappell
Harrison, David Ellsworth
Kent, Barbara
Knoll, Henry Albert
Labov, Jay Brian
Norton, James Michael
Rand, Peter W
Sidell, Bruce David
Spirito, Carl Peter

MARYLAND
Allen, Willard M
Amende, Lynn Meridith
Anderson, David Everett
Anderson, Larry Douglas
Anderson, Robert Simpers
Angelone, Luis
Augustine, Patricia C
Ballard, Kathryn Wise
Bareis, Donna Lynn
Barker, Jeffery Lange
Barraclough, Charles Arthur
Bass, Eugene L
Batson, David Banks
Baust, John G
Bean, Barbara Louise
Beisel, William R
Benevento, Louis Anthony
Bergey, Gregory Kent
Berman, Michael Roy
Blaustein, Mordecai P
Bodammer, Joel Edward
Bodian, David
Bolt, Douglas John
Boucher, John H
Brewer, Nathan Ronald
Brock, Mary Anne
Brownstein, Michael Jay
Buck, John Bonner
Bunger, Rolf
Burke, Robert Emmett
Burt, David Reed
Burton, Dennis Thorpe
Calvert, Richard John
Carlson, Drew E
Cecil, Helene Carter
Cerveny, Thelma Jannette
Chapin, John Ladner
Chitwood, David Joseph
Choudary, Jasti Bhaskararao
Clarke, David Harrison
Colombini, Marco
Commissiong, John Wesley
Contrera, Joseph Fabian
Coulombe, Harry N
Cronin, Thomas Wells
Dahlen, Roger W
Delahunty, George
De Monasterio, Francisco M
Denniston, Joseph Charles
Dhindsa, Dharam Singh
Donaldson, Sue Karen
Dubin, Norman H
Dubner, Ronald
Dubois, Andre T
Dudley, Peter Anthony
Ehrlich, Walter
Eipper, Betty Anne
Elders, Minnie Joycelyn
Engel, Bernard Theodore
Estienne, Mark Joseph
Ewing, Larry Larue
Fajer, Abram Bencjan
Fambrough, Douglas McIntosh
Farrell, Robert Edmund, Jr
Ferguson, Earl Wilson
Fitch, Kenneth Leonard
Fitzgerald, Robert Schaefer
Foster, Giraud Vernam

Fowler, Arnold K
Frank, Martin
Franklin, Renty Benjamin
Freed, Arthur Nelson
Freed, Michael Abraham
Gainer, Harold
Gann, Donald Stuart
Gatti, Philip John
Gold, Armand Joel
Goldman, Stephen Shepard
Graff, Morris Morse
Graham, Charles Raymond, Jr
Greulich, Richard Curtice
Gruber, Kenneth Allen
Haddy, Francis John
Hanig, Joseph Peter
Hansen, Barbara Caleen
Hansford, Richard Geoffrey
Hanson, Frank Edwin
Hardy, Lester B
Heath, Martha Ellen
Helke, Cinda Jane
Hempel, Franklin Glenn
Hill, Elwood Fayette
Hirshman, Carol A
Hobbs, Ann Snow
Homer, Louis David
Hopkins, Thomas Franklin
Hoversland, Arthur Stanley
Jasper, Robert Lawrence
Jenkins, Mamie Leah Young
Jenkins, Melvin Earl
Jordan, Alexander Walker, III
Kafka, Marian Stern
Kayar, Susan Rennie
Kennedy, Thomas James, Jr
Khan, Mushtaq Ahmad
Kiang, Juliann G
Kidd, Bernard Sean Langford
Kinnard, Matthew Anderson
Kowarski, A Avinoam
Kundig, Fredericka Dodyk
Kusano, Kiyoshi
Lange, Gordon David
Larrabee, Martin Glover
Lazar-Wesley, Eliane M
Levitan, Herbert
Lin, Diane Chang
Liu, Ching-Tong
Livengood, David Robert
Loeb, Marcia Joan
Lowensohn, Howard Stanley
Lymangrover, John R
MacCanon, Donald Moore
McGowan, Joan A
McIndoe, Darrell W
McLaughlin, Alan Charles
MacVittie, Thomas Joseph
Marban, Eduardo
Margolis, Ronald Neil
Marx, Stephen John
Mather, Ian Heywood
Mattson, Margaret Ellen
Meszler, Richard M
Migeon, Claude Jean
Milnor, William Robert
Monjan, Andrew Arthur
Munson, Paul Lewis
Murray, Gary Joseph
Murray, George Cloyd
Myslinski, Norbert Raymond
Nelson, Jay Arlen
Nelson, Phillip Gillard
Nelson, Ralph Francis
O'Rangers, John Joseph
Pancella, John Raymond
Pare, William Paul
Pellmar, Terry C
Permutt, Solbert
Plato, Chris C
Poggio, Gian Franco
Provine, Robert Raymond
Pubols, Lillian Menges
Raina, Ashok K
Rall, Wilfrid
Ramey, Estelle R
Rattner, Barnett Alvin
Rechcigl, Miloslav, Jr
Reed, Randall R
Revoile, Sally Gates
Robbins, Jacob
Robinson, David Lee
Roesijadi, Guritno
Roth, George Stanley
Saudek, Christopher D
Schramm, Lawrence Peter
Schuetz, Allen W
Scow, Robert Oliver
Sexton, Thomas John
Shapiro, Bert Irwin
Sherins, Richard J
Smith, Lewis Wilbert
Smith, Thomas Graves, Jr
Snipes, Charles Andrew
Spector, Novera Herbert
Spooner, Peter Michael
Steinmetz, Michael Anthony
Stewart, Doris Mae
Stolzenberg, Sidney Joseph
Straile, William Edwin
Stringfellow, Frank
Svoboda, James Arvid
Tasaki, Ichiji
Tearney, Russell James

Terris, James Murray
Theodore, Theodore Spiros
Traystman, Richard J
Trout, David Linn
Turner, R James
Tyler, Bonnie Moreland
Umberger, Ernest Joy
Vener, Kirt J
Vinores, Stanley Anthony
Wagner, Henry George
Watkins, Clyde Andrew
Watson, John Thomas
Weight, Forrest F
Weinreich, Daniel
Weintraub, Bruce Dale
Weisfeldt, Myron Lee
Whidden, Stanley John
Whitehorn, William Victor
Wildt, David Edwin
Williams, Walter Ford
Wintercorn, Eleanor Stiegler
Wolff, Jan
Wollman, Seymour Horace
Woods, Lewis Curry, III
Wurtz, Robert Henry
Yellin, Herbert
Yoshinaga, Koji
Zimbelman, Robert George

MASSACHUSETTS
Adolph, Alan Robert
Alpers, Joseph Benjamin
Ames, Adelbert, III
Anderson, Rosalind Coogan
Anderson-Olivo, Margaret
Askew, Eldon Wayne
Atema, Jelle
Axelrod, Lloyd
Banzett, Robert B
Barber, Saul Benjamin
Bengele, Howard Henry
Berman, Marlene Oscar
Bird, Stephanie J(ean)
Black, Donald Leighton
Bougas, James Andrew
Bridges, Robert Stafford
Bruno, Merle Sanford
Burggren, Warren William
Butler, James Preston
Callard, Gloria Vincz
Cameron, John Stanley
Camougis, George
Capuzzo, Judith M
Carey, Francis G
Chang, Joseph Yoon
Chappel, Scott Carlton
Chattoraj, Sati Charan
Clemens, Daniel Theodore
Corkin, Suzanne Hammond
Coyne, Mary Downey
Damassa, David Allen
Deegan, Linda Ann
Delaney, Patrick Francis, Jr
Dobson, James Gordon, Jr
Douglas, Pamela Susan
Dowling, John Elliott
Duby, Robert T
Duffy, Frank Hopkins
Durkot, Michael John
Feig, Larry Allen
Furshpan, Edwin Jean
George, Stephen Anthony
Goldberg, Alfred L
Goodman, Henry Maurice
Graves, William Earl
Griffin, Donald R(edfield)
Grossberg, Stephen
Grossman, William
Guimond, Robert Wilfrid
Gustafson, Alvar Walter
Habener, Joel Francis
Hales, Charles A
Hall, Robert Dilwyn
Hartline, Peter Haldan
Hartner, William Christopher
Hichar, Joseph Kenneth
Hillman, Robert Edward
Hinrichs, Katrin
Hobson, John Allan
Hoffmann, Joan Carol
Hollenberg, Norman Kenneth
Hubel, David Hunter
Irving, James Tutin
Kahn, Carl Ronald
Kaltenbach-Townsend, Jane
Kazemi, Homayoun
Kiang, Nelson Yuan-Sheng
Kunkel, Joseph George
Kunz, Thomas Henry
Lambert, Helen Haynes
Landowne, Milton
LeFevre, Paul Green
Lessie, Thomas Guy
Liss, Robert H
Long, Nancy Carol
Longcope, Christopher
Lowenstein, Edward
Lynch, Harry James
McCarley, Robert William
Macchi, I Alden
McCormick, Stephen Daniel
McCracken, John Aitken
Mandels, Mary Hickox
Maran, Janice Wengerd

Marieb, Elaine Nicpon
Marks, Leon Joseph
Marsh, Richard L
Millard, William James
Miller, James Albert, Jr
Mitchell, David Hillard
Moore-Ede, Martin C
Mordes, John Peter
Morgan, James Philip
Morgane, Peter J
Morin, Walter Arthur
Moskowitz, Michael Arthur
Mountain, David Charles, Jr
Murphey, Rodney Keith
Naqvi, S Rehan Hasan
Neer, Robert M
O'Malley, Alice T
Palmer, John Derry
Payne, Bertram R
Pfister, Richard Charles
Pober, Zalmon
Pontoppidan, Henning
Poon, Chi-Sang
Potter, David Dickinson
Powers, J Bradley
Prestwich, Kenneth Neal
Remmel, Ronald Sylvester
Reppert, Steve Marion
Richardson, George S
Riggi, Stephen Joseph
Rigney, David Roth
Ross, James Neil, Jr
Ryan, Kenneth John
Saide, Judith Dana
Saper, Clifford B
Schaub, Robert George
Schneeberger, Eveline E
Schomer, Donald Lee
Senior, Boris
Shepro, David
Singer, Joshua J
Skavenski, Alexander Anthony
Skrinar, Gary Stephen
Smith, Peter John Shand
Smith, Wendy Anne
Snyder, Benjamin Willard
Solomon, Jolane Baumgarten
Stein, Diana B
Talamo, Barbara Lisann
Thies, Robert Scott
Torda, Clara
Treistman, Steven Neal
Trinkaus-Randall, Vickery E
Villee, Dorothy Balzer
Waloga, Geraldine
Webb, Helen Marguerite
Weinberger, Steven Elliott
Weiss, Earle Burton
Wenger, Christian Bruce
Widmaier, Eric Paul
Wiegner, Allen W
Woodhull-McNeal, Ann P
Wyse, Gordon Arthur
Zapol, Warren Myron
Zeldin, Michael Hermen
Zottoli, Steven Jaynes

MICHIGAN
Alpern, Mathew
Anderson, Gordon Frederick
Aulerich, Richard J
Barman, Susan Marie
Barney, Christopher Carroll
Beitins, Inese Zinta
Bergen, Werner Gerhard
Bernard, Rudy Andrew
Bohr, David Francis
Borer, Katarina Tomljenovic
Bowker, Richard George
Bradley, Robert Martin
Brooks, Samuel Carroll
Campbell, Kenneth Lyle
Casey, Kenneth L(yman)
Chou, Ching-Chung
Courtney, Gladys (Atkins)
Coyle, Peter
Cronkite, Donald Lee
Cunningham, James Gordon
D'Amato, Constance Joan
Dambach, George Ernest
Davenport, Horace Willard
Davis, Lynne
Dolan, David Francis
Dunbar, Joseph C
England, Barry Grant
Fenstermacher, Joseph Don
Fernandez, Hector R C
Foster, Douglas Layne
Frank, Robert Neil
Friar, Robert Edsel
Froiland, Thomas Gordon
Fromm, David
Gala, Richard R
Garlick, Robert L
Gebber, Gerard L
Gossain, Ved Vyas
Goudsmit, Esther Marianne
Gratz, Ronald Karl
Green, Daniel G
Guthe, Karl Frederick
Hajra, Amiya Kumar
Hall, Edward Dallas
Halter, Jeffrey Brian
Hawkins, Joseph Elmer

Physiology, Animal (cont)

Heisey, S(amuel) Richard
Henneman, Harold Albert
Hill, Richard William
Hoch, Frederic Louis
Hsu, Chen-Hsing
Ireland, James
Johnson, Herbert Gardner
Karsch, Fred Joseph
Keyes, Paul Landis
Kimball, Frances Adrienne
Klivington, Kenneth Albert
Kostyo, Jack Lawrence
Kurta, Allen
Lancaster, Cleo
Lauderdale, James W, Jr
Lawson, David Michael
Lehman, Grace Church
Mccorkle, Fred Miller
Mather, Edward Chantry
Midgley, A Rees, Jr
Mistretta, Charlotte Mae
Mitchell, Jerald Andrew
Moore, Thomas Edwin
Morrow, Thomas John
Nachreiner, Raymond F
Nag, Asish Chandra
Neudeck, Lowell Donald
Nuttall, Alfred L
Oakley, Bruce
Pavgi, Sushama
Penney, David George
Person, Steven John
Phillis, John Whitfield
Piercey, Montford F
Pittman, Robert Preston
Polin, Donald
Puro, Donald George
Rapundalo, Stephen T
Reynolds, Orland Bruce
Riegle, Gail Daniel
Rillema, James Alan
Rivera, Evelyn Margaret
Robert, Andre
Roberts, John Stephen
Robinson, Norman Edward
Rovner, David Richard
Rust, Steven Ronald
Schreur, Peggy Jo Korty Dobry
Sedensky, James Andrew
Shappirio, David Gordon
Sloane, Bonnie Fiedorek
Sparks, Harvey Vise
Spielman, William Sloan
Spilman, Charles Hadley
Stebbins, William Cooper
Stephenson, Robert Bruce
Stephenson, Robert Storer
Strand, James Cameron
Stromsta, Courtney Paul
Stucki, Jacob Calvin
Studier, Eugene H
Tannen, Richard L
Tigchelaar, Peter Vernon
Tosi, Oscar I
Tucker, Herbert Allen
Valenstein, Elliot Spiro
Vander, Arthur J
Webb, Paul
Webb, R Clinton
Wiley, John W
Wolterink, Lester Floyd

MINNESOTA

Alsum, Donald James
Ankley, Gerald Thomas
Baker, Shirley Marie
Barker, Laren Dee Stacy
Barnwell, Franklin Hershel
Bauer, Gustav Eric
Brimijoin, William Stephen
Daggs, Ray Gilbert
Di Salvo, Joseph
Duke, Gary Earl
Dziuk, Harold Edmund
Ebner, Timothy John
Eisenberg, Richard Martin
Engeland, William Charles
Fohlmeister, Jurgen Fritz
Forbes, Donna Jean
From, Arthur Harvey Leigh
Gallaher, Daniel David
Good, A L
Gorman, Colum A
Haller, Edwin Wolfgang
Hanson, Robert C
Hedgecock, Le Roy Darien
Heller, Lois Jane
Hooper, Alan Bacon
Hunter, Alan Graham
Iadecola, Costantino
Jacobson, Joan
Kallok, Michael John
Keshaviah, Prakash Ramnathpur
Klicka, John Kenneth
Knox, Charles Kenneth
Kromann, Rodney P
Lambert, Edward Howard
Lennon, Vanda Alice
Low, Walter Cheney
McCaffrey, Thomas Vincent
McDermott, Richard P
Marx, George Donald

Mayberry, William Eugene
Ogburn, Phillip Nash
Oppenheimer, Jack Hans
Phillips, Richard Edward
Polzin, David J
Poppele, Richard E
Purple, Richard L
Robinette, Martin Smith
Schmalz, Philip Frederick
Schwartz, Harold Leon
Sehe, Charles Theodore
Sicard, Raymond E(dward)
Smith, Thomas Jay
Swift, Michael Crane
Szurszewski, Joseph Henry
Thompson, Edward William
Tindall, Donald J
Ungar, Frank
Weir, Edward Kenneth
Wheaton, Jonathan Edward
Wilson, Michael John

MISSISSIPPI

Anderson, Gary
Biesiot, Patricia Marie
Bloom, Sherman
Bradley, Doris P
Buddington, Randal K
Chambers, Janice E
Coleman, Thomas George
Dickens, Joseph Clifton
Fuquay, John Wade
Ikenberry, Roy Dewayne
Larsen, James Bouton
Long, Billy Wayne
Lynch, James Carlyle
Montani, Jean-Pierre
Nelson, Norman Crooks
Norman, Roger Atkinson, Jr
Ramaswamy, Sonny B
Read, Virginia Hall
Smith, Manis James, Jr
Thaxton, James Paul

MISSOURI

Ackerman, Joseph John Henry
Ahmed, Mahmoud S
Armbrecht, Harvey James
Avioli, Louis
Baer, Robert W
Bergmann, Steven R
Blaine, Edward Homer
Breitenbach, Robert Peter
Buonomo, Frances Catherine
Burns, Thomas Wade
Burton, Harold
Byatt, John Christopher
Carrel, James Elliott
Coles, Richard Warren
Collier, Robert Joseph
Coyer, Philip Exton
Dale, Homer Eldon
Davis, Hallowell
Feir, Dorothy Jean
Fredrickson, John Murray
Gardner, Jerry David
Geisbuhler, Timothy Paul
Gerhardt, H Carl, Jr
Gibbs, Finley P
Grunt, Jerome Alvin
Hammerman, Marc R
Highstein, Stephen Morris
Hunt, Carlton Cuyler
Johnson, J(oseph) Alan
Kipnis, David Morris
Kohlmeier, Ronald Harold
Luxon, Bruce Arlie
Martin, Charles Everett
Mercer, Robert William
Mills, Steven Harlon
Mitchell, Henry Andrew
Molnar, Charles Edwin
Morton, Phillip A
Nickols, G Allen
Packman, Paul Michael
Parker, Mary Langston
Paull, Willis K, Jr
Peck, Carol King
Peck, William Arno
Peters, Ralph I
Peterson, Donald Frederick
Premachandra, Bhartur N
Quay, Wilbur Brooks
Rash, Jay Justen
Reidinger, Russell Frederick, Jr
Rovainen, Carl (Marx)
Rowe, Vernon Dodds
Rubin, Bruce Kalman
Ruh, Mary Frances
St Omer, Vincent Victor
Salkoff, Lawrence Benjamin
Savery, Harry P
Schonfeld, Gustav
Sharp, John Roland
Spiers, Donald Ellis
Starling, Jane Ann
Stein, Paul S G
Tannenbaum, Michael Glen
Thompson, Daniel James
Thomsen, Robert Harold
Tolbert, Daniel Lee
Tomasi, Thomas Edward
Tumosa, Nina Jean
Van Essen, David Clinton

Veenhuizen, Jeffrey J
Vineyard, Billy Dale
Volkert, Wynn Arthur
Warren, James C
Welply, Joseph Kevin
Weppelman, Roger Michael
Whitten, Elmer Hammond
Wilkens, Lon Allan
York, Donald Harold

MONTANA

Bradbury, James Thomas
Chaney, Robert Bruce, Jr
Coe, John Emmons
Dratz, Edward Alexander
Fevold, Harry Richard
Foresman, Kerry Ryan
Hull, Maurice Walter
Kilgore, Delbert Lyle, Jr
Kirkpatrick, Jay Franklin
McFeters, Gordon Alwyn
McMillan, James Alexander
Parker, Charles D
Patent, Gregory Joseph
Tietz, William John, Jr

NEBRASKA

Barnawell, Earl B
Carmines, Pamela K(ay)
Clark, Francis John
Clemens, Edgar Thomas
Cornish, Kurtis George
DeGraw, William Allen
Ellingson, Robert James
Ellington, Earl Franklin
Farr, Lynne Adams
Ford, Johny Joe
Fougeron, Myron George
Geluso, Kenneth Nicholas
Grotjan, Harvey Edward, Jr
Jones, Timothy Arthur
Larson, Larry Lee
Laster, Danny Bruce
Littledike, Ernest Travis
Pancoe, William Louis, Jr
Pekas, Jerome Charles
Roberts, Jane Carolyn
Rodriguez-Sierra, Jorge F
Tharp, Gerald D
Yen, Jong-Tseng
Zucker, Irving H

NEVADA

Anderson, Bernard A
Garner, Duane LeRoy
Marlow, Ronald William
Nellor, John Ernest
Ort, Carol
Potter, Gilbert David
Sanders, Kenton M
Starkweather, Peter Lathrop
Turner, John K

NEW HAMPSHIRE

Bogdonoff, Philip David, Jr
Brink-Johnsen, Truls
Cahill, George Francis, Jr
Chabot, Christopher Cleaves
Crandell, Walter Bain
Daubenspeck, John Andrew
Edwards, Brian Ronald
Galton, Valerie Anne
Gosselin, Robert Edmond
Hermann, Howard T
Langford, George Malcolm
Lesser, Michael Patrick
Munck, Allan Ulf
Musiek, Frank Edward
Nattie, Eugene Edward
St John, Walter McCoy
Sokol, Hilda Weyl
Sower, Stacia Ann
Stanton, Bruce Alan
Tokay, F Harry

NEW JERSEY

Ahn, Ho-Sam
Arbeeny, Cynthia M
Arthur, Alan Thorne
Bagnell, Carol A
Beyer-Mears, Annette
Bird, John William Clyde
Boyle, Joseph, III
Brooks, Jerry R
Bullock, John
Campfield, L Arthur
Carlson, Richard P
Chinard, Francis Pierre
Collins, Elliott Joel
Conn, Hadley Lewis, Jr
Connor, John Arthur
Convey, Edward Michael
Cornell, James S
Curcio, Lawrence Nicholas
Curro, Frederick A
Deetz, Lawrence Edward, III
Denhardt, David Tilton
Duran, Walter Nunez
Edelman, Norman H
Edinger, Henry Milton
Egan, Robert Wheeler
Emmers, Raimond
Ertel, Norman H
Farmanfarmaian, Allahverdi

Farnsworth, Patricia Nordstrom
Foley, James E
Goldsmith, Laura Tobi
Gona, Amos G
Gona, Ophelia Delaine
Goodman, Frank R
Grover, Gary James
Gruszczyk, Jerome Henry
Hafs, Harold David
Hahn, DoWon
Hall, James Conrad
Hall, Joseph L
Halmi, Nicholas Stephen
Hendler, Edwin
Hockel, Gregory Martin
Hofmann, Frederick Gustave
Hruza, Zdenek
Hubbard, John W
Ianuzzo, C David
Julesz, Bela
Kapoor, Inder Prakash
Katz, George Maxim
Katz, Larry Steven
Kirschner, Marvin Abraham
Kumbaraci-Jones, Nuran Melek
Kung, Ted Teshih
Lenz, Paul Heins
Lettvin, Jerome Y
Levenstein, Irving
Levin, Barry Edward
Levy, David Edward
Lisk, Robert Douglas
Lowndes, Herbert Edward
McArdle, Joseph John
Martinez del Rio, Carlos
Mellin, Theodore Nelson
Merrill, Gary Frank
Meyers, Kenneth Purcell
Mittler, James Carlton
Mohammed, Kasheed
Moreland, Suzanne
Mowles, Thomas Francis
Mueller, Stephen Neil
Newton, Paul Edward
Nocenti, Mero Raymond
Oddis, Leroy
Page, Charles Henry
Paterniti, James R, Jr
Perryman, James Harvey
Pitts, Barry James Roger
Ponessa, Joseph Thomas
Rabii, Jamshid
Royce, Paul C
Ruknudin, Abdul Majeed
Rumsey, William LeRoy
Saini, Ravinder Kumar
Salans, Lester Barry
Sargent, William Quirk
Siegel, Allan
Solomon, Thomas Allan
Steele, Ronald Edward
Stein, Donald Gerald
Steiner, Kurt Edric
Steinetz, Bernard George, Jr
Stevenson, Nancy Roberta
Swenson, Theresa Lynn
Swislocki, Norbert Ira
Szewczak, Mark Russell
Szot, Robert Joseph
Tomeo-Dymkowski, Aline Claire
Virkar, Raghunath Atmaram
Wallace, Edith Winchell
Watnick, Arthur Saul
Weiss, Gerson
Wilbur, Robert Daniel
Zbuzek, Vlasta Kmentova
Zulalian, Jack

NEW MEXICO

Bernstein, Marvin Harry
Botsford, James L
Buss, William Charles
Dressendorfer, Rudy
Duszynski, Donald Walter
Easley, Stephen Phillip
Hallford, Dennis Murray
Herman, Ceil Ann
Luft, Ulrich Cameron
McNamara, Mary Colleen
Malvin, Gary M
Matchett, William H
Mauderly, Joe L
Mauro, Jack Anthony
Mills, David Edward
Muggenburg, Bruce Al
Omdahl, John L
Partridge, L Donald
Priola, Donald Victor
Ratliff, Floyd
Ratner, Albert
Seagrave, JeanClare

NEW YORK

Adkins-Regan, Elizabeth Kocher
Adolph, Edward Frederick
Aldrich, Thomas K
Alexander, Robert Spence
Altszuler, Norman
Amassian, Vahe Eugene
Antzelevitch, Charles
April, Ernest W
Aronson, Ronald Stephen
Asanuma, Hiroshi
Atlas, Steven Alan

BIOLOGICAL SCIENCES / 151

Aull, Felice
Baizer, Joan Susan
Baker, Robert George
Barac-Nieto, Mario
Beermann, Donald Harold
Bender, David Bowman
Bennett, Michael Vander Laan
Benzo, Camillo Anthony
Bergman, Emmett Norlin
Berlyne, Geoffrey Merton
Bickerman, Hylan A
Bienvenue, Gordon Raymond
Bito, Laszlo Z
Blair, Martha Longwell
Blanck, Thomas Joseph John
Bleicher, Sheldon Joseph
Bookchin, Robert M
Boozer, Carol Sue Neely
Borowsky, Betty Marian
Brachfeld, Norman
Brown, Oliver Monroe
Brown, Patricia Stocking
Brown, Stephen Clawson
Browner, Robert Herman
Brust, Manfred
Bukovsan, William
Burchfiel, James Lee
Bushinsky, David Allen
Butler, Robert Neil
Cabot, John Boit, II
Canfield, Robert E
Canfield, William H
Carlson, Albert Dewayne, Jr
Carr, Ronald E
Catapane, Edward John
Center, Sharon Anne
Centola, Grace Marie
Cerny, Frank J
Chan, Stephen
Chang, Chin-Chuan
Christy, Nicholas Pierson
Claus, Thomas Harrison
Clitheroe, H John
Cohan, Christopher Scott
Cohen, Avis Hope
Cohen, Bernard
Cohen, Morton Irving
Cooper, George William
Corradino, Robert Anthony
Crain, Stanley M
Crandall, David L
Davies, Kelvin J A
Diakow, Carol
Dickson, Stanley
Dobson, Alan
Doering, Charles Henry
Dow, Bruce MacGregor
Duffey, Michael Eugene
DuFrain, Russell Jerome
Durkovic, Russell George
Egan, Edmund Alfred
Engbretson, Gustav Alan
Erasmus, Beth de Wet (Fleming)
Farhi, Leon Elie
Federico, Olga Maria
Felig, Philip
Fetcho, Joseph Robert
Finkelstein, Jacob Noah
Fletcher, Paul Wayne
Font, Cecilio Rafael
Foote, Robert Hutchinson
Fortune, Joanne Elizabeth
Fourtner, Charles Russell
Fox, Kevin A
Frankel, Arthur Irving
Freed, Simon
Freedman, Michael Lewis
French, Alan Raymond
Frumkes, Thomas Eugene
Gapp, David Alger
Gardner, Daniel
Gardner, Esther Polinsky
Gasteiger, Edgar Lionel
Geary, Norcross D
Gershengorn, Marvin Carl
Gershon, Michael David
Gibson, Kenneth David
Gil, Joan
Goldberg, Louis J
Goldfarb, David S
Goldfarb, Roy David
Gonzalez, Eulogio Raphael
Gootman, Norman Lerner
Gootman, Phyllis Myrna Adler
Greenberg, Danielle
Grew, John C
Griswold, Joseph Garland
Grota, Lee J
Habicht, Gail Sorem
Hainline, Louise
Halpern, Bruce Peter
Halpern, Mimi
Halpryn, Bruce
Han, Jaok
Harkavy, Allan Abraham
Harris, C(harles) Leon
Harris-Warrick, Ronald Morgan
Harth, Erich Martin
Hawkins, Robert Drake
Hayes, Carol J
Hechemy, Karim E
Henneberry, Richard Christopher
Higashiyama, Tadayoshi
Hillman, Dean Elof

Hirsch, Helmut V B
Hochberg, Irving
Hoffman, Roger Alan
Holtzman, Eric
Holtzman, Seymour
Hood, Donald C
Horn, Eugene Harold
Horvath, Fred Ernest
Hotchkiss, Sharon K
Houpt, Katherine Albro
Houpt, Thomas Richard
Howland, Howard Chase
Hudecki, Michael Stephen
Hughes, Patrick Richard
Hyde, Richard Witherington
Isseroff, Hadar
Jacklet, Jon Willis
Jacobs, Laurence Stanton
Jacobson, Herbert (Irving)
Jahan-Parwar, Behrus
Jeffrey, John J
Johnson, Patricia R
Kandel, Eric Richard
Kao, Chien Yuan
Kaplan, John Ervin
Karl, Peter I
Kashin, Philip
Kass, Robert S
Katsel, Pavel Leon
Kaufman, Albert Irving
Kerlan, Joel Thomas
Khanna, Shyam Mohan
Kimmich, George Arthur
Kincl, Fred Allan
King, Thomas K C
Klocke, Robert Albert
Knigge, Karl Max
Koenig, Edward
Kohn, Michael
Koizumi, Kiyomi
Kostreva, David Robert
Krasney, John Andrew
Krey, Lewis Charles
Krulwich, Terry Ann
Kupfer, Sherman
Kupperman, Herbert Spencer
Kydd, David Mitchell
Lang, Charles H
Lawton, Richard Woodruff
Lee, James B
Lengemann, Frederick William
Levey, Harold Abram
Levitz, Mortimer
Liebman, Frederick Melvin
Liebow, Charles
Llinas, Rodolfo
Lnenicka, Gregory Allen
Loew, Ellis Roger
Loretz, Christopher Alan
Lutton, John D
MacGillivray, Margaret Hilda
McKenna, Olivia Cleveland
Mahler, Richard Joseph
Mahoney, Robert Patrick
Makous, Walter L
Malhotra, Ashwani
Mantel, Linda Habas
Maple, William Thomas
Martin, Constance R
Mason, Elliott Bernard
Maurice, David Myer
Medici, Paul T
Mehaffey, Leathem, III
Mellins, Robert B
Menguy, Rene
Messina, Edward Joseph
Michaelson, Solomon M
Miller, Myron
Miller, Richard Avery
Mindich, Leonard Eugene
Montefusco, Cheryl Marie
Morck, Timothy Anton
Morris, Marilyn Emily
Mueller, August P
Munro, Donald W, Jr
Murphy, Michael Joseph
Nahas, Gabriel Georges
Nathan, Marc A
Nelson, Margaret Christina
Newell, Jonathan Clark
Nicholson, J(ohn) Charles (Godfrey)
Nickerson, Peter Ayers
Notides, Angelo C
Oberdorster, Gunter
O'Connor, John Francis
Oesterreich, Roger Edward
Ornt, Daniel Burrows
Overweg, Norbert I A
Pagala, Murali Krishna
Pandit, Hemchandra M
Pasik, Pedro
Pasik, Tauba
Peters, Theodore, Jr
Pfaff, Donald Wells
Pfaffmann, Carl
Pollard, Jeffrey William
Potts, Gordon Oliver
Poulos, Dennis A
Primack, Marshall Philip
Purpura, Dominick Paul
Reilly, Margaret Anne
Reimers, Thomas John
Richman, Robert Alan
Rifkind, Arleen B

Ringo, James Lewis
Robertson, Douglas Reed
Rockwood, William Philip
Rogers, Philip Virgilius
Rosen, John Friesner
Rosenberg, Warren L
Rosensweig, Norton S
Rosoff, Betty
Rubin, Ronald Philip
Sack, Robert A
St Leger, Raymond John
Sakitt, Barbara
Salpeter, Miriam Mirl
Saxena, Brij B
Scalia, Frank
Scarpelli, Emile Michael
Schiff, Joel D
Schlafer, Donald H
Schlesinger, Richard B
Schmidt, John Thomas
Schoenfeld, Cy
Schreibman, Martin Paul
Schryver, Herbert Francis
Sellers, Alvin Ferner
Sharma, Sansar C
Shellabarger, Claire J
Shepherd, Julian Granville
Sherman, Frederick George
Sherman, Samuel Murray
Sherwood, Louis Maier
Shumway, Sandra Elisabeth
Singh, Toolsee J
Skinner, Kathleen Mary
Sonenberg, Martin
Sorrentino, Sandy Jr
Southwick, Edward E
Sparks, Charles Edward
Spaulding, Stephen Waasa
Spray, David Conover
Stagg, Ronald M
Steffen, Daniel G
Stein, Theodore Anthony
Stephenson, John Leslie
Stoner, Larry Clinton
Stouter, Vincent Paul
Strand, Fleur Lillian
Sturr, Joseph Francis
Sundaram, Kalyan
Talhouk, Rabih Shakib
Tamir, Hadassah
Tapper, Daniel Naphtali
Thau, Rosemarie B Zischka
Thornborough, John Randle
Tobach, Ethel
Tsan, Min-Fu
Udin, Susan Boymel
Van Liew, Hugh Davenport
Van Tienhoven, Ari
Ventura, William Paul
Vogel, Henry
Vroman, Leo
Wallman, Joshua
Walters, Deborah K W
Waterhouse, Joseph Stallard
Weiss, Klaudiusz Robert
Weitzman, Elliot D
White, James Patrick
Wilson, Victor Joseph
Wit, Andrew Lewis
Witkovsky, Paul
Wittig, Kenneth Paul
Wulff, Verner John
Wyman, Richard L
Yao, Alice C
Yarns, Dale A
Yasumura, Seiichi
Yazulla, Stephen
Yellin, Edward L
Yu, Ts'ai-Fan
Zablow, Leonard
Zimmerman, Jay Alan

NORTH CAROLINA
Albright, Bruce Calvin
Aldridge, David William
Anderson, Nels Carl, Jr
Anderson, Roger Fabian
Arendshorst, William John
Beckman, David Lee
Bentzel, Carl Johan
Black, Betty Lynne
Bo, Walter John
Bredeck, Henry E
Breese, George Richard
Brinn, Jack Elliott, Jr
Burden, Hubert White
Carroll, Robert Graham
Caruolo, Edward Vitangelo
Casseday, John Herbert
Clark, Martin Ralph
Clemmons, David Roberts
Coffey, James Cecil, Jr
Corless, Joseph Michael James
Cornwell, John Calhoun
Coulter, Norman Arthur, Jr
Croom, Warren James, Jr
Daw, John Charles
Di Giulio, Richard Thomas
Dimmick, John Frederick
Dreyer, Duane Arthur
Dusseau, Jerry William
Edens, Frank Wesley
Eldridge, Frederic L
Eldridge, John Charles

Erickson, Robert Porter
Faber, James Edward
Fail, Patricia A
Fairchild, Homer Eaton
Fletcher, Donald James
Folinsbee, Lawrence John
Furth, Eugene David
Garren, Henry Wilburn
Gatten, Robert Edward, Jr
Gilmore, Stuart Irby
Glasser, Richard Lee
Glassman, Edward
Goetsch, Dennis Donald
Goode, Lemuel
Gordon, Christopher John
Gottschalk, Carl William
Green, Harold D
Greenfield, Joseph C, Jr
Hazzard, William Russell
Hedlund, Laurence William
Henson, Anna Miriam (Morgan)
Hicks, T Philip
Hodson, Charles Andrew
Javel, Eric
Kaufman, William
Kizer, John Stephen
Klitzman, Bruce
Korach, Kenneth Steven
Lassiter, William Edmund
Lawson, Edward Earle
Lieberman, Melvyn
Little, William C
Lobaugh, Bruce
Louis, Thomas Michael
Lund, Pauline Kay
McManus, Thomas (Joseph)
Malindzak, George S, Jr
Means, Anthony R
Meissner, Gerhard
Meyer, Ralph A, Jr
Mikat, Eileen M
Miller, David S
Miller, William Laubach
Millstein, Lloyd Gilbert
Mochrie, Richard D
Moore, John Wilson
Myers, Robert Durant
Nayfeh, Shihadeh Nasri
O'Brien, Deborah A
Ontjes, David A
Parker, John Curtis
Pasipoularides, Ares D
Perlmutt, Joseph Hertz
Pritchard, John B
Reel, Jerry Royce
Robinson, Jerry Allen
Roer, Robert David
Rustioni, Aldo
Schanberg, Saul M
Schmidt, Stephen Paul
Schmidt-Nielsen, Knut
Shaw, Helen Lester Anderson
Simpson, Everett Coy
Siopes, Thomas David
Smith, Donald Eugene
Smith, Frank Houston
Smith, Ned Allan
Sonntag, William Edmund
Stevens, Charles Edward
Strandhoy, Jack W
Stuart, Ann Elizabeth
Stumpf, Walter Erich
Taylor, Martha Loeb
Thubrikar, Mano J
Toverud, Svein Utheim
Turner, James Eldridge
Tyrey, Lee
Underwood, Herbert Arthur, Jr
Underwood, Louis Edwin
Vandemark, Noland Leroy
Vandenbergh, John Garry
Van Wyk, Judson John
Wallace, Andrew Grover
Watts, John Albert, Jr
Whitsel, Barry L
Williams, Redford Brown, Jr
Wise, George Herman
Wolcott, Thomas Gordon
Yarger, William E

NORTH DAKOTA
Aschbacher, Peter William
Fivizzani, Albert John, Jr
Haunz, Edgar Alfred
Joshi, Madhusudan Shankarrao
Leopold, Roger Allen
Reynolds, Lawrence P
Roseland, Craig R
Saari, Jack Theodore
Samson, Willis Kendrick
Sheridan, Mark Alexander
Stinnett, Henry Orr
Zogg, Carl A

OHIO
Adams, Walter C
Akeson, Richard Allan
Arlian, Larry George
Baker, Saul Phillip
Barnes, Karen Louise
Behr, Stephen Richard
Berger, Kenneth Walter
Biagini, Raymond E
Billman, George Edward

Physiology, Animal (cont)

Birkhahn, Ronald H
Black, Craig Patrick
Bohrman, Jeffrey Stephen
Bolls, Nathan J, Jr
Boyer, Jere Michael
Butz, Andrew
Chen, Wen Yuan
Chester, Edward Howard
Clanton, Thomas L
Clark, Kenneth Edward
Claussen, Dennis Lee
Coleman, Marilyn A
Cooke, Helen Joan
Cooper, Gary Pettus
Cornhill, John Fredrick
Costanzo, Jon P
Costello, Walter James
Cruce, William L R
Curry, John Joseph, III
Dabrowski, Konrad
Dage, Richard Cyrus
Decker, Clarence Ferdinand
Dedman, John Raymond
DeVillez, Edward Joseph
Donnelly, Kenneth Gerald
Dun, Nae Jiuum
Dutta, Hiran M
Faber, Lee Edward
Faiman, Charles
Fondacaro, Joseph D
Foreman, Darhl Lois
Freed, James Melvin
Fritz, Marc Anthony
Genuth, Saul M
Glaser, Roger Michael
Goldstein, David Louis
Greenstein, Julius S
Grossie, James Allen
Grossman, Charles Jerome
Gwinn, John Frederick
Hamlin, Robert Louis
Harder, John Dwight
Harrison, Donald C
Haxhiu, Musa A
Hebbard, Frederick Worthman
Heesch, Cheryl Miller
Henschel, Austin
Hill, Richard M
Hinman, Channing L
Hopfer, Ulrich
Jegla, Thomas Cyril
Johnson, Melvin Andrew
Johnston, John O'Neal
Jones, Stephen Wallace
Kaiserman-Abramof, Ita Rebeca
Kern, Michael Don
Kogut, Maurice D
Kolenbrander, Harold Mark
Kolodziej, Bruno J
LaManna, Joseph Charles
Landis, Story Cleland
Lasek, Raymond J
Latshaw, J David
Lemon, Peter Willian Reginald
Loewenstein, Joseph Edward
Malhotra, Om P
Martin, Elden William
Meserve, Lee Arthur
Metz, David A
Michal, Edwin Keith
Millard, Ronald Wesley
Moulton, Bruce Carl
Mudry, Karen Michele
Murray, Finnie Ardrey, Jr
Needham, Glen Ray
Nishikawara, Margaret T
Nodar, Richard (H)enry
Nussbaum, Noel Sidney
Orcutt, Frederic Scott, Jr
Orosz, Charles George
Paradise, Norman Francis
Patterson, Michael Milton
Pettegrew, Raleigh K
Pilati, Charles Francis
Pinheiro, Marilyn Lays
Quinn, David Lee
Radabaugh, Dennis Charles
Rall, Jack Alan
Ray, Richard Schell
Replogle, Clyde R
Robbins, Norman
Robinson, Michael K
Rogers, Richard C
Rothchild, Irving
Sabourin, Thomas Donald
Saiduddin, Syed
Schoessler, John Paul
Sherman, Robert George
Shertzer, Howard Grant
Shipley, Michael Thomas
Smith, Clifford James
Smith, D Edwards
Smith, David Varley
Srivastava, Laxmi Shanker
Stevenson, J(oseph) Ross
Stokes, Bradford Taylor
Stuesse, Sherry Lynn
Tassava, Roy A
Teyler, Timothy James
Treick, Ronald Walter
Turner, John W, Jr
Tzagournis, Manuel

Van Ummersen, Claire Ann
Vardaris, Richard Miles
Vogel, Thomas Timothy
Voneida, Theodore J
Wagner, Thomas Edwards
Weiss, Harold Samuel
Wiley, Ronald Lee
Willett, Lynn Brunson
Wilson, David Franklin
Wolin, Lee Roy
Yocum, George David
Zigmond, Richard Eric

OKLAHOMA
Baird, Troy Alan
Bastian, Joseph
Bourdeau, James Edward
Burgus, Roger Cecil
Chung, Kyung Won
Clark, James Bennett
Dormer, Kenneth John
Farber, Jay Paul
Faulkner, Lloyd (Clarence)
Foreman, Robert Dale
Gass, George Hiram
Gunn, Chesterfield Garvin, Jr
Haines, Howard Bodley
Houlihan, Rodney T
Hurst, Jerry G
Hutchison, Victor Hobbs
Jones, Sharon Lynn
Kinasewitz, Gary Theodore
King, Mary Margaret
Kling, Ozro Ray
Martin, Loren Gene
Meier, Charles Frederick, Jr
Miner, Gary David
Pento, Joseph Thomas
Robertson, William G
Sauer, John Robert
Schaefer, Carl Francis
Scherlag, Benjamin J
Shirley, Barbara Anne
Simpson, Ocleris C
Stith, Rex David
Thies, Roger E
Trachewsky, Daniel
Turman, Elbert Jerome
Van Thiel, David H
Wettemann, Robert Paul

OREGON
Anderson, Amy Elin
Arch, Stephen William
Bandick, Neal Raymond
Barmack, Neal Herbert
Bethea, Cynthia Louise
Bradshaw, William Emmons
Brookhart, John Mills
Brown, Arthur Charles
Brownell, Philip Harry
Carlisle, Kay Susan
Chilgren, John Douglas
Christensen, Ned Jay
Church, David Calvin
Clark, Mary Eleanor
Coleman, Ralph Orval, Jr
Davis, Steven Lewis
Grimm, Robert John
Gunberg, David Leo
Gwilliam, Gilbert Franklin
Harris, Nellie Robbins
Hillman, Stanley Severin
Hisaw, Frederick Lee, Jr
Keenan, Edward James
Kohler, Peter
Levine, Leonard
Lilly, David J
Ludlam, William Myrton
McCollum, Gin
Maksud, Michael George
Marrocco, Richard Thomas
Mason, Robert Thomas
Maurer, Richard Allen
Meikle, Mary B
Montagna, William
Moore, Frank Ludwig
Novy, Miles Joseph
Oginsky, Evelyn Lenore
Ojeda, Sergio Raul
Osgood, David William
Phoenix, Charles Henry
Pritchard, Austin Wyatt
Ramaley, Judith Aitken
Rasmussen, Lois E Little
Roberts, Michael Foster
Roselli, Charles Eugene
Roth, Niles
Ruwe, William David
Soderwall, Arnold Larson
Spies, Harold Glen
Stormshak, Fredrick
Stouffer, Richard Lee
Swanson, Lloyd Vernon
Swanson, Robert E
Taylor, Mary Lowell Branson
Taylor, Walter Herman, Jr
Terwilliger, Nora Barclay
Thorson, Thomas Bertel
Wilson, Marlene Moore
Young, Norton Bruce
Zimmerman, Earl Abram

PENNSYLVANIA
Angstadt, Robert B
Aston-Jones, Gary Stephen
Bex, Frederick James
Binkley, Sue Ann
Borle, Andre Bernard
Botelho, Stella Yates
Bradley, Stanley Edward
Brooks, David Patrick
Buckelew, Thomas Paul
Burgi, Ernest, Jr
Buskirk, Elsworth Robert
Cameron, William Edward
Castell, Donald O
Clark, James Michael
Colby, Howard David
Compher, Marvin Keen, Jr
Corbin, Alan
Cowan, Alan
Davies, Richard Oelbaum
Davis, Robert Harry
DeCaro, Thomas F
De Groat, William C, Jr
DeMartinis, Frederick Daniel
Demers, Laurence Maurice
Di George, Angelo Mario
Dixit, Rakesh
Dodgson, Susanna Jane
Ertel, Robert James
Faber, Donald Stuart
Fish, Frank Eliot
Flemister, Sarah C
Florant, Gregory Lester
Flynn, John Thomas
Frazier, James Lewis
Fritz, George Richard, Jr
Fromm, Gerhard Hermann
Fuchs, Franklin
Fuller, Ellen Oneil
Gerstein, George Leonard
Glazier, Douglas Stewart
Goodman, David Barry Poliakoff
Gregg, Christine M
Grosvenor, Clark Edward
Grunstein, Michael M
Guida, Vincent George
Hagen, Daniel Russell
Hale, Creighton J
Harris, Dorothy Virginia
Hartley, Harold V, Jr
Henderson, Linda Shlatz
Herlyn, Dorothee Maria
Holliday, Charles Walter
Hook, Jerry B
Horn, Lyle William
Horner, George John
Hymer, Wesley C
Idzkowsky, Henry Joseph
Isom, Harriet C
Jacobs, George Joseph
Jacobson, Frank Henry
Jacoby, Henry I
Joos, Barbara
Kelsey, Ruben Clifford
Kennedy, Ann Randtke
Kennedy, Michael Craig
Kidawa, Anthony Stanley
Kinter, Lewis Boardman
Kivlighn, Salah Dean
Klinger, Thomas Scott
Kornfield, A(lfred) T(heodore)
Kozak, Wlodzimierz M
Krause, Stephen Myron
Kwiecinski, Gary George
Lamperti, Albert A
Lankford, Edward B
Leach, Roland Melville, Jr
Leavitt, Marc Laurence
Lefer, Allan Mark
Levey, Gerald Saul
Liebman, Paul Arno
Lindgren, Clark Allen
McElligott, James George
McKinstry, Donald Michael
McNair, Dennis M
Martin, John Samuel
Mashaly, Magdi Mohamed
Mendelson, Emanuel Share
Miller, Kirk
Minsker, David Harry
Miselis, Richard Robert
Molt, James Teunis
Monson, Frederick Carlton
Mote, Michael Isnardi
Motoyama, Etsuro K
Mullin, James Michael
Murray, Marion
Nadelhaft, Irving
Orr, Nancy Hoffner
Palmer, Larry Alan
Panuska, Joseph Allan
Peaslee, Margaret H
Persky, Harold
Pirone, Louis Anthony
Pitkin, Ruthanne B
Pitkow, Howard Spencer
Powell, William John, Jr
Pruett, Esther D R
Rao, Kalipatnapu Narasimha
Rapp, Paul Ernest
Redgate, Edward Stewart
Robertson, Robert James
Rodan, Gideon Alfred
Ross, Leonard Lester

Salzberg, Brian Matthew
Sandler, Rivka Black
Savage, Carl Richard, Jr
Scherer, Peter William
Scott, Donald, Jr
Sekerka, Robert Floyd
Shank, Richard Paul
Shaw, Ralph Arthur
Short, John Albert
Sidie, James Michael
Siegfried, John Barton
Skinner, James Ernest
Smith, M Susan
Sojka, Gary Allan
Sperling, Mark Alexander
Steele, Craig William
Strauss, Jerome Frank, III
Stricker, Edward Michael
Studer, Rebecca Kathryn
Summy-Long, Joan Yvette
Surmacz, Cynthia Ann
Takashima, Shiro
Taylor, Rhoda E
Thomas, Steven P
Tobin, Thomas Vincent
Trelka, Dennis George
Trevino, Daniel Louis
Trippodo, Nick Charles
Tulenko, Thomas Norman
Vargo, Steven William
Vasilatos-Younken, Regina
Vickers, Florence Foster
Vollmer, Regis Robert
Vomachka, Archie Joel
Weiss, Michael Stephen
Weisz, Judith
Wickersham, Edward Walker
Wideman, Robert Frederick, Jr
Winchester, Richard Albert
Young, In Min
Yushok, Wasley Donald
Zabara, Jacob
Zaccaria, Robert Anthony
Zelis, Robert Felix
Zimmerman, Irwin David

RHODE ISLAND
Arnold, Mary B
Durfee, Wayne King
Ebner, Ford Francis
Elbaum, Charles
Hai, Chi-Ming
Hammen, Carl Schlee
Hammen, Susan Lum
Harrison, Robert William
Marsh, Donald Jay
Morris, David Julian
Pilson, Michael Edward Quinton
Rice, Michael Alan
Richardson, Peter Damian

SOUTH CAROLINA
Abel, Francis Lee
Baxter, Ann Webster
Beck, Ronald Richard
Bell, Norman H
Birrenkott, Glenn Peter, Jr
Blackburn, John Gill
Blake, Charles Albert
Campbell, Gary Thomas
Cheng, Thomas Clement
Coleman, James Roland
Congdon, Justin D
Cordova-Salinas, Maria Asuncion
DeCoursey, Patricia Alice Jackson
Dickey, Joseph Freeman
Finlay, Mary Fleming
Gibson, Mary Morton
Glick, Bruce
Hays, Sidney Brooks
Henricks, Donald Maurice
Jagoe, Charles Henry
Katz, Sidney
McFarland, Kay Flowers
Ondo, Jerome G
Powell, Donald Ashmore
Powell, Harold
Roel, Lawrence Edmund
Weymouth, Richard J
Wilbur, Donald Lee
Wise, William Curtis

SOUTH DAKOTA
Heisinger, James Fredrick
Johnson, Jeffery Lee
Johnson, Leland Gilbert
Langworthy, Thomas Allan
Martin, Douglas Stuart
Pengra, Robert Monroe
Roller, Michael Harris
Schlenker, Evelyn Heymann
Slyter, Arthur Lowell
Westby, Carl A
Zawada, Edward T, Jr
Zeigler, David Wayne

TENNESSEE
Asp, Carl W
Banerjee, Mukul Ranjan
Bealer, Steven Lee
Beaty, Orren, III
Biaggioni, Italo
Blatteis, Clark Martin
Bogitsh, Burton Jerome

Brigham, Kenneth Larry
Camacho, Alvro Manuel
Carter, Clint Earl
Cherrington, Alan Douglas
Chytil, Frank
Cook, Robert James
Davis, Kenneth Bruce, Jr
Davis, William James
Dooley, Elmo S
Durham, Ross M
Eiler, Hugo
Fleischer, Sidney
Foreman, Charles William
Freeman, John A
Huggins, James Anthony
Law, Peter Koi
Lee, Kai-Lin
Leffler, Charles William
Levene, John Reuben
Levine, Jon Howard
Mahajan, Satish Chander
Mallette, John M
Martinez-Hernandez, Antonio
Mrotek, James Joseph
Norden, Jeanette Jean
Norton, Virginia Marino
Orgebin-Crist, Marie-Claire
Peterson, Darwin Wilson
Reddy, Churku Mohan
Regen, David Marvin
Richmond, Chester Robert
Rieke, Garl Kalman
Sander, Linda Dian
Sanders, Jay W
Shirley, Herschel Vincent, Jr
Snapper, James Robert
Solomon, Solomon Sidney
Steen, R Grant
Stone, Robert Edward, Jr
Taylor, Robert Emerald, Jr
Turner, Barbara Bush
Weinstein, Ira
Welch, Hugh Gordon
Wondergem, Robert

TEXAS
Achilles, Robert F
Alexander, William Carter
Alkadhi, Karim A
Allen, Edward Patrick
Alter, William A, III
Amoss, Max St Clair
Ashworth, Robert David
Azizi, Sayed Ausim
Banerji, Tapan Kumar
Barker, Kenneth Leroy
Barnes, George Edgar
Bellinger, Larry Lee
Bertrand, Helen Anne
Birnbaumer, Lutz
Blankenship, James Emery
Bockholt, Anton John
Bonner, Hugh Warren
Booth, Frank
Boudreau, James Charles
Breen, Gail Anne Marie
Brooks, Barbara Alice
Burks, James Kenneth
Burns, John Mitchell
Burton, Karen Poliner
Burton, Russell Rohan
Butler, Bruce David
Caffrey, James Louis
Cameron, James N
Campbell, Bonnalie Oetting
Cartwright, Aubrey Lee, Jr
Castracane, V Daniel
Cate, Rodney Lee
Chick, Thomas Wesley
Christensen, Burgess Nyles
Chubb, Curtis Evans
Clark, James Henry
Clark, James Richard
Claus-Walker, Jacqueline Lucy
Clench, Mary Heimerdinger
Coats, Alfred Cornell
Convertino, Victor Anthony
Cooper, Cary Wayne
Crass, Maurice Frederick, III
Crawford, Isaac Lyle
Crews, David Pafford
Dafny, Nachum
Dahl, Roy Dennis
Dalterio, Susan Linda
Danhof, Ivan Edward
Davis, James Royce
DeFrance, Jon Fredric
DiMichele, Leonard Vincent
Downey, H Fred
Eldridge, David Wyatt
Entman, Mark Lawrence
Espey, Lawrence Lee
Evans, James Warren
Falck, Frank James
Fawcett, Colvin Peter
Fisher, Frank M, Jr
Foote, Wilford Darrell
Foster, Daniel W
Fox, Donald A
Frasher, Wallace G, Jr
Frazier, Loy William, Jr
Fremming, Benjamin DeWitt
Friend, Ted H
Frigyesi, Tamas L
Garfield, Robert Edward
Gaugl, John F
Gerken, George Manz
German, Dwight Charles
Glantz, Raymon M
Glasser, Stanley Richard
Goldstein, Margaret Ann
Gonzalez-Lima, Francisco
Gordon, Wayne Lecky
Gottesfeld, Zehava
Gotto, Antonio Marion, Jr
Guest, Maurice Mason
Guo, Yan-Shi
Haensly, William Edward
Hancock, Michael B
Harper, Michael John Kennedy
Hartley, Craig Jay
Hartsfield, Sandee Morris
Hazelwood, Robert Leonard
Hazlett, David Richard
Henning, Susan June
Higgins, Linden Elizabeth
Hill, Ronald Stewart
Holman, Gerald Hall
Hotchkiss, Julane
Hudspeth, Albert James
Hupp, Eugene Wesley
Irwin, Louis Neal
Jauchem, James Robert
Johnson, Bryan Hugh
Johnson, Larry
Kasschau, Margaret Ramsey
Keating, Robert Joseph
Kenny, Alexander Donovan
King, Richard Joe
Knobil, Ernst
Konecci, Eugene B
Krock, Larry Paul
Kubena, Leon Franklin
Kunze, Diana Lee
Kuo, Lih
Larimer, James Lynn
Lawrence, Addison Lee
Leach-Huntoon, Carolyn S
Lebovitz, Robert Mark
Lees, George Edward
Lifschitz, Meyer David
Lucas, Edgar Arthur
McCann, Samuel McDonald
McCrady, James David
McGrath, James Joseph
McMahon, Robert Francis, III
Madison, Leonard Lincoln
Mailman, David Sherwin
Margolin, Solomon B
Martin, Frederick N
Martinez, J Ricardo
Masoro, Edward Joseph
Mattingly, Stephen Joseph
Maxwell, Leo C
Meola, Roger Walker
Michael, Lloyd Hal
Mitchell, Jere Holloway
Moldawer, Marc
Moore, Lee E
Morrow, Dean Huston
Moss, Robert L
Nathan, Richard D
Norman, Reid Lynn
Nusynowitz, Martin Lawrence
O'Benar, John DeMarion
Olson, John S
O'Malley, Bert W
Owens, David William
Pak, Charles Y
Parkening, Terry Arthur
Parker, Nick Charles
Peake, Robert Lee
Pitts, Donald Graves
Poluhowich, John Jacob
Pond, Wilson Gideon
Powell, Michael Robert
Preslock, James Peter
Proppe, Duane W
Redburn, Dianna Ammons
Reeves, T Joseph
Reid, Michael Baron
Reiter, Russel Joseph
Reuss, Luis
Richards, Joanne S
Riehl, Robert Michael
Rock, Michael Keith
Roeser, Ross Joseph
Roffwarg, Howard Philip
Rosenfeld, Charles Richard
Rudenberg, Frank Hermann
Samaan, Naguib A
Sant'Ambrogio, Giuseppe
Schafer, Rollie R
Scheel, Konrad Wolfgang
Schindler, William Joseph
Schrader, William Thurber
Scurry, Murphy Townsend
Senseman, David Michael
Shade, Robert Eugene
Silverthorn, Dee Unglaub
Simmons, David J
Smatresk, Neal Joseph
Snyder, Gary Dean
Soloff, Melvyn Stanley
Soriero, Alice Ann
Spence, Dale William
Sperling, Harry George
Sponsel, William Eric
Stacy, David Lowell
Stancel, George Michael
Stein, Jerry Michael
Storey, Arthur Thomas
Stubbs, Donald William
Stull, James Travis
Suddick, Richard Phillips
Tate, Charlotte Anne
Taubert, Kathryn Anne
Tsin, Andrew Tsang Cheung
Vanatta, John Crothers, III
Varma, Surendra K
Vaughan, Mary Kathleen
Vick, Robert Lore
Villa, Vicente Domingo
Vinson, S Bradleigh
Ward, Walter Frederick
Warner, Marlene Ryan
Waymire, Jack Calvin
Wayner, Matthew John
Weems, William Arthur
White, Fred Newton
Whitmore, Donald Herbert, Jr
Whitsett, Johnson Mallory, II
Wilkerson, James Edward
Williams, Charles Herbert
Williams, Gary Lynn
Willis, William Darrell, Jr
Wilson, Everett D
Wilson, L Britt
Witzel, Donald Andrew
Wolfe, James Wallace
Wolfenberger, Virginia Ann
Wordinger, Robert James
Worthy, Graham Anthony James
Yorio, Thomas
Yu, Byung Pal

UTAH
Albertine, Kurt H
Baranowski, Robert Louis
Bland, Richard David
Brown, Harold Mack, Jr
Burgess, Paul Richards
Conlee, Robert Keith
Ellis, LeGrande Clark
Eyzaguirre, Carlos
Foote, Warren Christopher
Galster, William Allen
Gessaman, James A
Grosser, Bernard Irving
Hansen, Afton M
Hsiao, Ting Huan
Jensen, David
Kesner, Raymond Pierre
Lasater, Eric Martin
Lighton, John R B
Lords, James Lafayette
McKenzie, Jess Mack
Mecham, Merlin J
Nelson, Don Harry
Normann, Richard A
Rhees, Reuben Ward
Rodin, Martha Kinscher
Roeder, Beverly Louise
Ruhmann Wennhold, Ann Gertrude
Shumway, Richard Phil
Tuckett, Robert P
Wilson, Dana E
Winder, William W
Wurst, Gloria Zettle
Yoshikami, Doju

VERMONT
Bartel, Lavon L
Evans, John N
Otter, Timothy
Slack, Nelson Hosking
Webb, George Dayton
Welsh, George W, III

VIRGINIA
Adams, James Milton
Andrykovitch, George
Bass, Michael Lawrence
Berne, Robert Matthew
Blackard, William Griffith
Blatt, Elizabeth Kempske
Block, Gene David
Burr, Helen Gunderson
Ching, Melvin Chung Hing
Clarke, Gary Anthony
Colmano, Germille
Corwin, Jeffrey Todd
Costanzo, Linda Schupper
Dane, Charles Warren
Davis, Joel L
Denbow, Donald Michael
Dendinger, James Elmer
Desjardins, Claude
Duling, Brian R
Dunn, John Thornton
Edwards, Leslie Erroll
Fariss, Bruce Lindsay
Feher, Joseph John
Fenner-Crisp, Penelope Ann
Fine, Michael Lawrence
Friesen, Wolfgang Otto
Gray, Faith Harriet
Grider, John Raymond
Gwazdauskas, Francis Charles
Hackett, John Taylor
Hall, Richard Eugene
Hanna, George R
Hardie, Edith L
Harper, Laura Jane
Heath, Alan Gard
Herbein, Joseph Henry, Jr
Hickman, Cleveland Pendleton, Jr
Hodgen, Gary Dean
Johnston, David Ware
Jones, John Evan
Jones, Joyce Howell
Katz, Alan Charles
Kirkpatrick, Roy Lee
Kramp, Robert Charles
Kripke, Bernard Robert
Kronfeld, David Schultz
Lloyd, John Willie, III
McCarty, Richard Charles
McCowen, Sara Moss
Manson, Nancy Hurt
Martin, James Henry, III
Mayer, David Jonathan
Mellon, DeForest, Jr
Nolan, Stanton Peelle
Olsen, Kathie Lynn
Owen, John Atkinson, Jr
Owens, Gary
Paulsen, Elsa Proehl
Pepe, Gerald Joseph
Pittman, Roland Nathan
Pitts, Grover Cleveland
Povlishock, John Theodore
Prewitt, Russell Lawrence, Jr
Price, Steven
Rogol, Alan David
Rubio, Rafael
Ruhling, Robert Otto
Schwab, Walter Edwin
Scott, David Evans
Shukla, Kamal Kant
Somlyo, Avril Virginia
Stein, Barry Edward
Stewart, Jennifer Keys
Swanson, Robert James
Talbert, George Brayton
Umminger, Bruce Lynn
Walsh, Scott Wesley
Wei, Enoch Ping
Wells, John Morgan, Jr
Williams, Gerald Albert
Williams, Patricia Bell
Wingard, Christopher Jon
Wolfe, Barbara Ann
Wolford, John Henry

WASHINGTON
Adkins, Ronald James
Ammann, Harriet Maria
Artru, Alan Arthur
Avner, Ellis David
Baksi, Samarendra Nath
Barnes, Charles Dee
Beck, Thomas W
Berger, Albert Jeffrey
Bierman, Edwin Lawrence
Binder, Marc David
Blinks, John Rogers
Bowden, Douglas Mchose
Bremner, William John
Brengelmann, George Leslie
Canfield, Robert Charles
Carlberg, Karen Ann
Crampton, George H
Dauble, Dennis Deene
Deyrup-Olsen, Ingrith Johnson
Dickhoff, Walton William
Dickson, William Morris
Duncan, Gordon W
Ensinck, John William
Estergreen, Victor Line
Gibson, Flash
Goodner, Charles Joseph
Gordon, Albert McCague
Halbert, Sheridan A
Hampton, John Kyle, Jr
Harding, Joseph Warren, Jr
Harper, Alfred Edwin
Heitkemper, Margaret M
Hicks, David L
Hildebrandt, Jacob
Huey, Raymond Brunson
Kahan, Linda Beryl
Kaufman, William Carl, Jr
Kenagy, George James
Kennedy, Thelma Temy
Landau, Barbara Ruth
Leung, Frederick C
Martin, Arthur Wesley
Mendelson, Martin
Moffett, David Franklin, Jr
Morgan, Kenneth Robb
Morrison, Peter Reed
Murphy, Mary Eileen
Nathanson, Neil Marc
Palka, John Milan
Palmer, John M
Paulsen, Charles Alvin
Prinz, Patricia N
Prusch, Robert Daniel
Reeves, Jerry John
Reh, Thomas Andrew
Reid, Donald House
Rich, Clayton
Riddiford, Lynn Moorhead
Rodieck, Robert William
Rubel, Edwin W

Physiology, Animal (cont)

Ruvalcaba, Rogelio H A
Santisteban, George Anthony
Schneider, David Edwin
Simpson, John Barclay
Smith, Lynwood S
Smith, Orville Auverne
Spencer, Merrill Parker
Stanek, Karen Ann
Steiner, Robert Alan
Taborsky, Gerald J, Jr
Teller, Davida Young
Toivola, Pertti Toivo Kalevi
Van Citters, Robert L
White, Fredric Paul
Whiteley, Helen Riaboff
Willows, Arthur Owen Dennis
Yancey, Paul Herbert
Youmans, William Barton

WEST VIRGINIA
Brown, David E
Brown, Paul B
Castranova, Vincent
Chambers, John William
Cochrane, Robert Lowe
Collins, William Edgar
Connors, John Michael
Culberson, James Lee
Dailey, Robert Arthur
Einzig, Stanley
Fisher, Martin Joseph
Foster, Joyce Geraldine
Gladfelter, Wilbert Eugene
Inskeep, Emmett Keith
Johnson, Michael D
Mugaas, John Nels
Overbeck, Henry West
Peterson, Ronald A
Reilly, Frank Daniel
Weber, Kenneth C

WISCONSIN
Anand, Amarjit Singh
Attie, Alan D
Bach-Y-Rita, Paul
Bavister, Barry Douglas
Benjamin, Robert Myles
Bergtrom, Gerald
Bisgard, Gerald Edwin
Bownds, M Deric
Boyeson, Michael George
Bremel, Robert Duane
Chi, Che
Coon, Robert L
Cowley, Allen Wilson, Jr
Davis, Larry Dean
Dierschke, Donald Joe
Ehle, Fred Robert
Fahning, Melvyn Luverne
Flax, Stephen Wayne
Folts, John D
Gilboe, David Dougherty
Goldberg, Alan Herbert
Gorski, Jack
Hager, Steven Ralph
Hamilton, Lyle Howard
Harrington, Sandra Serena
Hetzler, Bruce Edward
Howard, Thomas Hyland
Hutz, Reinhold Josef
Jackson, Meyer B
Keesey, Richard E
Kemnitz, Joseph William
Knoll, Jack
Lalley, Peter Michael
Lanphier, Edward Howell
Ledwitz-Rigby, Florence Ina
Lombard, Julian H
Lorton, Steven Paul
Marsh, Benjamin Bruce
Minnich, John Edwin
Nagle, Francis J
Nichols, Roy Elwyn
Piacsek, Bela Emery
Plotka, Edward Dennis
Raff, Hershel
Reneau, John
Rhode, William Stanley
Rose, Jerzy Edwin
Rouse, Thomas C
Schlough, James Sherwyn
Sidky, Younan Abdel Malik
Smith, Dean Orren
Spurr, Gerald Baxter
Staszak, David John
Steele, Robert Darryl
Stekiel, William John
Stevens, Richard Joseph
Stowe, David F
Stretton, Antony Oliver Ward
Struble, Craig Bruce
Sullivan, John Joseph
Terasawa, EI
Walgenbach-Telford, Susan Carol
Weil, Michael Ray
Wentworth, Bernard C
Will, James Arthur
Wolf, Richard Clarence
Woolsey, Clinton Nathan
Yin, Tom Chi Tien
Zuperku, Edward John

WYOMING
Dellenback, Robert Joseph
Dunn, Thomas Guy
Rose, James David
Smith, William Kirby

PUERTO RICO
Afanador, Arthur Joseph
Del Castillo, Jose
Haddock, Lillian
Kuffler, Damien Paul
Opava-Stitzer, Susan C
Orkand, Richard K
Santacana, Guido E

ALBERTA
Bland, Brian Herbert
Campenot, Robert Barry
Coulter, Glenn Hartman
Dawson, J W
Hohn, Emil Otto
Hutchison, Kenneth James
Jaques, Louis Barker
Jones, Richard Lee
LeBlanc, Francis Ernest
Lukowiak, Kenneth Daniel
Lysyk, Timothy James
Mackay, William Charles
McMahon, Brian Robert
Mears, Gerald John
Molnar, George D
Moore, Graham John
Neuwirth, Maria
Pang, Peter Kai To
Pearson, Keir Gordon
Peter, Richard Ector
Pittman, Quentin J
Russell, James Christopher
Smith, Richard Sidney
Spencer, Andrew Nigel
Stein, Richard Bernard
Stell, William Kenyon
Tyberg, John Victor
Walsh, Michael Patrick
Wang, Lawrence Chia-Huang
Whitelaw, William Albert
Wilkens, Jerrel L
Wilson, Frank B
Winter, Jeremy Stephen Drummond

BRITISH COLUMBIA
Bailey, Charles Basil Mansfield
Bruchovsky, Nicholas
Clarke, William Craig
Copp, Douglas Harold
Cottle, Merva Kathryn Warren
Cynader, Max Sigmund
Domenici, Paolo
Donaldson, Edward Mossop
Fankboner, Peter Vaughn
Freeman, Hugh J
Gosline, John M
Haunerland, Norbert Heinrich
Hayward, John S
Hoffer, J(oaquin) A(ndres)
Hughes, Maryanne Robinson
Jones, David Robert
Kennedy, Christopher Jesse
Kraintz, Leon
Krishnamurti, Cuddalore Rajagopal
Lederis, Karolis (Karl)
McKeown, Brian Alfred
Milsom, William Kenneth
Rennie, Paul Steven
Schwarz, Dietrich Walter Friedrich
Steeves, John Douglas
Vincent, Steven Robert
Ward, David Mercer

MANITOBA
Anthonisen, Nicholas R
Bihler, Ivan
Dyck, G(Gerry) W
Foster, Charles David Owen
Giles, Michael Arthur
Greenway, Clive Victor
Hara, Toshiaki J
Huebner, Judith Dee
Nance, Dwight Maurice
Polimeni, Philip Iniziato
Singal, Pawan Kumar
Swierstra, Ernest Emke

NEW BRUNSWICK
Anderson, John Murray
Beninger, Peter Gerard
Paim, Uno
Seabrook, William Davidson
Waiwood, Kenneth George

NEWFOUNDLAND
Brosnan, John Thomas
Fletcher, Garth L
Hillman, Donald Arthur
Kao, Ming-hsiung
Morris, Claude C
Rusted, Ian Edwin L H
Steele, Vladislava Julie
Virgo, Bruce Barton

NOVA SCOTIA
Dudar, John Douglas
Ferrier, Gregory R
Finley, John P

Horrobin, David Frederick
Leslie, Ronald Allan
McDonald, Terence Francis
Marshall, William Smithson
O'Dor, Ronald Keith
Rasmusson, Douglas Dean
Semba, Kazue
Szerb, John Conrad
Toews, Daniel Peter
Weld, Charles Beecher

ONTARIO
Abrahams, Vivian Cecil
Ackermann, Uwe
Adamson, S Lee
Ainsworth, Louis
Alikhan, Muhammad Akhtar
Atwood, Harold Leslie
Ballantyne, James Stuart
Bayliss, Colin Edward
Begin-Heick, Nicole
Biro, George P
Bisby, Mark A
Bols, Niels Christian
Broughton, Roger James
Buchanan, George Dale
Burton, John Heslop
Butler, Richard Gordon
Cafarelli, Enzo Donald
Carlen, Peter Louis
Challis, John Richard George
Chapler, Christopher Keith
Ciriello, John
Coceani, Flavio
Cohen, Saul Louis
Cooke, J David
Cunnane, Stephen C
Davey, Kenneth George
Davis, Alan
De Bold, Adolfo J
Depocas, Florent
Downer, Roger George Hamill
Etches, Robert J
Fenwick, James Clarke
Ferguson, Alastair Victor
Flanagan, Peter Rutledge
Foskett, J Kevin
Friesen, Henry George
Gardner, David R
Graham, Arthur Renfree
Hetenyi, Geza Joseph
Hobkirk, Ronald
Horner, Alan Alfred
Houston, Arthur Hillier
Ihssen, Peter Edowald
Karmazyn, Morris
King, Cheryl E
King, Gordon James
Kinson, Gordon A
Korecky, Borivoj
Kraicer, Jacob
Kuroski-De Bold, Mercedes Lina
Laidlaw, John Coleman
Langille, Brian Lowell
Leatherland, John F
Lee, Robert Maung Kyaw Win
Leung, Lai-Wo Stan
Ling, Daniel
Loeb, Gerald Eli
Lucis, Ojars Janis
Lucis, Ruta
McLachlan, Crapper Donald Raymond
McLachlan, Richard Scott
Milligan, John Vorley
Millman, Barry Mackenzie
Montemurro, Donald Gilbert
Moon, Thomas William
Murphy, John Thomas
Osmond, Daniel Harcourt
Pang, Cho Yat
Parker, John Orval
Perrault, Marcel Joseph
Rangachari, Patanji Srinivasa Kumar
Renaud, Leo P
Rixon, Raymond Harwood
Schimmer, Bernard Paul
Schneiderman, Jacob Harry
Seguin, Jerome Joseph
Sessle, Barry John
Sivak, Jacob Gershon
Slutsky, Arthur
Smith, Jonathan Jeremy Berkeley
Smith, Lorraine Catherine
Song, Seh-Hoon
Steel, Colin Geoffrey Hendry
Steele, John Earle
Stevens, Ernest Donald
Storey, Kenneth Bruce
Sullivan, Charlotte Murdoch
Tait, James Simpson
Tepperman, Barry Lorne
Toews, Cornelius J
Tsang, Charles Pak Wai
Vranic, Mladen
Weatherley, Alan Harold
Webb, Rodney A
Weick, Richard Fred
Wellman, Angela Myra
Wiebe, John Peter
Wood, Christopher Michael
Yip, Cecil Cheung-Ching
Zamel, Noe
Zarzecki, Peter

PRINCE EDWARD ISLAND
Amend, James Frederick

QUEBEC
Albert, Paul Joseph
Anctil, Michel
Barden, Nicholas
Belanger, Alain
Bellabarba, Diego
Besner, Michel
Billette, Jacques
Boisclair, Daniel
Borgeat, Pierre
Brouillette, Robert T
Chretien, Michel
Descarries, Laurent
Digby, Peter Saki Bassett
Dubé, François
Dupont, Andre Guy
Dykes, Robert William
Farmer, Chantal
Ford-Hutchinson, Anthony W
Garcia, Raul
Gariepy, Claude
Genest, Jacques
Gold, Allen
Guderley, Helga Elizabeth
Johnston, C Edward
Jones, Barbara Ellen
Jones, Geoffrey Melvill
Kalant, Norman
Kallai-Sanfacon, Mary-Ann
Krnjevic, Kresimir
Labrie, Fernand
Landry, Fernand
Lapp, Wayne Stanley
Larochelle, Jacques
Lehoux, Jean-Guy
Lemay, Jean-Paul
Letarte, Jacques
MacIntosh, Frank Campbell
Mackey, Michael Charles
Macklem, Peter Tiffany
Manjunath, Puttaswamy
Marsden, Joan Chauvin
Mortola, Jacopo Prospero
Mulay, Shree
Nadeau, Reginald Antoine
Nadler, Norman Jacob
Outerbridge, John Stuart
Parent, Andre
Pelletier, Georges H
Percy, Jonathan Arthur
Peronnet, Francois R R
Roberts, Kenneth David
Rochefort, Joseph Guy
Sandor, Thomas
Schiffrin, Ernesto Luis
Srivastava, Ashok Kumar
Srivastava, Prakash Narain
Tannenbaum, Gloria Shaffer
Tenenhouse, Alan M
Vinay, Patrick

SASKATCHEWAN
Patience, John Francis
Thornhill, James Arthur
Weisbart, Melvin

OTHER COUNTRIES
Allessie, Maurits
Alvarez-Buylla, Ramon
Bhatnagar, Ajay Sahai
Biederman-Thorson, Marguerite Ann
Bronk, John Ramsey
Buttemer, William Ashley
Chang, H K
Chen, Jan-Kan
Clamann, H Peter
Comfort, Alexander
Criss, Wayne Eldon
De Troyer, Andre Jules
Eik-Nes, Kristen Borger
Garcia-Sainz, J Adolfo
Gontier, Jean Roger
Haight, John Richard
Hall, Peter Francis
Hedglin, Walter L
Hughes, Abbie Angharad
Ichikawa, Shuichi
Ishida, Yukisato
Jarvis, Simon Michael
Josenhans, William T
Kaye, Alvin Maurice
Koella, Werner Paul
Kraicer, Peretz Freeman
Kuo, Ching-Ming
Lammers, Wim
Laurent, Pierre
Lekeux, Pierre Marie
Lieb, William Robert
Livett, Bruce G
Lu, Guo-Wei
Mezquita, Cristobal
Miyasaka, Kyoko
Phillips, George Douglas
Quintanllha, Alexandre Tiedtke
Racotta, Radu Gheorghe
Rettori, Ovidio
Richmond, Robert H
Sabry, Ismail
Santidrian, Santiago
Schliwa, Manfred
Schmeling, Sheila Kay

Sellin, Lawrence C
Smokovitis, Athanassios A
Srygley, Robert Baxter
Suga, Hiroyuki
Thorson, John Wells
Werman, Robert
Westerhof, Nicolaas
Withers, Philip Carew
Young, Bruce Arthur
Zapata, Patricio
Zbuzek, Vratislav

Physiology, General

ALABAMA
Al-lami, Fadhil
Barras, Donald J
Benos, Dale John
Besse, John C
Burns, Moore J
Cain, Stephen Malcolm
Carlo, Waldemar Alberto
Carpenter, Frank Grant
Carter, Charlotte
Cline, George Bruce
Davis, Elizabeth Young
Gibbons, Ashton Frank Eleazer
Gottlieb, Sheldon F
Hageman, Gilbert Robert
Hartley, Marshall Wendell
Hudson, Jack William, Jr
Lehman, Robert Harold
Lisano, Michael Edward
McDaniel, Gayner Raiford
Mitchell, Joseph Christopher
Neill, Jimmy Dyke
Nelson, Karl Marcus
Pace, Caroline S
Prejean, Joe David
Schneyer, Charlotte A
Schoepfle, Gordon Marcus
Shoemaker, Richard Leonard
Singh, Jarnail
Sparks, David Lee
Taylor, Aubrey Elmo
Ward, Coleman Younger
Warnock, David Gene
Wiggins, Earl Lowell

ALASKA
Behrisch, Hans Werner
Dilmino, Michael Joseph
Elsner, Robert

ARIZONA
Bernays, Elizabeth Anna
Bogner, Phyllis Holt
Bowman, Douglas Clyde
Braun, Eldon John
Burt, Janis Mae
Dantzler, William Hoyt
Gruener, Raphael P
Hadley, Neil F
Halonen, Marilyn Jean
Hazel, Jeffrey Ronald
Johnson, David Gregory
Johnson, Oliver William
Johnson, Paul Christian
Kaltenbach, Carl Colin
Koldovsky, Otakar
Markle, Ronald A
Morgan, Wayne Joseph
Russell, Findlay Ewing
Schmid, Jack Robert
Schopp, Robert Thomas
Shields, Jimmie Ken
Skinner, James Stanford
Stuart, Douglas Gordon
Tipton, Charles M
Varnell, Thomas Raymond
Witten, Mark Lee

ARKANSAS
Chowdhury, Parimal
Conaway, Howard Herschel
Demarest, Jeffrey R
Doyle, Lee Lee
Flanigan, William J
Griffin, Edmond Eugene
Koike, Thomas Isao
Lindley, Barry Drew
Martin, Duncan Willis
Martin, Richard Harvey
Molnar, George William
Moss, Alfred Jefferson, Jr
Rayford, Phillip Leon
Webber, Richard John

CALIFORNIA
Adorante, Joseph S
Agnew, William Finley
Alpen, Edward Lewis
Applegarth, Adrienne P
Arcadi, John Albert
Arendt, Kenneth Albert
Arnaud, Claude Donald, Jr
Bainton, Cedric R
Bair, Thomas De Pinna
Balchum, Oscar Joseph
Baldwin, Kenneth M
Balon, Thomas William
Bancroft, Richard Wolcott
Baum, David

Baxter, Claude Frederick
Beaver, W(illiam) L(awrence)
Behrman, Richard Elliot
Bejar, Ezra
Berger, Ralph Jacob
Bergman, Richard N
Bernauer, Edmund Michael
Berry, Christine Albachten
Berry, Stephen J
Bianchi, Donald Ernest
Billings, Charles Edgar, Jr
Birren, James E(mmett)
Bloor, Colin Mercer
Blume, Frederick Duane
Boda, James Marvin
Bradley, JoAnn D
Brandt, Charles Lawrence
Brokaw, Charles Jacob
Brooks, George Austin
Brown, Kenneth Taylor
Bryson, George Gardner
Buffett, Rita Frances
Burri, Betty Jane
Burwen, Susan Jo
Bushnell, James Judson
Cala, Peter M
Callantine, Merritt Reece
Carregal, Enrique Jose Alvarez
Carsten, Mary E
Case, James Frederick
Catran, Jack
Chaffee, Rowand R J
Chan, Timothy M
Chapman, Lloyd William
Charles, M Arthur
Chien, Shu
Childress, James J
Chou, Chung-Kwang
Christensen, Halvor Niels
Clegg, James S
Clements, John Allen
Close, Perry
Cobb, Jewel Plummer
Code, Charles Frederick
Cogan, Martin
Cohn, Stanton Harry
Collier, Clarence Robert
Cons, Jean Marie Abele
Contreras, Thomas Jose
Cooper, Allen D
Coppenger, Claude Jackson
Cornelius, Charles Edward
Courtney, Kenneth Randall
Crandall, Edward D
Crowe, John H
Curry, Fitz-Roy Edward
Curtis, Dwayne H
Davis, James A
Dawson, John E
Diamond, Jared Mason
Dole, William Paul
Dunn, Arnold Samuel
Ernst, Ralph Ambrose
Estrada, Norma Ruth
Fagin, Katherine Diane
Fallon, James Harry
Feldmeth, Carl Robert
Fiorindo, Robert Philip
Fisher, Delbert A
Flacke, Joan Wareham
Flacke, Werner Ernst
Flaim, Kathryn Erskine
Flaim, Stephen Frederick
Foley, Duane H
Forte, John Gaetano
Fortes, George (Peter Alexander)
Fox, Joan Elizabeth Bothwell
Frank, Joy Sopis
Fronek, Arnost
Galambos, Robert
Galloway, Raymond Alfred
Gibor, Ahron
Gietzen, Dorothy Winter
Goldstein, Bernard
Goodman, Joan Wright (Mrs Charles D)
Gordon, David Buddy
Gough, David Arthur
Greenleaf, John Edward
Guillemin, Roger (Charles Louis)
Haas, Gustav Frederick
Hackett, Nora Reed
Hackney, Jack Dean
Hannon, John Patrick
Hansen, James E
Harrison, Florence Louise
Harrison, Michael R
Hastings, David Frank
Hatton, Glenn Irwin
Heber, David
Heinrich, Milton Rollin
Heller, Horace Craig
Heusner, Alfred August
Hicks, Jimmie Lee
Hill, Robert
Holley, Daniel Charles
Holmsen, Theodore Waage
Homsher, Earl Edwin, II
Honrubia, Vicente
Hopf, Harriet Dudey Williams
Horres, Charles Russell, Jr
Horton, Richard
Horvath, Steven Michael
Hsueh, Aaron Jen Wang
Huszczuk, Andrew Huszcza

Hutton, Kenneth Earl
Jevning, Ron
Jolley, Weldon Bosen
Jones, James Henry
Josephson, Robert Karl
Kappagoda, C Tissa
Kapur, Krishan Kishore
Karuza, Sarunas Kazys
Kaufman, Marc P
Keens, Thomas George
Keil, Lanny Charles
Kellogg, Ralph Henderson
Khatra, Balwant Singh
Kim, Kwang-Jin
Kleeman, Charles Richard
Kloner, Robert A
Klouda, Mary Ann Aberle
Koziol, Brian Joseph
Kroc, Robert Louis
Kun, Ernest
Lance, Valentine A
Laris, Philip Charles
Lasley, Bill Lee
Lee, Jui Shuan
Leon, Henry A
Levy, Joseph Victor
Lianides, Sylvia Panagos
Linde, Leonard M
Linker-Israeli, Mariana
Lobl, Thomas Jay
Lolley, Richard Newton
Longo, Lawrence Daniel
Lorber, Victor
Lukin, L(arissa) S(kvortsov)
Lutt, Carl J
Lyon, Irving
McBlair, William
McClintic, Joseph Robert
Macey, Robert Irwin
McGuigan, Frank Joseph
Maffly, Roy Herrick
Martinson, Ida Marie
Meehan, John Patrick
Meerbaum, Samuel
Michael, Ernest Denzil, Jr
Mitchell, Robert A
Mommaerts, Wilfried
Morey-Holton, Emily Rene
Morse, John Thomas
Morton, Martin Lewis
Muscatine, Leonard
Newsom, Bernard Dean
Nicoll, Charles S
Nolan, Janiece Simmons
Ogasawara, Frank X
Ogden, Thomas E
Omid, Ahmad
Oyama, Jiro
Pace, Nello
Paolini, Paul Joseph
Parlow, Albert Francis
Pearl, Ronald G
Pengelley, Eric T
Peper, Erik
Peterson, Charles Marquis
Pickwell, George Vincent
Pilgeram, Laurence Oscar
Pollard, James Edward
Povzhitkov, Moysey Michael
Powanda, Michael Christopher
Power, Gordon G
Pratley, James Nicholas
Rabinowitz, Lawrence
Ralston, Henry James
Reid, Ian Andrew
Renkin, Eugene Marshall
Reynolds, Robert Williams
Rhode, Edward A, Jr
Ridley, Peter Tone
Robin, Eugene Debs
Rosenfeld, Sheldon
Ross, Gordon
Rothman, Stephen Sutton
Rubas, Werner
Rubinstein, Eduardo Hector
Rudolph, Abraham Morris
Rudy, Paul Passmore, Jr
Runion, Howell Irwin
Russell, Sharon May
Sanui, Hisashi
Scheer, Bradley Titus
Schein, Arnold Harold
Schmid-Schoenbein, Geert W
Schneidkraut, Marlowe J
Schooley, John C
Schwab, Robert G
Schweisthal, Michael Robert
Scobey, Robert P
Seegers, Walter Henry
Sellers, Alvin Louis
Setler, Paulette Elizabeth
Shaw, Jane E
Shellock, Frank G
Shore, Virgie Guinn
Shvartz, Esar
Simmons, Daniel Harold
Simmons, Francis Blair
Sirota, Jonas H
Smith, Arthur Hamilton
Smith, Robert Elphin
Smulders, Anthony Peter
Sobin, Sidney S
Sonnenschein, Ralph Robert
Stabenfeldt, George H

Starr, Arnold
Staub, Norman Croft
Stein, Myron
Stein, Seymour Norman
Stiffler, Daniel F
Strautz, Robert Lee
Sugar, Oscar
Swan, Harold James Charles
Swanson, George D
Talbot, Prudence
Taylor, Anna Newman
Thomas, Robert E
Thrasher, Terry Nicholas
Tierney, Donald Frank
Timiras, Paola Silvestri
Torre-Bueno, Jose Rollin
Trei, John Earl
Tremor, John W
Tullis, Richard Eugene
Turgeon, Judith Lee
Turley, Kevin
Tuttle, Ronald Ralph
Vidoli, Vivian Ann
Vomhof, Daniel William
Wagner, Jeames Arthur
Wasserman, Karlman
Wasterlain, Claude Guy
Weathers, Wesley Wayne
Weidner, William Jeffrey
Weissberg, Robert Murray
Wells, J Gordon
Wendel, Otto Theodore, Jr
Westfahl, Pamela Kay
White, Stephen Halley
White, Timothy P
Wilson, Archie Fredric
Wilson, Barry William
Woolley, Dorothy Elizabeth Schumann
Wright, Ernest Marshall
Wright, Richard Kenneth
Yates, Francis Eugene
Young, Ho Lee
Younoszai, Rafi
Yu, Jen
Zieg, Roger Grant
Zweifach, Benjamin William

COLORADO
Audesirk, Teresa Eck
Balke, Bruno
Battaglia, Frederick Camillo
Beam, Kurt George, Jr
Best, Jay Boyd
Betz, William J
Brockway, Alan Priest
Burke, Edmund R
Burke, Thomas Joseph
Carey, Cynthia
Draznin, Boris
Fuller, Barbara Fink Brockway
Gotshall, Robert William
Gray, John Stephens
Hand, Steven Craig
Harold, Franklin Marcel
Irish, James McCredie, III
Ishii, Douglas Nobuo
Korr, Irvin Morris
Krausz, Stephen
Levine, Simon Rock
McMurtry, Ivan Fredrick
Meschia, Giacomo
Neville, Margaret Cobb
Pautler, Eugene L
Pickett, Bill Wayne
Quissell, David Olin
Ralph, Charles Leland
Reeve, Ernest Basil
Rich, Royal Allen
Richards, Edmund A
Roberts, P Elaine
Roper, Stephen David
Sexton, Alan William
Snyder, Gregory Kirk
Takeda, Yasuhiko
Trapani, Ignatius Louis
Ulvedal, Frode
Vader-Lindholm, Connie
Wilber, Charles Grady
Williams, Ronald Lee

CONNECTICUT
Agulian, Samuel Kevork
Anderson, Rebecca Jane
Aronson, Peter S
Baird, James Leroy, Jr
Behrman, Harold R
Berliner, Robert William
Bigland-Ritchie, Brenda
Borgia, Julian Frank
Boulpaep, Emile L
Bronner, Felix
Chandler, William Knox
Clark, Nancy Barnes
Constantine, Jay Winfred
Crean, Geraldine L
Dawson, Margaret Ann
Dumont, Allan E
Edwards, Lawrence Jay
Fiore, Carl
Fleming, James Stuart
Gagge, Adolf Pharo
Gee, J Bernard L
Giebisch, Gerhard Hans
Gwynn, Robert H

Physiology, General (cont)

Hayslett, John P
Hirsh, Eva Maria Hauptmann
Hoffman, Joseph Frederick
Jaffe, Laurinda A
Jungas, Robert Leando
Katz, Arnold Martin
Koeppen, Bruce Michael
Landmesser, Lynn Therese
Lee, Thomas W
Mack, Gary W
Matovcik, Lisa M
Naftolin, Frederick
Oates, Peter Joseph
Patterson, John Ward
Rajendran, Vazhaikkurichi M
Rawson, Robert Orrin
Renfro, J Larry
Riblet, Leslie Alfred
Ritchie, Joseph Murdoch
Roos, Henry
Schnitzler, Ronald Michael
Schwartz, Tobias Louis
Sha'afi, Ramadan Issa
Siegel, Norman Joseph
Sikand, Rajinder S
Slayman, Clifford L
Sobol, Bruce J
Stitt, John Thomas
Taylor, Kenneth J W
Turnipseed, Marvin Roy
Verses, Christ James
Waterman, Talbot H(owe)
Weissman, Sherman Morton
Woods, Joseph James
Wright, Hastings Kemper
Zaret, Barry L

DELAWARE
Skopik, Steven D
South, Frank E
Turlapaty, Prasad
Wheeler, Allan Gordon

DISTRICT OF COLUMBIA
Atkins, James Lawrence
Bellamy, Ronald Frank
Burger, Edward James, Jr
Carpentier, Robert George
Cassidy, Marie Mullaney
Chaput, Raymond Leo
Coleman, Bernell
Freygang, Walter Henry, Jr
Geelhoed, Glenn William
Greenfield, Wilbert
Hamosh, Margit
Hamosh, Paul
Haskins, Caryl Parker
Johnson, Thomas F
Kasbekar, Dinkar Kashinath
Koering, Marilyn Jean
Lakshman, M Raj
Lee, Cheng-Chun
Leto, Salvatore
Lilienfield, Lawrence Spencer
Littler, Diane Sullion
Nardone, Roland Mario
Nath, Jayasree
Orlans, F Barbara
Pearce, Frederick James
Ramwell, Peter William
Raub, William F
Ronkin, R(aphael) R(ooser)
Rose, John Charles
Sinkford, Jeanne C
Starr, Matthew C
Sudia, Theodore William
Sulzman, Frank Michael
Trusal, Lynn R
Verrier, Richard Leonard
Watkins, Don Wayne
Wong, Harry Yuen Chee
Yamamoto, William Shigeru

FLORIDA
Abraham, William Michael
Abrams, Robert Marlow
Anderson, William McDowell
Baiardi, John Charles
Balis, John Ulysses
Barkalow, Derek Talbot
Barrera, Frank
Battista, Sam P
Beidler, Lloyd M
Belardinelli, Luiz
Berkley, Karen J
Bressler, Steven L
Briscoe, Anne M
Brown, John Lott
Bundy, Roy Elton
Buss, Daryl Dean
Chaet, Alfred Bernard
Cody, D Thane
Cousins, Robert John
Davis, Darrell Lawrence
Dietz, John R
Drummond, Willa H
Dudley, Gary A
Ebbs, Jane Cotton
Edwards, Charles
Eitzman, Donald V
Ekberg, Donald Roy
Ellington, William Ross

Elmadjian, Fred
Epstein, Murray
Evans, David Hudson
Ferguson, John Carruthers
Fregly, Melvin James
Genthner, Barbara Robyn Sharak
Gerencser, George A
Gilmore, Joseph Patrick
Glennon, Joseph Anthony
Hayashi, Teru
Hays, Elizabeth Teuber
Ho, Ren-jye
Howell, David Sanders
Hurst, Josephine M
Jacobson, Howard Newman
Kemph, John Patterson
Kerrick, Wallace Glenn Lee
Kessler, Richard Howard
Kissen, Abbott Theodore
Landowne, David
Lawrence, John M
Lawrence, Merle
Legler, Donald Wayne
Mallery, Charles Henry
Mayrovitz, Harvey N
Nagel, Joachim Hans
Natzke, Roger Paul
Nelson, Eldon Lane, Jr
Nelson, Thomas Eustis, Jr
Nicolosi, Gregory Ralph
Nicosia, Santo Valerio
Nolasco, Jesus Bautista
Nunez, German
Olsson, Ray Andrew
Penhos, Juan Carlos
Posner, Philip
Raizada, Mohan K
Rao, Krothapalli Ranga
Rao, Papineni Seethapathi
Reynafarje, Baltazar
Reynolds, Orr Esrey
Rhamy, Robert Keith
Robbins, Ralph Compton
Rockstein, Morris
Roeder, Martin
Rose, Birgit
Rosenthal, Myron
Samis, Harvey Voorhees, Jr
Schwassman, Horst Otto
Shireman, Rachel Baker
Shiverick, Kathleen Thomas
Sick, Thomas J
Silverman, Sol Richard
Soliman, Magdi R I
Sparkman, Marjorie Frances
Speckmann, Elwood W
Stevens, Bruce Russell
Sturbaum, Barbara Ann
Thatcher, William Watters
Vogh, Betty Pohl
Wells, Charles Henry
Wilcox, Christopher Stuart
Williams, Clyde Michael
Williams, Joseph Francis
Wilson, Henry R
Wood, Charles Evans
Woodside, Kenneth Hall

GEORGIA
Abercrombie, Ronald Ford
Armstrong, Robert Beall
Ashley, Doyle Allen
Balcer-Brownstein, Josefine P
Blum, Murray Sheldon
Boyd, Louis Jefferson
Caldwell, Robert William
Chew, Catherine S
Davidson, John Keay, III
De Felice, Louis John
DeHaan, Robert Lawrence
Edelhauser, Henry F
Eriksson, Karl-Erik Lennart
Geber, William Frederick
Ginsburg, Jack Martin
Godt, Robert Eugene
Green, Keith
Guinan, Mary Elizabeth
Gunn, Robert Burns
Hendrich, Chester Eugene
Hersey, Stephen J
Hockman, Charles Henry
Huber, Thomas Lee
Hunter, Frissell Roy
Innes, David Lyn
Jackson, Richard Thomas
Khan, Mohammad Iqbal
Kolbeck, Ralph Carl
Mann, David R
Manning, John W
Martin, David Edward
Matsumoto, Yorimi
Moriarty, C Michael
Nabrit, Samuel Milton
Nichols, J Wylie
Pashley, David Henry
Pooler, John Preston
Popovic, Vojin
Porterfield, Susan Payne
Reichard, Sherwood Marshall
Rinard, Gilbert Allen
Steinbeck, Klaus
Taylor, Andrew T, Jr
Taylor, Robert Clement
Toney, Thomas Wesley

Tuttle, Elbert P, Jr
Van De Water, Joseph M
Waitzman, Morton Benjamin
Weatherred, Jackie G
White, John Francis
Whitford, Gary M
Wiedmeier, Vernon Thomas
Willis, John Steele
Woods, Wendell David
Young, Leona Graff

HAWAII
Ahearn, Gregory Allen
Chang, Franklin
Druz, Walter S
Kodama, Arthur Masayoshi
Lin, Yu-Chong
Read, George Wesley
Reed, Stuart Arthur
Rogers, Terence Arthur
Segal, Earl
Smith, Richard Merrill

IDAHO
Bunde, Daryl E
Christian, Ross Edgar
Dahmen, Jerome J
Ferguson, James Homer
Laskowski, Michael Bernard
McKean, Thomas Arthur
Seeley, Rod R

ILLINOIS
Abraira, Carlos
Al-Bazzaz, Faiq J
Albrecht, Ronald Frank
Allgood, Joseph Patrick
Anderson, John Denton
Aprison, Barry Steven
Bacus, James William
Bailie, Michael David
Barany, Kate
Barr, Lloyd
Bartke, Andrzej
Becker, John Henry
Beecher, Christopher W W
Bell, Richard Dennis
Blivaiss, Ben Burton
Browning, Ronald Anthony
Buchholz, Robert Henry
Buetow, Dennis Edward
Burhop, Kenneth Eugene
Carlson, Gerald Eugene
Catchpole, Hubert Ralph
Chan, Yun Lai
Cralley, John Clement
Cromer, John A
Cureton, Thomas (Kirk), Jr
Curtis, Brian Albert
Dawe, Albert Rolke
Dawson, M Joan
DeClue, Jim W
Denbo, John Russell
Devries, Arthur Leland
Domroese, Kenneth Arthur
Ducoff, Howard S
Dunagan, Tommy Tolson
Dziuk, Philip J
Ebert, Paul Allen
Ecker, Richard Eugene
Eder, Douglas Jules
Eisenberg, Robert S
Ellert, Martha Schwandt
Ernest, J Terry
Filkins, James P
Fozzard, Harry A
Frehn, John
Garthwaite, Susan Marie
Gartner, Lawrence Mitchel
Gibbons, Larry V
Gibori, Geula
Gillette, Martha Ulbrick
Glaviano, Vincent Valentino
Green, Jonathan P
Greenberg, Ruven
Grimm, Arthur F
Gu, Yan
Haas, Mark
Harris, Ruth B S
Harris, Stanley Cyril
Hassan, Aslam Sultan
Hast, Malcolm Howard
Hawkins, Richard Albert
Hawley, Philip Lines
Heath, James Edward
Hechter, Oscar Milton
Helman, Sandy I
Hentges, Lynnette S W
Houk, James Charles
Humphrey, Gordon Laird
Hunter, William Sam
Hunzicker-Dunn, Mary
Jackson, Gary Loucks
Jacobs, H Kurt
Jakobsson, Eric Gunnar, Sr
Jung, Frederic Theodore
Kaplan, Harold M
Karczmar, Alexander George
Kelso, Albert Frederick
Khan, Mohammed Nasrullah
Kidder, George Wallace, III
Kim, Mi Ja
Klein, Diane M
Koblick, Daniel Cecil

Koushanpour, Esmail
Kyncl, J Jaroslav
Lax, Louis Carl
Leven, Robert Maynard
Lin, James Chih-I
Lindheimer, Marshall D
Lipsius, Stephen Lloyd
Lit, Alfred
Lodge, James Robert
Loeb, Jerod M
Loizzi, Robert Francis
Lopez-Majano, Vincent
Lorand, Laszlo
McCormack, Charles Elwin
Madsen, David Christy
Marotta, Sabath Fred
Masken, James Frederick
Mason, George Robert
Metz, John Thomas
Miller, Donald Morton
Morrison, Shaun Francis
Myers, James Hurley
Nelson, Douglas O
Nelson, Ralph A
Nicolette, John Anthony
Niles, Walter Dulany, II
Nyhus, Lloyd Milton
Palincsar, Edward Emil
Pederson, Vernon Clayton
Peiss, Clarence Norman
Perkins, William Eldredge
Perlman, Robert
Perry, Harold Tyner, Jr
Preston, Robert Leslie
Prosser, Clifford Ladd
Proudfit, Carol Marie
Rabuck, David Glenn
Rakowski, Robert F
Ratzlaff, Kermit O
Ringham, Gary Lewis
Ripps, Harris
Rose, Richard Carrol
Russell, Brenda
Sanderson, Glen Charles
Sayeed, Mohammed Mahmood
Schreider, Bruce David
Scommegna, Antonio
Simmons, William Howard
Singh, Sant Parkash
Sleator, William Warner, Jr
Snarr, John Frederic
Solaro, R John
Sukowski, Ernest John
Tallitsch, Robert Boyde
Taylor, Ardell Nichols
Ten Eick, Robert Edwin
Thilenius, Otto G
Thomas, John Xenia, Jr
Thompson, Leif Harry
Toback, F(rederick) Gary
Tobin, Martin John
Truong, Xuan Thoai
Ts'ao, Chung-Hsin
Turnquist, Richard Lee
Van de Kar, Louis David
Vesselinovitch, Stan Dushan
Walsh, Raymond Robert
Wessel, Hans U
Whisler, Kenneth Eugene
Wilkinson, Harold L
Zak, Radovan Hynek

INDIANA
Alliston, Charles Walter
Asano, Tomoaki
BeMiller, Paraskevi Mavridis
Blasingham, Mary Cynthia
Bottoms, Gerald Doyle
Brett, William John
Christ, Daryl Dean
Dantzig, Anne H
Dungan, Kendrick Webb
Echtenkamp, Stephen Frederick
Elharrar, Victor
Friedman, Julius Jay
Fuller, Forst Donald
Gaddis, Monica Louise
Garton, David Wendell
Geddes, LaNelle Evelyn
Grabowski, Sandra Reynolds
Greenspan, Kalman
Gunst, Susan Jane
Heath, Gordon Glenn
Henzlik, Raymond Eugene
Hertzler, Emanuel Cassel
Hinds, Marvin Harold
Howard, David K
Hui, Chiu Shuen
Hunt, Linda Margaret
Hurst, Robert Nelson
Hyslop, Paul A
Knoebel, Leon Kenneth
Kurz, Kenneth D
Lane, Ardelle Catherine
Lin, Tsung-Min
Maass-Moreno, Roberto
Marquis, Norman Ronald
Mays, Charles Edwin
Meiss, Richard Alan
Meyers, Michael Clinton
Nunn, Arthur Sherman, Jr
Randall, Barbara Feucht
Randall, James Edwin
Randall, Walter Clark

Renzi, Alfred Arthur
Rhoades, Rodney A
Rinkema, Lynn Ellen
Schaeffer, John Frederick
Schauf, Charles Lawrence
Selkurt, Ewald Erdman
Shea, Philip Joseph
Silbaugh, Steven A
Singleton, Wayne Louis
Sisson, George Maynard
Stephenson, William Kay
Strickholm, Alfred
Suthers, Roderick Atkins
Tamar, Henry
Wilkin, Peter J
William, James C, Jr
Witzmann, Frank A
Wolff, Ronald Keith

IOWA
Bergman, Ronald Arly
Bishop, Stephen Hurst
Christensen, James
Deavers, Daniel Ronald
Dunham, Jewett
Engen, Richard Lee
Folk, George Edgar, Jr
Gisolfi, Carl Vincent
Guest, Mary Frances
Gurll, Nelson
Horst, Ronald Lee
Huiatt, Ted W
Imig, Charles Joseph
Kirkland, Willis L
Kolder, Hansjoerg E
Masat, Robert James
Mennega, Aaldert
Norcia, Leonard Nicholas
Osborne, James William
Rulon, Russell Ross
Searle, Gordon Wentworth
Siebes, Maria
Spaziani, Eugene
Welsh, Michael James
Woolley, Donald Grant
Wunder, Charles C(ooper)

KANSAS
Alper, Richard H
Balfour, William Mayo
Call, Edward Prior
Cheney, Paul David
Clancy, Richard L
Gillespie, Jerry Ray
Greenberger, Norton Jerald
Greenwald, Gilbert Saul
Hermreck, Arlo Scott
Longley, William Joseph
Loudon, Catherine
Morrissey, J Edward
Norris, Patricia Ann
Nusser, Wilford Lee
Ringle, David Allan
Shirer, Hampton Whiting
Smith, Joseph Emmitt
Trank, John W
Upson, Dan W
Urban, James Edward
Valenzeno, Dennis Paul
Yochim, Jerome M

KENTUCKY
Anderson, Gary Lee
Bailey, Donald Wycoff
Bennett, Thomas Edward
Bowen, Thomas Earle, Jr
Chappell, Guy Lee Monty
Collins, Delwood C
Crawford, Eugene Carson, Jr
Diana, John N
Douglas, Robert Hazard
Dowell, Russell Thomas
Fell, Ronald Dean
Frazier, Donald Tha
Gailey, Franklin Bryan
Getchell, Thomas Vincent
Gillilan, Lois Adell
Gordon, Helmut Albert
Gross, David Ross
Huang, Kee-Chang
Humphreys, Wallace F
Jimenez, Agnes E
Jokl, Ernst
Kraman, Steve Seth
Lai, Yih-Loong
Lee, Lu-Yuan
Legan, Sandra Jean
Moore, James Carlton
Musacchia, X J
Parkins, Frederick Milton
Passmore, John Charles
Peretz, Bertram
Randall, David Clark
Richardson, Daniel Ray
Schilb, Theodore Paul
Still, Eugene Updike
Thornton, Paul A
Walker, Sheppard Matthew
Wekstein, David Robert
Wiegman, David L
Wyssbrod, Herman Robert
Zechman, Frederick William, Jr

LOUISIANA
Arimura, Akira
Beard, Elizabeth L
Bhattacharyya, Ashim Kumar
Burns, Alastair H
Cullen, John Knox, Jr
Davis, George Diament
Falcon, Carroll James
Flournoy, Robert Wilson
Franklin, Beryl Cletis
George, Ronald Baylis
Granger, D Niel
Harrison, Lura Ann
Hayes, Donald H
Heneghan, James Beyer
Hyman, Edward Sidney
Kalogeris, Therodore J
Klyce, Stephen Downing
Kreider, Jack Leon
Kreisman, Norman Richard
Kvietys, Peter R
Liles, Samuel Lee
McAfee, Robert Dixon
McDonough, Kathleen H
Mehendale, Harihara Mahadeva
Meier, Albert Henry
Miller, William Wadd, III
Morrissette, Maurice Corlette
Nance, Francis Carter
Newman, Wiley Clifford, Jr
Olmsted, Clinton Albert
Petri, William Henry, III
Porter, Johnny Ray
Randall, Howard M
Ricks, Beverly Lee
Robinson, George Edward, Jr
Robinson, Roy Garland, Jr
Roheim, Paul Samuel
Roussel, Joseph Donald
Shiau, Yih-Fu
Sloop, Charles Henry
Spanier, Arthur M
Spitzer, John J
Vela, Adan Richard
Wallin, John David
Watt, Edward William
Weill, Hans
Welbourne, Tomas C
Yeh, Kwo-Yih

MAINE
Bennett, Miriam Frances
Bredenberg, Carl Eric
Chute, Robert Maurice
Gelinas, Douglas Alfred
Hiatt, Robert Burritt
Hogben, Charles Adrian Michael
Logan, Rowland Elizabeth
Lycette, R(ichard) (Milton)
Major, Charles Walter
Twarog, Betty Mack

MARYLAND
Abbrecht, Peter H
Ahrens, Richard August
Alexander, Nancy J
Altland, Paul Daniel
Anderson, James Howard
Anderson, Norman Gulack
Baker, George Thomas, III
Baker, Houston Richard
Balaban, Robert S
Baum, Siegmund Jacob
Beall, James Robert
Benjamin, Fred Berthold
Benton, Allen William
Biersner, Robert John
Blackman, Marc Roy
Bodammer, Joel Edward
Bolt, Douglas John
Bricker, Jerome Gough
Bromberger-Barnea, Baruch (Berthold)
Brunner, Martha J
Brusilow, Saul W
Canonico, Peter Guy
Chaudhari, Anshumali
Choudary, Jasti Bhaskararao
Cipriano, Leonard Francis
Coble, Anna Jane
Covell, David Gene
Craig, Francis Northrop
Critz, Jerry B
Cummings, Edmund George
Dalton, John Charles
Darby, Eleanor Muriel Kapp
Dasler, Adolph Richard
Delahunty, George
Dermody, William Christian
Donaldson, Sue Karen
Eichholz, Alexander
Ewing, Larry Larue
Fajer, Abram Bencjan
Fitzgerald, Robert Schaefer
Freas, William
Gallin, Elaine K
Gamble, James Lawder, Jr
Gilbert, Daniel Lee
Glaser, Edmund M
Gonzalez-Fernandez, Jose Maria
Greenbaum, Leon J, Jr
Greisman, Sheldon Edward
Griffo, Zora Jasincuk
Grollman, Sigmund
Hajdu, Stephen
Hallfrisch, Judith
Hamilton, Clara Eddy
Hammer, John A
Hassett, Charles Clifford
Hegyeli, Ruth I E J
Heineman, Frederick W
Hellman, Alfred
Hertz, Roy
Higgins, William Joseph
Himwich, Williamina (Elizabeth) Armstrong
Hoffman, Donald J
Horton, Richard Greenfield
Hoskin, George Perry
Jackson, Michael J
Jaeger, James J
Johnson, Lawrence Arthur
Jordan, Alexander Walker, III
Joy, Robert John Thomas
Jurf, Amin N
Kafka, Marian Stern
Kalberer, John Theodore, Jr
Kaplan, Ann Esther
Kenimer, James G
Ketchel, Melvin M
Kety, Seymour S
Kiley, James P
King, Gregory L
Kinnamon, Kenneth Ellis
Koehler, Raymond Charles
Kunos, George
Lall, Abner Bishamber
Lederer, William Jonathan
Lee-Ham, Doo Young
LeRoith, Derek
Levinson, John Z
Levy, Alan C
Linberg, Steven E
Lincicome, David Richard
Litten, Raye Z, III
Lo, Chu Shek
Lorber, Mortimer
Lowy, R Joel
Lymn, Richard Wesley
McCormick, Kathleen Ann
Maciag, Thomas Edward
Maddox, Yvonne T
Maloney, Peter Charles
Marban, Eduardo
Marwah, Joe
Mercado, Teresa I
Meyer, Richard Arthur
Mitchell, Thomas George
Monti, John Anthony
Moran, Walter Harrison, Jr
Morgan, Raymond P
Mountcastle, Vernon Benjamin
Mullins, Lorin John
Murano, Genesio
Nekola, Mary Virginia
Newball, Harold Harcourt
Orahovats, Peter Dimiter
Ortmeyer, Heidi Karen
Paape, Max J
Packard, Barbara B K
Pamnani, Motilal Bhagwandas
Patanelli, Dolores J
Phelps, Harriette Longacre
Pierce, Sidney Kendrick
Pinter, Gabriel George
Podolsky, Richard James
Pursel, Vernon George
Quimby, Freeman Henry
Rall, Joseph Edward
Ramey, Estelle R
Rapoport, Stanley I
Rawlings, Samuel Craig
Redick, Thomas Ferguson
Rexroad, George Earle, Jr
Robinson, David Adair
Rosenstein, Laurence S
Rovelstad, Gordon Henry
Salcman, Michael
San Antonio, James Patrick
Sastre, Antonio
Scherbenske, M James
Schmidt, Edward Matthews
Schoenberg, Mark
Schweizer, Malvina
Scott, William Wallace
Scow, Robert Oliver
Shamoo, Adil E
Siebens, Arthur Alexandre
Sokoloff, Louis
Solomon, Neil
Specht, Heinz
Stojilkovic, Stanko S
Stolzenberg, Sidney Joseph
Striker, G E
Triantaphyllopoulos, Demetrios
Vargas, Fernando Figueroa
Varma, Shambhu D
Vollmer, Erwin Paul
Wade, James B
Waltrup, Paul John
Weinberg, Robert P
Wier, Withrow Gil
Williams, Robert Jackson
Yau, King-Wai
Yen-Koo, Helen C
Zaharko, Daniel Samuel
Zierler, Kenneth

MASSACHUSETTS
Adams, Jack H(erbert)
Adelman, William Joseph, Jr
Armett-Kibel, Christine
Arvan, Peter
Bade, Maria Leipelt
Barger, A(braham) Clifford
Barkman, Robert Cloyce
Bass, David Eli
Beasley, Debbie Sue
Berman, Herbert Joshua
Biggers, John Dennis
Biswas, Debajit K
Bizzi, Emilio
Blake, Thomas R
Bloomquist, Eunice
Boisse, Norman Robert
Botticelli, Charles Robert
Brain, Joseph David
Brown, Robert Glenn
Bruce, John Irvin
Burggren, Warren William
Burse, Richard Luck
Cantiello, Horacio Fabio
Chidsey, Jane Louise
Cochrane, David Earle
Cole, Edward Anthony
Covino, Benjamin Gene
Cymerman, Allen
Czeisler, Charles Andrew
Dane, Benjamin
Dayal, Yogeshwar
Deen, William Murray
Epstein, Franklin Harold
Essig, Alvin
Fay, Fredric S
Freadman, Marvin Alan
Gaensler, Edward Arnold
Ganley, Oswald Harold
Gifford, Cameron Edward
Goldberg, Alfred L
Goldman, David Eliot
Goldman, Ralph Frederick
Gonzalez, Richard Rafael
Goodman, Henry Maurice
Graves, William Earl
Green, Howard
Guimond, Robert Wilfrid
Gullans, Steven R
Hardy, William Lyle
Haschemeyer, Audrey Elizabeth Veazie
Hastings, John Woodland
Hedley-Whyte, John
Heglund, Norman C
Hilden, Shirley Ann
Hodgson, Edward Shilling
Horowitz, Arie
Howe, George R
Ingram, Roland Harrison
Irwin, Richard Stephen
Ishikawa, Sadamu
Jackson, Benjamin T
Jaffe, Lionel F
Jain, Rakesh Kumar
Johnson, Elsie Ernest
Kamen, Gary P
Kaminer, Benjamin
Kayne, Herbert Lawrence
Kazemi, Homayoun
Keeney, Clifford Emerson
Kolka, Margaret A
Lazarte, Jaime Esteban
Leavis, Paul Clifton
Leeman, Susan Epstein
LeFevre, Marian E Willis
LeFevre, Paul Green
Li, Jeanne B
Libby, Peter
Lipinski, Boguslaw
Liscum, Laura
Loewenfeld, Irene Elizabeth
Loring, Stephen H
Marcus, Elliot M
Mayer, Jean
Meldon, Jerry Harris
Milburn, Nancy Stafford
Millard, William James
Miller, Andrew Leitch
Morgan, Kathleen Greive
Mulvey, Philip Francis, Jr
O'Connor, William Brian
Orlidge, Alicia
Pandolf, Kent Barry
Pappenheimer, John Richard
Patton, John F
Pearincott, Joseph V
Prestwich, Kenneth Neal
Previte, Joseph James
Pryor, Marilyn Ann Zirk
Reichlin, Seymour
Resnick, Oscar
Ricci, Benjamin
Roberts, John Lewis
Rock, Paul Bernard
Romanoff, Elijah Bravman
Rosemberg, Eugenia
Scheid, Cheryl Russell
Schnitzer, Jan Eugeneusz
Schwartz, Bernard
Schwartz, John H
Shamgochian, Maureen Dowd
Shipley, Reginald A
Slechta, Robert Frank
Smith, Curtis Griffin

Physiology, General (cont)

Snedecor, James George
Solomon, Jolane Baumgarten
Stephenson, Lou Ann
Szlyk-Modrow, Patricia Carol
Tang, Shiow-Shih
Taylor, Charles Richard
Taylor, Robert E
Tomera, John F
Tong, Edmund Y
Tosteson, Daniel Charles
Trimmer, Barry Andrew
Ullrick, William Charles
Villars, Felix Marc Hermann
Vogel, James Alan
Waggener, Thomas Barrow
Walkowiak, Edmund Francis
Walsh, John V
Wilgram, George Friederich
Wilson, Thomas Hastings
Yellen, Gary
Zetter, Bruce Robert

MICHIGAN
Adams, Thomas
Aiken, James Wavell
Barnhart, Marion Isabel
Barraco, I Robin A
Be Ment, Spencer L
Bernard, Rudy Andrew
Blackwell, Leo Herman
Bradley, Robert Martin
Briggs, Josephine P
Buhl, Allen Edwin
Burlington, Roy Frederick
Butterworth, Francis M
Carter-Su, Christin
Chimoskey, John Edward
Christiansen, Richard Louis
Churchill, Paul Clayton
Collins, Robert James
Cooper, Theodore
D'Alecy, Louis George
Dawson, David Charles
Dawson, William Ryan
Derksen, Frederick Jan
Deuben, Roger R
Dillon, Patrick Francis
Dukelow, W Richard
England, Barry Grant
Ernst, Stephen Arnold
Faulkner, John A
Fenstermacher, Joseph Don
Ferguson-Miller, Shelagh Mary
Fernandez-Madrid, Felix
Fernandez y Cossio, Hector Rafael
Foa, Piero Pio
Freedman, Robert Russell
Frohman, Charles Edward
Fromm, Paul Oliver
Gala, Richard R
Garvin, Jeffrey Lawrence
Gelderloos, Orin Glenn
Gerritsen, George Contant
Gordon, Sheldon Robert
Hansen-Smith, Feona May
Heffner, Thomas G
Henry, Raymond Leo
Horowitz, Samuel Boris
Howatt, William Frederick
Hsu, Chen-Hsing
Jackson, William F
Jacquez, John Alfred
Jochim, Kenneth Erwin
Johnson, Shirley Mae
Johnston, Raymond F
Kaldor, George
Kootsey, Joseph Mailen
Kostyo, Jack Lawrence
Kratochvil, Clyde Harding
Lauderdale, James W, Jr
Lewis, Benjamin Marzluff
Lindemann, Charles Benard
McCoy, Lowell Eugene
Malvin, Richard L
Marshall, Norman Barry
Meites, Joseph
Menge, Alan C
Meyer, Ronald Anthony
Mitchell, Jerald Andrew
Morrow, Thomas John
Neudeck, Lowell Donald
Pax, Ralph A
Penney, David George
Piercey, Montford F
Ram, Jeffrey L
Reynolds, Orland Bruce
Riegle, Gail Daniel
Rillema, James Alan
Ringer, Robert Kosel
Roberts, John Stephen
Schwartz, Jessica
Shansky, Michael Steven
Shepard, Robert Stanley
Sherman, James H
Silbergleit, Allen
Steiman, Henry Robert
Stockwell, Charles Warren
Stout, John Frederick
Swanson, Curtis James
Van Harn, Gordon L
Walz, Daniel Albert
Weg, John Gerard
Whitten, Bertwell Kneeland
Williams, John Andrew
Williams, William James
Winbury, Martin M
Winkler, Barry Steven
Wittle, Lawrence Wayne
Wood, Jack Sheehan

MINNESOTA
Alsum, Donald James
Bacaner, Marvin Bernard
Bache, Robert James
Beck, Kenneth Charles
Benson, Katherine Alice
Carter, Earl Thomas
Delaney, John P
Dennis, Clarence
Dousa, Thomas Patrick
Eaton, John Wallace
Frederick, Edward C
Gallant, Esther May
Ganguli, Mukul Chandra
Gebhard, Roger Lee
Georgopoulos, Apostolos P
Graham, Edmund F
Graubard, Mark Aaron
Grim, Eugene
Haas, John Arthur
Haller, Edwin Wolfgang
Hancock, Peter Adrian
Housmans, Philippe Robert H P
Humphrey, Edward William
Hyatt, Robert Eliot
Iaizzo, Paul Anthony
Jankus, Edward Francis
Johnson, Ivan M
Johnson, Vincent Arnold
Kane, William J
Kaye, Michael Peter
Keys, Ancel (Benjamin)
Khraibi, Ali A
Knox, Franklyn G
Levitt, David George
Lewis, Debra A
Mabry, Paul Davis
Meyer, Maurice Wesley
Michels, Lester David
Miller, Robert F
Purple, Richard L
Rehder, Kai
Robb, Richard A
Sammak, Paul J
Shepherd, John Thompson
Taylor, Stuart Robert
Trachte, George J
Wallace, Kendall B
Wangensteen, Ove Douglas
Welter, Alphonse Nicholas
Wood, Earl Howard

MISSISSIPPI
Adair, Thomas H
Ashburn, Allen David
Bearden, Henry Joe
Brewer, Franklin Douglas
Douglas, Ben Harold
Dzielak, David J
Granger, Joey Paul
Guyton, Arthur Clifton
Hall, John Edward
Hester, Robert Leslie
Jennings, David Phipps
Manning, R Davis, Jr
Pace, Henry Buford
Russell, Raymond Alvin
Thompson, Walter Rolph
Turner, Manson Don
Walker, James Frederick
Young, David Bruce

MISSOURI
Baile, Clifton Augustus, III
Biellier, Harold Victor
Blount, Don Houston
Boyarsky, Saul
Brown, Herbert Ensign
Collier, Robert Joseph
Cook, Mary Rozella
Cornell, Creighton N
Coyer, Philip Exton
Dahms, Thomas Edward
Davis, James Othello
Eldredge, Donald Herbert
Ellsworth, Mary Litchfield
Fickess, Douglas Ricardo
Forker, E Lee
Graham, Charles
Griggs, Douglas M, Jr
Holloszy, John O
Howlett, Allyn C
Jen, Philip Hungsun
Johnson, Harold David
Jones, Allan W
Keane, John Francis, Jr
Kim, Hyun Dju
Krukowski, Marilyn
Krum, Alvin A
Liu, Maw-Shung
Mehrle, Paul Martin, Jr
Merrick, Arthur West
Mock, Orin Bailey
Morrill, Callis Gary
Mueckler, Mike Max
Peissner, Lorraine C
Platner, Wesley Stanley
Polakoski, Kenneth Leo
Roos, Albert
Rovetto, Michael Julien
Rubin, Maryiln Bernice
Ruh, Mary Frances
Russell, Robert Lee
Sage, Martin
Savery, Harry P
Schadt, James C
Senay, Leo Charles, Jr
Stahl, Philip Damien
Suga, Nobuo
Thomsen, Robert Harold
Van Beaumont, Karel William
Wang, Bin Ching
Welling, Larry Wayne
Winter, Henry Frank, Jr
Zatzman, Marvin Leon
Zepp, Edwin Andrew

MONTANA
Bintz, Gary Luther
Burfening, Peter J
Butler, Hugh C
Ericsson, Ronald James
Harrington, Joseph D
Rutledge, Lester T
Visscher, Saralee Neumann

NEBRASKA
Abel, Peter William
Adrian, Thomas E
Andrews, Richard Vincent
Belknap, Robert Wayne
Brody, Alfred Walter
Fougeron, Myron George
Gale, Henry H
Hill, Marvin Francis
Landolt, Paul Albert
Magee, Donal Francis
Murthy, Veeraraghavan Krishna
Scholes, Norman W
Shaw, David Harold
Ware, Frederick
Weeks, Donald Paul

NEVADA
Hillyard, Stanley Donald
Nellor, John Ernest
Yousef, Mohamed Khalil
Zanjani, Esmail Dabaghchian

NEW HAMPSHIRE
Bartlett, Donald, Jr
Berndtson, William Everett
Black, Wallace Gordon
DeMaggio, Augustus Edward
Detar, Reed L
Dixon, Wallace Clark, Jr
Dunlap, Jay Clark
Forster, Roy Philip
McCann, Frances Veronica
Naitove, Arthur
Ou, Lo-Chang
Rochelle, Lori Gatzy
Sasner, John Joseph, Jr
Tenney, Stephen Marsh
Valtin, Heinz

NEW JERSEY
Ahmed, S Sultan
Auletta, Carol Spence
Barker, June Northrop
Bauman, John W, Jr
Bianchi, John J
Blaiklock, Robert George
Blumenthal, Rosalyn D
Byrne, Jeffrey Edward
Carlin, Ronald D
Carlson, Gerald M
Cevallos, William Hernan
Choi, Sook Y
Cizek, Louis Joseph
Convey, Edward Michael
Crane, Robert Kellogg
DeFouw, David O
Diecke, Friedrich Paul Julius
Duran, Walter Nunez
Edinger, Henry Milton
England, Sandra J
Farnsworth, Patricia Nordstrom
Feinblatt, Joel Daniel
Fenichel, Richard Lee
Firschein, Hilliard E
Frankel, Harry Meyer
Frascella, Daniel W
Freund, Matthew J
Friedman, Kenneth Joseph
Garrick, Rita Anne
Gaudino, Mario
Geczik, Ronald Joseph
Goggins, Jean A
Goldenberg, Marvin M
Green, James Weston
Hafs, Harold David
Hageman, William E
Hall, James Conrad
Hassert, G(eorge) Lee, Jr
Heaslip, Richard Joseph
Henry, Sydney Mark
Hessler, Jack Ronald
Hey, John Anthony
Hruza, Zdenek
Isaacson, Allen
Ivatt, Raymond John
Kahn, Arthur Jole
Kleinberg, William
Krauthamer, George Michael
Kuna, Samuel
Lenz, Paul Heins
Lisk, Robert Douglas
Lodge, Nicholas John
Long, James Frantz
Lourenco, Ruy Valentim
Macdonald, Gordon J
McElligott, Mary Ann
McIlreath, Fred J
Mason, Richard Canfield
Mele, Frank Michael
Mittler, James Carlton
Moe, Robert Anthony
Opdyke, David F
Osborne, Melville
Phillips, Joy Burcham
Poutsiaka, John William
Raab, Jacob Lee
Reeves, John Paul
Reiser, H Joseph
Ress, Rudyard Joseph
Ritchie, David Malcolm
Rowley, George Richard
Rumsey, William LeRoy
Saldarini, Ronald John
Scott, Mary Celine
Shelden, Robert Merten
Shellenberger, Carl H
Siegel, John H
Sinclair, J Cameron
Sinha, Arabinda Kumar
Smith, Amelia Lillian
Spoerlein, Marie Teresa
Steele, Ronald Edward
Stephenson, Elizabeth Weiss
Stout, Marguerite Annette
Sturkie, Paul David
Sutherland, Donald James
Van Inwegen, Richard Glen
Virkar, Raghunath Atmaram
Vogel, W Mark
Waggoner, William Charles
Wang, Hsueh-Hwa
Wang, Shih Chun
Weiss, Harvey Richard
Wilson, Walter LeRoy
Winters, Robert Wayne
Wong, Keith Kam-Kin
You, Kwan-sa
Zambraski, Edward K
Zbuzek, Vlasta Kmentova

NEW MEXICO
Bachrach, Arthur Julian
Bolie, Victor Wayne
Galey, William Raleigh
Hallford, Dennis Murray
Hurwitz, Leon
Kluger, Matthew Jay
Leach, John Kline
Loeppky, Jack Albert
Partridge, L Donald
Ratner, Albert
Rich, Travis Dean
Riedesel, Marvin LeRoy
Smoake, James Alvin
Solomon, Sidney
Tombes, Averett S
Venters, Michael Dyar
Whalen, William James
Wood, Stephen Craig

NEW YORK
Acara, Margaret A
Adolph, Edward Frederick
Al-Awqati, Qais
Altura, Bella T
Altura, Burton Myron
Andersen, Olaf Sparre
Axen, Kenneth
Baird, Malcolm Barry
Barac-Nieto, Mario
Baruch, Sulamita B
Batt, Ellen Rae
Beal, Myron Clarence
Bedford, John Michael
Beer, Bernard
Bellve, Anthony Rex
Bentley, Barbara Lee
Bergofsky, Edward Harold
Bernardis, Lee L
Bernstein, Alvin Stanley
Beyenbach, Klaus Werner
Bishop, Beverly Petterson
Bisson, Mary A
Blank, Martin
Blanpied, George David
Borden, Edward B
Borgese, Thomas A
Bortoff, Alexander
Brandt, Philip Williams
Breed, Ernest Spencer
Broadway, Roxanne Meyer
Brodsky, William Aaron
Brunson, John Taylor
Bryan, John Kent
Buchholz, R Alan
Buckley, Nancy Margaret
Burkhoff, Daniel

Burlington, Harold
Cabot, John Boit, II
Camporesi, Enrico M
Candia, Oscar A
Case, Robert B
Catapane, Edward John
Chang, Chin-Chuan
Cockett, Abraham Timothy K
Cohen, Julius Jay
Cohen, Stephen Robert
Cohn, Deirdre Arline
Copley, Alfred Lewin
Craig, Albert Burchfield, Jr
Cranefield, Paul Frederic
Cunningham, Dorothy J
Deane, Norman
Del Guercio, Louis Richard M
Dell, Ralph Bishop
Di Salvo, Nicholas Armand
Downey, John A
Dunham, Philip Bigelow
Durkin, Patrick Ralph
Duta, Purabi
Dutton, Robert Edward, Jr
Egan, Edmund Alfred
Eggena, Patrick
Ellis, Keith Osborne
El-Sherif, Nabil
Farber, Jorge
Farhi, Leon Elie
Feder, Harvey Herman
Feld, Leonard Gary
Field, Michael
Finberg, Laurence
Finster, Mieczyslaw
Firriolo, Domenic
Fischbarg, Jorge
Fisher, Vincent J
Frankel, Arthur Irving
Freedman, Jeffrey Carl
Fried, George H
Friedman, Ronald Marvin
Fuchs, Anna-Riitta
Fusco, Madeline M
Gambert, Steven Ross
Gatto, Louis Albert
Gerst, Paul Howard
Gibson, Quentin Howieson
Gidari, Anthony Salvatore
Giering, John Edgar
Glaser, Warren
Godec, Ciril J
Gordon, Albert Saul
Gouras, Peter
Grassl, Steven Miller
Greif, Roger Louis
Grosso, Leonard
Grumbach, Leonard
Gupta, Raj K
Gupta, Sanjeev
Gurtner, Gail H
Haldar, Jaya
Hamilton, Robert William
Hammerman, David Lewis
Hartenstein, Roy
Hays, Richard Mortimer
Hazen, Annette
Henshaw, Robert Eugene
Herreid, Clyde F, II
Higgins, James Thomas, Jr
Hinkle, Patricia M
Hintze, Thomas Henry
Hirsch, Judith Ann
Hong, Suk Ki
Honig, Carl Robert
Hornung, David Eugene
Horowicz, Paul
Howell, Barbara Jane
Iovino, Anthony Joseph
Jacoby, Jean
Jalife, Jose
Javitt, Norman B
Jean-Baptiste, Emile
Johnson, Arnold
Johnson, Roger A
Jones, Susan Muriel
Jung, Chan Yong
Kaley, Gabor
Kamm, Donald E
Kanter, Gerald Sidney
Kao, Chien Yuan
Kao, Frederick
Kaplan, Ehud
Kaplan, Martin L
Karmali, Rashida A
Karpatkin, Simon
Karr, James Presby
Kashin, Philip
Kavaler, Frederic
Keller, Dolores Elaine
Kiely, Lawrence J
Kirchberger, Madeleine
Kleinman, Leonard I
Koester, John D
Koizumi, Kiyomi
Kow, Lee-Ming
Kraus, Shirley Ruth
Kriebel, Mahlon E
Lai, Fong M
Lalezari, Parviz
Lambert, Francis Lincoln
Laragh, John Henry
Lawton, Richard Woodruff
Lee, Chin Ok

Levy, Harvey Merrill
Lindsay, William Germer, Jr
Lippes, Jack
LoBue, Joseph
Lockshin, Richard Ansel
Loeb, John Nichols
Loegering, Daniel John
Lundgren, Claes Erik Gunnar
Maitra, Subir R
Marin, Matthew Gruen
Mattfeld, George Francis
Maturo, Joseph Martin, III
Michielli, Donald Warren
Minatoya, Hiroaki
Mizejewski, Gerald Jude
Morris, Thomas Wilde
Mozell, Maxwell Mark
Murphy, Michael Joseph
Murrish, David Earl
Nace, Paul Foley
Nagler, Arnold Leon
Natke, Ernest, Jr
Neidle, Enid Anne
Nieman, Gary Frank
Oronsky, Arnold Lewis
Paganelli, Charles Victor
Pappas, George Stephen
Parsons, Robert Hathaway
Pearl, William
Peirce, Edmund Converse, II
Pendergast, David R
Pentyala, Srinivas Narasimha
Pereira, Martin Rodrigues
Phipps, Roger John
Pilkington, Lou Ann
Pilkis, Simon J
Poston, Hugh Arthur
Pruitt, Nancy L
Purandare, Yeshwant K
Quaroni, Andrea
Ramazzotto, Louis John
Ranu, Harcharan Singh
Rasweiler, John Jacob, IV
Rayson, Barbara M
Redisch, Walter
Reeves, Robert Blake
Rennie, Donald Wesley
Rockwood, William Philip
Rosendorff, Clive
Rosenthal, William S
Saba, Thomas Maron
Sachs, John Richard
Satir, Birgit H
Scheuer, James
Schneider, Bruce Solomon
Schneiweiss, Jeannette W
Schreiber, Sidney S
Schwartz, Irving Leon
Shrager, Peter George
Shreeve, Walton Wallace
Siegel, Irwin Michael
Skulan, Thomas William
Sloviter, Robert Seth
Smith, Ora Kingsley
Socolar, Sidney Joseph
Spoor, Ryk Peter
Spotnitz, Henry Michael
Stagg, Ronald M
Stein, Philip
Stoner, Larry Clinton
Sullivan, W(alter) James
Susca, Louis Anthony
Taggart, John Victor
Taub, Mary L
Tedeschi, Henry
Ter Kuile, Benno Herman
Trimble, Mary Ellen
Tullson, Peter C
Tunik, Bernard D
Turinsky, Jiri
Tuttle, Richard Suneson
Van der Kloot, William George
Van Liew, Judith Bradford
Vassalle, Mario
Volavka, Jan
Wallenstein, Martin Cecil
Wasserman, Robert Harold
Weiner, Richard
Weinstein, Harel
Williams, Marshall Henry, Jr
Windhager, Erich E
Wittenberg, Beatrice A
Wittenberg, Jonathan B
Wittner, Murray
Wolf, Robert Lawrence
Wolin, Michael Stuart
Yablonski, Michael Eugene
Yao, Alice C
York, James Lester
Yozawitz, Allan
Zadunaisky, Jose Atilio
Ziegler, Frederick Dixon
Zorzoli, Anita
Zucker, Marjorie Bass

NORTH CAROLINA
Adler, Kenneth B
Amen, Ralph DuWayne
Anderson, John Joseph Baxter
Arendshorst, William John
Argenzio, Robert Alan
Barrow, Emily Mildred Stacy
Bawden, James Wyatt
Bennett, Peter Brian

Bentley, Peter John
Blanchard, Susan M
Blum, Jacob Joseph
Brodish, Alvin
Bull, Leonard Seth
Chan, Po Chuen
Crenshaw, Miles Aubrey
Curtin, Terrence M
Dohm, Gerald Lynis
Eastin, William Clarence, Jr
Ennis, Ella Gray Wilson
Faust, Robert Gilbert
Ferguson, John Howard
Finn, Arthur Leonard
Garren, Henry Wilburn
Garrett, Ruby Joyce Burriss
Glenn, Thomas M
Glower, Donald D, Jr
Gutknecht, John William
Hackney, Anthony C
Hayek, Dean Harrison
Humm, Douglas George
Hutchins, Phillip Michael
Jobsis, Frans Frederik
King, Theodore Matthew
Kokas, Eszter B
Kylstra, Johannes Arnold
Lazarus, Lawrence H
Lenhard, James M
Lieberman, Edward Marvin
Lieberman, Melvyn
Longmuir, Ian Stewart
Louie, Dexter Stephen
Mandel, Lazaro J
Menhinick, Edward Fulton
Mikat, Eileen M
Miller, Augustus Taylor, Jr
Miller, Inglis J, Jr
Mills, Elliott
Morse, Roy E
Mouw, David Richard
Murdock, Harold Russell
Noe, Frances Elsie
Olson, Howard H
Pallotta, Barry S
Parker, John Curtis
Patterson, David Thomas
Perl, Edward Roy
Proffit, William R
Ramp, Warren Kibby
Robinson, Jerry Allen
Robison, Odis Wayne
Rose, James C
Sar, Madhabananda
Sargent, Frank Dorrance
Schmidt-Nielsen, Knut
Shinkman, Paul G
Skinner, Newman Sheldon, Jr
Smith, Thomas Lowell
Somjen, George G
Swan, Algernon Gordon
Talmage, Roy Van Neste
Thurber, Robert Eugene
Townes, Mary McLean
Tucker, Vance Alan
Ulberg, Lester Curtiss
Van Horn, Diane Lillian
Vogel, Steven
Waugh, William Howard
Welsch, Frank
Wiggers, Harold Carl
Wilbur, Karl Milton
Wolgemuth, Richard Lee
Wooles, Wallace Ralph
Yonce, Lloyd Robert

NORTH DAKOTA
Brumleve, Stanley John
Davison, Kenneth Lewis
Duerr, Frederick G
Gerst, Jeffery William
Joshi, Madhusudan Shankarrao
Lukaski, Henry Charles
Saari, Jack Theodore

OHIO
Adams, Walter C
Adragna, Norma C
Alway, Stephen Edward
Appert, Hubert Ernest
Bajpai, Praphulla K
Banks, Robert O
Banwell, John G
Barlow, George
Barr, Harry L
Bhattacharya, Amar Nath
Bhattacharya, Amit
Bozler, Emil
Brand, Paul Hyman
Bryant, Shirley Hills
Budd, Geoffrey Colin
Burns, Elizabeth Mary
Cerroni, Rose E
Chakraborty, Joana J(yotsna)
Clark, David Lee
Connors, Alfred F, Jr
Cooke, Helen Joan
Corson, Samuel Abraham
Daroff, Robert Barry
Dujardin, Jean-Pierre L
Ely, Daniel Lee
Faiman, Charles
Ferguson, Marion Lee
Ferrario, Carlos Maria

Foulkes, Ernest Charles
Freed, James Melvin
Friedlander, Ira Ray
Fry, Donald Lewis
Geha, Alexander Salim
Greenwald, Lewis
Grossie, James Allen
Grupp, Gunter
Grupp, Ingrid L
Hall, Philip Wells, III
Halm, Dan Robert
Handwerger, Stuart
Hanson, Kenneth Marvin
Hassler, Craig Reinhold
Holtkamp, Dorsey Emil
Horseman, Nelson Douglas
Hoshiko, Tomuo
Howell, John N
Johnson, Dale Richard
Jones, Patricia H
Jones, Richard Dell
Jung, Dennis William
Khairallah, Philip Asad
Kirby, Albert Charles
Kleinerman, Jerome
Kline, Daniel Louis
Knox, Francis Stratton, III
Krontiris-Litowitz, Johanna Kaye
Kunz, Albert L
Lauf, Peter Kurt
Lessler, Milton A
Levy, Matthew Nathan
Lewis, Lena Armstrong
Lipsky, Joseph Albin
Lotz, W Gregory
Lustick, Sheldon Irving
McGrady, Angele Vial
Macht, Martin Benzyl
Maguire, Michael Ernest
Maron, Michael Brent
Martin, Paul Joseph
Meserve, Lee Arthur
Metting, Patricia J
Mostardi, Richard Albert
Naito, Herbert K
Nathan, Paul
Neiman, Gary Scott
Nelson, Leonard
Newman, Howard Abraham Ira
Nielson, Read R
Noyes, David Holbrook
Nussbaum, Noel Sidney
Olson, Lynne E
Patil, Popat N
Paul, Lawrence Thomas
Paul, Richard Jerome
Peterjohn, Glenn William
Phillips, Chandler Allen
Recknagel, Richard Otto
Reiser, Peter Jacob
Robbins, David O
Rudy, Yoram
Saiduddin, Syed
Scarpa, Antonio
Schinagl, Erich F
Senturia, Jerome B(asil)
Slonim, Arnold Robert
Sperelakis, Nicholas
Stokes, Robert Mitchell
Stone, Kathleen Sexton
Strohl, Kingman P
Stuesse, Sherry Lynn
Sutherland, James McKenzie
Thomson, Dale S
Tomashefski, Joseph Francis
Travis, Randall Howard
Walenga, Ronald W
Webb, Paul
Wideman, Cyrilla Helen
Wilson, James Albert
Wilson, Kenneth Glade
Wood, Jackie Dale

OKLAHOMA
Beames, Calvin G, Jr
Beesley, Robert Charles
Benyajati, Siribhinya
Blair, Robert William
Bogenschutz, Robert Parks
Cooper, Richard Grant
Cox, Beverley Lenore
DeLacerda, Fred G
Harrison, Aix B
Higgins, E Arnold
Hinshaw, Lerner Brady
Houlihan, Rodney T
Hurst, Jerry G
Johnson, Becky Beard
Johnson, Gerald, III
Lovallo, William Robert
McDonald, Leslie Ernest
Massion, Walter Herbert
Maton, Paul Nicholas
Mobley, Bert A
Newcomer, Wilbur Stanley
Olson, Robert Leroy
Oyler, J Mack
Shirley, Barbara Anne
Thomas, Sarah Nell
Wickham, M Gary

OREGON
Anderson, Debra F
Bennett, Robert M

Physiology, General (cont)

Carleton, Blondel Henry
Chilgren, John Douglas
Conte, Frank Philip
Crawshaw, Larry Ingram
Debons, Albert Frank
Dejmal, Roger Kent
Dow, Robert Stone
Faber, Jan Job
Hardt, Alfred Black
Hermsmeyer, Ralph Kent
Hillman, Stanley Severin
Hohimer, A Roger
Hoskins, Dale Douglas
Johnson, Larry Reidar
Keyes, Jack Lynn
Kleinholz, Lewis Hermann
Levine, Leonard
Lucas, Oscar Nestor
Mela-Riker, Leena Marja
Peterson, Clare Gray
Rampone, Alfred Joseph
Resko, John A
Rutten, Michael John
Shannon, James Augustine
Smith, Charles Welstead
Somero, George Nicholls
Stones, Robert C
Tanz, Ralph
Van Hassel, Henry John
Von Dreele, Margaret M
Weimar, Virginia Lee
Worrest, Robert Charles
Yatvin, Milton B
Zauner, Christian Walter

PENNSYLVANIA
Agersborg, Helmer Pareli Kjerschow, Jr
Agus, Zalman S
Ambromovage, Anne Marie
Angelakos, Evangelos Theodorou
Anthony, Adam
Armstead, William M
Armstrong, Clay M
Atwood, Mark Wyllie
Baechler, Charles Albert
Becker, Anthony J, Jr
Bianchi, Carmine Paul
Bond, Judith
Borle, Andre Bernard
Brobeck, John Raymond
Butler, Thomas Michael
Civan, Mortimer M
Coburn, Ronald F
Cox, Robert Harold
Cristofalo, Vincent Joseph
Cutler, Winnifred Berg
Davies, Robert Ernest
DeBias, Domenic Anthony
Detweiler, David Kenneth
De Weer, Paul Joseph
DiPasquale, Gene
Dodgson, Susanna Jane
Doghramji, Karl
Drees, John Allen
Driska, Steven P
Dunson, William Albert
Edwards, McIver Williamson, Jr
English, Leigh Howard
Fawley, John Philip
Fischer, Grace Mae
Fisher, Aron Baer
Fishman, Alfred Paul
Flemister, Launcelot Johnson
Flemister, Sarah C
Fluck, Richard Allen
Forster, Robert E, II
Fox, Karl Richard
Freeman, Alan R
Friedman, M H
Fritz, George Richard, Jr
Fuchs, Franklin
Gay, Carol Virginia Lovejoy
Gee, William
Gellai, Miklos
Glauser, Elinor Mikelberg
Goldberg, Martin
Goldman, Yale E
Greene, Charlotte Helen
Grego, Nicholas John
Hammerstedt, Roy H
Hart, Robert Gerald
Hartman, Arthur Dalton
Hasler, Marilyn Jean
Hausberger, Franz X
Henderson, George Richard
Herbison, Gerald J
Hitner, Henry William
Holliday, Charles Walter
Hollis, Theodore M
Horn, Lyle William
Idell-Wenger, Jane Arlene
Jefferson, Leonard Shelton
Johnson, Ernest Walter
Kamon, Eliezer
Kauffman, Gordon Lee, Jr
Kenney, William Lawrence
Killian, Gary Joseph
Kimbel, Philip
Knuttgen, Howard G
Kovach, Arisztid G B
Lahiri, Sukhamay
Langan, William Bernard
Lentini, Eugene Alfred Anthony
Levin, Sidney Seamore
Levine, Elliot Myron
Lindgren, Clark Allen
Ling, Gilbert Ning
Loeb, Alex Lewis
Loewy, Ariel Gideon
Lynch, Peter Robin
Magno, Michael Gregory
Martin, John Samuel
Mashaly, Magdi Mohamed
Meriney, Stephen D
Michelson, Eric L
Miles, Daniel S
Minsker, David Harry
Mir, Ghulam Nabi
Mitchell, Robert Bruce
Moore, E(arl) Neil
Morgan, Howard E
Mortimore, Glenn Edward
Neff, William H
Negro-Vilar, Andres F
Newell, Allen
Peachey, Lee DeBorde
Pegg, Anthony Edward
Pelleg, Amir
Piscitelli, Joseph
Pitt, Bruce R
Polgar, George
Post, Robert Lickely
Puri, Rajinder N
Rannels, Donald Eugene, Jr
Rattan, Satish
Reibel-Shinfeld, Diane Karen
Reinking, Larry Norman
Roberts, Shepherd (Knapp de Forest)
Robishaw, Janet D
Rosenfeld, Leonard M
Russell, John McCandless
Salama, Guy
Sather, Bryant Thomas
Schauer, Richard C
Senft, Joseph Philip
Shaffer, Thomas Hillard
Shapiro, Herbert
Sidie, James Michael
Siegfried, John Barton
Siegman, Marion Joyce
Smith, M Susan
Spear, Joseph Francis
Tansy, Martin F
Taylor, Paul M
Thomas, Steven P
Tong, Winton
Torchiana, Mary Louise
Torres, Joseph Charles
Trainer, David Gibson
Tuma, Ronald F
Upton, G Virginia
Vagnucci, Anthony Hillary
Waterhouse, Barry D
Watrous, James Joseph
Weber, Annemarie
Weiser, Philip Craig
Weisz, Judith
Wesson, Laurence Goddard, Jr
Whinnery, James Elliott
Whitfield, Carol F(aye)
Wickersham, Edward Walker
Winegrad, Saul
Wing, Rena R
Wolf, Stewart George, Jr
Wolfersberger, Michael Gregg
Wolken, Jerome Jay
Woychik, John Henry
Zabara, Jacob
Zavodni, John J

RHODE ISLAND
Constantine, Herbert Patrick
Cserr, Helen F
Degnan, Kevin John
Dolyak, Frank
Dowben, Robert Morris
Goldstein, Leon
Harrison, Robert William
Hill, Robert Benjamin
Jackson, Donald Cargill
Marshall, Jean McElroy
Stewart, Peter Arthur

SOUTH CAROLINA
Allen, Thomas Charles
Bond, Robert Franklin
Brodoff, Bernard Noah
Burnett, Louis E
Cole, Benjamin Theodore
Colwell, John Amory
Cook, James Arthur
Cooper, George, IV
Cromer, Jerry Haltiwanger
Fairbanks, Gilbert Wayne
Frayser, Katherine Regina
Fredericks, Christopher M
Fulton, George P(earman)
Hays, Ruth Lanier
Hempling, Harold George
Horres, Alan Dixon
Hughes, Buddy Lee
Katz, Sidney
Kinard, Fredrick William
Knight, Anne Bradley
Leonard, Walter Raymond
McCutcheon, Ernest P
McNamee, James Emerson
Pharr, Pamela Northington
Pivorun, Edward Broni
Runey, Gerald Luther
Sias, Fred R, Jr
Smiley, James Watson
Taber, Elsie
Tempel, George Edward
Watson, Philip Donald
Wheeler, Alfred Portius
Wheeler, Darrell Deane
Wolf, Matthew Bernard

SOUTH DAKOTA
Diggins, Maureen Rita
Goodman, Barbara Eason
Hietbrink, Bernard Edward
Schlenker, Evelyn Heymann
Schramm, Mary Arthur

TENNESSEE
Ahokas, Robert A
Arnold, William Archibald
Bagby, Roland Mohler
Banerjee, Mukul Ranjan
Beard, James David
Chen, Thomas Tien
Coburn, Corbett Benjamin, Jr
Cook, John Samuel
Corbin, Jack David
Coulson, Patricia Bunker
Davis, William James
Dix, John Willard
Fitzgerald, Laurence Rockwell
Friesinger, Gottlieb Christian
Fry, Richard Jeremy Michael
Gan, Rong Zhu
Ginski, John Martin
Groer, Maureen
Hall, Hugh David
Hardman, Joel G
Harris, Thomas R(aymond)
Jackowski, Suzanne
Johnson, Leonard Roy
Joyner, William Lyman
Kant, Kenneth James
Kao, Race Li-Chan
Kono, Tetsuro
Lawrence, William Homer
Leffler, Charles William
Liles, James Neil
MacFadden, Donald Lee
McGuinness, Owen P
Mahajan, Satish Chander
Manley, Emmett S
Manthey, Arthur Adolph
Meng, H C
Moses, Henry A
Neff, Robert Jack
Park, Charles Rawlinson
Rasch, Robert
Reynolds, Leslie Boush, Jr
Robert, Timothy Albert
Sander, Linda Dian
Schneider, Edward Greyer
Share, Leonard
Stahlman, Mildred
Stiles, Robert Neal
Tardif, Suzette Davis
Thakar, Jay H
Thomason, Donald Brent
Van Middlesworth, Lester
Williams, Carole A
Wiser, Cyrus Wymer
Wondergem, Robert
Wood, Henderson Kingsberry
Wood, William Booth
Zemel, Michael Barry

TEXAS
Allen, Julius Cadden
Allen, Thomas Hunter
Angelides, Kimon Jerry
Armstrong, George Glaucus, Jr
Ashton, Juliet H
Baker, Lee Edward
Baker, Robert David
Barrow, Robert E
Baumgardner, F Wesley
Bazer, Fuller Warren
Beard, James B
Bishop, Jack Garland
Bishop, Vernon S
Blomqvist, Carl Gunnar
Bristol, Bernard Noah
Brock, Tommy A
Brodwick, Malcolm Stephen
Bronson, Franklin Herbert
Brown, Arthur Morton
Brown, Jack Harold Upton
Buderer, Melvin Charles
Bullard, Truman Robert
Burns, John Mitchell
Burns, John W
Butcher, Reginald William
Butler, Bruce David
Byrne, John Howard
Campbell, Bonnalie Oetting
Carnes, David Lee, Jr
Castracane, V Daniel
Castro, Gilbert Anthony
Clark, James Henry
Cooper, William Anderson
Csaky, Tihamer Zoltan
Dietlein, Lawrence Frederick
Dill, Russell Eugene
Eddy, Carlton Anthony
Elizondo, Reynaldo S
Fife, William Paul
Fishman, Harvey Morton
Fletcher, John Lynn
Flickinger, George Latimore, Jr
Franklin, Thomas Doyal, Jr
Frishman, Laura J(ean)
Glantz, Raymon M
Gothelf, Bernard
Granger, Harris Joseph
Green, Gary Miller
Grundy, Scott Montgomery
Guentherman, Robert Henry
Guo, Yan-Shi
Gutierrez, Guillermo
Hall, Charles Eric
Hazlewood, Carlton Frank
Herd, James Alan
Herlihy, Jeremiah Timothy
Hester, Richard Kelly
Hettinger, Deborah D R
Hightower, Nicholas Carr, Jr
Hill, Ronald Stewart
Hoage, Terrell Rudolph
Holly, Frank Joseph
Horger, Lewis Milton
Horne, Francis R
Hughes, Maysie J H
Humphrey, Ronald DeVere
Hutchens, John Oliver
Ivy, John L
James, Harold Lee
Jaweed, Mazher
Jennings, Michael Leon
Johnson, John Marshall
Johnson, Robert Lee
Jones, Carl E
Jordan, Paul H, Jr
Kallus, Frank Theodore
Kalu, Dike Ndukwe
Kasschau, Margaret Ramsey
Keeley, Larry Lee
Kellaway, Peter
Knobil, Ernst
Koke, Joseph R
Kraemer, Duane Carl
Kraus-Friedmann, Naomi
Krise, George Martin
Krock, Larry Paul
Kronenberg, Richard Samuel
Langston, Jimmy Byrd
Leach-Huntoon, Carolyn S
Lee, Patrice Anne
Lehmkuhl, L Don
Leroy, Robert Frederick
Lewis, Simon Andrew
Liepa, George Uldis
Lipton, James Matthew
Little, Perry L
Livingston, Linda
Loftus, Joseph P, Jr
Lott, James Robert
Loubser, Paul Gerhard
Lutherer, Lorenz O
McCarter, Roger John Moore
McCarthy, John Lawrence, Jr
McDonald, Harry Sawyer
McGrath, James Joseph
Maclean, Graeme Stanley
Masoro, Edward Joseph
Maxwell, Leo C
Meininger, Gerald A
Miller, Thomas Allen
Moldenhauer, Ralph Roy
Montgomery, Edward Harry
Mukherjee, Amal
Musgrave, F Story
Norris, William Elmore, Jr
Norwood, James S
O'Brien, Larry Joe
Ordway, George A
Orem, John
Peterson, Lysle Henry
Poenie, Martin Francis
Porter, John Charles
Pratt, David R
Preslock, James Peter
Raven, Peter Bernard
Reid, Michael Baron
Reuss, Luis
Rogers, Walter Russell
Rosborough, John Paul
Sallee, Verney Lee
Sant'Ambrogio, Giuseppe
Schrank, Auline Raymond
Schuhmann, Robert Ewald
Schultz, Stanley George
Schwalm, Fritz Ekkehardt
Shepherd, Albert Pitt, Jr
Sikes, James Klingman
Siler-Khodr, Theresa M
Simmons, David J
Slack, Jim Marshall
Smatresk, Neal Joseph
Smith, Edwin Lee
Smith, Thomas Caldwell
Sordahl, Louis A
Soriero, Alice Ann
Srebro, Richard
Sybers, Harley D
Szilagyi, Julianna Elaine

Taurog, Alvin
Taylor, Alan Neil
Templeton, Gordon Huffine
Thompson, James Charles
Traber, Daniel Lee
Triano, John Joseph
Valentich, John David
Vanatta, John Crothers, III
Wall, Malcolm Jefferson, Jr
Ward, Walter Frederick
Wauquier, Albert
Weisbrodt, Norman William
Weitlauf, Harry
White, Fred Newton
Widner, William Richard
Wildenthal, Kern
Williams, Fred Eugene
Williams, Robert Sanders
Williams, W Jon-Allan
Wills, Nancy Kay
Wilson, Everett D
Wilson, Forest Ray, II
Wohlman, Alan
Wong, Brendan So
Woodward, Donald Jay
Wright, Kenneth C
Young, Margaret Claire

UTAH
Bloxham, Don Dee
Clark, C Elmer
Conlee, Robert Keith
Dixon, John Aldous
Graff, Darrell Jay
Heninger, Richard Wilford
Hibbs, John Burnham
Jaussi, August Wilhelm
Johnston, Stephen Charles
Kuida, Hiroshi
Parkin, James Lamar
Renzetti, Attilio D, Jr
Sanders, Raymond Thomas
Sernka, Thomas John
Sharp, Joseph C(ecil)
Stringham, Reed Millington, Jr
Urry, Ronald Lee
Walker, John Lawrence, Jr
Wardell, Joe Russell, Jr
Warner, Homer R
Windham, Carol Thompson
Woodbury, John Walter

VERMONT
Alpert, Norman Roland
Bevan, Rosemary D
Chambers, Alfred Hayes
Davison, John (Amerpohl)
Fahey, John Vincent
Foss, Donald C
Gennari, F John
Halpern, William
Hanson, John Sherwood
Heinrich, Bernd
Hendley, Edith Di Pasquale
Johnson, Robert Eugene
King, Patricia A
Low, Robert Burnham
McCrorey, Henry Lawrence
Mulieri, Berthann Scubon
Mulieri, Louis A
Nye, Robert Eugene, Jr
Parsons, Rodney Lawrence

VIRGINIA
Baker, Donald Granville
Barrett, Richard John
Beach, James M
Biber, Thomas U L
Boadle-Biber, Margaret Clare
Bornmann, Robert Clare
Bowman, Edward Randolph
Briggs, Fred Norman
Brown, Barry Lee
Buikema, Arthur L, Jr
Chevalier, Robert Louis
Costanzo, Richard Michael
Daniel, Joseph Car, Jr
De Long, Chester Wallace
DeSimone, John A
Desjardins, Claude
Edwards, Leslie Erroll
Ford, George Dudley
Gewirtz, David A
Gourley, Desmond Robert Hugh
Hart, Jayne Thompson
Hinton, Barry Thomas
Hoptman, Julian
Howards, Stuart S
Howes, Cecil Edgar
Hundley, Louis Reams
Kalimi, Mohammed Yahya
Kaul, Sanjiv
Kelleher, Dennis L
Kimbrough, Theo Daniel, Jr
Knight, James William
Kontos, Hermes A
Kory, Ross Conklin
Kutchai, Howard C
Leighton, Alvah Theodore, Jr
McNabb, Roger Allen
Maio, Domenic Anthony
Makhlouf, Gabriel Michel
Mangum, Charlotte P
Menaker, Michael

Mengebier, William Louis
Mikulecky, Donald C
Murphy, Richard Alan
Osborne, Paul James
Peach, Michael Joe
Pepe, Gerald Joseph
Pinschmidt, Mary Warren
Poland, James Leroy
Price, Steven
Ridgway, Ellis Branson
Rochester, Dudley Fortescue
Rosenblum, William I
Saacke, Richard George
Scanlon, Patrick Francis
Schreiner, George E
Sellers, Cletus Miller, Jr
Somlyo, Andrew Paul
Stillwell, Edgar Feldman
Suter, Daniel B
Swain, David P
Szabo, Gabor
Talbot, Richard Burritt
Tolles, Walter Edwin
Turner, Terry Tomo
Tuttle, Jeremy Ballou
Van Krey, Harry P
Watlington, Charles Oscar
Wei, Enoch Ping
Weltman, Arthur
Wilson, John Drennan
Witorsch, Raphael Jay

WASHINGTON
Akers, Thomas Kenny
Anderson, Marjorie Elizabeth
Bassingthwaighte, James B
Bhansali, Praful V
Blinks, John Rogers
Bredahl, Edward Arlan
Burnell, James McIndoe
Campbell, Kenneth B
Catterall, William A
Cohen, Arthur LeRoy
Detwiler, Peter Benton
Dickhoff, Walton William
Feigl, Eric O
Free, Michael John
Gale, Charles C, Jr
Gellert, Ronald J
Glomset, John A
Gordon, Albert McCague
Hanegan, James L
Hegyvary, Csaba
Heinle, Donald Roger
Hille, Bertil
Hornbein, Thomas F
Jackson, Kenneth Lee
Kaplan, Alex
Kastella, Kenneth George
Kirschner, Leonard Burton
Koch, Alan R
Koenig, Jane Quinn
Martin, Arthur Wesley
Matlock, Daniel Budd
Patton, Harry Dickson
Phillips, Richard Dean
Plisetskaya, Erika Michael
Pollack, Gerald H
Porter, Charles Warren
Reid, Donald House
Rowell, Loring B
Sandler, Harold
Scher, Allen Myron
Schoene, Robert B
Stien, Howard M
Stirling, Charles E
Su, Judy Ya-Hwa Lin
Taylor, Eugene M
Towe, Arnold Lester
Went, Hans Adriaan
Whatmore, George Bernard
White, Ronald Jerome
Woods, Stephen Charles

WEST VIRGINIA
Aulick, Louis H
Bond, Guy Hugh
Castranova, Vincent
Cochrane, Robert Lowe
Dailey, Robert Arthur
Franz, Gunter Norbert
Hedge, George Albert
Larson, Gary Eugene
Lee, Ping
Lindsay, Hugh Alexander
Stauber, William Taliaferro

WISCONSIN
Bass, Paul
Benjamin, Hiram Bernard
Bittar, Evelyn Edward
Brittain, David B
Brooks, Jack Carlton
Brugge, John F
Carlson, Stanley David
Cartee, Gregory D
ColAs, Antonio E
Condon, Robert Edward
Cowley, Allen Wilson, Jr
Craig, Elizabeth Anne
Cvancara, Victor Alan
Dawson, Christopher A
Dennis, Warren Howard
Ebert, Thomas Jay

Effros, Richard Matthew
First, Neal L
Forster, Hubert Vincent
Gaspard, Kathryn Jane
Hall, Kent D
Harder, David Rae
Hofmann, Lorenz M
Holden, James Edward
Kampine, John P
Kendrick, John Edsel
Kleinman, Jack G
Klitgaard, Howard Maynard
Kochan, Robert George
Lauson, Henry Dumke
Light, Douglas B
Lombard, Julian H
Macintyre, Bruce Alexander
Madden, Jane A
Moss, Richard
Olsen, Ward Alan
Osborn, Jeffrey L
Pace, Marvin M
Plotka, Edward Dennis
Porth, Carol Mattson
Rankin, John Horsley Grey
Reddan, William Gerald
Rickaby, David A
Rieselbach, Richard Edgar
Roman, Richard J
Sarna, Sushil K
Sheldahl, Lois Marie
Smith, Curtis Alan
Smith, James John
Smith, Luther Michael
Smith, Milton Reynolds
Snyder, Ann C
Sothmann, Mark Steven
Staszak, David John
Stone, William Ellis
Stowe, David F
Terry, Leon Cass
Thibodeau, Gary A
Wen, Sung-Feng
Wolf, Richard Clarence

PUERTO RICO
De Mello, W Carlos
Fernandez-Repollet, Emma D
Garcia-Castro, Ivette
Santos-Martinez, Jesus

ALBERTA
Baumber, John Scott
Beatty, David Delmar
Birdsell, Dale Carl
Cheeseman, Christopher Ian
Cooper, Keith Edward
Doornenbal, Hubert
Famiglietti, Edward Virgil, Jr
Frank, George Barry
Hammond, Brian Ralph
Henderson, Ruth McClintock
Kaufman-Jacobs, Susan E
Lauber, Jean Kautz
MacIntosh, Brian Robert
Mears, Gerald John
Pang, Peter Kai To
Poznansky, Mark Joab
Schachter, Melville
Thomas, Norman Randall
Veale, Warren Lorne

BRITISH COLUMBIA
Bates, David Vincent
Belton, Peter
Bressler, Bernard Harvey
Cottle, Walter Henry
Cramer, Carl Frederick
Cynader, Max Sigmund
Dehnel, Paul Augustus
Dorchester, John Edmund Carleton
Hahn, Peter
Hoar, William Stewart
Keeler, Ralph
Ledsome, John R
Leung, So Wah
Lioy, Franco
McLennan, Hugh
Palaty, Vladimir
Perks, Anthony Manning
Phillips, John Edward
Quastel, D M J
Randall, David John
Sanders, Harvey David
Skala, Josef Petr
Thurlbeck, William Michael
Webber, William A
Wong, Norman L M

MANITOBA
Bodnaryk, Robert Peter
Carter, Stefan A
Chernick, Victor
Dandy, James William Trevor
Dhalla, Naranjan Singh
Gaskell, Peter
Gerber, George Hilton
Jordan, Larry
Kepron, Wayne
Lautt, Wilfred Wayne
Moorhouse, John A
Naimark, Arnold
Stephens, Newman Lloyd
Younes, Magdy K

Zahradka, Peter

NEW BRUNSWICK
Cowan, F Brian M
Driedzic, William R
Saunders, Richard Lee

NEWFOUNDLAND
Mookerjea, Sailen
Nolan, Richard Arthur
Payton, Brian Wallace

NOVA SCOTIA
Armour, John Andrew
Cook, Robert Harry
Dolphin, Peter James
Freeman, Harry Cleveland
Guernsey, Duane L
Hanna, Brian Dale
Hatcher, James Donald
Howlett, Susan Ellen
Issekutz, Bela, Jr
Rautaharju, Pentti M
Wong, Alan Yau Kuen

ONTARIO
Ackles, Kenneth Norman
Andrew, George McCoubrey
Ashwin, James Guy
Atwood, Harold Leslie
Barclay, Jack Kenneth
Buick, Fred, Jr
Calaresu, Franco Romano
Campbell, Edward J Moran
Campbell, James
Cowan, John
Cummins, W(illiam) Raymond
Cunningham, David A
DesMarais, Andre
Dirks, John Herbert
Downie, Harry G
Ettinger, George Harold
Farrar, John Keith
Fiser, Paul S(tanley)
Fraser, Donald
Froese, Alison Barbara
Frost, Barrie James
George, John Caleekal
Girvin, John Patterson
Grayson, John
Grinstein, Sergio
Groom, Alan Clifford
Harding, Paul George Richard
Harrison, Robert Victor
Hoffman-Goetz, Laurie
Hoffstein, Victor
Huang, Bing Shaun
Jacobson, Stuart Lee
Jennings, Donald B
Johnston, Miles Gregory
Jones, Douglas L
Karmazyn, Morris
Kinson, Gordon A
Kooh, Sang Whay
Kraicer, Jacob
Lefcoe, Neville
Logothetopoulos, J
Lynn, Denis Heward
Machin, J
Martin, Julio Mario
Mercer, Paul Frederick
Mettrick, David Francis
Money, Kenneth Eric
Morris, Catherine Elizabeth
Napke, Edward
Norwich, Kenneth Howard
O'Hea, Eugene Kevin
Phillipson, Eliot Asher
Philogene, Bernard J R
Rakusan, Karel Josef
Rappaport, Aron M
Roslycky, Eugene Bohdan
Roth, René Romain
Sen, Amar Kumar
Shah, Bhagwan G
Shephard, Roy Jesse
Sirek, Anna
Sirek, Otakar Victor
Stinson, Michael Roy
Sun, Anthony Mein-Fang
Talesnik, Jaime
Trevors, Jack Thomas
Vallerand, Andre L
Veliky, Ivan Alois
Vranic, Mladen
Weaver, Lynne C
Younglai, Edward Victor
Zimmerman, Arthur Maurice

QUEBEC
Anand-Srivastava, Madhu Bala
Barcelo, Raymond
Bates, Jason H T
Bergeron, Georges Albert
Bergeron, Michel
Billette, Jacques
Bolte, Edouard
Briere, Normand
Bukowiecki, Ludwik J
Burgess, John Herbert
Capek, Radan
Castellucci, Vincent F
Collier, Brian
Cruess, Richard Leigh

Physiology, General (cont)

De Champlain, Jacques
DeRoth, Laszlo
Digby, Peter Saki Bassett
Dufour, Jacques John
Dugal, Louis Paul
Dunnigan, Jacques
Frojmovic, Maurice Mony
Galeano, Cesar
Gjedde, Albert
Goltzman, David
Kapoor, Narinder N
Larochelle, Jacques
Lavoie, Jean-Marc
Leblanc, Jacques Arthur
Lemonde, Andre
MacIntosh, Frank Campbell
Milic-Emili, Joseph
Mortola, Jacopo Prospero
Olivier, Andre
Payet, Marcel Daniel
Pearson Murphy, Beverly Elaine
Polosa, Canio
Ponka, Premysl
Potvin, Pierre
Quillen, Edmond W, Jr
Ruiz-Petrich, Elena
Schanne, Otto F
Tenenhouse, Harriet Susie
Tremblay, Angelo
Van Gelder, Nico Michel
Viereck, Christopher
Wolsky, Alexander

SASKATCHEWAN
Murphy, Bruce Daniel
Sulakhe, Prakash Vinayak
Thornhill, James Arthur
Williams, Charles Melville

OTHER COUNTRIES
Alivisatos, Spyridon Gerasimos Anastasios
Almers, Wolfhard
Binder, Bernd R
Bligh, John
Boullin, David John
Braquet, Pierre G
Chertok, Robert Joseph
De Alba Martinez, Jorge
Delahayes, Jean
De Troyer, Andre Jules
Dikstein, Shabtay
Fawaz, George
Fleckenstein, Albrecht
Fritz, Irving Bamdas
Garcia Ramos, Juan
Gatz, Randall Neal
Gontier, Jean Roger
Hall, Peter
Hall, Peter Francis
Hazeyama, Yuji
Heiniger, Hans-Jorg
Heinz, Erich
Hisada, Mituhiko
Huxley, Andrew (Fielding)
Itiaba, Kibe
Josenhans, William T
Kawamura, Hiroshi
King, Dorothy Wei (Cheng)
Kordan, Herbert Allen
Kraicer, Peretz Freeman
Lu, Guo-Wei
McBroom, Marvin Jack
Machne, Xenia
McMillan, Joseph Patrick
Meli, Alberto L G
Mezquita, Cristobal
Monoš, Emil
Radford, Edward Parish
Reuter, Harald
Rodahl, Kaare
Sampson, Sanford Robert
Schneider, Wolfgang Johann
Schonbaum, Eduard
Shiraki, Keizo
Smolander, Martti Juhani
Torun, Benjamin
Tung, Che-Se
Valencia, Mauro Eduardo
Van Driessche, Willy
Vane, John Robert
Veicsteinas, Arsenio
Ward, Susan A
Watanabe, Yoshio
Weidmann, Silvio
Welty, Joseph D
Whipp, Brian James
Zapata, Patricio

Physiology, Plant

ALABAMA
Allen, Seward Ellery
Backman, Paul Anthony
Biswas, Prosanto K
Cherry, Joe H
Davis, Donald Echard
Elgavish, Ada S
Eyster, Henry Clyde
Gilliam, Charles Homer
Henderson, James Henry Meriwether

Jones, Daniel David
Jones, Jeanette
Locy, Robert Donald
O'Kelley, Joseph Charles
Peterson, Curtis Morris
Rajanna, Bettaiya
Rogers, Hugo H, Jr
Singh, Bharat
Teyker, Robert Henry
Truelove, Bryan
Weete, John Donald
Whatley, Booker Tillman

ALASKA
Stekoll, Michael Steven

ARIZONA
Allen, Stephen Gregory
Bartels, Paul George
Davis, Edwin Alden
Erickson, Eric Herman, Jr
Fowler, Dona Jane
Glenn, Edward Perry
Guinn, Gene
Hammond, H David
Harris, Leland
Hendrix, Donald Louis
Hull, Herbert Mitchell
Katterman, Frank Reinald Hugh
Kimball, Bruce Arnold
Larkins, Brian Allen
Lipke, William G
McDaniel, Robert Gene
Mellor, Robert Sydney
Morton, Howard LeRoy
Newman, David William
O'Leary, James William
Radin, John William
Reid, Charles Phillip Patrick
Tinus, Richard Willard
Towill, Leslie Ruth
Trelease, Richard Norman
Upchurch, Robert Phillip
Verbeke, Judith Ann
Wright, Daniel Craig

ARKANSAS
Einert, Alfred Erwin
Evans, Raymond David
Lane, Forrest Eugene
McMasters, Dennis Wayne
Morris, Justin Roy
Nickell, Louis G
Oosterhuis, Derrick M
Reed, Hazell
Roberson, Ward Bryce
Stewart, James McDonald
Stutte, Charles A
Turnipseed, Glyn D
West, Charles Patrick
Wickliff, James Leroy

CALIFORNIA
Addicott, Fredrick Taylor
Anderson, Lars William James
Arditti, Joseph
Ashton, Floyd Milton
Bailey-Serres, Julia
Baldy, Richard Wallace
Bayer, David E
Beevers, Harry
Bell, Charles W
Benson, Andrew Alm
Bils, Robert F
Bird, Harold L(eslie), Jr
Blakely, Lawrence Mace
Bledsoe, Caroline Shafer
Bolar, Marlin L
Bonner, Bruce Albert
Brandon, David Lawrence
Bray, Elizabeth Ann
Brecht, Patrick Ernest
Breidenbach, Rowland William
Briggs, Winslow Russell
Buchanan, Bob Branch
Campbell, Bruce Carleton
Castelfranco, Paul Alexander
Catlin, Peter Bostwick
Chrispeels, Maarten Jan
Coggins, Charles William, Jr
Conn, Eric Edward
Cooper, James Burgess
Corcoran, Mary Ritzel
Corse, Joseph Walters
Cronshaw, James
Currier, Herbert Bashford
Dalton, Francis Norbert
Davies, Huw M
Downing, Michael Richard
Dugger, Willie Mack, Jr
Duniway, John Mason
Du Pont, Frances Marguerite
Durzan, Donald John
Eaks, Irving Leslie
Einset, John William
Elkin, Lynne Osman
Epstein, Emanuel
Erickson, Louis Carl
Estermann, Eva Frances
Finkle, Bernard Joseph
Finn, James Crampton, Jr
Fong, Franklin
Fork, David Charles
Fosket, Daniel Elston

French, Charles Stacy
Fuller, Glenn
Gerwick, Ben Clifford, III
Grantz, David Arthur
Gray, Reed Alden
Greene, Richard Wallace
Grieve, Catherine Macy
Grimes, Donald Wilburn
Grossenbacher, Karl A(lbert)
Haard, Norman F
Hall, Anthony Elmitt
Haxo, Francis Theodore
Heath, Robert Louis
Hess, Charles
Hess, Frederick Dan
Hewitt, Allan A
Hirsch, Ann Mary
Holm-Hansen, Osmund
Houpis, James Louis Joseph
Howe, George Franklin
Hsiao, Theodore Ching-Teh
Huang, Anthony Hwoon Chung
Huffaker, Ray C
Hurkman, William James
Jacobson, Louis
Johnson, Kenneth Duane
Jones, Russell Lewis
Jordan, Lowell Stephen
Kader, Adel Abdel
Kemp, Paul Raymond
Kester, Dale Emmert
Ketellapper, Hendrik Jan
Kliewer, Walter Mark
Koehler, Don Edward
Kohl, Harry Charles, Jr
Kozlowski, Theodore Thomas
Labanauskas, Charles K
Labavitch, John Marcus
Laties, George Glushanok
Leonard, Robert Thomas
Lewis, Lowell N
Libby, William John, (Jr)
Liebhardt, William C
Lincoln, Richard G
Lipton, Werner Jacob
Long, Sharon Rugel
Loomis, Robert Simpson
Lovatt, Carol Jean
Lovelace, C James
Lucas, William John
Lyons, James Martin
Maas, Eugene Vernon
Maas, Stephen Joseph
MacDonald, James Douglas
McNairn, Robert Blackwood
Madore, Monica Agnes
Martineau, Belinda M
Mazelis, Mendel
Melis, Anastasios
Merchant, Sabeeha
Morris, Leonard Leslie
Murashige, Toshio
Neel, James William
Nevins, Donald James
Ngo, That Tjien
Nieman, Richard Hovey
Nobel, Park S
Norlyn, Jack David
Norris, Robert Francis
Nothnagel, Eugene Alfred
Oechel, Walter C
Ow, David Wing
Park, Roderic Bruce
Percival, Frank William
Phinney, Bernard Orrin
Pierce, Wayne Stanley
Porter, Clark Alfred
Pratt, Harlan Kelley
Purves, William Kirkwood
Rappaport, Lawrence
Ray, Peter Martin
Rayle, David Lee
Rendig, Victor Vernon
Ritenour, Gary Lee
Roberts, Stephen Winston
Rogers, Bruce Joseph
Romani, Roger Joseph
Rubatzky, Vincent E
Ryugo, Kay
Sacher, Robert Francis
Sachs, Roy M
Saltveit, Mikal Endre, Jr
Sayre, Robert Newton
Schiefferstein, Robert Harold
Shih, Ching-Yuan G
Shugarman, Peter Melvin
Silberstein, Otmar Otto
Somerville, Christopher Roland
Sommer, Noel Frederick
Sparks, Steven Richard
Stehsel, Melvin Louis
Stemler, Alan James
Stocking, Clifford Ralph
Stone, Edward Curry
Taiz, Lincoln
Tanada, Takuma
Terry, Norman
Thornton, Robert Melvin
Ting, Irwin Peter
Tobin, Elaine Munsey
Uriu, Kiyoto
Vanderhoef, Larry Neil
Vinters, Harry Valdis
Vreman, Hendrik Jan

Wallace, Arthur
Wallace, Joan M
Weatherspoon, Charles Phillip
Weaver, Ellen Cleminshaw
Weaver, Robert John
Weber, John R
Weimberg, Ralph
Wiley, Lorraine
Wilkins, Harold
Willemsen, Roger Wayne
Wu, Jia-Hsi
Yamaguchi, Masatoshi
Yang, Shang Fa
Zavala, Maria Elena
Zscheile, Frederick Paul, Jr

COLORADO
Alford, Donald Kay
Barney, Charles Wesley
Basham, Charles W
Bonde, Erik Kauffmann
Brown, Lewis Marvin
Dever, John E, Jr
Eley, James H
Fly, Claude Lee
Gough, Larry Phillips
Hanan, Joe John
Hendrix, John Edwin
Holm, David Garth
Kaufmann, Merrill R
Linck, Albert John
Nabors, Murray Wayne
Pollock, Bruce McFarland
Ross, Cleon Walter
Ryan, Michael G
Schweizer, Edward E
Seibert, Michael
Sowell, John Basil
Storey, Richard Drake
Stushnoff, Cecil
Workman, Milton

CONNECTICUT
Ahrens, John Frederick
Ashton, Peter Mark Shaw
Berlyn, Graeme Pierce
Crain, Richard Cullen
Galston, Arthur William
Geballe, Gordon Theodore
Goldsmith, Mary Helen Martin
Gordon, John C
Kennard, William Crawford
Koning, Ross E
Parker, Johnson
Peterson, Richard Burnett
Rother, Ana
Satter, Ruth
Slayman, Clifford L
Stowe, Bruce Bernot
Wargo, Philip Matthew
Warren, Richard Scott
Wetherell, Donald Francis
Zelitch, Israel

DELAWARE
Beyer, Elmo Monroe, Jr
Boyer, John Strickland
Bozarth, Gene Allen
Gatenby, Anthony Arthur
Giaquinta, Robert T
Green, Jerome
Hodson, Robert Cleaves
Knowles, Francis Charles
Lichtner, Francis Thomas, Jr
Lin, Willy
Matlack, Albert Shelton
Reasons, Kent M
Riggleman, James Dale
Seliskar, Denise Martha
Somers, George Fredrick, Jr
Wittenbach, Vernon Arie

DISTRICT OF COLUMBIA
Allen, James Ralston
Chen, H R
Donaldson, Robert Paul
Gordon, William Ransome
Halstead, Thora Waters
Islam, Nurul
Keitt, George Wannamaker, Jr
Littler, Mark Masterton
Rabson, Robert
Shropshire, Walter, Jr
Sulzman, Frank Michael
Tompkins, Daniel Reuben
Wiggans, Samuel Claude

FLORIDA
Albrigo, Leo Gene
Ayers, Alvin Dearing
Bausher, Michael George
Bennette, Jerry Mac
Biggs, Robert Hilton
Boss, Manley Leon
Bowes, George Ernest
Bryan, Herbert Harris
Burdine, Howard William
Buslig, Bela Stephen
Calvert, David Victor
Campbell, Carl Walter
Cantliffe, Daniel James
Chen, Tsong Meng
Cooper, William Cecil
Cropper, Wendell Parker, Jr

BIOLOGICAL SCIENCES / 163

Davenport, Thomas Lee
Dusky, Joan Agatha
Fischler, Drake Anthony
Fitzpatrick, George
Fritz, George John
Gaskins, Murray Hendricks
Gilreath, James Preston
Grunwald, Claus Hans
Hall, Chesley Barker
Haller, William T
Homann, Peter H
Huber, Donald John
Jackson, William Thomas
Karsten, Kenneth Stephen
Kossuth, Susan
Lee, David Webster
Mecklenburg, Roy Albert
Mullins, John Thomas
Orgell, Wallace Herman
Owens, Clarence Burgess
Parsons, Lawrence Reed
Roberts, Donald Ray
Ruelke, Otto Charles
Sager, John Clutton
Schroder, Vincent Nils
Silverman, David Norman
Sinclair, Thomas Russell
Smith, Paul Frederick
Smith, Richard Clark
Steward, Kerry Kalen
Stewart, Herbert
Sweet, Haven C
Thompson, Neal Philip
Vasil, Vimla
Verkade, Stephen Dunning
Weigel, Russell C(ornelius), Jr
Wells, Gary Neil
West, Sherlie Hill
Wheaton, Thomas Adair
White, Timothy Lee
Wilcox, Merrill
Wilfret, Gary Joe
Williams, Robert Haworth
Wilson, William Curtis
Wiltbank, William Joseph
Yelenosky, George

GEORGIA
Burns, Robert Emmett
Cutler, Horace Garnett
Dashek, William Vincent
Dekazos, Elias Demetrios
Dinus, Ronald John
Donoho, Clive Wellington, Jr
Duncan, Ronny Rush
Eastin, Emory Ford
Espelie, Karl Edward
Hendrix, Floyd Fuller, Jr
Hewlett, John David
Key, Joe Lynn
McLaurin, Wayne Jefferson
Michel, Burlyn Everett
Mixon, Aubrey Clifton
Nes, William David
Ohki, Kenneth
Palevitz, Barry Allan
Phatak, Sharad Chintaman
Pratt, Lee Herbert
Pressey, Russell
Reger, Bonnie Jane
Reilly, Charles Conrad
Robinson, Margaret Chisolm
Romanovicz, Dwight Keith
Sansing, Norman Glenn
Schmidt, Gregory Wayne
Smith, Albert Ernest, Jr
Smith, Morris Wade
Smittle, Doyle Allen
Sommer, Harry Edward
Sparks, Darrell
Vencill, William Keith
Vines, Herbert Max
Walker, Alma Toevs
Wilkinson, Robert E
Wilkinson, Robert Eugene
Wood, Bruce Wade

HAWAII
Akamine, Ernest Kisei
Albert, Henrick Horst
Cooil, Bruce James
De la Pena, Ramon Serrano
Demanche, Edna Louise
Fitch, Maureen Meiko Masuda
Kefford, Noel Price
Krauss, Beatrice Hilmer
Moore, Paul Harris
Patterson, Gregory Matthew Leon
Rauch, Fred D
Sakai, William Shigeru
Sanford, Wallace Gordon
Skolmen, Roger Godfrey
Sun, Samuel Sai-Ming
Tanabe, Michael John
Teramura, Alan Hiroshi
Vennesland, Birgit
Webb, David Thomas

IDAHO
Bowmer, Richard Glenn
Dwelle, Robert Bruce
Eberlein, Charlotte
Kleinkopf, Gale Eugene
Kochan, Walter J

LeTourneau, Duane John
Muzik, Thomas J
Oliver, David John
Roberts, Lorin Watson
Sojka, Robert E
Spomer, George Guy
Thill, Donald Cecil

ILLINOIS
Ashby, William Clark
Ausich, Rodney L
Blair, Louis Curtis
Briskin, Donald Phillip
Brown, Lindsay Dietrich
Chappell, Dorothy Field
Chen, Shepley S
Cordes, William Charles
Cross, John W
Dove, Lewis Dunbar
Edwards, Harold Herbert
Eskins, Kenneth
Feder, Martin Elliott
Galsky, Alan Gary
Gardner, Harold Wayne
Gassman, Merrill Loren
Gould, John Michael
Govindjee, M
Hageman, Richard Harry
Hanson, John Bernard
Harper, James Eugene
Hauptmann, Randal Mark
Heichel, Gary Harold
Heller, Alfred
Hesketh, J D
Hinchman, Ray Richard
Holbrook, Gabriel Peter
Holt, Donald Alexander
Howell, Robert Wayne
Huck, Morris Glen
Leffler, Harry Rex
Lippincott, Barbara Barnes
Lippincott, James Andrew
Lorenz, Ralph William
McCracken, Derek Albert
McIlrath, Wayne Jackson
Melhado, L(ouisa) Lee
Meyer, Martin Marinus, Jr
Miernyk, Jan Andrew
Ogren, William Lewis
Ort, Donald Richard
Pappelis, Aristotel John
Player, Mary Anne
Portis, Archie Ray, Jr
Portz, Herbert Lester
Rebeiz, Constantin Anis
Rinne, Robert W
Ruddat, Manfred
Sachs, Martin M
Skirvin, Robert Michael
Skok, John
Smiciklas, Kenneth Donald
Splittstoesser, Walter E
Spomer, Louis Arthur
Stoller, Edward W
Tindall, Donald R
Van Sambeek, Jerome William
Verduin, Jacob
Weber, Evelyn Joyce
Weidner, Terry Mohr
Widholm, Jack Milton
Williams, David James
Wilson, Richard Hansel
Woolley, Joseph Tarbet
Yopp, John Herman

INDIANA
Alder, Edwin Francis
Barr, Rita
Bauman, Thomas Trost
Bhattacharya, Pradeep Kumar
Boyd, Frederick Mervin
Brenneman, James Alden
Bucholtz, Dennis Lee
Castleberry, Ron M
Chaney, William R
Coolbaugh, Ronald Charles
Crouch, Martha Louise
Dilley, Richard Alan
Dunkle, Larry D
Foley, Michael Edward
Gentile, Arthur Christopher
Hagen, Charles William, Jr
Hodges, Thomas Kent
Housley, Thomas Lee
Kapustka, Lawrence A
Keck, Robert William
Manthey, John August
Miller, Carlos Oakley
Mitchell, Cary Arthur
Nichols, Kenneth E
Ockerse, Ralph
Polley, Lowell David
Raveed, Dan
Reazin, George Harvey, Jr
Rhykerd, Charles Loren
Ruesink, Albert William
Schreiber, Marvin Mandel
Smith, Charles Edward, Jr
Stiller, Mary Louise
Togasaki, Robert K
Tsai, Chia-Yin
Waldrep, Thomas William
Watson, Maxine Amanda
Webb, Mary Alice

Williams, Gene R
Williams, James Lovon, Jr
Wolt, Jeffrey Duaine
Woodson, William Randolph
Wright, William Leland

IOWA
Anderson, Irvin Charles
Berg, Virginia Seymour
Burris, Joseph Stephen
Cavalieri, Anthony Joseph, II
Chaplin, Michael H
Czarnecki, David Bruce
Durkee, Lenore T
Ford, Clark Fugier
George, John Ronald
Imsande, John
LaMotte, Clifford Elton
Lillehoj, Eivind B
Muir, Robert Mathew
Pearce, Robert Brent
Shibles, Richard Marwood
Showers, William Broze, Jr
Sjölund, Richard David
Smith, Frederick George
Stewart, Cecil R
Svec, Leroy Vernon
Tjostem, John Leander
Wendel, Jonathan F
Yager, Robert Eugene

KANSAS
Borchert, Rolf
Coyne, Patrick Ivan
Davis, Lawrence Clark
Guikema, James Allen
Jennings, Paul Harry
Johnson, Lowell Boyden
Kirkham, M B
Li, Yi
Murphy, John Joseph
Murphy, Larry S
Paulsen, Gary Melvin
Wiest, Steven Craig
Wong, Peter P

KENTUCKY
Bush, Lowell Palmer
Dunham, Valgene Loren
Dyar, James Joseph
Harmet, Kenneth Herman
Hiatt, Andrew Jackson
Kasperbauer, Michael J
Lacefield, Garry Dale
Lambert, Roger Gayle
Lasheen, Aly M
Leggett, James Everett
Lowe, Richie Howard
Roberts, Clarence Richard
Straley, Susan Calhoon
Toman, Frank R
Wagner, George Joseph

LOUISIANA
Alam, Jawed
Baker, John Bee
Barber, John Threlfall
Barnett, James P
Benda, Gerd Thomas Alfred
Bhatnagar, Deepak
Board, James Ellery
Carpenter, Stanley Barton
Carter, Mason Carlton
Cohn, Marc Alan
Faw, Wade Farris
Gosselink, James G
Hough, Walter Andrew
Jacks, Thomas Jerome
Longstreth, David J
Lorio, Peter Leonce, Jr
Martin, Freddie Anthony
Moore, Thomas Stephen, Jr
Musgrave, Mary Elizabeth
Nair, Pankajam K
Pezeshki, S Reza
Rogers, Robert Larry
Standifer, Leonides Calmet, Jr
Stevenson, Enola L
Triplett, Barbara Ann
Webert, Henry S
Yatsu, Lawrence Y

MAINE
Brawley, Susan Howard
Clapham, William Montgomery
Eggert, Franklin Paul
Laber, Larry Jackson
Schwintzer, Christa Rose
Thomas, Robert James
Tjepkema, John Dirk

MARYLAND
Bandel, Vernon Allan
Barnett, Neal Mason
Britz, Steven J
Bruns, Herbert Arnold
Bunce, James Arthur
Butcher, Henry Clay, IV
Chappelle, Emmett W
Chitwood, David Joseph
Christiansen, Meryl Naeve
Christy, Alfred Lawrence
Cleland, Charles Frederick
Deitzer, Gerald Francis

Falkenstein, Kathy Fay
Filner, Barbara
Foy, Charles Daley
Freiberg, Samuel Robert
French, Richard Collins
Gantt, Elisabeth
Gouin, Francis R
Gross, Kenneth Charles
Habermann, Helen M
Herman, Eliot Mark
Herrett, Richard Allison
Hill, Jane Virginia Foster
Hirsh, Allen Gene
Hruschka, Howard Wilbur
Hurtt, Woodland
Isensee, Allan Robert
Josephs, Melvin Jay
Krauss, Robert Wallfar
Krizek, Donald Thomas
Lee, Edward Hsien-Chi
Liverman, James Leslie
Long, James Delbert
Margulies, Maurice
Matthews, Benjamin F
Miura, George Akio
Mulchi, Charles Lee
Owens, Lowell Davis
Patterson, Glenn Wayne
Racusen, Richard Harry
Ramagopal, Subbanaidu
Ridley, Esther Joanne
Romberger, John Albert
St John, Judith Brook
Saunders, James Allen
Shear, Cornelius Barrett
Simpson, Marion Emma
Slocum, Robert Dale
Sloger, Charles
Smith, William Owen
Stanley, Ronald Alwin
Steffens, George Louis
Sterrett, John Paul
Sullivan, Joseph Hearst
Sze, Heven
Tester, Cecil Fred
Vigil, Eugene Leon
Wadleigh, Cecil Herbert
Wang, Chien Yi
Warmbrodt, Robert Dale
Watada, Alley E
Williams, Robert Jackson
Woodstock, Lowell Willard
Youle, Richard James
Zacharius, Robert Marvin
Zimmerman, Richard Hale

MASSACHUSETTS
Alpert, Peter
Barker, Allen Vaughan
Bazzaz, Maarib Bakri
Bogorad, Lawrence
Burris, John Edward
Craker, Lyle E
Dacey, John W H
Devlin, Robert M
DiLiddo, Rebecca McBride
Feinleib, Mary Ella (Harman)
Frederick, Sue Ellen
Freeberg, John Arthur
Gibbs, Martin
Gruber, Peter Johannes
Hepler, Peter Klock
Holmes, Francis W(illiam)
Howe, Kenneth Jesse
Kadkade, Prakash Gopal
Kamien, Ethel N
Klein, Attila Otto
Melan, Melissa A
Miller, Andrew Leitch
Offner, Gwynneth Davies
Orme-Johnson, William H
Papastoitsis, Gregory
Reid, Philip Dean
Rivera, Ezequiel Ramirez
Schiff, Jerome A
Stern, Arthur Irving
Tattar, Terry Alan
Thomas, Aubrey Stephen, Jr
Torrey, John Gordon
Whitaker, Ellis Hobart

MICHIGAN
Adams, Paul Allison
Adler, John Henry
Bukovac, Martin John
Campbell, Wilbur Harold
Christenson, Donald Robert
Dennis, Frank George, Jr
Dilley, David Ross
Good, Norman Everett
Halloin, John McDonell
Hare, Leonard N
Henry, Egbert Winston
Hough, Richard Anton
Howell, Gordon Stanley, Jr
Ikuma, Hiroshi
Inselberg, Edgar
Isleib, Donald Richard
Johnston, Taylor Jimmie
Kaufman, Peter Bishop
Kende, Hans Janos
Kivilaan, Aleksander
Kubiske, Mark E
Lang, Anton

Physiology, Plant (cont)

Loescher, Wayne Harold
McIntosh, Lee
Mullison, Wendell Roxby
Nadler, Kenneth David
Nooden, Larry Donald
Podila, Gopi Krishna
Preiss, Jack
Putnam, Alan R
Ries, Stanley K
Schneider, Michael J
Stein, Howard Jay
Vander Beek, Leo Cornelis
Yocum, Conrad Schatte
Zeevaart, Jan Adriaan Dingenis

MINNESOTA
Ahlgren, George E
Andersen, Robert Neils
Behrens, Richard
Brenner, Mark
Bushnell, William Rodgers
Choe, Hyung Tae
Crookston, Robert Kent
Frenkel, Albert W
Gengenbach, Burle Gene
Hackett, Wesley P
Jonas, Herbert
Klemer, Andrew Robert
Kraft, Donald J
Li, Pen H (Paul)
Pratt, Douglas Charles
Rehwaldt, Charles A
Silflow, Carolyn Dorothy
Singer, Susan Rundell
Smith, Lawrence Hubert
Soulen, Thomas Kay
Stadelmann, Eduard Joseph
Stadtherr, Richard James
Sucoff, Edward Ira
Sullivan, Timothy Paul
Tautvydas, Kestutis Jonas
Zeyen, Richard John

MISSISSIPPI
Creech, Roy G
Duke, Stephen Oscar
Elmore, Carroll Dennis
Friend, Alexander Lloyd
Heatherly, Larry G
Hoagland, Robert Edward
Hodges, Harry Franklin
Hodges, John Deavours
Kurtz, Mark Edward
McChesney, James Dewey
McWhorter, Chester Gray
Roark, Bruce (Archibald)
Spiers, James Monroe
Wills, Gene David

MISSOURI
Backus, Elaine Athene
Blevins, Dale Glenn
Carpenter, Will Dockery
Donald, William Waldie
Duncan, David Robert
Einhellig, Frank Arnold
Field, Christopher Bower
Fox, J Eugene
Gay, Don Douglas
George, Milon Fred
Hayes, Alice Bourke
Houghton, John M
Kohl, Daniel Howard
Lambeth, Victor Neal
Mertz, Dan
Miles, Charles Donald
Montague, Michael James
Nelson, Curtis Jerome
Pallardy, Stephen Gerard
Paul, Kamalendu Bikash
Pettit, Robert Eugene
Pickard, Barbara G
Pikaard, Craig Stuart
Schumacher, Richard William
Sells, Gary Donnell
Symington, Janey Studt
Varner, Joseph Elmer
Vogt, Albert R

MONTANA
Behan, Mark Joseph
Bilderback, David Earl
Carpenter, Bruce H
Mills, Ira Kelly
Sheridan, Richard P
Stallknecht, Gilbert Franklin

NEBRASKA
Chollet, Raymond
Daly, Joseph Michael
Davies, Eric
Eastin, John A
Elthon, Thomas Eugene
Higley, Leon George
Kinbacher, Edward John
Mason, Stephen Carl
Partridge, James Enoch
Read, Paul Eugene
Roeth, Frederick Warren
Rowland, Neil Wilson
Schwartzbach, Steven Donald
Specht, James Eugene

Spreitzer, Robert Joseph

NEVADA
Fox, Carl Alan
Howland, Joseph E(mery)
Taylor, George Evans, Jr

NEW HAMPSHIRE
Eggleston, Patrick Myron
Lesser, Michael Patrick
Minocha, Subhash C

NEW JERSEY
Bahr, James Theodore
Brown, Thomas Edward
Bruno, Stephen Francis
Carr, Richard J
Clark, Harold Eugene
Davis, Robert Foster, Jr
Durkin, Dominic J
Gaynor, John James
Ghai, Geetha R
Gramlich, James Vandle
Grant, Neil George
Greenfield, Sydney Stanley
Gruenhagen, Richard Dale
Hu, Ching-Yeh
Marrese, Richard John
Nelson, Nathan
Reid, Hay Bruce, Jr
Vasconcelos, Aurea C
Zilinskas, Barbara Ann

NEW MEXICO
Carroll, Arthur Paul
Cotter, Donald James
Essington, Edward Herbert
Fisher, James Thomas
Fowler, James Lowell
Gosz, James Roman
Gutschick, Vincent Peter
Johnson, Gordon Verle
Mexal, John Gregory
Phillips, Gregory Conrad
Throneberry, Glyn Ogle

NEW YORK
Alves, Leo Manuel
Ammirato, Philip Vincent
Bing, Arthur
Birecka, Helena M
Bisson, Mary A
Brody, Marcia
Buckley, Edward Harland
Budd, Thomas Wayne
Carroll, Robert Buck
Chabot, Brian F
Collins, Carol Desormeau
Creasy, Leroy L
Davies, Peter John
Dietert, Margaret Flowers
Domozych, David S
Earle, Elizabeth Deutsch
Ecklund, Paul Richard
Edgerton, Louis James
Evans, Lance Saylor
Ewing, Elmer Ellis
Fisher, Nicholas Seth
Gallagher, Jane Chispa
Gerard, Valrie Ann
Griffin, David H
Gussin, Arnold E S
Haber, Alan Howard
Hind, Geoffrey
Jacobson, Jay Stanley
Jagendorf, Andre Tridon
Khan, Anwar Ahmad
Klingensmith, Merle Joseph
Krikorian, Abraham D
Lakso, Alan Neil
LaRue, Thomas A
Law, David Martin
Leopold, Aldo Carl
Linkins, Arthur Edward
Lyman, Harvard
Ma, Hong
McCune, Delbert Charles
McDaniel, Carl Nimitz
MacLean, David Cameron
Mancinelli, Alberto L
Mantai, Kenneth Edward
Marra, John Frank
Miller, John Henry
Obendorf, Ralph Louis
Pietraface, William John
Pool, Robert Morris
Posner, Herbert Bernard
Powell, Loyd Earl, Jr
Rana, Mohammad A
Reeves, Stuart Graham
Rhee, G-Yull
Robinson, Beatrice Letterman
Robinson, Terence Lee
Schaedle, Michail
Shrift, Alex
Siegelman, Harold William
Simon, Robert David
Sirois, David Leon
Spanswick, Roger Morgan
Steponkus, Peter Leo
Tan-Wilson, Anna L(i)
Titus, John Elliott
Truscott, Frederick Herbert
Weinstein, Leonard Harlan

Welch, Ross Maynard
Wilcox, Hugh Edward
Wilson, Karl A

NORTH CAROLINA
Allen, Nina Stromgren
Bird, Kimon T
Burns, Joseph Charles
Camp Hay, Pamela Jean
Carter, Thomas Edward, Jr
Collins, Henry A
Cooke, Anson Richard
Corbin, Frederick Thomas
Cowett, Everett R
Danielson, Loran Leroy
Darling, Marilyn Stagner
Davis, Daniel Layten
Davis, Graham Johnson
De Hertogh, August Albert
DeJong, Donald Warren
Downs, Robert Jack
DuBay, Denis Thomas
Ellenson, James L
Fiscus, Edwin Lawson
Friedrich, James Wayne
Gross, Harry Douglass
Haun, Joseph Rhodes
Heck, Walter Webb
Hellmers, Henry
Holm, Robert E
Huang, Jeng-Sheng
Jaffe, Mordecai J
Jividen, Gay Melton
Kamykowski, Daniel
Knauft, David A
Kramer, Paul Jackson
Long, Raymond Carl
Maier, Robert Hawthorne
Manning, David Treadway
Matthysse, Ann Gale
Miller, Conrad Henry
Miller, Joseph Edwin
Miller, Robert James, II
Moll, Robert Harry
Moreland, Donald Edwin
Morrison, Ralph M
Mott, Ralph Lionel
Naylor, Aubrey Willard
Nelson, Paul Victor
O'Neal, Thomas Denny
Pattee, Harold Edward
Peet, Mary Monnig
Pirrung, Michael Craig
Ramus, Joseph S
Richardson, Curtis John
Rogerson, Asa Benjamin
Sanders, Douglas Charles
Scofield, Herbert Temple
Scott, Tom Keck
Sehgal, Prem P
Seltmann, Heinz
Shafer, Thomas Howard
Sheets, Thomas Jackson
Siedow, James N
Sisler, Edward C
Spiker, Steven L
Stier, Howard Livingston
Sunda, William George
Terborgh, John J
Thompson, William Francis, III
Troyer, James Richard
Volk, Richard James
Wallace, James William, Jr
Webb, Burleigh C
Weintraub, Robert Louis
Williamson, Ralph Edward
Worsham, Arch Douglas
Zublena, Joseph Peter

NORTH DAKOTA
Davis, David G
Doney, Devon Lyle
Duysen, Murray E
Frank, Albert Bernard
Freeman, Thomas Patrick
Galitz, Donald S
Greenwald, Stephen Mark
Metzger, James David
Ozbun, Jim L
Scholz, Earl Walter
Seiler, Gerald Joseph
Shimabukuro, Richard Hideo
Starks, Thomas Leroy
Suttle, Jeffrey Charles
Zimmerman, Don Charles

OHIO
Aboul-Ela, Mohamed Mohamed
Baker, David Bruce
Bendixen, Leo E
Brooks, James Reed
Cline, Morris George
Cohn, Norman Stanley
Cullis, Christopher Ashley
Darrow, Robert A
De La Fuente, Rollo K
Dollwet, Helmar Hermann Adolf
Edwards, Kathryn Louise
Evans, Michael Leigh
Francko, David Alex
Geiger, Donald R
Graves, Anne Carol Finger
Gross, Elizabeth Louise
Hangarter, Roger Paul

Henderlong, Paul Robert
Honda, Shigeru Irwin
Jegla, Dorothy Eldredge
Jyung, Woon Heng
Ku, Han San
Larson, Laurence Arthur
Liu, Ting-Ting Y
Madden, L V
Mayne, Berger C
Noble, Reginald Duston
Perley, James E
Pickering, Ed Richard
Platt, Robert Swanton, Jr
Richardson, Keith Erwin
Roberts, Bruce R
Sack, Fred David
Salminen, Seppo Ossian
Schaff, John Franklin
Servaites, Jerome Casimer
Silvius, John Edward
Streeter, John Gemmil
Swanson, Carroll Arthur

OKLAHOMA
Basler, Eddie, Jr
Beevers, Leonard
Couch, Richard Wesley
Dickinson, David Budd
Fletcher, John Samuel
Hawxby, Keith William
Ketring, Darold L
Mort, Andrew James
Ownby, James Donald
Schubert, Karel Ralph

OREGON
Amundson, Robert Gale
Armstrong, Donald James
Arp, Daniel James
Bishop, Norman Ivan
Bullock, Richard Melvin
Chilcote, David Owen
Ching, Te May
Couey, H Melvin
Daley, Laurence Stephen
Dalton, David Andrews
Fuchigami, Leslie H
Hannaway, David Bryon
Ho, Iwan
Kelsey, Rick G
Klepper, Elizabeth Lee (Betty)
Lagerstedt, Harry Bert
McClendon, John Haddaway
Mielke, Eugene Albert
Mok, Machteld Cornelia
Moore, Thomas Carrol
Moss, Dale Nelson
Nitsos, Ronald Eugene
Proebsting, William Martin
Rasmussen, Reinhold Albert
Rottink, Bruce Allan
Schreiner, Roger Paul
Scott, Eion George
Scott, Peter Carlton
Smith, Orrin Ernest
Snow, Michael Dennis
Stafford, Helen Adele
Stark, Nellie May
Stein, William Ivo
Tingey, David Thomas
Tocher, Richard Dana
Toumadje, Arazdordi
Weber, James Alan
Weiser, Conrad John
Westwood, Melvin (Neil)
Zaerr, Joe Benjamin

PENNSYLVANIA
Barr, Richard Arthur
Bayer, Margret H(elene) Janssen
Black, Robert Corl
Blumenfield, David
Boutros, Osiris Wahba
Brennan, Thomas Michael
Brunkard, Kathleen Marie
Burley, J William Atkinson
Coleman, Robert E
Cosgrove, Daniel Joseph
Dietrich, William Edward
Doner, Landis Willard
Evensen, Kathleen Brown
Fales, Steven Lewis
Fett, William Frederick
Fluck, Richard Allen
Hamilton, Robert Hillery, Jr
Harding, Roy Woodrow, Jr
Hill, Kenneth Lee
Husic, Harold David
Jacobs, Mark
Jacobsohn, Myra K
Jung, Gerald Alvin
Kendall, William Anderson
Knievel, Daniel Paul
Koide, Roger Tai
Krueger, Charles Robert
List, Albert, Jr
Lockard, Raymond G
Lowenhaupt, Benjamin
Maksymowych, Roman
Mickle, Ann Marie
Moreau, Robert Arthur
Partanen, Carl Richard
Pike, Carl Stephen
Robinson, Curtis

Schultz, Jack C
Shannon, Jack Corum
Snyder, Freeman Woodrow
Stephenson, Andrew George
Steucek, Guy Linsley
Thimann, Kenneth Vivian
Wallner, Stephen John
Warner, Harlow Lester
Williams, Stephen Edward
Witham, Francis H
Zimmerer, Robert P

RHODE ISLAND
Albert, Luke Samuel
Beale, Samuel I
Holowinsky, Andrew Wolodymyr
Hull, Richard James

SOUTH CAROLINA
Camper, Nyal Dwight
Carter, George Emmitt, Jr
Coston, Donald Claude
Gossett, Billy Joe
Hook, Donal D
Kerstetter, Rex E
Lincoln, David Erwin
McClain, Eugene Fredrick
McGregor, William Henry Davis
Morris, James T
Nelson, Eric Alan
Singh, Raghbir
Wallace, Susan Ulmer
Whitney, John Barry, Jr

SOUTH DAKOTA
Dybing, Clifford Dean
Ferguson, Michael William
Gaines, Jack Raymond
Moore, Raymond A
Nelson, Donald Carl
Tatina, Robert Edward
Tieszen, Larry L

TENNESSEE
Davenport, James Whitcomb
Dettbarn, Wolf Dietrich
Eskew, David Lewis
Foutch, Harley Wayne
Hayes, Robert M
Henson, James Wesley
Holton, Raymond William
Howell, Golden Leon
Jarrett, Harry Wellington, III
Jones, Larry Hudson
Jones, Larry Warner
Luxmoore, Robert John
McLaughlin, Samuel Brown
Mullin, Beth Conway
Naylor, Gerald Wayne
Norby, Richard James
Reynolds, John Horace
Sams, Carl Earnest
Schwarz, Otto John
Shipp, Oliver Elmo
Stone, Benjamin P
Trigiano, Robert Nicholas

TEXAS
Allen, Randy Dale
Arntzen, Charles Joel
Bell, Alois Adrian
Benedict, C R
Bovey, Rodney William
Brand, Jerry Jay
Briske, David D
Chadwick, Arthur Vorce
Cook, Charles Garland
Coulter, Murray W
Davies, Frederick T, Jr
Duble, Richard Lee
Dunton, Kenneth Harlow
Durham, James Ivey
Fucik, John Edward
Funkhouser, Edward Allen
Garay, Andrew Steven
Goodin, Joe Ray
Griffing, Lawrence Randolph
Hall, Timothy Couzens
Hall, Wayne Clark
Hopper, Norman Wayne
Irvine, James Estill
Joham, Howard Ernest
Johanson, Lamar
Jordan, Wayne Robert
Kennedy, Robert Alan
Krieg, Daniel R
Lineberger, Robert Daniel
Lingle, Sarah Elizabeth
Lipe, John Arthur
McBee, George Gilbert
McCully, Wayne Gunter
McKnight, Thomas David
McLemore, Bobbie Frank
Meyer, Robert Earl
Miller, Julian Creighton, Jr
Mills, William Ronald
Morgan, Page Wesley
Nguyen, Henry Thien
Norris, William Elmore, Jr
Poulsen, Lawrence Leroy
Rizzo, Peter Jacob
Rogers, Suzanne M Dethier
Sij, John William
Skelton, Bobby Joe

Smith, Don Wiley
Smith, James Douglas
Smith, Roberta Hawkins
Smith, Russell Lamar
Soltes, Edward John
Spear, Irwin
Storey, James Benton
Taylor, Richard Melvin
Van Alfen, Neal K
Vance, Benjamin Dwain
Van Overbeek, Johannes
Vietor, Donald Melvin
Whitenberg, David Calvin
Wiegand, Oscar Fernando
Williams, Ralph Edward
Winter, Steven Ray

UTAH
Bennett, Jesse Harland
Caldwell, Martyn Mathews
Campbell, William Frank
Chatterton, Norman Jerry
Ehleringer, James Russell
Harrison, Bertrand Fereday
Johnson, Douglas Allan
McNulty, Irving Bazil
Miller, Gene Walker
Openshaw, Martin David
Rasmussen, Harry Paul
Salisbury, Frank Boyer
Smith, Bruce Nephi
Tilton, Varien Russell
Walker, Edward Bell Mar
Walser, Ronald Herman
Wang, Richard Ruey-Chyi
Weber, Darrell J
Welkie, George William
Williams, M Coburn

VERMONT
Currier, William Wesley
Etherton, Bud
Gregory, Robert Aaron
Klein, Deana Tarson
Klein, Richard M
Landgren, Craig Randall
Spearing, Ann Marie
Tabor, Christopher Alan
Tyree, Melvin Thomas

VIRGINIA
Alberte, Randall Sheldon
Alscher, Ruth
Bingham, Samuel Wayne
Brown, Gregory Neil
Burns, Russell MacBain
Chevone, Boris Ivan
Coartney, James S
Cowles, Joe Richard
Dawson, Murray Drayton
Drake, Charles Roy
Foy, Chester Larrimore
Fuller, Stephen William
Hale, Maynard George
Hatzios, Kriton Kleanthis
Hayashi, Fumihiko
Hess, John Lloyd
Hortick, Harvey J
Jones, Johnnye M
Kreh, Richard Edward
Lacy, George Holcombe
Marcy, Joseph Edwin
Mathes, Martin Charles
Nielsen, Peter Tryon
Orcutt, David Michael
Parrish, David Joe
Peters, Gerald Alan
Roshal, Jay Yehudie
Rosinski, Joanne
Stepka, William
Stetler, David Albert
Webb, Kenneth L
Young, Donald Raymond

WASHINGTON
Brewer, Howard Eugene
Cataldo, Dominic Anthony
Cellarius, Richard Andrew
Chevalier, Peggy
Clark, James Richard
Cleland, Robert E
Dean, Bill Bryan
DePuit, Edward J
Edwards, Gerald Elmo
Elfving, Donald Carl
Evans, David W
Fisher, Donald B
George, Donald Wayne
Greenblatt, Gerald A
Hatheway, William Howell
Heebner, Charles Frederick
Hiller, Larry Keith
Iritani, W M
Irving, Patricia Marie
Ketchie, Delmer O
Loewus, Frank A
Lopushinsky, William
McFadden, Bruce Alden
Meeuse, Bastiaan J D
Ogg, Alex Grant, Jr
Pack, Merrill Raymond
Patterson, Max E
Perry, Mary Jane
Peterson, John Carl

Poovaiah, Bachettira Wthappa
Radwan, Mohamed Ahmed
Raese, John Thomas
Schrader, Lawrence Edwin
Shortess, David Keen
Thornton, Robert Kim
Tukey, Harold Bradford, Jr
Warner, Robert Lewis
Williams, Max W
Withner, Carl Leslie, Jr
Yu, Ming-Ho
Zimmermann, Charles Edward

WEST VIRGINIA
Biggs, Alan Richard
Blaydes, David Fairchild
Clark, Ralph B
Day, Thomas Arthur
Elkins, John Rush
Foster, Joyce Geraldine
Horton, Billy D
Ingle, L Morris
Leather, Gerald Roger
Mills, Howard Leonard

WISCONSIN
Albrecht, Kenneth Adrian
Amasino, Richard M
Armstrong, George Michael
Balke, Nelson Edward
Barton, Kenneth Allen
Beck, Gail Edwin
Becker, Wayne Marvin
Cao, Weixing
Daie, Jaleh
Dickson, Richard Eugene
Duke, Stanley Houston
Dutton, Herbert Jasper
Feirer, Russell Paul
Gerloff, Gerald Carl
Harris, Joseph Belknap
Heggestad, Howard Edwin
Helgeson, John Paul
Holm, LeRoy George
Johnson, Morris Alfred
Keegstra, Kenneth G
Knous, Ted R
Maravolo, Nicholas Charles
Millington, William Frank
Nelson, Neil Douglas
Nelson, Oliver Evans, Jr
Norman, John Matthew
Paulson, William H
Peterson, David Maurice
Quirk, John Thomas
Schwartz, Leander Joseph
Sharkey, Thomas D
Skoog, Folke
Sussman, Michael R
Thiesfeld, Virgil Arthur
Tibbitts, Theodore William
Weeks, Thomas F
Wolter, Karl Erich
Wyse, Roger Earl

WYOMING
Fisser, Herbert George
Smith, William Kirby

PUERTO RICO
Jordan-Molero, Jaime E
Lugo, Herminio Lugo

ALBERTA
Allan, John R
Briggs, Keith Glyn
Cossins, Edwin Albert
Feng, Yongsheng
Gorham, Paul Raymond
Laroche, Andre
Lynch, Dermot Roborg
Pharis, Richard Persons
Reid, David Mayne
Roberts, David Wilfred Alan
Swanson, Robert Harold
Thorpe, Trevor Alleyne
Vanden Born, William Henry
Whitehouse, Ronald Leslie S
Zalik, Saul

BRITISH COLUMBIA
Ballantyne, David John
Cumming, Bruce Gordon
Eastwell, Kenneth Charles
Fisher, Francis John Fulton
Lavender, Denis Peter
Looney, Norman E
Pollard, Douglas Frederick William
Prasad, Raghubir (Raj)
Runeckles, Victor Charles
Setterfield, George Ambrose
Srivastava, Lalit Mohan
Stout, Darryl Glen
Taylor, Iain Edgar Park
Vidaver, William Elliott
Warrington, Patrick Douglas
Waygood, Ernest Roy
Whyte, John Nimmo Crosbie

MANITOBA
LaCroix, Lucien Joseph
McVetty, Peter Barclay Edgar
Staniforth, Richard John
Vessey, Joseph Kevin

NEW BRUNSWICK
Bonga, Jan Max
Chopin, Thierry Bernard Raymond
Coleman, Warren Kent
Fensom, David Strathern
Little, Charles Harrison Anthony
Yoo, Bong Yul

NEWFOUNDLAND
Hampson, Michael Chisnall

NOVA SCOTIA
Bidwell, Roger Grafton Shelford
Blatt, Carl Roger
Chen, Lawrence Chien-Ming
Cullen, John Joseph
Kimmins, Warwick Charles
Laycock, Maurice Vivian

ONTARIO
Arnison, Paul Grenville
Barker, William George
Bewley, John Derek
Birmingham, Brendan Charles
Canvin, David T
Chang, Fa Yan
Colman, Brian
Colombo, Stephen John
Cummins, W(illiam) Raymond
Czuba, Margaret
Dumbroff, Erwin Bernard
Effer, W R
Farmer, Robert E, Jr
Ferrier, Jack Moreland
Fletcher, Ronald Austin
Forsyth, Frank Russell
Greenberg, Bruce Matthew
Heath, Michele Christine
Higgins, Verna Jessie
Hofstra, Gerald Gerrit
Hogan, Gary D
Hope, Hugh Johnson
Hopkins, William George
Horton, Roger Francis
Ingratta, Frank Jerry
Israelstam, Gerald Frank
Johnsen, Kurt H
Joy, Kenneth Wilfred
Lee, Tsung Ting
Leppard, Gary Grant
Liptay, Albert
Mack, Alexander Ross
Minshall, William Harold
Nozzolillo, Constance
Oaks, B Ann
Ormrod, Douglas Padraic
O'Sullivan, John
Peirson, David Robert
Pillay, Dathathry Trichinopoly Natraj
Proctor, John Thomas Arthur
Rauser, Wilfried Ernst
Riekels, Jerald Wayne
Rosa, Nestor
St Pierre, Jean Claude
Siminovitch, David
Stokes, Pamela Mary
Switzer, Clayton Macfie
White, Gordon Allan
Wightman, Frank
Winget, Carl Henry

PRINCE EDWARD ISLAND
Beauregard, Marc Daniel
MacQuarrie, Ian Gregor
Suzuki, Michio

QUEBEC
Bigras, Francine Jeanne
Bolduc, Reginald J
Boll, William George
Cormier, Francois
Furlan, Valentin
Gibbs, Sarah Preble
Girouard, Ronald Maurice
Levasseur, Maurice Edgar
Maclachlan, Gordon Alistair
Margolis, Hank A
Phan, Chon-Ton
Poole, Ronald John
Sarhan, Fathey
Willemot, Claude

SASKATCHEWAN
Clarke, John Mills
Gusta, Lawrence V
Kartha, Kutty Krishnan
King, John
Kurz, Wolfgang Gebhard Walter
Quick, William Andrew
Redmann, Robert Emanuel
Tanino, Karen Kikumi
Turel, Franziska Lili Margarete
Walther, Alina

OTHER COUNTRIES
Dainty, Jack
Delmer, Deborah P
Dreosti, Ivor Eustace
Edelman, Marvin
Gressel, Jonathan Ben
Maass, Wolfgang Siegfried Gunther
MacCarthy, Jean Juliet
Medina, Ernesto Antonio
Rakowski, Krzysztof J

166 / DISCIPLINE INDEX

Physiology, Plant (cont)
Shubert, L Elliot
Steer, Martin William
Thompson, Peter Allan
Todd, Glenn William
Zenk, Meinhart Hans Christian

Protein Science

ALABAMA
Kilpatrick, John Michael
Paxton, Ralph
Urry, Dan Wesley

CALIFORNIA
Burns, John William
Cecchini, Gary
Cleland, Jeffrey Lynn
Harford, Joe Bryan
Heeb, Mary Jo
Hora, Maninder Singh
Houston, L L
Hurkman, William James
Jarnagin, Kurt
Lasky, Richard David
Ling, Nicholas Chi-Kwan
Marcus, Frank
Nguyen, Binh Trong
Paech, Christian
Rosenquist, Grace Link
Sliwkowski, Mary Burke
Sy, Jose
Treat-Clemons, Lynda George
Unver, Ercan
Yang, Heechung
Zaidi, Iqbal Mehdi

COLORADO
Lin, Leu-Fen Hou

CONNECTICUT
Deutscher, Murray Paul
Schepartz, Alanna

DISTRICT OF COLUMBIA
Phillips, Terence Martyn

GEORGIA
Brinton, Margo A
Gerschenson, Mariana
Hargrove, James Lee
Kiefer, Charles R(andolph)

ILLINOIS
Ahmad, Sarfraz
Budzik, Gerald P
Eisenhauer, Donald Alan
Kapsalis, Andreas A
Miyazaki, John H
Musa, Wafaa Arafat
Myatt, Elizabeth Anne
Schultz, Richard Michael
Spangler, Brenda Dolgin
Winkler, Martin Alan

INDIANA
Hurrell, John Gordon

IOWA
Metzler, David Everett

KANSAS
Gegenheimer, Peter Albert
Lookhart, George LeRoy

KENTUCKY
Jayaram, Beby

LOUISIANA
Dionigi, Christopher Paul
Ullah, Abul Hasnat

MARYLAND
Ahmed, Syed Ashrafuddin
Allnutt, F C Thomas
Chen, Hao-Chia
Cheng, Tu-chen
Davis, Julien Sinclair
Kole, Hemanta Kumar
Mihalyi, Elemer
Mostwin, Jacek Lech
Noguchi, Constance Tom
Pattabiraman, Nagarajan
Pedersen, Peter L
Pinkus, Lawrence Mark
Poljak, Roberto Juan
Robey, Frank A
Rose, George David
Wirth, Peter James

MASSACHUSETTS
Cantiello, Horacio Fabio
Cohen, Jonathan Brewer
Lu, Frederick Ming
Margolin, Alexey L
Poteete, Anthony Robert
Saidha, Tekchand
Sauer, Robert Thomas
Taubman, Martin Arnold
Yang, Shiow-shong
Yellen, Gary
Zabrecky, James R

MINNESOTA
Coleman, Patrick Louis
LaPorte, David Coleman

MISSISSIPPI
Bateman, Robert Carey, Jr
Wellman, Susan Elizabeth

MISSOURI
Gorski, Jeffrey Paul

NEBRASKA
Chollet, Raymond

NEW HAMPSHIRE
Harbury, Henry Alexander

NEW JERSEY
Krystek, Stanley R, Jr
Logdberg, Lennart Erik
Lunn, Charles Albert
Martinez del Rio, Carlos
Nalin, Carlo M
Raman, Jay Ananth
Sheriff, Steven

NEW YORK
Bishop, David Franklin
Burke, G Thompson
Lepp, Cyrus Andrew
Maley, Gladys Feldott
Paine, Philip Lowell
Payne, Michael Ross
Prasthofer, Thomas W
Ter Kuile, Benno Herman

NORTH CAROLINA
Edgell, Marshall Hall
Horton, Horace Robert
Privalle, Christopher Thomas

OHIO
Blumenthal, Robert Martin
Brooks, James Reed
Francis, Joseph William
Gronostajski, Richard M
Jentoft, Joyce Eileen

PENNSYLVANIA
Alonso-Caplen, Firelli V
Bjornsti, Mary-Ann
Brinigar, William Seymour, Jr
Kronick, Paul Leonard
Shiman, Ross
Weisel, John Winfield

RHODE ISLAND
Biggins, John
Yoon, Poksyn Song

TENNESSEE
Chunduru, Srinivas Kishore
Clauberg, Martin
Conary, Jon Thurston
Davenport, James Whitcomb
Nishimoto, Satoru Kenneth
Niyogi, Salil Kumar

TEXAS
Kurosky, Alexander
Rainwater, David Luther
Wold, Finn

VIRGINIA
Bauerle, Ronald H
Biben, Maxeen G
Jones, Daniel David
Sayala, Chhaya

WASHINGTON
Neurath, Hans

WISCONSIN
Burgess, Richard Ray
Hefle, Susan Lynn

BRITISH COLUMBIA
Richards, William Reese

ONTARIO
Capone, John Pasquale
Goldberg, Gary Steven
Hope, Hugh Johnson
Huber, Carol (Saunderson)
Lepock, James Ronald
Neelin, James Michael
Plaxton, William Charles
Rand, Richard Peter
Wasi, Safia
Wood, Janet Marion

PRINCE EDWARD ISLAND
Beauregard, Marc Daniel

QUEBEC
Guillemette, Gaetan
Lagace, Lisette
Lin, Sheng-Xiang
Sairam, Malur Ramaswamy
Yakunin, Alexander F

OTHER COUNTRIES
Argos, Patrick
Said, Mostafa Mohamed

Toxicology

ALABAMA
Coker, Samuel Terry
Hood, Ronald David
Liu, Ray Ho
Yarbrough, James David

ALASKA
Kennish, John M

ARIZONA
Halpert, James Robert
Witten, Mark Lee

ARKANSAS
Allaben, William Thomas
Badger, Thomas Mark
Carter, Charleata A
Casciano, Daniel Anthony
Doerge, Daniel Robert
Hansen, Deborah Kay
Hinson, Jack Allsbrook
Holson, Ralph Robert
Jackson, Carlton Darnell
Leakey, Julian Edwin Arundell
Paule, Merle Gale
Roberts, Dean Winn, Jr
Schieferstein, George Jacob
Sheehan, Daniel Michael
Soderberg, Lee Stephen Freeman
Wessinger, William David
Young, John Falkner

CALIFORNIA
Anderson, Geoffrey Robert
Bailey, David Nelson
Baldwin, Robert Charles
Ballard, Ralph Campbell
Bendix, Selina (Weinbaum)
Bjeldanes, Leonard Frank
Buck, Alan Charles
Buckpitt, Alan R
Budny, John Arnold
Casida, John Edward
Collins, James Francis
Coughlin, James Robert
Crosby, Donald Gibson
Danse, Ilene H Raisfeld
Davis, Brian Kent
Davis, William Ellsmore, Jr
DePass, Linval R
Eastmond, David Albert
Esposito, Michael Salvatore
Farber, Sergio Julio
Flegal, Arthur Russell, Jr
Froebe, Larry R
Furst, Arthur
Goldman, Marvin
Hackney, Jack Dean
Hochstein, Paul Eugene
Hoke, Glenn Dale
Hollinger, Mannfred Alan
Howd, Robert A
Hughes, James Paul
Hutchin, Maxine E
Kassakhian, Garabet Haroutioun
Kelman, Bruce Jerry
Kland, Mathilde June
Koschier, Francis Joseph
Landolph, Joseph Richard, Jr
Lapota, David
Last, Jerold Alan
Leung, Peter
Malone, Marvin Herbert
Maxwell, Kenneth Eugene
Mehra, Rajesh Kumar
Newell, Gordon Wilfred
Ohlendorf, Harry Max
Parker, Kenneth D
Paustenbach, Dennis J
Rabovsky, Jean
Repique, Eliseo
Rodriguez, Eloy
Sanders, Brenda Marie
Schneider, Meier
Short, Jay M
Smith, Martyn Thomas
Sunshine, Irving
Tune, Bruce Malcolm
Victery, Winona Whitwell
Wilson, Barry William
Witham, Clyde Lester
Witschi, Hanspeter R

COLORADO
Aldrich, Franklin Dalton
Brooks, Bradford O
Corby, Donald G
Hall, Alan H
Hesterberg, Thomas William
Johnsen, Richard Emanuel
Rumack, Barry H
Shopp, George Milton, Jr
Sokol, Ronald Jay
Wilber, Charles Grady
Wilson, Vincent L
Wingeleth, Dale Clifford

CONNECTICUT
Amacher, David E
Berdick, Murray
Buck, Marion Gilmour
Cohen, Steven Donald

Dawson, Margaret Ann
Kowolenko, Michael D
Lacouture, Peter George
McConnell, William Ray
Milzoff, Joel Robert
Monro, Alastair Macleod
Pfeifer, Richard Wallace
Rosenberg, Philip
Schurig, John Eberhard
Snellings, William Moran
Stadnicki, Stanley Walter, Jr
Tyler, Tipton Ransom

DELAWARE
Blum, Lee M
Daly, John Joseph, Jr
Malley, Linda Angevine
Pierson, Keith Bernard
Scala, Robert Andrew
Staples, Robert Edward

DISTRICT OF COLUMBIA
Abernathy, Charles Owen
April, Robert Wayne
Arcos, Joseph (Charles)
Bolger, P Michael
Byrd, Daniel Madison, III
Elsayed, Nabil M
Feeney, Gloria Comulada
Gunn, John William, Jr
Irausquin, Hiltje
Lai, David Ying-lun
Mac, Michael John
McMahon, Timothy F
Milstein, Stanley Richard
Moore, John Arthur
Nadolney, Carlton H
Page, Samuel William
Putzrath, Resha Mae
Raslear, Thomas G
Rhoden, Richard Allan
Saslaw, Leonard David
Schwartz, Sorell Lee
Silver, Francis
Sobotka, Thomas Joseph
Thomas, Richard Dean
Todhunter, John Anthony
Zeeman, Maurice George

FLORIDA
Alam, Mohammed Khurshid
Barron, Mace Gerald
Boxill, Gale Clark
Frank, H Lee
Gilbert-Barness, Enid F
James, Margaret Olive
Levy, Richard
Lu, Frank Chao
McKenney, Charles Lynn, Jr
Mayer, Foster Lee, Jr
Milton, Robert Mitchell
Murchison, Thomas Edgar
Porvaznik, Martin
Radomski, Jack London
Rao, Krothapalli Ranga
Spieler, Richard Earl
Teaf, Christopher Morris

GEORGIA
Bates, Harold Brennan, Jr
Berg, George G
Catravas, John D
Curley, Winifred H
David, Robert Michael
Denson, Donald D
Dusenbery, David Brock
Hargrove, Robert John
Holt, Peter Stephen
Krolak, John Michael
Lehner, Andreas Friedrich
McGrath, William Robert
Sheppard, David Craig
Snell, Terry Wayne
White, Donald Henry
Winger, Parley Vernon
Wolven-Garrett, Anne M

HAWAII
Hokama, Yoshitsugi
Kodama, Arthur Masayoshi
Liu, Yong-Biao
Seifert, Josef
Sherman, Martin

ILLINOIS
Ambre, John Joseph
Benkendorf, Carol Ann
Bhattacharyya, Maryka Horsting
Brockman, Herman E
Brown, Daniel Joseph
Chapman, John R
Feldbush, Thomas Lee
Fitzloff, John Frederick
Foreman, Ronald Louis
Francis, Bettina Magnus
Garvin, Paul Joseph, Jr
Gu, Yan
Hansen, Larry George
Hermann, Edward Robert
Kaistha, Krishan K
Kaminski, Edward Jozef
Kethley, John Bryan
Khan, Mohammed Abdul Quddus
McConnachie, Peter Ross

McCormick, David Loyd
Miller, Gregory Duane
Musa, Wafaa Arafat
Narahashi, Toshio
Northup, Sharon Joan
Patterson, Douglas Reid
Reilly, Christopher Aloysius, Jr
Sherwood, Robert Lawrence
Taylor, Julius David
Tindall, Donald R
Visek, Willard James
White, Randy D
Yeates, Donovan B

INDIANA
Amundson, Merle E
Besch, Henry Roland, Jr
Born, Gordon Stuart
Borowitz, Joseph Leo
Caplis, Michael E
Gehring, Perry James
Lindstrom, Terry Donald
Litov, Richard Emil
Maickel, Roger Philip
Meyers, Donald Bates
Paladino, Frank Vincent
Pierce, William Meredith, Jr
Schulze, Gene Edward
Vodicnik, Mary Jo
Wheeler, Ralph John
Wolff, Ronald Keith

IOWA
Anderson, Todd Alan
Baron, Jeffrey
Coats, Joel Robert
McDonald, Donald Burt
Stahr, Henry Michael
Tephly, Thomas R

KANSAS
Browne, Ronald K
Li, Jonathan J
Oehme, Frederick Wolfgang
Parkinson, Andrew
Pazdernik, Thomas Lowell
Pickrell, John A
Pierce, John Thomas
Schroeder, Robert Samuel
Walaszek, Edward Joseph

KENTUCKY
Chen, Theresa S
Glauert, Howard Perry
Gupta, Ramesh C
Hurst, Harrell Emerson
Jarboe, Charles Harry
Kasarskis, Edward Joseph
Myers, Steven Richard
Nerland, Donald Eugene
Panemangalore, Myna
Volp, Robert Francis
Yokel, Robert Allen

LOUISIANA
Carter, Mary Kathleen
Fingerman, Milton
Hughes, Janice S
Komiskey, Harold Louis
Lee, Jordan Grey
Manno, Barbara Reynolds
Mehendale, Harihara Mahadeva
Olubadewo, Joseph Olanrewaju
Parent, Richard Alfred
Pryor, William Austin
Smith, Robert Leonard
Swenson, David Harold

MARYLAND
Alleva, Frederic Remo
Anderson, Robert Simpers
Barbehenn, Elizabeth Kern
Barbour, Michael Thomas
Bass, Eugene L
Beall, James Robert
Beyer, W Nelson
Bigger, Cynthia Anita Hopwood
Burton, Dennis Thorpe
Caplan, Yale Howard
Cerveny, Thelma Jannette
Chatterjee, Subroto
Choudary, Jasti Bhaskararao
Cone, Edward Jackson
Creasia, Donald Anthony
Dacre, Jack Craven
Fechter, Laurence David
Finch, Robert Allen
Fitzgerald, Glenna Gibbs (Cady)
Frazier, John Melvin
Friedman, Leonard
Gift, James J
Glocklin, Vera Charlotte
Glowa, John Robert
Goldman, Dexter Stanley
Gori, Gio Batta
Green, Martin David
Gryder, Rosa Meyersburg
Gupta, Gian Chand
Habig, William Henry
Hall, Richard Leland
Hathcock, John Nathan
Herman, Eugene H
Hill, Elwood Fayette
Horakova, Zdenka

Hsia, Mong Tseng Stephen
Jacobs, Abigail Conway
Jacobson-Kram, David
Joshi, Sewa Ram
Kiang, Juliann G
Kokoski, Charles Joseph
Kraybill, Herman Fink
Landauer, Michael Robert
Lee, Fang-Jen Scott
Lee, I P
Lijinsky, William
Melancon, Mark J
Milman, Harry Abraham
Misra, Rohini Rita
Morcock, Robert Edward
Morton, Joseph James Pandozzi
Mufson, R Allan
Nuite-Belleville, Jo Ann
Osawa, Yoichi
Palmer, Winifred G
Platz, Robert Dale
Prasanna, Hullahalli Rangaswamy
Putman, Donald Lee
Ragsdale, Nancy Nealy
Rao, Prasad Yarlagadda
Rattner, Barnett Alvin
Roddy, Martin Thomas
Roesijadi, Guritno
Rubin, Robert Jay
Saffiotti, Umberto
Sagripanti, Jose-Luis
Sahu, Saura Chandra
Shih, Tsung-Ming Anthony
Snipes, Charles Andrew
Steele, Vernon Eugene
Umbreit, Thomas Hayden
Vocci, Frank Joseph
Wannemacher, Robert, Jr
Weisburger, Elizabeth Kreiser
Whaley, Wilson Monroe
Whiting, John Dale, Jr
Whittaker, Paul
Woolverton, Christopher Jude
Wykes, Arthur Albert
Yager, James Donald, Jr
Yang, Shen Kwei

MASSACHUSETTS
Adams, Richard A
Amdur, Mary Ochsenhirt
Anderson, Rosalind Coogan
Busby, William Fisher, Jr
Chou, Iih-Nan (George)
Demple, Bruce
Edwards, Gordon Stuart
Gallo, Richard Louis
Goldmacher, Victor S
Graffeo, Anthony Philip
Hoar, Richard Morgan
Hoffmann, George Robert
Nanji, Amin Akbarali
Newcombe, David S
Robinson, William Edward
Rohovsky, Michael William
Sivak, Andrew
Slapikoff, Saul Abraham
Smith, Dennis Matthew
Smuts, Mary Elizabeth
Wogan, Gerald Norman

MICHIGAN
Aaron, Charles Sidney
Aulerich, Richard J
Beaudoin, Allan Roger
Bedford, James William
Bernstein, Isadore A
Biermann, Janet Sybil
Braselton, Webb Emmett, Jr
Butterworth, Francis M
Chang, Chia-Cheng
Chopra, Dharam Pal
Cochran, Kenneth William
De La Iglesia, Felix Alberto
Dutta, Saradindu
Dwivedi, Rama S
Elferink, Cornelius Johan
Elliott, George Algimon
Frade, Peter Daniel
Garg, Bhagwan D
Giesy, John Paul, Jr
Hollenberg, Paul Frederick
Hult, Richard Lee
Kamrin, Michael Arnold
Krishna, Gopala
Kruse, James Alexander
Lalwani, Narendra Dhanraj
Landrum, Peter Franklin
Lindblad, William John
Mccorkle, Fred Miller
MacDonald, John Robert
Mitchell, Jerry R
Moolenaar, Robert John
Novak, Raymond Francis
Passino, Dora R May
Piper, Walter Nelson
Reeves, Andrew Louis
Reiners, John Joseph, Jr
Richardson, Rudy James
Roth, Robert Andrew, Jr
Sikarskie, James Gerard
Sinsheimer, Joseph Eugene
Sleight, Stuart Duane
Theiss, Jeffrey Charles

Van Dyke, Russell Austin

MINNESOTA
Ankley, Gerald Thomas
Chesney, Charles Frederic
Goble, Frans Cleon
McNaught, Donald Curtis
Messing, Rita Bailey
Nagasawa, Herbert Tsukasa
Roy, Robert Russell
Smith, Thomas Jay
Willard, Paul W

MISSISSIPPI
Benson, William Hazlehurst
Chambers, Janice E
Hall, Margaret (Margot) Jean
Heitz, James Robert
Hwang, Huey-Min
Larsen, James Bouton

MISSOURI
Badr, Mostafa
Briner, Robert C
Brown, Olen Ray
Howlett, Allyn C
Johannsen, Frederick Richard
Kuhns, John Farrell
Lau, Brad W C
Mehrle, Paul Martin, Jr
Mills, Steven Harlon
Schoettger, Richard A
Su, Kwei Lee
Thompson, Daniel James
Wilkinson, Ralph Russell
Yanders, Armon Frederick

NEBRASKA
Berndt, William O
Cohen, Samuel Monroe
Corbett, Michael Dennis
Nipper, Henry Carmack
Schneider, Norman Richard
Scholes, Norman W
Taylor, Stephen Lloyd

NEVADA
Wiemeyer, Stanley Norton

NEW HAMPSHIRE
Chabot, Christopher Cleaves
Gosselin, Robert Edmond
Rochelle, Lori Gatzy

NEW JERSEY
Abdel-Rahman, Mohamed Shawky
Aberman, Harold Mark
Abrutyn, Donald
Amemiya, Kenjie
Arthur, Alan Thorne
Auletta, Carol Spence
Baker, Thomas
Bogden, John Dennis
Burger, Joanna
Cooper, Keith Raymond
Curcio, Lawrence Nicholas
Dairman, Wallace M
De Salva, Salvatore Joseph
Durham, Stephen K
Farmanfarmaian, Allahverdi
Frenkel, Gerald Daniel
Gertner, Sheldon Bernard
Ghosh, Arati
Goel, Mahesh C
Hayes, Terence James
Hite, Mark
Jaeger, Rudolph John
Kapp, Robert Wesley, Jr
Katz, Laurence Barry
Levin, Wayne
Lewis, Neil Jeffrey
Lewis, Steven Craig
Lin, Yi-Ching
Lipman, Jack M
Lukacsko, Alison B
McKenzie, Basil Everard
Mattes, William Bustin
Mehlman, Myron A
Mitala, Joseph Jerrold
Moral, Josefa Liduvina
Rothstein, Edwin C(arl)
Scanes, Colin G
Schrankel, Kenneth Reinhold
Schreiner, Ceinwen Ann
Stark, Dennis Michael
Swenson, Christine Erica
Szot, Robert Joseph
Tierney, William John
Weis, Peddrick
Willson, John Ellis

NEW MEXICO
Bechtold, William Eric
Belinsky, Steven Alan
Corcoran, George Bartlett, III
Dahl, Alan Richard
Finch, Gregory Lee
Hadley, William Melvin
Henderson, Rogene Faulkner
Hoover, Mark Douglas
Johnson, Neil Francis
Lechner, John Fred
Mauderly, Joe L
Mewhinney, James Albert

North-Root, Helen May
Smith, Garmond Stanley
Thomassen, David George
Thompson, Jill Charlotte
Tillery, Marvin Ishmael

NEW YORK
Abraham, Rajender
Aschner, Michael
Astill, Bernard Douglas
Babich, Harvey
Barber, Eugene Douglas
Bigelow, Sanford Walker
Bloom, Stephen Earl
Bogoroch, Rita
Bora, Sunder S
Borenfreund, Ellen
Boyd, Juanell N
Brown, John Francis, Jr
Carpenter, David O
Castellion, Alan William
Chauhan, Ved P S
Chishti, Muhammad Azhar
Christie, Nelwyn T
Cronkite, Eugene Pitcher
David, Oliver Joseph
Davies, Kelvin J A
Drobeck, Hans Peter
Durkin, Patrick Ralph
Elahi, Nasik
Ely, Thomas Sharpless
Ferin, Juraj
Finkelstein, Jacob Noah
Font, Cecilio Rafael
Gates, Allen H(azen), Jr
Gordon, Harry William
Hard, Gordon Charles
Hoffman, Donald Bertrand
Iden, Charles R
Kaplan, Joel Howard
Kostyniak, Paul J
Krasavage, Walter Joseph
Lagunowich, Laura Andrews
Latimer, Clinton Narath
Lin, George Hung-Yin
McCann, Michael Francis
Maines, Mahin D
Mantel, Linda Habas
Miller, Richard Kermit
Milmore, John Edward
Neal, Michael William
Oberdorster, Gunter
O'Donoghue, John Lipomi
Pinto, John
Pool, William Robert
Richie, John Peter, Jr
Salwen, Martin J
Smith, Thomas Harry Francis
Sundaram, Kalyan
Thysen, Benjamin
Trombetta, Louis David
Ventura, William Paul
Warner, Garvin L
Weinstein, Leonard Harlan
Weisburger, John Hans
Weiss, Bernard
Zavon, Mitchell Ralph
Zedeck, Morris Samuel

NORTH CAROLINA
Abbott, Barbara D
Adler, Kenneth B
Aldridge, David William
Allis, John W
Andersen, Melvin Ernest
Anderson, Roger Fabian
Bishop, Jack Belmont
Bond, James Anthony
Boreiko, Craig John
Boyd, Jeffrey Allen
Burkhart, James Gaylord
Casanova, Mercedes
Chan, Po Chuen
Claxton, Larry Davis
Damstra, Terri
Dauterman, Walter Carl
Devlin, Robert B
Dieter, Michael Phillip
Di Giulio, Richard Thomas
Dunnick, June K
Durham, William Fay
Eastin, William Clarence, Jr
Fail, Patricia A
Fouts, James Ralph
Gardner, Donald Eugene
Genter, Mary Beth
Goyer, Robert Andrew
Graham, Doyle Gene
Hart, Larry Glen
Hassan, Hosni Moustafa
Heck, Henry d'Arcy
Heck, Jonathan Daniel
Hickey, Anthony James
Hodgson, Ernest
Johnson, Franklin M
Kodavanti, Prasada Rao S
Kodavanti, Urmila P
Krasny, Harvey Charles
Lawton, Michael P
Lewis, Susan Erskine
Little, Patrick Joseph
McBay, Arthur John
MacPhail, Robert C
Massaro, Edward Joseph

Toxicology (cont)

Matthews, Hazel Benton, Jr
Mennear, John Hartley
Mercer, Robert R
Mitchell, Ann Denman
Moreland, Donald Edwin
Morgan, Kevin Thomas
Phelps, Allen Warner
Popp, James Alan
Portier, Christopher Jude
Riviere, Jim Edmond
Roop, Robert Kenneth
Saunders, Donald Roy
Schiller, Carol Masters
Schwetz, Bernard Anthony
Steinhagen, William Herrick
Stoskopf, Michael Kerry
Suk, William Alfred
Sumner, Darrell Dean
Tice, Raymond Richard
Tilson, Hugh Arval
Toews, Arrel Dwayne
Tucker, Walter Eugene, Jr
Turner, Alvis Greely
Walters, Douglas Bruce
Waters, Michael Dee
Welsch, Frank

NORTH DAKOTA
Hein, David William
Messiha, Fathy S
Schnell, Robert Craig

OHIO
Alden, Carl L
Banning, Jon Willroth
Baumann, Paul C
Benedict, James Harold
Bertram, Timothy Allyn
Biagini, Raymond E
Bingham, Eula
Brain, Devin King
Carmichael, Wayne William
Carpenter, Robert Leland
Carter, R Owen, Jr
Chengelis, Christopher P
Collins, William John
Coots, Robert Herman
Dodd, Darol Dennis
Dougherty, John A
Homer, George Mohn
Imondi, Anthony Rocco
Klemm, Donald J
Langford, Roland Everett
Langley, Albert E
Lewis, Trent R
Olson, Cerl Thomas
Palmer, Brent David
Pavelic, Zlatko Paul
Pereira, Michael Alan
Pesce, Amadeo J
Peterle, Tony J
Pfohl, Ronald John
Robinson, Michael K
Rynbrandt, Donald Jay
Sidman, Charles L
Skelly, Michael Francis
Staubus, Alfred Elsworth
Tabor, Marvin Wilson
Taylor, Douglas Hiram
Thomas, Craig Eugene
Tong, James Ying-Peh
Toraason, Mark
Vorhees, Charles V
Willett, Lynn Brunson
Wong, Laurence Cheng Kong
Woodward, James Kenneth
Yeary, Roger A
Ziller, Stephen A, Jr

OKLAHOMA
Bantle, John Albert
Chaturvedi, Arvind Kumar
Coleman, Nancy Pees
Coleman, Ronald Leon
Crane, Charles Russell
Dubowski, Kurt M(ax)
Fletcher, John Samuel
Floyd, Robert A
Reinke, Lester Allen
Rikans, Lora Elizabeth
Rolf, Lester Leo, Jr

OREGON
Blythe, Linda L
Chandler, David B
Ferraro, Steven Peter
Garton, Ronald Ray
Henny, Charles Joseph
Kennedy, Margaret Wiener
Koller, Loren D
Miller, Terry Lee
Moldenke, Alison Feerick
Sinnhuber, Russell Otto

PENNSYLVANIA
Alarie, Yves
Alperin, Richard Junius
Carpenter, Charles Patten
Coon, Julius Mosher
Dean, Jack Hugh
Deckert, Fred W
Dixit, Rakesh
English, Leigh Howard
Fogleman, Ralph William
Forman, Debra L
Goldstein, Robin Sheryl
Hajjar, Nicolas Philippe
Hurt, Susan Schilt
Johnson, Elmer Marshall
Karol, Meryl Helene
Kinter, Lewis Boardman
Lazo, John Stephen
Linzey, Alicia Vogt
McKinstry, Donald Michael
Majumdar, Shyamal K
Mattison, Donald Roger
Miller, Arthur James
Oldham, James Warren
Saporito, Michael S
Schwartz, Edward
Seligson, Frances Hess
Sherman, Larry Ray
Smith, Roger Bruce
Sprince, Herbert
Steele, Craig William
Stong, David Bruce
Sun, James Dean
Thoman, Charles James
Urbach, Frederick
Weil, Carrol S
Yankell, Samuel L
Zacchei, Anthony Gabriel
Zaleski, Jan F
Zarkower, Arian
Zemaitis, Michael Alan

RHODE ISLAND
Brungs, William Aloysius
Ho, Kay T
Shaikh, Zahir Ahmad
Yevich, Paul Peter

SOUTH CAROLINA
Best, Robert Glen
Gard, Nicholas William
Jagoe, Charles Henry
Jollow, David J
Woodard, Geoffrey Dean Leroy

SOUTH DAKOTA
Evenson, Donald Paul
Hamilton, Steven J

TENNESSEE
Briggs, Robert Chester
Chung, King-Thom
Churchill, John Alvord
Dees, Craig
Dettbarn, Wolf Dietrich
Draughon, Frances Ann
Gerwe, Roderick Daniel
Lawrence, William Homer
Lyman, Beverly Ann
Maier, Kurt Jay
Marnett, Lawrence Joseph
Miller, Mark Steven
Miyamoto, Michael Dwight
Robert, Timothy Albert
Shugart, Lee Raleigh
Skalko, Richard G(allant)
Suter, Glenn Walter, II
Tardif, Suzette Davis
White, David Cleaveland
Wilson, Benjamin James

TEXAS
Ahmed, Ahmed Elsayed
Ansari, Guhlam Ahmad Shakeel
Arenaz, Pablo
Bailey, Everett Murl, Jr
Bjorndahl, Jay Mark
Bost, Robert Orion
Burghardt, Robert Casey
Busbee, David L
Chanh, Tran C
Cody, John T
Garcia, Hector D
Gingell, Ralph
Gothelf, Bernard
Harvey, Roger Bruce
Heinze, John Edward
Hsie, Abraham Wuhsiung
Jaweed, Mazher
Kennedy, James H
Kim, Hyeong Lak
Kokkinakis, Demetrius Michael
Kuhn, Janice Oseth
Lal, Harbans
McGrath, James Joseph
Meistrich, Marvin Lawrence
Meltz, Martin Lowell
Merrill, Gerald Alan
Moreland, Ferrin Bates
Nechay, Bohdan Roman
Randerath, Kurt
Ray, Allen Cobble
Reagor, John Charles
Wallace, Jack E
Ward, Jonathan Bishop, Jr
Wong, Shan Shekyuk

UTAH
Cheney, Carl D
Galster, William Allen
Moody, David Edward
Segelman, Alvin Burton
Sharma, Raghubir Prasad

VERMONT
Kneip, Theodore Joseph
Nicklas, Janice A
Schaeffer, Warren Ira
Sinclair, Peter Robert

VIRGINIA
Allison, Trenton B
Atrakchi, Aisar Hasan
Bass, Michael Lawrence
Batra, Karam Vir
Bleiberg, Marvin Jay
Borzelleca, Joseph Francis
Bradley, Sterling Gaylen
Cairns, John, Jr
Chinnici, Joseph (Frank) Peter
Cifone, Maria Ann
Donohue, Joyce Morrissey
Douglas, Jocelyn Fielding
Du, Julie (Yi-Fang) Tsai
Egle, John Lee, Jr
Gregory, Arthur Robert
Gross, Stanley Burton
Haber, Lynne Tracey
Hess, John Lloyd
Hill, Jim T
Katz, Alan Charles
Lipnick, Robert Louis
Locke, Krystyna Kopaczyk
Locke, Raymond Kenneth
Macdonald, Timothy Lee
McWright, Cornelius Glen
Munson, Albert Enoch
Nagarkatti, Mitzi
Nagarkatti, Prakash S
Pages, Robert Alex
Reno, Frederick Edmund
Sahli, Brenda Payne
Wands, Ralph Clinton
Wilkinson, Christopher Foster

WASHINGTON
Booth, Pieter Nathaniel
Bull, Richard J
Dauble, Dennis Deene
Karagianes, Manuel Tom
Long, Edward R
Mahlum, Daniel Dennis
Meadows, Gary Glenn
Sasser, Lyle Blaine
Sikov, Melvin Richard
Way, Jon Leong
Wehner, Alfred Peter
Whelton, Bartlett David
Yu, Ming-Ho

WEST VIRGINIA
Barnett, John Brian
Castranova, Vincent
Rankin, Gary O'Neal
Reasor, Mark Jae

WISCONSIN
Balke, Nelson Edward
Brooks, Arthur S
Cook, Mark Eric
Dasgupta, Bibhuti R
Dickerson, Charlesworth Lee
Dukenschein, Jeanne Therese
Fahl, William Edwin
Gallenberg, Loretta A
Ganther, Howard Edward
Haasch, Mary Lynn
Hake, Carl (Louis)
Hefle, Susan Lynn
Kaplan, Stanley
Kasper, Charles Boyer
Kitzke, Eugene David
Miller, James Alexander
Pariza, Michael Willard
Saryan, Leon Aram

WYOMING
Lockwood, Jeffrey Alan

PUERTO RICO
Santacana, Guido E

ALBERTA
Cook, David Alastair
Coppock, Robert Walter
Gorham, Paul Raymond
Lorscheider, Fritz Louis

BRITISH COLUMBIA
Autor, Anne Pomeroy
Banister, Eric Wilton
Bellward, Gail Dianne
Chapman, Peter Michael
Davison, Allan John
Dost, Frank Norman
Kafer, Etta (Mrs E R Boothroyd)
Kennedy, Christopher Jesse
Lauzier, Raymond B
Law, Francis C P
McKeown, Brian Alfred
Warrington, Patrick Douglas

MANITOBA
Giles, Michael Arthur
Glavin, Gary Bertrun
Pip, Eva
Sheppard, Stephen Charles
Szekely, Joseph George

NEW BRUNSWICK
Fairchild, Wayne Lawrence
Haya, Katsuji

NEWFOUNDLAND
Virgo, Bruce Barton

NOVA SCOTIA
Cook, Robert Harry
Gilgan, Michael Wilson
MacRae, Thomas Henry
Stratton, Glenn Wayne
Vandermeulen, John Henri

ONTARIO
Appanna, Vasu Dhananda
Bray, Tammy M
Bunce, Nigel James
Burnison, Bryan Kent
Chappel, Clifford
Cherian, M George
Chettle, David Robert
Cummins, Joseph E
Ewins, Peter J
Feuer, George
Graham, Richard Charles Burwell
Greenberg, Bruce Matthew
Harris, Charles Ronald
Hebert, Paul David Neil
Hollbach, Natasha Coffin
Kalow, Werner
Kelso, John Richard Murray
Khanna, Jatinder Mohan
Liu, Dickson Lee Shen
Logan, David Mackenzie
McCalla, Dennis Robert
Mayer, Joel
Moon, Thomas William
O'Brien, Peter J
Robinson, Gerald Arthur
Roy, Marie Lessard
Schneider, Henry
Smith, Lorraine Catherine
Trenholm, Harold Locksley
Weseloh, D V (Chip)

QUEBEC
Campbell, Peter G C
Germain, Lucie
La Roche, Gilles
Plaa, Gabriel Leon
Tessier, Andre

SASKATCHEWAN
Blakley, Barry Raymond
Huang, P M
Irvine, Donald Grant
Khachatourians, George G

OTHER COUNTRIES
Aoyama, Isao
Back, Kenneth Charles
Bernstein, David Maier
Gustafsson, Jan-Ake
Ide, Hiroyuki
Jondorf, W Robert
Mølhave, Lars
Nabeshima, Toshitaka
Nakagawa, Shizutoshi
Orrenius, Sten
Said, Mostafa Mohamed
Stoppani, Andres Oscar Manuel
Sung, Chen-Yu
Van Loveren, Henk

Zoology

ALABAMA
Adams, Curtis H
Arrington, Richard, Jr
Bacon, Arthur Lorenza
Best, Troy Lee
Beyers, Robert John
Boozer, Reuben Bryan
Boschung, Herbert Theodore
Brown, Jack Stanley
Carter, Howard Payne
Davis, David Gale
Dixon, Carl Franklin
Dobson, F(rederick) Stephen
Dusi, Julian Luigi
Fincher, John Albert
Fitzpatrick, J(oseph) F(erris), Jr
Folkerts, George William
Freeman, John A
Gilliland, Floyd Ray, Jr
Guyer, Craig
Hajduk, Stephen Louis
Hartley, Marshall Wendell
Hemphill, Andrew Frederick
Holler, Nicholas Robert
Holliman, Dan Clark
Jenkins, Ronald Lee
Loop, Michael Stuart
McClintock, James Bruce
Mangat, Baldev Singh
Mason, William Hickmon
Mitchell, Joseph Christopher
Modlin, Richard Frank
Mount, Robert Hughes

Nelson, David Herman
Sanford, L G
Sharma, Udhishtra Deva
Shipp, Robert Lewis
Wolfe, James Leonard
Yokley, Paul, Jr

ALASKA
Fay, Francis Hollis
Feist, Dale Daniel
Hameedi, Mohammad Jawed
Hatch, Scott Alexander
Highsmith, Raymond C
Kessel, Brina
Klein, David Robert
Kudenov, Jerry David
Peyton, Leonard James
Schoen, John Warren
Weeden, Robert Barton
West, George Curtiss
Williamson, Francis Sidney Lanier

ARIZONA
Alvarado, Ronald Herbert
Anderson, Glenn Arthur
Bradshaw, Gordon Van Rensselaer
Brown, Bryan Turner
Clark, Clarence Floyd
Cockrum, Elmer Lendell
Cohen, Andrew Scott
Cupp, Eddie Wayne
Davis, Russell Price
Dunlap, Donald Gene
Faeth, Stanley Herman
Fouquette, Martin John, Jr
Grim, J(ohn)Norman
Hadley, Mac Eugene
Hendrickson, John Roscoe
Johnson, Oliver William
Johnson, Robert Andrew
Jollie, Malcolm Thomas
Kay, Fenton Ray
Krutzsch, Philip Henry
Landers, Earl James
Lowe, Charles Herbert, Jr
McClure, Michael Allen
Maienschein, Jane Ann
Mead, Albert Raymond
Mead, James Irving
Minckley, Wendell Lee
Montgomery, Willson Linn
Ohmart, Robert Dale
Papaj, Daniel Richard
Patterson, Robert Allen
Pough, Frederick Harvey
Rosenzweig, Michael Leo
Russell, Stephen Mims
Smith, Andrew Thomas
Spofford, Sally Hoyt
Spofford, Walter Richardson, II
Van Riper, Charles, III
Vaughan, Terry Alfred
Vial, James Leslie
Walsberg, Glenn Eric
Welch, Claude Alton
White, John Anderson
Wilkes, Stanley Northrup
Woodin, William Hartman, III
Zimmerman, James Roscoe
Zweifel, Richard George

ARKANSAS
Bailey, Claudia F
Baker, Frank Hamon
Beadles, John Kenneth
Carter, Charleata A
Corkum, Kenneth C
Couser, Raymond Dowell
Dorris, Peggy Rae
Fribourgh, James H
Hanebrink, Earl L
Harp, George Lemaul
Heidt, Gary A
Kilambi, Raj Varad
Kraemer, Louise Russert
McDaniel, Van Rick
Morgans, Leland Foster
Plummer, Michael V(an)
Rakes, Jerry Max
Robison, Henry Welborn
Schmitz, Eugene H
Sealander, John Arthur, Jr
Smith, Kimberly Gray
Walker, James Martin
Wennerstrom, David E

CALIFORNIA
Alberch, Pere
Alexander, Claude Gordon
Alvarino, Angeles
Anderson, Bertin W
Andrews, Fred Gordon
Andrus, William DeWitt, Jr
Anthony, Ronald Lewis
Antipa, Gregory Alexis
Armstrong, Peter Brownell
Arnold, John Ronald
Arora, Harbans Lall
Arp, Alissa Jan
Austin, Joseph Wells
Bair, Thomas De Pinna
Baker, Mary Ann
Baldwin, James Gordon
Balgooyen, Thomas Gerrit
Band, Rudolph Neal
Barlow, George Webber
Bartholomew, George Adelbert
Beatty, Kenneth Wilson
Beck, Albert J
Beeman, Robert D
Bennett, Albert Farrell
Bern, Howard Alan
Berrend, Robert E
Bertsch, Hans
Beuchat, Carol Ann
Black, John David
Bolaffi, Janice Lerner
Boolootian, Richard Andrew
Bowers, Darl Eugene
Bowers, Roger Raymond
Boyd, Milton John
Bradbury, Margaret G
Bradley, Timothy Jud
Brand, Leonard Roy
Brattstrom, Bayard Holmes
Brittan, Martin Ralph
Broadbooks, Harold Eugene
Brownell, Robert Leo, Jr
Brusca, Gary J
Buchsbaum, Ralph
Buth, Donald George
Cailliet, Gregor Michel
Calarco, Patricia G
Callison, George
Campbell, James L
Carpenter, Roger Edwin
Castro, Peter
Cech, Joseph Jerome, Jr
Chamberlain, Dilworth Woolley
Chang, Ernest Sun-Mei
Chappell, Mark Allen
Chen, Lo-Chai
Cheng, Lanna
Christianson, Lee (Edward)
Church, Ronald L
Cliff, Frank Samuel
Coan, Eugene Victor
Cody, Martin L(eonard)
Cogswell, Howard Lyman
Cohen, Anne Carolyn Constant
Cohen, Daniel Morris
Cohen, Nathan Wolf
Coleman, Ronald Murray
Collias, Elsie Cole
Collias, Nicholas Elias
Collier, Gerald
Collins, Charles Thompson
Connelly, Thomas George
Costa, Daniel Paul
Courchesne, Eric
Cowles, David Lyle
Cox, George W
Crane, Jules M, Jr
Crouch, James Ensign
DeMartini, John
Desharnais, Robert Anthony
DeWolfe, Barbara Blanchard Oakeson
Dowell, Armstrong Manly
Drewes, Robert Clifton
Durrant, Barbara Susan
Eakin, Richard Marshall
Ebeling, Alfred W
Edmunds, Peter James
Edwards, J Gordon
Endler, John Arthur
Enright, James Thomas
Eschmeyer, William Neil
Estes, James Allen
Etheridge, Richard Emmett
Evans, Kenneth Jack
Federici, Brian Anthony
Feldmeth, Carl Robert
Fell, Howard Barraclough
Ferguson, William E
Fierstine, Harry Lee
Filice, Francis P
Fiorindo, Robert Philip
Fisler, George Frederick
Flaim, Francis Richard
Follett, Wilbur Irving
Fox, Laurel R
Fusaro, Craig Allen
Gabel, James Russel
Galt, Charles Parker, Jr
Garth, John Shrader
Gascoigne, Nicholas Robert John
Gaudin, Anthony J
Ghiselin, Michael Tenant
Goldberg, Stephen Robert
Goldstein, Bernard
Gorman, Cornelia M
Gorman, George Charles
Gould, Douglas Jay
Gray, Constance Helen
Greene, Richard Wallace
Guthrie, Daniel Albert
Haas, Richard
Hammar, Allan H
Hand, Cadet Hammond, Jr
Hanson, William Roderick
Hardy, Cecil Ross
Harmon, Wallace Morrow
Harris, Lester Earle, Jr
Hemingway, George Thomson
Hendler, Gordon Lee
Hermans, Colin Olmsted
Hessler, Robert Raymond
Hildebrand, Milton
Hill, Ray Allen
Ho, Ju-Shey
Hobson, Edmund Schofield
Holland, Nicholas Drew
Hooper, Emmet Thurman, Jr
Horn, Michael Hastings
Houston, Roy Seamands
Howell, Thomas Raymond
Hoyt, Donald Frank
Hsueh, Aaron Jen Wang
Huckaby, David George
Hunsaker, Don, II
Iwamoto, Tomio
Jacobs, John Allen
James, Thomas William
Jameson, Everett Williams, Jr
Jennrich, Ellen Coutlee
Jessop, Nancy Meyer
Johnson, Eric Van
Johnson, Ned Keith
Jones, Ira
Jones, James Henry
Jones, Ronald McClung
Jorgenson, Edsel Carpenter
Keller, Raymond E
Kenk, Vida Carmen
Kitting, Christopher Lee
Kooyman, Gerald Lee
Krejsa, Richard Joseph
Lambert, Charles Calvin
Lapota, David
Lawrence, George Edwin
Lee, Sue Ying
Leibovitz, Brian
Leitner, Philip
Lenhoff, Howard Maer
Leviton, Alan Edward
Lewis, Cynthia Lucille
Licht, Paul
Lidicker, William Zander, Jr
Lindberg, David Robert
Loher, Werner J
Longhurst, John Charles
Loomis, Richard Biggar
Loomis, Robert Henry
Lovich, Jeffrey Edward
Lownsbery, Benjamin Ferris
Luckock, Arlene Suzanne
Lyke, Edward Bonsteel
McClure, Howe Elliott
McCosker, John E
Mace, John Weldon
McFarland, William Norman
McLaughlin, Charles Albert
McLean, James H
McLean, Norman, Jr
McLeod, Samuel Albert
MacMillen, Richard Edward
Maggenti, Armand Richard
Main, Robert Andrew
Mankau, Sarojam Kurudamannil
Marler, Peter
Martin, Joel William
Mayhew, Wilbur Waldo
Mazia, Daniel
Mead, Giles Willis
Mercer, Edward King
Metzner, Walter
Morin, James Gunnar
Morris, Robert Wharton
Morton, Martin Lewis
Mossman, Archie Stanton
Moyle, Peter Briggs
Mulligan, Timothy James
Mulloney, Brian
Murphy, Ted Daniel
Muscatine, Leonard
Myrick, Albert Charles, Jr
Nagy, Kenneth Alex
Nandi, Satyabrata
Nelson, Keith
Newberry, Andrew Todd
Noffsinger, Ella Mae
Norris, Kenneth Stafford
Nybakken, James W
Ohlendorf, Harry Max
Orr, Robert Thomas
Padian, Kevin
Park, Taisooe
Parrish, Richard Henry
Patton, James Lloyd
Pearse, Vicki Buchsbaum
Perrin, William Fergus
Petersen, Bruce Wallace
Phleger, Charles Frederick
Pitelka, Frank Alois
Pitelka, Louis Frank
Platzer, Edward George
Plopper, Charles George
Pohlo, Ross
Poinar, George O, Jr
Potts, Donald Cameron
Power, Dennis Michael
Presch, William Frederick
Price, Mary Vaughan
Prothero, Donald Ross
Randall, Janet Ann
Raski, Dewey John
Reisen, William Kenneth
Reish, Donald James
Risser, Arthur Crane, Jr
Roche, Edward Towne
Roe, Pamela
Roest, Aryan Ingomar
Rogoff, William Milton
Rosenblatt, Richard Heinrich
Rudd, Robert L
Ruibal, Rodolfo
Saffo, Mary Beth
Sassaman, Clay Alan
Scherba, Gerald Marron
Schooley, Caroline Naus
Scott, Michael David
Scott, Norman Jackson, Jr
Sekhon, Sant Singh
Shellhammer, Howard Stephen
Sherman, Irwin William
Sibley, Charles Gald
Simpson, Larry P
Smith, Kenneth Lawrence, Jr
Smith, Ralph Ingram
Soltz, David Lee
Soule, John Dutcher
Spieler, Richard Arno
Spieth, Herman Theodore
Standing, Keith M
Stanton, Toni Lynn
Starrett, Andrew
Stebbins, Robert Cyril
Steele, Arnold Edward
Stephens, Grover Cleveland
Stewart, Glenn Raymond
Stiffler, Daniel F
Stohler, Rudolf
Stott, Kenhelm Welburn, Jr
Strohman, Richard Campbell
Swift, Camm Churchill
Sydeman, William J
Taylor, Dwight Willard
Taylor, Leighton Robert, Jr
Thomas, Barry
Thomas, Bert O
Thueson, David Orel
Tisserat, Brent Howard
Tjioe, Djoe Tjhoo
Tomlinson, Jack Trish
Torch, Reuben
Torell, Donald Theodore
Trapp, Gene Robert
Tribbey, Bert Allen
Triplett, Edward Lee
Turner, Frederick Brown
Udvardy, Miklos Dezso Ferenc
Vehrencamp, Sandra Lee
Viglierchio, David Richard
Wake, Marvalee H
Warter, Stuart L
Washburn, Sherwood L
Waters, James Frederick
Welsh, James Francis
Wenner, Adrian Manley
Westervelt, Clinton Albert, Jr
Westfahl, Pamela Kay
Weston, Henry Griggs, Jr
Wickler, Steven John
Widmer, Elmer Andreas
Williams, Daniel Frank
Wilson, Barry William
Wirtz, William Otis, II
Wood, Forrest Glenn
Wood, Richard Lyman
Woods, Geraldine Pittman
Woodwick, Keith Harris
Wright, Richard Kenneth
Yarnall, John Lee
Zieg, Roger Grant
Zimmer, Russel Leonard
Zuk, Marlene

COLORADO
Angell, Robert Walker
Apley, Martyn Linn
Armstrong, David M(ichael)
Audesirk, Teresa Eck
Belden, Don Alexander, Jr
Bock, Carl E
Bushnell, John Horace
Capen, Ronald L
Carnathan, Gilbert William
Cassel, J(oseph) Frank(lin)
Cohen, Robert Roy
Cruz, Alexander
Dudek, F Edward
Enderson, James H
Erickson, James George
Fall, Michael William
Fausch, Kurt Daniel
Finger, Thomas Emanuel
Finley, Robert Byron, Jr
Fuller, Barbara Fink Brockway
Green, Jeffrey Scott
Hall, Joseph Glenn
Hanken, James
Hansen, Richard M
Harriss, Thomas T
Herrmann, Scott Joseph
Hibler, Charles Phillip
Janes, Donald Wallace
Jones, Richard Evan
Lawrence, Robert G
Lehner, Philip Nelson
Lesh-Laurie, Georgia Elizabeth
Linder, Allan David
McClellan, John Forbes
McLean, Robert George
Marquardt, William Charles
Moury, John David
Nash, Donald Joseph

Zoology (cont)

Packard, Gary Claire
Pennak, Robert William
Pettus, David
Plakke, Ronald Keith
Reynolds, Richard Truman
Schmidt, Gerald D
Seilheimer, Jack Arthur
Smith, Hobart Muir
Spencer, Albert William
Sublette, James Edward
Vohs, Paul Anthony, Jr
Voth, David Richard
Ward, Gerald Madison
Woolley, Tyler Anderson
Wunder, Bruce Arnold
Zeiner, Frederick Neyer

CONNECTICUT

Aylesworth, Thomas Gibbons
Bernard, Richard Fernand
Blackburn, Daniel Glenn
Booth, Charles E
Brewer, Robert Hyde
Brush, Alan Howard
Caira, Janine Nicole
Carlton, James Theodore
Chichester, Lyle Franklin
Clark, George Alfred, Jr
Cohen, Melvin Joseph
DeCoursey, Russell Myles
Fiore, Carl
Gable, Michael F
Goodall, Jane
Grimm-Jorgensen, Yvonne
Hanks, James Elden
Hartman, Willard Daniel
James, Hugo A
Jokinen, Eileen Hope
Kent, John Franklin
Loomis, Stephen Henry
Schake, Lowell Martin
Simpson, Tracy L
Singletary, Robert Lombard
Swain, Elisabeth Ramsay
Taigen, Theodore Lee
Thurberg, Frederick Peter
Wagner, Gunter Paul
Wells, Kentwood David
Wheeler, Bernice Marion

DELAWARE

Boord, Robert Lennis
Carriker, Melbourne Romaine
Curtis, Lawrence Andrew
Ferguson, Thomas
Gaffney, Patrick M
Karlson, Ronald Henry
Koeppe, Mary Kolean
Pierson, Keith Bernard
Roth, Roland Ray
Targett, Timothy Erwin
Tripp, Marenes Robert

DISTRICT OF COLUMBIA

Banks, Richard Charles
Banks, William Michael
Banta, William Claude
Barnard, Jerry Laurens
Bayer, Frederick Merkle
Beehler, Bruce McPherson
Bellmer, Elizabeth Henry
Bernor, Raymond Louis
Bowman, Thomas Elliot
Burns, John McLauren
Bushman, John Branson
Cairns, Stephen Douglas
Collette, Bruce Baden
Cooper, George Everett
De Queiroz, Kevin
Dimond, Marie Therese
Domning, Daryl Paul
Fauchald, Kristian
Feeney, Gloria Comulada
Gladue, Brian Anthony
Goldberg, Aaron
Hair, Jay Dee
Hobbs, Horton Holcombe, Jr
Hope, William Duane
Kapur, Shakti Prakash
Knapp, Leslie W
Knowlton, Robert Earle
Lachner, Ernest Albert
Lovejoy, Thomas E
McDiarmid, Roy Wallace
Manning, Raymond B
Mathis, Wayne Neilsen
Merchant, Henry Clifton
Nickle, David Allan
Nickum, John Gerald
Olson, Storrs Lovejoy
Paris, Oscar Hall
Pawson, David Leo
Perez-Farfante, Isabel Cristina
Pettibone, Marian Hope
Puglia, Charles Raymond
Ralls, Katherine Smith
Rehder, Harald Alfred
Ripley, Sidney Dillon, II
Robinson, Michael Hill
Roper, Clyde Forrest Eugene
Springer, Victor Gruschka
Tyler, James Chase

Urban, Edward Robert, Jr
Van Arsdel, William Campbell, III
Waller, Thomas Richard
Watson, George E, III
Weitzman, Stanley Howard
West, William Lionel
Williams, Austin Beatty
Williams, Jeffrey Taylor
Wilson, Don Ellis
Zug, George R

FLORIDA

Adams, Roger Omar
Alam, Mohammed Khurshid
Allen, Ted Tipton
Anderson, Peter Alexander Vallance
Auffenberg, Walter
Berner, Lewis
Bick, George Herman
Bortone, Stephen Anthony
Branch, Lyn Clarke
Britt, N Wilson
Brockmann, Helen Jane
Brodkorb, Pierce
Caldwell, Melba Carstarphen
Carpenter, James Woodford
Castner, James Lee
Clark, Kerry Bruce
Colwin, Arthur Lentz
Colwin, Laura Hunter
Cooley, Nelson Reede
Courtenay, Walter Rowe, Jr
Courtney, Charles Hill
De Witt, Robert Merkle
Dickinson, Joshua Clifton, Jr
Dickson, Donald Ward
Dietz, John R
Ehrhart, Llewellyn McDowell
Ellis, Leslie Lee, Jr
Evans, David Hudson
Feddern, Henry A
Feldmesser, Julius
Ferguson, John Carruthers
Fitzpatrick, John Weaver
Fleming, Theodore Harris
Flowers, Ralph Wills
Forrester, Donald J
Fournie, John William
Frederick, Peter Crawford
Giblin-Davis, Robin Michael
Gilbert, Carter Rowell
Gilbert, Perry Webster
Gleeson, Richard Alan
Goldberg, Stephen
Goldberg, Walter M
Gordon, Kenneth Richard
Grabowski, Casimer Thaddeus
Greenberg, Michael John
Grobman, Arnold Brams
Groupe, Vincent
Gude, Richard Hunter
Gupta, Virendra K
Hardy, John William
Hays, Elizabeth Teuber
Heard, William Herman
Heckerman, Raymond Otto
Heckrotte, Carlton
Hinsch, Gertrude Wilma
Holling, Crawford Stanley
Hopkins, Thomas Lee
Hribar, Lawrence Joseph
Kale, Herbert William, II
Kaufmann, John Henry
Kerr, John Polk
Kinloch, Robert Armstrong
Kirkwood, James Benjamine
Kruczynski, William Leonard
Layne, James Nathaniel
Lessios, Harilaos Angelou
Levey, Douglas J
Levy, Richard
Lillywhite, Harvey B
McKenney, Charles Lynn, Jr
McNab, Brian Keith
McSorley, Robert
McSweeny, Edward Shearman
Mariscal, Richard North
Maturo, Frank J S, Jr
Means, D(onald) Bruce
Mendez, Eustorgio
Meryman, Charles Dale
Michel, Harding B
Miyamoto, Michael Masao
Moore, Donald Richard
Moore, Joseph Curtis
Mushinsky, Henry Richard
Myrberg, Arthur August, Jr
Nordlie, Frank Gerald
O'Bannon, John Horatio
Odell, Daniel Keith
Overman, Amegda Jack
Patterson, Richard Sheldon
Peckham, Richard Stark
Perry, Vernon G
Porter, James Armer, Jr
Porter, Sanford Dee
Povar, Morris Leon
Reynolds, John Elliott, III
Rhoades, Harlan Leon
Rice, Mary Esther
Rice, Stanley Alan
Richards, William Joseph
Riggs, Carl Daniel
Roberts, Larry Spurgeon

Robinson, Gerald Garland
Ruffer, David G
Savage, Jay Mathers
Sayles, Everett Duane
Schwartz, Albert
Simon, Joseph Leslie
Smith, Sharon Louise
Snelson, Franklin F, Jr
Spieler, Richard Earl
Stafford, Robert Oppen
Stevenson, Henry Miller
Stokes, Donald Eugene
Sturbaum, Barbara Ann
Tarjan, Armen Charles
Taylor, George Thomas
Taylor, Walter Kingsley
Telford, Sam Rountree, Jr
Thommes, Robert Charles
Thompson, Fred Gilbert
Thurber, Walter Arthur
Tihen, Joseph Anton
Travis, Joseph
Tripp, John Rathbone
Wallbrunn, Henry Maurice
Webb, Sawney David
White, Arlynn Quinton, Jr
Wing, Elizabeth S
Wireman, Kenneth
Witham, P(hilip) Ross
Witherington, Blair Ernest
Wolff, Ronald Gilbert
Woodruff, Robert Eugene
Woods, Charles Arthur
Woolfenden, Glen Everett
Wyneken, Jeanette
Yerger, Ralph William
Young, Craig Marden
Yunker, Conrad Erhardt

GEORGIA

Adkison, Daniel Lee
Amerson, Grady Malcolm
Avise, John C
Bates, Harold Brennan, Jr
Briggs, John Carmon
Brown, Paul Lopez
Burbanck, William Dudley
Caldwell, Sloan Daniel
Comeau, Roger William
Cutler, Horace Garnett
Davenport, Leslie Bryan, Jr
English, Arthur William
Fechheimer, Marcus
Fitt, William K
Green, Edwin Alfred
Hanson, William Lewis
Hicks, Heraline Elaine
Hunter, Frissell Roy
Husain, Ansar
Johnson, Alva William
Keeler, Clyde Edgar
Klompen, J S H
Kneib, Ronald Thomas
Laerm, Joshua
McKeever, Sturgis
Mapp, Frederick Everett
Miller, George C
Mills, James Norman
Murray, Joan Baird
Nadler, Ronald D
Odum, Eugene Pleasants
Patterson, Rosalyn Mitchell
Paulin, Jerome John
Porter, James W
Provost, Ernest Edmund
Relyea, Kenneth George
Saladin, Kenneth S
Scott, Donald Charles
Seerley, Robert Wayne
Sharp, Homer Franklin, Jr
Shibley, John Luke
Shure, Donald Joseph
Sridaran, Rajagopala
Stirewalt, Harvey Lee
Taylor, Robert Clement
Tietjen, William Leighton
Urban, Emil Karl
Urso, Paul
Utley, Philip Ray
Volpe, Erminio Peter
White, Donald Henry
Willis, John Steele

HAWAII

Ahearn, Gregory Allen
Arnold, John Miller
Bailey-Brock, Julie Helen
Berg, Carl John, Jr
Berger, Andrew John
DeMartini, Edward Emile
Eddinger, Charles Robert
Eldredge, Lucius G
Freed, Leonard Alan
Grau, Edward Gordon
Greenfield, David Wayne
Grove, John Sinclair
Hadfield, Michael Gale
Haley, Samuel Randolph
Hunter, Cynthia L
Kamemoto, Fred Isamu
Kay, Elizabeth Alison
Kinzie, Robert Allen, III
Koshi, James H
Little, Harold Franklin

Losey, George Spahr, Jr
Polovina, Jeffrey Joseph
Pyle, Robert Lawrence
Reed, Stuart Arthur
Reimer, Diedrich
Roderick, George Karlsson
Tinker, Spencer Wilkie
Tomich, Prosper Quentin
Vargas, Roger I
Wyban, James A

IDAHO

Berrett, Delwyn Green
Bunde, Daryl E
Cade, Thomas Joseph
Fritchman, Harry Kier, II
Goddard, Stephen
Hibbert, Larry Eugene
Johnson, James Blakeslee
Kelley, Fenton Crosland
Mead, Rodney A
Reno, Harley W
Schell, Stewart Claude
Trost, Charles Henry
Vance, Velma Joyce
Wootton, Donald Merchant

ILLINOIS

Alger, Nelda Elizabeth
Altmann, Jeanne
Andrews, Richard D
Axtell, Ralph William
Bartell, Marvin H
Beck, Sidney L
Beecher, William John
Bennett, Cecil Jackson
Berry, James Frederick
Bettice, John Allen
Bieler, Rudiger
Binford, Laurence Charles
Blake, Emmet Reid
Bolt, John Ryan
Brand, Raymond Howard
Brandon, Ronald Arthur
Brown, Lauren Evans
Buhse, Howard Edward, Jr
Burhop, Kenneth Eugene
Burr, Brooks Milo
Carr, Tommy Russell
Cracraft, Joel Lester
Damaskus, Charles William
Dinsmore, Charles Earle
Dudkiewicz, Alan Bernard
Dunn, Robert Bruce
Dybas, Linda Kathryn
Edwards, Dale Ivan
Fanslow, Don J
Feder, Martin Elliott
Fooden, Jack
Forman, G Lawrence
Foster, Merrill W
Foulkes, Robert Hugh
Franks, Edwin Clark
Freiburg, Richard Eighme
Funk, Richard Cullen
Garoian, George
Garrigus, Upson Stanley
George, William
Giere, Frederic Arthur
Girard, G Tanner
Goldberg, Robert Jack
Goodrich, Michael Alan
Graber, Richard Rex
Gramza, Anthony Francis
Green, Jonathan P
Griffiths, Thomas Alan
Hausler, Carl Louis
Heaney, Lawrence R
Heath, James Edward
Hetzel, Howard Roy
Hoffmeister, Donald Frederick
Holmes, E(dward) Bruce
Hunt, Lawrence Barrie
Jablonski, David
Jensen, Donald Reed
Johnson-Wint, Barbara Paule
Kanatzar, Charles Leplie
KerbisPeterhans, Julian C
Ketterer, John Joseph
Kruse, Kipp Colby
Lanyon, Scott Merril
Larson, Ingemar W
Le Febvre, Eugene Allen
Lehmann, Wilma Helen
Levine, Norman Dion
Lima, Gail M
Lombard, Richard Eric
Lutsch, Edward F
Mackal, Roy Paul
McQuistion, Thomas Evin
Malek, Richard Barry
Marx, Hymen
Mehta, Rajendra G
Meseth, Earl Herbert
Moll, Edward Owen
Morgan, Juliet
Munyer, Edward Arnold
Myer, Donal Gene
Nadler, Charles Fenger
Naples, Virginia L
Nielsen, Peter James
Page, Lawrence Merle
Peterson, Darryl Ronnie
Phillips, Carleton Jaffrey

Phillips, Christopher Alan
Pruett-Jones, Stephen Glen
Rabb, George Bernard
Rabinowitch, Victor
Ratzlaff, Kermit O
Reed, Charles Allen
Ridgeway, Bill Tom
Robertson, Hugh Mereth
Rosenthal, Gerson Max, Jr
Roth, Allan Charles
Sanderson, Glen Charles
Sather, J Henry
Shepherd, Benjamin Arthur
Shomay, David
Sprugel, George, Jr
Stains, Howard James
Thiruvathukal, Kris V
Thomerson, Jamie E
Thompson, Charles Frederick
Thonar, Eugene Jean-Marie
Throckmorton, Lynn Hiram
Thurow, Gordon Ray
Traylor, Melvin Alvah, Jr
Troll, Ralph
Uzzell, Thomas
Vetter, Richard L
Voight, Janet Ruth
Von Zellen, Bruce Walfred
Voris, Harold K
Waltenbaugh, Carl
Waring, George Houstoun, IV
Warnock, John Edward
Weigel, Robert David
Whelan, Christopher John
Whitt, Dixie Dailey
Williams, Thomas Alan
Wittig, Gertraude Christa
Wolff, Robert John
Zalisko, Edward John

INDIANA
Amlaner, Charles Joseph, Jr
Brodman, Robert David
Burkholder, Timothy Jay
Cable, Raymond Millard
Cooper, William Edgar, Jr
Corrigan, John Joseph
Crowell, (Prince) Sears, Jr
Current, William L
Durflinger, Elizabeth Ward
Dustman, John Henry
Eberly, William Robert
Ferris, John Mason
Ferris, Virginia Rogers
Frey, David Grover
Garton, David Wendell
Gidda, Jaswant Singh
Goetz, Frederick William, Jr
Harmon, Bud Gene
Harrington, Rodney B
Hopp, William Beecher
Iverson, John Burton
Jacobs, Merle Emmor
Johnson, Willis Hugh
Jones, Duvall Albert
Judge, Max David
Kelley, George W, Jr
Koritnik, Donald Raymond
Krekeler, Carl Herman
Lamberti, Gary Anthony
List, James Carl
Lodge, David Michael
Lucas, Jeffrey Robert
McCafferty, William Patrick
Maloney, Michael Stephen
Mays, Charles Edwin
Meyer, Frederick Richard
Minton, Sherman Anthony
Nelson, Craig Eugene
Olson, John Bennet
Ott, Karen Jacobs
Pachut, Joseph F(rancis), Jr
Perrill, Stephen Arthur
Platt, Thomas Reid
Preer, John Randolph, Jr
Roales, Robert R
Rowland, William Joseph
Santerre, Robert Frank
Sever, David Michael
Singleton, Wayne Louis
Stob, Martin
Tamar, Henry
Tweedell, Kenyon Stanley
Webster, Jackson Dan
Werth, Robert Joseph
Williams, Eliot Churchill
Wise, Charles Davidson
Zinsmeister, William John

IOWA
Ackerman, Ralph Austin
Adams, Donald Robert
Beachy, Christopher King
Beams, Harold William
Bowles, John Bedell
Buttrey, Benton Wilson
Cherian, Sebastian K
Ciochon, Russell Lynn
Cook, Kenneth Marlin
Dawson, David Lynn
Dolphin, Warren Dean
Dunham, Jewett
Hicks, Ellis Arden
Jahn, J Russell

Kessel, Richard Glen
Klaas, Erwin Eugene
Krafsur, Elliot Scoville
Laffoon, John
Lewis, Charles J
Lewis, Robert Earl
Lyon, David Louis
Menzel, Bruce Willard
Mertins, James Walter
Norton, Don Carlos
Parrish, Frederick Charles, Jr
Reitan, Phillip Jennings
Robson, Richard Morris
Rogers, Frances Arlene
Rogers, Thomas Edwin
Rundell, Harold Lee
Solursh, Michael
Stevermer, Emmett J
Sullivan, Charles Henry
Ulrich, Merwyn Gene
Welshons, William John
Wilson, Nixon Albert
Zmolek, William G

KANSAS
Allen, Deloran Matthew
Ashe, James S
Besharse, Joseph Culp
Boyd, Roger Lee
Boyer, Don Raymond
Burkholder, John Henry
Burton, Paul Ray
Choate, Jerry Ronald
Cink, Calvin Lee
Clark, George Richmond, II
Clarke, Robert Francis
Cross, Frank Bernard
Cutler, Bruce
Dehner, Eugene William
Duellman, William Edward
Ely, Charles Adelbert
Fautin, Daphne Gail
Fleharty, Eugene
Franzen, Dorothea Susanna
Good, Don L
Greenfield, Michael Dennis
Hays, Horace Albennie
Humphrey, Philip Strong
Jenkinson, Marion Anne
Johnson, John Christopher, Jr
Johnston, Richard Fourness
Kaufman, Donald Wayne
Klemm, Robert David
Leavitt, Wendell William
McCrone, John David
Michener, Charles Duncan
Morrissey, J Edward
Nelson, Victor Eugene
Neufeld, Gaylen Jay
Orr, James Anthony
Orr, Orty Edwin
Owen, Bernard Lawton
Platt, Dwight Rich
Prophet, Carl Wright
Reichman, Omer James
Robins, Charles Richard
Stockhammer, Karl Adolf
Thompson, Max Clyde
Upton, Steve Jay
Walker, Richard Francis
Westfall, Jane Anne
Wiley, E(dward) O(rlando), III

KENTUCKY
Barbour, Roger William
Birge, Wesley Joe
Branson, Branley Allan
Bryant, William Stanley
Budde, Mary Laurence
Cole, Evelyn
Crawford, Eugene Carson, Jr
Cupp, Paul Vernon, Jr
Davis, Wayne Harry
Derrickson, Charles M
Ely, Donald Gene
Ferner, John William
Ferrell, Blaine Richard
Hamon, J Hill
Hilton, Frederick Kelker
Howell, Jerry Fonce, Jr
Hoyt, Robert Dan
Just, John Josef
Kuehne, Robert Andrew
Lindsay, Dwight Marsee
Monroe, Burt Leavelle, Jr
Musacchia, X J
Oetinger, David Frederick
Pearson, William Dean
Prins, Rudolph
Rawls, John Marvin, Jr
Shadowen, Herbert Edwin
Sih, Andrew
Spoor, William Arthur
Stephens, Noel, Jr
Tamburro, Kathleen O'Connell
Thorp, James Harrison, III
Tucker, Ray Edwin
Uglem, Gary Lee
Webster, Carl David
Westneat, David French, Jr
Whiteker, McElwyn D
Wilder, Cleo Duke
Williams, John C

LOUISIANA
Avent, Robert M
Bamforth, Stuart Shoosmith
Bauer, Raymond Thomas
Black, Joe Bernard
Boertje, Stanley
Bryan, Charles F
Christian, Frederick Ade
Collins, Richard Lapointe
Cordes, Carroll Lloyd
Davis, Billy J
Dessauer, Herbert Clay
Douglas, Neil Harrison
Dundee, Harold A
Etheridge, Albert Louis
Eyster, Marshall Blackwell
Fairbanks, Laurence Dee
Felder, Darryl Lambert
Fleeger, John Wayne
Gassie, Edward William
Goertz, John William
Green, Jeffrey David
Grodner, Robert Maynard
Hardberger, Florian Max
Hardy, Laurence McNeil
Harman, Walter James
Harrison, Richard Miller
Hastings, Robert Wayne
Heins, David Carl
Herke, William Herbert
Homberger, Dominique Gabrielle
Hughes, Janice S
Kee, David Thomas
Khalaf, Kamel T
Kinn, Donald Norman
Kreider, Jack Leon
Lane, James Dale
Lanners, H Norbert
Loden, Michael Simpson
Meier, Albert Henry
Nair, Pankajam K
Norris, William Warren
Platt, William Joshua, III
Poirrier, Michael Anthony
Remsen, James Vanderbeek, Jr
Ricks, Beverly Lee
Rossman, Douglas Athon
Seigel, Richard Allyn
Shaw, Richard Francis
Shepherd, David Preston
Silverman, Harold
Smalley, Alfred Evans
Stewart, T Bonner
Suttkus, Royal Dallas
Toscano, William Agostino, Jr
Tso, Patrick Po-Wing
Turner, Hugh Michael
Wakeman, John Marshall
Williams, Kenneth L

MAINE
Allen, Kenneth William
Bell, Allen L
Borei, Hans Georg
Butler, Ronald George
Dean, David
Dearborn, John Holmes
De Witt, Hugh Hamilton
Eckelbarger, Kevin Jay
Francq, Edward Nathaniel Lloyd
Gainey, Louis Franklin, Jr
Glanz, William Edward
Greenwood, Paul Gene
Huntington, Charles Ellsworth
Larsen, Peter Foster
McCleave, James David
Martin, Robert Lawrence
Mazurkiewicz, Michael
Palmer, Ralph Simon
Parker, James Willard
Tyler, Seth
Watling, Les
Yuhas, Joseph George

MARYLAND
Adams, Lowell William
Ahl, Alwynelle S
Albright, Joseph Finley
Alleva, Frederic Remo
Altland, Paul Daniel
Altman, Philip Lawrence
Anderson, Lucy Macdonald
Ball, Gregory Francis
Barbour, Michael Thomas
Barnett, Audrey
Barry, Ronald Everett, Jr
Bass, Eugene L
Beyer, W Nelson
Buck, John Bonner
Burton, Dennis Thorpe
Cammen, Leon Matthew
Cargo, David Garrett
Carney, William Patrick
Chace, Fenner Albert, Jr
Chen, T R
Clark, Eugenie
Coulombe, Harry N
Creighton, Phillip David
Cronin, Thomas Wells
Dahl, Carol A
Dalton, John Charles
Delahunty, George
Dwyer, Dennis Michael
Eckardt, Michael Jon

Elwood, Jerry William
Endo, Burton Yoshiaki
Erickson, Howard Ralph
Fayer, Ronald
Forester, Donald Charles
Freed, Arthur Nelson
Futcher, Anthony Graham
Gillespie, Walter Lee
Golden, Alva Morgan
Goldsmith, Paul Kenneth
Graham, Charles Raymond, Jr
Green, Marie Roder
Hairstone, Marcus A
Hamilton, Clara Eddy
Heath, Martha Ellen
Highton, Richard
Hines, Anson Hemingway
Hoberg, Eric Paul
Hocutt, Charles H
Hoskin, George Perry
Hotton, Nicholas, III
Howe, Marshall Atherton
Huheey, James Edward
Kayar, Susan Rennie
Khan, Hameed A
Kirby, Paul Edward
Kloetzel, John Arthur
Koval, Thomas Michael
Lall, Abner Bishamber
Landauer, Michael Robert
Leach, Berton Joe
Leedy, Daniel Loney
Lewis, Herman William
Linder, Harris Joseph
McAtee, Lloyd Thomas
McVey, James Paul
Marsden, Halsey M
Maslow, David E
Merlino, Glenn T
Messersmith, Donald Howard
Meszler, Richard M
Michelson, Edward Harlan
Miller, Dorothea Starbuck
Munson, Donald Albert
Muul, Illar
Nelson, Jay Arlen
Nichols, James Dale
Nickle, William R
Oberdorfer, Michael Douglas
O'Grady, Richard Terence
Osman, Richard William
Palmer, Timothy Trow
Pancella, John Raymond
Park, Lee Crandall
Petty, William Clayton
Phillips, William George
Piavis, George Walter
Pierson, Bernice Frances
Pilitt, Patricia Ann
Platz, Robert Dale
Popper, Arthur N
Potter, Jane Huntington
Powers, Kendall Gardner
Provenza, Dominic Vincent
Reddy, Gunda
Remondini, David Joseph
Rice, Clifford Paul
Robbins, Chandler Seymour
Roesijadi, Guritno
Rose, Kenneth David
Schiff, Nathan Mark
Scully, Erik Paul
Shafer, W Sue
Simons, Daniel J
Small, Eugene Beach
Smith, F Harrell
Song, Jiakun
Sparling, Donald Wesley, Jr
Steinmetz, Michael Anthony
Stewart, Doris Mae
Stoecker, Diane Kastelowitz
Stringfellow, Frank
Swinebroad, Jeff
Tarshis, Irvin Barry
Thorington, Richard Wainwright, Jr
Trauger, David Lee
Vener, Kirt J
Weisberg, Stephen Barry
Wildt, David Edwin
Wiley, Martin Lee
Woods, Lewis Curry, III
Yoder, Wayne Alva

MASSACHUSETTS
Ahlberg, Henry David
Anderson-Olivo, Margaret
Atkinson-Templeman, Kristine Hofgren
Barber, Saul Benjamin
Bond, George Walter
Booke, Henry Edward
Borton, Anthony
Boss, Kenneth Jay
Bowdan, Elizabeth Segal
Bridges, Robert Stafford
Burggren, Warren William
Butman, Cheryl Ann
Capuzzo, Judith M
Castro, Gonzalo
Chlapowski, Francis Joseph
Clemens, Daniel Theodore
Cornman, Ivor
Dasgupta, Arijit M
Deegan, Linda Ann
Duhamel, Raymond C

Zoology (cont)

Duncan, Stewart
Eschenberg, Kathryn (Marcella)
Ezzell, Robert Marvin
Foster, William Burnham
Furth, David George
Gibbons, Michael Francis, Jr
Grant, William Chase, Jr
Griffin, Donald R(edfield)
Gustafson, Alvar Walter
Harbison, G Richard
Healy, William Ryder
Hichar, Joseph Kenneth
Hillman, Robert Edward
Hoff, James Gaven
Honigberg, Bronislaw Mark
Horner, B Elizabeth
Jahoda, John C
Jearld, Ambrose, Jr
Jones, Gwilym Strong
Klingener, David John
Kroodsma, Donald Eugene
Kunz, Thomas Henry
Laprade, Mary Hodge
Liem, Karel F
Lovejoy, David Arnold
McCormick, Stephen Daniel
McLaren, J Philip
Millard, William James
Moffett, Mark William
Moir, Ronald Brown, Jr
Morse, M Patricia
Muckenthaler, Florian August
Mulcare, Donald J
Naqvi, S Rehan Hasan
Paracer, Surindar Mohan
Payne, Bertram R
Paynter, Raymond Andrew, Jr
Pearce, John Bodell
Pearincott, Joseph V
Pechenik, Jan A
Perry, Alfred Eugene
Prescott, John Hernage
Prestwich, Kenneth Neal
Pyle, Robert Wendell
Rafferty, Nancy S
Rauch, Harold
Riser, Nathan Wendell
Robinson, William Edward
Sanders, Howard L(awrence)
Sargent, Theodore David
Scheltema, Amelie Hains
Smith, Frederick Edward
Smith, Wendy Anne
Snow, Beatrice Lee
Snyder, Dana Paul
Telford, Sam Rountree, III
Tilley, Stephen George
Trimmer, Barry Andrew
Trinkaus-Randall, Vickery E
Turner, Jefferson Taylor
Turner, Ruth Dixon
Vankin, George Lawrence
Waters, Joseph Hemenway
Werntz, Henry Oscar
Wichterman, Ralph
Widmayer, Dorothea Jane
Williams, Ernest Edward
Williamson, Peter George
Wright, Mary Lou
Wyse, Gordon Arthur
Zinn, Donald Joseph
Zottoli, Robert

MICHIGAN
Aggarwal, Surinder K
Alexander, Richard Dale
Atkinson, James William
Balaban, Martin
Blankespoor, Harvey Dale
Bowen, Stephen Hartman
Bowker, Richard George
Brady, Allen Roy
Brewer, Richard (Dean)
Buhl, Allen Edwin
Burch, John Bayard
Butsch, Robert Stearns
Caswell, Herbert Hall, Jr
Cather, James Newton
Chobotar, Bill
Conder, George Anthony
Cook, David Russell
Cooper, William E
Courtney, Gladys (Atkins)
Dillery, Dean George
Distler, Jack, Jr
Edgar, Arlan Lee
Eichler, Victor B
Engemann, Joseph George
Eyer, Lester Emery
Fennel, William Edward
Fink, William Lee
Fleming, Richard Cornwell
Frantz, William Lawrence
Gangwere, Stanley Kenneth
Gans, Carl
Gillingham, James Clark
Grudzien, Thaddeus Arthur, Jr
Hill, Richard William
Hill, Susan Douglas
Hooper, Frank Fincher
Horowitz, Samuel Boris
Houts, Larry Lee
Hudson, Roy Davage
Kemp, Norman Everett
Kerfoot, Wilson Charles
King, John Arthur
Kluge, Arnold Girard
Kurta, Allen
Lamberts, Austin E
Lauff, George Howard
Lehman, Grace Church
Lehnert, James Patrick
Lenski, Richard E
Lew, Gloria Maria
Lunk, William Allan
McCrimmon, Donald Alan, Jr
Martin, Joseph Patrick, Jr
Merrill, Dorothy
Miller, Brinton Marshall
Miller, Robert Rush
Myers, Philip
Nussbaum, Ronald Archie
Pavgi, Sushama
Payne, Robert B
Pelzer, Charles Francis
Person, Steven John
Peters, Lewis Ernest
Porter, Calvin Anthon
Porter, Thomas Wayne
Ritland, Richard Martin
Rivera, Evelyn Margaret
Rusch, Wilbert H, Sr
Sacco, Anthony G
Saltiel, Alan Robert
Scarborough, Charles Spurgeon
Smith, Gerald Ray
Smith, P(aul) Dennis
Smith, R Jay
Speare, Edward Phelps
Stehr, Frederick William
Storer, Robert Winthrop
Tenbroek, Bernard John
Thompson, William Lay
Thoresen, Asa Clifford
Vatsis, Kostas Petros
Waffle, Elizabeth Lenora
Walker, Glenn Kenneth
Weiss, Mark Lawrence
Werner, Earl Edward
Whiteman, Eldon Eugene
Winterstein, Scott Richard

MINNESOTA
Arthaud, Raymond Louis
Arthur, John W
Baker, Shirley Marie
Ballard, Neil Brian
Barnwell, Franklin Hershel
Barrett, James Martin
Bartel, Monroe H
Birney, Elmer Clea
Breckenridge, Walter John
Collins, Hollie L
Corbin, Kendall Wallace
Dapkus, David Conrad
Dearden, Douglas Morey
Downing, William Lawrence
Dropkin, Victor Harry
Erickson, James Eldred
Foose, Thomas John
Ford, Norman Lee
Frederick, Edward C
Frydendall, Merrill J
Gilbertson, Donald Edmund
Goble, Frans Cleon
Grewe, Alfred H, Jr
Hanson, Lester Eugene
Hazard, Evan Brandao
Heist, Herbert Ernest
Herman, William S
Heuschele, Ann
Hofslund, Pershing Benard
Hoppe, David Matthew
Jannett, Frederick Joseph, Jr
Johnson, Ivan M
Johnson, Vincent Arnold
Lukens, Paul W, Jr
McCann, Lester J
Mech, Lucyan David
Murdock, Gordon Robert
Noetzel, David Martin
Noland, Wayland Evan
Olson, Magnus
Preiss, Frederick John
Pusey, Anne E
Rathbun, William B
Saccoman, Frank (Michael)
Shoger, Ross L
Sinha, Akhouri Achyutanand
Sulerud, Ralph L
Swift, Michael Crane
Tordoff, Harrison Bruce
Underhill, James Campbell
Wagenbach, Gary Edward
Warner, Dwain Willard
Weibust, Robert Smith
Williams, Steven Frank
Wyatt, Ellis Junior
Zischke, James Albert

MISSISSIPPI
Anderson, Gary
Boyd, Leroy Houston
Boykins, Ernest Aloysius, Jr
Cliburn, Joseph William
Cooper, Charles Morris
Goddard, Jerome
Gunter, Gordon
Helms, Thomas Joseph
Ikenberry, Roy Dewayne
Jackson, Jerome Alan
Keiser, Edmund Davis, Jr
Longest, William Douglas
Ramaswamy, Sonny B
Randolph, Kenneth Norris
Ross, Stephen Thomas
Spann, Charles Henry
Sutton, William Wallace
Walley, Willis Wayne

MISSOURI
Aldridge, Robert David
Belshe, John Francis
Blaine, Edward Homer
Bolla, Robert Irving
Breitenbach, Robert Peter
Buchanan, Bryant W
Calabrese, Diane M
Cheverud, James Michael
Clare, Stewart
Coles, Richard Warren
Cook, Nathan Howard
Davis, Jerry Collins
Elliott, Dana Ray
Fickess, Douglas Ricardo
Gerdes, Charles Frederick
Gerhardt, H Carl, Jr
Gersbacher, Willard Marion
Gill, James Wallace
Goodge, William Russell
Hawksley, Oscar
Hazelwood, Donald Hill
Ignoffo, Carlo Michael
Ingersol, Robert Harding
Larson, Allan
Losos, Jonathan B
Luecke, Richard H
McCollum, Clifford Glenn
Mathis, Sharon Alicia
Metter, Dean Edward
Miles, Charles David
Mitchell, Henry Andrew
Mock, Orin Bailey
Momberg, Harold Leslie
Myers, Richard F
Pflieger, William Leo
Rasmussen, David Tab
Sage, Martin
Schreiweis, Donald Otto
Sexton, Owen James
Sharp, John Roland
Smith, Nathan Elbert
Smith, Richard Jay
Sorenson, Marion W
Spiers, Donald Ellis
Taber, Charles Alec
Tannenbaum, Michael Glen
Train, Carl T
Twente, John W
Williams, Henry Warrington
Williamson, James Lawrence
Wilson, Stephen W

MONTANA
Bakken, Arnold
Ball, Irvin Joseph
Brunson, Royal Bruce
Foresman, Kerry Ryan
Harrington, Joseph D
Johnson, Oscar Walter
Kaya, Calvin Masayuki
Kendall, Katherine Clement
Kilgore, Delbert Lyle, Jr
Mitchell, Lawrence Gustave
O'Gara, Bart W
Pfeiffer, Egbert Wheeler
Picton, Harold D
Sheldon, Andrew Lee
Tibbs, John Francisco
Weisel, George Ferdinand, Jr
Werner, John Kirwin
Wright, Philip Lincoln

NEBRASKA
Adams, Charles Henry
Ballinger, Royce Eugene
Bliese, John C W
Boernke, William E
Calvert, Jay Gregory
Dalley, Arthur Frederick, II
Fritschen, Robert David
Geluso, Kenneth Nicholas
Genoways, Hugh Howard
Gunderson, Harvey Lorraine
Hergenrader, Gary Lee
Janovy, John, Jr
Johnsgard, Paul Austin
Jones, John Ackland
Kutler, Benton
Lynch, John Douglas
McGaugh, John Wesley
Peters, Edward James
Pritchard, Mary (Louise) Hanson
Sharpe, Roger Stanley

NEVADA
Benes, Elinor Simson
Douglas, Charles Leigh
Garner, Duane LeRoy
Mahadeva, Madhu Narayan
O'Farrell, Michael John
O'Farrell, Thomas Paul
Parmelee, David Freeland
Ryser, Fred A, Jr
Winokur, Robert Michael

NEW HAMPSHIRE
Bergman, Kenneth David
Borror, Arthur Charles
Dingman, Jane Van Zandt
Eggleston, Patrick Myron
Gilbert, John Jouett
Harrises, Antonio Efthemios
Lavoie, Marcel Elphege
Sasner, John Joseph, Jr
Spencer, Larry T
Stettenheim, Peter
Strout, Richard Goold
Tyson, Greta E
Walker, Warren Franklin, Jr
Wu, Lin
Wurster-Hill, Doris Hadley

NEW JERSEY
Agnish, Narsingh Dev
Ahn, Ho-Sam
Beer, Colin Gordon
Brattsten, Lena B
Brubaker, Paul Eugene, II
Campbell, William Cecil
Chizinsky, Walter
Collier, Marjorie McCann
Cook, Marie Mildred
Crossner, Kenneth Alan
Feldman, Susan C
Ford, Daniel Morgan
Gardiner, Lion Frederick
Gaumer, Albert Edwin Hellick
Gittleson, Stephen Mark
Grant, Peter Raymond
Grassle, Judith Payne
Griffo, James Vincent, Jr
Grinberg-Funes, Ricardo A
Gupta, Godaveri Rawat
Hall, James Conrad
Halpern, Myron Herbert
Hu, Yaping
Isquith, Irwin R
Jenkins, William Robert
Johnson, Leslie Kilham
Jolly, Clifford J
Joslyn, Dennis Joseph
Kantor, Sidney
Kent, George Cantine, Jr
Kiley, Charles Walter
Koepp, Stephen John
Leck, Charles Frederick
Levine, Donald Martin
Lutz, Richard Arthur
McGhee, George Rufus, Jr
Martin, Robert Allen
Massa, Tobias
Petriello, Richard P
Pollock, Leland Wells
Power, Harry W, III
Sacks, Martin
Saiff, Edward Ira
Sellmer, George Park
Stearns, Donald Edison
Van Gelder, Richard George
Virkar, Raghunath Atmaram
Wallace, Edith Winchell
Weiss, Mitchell Joseph
Wilhoft, Daniel C
Willis, Jacalyn Giacalone

NEW MEXICO
Brown, James Hemphill
Conant, Roger
Cooch, Frederick Graham
Corliss, John Ozro
Crissman, Harry Allen
Degenhardt, William George
Duncan, Irma W
Findley, James Smith
Hayward, Bruce Jolliffe
Herman, Ceil Ann
Hubbard, John Patrick
Jones, Kirkland Lee
La Pointe, Joseph L
Ligon, James David
Martignoni, Mauro Emilio
Raitt, Ralph James, Jr
Richman, David Bruce
Stallcup, William Blackburn, Jr
Stricker, Stephen Alexander
Thaeler, Charles Schropp, Jr
Whitmore, Mary (Elizabeth) Rowe
Wolberg, Donald Lester
Zimmerman, Dale A

NEW YORK
Abbatiello, Michael James
Adler, Kraig (Kerr)
Alexander, A Allan
Anderson, Sydney
Andrle, Robert Francis
Ayres, Jose Marcio Correa
Baker-Cohen, Katherine France
Battin, William T
Beason, Robert Curtis
Behrens, Mildred Esther
Bell, Michael Allen
Bell, Robin Graham

Benton, Allen Haydon
Bernardis, Lee L
Blackler, Antonie W C
Blake, Robert Wesley
Bock, Walter Joseph
Borowsky, Richard Lewis
Bradbury, Michael Wayne
Brannigan, David
Breed, Helen Illick
Brett, Betty Lou Hilton
Brett, Carlton Elliot
Britten, Bryan Terrence
Broseghini, Albert L
Brothers, Edward Bruce
Brown, Patricia Stocking
Brown, Stephen Clawson
Bruning, Donald Francis
Chamberlain, James Luther
Clark, Ralph M
Conover, David Olmstead
Crocker, Denton Winslow
Crowell, Robert Merrill
Cutler, Edward Bayler
Daniels, Robert Artie
De Gennaro, Louis D
Delson, Eric
Deubler, Earl Edward, Jr
Dowling, Herndon Glenn, (Jr)
Dudley, Patricia
Easton, Douglas P
Eaton, Stephen Woodman
Eisner, Thomas
Elrod, Joseph Harrison
Emerson, William Keith
Engbretson, Gustav Alan
Factor, Jan Robert
Fetcho, Joseph Robert
Finks, Robert Melvin
Finlay, Peter Stevenson
Fowler, James A
Frankel, Arthur Irving
French, Alan Raymond
Geisler, Grace Sr
Gering, Robert Lee
Goodwin, Robert Earl
Graham, William Joseph
Greenhall, Arthur Merwin
Greenlaw, Jon Stanley
Grove, Patricia A
Haefner, Paul Aloysius, Jr
Hainsworth, Fenwick Reed
Hardy, Matthew Phillip
Haresign, Thomas
Harman, Willard Nelson
Harris, C(harles) Leon
Haugh, John Richard
Hecht, Max Knobler
Heidenthal, Gertrude Antoinette
Hertz, Paul Eric
Holtzman, Seymour
Horst, G Roy
Howland, Howard Chase
Hughes, Stephen Edward
Hutt, Frederick Bruce
Johansson, Toge (Tage) Sigvard Kjell
Johnson, Lawrence Lloyd
Kallen, Frank Clements
Kallman, Klaus D
Kanzler, Walter Wilhelm
Keen, William Hubert
Keithly, Janet Sue
Klein, Harold George
Kolmes, Steven Albert
Koopman, Karl Friedrich
Krause, David Wilfred
Krieg, David Charles
Krishna, Kumar
Lackey, James Alden
Landing, Ed
Landry, Stuart Omer, Jr
Lanyon, Wesley Edwin
LaRow, Edward J
Lasker, Howard Robert
Laychock, Suzanne Gale
Levin, Norman Lewis
Lipke, Peter Nathan
LoGerfo, John J
Lorch, Joan
Loretz, Christopher Alan
McCourt, Robert Perry
McCune, Amy Reed
McDowell, Sam Booker
McGraw, James Carmichael
MacIntyre, Giles T
McIntyre, Judith Watland
Madison, Dale Martin
Mantel, Linda Habas
Maple, William Thomas
Maxwell, George Ralph, II
Menon, Gopinathan K(unnariath)
Meyer, Axel
Miller, Sue Ann
Monaco, Lawrence Henry
Mueller, Justus Frederick
Muller-Schwarze, Dietland
Murphy, Michael Joseph
Myers, Charles William
Nelson, Sigurd Oscar, Jr
Norton, Roy Arnold
Nur, Uzi
Oaks, Emily Caywood Jordan
Organ, James Albert
Ortman, Robert A
Osterberg, Donald Mouretz

Payne, Harrison H
Peckarsky, Barbara Lynn
Pentyala, Srinivas Narasimha
Perlmutter, Alfred
Phillips-Quagliata, Julia Molyneux
Platnick, Norman I
Pokorny, Kathryn Stein
Pough, Richard Hooper
Rasweiler, John Jacob, IV
Rausch, James Peter
Reilly, Marguerite
Reisman, Howard Maurice
Rich, Abby M
Richmond, Milo Eugene
Rieder, Conly LeRoy
Rivest, Brian Roger
Roecker, Robert Maar
Rosi, David
Rough, Gaylord Earl
Rudzinska, Maria Anna
Ryan, Richard Alexander
Satir, Peter
Schaller, George Beals
Schneider, Kathryn Claire (Johnson)
Schreibman, Martin Paul
Schuster, Frederick Lee
Shields, William Michael
Short, Lester Le Roy
Siegfried, Clifford Anton
Skinner, Kathleen Mary
Slobodkin, Lawrence Basil
Smith, Clarence Lavett
Sohacki, Leonard Paul
Stewart, Kenton M
Stewart, Margaret McBride
Stouffer, James Ray
Stouter, Vincent Paul
Strayer, David Lowell
Summers, Robert Gentry, Jr
Swartz, Gordon Elmer
Swartzendruber, Donald Clair
Szabo, Piroska Ludwig
Tattersall, Ian
Thompson, Steven Risley
Thornborough, John Randle
Titus, John Elliott
Tobach, Ethel
VanDruff, Larry Wayne
Vasey, Carey Edward
Vuilleumier, Francois
Watanabe, Myrna Edelman
Wells, John West
Wemyss, Courtney Titus, Jr
Wenk, Eugene J
Werner, Robert George
Whiting, Anne Margaret
Williams, George Christopher
Windsor, Donald Arthur
Wolf, Larry Louis
Wood, Raymond Arthur
Wright, Margaret Ruth
Wysolmerski, Theresa
Zimmerman, Jay Alan

NORTH CAROLINA
Aldridge, David William
Allen, Charles Marvin
Ankel-Simons, Friderun Annursel
Bailey, Joseph Randle
Baker, Elizabeth McIntosh
Barrick, Elliott Roy
Biggs, Walter Clark, Jr
Boliek, Irene
Bookhout, Cazlyn Green
Bradbury, Phyllis Clarke
Brown, Richard Dean
Bruce, Richard Conrad
Butts, Jeffrey A
Colvard, Dean Wallace
Costlow, John DeForest
Coyle, Frederick Alexander
Darling, Marilyn Stagner
Derrick, Finnis Ray
Dimock, Ronald Vilroy, Jr
Edwards, James Wesley
Esch, Gerald Wisler
Feducia, John Alan
Fehon, Jack Harold
Freeman, John Alderman
Gilbert, Lawrence Irwin
Gordon, Christopher John
Govoni, John Jeffrey
Gregg, John Richard
Grosch, Daniel Swartwood
Guyselman, J(ohn) Bruce
Hairston, Nelson George
Hampton, Carolyn Hutchins
Harrison, Frederick Williams
Heatwole, Harold Franklin
Hedlund, Laurence William
Hendrickson, Herbert T
Higgins, Robert Price
Hylander, William Leroy
Jones, Claiborne Stribling
Kalmus, Gerhard Wolfgang
Klopfer, Peter Hubert
Komma, Donald Jerry
Lassiter, Charles Albert
Lay, Douglas M
Lee, Greta Marlene
Leise, Esther M
Lessler, Judith Thomasson
Lindquist, David Gregory
Link, Garnett William, Jr

Lutz, Paul E
Lynn, William Gardner
Lytle, Charles Franklin
McCrary, Anne Bowden
McDaniel, Susan Griffith
McDowell, Robert E, Jr
McLeod, Michael John
McMahan, Elizabeth Anne
Manooch, Charles Samuel, III
Meyer, John Richard
Miller, Grover Cleveland
Moore, Allen Murdoch
Moseley, Lynn Johnson
Mueller, Helmut Charles
Mueller, Nancy Schneider
O'Hara, Robert James
Parnell, James Franklin
Peterson, Charles Henry
Poulton, Bruce R
Powell, Roger Allen
Reice, Seth Robert
Reynolds, Joshua Paul
Rittschof, Daniel
Roberts, John Fredrick
Roer, Robert David
Schwartz, Frank Joseph
Sharer, Archibald Wilson
Simons, Elwyn L
Simpson, Everett Coy
Smith, Donald Eugene
Smith, Ned Allan
Spuller, Robert L
Staddon, John Eric Rayner
Steinhagen, William Herrick
Stewart, Paul Alva
Stroud, Richard Hamilton
Sulik, Kathleen Kay
Sullivan, James Bolling
Sutcliffe, William Humphrey, Jr
Thomas, Mary Beth
Trainer, John Ezra, Jr
Triantaphyllou, Hedwig Hirschmann
Vogel, Steven
Wainwright, Stephen Andrew
Ward, Robert T
Webster, William David
Williams, Ann Houston
Wolk, Robert George
Yarbrough, Charles Gerald
Yongue, William Henry

NORTH DAKOTA
Duerr, Frederick G
Erickson, Duane Otto
Gerst, Jeffery William
Grier, James William
Kannowski, Paul Bruno
Lieberman, Milton Eugene
Lokemoen, John Theodore
Riemann, John G
Seabloom, Robert W
Wrenn, William J

OHIO
Alicino, Nicholas J
Baker, Peter C
Banning, Jon Willroth
Barbour, Clyde D
Barnes, Herbert M
Barrow, James Howell, Jr
Beal, Kathleen Grabaskas
Berra, Tim Martin
Birky, Carl William, Jr
Brent, Morgan McKenzie
Burns, Robert David
Burtt, Edward Howland, Jr
Case, Denis Stephen
Claussen, Dennis Lee
Clise, Ronald Leo
Cooke, Helen Joan
Costanzo, Jon P
Crites, John Lee
Culver, David Alan
Dabrowski, Konrad
Daniel, Paul Mason
Delphia, John Maurice
Dexter, Ralph Warren
Dickerman, Richard Curtis
Elfner, Lynn Edward
Etges, Frank Joseph
Faber, Lee Edward
Ferguson, Marion Lee
Foreman, Darhl Lois
Fry, Anne Evans
Gates, Michael Andrew
Gatz, Arthur John, Jr
Gaunt, Abbot Stott
Gilloteaux, Jacques Jean-Marie A
Goldstein, David Louis
Gottschang, Jack Louis
Grassmick, Robert Alan
Greenstein, Julius S
Guttman, Sheldon
Hahnert, William Franklin
Hintz, Howard William
Hoagstrom, Carl William
Hoff, Kenneth Michael
Hubschman, Jerry Henry
Hummon, William Dale
Ingersoll, Edwin Marvin
Isler, Gene A
Jackson, Dale Latham
Keiser, Terry Dean
Kern, Michael Don

Klemm, Donald J
Kopchick, John J
Lane, Roger Lee
McLean, Edward Bruce
Madenjian, Charles Paul
Martin, Scott McClung
Mayfield, Harold Ford
Miller, John Wesley, Jr
Moore, Nelson Jay
Moore, Randy
Myser, Willard C
Orcutt, Frederic Scott, Jr
Palmer, Brent David
Pannabecker, Richard Floyd
Parrish, Wayne
Patton, Wendell Keeler
Peterjohn, Glenn William
Peterle, Tony J
Pettegrew, Raleigh K
Pilati, Charles Francis
Plimpton, Rodney F, Jr
Przybylski, Ronald J
Putnam, Loren Smith
Rab, Paul Alexis
Rothenbuhler, Walter Christopher
Rovner, Jerome Sylvan
Rubin, David Charles
St John, Fraze Lee
Saksena, Vishnu P
Scott, John Paul
Shaffer, Jacquelin Bruning
Sherman, Robert George
Shetlar, David John
Stansbery, David Honor
Stein, Carol B
Stein, Roy Allen
Stoffer, Richard Lawrence
Tandler, Bernard
Taylor, Douglas Hiram
Thomas, Cecil Wayne
Trautman, Milton Bernhard
Uetz, George William
Valentine, Barry Dean
Vaughn, Charles Melvin
Venard, Carl Ernest
Weis, Dale Stern
White, Andrew Michael
Wilson, Richard Ferrin
Wood, Jackie Dale
Wood, Timothy Smedley

OKLAHOMA
Abram, James Baker, Jr
Baird, Troy Alan
Black, Jeffrey Howard
Brown, Harley Procter
Brown, Marie Jenkins
Carpenter, Charles Congden
Carter, William Alfred
Chapman, Brian Richard
Clemens, Howard Paul
Cox, Beverley Lenore
Fox, Stanley Forrest
Glass, Bryan Pettigrew
Haines, Howard Bodley
Hellack, Jenna Jo
Hill, Loren Gilbert
Hutchison, Victor Hobbs
Korstad, John Edward
Lindsay, Hague Leland, Jr
Mares, Michael Allen
Matthews, William John
Miller, Helen Carter
Miller, Rudolph J
Nelson, John Marvin, Jr
Ownby, Charlotte Ledbetter
Radke, William John
Russell, Charles Clayton
Schnell, Gary Dean
Seto, Frank
Shorter, Daniel Albert
Sonleitner, Frank Joseph
Talent, Larry Gene
Thornton, John William
Tyler, Jack D
Vestal, Bedford Mather
Vitt, Laurie Joseph
Walters, Lowell Eugene
Young, Sharon Clairene

OREGON
Anderson, Amy Elin
Beatty, Joseph John
Bethea, Cynthia Louise
Bierzychudek, Paulette F
Boddy, Dennis Warren
Bond, Carl Eldon
Callahan, Clarence Arthur
Carter, Richard Thomas
Cheeke, Peter Robert
Cimberg, Robert Lawrence
Coffey, Marvin Dale
Crawford, John Arthur
Cross, Stephen P
Doudoroff, Peter
Fahrenbach, Wolf Henrich
Field, Katharine G
Forbes, Richard Bryan
Gonor, Jefferson John
Gunberg, David Leo
Gwilliam, Gilbert Franklin
Henny, Charles Joseph
Hillman, Stanley Severin
Hisaw, Frederick Lee, Jr

Zoology (cont)

Hixon, Mark A
Hughes, Robert Mason
Jensen, Harold James
Lubchenco, Jane
Markle, Douglas Frank
Maser, Chris
Mix, Michael Cary
Neiland, Kenneth Alfred
Olson, Robert Eldon
Osgood, David William
Owczarzak, Alfred
Quast, Jay Charles
Radovsky, Frank Jay
Roberts, Michael Foster
Simpson, Leonard
Spencer, Peter Simner
Storm, Robert MacLeod
Taylor, Jocelyn Mary
Thorson, Thomas Bertel
Walters, Roland Dick
Warren, Charles Edward
Weaver, Morris Eugene
Wirtz, John Harold

PENNSYLVANIA
Alperin, Richard Junius
Anderson, David Robert
Barnes, Robert Drane
Barnett, Leland Bruce
Beach, Neil William
Bell, Edwin Lewis, II
Bellis, Edward David
Bergstresser, Kenneth A
Billingsley, Melvin Lee
Buckelew, Thomas Paul
Bursey, Charles Robert
Carpenter, Esther
Catalano, Raymond Anthony
Chase, Robert Silmon, Jr
Clark, Richard James
Cole, James Edward
Cundall, David Langdon
Cunningham, Harry N, Jr
Dalby, Peter Lenn
Davis, George Morgan
Debouck, Christine Marie
Denoncourt, Robert Francis
Dollahon, Norman Richard
Dunson, William Albert
Eggert, Robert Glenn
Fabian, Michael William
Farber, Phillip Andrew
Feuer, Robert Charles
Fish, Frank Eliot
Foor, W Eugene
Franz, Craig Joseph
Fredrickson, Richard William
Gallati, Walter William
Gibley, Charles W, Jr
Gill, Frank Bennington
Glazier, Douglas Stewart
Grosvenor, Clark Edward
Guida, Vincent George
Halbrendt, John Marthon
Hall, John Sylvester
Harclerode, Jack E
Hayes, Wilbur Frank
Henderson, Alex
Hendrix, Sherman Samuel
Hill, Frederick Conrad
Holliday, Charles Walter
Hulse, Arthur Charles
Jacobs, George Joseph
Jeffries, William Bowman
Joos, Barbara
Kamon, Eliezer
Karlson, Eskil Leannart
Kesner, Michael H
Kimmel, William Griffiths
Kirkland, Gordon Laidlaw, Jr
Klinger, Thomas Scott
Krishtalka, Leonard
Kwiecinski, Gary George
Legge, Thomas Nelson
Linzey, Alicia Vogt
McCoy, Clarence John, Jr
McDermott, John Joseph
McKinstry, Donald Michael
Mann, Alan Eugene
Meyer, R Peter
Miller, Richard Lee
Neff, William H
O'Connor, John Dennis
Ogren, Robert Edward
Ostrovsky, David Saul
Otte, Daniel
Parkes, Kenneth Carroll
Pearson, David Leander
Pitkin, Ruthanne B
Porter, Keith Roberts
Prezant, Robert Steven
Raikow, Robert Jay
Reif, Charles Braddock
Robertson, Robert
Rockwell, Kenneth H
Rogers, William Edwin
Rosenberg, Gary
Schlitter, Duane A
Selander, Robert Keith
Sherritt, Grant Wilson
Shostak, Stanley
Siegel, Michael Ian

Sillman, Emmanuel I
Silverman, Jerald
Snyder, Donald Benjamin
Stauffer, Jay Richard, Jr
Streu, Herbert Thomas
Taylor, D Lansing
Thomas, Roger David Keen
Thompson, Roger Kevin Russell
Thomson, Keith Stewart
Towne, William F
Trainer, John Ezra, Sr
Turoczi, Lester J
Twiest, Gilbert Lee
Uricchio, William Andrew
Vomachka, Archie Joel
Webb, Glenn R
West, Keith P
Williams, Russell Raymond
Williams, Timothy C
Wilson, Lowell L
Winkelmann, John Roland
Yahner, Richard Howard
Zaccaria, Robert Anthony
Zenisek, Cyril James

RHODE ISLAND
Dolyak, Frank
Goslow, George E, Jr
Hammen, Carl Schlee
Harrison, Robert William
Heppner, Frank Henry
Howe, Robert Johnston
Hyland, Kerwin Ellsworth, Jr
Jeffries, Harry Perry
Lazell, James Draper
Miller, Don Curtis
Rand, David McNear
Wahle, Richard Andreas

SOUTH CAROLINA
Anderson, William Dewey, Jr
Biernbaum, Charles Knox
Bildstein, Keith Louis
Brisbin, I Lehr, Jr
Browdy, Craig Lawrence
Brusca, Richard Charles
Carew, James L
Forsythe, Dennis Martin
Fox, Richard Shirley
Gauthreaux, Sidney Anthony
Gibbons, J Whitfield
Guram, Malkiat Singh
Harrison, Julian R, III
Helms, Carl Wilbert
Jagoe, Charles Henry
Johnson, Albert Wayne
Johnson, Robert Karl
Kelly, Robert Withers
Labanick, George Michael
Lacher, Thomas Edward, Jr
Leonard, Walter Raymond
Lewis, Stephen Albert
Mulvey, Margaret
Newman, Michael Charles
Pivorun, Edward Broni
Roache, Lewie Calvin
Sandifer, Paul Alan
Schindler, James Edward
Scott, Jaunita Simons
Spooner, John D
Stewart, Shelton E
Swallow, Richard Louis
Teska, William Reinhold
Watabe, Norimitsu
Weeks, Stephen Charles
Wheeler, Alfred Portius

SOUTH DAKOTA
Dieter, Charles David
Dillon, Raymond Donald
Haertel, John David
Harrell, Byron Eugene
Martin, James Edward
Parke, Wesley Wilkin
Romans, John Richard
Smolik, James Darrell
Tatina, Robert Edward
Wahlstrom, Richard Carl
Zeigler, David Wayne

TENNESSEE
Alsop, Frederick Joseph, III
Amy, Robert Lewis
Auerbach, Stanley Irving
Benz, George William
Bernard, Ernest Charles
Brinkhurst, Ralph O
Brode, William Edward
Carlton, Robert Austin
Chance, Charles Jackson
Chandler, Clay Morris
Coutant, Charles Coe
Cushing, Bruce S
Dennison, Clifford C
Dumont, James Nicholas
Dyer, Melvin I
Echternacht, Arthur Charles
Eddlemon, Gerald Kirk
Elliott, Alice
Etnier, David Allen
Foreman, Charles William
Freeman, John Richardson
Greenberg, Neil
Gutzke, William H N

Harrison, Robert Edwin
Hochman, Benjamin
James, Ted Ralph
Kathman, R Deedee
Kennedy, Michael Lynn
Lawson, James Everett
Loar, James M
McCarthy, John F
McGavock, Walter Donald
McGhee, Charles Robert
Millemann, Raymond Eagan
Murphy, George Graham
Nelson, Diane Roddy
Nunnally, David Ambrose
Parchment, John Gerald
Parmalee, Paul Woodburn
Payne, James
Polis, Gary Allan
Richardson, Don Orland
Riechert, Susan Elise
Schultz, Terry Wayne
Scott, Mack Tommie
Snyder, David Hilton
Staub, Robert J
Vogel, Howard H, Jr
Voorhees, Larry Donald
Wallace, Gary Oren
Welch, Hugh Gordon
Wilhelm, Walter Eugene
Wilson, James Lester
Yeatman, Harry Clay

TEXAS
Adams, Clark Edward
Aggarwal, Bharat Bhushan
Allen, Archie C
Applegate, Richard Lee
Arnold, Keith Alan
Baker, James Haskell
Baker, Rollin Harold
Banerji, Tapan Kumar
Barth, Robert Hood, Jr
Bassett, James Wilbur
Bazer, Fuller Warren
Beasley, Clark Wayne
Bedinger, Charles Arthur, Jr
Belk, Gene Denton
Bickham, John W
Bischoff, Harry William
Bramblett, Claud Allen
Cameron, Ivan Lee
Campbell, Carlos Boyd Godfrey
Carpenter, Zerle Leon
Chaney, Allan Harold
Chrapliwy, Peter Stanley
Clark, Donald Ray, Jr
Clayton, Dale Leonard
Clench, Mary Heimerdinger
Cooper, William Anderson
Crandall, Keith Alan
Crawford, Isaac Lyle
Crews, David Pafford
Dalquest, Walter Woelber
Davis, Gordon Wayne
Dixon, James Ray
Dunbar, Bonnie Sue
Dunkle, Sidney Warren
Faulkner, Russell Conklin, Jr
Flury, Alvin Godfrey
Formanowicz, Daniel Robert, Jr
Garner, Herschel Whitaker
Glenn, William Grant
Greding, Edward J, Jr
Green, George G
Greenbaum, Ira Fred
Greenberg, Les Paul
Hanlon, Roger Thomas
Hannan, Herbert Herrick
Harris, Arthur Horne
Harry, Harold William
Higgins, Linden Elizabeth
Hillis, David Mark
Honeycutt, Rodney Lee
Hubbs, Clark
Hudson, Frank Alden
Huggins, Sara Espe
Ingold, Donald Alfred
Jacobson, Antone Gardner
Johnston, John Spencer
Jones, Clyde J
Jones, J Knox, Jr
Judd, Frank William
Kalthoff, Klaus Otto
Kammerer, Candace Marie
Kennedy, James H
Killebrew, Flavius Charles
Kirkpatrick, Mark Adams
Klaus, Ewald Fred, Jr
Kuntz, Robert Elroy
Lewis, Simon Andrew
Lieb, Carl Sears
Long, James Duncan
McAllister, Chris Thomas
McCarley, Wardlow Howard
McDonald, Harry Sawyer
McEachran, John D
McMahon, Robert Francis, III
Martin, Robert Frederick
Meacham, William Ross
Meade, Thomas Gerald
Mecham, John Stephen
Metcalf, Artie Lou
Montagna, Paul A
Murray, Harold Dixon

Neill, Robert Lee
Newman, George Allen
Nichols, James Ross
Owens, David William
Parker, Nick Charles
Parker, Robert Hallett
Pettingill, Olin Sewall, Jr
Pierce, Benjamin Allen
Pierce, Jack Robert
Pyburn, William F
Ramsey, Jed Junior
Robertson, Walter Volley
Robinson, Arin Forest
Roller, Herbert Alfred
Rose, Francis L
Rubink, William Louis
Rylander, Michael Kent
Samollow, Paul B
Sanders, Ottys E
Schlueter, Edgar Albert
Schmidly, David James
Schwalm, Fritz Ekkehardt
Scudday, James Franklin
Self, Hazzle Layfette
Shake, Roy Eugene
Sissom, Stanley Lewis
Smatresk, Neal Joseph
Smith, William H
Smith, William Russell
Steele, David Gentry
Strawn, Robert Kirk
Tamsitt, James Ray
Telfair, Raymond Clark, II
Thames, Walter Hendrix, Jr
Throckmorton, Gaylord Scott
Turco, Charles Paul
Tuttle, Merlin Devere
Vaughan, Mary Kathleen
Warner, Marlene Ryan
Wasserman, Aaron Osias
Webb, Robert G
Wellso, Stanley Gordon
Wheeler, Marshall Ralph
Wicksten, Mary Katherine
Wilson, William Thomas
Wohlschlag, Donald Eugene
Worthington, Richard Dane
Wursig, Bernd Gerhard
Yaden, Senka Long
Zimmerman, Earl Graves

UTAH
Aird, Steven Douglas
Bahler, Thomas Lee
Balph, Martha Hatch
Behle, William Harroun
Bissonette, John Alfred
Bosakowski, Thomas
Cuellar, Orlando
Dixon, Keith Lee
Evans, Frederick Read
Fitzgerald, Paul Ray
Frost, Herbert Hamilton
Gaufin, Arden Rupert
George, Sarah B(rewster)
Graff, Darrell Jay
Griffin, Gerald D
Hansen, Afton M
Harris, John Michael
Hulet, Clarence Veloid
Jeffery, Duane Eldro
Jensen, Emron Alfred
Johnston, Stephen Charles
Legler, John Marshall
Liddell, William David
MacMahon, James A
Maurer, Brian Alan
Messina, Frank James
Murphy, Joseph Robison
Negus, Norman Curtiss
Rickart, Eric Allan
Schoenwolf, Gary Charles
Sites, Jack Walter, Jr
Smith, Howard Duane
Smith, Nathan McKay
Van De Graaff, Kent Marshall
White, Clayton M

VERMONT
Bell, Ross Taylor
Brodie, Edmund Darrell, Jr
Chipman, Robert K
Detwyler, Robert
Hitchcock, Harold Bradford
Kilpatrick, Charles William
Schall, Joseph Julian
Stevens, Dean Finley

VIRGINIA
Adkisson, Curtis Samuel
Alden, Raymond W, III
Andrews, Jay Donald
Atkins, David Lynn
Barton, Alexander James
Birdsong, Ray Stuart
Blood, Benjamin Donald
Briggs, Jeffrey L
Brown, Luther Park
Burreson, Eugene M
Bush, Francis M
Byrd, Mitchell Agee
Carico, James Edwin
Chipley, Robert MacNeill
Corwin, Jeffrey Todd

Cranford, Jack Allen
Cruthers, Larry Randall
Davis, John Edward, Jr
Dent, James (Norman)
Diehl, Fred A
Doherty, John Douglas
Eckerlin, Ralph Peter
Eller, Arthur L, Jr
Emsley, Michael Gordon
Ernst, Carl Henry
Fenner-Crisp, Penelope Ann
Gaines, Gregory
Gourley, Eugene Vincent
Gray, Faith Harriet
Heath, Alan Gard
Hickman, Cleveland Pendleton, Jr
Hoffman, Richard Lawrence
Holsinger, John Robert
Holt, Perry Cecil
Jahn, Laurence R
Johnson, Raymond Earl
Johnson, Rose Mary
Jopson, Harry Gorgas Michener
Karowe, David Nathan
Keinath, John Allen
Kornegay, Ervin Thaddeus
Kornicker, Louis Sampson
McKitrick, Mary Caroline
McNabb, Roger Allen
Mangum, Charlotte P
Marcellini, Dale Leroy
Maroney, Samuel Patterson, Jr
Mehner, John Frederick
Moncrief, Nancy D
Muncy, Robert Jess
Murray, Joseph James, Jr
Musick, John A
Neves, Richard Joseph
Opell, Brent Douglas
Osborne, Paul James
Pinschmidt, William Conrad, Jr
Pugh, Jean Elizabeth
Ray, G Carleton
Rebhun, Lionel Israel
Russell, Catherine Marie
Sandow, Bruce Arnold
Savitzky, Alan Howard
Shugart, Herman Henry, Jr
Shuster, Carl Nathaniel, Jr
Simpson, Margaret
Sonenshine, Daniel E
Spangenberg, Dorothy Breslin
Stanley, Melissa Sue Millam
Stevens, Robert Edward
Turner, Bruce Jay
Turney, Tully Hubert
Umminger, Bruce Lynn
Vaughan, Michael Ray
Wells, Ouida Carolyn
West, Warwick Reed, Jr
Wilbur, Henry Miles
Wilson, John William, III
Winston, Judith Ellen
Wood, Robertson Harris Langley
Woolcott, William Starnold

WASHINGTON
Baumel, Julian Joseph
Boersma, P Dee
Braham, Howard Wallace
Brenchley, Gayle Anne
Broad, Alfred Carter
Brown, Herbert Allen
Brown, Robert Harrison
Clark, Glen W
Crabtree, David Melvin
deRoos, Roger McLean
Dickhoff, Walton William
Drabek, Charles Martin
Dumas, Philip Conrad
Eastlick, Herbert Leonard
Edwards, John S
Ford, James
Gibson, Flash
Gilbert, Frederick Franklin
Gorbman, Aubrey
Harrington, Edward James
Hayes, Murray Lawrence
Henry, Dora Priaulx
Jackson, Crawford Gardner, Jr
Johnson, Murray Leathers
Johnson, Phyllis Truth
Johnson, Richard Evan
Karlstrom, Ernest Leonard
Kohn, Alan Jacobs
Kozloff, Eugene Nicholas
Laird, Charles David
Landis, Wayne G
Larsen, John Herbert, Jr
Long, Edward R
Loughlin, Thomas Richard
Luchtel, Daniel Lee
McCloskey, Lawrence Richard
McLaughlin, Patsy Ann
Martin, Arthur Wesley
Martin, Dennis John
Miller, Elwood Morton
Mills, Claudia Eileen
Moffett, David Franklin, Jr
Morgan, Kenneth Robb
Murphy, Mary Eileen
Napp, Jeffrey M
Paine, Robert T
Paulson, Dennis R

Pietsch, Theodore Wells
Plisetskaya, Erika Michael
Porter, Charles Warren
Reischman, Placidus George
Rempel, Arthur Gustav
Rice, Dale Warren
Ross, June Rosa Pitt
Scheffer, Victor B
Schroeder, Michael Allen
Schroeder, Paul Clemens
Senger, Clyde Merle
Stien, Howard M
Thompson, Christopher William
Turner, William Junior
VanBlaricom, Glenn R
Wakimoto, Barbara Toshiko
Way, Jon Leong
White, Ronald Jerome
Yao, Meng-Chao

WEST VIRGINIA
Hall, George Arthur, Jr
Hall, John Edgar
Mugaas, John Nels
Pauley, Thomas Kyle
Reyer, Randall William
Seidel, Michael Edward
Shalaway, Scott D
Tarter, Donald Cain
Taylor, Ralph Wilson

WISCONSIN
Abramoff, Peter
Baylis, Jeffrey Rowe
Beck, Stanley Dwight
Bloom, Alan S
Brynildson, Oscar Marius
Crowe, David Burns
Curtin, Charles Gilbert
Emlen, John Thompson, Jr
Fraser, Lemuel Anderson
Gottfried, Bradley M
Grittinger, Thomas Foster
Guilford, Harry Garrett
Hasler, Arthur Davis
Hodgson, James Russell
Hoffman, William F
Jeanne, Robert Lawrence
Johnson, Wendel J
Lai, Ching-San
Leutenegger, Walter
Light, Douglas B
Lombard, Julian H
Long, Charles Alan
Long, Claudine Fern
Mahmoud, Ibrahim Younis
Michaud, Ted C
Minock, Michael Edward
Norden, Carroll Raymond
North, Charles A
Passano, Leonard Magruder
Piacsek, Bela Emery
Riegel, Ilse Leers
Ruffolo, John Joseph, Jr
Rusch, Donald Harold
Rust, Charles Chapin
Seale, Dianne B
Staszak, David John
Struble, Craig Bruce
Tappen, Neil Campbell
Temple, Stanley A
Weil, Michael Ray
Weise, Charles Martin
Wilson, Richard Howard
Wittrock, Darwin Donald
Young, Howard Frederick

WYOMING
Botkin, Merwin P
Diem, Kenneth Lee
Dolan, Joseph Edward
George, Robert Porter
Hubert, Wayne Arthur
Kingston, Newton
Nunamaker, Richard Allan
Parker, Michael
Tabachnick, Walter J

PUERTO RICO
Cutress, Charles Ernest
Rivero, Juan Arturo
Roman, Jesse
Santacana, Guido E
Squire, Richard Douglas
Thomas, (John) (Paul) Richard
Voltzow, Janice

ALBERTA
Cavey, Michael John
Chia, Fu-Shiang
Clifford, Hugh Fleming
Currie, Philip John
Dixon, Elisabeth Ann
Geist, Valerius
Holmes, John Carl
Lauber, Jean Kautz
Macpherson, Andrew Hall
Mahrt, Jerome L
Nelson, Joseph Schieser
Neuwirth, Maria
Pittman, Quentin J
Rosenberg, Herbert Irving
Russell, Anthony Patrick
Scrimgeour, Garry Joseph

Seghers, Benoni Hendrik
Spencer, Andrew Nigel
Steiner, Andre Louis
Stirling, Ian G
Strobeck, Curtis
Wang, Lawrence Chia-Huang
Wilson, Mark Vincent Hardman

BRITISH COLUMBIA
Adamson, Martin Leif
Bandy, Percy John
Booth, Amanda Jane
Bousfield, Edward Lloyd
Bower, Susan Mae
Burke, Robert D
Clarke, William Craig
Cowan, Ian McTaggart
Dehnel, Paul Augustus
Dill, Lawrence Michael
Fontaine, Arthur Robert
Galindo-Leal, Carlos Enrique
Gosline, John M
Jamieson, Glen Stewart
Jones, David Robert
Kasinsky, Harold Edward
Lewis, Alan Graham
Liley, Nicholas Robin
Lindsey, Casimir Charles
Mackie, George Owen
McLean, John Alexander
Marliave, Jeffrey Burton
Milsom, William Kenneth
Newman, Murray Arthur
Nordan, Harold Cecil
Nursall, John Ralph
Randall, David John
Reimchen, Thomas Edward
Stringam, Elwood Williams
Tunnicliffe, Verena Julia
Ward, David Mercer
Wilimovsky, Norman Joseph

MANITOBA
Abrahams, Mark Vivian
Ayles, George Burton
Bodaly, Richard Andrew
Braekevelt, Charlie Roger
Clayton, James Wallace
Dandy, James William Trevor
Hara, Toshiaki J
Huebner, Judith Dee
Iverson, Stuart Leroy
Laale, Hans W
Novak, Marie Marta
Preston, William Burton
Pruitt, William O(badiah), Jr
Smith, Roger Francis Cooper
White, Noel David George
Zach, Reto

NEW BRUNSWICK
Beninger, Peter Gerard
Cowan, F Brian M
Eveleigh, Eldon Spencer
Hanson, John Mark
Pohle, Gerhard Werner
Scott, William Beverley

NEWFOUNDLAND
Aldrich, Frederick Allen
Bennett, Gordon Fraser
Davis, Charles (Carroll)
Kao, Ming-hsiung
Khan, Rasul Azim
Mercer, Malcolm Clarence
Morris, Claude C
Steele, Donald Harold
Steele, Vladislava Julie
Threlfall, William

NOVA SCOTIA
Bleakney, John Sherman
Brown, Richard George Bolney
Daborn, Graham Richard
Dadswell, Michael John
Ferrier, Gregory R
Garside, Edward Thomas
Leslie, Ronald Allan
O'Dor, Ronald Keith
Rojo, Alfonso

ONTARIO
Atwood, Harold Leslie
Baker, Allan John
Ballantyne, James Stuart
Balon, Eugene Kornel
Banfield, Alexander William Francis
Barlow, Jon Charles
Barr, David Wallace
Beamish, Frederick William Henry
Berger, Jacques
Beverley-Burton, Mary
Bols, Niels Christian
Bond, Edwin Joshua
Buchanan-Smith, Jock Gordon
Butler, Richard Gordon
Cade, William Henry
Calder, Dale Ralph
Cameron, Duncan MacLean, Jr
Carlisle, David Brez
Casselman, John Malcolm
Chant, Donald A
Chapman, David MacLean
Chengalath, Rama

Ciriello, John
Coad, Brian William
Cook, David Greenfield
Crossman, Edwin John
Darling, Donald Christopher
Desser, Sherwin S
Dodson, Edward O
Emery, Alan Roy
Engstrom, Mark Douglas
Ewins, Peter J
Foskett, J Kevin
George, John Caleekal
Gill, Bruce Douglas
Godfrey, William Earl
Govind, Choonilal Keshav
Gray, David Robert
Harington, Charles Richard
Hebert, Paul David Neil
Helmuth, Hermann Siegfried
Keenleyside, Miles Hugh Alston
Kelso, John Richard Murray
Kott, Edward
Kovacs, Kit M
Lai-Fook, Joan Elsa I-Ling
Lavigne, David M
Leach, Joseph Henry
Leatherland, John F
Lee, David Robert
Lingwood, Clifford Alan
Lynn, Denis Heward
McAllister, Donald Evan
McMillan, Donald Burley
Middleton, Alex Lewis Aitken
Morris, Catherine Elizabeth
Muir, Barry Sinclair
Munroe, Eugene Gordon
Noakes, David Lloyd George
Olthof, Theodorus Hendrikus Antonius
Ouellet, Henri
Pang, Cho Yat
Parsons, Thomas Sturges
Petras, Michael Luke
Rising, James David
Robertson, Raleigh John
Ronald, Keith
Roots, Betty Ida
Rossant, Janet
Saleuddin, Abu S
Scadding, Steven Richard
Scott, David Maxwell
Seligy, Verner Leslie
Shih, Chang-Tai
Shivers, Richard Ray
Singal, Dharam Parkash
Singh, Ripu Daman
Smetana, Ales
Smith, Donald Alan
Smith, Jonathan Jeremy Berkeley
Smol, John Paul
Solman, Victor Edward Frick
Steele, John Earle
Stevens, Ernest Donald
Stone, John Bruce
Tomlin, Alan David
Watson, Wynnfield Young
Weseloh, D V (Chip)
Winterbottom, Richard
Youson, John Harold

PRINCE EDWARD ISLAND
Drake, Edward Lawson

QUEBEC
Ali, Mohamed Ather
Anctil, Michel
Boisclair, Daniel
Browman, Howard Irving
Brunel, Pierre
Castonguay, Martin
Chase, Ronald
Darveau, Marcel
Ferron, Jean H
Filteau, Gabriel
Grant, James William Angus
Green, David M(artin)
Guderley, Helga Elizabeth
Johnston, C Edward
Kapoor, Narinder N
McNeil, Raymond
Magnan, Pierre
Nishioka, Yutaka
Prescott, Jacques
Reiswig, Henry Michael
Sainte-Marie, Bernard
Sanborne, Paul Michael
Srivastava, Prakash Narain

SASKATCHEWAN
Brandell, Bruce Reeves
Brigham, R Mark
Forsyth, Douglas John
Gilmour, Thomas Henry Johnstone
Gratto-Trevor, Cheri Lynn
Hobson, Keith Alan
Makowski, Roberte Marie Denise
Mitchell, George Joseph
Oliphant, Lynn Wesley
Secoy, Diane Marie

OTHER COUNTRIES
Buttemer, William Ashley
Craig, John Frank
Haight, John Richard
Harbach, Ralph Edward

176 / DISCIPLINE INDEX

Zoology (cont)

Hisada, Mituhiko
Jahn, Ernesto
Kaiser, Hinrich
Komers, Petr E
Laurent, Pierre
Lochhead, John Hutchison
McMillan, Joseph Patrick
Marsh, James Alexander, Jr
Myres, Miles Timothy
O'Neill, Patricia Lynn
Paulay, Gustav
Phillips, Allan Robert
Rich, Thomas Hewitt
Richmond, Robert H
Rodriguez, Gilberto
Rubinoff, Ira
Sabry, Ismail
Saether, Ole Anton
Schliwa, Manfred
Schmidt-Koenig, Klaus
Withers, Philip Carew
Wynne-Edwards, Vero Copner
Yosef, Reuven
Zhang, Zhi-Qiang

Other Biological Sciences

ALABAMA
Bernstein, Maurice Harry
Bradley, James T
Guyer, Craig
Lewis, Marian L Moore
Magargal, Wells Wrisley, II
Singh, Jarnail
Smith-Somerville, Harriett Elizabeth
Strada, Samuel Joseph
Urry, Dan Wesley
Wilson, G Dennis

ALASKA
Proenza, Luis Mariano
Rockwell, Julius, Jr

ARIZONA
Allen, John Jeffery
Berens, Michael E
Cheal, MaryLou
Figueredo, Aurelio Jose
Flint, Hollis Mitchell
Gendler, Sandra J
George, M Colleen
Gerner, Eugene Willard
Gilkey, John Clark
Hammond, H David
Henshaw, Paul Stewart
Reitan, Ralph Meldahl
Soren, David
Taylor, Richard G
Villanueva, Antonio R
Williams, Stuart K, II

ARKANSAS
Badger, Thomas Mark
Baker, Max Leslie
Epstein, Joshua
Griffin, Edmond Eugene
Hart, Ronald Wilson
Leone, Charles Abner
Nagle, William Arthur
Sherman, Jerome Kalman

CALIFORNIA
Abramson, Stephan B
Ainsworth, Earl John
Alpen, Edward Lewis
Amerine, Maynard Andrew
Ashe, John Herman
Babcock, Gary G
Bagshaw, Malcolm A
Bailey, Ian L
Baptista, Luis Felipe
Batchelder, William Howard
Belisle, Barbara Wolfanger
Bennett, John Francis
Bentley, David R
Billingham, John
Birren, James E(mmett)
Blakely, Eleanor Alice
Bonura, Thomas
Bradbury, Margaret G
Brown, J Martin
Bryant, Peter James
Burgess, Teresa Lynn
Campbell, Kenneth Eugene, Jr
Castro, Joseph Ronald
Chao, Fu-chuan
Cobb, Fields White, Jr
Cohn, Stanton Harry
Cox, Cathleen Ruth
Cramer, Donald V
Crandall, Edward D
Crane, Jules M, Jr
Cross, Carroll Edward
Deen, Dennis Frank
Dement, William Charles
Dewey, William
Dillman, Robert O
Dobson, R Lowry
Ensminger, Marion Eugene
Epstein, John Howard
Erickson, Jon W

Fox, Joan Elizabeth Bothwell
Fuller, Thomas Charles
Gish, Duane Tolbert
Glick, David
Go, Vay Liang
Goeddel, David V
Gold, Daniel P
Goldfine, Ira D
Goldstein, Norman N
Goldyne, Marc E
Gondos, Bernard
Graves, Joseph L
Hager, E Jant
Hamilton, Robert L, Jr
Hammel, Eugene A
Hanggi, Evelyn Betty
Harris, John Wayne
Hoch, Paul Edwin
Hoffman, Robert M
Holley, Daniel Charles
Holmquist, Walter Richard
Howard, Walter Egner
Innerarity, Tom L
Jerison, Harry Jacob
Jones, Claris Eugene, Jr
Jukes, Thomas Hughes
Kallman, Robert Friend
Kodira, Umesh Chengappa
Kohn, Fred R
Kraft, Lisbeth Martha
Kurohara, Samuel S
Kutas, Marta
Lanier, Lewis L
Lawyer, Arthur L
Lee, Nancy L
Leon, Michael Allan
Lewis, Robert Allen
Li, Thomas M
Liburdy, Robert P
Lorence, Matthew C
McDonnel, Gerald M
McEwen, Joan Elizabeth
McGuigan, Frank Joseph
Marcus, Carol Silber
Marini, Mario A
Matthes, Michael Taylor
Meiselman, Herbert Joel
Meredith, Carol N
Miledi, Ricardo
Morse, Michael S
Mussen, Eric Carnes
Nichols, Barbara Ann
Nissenson, Robert A
Osborne, Michael Andrew
Padian, Kevin
Painter, Robert Blair
Peterson, Andrew R
Phillips, Theodore Locke
Ralph, Peter
Rao, Tadimeti Seetapati
Rees, Rees Bynon
Ribak, Charles Eric
Rivier, Catherine L
Robles, Laura Jeanne
Rosenzweig, Mark Richard
Ryder, Oliver A
Schonewald-Cox, Christine Micheline
Seto, Joseph Tobey
Shapiro, Lucille
Smith, Charles G
Snape, William J
Sondhaus, Charles Anderson
Sperling, George
Squire, Larry Ryan
Starrett, Priscilla Holly
Tamerius, John
Trapani, Robert Vincent
Umiel, Tehila
Urch, Umbert Anthony
Varon, Myron Izak
Vinters, Harry Valdis
Vlasuk, George P
Washburn, Sherwood L
Weil, Jon David
Weinstein, Irwin M
Weissman, Irving L
Wenzel, Bernice Martha
Wyrwicka, Wanda
Yerram, Nagender Rao

COLORADO
Bedford, Joel S
Conger, John Douglas
Cook, James L
Elkind, Mortimer M
Gillette, Edward LeRoy
Harold, Ruth L
Hilmas, Duane Eugene
Johnson, Thomas Eugene
Johnson, Timothy John Albert
Kloppel, Thomas Mathew
Lett, John Terence
Liechty, Richard Dale
Parton, William Julian, Jr
Prasad, Kedar N
Rausch, Steven K
Stith, Bradley James
Verdeal, Kathey Marie
Voss, James Leo
Watkins, Linda Rothblum
Whicker, Floyd Ward

CONNECTICUT
Badoyannis, Helen Litman

Coe, Michael Douglas
Conklin, Harold Colyer
Eustice, David Christopher
Farmer, Mary Winifred
Fischer, James Joseph
Klein, Norman W
Levenson, Robert
Lurie, Alan Gordon
Morrow, Jon S
Pooley, Alan Setzler
Rouse, Irving
Smilowitz, Henry Martin
Trumbore, Mark W
Walton, Alan George
Wescott, Roger Williams

DELAWARE
Melby, James Michael
Ray, Thomas Shelby

DISTRICT OF COLUMBIA
Ampy, Franklin
Bednarek, Jana Marie
Carlson, William Dwight
Cuatrecasas, Jose
Douple, Evan Barr
Field, Ruth Bisen
Green, Rayna Diane
Halstead, Thora Waters
Henderson, Ellen Jane
Hollinshead, Ariel Cahill
Lakshman, M Raj
Lavine, Robert Alan
McCarty, Gale A
Malone, Thomas E
Milbert, Alfred Nicholas
Miles, Corbin I
Norrbom, Allen Lee
Parasuraman, Raja
Robinette, Charles Dennis
Sarin, Prem S
Saslaw, Leonard David
Smith, David Allen
Sparrowe, Rollin D
Stanford, Dennis Joe
Tangney, John Francis
Vietmeyer, Noel Duncan
Wassef, Nabila M

FLORIDA
Achey, Phillip M
Anderson, William McDowell
Armelagos, George John
Bloom, Stephen Allen
Bruner, Harry Davis
Chavin, Walter
Christie, Richard G
Cromroy, Harvey Leonard
Dribin, Lori
Frederick, Peter Crawford
Gilbert-Barness, Enid F
Hope, George Marion
Keane, Robert W
Litosch, Irene
Lombardi, Max H
Patterson, Richard Sheldon
Raizada, Mohan K
Robitaille, Henry Arthur
Routh, Donald Kent
Silver, Warren Seymour
Smith, Lawton Harcourt
Stroud, Robert Michael
Teitelbaum, Philip
Thuning-Roberson, Claire Ann
Tumer, Nihal
Undeen, Albert Harold
West, Richard Lowell
Wheeler, George Carlos
Wheeler, Jeanette Norris

GEORGIA
Ades, Edwin W
Barfuss, Delon Willis
Bryan, John Henry Donald
Cureton, Kirk J
Eriksson, Karl-Erik Lennart
Gould, Kenneth G
Isaac, Walter
Karow, Armand M(onfort), Jr
Kaufman, Leo
Kubica, George P
Proctor, Charles Darnell
Reichard, Sherwood Marshall
Romanovicz, Dwight Keith
Tzianabos, Theodore
Urso, Paul
Zaia, John Anthony

HAWAII
Carlson, John Gregory
Fast, Arlo Wade
Fok, Agnes Kwan
Howarth, Francis Gard
Kaneshiro, Kenneth Yoshimitsu
Lee, Cheng-Sheng
Morin, Marie Patricia
Schmitt, Donald Peter
Townsley, Sidney Joseph
Walsh, William Arthur

IDAHO
Roberts, Lorin Watson

ILLINOIS
Al-Ubaidi, Muayyad R
Barr, Susan Hartline
Becker, John Henry
Bevan, William
Capone, James J
Coulson, Richard L
Dawson, M Joan
Desai, Parimal R
DiMuzio, Michael Thomas
Dudley, Horace Chester
Feder, Martin Elliott
Frenkel, Niza B
Fritz, Thomas Edward
Grahn, Douglas
Heaney, Lawrence R
Heller, Alfred
Holtzman, Richard Beves
Juraska, Janice Marie
Kaine, Brian Paul
King, David George
Klass, Michael R
Knospe, William H
Kundu, Samar K
Landsberg, Lewis
McQuistion, Thomas Evin
Mathews, Martin Benjamin
Mednieks, Maija
Mewissen, Dieudonne Jean
Moretz, Roger C
Nadler, Steven Anthony
Naples, Virginia L
Nicholas, Ralph Wallace
Peak, Meyrick James
Rabb, George Bernard
Ratajczak, Helen Vosskuhler
Rechtschaffen, Allan
Reichmann, Manfred Eliezer
Rips, Lance Jeffrey
Rosner, Marsha R
Rundo, John
Simpson, Sidney Burgess, Jr
Sinclair, James Burton
Sitrin, Michael David
Smith, Douglas Calvin
Stefani, Stefano
Stehney, Andrew Frank
Thomson, John Ferguson
Voight, Janet Ruth
Wencel-Drake, June D
White, Randy D
Zimmerman, Roger Paul

INDIANA
Chandrasekhar, S
Chernoff, Ellen Ann Goldman
Daily, Fay Kenoyer
Ferris, Virginia Rogers
Green, Morris
Guth, S(herman) Leon
Kwon, Byoung Se
Wagner, Morris
Zinsmeister, William John

IOWA
Beachy, Christopher King
Dutton, Gary Roger
Isely, Duane
LaBrecque, Douglas R
Laffoon, John
Osborne, James William
Riley, Edgar Francis, Jr
Ruth, Royal Francis

KANSAS
Cho, Cheng T
Choate, Jerry Ronald
Cutler, Bruce
Dass, David
Frost-Mason, Sally Kay
Hays, Horace Albennie
Kimler, Bruce Franklin
Padmanabhan, Radhakrishnan
Shaw, Edward Irwin
Uyeki, Edwin M

KENTUCKY
Bonner, Philip Hallinder
Calkins, John
Carter, Julia H
Crooks, Peter Anthony
Feola, Jose Maria
Friedler, Robert M
Giammara, Beverly L Turner Sites
Urano, Muneyasu

LOUISIANA
Hughes, Janice S
Knaus, Ronald Mallen
McGrath, Hugh, Jr
Musgrave, Mary Elizabeth
Olson, Gayle A
Shaw, Richard Francis
Swenson, David Harold

MAINE
Harrison, David Ellsworth
Lovett, Edmund J, III
Musgrave, Stanley Dean

MARYLAND
Anderson, Mauritz Gunnar
Atwell, Constance Woodruff
Baker, Carl Gwin

Baker, George Thomas, III
Bank-Schlegel, Susan Pamela
Baum, Siegmund Jacob
Bick, Katherine Livingstone
Blithe, Diana Lynn
Brook, Itzhak
Burton, Dennis Thorpe
Chen, Raymond F
Chou, Janice Y
Crampton, Theodore Henry Miller
Dixon-Holland, Deborah Ellen
Doubt, Thomas J
Driscoll, Bernard F
Earnshaw, William C
Eddy, Hubert Allen
Ederer, Fred
Fahy, Gregory Michael
Fozard, James Leonard
Gallin, John I
Geller, Ronald G
Gill, John Russell, Jr
Golding, Hana
Gonzalez, Frank J
Gorden, Phillip
Goroff, Diana K
Gray, Paulette S
Harrison, George H
Hawkins, Michael John
Hendrix, Thomas Russell
Henrich, Curtis J
Hochstein, Herbert Donald
Hodge, Frederick Allen
Hubbard, Ann Louise
James, Stephen P
Kammula, Raju G
Khachaturian, Zaven Setrak
Kimes, Brian William
Kinnamon, Kenneth Ellis
Kloetzel, John Arthur
Kraemer, Kenneth H
Krusberg, Lorin Ronald
Kung, Hsiang-Fu
Kuo, Scot Charles
Lakatta, Edward G
Larsen, Paul M
Leach, William Matthew
Leef, James Lewis
Leong, Kam W
Ley, Herbert L, Jr
Liau, Gene
Lipman, David J
Lowder, James N
Lowy, R Joel
McDonald, Lee J
Margolis, Ronald Neil
Marino, Pamela A
Marzella, Louis
Meryman, Harold Thayer
Michelsen, Arve
Mitsuya, Hiroaki
Morhardt, Sia S
Neville, David Michael, Jr
O'Grady, Richard Terence
Oliver, Constance
Ozato, Keiko
Pagano, Richard Emil
Page, Norbert Paul
Peterson, Jane Louise
Pienta, Roman Joseph
Pilitt, Patricia Ann
Plato, Chris C
Purcell, Jennifer Estelle
Rall, William Frederick
Riesz, Peter
Rifkind, Basil M
Robinson, David Mason
Shulman, N Raphael
Skidmore, Wesley Dean
Sobocinski, Philip Zygmund
Spector, Novera Herbert
Spiegel, Allen M
Stratmeyer, Melvin Edward
Top, Franklin Henry, Jr
Vande Woude, George
Wachholz, Bruce William
Weiss, Joseph Francis
Wright, William W
Zbar, Berton
Zoon, Kathryn C

MASSACHUSETTS
Adams, Richard A
Allen, Philip G
Alt, Frederick W
Baldwin, William Russell
Barone, Leesa M
Bell, Eugene
Bennett, Holly Vander Laan
Bikoff, Elizabeth K
Bynum, T E
Carney, Walter Patrick
Caspar, Donald Louis Dvorak
Colt, LeBaron C, Jr
Davis, William Edwin, Jr
Eil, Charles
Ennis, Francis A
Estes, William K
Ezzell, Robert Marvin
Fimian, Walter Joseph, Jr
Gardner, Howard Earl
Geisterfer-Lowrance, Anja A T
Giffin, Emily Buchholtz
Greene, Reginald
Grigg, Peter

Hagen, Susan James
Halle, Morris
Hartwig, John
Hebben, Nancy
Henneman, Elwood
Hu, Jing-Shan
Jacobs, Jerome Barry
Jones, Rosemary
Kung, Patrick C
Lee, John K
McIntire, Kenneth Robert
Marcus, Elliot M
Mulvey, Philip Francis, Jr
Naqvi, S Rehan Hasan
Ofner, Peter
Orlidge, Alicia
Paul, Benoy Bhushan
Raviola, Elio
Saper, Clifford B
Sekuler, Robert W
Sommers Smith, Sally K
Stein, Janet Lee Swinehart
Sullivan, William Daniel
Swann, David A
Tannenbaum, Steven Robert
Tilley, Stephen George
Volkmann, Frances Cooper
Woodbury, Richard Benjamin
Wrangham, Richard Walter
Ziomek, Carol A

MICHIGAN
Allen, Edward David
Barald, Kate Francesca
Berent, Stanley
Brawn, Mary Karen
Bursian, Steven John
Dajani, Adnan S
Erickson, Laurence A
Fisher, Leslie John
Goustin, Anton Scott
Hansen-Smith, Feona May
Hirata, Fusao
Jensen, James Burt
Kaufman, Donald Barry
Kawooya, John Kasajja
Keyes, Paul Landis
Klomparens, Karen L
Kurachi, Kotoku
Laborde, Alice L
Ledbetter, Steven R
Lindsay, Robert Kendall
Lum, Lawrence
Lynne-Davies, Patricia
McManus, Myra Jean
Marcelo, Cynthia Luz
Maruyama, Yosh
Miller, Josef Mayer
Motiff, James P
Moudgil, Virinder Kumar
Porter, Calvin Anthon
Radin, Eric Leon
Raub, Thomas Jeffrey
Roberts, Joan Marie
Smith, David I
Solomon, Allen M
Weiss, Bernard
Zacks, James Lee

MINNESOTA
Bishop, Jonathan S
Brambl, Robert Morgan
Breckenridge, Walter John
Gores, Gregory J
Graham, Edmund F
Hamilton, David Whitman
Jenkins, Robert Brian
Kornblith, Carol Lee
La Russo, Nicholas F
Lindgren, Alice Marilyn Lindell
Mabry, Paul Davis
Mount, Donald I
Ordway, Ellen
Percich, James Angelo
Rohrbach, Michael Steven
Rosenberg, Andreas
Salisbury, Jeffrey L
Song, Chang Won
Swingle, Karl Frederick
Vetter, Richard J
Yellin, Absalom Moses

MISSISSIPPI
Alexander, William Nebel
Lloyd, Edwin Phillips
McClurkin, Iola Taylor

MISSOURI
Alexander, Stephen
Burchard, Jeanette
Connolly, Daniel Thomas
Cook, Mary Rozella
Dilley, William G
Glenn, Kevin Challon
Kenny, Michael Thomas
Leimgruber, Richard M
Mecham, Robert P
Medoff, Judith
Menz, Leo Joseph
Monzyk, Bruce Francis
Morris, Stephen Jon
Morton, Phillip A
Moulton, Robert Henry
Roloff, Marston Val

Tolmach, L(eonard) J(oseph)

MONTANA
Bergman, Robert K
Grund, Vernon Roger
Rumely, John Hamilton
Rutledge, Lester T

NEBRASKA
Dalrymple, Glenn Vogt
Keeler, Kathleen Howard
Kolar, Joseph Robert, Jr
Morris, Mary Rosalind

NEVADA
Franson, Raymond Lee
Hoelzer, Guy Andrew

NEW HAMPSHIRE
Easstman, Alan
Fagerberg, Wayne Robert
McPeek, Mark Alan
Normandin, Robert F
Swartz, Harold M
Wurster-Hill, Doris Hadley

NEW JERSEY
Alfano, Michael Charles
Black, Ira B
Blood, Christine Haberern
Carlson, Gerald M
DeForrest, Jack M
DeMaio, Donald Anthony
Dubin, Donald T
Dugan, Gary Edwin
Founds, Henry William
Frances, Saul
Heaslip, Richard Joseph
Hirsch, Robert L
Huber, Ivan
Jacobus, David Penman
Koster, William Henry
Levine, O Robert
Lewis, Steven Craig
Manne, Veeraswamy
Mele, Frank Michael
Naro, Paul Anthony
Riley, David
Schrankel, Kenneth Reinhold
Sherrick, Carl Edwin
Stern, William
Thompson, Robert L
Tiku, Moti
Tischio, John Patrick
Triemer, Richard Ernest

NEW MEXICO
Dressendorfer, Rudy
Guilmette, Raymond Alfred
Lechner, John Fred
Ley, Kenneth D
Lundgren, David L(ee)
Mewhinney, James Albert
Pena, Hugo Gabriel
Richman, David Bruce
Thomas, Kimberly W
Thomassen, David George
Trotter, John Allen

NEW YORK
Adesnik, Milton
Altar, Charles Anthony
Altman, Kurt Ison
Bachrach, Howard L
Banerjee, Debendranath
Baroni, Timothy J
Bateman, John Laurens
Bell, Michael Allen
Berkowitz, Jesse M
Bernstein, Alvin Stanley
Beyer, Carl Fredrick
Black, Lindsay MacLeod
Boutjdir, Mohamed
Bruce, Alan Kenneth
Bullough, Vern L
Carsten, Arland L
Casarett, Alison Provoost
Center, Sharon Anne
Chasalow, Fred I
Chou, Shyan-Yih
Colman, David Russell
Concannon, Joseph N
Corradino, Robert Anthony
Cronkite, Eugene Pitcher
Crouch, Billy G
Davis, Thomas Richard
Deschner, Eleanor Elizabeth
DiCioccio, Richard A
DiGaudio, Mary Rose
Domozych, David S
Dougherty, Thomas John
Dounce, Alexander Latham
Duff, Ronald George
Erlij, David
Falk, Dean
Ferris, Steven Howard
Fischman, Marian Weinbaum
Godson, Godfrey Nigel
Greene, Lloyd A
Grove, Patricia A
Hall, Eric John
Hanafusa, Hidesaburo
Hashim, George A
Haycock, Dean A

Hopkins, Betty Jo Henderson
Horst, Ralph Kenneth
Janakidevi, K
Kaplan, Ehud
Kilpper, Robert William
Kim, Jae Ho
Knauf, Philip A
Lau, Joseph T Y
Law, Paul Arthur
Lawrence, Christopher William
Lazarow, Paul B
Lee, Soohee
Lennarz, William J
Levine, Louis David
McGarry, Michael P
McOsker, Charles C
Macris, Nicholas T
Maillie, Hugh David
Makous, Walter L
Masur, Sandra Kazahn
Menon, Gopinathan K(unnariath)
Meruelo, Daniel
Miller, Morton W
Movshon, J Anthony
Norell, Mark Allen
Norton, Larry
Osterberg, Donald Mouretz
Padawer, Jacques
Patil, Jaygonda
Peerschke, Ellinor Irmgard Barbara
Pfaffmann, Carl
Pogo, Beatriz G T
Powell, Sharon Kay
Redmond, Billy Lee
Reichert, Leo E, Jr
Reiss, Diana
Repasky, Elizabeth Ann
Ritch, Robert
Rivenson, Abraham S
Rosenberger, David A
Rosendorff, Clive
Rowe, Arthur W(ilson)
Sackeim, Harold A
Sanders, F Kingsley
Schmidt, John Thomas
Serafy, D Keith
Shellabarger, Claire J
Siddiqui, M A Q
Sobel, Jael Sabina
Stross, Raymond George
Tedlock, Dennis
Treser, Gerhard
Vladutiu, Georgirene Dietrich
Wiesel, Torsten Nils
Wigler, Michael H
Woodard, Helen Quincy
Worrall, James Joseph
Wurster, Charles F
Wysocki, Annette Bernadette
Yang, Shung-Jun
Yazulla, Stephen

NORTH CAROLINA
Allen, Nina Stromgren
Barrett, J(ames) Carl
Bennett, William D
Branch, Clarence Joseph
Butts, Jeffrey A
Curnow, Randall T
Esch, Gerald Wisler
Hall, Warren G
Horstman, Donald H
Hoss, Donald Earl
Jameson, Charles William
Jones(Stover), Betsy
Kovacs, Charles J
Kunkel, Thomas A
Kwock, Lester
Lenhard, James M
Miwa, Gerald T
Morgan, Kevin Thomas
Nettesheim, Paul
Reed, William
Rotman, Harold H
Rublee, Parke Alstan
Wheeler, Kenneth Theodore, Jr
Wise, Ernest George
Zimmerman, Thomas Paul

NORTH DAKOTA
Kelleher, James Joseph
Uthus, Eric O

OHIO
Bank, Harvey L
Bhattacharya, Amit
Boissy, Raymond E
Britton, Steven Loyal
Brunner, Robert Lee
Chen, I-Wen
Coots, Robert Herman
Fanaroff, Avroy A
Francko, David Alex
Fucci, Donald James
Goldblatt, Peter Jerome
Grossman, Charles Jerome
Harding, Clifford Vincent, III
Hiles, Maurice
Jegla, Dorothy Eldredge
Kvalnes, Kalla K
Lacey, Beatrice Cates
Lee, Harold Hon-Kwong
Leis, Jonathan Peter
Lieberman, Michael A

Other Biological Sciences (cont)

McCarthy, F D
Mendenhall, Charles L
Nixon, Charles William
Nordlund, James John
Nygaard, Oddvar Frithjof
Oliver, Joel Day
Olsen, Richard George
Patterson, Michael Milton
Pribor, Donald B
Rein, Diane Carla
Thomas, Craig Eugene
Varnes, Marie Elizabeth
Yang, Tsanyen
Yates, Allan James

OKLAHOMA
Elisens, Wayne John
Friedberg, Wallace
Gimble, Jeffrey M
Holloway, Frank A
Mares, Michael Allen
Simpson, Ocleris C
Sinclair, James Lewis
Talent, Larry Gene
Thomas, Sarah Nell

OREGON
Bethea, Cynthia Louise
Brownell, Philip Harry
Burry, Kenneth A
Callahan, Clarence Arthur
Cereghino, James Joseph
Clinton, Gail M
Cross, Stephen P
Denison, William Clark
Hughes, Robert Mason
Koong, Ling-Jung
Liss, William John
Pizzimenti, John Joseph
Straube, Robert Leonard
Yatvin, Milton B

PENNSYLVANIA
Agarwal, Jai Bhagwah
Andes, Charles Lovett
Baldino, Frank, Jr
Barchi, Robert Lawrence
Bex, Frederick James
Biaglow, John E
Boggs, Sallie Patton Slaughter
Brennan, James Thomas
Clark, T(helma) K
Conger, Alan Douglas
Courtney, Richard James
Cox, Malcolm
Dinges, David Francis
Edelstone, Daniel I
Emberton, Kenneth C
Ewing, David Leon
Foley, Thomas Preston, Jr
Forbes, Paul Donald
Fussell, Catharine Pugh
Glick, Mary Catherine
Gollin, Susanne Merle
Gombar, Charles T
Greene, Mark Irwin
Halbrendt, John Marthon
Hauptman, Stephen Phillip
Hsu, Kuo-Hom Lee
Isom, Harriet C
Jacobs, George Joseph
Kennedy, Ann Randtke
Kim, Ke Chung
Kunapuli, Satya P
Ladman, Aaron J(ulius)
Leeper, Dennis Burton
Leon, Shalom A
Lynn, Robert K
Maloy, W Lee
Mason, James Russell
Maul, Gerd G
Maxson, Linda Ellen R
Moyer, Robert (Findley)
Mullin, James Michael
Olds-Clarke, Patricia Jean
Olexa, Stephanie A
Pruett, Esther D R
Saunders, William Bruce
Sax, Martin
Ubels, John L
Watson, Joseph Alexander
Wing, Rena R
Zigmond, Sally H

RHODE ISLAND
Biancani, Piero
Bibb, Harold David
Goos, Roger Delmon
McMillan, Paul N
Waage, Jonathan King

SOUTH CAROLINA
Alley, Thomas Robertson
Appel, James B
Creek, Kim E
Ghaffar, Abdul
Hahn, Amy Beth
Horn, Charles Norman
Mitchell, Hugh Bertron
Powers, Edward Lawrence

Rembert, David Hopkins, Jr
Weeks, Stephen Charles

TENNESSEE
Allen, William Richard
Alsop, Frederick Joseph, III
Carlson, James Gordon
Clapp, Neal K
Coons, Lewis Bennion
Dix, John Willard
Fry, Richard Jeremy Michael
Gutzke, William H N
Hadden, Charles Thomas
Hubner, Karl Franz
Isom, Billy Gene
Keim, Robert Gerald
Limbird, Lee Eberhardt
Mazur, Peter
Monaco, Paul J
Noonan, Thomas Robert
Reddick, Bradford Beverly
Rushton, Priscilla Strickland
Schlarbaum, Scott E
Sharp, Aaron John
Storer, John B
Sun, Deming
Walne, Patricia Lee
Whitt, Michael A

TEXAS
Alhossainy, Effat M
Attanasio, Roberta
Aumann, Glenn D
Baron, Samuel
Boldt, David H
Bryant, Vaughn Motley, Jr
Byrne, John Howard
Christ, John Ernest
Ciomperlik, Matthew A
Cooper, Ronda Fern
Cox, Ann Bruger
Draper, Rockford Keith
Dung, H C
Ford, George Peter
Frishman, Laura J(ean)
Gaulden, Mary Esther
Gogol, Edward Peter
Granger, Harris Joseph
Gyorkey, Ferenc
Haye, Keith R
Howe, William Edward
Jenkins, Vernon Kelly
Kennedy, Joseph Patrick
Klaus, Ewald Fred, Jr
Ku, Bernard Siu-Man
Kunau, Robert, Jr
Lachman, Lawrence B
Lapeyre, Jean-Numa
Levine, Alan E
Lopez-Berestein, Gabriel
Meltz, Martin Lowell
Moller, Peter C
Nachtwey, David Stuart
Oujesky, Helen Matusevich
Pennebaker, James W
Prasad, Naresh
Rodarte, Joseph Robert
Sauer, Helmut Wilhelm
Schneider, Nancy Reynolds
Schonbrunn, Agnes
Soltes, Edward John
Steffy, John Richard
Steinberger, Anna
Thalmann, Robert H
Tomasovic, Stephen Peter
Ullrich, Robert Leo
Wiederhold, Michael L
Wilmore, Jack H

UTAH
Barnett, Bill B
Dethlefsen, Lyle A
Mahoney, Arthur W
Ramachandran, Chittoor Krishna
Stevens, Walter
Warters, Raymond Leon

VERMONT
Watters, Christopher Deffner

VIRGINIA
Barak, Eve Ida
Barranco, Sam Christopher, III
Bonvillian, John Doughty
Braciale, Vivian Lam
Carson, Keith Alan
Courtney, Mark William
Davis, Joel L
De Haan, Henry J
Edwards, Ernest Preston
Fuller, Stephen William
Hilu, Khidir Wanni
Holsinger, John Robert
Homsher, Paul John
Hymes, Dell Hathaway
Jones, Johnnye M
Knauer, Thomas E
Looney, William Boyd
McGowan, John Joseph
McKitrick, Mary Caroline
Martin, William Wallace
Moncrief, Nancy D
Nelson, Neal Stanley
Olsen, Kathie Lynn

Rozzell, Thomas Clifton
Senti, Frederic R(aymond)
Sirica, Alphonse Eugene
Stevens, Robert Edward
Townsend, Lawrence Willard
Wilson, John Drennan
Winston, Judith Ellen

WASHINGTON
Abrass, Christine K
Bair, William J
Bonham, Kelshaw
Branca, Andrew A
Christensen, Gerald M
Colby, Susan Melanie
Crill, Wayne Elmo
Dale, Beverly A
Edwards, John S
Greenberg, Philip D
Greenlee, Theodore K
Heiserman, Gary
Hungate, Frank Porter
McDougall, James K
Mackensie, Alan P
Martin, Carroll James
Moffett, Stacia Brandon
Monan, Gerald E
Norvell, Lorelei Lehwalder
Rasey, Janet Sue
Sanders, Charles Leonard, Jr
Sieker, Larry Charles
Singer, Jack W
Teller, Davida Young
Thompson, Roy Charles, Jr
Walters, Ronald Arlen
Wolf, Norman Sanford

WEST VIRGINIA
Reid, Robert Leslie

WISCONSIN
Amrani, David L
Bosnjak, Zeljko J
Cartee, Gregory D
Clifton, Kelly Hardenbrook
Cummings, John Albert
Curtin, Charles Gilbert
Duewer, Elizabeth Ann
Goldstein, Robert
Gould, Michael Nathan
Greiff, Donald
Grunewald, Ralph
Iltis, Hugh Hellmut
Kostenbader, Kenneth David, Jr
Kowal, Robert Raymond
Mackie, Thomas Rockwell
Mandel, Neil
Norris, William Penrod
Oyen, Ordean James
Schwartz, Bradford S
Sieber, Fritz
Tulunay-Keesey, Ulker

WYOMING
Doerges, John E
Humburg, Neil Edward

PUERTO RICO
Jordan-Molero, Jaime E
Squire, Richard Douglas
Voltzow, Janice

ALBERTA
Bidgood, Bryant Frederick
Church, Robert Bertram
Famiglietti, Edward Virgil, Jr
McGann, Locksley Earl
Sohal, Parmjit S
Wyse, David George

BRITISH COLUMBIA
Alfaro, Rene Ivan
Campbell, Alan
Freeman, Hugh J
Hoffmann, Geoffrey William
Sinclair, John G
Skarsgard, Lloyd Donald
Tingle, Aubrey James

MANITOBA
Chong, James York
Gerrard, Jonathan M
Gill, Clifford Cressey

NEW BRUNSWICK
Waiwood, Kenneth George

NOVA SCOTIA
Connolly, John Francis
Horackova, Magda
Scaratt, David J
Stobo, Wayne Thomas

ONTARIO
Basu, Prasanta Kumar
Burba, John Vytautas
Calder, Dale Ralph
Capone, John Pasquale
Carstens, Eric Bruce
Chadwick, June Stephens
Chang, Patricia Lai-Yung
Clark, Gordon Murray
Cosmos, Ethel
Cutz, Ernest

Czuba, Margaret
Dalpe, Yolande
DesMarais, Andre
Elinson, Richard Paul
Forer, Arthur H
Fourney, Ronald Mitchell
Goldberg, Gary Steven
Hare, William Currie Douglas
Hill, Richard Peter
Inch, William Rodger
Kean, Vanora Mabel
Kovacs, Kit M
Kruuv, Jack
Lau, Catherine Y
Leibo, Stanley Paul
Lingwood, Clifford Alan
McKneally, Martin F
Redhead, Scott Alan
Schneider, Bruce Alton
Tannock, Ian Frederick
Weseloh, D V (Chip)
Whitmore, Gordon Francis

QUEBEC
Antakly, Tony
Ferdinandi, Eckhardt Stevan
Galbraith, David Allan
Legendre, Pierre
Lehnert, Shirley Margaret
Verdy, Maurice

SASKATCHEWAN
Lapp, Martin Stanley
Ripley, Earle Allison

OTHER COUNTRIES
Acres, Robert Bruce
Black, James
Brondz, Boris Davidovich
Bye, Robert Arthur
Campbell, Virginia Wiley
Farid, Nadir R
Ganchrow, Donald
Hahn, Eric Walter
Jarvis, Simon Michael
Jondorf, W Robert
Kang, Kewon
Lin, Chin-Tarng
Lwoff, Andre Michel
MacCarthy, Jean Juliet
Miquel, Jaime
Oxnard, Charles Ernest
Quastel, Michael Reuben
Shimaoka, Katsutaro
Snyder, Allan Whitnack
Srygley, Robert Baxter
Thompson, Peter Allan
Wynne-Edwards, Vero Copner
Zenk, Meinhart Hans Christian

CHEMISTRY

Agricultural & Food Chemistry

ALABAMA
Ammerman, Gale Richard
Rymal, Kenneth Stuart
Small, Robert James

ALASKA
Kennish, John M

ARIZONA
Emery, Donald F
Hendrix, Donald Louis
Hoffmann, Joseph John
Kline, Ralph Willard
Shaw, Ellsworth
Steffen, Albert Harry

ARKANSAS
Conn, James Frederick
Crandall, Philip Glen
Doerge, Daniel Robert
Hood, Robert L
Horan, Francis E

CALIFORNIA
Baker, Frederick Charles
Berhow, Mark Alan
Bernhard, Richard Allan
Bolin, Harold R
Chen, Ming Sheng
Chen, Tung-Shan
Chou, Tsong-Wen
Cooper, Geoffrey Kenneth
Coughlin, James Robert
Cramer, Archie Barrett
Crosby, Donald Gibson
Dutra, Ramiro Carvalho
Eisenberg, Sylvan
Fancher, Llewellyn W
Farmer, Walter Joseph
Feeney, Robert Earl
Fleming, Sharon Elayne
Fung, Steven
Giang, Benjamin Yunwen
Haddon, William F, (Jr)
Hamaker, John Warren
Hammock, Bruce Dupree
Heil, John F, Jr
Henika, Richard Grant
Henrick, Clive Arthur

Hill, Fredric William
Hokama, Takeo
Howe, Robert Kenneth
Ja, William Yin
Jen, Joseph Jwu-Shan
Johnson, Phyllis Elaine
Josephson, Ronald Victor
Kean, Chester Eugene
Khayat, Ali
Kinsella, John Edward
L'Annunziata, Michael Frank
Lee, David Louis
Leo, Albert Joseph
Lewis, Sheldon Noah
Liang, Yola Yueh-o
Luh, Bor Shiun
Lundeen, Glen Alfred
McNall, Lester R
Maier, V(incent) P(aul)
Masri, Merle Sid
Mayer, William Joseph
Mihailovski, Alexander
Millar, Jocelyn Grenville
Montecalvo, Joseph, Jr
Nielsen, Milo Alfred
Patton, Stuart
Petrowski, Gary E
Quistad, Gary Bennet
Rendig, Victor Vernon
Riffer, Richard
Rockland, Louis B
Rodriguez, Eloy
Roitman, James Nathaniel
Russell, Gerald Frederick
Rynd, James Arthur
Samuels, Robert Bireley
Sayre, Robert Newton
Schaefer, Charles Herbert
Singleton, Vernon LeRoy
Smith, Lloyd Muir
Spatz, David Mark
Thomson, John Ansel Armstrong
Thomson, Tom Radford
Thornton, John Irvin
Trumble, John Thomas
Vreeland, Valerie Jane
Walker, Howard David
Weast, Clair Alexander
Webb, Albert Dinsmoor
Williams, John Wesley
Winterlin, Wray Laverne
Young, Donald C
Yuen, Wing
Zavarin, Eugene

COLORADO
Bower, Nathan Wayne
Cannon, William Nathaniel
Cleary, Michael E
Heberlein, Douglas G(aravel)
Kimoto, Walter Iwao
Long, Austin Richard
Lorenz, Klaus J
Mortvedt, John Jacob
Mosier, Arvin Ray
Nachtigall, Guenter Willi
Norstadt, Fred A
Richardson, Thomas
Turner, James Howard
Verdeal, Kathey Marie
Wright, Terry L

CONNECTICUT
Agarwal, Vipin K
Auerbach, Michael Howard
Banijamali, Ali Reza
Bibeau, Thomas Clifford
Davis, Robert Glenn
Hankin, Lester
Mattina, Mary Jane Incorvia
Pierce, James Benjamin
Putterman, Gerald Joseph
Relyea, Douglas Irving
Schaaf, Thomas Ken

DELAWARE
Anderson, Jeffrey John
Austin, Paul Rolland
Beestman, George Bernard
Dubas, Lawrence Francis
Jelinek, Arthur Gilbert
Koeppe, Mary Kolean
May, Ralph Forrest
Metzger, James Douglas
Moberg, William Karl
Montague, Barbara Ann
Morrissey, Bruce William
Rasmussen, Arlette Irene
Roe, David Christopher
Saegebarth, Klaus Arthur
Sandell, Lionel Samuel
Taves, Milton Arthur
Thoroughgood, Carolyn A
Wommack, Joel Benjamin, Jr
Woods, Thomas Stephen

DISTRICT OF COLUMBIA
Flora, Lewis Franklin
Hammer, Charles F
Kuznesof, Paul Martin
Morehouse, Kim Matthew
Page, Samuel William
Sarmiento, Rafael Apolinar
Schneider, Bernard Arnold
Staruszkiewicz, Walter Frank, Jr
Treichler, Ray

FLORIDA
Attaway, John Allen
Berry, Robert Eddy
Boehme, Werner Richard
Calvert, David Victor
Cort, Winifred Mitchell
Gash, Virgil Walter
Gregory, Jesse Forrest, III
Grimm, Charles Henry
Grunwald, Claus Hans
Hagenmaier, Robert Doller
Hill, Kenneth Richard
Hunter, George L K
Jackel, Simon Samuel
Kilsheimer, John Robert
Knapp, Joseph Leonce, Jr
Matthews, Richard Finis
Nagy, Steven
Nakashima, Tadayoshi
Nanz, Robert Augustus Rollins
Newsom, William S, Jr
Ross, Lynne Fischer
Shuey, William Carpenter
Trelease, Richard Davis
Wheeler, Willis Boly
Wolf, Benjamin
Yu, Simon Shyi-Jian

GEORGIA
Barton, Franklin Ellwood, II
Brown, David Smith
Chortyk, Orestes Timothy
Churchill, Frederick Charles
Clark, Benjamin Cates, Jr
Colby, David Anthony
Gaines, Tinsley Powell
Iacobucci, Guillermo Arturo
Jain, Anant Vir
Leffingwell, John C
Lillard, Dorris Alton
Loewenstein, Morrison
Myers, Dirck V
Smit, Christian Jacobus Bester
Wilson, David Merl

HAWAII
Hilton, H Wayne
Moritsugu, Toshio
Sherman, Martin
Young, Hong Yip

IDAHO
Davidson, Philip Michael
Hibbs, Robert A
Lehrer, William Peter, Jr
Muneta, Paul
Natale, Nicholas Robert

ILLINOIS
Abbott, Thomas Paul
Berenbaum, May Roberta
Bernetti, Raffaele
Bietz, Jerold Allen
Blair, Louis Curtis
Bookwalter, George Norman
Buchman, Russell
Burkwall, Morris Paton, Jr
Cheryan, Munir
Christianson, Donald Duane
Cote, Gregory Lee
Cross, John W
Crovetti, Aldo Joseph
Donnelly, Thomas Henry
Eckner, Karl Friedrich
Fahey, George Christopher, Jr
Fink, Kenneth Howard
Germino, Felix Joseph
Halaby, George Anton
Hamer, Martin
Hefley, Alta Jean
Heydanek, Menard George, Jr
Hoepfinger, Lynn Morris
Honig, David Herman
Hou, Ching-Tsang
Hynes, John Thomas
Inglett, George Everett
Kinghorn, Alan Douglas
Kite, Francis Ervin
Klein, Barbara P
Knake, Ellery Louis
Krishnamurthy, Ramanathapur Gundachar
Larson, Bruce Linder
Levin, Alfred A
Lundquist, Burton Russell
McGorrin, Robert Joseph
Maher, George Garrison
Moser, Kenneth Bruce
Mounts, Timothy Lee
Nielsen, Harald Christian
Nishida, Toshiro
Pai, Sadanand V
Parrill, Irwin Homer
Rackis, Joseph John
Rankin, John Carter
Sadler, George D
Schaefer, Wilbur Carls
Schanefelt, Robert Von
Scheid, Harold E
Sebree, Bruce Randall
Sessa, David Joseph
Sheldon, Victor Lawrence
Singh, Laxman
Soucie, William George
Stanford, Marlene A
Tiemstra, Peter J
Traxler, James Theodore
Van der Kloot, Albert Peter
Wang, Pie-Yi
Weber, Evelyn Joyce
Whitney, Robert McLaughlin
Williams, Joseph Lee
Woolson, Edwin Albert
Yates, Shelly Gene

INDIANA
Alter, John Emanuel
Axelrod, Bernard
Beck, James Richard
Butler, Larry G
Corrigan, John Joseph
Glover, David Val
Hanks, Alan Ray
Katz, Frances R
Kossoy, Aaron David
Michel, Karl Heinz
Mihaliak, Charles Andrew
Pratt, Dan Edwin
Rogers, Richard Brewer
Walker, Jerry C

IOWA
Barthel, William Frederick
Coats, Joel Robert
Johnson, Lawrence Alan
Murphy, Patricia A
Rebstock, Theodore Lynn
Stahr, Henry Michael

KANSAS
Chung, Okkyung Kim
Hooker, Mark L
Hoseney, Russell Carl
Krum, Jack Kern
Oppert, Brenda
Schroeder, Robert Samuel
Seitz, Larry Max
Shogren, Merle Dennis
Stutz, Robert L
Von Rumker, Rosmarie

KENTUCKY
Gird, Steven Richard
Knapp, Frederick Whiton
Luzzio, Frederick Anthony
Massik, Michael
Wong, John Lui

LOUISIANA
Bayer, Arthur Craig
Blouin, Florine Alice
Connick, William Joseph, Jr
Domelsmith, Linda Nell
Fischer, Nikolaus Hartmut
Gonzales, Elwood John
Liuzzo, Joseph Anthony
Magin, Ralph Walter
Marshall, Wayne Edward
Palmgren, Muriel Signe
St Angelo, Allen Joseph
Settoon, Patrick Delano
Spanier, Arthur M
Vercellotti, John R

MAINE
Bushway, Rodney John
Pierce, Arleen Cecilia

MARYLAND
Anderson, David Leslie
Argauer, Robert John
Baker, Doris
Beroza, Morton
Brause, Allan R
Caret, Robert Laurent
Christy, Alfred Lawrence
Fulger, Charles V
Galetto, William George
Gilmore, Thomas Meyer
Hall, Richard Leland
Herner, Albert Erwin
Horwitz, William
Inscoe, May Nilson
Jacobson, Martin
Kaufman, Donald DeVere
Kleier, Daniel Anthony
Lee, Yuen San
Liang, Shu-Mei
Mandava, Naga Bhushan
Menn, Julius Joel
Moats, William Alden
Modderman, John Philip
Mokhtari-Rejali, Nahld
Nelson, Judd Owen
Porzio, Michael Anthony
Ragsdale, Nancy Nealy
Rice, Clifford Paul
Salwin, Harold
Schechter, Milton Seymour
Skidmore, Wesley Dean
Sphon, James Ambrose
Veitch, Fletcher Pearre, Jr
Wong, Noble Powell
Zeleny, Lawrence

MASSACHUSETTS
Angelini, Pio
Devlin, Robert M
Ehntholt, Daniel James
Fagerson, Irving Seymour
Gorfien, Harold
Hagopian, Miasnig
Jarboe, Jerry K(ent)
Lapuck, Jack Lester
Learson, Robert Joseph
Mabrouk, Ahmed Fahmy
Marquis, Judith Kathleen
Nawar, Wassef W
Salant, Abner Samuel
Tan, Barrie
Taub, Irwin A(llen)

MICHIGAN
Arrington, Jack Phillip
Bjork, Carl Kenneth, Sr
Brunner, Jay Robert
Budde, Paul Bernard
Carlotti, Ronald John
Ferguson-Miller, Shelagh Mary
Frawley, Nile Nelson
Gilbertson, Terry Joel
Gray, Ian
Hightower, Kenneth Ralph
Hollingsworth, Cornelia Ann
Hollingworth, Robert Michael
Jensen, David James
Laks, Peter Edward
Landrum, Peter Franklin
Leng, Marguerite Lambert
Mergentime, Max
Muentener, Donald Arthur
Pynnonen, Bruce W
Shelef, Leora Aya
Skarsaune, Sandra Kaye
Smith, Denise Myrtle
Walters, Stephen Milo

MINNESOTA
Anderson, George Robert
Atwell, William Alan
Chandan, Ramesh Chandra
Dahle, Leland Kenneth
Fulmer, Richard W
Jezeski, James John
Johnson, Guy Henry
Lonergan, Dennis Arthur
Reinhart, Richard D
Roach, J Robert
Standing, Charles Nicholas
Thompson, Richard David
Throckmorton, James Rodney
Van Valkenburg, Jeptha Wade, Jr

MISSISSIPPI
Alley, Earl Gifford
Cardwell, Joe Thomas
Collins, Johnnie B
Hedin, Paul A
Heitz, James Robert
Minyard, James Patrick

MISSOURI
Bailey, Milton (Edward)
Bradley, Guy Martin
Dutra, Gerard Anthony
Erickson, David R
Franz, John Edward
Freeman, Robert Clarence
Hammann, William Curl
Heymann, Hildegarde
Khan, Adam
Klein, Andrew John
Koenig, Karl E
Moedritzer, Kurt
Newell, Jon Albert
Pfander, William Harvey
Radke, Rodney Owen
Schleppnik, Alfred Adolf
Schumacher, Richard William
Seymour, Keith Goldin
Sharp, Dexter Brian
Shaver, Kenneth John
Stalling, David Laurence
Stenseth, Raymond Eugene
Stout, Edward Irvin
Waggle, Doyle H

MONTANA
McGuire, Charles Francis
Richards, Geoffrey Norman

NEBRASKA
Corbett, Michael Dennis
McClurg, James Edward
Taylor, Stephen Lloyd
Taylor, Steve L
Wallen, Stanley Eugene

NEW HAMPSHIRE
Feliciotti, Enio
Finn, James Walter
Ikawa, Miyoshi

NEW JERSEY
Addor, Roger Williams
Barringer, Donald F, Jr
Borenstein, Benjamin
Carlucci, Frank Vito
Cavender, Patricia Lee

Agricultural & Food Chemistry (cont)

Cevasco, Albert Anthony
Chang, Hsien-Hsin
Chang, Stephen Szu Shiang
Chiang, Bin-Yea
Christenson, Philip A
Coggon, Philip
Crosby, Guy Alexander
Focella, Antonino
Fost, Dennis L
Gatterdam, Paul Esch
Giacobbe, Thomas Joseph
Giddings, George Gosselin
Glass, Michael
Gruenhagen, Richard Dale
Halfon, Marc
Harding, Maurice James Charles
Hatch, Charles Eldridge, III
Ho, Chi-Tang
Hsu, Kenneth Hsuehchia
Jacobson, Glen Arthur
Jennings, Carl Anthony
Johnson, Bobby Ray
Klemann, Lawrence Paul
Labows, John Norbert, Jr
Ladner, David William
Lavanish, Jerome Michael
Lee, Tung-Ching
Leeder, Joseph Gorden
Levine, Harry
Lutz, Albert William
Lynch, Charles Andrew
Manley, Charles Howland
Maravetz, Lester L
Maulding, Donald Roy
Miller, Albert Thomas
Mustafa, Shams
Naipawer, Richard Edward
Pearce, David Archibald
Quinn, James Allen
Ramsey, Arthur Albert
Rein, Alan James
Rinehart, Jay Kent
Rosen, Joseph David
Schaich, Karen Marie
Scharpf, Lewis George, Jr
Stults, Frederick Howard
Sullivan, Andrew Jackson
Tway, Patricia C
Vallese, Frank M
Vanden Heuvel, William John Adrian, III
Van Saun, William Arthur
Volpp, Gert Paul Justus
Waltking, Arthur Ernest
Wislocki, Peter G
Yoder, Donald Maurice

NEW MEXICO
Del Valle, Francisco Rafael
Smith, Garmond Stanley

NEW YORK
Abayev, Michael
Acree, Terry Edward
Aten, Carl Faust, Jr
Bourke, John Butts
Bourne, Malcolm Cornelius
Brenner, Mortimer W
Buchel, Johannes A
Cante, Charles John
Dowling, Joseph Francis
Ehmann, Edward Paul
Kahan, Sidney
Lewis, Bertha Ann
Lichtman, Irwin A
Lisk, Donald J
Luther, Herbert George
Mackay, Donald Alexander Morgan
Mansfield, Kevin Thomas
Marov, Gaspar J
Mirviss, Stanley Burton
Reich, Ismar M(eyer)
Roelofs, Wendell L
Roth, Howard
Shallenberger, Robert Sands
Smith, Laura Lee Weisbrodt
Walter, Reginald Henry

NORTH CAROLINA
Anderson, Thomas Ernest
Apperson, Charles Hamilton
Beck, Charles I
Bell, Jimmy Holt
Chung, Henry Hsiao-Liang
Cooke, Anson Richard
Engle, Thomas William
Gemperline, Margaret Mary Cetera
Hermanson, Harvey Philip
Hinze, Willie Lee
Holm, Robert E
Kuhr, Ronald John
Kurtz, A Peter
LeBaron, Homer McKay
Lineback, David R
McDaniel, Roger Lanier, Jr
Manning, David Treadway
Marable, Nina Louise
Maselli, John Anthony
Moates, Robert Franklin
Neumann, Calvin Lee

Patil, Arvind Shankar
Patterson, Ronald Brinton
Perfetti, Patricia F
Reed, Gerald
Schwindeman, James Anthony
Senzel, Alan Joseph
Sheets, Thomas Jackson
Worsham, Arch Douglas

NORTH DAKOTA
D'Appolonia, Bert Luigi
Davison, Kenneth Lewis
Donnelly, Brendan James
McDonald, Clarence Eugene
Satterlee, Lowell Duggan
Stolzenberg, Gary Eric
Walsh, David Ervin

OHIO
Battershell, Robert Dean
Buck, Keith Taylor
Cotton, Wyatt Daniel
Dolfini, Joseph E
Dollimore, David
Eisenhardt, William Anthony, Jr
Flautt, Thomas Joseph, Jr
Greenlee, Kenneth William
Heckert, David Clinton
Jackson, Earl Rogers
Kallenberger, Waldo Elbert
Litchfield, John Hyland
Magee, Thomas Alexander
Marks, Alfred Finlay
Meinert, Walter Theodore
Mohlenkamp, Marvin Joseph, Jr
Myhre, David V
Ockerman, Herbert W
Peng, Andrew Chung Yen
Powers, Larry James
Schafer, Mary Louise
Schweitzer, Mark Glenn
Sevenants, Michael R
Strobel, Rudolf G K
Watson, Maurice E
Watson, Stanley Arthur
Yamazaki, William Toshi
Youngquist, R(udolph) William

OKLAHOMA
Abbott, Donald Clayton
Bingham, Robert J
Fish, Wayne William
Seideman, Walter E

OREGON
Arnold, Roy Gary
Fang, Sheng Chung
Laver, Murray Lane
Norris, Logan Allen
Swisher, Horton Edward
Wrolstad, Ronald Earl

PENNSYLVANIA
Albright, Fred Ronald
Carter, Linda G
Cerbulis, Janis
Chen, Audrey Weng-Jang
Cherry, John P
Di Cuollo, C John
Doner, Landis Willard
Dyott, Thomas Michael
Farrell, Harold Maron, Jr
Fishman, Marshall Lewis
Foglia, Thomas Anthony
Gluntz, Martin L
Hardesty, Patrick Thomas
Harris, James Joseph
Herzog, David Paul
Hess, Earl Hollinger
Hicks, Kevin B
Holsinger, Virginia Harris
Hood, Lamartine Frain
Hunt, David Allen
Hurst, William Jeffrey
Hurt, Susan Schilt
Johnson, Wayne Orrin
Kroger, Manfred
Kronick, Paul Leonard
Kurtz, David Allan
Lopez, R C Gerald
Lyman, William Ray
McCurry, Patrick Matthew, Jr
Maerker, Gerhard
Marmer, William Nelson
Moreau, Robert Arthur
Mozersky, Samuel M
Neubeck, Clifford Edward
Newirth, Terry L
Ocone, Luke Ralph
Pfeffer, Philip Elliot
Rogerson, Thomas Dean
Rothbart, Herbert Lawrence
Rothman, Alan Michael
Schwing, Gregory Wayne
Stinson, William Sickman, Jr
Strohm, Paul F
Strong, Frederick Carl, III
Talley, Eugene Alton
Thayer, Donald Wayne
Tunick, Michael Howard
Wasserman, Aaron E
Wolff, Ivan A
Yih, Roy Yangming
Zaika, Laura Larysa

RHODE ISLAND
Olney, Charles Edward
Rand, Arthur Gorham, Jr

SOUTH CAROLINA
Brooks, Thomas
Dellicolli, Humbert Thomas
Hunter, George William
Morr, Charles Vernon
Surak, John Godfrey

SOUTH DAKOTA
Halverson, Andrew Wayne
Parsons, John G

TENNESSEE
Campbell, Ada Marie
Demott, Bobby Joe
Woods, Alvin Edwin

TEXAS
Beier, Ross Carlton
Brown, Dale Gordon
Chen, Anthony Hing
Elder, Vincent Allen
Goforth, Deretha Rainey
Hendrickson, Constance McRight
Kendall, John Hugh
Mills, William Ronald
Mitchell, Roy Ernest
Parker, Harry W(illiam)
Ray, Allen Cobble
Sand, Ralph E
Schlameus, Herman Wade
Stoner, Graham Alexander
Varsel, Charles John
Will, Paul Arthur

UTAH
Balandrin, Manuel F
DelMar, Eric G
Ernstrom, Carl Anthon
Johnson, John Hal
Richardson, Gary Haight
Segelman, Florence H
Skujins, John Janis
Van Wagenen, Bradford Carr
Wallace, Volney

VIRGINIA
Bishop, John Russell
Dockter, Kenneth Wylie
Evans, Herbert John
Felton, Staley Lee
Ferguson, Robert Nicholas
Glasser, Wolfgang Gerhard
Harowitz, Charles Lichtenberg
Harvey, William Ross
Huang, H(sing) T(sung)
Ikeda, Robert Mitsuru
Marcy, Joseph Edwin
Palmer, James Kenneth
Phillips, Jean Allen
Rickett, Frederic Lawrence
Rippen, Thomas Edward
Sprinkle, Robert Shields, III
Thompson, Mark Ewell

WASHINGTON
Brown, David Francis
Chinn, Clarence Edward
Ericsson, Lowell Harold
Gruger, Edward H, Jr
Ingalsbe, David Weeden
Lewis, Norman G
Pigott, George M
Pomeranz, Yeshajahu
Powers, Joseph Robert
Rasco, Barbara A
Rigby, F Lloyd
Rubenthaler, Gordon Lawrence
Spinelli, John
Stansby, Maurice Earl
Stout, Virginia Falk
Sullivan, John W
Swanson, Barry Grant
Wekell, Marleen Marie
Worthington, Ralph Eric
Zimmermann, Charles Edward

WEST VIRGINIA
Barnes, Robert Keith
Foster, Joyce Geraldine

WISCONSIN
Balke, Nelson Edward
Ballantine, Larry Gene
Bartholomew, Darrell Thomas
Besch, Gordon Otto Carl
Boggs, Lawrence Allen
Broderick, Glen Allen
Bundy, Larry Gene
Chicoye, Etzer
Cooper, Elmer James
Hollenberg, David Henry
Lindsay, Robert Clarence
Olson, Norman Fredrick
Peterson, David Maurice
Priest, Matthew A
Sanderson, Gary Warner
Seitz, Eugene W
Springer, Edward L(ester)
Struble, Craig Bruce
Wall, Joseph Sennen

WYOMING
Katta, Jayaram Reddy
Miller, Glenn Joseph
Rohwer, Robert G
Vance, George Floyd

ALBERTA
Campbell, John Stewart
Hawrysh, Zenia Jean
Hill, Bernard Dale
Kalra, Yash Pal

BRITISH COLUMBIA
Anderson, John Ansel
Li-Chan, Eunice
Majak, Walter
Nakai, Shuryo

MANITOBA
Bushuk, Walter
Kovacs, Miklos I P
Kruger, James Edward
Lukow, Odean Michelin
MacGregor, Alexander William
Noll, John Stephen
Singh, Harwant
Tipples, Keith H
Tkachuk, Russell

NOVA SCOTIA
Ackman, Robert George
Gilgan, Michael Wilson
Robinson, Arthur Robin
Warman, Philip Robert

ONTARIO
Belanger, Jacqueline M R
Brewer, Arthur David
Campbell, James Alexander
Campbell, Kenneth Wilford
Collins, Frank William, Jr
Court, William Arthur
Cunningham, Hugh Meredith
DeMan, John Maria
Ferrier, Leslie Kenneth
Fortin, Andre Francis
Greenhalgh, Roy
Harris, Charles Ronald
Hay, George William
Khan, Shahamat Ullah
Loughheed, Thomas Crossley
Pare, J R Jocelyn
Ratnayake, Walisundera Mudiyanselage Nimal
Rubin, Leon Julius
Sarwar, Ghulam
Scott, Fraser Wallace
Scott, Peter Michael
Siminovitch, David
Spencer, Elvins Yuill
Stavric, Bozidar
Thompson, Lilian Umale
Trenholm, Harold Locksley
Tu, Chin Ming
Wigfield, Yuk Yung
Yaguchi, Makoto
Young, James Christopher F

QUEBEC
Britten, Michel
Common, Robert Haddon
Panalaks, Thavil
Sood, Vijay Kumar

SASKATCHEWAN
Grant, Donald R
Grover, Rajbans
Julien, Jean-Paul
Smith, Allan Edward
Sosulski, Frank Walter

OTHER COUNTRIES
Brown, Barker Hastings
Buckmire, Reginald Eugene
Buckwold, Sidney Joshua
Chan, Hak-Foon
Gregory, Peter
Kurobane, Itsuo
Lee, Men Hui
Malherbe, Roger F
Perry, Albert Solomon

Analytical Chemistry

ALABAMA
Arendale, William Frank
Askew, William Crews
Aull, John Louis
Baker, June Marshall
Barrett, William Jordan
Beeman, Curt Pletcher
Bertsch, Wolfgang
Campbell, Robert Terry
Daniel, Robert Eugene
Emerson, Merle T
Garmon, Ronald Gene
Geppert, Gerard Allen
Gerdom, Larry E
Hamer, Justin Charles
Harris, Albert Zeke
Ho, Mat H
Hung, George Wen-Chi
Johnson, Frank Junior

Lambert, James LeBeau
Liu, Ray Ho
Lloyd, Nelson Albert
McNutt, Ronald Clay
Makhija, Suraj Parkash
Miller, Herbert Crawford
Mountcastle, William R, Jr
Retief, Daniel Hugo
Rymal, Kenneth Stuart
Sides, Gary Donald
Small, Robert James
Smith, Frederick Paul
Strohl, John Henry
Tan, Boen Hie
Timkovich, Russell
Toren, Eric Clifford, Jr
Veazey, Thomas Mabry
Ward, Edward Hilson
Weinberg, David Samuel
Woodham, Donald W

ALASKA
Kennish, John M
LaPerriere, Jacqueline Doyle
Stolzberg, Richard Jay

ARIZONA
Ambrose, Robert T
Andreen, Brian H
Blouin, Leonard Thomas
Boynton, William Vandegrift
Burke, Michael Francis
Cline, Athol
Denton, M Bonner
Fernando, Quintus
Freiser, Henry
Fuchs, Jacob
George, Thomas D
Gilbert, Don Dale
Hildebrandt, Wayne Arthur
Juvet, Richard Spalding, Jr
Karasek, Francis Warren
Kirby, Robert Emmet
Knowlton, Gregory Dean
Liu, Chui Hsun
Mosher, Richard Arthur
Moss, Rodney Dale
Muggli, Robert Zeno
Swinehart, Bruce Arden
Towle, Louis Wallace
Valenty, Steven Jeffrey
Whaley, Thomas Patrick
Whitcomb, Donald Leroy
Wright, Daniel Craig
Yates, Ann Marie

ARKANSAS
Bobbitt, Donald Robert
Bonner, Errol Marcus
Bowman, Leo Henry
Commerford, John D
Fortner, Wendell Lee
Fu, Peter Pi-Cheng
Gosnell, Aubrey Brewer
Hood, Robert L
Howick, Lester Carl
Jimerson, George David
Ku, Victoria Feng
Mitchell, Richard Sibley
Nisbet, Alex Richard
Nix, Joe Franklin
Parks, Albert Fielding
Senseman, Scott Allen
Strong, Robert Stanley

CALIFORNIA
Abbott, Seth R
Abdel-Baset, Mahmoud B
Aberth, William H
Adinoff, Bernard
Ahmed, Faizy
Akmal, Naim
Amoore, John Ernest
Andresen, Brian Dean
Anson, Fred (Colvig)
Bailey, David Nelson
Baker, Frederick Charles
Barrall, Edward Martin, II
Beilby, Alvin Lester
Belsky, Theodore
Bentley, Kenton Earl
Bernhard, Richard Allan
Bilow, Norman
Boyle, Walter Gordon, Jr
Braun, Donald E
Bremner, Raymond Wilson
Brookman, David Joseph
Brundle, Christopher Richard
Burlingame, Alma L
Burnham, Alan Kent
Burtner, Dale Charles
Bushey, Albert Henry
Buttrill, Sidney Eugene, Jr
Bystroff, Roman Ivan
Calkins, Russel Crosby
Camenzind, Mark J
Cape, John Anthony
Caputi, Roger William
Carle, Glenn Clifford
Carr, Edward Mark
Carr, Robert J(oseph)
Chaney, Charles Lester
Chang, Ming-Houng (Albert)
Chatfield, Dale Alton

Chidsey, Christopher E
Chodos, Arthur A
Christoffersen, Donald John
Ciaramitaro, David A
Clark, Dennis Richard
Clarkson, Jack E
Cocks, George Gosson
Colovos, George
Cotts, Patricia Metzger
Crawford, Richard Whittier
Creek, Jefferson Louis
Crichton, David
Crutchfield, Charlie
Current, Jerry Hall
Curry, Bo(stick) U
Czuha, Michael, Jr
Davison, John Blake
Day, Robert James
Dea, Phoebe Kin-Kin
DeJongh, Don C
Delker, Gerald Lee
Deuble, John Lewis, Jr
Donnelly, Timothy C
Doyle, Walter M
Duffield, Jack Jay
Edgerley, Dennis A
Epstein, Barry D
Estill, Wesley Boyd
Evans, Charles Andrew, Jr
Falick, Arnold M
Farrington, Paul Stephen
Fassel, Velmer Arthur
Fedder, Steven Lee
Feldman, Fredric J
Fenimore, David Clarke
Fernandez, Alberto Antonio
Fetzer, John Charles
Fine, Dwight Albert
Fischer, Robert Blanchard
Fisher, Richard Paul
Fletcher, Aaron Nathaniel
Fraser, James Mattison
Frye, Herschel Gordon
Fujikawa, Norma Sutton
Gallegos, Emilio Juan
Gantzel, Peter Kellogg
Gardiner, Kenneth William
Gaston, Lyle Kenneth
Giang, Benjamin Yunwen
Gillen, Keith Thomas
Goh, Kean S
Goldberg, Ira Barry
Gordon, Benjamin Edward
Gordon, Joseph Grover, II
Gordon, Robert Julian
Gouw, T(an) H(ok)
Grant, Patrick Michael
Green, Robert Bennett
Greig, Douglas Richard
Grubbs, Charles Leslie
Grunbaum, Benjamin Wolf
Gump, Barry Hemphill
Haddon, William F, (Jr)
Hanneman, Walter W
Harrar, Jackson Elwood
Harris, Daniel Charles
Hartigan, Martin Joseph
Haugen, Gilbert R
Hausmann, Werner Karl
Hayes, Janan Mary
Helmer, John
Hem, John David
Hessel, Donald Wesley
Hobart, David Edward
Holmes, Ivan Gregory
Hovanec, B(ernard) Michael
Hubert, Jay Marvin
Hurd, Ralph Eugene
Ja, William Yin
Jacob, Peyton, III
Jacob, Robert Allen
Jain, Naresh C
Jakob, Fredi
Jankowski, Stanley John
Janota, Harvey Franklin
Jardine, Ian
Jayne, Jerrold Clarence
Jenden, Donald James
Jennings, Walter Goodrich
Jensen, James Leslie
Jentoft, Ralph Eugene, Jr
Johanson, Robert Gail
Johnson, LeRoy Franklin
Johnson, Phyllis Elaine
Judd, Stanley H
Kabra, Pokar Mal
Kang, Tae Wha
Kassakhian, Garabet Haroutioun
Katekaru, James
Kehoe, Thomas J
Kennedy, John Harvey
Kim, Ki Hong
Kirkpatrick, James W
Kissel, Charles Louis
Kland, Mathilde June
Klein, David Henry
Kong, Eric Siu-Wai
Kothny, Evaldo Luis
Koziol, Brian Joseph
Kuhlmann, Karl Frederick
Kuwahara, Steven Sadao
Landgraf, William Charles
Laub, Richard J
Lawless, James George

Lee, Alfred Tze-Hau
Liang, Yola Yueh-o
Liberman, Martin Henry
Liggett, Lawrence Melvin
Lin, Jiann-Tsyh
Lincoln, Kenneth Arnold
Lindner, Elek
Loda, Richard Thomas
Lord, Harry Chester, III
Lucci, Robert Dominick
Lundin, Robert Enor
Lysyj, Ihor
McGown, Evelyn L
Mach, Martin Henry
McIver, Robert Thomas, Jr
McKaveney, James P
McKinney, Ted Meredith
MacWilliams, Dalton Carson
Mallett, William Robert
Mandelin, Dorothy Jane Bearcroft
Marantz, Laurence Boyd
Markowitz, Samuel Solomon
Marshall, Donald D
Mason, James Willard
Matsuyama, George
Medrud, Ronald Curtis
Meinhard, James Edgar
Miller, George E
Moody, Kenton J
Mooney, John Bernard
Mooney, Larry Albert
Moore, Gerald L
Morris, Jerry Lee, Jr
Mosen, Arthur Walter
Neptune, John Addison
Neufeld, Jerry Don
Nilsson, William A
Nowak, Anthony Victor
Oh, Chan Soo
Okamura, Judy Paulette
Ozari, Yehuda
Palmer, Thomas Adolph
Pappatheodorou, Sofia
Parrish, William
Parsons, Michael L
Paselk, Richard Alan
Passchier, Arie Anton
Passell, Thomas Oliver
Pearson, Robert Melvin
Pease, Burton Frank
Penton, Zelda Eve
Perkins, Willis Drummond
Perone, Samuel Patrick
Perry, Dale Lynn
Pesek, Joseph Joel
Pierce, Matthew Lee
Pollack, Louis Rubin
Potter, John Clarkson
Preus, Martin William
Putz, Gerard Joseph
Que Hee, Shane Stephen
Rabenstein, Dallas Leroy
Raby, Bruce Alan
Rainis, Andrew
Rao, Tadimeti Seetapati
Ratcliff, Milton, Jr
Rhodes, David R
Rider, Benjamin Franklin
Rivier, Jean E F
Roberts, Julian Lee, Jr
Robinson, Merton Arnold
Rocklin, Roy David
Russ, Guston Price, III
Rynd, James Arthur
Sahbari, Javad Jabbari
Sanborn, Russell Hobart
Saperstein, David Dorn
Scattergood, Thomas W
Schaleger, Larry L
Schelar, Virginia Mae
Schwall, Richard Joseph
Seim, Henry Jerome
Selig, Walter S
Shields, Loran Donald
Short, Michael Arthur
Silverman, Herbert Philip
Singh, Jai Prakash
Skoog, Douglas Arvid
Skowronski, Raymund Paul
Sloane, Howard J
Smith, Charles Aloysius
Smith, James Hart
Smith, Leverett Ralph
Solomon, Malcolm David
Stevens, William George
Stevenson, Robert Lovell
Strahl, Erwin Otto
Suffet, I H (Mel)
Sunshine, Irving
Suryaraman, Maruthuvakudi Gopalasastri
Sussman, Howard H
Teeter, Richard Malcolm
Thomas, Richard Sanborn
Thompson, John Michael
Thornton, John Irvin
Throop, Lewis John
Ting, Chih-Yuan Charles
Tsao, Constance S
Tseng, Chien Kuei
Tuffly, Bartholomew Louis
VanAntwerp, Craig Lewis
Vedvick, Thomas Scott
Vomhof, Daniel William
Vreman, Hendrik Jan

Walker, Joseph
Wallcave, Lawrence
Walz, Alvin Eugene
Warren, Paul Horton
Wauchope, Robert Donald
Weiss, Fred Toby
Weiss, Harold Gilbert
Weiss, Roger Harvey
Wensley, Charles Gelen
West, Charles David
West, Donald Markham
Westover, Lemoyne Byron
Whatley, Thomas Alvah
Whiteker, Roy Archie
Wiesendanger, Hans Ulrich David
Wiley, Michael David
Wilkins, Charles Lee
Willis, William Van
Wilson, Charles Oren
Winterlin, Wray Laverne
Worden, Earl Freemont, Jr
Wright, John Marlin
Wu, Jiann-Long
Yang, Yan-Bo
Yavrouian, Andre
Young, Donald Charles
Young, Ho Lee
Yuen, Wing
Zabin, Burton Allen
Zander, Andrew Thomas
Zellmer, David Louis
Ziegler, Carole L

COLORADO
Albert, Harrison Bernard
Allen, Marvin Carrol
Barford, Robert A
Barisas, Bernard George, Jr
Birks, John William
Bower, Nathan Wayne
Bruno, Thomas J
Chao, Tsun Tien
Chisholm, James Joseph
Christie, Joseph Herman
Cleary, Michael E
Edwards, Kenneth Ward
Elliott, Cecil Michael
Erdmann, David E
Espoy, Henry Marti
Fennessey, Paul V
Fishman, Marvin Joseph
Fitzpatrick, Francis Anthony
Herrmann, Scott Joseph
Hill, Walter Edward, Jr
Hitchcock, Eldon Titus
Hutto, Francis Baird, Jr
Janes, Donald Wallace
Jones, Berwyn E
Kinsinger, James A
Kniebes, Duane Van
Koval, Carl Anthony
Long, Austin Richard
McAuliffe, Clayton Doyle
MacCarthy, Patrick
Maciel, Gary Emmet
Mahan, Kent Ira
Martin, Charles R
Mehs, Doreen Margaret
Micheli, Roger Paul
Morrison, Charles Freeman, Jr
Murphy, Robert Carl
Norton, Daniel Remsen
Petersen, Joseph Claine
Rhoads, William Denham
Richardson, Albert Edward
Ritchey, John Michael
Rogerson, Peter Freeman
Schalge, Alvin Laverne
Sievers, Robert Eugene
Skogen, Haven Sherman
Skogerboe, Rodney K
Smith, Dwight Morrell
Smith, John Elvans
Spence, Jack Taylor
Struempler, Arthur W
Stull, Dean P
Tackett, James Edwin, Jr
Thompson, Ronald G
Turner, James Howard
Van Vorous, Ted
Vejvoda, Edward
Voorhees, Kent Jay
Walton, Harold Frederic
Webb, John Day
Wildeman, Thomas Raymond
Wingeleth, Dale Clifford
Woerner, Dale Earl

CONNECTICUT
Agarwal, Vipin K
Alpert, Nelson Leigh
Anderson, Carol Patricia
Andrade, Manuel
Auerbach, Michael Howard
Baldwin, Ronald Martin
Banijamali, Ali Reza
Barney, James Earl, II
Buck, Marion Gilmour
Burnett, Robert Walter
Campbell, Bruce Henry
Chambers, William Edward
Chang, Ted T
Cincotta, Joseph John
Clarke, George

Analytical Chemistry (cont)

Cohen, Edward Morton
Cooper, James William
Coulter, Paul David
Csejka, David Andrew
Curley, James Edward
Ettre, Leslie Stephen
Golden, Gerald Seymour
Gordon, Philip N
Gray, Edward Theodore, Jr
Greenhouse, Steven Howard
Greenough, Ralph Clive
Groth, Joyce Lorraine
Henderson, David Edward
Hiskey, Clarence Francis
I, Ting-Po
Johnson, Bruce McDougall
Kavarnos, George James
Kluender, Harold Clinton
Kozak, Gary S
Krol, George J
Kuck, Julius Anson
Lamm, Foster Philip
Levy, Gabor Bela
MacDonald, John Chisholm
McMurray, Walter Joseph
McNally, John G
Martoglio, Pamela Anne
Mattina, Mary Jane Incorvia
Mayell, Jaspal Singh
Michel, Robert George
Miller, Jerry K
Murai, Kotaro
Page, Edgar J(oseph)
Park, George Bennet
Paul, Jeddeo
Reffner, John A
Rennhard, Hans Heinrich
Rook, Harry Lorenz
Rusling, James Francis
Santacana-Nuet, Francisco
Sarneski, Joseph Edward
Savitzky, Abraham
Sease, John William
Shalvoy, Richard Barry
Shaw, Brenda Roberts
Stock, John Thomas
Stuart, James Davies
Suib, Steven L
Swanson, Donald Leroy
Tom, Glenn McPherson
Voorhies, John Davidson
Ziegler, William Arthur

DELAWARE

Abrahamson, Earl Arthur
Baaske, David Michael
Barth, Howard Gordon
Bauchwitz, Peter S
Benson, Richard Edward
Bly, Donald David
Brame, Edward Grant, Jr
Buchta, Raymond Charles
Carberry, Judith B
Chase, David Bruce
Chiu, Jen
Clarke, John Frederick Gates, Jr
Conner, Albert Z
Crecely, Roger William
Crippen, Raymond Charles
Dippel, William Alan
Durbin, Ronald Priestley
Dwivedi, Anil Mohan
Erdmann, Duane John
Evans, Dennis Hyde
Firment, Lawrence Edward
Fleming, Sydney Winn
Gardiner, John Alden
Gillow, Edward William
Gold, Harvey Saul
Graham, John Alan
Haq, Mohammad Zamir-ul
Harlow, Richard Leslie
Hartzell, Charles Ross, III
Harvey, John, Jr
Hobgood, Richard Troy, Jr
Hoegger, Erhard Fritz
Jack, John James
Johnson, Donald Richard
Kaiser, Mary Agnes
Ketterer, Paul Anthony
Kirkland, Joseph Jack
Kissa, Erik
Kiviat, Fred E
Koeppe, Mary Kolean
Lam, Gilbert Nim-Car
Larsen, Barbara Seliger
Lee, Shung-Yan Luke
Levy, Paul F
McCurdy, Wallace Hutchinson, Jr
McEwen, Charles Nehemiah
Majors, Ronald E
Mathre, Owen Bertwell
Memeger, Wesley, Jr
Milian, Alwin S, Jr
Mitchell, John, Jr
Mollica, Joseph Anthony
Monagle, Daniel J
Moore, Charles B(ernard)
Moore, Ralph Bishop
Moser, Glenn Allen
Mowery, Richard Allen, Jr
Munson, Burnaby

Narvaez, Richard
Neal, Thomas Edward
Neff, Bruce Lyle
Nichols, James Randall
Paris, Jean Philip
Patterson, Gordon Derby, Jr
Pierson, Keith Bernard
Quarry, Mary Ann
Quon, Check Yuen
Reuben, Jacques
Salzman, Steven Kerry
Sammak, Emil George
Sauerbrunn, Robert Dewey
Scott, James Alan
Stelting, Kathleen Marie
Stockburger, George Joseph
Stuper, Andrew John
Swann, Charles Paul
Tise, Frank P
Twelves, Robert Ralph
Ward, George A
Wetlaufer, Donald Burton
Whitney, Charles Candee, Jr
Williams, Reed Chester

DISTRICT OF COLUMBIA

Blackburn, Thomas Roy
Breen, Joseph John
Cash, Gordon Graham
Collat, Justin White
De Levie, Robert
Frank, Richard Stephen
Girard, James Emery
Gunn, John William, Jr
Hammer, Charles F
Hirsch, Roland Felix
Jett-Tilton, Marti
Karle, Jean Marianne
Miller, David Jacob
Morehouse, Kim Matthew
Morris, Joseph Burton
Nesheim, Stanley
Nordquist, Paul Edgard Rudolph, Jr
O'Grady, William Edward
Paabo, Maya
Page, Samuel William
Plost, Charles
Poon, Bing Toy
Preer, James Randolph
Sarmiento, Rafael Apolinar
Schmidt, William Edward
Scott, Kenneth Richard
Shirk, James Siler
Staruszkiewicz, Walter Frank, Jr
Tanner, James Thomas
Vertes, Akos
Wade, Clarence W R
Wang, Jin Tsai
Wilkinson, John Edwin
Wyatt, Jeffrey Renner

FLORIDA

Adams, Martha Lovell
Attaway, John Allen
Bates, Roger Gordon
Baumann, Arthur Nicholas
Bledsoe, James O, Jr
Bowman, Ray Douglas
Brajter-Toth, Anna F
Braman, Robert Steven
Carlson, David Arthur
Cohen, Martin Joseph
Coolidge, Edwin Channing
Davidson, Mark Rogers
Debnath, Sadhana
Delfino, Joseph John
Dinsmore, Howard Livingstone
Drake, Robert Firth
Dubravcic, Milan Frane
Foner, Samuel Newton
Francis, Stanley Arthur
Frediani, Harold Arthur
Fuller, Harold Wayne
Garn, Paul Donald
Gnaedinger, Richard H
Goldring, Lionel Solomon
Gordon, Saul
Graves, Robert Joseph
Green, Floyd J
Gropp, Armin Henry
Hamill, James Junior
Hardman, Bruce Bertolette
Harrison, Willard Wayne
Hartley, Arnold Manchester
Heidner, Robert Hubbard
Hill, Kenneth Richard
Hormats, Ellis Irving
Huber, Joseph William, III
Hunter, George L K
Jackson, George Frederick, III
Jackson, Harold Woodworth
James, Margaret Olive
Keirs, Russell John
Kelley, Myron Truman
Kimble, Allan Wayne
King, Roy Warbrick
Krc, John, Jr
Kroger, Hanns H
Kurtz, George Wilbur
Lacoste, Rene John
Laitinen, Herbert August
Li, San
Loder, Edwin Robert

Long, Calvin H
McGee, William Walter
McKinney, Robert Wesley
Mann, Charles Kenneth
Marshall, Alan George
Martin, Barbara Bursa
Martin, Dean Frederick
Maskal, John
Masters, Larry William
Meisels, Gerhard George
Merrill, Jerald Carl
Miles, Carl J
Miller, William Knight
Mounts, Richard Duane
Moye, Hugh Anson
Munk, Miner Nelson
Myers, Richard Lee
Normile, Hubert Clarence
Olsen, Eugene Donald
Paterson, Arthur Renwick
Perchalski, Robert
Poet, Raymond B
Resnik, Frank Edward
Riedhammer, Thomas M
Roach, Don
Rogers, Lewis Henry
Ross, Lynne Fischer
Ryznar, John W
Schilt, Alfred Ayars
Schmid, Gerhard Martin
Schulman, Stephen Gregory
Searle, Norma Zizmer
Senftleber, Fred Carl
Shaw, Philip Eugene
Spiegelhalter, Roland Robert
Stewart, Burch Byron
Strunk, Duane H
Swartz, William Edward, Jr
Szonntagh, Eugene L(eslie)
Urone, Paul
Vickers, Thomas J
Wallace, Gerald Wayne
Wang, Tsen Chen
Warshowsky, Benjamin
Wheeler, Willis Boly
Wiebush, Joseph Roy
Willard, Thomas Maxwell
Winefordner, James D
Xu, Renliang
Yost, Richard A
Zuo, Yuegang

GEORGIA

Anderson, James Leroy
Baughman, George Larkins
Bayer, Charlene Warres
Black, Billy C, II
Blum, Murray Sheldon
Bottomley, Lawrence Andrew
Boudinot, Frank Douglas
Brewer, John Gilbert
Brooks, John Bill
Busch, Kenneth Louis
Cadwallader, Donald Elton
Chandler, James Harry, III
Chortyk, Orestes Timothy
Churchill, Frederick Charles
Clark, Benjamin Cates, Jr
Colby, David Anthony
Cunningham, Alice Jeanne
Day, Reuben Alexander, Jr
Doetsch, Paul William
Donaldson, William Twitty
Ellington, James Jackson
Flaschka, Hermenegild Arved
Freese, William P, II
Frierson, William Joe
Furse, Clare Taylor
Gaines, Tinsley Powell
Garrison, Arthur Wayne
Gollob, Lawrence
Guerrant, Gordon Owen
Gunter, Bobby J
Hicks, Donald Gail
Hodges, Linda Carol
Holzer, Gunther Ulrich
Hoover, Thomas Burdett
Husa, William John, Jr
Isaac, Robert A
Jackson, William Morrison
Jain, Anant Vir
James, Franklin Ward
James, Jeffrey
Karliner, Jerrold
Koelsche, Charles L
Kohl, Paul Albert
Krajca, Kenneth Edward
Lehner, Andreas Friedrich
Lindauer, Maurice William
Love, Jimmy Dwane
McBride, Clifford Hoyt
McGuire, John Murray
Matthews, Edward Whitehouse
Miles, James William
Nicolson, Paul Clement
O'Neal, Floyd Breland
Parker, Lloyd Robinson, Jr
Pyle, John Tillman
Rhoades, James Lawrence
Robbins, Wayne Brian
Robinson, George Waller
Rogers, Lockhart Burgess
Schweri, Margaret Mary
Simmons, George Allen

Steele, Jack
Stratton, Cedric
Sturrock, Peter Earle
Thruston, Alfred Dorrah, Jr
Trawick, William George
Tsao, Mark Fu-Pao
Vander Velde, George
Williams, John Frederick
Wilson, David Merl
Woodford, James

HAWAII

Crane, Sheldon Cyr
Forster, William Owen
Grothaus, Paul Gerard
Hertlein, Fred, III
Huebert, Barry Joe
Kemp, Paul James
Malmstadt, Howard Vincent
Moritsugu, Toshio
Pecsok, Robert Louis
Roberts, Robert Russell
Sansone, Francis Joseph
Seifert, Josef

IDAHO

Arcand, George Myron
Delmastro, Ann Mary
Delmastro, Joseph Raymond
Farwell, Sherry Owen
Griffiths, Peter Roughley
Hatcher, Herbert John
Hibbs, Robert A
Kirchmeier, Robert L
Lewis, Leroy Crawford
Mincher, Bruce J
Sill, Claude Woodrow
Sutton, John Curtis
Wright, Kenneth James

ILLINOIS

Adlof, Richard Otto
Afremow, Leonard Calvin
Altpeter, Lawrence L, Jr
Anderson, Arnold Lynn
Atoji, Masao
Babcock, Robert Frederick
Bailey, David Newton
Bath, Donald Alan
Bekersky, Ihor
Bernath, Tibor
Bernetti, Raffaele
Bingham, Carleton Dille
Bloemer, William Louis
Bohn, Paul William
Brezinski, Darlene Rita
Brooks, Kenneth Conrad
Brubaker, Inara Mencis
Buko, Alexander Michael
Burns, Richard Price
Carnahan, Jon Winston
Caskey, Albert Leroy
Chadde, Frank Ernest
Chandler, Dean Wesley
Chao, Sherman S
Chipman, Gary Russell
Chou, Mei-In Melissa Liu
Connolly, James D
Cummings, Thomas Fulton
Davis, Andrew Morgan
DeFord, Donald Dale
Dickerson, Donald Robert
Dietz, Mark Louis
Domsky, Irving Isaac
Doody, Marijo
Erickson, Mitchell Drake
Eskins, Kenneth
Fairman, William Duane
Faulkner, Larry Ray
Ferrara, Louis W
Ferren, Larry Gene
Filson, Don P
Firsching, Ferdinand Henry
Fitch, Alanah
Fitzloff, John Frederick
Ganchoff, John Christopher
Gibbons, Larry V
Goss, George Robert
Gray, Linsley Shepard, Jr
Greene, John Philip
Grove, Ewart Lester
Guyon, John Carl
Haberfeld, Joseph Lennard
Halfman, Clarke Joseph
Haugen, David Allen
Hazdra, James Joseph
Hefley, Alta Jean
Hime, William Gene
Hockman, Deborah C
Hoepfinger, Lynn Morris
Hollins, Robert Edward
Holt, Ben Dance
Homeier, Edwin H, Jr
Hughes, Benjamin G
Inskip, Ervin Basil
Jakubiec, Robert Joseph
Janghorbani, Morteza
Jarke, Frank Henry
Jaselskis, Bruno
Jiu, James
Johari, Om
Johnson, John Harold
Johnson, Paul Lorentz
Justen, Lewis Leo

Kaplan, Ephraim Henry
Kennedy, Albert Joseph
Khattab, Ghazi M A
Kieft, Richard Leonard
Klemm, Waldemar Arthur, Jr
Kornel, Ludwig
Koropchak, John
Krawetz, Arthur Altshuler
Ku, Yi-Yin
Lambert, Glenn Frederick
Lambert, Joseph B
Lane, William James
Lasley, Stephen Michael
Layloff, Thomas
Levenberg, Milton Irwin
Levi-Setti, Riccardo
Lewin, Seymour Z
Li, Yao-En
Linde, Harry Wight
Linder, Louis Jacob
Lucchesi, Claude A
McCrone, Walter C
McGorrin, Robert Joseph
MacKichan, Janis Jean
Maclay, G Jordan
Made-Gowda, Netkal M
Markunas, Peter Charles
Markuszewski, Richard
Marley, Nancy Alice
Marquart, John R
Martin, Ronald LeRoy
Matulis, Raymond M
Melendres, Carlos Arciaga
Melford, Sara Steck
Melton, Marilyn Anders
Metcalfe, Lincoln Douglas
Minear, Roger Allan
Moore, Carl Edward
Mounts, Timothy Lee
Nagy, Zoltan
Neas, Robert Edwin
Newcome, Marshall Millar
Newhart, M(ary) Joan
Nieman, Timothy Alan
Norman, Richard Daviess
Osten, Donald Edward
Pankratz, Ronald Ernest
Pasterczyk, William Robert
Peck, Theodore Richard
Petro, Peter Paul, Jr
Pietri, Charles Edward
Piwoni, Marvin Dennis
Prasad, Ramanujam Srinivasa
Rein, James Earl
Rinehart, Kenneth Lloyd
Rohwedder, William Kenneth
Ruch, Rodney R
Scheeline, Alexander
Scholfield, Charles Rexel
Scholz, Robert George
Schroeer, Juergen Max
Sedlet, Jacob
Sellers, Donald Roscoe
Sennello, Lawrence Thomas
Shapiro, Rubin
Sheft, Irving
Sherren, Anne Terry
Sibbach, William Robert
Singh, Laxman
Steele, Ian McKay
Stetter, Joseph Robert
Stoffer, Robert Llewellyn
Stubblefield, Robert Douglas
Surgi, Marion Rene
Sutton, Lewis McMechan
Sytsma, Louis Frederick
Szpunar, Carole Bryda
Tomkins, Marion Louise
Trent, John Ellsworth
Vandeberg, John Thomas
Van Duyne, Richard Palmer
Vanysek, Petr
Washburn, William H
Wei, Lester Yeehow
Wenzel, Bruce Erickson
Wilks, Alan Delbert
Wimer, David Carlisle
Winans, Randall Edward
Wingender, Ronald John
Woolson, Edwin Albert
Young, Austin Harry
Young, David W

INDIANA
Alter, John Emanuel
Amundson, Merle E
Amy, Jonathan Weekes
Bair, Edward Jay
Baird, William McKenzie
Bambenek, Mark A
Becker, Elizabeth Ann (White)
Bishara, Rafik Hanna
Boaz, Patricia Anne
Boguslaski, Robert Charles
Bottei, Rudolph Santo
Bowers, Larry Donald
Burden, Stanley Lee, Jr
Chernoff, Donald Alan
Contario, John Joseph
Day, Edgar William, Jr
DeLong, Allyn F
Demkovich, Paul Andrew
Diamond, Steven Elliot
Doeden, Gerald Ennen

Dube, David Gregory
Dubin, Paul Lee
Dyer, Rolla McIntyre, Jr
Elkin, Robert Glenn
Ellefsen, Paul
Fox, Owen Forrest
Free, Helen M
Fricke, Gordon Hugh
Gainer, Frank Edward
Gallo, Duane Gordon
Gunter, Claude Ray
Guthrie, Frank Albert
Hanks, Alan Ray
Hayes, John Michael
Hemmes, Paul Richard
Heydegger, Helmut Roland
Hieftje, Gary Martin
Hites, Ronald Atlee
Huitink, Geraldine M
Hulbert, Matthew H
Jansing, Jo Ann
Kamat, Prashant V
Kennedy, Edward Earl
Kissinger, Peter Thomas
Kramer, William J
Kuzel, Norbert R
Lagu, Avinash L
Lee, Linda Shahrabani
Lytle, Fred Edward
McGinness, James Donald
McLafferty, John J, Jr
McMasters, Donald L
Margerum, Dale William
Marsh, Max Martin
Mazac, Charles James
Merritt, Lynne Lionel, Jr
Meyerson, Seymour
Michel, Karl Heinz
Mihaliak, Charles Andrew
Morris, David Alexander Nathaniel
Murray, Lionel Philip, Jr
Nagel, Edgar Herbert
Niss, Hamilton Frederick
Occolowitz, John Lewis
Pacer, Richard A
Pardue, Harry L
Peters, Dennis Gail
Pierce, William Meredith, Jr
Pribush, Robert A
Quinney, Paul Reed
Reilly, James Patrick
Reuland, Donald John
Rickard, Eugene Clark
Rivers, Paul Michael
Schaap, Ward Beecher
Schall, Elwyn DeLaurel
Schultz, Franklin Alfred
Shaw, Vernon Reed
Shockey, William Lee
Shultz, Clifford Glen
Siefker, Joseph Roy
Smith, David Lee
Smith, Jean Blair
Smucker, Arthur Allan
Stevenson, William Campbell
Stroube, William Bryan, Jr
Sutula, Chester Louis
Tennyson, Richard Harvey
Timma, Donald Lee
Weaver, Michael John
Wheeler, Ralph John
Wolszon, John Donald
Wozniak, Timothy James
Yan, Sau-Chi Betty
Zimmerman, John F

IOWA
Baetz, Albert L
Barua, Arun B
Bastiaans, Glenn John
Binz, Carl Michael
Buchanan, Edward Bracy, Jr
Cotton, Therese Marie
Diehl, Harvey
Edelson, Martin Charles
Fritz, James Sherwood
Hammerstrom, Harold Elmore
Houk, Robert Samuel
Jacob, Fielden Emmitt
Keiser, Jeffrey E
Kniseley, Richard Newman
Koerner, Theodore Alfred William, Jr
Kracher, Alfred
McClelland, John Frederick
Malone, Diana
Pflaum, Ronald Trenda
Schmidt, Reese Boise
Stahr, Henry Michael
Svec, Harry John
Swartz, James E
Watkins, Stanley Read
Weeks, Stephan John
Yeung, Edward Szeshing

KANSAS
Adams, Ralph Norman
Davis, Lawrence Clark
Decedue, Charles Joseph
Hammaker, Robert Michael
Hawley, Merle Dale
Hiebert, Allen G
Isenhour, Thomas Lee
Iwamoto, Reynold Toshiaki
Judson, Charles Morrill

Kuwana, Theodore
Lambert, Jack Leeper
Landis, Arthur Melvin
Lehman, Thomas Alan
Li, Yi
Lindenbaum, Siegfried
Lookhart, George LeRoy
Lunte, Craig Edward
Macke, Gerald Fred
Marshall, Delbert Allan
Meloan, Clifton E
Pierce, John Thomas
Schorno, Karl Stanley
Seitz, Larry Max
Sellers, Douglas Edwin
Sherwood, Peter Miles Anson
Singhal, Ram P
Sobczynski, Dorota
Soodsma, James Franklin
Sunderman, Herbert D
Swanson, Lynn Allen
Wilson, George Spencer
Wright, Charles Hubert

KENTUCKY
Byrn, Ernest Edward
Byrne, Francis Patrick
Clark, Howell R
Cooke, Samuel Leonard
Davidson, John Edwin
Fabricant, Barbara Louise
Gird, Steven Richard
Hartman, David Robert
Helms, Boyce Dewayne
Hessley, Rita Kathleen
Holler, Floyd James
Hurst, Harrell Emerson
Johnson, Lawrence Robert
Keely, William Martin
Klingenberg, Joseph John
Lauterbach, John Harvey
McClellan, Bobby Ewing
Moldoveanu, C Serban
Moorhead, Edward Darrell
O'Reilly, James Emil
Pan, Wei-Ping
Phillips, John Perrow
Reed, Kenneth Paul
Riley, John Thomas
Robertson, John David
Rosenberg, Alexander F
Schulz, William
Shank, Lowell William
Smith, Walter Thomas, Jr
Thio, Alan Poo-An
Wong, John Lui

LOUISIANA
Bailey, George William
Berg, Eugene Walter
Berni, Ralph John
Braun, Robert Denton
Brown, Lawrence E(ldon)
Carver, James Clark
Chrastil, Joseph
Clarke, Margaret Alice
Colgrove, Steven Gray
Connick, William Joseph, Jr
DeLeon, Ildefonso R
Drushel, Harry (Vernon)
Evilia, Ronald Frank
Fischer, Nikolaus Hartmut
Fitzpatrick, Jimmie Doile
Fontenot, Martin Mayance, Jr
Gale, Robert James
Geissler, Paul Robert
Gibson, David Michael
Greco, Edward Carl
Ham, Russell Allen
Hankins, B(obby) E(ugene)
Hargis, Larry G
Imhoff, Donald Wilbur
Kuan, Shia Shiong
Lee, John Yuchu
Mark, Harold Wayne
Merrill, Howard Emerson
Montalvo, Joseph G, Jr
Morris, Cletus Eugene
Morris, Nancy Mitchell
Moseley, Patterson B
Oberding, Dennis George
Overton, Edward Beardslee
Peard, William John
Riddick, John Allen
Rimes, William John
Robinson, James William
Seidler, Rosemary Joan
Shih, Frederick F
Slaven, Robert Walter
Smith, Isaac Litton
Smith, Robert Leonard
Thomas, Samuel Gabriel
Vercellotti, John R
Vidaurreta, Luis E
Walters, Fred Henry
Warner, Isiah Manuel
West, Philip William
White, June Broussard

MAINE
Bushway, Rodney John
Cronn, Dagmar Rais
Garside, Christopher
Machemer, Paul Ewers

Taylor, Anthony Boswell
Whitten, Maurice Mason

MARYLAND
Adams, James Miller
Alexander, Thomas Goodwin
Anderson, David Leslie
Argauer, Robert John
Babrauskas, Vytenis
Baker, Doris
Barnes, Ira Lynus
Becker, Donald Arthur
Beecher, Gary Richard
Behar, Marjam Gojchlerner
Benson, Walter Roderick
Beroza, Morton
Block, Jacob
Bryden, Wayne A
Callahan, John Joseph
Callahan, Mary Vincent
Caplan, Yale Howard
Cassidy, James Edward
Chakrabarti, Siba Gopal
Christensen, Richard G
Clark, Patricia
Cone, Edward Jackson
Cotter, Robert James
Darneal, Robert Lee
Davis, Henry McRay
DeToma, Robert Paul
Devoe, James Rollo
Dillon, James Greenslade
Downing, Robert Gregory
Duignan, Michael Thomas
Eng, Leslie
Fassett, John David
Fauth, Mae Irene
Fenselau, Catherine Clarke
Ferretti, James Alfred
Feyns, Liviu Valentin
Fisher, Dale John
Flora, Karl Philip
Freeman, David Haines
Gajan, Raymond Joseph
Gilfrich, John Valentine
Goldin, Abraham Samuel
Gordon, Glen Everett
Grady, Lee Timothy
Graham, Joseph H
Groopman, John Davis
Hackert, Raymond L
Haenni, Edward Otto
Hartstein, Arthur M
Heinrich, Kurt Francis Joseph
Herner, Albert Erwin
Hertz, Harry Steven
Hoffman, Michael K
Hornstein, Irwin
Horwitz, William
Hsu, Chen C
Hsu, Stephen M
Issaq, Haleem Jeries
Jones, Donald Eugene
Jovancicevic, Vladimir
Kalasinsky, Victor Frank
Kasler, Franz Johann
Kaufman, Samuel
Kawalek, Joseph Casimir, Jr
Kelly, William Robert
Khare, Mohan
Kirkendall, Thomas Dodge
Klein, Michael
Kline, Jerry Robert
Koch, Thomas Richard
Koch, William Frederick
Kroll, Martin Harris
Leapman, Richard David
Lepley, Arthur Ray
Lerner, Melvin
Lerner, Pauline
Liang, Shoudeng
Lijinsky, William
Linder, Seymour Martin
Lusby, William Robert
McCandliss, Russell John
McCurley, Marian Frances
McGuire, Francis Joseph
Mahle, Nels H
Margalit, Nehemiah
Marinenko, George
Martinez, Richard Isaac
May, Irving
May, Joan Christine
May, Willie Eugene
Mead, Marshall Walter
Melancon, Mark J
Miller, C David
Miziolek, Andrzej Wladyslaw
Modderman, John Philip
Mokhtari-Rejali, Nahld
Moody, John Robert
Murray, George Milton
Muschik, Gary Mathew
Muser, Marc
Musselman, Nelson Page
Newbury, Dale Elwood
Novak, Thaddeus John
O'Haver, Thomas Calvin
O'Rangers, John Joseph
Pannu, Sardul S
Paulsen, Paul
Peterson, John Ivan
Piotrowicz, Stephen R
Pomerantz, Irwin Herman

Analytical Chemistry (cont)

Prouty, Richard Metcalf
Rakhit, Gopa
Rasberry, Stanley Dexter
Rice, Clifford Paul
Rice, James K
Risby, Terence Humphrey
Roller, Peter Paul
Rollins, Orville Woodrow
Rotherham, Jean
St John, Peter Alan
Salwin, Harold
Samuel, Aryeh Hermann
Schaffer, Robert
Schifreen, Richard Steven
Schwartz, Robert Saul
Smardzewski, Richard Roman
Snow, Milton Leonard
Snow, Philip Anthony
Sphon, James Ambrose
Stadlbauer, John Mannix
Stalling, David L
Stuntz, Calvin Frederick
Taylor, John Keenan
Theimer, Edgar E
Thompson, Donald Leroy
Tompa, Albert S
Topping, Joseph John
Turk, Gregory Chester
Vance, Hugh Gordon
Vanderslice, Joseph Thomas
Veillon, Claude
Wexler, Arthur Samuel
Whiting, John Dale, Jr
Willeboordse, Friso
Williams, Frederick Wallace
Wolf, Wayne Robert
Yergey, Alfred L, III
Young, Harold Henry
Yurow, Harvey Warren
Zacharius, Robert Marvin

MASSACHUSETTS

Adler, Norman
Ahearn, James Joseph, Jr
Ammlung, Richard Lee
Atkins, Jaspard Harvey
Baratta, Edmond John
Barnes, Ramon M
Berera, Geetha Poonacha
Berlandi, Francis Joseph
Berry, Vern Vincent
Bessette, Russell R
Bidlingmeyer, Brian Arthur
Biemann, Klaus
Billo, Edward Joseph
Boehm, Paul David
Bracco, Donato John
Brauner, Phyllis Ambler
Buono, John Arthur
Carey, George H
Cembrola, Robert John
Chung, Frank H
Clarke, Michael J
Coleman, Geoffry N
Coppola, Elia Domenico
Cormier, Alan Dennis
Cosgrove, James Francis
Costello, Catherine E
Crabtree, Donald V
Curran, David James
David, Donald J
Davies, Geoffrey
Dehal, Shangara Singh
DiNardi, Salvatore Robert
Donoghue, John Timothy
Ehntholt, Daniel James
Ellingboe, James
Ellis, David Wertz
Ezrin, Myer
Flagg, John Ferard
Fletcher, Kenneth Steele, III
Fong, Godwin Wing-Kin
Forman, Earl Julian
Frant, Martin S
Fritzsche, Alfred Keith
Gersh, Michael Elliot
Gershman, Louis Leo
Gerteisen, Thomas Jacob
Gibb, Thomas Robinson Pirie
Gilbert, Theodore William, Jr
Gilbert, Thomas Rexford
Girard, Francis Henry
Glajch, Joseph Louis
Gopikanth, M L
Gore, William Earl
Graffeo, Anthony Philip
Griffin, Paul Joseph
Haas, John William, Jr
Hammond, Sally Katharine
Hanselman, Raymond Bush
Hayden, Thomas Day
Heyn, Arno Harry Albert
Hobbs, John Robert
Hochella, Norman Joseph
Jankowski, Conrad M
Jarboe, Jerry K(ent)
Kaplan, David Lee
Karger, Barry Lloyd
Klanfer, Karl
Kliem, Peter O
Kocon, Richard William
Krull, Ira Stanley
Langmuir, Margaret Elizabeth Lang
Learson, Robert Joseph
Legg, Kenneth Deardorff
Lembo, Nicholas J
Leonard, Edward H
Lester, Joseph Eugene
Li, Kuang-Pang
Li, Rounan
Licht, Stuart Lawrence
Light, Truman S
Lingane, James Joseph
Little, James Noel
Livingston, Hugh Duncan
Lublin, Paul
Macaione, Domenic Paul
McGarry, Margaret
Malenfant, Arthur Lewis
Mattina, Charles Frederick
Merritt, Margaret Virginia
Mowery, Dwight Fay, Jr
Neue, Uwe Dieter
Olmez, Ilhan
Olver, John Walter
Pinkus, Jack Leon
Plankey, Francis William, Jr
Pojasek, Robert B
Ram, Neil Marshall
Reckhow, David Alan
Robbat, Albert, Jr
Robertson, Donald Hubert
Ross, James William, Jr
Rubin, Leon E
Rupich, Martin Walter
Serafin, Frank G
Serra, Jerry M
Shepardson, John U
Siegal, Bernard
Siggia, Sidney
Sudmeier, James Lee
Takman, Bertil Herbert
Thomas, Martha Jane Bergin
Thorstensen, Thomas Clayton
Tramondozzi, John Edmund
Turnquist, Carl Richard
Tyson, Julian Fell
Uden, Peter Christopher
Varco-Shea, Theresa Camille
Vouros, Paul
Wallace, David H
Wallace, Frederic Andrew
Wechter, Margaret Ann

MICHIGAN

Allison, John
Anderson, Charles Thomas
Armentrout, David Noel
Bauer, William Eugene
Benson, Edmund Walter
Bernius, Mark Thomas
Betso, Stephen Richard
Bhatt, Padmamabh P
Bone, Larry Irvin
Bowman, Phil Bryan
Braselton, Webb Emmett, Jr
Bredeweg, Robert Allen
Butts, Susan Beda
Cadle, Steven Howard
Cantor, David Milton
Chance, Robert L
Chrepta, Stephen John
Chulski, Thomas
Clipper, Scott Alan
Coburn, Joel Thomas
Coleman, David Manley
Cope, Virgil W
Corrigan, Dennis Arthur
Crouch, Stanley Ross
Crow, Frank Warren
Crummett, Warren B
D'Itri, Frank M
Dragun, James
Dukes, Gary Rinehart
Elving, Philip Juliber
Erlich, Ronald Harvey
Fawcett, Timothy Goss
Floutz, William Vaughn
Forist, Arlington Ardeane
Frade, Peter Daniel
Frawley, Nile Nelson
Gaarenstroom, Stephen William
Gill, Harold Hatfield
Goetz, Rudolph W
Goodwin, Jesse Francis
Gordus, Adon Alden
Gulick, Wilson M, Jr
Gump, J R
Hajratwala, Bhupendra R
Hamlin, William Earl
Hart, Donald John
Heeschen, Jerry Parker
Herrinton, Paul Matthew
Holcomb, Ira James
Holkeboer, Paul Edward
Hood, Robin James
Howell, James Arnold
Hutchinson, Kenneth A
Ikuma, Hiroshi
Jackson, Larry Lynn
Jensen, David James
Johnson, Jack (Lamar)
Johnson, Ray Leland
Kaiser, David Gilbert
Kallos, George J
Kang, Uan Gen
Kelly, Nelson Allen
King, Stanley Shih-Tung
Kolat, Robert S
Kuo, Mingshang
Landrum, Peter Franklin
Langvardt, Patrick William
Larson, John Grant
Leenheer, Mary Janeth
Lewis, Lynn Loraine
Lindblad, William John
Lipton, Michael Forrester
Lorch, Steven Kalman
Lubman, David Mitchell
McEwen, David John
McLean, James Dennis
Maheswari, Shyam P
Marquardt, Roland Paul
Meyerhoff, Mark Elliot
Morris, Michael D
Mosher, Robert Eugene
Munson, James William
Murie, Richard A
Myers, Harvey Nathaniel
Nader, Bassam Salim
Nestrick, Terry John
Neumann, Fred William
Nevius, Timothy Alfred
Parker, Gordon Arthur
Petersen, Donald Ralph
Petzold, Edgar
Pfeiffer, Curtis Dudley
Poole, Colin Frank
Popov, Alexander Ivan
Potter, Noel Marshall
Price, Harold Anthony
Prostak, Arnold S
Przybytek, James Theodore
Rainey, Mary Louise
Ray, Jesse Paul
Reim, Robert E
Rengan, Krishnaswamy
Roberts, Charles Brockway
Robertson, John Harvey
Rorabacher, David Bruce
Rulfs, Charles Leslie
Schuetzle, Dennis
Seymour, Michael Dennis
Shah, Shirish A
Shimp, Neil Frederick
Siegl, Walter Otto
Sinsheimer, Joseph Eugene
Skelly, Norman Edward
Slywka, Gerald William Alexander
Small, Hamish
Smart, James Blair
Smith, Kenneth Edward
Smith, Ralph G
Smith, Richard Harding
Spang, Arthur William
Spurlock, Carola Henrich
Steinhaus, Ralph K
Stenger, Vernon Arthur
Struck, William Anthony
Suggs, William Terry
Swarin, Stephen John
Swartz, Grace Lynn
Swathirajan, S
Swovick, Melvin Joseph
Szutka, Anton
Taraszka, Anthony John
Thomas, Kenneth Eugene, III
Timnick, Andrew
Tou, James Chieh
Tsuji, Kiyoshi
Turley, June Williams
Vanderwielen, Adrianus Johannes
VanEffen, Richard Michael
Van Hall, Clayton Edward
Walters, Stephen Milo
Warren, H(erbert) Dale
Watson, Jack Throck
Weber, Dennis Joseph
Whitfield, Richard George
Wong, Peter Alexander
Yang, Victor Chi-Min
Zak, Bennie

MINNESOTA

Adams, Ernest Clarence
Andria, George D
Atwell, William Alan
Baude, Frederic John
Brenner, Mark
Brezonik, Patrick Lee
Bydalek, Thomas Joseph
Carr, Peter William
Clapp, C(harles) Edward
DeRoos, Fred Lyn
Dittmar, Rebecca May
Duerst, Richard William
Durnick, Thomas Jackson
Ellefson, Ralph Donald
Evans, John Fenton
Farm, Raymond John
Farnum, Sylvia A
Forrette, John Elmer
Freier, Herbert Edward
Fritsch, Carl Walter
Gyberg, Arlin Enoch
Haddad, Louis Charles
Hagen, Donald Frederick
Henney, Robert Charles
Holden, John B, Jr
Jensen, Richard Erling
Kariv-Miller, Essie
Katz, William
Kolthoff, Izaak Maurits
Kovacic, Joseph Edward
Latterell, Joseph J
Lodge, Timothy Patrick
Lonnes, Perry Bert
MacKellar, William John
McKenna, Jack F(ontaine)
McMullen, James Clinton
Marsh, Frederick Leon
Marshall, John Clifford
Mishmash, Harold Edward
Newmark, Richard Alan
Olsen, Rodney L
Oxborrow, Gordon Squires
Poe, Donald Patrick
Potts, Lawrence Walter
Ramette, Richard Wales
Rejto, Peter A
Thatcher, Walter Eugene
Thompson, Richard David
Tiers, George Van Dyke
Todd, Jerry William
Toren, Paul Edward
Walters, John Philip

MISSISSIPPI

Alley, Earl Gifford
Carter, Fairie Lyn
Doumit, Carl James
ElSohly, Mahmoud Ahmed
Fawcett, Newton Creig
Hussey, Charles Logan
Kopfler, Frederick Charles
Luker, William Dean
Minyard, James Patrick
Monts, David Lee
Tai, Han

MISSOURI

Armstrong, Daniel Wayne
Arneson, Dora Williams
Beaver, Earl Richard
Bier, Dennis Martin
Bohanon, Joseph Terril
Brackmann, Richard Theodore
Brescia, Vincent Thomas
Briner, Robert C
Chen, Yih-Wen
Cheng, Kuang Lu
Clowers, Churby Conrad, Jr
Cole, Douglas L
Conkin, Robert A
Coria, Jose Conrado
Craver, Clara Diddle (Smith)
Dahl, William Edward
Das, Manik Lal
Drew, Henry D
Emery, Edward Mortimer
Field, Byron Dustin
Frankoski, Stanley P
Freeland, Max
Freeman, John Jerome
Gay, Don Douglas
Geisman, Raymond August, Sr
Gerhardt, Klaus Otto
Gibbons, James Joseph
Glascock, Michael Dean
Grayson, Michael A
Greenlief, Charles Michael
Gustafson, Mark Edward
Haggerty, William Joseph, Jr
Hallas, Laurence Edward
Hardtke, Fred Charles, Jr
Haynes, William Miller
Housmyer, Carl Leonidas
Jungclaus, Gregory Alan
Kaelble, Emmett Frank
Kaley, Robert George, II
Keller, Robert Ellis
Kenyon, Allen Stewart
Kitchin, Robert Walter
Klein, Andrew John
Kluba, Richard Michael
Koirtyohann, Samuel Roy
Korotev, Randall Lee
Ladenson, Jack Herman
Larson, Wilbur John
Leventis, Nicholas
Levy, Ram Leon
Lott, Peter F
Ludwig, Frederick John, Sr
Macias, Edward S
Malik, Joseph Martin
Mori, Erik Jun
Moulton, Robert Henry
Neau, Steven Henry
Newell, Jon Albert
Ogilvie, James Louis
Pickett, Edward Ernest
Plimmer, Jack Reynolds
Podosek, Frank A
Rath, Nigam Prasad
Roy, Rabindra (Nath)
Rucppel, Melvin Leslie
Schaeffer, Harold F(ranklin)
Sharp, Dexter Brian
Sherman, William Reese
Smith, David Warren
Smith, Robert Eugene
Smith, Ronald Gene
Sotiriou-Leventis, Chariklia
Soundararajan, Rengarajan

Stout, Edward Irvin
Sunde, Roger Allan
Talbott, Ted Delwyn
Thielmann, Vernon James
Thompson, Richard Michael
Tuthill, Samuel Miller
Wilcox, Harold Kendall
Woodhouse, Edward John
Worley, Jimmy Weldon

MONTANA
Amend, John Robert
Layman, Wilbur A
McCauley, Gerald Brady
Volborth, Alexis

NEBRASKA
Blickensderfer, Peter W
Carr, James David
Freidline, Charles Eugene
Gross, Michael Lawrence
Heath, Eugene Cartmill
Howell, Daniel Bunce
Johar, J(ogindar) S(ingh)
Kaplan, Sanford Sandy
Kenkel, John V
Kolade, Alabi E
Mattern, Paul Joseph
Mattes, Frederick Henry
Nipper, Henry Carmack
Phelps, George Clayton
Solsky, Joseph Fay

NEVADA
Betowski, Leon Donnelly, Jr
Billingham, Edward J, Jr
Eastwood, DeLyle
Gerard, Jesse Thomas
Johnson, Brian James
Klainer, Stanley M
Miles, Maurice Jarvis
Pierson, William R
Rhees, Raymond Charles
Schooley, David Allan
Sovocool, G Wayne
Tanner, Roger Lee
Ulrich, William Frederick
Vincent, Harold Arthur
Wakayama, Edgar Junro

NEW HAMPSHIRE
Damour, Paul Lawrence
Finn, James Walter
Foglesong, Wavell Wainwright
Hebert, Normand Claude
Illian, Carl Richard
Leipziger, Fredric Douglas
McGinness, James E
Merritt, Charles, Jr
Neil, Thomas C
Pfluger, Clarence Eugene
Roberts, John Edwin
Seitz, William Rudolf
Stepenuck, Stephen Joseph, Jr

NEW JERSEY
Abdou, Hamed M
Achari, Raja Gopal
Aczel, Thomas
Adler, Seymour Jacob
Aldrich, Haven Scott
Altman, Lawrence Jay
Andrade, John Robert
Anthony, Linda J
Ardelt, Wojciech Joseph
Aronovic, Sanford Maxwell
Ashworth, Harry Arthur
Atwater, Beauford W
Banick, William Michael, Jr
Barnes, Robert Lee
Bathala, Mohinder S
Beck, John Louis
Beispiel, Myron
Bentz, Bryan Lloyd
Benz, Wolfgang
Bhattacharyya, Pranab K
Boczkowski, Ronald James
Bodin, Jerome Irwin
Bornstein, Alan Arnold
Bornstein, Michael
Borysko, Emil
Bose, Ajay Kumar
Bozzelli, Joseph William
Bradstreet, Raymond Bradford
Brand, William Wayne
Brody, Stuart Martin
Brofazi, Frederick R
Brown, John Angus
Buch, Robert Michael
Burnett, Bruce Burton
Carlucci, Frank Vito
Carter, James Evan
Chandrasekhar, Prasanna
Chaudhri, Safee U
Christenson, Philip A
Ciaccio, Leonard Louis
Citron, Irvin Meyer
Cohen, Allen Irving
Cone, Conrad
Conley, Jack Michael
Cronin, John
Daly, Robert E
Dean, Donald E
DeCastro, Arthur

De Silva, John Arthur F
Diegnan, Glenn Alan
Downing, George V, Jr
Duerr, J Stephen
Dunham, John Malcolm
Dunn, William Howard
Dziedzic, Joseph
Eastman, David Willard
Edelson, Edward Harold
Edelstein, Harold
Egan, Richard Stephen
Egli, Peter
Ehrlich, Julian
Eider, Norman George
Fawzi, Ahmad B
Feldman, Nicholas
Feng, Rong
Ferraro, Charles Frank
Fink, David Warren
Finston, Harmon Leo
Fitchett, Gilmer Trower
Fix, Kathleen A
Flanders, Clifford Auten
Fong, Jones W
Frantz, Beryl May
Ganapathy, Ramachandran
Geller, Milton
Gerecht, J Fred
Gierer, Paul L
Glass, John Richard
Goffman, Martin
Granchi, Michael Patrick
Grandolfo, Marian Carmela
Grey, Peter
Gullo, Vincent Philip
Hackman, Martin Robert
Hagel, Robert B
Hall, Gene Stephen
Hall, Gretchen Randolph
Halpern, Donald F
Hammond, Willis Burdette
Hanley, Arnold V
Hanna, Samir A
Hansen, Holger Victor
Hansen, Ralph Holm
Hartkopf, Arleigh Van
Heacock, Craig S
Heveran, John Edward
Hills, Stanley
Ho, Chi-Tang
Hobart, Everett W
Hoffman, Clark Samuel, Jr
Hoffman, Henry Tice, Jr
Hokanson, Gerard Clifford
Huger, Francis P
Humphreys, Robert William Riley
Iorns, Terry Vern
Jackson, Thomas A J
Jacobs, Morton Howard
Jacobson, Harold
Jain, Nemichand B
Janssen, Richard William
Javick, Richard Anthony
Jemal, Mohammed
Jensen, Norman P
Johnson, Hilding Reynold
Joseph, John Mundancheril
Kabadi, Balachandra N
Kadin, Harold
Kagan, Michael Z
Kallmann, Silve
Kaplan, Gerald
Katz, Sidney A
Kender, Donald Nicholas
Kern, Werner
Kikta, Edward Joseph, Jr
Kim, Benjamin K
Kirschbaum, Joel Jerome
Kline, Berry James
Kobrin, Robert Jay
Kohout, Frederick Charles, III
Kolis, Stanley Joseph
Korfmacher, Walter Averill
Krey, Philip W
Kumari, Durga
Kuritzkes, Alexander Mark
Kwon, Joon Taek
Labows, John Norbert, Jr
Lalancette, Roger A
Laughlin, Alice
Lee, Catherine Coyle
Letterman, Herbert
Lewis, Arnold D
Liebowitz, Stephen Marc
Lin, Denis Chung Kam
Lofstrom, John Gustave
Lorenz, Patricia Ann
Love, L J Cline
MacDonald, Alexander, Jr
McGuire, David Kelty
McMahon, David Harold
McSharry, William Owen
Maldacker, Thomas Anton
Maloy, Joseph T
Medwick, Thomas
Melveger, Alvin Joseph
Mergens, William Joseph
Miller, James Monroe
Miller, Warren Victor
Milliman, George Elmer
Mitchell, James Winfield
Montana, Anthony J
Montgomery, Richard Millar
Moros, Stephen Andrew

Mowitz, Arnold Martin
Mukai, Cromwell Daisaku
Murthy, Vadiraja Venkatesa
Mussinan, Cynthia June
Mustafa, Shams
Nadkarni, Ramachandra Anand
Naik, Datta Vittal
Nicolau, Gabriela
Niedermayer, Alfred O
O'Connor, Joseph Michael
O'Connor, Matthew James
Okinaka, Yutaka
Opila, Robert L, Jr
Oppenheimer, Larry Eric
Ortiz-Martinez, Aury
Oyler, Alan Richard
Pamer, Treva Louise
Papastephanou, Constantin
Patrick, James Edward
Phillips, Wendell Francis
Pilkiewicz, Frank George
Platt, Thomas Boyne
Pluscec, Josip
Rabel, Fredric M
Rau, Eric
Redden, Patricia Ann
Reents, William David, Jr
Reissmann, Thomas Lincoln
Rigler, Neil Edward
Roberts, Ronald Frederick
Rodgers, Robert Stanleigh
Rolle, F Robert
Romano, Salvatore James
Rose, Ira Marvin
Rose, Stuart Alan
Rosen, Joseph David
Rosen, Robert T
Rosenberg, Ira Edward
Rothstein, Edwin C(arl)
Roy, Ram Babu
Sattur, Theodore W
Schwartzkopf, George, Jr
Scott, Mary Celine
Shearer, Charles M
Shinkai, Ichiro
Sibilia, John Philip
Sieh, David Henry
Sinclair, James Douglas
Singleton, Bert
Smyers, William Hays
Somkaite, Rozalija
Spohn, Ralph Joseph
Sterbenz, Francis Joseph
Stillman, John Edgar
Stober, Henry Carl
Szap, Peter Charles
Szyper, Mira
Taylor, John William
Tibbetts, Merrick Sawyer
Toome, Voldemar
Trewella, Jeffrey Charles
Triglia, Emil J
Tse, Francis Lai-Sing
Turi, Paul George
Turse, Richard S
Tway, Patricia C
Vallese, Frank M
Venturella, Vincent Steven
Vickroy, David Gill
Virgili, Luciano
Vogel, Veronica Lee
Waltking, Arthur Ernest
Westerdahl, Carolyn Ann Lovejoy
Westerdahl, Raymond P
Weston, Charles Alvin
Whigan, Daisy B
Wilson, Mabel F
Witkowski, Mark Robert
Wittick, James John
Wolf, Thomas
Woodruff, Hugh Boyd
Zaim, Semih
Zanzucchi, Peter John
Zenchelsky, Seymour Theodore

NEW MEXICO
Apel, Charles Turner
Balagna, John Paul
Beattie, Willard Horatio
Bechtold, William Eric
Bentley, Glenn E
Binder, Irwin
Bowen, Scott Michael
Burns, Frank Bernard
Campbell, George Melvin
Caton, Roy Dudley, Jr
Clark, Robert Paul
Cunningham, Paul Thomas
Enke, Christie George
Essington, Edward Herbert
Ewing, Galen Wood
Fritz, Georgia T(homas)
George, Raymond S
Haaland, David Michael
Hakkila, Eero Arnold
Heaton, Richard Clawson
Jackson, Darryl Dean
Keller, Roy Alan
Kelly, Clark Andrew
Kenna, Bernard Thomas
Loughran, Edward Dan
Matlack, George Miller
Mead, Richard Wilson
Morales, Raul

Nielsen, Stuart Dee
Niemczyk, Thomas M
Nogar, Nicholas Stephen
Ohline, Robert Wayne
Ott, Donald George
Park, Su-Moon
Patterson, James Howard
Rayson, Gary D
Renschler, Clifford Lyle
Rogers, Raymond N
Sedlacek, William Adam
Sherman, Robert Howard
Shore, Fred L
Smith, James Lewis
Smith, Maynard E
Smith, Wayne Howard
Spall, Walter Dale
Stanbro, William David
Trujillo, Patricio Eduardo
Vanderborgh, Nicholas Ernest
Walton, George
Wang, Joseph
Weissman, Suzanne Heisler
West, Mike Harold
Williams, Mary Carol
Yasuda, Stanley K

NEW YORK
Abraham, Jerrold L
Abruna, Hector D
Ahuja, Satinder
Aikens, David Andrew
Alben, Katherine Turner
Alliet, David F
Anderson, Frank Wallace
Andrews, John Parray
Andrews, Mark Allen
Angelino, Norman J
Antonucci, Frank Ralph
Apai, Gustav Richard, II
Attygalle, Athula B
Baden, Harry Christian
Beach, David H
Becker, Harry Carroll
Berger, Selman A
Billmeyer, Fred Wallace, Jr
Birke, Ronald Lewis
Bixler, John Wilson
Blackwell, Crist Scott
Boettger, Susan D
Bonvicino, Guido Eros
Bottger, Gary Lee
Bowser, James Ralph
Brown, Oliver Monroe
Bruckenstein, Stanley
Burlitch, James Michael
Bush, David Graves
Campion, James J
Cappel, C Robert
Capretta, Umberto
Cardenas, Raúl R, Jr
Carnahan, James Claude
Carpenter, Thomas J
Carson, Chester Carrol
Chang, Jack Che-Man
Charola, Asuncion Elena
Conway, Walter Donald
Cooke, William Donald
Cooper, Aaron David
Corth, Richard
Cover, Richard Edward
Cratty, Leland Earl, Jr
Creasy, William Russel
D'Angelo, Gaetano
Daves, Glenn Doyle, Jr
Davis, Abram
DeForest, Peter Rupert
Dehm, Richard Lavern
DeStefano, Anthony Joseph
Dexter, Theodore Henry
Dietz, Edward Albert, Jr
Dietz, Russell Noel
Dinan, Frank J
DiNunzio, James E
Douglass, Pritchard Calkins
Dumoulin, Charles Lucian
Durst, Richard Allen
Dutta, Shib Prasad
Dwyer, Robert Francis
Eadon, George Albert
Edsberg, Robert Leslie
Fay, Homer
Fiala, Emerich Silvio
Forcier, George Arthur
Froelich, Philip Nissen
Gardella, Joseph Augustus, Jr
Gaver, Robert Calvin
Gilbert, Jack Pittard
Glickstein, Joseph
Gluck, Ronald Monroe
Goddard, John Burnham
Goldman, James Allan
Gordon, Harry William
Greizerstein, Hebe Beatriz
Grossman, William Elderkin Leffingwell
Hagan, William John, Jr
Hall, Ernest Leroy
Halpern, Mordecai Joseph
Hankes, Lawrence Valentine
Hanson, Jonathan C
Hardy, Ralph Wilbur Frederick
Heininger, Clarence George, Jr
Heintz, Edward Allein
Hopke, Philip Karl

CHEMISTRY / 185

Analytical Chemistry (cont)

Huie, Carmen Wah-Kit
Iden, Charles R
Ilmet, Ivor
Irsa, Adolph Peter
Jackson, Kenneth William
Jacobson, Jay Stanley
Jaffe, Marvin Richard
Janauer, Gilbert E
Jespersen, Neil David
Johnson, Raymond Nils
Jones, Stanley Leslie
Julian, Donald Benjamin
Karp, Stewart
Keesee, Robert George
Kesner, Leo
Kho, Boen Tong
Killian, Carl Stanley
Kingston, Charles Richard
Kissel, Thomas Robert
Klingele, Harold Otto
Kross, Robert David
Kruegel, Alice Virginia
Lam, Stanley K
Lane, Keith Aldrich
Launer, Philip Jules
Leff, Judith
Lerman, Steven I
Lessor, Edith Schroeder
Levinson, Steven R
Levy, Arthur Louis
Levy, George Charles
Liang, Charles C
Liao, Hsueh-Liang
Ligon, Woodfin Vaughan, Jr
Loach, Kenneth William
Locke, David Creighton
Lorenzo, George Albert
Loscalzo, Anne Grace
Lovecchio, Frank Vito
Luders, Richard Christian
Ma, Maw-Suen
Macero, Daniel Joseph
McGriff, Richard Bernard
McHugh, James Anthony, Jr
McLafferty, Fred Warren
McLennan, Scott Mellin
Mahony, John Daniel
Manche, Emanuel Peter
Margoshes, Marvin
Marov, Gaspar J
Marr, David Henry
Matthews, Dwight Earl
Meloon, Daniel Thomas, Jr
Mennitt, Philip Gary
Merritt, Paul Eugene
Meyer, John Austin
Michiels, Leo Paul
Molina, John Francis
Molloy, Andrew A
Morrison, George Harold
Mourning, Michael Charles
Muller, Olaf
Murphy, Kenneth Robert
Mylroie, Victor L
Nathanson, Benjamin
Nealy, Carson Louis
Nelson, Kurt Herbert
Neuborт, Shimon
Noel, Dale Leon
Norton, Elinor Frances
Oberholtzer, James Edward
O'Donnell, Raymond Thomas
O'Keefe, Patrick William
Orna, Mary Virginia
Orzech, Chester Eugene, Jr
Owens, Patrick M
Padmanabhan, G R
Parker, Frank S
Parsons, Patrick Jeremy
Pearse, George Ancell, Jr
Pittman, Kenneth Arthur
Pontius, Dieter J J
Potter, George Henry
Prigot, Melvin
Przybylowicz, Edwin P
Rand, Salvatore John
Reardon, Joseph Daniel
Redalieu, Elliot
Reddy, Thomas Bradley
Reeves, Stuart Graham
Reinmuth, William Henry
Reuter, Wilhad
Richtol, Herbert H
Roboz, John
Rogers, Donald Warren
Rosenthal, Donald
Ross, Malcolm S F
Rothchild, Robert
Russell, Virginia Ann
Sage, Gloria W
Salotto, Anthony W
Sandifer, James Roy
Sauer, Charles William
Sayegh, Joseph Frieh
Schaefer, Robert William
Schmeltz, Irwin
Schneider, Frank L
Schottmiller, John Charles
Schucker, Gerald D
Schupp, Orion Edwin, III
Schwartz, Herbert Mark
Segatto, Peter Richard
Sharkey, John Bernard
Sharma, Minoti
Shilman, Avner
Silver, Herbert Graham
Singh, Surjit
Solomon, Jerome Jay
Spielholtz, Gerald I
Spink, Charles Harlan
Strojny, Norman
Stryker, Martin H
Su, Yao Sin
Swartz, James Lawrence
Sweet, Richard Clark
Takeuchi, Esther Sans
Talley, Charles Peter
Teitelbaum, Charles Leonard
Thumm, Byron Ashley
Tong, Stephen S C
Triplett, Kelly B
Underkofler, William Leland
Van Geet, Anthony Leendert
Vitus, Carissima Marie
Wandass, Joseph Henry
Wapnir, Raul A
Weisler, Leonard
Werner, Thomas Clyde
West, Kenneth Calvin
Whitlock, L Ronald
Wiberley, Stephen Edward
Wong, Lan Kan
Wood, David
Wright, Charles Joseph
Wright, Robert W
Wu, Konrad T
Wu, Mingdan
Youker, John
Yu, Ming Lun
Zuehlke, Carl William
Zuman, Petr

NORTH CAROLINA
Adcock, Louis Henry
Alam, Mohammed Ashraful
Anderegg, Robert James
Bell, Jimmy Holt
Boss, Charles Ben
Bryan, Horace Alden
Buck, Richard Pierson
Burnett, John Nicholas
Bursey, Joan Tesarek
Bursey, Maurice Moyer
Chou, David Yuan Pin
Chung, Henry Hsiao-Liang
Clapp, William Lee
Clements, John B(elton)
Cochran, George Thomas
Collier, Herman Edward, Jr
Coury, Louis
Cross, Robert Edward
Crumpler, Thomas Bigelow
Culbreth, Judith Elizabeth
Cundiff, Robert Hall
Decker, Clifford Earl, Jr
Dobbins, James Talmage, Jr
Forman, Donald T
Gangwal, Santosh Kumar
Ganz, Charles Robert
Gemperline, Margaret Mary Cetera
Gemperline, Paul Joseph
Gibson, Robert Harry
Gillikin, Jesse Edward, Jr
Green, Charles Raymond
Hadzija, Bozena Wesley
Haile, Clarence Lee
Hanck, Kenneth William
Harrell, T Gibson
Harrison, Stanley L
Hass, James Ronald
Heck, Henry d'Arcy
Heckman, Robert Arthur
Heintzelman, Richard Wayne
Herman, Harvey Bruce
Hinze, Willie Lee
Hong, Donald David
Hurlbert, Bernard Stuart
Jackman, Donald Coe
Jayanty, R K M
Jezorek, John Robert
Johnson, Delwin Phelps
Johnson, J(ames) Donald
Jorgenson, James Wallace
Kersey, Robert Lee, Jr
Kirby, James Ray
Knecht, Laurance A
Leidy, Ross Bennett
Levine, Solomon Leon
Lewis, Claude Irenius
Li, Chia-Yu
Linton, Richard William
Ljung, Harvey Albert
Lochmuller, Charles Howard
Lueck, Charles Henry
Lunney, David Clyde
Ma, Tsu Sheng
McBay, Arthur John
McCarty, Billy Dean
McDaniel, Roger Lanier, Jr
Martin, G(uy) William, Jr
Mays, Rolland Lee
Murray, Royce Wilton
O'Connor, Lila Hunt
Olander, Donald Paul
Osteryoung, Robert Allen
Padmore, Joel M
Panek, Edward John
Parker, Carol Elaine Greenberg
Parks, Ross Lombard
Patterson, Ronald Brinton
Pellizzari, Edo Domenico
Perfetti, Thomas Albert
Pickett, John Harold
Pobiner, Harvey
Ramaswamy, H N
Remington, Lloyd Dean
Robarge, Wayne
Rochow, Theodore George
Sawardeker, Jawahar Sazro
Schickedantz, Paul David
Scott, Donald Ray
Senzel, Alan Joseph
Shackelford, Walter McDonald
Sneade, Barbara Herbert
Steinhagen, William Herrick
Stejskal, Edward Otto
Strobel, Howard Austin
Stubblefield, Charles Bryan
Thomas, Elizabeth Wadsworth
Tomer, Kenneth Beamer
Tyndall, John Raymond
Upton, Ronald P
Volk, Richard James
Wani, Mansukhlal Chhaganlal
Whitehurst, Garnett Brooks
Willey, Joan DeWitt
Williams, Jean Paul
Wilson, Nancy Keeler
Woosley, Royce Stanley
Yeowell, David Arthur
Youmans, Hubert Lafay

NORTH DAKOTA
Johnson, Arnold Richard, Jr
Khalil, Shoukry Khalil Wahba
Metzger, James David
Stolzenberg, Gary Eric
Tallman, Dennis Earl
Vogel, Gerald Lee

OHIO
Allenson, Douglas Rogers
Ampulski, Robert Stanley
Anderson, Larry Bernard
Attalla, M Albert
Austern, Barry M
Beres, John Joseph
Bimber, Russell Morrow
Black, Arthur Herman
Booth, David
Brain, Devin King
Brandt, Manuel
Bromund, Richard Hayden
Bromund, Werner Hermann
Budde, William L
Budke, Clifford Charles
Buell, Glen R
Burg, William Robert
Burrows, Kerilyn Christine
Butkus, Antanas
Campbell, Donald R
Chambers, Lee Mason
Chang, Jung-Ching
Chester, Thomas Lee
Chong, Clyde Hok Heen
Christian, John B
Cole, Jerry Joe
Cordell, Richard William
Couri, Daniel
Cox, James Allan
Crable, John Vincent
Cryberg, Richard Lee
Danielson, Neil David
Dehne, George Clark
Dewald, Howard Dean
Diem, Hugh E(gbert)
Dobbelstein, Thomas Norman
Duff, Robert Hodge
Durand, Edward Allen
Eisentraut, Kent James
Fabris, Hubert
Fike, Winston
Folk, Theodore Lamson
Fraser, Alex Stewart
Freeberg, Fred E
Gerlach, Edward Rudolph
Gilpin, Roger Keith
Gordon, Gilbert
Gordon, Sydney Michael
Gorse, Joseph
Grasselli, Jeanette Gecsy
Greenberg, Mark Shiel
Greene, Arthur Frederick, Jr
Greenlee, Kenneth William
Greinke, Ronald Alfred
Grieshammer, Lawrence Louis
Gurley, Thomas Wood
Gustafson, David Harold
Gustafson, Terry Lee
Hall, Ronald Henry
Harmon, Dale Joseph
Heineman, William Richard
Henry, William Mellinger
Hess, George G
Hibbits, James Oliver, Jr
Hilton, Ashley Stewart
Hively, Robert Arland
Hoffman, William Andrew, Jr
Holbert, Gene W(arwick)
Holtman, Mark Steven
Holubec, Zenowie Michael
Homan, Ruth Elizabeth
Howell, Norman Gary
Hubbard, Arthur T
Hutchison, William Marwick
Ikenberry, Luther Curtis
Jakobsen, Robert John
Jensen, Adolph Robert
Johnson, David Barton
Kallenberger, Waldo Elbert
Karweik, Dale Herbert
Katon, John Edward
Katovic, Vladimir
Kay, Peter Steven
Keily, Hubert Joseph
Knowlton, David A
Koch, Ronald Joseph
Koehler, Mark E
Koknat, Friedrich Wilhelm
Kornbrekke, Ralph Erik
Krishen, Anoop
Krivis, Alan Frederick
Krochta, William G
Kuemmel, Donald Francis
Kurz, David W
Lamb, Robert Edward
Lang, James Frederick
Laning, Stephen Henry
Lattimer, Robert Phillips
Latz, Howard W
Leussing, Daniel, Jr
Liao, Shu-Chung
Link, William Edward
Lott, John Alfred
Lucas, Kenneth Ross
Lucke, William E
McCreery, Richard Louis
McDonald, John William
McFarland, Charles Warren
MacGee, Joseph
Mao, Hsiang-Kuen Mark
Mark, Harry Berst, Jr
Marks, Alfred Finlay
Mateescu, Gheorghe D
Meal, Larie L
Megargle, Robert G
Miller, Barry
Miller, Theodore Lee
Misono, Kunio Shiraishi
Mitchum, Ronald Kem
Myers, Ronald Eugene
Neveu, Darwin D
Nicholson, D Allan
Oertel, Richard Paul
Ogle, Pearl Rexford, Jr
Olson, Carter LeRoy
Olynyk, Paul
Pacey, Gilbert E
Pappas, Leonard Gust
Pappenhagen, James Meredith
Patel, Siddharth Manilal
Pausch, Jerry Bliss
Porter, Leo Earle
Raynie, Douglas Edward
Riechel, Thomas Leslie
Rinaldi, Peter L
Rubinson, Judith Faye
Rubinson, Kenneth A
Rynasiewicz, Joseph
Saltzman, Bernard Edwin
Sams, Richard Alvin
Saraceno, Anthony Joseph
Schroeder, Friedhelm
Schumacher, Roy Joseph
Schweitzer, Mark Glenn
Seabaugh, Pyrtle W
Shaw, Elwood R
Silver, Gary Lee
Smith, Francis White
Smith, Jerome Paul
Spokane, Robert Bruce
Sporek, Karel Frantisek
Srinivasan, Vakula S
Stadler, Louis Benjamin
Stephens, Marvin Wayne
Stotz, Robert William
Sympson, Robert F
Tabor, Marvin Wilson
Tan, Henry S I
Taylor, Michael Lee
Thomas, Quentin Vivian
Thompson, Robert Quinton
Tiernan, Thomas Orville
Tong, James Ying-Peh
Trivisonno, Charles F(rancis)
Turnquist, Truman Dale
Tyler, Willard Philip
Ullman, Alan Howard
Uscheek, Dave Petrovich
Wadelin, Coe William
Watson, Maurice E
Weeks, Thomas Joseph, Jr
Westneat, David French
Wharton, H(arry) Whitney
Wilde, Bryan Edmund
Williams, Robert Calvin
Williams, Theodore Roosevelt
Wilson, Larry Eugene
Wright, George Joseph
Yanko, William Harry
Yellin, Wilbur
Zakriski, Paul Michael
Zaye, David F

CHEMISTRY / 187

OKLAHOMA
Al-Shaieb, Zuhair Fouad
Anderson, Paul Dean
Battiste, David Ray
Beckett, James Reid
Brown, Kenneth Henry
Cabbiness, Dale Keith
Carubelli, Raoul
Coffman, Harold H
Cowley, Thomas Gladman
DiFeo, Daniel Richard, Jr
Dryhurst, Glenn
Evens, F Monte
Fish, Wayne William
Frost, Jackie Gene
Greenwood, Gil Jay
Grigsby, Ronald Davis
Hamming, Mynard C
Leslie, Wallace Dean
Linder, Donald Ernst
Maddin, Charles Milford
Miller, John Walcott
Moczygemba, George A
Monn, Donald Edgar
Mottola, Horacio Antonio
Nalley, Elizabeth Ann
Natowsky, Sheldon
Paxson, John Ralph
Puls, Robert W
Reinbold, Paul Earl
Robinson, Jack Landy
Shew, Delbert Craig
Shioyama, Tod Kay
Van De Steeg, Garet Edward
Varga, Louis P
Wharry, Stephen Mark
Worthington, James Brian

OREGON
Adams, Frank William
Binder, Bernhard
Brinkley, John Michael
Church, Larry B
Clark, James Orie, II
DuFresne, Ann
Elia, Victor John
Freund, Harry
Gerke, John Royal
Goodney, David Edgar
Hackleman, David E
Hawkes, Stephen J
Humphrey, J Richard
Ingle, James Davis, Jr
Kay, Michael Aaron
Ko, Hon-Chung
Long, James William
Matthes, Steven Allen
Perlich, Robert Willard
Piepmeier, Edward Harman
Roe, David Kelmer
Szalecki, Wojciech Jozef
Williams, Evan Thomas

PENNSYLVANIA
Achey, Frederick Augustus
Adler, Irving Larry
Albright, Fred Ronald
Allen, Eugene (Murray)
Almond, Harold Russell, Jr
Angeloni, Francis M
Anspon, Harry Davis
Asher, Sanford Abraham
Baker, Harold Weldon
Bandy, Alan Ray
Barnett, Herald Alva
Barnhart, Barry B
Baum, Harry
Beitchman, Burton David
Boggs, William Emmerson
Bouis, Paul Andre
Box, Larry
Brenner, Gerald Stanley
Brooks, Marvin Alan
Brown, Neil Harry
Campbell, Iain Malcolm
Cardarelli, Joseph S
Carlson, Dana Peter
Carlson, Gerald Leroy
Caruso, Sebastian Charles
Cavanaugh, James Richard
Chen, Audrey Weng-Jang
Cheng, Cheng-Yin
Cheng, Hung-Yuan
Christie, Michael Allen
Clavan, Walter
Coetzee, Johannes Francois
Connelly, Carolyn Thomas
Cook, Ronald Frank
Cornell, Donald Gilmore
Cryer, Dennis Robert
Danchik, Richard S
Dedinas, Jonas
De Jong, Gary Joel
Derby, James Victor
Diorio, Alfred Frank
Doner, Landis Willard
Dulka, Joseph John
Duswalt, Allen Ainsworth, Jr
Einhorn, Philip A
Elder, James Franklin, Jr
Elson, Jesse
Ewing, Andrew Graham
Fairchild, Edward H
Farlee, Rodney Dale

Felty, Wayne Lee
Fiddler, Walter
Finseth, Dennis Henry
Foglia, Thomas Anthony
Follweiler, Douglas MacArthur
Fox, Jay B, Jr
Frankel, Lawrence (Stephen)
Franz, David Alan
Freedman, Robert Wagner
Frohliger, John Owen
Funke, Phillip T
Gergova, Katia M
Giannovario, Joseph Anthony
Godfrey, Robert Allen
Golton, William Charles
Gottlieb, Irvin M
Grinstein, Reuben H
Grob, Robert Lee
Guthrie, Joseph D
Hafford, Bradford C
Hardesty, Patrick Thomas
Harrington, George William
Hedrick, Jack LeGrande
Hercules, David Michael
Herman, Richard Gerald
Herzog, David Paul
Herzog, Leonard Frederick, II
Hicks, Kevin B
Hinkel, Robert Dale
Ho, Floyd Fong-Lok
Hoberman, Alfred Elliott
Holifield, Charles Leslie
Holm, Reimer
Honig, Richard Edward
Hopson, Kevin Matthew
Hughes, Michael Charles
Hurst, William Jeffrey
Hurwitz, Jan Krosst
Janicki, Casimir A
Johnston-Feller, Ruth M
Jordan, Joseph
Judd, Jane Harter
Jurs, Peter Christian
Kanzelmeyer, James Herbert
Karp, Howard
Kelly, Ernest L
Khan, Shakil Ahmad
Khosah, Robinson Panganai
Kieft, Lester
Kimlin, Mary Jayne
King, Richard Warren
Kleinman, Roberta Wilma
Kramer, Raymond Arthur
Krapf, George
Krzeminski, Stephen F
Kuebler, John Ralph, Jr
Kugler, George Charles
Kurtz, David Allan
Larkin, Robert Hayden
Latshaw, David Rodney
Lazar, Anna
Lemmon, Donald H
Leyon, Robert Edward
Liberti, Paul A
Locke, Harold Ogden
Louie, Ming
Lovell, Harold Lemuel
Lyman, William Ray
McCallum, Keith Stuart
MacDonald, Hubert C, Jr
McKay, James Brian
Maerker, Gerhard
Mainieri, Robert
Mair, Robert Dixon
Manka, Dan P
Markham, James J
Marmer, William Nelson
Maroulis, Peter James
Martin, Aaron J
Martin, John Robert
Melnick, Aryeh M(enashe)
Mifflin, Theodore Edward
Minard, Robert David
Mozersky, Samuel M
Nagy, Dennis J
Nikelly, John G
Noceti, Richard Paul
Obbink, Russell C
Ohnesorge, William Edward
Ottenstein, Daniel
Owens, Kevin Glenn
Pacer, John Charles
Parees, David Marc
Perry, Mary Hertzog
Peterson, James Oliver
Phifer, Lyle Hamilton
Phillips, John Spencer
Pinschmidt, Robert Krantz, Jr
Preti, George
Prohaska, Charles Anton
Reiff, Harry Elmer
Retcofsky, Herbert L
Robinson, Douglas Walter
Rodell, Michael Byron
Rodgers, Sheridan Joseph
Roesmer, Josef
Rogers, Horace Elton
Romberger, Karl Arthur
Ross, Stephen T
Rothbart, Herbert Lawrence
Rothman, Alan Michael
Rutgers, Jay G
Sadtler, Philip
Sanders, Charles Irvine

Scheirer, James E
Schobert, Harold Harris
Schroeder, Thomas Dean
Schultz, Hyman
Schweighardt, Frank Kenneth
Sheffield, Ann Elizabeth
Shepherd, Rex E
Shergalis, William Anthony
Sherma, Joseph A
Sherman, Anthony Michael
Sherman, Larry Ray
Shive, Donald Wayne
Siegel, Melvin Walter
Siegel, Richard C
Simmons, Daryl Michael
Simon, Joseph Matthew
Smith, Graham Monro
Smith, James Stanley
Smith, Stewart Edward
Snider, Albert Monroe, Jr
Spritzer, Michael Stephen
Stahl, John Wendell
State, Harold M
Stiller, Richard L
Stone, Herman
Straub, William Albert
Streuli, Carl Arthur
Strong, Frederick Carl, III
Sylvester, James Edward
Syty, Augusta
Tackett, Stanford L
Talley, Eugene Alton
Taylor, David Cobb
Taylor, Robert Morgan
Thornton, Donald Carlton
Tingle, William Herbert
Toberman, Ralph Owen
Tunick, Michael Howard
Turner, William Richard
Umbreit, Gerald Ross
Urenovitch, Joseph Victor
Varma, Asha
Veening, Hans
Vickers, Stanley
Vinson, Joe Allen
Ward, Laird Gordon Lindsay
Warren, Richard Joseph
Weber, Frank L
Weiss, Paul Storch
Westmoreland, David Gray
Williams, John Roderick
Winograd, Nicholas
Wittle, John Kenneth
Yohe, Thomas Lester
Young, Irving Gustav
Youngstrom, Richard Earl
Zacchei, Anthony Gabriel
Zaika, Laura Larysa
Zalipsky, Jerome Jaroslaw
Zinser, Edward John

RHODE ISLAND
Annino, Raymond
Brown, Christopher W
Cruickshank, Alexander Middleton
Fasching, James Le Roy
Gearing, Juanita Newman
Kirschenbaum, Louis Jean
Kreiser, Ralph Rank
Martin, Horace F
Morris, George V
Moyerman, Robert Max
Quinn, James Gerard
Smith, Alan Jerrard
Swift, Dorothy Garrison
Walsh, John Thomas
Zuehlke, Richard William

SOUTH CAROLINA
Aseleon, Gary Lee
Baumann, Elizabeth Wilson
Blood, Elizabeth Reid
Clayton, Fred Ralph, Jr
Cogswell, George Wallace
Davis, Joseph B
Davis, Raymond F
Deanhardt, Marshall Lynn
Edwards, John C
Elzerman, Alan William
Emrick, Edwin Roy
Farmer, Larry Bert
Fulda, Myron Oscar
Goode, Julia Pratt
Goode, Scott Roy
Gustin, Vaughn Kenneth
Hochel, Robert Charles
Hofstetter, Kenneth John
Holcomb, Herman Perry
Hopper, Michael James
Hunter, George William
Jacobs, William Donald
Kendall, David Nelson
Kinard, W Frank
King, Lee Curtis
Knapp, Daniel Roger
Lincoln, David Erwin
Lovins, Robert E
McFarren, Earl Francis
McGill, Julian Edward
Maier, Herbert Nathaniel
Malstrom, Robert Arthur
Morrow, William Scot
Oswald, Edward Odell
Parker, Julian E, III

Peterson, Stephen Frank
Philp, Robert Herron, Jr
Pike, LeRoy
Rains, Theodore Conrad
Sargent, Roger N
Stampf, Edward John, Jr
Waldman, Barbara Criscuolo
Wilhite, Elmer Lee
Wynn, James Elkanah

SOUTH DAKOTA
Hilderbrand, David Curtis
Looyenga, Robert William

TENNESSEE
Bhattacharya, Syamal Kanti
Burtis, Carl A, Jr
Caflisch, Edward George
Caflisch, George Barrett
Cain, Carl, Jr
Chambers, James Q
Cheng, Meng-Dawn
Christie, Warner Howard
Cook, Kelsey Donald
Crane, Laura Jane
Daniel, Douglas
Dean, John Aurie
Desiderio, Dominic Morse, Jr
Dillard, James William
Dilts, Robert Voorhees
Duncan, Budd Lee
Dyer, Frank Falkoner
Engel, Adolph James
Ferguson, Robert Lynn
Franklin, James Curry
Garber, Robert William
Guerin, Michael Richard
Hahn, Richard Balser
Hall, Larry Cully
Hicks, Jackson Earl
Honaker, Carl Boggess
Horton, Charles Abell
Jarrett, Harry Wellington, III
Johnson, Eric Robert
Klatt, Leon Nicholas
Laboda, Henry M
Lynch, John August
Lyons, Harold
McDowell, William Jackson
McDuffie, Bruce
McNeely, Robert Lewis
Mahlman, Harvey Arthur
Mamantov, Gleb
Maya, Leon
Miller, Francis Joseph
Morie, Gerald Prescott
Morrow, Roy Wayne
Mueller, Theodore Rolf
Murphy, William R
Nicely, Vincent Alvin
Nyssen, Gerard Allan
Osborne, Charles Edward
Otis, Marshall Voigt
Patterson, Truett Clifton
Payne, DeWitt Allen
Raasch, Lou Reinhart
Redfearn, Richard Daniel
Rice, Walter Wilburn
Ross, Harley Harris
Rutenberg, Aaron Charles
Scott, Dan Dryden
Scroggie, Lucy E
Shults, Wilbur Dotry, II
Smith, David Huston
Sweetman, Brian Jack
Taylor, Ellison Hall
Taylor, Kirman
Todd, Peter Justin
Tomkins, Bruce
Torrey, Rubye Prigmore
Uziel, Mayo
Vo-Dinh, Tuan
Weber, Charles William
Wehry, Earl L, Jr
Whetsel, Kermit Bazil
White, David Cleaveland
White, James Carl
Wignall, George Denis
Wiser, James Eldred
Wohlfort, Sam Willis
Wojciechowski, Norbert Joseph
Yoakum, Anna Margaret
Young, Jack Phillip

TEXAS
Adkins, John Earl, Jr
Allison, Jean Batchelor
Alsop, John Henry, III
Anselmo, Vincent C
Ayres, Gilbert Haven
Baker, Charles Taft
Banta, Marion Calvin
Bard, Allen Joseph
Bartsch, Richard Allen
Batten, Charles Francis
Beall, Charles
Beck, Benny Lee
Beier, Ross Carlton
Benson, Royal H
Bhatia, Kishan
Blay, George Albert
Botto, Robert Irving
Braterman, Paul S
Busch, Kenneth Walter

Analytical Chemistry (cont)

Busch, Marianna Anderson
Bushey, Michelle Marie
Campbell, Dan Norvell
Caprioli, Richard Michael
Cates, Vernon E
Chamberlain, Nugent Francis
Chen, Edward Chuck-Ming
Cocke, David Leath
Cogswell, Howard Winwood, Jr
Cole, Larry Lee
Cruser, Stephen Alan
Cummiskey, Charles
Daugherty, Kenneth E
De Groot, Peter Bernard
Deming, Stanley Norris
Deviney, Marvin Lee, Jr
DiFoggio, Rocco
Dillin, Dennis
Drake, Edgar Nathaniel, II
Dunn, Danny Leroy
Elder, Vincent Allen
Elthon, Donald L
Emerson, David Edwin
Faris, Sam Russell
Fitzgerald, Jerry Mack
Floyd, Willis Waldo
Foster, Norman George
Fowler, Robert McSwain
Frazee, Jerry D
Fritsche, Herbert Ahart, Jr
Fuller, Martin Emil
Gerlach, John Louis
Ghowsi, Kiumars
Gibson, Everett Kay, Jr
Grant, Clarence Lewis
Grogan, Michael John
Gupta, Vishnu Das
Hall, Randall Clark
Harris, Edward Lyndol
Hart, Haskell Vincent
Harvey, Mack Creede
Hausler, Rudolf H
Hayman, Alan Conrad
Holcombe, James Andrew
Horning, Evan Charles
Howard, Charles
Humphrey, Ray Eicken
Hunt, Richard Henry
Iddings, Frank Allen
Iskander, Felib Youssef
Johnson, David Russell
Johnson, Ralph Alton
Jones, Lawrence Ryman
Jones, Llewellyn Claiborne, Jr
Kadish, Karl Mitchell
Karchmer, Jean Herschel
Kaye, Howard
Keith, Lawrence H
Kenner, Charles Thomas
Keyworth, Donald Arthur
King, Thomas Scott
Kokkinakis, Demetrius Michael
Lee, George H, II
Lewis, Donald Richard
Liehr, Joachim G
Linowski, John Walter
Lin-Vien, Daimay
Luttrell, George Howard
McDougall, Robert I
Malloy, Thomas Bernard, Jr
Mansfield, Clifton Tyler
Markwell, Dick Robert
Maute, Robert Lewis
Melton, James Ray
Mendez, Victor Manuel
Middleditch, Brian Stanley
Millar, John David
Miller, James Franklin
Moltzan, Herbert John
Moreland, Ferrin Bates
Mosier, Benjamin
Muhs, Merrill Arthur
Newton, Robert Andrew
Patel, Bhagwandas Mavjibhai
Patton, Leo Wesley
Peurifoy, Paul Vastine
Poe, Richard D
Pogue, Randall F
Posey, Daniel Earl
Rajeshwar, Krishnan
Ray, Allen Cobble
Reed, John J R
Reinecke, Manfred Gordon
Rekers, Robert George
Rhodes, Robert Carl
Ricca, Paul Joseph
Rivera, William Henry
Robinson, J(ames) Michael
Robinson, Robert Eugene
Rodriguez, Charles F
Rollwitz, William Lloyd
Saleh, Farida Yousry
Sand, Ralph E
Sawyer, Donald Turner, Jr
Schwartz, Robert Donald
Schweikert, Emile Alfred
Scott, Robert Edward
Shaer, Elias Hanna
Shelly, Dennis C
Shukla, Shyam Swaroop
Sikes, James Klingman
Snapp, Thomas Carter, Jr
Southern, Thomas Martin
Spallholz, Julian Ernest
Starks, Aubrie Neal, Jr
Stoner, Graham Alexander
Stubbeman, Robert Frank
Thompson, Richard John
Tibbals, Harry Fred, III
Van Dreal, Paul Arthur
Varsel, Charles John
Vien, Steve Hung
Walters, Lester James, Jr
Wang, Ting-Tai Helen
Wendel, Carlton Tyrus
Wendlandt, Wesley W
Wentworth, Wayne
Whiteside, Charles Hugh
Wilson, Ray Floyd
Windham, Ronnie Lynn
Winfrey, J C
Wong, William Wai-Lun
Yau, Wallace Wen-Chuan
Yerick, Roger Eugene
Zlatkis, Albert

UTAH

Adler, Robert Garber
Brauner, Kenneth Martin
Bunger, James Walter
Burdett, Lorenzo Worth
Burton, Frederick Glenn
Butler, Eliot Andrew
Comeford, John J(ack)
Eatough, Delbert J
Eyring, Edward M
Farnsworth, Paul Burton
Garrard, Verl Grady
Goates, Steven Rex
Grey, Gothard C
Hall, David Warren
Harris, Joel Mark
James, Helen Jane
McKenzie, Jess Mack
Mangelson, Nolan Farrin
Marcus, Mark
Myers, Marcus Norville
Phillips, Lee Revell
Pyper, James William
Tuddenham, W(illiam) Marvin
Walker, Donald I
Wallace, Volney
Woolley, Earl Madsen

VERMONT

Abajian, Paul G
Geiger, William Ebling, Jr
Keenan, Robert Gregory
Kellner, Stephan Maria Eduard
Kneip, Theodore Joseph
Pipenberg, Kenneth James
Pool, Edwin Lewis
Provost, Ronald Harold
Tremmel, Carl George

VIRGINIA

Allen, Ralph Orville, Jr
Anderson, Carl William (Bill)
Anderson, Samuel
Armstrong, Alfred Ringgold
Baedecker, Philip A
Bartschmid, Betty Rains
Beyad, Mohammed Hossain
Boots, Sharon G
Brandt, Richard Bernard
Breder, Charles Vincent
Burton, Willard White
Cantu, Antonio Arnoldo
Chandler, Carl Davis, Jr
Clarke, Alan R
Cutter, Gregory Allan
Dalton, Roger Wayne
Decorpo, James Joseph
Demas, James Nicholas
Dessy, Raymond Edwin
Dominey, Raymond Nelson
Dyer, Randolph H
Einolf, William Noel
Fabbi, Brent Peter
Feinstein, Hyman Israel
Ferguson, Robert Nicholas
Fischbach, Henry
Fonong, Tekum
Frodyma, Michael Mitchell
Garrett, Benjamin Caywood
Glanville, James Oliver
Greene, Virginia Carvel
Grossman, Jeffrey N
Grunder, Fred Irwin
Gushee, Beatrice Eleanor
Harvey, William Ross
Hassler, William Woods
Hawkridge, Fred Martin
Howard, John William
Huggett, Robert James
Hunt, Donald F
Kuhn, William Frederick
Kutsher, George Samuel
Le Febvre, Edward Ellsworth
Lewis, Cornelius Crawford
Leyden, Donald E
Link, William B
Macko, Stephen Alexander
McNair, Harold Monroe
Mahoney, Bernard Launcelot, Jr
Marcy, Joseph Edwin
Martin, Albert Edwin
Mason, John Grove
Mefford, David Allen
Meites, Louis
Morgan, Evan
Morgan, John Walter
Olin, Jacqueline S
Osteryoung, Janet G
Palmer, Curtis Allyn
Parsons, James Sidney
Powell, William Allan
Price, Byron Frederick
Rudat, Martin August
Saalfeld, Fred Erich
Schnepfe, Marian Moeller
Settle, Frank Alexander, Jr
Shaheen, Donald G
Simon, Frederick Otto
Sobol, Stanley Paul
Spies, Joseph Reuben
Sprinkle, Robert Shields, III
Stewart, Kent Kallam
Tiedemann, Albert William, Jr
Turner, Anne Halligan
Van Norman, John Donald
Van't Riet, Bartholomeus
Vilcins, Gunars
Walters, Allan N
Watson, Duane Craig
Whidby, Jerry Frank
Wickham, James Edgar, Jr
Will, Fritz, III
Wood, George Marshall

WASHINGTON

Alberts, Gene S
Anderson, Donald Hervin
Ballou, Nathan Elmer
Barnes, Edwin Ellsworth
Behm, Roy
Berry, Keith O
Briggs, William Scott
Butler, Lillian Ida
Butts, William Cunningham
Callis, James Bertram
Campbell, Milton Hugh
Cheng, Xueheng
Christian, Gary Dale
Cormack, James Frederick
Daniel, J(ack) Leland
Diebel, Robert Norman
Duncan, Leonard Clinton
Easty, Dwight Buchanan
Ericsson, Lowell Harold
Felicetta, Vincent Frank
Felton, Samuel Page
Filby, Royston Herbert
Fordyce, David Buchanan
Frank, Andrew Julian
Gahler, Arnold Robert
Godar, Edith Marie
Goheen, Steven Charles
Groten, Barney
Healy, Michael L
Hill, Herbert Henderson, Jr
Huestis, Laurence Dean
Ingalsbe, David Weeden
Janata, Jiri
Jones, Jerry Lynn
Jones, Thomas Evan
Kalkwarf, Donald Riley
Kaye, James Herbert
King, Donald M
Koehmstedt, Paul Leon
Kowalski, Bruce Richard
Lee, Steven Hunter
McCown, John Joseph
Malik, Sohail
Mehlhaff, Leon Curtis
Mopper, Kenneth
Nelson, Ivory Vance
Noyes, Claudia Margaret
Pool, Karl Hallman
Radziemski, Leon Joseph
Rasco, Barbara A
Robinson, Rex Julian
Scott, Frederick Arthur
Serne, Roger Jeffrey
Shull, Charles Morell, Jr
Sklarew, Deborah S
Smith, Richard Dale
Smith, Robert Victor
Spinelli, John
Stevens, Vernon Lewis
Stout, Virginia Falk
Stromatt, Robert Weldon
Symonds, Robert B
Taylor, Murray East
Warner, John Scott
Warner, Ray Allen
Weyh, John Arthur
Zimmerman, Gary Alan

WEST VIRGINIA

Bhasin, Madan M
Chisholm, William Preston
Das, Kamalendu
Dawson, Thomas Larry
Dunphy, James Francis
Elliston, John E
Finklea, Harry Osborn
Fisher, John F
Glick, Charles Frey
Kalb, G William
Lamey, Steven Charles
Macnaughtan, Donald, Jr
Martin, John Perry, Jr
Ode, Richard Herman
Pribble, Mary Jo
Richter, G Paul
Sandridge, Robert Lee
Smith, James Allbee
Sperati, Carleton Angelo

WISCONSIN

Alvarez, Robert
Baetke, Edward A
Bayer, Richard Eugene
Berge, Douglas G
Beyerlein, Floyd Hilbert
Blaedel, Walter John
Broderick, Glen Allen
Casper, Lawrence Allen
Clark, Harlan Eugene
Conigliaro, Peter James
Connors, Kenneth A
Constant, Marc Duncan
Doerr, Robert George
Drexler, Edward James
Eckert, Alfred Carl, Jr
Eggert, Arthur Arnold
Evenson, Merle Armin
Ewald, Fred Peterson, Jr
Feinberg, Benjamin Allen
Ferguson, John Allen
Hansen, Robert Conrad
Hassinger, Mary Colleen
Hoffman, Norman Edwin
Hynek, Robert James
Kao, Wen-Hong
Kehres, Paul W(illiam)
Kincaid, James Robert
Knight, Homer Talcott
Landucci, Lawrence L
Lanterman, Elma
Larson, Wilbur S
Lemm, Arthur Warren
Lindsay, Robert Clarence
Markley, John Lute
Parnell, Donald Ray
Peace, George Earl, Jr
Polcyn, Daniel Stephen
Priest, Matthew A
Propp, Jacob Henry
Qureshi, Nilofer
Rainville, David Paul
Ranganathan, Brahmanpalli Narasimhamurthy
Ritter, Garry Lee
Rosenthal, Jeffrey
Scheppers, Gerald J
Schwab, Helmut
Scott, Lawrence William
Shain, Irving
Smith, Lloyd Michael
Sommers, Raymond A
Stone, William Ellis
Taylor, James Welch
Taylor, Paul John
Trischan, Glenn M
West, Kevin James
Wiersma, James H
Wright, John Curtis
Wright, Steven Martin
Zinkel, Duane Forst

WYOMING

Archer, Vernon Shelby
Dorrence, Samuel Michael
Duvall, John Joseph
Hurtubise, Robert John
Netzel, Daniel Anthony
Poulson, Richard Edwin
Schabron, John F
Sweet, David Paul

PUERTO RICO

ALZERRECA, ARNALDO
Carrasquillo, Arnaldo
Infante, Gabriel A
Stephens, William Powell

ALBERTA

Birss, Viola Ingrid
Birss, Viola Ingrid
Cantwell, Frederick Francis
Codding, Edward George
Coutts, Ronald Thomson
D'Agostino, Paul A
Fedorak, Phillip Michael
Feng, Joseph C
Harris, Walter Edgar
Hill, Bernard Dale
Hogg, Alan Mitchell
Horlick, Gary
Kratochvil, Byron
Plambeck, James Alan
Yamdagni, Raghavendra
Yeager, Howard Lane

BRITISH COLUMBIA

Anderson, John Ansel
D'Auria, John Michael
Hocking, Martin Blake
Law, Francis C P
McAuley, Alexander
Oliver, Barry Gordon
Wade, Adrian Paul

Whyte, John Nimmo Crosbie
Yunker, Mark Bernard

MANITOBA
Attas, Ely Michael
Bigelow, Charles C
Chow, Arthur
Kovacs, Miklos I P
Lange, Bruce Ainsworth
Muir, Derek Charles G
Saluja, Preet Pal Singh
Stewart, Reginald Bruce
Tkachuk, Russell
Vandergraaf, Tjalle T
Williams, Philip Carslake

NEW BRUNSWICK
Brewer, Douglas G
Grant, Douglas Hope
Mallet, Victorin Noel

NEWFOUNDLAND
Georghiou, Paris Elias
Longerich, Henry Perry
Stein, Allan Rudolph

NOVA SCOTIA
Ackman, Robert George
Addison, Richard Frederick
Aue, Walter Alois
Bridgeo, William Alphonsus
Chatt, Amares
Ellenberger, Herman Albert
Gilgan, Michael Wilson
Holzbecher, Jiri
Jamieson, William David
Odense, Paul Holger
Ramaley, Louis
Ryan, Douglas Earl
Stiles, David A
Wentzell, Peter Dale

ONTARIO
Afghan, Baderuddin Khan
Arsenault, Guy Pierre
Atkinson, George Francis
Bannard, Robert Alexander Brock
Baptista, Jose Antonio
Belanger, Jacqueline M R
Berman, Shier
Bhatnagar, Dinech C
Bidleman, Terry Frank
Bioleau, Luc J R
Boorn, Andrew William
Chakrabarti, Chuni Lal
Chau, Alfred Shun-Yuen
Chiba, Mikio
Connelly, Philip Walter
Corsini, A
Court, William Arthur
Crocker, Iain Hay
Cross, Charles Kenneth
Currah, Jack Ellwood
Dick, James Gardiner
Dumbroff, Erwin Bernard
Dunford, Raymond A
Foster Roberts, M Glenys
French, J(ohn) Barry
Gordon, Myra
Graham, Ronald Powell
Greenhalgh, Roy
Guevremont, Roger M
Gulens, Janis
Harrison, Alexander George
Hay, George William
Heggie, Robert Murray
Hileman, Orville Edwin, Jr
Hill, Martha Adele
Hogan, Gary D
Holland, William John
Holloway, Clive Edward
Ihnat, Milan
Jardine, John McNair
Kaiser, Klaus L(eo) E(duard)
Krull, Ulrich Jorg
Lawrence, John
Lesage, Suzanne
Logan, James Edward
Loughheed, Thomas Crossley
Lucas, Douglas M
McLaren, James Walker
Makhija, Ramesh C
Marshall, Heather
Martin, Trevor Ian
Maynes, Albion Donald
Meresz, Otto
Miller, Jack Martin
Onuska, Francis Ivan
Pang, Henrianna Yicksing
Pare, J R Jocelyn
Perrin, Carrol Hollingsworth
Ramachandran, Vangipuram S
Reffes, Howard Allen
Russell, Douglas Stewart
Sahasrabudhe, Madhu R
Sandler, Samuel
Schroeder, William Henry
Scott, Peter Michael
Sekerka, Ivan
Shah, Bhagwan G
Shearer, Duncan Allan
Singer, Eugen
Sowa, Walter
Stairs, Robert Ardagh

Stillman, Martin John
Sturgeon, Ralph Edward
Subramanian, Kunnath Sundara
Tang, You-Zhi
Thibert, Roger Joseph
Ulpian, Carla
Westaway, Kenneth C
Westland, Alan Duane
Wigfield, Yuk Yung
Wolkoff, Aaron Wilfred
Wood, Gordon Walter
Wright, Maurice Morgan
Yang, Paul Wang
Young, James Christopher F
Zakaib, Daniel D

QUEBEC
Allen, Lawrence Harvey
Belanger, Patrice Charles
Borgeat, Pierre
Campbell, Peter G C
Cormier, Francois
Desjardins, Claude W
Dorris, Gilles Marcel
Douek, Maurice
Fresco, James Martin
Goldstein, Sandu M
Gurudata, Neville
Lasia, Andrzej
Lennox, Robert Bruce
Lepine, Francois
Panalaks, Thavil
Pelletier, Omer
Powell, William St John
Power, Joan F
Purdy, William Crossley
Roy, Jean-Claude
Verschingel, Roger H C
Yates, Claire Hilliard
Zienius, Raymond Henry

SASKATCHEWAN
Cassidy, Richard Murray
Durden, David Alan
Korsrud, Gary Olaf
Larson, D Wayne
Smith, Allan Edward

OTHER COUNTRIES
Andreae, Meinrat Otto
Balakrishnan, Narayana Swamy
Bess, Robert Carl
Brown, Charles Eric
Buckwold, Sidney Joshua
Chang, Shuya
Cheung, Mo-Tsing Miranda
Collins, Carol Hollingworth
Collins, Kenneth Elmer
Edmonds, James W
Finlayson, James Bruce
Guerrero, Ariel Heriberto
Haegele, Klaus D
Hasty, Robert Armistead
Irgolic, Kurt Johann
Lewis, Arnold Leroy, II
Lingane, Peter James
Loeliger, David A
Lonsdale-Eccles, John David
Milner, David
Newlands, Michael John
Perry, John Arthur
Raveendran, Ekarath
Sinha, Asru Kumar
Steichen, Richard John
Wenclawiak, Bernd Wilhelm

Biochemistry

ALABAMA
Bennett, Leonard Lee, Jr
Bertsch, Wolfgang
Breithaupt, Lea Joseph, Jr
Cheung, Herbert Chiu-Ching
Dorai-Raj, Diana Glover
Friedman, Michael E
Funkhouser, Jane D
Furuto, Donald K
Ho, Mat H
Krishna, N Rama
McCarthy, Dennis Joseph
Miller, George Paul
Nordlund, Thomas Michael
Parks, Paul Franklin
Prince, Charles William
Pritchard, David Graham
Rahn, Ronald Otto
Sakai, Ted Tetsuo
Secrist, John Adair, III
Tan, Boen Hie
Taylor, Kenneth Boivin
Timkovich, Russell
Truelove, Bryan
Weete, John Donald
Wilken, Leon Otto, Jr

ALASKA
Fink, Thomas Robert
Kennish, John M
Lane, Robert Harold

ARIZONA
Bates, Robert Wesley
Blankenship, Robert Eugene

Brown, Michael Frederick
Chvapil, Milos
Cronin, John Read
Diehn, Bodo
Foltz, Thomas Roberts, Jr
Gunderson, Hans Magelssen
Harris, Leland
Lohr, Dennis Evan
Morton, Howard LeRoy
Rose, Seth David
Sander, Eugene George
Smith, Cecil Randolph, Jr
Tollin, Gordon
Valenty, Vivian Briones
Yamamura, Henry Ichiro

ARKANSAS
Beranek, David T
Doerge, Daniel Robert
Doyle, Miles Lawrence
Everett, Wilbur Wayne
Fu, Peter Pi-Cheng
Geren, Collis Ross
Goodwin, James Crawford
Jones, Joe Maxey
Koeppe, Roger E, II
Ku, Victoria Feng
Thoma, John Anthony
Tollett, James Terrell
Winter, Charles Gordon

CALIFORNIA
Abraham, Sandy
Albert, Jerry David
Allison, William S
Amoore, John Ernest
Ansari, Ali
Atwood, Linda
Baker, Frederick Charles
Baldwin, Robert Lesh
Baptist, Victor Harry
Barton, Jacqueline K
Bernardin, John Emile
Bigler, William Norman
Bird, Harold L(eslie), Jr
Blank, Gregory Scott
Bond, Martha W(illis)
Bovard, Freeman Carroll
Boxer, Steven George
Boyer, Paul Delos
Brooks, George Austin
Brostoff, Steven Warren
Brown, James Edward
Bruice, Thomas Charles
Budny, John Arnold
Burri, Betty Jane
Butler, Alison
Cabot, Myles Clayton
Campagnoni, Anthony Thomas
Cantor, Charles Robert
Chan, Bock G
Chan, David S
Chan, Sunney Ignatius
Chang, George Washington
Chang, Yi-Han
Cheng, Sze-Chuh
Chenoweth, Dennis Edwin
Choong, Hsia Shaw-Lwan
Choudary, Prabhakara Velagapudi
Christianson, Michael Lee
Chuong, Cheng-Ming
Codrington, Robert Smith
Cole, Roger David
Conover, Woodrow Wilson
Coppedge, James
Coyle, Bernard Andrew
Craik, Charles S
Cronin, Michael John
Dafforn, Geoffrey Alan
Dahms, Arthur Stephen
Davis, Ward Benjamin
Dawson, Marcia Ilton
DeLange, Robert J
Delk, Ann Stevens
Dennis, Edward A
Deonier, Richard Charles
Dickerson, Richard Earl
Dunaway, Marietta
Eiduson, Samuel
Einstein, Elizabeth Roboz
Eisenberg, David
Ernst, Roberta Dorothea
Feeney, Robert Earl
Felgner, Philip Louis
Fink, Anthony Lawrence
Fleischer, Everly B
Fowler, Audree Vernee
Freedland, Richard A
Fukuda, Minoru
Fuller, Glenn
Gagne, Robert Raymond
Galaway, Ronald Alvin
Garrison, Warren Manford
Geller, Edward
George-Nascimento, Carlos
Gerig, John Thomas
Gilleland, Martha Jane
Gish, Duane Tolbert
Glaser, Robert J
Glick, David
Gober, James William
Grassetti, Davide Riccardo
Green, Donald Eugene
Grimes, Carol Jane Galles

Gum, James Raymond, Jr
Hagler, Arnold T
Hajdu, Joseph
Hamilton, Carole Lois
Haney, David N
Hansen, Robert John
Hearst, John Eugene
Heath, Robert Louis
Hill, Terrell Leslie
Hinkson, Jimmy Wilford
Hoke, Glenn Dale
Hong, Keelung
Hooker, Thomas M, Jr
Horowitz, Sylvia Teich
Hugli, Tony Edward
Hurd, Ralph Eugene
Hutchin, Maxine E
Iacono, James M
Jackson, Craig Merton
Jacob, Peyton, III
Jacobson, Ralph Allen
James, Thomas Larry
Jardine, Ian
Jensen, Ronald Harry
Jewett, Sandra Lynne
Johnson, Eric F
Johnson, Herman Leonall
Jones, Richard Elmore
Kasarda, Donald David
Kavenoff, Ruth
Keelung, Hong
Kenyon, George Lommel
Khanna, Pyare Lal
Klinman, Judith Pollock
Kosow, David Phillip
Koziol, Brian Joseph
Kraut, Joseph
Kuhlmann, Karl Frederick
Kuwahara, Steven Sadao
Langerman, Neal Richard
Lee, David Louis
Leighton, Terrance J
Leonard, Nelson Jordan
Levy, Daniel
Lin, Grace Woan-Jung
Lin, Jiann-Tsyh
Linder, Maria C
Lindner, Elek
Lippmann, Wilbur
Loew, Gilda Harris
Lohrmann, Rolf
Lorance, Elmer Donald
McConnell, Harden Marsden
McGee, Lawrence Ray
McKenna, Charles Edward
Maggio, Edward Thomas
Maier, V(incent) P(aul)
Maki, August Harold
Mathewson, James H
Matthews, David Allan
Megraw, Robert Ellis
Metzger, Robert P
Miller, Jon Philip
Miller, Sol
Minch, Michael Joseph
Mitoma, Chozo
Miyada, Don Shuso
Mosher, Carol Walker
Nagel, Glenn M
Nambiar, Krishnan P
Neil, Gary Lawrence
Neilands, John Brian
Ngo, That Tjien
Ning, Robert Y
Nitecki, Danute Emilija
Osborn, Terry Wayne
Pan, Richard Y
Parsons, Stanley Monroe
Paselk, Richard Alan
Payne, Philip Warren
Pellegrini, Maria C
Peter, James Bernard
Peters, Marvin Arthur
Petryka, Zbyslaw Jan
Pettit, David J
Philo, John Sterner
Pigiet, Vincent P
Poling, Stephen Michael
Reilly, Dorothea Eleanor
Riffer, Richard
Riley, Richard Fowble
Ritzmann, Ronald Fred
Roberts, Sidney
Robins, Roland Kenith
Roche, George William
Rosenberg, Lawson Lawrence
Rosenthal, Allan Lawrence
Rynd, James Arthur
Sanchez, Robert A
Sartoris, David John
Sauer, Kenneth
Sayre, Robert Newton
Schein, Arnold Harold
Schmid, Carl William
Schmidt, Paul Gardner
Schumaker, Verne Norman
Scogin, Ron Lynn
Shafer, Richard Howard
Shaffer, Patricia Marie
Shen, Wei-Chiang
Siegel, Brock Martin
Silberstein, Otmar Otto
Slattery, Charles Wilbur
Sliwkowski, Mary Burke

Biochemistry (cont)

Solomon, Edward I
Starr, Mortimer Paul
Stumph, William Edward
Swanson, Anne Barrett
Swinehart, James Herbert
Thomas, Charles Allen, Jr
Thorner, Jeremy William
Tinoco, Joan W H
Tomich, John Matthew
Treat-Clemons, Lynda George
Trevor, Anthony John
Tutupalli, Lohit Venkateswara
Tyrrell, David John
Ullman, Edwin Fisher
Valenzuela, Pablo
Vandlen, Richard Lee
Varon, Silvio Salomone
Vedvick, Thomas Scott
Vinogradoff, Anna Patricia
Ward, Raymond Leland
Waskell, Lucy A
Weber, Bruce Howard
Weiss, Theodore Joel
Weissman, Norman
Westcott, Keith R
Wikman-Coffelt, Joan
Willhite, Calvin Campbell
Wilson, Charles Oren
Winkler, Marjorie Everett
Wisnieski, Bernadine Joann
Wong, Chi-Huey
Wright, Ernest Marshall
Wu, Sing-Yung
Wynston, Leslie K
Yang, Heechung
Yang, Jen Tsi
Young, Ho Lee
Zahnley, James Curry
Zimm, Bruno Hasbrouck

COLORADO
Allen, Kenneth G D
Barisas, Bernard George, Jr
Beck, Steven R
Bourquin, Al Willis J
Cann, John Rusweiler
Cech, Carol Martinson
Cech, Thomas Robert
Collins, Allan Clifford
Dewey, Thomas Gregory
Dynan, William Shelley
Elliott, Cecil Michael
Fitzpatrick, Francis Anthony
Freed, John Howard
Gal, Joseph
Gill, Stanley Jensen
Jones, Steven Wayne
Levine, Simon Rock
Martin, Susan Scott
Miller, William Theodore
Musick, James R
Petersen, Gene
Pitts, Malcolm John
Repine, John E
Schonbeck, Niels Daniel
Stushnoff, Cecil
Taber, Richard Lawrence
Uhlenbeck, Olke Cornelis
Ulvedal, Frode
Vigers, Alison J
Woody, A-Young Moon
York, Sheldon Stafford

CONNECTICUT
Adler, Alan David
Armitage, Ian MacLeod
Beardsley, George Peter
Boell, Edgar John
Branchini, Bruce Robert
Braswell, Emory Harold
Brudvig, Gary Wayne
Caradonna, John Philip
Celmer, Walter Daniel
Chaturvedi, Rama Kant
Chello, Paul Larson
Crothers, Donald M
Doyle, Terrence William
Eagar, Robert Gouldman, Jr
Geiger, Edwin Otto
Gent, Martin Paul Neville
Godchaux, Walter, III
Gomez-Cambronero, Julian
Grattan, James Alex
Haake, Paul
Haakonsen, Harry Olav
Harwood, Harold James, Jr
Haubrich, Dean Robert
Hook, Derek John
I, Ting-Po
Knox, James Russell, Jr
Lamm, Foster Philip
Langner, Ronald O
Lee, Henry C
McMurray, Walter Joseph
Phillips, Alvah H
Prestegard, James Harold
Prigodich, Richard Victor
Putterman, Gerald Joseph
Richards, Frederic Middlebrook
Rooney, Seamus Augustine
Ross, Donald Joseph
Sartorelli, Alan Clayton
Schepartz, Alanna
Schuster, Todd Mervyn
Skinner, H Catherine W
Sturtevant, Julian Munson
Taylor, Duncan Paul
Verdi, James L

DELAWARE
Anderson, Jeffrey John
Beer, John Joseph
Beyer, Elmo Monroe, Jr
Billheimer, Jeffrey Thomas
Colman, Roberta F
Davis, Leonard George
DeGrado, William F
Domaille, Peter John
Duke, Jodie Lee, Jr
Ehrlich, Robert Stark
Galbraith, William
Hartzell, Charles Ross, III
Herron, Norman
Hodges, Charles Thomas
Johnson, Donald Richard
Kettner, Charles Adrian
Knowles, Francis Charles
Larsen, Barbara Seliger
Lerman, Charles Lew
Link, Bernard Alvin
Malley, Linda Angevine
Melby, James Michael
Moe, Gregory Robert
Morrissey, Bruce William
Nelson, Mark James
Sariaslani, Sima
Sharp, Jonathan Hawley
Taves, Milton Arthur
Temple, Stanley
Thompson, Jeffery Scott
Thoroughgood, Carolyn A
Thorpe, Colin
Wetlaufer, Donald Burton
Whitney, Charles Candee, Jr

DISTRICT OF COLUMBIA
Bednarek, Jana Marie
Doctor, Bhupendra P
El Khadem, Hassan S
Field, Ruth Bisen
Gaber, Bruce Paul
Garavelli, John Stephen
Hare, Peter Edgar
Heman-Ackah, Samuel Monie
Isbell, Horace Smith
Johnston, Laurance S
Karle, Jean Marianne
Khanna, Krishan L
Kowalsky, Arthur
Ladisch, Stephan
May, Leopold
Miles, Corbin I
Mog, David Michael
Mullin, Brian Robert
O'Leary, Timothy Joseph
Perros, Theodore Peter
Pierce, Dorothy Helen
Plost, Charles
Rabin, Robert
Rogers, Senta S(tephanie)
Rosenberg, Robert Charles
Saslaw, Leonard David
Smith, Thomas Elijah
Stiehler, Robert D(aniel)
Walsh, Stephen G
Yang, David Chih-Hsin

FLORIDA
Ahmad, Fazal
Bash, Paul Anthony
Bernstein, Sheldon
Bertholf, Roger L
Blossey, Erich Carl
Bowman, Ray Douglas
Chapman, Michael S
Chun, Paul W
Cross, Timothy Albert
Davison, Clarke
Denslow, Nancy Derrick
Dunn, Ben Monroe
Eichler, Duane Curtis
Elfenbein, Gerald Jay
Ferl, Robert Joseph
Finger, Kenneth F
Frishman, Daniel
Fuller, Harold Wayne
Garrett, Edward Robert
Gilmer, Penny Jane
Goldstein, Arthur Murray
Gorman, Marvin
Grodberg, Marcus Gordon
Grossman, Steven Harris
Hackney, John Franklin
Halpern, Ephriam Philip
Hanks, Robert William
Henson, Carl P
Holt, Thomas Manning
Horn, Joanne Marie
Huber, Joseph William, III
James, Margaret Olive
Katz, Albert Barry
Keller, R Kennedy
Lee, Henry Joung
Lee, Ted C K
Leibman, Kenneth Charles
Makowski, Lee
Man, Eugene H
Marshall, Alan George
Meloche, Henry Paul
Miles, Richard David
Nagy, Steven
Olsen, Eugene Donald
Owen, Terence Cunliffe
Price, David Alan
Pritham, Gordon Herman
Rill, Randolph Lynn
Sang, Qing-Xiang
Schleyer, Walter Leo
Schoor, W Peter
Schultz, Duane Robert
Shiverick, Kathleen Thomas
Smith, Dennis Eugene
Stewart, Ivan
Teal, Peter E A
Trefonas, Louis Marco
Vander Meer, Robert Kenneth
Wheeler, Willis Boly
White, Frederick Howard, Jr
Williams, Joseph Francis
Winkler, Bruce Conrad
Yu, Simon Shyi-Jian

GEORGIA
Abdel-Latif, Ata A
Anderson, James Leroy
Bain, James Arthur
Batra, Gopal Krishan
Colby, David Anthony
Darvill, Alan G
David, Robert Michael
Dean, Jeffrey
DerVartanian, Daniel Vartan
Etzler, Frank M
Evans, Donald Lee
Habte-Mariam, Yitbarek
Hodges, Linda Carol
Holzer, Gunther Ulrich
Iacobucci, Guillermo Arturo
Jain, Anant Vir
Kuck, John Frederick Read, Jr
Lawton, Richard G
Mahesh, Virendra B
Martinek, Robert George
May, Sheldon William
Mickelsen, Olaf
Moinuddin, Jessie Fischer
Morse, Stephen Allen
Nes, William David
Newell, Steven Young
Pohl, Douglas George
Powers, James Cecil
Ray, Apurba Kanti
Reiss, Errol
Rhoades, James Lawrence
Sanford, Gary L
Scott, Robert Allen
Shapira, Raymond
Steele, Jack
Torres, Lourdes Maria
Trawick, William George
Van Halbiek, Herman
Wilkinson, Robert E
Wilson, William David
Worthington, Robert Earl
Yu, Nai-Teng

HAWAII
Ako, Harry Mu Kwong Ching
Loo, Yen-Hoong
Moritsugu, Toshio
Seifert, Josef
Tang, Chung-Shih
Vennesland, Birgit

IDAHO
Dreyfus, Pierre Marc
Eberlein, Charlotte
Elmore, John Jesse, Jr
Goettsch, Robert
Keay, Leonard
Lehrer, William Peter, Jr

ILLINOIS
Anderson, Robert Lewis
Bachrach, Joseph
Baich, Annette
Bietz, Jerold Allen
Bolen, David Wayne
Brubaker, George Randell
Cannon, John Burns
Chen, Wen Sherng
Chignell, Derek Alan
Cifonelli, Joseph Anthony
Cote, Gregory Lee
Cullum, Malford Eugene
Dikeman, Roxane Norris
Doane, William M
Drengler, Keith Allan
Druse-Manteuffel, Mary Jeanne
Dybel, Michael Wayne
Emken, Edward Allen
Erman, James Edwin
Eskins, Kenneth
Farnsworth, Norman R
Ferren, Larry Gene
Foster, George A, Jr
Friedman, Robert Bernard
Gardner, Harold Wayne
Gennis, Robert Bennett
Gould, John Michael
Griffin, Harold Lee
Grotefend, Alan Charles
Groziak, Michael Peter
Gumport, Richard I
Hagar, Lowell Paul
Harris, Donald Wayne
Haselkorn, Robert
Hathaway, Robert J
Held, Irene Rita
Heller, Michael
Henkin, Jack
Herting, Robert Leslie
Hodge, John Edward
Hoepfinger, Lynn Morris
Hynes, John Thomas
Iqbal, Zafar
Jones, Marjorie Ann
Kass, Guss Sigmund
Katzenellenbogen, Benita Schulman
Katzenellenbogen, John Albert
Klotz, Irving Myron
Koritala, Sanbasivaroa
Kornel, Ludwig
Kresheck, Gordon C
Kumar, Sudhir
Lambert, Mary Pulliam
Lash, Timothy David
Lee, Charlotte
Lehman, Dennis Dale
Lerner, Louis L(eonard)
Lipka, James J
Loach, Paul A
Lopatin, William
Makinen, Marvin William
Margoliash, Emanuel
Matthews, Clifford Norman
Mimnaugh, Michael Neil
Miyazaki, John H
Molotsky, Hyman Max
Moore, Edwin Granville
Moser, Kenneth Bruce
Mota de Freitas, Duarte Emanuel
Needleman, Saul Ben
Neet, Kenneth Edward
Olsen, Kenneth Wayne
Ostrow, Jay Donald
Panasenko, Sharon Muldoon
Papaioannou, Stamatios E
Parker, Helen Meister
Parsons, Carl Michael
Paschall, Eugene F
Paulson, Glenn
Peterson, Melbert Eugene
Portnoy, Norman Abbye
Pryde, Everett Hilton
Reynolds, Rosalie Dean (Sibert)
Rinehart, Kenneth Lloyd
Rotenberg, Keith Saul
Rothfus, John Arden
Salamon, Ivan Istvan
Sawinski, Vincent John
Seidman, Martin
Sennello, Lawrence Thomas
Sessa, David Joseph
Sherman, Warren V
Shriver, John William
Siedler, Arthur James
Silverman, Richard Bruce
Simmons, William Howard
Smittle, Richard Baird
Snyder, William Robert
Soucie, William George
Spangler, Brenda Dolgin
Subbaiah, Papasani Venkata
Szuchet, Sara
Tao, Mariano
Treadway, William Jack, Jr
Tsang, Joseph Chiao-Liang
Van Cleve, John Woodbridge
Vanderkooi, Garret
Van Fossan, Donald Duane
Van Kley, Harold
Van Lanen, Robert Jerome
Van Winkle, Lon J
Wagner, Gerald C
Walhout, Justine I Simon
Wargel, Robert Joseph
Weber, Evelyn Joyce
Westley, John Leonard
Whisler, Kenneth Eugene
Williams, Michael
Yeung, Tin-Chuen
Young, Jay Maitland
Zeffren, Eugene

INDIANA
Alter, John Emanuel
Baird, William McKenzie
Beckman, Jean Catherine
BeMiller, James Noble
Boguslaski, Robert Charles
Boschmann, Erwin
Bosron, William F
Bowers, Larry Donald
Burck, Philip John
Caplis, Michael E
Chandrasekaran, Rengaswami
Coburn, Stephen Putnam
Cox, David Jackson
Davis, Eldred Jack
Diamond, Steven Elliot
Fife, Wilmer Krafft
Fox, Owen Forrest
Gantzer, Mary Lou

Gin, Jerry Ben
Gorenstein, David George
Gunter, Claude Ray
Hamill, Robert L
Hanks, Alan Ray
Haslanger, Martin Frederick
Hodes, Marion Edward
Huber, Paul William
Jenkins, Winborne Terry
Johnson, Eric Richard
Kinzel, Jerry J
Kleber, John William
Kory, Mitchell
Kreiser, Thomas H(arry)
Lagu, Avinash L
Laskowski, Michael, Jr
Leoschke, William Leroy
Long, Eric Charles
McBride, William Joseph
Michel, Karl Heinz
Mihaliak, Charles Andrew
Nicholson, Ralph Lester
Nowak, Thomas
Patwardhan, Bhalchandra H
Rainey, Donald Paul
Ray, William Jackson, Jr
Richardson, John Paul
Roach, Peter John
Robertson, Donald Edwin
Rodia, Jacob Stephen
Roeske, Roger William
Ryder, Kenneth William, Jr
Serianni, Anthony Stephan
Slavik, Nelson Sigman
Smith, Gerald Floyd
Smith, Jean Blair
Smucker, Arthur Allan
Sullivan, John Lawrence
Szuhaj, Bernard F
Tunnicliff, Godfrey
Umbarger, H Edwin
Van Etten, Robert Lee
Vlahos, Chris John
Weller, Lowell Ernest
Wendel, Samuel Reece
Widlanski, Theodore Solomon
Wild, Gene Muriel
Wildfeuer, Marvin Emanuel
Wright, Walter Eugene
Yan, Sau-Chi Betty

IOWA
Abadi, Djahanguir M
Baetz, Albert L
Barua, Arun B
Cotton, Therese Marie
Foss, John G(erald)
Goff, Harold Milton
Hampton, David Clark
Hanson, Austin Moe
Kostic, Nenad M
Lim, Ramon (Khe Siong)
Linhardt, Robert John
Marshall, William Emmett
Maruyama, George Masao
Masat, Robert James
Meints, Clifford Leroy
Murphy, Patricia A
Nair, Vasu
Newton, John Marshall
Rosazza, John N
Speckard, David Carl
Stageman, Paul Jerome
Uhlenhopp, Elliott Lee
Verkade, John George

KANSAS
Adams, Ralph Norman
Barton, Janice Sweeny
Bates, Lynn Shannon
Borchardt, Ronald T
Clarenburg, Rudolf
Colen, Alan Hugh
Decedue, Charles Joseph
Grunewald, Gary Lawrence
Hanzlik, Robert Paul
Kastner, Curtis Lynn
Kramer, Karl Joseph
Lookhart, George LeRoy
Lunte, Craig Edward
Mueller, Delbert Dean
Murdock, Archie Lee
Muthukrishnan, Subbaratnam
Nicholson, Larry Michael
Schloss, John Vinton
Schowen, Richard Lyle
Seitz, Larry Max

KENTUCKY
Butterfield, David Allan
Diedrich, Donald Frank
Dowell, Russell Thomas
Foster, Thomas Scott
Hartman, David Robert
Kohler, Heinz
Petering, Harold George
Porter, William Hudson
Spatola, Arno F
Teller, David Norton
Volp, Robert Francis
Wong, 'ohn Lui

LOUISIANA
Alworth, William Lee

Barker, Louis Allen
Bazan, Nicolas Guillermo
Bernofsky, Carl
Biersmith, Edward L, III
Chrastil, Joseph
Doomes, Earl
George, William Jacob
Goldberg, Stanley Irwin
Gormus, Bobby Joe
Haycock, John Winthrop
Kadan, Ranjit Singh
Kuan, Shia Shiong
Liuzzo, Joseph Anthony
Maverick, Andrew William
Neucere, Joseph Navin
Pfeifer, Gerard David
Rhoads, Robert E
St Angelo, Allen Joseph
Shepherd, Raymond Edward
Shih, Frederick F
Sinha, Sudhir K
Snow, Lloyd Dale
Taylor, Eric Robert
Toscano, William Agostino, Jr
Weidig, Charles F
Wiechelman, Karen Janice

MAINE
Blake, Richard D
Guiseley, Kenneth B
Nagle, John David
Page, David Sanborn
Pierce, Arleen Cecilia
Renn, Donald Walter
Sherblom, Anne P

MARYLAND
Abrell, John William
Albers, Robert Wayne
Andres, Scott Fitzgerald
Ashcom, James D
Aszalos, Adorjan
Axelrod, Julius
Bax, Ad
Beckner, Suzanne K
Beecher, Gary Richard
Bennett, David Arthur
Berzofsky, Jay Arthur
Bicknell-Brown, Ellen
Bills, Donald Duane
Bis, Richard F
Bleecker, Margit
Bohr, Vilhelm Alfred
Braatz, James Anthony
Broomfield, Clarence A
Bruck, Stephen Desiderius
Bucci, Enrico
Burton, Lester Percy Joseph
Bush, C Allen
Callahan, Mary Vincent
Cantoni, Giulio L
Cassidy, James Edward
Chang, Henry
Cheh, Albert Mei-Chu
Chen, Hao
Chen, Joseph Ke-Chou
Ching, Wei-Mei
Chmurny, Alan Bruce
Chmurny, Gwendolyn Neal
Chuang, De-Maw
Cohen, Jack Sidney
Cohen, Louis Arthur
Collins, John Henry
Copeland, Edmund Sargent
Cross, David Ralston
Daly, John William
Dawidowicz, Eliezar A
Dighe, Shrikant Vishwanath
Eanes, Edward David
Eaton, William Allen
Eby, Denise
Eichhorn, Gunther Louis
Eisenberg, Frank, Jr
Elin, Ronald John
Fales, Henry Marshall
Fenselau, Catherine Clarke
Fox, Barbara Saxton
Fraser, Blair Allen
Freimuth, Henry Charles
Friedman, Fred K
Fuhr, Irvin
Ganguly, Pankaj
Garland, Donita L
Garner, Daniel Dee
Gillette, James Robert
Glaudemans, Cornelis P
Goldberg, Michael Ian
Goldman, Dexter Stanley
Goldstein, Jorge Alberto
Hackley, Brennie Elias, Jr
Hairstone, Marcus A
Hanover, John Allan
Herner, Albert Erwin
Hinton, Deborah M
Holmes, Kathryn Voelker
Holt, James Allen
Hosmane, Ramachandra Sadashiv
Hubbard, Donald
Ingham, Kenneth Culver
Inman, John Keith
Irving, George Washington, Jr
Jenkins, Mamie Leah Young
Josephs, Steven F
Kador, Peter Fritz

Kaplan, Ann Esther
Karpel, Richard Leslie
Kayser, Robert Helmut
Kelsey, Morris Irwin
Kleinman, Hynda Karen
Kohn, Leonard David
Krakauer, Henry
Lamy, Peter Paul
Landsman, David
Langone, John Joseph
Lee-Ham, Doo Young
Lerner, Pauline
Litterst, Charles Lawrence
Liu, Darrell T
Lo, Theresa Nong
London, Edythe D
Luborsky, Samuel William
Ludden, Thomas Marcellus
Maciag, Thomas Edward
McKenna, Mary Catherine
MacKerell, Alexander Donald, Jr
McPhie, Peter
Marquardt, Warren William
Marquez, Victor Esteban
Martenson, Russell Eric
Melville, Robert S
Menn, Julius Joel
Mickel, Hubert Sheldon
Miles, Edith Wilson
Miller, Paul Scott
Miller, Stephen P F
Minton, Allen Paul
Murray, Gary Joseph
Myers, Charles
Ness, Robert Kiracofe
Paskins-Hurlburt, Andrea Jeanne
Passwater, Richard Albert
Pedersen, Peter L
Podleckis, Edward Vidas
Pollack, Ralph Martin
Ponnamperuma, Cyril Andrew
Poston, John Michael
Prasanna, Hullahalli Rangaswamy
Quarles, Richard Hudson
Ralapati, Suresh
Reddi, A Hari
Repaske, Roy
Rifkind, Joseph Moses
Roberts, Anita Bauer
Roberts, David Duncan
Robey, Frank A
Robinson-White, Audrey Jean
Rose, George David
Rosenstein, Robert William
Sahu, Saura Chandra
Sampugna, Joseph
Sass, Neil Leslie
Sayer, Jane M
Schaffer, Robert
Schechter, Alan Neil
Schifreen, Richard Steven
Schneider, Walter Carl
Seifried, Adele Susan Corbin
Seifried, Harold Edwin
Sharrett, A Richey
Shore, Moris Lawrence
Skidmore, Wesley Dean
Snader, Kenneth Means
Speedie, Marilyn Kay
Spooner, Peter Michael
Steiner, Robert Frank
Surrey, Kenneth
Suskind, Sigmund Richard
Swann, Madeline Bruce
Tallent, William Hugh
Taniuchi, Hiroshi
Titus, Elwood Owen
Townsend, Craig Arthur
Townsley, John D
Ts'o, Paul On Pong
Tso, Tien Chioh
Tung, Ming Sung
Varner, Hugh H
Veitch, Fletcher Pearre, Jr
Vijay, Inder Krishan
Walker, Mary Clare
Walter, Eugene LeRoy, Jr
Watkins, Paul Allan
Weinstein, Howard
Welch, Arnold D(eMerritt)
Whiting, John Dale, Jr
Wickner, Sue
Williams, Wesley M
Williamson, Charles Elvin
Wilson, Maureen O
Wolff, John B
Wu, Roy Shih-Shyong
Wubbels, Gene Gerald
Wykes, Arthur Albert
Yeh, Lai-Su Lee
Yoho, Clayton W

MASSACHUSETTS
Adler, Alice Joan
Agrawal, Sudhir
Ahern, David George
Ajami, Alfred Michel
Akkara, Joseph A
Askew, Eldon Wayne
Auer, Henry Ernest
Babayan, Vigen Khachig
Ball, Derek Harry
Bass, Norman Herbert
Berlowitz Tarrant, Laurence

Biemann, Klaus
Bing, David H
Browne, Douglas Townsend
Campbell, Mary Kathryn
Cantley, Lewis Clayton
Cantor, Charles R
Cathou, Renata Egone
Champion, Paul Morris
Chase, Arleen Ruth
Chipman, Wilmon B
Clarke, Michael J
Cohen, Saul G
Crabtree, Donald V
DeArruda, Monika Vicira
Dehal, Shangara Singh
Demain, Arnold Lester
Dickinson, Leonard Charles
DiLiddo, Rebecca McBride
Doctrow, Susan R
Downer, Nancy Wuerth
Ehlers, Mario Ralf Werner
Fossel, Eric Thor
Foye, William Owen
Friedman, Orrie Max
Furbish, Francis Scott
Gallop, Paul Myron
Galper, Jonas Bernard
Girard, Francis Henry
Gounaris, Anne Demetra
Greenaway, Frederick Thomas
Hansen, David Elliott
Harrison, Stephen Coplan
Hauser, George
Hellman, Kenneth P
Hobey, William David
Hochella, Norman Joseph
Hodge, Richard Paul
Holick, Sally Ann
Jahngen, Edwin Georg Emil, Jr
Jarboe, Jerry K(ent)
Jungalwala, Firoze Bamanshaw
Kantrowitz, Evan R
Kaplan, David Lee
King, Jonathan (Alan)
Klibanov, Alexander M
Klotz, Lynn Charles
Knowles, Aileen Foung
Knowles, Jeremy Randall
Kobayashi, Kazumi
Kocon, Richard William
Koltun, Walter Lang
Kominz, David Richard
Kropf, Allen
Kupfer, David
Learson, Robert Joseph
Lees, Marjorie Berman
Libby, Peter
Lin, Gloria C
Lippard, Stephen J
Lovett, Charles McVey
Lowey, Susan
Marquis, Judith Kathleen
Marx, Kenneth Allan
Nauss, Kathleen Minihan
Nickerson, Richard G
O'Brien, Richard Desmond
Ocain, Timothy Donald
Olsen, Bjorn Reino
Pinkus, Jack Leon
Plocke, Donald J
Pluskal, Malcolm Graham
Redfield, Alfred Guillou
Riordan, James F
Robbat, Albert, Jr
Rosenkrantz, Harris
Ruelius, Hans Winfried
Shuster, Louis
Silver, Marc Stamm
Simons, Elizabeth Reiman
Sipe, Jean Dow
Small, Donald MacFarland
Stephens, Raymond Edward
Tan, Barrie
Thomas, Peter
Timasheff, Serge Nicholas
Tolan, Dean Richard
Torchilin, Vladimir Petrovich
Umans, Robert Scott
Vellaccio, Frank
Warren, Christopher David
Whitesides, George McClelland
Williamson, Kenneth L(ee)
Wineman, Robert Judson
Winkelman, James W
Wolfson, Adele Judith
Worden, Patricia Barron
Zannis, Vassilis I

MICHIGAN
Ahn, Kyunghye
Akil, Huda
Babcock, Gerald Thomas
Benjamins, Joyce Ann
Biermann, Janet Sybil
Brown, Ray Kent
Callewaert, Denis Marc
Campbell, Wilbur Harold
Caspers, Mary Lou
Chang, Chi Kwong
Cook, Paul Laverne
Coward, James Kenderdine
Craig, Winston John
Crippen, Gordon Marvin
Czarnik, Anthony William

Biochemistry (cont)

Dekker, Eugene Earl
Drach, John Charles
Edwards, Brian F
Eliezer, Naomi
Fernandez-Madrid, Felix
Fischer, Lawrence J
Forist, Arlington Ardeane
Garlick, Robert L
Hakim, Margaret Heath
Halvorson, Herbert Russell
Herzig, David Jacob
Holland, Jack Calvin
Hultquist, Donald Elliott
Janik, Borek
Kaiser, David Gilbert
Kang, Uan Gen
Kiechle, Frederick Leonard
Klein, Ronald Don
Kuo, Mingshang
La Du, Bert Nichols, Jr
Laks, Peter Edward
Lindblad, William John
Lorch, Steven Kalman
McBean, Lois D
McCarville, Michael Edward
McCullough, Willard George
McGinnis, Gary David
McIntosh, Lee
McNair, Ruth Davis
Marletta, Michael Anthony
Marshall, Charles Wheeler
Matthews, Rowena Green
Miller, Susan Mary
Mizukami, Hiroshi
Moore, Kenneth Edwin
Mostafapour, M Kazem
Moyer, Carl Edward
O'Connell, Paul William
Ogilvie, Marvin Lee
Parikh, Jekishan R
Penner-Hahn, James Edward
Peterson-Kennedy, Sydney Ellen
Petzold, Edgar
Phan, Sem Hin
Randinitis, Edward J
Reeves, Andrew Louis
Root-Bernstein, Robert Scott
Roth, Robert Andrew, Jr
Salmond, William Glover
Schacht, Jochen (Heinrich)
Schullery, Stephen Edmund
Sellinger, Otto Zivko
Servis, Robert Eugene
Shore, Joseph D
Showalter, Howard Daniel Hollis
Smart, James Blair
Smith, Richard Harding
Stach, Robert William
Thoene, Jess Gilbert
Tormanen, Calvin Douglas
Tuls, Jody Lynn Foy
Tung, Fred Fu
Uhler, Michael David
Vaughn, Clarence Benjamin
Verrill, Harland Lester
Watson, Jack Throck
Welsch, Clifford William, Jr
Wilson, John Edward
Woster, Patrick Michael
Yancey, Robert John, Jr
Yang, Victor Chi-Min
Yurewicz, Edward Charles
Zand, Robert

MINNESOTA

Adams, Ernest Clarence
Adiarte, Arthur Lardizabal
Allewell, Norma Mary
Ames, Matthew Martin
Barany, George
Baumann, Wolfgang J
Bloomfield, Victor Alfred
Brockman, Howard Lyle, Jr
Brown, Rhoderick Edmiston, Jr
Carter, John Vernon
Conard, Gordon Joseph
Csallany, Agnes Saari
Drewes, Lester Richard
Farnum, Sylvia A
Freier, Esther Fay
Fulmer, Richard W
Gray, Gary Ronald
Haddad, Louis Charles
Hartman, Boyd Kent
Jemmerson, Ronald Renomer Weaver
Kluetz, Michael David
Kyle, Robert Arthur
Lawrenz, Frances Patricia
Lin, Sping
Lovrien, Rex Eugene
Malewicz, Barbara Maria
Nicol, Susan Elizabeth
Perisho, Clarence R
Que, Lawrence, Jr
Rao, Gundu Hirisave Rama
Schmidt, Jane Ann
Stein, Paul John
Swaiman, Kenneth F
Tsong, Tian Yow
Ugurbil, Kamil
Vatassery, Govind T
Wei, Guang-Jong Jason

MISSISSIPPI

Boyle, John Aloysius, Jr
Brown, Kenneth Lawrence
Case, Steven Thomas
Chaires, Jonathan Bradford
Fawcett, Newton Creig
Graves, David E
Hall, Margaret (Margot) Jean
Hedin, Paul A
Ho, Ing Kang
Kopfler, Frederick Charles
McCormick, Charles Lewis, III
Peterson, Harold LeRoy
Wellman, Susan Elizabeth
Williamson, John S

MISSOURI

Ackers, Gary K
Adams, Steven Paul
Black, Wayne Edward
Briner, Robert C
Brown, Barbara Illingworth
Burke, William Joseph
Burleigh, Bruce Daniel, Jr
Byatt, John Christopher
Byington, Keith H
Cannon, John Francis
Chiappinelli, Vincent A
Cicero, Theodore James
Coudron, Thomas A
Covey, Douglas Floyd
Das, Manik Lal
De Blas, Angel Luis
Delaware, Dana Lewis
Feder, Joseph
Ferrendelli, James Anthony
Fraley, Robert Thomas
Francis, Faith Ellen
Geison, Ronald Leon
Green, Vernon Albert
Hertelendy, Frank
Holtzer, Alfred Melvin
Hood, Larry Lee
Howard, Susan Carol Pearcy
Janski, Alvin Michael
Lau, Brad W C
Lipkin, David
Lohman, Timothy Michael
Lozeron, Homer A
McCormick, John Pauling
MacQuarrie, Ronald Anthony
Marshall, Lucia Garcia-Iniguez
Marx, Jon William
Mayer, Dennis T
Mortensen, Harley Eugene
Moscatelli, Ezio Anthony
Newell, Jon Albert
Piepho, Robert Walter
Ramsey, Robert Bruce
Rogic, Milorad Mihailo
Rosenthal, Harold Leslie
Rowe, Elizabeth Snow
Schleppnik, Alfred Adolf
Schmitz, Kenneth Stanley
Schuck, James Michael
Shukla, Shivendra Dutt
Sikorski, James Alan
Slatopolsky, Eduardo
Smith, Carl Hugh
Smith, Robert Eugene
Sun, Albert Yung-Kwang
Turner, John T
Tyagi, Suresh C
Vineyard, Billy Dale
Wheeler, James Donlan
Whittle, Philip Rodger
Wilkinson, Ralph Russell
Wood, David Collier
Yarbro, John Williamson
Yunger, Libby Marie
Zahler, Warren Leigh

MONTANA

Dooley, David Marlin
Parker, Keith Krom
Richardson, Charles
Thompson, Holly Ann

NEBRASKA

Casey, Carol A
Ford, Johny Joe
Gold, Barry Irwin
Harbison, Gerard Stanislaus
Hulce, Martin R
Lockridge, Oksana Maslivec
MacDonald, Richard G
Nagel, Donald Lewis
Parkhurst, Lawrence John
Rasmussen, Russell Lee
Smith, David Hibbard
Yen, Jong-Tseng

NEVADA

Dreiling, Charles Ernest
Jay, James Monroe
Lightner, David A
Peck, Ernest James, Jr
Wakayama, Edgar Junro
Wolf, Edward Charles

NEW HAMPSHIRE

Chang, Ta-Yuan
Chasteen, Norman Dennis
Ikawa, Miyoshi

Kull, Fredrick J
Maudsley, David V
Wetterhahn, Karen E

NEW JERSEY

Alberts, Alfred W
Ardelt, Wojciech Joseph
Bagnell, Carol A
Bathala, Mohinder S
Becker, Joseph Whitney
Bernholz, William Francis
Bhattacharjee, Himangshu Ranjan
Boskey, Adele Ludin
Breslow, Esther M G
Bryant, Robert Wesley
Buch, Robert Michael
Burbaum, Beverly Wolgast
Caldwell, Carlyle Gordon
Chaikin, Philip
Chiao, Yu-Chih
Chinard, Francis Pierre
Chovan, James Peter
Comai-Fuerherm, Karen
Cuffari, Gilbert Luke
Cushman, David Wayne
Dashman, Theodore
DeBari, Vincent A
Deetz, Lawrence Edward, III
Denhardt, David Tilton
Desjardins, Raoul
Diegnan, Glenn Alan
Dunikoski, Leonard Karol, Jr
Durso, Donald Francis
Dvornik, Dushan Michael
Ehrlich, Julian
Ellenbogen, Leon
Erenrich, Eric Howard
Erenrich, Evelyn Schwartz
Felix, Arthur M
Feng, Rong
Ferguson, Karen Anne
Fernandes, Prabhavathi Bhat
Fordham, William David
Frederick, James R
Fu, Shou-Cheng Joseph
Gagna, Claude Eugene
Garfinkel, Harmon Mark
Georgopapadakou, Nafsika Eleni
Giddings, George Gosselin
Gidwani, Ram N
Gold, Barry Ira
Grosso, John A
Hagan, James Joyce
Hall, Stan Stanley
Harkins, Robert W
Harrison-Johnson, Yvonne E
Heacock, Craig S
Heimer, Edgar P
Hensens, Otto Derk
Ivatt, Raymond John
Jackson, Richard Lee
Jacobs, Morton Howard
Jacobson, Martin Michael
Jakubowski, Hieronim Zbigniew
Jones, Martha Ownbey
Jordan, Frank
Joseph, John Mundancheril
Kaczorowski, Gregory John
Kamm, Jerome J
Kauzmann, Walter (Joseph)
Kirschbaum, Joel Jerome
Kline, Toni Beth
Kolis, Stanley Joseph
Koonce, Samuel David
Kripalani, Kishin J
Kristol, David Sol
Lam, Yiu-Kuen Tony
Lee, Catherine Coyle
Lee, Tung-Ching
Leeds, Norma S
Lin, Chin-Chung
McCaslin, Darrell
MacCoss, Malcolm
Mackerer, Carl Robert
Madison, Vincent Stewart
Manning, Gerald Stuart
Manowitz, Paul
Margolis, Frank L
Mark, David Fu-Chi
Mendelsohn, Richard
Montville, Thomas Joseph
Murthy, Vadiraja Venkatesa
Naipawer, Richard Edward
Nelson, Nathan
Ofengand, Edward James
Olson, Wilma King
OSullivan, Joseph
Palmer, Keith Henry
Patel, Ramesh N
Pittz, Eugene P
Quinn, Gertrude Patricia
Ramanarayanan, Madhava
Roberts, Ronald C
Roland, Dennis Michael
Rothberg, Lewis Josiah
Saferstein, Richard
Salowe, Scott P
Schaich, Karen Marie
Scharpf, Lewis George, Jr
Schrier, Eugene Edwin
Schugar, Harvey
Schwenker, Robert Frederick, Jr
Shefer, Sarah
Sheikh, Maqsood A

Sheriff, Steven
Singh, Vishwa Nath
Smith, Elizabeth Melva
Snyder, Robert
Spiro, Thomas George
Stillman, John Edgar
Strauss, George
Stults, Frederick Howard
Symchowicz, Samson
Takahashi, Mark T
Tindall, Charles Gordon, Jr
Tolman, Richard Lee
Trifan, Daniel Siegfried
Tsong, Yun Yen
Vandegrift, Vaughn
Villafranca, Joseph John
Wassarman, Paul Michael
Weill, Carol Edwin
Wennogle, Lawrence
White, Ronald E
Wildnauer, Richard Harry
Wislocki, Peter G
Yarmush, Martin Leon

NEW MEXICO

Birnbaum, Edward Robert
Darnall, Dennis W
Fee, James Arthur
Gurd, Ruth Sights
Guziec, Frank Stanley, Jr
Guziec, Lynn Erin
Jones, Peter Frank
Matwiyoff, Nicholas Alexander
Thompson, Jill Charlotte
Trujillo, Ralph Eusebio
Wild, Gaynor (Clarke)
Woodfin, Beulah Marie

NEW YORK

Abramowicz, Daniel Albert
Abramson, Morris Barnet
Andrews, John Parray
Angelino, Norman J
Bachrach, Howard L
Baird, Barbara A
Bardos, Thomas Joseph
Baum, Stuart J
Bello, Jake
Benuck, Myron
Berl, Soll
Bettelheim, Frederick A
Bitcover, Ezra Harold
Bittman, Robert
Blank, Robert H
Blei, Ira
Blume, Arthur Joel
Bofinger, Diane P
Bonvicino, Guido Eros
Bookchin, Robert M
Borer, Philip N
Brewer, Curtis Fred
Brodie, Jonathan D
Broekman, Marinus Johan
Brown, John Francis, Jr
Callender, Robert
Carpenter, Barry Keith
Cavalieri, Liebe Frank
Cerny, Laurence Charles
Charton, Marvin
Chiulli, Angelo Joseph
Cody, Vivian
Cohen, Gerald
Cohen, Leonard A
Cohen, Stephen Robert
Conklin, John Douglas, Sr
Corfield, Peter William Reginald
Coupet, Joseph
Cowburn, David
Crandall, David L
D'Angelo, Gaetano
Datta, Ranajit Kumar
Day, Loren A
DeForest, Peter Rupert
Denning, George Smith, Jr
Dick, William Edwin, Jr
Diem, Max
Dolak, Terence Martin
Elahi, Nasik
Fabry, Mary E Riepe
Fabry, Thomas Lester
Farina, Peter R
Fenton, Sandra Sulikowski
Finkelstein, Jacob Noah
Finlay, Thomas Hiram
Fleisher, Martin
Fletcher, Paul Wayne
Freedman, Lewis Simon
Gerolimatos, Barbara
Gessner, Teresa
Ginsberg-Fellner, Fredda Vita
Gioannini, Theresa Lee
Goldberg, Arthur H
Grand, Stanley
Greene, Lloyd A
Gurel, Demet
Guttenplan, Joseph B
Hagan, William John, Jr
Hajjar, David Phillip
Hakala, Maire Tellervo
Hamilton, Mary Jane Gill
Hamsher, James J
Henn, Fritz Albert
Hilborn, David Alan
Hoffman, Linda M

Hogg, James Felter
Holowka, David Allan
Hommes, Frits A
Horowitz, Marilyn Stephens
Hough, Jane Linscott
Howard, Irmgard Matilda Keeler
Hui, Koon-Sea
Joshi, Sharad Gopal
Kaiserman, Howard Bruce
Kallenbach, Neville R
Kalman, Thomas Ivan
Kapuscinski, Jan
Kemphues, Kenneth J
Killian, Carl Stanley
King, Te Piao
Kingston, Charles Richard
Kirchberger, Madeleine
Krugh, Thomas Richard
Kurylo-Borowska, Zofia
Landsberger, Frank Robbert
Lang, Helga M (Sr Therese)
Lapin, Evelyn P
Lau-Cam, Cesar A
Lee, Ching-Li
Lee, May D-Ming (Lu)
Lee, Tzoong-Chyh
Leifer, Zer
Lepp, Cyrus Andrew
Li, Lu Ku
Lieberman, Seymour
Lin, Yong Yeng
Lobo, Angelo Peter
Luine, Victoria Nall
McLendon, George L
Mayers, George Louis
Meinwald, Yvonne Chu
Miller, David Lee
Miller, Herbert Kenneth
Milmore, John Edward
Mitacek, Eugene Jaroslav
Monaco, Regina R
Morris, John Emory
Morrow, Janet Ruth
Mundorff-Shrestha, Sheila Ann
Murphy, Randall Bertrand
Myer, Yash Paul
Napoli, Joseph Leonard
Neenan, John Patrick
Nemethy, George
Nisselbaum, Jerome Seymour
Noble, Robert Warren, Jr
Oreskes, Irwin
Page, Gregory Vincent
Philipp, Manfred (Hans)
Phillips, Gerald B
Pittman, Kenneth Arthur
Plane, Robert Allen
Powers, John Clancey, Jr
Premuzic, Eugene T
Purandare, Yeshwant K
Raventos-Suaraz, Carmen Elvira
Reich, Marvin Fred
Roboz, John
Roll, David E
Rowlett, Roger Scott
Rubin, Byron Herbert
Russell, Charlotte Sananes
Sabban, Esther Louise
Samuels, Stanley
Schayer, Richard William
Scheraga, Harold Abraham
Schliselfeld, Louis Harold
Schneck, Larry
Schneider, Arthur Lee
Schulz, Horst H
Schwartz, Ernest
Shiao, Daniel Da-Fong
Shultz, Walter
Siew, Ernest L
Simon, Eric Jacob
Smith, Frank Ackroyd
Solomon, Jerome Jay
Sparks, Charles Edward
Springer, Charles Sinclair, Jr
Staples, Basil George
Stark, John Howard
Starzak, Michael Edward
Strekas, Thomas C
Stryker, Martin H
Sufrin, Janice Richman
Tabachnick, Milton
Tepperman, Helen Murphy
Tieckelmann, Howard
Turner, Douglas Hugh
Usher, David Anthony
Wajda, Isabel
Warden, Joseph Tallman
Weisburger, John Hans
Wetmur, James Gerard
Williams, Noreen
Wolin, Michael Stuart
Wootton, John Francis
Wu, Mingdan
Yeagle, Philip L
Yu, Andrew B C
Zilversmit, Donald Berthold
Zipfel, Warren Roger
Zobel, C(arl) Richard

NORTH CAROLINA
Abernethy, John Leo
Allis, John W
Anderegg, Robert James
Aune, Kirk Carl

Bonaventura, Celia Jean
Bonaventura, Joseph
Bredeck, Henry E
Burrous, Stanley Emerson
Burt, Charles Tyler
Buttke, Thomas Martin
Campbell, Robert Joseph
Cavallito, Chester John
Chapman, John Franklin, Jr
Chazotte, Brad Nelson
Cheng, Pi-Wan
Clendenon, Nancy Ruth
Cowling, Ellis Brevier
Di Giulio, Richard Thomas
D'Silva, Themistocles Damasceno Joaquim
Ellenson, James L
Ferris, Robert Monsour
Fletcher, Paul Litton, Jr
Gilbert, Lawrence Irwin
Glinski, Ronald P
Harpold, Michael Alan
Hastings, Felton Leo
Henkens, Robert William
Hiskey, Richard Grant
Hodges, Helen Leslie
Hooper, Irving R
Horton, Horace Robert
Huang, Jamin
Kepler, Carol R
Knopp, James A
LeBaron, Homer McKay
Lee, Martin Jerome
Lentz, Barry R
Lineback, David R
Lister, Mark David
London, Robert Elliot
McIlwain, David Lee
Maier, Robert Hawthorne
Marco, Gino Joseph
Marzluff, William Frank, Jr
Matthews, Hazel Benton, Jr
Millen, Jane
Mold, James Davis
Mortenson, Leonard Earl
Nelson, Donald J
Nesnow, Stephen Charles
Nozaki, Yasuhiko
Parikh, Indu
Pirrung, Michael Craig
Raleigh, James Arthur
Renuart, Adhemar William
Risley, John Marcus
Schroeder, David Henry
Schuh, Merlyn Duane
Shaw, Barbara Ramsay
Smith, Gary Keith
Spremulli, Linda Lucy
Straub, Karl David
Struve, William George
Suzuki, Kunihiko
Switzer, Mary Ellen Phelan
Thompson, Nancy Lynn
Tucker, Charles Leroy, Jr
VanMiddlesworth, Frank L
Wani, Mansukhlal Chhaganlal
Weintraub, Robert Louis
Welch, Richard Martin
Weybrew, Joseph Arthur
Woolridge, Edward Daniel

NORTH DAKOTA
Ary, Thomas Edward
Frear, Donald Stuart
Metzger, James David
Milne, David Bayard
Stolzenberg, Gary Eric

OHIO
Adragna, Norma C
Baur, Fredric John, Jr
Berliner, Lawrence J
Blazyk, Jack
Bobst, Albert M
Bohinski, Robert Clement
Borders, Charles LaMonte, Jr
Brody, Richard Simon
Brueggemeier, Robert Wayne
Burke, Morris
Campbell, Jeptha Edward, Jr
Chengelis, Christopher P
Cummings, Sue Carol
Decker, Clarence Ferdinand
Decker, Lucile Ellen
Dedman, John Raymond
DiCorleto, Paul Eugene
Ekvall, Shirley W
Farrier, Noel John
Fuchsman, William Harvey
Glende, Eric A, Jr
Green, Ralph
Grooms, Thomas Albin
Haddox, Charles Hugh, Jr
Hagerman, Ann Elizabeth
Halsall, H(allen) Brian
Harpst, Jerry Adams
Heineman, William Richard
Hickson, John LeFever
Hoff, Kenneth Michael
Horrocks, Lloyd Allen
Hoss, Wayne Paul
Jacobsen, Donald Weldon
Jakobsen, Robert John
Johnson, David Barton

Klopman, Gilles
Knowlton, David A
Kolattukudy, P E
Koo, Peter H
Kreishman, George Paul
Lang, James Frederick
Lateef, Abdul Bari
Lechner, Joseph H
Levin, Jerome Allen
Lutton, John Kazuo
McClure, Jerry Weldon
McLoughlin, Daniel Joseph
Matlin, Albert R
Mattice, Wayne Lee
Meites, Samuel
Miller, Carl Henry, Jr
Morgan, Marjorie Susan
Myhre, David V
Nee, Michael Wei-Kuo
Pereira, Michael Alan
Richardson, Keith Erwin
Richter, Helen Wilkinson
Rosenberry, Terrone Lee
Ross, Robert Talman
Rubinson, Judith Faye
Schweitzer, Mark Glenn
Scovell, William Martin
Skelly, Michael Francis
Spokane, Robert Bruce
Stephens, Marvin Wayne
Strobel, Rudolf G K
Stuehr, John Edward
Stukus, Philip Eugene
Tabor, Marvin Wilson
Thompson, Robert Quinton
Viola, Ronald Edward
Walker, Thomas Eugene
Warner, Ann Marie
Watson, Barry
Weber, George Russell
White, Robert J
Wiginton, Dan Allen
Wilson, Gustavus Edwin

OKLAHOMA
Banschbach, Martin Wayne
Crane, Charles Russell
Decker, Rolan Van
DiFeo, Daniel Richard, Jr
Dunlap, William Joe
Ford, Sharon R
Gorin, George
Johnson, Arthur Edward
Lee, Kyung No
McGurk, Donald J
Matsumoto, Hiroyuki
Mort, Andrew James
Nelson, Eldon Carl
O'Neal, Steven George
Patterson, Manford Kenneth, Jr
Pomes, Adrian Francis
Prabhu, Vilas Anandrao
Reinhart, Gregory Duncan
Sachdev, Goverdhan Pal
Snider, Theodore Eugene
Spivey, Howard Olin
Wright, John Ricken

OREGON
Ayres, James Walter
Bartos, Frantisek
Brinkley, John Michael
Buhler, Donald Raymond
Dahlquist, Frederick Willis
Ehler, Kenneth Walter
Gardner, Sara A
Gold, Michael Howard
Gould, Steven James
Griffith, O Hayes
Halko, David Joseph
Hays, John Bruce
Horton, Aaron Wesley
Johnson, Walter Curtis, Jr
Jost, Patricia Cowan
Loehr, Thomas Michael
McClard, Ronald Wayne
McQuate, Robert Samuel
Nitsos, Ronald Eugene
Peticolas, Warner Leland
Robinson, Arthur B
Sandifer, Ronda Margaret
Seaman, Geoffrey Vincent F
Van Holde, Kensal Edward
Von Hippel, Peter Hans
Walters, Roland Dick
Zimmerman, Earl Abram

PENNSYLVANIA
Albright, Fred Ronald
Alhadeff, Jack Abraham
Anderson, Lewis L
Babu, Uma Mahesh
Barr, Richard Arthur
Basch, Jay Justin
Bednar, Rodney Allan
Behe, Michael Joseph
Benedict, Robert Curtis
Bering, Charles Lawrence
Bossard, Mary Jeanette
Bothner-By, Aksel Arnold
Brown, Eleanor Moore
Burgess, David Ray
Callahan, Hugh James
Castric, Peter Allen

Chapman, Toby Marshall
Clapp, Charles H
Cohen, Margo Nita Panush
Cohn, Mildred
Constantinides, Panayiotis Pericleous
Cordova, Vincent Frank
Cottrell, Ian William
Crissman, Jack Kenneth, Jr
Dalton, Colin
Darke, Paul
Davis, Hugh Michael
Deckert, Fred W
Doner, Landis Willard
Drake, B(illy) Blandin
Dreisbach, Joseph Herman
Dreyfuss, Gideon
Duggan, Daniel Edward
Dvonch, William
Egolf, Roger
Eisner, Robert Linden
Erhan, Semih M
Ewing, Andrew Graham
Farrar, John J
Feairheller, Stephen Henry
Fisher, Tom Lyons
Fishman, Marshall Lewis
George, Philip
Giuliano, Robert Michael
Gleim, Robert David
Glover, George Irvin
Glusker, Jenny Pickworth
Gold, Martin
Hackney, David Daniel
Haines, William Joseph
Hajjar, Nicolas Philippe
Hamilton, Gordon Andrew
Hendry, Richard Allan
Herskovitz, Thomas
Hicks, Kevin B
Higman, Henry Booth
Hoek, Joannes (Jan) Bernardus
Horrocks, William DeWitt, Jr
Hurst, William Jeffrey
Husic, Harold David
Jaffe, Eileen Karen
Janssen, Frank Walter
Johnson, Kenneth Allen
Johnson, Richard William
Kahn, Leo David
Keen, James H
Knievel, Daniel Paul
Kopple, Kenneth D(avid)
Kruse, Lawrence Ivan
Kuo, Lawrence C
Layne, Porter Preston
Lazo, John Stephen
Libby, Carol Baker
Libby, R Daniel
Linask, Kersti Katrin
Llinas, Miguel
Lublin, Fred D
Malin, Edyth
Markham, George Douglas
Matthews, Charles Robert
Metcalf, Brian Walter
Metrione, Robert M
Mooney, David Samuel
Moreau, Robert Arthur
Motsavage, Vincent Andrew
Mottur, George Preston
Mozersky, Samuel M
Na, George Chao
Neubeck, Clifford Edward
Newirth, Terry L
Niewiarowski, Stefan
Ochiai, Ei-Ichiro
Ocone, Luke Ralph
Parker, Leslie
Perlis, Irwin Bernard
Phillips, Michael Canavan
Prescott, David Julius
Price, Charles Coale
Ritter, Carl A
Rosenberg, Jerome Laib
Russell, Alan James
Sax, Martin
Schengrund, Cara-Lynne
Schray, Keith James
Selinsky, Barry Steven
Seybert, David Wayne
Shaw, Leslie M J
Sideropoulos, Aris S
Sitrin, Robert David
Sloviter, Henry Allan
Smith, John Bryan
Soriano, David S
Southwick, Philip Lee
Spangler, Martin Ord Lee
Stevens, Charles Le Roy
Storey, Bayard Thayer
Straub, Thomas Stuart
Takashima, Shiro
Templer, David Allen
Thoman, Charles James
Thornton, Edward Ralph
Thornton, Elizabeth K
Tomezsko, Edward Stephen John
Tu, Shu-i
Tzodikov, Nathan Robert
Vickers, Stanley
Voet, Donald Herman
Wedler, Frederick Charles Oliver, Jr
Wetzel, Ronald Burnell
Wickstrom, Eric

Biochemistry (cont)

Williams, Taffy J
Witham, Francis H
Witting, Lloyd Allen
Wolfersberger, Michael Gregg
Wolken, Jerome Jay
Yen, Tim J
Yoon, Kyonggeun
Zacchei, Anthony Gabriel

RHODE ISLAND
Brautigan, David L
Cane, David Earl
Hartman, Karl August
Hawrot, Edward

SOUTH CAROLINA
Amma, Elmer Louis
Banik, Narendra Lal
Dawson, John Harold
Evans, David Wesley
Greene, George C, III
Greenfield, Seymour
Hunter, George William
Kinard, Fredrick William
Lindenmayer, George Earl
Lundstrom, Ronald Charles
Wang, Huei-Hsiang Lisa

SOUTH DAKOTA
Cook, David Edgar
Evenson, Donald Paul
Marshall, Finley Dee
West, Thomas Patrick

TENNESSEE
Baldwin, Charles M
Bell, Helen
Benton, Charles Herbert
Bhattacharya, Syamal Kanti
Crane, Laura Jane
Egan, B Zane
Eyring, Edward J
Farrington, Joseph Kirby
Fridland, Arnold
Gerwe, Roderick Daniel
Goertz, Grayce Edith
Greenbaum, Elias
Griffin, Guy David
Harris, Durward Smith
Hartman, Frederick Cooper
Hossler, Fred E
Incardona, Antonino L
Jarrett, Harry Wellington, III
Jernigan, Howard Maxwell, Jr
Johnson, David Andrew
Kelley, Jim Lee
Laboda, Henry M
Legg, Ivan
Lowe, James N
Reddick, Bradford Beverly
Roberts, DeWayne
Smith, Jerry Howard
Staros, James Vaughan
Thakar, Jay H
Turner, Rex Howell
Uziel, Mayo
Waddell, Thomas Groth
Wakefield, Troy, Jr
Waterman, Michael Roberts
Williams, Robley Cook, Jr
Worsham, Lesa Marie Spacek

TEXAS
Adams, Emory Temple, Jr
Albrecht, Thomas Blair
Allen, Robert Charles
Alvarez-Gonzalez, Rafael
Arntzen, Charles Joel
Ball, M Isabel
Bates, George Winston
Besch, Paige Keith
Brown, Richard Malcolm, Jr
Brown, Wanda Lois
Burzynski, Stanislaw Rajmund
Bushey, Michelle Marie
Cate, Rodney Lee
Cody, John T
Cook, Paul Fabyan
Crookshank, Herman Robert
Czerwinski, Edmund William
Davis, Donald Robert
Dean, David Devereaux
Dear, Robert E A
Deloach, John Rooker
Dempsey, Walter B
Dwyer, Lawrence Arthur
Eakin, Richard Timothy
Everse, Johannes
Fleenor, Marvin Bension
Funkhouser, Edward Allen
Furlong, Norman Burr, Jr
Gierasch, Lila Mary
Gillard, Baiba Kurins
Gilman, Alfred G
Gray, Donald Melvin
Gray, Horace Benton, Jr
Gutsche, Carl David
Guynn, Robert William
Hackert, Marvin LeRoy
Heffley, James D
Hendrickson, Constance McRight
Ho, Begonia Y

Ho, Beng Thong
Ho, Dah-Hsi
Jennings, Michael Leon
Joshi, Vasudev Chhotalal
Kellems, Rodney E
Kohn, Harold Lewis
Kraus-Friedmann, Naomi
Kroschewsky, Julius Richard
Lupski, James Richard
Mabry, Tom Joe
McAdoo, David J
McConathy, Walter James
Macfarlane, Ronald Duncan
Massey, John Boyd
Masters, Bettie Sue Siler
Modak, Arvind T
Moreland, Ferrin Bates
Nachman, Ronald James
Nordyke, Ellis Larrimore
Olson, John S
Oray, Bedii
Pace, Carlos Nick
Parry, Ronald John
Piziak, Veronica Kelly
Ranney, David Francis
Rassin, David Keith
Raval, Dilip N
Ray, Allen Cobble
Renthal, Robert David
Robberson, Donald Lewis
Robinson, Neal Clark
Robison, George Alan
Rogers, Lorene Lane
Romeo, Tony
Saunders, Priscilla Prince
Scott, Alastair Ian
Shanley, Mark Stephen
Shaw, Robert Wayne
Sherry, Allan Dean
Siede, Wolfram
Skelley, Dean Sutherland
Slaga, Thomas Joseph
Sloan, Tod Burns
Sponsel, William Eric
Stone, Irving Charles, Jr
Teng, Jon Ie
Waldbillig, Ronald Charles
Weigel, Paul H(enry)
Wendt, Theodore Mil
Widger, William Russell
Williams, Charles Herbert
Williams, Mary Carr
Wilson, Carol Maggart
Wilson, Lon James
Wong, Shan Shekyuk
Woo, Savio L C
Wright, Stephen E
Zardeneta, Gustavo
Ziegler, Miriam Mary

UTAH
Barnett, Bill B
Gesteland, Raymond Frederick
Jensen, David
Johnston, Stephen Charles
Kemp, John Wilmer
Knapp, Gayle
Linker, Alfred
Moody, David Edward
Pitt, Charles H
Prescott, Stephen M
Ramachandran, Chittoor Krishna
Robins, Morris Joseph
Rothenberg, Mortimer Abraham
Scouten, William Henry
Seeley, Schuyler Drannan
Segelman, Alvin Burton
Smith, Marvin Artell
Walker, Edward Bell Mar
Windham, Carol Thompson

VERMONT
Crooks, George Chapman
Mitchell, William Alexander
Moyer, Walter Allen, Jr
Rinse, Jacobus
Tritton, Thomas Richard
Walters, Carol Price
Wurzburg, Otto Bernard

VIRGINIA
Allison, Trenton B
Averill, Bruce Alan
Barnett, Lewis Brinkley
Biltonen, Rodney Lincoln
Blackmore, Peter Fredrick
Blair, Barbara Ann
Brunelle, Richard Leon
Chandler, Jerry LeRoy
Cole, Patricia Ellen
Collins, Galen Franklin
Cone, Clarence Donald, Jr
Dannenburg, Warren Nathaniel
Donohue, Joyce Morrissey
Easter, Donald Philips
Esen, Asim
Ferguson, Robert Nicholas
Fonong, Tekum
Gander, George William
Gandour, Richard David
Gaskin, Felicia
Hall, Philip Layton
Hunt, Donald F
Jacobs, Maryce Mercedes

Johnston, Pauline Kay
Macdonald, Timothy Lee
Martin, Robert Bruce
Monroe, Stuart Benton
Olin, Jacqueline S
Pages, Robert Alex
Perillo, Benjamin Anthony
Pierson, Merle Dean
Place, Janet Dobbins
Rodig, Oscar Rudolf
Schroeder, Walter Adolph
Shipe, James R, Jr
Sirica, Alphonse Eugene
Skelly, Jerome Philip, Sr
Smibert, Robert Merrall, II
Smith, Harold Linwood
Stucky, Gary Lee
Taylor, Ronald Paul
Voige, William Huntley
Welt, Isaac Davidson
Whitney, George Stephen
Willett, James Delos
Wishner, Lawrence Arndt
Yu, Robert Kuan-Jen
Zaborsky, Oskar Rudolf

WASHINGTON
Anderson, Larry Ernest
Atkins, William M
Baillie, Thomas Allan
Benditt, Earl Philip
Bland, Jeffrey S
Cheng, Xueheng
Deming, Jody W
Duncan, James Byron
Ericsson, Lowell Harold
Gray, Patrick William
Jensen, Lyle Howard
Lewis, Norman G
Lybrand, Terry Paul
Medcalf, Darrell Gerald
Meeuse, Bastiaan J D
Palmiter, Richard
Patt, Leonard Merton
Phillips, Steven J
Read, David Hadley
Rees, Allan W
Reeves, Raymond
Roubal, William Theodore
Russo, Salvatore Franklin
Ryan, Clarence Augustine, Jr
Satterlee, James Donald
Smith, Richard Dale
Teller, David Chambers
Wekell, Marleen Marie
Whelton, Bartlett David
Zimmerman, Gary Alan

WEST VIRGINIA
Elkins, John Rush
Lotspeich, Frederick Jackson
Roberts, Joseph Linton
Wiggins, Richard Calvin

WISCONSIN
Anderson, Laurens
Antholine, William E
Boggs, Lawrence Allen
Cleland, William Wallace
Cohen, Philip Pacy
Craig, Elizabeth Anne
Cushing, Merchant Leroy
Dwyer-Hallquist, Patricia
Hake, Carl (Louis)
Hoffman, Norman Edwin
Johns, Philip Timothy
Johnson, Morris Alfred
Johnson, Warren Victor
Jones, Berne Lee
Kirk, T Kent
Klaassen, Dwight Homer
Krahnke, Harold C
Markley, John Lute
Miller, Joyce Fiddick
Petering, David Harold
Raines, Ronald T
Record, M Thomas, Jr
Rich, Daniel Hulbert
Romans-Hess, Alice Yvonne
Rosenberg, Robert Melvin
Runquist, Alfonse William
Sanderson, Gary Warner
Saryan, Leon Aram
Shaw, C Frank, III
Sih, Charles John
Skatrud, Thomas Joseph
Smith, Milton Reynolds
Soerens, Dave Allen
Struble, Craig Bruce
Twining, Sally Shinew
Ward, Kyle, Jr
Westler, William Milo
Worley, John David

WYOMING
Lewis, Randolph Vance

PUERTO RICO
Habeeb, Ahmed Fathi Sayed Ahmed
Infante, Gabriel A
Preston, Alan Martin
Ramirez, J Roberto
Salas-Quintana, Salvador

ALBERTA
Bide, Richard W
Dunford, Hugh Brian
Hill, Bernard Dale
Hodges, Robert Stanley
James, Michael G
Kotovych, George
Moore, Graham John
Roche, Rodney Sylvester
Stiles, Michael Edgecombe
Sykes, Brian Douglas
Van De Sande, Johan Hubert
Vederas, John Christopher
Wiebe, Leonard Irving

BRITISH COLUMBIA
Anderson, John Ansel
Cushley, Robert John
Dolphin, David Henry
Ekramoddoullah, Abul Kalam M
Haunerland, Norbert Heinrich
Irvine, George Norman
Mauk, Arthur Grant
Oehlschlager, Allan Cameron
Smith, Michael
Withers, Stephen George

MANITOBA
Duckworth, Henry William (Harry)
Forrest, Bruce James
Hasinoff, Brian Brennen
Hougen, Frithjof W
Jamieson, James C
McDonald, Bruce Eugene
MacGregor, Elizabeth Ann
Mantsch, Henry H
Rosenmann, Edward A
Stevens, Frits Christiaan
Vitti, Trieste Guido

NEWFOUNDLAND
Shaw, Derek Humphrey

NOVA SCOTIA
Dolphin, Peter James
Freeman, Harry Cleveland
Gilgan, Michael Wilson
Odense, Paul Holger
O'Dor, Ronald Keith
Verpoorte, Jacob A
Vining, Leo Charles
White, Robert Lester
White, Thomas David

ONTARIO
Brisbin, Doreen A
Burton, Graham William
Callahan, John William
Caplan, Donald
Chua, Kian Eng
Clarke, Anthony John
Cupp, Calvin R
Currie, Violet Evadne
Deslauriers, Roxanne Marie Lorraine
Dorrell, Douglas Gordon
Epand, Richard Mayer
Ferrier, Leslie Kenneth
Feuer, George
Gibson, Rosalind Susan
Goldsack, Douglas Eugene
Guthrie, James Peter
Janzen, Edward George
Kates, Morris
Khanna, Jatinder Mohan
Liew, Choong-Chin
Logan, James Edward
Macdonald, Peter Moore
Marks, Charles Frank
Mickle, Donald Alexander
Rodrigo, Russell Godfrey
Romans, Robert Gordon
Sarwar, Ghulam
Spencer, John Hedley
Spenser, Ian Daniel
Stancer, Harvey C
Stillman, Martin John
Thibert, Roger Joseph
Topp, Edward
Trenholm, Harold Locksley
Tsang, Charles Pak Wai
Wiebe, John Peter
Wood, Peter John
Wu, Tai Wing
Yaguchi, Makoto
Yang, Paul Wang
Yeung, David Lawrence

QUEBEC
Aubry, Muriel
Bateman, Andrew
Belanger, Patrice Charles
Berlinguet, Louis
Boulet, Marcel
Braun, Peter Eric
Chenevert, Robert (Bernard)
Ecobichon, Donald John
Escher, Emanuel
Jurasek, Lubomir
Khan, Masood
Lennox, Robert Bruce
Levy, Samuel Wolfe
Masut, Remo Antonio
Paice, Michael
Pezolet, Michel

Portelance, Vincent Damien
Roughley, Peter James
Schiller, Peter Wilhelm
Sikorska, Hanna
Srivastava, Uma Shanker
Van Calsteren, Marie-Rose
Vinay, Patrick
Wolfe, Leonhard Scott
Zamir, Lolita Ora

SASKATCHEWAN
Blair, Donald George Ralph
Boulton, Alan Arthur
Denford, Keith Eugene
Howarth, Ronald Edward
Khandelwal, Ramji Lal
McLennan, Barry Dean
Wood, James Douglas

OTHER COUNTRIES
Bar-Shavit, Rachel
Carmichael, David James
Chipman, David Mayer
Diederich, Francois Nico
Glick, David M
Gustafsson, Jan-Ake
Hazeyama, Yuji
Hebborn, Peter
Hopkins, Colin Russell
Jensen, Elwood Vernon
Judis, Joseph
Kerkay, Julius
Kroon, Paulus Arie
Kumar, S Anand
Kuriyama, Kinya
Kuttab, Simon Hanna
Ladinsky, Herbert
Lamoyi, Edmundo
Liao, Ta-Hsiu
Maragoudakis, Michael E
Markert, Claus O
Michaelsen, Terje E
Mitruka, Brij Mohan
Muschek, Lawrence David
Neuse, Eberhard Wilhelm
Rinehart, Frank Palmer
Sanger, Frederick
Santoro, Ferrucio Fontes
Schwartz, Alan William
Shapiro, Stuart
Sinha, Asru Kumar
Spiess, Joachim
Trentham, David R
Wan, Abraham Tai-Hsin
Wenclawiak, Bernd Wilhelm

Chemical Dynamics

ALABAMA
Bakker, Martin Gerard
Leslie, Thomas M
Meagher, James Francis
Stanbury, David McNeil
Williamson, Ashley Deas

ARIZONA
Lin, Sheng Hsien
Smith, Mark Alan

ARKANSAS
Becker, Ralph Sherman
Fry, Arthur James

CALIFORNIA
Atkinson, Roger
Auerbach, Daniel J
Benson, Sidney William
Berry, Michael James
Bowers, Michael Thomas
Brauman, John I
Brown, Nancy J
Bunton, Clifford A
Chidsey, Christopher E
Connick, Robert Elwell
Curtis, Earl Clifton, Jr
Delker, Gerald Lee
Dows, David Alan
Fisk, George Ayrs
Garrison, Warren Manford
Gelinas, Robert Joseph
Golden, David Mark
Goldstein, Elisheva
Green, Robert Bennett
Hajdu, Joseph
Harris, Charles Bonner
Haugen, Gilbert R
Head-Gordon, Martin Paul
Houk, Kendall N
Huestis, David Lee
Janda, Kenneth Carl
Jeffries, Jay B
Kibby, Charles Leonard
Kulander, Kenneth Charles
Lindner, Duane Lee
Ludwig, Frank Arno
McIver, Robert Thomas, Jr
Martens, Craig Colwell
Martin, L(aurence) Robbin
Mason, D(avid) M(alcolm)
Miller, William Hughes
Murov, Steven Lee
Okumura, Mitchio
Pearson, Michael J

Philo, John Sterner
Reisler, Hanna
Simon, John Douglas
Skolnick, Jeffrey
Syage, Jack A
True, Nancy S
Weinberg, William Henry
Whaley, Katharine Birgitta
Williamson, David Gadsby
Zimmerman, Ivan Harold

COLORADO
Anderson, Larry Gene
Bierbaum, Veronica Marie
Cristol, Stanley Jerome
Frank, Arthur Jesse
Giulianelli, James Louis
Leone, Stephen Robert
Lucas, George Bond
Martin, Charles R
Parson, Robert Paul

CONNECTICUT
Haake, Paul
Hinchen, John J(oseph)
Markham, Claire Agnes
Saunders, Martin
Seery, Daniel J

DELAWARE
Coulson, Dale Robert
Dixon, David Adams
Doren, Douglas James
Gay, Frank P
Gentzler, Robert E
Mills, Patrick Leo, Sr
Monroe, Bruce Malcolm
Ridge, Douglas Poll
Sparks, Donald

DISTRICT OF COLUMBIA
Earley, Joseph Emmet
Hancock, Kenneth George
Laufer, Allan Henry

FLORIDA
Babich, Michael Wayne
Brucat, Philip John
Hanrahan, Robert Joseph
Myerson, Albert Leon
Rhodes, William Clifford
Safron, Sanford Alan
Stoufer, Robert Carl
Young, Frank Glynn

GEORGIA
Dixon, Dabney White
Gole, James Leslie
Kutal, Charles Ronald
Legare, Richard J
Lin, Ming Chang
Nelson, Robert Norton
Uzer, Ahmet Turgay
Whetten, Robert Lloyd

IDAHO
Christian, Jerry Dale

ILLINOIS
Astumian, Raymond Dean
Boldridge, David William
Boyd, Mary K
Burrell, Elliott Joseph, Jr
Fleming, Graham Richard
Gordon, Sheffield
Gruebele, Martin
Henry, Patrick M
Kiefer, John Harold
Lambert, Joseph B
Makri, Nancy
Michael, Joe Victor
Pratt, Stephen Turnham
Sachtler, Wolfgang Max Hugo
Schatz, George Chappell
Scheeline, Alexander
Sibener, Steven Jay
Spears, Kenneth George
Suslick, Kenneth Sanders
Ulrey, Stephen Scott
Warf, C Cayce, Jr
Young, Charles Edward
Young, Linda

INDIANA
Adelman, Steven A
Grant, Edward R
Malik, David Joseph
Peters, Dennis Gail
Pimblott, Simon Martin
Racke, Kenneth David
Serianni, Anthony Stephan
Shortridge, Robert Glenn, Jr
Thomas, John Kerry

IOWA
DePristo, Andrew Elliott
Swartz, James E

KANSAS
Chu, Shih I
Fujimoto, Gordon Takeo
Hierl, Peter Marston
Lehman, Thomas Alan

KENTUCKY
Clouthier, Dennis James
Farina, Robert Donald
Moorhead, Edward Darrell
Slocum, Donald Warren

LOUISIANA
Kurtz, Richard Leigh
Witriol, Norman Martin

MARYLAND
Beri, Avinash Chandra
Bowen, Kit Hansel
Cavanagh, Richard Roy
Cody, Regina Jacqueline
Dagdigian, Paul J
Flynn, Joseph Henry
Gann, Richard George
Heimerl, Joseph Mark
Huang, Charles Y
Huie, Robert Elliott
King, David S(cott)
Kurylo, Michael John, III
Martinez, Richard Isaac
Miziolek, Andrzej Wladyslaw
Neta, Pedatsur
Stein, Stephen Ellery
Stief, Louis J
Wubbels, Gene Gerald

MASSACHUSETTS
Bowers, Peter George
Brenner, Howard
Doering, William von Eggers
Epstein, Irving Robert
Freedman, Andrew
Gardner, James A
Gersh, Michael Elliot
Hornig, Donald Frederick
Kolb, Charles Eugene, Jr
Lichtin, Norman Nahum
Marshall, Mark David
Molina, Mario Jose
Rawlins, Wilson Terry
Rose, Timothy Laurence
Stein, Samuel H
Sun, Yan
Van Eikeren, Paul

MICHIGAN
Barker, John Roger
Bartell, Lawrence Sims
Berndt, Donald Carl
Chou, Kuo Chen
Crittenden, John Charles
Hase, William Louis
LeBel, Norman Albert
Lubman, David Mitchell
Mutch, George William
Newcomb, Martin
Polik, William Frederick
Rothe, Erhard William
Rothschild, Walter Gustav
Wu, Ching-Hsong George

MINNESOTA
Atwell, William Alan
Bates, Frank S
Carr, Robert Wilson, Jr
Eades, Robert Alfred
Kreevoy, Maurice M

MISSOURI
Ackerman, Joseph John Henry
Bradburn, Gregory Russell
Frederick, Raymond H
Murray, Robert Wallace
Thomas, Timothy Farragut
Whitefield, Philip Douglas

MONTANA
Field, Richard Jeffrey

NEBRASKA
Provencher, Gerald Martin

NEW HAMPSHIRE
Bel Bruno, Joseph James
Hughes, Russell Profit
Winn, John Sterling

NEW JERSEY
Bahr, Charles Chester
Bernasek, Steven Lynn
Brus, Louis Eugene
Calcote, Hartwell Forrest
Celiano, Alfred
Choi, Byung Chang
Church, John Armistead
Ciaccio, Leonard Louis
Fisanick, Georgia Jeanne
Fu, Shou-Cheng Joseph
Gottscho, Richard Alan
Harris, Alexander L
McVey, Jeffrey King
Schoenfeld, Theodore Mark
Slagg, Norman

NEW MEXICO
Bellum, John Curtis
Cross, Jon Byron
Frauenfelder, Hans
Granoff, Barry
Greiner, Norman Roy

Hoffbauer, Mark Arles
Lyman, John L
Nevitt, Thomas D
Nogar, Nicholas Stephen
Pack, Russell T
Smith, Wayne Howard
Streit, Gerald Edward
Walters, Edward Albert

NEW YORK
Agosta, William Carleton
Atwood, Jim D
Bauer, Simon Harvey
Bersohn, Richard
Brown, Eric Richard
Brunschwig, Bruce Samuel
Buff, Frank Paul
Chapman, Sally
Chu, Liang Tung
Cole, David Le Roy
Creasy, William Russel
Ezra, Gregory Sion
Fontijn, Arthur
Garvey, James F
Gurel, Okan
Haim, Albert
Hehre, Edward James
Hudson, John B(alch)
Kallen, Thomas William
Kissel, Thomas Robert
Kuivila, Henry Gabriel
Lees, Alistair John
McGee, Thomas Howard
McLendon, George L
Malerich, Charles
Myers, Anne Boone
Ockman, Nathan
Philipp, Manfred (Hans)
Rowlett, Roger Scott
Schwarz, Harold A
Shiao, Daniel Da-Fong
Simonelli, Anthony Peter
Srinivasan, Rangaswamy
Sutin, Norman
Takacs, Gerald Alan
Van Verth, James Edward
Warden, Joseph Tallman
Weston, Ralph E, Jr
White, Michael George
Wiesenfeld, John Richard
Wu, Konrad T
Zhang, John Zeng Hui

NORTH CAROLINA
Carmichael, Halbert Hart
MacPhail, Richard Allyn
Martin, Ned Harold
Miller, Roger Ervin

NORTH DAKOTA
Tallman, Dennis Earl

OHIO
Belles, Frank Edward
Brilliant, Howard Michael
Dorfman, Leon Monte
Edwards, Jimmie Garvin
Fullmer, June Z(immerman)
Gordon, Gilbert
Gustafson, Terry Lee
Herbst, Eric
Laali, Kenneth Khosrow
Meckstroth, Wilma Koenig
Mortensen, Earl Miller
Platz, Matthew S
Richter, Helen Wilkinson
Rodgers, Michael A J
Rose, Mitchell
Schildcrout, Steven Michael
Snavely, Deanne Lynn
Taylor, Jay Eugene
Thayer, John Stearns
Warshay, Marvin

OKLAHOMA
Geibel, Jon Frederick

OREGON
Balko, Barbara Ann
Fleming, Bruce Ingram
Girardeau, Marvin Denham, Jr
Grover, James Robb
Haygarth, John Charles
Noyes, Richard Macy
Tyler, David Ralph

PENNSYLVANIA
Aviles, Rafael G
Bednar, Rodney Allan
Castleman, Albert Welford, Jr
Chay, Teresa R
Dai, Hai-Lung
Felty, Wayne Lee
Grabowski, Joseph J
Grinstein, Reuben H
Helfferich, Friedrich G
Hofer, Lawrence John Edward
Marcelin, George
Shepherd, Rex E
Simonaitis, Romualdas
Staley, Stuart Warner
Straub, Thomas Stuart
Sturm, James Edward
Tennent, Howard Gordon

Chemical Dynamics (cont)

Wood, Kurt Arthur
Yuan, Jian-Min

RHODE ISLAND
Kirschenbaum, Louis Jean

SOUTH CAROLINA
Miley, John Wulbern

TENNESSEE
Carman, Howard Smith, Jr
Chen, Chung-Hsuan
Crawford, Oakley H
Datz, Sheldon
Feigerle, Charles Stephen
Klots, Cornelius E
Metcalf, David Halstead
Miller, John Cameron
Taylor, Ellison Hall
Trowbridge, Lee Douglas

TEXAS
Albright, John Grover
Brooks, Philip Russell
Cares, William Ronald
DeBerry, David Wayne
De Groot, Peter Bernard
Dobson, Gerard Ramsden
Eaker, Charles William
Holwerda, Robert Alan
Johnson, James Elver
Marshall, Paul
Schmalz, Thomas G
Tibbals, Harry Fred, III

UTAH
Beishline, Robert Raymond
Bentrude, Wesley George
Czerlinski, George Heinrich
Goates, Steven Rex
Shirts, Randall Brent
Wahrhaftig, Austin Levy

VIRGINIA
Bogan, Denis John
Bryant, Robert George
DeGraff, Benjamin Anthony
Demas, James Nicholas
Desjardins, Steven G
Fisher, Farley
Hsu, David Shiao-Yo
Meites, Louis
Palmer, Curtis Allyn
Sanzone, George
Schiavelli, Melvyn David
Turner, Anne Halligan
Tyson, John Jeanes

WASHINGTON
Baughcum, Steven Lee
Duvall, George Evered
Root, John Walter
Satterlee, James Donald
Tabbutt, Frederick Dean
Windsor, Maurice William

WISCONSIN
Crim, F(orrest) Fleming
Megahed, Sid A
Moore, John Ward
Nathanson, Gilbert M
Viswanathan, Ramaswami
Zimmerman, Howard Elliot

WYOMING
Jacobson, Irven Allan, Jr

PUERTO RICO
Infante, Gabriel A

ALBERTA
Birss, Viola Ingrid
Dunford, Hugh Brian
Freeman, Gordon Russel
Sanger, Alan Rodney
Tschuikow-Roux, Eugene

BRITISH COLUMBIA
Basco, N

MANITOBA
Hasinoff, Brian Brennen
McFarlane, Joanna

NEW BRUNSWICK
Reinsborough, Vincent Conrad

NOVA SCOTIA
Pacey, Philip Desmond
Roscoe, John Miner

ONTARIO
Anlauf, Kurt Guenther
Back, Robert Arthur
Bolton, James R
Brumer, Paul William
Buncel, Erwin
Carrington, Tucker
Halfon, Efraim
Kresge, Alexander Jerry
Liu, Wing-Ki
Penner, Glenn H

Sadowski, Chester M

QUEBEC
Serpone, Nick
Srivastava, Uma Shanker

SASKATCHEWAN
Steer, Ronald Paul

OTHER COUNTRIES
Balakrishnan, Narayana Swamy
Bondybey, Vladimir E
Forst, Wendell
Hayon, Elie M
Kay, Kenneth George

Chemistry, General

ALABAMA
Askew, William Crews
Aull, John Louis
Baker, June Marshall
Barker, Samuel Lamar
Blanchard, Richard Lee
Carpenter, Charles H
Cataldo, Charles Eugene
Dorai-Raj, Diana Glover
Ellenburg, Janus Yentsch
Fendley, Ted Wyatt
Frings, Christopher Stanton
Garner, Robert Henry
Gill, Piara Singh
Govil, Narendra Kumar
Hardin, Ian Russell
Hughes, Edwin R
Kennedy, Frank Metler
Lammertsma, Koop
Lieberman, Robert
Ludvigsen, F J B
Marano, Gerald Alfred
Metzger, Robert Melville
Mishra, Satya Narayan
Pan, Chai-Fu
Scheiner, Bernard James
Sheridan, Richard Collins
Shoemaker, John Daniel, Jr
Smith, Lewis Taylor
Stevens, Frank Joseph
Summerlin, Lee R
Tamburin, Henry John
Toffel, George Mathias
Walia, Amrik Singh
Warfield, Carol Larson
Watkins, Charles Lee
Whitehead, Fred
Yee, Tin Boo
Ziegler, Paul Fout

ALASKA
Lokken, Donald Arthur

ARIZONA
Allen, John Rybolt
Alley, Starling Kessler, Jr
Barr, Charles (Robert)
Brown, Paul Wayne
Connor, Ralph (Alexander)
Cox, Fred Ward, Jr
Davis, Thomas P
Delbecq, Charles Jarchow
Fancher, Otis Earl
Glaunsinger, William Stanley
Helbert, John N
Kierbow, Julie Van Note Parker
King, William Mattern
Kriege, Owen Hobbs
Larson, Lester Mikkel
Miller, Glenn Harry
Nordstrom, Brian Hoyt
Robinette, Hillary, Jr
Rogers, Alan Barde
Shafer, M(errill) W(ilbert)
Tam, Wing Yim
Tamborski, Christ
Thomas, Charles L(amar)
Urry, Wilbert Herbert
Vandenberg, Edwin James
Wise, Edward Nelson

ARKANSAS
Geren, Collis Ross
Koeppe, Roger E, II
Ku, Victoria Feng
Obenland, Clayton O
Poirier, Lionel Albert
Shook, Thomas Eugene

CALIFORNIA
Abrams, Irving M
Adams, George Baker
Ahuja, Jagan N
Alire, Richard Marvin
Allen, Esther C(ampbell)
Altgelt, Klaus H
Altman, David
Anderson, Robert Emra
Andreas, John M(oore)
Andrews, Lawrence James
Ansari, Ali
Aragon, Sergio Ramiro
Arias, Jose Manuel
Arnold, James Richard
Atkins, Don Carlos, Jr

Atkinson, Roger
Avera, Fitzhugh Lee
Barusch, Maurice R
Battles, Willis Ralph
Beardslee, Ronald Allen
Becker, Joseph F
Beckman, Arnold Orville
Behr, Inga
Beilby, Alvin Lester
Berman, Horace Aaron
Bishop, John William
Bishop, Keith C, III
Bondar, Richard Jay Laurent
Bonner, Lyman Gaylord
Bonner, William Andrew
Bradbury, E Morton
Brauman, John I
Brink, David Liddell
Brouillard, Robert Ernest
Bruenner, Rolf Sylvester
Buck, Douglas Earl
Budny, John Arnold
Byall, Elliott Bruce
Campbell, George Washington, Jr
Chaikin, Saul William
Chou, Yungnien John
Clark, Leigh Bruce
Cliath, Mark Marshall
Condit, Paul Carr
Condit, Ralph Howell
Cooper, Wilson Wayne
Cosby, Philip C
Cowperthwaite, Michael
Coyner, Eugene Casper
Craik, Charles S
Csicsery, Sigmund Maria
Cubicciotti, Daniel David
Current, Steven P
Cusumano, James A
Cutforth, Howard Glen
Deck, Joseph Francis
Deleray, Arthur Loyd
Deshpande, Shirkant V
De Vries, John Edward
Domeniconi, Michael John
Drewes, Patricia Ann
Eastman, John W
Eikrem, Lynwood Olaf
Etzler, Dorr Homer
Feiler, William A, Jr
Feng, Da-Fei
Fenton, Donald Mason
Feramisco, James Robert
Ferguson, Lloyd Noel
Fletcher, Peter C
Foehr, Edward Gotthard
Frank, Joan Patricia
Frankel, Richard Barry
Frazer, Jack Winfield
Furby, Neal Washburn
Gamble, Fred Ridley, Jr
Gardner, Marjorie Hyer
Gerstein, Melvin
Gilbert, Francis Evalo
Goble, James H, Jr
Gold, Marvin B
Goldberg, Sabine Ruth
Gordon, Benjamin Edward
Groot, Cornelius
Haimes, Florence Catherine
Hale, Ron L
Halstead, Bruce W
Hamlin, Kenneth Eldred, Jr
Hansch, Corwin Herman
Harding-Barlow, Ingeborg
Hardy, Edgar Erwin
Harrar, Jackson Elwood
Hathaway, Gary Michael
Havlicek, Stephen
Hawkes, Wayne Christian
Hayter, Roy G
Heinemann, Heinz
Hendricks, Grant Walstein
Hicks, Harry Gross
Hickson, Donald Andrew
Hill, Russell John
Hinkson, Jimmy Wilford
Hobart, David Edward
Hochberg, Melvin
Hoey, Danny Lee
Holmquist, Walter Richard
Holzmann, Richard Thomas
Hooper, Catherine Evelyn
Horn, Christian Friedrich
Hydock, Joseph J
Hygh, Earl Hampton
Ja, William Yin
Janda, Kim D
Janus, Alan Robert
Johnson, LeRoy Franklin
Johnson, Richard D
Jones, Francis Tucker
Jones, Rebecca Anne
Jorgensen, Paul J
Kabra, Pokar Mal
Kamp, David Allen
Kane, Stephen Shimmon
Kaska, William Charles
Katsumoto, Kiyoshi
Kauffman, George Bernard
Keilin, Bertram
Kelley, Michael J
Kempter, Charles Prentiss
Keyzer, Hendrik

Kharasch, Norman
Kirsch, Milton
Klein, August S
Koenig, Nathan Hart
Kohn, Gustave K
Kolbezen, Martin (Joseph)
Kray, Louis Robert
Krikorian, Oscar Harold
Krummenacher, Daniel
Kuo, Harng-Shen
Kurnick, Allen Abraham
La Ganga, Thomas S
Landolph, Joseph Richard, Jr
Lawton, Emil Abraham
Lee, Vin-Jang
Lee, Young-Jin
Leider, Herman R
Lemmon, Richard Millington
Leyden, Richard Noel
Linder, Maria C
Liska, Kenneth J
Llenadol, Ramon
Loomis, Albert Geyer
Lukens, Herbert Richard, Jr
LuValle, James Ellis
Maas, Keith Allan
McCloskey, Allen Lyle
McKoy, Vincent
Madix, Robert James
Mar, Raymond W
Marantz, Laurence Boyd
Marquart, Ronald Gary
Marquis, Marilyn A
Meiners, Henry C(ito)
Meinhard, James Edgar
Melgard, Rodney
Metzger, Robert P
Michael, Leslie William
Michels, Lloyd R
Miller, Stanley Lloyd
Moffitt, Robert Allan
Monson, Richard Stanley
Moore, C Bradley
Morgal, Paul Walter
Morgan, James John
Morris, Jerry Lee, Jr
Mosen, Arthur Walter
Mullis, Kary B
Murov, Steven Lee
Mysels, Estella Katzenellenbogen
Naylor, Benjamin Franklin
Needles, Howard Lee
Neesby, Torben Emil
Neff, Loren Lee
Neher, Maynard Bruce
Neu, Ernest Ludwig
Newsam, John M
Newsom, Herbert Charles
Newton, Amos Sylvester
Nicolaou, Kyriacos Costa
Nixon, Alan Charles
Nozaki, Kenzie
Oestreicher, Hans
O'Konski, Chester Thomas
Owen, Walter Wycliffe
Ozari, Yehuda
Pallos, Ferenc M
Pathania, Rajeshwar S
Pattabhiraman, Tammanur R
Pauling, Linus Carl
Pearson, Michael J
Pearson, Ralph Gottfrid
Peeters, Randall Louis
Perrino, Charles T
Petrucci, Ralph Herbert
Pettijohn, Richard Robert
Phipps, Peter Beverley Powell
Pileggi, Vincent Joseph
Pokras, Harold Herbert
Pollack, Louis Rubin
Potter, John Clarkson
Ralls, Jack Warner
Recht, Howard Leonard
Reed, Russell, Jr
Rempel, Herman G
Repique, Eliseo
Riley, John Francis
Rocklin, Albert Louis
Rogoff, William Milton
Roukes, Michael L
Rowland, F Sherwood
Rudy, Thomas Philip
Savedoff, Lydia Goodman
Schall, Roy Franklin, Jr
Schick, Lloyd Alan
Schlatter, James Cameron
Schneider, Meier
Schugart, Kimberly A
Schultz, Peter G
Schumacher, Joseph Charles
Schweiker, George Christian
Searcy, A(lan) W(inn)
Sedat, John William
Sellas, James Thomas
Shackle, Dale Richard
Sharts, Clay Marcus
Simon, John Douglas
Simpson, William Tracy
Singer, Stanley
Sloane, Howard J
Slota, Peter John, Jr
Smith, Charles Francis, Jr
Smith, Elbert George
Snyder, Lloyd Robert

CHEMISTRY / 197

Spangler, John Allen
Spitze, LeRoy Alvin
Sridhar, Rajagopalan
Stevenson, David P(aul)
Stosick, Arthur James
Strouse, Charles Earl
Subramani, Suresh
Suess, Hans Eduard
Sweeney, William Alan
Thomas, Gerald Andrew
Tiedcke, Carl Heinrich Wilhelm
Ting, Shih-Fan
Toy, Madeline Shen
Traylor, Patricia Shizuko
Trogler, William C
Truesdell, Alfred Hemingway
Tsao, Makepeace Uho
Tsien, Roger Yonchien
Turnlund, Judith Rae
Vagnini, Livio L
Valentekovich, Marija Nikoletic
Vedvick, Thomas Scott
Vogel, Roger Frederick
Wadman, W Hugh
Walberg, Clifford Bennett
Walker, Jimmy Newton
Walwick, Earle Richard
Ward, John F
Weare, John H
Weinberg, William Henry
Weiss, Herbert V
Weissman, Earl Bernard
Westberg, Karl Rogers
Wheaton, Robert Miller
Whiteker, Roy Archie
Wieland, Bruce Wendell
Wiktorowicz, John Edward
Wiley, Richard Haven
Wilkes, John Barker
Williams, Lewis David
Williamson, Stanley Morris
Wilson, Martin
Wofsy, Leon
Wolf, Kathleen A
Wood, Peter Douglas
Woodward, Ervin Chapman, Jr
Wright, Ernest Marshall
Yaffe, Ruth Powers
Yavrouian, Andre
Yerram, Nagender Rao
Young, Ho Lee
Zerez, Charles Raymond
Zukoski, Edward Edom
Zwicker, Benjamin M G

COLORADO
Argabright, Perry A
Bedford, Joel S
Beren, Sheldon Kuciel
Catalano, Anthony William
Conant, Dale Holdrege
Daunora, Louis George
Dinneen, Gerald Uel
Eliot, Robert S
Ellison, Gayfree Barney
Estler, Ron Carter
Fall, R Ray
Ferguson, William Sidney
Giulianelli, James Louis
Gottschall, W Carl
Greyson, Jerome
Guinn, Denise Eileen
Harrison, Merle E(dward)
Heikkinen, Henry Wendell
Hersh, Sylvan David
Kano, Adeline Kyoko
Kennerly, George Warren
Kremers, Howard Earl
Lefever, Robert Allen
Lindquist, Robert Marion
Michel, Lester Allen
Michl, Josef
Navratil, James Dale
Pearson, John Richard
Peterson, Alan Herbert
Puleston, Harry Samuel
Quissell, David Olin
Richardson, Albert Edward
Schwartz, Donald
Sharma, Brahma Dutta
Skogen, Haven Sherman
Starr, Robert I
Stewart, James Joseph Patrick
Tosch, William Conrad
Wingeleth, Dale Clifford

CONNECTICUT
Anderson, Carol Patricia
Anderson, Rebecca Jane
Andrade, Manuel
Calbo, Leonard Joseph
Casberg, John Martin
Cooke, Theodore Frederic
Corbett, John Frank
Covey, Irene Mabel
Crothers, Donald M
Danishefsky, Samuel
Dooley, Joseph Francis
Essenfeld, Amy
Faust, John Philip
Feinland, Raymond
Fenn, John Bennett
Frank, Simon
Fried, John H

Galasso, Francis Salvatore
Gans, Eugene Howard
Garcia, Mario Leopoldo
Gingold, Kurt
Goodstein, Madeline P
Halverson, Frederick
Hekal, Ihab M
Hershenson, Herbert Malcolm
Hofrichter, Charles Henry
Humphrey, Bingham Johnson
Janeway, Charles Alderson, Jr
Johnstone, John William, Jr
Joyce, William
Knollmueller, Karl Otto
Kruse, Jurgen M
Leavitt, Julian Jacob
Levine, Leon
Loft, John T
Lybeck, A(lvin) H(iggins)
Maizell, Robert Edward
Man, Evelyn Brower
Martoglio, Pamela Anne
Masterton, William Lewis
Mayell, Jaspal Singh
Menkart, John
Mesrobian, Robert Benjamin
Mroczkowski, Stanley
Nielsen, John Merle
Noack, Manfred Gerhard
Norton, Louis Arthur
Oberstar, Helen Elizabeth
O'Neill, James F
Piehl, Donald Herbert
Prabulos, Joseph J, Jr
Pulito, Aldo Martin
Reed, Charles E(li)
Savinelli, Emilio A
Scardera, Michael
Schlessinger, Gert Gustav
Schoenbrunn, Erwin F(rederick)
Schroeder, Hansjuergen Alfred
Scola, Daniel Anthony
Scott, Robert Neal
Seefried, Carl G, Jr
Sieckhaus, John Francis
Steingiser, Samuel
Stewart, Albert Clifton
Sveda, Michael
Thomas, Arthur L
Todd, Harold David
Toralballa, Gloria C
Toy, Arthur Dock Fon
Tranner, Frank
Vaughan, Wyman Ristine
Welcher, Richard Parke
Willette, Gordon Louis
Wojtowicz, John Alfred
Woronick, Charles Louis
Wuskell, Joseph P

DELAWARE
Applegate, Lynn E
Auspos, Lawrence Arthur
Beestman, George Bernard
Birkenhauer, Robert Joseph
Borchardt, Hans J
Brandner, John David
Brehm, Warren John
Burch, Robert Ray, Jr
Burdick, Charles Lalor
Calkins, William Harold
Chang, Catherine Teh-Lin
Chi, Minn-Shong
Clement, Robert Alton
Cook, Leslie G(ladstone)
Craig, Alan Daniel
Craig, Raymond Allen
D'Amore, Michael Brian
Dasgupta, Sunil Priya
Dunlop, Edward Clarence
Earle, Ralph Hervey, Jr
Erdmann, Duane John
Fahl, Roy Jackson, Jr
Ford, Thomas Aven
Freerksen, Deborah Lynne (Chalmers)
Gardiner, John Alden
Gemmell, Gordon D(ouglas)
Gerlach, Howard G, Jr
Gilbert, Walter Wilson
Gloor, Walter Ervin
Hamilton, Jefferson Merritt, Jr
Hardham, William Morgan
Herkes, Frank Edward
Hershkowitz, Robert L
Hill, Frederick Burns, Jr
Hume, Harold Frederick
Ivett, Reginald William
Johnson, Kenneth Earl
Klein, Michael Tully
Koch, Theodore Augur
Kolber, Harry John
Kosak, John R
Leibu, Henry J
Litchfield, William John
Maclachlan, Alexander
Mahler, Walter
Malley, Linda Angevine
Mallonee, James Edgar
Mandel, Zoltan
Martin, Arthur Francis
Mathre, Owen Bertwell
Maynard, Carl Wesley, Jr
Mighton, Harold Russell
Miles, James Lowell

Milian, Alwin S, Jr
Neal, Thomas Edward
Niedzielski, Edmund Luke
Pace, Salvatore Joseph
Parshall, George William
Podlas, Thomas Joseph
Quon, Check Yuen
Resnick, Paul R
Rexford, Dean R
Riches, Wesley William
Rondestvedt, Christian Scriver, Jr
Rudolph, Jeffrey Stewart
Saffer, Henry Walker
Sands, Seymour
Scalfarotto, Robert Emil
Schenker, Henry Hans
Shoaf, Charles Jefferson
Sloan, Martin Frank
Solenberger, John Carl
Sonnichsen, George Carl
Stewart, Edward William
Stiles, A(lvin) B(arber)
Stockburger, George Joseph
Streicher, Michael A(lfred)
Sturgis, Bernard Miller
Summers, John Clifford
Takvorian, Kenneth Bedrose
Theopold, Klaus Hellmut
Tullio, Victor
Valdsaar, Herbert
Weeks, Gregory Paul
Williams, Richard Anderson
Wu, Ting Kai
Yamamoto, Y Stephen
Young, Charles Albert

DISTRICT OF COLUMBIA
Bannerman, Douglas George
Bertsch, Charles Rudolph
Boron, David J
Bowen, David Hywel Michael
Bultman, John D
Cohen, Alex
Di Carlo, Frederick Joseph
Frattali, Victor Paul
Good, Mary Lowe
Henry, Richard Lynn
Johnson, Martin R
McCafferty, Edward
Nandedkar, Arvindkumar Narhari
Newman, Pauline
Nicholson, Richard Selindh
Perros, Theodore Peter
Pierce, Dorothy Helen
Raber, Douglas John
Ramsey, Jerry Warren
Rhoads, Allen R
Sarmiento, Rafael Apolinar
Vanderveen, John Edward
Wakelyn, Phillip Jeffrey
Wyatt, Jeffrey Renner
Yang, David Chih-Hsin

FLORIDA
Andregg, Charles Harold
Billica, Harry Robert
Bixler, Dean A
Bjorksten, Johan Augustus
Block, Seymour Stanton
Boer, F Peter
Borovsky, Dov
Browning, Joe Leon
Carter, James Harrison, II
Christenson, Roger Morris
Clark, Ralph O
Connors, William Matthew
Coppoc, William Joseph
Crentz, William Luther
Cross, Timothy Albert
Croxall, Willard (Joseph)
Dasher, Paul James
Davis, Dean Frederick
Davis, John Armstrong
Dean, Robert Reed
DeRosset, Armand John
Detrick, Robert Sherman
Dewar, Michael James Steuart
Drake, Robert Firth
Dresdner, Richard David
Duranceau, Steven Jon
Filson, Malcolm Harold
Fox, Gerald
Foy, Robert Bastian
Frazier, Stephen Earl
Frishman, Daniel
Gerwe, Raymond Daniel
Grafstein, Daniel
Gross, Malcolm Edmund
Halsey, John Joseph
Hamill, James Junior
Hanley, James Richard, Jr
Harrold, Gordon Coleson
Haug, Arthur John
Head, Ronald Alan
Hein, Richard Earl
Hellwege, Herbert Elmore
Henkin, Hyman
Hieserman, Clarence Edward
Holub, Fred F
Hormats, Ellis Irving
Howsmon, John Arthur
Ingwalson, Raymond Wesley
Jackson, George Richard
Jackson, William G(ordon)

Karsten, Kenneth Stephen
Katritzky, Alan R
Lacoste, Rene John
Lee, Harley Clyde
Leffler, John Edward
Leonard, Reid Hayward
Leverenz, Humboldt Walter
Lieberman, Samuel Victor
Lippe, Robert Lloyd
Mador, Irving Lester
Magee, John Robert
Miles, Delbert Howard
Mizell, Louis Richard
Morejon, Clara Baez
Moroni, Eneo C
Moskowitz, Mark Lewis
Pande, Gyan Shanker
Pappas, Anthony John
Parks, Kenneth Lee
Parrish, Clyde Franklin
Radomski, Jack London
Ramaswamy, C
Ray, Robert Allen
Reed, Sherman Kennedy
Rocco, Gregory Gabriel
Rochow, Eugene George
Ropp, Walter Shade
Sackett, William Malcolm
Salsbury, Jason Melvin
Sathe, Shridhar Krishna
Shane, Robert S
Sharkey, William Henry
Shaw, Philip Eugene
Singley, John Edward
Sisler, Harry Hall
Smith, Paul Vergon, Jr
Spielberger, Charles Donald
Stang, Louis George
Stoufer, Robert Carl
Stump, Eugene Curtis, Jr
Tarrant, Paul
Thomas, Vera
Walton, Charles William
Wenzinger, George Robert
Werner, Rudolf
Wiles, Robert Allan
Wissow, Lennard Jay
Wittbecker, Emerson LaVerne
Witte, Michael
Wood, James Brent, III
Young, Frank Glynn

GEORGIA
Ali, Monica McCarthy
Allison, Jerry David
Aseff, George V
Block, Toby Fran
Boone, Donald Joe
Burrows, Walter Herbert
Chen, Chia Ming
Crawford, Van Hale
Danzer, Laurence Alfred
DeLorenzo, Ronald Anthony
Dooley, William Paul
Fales, Frank Weck
Fay, Alice D Awtrey
Fineman, Manuel Nathan
Fitzwater, Robert N
Friedman, Harold Bertrand
Garst, John Fredric
Gilbert, Frederick Emerson, Jr
Hill, Craig Livingston
Hodges, Linda Carol
Jones, William Henry, Jr
Kollig, Heinz Philipp
Lewis, Jasper Phelps
Lin, Ming Chang
Lohuis, Delmont John
McBay, Henry Cecil
McBride, Clifford Hoyt
Martinek, Robert George
Middleton, Henry Moore, III
Montgomery, Daniel Michael
Morley, Robert
Palmer, Richard Carl
Robbins, Wayne Brian
Schultz, Donald Paul
Sgoutas, Demetrios Spiros
Sizemore, Douglas Reece
Szostak, Rosemarie
Watts, Sherrill Glenn
Whitten, Kenneth Wayne
Young, C(larence) B(ernard) F(ehrler)

HAWAII
Divis, Allan Francis
Kaye, Wilbur (Irving)
Seo, Stanley Toshio
Tu, Chen Chuan
Zaleski, Halina Maria

IDAHO
Gustafson, Donald Arvid
Wright, Kenneth James

ILLINOIS
Adelson, Bernard Henry
Anders, Edward
Anderson, David John
Anderson, Kenning M
Anderson, Scott
Atcher, Robert
Baitinger, William F, Jr
Barrall, Raymond Charles

Chemistry, General (cont)

Benkendorf, Carol Ann
Berkman, Michael G
Bermes, Edward William, Jr
Bingham, Carleton Dille
Blackburn, Paul Edward
Brace, Neal Orin
Brooks, Kenneth Conrad
Burney, Donald Eugene
Burwell, Robert Lemmon, Jr
Carnahan, Jon Winston
Chao, Sherman S
Chignell, Derek Alan
Cohen, Donald
Cohen, Harry
Colton, Frank Benjamin
Curtis, Veronica Anne
Damusis, Adolfas
De Pasquali, Giovanni
Dinerstein, Robert Alvin
Drickamer, Harry George
Dybalski, Jack Norbert
Elkins, Robert Hiatt
Epstein, Morton Batlan
Field, Kurt William
Fiess, Harold Alvin
Foster, Harold Marvin
Frame, Robert Roy
Friedland, Waldo Charles
Friedman, Bernard Samuel
Goran, Morris
Goss, George Robert
Green, Frank Orville
Greenberg, Elliott
Grosz, Oliver
Hadley, Elbert Hamilton
Hampel, Clifford Allen
Harper, Jon Jay
Harrington, Joseph Anthony
Harris, Samuel William
Harrison, Ben
Hartlage, James Albert
Henning, Lester Allan
Henry, Robert Edwin
Hirsch, Arnold
Holtzman, Richard Beves
Ichniowski, Thaddeus Casimir
Inskip, Ervin Basil
Iveson, Herbert Todd
Jason, Emil Fred
Johnson, Richard Joseph
Kalinowski, Mathew Lawrence
Kaufman, Priscilla C
Kirkpatrick, Joel Lee
Kleiman, Morton
Konopinski, Virgil J
Laird, Don M
Lee, Anthony L
Lerner, Louis L(eonard)
Lester, George Ronald
Lillwitz, Lawrence Dale
McDonald, J Douglas
Magnus, George
Mahajan, Om Prakash
Maltenfort, George Gunther
Marks, Tobin Jay
Mason, Donald Frank
Matson, Howard John
Maynert, Everett William
Melton, Marilyn Anders
Miller, John Robert
Nash, Ralph Glen
Nelson, Arthur Kendall
Neuzil, Richard William
Newhart, M(ary) Joan
Norris, James Rufus, Jr
Parent, Joseph D(ominic)
Penrose, William Roy
Perlow, Mina Rea Jones
Porsche, Jules D(ownes)
Princen, Lambertus Henricus
Regunathan, Perialwar
Richmond, Patricia Ann
Riley, Reed Farrar
Rosenberg, Saul Howard
Ryan, Julian Gilbert
Salutsky, Murrell Leon
Sawinski, Vincent John
Sehgal, Lakshman R
Shida, Mitsuzo
Silvernail, Walter Lawrence
Simmons, William Howard
Simon, Wilbur
Singh, Mohinder
Slama, Francis J
Snow, Adolph Isaac
Stavrolakis, J(ames) A(lexander)
Steele, Ian McKay
Steele, Sidney Russell
Stehney, Andrew Frank
Steindler, Martin Joseph
Stipanovic, Bozidar J
Stuart, Alfred Herbert
Stucker, Joseph Bernard
Studier, Martin Herman
Sutton, Russell Paul
Sytsma, Louis Frederick
Thompson, John
Treadway, William Jack, Jr
Trimble, Russell Fay
Walhout, Justine I Simon
Warne, Thomas Martin
Welsh, Thomas Laurence
Wing, Robert Edward
Woolson, Edwin Albert
Wysocki, Allen John
Yang, Nien-Chu
Yuster, Philip Harold

INDIANA

Abdulla, Riaz Fazal
Amy, Jonathan Weekes
Baker, Timothy Stanley
Bodner, George Michael
Booe, J(ames) M(arvin)
Boyd, Frederick Mervin
Chisholm, Malcolm Harold
Chmiel, Chester T
Christou, George
Demkovich, Paul Andrew
Diamond, Sidney
Diamond, Steven Elliot
Edgell, Walter Francis
Felger, Maurice Monroe
Fricke, Gordon Hugh
Gardner, David Arnold
Getzendaner, Milton Edmond
Hellman, Henry Martin
Johnston, Katharine Gentry
Kampen, Emerson
Kissinger, Peter Thomas
Kroon, James Lee
Magnus, Philip Douglas
Martin, Jerome
Mead, Darwin James
Meyerson, Seymour
Montieth, Richard Voorhees
Moorehead, Wells Rufus
Morrison, William Alfred
Natarajan, Viswanathan
Novotny, Milos V
Proksch, Gary J
Rapkin, Myron Colman
Ross, Alberta Barkley
Scherer, George Allen
Shultz, Clifford Glen
Shupe, Robert Eugene
Sieloff, Ronald F
Sowers, Edward Eugene
Stratton, Wilmer Joseph
Stuckwisch, Clarence George
Swain, Richard Russell
Taylor, Harold Mellon
Timma, Donald Lee
Turner, James A
Walsh, Patrick Noel
Wang, Jin-Liang
Wehrmeister, Herbert Louis
Wilks, Louis Phillip
Williams, Leamon Dale
Yordy, John David

IOWA

Carr, Duane Tucker
Firstenberger, B(urnett) G(eorge)
Hanak, Joseph J
Meints, Clifford Leroy
Reuland, Robert John

KANSAS

Baranczuk, Richard John
Beadle, Buell Wesley
Beard, William Quinby, Jr
Beechan, Curtis Michael
Bellet, Eugene Marshall
Berg, J(ohn) Robert
Boyle, Kathryn Moyne Ward Dittemore
Bricker, Clark Eugene
Choguill, Harold Samuel
Christian, Robert Vernon, Jr
Curless, William Toole
Dreschhoff, Gisela Auguste-Marie
Finney, Karl Frederick
Glick, John Henry, Jr
Hefferren, John James
Hiebert, Allen G
Hirschmann, Robert P
Lanning, Francis Chowing
Lehman, Thomas Alan
McCormick, Bailie Jack
Marshall, Delbert Allan
Reynolds, Charles Albert
Rumpel, Max Leonard
Schrenk, William George
Shaver, Lee Alan

KENTUCKY

Beck, Lloyd Willard
Bhattacharya-Chatterjee, Malaya
Doderer, George Charles
Gird, Steven Richard
Helms, Boyce Dewayne
Hettinger, William Peter, Jr
Hunter, Norman W
Klingenberg, Joseph John
Knopf, Daniel Peter
McDermott, Dana Paul
Plucknett, William Kennedy
Wagner, William Frederick
Wheeler, Thomas Jay

LOUISIANA

Andrews, Bethlehem Kottes
Baird, William C, Jr
Beeler, Myrton Freeman
Blanchard, Eugene Joseph
Broeg, Charles Burton
Bromberg, Milton Jay
Brown, Jerome Engel
Brown, Lawrence E(ldon)
Calamari, Timothy A, Jr
Cook, Shirl Eldon
Daigle, Donald J
Danti, August Gabriel
Dearth, James Dean
DeMonsabert, Winston Russel
Drawe, Sister M Veronica
Gormus, Bobby Joe
Greco, Edward Carl
Halbert, Thomas Risher
Hamer, Jan
Harvey, Clarence Charles, (Jr)
Hocart, Simon
Kennedy, Frank Scott
Laurent, Sebastian Marc
Mangham, Jesse Roger
Merrill, Howard Emerson
Morris, Cletus Eugene
Moseley, Harry Edward
Plonsker, Larry
Reeves, Wilson Alvin
Reichle, Alfred Douglas
Reid, John David
Risinger, Gerald E
Roberts, Earl John
Roberts, Reginald Francis
Sachdev, Sham L
Siddall, Thomas Henry, III
Sinha, Sudhir K
Smith, Charles Hooper
Srinivasan, Sathanur Ramachandran
Stahly, Glenn Patrick
Vingiello, Frank Anthony
Warner, Isiah Manuel
Watson, Dennis Ronald
Yeadon, David Allou

MAINE

Bragdon, Robert Wright
Cartier, George Thomas
Mundy, Bradford Philip
Nagle, Jeffrey Karl
Nahabedian, Kevork Vartan
Rundell, Clark Ace
Sottery, Theodore Walter
Stanley, Norman Francis
Taylor, Anthony Boswell

MARYLAND

Affens, Wilbur Allen
Alayash, Abdu I
Barrack, Carroll Marlin
Baummer, J Charles, Jr
Berch, Julian
Berg, Jeremy M
Berkowitz, Sidney
Black, Simon
Braude, George Leon
Callahan, Mary Vincent
Carr, Charles Jelleff
Caswell, Robert Little
Chan, Daniel Wan-Yui
Cheng, Sheue-yann
Cohen, Howard Joseph
Collins, John Henry
Conrad, Edward Ezra
Cookson, John Thomas, Jr
Crisler, Joseph Presley
Currie, Lloyd Arthur
Dingell, James V
Edinger, Stanley Evan
Ellis, Rex
Ernest, Michael Vance, Sr
Fields, Richard Joel
Fitch, Steven Joseph
Fowler, Emil Eugene
Frank, Victor Samuel
Freimuth, Henry Charles
Fristrom, Robert Maurice
Gann, Richard George
Ghoshtagore, Rathindra Nath
Gibian, Thomas George
Goldenson, Jerome
Grant, David Graham
Haffner, Richard William
Hanford, William Edward
Heath, George A(ugustine)
Heimerl, Joseph Mark
Horney, Amos Grant
Hoster, Donald Paul
Houk, Albert Edward Hennessee
Howell, Barbara Fennema
Hutchins, Clyde
Ibrahim, A Mahammad
James, John Cary
Jones, Owen Lloyd
Jones, Thomas Oswell
Kappe, David Syme
Koch, William Frederick
Kolobielski, Marjan
Krynitsky, John Alexander
Landgrebe, Albert R
Lee, Theresa
Lepley, Arthur Ray
Lever, John
Levine, Alan Stewart
Levy, Robert
Lijinsky, William
McCandliss, Russell John
McCormick, Anna M
Marans, Nelson Samuel
Margalit, Nehemiah
Mariano, Patrick S
Massie, Samuel Proctor
Mazumder, Bibhuti R
Melville, Robert S
Merz, Kenneth M(alcolm), Jr
Miller, Hugh Hunt
Milne, George William Anthony
Misra, Renuka
Montgomery, Stewart Robert
Mowry, David Thomas
Murray, Gary Joseph
Naibert, Zane Elvin
Ondov, John Michael
Ordway, Fred
Osawa, Yoichi
Panayappan, Ramanathan
Peiser, Herbert Steffen
Phelan, Earl Walter
Pick, Robert Orville
Placious, Robert Charles
Powell, Francis X
Prasanna, Hullahalli Rangaswamy
Priest, Homer Farnum
Rice, Rip G
Richards, Joseph Dudley
Richman, Robert Michael
Rochlin, Phillip
Rosenblatt, David Hirsch
Rosenblum, Annette Tannenholz
Rostenbach, Royal E(dwin)
Roswell, David Frederick
Rush, Cecil Archer
Sanders, James Grady
Schattner, Robert I
Schifreen, Richard Steven
Shah, Shirish
Siatkowski, Ronald E
Sieckmann, Donna G
Silverman, Joseph
Skalny, Jan Peter
Smith, Betty F
Smith, Ieuan Trevor
Strauss, Simon Wolf
Takagi, Shozo
Taylor, Harold Nathaniel
Trimmer, Robert Whitfield
Trus, Benes L
Tullius, Thomas D
Vennos, Mary Susannah
Ward, Joseph Richard
Watters, Robert Lisle
Wells, James Robert
Weser, Don Benton
White, Blanche Babette
Wiggin, Edwin Albert
Wolford, Richard Kenneth
Wright, William Wynn
Wubbels, Gene Gerald
Yeh, Kwan-Nan
Yeh, Lai-Su Lee
Young, Jay Alfred
Zamora, Antonio
Zdravkovich, Vera

MASSACHUSETTS

Agrawal, Sudhir
Anderson, James Gilbert
Backman, Keith Cameron
Bailey, Milton
Banderet, Louis Eugene
Baratta, Edmond John
Bartlett, Paul Doughty
Bazinet, Maurice L
Berlandi, Francis Joseph
Berry, Vern Vincent
Blank, Charles Anthony
Bronstein, Irena Y
Brown, Harold Hubley
Bump, Charles Kilbourne
Burkhardt, Alan Elmer
Chang, Raymond
Charkoudian, John Charles
Chase, Fred Leroy
Chen, Peter
Clemson, Harry C
Clydesdale, Fergus Macdonald
Coleman, George W(illiam)
Cook, Michael Miller
Coulter, Lowell Vernon
Crandlemere, Robert Wayne
Dasgupta, Arijit M
Deck, Joseph Charles
Deutsch, Marshall Emanuel
Doshi, Anil G
Elfbaum, Stanley Goodman
Evans, David A
Feinstein, Leonard Gordon
Fine, David H
Fischer, George A
Forchielli, Americo Lewis
Friedenstein, Hanna
Gibson, George
Giese, Roger Wallace
Giessen, Bill C(ormann)
Golubovic, Aleksandar
Gorfien, Harold
Green, Milton
Hadjian, Richard Albert
Hamilton, Charles William
Handy, Carleton Thomas
Hepler, Peter Klock
Hoffman, Donald Oliver
Holland, Andrew Brian

Hume, David Newton
Ives, Robert Southwick
Kaplan, Lawrence Jay
Kliem, Peter O
Korenstein, Ralph
Lin, Sin-Shong
MacIver, Donald Stuart
Markland, William R
Marshall, Mark David
Merken, Melvin
Morbey, Graham Kenneth
Naidus, Harold
Olson, Arthur Russell
O'Malley, Robert Francis
Panto, Joseph Salvatore
Pappalardo, Romano Giuseppe
Parkinson, R(obert) E(dward)
Peirent, Robert John
Peng, Fred Ming-Sheng
Raphael, Thomas
Raymond, Samuel
Ritt, Paul Edward, Jr
Rogers, Howard Gardner
Ross, Sidney David
Schaffel, Gerson Samuel
Shepp, Allan
Sigai, Andrew Gary
Sleeman, Richard Alexander
Smith, William Edward
Stein, Samuel H
Strimling, Walter Eugene
Suplinskas, Raymond Joseph
Swank, Thomas Francis
Taunton-Rigby, Alison
Taylor, Lloyd David
Trotz, Samuel Isaac
Vasilos, Thomas
Veidis, Mikelis Valdis
Wald, Fritz Veit
Waller, David Percival
Wechter, Margaret Ann
Weller, Paul Franklin
Wiener, Robert Newman
Wood, John Stanley
Wyman, John E
Zapsalis, Charles

MICHIGAN
Ahn, Kyunghye
Allan, David
Allenstein, Richard Van
Anders, Oswald Ulrich
Anderson, Melvin Lee
Anderson, Robert Hunt
Anderson, Theodore
Bartleson, John David
Blinn, Lorena Virginia
Boundy, Ray Harold
Broene, Herman Henry
Bruner, Leonard Bretz, Jr
Colingsworth, Donald Rudolph
Cook, Paul Laverne
Corrigan, Dennis Arthur
Davis, Ralph Anderson
Deck, Charles Francis
Deering, Carl F
Dirkse, Thedford Preston
Doyle, Daryl Joseph
Duggan, Helen Ann
Eldis, George Thomas
Epstein, Emanuel
Fearon, Frederick William Gordon
Felmlee, William John
Fortuna, Edward Michael, Jr
Frawley, Nile Nelson
Frevel, Ludo Karl
Frisch, Kurt Charles
Gayer, Karl Herman
Gendernalik, Sue Aydelott
Goliber, Edward William
Grochoski, Gregory T
Gruen, Fred Martin
Hallada, Calvin James
Hartman, Robert John
Hartwell, George E
Heyman, Laurel Elaine
Hinkamp, James Benjamin
Jacko, Michael George
Johnson, James Leslie
Kangas, Donald Arne
Kardos, Otto
Kennelly, William J
Kirschner, Stanley
Kolopajlo, Lawrence Hugh
Kummer, Joseph T
LaBarge, Robert Gordon
Leddy, James Jerome
LeGrow, Gary Edward
Lentz, Charles Wesley
Lustgarten, Ronald Krisses
McGraw, Leslie Daniel
Maheswari, Shyam P
Mance, Andrew Mark
Meyer, Heinz Friedrich
Miller, Thomas Lee
Moissides-Hines, Lydia Elizabeth
Moser, Frank Hans
Mullins, John A
Nelson, John Arthur
Northup, Melvin Lee
Peery, Clifford Young
Polmanteer, Keith Earl
Rabold, Gary Paul
Rand, Cynthia Lucille

Reim, Robert E
Ruof, Clarence Herman
Sardesai, Vishwanath M
Schaap, A Paul
Schneider, Eric West
Scott, Joseph Hurlong
Sheets, Donald Guy
Shelef, Mordecai
Singh, Lal Pratap S
Snyder, Dexter Dean
Solomon, David Eugene
Speier, John Leo, Jr
Stevens, Calvin Lee
Streiff, Anton Joseph
Strojny, Edwin Joseph
Swartz, Grace Lynn
Swathirajan, S
Tai, Julia Chow
Tasker, Clinton Waldorf
Terry, Samuel Matthew
Vanderkooi, William Nicholas
Vrieland, Gail Edwin
Wang, Yar-Ming
Willson, Philip James
Yeung, Patrick Pui-hang
Zabik, Matthew John

MINNESOTA
Agre, Courtland LeVerne
Baker, Michael Harry
Baldwin, Arthur Richard
Beebe, George Warren
Brink, Norman George
Carlson, Robert Leonard
Childs, William Ves
Cowles, Edward J
Di Gangi, Frank Edward
Durnick, Thomas Jackson
Eisenreich, Steven John
Finholt, Albert Edward
Frank, William Charles
Fraser, Judy Ann
Fridinger, Tomas Lee
Haase, Ashley Thomson
Hagen, Donald Frederick
Harrison, Stuart Amos
Holler, Albert Cochran
Iwasaki, Iwao
Janes, Donald Lucian
Jewsbury, Wilbur
Kleber, Eugene Victor
McKenna, Jack F(ontaine)
McMullen, James Clinton
Magnuson, Vincent Richard
Malzer, Gary Lee
Mayerle, James Joseph
Meehan, Edward Joseph
Miessler, Gary Lee
Nowlin, Duane Dale
Paterson, William Gordon
Paulson, Bradley
Pearlson, Wilbur H
Sahyun, Melville Richard Valde
Schwartz, A(lbert) Truman
Seibold, Carol Duke
Sherman, Patsy O'Connell
Sobieski, James Fulton
Swofford, Harold S, Jr
Throckmorton, James Rodney
Wertz, John Edward

MISSISSIPPI
Altenkirch, Robert Ames
Emerich, Donald Warren
Pinson, James Wesley
Russell, Joseph Louis
Wolverton, Billy Charles

MISSOURI
Bohanon, Joseph Terril
Bovy, Philippe R
Brew, William Barnard
Churchill, Ralph John
Craver, Clara Diddle (Smith)
Craver, John Kenneth
Crutchfield, Marvin Mack
Dyroff, David Ray
Eime, Lester Oscar
Elliott, Joseph Robert
Farnsworth, Marie
Festa, Roger Reginald
Fox, Dale Bennett
Hellerstein, Stanley
Hemphill, Louis
Hodges, Glenn R(oss)
Homeyer, August Henry
James, William Joseph
Janski, Alvin Michael
Kern, Roland James
Khalifah, Raja Gabriel
King, Perry, Jr
Kishore, Ganesh M
Ksycki, Mary Joecile
Kuhns, John Farrell
Lembke, Roger Roy
Long, Lawrence William
Lynch, Dan K
McConaghy, John Stead, Jr
McGinnes, Edgar Allan, Jr
Magruder, Willis Jackson
Mohrman, Harold W
Morse, Ronald Loyd
Nason, Howard King
Neau, Steven Henry

Orban, Edward
Palmer, John Frank, Jr
Perry, Randolph, Jr
Russell, Robert Raymond
Russo, Michael Eugene
Scallet, Barrett Lerner
Schwarz, Richard
Sherman, William Reese
Shri, Thanedar
Singh, Harbhajan
Stone, Bobbie Dean
Teaford, Margaret Elaine
Waldron, Harold Francis
Walsh, Robert Jerome
Wang, Maw Shiu
Welch, Michael John
Wilcox, Harold Kendall
Winicov, Murray William
Yeager, John Frederick
Zienty, Ferdinand B

MONTANA
Baker, Graeme Levo
Fessenden, Ralph James
Gloege, George Herman
Goering, Kenneth Justin
Julian, Gordon Ray
Parker, Keith Krom
White, John Greville

NEBRASKA
Garey, Carroll Laverne
Griswold, Norman Ernest
Harbison, Gerard Stanislaus
Heaney, Robert Proulx
Mattes, Frederick Henry
Nair, Chandra Kunju
Quigley, Herbert Joseph, Jr
Rack, Edward Paul
Sturgeon, George Dennis
Wang, Chin Hsien

NEVADA
Dean, John Gilbert
Dickson, Frank Wilson
Fox, Neil Stewart
Kuroda, Paul Kazuo
Nazy, John Robert
Thamer, B(urton) J(ohn)

NEW HAMPSHIRE
Cotter, Robert James
Custer, Michael
Roberts, John Edwin
Wolff, Nikolaus Emanuel

NEW JERSEY
Adler, Irwin L
Ager, John Winfrid
Albers-Schonberg, Georg
Allen, Leland Cullen
Alvarez, Vernon Leon
Andreatch, Anthony J
Antonsen, Donald Hans
Armamento, Eduardo T
Babson, Arthur Lawrence
Barile, George Conrad
Barnes, Robert Lee
Bassett, Alton Herman
Berk, Bernard
Binder, Michael
Blank, Zvi
Bluestein, Claire
Borowsky, Harry Herbert
Bosniack, David S
Brady, Thomas E
Brennan, James A
Brewer, Glenn A, Jr
Brody, Stuart Martin
Brown, George Lincoln
Brown, Stanley Monty
Bulusu, Suryanarayana
Butensky, Irwin
Campbell, Clement, Jr
Carroll, James Joseph
Carter, Richard John
Chang, Jun Hsin
Cheng, Kang
Cohen, Howard Melvin
Cohn, J Gunther
Comai-Fuerherm, Karen
Condon, Francis Edward
Crowell, John Patrick
Cruz, Mamerto Manahan, Jr
Curry, Michael Joseph
Curtis, Robert K
Cutler, Frank Allen, Jr
Dean, Donald E
Deen, Harold E(ugene)
DeMartino, Ronald Nicholas
Demko, Donald
Drelich, Arthur (Herbert)
Eachus, Spencer William
Eigen, Edward
Elden, Richard Edward
Feldman, Nicholas
Firth, William Charles, Jr
Fischer, Robert Leigh
Flexser, Leo Aaron
Frey, William Carl
Fung, Shun Chong
Gajewski, Fred John
Gambino, S(alvatore) Raymond
Gavini, Muralidhara B

Gawley, Irwin H, Jr
Gerecht, J Fred
Gershon, Sol D
Gilbert, Richard Lapham, Jr
Glass, Michael
Goodloe, Paul Miller, II
Grabowski, Edward Joseph John
Granito, Charles Edward
Gray, Russell Houston
Greenblatt, Martha
Grubman, Wallace Karl
Haft, Jacob I
Halfon, Marc
Hay, Peter Marsland
Heffron, Peter John
Herdklotz, John Key
Herrington, Kermit (Dale)
Herzog, Hershel Leon
Hollander, Max Leo
Huang, Dao Pei
Huger, Francis P
Hughes, O Richard
Idol, James Daniel, Jr
Jaruzelski, John Janusz
Jennings, Carl Anthony
Jeter, Hewitt Webb
Jones, Martha Ownbey
Kaplan, Michael
Karg, Gerhart
Katz, Sidney A
Kern, Werner
Knapp, Malcolm Hammond
Koonce, Samuel David
Kraus, Hubert Adolph
Kuebler, William Frank, Jr
Lafornara, Joseph Philip
Lam, Fuk Luen
LaPalme, Donald William
Laudise, Robert Alfred
Layng, Edwin Tower
Lederman, Sally Ann
Leong, William
Levinson, Sidney Bernard
Levy, Marilyn
Lin, Ruey Y
Lobunez, Walter
Los, Marinus
LoSurdo, Antonio
Louis, Kwok Toy
Luberoff, Benjamin Joseph
Luthy, Jakob Wilhelm
McGinnis, James Lee
McVey, Jeffrey King
Maleeny, Robert Timothy
Manganaro, James Lawrence
Manganelli, Raymond M(ichael)
Maso, Henry Frank
Mellberg, James Richard
Melveger, Alvin Joseph
Miale, Joseph Nicolas
Mitchell, Thomas Owen
Most, Joseph Morris
Nadkarni, Ramachandra Anand
Naglieri, Anthony N
Navrotsky, Alexandra
Obeji, John T
Pader, Morton
Parisse, Anthony John
Perrotta, James
Price, Alson K
Priesing, Charles Paul
Pugliese, Michael
Rankel, Lillian Ann
Rau, Eric
Reichle, Walter Thomas
Rinehart, Jay Kent
Robbins, Clarence Ralph
Robinson, Edwin Allin
Rolston, Charles Hopkins
Ropp, Richard C
Rosano, Henri Louis
Rosenberg, Howard Alan
Rosenthal, Fritz
Saldick, Jerome
Salerno, Alphonse
Sausville, Joseph Winston
Schlossman, Mitchell Lloyd
Schneider, Joseph
Schneider, Paul
Schrier, Melvin Henry
Schubert, Rudolf
Schuler, Mathias John
Schwartz, Newton
Scoles, Giacinto
Scott, William Edwin
Sekutowski, Dennis G
Shah, Atul A
Sherrick, Carl Edwin
Shulman, George
Sikder, Santosh K
Silvestri, Anthony John
Sincius, Joseph Anthony
Sivco, Deborah L
Soled, Stuart
Sonn, George Frank
Sorkin, Marshall
Spiegel, Herbert Eli
Spiegelman, Gerald Henry
Spiro, Thomas George
Staum, Muni M
Steiger, Fred Harold
Stempel, Arthur
Stern, Eric Wolfgang
Stingl, Hans Alfred

Chemistry, General (cont)

Stoy, William S
Stults, Frederick Howard
Subramanyam, Dilip Kumar
Suciu, George Dan
Sugam, Richard Jay
Sutman, Frank X
Swenson, Richard Waltner
Swenson, Theresa Lynn
Sykes, Donald Joseph
Szyper, Mira
Tatyrek, Alfred Frank
Thelin, Jack Horstmann
Thich, John Adong
Thomas, Robert Joseph
Thurman, Richard Gary
Torok, Andrew, Jr
Townley, Robert William
Tripathi, Uma Prasad
Turse, Richard S
Van de Castle, John F
Van Saun, William Arthur
Von, Isaiah
Voss, Kenneth Edwin
Vyas, Brijesh
Watt, William Russell
Webb, Philip Gilbert
Wehrli, Pius Anton
Weisgerber, George Austin
White, John Francis
Wilson, Robert G
Wood, Darwin Lewis
Woodbridge, Joseph Eliot
Young, Edmond Grove
Young, Lewis Brewster
Zaim, Semih
Zoss, Abraham Oscar

NEW MEXICO
Anderson, William Loyd
Attrep, Moses, Jr
Baker, Richard Dean
Balagna, John Paul
Baxman, Horace Roy
Bentley, Glenn E
Bravo, Justo Baladjay
Browne, Charles Idol
Bryant, Ernest Atherton
Caton, Roy Dudley, Jr
Chapman, Robert Dale
Cheavens, Thomas Henry
Cowan, George A
Cox, Lawrence Edward
Daniels, William Richard
Eller, Phillip Gary
Fisher, James Delbert
Ford, George Pratt
Giorgi, Angelo Louis
Groblewski, Gerald Eugene
Guenther, Arthur Henry
Hardy, Paul Wilson
Heldman, Julius David
Kahn, Milton
Kjeldgaard, Edwin Andreas
Kunz, Walter Ernest
Levine, Herman Saul
Luehr, Charles Poling
McKenzie, James Montgomery
Magnani, Nicholas J
Ott, Donald George
Patterson, James Howard
Powers, Dana Auburn
Sparks, Morgan
Standefer, Jimmy Clayton
Steinhaus, David Walter
Stubbs, Morris Frank
Summers, Donald Balch
Wolfsberg, Kurt
Zack, Neil Richard

NEW YORK
Adler, Stephen Fred
Albrecht, Andreas Christopher
Allen, Augustine Oliver
Allen, Donald Stewart
Allen, Gary William
Allentoff, Norman
Alliet, David F
Armour, Eugene Arthur
Auerbach, Clemens
Bacon, Egbert King
Bacon, Robert Elwin
Bahl, Om Parkash
Bard, Charleton Cordery
Baron, Arthur L
Barr, Donald Eugene
Bielski, Benon H J
Blatt, Sylvia
Booms, Robert Edward
Bove, John L
Brauer, Joseph B(ertram)
Brink, Gilbert Oscar
Brody, Bernard B
Brown, Eric Reeder
Brown, Pamela Ann McElroy
Brown, Robert Getman
Brust, David Philip
Bunce, Stanley Chalmers
Burge, Robert Ernest, Jr
Burtt, Benjamin Pickering
Cabeen, Samuel Kirkland
Cappel, C Robert
Caso, Marguerite Miriam
Chang, Hai-Won
Chen, Cindy Chei-Jen
Chen, Stephen P K
Chilcote, Max Eli
Chodroff, Saul
Clemens, Carl Frederick
Cofrancesco, Anthony J
Cohen, Jacob Isaac
Coll, Hans
Conant, Robert Henry
Condrate, Robert Adam
Coppens, Philip
Corbeels, Roger
Couch, Troy Lee
Dabrowiak, James Chester
Daniels, John Maynard
Deangelis, Thomas P
Dessauer, John Hans
Dille, Kenneth Leroy
Di Paola, Mario
Dixon, William Brightman
Drickamer, Kurt
Edgerton, Robert Flint
Edwards, Leila
Eibeck, Richard Elmer
Eirich, Frederick Roland
Feit, Irving N
Feitler, David
Feldman, Larry Howard
Fischer, Leewellyn C
Flank, William H
Fleming, James Charles
Follows, Alan Greaves
Forsyth, Paul Francis
Freeman, John Paul
Freeman, Richard Carl
Friberg, Stig E
Fromageot, Henri Pierre-Marcel
Garza, Cutberto
Gladstone, Matthew Theodore
Gloyer, Stewart Edward
Gregor, Harry Paul
Griffiths, Joan Martha
Grina, Larry Dale
Grubman, Marvin J
Gurel, Demet
Habig, Robert L
Harpp, David Noble
Hastings, Julius Mitchell
Hensler, Joseph Raymond
Hersh, Leroy S
Hohnadel, David Charles
Holder, Charles Burt, Jr
Holtz, Carl Frederick
Horvat, Robert Emil
Hunt, Heman Dowd
Hunt, Roy Edward
Ivashkiv, Eugene
Jaffe, Marvin Richard
Jelling, Murray
Johnson, Hollister, Jr
Jones, Stanley Leslie
Joseph, Solomon
Julian, Donald Benjamin
Kablaoui, Mahmoud Shafiq
Kallen, Thomas William
Kaufmann, Peter John
Keirns, Mary Hull
Keisch, Bernard
Kellogg, Lillian Marie
Kennard, Kenneth Clayton
Kirshenbaum, Isidor
Kishore, Gollamudi Sitaram
Klerer, Julius
Kurz, Richard Karl
Lai, Yuan-Zong
Lam, Stanley K
Larson, Allan Bennett
Law, Paul Arthur
LeBlanc, Jerald Thomas
Lee, Ching-Li
Lerman, Sidney
Levin, Robert Aaron
Lincoln, Lewis Lauren
Loscalzo, Anne Grace
Lum, Kin K
Mackles, Leonard
McLaen, Donald Francis
Martell, Michael Joseph, Jr
Marton, Renata
Mathad, Gangadhara S
Maurer, Gernant E
Mausner, Jack
Merkel, Paul Barrett
Merrigan, Joseph A
Miller, Melvin J
Model, Frank Steven
Morrison, John Agnew
Mourning, Michael Charles
Muenter, Annabel Adams
Mukamel, Shaul
Mullen, Patricia Ann
Mumbach, Norbert R
Murphy, Cornelius Bernard
Murphy, Daniel Barker
Mylroie, Victor L
Naider, Fred R
Nash, Robert Arnold
Nelson, Lawrence Barclay
Nerken, Albert
Neuberger, Dan
Nolan, John Thomas, Jr
Ober, Christopher Kemper
Oberender, Frederick G
Oja, Tonis
Osborn, H(arland) James
Osterholtz, Frederick David
Padalino, Stephen John
Pedroza, Gregorio Cruz
Pemrick, Raymond Edward
Pepe, Joseph Philip
Pepkowitz, Leonard Paul
Perlstein, Jerome Howard
Pesce, Michael A
Post, Howard William
Potter, George Henry
Purandare, Yeshwant K
Quinan, James Roger
Rimai, Donald Saul
Rocks, Lawrence
Roffman, Steven
Rothermel, Joseph Jackson
Saeva, Franklin Donald
Saletan, Leonard Timothy
Salvo, Joseph J
Scaife, Charles Walter John
Schaffrath, Robert Eben
Schallenberg, Elmer Edward
Schallhorn, Charles H
Schick, Jerome David
Schindler, Hans
Scholes, Samuel Ray, Jr
Schwartz, Morton K
Seanor, Donald A
Seus, Edward J
Shapiro, Raymond E
Sharbaugh, Amandus Harry
Sharma, Minoti
Sheeran, Stanley Robert
Shen, Samuel Yi-Wen
Silleck, Clarence Frederick
Simpson, William Henry
Slate, Floyd Owen
Slezak, Jane Ann
Smith, Gale Eugene
Sparberg, Esther Braun
Spingola, Frank
Spittler, Terry Dale
Springer, Dwight Sylvan
Stachel, Johanna
Stasiw, Roman Orest
Sterrett, Frances Susan
Stiel, Leonard Irwin
Stoll, William Russell
Sturges, Stuart
Sullivan, Michael Francis
Sutherland, George Leslie
Sweeney, William Mortimer
Swinehart, James Stephen
Tassinari, Silvio John
Texter, John
Thomson, Gerald Edmund
Tischer, Thomas Norman
Toribara, Taft Yutaka
Tuite, Robert Joseph
Urban, Joseph
Van Norman, Gilden Ramon
Via, Francis Anthony
Vincent, George Paul
Vogel, Alfred Morris
Wadlinger, Robert Louis Peter
Whittingham, M(ichael) Stanley
Winstrom, Leon Oscar
Wolfson, Leonard Louis
Wyatt, Benjamin Woodrow
Yao, Jerry Shi Kuang
Yourtee, Lawrence Karn
Zakkay, Victor
Zimar, Frank
Zingaro, Joseph S
Zinman, Walter George
Zwick, Daan Marsh
Zwicker, Walter Karl

NORTH CAROLINA
Andersen, Melvin Ernest
Apellaniz, Joseph E P
Aycock, Benjamin Franklin
Bauermeister, Herman Otto
Beaumont, Ralph Harrison, Jr
Beck, Keith Russell
Berry, George William
Bockstahler, Theodore Edwin
Bonk, James F
Bradley, Harris Walton
Bright, Gordon Stanley
Casey, James Patrick
Chang, Hou-Min
Chou, David Yuan Pin
Christman, Russell Fabrique
Cialdella, Cataldo
Clemens, Donald Faull
Clifford, Alan Frank
Copulsky, William
Crutcher, Harold L
Dawe, Richard Donald
Deal, Glenn W, Jr
DeSimone, Joseph M
Dixit, Ajit Suresh
Edwards, Ben E
Ellestad, Reuben B
Findley, William Robert
Fine, Jo-David
Flokstra, John Hilbert
Gallent, John Bryant
Gillespie, Arthur Samuel, Jr
Goldstein, Irving Solomon
Green, Charles Raymond
Groszos, Stephen Joseph
Guthrie, Donald Arthur
Hall, Seymour Gerald
Herrell, Astor Y
Hess, Daniel Nicholas
Hilliard, Roy C
Hoelzel, Charles Bernard
Hoffman, Doyt K, Jr
Hovis, Louis Samuel
Hurlbert, Bernard Stuart
Johnson, Ronald Sanders
Jones, Daniel Silas, Jr
Jones, Samuel O'Brien
Leake, Norman
McPhail, Andrew Tennent
Martens, Christopher Sargent
Morris, Leo Raymond
Moseman, Robert Fredrick
Mushak, Paul
Nowell, John William
Purchase, Earl Ralph
Senkus, Murray
Smith, Fred R, Jr
Smith, Susan T
Sookne, Arnold Maurice
Spremulli, Linda Lucy
Squibb, Samuel Dexter
Taylor, Barry Edward
Teague, Claude Edward, Jr
Tulagin, Vsevolod
Wang, Victor Kai-Kuo
Watts, Plato Hilton, Jr
Wehner, Philip
Wollin, Goesta
Woosley, Royce Stanley

NORTH DAKOTA
Abrahamson, Harmon Bruce
Bauer, Armand
Fleeker, James R
Rathmann, Franz Heinrich

OHIO
Addanki, Somasundaram
Antler, Morton
Atam-Alibeckoff, Galib-Bey
Bailey, John Clark
Ball, David Ralph
Bayless, Philip Leighton
Behrmann, Eleanor Mitts
Bhatti, Asif
Botros, Raouf
Brown, Glenn Halstead
Burrows, Kerilyn Christine
Campbell, Donald R
Cardon, Samuel Zelig
Carter, Carolyn Sue
Cobb, Thomas Berry
Coe, Kenneth Loren
Coffin, Frances Dunkle
Cole, John Oliver
Connor, Daniel S
Cook, William R, Jr
Croxton, Frank Cutshaw
Damian, Carol G
Darling, Samuel Mills
DeWitt, Bernard James
Dorfman, Leon Monte
Dove, Ray Allen
Eaker, Charles Mayfield
Finn, John Martin
Fischer, Mark Benjamin
Fortman, John Joseph
Gage, Frederick Worthington
Garascia, Richard Joseph
Garrett, Alfred Benjamin
Gibbins, Betty Jane
Gibbons, Louis Charles
Goetz, Richard W
Goodson, Alan Leslie
Graham, Paul Whitener Link
Grisaffe, Salvatore J
Guo, Hua
Hall, Ronald Henry
Hanes, Ronnie Michael
Hanks, Richard Donald
Harmony, Judith A K
Harris, Arlo Dean
Hart, David Joel
Hein, Richard William
Helminiak, Thaddeus Edmund
Hine, Jack
Holland, William Frederick
Innes, John Edwin
Ish, Carl Jackson
Johnston, Herbert Norris
Kawahara, Fred Katsumi
Kay, Peter Steven
Keily, Hubert Joseph
Kershner, Carl John
Khosla, Mahesh C
Klopman, Gilles
Knecht, Walter Ludwig
Knipple, Warren Russell
Kohl, Fred John
Koknat, Friedrich Wilhelm
Kornbrekke, Ralph Erik
Kozikowski, Barbara Ann
Kwiatek, Jack
Lateef, Abdul Bari
Lehn, William Lee
Lichtin, J Leon
Luoma, Ernie Victor
McCarty, Lewis Vernon

McCune, Homer Wallace
McKinnis, Charles Leslie
McLoughlin, Daniel Joseph
Mayer, Ramona Ann
Meckstroth, Wilma Koenig
Meites, Samuel
Mettee, Howard Dawson
Michael, William R
Miller, William Reynolds, Jr
Morton, Maurice
Muntean, Richard August
Muse, Joel, Jr
Myers, Elliot H
Nelson, Gilbert Harry
Newby, John R
Nygaard, Oddvar Frithjof
Oesterling, Thomas O
Olson, Walter T
Ouimet, Alfred J, Jr
Patel, Siddharth Manilal
Paynter, John, Jr
Peters, Lynn Randolph
Portfolio, Donald Carl
Rau, Allen H
Raynie, Douglas Edward
Renoll, Mary Wilhelmine
Rich, Ronald Lee
Roberts, Timothy R
Rubinson, Judith Faye
Rubinson, Kenneth A
Schaff, John Franklin
Schmidt, Donald L
Semon, Waldo Lonsbury
Shine, Daniel Phillip
Sievert, Carl Frank
Skees, Hugh Benedict
Skiest, Eugene Norman
Spencer, Walter William
Spittler, Ernest George
Spoehr, Albert Frederick
Standley, Paul Melvin
Stevens, Henry Conrad
Stevenson, Don R
Sullenger, Don Bruce
Suter, Robert Winford
Sweet, Thomas Richard
Taylor, Paul Duane
Thekdi, Arvind C
Thomas, George B
Toeniskoetter, Richard Henry
Towarnicky, J
Towns, Robert Lee Roy
Turnbull, Kenneth
Van Winkle, Quentin
Velenyi, Louis Joseph
Volk, Murray Edward
Walters, Martha I
Weidner, Bruce Van Scoyoc
Wiedenheft, Charles John
Wilkes, Charles Eugene
Yang, Philip Yung-Chin
Yang, Tsanyen
Yocum, Ronald Harris

OKLAHOMA
Ball, John Sigler
Beaver, W Don
Bills, John Lawrence
Bresson, Clarence Richard
Coffman, Harold H
Cowley, Thomas Gladman
Daake, Richard Lynn
Das, Paritosh Kumar
Dermer, Otis Clifford
Doss, Richard Courtland
Fenstermaker, Roger William
Frost, Jackie Gene
Hamm, Donald Ivan
Heasley, Gene
Hermann, John Alexander
Holtmyer, Marlin Dean
Lee, Diana Mang
Linder, Donald Ernst
Matson, Michael Steven
Matson, Ted P
Mills, King Louis, Jr
Motz, Kaye La Marr
Natowsky, Sheldon
Neptune, William Everett
Nesbitt, Stuart Stoner
Norell, John Reynolds
Nye, Mary Jo
O'Neal, Steven George
Passey, Richard Boyd
Pitchford, Armin Cloyst
Pritchard, James Edward
Purdue, Jack Olen
Richardson, Verlin Homer
Scott, Charles Edward
Varga, Louis P
Williamson, William Burton
Wimmer, Donn Braden
Witt, Donald Reinhold

OREGON
Arp, Daniel James
Barnum, Dennis W
Bartlett, James Kenneth
Bettega, Richard
Brandt, Howard Allen
Charlton, David Berry
Dittmer, Karl
Gerke, John Royal
Graham, Beardsley

Halko, Barbara Tomlonovic
Keedy, Curtis Russell
Larsen, Marlin Lee
Lochner, Janis Elizabeth
Long, James William
MacVicar, Robert William
Meredith, Robert E(ugene)
Pankow, James Frederick
Scott, Peter Carlton
Shoemaker, David Powell
Terraglio, Frank Peter
Torley, Robert Edward
Wang, Chih Hsing
Winniford, Robert Stanley
Woldegiorgis, Gebretateos

PENNSYLVANIA
Abrams, Ellis
Adams, Roy Melville
Andelman, Julian Barry
Ayers, Joseph W
Baker, Rees Terence Keith
Barker, Franklin Brett
Bartish, Charles Michael Christopher
Batzar, Kenneth
Beavers, Ellington McHenry
Bednar, Rodney Allan
Bell, Ian
Berkowitz, Harry Leo
Bissey, Luther Trauger
Bodamer, George Willoughby
Bogard, Andrew Dale
Bossard, Mary Jeanette
Bowman, Kenneth Aaron
Bowne, Samuel Winter, Jr
Brantley, Susan Louise
Buhle, Emmett Loren
Calkins, Charles Richard
Casassa, Ethel Zaiser
Castle, John Edwards
Cheng, Cheng-Yin
Cheng, Hung-Yuan
Cheng, Sheau-Mei
Chenot, Charles Frederic
Clark, T(helma) K
Clarke, Duane Grookett
Clement, Gerald Edwin
Coleman, Charles Mosby
Conroy, James Strickler
Constantinides, Panayiotis Pericleous
Cowen, William Frank
Cupper, Robert Alton
Dalal, Fram Rustom
Deischer, Claude Knauss
Dicciani, Nance Katherine
Diethorn, Ward Samuel
Dunkelberger, Tobias Henry
Einhorn, Philip A
Ernst, Richard Edward
Eror, Nicholas George, Jr
Fearing, Ralph Burton
Feeman, James Frederic
Feldman, Kenneth Scott
Ficher, Miguel
Fowkes, Frederick Mayhew
Frank, William Benson
Freedman, Arthur Jacob
Friedman, Arnold Carl
Fugger, Joseph
Gilbert, William Irwin
Gillman, Hyman David
Gogel, Germaine E
Gottscho, Alfred M(orton)
Greenberg, Charles Bernard
Greifer, Aaron Philip
Grundmann, Christoph Johann
Guild, Lloyd V
Gur, David
Hall, Gary R
Hall, W Keith
Hall, William Heinlen
Halpern, Benjamin David
Harkins, Thomas Regis
Harren, Richard Edward
Harrison, John William
Hauser, William P
Hoberman, Alfred Elliott
Hochstrasser, Robin
Hosler, Peter
Huff, George Franklin
Hunter, James Bruce
Johnston, Gordon Robert
Karo, Wolf
Kassis, Shouki
Kauffman, Joel Mervin
Khatami, Mahin
Kirklin, Perry William
Kirsch, Francis William
Kitazawa, George
Kleinman, Roberta Wilma
Kochar, Harvinder K
Kuebler, John Ralph, Jr
Lange, Barry Clifford
Lau, Kenneth W
Lazar, Anna
Libby, R Daniel
Lloyd, Thomas Blair
Lyman, William Ray
McCurry, Patrick Matthew, Jr
MacDiarmid, Alan Graham
McGovern, John Joseph
Martinez de Pinillos, Joaquin Victor
Marton, Joseph
Matocha, Charles K

Meadows, Geoffrey Walsh
Melamed, Nathan T
Mendelsohn, Morris A
Merner, Richard Raymond
Metzger, Sidney Henry, Jr
Meyer, Ellen McGhee
Millard, Frederick William
Mirhej, Michael Edward
Misra, Sudhan Sekher
Mortimer, Charles Edgar
Moscony, John Joseph
Myers, Raymond Reever
Nair, K Manikantan
Nath, Amar
Nemeth, Edward Joseph
Neubeck, Clifford Edward
Nishioka, Masaharu
Nuessle, Albert Christian
Peterman, Keith Eugene
Phillips, John Spencer
Pullukat, Thomas Joseph
Qazi, Mahmood A
Rao, V Udaya S
Renfrew, Edgar Earl
Roche, James Norman
Rogers, Ralph Loucks
Romovacek, George R
Rowe, Jay Elwood
Ruh, Edwin
Rushton, Brian Mandel
Saffer, Charles Martin, Jr
Sanders, Charles Irvine
Sankar, Suryanarayan G
Sax, Sylvan Maurice
Schnable, George Luther
Schramm, Robert Frederick
Schrum, Robert Wallace
Sexsmith, Frederick Hamilton
Shapiro, Zalman Mordecai
Sheeran, Patrick Jerome
Sieger, John S(ylvester)
Singh, Baldev
Sloat, Charles Allen
Smith, Carolyn Jean
Smith, Charles Lea
Smith, James S(terrett)
Smith, James Stanley
Stanley, Edward Livingston
Steiner, Russell Irwin
Stengle, William Bernard
SubbaRao, Saligrama C
Suyama, Yoshitaka
Suydam, Frederick Henry
Sykes, James Aubrey, Jr
Trachtman, Mendel
Tulk, Alexander Stuart
Van Dolah, Robert Wayne
Vaux, James Edward, Jr
Venuto, Paul B
Vick, Gerald Kieth
Wang, Ke-Chin
Warme, Paul Kenneth
Weldes, Helmut H
Werkman, Joyce
Windisch, Rita M
Wohleber, David Alan
Wootten, Michael John
Wu, Ching-Yong
Xu, Zhifu
Zaika, Laura Larysa
Zimmt, Werner Siegfried

RHODE ISLAND
Brown, Phyllis R
Caroselli, Remus Francis
Da Silva, Joseph
De Bethune, Andre Jacques
Di Pippo, Ascanio G
Gerritsen, Hendrik Jurjen
Griffiths, William C
Landskroener, Peter
MacKenzie, Scott, Jr
Storm, Carlyle Bell

SOUTH CAROLINA
Adams, Daniel Otis
Amidon, Roger Welton
Asleson, Gary Lee
Aspland, John Richard
Bailey, Carl Williams, III
Ball, Frank Jervery
Baynes, John William
Bibler, Ned Eugene
Bowman, Wilfred William
Buurman, Clarence Harold
Carnes, Richard Albert
Cathey, LeConte
Chase, Vernon Lindsay
Clayton, Fred Ralph, Jr
Culbertson, Edwin Charles
DesMarteau, Darryl D
Earl, Charles Riley
Ghaffar, Abdul
Goldstein, Herman Bernard
Guthrie, Roger Thackston
Hayes, David Wayne
Holcomb, Herman Perry
Hunter, George William
King, Lee Curtis
Kistler, Malathi K
LaFleur, Kermit Stillman
Liebman, Arnold Alvin
Macaulay, Andrew James
Metz, Clyde Raymond

Mishra, Nawin C
Moncrief, John William
Montenyohl, Victor Irl
Otto, Wolfgang Karl Ferdinand
Park, Conrad B
Pecka, James Thomas
Russell, Allen Stevenson
Sax, Karl Jolivette
Scott, Jaunita Simons
Spain, James Dorris, Jr
Spooner, George Hansford
Sproul, Gordon Duane
Weiss, Michael Karl
Wheat, Joseph Allen
Wise, John Thomas
Worsham, Walter Castine

SOUTH DAKOTA
Dible, William Trotter, Jr
Hanson, Milton Paul
Jensen, William Phelps
Johnson, Elmer Roger
Landborg, Richard John

TENNESSEE
Adams, Robert Edward
Anderson, Roger W(hiting)
Barker, Marvin
Bloor, John E
Brokaw, George Young
Butler, Thomas Arthur
Capdevila, H Jorge
Clark, Richard Bennett
Clemens, Robert Jay
Coleman, Charles F(ranklin)
Crane, Laura Jane
Davis, Phillip Howard
Davis, Wallace, Jr
Egan, B Zane
Elowe, Louis N
Embree, Norris Dean
Ghosh, Mriganka M(ouli)
Gleason, Geoffrey
Gupta, Brij Mohan
Hefferlin, Ray (Alden)
Henry, Jonathan Flake
Hibbs, Roger Franklin
Horton, Charles Abell
Hurst, G Samuel
Kammann, Karl Philip, Jr
Keim, Christopher Peter
Kelley, Jim Lee
Kelly, Minton J
Lappin, Gerald R
Lyon, William Southern, Jr
McDaniel, Edgar Lamar, Jr
Martin, James Cuthbert
Martin, Richard Blazo
Mattison, Louis Emil
Natelson, Samuel
Nunez, Loys Joseph
Nyssen, Gerard Allan
Pearson, Donald Emanual
Penner, Hellmut Philip
Perkins, James
Perry, Lloyd Holden
Pinkerton, Frank Henry
Poutsma, Marvin Lloyd
Rainey, Robert Hamric
Rudolph, Philip S
Schmitt, Charles Rudolph
Schreyer, James Marlin
Schulert, Arthur Robert
Simhan, Raj
Strehlow, Richard Alan
Stroud, Robert Wayne
Sublett, Robert L
Taylor, Kirman
Torrey, Rubye Prigmore
Van Wazer, John Robert

TEXAS
Adams, Charles Rex
Adams, Herman Ray
Adcock, Willis Alfred
Aguilo, Adolfo
Ahlberg, Dan Leander
Altmiller, Henry
Aquino, Dolores Catherine
Aslam, Mohammad
Asperger, Robert George, Sr
Bachmann, John Henry, Jr
Baker, Samuel I
Barton, Derek Harold Richard
Battista, Orlando Aloysius
Bellavance, David Walter
Benner, Gereld Stokes
Bennett, Richard Henry
Bernal, Ivan
Bradley, Charles
Broun, Thorowgood Taylor, Jr
Bungo, Michael William
Burkart, Leonard F
Bush, Warren Van Ness
Carbajal, Bernard Gonzales, III
Carlson, Kenneth Theodore
Castrillon, Jose P A
Cavitt, Stanley Bruce
Cleek, Given Wood
Crecelius, Robert Lee
Cronkright, Walter Allyn, Jr
Cruse, Robert Ridgely
Daugherty, Kenneth E
DeBerry, David Wayne

Chemistry, General (cont)

Dille, Roger McCormick
Dobrott, Robert D
Ducis, Ilze
Edwards, Gayle Dameron
Edwards, George
Ellzey, Marion Lawrence, Jr
Escue, Richard Byrd, Jr
Farhataziz, Mr
Ferraris, John Patrick
Fleenor, Marvin Bension
Floyd, Willis Waldo
Folkers, Karl August
Foster, Walter Edward
Goodwin, John Thomas, Jr
Grissom, David
Gryting, Harold Julian
Guenther, Frederick Oliver
Haberman, John Phillip
Hale, Cecil Harrison
Ham, Joe Strother
Hammond, James W
Harbordt, C(harles) Michael
Hardcastle, James Edward
Hart, Paul Robert
Hassell, Clinton Alton
Hausler, Rudolf H
Heller, Adam
Henery, James Daniel
Herbert, Stephen Aven
Hoefelmeyer, Albert Bernard
Horridge, Patricia Emily
Hosmane, Narayan Sadashiv
Idson, Bernard
Jeanes, Jack Kenneth
Johnson, Mary Lynn Miller
Jones, Jesse W
Jordan, John William
Keenan, Roy W
Kelso, Edward Albert
King, Edward Frazier
Koestler, Robert Charles
Kulkarni, Padmakar Venkatrao
Ladde, Gangaram Shivlingappa
Laity, John Lawrence
Lawson, Jimmie Brown
Levine, Duane Gilbert
Levy, Leon Bruce
Lew, Chel Wing
Lupski, James Richard
Lyle, Robert Edward, Jr
McAdoo, David J
Macaluso, Anthony, Sr
McCown, Joseph Dana
Machacek, Oldrich
McMillin, Jeanie
McPherson, Clinton Marsud
Manning, Harold Edwin
Meerbott, William Keddie
Merchant, Philip, Jr
Miller, James Richard
Moffat, James
Morgan, Bryan Edward
Morgan, Leon Owen
Moser, James Howard
Mosier, Benjamin
Moy, Mamie Wong
Naae, Douglas Gene
Norton, Scott J
O'Connor, Rod
Ostroff, Anton G
Otto, John B, Jr
Pannell, Richard Byron
Parker, Patrick LeGrand
Patil, Kashinath Z(iparu)
Phillips, Guy Frank
Pierce, James Kenneth
Reeves, Perry Clayton
Reisberg, Joseph
Riddle, David
Rigdon, Orville Wayne
Rodriguez, Charles F
Roehrig, Gerald Ralph
Rosenfeld, Daniel David
Safe, Stephen Harvey
Sartor, Albin Francis, Jr
Schroepfer, George John, Jr
Sheshtawy, Adel A
Shilstone, James Maxwell, Jr
Sicilio, Fred
Simonsen, Stanley Harold
Skarlos, Leonidas
Skelley, Dean Sutherland
Stewart, Frank Edwin
Stranges, Anthony Nicholas
Suttle, Andrew Dillard, Jr
Tang, Yi-Noo
Tanner, Alan Roger
Towne, Jack C
Unruh, Jerry Dean
Varma, Rajender S
Wagner, Lawrence Carl
Walbrick, Johnny Mac
Wald, Milton M
Waters, John Albert
Weigel, Paul H(enry)
Wendt, Theodore Mil
Willson, Carlton Grant
Woller, William Henry
Wurth, Thomas Joseph
Yao, Joe

UTAH
Alexander, Guy B
Ash, Kenneth Owen
Borrowman, S Ralph
Christensen, Edward Richards
Davidson, Thomas Ferguson
Gates, Henry Stillman
Hansen, Lee Duane
Haslem, William Joshua
Hill, George Richard
Hunter, Byron Alexander
Langheinrich, Armin P(aul)
McCloskey, James Augustus, Jr
Miller, Richard Roy
Oblad, Alex Golden
Poulter, Charles Dale
Stokes, Barry Owen
Stuart, David Marshall
Sudweeks, Walter Bentley
Ursenbach, Wayne Octave

VERMONT
Gilbert, Arthur Donald
Lang, William Harry
Preuss, Albert F
Steele, Richard
Vellturo, Anthony Francis
Wooding, William Minor

VIRGINIA
Alexander, Allen Leander
Bailey, Susan Goodman
Ball, Donald Lee
Balmforth, Dennis
Berry, Clark Green
Blair, Barbara Ann
Bliss, Laura
Blodgett, Robert Bell
Brignoli Gable, Carol
Casali, Liberty
Clark, Walter Ernest
Cole, James Webb, Jr
Conn, William David
Cool, Raymond Dean
De Marco, Ronald Anthony
DeVore, Thomas Carroll
Dodgen, Durward F
Durrill, Preston Lee
Finley, Arlington Levart
Fischbach, Henry
Garretson, Harold H
Goodson, Louie Aubrey, Jr
Greinke, Everett D
Gruemer, Hanns-Dieter
Haas, Carol Kressler
Hecht, Sidney Michael
Hemley, John Julian
Holmes, Joseph Charles
Huggett, Clayton (McKenna)
Jackson, Elizabeth Burger
Jenkins, Robert Walls, Jr
Keihn, Frederick George
Kuhn, William Frederick
Lapporte, Seymour Jerome
Loranger, William Farrand
McHenry, William Earl
Morrell, William Egbert
Moyers, Jarvis Lee
Mukherjee, Tapan Kumar
Murday, James Stanley
Murray, John Wolcott
Newman, Richard Holt
Palmer, Curtis Allyn
Pohlmann, Juergen Lothar Wolfgang
Rainer, Norman Barry
Rainis, Albert Edward
Ritz, Victor Henry
Rothstein, Lewis Robert
Savory, John
Scharpf, William George
Scully, Frank E, Jr
Segura, Gonzalo, Jr
Sobel, Robert Edward
Spindel, William
Stein, Richard Louis
Stolow, Nathan
Turner, Carlton Edgar
Twilley, Ian Charles
Urbanik, Arthur Ronald
Van Norman, John Donald
Watson, Robert Francis
Wayland, Rosser Lee, Jr
Wei, Enoch Ping
Wilber, Joe Casley, Jr
Zakrzewski, Thomas Michael

WASHINGTON
Atkins, William M
Barton, Gerald Blackett
Bennett, Clifton Francis
Berry, Keith O
Brouns, Richard John
Cady, George Hamilton
Carlson, Lewis John
Chinn, Clarence Edward
Coe, Richard Hanson
Crittenden, Alden La Rue
Diebel, Robert Norman
Evans, John Charles, Jr
Exarhos, Gregory James
Fournier, R E Keith
Gahler, Arnold Robert
Garland, John Kenneth
Goldschmid, Otto
Guion, Thomas Hyman
Hausenbuiller, Robert Lee
Henderson, Lawrence J
Ho, Lydia Su-yong
Hrutfiord, Bjorn F
Jones, Carl Trainer
Kaplan, Alex
Kelly, Jeffrey John
Kent, Ronald Allan
Killingsworth, Lawrence Madison
Kokta, Milan Rastislav
Lackey, Homer Baird
Liang, Shou Chu
Lo, Cheng Fan
Ludwig, Charles Heberle
McCown, John Joseph
Macklin, John Welton
Milberger, Ernest Carl
Morris, Daniel Luzon
Nowotny, Kurt A
Patt, Leonard Merton
Pool, Karl Hallman
Riehl, Jerry A
Schmidt, Eckart W
Sheppard, John Clarence
Tazuma, James Junkichi
Thomas, Berwyn Brainerd
Tinker, John Frank
Van Tuyl, Harold Hutchison
West, Martin Luther
Wilson, Archie Spencer
Worthington, Ralph Eric
Zimbrick, John David

WEST VIRGINIA
Bartley, William J
Bhasin, Madan M
Cunningham, Newlin Buchanan
Giza, Chester Anthony
Karr, Clarence, Jr
Longanbach, James Robert
MacPeek, Donald Lester
Manyik, Robert Michael
Matthews, Virgil Edison
Moffa, David Joseph
Osborn, Claiborn Lee
Richter, G Paul
Sherman, Paul Dwight, Jr
Smith, Percy Leighton
Stansbury, Harry Adams, Jr
Wilson, Thomas Putnam

WISCONSIN
Adams, James William
Adrian, Alan Patrick
Anderson, Stephen William
Andrews, Oliver Augustus
Brebrick, Robert Frank, Jr
Cherayil, George Devassia
Deutsch, Harold Francis
Doumas, Basil T
Downs, Martin Luther
Dunn, Stanley Austin
Ediger, Mark D
Ellis, Arthur Baron
Feist, William Charles
Glasoe, Paul Kirkwold
Gloyer, Stewart Wayne
Hellman, Nison Norman
Hill, John William
Holland, Dewey G
Hossain, Shafi Ul
Huber, Calvin
Ihde, Aaron John
Isenberg, Irving Harry
Jenny, Neil Allan
Kao, Wen-Hong
Klopotek, David L
Laessig, Ronald Harold
Long, Claudine Fern
Makela, Lloyd Edward
Millett, Merrill Albert
Moore, John Ward
Morton, Stephen Dana
Oehmke, Richard Wallace
Perry, Billy Wayne
Punwar, Jalamsinh K
Reinders, Victor A
Roth, Marie M
Sanyer, Necmi
Sasse, Edward Alexander
Savereide, Thomas J
Schopler, Harry A
Schwab, Helmut
Scott, Lawrence William
Shakhashiri, Bassam Zekin
Showalter, Donald Lee
Siegfried, Robert
Smith, Milton Reynolds
Sommers, Raymond A
Stevens, Michael Fred
Trytten, Roland Aaker
Weipert, Eugene Allen
West, Kevin James
Wright, Steven Martin
Young, Raymond A

WYOMING
Haines, William Emerson
Latham, DeWitt Robert
Noe, Lewis John
Robertson, Raymond E(liot)
Robinson, Wilbur Eugene
Seese, William Shober

PUERTO RICO
ALZERRECA, ARNALDO
Baus, Bernard V(illars)
Ramos, Lillian
Stephens, William Powell

ALBERTA
Bird, Gordon Winslow
Birss, Viola Ingrid
Davis, Stuart George
Hepler, Loren George
Holmes, Owen Gordon
Hyne, James Bissett
Jones, R Norman
Krouse, Howard Roy
Rauk, Arvi
Strausz, Otto Peter
Wieser, Helmut

BRITISH COLUMBIA
Becker, Edward Samuel
Gerry, Michael Charles Lewis
Godard, Hugh P(hillips)
James, Douglas Garfield Limbrey
Patey, Grenfell N
Polglase, William James
Spitzer, Ralph
Wiles, David M

MANITOBA
Biswas, Shib D
Letkeman, Peter
Muir, Derek Charles G
Singh, Ajit
Sunder, Sham
Tkachuk, Russell
Vandergraaf, Tjalle T

NEW BRUNSWICK
Mallet, Victorin Noel
Mehra, Mool Chand
Unger, Israel

NEWFOUNDLAND
Scott, John Marshall William

NOVA SCOTIA
Knop, Osvald
McAlduff, Edward J
McMillan, Alan F
Matthews, Frederick White

ONTARIO
Alper, Anne Elizabeth
Boorn, Andrew William
Brownstein, Sydney Kenneth
Buckler, Ernest Jack
Burgess, William Howard
Dyne, Peter John
Fingas, Merv Floyd
George, Albert El Deeb
Gillespie, Ronald James
Goring, David Arthur Ingham
Gough, Sidney Roger
Gould, William Douglas
Gulens, Janis
Heyding, Robert Donald
Hollbach, Natasha Coffin
Holloway, Clive Edward
Ingruber, Otto Vincent
Jacobs, Patrick W M
Joubin, Franc Renault
Koningstein, Johannes A
Kudo, Akira
Lawrence, John
Lean, David Robert Samuel
Lee, Lilian M
Lemire, Robert James
Leonard, John Alex
Levere, Trevor Harvey
Limerick, Jack McKenzie, Sr
Logan, Charles Donald
McBryde, William Arthur Evelyn
McIntosh, Alexander Omar
MacKay, Donald Douglas
Moo-Young, Murray
Morley, Harold Victor
Mutton, Donald Barrett
Nicolle, Francois Marcel Andre
Prince, Alan Theodore
Puddington, Ira E(dwin)
Ramachandran, Vangipuram S
Roovers, J
Ross, Robert Anderson
Ryan, Michael T
Sadler, Arthur Graham
Sadowski, Chester M
Saha, Jadu Gopal
Sanderson, H Preston
Schroeder, William Henry
Shurvell, Herbert Francis
Subramanian, Kunnath Sundara
Usselman, Melvyn Charles
Van Loon, Jon Clement
Wade, Robert Simson
Zakaib, Daniel D

QUEBEC
Ayroud, Abdul-Mejid
Beique, Rene Alexandre
Belanger, Alain
Bolker, Henry Irving
Cooper, David Gordon
Gurudata, Neville

Lau, Cheuk Kun
Onyszchuk, Mario
Pelletier, Gerard Eugene
Roy, Jean-Claude
Schucher, Reuben
Schwab, Andreas J
Tan, Ah-Ti Chu
Tomlinson, George Herbert
Tonks, David Bayard
Yaffe, Leo

SASKATCHEWAN
Cassidy, Richard Murray
Johnson, Keith Edward
Smith, Allan Edward
Spinks, John William Tranter
Stewart, John Wray Black
Sutherland, Ronald George
Woods, Robert James

OTHER COUNTRIES
Ache, Hans Joachim
Bergstrom, K Sune D
Caesar, Philip D
Cornforth, John Warcup
De Haën, Christoph
Dreosti, Ivor Eustace
Fey, George Ting-Kuo
Friedel, Jacques
Gorin, Philip Albert James
Gwilt, John Ruff
Hasty, Robert Armistead
Huber, Robert
Metzger, Gershon
Parker, David H
Pearson, Arthur David
Phillips, David Colin
Ramachandran, Venkataraman
Salk, Sung-Ho Suck
Sarkisov, Pavel Jebraelovich
Sleeter, Thomas David
Steadman, Robert George
Stumm, Werner
Tits, Jacques
Todd, Alexander Robertus
Webb, Cynthia Ann Glinert
Wilkinson, Geoffrey
Winn, Edward Barriere
Wittig, Georg Friedrich Karl

Inorganic Chemistry

ALABAMA
Beal, James Burton, Jr
Chastain, Benjamin Burton
Copeland, David Anthony
Gerdom, Larry E
Jackson, Margaret E
Johnson, Frederic Allan
Knockemus, Ward Wilbur
Krannich, Larry Kent
Lucas, William R(ay)
Ludwick, Larry Martin
McNutt, Ronald Clay
Makhija, Suraj Parkash
Marano, Gerald Alfred
Nikles, David Eugene
Perry, William Daniel
Retief, Daniel Hugo
Stanbury, David McNeil
Teggins, John E
Van Artsdalen, Ervin Robert
Vigee, Gerald S
Ward, Edward Hilson

ALASKA
Lane, Robert Harold
Lokken, Donald Arthur

ARIZONA
Beaumont, Randolph Campbell
Birk, James Peter
DeKorte, John Martin
Eichinger, Jack Waldo, Jr
Enemark, John Henry
Eyring, LeRoy
Feltham, Robert Dean
Glick, Milton Don
Heath, Roy Elmer
Jull, Anthony John Timothy
Keller, Philip Charles
Knowlton, Gregory Dean
Lichtenberger, Dennis Lee
Liu, Chui Hsun
Marks, Ronald Lee
O'Keeffe, Michael
Post, Roy G
Robertson, Frederick Noel
Roche, Thomas Stephen
Rund, John Valentine
Smith, Rodger Chapman
Whaley, Thomas Patrick
Willson, Donald Bruce
Zielen, Albin John

ARKANSAS
Cordes, Arthur Wallace
Dodson, B C
Gwinup, Paul D
Hood, Robert L
Jimerson, George David
Johnson, Dale A
Netherton, Lowell Edwin

Palmer, Bryan D
Pearson, Robert Stanley
Ransford, George Henry
Schafer, Lothar
Schultz, Donald Raymond
Stanitski, Conrad Leon
Stuckey, John Edmund
Teague, Marion Warfield
Trigg, William Walker
Wiggins, James William

CALIFORNIA
Adams, George Baker
Almond, Hy
Alvarez, Vincent Edward
Bailin, Lionel J
Baisden, Patricia Ann
Balch, Alan Lee
Bartlett, Neil
Barton, Jacqueline K
Bau, Robert
Beck, Roland Arthur
Bennett, Larry E
Bercaw, John Edward
Blair, George Richard
Boardman, William Walter, Jr
Bonner, Norman Andrew
Bowen, Ruth Justice
Brown, William E(ric)
Burg, Anton Behme
Butler, Alison
Callahan, Kenneth Paul
Calvin, Melvin
Camenzind, Mark J
Canning, T(homas) F
Caputi, Roger William
Chock, Ernest Phaynan
Ciaramitaro, David A
Cihonski, John Leo
Cobble, James Wikle
Cohn, Kim
Collman, James Paddock
Connick, Robert Elwell
Constant, Clinton
Covey, William Danny
Coyle, Bernard Andrew
Crutchfield, Charlie
Cubicciotti, Daniel David
Current, Steven P
Czuha, Michael, Jr
Darnell, Alfred Jerome
Davis, Clyde Edward
Davison, John Blake
De Haan, Frank P
Delker, Gerald Lee
Dines, Martin Benjamin
Doedens, Robert John
Ecker, David John
Evans, William John
Feher, Frank J
Feiler, William A, Jr
Feldman, Fredric J
Fine, Dwight Albert
Fleischauer, Paul Dell
Fleischer, Everly B
Ford, Peter Campbell
Frenzel, Lydia Ann Melcher
Gagne, Robert Raymond
Gerlach, John Norman
Ginell, William Seaman
Goetschel, Charles Thomas
Gold, Marvin B
Goldberg, Sabine Ruth
Goldwhite, Harold
Gordon, Joseph Grover, II
Gorman, Melville
Grant, Louis Russell, Jr
Grantham, Leroy Francis
Gray, Harry B
Greenstadt, Melvin
Grubbs, Robert Howard
Grunthaner, Frank John
Hall, Kenneth Lynn
Hamilton, Hobart Gordon, Jr
Hardcastle, Kenneth Irvin
Harris, Daniel Charles
Harris, Gordon McLeod
Hawthorne, Marion Frederick
Hempel, Judith Cato
Hendrickson, David Norman
Herman, Zelek Seymour
Hiller, Frederick W
Hobart, David Edward
Hodgson, Keith Owen
Hoffman, Charles John
Hoffmann, Michael Robert
Holzmann, Richard Thomas
Hornig, Howard Chester
Hubred, Gale L
Hulet, Ervin Kenneth
Hutchinson, Bennett Buckley
Ivanov, Igor C
Jaecker, John Alvin
James, Dean B
Jayne, Jerrold Clarence
Johansson, Robert Gail
Johnson, Oliver
Jolly, William Lee
Jones, Patrick Ray
Kaesz, Herbert David
Katzin, Leonard Isaac
Kauffman, George Bernard
Kay, Eric
Kehoe, Thomas J

Kennedy, James Vern
Kieft, John A
King, William Robert, Jr
Knobler, Carolyn Berk
Korst, William Lawrence
Kothny, Evaldo Luis
Kratzer, Reinhold
Kubota, Mitsuru
Labinger, Jay Alan
Landesman, Herbert
Landis, Vincent J
Lawton, Emil Abraham
Lee, Yat-Shir
Lewis, Nathan Saul
Lind, Carol Johnson
Lustig, Max
McBride, William Robert
McKaveney, James P
Malik, Jim Gorden
Malouf, George M
Mandelin, Dorothy Jane Bearcroft
Markowitz, Samuel Solomon
Marrocco, Matthew Louis
Marshall, Donald D
Masters, Burton Joseph
Matheson, Arthur Ralph
Maya, Walter
Meyer, Carl Beat
Michlmayr, Manfred
Mode, V(incent) Alan
Moe, George
Monchamp, Roch Robert
Mosen, Arthur Walter
Murbach, Earl Wesley
Nathan, Lawrence Charles
Nelson, Norvell John
Newkirk, Herbert William
Newsam, John M
Ogimachi, Naomi Neil
Palmer, Thomas Adolph
Panzer, Richard Earl
Pappatheodorou, Sofia
Pearson, Michael J
Perry, Dale Lynn
Phipps, Peter Beverley Powell
Po, Henry N
Potter, Norman D
Potts, John Calvin
Raby, Bruce Alan
Rard, Joseph Antoine
Raymond, Kenneth Norman
Reed, Christopher Alan
Reinhardt, Richard Alan
Rhein, Robert Alden
Ring, Morey Abraham
Robinson, Paul Ronald
Rogers, Howard H
Rosenberg, Sanders David
Rossman, George Robert
Rusch, Peter Frederick
Russell, John Blair
Rustad, Douglas Scott
Sailor, Michael Joseph
Salot, Stuart Edwin
Sangster, Raymond Charles
Schack, Carl J
Schrauzer, Gerhard N
Schugart, Kimberly A
Sellas, James Thomas
Sharma, Prasanta
Sharpless, K Barry
Shoemaker, Carlyle Edward
Siegel, Bernard
Silber, Herbert Bruce
Smith, Robert Alan
Solomon, Edward I
Sprague, Robert W
Steinman, Robert
Stucky, Galen Dean
Swinehart, James Herbert
Taube, Henry
Tillay, Eldrid Wayne
Tilley, T(erry) Don
Toney, Joe David
Trogler, William C
Valentine, Joan Selverstone
Van Alten, Lloyd
Vander Wall, Eugene
Van Hecke, Gerald Raymond
Vernon, Gregory Allen
Vogel, Roger Frederick
Wadley, Margil Warren
Wagner, Ross Irving
Waldo, Willis Henry
Warf, James Curren
Weaver, Henry D, Jr
Weber, Carl Joseph
Webster, Clyde Leroy, Jr
Weiss, Harold Gilbert
Weisz, Robert Stephen
Wickham, Donald G
Wilk, William David
Williams, Colin James
Willis, William Van
Winchell, Robert E
Winkler, Marjorie Everett
Wong, Hans Kuomin
Wong, Joe
Young, Donald C
Zabin, Burton Allen
Zink, Jeffrey Irve

COLORADO
Anderson, Oren P

Barry, Henry F
Blake, Daniel Melvin
Clough, Francis Bowman
Conner, Jack Michael
Cummins, Jack D
Daugherty, Ned Arthur
Dickerhoof, Dean W
Eaton, Gareth Richard
Eaton, Sandra Shaw
Erdmann, David E
Fields, Clark Leroy
Fleming, Michael Paul
Frye, James Sayler
Geller, Seymour
Gerteis, Robert Louis
Gottschall, W Carl
Haas, Frank C
Hadzeriga, Pablo
Hall, James Louis
Heberlein, Douglas G(aravel)
Hutto, Francis Baird, Jr
Hyatt, David Ernest
Keder, Wilbert Eugene
King, Edward Louis
Koval, Carl Anthony
Legal, Casimer Claudius, Jr
Lewis, Clifford Jackson
Macalady, Donald Lee
Mehs, Doreen Margaret
Miner, Frend John
Moody, David Coit, III
Muscatello, Anthony Curtis
Norman, Arlan Dale
Parkinson, Bruce Alan
Pierpont, Cortlandt Godwin
Pundsack, Frederick Leigh
Ritchey, John Michael
Sanderson, Robert Thomas
Schaeffer, Riley
Sievers, Robert Eugene
Spence, Jack Taylor
Splittgerber, George H
Thompson, Gary Haughton
Thompson, Ronald G
Watkins, Kay Orville
Yang, In Che

CONNECTICUT
Alfieri, Charles C
Brudvig, Gary Wayne
Caradonna, John Philip
Carrano, Salvatore Andrew
Chamberland, Bertrand Leo
Crabtree, Robert H
Eppler, Richard A
Faller, John William
Galasso, Francis Salvatore
Golden, Gerald Seymour
Grayson, Martin
Haake, Paul
Hiskey, Clarence Francis
Horrigan, Philip Archibald
Kilbourn, Barry T
Kostiner, Edward S
Kozlowski, Adrienne Wickenden
Kuck, Julius Anson
Leavitt, Julian Jacob
McKeon, James Edward
Moeller, Carl William, Jr
Moore, Robert Earl
Moyer, Ralph Owen, Jr
Noack, Manfred Gerhard
Prigodich, Richard Victor
Roscoe, John Stanley, Jr
Sarada, Thyagaraja
Sarneski, Joseph Edward
Schiessl, H(enry) W(illiam)
Shaw, Brenda Roberts
Suib, Steven L
Tanaka, John
Tom, Glenn McPherson
Yang, Darchun Billy

DELAWARE
Baker, Ralph Thomas
Bissot, Thomas Charles
Boughton, John Harland
Braun, Juergen Hans
Brill, Thomas Barton
Bulkowski, John Edmund
Burmeister, John Luther
Calabrese, Joseph C
Chowdhry, Uma
Corbin, David Richard
Domaille, Peter John
Drinkard, William Charles, Jr
Falletta, Charles Edward
Foley, Henry Charles
Glaeser, Hans Hellmut
Hertzenberg, Elliot Paul
Hess, Richard William
Hoppenjans, Donald William
Ittel, Steven Dale
Kolski, Thaddeus L(eonard)
Kruse, Walter M
Ling, Harry Wilson
Longhi, Raymond
Luther, George William, III
Lyon, Donald Wilkinson
McClelland, Alan Lindsey
McGinnis, William Joseph
McLain, Stephan James
Mahler, Walter
Mead, Edward Jairus

Inorganic Chemistry (cont)

Meyer, James Melvin
Moran, Edward Francis, Jr
Moser, Glenn Allen
Phillips, Brian Ross
Potrafke, Earl Mark
Roe, David Christopher
Rogers, Donald B
Shannon, Robert Day
Sharp, Kenneth George
Sopp, Samuel William
Spaulding, Len Davis
Theopold, Klaus Hellmut
Thompson, Jeffery Scott
Thorpe, Colin
Tolman, Chadwick Alma
Tomic, Ernst Alois
Wasfi, Sadiq Hassan

DISTRICT OF COLUMBIA
Baker, Louis Coombs Weller
Bertsch, Charles Rudolph
Burnett, John Lambe
Butcher, Raymond John
Butter, Stephen Allan
Crum, John Kistler
Earley, Joseph Emmet
Erstfeld, Thomas Ewald
Good, Mary Lowe
Henry, Richard Lynn
Kuznesof, Paul Martin
Larsen, Lynn Alvin
McCormack, Mike
Marianelli, Robert Silvio
Nordquist, Paul Edgard Rudolph, Jr
Ondik, Helen Margaret
Plost, Charles
Pope, Michael Thor
Preer, James Randolph
Rosenberg, Robert Charles
Rowley, David Alton
Shirk, Amy Emiko
Stern, Kurt Heinz
Varga, Gideon Michael, Jr
Wang, Jin Tsai
Webb, Alan Wendell
White, David Gover
Yesinowski, James Paul

FLORIDA
Adisesh, Setty Ravanappa
Ahmann, Donald H(enry)
Austin, Alfred Ells
Babich, Michael Wayne
Baumann, Arthur Nicholas
Bettinger, Donald John
Betzer, Peter Robin
Birdwhistell, Ralph Kenton
Burbage, Joseph James
Callahan, James Louis
Carlson, Gordon Andrew
Choppin, Gregory Robert
Clark, Ronald Jene
Clausen, Chris Anthony
Crompton, Charles Edward
DeLap, James Harve
Dewanjee, Mrinal K
Drago, Russell Stephen
Drake, Robert Firth
Everett, Kenneth Gary
Frazier, Stephen Earl
Goedken, Virgil Linus
Gu, Binhe
Halperin, Joseph
Hare, Curtis R
Hellwege, Herbert Elmore
Iloff, Phillip Murray, Jr
Jungbauer, Mary Ann
King, Walter Bernard
Kirshenbaum, Abraham David
Kroger, Hanns H
Lee, Kah-Hock
Lewis, Nita Aries
McFarlin, Richard Francis
Martin, Barbara Bursa
Martin, Dean Frederick
Maskal, John
Mellon, Edward Knox, Jr
Neithamer, Richard Walter
Palenik, Gus J
Palmer, Jay
Penner, Siegfried Edmund
Perumareddi, Jayarama Reddi
Remmel, Randall James
Roe, William P(rice)
Rudd, DeForest Porter
Ryschkewitsch, George Eugene
Sheard, John Leo
Sisler, Harry Hall
Stanko, Joseph Anthony
Stoufer, Robert Carl
Whitaker, Robert Dallas
Whitney, Ellsworth Dow
Willard, Thomas Maxwell
Worrell, Jay H

GEORGIA
Anderson, Bruce Martin
Ashby, Eugene Christopher
Barefield, Edward Kent
Bertrand, Joseph Aaron
Block, Toby Fran
Bottomley, Lawrence Andrew

Brandau, Betty Lee
Centofanti, Louis F
Crawford, Van Hale
Daane, Adrian Hill
DeLorenzo, Ronald Anthony
Griffiths, James Edward
Henneike, Henry Fred
Hicks, Donald Gail
Hunt, Gary W
Hunt, Harold Russell, Jr
Islam, M Safizul
James, Jeffrey
Johnson, Ronald Carl
King, Robert Bruce
Koelsche, Charles L
Kutal, Charles Ronald
Lockhart, William Lafayette
Netzel, Thomas Leonard
Neumann, Henry Matthew
Rees, William Smith, Jr
Richardson, Susan D
Royer, Donald Jack
Ruff, John K
Scott, Robert Allen
Sears, Curtis Thornton, Jr
Spencer, Jesse G
Steele, Jack
Stone, John Austin
Stratton, Cedric
Thomas, Frank Harry
Torres, Lourdes Maria
Waggoner, William Horace
Whitten, Kenneth Wayne
Williams, Daniel James
Williams, George Nathaniel
Young, Raymond Hinchcliffe, Jr

HAWAII
Cramer, Roger Earl
Gilje, John
Hirayama, Chikara
Kemp, Paul James
Radtke, Richard Lynn
Waugh, John Lodovick Thomson
Wrathall, Jay W

IDAHO
Arcand, George Myron
Baker, John David
Batey, Harry Hallsted, Jr
Benson, Ernest Phillip, Jr
Bills, Charles Wayne
Carter, Loren Sheldon
Grahn, Edgar Howard
Hammer, Robert Russell
Kirchmeier, Robert L
Lewis, Leroy Crawford
Miller, Norman E
Mincher, Bruce J
Nagel, Terry Marvin
Olander, James Alton
Shreeve, Jean'ne Marie
Slansky, Cyril M
Strommen, Dennis Patrick
Tracy, Joseph Walter
Wright, Kenneth James

ILLINOIS
Adams, Max Dwain
Allred, Albert Louis
Appelman, Evan Hugh
Bailar, John Christian, Jr
Basile, Louis Joseph
Basolo, Fred
Beese, Ronald Elroy
Belford, R Linn
Berntsen, Robert Andyv
Bertolacini, Ralph James
Broach, Robert William
Brooks, Kenneth Conrad
Brown, Theodore Lawrence
Brubaker, George Randell
Brubaker, Inara Mencis
Bunting, Roger Kent
Burdett, Jeremy Keith
Burns, Richard Price
Cafasso, Fred A
Carlin, Richard Lewis
Carnahan, Jon Winston
Carrado, Kathleen Anne
Chang, Chin Hsiung
Chao, Sherman S
Chellew, Norman Raymond
Coley, Ronald Frank
Court, Anita
Curtiss, Larry Alan
Danzig, Morris Juda
De Pasquali, Giovanni
Dietz, Mark Louis
Doody, Marijo
Duttaahmed, A
Fischer, Albert Karl
Freeman, Wade Austin
Funck, Larry Lehman
Ganchoff, John Christopher
Geiser, Urs Walter
Girolami, Gregory Scott
Hakewill, Henry, Jr
Halpern, Jack
Hamerski, Julian Joseph
Hanson, John Elbert
Harris, Ronald L
Henry, Patrick M
Herlinger, Albert William

Hess, Wendell Wayne
Hoffman, Alan Bruce
Hoffman, Brian Mark
Hohnstedt, Leo Frank
Hollins, Robert Edward
Homeier, Edwin H, Jr
Hopkins, Paul Donald
Horwitz, Earl Philip
House, James Evan, Jr
House, Kathleen Ann
Huang, Leo W
Hughes, Benjamin G
Huston, John Lewis
Ibers, James Arthur
Jaffe, Philip Monlane
Jaselskis, Bruno
Karayannis, Nicholas M
Karraker, Robert Harreld
Keiter, Ellen Ann
Keiter, Richard Lee
Kennelly, Mary Marina
Kieft, Richard Leonard
Klemm, Waldemar Arthur, Jr
Klemperer, Walter George
Kokalis, Soter George
Kuhajek, Eugene James
Leibowitz, Leonard
Li, Yao-En
Lipka, James J
Liu, Chui Fan
Lofquist, Marvin John
Made-Gowda, Netkal M
Maroni, Victor August
Mason, W Roy, III
Melford, Sara Steck
Mendelsohn, Marshall H
Mirkin, Chad Alexander
Mota de Freitas, Duarte Emanuel
Nash, Kenneth Laverne
Nebgen, John William
Nicolae, George G
Nilges, Mark J
Partenheimer, Walter
Pavkovic, Stephen F
Perkins, Alfred J
Perlow, Mina Rea Jones
Poeppelmeier, Kenneth Reinhard
Pohlmann, Hans Peter
Poskozim, Paul Stanley
Postmus, Clarence, Jr
Rathke, Jerome William
Rauchfuss, Thomas Bigley
Reagan, William Joseph
Reidies, Arno H
Rogers, Robin Don
Rotella, Frank
Schmulbach, Charles David
Schreiner, Felix
Schug, Kenneth
Schultz, Arthur Jay
Schulz, Charles Emil
Sheft, Irving
Shriver, Duward F
Stein, Lawrence
Steunenberg, Robert Keppel
Suslick, Kenneth Sanders
Thompson, Arthur Robert
Thompson, Martin Leroy
Treptow, Richard S
Trevorrow, Laverne Everett
Trimble, Russell Fay
Vaughn, Joe Warren
Warf, C Cayce, Jr
Weil, Thomas Andre
West, Douglas Xavier
White, Jesse Edmund
Williams, Jack Marvin
Wimer, David Carlisle
Woods, Mary
Zimmerman, Donald Nathan

INDIANA
Bottei, Rudolph Santo
Brown, Herbert Charles
Case, Vernon Wesley
Chisholm, Malcolm Harold
Christou, George
Davenport, Derek Alfred
DeSantis, John Louis
Fehlner, Thomas Patrick
Ferraudi, Guillermo Jorge
Friedel, Arthur W
Garber, Lawrence L
George, James E
Green, Mark Alan
Hofman, Emil Thomas
Holowaty, Michael O
Joyner, Ralph Delmer
Kubiak, Clifford P
Landreth, Ronald Ray
Lipschutz, Michael Elazar
Long, Eric Charles
McMillin, David Robert
Margerum, Dale William
Morrison, William Alfred
Phillips, John R
Pilger, Richard Christian, Jr
Pitha, John Joseph
Pribush, Robert A
Reuland, Donald John
Robinson, William Robert
Schaap, Ward Beecher
Scheidt, Walter Robert
Schwan, Theodore Carl

Siefker, Joseph Roy
Storhoff, Bruce Norman
Swartz, Marjorie Louise
Todd, Lee John
Voorhoeve, Rudolf Johannes Herman
Walter, Joseph L
Wentworth, Rupert A D
Wyma, Richard J

IOWA
Angelici, Robert Joe
Bakac, Andreja
Bennett, William Earl
Chang, James C
Corbett, John Dudley
Deskin, William Arna
Doyle, John Robert
Eaton, David Leo
Elsbernd, Helen
Espenson, James Henry
Eyman, Darrell Paul
Goff, Harold Milton
Hammerstrom, Harold Elmore
Hutton, Wilbert, Jr
Kostic, Nenad M
McCarley, Robert Eugene
Martin, Don Stanley, Jr
Messerle, Louis
Miller, Gordon James
Powell, Jack Edward
Quass, La Verne Carl
Rila, Charles Clinton
Swartz, James E
Woods, Joe Darst

KANSAS
Boyle, Kathryn Moyne Ward Dittemore
Busch, Daryle Hadley
Cohen, Sheldon H
Frazier, A Joel
Gimple, Glenn Edward
Greene, Frank T
Griswold, Ernest
Gusenius, Edwin Mauritz
Hiebert, Allen G
Johnson, William Jacob
Klabunde, Kenneth John
Kleinberg, Jacob
Lambert, Jack Leeper
Landis, Arthur Melvin
McCormick, Bailie Jack
McElroy, Albert Dean
Mertes, Kristin Bowman
Potts, Melvin Lester
Purcell, Keith Frederick
Rumpel, Max Leonard
Sullivan, Mary Louise

KENTUCKY
Beall, Gary Wayne
Boggess, Gary Wade
Buchanan, Robert Martin
Chamberlin, John MacMullen
Clark, Howell R
Crawford, Thomas H
Daly, John Matthew
Davidson, John Edwin
Henrickson, Charles Henry
Hunt, Richard Lee
Keely, William Martin
Kiser, Robert Wayne
Klingenberg, Joseph John
McDermott, Dana Paul
Magnuson, Winifred Lane
Niedenzu, Kurt
Powell, Howard B
Richter, Edward Eugene
Sarma, Atul C
Selegue, John Paul
Sweeny, Daniel Michael
Tharp, A G
Vance, Robert Floyd

LOUISIANA
Anex, Basil Gideon
Bains, Malkiat Singh
Berni, Ralph John
Blanchard, Eugene Joseph
Boudreaux, Edward A
Brown, Leo Dale
Carmichael, J W, Jr
Day, Marion Clyde, Jr
Doomes, Earl
Field, Jack Everett
Frey, Frederick Wolff, Jr
Gale, Robert James
Halbert, Thomas Risher
Ham, Russell Allen
Jonassen, Hans Boegh
Kennedy, Frank Scott
Klobucar, William Dirk
Lanoux, Sigred Boyd
Mague, Joel Tabor
Maverick, Andrew William
Overton, Edward Beardslee
Peard, William John
Raisen, Elliott
Robson, Harry Edwin
Schweizer, Albert Edward
Seidler, Rosemary Joan
Selbin, Joel
Sen, Buddhadev
Siriwardane, Upali
Stanley, George Geoffrey

Ter Haar, Gary L
Watkins, Steven F
Weidig, Charles F
White, June Broussard
Wiewiorowski, Tadeusz Karol

MAINE
Gordon, Nancy Rowan
Nagle, Jeffrey Karl
Patterson, Howard Hugh
Russ, Charles Roger
Smith, Wayne Lee
Wheatland, David Alan

MARYLAND
Albinak, Marvin Joseph
Barnes, James Alford
Biddle, Richard Albert
Block, Jacob
Bonnell, David William
Bonsack, James Paul
Boucher, Laurence James
Boyd, Alfred Colton, Jr
Brinckman, Frederick Edward, Jr
Burton, Lester Percy Joseph
Carter, John Paul
Chamberlain, David Leroy, Jr
Coleman, James Stafford
Collins, Alva LeRoy, Jr
Cooperstein, Raymond
Coyle, Thomas Davidson
Debye, Nordulf Wiking Gerud
DeCraene, Denis Fredrick
Eanes, Edward David
Eichhorn, Gunther Louis
Freeman, Horatio Putnam
Ginther, Robert J
Griffo, Joseph Salvatore
Grim, Samuel Oram
Gryder, John William
Hastie, John William
Hsu, Chen C
Huheey, James Edward
Jaquith, Richard Herbert
Joedicke, Ingo Bernd
Josephs, Steven F
Karlin, Kenneth Daniel
Kask, Uno
Kayser, Richard Francis
Koubek, Edward
Lussier, Roger Jean
McCawley, Frank X(avier)
McMurdie, Howard Francis
Magee, John Storey
Margalit, Nehemiah
Maselli, James Michael
Mathew, Mathai
May, Joan Christine
Murch, Robert Matthews
Murray, George Milton
Muser, Marc
Pace, Judith G
Poli, Rinaldo
Pribula, Alan Joseph
Richman, Robert Michael
Ritter, Joseph John
Roethel, David Albert Hill
Rollins, Orville Woodrow
Shade, Joyce Elizabeth
Swann, Madeline Bruce
Tarshis, Irvin Barry
Thompson, Joseph Kyle
Tuve, Richard Larsen
Wason, Satish Kumar
Zimmerman, John Gordon
Zuckerbrod, David

MASSACHUSETTS
Ammlung, Richard Lee
Archer, Ronald Dean
Auld, David Stuart
Baird, Donald Heston
Bates, Carl H
Beall, Herbert
Berka, Ladislav Henry
Bessette, Russell R
Billo, Edward Joseph
Brauner, Phyllis Ambler
Burgess, Thomas Edward
Carey, George H
Caslavska, Vera Barbara
Caslavska, Jaroslav Ladislav
Chin, David
Clarke, Michael J
Cohen, Saul Mark
Coleman, William Fletcher
Comer, Joseph John
Condike, George Francis
Croft, William Joseph
Davies, Geoffrey
Deckert, Cheryl A
Donoghue, John Timothy
Duecker, Heyman Clarke
Ekman, Carl Frederick W
Feakes, Frank
Foxman, Bruce Mayer
Goldberg, Gershon Morton
Greenaway, Frederick Thomas
Greenblatt, David J
Haas, Terry Evans
Higgins, Dorothy
Hoffman, Morton Z
Hollocher, Thomas Clyde, Jr
Holm, Richard H
Holmes, Robert Richard
Hooper, Robert John
Howard, Wilmont Frederick, Jr
Kaczmarczyk, Alexander
Knox, Kerro
Korenstein, Ralph
Kowalak, Albert Douglas
Kustin, Kenneth
Langer, Horst Günter
Li, Rounan
Linck, Robert George
Lippard, Stephen J
Loretz, Thomas J
McGarry, Margaret
Marganian, Vahe Mardiros
Menashi, Jameel
Milburn, Ronald McRae
Minne, Ronn N
Moore, Robert Edmund
Natansohn, Samuel
Palilla, Frank C
Powers, Donald Howard, Jr
Quinlan, Kenneth Paul
Reiff, William Michael
Reis, Arthur Henry, Jr
Ricci, John Silvio, Jr
Richason, George R, Jr
Riordan, James F
Schlaikjer, Carl Roger
Schroch, Richard Royce
Seyferth, Dietmar
Shinn, Dennis Burton
Smith, Joseph Harold
Sosinsky, Barrie Alan
Stuhl, Louis Sheldon
Sullivan, Edward Augustine
Suplinskas, Raymond Joseph
Swank, Thomas Francis
Szafran, Zvi
Taub, Irwin A(llen)
Thomas, Martha Jane Bergin
Tulip, Thomas Hunt
Urry, Grant Wayne
Wang, Robert T
Weibrecht, Walter Eugene
Weller, Paul Franklin
Wheelock, Kenneth Steven
Wilson, Linda S (Whatley)
Wreford, Stanley S
Wrighton, Mark Stephen
Yang, Julie Chi-Sun
Zompa, Leverett Joseph

MICHIGAN
Anderson, Charles Thomas
Anderson, Melvin Lee
Baney, Ronald Howard
Bennett, Stephen Lawrence
Benson, Edmund Walter
Brault, Margaret A
Brenner, Alan
Brewer, Stephen Wiley, Jr
Brown, Donald John
Brubaker, Carl H, Jr
Butts, Susan Beda
Carpenter, Michael Kevin
Chance, Robert L
Chandra, Grish
Collins, Ronald William
Cooke, Dean William
Coucouvanis, Dimitri N
Curtis, Myron David
Davidovits, Joseph
Dawson, Gladys Quinty
Deck, Charles Francis
DeKock, Roger Lee
Dennis, Mary
Dukes, Gary Rinehart
Eastland, George Warren, Jr
Eick, Harry Arthur
Eliezer, Isaac
Endicott, John F
Evans, B J
Fisher, Galen Bruce
Fleming, Suzanne M
Flynn, James Patrick
Frey, John Erhart
Galloway, Gordon Lynn
Gaswick, Dennis C
Gump, J R
Hamilton, James Beclone
Hammer, Robert Nelson
Harmon, Kenneth Millard
Harris, Paul Jonathan
Hartwell, George E
Hutchison, James Robert
Jekel, Eugene Carl
Johnson, David Alfred
Johnson, Wayne Douglas
Kanamueller, Joseph M
Kanatzidis, Mercouri G
Kirschner, Stanley
Knop, Charles Philip
Kopperl, Sheldon Jerome
Kuczkowski, Robert Louis
Lane, George Ashel
Latham, Ross, Jr
Lehman, Duane Stanley
Lintvedt, Richard Lowell
Lipowitz, Jonathan
Luehrs, Dean C
McCollough, Fred, Jr
McLean, John A, Jr
Magnell, Kenneth Robert
Mance, Andrew Mark
Moyer, John Raymond
Murchison, Craig Brian
Oei, Djong-Gie
Oliver, John Preston
Pecoraro, Vincent L
Penner-Hahn, James Edward
Peters, Till Justus Nathan
Peterson-Kennedy, Sydney Ellen
Pinnavaia, Thomas J
Popov, Alexander Ivan
Potts, Richard Allen
Roberts, Charles Brockway
Rorabacher, David Bruce
Rulfs, Charles Leslie
Rycheck, Mark Rule
Seefurth, Randall N
Skelcey, James Stanley
Spees, Steven Tremble, Jr
Spike, Clark Ghael
Squattrito, Philip John
Stowe, Robert Allen
Swartz, Grace Lynn
Tamres, Milton
Taylor, Robert Craig
Taylor, Stephen Keith
Turner, Almon George, Jr
Van Doorne, William
Vukasovich, Mark Samuel
Warren, H(erbert) Dale
Williams, Donald Howard
Wilson, Laurence Edward
Yamauchi, Masanobu

MINNESOTA
Abeles, Tom P
Alich, Agnes Amelia
Alton, Earl Robert
Baum, Burton Murry
Bolles, Theodore Frederick
Britton, William Giering
Bydalek, Thomas Joseph
Carberry, Edward Andrew
Carlson, Robert Leonard
Carpenter, John Harold
Chamberlain, Craig Stanley
Dendinger, Richard Donald
Dinga, Gustav Paul
Dreyer, Kirt A
Ellis, John Emmett
Finholt, James E
Fleming, Peter B
Foss, Frederick William, Jr
Gabor, Thomas
Gebelt, Robert Eugene
Gladfelter, Wayne Lewis
Graham, Robert Leslie
Haas, Larry Alfred
Heino, Walden Leo
Hill, Brian Kellogg
Howell, Peter Adam
Johnson, Bryce Vincent
King, Reatha Clark
McKenna, Jack F(ontaine)
Magnuson, Vincent Richard
Mathiason, Dennis R
Mayerle, James Joseph
Mentone, Pat Francis
Miessler, Gary Lee
Mueting, Ann Marie
Neuvar, Erwin W
Oberle, Thomas M
Owens, Kenneth Eugene
Pignolet, Louis H
Plovnick, Ross Harris
Potts, Lawrence Walter
Que, Lawrence, Jr
Reynolds, Warren Lind
Siedle, Allen R
Stadtherr, Leon
Stein, Paul John
Tarr, Donald Arthur
Thompson, Larry Clark
Thompson, Mary E
Wolsey, Wayne C
Wood, John Martin

MISSISSIPPI
Ahmed, Ismail Yousef
Bishop, Allen David, Jr
Brown, Kenneth Lawrence
Deshpande, Krishnanath Bhaskar
Doumit, Carl James
Howell, John Emory
Klingen, Theodore James
McMahan, William H
Wertz, David Lee

MISSOURI
Anderson, Harold D
Bahn, Emil Lawrence, Jr
Barton, Lawrence
Bauman, John E, Jr
Bleeke, John Richard
Brammer, Lee
Chen, Yih-Wen
Connolly, John W
Corey, Eugene R
Corey, Joyce Yagla
Coria, Jose Conrado
Craddock, John Harvey
Crutchfield, Marvin Mack
Day, D(elbert) E(dwin)
Dean, Walter Keith
Deutsch, Edward Allen
Donovan, John Richard
Droll, Henry Andrew
Ebner, Jerry Rudolph
Felthouse, Timothy R
Forsberg, John Herbert
Forster, Denis
French, James Edwin
Gard, David Richard
Grindstaff, Wyman Keith
Groenweghe, Leo Carl Denis
Haymore, Barry Lant
Heitsch, Charles Weyand
Johnson, Mary Frances
Katti, Kattesh V
King, Thomas Morgan
Lang, Gerhard Paul
Lannert, Kent Philip
Lawless, Edward William
Leifield, Robert Francis
Long, Gary John
McDonald, Hector O
Malone, Leo Jackson, Jr
Moedritzer, Kurt
Monzyk, Bruce Francis
Moore, Edward Lee
Murmann, Robert Kent
O'Brien, James Francis
Readnour, Jerry Michael
Riley, Dennis Patrick
Ross, Frederick Keith
Schlemper, Elmer Otto
Schwarz, Richard
Solodar, Arthur John
Staniforth, Robert Arthur
Stary, Frank Edward
Stearns, Robert Inman
Stone, Bobbie Dean
Switzer, Jay Alan
Tappmeyer, Wilbur Paul
Teicher, Harry
Thompson, Richard Claude
Tompkins, Jay Allen
Tyagi, Suresh C
Ucko, David A
Wehrmann, Ralph F(rederick)
Wilcox, Harold Kendall
Zetlmeisl, Michael Joseph

MONTANA
Abbott, Edwin Hunt
Dooley, David Marlin
Elliott, Eugene Willis
Emerson, Kenneth
Jennings, Paul W
Osterheld, Robert Keith
Rosenberg, Edward
Thomas, Forrest Dean, II
Van Meter, Wayne Paul

NEBRASKA
Freidline, Charles Eugene
George, T Adrian
Griswold, Norman Ernest
Harris, Holly Ann
Harris, Robert Hutchison
Holtzclaw, Henry Fuller, Jr
Howell, Daniel Bunce
Mattes, Frederick Henry
Rieke, Reuben Dennis
Sterner, Carl D

NEVADA
Grenda, Stanley C
LeMay, Harold E, Jr
MacDonald, David J
Nelson, John Henry
Rhees, Raymond Charles
Ulrich, William Frederick
Williams, Loring Rider

NEW HAMPSHIRE
Chasteen, Norman Dennis
Dodson, Vance Hayden, Jr
Fogleman, Wavell Wainwright
Haendler, Helmut Max
Hoag, Roland Boyden, Jr
Hughes, Russell Profit
Mooney, Richard Warren
Roberts, John Edwin
Soderberg, Roger Hamilton
Springsteen, Kathryn Rose Mooney
Wetterhahn, Karen E
Wong, Edward Hou

NEW JERSEY
Arbuckle, Georgia Ann
Bamberger, Curt
Barnes, Robert Lee
Benedict, Joseph T
Biagetti, Richard Victor
Bocarsly, Andrew B
Bogden, John Dennis
Brill, Yvonne Claeys
Brous, Jack
Buckley, R Russ
Castor, William Stuart, Jr
Celiano, Alfred
Chandrasekhar, Prasanna
Chester, Arthur Warren
Chianelli, Russell Robert
Clarke, Lilian A
Coakley, Mary Peter
Cohen, Paul

Inorganic Chemistry (cont)

Crerar, David Alexander
Cullen, Glenn Wherry
Dentai, Andrew G
Dess, Howard Melvin
Dreeben, Arthur B
Dzierzanowski, Frank John
Ebert, Lawrence Burton
Ekstrom, Lincoln
Falconer, Warren Edgar
Falk, Charles David
Farrauto, Robert Joseph
Finston, Harmon Leo
Fix, Kathleen A
Forbes, Charles Edward
Garwood, William Everett
Ginsberg, Alvin Paul
Gore, Ernest Stanley
Grams, Gary Wallace
Granchi, Michael Patrick
Grandolfo, Marian Carmela
Greenblatt, Martha
Haden, Walter Linwood, Jr
Hall, Gretchen Randolph
Hall, Richard Eugene
Hayles, William Joseph
Hills, Stanley
Huchital, Daniel H
Humiec, Frank S, Jr
Inniss, Daryl
Isied, Stephan Saleh
Jackson, Julius
Jacobson, Stephen Ernest
Johnson, Jack Wayne
Joy, George Cecil, III
Kasper, Horst Manfred
Kern, Werner
Kestner, Mark Otto
Klein, Richard M
Kluiber, Rudolph W
Kowalski, Stephen Wesley
Kramer, George Mortimer
Krueger, Paul Carlton
Kuehl, Guenter Hinrich
Kwon, Joon Taek
Larkin, William Albert
Larsen, Marilyn Ankeney
Le Duc, J-Adrien Maher
Lee, Catherine Coyle
Libowitz, George Gotthart
Lipp, Steven Alan
Lynde, Richard Arthur
McCarroll, William Henry
McClure, Donald Stuart
Magder, Jules
Marezio, Massimo
Matsuguma, Harold Joseph
Miller, Arthur
Miller, Warren Victor
Morrow, Scott
Munday, Theodore F
Murray, Leo Thomas
Naik, Datta Vittal
Nelson, Gregory Victor
Nelson, Roger Edwin
Pinch, Harry Louis
Potenza, Joseph Anthony
Rabinovich, Eliezer M
Radimer, Kenneth John
Rankel, Lillian Ann
Reichle, Walter Thomas
Rivela, Louis John
Robbins, Murray
Rollmann, Louis Deane
Ropp, Richard C
Rosan, Alan Mark
Rubino, Andrew M
Schneemeyer, Lynn F
Schoenberg, Leonard Norman
Schugar, Harvey
Sekutowski, Dennis G
Sherry, Howard S
Simplicio, Jon
Sinclair, William Robert
Slepetys, Richard Algimantas
Spiro, Thomas George
Stange, Hugo
Stiefel, Edward
Strange, Ronald Stephen
Suchow, Lawrence
Szyper, Mira
Tannenbaum, Stanley
Tauber, Arthur
Tecotzky, Melvin
Thich, John Adong
Tiethof, Jack Alan
Tripathi, Uma Prasad
Vandegrift, Vaughn
Vanderspurt, Thomas Henry
Van Uitert, LeGrand G(erard)
Vickroy, David Gill
Weschler, Charles John
Wheaton, Gregory Alan
Whigan, Daisy B
White, Lawrence Keith
Wise, John James
Wong, Ching-Ping
Yocom, Perry Niel

NEW MEXICO

Alexander, Martin Dale
Asprey, Larned Brown
Baca, Glenn
Behrens, Robert George
Birnbaum, Edward Robert
Bowen, Scott Michael
Bravo, Justo Baladjay
Chapman, Robert Dale
Dahl, Alan Richard
DeArmond, M Keith
DuBois, Frederick Williamson
Eckert, Juergen
George, Raymond S
Harrill, Robert W
Heaton, Richard Clawson
Jackson, Darryl Dean
Kelber, Jeffry Alan
Kubas, Gregory Joseph
Levy, Samuel C
Maier, William Bryan, II
Mason, Caroline Faith Vibert
Newton, Thomas William
Onstott, Edward Irvin
Paine, Robert Treat, Jr
Penneman, Robert Allen
Peterson, Charles Leslie
Peterson, Eugene James
Plymale, Donald Lee
Popp, Carl John
Powers, Dana Auburn
Quinn, Rod King
Schulz, Wallace Wendell
Stinecipher, Mary Margaret
Swanson, Basil Ian
Tapscott, Robert Edwin
Thomas, Kimberly W
West, Mike Harold
Yalman, Richard George
Zack, Neil Richard

NEW YORK

Abruna, Hector D
Adin, Anthony
Allenbach, Charles Robert
Amero, Bernard Alan
Anbar, Michael
Anthony, Thomas Richard
Ashen, Philip
Atwood, Jim D
Bailey, Ronald Albert
Banks, Ephraim
Baum, Parker Bryant
Baum, Stuart J
Beachley, Orville Theodore, Jr
Benzinger, William Donald
Birnbaum, Ernest Rodman
Bogucki, Raymond Francis
Bonner, Francis Truesdale
Bowser, James Ralph
Brady, James Edward
Brunschwig, Bruce Samuel
Bryan, Philip Steven
Bullock, R Morris
Bulloff, Jack John
Burlitch, James Michael
Campisi, Louis Sebastian
Chamberlain, Phyllis Ione
Chan, Shu Fun
Churchill, Melvyn Rowen
Cohen, Irwin A
Cole, Roger M
Cook, Edward Hoopes, Jr
Corfield, Peter William Reginald
Cratty, Leland Earl, Jr
Crayton, Philip Hastings
Creutz, Carol
Dabrowiak, James Chester
Danielson, Paul Stephen
DeCrosta, Edward Francis, Jr
Deiters, Joan A
Dexter, Theodore Henry
D'Heurle, Francois Max
Dillard, Clyde Ruffin
DiSalvo, Francis Joseph, Jr
Dodge, Cleveland John
Donovan, Thomas Arnold
Dumbaugh, William Henry, Jr
Eachus, Raymond Stanley
Eberts, Robert Eugene
Eibeck, Richard Elmer
Eisenberg, Max
Eisenberg, Richard
Fay, Robert Clinton
Featherstone, John Douglas Bernard
Feldman, Isaac
Flanigen, Edith Marie
Flank, William H
Frost, Robert Edwin
Gambino, Richard Joseph
Gancy, Alan Brian
Geiger, David Kenneth
Gentile, Philip
Gerace, Paul Louis
Goddard, John Burnham
Goldberg, David Elliott
Goldberg, Stephen Zalmund
Gortsema, Frank Peter
Gysling, Henry J
Hahn, Richard Leonard
Haim, Albert
Haitko, Deborah Ann
Halliday, Robert William
Hares, George Bigelow
Holland, Hans J
Holtzberg, Frederic
Hriljac, Joseph A
Hubbard, Richard Alexander, II
Hyde, Kenneth E
Interrante, Leonard V
Jaffe, Marvin Richard
Jones, Stanley Leslie
Jones, Wayne E
Jones, William Davidson
Kallen, Thomas William
Keeler, Robert Adolph
Keister, Jerome Baird
Kirchner, Richard Martin
Kivlighn, Herbert Daniel, Jr
Klein, Donald Lee
Klinck, Ross Edward
Koch, Stephen Andrew
Koetzle, Thomas F
Kokoszka, Gerald Francis
Kotz, John Carl
Kozlowski, Theodore R
Kwitowski, Paul Thomas
Larach, Simon
Lashewycz-Rubycz, Romana A
Lavallee, David Kenneth
Lazarus, Marc Samuel
Lees, Alistair John
Lefkowitz, Stanley A
Lichtman, Irwin A
Liu, Chong Tan
Luckey, George William
McCarthy, Paul James
MacDowell, John Fraser
McGrath, John F
Mackay, Raymond Arthur
McLendon, George L
Madan, Stanley Krishen
Malerich, Charles
Mathad, Gangadhara S
Matienzo, Luis J
Mayr, Andreas
Meloon, Daniel Thomas, Jr
Meloon, David Rand
Michiels, Leo Paul
Milton, Kirby Mitchell
Model, Frank Steven
Morrow, Jack I
Morrow, Janet Ruth
Muller, Olaf
Mullhaupt, Joseph Timothy
Myers, Clifford Earl
Nancollas, George H
Parkin, Gerard Francis Ralph
Parsons, Patrick Jeremy
Patmore, Edwin Lee
Patton, Robert Lyle
Pence, Harry Edmond
Pizer, Richard David
Poncha, Rustom Pestonji
Popp, Gerhard
Porter, Richard Francis
Price, Harry James
Puttlitz, Karl Joseph
Raider, Stanley Irwin
Robinson, Martin Alvin
Rossi, Miriam
Rupp, John Jay
Russell, Virginia Ann
Scaife, Charles Walter John
Schmeckenbecher, Arnold F
Schottmiller, John Charles
Scott, Bruce Albert
Shelby, James Elbert
Sheridan, Peter Sterling
Shineman, Richard Shubert
Shoup, Robert D
Silver, Herbert Graham
Srivastava, Suresh Chandra
Stephen, Keith H
Strekas, Thomas C
Sturmer, David Michael
Sujishi, Sei
Sutin, Norman
Swanson, Alan Wayne
Syed, Ashfaquzzaman
Szalda, David Joseph
Thompson, David Allen
Thurner, Joseph John
Triplett, Kelly B
Vaska, Lauri
Villa, Juan Francisco
Vinal, Richard S
Vogel, Glenn Charles
Von Winbush, Samuel
Walsh, Edward Nelson
Walter, Paul Hermann Lawrence
Wamser, Christian Albert
Weick, Charles Frederick
Wexell, Dale Richard
Wiedemeier, Heribert
Wilkes, Glenn Richard
Williams, Jimmie Lewis
Wiseman, George Edward
Wyman, Donald Paul
Xue, Yongpeng
Zeldin, Martel
Ziolo, Ronald F
Zipp, Arden Peter
Zubieta, Jon Andoni

NORTH CAROLINA

Allen, Harry Clay, Jr
Bereman, Robert Deane
Bottjer, William George
Bowkley, Herbert Louis
Bryan, Horace Alden
Buchanan, James Wesley
Burrus, Robert Tilden
Carr, Dodd S(tewart)
Clemens, Donald Faull
Clifford, Alan Frank
Cochran, George Thomas
Collier, Francis Nash, Jr
Collier, Herman Edward, Jr
Dilts, Joseph Alstyne
DuBois, Thomas David
Epperson, E(dward) Roy
Farona, Michael F
Frieser, Rudolf Gruenspan
Garrou, Philip Ernest
Gillikin, Jesse Edward, Jr
Gray, Walter C(larke)
Grimley, Eugene Burhans, III
Hammaker, Geneva Sinquefield
Hatfield, William E
Hentz, Forrest Clyde, Jr
Hicks, Kenneth Ward
Horner, William Wesley
Huang, Jamin
Hugus, Z Zimmerman, Jr
Jackels, Susan Carol
Jackman, Donald Coe
Jezorek, John Robert
Levy, Newton, Jr
Little, William Frederick
Long, George Gilbert
McAllister, Warren Alexander
McDonald, Daniel Patrick
MacInnes, David Fenton, Jr
May, Walter Ruch
Melson, Gordon Anthony
Meyer, Thomas J
Murray, Royce Wilton
Noftle, Ronald Edward
Palmer, Richard Alan
Pruett, Roy L
Ramaswamy, H N
Reisman, Arnold
Rhyne, Thomas Crowell
Richter, Harold Gene
Schreiner, Anton Franz
Sink, Donald Woodfin
Spielvogel, Bernard Franklin
Switzer, Mary Ellen Phelan
Templeton, Joseph Leslie
Walsh, Edward John
Wasson, John R
Weaver, Ervin Eugene
Whangbo, Myung Hwan
Wilson, Lauren R
Zumwalt, Lloyd Robert

NORTH DAKOTA

Abrahamson, Harmon Bruce
Boudjouk, Philip Raymond
Broberg, Joel Wilbur
Garvey, Roy George
Hoggard, Patrick Earle
McCarthy, Gregory Joseph
Morris, Melvin L
Young, Clifton A

OHIO

Ackermann, Martin Nicholas
Alexander, John J
Attig, Thomas George
Bacon, Frank Rider
Bacon, William Edward
Barnett, Kenneth Wayne
Bendure, Robert J
Bernays, Peter Michael
Bimber, Russell Morrow
Blinn, Elliott L
Blum, Patricia Rae
Brown, William Anderson
Burow, Duane Frueh
Bursten, Bruce Edward
Buttlar, Rudolph O
Carfagno, Daniel Gaetano
Carrabine, John Anthony
Chong, Clyde Hok Heen
Cooley, William Edward
Cummings, Sue Carol
Curry, John D
Dahlgren, Richard Marc
Darlington, William Bruce
Davies, Julian Anthony
Dowell, Michael Brendan
Duffy, Norman Vincent, Jr
Durand, Edward Allen
Edwards, Jimmie Garvin
Elder, Richard Charles
Eley, Richard Robert
Erbacher, John Kornel
Everson, Howard E
Fischer, Mark Benjamin
Fortman, John Joseph
Fox, Robert Kriegbaum
Fryxell, Robert Edward
Gage, Frederick Worthington
Gallagher, Patrick Kent
Gaus, Paul Louis
Ghose, Hirendra M
Gilbert, George Lewis
Gordon, Gilbert
Gould, Edwin Sheldon
Greene, Arthur Frederick, Jr
Harrington, Roy Victor
Harris, Arlo Dean
Hendricker, David George
Hill, Ann Gertrude

Hohman, William H
Holtman, Mark Steven
Houk, Clifford C
Huff, Norman Thomas
Hunt, Dominic Joseph
Hunt, Jerry Donald
Hunter, James Charles
Hurley, Forrest Reyburn
Hurst, Peggy Morison
Jache, Albert William
Jaeger, Ralph R
Johnson, Gordon Lee
Karipides, Anastas
Katovic, Vladimir
Kenney, Malcolm Edward
Kline, Robert Joseph
Koehler, Mark E
Kokenge, Bernard Russell
Kozikowski, Barbara Ann
Krause, Horatio Henry
Kroenke, William Joseph
Kupel, Richard E
Kurz, David W
Livigni, Russell Anthony
Lonadier, Frank Dalton
Love, Calvin Miles
McDonald, John William
Malkin, Irving
Mast, Roy Clark
Meek, Devon Walter
Mezey, Eugene Julius
Middaugh, Richard Lowe
Molnar, Stephen P
Movius, William Gust
Muntean, Richard August
Myers, R(alph) Thomas
Myers, Ronald Eugene
Nicholson, D Allan
Nikodem, Robert Bruce
Ogle, Pearl Rexford, Jr
Olszanski, Dennis John
O'Neill, Richard Thomas
Owen, James Emmet
Pappas, Leonard Gust
Paynter, John, Jr
Pinkerton, A Alan
Pirooz, Perry Parviz
Place, Robert Daniel
Porter, Spencer Kellogg
Reed, A Thomas
Reed, William Robert
Rich, Ronald Lee
Riechel, Thomas Leslie
Rogers, Donald Richard
Sadurski, Edward Alan
Saraceno, Anthony Joseph
Schildcrout, Steven Michael
Schneider, Wolfgang W
Schram, Eugene P
Seufzer, Paul Richard
Shaw, Wilfrid Garside
Shore, Sheldon Gerald
Silver, Gary Lee
Slater, James Louis
Slotter, Richard Arden
Smith, Robert Kingston
Smithson, George Raymond, Jr
Snyder, Donald Lee
Stambaugh, Edgel Pryce
Stout, Barbara Elizabeth
Strecker, Harold Arthur
Swinehart, Carl Francis
Tessier, Claire Adrienne
Thayer, John Stearns
Thomas, Terence Michael
Thompson, Lancelot Churchill Adalbert
Toeniskoetter, Richard Henry
Uebele, Curtis Eugene
Urbach, Frederick Lewis
Vallee, Richard Earl
Velenyi, Louis Joseph
Viola, Ronald Edward
Wallace, William J
Weeks, Thomas Joseph, Jr
Wilhelm, Dale Leroy
Williams, John Paul
Wojcicki, Andrew
Yellin, Wilbur
Yingst, Ralph Earl
Young, Patrick Henry

OKLAHOMA
Atkinson, Gordon
Castleberry, George E
Conley, Francis Raymond
Daake, Richard Lynn
Davis, Robert Elliott
Dill, Edward D
Eastman, Alan D
Ellis, David Allen
Eubanks, Isaac Dwaine
Fodor, Lawrence Martin
Fogel, Norman
Frame, Harlan D
Gabriel, Henry
Geibel, Jon Frederick
Hagen, Arnulf Peder
Holt, Elizabeth Manners
Holt, Smith Lewis, Jr
Johnston, Harlin Dee
Kucera, Clare H
Lindahl, Charles Blighe
Matson, Michael Steven
Meshri, Dayaldas Tanumal

Moczygemba, George A
Moore, Thomas Edwin
Nicholas, Kenneth M
Noll, Leo Albert
Paxson, John Ralph
Purdie, Neil
Reinbold, Paul Earl
Robertson, Wilbert Joseph, Jr
Schmidt, Donald Dean
Seeney, Charles Earl
Shioyama, Tod Kay
Summers, Jerry C
Welch, Melvin Bruce
Whitfill, Donald Lee
Williamson, William Burton
Witt, Donald Reinhold

OREGON
Binder, Bernhard
Brown, Bruce Willard
De Kock, Carroll Wayne
Duell, Paul M
Dunne, Thomas Gregory
Ferrante, Michael John
Gard, Gary Lee
Halko, Barbara Tomlonovic
Halko, David Joseph
Krueger, James Harry
Loehr, Thomas Michael
Norris, Thomas Hughes
Parsons, Theran Duane
Pond, Judson Samuel
Ryan, Robert Reynolds
Shoemaker, Clara Brink
Silverman, Morris Bernard
Sleight, Arthur William
Tyler, David Ralph
Yoke, John Thomas

PENNSYLVANIA
Addison, Anthony William
Anderson, Wayne Philpott
Armor, John N
Barnhart, Barry B
Bartish, Charles Michael Christopher
Bassner, Sherri Lynn
Baurer, Theodore
Beichl, George John
Belitskus, David
Berman, David Alvin
Berry, Robert Walter
Birdsall, William John
Bobonich, Harry Michael
Breyer, Arthur Charles
Brixner, Lothar Heinrich
Brown, Patrick Michael
Brown, Paul Edmund
Burch, Mary Kappel
Burness, James Hubert
Byler, David Michael
Chiola, Vincent
Chu, Vincent H(ao) K(wong)
Chung, Tze-Chiang
Clayton, John Charles (Hastings)
Coe, Charles Gardner
Cohen, Alvin Jerome
Cooper, John Neale
Cooper, Norman John
Cordes, Eugene H
Cornelius, Richard Dean
Crocket, David Scott
Douglas, Bodie E
Dreibelbis, John Adam
Durante, Vincent Anthony
Elkind, Michael John
Esposito, John Nicholas
Falcone, James Salvatore, Jr
Faltynek, Robert Allen
Faut, Owen Donald
Feely, Wayne E
Felty, Wayne Lee
Fink, Colin Ethelbert
Fleischmann, Charles Werner
Forchheimer, Otto Louis
Franco, Nicholas Benjamin
Friedman, Lawrence Boyd
Garbarini, Victor C
Gardner, David Milton
Garrett, Michael Benjamin
Gebhardt, Joseph John
Gonick, Ely
Goochee, Herman Francis
Gottlieb, Irvin M
Greifer, Aaron Philip
Gustison, Robert Abdon
Haas, Charles Gustavus, Jr
Hafford, Bradford C
Harmer, Martin Paul
Harris, James Joseph
Herman, Richard Gerald
Holifield, Charles Leslie
Horrocks, William DeWitt, Jr
Hudson, Robert McKim
Hultman, Carl A
Hutchins, MaryGail Kinzer
Jackovitz, John Franklin
Jenkins, Wilmer Atkinson, II
Johnson, Richard William
Johnston, William Dwight
Kay, Jack Garvin
Khan, Shakil Ahmad
Kifer, Edward W
King, James P
Kriner, William Arthur

Krivak, Thomas Gerald
Kuebler, John Ralph, Jr
Kunesh, Charles Joseph
Leffler, Amos J
Long, Kenneth Maynard
Lu, Guangquan
MacDiarmid, Alan Graham
MacDonald, Hubert C, Jr
MacInnis, Martin Benedict
Maricondi, Chris
Meyer, Ellen McGhee
Mikulski, Chester Mark
Misra, Sudhan Sekher
Mitra, Grihapati
Moss, Herbert Irwin
Mulay, Laxman Nilakantha
Murray, Edward Conley
Ochiai, Ei-Ichiro
Ocone, Luke Ralph
Opitz, Herman Ernest
Parker, William Evans
Pasternack, Robert Francis
Peterman, Keith Eugene
Pez, Guido Peter
Phillips, John Spencer
Poss, Stanley M
Prados, Ronald Anthony
Pytlewski, Louis Lawrence
Radzikowski, M St Anthony
Ravi, Natarajan
Reeder, Ray R
Resnik, Robert Kenneth
Ristey, William J
Rodgers, Glen Ernest
Root, Charles Arthur
Rosenberg, Richard Martin
Schaeffer, Charles David
Schnable, George Luther
Shepherd, Rex E
Shim, Benjamin Kin Chong
Shozda, Raymond John
Sloat, Charles Allen
Smid, Robert John
Smith, Herbert L
Smyth, Donald Morgan
State, Harold M
Stehly, David Norvin
Steinbrecher, Lester
Stetson, Harold W(ilbur)
Steward, Omar Waddington
Straub, Darel K
Stubican, Vladimir S(tjepan)
Swift, Harold Eugene
Tennent, Howard Gordon
Thibeault, Jack Claude
Titus, Donald Dean
Torop, William
Van Dyke, Charles H
Vernon, William Wallace
Wade, Charles Gary
Ward, Laird Gordon Lindsay
Warren, Alan
Wartik, Thomas
Watterson, Kenneth Franklin
Wayland, Bradford B
Weiss, Gerald S
Welch, Cletus Norman
Weldes, Helmut H
Wicker, Robert Kirk
Willeford, Bennett Rufus, Jr
Witkowski, Robert Edward
Wittle, John Kenneth
Yoder, Claude H
Zatko, David A
Zeigler, A(lfred) G(eyer)

RHODE ISLAND
Cruickshank, Alexander Middleton
Edwards, John Oelhaf
Greene, David Lee
Kirschenbaum, Louis Jean
Kreiser, Ralph Rank
Laferriere, Arthur L
Nelson, Wilfred H
Rieger, Philip Henri
Risen, William Maurice, Jr
Smith, Alan Jerrard
Sprowles, Jolyon Charles
Wold, Aaron

SOUTH CAROLINA
Allen, Joe Frank
Amma, Elmer Louis
Bach, Ricardo O
Clayton, Fred Ralph, Jr
Culbertson, Edwin Charles
DesMarteau, Darryl D
Dukes, Michael Dennis
Fanning, James Collier
Fauth, David Jonathan
Fowler, John Rayford
Gouge, Edward Max
Groce, William Henry (Bill)
Hahs, Sharon K
Hochel, Robert Charles
Horner, Sally Melvin
Hummers, William Strong, Jr
Jeremias, Charles George
Kane-Maguire, Noel Andrew Patrick
Malstrom, Robert Arthur
Mercer, Edward Everett
Moore, Lawrence Edward
Myrick, Michael Lenn
Odom, Jerome David

Orebaugh, Errol Glen
Owen, John Harding
Petersen, John David
Rodesiler, Paul Frederick
Sproul, Gordon Duane
Stampf, Edward John, Jr
Thompson, Major Curt
Weiss, Michael Karl

SOUTH DAKOTA
Gehrke, Henry
Jonte, John Haworth
Viste, Arlen E

TENNESSEE
Baes, Charles Frederick, Jr
Bamberger, Carlos Enrique Leopoldo
Barber, Eugene John
Barnes, Craig Eliot
Begun, George Murray
Billington, Douglas S(heldon)
Boston, Charles Ray
Brown, Gilbert Morris
Bull, William Earnest
Caflisch, Edward George
Campbell, David Owen
Davis, Jimmy Henry
Davis, Phillip Howard
Dean, Walter Lee
Dieck, Ronald Lee
Doty, Mitchell Emerson
Egan, B Zane
Gibson, John Knight
Guenther, William Benton
Haas, Paul Arnold
Hanusa, Timothy Paul
Hoffman, Kenneth Wayne
Houk, Larry Wayne
Jeter, David Yandell
Joesten, Melvin D
Jones, Mark Martin
Keenan, Charles William
Kelmers, Andrew Donald
Larkins, Thomas Hassell, Jr
Legg, Ivan
Lloyd, Milton Harold
Lukehart, Charles Martin
McDowell, William Jackson
McGill, Robert Mayo
Mamantov, Gleb
Marshall, Robert Herman
Maya, Leon
Metcalf, David Halstead
Miller, Paul Thomas
Moyer, Bruce A
Nehls, James Warwick
Nyssen, Gerard Allan
Obear, Frederick W
O'Kelley, Grover Davis
Okrasinski, Stanley John
Patterson, Truett Clifton
Petersen, Harold, Jr
Peterson, Joseph Richard
Sachleben, Richard Alan
Schweitzer, George Keene
Sears, Mildred Bradley
Shor, Arthur Joseph
Silverman, Meyer David
Stites, Joseph Gant, Jr
Summers, James Thomas
Sworski, Thomas John
Taylor, Kirman
Thoma, Roy E
Wardeska, Jeffrey Gwynn
Wood, James Lee
Woods, Clifton, III
Worsham, Lesa Marie Spacek
Ziegler, Robert G

TEXAS
Anderson, Herbert Hale
Ashley, Kenneth R
Baker, Willie Arthur, Jr
Bear, John L
Bellavance, David Walter
Benner, Gereld Stokes
Braterman, Paul S
Burke, John A
Busch, Marianna Anderson
Cannon, Dickson Y
Carlisle, Gene Ozelle
Carlson, Robert Kenneth
Cates, Vernon E
Clearfield, Abraham
Cotton, Frank Albert
Cowley, Alan H
Cude, Willis Augustus, Jr
Cummiskey, Charles
Darensbourg, Donald Jude
Darensbourg, Marcetta York
Dillin, Dennis
Dobson, Gerard Ramsden
Duncan, Walter Marvin, Jr
Elward-Berry, Julianne
Evans, Wayne Errol
Fackler, John Paul
Frisch, P Douglas
Gatti, Anthony Roger
Geanangel, Russell Alan
Gerlach, John Louis
Gingerich, Karl Andreas
Girardot, Peter Raymond
Hall, Michael Bishop
Heller, Adam

Inorganic Chemistry (cont)

Holwerda, Robert Alan
Hoover, William L
Horton, Robert Louis
Hosmane, Narayan Sadashiv
Howard, Charles
Howatson, John
Huege, Fred Robert
Hurlburt, H(arvey) Zeh
Iyer, Ramakrishnan S
Jacobson, Allan Joseph
Job, Robert Charles
Johnson, Malcolm Pratt
Kelly, Henry Curtis
King, Edward Frazier
Klassen, David Morris
Koehler, William Henry
Kust, Roger Nayland
Lagowski, Joseph John
Lattman, Michael
Li, Shu
Longo, John M
Lucid, Michael Francis
Mango, Frank Donald
Margrave, John Lee
Martell, Arthur Earl
Martin, Donald Ray
Messenger, Joseph Umlah
Meyer, W(illiam) Keith
Meyers, M Douglas
Mills, Jerry Lee
Mosier, Benjamin
Muir, Mariel Meents
Neilson, Robert Hugh
Ortego, James Dale
Pennington, David Eugene
Porter, Vernon Ray
Puligandla, Viswanadham
Riggs, Charles Lee
Rosenthal, Michael R
Sager, Ray Stuart
Sams, Lewis Calhoun, Jr
Sawyer, Donald Turner, Jr
Schultz, Linda Dalquest
Seaton, Jacob Alif
Sessler, Jonathan Lawrence
Sicilio, Fred
Singleton, David Michael
Slinkard, William Earl
Steinfink, Hugo
Stranges, Anthony Nicholas
Theriot, Leroy James
Thompson, Michael McCray
Walmsley, Frank
Walmsley, Judith Abrams
Wendlandt, Wesley W
Whitmire, Kenton Herbert
Whittlesey, Bruce R
Williams, Rickey Jay
Wilson, Bobby L
Wilson, Lon James
Wilson, Peggy Mayfield Dunlap
Wisian-Neilson, Patty Joan
Witmer, William Byron
Zingaro, Ralph Anthony

UTAH
Adler, Robert Garber
Bills, James LaVar
Breckenridge, William H
Cannon, John Francis
Ernst, Richard Dale
Eyring, Edward M
Izatt, Reed McNeil
Kelsey, Stephen Jorgensen
Kodama, Goji
Malouf, Emil Edward
Miller, Joel Steven
Morse, Karen W
Nordmeyer, Francis R
Parry, Robert Walter
Ragsdale, Ronald O
Richmond, Thomas G
Sauer, Dennis Theodore
Stoker, Howard Stephen
Waddell, Kidd M
Welch, Garth Larry
Wilson, Byron J

VERMONT
Allen, Christopher Whitney
Friihauf, Edward Joe
Jasinski, Jerry Peter
Keenan, Robert Gregory
Krause, Ronald Alfred
Schuele, William John

VIRGINIA
Addamiano, Arrigo
Anderson, Samuel
Averill, Bruce Alan
Bailey, Susan Goodman
Beach, Harry Lee, Jr
Beck, James Donald
Berg, John Richard
Calle, Luz Marina
Cavanaugh, Margaret Anne
Crissman, Judith Anne
DeFotis, Gary Constantine
Demas, James Nicholas
Dominey, Raymond Nelson
Evans, Howard Tasker, Jr
Finkelman, Robert Barry
Finley, Arlington Levart
Fisher, James Russell
Fox, William B
Galloway, James Neville
Garrett, Barry B
Glanville, James Oliver
Goehring, John Brown
Grimes, Russell Newell
Gushee, Beatrice Eleanor
Hazlehurst, David Anthony
Hufstedler, Robert Sloan
Hunt, John Baker
Hutchison, David Allan
Joebstl, Johann Anton
Liss, Ivan Barry
Malin, John Michael
Martin, Robert Bruce
Matuszko, Anthony Joseph
Merola, Joseph Salvatore
Morgan, Evan
Mosbo, John Alvin
Myers, William Howard
Naeser, Charles Rudolph
Newton, William Edward
Pasko, Thomas Joseph, Jr
Porterfield, William Wendell
Quagliano, James Vincent
Reed, Joseph
Richardson, Frederick S
Saalfeld, Fred Erich
Sacks, Lawrence J
St Clair, Anne King
Schnepfe, Marian Moeller
Schreiber, Henry Dale
Smith, Jerry Joseph
Stagg, William Ray
Taylor, Larry Thomas
Thompson, David Wallace
Topich, Joseph
Venezky, David Lester
Watt, William Joseph
Welch, Raymond Lee
Wickham, James Edgar, Jr
Workman, Marcus Orrin

WASHINGTON
Axtell, Darrell Dean
Ballou, Nathan Elmer
Barney, Gary Scott
Bates, J(unior) Lambert
Berry, Keith O
Boone, James Lightholder
Brehm, William Frederick
Burger, Leland Leonard
Cady, George Hamilton
Chinn, Clarence Edward
Cox, John Layton
Crook, Joseph Raymond
Crosby, Glenn Arthur
Crouthamel, Carl Eugene
Dreyfuss, Robert George
Duncan, Leonard Clinton
Eddy, Lowell Perry
Frankart, William A
Gibbins, Sidney Gore
Gregory, Norman Wayne
Grieb, Merland William
Haight, Gilbert Pierce, Jr
Hill, Orville Farrow
Hunt, John Philip
Jones, Thomas Evan
Kent, Ronald Allan
Koehmstedt, Paul Leon
Koerker, Frederick William
Ladd, Kaye Victoria
Macklin, John Welton
Millard, George Buente
Moore, Raymond H
Morrey, John Rolph
Morse, Joseph Grant
Neal, John Alexander
Norman, Joe G, Jr
Nowotny, Kurt A
Nyman, Carl John
O'Brien, Thomas Doran
Partridge, Jerry Alvin
Ryan, Jack Lewis
Sheppard, John Clarence
Templeton, John Charles
Terada, Kazuji
Weyh, John Arthur
Wicholas, Mark L
Wilson, Archie Spencer
Yunker, Wayne Harry

WEST VIRGINIA
Babb, Daniel Paul
Bhasin, Madan M
Bond, Stephon Thomas
Chakrabarty, Manoj R
Das, Kamalendu
Das Sarma, Basudeb
Dean, Warren Edgell
Hall, James Lester
Jones, Wilber Clark
Nakon, Robert Steven
Petersen, Jeffrey Lee
Peterson, James Lowell
Pribble, Mary Jo
Richter, G Paul
Smith, James Allbee
Swiger, Elizabeth Davis

WISCONSIN
Antholine, William E
Bayer, Richard Eugene
Becker, Edward Brooks
Campbell, Donald L
Colton, Ervin
Dahl, Lawrence Frederick
Discher, Clarence August
Evans, James Stuart
Ferguson, John Allen
Fraser, Margaret Shirley
Gaines, Donald Frank
Hansen, Robert Conrad
Haworth, Daniel Thomas
Kistner, Clifford Richard
Larsen, Edwin Merritt
Larson, Wilbur S
Marking, Ralph H
Moore, John Ward
Post, Elroy Wayne
Radtke, Douglas Dean
Rainville, David Paul
Roubal, Ronald Keith
Scott, Earle Stanley
Shaw, C Frank, III
Siebring, Barteld Richard
Sinko, John
Spencer, Brock
Treichel, Paul Morgan, Jr
Utke, Allen R
Watters, Kenneth Lynn
Wright, Steven Martin

WYOMING
Coates, Geoffrey Edward
Hodgson, Derek John
Morgan, George L
Schuman, Gerald E
Sullivan, Brian Patrick

PUERTO RICO
Lamba, Ram Sarup

ALBERTA
Boorman, Philip Michael
Cavell, Ronald George
Chivers, Tristram
Cowie, Martin
Graham, William Arthur Grover
Jack, Thomas Richard
Langford, Cooper Harold, III
Sanger, Alan Rodney

BRITISH COLUMBIA
Bushnell, Gordon William
Chaklader, Asoke Chandra Das
Cullen, William Robert
Dixon, Keith R
Einstein, Frederick W B
Howard, John
James, Brian Robert
Kirk, Alexander David
McAuley, Alexander
Orvig, Chris
Sutton, Derek

MANITOBA
Baldwin, W George
Janzen, Alexander Frank
Lange, Bruce Ainsworth
Taylor, Peter
Westmore, John Brian

NEW BRUNSWICK
Bottomley, Frank
Brewer, Douglas G
Fraser, Alan Richard
Mehra, Mool Chand
Passmore, Jack
Whitla, William Alexander

NEWFOUNDLAND
Clase, Howard John
Gogan, Niall Joseph
Haines, Robert Ivor
Lucas, Colin Robert
Rayner-Canham, Geoffrey William

NOVA SCOTIA
Burford, Neil
Cameron, T Stanley
Clark, Howard Charles
Faught, John Brian
Knop, Osvald
Peach, Michael Edwin
Ryan, Douglas Earl
Whiteway, Stirling Giddings

ONTARIO
Ashbrook, Allan William
Baldwin, Howard Wesley
Bancroft, George Michael
Barrett, Peter Fowler
Bhatnagar, Dinech C
Bioleau, Luc J R
Boorn, Andrew William
Brown, Ian David
Cale, William Robert
Carty, Arthur John
Chakrabarti, Chuni Lal
Cocivera, Michael
Dean, Philip Arthur Woodworth
Detellier, Christian
Drake, John Edward
Ettel, Victor Alexander
Fyfe, William Sefton
Gillespie, Ronald James
Greedan, John Edward
Haines, Roland Arthur
Hair, Michael L
Hartman, John Stephen
Hemmings, Raymond Thomas
Kaiser, Klaus L(eo) E(duard)
Kidd, Robert Garth
Lever, Alfred B P
Lister, Maurice Wolfenden
Lock, Colin James Lyne
Mackiw, Vladimir Nicolaus
Makhija, Ramesh C
Miller, Jack Martin
Norris, A R
Payne, Nicholas Charles
Pietro, William Joseph
Poe, Anthony John
Poulton, Donald John
Puddephatt, Richard John
Rempel, Garry Llewellyn
Richardson, Mary Frances
Ritcey, Gordon M
Rumfeldt, Robert Clark
Sadana, Yoginder Nath
Senoff, Caesar V
Stairs, Robert Ardagh
Stillman, Martin John
Thompson, James Charlton
Tomlinson, Richard Howden
Tuck, Dennis George
Walker, Alan
Walker, Ian Munro
Westland, Alan Duane
Wiles, Donald Roy
Willis, Christopher John

QUEBEC
Andrews, Mark Paul
Couillard, Denis
Hogan, James Joseph
Jabalpurwala, Kaizer E
Lalancette, Jean-Marc
Lau, Cheuk Kun
Onyszchuk, Mario
Serpone, Nick
Theophanides, Theophile

SASKATCHEWAN
Senior, John Brian
Strathdee, Graeme Gilroy
Waltz, William Lee

OTHER COUNTRIES
Bulkin, Bernard Joseph
Collins, Kenneth Elmer
El-Awady, Abbas Abbas
Grunwald, John J
Housecroft, Catherine Elizabeth
Irgolic, Kurt Johann
Loeliger, David A
Lonsdale-Eccles, John David
McGuire, George
McRae, Wayne A
Newlands, Michael John
Sarkisov, Pavel Jebraelovich
Schleyer, Paul von Rague
Truex, Timothy Jay
Walton, Wayne J A, Jr
Wilkins, Ralph G
Wilkinson, Geoffrey

Nuclear Chemistry

ALABAMA
Lecklitner, Myron Lynn

ARIZONA
Boynton, William Vandegrift
Burgus, Warren Harold
Jull, Anthony John Timothy
Stamm, Robert Franz
Whitehurst, Harry Bernard

ARKANSAS
Day, James Meikle
Palmer, Bryan D

CALIFORNIA
Baisden, Patricia Ann
Bonner, Norman Andrew
Camp, David Conrad
Carr, Robert J(oseph)
Chaubal, Madhukar Gajanan
Croft, Paul Douglas
Diamond, Richard Martin
Donovan, Paul F(rancis)
Dougan, Arden Diane
Gardner, Donald Glenn
Goishi, Wataru
Grant, Patrick Michael
Grantham, Leroy Francis
Hawkes, Wayne Christian
Hicks, Harry Gross
Hobart, David Edward
Hoff, Richard William
Hoffman, Darleane Christian
Hollander, Jack M
Holmes, Donald Eugene
Holmes, Ivan Gregory
Hulet, Ervin Kenneth

Ide, Roger Henry
King, William Robert, Jr
Knox, William Jordan
Kocol, Henry
Kruger, Paul
Lederer, C Michael
Leich, Douglas Albert
Leventhal, Leon
Levy, Harris Benjamin
Ling, Alan Campbell
Lugmair, Guenter Wilhelm
MacDonald, Norman Scott
Markowitz, Samuel Solomon
Marti, Kurt
Martz, Harry Edward, Jr
Masters, Burton Joseph
Metzger, Albert E
Michels, Lloyd R
Miller, George E
Miskel, John Albert
Mode, V(incent) Alan
Moody, Kenton J
Najafi, Ahmad
Nethaway, David Robert
Niemeyer, Sidney
Nurmia, Matti Juhani
Parker, Winifred Ellis
Pehl, Richard Henry
Petryka, Zbyslaw Jan
Pettijohn, Richard Robert
Poskanzer, Arthur M
Raby, Bruce Alan
Rasmussen, John Oscar, Jr
Ruiz, Carl P
Russ, Guston Price, III
Shotts, Wayne J
Smith, Charles Francis, Jr
Stewart, Donald Charles
Struble, Gordon Lee
Sugihara, Thomas Tamotsu
Swanson, David G, Jr
Tewes, Howard Allan
Van Hise, James R
Varteressian, K(egham) A(rshavir)
Wasson, John Taylor
Wiesendanger, Hans Ulrich David
Wild, John Frederick
Wolf, Walter
Wu, Jiann-Long
Wunderly, Stephen Walker

COLORADO
Gottschall, W Carl
Harlan, Ronald A
Heberlein, Douglas G(aravel)
Keder, Wilbert Eugene
Lucas, George Bond
Martell, Edward A
Muscatello, Anthony Curtis
Richardson, Albert Edward
Vejvoda, Edward
Yang, In Che

CONNECTICUT
Soufer, Robert
Ziegler, Theresa Frances

DISTRICT OF COLUMBIA
Burnett, John Lambe
Houck, Frank Scanland
McCormack, Mike

FLORIDA
Crompton, Charles Edward
Halperin, Joseph
Hertel, George Robert
Ostlund, H Gote
Stang, Louis George
Wethington, John A(bner), Jr

GEORGIA
Jackson, William Morrison
Kahn, Bernd
Neumann, Henry Matthew
Reif, Donald John
Stone, John Austin

IDAHO
Baker, John David
Gehrke, Robert James

ILLINOIS
Ahmad, Irshad
Bingham, Carleton Dille
Burrell, Elliott Joseph, Jr
Coley, Ronald Frank
Finn, Patricia Ann
Flynn, Kevin Francis
Horwitz, Earl Philip
Johnson, Irving
Kaufman, Sheldon Bernard
Kennedy, Albert Joseph
Liaw, Jye Ren
Mason, Donald Frank
Meadows, James Wallace, Jr
Melhado, L(ouisa) Lee
Nash, Kenneth Laverne
Pietri, Charles Edward
Sedlet, Jacob
Steinberg, Ellis Philip
Steindler, Martin Joseph
Yule, Herbert Phillip

INDIANA
Aprahamian, Ani
Contario, John Joseph
Kolata, James John
Lipschutz, Michael Elazar
Reuland, Donald John
Shaw, Stanley Miner
Stroube, William Bryan, Jr
Viola, Victor E, Jr

IOWA
Hill, John Christian
Martin, Don Stanley, Jr
Voigt, Adolf F
Watkins, G(ordon) Leonard

KENTUCKY
Ehmann, William Donald
Nathan, Richard Arnold
Robertson, John David
Yates, Steven Winfield

LOUISIANA
Keeley, Dean Francis

MARYLAND
Anderson, David Leslie
Becker, Donald Arthur
Buchanan, John Donald
Colle, Ronald
Cooperstein, Raymond
Del Grosso, Vincent Alfred
Downing, Robert Gregory
Friedman, Abraham S(olomon)
Gause, Evelyn Pauline
Gordon, Glen Everett
Guinn, Vincent Perry
Lamaze, George Paul
Leak, John Clay, Jr
Mignerey, Alice Cox
Poochikian, Guiragos K
Roche, Lidia Alicia
Roethel, David Albert Hill
Rzeszotarski, Waclaw Janusz
Skrabek, Emanuel Andrew
Smith, Elmer Robert
Stalling, David L
Walters, William Ben
Weber, Clifford E
Wing, James

MASSACHUSETTS
Brenner, Daeg Scott
Carey, George H
Davis, Michael Allan
Glajch, Joseph Louis
Huang, Shaw-Guang
Kafalas, Peter
Lis, Steven Andrew
Livingston, Hugh Duncan
Olmez, Ilhan
Richason, George R, Jr
Tulip, Thomas Hunt

MICHIGAN
Anders, Oswald Ulrich
Griffin, Henry Claude
Griffioen, Roger Duane
Huang, Che C
Morrissey, David Joseph
Rengan, Krishnaswamy
Robbins, John Alan
Schneider, Eric West
Thomas, Kenneth Eugene, III
Thompson, Paul Woodard

MINNESOTA
Carpenter, John Harold

MISSISSIPPI
Trehan, Rajender

MISSOURI
Ames, Donald Paul
Black, Wayne Edward
Freeman, Robert Clarence
Glascock, Michael Dean
Macias, Edward S
Manuel, Oliver K
Thompson, Clifton C
Troutner, David Elliott
Volkert, Wynn Arthur

NEBRASKA
Blotcky, Alan Jay
Borchert, Harold R

NEVADA
Pierson, William R

NEW HAMPSHIRE
Blum, Joel David

NEW JERSEY
Armamento, Eduardo T
Cadet, Gardy
Egli, Peter
Ganapathy, Ramachandran
Gavini, Muralidhara B
Hall, Gene Stephen
Kramer, George Mortimer
Krey, Philip W
Nicholls, Gerald P
Pickering, Miles Gilbert

Siegel, Jeffry A

NEW MEXICO
Apt, Kenneth Ellis
Barr, Donald Westwood
Bild, Richard Wayne
Binder, Irwin
Bowen, Scott Michael
Bryant, Ernest Atherton
Butler, Gilbert W
Catlett, Duane Stewart
Erdal, Bruce Robert
Fowler, Malcolm McFarland
Gancarz, Alexander John
Giesler, Gregg Carl
Hakkila, Eero Arnold
Jackson, Darryl Dean
Kiley, Leo Austin
Marlow, Keith Winton
Mason, Allen Smith
Molecke, Martin A
Norris, Andrew Edward
Orth, Charles Joseph
Paxton, Hugh Campbell
Penneman, Robert Allen
Pillay, K K Sivasankara
Reedy, Robert Challenger
Sedlacek, William Adam
Thomas, Charles Carlisle, Jr
Thomas, Kimberly W
Vieira, David John
Wahl, Arthur Charles
Wilhelmy, Jerry Barnard
Williams, Robert Allen
Wolfsberg, Kurt
Yates, Mary Anne

NEW YORK
Alexander, John Macmillan, Jr
Chan, Shu Fun
Chrien, Robert Edward
Chu, Yung Yee
Clark, Herbert Mottram
Cooper, Arthur Joseph L
Cumming, James B
Dahl, John Robert
Davis, Raymond, Jr
Debiak, Ted Walter
Duek, Eduardo Enrique
Finn, Ronald Dennet
Foreman, Bruce Milburn, Jr
Friedlander, Gerhart
Gill, Ronald Lee
Hahn, Richard Leonard
Hartmann, Francis Xavier
Haustein, Peter Eugene
Hoff, Wilford J, Jr
Holden, Norman Edward
Hopke, Philip Karl
Huizenga, John Robert
Husain, Liaquat
Katcoff, Seymour
Komorek, Michael Joesph, Jr
Ma, Maw-Suen
McHugh, James Anthony, Jr
Mackay, Raymond Arthur
MacKenzie, Donald Robertson
Matuszek, John Michael, Jr
Mausner, Leonard Franklin
Meyer, John Austin
Nealy, Carson Louis
Schroder, Wolf-Udo
Srivastava, Suresh Chandra
Stark, Ruth E
Tassinari, Silvio John
Ward, Thomas Edmund
Wolf, Alfred Peter

NORTH CAROLINA
Conn, Paul Kohler
Kalbach, Constance
Spicer, Leonard Dale
Wyrick, Steven Dale

OHIO
Bogar, Louis Charles
Brandhorst, Henry William, Jr
Chong, Clyde Hok Heen
Decker, Clarence Ferdinand
Hummon, William Dale
John, George
Madia, William J
Muntean, Richard August
Parikh, Sarvabhaum Sohanlal
Patton, Finis S, Jr
Poggenburg, John Kenneth, Jr
Rudy, Clifford R
Saha, Gopal Bandhu
Stout, Barbara Elizabeth
Washburn, Lee Cross
Yanko, William Harry

OREGON
Church, Larry B
DeMent, Jack (Andrew)
Goles, Gordon George
Grover, James Robb
Huh, Chih-An
Kay, Michael Aaron
Loveland, Walter (David)
Schmitt, Roman A
Williams, Evan Thomas

PENNSYLVANIA
Bogard, Andrew Dale
Bouis, Paul Andre
Brown, Paul Edmund
Catchen, Gary Lee
Chubb, Walston
Heys, John Richard
Hogan, Aloysius Joseph, Jr
Jaffe, Eileen Karen
Jester, William A
Kahler, Albert Comstock, III
Kaplan, Morton
Karol, Paul J(ason)
Kay, Jack Garvin
Kohman, Truman Paul
Pacer, John Charles
Roesmer, Josef
Sturm, James Edward
Tzodikov, Nathan Robert

SOUTH CAROLINA
Allen, Joe Frank
Bibler, Ned Eugene
Boyd, George Edward
Cheng, Kenneth Tat-Chiu
Fowler, John Rayford
Groh, Harold John
Hawkins, Richard Horace
Hochel, Robert Charles
Hofstetter, Kenneth John
Holcomb, Herman Perry
Kinard, W Frank
McDonell, William Robert
Moore, Willard S
Plodinec, Matthew John
Wilhite, Elmer Lee

TENNESSEE
Anderson, Roger W(hiting)
Bemis, Curtis Elliot, Jr
Bhattacharya, Syamal Kanti
Bigelow, John E(aly)
Campbell, David Owen
Dyer, Frank Falkoner
Ferguson, Robert Lynn
Haas, Paul Arnold
Hahn, Richard Balser
Hamilton, Joseph H, Jr
Johnson, Noah R
Lindemer, Terrence Bradford
McDowell, William Jackson
Moyer, Bruce A
Plasil, Franz
Ricci, Enzo
Schweitzer, George Keene
Taylor, Kirman
Young, Jack Phillip

TEXAS
Conway, Dwight Colbur
Farhataziz, Mr
Gnade, Bruce E
Haenni, David Richard
Hassell, Clinton Alton
Hunt, Richard Henry
Iskander, Felib Youssef
Lamb, James Francis
Ludwig, Allen Clarence, Sr
Miller, James Richard
Patel, Bhagwandas Mavjibhai
Tilbury, Roy Sidney
Walters, Lester James, Jr
Wright, James Foley

UTAH
Mangelson, Nolan Farrin

VIRGINIA
Allen, Ralph Orville, Jr
Baedecker, Philip A
Baker, Paul, Jr
Chulick, Eugene Thomas
Davidson, Charles Nelson
Grossman, Jeffrey N
Morgan, John Walter
Morrow, Richard Joseph
Owens, Charles Wesley
Perry, Dennis Gordon
Sprinkle, Robert Shields, III
Wong, George Tin Fuk

WASHINGTON
Anderson, Harlan John
Ballou, Nathan Elmer
Biskupiak, Joseph E
Brodzinski, Ronald Lee
Brown, Donald Jerould
Chinn, Clarence Edward
Clark, Donald Eldon
Daniel, J(ack) Leland
Fisher, Darrell R
Garland, John Kenneth
Hendrickson, Waldemar Forrsel
Hill, Orville Farrow
Kaye, James Herbert
Krohn, Kenneth Albert
Pillay, Gautam
Reeder, Paul Lorenz
Tingey, Garth Leroy
Wheelwright, Earl J
Wilbur, Daniel Scott
Wogman, Ned Allen

Nuclear Chemistry (cont)

WYOMING
Adams, Richard Melverne

PUERTO RICO
Souto Bachiller, Fernando Alberto

ALBERTA
Apps, Michael John

BRITISH COLUMBIA
D'Auria, John Michael

MANITOBA
Attas, Ely Michael
Iverson, Stuart Leroy
Sunder, Sham
Sze, Yu-Keung
Tait, John Charles
Wikjord, Alfred George

NOVA SCOTIA
Chatt, Amares
Holzbecher, Jiri

ONTARIO
Ball, Gordon Charles
Crocker, Iain Hay
Davies, John Arthur
Eastwood, Thomas Alexander
Hay, George William
Horn, Dag
Jardine, John McNair
Jervis, Robert E
Marshall, Heather
Thode, Henry George

QUEBEC
Diksic, Mirko
Hogan, James Joseph
Yaffe, Leo

SASKATCHEWAN
Cassidy, Richard Murray

OTHER COUNTRIES
Collins, Kenneth Elmer
Menon, Manchery Prabhakara
Qureshi, Iqbal Hussain
Yellin, Joseph

Organic Chemistry

ALABAMA
Atwood, Jerry Lee
Ayers, Orval Edwin
Beasley, James Gordon
Benington, Frederick
Byrd, James Dotson
Caine, Drury Sullivan, III
Cava, Michael Patrick
Chou, Libby Wang
Christian, Samuel Terry
Coburn, William Carl, Jr
Compton, Jack
Cooper, Emerson Amenhotep
Dean, David Lee
Elliott, Robert Daryl
Evanochko, William Thomas
Feazel, Charles Elmo, Jr
Hamer, Justin Charles
Hargis, J Howard
Hart, David R
Huskins, Chester Walker
Ingram, Sammy Walker, Jr
Isbell, Raymond Eugene
Jackson, Thomas Gerald
Kearley, Francis Joseph, Jr
Kiely, Donald Edward
Lambert, James LeBeau
Langston, James Horace
Leslie, Thomas M
Livant, Peter David
Ludwick, Adriane Gurak
McManus, Samuel Plyler
Miller, Nathan C
Montgomery, John Atterbury
Murray, Thomas Pinkney
Nair, Madhavan G
Parish, Edward James
Patterson, William Jerry
Penn, Benjamin Grant
Piper, James Robert
Renard, Jean Joseph
Retief, Daniel Hugo
Richmond, Charles William
Riley, Thomas N
Sani, Brahma Porinchu
Sayles, David Cyril
Secrist, John Adair, III
Shealy, Y(oder) Fulmer
Shevlin, Philip Bernard
Small, Robert James
Squillacote, Michael Edward
Stephens, William D
Tamburin, Henry John
Temple, Carroll Glenn
Thomas, Hazel Jeanette
Thomas, Joseph Calvin
Thompson, Wynelle Doggett
Thorpe, Martha Campbell
Veazey, Thomas Mabry

Walsh, David Allan
Weinberg, David Samuel
Wingard, Robert Eugene, Jr
Worley, S D
Youngblood, Bettye Sue

ALASKA
Collins, Jeffery Allen
Reppond, Kermit Dale

ARIZONA
Adamson, Albert S, Jr
Adickes, H Wayne
Allen, James Durwood
Anderson, Floyd Edmond
Anthes, John Allen
Barry, Arthur John
Barstow, Leon E
Bates, Robert Brown
Berry, James Wesley
Burgoyne, Edward Eynon
Caple, Gerald
Carter, Herbert Edmund
Coe, Beresford
Deutschman, Archie John, Jr
Doubek, Dennis Lee
Eskelson, Cleamond D
Gardlund, Zachariah Gust
Glass, Richard Steven
Greenfield, Harold
Gust, J Devens, Jr
Harris, Leland
Heaton, Charles Daniel
Heinle, Preston Joseph
Hinman, Charles Wiley
Houtman, Thomas, Jr
Hoyt, Earle B, Jr
Hruby, Victor J
Huffman, Robert Wesly
Jungermann, Eric
Kelley, Alec Ervin
Kelley, Maurice Joseph
Kellman, Raymond
Liddell, Robert William, Jr
Mariella, Raymond Peel
Moore, Ana M L
Moss, Rodney Dale
Moyer, Patricia Helen
Muggli, Robert Zeno
Mulvaney, James Edward
Pettit, George Robert
Remers, William Alan
Rose, Seth David
Smith, Delmont K
Steelink, Cornelius
Tamborski, Christ
Valenty, Steven Jeffrey
Valenty, Vivian Briones

ARKANSAS
Black, Howard Charles
Commerford, John D
De Luca, Donald Carl
Doerge, Daniel Robert
Fry, Arthur James
Goodwin, James Crawford
Goodwin, Thomas Elton
Gosnell, Aubrey Brewer
Ku, Victoria Feng
Manion, Jerald Monroe
Meyer, Walter Leslie
Nave, Paul Michael
Netherton, Lowell Edwin
Odell, Norman Raymond
Ransford, George Henry
Setliff, Frank Lamar
Siegel, Samuel
Sifford, Dewey H
Swindell, Robert Thomas
Wright, Joe Carrol
Yang, Dominic Tsung-Che
York, John Lyndal

CALIFORNIA
Abdel-Baset, Mahmoud B
Abegg, Victor Paul
Abell, Jared
Abrash, Henry I
Adams, John Howard
Ainsworth, Cameron
Akawie, Richard Isidore
Akmal, Naim
Albert, Anthony Harold
Alexander, Edward Cleve
Allen, Charles Freeman
Allen, William Merle
Anderson, Clyde Lee
Anderson, Curtis Benjamin
Andresen, Brian Dean
Andrews, Lawrence James
Anet, Frank Adrien Louis
Assony, Steven James
Athey, Robert Douglas, Jr
Atkins, Ronald Leroy
Aue, Donald Henry
Bacha, John D
Bacskai, Robert
Baker, Don Robert
Ballinger, Peter Richard
Banigan, Thomas Franklin, Jr
Barnum, Emmett Raymond
Baum, John William
Beattie, Thomas Robert
Belloli, Robert Charles

Belmares, Hector
Benson, Harriet
Benson, Robert Franklin
Bergman, Robert George
Bernasconi, Claude Francois
Bernhard, Richard Allan
Berteau, Peter Edmund
Biggerstaff, Warren Richard
Billig, Franklin A
Bishop, Keith C, III
Bissell, Eugene Richard
Blair, Charles Melvin, Jr
Block, Michael Joseph
Bluestone, Henry
Boger, Dale L
Bolt, Robert O'Connor
Bond, Frederick Thomas
Bonner, William Andrew
Bottini, Albert Thomas
Bozak, Richard Edward
Braithwaite, Charles Henry, Jr
Brault, Robert George
Brauman, John I
Brauman, Sharon K(ruse)
Brotherton, Robert John
Brown, Charles Allan
Brown, James Edward
Brown, Stuart Houston
Bruice, Thomas Charles
Bublitz, Donald Edward
Buck, Alan Charles
Buckles, Robert Edwin
Bunnett, Joseph Frederick
Bunton, Clifford A
Burke, Richard Lerda
Burlingame, Alma L
Byrd, Norman Robert
Calvin, Melvin
Camenzind, Mark J
Campbell, Thomas Cooper
Carty, Daniel T
Casanova, Joseph
Caserio, Marjorie C
Cason, James, Jr
Castro, Albert Joseph
Castro, Anthony J
Castro, Charles E
Chambers, Robert Rood
Chang, Ding
Chang, Frederic Chewming
Chapman, Orville Lamar
Chau, Michael Ming-Kee
Chauffe, Leroy
Chiddix, Max Eugene
Chin, Hsiao-Ling M
Ching, Ta Yen
Cho, Arthur Kenji
Choong, Hsia Shaw-Lwan
Ciaramitaro, David A
Ciula, Richard Paul
Clark, David Ellsworth
Cleland, George Horace
Cobb, Emerson Gillmore
Cole, Thomas Ernest
Collman, James Paddock
Colwell, William Tracy
Comer, William Timmey
Cooper, Douglas Elhoff
Cooper, Geoffrey Kenneth
Correia, John Sidney
Craig, John Cymerman
Craig, John Horace
Cram, Donald James
Cromartie, Thomas Houston
Cross, Alexander Dennis
Csendes, Ernest
Currell, Douglas Leo
Current, Steven P
Dafforn, Geoffrey Alan
Dahl, Klaus Joachim
Dalman, Gary
DaMassa, Al John
Daub, Guido William
Dauben, William Garfield
David, Gary Samuel
Davis, Jay C
Dawson, Daniel Joseph
DeAcetis, William
DeJongh, Don C
Dennis, Edward A
Dervan, Peter Brendan
Deshpande, Shirkant V
Detert, Francis Lawrence
Deutsch, Daniel Harold
Dev, Vasu
DeVenuto, Frank
Dhami, Kewal Singh
Diaz, Arthur Fred
Diekman, John David
DiGiorgio, Joseph Brun
Dills, Charles E
Dines, Martin Benjamin
Djerassi, Carl
Docks, Edward Leon
Dougherty, Dennis A
Drisko, Richard Warren
Dugaiczyk, Achilles
Duggins, William Edgar
Durham, Lois Jean
Dutra, Ramiro Carvalho
Eastman, Richard Hallenbeck
Eck, David Lowell
Edamura, Fred Y
Embree, Harland Dumond

Endres, Leland Sander
Engler, Edward Martin
Erdman, Timothy Robert
Fahey, Robert C
Fancher, Llewellyn W
Fang, Fabian Tien-Hwa
Faulkner, D John
Feher, Frank J
Felix, Raymond Anthony
Field, George Francis
Fife, Thomas Harley
Fischback, Bryant C
Fish, Richard Wayne
Fisher, Richard Paul
Flood, Thomas Charles
Fohlen, George Marcel
Foote, Christopher S
Forkey, David Medrick
Foscante, Raymond Eugene
Frankel, Edwin N
Freeman, Fillmore
Frey, Thomas G
Fried, John
Friedman, Mendel
Friedrich, Edwin Carl
Friess, Seymour Louis
Frost, Kenneth Almeron, Jr
Fuhrman, Albert William
Fukuto, Tetsuo Roy
Fuller, Glenn
Fung, Steven
Gaffield, William
Gale, Laird Housel
Gall, Walter George
Gallegos, Emilio Juan
Gandler, Joseph Rubin
Garber, John Douglas
Gash, Kenneth Blaine
Giang, Benjamin Yunwen
Giants, Thomas W
Gibbons, Loren Kenneth
Gish, Duane Tolbert
Glaser, Charles Barry
Glaser, Robert J
Glover, Leon Conrad, Jr
Goetschel, Charles Thomas
Gold, Marvin Harold
Goldish, Dorothy May (Bowman)
Goodman, Murray
Goodrich, Judson Earl
Gordon, Chester Duncan
Gordon, Robert Julian
Grant, Barbara Dianne
Green, David Claude
Greene, Charles Richard
Grethe, Guenter
Gross, Paul Hans
Grubbs, Charles Leslie
Grubbs, Edward
Grubbs, Robert Howard
Gschwend, Heinz W
Guner, Osman Fatih
Gupta, Amitava
Hajdu, Joseph
Hale, Ron L
Hall, Thomas Kenneth
Hamel, Edward E
Hamilton, Carole Lois
Hammond, James Alexander, Jr
Han, Yuri Wha-Yul
Hanneman, Walter W
Harris, Edwin Randall
Harris, Francis Laurie
Harris, William Merl
Havlicek, Stephen
Haxo, Henry Emile, Jr
Hearst, Peter Jacob
Heasley, Victor Lee
Heathcock, Clayton Howell
Heller, Jorge
Helmkamp, George Kenneth
Hendrickson, Yngve Gust
Henrick, Clive Arthur
Henry, David Weston
Henry, Ronald Andrew
Hershey, John William Baker
Hess, Patrick Henry
Hessel, Donald Wesley
Hill, Marion Elzie
Hobbs, M Floyd
Hoch, Paul Edwin
Hogen-Esch, Thieo Eltjo
Hokama, Takeo
Holtz, David
Horowitz, Robert Miller
Houk, Kendall N
Hounshell, William Douglas
Howe, Robert Kenneth
Howell, Edward Tillson
Hsu, Ming-Ta Sung
Hughes, John Lawrence
Hullar, Theodore Lee
Ingraham, Lloyd Lewis
Isensee, Robert William
Ja, William Yin
Jacob, Peyton, III
Jacobs, Thomas Lloyd
Jaffe, Annette Bronkesh
James, Philip Nickerson
Jensen, Harbo Peter
Jensen, James Leslie
Johanson, Robert Gail
Johnson, Harry William, Jr
Johnson, Kenneth

CHEMISTRY / 211

Johnson, LeRoy Franklin
Johnson, Michael Ross
Johnson, William K
Johnson, William Summer
Jones, Gordon Henry
Jones, Todd Kevin
Judson, Horace Augustus
Jules, Leonard Herbert
Jung, Michael Ernest
Juster, Norman Joel
Kakis, Frederic Jacob
Kamp, David Allen
Kaneko, Thomas Motomi
Kaufman, Daniel
Kaye, Samuel
Keeffe, James Richard
Kenyon, George Lommel
Kepner, Richard Edwin
Ketcham, Roger
Khanna, Pyare Lal
Kiely, John Steven
Kilday, Warren D
Kim, Chung Sul (Sue)
King, John Mathews
Kissel, Charles Louis
Klager, Karl
Kland, Mathilde June
Klinedinst, Paul Edward, Jr
Klinman, Judith Pollock
Kolyer, John M
Kong, Eric Siu-Wai
Konigsberg, Moses
Korte, William David
Koshland, Daniel Edward, Jr
Kovacs (Nagy), Hanna
Kozlowski, Robert H
Krantz, Allen
Krantz, Reinhold John
Kratzer, Reinhold
Krbechek, Leroy O
Krubiner, Alan Martin
Kryger, Roy George
Kumamoto, Junji
Kumli, Karl F
Kurihara, Norman Hiromu
Kurkov, Victor Peter
Kushner, Arthur S
Labinger, Jay Alan
Lamb, Sandra Ina
Lambert, Frank Lewis
Landrum, Billy Frank
Langford, Robert Bruce
Larrabee, Richard Brian
Lawson, Daniel David
Lee, Alfred Tze-Hau
Lee, David Louis
Lee, Stuart M(ilton)
Lee, William Wei
Lee, Young-Jin
Leonard, Nelson Jordan
Levy, M(orton) Frank
Lewis, Sheldon Noah
Leyden, Richard Noel
Liang, Yola Yueh-o
Licari, James John
Liggett, Lawrence Melvin
Lin, Jiann-Tsyh
Lin, Tz-Hong
Lind, Carol Johnson
Lindquist, Robert Nels
Lipson, Melvin Alan
Little, John Clayton
Little, Raymond Daniel
Livingston, Douglas Alan
Lorensen, Lyman Edward
Low, Hans
Lowe, Warren
Lowy, Peter Herman
Lucci, Robert Dominick
Luibrand, Richard Thomas
Lyon, Cameron Kirby
Mabey, William Ray
McCaleb, Kirtland Edward
McCloskey, Chester Martin
Mach, Martin Henry
MacKay, Kenneth Donald
McKenna, Charles Edward
McNall, Lester R
Manatt, Stanley L
Mansfield, Joseph Victor
Maricich, Tom John
Marmor, Solomon
Marshall, Henry Peter
Marsi, Kenneth Larue
Marten, David Franklin
Martin, Eugene Christopher
Matthews, Gary Joseph
Matuszak, Charles A
Maya, Walter
Mayer, William Joseph
Mayfield, Darwin Lyell
Meader, Arthur Lloyd, Jr
Merrill, Ronald Eugene
Metzner, Ernest Kurt
Meyer, Gregory Carl
Meyers, Robert Allen
Micheli, Robert Angelo
Midland, Michael Mark
Mihailovski, Alexander
Mill, Theodore
Millar, Jocelyn Grenville
Miller, Leroy Jesse
Miller, Robert Dennis
Miller, Russell Bryan

Minch, Michael Joseph
Mirza, John
Mitchell, Alexander Rebar
Moffatt, John Gilbert
Molyneux, Russell John
Montana, Andrew Frederick
Moore, Harold W
Moos, Walter Hamilton
Morgan, Alan Raymond
Morgan, Thomas Kenneth, Jr
Mori, Raymond I
Morton, Thomas Hellman
Mosher, Harry Stone
Moye, Anthony Joseph
Muchowski, Joseph Martin
Munger, Charles Galloway
Murov, Steven Lee
Musker, Warren Kenneth
Musser, John H
Myhre, Philip C
Najafi, Ahmad
Nambiar, Krishnan P
Nebenzahl, Linda Levine
Neher, Maynard Bruce
Neil, Gary Lawrence
Nelson, Norvell John
Nestor, John Joseph, Jr
Neuman, Robert C, Jr
Newey, Herbert Alfred
Newsom, Herbert Charles
Ngo, That Tjien
Nielsen, Arnold Thor
Nilsson, William A
Ning, Robert Y
Nitecki, Danute Emilija
Noble, Paul, Jr
Northington, Dewey Jackson, Jr
Novak, Bruce Michael
Noyce, Donald Sterling
Nozaki, Kenzie
Nugent, Maurice Joseph, Jr
Nussenbaum, Siegfried Fred
Nutt, Ruth Foelsche
Nyberg, David Dolph
Nyquist, Harlan LeRoy
Oakes, John Morgan
Ogimachi, Naomi Neil
Oh, Chan Soo
Okamura, William H
Olah, George Andrew
Olah, Judith Agnes
Olivier, Kenneth Leo
Olsen, Carl John
Onopchenko, Anatoli
Orgel, Leslie E
Ottke, Robert Crittenden
Overman, Larry Eugene
Oyama, Vance I
Ozari, Yehuda
Palchak, Robert Joseph Francis
Pallos, Ferenc M
Pan, Richard Y
Pandell, Alexander Jerry
Pappatheodorou, Sofia
Patterson, Dennis Bruce
Paulson, Donald Robert
Paulson, Suzanne Elizabeth
Pellegrini, Maria C
Pennington, Frank Cook
Percival, Douglas Franklin
Perlmutter, Milton Manuel
Perrin, Charles Lee
Perry, Robert Hood, Jr
Peters, Howard McDowell
Phillips, Donald David
Phillips, Lee Vern
Phillips, Robert Edward
Pinnell, Robert Peyton
Prakash, Surya G K
Preus, Martin William
Putz, Gerard Joseph
Quistad, Gary Bennet
Raley, John Howard
Rapoport, Henry
Ratcliff, Milton, Jr
Reardon, Edward Joseph, Jr
Recsei, Andrew A
Reed, Russell, Jr
Reese, Floyd Ernest
Reifschneider, Walter
Reinheimer, John David
Reist, Elmer Joseph
Replogle, Lanny Lee
Reyes, Zoila
Richardson, Wallace Lloyd
Rickborn, Bruce Frederick
Rife, William C
Riffer, Richard
Rinderknecht, Heinrich
Ripka, William Charles
Roberts, John D
Robertson, David W
Robertson, David Wayne
Robins, Roland Kenith
Rocklin, Albert Louis
Rodemeyer, Stephen A
Rodgers, James Edward
Ross, David Samuel
Ross, Donald Lewis
Rossiter, Bryant William
Rubin, Carol Marie
Rudy, Thomas Philip
Rusay, Ronald Joseph
Russell, John George

Ryan, Patrick Walter
Rynd, James Arthur
Saltman, William Mose
Saltonstall, Clarence William, Jr
Sanchez, Robert A
Santi, Daniel V
Savitz, Maxine Lazarus
Schaleger, Larry L
Schaumberg, Gene David
Schore, Neil Eric
Schwartz, Joseph Robert
Schwarzer, Carl G
Seapy, Dave Glenn
Selassie, Cynthia R
Sellas, James Thomas
Selover, James Carroll
Selter, Gerald A
Selwitz, Charles Myron
Servis, Kenneth L
Seubold, Frank Henry, Jr
Shackelford, Scott Addison
Shapiro, Bernard Lyon
Sharpless, K Barry
Sharts, Clay Marcus
Shelden, Harold Raymond, II
Shellhamer, Dale Francis
Shelton, John C
Shen, Kelvin Kei-Wei
Shen, Yvonne Feng
Siegel, Brock Martin
Siegfried, William
Silva, Ricardo
Simpson, John Ernest
Sims, James Joseph
Singaram, Bakthan
Singer, Lawrence Alan
Singh, Jai Prakash
Singh, Prithipal
Sipos, Frank
Skowronski, Raymund Paul
Smith, Kevin Malcolm
Smith, Leverett Ralph
Smith, Louis Charles
Smith, Robert Alan
Smith, Thomas Henry
Sobolev, Igor
Solomon, Malcolm David
Soloway, S Barney
Sommer, Leo Harry
Spatz, David Mark
Spenger, Robert E
Spray, Clive Robert
Stanley, William Lyons
Stanton, Garth Michael
Stehsel, Melvin Louis
Steinberg, Gunther
Steinberg, Howard
Steinman, Robert
Sterling, Robert Fillmore
Still, Gerald G
Stone, Herbert
Stonebraker, Peter Michael
Stout, Charles Allison
Stoutamire, Donald Wesley
Straus, William R
Streitwieser, Andrew, Jr
Suzuki, Shigeto
Sweeney, William Alan
Swidler, Ronald
Szabo, Emery D
Taagepera, Mare
Taft, Robert Wheaton, Jr
Tanabe, Masato
Teach, Eugene Gordon
Teeter, Richard Malcolm
Teranishi, Roy
Tess, Roy William Henry
Thelen, Charles John
Thompson, Evan M
Thomson, Tom Radford
Thorsett, Eugene Deloy
Tilles, Harry
Ting, Chih-Yuan Charles
Tokes, Laszlo Gyula
Tomich, John Matthew
Tong, Yulan Chang
Toy, Madeline Shen
Traylor, Teddy G
Trost, Barry M
Trowbridge, Dale Brian
Tsao, Constance S
Ullman, Edwin Fisher
Untch, Karl G(eorge)
Urbach, Karl Frederic
VanAntwerp, Craig Lewis
Van Tamelen, Eugene Earl
Varteressian, K(egham) A(rshavir)
Verbiscar, Anthony James
Verheyden, Julien P H
Vinogradoff, Anna Patricia
Virnig, Michael Joseph
Vollhardt, K Peter C
Vollmar, Arnulf R
Von HUngen, Kern
Wade, Robert Harold
Waiss, Anthony C, Jr
Walba, Harold
Walker, Francis H
Walker, Joseph
Wall, Robert Gene
Wallace, Edwin Garfield
Wallace, Robert Allan
Webb, William Paul
Weck, Friedrich Josef

Wedegaertner, Donald K
Wegner, Patrick Andrew
Weisgraber, Karl Heinrich
Welch, Frank Joseph
Weliky, Norman
Wender, Paul Anthony
Wenkert, Ernest
Wensley, Charles Gelen
Westover, James Donald
Wight, Hewitt Glenn
Wikman-Coffelt, Joan
Wiley, Michael David
Wilgus, Donovan Ray
Williams, John Wesley
Williams, Morris Evan
Williams, Peter M
Willson, Clyde D
Wilson, Elwood Justin, Jr
Winterlin, Wray Laverne
Wipke, W Todd
Wolfe, James Frederick
Wong, Chi-Huey
Wong, Shi-Yin
Woo, Gar Lok
Wood, William Fulton
Woods, William George
Wright, John Marlin
Wudl, Fred
Wunderly, Stephen Walker
Yamada, Yoshikazu
Yamagishi, Frederick George
Yang, Meiling T
Yang, Yan-Bo
Yee, Tucker Tew
Yen, Teh Fu
Youngman, Edward August
Zimmerman, Mary Prislopski
Zon, Gerald
Zweifel, George

COLORADO
Bailey, David Tiffany
Barrett, Anthony Gerard Martin
Bean, Gerritt Post
Beel, John Addis
Bierbaum, Veronica Marie
Blake, Daniel Melvin
Cannon, William Nathaniel
Caruthers, Marvin Harry
Champion, William (Clare)
Cleary, Michael E
Cook, William Boyd
Coomes, Richard Merril
Cristol, Stanley Jerome
Daughenbaugh, Randall Jay
DeBruin, Kenneth Edward
De Puy, Charles Herbert
Dewey, Fred McAlpin
DoMinh, Thap
Druelinger, Melvin L
Duke, Roy Burt, Jr
Dyckes, Douglas Franz
Ellison, Gayfree Barney
Evans, David Lane
Fennessey, Paul V
Finke, Richard Gerald
Fleming, Michael Paul
Gabel, Richard Allen
Gentry, Willard Max, Jr
Goren, Mayer Bear
Gramera, Robert Eugene
Guinn, Denise Eileen
Guss, Cyrus Omar
Hammer, Charles Rankin
Hegedus, Louis Stevenson
Hirs, Christophe Henri Werner
Hornback, Joseph Michael
Hultquist, Martin Everett
Hyatt, David Ernest
Jones, Harold Lester
Kice, John Lord
Klundt, Irwin Lee
Koch, Tad H
Langworthy, William Clayton
Lehman, Joe Junior
Levin, Ronald Harold
Lindquist, Robert Marion
Lodewijk, Eric
Lucas, George Bond
Meek, John Sawyers
Meyers, Albert Irving
Micheli, Roger Paul
Miller, Arnold Reed
Miller, William Theodore
Murphy, Robert Carl
Nachtigall, Guenter Willi
Nightingale, Dorothy Virginia
Ozog, Francis Joseph
Parungo, Farn Pwu
Patton, James Winton
Peters, Kevin Scott
Petersen, Joseph Claine
Pyler, Richard Ernst
Robertson, Jerold C
Schoffstall, Allen M
Schreck, James O(tto)
Schroeder, Herbert A(ugust)
Severson, Roland George
Shelton, Robert Wayne
Spraggins, Robert Lee
Stermitz, Frank
Stickler, William Carl
Stille, John Kenneth
Stribley, Rexford Carl

Organic Chemistry (cont)

Tomasi, Gordon Ernest
Turner, James Howard
Voorhees, Kent Jay
Walba, David Mark
Weliky, Irving
Wilkes, John Stuart
Ziebarth, Timothy Dean

CONNECTICUT
Alexanian, Vazken Arsen
Alfieri, Charles C
Amundsen, Lawrence Hardin
Andrews, Glenn Colton
Arif, Shoaib
Baccei, Louis Joseph
Bailey, William Francis
Baker, Leonard Morton
Baldwin, Ronald Martin
Banijamali, Ali Reza
Banucci, Eugene George
Barnes, Carl E
Batorewicz, Wadim
Bentz, Alan P(aul)
Berson, Jerome Abraham
Bindra, Jasjit Singh
Bloom, Barry Malcolm
Bobko, Edward
Bollyky, L(aszlo) Joseph
Bongiorni, Domenic Frank
Borden, George Wayne
Bordner, Jon D B
Branchini, Bruce Robert
Brittelli, David Ross
Brown, Keith Charles
Buchholz, Allan C
Byrne, Kevin J
Calbo, Leonard Joseph
Chandler, Roger Eugene
Chang, Pauline (Wuai) Kimm
Chu, Hsien-Kun
Church, Robert Fitz (Randolph)
Conover, Lloyd Hillyard
Corbett, John Frank
Covey, Irene Mabel
Covey, Rupert Alden
Crabtree, Robert H
Crast, Leonard Bruce, Jr
Crawford, Thomas Charles
Cronin, Timothy H
Cue, Berkeley Wendell, Jr
Czuba, Leonard J
Dailey, Joseph Patrick
Davis, Robert Glenn
Desio, Peter John
Dirlam, John Philip
Dominy, Beryl W
Dow, Robert Lee
Doyle, Terrence William
Drucker, Arnold
Eagar, Robert Gouldman, Jr
Elder, John William
Epling, Gary Arnold
Essenfeld, Amy
Evans, Charles P
Faller, John William
Fenton, Wayne Alexander
Fine, Leonard W
Fiore, Joseph Vincent
Firestone, Raymond A
Franck, Richard W
Franks, Neal Edward
Frauenglass, Elliott
Friedlander, Henry Z
Fry, Albert Joseph
Gavin, David Francis
Gewanter, Herman Louis
Giuffrida, Robert Eugene
Grayson, Martin
Harbert, Charles A
Heeren, James Kenneth
Henderson, William Arthur, Jr
Hendrick, Michael Ezell
Henkel, James Gregory
Herz, Jack L
Hess, Hans-Jurgen Ernst
Hines, Paul Steward
Hinman, Richard Leslie
Holland, Gerald Fagan
Hopper, Paul Frederick
Horrocks, Robert H
Huang, Samuel J
Humphrey, Bingham Johnson
Husband, Robert Murray
Ives, Jeffrey Lee
Jacobi, Peter Alan
Jorgensen, William L
Katz, Leon
Kavarnos, George James
Kerridge, Kenneth A
Klemarczyk, Philip Thaddeus
Klemchuk, Peter Paul
Kluender, Harold Clinton
Knollmueller, Karl Otto
Koe, B Kenneth
Konigsbacher, Kurt S
Korst, James Joseph
Kuck, Julius Anson
Kuryla, William C
LaMattina, John Lawrence
Lande, Saul
Leavitt, Julian Jacob
Leis, Donald George

Leonard, Robert F
Lin, Tai-Shun
Lipinski, Christopher Andrew
Lipka, Benjamin
Loew, Leslie Max
Lombardino, Joseph George
Luh, Yuhshi
McArthur, Richard Edward
McBride, James Michael
McClenachan, Ellsworth C
McFarland, James William
McGregor, Donald Neil
McGuinness, James Anthony
McKeon, James Edward
McManus, James Michael
McMurray, Walter Joseph
Malofsky, Bernard Miles
Marfat, Anthony
Matsuda, Ken
Meierhoefer, Alan W
Melvin, Lawrence Sherman, Jr
Meriwether, Lewis Smith
Miller, Audrey
Miller, Wilbur Hobart
Moppett, Charles Edward
Mosby, William Lindsay
Murai, Kotaro
Nielsen, John Merle
Noland, James Sterling
O'Connell, Edmond J, Jr
O'Shea, Francis Xavier
Panicci, Ronald J
Panzer, Hans Peter
Patterson, Elizabeth Chambers
Pegolotti, James Alfred
Perry, Clark William
Philips, Judson Christopher
Phillips, Benjamin
Pierce, James Benjamin
Pike, Roscoe Adams
Pinson, Ellis Rex, Jr
Poindexter, Graham Stuart
Prokai, Bela
Proverb, Robert Joseph
Rauch, Albert Lee
Rauhut, Michael McKay
Rennhard, Hans Heinrich
Ressler, Charlotte
Rich, Richard Douglas
Roos, Leo
Rosati, Robert Louis
Rose, James Stephenson
Rosenman, Irwin David
Rothenberg, Alan S
Ryan, Richard Patrick
Salce, Ludwig
Sarges, Reinhard
Saunders, Martin
Scardera, Michael
Schepartz, Alanna
Schmir, Gaston L
Schnur, Rodney Caughren
Sciarini, Louis J(ohn)
Sedlak, John Andrew
Shalaby, Shalaby W
Shine, Timothy D
Singh, Ajaib
Smith, Curtis Page
Smith, Douglas Stewart
Smith, Frederick Albert
Stevens, Malcolm Peter
Stott, Paul Edwin
Strunk, Richard John
Stuber, Fred A
Suh, John Taiyoung
Tate, Bryce Eugene
Thomas, Paul David
Turk, Amos
Ulrich, Henri
Valentine, Donald H, Jr
Vinick, Fredric James
Volkmann, Robert Alfred
Wasserman, Harry H
Wassmundt, Frederick William
Welch, Willard McKowan, Jr
Wharton, Peter Stanley
Whelan, William Paul, Jr
Wiberg, Kenneth Berle
Willner, David
Witczak, Zbigniew J
Wright, Alan Carl
Wuskell, Joseph P
Wynn, Charles Martin, Sr
Wystrach, Vernon Paul
Yevich, Joseph Paul
Ziegler, Frederick Edward

DELAWARE
Agnello, Eugene Joseph
Ainbinder, Zarah
Aldrich, Paul E
Ali-Khan, Ausat
Anderson, Albert Gordon
Anderson, Burton Carl
Angerer, J(ohn) David
Ansul, Gerald R
Armitage, John Brian
Athey, Robert Jackson
Austin, Paul Rolland
Autenrieth, John Stork
Bankert, Ralph Allen
Barker, Harold Clinton
Bartels, George William, Jr
Bauchwitz, Peter S

Baylor, Charles, Jr
Beare, Steven Douglas
Beer, John Joseph
Bellina, Russell Frank
Benson, Frederic Rupert
Bjornson, August Sven
Black, Donald K
Blomstrom, Dale Clifton
Boden, Herbert S
Bowers, George Henry, III
Breazeale, Almut Frerichs
Brehm, Warren John
Breslow, David Samuel
Brodoway, Nicolas
Brown, Robert Raymond
Burg, Marion
Buser, Kenneth Rene
Cairncross, Allan
Carboni, Rudolph A
Carlson, Bruce Arne
Carlson, Norman Arthur
Carnahan, James Elliot
Cavanaugh, Robert J
Chambers, Vaughan Crandall, (Jr)
Cherkofsky, Saul Carl
Chorvat, Robert John
Citron, Joel David
Class, Jay Bernard
Clayton, Anthony Broxholme
Clement, Robert Alton
Cone, Michael McKay
Confalone, Pat N
Cook, Gordon Smith
Cords, Donald Philip
Corner, James Oliver
Coulson, Dale Robert
Cripps, Harry Norman
Crouse, Dale McClish
Crow, Edwin Lee
Daly, John Joseph, Jr
D'Amore, Michael Brian
Darby, Robert Albert
David, Israel A
DeDominicis, Alex John
DeGrado, William F
Dessauer, Rolf
Dishart, Kenneth Thomas
Diveley, William Russell
Dyer, Elizabeth
Eaton, David Fielder
Ehrich, F(elix) Frederick
Eleuterio, Herbert Sousa
Engelhardt, Vaughn Arthur
Espy, Herbert Hastings
Evans, Franklin James, Jr
Fawcett, Mark Stanley
Feinberg, Stewart Carl
Feltzin, Joseph
Finizio, Michael
Finlay, Joseph Burton
Foldi, Andrew Peter
Ford, Thomas Aven
Foss, Robert Paul
Frankenburg, Peter Edgar
Franklin, Richard Crawford
Frey, William Adrian
Fuchs, Julius Jakob
Fukunaga, Tadamichi
Fullhart, Lawrence, Jr
Gano, Robert Daniel
Ganti, Venkat Rao
Garland, Charles E
Garrison, William Emmett, Jr
Gervay, James Edmund
Gibbs, Hugh Harper
Gilbert, Walter Wilson
Gorski, Robert Alexander
Gosser, Lawrence Wayne
Grot, Walther Gustav Fredrich
Gruber, Wilhelm F
Guggenberger, Lloyd Joseph
Gumprecht, William Henry
Hanzel, Robert Stephen
Haq, Mohammad Zamir-ul
Harmuth, Charles Moore
Harris, George Christe
Harris, John Ferguson, Jr
Hartzler, Harris Dale
Hasty, Noel Marion
Haubein, Albert Howard
Hayek, Mason
Hays, John Thomas
Heck, Richard Fred
Heldt, Walter Z
Herkes, Frank Edward
Hilfiker, Franklin Roberts
Hill, Frederick Burns, Jr
Hirsty, Sylvain Max
Hirwe, Ashalata Shyamsunder
Hoegger, Erhard Fritz
Hoehn, Harvey Herbert
Hoeschele, Guenther Kurt
Hoh, George Lok Kwong
Holfeld, Winfried Thomas
Holmes, David Willis
Holyoke, Caleb William, Jr
Honsberg, Wolfgang
Hood, Horace Edward
Hoover, Fred Wayne
Hornberger, Carl Stanley, Jr
Howe, King Lau
Hummel, Donald George
Immediata, Tony Michael
Inskip, Harold Kirkwood

Jackson, Harold Leonard
Jaffe, Edward E
Jamison, Joel Dexter
Jelinek, Arthur Gilbert
Jenner, Edward L
Johnson, Alexander Lawrence
Kaplan, Ralph Benjamin
Kassal, Robert James
Keating, James T
Kegelman, Matthew Roland
Kettner, Charles Adrian
Kittila, Richard Sulo
Knipmeyer, Hubert Elmer
Kohan, Melvin Ira
Kosak, John R
Kramer, Brian Dale
Krespan, Carl George
Kwok, Wo Kong
Landerl, Harold Paul
Langsdorf, William Philip
Lann, Joseph Sidney
Lazaridis, Christina Nicholson
LeClaire, Claire Dean
Lee, Shung-Yan Luke
Leser, Ernst George
Levitt, George
Libbey, William Jerry
Linsay, Ernest Charles
Lipscomb, Robert DeWald
Logullo, Francis Mark
Loken, Halvar Young
Longhi, Raymond
Loomis, Gary Lee
Lorenz, Carl Edward
Lukach, Carl Andrew
McCormack, William Brewster
McCoy, V Eugene, Jr
McGonigal, William E
McKay, Sandra J
McLain, Stephan James
Mallonee, James Edgar
Manos, Philip
Matlack, Albert Shelton
Maury, Lucien Garnett
Maynes, Gordon George
Mazur, Stephen
Mekler, Arlen B
Memeger, Wesley, Jr
Metzger, James Douglas
Monagle, Daniel J
Moncure, Henry, Jr
Monroe, Bruce Malcolm
Monroe, Elizabeth McLeister
Moore, James Alexander
Moore, Ralph Bishop
Mori, Peter Taketoshi
Moses, Francis Guy
Mrowca, Joseph J
Munn, George Edward
Murray, Roger Kenneth, Jr
Nader, Allan E
Naylor, Marcus A, Jr
Nelson, Jerry Allen
Newby, William Edward
Niedzielski, Edmund Luke
Norman, Oscar Loris
Norton, Lilburn Lafayette
Norton, Richard Vail
Ojakaar, Leo
Pagano, Alfred Horton
Pappas, Nicholas
Park, Chung Ho
Patterson, George Harold
Paulshock, Marvin
Pelosi, Lorenzo Fred
Pensak, David Alan
Percival, William Colony
Petersen, Wallace Christian
Plambeck, Louis, Jr
Porter, Hardin Kibbe
Potrafke, Earl Mark
Pretka, John E
Prosser, Robert M
Prosser, Thomas John
Pruckmayr, Gerfried
Putzig, Donald Edward
Quisenberry, Richard Keith
Raasch, Maynard Stanley
Rajagopalan, Parthasarathi
Ramler, Edward Otto
Rave, Terence William
Raynolds, Stuart
Read, Robert E
Reardon, Joseph Edward
Reid, Donald Eugene
Remington, William Roscoe
Richardson, Graham McGavock
Richardson, Paul Noel
Rombach, Louis
Rondestvedt, Christian Scriver, Jr
Saegebarth, Klaus Arthur
Sauers, Richard Frank
Sausen, George Neil
Schadt, Frank Leonard, III
Schappell, Frederick George
Scherer, Kirby Vaughn, Jr
Schmiegel, Walter Werner
Schroeder, Herman Elbert
Schwartz, Jerome Lawrence
Schweitzer, Carl Earle
Schweizer, Edward Ernest
Senkler, George Henry, Jr
Shealy, Otis Lester
Simmons, Howard Ensign, Jr

Simms, John Alvin
Simpson, David Alexander
Singh, Gurdial
Skolnik, Herman
Sloan, Gilbert Jacob
Sloan, Martin Frank
Smart, Bruce Edmund
Smat, Robert Joseph
Smiley, Robert Arthur
Smith, Claibourne Davis
Spaulding, Len Davis
Sroog, Cyrus Efrem
Stahl, Roland Edgar
Stanton, William Alexander
Steller, Kenneth Eugene
Stockburger, George Joseph
Strobach, Donald Roy
Subramanian, Pallatheri Manackal
Sundelin, Kurt Gustav Ragnar
Swerlick, Isadore
Taber, Douglass Fleming
Tan, Henry Harry
Targett, Nancy McKeever
Tatum, William Earl
Taves, Milton Arthur
Taylor, Robert Burns, Jr
Temple, Stanley
Terss, Robert H
Thompson, Robert Gene
Thornton, Roger Lea
Tise, Frank P
Tolman, Chadwick Alma
Trainor, George L
Traumann, Klaus Friedrich
Trost, Henry Biggs
Twelves, Robert Ralph
Un, Howard Ho-Wei
Van Gulick, Norman Martin
Venkatachalam, Taracad Krishnan
Wagner, Hans
Wallenberger, Frederick Theodore
Walter, Henry Clement
Waring, Derek Morris Holt
Wasserman, Edel
Wat, Edward Koon Wah
Webers, Vincent Joseph
Webster, Owen Wright
Wiley, Douglas Walker
Williams, Ebenezer David, Jr
Witterholt, Vincent Gerard
Wojtkowski, Paul Walter
Woods, Thomas Stephen
Wright, Everett James
Wright, Leon Wendell
Wuonola, Mark Arvid
Yamamoto, Y Stephen
Young, Charles Albert

DISTRICT OF COLUMBIA
Alexander, Benjamin H
Bednarek, Jana Marie
Cantrell, Thomas Samuel
Caress, Edward Alan
Carson, Frederick Wallace
Cash, Gordon Graham
Cohen, Alex
Cotruvo, Joseph Alfred
Crist, DeLanson Ross
El Khadem, Hassan S
Engler, Reto Arnold
Fearn, James Ernest
Feldman, Martin Robert
Filipescu, Nicolae
Fleming, Patrick John
Gist, Lewis Alexander, Jr
Graminski, Edmond Leonard
Hammer, Charles F
Hancock, Kenneth George
Hare, Peter Edgar
Hirzy, John William
Horak, Vaclav
Horton, Derek
Hudrlik, Anne Marie
Hudrlik, Paul Frederick
Keller, Teddy Monroe
King, Michael M
Klayman, Daniel Leslie
Kordoski, Edward William
Kulawiec, Robert Joseph
Lourie, Alan David
Merrifield, D Bruce
Metz, Fred L
Milstein, Stanley Richard
Mog, David Michael
Panar, Manuel
Perfetti, Randolph B
Phillips, Don Irwin
Pierce, Dorothy Helen
Ponaras, Anthony A
Poon, Bing Toy
Powell, Justin Christopher
Rogers, Senta S(tephanie)
Roscher, Nina Matheny
Spurlock, Langley Augustine
Sunderlin, Charles Eugene
Vanderhoek, Jack Yehudi
Wade, Clarence W R
Weiss, Richard Gerald
White, Philip Cleaver
Wise, Hugh Edward, Jr

FLORIDA
Alderson, Thomas
Arntzen, Clyde Edward
Astrologes, Gary William
Attaway, John Allen
Baker, Earl Wayne
Barnes, Garrett Henry, Jr
Batich, Christopher David
Battiste, Merle Andrew
Beiler, Theodore Wiseman
Bieber, Theodore Immanuel
Bishop, Charles Anthony
Black, William Bruce
Bledsoe, James O, Jr
Boehme, Werner Richard
Bonnell, James Monroe
Bordenca, Carl
Brown, Henry Clay
Brown, Henry Clay, III
Burney, Curtis Michael
Busse, Robert Franklyn
Butler, George Bergen
Callen, Joseph Edward
Cardenas, Carlos Guillermo
Carlson, David Arthur
Carnes, Joseph John
Castle, Raymond Nielson
Chang, Clifford Wah Jun
Chapman, Richard David
Cioslowski, Jerzy
Clark, Samuel Friend
Cline, Edward Terry
Coke, C(hauncey) Eugene
Conrad, Walter Edmund
Coolidge, Edwin Channing
Couch, Margaret Wheland
Curry, Thomas Harvey
Dalton, Philip Benjamin
Davis, Curry Beach
Denny, George Hutcheson
Derfer, John Mentzer
DeTar, DeLos Fletcher
Deyrup, James Alden
Dolbier, William Read, Jr
Dougherty, Ralph C
Dunn, Ben Monroe
Duranceau, Steven Jon
Ebetino, Frank Frederick
Eldred, Nelson Richards
Ellis, Leonard Culberth
Engel, John Hal, Jr
Fernandez, Jack Eugene
Figdor, Sanford Kermit
Fisher, George Harold, Jr
Fishman, Jack
Fozzard, George Broward
Freeman, Kenneth Alfrey
Friedlander, Herbert Norman
Garbrecht, William Lee
Gash, Virgil Walter
Gawley, Robert Edgar
Giner-Sorolla, Alfredo
Gordon, Joseph R
Gough, Robert George
Gourse, Jerome Allen
Green, Floyd J
Gupton, John
Gurst, Jerome E
Hahn, Elliot F
Hammer, Richard Hartman
Hanley, James Richard, Jr
Hardman, Bruce Bertolette
Harvey, George Ranson
Heberling, Jack Waugh, Jr
Hechenbleikner, Ingenuin Albin
Helling, John Frederic
Herriott, Arthur W
Herz, Werner
Heying, Theodore Louis
Hicks, Elija Maxie, Jr
Hill, Kenneth Richard
Hisey, Robert Warren
Hoeppner, Conrad Henry
Hoffman, Warren E
Holmer, Donald A
Huba, Francis
Huber, Joseph William, III
Hunt, William Cecil
Hunter, George L K
Ihndris, Raymond Will
Iloff, Phillip Murray, Jr
Ingwalson, Raymond Wesley
Ireland, Robert Ellsworth
Jackson, George Richard
Jaeger, Herbert Karl
Jones, William Maurice
Joyce, Robert Michael
Jurch, George Richard, Jr
Kagan, Benjamin
Kane, Bernard James
Kane, Howard L
King, Roy Warbrick
Klacsmann, John Anthony
Kleinschmidt, Albert Willoughby
Kowkabany, George Norman
Krafft, Marie Elizabeth
Kreider, Henry Royer
Kunisi, Venkatasubban S
Kurchacova, Elva S
Kutner, Abraham
Lawrence, Franklin Isaac Latimer
Leffler, John Edward
Levis, William Walter, Jr
Levy, Joseph
Lewis, George Edwin
Light, Robley Jasper
Lillien, Irving
LoCicero, Joseph Castelli
Lombardo, Anthony
McBride, Joseph James, Jr
McCall, Marvin Anthony
McMullen, Warren Anthony
Mandell, Leon
Mason, Kenneth M
Mattson, Guy C
Mehta, Nariman Bomanshaw
Miles, David H
Millor, Steven Ross
Mitch, Frank Allan
Moroni, Eneo C
Moskowitz, Mark Lewis
Murphy, James Gilbert
Nersasian, Arthur
Newkome, George Richard
Noller, David Conrad
Nordby, Harold Edwin
Oberst, Fred William
Odioso, Raymond C
Offenhauer, Robert Dwight
Orloff, Harold David
Osbahr, Albert J, Jr
Owen, Gwilym Emyr, Jr
Owen, Terence Cunliffe
Parkanyi, Cyril
Parker, Earl Elmer
Parmerter, Stanley Marshall
Payton, Albert Levern
Penner, Siegfried Edmund
Peterson, Robert Hampton
Pickard, Porter Louis, Jr
Pinder, Albert Reginald
Plapinger, Robert Edwin
Pop, Emil
Porter, Lee Albert
Postelnek, William
Ramsey, Brian Gaines
Rice, Frederick Anders Hudson
Richards, Marvin Sherrill
Rider, Don K(eith)
Ritchie, Calvin Donald
Rogan, John B
Rokach, Joshua
Ross, Fred Michael
Royals, Edwin Earl
Sager, William Frederick
Salsbury, Jason Melvin
Saltiel, Jack
Saunders, James Henry
Schneller, Stewart Wright
Schwartz, Martin Alan
Segal, Alvin
Sellers, John William
Shabica, Anthony Charles, Jr
Sharkey, William Henry
Shaw, Philip Eugene
Sheffer, Howard Eugene
Shepard, John Reed
Sherman, Albert Herman
Sherman, Edward
Smart, William Donald
Smith, William Mayo
Snyder, Carl Henry
Sollman, Paul Benjamin
Soto, Aida R
Spatz, Sydney Martin
Spayd, Richard W
Spencer, John Lawrence
Sperley, Richard Jon
Stahl, Joel S
Starke, Albert Carl, Jr
Steadman, Thomas Ree
Stolberg, Marvin Arnold
Stump, Eugene Curtis, Jr
Sullivan, Lloyd John
Surrey, Alexander Robert
Sweet, Ronald Lancelot
Szmant, Herman Harry
Travnicek, Edward Adolph
Urbas, Branko
Van Handel, Emile
Walborsky, Harry M
Walter, Charles Robert, Jr
Wander, Joseph Day
Warren, Harold Hubbard
Watkins, Spencer Hunt
Webb, Robert Lee
Wiles, Robert Allan
Williams, Robert Hackney
Wise, Raleigh Warren
Witte, Michael
Wood, James Brent, III
Woodward, Fred Erskine
Zimmer, William Frederick, Jr
Zoltewicz, John A
Zook, Harry David

GEORGIA
Allinger, Norman Louis
Allison, John P
Anderson, Gloria Long
Ashby, Eugene Christopher
Baarda, David Gene
Barbas, John Theophani
Barefield, Edward Kent
Barreras, Raymond Joseph
Barton, Franklin Ellwood, II
Bayer, Charlene Warres
Bergman, Elliot
Blanton, Charles DeWitt, Jr
Blomquist, Richard Frederick
Boschan, Robert Herschel
Boxer, Robert Jacob
Boykin, David Withers, Jr
Bralley, James Alexander
Burgess, Edward Meredith
Carter, Mary Eddie
Caughman, Henry Daniel
Chandler, James Harry, III
Chortyk, Orestes Timothy
Chu, Chung K
Churchill, Frederick Charles
Clark, Benjamin Cates, Jr
Clendinning, Robert Andrew
Cole, Thomas Winston, Jr
Counts, Wayne Boyd
Derrick, Mildred Elizabeth
Dimmel, Donald R
Dinwiddie, Joseph Gray, Jr
Dixon, Dabney White
DuBois, Grant Edwin
Duvall, Harry Marean
Ellington, James Jackson
Ennor, Kenneth Stafford
Esslinger, William Glenn
Fay, Alice D Awtrey
Freese, William P, II
Garrett, David William
Gianturco, Maurizio
Girardot, Jean Marie Denis
Glassick, Charles Etzweiler
Goldsmith, David Jonathan
Gollob, Lawrence
Grovenstein, Erling, Jr
Hargrove, Robert John
Harris, Henry Earl
Heyd, Charles E
Hicks, Arthur M
Hill, Richard Keith
Hodges, Linda Carol
Holtkamp, Freddy Henry
House, Herbert Otis
Howard, John Charles
Hung, William Mo-Wei
Hurd, Phillip Wayne
Iacobucci, Guillermo Arturo
Iannicelli, Joseph
Jen, Yun
Johnson, Robert William, Jr
Jones, Ronald Goldin
Karliner, Jerrold
Kellogg, Craig Kent
Khan, Ishrat Mahmood
Knight, James Albert, Jr
Knopka, W N
Krajca, Kenneth Edward
Lawton, Richard G
Lester, Charles Turner
Lewis, Silas Davis
Liebeskind, Lanny Steven
Liotta, Dennis C
Lomax, Eddie
Love, Jimmy Dwane
McCarthy, Neil Justin, Jr
McKinney, Roger Minor
Mahesh, Virendra B
Marascia, Frank Joseph
Mathews, Walter Kelly
Matthews, Edward Whitehouse
Menger, Fred M
Mickelsen, Olaf
Newton, Melvin Gary
Pacifici, James Grady
Padwa, Albert
Pelletier, S William
Pepoy, Louis John
Perdue, Edward Michael
Pohl, Douglas George
Polk, Malcolm Benny
Ponder, Billy Wayne
Powers, James Cecil
Proctor, Charles Darnell
Radford, Terence
Raut, Kamalakar Balkrishna
Rees, William Smith, Jr
Robbins, Paul Edward
Rodriguez, Augusto
Roobol, Norman R
Ruenitz, Peter Carmichael
Sandri, Joseph Mario
Schmalz, Alfred Chandler
Schwerzel, Robert Edward
Sears, Curtis Thornton, Jr
Shapira, Raymond
Sommers, Jay Richard
Spence, Gavin Gary
Spriggs, Alfred Samuel
Stammer, Charles Hugh
Stanfield, James Armond
Su, Helen Chien-Fan
Thompson, Bobby Blackburn
Tolbert, Laren Malcolm
Trumbull, Elmer Roy, Jr
Tucker, Willie George
Turbak, Albin Frank
Tyler, Jean Mary
Walker, John J
Walker, Russell Wagner
Wiesboeck, Robert A
Williams, George Nathaniel
Wolf, Monte William
Yost, Robert Stanley
Young, Raymond Hinchcliffe, Jr
Zalkow, Leon Harry
Zepp, Richard Gardner
Zvejnieks, Andrejs

Organic Chemistry (cont)

HAWAII
Antal, Michael Jerry, Jr
Dority, Guy Hiram
Hiu, Dawes Nyukleu
Howton, David Ronald
Kiefer, Edgar Francis
Kloetzel, Milton Carl
Larson, Harold Olaf
Liu, Robert Shing-Hei
Moore, Richard E
Mower, Howard Frederick
Norton, Ted Raymond
Scheuer, Paul Josef

IDAHO
Banks, Richard C
Braun, Loren L
Bush, David Clair
Castle, Lyle William
Cooley, James Hollis
Dalton, Jack L
Hatcher, Herbert John
Imel, Arthur Madison
Isaacson, Eugene I
Matjeka, Edward Ray
Mosher, Michael David
Natale, Nicholas Robert
Ronald, Bruce Pender
Sarett, Lewis Hastings
Sutton, John Curtis
Wiegand, Gayl
Winkel, Cleve R

ILLINOIS
Adlof, Richard Otto
Ager, David John
Allen, John Kay
Anderson, Arnold Lynn
Andrews, Eugene Raymond
Applequist, Douglas Einar
Arendsen, David Lloyd
Argoudelis, Chris J
Armstrong, P(aul) Douglas
Arnold, Richard Thomas
Ashley, Warren Cotton
Babler, James Harold
Bagby, Marvin Orville
Baldoni, Andrew Ateleo
Baran, John Stanislaus
Bauer, Ludwig
Baumgarten, Ronald J
Beak, Peter
Bechara, Ibrahim
Beck, Karl Maurice
Berger, Daniel Richard
Bergstrom, Clarence George
Berman, Lawrence Uretz
Bernath, Tibor
Bernetti, Raffaele
Bernhardt, Randal Jay
Beyler, Roger Eldon
Blaha, Eli William
Bordwell, Frederick George
Boshart, Gregory Lew
Bowles, William Allen
Boyd, Mary K
Brace, Neal Orin
Breitbeil, Fred W, III
Brooks, Dee W
Brown, George Earl
Bryant, Rhys
Buchanan, David Hamilton
Buchman, Russell
Buntrock, Robert Edward
Burgauer, Paul David
Burney, Donald Eugene
Cammarata, Peter S
Carr, Lawrence John
Carrick, Wayne Lee
Cengel, John Anthony
Cerefice, Steven A
Chao, Tai Siang
Cheng, Shu-Sing
Chorghade, Mukund Shankar
Chou, Mei-In Melissa Liu
Churchill, Constance Louise
Clark, Stephen Darrough
Closs, Gerhard Ludwig
Coates, Robert Mercer
Cochrane, Chappelle Cecil
Cordell, Geoffrey Alan
Cosper, David Russell
Crovetti, Aldo Joseph
Crumrine, David Shafer
Cunico, Robert Frederick
Currie, Bruce LaMonte
Curtin, David Yarrow
Curtis, Veronica Anne
Danzig, Morris Juda
DeBoer, Edward Dale
Debus, Allen George
Dellaria, Joseph Fred, Jr
DeYoung, Edwin Lawson
Dixit, Saryu N
Doane, William M
Drengler, Keith Allan
Dybalski, Jack Norbert
Eachus, Alan Campbell
Eaton, Philip Eugene
Eby, Lawrence Thornton
Economy, James
Ellis, Jerry William
Empen, Joseph A
Erby, William Arthur
Erickson, Mitchell Drake
Evans, Robert John
Falk, John Carl
Fanta, George Frederick
Fanta, Paul Edward
Fassnacht, John Hartwell
Faubl, Hermann
Feinstein, Allen Irwin
Field, Kurt William
Fields, Ellis Kirby
Filler, Robert
Foote, Carlton Dan
Foster, George A, Jr
Frame, Robert Roy
Frank, Forrest Jay
Frey, David Allen
Fried, Josef
Gaffney, Jeffrey Steven
Garland, Robert Bruce
Garven, Floyd Charles
Gasser, William
Gast, Lyle Everett
Gavlin, Gilbert
Gearien, James Edward
Gebauer, Peter Anthony
Gehring, Harvey Thomas
Gemmer, Robert Valentine
Gershbein, Leon Lee
Gibbs, James Albert
Ginger, Leonard George
Giori, Claudio
Glaser, Milton Arthur
Gleeson, James Newman
Glover, Allen Donald
Goeckner, Norbert Anthony
Golinkin, Herbert Sheldon
Goretta, Louis Alexander
Grotefend, Alan Charles
Groziak, Michael Peter
Gupta, Goutam
Gutberlet, Louis Charles
Hac, Lucile R
Hakewill, Henry, Jr
Hall, J Herbert
Hamer, Martin
Hansen, Donald Willis, Jr
Hansen, John Frederick
Harper, Jon Jay
Hartlage, James Albert
Harvey, Ronald Gilbert
Haugen, David Allen
Hauptschein, Murray
Henkin, Jack
Henning, Lester Allan
Henry, Patrick M
Hoeg, Donald Francis
Hofreiter, Bernard T
Hogan, Philip
Hopkins, Paul Donald
Huang, Leo W
Huang, Shu-Jen Wu
Huffman, George Wallen
Hughes, Robert David
Hutchcroft, Alan Charles
Ika, Prasad Venkata
Illingworth, George Ernest
James, David Eugene
Janssen, Jerry Frederick
Jezl, James Louis
Johnson, Calvin Keith
Johnson, Carl Edwin
Johnson, Dale Howard
Johnson, John Harold
Jones, David A, Jr
Jones, Marjorie Ann
Juenge, Eric Carl
Jungmann, Richard A
Kabara, Jon Joseph
Kagan, Jacques
Karll, Robert E
Katzenellenbogen, John Albert
Kaufman, Priscilla C
Kennelly, Mary Marina
Kerdesky, Francis A J
Kerkman, Daniel Joseph
Kester, Dennis Earl
Kevill, Dennis Neil
Kharas, Gregory B
Khattab, Ghazi M A
Kim, Ki Hwan
King, Lafayette Carroll
Kipp, James Edwin
Kirkpatrick, Joel Lee
Kissel, William John
Kleiman, Morton
Klein, Larry L
Klemm, Richard Andrew
Klimstra, Paul D
Knaggs, Edward Andrew
Knobloch, James Otis
Knutson, Clarence Arthur, Jr
Kolb, Doris Kasey
Kolb, Kenneth Emil
Krajewski, John J
Kretchmer, Richard Allan
Kruse, Carl William
Ku, Yi-Yin
Kucera, Thomas J
Kuceski, Vincent Paul
Kuhajek, Eugene James
Kuhlmann, George Edward
Kurath, Paul
Kurz, Michael E
Lambert, Joseph B
Lan, Ming-Jye
Larson, Richard Allen
Lash, Timothy David
Lauterbach, Richard Thomas
Lee, Charlotte
Lee, Cheuk Man
Lemper, Anthony Louis
Letsinger, Robert Lewis
Levin, Alfred A
Lewis, Frederick D
Lewis, Morton
Lillwitz, Lawrence Dale
Lindbeck, Wendell Arthur
Lindberg, Steven Edward
Lira, Emil Patrick
Little, Randel Quincy, Jr
Loire, Norman Paul
Lucas, Glennard Ralph
Lynch, Darrel Luvene
Lynch, Don Murl
McAlpine, James Bruce
McCollum, John David
McLaughlin, Robert Lawrence
Magnus, George
Maher, George Garrison
Mallia, Anantha Krishna
Malloy, Thomas Patrick
Matson, Howard John
Matta, Michael Stanley
Matthews, Clifford Norman
Mazur, Robert Henry
Meguerian, Garbis H
Melhado, L(ouisa) Lee
Meyer, Delbert Henry
Meyers, Cal Yale
Mihina, Joseph Stephen
Mikulec, Richard Andrew
Miller, Francis Marion
Miller, Sidney Israel
Mimnaugh, Michael Neil
Mock, William L
Mohan, Prem
Morello, Edwin Francis
Moriarty, Robert M
Moser, Kenneth Bruce
Mounts, Timothy Lee
Muffley, Harry Chilton
Murphy, Thomas Joseph
Narske, Richard Martin
Neumiller, Harry Jacob, Jr
Noren, Gerry Karl
Novak, Robert William
Nuzzo, Ralph George
Oono, Yoshitsugu
Osman, Elizabeth Mary
Otey, Felix Harold
Padrta, Frank George
Pai, Sadanand V
Paisley, David M
Patinkin, Seymour Harold
Patrick, Timothy Benson
Paul, Iain C
Perkins, Edward George
Perun, Thomas John
Peterson, Melbert Eugene
Piatak, David Michael
Piehl, Frank John
Pines, Herman
Pirkle, William H
Portis, Archie Ray, Jr
Portnoy, Norman Abbye
Potempa, Lawrence Albert
Poulos, Nicholas A
Prout, Franklin Sinclair
Pryde, Everett Hilton
Radford, Herschel Donald
Raheel, Mastura
Rakoff, Henry
Rao, Yedavalli Shyamsunder
Rathke, Jerome William
Rausch, David John
Rawlinson, David John
Rees, Thomas Charles
Reidies, Arno H
Reily, William Singer
Rentmeester, Kenneth R
Reynolds, Rosalie Dean (Sibert)
Richmond, James M
Rinehart, Kenneth Lloyd
Robin, Burton Howard
Rocek, Jan
Roderick, William Rodney
Rorig, Kurt Joachim
Rosen, Bruce Irwin
Rosenberg, Saul Howard
Rosenbrook, William, Jr
Sacco, Louis Joseph, Jr
Sample, James Halverson
Sause, H William
Schaap, Luke Anthony
Schaeffer, David Joseph
Schmukler, Seymour
Scholfield, Charles Rexel
Schriesheim, Alan
Schuster, Gary Benjamin
Schwab, Arthur William
Scouten, Charles Ervin
Seigler, David Stanley
Shiner, Edward Arnold
Shipchandler, Mohammed Tyebji
Shoffner, James Priest
Shone, Robert L
Shriner, Ralph Lloyd
Shulman, Sol
Silverman, Richard Bruce
Sinclair, Henry Beall
Smith, Gerard Vinton
Smith, Homer Alvin, Jr
Smith, Robert Johnson
Smith, Stanley Glen
Snyder, Gary James
Snyder, William Robert
Spangler, Charles William
Spitzer, William Carl
Stephens, James Regis
Stock, Leon M
Stowell, James Kent
Sytsma, Louis Frederick
Szpunar, Carole Bryda
Tiefenthal, Harlan E
Tomomatsu, Hideo
Tortorello, Anthony Joseph
Traxler, James Theodore
Trevillyan, Alvin Earl
Turner, Fred Allen
Tyner, David Anson
Ulrey, Stephen Scott
Unglaube, James M
Van Lanen, Robert Jerome
Van Strien, Richard Edward
Vasiliauskas, Edmund
Verbanac, Frank
Voedisch, Robert W
Vosti, Donald Curtis
Walhout, Justine I Simon
Walter, Robert Irving
Walton, Henry Miller
Wang, Gary T
Warne, Thomas Martin
Wasielewski, Michael Roman
Weatherbee, Carl
Webber, Gayle Milton
Weier, Richard Mathias
Wentworth, Gary
Weston, Arthur Walter
Weyna, Philip Leo
Wiedenmann, Lynn G
Wilke, Robert Nielsen
Wilt, James William
Wimer, David Carlisle
Winans, Randall Edward
Witt, John, Jr
Wolf, Leslie Raymond
Wolf, Richard Eugene
Wolff, William Francis
Wotiz, John Henry
Wysocki, Allen John
Yamamoto, Diane M
Yokelson, Howard Bruce
Yonan, Edward E
Young, Austin Harry
Young, David W
Young, Harland Harry
Zaugg, Harold Elmer
Zeffren, Eugene
Zletz, Alex

INDIANA
Archer, Robert Allen
Arndt, Henry Clifford
Bailey, Thomas Daniel
Bakker, Gerald Robert
Barnett, Charles Jackson
Bays, James Philip
Beck, James Richard
Becker, Elizabeth Ann (White)
Beckman, Jean Catherine
Belagaje, Rama M
Benjaminov, Benjamin S
Benkeser, Robert Anthony
Benton, Francis Lee
Bergstrom, Donald Eugene
Blickenstaff, Robert Theron
Boguslaski, Robert Charles
Bolton, Benjamin A
Borchert, Peter Jochen
Bosin, Talmage R
Bostick, Edgar E
Bottorff, Edmond Milton
Bradway, Keith E
Brandt, Karl Garet
Brannon, Donald Ray
Brewster, James Henry
Brinkmeyer, Raymond Samuel
Brooker, Robert Munro
Brown, Herbert Charles
Burkett, Howard (Benton)
Burow, Kenneth Wayne, Jr
Butler, John Mann
Byrn, Stephen Robert
Carlson, Merle Winslow
Carmack, Marvin
Carnahan, Robert Edward
Chateauneuf, John Edward
Cook, Addison Gilbert
Cook, Donald Jack
Cook, Kenneth Emery
Cooper, Robin D G
Corey, Paul Frederick
Crandall, Jack Kenneth
Cushman, Mark
Cutshall, Theodore Wayne
Danehy, James Philip
Debono, Manuel
DeSantis, John Louis
Dominianni, Samuel James

CHEMISTRY / 215

Donner, David Bruce
Dorman, Douglas Earl
Dunn, Howard Eugene
Dykstra, Stanley John
Easton, Nelson Roy
Egly, Richard S(amuel)
Ellingson, Rudolph Conrad
Emmick, Thomas Lynn
Farkas, Eugene
Feigl, Dorothy M
Feuer, Henry
Fieldhouse, John W
Fife, Wilmer Krafft
Flaugh, Michael Edward
Flynn, John Joseph, Jr
Fornefeld, Eugene Joseph
Freeman, Jeremiah Patrick
Frump, John Adams
Gajewski, Joseph J
Gallucci, Robert Russell
Gerns, Fred Rudolph
Gilham, Peter Thomas
Goe, Gerald Lee
Goossens, John Charles
Grieco, Paul Anthony
Grutzner, John Brandon
Gutowski, Gerald Edward
Harper, Edwin T
Harper, Richard Waltz
Haslanger, Martin Frederick
Helquist, Paul M
Hennion, George Felix
Herron, David Kent
Hsiung, Hansen M
Hull, Clarence Joseph
Ingraham, Joseph Sterling
Jackson, Bill Grinnell
Jacobs, Martin John
Johnston, Katharine Gentry
Kaslow, Christian Edward
Kelly, Walter James
Kilsheimer, Sidney Arthur
Kirst, Herbert Andrew
Kjonaas, Richard A
Koller, Charles Richard
Koppel, Gary Allen
Kornblum, Nathan
Kovacic, Peter
Kraemer, John Francis
Kramer, William J
Kreider, Leonard C(ale)
Kress, Thomas Joseph
Lantero, Oreste John, Jr
Lavagnino, Edward Ralph
Leonard, Jack E
Lewis, Dennis Allen
Lipkowitz, Kenny Barry
Long, Eric Charles
Longroy, Allan Leroy
Lorentzen, Keith Eden
Loudon, Gordon Marcus
Lutz, Wilson Boyd
McFarland, John William
McIntyre, Thomas William
Majewski, Robert Francis
Markley, Lowell Dean
Marshall, Frederick J
Matsumoto, Ken
Meyerson, Seymour
Miesel, John Louis
Mikolasek, Douglas Gene
Miller, Edward George
Miller, Roy Glenn
Miranda, Thomas Joseph
Montgomery, Lawrence Kernan
Morris, John F
Morrison, Harry
Myerholtz, Ralph W, Jr
Negishi, Ei-ichi
Nelson, Nils Keith
O'Doherty, George Oliver-Plunkett
O'Donnell, Martin James
Osborne, David Wendell
Pasto, Daniel Jerome
Patwardhan, Bhalchandra H
Porter, Herschel Donovan
Pyle, James L
Quinn, Michael H
Rand, Leon
Rivers, Paul Michael
Robertson, Donald Edwin
Robey, Roger Lewis
Ross, Joseph Hansbro
Safdy, Max Errol
Salerni, Oreste Leroy
Schroeder, Juel Pierre
Schwan, Theodore Carl
Serianni, Anthony Stephan
Shankland, Rodney Veeder
Shiner, Vernon Jack, Jr
Siebert, John
Sieloff, Ronald F
Smith, Gerald Floyd
Smith, Lewis Oliver, Jr
Soper, Quentin F(rancis)
Sousa, Lynn Robert
Sowers, Edward Eugene
Stamper, Martha C
Stehouwer, David Mark
Sullivan, Hugh R, Jr
Sybert, Paul Dean
Tennyson, Richard Harvey
Thibault, Thomas Delor
Thompson, Gerald Lee

Toomey, Joseph Edward
Trinler, William A
Trischler, Floyd D
Trott, Gene F
Trozzolo, Anthony Marion
Truce, William Everett
Turner, James A
Uloth, Robert Henry
Van Dyke, John William, Jr
Van Etten, Robert Lee
Van Heyningen, Earle Marvin
Vaughn, Thomas Hunt
Wang, Jin-Liang
Webber, J Alan
Weiner, Henry
Welch, Zara D
Weller, Lowell Ernest
Wendel, Samuel Reece
White, Harold Keith
Widlanski, Theodore Solomon
Winslow, Alfred Edwards
Wiseman, Park Allen
Wolinsky, Joseph
Wright, Ian Glaisby
Wu, Yao Hua
Yordy, John David

IOWA
Ault, Addison
Barton, Thomas J
Bearce, Winfield Hutchinson
Burton, Donald Joseph
Carlson, Emil Herbert
Czarny, Michael Richard
Docken, Adrian (Merwin)
Doorenbos, Harold E
Downing, Donald Talbot
Franklin, Robert Louis
Geels, Edwin James
Graybill, Bruce Myron
Hampton, David Clark
Kercheval, James William
Kraus, Kenneth Wayne
Larock, Richard Craig
Lindberg, James George
Linhardt, Robert John
McGrew, Leroy Albert
MacMillan, James G
Nair, Vasu
Neumann, Marguerite
Osuch, Carl
Rila, Charles Clinton
Russell, Glen Allan
Summers, William Allen, Jr
Swartz, James E
Trahanovsky, Walter Samuel
Verkade, John George
Welch, Dean Earl
Wertz, Philip Wesley
Wiemer, David F

KANSAS
Baumgartner, George Julius
Beechan, Curtis Michael
Boyle, Kathryn Moyne Ward Dittemore
Breed, Laurence Woods
Briles, George Herbert
Burgstahler, Albert William
Cahoy, Roger Paul
Carlson, Robert Gideon
Chappelow, Cecil Clendis, Jr
Cheng, Chia-Chung
Choguill, Harold Samuel
Clevenger, Richard Lee
Crandall, Elbert Williams
Duncan, William Perry
Engler, Thomas A
Englund, Charles R
Fuhlhage, Donald Wayne
Givens, Richard Spencer
Glazier, Robert Henry
Grunewald, Gary Lawrence
Hanley, Wayne Stewart
Hua, Duy Huu
Huyser, Earl Stanley
Johnson, John Webster, Jr
Klabunde, Kenneth John
Landgrebe, John A
Lenhert, Anne Gerhardt
Levy, Edward Robert
McDonald, Richard Norman
Marx, Michael
Moyer, Melvin Isaac
Paukstelis, Joseph V
Schowen, Richard Lyle
Schroeder, Robert Samuel
Seib, Paul A
Slater, Carl David
Stutz, Robert L
Talaty, Erach R
Wiley, James C, Jr

KENTUCKY
Bendall, Victor Ivor
Brown, Ellis Vincent
Fields, Donald Lee
Flachsman, Robert Louis, Jr
Gibson, Dorothy Hinds
Givens, Edwin Neil
Glenn, Furman Eugene
Graver, Richard Byrd
Guthrie, Robert D
Hamilton-Kemp, Thomas Rogers
Hendon, Joseph C

Hessley, Rita Kathleen
Hussung, Karl Frederick
Huttenlocher, Dietrich F
Johnson, Robert Reiner
Kadaba, Pankaja Kooveli
Kargl, Thomas E
Kennedy, John Elmo, Jr
Kornet, Milton Joseph
Lloyd, William Gilbert
Lukes, Robert Michael
Magid, Linda Lee Jenny
Masters, John Edward
Meier, Mark Stephan
Meisenheimer, John Long
Nathan, Richard Arnold
Owen, Robert Allan
Patterson, John Miles
Price, Martin Burton
Reasoner, John W
Richard, John P
Richter, Alfred Eugene
Sagar, William Clayton
Sanford, Robert Alois
Schulz, William
Scott, George William
Selegue, John Paul
Shoemaker, Gradus Lawrence
Slocum, Donald Warren
Smith, Stanford Lee
Smith, Walter Thomas, Jr
Taylor, Kenneth Grant
Thio, Alan Poo-An
Tucker, Irwin William
Wilson, Gordon, Jr
Wilson, Joseph William

LOUISIANA
Avonda, Frank Peter
Babb, Robert Massey
Baird, William C, Jr
Bauer, Beverly Ann
Bayer, Arthur Craig
Bertoniere, Noelie Rita
Biersmith, Edward L, III
Blanchard, Eugene Joseph
Blouin, Florine Alice
Boozer, Charles (Eugene)
Boyer, Joseph Henry
Brackenridge, David Ross
Bromberg, Milton Jay
Byrd, David Shelton
Camp, Ronald Lee
Cartledge, Frank
Chen, Hoffman Hor-Fu
Clarke, Wilbur Bancroft
Colgrove, Steven Gray
Connick, William Joseph, Jr
Conrad, Franklin
Corkern, Walter Harold
Couvillion, John Lee
Daly, William Howard
Daul, George Cecil
Davenport, Tom Forest, Jr
Davis, Bryan Terence
Domelsmith, Linda Nell
Doomes, Earl
Edwards, Joseph D, Jr
Eisenbraun, Allan Alfred
Ellzey, Samuel Edward, Jr
Ensley, Harry Eugene
Everly, Charles Ray
Fischer, Nikolaus Hartmut
Fitzpatrick, Jimmie Doile
Frank, Arlen W(alker)
Franklin, William Elwood
Gallo, August Anthony
Gibson, David Michael
Goldberg, Stanley Irwin
Griffin, Gary Walter
Hanson, Marvin Wayne
Harper, Robert John, Jr
Holt, Robert Louis
Hornbaker, Edwin Dale
Ieyoub, Kalil Phillip
Impastato, Fred John
Jackisch, Philip Frederick
Jacks, Thomas Jerome
Kehoe, Lawrence Joseph
Kennedy, Frank Scott
Klobucar, William Dirk
Koenig, Paul Edward
Ledford, Thomas Howard
Lee, Burnell
Lee, John Yuchu
McGraw, Gerald Wayne
Magin, Ralph Walter
Mangham, Jesse Roger
Mark, Harold Wayne
Marvel, John Thomas
Mazzeno, Laurence William
Mollere, Phillip David
Moore, Richard Newton
Munchausen, Linda Lou
Neher, Clarence M
Nevill, William Albert
Parent, Richard Alfred
Pepperman, Armand Bennett, Jr
Petterson, Robert Carlyle
Pine, Lloyd A
Plonsker, Larry
Price, Leonard
Pryor, William Austin
Rabideau, Peter W
Reeves, Richard Edwin

Reich, Donald Arthur
Reinhardt, Robert Milton
Roberts, Donald Duane
Robinson, Gene Conrad
Rogers, Stearns Walter
Schexnayder, Mary Anne
Schoellmann, Guenther
Sevenair, John P
Shin, Kju Hi
Shubkin, Ronald Lee
Slaven, Robert Walter
Smalley, Arnold Winfred
Smith, Grant Warren, II
Smith, Robert Leonard
Stahly, Glenn Patrick
Stanonis, David Joseph
Starrett, Richmond Mullins
Steinmetz, Walter Edmund
Stocker, Jack H(ubert)
Stowell, John Charles
Sumrell, Gene
Theriot, Kevin Jude
Timberlake, Jack W
Traynham, James Gibson
Trisler, John Charles
Vail, Sidney Lee
Vercellotti, John R
Vigo, Tyrone Lawrence
Walia, Jasjit Singh
Walter, Thomas James
Walton, Warren Lewis
Weidig, Charles F
Welch, Clark Moore
Wells, Darthon Vernon
Willer, Rodney Lee
Wright, Oscar Lewis
Zietz, Joseph R, Jr

MAINE
Aft, Harvey
Bentley, Michael David
Bonner, Willard Hallam, Jr
Borror, Alan L
Bragdon, Robert Wright
Duncan, Charles Donald
Fort, Raymond Cornelius, Jr
Green, Brian
Jensen, Bruce L
McCarty, John Edward
Mattor, John Alan
Mayo, Dana Walker
Mueller, George Peter
Newton, Thomas Allen
Pierce, Arleen Cecilia
Reid, Evans Burton
Shepard, Robert Andrews
Spencer, Claude Franklin
Tedeschi, Robert James
Tewksbury, L Blaine
Welldon, Paul Burke
Winicov, Herbert

MARYLAND
Aaron, Herbert Samuel
Adolph, Horst Guenter
Aldridge, Mary Hennen
Ammon, Herman L
Andrulis, Peter Joseph, Jr
Bailey, William John
Banks, Harold Douglas
Barnes, Charlie James
Beisler, John Albert
Bellet, Richard Joseph
Berkowitz, Lewis Maurice
Berkowitz, Sidney
Beroza, Morton
Blum, Stanley Walter
Borkovec, Alexej B
Bosmajian, George
Brown, Alexandra Louise
Brown, Alfred Edward
Buras, Edmund Maurice
Burrows, Elizabeth Parker
Burton, Lester Percy Joseph
Bush, Richard Wayne
Cahnmann, Hans Julius
Campbell, Paul Gilbert
Caret, Robert Laurent
Catravas, George Nicholas
Cavagna, Giancarlo Antonio
Chamberlain, David Leroy, Jr
Chaykovsky, Michael
Chen, Hao
Chiu, Ching Ching
Chmurny, Alan Bruce
Clark, Patricia
Clarke, Frederic B, III
Coates, Arthur Donwell
Cogliano, Joseph Albert
Cohen, Louis Arthur
Cowan, Dwaine O
Cragg, Gordon Mitchell
Creighton, Donald John
Cullen, William Charles
Delaney, Harold
De Matte, Michael L
Dighe, Shrikant Vishwanath
Dillon, James Greenslade
DiNunno, Cecil Malmberg
Doukas, Harry Michael
Dragun, Henry L
Dubois, Ronald Joseph
Eng, Leslie
Engle, Robert Rufus

Organic Chemistry (cont)

Epstein, Joseph
Evanega, George R
Falci, Kenneth Joseph
Fales, Henry Marshall
Fatiadi, Alexander Johann
Feinberg, Robert Samuel
Feldman, Alfred Philip
Fenselau, Allan Herman
Ferretti, Aldo
Feyns, Liviu Valentin
Fike, Harold Lester
Fisher, E(arl) Eugene
Fleming, Paul Robert
Florin, Roland Eric
Fox, Robert Bernard
Frank, Victor Samuel
Fraser, Blair Allen
Gabrielsen, Bjarne
Gaidis, James Michael
Gajan, Raymond Joseph
Garman, John Andrew
Goldstein, Jorge Alberto
Gould, Jack Richard
Graham, Joseph H
Gray, Allan P
Guthrie, James Leverette
Hackley, Brennie Elias, Jr
Hamilton, William Lander
Harmon, Alan Dale
Hartwell, Jonathan Lutton
Haske, Bernard Joseph
Henery-Logan, Kenneth Robert
Herndon, James W
Herner, Albert Erwin
Herrick, Elbert Charles
Hertz, Harry Steven
Heumann, Karl Fredrich
Highet, Robert John
Hillery, Paul Stuart
Hillstrom, Warren W
Hoffman, Michael K
Hoffsommer, John C
Homberg, Otto Albert
Hornstein, Irwin
Hosmane, Ramachandra Sadashiv
Hoster, Donald Paul
Hsia, Mong Tseng Stephen
Hyatt, Asher Angel
Inscoe, May Nilson
Isaacs, Tami Yvette
Jacobson, Arthur E
Jacobson, Martin
James, John Cary
Jaouni, Taysir M
Jarvis, Bruce B
Jenkins, Mamie Leah Young
Jerina, Donald M
Karpetsky, Timothy Paul
Karten, Marvin J
Kayser, Richard Francis
Keefer, Larry Kay
Khan, Hameed A
Kleinspehn, George Gehret
Kovach, Eugene George
Kreysa, Frank Joseph
Krutzsch, Henry C
Kutik, Leon
Leak, John Clay, Jr
Lednicer, Daniel
Levy, Joseph Benjamin
Lewis, Cameron David
Liebman, Joel Fredric
Linder, Seymour Martin
Longenecker, William Hilton
McElroy, Wilbur Renfrew
McGuire, Francis Joseph
Markey, Sanford Philip
Marquez, Victor Esteban
Martenson, Russell Eric
Massie, Samuel Proctor
May, Willie Eugene
Mazzocchi, Paul Henry
Michejda, Christopher Jan
Mills, Frank D
Misra, Renuka
Morgan, Charles Robert
Moschel, Robert Carl
Murch, Robert Matthews
Murr, Brown L, Jr
Muschik, Gary Mathew
Naff, Marion Benton
Neta, Pedatsur
Nickon, Alex
Novak, Thaddeus John
Oliverio, Vincent Thomas
Paull, Kenneth Dywain
Pitha, Josef
Pohland, Albert
Pollack, Ralph Martin
Pomerantz, Irwin Herman
Posner, Gary Herbert
Preusch, Peter Charles
Raksis, Joseph W
Ralapati, Suresh
Reynolds, Kevin A
Rice, Kenner Cralle
Ridgway, Robert Worrell
Robinson, Cecil Howard
Rodewald, Lynn B
Rosenblatt, David Hirsch
Rosenblum, Annette Tannenholz
Rotherham, Jean
Rowell, Charles Frederick
Rzeszotarski, Waclaw Janusz
Saroff, Harry Arthur
Sayer, Jane M
Schechter, Milton Seymour
Schiffmann, Elliot
Schroeder, Michael Allan
Schwarz, Meyer
Shapiro, Robert Howard
Silversmith, Ernest Frank
Simmons, Thomas Carl
Simons, Samuel Stoney, Jr
Smedley, William Michael
Smith, Betty F
Smith, William Owen
Snader, Kenneth Means
Spande, Thomas Frederick
Sphon, James Ambrose
Sripada, Pavanaram Kameswara
Stalling, David L
Steck, Edgar Alfred
Stein, Harvey Philip
Steinman, Harry Gordon
Strier, Murray Paul
Sweeting, Linda Marie
Talbert, Preston Tidball
Tamminga, Carol Ann
Terry, Paul H
Thompson, Malcolm J
Timony, Peter Edward
Titus, Elwood Owen
Tolgyesi, Eva
Torrence, Paul Frederick
Townsend, Craig Arthur
Trimmer, Robert Whitfield
Tsai, Lin
Valega, Thomas Michael
Veitch, Fletcher Pearre, Jr
Vestling, Martha Meredith
Vitullo, Victor Patrick
Waters, Rolland Mayden
Watthey, Jeffrey William Herbert
Weinberger, Lester
Werber, Frank Xavier
Weser, Don Benton
Whaley, Wilson Monroe
Wheeler, James William, Jr
White, Emil Henry
Willette, Robert Edmond
Winestock, Claire Hummel
Wingrove, Alan Smith
Wirth, Peter James
Witkop, Bernhard
Wood, Louis L
Woods, Charles William
Wubbels, Gene Gerald
Yagi, Haruhiko
Yarbrough, Arthur C, Jr
Yeh, Lai-Su Lee
Yoho, Clayton W
Zaczek, Norbert Marion
Zdravkovich, Vera
Zehner, Lee Randall
Ziffer, Herman

MASSACHUSETTS

Aberhart, Donald John
Ahearn, James Joseph, Jr
Ahern, David George
Allen, Duff Shederic, Jr
Allen, Malwina I
Anselme, Jean-Pierre L M
Atkinson, Edward Redmond
Auld, David Stuart
Bader, Henry
Bannister, William Warren
Basdekis, Costas H
Bearden, William Harlie
Becker, Ernest I
Bender, Howard Sanford
Bennett, Ovell Francis
Berchtold, Glenn Allen
Berman, Elliot
Biemann, Klaus
Bluhm, Aaron Leo
Boehm, Paul David
Bornstein, Joseph
Bragole, Robert A
Browne, Douglas Townsend
Browne, Sheila Ewing
Buchi, George
Buckler, Sheldon A
Burson, Sherman Leroy, Jr
Byrnes, Eugene William
Carpino, Louis A
Caspi, Eliahu
Chang, Frank C
Chang, Kuang-Chou
Chapin, Earl Cone
Cheema, Zafarullah K
Chipman, Wilmon B
Clagett, Donald Carl
Clapp, Richard Crowell
Coffin, Perley Andrews
Cohen, Merrill
Cohen, Saul G
Cohen, Saul Mark
Comeford, Lorrie Lynn
Cook, Michael Miller
Coplan, Myron J
Corey, Elias James
Coull, James M
Coury, Arthur Joseph
Crawford, Jean Veghte
Danheiser, Rick Lane
Davis, Robert Bernard
Dehal, Shangara Singh
Demember, John Raymond
Dixon, Brian Gilbert
Doering, William von Eggers
Donoghue, John Timothy
Drumm, Manuel Felix
Dunny, Stanley
Ehntholt, Daniel James
Ehret, Anne
Erickson, Karen Louise
Fessler, William Andrew
Fetscher, Charles Arthur
Figueras, John
Filer, Crist Nicholas
Finkelstein, Manuel
Foye, William Owen
Frank, Jean Ann
Freedman, Harold Hersh
Garner, Albert Y
Georgian, Vlasios
Gerteisen, Thomas Jacob
Giering, Warren Percival
Girard, Francis Henry
Goldberg, Gershon Morton
Gomez, Ildefonso Luis
Gorbunoff, Marina J
Gore, William Earl
Gounaris, Anne Demetra
Grasshoff, Jurgen Michael
Gray, Nancy M
Greene, Frederick Davis, II
Haber, Stephen B
Haensel, Vladimir
Hagopian, Miasnig
Hall, George E
Hansen, David Elliott
Hathaway, Susan Jane
Hearn, Michael Joseph
Hendrickson, James Briggs
Herz, Matthew Lawrence
Hixson, Stephen Sherwin
Hodes, William L
Hodgdon, Russell Bates
Holick, Sally Ann
Hoover, M Frederick
Howell, David Moore
Hurd, Richard Nelson
Idelson, Martin
Isaks, William H
Israel, Stanley C
Jachimowicz, Felek
Jahngen, Edwin Georg Emil, Jr
Jarret, Ronald Marcel
Jennings, Bojan Hamlin
Jiang, Jack Bau-Chien
Jones, Elmer Everett
Jones, Guilford, II
Kafrawy, Adel
Kaller, Brian Francis
Kantor, Simon William
Keehn, Philip Moses
Kelly, Thomas Ross
Kemp, Daniel Schaeffer
Kenley, Richard Alan
Keough, John Henry
Khorana, Har Gobind
Kishi, Yoshito
Klibanov, Alexander M
Knapczyk, Jerome Walter
Knapp, Robert Lester
Kramer, Charles Edwin
Krieger, Jeanne Kann
Krueger, Robert A
Krull, Ira Stanley
Lada, Arnold
Lapkin, Milton
LaSala, Edward Francis
Laufer, Daniel A
Lee, Kang In
Lenz, George Richard
Lenz, Robert William
Le Quesne, Philip William
Lichtin, Norman Nahum
Lillya, Clifford Peter
Long, Alan K
Lowry, Nancy
Lowry, Thomas Hastings
Macaione, Domenic Paul
McEwen, William Edwin
McGrath, Michael Glennon
McMaster, Paul D
Macnair, Richard Nelson
McWhorter, Earl James
Magarian, Charles Aram
Maniscalco, Ignatius Anthony
Margolin, Alexey L
Markees, Diether Gaudenz
Markgraf, J(ohn) Hodge
Masamune, Satoru
Medalia, Avrom Izak
Mehta, Avinash C
Mehta, Mahendra
Miliora, Maria Teresa
Miller, Bernard
Miller, Bernard
Moser, William Ray
Moses, Ronald Elliot
Mowery, Dwight Fay, Jr
Nickerson, Richard G
Nussbaum, Alexander Leopold
Obermayer, Arthur S
Ocain, Timothy Donald
Ofner, Peter
O'Rell, Dennis Dee
Orphanos, Demetrius George
Pandya, Ashish
Parham, Marc Ellous
Pars, Harry George
Pavlik, James William
Pavlos, John
Pearson, Myrna Schmidt
Perkins, Janet Sanford
Peterson, Janet Brooks
Pian, Charles Hsueh Chien
Pike, Ronald Marston
Pinkus, Jack Leon
Piper, James Underhill
Powers, Donald Howard, Jr
Quin, Louis DuBose
Rademacher, Leo Edward
Rausch, Marvin D
Razdan, Raj Kumar
Rebek, Julius, Jr
Refojo, Miguel Fernandez
Renfroe, Harris Burt
Rivin, Donald
Roberts, Francis Donald
Rogers, Howard Gardner
Rosenberg, Joseph
Rosenblum, Myron
Rosenfeld, Stuart Michael
Rosowsky, Andre
Rossitto, Conrad
Rowe, Paul E
Rubinstein, Harry
Sahatjian, Ronald Alexander
Salamone, Joseph C
Samour, Carlos M
Santer, James Owen
Sardella, Dennis Joseph
Scanio, Charles John Vincent
Schlaikjer, Carl Roger
Schlein, Herbert
Schrock, Richard Royce
Schuler, Robert Frederick
Scott, Lawrence Tressler
Serra, Jerry M
Sheehan, John Clark
Shuford, Richard Joseph
Simon, Myron Sydney
Smith, William Edward
Snider, Barry B
Soffer, Milton David
Sosinsky, Barrie Alan
Stark, James Cornelius
Stein, Samuel H
Stemniski, John Roman
Stillwell, Richard Newhall
Stolow, Robert David
Strem, Michael Edward
Stuhl, Louis Sheldon
Sullivan, Charles Irving
Takman, Bertil Herbert
Taylor, Lloyd David
Thomas, George Richard
Thomas, Peter
Towns, Donald Lionel
Trachtenberg, Edward Norman
Tramondozzi, John Edmund
Trotz, Samuel Isaac
Truesdale, Larry Kenneth
Van der Burg, Sjirk
Van Eikeren, Paul
Vanelli, Ronald Edward
Viola, Alfred
Vogel, George
Vouros, Paul
Waller, David Percival
Walter, Henry Alexander
Wang, Chi-Hua
Wang, Nancy Yang
Warner, John Charles
Warner, Philip Mark
Watterson, Arthur C, Jr
Webster, Eleanor Rudd
Weininger, Stephen Joel
Wentworth, Stanley Earl
Westheimer, Frank Henry
Whitesides, George McClelland
Whitney, Robert Byron
Whitney, Thomas Allen
Wick, Emily Lippincott
Williams, John Russell
Williamson, Kenneth L(ee)
Wilson, Angus
Wotiz, Herbert Henry
Young, Richard L
Zavisza, Daniel Maxmillian
Ziomek, Carol A

MICHIGAN

Ahn, Kyunghye
Anderson, Amos Robert
Anderson, Hugh Verity
Anderson, Theodore
Andrewes, Arthur George
Arrington, Jack Phillip
Ashe, Arthur James, III
Atwell, William Henry
Augelli-Szafran, Corinne E
Bailey, Donald Leroy
Bajzer, William Xavier
Baker, Robert Henry
Bannister, Brian
Barry, Roger Donald
Beal, Philip Franklin, III

Belmont, Daniel Thomas
Berman, Ellen Myra
Berndt, Donald Carl
Blair, Etcyl Howell
Blankespoor, Ronald Lee
Blankley, Clifton John
Blatchford, John Kerslake
Blecker, Harry Herman
Boyer, Rodney Frederick
Braidwood, Clinton Alexander
Bremmer, Bart J
Brieger, Gottfried
Brodasky, Thomas Francis
Brubaker, Carl H, Jr
Brust, Harry Francis
Bundy, Gordon Leonard
Burgert, Bill E
Butler, Donald Eugene
Caron, E(dgar) Louis
Chang, Chi Kwong
Chen, Huai Gu
Chen, Ming Chih
Cobler, John George
Cody, Wayne Livingston
Compere, Edward L, Jr
Connor, David Thomas
Cook, Paul Laverne
Cook, Richard James
Coward, James Kenderdine
Cox, David Buchtel
Craig, Winston John
Creger, Paul LeRoy
Crump, John William
Culbertson, Townley Payne
Curtis, Myron David
Czarnik, Anthony William
Darby, Nicholas
Davis, Pauls
DeCamp, Mark Rutledge
Deering, Carl F
Delia, Thomas J
De Wall, Gordon
Dewitt, Sheila Hobbs
DeYoung, Jacob J
Dilling, Wendell Lee
Domagala, John Michael
Dorman, Linneaus Cuthbert
Dreyfuss, M(ax) Peter
Dugan, LeRoy, Jr
Dusenbery, Ruth Lillian
Ege, Seyhan Nurettin
Elrod, David Wayne
Farnum, Donald G
Ferrell, William James
Fisher, Jed Freeman
Ford, Dwain
Forsblad, Ingemar Bjorn
Frawley, Nile Nelson
French, James C
Frye, Cecil Leonard
Furey, Robert Lawrence
Gadwood, Robert Charles
Gammill, Ronald Bruce
Ginsburg, David
Goel, Om Prakash
Goetz, Rudolph W
Gorden, Berner J
Grabiel, Charles Edward
Graham, John
Greenfield, John Charles
Gregg, David Henry
Grostic, Marvin Ford
Hall, Charles Mack
Hall, Richard Harold
Hamlin, William Earl
Harmon, Robert E
Hart, Donald John
Hart, Harold
Hartgerink, Ronald Lee
Hartman, Robert John
Hartwell, George E
Hauxwell, Ronald Earl
Havel, James Joseph
Hays, Donald R
Heinrikson, Robert L
Heins, Conrad F
Hennis, Henry Emil
Herbener, Roland Eugene
Herr, Ross Robert
Herrinton, Paul Matthew
Hester, Jackson Boling, Jr
Heyd, William Ernst
Heyman, Duane Allan
Hinkamp, Paul Eugene, (II)
Hoeksema, Herman, Jr
Holubka, Joseph Walter
Horwitz, Jerome Philip
Hosler, John Frederick
Howe, Norman Elton, Jr
Howe, William Jeffrey
Howell, Bob A
Hsi, Richard S P
Huang, Che C
Hubbard, James Stuart
Huber, Joel E
Hupe, Donald John
Hylander, David Peter
Iffland, Don Charles
Jackson, James Edward
Jackson, Larry Lynn
Johnson, Carl Randolph
Johnson, Roy Allen
Jones, Frank Norton
Jones, Giffin Denison

Kagan, Fred
Kaltenbronn, James S
Kan, Peter Tai Yuen
Kang, Uan Gen
Karabatsos, Gerasimos J
Kelly, Robert Charles
Kenny, David Herman
Kim, Yungki
Knapp, Gordon Grayson
Komarmy, Julius Michael
Korcek, Stefan
Kurtz, Richard Robert
Labana, Santokh Singh
Laine, Richard Mason
Langdon, William Keith
Larsen, Eric Russell
Larson, G(ustav) Olof
LeBel, Norman Albert
LeGoff, Eugene
Lewandos, Glenn S
Lewenz, George F
Logue, Marshall Woford
Longone, Daniel Thomas
Lorand, John Peter
Love, Jim
Lustgarten, Ronald Krisses
Lyster, Mark Allan
McCall, John Michael
McGregor, Stanley Dane
Maddox, V Harold, Jr
Malchick, Sherwin Paul
Manhart, Joseph Heritage
Marino, Joseph Paul
Marquardt, Roland Paul
Massingill, John Lee, Jr
Mathison, Ian William
Mattano, Leonard August
Meister, Peter Dietrich
Mendenhall, George David
Messing, Sheldon Harold
Mich, Thomas Frederick
Miller, Susan Mary
Mills, Jack F
Mixan, Craig Edward
Moffett, Robert Bruce
Moll, Harold Wesbrook
Moore, Richard Anthony
Morrison, Glenn C
Mungall, William Stewart
Muntz, Ronald Lee
Myers, Donald Royal
Nader, Bassam Salim
Nagler, Robert Carlton
Narayan, Ramani
Narayan, Tv Lakshmi
Nestrick, Terry John
Neumann, Fred William
Newcomb, Martin
Nienhouse, Everett J
Nordin, Ivan Conrad
Nordstrom, J David
Nowak, Robert Michael
Nummy, William Ralph
Otterbacher, Eric Wayne
Ouderkirk, John Thomas
Overberger, Charles Gilbert
Parks, Terry Everett
Patton, John Thomas, Jr
Pearlman, Bruce A
Pearson, William H
Peery, Clifford Young
Pence, Leland Hadley
Petersen, Quentin Richard
Pews, Richard Garth
Pino, Lewis Nicholas
Pizzini, Louis Celeste
Platt, Alan Edward
Prakash, D Kash
Przybytek, James Theodore
Raban, Morton
Raley, Charles Francis, Jr
Ramsay, Ogden Bertrand
Rand, Cynthia Lucille
Rathke, Michael William
Rausch, Douglas Alfred
Reid, Stanley Lyle
Reusch, William Henry
Roth, Jerome A
Ryan, John William
Rycheck, Mark Rule
Ryntz, Rose A
Sacks, Clifford Eugene
Salmond, William Glover
Sandel, Vernon Ralph
Sarge, Theodore William
Schaaf, Robert Lester
Schaap, A Paul
Scheidt, Francis Matthew
Schmidt, Donald L
Schneider, John Arthur
Schroeder, William, Jr
Schumann, Edward Lewis
Scott, Allen
Shirley, Robert Louis
Skorcz, Joseph Anthony
Smith, Harry Andrew
Smith, Peter A(lan) S(omervail)
Staklis, Andris A
Stern, Robert Louis
Stevens, Violete L
Stille, John Robert
Stimson, Miriam Michael
Stratton, Charlotte Dianne
Stright, Paul Leonard

Strom, Robert Michael
Su, George Chung-Chi
Sullivan, John M
Sweeley, Charles Crawford
Swisher, Joseph Vincent
Swovick, Melvin Joseph
Tabor, Theodore Emmett
Tanis, Steven Paul
Taylor, Stephen Keith
Thomas, Richard Charles
Timko, Joseph Michael
Tomalia, Donald Andrew
Topliss, John G
Turner, Robert Lawrence
Tyler, Leslie J
Van Rheenen, Verlan H
Wagner, John Garnet
Wagner, Peter J
Walker, Jerry Arnold
Walles, Wilhelm Egbert
Warner, Donald Theodore
Warpehoski, Martha Anna
Weiner, Steven Allan
Weisbach, Jerry Arnold
Werbel, Leslie Morton
Westland, Roger D(ean)
Whaley, Howard Arnold
White, David Raymond
White, Jerry Eugene
Wickrema Sinha, Asoka J
Wierenga, Wendell
Williamson, Jerry Robert
Wise, Lawrence David
Wiseman, John R
Woo, P(eter) W(ing) K(ee)
Work, Stewart W
Woster, Patrick Michael
Yankee, Ernest Warren
Yeung, Patrick Pui-hang
Young, David Caldwell
Youngdale, Gilbert Arthur
Zabriskie, John Lansing, Jr
Zemlicka, Jiri

MINNESOTA
Abraham, Robert Thomas
Abul-Hajj, Jusuf J
Agre, Courtland LeVerne
Ames, Matthew Martin
Andrus, Milton Henry, Jr
Banitt, Elden Harris
Barks, Paul Allan
Baude, Frederick John
Baum, Burton Murry
Beebe, George Warren
Benham, Judith Laureen
Bird, Charles Norman
Boaz, David Paul
Brunelle, Thomas E
Cadotte, John Edward
Caple, Ronald
Carberry, Edward Andrew
Carlin, Charles Herrick
Carlson, Janet Lynn
Carlson, Robert M
Chaffin, Tommy L
Christian, Curtis Gilbert
Clemens, Lawrence Martin
Cook, Jack E
Crooks, Stephen Lawrence
Davis, Eugenia Asimakopoulos
Dickie, John Peter
Dodson, Raymond Monroe
Dybvig, Douglas Howard
Egberg, David Curtis
Eian, Gilbert Lee
Ellefson, Ralph Donald
Ellis, John Emmett
Erickson, John Gerhard
Esmay, Donald Levern
Etter, Margaret Cairns
Evensen, Thomas James
Farnum, Bruce Wayne
Farnum, Sylvia A
Fenton, Stuart William
Frank, William Charles
Fridinger, Tomas Lee
Fuchsman, Charles H(erman)
Fulmer, Richard W
Gassman, Paul G
Goon, David James Wong
Gray, Gary Ronald
Guehler, Paul Frederick
Harriman, Benjamin Ramage
Holmen, Raymond Emanuel
Holum, John Robert
Horner, James William, Jr
Hoye, Thomas Robert
Hughes, Mark
Jochman, Richard Lee
Johnston, Manley Roderick
Kariv-Miller, Essie
Klein, William Arthur
Kovacic, Joseph Edward
Kowanko, Nicholas
Krogh, Lester Christensen
Kropp, James Edward
Langsjoen, Arne Nels
Larson, Eric George
Leete, Edward
Li, Wu-Shyong
Lockwood, Robert Greening
Lu, Shih-Lai
Lucast, Donald Hurrell

Meeks, Benjamin Spencer, Jr
Mendel, Arthur
Mohrig, Jerry R
Morath, Richard Joseph
Mutsch, Edward L
Nelson, John Daniel
Newman, Norman
Noland, Wayland Evan
Opie, Joseph Wendell
Ostercamp, Daryl Lee
Patel, Kalyanji U
Pearson, Wesley A
Petersen, Robert J
Pletcher, Wayne Albert
Punderson, John Oliver
Ree, Buren Russel
Reever, Richard Eugene
Reich, Charles
Reid, Thomas S
Riehl, Mary Agatha
Rislove, David Joel
Roach, J Robert
Robertson, Jerry Earl
Robins, Janis
Runquist, Olaf A
Scherrer, Robert Allan
Schlenk, Hermann
Schultz, Robert John
Seffl, Raymond James
Shier, Wayne Thomas
Sjolander, John Rogers
Skoog, Ivan Hooglund
Snow, John Elbridge
Snyder, William Richard
Sobieski, James Fulton
Sorensen, David Perry
Spessard, Gary Oliver
Spliethoff, William Ludwig
Stepan, Alfred Henry
Stephens, Dale Nelson
Stocker, Fred Butler
Stricklin, Buck
Sullivan, Betty J
Talbott, Richard Lloyd
Taylor, Charles William
Throckmorton, James Rodney
Tiers, George Van Dyke
Toren, George Anthony
Trepka, Robert Dale
Wear, Robert Lee
Weberg, Berton Charles
Weiss, Douglas Eugene
Wen, Richard Yutze
Werth, Richard George
White, Joe Wade
Williams, Edmond Brady
Wittman, William F
Wollner, Thomas Edward
Zollinger, Joseph LaMar

MISSISSIPPI
Alley, Earl Gifford
Bauer, Stewart Thomas
Bedenbaugh, Angela Lea Owen
Bedenbaugh, John Holcombe
Behr, Lyell Christian
Berry, Roy Alfred, Jr
Borne, Ronald Francis
Brent, Charles Ray
Brown, Kenneth Lawrence
Cain, Charles Eugene
Creed, David
Duffey, Donald Creagh
Elakovich, Stella Daisy
Fisher, Thomas Henry
Germany, Archie Herman
Greever, Joe Carroll
Hedin, Paul A
Heimer, Norman Eugene
Johnson, Carl Erick
McCormick, Charles Lewis, III
Mathias, Lon Jay
Minyard, James Patrick
Noe, Eric Arden
Panetta, Charles Anthony
Pittman, Charles U, Jr
Scott, Robert Blackburn, Jr
Seymour, Raymond B(enedict)
Sindelar, Robert D
Stefani, Andrew Peter
Thames, Shelby Freland
Thompson, Alonzo Crawford
Van Aller, Robert Thomas
Wescott, Lyle DuMond, Jr
Wierengo, Cyril John, Jr

MISSOURI
Adams, Kenneth Howard
Adams, Steven Paul
Ahmad, Moghisuddin
Alt, Gerhard Horst
Anagnostopoulos, Constantine E
Armbruster, Charles William
Armstrong, Daniel Wayne
Bachman, Gerald Lee
Baker, Joseph Willard
Balthazor, Terrell Mack
Bechtle, Gerald Francis
Bleeke, John Richard
Bodine, Richard Shearon
Bohanon, Joseph Terril
Bovy, Philippe R
Briner, Robert C
Brodmann, John Milton

Organic Chemistry (cont)

Carpenter, Sammy
Caughlan, John Arthur
Chapman, Douglas Wilfred
Chen, Yih-Wen
Chickos, James S
Clark, Frank S
Clark, Sidney Gilbert
Cole, Douglas L
Conkin, Robert A
Covey, Douglas Floyd
Dahl, William Edward
Dale, Wesley John
Darby, Joseph Raymond
Delaware, Dana Lewis
Dias, Jerry Ray
DiFate, Victor George
Dill, Dale Robert
Diuguid, Lincoln Isaiah
Dixon, Marvin Porter
Dub, Michael
Duchek, John Robert
Dutra, Gerard Anthony
Eime, Lester Oscar
Elliott, William H
Engel, James Francis
Faulk, Dennis Derwin
Freeman, Robert Clarence
Froemsdorf, Donald Hope
Gaertner, Van Russell
Garin, David L
Gaspar, Peter Paul
Gerhardt, Klaus Otto
Giarrusso, Frederick Frank
Glaser, Rainer
Glaspie, Peyton Scott
Godt, Henry Charles, Jr
Gokel, George William
Gordon, Annette Waters
Guthrie, David Burrell
Hammann, William Curl
Hancock, Anthony John
Hedrick, Ross Melvin
Heininger, S(amuel) Allen
Herber, John Frederick
Hobbs, Charles Floyd
Holm, Myron James
Hortmann, Alfred Guenther
Hyndman, Harry Lester
Jamieson, Norman Clark
Jason, Mark Edward
Kaiser, Edwin Michael
Kalota, Dennis Jerome
Keller, William John
Knox, Walter Robert
Koeltzow, Donald Earl
Koenig, Karl E
Krueger, Paul A
Ku, Audrey Yeh
Kurz, Joseph Louis
Lanson, Herman Jay
Lennon, Patrick James
Leventis, Nicholas
Li, Tao Ping
Liao, Tsung-Kai
Lipkin, David
Loeppky, Richard N
Lovett, Eva G
Ludwig, Frederick John, Sr
Lusskin, Robert Miller
McConaghy, John Stead, Jr
McCormick, John Pauling
McCoy, Layton Leslie
McHugh, Kenneth Laurence
MacQuarrie, Ronald Anthony
Malik, Joseph Martin
Malzahn, Ray Andrew
Mange, Franklin Edwin
Markowski, Henry Joseph
Matzner, Edwin Arthur
Moedritzer, Kurt
Moeller, Kevin David
Morris, Donald Eugene
Morse, Ronald Loyd
Mosher, Melvyn Wayne
Mottus, Edward Hugo
Mount, Ramon Albert
Munch, John Howard
Murray, Robert Wallace
Murrill, Evelyn A
Oftedahl, Marvin Loren
Owsley, Dennis Clark
Palmer, John Frank, Jr
Paulik, Frank Edward
Peacock, Val Edward
Phillips, Bruce Edwin
Pivonka, William
Plimmer, Jack Reynolds
Podrebarac, Eugene George
Popp, Frank Donald
Rabjohn, Norman
Raizman, Paula
Rath, Nigam Prasad
Ratts, Kenneth Wayne
Redmore, Derek
Richards, Charles Norman
Robinson, Donald Alonzo
Rogic, Milorad Mihailo
Rothrock, Thomas Stephenson
Rueppel, Melvin Leslie
Sabacky, M Jerome
Salivar, Charles Joseph
Sathe, Sharad Somnath
Schisla, Robert M
Schuck, James Michael
Schultz, John E
Schultz, Robert George
Schumacher, Ignatius
Schwarz, Richard
Searles, Scott, Jr
Sears, J Kern
Sharp, Dexter Brian
Sikorski, James Alan
Singleton, Tommy Clark
Slocombe, Robert Jackson
Smith, Lowell R
Smith, Ronald Gene
Snyder, Joseph Quincy
Solodar, Arthur John
Sotiriou-Leventis, Chariklia
Stary, Frank Edward
Stenseth, Raymond Eugene
Stout, Edward Irvin
Sweet, Frederick
Swisher, Robert Donald
Talbott, Ted Delwyn
Teng, James
Tompkins, Jay Allen
Tonkyn, Richard George
Vineyard, Billy Dale
Wagenknecht, John Henry
Wasson, Richard Lee
Watkins, Darrell Dwight, Jr
Whittle, Philip Rodger
Wilbur, James Myers, Jr
Wildi, Bernard Sylvester
Williams, Byron Lee, Jr
Wilson, James Dennis
Winicov, Murray William
Winter, Rudolph Ernst Karl
Woodbrey, James C
Worley, Jimmy Weldon
Wulfman, David Swinton
Zee-Cheng, Robert Kwang-Yuen
Zey, Robert L
Zienty, Ferdinand B

MONTANA

Craig, Arnold Charles
Jennings, Paul W
Juday, Richard Evans
Stewart, John Mathews

NEBRASKA

Baumgarten, Henry Ernest
Cavalieri, Ercole Luigi
Clark, Ronald David
Corbett, Michael Dennis
Cromwell, Norman Henry
Demuth, John Robert
George, Anne Denise
Gold, Barry Irwin
Gross, Michael Lawrence
Holmquist, Barton
Hulce, Martin R
Johnson, John Arnold
Jones, Lee Bennett
Kaufman, Don Allen
Kingsbury, Charles Alvin
Klein, Francis Michael
Laursen, Paul Herbert
Linn, Carl Barnes
Linstromberg, Walter William
Looker, James Howard
Nagel, Donald Lewis
O'Leary, Marion Hugh
Raha, Chitta Ranjan
Rasmussen, Russell Lee
Rieke, Reuben Dennis
Roark, James L
Ruyle, William Vance
Shearer, Greg Otis
Smith, David Hibbard
Takemura, Kaz H(orace)
Vennerstrom, Jonathan Lee
Wheeler, Desmond Michael Sherlock
Wood, James Kenneth

NEVADA

Emerson, David Winthrop
Fox, Neil Stewart
Kemp, Kenneth Courtney
Lightner, David A
Nazy, John Robert
Nelson, John Henry
Poziomek, Edward John
Rose, Charles Buckley
Seiber, James N
Smith, Robert Bruce
Sovocool, G Wayne
Titus, Richard Lee

NEW HAMPSHIRE

Andersen, Kenneth K
Balmer, Clifford Earl
Beck, Mae Lucille
Bohm, Howard A
Coburn, Everett Robert
Cotter, Robert James
Creagh-Dexter, Linda T
Curphey, Thomas John
Detweiler, W Kenneth
Donaruma, L Guy
English, Jackson Pollard
Gartner, Edward A
Gorham, William Franklin
Gribble, Gordon W
Harris, Thomas David
Hughes, Russell Profit
Jones, Paul Raymond
Lemal, David M
Martin, William Butler, Jr
Morrison, James Daniel
Neil, Thomas C
Ort, Morris Richard
Pratt, Richard J
Shafer, Paul Richard
Spencer, Thomas A
Springsteen, Kathryn Rose Mooney
Stewart, Roberta A
Tewksbury, Charles Isaac
Upham, Roy Herbert
Walling, Cheves (Thomson)
Weisman, Gary Raymond
Woodbury, Richard Paul
Worman, James John
Zager, Ronald

NEW JERSEY

Aaronoff, Burton Robert
Abramovici, Miron
Achari, Raja Gopal
Adams, Phillip
Adams, Richard Ernest
Addor, Roger Williams
Afonso, Adriano
Aguiar, Adam Martin
Alekman, Stanley L
Allen, Richard Charles
Altman, Lawrence Jay
Anderson, Lowell Ray
Anderson, Paul LeRoy
Anderson, Robert Christian
Andose, Joseph D
Andrade, John Robert
Angel, Henry Seymour
Angier, Robert Bruce
Ariyan, Zaven S
Armbruster, David Charles
Asato, Goro
Ashby, Bruce Allan
Ashcraft, Arnold Clifton, Jr
Atwater, Beauford W
Atwater, Norman Willis
Auerbach, Andrew Bernard
Auerbach, Victor
Augustine, Robert Leo
Bach, Frederick L
Badin, Elmer John
Bagli, Jenanbux Framroz
Bair, Kenneth Walter
Ballina, Rudolph August
Bamberger, Curt
Barclay, Robert, Jr
Barile, George Conrad
Barnabeo, Austin Emidio
Barnett, Ronald E
Barringer, Donald F, Jr
Batcho, Andrew David
Battisti, Angelo James
Bauer, Frederick William
Bauman, Robert Andrew
Bavley, Abraham
Baylouny, Raymond Anthony
Beinfest, Sidney
Beispiel, Myron
Bellis, Harold E
Berg, Jeffrey Howard
Berkelhammer, Gerald
Bernholz, William Francis
Berry, William Lee
Bertelo, Christopher Anthony
Bertz, Steven Howard
Besso, Michael M
Bezwada, Rao Srinivasa
Blake, Jules
Bloom, Allen
Bluestein, Allen Channing
Bluestein, Bernard Richard
Bluestein, Claire
Bohrer, James Calvin
Boikess, Robert S
Boothe, James Howard
Borah, Kripanath
Borowitz, Grace Burchman
Bose, Ajay Kumar
Bosniack, David S
Bouboulis, Constantine Joseph
Bowman, Lewis Wilmer
Bowman, Robert Mathews
Boyle, Richard James
Boyle, William Johnston, Jr
Bradstreet, Raymond Bradford
Brady, Thomas E
Brand, William Wayne
Brandman, Harold A
Brennan, James A
Breslow, Ronald
Brill, William Franklin
Brons, Cornelius Hendrick
Brooks, Robert Alan
Brown, John Angus
Brown, Richard Emery
Buck, Carl John
Bullock, Francis Jeremiah
Bulusu, Suryanarayana
Burbaum, Beverly Wolgast
Burlant, William Jack
Burton, Gilbert W
Cardis, Angeline Baird
Carrock, Frederick E
Cavender, Patricia Lee
Cevasco, Albert Anthony
Chamberlin, Earl Martin
Chandross, Edwin A
Chang, Hsien-Hsin
Chapas, Richard Bernard
Charbonneau, Larry Francis
Chaudhri, Safee U
Chen, Shuhchung Steve
Chiao, Wen Bin
Chibnik, Sheldon
Chien, Ping-Lu
Christenson, Philip A
Church, John Armistead
Coan, Stephen B
Cohen, Murray Samuel
Cohen, Noal
Colicelli, Elena Jeanmarie
Comai-Fuerherm, Karen
Conciatori, Anthony Bernard
Condon, Michael Edward
Cone, Conrad
Controulis, John
Cook, Alan Frederick
Cooke, Robert Sanderson
Coombs, Robert Victor
Cordon, Martin
Crosby, Guy Alexander
Cruickshank, P A
Cummins, Richard Williamson
Cunningham, Earlene Brown
Cutler, Frank Allen, Jr
Dain, Jeremy George
Davies, Richard Edgar
Dawson, Arthur Donovan
Dayan, Jason Edward
Deem, Mary Lease
Degginger, Edward R
Delaney, Edward Joseph
Delmonte, David William
Denney, Donald Berend
Derieg, Michael E
DeStevens, George
Diassi, Patrick Andrew
DiBella, Eugene Peter
Diegnan, Glenn Alan
Draper, Richard William
Dreby, Edwin Christian, III
Dunkel, Morris
Durante, Anthony Joseph
Durette, Philippe Lionel
Dursch, Friedrich
Eastman, David Willard
Eby, John Martin
Eck, John Clifford
Eckler, Paul Eugene
Edelson, Edward Harold
Edelstein, Harold
Egli, Peter
Ehrenfeld, Robert Louis
Ehrlich, Robert
Emert, Jack Isaac
Essery, John M
Evers, William John
Fagan, Paul V
Fahey, John Leonard
Fahrenholtz, Kenneth Earl
Fahrenholtz, Susan Roseno
Fand, Theodore Ira
Farina, Thomas Edward
Farrar, Grover Louis
Fath, Joseph
Feig, Gerald
Feldman, Martin Louis
Ferstandig, Louis Lloyd
Finkelstein, Jacob
Finley, Joseph Howard
Firth, William Charles, Jr
Fishman, David H
Fishman, Morris
Fitchett, Gilmer Trower
Fitzgerald, Patrick Henry
Florey, Klaus
Fobare, William Floyd
Fono, Andrew
Ford, Neville Finch
Forster, Warren Schumann
Fost, Dennis L
Francis, John Elsworth
Frankenfeld, John William
Franz, Curtis Allen
Frey, Sheldon Ellsworth
Fry, John Sedgwick
Fryer, Rodney Ian
Gaffney, Barbara Lundy
Gal, George
Gall, Martin
Gallopo, Andrew Robert
Gander, Robert Johns
Ganguly, Ashit K
Gans, Manfred
Garty, Kenneth Thomas
Garwood, William Everett
Genzer, Jerome Daniel
Gerber, Samuel Michael
Gerecht, J Fred
Gessler, Albert Murray
Giacobbe, Thomas Joseph
Gierer, Paul L
Gilbert, Allan Henry
Gilman, Norman Washburn
Girijavallabhan, Viyyoor Moopil
Gladstone, Harold Maurice
Glamkowski, Edward Joseph

Glickman, Samuel Arthur
Glidden, Richard Mills
Goeke, George Leonard
Gold, Elijah Herman
Goldstein, Albert
Gorbaty, Martin Leo
Gottesman, Roy Tully
Gould, Kenneth Alan
Grabowski, Edward Joseph John
Graham, Harold Nathaniel
Grams, Gary Wallace
Granito, Charles Edward
Gray, Frederick William
Green, Joseph
Green, Michael John
Greenberg, Arthur
Grenda, Victor J
Griffin, William Dallas
Griffith, Martin G
Gross, Leonard
Grosso, John A
Groves, John Taylor, III
Guida, Wayne Charles
Guiducci, Mariano A
Gullo, Vincent Philip
Gunberg, Paul F
Haag, Werner O
Haddon, Robert Cort
Hagmann, William Kirk
Halfon, Marc
Halgren, Thomas Arthur
Hall, John B
Hall, Luther Axtell Richard
Hall, Stan Stanley
Halpern, Donald F
Hammond, Willis Burdette
Hannah, John
Hansen, Holger Victor
Hansen, Ralph Holm
Hanson, James Edward
Harding, Maurice James Charles
Hardtmann, Goetz E
Harris, Guy H
Harvey, Robert Joseph
Hatch, Charles Eldridge, III
Hazen, George Gustave
Heffron, Peter John
Heiba, El-Ahmadi Ibrahim
Helsley, Grover Cleveland
Herman, Daniel Francis
Hetzel, Donald Stanford
Heyman, Karl
High, LeRoy Bertolet
Hirsch, Arthur
Hirsch, Jerry Allan
Hirshman, Justin Leonard
Hlavacek, Robert John
Ho, Chi-Tang
Hochstetler, Alan Ray
Hoffman, Dorothea Heyl
Hoffman, Roger Allen
Horowitz, Hugh H(arris)
Hort, Eugene Victor
Horton, Robert Louis
Houlihan, William Joseph
Huber, Melvin Lefever
Humber, Leslie George
Humphreys, Robert William Riley
Hundert, Murray Bernard
Hung, Paul Ling-Kong
Hymans, William E
Idol, James Daniel, Jr
Inglessis, Criton George S
Jacob, Theodore August
Jacobson, Stephen Ernest
Jaruzelski, John Janusz
Jenkins, Alfred Martin
Jennings, Carl Anthony
Jensen, Norman P
Jernow, Jane L
Ji, Sungchul
Jirkovsky, Ivo
Jones, Howard
Jones, Maitland G
Jones, Martha Ownbey
Jones, Roger Alan
Jones, William Howry
Kaback, Stuart Mark
Kaeding, Warren William
Kahn, Donald Jay
Kalm, Max John
Kaminski, James Joseph
Kaminski, Joan M
Kaniecki, Thaddeus John
Kaplan, Martin L
Karady, Sandor
Karl, Curtis Lee
Karmas, George
Kasparek, Stanley Vaclav
Kaufman, Harold Alexander
Keaveney, William Patrick
Kees, Kenneth Lewis
Keith, Dennis Dalton
Kellgren, John
Keogh, Michael John
Kern, Werner
Kierstead, Richard Wightman
Kim, Benjamin K
King, John A(lbert)
Klaas, Nicholas Paul
Klaubert, Dieter Heinz
Klein, Richard M
Klemann, Lawrence Paul
Klingsberg, Erwin

Klinke, David J
Knapp, Malcolm Hammond
Kolc, Jaroslav Jerry F
Konecky, Milton Stuart
Konort, Mark D
Konzelman, Leroy Michael
Koonce, Samuel David
Kozikowski, Alan Paul
Kramer, George Mortimer
Kreeger, Russell Lowell
Kreft, Anthony Frank, III
Krishnamurthy, Subrahmanya
Krishnan, Gopal
Kristol, David Sol
Kroll, Emanuel
Kruh, Daniel
Kuder, James Edgar
Kumar, Suriender
Kumari, Durga
Kuntz, Irving
Kuo, Chan-Hwa
Kuritzkes, Alexander Mark
Kwartler, Charles Edward
Labows, John Norbert, Jr
Ladner, David William
Lakritz, Julian
Lam, Yiu-Kuen Tony
Lang, Philip Charles
Langley, Robert Charles
Largman, Theodore
Lau, Roland
Laufer, Robert J
Lazarus, Allan Kenneth
Leal, Joseph Rogers
Leary, Ralph John
Lee, Do-Jae
Lee, Shui Lung
Leeds, Morton W
Leeds, Norma S
Lefar, Morton Saul
Lehne, Richard Karl
Leong, William
Lerner, Lawrence Robert
Levi, Elliott J
Levine, Aaron William
Levy, Alan B
Lewis, Neil Jeffrey
Liang, Wei Chuan
Liauw, Koei-Liang
Liberles, Arno
Lies, Thomas Andrew
Liesch, Jerrold Michael
Linn, Bruce Oscar
Liu, Yu-Ying
Lo, Elizabeth Shen
Loffelman, Frank Fred
Login, Robert Bernard
Lohner, Donald J
Lorenz, Donald H
Love, George M
Luk, Kin-Chun C
Lyding, Arthur R
Lynch, Charles Andrew
McCaully, Ronald James
MacCoss, Malcolm
McDowell, Wilbur Benedict
McGahren, William James
McGinnis, James Lee
McKay, Donald Edward
McVey, Jeffrey King
Manhas, Maghar Singh
Marburg, Stephen
Marder, Herman Lowell
Mares, Frank
Margerison, Richard Bennett
Martin, Lawrence Leo
Masuelli, Frank John
Mathew, Chempolil Thomas
Matthews, Demetreos Nestor
Matzner, Markus
Maulding, Donald Roy
Meienhofer, Johannes Arnold
Meisel, Seymour Lionel
Meislich, Herbert
Melillo, David Gregory
Meyer, Victor Bernard
Miale, Joseph Nicolas
Milionis, Jerry Peter
Miner, Robert Scott, Jr
Mislow, Kurt Martin
Misner, Robert E
Mitchell, Thomas Owen
Mohacsi, Erno
Mohan, Arthur G
Moroni, Antonio
Moss, Robert Allen
Mrozik, Helmut
Mukai, Cromwell Daisaku
Munsell, Monroe Wallwork
Murray, John Joseph
Murray, Leo Thomas
Murty, Dasika Radha Krishna
Nafissi, Mehdi
Nager, Urs Felix
Naglieri, Anthony N
Naipawer, Richard Edward
Napier, Roger Paul
Naro, Paul Anthony
Nebel, Carl Walter
Neiswender, David Daniel
Nelson, John Archibald
Neustadt, Bernard Ray
Newland, Robert Joe
Newmark, Harold Leon

Noshay, Allen
Olechowski, Jerome Robert
Olin, Arthur David
Oliveto, Eugene Paul
Olmstead, William N(eergaard)
Olson, Gary Lee
Ondetti, Miguel Angel
Oppelt, John Christian
Osuch, Christopher Erion
Owens, Clifford
Oyler, Alan Richard
Palermo, Felice Charles
Palmere, Raymond M
Pandey, Ramesh Chandra
Panson, Gilbert Stephen
Papaioannou, Christos George
Pappalardo, Leonard Thomas
Pappas, James John
Parisek, Charles Bruce
Parker, William Lawrence
Patrick, James Edward
Paul, Rolf
Paustian, John Earle
Pavlin, Mark Stanley
Peake, Clinton J
Perkel, Robert Jules
Perlmutter, Howard D
Perry, Donald Dunham
Petrillo, Edward William
Pheasant, Richard
Pierson, William Grant
Pilkiewicz, Frank George
Pines, Seemon H
Pinto, Frank G
Pollack, Maxwell Aaron
Pollak, Kurt
Popkin, A(lexander) H
Popper, Thomas Leslie
Powell, Burwell Frederick
Prapas, Aristotle G
Price, Alson K
Priesing, Charles Paul
Prince, Martin Irwin
Puar, Mohindar S
Pugliese, Michael
Ramsden, Hugh Edwin
Rashidbaigi, Abbas
Raymond, Maurice A
Rebenfeld, Ludwig
Regna, Peter P
Reichle, Walter Thomas
Reichmanis, Elsa
Reider, Paul Joseph
Reilly, Eugene Patrick
Reimann, Hans
Riccobono, Paul Xavier
Rieger, Martin Max
Rigler, Neil Edward
Rinehart, Jay Kent
Robb, Ernest Willard
Robbins, Clarence Ralph
Roberts, Floyd Edward, Jr
Roberts, William John
Robertson, Dale Norman
Robin, Michael
Rodewald, Paul Gerhard, Jr
Rolston, Charles Hopkins
Roper, Robert
Rosan, Alan Mark
Rose, Ira Marvin
Rosegay, Avery
Rosen, Joseph David
Rosen, Marvin
Rosen, Perry
Rosen, William Edward
Rosenberg, Ira Edward
Rosenthal, Arnold Joseph
Ross, Lawrence James
Rossi, Robert Daniel
Roth, Heinz Dieter
Rovnyak, George Charles
Roy, Ram Babu
Ruby, Philip Randolph
Ruddy, Arlo Wayne
Rutenberg, Morton Wolf
Rutledge, Thomas Franklin
Ryzlak, Maria Teresa
Salvesen, Robert H
San Filippo, Joseph, Jr
Sauers, Ronald Raymond
Scanley, Clyde Stephen
Schipper, Edgar
Schlosberg, Richard Henry
Schmidle, Claude Joseph
Schonfeld, Edward
Schreiber, William Lewis
Schueler, Paul Edgar
Schulz, Donald Norman
Schwartz, Herbert
Schwartz, Jeffrey
Schwartzkopf, George, Jr
Schwarz, Hans Jakob
Schwarz, John Samuel Paul
Scott, Donald Albert
Scott, Eric James Young
Sepkoski, Joseph John, Sr
Sestanj, Kazimir
Sexsmith, David Randal
Shearer, Charles M
Sheats, John Eugene
Shechter, Leon
Sheikh, Maqsood A
Sherr, Allan Ellis
Shih, Jenn-Shyong

Shine, Robert John
Shinkai, Ichiro
Shulman, George
Shutske, Gregory Michael
Sidi, Henri
Siegel, Maurice L
Sieh, David Henry
Sifniades, Stylianos
Silvestri, Anthony John
Simmons, Jean Elizabeth Margaret
Simonoff, Robert
Singleton, Bert
Sisenwine, Samuel Fred
Siskin, Michael
Skotnicki, Jerauld S
Skoultchi, Martin Milton
Skovronek, Herbert Samuel
Skrypa, Michael John
Slade, Joel S
Slates, Harry Lovell
Slezak, Frank Bier
Slusarchyk, William Allen
Smith, Eileen Patricia
Smith, Elizabeth Melva
Soldati, Gianluigi
Spano, Francis A
Speers, Louise (Mrs Henry Croix)
Speert, Arnold
Spiegelman, Gerald Henry
Spohn, Ralph Joseph
Sprague, Peter Whitney
Stanaback, Robert John
Stange, Hugo
Stefancsik, Ernest Anton
Stein, Charles W C
Stein, Reinhardt P
Steinbach, Leonard
Steinberg, Eliot
Steinman, Martin
Steltenkamp, Robert John
Sternbach, Leo Hynryk
Stingl, Hans Alfred
Stockton, John Richard
Stone, Edward
Stone, Edward John
Storfer, Stanley J
Stork, Gilbert (Josse)
Struve, William Scott
Sturzenegger, August
Suciu, George Dan
Sugathan, Kanneth Kochappan
Sulzberg, Theodore
Susi, Peter Vincent
Taub, David
Taylor, Edward Curtis
Taylor, Gary N
Tencza, Thomas Michael
Tessler, Martin Melvyn
Thaler, Warren Alan
Theilheimer, William
Thompson, Hugh Walter
Tibbetts, Merrick Sawyer
Tilley, Jefferson Wright
Tolman, Richard Lee
Tomcufcik, Andrew Stephen
Tracy, David J
Trewella, Jeffrey Charles
Trifan, Daniel Siegfried
Tseng, Shin-Shyong
Tsong, Yun Yen
Turro, Nicholas John
Valiaveedan, George Devasia
Van Der Veen, James Morris
Van Order, Robert Bruce
Van Saun, William Arthur
Venturella, Vincent Steven
Vial, Theodore Merriam
Virgili, Luciano
Vlattas, Isidoros
Volante, Ralph Paul
Von, Isaiah
Vona, Joseph Albert
Vukov, Rastko
Wagner, Arthur Franklin
Wagner, Frank A, Jr
Walsh, John Paul
Wanat, Stanley Frank
Warfield, Peter Foster
Warren, Craig Bishop
Watt, William Russell
Watts, Daniel Jay
Webb, Philip Gilbert
Wei, Chung-Chen
Weigmann, Hans-Dietrich H
Weinberger, Harold
Weintraub, Leonard
Weintraub, Lester
Wellman, William Edward
Werner, Lincoln Harvey
Weston, Charles Alvin
Wheaton, Gregory Alan
Whitehurst, Darrell Duayne
Wilkinson, Raymond George
Willard, Paul Edwin
Williams, Thomas Henry
Willits, Charles Haines
Wilson, Armin Guschel
Wilson, Evelyn H
Wilson, Harold Frederick
Winslow, Field Howard
Witkowski, Joseph Theodore
Wittekind, Raymond Richard
Witz, Gisela
Wohl, Ronald A

Organic Chemistry (cont)

Wolf, Frank James
Wolf, Philip Frank
Wolff, Steven
Wong, Ching-Ping
Woodworth, Curtis Wilmer
Wu, Joseph Woo-Tien
Wu, Mu Tsu
Wu, Tse Cheng
Yale, Harry Louis
You, Kwan-sa
Young, Lewis Brewster
Young, Sanford Tyler
Zalay, Andrew W(illiam)
Zambito, Arthur Joseph
Zask, Arie
Ziegler, John Benjamin
Zinnes, Harold
Zoss, Abraham Oscar

NEW MEXICO
Amai, Robert Lin Sung
Auerbach, Irving
Benicewicz, Brian Chester
Brower, Kay Robert
Burns, Frank Bernard
Cahill, Paul A
Chapman, Robert Dale
Cheavens, Thomas Henry
Clark, Ronald Duane
Clough, Roger Lee
Coburn, Michael Doyle
Corcoran, George Bartlett, III
Corrigan, John Raymond
Cowan, John C
Curtice, Jay Stephen
Dahl, Alan Richard
Danen, Wayne C
Davis, Dennis Duval
Dorko, Ernest A
Evans, Latimer Richard
George, Raymond S
Guziec, Frank Stanley, Jr
Guziec, Lynn Erin
Hatch, Melvin (Jay)
Hoffman, Robert Vernon
Hollstein, Ulrich
Kjeldgaard, Edwin Andreas
Larson, Thomas E
Loftfield, Robert Berner
Lwowski, Walter Wilhelm Gustav
Marcy, Willard
Morton, John West, Jr
Nevin, Robert Stephen
Nielsen, Stuart Dee
Ogilby, Peter Remsen
Ott, Donald George
Papadopoulos, Eleftherios Paul
Shore, Fred L
Stinecipher, Mary Margaret
Wewerka, Eugene Michael
Whaley, Thomas Williams
Williams, Joel Mann, Jr
Zeigler, John Martin

NEW YORK
Abraham, Carl J
Acerbo, Samuel Nicholas
Ackerman, Hervey Winfield, Jr
Adduci, Jerry M
Addy, John Keith
Adler, George
Aftergut, Siegfried
Agosta, William Carleton
Alaimo, Robert J
Albrecht, Frederick Xavier
Albright, Jay Donald
Allen, Gary William
Allen, Lewis Edwin
Althuis, Thomas Henry
Altland, Henry Wolf
Altschul, Rolf
Anderson, Wayne Keith
Andrews, Mark Allen
Angelino, Norman J
Arcesi, Joseph A
Archie, William C, Jr
Arkell, Alfred
Armour, Eugene Arthur
Armstrong, William Lawrence
Arnowich, Beatrice
Attygalle, Athula B
Aviram, Ari
Axelrad, George
Axenrod, Theodore
Ayral-Kaloustian, Semiramis
Badding, Victor George
Baechler, Raymond Dallas
Bak, David Arthur
Balassa, Leslie Ladislaus
Baldwin, John E
Banay-Schwartz, Miriam
Bank, Shelton
Barany, Francis
Barkey, Kenneth Thomas
Baron, Arthur L
Barr, Donald Eugene
Barrett, Edward Joseph
Barringer, William Charles
Bass, Jon Dolf
Baum, George
Baum, Martin David
Baumgarten, Reuben Lawrence

Beavers, Dorothy (Anne) Johnson
Beavers, Leo Earice
Beck, Curt Werner
Bell, Malcolm Rice
Bell, Thomas Wayne
Bennett, James Gordy, Jr
Bent, Richard Lincoln
Benz, George William
Berdahl, Donald Richard
Berger, S Edmund
Bergmark, William R
Berry, E Janet
Bieron, Joseph F
Billman, John Henry
Bitha, Panayota
Blackman, Samuel William
Block, Eric
Blood, Charles Allen, Jr
Bluestein, Ben Alfred
Boeckman, Robert K, Jr
Boettger, Susan D
Bolon, Donald A
Bonvicino, Guido Eros
Booth, Robert Edwin
Borch, Richard Frederic
Borders, Donald B
Borowitz, Irving Julius
Boudakian, Max Minas
Box, Vernon G S
Brabander, Herbert Joseph
Bradley, Arthur
Bradlow, Herbert Leon
Brewer, Curtis Fred
Brown, John Francis, Jr
Brust, David Philip
Bullock, R Morris
Bunce, Stanley Chalmers
Burge, Robert Ernest, Jr
Burgmaier, George John
Burrows, Cynthia Jane
Cahn, Arno
Campbell, Bruce (Nelson), Jr
Campbell, Gerald Allan
Cappel, C Robert
Carabateas, Philip M
Carle, Kenneth Roberts
Carnahan, James Claude
Carpenter, Barry Keith
Carpenter, James William
Carpenter, Thomas J
Carr, Russell L K
Carroll, Robert Baker
Cathcart, John Almon
Ceprini, Mario Q
Chafetz, Harry
Chapman, Derek D
Charton, Marvin
Chen, Chin Hsin
Chen, Shaw-Horng
Childers, Robert Lee
Christensen, Larry Wayne
Christiansen, Robert George
Christman, David R
Chu, Joseph Yung-Chang
Clardy, Jon Christel
Clark, Charles Austin
Closson, William Deane
Coburn, Robert A
Cochran, John Charles
Cofrancesco, Anthony J
Cohen, Hyman L
Cohen, Sidney
Collins, Joseph Charles, Jr
Collum, David Boshart
Condit, Paul Brainard
Conway, Walter Donald
Cooper, Arthur Joseph L
Copelin, Harry B
Crivello, James V
Cunningham, Michael Paul
Curran, William Vincent
Cutler, Louise Marie
Cutler, Royal Anzly, Jr
Dagostino, Vincent F
D'Angelo, Gaetano
Daniel, Daniel S
Daniels, Ralph
Danishefsky, Isidore
Dannenberg, Joseph
Darlak, Robert
Daum, Sol Jacob
Daves, Glenn Doyle, Jr
Davis, Horace Raymond
Davis, Marvin Lester
Day, Jack Calvin
Dehn, Joseph William, Jr
Delton, Mary Helen
Denk, Ronald H
De Selms, Roy Charles
Dickinson, William Borden
Dieterich, David Allan
Dille, Kenneth Leroy
Dinan, Frank J
Dittmer, Donald Charles
Dixon, James Edward
Doubleday, Charles E, Jr
Dougherty, Charles Michael
Dutta, Shib Prasad
Eadon, George Albert
Earing, Mason Humphry
Eberhard, Anatol
Eckroth, David Raymond
Edelman, Robert
Ehrenson, Stanton Jay

Eikenberry, Jon Nathan
Eisch, John Joseph
Ellestad, George A
Elwood, James Kenneth
Engebrecht, Ronald Henry
Engel, Robert Ralph
Epstein, Joseph William
Erickson, Wayne Francis
Esse, Robert Carlyle
Estes, John H
Evans, Francis Eugene
Evert, Henry Earl
Factor, Arnold
Fanshawe, William Joseph
Farb, Edith H
Fedor, Leo Richard
Feigenson, Gerald William
Feit, Irving N
Feitler, David
Felty, Evan J
Fernandez, Jose Martin
Ferraro, John J
Ferris, James Peter
Fields, Thomas Lynn
Filandro, Anthony Salvatore
Finkbeiner, Herman Lawrence
Finley, Kay Thomas
Finzel, Rodney Brian
Fix, Delbert Dale
Ford, John Albert, Jr
Fowler, Frank Wilson
Fowler, Joanna S
Fox, Adrian Samuel
Francis, Arokiasamy Joseph
Frazza, Everett Joseph
Frechet, Jean M J
Frenkel, Krystyna
Friedman, Paul
Friedman, Seymour K
Friedrich, Louis Elbert
Frye, Robert Bruce
Fuerst, Adolph
Gage, Clarke Lyman
Ganem, Bruce
Gao, Jiali
Gates, Marshall DeMotte, Jr
Geering, Emil John
George, Philip Donald
Georgiev, Vassil St
Gershon, Herman
Gettler, Joseph Daniel
Giannotti, Ralph Alfred
Giguere, Raymond Joseph
Gilman, Robert Edward
Gilmont, Ernest Rich
Ginos, James Zissis
Gioannini, Theresa Lee
Glaros, George Raymond
Gletsos, Constantine
Gloyer, Stewart Edward
Gold, Allen Morton
Goldberg, Joseph Louis
Goldschmidt, Eric Nathan
Gordon, Irving
Gortler, Leon Bernard
Gray, D Anthony
Greco, Claude Vincent
Green, Mark M
Greizerstein, Walter
Grina, Larry Dale
Grubman, Marvin J
Gruenbaum, William Tod
Gruett, Monte Deane
Grushkin, Bernard
Gugig, William
Gurel, Demet
Gurien, Harvey
Gysling, Henry J
Haase, Jan Raymond
Haberfield, Paul
Hagan, William John, Jr
Hahn, Roger C
Haitko, Deborah Ann
Halm, James Maurice
Hamb, Fredrick Lynn
Hamilton, Lewis R
Hammer, Richard Benjamin
Hammond, George Simms
Hargreaves, Ronald Thomas
Harris, Robert L
Harrison, James Beckman
Hawks, George H, III
Hecht, Stephen Samuel
Hellmuth, Walter Wilhelm
Henion, Richard S
Henzel, Richard Paul
Hepfinger, Norbert Francis
Herbrandson, Harry Fred
Herbstman, Sheldon
Herdle, Lloyd Emerson
Hickernell, Gary L
Hileman, Kenneth Leroy
Hill, John Hamon Massey
Hillman, Manny
Hindersinn, Raymond Richard
Hirschberg, Albert I
Hlavka, Joseph John
Hoffman, Linda M
Hoffman, Theodore P
Hoffmann, Dietrich
Honig, Milton Leslie
Hough, Lindsay B
Howard, Philip Hall
Hoyte, Robert Mikell

Hull, Carl Max
Indictor, Norman
Inman, Charles Gordon
Isaacson, Henry Verschay
Izzo, Patrick Thomas
Jaffe, Fred
Jahn, Edwin Cornelius
Jain, Rakesh Kumar
James, Daniel Shaw
Jean, George Noel
Jelinski, Lynn W
Jelling, Murray
Jenkins, Philip Winder
Jesaitis, Raymond G
Jochsberger, Theodore
Joffee, Irving Brian
Johansson, Sune
Johns, William Francis
Johnson, Bruce Fletcher
Johnson, Francis
Johnson, Thomas Lynn
Jones, Gerald Walter
Jones, William Davidson
Joshua, Henry
Kalenda, Norman Wayne
Kalman, Thomas Ivan
Kapecki, Jon Alfred
Kaplan, Melvin
Kapuscinski, Jan
Katz, Thomas Joseph
Kaufman, Frank B
Keister, Jerome Baird
Kelly, Kenneth William
Kende, Andrew S
Kerber, Robert Charles
Kestner, Melvin Michael
Khan, Jamil Akber
Kinnel, Robin Bryan
Kirdani, Rashad Y
Klanderman, Bruce Holmes
Klein, Bernard
Klijanowicz, James Edward
Klingele, Harold Otto
Klose, Thomas Richard
Knauer, Bruce Richard
Koch, Heinz Frank
Koeng, Fred R
Kofron, James Thomas, Jr
Kohlbrenner, Philip John
Kosak, Alvin Ira
Krabbenhoft, Herman Otto
Krause, Josef Gerald
Kresse, Jerome Thomas
Krishnamurthy, Sundaram
Kruegel, Alice Virginia
Krueger, James Elwood
Krueger, William E
Kubisen, Steven Joseph, Jr
Kudzin, Stanley Francis
Kuivila, Henry Gabriel
Kullnig, Rudolph K
Kumler, Philip L
Kupchik, Eugene John
Labianca, Dominick A
Laganis, Deno
Lai, Yu-Chin
LaLonde, Robert Thomas
Landesberg, Joseph Marvin
Lang, Stanley Albert, Jr
Lansbury, Peter Thomas
Lau, Philip T S
Laubach, Gerald D
LeBlanc, Jerald Thomas
Lee, Lieng-Huang
Lee, May D-Ming (Lu)
Lee, Tzoong-Chyh
Lee, Ving Jick
Leibman, Lawrence Fred
Lengyel, Istvan
Le Noble, William Jacobus
Leone, Ronald Edmund
Leubner, Gerhard Walter
Levine, Stephen Alan
Ligon, Woodfin Vaughan, Jr
Limburg, William W
Lin, Mow Shiah
Lobo, Angelo Peter
Loev, Bernard
Lok, Roger
Looker, Jerome J
Lorenzo, George Albert
MacAvoy, Thomas Coleman
McClure, Judson P
McConnell, James Francis
McGinn, Clifford
McGriff, Richard Bernard
Machiele, Delwyn Earl
McKelvie, Neil
McLaen, Donald Francis
MacLaury, Michael Risley
MacLeay, Ronald E
McMurry, John Edward
McNelis, Edward Joseph
Malan, Rodwick LaPur
Mansfield, Kevin Thomas
Marcelli, Joseph F
March, Jerry
Mark, Robert Vincent
Markovitz, Mark
Martellock, Arthur Carl
Maul, James Joseph
Mayr, Andreas
Meinwald, Jerrold

CHEMISTRY / 221

Meinwald, Yvonne Chu
Menapace, Lawrence William
Meyers, Marian Bennett
Mihajlov, Vsevolod S
Miller, David Lee
Miller, James Robert
Miller, Theodore Charles
Miller, William Taylor
Milton, Kirby Mitchell
Mirviss, Stanley Burton
Misra, Anand Lal
Mitacek, Eugene Jaroslav
Modic, Frank Joseph
Mooberry, Jared Ben
Moon, Sung
Moore, James Alfred
Moran, Juliette May
Morduchowitz, Abraham
Morrill, Terence Clark
Mourning, Michael Charles
Murdock, Keith Chadwick
Murphy, Daniel Barker
Mustacchi, Henry
Mylroie, Victor L
Nealey, Richard H
Neeman, Moshe
Neill, Alexander Bold
Nettleton, Donald Edward, Jr
Neumann, Stephen Michael
Neveu, Maurice C
Norcross, Bruce Edward
Ober, Christopher Kemper
Oberender, Frederick G
O'Brien, Anne T
Ojima, Iwao
Olafsson, Patrick Gordon
Oliver, Gene Leech
Orlando, Charles M
Osawa, Yoshio
Osborn, H(arland) James
Osterholtz, Frederick David
Pachter, Irwin Jacob
Pankiewicz, Krzysztof Wojciech
Parikh, Hemant Bhupendra
Parsons, Timothy F
Partch, Richard Earl
Patmore, Edwin Lee
Pavlisko, Joseph Anthony
Pedroza, Gregorio Cruz
Pellegrini, Frank C
Pelosi, Evelyn Tyminski
Pelosi, Stanford Salvatore, Jr
Pemrick, Raymond Edward
Penn, Lynn Sharon
Pepe, Joseph Philip
Personeus, Gordon Rowland
Petersen, Ingo Hans
Petersen, Kenneth C
Petropoulos, Constantine Chris
Phillips, Donald Kenney
Pickett, James Edward
Pier, Harold William
Piper, Douglas Edward
Plue, Arnold Frederick
Pollart, Dale Flavian
Ponticello, Ignazio Salvatore
Pontius, Dieter J J
Portlock, David Edward
Poslusny, Jerrold Neal
Poss, Andrew Joseph
Post, Howard William
Potter, George Henry
Potts, Kevin T
Powers, John Clancey, Jr
Prigot, Melvin
Pucknat, John Godfrey
Purandare, Yeshwant K
Rauch, Emil B(runo)
Ravindran, Nair Narayanan
Reckhow, Warren Addison
Redalieu, Elliot
Rees, William Wendell
Regan, Thomas Hartin
Rehfuss, Mary
Reich, Marvin Fred
Reidlinger, Anthony A
Rhodes, Yorke Edward
Rice, David E
Richards, Jack Lester
Riecke, Edgar Eric K
Roelofs, Wendell Lee
Rosen, Milton Jacques
Rosenfeld, Jerold Charles
Ross, Malcolm S F
Ross, Robert Edward
Rothchild, Robert
Rousseau, Viateur
Rush, Kent Rodney
Russell, Charlotte Sananes
Rutkin, Philip
Ryan, Thomas John
Sanchez, Jose
Sandhu, Mohammad Akram
Sands, Richard Dayton
Sapino, Chester, Jr
Sauer, Charles William
Saunders, William Hundley, Jr
Sauter, Frederick Joseph
Savage, Dennis Jeffrey
Schaeffer, James Robert
Schaffrath, Robert Eben
Scheiner, Peter
Schleigh, William Robert
Schlessinger, Richard H

Schlicht, Raymond Charles
Schmeltz, Irwin
Schmitz, Henry
Schuerch, Conrad
Schulenberg, John William
Schultz, Arthur George
Schultz, William Clinton
Schulz, Horst H
Schuster, David Israel
Schwab, Linda Sue
Schwan, Thomas James
Schwartz, Leonard H
Searle, Roger
Searles, Arthur Langley
Seltzer, Stanley
Semmelhack, Martin F
Seus, Edward J
Shaath, Nadim Ali
Shafer, Sheldon Jay
Shannon, Frederick Dale
Shapiro, Robert
Sharma, Moheswar
Sheppard, Chester Stephen
Shields, Joan Esther
Siegart, William Raymond
Siew, Ernest L
Silberman, Robert G
Silleck, Clarence Frederick
Silveira, Augustine, Jr
Sims, Rex J
Siuta, Gerald Joseph
Sleezer, Paul David
Slusarczuk, George Marcelius Jaremias
Smith, Richard Frederick
Smith, Thomas Woods
Snyder, Harry Raymond, Jr
Sogah, Dotsevi Yao
Sojka, Stanley Anthony
Solo, Alan Jere
Soloway, Harold
Soloway, Saul
Sorkin, Howard
Sorter, Peter F
Sowa, John Robert
Spangler, Fred Walter
Sparapany, John Joseph
Spittler, Terry Dale
Sprung, Joseph Asher
Srinivasan, Rangaswamy
Staples, Jon T
Starkey, Frank David
Stecher, Emma Dietz
Steinberg, David H
Stephens, Lawrence James
Stern, Max Herman
Sticker, Robert Earl
Still, W Clark, Jr
Stohrer, Gerhard
Stone, Joe Thomas
Stoner, George Green
Stradling, Samuel Stuart
Sturmer, David Michael
Su, Tah-Mun
Sullivan, Michael Francis
Sutherland, George Leslie
Svoronos, Paris D N
Swicklik, Leonard Joseph
Swinehart, James Stephen
Syed, Ashfaquzzaman
Takeuchi, Esther Sans
Teasdale, William Brooks
Teegarden, David Morrison
Tesoro, Giuliana C
Thomas, Harold Todd
Thomas, Telfer Lawson
Timell, Tore Erik
Tokoli, Emery G
Tomasz, Maria
Tometsko, Andrew M
Troll, Walter
Trust, Ronald I
Tufariello, Joseph James
Tuites, Richard Clarence
Turek, William Norbert
Turnblom, Ernest Wayne
Uebel, Jacob John
Ursino, Joseph Anthony
Van Duuren, Benjamin Louis
Van Verth, James Edward
Verter, Herbert Sigmund
Vogel, Philip Christian
Von Strandtmann, Maximillian
Vratsanos, Spyros M
Vullo, William Joseph
Wagner, Gerald Roy
Wallis, Thomas Gary
Walsh, Edward Nelson
Ward, Frank Kernan
Warren, James Donald
Wasacz, John Peter
Watanabe, Kyoichi A
Webster, Francis X
Weetman, David G
Weil, Edward David
Weisler, Leonard
Weiss, David Steven
Weiss, Irma Tuck
Weiss, Martin Joseph
Weltman, Clarence A
White, Dwain Montgomery
White, Ralph Lawrence, Jr
Whitten, David G
Wiegert, Philip E
Wilcox, Charles Frederick, Jr

Wilen, Samuel Henry
Wilson, Burton David
Wilson, Stephen Ross
Windholz, Thomas Bela
Winter, Roland Arthur Edwin
Wiseman, George Edward
Wissner, Allan
Wittcoff, A Harold
Wolf, Alfred Peter
Wolfarth, Eugene F
Wolfe, Roger Thomas
Wolinski, Leon Edward
Wolny, Friedrich Franz
Wood, David
Wu, Mingdan
Wyman, Donald Paul
Xu, Zhenchun
Yaffe, Roberta
Yourtee, Lawrence Karn
Yu, Chia-Nien
Yu, Shiu Yeh
Yunick, Robert P
Yuska, Henry
Zalay, Ethel Suzanne
Zalewski, Edmund Joseph
Zavitsas, Andreas Athanasios
Zawadzki, Joseph Francis
Zerwekh, Charles Ezra, Jr
Zieger, Herman Ernst
Zimmerman, Barry
Zuckerman, Samuel
Zuman, Petr

NORTH CAROLINA
Abbey, Kirk J
Alston, Peter Van
Arnett, Edward McCollin
Arnold, Allen Parker
Arnold, Luther Bishop, Jr
Ayers, Paul Wayne
Barborak, James Carl
Bauermeister, Herman Otto
Bealor, Mark Dabney
Beauchamp, Lilia Marie
Bell, Jimmy Holt
Bockstahler, Theodore Edwin
Boswell, Donald Eugene
Bradsher, Charles Kilgo
Brieaddy, Lawrence Edward
Bristol, Douglas Walter
Brookhart, Maurice S
Bryan, Carl Eddington
Bumgardner, Carl Lee
Bursey, Joan Tesarek
Bushey, Dean Franklin
Carroll, F Ivy
Carroll, Felix Alvin, Jr
Carson, Bonnie L Bachert
Cavallito, Chester John
Chaney, David Webb
Chignell, Colin Francis
Chilton, William Scott
Chopra, Naiter Mohan
Cialdella, Cataldo
Clark, Russell Norman
Cline, Warren Kent
Coke, James Logan
Cook, Clarence Edgar
Cook, Lawrence C
Cote, Philip Norman
Cross, Robert Edward
Cubberley, Adrian H
Culberson, Chicita Frances
Cunningham, Carol Clem
Danieley, Earl
Daumit, Gene Philip
Davis, Roman
Dixit, Ajit Suresh
D'Silva, Themistocles Damasceno Joaquim
Dudley, Kenneth Harrison
Dufresne, Richard Frederick
Durden, John Apling, Jr
Dye, William Thomson, Jr
Edwards, Ben E
Eliel, Ernest Ludwig
Engel, John Francis
Erickson, Bruce Wayne
Etter, Robert Miller
Evans, Evan Franklin
Fabricius, Dietrich M
Farrissey, William Joseph, Jr
Felten, John James
Forrester, Sherri Rhoda
Frank, Robert Loeffler
Fraser-Reid, Bertram Oliver
Fredericksen, James Monroe
Freedman, David
Freeman, Harold Stanley
Fu, Wallace Yamtak
Fytelson, Milton
Gains, Lawrence Howard
Ganz, Charles Robert
Ghirardelli, Robert George
Giles, Jesse Albion, III
Gillikin, Jesse Edward, Jr
Goldstein, Irving Solomon
Graham, Jack Raymond
Greenspan, Frank Philip
Guthrie, Donald Arthur
Haile, Clarence Lee
Harfenist, Morton
Heckman, Robert Arthur
Heintzelman, Richard Wayne

Henderson, Thomas Ranney
Herbst, Robert Max
Higgins, Robert H
Hiskey, Richard Grant
Hix, Homer Bennett
Hoelzel, Charles Bernard
Huang, Jamin
Huber, W(ilson) Frederick
Huffman, K(enneth) Robert
Hurwitz, Melvin David
Husk, George Ronald
Izydore, Robert Andrew
Jameson, Charles William
Jeffs, Peter W
Johnston, John
Jones, Rufus Sidney
Jones, Stephen Thomas
Jones, William Jonas, Jr
Jung, James Moser
Kasperek, George James
Kelley, James Leroy
King, Henry Lee
Knight, David Bates
Kropp, Paul Joseph
Kupstas, Edward Eugene
Kuyper, Lee Frederick
Laganis, Evan Dean
Lamb, Robert Charles
Lee, Yue-Wei
Leiserson, Lee
Levine, Samuel Gale
Levy, Jack Benjamin
Levy, Louis A
Lewin, Anita Hana
Lewis, Robert Glenn
Little, William Frederick
Loeffler, Larry James
Loeppert, Richard Henry
Long, Robert Allen
Lynn, John Wendell
Ma, Tsu Sheng
McDaniel, Roger Lanier, Jr
McKay, Jerry Bruce
Majewski, Theodore E
Manning, Robert Joseph
Marco, Gino Joseph
Martin, Gary Edwin
Martin, Ned Harold
Martin, Richard Hugo
Martin, William Royall, Jr
Mathewes, David A
Mechanic, Gerald
Miles, George Benjamin
Miles, Marion Lawrence
Miller, Harry Brown
Miller, Walter Peter
Moates, Robert Franklin
Morris, Gene Franklin
Morris, Leo Raymond
Morrison, Robert W, Jr
Myers, John Albert
Narasimhachari, Nedathur
Neale, Robert S
Neumann, Calvin Lee
Newell, Marjorie Pauline
Nielsen, Lawrence Arthur
Norris, Terry Orban
Panek, Edward John
Partridge, John Joseph
Patterson, Ronald Brinton
Peck, David W
Perfetti, Thomas Albert
Phillips, Arthur Page
Pienta, Norbert J
Pirrung, Michael Craig
Porter, Ned Allen
Powers, Edward James
Preston, Jack
Price, Howard Charles
Proffitt, Thomas Jefferson, Jr
Pruett, Roy L
Purrington, Suzanne T
Raleigh, James Arthur
Ranieri, Richard Leo
Rapoport, Lorence
Reese, Millard Griffin, Jr
Rehberg, Chessie Elmer
Reid, Jack Richard
Richardson, Stephen Giles
Rideout, Janet Litster
Riemann, James Michael
Risley, John Marcus
Rivers, Jessie Markert
Rodgman, Alan
Roth, Roy William
Russell, Henry Franklin
Scharver, Jeffrey Douglas
Schultz, Frederick John
Schumacher, Joseph Nicholas
Schwindeman, James Anthony
Sherer, James Pressly
Short, Franklin Willard
Soeder, Robert W
Sparacino, Charles Morgan
Spears, Alexander White, III
Sprague, Robert Hicks
Squibb, Samuel Dexter
Sternbach, Daniel David
Swaringen, Roy Archibald, Jr
Teague, Harold Junior
Thomas, Alford Mitchell
Thompson, Crayton Beville
Tinsley, Samuel Weaver
Tomer, Kenneth Beamer

Organic Chemistry (cont)

Tucker, William Preston
Tyndall, John Raymond
VanMiddlesworth, Frank L
Waite, Moseley
Walsh, Thomas David
Wani, Mansukhlal Chhaganlal
Waters, James Augustus
Wehner, Philip
Wetmore, David Eugene
Whangbo, Myung Hwan
Wheeler, Thomas Neil
Wilder, Pelham, Jr
Wilson, Nancy Keeler
Wiseman, Jeffrey Stewart
Woosley, Royce Stanley
Wyrick, Steven Dale

NORTH DAKOTA
Boudjouk, Philip Raymond
Donnelly, Brendan James
Jasperse, Craig Peter
Johnson, A William
Lambeth, David Odus
Olson, Edwin S
Pomonis, James George
Rudesill, James Turner
Shelver, William H
Stenberg, Virgil Irvin
Sugihara, James Masanobu
Tanaka, Fred Shigeru
Woolsey, Neil Franklin

OHIO
Abraham, Tonson
Albright, James Andrew
Alford, Harvey Edwin
Ambelang, Joseph Carlyle
Andrist, Anson Harry
Arora, Sardari Lal
Attig, Thomas George
Austin, Robert Andrae
Averill, Seward Junior
Baclawski, Leona Marie
Ball, David Ralph
Baltazzi, Evan S
Bann, Robert (Francis)
Barnett, Kenneth Wayne
Barrer, Daniel Edward
Battershell, Robert Dean
Bauer, Richard G
Baughman, Glenn Laverne
Baur, Fredric John, Jr
Behrmann, Eleanor Mitts
Bender, Daniel Frank
Berger, Mitchell Harvey
Bergomi, Angelo
Bernstein, Stanley Carl
Bertelson, Robert Calvin
Bertsch, Robert Joseph
Bethea, Tristram Walker, III
Bimber, Russell Morrow
Binkley, Roger Wendell
Blewett, Charles William
Blum, Patricia Rae
Bockhoff, Frank James
Bond, William Bradford
Borgnaes, Dan
Brady, Leonard Everett
Brain, Devin King
Brannen, William Thomas, Jr
Brittain, William Joseph
Broaddus, Charles D
Brodhag, Alex Edgar, Jr
Brody, Richard Simon
Bromund, Werner Hermann
Brueggemeier, Robert Wayne
Brundage, Donald Keith
Bruno, David Joseph
Buck, Keith Taylor
Buell, Glen R
Bufkin, Billy George
Bumpus, Francis Merlin
Burke, Morris
Burt, Gerald Dennis
Campbell, Robert Wayne
Canfield, James Howard
Carlock, John Timothy
Carr, Albert A
Cassady, John Mac
Chamberlin, William Bricker, III
Clark, Alan Curtis
Clemans, George Burtis
Clement, William H
Cohen, Irwin
Coleman, John Franklin
Coleman, Lester Earl, (Jr)
Colvin, Howard Allen
Compton, Ell Dee
Conley, Robert T
Connor, Daniel S
Convery, Robert James
Coran, Aubert Y
Corbin, James Lee
Cosgrove, Stanley Leonard
Costanza, Albert James
Cotton, Wyatt Daniel
Cryberg, Richard Lee
Culbertson, Billy Muriel
Curry, Howard Millard
Dalton, James Christopher
Dana, Robert Clark
Dannley, Ralph Lawrence
Davies, Julian Anthony
Davis, Edward Melvin
Davis, James Allen
Delcamp, Robert Mitchell
Delvigs, Peter
DeMarco, John Gregory
Denham, Joseph Milton
Deters, Donald W
Dial, William Richard
Di Biase, Stephen Augustine
Dietrich, Joseph Jacob
Dimond, Harold Lloyd
Dinbergs, Kornelius
Dolfini, Joseph E
Doran, Thomas J, Jr
Doyle, Richard Robert
Duddey, James E
Duke, June Temple
Dunn, Horton, Jr
Dunning, John Walcott
Edman, James Richard
Eichel, Herman Joseph
Eicher, John Harold
Eisenhardt, William Anthony, Jr
Elmer, Otto Charles
Emrick, Donald Day
Erman, William F
Essig, Henry J
Eveslage, Sylvester Lee
Fabris, Hubert
Farkas, Julius
Fayter, Richard George, Jr
Fechter, Robert Bernard
Feld, William Adam
Feldman, Julian
Fendler, Eleanor Johnson
Fentiman, Allison Foulds, Jr
Fineberg, Herbert
Finelli, Anthony Francis
Fishel, Derry Lee
Flechtner, Thomas Welch
Fleischman, Darrell Eugene
Flynn, Gary Alan
Foldvary, Elmer
Ford, Emory A
Fox, Bernard Lawrence
Fraenkel, Gideon
Frank, Charles Edward
Franks, Allen P
Franks, John Anthony, Jr
Fry, James Leslie
Fu, Yun-Lung
Furry, Benjamin K
Galloway, Ethan Charles
Gano, James Edward
Gaul, Richard Joseph
Germann, Richard P(aul)
Gibson, Thomas William
Gifford, David Stevens
Gilbert, Eugene Charles
Gilde, Hans-Georg
Gillis, Bernard Thomas
Ginning, P(aul) R(oll)
Glasgow, David Gerald
Gleim, Clyde Edgar
Gloth, Richard Edward
Goebel, Charles Gale
Goetz, Richard W
Goodson, Alan Leslie
Gordon, John Edward
Gosselink, Eugene Paul
Gould, Edwin Sheldon
Graef, Walter L
Gray, Don Norman
Greenbaum, Sheldon Boris
Greene, Hoke Smith
Greene, Janice L
Griffin, Claibourne Eugene, Jr
Grimm, Robert Arthur
Grose, Herschel Gene
Grotta, Henry Monroe
Gustafson, David Harold
Hanes, Ronnie Michael
Harrington, Roy Victor
Harris, Richard Lee
Hart, David Joel
Hausch, Walter Richard
Hawbecker, Byron L
Hayes, Jeffrey Charles
Haynes, LeRoy Wilbur
Hays, Byron G
Hazy, Andrew Christopher
Heckert, David Clinton
Hein, Richard William
Helmick, Larry Scott
Hergenrother, William Lee
Herliczek, Siegfried H
Herold, Robert Johnston
Hess, George G
Hickman, Howard Minor
Hirschmann, Hans
Hoke, Donald I
Holubec, Zenowie Michael
Horgan, Stephen William
Horne, Samuel Emmett, Jr
Hosansky, Norman Leon
Houser, John J
Howald, Jeremiah Mark
Hunt, Jerry Donald
Huntsman, William Duane
Hutchison, Robert B
Ingham, Robert Kelly
Innes, John Edwin
Jacobs, Harvey
Jacobs, Richard Lee
Jarvi, Esa Tero
Johnson, David Barton
Johnson, Robert Gudwin
Jones, Winton D, Jr
Kagen, Herbert Paul
Kane, James Joseph
Kaplan, Fred
Katz, Morton
Katzen, Raphael
Kay, Edward Leo
Kay, Peter Steven
Ketcha, Daniel Michael
Kinsman, Donald Vincent
Kinstle, James Francis
Kinstle, Thomas Herbert
Knowlton, David A
Kofron, William G
Kolattukudy, P E
Korach, Malcolm
Krabacher, Bernard
Kreuz, John Anthony
Kriens, Richard Duane
Kupel, Richard E
Kurtz, David Williams
Kurz, David W
Kwiatek, Jack
Laali, Kenneth Khosrow
Lang, James Frederick
Laughlin, Robert Gene
Lawson, David Francis
Layer, Robert Wesley
Leckonby, Roy Alan
Leffler, Marlin Templeton
Letton, James Carey
Levin, Robert Harold
Lewis, Irwin C
Lipinsky, Edward Solomon
Litt, Morton Herbert
Livigni, Russell Anthony
Loening, Kurt L
Logan, Ted Joe
Losekamp, Bernard Francis
Loughran, Gerard Andrew, Sr
Lucier, John J
Lukin, Marvin
McCain, George Howard
McCarty, Frederick Joseph
McFarland, Charles Warren
McLoughlin, Daniel Joseph
McMillan, Robert McKee
McRowe, Arthur Watkins
Maender, Otto William
Magee, Thomas Alexander
Mao, Hsiang-Kuen Mark
Mao, Simon Jen-Tan
Marshall, Richard Allen
Mathers, Alexander Pickens
Matlin, Albert R
Mattson, Raymond Harding
Mayor, Rowland Herbert
Meinert, Walter Theodore
Meinhardt, Norman Anthony
Meloy, Carl Ridge
Michaelis, Carl I
Michelman, John S
Milone, Charles Robert
Molnar, Stephen P
Mouk, Robert Watts
Mundell, Percy Meldrum
Munson, H Randall, Jr
Murdoch, Arthur
Murphy, Walter Thomas
Muse, Joel, Jr
Naples, Felix John
Nardi, John Christopher
Neckers, Douglas
Nee, Michael Wei-Kuo
Nelson, Wayne Franklin
Newman, Melvin Spencer
Nicholas, Paul Peter
Nicholson, D Allan
Nobis, John Francis
Oberster, Arthur Eugene
O'Connor, David Evans
Orchin, Milton
Ostrum, G(eorge) Kenneth
Ouellette, Robert J
Oziomek, James
Pace, Henry Alexander
Palopoli, Frank Patrick
Paquette, Leo Armand
Parish, Darrell Joe
Parker, Richard Ghrist
Peet, Norton Paul
Percec, Virgil
Peters, Lynn Randolph
Peterson, Donald J
Petrarca, Anthony Edward
Pialet, Joseph William
Platau, Gerard Oscar
Platz, Matthew S
Portoghese, Philip S...

Portfolio, Donald Carl
Powell, Warren Howard
Powers, Larry James
Price, John Avner
Pritchett, Ervin Garrison
Prudence, Robert Thomas
Pynadath, Thomas I
Raciszewski, Zbigniew
Rader, Charles Phillip
Ramey, Chester Eugene
Ramp, Floyd Lester
Rau, Allen H
Reilly, Charles Bernard
Rhinesmith, Herbert Silas
Richardson, Alfred, Jr
Richardson, Kathleen Schueller
Rinaldi, Peter L
Rizzi, George Peter
Roberts, Durward Thomas, Jr
Robins, Richard Dean
Robinson, Kenneth Robert
Roha, Max Eugene
Rosen, Irving
Ryan, Anne Webster
Ryerson, George Douglas
Sadurski, Edward Alan
Sampson, Paul
Sarbach, Donald Victor
Sartoris, Nelson Edward
Scalzi, Francis Vincent
Scanlan, Mary Ellen
Scheve, Bernard Joseph
Schilling, Curtis Louis, Jr
Schlossman, Irwin S
Schmidt, Francis Henry
Schmidt, Paul J
Schneider, Wolfgang W
Schnettler, Richard Anselm
Schollenberger, Charles Sundy
Schroll, Gene E
Schulze, William Eugene
Schumacher, Roy Joseph
Scozzie, James Anthony
Sebastian, John Francis
Selden, George
Selker, Milton Leonard
Serve, Munson Paul
Shanklin, James Robert, Jr
Shechter, Harold
Sheibley, Fred Easly
Shelton, James Reid
Sheppard, William James
Shertzer, Howard Grant
Siedschlag, Karl Glenn, Jr
Siefken, Mark William
Siklosi, Michael Peter
Sill, Arthur DeWitt
Sinner, Donald H
Smith, Daniel John
Smith, Douglas Alan
Spanninger, Philip Andrew
Spessard, Dwight Rinehart
Spialter, Leonard
Spitzer, Jeffrey Chandler
Sporek, Karel Frantisek
Staker, Donald David
Stansfield, Roger Ellis
Stein, Daryl Lee
Sternfeld, Marvin
Stevens, Henry Conrad
Stevenson, Don R
Stoldt, Stephen Howard
Surbey, Donald Lee
Swartzentruber, Paul Edwin
Swenton, John Stephen
Tarvin, Robert Floyd
Taschner, Michael J
Tate, David Paul
Taylor, Richard Timothy
Teller, Daniel Myron
Temple, Robert Dwight
Theis, Richard James
Theisen, Cynthia Theres
Thomas, Alexander Edward, III
Thomas, Lewis Edward
Thomas, McCalip Joseph
Thompson, James Edwin
Throckmorton, Peter E
Traynor, Lee
Trivisonno, Charles F(rancis)
Tsai, Chung-Chieh
Tsai, Ming-Daw
Turkel, Rickey M(artin)
Turnbull, Kenneth
Undeutsch, William Charles
Uscheek, Dave Petrovich
Vogel, Paul William
Volk, Murray Edward
Von Ostwalden, Peter Weber
Wade, Peter Cawthorn
Wagner, Melvin Peter
Walker, Thomas Eugene
Walsh, James Aloysius
Washburn, Lee Cross
Weaver, William Michael
Webb, Thomas Howard
Webster, James Albert
Weinstein, Arthur Howard
Weintraub, Philip Marvin
Weisgerber, David Wendelin
Werner, Raymond Edmund
Westerman, Ira John
Westerman, Philip William
Wideman, Lawson Gibson
Williger, Ervin John
Wilson, Gustavus Edwin
Winkler, Robert Randolph
Wise, Richard Melvin
Witiak, Donald T
Woo, James T K
Wright, Robert L
Yanko, William Harry
Yocum, Ronald Harris
York, Owen, Jr
Youngs, Wiley Jay
Zaremsky, Baruch

Zilch, Karl T
Zimmer, Hans

OKLAHOMA
Alaupovic, Petar
Baldwin, Roger Allan
Banasiak, Dennis Stephen
Battiste, David Ray
Beckett, James Reid
Berlin, Kenneth Darrell
Bost, Howard William
Bunce, Richard Alan
Burlett, Donald James
Card, Roger John
Chaturvedi, Arvind Kumar
Clapper, Thomas Wayne
Craig, Louis Elwood
Crain, Donald Lee
Das, Paritosh Kumar
Dermer, Otis Clifford
Durr, Albert Matthew, Jr
Eby, Harold Hildenbrandt
Eddington, Carl Lee
Edmison, Marvin Tipton
Efner, Howard F
Eisenbraun, Edmund Julius
Fahey, Darryl Richard
Farrar, Ralph Coleman
Geibel, Jon Frederick
Greenwood, Gil Jay
Hartzfeld, Howard Alexander
Hertzler, Donald Vincent
Hill, James Wagy
Hodnett, Ernest Matelle
Holt, Elizabeth Manners
Johnson, Wallace Delmar
Kleinschmidt, Roger Frederick
Kokesh, Fritz Carl
Kucera, Clare H
Lehr, Roland E
McGuire, Stephen Edward
McGurk, Donald J
Magarian, Robert Armen
Matson, Michael Steven
Minor, John Threecivelous
Moore, Donald R
Motz, Kaye La Marr
Natowsky, Sheldon
Nicholas, Kenneth M
Nye, Mary Jo
Pober, Kenneth William
Riggs, Olen Lonnie, Jr
Schiff, Sidney
Seeney, Charles Earl
Selman, Charles Melvin
Shotts, Adolph Calveran
Snider, Theodore Eugene
Thyvelikakath, George
Tillman, Richard Milton
Trepka, William James
Wang, Binghe
Wharry, Stephen Mark
White, Harold McCoy
Wurth, Michael John
Yu, Chang-An
Zuech, Ernest A

OREGON
Autrey, Robert Luis
Badger, Rodney Allan
Becker, Robert Richard
Bloomfield, Jordan Jay
Boekelheide, Virgil Carl
Brinkley, John Michael
Christensen, Bert Einar
Civelli, Oliver
Cronyn, Marshall William
Currie, James Orr, Jr
Deinzer, Max Ludwig
Dolby, Lloyd Jay
Duncan, James Alan
Ehler, Kenneth Walter
Elia, Victor John
Firkins, John Lionel
Fleming, Bruce Ingram
Freeman, Peter Kent
Gleicher, Gerald Jay
Gould, Steven James
Grettie, Donald Pomeroy
Haugland, Richard Paul
Hauser, Frank Marion
Hudak, Norman John
Jaeger, Charles Wayne
Keana, John F W
Keevil, Thomas Alan
Klemm, LeRoy Henry
Klopfenstein, Charles E
Knott, Donald Macmillan
Koenig, Thomas W
Laver, Murray Lane
Levinson, Alfred Stanley
Lutz, Raymond Paul
Merz, Paul Louis
Oatfield, Harold
Paudler, William W
Pawlowski, Norman E
Postl, Anton
Rose, Norman Carl
Sandifer, Ronda Margaret
Schink, Chester Albert
Shilling, Wilbur Leo
Szalecki, Wojciech Jozef
Thies, Richard William
Wamser, Carl Christian

Weisenborn, Frank L
White, James David
Whittemore, Charles Alan

PENNSYLVANIA
Abou-Gharbia, Magid
Ackermann, Guenter Rolf
Adler, Irving Larry
Albright, Robert Lee
Almond, Harold Russell, Jr
Alvino, William Michael
Anspon, Harry Davis
Austin, Thomas Howard
Baird, Merton Denison
Bakule, Ronald David
Bare, Thomas M
Barie, Walter Peter, Jr
Barnes, David Kennedy
Bauer, William, Jr
Baumann, Jacob Bruce
Bayer, Horst Otto
Beck, Paul Edward
Becker, Robert Hugh
Beckley, Ronald Scott
Beitchman, Burton David
Bell, Ian
Bell, Stanley C
Benfey, Otto Theodor
Benkovic, Stephen J
Bennett, Richard Bond
Benson, Barrett Wendell
Berges, David Alan
Bergo, Conrad Hunter
Berkheimer, Henry Edward
Berliner, Ernst
Berliner, Frances (Bondhus)
Bidlack, Verne Claude, Jr
Bihovsky, Ron
Bingham, Richard Charles
Birnbaum, Hermann
Bissing, Donald Eugene
Blank, Benjamin
Blommers, Elizabeth Ann
Bluhm, Harold Frederick
Bock, Mark Gary
Bockrath, Bradley Charles
Bohen, Joseph Michael
Bolgiano, Nicholas Charles
Bolhofer, William Alfred
Bondinell, William Edward
Bortnick, Newman Mayer
Bosso, Joseph Frank
Bouis, Paul Andre
Bower, George Myron
Bowman, Robert Samuel
Brady, Stephen Francis
Braunstein, David Michael
Breazeale, Robert David
Brenner, Gerald Stanley
Brockington, James Wallace
Brownell, George L
Browning, Daniel Dwight
Brutcher, Frederick Vincent, Jr
Buhle, Emmett Loren
Bulbenko, George Fedir
Burke, James David
Burns, H Donald
Cairns, Theodore L
Cann, Michael Charles
Canter, Neil M
Capaldi, Eugene Carmen
Caputo, Joseph Anthony
Carlin, Robert Burnell
Carlson, Dana Peter
Carlson, Glenn Richard
Carter, Linda G
Caruso, Sebastian Charles
Casey, Adria Catala
Castle, John Edwards
Cawley, John Joseph
Cenci, Harry Joseph
Chakrabarti, Paritosh M
Chang, Charles Hung
Chang, Ching-Jen
Chang, Wen-Hsuan (Wayne)
Childress, Scott Julius
Chiola, Vincent
Chong, Berni Patricia
Chong, Joshua Anthony
Chow, Alfred Wen-Jen
Christie, Michael Allen
Christie, Peter Allan
Clemens, David Henry
Clovis, James S
Cohen, Gordon Mark
Cohen, Theodore
Condo, Albert Carman, Jr
Connor, James Edward, Jr
Conte, John Salvatore
Cook, Richard Sherrard
Coraor, George Robert
Cramer, Richard (David)
Crano, John Carl
Crosby, Alan Hubert
Cunningham, Howard
Cupper, Robert Alton
Curran, Dennis Patrick
Dalton, Augustine Ivanhoe, Jr
Dalton, David Robert
Damle, Suresh B
Davie, William Raymond
Davis, Brian Clifton
Davis, Franklin A
Deno, Norman C

Denoon, Clarence England, Jr
De Tommaso, Gabriel Louis
DeWald, Horace Albert
De Witt, Hobson Dewey
Dickstein, Jack
Dimmig, Daniel Ashton
Dixon, Joseph Ardiff
Djuric, Stevan Wakefield
Dohany, Julius Eugene
Donald, Dennis Scott
Dowbenko, Rostyslaw
Dowd, Paul
Dowd, Susan Ramseyer
Downs, James Joseph
Dreibelbis, John Adam
Dresden, Carlton F
Dressler, Hans
Dromgold, Luther D
Dunlap, Lawrence H
Dunn, George Lawrence
Dupont, Paul Emile
Durand, Marc L
Durrell, William S
Dymicky, Michael
Ebling, Irving Nelson
Ellis, Jeffrey Raymond
Emling, Bertin Leo
Emmons, William David
Engelhardt, Edward Louis
Erb, Dennis J
Ertelt, Henry Robinson
Evans, Ben Edward
Farcasiu, Dan
Feairheller, Stephen Henry
Fearing, Ralph Burton
Feely, Wayne E
Feeman, James Frederic
Fehnel, Edward A(dam)
Feller, Robert Livingston
Felley, Donald Louis
Fenoglio, Richard Andrew
Ferren, Richard Anthony
Field, Nathan David
Fishel, John B
Flitter, David
Foglia, Thomas Anthony
Follweiler, Joanne Schaaf
Foltz, George Edward
Ford, Michael Edward
Francis, Peter Schuyler
Franko-Filipasic, Borivoj Richard Simon
Freed, Meier Ezra
Freedman, Robert Wagner
Freeman, James Harrison
Freidinger, Roger Merlin
Freyermuth, Harlan Benjamin
Friedman, Sidney
Funke, Phillip T
Fusco, Gabriel Carmine
Gadek, Frank Joseph
Gallagher, George Arthur
Gardner, David Milton
Geoffroy, Gregory Lynn
Gergova, Katia M
Gerhardt, George William
Giuliano, Robert Michael
Glover, George Irvin
Gluntz, Martin L
Godfrey, John Carl
Goldman, Theodore Daniel
Goldsby, Arthur Raymond
Goldstein, Theodore Philip
Gormley, William Thomas
Grabowski, Joseph J
Greenberg, Herman Samuel
Grey, Roger Allen
Grezlak, John Henry
Griffith, Michael Grey
Grillot, Gerald Francis
Grinstein, Reuben H
Gruber, Gerald William
Hager, Robert B
Haggard, Richard Allan
Hamel, Coleman Rodney
Hammons, James Hutchinson
Hanauer, Richard
Harrison, Aline Margaret
Harrison, Ernest Augustus, Jr
Hart, William Forris
Hartline, Richard
Hartman, John Alan
Hartman, Kenneth Eugene
Hartranft, George Robert
Hausser, Jack W
Hedenburg, John Frederick
Heiberger, Philip
Heindel, Ned Duane
Heine, Harold Warren
Henderson, Richard Elliott Lee
Henrichs, Paul Mark
Hergert, Herbert L
Herman, Frederick Louis
Herskovitz, Thomas
Hertler, Walter Raymond
Hess, Ronald Eugene
Heys, John Richard
Hirsch, Stephen Simeon
Hirschmann, Ralph Franz
Hlasta, Dennis John
Hockswender, Thomas Richard
Hofferth, Burt Frederick
Hoffman, Jacob Matthew, Jr
Hoiness, David Eldon
Holden, Kenneth George

Hoops, Stephen C
Hotelling, Eric Bell
Howard, Edgar, Jr
Huffman, Allan Murray
Humer, Philip Wilson
Hummer, James Knight
Hunt, David Allen
Hunter, Wood E
Hutchins, MaryGail Kinzer
Hutchins, Robert Owen, Sr
Hutton, Thomas Watkins
Hutzler, John R
Hwang, Bruce You-Huei
Ingram, Alvin Richard
Irwin, Philip George
Jabloner, Harold
Jackman, Lloyd Miles
Jacobson, Richard Martin
Johnson, Leroy Dennis
Johnson, Richard William
Johnson, Wayne Orrin
Johnston, Gordon Robert
Johnston, Jean Vance
Jones, James Holden
Jones, Jennings Hinch
Joullie, Madeleine M
Karo, Wolf
Kauer, James Charles
Kauffman, Joel Mervin
Kehr, Clifton Leroy
Kennedy, Flynt
Kerst, A(l) Fred
Khan, Shabbir Ahmed
Khosah, Robinson Panganai
Kidwell, Roger Lynn
Kim, Jean Bartholomew
Kingsbury, William Dennis
Kirch, Lawrence S
Klapproth, William Jacob, Jr
Kleinman, Roberta Wilma
Koch, Walter Theodore
Koob, Robert Philip
Kopchik, Richard Michael
Kopple, Kenneth D(avid)
Kowalski, Conrad John
Krackov, Mark Harry
Kress, Bernard Hiram
Kronberger, Karlheinz
Krow, Grant Reese
Kruse, Lawrence Ivan
Kukla, Michael Joseph
Kulp, Stuart S
Kuzmak, Joseph Milton
Labosky, Peter, Jr
LaCount, Robert Bruce
Laemmle, Joseph Thomas
Lange, Barry Clifford
Larsen, John W
Larson, Gerald Louis
Lavoie, Alvin Charles
Leake, William Walter
Leber, Phyllis Ann
Lee, Robert William
Leeper, Robert Walz
Leiby, Robert William
Lenox, Ronald Sheaffer
Leonard, John Joseph
Leston, Gerd
Lever, Cyril, Jr
Levesque, Charles Louis
Levine, Robert
Libby, Carol Baker
Libby, R Daniel
Lieberman, Hillel
Lindsey, William B
Linn, William Joseph
Lipsitz, Paul
Liu, Andrew Tze-chiu
Livengood, Samuel Miller
Lopez, R C Gerald
Lorenz, Roman R
Lovald, Roger Allen
Lovett, John Robert
Lugar, Richard Charles
Lumma, William Carl, Jr
Luskin, Leo Samuel
Lyons, James Edward
McCarthy, John Randolph
McGrath, Thomas Frederick
Machleder, Warren Harvey
McInerney, Eugene F
McKeever, Charles H
McKelvey, Donald Richard
Madhav, Ram
Magill, Joseph Henry
Mallory, Clelia Wood
Mallory, Frank Bryant
Manger, Martin C
Manka, Dan P
Mantell, Gerald Jerome
Mao, Chung-Ling
Marmer, William Nelson
Maryanoff, Bruce Eliot
Maryanoff, Cynthia Anne Milewski
Masciantonio, Philip
Mascioli, Rocco Lawrence
Maslak, Przemyslaw Boleslaw
Matyjaszewski, Krzysztof
Mendelson, Wilford Lee
Merner, Richard Raymond
Messer, Wayne Ronald
Middleton, William Joseph
Milakofsky, Louis
Miller, Thomas Gore

Organic Chemistry (cont)

Minard, Robert David
Mokotoff, Michael
Monsimer, Harold Gene
Montgomery, Ronald Eugene
Mooney, David Samuel
Moore, Gordon George
Morse, Lewis David
Mortimer, Charles Edgar
Mount, Lloyd Gordon
Muck, Darrel Lee
Murphy, Clarence John
Musser, David Musselman
Myers, Earl Eugene
Myron, Thomas L(eo)
Naegele, Edward Wister, Jr
Naples, John Otto
Nedwick, John Joseph
Neidig, Howard Anthony
Nelson, George Leonard
Nemec, Joseph William
Newirth, Terry L
Newitt, Edward James
Niederhauser, Warren Dexter
Noceti, Richard Paul
Nodiff, Edward Albert
Novak, Ronald William
Novello, Frederick Charles
Noyes, Paul R
Nyi, Kayson
Obbink, Russell C
Ocone, Luke Ralph
Olofson, Roy Arne
Opie, Thomas Ranson
Orphanides, Gus George
Orr, Robert S
Osei-Gyimah, Peter
Owens, Frederick Hammann
Packman, Albert M
Papanikolaou, Nicholas E
Parish, Roger Cook
Patsiga, Robert A
Patterson, Dennis Ray
Paulson, Mark Clements
Peck, Richard Merle
Peffer, John Roscoe
Pelczar, Francis A
Pellegrini, John P, Jr
Perchonock, Carl David
Perry, Mary Hertzog
Peterson, James Oliver
Petronio, Marco
Phillips, Marshall
Piccolini, Richard John
Pinschmidt, Robert Krantz, Jr
Plant, William J
Ponticello, Gerald S
Poos, George Ireland
Potter, Neil H
Precopio, Frank Mario
Press, Jeffery Bruce
Press, Linda Seghers
Preti, George
Price, Charles Coale
Price, Edward Hector
Quinn, Edwin John
Rabiger, Dorothy June
Ramachandran, Subramania
Rapp, Robert Dietrich
Rauch, Stewart Emmart, Jr
Reider, Malcolm John
Reifenberg, Gerald H
Reiff, Harry Elmer
Reingold, I(ver) David
Reitz, Allen Bernard
Reitz, Robert Rex
Remar, Joseph Francis
Remy, David Carroll
Renfrew, Edgar Earl
Resconich, Samuel
Reuwer, Joseph Francis, Jr
Richey, Herman Glenn, Jr
Roberts, Bryan Wilson
Robinson, Donald Nellis
Rogers, Janet
Rogers, Ralph Loucks
Rogerson, Thomas Dean
Rosenburg, Dale Weaver
Rosenthal, Rudolph
Ross, Stephen T
Rowe, Jay Elwood
Rowland, Alex Thomas
Rush, James E
Russell, Peter Byrom
Russey, William Edward
Russo, Thomas Joseph
Saari, Walfred Spencer
Saggiomo, Andrew Joseph
Salvino, Joseph Michael
Samanen, James Martin
Santilli, Arthur A
Sasin, Richard
Sattsangi, Prem Das
Sawicki, John Edward
Sawyer, David W(illiam)
Scala, Luciano Carlo
Schaeffer, Lee Allen
Schauble, J Herman
Schearer, William Richard
Schiff, Paul L, Jr
Schneider, Henry Joseph
Scholnick, Frank
Schreiber, Kurt Clark
Schreyer, Ralph Courtenay
Schrof, William Ernst John
Schuster, Ingeborg I M
Schwing, Gregory Wayne
Segall, Stanley
Seybert, David Wayne
Shalit, Harold
Shaw, John Thomas
Shepard, Kenneth LeRoy
Sherman, Anthony Michael
Shiue, Chyng-Yann
Shozda, Raymond John
Sidler, Jack D
Signorino, Charles Anthony
Silbert, Leonard Stanton
Singh, Baldev
Sinotte, Louis Paul
Sircar, Ila
Sircar, Jagadish Chandra
Sittenfield, Marcus
Skell, Philip S
Smart, G N Russell
Smith, Amos Brittain, III
Smith, Francis Xavier
Smith, Graham Monro
Smith, Howard Leroy
Smith, James Stanley
Smith, Perrin Gary
Smith, Robert Lawrence
Smith, W Novis, Jr
Sollott, Gilbert Paul
Solodar, Warren E
Solomon, M Michael
Somkuti, George A
Sonnet, Philip E
Soriano, David S
Southwick, Philip Lee
Spangler, Martin Ord Lee
Spencer, Ralph Donald
Spindt, Roderick Sidney
Staiger, Roger Powell
Staley, Stuart Warner
Statton, Gary Lewis
Stedman, Robert John
Steiner, Russell Irwin
Stevens, Travis Edward
Stevenson, Robert William
Stewart, Thomas
Stine, William R
Stinson, Edgar Erwin
Stone, Herman
Stong, David Bruce
Straub, Thomas Stuart
Strike, Donald Peter
Strohm, Paul F
Sundeen, Joseph Edward
Suter, Stuart Ross
Swamer, Frederic Wurl
Swift, Graham
Takeshita, Tsuneichi
Tatum, Charles Maris
Tekel, Ralph
Templer, David Allen
Thoman, Charles James
Thompson, Wayne Julius
Thornton, Edward Ralph
Thornton, Elizabeth K
Tomezsko, Edward Stephen John
Touchstone, Joseph Cary
Traynor, Sean G
Trimitsis, George B
Tse, Rose (Lou)
Turner, Andrew B
Tzodikov, Nathan Robert
Ullyot, Glenn Edgar
Vanderwerff, William D
Van Dyke, Charles H
Van Gemert, Barry
Van Horn, Ruth Warner
Varkey, Thankamma Eapen
Vaux, James Edward, Jr
Veber, Daniel Frank
Venuto, Paul B
Vernon, John Ashbridge
Vick, Gerald Kieth
Vickers, Stanley
Vladutz, George E
Vogel, Martin
Waddell, Walter Harvey
Wade, Charles Gary
Wade, Peter Allen
Wagner, Robert Edwin
Walker, Frederick
Walker, Ruth Angelina
Waller, Francis Joseph
Walsh, Edward Joseph, Jr
Walters, Lee Rudyard
Warfel, David Ross
Warrick, Percy, Jr
Washburne, Stephen Shepard
Watterson, Kenneth Franklin
Weaver, Leo James
Weese, Richard Henry
Wei, Yen
Weiler, Ernest Dieter
Weinreb, Steven Martin
Weinstock, Joseph
Weiss, Benjamin
Weiss, James Allyn
Weldes, Helmut H
Welsh, David Albert
Wempe, Lawrence Kyran
Wender, Irving
Wentland, Mark Philip
Werner, Ervin Robert, Jr
Westley, John William
White, Harry Joseph
Whitney, Joel Gayton
Widdowson, Katherine Louisa
Wilkins, Cletus Walter, Jr
Wilkins, Raymond Leslie
Williams, Donald Robert
Williams, John Roderick
Williamson, Hugh A
Wineholt, Robert Leese
Winstead, Meldrum Barnett, Jr
Wintner, Claude Edward
Witschard, Gilbert
Woodward, David Willcox
Wu, Ching-Yong
Wuchter, Richard B
Wunz, Paul Richard, Jr
Yeager, Sandra Ann
Yost, John Franklin
Younes, Usama E
Young, Thomas Edwin
Zacharias, David Edward
Zajacek, John George
Zanger, Murray
Zell, Howard Charles

RHODE ISLAND

Abell, Paul Irving
Casey, John Edward, Jr
Cheer, Clair James
Di Pippo, Ascanio G
Elango, Varadaraj
Forbes, Malcolm Holloway
Galkowski, Theodore Thaddeus
Goodman, Leon
Kroll, Harry
Laferriere, Arthur L
Lawler, Ronald George
MacKay, Francis Patrick
Magyar, Elaine Stedman
Magyar, James George
Meschino, Joseph Albert
Morris, David Julian
Moyerman, Robert Max
Nace, Harold Russ
Nugent, James F
Parker, Kathlyn Ann
Perry, Edward Mahlon
Quinn, James Gerard
Rerick, Mark Newton
Rieger, Anne Lloyd
Rosen, William M
Saltzman, Martin D
Slater, Schuyler G
Stokes, William Moore
Suggs, John William
Tien, Rex Yuan
Turcotte, Joseph George
Von Riesen, Daniel Dean
Williams, John Collins, Jr
Williard, Paul Gregory

SOUTH CAROLINA

Abramovitch, Rudolph Abraham Haim
Bailey, Roy Horton, Jr
Ball, Frank Jervery
Ballentine, Alva Ray
Barkley, Lloyd Blair
Bauknight, Charles William, Jr
Beam, Charles Fitzhugh, Jr
Berger, Richard S
Bishop, Muriel Boyd
Bliss, Arthur Dean
Bly, Robert Stewart
Brooks, Thomas
Buurman, Clarence Harold
Cantrill, James Egbert
Carnes, Richard Albert
Carter, Kenneth Nolon
Cavin, William Pinckney
Cogswell, George Wallace
Crounse, Nathan Norman
DeBrunner, Ralph Edward
Dieter, Janice Wong
Dieter, Richard Karl
Eastes, Frank Elisha
Evans, David Wesley
Falkehag, S Ingemar
Farmer, Larry Bert
Fauth, David Jonathan
Gillespie, Robert Howard
Glymph, Eakin Milton
Goldstein, Herman Bernard
Griffith, Elizabeth Ann Hall
Groce, William Henry (Bill)
Guthrie, Roger Thackston
Hardwicke, James Ernest, Jr
Hartz, Roy Eugene
Henderson, Richard Wayne
Hill, Arthur Joseph, Jr
Holsten, John Robert
Huffman, John William, Jr
Hutter, George Frederick
Hynes, John Barry
Jeremias, Charles George
Jones, Edward Stephen
Kelly, Robert James
Knapp, Daniel Roger
Knowles, Cecil Martin
Koli, Andrew Kaitan
Krantz, Karl Walter
Kubler, Donald Gene
Kuhn, Hans Heinrich
Leopold, Robert Summers
Liebman, Arnold Alvin
Lindemann, Martin K
Machell, Greville
Maricq, John
Marshall, James Arthur
Martin, Tellis Alexander
Marullo, Nicasio Philip
Miley, John Wulbern
Moore, Alexander Mazyck
Odom, Homer Clyde, Jr
Parker, Julian E, III
Pawson, Beverly Ann
Peterson, Paul E
Porter, John J
Posey, Robert Giles
Preiss, Donald Merle
Robinson, Myrtle Tonne
Robinson, Robert Earl
Ropp, Gus Anderson
Ross, Stanley Elijah
Rothrock, George Moore
Rowlett, Russell Johnston, Jr
Sanderfer, Paul Otis
Schaefer, Frederic Charles
Schneider, William Paul
Scruggs, Jack G
Shah, Hamish V
Sheehan, William C
Sick, Lowell Victor
Sorell, Henry P
Teague, Peyton Clark
Theuer, William John
Tour, James Mitchell
Von Rosenberg, Joseph Leslie, Jr
Wagner, William Sherwood
Wearn, Richard Benjamin
Weinstock, Leonard M
Weiss, James Owen
Weiss, Michael Karl
Wishman, Marvin

SOUTH DAKOTA

Hanson, Milton Paul
Kintner, Robert Roy
Lewis, David Edwin
Scott, George Prescott
Shryock, Gerald Duane
Stoner, Marshall Robert
Wadsworth, William Steele, Jr
Wagner, Charles Roe

TENNESSEE

Adcock, James Luther
Alexandratos, Spiro
Anjaneyulu, P S R
Bahner, Carl Tabb
Baillargeon, Victor Paul
Baldwin, Charles M
Ball, Frances Louise
Barron, Eugene Roy
Barton, Kenneth Ray
Benjamin, Ben Monte
Benton, Charles Herbert
Bentz, Ralph Wagner
Blackstock, Silas Christopher
Boone, James Ronald
Bowman, Newell Stedman
Brennan, Michael Edward
Bucovaz, Edsel Tony
Caflisch, Edward George
Chitwood, James Leroy
Clark, Richard Bennett
Clemens, Robert Jay
Cleveland, James Perry
Cliffton, Michael Duane
Collins, Jerry Dale
Coover, Harry Wesley, Jr
Dale, John Irvin, III
Daniel, Douglas
Dean, Walter Lee
Dombroski, John Richard
Eastham, Jerome Fields
Elliott, Irvin Wesley
Fagerburg, David Richard
Fenyes, Joseph Gabriel Egon
Field, Lamar
Finch, Gaylord Kirkwood
Foster, Charles Howard
Gagen, James Edwin
Galbraith, Harry Wilson
Germroth, Ted Calvin
Gerwe, Roderick Daniel
Gibson, Gerald W
Gilliom, Richard D
Gilow, Helmuth Martin
Gleason, Edward Hinsdale, Jr
Griffin, Guy David
Griscom, Richard William
Gross, Benjamin Harrison
Harding, Charles Enoch
Harris, Thomas Munson
Harwell, Kenneth Elzer
Hasek, Robert Hall
Hayes, Robert M
Hess, Bernard Andes, Jr
Ho, Patience Ching-Ru
Hoffman, Kenneth Wayne
Holmes, Jerry Dell
Hudnall, Phillip Montgomery
Hutchinson, James Herbert, Jr
Hyatt, John Anthony
Irick, Gether, Jr
Jenkins, James William

Jones, Glenn Clark
Kabalka, George Walter
Kammann, Karl Philip, Jr
Kennedy, Robert Wilson
Kirchner, Frederick Karl
Knee, Terence Edward Creasey
Kreh, Donald Willard
Krutak, James John, Sr
Kuo, Chung-Ming
Lane, Charles A
Langford, Paul Brooks
Lasslo, Andrew
Lowe, James N
Lowe, James Urban, II
Lu, Mary Kwang-Ruey Chao
Lura, Richard Dean
McCarthy, John F
McCollum, Anthony Wayne
McCombs, Charles Allan
McConnell, Richard Leon
McDaniel, Edgar Lamar, Jr
Magid, Ronald
Mani, Rama I
Marnett, Lawrence Joseph
Martin, James Cullen
Martin, James Cuthbert
Martin, Richard Blazo
Meen, Ronald Hugh
Miller, James L
Miller, Robert Witherspoon
Nagel, Fritz John
Newland, Gordon Clay
O'Connor, Timothy Edmond
Pagni, Richard
Parish, Harlie Albert, Jr
Pearson, Donald Emanual
Penner, Hellmut Philip
Pera, John Dominic
Perry, Lloyd Holden
Peterson, William Roger
Pond, David Martin
Poutsma, Marvin Lloyd
Puerckhauer, Gerhard Wilhelm Richard
Raaen, Vernon F
Rash, Fred Howard
Redfearn, Richard Daniel
Reynolds, Brian Edgar
Robinson, Charles Nelson
Rutenberg, Aaron Charles
Sachleben, Richard Alan
Schreiber, Eric Christian
Smith, Howard E
Smith, Jerry Howard
Snell, Robert L
Solomons, William Ebenezer
Spadafino, Leonard Peter
Stites, Joseph Gant, Jr
Sublett, Bobby Jones
Sweetman, Brian Jack
Tarbell, Dean Stanley
Thiessen, William Ernest
Thweatt, John G
Todd, Peter Justin
Tuleen, David L
Turner, S Richard
Tyczkowski, Edward Albert
Vachon, Raymond Normand
Van Sickle, Dale Elbert
Volpe, Angelo Anthony
Waddell, Thomas Groth
Wang, Richard Hsu-Shien
Warren, Mitchum Ellison, Jr
Washington, Arthur Clover
Watts, Exum DeVer
Webb, James L A
Wicker, Thomas Hamilton, Jr
Wilkin, Louis Alden
Willcott, Mark Robert, III
Witzeman, Jonathan Stewart
Wyse, B D
Zoeller, Joseph Robert

TEXAS
Adam, Klaus
Adamcik, Joe Alfred
Adams, Leon Milton
Albach, Roger Fred
Allen, Robert Paul
Allen, Robert Ray
Ansari, Guhlam Ahmad Shakeel
Applewhite, Thomas H
Arganbright, Robert Philip
Ashton, Joseph Benjamin
Aslam, Mohammad
Aspelin, Gary B
Aufdermarsh, Carl Albert, Jr
Bailey, Philip Sigmon
Bartsch, Richard Allen
Bauer, Ronald Sherman
Bauld, Nathan Louis
Beede, Charles Herbert
Beier, Ross Carlton
Belew, John Seymour
Benitez, Francisco Manuel
Bennett, Robert Putnam
Benson, Royal H
Bergbreiter, David Edward
Besozzi, Alfio Joseph
Bhatia, Kishan
Biehl, Edward Robert
Billups, W Edward
Birney, David Martin
Borchardt, John Keith
Bost, Robert Orion

Brader, Walter Howe, Jr
Brady, Donnie Gayle
Brady, William Thomas
Brandenberger, Stanley George
Brown, Dale Gordon
Busby, Hubbard Taylor, Jr
Bush, Warren Van Ness
Cabanes, William Ralph, Jr
Caldwell, Richard A
Cameron, Margaret Davis
Cargill, Robert Lee, Jr
Carlton, Donald Morrill
Castrillon, Jose P A
Caswell, Lyman Ray
Cate, Rodney Lee
Cates, Lindley A
Cavender, James Vere, Jr
Ciufolini, Marco A
Cobb, Raymond Lynn
Cogdell, Thomas James
Cole, Larry Lee
Condray, Ben Rogers
Cook, Paul Fabyan
Cox, James Reed, Jr
Cragoe, Edward Jethro, Jr
Crawford, James Dalton
Crawford, James Worthington
Croce, Louis J
Croft, Thomas Stone
Cross, John Parson
Cupples, Barrett L(eMoyne)
Cuscurida, Michael
Cywinski, Norbert Francis
Damodaran, Kalyani Muniratnam
Darensbourg, Donald Jude
Darensbourg, Marcetta York
Darlage, Larry James
Davenport, Kenneth Gerald
Dawes, John Leslie
Dear, Robert E A
De La Mare, Harold Elison
Denekas, Milton Oliver
Dill, Charles William
Dillard, Robert Garing, Jr
Dimitroff, Edward
Doyle, Michael P
Dukatz, Ervin L, Jr
Dunn, Danny Leroy
Duty, Robert C
Dvoretzky, Isaac
Dwyer, Lawrence Arthur
Edmondson, Morris Stephen
Ekerdt, John Gilbert
Ellison, Robert Hardy
Elsenbaumer, Ronald Lee
Engel, Paul Sanford
Etter, Raymond Lewis, Jr
Everse, Johannes
Fain, Robert C
Fan, Joyce Wang
Fawcett, Colvin Peter
Ferraris, John Patrick
Fitch, John William, III
Floyd, Joseph Calvin
Floyd, Willis Waldo
Fodor, George E
Fonken, Gerhard Joseph
Ford, George Peter
Fox, Marye Anne
Francis, William Connett
Franzl, Robert E
Fuchs, Richard
Fukuyama, Tohru
Garrett, James M
Gerlach, John Louis
Giam, Choo-Seng
Gilbert, John Carl
Goldstein, Herbert Jay
Gordon, Wayne Lecky
Graves, Robert Earl
Gryting, Harold Julian
Guerrant, William Barnett, Jr
Guidry, Carlton Levon
Gum, Wilson Franklin, Jr
Gutsche, Carl David
Gwynn, Donald Eugene
Ham, George Edward
Hamilton, Janet V
Harding, Kenn E
Harlan, Horace David
Hautala, Richard Roy
Hayman, Alan Conrad
Haynes, George Rufus
Headley, Allan Dave
Hedrich, Loren Wesley
Herndon, William Cecil
Hickner, Richard Allan
Hobbs, Charles Clifton, Jr
Hoblit, Louis Douglas
Hogg, John Leslie
Holst, Edward Harland
Horeczy, Joseph Thomas
Houston, James Grey
Huang, Charles T L
Hunt, Richard Henry
Hurdis, Everett Cushing
Idoux, John Paul
Imhoff, Michael Andrew
Isbell, Arthur Furman
Iyer, Ramakrishnan S
Jacobus, Otha John
Jaszberenyi, Joseph C
Jeskey, Harold Alfred
Johnson, Fred Lowery, Jr

Johnson, James Elver
Johnson, Malcolm Pratt
Jones, Paul Ronald
Journeay, Glen Eugene
Kamego, Albert Amil
Kaye, Howard
Keith, Lawrence H
Kelsey, Donald Ross
Kerr, Ralph Oliver
Kirk, James Curtis
Kmiecik, James Edward
Knapp, Roger Dale
Kochi, Jay Kazuo
Kohn, Harold Lewis
Koons, Charles Bruce
Kraychy, Stephen
Kyba, Evan Peter
Lakshmanan, P R
Landolt, Robert George
Larkin, John Michael
Leftin, Harry Paul
Levy, Leon Bruce
Lewis, Edward Sheldon
Liehr, Joachim G
Lin-Vien, Daimay
Lloyd, Winston Dale
Lonzetta, Charles Michael
Lowrey, Charles Boyce
Lundeen, Allan Jay
Lyle, Gloria Gilbert
Lyle, Robert Edward, Jr
Mabry, Tom Joe
McCown, Joseph Dana
McCoy, David Ross
McCullough, James Douglas, Jr
McCullough, Thomas F
McGirk, Richard Heath
Malpass, Dennis B
Mango, Frank Donald
Mangold, Donald J
Manning, Harold Edwin
Marchand, Alan Philip
Marquis, Edward Thomas
Marshall, James Lawrence
Martin, Charles William
Martin, Stephen Frederick
Marx, John Norbert
Mason, Perry Shipley, Jr
May, James Aubrey, Jr
Meinschein, Warren G
Melville, Marjorie Harris
Mendez, Victor Manuel
Miller, Emery B
Mills, Nancy Stewart
Moltzan, Herbert John
Monti, Stephen Arion
Muhs, Merrill Arthur
Mulvey, Dennis Michael
Musser, Michael Tuttle
Naae, Douglas Gene
Nachman, Ronald James
Nagyvary, Joseph
Nash, William Donald
Naylor, Carter Graham
Newsom, Raymond A
Newton, Robert Andrew
Oakes, Billy Dean
O'Brien, Daniel H
O'Farrell, Charles Patrick
O'Neal, Hubert Ronald
Otken, Charles Clay
Padgett, Algie Ross
Patil, Kashinath Z(iparu)
Patton, Leo Wesley
Patton, Tad LeMarre
Peck, Merlin Larry
Piziak, Veronica Kelly
Plank, Don Allen
Platte, Howard
Plummer, Benjamin Frank
Poe, Richard D
Pomerantz, Martin
Proops, William Robert
Pullig, Tillman R(upert)
Ramanujam, V M Sadagopa
Rao, Pemmaraju Narasimha
Razniak, Stephen L
Reeder, Charles Edgar
Reeves, W Preston
Reid, James Cutler
Reinecke, Manfred Gordon
Rhodes, Robert Carl
Richardson, Arian Gilbert
Richter, Reinhard Hans
Rigdon, Orville Wayne
Roberts, Royston Murphy
Roberts, Thomas David
Robinson, Alfred Green
Robinson, J(ames) Michael
Roehrig, Gerald Ralph
Rosenquist, Edward P
Rowton, Richard Lee
Russell, Thomas Webb
Sample, Thomas Earl, Jr
Sand, Ralph E
Schimelpfenig, Clarence William
Schnizer, Arthur Wallace
Schueler, Bruno Otto Gottfried
Scott, Alastair Ian
Sessler, Jonathan Lawrence
Shine, Henry Joseph
Shumate, Kenneth McClellan
Simpson, Billy Doyle
Simpson, James Urban

Singleton, David Michael
Skarlos, Leonidas
Skinner, Charles Gordon
Smith, Curtis William
Smith, Leland Leroy
Smith, William Burton
Smolinsky, Gerald
Smutny, Edgar Josef
Snapp, Thomas Carter, Jr
Soltes, Edward John
Sonntag, Norman Oscar Victor
Sonntag, Roy Windham
Soulen, Robert Lewis
Speranza, George Phillip
Steinhardt, Charles Kendall
Stephenson, Danny Lon
Strom, E(dwin) Thomas
Sullivan, Thomas Allen
Sund, Eldon H
Supple, Jerome Henry
Tanner, Alan Roger
Ternay, Andrew Louis, Jr
Thompson, Michael McCray
Thyagarajan, B S
Trotter, John Wayne
Tweedie, Virgil Lee
Unruh, Jerry Dean
Vanderbilt, Jeffrey James
Van Dijk, Christiaan Pieter
Varma, Rajender S
Venier, Clifford George
Vogelfanger, Elliot Aaron
Wagner, Frank S, Jr
Walbrick, Johnny Mac
Walter, Reuben
Wang, Ting-Tai Helen
Warner, Charles D
Wentland, Stephen Henry
Wheeler, Edward Norwood
Whitesell, James Keller
Whittle, John Antony
Williams, Charles Herbert
Willson, Carlton Grant
Winfrey, J C
Witt, Enrique Roberto
Wood, Randall Dudley
Woodyard, James Douglas
Workman, Wesley Ray
Wulfers, Thomas Frederick
Yager, Billy Joe
Yarian, Dean Robert
Yeakey, Ernest Leon
Zlatkis, Albert

UTAH
Adams, Michael Curtis
Aldous, Duane Leo
Allred, Evan Leigh
Alvord, Donald C
Beishline, Robert Raymond
Bentrude, Wesley George
Blackham, Angus Udell
Bowman, Carlos Morales
Bradshaw, Jerald Sherwin
Broadbent, Hyrum Smith
Broom, Arthur Davis
Dehm, Henry Christopher
DelMar, Eric G
Elmslie, James Stewart
Epstein, William Warren
Foltz, Rodger L
Gladysz, John A
Guillot, David G
Hall, David Warren
Hawkins, Richard Thomas
Hinshaw, Jerald Clyde
Horton, Walter James
Hunter, Byron Alexander
Johnson, John Hal
Johnson, LaVell R
Nelson, Kay LeRoi
Nielsen, Donald R
Olsen, Richard Kenneth
Paul, Edward Gray
Phillips, Lee Revell
Robins, Morris Joseph
Smith, Grant Gill
Stang, Peter John
Sudweeks, Walter Bentley
Thompson, Grant
Van Orden, Harris O
Weinshenker, Ned Martin

VERMONT
Bushweller, Charles Hackett
Gleason, Robert Willard
Harnest, Grant Hopkins
Hawthorne, Robert Montgomery, Jr
Jewett, John Gibson
Kaplow, Leonard Samuel
Krapcho, Andrew Paul
Kuehne, Martin Eric
McCormack, John Joseph, Jr
Magnien, Ernest
Murray, James Gordon
Torkelson, Arnold
White, William North
Woodworth, Robert Cummings
Worrall, Winfield Scott

VIRGINIA
Abraham, Donald James
Atkins, Robert Charles
Augl, Joseph Michael

Organic Chemistry (cont)

Baedecker, Mary Jo
Bass, Robert Gerald
Bell, Charles E, Jr
Bertram, Leon Leroy
Beyad, Mohammed Hossain
Bikales, Norbert M
Boatman, Sandra
Breder, Charles Vincent
Brownlee, Paula Pimlott
Burke, Hanna Suss
Burton, Willard White
Carey, Francis Arthur
Castagnoli, Neal, Jr
Chamot, Dennis
Chlebowski, Jan F
Chong, Shuang-Ling
Christian, Jack G
Clark, Allen Keith
Clough, Stuart Chandler
Cox, Richard Harvey
Crowell, Thomas Irving
Dahlgard, Muriel Genevieve
Dardoufas, Kimon C
Deinet, Adolph Joseph
DeVries, George H
Dillard, John Gammons
Easter, Donald Philips
Eby, Charles J
Edwards, William Brundige, III
Einolf, William Noel
Fisher, Charles Harold
Fitzgerald, James Allen
Fonong, Tekum
Gager, Forrest Lee, Jr
Gal, Andrew Eugene
Gamble, Dean Franklin
Gandour, Richard David
Gerow, Clare William
Gibson, Harry William
Gillespie, Jesse Samuel, Jr
Glassco, William Shoemaker
Goller, Edwin John
Gould, David Huntington
Gratz, Roy Fred
Griffiths, David Warren
Grivsky, Eugene Michael
Haas, Carol Kressler
Hall, Philip Layton
Hammer, Gary G
Hansrote, Charles Johnson, Jr
Harowitz, Charles Lichtenberg
Hartzler, Jon David
Harvey, William Ross
Hassler, William Woods
Hazlett, Robert Neil
Heisey, Lowell Vernon
Henderson, Ulysses Virgil, Jr
Henson, Paul D
Heuberger, Oscar
Hill, Carl McClellan
Hill, Trevor Bruce
Hudlicky, Milos
Hudlicky, Tomas
Hudson, Frederick Mitchell
Hunt, Donald F
Hutchison, David Allan
Ihrman, Kryn George
Jelinek, Charles Frank
Jensen, Arnold William
Johnson, Fatima Nunes
Johnston, Byron E
Jonas, John Joseph
Kallianos, Andrew George
Kauffman, Glenn Monroe
Kingston, David George Ian
Klingensmith, George Bruce
Kondo, Norman Shigeru
Konizer, George Burr
Lambert, Rogers Franklin
Landis, Phillip Sherwood
Lapporte, Seymour Jerome
Leake, Preston Hildebrand
Lefebvre, Yvon
Link, William B
Lipnick, Robert Louis
Little, Edwin Demetrius
McClenon, John R
Macdonald, Timothy Lee
McEntee, Thomas Edwin
McGrath, James Edward
McHenry, William Earl
Mateer, Richard Austin
Matuszko, Anthony Joseph
Melton, Thomas Mason
Mengenhauser, James Vernon
Merkel, Timothy Franklin
Meyer, Leo Francis
Monroe, Stuart Benton
Moore, Theron Langford
Nealy, David Lewis
O'Brien, Michael Harvey
Ogilvie, James William, Jr
Ogliaruso, Michael Anthony
Ottenbrite, Raphael Martin
Oyama, Shigeo Ted
Patrick, James Burns
Perozzi, Edmund Frank
Pine, Stanley H
Rainer, Norman Barry
Ramirez, Fausto
Richardson, William Harry
Rim, Yong Sung

Roberts, David Craig
Rodig, Oscar Rudolf
Roscher, David Moore
Rosenberg, Murray David
Ross, Alexander
Roth, Ronald John
Rubin, Alan Barry
Rubottom, George M
Sahli, Muhammad S
Sastri, Vinod Ram
Sayala, Chhaya
Scharpf, William George
Schiavelli, Melvyn David
Schirch, Laverne Gene
Schulz, Johann Christoph Friedrich
Seligman, Robert Bernard
Sheehan, Desmond
Shen, Tsung Ying
Shillington, James Keith
Showell, John Sheldon
Slayden, Suzanne Weems
Smith, Craig La Salle
Smith, James Doyle
Southwick, Everett West
Stalick, Wayne Myron
Starkovsky, Nicolas Alexis
Starnes, William Herbert, Jr
Stubbins, James Fiske
Stubblefield, Frank Milton
Sundberg, Richard J
Taylor, William Irving
Teng, Lina Chen
Thompson, David Wallace
Thompson, Mark Ewell
Turer, Jack
Urbanik, Arthur Ronald
Uwaydah, Ibrahim Musa
Verell, Ruth Ann
Walker, Grayson Watkins
Walton, Theodore Ross
Wasti, Khizar
Wayland, Rosser Lee, Jr
Welch, Raymond Lee
Wetmore, Stanley Irwin, Jr
Whitney, George Stephen
Wilkinson, William Kenneth
Willett, James Delos
Williams, Roy Lee
Wilson, Donald Richard
Winters, Lawrence Joseph
Wolfe, James F
Woo, Yin-tak
Yost, William Lassiter
Zaborsky, Oskar Rudolf

WASHINGTON

Alberts, Arnold A
Allan, George Graham
Andersen, Niels Hjorth
Anderson, Arthur G, Jr
Anderson, Charles Dean
Babad, Harry
Baillie, Thomas Allan
Barrueto, Richard Benigno
Beelik, Andrew
Beug, Michael William
Blake, James J
Bocksch, Robert Donald
Borden, Weston Thatcher
Briggs, William Scott
Chambers, James Richard
Cheng, Xueheng
Conca, Romeo John
Cooke, Manning Patrick, Jr
Dann, John Robert
Duhl-Emswiler, Barbara Ann
Fletcher, Thomas Lloyd
Frankart, William A
Franz, James Alan
Giddings, William Paul
Godar, Edith Marie
Goheen, David Wade
Hansen, Michael Roy
Hedges, John Ivan
Herrick, Franklin Willard
Hine, John Maynard
Hoblitt, Richard Patrick
Holzman, George
Hopkins, Paul Brink
Huestis, Laurence Dean
Jacoby, Lawrence John
Johnson, Carl Arnold
Johnson, Donald Curtis
Kaplan, Phyllis Deen
Kreibich, Roland
Kriz, George Stanley
Kruger, Albert Aaron
Lago, James
Lampman, Gary Marshall
Latourette, Harold Kenneth
Leonard, John Edward
Lepse, Paul Arnold
McDonough, Leslie Marvin
McMillan, Kirk Dugald
Matteson, Donald Stephen
Mayer, Julian Richard
Meyer, Rich Bakke, Jr
Minckler, Leon Sherwood, Jr
Nelson, Randall Bruce
Orth, George Otto, Jr
Patt, Leonard Merton
Pavia, Donald Lee
Pocker, Yeshayau
Raucher, Stanley

Read, David Hadley
Reintjes, Marten
Ronald, Robert Charles
Rowland, Stanley Paul
Schubert, Wolfgang Manfred
Sears, Karl David
Senear, Allen Eugene
Short, William Arthur
Stacy, Gardner W
Steckler, Bernard Michael
Stout, Virginia Falk
Trager, William Frank
Varanasi, Usha
Vessel, Eugene David
Wade, Leroy Grover, Jr
Walecka, Jerrold Alberts
Warner, John Scott
Wasserman, William Jack
Wilbur, Daniel Scott
Wirth, Joseph Glenn
Wither, Ross Plummer
Wollwage, Paul Carl
Wong, Chun-Ming
Young, Robert Hayward
Zimmerman, Gary Alan

WEST VIRGINIA

Alm, Robert M
Anderson, Gary Don
Barnes, Robert Keith
Bartley, William J
Bryant, David
Campbell, Clyde Del
Capstack, Ernest
Digman, Robert V
Douglass, James Edward
Doumaux, Arthur Roy, Jr
Draper, John Daniel
Driesch, Albert John
Dunphy, James Francis
Elkins, John Rush
Fodor, Gabor
Giza, Chester Anthony
Giza, Yueh-Hua Chen
Harrison, Arnold Myron
Hess, Lawrence George
Hubbard, John Lewis
Hussey, Edward Walter
Johnson, George Frederick
Johnson, Richard Lawrence
Kaplan, Leonard
Knopf, Robert John
Knowles, Richard N
Kurland, Jonathan Joshua
Lamey, Steven Charles
Lemke, Thomas Franklin
McCain, James Herndon
MacDowell, Denis W H
MacPeek, Donald Lester
Markiw, Roman Teodor
Matthews, Virgil Edison
Moore, William Robert
Muth, Chester William
Ode, Richard Herman
Osborn, Claiborn Lee
Papa, Anthony Joseph
Peascoe, Warren Joseph
Pianfetti, John Andrew
Rankin, Gary O'Neal
Ream, Bernard Claude
Reid, C Glenn
Robson, John Howard
Ruoff, William (David)
Sandridge, Robert Lee
Sherman, Paul Dwight, Jr
Smith, James Allbee
Speck, Rhoads McClellan
Steiner, Werner Douglas
Steinle, Edmund Charles, Jr
Volker, Eugene Jeno

WISCONSIN

Anderson, Stephen William
Baetke, Edward A
Bender, Margaret McLean
Biester, John Louis
Boggs, Lawrence Allen
Boye, Frederick C
Brown, Kenneth Howard
Brown, William Henry
Burke, Steven Douglas
Calvanico, Nickolas Joseph
Casey, Charles P
Chenier, Philip John
Chitharanjan, D
Cleary, James William
Conigliaro, Peter James
Cremer, Sheldon E
Crimmins, Timothy Francis
Dalrymple, David Lawrence
Dickerson, Charlesworth Lee
Doerr, Robert George
D'Orazio, Vincent T
Dwyer, Sean G
Dwyer-Hallquist, Patricia
Farnsworth, Carl Leon
Feist, William Charles
Friedlander, William Sheffield
Garmaise, David Lyon
Gellman, Samuel Helmer
Goering, Harlan Lowell
Greene, Charles Edwin
Griffith, Owen W
Hamm, Kenneth Lee

Harkin, John McLay
Hart, Phillip A
Hedden, Gregory Dexter
Helms, John F
Hill, Elgin Alexander
Hoffman, Norman Edwin
Hollenberg, David Henry
Horton, Joseph William
Isenberg, Norbert
Johnson, Hal G(ustav)
Klink, Joel Richard
Klopotek, David L
Kolb, Vera
Kyung, Jai Ho
Landucci, Lawrence L
Lee, Kathryn Adele Bunding
Lepeska, Bohumir
Lin, Stephen Y
Lokensgard, Jerrold Paul
Losin, Edward Thomas
McClenahan, William St Clair
McKelvey, Ronald Deane
Magnuson, Eugene Robert
Manly, Donald G
Mathews, Frederick John
Megahed, Sid A
Moore, Leonard Oro
Nelsen, Stephen Flanders
Nitz, Otto William Julius
Ochrymowycz, Leo Arthur
Olson, Melvin Martin
Pearl, Irwin Albert
Perlman, Kato (Katherine) Lenard
Post, Elroy Wayne
Posvic, Harvey Walter
Priest, Matthew A
Puhl, Richard James
Raines, Ronald T
Rainville, David Paul
Rausch, Gerald
Reich, Hans Jurgen
Reich, Ieva Lazdins
Reichenbacher, Paul H
Roberts, Richard W
Roth, Marie M
Rowe, John Westel
Runquist, Alfonse William
Savereide, Thomas J
Scamehorn, Richard Guy
Schnack, Larry G
Schnoes, Heinrich Konstantin
Sedor, Edward Andrew
Shaver, Roy Allen
Sneen, Richard Allen
Soerens, Dave Allen
Sosnovsky, George
Splies, Robert Glenn
Stackman, Robert W
Stevens, Michael Fred
Theine, Alice
Thompson, Norman Storm
Thurmaier, Roland Joseph
Tonnis, John A
Tsao, Francis Hsiang-Chian
Turner, Robert James
Vedejs, Edwin
Ward, Kyle, Jr
Weipert, Eugene Allen
Wendland, Ray Theodore
West, Robert C
Westler, William Milo
Whyte, Donald Edward
Wilds, Alfred Lawrence
Zimmerman, Howard Elliot
Zinkel, Duane Forst

WYOMING

Coates, Geoffrey Edward
Decora, Andrew Wayne
Dorrence, Samuel Michael
Duvall, John Joseph
Harnsberger, Paul Michael
Jaeger, David Allen
Kelly, Floyd W, Jr
McDonald, Francis Raymond
Maurer, John Edward
Nelson, David Alan
Northcott, Jean
Raulins, Nancy Rebecca
Rhoads, Sara Jane
Robertson, Raymond E(liot)
Schultz, Harry Pershing
Seese, William Shober
Speight, James G

PUERTO RICO

Carrasquillo, Arnaldo
Eberhardt, Manfred Karl
Gonzalez De Alvarez, Genoveva
Lamba, Ram Sarup
Souto Bachiller, Fernando Alberto

ALBERTA

Ayer, William Alfred
Bachelor, Frank William
Blackburn, Edward Victor
Brown, Robert Stanley
Cook, David Alastair
Crawford, Robert James
Creighton, Stephen Mark
Dixon, Elisabeth Ann
Elofson, Richard Macleod
Hooz, John
Kopecky, Karl Rudolph

CHEMISTRY / 227

Lemieux, Raymond Urgel
Lown, James William
Micetich, Ronald George
Moschopedis, Speros E
Robertson, Ross Elmore
Sorensen, Theodore Strang
Tanner, Dennis David
Wayman, Morris
Williams, Jack L R

BRITISH COLUMBIA
Chalmers, William
Chow, Yuan Lang
Dutton, Guy Gordon Studdy
Freter, Kurt Rudolf
Gardner, Joseph Arthur Frederick
Hayward, Lloyd Douglas
Hocking, Martin Blake
Howard, John
Kiehlmann, Eberhard
Kutney, James Peter
Manville, John Fieve
Mitchell, Reginald Harry
Piers, Edward
Pincock, Richard Earl
Preston, Caroline Margaret
Rosenthal, Alex
Sams, John Robert, Jr
Scheffer, John R
Singh, Raj Kumari
Slessor, Keith Norman
Stewart, Ross
Struble, Dean L
Tener, Gordon Malcolm
Weiler, Lawrence Stanley
Withers, Stephen George
Yates, Keith

MANITOBA
Charlton, James Leslie
Chow, Donna Arlene
Hunter, Norman Robert
McKinnon, David M
Wong, Chiu Ming

NEW BRUNSWICK
Adams, Kenneth Allen Harry
Barclay, Lawrence Ross Coates
Findlay, John A
Jankowski, Christopher K
Kelly, Ronald Burger
Strunz, G(eorge) M(artin)
Stuart, Ronald S
Valenta, Zdenek

NEWFOUNDLAND
Anderson, Hugh John
Orr, James Cameron
Shahidi, Fereidoon
Stein, Allan Rudolph

NOVA SCOTIA
Ackman, Robert George
Arnold, Donald Robert
Arseneau, Donald Francis
Aue, Walter Alois
Bunbury, David Leslie
Grindley, T Bruce
Grossert, James Stuart
Hooper, Donald Lloyd
Leffek, Kenneth Thomas
Lynch, Brian Maurice
McCulloch, Archibald Wilson
Ogilvie, Kelvin Kenneth
Piorko, Adam M
White, Robert Lester

ONTARIO
Akhtar, Mohammad Humayoun
Alper, Anne Elizabeth
Alper, Howard
ApSimon, John W
Arsenault, Guy Pierre
Baer, Hans Helmut
Baines, Kim Marie
Bannard, Robert Alexander Brock
Bell, Russell A
Bharucha, Keki Rustomji
Birnbaum, George I
Bohme, Diethard Kurt
Bourns, Arthur N
Brook, Adrian Gibbs
Buchanan, Gerald Wallace
Bunce, Nigel James
Buncel, Erwin
Bunting, John William
Burton, Graham William
Butler, Douglas Neve
Cann, Malcolm Calvin
Childs, Ronald Frank
Clark, Ferrers Robert Scougall
Court, William Arthur
Detellier, Christian
Dolenko, Allan John
Duff, James McConnell
Dunn, John Robert
Durst, Tony
Edwards, Douglas Cameron
Edwards, Oliver Edward
Eisenhauer, Hugh Ross
Fallis, Alexander Graham
Fitt, Peter Stanley
Fraser, Robert Rowntree
Gingras, Bernard Arthur

Gordon, Myra
Graham, Bruce Allan
Greenberg, Bruce Matthew
Guthrie, James Peter
Hach, Vladimir
Harris, John William
Harrison, Alexander George
Harrison, William Ashley
Heacock, Ronald A
Heggie, Robert Murray
Holland, Herbert Leslie
Hopkins, Clarence Yardley
Howard, James Anthony
Hughes, David William
Ingold, Keith Usherwood
Janzen, Edward George
Jones, John Bryan
Jones, Maurice Harry
Kaiser, Klaus L(eo) E(duard)
Kazmaier, Peter Michael
King, James Frederick
Kluger, Ronald H
Kramer, John Karl Gerhard
Krepinsky, Jiri J
Kresge, Alexander Jerry
Lange, Gordon Lloyd
Laughton, Paul MacDonell
Lautens, Mark
Lautenschlaeger, Friedrich Karl
Lee-Ruff, Edward
Leitch, Leonard Christie
Lesage, Suzanne
Lewars, Errol George
Leznoff, Clifford Clark
McArthur, Colin Richard
McIntosh, John McLennan
McKague, Allan Bruce
MacLean, David Bailey
McLean, Stewart
Marks, Gerald Samuel
Martin, Trevor Ian
Massiah, Thomas Frederick
Mayo, De Paul
Meresz, Otto
Moir, Robert Young
Morand, Peter
Morita, Hirokazu
Narang, Saran A
Onuska, Francis Ivan
Prokipcak, Joseph Michael
Puddephatt, Richard John
Rakhit, Sumanas
Ratnayake, Walisundera Mudiyanselage Nimal
Rees, Alun Hywel
Reid, Sidney George
Reynolds-Warnhoff, Patricia
Ritcey, Gordon M
Rodrigo, Russell Godfrey
Rutherford, Kenneth Gerald
Sanderson, Edwin S
Schmid, George Henry
Scott, Andrew Edington
Scott, Peter Michael
Serdarevich, Bogdan
Siddiqui, Iqbal Rafat
Sniekus, Victor A
Sowa, Walter
Spencer, Elvins Yuill
Still, Ian William James
Stothers, John Bailie
Strachan, William Michael John
Szabo, Arthur Gustav
Szarek, Walter Anthony
Tidwell, Thomas Tinsley
Tuck, Dennis George
Usselman, Melvyn Charles
Vijay, Hari Mohan
Warkentin, John
Warnhoff, Edgar William
Weedon, Alan Charles
Werstiuk, Nick Henry
Westaway, Kenneth C
Wigfield, Donald Compston
Wigfield, Yuk Yung
Wightman, Robert Harlan
Winnik, Francoise Martine
Winnik, Mitchell Alan
Winthrop, Stanley Oscar
Wood, Gordon Walter
Woolford, Robert Graham
Yates, Peter
Young, James Christopher F
Ziegler, Peter

PRINCE EDWARD ISLAND
Liu, Michael T H
Palmer, Glenn Earl
Rigney, James Arthur

QUEBEC
Atkinson, Joseph George
Belanger, Patrice Charles
Brown, Gordon Manley
Burnell, Robert H
Canonne, Persephone
Chan, Tak-Hang
Chenevert, Robert (Bernard)
Chubb, Francis Learmonth
Clayton, David Walton
Colebrook, Lawrence David
Cook, Robert Douglas
Daessle, Claude
Darling, Graham Davidson

Deans, Sidney Alfred Vindin
DeMedicis, E M J A
Deslongchamps, Pierre
Diksic, Mirko
Doughty, Mark
Dugas, Hermann
Edward, John Thomas
Engel, Charles Robert
Favre, Henri Albert
Ferdinandi, Eckhardt Stevan
Fliszar, Sandor
Garneau, Francois Xavier
Gaudry, Roger
Gravel, Denis Fernand
Gurudata, Neville
Hamlet, Zacharias
Hanessian, Stephen
Hay, Allan Stuart
Heisler, Seymour
Heitner, Cyril
Joly, Louis Philippe
Just, George
Khalil, Michel
Lau, Cheuk Kun
Lennox, Robert Bruce
Lepine, Francois
Lessard, Jean
Paice, Michael
Perlin, Arthur Saul
Perron, Yvon G
Portelance, Vincent Damien
Powell, William St John
Richer, Jean-Claude
Roughley, Peter James
Vincent, Donald Leslie
Wuest, James D
Yeats, Ronald Bradshaw
Zamir, Lolita Ora

SASKATCHEWAN
Chandler, William David
Gear, James Richard
Grant, Donald R
Lee, Donald Garry
Martin, Robert O
Mezey, Paul G
Pepper, James Morley
Slater, George P
Steck, Warren Franklin

OTHER COUNTRIES
Baddiley, James
Baker, Frank Weir
Barnes, Roderick Arthur
Bhatnagar, Ajay Sahai
Boffa, Lidia C
Booth, Gary Edwin
Boustany, Kamel
Brown, Peter
Bussert, Jack Francis
Carter, Robert Everett
Chan, Hak-Foon
Chang, Shuya
Choi, Yong Chun
Collins, Carol Hollingworth
Diederich, Francois Nico
Edmonds, James W
Evleth, Earl Mansfield
Fersht, Alan Roy
Ghosh, Anil Chandra
Goldstein, Melvin Joseph
Graber, Robert Philip
Griffin, Anselm Clyde, III
Grisar, Johann Martin
Hajos, Zoltan George
Hassner, Alfred
Hummel, Hans Eckhardt
Hutchison, John Joseph
Kellogg, Richard Morrison
Kent, Stephen Brian Henry
Kim, Dong Han
Kosower, Edward Malcolm
Krakower, Gerald W
Kuttab, Simon Hanna
Levand, Oscar
Levin, Gideon
Little, John Stanley
Lonsdale-Eccles, John David
Low, Chow-Eng
Low, Teresa Lingchun Kao
Lowrance, William Wilson, Jr
McCullough, John James
Malherbe, Roger F
Meyers, Martin Bernard
Millar, John Robert
Milligan, Barton
Paik, Young-Ki
Pohland, Hermann W
Prelog, Vladimir
Schleyer, Paul von Rague
Siegel, Herbert
Slagel, Robert Clayton
Solomons, Thomas William Graham
Spaght, Monroe Edward
Strong, Jerry Glenn
Trentham, David R
Trofimov, Boris Alexandrovich
Vasilakos, Nicholas Petrou
Von Stryk, Frederick George
Ward, John Edward

Pharmaceutical Chemistry

ALABAMA
Beasley, James Gordon
Darling, Charles Milton
Elliott, Robert Daryl
Hocking, George Macdonald
Kilpatrick, John Michael
Montgomery, John Atterbury
Morin, Richard Dudley
Nair, Madhavan G
Parish, Edward James
Ravis, William Robert
Riley, Thomas N
Sakai, Ted Tetsuo
Shealy, Y(oder) Fulmer
Struck, Robert Frederick
Tan, Boen Hie
Temple, Carroll Glenn
Thomas, Hazel Jeanette
Vacik, James P
Walsh, David Allan
Wilken, Leon Otto, Jr

ARIZONA
Clinton, Raymond Otto
Cole, Jack Robert
Granatek, Alphonse Peter
Hoffmann, Joseph John
Martin, Arnold R
Parks, Lloyd McClain
Reed, Fred DeWitt, Jr
Remers, William Alan
Shively, Charles Dean
Yalkowsky, Samuel Hyman

ARKANSAS
Lattin, Danny Lee
Mittelstaedt, Stanley George
Sorenson, John R J
Tucker, Robert Gene

CALIFORNIA
Ainsworth, Cameron
Bartlett, Janeth Marie
Beattie, Thomas Robert
Benet, Leslie Z
Bernstein, Lawrence Richard
Biles, John Alexander
Boger, Dale L
Bondar, Richard Jay Laurent
Brochmann-Hanssen, Einar
Burlingame, Alma L
Carter, Kenneth
Chan, Kenneth Kin-Hing
Chaubal, Madhukar Gajanan
Chinn, Leland Jew
Chou, Yungnien John
Cleland, Jeffrey Lynn
Colwell, William Tracy
Craig, John Cymerman
Craik, Charles S
Cureton, Glen Lee
Degraw, Joseph Irving, Jr
Dev, Vasu
Felgner, Philip Louis
Fried, John
Friedman, Alan E
Fries, David Samuel
Fung, Steven
Gerber, Bernard Robert
Gordon, Eric Michael
Gordon, Samuel Morris
Goyan, Jere Edwin
Green, Donald Eugene
Gschwend, Heinz W
Guner, Osman Fatih
Hamilton, J(ames) Hugh
Hamor, Glenn Herbert
Haney, David N
Henry, David Weston
Hoener, Betty-ann
Hooker, Thomas M, Jr
Hora, Maninder Singh
Hostetler, Karl Yoder
Hughes, John Lawrence
Jacob, Peyton, III
Johnson, Howard (Laurence)
Jones, Richard Elmore
Jones, Todd Kevin
Katz, Martin
Kenyon, George Lommel
Kertesz, Jean Constance
Kiely, John Steven
Klein, David Henry
Koda, Robert T
Kollman, Peter Andrew
Kong, Eric Siu-Wai
Krezanoski, Joseph Z
Krstenansky, John Leonard
Kryger, Roy George
Lad, Pramod Madhusudan
Landgraf, William Charles
Lasky, Richard David
Lawless, Gregory Benedict
Lazo-Wasem, Edgar A
Lee, David Louis
Leo, Albert Joseph
Lien, Eric Jung-Chi
Lin, Tz-Hong
Ling, Nicholas Chi-Kwan
Livingston, Douglas Alan
Lobl, Thomas Jay
Lucci, Robert Dominick

Pharmaceutical Chemistry (cont)

McGee, Lawrence Ray
McKenna, Charles Edward
Matin, Shaikh Badarul
Matuszak, Alice Jean Boyer
Mayer, William Joseph
Meehan, Thomas (Dennis)
Mitchner, Hyman
Mufson, Daniel
Musser, John H
Najafi, Ahmad
Nematollahi, Jay
Nestor, John Joseph, Jr
Nguyen, Tue H
Nikoui, Nik
Ning, Robert Y
Nutt, Ruth Foelsche
Okamura, William H
Ortiz de Montellano, Paul Richard
Pearlman, Rodney
Perlmutter, Milton Manuel
Pierschbacher, Michael Dean
Pigiet, Vincent P
Rao, Tadimeti Seetapati
Reifenrath, William Gerald
Ripka, William Charles
Robertson, David Wayne
Roscoe, Charles William
Rucker, Robert Blain
Safir, Sidney Robert
Sargent, Thornton William, III
Selassie, Cynthia R
Shackelford, Scott Addison
Shell, John Weldon
Shen, Wei-Chiang
Singaram, Bakthan
Sorby, Donald Lloyd
Tozer, Thomas Nelson
Tretter, James Ray
Untch, Karl G(eorge)
Vollhardt, K Peter C
Wechter, William Julius
Weers, Jeffry Greg
Wender, Paul Anthony
Whitlow, Marc David
Williams, Lewis David
Wolf, Walter
Wolff, Manfred Ernst
Wu, Jiann-Long
Zderic, John Anthony

COLORADO
Barrett, Anthony Gerard Martin
Fitzpatrick, Francis Anthony
Guinn, Denise Eileen
Micheli, Roger Paul
Nachtigall, Guenter Willi
Ruth, James Allan
Seely, Robert J
Thompson, John Alec
Wingeleth, Dale Clifford
Wright, Terry L

CONNECTICUT
Baldwin, Ronald Martin
Banijamali, Ali Reza
Bongiorni, Domenic Frank
Brittelli, David Ross
Church, Robert Fitz (Randolph)
Cohen, Edward Morton
Crast, Leonard Bruce, Jr
Cue, Berkeley Wendell, Jr
Davis, Robert Glenn
Dirlam, John Philip
Doyle, Terrence William
Epling, Gary Arnold
Forssen, Eric Anton
Gauthier, George James
Ghosh, Dipak K
Harbert, Charles A
Harwood, Harold James, Jr
Henkel, James Gregory
Hess, Hans-Jurgen Ernst
Heyd, Allen
Hiskey, Clarence Francis
Hite, Gilbert J
Hofmann, Corris Mabelle
Holland, Gerald Fagan
Horhota, Stephen Thomas
Ives, Jeffrey Lee
Johnson, Bruce McDougall
Kanig, Joseph Louis
Kerridge, Kenneth A
Kluender, Harold Clinton
Kotick, Michael Paul
Kreighbaum, William Eugene
Kriesel, Douglas Clare
LaMattina, John Lawrence
Lawson, John Edward
Lin, Tai-Shun
Luh, Yuhshi
McGregor, Donald Neil
McLamore, William Merrill
McManus, James Michael
Makriyannis, Alexandros
Marfat, Anthony
Mayell, Jaspal Singh
Melvin, Lawrence Sherman, Jr
Milne, George McLean, Jr
Monro, Alastair Macleod
Moppett, Charles Edward
Nieforth, Karl Allen
Ozols, Juris
Rosati, Robert Louis
Ryan, Richard Patrick
Sarges, Reinhard
Schnur, Rodney Caughren
Shapiro, Warren Barry
Suh, John Taiyoung
Tamorria, Christopher Richard
Temple, Davis Littleton, Jr
Thomas, Paul David
Vida, Julius
Volkmann, Robert Alfred
Witczak, Zbigniew J
Yevich, Joseph Paul
Zalucky, Theodore B

DELAWARE
Aldrich, Paul E
Aungst, Bruce J
Baaske, David Michael
Boswell, George A
Carr, John B
Chaudhari, Bipin Bhudharlal
Cherkofsky, Saul Carl
Chorvat, Robert John
Confalone, Pat N
Crow, Edwin Lee
Dubas, Lawrence Francis
Finizio, Michael
Gakenheimer, Walter Christian
Hesp, B
Holyoke, Caleb William, Jr
Huang, Chin Pao
Jackson, Thomas Edwin
Johnson, Alexander Lawrence
Kettner, Charles Adrian
Lam, Gilbert Nim-Car
Lee, Shung-Yan Luke
Lerman, Charles Lew
Mollica, Joseph Anthony
Park, Chung Ho
Quarry, Mary Ann
Rajagopalan, Parthasarathi
Rifino, Carl Biaggio
Sundelin, Kurt Gustav Ragnar
Turlapaty, Prasad
Wuonola, Mark Arvid

DISTRICT OF COLUMBIA
Hajiyani, Mehdi Hussain
Horak, Vaclav
Klayman, Daniel Leslie
La Pidus, Jules Benjamin
Milbert, Alfred Nicholas
Miller, David Jacob
Milstein, Stanley Richard
Poon, Bing Toy
Scott, Kenneth Richard
Telang, Vasant G
Thomas, Richard Dean
Weber, John Donald

FLORIDA
Bodor, Nicholas Stephen
Child, Ralph Grassing
Curlin, Lemuel Calvert
Ebert, William Robley
Fitzgerald, Thomas James
Garrett, Edward Robert
Goldberg, Eugene P
Halpern, Ephriam Philip
Hammer, Henry Felix
Hammer, Richard Hartman
Howes, John Francis
Lee, Henry Joung
Linfield, Warner Max
Mehta, Nariman Bomanshaw
Millor, Steven Ross
Nakashima, Tadayoshi
Pop, Emil
Riedhammer, Thomas M
Rokach, Joshua
Schwartz, Michael Averill
Wallace, Gerald Wayne
Wheeler, Frank Carlisle
Yost, Richard A
Young, Michael David

GEORGIA
Blanton, Charles DeWitt, Jr
Boudinot, Frank Douglas
Brennan, John Joseph
Chu, Chung K
Churchill, Frederick Charles
Doetsch, Paul William
Ellington, James Jackson
Honigberg, Irwin Leon
Israili, Zafar Hasan
La Rocca, Joseph Paul
Liotta, Dennis C
Lopez, Antonio Vincent
Matthews, Hewitt William
Pelletier, S William
Powers, James Cecil
Proctor, Charles Darnell
Rhodes, Robert Allen
Ruenitz, Peter Carmichael
Stewart, James T
Thompson, Bobby Blackburn
Tsao, Mark Fu-Pao
Waters, Kenneth Lee
Woodford, James

HAWAII
Grothaus, Paul Gerard

IDAHO
Goettsch, Robert
Isaacson, Eugene I
Natale, Nicholas Robert

ILLINOIS
Alam, Abu Shafiul
Amsel, Lewis Paul
Arendsen, David Lloyd
Armstrong, P(aul) Douglas
Bauer, Ludwig
Borgman, Robert John
Brigham, Robert B
Brooks, Dee W
Buchman, Russell
Burgauer, Paul David
Cannon, John Burns
Chen, Wen Sherng
Chorghade, Mukund Shankar
Collins, Paul Waddell
Currie, Bruce LaMonte
Debus, Allen George
Dellaria, Joseph Fred, Jr
DeRose, Anthony Francis
Dunn, William Joseph
Eachus, Alan Campbell
Fitzloff, John Frederick
Garland, Robert Bruce
Gearien, James Edward
Hansen, Donald Willis, Jr
Hardwidge, Edward Albert
Jenkins, William Wesley
Jiu, James
Johnson, Dale Howard
Jones, Peter Hadley
Jones, Ralph William
Joslin, Robert Scott
Karim, Aziz
Kerdesky, Francis A J
Kerkman, Daniel Joseph
Kim, Ki Hwan
Kipp, James Edwin
Klimstra, Paul D
Krafft, Grant A
Krimmel, C Peter
Lee, Cheuk Man
Lu, Matthias Chi-Hwa
Martin, Yvonne Connolly
Mohan, Prem
Moon, Byong Hoon
Nysted, Leonard Norman
Pariza, Richard James
Piatak, David Michael
Possley, Leroy Henry
Prasad, Ramanujam Srinivasa
Propst, Catherine Lamb
Ranade, Vinayak Vasudeo
Rao, Gopal Subba
Roderick, William Rodney
Rorig, Kurt Joachim
Roseman, Theodore Jonas
Rosenberg, Saul Howard
Sause, H William
Seidenfeld, Jerome
Sennello, Lawrence Thomas
Shipchandler, Mohammed Tyebji
Shone, Robert L
Siegel, Sheldon
Silverman, Richard Bruce
Singiser, Robert Eugene
Smith, Homer Alvin, Jr
Sommers, Armiger Henry
Stout, David Michael
Tsang, Joseph Chiao-Liang
Ulrey, Stephen Scott
Weston, Arthur Walter
Wimer, David Carlisle
Winn, Martin
Woroch, Eugene Leo
Yamamoto, Diane M
Young, David W
Zimmer, Arthur James

INDIANA
Amundson, Merle E
Bergstrom, Donald Eugene
Boenigk, John William
Born, Gordon Stuart
Boyd, Donald Bradford
Campaigne, Ernest Edward
Chang, Ching-Jer
Childers, Ray Fleetwood
Conine, James William
Cooper, Robin D G
Corey, Paul Frederick
Cushman, Mark
Cwalina, Gustav Edward
DeLong, Allyn F
Dinner, Alan
Dorman, Douglas Earl
Farkas, Eugene
Fricke, Gordon Hugh
Grant, Ernest Walter
Green, Mark Alan
Gutowski, Gerald Edward
Harper, Richard Waltz
Haslanger, Martin Frederick
Havel, Henry Acken
Hulbert, Matthew H
Indelicato, Joseph Michael
Jackson, Bill Grinnell
Kaufman, Karl Lincoln
Kessler, Wayne Vincent
Knevel, Adelbert Michael
Kornfeld, Edmund Carl
Kossoy, Aaron David
Lappas, Lewis Christopher
Larsen, Aubrey Arnold
McIntyre, Thomas William
McKeehan, Charles Wayne
McLaughlin, Jerry Loren
Marshall, Winston Stanley
Massey, Eddie H
Matsumoto, Ken
Mikolasek, Douglas Gene
Morrow, Duane Francis
Murphy, Charles Franklin
Nichols, David Earl
Nirschl, Joseph Peter
Park, Kinam
Patwardhan, Bhalchandra H
Pikal, Michael Jon
Pioch, Richard Paul
Rickard, Eugene Clark
Robertson, Donald Edwin
Safdy, Max Errol
Salerni, Oreste Leroy
Sanchez, Ignacio
Shaw, Stanley Miner
Stiver, James Frederick
Stroube, William Bryan, Jr
Su, Kenneth Shyan-Ell
Thompkins, Leon
Wheeler, William Joe

IOWA
Cannon, Joseph G
Linhardt, Robert John
Neumann, Marguerite
Parrott, Eugene Lee
Poust, Rolland Irvin
Watkins, G(ordon) Leonard
Wiley, Robert A
Wurster, Dale Eric, Jr

KANSAS
Beechan, Curtis Michael
Borchardt, Ronald T
Cheng, Chia-Chung
Gillingham, James M
Grunewald, Gary Lawrence
Hanzlik, Robert Paul
Haslam, John Lee
Lindenbaum, Siegfried
Lunte, Craig Edward
Marx, Michael
Owensby, Clenton Edgar
Pazdernik, Thomas Lowell
Rork, Gerald Stephen
Rytting, Joseph Howard
Stella, Valentino John
Stutz, Robert L
Walaszek, Edward Joseph

KENTUCKY
Crooks, Peter Anthony
DeLuca, Patrick Phillip
Digenis, George A
Kadaba, Pankaja Kooveli
Kornet, Milton Joseph
Luzzio, Frederick Anthony
Smith, Walter Thomas, Jr
Thio, Alan Poo-An

LOUISIANA
Bauer, Dennis Paul
Daigle, Josephine Siragusa
Davenport, Tom Forest, Jr
Gallo, August Anthony
Hornbaker, Edwin Dale
Weidig, Charles F

MAINE
Bragdon, Robert Wright
Dice, John Raymond
Tedeschi, Robert James

MARYLAND
Alexander, Thomas Goodwin
Barnstein, Charles Hansen
Benson, Walter Roderick
Blum, Stanley Walter
Callery, Patrick Stephen
Carr, Charles Jelleff
Carter, John Paul, Jr
Chang, Charles Yu-Chun
Chang, Henry
Chiu, Tony Man-Kuen
Cohen, Louis Arthur
Driscoll, John Stanford
Ellin, Robert Isadore
Feyns, Liviu Valentin
Fleming, Paul Robert
Flora, Karl Philip
Gelberg, Alan
Grady, Lee Timothy
Graham, Joseph H
Gray, Allan P
Hackley, Brennie Elias, Jr
Henery-Logan, Kenneth Robert
Hillery, Paul Stuart
Hollenbeck, Robert Gary
Hosmane, Ramachandra Sadashiv
Jacobson, Kenneth Alan
Kador, Peter Fritz

Kaiser, Carl
Keefer, Larry Kay
Klein, Michael
Klioze, Oscar
Kochhar, Man Mohan
Kramer, Stanley Phillip
Kumkumian, Charles Simon
Leak, John Clay, Jr
Leslie, James
Levine, Alan Stewart
Lo, Theresa Nong
McShefferty, John
Marquez, Victor Esteban
Massie, Samuel Proctor
Misra, Renuka
Mokhtari-Rejali, Nahld
Nemec, Josef
Oxman, Michael Allan
Pitha, Josef
Poochikian, Guiragos K
Quinn, Frank Russell
Rapaka, Rao Sambasiva
Rice, Kenner Cralle
Rodgers, Imogene Sevin
Roller, Peter Paul
Rzeszotarski, Waclaw Janusz
Schattner, Robert I
Schwartz, Paul
Schwartz, Samuel Meyer
Scovill, John Paul
Sharma, Dinesh C
Sheinin, Eric Benjamin
Short, James Harold
Shroff, Arvin Pranlal
Sinha, Birandra Kumar
Smith, Edward
Snow, Philip Anthony
Theimer, Edgar E
Watthey, Jeffrey William Herbert
Weiss, Peter Joseph
Whaley, Wilson Monroe
Whiting, John Dale, Jr
Willette, Robert Edmond
Wolters, Robert John
Wykes, Arthur Albert
Yeh, Shu-Yuan
Zenitz, Bernard Leon
Zenker, Nicolas

MASSACHUSETTS
Ajami, Alfred Michel
Auer, Henry Ernest
Bessette, Russell R
Cantley, Lewis Clayton
Cheema, Zafarullah K
Davis, Michael Allan
Driedger, Paul Edwin
Filer, Crist Nicholas
Fong, Godwin Wing-Kin
Foye, William Owen
Gerteisen, Thomas Jacob
Glajch, Joseph Louis
Granchelli, Felix Edward
Gray, Nancy M
Gund, Peter Herman Lourie
Jiang, Jack Bau-Chien
Kafrawy, Adel
Kobayashi, Kazumi
LaSala, Edward Francis
Margulis, Thomas N
Mickles, James
Mueller, Robert Kirk
Neumeyer, John L
Ocain, Timothy Donald
Parham, Marc Ellous
Razdan, Raj Kumar
Renfroe, Harris Burt
Schomer, Donald Lee
Siegal, Bernard
Smith, Pierre Frank
Snider, Barry B
Takman, Bertil Herbert
Torchilin, Vladimir Petrovich
Tuckerman, Murray Moses
Van Eikeren, Paul
Warner, John Charles
Williams, David Allen
Williams, David Lloyd
Woodman, Peter William
Wright, George Edward
Yesair, David Wayne

MICHIGAN
Abramson, Hanley N
Albertson, Noel Frederick
Augelli-Szafran, Corinne E
Axen, Udo Friedrich
Barksdale, Charles Madsen
Belmont, Daniel Thomas
Berman, Ellen Myra
Bhatt, Padmamabh P
Capps, David Bridgman
Chen, Huai Gu
Cody, Wayne Livingston
Counsell, Raymond Ernest
Coward, James Kenderdine
Creger, Paul LeRoy
Culbertson, Townley Payne
Czarnik, Anthony William
Deering, Carl F
Delia, Thomas J
Domagala, John Michael
Dorman, Linneaus Cuthbert
Drach, John Charles

Dukes, Gary Rinehart
Dunbar, Joseph Edward
Dunker, Melvin Frederick William
Elslager, Edward Faith
Gadwood, Robert Charles
Goel, Om Prakash
Gregg, David Henry
Guttendorf, Robert John
Harmon, Robert E
Heinrikson, Robert L
Herrinton, Paul Matthew
Heyd, William Ernst
Hiestand, Everett Nelson
Hoefle, Milton Louis
Huang, Che C
Jacoby, Ronald Lee
Jones, Eldon Melton
Joshi, Mukund Shankar
Kagan, Fred
Kaiser, David Gilbert
Klutchko, Sylvester
Koch, Melvin Vernon
Koshy, K Thomas
Laks, Peter Edward
Lintner, Carl John, Jr
Lipton, Michael Forrester
McCall, John Michael
McGovren, James Patrick
Magerlein, Barney John
Marletta, Michael Anthony
Mathison, Ian William
Maxey, Brian William
Meyer, Robert F
Moffett, Robert Bruce
Moissides-Hines, Lydia Elizabeth
Morozowich, Walter
Nagler, Robert Carlton
Nelson, Norman Allan
Pfeiffer, Curtis Dudley
Philipsen, Judith Lynne Johnson
Rand, Cynthia Lucille
Reid, William Bradley
Schumann, Edward Lewis
Schut, Robert N
Sinkula, Anthony Arthur
Sinsheimer, Joseph Eugene
Stanley, John Pearson
Suggs, William Terry
Taraszka, Anthony John
Thomas, Richard Charles
Tinney, Francis John
Tobey, Stephen Winter
Topliss, John G
Townsend, Leroy B
Unangst, Paul Charles
Valvani, Shri Chand
Van Rheenen, Verlan H
Wagner, John Garnet
Weisbach, Jerry Arnold
Westland, Roger D(ean)
Wilkinson, Paul Kenneth
Woo, P(eter) W(ing) K(ee)
Wormser, Henry C
Woster, Patrick Michael
Yankee, Ernest Warren
Ye, Qizhuang

MINNESOTA
Ames, Matthew Martin
Bunker, James Edward
Crooks, Stephen Lawrence
Erickson, Edward Herbert
Farber, Elliott
Grant, David James William
Hanna, Patrick E
Jochman, Richard Lee
Kolthoff, Izaak Maurits
Mikhail, Adel Ayad
Nagasawa, Herbert Tsukasa
Portoghese, Philip S
Reever, Richard Eugene
Rippie, Edward Grant
Robertson, Jerry Earl
Thompson, Richard David
Vince, Robert
Yapel, Anthony Francis, Jr

MISSISSIPPI
Baker, John Keith
Borne, Ronald Francis
Hufford, Charles David
Nobles, William Lewis
Sam, Joseph
Thompson, Alonzo Crawford
Waller, Coy Webster

MISSOURI
Arneson, Dora Williams
Asrar, Jawed
Baker, Joseph Willard
Banakar, Umesh Virupaksh
Bockserman, Robert Julian
Bovy, Philippe R
Chafetz, Lester
Cheng, Lawrence Kar-Hiu
Delaware, Dana Lewis
Dill, Dale Robert
Drew, Henry D
Eibert, John, Jr
Gardner, Jerry David
Godt, Henry Charles, Jr
Gustafson, Mark Edward
Johnson, Richard Dean
Kenyon, Allen Stewart

Krueger, Paul A
Lennon, Patrick James
Liao, Tsung-Kai
Marshall, Lucia Garcia-Iniguez
Moeller, Kevin David
Neau, Steven Henry
Nuessle, Noel Oliver
Podrebarac, Eugene George
Popp, Frank Donald
Raffelson, Harold
Rost, William Joseph
Sabacky, M Jerome
Sathe, Sharad Somnath
Schleppnik, Alfred Adolf
Schumacher, Ignatius
Sikorski, James Alan
Stenseth, Raymond Eugene
Volkert, Wynn Arthur
Wasson, Richard Lee
Wheeler, James Donlan
Woodhouse, Edward John
Zee-Cheng, Robert Kwang-Yuen
Zienty, Ferdinand B
Zuzack, John W

MONTANA
Pettinato, Frank Anthony

NEBRASKA
Gold, Barry Irwin
Harris, Lewis Eldon
Lee, Jean Chor-Yin Wong
Mauger, John William
Murray, Wallace Jasper
Nagel, Donald Lewis
Roche, Edward Browining
Ruyle, William Vance
Saski, Witold
Small, LaVerne Doreyn
Vennerstrom, Jonathan Lee

NEW HAMPSHIRE
Smith, Terry Douglas

NEW JERSEY
Abdou, Hamed M
Adams, Richard Ernest
Andrade, John Robert
Ardelt, Wojciech Joseph
Aronovic, Sanford Maxwell
Asano, Akira
Bailey, Leonard Charles
Beck, John Louis
Berger, Joel Gilbert
Bernady, Karel Francis
Blackburn, Dale Warren
Bodin, Jerome Irwin
Bollinger, Frederick W(illiam)
Borah, Kripanath
Borris, Robert P
Bose, Ajay Kumar
Breslauer, Kenneth John
Brine, Charles James
Brody, Stuart Martin
Brofazi, Frederick R
Bryan, Wilbur Lowell
Buch, Robert Michael
Burbaum, Beverly Wolgast
Butensky, Irwin
Carlucci, Frank Vito
Carter, James Evan
Cavender, Patricia Lee
Chen, James L
Chertkoff, Marvin Joseph
Chien, Yie W
Clarke, Frank Henderson
Clayton, John Mark
Colaizzi, John Louis
Controulis, John
Cook, Alan Frederick
Crawley, Lantz Stephen
Cross, John Milton
Dean, Donald E
Delaney, Edward Joseph
Denton, John Joseph
De Silva, John Arthur F
Diamond, Julius
DiFazio, Louis T
Draper, Richard William
Dugan, Gary Edwin
Dunn, William Howard
Durette, Philippe Lionel
Egli, Peter
Elder, John Philip
Erhardt, Paul William
Essery, John M
Fahey, John Leonard
Fahrenholtz, Kenneth Earl
Fand, Theodore Ira
Farng, Richard Kwang
Faust, Richard Edward
Finkelstein, Jacob
Fitchett, Gilmer Trower
Focella, Antonino
Fong, Jones W
Fryer, Rodney Ian
Garland, William Arthur
Geller, Milton
Gershon, Sol D
Gimelli, Salvatore Paul
Girijavallabhan, Viyyoor Moopil
Gold, Daniel Howard
Goldemberg, Robert Lewis
Gordziel, Steven A

Grier, Nathaniel
Groeger, Theodore Oskar
Gyan, Nanik D
Hager, Douglas Francis
Hagmann, William Kirk
Halpern, Donald F
Hanna, Samir A
Hardtmann, Goetz E
Hatzenbuhler, Douglas Albert
Heck, James Virgil
Heilman, William Paul
Helsley, Grover Cleveland
Hirsch, Allen Frederick
Hoff, Dale Richard
Hoffman, Allan Jordan
Hokanson, Gerard Clifford
Houlihan, William Joseph
Howe, Eugene Everett
Howell, Charles Frederick
Ikeda, Tatsuya
Infeld, Martin Howard
Jacobs, Allen Leon
Jacobson, Harold
Jaffe, James Mark
Jain, Nemichand B
Jensen, Norman P
Jirkovsky, Ivo
Jones, Howard
Kabadi, Balachandra N
Kaczorowski, Gregory John
Kagan, Michael Z
Kalm, Max John
Kaplan, Leonard Louis
Kasparek, Stanley Vaclav
Katz, Irwin Alan
Kees, Kenneth Lewis
Kim, Benjamin K
Kirsch, Nathan Carl
Klaubert, Dieter Heinz
Kline, Berry James
Kline, Toni Beth
Koster, William Henry
Kreft, Anthony Frank, III
Lehr, Hanns H
Lewis, Neil Jeffrey
Lieberman, Herbert A
Liebowitz, Stephen Marc
Lin, Yi-Jong
Linn, Bruce Oscar
Lippmann, Irwin
Liu, Yu-Ying
Love, L J Cline
Luk, Kin-Chun C
Lukas, George
McCaully, Ronald James
MacCoss, Malcolm
McDowell, Wilbur Benedict
McKearn, Thomas Joseph
Marlowe, Edward
Martin, Lawrence Leo
Mehta, Atul Mansukhbhai
Merkle, F Henry
Mezick, James Andrew
Michel, Gerd Wilhelm
Mikes, John Andrew
Miller, Larry Gene
Miner, Robert Scott, Jr
Moros, Stephen Andrew
Nessel, Robert J
Nicolau, Gabriela
O'Connor, Joseph Michael
Olson, Gary Lee
Omar, Mostafa M
Ortiz-Martinez, Aury
Parisse, Anthony John
Patchett, Arthur Allan
Perrotta, James
Pflug, Gerald Ralph
Pilkiewicz, Frank George
Puri, Surendra Kumar
Rasmusson, Gary Henry
Regna, Peter P
Reif, Van Dale
Rieger, Martin Max
Roland, Dennis Michael
Rosenberg, Howard Alan
Rosenthal, Murray William
Rovnyak, George Charles
Schaffner, Carl Paul
Schnaare, Roger L
Sciarrone, Bartley John
Semenuk, Nick Sarden
Sestanj, Kazimir
Shah, Atul A
Sieh, David Henry
Simonoff, Robert
Skotnicki, Jerauld S
Slade, Joel S
Slocum, Donald Hillman
Slusarchyk, William Allen
Solis-Gaffar, Maria Corazon
Somkaite, Rozalija
Spitznagle, Larry Allen
Staum, Muni M
Stein, Reinhardt P
Steinberg, Eliot
Steinman, Martin
Stober, Henry Carl
Taylor, John William
Tolman, Richard Lee
Turse, Richard S
Tway, Patricia C
Vanden Heuvel, William John Adrian, III

Pharmaceutical Chemistry (cont)

Venturella, Vincent Steven
Villani, Frank John, Sr
Virgili, Luciano
Vlattas, Isidoros
Walser, Armin
Watts, Daniel Jay
Weintraub, Herschel Jonathan R
Weintraub, Leonard
Weiss, Marvin
Whigan, Daisy B
White, Ronald E
Wilkinson, Raymond George
Williams, Jeffrey Walter
Willis, Carl Raeburn, Jr
Wilson, Armin Guschel
Witkowski, Joseph Theodore
Wohl, Ronald A
Wright, William Blythe, Jr
Wu, Mu Tsu
Yakubik, John
You, Kwan-sa
Zask, Arie

NEW MEXICO
Garst, John Eric
Guziec, Frank Stanley, Jr
Kelly, Clark Andrew
Saenz, Reynaldo V

NEW YORK
Ackerman, James Howard
Agnew-Marcelli, G(ladys) Marie
Albright, Jay Donald
Alks, Vitauts
Allen, George Rodger, Jr
Anderson, Wayne Keith
Archer, Sydney
Balassa, Leslie Ladislaus
Bardos, Thomas Joseph
Barringer, William Charles
Bitha, Panayota
Bizios, Rena
Bonvicino, Guido Eros
Burger, Richard Melton
Carabateas, Philip M
Chandran, V Ravi
Charton, Marvin
Chheda, Girish B
Christman, David R
Coburn, Robert A
Cohen, Elliott
Cooper, Aaron David
Crenshaw, Ronnie Ray
Curran, William Vincent
Dabrowiak, James Chester
Daum, Sol Jacob
Dutta, Shib Prasad
Feit, Irving N
Fields, Thomas Lynn
Finn, Ronald Dennet
Fliedner, Leonard John, Jr
Fondy, Thomas Paul
Fung, Ho-Leung
Gaver, Robert Calvin
Georgiev, Vassil St
Gershon, Herman
Ginos, James Zissis
Goldschmidt, Eric Nathan
Gordon, Maxwell
Gottesman, Elihu
Gottstein, William J
Gringauz, Alex
Haberfield, Paul
Hong, Chung Il
Hoyte, Robert Mikell
Ivashkiv, Eugene
Jarowski, Charles I
Johns, William Francis
Johnson, Robert Ed
Kalman, Thomas Ivan
Kapoor, Amrit Lal
Kapuscinski, Jan
Khan, Jamil Akber
Klein, Harvey Gerald
Kohlbrenner, Philip John
Krueger, James Elwood
Kushner, Samuel
Larson, Allan Bennett
Laubach, Gerald D
Lee, May D-Ming (Lu)
Loev, Bernard
Mackles, Leonard
Mercier, Gustavo Alberto, Jr
Meyer, Walter Edward
Miller, Theodore Charles
Misra, Anand Lal
Monaco, Regina R
Moore, Maurice Lee
Morris, Marilyn Emily
Murdock, Keith Chadwick
Nash, Robert Arnold
Neenan, John Patrick
Novotny, Jaroslav
O'Brien, Anne T
Ojima, Iwao
Orzech, Chester Eugene, Jr
Paikoff, Myron
Pankiewicz, Krzysztof Wojciech
Partch, Richard Earl
Portmann, Glenn Arthur
Poss, Andrew Joseph
Redalieu, Elliot
Reich, Marvin Fred
Rich, Arthur Gilbert
Ross, Malcolm S F
Rossi, Miriam
Rubin, Byron Herbert
Sayegh, Joseph Frieh
Sellstedt, John H
Shaath, Nadim Ali
Sharma, Ram Ashrey
Siggins, James Ernest
Simonelli, Anthony Peter
Siuta, Gerald Joseph
Snyder, Harry Raymond, Jr
Solo, Alan Jere
Soloway, Harold
Spiegel, Allen J
Strojny, Norman
Taub, Abraham
Triggle, David J
Trust, Ronald I
Upeslacis, Janis
Urdang, Arnold
Venkataraghavan, R
Von Strandtmann, Maximillian
Warner, Paul Longstreet, Jr
Warren, James Donald
Webb, William Gatewood
Weiss, Martin Joseph
Welles, Harry Leslie
White, Ralph Lawrence, Jr
Wolf, Alfred Peter
Wong, Lan Kan
Wood, David
Xu, Zhenchun
Yu, Andrew B C
Yudelson, Joseph Samuel

NORTH CAROLINA
Beverung, Warren Neil, Jr
Bigham, Eric Cleveland
Blumenkopf, Todd Andrew
Boettner, Fred Easterday
Booth, Raymond George
Boyd, Robert Edward, Sr
Brieaddy, Lawrence Edward
Brown, Horace Dean
Carl, Philip Louis
Carr, Fred K
Cavallito, Chester John
Chae, Kun
Chung, Henry Hsiao-Liang
Clarke, Robert LaGrone
Cocolas, George Harry
Cory, Michael
Davis, Roman
Dixit, Ajit Suresh
Doak, George Osmore
Durden, John Apling, Jr
Engle, Thomas William
Fernandes, Daniel James
Foernzler, Ernest Carl
Freeman, Harold Stanley
Gemperline, Paul Joseph
Hadzija, Bozena Wesley
Hager, George Philip, Jr
Hickey, Anthony James
Higgins, Robert H
Hinze, Willie Lee
Hong, Donald David
Hooper, Irving R
Hurlbert, Bernard Stuart
Ishaq, Khalid Sulaiman
Izydore, Robert Andrew
Kelley, James Leroy
Kelsey, John Edward
Kuyper, Lee Frederick
Lee, Kuo-Hsiung
Lee, Yue-Wei
Long, Robert Allen
McCutcheon, Rob Stewart
McDermed, John Dale
Martin, Gary Edwin
Millen, Jane
Misek, Bernard
Narasimhachari, Nedathur
Nielsen, Lawrence Arthur
Piantadosi, Claude
Price, Kenneth Elbert
Reid, Jack Richard
Richardson, Stephen Giles
Rideout, Janet Litster
Schaeffer, Howard John
Semeniuk, Fred Theodor
Short, Franklin Willard
Sigel, Carl William
Spielvogel, Bernard Franklin
Swarbrick, James
Thomas, Elizabeth Wadsworth
Waters, James Augustus
Wyrick, Steven Dale
Yeowell, David Arthur

NORTH DAKOTA
Ary, Thomas Edward
Jasperse, Craig Peter
Magarian, Edward O
Shelver, William H
Vincent, Muriel C

OHIO
Billups, Norman Fredrick
Bope, Frank Willis
Brueggemeier, Robert Wayne
Carr, Albert A
Cassady, John Mac
Didchenko, Rostislav
Dines, Allen I
Dolfini, Joseph E
Eisenhardt, William Anthony, Jr
Erman, William F
Fall, Harry H
Feld, William Adam
Fentiman, Allison Foulds, Jr
Fleming, Robert Willerton
Flynn, Gary Alan
Frank, Sylvan Gerald
Freedman, Jules
Germann, Richard P(aul)
Gilpin, Roger Keith
Graef, Walter L
Greenlee, Kenneth William
Holbert, Gene W(arwick)
Homan, Ruth Elizabeth
Horgan, Stephen William
Huber, Harold E
Jarvi, Esa Tero
Jordan, Freddie L
Keily, Hubert Joseph
Letton, James Carey
Leyda, James Perkins
Lichtin, J Leon
Lutton, John Kazuo
McCarty, Frederick Joseph
McVean, Duncan Edward
Messer, William Sherwood, Jr
Morgan, Stanley L
Munson, H Randall, Jr
Myhre, David V
Norris, Paul Edmund
Parker, Roger A
Peet, Norton Paul
Powers, Larry James
Rau, Allen H
Sampson, Paul
Schnettler, Richard Anselm
Shanklin, James Robert, Jr
Sharma, Rameshwar Kumar
Soloway, Albert Herman
Stadler, Louis Benjamin
Stansloski, Donald Wayne
Staubus, Alfred Ellsworth
Stotz, Robert William
Tan, Henry S I
Taschner, Michael J
Wade, Peter Cawthorn
Wagner, Eugene Ross
Warner, Ann Marie
Warner, Victor Duane
Washburn, Lee Cross
Webb, Norval Ellsworth, Jr
Witiak, Donald T
Zoglio, Michael Anthony

OKLAHOMA
Berlin, Kenneth Darrell
Bunce, Richard Alan
Carubelli, Raoul
Dryhurst, Glenn
Harris, Loyd Ervin
Magarian, Robert Armen
Ortega, Gustavo Ramon
Prabhu, Vilas Anandrao
Ratto, Peter Angelo
Shough, Herbert Richard
Wang, Binghe

OREGON
Block, John Harvey
Hauser, Frank Marion
Schultz, Harry Wayne

PENNSYLVANIA
Abou-Gharbia, Magid
Ackermann, Guenter Rolf
Almond, Harold Russell, Jr
Bailey, Denis Mahlon
Baird, Michael Jefferson
Bihovsky, Ron
Blank, Benjamin
Bock, Mark Gary
Borke, Mitchell Louis
Brady, Stephen Francis
Brenner, Gerald Stanley
Brown, Neil Harry
Burman, Sudhir
Casey, Adria Catala
Cheng, Hung-Yuan
Chow, Alfred Wen-Jen
Cooperman, Barry S
Deist, Robert Paul
Dempski, Robert E
Djuric, Stevan Wakefield
Douglas, Bryce
Egolf, Roger
Elson, Jesse
Evans, Ben Edward
Faiferman, Isidore
Falkiewicz, Michael Joseph
Faltynek, Robert Allen
Feldman, Joseph Aaron
Fong, Kei-Lai Lau
Freidinger, Roger Merlin
Galinsky, Alvin M
Gangjee, Aleem
Gardella, Libero Anthony
Gennaro, Alfonso Robert
Grebow, Peter Eric
Grim, Wayne Martin
Grindel, Joseph Michael
Gunther, Wolfgang Hans Heinrich
Haines, William Joseph
Harrison, Aline Margaret
Hartman, Kenneth Eugene
Heindel, Ned Duane
Herzog, Karl A
Hlasta, Dennis John
Hoffman, Jacob Matthew, Jr
Hoover, John Russel Eugene
Howard, Stephen Arthur
Hwang, Bruce You-Huei
Janicki, Casimir A
Kade, Charles Frederick, Jr
Kauffman, Joel Mervin
Kelly, Ernest L
King, Robert Edward
Kingsbury, William Dennis
Kowalski, Conrad John
Kowarski, Chana R
Kruse, Lawrence Ivan
Kwan, King Chiu
Lapidus, Milton
Larson, Gerald Louis
Lazar, Anna
Lee, Robert William
Lorenz, Roman R
Lumma, William Carl, Jr
Mackowiak, Elaine DeCusatis
Martin, Bruce Douglas
Maryanoff, Bruce Eliot
Maryanoff, Cynthia Anne Milewski
Melamed, Sidney
Melander, Wayne Russell
Mishra, Dinesh S
Mokotoff, Michael
Morse, Lewis David
Motsavage, Vincent Andrew
ONeill, Joseph Lawrence
Packman, Albert M
Parish, Roger Cook
Poole, John William
Press, Jeffery Bruce
Price, Charles Coale
Prugh, John Drew
Rasmussen, Chris Royce
Raymond, Matthew Joseph
Reiff, Harry Elmer
Reitz, Allen Bernard
Russo, Emanuel Joseph
Saggiomo, Andrew Joseph
Salvino, Joseph Michael
Santilli, Arthur A
Santora, Norman Julian
Sarantakis, Dimitrios
Scheindlin, Stanley
Schwartz, Joseph Barry
Shepard, Kenneth LeRoy
Shiue, Chyng-Yann
Singh, Baldev
Sinotte, Louis Paul
Sklar, Stanley
Smith, Robert Lawrence
Southwick, Philip Lee
Strike, Donald Peter
Studt, William Lyon
Sunshine, Warren Lewis
Sutton, Blaine Mote
Templer, David Allen
Tice, Linwood Franklin
Ullyot, Glenn Edgar
Urenovitch, Joseph Victor
Wei, Peter Hsing-Lien
Weinkam, Robert Joseph
Weinstock, Joseph
Wentland, Mark Philip
Wetzel, Ronald Burnell
Widdowson, Katherine Louisa
Williams, John Roderick
Wilson, James William
Wineholt, Robert Leese
Woltersdorf, Otto William, Jr
Wood, Kurt Alan
Yeager, Sandra Ann
Yellin, Tobias O
Youngstrom, Richard Earl
Yu, Ruey Jiin
Zanowiak, Paul

RHODE ISLAND
Abushanab, Elie
Elango, Varadaraj
Sapolsky, Asher Isadore
Smith, Alan Jerrard
Smith, Charles Irvel
Turcotte, Joseph George

SOUTH CAROLINA
Bauguess, Carl Thomas, Jr
Beamer, Robert Lewis
Dodds, Alvin Franklin
Haskell, Theodore Herbert, Jr
Martin, Tellis Alexander
Moore, Alexander Mazyck
Sadik, Farid
Schmidt, Gilbert Carl
Scruggs, Jack G
Shah, Hamish V
Walter, Wilbert George
Wang, Huei-Hsiang Lisa
Wynn, James Elkanah

SOUTH DAKOTA
Bailey, Harold Stevens
Omodt, Gary Wilson

TENNESSEE
Avis, Kenneth Edward
Gilliom, Richard D
Gollamudi, Ramachander
Handorf, Charles Russell
Kabalka, George Walter
Laboda, Henry M
Lasslo, Andrew
Lyman, Beverly Ann
Miller, Duane Douglas
Murphy, William R
Otis, Marshall Voigt
Parish, Harlie Albert, Jr
Patel, Tarun R
Pinajian, John Joseph
Reynolds, Brian Edgar
Sastry, Bhamidipaty Venkata Rama
Sheth, Bhogilal
Smith, Howard E
Solomons, William Ebenezer
Tam, James Pingkwan
Wells, Jack Nulk
Wigler, Paul William

TEXAS
Alam, Maktoob
Aslam, Mohammad
Bapatla, Krishna M
Bikin, Henry
Boblitt, Robert LeRoy
Cates, Lindley A
Chen, Shih-Fong
Ciufolini, Marco A
Cragoe, Edward Jethro, Jr
Delgado, Jaime Nabor
Doluisio, James Thomas
Harrell, William Broomfield
Hedrich, Loren Wesley
Hogan, Michael Edward
Hurley, Laurence Harold
Iskander, Felib Youssef
Jaszberenyi, Joseph C
Jones, Lawrence Ryman
Kasi, Leela Peshkar
Lemke, Thomas Lee
Marchand, Alan Philip
Marshall, Gailen D, Jr
Monk, Clayborne Morris
Mosier, Benjamin
Mulvey, Dennis Michael
Quintana, Ronald Preston
Reinecke, Manfred Gordon
Robinson, J(ames) Michael
Schlameus, Herman Wade
Senkowski, Bernard Zigmund
Stavchansky, Salomon Ayzenman
Szczesniak, Raymond Albin
Ternay, Andrew Louis, Jr
Webber, Marion George

UTAH
Aldous, Duane Leo
Anderson, Bradley Dale
Balandrin, Manuel F
Broadbent, Hyrum Smith
Broom, Arthur Davis
DelMar, Eric G
Higuchi, William Iyeo
Kim, Sung Wan
Kopecek, Jindrich
Mason, Robert C
Moe, Scott Thomas
Petersen, Robert Virgil
Roll, David Byron
Segelman, Alvin Burton
Van Wagenen, Bradford Carr
Yost, Garold Steven

VERMONT
McCormack, John Joseph, Jr

VIRGINIA
Abraham, Donald James
Andrako, John
Boots, Marvin Robert
Boots, Sharon G
Burger, Alfred
Cale, Albert Duncan, Jr
Casola, Armand Ralph
Castagnoli, Neal, Jr
Collins, Galen Franklin
De Camp, Wilson Hamilton
Dvorchik, Barry Howard
Fedrick, James Love
Feldmann, Edward George
Glassco, William Shoemaker
Greene, Virginia Carvel
Haines, Bernard A
Harris, Paul Robert
Hudlicky, Milos
Johnson, Fatima Nunes
Johnston, Byron E
Kier, Lemont Burwell
Lunsford, Carl Dalton
Martin, Albert Edwin
May, Everette Lee
Murphey, Robert Stafford
Neumann, Helmut Carl
Patel, Appasaheb Raojibhai
Perillo, Benjamin Anthony

Salley, John Jones, Jr
Stubbins, James Fiske
Sweeney, Thomas Richard
Teng, Lina Chen
Uwaydah, Ibrahim Musa
Varma, Ravi Kannadikovilakom
Wasti, Khizar
Weaver, Warren Eldred
Welstead, William John, Jr

WASHINGTON
Abdel-Monem, Mahmoud Mohamed
Baillie, Thomas Allan
Bhat, Venkatramana Kakekochi
Biskupiak, Joseph E
Brady, Lynn R
Duhl-Emswiler, Barbara Ann
Dunbar, Philip Gordon
Fullerton, Dwight Story
Galpin, Donald R
Huitric, Alain Corentin
Kaplan, Phyllis Deen
Lybrand, Terry Paul
McCarthy, Walter Charles
Malik, Sohail
Meyer, Rich Bakke, Jr
Nelson, Wendel Lane
Smith, Robert Victor
Staiff, Donald C
Trager, William Frank
Whelton, Bartlett David
Widmaier, Robert George
Wilbur, Daniel Scott

WEST VIRGINIA
Fodor, Gabor
Gilmore, William Franklin
Rankin, Gary O'Neal

WISCONSIN
Burke, Steven Douglas
Chenier, Philip John
Connors, Kenneth A
Cook, James Minton
Griffith, Owen W
Gross, Herbert Michael
Heath, Timothy Douglas
Hollenberg, David Henry
Kolb, Vera
Priest, Matthew A
Robinson, Joseph Robert
Shaw, C Frank, III
Zografi, George

WYOMING
Brunett, Emery W
Nelson, Kenneth Fred

PUERTO RICO
Gonzalez De Alvarez, Genoveva

ALBERTA
Coutts, Ronald Thomson
Knaus, Edward Elmer
Locock, Robert A
Micetich, Ronald George
Moskalyk, Richard Edward
Rogers, James Albert
Wiebe, Leonard Irving

BRITISH COLUMBIA
Chatten, Leslie George
Hayward, Lloyl Douglas

MANITOBA
Forrest, Bruce James
Hasinoff, Brian Brennen
Hunter, Norman Robert
Kim, Ryung-Soon (Song)
Lange, Bruce Ainsworth
Steele, John Wiseman

NOVA SCOTIA
Farmer, Patrick Stewart
Mezei, Michael

ONTARIO
Bend, John Richard
Bhavnani, Bhagu R
Butler, Douglas Neve
Dunford, Raymond A
Hach, Vladimir
Howard-Lock, Helen Elaine
Kaiser, Klaus L(eo) E(duard)
Leznoff, Clifford Clark
Lock, Colin James Lyne
Lovering, Edward Gilbert
Lucas, Douglas M
Nogrady, Thomas
Pang, Henrianna Yicksing
Rees, Alun Hywel
Shah, Ashok Chandulal
Szarek, Walter Anthony
Teare, Frederick Wilson

QUEBEC
Caille, Gilles
Chenevert, Robert (Bernard)
Demerson, Christopher
Desjardins, Claude W
Engel, Charles Robert
Ferdinandi, Eckhardt Stevan
Gleason, Clarence Henry
Goldstein, Sandu M

Joly, Louis Philippe
Levi, Irving
Perron, Yvon G
Regoli, Domenico
Salvador, Romano Leonard
Simon, David Zvi
Yates, Claire Hilliard
Young, Robert Norman

SASKATCHEWAN
Dimmock, Jonathan Richard
Gupta, Vidya Sagar
Mikle, Janos J
Quail, John Wilson
Zuck, Donald Anton

OTHER COUNTRIES
Braquet, Pierre G
Ghosh, Anil Chandra
Grisar, Johann Martin
Hajos, Zoltan George
Hassner, Alfred
Kim, Dong Han
Kreider, Eunice S
Kuttab, Simon Hanna
Lambrecht, Richard Merle
Lawson, William Burrows
Nair, Vijay
Sung, Chen-Yu

Physical Chemistry

ALABAMA
Allan, Barry David
Arendale, William Frank
Atwood, Jerry Lee
Ayers, Orval Edwin
Baird, James Kern
Bakker, Martin Gerard
Bartlett, James Holly
Beal, James Burton, Jr
Brandt, Luther Warren
Bugg, Charles Edward
Cappas, C
Coburn, William Carl, Jr
Cohn, Ernst M
Cole, George David
Copeland, David Anthony
Dismukes, Edward Brock
Dorai-Raj, Diana Glover
Edwards, Oscar Wendell
Ellis, Richard Bassett
Emerson, Merle T
Fredericks, William John
Gant, Fred Allan
Gerdom, Larry E
Gill, Piara Singh
Hand, Clifford Warren
Harris, Albert Zeke
Hatfield, John Dempsey
Hazlegrove, Leven S
Hirko, Ronald John
Huffman, Ernest Otto
Hung, George Wen-Chi
Jackson, Margaret E
Johnson, Frederic Allan
Khaled, Mohammad Abu
Kispert, Lowell Donald
Laliberte, Laurent Hector
Livant, Peter David
Maclean, Donald Isadore
Meagher, James Francis
Metzger, Robert Melville
Miller, Meredith
Miyagawa, Ichiro
Mountcastle, William R, Jr
Neely, William Charles
Pan, Chai-Fu
Rahn, Ronald Otto
Renard, Jean Joseph
Richards, Nolan Earle
Riley, Clyde
Rosenberger, Franz
Sides, Gary Donald
Smith, Anthony James
Smith, Donald Foss
Smith, Richard Neilson
Smith, Steven Patrick Decland
Squillacote, Michael Edward
Stanbury, David McNeil
Stein, Dale Franklin
Thorpe, Martha Campbell
Timkovich, Russell
Van Artsdalen, Ervin Robert
Vigee, Gerald S
Ward, Charlotte Reed
Ward, Curtis Howard
Wharton, Walter Washington
Whorton, Rayburn Harlen
Williamson, Ashley Deas
Workman, Gary Lee
Worley, S D

ALASKA
Douthat, Daryl Allen
Evans, William Harrington
Godbey, William Givens
Hoskins, Leo Claron
Reppond, Kermit Dale
Roberts, Thomas D
Rutledge, Gene Preston
Wentink, Tunis, Jr

ARIZONA
Angell, C A
Bailey, Samuel David
Balasubramanian, Krishnan
Barfield, Michael
Barry, Arthur John
Bashkin, Stanley
Brathovde, James Robert
Brown, Duane
Brown, Michael Frederick
Carr, Clide Isom
Carruthers, Lucy Marston
Dodd, Charles Gardner
Eastman, Michael Paul
Engleman, Rolf, Jr
Eyring, LeRoy
Fackler, Walter Valentine, Jr
Feltham, Robert Dean
Forster, Leslie Stewart
Fox, Michael Jean
Frost, Arthur Atwater
Gibbs, Robert John
Gladrow, Elroy Merle
Glaunsinger, William Stanley
Glick, Milton Don
Golden, Sidney
Guilbert, John M
Harris, Leland
Helbert, John N
Jones, Samuel Stimpson
Kierbow, Julie Van Note Parker
Knowlton, Gregory Dean
Kohler, Sigurd H
Krasnow, Marvin Ellman
Kukolich, Stephen George
Lin, Sheng Hsien
Marzke, Robert Franklin
Meibuhr, Stuart Gene
Milberg, Morton Edwin
Milner, Paul Chambers
Mosley, John Ross
Post, Roy G
Roche, Thomas Stephen
Rupley, John Allen
Sears, Donald Richard
Seely, Gilbert Randall
Shoemaker, Richard Lee
Smith, Mark Alan
Stamm, Robert Franz
Steimle, Timothy C
Swanson, John William
Swink, Laurence N
Tompkin, Gervaise William
Urbain, Walter Mathias
Walker, Walter W(yrick)
Winfree, Arthur T
Wyckoff, Ralph Walter Graystone
Young, Robert A
Zielen, Albin John

ARKANSAS
Anderson, Robbin Colyer
Becker, Ralph Sherman
Blyholder, George Donald
Bolmer, Perce W
Broach, Wilson J
Evans, Frederick Earl
Gildseth, Wayne
Glendening, Norman Willard
Gwinup, Paul D
Hinton, James Faulk
Johnson, Dale A
Kraemer, Louise Margaret
Krause, Paul Frederick
McCarty, Clark William
Netherton, Lowell Edwin
Pryor, Joseph Ehrman
Pulay, Peter
Schafer, Lothar
Steinmeier, Robert C
Stuckey, John Edmund
Teague, Marion Warfield
Wagner, George Hoyt
Wear, James Otto
Williams, William Donald
Wilson, Edmond Woodrow, Jr
Woodland, Dorothy Jane

CALIFORNIA
Abrams, Marvin Colin
Acrivos, Juana Luisa Vivo
Adam, Randall Edward
Adams, George Baker
Adamson, Arthur Wilson
Addy, Tralance Obuama
Adinoff, Bernard
Ahlers, Guenter
Aklonis, John Joseph
Alivisatos, A Paul
Allamandola, Louis John
Allen, Thomas Lofton
Alter, Henry Ward
Altgelt, Klaus H
Altman, Robert Leon
Amey, Ralph Leonard
Amis, Eric J
Amster, Adolph Bernard
Andersen, Hans Christian
Andersen, Wilford Hoyt
Anderson, Roger E
Anderson, Roger W
Andrews, Frank Clinton
Arias, Jose Manuel
Arslancan, Ahmet N

Physical Chemistry (cont)

Athey, Robert Douglas, Jr
Atkinson, Roger
Atkinson, Russell H
Auerbach, Daniel J
Augood, Derek Raymond
Bailin, Lionel J
Baisden, Patricia Ann
Baker, Bernard Ray
Baker, Richard William
Baldeschwieler, John Dickson
Batzel, Roger Elwood
Baugh, Ann Lawrence
Baumgartner, Werner Andreas
Baur, Mario Elliott
Bayes, Kyle D
Beauchamp, Jesse Lee
Beaudet, Robert A
Beck, Roland Arthur
Beck, Steven Michael
Becker, Edwin Norbert
Beek, John
Belmares, Hector
Beni, Gerardo
Benson, Sidney William
Berlad, Abraham Leon
Bernasconi, Claude Francois
Berry, Michael James
Bezman, Richard David
Bittner, Harlan Fletcher
Blair, Charles Melvin, Jr
Blann, H Marshall
Boardman, William Walter, Jr
Boehm, Felix H
Bonner, Norman Andrew
Borg, Richard John
Boudart, Michel
Bowers, Michael Thomas
Bowman, Robert Clark, Jr
Boxer, Steven George
Bragin, Joseph
Brauman, John I
Brauman, Sharon K(ruse)
Braun, Robert Leore
Brewer, Leo
Brodd, Ralph James
Broido, Abraham
Bromley, Le Roy Alton
Brown, Nancy J
Brownscombe, Eugene Russell
Brundle, Christopher Richard
Bryden, John Heilner
Buckles, Robert Edwin
Burland, Donald Maxwell
Burnham, Alan Kent
Bushey, Albert Henry
Cadogan, Kevin Denis
Cahill, Richard William
Cairns, Elton James
Callahan, Kenneth Paul
Campbell, John Hyde
Campbell, Thomas Cooper
Cannon, Peter
Cantow, Manfred Josef Richard
Caprioglio, Giovanni
Caputi, Roger William
Carniglia, Stephen C(harles)
Carr, Robert J(oseph)
Carruth, Willis Lee
Carter, Emily Ann
Case, David Andrew
Castro, George
Catsiff, Ephraim Herman
Chackerian, Charles, Jr
Chandler, David
Chang, Elfreda Te-Hsin
Chang, Shih-Ger
Chau, Michael Ming-Kee
Chen, Timothy Shieh-Sheng
Cher, Mark
Chesick, John Polk
Chesnut, Dwayne A(llen)
Cheung, Jeffrey Tai-Kin
Chiang, Anne
Chidsey, Christopher E
Chock, Ernest Phaynan
Chompff, Alfred J(ohan)
Christe, Karl Otto
Christofferson, Glen Davis
Cichowski, Robert Stanley
Clarkson, Jack E
Cobble, James Wikle
Coffey, Dewitt, Jr
Cohen, Norman
Cole, Richard
Cole, Terry
Coleman, Charles Clyde
Colmenares, Carlos Adolfo
Colodny, Paul Charles
Compton, Leslie Ellwyn
Cook, Glenn Melvin
Cook, Robert Crossland
Coppel, Claude Peter
Cordes, Herman Fredrick
Correia, John Sidney
Cotts, David Bryan
Cotts, Patricia Metzger
Covey, William Danny
Creek, Jefferson Louis
Croll, Ian Murray
Crosley, David Risdon
Csicsery, Sigmund Maria
Cummings, John Patrick
Current, Jerry Hall
Curtis, Earl Clifton, Jr
Cusumano, James A
Czuha, Michael, Jr
Dalla Betta, Ralph A
Dalton, Larry Raymond
Darnell, Alfred Jerome
Davenport, John Eaton
Davies, David Huw
Dawson, Kenneth Adrian
Day, Robert J(ames)
Day, Robert James
Dea, Phoebe Kin-Kin
Deal, Bruce Elmer
Dee, Diana
De Haan, Frank P
DeHollander, William Roger
Detz, Clifford M
Deutsch, Daniel Harold
DiGiorgio, Joseph Brun
Dill, Kenneth Austin
Dills, Charles E
Dobyns, Leona Danette
Dodge, Richard Patrick
Dodson, Charles Leon, Jr
Dodson, Richard Wolford
Dolbear, Geoffrey Emerson
Domeniconi, Michael John
Donnelly, Timothy C
Donovan, John W
Dorough, Gus Downs, Jr
Dows, David Alan
Duff, Russell Earl
Duisman, Jack Arnold
Dulmage, William James
Dyck, Rudolph Henry
Dzieciuch, Matthew Andrew
Eatough, Norman L
Eckstrom, Donald James
Eichelberger, Robert Leslie
Ellis, Elizabeth Carol
El-Sayed, Mostafa Amr
Endres, Leland Sander
Engler, Edward Martin
English, Gerald Alan
Epstein, Barry D
Epstein, Leo Francis
Erickson, Jon W
Fadley, Charles Sherwood
Falick, Arnold M
Farber, Joseph
Farber, Milton
Farone, William Anthony
Fass, Richard A
Fassel, Velmer Arthur
Faulkner, Thomas Richard
Fayer, Michael David
Fergin, Richard Kenneth
Feuerstein, Seymour
Fife, Thomas Harley
Fineman, Morton A
Fink, William Henry
Fisher, Robert Amos, Jr
Fisk, George Ayrs
Fleischauer, Paul Dell
Fleming, Edward Homer, Jr
Fletcher, Aaron Nathaniel
Fraser, James Mattison
Frashier, Loyd Dola
Fratiello, Anthony
Freer, Stephan T
Freiling, Edward Clawson
French, Edward P(erry)
Fritz, James John
Gale, Laird Housel
Gallegos, Emilio Juan
Galperin, Irving
Gantzel, Peter Kellogg
Gardiner, Kenneth William
Gardner, Donald Glenn
Gardner, Phillip John
Garrison, Warren Manford
Gatlin, Lila L
Gay, Charles Francis
Gay, Richard Leslie
Gaynor, Joseph
Gee, Allen
Gelbart, William M
Gerber, Bernard Robert
Ghandehari, Mohammad Hossein
Gibson, Thomas Alvin, Jr
Gillen, Keith Thomas
Gilmore, John T
Ginell, William Seaman
Givens, William Geary
Godycki, Ludwig Edward
Goerke, Jon
Gofman, John William
Goldberg, Ira Barry
Goldberg, Sabine Ruth
Goldish, Elihu
Goldsmith, Henry Arnold
Goldstein, Elisheva
Golub, Morton Allan
Gordon, Alvin S
Gordon, Joseph Grover, II
Graf, Peter Emil
Grantham, Leroy Francis
Green, John William
Green, Michael Philip
Green, Robert Bennett
Greenberg, Sidney Abraham
Greene, Charles Richard
Grieman, Frederick Joseph
Gross, Paul Hans
Grossman, Jack Joseph
Grubbs, Edward
Gruhn, Thomas Albin
Grunthaner, Frank John
Gunnink, Raymond
Gupta, Amitava
Gustavson, Marvin Ronald
Gwinn, William Dulaney
Haag, Robert Marlay
Hadley, Steven George
Haendler, Blanca Louise
Haffner, James Wilson
Hahn, Harold Thomas
Hall, James Timothy
Hall, Kenneth Lynn
Hammond, R Philip
Handy, Lyman Lee
Hansford, Rowland Curtis
Hanson, Carl Veith
Hanson, Mervin Paul
Harkins, Carl Girvin
Harnsberger, Hugh Francis
Harris, David Owen
Harris, Gordon McLeod
Harris, Robert A
Harrison, George Conrad, Jr
Harrison, Jonas P
Haskell, Vernon Charles
Haugen, Gilbert R
Hauk, Peter
Hayes, Claude Q C
Hazi, Andrew Udvar
Head-Gordon, Martin Paul
Heinemann, Heinz
Heldman, Morris J
Helgeson, Harold Charles
Hem, John David
Hemminger, John Charles
Hertzler, Barry Lee
Hickson, Donald Andrew
Hidy, George Martel
Hiemenz, Paul C
Hill, Russell John
Hiller, Frederick W
Hills, Graham William
Hiraoka, Hiroyuki
Hisatsune, Isamu Clarence
Hobbs, M Floyd
Hoffman, Charles John
Hoffman, Darleane Christian
Hollenberg, J Leland
Holmes, Donald Eugene
Holzmann, Richard Thomas
Hong, Keelung
Hoover, William Graham
Hornig, Howard Chester
Hornung, Erwin William
Horsley, John Anthony
Horsma, David August
Hostettler, John Davison
Houle, Frances Anne
Howe, John A
Hsia, Yu-Ping
Hsieh, Paul Yao Tong
Hubert, Jay Marvin
Hunziker, Heinrich Erwin
Hutchinson, Bennett Buckley
Hutchinson, Eric
Iglesia, Enrique
Imhof, William Lowell
Ingraham, Lloyd Lewis
Innes, William Beveridge
Irani, Riyad Ray
Jackson, William Morgan
Jaecker, John Alvin
Jaffe, Annette Bronkesh
Jaffe, Sigmund
James, Dean B
James, Edward, Jr
Janda, Kenneth Carl
Jeffries, Jay B
Jendresen, Malcolm Dan
Jennings, Harley Young, Jr
Jensen, James Leslie
Jentoft, Ralph Eugene, Jr
Johnson, Carl Emil, Jr
Johnson, LeRoy Franklin
Johnston, Harold Sledge
Jones, Dane Robert
Jones, Patrick Ray
Jones, Richard Elmore
Kaelble, David Hardie
Kakis, Frederic Jacob
Kallo, Robert Max
Kamen, Martin David
Kaplan, David Gilbert
Kaplan, Selig N(eil)
Kardonsky, Stanley
Karukstis, Kerry Kathleen
Katekaru, James
Katzin, Leonard Isaac
Kay, Eric
Kearns, David R
Keefer, Raymond Marsh
Keeling, Charles David
Kcilin, Bertram
Keizer, Joel Edward
Kellerman, Martin
Kempter, Charles Prentiss
Kendig, William Milliam
Keneshea, Francis Joseph
Kennedy, James Vern
Kennedy, John Harvey
Keys, Richard Taylor
Keyzer, Hendrik
Kibby, Charles Leonard
Kieffer, William Franklin
Kieft, John A
Kinderman, Edwin Max
King, James, Jr
King, William Robert, Jr
Kinney, Gilbert Ford
Kinoshita, Kimio
Kirtman, Bernard
Kivelson, Daniel
Kliger, David Saul
Klinman, Judith Pollock
Knipe, Richard Hubert
Knobler, Charles Martin
Kohler, Bryan Earl
Kollman, Peter Andrew
Konowalow, Daniel Dimitri
Koob, Robert Duane
Korst, William Lawrence
Kosiba, Walter Louis
Kozawa, Akiya
Kramer, Henry Herman
Krikorian, Oscar Harold
Kroger, F(erdinand) A(nne)
Kropp, John Leo
Kumamoto, Junji
Kunc, Joseph Anthony
Kuntz, Irwin Douglas, Jr
Kuppermann, Aron
Lam, Kai Shue
Landel, Robert Franklin
Lang, Neil Charles
Lang, Valerie Ilona
Langer, Sidney
Langlois, Gordon Ellerby
Lapple, Charles E
Laub, Richard J
Laudenslager, James Bishop
Lee, Yuan Tseh
Leffler, Esther Barbara
Leiga, Algird George
Leonard, William J, Jr
Lepie, Albert Helmut
Lerner, Narcinda Reynolds
Lester, William Alexander, Jr
Levine, Charles (Arthur)
Levine, Howard Bernard
Levy, Ricardo Benjamin
Lewis, Nathan Saul
Lewis, Sheldon Noah
Liberman, Martin Henry
Lincoln, Kenneth Arnold
Lind, Maurice David
Linde, Peter Franz
Lindenberg, Katja
Lindner, Duane Lee
Lindner, Manfred
Lindquist, Robert Henry
Ling, Alan Campbell
Littauer, Ernest Lucius
Liu, Ming-Biann
Live, David H
Lo, George Albert
Loda, Richard Thomas
Long, Franklin A
Lord, Harry Chester, III
Lorenz, Max R(udolph)
Ludwig, Frank Arno
Lundin, Robert Enor
Luther, Marvin L
LuValle, James Ellis
Lyon, David N
Maas, Keith Allan
McCalley, Roderick Canfield
McCarty, Jon Gilbert
McClellan, Aubrey Lester
McConnell, Harden Marsden
McIver, Robert Thomas, Jr
McKenzie, Donald Edward
Mackenzie, John D(ouglas)
McKinney, Ted Meredith
MacLaren, Richard Oliver
McMillan, William George
MacWilliams, Dalton Carson
Magde, Douglas
Magee, John Lafayette
Malik, Jim Gorden
Maloney, Kenneth Long
Manasevit, Harold Murray
Mandelin, Dorothy Jane Bearcroft
Mansfeld, Florian Berthold
Mantei, Kenneth Alan
Maple, M Brian
Marcus, Rudolph Arthur
Marcus, Rudolph Julius
Margerum, John David
Margolis, Jack Selig
Markowitz, Samuel Solomon
Marsden, Sullivan S(amuel), Jr
Marsh, Richard Edward
Marshall, Henry Peter
Marsi, Kenneth Larue
Martin, L(aurence) Robbin
Martin, Richard McKelvy
Martin, Stanley Buel
Marx, Paul Christian
Mason, Harold Frederick
Mathews, Larry Arthur
Mayer, Stanley Wallace
Meads, Philip F
Meares, Claude Francis
Medrud, Ronald Curtis

CHEMISTRY / 233

Menefee, Emory
Merrin, Seymour
Merten, Ulrich
Meschi, David John
Metiu, Horia I
Meyer, Paul
Miles, Melvin Henry
Miller, Arnold
Miller, Donald Gabriel
Miller, George E
Miller, James Angus
Miller, William Hughes
Millikan, Roger Conant
Minch, Michael Joseph
Mishuck, Eli
Mlodozeniec, Arthur Roman
Moe, George
Moerner, W(illiam) E(sco)
Molinari, Robert James
Moore, Gordon Earle
Moretto, Luciano G
Morris, Jerry Lee, Jr
Muller, Rolf Hugo
Mustafa, Mohammed G
Myers, Benjamin Franklin, Jr
Myers, Rollie John, Jr
Mysels, Karol J(oseph)
Nadler, Melvin Philip
Namboodiri, Madassery Neelakantan
Nanes, Roger
Nash, Charles Presley
Nash, James Richard
Naylor, Benjamin Franklin
Nebenzahl, Linda Levine
Neti, Radhakrishna Murty
Neuman, Robert C, Jr
Newkirk, Herbert William
Newman, John Scott
Nichols, Ambrose Reuben, Jr
Nicholson, Daniel Elbert
Nicholson, Margie May
Nicol, Malcolm Foertner
Niday, James Barker
Nimer, Edward Lee
Nishibayashi, Masaru
Noble, Paul, Jr
Noring, Jon Everett
Norman, John Harris
Notley, Norman Thomas
Nugent, Leonard James
Nunn, Roland Cicero
Oakes, John Morgan
Obremski, Robert John
Offen, Henry William
Okumura, Mitchio
O'Loane, James Kenneth
O'Neal, Harry E
Orcutt, John Arthur
Orttung, William Herbert
Ory, Horace Anthony
Otto, Roland John
Otvos, John William
Panzer, Richard Earl
Parks, Norris Jim
Passchier, Arie Anton
Passell, Thomas Oliver
Payton, Patrick Herbert
Pearson, Michael J
Pearson, Robert Melvin
Pecora, Robert
Peller, Leonard
Peterson, Donald Bruce
Peterson, Donald Lee
Petrucci, Ralph Herbert
Petruska, John Andrew
Pettijohn, Richard Robert
Phillips, Norman Edgar
Phillips, Roger Winston
Philpott, Michael Ronald
Pickett, Herbert McWilliams
Pines, Alexander
Pitt, Colin Geoffrey
Pitzer, Kenneth Sanborn
Plock, Richard James
Po, Henry N
Potter, Norman D
Potts, John Calvin
Prenzlow, Carl Frederick
Pritchard, Glyn O
Pritchett, Thomas Ronald
Prussin, Stanley Gerald
Pye, Earl Louis
Quinlan, John Edward
Rainis, Andrew
Raleigh, Douglas Overholt
Raley, John Howard
Rard, Joseph Antoine
Rast, Howard Eugene, Jr
Razouk, Rashad Elias
Recht, Howard Leonard
Reichman, Sandor
Reisler, Hanna
Reiss, Howard
Rentzepis, Peter M
Rescigno, Thomas Nicola
Rhee, Jay Jea-yong
Rhim, Won-Kyu
Rhodes, David R
Richards, L(orenzo) Willard
Richardson, Jeffery Howard
Richter, Herbert Peter
Ritzman, Robert L
Roberts, Edwin Kirk
Rocklin, Albert Louis

Rodemeyer, Stephen A
Rodgers, James Edward
Roeder, Stephen Bernhard Walter
Rosa, Eugene John
Rosenblatt, Gerd Matthew
Rosenblum, Stephen Saul
Ross, David Samuel
Ross, John
Ross, Philip Norman, Jr
Roth, Walter
Rowland, Richard Lloyd
Roy, Prodyot
Rudolph, Raymond Neil
Ruiz, Carl P
Rusch, Peter Frederick
Russell, Kenneth Homer
Russell, Thomas Paul
Rustad, Douglas Scott
Ryason, Porter Raymond
Salovey, Ronald
Sanborn, Russell Hobart
Sancier, Kenneth Martin
Saperstein, David Dorn
Savitz, Maxine Lazarus
Saxon, Roberta Pollack
Scharber, Samuel Robert, Jr
Scharpen, LeRoy Henry
Scher, Herbert Benson
Scherer, James R
Schimitschek, Erhard Josef
Schleich, Thomas W
Schmidt, Hartland H
Schmidt, Raymond LeRoy
Schnepp, Otto
Schremp, Frederic William
Schugart, Kimberly A
Schwartz, Joseph Robert
Schwartz, Larry L
Schwartz, Robert Nelson
Scott, Gary Walter
Scott, Robert Lane
Seaborg, Glenn Theodore
Seki, Hajime
Senozan, Nail Mehmet
Servis, Kenneth L
Shackelford, Scott Addison
Shaka, Athan James
Sharp, Terry Earl
Shelby, Robert McKinnon
Shetlar, Martin David
Shin, Soo H
Shoolery, James Nelson
Shreve, George Wilcox
Shudde, Rex Hawkins
Shykind, David
Siebert, Eleanor Dantzler
Siegel, Bernard
Siegel, Irving
Silva, Robert Joseph
Silverman, Jacob
Sime, Rodney J
Sime, Ruth Lewin
Simon, John Douglas
Simpson, Howard Douglas
Singh, Hakam
Sinton, Steven Williams
Skolnick, Jeffrey
Sly, William Glenn
Smith, James Hart
Smith, Louis Charles
Snyder, Robert Gene
Solarz, Richard William
Solomon, Edward I
Somorjai, Gabor Arpad
Spilburg, Curtis Allen
Sporer, Alfred Herbert
Sposito, Garrison
Sprokel, Gerard J
Staudhammer, Peter
Steele, Warren Cavanaugh
Steffgen, Frederick Williams
Steinberg, Gunther
Steinberg, Martin
Steinmetz, Wayne Edward
Stephens, Frank Samuel
Stern, John Hanus
Stern, Richard Cecil
Stevens, William George
Stewart, Robert Daniel
Strauss, Herbert L
Streitwieser, Andrew, Jr
Stross, Fred Helmut
Struble, Gordon Lee
Sullivan, Richard Frederick
Suryaraman, Maruthuvakudi Gopalasastri
Sutton, David George
Swalen, Jerome Douglas
Swanson, David G, Jr
Sweeney, Michael Anthony
Switkes, Eugene
Syage, Jack A
Symons, Philip Charles
Szwarc, Michael
Taagepera, Mare
Taft, Robert Wheaton, Jr
Tang, Stephen Shien-Pu
Tannenbaum, Irving Robert
Templeton, David Henry
Tewes, Howard Allan
Theeuwes, Felix
Theodorou, Doros Nicolas
Thomas, John Richard
Thompson, Douglas Stuart
Thompson, Evan M

Thomson, Tom Radford
Tice, Russell L
Tietjen, James Joseph
Tinoco, Ignacio, Jr
Tinti, Dino S
Tobias, Charles W
Toensing, C(larence) H(erman)
Topol, Leo Eli
Trotter, Philip James
True, Nancy S
Trueblood, Kenneth Nyitray
Tuck, Leo Dallas
Tuffly, Bartholomew Louis
Tully, Frank Paul
Valletta, Robert M
Vander Wall, Eugene
Van Hecke, Gerald Raymond
Van Thiel, Mathias
Van Wart, Harold Edgar
Varteressian, K(egham) A(rshavir)
Vermeulen, Theodore (Cole)
Vilker, Vincent Lee
Viola, John Thomas
Vogel, Richard Clark
Vold, Marjorie Jean
Vold, Regitze Rosenorn
Volman, David H
Vroom, David Archie
Wade, Charles Gordon
Wall, Frederick Theodore
Walz, Alvin Eugene
Wang, John Ling-Fai
Ward, John William
Warshel, Arieh
Wauchope, Robert Donald
Weaver, Henry D, Jr
Weck, Friedrich Josef
Weed, H(omer) C(lyde)
Weers, Jeffry Greg
Weinberg, William Henry
Weir, William David
Weiss, Harold Gilbert
Wemmer, David Earl
Wensley, Charles Gelen
Wessel, John Emmit
Wettack, F Sheldon
Whaley, Katharine Birgitta
Whatley, Thomas Alvah
Wheeler, John C
Whittam, James Henry
Wiesendanger, Hans Ulrich David
Wilemski, Gerald
Williams, Colin James
Williams, Richard Stanley
Willis, Grover C, Jr
Wilson, Kent Raymond
Winston, Harvey
Wolf, Kathleen A
Wolfsberg, Max
Wong, Dorothy Pan
Wong, Hans Kuomin
Wong, Joe
Wong, K(wee) C
Wood, David Wells
Woods, William George
Woodson, John Hodges
Worden, Earl Freemont, Jr
Wunderly, Stephen Walker
Wydeven, Theodore
Yablonovitch, Eli
Yang, Meiling T
Yanick, Nicholas Samuel
Yannoni, Costantino Sheldon
Yates, Robert Edmunds
Yayanos, A Aristides
Yee, Rena
Yeh, Yin
Young, David A
Zare, Richard Neil
Zewail, Ahmed H
Zink, Jeffrey Irve
Zwerdling, Solomon

COLORADO

Anderson, Larry Gene
Ashby, Carl Toliver
Barford, Robert A
Bernstein, Elliot R
Bierbaum, Veronica Marie
Birks, John William
Blakeslee, A Eugene
Bruno, Thomas J
Cadle, Richard Dunbar
Callanan, Jane Elizabeth
Calvert, Jack George
Catalano, Anthony William
Chiou, Cary T(sair)
Christoffersen, Ralph Earl
Clough, Francis Bowman
Conner, Jack Michael
Connolly, John Stephen
Daughenbaugh, Randall Jay
de Heer, Joseph
DePoorter, Gerald Leroy
Dewey, Thomas Gregory
DoMinh, Thap
Douglas, Larry Joe
Eberhart, James Gettins
Edwards, Harry Wallace
Edwards, Kenneth Ward
Ely, James Frank
Estler, Ron Carter
Fischer, William Henry
Frank, Arthur Jesse

Frye, James Sayler
Gibbons, David Louis
Giulianelli, James Louis
Greene, Kenneth Titsworth
Greyson, Jerome
Hanley, Howard James Mason
Howard, Carleton James
James, MarLynn Rees
Keder, Wilbert Eugene
Kennedy, George Hunt
King, Lowell Alvin
Koch, William George
Ladanyi, Branka Maria
Lee, Raymond Curtis
Leone, Stephen Robert
Lineberger, William Carl
Macalady, Donald Lee
Maciel, Gary Emmet
Mahan, Kent Ira
Martin, Charles R
Mills, James Wilson
Milne, Thomas Anderson
Montgomery, Robert L
Noufi, Rommel
Nozik, Arthur Jack
Oetting, Franklin Lee
Olson, George Gilbert
Overend, Ralph Phillips
Ozog, Francis Joseph
Parkinson, Bruce Alan
Parungo, Farn Pwu
Peters, Kevin Scott
Phillipson, Paul Edgar
Presley, Cecil Travis
Rainwater, James Carlton
Riter, John Randolph, Jr
Robertson, Jerold C
Schwenz, Richard William
Sharma, Brahma Dutta
Sherrill, Bette Cecile Benham
Smith, Dwight Morrell
Solie, Thomas Norman
Solomon, Susan
Spain, Ian L
Stoner, Allan Wilbur
Strickler, Stewart Jeffery
Thompson, Gary Haughton
Tosch, William Conrad
Turner, James Howard
Vaughan, John Dixon
Watkins, Kay Orville
Watkins, Kenneth Walter
Webb, John Day
Weiner, Eugene Robert
Whatley, Alfred T
Wilkes, John Stuart
Witters, Robert Dale
Woody, Robert Wayne
Wright, Terry L
Yarar, Baki
Yates, Albert Carl

CONNECTICUT

Adler, Alan David
Arnold, John Richard
Bandes, Herbert
Barber, William Austin
Barrante, James Richard
Benson, Loren Allen
Berets, Donald Joseph
Beveridge, David L
Boggio, Joseph E
Bohn, Robert K
Bongiorni, Domenic Frank
Brinen, Jacob Solomon
Brooks, Clyde S
Brown, Charles Thomas
Brown, Keith Charles
Brown, Oliver Leonard Inman
Brudvig, Gary Wayne
Buchholz, Allan C
Burford, M Gilbert
Camp, Eldridge Kimbel
Casper, John Matthew
Castonguay, Richard Norman
Chang, Ted T
Cheng, Francis Sheng-Hsiung
Chupka, William Andrew
Clarke, George A
Corning, Mary Elizabeth
Csejka, David Andrew
David, Carl Wolfgang
DeDecker, Hendrik Kamiel Johannes
DePhillips, Henry Alfred, Jr
Dobbs, Gregory Melville
Douville, Phillip Raoul
Emond, George T
Epling, Gary Arnold
Gallivan, James Bernard
Garcia, Mario Leopoldo
Glasel, Jay Arthur
Goldstein, Marvin Sherwood
Gray, Thomas James
Gum, Mary Lou
Haller, Gary Lee
Hekal, Ihab M
Herbette, Leo G
Herron, Michael Myrl
Herz, Jack L
Hinchen, John J(oseph)
Hiskey, Clarence Francis
Horhota, Stephen Thomas
Hou, Kenneth C
I, Ting-Po

Physical Chemistry (cont)

Katz, Lewis
Kaufman, Ernest D
Kontos, Emanuel G
Kozak, Gary S
Krol, George J
Landsman, Douglas Anderson
Lasaga, Antonio C
Leniart, Daniel Stanley
Linke, William Finan
Lisman, Frederick Louis
Lowry, Eric G
Lumb, Ralph F
Lyons, Philip Augustine
McBride, James Michael
McKeon, Mary Gertrude
Markham, Claire Agnes
Martoglio, Pamela Anne
Mellor, John
Meriwether, Lewis Smith
Michels, H(orace) Harvey
Moore, Robert Earl
Morehead, Frederick Ferguson, Jr
Murai, Kotaro
Noack, Manfred Gerhard
Novick, Stewart Eugene
O'Connell, Edmond J, Jr
Orloff, Malcolm Kenneth
Orr, William Campbell
Papazian, Louis Arthur
Patterson, Andrew, Jr
Petersson, George A
Phillips, George Wygant, Jr
Piehl, Donald Herbert
Plessy, Boake Lucien
Prestegard, James Harold
Prigodich, Richard Victor
Pringle, Wallace C, Jr
Richards, Frederic Middlebrook
Roos, Leo
Roscoe, John Stanley, Jr
Rozelle, Lee T
Rusling, James Francis
Sanders, William Albert
Sarada, Thyagaraja
Schmitt, Joseph Lawrence, Jr
Schuster, Todd Mervyn
Scott, Raymond Peter William
Seefried, Carl G, Jr
Sethi, Dhanwant S
Shaw, Montgomery Throop
Singh, Ajaib
Skewis, John David
Skutnik, Bolesh Joseph
Smellie, Robert Henderson, Jr
Smith, Frederick Albert
Smith, Sidney Ruven
Smith, Trudy Enzer
Smith, Wendell Vandervort
Steingiser, Samuel
Stewart, Gerald Walter
Stuber, Fred A
Stwalley, William Calvin
Swanson, Donald Leroy
Tobin, Marvin Charles
Tucci, James Vincent
Vaisnys, Juozas Rimvydas
Valentine, Donald H, Jr
Voorhies, John Davidson
Wasser, Richard Barkman
Wellman, Russel Elmer
Wheeler, George Lawrence
Whipple, Earl Bennett
Wolfram, Leszek January
Yang, Darchun Billy
Young, George Jamison
Ziegler, Theresa Frances

DELAWARE

Abrams, Lloyd
Alms, Gregory Russell
Andersen, Donald Edward
Arrington, Charles Hammond, Jr
Baidins, Andrejs
Barteau, Mark Alan
Barton, Randolph, Jr
Bauchwitz, Peter S
Beare, Steven Douglas
Becher, Paul
Becker, Aaron Jay
Bissot, Thomas Charles
Blaker, Robert Hockman
Bond, Arthur Chalmer
Boyd, Robert Henry
Brandner, John David
Braun, Juergen Hans
Brill, Thomas Barton
Brown, Charles Julian, Jr
Buck, Warren Howard
Cella, Richard Joseph, Jr
Chang, Catherine Teh-Lin
Chase, David Bruce
Chowdhry, Uma
Chu, Sung Gun
Cohen, Edward David
Coleman, Marcia Lepri
Cone, Michael McKay
Cooney, John Leo
Corbin, David Richard
Culberson, Charles Henry
Dahl, Alton
Dam, Rudy Johan
Davison, Robert Wilder

Day, Herman O'Neal, Jr
Dettre, Robert Harold
Dixon, David Adams
Domaille, Peter John
Drake, Arthur Edwin
Dybowski, Cecil Ray
Eaton, David Fielder
Ehrich, F(elix) Frederick
English, Alan Dale
Espy, Herbert Hastings
Evans, Dennis Hyde
Falletta, Charles Edward
Ferguson, Raymond Craig
Firment, Lawrence Edward
Fleming, Sydney Winn
Foley, Henry Charles
Foss, Robert Paul
Frensdorff, H Karl
Futrell, Jean H
Galli, Paolo
Gierke, Timothy Dee
Gillow, Edward William
Glasgow, Louis Charles
Gold, Harvey Saul
Gore, Wilbert Lee
Gorski, Robert Alexander
Gosser, Lawrence Wayne
Grant, David Evans
Hayek, Mason
Held, Robert Paul
Heldt, Walter Z
Hembree, George Hunt
Herglotz, Heribert Karl Josef
Hiller, Dale Murray
Hoh, George Lok Kwong
Hoppenjans, Donald William
Howe, King Lau
Hsiao, Benjamin S
Huntsberger, James Robert
Ingersoll, Henry Gilbert
Ireland, Carol Beard
Johnson, Rulon Edward, Jr
Kegelman, Matthew Roland
Keidel, Frederick Andrew
Keller, Philip Joseph
Kissa, Erik
Kobsa, Henry
Koch, Theodore Augur
Kolber, Harry John
Kramer, Brian Dale
Kruse, Walter M
Krusic, Paul Joseph
Langsdorf, William Philip
Larry, John Robert
Lauzon, Rodrigue Vincent
Lund, John Turner
Lyons, Peter Francis
McCoy, V Eugene, Jr
McCullough, Roy Lynn
McGinnis, William Joseph
Malone, Creighton Paul
Mannis, Fred
Marshall, William Joseph
Mastrangelo, Sebastian Vito Rocco
Maury, Lucien Garnett
Merrifield, Richard Ebert
Mills, George Alexander
Moore, Ralph Bishop
Morrissey, Bruce William
Moses, Francis Guy
Munson, Burnaby
Neff, Bruce Lyle
Newby, William Edward
Nichols, James Randall
Niedzielski, Edmund Luke
Noggle, Joseph Henry
Odiorne, Truman J
Ovenall, Derick William
Overman, Joseph DeWitt
Palmer, Alan Blakeslee
Paris, Jean Philip
Pariser, Rudolph
Parkins, John Alexander
Paulson, Charles Maxwell, Jr
Perri, Joseph Mark
Phillips, Brian Ross
Pierce, Marion Armbruster
Pierce, Robert Henry Horace, Jr
Pieski, Edwin Thomas
Podlas, Thomas Joseph
Pretka, John E
Reddy, Gade Subbarami
Restaino, Alfred Joseph
Reuben, Jacques
Ridge, Douglas Poll
Roe, David Christopher
Ross, William D(aniel)
Saffer, Henry Walker
St John, Daniel Shelton
Sandell, Lionel Samuel
Sarner, Stanley Frederick
Schmude, Keith E
Senkler, George Henry, Jr
Sheer, M Lana
Simpson, David Alexander
Slade, Arthur Laird
Sloan, Gilbert Jacob
Smart, Bruce Edmund
Smith, Kenneth McGregor
Smith, Ronald W
Sopp, Samuel William
Sparks, Donald
Spurlin, Harold Morton
Staikos, Dimitri Nickolas

Staley, Ralph Horton
Starkweather, Howard Warner, Jr
Tanikella, Murty Sundara Sitarama
Thayer, Chester Arthur
Thompson, Robert Gene
Trumbore, Conrad Noble
Van Dyk, John William
Walsh, Robert Michael
Webb, Charles Alan
Weiher, James F
Wendt, Robert Charles
Wetlaufer, Donald Burton
Whitehouse, Bruce Alan
Wiley, Douglas Walker
Wolfe, William Ray, Jr
Wood, Robert Hemsley
Wopschall, Robert Harold
Yiannos, Peter N
Zimmerman, Joseph

DISTRICT OF COLUMBIA

Abelson, Philip Hauge
Ali, Mahamed Akbar
Bates, Richard Doane, Jr
Bednarek, Jana Marie
Bell, Jerry Alan
Berkowitz, Joan B
Blackburn, Thomas Roy
Brice, Luther Kennedy
Burley, Gordon
Burnett, John Lambe
Butler, James Ehrich
Callanan, Margaret Joan
Chang, Eddie Li
Chiu, Ying-Nan
Corsaro, Robert Dominic
Crist, DeLanson Ross
De Levie, Robert
Donahue, D Joseph
Earley, Joseph Emmet
Fearn, James Ernest
Graminski, Edmond Leonard
Gress, Mary Edith
Griscom, David Lawrence
Gutman, David
Halpern, Joshua Baruch
Hancock, Kenneth George
Hicks, Janice Marie
Hoering, Thomas Carl
Hughes, Robert Edward
Jawed, Inam
Karle, Jerome
Klein, Philipp Hillel
Laufer, Allan Henry
Lee, Robert E, Jr
Lutz, George John
McCormack, Mike
McGrory, Joseph Bennett
Mack, Julius L
McNeal, Robert Joseph
Maisch, William George
Martire, Daniel Edward
May, Leopold
Merrifield, D Bruce
Morehouse, Kim Matthew
Munson, Ronald Alfred
O'Grady, William Edward
Okabe, Hideo
O'Leary, Timothy Joseph
Panar, Manuel
Pierce, Dorothy Helen
Poranski, Chester F, Jr
Price, Elton
Raizen, Senta Amon
Ramaker, David Ellis
Roscher, Nina Matheny
Rowley, David Alton
Schnur, Joel Martin
Shirk, James Siler
Smith, Leslie E
Stern, Kurt Heinz
Sterrett, Kay Fife
Sutter, John Ritter
Tanczos, Frank I
Tolles, William Marshall
Trzaskoma, Patricia Povilitis
Tsang, Tung
Turner, Noel Hinton
Vertes, Akos
Webb, Alan Wendell
Weiss, Richard Gerald
Yesinowski, James Paul

FLORIDA

Adisesh, Setty Ravanappa
Allen, Eric Raymond
Ambrose, John Russell
Austin, Alfred Ells
Babich, Michael Wayne
Bailey, Thomas L, III
Banter, John C
Barron, Saul
Bartlett, Rodney Joseph
Batich, Christopher David
Baum, J(ames) Clayton
Binford, Jesse Stone, Jr
Birdwhistell, Ralph Kenton
Bishop, Charles Anthony
Block, Ronald Edward
Boehnke, David Neal
Boonstra, Bram B
Brey, Wallace Siegfried, Jr
Bright, Willard Mead
Brucat, Philip John

Carraher, Charles Eugene, Jr
Choppin, Gregory Robert
Cioslowski, Jerzy
Colgate, Samuel Oran
Criss, Cecil M
Crompton, Charles Edward
Davis, Jefferson Clark, Jr
Debnath, Sadhana
DeLap, James Harve
Diamond, Jacob J(oseph)
Dibeler, Vernon Hamilton
Dobbins, Robert Joseph
Dougherty, Ralph C
Drake, Robert Firth
Drost-Hansen, Walter
Duedall, Iver Warren
Edelson, David
Ellis, William Hobert
Ernsberger, Fred Martin
Eubank, William Roderick
Eyler, John Robert
Foner, Samuel Newton
Francis, Stanley Arthur
Fried, Vojtech
Friedlander, Herbert Norman
Friedman, Raymond
Fulton, Robert Lester
Garn, Paul Donald
Garrett, Richard Edward
Glick, Richard Edwin
Glover, Roland Leigh
Goldring, Lionel Solomon
Goll, Robert John
Gropp, Armin Henry
Gruntfest, Irving James
Gurry, Robert Wilton
Halperin, Joseph
Hanrahan, Robert Joseph
Hardman, Bruce Bertolette
Hare, Curtis R
Harrow, Lee Salem
Haugh, Eugene (Frederick)
Henkin, Hyman
Herczog, Andrew
Hertel, George Robert
Hirt, Thomas J(ames)
Hoeve, Cornelis Abraham Jacob
Huba, Francis
Hudson, Alice Peterson
Hudson, Reggie Lester
Hull, Harry H
Jackson, George Frederick, III
Johnsen, Russell Harold
Johnston, Milton Dwynell, Jr
Jones, Walter H(arrison)
Julien, Hiram Paul
Jurch, George Richard, Jr
Kasha, Michael
Kennelley, James A
Kerker, Milton
Kerr, Robert Lowell
Kirshenbaum, Abraham David
Krakow, Burton
Kreider, Henry Royer
Leblanc, Roger M
Lee, Ted C K
Leidheiser, Henry, Jr
Li, San
Linder, Bruno
Llewellyn, J(ohn) Anthony
Marshall, Alan George
Maskal, John
Maybury, Paul Calvin
Meisels, Gerhard George
Merrill, Jerald Carl
Micha, David Allan
Millero, Frank Joseph, Jr
Mills, Alfred Preston
Milton, Robert Mitchell
Moody, Leroy Stephen
Morehouse, Clarence Kopperl
Muga, Marvin Luis
Muschlitz, Earle Eugene, Jr
Myers, Gardiner Hubbard
Myerson, Albert Leon
Newman, Roger
Ors, Jose Alberto
Osipow, Lloyd Irving
Palmer, Jay
Parkanyi, Cyril
Parker, John Hilliard
Parreira, Helio Correa
Parrish, Clyde Franklin
Person, Willis Bagley
Perumareddi, Jayarama Reddi
Pierce, James Bruce
Pitt, Donald Alfred
Pittman, Melvin Amos
Plescia, Otto John
Polejes, J(acob) D
Ramaswamy, C
Ramsey, Brian Gaines
Reilly, Charles Austin
Riedhammer, Thomas M
Rikvold, Per Arne
Roth, James Frank
Ryznar, John W
Sadler, Monroe Scharff
Saffer, Alfred
Safron, Sanford Alan
Sakhnovsky, Alexander Alexandrovitch
Schmidt, Klaus H
Schulman, Stephen Gregory
Searle, Norma Zizmer

CHEMISTRY / 235

Seiger, Harvey N
Sheldon, John William
Snyder, Patricia Ann
Steadman, Thomas Ree
Stevens, Brian
Stewart, Burch Byron
Swartz, William Edward, Jr
Szczepaniak, Krystyna
Turner, Ralph Waldo
Urone, Paul
Vala, Martin Thorvald, Jr
Van Zee, Richard Jerry
Vergenz, Robert Allan
Wallace, Gerald Wayne
Wei-Berk, Caroline
Weltner, William, Jr
Wenger, Franz
Whitney, Ellsworth Dow
Wolsky, Sumner Paul
Woodward, Fred Erskine
Worrell, Jay H
Wu, Jin Zhong
Xu, Renliang
Young, Frank Glynn
Zaukelies, David Aaron
Zerner, Michael Charles

GEORGIA
Anderson, Bruce Martin
Anderson, James Leroy
Barr, John Baldwin
Barton, Franklin Ellwood, II
Batten, George L, Jr
Block, Toby Fran
Borkman, Raymond Francis
Brandau, Betty Lee
Buzzelli, Edward S
Camp, Ronnie Wayne
Carreira, Lionel Andrade
Cheng, Wu-Chieh
Clever, Henry Lawrence
Colvin, Clair Ivan
Combs, Leon Lamar, III
Cooper, Charles Dewey
Cooper, Gerald Rice
Crider, Fretwell Goer
Cummings, Frank Edson
Danzer, Laurence Alfred
DeLorenzo, Ronald Anthony
Derrick, Mildred Elizabeth
Dever, David Francis
Eberhardt, William Henry
Etzler, Frank M
Felton, Ronald H
Friedman, Harold Bertrand
Garmon, Lucille Burnett
Garrett, Bowman Staples
Gayles, Joseph Nathan, Jr
Goldstein, Jacob Herman
Gole, James Leslie
Grenga, Helen E(va)
Griffiths, James Edward
Habte-Mariam, Yitbarek
Harmer, Don Stutler
Hart, Raymond Kenneth
Heric, Eugene LeRoy
Hopkins, Harry P, Jr
James, Jeffrey
Johnston, Francis J
Jones, Alan Richard
Karickhoff, Samuel Woodford
Kaufman, Myron Jay
Keller, Oswald Lewin
Kohl, Paul Albert
Kotliar, Abraham Morris
Kromann, Paul Roger
Lee, John William
Lin, Ming Chang
Lindauer, Maurice William
Liu, Fred Wei Jui
Lokken, Stanley Jerome
MacMillan, Joseph Edward
Macur, George J
Manson, Steven Trent
Melton, Charles Estel
Miller, George Alford
Moran, Thomas Francis
Nelson, Robert Norton
Netzel, Thomas Leonard
Nickerson, John David
Orofino, Thomas Allan
Pacifici, James Grady
Perdue, Edward Michael
Peterson, Keith L
Pierotti, Robert Amadeo
Pignocco, Arthur John
Ray, Apurba Kanti
Reif, Donald John
Rhoades, James Lawrence
Richardson, Susan D
Rouvray, Dennis Henry
Samuels, Robert Joel
Sanders, T H, Jr
Schwerzel, Robert Edward
Spauschus, Hans O
Spencer, Jesse G
Starr, Thomas Louis
Steele, Jack
Stone, John Austin
Story, Troy L, Jr
Sturrock, Peter Earle
Teja, Amyn Sadruddin
Thomas, Tudor Lloyd
Tincher, Wayne Coleman
Trawick, William George
Walker, Russell Wagner
Walker, William Comstock
Walters, John Philip
Watts, Sherrill Glenn
Whetten, Robert Lloyd
Wilson, William David
Winnick, Jack
Zepp, Richard Gardner

HAWAII
Bopp, Thomas Theodore
Brantley, Lee Reed
Chu, Victor Fu Hua
Dorrance, William Henry
Frystak, Ronald Wayne
Hanna, Melvin Wesley
Hirayama, Chikara
Huebert, Barry Joe
Ihrig, Judson La Moure
Inskeep, Richard Guy
McDonald, Ray Locke
Mader, Charles Lavern
Muenow, David W
Rollins, Ronald Roy
Seff, Karl
Sood, Satya P
Waugh, John Lodovick Thomson

IDAHO
Baker, John David
Berreth, Julius R
Carter, Loren Sheldon
Castle, Lyle William
Christian, Jerry Dale
Hagen, Jack Ingvald
Hammer, Robert Russell
Harmon, J Frank
Heckler, George Earl
Lewis, Leroy Crawford
Miller, Richard Lloyd
Nagel, Terry Marvin
Ramshaw, John David
Renfrew, Malcolm MacKenzie
Schuman, Robert Paul
Schutte, William Calvin
Tallman, Richard Louis
Tanner, John Eyer, Jr
Wai, Chien Moo

ILLINOIS
Abraham, Bernard M
Albrecht, William Lloyd
Anysas, Jurgis Arvydas
Arnold, Richard Thomas
Astumian, Raymond Dean
Atoji, Masao
Bader, Samuel David
Baker, Louis, Jr
Bartels, David M
Beese, Ronald Elroy
Begala, Arthur James
Belford, R Linn
Berkowitz, Joseph
Berry, Richard Stephen
Blander, Milton
Bloemer, William Louis
Blumberg, Avrom Aaron
Boldridge, David William
Bonham, Russell Aubrey
Brubaker, Inara Mencis
Budrys, Rimgaudas S
Burdett, Jeremy Keith
Burns, Richard Price
Burrell, Elliott Joseph, Jr
Burwell, Robert Lemmon, Jr
Cafasso, Fred A
Carlson, Keith Douglas
Carnahan, Jon Winston
Carnall, William Thomas
Cengel, John Anthony
Ceperley, David Matthew
Chandler, Dean Wesley
Chang, Chin Hsiung
Chen, Juh Wah
Clardy, LeRoy
Clarkson, Robert Breck
Coley, Ronald Frank
Coutts, John Wallace
Crespi, Henry Lewis
Crumrine, David Shafer
Cummings, Thomas Fulton
Cunningham, George Lewis, Jr
Cutnell, John Daniel
Dehmer, Joseph L(eonard)
Dehmer, Patricia Moore
DeVault, Don Charles
Diamond, Herbert
Dickinson, Helen Rose
Didwania, Hanuman Prasad
Dlott, Dana D
Dobbs, Frank W
Dobry, Alan (Mora)
Duttaahmed, A
Ebdon, David William
Ehrlich, Gert
Eissler, Robert L
Eliason, Morton A
El Saffar, Zuhair M
Faber, Roger Jack
Falkenstein, Gary Lee
Faulkner, Larry Ray
Feng, Paul Yen-Hsiung
Fields, Paul Robert
Fildes, John M
Filson, Don P
Fischer, Albert Karl
Fitch, Alanah
Fleming, Graham Richard
Frederick, Kenneth Jacob
Freed, Karl F
Gaffney, Jeffrey Steven
Geiser, Urs Walter
Gemmer, Robert Valentine
Gindler, James Edward
Gislason, Eric Arni
Gleeson, James Newman
Glogovsky, Robert L
Glonek, Thomas
Golinkin, Herbert Sheldon
Gomer, Robert
Gordon, Robert Jay
Gordon, Sheffield
Granick, Steve
Gray, Linsley Shepard, Jr
Green, David William
Griffin, Harold Lee
Grove, Ewart Lester
Gruebele, Martin
Gruen, Dieter Martin
Gutfreund, Kurt
Gutowsky, H(erbert) S(ander)
Hadley, Fred Judson
Hakala, Reino William
Hardwidge, Edward Albert
Henderson, George Asa
Henderson, Giles Lee
Hersh, Herbert N
Hinckley, Conrad Cutler
Hinks, David George
Hoeg, Donald Francis
Hoffman, Alan Bruce
Hoffman, Brian Mark
Holcomb, David Nelson
Hollins, Robert Edward
Homeier, Edwin H, Jr
Hopkins, Paul Donald
House, James Evan, Jr
Huang, Leo W
Huang, Shu-Jen Wu
Hummel, John Philip
Huston, John Lewis
Hutchison, Clyde Allen, Jr
Ika, Prasad Venkata
Illingworth, George Ernest
Iton, Lennox Elroy
Jaffey, Arthur Harold
Jameson, A Keith
Johnson, Irving
Johnson, Marvin Francis Linton
Johnson, Robert E
Jonah, Charles D
Jonas, Jiri
Jones, Richard Evan, Jr
Jones, Thomas Hubbard
Josefson, Clarence Martin
Justen, Lewis Leo
Kanofsky, Jeffrey Ronald
Kaplan, Sam H
Kasner, Fred E
Katz, Joseph J
Katz, Sidney
Kaufman, Sheldon Bernard
Keiderling, Timothy Allen
Keiter, Ellen Ann
Kevill, Dennis Neil
Kiefer, John Harold
Kipp, James Edwin
Klein, Robert L
Kleppa, Ole Jakob
Kooser, Robert Galen
Koster, David F
Kotin, Leonard
Krawetz, Arthur Altshuler
Krueger, Robert Harold
Kuhlmann, George Edward
Kung, Harold Hing Chuen
Lam, Daniel J
Lang, Robert Phillip
LaPlanche, Laurine A
Larsen, Robert Peter
Lauterbur, Paul Christian
LeBreton, Pierre Robert
Leibowitz, Leonard
Leland, Frances E(lbridge)
Lerman, Zafra Margolin
Lester, George Ronald
Levy, Donald Harris
Lewin, Seymour Z
Lewis, Frederick D
Light, John Caldwell
Lisy, James Michael
Longworth, James W
Lucchesi, Claude A
Lustig, Stanley
McDonald, Hugh Joseph
Mahajan, Om Prakash
Makinen, Marvin William
Makowski, Mieczyslaw Paul
Makri, Nancy
Maroni, Victor August
Marquart, John R
Marvin, Henry Howard, Jr
Masel, Richard Isaac
Mason, Donald Frank
Meadows, James Wallace, Jr
Meguerian, Garbis H
Meisel, Dan
Melendres, Carlos Arciaga
Mendelsohn, Marshall H
Metz, Florence Irene
Meyer, Edwin F
Michael, Joe Victor
Miller, John Robert
Mooring, Francis Paul
Nachtrieb, Norman Harry
Nagy, Zoltan
Nandi, Satyendra Prosad
Nash, Kenneth Laverne
Nebgen, John William
Nieman, George Carroll
Nilges, Mark J
Nordine, Paul Clemens
Nuzzo, Ralph George
Oka, Takeshi
O'Neal, Harry Roger
Osten, Donald Edward
Ostrow, Jay Donald
Oxtoby, David W
Padrta, Frank George
Parks, Eric K
Parrill, Irwin Homer
Paul, Iain C
Pearlstein, Arne Jacob
Perkins, Alfred J
Pertel, Richard
Pohlmann, Hans Peter
Poppe, Wassily
Pratt, Stephen Turnham
Primak, William L
Rakowsky, Frederick William
Ratner, Mark A
Rayudu, Garimella V S
Renner, Terrence Alan
Rice, Stuart Alan
Richards, R Ronald
Ries, Herman Elkan, Jr
Riley, Stephen James
Rocek, Jan
Rothman, Alan Bernard
Russell, Morley Egerton
Saboungi, Marie-Louise Jean
Sauer, Myran Charles, Jr
Schatz, George Chappell
Scheiner, Steve
Schlossman, Mark Loren
Schrage, Samuel
Schreiner, Felix
Schweizer, Kenneth Steven
Scouten, Charles George
Secrest, Donald H
Sedlet, Jacob
Seebauer, Edmund Gerard
Sessa, David Joseph
Shain, Albert Leopold
Sherman, Warren V
Shriver, John William
Sibener, Steven Jay
Sinha, Shome Nath
Smith, Darwin Waldron
Smith, Gerard Vinton
Snelson, Alan
Snyder, Gary James
Spangler, Charles William
Spears, Kenneth George
Sproul, William Dallas
Stair, Peter Curran
Stanford, Marlene A
Stetter, Joseph Robert
Stout, John Willard
Streets, David George
Strehlow, Roger Albert
Sugarman, Nathan
Tetenbaum, Marvin
Thurnauer, Marion Charlotte
Trevorrow, Laverne Everett
Trifunac, Alexander Dimitrije
Turkevich, Anthony
Unger, Lloyd George
Vandeberg, John Thomas
Van Duyne, Richard Palmer
Van Lente, Kenneth Anthony
Vanysek, Petr
Vaughn, Joe Warren
Veis, Arthur
Veleckis, Ewald
Viehland, Larry Alan
Wallace, Thomas Patrick
Walter, Robert Irving
Warf, C Cayce, Jr
Washington, Elmer L
Wasielewski, Michael Roman
Weisleder, David
Weitz, Eric
Wenzel, Bruce Erickson
Wesolowski, Wayne Edward
Wharton, Lennard
White, Jesse Edmund
Williams, Lesley Lattin
Wilson, Robert Steven
Wingender, Ronald John
Woo, Lecon
Wozniak, Wayne Theodore
Wu, Ying Victor
Young, Austin Harry
Young, Charles Edward
Young, Linda
Zaromb, Solomon
Zimmerman, Donald Nathan
Zletz, Alex

Physical Chemistry (cont)

INDIANA
Adelman, Steven A
Allerhand, Adam
Alter, John Emanuel
Bair, Edward Jay
Balajee, Shankverm R
Bangs, Leigh Buchanan
Bischoff, Robert Francis
Boaz, Patricia Anne
Booe, J(ames) M(arvin)
Borden, Kenneth Duane
Bostick, Edgar E
Bryan, William Phelan
Byrn, Stephen Robert
Chateauneuf, John Edward
Chernoff, Donald Alan
Daly, Patrick Joseph
Desai, Pramod D
Droege, John Walter
Dubin, Paul Lee
Dykstra, Clifford Elliot
Ewing, George Edward
Ferraudi, Guillermo Jorge
Fessenden, Richard Warren
Fidelle, Thomas Patrick
Flick, Cathy
Fong, Francis K
Fong, Shao-Ling
Genshaw, Marvin Alden
Grant, Edward R
Grimley, Robert Thomas
Hagstrom, Stanley Alan
Hall, David Alfred
Halpern, Arthur Merrill
Hamill, William Henry
Havel, Henry Acken
Hayes, Robert Green
Helman, William Phillip
Hemmes, Paul Richard
Heydegger, Helmut Roland
Hubble, Billy Ray
Hug, Gordon L
Hulbert, Matthew H
Hunt, Ann Hampton
Jones, Noel Duane
Kamat, Prashant V
Kelly, Edward Joseph
King, Gayle Nathaniel
Kinsey, Philip A
Kirsch, Joseph Lawrence, Jr
Kosman, Warren Melvin
Kossoy, Aaron David
Kuo, Charles C Y
Landreth, Ronald Ray
Langer, Lawrence Marvin
Langhoff, Peter Wolfgang
Larter, Raima
Laskowski, Michael, Jr
Leonard, Jack E
Lorentzen, Keith Eden
Luerssen, Frank W
McKinney, Paul Caylor
Madden, Keith Patrick
Malik, David Joseph
Mark, Earl Larry
Marsh, Max Martin
Mazac, Charles James
Meiser, John H
Meyerson, Seymour
Montgomery, Lawrence Kernan
Moore, Walter John
Morris, David Alexander Nathaniel
Mozumder, Asokendu
Muller, Norbert
Murray, Lionel Philip, Jr
Ogren, Paul Joseph
Onwood, David P
Ostroy, Sanford Eugene
Parmenter, Charles Stedman
Patterson, Larry K
Pearlstein, Robert Milton
Peters, Dennis Gail
Pikal, Michael Jon
Pilger, Richard Christian, Jr
Pimblott, Simon Martin
Pinkham, Chester Allen, III
Porile, Norbert Thomas
Reilly, James Patrick
Richardson, James Wyman
Ricketts, John Adrian
Schuler, Robert Hugo
Schultz, Franklin Alfred
Shiner, Vernon Jack, Jr
Shoup, Charles Samuel, Jr
Shuman, Cornwell A
Simms, Paul C
Smith, David Lee
Smith, Gerald Duane
Sousa, Lynn Robert
Steinrauf, Larry King
Stevenson, Kenneth Lee
Streib, William E
Strieder, William
Subramanian, Sethuraman
Sutula, Chester Louis
Tatum, James Patrick
Tensmeyer, Lowell George
Thomas, John Kerry
Toomey, Joseph Edward
Tripathi, Gorakh Nath Ram
Trozzolo, Anthony Marion
Vest, Robert W(ilson)
Voorhoeve, Rudolf Johannes Herman
Wagner, Eugene Stephen
Weaver, Michael John
Williams, Edward James
Wyma, Richard J

IOWA
Applequist, Jon Barr
Baenziger, Norman Charles
Baker, James LeRoy
Breuer, George Michael
Buettner, Garry Richard
Cater, Earle David
Chang, James C
Coffman, Robert Edgar
Cotton, Therese Marie
DePristo, Andrew Elliott
Erickson, Luther E
Franzen, Hugo Friedrich
Friedrich, Bruce H
Gerstein, Bernard Clemence
Gordon, Mark Stephen
Gschneidner, Karl A(lbert), Jr
Hansen, Peter Jacob
Hansen, Robert Suttle
Hoekstra, John Junior
Jackson, Herbert Lewis
Jacob, Fielden Emmitt
Jacobson, Robert Andrew
Jordan, Truman H
Kintanar, Agustin
Kozak, John Joseph
Lutz, Robert William
Maatman, Russell Wayne
Martin, Don Stanley, Jr
Martin, S W
Moriarty, John Lawrence, Jr
Mottley, Carolyn
Powell, Jack Edward
Reuland, Robert John
Rider, Paul Edward, Sr
Sims, Leslie Berl
Smith, John F(rancis)
Smith, Karen Ann
Struve, Walter Scott
Svec, Harry John
Swenson, Charles Allyn
Wilhelm, Harley A
Wurster, Dale Eric, Jr
Yeung, Edward Szeshing
Zemke, Warren T

KANSAS
Allen, Anneke S
Argersinger, William John, Jr
Copeland, James Lewis
Fujimoto, Gordon Takeo
Funderburgh, James Louis
Gilles, Paul Wilson
Greene, Frank T
Greenlief, Charles M
Hammaker, Robert Michael
Harmony, Marlin D
Hiebert, Allen G
Ho, James Chien Ming
Judson, Charles Morrill
Kruh, Robert Frank
Kuwana, Theodore
Latschar, Carl Ernest
Lehman, Thomas Alan
Lindenbaum, Siegfried
Lookhart, George LeRoy
McElroy, Albert Dean
Moser, Herbert Charles
Noelken, Milton Edward
Potts, Melvin Lester
Renich, Paul William
Rumpel, Max Leonard
Rytting, Joseph Howard
Setser, Donald W
Shearer, Edmund Cook
Sobczynski, Dorota
Sobczynski, Radek
Sorensen, Christopher Michael
Talaty, Erach R
Tran, Loc Binh
Wahlbeck, Phillip Glenn

KENTUCKY
Armendarez, Peter X
Beall, Gary Wayne
Beyer, Louis Martin
Bujake, John Edward, Jr
Butterfield, David Allan
Byrn, Ernest Edward
Chamberlin, John MacMullen
Clouthier, Dennis James
Companion, Audrey (Lee)
Conley, Harry Lee, Jr
Davis, Burtron H
Draper, Arthur Lincoln
Du Pre, Donald Bates
Farrar, David Turner
Field, Jay Ernest
Goren, Alan Charles
Guthrie, Robert D
Henley, Melvin Brent, Jr
Hettinger, William Peter, Jr
Johnson, Lawrence Robert
Keely, William Martin
Klein, Elias
Kramer, Jerry Martin
McDermott, Dana Paul
Magid, Linda Lee Jenny
Mapes, William Henry
Mitchell, Maurice McClellan, Jr
Moorhead, Edward Darrell
Mueller, Sr Rita Marie
Pan, Wei-Ping
Parekh, Bhupendra Kumar
Pearson, Earl F
Plucknett, William Kennedy
Porter, Richard A
Price, Martin Burton
Reucroft, Philip J
Sands, Donald Edgar
Smiley, Harry M
Smith, Terry Edward
Tabibian, Richard
Taylor, Morris D
Thompson, Ralph J
Trapp, Charles Anthony
Wilkins, Curtis C
Williams, Donald Elmer

LOUISIANA
Allen, Susan Davis
Anex, Basil Gideon
Arthur, Jett Clinton, Jr
Bains, Malkiat Singh
Benerito, Ruth Rogan
Berni, Ralph John
Bissell, Charles Lynn
Boudreaux, Edward A
Bursh, Talmage Poutau
Byrd, David Shelton
Camp, Ronald Lee
Carmichael, J W, Jr
Carpenter, Dewey Kenneth
Chrastil, Joseph
Clarke, Margaret Alice
Craig, James Porter, Jr
Crowder, Gene Autrey
Day, Marion Clyde, Jr
Fagley, Thomas Fisher
Fischer, David John
Franklin, William Elwood
French, Alfred Dexter
Gale, Robert James
Gonzalez, Richard Donald
Hamori, Eugene
Hanlon, Thomas Lee
Jones, Daniel Elven
Jung, Hilda Ziifle
Kaldor, Andrew
Keeley, Dean Francis
Kern, Ralph Donald, Jr
Kestner, Neil R
Klasinc, Leo
Kumar, Devendra
La Rochelle, John Hart
Looney, Ralph William
McGlynn, Sean Patrick
Mark, Harold Wayne
Miller, Kenneth Jay
Moseley, Harry Edward
Nauman, Robert Vincent
Nelson, Mary Lockett
Oliver, James Russell
Perkins, Richard Scott
Phillips, Travis J
Ponter, Anthony Barrie
Pullen, Bailey Price
Raisen, Elliott
Ratchford, Robert James
Riddick, Frank Adams, Jr
Roberts, Donald Duane
Robinson, Press L
Robson, Harry Edwin
Runnels, Kelli
Schexnayder, Mary Anne
Schroeder, Rudolph Alrud
Schucker, Robert Charles
Scott, John Delmoth
Sen, Buddhadev
Smith, Charles Hooper
Smith, Martin Bristow
Sulkes, Mark
Thompson, Ronald Hobart
Valsaraj, Kalliat Thazhathuveetil
Vickroy, Virgil Vester, Jr
Von Bodungen, George Anthony
Wan, Peter J
Ward, Truman L
Wendt, Richard P
Wharton, James Henry
Williams, Hulen Brown
Wood, James Manley, Jr
Wright, Oscar Lewis
Zagar, Walter T

MAINE
Allen, Roger Baker
Batha, Howard Dean
Birkett, James Davis
Boyles, James Glenn
Butcher, Samuel Shipp
Duncan, Charles Donald
Dunlap, Robert D
Fort, Raymond Cornelius, Jr
Goodfriend, Paul Louis
Harris, Robert Laurence
Holden, James Richard
Kirkpatrick, Francis Hubbard
Lauterbach, George Ervin
Merrifield, Paul Elliott
Nagle, Jeffrey Karl
Patterson, Howard Hugh
Rasaiah, Jayendran C
Shattuck, Thomas Wayne
Simard, Gerald Lionel
Smith, Wayne Lee
Sottery, Theodore Walter
Stauffer, Charles Henry
Stebbins, Richard Gilbert
Wilson, Thomas Lee

MARYLAND
Abramowitz, Stanley
Adrian, Frank John
Albers, Edwin Wolf
Aranow, Ruth Lee Horwitz
Avery, William Hinckley
Bardo, Richard Dale
Barnes, Charlie James
Barnes, James Alford
Bax, Ad
Becker, Edwin Demuth
Benson, Richard C
Berl, Walter G(eorge)
Bernecker, Richard Rudolph
Bertocci, Ugo
Bestul, Alden Beecher
Bicknell-Brown, Ellen
Blankenship, Floyd Allen
Block, Jacob
Block, Stanley
Blumenthal, Robert Paul
Bonnell, David William
Bonsack, James Paul
Bowen, Kit Hansel
Brenner, Abner
Brown, Samuel Heffner
Brown, Walter Eric
Bryden, Wayne A
Callahan, John Joseph
Callahan, Mary Vincent
Campbell, William Joseph
Casassa, Michael Paul
Castellan, Gilbert William
Castle, Peter Myer
Cavanagh, Richard Roy
Cave, William Thompson
Chang, Ren-Fang
Chang, Shu-Sing
Chapin, Douglas Scott
Charney, Elliot
Chen, Ching-Nien
Chmurny, Gwendolyn Neal
Chow, Laurence Chung-Lung
Clarke, Frederic B, III
Coates, Arthur Donwell
Cody, Regina Jacqueline
Cohen, Gerson H
Coleman, James Stafford
Colle, Ronald
Colwell, Jack Harold
Coplan, Michael Alan
Coriell, Sam Ray
Coursey, Bert Marcel
Cross, David Ralston
Cunniff, Patricia A
Currie, Lloyd Arthur
Dagdigian, Paul J
Darwent, Basil de Baskerville
Dauber, Edwin George
Davis, George Thomas
Davis, Henry McRay
Davis, Julien Sinclair
Debye, Nordulf Wiking Gerud
Dehn, James Theodore
Deitz, Victor Reuel
DeToma, Robert Paul
Deutch, John Mark
DeVoe, Howard Josselyn
Dickens, Brian
Diness, Arthur M(ichael)
Doan, Arthur Sumner, Jr
Dragun, Henry L
Duignan, Michael Thomas
Dunning, Herbert Neal
Eanes, Edward David
Eden, Murray
Edinger, Stanley Evan
Egan, William Michael
Egelhoff, William Frederick, Jr
Eng, Leslie
Epstein, Joseph
Ewing, June Swift
Ferretti, James Alfred
Fifer, Robert Alan
Flynn, Joseph Henry
Forziati, Alphonse Frank
Fox, Robert Bernard
Fraser, Gerald Timothy
Freedman, Eli (Hansell)
Freeman, David Haines
Frey, Douglas D
Frohndsdorff, Geoffrey James Carl
Fulmer, Glenn Elton
Gann, Richard George
Garvin, David
Gevantman, Lewis Herman
Goldberg, Robert Nathan
Goldenberg, Neal
Gornick, Fred
Gottlieb, Melvin Harvey
Grant, Warren Herbert
Gravatt, Claude Carrington, Jr
Green, John Arthur Savage
Greenhouse, Harold Mitchell
Greer, Sandra Charlene

Griffo, Joseph Salvatore
Gryder, John William
Guruswamy, Vinodhini
Hadermann, Albert Felix
Haller, Wolfgang Karl
Hamer, Walter Jay
Hampson, Robert F, Jr
Hartman, Kenneth Owen
Hartstein, Arthur M
Hashmall, Joseph Alan
Hastie, John William
Hedman, Fritz Algot
Herrick, Claude Cummings
Herron, John Thomas
Heydemann, Peter Ludwig Martin
Hiatt, Caspar Wistar, III
Hillstrom, Warren W
Hirsh, Allen Gene
Hoffman, Michael K
Hofrichter, Harry James
Hollandsworth, Clinton E
Hougen, Jon T
Houseman, Barton L
Howell, Barbara Fennema
Hsu, Chen C
Huang, Charles Y
Hunt, Paul Payson
Huntress, Wesley Theodore, Jr
Iwasa, Kuni H
Jackson, Jo-Anne Alice
Jacox, Marilyn Esther
Jaffe, Harold
James, Stanley D
Jarvis, Neldon Lynn
Jernigan, Robert Lee
Jonas, Leonard Abraham
Jones, Owen Lloyd
Jovancicevic, Vladimir
Julienne, Paul Sebastian
Kalasinsky, Victor Frank
Kandel, Richard Joshua
Kappe, David Syme
Katz, Joseph L
Kaufman, Joyce J
Kaufman, Samuel
Kelsh, Dennis J
Ketley, Arthur Donald
Khare, Mohan
Kippenberger, Donald Justin
Kirchhoff, William Hayes
Kirkien-Rzeszotarski, Alicja M
Kleier, Daniel Anthony
Klein, Nathan
Koski, Walter S
Kossiakoff, Alexander
Kreis, Ronald W
Krisher, Lawrence Charles
Krumbein, Simeon Joseph
Kumins, Charles Arthur
Kurylo, Michael John, III
Kushner, Lawrence Maurice
Lafferty, Walter J
Land, William Everett
Larkin, David
Larson, Clarence Edward
Lee, Byungkook
Lee, Frederick Strube
Lepley, Arthur Ray
LeRoy, André François
Levin, Ira William
Levin, Irvin
Levy, Joseph Benjamin
Liang, Shoudeng
Lide, David Reynolds, Jr
Linevsky, Milton Joshua
Linzer, Melvin
Loebenstein, William Vaille
Long, Charles Anthony
Luborsky, Samuel William
Lyons, John Winship
McDaniel, Carl Vance
McDonald, Jimmie Reed
McDowell, Hershel
McLaughlin, Alan Charles
McLaughlin, William Lowndes
McNesby, James Robert
Magee, John Storey
Maher, Philip Kenerick
Malarkey, Edward Cornelius
Malmberg, Marjorie Schooley
Margalit, Nehemiah
Marinenko, George
Marqusee, Jeffrey Alan
Marshall, Walter Lincoln
Martinez, Richard Isaac
May, Willie Eugene
Maycock, John Norman
Mazumder, Bibhuti R
Meyer, Richard Adlin
Michejda, Christopher Jan
Mies, Frederick Henry
Miller, Gerald Ray
Miller, Melvin P
Miller, Raymond Earl
Miller, Walter E
Minton, Allen Paul
Misra, Dwarika Nath
Miziolek, Andrzej Wladyslaw
Monchick, Louis
Moore, John Hays
Mopsik, Frederick Israel
Moses, Saul
Munn, Robert James
Murr, Brown L, Jr

Myers, Lawrence Stanley, Jr
Neta, Pedatsur
Neumann, Dan Alan
Novotny, Donald Bob
Offutt, William Franklin
O'Hare, Patrick A G
Olson, William Bruce
Ondov, John Michael
Orlick, Charles Alex
Ostrofsky, Bernard
Parsegian, Vozken Adrian
Paule, Robert Charles
Pawlowski, Anthony T
Peters, Alan Winthrop
Petersen, Raymond Carl
Peterson, Miller Harrell
Peterson, Norman Cornelius
Pierce, Elliot Stearns
Piermarini, Gasper J
Plumlee, Karl Warren
Price, Donna
Rader, Charles Allen
Rakhit, Gopa
Raksis, Joseph W
Rao, Gopalakrishna M
Reckley, John
Redmon, Michael James
Reilly, Hugh Thomas
Riesz, Peter
Rifkind, Joseph Moses
Roberts, Ralph
Robertson, Baldwin
Robey, Frank A
Robinson, Dean Wentworth
Roche, Lidia Alicia
Romans, James Bond
Rosenberg, Arnold Morry
Rosenblatt, David Hirsch
Rosenfield, Joan E
Rotariu, George Julian
Rothman, Sam
Rowell, Charles Frederick
Ruby, Stanley
Rush, John Joseph
Ruth, John Moore
Rutner, Emile
Saba, William George
Sahu, Saura Chandra
Salwin, Arthur Elliott
Samuel, Aryeh Hermann
Satkiewicz, Frank George
Scarbrough, Frank Edward
Scheer, Milton David
Schroeder, LeRoy William
Schroeder, Michael Allan
Schuldiner, Sigmund
Schultz, John Wilfred
Sheinson, Ronald Swiren
Silverstone, Harris Julian
Silverton, James Vincent
Simic, Michael G
Simmons, Joe Denton
Smardzewski, Richard Roman
Snow, Milton Leonard
Sondergaard, Neal Albert
Spangler, Glenn Edward
Stadlbauer, John Mannix
Stein, Stephen Ellery
Stephenson, John Carter
Stimler, Suzanne Stokes
Strauss, Mary Jo
Strier, Murray Paul
Stromberg, Robert Remson
Strong, Laurence Edward
Suenram, Richard Dee
Swann, Madeline Bruce
Sweeting, Linda Marie
Tannenbaum, Harvey
Taylor, John Keenan
Tevault, David Earl
Thirumalai, Devarajan
Thompson, Donald Leroy
Thompson, Joseph Kyle
Thompson, Robert John, Jr
Thompson, Warren Elwin
Tompa, Albert S
Tompkins, Robert Charles
Trus, Benes L
Tung, Ming Sung
Tupper, Kenneth Joseph
Tuve, Richard Larsen
Urbach, Herman B
Vander Hart, David Lloyd
Vargas, Fernando Figueroa
Verdier, Peter Howard
Vincent, James Sidney
Wang, Francis Wei-Yu
Wason, Satish Kumar
Weber, Alfons
Weber, Leon
Weber, Mark
Weiner, John
Weiss, Charles, Jr
Williams, Ellen D
Willis, Phyllida Mave
Wilmot, George Barwick
Wing, James
Wu, En Shinn
Wu, Yung-Chi
Yagi, Haruhiko
Yarkony, David R
Yeh, Herman Jia-Chain
Young, Jay Alfred
Zimmerman, John Gordon

Zimmerman, Steven B
Zuckerbrod, David
Zwanzig, Robert Walter

MASSACHUSETTS
Ackerman, Jerome Leonard
Adelman, Albert H
Adler, Alice Joan
Adler, Norman
Alberty, Robert Arnold
Allen, Malwina I
Armington, Alton
Atkins, Jaspard Harvey
Baboian, Robert
Baglio, Joseph Anthony
Baird, Donald Heston
Bald, Kenneth Charles
Bannister, William Warren
Bearden, William Harlie
Berka, Ladislav Henry
Bernard, Walter Joseph
Bethune, John Lemuel
Birstein, Seymour J
Blank, Charles Anthony
Boedtker Doty, Helga
Bonner, Francis Joseph
Bowers, Peter George
Bracco, Donato John
Brandts, John Frederick
Bridgman, Wilbur Benjamin
Butler, James Newton
Caledonia, George Ernest
Carey, George H
Cathou, Renata Egone
Cembrola, Robert John
Cerankowski, Leon Dennis
Chance, Kelly Van
Chang, Kuang-Chou
Chang, Raymond
Charkoudian, John Charles
Chien, James C W
Chiu, Tak-Ming
Chou, Kuo-Chih
Chung, Frank H
Clough, Stuart Benjamin
Cobb, Carolus M
Coleman, William Fletcher
Comeford, Lorrie Lynn
Connors, Robert Edward
Cormier, Alan Dennis
Coulter, Lowell Vernon
Daley, Henry Owen, Jr
D'Amato, Richard John
Dampier, Frederick Walter
Das, Pradip Kumar
Davidovits, Paul
Davison, Peter Fitzgerald
DeCosta, Peter F(rancis)
Dewald, Robert Reinhold
Dickinson, Alan Charles
Dickinson, Leonard Charles
DiNardi, Salvatore Robert
Donoghue, John Timothy
Dorain, Paul Brendel
Druy, Mark Arnold
Ehret, Anne
Elliott, John Frank
Engelke, John Leland
Epstein, Irving Robert
Eschenroeder, Alan Quade
Feakes, Frank
Fennelly, Paul Francis
Field, Robert Warren
Fink, Richard David
Fleck, George Morrison
Flynn, George P(atrick)
Freedman, Andrew
Friend, Cynthia Marie
Fritzsche, Alfred Keith
Galligan, John D(onald)
Gardner, Donald Murray
Garland, Carl Wesley
George, James Henry Bryn
Gersh, Michael Elliot
Gibbard, H Frank
Giner, Jose D
Glazman, Yuli M
Gopikanth, M L
Gordon, George Selbie
Gordon, Roy Gerald
Green, Byron David
Greenaway, Frederick Thomas
Greif, Mortimer
Griesinger, David Hadley
Grunwald, Ernest Max
Haas, John William, Jr
Hagnauer, Gary Lee
Hall, Lowell Headley, II
Hamilton, James Arthur
Hanselman, Raymond Bush
Harrison, Anna Jane
Harvey, Walter William
Hayden, Thomas Day
Henchman, Michael J
Herschbach, Dudley Robert
Herz, Matthew Lawrence
Herzfeld, Judith
Hess, John Monroe Converse
Hobbs, John Robert
Hodgdon, Russell Bates
Hoffman, Morton Z
Hollister, Charlotte Ann
Hopkins, Esther Arvilla Harrison
Hornig, Donald Frederick

Huang, Jan-Chan
Huffman, Robert Eugene
Illinger, Joyce Lefever
Illinger, Karl Heinz
Ingle, George William
Isaks, Martin
Jachimowicz, Felek
Johnson, Keith Huber
Johnston, A Sidney
Jones, Guilford, II
Jost, Ernest
Karplus, Martin
Keat, Paul Powell
Kellner, Jordan David
Klemperer, William
Klinedinst, Keith Allen
Kolodny, Nancy Harrison
Kopf, Peter W
Kowalak, Albert Douglas
Krieger, Jeanne Kann
Kushick, Joseph N
Kustin, Kenneth
Lamola, Angelo Anthony
Langmuir, Margaret Elizabeth Lang
Lebowitz, Elliot
Lester, Joseph Eugene
Levine, Oscar
Levy, Boris
Lewis, Armand Francis
Licht, Stuart Lawrence
Lichtin, Norman Nahum
Lindsay, William Tenney, Jr
Linschitz, Henry
Lipscomb, William Nunn, Jr
Loehlin, James Herbert
Lowey, Susan
Lowry, Nancy
Luft, Ludwig
McFadden, David Lee
Mac Knight, William John
Maher, Galeb Hamid
Malin, Murray Edward
Manchester, Kenneth Edward
Mandl, Alexander Ernst
Mariani, Henry A
Marrero, Hector G
Marshall, Mark David
Meal, Janet Hawkins
Medalia, Avrom Izak
Mehta, Mahesh J
Menashi, Jameel
Messer, Charles Edward
Meyer, Vincent D
Millard, Richard James
Miller, Keith Wyatt
Molina, Mario Jose
Monson, Peter A
Morris, Robert Alan
Murad, Edmond
Nash, Leonard Kollender
Nelson, David Robert
Nelson, Keith A
Nemeth, Ronald Louis
Neue, Uwe Dieter
Norment, Hillyer Gavin
Olderman, Jerry
Oppenheim, Irwin
Pal, Uday Bhanu
Pandolfe, William David
Papaefthymiou, Georgia Christou
Patrick, Richard Montgomery
Paulsen, Duane E
Paulson, John Frederick
Pemsler, J(oseph) Paul
Peri, John Bayard
Perkins, Janet Sanford
Phillies, George David Joseph
Plumb, Robert Charles
Powell, Arnet L
Prock, Alfred
Pyun, Chong Wha
Ragle, John Linn
Ramsley, Alvin Olsen
Rauh, Robert David, Jr
Rawlins, Wilson Terry
Ricci, John Silvio, Jr
Rivin, Donald
Roberts, Mary Fedarko
Rock, Elizabeth Jane
Roebber, John Leonard
Rose, Timothy Laurence
Rosenthal, Lois C
Ross, James William, Jr
Rowe, Paul E
Rowell, Robert Lee
Rupich, Martin Walter
Scala, Alfred Anthony
Schempp, Ellory
Schneider, Maxyne Dorothy
Schonhorn, Harold
Schwartz, Lowell Melvin
Serafin, Frank G
Shimizu, Nobumichi
Silbey, Robert James
Simons, Harold Lee
Slater, Richard Craig
Smith, Joseph Harold
Soltzberg, Leonard Jay
Spitler, Mark Thomas
Sprengnether, Michele M
Steel, Colin
Stein, Samuel H
Steinfeld, Jeffrey Irwin
Stengle, Thomas Richard

Physical Chemistry (cont)

Stidham, Howard Donathan
Swank, Thomas Francis
Szafran, Zvi
Taub, Irwin A(llen)
Tauer, Kenneth J
Taylor, Lloyd David
Taylor, Raymond L
Thomas, Martha Jane Bergin
Torre, Frank John
Towns, Donald Lionel
Tsuk, Andrew George
Turnbull, David
Tuttle, Thomas R, Jr
Tykodi, Ralph John
Uhlig, Herbert H(enry)
Ukleja, Paul Leonard Matthew
Umans, Robert Scott
Underwood, Donald Lee
Vanderslice, Thomas Aquinas
Vanpee, Marcel
Viggiano, Albert
Waack, Richard
Wallace, Frederic Andrew
Wang, Chih-Lueh Albert
Wang, Robert T
Waugh, John Stewart
Weaver, Edwin Snell
Weininger, Stephen Joel
Weiss, Karl H
Wen, Wen-Yang
Westmoreland, Phillip R
Wilde, Anthony Flory
Wilson, Edgar Bright
Wrathall, Donald Prior
Wright, David Franklin
Wrighton, Mark Stephen
Yannas, Ioannis Vassilios
Yannoni, Nicholas
Yoshino, Kouichi

MICHIGAN
Abu-Isa, Ismat Ali
Adams, Wade J
Anders, Oswald Ulrich
Anderson, James E
Bartell, Lawrence Sims
Bates, John Bertram
Bauer, David Robert
Ben, Manuel
Bennett, Stephen Lawrence
Berry, Myron Garland
Bettman, Max
Blint, Richard Joseph
Blurton, Keith F
Bone, Larry Irvin
Brenner, Alan
Brodasky, Thomas Francis
Butts, Susan Beda
Carpenter, Michael Kevin
Chao, Mou Shu
Chessin, Hyman
Cho, Byong Kwon
Clarke, Richard Penfield
Cochrane, Hector
Cogan, Harold Louis
Coleman, David Manley
Cook, Richard James
Cornilsen, Bahne Carl
Craig, Robert George
Cratin, Paul David
Crippen, Gordon Marvin
Curnutt, Jerry Lee
Dahlgren, George
Dahm, Donald B
Deal, Ralph Macgill
Diamond, Howard
Dickie, Ray Alexander
Diesen, Ronald W
Dirkse, Thedford Preston
Downey, Joseph Robert, Jr
Doyle, Daryl Joseph
Drake, Michael Cameron
Duchamp, David James
Dunn, Thomas M
Dye, James Louis
Eastland, George Warren, Jr
Ebbing, Darrell Delmar
Eick, Harry Arthur
Eliezer, Isaac
Eliezer, Naomi
Ellis, Thomas Stephen
Endicott, John F
Filisko, Frank Edward
Fisher, Edward Richard
Fisher, Galen Bruce
Flynn, James Patrick
Foister, Robert Thomas
Forist, Arlington Ardeane
Francis, Anthony Huston
Francisco, Joseph Salvadore, Jr
Freyberger, Wilfred L(awson)
Gagnon, Eugene Gerald
Gandhi, Harendra Sakarlal
Garrett, David L, Jr
Gendernalik, Sue Aydelott
Gland, John Louis
Gorse, Robert August, Jr
Graves, Bruce Bannister
Greene, Bettye Washington
Griffin, Henry Claude
Gulick, Wilson M, Jr
Gunther, Ronald George

Hahne, Rolf Mathieu August
Hajratwala, Bhupendra R
Halberstadt, Marcel Leon
Hansen, Robert Douglas
Harris, Stephen Joel
Harrison, James Francis
Hase, William Louis
Heeschen, Jerry Parker
Heller, Hanan Chonon
Herk, Leonard Frank
Hillig, Kurt Walter, II
Hoare, James Patrick
Hong, Kuochih
Hood, Robin James
Houser, Thomas J
Howe, Norman Elton, Jr
Huettner, David Joseph
Hutchison, James Robert
Iyengar, Doreswamy Raghavachar
Jacko, Michael George
Jacobs, Gerald Daniel
Johnson, David Alfred
Johnson, Ray Leland
Johnson, Richard Allen
Julien, Larry Marlin
Kagel, Ronald Oliver
Kaiser, Edward William, Jr
Kawooya, John Kasajja
Kelly, Nelson Allen
Kenney, Donald J
Killgoar, Paul Charles, Jr
King, Peter Foster
King, Stanley Shih-Tung
Komarmy, Julius Michael
Kopelman, Raoul
Kuczkowski, Robert Louis
Kunz, Albert Barry
Kuo, Mingshang
Lane, George Ashel
Larsen, Eric Russell
Larson, John Grant
Lee, Chuan-Pu
Lee, Wei-Ming
Leffert, Charles Benjamin
Leifer, Leslie
Leroi, George Edgar
Levine, Samuel
Li, Chi-Tang
Lindfors, Karl Russell
Lowry, George Gordon
Lubman, David Mitchell
Macarthur, Donald M
McDonald, Charles Joseph
McHarris, William Charles
McTague, John Paul
Mansfield, Marc L
Mayer, William John
Mendenhall, George David
Miller, Philip Joseph
Miller, Steven Ralph
Moore, Carl
Morgan, Roger John
Morrissey, David Joseph
Moyer, John Raymond
Murchison, Craig Brian
Murchison, Pamela W
Mutch, George William
Nazri, Gholam-Abbas
Neely, Brock Wesley
Niekamp, Carl William
Nordman, Christer Eric
Novak, Raymond Francis
Oei, Djong-Gie
O'Neil, James R
Otto, Klaus
Ouderkirk, John Thomas
Passino, Dora R May
Penner-Hahn, James Edward
Peters, James
Petersen, Donald Ralph
Peterson-Kennedy, Sydney Ellen
Petrella, Ronald Vincent
Platt, Alan Edward
Plaush, Albert Charles
Polik, William Frederick
Powell, Ralph Robert
Ramamurthy, Amurthur C
Reck, Gene Paul
Revzin, Arnold
Riehl, James Patrick
Rieke, James Kirk
Riley, Bernard Jerome
Rogers, Jerry Dale
Rothe, Erhard William
Rothschild, Walter Gustav
Russell, Joel W
Sandel, Vernon Ralph
Sarge, Theodore William
Sarkar, Nitis
Saunders, Frank Linwood
Schlick, Shulamith
Schnitker, Jurgen H
Schullery, Stephen Edmund
Schwartz, Shirley E
Schwendeman, Richard Henry
Seefurth, Randall N
Sell, Jeffrey Alan
Sevilla, Michael Douglas
Sharma, Ram Autar
Sharp, Robert Richard
Sheetz, David P
Skelly, Norman Edward
Skochdopole, Richard E
Sloane, Thompson Milton

Slomp, George
Small, Hamish
Smith, Arthur Gerald
Snyder, Dexter Dean
Solc, Karel
Soulen, John Richard
Spurgeon, William Marion
Stark, Forrest Otto
Stowe, Robert Allen
Summitt, W(illiam) Robert
Swathirajan, S
Swets, Don Eugene
Tai, Julia Chow
Tamres, Milton
Taylor, Kathleen C
Taylor, Robert Cooper
Thacker, Raymond
Tischer, Ragnar P(ascal)
Tomalia, Donald Andrew
Tou, James Chieh
Tuesday, Charles Sheffield
Ullman, Robert
Vanderwielen, Adrianus Johannes
Vukasovich, Mark Samuel
Wahl, Werner Henry
Wallington, Timothy John
Warrick, Earl Leathen
Warrington, Terrell L
Watenpaugh, Keith Donald
Weber, Dennis Joseph
Weiner, Steven Allan
Weissman, Eugene Y(ehuda)
Wessling, Ritchie A
Westrum, Edgar Francis, Jr
Whitfield, Richard George
Wicke, Brian Garfield
Willermet, Pierre Andre
Williams, Ronald Lloyde
Wims, Andrew Montgomery
Woehler, Scott Edwin
Wong, Peter Alexander
Yee, Albert Fan
Zanini-Fisher, Margherita

MINNESOTA
Adams, Roger James
Agarwal, Som Prakash
Allen, Martin
Anderson, George Robert
Anderson, Ronald Keith
Barany, George
Barden, Roland Eugene
Beetch, Ellsworth Benjamin
Bohon, Robert Lynn
Bolles, Theodore Frederick
Bonne, Ulrich
Britton, William Giering
Brom, Joseph March, Jr
Bryan, Thomas T
Carr, Robert Wilson, Jr
Carter, Orwin Lee
Childs, William Ves
Crawford, Bryce (Low), Jr
Dahler, John S
Davis, Howard Ted
Dempsey, John Nicholas
Dendinger, Richard Donald
Dickson, Arthur Donald
Dierenfeldt, Karl Emil
Drew, Bruce Arthur
Dreyer, Kirt A
Duerst, Richard William
Durnick, Thomas Jackson
Erickson, John M
Erickson, Randall L
Errede, Louis A
Evans, Douglas Fennell
Evans, John Fenton
Farnum, Sylvia A
Fester, Keith Edward
Finholt, James E
Gabor, Thomas
Gavin, Robert M, Jr
Gentry, William Ronald
Giese, Clayton
Graham, Robert Leslie
Grant, David James William
Grundmeier, Ernest Winston
Gyberg, Arlin Enoch
Haas, Larry Alfred
Hanson, Allen Louis
Hardgrove, George Lind, Jr
Harriss, Donald K
Hexter, Robert Maurice
Hogle, Donald Hugh
Howell, Peter Adam
Jones, Lester Tyler
Khalafalla, Sanaa E
King, Reatha Clark
Kolthoff, Izaak Maurits
Kreevoy, Maurice M
Labuza, Theodore Peter
Lai, Juey Hong
Leopold, Doreen Geller
Lesikar, Arnold Vincent
Lumry, Rufus Worth, II
McBrady, John J
McKenna, Jack F(ontaine)
McKeown, James John
Marsh, Frederick Leon
Mentone, Pat Francis
Miessler, Gary Lee
Montgomery, Peter Williams
Moore, Francis Bertram

Moscowitz, Albert
Nichol, James Charles
Olsen, Douglas Alfred
Oriani, Richard Anthony
Owens, Kenneth Eugene
Peterson, Lowell E
Pocius, Alphonsus Vytautas
Polnaszek, Carl Francis
Prager, Stephen
Ramette, Richard Wales
Robins, Janis
Robinson, Glen Moore, III
Roquitte, Bimal C
Rose, Wayne Burl
Roska, Fred James
Rutledge, Robert L
Ryan, John Peter
Salmon, Oliver Norton
Schmidt, Lanny D
Schwartz, A(lbert) Truman
Slowinski, Emil J, Jr
Smyrl, William Hiram
Sobieski, James Fulton
Stankovich, Marian Theresa
Stoesz, James Darrel
Strong, Judith Ann
Suryanarayanan, Raj Gopalan
Tamsky, Morgan Jerome
Thompson, Herbert Bradford
Thompson, Mary E
Tiers, George Van Dyke
Truhlar, Donald Gene
Tsai, Bilin Paula
Ugurbil, Kamil
White, Joe Wade
Yapel, Anthony Francis, Jr

MISSISSIPPI
Baker, John Keith
Bass, Henry Ellis
Bishop, Allen David, Jr
Boyle, John Aloysius, Jr
Brady, Ruth Mary
Brent, Charles Ray
Calloway, E Dean
Cooksy, Andrew Lloyd
Crawford, Crayton McCants
Deshpande, Krishnanath Bhaskar
Duffey, Donald Creagh
Fisher, Thomas Henry
Klingen, Theodore James
Legg, John Wallis
Little, Brenda Joyce
McCain, Douglas C
McCormick, Charles Lewis, III
Miller, Donald Piguet
Monts, David Lee
Myers, Richard Showse
Pinson, James Wesley
Pojman, John Anthony
Posey, Franz Adrian
Stefani, Andrew Peter
Trehan, Rajender
Weissman, William
Wertz, David Lee
West, Rose Gayle
Wilson, Wilbur William
Yun, Kwang-Sik

MISSOURI
Ackerman, Joseph John Henry
Ackers, Gary K
Ames, Donald Paul
Bahn, Emil Lawrence, Jr
Baiamonte, Vernon D
Bard, James Richard
Baughman, Russell George
Beaver, Earl Richard
Beistel, Donald W
Benjamin, Philip Palamoottil
Bertrand, Gary Lane
Blum, Frank D
Bornmann, John Arthur
Bradburn, Gregory Russell
Bradley, Guy Martin
Brown, Robert Eugene
Burke, James Joseph
Carpenter, James Franklin
Chen, Yih-Wen
Conradi, Mark Stephen
Crutchfield, Marvin Mack
Dias, Jerry Ray
Dixon, Marvin Porter
Drew, Henry D
Frederick, Raymond H
Freeman, John Jerome
Gard, Janice Koles
Gaspar, Peter Paul
Gerfen, Charles Otto
Gibbons, James Joseph
Glaser, Rainer
Glaspie, Peyton Scott
Gleaves, John Thompson
Greenlief, Charles Michael
Griffith, Edward Jackson
Hansen, Richard Lee
Harris, Harold H
Henis, Jay Myls Stuart
Hinkebein, John Arnold
Holloway, Thomas Thornton
Holsen, James N(oble)
Holtzer, Marilyn Emerson
James, William Joseph
Jason, Mark Edward

Kelley, John Daniel
Killingbeck, Stanley
King, Thomas Morgan
Kitchin, Robert Walter
Koerner, William Elmer
Kowert, Bruce Arthur
Kuntz, Robert Roy
Kurz, James Eckhardt
Kurz, Joseph Louis
Larsen, David W
Lee, Emerson Howard
Lembke, Roger Roy
Leventis, Nicholas
Lilenfeld, Harvey Victor
Lin, Tien-Sung Tom
Lipkin, David
Lissant, Kenneth Jordan
Loeppky, Richard N
Lohman, Timothy Michael
Long, Gary John
Lott, Peter F
Lund, Louis Harold
McDonald, Hector O
McGraw, Robert Leonard
Macias, Edward S
Mosher, Melvyn Wayne
Munch, Ralph Howard
O'Brien, James Francis
Palamand, Suryanarayana Rao
Piper, Roger D
Raw, Cecil John Gough
Readnour, Jerry Michael
Rice, Bernard
Richards, Charles Norman
Roach, Donald Vincent
Robaugh, David Allan
Robertson, Bobby Ken
Roy, Rabindra (Nath)
Sarantites, Demetrios George
Schaefer, Jacob Franklin
Schaeffer, Harold F(ranklin)
Schmitt, John Leigh
Seeley, Robert D
Seymour, Keith Goldin
Sheets, Ralph Waldo
Stary, Frank Edward
Streng, William Harold
Takano, Masaharu
Thomas, George Joseph, Jr
Thomas, Timothy Farragut
Thompson, Clifton C
Tria, John Joseph, Jr
Troutner, David Elliott
Unland, Mark Leroy
Wagenknecht, John Henry
Weissman, Samuel Isaac
Welsh, William James
Whitefield, Philip Douglas
Wilkinson, Ralph Russell
Winicov, Murray William
Wolf, Clarence J
Wong, Tuck Chuen
Woodbrey, James C
Yaris, Robert
Yasuda, Hirotsugu
Zetlmeisl, Michael Joseph

MONTANA
Barnhart, David M
Bradley, Daniel Joseph
Callis, Patrik Robert
Caughlan, Charles Norris
Conn, Paul Joseph
Craig, Arnold Charles
Emerson, Kenneth
Graham, Raymond
Howald, Reed Anderson
Layman, Wilbur A
Osterheld, Robert Keith
Rice, Richard Eugene
Valencich, Trina J
Woodbury, George Wallis, Jr
Yates, Leland Marshall
Zelezny, William Francis

NEBRASKA
Coleman, George Hunt
Diestler, Dennis Jon
Eckhardt, Craig Jon
Garey, Carroll Laverne
Howell, Daniel Bunce
King, Delbert Leo
Kuecker, John Frank
Rack, Edward Paul
Scholz, John Joseph, Jr
Shearer, Greg Otis
Skopp, Joseph Michael
Snipp, Robert Leo
Sonderegger, Theo Brown
Sterner, Carl D
Swanson, Jack Lee
Swanson, James A
Vanderzee, Cecil Edward
Zebolsky, Donald Michael

NEVADA
Blincoe, Clifton (Robert)
Burkhart, Richard Delmar
Earl, Boyd L
Eastwood, DeLyle
Farley, John William
Gerard, Jesse Thomas
Harrington, Rodney E
Jones, Denny Alan

Kemp, Kenneth Courtney
Klainer, Stanley M
McCarroll, Bruce
MacDonald, David J
Pierson, William R
Shin, Hyung Kyu
Smith, Ross W
Sovocool, G Wayne

NEW HAMPSHIRE
Amell, Alexander Renton
Bel Bruno, Joseph James
Bertrand, Rene Robert
Braun, Charles Louis
Chasteen, Norman Dennis
Cleland, Robert Lindbergh
Creagh-Dexter, Linda T
Damour, Paul Lawrence
Dancy, Terence E(rnest)
Ehrlich, Paul
Ellis, Samuel Benjamin
Greenwald, Harold Leopold
Hornig, James Frederick
Hubbard, Colin D
Kegeles, Gerson
Kelemen, Denis George
Martin, William Butler, Jr
Mayne, Howard R
Mooney, Richard Warren
Nadeau, Herbert Gerard
Pilar, Frank Louis
Rice, Dale Wilson
Stepenuck, Stephen Joseph, Jr
Stockmayer, Walter H(ugo)
Strazdins, Edward
Swift, Robinson Marden
Walling, Cheves (Thomson)
Weisman, Gary Raymond
Winn, John Sterling

NEW JERSEY
Aldrich, Haven Scott
Alekman, Stanley L
Alvarez, Vernon Leon
Alyea, Hubert Newcombe
Amron, Irving
Ander, Paul
Andrews, Rodney Denlinger, Jr
Appleby, Alan
Armstrong, William David
Auborn, James John
Axtmann, Robert Clark
Bacher, Frederick Addison
Bachman, Kenneth Charles
Badin, Elmer John
Bahr, Charles Chester
Bair, Harvey Edward
Baker, William Oliver
Ban, Vladimir Sinisa
Barnard, William Sprague
Barnett, Ronald E
Bartok, William
Behl, Wishvender K
Bender, Max
Bendure, Raymond Lee
Bergh, Arpad A
Bernasek, Steven Lynn
Bernholz, William Francis
Bhattacharjee, Himanshu Ranjan
Bhattacharyya, Pranab K
Biagetti, Richard Victor
Binder, Michael
Bird, George Richmond
Blanc, Joseph
Bloom, Allen
Bocarsly, Andrew B
Bonacci, John C
Booman, Keith Albert
Boskey, Adele Ludin
Bozzelli, Joseph William
Breslauer, Kenneth John
Brous, Jack
Brown, Joe Ned, Jr
Brown, Stanley Monty
Brugger, John Edward
Brus, Louis Eugene
Buck, Thomas M
Buckley, Denis Noel
Buckley, R Russ
Buhks, Ephraim
Bulas, Romuald
Burley, David Richard
Burwasser, Herman
Calcote, Hartwell Forrest
Cardillo, Mark J
Carlucci, Frank Vito
Carrock, Frederick E
Castor, William Stuart, Jr
Chandross, Edwin A
Chatterjee, Pronoy Kumar
Chaudhri, Safee U
Chiao, Yu-Chih
Chiu, Tin-Ho
Cho, Kon Ho
Church, John Armistead
Ciaccio, Leonard Louis
Coakley, Mary Peter
Cohen, Abraham Bernard
Cohen, George Lester
Cotter, Martha Ann
Couchman, Peter Robert
Cruice, William James
Cummin, Alfred Samuel
Dahlstrom, Bertil Philip, Jr

Davidson, Theodore
Davis, Stanley Gannaway
Dean, Anthony Marion
Debenedetti, Pablo Gaston
De Korte, Aart
DiMasi, Gabriel Joseph
Dismukes, Gerard Charles
Donovan, Sandra Steranka
Douglass, D C
Downing, George V, Jr
Dreby, Edwin Christian, III
Dreeben, Arthur B
Dunbar, Phyllis Marguerite
Dunn, William Howard
Dzierzanowski, Frank John
Ebert, Lawrence Burton
Ekstrom, Lincoln
Elder, John Philip
Erenrich, Eric Howard
Ermler, Walter Carl
Evans, Warren William
Falconer, Warren Edgar
Falk, Charles David
Farrington, Thomas Allan
Farrow, Leonilda Altman
Felder, William
Feldstein, Nathan
Ferrigno, Thomas Howard
Fetters, Lewis
Filas, Robert William
Fisanick, Georgia Jeanne
Fitzgerald, Patrick Henry
Fix, Kathleen A
Flato, Jud B
Forbes, Charles Edward
Forster, Eric Otto
Foy, Walter Lawrence
Frankenthal, Robert Peter
Fraser, Donald Boyd
Freund, Howard John
Fuller, Everett J
Fung, Shun Chong
Gagliardi, L John
Gale, Paula Jane
Gans, Manfred
Garfinkel, Harmon Mark
Garik, Vladimir L
Gelfand, Jack Jacob
Gethner, Jon Steven
Glarum, Sivert Herth
Glassman, Irvin
Gold, Daniel Howard
Goldblatt, Irwin Leonard
Goldstein, Martin
Goldstein, Robert Lawrence
Goodkin, Jerome
Goodman, Lionel
Gore, Ernest Stanley
Gottscho, Richard Alan
Grasselli, Robert Karl
Grelecki, Chester
Griffith, Martin G
Haag, Werner O
Haden, Walter Linwood, Jr
Hager, Douglas Francis
Halaby, Sami Assad
Halgren, Thomas Arthur
Harpell, Gary Allan
Harris, Alexander L
Harris, Leonce Everett
Hatzenbuhler, Douglas Albert
Hayles, William Joseph
Healey, Frank Henry
Heilweil, Israel Joel
Hen, John
Herber, Rolfe H
Herdklotz, John Key
Herzog, Gregory F
Hills, Stanley
Hobson, Melvin Clay, Jr
Hoffman, Henry Tice, Jr
Horowitz, Hugh H(arris)
House, Edward Holcombe
Hsu, Edward Ching-Sheng
Huang, John S
Hung, John Hui-Hsiung
Hunger, Herbert Ferdinand
Ikeda, Dennis Noel
Ilardi, Joseph Michael
Inniss, Daryl
Isaac, Peter Ashley Hammond
Isbrandt, Lester Reinhardt
Izod, Thomas Paul John
Jacobson, Harold
Jain, Nemichand B
Joffe, Joseph
Jones, Francis Thomas
Jordan, Andrew Stephen
Kagann, Robert Howard
Kamath, Yashavanth Katapady
Kaplan, Martin L
Kaplan, Michael
Karg, Gerhart
Kasper, Horst Manfred
Kaufmann, Kenneth James
Kauzmann, Walter (Joseph)
Kenkare, Divaker B
Kennedy, Anthony John
Kerr, George Thomson
Kimmel, Elias
Kimmel, Howard S
Kleiner, Walter Bernhard
Kohout, Frederick Charles, III
Konde, Anthony Joseph

Kornblum, Saul S
Kosel, George Eugene
Kowalski, Ludwik
Kramer, George Mortimer
Krenos, John Robert
Krueger, Paul Carlton
Kunin, Robert
Kunzler, John Eugene
Kydd, Paul Harriman
Lai, Kuo-Yann
Laity, Richard Warren
Lang, Frank Theodore
Lanzerotti, Mary Yvonne DeWolf
LaPalme, Donald William
Larsen, Marilyn Ankeney
Laurie, Victor William
Le Duc, J-Adrien Maher
Lee, Do-Jae
Levi, Elliott J
Levine, Harry
Levine, Richard S
Li, Hong
Liebe, Donald Charles
Link, Gordon Littlepage
Loh, Roland Ru-loong
Lorenz, Patricia Ann
LoSurdo, Antonio
Lowell, A(rthur) I(rwin)
Lucchesi, Peter J
Lui, Yiu-Kwan
Lyon, Richard Kenneth
McAfee, Kenneth Bailey, Jr
McCall, David Warren
McCauley, James A
McCleary, Harold Russell
McClure, Donald Stuart
McCullough, John Price
MacFarlane, Robert, Jr
McKnight, Lee Graves
McMahon, Paul E
McNevin, Susan Clardy
Madey, Theodore Eugene
Magder, Jules
Marra, Dorothea Catherine
Maulding, Hawkins Valliant, Jr
Meiboom, Saul
Mendelsohn, Richard
Meyer, Albert William
Miller, Arthur
Mohacsi, Erno
Monse, Ernst Ulrich
Montana, Anthony J
Mraw, Stephen Charles
Mucenieks, Paul Raimond
Mukai, Cromwell Daisaku
Mullen, Robert Terrence
Murarka, Shyam Prasad
Nace, Donald Miller
Naro, Paul Anthony
Nelson, Roger Edwin
Newman, Stephen Alexander
Nilsen, Walter Grahn
O'Connor, Matthew James
Okinaka, Yutaka
Olechowski, Jerome Robert
Olmstead, William N(eergaard)
Olson, David Harold
Opila, Robert L, Jr
Oppenheimer, Larry Eric
Osuch, Christopher Erion
Owens, Clifford
Owens, Frank James
Panish, Morton B
Panson, Gilbert Stephen
Parisi, George I
Parker, Richard C
Patel, Gordhanbhai Nathalal
Paul, Edward W
Perkins, Harolyn King
Pethica, Brian Anthony
Pierce, Robert Charles
Pierson, William Grant
Pilla, Arthur Anthony
Pinch, Harry Louis
Poe, Martin
Poindexter, Edward Haviland
Polestak, Walter John S
Potenza, Joseph Anthony
Princen, Henricus Mattheus
Puar, Mohindar S
Pugliese, Michael
Rabitz, Herschel Albert
Raghavachari, Krishnan
Raich, Henry
Raines, Thaddeus Joseph
Ramachandran, Pallassana N
Ramaprasad, K R (Ram)
Randall, James Carlton, Jr
Raveche, Harold Joseph
Raynor, Susanne
Readdy, Arthur F, Jr
Rebick, Charles
Redden, Patricia Ann
Reents, William David, Jr
Regna, Peter P
Reilly, Eugene Patrick
Rein, Alan James
Rennert, Joseph
Richlin, Jack
Rieger, Martin Max
Roberts, Ronald Frederick
Robertson, Nat Clifton
Robins, Jack
Rollino, John

Physical Chemistry (cont)

Rome, Martin
Rosenberg, Allan (Herbert)
Rosenberg, Howard Alan
Ross, Lawrence James
Roth, Heinz Dieter
Rousseau, Denis Lawrence
Rowe, Carleton Norwood
Rubin, Herbert
Rubino, Andrew M
Sagal, Matthew Warren
Saldick, Jerome
Salomon, Mark
Sanderson, Benjamin S
Sandus, Oscar
Sayres, Alden R
Schlegel, James M
Schueler, Paul Edgar
Schwartz, Bertram
Schwartzkopf, George, Jr
Scott, Eric James Young
Shah, Atul A
Shallcross, Frank V(an Loon)
Shanefield, Daniel J
Shaw, Henry
Sherry, Howard S
Shewmaker, James Edward
Shombert, Donald James
Siano, Donald Bruce
Sibilia, John Philip
Sifniades, Stylianos
Sinclair, William Robert
Sinfelt, John Henry
Slagg, Norman
Smith, Donald Eugene
Smith, Eileen Patricia
Smith, George Byron
Smith, Harlan Millard
Smith, Richard Pearson
Soos, Zoltan Geza
Speert, Arnold
Spiro, Thomas George
Stillinger, Frank Henry
Stivala, Salvatore Silvio
Strauss, George
Strauss, Ulrich Paul
Suchow, Lawrence
Szyper, Mira
Tanenbaum, Morris
Thornton, C G
Tiethof, Jack Alan
Toby, Sidney
Toome, Voldemar
Townsend, Palmer W
Trambarulo, Ralph
Trewella, Jeffrey Charles
Trumbore, Forrest Allen
Tsonopoulos, Constantine
Tucci, Edmond Raymond
Tully, John Charles
Tunc, Deger Cetin
Turkevich, John
Tway, Patricia C
Valentini, James Joseph
Vanderspurt, Thomas Henry
Van Hook, James Paul
Vasile, Michael Joseph
Virgili, Luciano
Vogel, Veronica Lee
Vyas, Brijesh
Wagner, Sigurd
Waltking, Arthur Ernest
Wang, Chih Chun
Warren, Craig Bishop
Weakliem, Herbert Alfred, Jr
Wei, Chung-Chen
Wenzel, John Thompson
Weschler, Charles John
Westerdahl, Carolyn Ann Lovejoy
Westerdahl, Raymond P
Wheaton, Gregory Alan
White, Lawrence Keith
Wijnen, Joseph M H
Williams, Bernard Leo
Williams, Dale Gordon
Williams, Richard
Wise, John James
Wolf, Philip Frank
Woodbridge, Joseph Eliot
Wu, Ellen Lem
Yablonsky, Harvey Allen
Yardley, James Thomas, III

NEW MEXICO

Alam, Mansoor
Ames, Lynford Lenhart
Asprey, Larned Brown
Auerbach, Irving
Baca, Glenn
Bard, Richard James
Bayhurst, Barbara P
Bechtold, William Eric
Behrens, Robert George
Blais, Normand C
Brabson, George Dana, Jr
Brandvold, Donald Keith
Breshears, Wilbert Dale
Butler, Michael Alfred
Cady, Howard Hamilton
Campbell, George Melvin
Carlson, Gary Alden
Catlett, Duane Stewart
Clark, Robert Paul

Clifton, David Geyer
Coppa, Nicholas V
Cromer, Don Tiffany
Cross, Jon Byron
Cunningham, Paul Thomas
Curtice, Jay Stephen
Cuthrell, Robert Eugene
Danen, Wayne C
DeArmond, M Keith
Dinegar, Robert Hudson
Doddapaneni, Narayan
Dorko, Ernest A
Dropesky, Bruce Joseph
Duffy, Clarence John
Eckert, Juergen
Ellis, Walton P
Engelke, Raymond Pierce
Enke, Christie George
Erickson, Kenneth Lynn
Erikson, Jay Arthur
Ewing, Galen Wood
Ewing, Gordon J
Eyster, Eugene Henderson
Fowler, Malcolm McFarland
Fries, James Andrew
Fries, Ralph Jay
Grisham, Genevieve Dwyer
Gutschick, Vincent Peter
Hammel, Edward Frederic
Hartley, Danny L
Heaton, Maria Malachowski
Hoehn, Martha Vaughan
Hoffbauer, Mark Arles
Hughes, Robert Clark
Jennings, Charles Warren
Jensen, Reed Jerry
Jones, Peter Frank
Jones, Wesley Morris
Keizer, Clifford Richard
Keller, Richard Alan
Kenna, Bernard Thomas
Kerley, Gerald Irwin
Krupka, Milton Clifford
Laquer, Henry L
Larson, Thomas E
Levine, Herman Saul
Levy, Samuel C
Lieberman, Morton Leonard
Lyman, John L
Lynch, Richard Wallace
McGuire, Joseph Clive
McLaughlin, Donald Reed
Mann, Joseph Bird, (Jr)
Martin, James Ellis
Mason, Caroline Faith Vibert
Mills, Robert Leroy
Mitchell, David Wesley
Morosin, Bruno
Mulford, Robert Neal Ramsay
Nathan, Charles C(arb)
Newton, Thomas William
Niemczyk, Thomas M
Nogar, Nicholas Stephen
Northrop, David A(mos)
O'Brien, Harold Aloysious, Jr
Ogard, Allen E
Ogilby, Peter Remsen
Olson, Douglas Bernard
Olson, William Marvin
Onstott, Edward Irvin
Orth, Charles Joseph
Ortiz, Joseph Vincent
Park, Su-Moon
Peden, Charles H F
Peek, H(arry) Milton
Peterson, Charles Leslie
Peterson, Dean Everett
Purser, Fred O
Quinn, Rod King
Rabideau, Sherman Webber
Ramsay, John Barada
Renschler, Clifford Lyle
Rice, James Kinsey
Robinson, C Paul
Romig, Alton Dale, Jr
Sattizahn, James Edward, Jr
Schaefer, Dale Wesley
Schott, Garry Lee
Schufle, Joseph Albert
Sedlacek, William Adam
Seegmiller, David W
Sheehan, William Francis
Shepard, Joseph William
Sherman, Robert Howard
Smith, Gordon Meade
Smith, James Lewis
Sorem, Michael Scott
Stanbro, William David
Stark, Walter Alfred, Jr
Stinecipher, Mary Margaret
Streit, Gerald Edward
Sullivan, John Henry
Taber, Joseph John
Tapscott, Robert Edwin
Taylor, Gene Warren
Vier, Dwayne Trowbridge
Walker, Robert Bridges
Wallace, Terry Charles, Sr
Walters, Edward Albert
Walton, George
Walton, Roddy Burke
Wampler, Fred Benny
Ward, John William
Werbelow, Lawrence Glen

Werkema, George Jan
Wewerka, Eugene Michael
Williams, David Cary
Wolfsberg, Kurt
Wood, Gerry Odell
Wu, Adam Yu
Young, Ainslie Thomas, Jr
Zeltmann, Alfred Howard

NEW YORK

Abramowicz, Daniel Albert
Abruna, Hector D
Addy, John Keith
Adelstein, Peter Z
Adin, Anthony
Adler, George
Alben, Katherine Turner
Albrecht, Andreas Christopher
Albrecht, Frederick Xavier
Albrecht, William Melvin
Alexander, John Macmillan, Jr
Alfieri, Gaetano T
Allen, Augustine Oliver
Allenbach, Charles Robert
Altschul, Rolf
Amborski, Leonard Edward
Amero, Bernard Alan
Anbar, Michael
Anderson, Frank Wallace
Anderson, Herbert Rudolph, Jr
Apai, Gustav Richard, II
Araujo, Roger Jerome
Archie, William C, Jr
Arendt, Ronald H
Arents, John (Stephen)
Armanini, Louis Anthony
Arnowich, Beatrice
Aronson, Seymour
Arquette, Gordon James
Aten, Carl Faust, Jr
Atwood, Gilbert Richard
Aven, Manuel
Aviram, Ari
Avouris, Phaedon
Axe, John Donald
Baer, Norbert Sebastian
Baetzold, Roger C
Bagchi, Pranab
Bahary, William S
Baier, Robert Edward
Baird, Barbara A
Bak, David Arthur
Bakhru, Hassaram
Baldwin, John E
Bank, Shelton
Barile, Raymond Conrad
Bartholomew, Roger Frank
Bauer, Simon Harvey
Baum, Parker Bryant
Baumgartner, Reuben Lawrence
Beer, Sylvan Zavi
Belfort, Georges
Bent, Brian E
Benzinger, William Donald
Bettelheim, Frederick A
Beuhler, Robert James, Jr
Bevak, Joseph Perry
Bigeleisen, Jacob
Birge, Robert Richards
Birke, Ronald Lewis
Bishop, Marvin
Blackwell, Crist Scott
Bloch, Aaron Nixon
Bluhm, Terry Lee
Blum, Samuel Emil
Bobka, Rudolph J
Bodi, Lewis Joseph
Bommaraju, Tilak V
Bonner, Francis Truesdale
Bottger, Gary Lee
Brady, George W
Bragg, John Kendal
Bramwell, Fitzgerald Burton
Bray, Norman Francis
Brenner, Henry Clifton
Briehl, Robin Walt
Browall, Kenneth Walter
Brown, Eric Richard
Bruckenstein, Stanley
Brumbaugh, Donald Verwey
Brumberger, Harry
Brunschwig, Bruce Samuel
Buff, Frank Paul
Buhsmer, Charles P
Bulloff, Jack John
Burtt, Benjamin Pickering
Cadenhead, David Allan
Campion, James J
Cante, Charles John
Capwell, Robert J
Carle, Kenneth Roberts
Carroll, Harvey Franklin
Carroll, Robert William
Chaffee, Eleanor
Chandler, Horace W
Chang, Chin-An
Chang, Joseph Yung
Chang, Shih-Yung
Chapman, Sally
Charola, Asuncion Elena
Charton, Marvin
Chen, Chang-Hwei
Chen, Cindy Chei-Jen
Cherin, Paul

Chiang, Joseph Fei
Chiang, Yuen-Sheng
Chin, Der-Tau
Chou, Chung-Chi
Chou, Tzi Shan
Chu, Benjamin Peng-Nien
Chu, Joseph Yung-Chang
Chu, Liang Tung
Chu, Yung Yee
Clark, Alfred
Cloney, Robert Dennis
Cohen, Stephen Robert
Cole, David Le Roy
Collier, Susan S
Conan, Robert James, Jr
Cook, Edward Hoopes, Jr
Cooke, Derry Douglas
Cool, Terrill A
Cooper, Walter
Coppola, Patrick Paul
Corth, Richard
Cratty, Leland Earl, Jr
Creasy, William Russel
Cunningham, Michael Paul
Curme, Henry Garrett
Dailey, Benjamin Peter
Daniel, Daniel S
Dannenberg, Joseph
Dannhauser, Walter
Darrow, Frank William
Davenport, Lesley
Davis, William Donald
DeCrosta, Edward Francis, Jr
Dehn, Joseph William, Jr
Demerjian, Kenneth Leo
DeStefano, Anthony Joseph
Deutsch, John Ludwig
Diem, Max
Dieterich, David Allan
Dill, Aloys John
Dingledy, David Peter
Disch, Raymond L
Dixon, William Brightman
Doetschman, David Charles
Doigan, Paul
Dondes, Seymour
Donoian, Haig Cadmus
Doremus, Robert Heward
Doubleday, Charles E, Jr
Dumbaugh, William Henry, Jr
Duncan, Thomas Michael
Dwyer, Robert Francis
Eachus, Raymond Stanley
Eberts, Robert Eugene
Egan, James John
Ehrenson, Stanton Jay
Ehrlich, Sanford Howard
Eidinoff, Maxwell Leigh
Eikenberry, Jon Nathan
Eisenberg, Max
Elder, Fred A
Ells, Victor Raymond
Elmore, Glenn Van Ness
Everhard, Martin Edward
Factor, Arnold
Fajer, Jack
Farb, Edith H
Farrar, James Martin
Fay, Homer
Featherstone, John Douglas Bernard
Fehlner, Francis Paul
Feigenson, Gerald William
Feldberg, Stephen William
Feldman, Isaac
Feldman, Larry Howard
Felty, Evan J
Filbert, Augustus Myers
Finkbeiner, Herman Lawrence
Finley, Kay Thomas
Finzel, Rodney Brian
Fishman, Jerry Haskel
Flank, William H
Flannery, John B, Jr
Flom, Donald Gordon
Flynn, George William
Folan, Lorcan Michael
Fontijn, Arthur
Fox, William
Fraenkel, George Kessler
Freed, Jack H
Freedman, Jeffrey Carl
Friedlander, Gerhart
Friedman, Harold Leo
Friedman, Joel Mitchel
Friedman, Lewis
Friedman, Seymour K
Frisch, Harry Lloyd
Fuerst, Adolph
Gaines, George Loweree, Jr
Gancy, Alan Brian
Gans, Paul Jonathan
Gardner, William Lee
Garetz, Bruce Allen
Garvey, James F
Gaudioso, Stephen Lawrence
Gawer, Albert Henry
Geacintov, Nicholas
Gentner, Robert F
Gershinowitz, Harold
Gettler, Joseph Daniel
Ghezzo, Mario
Gill, Robert Anthony
Gill, Ronald Lee
Gillies, Charles Wesley

CHEMISTRY / 241

Gilman, Paul Brewster, Jr
Gilmour, Hugh Stewart Allen
Ginell, Robert
Goldfarb, Theodore D
Gollob, Fred
Good, Robert James
Goodisman, Jerry
Goodman, Jerome
Goodman, Seymour
Gordon, Barry Maxwell
Gould, Robert Kinkade
Green, Michael Enoch
Greenbaum, Steven Garry
Greenspan, Joseph
Griffel, Maurice
Griffin, Jane Flanigen
Grubb, Willard Thomas
Grushkin, Bernard
Haas, David Jean
Haas, Werner E L
Hach, Edwin E, Jr
Hall, Richard Travis
Haller, Ivan
Halpern, Mordecai Joseph
Hanson, David M
Hanson, Jonathan C
Hanson, Louise I Karle
Harbottle, Garman
Hargitay, Bartholomew
Harrison, Shirley Wanda
Harwood, Colin Frederick
Hecht, Charles Edward
Heininger, Clarence George, Jr
Helfgott, Cecil
Hendricks, Robert William
Herkstroeter, William G
Herley, Patrick James
Herman, Irving Philip
Herskovits, Theodore Tibor
Hertl, William
Herz, Arthur H
Heston, William May, Jr
Hickmott, Thomas Ward
Hill, Cliff Otis
Hill, Derek Leonard
Hillig, William Bruno
Hillman, Manny
Hodge, Ian Moir
Holden, Norman Edward
Holleran, Eugene Martin
Hollinger, Henry Boughton
Holmes, Curtis Frank
Holroyd, Richard Allan
Hopke, Philip Karl
Houston, Paul Lyon
Howell, James MacGregor
Howery, Darryl Gilmer
Hsu, Ming-Ta
Htoo, Maung Shwe
Huang, Suei-Rong
Hubbard, Richard Alexander, II
Huizenga, John Robert
Hull, Michael Neill
Iden, Charles R
Ilmet, Ivor
Imperial, George Romero
Innes, Kenneth Keith
Irani, N F
Irsa, Adolph Peter
Isaacs, Hugh Solomon
Ishida, Takanobu
Jach, Joseph
Janauer, Gilbert E
Janz, George John
Jelinski, Lynn W
Jellinek, Hans Helmut Gunter
Jochsberger, Theodore
Joffee, Irving Brian
Johansson, Sune
Johnson, Myrle F
Johnson, Philip M
Johnston, Don Richard
Jones, Wayne E
Jorgensen, Helmuth Erik Milo
Jorne, Jacob
Joshi, Bhairav Datt
Kaarsberg, Ernest Andersen
Kaiserman, Howard Bruce
Kambour, Roger Peabody
Kapecki, Jon Alfred
Kasper, John Simon
Katcoff, Seymour
Kaufman, Frank B
Keeler, Robert Adolph
Keesee, Robert George
Keller, D Steven
Keller, Douglas Vern, Jr
Kent, Henry Johann
Khare, Bishun Narain
Kim, Sang Hyung
Kiphart, Kerry
Kiss, Klara
Kittelberger, John Stephen
Kivlighn, Herbert Daniel, Jr
Klerer, Julius
Klinck, Ross Edward
Koch, Heinz Frank
Koeppl, Gerald Walter
Koetzle, Thomas F
Kohrt, Carl Fredrick
Kokoszka, Gerald Francis
Korinek, George Jiri
Kratohvil, Josip
Krause, Sonja

Kreilick, Robert W
Kronman, Martin Jesse
Kubisen, Steven Joseph, Jr
Kuki, Atsuo
Kullnig, Rudolph K
Kumbar, Mahadevappa M
Kwei, Ti-Kang
Ladd, John Herbert
Lane, Richard L
Lange, Christopher Stephen
LaPietra, Joseph Richard
Larach, Simon
Lasker, Sigmund E
Laurenzi, Bernard John
Lazarus, Marc Samuel
LeBlanc, Oliver Harris, Jr
Lebowitz, Jacob Mordecai
Lee, Lieng-Huang
Lees, Alistair John
Lefkowitz, Stanley A
LeGrand, Donald George
Leone, James A
Leubner, Ingo Herwig
Leung, Pak Sang
Levine, Ira Noel
Levine, Samuel W
Levkov, Jerome Stephen
Levy, Arthur Louis
Lewis, David Kenneth
Liang, Charles C
Liang, Kai
Liao, Martha
Lichtman, Irwin A
Liebeskind, Herbert
Lindholm, Robert D
Lindsay, Derek Michael
Lipsig, Joseph
Loebl, Ernest Moshe
Lombardi, John Rocco
Long, Michael Edgar
Longhi, John
Lorenzen, Jerry Alan
Loring, Roger Frederic
Lounsbury, John Baldwin
Low, Manfred Josef Dominik
Luborsky, Fred Everett
Luckey, George William
Ludlum, Kenneth Hills
Luner, Philip
Macaluso, Pat
McBreen, James
McCarthy, Paul James
MacColl, Robert
McDonald, Robert Skillings
McElligott, Peter Edward
McGee, Thomas Howard
McGrath, John F
Mackay, Raymond Arthur
McKee, Douglas William
McKelvey, John Murray
MacKenzie, Donald Robertson
McLaren, Eugene Herbert
Mahony, John Daniel
Malerich, Charles
Mancuso, Richard Vincent
Marchetti, Alfred Paul
Marino, Robert Anthony
Marion, Alexander Peter
Mark, Herman Francis
Martin, Francis W
Martini, Catherine Marie
Massa, Dennis Jon
Massa, Louis
Matijevic, Egon
May, John Walter
Meloon, David Rand
Mendel, John Richard
Menger, Eva L
Mennitt, Philip Gary
Merkel, Paul Barrett
Metz, Donald J
Meyerson, Bernard Steele
Mezei, Mihaly
Michiels, Leo Paul
Mihajlov, Vsevolod S
Miller, Harold A
Miller, Richard J
Miller, Warren James
Milner, Clifford E
Mitacek, Paul, Jr
Mitchell, John Jacob
Mittal, Kashmiri Lal
Model, Frank Steven
Modic, Frank Joseph
Molina, John Francis
Monahan, Alan Richard
Montano, Pedro Antonio
Morawetz, Herbert
Morrison, Ian Douglas
Morrow, Jack I
Moskowitz, Jules Warren
Moynihan, Cornelius Timothy
Mucci, Joseph Francis
Muenter, Annabel Adams
Muenter, John Stuart
Mullen, Patricia Ann
Mullhaupt, Joseph Timothy
Murphy, Daniel Barker
Murphy, James A
Myers, Anne Boone
Myers, Clifford Earl
Nafie, Laurence Allen
Nancollas, George H
Nash, Robert Joseph

Needham, Charles D
Nemethy, George
Neubert, Theodore John
Neugebauer, Constantine Aloysius
Niedrach, Leonard William
Norcross, Bruce Edward
O'Donnell, Raymond Thomas
Orem, Michael William
Oreskes, Irwin
Orna, Mary Virginia
Osborg, Hans
Osgood, Richard Magee, Jr
Oster, Gerald
Osterhoudt, Hans Walter
Ott, Henry C(arl)
Owens, Patrick M
Padmanabhan, G R
Padnos, Norman
Pan, Kee-Chuan
Panagiotopoulos, Athanassios Z
Paraszczak, Jurij Rostyslan
Parravano, Carlo
Pasfield, William Horton
Patsis, Angelos Vlasios
Patton, Elizabeth VanDyke
Pavlisko, Joseph Anthony
Pearson, James Murray
Penn, Lynn Sharon
Perry, Edmond S
Petrakis, Leonidas
Petro, Anthony James
Petrucci, Sergio
Philips, Laura Alma
Piersma, Bernard J
Plummer, William Allan
Pochan, John Michael
Poncha, Rustom Pestonji
Porter, Richard Francis
Porter, Richard Needham
Poslusny, Jerrold Neal
Posner, Aaron Sidney
Powers, Dale Robert
Powers, Robert William
Praissman, Melvin
Prasad, Paras Nath
Preiss, Ivor Louis
Preses, Jack Michael
Price, Harry James
Proskauer, Eric S
Rand, Salvatore John
Ransom, Bruce Davis
Rashkin, Jay Arthur
Raymonda, John Warren
Reardon, Joseph Daniel
Reddy, Thomas Bradley
Reeves, Robert R
Rein, Robert
Remsberg, Louis Philip, Jr
Resing, Henry Anton
Rhodin, Thor Nathaniel, Jr
Ricci, John Ettore
Richtol, Herbert H
Ristow, Bruce W
Roepe, Paul David
Rogers, Donald Warren
Romankiw, Lubomyr Taras
Ronn, Avigdor Meir
Rose, Philip I
Rosen, Milton Jacques
Rosenkrantz, Marcy Ellen
Rosenthal, Donald
Rosman, Howard
Rosoff, Morton
Ross, John Brandon Alexander
Ross, Sydney
Rossington, David Ralph
Ruckman, Mark Warren
Ruoff, Arthur Louis
Rzad, Stefan Jacek
Sage, Gloria W
Sakano, Theodore K
Saleeb, Fouad Zaki
Salotto, Anthony W
Saltsburg, Howard Mortimer
Salzberg, Hugh William
Samworth, Eleanor A
Sandifer, James Roy
Satir, Birgit H
Saturno, Antony Fidelas
Saunders, William Hundley, Jr
Scaringe, Raymond Peter
Schick, Martin J
Schottmiller, John Charles
Schreurs, Jan W H
Schulman, Jerome M
Schwartz, Geraldine Cogin
Schwartz, Stephen Eugene
Schwarz, Harold A
Schwarz, William Merlin, Jr
Scott, Bruce Albert
Segatto, Peter Richard
Seltzer, Stanley
Shahin, Michael M
Sheats, George Frederic
Shelby, James Elbert
Shiao, Daniel Da-Fong
Shilman, Avner
Shoup, Robert D
Siew, Ernest L
Silver, Herbert Graham
Simonelli, Anthony Peter
Singh, Surjit
Skarulis, John Anthony
Skelly, David W

Smid, Johannes
Smith, Gale Eugene
Smith, George David
Smith, Kenneth Judson, Jr
Smith, Michael
Smith, Norman Obed
Snyder, Grace
Snyder, Lawrence Clement
Snyder, Robert Lyman
Sojka, Stanley Anthony
Solomon, Frank I
Solomon, Jack
Solomon, Jerome Jay
Somasundaran, P(onisseril)
Sorkin, Howard
Sparapany, John Joseph
Spink, Charles Harlan
Stanton, Richard Edmund
Stauffer, Robert Eliot
Sterman, Melvin David
Sterman, Samuel
Stern, Silviu Alexander
Stone, Joe Thomas
Stookey, Stanley Donald
Strekas, Thomas C
Sturmer, David Michael
Suggitt, Robert Murray
Sun, Siao Fang
Sundberg, Michael William
Sutin, Norman
Sweet, Richard Clark
Swingley, Charles Stephen
Szebenyi, Doletha M E
Szekely, Andrew Geza
Takacs, Gerald Alan
Tan, Julia S
Tan, Yen T
Tang, Ignatius Ning-Bang
Taylor, Dale Frederick
Testa, Anthony Carmine
Texter, John
Thomas, Harold Todd
Thomas, Robert
Tomkiewicz, Micha
Tong, Stephen S C
Transue, Laurence Frederick
Trevoy, Donald James
Tseng, Linda
Van Geet, Anthony Leendert
Vinciguerra, Michael Joseph
Vincow, Gershon
Violante, Michael Robert
Vitus, Carissima Marie
Vogel, Philip Christian
Vonnegut, Bernard
Von Winbush, Samuel
Wait, Samuel Charles, Jr
Waitkins, George Raymond
Walker, Michael Sidney
Walnut, Thomas Henry, Jr
Wandass, Joseph Henry
Ward, Anthony Thomas
Warden, Joseph Tallman
Weiss, David Steven
Weiss, Jerome
Weller, S(ol) W(illiam)
Weltman, Clarence A
Wentorf, Robert H, Jr
Weston, Ralph E, Jr
White, Michael George
Whitten, David G
Widom, Benjamin
Wieder, Grace Marilyn
Wiegert, Philip E
Wiesenfeld, John Richard
Wilcox, Charles Frederick, Jr
Will, Frit Gustav
Williams, David James
Wilson, James M
Winstrom, Leon Oscar
Winter, William Thomas
Wiswall, Richard H, Jr
Wolfarth, Eugene F
Woodward, Arthur Eugene
Wotherspoon, Neil
Wright, John Fowler
Wrobel, Joseph Jude
Wu, Konrad T
Xue, Yongpeng
Yeagle, Philip L
Yencha, Andrew Joseph
Youker, John
Young, Ralph Howard
Yudelson, Joseph Samuel
Zavitsas, Andreas Athanasios
Zeltmann, Eugene W
Zevos, Nicholas
Ziegler, James Francis
Ziemniak, Stephen Eric
Zollweg, John Allman
Zuman, Petr

NORTH CAROLINA
Abbey, Kathleen Mary Kyburz
Ahrens, Rolland William
Alam, Mohammed Ashraful
Allen, Harry Clay, Jr
Allis, John W
Alston, Peter Van
Ambrosiani, Vincent F
Andrady, Anthony Lakshman
Arnett, Edward McCollin
Arnold, Luther Bishop, Jr
Ayers, Caroline LeRoy

Physical Chemistry (cont)

Ayers, Paul Wayne
Baer, Tomas
Bassett, David R
Blackman, Carl F
Bosley, David Emerson
Bowen, Lawrence Hoffman
Buchanan, James Wesley
Buck, Richard Pierson
Burrus, Robert Tilden
Bush, Stewart Fowler
Cahill, Charles L
Carroll, Felix Alvin, Jr
Cassen, Thomas Joseph
Chesnut, Donald Blair
Chou, David Yuan Pin
Coots, Alonzo Freeman
Daniels, William Ward
Davis, Henry Mauzee
Davis, Thomas Wilders
Dayton, Benjamin Bonney
Dearman, Henry Hursell
Echols, Joseph Todd, Jr
Elleman, Thomas Smith
Ellenson, James L
Ellison, Alfred Harris
Frieser, Rudolf Gruenspan
Gable, Ralph William
Getzen, Forrest William
Gillikin, Jesse Edward, Jr
Gillooly, George Rice
Goodwin, Frank Erik
Gross, Paul Magnus, Jr
Hall, Carol Klein
Hall, William Earl
Hamilton, Willard Charlson
Hartzog, James Victor
Haseley, Edward Albert
Hentz, Forrest Clyde, Jr
Hermans, Jan Joseph
Hickman, James Joseph
Hicks, Kenneth Ward
Hill, Eric Stanley
Hobbs, Marcus Edwin
Hornack, Frederick Mathew
Howell, James Milton
Hutchins, John R(ichard), III
Irvine, James Bosworth
Jackels, Charles Frederick
Jackson, Earl Graves
Jarnagin, Richard Calvin
Johnson, Charles Sidney, Jr
Johnson, James Edwin
Jones, Daniel Silas, Jr
Kirk, David Clark
Knecht, Laurance A
Kornfeil, Fred
Krigbaum, William Richard
Kuppers, James Richard
Leiserson, Lee
Lentz, Barry R
Lewin, Anita Hana
Lin, Stephen Fang-Maw
Loeppert, Richard Henry
McDonald, Daniel Patrick
MacPhail, Richard Allyn
Manning, Robert Joseph
Martin, G(uy) William, Jr
Martin, Richard Hugo
May, Walter Ruch
Meelheim, Richard Young
Miller, Robert L
Miller, Roger Ervin
Montet, George Louis
Moreland, Charles Glen
Neece, George A
Neumann, Calvin Lee
Owens, Boone Bailey
Park, Kisoon
Peck, David W
Pedersen, Lee G
Pobiner, Harvey
Poirier, Jacques Charles
Raleigh, James Arthur
Reisman, Arnold
Reynolds, John Hughes, IV
Rhyne, Thomas Crowell
Rochow, Theodore George
Samulski, Edward Thaddeus
Schuh, Merlyn Duane
Scott, Donald Ray
Simon, Dorothy Martin
Smith, Peter
Spears, Alexander White, III
Spicer, Leonard Dale
Stejskal, Edward Otto
Stellner, Kevin
Stevens, John Gehret
Strobel, Howard Austin
Sunda, William George
Swarbrick, James
Swofford, Robert Lewis
Tanford, Charles
Thompson, Nancy Lynn
Vail, Charles Brooks
Vanselow, Clarence Hugo
Wang, Victor Kai-Kuo
Wasson, John R
Watts, Plato Hilton, Jr
Wertz, Dennis William
Wilfong, Robert Edward
Wilhite, Douglas Lee
Wilson, Nancy Keeler
Yang, Weitao
Young, William Anthony
Zeldes, Henry
Zumwalt, Lloyd Robert

NORTH DAKOTA
Abrahamson, Harmon Bruce
Bierwagen, Gordon Paul
Gillispie, Gregory David
Hoggard, Patrick Earle
Kulevsky, Norman
Treumann, William Borgen
Vogel, Gerald Lee
Willson, Warrack Grant

OHIO
Abel, Alan Wilson
Abraham, Tonson
Adams, Harold Elwood
Allenson, Douglas Rogers
Amata, Charles David
Amjad, Zahid
Ampulski, Robert Stanley
Andrist, Anson Harry
Attalla, Albert
Aukrust, Egil
Bacon, Frank Rider
Baczek, Stanley Karl
Bailey, John Clark
Baker, Charles Edward
Baranwal, Krishna Chandra
Batt, Russell Howard
Battino, Rubin
Bauman, Richard Gilbert
Belles, Frank Edward
Bender, Daniel Frank
Beres, John Joseph
Bernays, Peter Michael
Biefeld, Paul Franklin
Bierlein, James A(llison)
Binkley, Roger Wendell
Binning, Robert Christie
Birkhahn, Ronald H
Bittker, David Arthur
Blocher, John Milton, Jr
Blomgren, George Earl
Bockhoff, Frank James
Bond, William Bradford
Briggs, Thomas N
Broge, Robert Walter
Bulgrin, Vernon Carl
Bundschuh, James Edward
Burow, Duane Frueh
Cameron, David George
Cantrell, Joseph Sires
Carfagno, Daniel Gaetano
Carlton, Terry Scott
Carman, Charles Jerry
Carter, R Owen, Jr
Cassim, Joseph Yusuf Khan
Ceasar, Gerald P
Chambers, Lee Mason
Chang, Jung-Ching
Chrysochoos, John
Coffin, Frances Dunkle
Collins, Edward A
Courchene, William Leon
Craig, Norman Castleman
Culbertson, Billy Muriel
Dahlgren, Richard Marc
Dalton, James Christopher
Darlington, William Bruce
Davies, Julian Anthony
Dawson, Steven Michael
Day, Jesse Harold
Den Besten, Ivan Eugene
De Vries, Adriaan
Dewhurst, Harold Ainslie
Dietrick, Harry Joseph
Dollimore, David
Drauglis, Edmund
Dunbar, Robert Copeland
Dymerski, Paul Peter
Eckert, George Frank
Eckstein, Bernard Hans
Edwards, Jimmie Garvin
Eisenhardt, William Anthony, Jr
Eley, Richard Robert
Endres, Paul Frank
Erickson, David Leland
Everson, Howard E
Fendler, Eleanor Johnson
Firestone, Richard Francis
Flautt, Thomas Joseph, Jr
Flechtner, Thomas Welch
Fleming, Paul Daniel, III
Foos, Raymond Anthony
Ford, William Ellsworth
Fordyce, James Stuart
Foreman, Dennis Walden, Jr
Foster, Alfred Field
Fox, Robert Kriegbaum
Fraenkel, Gideon
Frank, Sylvan Gerald
Frederick, John Edgar
Fryxell, Robert Edward
Fullmer, June Z(immerman)
Gage, Frederick Worthington
Gardner-Chavis, Ralph Alexander
Garner, Harry Richard
Ghose, Hirendra M
Goldman, Stephen Allen
Gordon, John Edward
Gray, John Augustus, III
Greenberg, Daniel
Greene, Hoke Smith
Grieshammer, Lawrence Louis
Griffin, John Leander
Griffith, Cecil Baker
Gupta, Vijay Kumar
Haas, Trice Walter
Hargis, I Glen
Harpst, Jerry Adams
Hassell, John Allen
Hatfield, Marcus Rankin
Hays, Byron G
Herbst, Eric
Herrmann, Kenneth Walter
Hickok, Robert Lee
Hill, Ann Gertrude
Hirst, Robert Charles
Hobrock, Don Leroy
Howell, Norman Gary
Hubbard, Arthur T
Huddleston, George Richmond, Jr
Hunter, James Charles
Hwang, Soon Muk
Hyndman, John Robert
Ishida, Hatsuo
Jackobs, John Joseph
Jacobs, Harvey
Jaffe, Hans
Jakobsen, Robert John
Jandacek, Ronald James
Jendrek, Eugene Francis, Jr
Jenkins, Robert George
Johnson, Gordon Lee
Johnson, Harlan Bruce
Jones, Edward Grant
Joyce, Blaine R
Karipides, Anastas
Katon, John Edward
Kaufman, Miron
Keil, Robert Gerald
Kershner, Carl John
Kircher, John Frederick
Koehler, Mark E
Kohl, Fred John
Kollen, Wendell James
Kornbrekke, Ralph Erik
Kozikowski, Barbara Ann
Krabacher, Bernard
Krieger, Irvin Mitchell
Kruse, Ferdinand Hobert
Kuska, Henry (Anton)
Lattimer, Robert Phillips
Laughlin, Robert Gene
Lauver, Richard William
Leckonby, Roy Alan
Lee, Min-Shiu
Lemmerman, Karl Edward
Lewis, Irwin C
Lewis, Richard Thomas
Liao, Shu-Chung
Lim, Edward C
Litt, Morton Herbert
Livigni, Russell Anthony
Livingston, Daniel Isadore
Loening, Kurt L
Lonadier, Frank Dalton
Love, Calvin Miles
Lucas, Kenneth Ross
Luebbe, Ray Henry, Jr
Luoma, John Robert Vincent
McFarland, Charles Warren
McMillan, Garnett Ramsay
McQuigg, Robert Duncan
Malkin, Irving
Mark, Harry Berst, Jr
Marker, Leon
Martinsons, Aleksandrs
Mast, Roy Clark
Mateescu, Gheorghe D
Matesich, Mary Andrew
Mathews, A L
Mathews, Collis Weldon
Mattice, Wayne Lee
Meal, Larie L
Meckstroth, Wilma Koenig
Meeks, Frank Robert
Metcalfe, Joseph Edward, III
Meyer, Norman James
Middaugh, Richard Lowe
Miller, Barry
Miller, James Roland
Miller, Terry Alan
Miller, William Reynolds, Jr
Mochel, Virgil Dale
Moon, Tag Young
Morin, Brian Gerald
Mortensen, Earl Miller
Mosser, John Snavely
Muchow, Gordon Mark
Mulligan, Bernard
Myers, R(alph) Thomas
Nardi, John Christopher
Neckers, Douglas
Neff, Vernon Duane
Nelson, Wayne Franklin
Newman, David S
Nikodem, Robert Bruce
Noda, Isao
Oertel, Richard Paul
Ogle, Pearl Rexford, Jr
Oliver, Joel Day
Olszanski, Dennis John
O'Neill, Richard Thomas
Palik, Emil Samuel
Parson, John Morris
Petschek, Rolfe George
Phillips, David Berry
Pinkerton, A Alan
Place, Robert Daniel
Platz, Matthew S
Pobereskin, Meyer
Porter, Raymond P
Porter, Spencer Kellogg
Post, Robert Elliott
Potter, Ralph Miles
Powell, David Lee
Poynter, James William
Provder, Theodore
Prusaczyk, Joseph Edward
Ramp, Floyd Lester
Reardon, Joseph Patrick
Reed, Allan Hubert
Reed, William Robert
Reno, Martin A
Reuter, Robert A
Rich, Joseph William
Rich, Ronald Lee
Richter, Helen Wilkinson
Riga, Alan
Rightmire, Robert
Ritchey, William Michael
Ritter, Hartien Sharp
Rodgers, Michael A J
Roe, Ryong-Joon
Rogers, Charles Edwin
Ross, Robert Talman
Ruch, Richard Julius
Russell, Leonard Nelson
Schenz, Anne Filer
Schenz, Timothy William
Schildcrout, Steven Michael
Schneider, Wolfgang W
Schochet, Melvin Leo
Schoonmaker, Richard Clinton
Schumm, Brooke, Jr
Scott, Roy Albert, III
Sebastian, John Francis
Selker, Milton Leonard
Selover, Theodore Britton, Jr
Seybold, Paul Grant
Shaw, Wilfrid Garside
Sicka, Richard Walter
Siebert, Alan Roger
Sieglaff, Charles Lewis
Silver, Gary Lee
Simha, Robert
Singer, Leonard Sidney
Singer, Sherwin
Smith, Warren Harvey
Snavely, Deanne Lynn
Snyder, Donald Lee
Sodd, Vincent J
Sokoloski, Theodore Daniel
Spencer, Harry Edwin
Sprague, Estel Dean
Srinivasan, Vakula S
Stansbrey, John Joseph
Sternlicht, Himan
Stoner, Elaine Carol Blatt
Strobel, Rudolf G K
Sullenger, Don Bruce
Swift, Terrence James
Sympson, Robert F
Taylor, William Johnson
Temple, Robert Dwight
Thomas, Lazarus Daniel
Thomas, Terence Michael
Tiernan, Thomas Orville
Tong, James Ying-Peh
Townley, Charles William
Tsai, Chun-Che
Tuan, Debbie Fu-Tai
Turnbull, Bruce Felton
Ulmer, Richard Clyde
Uys, Johannes Marthinus
Vallee, Richard Earl
Van Lier, Jan Antonius
Vassiliades, Anthony E
Verhoek, Frank Henry
Wagner, Melvin Peter
Walsh, James Aloysius
Westenbarger, Gene Arlan
Whitman, Donald Ray
Wilde, Bryan Edmund
Williams, Elmer Lee
Williams, Josephine Louise
Williams, Robert Calvin
Wolken, George, Jr
Woltz, Frank Earl
Wu, Richard Li-Chuan
Yeager, Ernest Bill
Yellin, Wilbur
Zoglio, Michael Anthony
Zubler, Edward George

OKLAHOMA
Ackerson, Bruce J
Andersen, Terrell Neils
Atkinson, Gordon
Baldwin, Bernard Arthur
Brown, Kenneth Henry
Burchfield, Thomas Elwood
Burr, John Green
Canham, Richard Gordon
Chatenever, Alfred
Christian, Sherril Duane
Cook, Charles Falk
Cunningham, Clarence Marion

Das, Paritosh Kumar
Devlin, Joseph Paul
Dill, Edward D
Dwiggins, Claudius William, Jr
Eastman, Alan D
Ellis, David Allen
Fenstermaker, Roger William
Finch, Jack Norman
Fogel, Norman
Frame, Harlan D
Frech, Roger
Freeman, Robert David
Gall, James William
Gregory, M Duane
Grigsby, Ronald Davis
Grotheer, Morris Paul
Growcock, Frederick Bruce
Guillory, Jack Paul
Hall, Colby D(ixon), Jr
Harris, Jesse Ray
Harrison, R(oland) H(enry)
Harwell, Jeffrey Harry
Heckelsberg, Louis Fred
Hobbs, Anson Parker
Howard, Robert Ernest
Janzen, Jay
Johnson, Marvin M
Johnson, Timothy Walter
Kemp, Marwin K
Lanning, William Clarence
Lorenz, Philip Boalt
Lyon, William Graham
McDaniel, Max Paul
Mains, Gilbert Joseph
Matson, Ted P
Meshri, Dayaldas Tanumal
Murphy, George Washington
Neely, Stanley Carrell
Neptune, William Everett
Nolan, George Junior
Noll, Leo Albert
Nye, Mary Jo
Owen, James Robert
Palmer, Bruce Robert
Parrott, Stephen Laurent
Purdie, Neil
Purdue, Jack Olen
Riggs, Olen Lonnie, Jr
Schaffer, Arnold Martin
Schmidt, Donald Dean
Shell, Francis Joseph
Shock, D'Arcy Adriance
Spivey, Howard Olin
Stratton, Charles Abner
Sundberg, Kenneth Randall
Thomason, William Hugh
Thompson, Donald Leo
Thomson, George Herbert
Van der Helm, Dick
Van De Steeg, Garet Edward
Vinatieri, James Edward
Whitfill, Donald Lee
Williamson, William Burton
Wright, Paul McCoy

OREGON
Abbott, Andrew Doyle
Arthur, John Read, Jr
Balko, Barbara Ann
Broide, Michael Lynn
Chittick, Donald Ernest
Daniels, Malcolm
Decius, John Courtney
Dunne, Thomas Gregory
Durham, George Stone
Dyke, Thomas Robert
Engelking, Paul Craig
Evans, Glenn Thomas
Faler, Kenneth Turner
Feldman, Milton H
Firkins, John Lionel
Gatz, Carole R
Grover, James Robb
Hackleman, David E
Hamby, Drannan Carson
Hannum, Steven Earl
Hard, Thomas Michael
Hardwick, John Lafayette
Haygarth, John Charles
Hedberg, Kenneth Wayne
Hermens, Richard Anthony
Horne, Frederick Herbert
Hower, Charles Oliver
Keedy, Curtis Russell
Klopfenstein, Charles E
Ko, Hon-Chung
Loehr, Thomas Michael
Lonsdale, Harold Kenneth
Loveland, Walter (David)
Lutz, Raymond Paul
McClure, David Warren
Maze, Robert Craig
Mazo, Robert Marc
Mickelsen, John Raymond
Myers, George E
Nafziger, Ralph Hamilton
Nibler, Joseph William
Nielsen, Lawrence Ernie
Norris, Thomas Hughes
Noyes, Richard Macy
Payton, Arthur David
Photinos, Panos John
Prentice, Jack L
Robinson, Arthur B

Roe, David Kelmer
Schellman, John Anthony
Seevers, Robert Edward
Skinner, Gordon Bannatyne
Swanson, Lynwood Walter
Tenney, Agnes
Thomas, Thomas Darrah
Tyler, David Ralph
Warren, William Willard, Jr
White, Horace Frederick
Winniford, Robert Stanley

PENNSYLVANIA
Allara, David Lawrence
Ambs, William Joseph
Anderson, James B
Anderson, Jay Martin
Anderson, John Howard
Ascah, Ralph Gordon
Asher, Sanford Abraham
Aviles, Rafael G
Backus, John King
Bakule, Ronald David
Bandy, Alan Ray
Banks, Grace Ann
Barnartt, Sidney
Barnes, Mary Westergaard
Barnett, Herald Alva
Bass, Jonathan Langer
Basseches, Harold
Batdorf, Robert Ludwig
Baurer, Theodore
Bechtold, Max Frederick
Beck, Paul W
Beichl, George John
Belitskus, David
Benkovic, Stephen J
Bent, Henry Albert
Bergo, Conrad Hunter
Berkheimer, Henry Edward
Berry, Robert Walter
Birks, Neil
Bivens, Richard Lowell
Blaustein, Bernard Daniel
Bock, Charles Walter
Boggs, William Emmerson
Bohning, James Joel
Bonewitz, Robert Allen
Borkowski, Raymond P
Bothner-By, Aksel Arnold
Bowman, Kenneth Aaron
Bowman, Robert Samuel
Brendley, William H, Jr
Breyer, Arthur Charles
Bromberg, J Philip
Brown, Patrick Michael
Brown, Paul Edmund
Bugosh, John
Butera, Richard Anthony
Cardinal, John Robert
Caretto, Albert A, Jr
Carlson, Gerald Leroy
Carroll, Robert J
Cartier, Peter G
Casassa, Edward Francis
Castellano, Salvatore, Mario
Castleman, Albert Welford, Jr
Catchen, Gary Lee
Cavanaugh, James Richard
Cawley, John Joseph
Chaiken, Robert Francis
Chakrabarti, Paritosh M
Chase, Grafton D
Chaudhury, Manoj Kumar
Cheng, Sheau-Mei
Chiu, Tai-Woo
Chu, Vincent H(ao) K(wong)
Chuang, Strong Chieu-Hsiung
Chung, Tze-Chiang
Clayton, John Charles (Hastings)
Clovis, James S
Cochran, Charles Norman
Cohn, Mildred
Cole, Milton Walter
Conroy, James Strickler
Cook, Donald Bowker
Cooper, John Neale
Corneliussen, Roger DuWayne
Cornell, Donald Gilmore
Courtney, Welby Gillette
Craig, Raymond S
Dai, Hai-Lung
Davis, Brian Clifton
Davis, Burl Edward
Dedinas, Jonas
Deno, Norman C
Dent, Anthony L
DiCarlo, Ernest Nicholas
Diehl, Renee Denise
Diorio, Alfred Frank
Dougherty, Eugene P
Downey, Bernard Joseph
Downs, James Joseph
Dreibelbis, John Adam
Duck, William N, Jr
Dunkelberger, Tobias Henry
Durante, Vincent Anthony
Dux, James Philip
Dymicky, Michael
Dzombak, William Charles
El-Aasser, Mohamed S
Ellison, Frank Oscar
Elson, Jesse
Epstein, Lawrence Melvin

Erb, Robert Allan
Fabish, Thomas John
Falcone, James Salvatore, Jr
Falkiewicz, Michael Joseph
Farlee, Rodney Dale
Farrington, Gregory Charles
Feighan, Maria Josita
Feller, Robert Livingston
Ferrell, D Thomas, Jr
Ferren, Richard Anthony
Fetsko, Jacqueline Marie
Finseth, Dennis Henry
Fischer, John Edward
Fishman, Marshall Lewis
Fitts, Donald Dennis
Flattery, David Kevin
Forchheimer, Otto Louis
Forsman, William C(omstock)
Fortnum, Donald Holly
Foster, Perry Alanson, Jr
Francis, Peter Schuyler
Franco, Nicholas Benjamin
Frank, William Benson
Frankel, Lawrence (Stephen)
Frenklach, Michael Y
Fresia, Elmo James
Frick, Neil Huntington
Fritsch, Arnold Rudolph
Fuget, Charles Robert
Furrow, Stanley Donald
Gao, Quanyin
Garbarini, Victor C
Garber, Charles A
Gardner, David Milton
Garrett, Michael Benjamin
Garrett, Robert Roth
Garrett, Thomas Boyd
Garrison, Barbara Jane
Gaughan, Renata Rysnik
Gerasimowicz, Walter Vladimir
Gittler, Franz Ludwig
Glandt, Eduardo Daniel
Gledhill, Ronald James
Gold, Lewis Peter
Goodwin, James Gordon, Jr
Gordon, William Edwin
Grabowski, Joseph J
Grady, Harold Roy
Graham, Arthur H(ughes)
Grant, Richard J
Greeley, Richard Stiles
Greenshields, John Bryce
Greifer, Aaron Philip
Greiner, Richard William
Grey, James Tracy, Jr
Griffiths, Robert Budington
Gulbransen, Earl Alfred
Hall, W Keith
Haltner, Arthur John
Hansen, Gerald Delbert, Jr
Hardman, Carl Charles
Harju, Philip Herman
Hart, Maurice I, Jr
Hatch, Richard C
Hazeltine, James Ezra, Jr
Heald, Emerson Francis
Henrichs, Paul Mark
Hertzberg, Martin
Hillner, Edward
Hiltz, Arnold Aubrey
Hochstrasser, Robin
Hofer, Lawrence John Edward
Hogan, Aloysius Joseph, Jr
Hollibaugh, William Calvert
Hollingsworth, Charles Alvin
Holm, Reimer
Hudson, Robert McKim
Hultman, Carl A
Hutta, Paul John
Irwin, Philip George
Jackovitz, John Franklin
James, Laylin Knox, Jr
Jenkins, Wilmer Atkinson, II
Jhon, Myung S
Johnson, Edwin Wallace
Johnson, Kenneth Allen
Jordan, Kenneth David
Kakar, Anand Swaroop
Kampas, Frank James
Kaplan, Morton
Katsanis, Eleftherios P
Kay, Jack Garvin
Kay, Robert Leo
Keller, Rudolf
Kelly, Ernest L
Kenson, Robert Earl
Kerst, A(l) Fred
Kifer, Edward W
Kim, Tai Kyung
Kimlin, Mary Jayne
King, James P
King, Richard Warren
Klein, Michael Lawrence
Kleinsteuber, Tilmann Christoph Werner
Klier, Kamil
Kliman, Harvey Louis
Koller, Robert Dene
Kraitchman, Jerome
Kronick, Paul Leonard
Krzeminski, Stephen F
Kumosinski, Thomas Francis
Kunesh, Charles Joseph
Kuzmak, Joseph Milton
Labes, Mortimer Milton

Lampe, Frederick Walter
Lancet, Michael Savage
Lange, Klaus Robert
Larsen, John W
Lautenberger, William J
Leader, Gordon Robert
Lee, Robert William
Leffler, Amos J
Leininger, Paul Miller
Leister, Harry M
Leonard, John Joseph
Leston, Gerd
Lewis, Bernard
Lewis, Charles William
Li, Norman Chung
Liberti, Paul A
Lloyd, Thomas Blair
Locke, Harold Ogden
Longo, Frederick R
Loprest, Frank James
Lovejoy, Roland William
Lowe, John Philip
Lu, Guangquan
Ludwig, Howard C
Ludwig, Oliver George
McCarthy, Raymond Lawrence
Macdonald, Digby Donald
McDowell, Maurice James
Mack, Lawrence Lloyd
MacKay, Colin Francis
McKelvey, Donald Richard
Maclay, William Nevin
Magill, Joseph Henry
Maguire, Mildred May
Mair, Robert Dixon
Mandelcorn, Lyon
Manes, Milton
Manley, Thomas Clinton
Marcelin, George
Martin, Aaron J
Martin, Edward Shaffer
Martinez de Pinillos, Joaquin Victor
Masciantonio, Philip
Mason, Charles Morgan
Matyjaszewski, Krzysztof
Mayer, George Emil
Mayo, Ralph Elliott
Meakin, Paul
Meeker, Thrygve Richard
Melander, Wayne Russell
Messer, Wayne Ronald
Micale, Fortunato Joseph
Michaels, Adlai Eldon
Miller, Alfred Charles
Miller, Eugene D
Miller, John George
Milz, Wendell Collins
Minn, Fredrick Louis
Mishra, Dinesh S
Mitchell, Donald J
Montjar, Monty Jack
Morris, Marlene Cook
Motsavage, Vincent Andrew
Mozersky, Samuel M
Muck, Darrel Lee
Mulay, Laxman Nilakantha
Music, John Farris
Myers, Howard
Na, George Chao
Nagy, Dennis J
Nath, Amar
Naylor, Robert Ernest, Jr
Neddenriep, Richard Joe
Neidig, Howard Anthony
Newitt, Edward James
Nishioka, Masaharu
Nixon, Eugene Ray
Norris, Wilfred Glen
Nylund, Robert E
Ochjai, Ei-Ichiro
Ohlberg, Stanley Miles
O'Mara, James Herbert
Opella, Stanley Joseph
Owens, Kevin Glenn
Pacer, John Charles
Palmer, Howard Benedict
Parchen, Frank Raymond, Jr
Patterson, Gary David
Perzak, Frank John
Pessen, Helmut
Peterson, Jack Kenneth
Pettit, Frederick S
Pez, Guido Peter
Phillips, John Spencer
Piccolini, Richard John
Pickering, Howard W
Pierce, Percy Everett
Pierce, Thomas Henry
Pinschmidt, Robert Krantz, Jr
Plazek, Donald John
Poole, John Anthony
Porter, Sydney W, Jr
Prados, Ronald Anthony
Pratt, David Wixon
Prohaska, Charles Anton
Radspinner, John Asa
Randall, William Carl
Ravi, Natarajan
Recktenwald, Gerald William
Reeder, Ray R
Retcofsky, Herbert L
Reuwer, Joseph Francis, Jr
Richardson, Ralph J
Richey, Willis Dale

Physical Chemistry (cont)

Riethof, Thomas Robert
Ristey, William J
Roesmer, Josef
Rogers, Horace Elton
Romberger, Karl Arthur
Roper, Gerald C
Rosenbaum, Eugene Joseph
Rosenblum, Charles
Rothbart, Herbert Lawrence
Rozelle, Ralph B
Ruppel, Thomas Conrad
Sallavanti, Robert Armando
Salomon, Robert Ephriam
Samuelson, H Vaughn
Schaeffer, William Dwight
Scheirer, James E
Schell, William R
Schettler, Paul Davis, Jr
Schiesser, Robert H
Schiffman, Louis F
Schott, Hans
Schreiber, Kurt Clark
Schuster, Ingeborg I M
Schweighardt, Frank Kenneth
Shapiro, Zalman Mordecai
Sheasley, W(illiam) David
Shergalis, William Anthony
Shim, Benjamin Kin Chong
Shirley, David Arthur
Shoaf, Mary La Salle
Shozda, Raymond John
Signorino, Charles Anthony
Silbert, Leonard Stanton
Simmons, Gary Wayne
Simonaitis, Romualdas
Singleton, Jack Howard
Siska, Peter Emil
Sloat, Charles Allen
Smith, Allan Laslett
Smith, David
Smith, Stewart Edward
Sollott, Gilbert Paul
Solomon, Joseph Alvin
Spear, Jo-Walter
Spell, Aldenlee
Spencer, James Nelson
Spinnler, Joseph F
Staskiewicz, Bernard Alexander
Staut, Ronald
Steele, William A
Steiger, Roger Arthur
Stewart, Robert F
Stokes, Charles Sommers
Stryker, Lynden J
Sturm, James Edward
Swain, Howard Aldred, Jr
Swarts, Elwyn Lowell
Swift, Harold Eugene
Szabo, Miklos Tamas
Szamosi, Janos
Takeshita, Tsuneichi
Taylor, Robert Morgan
Taylor, William Daniel
Tenney, Albert Seward, III
Tepper, Frederick
Tevebaugh, Arthur David
Thibeault, Jack Claude
Thomas, H Ronald
Thompson, Peter Trueman
Thornton, Elizabeth K
Tomezsko, Edward Stephen John
Trachtman, Mendel
Tristram-Nagle, Stephanie Ann
Tubbs, Robert Kenneth
Vanderhoff, John W
Vastola, Francis J
Venable, Emerson
Voet, Donald Herman
Voltz, Sterling Ernest
Wahnsiedler, Walter Edward
Walker, Robert Winn
Wallace, Paul Francis
Warrick, Percy, Jr
Watterson, Kenneth Franklin
Wefers, Karl
Weismann, Theodore James
Weiss, Paul Storch
Welch, Cletus Norman
Wells, Adoniram Judson
Westmoreland, David Gray
White, David
White, James Russell
White, Malcolm Lunt
White, Norman Edward
Williams, Ardis Mae
Williams, James Earl, Jr
Wilson, John Randall
Winters, Earl D
Wismer, Robert Kingsley
Witkowski, Robert Edward
Wojcik, John F
Wolke, Robert Leslie
Wootten, Michael John
Work, James Leroy
Wrledt, Henry Anderson
Wunderlich, Francis J
Yang, Arthur Jing-Min
Yao, Shang Jeong
Yates, John Thomas, Jr
Yeh, George Chiayou
Young, Harrison Hurst, Jr
Young, Irving Gustav

Zanger, Murray
Zelley, Walter Gauntt
Zeroka, Daniel
Ziegler, George William, Jr
Zimmerman, George Landis
Zordan, Thomas Anthony

RHODE ISLAND

Annino, Raymond
Baird, James Clyde
Calo, Joseph Manuel
Carpenter, Gene Blakely
Diebold, Gerald Joseph
Doll, Jimmie Dave
Estrup, Peder Jan Z
Greene, Edward Forbes
Hartman, Karl August
Johnson, Gordon Carlton
Kirschenbaum, Louis Jean
Kreiser, Ralph Rank
Kroll, Harry
Lawler, Ronald George
Marzzacco, Charles Joseph
Mason, Edward Allen
Morris, George V
Risen, William Maurice, Jr
Sprowles, Jolyon Charles
Stratt, Richard Mark
Williams, John Collins, Jr
Zuehlke, Richard William

SOUTH CAROLINA

Ashy, Peter Jawad
Avgeropoulos, George N
Bach, Ricardo O
Bailey, Charles Edward
Bailey, Roy Horton, Jr
Ballentine, Alva Ray
Bly, Robert Stewart
Bonner, Oscar Davis
Boyd, George Edward
Breazeale, William Horace, Jr
Browne, Colin Lanfear
Clark, Hugh Kidder
Davis, Joseph B
Dellicolli, Humbert Thomas
Diefendorf, Russell Judd
Durig, James Robert
Dusenbury, Joseph Hooker
Dworjanyn, Lee O(leh)
Falkehag, S Ingemar
Farewell, John P
Faust, John William, Jr
Force, Carlton Gregory
Gibbs, Ann
Gilkerson, William Richard
Griffith, Elizabeth Ann Hall
Groh, Harold John
Harbour, John Richard
Heberger, John M
Hennelly, Edward Joseph
Henderson, Richard Wayne
Herdklotz, Richard James
Hochel, Robert Charles
Hopper, Michael James
Hyder, Monte Lee
Jacober, William John
Jarvis, Christine Woodruff
Jordan, Kenneth Gary
Jumper, Charles Frederick
Kendall, David Nelson
Knight, Jere Donald
Knight, Lon Bishop, Jr
Likes, Carl James
Longfield, James Edgar
Lowry, Bright Anderson
Malstrom, Robert Arthur
Mercer, Edward Everett
Metz, Clyde Raymond
Myrick, Michael Lenn
Nanney, Thomas Ray
Orebaugh, Errol Glen
Otto, Wolfgang Karl Ferdinand
Owen, John Harding
Patterson, William Alexander
Perkins, William Clopton
Piper, John
Plodinec, Matthew John
Porter, John J
Sargent, Roger N
Savitsky, George Boris
Sharpe, Louis Haughton
Simon, Frederick Tyler
Simonsen, David Raymond
Spencer, Harold Garth
Stearns, Edwin Ira
Taylor, Peter Anthony
Tolbert, Thomas Warren
Varin, Roger Robert
Woolsey, Gerald Bruce

SOUTH DAKOTA

Coker, Earl Howard, Jr
Estee, Charles Remington
Hecht, Harry George
Martin, Willard John
Rue, Rolland R
Scott, George Prescott
Spinar, Leo Harold

TENNESSEE

Adams, Robert Edward
Airee, Shakti Kumar
Allen, Vernon R

Anderson, Richard Louis
Anderson, Roger W(hiting)
Annis, Brian Kitfield
Baes, Charles Frederick, Jr
Bamberger, Carlos Enrique Leopoldo
Barber, Eugene John
Bates, John Bryant
Baxter, John Edwards
Beahm, Edward Charles
Bedoit, William Clarence, Jr
Begun, George Murray
Bell, Jimmy Todd
Bentz, Ralph Wagner
Blanck, Harvey F, Jr
Bleier, Alan
Bond, Walter D
Boston, Charles Ray
Braunstein, Jerry
Brooks, Alfred Austin, Jr
Brown, Gilbert Morris
Brown, Lloyd Leonard
Browning, Horace Lawrence, Jr
Bullock, Jonathan S, IV
Bunick, Gerard John
Burns, John Howard
Busey, Richard Hoover
Busing, William Richard
Caflisch, George Barrett
Campbell, David Owen
Cantor, Stanley
Carman, Howard Smith, Jr
Cathcart, John Varn
Clarke, Ann Neistadt
Clarke, James Harold
Coleman, Charles F(ranklin)
Compere, Edgar Lattimore
Condon, James Benton
Daunt, Stephen Joseph
Davis, Jimmy Henry
Davis, Montie Grant
Davis, Phillip Howard
Dillard, James William
Doody, John Edward
Dudney, Nancy Johnston
Duncan, Budd Lee
Egan, B Zane
Evans, Joe Smith
Fagerburg, David Richard
Farrar, Robert Lynn, Jr
Feigerle, Charles Stephen
Field, Frank Henry
Fletcher, William Henry
Fort, Tomlinson
Franceschetti, Donald Ralph
Frederiksen, Dixie Ward
Fuzek, John Frank
Gangwer, Thomas E
Germroth, Ted Calvin
Gerwe, Roderick Daniel
Gibson, John Knight
Glasstone, Samuel
Gray, Allen G(ibbs)
Guenther, William Benton
Gustafson, Bruce LeRoy
Habenschuss, Anton
Hall, Larry Cully
Harding, Charles Enoch
Hayter, John Bingley
Hinds, Nancy Webb
Ho, Patience Ching-Ru
Hochanadel, Clarence Joseph
Hoffman, Everett John
Holmes, Howard Frank
Huang, Thomas Tao Shing
Ijams, Charles Carroll
Jenks, Glenn Herbert
Johnson, Elijah
Johnston, David Owen
Judish, John Paul
Judkins, Roddie Reagan
Ketchen, Eugene Earl
Kocher, David Charles
Kovac, Jeffrey Dean
Krause, Herbert Francis
Krause, Manfred Otto
Langford, Paul Brooks
Leed, Russell Ernest
Li, Ying Sing
Lietzke, Milton Henry
Lindemer, Terrence Bradford
Livingston, Ralph
Longmire, Martin Shelling
Lura, Richard Dean
McConnell, Richard Leon
McDowell, William Jackson
McGill, Robert Mayo
McGraw, Gary Earl
Malinauskas, Anthony Peter
Marshall, William Leitch
Martin, Thomas Waring
Maya, Leon
Metcalf, David Halstead
Miceli, Angelo Sylvestro
Miller, John Cameron
Moore, George Edward
Moore, Louis Doyle, Jr
Mortimer, Robert George
Moyer, Bruce A
Naddy, Badie Ihrahim
Newby, Frank Armon, Jr
Nicely, Vincent Alvin
Noid, Donald William
Nyssen, Gerard Allan
Overton, James Ray

Parkinson, William Walker, Jr
Patton, Hugh Wilson
Pearson, R(ay) L(eon)
Petersen, Harold, Jr
Peterson, Joseph Richard
Petke, Frederick David
Phillips, Paul J
Pinnaduwage, Lal Ariyaratna
Polavarapu, Prasad Leela
Pond, David Martin
Poutsma, Marvin Lloyd
Powell, George Louis
Quist, Arvin Sigvard
Rauscher, Grant K
Reynolds, Jefferson Wayne
Robbins, Gordon Daniel
Roesler, Joseph Frank
Rudolph, Philip S
Rush, Richard Marion
Russin, Nicholas Charles
Rutenberg, Aaron Charles
Schaad, Lawrence Joseph
Silverman, Meyer David
Smith, George Pedro
Soldano, Benny A
Staats, Percy Anderson
Starr, Duane Frank
Stoughton, Raymond Woodford
Stubbs, Gerald James
Sworski, Thomas John
Tarpley, Anderson Ray, Jr
Taylor, Ellison Hall
Tellinghuisen, Joel Barton
Thiessen, William Ernest
Thoma, Roy E
Tirman, Alvin
Toth, Louis McKenna
Trowbridge, Lee Douglas
Van Hook, William Alexander
Van Sickle, Dale Elbert
Vasofsky, Richard William
Waggener, William Cole
Ward, James Andrew
Watson, Marshall Tredway
Williams, Thomas Ffrancon
Wilson, David J
Woods, Clifton, III
Ying, Zhiqiang Charles
Young, Frederick Walter, Jr
Young, Howard Seth
Zuber, William Henry, Jr

TEXAS

Abrams, Albert
Ahlberg, Dan Leander
Albright, John Grover
Ali, Yusuf
Allen, Roland Emery
Allison, Jean Batchelor
Altmiller, Henry
Anselmo, Vincent C
Armstrong, Andrew Thurman
Baker, Charles Taft
Bannerman, James Knox
Banta, Marion Calvin
Bard, Allen Joseph
Barieau, Robert (Eugene)
Barna, Gabriel George
Basila, Michael Robert
Batten, Charles Francis
Benson, Herbert Linne, Jr
Berg, Mark Alan
Birney, David Martin
Blytas, George Constantin
Bockris, John O'Mara
Boggs, James Ernest
Brady, William Thomas
Broodo, Archie
Brooks, Philip Russell
Brown, James Michael
Brown, Robert Wade
Brusie, James Powers
Bullock, Kathryn Rice
Caldwell, Richard A
Callender, Wade Lee
Cares, William Ronald
Carlson, Douglas W
Caudle, Danny Dearl
Celii, Francis Gabriel
Chao, Jing
Chen, Edward Chuck-Ming
Chen, Michael Chia-Chao
Chu, Ting Li
Clark, Ronald Keith
Clingman, William Herbert, Jr
Closmann, Philip Joseph
Cocke, David Leath
Cole, David F
Collins, Francis Allen
Collins, Russell Lewis
Conway, Dwight Colbur
Cook, Evin Lee
Cook, Paul Fabyan
Cooley, Stone Deavours
Crawford, James Worthington
Croce, Louis J
Cross, Virginia Rose
Cruser, Stephen Alan
Cummiskey, Charles
Curl, Robert Floyd
Cusachs, Louis Chopin
Custard, Herman Cecil
Davis, Donald Robert
Davis, Raymond E

DeBerry, David Wayne
Denison, Jack Thomas
Desiderato, Robert, Jr
Deviney, Marvin Lee, Jr
Dimitroff, Edward
Dobson, Gerard Ramsden
Dorris, Kenneth Lee
Drake, Edgar Nathaniel, II
Dufner, Douglas Carl
Duncan, Walter Marvin, Jr
Duty, Robert C
Dyer, Lawrence D
Dzidic, Ismet
Eaker, Charles William
Ekerdt, John Gilbert
Ellzey, Marion Lawrence, Jr
Elward-Berry, Julianne
Engel, Paul Sanford
Fackler, John Paul
Farhataziz, Mr
Faris, Sam Russell
Faubion, Billy Don
Floyd, Willis Waldo
Forest, Edward
Foster, Norman George
Fox, Marye Anne
Frank, Steven Neil
Franklin, Thomas Chester
Frazee, Jerry D
Fredin, Leif G R
Freeman, Maynard Lloyd
Friedman, Robert Harold
Fuller, Martin Emil
Furlong, Norman Burr, Jr
Gardiner, William Cecil, Jr
Garza, Cesar Manuel
Gerlach, John Louis
Ghowsi, Kiumars
Giam, Choo-Seng
Gilbert, John Carl
Gingerich, Karl Andreas
Glass, Graham Percy
Gleim, Paul Stanley
Gnade, Bruce E
Goodman, David Wayne
Graves, Robert Earl
Grogan, Michael John
Hackerman, Norman
Hackert, Marvin LeRoy
Hamilton, Janet V
Hamilton, Walter S
Hance, Robert Lee
Harlan, Horace David
Hart, Haskell Vincent
Harwood, William H
Hautala, Richard Roy
Hedges, Richard Marion
Heller, Adam
Herndon, William Cecil
Hightower, Joe W(alter)
Hilton, Ray
Hinkson, Thomas Clifford
Holly, Frank Joseph
Hopkins, Robert Charles
Horowitz, Paul Martin
Horton, Robert Louis
Houston, James Grey
Huckaby, Dale Alan
Hungerford, Ed Vernon, III
Ivey, Robert Charles
Jackson, Roy Joseph
Jackson, William Roy, Jr
Johnson, David Russell
Johnson, Fred Lowery, Jr
Johnson, Grover Leon
Johnstone, C(harles) Wilkin
Jones, Lawrence Ryman
Jones, Marvin Thomas
Jones, Morton Edward
Kalfoglou, George
Kamego, Albert Amil
Keenan, Joseph Aloysius
Kellerhals, Glen E
Kennelley, Kevin James
Kevan, Larry
Kinsey, James Lloyd
Koenig, Charles Louis
Kohn, Erwin
Konrad, Dusan
Kulkarni, Padmakar Venkatrao
Kust, Roger Nayland
Laane, Jaan
Ladner, Sidney Jules
Lamb, James Francis
Lang, John Calvin, Jr
Larsen, Russell D
Leftin, Harry Paul
Lewis, Donald Richard
Lewis, Paul Herbert
Linowski, John Walter
Lin-Vien, Daimay
Lippmann, David Zangwill
Loos, Karl Rudolf
Lowrey, Charles Boyce
Lunsford, Jack Horner
Lurix, Paul Leslie, Jr
McAtee, James Lee, Jr
McCammon, James Andrew
McCullough, James Douglas, Jr
McDougall, Robert I
McDowell, Harding Keith
McNamara, Edward P(aul)
Mahendroo, Prem P
Malloy, Thomas Bernard, Jr

Marcotte, Ronald Edward
Margrave, John Lee
Marsh, Kenneth Neil
Marshall, Paul
Mattax, Calvin Coolidge
Melrose, James C
Melton, Lynn Ayres
Mendelson, Robert Allen
Menon, Venugopal Balakrishna
Mertens, Frederick Paul
Messenger, Joseph Umlah
Meyer, W(illiam) Keith
Miller, Richard Edward
Milliken, Spencer Rankin
Moore, Robert J(ames)
Muir, Mariel Meents
Munk, Petr
Murphy, Ruth Ann
Naistat, Samuel Solomon
Natowitz, Joseph Bernard
Neeley, Charles Mack
Neff, Laurence D
Newton, Robert Andrew
Nordlander, Peter Jan Arne
O'Keefe, Dennis Robert
Olson, John S
Otken, Charles Clay
Pak, Charles Y
Palmer, H(arold) A(rthur)
Pannell, Richard Byron
Patil, Kashinath Z(iparu)
Patsch, Wolfgang
Pauley, James L
Perry, Reeves Baldwin
Petersen, Donald H
Pettitt, Bernard Montgomery
Plank, Don Allen
Pogue, Randall F
Powers, Robert S(inclair), Jr
Prigogine, Ilya
Puligandla, Viswanadham
Rabalais, John Wayne
Rawls, Henry Ralph
Redding, Rogers Walker
Redington, Richard Lee
Reeder, Charles Edgar
Reynolds, William Walter
Riggs, Charles Lee
Robinson, George Wilse
Robinson, Joseph Dewey
Rodgers, Alan Shortridge
Roehrig, Gerald Ralph
Rogers, Jesse Wallace
Roof, Jack Glyndon
Rosenquist, Edward P
Rosenthal, Michael R
Rossky, Peter Jacob
Rulon, Richard M
Sager, Ray Stuart
Sample, Thomas Earl, Jr
Sanchez, Isaac Cornelius
Sardisco, John Baptist
Sass, Ronald L
Schelly, Zoltan Andrew
Schmalz, Thomas G
Shaer, Elias Hanna
Shanfield, Henry
Shaw, Don W
Shipsey, Edward Joseph
Sloan, Tod Burns
Smalley, Richard Errett
Smith, Joseph Patrick
Smith, Larry
Smith, Michael Lew
Snavely, Earl Samuel, Jr
Spenadel, Lawrence
Spencer, John Edward
Stewart, George Hudson
Streetman, John Robert
Strieter, Frederick John
Strom, E(dwin) Thomas
Stubbeman, Robert Frank
Sudbury, John Dean
Sullivan, James Thomas, Jr
Sward, Edward Lawrence, Jr
Synek, Miroslav (Mike)
Tang, Yi-Noo
Tannahill, Mary Margaret
Taylor, Edward Donald
Teal, Gordon Kidd
Templeton, Charles Clark
Thomas, Estes Centennial, III
Tiedemann, Herman Henry
Timmons, Richard B
Tomson, Mason Butler
Tong, Youdong
Tucker, Woodson Coleman, Jr
Van Dijk, Christiaan Pieter
Vaudo, Anthony Frank
Vernon, Lonnie William
Vos, Kenneth Dean
Wade, William H
Walker, Bennie Frank
Walmsley, Judith Abrams
Ward-McLemore, Ethel
Warren, Kenneth Wayne
Waters, John Albert
Watson, William Harold, Jr
Weatherford, W(illiam) D(ewey), Jr
Webb, Allen Nystrom
Webber, Stephen Edward
Weisman, R(obert) Bruce
Wentworth, Wayne
Whalen, James William

White, John Michael
Wilde, Kenneth Alfred
Wilde, Richard Edward, Jr
Wiley, John Robert
Wilhoit, Eugene Dennis
Williams, Calvit Herndon, Jr
Williams, Rickey Jay
Williamson, Luther Howard
Wilson, Joseph Edward
Wilson, Peggy Mayfield Dunlap
Wilson, Ray Floyd
Witmer, William Byron
Woessner, Donald Edward
Wood, Scott Emerson
Yager, Billy Joe
Zwolinski, Bruno John

UTAH
Anderson, Bradley Dale
Anderson, Keith Phillips
Bartholomew, Calvin Henry
Bascom, Willard D
Batty, Joseph Clair
Blackham, Angus Udell
Bodily, David Martin
Breckenridge, William H
Caldwell, Dennis
Cannon, John Francis
Carey, John Hugh
Comeford, John J(ack)
Cook, Melvin Alonzo
Eatough, Delbert J
Eyring, Edward M
Garrard, Verl Grady
Giddings, John Calvin
Goates, James Rex
Goates, Steven Rex
Grant, David Morris
Guillory, William Arnold
Guillot, David G
Hansen, Wilford Nels
Harris, Frank Ephraim, Jr
Harris, Joel Mark
Izatt, Reed McNeil
Kratochvil, Jiri
McAdams, Robert Eli
McIntyre, James Douglass Edmonson
Miller, Richard Roy
Miner, Bryant Albert
Moore, William Marshall
Nelson, Kay LeRoi
Ott, J Bevan
Oyler, Dee Edward
Pugmire, Ronald J
Pyper, James William
Rowley, Richard L
Seager, Spencer Lawrence
Simons, John Peter
Snow, Richard L
Thorne, James Meyers
Wahrhaftig, Austin Levy
Wiser, Wendell H(aslam)
Woolley, Earl Madsen

VERMONT
Berens, Alan Robert
Brokaw, Richard Spohn
Bushweller, Charles Hackett
Dasher, George Franklin, Jr
Flanagan, Ted Benjamin
Furukawa, Toshiharu
Jasinski, Jerry Peter
Kellner, Stephan Maria Eduard
Naumann, Robert Alexander
Provost, Ronald Harold
Rinse, Jacobus
Smith, Thomas David
Tyree, Melvin Thomas
Weimann, Ludwig Jan
Westenberg, Arthur Ayer
White, William North
Woodworth, Robert Cummings

VIRGINIA
Adams, Alayne A(mercier)
Addamiano, Arrigo
Andrews, William Lester Self
Barber, Patrick George
Barkin, Stanley M
Bartis, James Thomas
Bell, Charles E, Jr
Berg, John Richard
Birely, John H
Boldridge, William Franklin
Bryant, Robert George
Burns, Allan Fielding
Bushey, Gordon Lake
Calle, Luz Marina
Cantu, Antonio Arnoldo
Cardwell, Paul H(avens)
Chakravarti, Kalidas
Chong, Shuang-Ling
Collins, Frank Charles
Cover, Herbert Lee
Cozzens, Robert F
Croft, Barbara Yoder
Crowell, Thomas Irving
Dalton, Roger Wayne
Davidson, Robert Bellamy
De Camp, Wilson Hamilton
Decorpo, James Joseph
DeFotis, Gary Constantine
Demas, James Nicholas
Desjardins, Steven G

DeVore, Thomas Carroll
Dillard, John Gammons
Dominey, Raymond Nelson
Easter, Donald Philips
Erickson, Wayne Douglas
Evans, John Stanton, Jr
Field, Paul Eugene
Foley, Robert Thomas
Foster, Wilfred John Daniel
Fox, William B
Garrett, Barry B
Gibbs, William Eugene
Gillmor, R(obert) N(iles)
Graybeal, Jack Daniel
Greer, William Louis
Griffiths, David Warren
Gulrich, Leslie William, Jr
Hammer, Gary G
Hansrote, Charles Johnson, Jr
Harney, Brian Michael
Harris, William Charles
Hartung, Homer Arthur
Hazlehurst, David Anthony
Hazlett, Robert Neil
Herm, Ronald Richard
Herr, Frank Leaman, Jr
Hess, LaVerne Derryl
Hilderbrandt, Richard L
Hsu, David Shiao-Yo
Huddle, Benjamin Paul, Jr
Jacknow, Joel
Jacobs, Theodore Alan
Joebstl, Johann Anton
Jordan, Wade H(ampton), Jr
Julian, Maureen M
Kauffman, Glenn Monroe
Kershenstein, John Charles
Kiefer, Richard L
Kirby, Andrew Fuller
Klingensmith, George Bruce
Klobuchar, Richard Louis
Knudsen, Dennis Ralph
Konizer, George Burr
Kranbuehl, David Edwin
Kubu, Edward Thomas
Laszlo, Tibor S
Layson, William M(cIntyre)
Leung, Wing Hai
Lim, Young Woon (Peter)
McCoy, Joseph Hamilton
Macek, Andrej
McGee, Henry A(lexander), Jr
McHale, Edward Thomas
Maggio, Bruno
Mahoney, Bernard Launcelot, Jr
Malin, John Michael
Malloy, Alfred Marcus
Marron, Michael Thomas
Marshall, Maryan Lorraine
Mateer, Richard Austin
Meites, Louis
Mengenhauser, James Vernon
Meyers, Earl Lawrence
Moore, Edward Weldon
Morrow, Richard Joseph
Nakadomari, Hisamitsu
Nelson, David Lynn
Normansell, David E
Ogliaruso, Michael Anthony
Orwoll, Robert Arvid
Owens, Charles Wesley
Oyama, Shigeo Ted
Piepho, Susan Brand
Richard, Alfred Joseph
Richardson, Frederick S
Rogowski, Robert Stephen
Rony, Peter R(oland)
Roscher, David Moore
Rudat, Martin August
Saalfeld, Fred Erich
Sands, George Dewey
Sanzone, George
Satterthwaite, Cameron B
Saunders, Peter Reginald
Sayre, Edward Vale
Schatz, Paul Namon
Schmidt, Parbury Pollen
Schug, John Charles
Selwyn, Philip Alan
Shaub, Walter M
Shuler, Robert Lee
Shull, Don Louis
Shultz, Allan R
Sipe, Herbert James, Jr
Smith, Jerry Joseph
Spencer, Hugh Miller
Spokes, G(ilbert) Neil
Stagg, William Ray
Stoner, Glenn Earl
Stump, Billy Lee
Sumerlin, Neal Gordon
Sumner, Barbara Elaine
Terner, James
Tichenor, Robert Lauren
Ticknor, Leland Bruce
Troxell, Terry Charles
Turer, Jack
Viers, Jimmy Wayne
Vold, Robert Lawrence
Wagner-Bartak, Claus Gunter Johann
Wakeham, Helmut
Walker, Grayson Watkins
Watt, William Stewart
Way, Kermit R

Physical Chemistry (cont)

Whidby, Jerry Frank
Wickman, Herbert Hollis
Wightman, James Pinckney
Wilson, Mathew Kent
Wodarczyk, Francis John
Woolfolk, Robert William
Yankwich, Peter Ewald
Yap, William Tan
Zwolenik, James J

WASHINGTON
Asher, William Edward
Babad, Harry
Barney, Gary Scott
Baughcum, Steven Lee
Benveniste, Jacob
Bierlein, Theo Karl
Bierman, Sidney Roy
Bridgforth, Robert Moore, Jr
Burger, Leland Leonard
Callis, James Bertram
Center, Robert E
Chang, Yew Chun
Cheng, Xueheng
Clark, Donald Eldon
Colson, Steven Douglas
Cotton, John Edward
Cox, John Layton
Crosby, Glenn Arthur
Dodgen, Harold Warren
Doepker, Richard DuMont
Douglas, John Edward
Dunning, Thomas Harold, Jr
Eggers, David Frank, Jr
Engel, Thomas Walter
Evans, Thomas Walter
Ewing, J J
Exarhos, Gregory James
Fordyce, David Buchanan
Franz, James Alan
Frasco, David Lee
Friedrich, Donald Martin
Gammon, Richard Harriss
Gerhold, George A
Gouterman, Martin (Paul)
Greager, Oswald Herman
Gregory, Norman Wayne
Halsey, George Dawson, Jr
Hamm, Randall Earl
Hendrickson, Waldemar Forrsel
Hipps, Kerry W
Hunt, John Philip
Hurst, James Kendall
Janata, Jiri
Johnson, A(lfred) Burtron, Jr
Johnson, Wilbur Vance
Kalkwarf, Donald Riley
Kazanjian, Armen Roupen
Keaton, Clark M
Kells, Lyman Francis
Kells, Milton Carlisle
Kissin, G(erald) H(arvey)
Koehmstedt, Paul Leon
Krier, Carol Alnoth
Kriz, George Stanley
Krohn, Kenneth Albert
Kruger, Albert Aaron
Kwiram, Alvin L
Ladd, Kaye Victoria
Lambert, Maurice C
Leitz, Fred John, Jr
Leonard, John Edward
Lindenmeyer, Paul Henry
Lingren, Wesley Earl
McDowell, Robin Scott
Maki, Arthur George, Jr
Miller, John Howard
Minor, James E(rnest)
Moore, Emmett Burris, Jr
Moore, Robert Lee
Morrey, John Rolph
Moseley, W(illiam) David
Murdock, Gordon Alfred
Neuzil, Edward F
Nickerson, Robert Fletcher
Nightingale, Richard Edwin
Ohta, Masao
Pavia, Donald Lee
Person, James Carl
Pocker, Yeshayau
Poole, Donald Ray
Rabinovitch, B(enton) S(eymour)
Read, David Hadley
Reinhardt, William Parker
Roake, William Earl
Root, John Walter
Ryan, Jack Lewis
Satterlee, James Donald
Schomaker, Verner
Schurr, John Michael
Sherman, David Michael
Shull, Charles Morell, Jr
Sloane, Christine Scheid
Slutsky, Leon Judah
Smith, Francis Marlon
Smith, Richard Dale
Spencer, Arthur Milton, Jr
Strong, Robert Lyman
Stuve, Eric Michael
Styris, David Lee
Tabbutt, Frederick Dean
Tang, Kwong-Tin
Thorsell, David Linden
Tingey, Garth Leroy
Tobiason, Frederick Lee
Vandenbosch, Robert
Wei, Pax Samuel Pin
Wheelwright, Earl J
Whitmer, John Charles
Willett, Roger
Windsor, Maurice William
Wogman, Ned Allen
Yan, Johnson Faa
Young, Robert Hayward
Yunker, Wayne Harry
Zoller, William H

WEST VIRGINIA
Abel, William T
Bhasin, Madan M
Chakrabarty, Manoj R
Coleman, James Edward
Dalal, Nar S
Das, Kamalendu
Dawson, Thomas Larry
Dean, Warren Edgell
Derderian, Edmond Joseph
Dumke, Warren Lloyd
Finklea, Harry Osborn
Fisher, John F
Grindstaff, Teddy Hodge
Hager, Stanley Lee
Hall, George Arthur, Jr
Hall, James Lester
Hanrahan, Edward S
Harrison, Arnold Myron
Hickman, James Blake
Hoback, John Holland
Humphrey, George Louis
Kurland, Jonathan Joshua
Lim, James Khai-Jin
Marcinkowsky, Arthur Ernest
Martin, John Perry, Jr
Murrmann, Richard P
Naumann, Alfred Wayne
Petersen, Jeffrey Lee
Seshadri, Kalkunte S
Showalter, Kenneth
Stewart, Robert Francis
Vucich, M(ichael) G(eorge)
Wallace, William James Lord
Williamson, Kenneth Dale
Woodland, William Charles

WISCONSIN
Alvarez, Robert
Anderson, Harold J
Bahe, Lowell W
Barr, Tery Lynn
Beatty, James Wayne, Jr
Bender, Paul J
Blair, James Edward
Bondeson, Stephen Ray
Brandt, Werner Wilfried
Brebrick, Robert Frank, Jr
Bryce, Hugh Glendinning
Casper, Lawrence Allen
Cataldi, Horace A(nthony)
Chen, Franklin M
Ciriacks, John A(lfred)
Clark, Harlan Eugene
Cornwell, Charles Daniel
Coward, Nathan A
Crim, F(orrest) Fleming
Dahl, Lawrence Frederick
Discher, Clarence August
Draeger, Norman Arthur
Duwell, Ernest John
Dwyer, Sean G
Ehlert, Thomas Clarence
Evans, James Stuart
Farrar, Thomas C
Fenrick, Harold William
Fripiat, Jose J
Greenler, Robert George
Grotz, Leonard Charles
Haeberli, Willy
Hall, Lyle Clarence
Hansen, Paul Vincent, Jr
Harriman, John E
Hassinger, Mary Colleen
Hauser, Edward Russell
Holt, Matthew Leslie
Hyde, James Stewart
Inman, Ross
Kao, Wen-Hong
Keulks, George William
Kittsley, Scott Loren
Kotvis, Peter Van Dyke
Kresch, Alan J
Lang, Conrad Marvin
Lee, Kathryn Adele Bunding
Lo, Mike Mei-Kuo
Losin, Edward Thomas
McAllister, Jerome Watt
McCoy, Charles Ralph
Mallmann, Alexander James
Miller, Arild Justesen
Morton, Stephen Dana
Mukerjee, Pasupati
Nakamoto, Kazuo
Nathanson, Gilbert M
Norman, Jack C
Pollnow, Gilbert Frederick
Prall, Bruce Randall
Radtke, Douglas Dean

Record, M Thomas, Jr
Romary, John Kirk
Rosenberg, Robert Melvin
Roskos, Roland R
St Louis, Robert Vincent
Schrader, David Martin
Schwab, Helmut
Sheppard, Erwin
Spencer, Brock
Steiner, John F
Su, Lao-Sou
Subach, Daniel James
Tcheurekdjian, Noubar
Thomas, Howard Major
Tiedemann, William Harold
Trischan, Glenn M
Vanselow, Ralf W
Vaughan, Worth E
Viswanathan, Ramaswami
Wang, Pao-Kuan
Wendricks, Roland N
Westler, William Milo
Willard, John Ela
Woods, Robert Claude
Wright, John Curtis
Yu, Hyuk
Zimmerman, Howard Elliot
Zografi, George

WYOMING
Adams, Richard Melverne
Decora, Andrew Wayne
Duvall, John Joseph
Edmiston, Clyde
Hiza, Michael John, Jr
Jacobson, Irven Allan, Jr
McColloch, Robert James
Meyer, Edmond Gerald
Netzel, Daniel Anthony
Northcott, Jean
Poulson, Richard Edwin
Ryan, Victor Albert

PUERTO RICO
Blum, Lesser
Bode, James Daniel
Schwartz, Abraham
Siegel, George G
Souto Bachiller, Fernando Alberto

ALBERTA
Ali, Shahida Parvin
Armstrong, David Anthony
Bertie, John E
Birss, Viola Ingrid
Birss, Viola Ingrid
Blades, Arthur Taylor
Brown, Robert Stanley
Codding, Penelope Wixson
Creighton, Stephen Mark
Davis, Stuart George
Dunford, Hugh Brian
Freeman, Gordon Russel
Gunning, Harry Emmet
Harvey, Ross Buschlen
Heimann, Robert B(ertram)
Henderson, John Frederick
Hepler, Loren George
Hodgson, Gordon Wesley
Jacobson, Ada Leah
Kalantar, Alfred Husayn
Kebarle, Paul
Khanna, Faqir Chand
Kotovych, George
Krueger, Peter J
Kuntz, Garland Parke Paul
Langford, Cooper Harold, III
Lown, James William
McClung, Ronald Edwin Dawson
McCurdy, Keith G
Martin, John Scott
Muir, Donald Ridley
Overfield, Robert Edward
Plambeck, James Alan
Robertson, Ross Elmore
Russell, James Christopher
Sommerfeldt, Theron G
Sorensen, Theodore Strang
Strausz, Otto Peter
Tollefson, Eric Lars
Tschuikow-Roux, Eugene
Wieser, Helmut
Yamdagni, Raghavendra
Yeager, Howard Lane

BRITISH COLUMBIA
Balfour, Walter Joseph
Barrow, Gordon M
Bell, Thomas Norman
Benston, Margaret Lowe
Bree, Alan V
Brion, Christopher Edward
Burnell, Edwin Elliott
Dingle, Thomas Walter
Dunell, Basil Anderson
Farmer, James Bernard
Forgacs, Otto Lionel
Frost, David Cregreen
Funt, B Lionel
Gerry, Michael Charles Lewis
Gillis, Hugh Andrew
Harrison, Lionel George
Hatton, John Victor
Hayward, Lloyl Douglas

Kirk, Alexander David
Korteling, Ralph Garret
Leja, J(an)
Lower, Stephen K
McAuley, Alexander
McDowell, Charles Alexander
Malli, Gulzari Lal
Morrison, Stanley Roy
Murray, Francis E
O'Brien, Robert Neville
Ogryzlo, Elmer Alexander
Peters, E(rnest)
Porter, Gerald Bassett
Rieckhoff, Klaus E
Sams, John Robert, Jr
Sherwood, A Gilbert
Spratley, Richard Denis
Stewart, Ross
Trotter, James
Voigt, Eva-Maria
Walker, David Crosby
Wells, Edward Joseph
Yates, Keith
Zwarich, Ronald James

MANITOBA
Bigelow, Charles C
Bock, Ernst
Burchill, Charles Eugene
Forrest, Bruce James
Gesser, Hyman Davidson
Hutton, Harold M
Kartzmark, Elinor Mary
McFarlane, Joanna
MacGregor, Elizabeth Ann
Mantsch, Henry H
Sagert, Norman Henry
Saluja, Preet Pal Singh
Schaefer, Theodore Peter
Secco, Anthony Silvio
Singh, Ajit
Sunder, Sham
Sze, Yu-Keung
Tait, John Charles
Tomlinson, Michael
Torgerson, David Franklyn
Trick, Gordon Staples
Vikis, Andreas Charalambous
Westmore, John Brian

NEW BRUNSWICK
Brooks, Wendell V F
Grant, Douglas Hope
Grein, Friedrich
LeBel, Roland Guy
Read, John Frederick
Reinsborough, Vincent Conrad
Semeluk, George Peter
Thakkar, Ajit Jamnadas

NEWFOUNDLAND
Shahidi, Fereidoon
Smith, Frank Roylance
Stein, Allan Rudolph
Tremaine, Peter Richard

NOVA SCOTIA
Arnold, Donald Robert
Bunbury, David Leslie
Cooke, Robert Clark
Davies, Donald Harry
Falk, Michael
Hayes, Kenneth Edward
Jamieson, William David
Kreuzer, Hans Jurgen
Kusalik, Peter Gerard
Kwak, Jan C T
Leffek, Kenneth Thomas
Linton, Everett Percival
Lynch, Brian Maurice
MacDonald, John James
Pacey, Philip Desmond
Secco, Etalo Anthony
Wasylishen, Roderick Ernest
White, Mary Anne
Whiteway, Stirling Giddings

ONTARIO
Allnatt, Alan Richard
Amenomiya, Yoshimitsu
Ananthanarayanan, Vettaikkoru S
Anlauf, Kurt Guenther
Atack, Douglas
Aziz, Philip Michael
Aziz, Ronald A
Back, Margaret Helen
Baker, Robert G
Bancroft, George Michael
Bardwell, Jennifer Ann
Barradas, Remigio Germano
Barrie, Leonard Arthur
Barton, Stuart Samuel
Benson, George Campbell
Bernath, Peter Francis
Bharucha, Nana R
Bidinosti, Dino Ronald
Bohme, Diethard Kurt
Bolton, James R
Bourns, Arthur N
Brown, James Douglas
Brown, Richard Julian Challis
Brumer, Paul William
Bulani, Walter
Burton, Graham William

CHEMISTRY / 247

Carey, Paul Richard
Carrington, Tucker
Chan, Raymond Kai-Chow
Cherniak, Eugene Anthony
Childs, Ronald Frank
Cocivera, Michael
Conard, Bruce R
Conway, Brian Evans
Dacey, John Robert
Dawes, David Haddon
Dawson, Peter Henry
Dawson, Peter Thomas
Deckers, Jacques (Marie)
De Kimpe, Christian Robert
Detellier, Christian
Diaz, Carlos Manuel
Dick, James Gardiner
Dickson, Lawrence William
Dignam, Michael John
Dixon, William Rossander
Dugan, Charles Hammond
Dunn, John Robert
Eastwood, Thomas Alexander
Eaton, Donald Rex
Elias, Lorne
Filseth, Stephen V
Ford, Richard Westaway
Gilbert, John Barry
Glew, David Neville
Goodings, John Martin
Gordon, Robert Dixon
Gough, Sidney Roger
Graham, Michael John
Graydon, William Frederick
Greenstock, Clive Lewis
Guillet, James Edwin
Hackett, Peter Andrew
Hair, Michael L
Hammerli, Martin
Handa, Paul
Hanna, Geoffrey Chalmers
Harrison, Alexander George
Henry, Bryan Roger
Hines, William Grant
Hitchcock, Adam Percival
Holmes, John Leonard
Holtslander, William John
Howard, James Anthony
Hunter, Geoffrey
Ingold, Keith Usherwood
Ingraham, Thomas Robert
Ingruber, Otto Vincent
Irish, Donald Edward
Jackman, Thomas Edward
Jacobs, Patrick W M
James, Herbert I
Janzen, Edward George
Jones, Alister Vallance
Jones, Maurice Harry
Jones, William Ernest
Kapral, Raymond Edward
Kavassalis, Tom A
Kenney-Wallace, Geraldine Anne
Kilp, Toomas
King, Gerald Wilfrid
Klassen, Norman Victor
Klug, Dennis Dwayne
Koningstein, Johannes A
Kresge, Alexander Jerry
Kruus, Peeter
Kumaran, Mavinkal Kizhakkeveettil
Kuriakose, Areekattuthazhayil
Laidler, Keith James
Laposa, Joseph David
Leaist, Derek Gordon
Le Surf, Joseph Eric
Litvan, Gerard Gabriel
Lockwood, David John
Lorimer, John William
Lossing, Frederick Pettit
Loutfy, Rafik Omar
Lu, Wei-Kao
Lundell, O Robert
McAdie, Henry George
MacArthur, John Duncan
McCourt, Frederick Richard Wayne
Macdonald, Peter Moore
McGarvey, Bruce Ritchie
Mackay, Gervase Ian
McKenney, Donald Joseph
Mackiw, Vladimir Nicolaus
Maguire, Robert James
March, Raymond Evans
Marchessault, Robert Henri
Mayo, De Paul
Meath, William John
Moffat, John Blain
Morrow, Barry Albert
Moskovits, Martin
Moule, David
Murphy, William Frederick
Nesheim, Michael Ernest
Niki, Hiromi
Norris, A R
Oldham, Keith Bentley
Paraskevopoulos, George
Penner, Glenn H
Pietro, William Joseph
Platford, Robert Frederick
Polanyi, John Charles
Preston, Keith Foncell
Pritchard, Huw Owen
Ramachandran, Vangipuram S
Rand, Richard Peter

Reddoch, Allan Harvey
Reeves, Leonard Wallace
Ripmeester, John Adrian
Ross, Robert Anderson
Rumfeldt, Robert Clark
Rummery, Terrance Edward
Sadowski, Chester M
Sandler, Samuel
Sarkar, Bibudhendra
Sawicka, Barbara Danuta
Schiff, Harold Irvin
Schneider, William George
Sharp, James H
Shigeishi, Ronald A
Shurvell, Herbert Francis
Siebrand, Willem
Singleton, Donald Lee
Smith, Donald Reed
Southam, Frederick William
Stairs, Robert Ardagh
Stewart, James Allen
Sukava, Armas John
Symons, Edward Allan
Szabo, Arthur Gustav
Thode, Henry George
Tinker, David Owen
Tomlinson, Richard Howden
Wan, Jeffrey Kwok-Sing
Ware, William Romaine
Weir, Ronald Douglas
Werstiuk, Nick Henry
Westaway, Kenneth C
Whalley, Edward
Williams, Harry Leverne
Willis, Clive
Winnik, Francoise Martine
Wright, Maurice Morgan

PRINCE EDWARD ISLAND
Liu, Michael T H

QUEBEC
Alince, Bohumil
Allen, Lawrence Harvey
Allen, Sandra Lee
Andrews, Mark Paul
Cabana, Aldee
Claessens, Pierre
Colebrook, Lawrence David
Daoust, Hubert
Davis, Herbert John
Desnoyers, Jacques Edouard
Dorris, Gilles Marcel
Edward, John Thomas
Fliszar, Sandor
Gray, Derek Geoffrey
Grosser, Arthur Edward
Harrod, John Frank
Heitner, Cyril
Herman, Jan Aleksander
Holden, Harold William
Jabalpurwala, Kaizer E
Kimmerle, Frank
Lam, Vinh-Te
Lamarche, François
Lasia, Andrzej
Lau, Cheuk Kun
Leigh, Charles Henry
Leonard, Jacques Walter
LeRoy, Rodney Lash
Ouellet, Ludovic
Pezolet, Michel
Power, Joan F
Robertson, Alexander Allen
Sacher, Edward
Salahub, Dennis Russell
Sandorfy, Camille
Savoie, Rodrigue
Schreiber, Henry Peter
Sicotte, Yvon
Theophanides, Theophile
Thompson, Allan Lloyd
Van De Van, Theodorus Gertrudus Maria
Whitehead, Michael Anthony
Yong, Raymond N

SASKATCHEWAN
Bardwell, John Alexander Eddie
Barton, Richard J
Chandler, William David
Eager, Richard Livingston
Hontzeas, S
Knight, Arthur Robert
Kybett, Brian David
Larson, D Wayne
Lee, Donald Garry
Lowden, J Alexander
McCallum, Kenneth James
Rummens, F H A
Steer, Ronald Paul
Strathdee, Graeme Gilroy
VanCleave, Allan Bishop
Verrall, Ronald Ernest
Waltz, William Lee
Weil, John A

OTHER COUNTRIES
Antoniou, Antonios A
Arnell, John Carstairs
Balakrishnan, Narayana Swamy
Ballhausen, Lennart
Band, Yehuda Benzion
Bargon, Joachim

Barrett, Jerry Wayne
Bearman, Richard John
Belton, Geoffrey Richard
Berns, Donald Sheldon
Bess, Robert Carl
Bondybey, Vladimir E
Brown, Larry Robert
Buckingham, Amyand David
Bulkin, Bernard Joseph
Caron, Aimery Pierre
Carter, Robert Everett
Cartwright, Hugh Manning
Chang, Shuya
Chen, Chen-Tung Arthur
Collins, Carol Hollingworth
Collins, Kenneth Elmer
Delahay, Paul
Diederich, Francois Nico
Eland, John Hugh David
El-Awady, Abbas Abbas
El-Bayoumi, Mohamed Ashraf
Engelborghs, Yves
Ernst, Richard R
Forst, Wendell
Freeman, Raymond
French, William George
Friedel, Jacques
Grunwald, John J
Hall, Laurance David
Haymet, Anthony Douglas-John
Hayon, Elie M
Heckel, Edgar
Hiebert, Gordon Lee
Ilten, David Frederick
Ionescu, Lavinel G
Jortner, Joshua
Kan, Lou Sing
Kay, Kenneth George
Kendall, James Tyldesley
Klein, Ralph
Kordesch, Karl Victor
Levin, Gideon
Lingane, Peter James
McRae, Wayne A
Menon, Manchery Prabhakara
Mulas, Pablo Marcelo
Nowotny, Hans
Peatman, William Burling
Qian, Renyuan
Ree, Alexius Taikyue
Safran, Samuel A
Sarkisov, Pavel Jebraelovich
Saupe, Alfred (Otto)
Schlag, Edward William
Spacil, Henry Stephen
Spaght, Monroe Edward
Stoeber, Werner
Tanny, Gerald Brian
Tedmon, Craig Seward, Jr
Trofimov, Boris Alexandrovich
Turrell, George Charles

Polymer Chemistry

ALABAMA
Ayers, Orval Edwin
Beindorff, Arthur Baker
Byrd, James Dotson
Dean, David Lee
Fox, Sidney Walter
Galbraith, Ruth Legg
Galil, Fahmy
Hall, David Michael
Hazlegrove, Leven S
Kispert, Lowell Donald
Lewis, Danny Harve
Ludwick, Adriane Gurak
Nikles, David Eugene
Patterson, William Jerry
Penn, Benjamin Grant
Sayles, David Cyril
Tice, Thomas Robert
Urry, Dan Wesley
Walsh, William K
Wingard, Robert Eugene, Jr

ARIZONA
Adickes, H Wayne
Allen, James Durwood
Burke, William James
Carr, Clide Isom
Gardlund, Zachariah Gust
Hall, Henry Kingston, Jr
Heinle, Preston Joseph
Helbert, John N
Hunter, William Leslie
Kelley, Maurice Joseph
Kellman, Raymond
Kleinschmidt, Eric Walker
Ramohalli, Kumar Nanjunda Rao
Ratkowski, Donald J
Robertson, Frederick Noel
Valenty, Steven Jeffrey
Valenty, Vivian Briones
Vandenberg, Edwin James

ARKANSAS
Gosnell, Aubrey Brewer
Keplinger, Orin Clawson
Netherton, Lowell Edwin

CALIFORNIA
Adicoff, Arnold

Akawie, Richard Isidore
Aklonis, John Joseph
Akmal, Naim
Amis, Eric J
Athey, Robert Douglas, Jr
Bacskai, Robert
Baker, Richard William
Banigan, Thomas Franklin, Jr
Barasch, Werner
Bateman, John Hugh
Baur, Mario Elliott
Beck, Henry Nelson
Belmares, Hector
Bilow, Norman
Brauman, Sharon K(ruse)
Brown, Mark William
Byrd, Norman Robert
Cahill, Richard William
Carty, Daniel T
Casella, John Francis
Castro, Anthony J
Chatfield, Dale Alton
Chen, Timothy Shieh-Sheng
Chiang, Anne
Chou, Yungnien John
Colodny, Paul Charles
Cook, Robert Crossland
Cope, Oswald James
Cotts, David Bryan
Cotts, Patricia Metzger
Dahl, Klaus Joachim
Dawson, Daniel Joseph
Dawson, Kenneth Adrian
Denn, Morton M(ace)
Dhami, Kewal Singh
Dyball, Christopher John
Eichinger, Bruce Edward
Epstein, George
Even, William Roy, Jr
Feay, Darrell Charles
Fedors, Robert Francis
Feher, Frank J
Fohlen, George Marcel
Fredrickson, Glenn Harold
Fuhrman, Albert William
Gall, Walter George
Gaynor, Joseph
Giants, Thomas W
Glover, Leon Conrad, Jr
Golub, Morton Allan
Grieve, Catherine Macy
Grubbs, Charles Leslie
Han, Yuri Wha-Yul
Hass, Robert Henry
Haxo, Henry Emile, Jr
Hertzler, Barry Lee
Hess, Patrick Henry
Hoffman, Dennis Mark
Hogen-Esch, Thieo Eltjo
Hokama, Takeo
Hora, Maninder Singh
Horn, Christian Friedrich
Hsieh, You-Lo
Hsu, Ming-Ta Sung
Jacobs, Thomas Lloyd
Johanson, Robert Gail
Kalfayan, Sarkis Hagop
Kamp, David Allen
Kaneko, Thomas Motomi
Kasai, Paul Haruo
Kim, Chung Sul (Sue)
Kissel, Charles Louis
Kolyer, John M
Kong, Eric Siu-Wai
Kornfield, Julia Ann
Kratzer, Reinhold
Kurkov, Victor Peter
Landis, Abraham L
Landrum, Billy Frank
Lawson, Daniel David
Lawton, Emil Abraham
Leeg, Kenton J(ames)
Leeper, Harold Murray
Legge, Norman Reginald
Lewis, Robert Allen
Lewis, Sheldon Noah
Liberman, Martin Henry
Live, David H
Lorensen, Lyman Edward
Lovelace, Alan Mathieson
MacWilliams, Dalton Carson
Margerum, John David
Marks, Burton Stewart
Marrocco, Matthew Louis
Martin, Eugene Christopher
Meader, Arthur Lloyd, Jr
Menefee, Emory
Miller, Leroy Jesse
Molau, Gunther Erich
Molinari, Robert James
Needles, Howard Lee
Neidlinger, Hermann H
Newey, Herbert Alfred
Nikoui, Nik
Notley, Norman Thomas
Novak, Bruce Michael
Nozaki, Kenzie
Nyberg, David Dolph
O'Loane, James Kenneth
Ozari, Yehuda
Paciorek, Kazimiera J L
Patterson, Dennis Bruce
Percival, Douglas Franklin
Philipson, Joseph

Polymer Chemistry (cont)

Plamthottam, Sebastian S
Potter, Norman D
Preus, Martin William
Rabolt, John Francis
Reardon, Edward Joseph, Jr
Regulski, Thomas Walter
Reiss, Howard
Rhein, Robert Alden
Richter, Herbert Peter
Riley, Robert Lee
Ruetman, Sven Helmuth
Russell, Thomas Paul
Ryan, Patrick Walter
Salovey, Ronald
Saltman, William Mose
Saltonstall, Clarence William, Jr
Seiling, Alfred William
Sellas, James Thomas
Shinohara, Makoto
Shrontz, John William
Siegfried, William
Singh, Hakam
Skolnick, Jeffrey
Smith, Gary Eugene
Smith, Thor Lowe
Snyder, Robert Gene
Sovish, Richard Charles
Steinberg, Gunther
Sterling, Robert Fillmore
Sundberg, John Edwin
Szabo, Emery D
Taft, David Dakin
Tess, Roy William Henry
Theodorou, Doros Nicolas
Thomson, Tom Radford
Tilley, T(erry) Don
Toy, Madeline Shen
Tschoegl, Nicholas William
Tung, Lu Ho
Volksen, Willi
Wade, Charles Gordon
Wade, Robert Harold
Weber, Carl Joseph
Weber, William Palmer
Wensley, Charles Gelen
Wheeler, John C
Williams, Morris Evan
Winkler, DeLoss Emmet
Winston, Philip Eldridge, Jr
Wismer, Marco
Wolfe, James Frederick
Wood, David Wells
Woods, William George
Wrasidlo, Wolfgang Johann
Wudl, Fred
Yang, Yan-Bo
Yavrouian, Andre
Yee, Rena
Yeh, Gene Horng C
Yoon, Do Yeung
Youngman, Edward August
Zavarin, Eugene
Zeronian, Sarkis Haig
Ziegler, Carole L
Zimm, Bruno Hasbrouck

COLORADO

Bachman, Bonnie Jean Wilson
DoMinh, Thap
Druelinger, Melvin L
Dunn, Richard Lee
Frank, Arthur Jesse
Hermann, Allen Max
Lucas, George Bond
Mackinney, Herbert William
Martin, Charles R
Miller, Ivan Keith
Petersen, Joseph Claine
Sharma, Brahma Dutta
Stille, John Kenneth
Strauss, Eric L
Thompson, Ronald G
Voorhees, Kent Jay
Webb, John Day

CONNECTICUT

Arnold, John Richard
Baccei, Louis Joseph
Baker, Leonard Morton
Banucci, Eugene George
Baum, Bernard
Berdick, Murray
Bergen, Robert Ludlum, Jr
Castonguay, Richard Norman
Cooke, Theodore Frederic
Cornell, Robert Joseph
Coscia, Anthony Thomas
Dobay, Donald Gene
Drucker, Arnold
Easterbrook, Eliot Knights
Franks, Neal Edward
Goldberg, A Jon
Hauser, Martin
Hsu, Nelson Nae-Ching
Huang, Samuel J
Hwa, Jesse Chia Hsi
Johnson, Julian Frank
Klemarczyk, Philip Thaddeus
Koberstein, Jeffrey Thomas
Koch, Stanley D
Kontos, Emanuel G
Korin, Amos

Lamm, Foster Philip
Leavitt, Julian Jacob
Leonard, Robert F
Levine, Leon
Luh, Yuhshi
Malofsky, Bernard Miles
Meierhoefer, Alan W
Noland, James Sterling
O'Shea, Francis Xavier
Page, Edgar J(oseph)
Panzer, Hans Peter
Pellon, Joseph
Philips, Judson Christopher
Plessy, Boake Lucien
Powers, Joseph
Proverb, Robert Joseph
Rie, John E
Roberts, George P
Schile, Richard Douglas
Schmitt, Joseph Michael
Schrage, F Eugene
Schwier, Chris Edward
Scola, Daniel Anthony
Sedlak, John Andrew
Seefried, Carl G, Jr
Shaw, Montgomery Throop
Shine, William Morton
Sprague, G(eorge) Sidney
Steingiser, Samuel
Stevens, Malcolm Peter
Stuber, Fred A
Suen, T(zeng) J(iueq)
Tanaka, John
Ulrich, Henri
Wasser, Richard Barkman
Wellman, Russel Elmer
Wheeler, Edward Stubbs
Whelan, William Paul, Jr
White, Leroy Albert
Wolfram, Leszek January
Wuskell, Joseph P
Wystrach, Vernon Paul
Yang, Darchun Billy

DELAWARE

Abernathy, Henry Herman
Adelman, Robert Leonard
Akeley, David Francis
Ali-Khan, Ausat
Alms, Gregory Russell
Andersen, Donald Edward
Anderson, Burton Carl
Angelo, Rudolph J
Angerer, J(ohn) David
Armitage, John Brian
Austin, Paul Rolland
Bair, Thomas Irvin
Baird, Richard Leroy
Bannister, Robert Grimshaw
Barker, Harold Clinton
Barney, Arthur Livingston
Barth, Howard Gordon
Bauchwitz, Peter S
Baylor, Charles, Jr
Bercaw, James Robert
Beresniewicz, Aleksander
Boettcher, F Peter
Breazeale, Almut Frerichs
Brehm, Warren John
Breslow, David Samuel
Bristowe, William Warren
Brown, Morton
Brown, Robert Raymond
Brown, Stewart Cliff
Buck, Warren Howard
Burg, Marion
Cairncross, Allan
Calkins, William Harold
Caywood, Stanley William, Jr
Cella, Richard Joseph, Jr
Chiu, Jen
Chu, Sung Gun
Citron, Joel David
Class, Jay Bernard
Cluff, Edward Fuller
Collette, John Wilfred
Cone, Michael McKay
Cooper, Stuart L
Cooper, Terence Alfred
Corner, James Oliver
Craig, Alan Daniel
David, Israel A
Dawson, Robert Louis
DeBrunner, Marjorie R
DeDominicis, Alex John
Di Giacomo, Armand
Elliott, John Habersham
English, Alan Dale
Feinberg, Stewart Carl
Feltzin, Joseph
Ferguson, Raymond Craig
Fleming, Richard Allan
Flexman, Edmund A, Jr
Fogiel, Adolf W
Foldi, Andrew Peter
Ford, Thomas Aven
Foss, Robert Paul
Frankenburg, Peter Edgar
Franta, William Alfred
Frensdorff, H Karl
Funck, Dennis Light
Garrison, William Emmett, Jr
Gay, Frank P
Geer, Richard P

Gold, Harvey Saul
Goodman, Albert
Grot, Walther Gustav Fredrich
Guggenberger, Lloyd Joseph
Hardie, William George
Harmuth, Charles Moore
Harrell, Jerald Rice
Harris, John Ferguson, Jr
Hasty, Noel Marion
Hatchard, William Reginald
Held, Robert Paul
Hill, Frederick Burns, Jr
Hirsty, Sylvain Max
Hoegger, Erhard Fritz
Hoehn, Harvey Herbert
Hoeschele, Guenther Kurt
Honsberg, Wolfgang
Howard, Edward George, Jr
Howe, King Lau
Hsiao, Benjamin S
Hurford, Thomas Rowland
Hurley, William Joseph
Immediata, Tony Michael
Ittel, Steven Dale
Jackson, Harold Leonard
Kassal, Robert James
Katz, Manfred
Keating, James T
Keown, Robert William
Ketterer, Paul Anthony
Killian, Frederick Luther
Kitson, Robert Edward
Knipmeyer, Hubert Elmer
Kogon, Irving Charles
Kohan, Melvin Ira
Kolber, Harry John
Krespan, Carl George
Kwok, Wo Kong
Kwolek, Stephanie Louise
Landoll, Leo Michael
Lauzon, Rodrigue Vincent
Lazaridis, Christina Nicholson
Lewis, John Raymond
Libbey, William Jerry
Lin, Kuang-Farn
Lipscomb, Robert DeWald
Logothetis, Anestis Leonidas
Logullo, Francis Mark
Loken, Halvar Young
Longhi, Raymond
Longworth, Ruskin
Loomis, Gary Lee
Lukach, Carl Andrew
Lyons, Peter Francis
MacDonald, Robert Neal
McLain, Stephan James
McTigue, Frank Henry
Mahler, Walter
Maloney, Daniel Edwin
Manos, Philip
Markiewitz, Kenneth Helmut
Martin, Wayne Holderness
Matheson, Robert
Maynes, Gordon George
Meyer, James Melvin
Monagle, Daniel J
Moncure, Henry, Jr
Monroe, Bruce Malcolm
Moore, Ralph Bishop
Moser, Glenn Allen
Narvaez, Richard
Naylor, Marcus A, Jr
Neff, Bruce Lyle
Nelb, Gary William
Norling, Parry McWhinnie
Norton, Lilburn Lafayette
Ojakaar, Leo
Pailthorp, John Raymond
Pappas, Nicholas
Patterson, Gordon Derby, Jr
Pelosi, Lorenzo Fred
Pieski, Edwin Thomas
Plambeck, Louis, Jr
Podlas, Thomas Joseph
Putnam, Stearns Tyler
Quisenberry, Richard Keith
Rave, Terence William
Raynolds, Stuart
Reardon, Joseph Edward
Repka, Benjamin C
Restaino, Alfred Joseph
Rodriguez-Parada, Jose Manuel
Saegebarth, Klaus Arthur
Sammak, Emil George
Sauerbrunn, Robert Dewey
Schadt, Frank Leonard, III
Schaefgen, John Raymond
Scheiber, David Hitz
Schiewetz, D(on) B(oyd)
Schmiegel, Walter Werner
Schroeder, Herman Elbert
Schwartz, Jerome Lawrence
Shambelan, Charles
Shih, Hsiang
Shoaf, Charles Jefferson
Simms, John Alvin
Slade, Arthur Laird
Smart, Bruce Edmund
Smook, Malcolm Andrew
Sroog, Cyrus Efrem
Starkweather, Howard Warner, Jr
Sundet, Sherman Archie
Sweet, Arthur Thomas, Jr
Takvorian, Kenneth Bedrose

Tanikella, Murty Sundara Sitarama
Tanner, David
Tarney, Robert Edward
Tise, Frank P
Tuites, Donald Edgar
Un, Howard Ho-Wei
Van Dyk, John William
Van Gulick, Norman Martin
Van Trump, James Edmond
Vassallo, Donald Arthur
Vriesen, Calvin W
Wagner, Richard Lloyd
Wallenberger, Frederick Theodore
Walter, Henry Clement
Webb, Charles Alan
Webers, Vincent Joseph
Webster, Owen Wright
Wendt, Robert Charles
Whitehouse, Bruce Alan
Williams, Ebenezer David, Jr
Williams, Richard Anderson
Wilson, Frank Charles
Wu, Souheng
Zimmerman, Joseph
Zussman, Melvin Paul

DISTRICT OF COLUMBIA

Bultman, John D
Hirzy, John William
Horak, Vaclav
Keller, Teddy Monroe
Metz, Fred L
Powell, Justin Christopher
Smith, Leslie E
Stiehler, Robert D(aniel)
Wade, Clarence W R

FLORIDA

Alderson, Thomas
Bach, Hartwig C
Beatty, Charles Lee
Billica, Harry Robert
Black, William Bruce
Blossey, Erich Carl
Carraher, Charles Eugene, Jr
Chapman, Richard David
Christenson, Roger Morris
Clark, Harold Arthur
Cline, Edward Terry
Coke, C(hauncey) Eugene
Dannenberg, E(li) M
Dunbar, Richard Alan
Engel, John Hal, Jr
Friedlander, Herbert Norman
Frishman, Daniel
Goldberg, Eugene P
Goldring, Lionel Solomon
Goldstein, Arthur Murray
Gordon, Joseph R
Gutbezahl, Boris
Hammer, Clarence Frederick, Jr
Hanley, James Richard, Jr
Hicks, Elija Maxie, Jr
Hopkins, George C
Hsu, Tsong-Han
Jones, Walter H(arrison)
Katchman, Arthur
Kauder, Otto Samuel
Kenney, James Franklin
Kline, Gordon Mabey
Kraiman, Eugene Alfred
Kutner, Abraham
Levy, Richard
Mandelkern, Leo
Mattson, Guy C
Menyhert, William Robert
Moskowitz, Mark Lewis
Nelson, Gordon Leigh
Nersasian, Arthur
Newkome, George Richard
Owen, Gwilym Emyr, Jr
Parker, Earl Elmer
Polejes, J(acob) D
Pollard, Robert Eugene
Potts, James Edward
Remmel, Randall James
Rider, Don K(eith)
Ridgway, James Stratman
Rogan, John B
Russell, Charles Addison
Russell, James
Saunders, James Henry
Searle, Norma Zizmer
Seelbach, Charles William
Sharkey, William Henry
Silver, Frank Morris
Skeist, Irving
Slade, Philip Earl, Jr
Smith, William Mayo
Stahl, Joel S
Steadman, Thomas Ree
Ting, Robert Yen-Ying
Ucci, Pompelio Angelo
Weddleton, Richard Francis
Wei-Berk, Caroline
Willwerth, Lawrence James
Wittbecker, Emerson LaVerne
Xu, Renliang
Zimmer, William Frederick, Jr

GEORGIA

Allison, John P
Baxter, Gene Francis
Boschan, Robert Herschel

CHEMISTRY / 249

Bottomley, Lawrence Andrew
Chen, Chia Ming
Clendinning, Robert Andrew
Freese, William P, II
Garrett, David William
Gollob, Lawrence
Gupta, Rakesh Kumar
Habte-Mariam, Yitbarek
Hawkins, Walter Lincoln
Hurd, Phillip Wayne
Islam, M Safizul
Khan, Ishrat Mahmood
Knopka, W N
Krajca, Kenneth Edward
Legare, Richard J
McCarthy, Neil Justin, Jr
Marshall, Donald Irving
Morman, Michael T
Orofino, Thomas Allan
Pohl, Douglas George
Polk, Malcolm Benny
Ray, Apurba Kanti
Reif, Donald John
Rodriguez, Augusto
Rouvray, Dennis Henry
Samoilov, Sergey Michael
Samuels, Robert Joel
Stratton, Robert Alan
Tincher, Wayne Coleman
Tolbert, Laren Malcolm
Tsao, Mark Fu-Pao
Walters, John Philip
Weber, Robert Emil
Yost, Robert Stanley
Young, Raymond Hinchcliffe, Jr

HAWAII
Kemp, Paul James
Sood, Satya P

IDAHO
Lambuth, Alan Letcher
Renfrew, Malcolm MacKenzie

ILLINOIS
Albertson, Clarence E
Allen, John Kay
Alsberg, Henry
Anderson, Arnold Lynn
Barenberg, Sumner
Begala, Arthur James
Berman, Lawrence Uretz
Blaha, Eli William
Blanks, Robert F
Blitstein, John
Bonsignore, Patrick Vincent
Bradley, Steven Arthur
Bridgeford, Douglas Joseph
Burrell, Elliott Joseph, Jr
Carr, Lawrence John
Carrick, Wayne Lee
Chang, Franklin Shih Chuan
Chang, Shu-Pei
Chipman, Gary Russell
Cosper, David Russell
Damusis, Adolfas
DeBoer, Edward Dale
Didwania, Hanuman Prasad
Doane, William M
Donnelly, Thomas Henry
Empen, Joseph A
Falk, John Carl
Fanta, George Frederick
Freed, Karl F
Frey, David Allen
Gaylord, Richard J
Gehring, Harvey Thomas
Geil, Phillip H
Giori, Claudio
Glaser, Milton Arthur
Gordon, Gerald Arthur
Granick, Steve
Griffith, James H
Grotefend, Alan Charles
Gupta, Goutam
Gutfreund, Kurt
Haberfeld, Joseph Lennard
Henning, Lester Allan
Hoff, Raymond E
Huang, Shu-Jen Wu
Ika, Prasad Venkata
Irvin, Howard H
Isbister, Roger John
Johnson, Calvin Keith
Jones, Thomas Hubbard
Kester, Dennis Earl
Kharas, Gregory B
Khattab, Ghazi M A
Kissel, William John
Krajewski, John J
Lan, Ming-Jye
Lauterbach, Richard Thomas
Lemper, Anthony Louis
Lewis, Morton
Lillwitz, Lawrence Dale
Lindberg, Steven Edward
Lucas, Glennard Ralph
Lucchesi, Claude A
Magnus, George
Maher, George Garrison
Matthews, Clifford Norman
Meyers, Cal Yale
Miller, Sidney Israel
Mirabella, Francis Michael, Jr

Mirkin, Chad Alexander
Morello, Edwin Francis
Muzyczko, Thaddeus Marion
Myatt, Elizabeth Anne
Noren, Gerry Karl
Paschke, Edward Ernest
Reed, George W, Jr
Reily, William Singer
Rowe, Charles David
Sadler, George D
Sample, James Halverson
Schmukler, Seymour
Schweizer, Kenneth Steven
Scouten, Charles George
Shain, Albert Leopold
Shaneyfelt, Duane L
Sharp, Louis James, IV
Shriver, Duward F
Shulman, Sol
Sibbach, William Robert
Spangler, Charles William
Spitzer, William Carl
Stephens, James Regis
Sternberg, Shmuel Mookie
Stowell, James Kent
Stupp, Samuel Isaac
Su, Cheh-Jen
Thielen, Lawrence Eugene
Thompson, Arthur Robert
Tortorello, Anthony Joseph
Tuomi, Donald
Vandeberg, John Thomas
Wallace, Thomas Patrick
Wentworth, Gary
Weyna, Philip Leo
Whittington, Wesley Herbert
Wiedenmann, Lynn G
Wolf, Leslie Raymond
Wolf, Richard Eugene
Wolff, Thomas E
Woo, Lecon
Yokelson, Howard Bruce
Young, Austin Harry
Young, David W

INDIANA
Benford, Arthur E
Bolton, Benjamin A
Brewster, James Henry
Buchanan, Russell Allen
Butler, John Mann
Chandrasekaran, Rengaswami
Dubin, Paul Lee
Dyer, Rolla McIntyre, Jr
Felger, Maurice Monroe
Fidelle, Thomas Patrick
Fife, Wilmer Krafft
Goossens, John Charles
Jacobs, Martin Irwin
Kamat, Prashant V
Kelly, Walter James
Koller, Charles Richard
Kraemer, John Francis
Kreider, Leonard C(ale)
Kubiak, Clifford P
Liberti, Frank Nunzio
McGinness, James Donald
Miranda, Thomas Joseph
Morris, John F
Myerholtz, Ralph W, Jr
Park, Kinam
Peppas, Nikolaos Athanassiou
Quinn, Michael H
Sannella, Joseph L
Schroeder, Juel Pierre
Schwan, Theodore Carl
Sybert, Paul Dean
Thomas, John Kerry
Trischler, Floyd D
Wang, Jin-Liang
Williams, Edward James

IOWA
Benton, Kenneth Curtis
Dordick, Jonathan Seth
Hauenstein, Jack David
Rethwisch, David Gerard
Rider, Paul Edward, Sr

KANSAS
Breed, Laurence Woods
Chappelow, Cecil Clendis, Jr
Macke, Gerald Fred
Scherrer, Joseph Henry
Wiley, James C, Jr

KENTUCKY
Bean, C Thomas, Jr
Buchanan, Robert Martin
Coe, Gordon Randolph
Crim, Gary Allen
Glenn, Furman Eugene
Graver, Richard Byrd
Lauterbach, John Harvey
Lipscomb, Nathan Thornton
Masters, John Edward
Price, Martin Burton
Schulz, William
Shank, Charles Philip
Smith, Terry Edward
Smith, Walter Thomas, Jr
Tabibian, Richard
Watkins, Nancy Chapman

LOUISIANA
Arthur, Jett Clinton, Jr
Beckman, Joseph Alfred
Blouin, Florine Alice
Carpenter, Dewey Kenneth
Chrastil, Joseph
Daly, William Howard
Fischer, David John
Franklin, William Elwood
Frazier, Claude Clinton, III
Haefner, A(lbert) J(ohn)
Hanlon, Thomas Lee
Harper, Robert John, Jr
Hornbaker, Edwin Dale
LeBlanc, Robert Bruce
Lee, Burnell
Lee, John Yuchu
Li, Hsueh Ming
Looney, Ralph William
McGraw, Gerald Wayne
Magin, Ralph Walter
Penton, Harold Roy, Jr
Samuels, Martin E(lmer)
Shih, Frederick F
Theriot, Kevin Jude
Vickroy, Virgil Vester, Jr
Vigo, Tyrone Lawrence
Wan, Peter J
Welch, Clark Moore
Wickson, Edward James
Zagar, Walter T
Zaweski, Edward F
Zietz, Joseph R, Jr

MAINE
Lepoutre, Pierre
Tedeschi, Robert James
Thompson, Edward Valentine
Wilson, Thomas Lee

MARYLAND
Adolph, Horst Guenter
Bailey, William John
Bicknell-Brown, Ellen
Block, Ira
Bosmajian, George
Bowen, Rafael Lee
Boyce, Richard Joseph
Braatz, James Anthony
Brady, Robert Frederick, Jr
Brauer, Gerhard Max
Bruck, Stephen Desiderius
Callahan, John Joseph
Chang, Shu-Sing
Chiang, Chwan K
Choi, Kyu-Yong
Christensen, Richard G
Clark, Joseph E(dward)
Davis, George Thomas
Dillon, James Greenslade
Ferington, Thomas Edwin
Florin, Roland Eric
Flynn, Joseph Henry
Fox, Robert Bernard
Fulmer, Glenn Elton
Garroway, Allen N
Gause, Evelyn Pauline
Goldstein, Jorge Alberto
Grant, Warren Herbert
Greer, Sandra Charlene
Guthrie, James Leverette
Guttman, Charles M
Hadermann, Albert Felix
Hamilton, William Lander
Han, Charles Chih-Chao
Hoguet, Robert Gerard
Horowitz, Emanuel
Hunston, Donald Lee
Ibrahim, A Mahammad
Jernigan, Robert Lee
Kasianowicz, John James
Ketley, Arthur Donald
McElroy, Wilbur Renfrew
Marqusee, Jeffrey Alan
Morgan, Charles Robert
Moses, Saul
Muller, Robert E
Murch, Robert Matthews
Muser, Marc
Raksis, Joseph W
Raskin, Betty Lou
Sheppard, Norman F, Jr
Silverman, Joseph
Stromberg, Robert Remson
Tolgyesi, Eva
Tung, Ming Sung
Vandegaer, Jan Edmond
Wagner, Herman Leon
Wang, Francis Wei-Yu
Warfield, Robert Welmore
Werber, Frank Xavier
Wood, Lawrence Arnell
Wu, Wen-Li

MASSACHUSETTS
Ackerman, Jerome Leonard
Archer, Ronald Dean
Auer, Henry Ernest
Basdekis, Costas H
Becker, Ernest I
Bender, Howard Sanford
Blumstein, Alexandre
Blumstein, Rita Blattberg
Brack, Karl

Bragole, Robert A
Browne, Sheila Ewing
Bush, John Burchard, Jr
Carton, Edwin Beck
Cembrola, Robert John
Chu, Nori Yaw-Chyuan
Chung, Frank H
Clough, Stuart Benjamin
Coffin, Perley Andrews
Cohen, Fredric Sumner
Cohen, Merrill
Cohen, Robert Edward
Cohen, Saul Mark
Coplan, Myron J
Corey, Albert Eugene
Cotten, George Richard
Coury, Arthur Joseph
Dahm, Donald J
Dale, William
Das, Pradip Kumar
Dasgupta, Arijit M
David, Donald J
Davis, Robert Bernard
Deanin, Rudolph D
Deets, Gary Lee
Dickinson, Leonard Charles
Dixon, Brian Gilbert
Doshi, Anil G
Druy, Mark Arnold
Ezrin, Myer
Fessler, William Andrew
Fetscher, Charles Arthur
Fettes, Edward Mackay
Fritzsche, Alfred Keith
Gardner, Donald Murray
Garrett, Paul Daniel
Goldberg, Gershon Morton
Grasshoff, Jurgen Michael
Greif, Mortimer
Guilbault, Lawrence James
Haas, Howard Clyde
Hagnauer, Gary Lee
Hiatt, Norman Arthur
Hill, Loren Wallace
Hodes, William
Hodgdon, Russell Bates
Hoover, M Frederick
Huang, Jan-Chan
Illinger, Joyce Lefever
Israel, Stanley C
Jones, Alan A
Kafrawy, Adel
Kantor, Simon William
Kingston, George C
Knapczyk, Jerome Walter
Kopf, Peter W
Kramer, Charles Edwin
Langley, Kenneth Hall
Lapkin, Milton
Lavin, Edward
Lee, Eric Kin-Lam
Lee, Kang In
Lenz, Robert William
Lerman, Leonard Solomon
Lillya, Clifford Peter
Macaione, Domenic Paul
Mac Knight, William John
Macnair, Richard Nelson
Magarian, Charles Aram
Marinaccio, Paul J
Martin, Richard Hadley, Jr
Medalia, Avrom Izak
Mehta, Mahendra
Merken, Henry
Milgrom, Jack
Moniz, William B
Mont, George Edward
Mueller, Robert Kirk
Muthukumar, Murugappan
Naidus, Harold
Nemeth, Ronald Louis
Nickerson, Richard G
O'Rell, Dennis Dee
Padwa, Allen Robert
Pandya, Ashish
Parham, Marc Ellous
Peng, Fred Ming-Sheng
Perkins, Janet Sanford
Phillips, Philip Wirth
Phillips, Yorke Peter
Porter, Roger Stephen
Powers, Donald Howard, Jr
Pyun, Chong Wha
Rao, Devulapalli V G L N
Refojo, Miguel Fernandez
Reisman, Abraham Joseph
Rossitto, Conrad
Salamone, Joseph C
Sawan, Samuel Paul
Schlaikjer, Carl Roger
Serlin, Irving
Serra, Jerry M
Seyferth, Dietmar
Shuford, Richard Joseph
Singler, Robert Edward
Smith, Donald Ross
Smith, William Edward
Sonnichsen, Harold Marvin
Spangler, Lora Lee
Stein, Richard Stephen
Sullivan, Charles Irving
Taylor, Lloyd David
Temin, Samuel Cantor
Torchilin, Vladimir Petrovich

Polymer Chemistry (cont)

Tripathy, Sukant K
Tsuk, Andrew George
Turnquist, Carl Richard
Udipi, Kishore
Van der Burg, Sjirk
Vartanian, Leo
Waack, Richard
Watterson, Arthur C, Jr
Wentworth, Stanley Earl
Wheelock, Kenneth Steven
Whitesides, George McClelland
Williams, David John
Zavisza, Daniel Maxmillian

MICHIGAN
Abu-Isa, Ismat Ali
Allan, David
Anders, Oswald Ulrich
Anderson, John Norton
Arends, Charles Bradford
Arrington, Jack Phillip
Bailey, Donald Leroy
Bajzer, William Xavier
Baney, Ronald Howard
Bauer, David Robert
Betso, Stephen Richard
Bhatt, Padmamabh P
Boyer, Raymond Foster
Bremmer, Bart J
Brewer, George E(ugene) F(rancis)
Brown, William Bernard
Chattha, Mohinder Singh
Clark, Peter David
Cobler, John George
Davidovits, Joseph
Dearlove, Thomas John
Dennis, Kent Seddens
Dewitt, Sheila Hobbs
Dickie, Ray Alexander
Dorman, Linneaus Cuthbert
Doyle, Daryl Joseph
Dreyfuss, M(ax) Peter
Dreyfuss, Patricia
Eichhorn, J(acob)
Elias, Hans Georg
Eliezer, Naomi
Ellis, Thomas Stephen
Fuentes, Ricardo, Jr
Gallagher, James A
Gardon, John Leslie
Garrett, David L, Jr
Graham, John
Gulari, Esin
Harris, Paul Jonathan
Heeschen, Jerry Parker
Heins, Conrad F
Herk, Leonard Frank
Heyman, Duane Allan
Hinkamp, Paul Eugene, (II)
Holubka, Joseph Walter
Howell, Bob A
Jackson, Larry Lynn
Jarvis, Lactance Aubrey
Johnston, Norman Wilson
Jones, Frank Norton
Kang, Uan Gen
Keinath, Steven Ernest
Killgoar, Paul Charles, Jr
Kim, Yungki
Klempner, Daniel
Koster, Robert Allen
Kresta, Jiri Erik
Krimm, Samuel
Labana, Santokh Singh
Langley, Neal Roger
Larsen, Eric Russell
Laverty, John Joseph
Lee, Do Ik
Lee, Wei-Ming
Li, Chi
McDonald, Charles Joseph
McKeever, L Dennis
McLean, John A, Jr
Malchick, Sherwin Paul
Mansfield, Marc L
Massingill, John Lee, Jr
Meier, Dale Joseph
Meister, John J
Miller, Frederick Arnold
Miller, Robert Llewellyn
Moore, Carl
Morgan, Roger John
Narayan, Ramani
Narayan, Tv Lakshmi
Newman, Seymour
Nordstrom, J David
Overberger, Charles Gilbert
Patton, John Thomas, Jr
Peters, James
Platt, Alan Edward
Prakash, D Kash
Pynnonen, Bruce W
Rasmussen, Paul G
Reynard, Kennard Anthony
Rutter, Edward Walter, Jr
Ryntz, Rose A
Sarkar, Nitis
Saunders, Frank Linwood
Schlick, Shulamith
Schmidt, Donald L
Schneider, John Arthur
Schwartz, Shirley E

Skochdopole, Richard E
Smith, Arthur Gerald
Smith, Harry Andrew
Snyder, Robert
Solc, Karel
Stark, Forrest Otto
Stevens, Violete L
Swarin, Stephen John
Turner, Robert Lawrence
Ullman, Robert
Walles, Wilhelm Egbert
Weiss, Philip
Wessling, Ritchie A
Weyenberg, Donald Richard
Willermet, Pierre Andre
Williams, Fredrick David
Williamson, Jerry Robert
Wims, Andrew Montgomery
Wyzgoski, Michael Gary
Yee, Albert Fan
Zand, Robert

MINNESOTA
Abere, Joseph Francis
Agre, Courtland LeVerne
Andrus, Milton Henry, Jr
Babu, Gaddam N
Benham, Judith Laureen
Cook, Jack E
Diedrich, James Loren
Eian, Gilbert Lee
Erickson, Randall L
Esmay, Donald Levern
Farber, Elliott
Farm, Raymond John
Fick, Herbert John
Forester, Ralph H
Gobran, Ramsis
Goken, Garold Lee
Hammar, Walton James
Hendricks, James Owen
Hill, Brian Kellogg
Hogle, Donald Hugh
Holmen, Reynold Emanuel
Johnson, Bryce Vincent
Johnson, Lennart Ingemar
Johnston, Manley Roderick
Kovacic, Joseph Edward
Kropp, James Edward
Kuder, Robert Clarence
Lai, Juey Hong
Langager, Bruce Allen
Larson, Eric George
Li, Wu-Shyong
Lockwood, Robert Greening
Lodge, Timothy Patrick
Lu, Shih-Lai
Lucast, Donald Hurrell
Miller, Wilmer Glenn
Penney, William Harry
Pletcher, Wayne Albert
Prager, Julianne Heller
Punderson, John Oliver
Robins, Janis
Roska, Fred James
Sandberg, Carl Lorens
Sherman, Patsy O'Connell
Smith, Samuel
Snell, John B
Snyder, William Richard
Talbott, Richard Lloyd
Tamsky, Morgan Jerome
Tiers, George Van Dyke
Wear, Robert Lee
Weiss, Douglas Eugene
Wen, Richard Yutze
Williams, Todd Robertson
Wollner, Thomas Edward
Wright, Charles Dean

MISSISSIPPI
McCormick, Charles Lewis, III
Mathias, Lon Jay
Pandey, Ras Bihari
Pittman, Charles U, Jr
Pojman, John Anthony
Seymour, Raymond B(enedict)
Thames, Shelby Freland

MISSOURI
Ames, Donald Paul
Anagnostopoulos, Constantine E
Asrar, Jawed
Bechtle, Gerald Francis
Blum, Frank D
Carter, Don E
Conradi, Mark Stephen
Craver, Clara Diddle (Smith)
Eime, Lester Oscar
Frushour, Bruce George
Gadberry, Howard M(ilton)
Greenley, Robert Z
Harland, Ronald Scott
Holtzer, Marilyn Emerson
Katti, Kattesh V
Keller, William John
Klein, Andrew John
Koenig, Karl E
Ku, Audrey Yeh
Kurz, James Eckhardt
Lennon, Patrick James
Leventis, Nicholas
Levy, Ram Leon
Markowski, Henry Joseph

Millich, Frank
Munch, John Howard
Sears, J Kern
Seeley, Robert D
Slocombe, Robert Jackson
Smith, Robert Eugene
Snyder, Joseph Quincy
Sonnino, Carlo Benvenuto
Stary, Frank Edward
Stoffer, James Osber
Stout, Edward Irvin
Thach, Robert Edwards
Tonkyn, Richard George
Twardowski, Zbylut Jozef
Wilbur, James Myers, Jr
Wohl, Martin H
Woodbrey, James C
Yasuda, Hirotsugu

NEBRASKA
Garey, Carroll Laverne
Schadt, Randall James

NEVADA
Borders, Alvin Marshall
Dix, James Seward
Herzlich, Harold Joel

NEW HAMPSHIRE
Bertozzi, Eugene R
Cotter, Robert James
Donaruma, L Guy
Finn, James Walter
Gorham, William Franklin
Greenwald, Harold Leopold
Ort, Morris Richard
Petrasek, Emil John
Strazdins, Edward
Wang, Victor S F
Webb, Richard Lansing

NEW JERSEY
Aharoni, Shaul Moshe
Akkapeddi, Murali Krishna
Amundson, Karl Raymond
Arbuckle, Georgia Ann
Arendt, Volker Dietrich
Ariemma, Sidney
Armbrecht, Frank Maurice, Jr
Armbruster, David Charles
Ashcraft, Arnold Clifton, Jr
Auerbach, Andrew Bernard
Auerbach, Victor
Bair, Harvey Edward
Baker, William Oliver
Barclay, Robert, Jr
Bardoliwalla, Dinshaw Framroze
Barnabeo, Austin Emidio
Baum, Gerald A(llan)
Bender, Howard L
Berardinelli, Frank Michael
Berenbaum, Morris Benjamin
Bertelo, Christopher Anthony
Besso, Michael M
Bezwada, Rao Srinivasa
Bhattacharjee, Himangshu Ranjan
Blank, Zvi
Bouboulis, Constantine Joseph
Bovey, Frank Alden
Bowden, Murrae John Stanley
Bowen, J Hartley, Jr
Brine, Charles James
Burton, Gilbert W
Cais, Rudolf Edmund
Canterino, Peter John
Chan, Maureen Gillen
Chance, Ronald Richard
Chandrasekhar, Prasanna
Chandross, Edwin A
Chapas, Richard Bernard
Charbonneau, Larry Francis
Chen, Catherine S H
Chen, Shuhchung Steve
Chen, William Kwo-Wei
Chenicek, Albert George
Chiao, Wen Bin
Chiao, Yu-Chih
Chimes, Daniel
Chlanda, Frederick P
Cipriani, Cipriano
Cohen, Abraham Bernard
Conger, Robert Perrigo
Cooke, Robert Sanderson
Curry, Michael Joseph
Davies, Richard Edgar
DeBona, Bruce Todd
Delmonte, David William
Deshpande, Achyut Bhalchandra
Douglass, D C
Dyer, John
Eastman, David Willard
Eby, John Martin
Eckler, Paul Eugene
Erenrich, Eric Howard
Evers, William L
Fahrenholtz, Susan Roseno
Farrington, Thomas Allan
Fetters, Lewis
Filas, Robert William
Firth, William Charles, Jr
Fishman, David H
Fong, Jones W
Forbes, Charles Edward
Fredericks, Robert Joseph

Gander, Robert Johns
Gardiner, John Brooke
Gaughan, Roger Grant
Gaylord, Norman Grant
Giacobbe, Thomas Joseph
Giddings, Sydney Arthur
Gladstone, Harold Maurice
Gold, Daniel Howard
Goldblatt, Irwin Leonard
Goldstein, Albert
Gorbaty, Martin Leo
Green, Joseph
Greer, Albert H
Griskey, R(ichard) G(eorge)
Grubman, Wallace Karl
Hager, Douglas Francis
Hale, Warren Frederick
Hammond, Willis Burdette
Hansen, Ralph Holm
Hanson, Harry Thomas
Hanson, James Edward
Harpell, Gary Allan
Hartkopf, Arleigh Van
Helfand, Eugene
Herdklotz, John Key
Herman, Daniel Francis
Hirsch, Arthur
Hoffman, Henry Tice, Jr
Hort, Eugene Victor
Hulyalkar, Ramchandra K
Humphreys, Robert William Riley
Hundert, Murray Bernard
Hung, Paul Ling-Kong
Isaacson, Robert B
Jackson, Thomas A J
Jaffe, Michael
Jirgensons, Arnold
Johnston, Christian William
Johnston, John Eric
Kaback, Stuart Mark
Kamath, Yashavanth Katapady
Kaplan, Michael
Karl, Curtis Lee
Karol, Frederick J
Keaveney, William Patrick
Kelley, Joseph Matthew
Kellgren, John
Keogh, Michael John
Kirshenbaum, Gerald Steven
Kolc, Jaroslav Jerry F
Kreeger, Russell Lowell
Kresge, Edward Nathan
Kronenthal, Richard Leonard
Kruh, Daniel
Kuntz, Irving
Kveglis, Albert Andrew
Larkin, William Albert
Larson, Eric Heath
Lasky, Jack Samuel
Leal, Joseph Rogers
Lem, Kwok Wai
Levi, David Winterton
Levine, Harry
Levy, Alan
Li, Hsin Lang
Loan, Leonard Donald
Login, Robert Bernard
Lohse, David John
Lorenz, Donald H
Lovinger, Andrew Joseph
Lowell, A(rthur) I(rwin)
Lundberg, Robert Dean
Lyding, Arthur R
McGary, Charles Wesley, Jr
McGinnis, James Lee
Magder, Jules
Manning, Gerald Stuart
Marder, Herman Lowell
Matsuoka, Shiro
Matthews, Demetreos Nestor
Matzner, Markus
Melveger, Alvin Joseph
Meyer, Victor Bernard
Mikes, John Andrew
Miller, Albert Thomas
Miskel, John Joseph, Jr
Montana, Anthony J
Moroni, Antonio
Muldrow, Charles Norment, Jr
Munger, Stanley H(iram)
Munsell, Monroe Wallwork
Nguyen, Hien Vu
Noshay, Allen
O'Malley, James Joseph
Oppenheimer, Larry Eric
Osuch, Christopher Erion
Otterburn, Michael Storey
Pacansky, Thomas John
Palermo, Felice Charles
Pappalardo, Leonard Thomas
Pappas, S Peter
Parisek, Charles Bruce
Perkel, Robert Jules
Perry, Donald Dunham
Peters, Timothy
Prapas, Aristotle G
Pugliese, Michael
Quan, Xina Shu-Wen
Quarles, Richard Wingfield
Ramharack, Roopram
Ray-Chaudhuri, Dilip K
Raymond, Maurice A
Register, Richard Alan
Reich, Leo

Reich, Murray H
Reichmanis, Elsa
Reilly, Eugene Patrick
Riccobono, Paul Xavier
Riley, David Waegar
Roberts, William John
Robins, Jack
Roper, Robert
Rosen, Marvin
Rosenthal, Arnold Joseph
Rossi, Robert Daniel
Sacher, Alex
Sacks, William
Saferstein, Lowell G
Salvesen, Robert H
Saxon, Robert
Scanley, Clyde Stephen
Schmidle, Claude Joseph
Schmitt, George Joseph
Schonfeld, Edward
Schulz, Donald Norman
Schwab, Frederick Charles
Schwartzkopf, George, Jr
Schwenker, Robert Frederick, Jr
Sekutowski, Dennis G
Shah, Hasmukh N
Shechter, Leon
Shih, Jenn-Shyong
Sibilia, John Philip
Sidi, Henri
Silver, Frederick Howard
Skoultchi, Martin Milton
Slocum, Donald Hillman
Smyers, William Hays
Song, Jung Hyun
Song, Won-Ryul
Sonn, George Frank
Sorenson, Wayne Richard
Spitsbergen, James Clifford
Spurr, Orson Kirk, Jr
Steiger, Fred Harold
Stone, Edward
Strauss, George
Sugathan, Kenneth Kochappan
Sulzberg, Theodore
Szymanski, Chester Dominic
Takahashi, Akio
Tamarelli, Alan Wayne
Tatyrek, Alfred Frank
Thich, John Adong
Thompson, Larry Flack
Tillson, Henry Charles
Tornqvist, Erik Gustav Markus
Townsend, Palmer W
Van de Castle, John F
Ver Strate, Gary William
Vincent, Gerald Glenn
Vukov, Rastko
Wang, Tsuey Tang
Wasserman, David
Watt, William Russell
Weintraub, Lester
Whelan, John Michael
Williams, Bernard Leo
Winslow, Field Howard
Wissbrun, Kurt Falke
Wu, Tse Cheng
Yourtee, John Boteler
Zaim, Semih
Zoss, Abraham Oscar

NEW MEXICO
Arnold, Charles, Jr
Assink, Roger Alyn
Auerbach, Irving
Benicewicz, Brian Chester
Benziger, Theodore Michell
Cahill, Paul A
Gillen, Kenneth Todd
Liepins, Raimond
Martin, James Ellis
Nevin, Robert Stephen
Nielsen, Stuart Dee
Ogilby, Peter Remsen
Park, Su-Moon
Schaefer, Dale Wesley
Trujillo, Ralph Eusebio
Van Deusen, Richard L
Wicks, Zeno W, Jr
Williams, Joel Mann, Jr
Young, Ainslie Thomas, Jr
Zeigler, John Martin

NEW YORK
Abate, Kenneth
Adduci, Jerry M
Adelstein, Peter Z
Alliet, David F
Altscher, Siegfried
Amborski, Leonard Edward
Arcesi, Joseph A
Bahary, William S
Baron, Arthur L
Barr, Donald Eugene
Benham, Craig John
Benzinger, James Robert
Benzinger, William Donald
Berdahl, Donald Richard
Bettelheim, Frederick A
Beyer, George Leidy
Bluestein, Ben Alfred
Bluhm, Terry Lee
Brust, David Philip
Cabasso, Israel

Campbell, Gerald Allan
Campbell, Gregory August
Carnahan, James Claude
Cathcart, John Almon
Ceprini, Mario Q
Chen, Cindy Chei-Jen
Chen, Shaw-Horng
Chen, Tsang Jan
Chow, Che Chung
Ciccarelli, Roger N
Craig, Daniel
Dagostino, Vincent F
DeCrosta, Edward Francis, Jr
Dehn, Joseph William, Jr
Doetschman, David Charles
Dowell, Flonnie
Earing, Mason Humphry
Edelman, Robert
Feger, Claudius
Felty, Evan J
Fox, Adrian Samuel
Frechet, Jean M J
Frisch, Harry Lloyd
Frye, Robert Bruce
Fuerniss, Stephen Joseph
Gardella, Joseph Augustus, Jr
Gardner, Sylvia Alice
Gardner, William Howlett
George, Philip Donald
Giannelis, Emmanuel P
Gladstone, Matthew Theodore
Godsay, Madhu
Green, Mark M
Grina, Larry Dale
Gruenbaum, William Tod
Hamb, Fredrick Lynn
Hargitay, Bartholomew
Harris, Robert L
Harrison, James Beckman
Helfgott, Cecil
Henry, Arnold William
Hepfinger, Norbert Francis
Herdle, Lloyd Emerson
Hindersinn, Raymond Richard
Horowitz, Carl
Immergut, Edmund H(einz)
Imperial, George Romero
Indictor, Norman
Isaacson, Henry Verschay
Jacobson, Homer
Jelinski, Lynn W
Jelling, Murray
Jenekhe, Samson A
Johansson, Sune
Johnson, Bruce Fletcher
Jones, Wayne E
Julinao, Peter C
Kagansky, Larisa
Kamath, Vasanth Rathnakar
Kaplan, Mark Steven
Khan, Jamil Akber
Khanna, Ravi
Kim, Ki-Soo
Kiss, Klara
Klein, Gerald Wayne
Klijanowicz, James Edward
Kolb, Frederick J(ohn), Jr
Kosky, Philip George
Krause, Sonja
Kronstein, Max
Kross, Robert David
Kumler, Philip L
Kwei, Ti-Kang
Labianca, Dominick A
Laganis, Deno
Lai, Yu-Chin
Lau, Philip T S
Launer, Philip Jules
Lee, Lieng-Huang
Lhila, Ramesh Chand
Limburg, William W
Lynn, Merrill
MacLaury, Michael Risley
Mansfield, Kevin Thomas
Markovitz, Mark
Martellock, Arthur Carl
Mason, John Hugh
Massa, Dennis Jon
Mathes, Kenneth Natt
Meinwald, Yvonne Chu
Mendel, John Richard
Mermelstein, Robert
Metz, Donald J
Meyer, John Austin
Mijovic, Jovan
Miller, William Taylor
Mirviss, Stanley Burton
Mizes, Howard Albert
Model, Frank Steven
Monahan, Alan Richard
Moore, James Alfred
Morduchowitz, Abraham
Nelson, Robert Andrew
Ober, Christopher Kemper
Odian, George G
Orlando, Charles M
Otocka, Edward Paul
Palladino, William Joseph
Partch, Richard Earl
Pavlisko, Joseph Anthony
Pearce, Eli M
Pearson, James Murray
Pemrick, Raymond Edward
Penn, Lynn Sharon

Petersen, Kenneth C
Petrie, Sarah Elaine
Petropoulos, Constantine Chris
Pillar, Walter Oscar
Pollart, Dale Flavian
Post, Howard William
Potter, George Henry
Pucknat, John Godfrey
Rebel, William J
Relles, Howard
Rice, David E
Riecke, Edgar Eric K
Roedel, George Frederick
Rosenfeld, Jerold Charles
Rubin, Isaac D
Sabia, Raffaele
Sandhu, Mohammad Akram
Sarko, Anatole
Schuerch, Conrad
Seltzer, Raymond
Shafer, Sheldon Jay
Shannon, Frederick Dale
Shaw, Richard Gregg
Silleck, Clarence Frederick
Smid, Johannes
Smith, Donald Arthur
Smith, Thomas Woods
Sogah, Dotsevi Yao
Sorkin, Howard
Staples, Jon T
Stein, Richard James
Sterman, Melvin David
Sterman, Samuel
Stoner, George Green
Sutherland, Judith Elliott
Tan, Julia S
Teegarden, David Morrison
Tesoro, Giuliana C
Thomas, Harold Todd
Totten, George Edward
Triplett, Kelly B
Tuites, Richard Clarence
Van Norman, Gilden Ramon
Vogl, Otto
Waltcher, Irving
Ward, Frank Kernan
Weiss, Jonas
White, Dwain Montgomery
Winter, Roland Arthur Edwin
Winter, William Thomas
Wolinski, Leon Edward
Wolny, Friedrich Franz
Woodward, Arthur Eugene
Wyman, Donald Paul
Xu, Zhenchun
Xue, Yongpeng
Yudelson, Joseph Samuel
Yunick, Robert P
Zalewski, Edmund Joseph

NORTH CAROLINA
Abbey, Kathleen Mary Kyburz
Abbey, Kirk J
Ambrose, Richard Joseph
Andrady, Anthony Lakshman
Barborak, James Carl
Bassett, David R
Bobalek, Edward G(eorge)
Bryan, Carl Eddington
Burrus, Robert Tilden
Cates, David Marshall
Clark, Howard Garmany
Cuculo, John Anthony
Drechsel, Paul David
Evans, Evan Franklin
Farona, Michael F
Farrissey, William Joseph, Jr
Gilbert, Richard Dean
Goldstein, Irving Solomon
Groszos, Stephen Joseph
Hamner, William Frederick
Hersh, Solomon Philip
Holzwarth, George Michael
Hsieh, Henry Lien
Huffman, K(enneth) Robert
Hurwitz, Melvin David
Johnston, John
Jones, Rufus Sidney
Jones, William Jonas, Jr
King, Henry Lee
Kusy, Robert Peter
Ledbetter, Harvey Don
Loeppert, Richard Henry
McKay, Jerry Bruce
McPeters, Arnold Lawrence
Magat, Eugene Edward
Majewski, Theodore E
Miller, Walter Peter
Murray, Royce Wilton
Nash, James Lewis, Jr
Powers, Edward James
Preston, Jack
Rapoport, Lorence
Redmond, John Peter
Reese, Cecil Everett
Rochow, Theodore George
Roth, Roy William
Samulski, Edward Thaddeus
Sargeant, Peter Barry
Schell, William John
Squibb, Samuel Dexter
Stannett, Vivian Thomas
Stejskal, Edward Otto
Theil, Michael Herbert

Tonelli, Alan Edward
Tucker, Paul Arthur
Walsh, Edward John
Williams, Joel Lawson
Wooten, Willis Carl, Jr
Work, Robert Wyllie

NORTH DAKOTA
Adams, David George
Bierwagen, Gordon Paul

OHIO
Abraham, Tonson
Adams, Harold Elwood
Aggarwal, Sundar Lal
Ambler, Michael Ray
Ampulski, Robert Stanley
Bacon, William Edward
Baczek, Stanley Karl
Baer, Eric
Ball, Lawrence Ernest
Baranwal, Krishna Chandra
Barlow, Anthony
Barsky, Constance Kay
Bauer, Richard G
Baughman, Glenn Laverne
Beres, John Joseph
Berger, Mitchell Harvey
Bertsch, Robert Joseph
Bethea, Tristram Walker, III
Bond, William Bradford
Bonner, David Calhoun
Borgnaes, Dan
Bradley, Ronald W
Brittain, William Joseph
Calderon, Nissim
Campbell, Robert Wayne
Carman, Charles Jerry
Carpenter, William Graham
Carter, R Owen, Jr
Causa, Alfredo G
Cheng, Stephen Zheng Di
Cinadr, Bernard F(rank)
Coleman, John Franklin
Coleman, Lester Earl, (Jr)
Collins, Edward A
Conley, Robert T
Costanza, Albert James
Covitch, Michael J
Cunningham, Robert Elwin
Delvigs, Peter
D'Ianni, James Donato
Diem, Hugh E(gbert)
Dinbergs, Kornelius
Divis, Roy Richard
Dodge, James Stanley
Dollimore, David
Duddey, James E
Duke, June Temple
Dunn, Horton, Jr
Edman, James Richard
Ells, Frederick Richard
England, Richard Jay
Epstein, Arthur Joseph
Erman, William F
Essig, Henry J
Evans, Robert Morton
Evers, Robert C
Fabris, Hubert
Fall, Harry H
Fechter, Robert Bernard
Feld, William Adam
Fielding-Russell, George Samuel
Finelli, Anthony Francis
Ford, Emory A
Forsyth, Thomas Henry
Franks, Allen P
Frederick, John Edgar
Fried, Joel Robert
Fu, Yun-Lung
Galloway, Ethan Charles
Gates, Raymond Dee
Gebelein, Charles G
Gerace, Michael Joseph
Giffen, William Martin, Jr
Gippin, Morris
Gleim, Clyde Edgar
Goldman, Stephen Allen
Gray, Don Norman
Griffin, Richard Norman
Grimm, Robert Arthur
Gromelski, Stanley John, Jr
Gruber, Elbert Egidius
Hall, Judd Lewis
Hamed, Gary Ray
Harris, Frank Wayne
Harris, Richard Lee
Hartsough, Robert Ray
Harwood, Harold James
Hassell, John Allen
Hayes, Robert Arthur
Heineman, William Richard
Hergenrother, William Lee
Herliczek, Siegfried H
Herold, Robert Johnston
Hiles, Maurice
Hoke, Donald I
Hollis, William Frederick
Horne, Samuel Emmett, Jr
Hyndman, John Robert
Ishida, Hatsuo
Jabarin, Saleh Abd El Karim
Jamieson, Alexander MacRae
Kanakkanatt, Antony

Polymer Chemistry (cont)

Katz, Morton
Keck, Max Hans
Kell, Robert M
Kelley, Frank Nicholas
Kennedy, Joseph Paul
Kinstle, James Francis
Koehler, Mark E
Koenig, Jack L
Kollen, Wendell James
Kreuz, John Anthony
Kuo, Cheng-Yih
Kurz, David W
Kwiatek, Jack
Lal, Joginder
Lawson, David Francis
Lehr, Marvin Harold
Leininger, Robert Irvin
Lewis, Irwin C
Li, George Su-Hsiang
Lipinsky, Edward Solomon
Litt, Morton Herbert
Livigni, Russell Anthony
Livingston, Daniel Isadore
Lohr, Delmar Frederick, Jr
Losekamp, Bernard Francis
Loughran, Gerard Andrew, Sr
Lucas, Kenneth Ross
McCain, George Howard
McDonel, Everett Timothy
McGinniss, Vincent Daniel
McIntyre, Donald
McIntyre, William Ernest, Jr
Mark, James Edward
Marshall, Richard Allen
Mattice, Wayne Lee
Meinhardt, Norman Anthony
Merritt, Robert Edward
Metanomski, Wladyslaw Val
Mikesell, Sharell Lee
Miller, James Roland
Miller, William Reynolds, Jr
Milone, Charles Robert
Morton, Maurice
Mosser, John Snavely
Mouk, Robert Watts
Murphy, Walter Thomas
Myers, Ronald Eugene
Nakajima, Nobuyuki
Nelson, Wayne Franklin
Niemann, Theodore Frank
Noda, Isao
Ofstead, Eilert A
O'Leary, Kevin Joseph
Ostrum, G(eorge) Kenneth
Oziomek, James
Pace, Henry Alexander
Pappas, Leonard Gust
Parish, Darrell Joe
Percec, Virgil
Petschek, Rolfe George
Piirma, Irja
Portfolio, Donald Carl
Pritchett, Ervin Garrison
Prudence, Robert Thomas
Prusaczyk, Joseph Edward
Purdon, James Ralph, Jr
Purvis, John Thomas
Pyle, James Johnston
Quirk, Roderic P
Raciszewski, Zbigniew
Rader, Charles Phillip
Ramp, Floyd Lester
Reardon, Joseph Patrick
Reilly, Charles Bernard
Rinaldi, Peter L
Roberts, Durward Thomas, Jr
Robins, Richard Dean
Roe, Ryong-Joon
Rogers, Charles Edwin
Roha, Max Eugene
Rosen, Irving
Sarbach, Donald Victor
Scheve, Bernard Joseph
Schilling, Curtis Louis, Jr
Schmidt, Francis Henry
Schneider, Wolfgang W
Schollenberger, Charles Sundy
Schuele, Donald Edward
Scott, Kenneth Walter
Shelton, James Reid
Siebert, Alan Roger
Siefken, Mark William
Sieglaff, Charles Lewis
Simha, Robert
Sinclair, Richard Glenn, II
Sircar, Anil Kumer
Sliemers, Francis Anthony, Jr
Sommer, John G
Standish, Norman Weston
Stephens, Howard L
Stevenson, Don R
Stickney, Palmer Blaine
Strauss, Carl Richard
Tarvin, Robert Floyd
Tate, David Paul
Tessier, Claire Adrienne
Theis, Richard James
Thomas, George B
Throckmorton, Morford C
Toth, William James
Traynor, Lee
Trewiler, Carl Edward
Tsai, Chung-Chieh
Uebele, Curtis Eugene
Updegrove, Louis B
Uscheek, Dave Petrovich
Vanderlind, Merwyn Ray
Vassiliades, Anthony E
Voigt, Charles Frederick
Wagner, Melvin Peter
Warner, Walter Charles
Weinstein, Arthur Howard
Westerman, Ira John
Wideman, Lawson Gibson
Wiff, Donald Ray
Williams, Robert Calvin
Williger, Ervin John
Woo, James T K
Yang, Philip Yung-Chin
Yocum, Ronald Harris
Young, Patrick Henry
Youngs, Wiley Jay
Zaremsky, Baruch

OKLAHOMA

Bailey, F Wallace
Das, Paritosh Kumar
Efner, Howard F
Fodor, Lawrence Martin
Ford, Warren Thomas
Geibel, Jon Frederick
Graves, Toby Robert
Highsmith, Ronald Earl
Hogan, John Paul
Holtmyer, Marlin Dean
Johnson, Timothy Walter
Jones, Faber Benjamin
Linder, Donald Ernst
Lindstrom, Merlin Ray
Miller, John Walcott
Moczygemba, George A
Moore, Donald R
Natowsky, Sheldon
Pober, Kenneth William
Rotenberg, Don Harris
Schmidt, Donald Dean
Seeney, Charles Earl
Selman, Charles Melvin
Short, James N
Shue, Robert Sidney
Sonnenfeld, Richard John
Stacy, Carl J
Trepka, William James
Welch, Melvin Bruce
Wharry, Stephen Mark
Witt, Donald Reinhold
Zelinski, Robert Paul

OREGON

Fleming, Bruce Ingram
Merz, Paul Louis
Myers, George E
Nielsen, Lawrence Ernie
Shilling, Wilbur Leo
Skiens, William Eugene
Smith, Kelly L
Thies, Richard William
Wamser, Carl Christian

PENNSYLVANIA

Allara, David Lawrence
Allcock, Harry R(ex)
Alvino, William Michael
Anspon, Harry Davis
Arkles, Barry Charles
Backus, John King
Bagley, George Everett
Bakule, Ronald David
Balaba, Willy Mukama
Barie, Walter Peter, Jr
Bartovics, Albert
Bartz, Warren F(rederick)
Bassner, Sherri Lynn
Batzar, Kenneth
Bauer, William, Jr
Bauman, Bernard D
Beckley, Ronald Scott
Berkheimer, Henry Edward
Berry, Guy C
Blommers, Elizabeth Ann
Bolgiano, Nicholas Charles
Bolstad, Luther
Bower, George Myron
Braunstein, David Michael
Breazeale, Robert David
Brendley, William H, Jr
Buck, Jean Coberg
Burch, Mary Kappel
Buzzell, John Gibson
Canter, Neil M
Cardinal, John Robert
Carlson, Dana Peter
Carlson, Gerald Leroy
Cartier, Peter G
Castle, John Edwards
Cenci, Harry Joseph
Chang, Ching-Jen
Chapman, Toby Marshall
Chaudhury, Manoj Kumar
Chen-Tsai, Charlotte Hsiao-yu
Chiola, Vincent
Chong, Berni Patricia
Christie, Peter Allan
Chung, Tze-Chiang
Claiborne, C Clair
Cohen, Gordon Mark
Condo, Albert Carman, Jr
Conyne, Richard Francis
Cook, Donald Bowker
Corneliussen, Roger DuWayne
Cornell, John Alston
Das, Surya Kumar
De Tommaso, Gabriel Louis
Dickstein, Jack
Dimmig, Daniel Ashton
Doak, Kenneth Worley
Dohany, Julius Eugene
Dougherty, Eugene P
Dunlap, Lawrence H
Durrell, William S
Ebert, Philip E
Ehrhart, Wendell A
Ehrig, Raymond John
El-Aasser, Mohamed S
Ellis, Jeffrey Raymond
Erhan, Semih M
Fellmann, Robert Paul
Fishman, Marshall Lewis
Fitzgerald, Maurice E
Ford, Michael Edward
Francis, Peter Schuyler
Freeman, James Harrison
Frick, Neil Huntington
Frost, Lawrence William
Garber, Charles A
Garrett, Robert Roth
Gaughan, Renata Rysnik
Gillis, Marina N
Goddu, Robert Fenno
Goldman, Theodore Daniel
Golton, William Charles
Goodman, Alan Lawrence
Goodman, Donald
Graham, Roger Kenneth
Greiner, Richard William
Grey, Roger Allen
Grezlak, John Henry
Grinstein, Reuben H
Gruber, Gerald William
Gruenwald, Geza
Gunther, Wolfgang Hans Heinrich
Haggard, Richard Allan
Hanauer, Richard
Harris, James Joseph
Hartman, Marvis Edgar
Hartranft, George Robert
Heiberger, Philip
Hertler, Walter Raymond
Hirsch, Stephen Simeon
Hockswender, Thomas Richard
Holm, Reimer
Holy, Norman Lee
Hutchins, MaryGail Kinzer
Ingram, Alvin Richard
Irwin, Philip George
Irwin, William Edward
Jabloner, Harold
Johnston-Feller, Ruth M
Jones, Roger Franklin
Judge, Joseph Malachi
Karo, Wolf
Kehr, Clifton Leroy
Kennedy, Flynt
Khosah, Robinson Panganai
Klapproth, William Jacob, Jr
Kleinschuster, Jacob John
Kline, Richard William
Koob, Robert Philip
Kopchik, Richard Michael
Korchak, Ernest I(an)
Koziar, Joseph Cleveland
Kronberger, Karlheinz
Kronick, Paul Leonard
Kuzmak, Joseph Milton
Lake, Robert D
Langsam, Michael
Lantos, P(eter) R(ichard)
Lavoie, Alvin Charles
Lee, Ying Kao
Lesko, Patricia Marie
Leston, Gerd
Lloyd, Thomas Blair
Louie, Ming
Lovald, Roger Allen
Luck, Russell M
Luskin, Leo Samuel
Lynch, Thomas John
McDowell, Maurice James
Mahlman, Bert H
Mair, Robert Dixon
Mantell, Gerald Jerome
Mao, Chung-Ling
Matyjaszewski, Krzysztof
Meier, Joseph Francis
Melamed, Sidney
Meyer, Ellen McGhee
Morgan, Paul Winthrop
Morse, Lewis David
Murphy, Clarence John
Nannelli, Piero
Naples, John Otto
Natoli, John
Novak, Ronald William
Noyes, Paul R
O'Mara, James Herbert
Orphanides, Gus George
Osei-Gyimah, Peter
Parchen, Frank Raymond, Jr
Patsiga, Robert A
Patterson, Gary David
Peffer, John Roscoe
Peterson, Jack Kenneth
Petrich, Robert Paul
Pinschmidt, Robert Krantz, Jr
Plant, William J
Prane, Joseph W(illiam)
Price, Charles Coale
Quinn, Edwin John
Reuwer, Joseph Francis, Jr
Ristey, William J
Robinson, Donald Nellis
Rogers, Janet
Samuelson, H Vaughn
Santamaria, Vito William
Scala, Luciano Carlo
Schaller, Edward James
Schimmel, Karl Francis
Schott, Hans
Schreyer, Ralph Courtenay
Schrof, William Ernst John
Schultz, Ray Karl
Seiner, Jerome Allan
Sheasley, W(illiam) David
Sherman, Anthony Michael
Shim, Benjamin Kin Chong
Slysh, Roman Stephan
Smith, James David Blackhall
Smith, Stewart Edward
Smoot, Charles Richard
Snider, Albert Monroe, Jr
Snyder, Harold Lee
Sollott, Gilbert Paul
Solomon, M Michael
Soriano, David S
Sperling, Leslie Howard
Statton, Gary Lewis
Stevens, Travis Edward
Stone, Herman
Stryker, Lynden J
Swift, Graham
Szamosi, Janos
Templer, David Allen
Tennent, Howard Gordon
Thibeault, Jack Claude
Thomas, H Ronald
Tweedie, Adelbert Thomas
Vanderhoff, John W
Victorius, Claus
Vijayendran, Bhima R
Vogel, Martin
Vorchheimer, Norman
Walker, Augustus Chapman
Walker, Frederick
Warfel, David Ross
Watson, William Martin, Jr
Weese, Richard Henry
Wei, Yen
Weisfeld, Lewis Bernard
Welsh, David Albert
Wempe, Lawrence Kyran
Werner, Ervin Robert, Jr
White, Malcolm Lunt
Whiteman, John David
Wilkins, Cletus Walter, Jr
Williams, Donald Robert
Work, James Leroy
Work, William James
Yanai, Hideyasu Steve
Younes, Usama E
Zimmt, Werner Siegfried
Zinser, Edward John

RHODE ISLAND

Bide, Martin John
Nadolink, Richard Hughes
Scott, Peter Hamilton

SOUTH CAROLINA

Beam, Charles Fitzhugh, Jr
Berger, Richard S
Bliss, Arthur Dean
Brooks, Thomas
Culbertson, Edwin Charles
Dieter, Janice Wong
Drews, Michael James
Farewell, John P
Guthrie, Roger Thackston
Hartz, Roy Eugene
Hendrix, James Easton
Hon, David Nyok-Sai
Hopkins, Allen John
Hudgin, Donald Edward
King, Lee Curtis
Krantz, Karl Walter
Lindemann, Martin K
Longtin, Bruce
Machell, Greville
Otto, Wolfgang Karl Ferdinand
Pike, LeRoy
Posey, Robert Giles
Robinson, Myrtle Tonne
Robinson, Robert Earl
Ross, Stanley Elijah
Rothrock, George Moore
Scruggs, Jack G
Spencer, Harold Garth
Terry, Stuart Lee
Tour, James Mitchell
Wagner, William Sherwood
Weiss, James Owen
Wishman, Marvin

SOUTH DAKOTA

Hanson, Milton Paul

TENNESSEE
Alexandratos, Spiro
Allen, Vernon R
Bagrodia, Shriram
Barron, Eugene Roy
Barton, Kenneth Ray
Bedoit, William Clarence, Jr
Brennan, Michael Edward
Brown, George Marshall
Browning, Horace Lawrence, Jr
Caflisch, George Barrett
Chambers, Ralph Arnold
Clark, Edward Shannon
Clemens, Robert Jay
Cliffton, Michael Duane
Collins, Jerry Dale
Coover, Harry Wesley, Jr
Davis, Burns
Dean, Walter Lee
Dieck, Ronald Lee
Dombroski, John Richard
Dorsey, George Francis
Fagerburg, David Richard
Fordham, J(ames) Lynn
Fort, Tomlinson
Germroth, Ted Calvin
Gilkey, Russell
Gleason, Edward Hinsdale, Jr
Gray, Theodore Flint, Jr
Habenschuss, Anton
Harmer, David Edward
Harwell, Kenneth Elzer
Hoffman, Kenneth Wayne
Hutchinson, James Herbert, Jr
Jackson, Winston Jerome, Jr
Jones, Glenn Clark
Kashdan, David Stuart
Knee, Terence Edward Creasey
Kovac, Jeffrey Dean
Krutak, James John, Sr
Kuo, Chung-Ming
Leonard, Edward Charles, Jr
McConnell, Richard Leon
McFarlane, Finley Eugene
McPherson, James Louis
Mayberry, Thomas Carlyle
Moore, Louis Doyle, Jr
Newland, Gordon Clay
Nicely, Vincent Alvin
Overton, James Ray
Parkinson, William Walker, Jr
Perry, Lloyd Holden
Phillips, Paul J
Raynolds, Peter Webb
Redfearn, Richard Daniel
Sublett, Bobby Jones
Tibbetts, Clark
Turner, S Richard
Vachon, Raymond Normand
Volpe, Angelo Anthony
Wang, Richard Hsu-Shien
White, Alan Wayne
Wicker, Thomas Hamilton, Jr
Wildman, Gary Cecil
Witzeman, Jonathan Stewart
Wunderlich, Bernhard
Yau, Cheuk Chung

TEXAS
Aufdermarsh, Carl Albert, Jr
Baker, Charles Taft
Bartsch, Richard Allen
Beede, Charles Herbert
Benitez, Francisco Manuel
Bergbreiter, David Edward
Borchardt, John Keith
Brady, Donnie Gayle
Brostow, Witold Konrad
Brown, James Michael
Bruins, Paul F(astenau)
Cabanes, William Ralph, Jr
Cannon, Dickson Y
Cassidy, Patrick Edward
Cavender, James Vere, Jr
Cross, John Parson
Cross, Virginia Rose
Daues, Gregory W, Jr
De La Mare, Harold Elison
Deviney, Marvin Lee, Jr
Edmondson, Morris Stephen
Ellison, Robert Hardy
Elsenbaumer, Ronald Lee
Elward-Berry, Julianne
Engle, Damon Lawson
Etter, Raymond Lewis, Jr
Feit, Eugene David
Fish, John G
Floyd, Joseph Calvin
Fuller, Martin Emil
Goldstein, Herbert Jay
Goodwin, John Thomas, Jr
Goodwyn, Jack Ray
Gryting, Harold Julian
Gum, Wilson Franklin, Jr
Handlin, Dale L
Hart, Paul Robert
Heilman, William Joseph
Hickner, Richard Allan
Hill, Robert William
Holmes, Larry A
Hosmane, Narayan Sadashiv
Jennings, Alfred Roy, Jr
Kalfoglou, George
Kaye, Howard
Kelsey, Donald Ross
Kennelley, Kevin James
Keskkula, Henno
Kochhar, Rajindar Kumar
Kohn, Erwin
Lakshmanan, P R
Lloyd, Douglas Roy
Lyle, Robert Edward, Jr
McCullough, James Douglas, Jr
McDougall, Robert I
McGirk, Richard Heath
Mack, Mark Philip
Malpass, Dennis B
Mangold, Donald J
Martin, Charles William
May, James Aubrey, Jr
Mendelson, Robert Allen
Moltzan, Herbert John
Munk, Petr
Naae, Douglas Gene
Neeley, Charles Mack
Nemphos, Speros Peter
Newton, Robert Andrew
Olson, Danford Harold
Olstowski, Franciszek
Park, Vernon Kee
Patton, Tad LeMarre
Paul, Donald Ross
Pomerantz, Martin
Powell, Richard James
Rawls, Henry Ralph
Reed, Thomas Freeman
Richter, Reinhard Hans
Roberts, Thomas David
Robinson, Alfred Green
Sample, Thomas Earl, Jr
Schimelpfenig, Clarence William
Simpson, James Urban
Smith, Curtis William
Spendel, Lawrence
Speranza, George Phillip
Stahl, Glenn Allan
Strom, E(dwin) Thomas
Sward, Edward Lawrence, Jr
Thompson, Michael McCray
Vogelfanger, Elliot Aaron
Webber, Stephen Edward
Willson, Carlton Grant
Wisian-Neilson, Patty Joan

UTAH
Dehm, Henry Christopher
Elmslie, James Stewart
Guillot, David G
Kopecek, Jindrich
Thompson, Grant

VERMONT
Allen, Christopher Whitney
Berens, Alan Robert
Murray, James Gordon
Weimann, Ludwig Jan

VIRGINIA
Armstrong, Robert G
Barber, Patrick George
Barker, Robert Henry
Barkin, Stanley M
Bass, Robert Gerald
Bell, Vernon Lee, Jr
Beyad, Mohammed Hossain
Bikales, Norbert M
Breder, Charles Vincent
Bryant, Robert George
Campbell, Francis James
Farago, John
Farmer, Barry Louis
Fisher, Charles Harold
Fitzgerald, James Allen
French, David Milton
Gerow, Clare William
Gibson, Harry William
Glasser, Wolfgang Gerhard
Gratz, Roy Fred
Gulrich, Leslie William, Jr
Hahn, Walter Leopold
Harowitz, Charles Lichtenberg
Hartzler, Jon David
Hazlehurst, David Anthony
Hewett, James Veith
Hodge, James Dwight
Huggett, Clayton (McKenna)
Hussamy, Samir
Hutchison, David Allan
Inskeep, George Esler
Jablonski, Werner Louis
Jensen, Arnold William
Johnson, William Randolph, Jr
Johnston, Norman Joseph
Kranbuehl, David Edwin
Lim, Young Woon (Peter)
Lodoen, Gary Arthur
Lupinski, John Henry
McGrath, James Edward
Merkel, Timothy Franklin
Meyer, Leo Francis
Milford, George Noel, Jr
Monroe, Stuart Benton
Orwoll, Robert Arvid
Patterson, Earl E(dgar)
Rainer, Norman Barry
Sahli, Muhammad S
St Clair, Anne King
St Clair, Terry Lee
Sastri, Vinod Ram
Schuurmans, Hendrik J L
Sherbeck, L Adair
Squire, David R
Starnes, William Herbert, Jr
Stiehl, R(oy) Thomas, Jr
Stump, Billy Lee
Thompson, Mark Ewell
Tichenor, Robert Lauren
Ticknor, Leland Bruce
Urbanik, Arthur Ronald
Van Ness, Kenneth E
Walton, Theodore Ross
Weedon, Gene Clyde
Wilkinson, Thomas Lloyd, Jr
Wilkinson, William Kenneth
Wilson, Donald Richard
Wynne, Kenneth Joseph

WASHINGTON
Ericsson, Lowell Harold
Eustis, William Henry
Exarhos, Gregory James
Felicetta, Vincent Frank
Groten, Barney
Hansen, Michael Roy
Hine, John Maynard
Kesting, Robert E
Kruger, Albert Aaron
Lindenmeyer, Paul Henry
Lyman, Donald Joseph
McCarthy, Joseph L(ePage)
Mehlhaff, Leon Curtis
Minckler, Leon Sherwood, Jr
Peterson, James Macon
Ratner, Buddy Dennis
Rowland, Stanley Paul
Subramanian, Ravanasamudram Venkatachalam
Tobiason, Frederick Lee
Vessel, Eugene David
Wasserman, William Jack
Wollwage, Paul Carl
Wong, Chun-Ming
Zollars, Richard Lee

WEST VIRGINIA
Bailey, Frederick Eugene, Jr
Barnes, Robert Keith
Bryant, George Macon
Dawson, Thomas Larry
Dunphy, James Francis
Giza, Yueh-Hua Chen
Hager, Stanley Lee
Knopf, Robert John
Koleske, Joseph Victor
Matthews, Virgil Edison
Novak, Ernest Richard
Osborn, Claiborn Lee
Peascoe, Warren Joseph
Reid, C Glenn
Rieck, James Nelson
Robson, John Howard
Smith, Joseph James
Sperati, Carleton Angelo
Winston, Anthony

WISCONSIN
Anderson, Stephen William
Atalla, Rajai Hanna
Berge, John Williston
Bringer, Robert Paul
Brown, Kenneth Howard
Cataldi, Horace A(nthony)
Caulfield, Daniel Francis
Chenier, Philip John
Christiansen, Alfred W
Cleary, James William
Denton, Denice Dee
Dickerson, Charlesworth Lee
Dwyer, Sean G
Fadner, Thomas Alan
Feist, William Charles
Ferry, John Douglass
Fischer, Richard Martin, Jr
Fitch, Robert McLellan
Gross, James Richard
Hauser, Edward Russell
Holland, Dewey G
Klopotek, David L
Kurath, Sheldon Frank
Lemm, Arthur Warren
Lepeska, Bohumir
Lin, Stephen Y
Morris, Marion Clyde
Randall, Francis James
Ranganathan, Brahmanpalli Narasimhamurthy
Reichenbacher, Paul H
Schrag, John L
Sedor, Edward Andrew
Sheppard, Erwin
Soerens, Dave Allen
Stackman, Robert W
Strause, Sterling Franklin
Svoboda, Glenn Richard
Thurmaier, Roland Joseph
Verbrugge, Calvin James
Ward, Kyle, Jr
West, Robert C
Wilkie, Charles Arthur
Yu, Hyuk

WYOMING
Hiza, Michael John, Jr
Lewis, Randolph Vance

ALBERTA
Fisher, Harold M
Henderson, John Frederick

BRITISH COLUMBIA
Cox, Lionel Audley
Davis, Gerald Gordon
Funt, B Lionel
Hardwicke, Norman Lawson
Hocking, Martin Blake

MANITOBA
MacGregor, Elizabeth Ann
Singh, Ajit
Sze, Yu-Keung

NEW BRUNSWICK
Grant, Douglas Hope

NOVA SCOTIA
Lynch, Brian Maurice
Piorko, Adam M

ONTARIO
Alexandru, Lupu
Ananthanarayanan, Vettaikkoru S
Asculai, Samuel Simon
Brash, John Law
Buckler, Ernest Jack
Cameron, Irvine R
Carlsson, David James
Dolenko, Allan John
Eliades, Theo I
Golemba, Frank John
Huang, Robert Y M
Kabayama, Michiomi Abraham
Kazmaier, Peter Michael
Kilp, Toomas
Lautens, Mark
Marchessault, Robert Henri
Martin, Trevor Ian
Noolandi, Jaan
O'Driscoll, Kenneth F(rancis)
Penlidis, Alexander
Pintar, M(ilan) Mik
Rempel, Garry Llewellyn
Roovers, J
Rudin, Alfred
Russell, Kenneth Edwin
Sanderson, Edwin S
Slater, Gary W
Sundararajan, Pudupadi Ranganathan
Whittington, Stuart Gordon
Williams, Harry Leverne
Winnik, Francoise Martine
Winnik, Mitchell Alan

PRINCE EDWARD ISLAND
Burch, George Nelson Blair

QUEBEC
Alince, Bohumil
Allen, Lawrence Harvey
Andrews, Mark Paul
Bartnikas, Ray
Darling, Graham Davidson
Dorris, Gilles Marcel
Eisenberg, Adi
Feldman, Dorel
Gray, Derek Geoffrey
Harrod, John Frank
Hay, Allan Stuart
Heitner, Cyril
Holden, Harold William
Kokta, Bohuslav Vaclav
Lamarche, François
Leonard, Jacques Walter
Manley, Rockliffe St John
Patterson, Donald Duke
Prud'Homme, Jacques
Prud'homme, Robert Emery
St Pierre, Leon Edward
Utracki, Lechoslaw Adam
Wright, Archibald Nelson

OTHER COUNTRIES
Ahmed, Syed Mahmood
Brown, Charles Eric
Brown, Larry Robert
Chang, Shuya
Davidson, Daniel Lee
Gray, Kenneth W
Griffin, Anselm Clyde, III
Grunwald, John J
Hutchison, John Joseph
Isaacs, Philip Klein
Lin, Otto Chui Chau
McRae, Wayne A
Malherbe, Roger F
Markovitz, Hershel
Matsuo, Keizo
Meier, James Archibald
Millar, John Robert
Neuse, Eberhard Wilhelm
Qian, Renyuan
Reid, William John
Slagel, Robert Clayton
Steichen, Richard John
Tanny, Gerald Brian
Trofimov, Boris Alexandrovich

Polymer Chemistry (cont)

Varga, Janos M
Weise, Jurgen Karl

Protein Science

ALABAMA
Fox, Sidney Walter

CALIFORNIA
Bond, Martha W(illis)
Chamow, Steven M
Danko, Stephen John
Deshpande, Shirkant V
Ely, Kathryn R
Falick, Arnold M
Gee, Allen
Kraut, Joseph
Krstenansky, John Leonard
Lowman, Henry
Nguyen, Tue H
Raitzer, Cynthia Carilli
Singh, Jai Prakash
Winkler, Marjorie Everett
Yang, Yan-Bo

COLORADO
Jones, Steven Wayne

CONNECTICUT
Ozols, Juris
Shansky, Albert

FLORIDA
Chapman, Michael S
Lee, Ted C K
Lewis, Nita Aries

ILLINOIS
Breillatt, Julian Paul, Jr
Buko, Alexander Michael
Donnelly, Thomas Henry
Gruebele, Martin
Jones, David A, Jr
Rothfus, John Arden
Sessa, David Joseph
Sorensen, Keld
Spangler, Brenda Dolgin
Treadway, William Jack, Jr

INDIANA
Lagu, Avinash L
Laskowski, Michael, Jr
Smith, Jean Blair

IOWA
Metzler, David Everett
Speckard, David Carl

KENTUCKY
Jayaram, Beby
Spatola, Arno F

LOUISIANA
Snow, Lloyd Dale

MARYLAND
Ahmed, Syed Ashrafuddin
Ashcom, James D
Krutzsch, Henry C
Lees, Andrew
Wickner, Sue

MASSACHUSETTS
Akkara, Joseph A
Cantor, Charles R
Humphreys, Robert Edward
Kemp, Daniel Schaeffer
Laursen, Richard Allan
Leavis, Paul Clifton
Liang, Jack N
Pluskal, Malcolm Graham
Riordan, James F
Sipe, Jean Dow
Tolan, Dean Richard
Venkatesh, Yeldur Padmanabha
Wolfson, Adele Judith

MICHIGAN
Marletta, Michael Anthony
Mizukami, Hiroshi
Smith, Denise Myrtle
Tuls, Jody Lynn Foy
Ueda, Tetsufumi

MINNESOTA
Butkowski, Ralph J

MISSOURI
Dahl, William Edward
Pikaard, Craig Stuart
Sharma, Krishna
Sotiriou-Leventis, Chariklia
Tompkins, Jay Allen

MONTANA
Richardson, Charles

NEVADA
Schooley, David Allan

NEW JERSEY
Glenn, Jeffrey K
Otterburn, Michael Storey
Salowe, Scott P
Zask, Arie

NEW MEXICO
Frauenfelder, Hans

NEW YORK
Fenton, Sandra Sulikowski
Horowitz, Marilyn Stephens
Oreskes, Irwin
Prestwich, Glenn Downes

NORTH CAROLINA
Alonso, William R
Risley, John Marcus
Smith, Gary Keith

OHIO
Flanagan, Margaret Ann
Morgan, Marjorie Susan

OKLAHOMA
Johnson, Arthur Edward
Spivey, Howard Olin

PENNSYLVANIA
Addison, Anthony William
Burman, Sudhir
Driska, Steven P
Lazar, Anna
Libby, R Daniel
Malin, Edyth
Raymond, Matthew Joseph
Sardana, Mohinder K
Tien, Ming

SOUTH CAROLINA
Wang, Huei-Hsiang Lisa

TENNESSEE
Ernst-Fonberg, Marylou
Stubbs, Gerald James

TEXAS
Gilbert, Hiram Frazier

VIRGINIA
Brill, Arthur Sylvan
Martin, Robert Bruce
Sayala, Chhaya

WISCONSIN
Gellman, Samuel Helmer

ALBERTA
Hodges, Robert Stanley

BRITISH COLUMBIA
Cushley, Robert John

ONTARIO
Kluger, Ronald H
Lee, Lilian M

QUEBEC
Sanctuary, Bryan Clifford

OTHER COUNTRIES
Kim, Dong Han

Quantum Chemistry

ALABAMA
McKee, Michael Leland

ARIZONA
Balasubramanian, Krishnan
Lichtenberger, Dennis Lee
Steimle, Timothy C

ARKANSAS
Becker, Ralph Sherman
Pulay, Peter

CALIFORNIA
Acrivos, Juana Luisa Vivo
Bagus, Paul Saul
Calabrese, Philip G
Carter, Emily Ann
Fink, William Henry
Goddard, William Andrew, III
Goldstein, Elisheva
Head-Gordon, Martin Paul
Herman, Frank
Herman, Zelek Seymour
Horsley, John Anthony
Huestis, David Lee
Karo, Arnold Mitchell
Keeports, David
Kollman, Peter Andrew
Konowalow, Daniel Dimitri
Lester, William Alexander, Jr
Miller, William Hughes
Morrison, Harry Lee
Sahbari, Javad Jabbari
Saxon, Roberta Pollack
Schwartz, Robert Nelson
Shull, Harrison
Simon, Barry Martin
Sovers, Ojars Juris

Taylor, Howard S
Van Hecke, Gerald Raymond
Wolf, Kathleen A
Wolfsberg, Max

COLORADO
Morgan, William Lowell
Stephens, Jeffrey Alan
Stewart, James Joseph Patrick

CONNECTICUT
Bohn, Robert K
Brittelli, David Ross
Clarke, George A
Petersson, George A
Smooke, Mitchell D

DELAWARE
Dixon, David Adams
Doren, Douglas James
Morgan, John Davis, III
Szalewicz, Krzysztof

DISTRICT OF COLUMBIA
Kertesz, Miklos
Wing, William Hinshaw

FLORIDA
Bartlett, Rodney Joseph
Bash, Paul Anthony
Baum, J(ames) Clayton
Cioslowski, Jerzy
Dash, Harriman Harvey
Jones, Walter H(arrison)
Monkhorst, Hendrik J
Pop, Emil
Rhodes, William Clifford
Snyder, Patricia Ann
Szczepaniak, Krystyna
Vergenz, Robert Allan
Zerner, Michael Charles

GEORGIA
Gole, James Leslie
Mead, Chester Alden

IDAHO
Switendick, Alfred Carl

ILLINOIS
Boldridge, David William
Bouman, Thomas David
Curtiss, Larry Alan
Goodman, Gordon Louis
Josefson, Clarence Martin
O'Donnell, Terence J
Sachtler, Wolfgang Max Hugo
Sanders, Frank Clarence, Jr
Scheiner, Steve

INDIANA
Bentley, John Joseph
Boyd, Donald Bradford
Kosman, Warren Melvin
Lipkowitz, Kenny Barry
Malik, David Joseph

IOWA
Kostic, Nenad M
Miller, Gordon James
Ruedenberg, Klaus

KANSAS
Purcell, Keith Frederick
Zandler, Melvin E

KENTUCKY
Clouthier, Dennis James
Plucknett, William Kennedy

LOUISIANA
Anex, Basil Gideon
Mollere, Phillip David
Scott, John Delmoth

MARYLAND
Beri, Avinash Chandra
Coplan, Michael Alan
Dehn, James Theodore
Kaufman, Joyce J
Liebman, Joel Fredric
McDiarmid, Ruth
Rosenfeld, Joan E
Ruffa, Anthony Richard
Silverstone, Harris Julian
Stevens, Walter Joseph
Tupper, Kenneth Joseph

MASSACHUSETTS
Chang, Edward Shi Tou
Chen, Maynard Ming-Liang
Harrison, Ralph Joseph
Hornig, Donald Frederick
Platt, John Rader
Weiss, Karl H
Winick, Jeremy Ross

MICHIGAN
Chang, Tai Yup
DeKock, Roger Lee
Ebbing, Darrell Delmar
Francisco, Joseph Salvadore, Jr
Hunt, Katharine Lois Clark
Jackson, James Edward

Polik, William Frederick
Schnitker, Jurgen H
Tai, Julia Chow
Weidman, Robert Stuart
Woehler, Scott Edwin

MINNESOTA
Eades, Robert Alfred

MISSOURI
Baughman, Russell George
Ching, Wai-Yim
Hansen, Richard Lee
Plummer, Patricia Lynne Moore
Thompson, Clifton C

MONTANA
Callis, Patrik Robert

NEBRASKA
Eckhardt, Craig Jon

NEW HAMPSHIRE
Fogleman, Wavell Wainwright

NEW JERSEY
Adams, William Henry
Eckert, John Andrew
Ermler, Walter Carl
Frishberg, Carol
Haddon, Robert Cort
Harris, Leonce Everett
Inniss, Daryl
McClure, Donald Stuart
Phillips, James Charles
Raghavachari, Krishnan
Raynor, Susanne
Sonn, George Frank
Stein, Reinhardt P
Stillinger, Frank Henry
Upton, Thomas Hallworth
Vogel, Veronica Lee
Weintraub, Herschel Jonathan R

NEW MEXICO
Alldredge, Gerald Palmer
Cahill, Paul A
Cartwright, David Chapman
Engelke, Raymond Pierce
Jennison, Dwight Richard
Martin, Richard Lee
Ortiz, Joseph Vincent
Pack, Russell T
Peterson, Otis G
Redondo, Antonio
Sheehan, William Francis
Walters, Edward Albert

NEW YORK
Avouris, Phaedon
Barnett, Michael Peter
Birge, Robert Richards
Chang, Shih-Yung
Ehrenson, Stanton Jay
Ezra, Gregory Sion
Feldman, Isaac
Finzel, Rodney Brian
Gao, Jiali
Hanson, Louise I Karle
Hattman, Stanley
Jain, Duli Chandra
Joshi, Bhairav Datt
Loebl, Ernest Moshe
Mehler, Ernest Louis
Padnos, Norman
Pflug, Donald Ralph
Rosenkrantz, Marcy Ellen
Silverman, Benjamin David
Weinstein, Harel
White, Michael George

NORTH CAROLINA
Caves, Thomas Courtney
Miller, Robert L
Navangul, Himanshoo Vishnu
Rabinowitz, James Robert
Rubloff, Gary W
Swofford, Robert Lewis
Wilhite, Douglas Lee
Yang, Weitao

OHIO
Bursten, Bruce Edward
Del Bene, Janet Elaine
Hayes, Edward Francis
Houser, John J
Klopman, Gilles
Kurtz, David Williams
Madia, William J
Smith, Douglas Alan
Taylor, William Johnson

OKLAHOMA
Lauffer, Donald Eugene
Neely, Stanley Carrell
Zetik, Donald Frank

OREGON
Engelking, Paul Craig

PENNSYLVANIA
Francl, Michelle Miller
Garrison, Barbara Jane
Jordan, Kenneth David

Ludwig, Oliver George
Pierce, Thomas Henry
Pratt, David Wixon
Ranck, John Philip
Ravi, Natarajan
Smith, Graham Monro
Staley, Stuart Warner
Zeroka, Daniel

SOUTH CAROLINA
Avegeropoulos, G
Larcom, Lyndon Lyle

SOUTH DAKOTA
Duffey, George Henry

TENNESSEE
Crawford, Oakley H
Fagerburg, David Richard
Kutzler, Frank William
Nicely, Vincent Alvin
Painter, Gayle Stanford
Tellinghuisen, Joel Barton

TEXAS
Birney, David Martin
Boggs, James Ernest
Conway, Dwight Colbur
Ford, George Peter
Hall, Michael Bishop
Harwood, William H
Ho, Paul Siu-Chung
Kelsey, Donald Ross
Marcotte, Ronald Edward
Marshall, Paul
Pettitt, Bernard Montgomery
Rossky, Peter Jacob
Scuseria, Gustavo Enrique

UTAH
Wahrhaftig, Austin Levy

VIRGINIA
Desjardins, Steven G
Hsu, David Shiao-Yo
Kelley, Ralph Edward
Phillips, Donald Herman
Piepho, Susan Brand
Smith, Bertram Bryan, Jr

WASHINGTON
Dunning, Thomas Harold, Jr
Gouterman, Martin (Paul)
Moseley, W(illiam) David
Norman, Joe G, Jr
Poshusta, Ronald D
Sherman, David Michael

WEST VIRGINIA
Maruca, Robert Eugene

WISCONSIN
Barr, Tery Lynn
Bondeson, Stephen Ray
England, Walter Bernard
King, Frederick Warren

ALBERTA
Ali, Shahida Parvin
Boere, Rene Theodoor
Grabenstetter, James Emmett
Klobukowski, Mariusz Andrzej

BRITISH COLUMBIA
Brion, Christopher Edward
Dingle, Thomas Walter

MANITOBA
Somorjai, Rajmund Lewis

NEW BRUNSWICK
Grein, Friedrich
Sichel, John Martin
Thakkar, Ajit Jamnadas

NOVA SCOTIA
Boyd, Russell Jaye

ONTARIO
Baird, Norman Colin
Davison, Sydney George
Kazmaier, Peter Michael
King, Gerald Wilfrid
Meath, William John
Paldus, Josef
Schlesinger, Mordechay

QUEBEC
Salahub, Dennis Russell
Sandorfy, Camille
Whitehead, Michael Anthony

SASKATCHEWAN
Mezey, Paul G

OTHER COUNTRIES
Balakrishnan, Narayana Swamy
Bunge, Carlos Federico
Kay, Kenneth George
Keller, Jaime
Lowdin, Per-Olov

Spectroscopy

ALABAMA
Bakker, Martin Gerard
Coburn, William Carl, Jr

ARIZONA
Blankenship, Robert Eugene
Brown, Michael Frederick
Christensen, Kenner Allen
Fuchs, Jacob
McMillan, Paul Francis

ARKANSAS
Komoroski, Richard Andrew

CALIFORNIA
Allamandola, Louis John
Anet, Frank Adrien Louis
Berry, Michael James
Cain, William F
Cosby, Philip C
Curry, Bo(stick) U
Davidson, Arthur L
Fassel, Velmer Arthur
Fetzer, John Charles
Giang, Benjamin Yunwen
Hovanec, B(ernard) Michael
Jeffries, Jay B
Kasai, Paul Haruo
Keeports, David
Knudtson, John Thomas
Lang, Valerie Ilona
Lieberman, Robert Arthur
Lundin, Robert Enor
Meinhard, James Edgar
Okumura, Mitchio
Parsons, Michael L
Perkins, Willis Drummond
Pertica, Alexander Jose
Ross, Philip Norman, Jr
Sahbari, Javad Jabbari
Schwartz, Robert Nelson
Shaka, Athan James
Sharma, Prasanta
Shuh, David Kelly
Shykind, David
Snyder, Robert Gene
Stebbins, Jonathan F
Stohr, Joachim
Weers, Jeffry Greg
Wilson, Charles Oren

COLORADO
Birks, John William
Strickler, Stewart Jeffery
Vejvoda, Edward
Webb, John Day

CONNECTICUT
Bentz, Alan P(aul)
Bohn, Robert K
Caradonna, John Philip
Dobbs, Gregory Melville
Greenhouse, Steven Howard
Martoglio, Pamela Anne
Reffner, John A
Siegel, Norman Joseph
Ultee, Casper Jan

DELAWARE
Barteau, Mark Alan
Brame, Edward Grant, Jr
Chase, David Bruce
Crawford, Michael Karl
Firment, Lawrence Edward
Graham, John Alan
Narvaez, Richard
Wiley, Douglas Walker

DISTRICT OF COLUMBIA
Hicks, Janice Marie
May, Leopold
Misra, Prabhakar
Poranski, Chester F, Jr
Shirk, James Siler
Vertes, Akos
Yesinowski, James Paul

FLORIDA
Baum, J(ames) Clayton
Block, Ronald Edward
Brucat, Philip John
Holloway, Paul Howard
Li, San
Stewart, Burch Byron
Weltner, William, Jr

GEORGIA
De Sa, Richard John
Love, Jimmy Dwane
Tincher, Wayne Coleman
Torres, Lourdes Maria
Van Assendelft, Onno Willem

IDAHO
Castle, Lyle William
Griffiths, Peter Roughley

ILLINOIS
Adlof, Richard Otto
Belford, R Linn
Buko, Alexander Michael
Cutnell, John Daniel

Dehmer, Patricia Moore
Erickson, Mitchell Drake
Ferraro, John Ralph
Glaser, Michael
Gruebele, Martin
Gruen, Dieter Martin
Horwitz, Alan Fredrick
House, Kathleen Ann
Jonas, Jiri
Koropchak, John
LaPlanche, Laurine A
Marley, Nancy Alice
Masel, Richard Isaac
Mota de Freitas, Duarte Emanuel
Sibbach, William Robert
Surgi, Marion Rene
Thompson, Arthur Robert
Wozniak, Wayne Theodore
Young, Charles Edward

INDIANA
Chernoff, Donald Alan
Fehlner, Thomas Patrick
Havel, Henry Acken
Hunt, Ann Hampton
Murray, Lionel Philip, Jr
Thomas, John Kerry
Tripathi, Gorakh Nath Ram

IOWA
Edelson, Martin Charles
Kintanar, Agustin
Koerner, Theodore Alfred William, Jr
Metzler, David Everett
Stahr, Henry Michael
Struve, Walter Scott

KANSAS
Carper, William Robert
Hammaker, Robert Michael
Harmony, Marlin D
Marshall, Delbert Allan
Purcell, Keith Frederick
Sherwood, Peter Miles Anson
Sobczynski, Dorota

KENTUCKY
Riley, John Thomas
Smith, Stanford Lee

LOUISIANA
Anex, Basil Gideon
Frazier, Claude Clinton, III
Kumar, Devendra
Maverick, Andrew William
Miceli, Michael Vincent
Robinson, James William
Smith, Charles Hooper

MAINE
Nagle, Jeffrey Karl

MARYLAND
Burrows, Elizabeth Parker
Fales, Henry Marshall
Fenselau, Catherine Clarke
Gilfrich, John Valentine
Grim, Samuel Oram
Houseman, Barton L
Lafferty, Walter J
Lovas, Francis John
McCurley, Marian Frances
McDiarmid, Ruth
Miller, Stephen P F
Murray, George Milton
Neumann, Dan Alan
Robinson, Dean Wentworth
Siatkowski, Ronald E
Suenram, Richard Dee
Thompson, Warren Elwin
Vander Hart, David Lloyd

MASSACHUSETTS
Aronson, James Ries
Brecher, Charles
Chiu, Tak-Ming
Chu, Nori Yaw-Chyuan
Dasari, Ramachandra R
Druy, Mark Arnold
Gardner, James A
Greenaway, Frederick Thomas
Howard, Wilmont Frederick, Jr
Lee, Roland Robert
Marshall, Mark David
Varco-Shea, Theresa Camille

MICHIGAN
Coburn, Joel Thomas
Epps, Dennis Earl
Francisco, Joseph Salvadore, Jr
Hunt, Katharine Lois Clark
Langhoff, Charles Anderson
Leroi, George Edgar
Nader, Bassam Salim
Polik, William Frederick
Prakash, D Kash
Rothschild, Walter Gustav
Smith, A(lbert) Lee
Woehler, Scott Edwin

MINNESOTA
Bodine, Peter Van Nest
Dittmar, Rebecca May
Frank, William Charles

Leopold, Doreen Geller
Newmark, Richard Alan
Potts, Lawrence Walter
Smith, David Philip

MISSOURI
Ames, Donald Paul
Blum, Frank D
Coria, Jose Conrado
Drew, Henry D
Field, Byron Dustin
Glaser, Rainer
Grayson, Michael A
Greenlief, Charles Michael
Kaelble, Emmett Frank
Kitchin, Robert Walter
Sotiriou-Leventis, Chariklia
Wong, Tuck Chuen

MONTANA
Rosenberg, Edward

NEBRASKA
Hulce, Martin R

NEVADA
Eastwood, DeLyle

NEW HAMPSHIRE
Bel Bruno, Joseph James
Winn, John Sterling

NEW JERSEY
Aronovic, Sanford Maxwell
Atwater, Beauford W
Bhattacharjee, Himangshu Ranjan
Borris, Robert P
Brody, Stuart Martin
Brons, Cornelius Hendrick
Brus, Louis Eugene
Cahn, Frederick
Feng, Rong
Fisanick, Georgia Jeanne
Gethner, Jon Steven
Goodman, Lionel
Harris, Alexander L
Harris, Leonce Everett
Hatzenbuhler, Douglas Albert
McClure, Donald Stuart
Omar, Mostafa M
Toome, Voldemar
Turse, Richard S
Williams, Thomas Henry
Witkowski, Mark Robert

NEW MEXICO
Beattie, Willard Horatio
Chapman, Robert Dale
Haaland, David Michael
McCulla, William Harvey
Nogar, Nicholas Stephen
Rayson, Gary D
Werbelow, Lawrence Glen
Williams, Mary Carol

NEW YORK
Angell, Charles Leslie
Attygalle, Athula B
Bauer, Simon Harvey
Billmeyer, Fred Wallace, Jr
Birke, Ronald Lewis
Blackwell, Crist Scott
Dixon, William Brightman
Doetschman, David Charles
Dumoulin, Charles Lucian
Duncan, Thomas Michael
Feigenson, Gerald William
Hsu, Donald K
Hurd, Jeffery L
Kagansky, Larisa
Lashewycz-Rubycz, Romana A
Launer, Philip Jules
Loring, Roger Frederic
MacColl, Robert
Marrone, Paul Vincent
Myers, Anne Boone
Nathanson, Benjamin
Orzech, Chester Eugene, Jr
Owens, Patrick M
Philips, Laura Alma
Rhodes, Yorke Edward
Schwartz, Herbert Mark
Siggins, James Ernest
Swinehart, James Stephen
Webster, Francis X
Wong, Lan Kan
Wu, Konrad T
Zipfel, Warren Roger

NORTH CAROLINA
Chung, Henry Hsiao-Liang
Culbreth, Judith Elizabeth
Noftle, Ronald Edward
Pickett, John Harold
Risley, John Marcus
Stejskal, Edward Otto
Wilson, Nancy Keeler

OHIO
Chester, Thomas Lee
Diem, Hugh E(gbert)
Ford, William Ellsworth
Gorse, Joseph
Laali, Kenneth Khosrow

Spectroscopy (cont)

Noda, Isao
Pausch, Jerry Bliss
Rinaldi, Peter L
Rubinson, Kenneth A
Stout, Barbara Elizabeth
Thompson, Robert Quinton
Wilson, Gustavus Edwin
Young, Patrick Henry

OKLAHOMA
Johnson, Arthur Edward

OREGON
Balko, Barbara Ann
Ferrante, Michael John
Loehr, Thomas Michael
Matthes, Steven Allen

PENNSYLVANIA
Asher, Sanford Abraham
Barnett, Herald Alva
Bernheim, Robert A
Byler, David Michael
Chang, Charles Hung
Chen, Audrey Weng-Jang
Dai, Hai-Lung
Dedinas, Jonas
Fairchild, Edward H
Gao, Quanyin
Gold, Lewis Peter
Jordan, Kenneth David
Lemmon, Donald H
Lu, Guangquan
Miller, Foil Allan
Nelson, George Leonard
Pfeffer, Philip Elliot
Pratt, David Wixon
Ravi, Natarajan
Selinsky, Barry Steven
Weiss, Paul Storch

SOUTH CAROLINA
Myrick, Michael Lenn

TENNESSEE
Begun, George Murray
Carman, Howard Smith, Jr
Close, David Matzen
Cook, Kelsey Donald
Ice, Gene Emery
Li, Ying Sing
Metcalf, David Halstead
Miller, John Cameron
Rutenberg, Aaron Charles

TEXAS
Berg, Mark Alan
Busch, Kenneth Walter
Busch, Marianna Anderson
Celii, Francis Gabriel
Fredin, Leif G R
Gnade, Bruce E
Jaszberenyi, Joseph C
Laane, Jaan
Lattman, Michael
McGuire, Anne Vaughan
Miller, James Richard
Patel, Bhagwandas Mavjibhai
Rabalais, John Wayne
Rawls, Henry Ralph
Redington, Richard Lee
Woessner, Donald Edward

VIRGINIA
Andrews, William Lester Self
Bell, Harold Morton
Bodnar, Robert John
Drew, Russell Cooper
Franzosa, Edward Sykes
Graybeal, Jack Daniel
Piepho, Susan Brand
Roberts, Catherine Harrison

WASHINGTON
Baer, Donald Ray
Crosby, Glenn Arthur
Friedrich, Donald Martin
Hill, Herbert Henderson, Jr
Lyman, Donald Joseph
McDowell, Robin Scott
Remmele, Richard L
Styris, David Lee

WEST VIRGINIA
Chisholm, William Preston
Seehra, Mohindar Singh

WISCONSIN
Barr, Tery Lynn
Hill, Charles Graham, Jr
Lemm, Arthur Warren
West, Kevin James
Wright, John Curtis
Yin, Jun-Jie

WYOMING
Netzel, Daniel Anthony

ALBERTA
Egerton, Raymond Frank
Overfield, Robert Edward

BRITISH COLUMBIA
Balfour, Walter Joseph
Brion, Christopher Edward
Mitchell, Reginald Harry
Preston, Caroline Margaret

MANITOBA
Attas, Ely Michael
Dong, Ronald Y
Ens, E(rich) Werner
McFarlane, Joanna
Sunder, Sham

NEW BRUNSWICK
Jankowski, Christopher K

NOVA SCOTIA
Coxon, John Anthony
Hooper, Donald Lloyd
McAlduff, Edward J
Wasylishen, Roderick Ernest

ONTARIO
Baines, Kim Marie
Belanger, Jacqueline M R
Brand, John C D
Carey, Paul Richard
Detellier, Christian
Henry, Bryan Roger
Irish, Donald Edward
Krull, Ulrich Jorg
LeRoy, Robert James
Macdonald, Peter Moore
McGarvey, Bruce Ritchie
Mackay, Gervase Ian
Murphy, William Frederick
Pare, J R Jocelyn
Rodrigo, Russell Godfrey
Shurvell, Herbert Francis
Sturgeon, Ralph Edward
Subramanian, Kunnath Sundara
Winnik, Francoise Martine

PRINCE EDWARD ISLAND
Liu, Michael T H

QUEBEC
Andrews, Mark Paul
Sanctuary, Bryan Clifford
Sandorfy, Camille
Serpone, Nick
Van Calsteren, Marie-Rose

OTHER COUNTRIES
Bondybey, Vladimir E
Brown, Charles Eric
Gray, Kenneth W
Hall, Laurance David
Trentham, David R
Turrell, George Charles

Structural Chemistry

ALABAMA
Cook, William Joseph
McKee, Michael Leland
Sakai, Ted Tetsuo
Sayles, David Cyril
Squillacote, Michael Edward
Tamburin, Henry John
Walsh, David Allan

ARIZONA
Hruby, Victor J
Kukolich, Stephen George
McMillan, Paul Francis
Stamm, Robert Franz
Steimle, Timothy C

ARKANSAS
Fry, Arthur James
Pulay, Peter

CALIFORNIA
Abdel-Baset, Mahmoud B
Bau, Robert
Burlingame, Alma L
Callahan, Kenneth Paul
Delker, Gerald Lee
Dickerson, Richard Earl
Dougherty, Dennis A
Fetzer, John Charles
George, Patricia Margaret
Giants, Thomas W
Glaser, Robert J
Guner, Osman Fatih
Haddon, William F, (Jr)
Hardcastle, Kenneth Irvin
Hodgson, Keith Owen
Hurd, Ralph Eugene
Jaecker, John Alvin
Janda, Kenneth Carl
Johnson, Oliver
Kim, Sung-Hou
Knobler, Carolyn Berk
La Mar, Gerd Neustadter
Leo, Albert Joseph
Lind, Maurice David
Medrud, Ronald Curtis
Patterson, Dennis Bruce
Peterson, Selmer Wilfred
Prakash, Surya G K
Raymond, Kenneth Norman
Samson, Sten
Schaefer, William Palzer
Scherer, James R
Shoolery, James Nelson
Simpson, Howard Douglas
Sims, James Joseph
Stohr, Joachim
Stucky, Galen Dean
Tokes, Laszlo Gyula
True, Nancy S
Trueblood, Kenneth Nyitray
Untch, Karl G(eorge)
Van Tamelen, Eugene Earl
Wong, Joe

COLORADO
Anderson, Oren P
Geller, Seymour

CONNECTICUT
Armitage, Ian MacLeod
Bohn, Robert K
Chang, Ted T
Katz, Lewis
Kilbourn, Barry T
Knox, James Russell, Jr
Rennhard, Hans Heinrich
Suib, Steven L

DELAWARE
Brame, Edward Grant, Jr
Calabrese, Joseph C
DeGrado, William F
Domaille, Peter John
Ferguson, Raymond Craig
Fukunaga, Tadamichi
Harlow, Richard Leslie
Herron, Norman
Ittel, Steven Dale
Lerman, Charles Lew
Reuben, Jacques
Shannon, Robert Day
Van Trump, James Edmond
Wallenberger, Frederick Theodore

DISTRICT OF COLUMBIA
Baker, Louis Coombs Weller
Butcher, Raymond John
Gilardi, Richard Dean
Gress, Mary Edith
Karle, Isabella Lugoski
Karle, Jean Marianne
Kertesz, Miklos
Skelton, Earl Franklin

FLORIDA
Babich, Michael Wayne
Copeland, Richard Franklin
DeLap, James Harve
Li, San
Pepinsky, Raymond
Person, Willis Bagley
Snyder, Patricia Ann
Stanko, Joseph Anthony
Trefonas, Louis Marco

GEORGIA
Pelletier, S William
Richardson, Susan D
Suddath, Fred LeRoy, (Jr)
Wilson, William David

HAWAII
Seff, Karl

IDAHO
Castle, Lyle William

ILLINOIS
Bouman, Thomas David
Broach, Robert William
Buko, Alexander Michael
Chang, Chong-Hwan
Curtin, David Yarrow
Geiser, Urs Walter
Guggenheim, Stephen
Keiderling, Timothy Allen
Lambert, Joseph B
LaPlanche, Laurine A
Levenberg, Milton Irwin
Magnus, George
Makinen, Marvin William
Maroni, Victor August
Mueller, Melvin H(enry)
Olsen, Kenneth Wayne
Pluth, Joseph John
Poeppelmeier, Kenneth Reinhard
Rankin, John Carter
Reynolds, Rosalie Dean (Sibert)
Ries, Herman Elkan, Jr
Rogers, Don
Schultz, Arthur Jay
Shriver, Duward F
Tuomi, Donald
Vandeberg, John Thomas
Wang, Andrew H-J
Williams, Jack Marvin

INDIANA
Chandrasekaran, Rengaswami
Christou, George
Dorman, Douglas Earl
Huffman, John Curtis
Hunt, Ann Hampton
Koch, Kay Frances
Kubiak, Clifford P
Lipkowitz, Kenny Barry
Scheidt, Walter Robert
Serianni, Anthony Stephan
Streib, William E
Toomey, Joseph Edward
Williams, Edward James

IOWA
Jacobson, Robert Andrew
Kraus, Kenneth Wayne
McCarley, Robert Eugene
Messerle, Louis

KANSAS
Fateley, William Gene
Harmony, Marlin D
Landis, Arthur Melvin
Mertes, Kristin Bowman

KENTUCKY
Brock, Carolyn Pratt
Buchanan, Robert Martin
McDermott, Dana Paul
Sands, Donald Edgar
Selegue, John Paul

LOUISIANA
Brown, Leo Dale
Fronczek, Frank Rolf
Lee, Burnell
Watkins, Steven F
Willer, Rodney Lee

MARYLAND
Amzel, L Mario
Block, Stanley
Bowen, Kit Hansel
Coulter, Charles L
D'Antonio, Peter
Fraser, Blair Allen
Frazer, Benjamin Chalmers
Lusby, William Robert
Mathew, Mathai
Nemec, Josef
Revesz, Akos George
Siatkowski, Ronald E
Suenram, Richard Dee
Takagi, Shozo
Trus, Benes L

MASSACHUSETTS
Baglio, Joseph Anthony
Burns, Roger George
Costello, Catherine E
Dahm, Donald J
Eriks, Klaas
Foxman, Bruce Mayer
Fritzsche, Alfred Keith
Gore, William Earl
Haas, Terry Evans
Hayden, Thomas Day
Holmes, Robert Richard
Kishi, Yoshito
Krull, Ira Stanley
Lippard, Stephen J
Loehlin, James Herbert
Margulis, Thomas N
Reis, Arthur Henry, Jr
Williamson, Kenneth L(ee)
Wilson, Linda S (Whatley)

MICHIGAN
Adams, Wade J
Bartell, Lawrence Sims
Bettman, Max
Cody, Wayne Livingston
Cornilsen, Bahne Carl
Duchamp, David James
Eick, Harry Arthur
Einspahr, Howard Martin
Fawcett, Timothy Goss
Heeschen, Jerry Parker
Hillig, Kurt Walter, II
Hohnke, Dieter Karl
Jackson, James Edward
Jacobs, Gerald Daniel
King, Stanley Shih-Tung
Kirschner, Stanley
Kuczkowski, Robert Louis
Meister, John J
Oliver, John Preston
Ovshinsky, Stanford R(obert)
Penner-Hahn, James Edward
Petersen, Donald Ralph
Prakash, D Kash
Rand, Cynthia Lucille
Rohrer, Douglas C
Ryntz, Rose A
Saper, Mark A
Schlick, Shulamith
Squattrito, Philip John
Tai, Julia Chow
Tanis, Steven Paul
Tulinsky, Alexander
Woo, P(eter) W(ing) K(ee)

MINNESOTA
Almlof, Jan Erik
Baumann, Wolfgang J
Britton, Doyle
Etter, Margaret Cairns
Gray, Gary Ronald

Mayerle, James Joseph
Pignolet, Louis H
Stebbings, William Lee
Thompson, Herbert Bradford

MISSOURI
Baughman, Russell George
Brammer, Lee
Felthouse, Timothy R
Freeman, John Jerome
Haymore, Barry Lant
Heitsch, Charles Weyand
Jason, Mark Edward
Munch, John Howard
Ramsey, Robert Bruce
Rath, Nigam Prasad
Ross, Frederick Keith
Talbott, Ted Delwyn
Thomas, George Joseph, Jr
Thompson, Clifton C
Tompkins, Jay Allen
Wong, Tuck Chuen
Yelon, William B

NEBRASKA
Eckhardt, Craig Jon
Harris, Holly Ann

NEVADA
Lightner, David A

NEW HAMPSHIRE
Haendler, Helmut Max
Pfluger, Clarence Eugene

NEW JERSEY
Bahr, Charles Chester
Beck, John Louis
Becker, Joseph Whitney
Boskey, Adele Ludin
Cohen, Allen Irving
Dreeben, Arthur B
Egan, Richard Stephen
Elk, Seymour B
Greenblatt, Martha
Inniss, Daryl
Klaubert, Dieter Heinz
Kleinschuster, Stephen J, III
Lalancette, Roger A
Lam, Yiu-Kuen Tony
Lawton, Stephen Latham
Liesch, Jerrold Michael
McCaslin, Darrell
Marezio, Massimo
Mikes, John Andrew
Mislow, Kurt Martin
Nelson, Gregory Victor
Pilkiewicz, Frank George
Raghavachari, Krishnan
Schlesinger, David H
Schneemeyer, Lynn F
Sibilia, John Philip
Suchow, Lawrence
Thomas, Kenneth Alfred, Jr
Vandenberg, Joanna Maria
Venturella, Vincent Steven
Williams, Thomas Henry

NEW MEXICO
Cromer, Don Tiffany
Eckert, Juergen
Kubas, Gregory Joseph
Lawson, Andrew Cowper, II
Mills, Robert Leroy
Morosin, Bruno
Penneman, Robert Allen
Peterson, Eugene James
Von Dreele, Robert Bruce

NEW YORK
Agarwal, Ramesh Chandra
Bauer, Simon Harvey
Bednowitz, Allan Lloyd
Bell, Thomas Wayne
Blackwell, Crist Scott
Blessing, Robert Harry
Bourne, Philip Eric
Clardy, Jon Christel
Clemans, Stephen D
Cody, Vivian
Cox, David Ernest
Diem, Max
Dodge, Cleveland John
Dorset, Douglas Lewis
Finzel, Rodney Brian
Flanigen, Edith Marie
Gillies, Charles Wesley
Goldberg, Stephen Zalmund
Hanson, Jonathan C
Hanson, Louise I Karle
Herley, Patrick James
Hoard, James Lynn
Hriljac, Joseph A
Kirchner, Richard Martin
Koetzle, Thomas F
Lashewycz-Rubycz, Romana A
Lee, May D-Ming (Lu)
Levy, George Charles
Low, Barbara Wharton
Morrow, Janet Ruth
Myers, Clifford Earl
Parkin, Gerard Francis Ralph
Quigley, Gary Joseph
Reeke, George Norman, Jr

Rossi, Miriam
Rudman, Reuben
Rupp, John Jay
Seeman, Nadrian Charles
Snyder, Robert Lyman
Steinberg, David H
Syed, Ashfaquzzaman
Szalda, David Joseph
Szebenyi, Doletha M E
Thomas, Robert
Williams, Grahame John Bramald
Winter, William Thomas
Ziolo, Ronald F

NORTH CAROLINA
Allen, Harry Clay, Jr
Baird, Herbert Wallace
Hanker, Jacob S
Higgins, Robert H
Jones, Daniel Silas, Jr
Martin, Gary Edwin
Posner, Herbert S
Templeton, Joseph Leslie
Wahl, George Henry, Jr

NORTH DAKOTA
Garvey, Roy George
Radonovich, Lewis Joseph

OHIO
Burow, Duane Frueh
Campbell, James Edward
Cohen, Irwin
Elder, Richard Charles
Gano, James Edward
Hunter, James Charles
Jendrek, Eugene Francis, Jr
Kinstle, Thomas Herbert
Koknat, Friedrich Wilhelm
Kroenke, William Joseph
Mateescu, Gheorghe D
Meek, Devon Walter
Misono, Kunio Shiraishi
Oertel, Richard Paul
Olszanski, Dennis John
Pinkerton, A Alan
Schweitzer, Mark Glenn
Shore, Sheldon Gerald
Sill, Arthur DeWitt
Sullenger, Don Bruce
Tsai, Chun-Che
Wolff, Gunther Arthur
Youngs, Wiley Jay

OKLAHOMA
Daake, Richard Lynn
Wharry, Stephen Mark

OREGON
Brown, Bruce Willard
Ryan, Robert Reynolds
Shoemaker, Clara Brink
Shoemaker, David Powell
Sleight, Arthur William

PENNSYLVANIA
Brady, Stephen Francis
Brixner, Lothar Heinrich
Craven, Bryan Maxwell
Gordon, Janice Taylor
Hutchins, MaryGail Kinzer
Kopple, Kenneth D(avid)
Morris, Marlene Cook
Nelson, George Leonard
Perrotta, Anthony Joseph
Pez, Guido Peter
Ristey, William J
Ross, Stephen T
Sax, Martin
Shepherd, Rex E
Smith, Douglas Lee
Staley, Stuart Warner
Thomas, H Ronald
Widdowson, Katherine Louisa
Zacharias, David Edward

RHODE ISLAND
Carpenter, Gene Blakely
Hartman, Karl August
Williard, Paul Gregory

SOUTH CAROLINA
Amma, Elmer Louis
Brown, Farrell Blenn
Kendall, David Nelson
Posey, Robert Giles
Sproul, Gordon Duane
Stampf, Edward John, Jr

TENNESSEE
Barnes, Craig Eliot
Brown, George Marshall
Habenschuss, Anton
Hanusa, Timothy Paul
Hingerty, Brian Edward
Howell, Elizabeth E
Krutak, James John, Sr
Lenhert, P Galen
Li, Ying Sing
Polavarapu, Prasad Leela
Quist, Arvin Sigvard
Woods, Clifton, III

TEXAS
Boggs, James Ernest
Chu, Shirley Shan-Chi
Cotton, Frank Albert
Czerwinski, Edmund William
Davis, Michael I
Frisch, P Douglas
Hall, Michael Bishop
Hance, Robert Lee
Howatson, John
Ivey, Robert Charles
Jacobson, Allan Joseph
Kevan, Larry
Knapp, Roger Dale
Malloy, Thomas Bernard, Jr
Meyer, Edgar F
Naae, Douglas Gene
Redington, Richard Lee
Steinfink, Hugo
Urdy, Charles Eugene
Watson, William Harold, Jr
Whitmire, Kenton Herbert
Whittlesey, Bruce R

UTAH
Allred, Evan Leigh
Goates, Steven Rex
Kodama, Goji
Powers, Linda Sue

VERMONT
Allen, Christopher Whitney
Jasinski, Jerry Peter

VIRGINIA
Barber, Patrick George
Bryan, Robert Finlay
De Camp, Wilson Hamilton
Evans, Howard Tasker, Jr
Gandour, Richard David
Graybeal, Jack Daniel
Grimes, Russell Newell
Huddle, Benjamin Paul, Jr
Jones, Tappey Hughes
Richardson, Frederick S

WASHINGTON
Adman, Elinor Thomson
Camerman, Arthur
Garland, John Kenneth
Hakomori, Sen-Itiroh
Lingafelter, Edward Clay, Jr
Lytle, Farrel Wayne
Macklin, John Welton
Maki, Arthur George, Jr
Moore, Emmett Burris, Jr
Reid, Brian Robert
Schomaker, Verner
Smith, Richard Dale
Stenkamp, Ronald Eugene
Whitmer, John Charles
Wilson, Archie Spencer

WISCONSIN
Gellman, Samuel Helmer
Landucci, Lawrence L
Qureshi, Nilofer
Schnoes, Heinrich Konstantin
Spencer, Brock

WYOMING
Hodgson, Derek John
Netzel, Daniel Anthony

ALBERTA
Boere, Rene Theodoor
Codding, Penelope Wixson
Cowie, Martin
Krueger, Peter J
Tschuikow-Roux, Eugene
Wieser, Helmut

BRITISH COLUMBIA
Bushnell, Gordon William
Dahn, Jeffery Raymond
Gerry, Michael Charles Lewis

MANITOBA
Cerny, Petr
Lange, Bruce Ainsworth
Mantsch, Henry H
Schaefer, Theodore Peter
Secco, Anthony Silvio
Wikjord, Alfred George

NEW BRUNSWICK
Whitla, William Alexander

NEWFOUNDLAND
Hellou, Jocelyne

NOVA SCOTIA
Burford, Neil
Cameron, T Stanley
McAlduff, Edward J
Wasylishen, Roderick Ernest

ONTARIO
Baer, Hans Helmut
Birnbaum, George I
Brown, Ian David
Buncel, Erwin
Butler, Douglas Neve
Carty, Arthur John

Drake, John Edward
Hitchcock, Adam Percival
Huber, Carol (Saunderson)
Kodama, Hideomi
Lock, Colin James Lyne
MacLean, David Bailey
Payne, Nicholas Charles
Penner, Glenn H
Przybylska, Maria
Tun, Zin

QUEBEC
Cedergren, Robert J
Lin, Sheng-Xiang
Onyszchuk, Mario
Sygusch, Jurgen
Theophanides, Theophile

SASKATCHEWAN
Barton, Richard J
Quail, John Wilson
Robertson, Beverly Ellis

OTHER COUNTRIES
Bayliss, Peter
Hirshfeld, Fred Lurie
Housecroft, Catherine Elizabeth
Newlands, Michael John
Nyburg, Stanley Cecil
Schleyer, Paul von Rague
Wuthrich, Kurt

Synthetic Inorganic & Organometallic Chemistry

ALABAMA
Gerdom, Larry E
Haak, Frederik Albertus
Marano, Gerald Alfred

ARIZONA
Bartocha, Bodo
Lichtenberger, Dennis Lee
Roche, Thomas Stephen
Tamborski, Christ
Valenty, Steven Jeffrey
Whaley, Thomas Patrick

CALIFORNIA
Allen, William Merle
Bau, Robert
Callahan, Kenneth Paul
Chen, Timothy Shieh-Sheng
Cole, Thomas Ernest
Current, Steven P
Davison, John Blake
Evans, William John
Feher, Frank J
Feigelson, Robert Saul
Fisher, Richard Paul
Gagne, Robert Raymond
Glaser, Robert J
Goldwhite, Harold
Grieve, Catherine Macy
Harris, Daniel Charles
Holzmann, Richard Thomas
Hutchinson, Bennett Buckley
Jaecker, John Alvin
Kratzer, Reinhold
Kurkov, Victor Peter
Labinger, Jay Alan
Lawton, Emil Abraham
Lustig, Max
Manasevit, Harold Murray
Mitchell, Dennis Keith
Nelson, Norvell John
Novak, Bruce Michael
Nozaki, Kenzie
Pappatheodorou, Sofia
Patterson, Dennis Bruce
Perry, Dale Lynn
Rainis, Andrew
Raymond, Kenneth Norman
Rosenberg, Sanders David
Shelton, Robert Neal
Singaram, Bakthan
Smith, Robert Alan
Stucky, Galen Dean
Tilley, T(erry) Don
Vollhardt, K Peter C
Weber, Carl Joseph
Wender, Paul Anthony
Wilson, Charles Oren

COLORADO
Anderson, Oren P
Barrett, Anthony Gerard Martin
Fleming, Michael Paul
Ritchey, John Michael
Thompson, Ronald G

CONNECTICUT
Crabtree, Robert H
Haas, Thomas J
Holland, Gerald Fagan
McKeon, James Edward
Nielsen, John Merle
Suib, Steven L
Yang, Darchun Billy

DELAWARE
Baker, Ralph Thomas
Breslow, David Samuel

Synthetic Inorganic & Organometallic Chemistry (cont)

Bulkowski, John Edmund
Calabrese, Joseph C
Corbin, David Richard
Foley, Henry Charles
Ford, Thomas Aven
Fukunaga, Tadamichi
Glaeser, Hans Hellmut
Herron, Norman
Hess, Richard William
Ittel, Steven Dale
Karel, Karin Johnson
Longhi, Raymond
McKinney, Ronald James
McLain, Stephan James
Mills, George Alexander
Moser, Glenn Allen
Niedzielski, Edmund Luke
Repka, Benjamin C
Rheingold, Arnold L
Rondestvedt, Christian Scriver, Jr
Soboczenski, Edward John
Taber, Douglass Fleming
Theopold, Klaus Hellmut
Tomic, Ernst Alois

DISTRICT OF COLUMBIA
Butcher, Raymond John
Butter, Stephen Allan
Hudrlik, Paul Frederick
Kulawiec, Robert Joseph
Nordquist, Paul Edgard Rudolph, Jr
Pope, Michael Thor

FLORIDA
Hahn, Elliot F
Jones, William Maurice
Kauder, Otto Samuel
Newkome, George Richard
Rochow, Eugene George
Sisler, Harry Hall
Stanko, Joseph Anthony
Stoufer, Robert Carl

GEORGIA
Allison, John P
Anderson, Bruce Martin
Barbas, John Theophani
Bottomley, Lawrence Andrew
Chandler, James Harry, III
Islam, M Safizul
James, Jeffrey
Liotta, Dennis C
Rees, William Smith, Jr

HAWAII
Cramer, Roger Earl
Seff, Karl

IDAHO
Baker, John David
Bitterwolf, Thomas Edwin

ILLINOIS
Arzoumanidis, Gregory G
Bernhardt, Randal Jay
Broach, Robert William
Fischer, Albert Karl
Gembicki, Stanley Arthur
Girolami, Gregory Scott
Gupta, Goutam
Halpern, Jack
Harris, Ronald L
Harvey, Ronald Gilbert
Henry, Patrick M
Herlinger, Albert William
Hollins, Robert Edward
Hopper, Steven Phillip
Keiter, Richard Lee
Kieft, Richard Leonard
Lehman, Dennis Dale
Lipka, James J
Mendelsohn, Marshall H
Menke, Andrew G
Mirkin, Chad Alexander
Nuzzo, Ralph George
Poeppelmeier, Kenneth Reinhard
Rathke, Jerome William
Rauchfuss, Thomas Bigley
Rogers, Robin Don
Rosen, Bruce Irwin
Shapley, John Roger
Shriver, Duward F
Suslick, Kenneth Sanders
Trevillyan, Alvin Earl
Weil, Thomas Andre
Wolff, Thomas E
Yokelson, Howard Bruce

INDIANA
Bergstrom, Donald Eugene
Chisholm, Malcolm Harold
Christou, George
Fehlner, Thomas Patrick
Green, Mark Alan
Helquist, Paul M
Kreider, Leonard C(ale)
Kubiak, Clifford P
Miller, Roy Glenn
Negishi, Ei-ichi

Scheidt, Walter Robert
Todd, Lee John

IOWA
Angelici, Robert Joe
Eyman, Darrell Paul
Hampton, David Clark
Kostic, Nenad M
McCarley, Robert Eugene
Messerle, Louis
Miller, Gordon James

KANSAS
Landis, Arthur Melvin
McCormick, Bailie Jack
Purcell, Keith Frederick

KENTUCKY
Brill, Joseph Warren
Buchanan, Robert Martin
Gibson, Dorothy Hinds
Henrickson, Charles Henry
Jayaram, Beby
Meier, Mark Stephan
Owen, David Allan
Selegue, John Paul
Slocum, Donald Warren

LOUISIANA
Conrad, Franklin
Doomes, Earl
Ehrlich, Kenneth Craig
Frazier, Claude Clinton, III
Gautreaux, Marcelian Francis
Halbert, Thomas Risher
Ham, Russell Allen
Harper, Robert John, Jr
Hornbaker, Edwin Dale
Klobucar, William Dirk
Lee, John Yuchu
Mague, Joel Tabor
Mangham, Jesse Roger
Maverick, Andrew William
Mollere, Phillip David
Plonsker, Larry
Robson, Harry Edwin
Schweizer, Albert Edward
Selbin, Joel
Shin, Kju Hi
Shubkin, Ronald Lee
Siriwardane, Upali
Stahly, Glenn Patrick
Stanley, George Geoffrey
Stuntz, Gordon Frederick
Theriot, Kevin Jude
Welch, Clark Moore
Zaweski, Edward F

MAINE
Bragdon, Robert Wright

MARYLAND
Andrulis, Peter Joseph, Jr
Argauer, Robert John
Barnes, James Alford
Coyle, Thomas Davidson
DeCraene, Denis Fredrick
Fleming, Paul Robert
Gause, Evelyn Pauline
Grim, Samuel Oram
Joedicke, Ingo Bernd
Poli, Rinaldo
Ritter, Joseph John
Schwartz, Paul
Shade, Joyce Elizabeth
Simon, Robert Michael
Viswanathan, C T

MASSACHUSETTS
Archer, Ronald Dean
Baglio, Joseph Anthony
Becker, Ernest I
Ciappenelli, Donald John
Clarke, Michael J
Evans, Michael Douglas
Foxman, Bruce Mayer
Goldberg, Gershon Morton
Howard, Wilmont Frederick, Jr
Idelson, Martin
Kramer, Charles Edwin
Lippard, Stephen J
Moser, William Ray
Parham, Marc Ellous
Rupich, Martin Walter
Seyferth, Dietmar
Snider, Barry B
Stuhl, Louis Sheldon
Sullivan, Edward Augustine
Taunton-Rigby, Alison
Tulip, Thomas Hunt
Urry, Grant Wayne
Wheelock, Kenneth Steven
Wreford, Stanley S

MICHIGAN
Anderson, Amos Robert
Anderson, Melvin Lee
Atwell, William Henry
Bailey, Donald Leroy
Bajzer, William Xavier
Baney, Ronald Howard
Brenner, Alan
Butts, Susan Beda
Chandra, Grish

Chen, Huai Gu
Harris, Paul Jonathan
Hartwell, George E
Li, Shin-Hwa
Lintvedt, Richard Lowell
Lipton, Michael Forrester
Mance, Andrew Mark
Moehs, Peter John
Moser, Frank Hans
Newcomb, Martin
Oliver, John Preston
Rasmussen, Paul G
Rutter, Edward Walter, Jr
Rycheck, Mark Rule
Salinger, Rudolf Michael
Smart, James Blair
Stark, Forrest Otto
Stille, John Robert
Weyenberg, Donald Richard

MINNESOTA
Esmay, Donald Levern
Gladfelter, Wayne Lewis
Johnson, Bryce Vincent
Miessler, Gary Lee
Mueting, Ann Marie
Pignolet, Louis H
Reich, Charles
Schmit, Joseph Lawrence

MISSISSIPPI
Howell, John Emory
Pittman, Charles U, Jr

MISSOURI
Bleeke, John Richard
Felthouse, Timothy R
Frederick, Raymond H
Haymore, Barry Lant
Heitsch, Charles Weyand
Katti, Kattesh V
Krueger, Paul A
Lennon, Patrick James
Long, Gary John
Markowski, Henry Joseph
Moedritzer, Kurt
Morris, Donald Eugene
Rath, Nigam Prasad
Riley, Dennis Patrick
Rogic, Milorad Mihailo
Sabacky, M Jerome
Wulfman, David Swinton

MONTANA
Rosenberg, Edward

NEBRASKA
George, T Adrian
Rieke, Reuben Dennis

NEW HAMPSHIRE
Danforth, Raymond Hewes
Haendler, Helmut Max
Hughes, Russell Profit
Meloni, Edward George
Ort, Morris Richard

NEW JERSEY
Armbrecht, Frank Maurice, Jr
Battisti, Angelo James
Bertelo, Christopher Anthony
Bertz, Steven Howard
Boyle, Richard James
Cullen, Glenn Wherry
Dreeben, Arthur B
Giddings, Sydney Arthur
Granchi, Michael Patrick
Hung, Paul Ling-Kong
Iorns, Terry Vern
Jacobson, Stephen Ernest
Johnson, Jack Wayne
Klemann, Lawrence Paul
Kline, Toni Beth
Kuehl, Guenter Hinrich
Kwon, Joon Taek
Larkin, William Albert
Leong, William
McGinnis, James Lee
Magder, Jules
Prapas, Aristotle G
Ramsden, Hugh Edwin
Reichle, Walter Thomas
Rosan, Alan Mark
Rossi, Robert Daniel
Schwartz, Herbert
Sekutowski, Dennis G
Spohn, Ralph Joseph
Thich, Jan Adong
Tiethof, Jack Alan
Tucci, Edmond Raymond
Voss, Kenneth Edwin
Weissman, Paul Morton
Wheaton, Gregory Alan
Yocom, Perry Niel

NEW MEXICO
Bowen, Scott Michael
Harrill, Robert W
Kubas, Gregory Joseph
Mullendore, A(rthur) W(ayne)
Peterson, Eugene James
Smith, Wayne Howard
West, Mike Harold
Zeigler, John Martin

NEW YORK
Amero, Bernard Alan
Andrews, Mark Allen
Atwood, Jim D
Beachley, Orville Theodore, Jr
Bowser, James Ralph
Bullock, R Morris
Burlitch, James Michael
Carpenter, Barry Keith
Dabrowiak, James Chester
Daves, Glenn Doyle, Jr
Dehn, Joseph William, Jr
Eschbach, Charles Scott
Flanigen, Edith Marie
Gancy, Alan Brian
Gardner, Sylvia Alice
Giannelis, Emmanuel P
Hriljac, Joseph A
Interrante, Leonard V
Jones, Wayne E
Jones, William Davidson
Kallen, Thomas William
Katz, Thomas Joseph
Kerber, Robert Charles
Koch, Stephen Andrew
Kotz, John Carl
Landon, Shayne J
Lashewycz-Rubycz, Romana A
Lau, Philip T S
Lees, Alistair John
Lemanski, Michael Francis
McKelvie, Neil
MacLaury, Michael Risley
Mayr, Andreas
Miller, William Taylor
Mirviss, Stanley Burton
Morrow, Janet Ruth
Ojima, Iwao
Parkin, Gerard Francis Ralph
Philipp, Manfred (Hans)
Poss, Andrew Joseph
Post, Howard William
Rupp, John Jay
Sogah, Dotsevi Yao
Sullivan, Michael Francis
Triplett, Kelly B
Vaska, Lauri
Ziolo, Ronald F

NORTH CAROLINA
Barborak, James Carl
Bereman, Robert Deane
Brookhart, Maurice S
Clemens, Donald Faull
Doak, George Osmore
Farona, Michael F
Gains, Lawrence Howard
Garrou, Philip Ernest
Gillooly, George Rice
Groszos, Stephen Joseph
Jackels, Susan Carol
Laganis, Evan Dean
McDonald, Daniel Patrick
Noftle, Ronald Edward
O'Connor, Lila Hunt
Panek, Edward John
Pruett, Roy L
Schwindeman, James Anthony
Spielvogel, Bernard Franklin
Switzer, Mary Ellen Phelan
Templeton, Joseph Leslie
Wasson, John R
Wilson, Lauren R

NORTH DAKOTA
Abrahamson, Harmon Bruce
Boudjouk, Philip Raymond
Garvey, Roy George
Jasperse, Craig Peter

OHIO
Ackermann, Martin Nicholas
Alexander, John J
Attig, Thomas George
Blum, Patricia Rae
Bradley, Ronald W
Chrysochoos, John
Conder, Harold Lee
Cummings, Sue Carol
Davies, Julian Anthony
Didchenko, Rostislav
Fechter, Robert Bernard
Feldman, Julian
Fischer, Mark Benjamin
Fraenkel, Gideon
Franks, John Anthony, Jr
Gibson, Thomas William
Hayes, Jeffrey Charles
Horne, Samuel Emmett, Jr
Jache, Albert William
Jacobsen, Donald Weldon
Johnson, Gordon Lee
Kang, Jung Wong
Katovic, Vladimir
Kenney, Malcolm Edward
Koknat, Friedrich Wilhelm
Kroenke, William Joseph
Litt, Morton Herbert
McDonald, John William
Meek, Devon Walter
Myers, Ronald Eugene
Nee, Michael Wei-Kuo
Nicholson, D Allan
Olszanski, Dennis John

Pinkerton, A Alan
Schilling, Curtis Louis, Jr
Shore, Sheldon Gerald
Simmons, Harry Dady, Jr
Tessier, Claire Adrienne
Throckmorton, Peter E
Turkel, Rickey M(artin)
Williams, John Paul
Wojcicki, Andrew
Youngs, Wiley Jay

OKLAHOMA
Daake, Richard Lynn
Efner, Howard F
Graves, Victoria
Highsmith, Ronald Earl
Lindahl, Charles Blighe
Nicholas, Kenneth M

OREGON
Halko, Barbara Tomlonovic
Humphrey, J Richard
Tyler, David Ralph
Yoke, John Thomas

PENNSYLVANIA
Adams, Roy Melville
Allcock, Harry R(ex)
Balaba, Willy Mukama
Bartish, Charles Michael Christopher
Bassner, Sherri Lynn
Collins, Terrence James
Cooper, Norman John
Dent, Anthony L
Durante, Vincent Anthony
Faltynek, Robert Allen
Ford, Michael Edward
Gitlitz, Melvin H
Grey, Roger Allen
Gunther, Wolfgang Hans Heinrich
Hunt, David Allen
Jordan, Robert Kenneth
Kennedy, Flynt
Knoth, Walter Henry, Jr
Kraihanzel, Charles S
Laemmle, Joseph Thomas
Larson, Gerald Louis
Lavoie, Alvin Charles
Mair, Robert Dixon
Maryanoff, Cynthia Anne Milewski
Mascioli, Rocco Lawrence
Matyjaszewski, Krzysztof
Murphy, Clarence John
Pez, Guido Peter
Richey, Herman Glenn, Jr
Rodgers, Sheridan Joseph
Sherman, Larry Ray
Smart, G N Russell
Smith, W Novis, Jr
Sollott, Gilbert Paul
Stehly, David Norvin
Steward, Omar Waddington
Urenovitch, Joseph Victor
Waller, Francis Joseph
Weisfeld, Lewis Bernard
Willeford, Bennett Rufus, Jr

SOUTH CAROLINA
Adams, Richard Darwin
Cogswell, George Wallace
Culbertson, Edwin Charles
Evans, David Wesley
Guthrie, Roger Thackston
Lee, Burtrand Insung
Posey, Robert Giles
Reger, Daniel Lewis
Robinson, Myrtle Tonne
Robinson, Robert Earl
Shah, Hamish V
Stampf, Edward John, Jr
Tour, James Mitchell

TENNESSEE
Adcock, James Luther
Barnes, Craig Eliot
Fenyes, Joseph Gabriel Egon
Gibson, John Knight
Hanusa, Timothy Paul
Okrasinski, Stanley John
Summers, James Thomas
White, Alan Wayne
Woods, Clifton, III
Zoeller, Joseph Robert

TEXAS
Bergbreiter, David Edward
Braterman, Paul S
Chen, Edward Chuck-Ming
Ciufolini, Marco A
Cotton, Frank Albert
Davenport, Kenneth Gerald
Edmondson, Morris Stephen
Ekerdt, John Gilbert
Evans, Wayne Errol
Frisch, P Douglas
Hosmane, Narayan Sadashiv
Jacobson, Allan Joseph
Johnson, Malcolm Pratt
Jones, Paul Ronald
Kelsey, Donald Ross
Kochhar, Rajinder Kumar
Laane, Jaan
Lattman, Michael
Malpass, Dennis B

Marchand, Alan Philip
Mills, Nancy Stewart
Pannell, Keith Howard
Sessler, Jonathan Lawrence
Singleton, David Michael
Thompson, Michael McCray
Unruh, Jerry Dean
Whitmire, Kenton Herbert
Whittlesey, Bruce R
Wisian-Neilson, Patty Joan

UTAH
Clark, David Neil
Ernst, Richard Dale
Kodama, Goji
Richmond, Thomas G
Van Wagenen, Bradford Carr

VERMONT
Allen, Christopher Whitney
Jasinski, Jerry Peter
Krause, Ronald Alfred

VIRGINIA
Averill, Bruce Alan
Bailey, Susan Goodman
Cavanaugh, Margaret Anne
Grimes, Russell Newell
Hutchison, David Allan
Lodoen, Gary Arthur
Merola, Joseph Salvatore
Myers, William Howard
St Clair, Anne King
Sayala, Chhaya
Topich, Joseph
Turner, Anne Halligan

WASHINGTON
Bennett, Clifton Francis
Smith, Victor Herbert

WEST VIRGINIA
Breneman, William C
Smith, James Allbee

WISCONSIN
Baetke, Edward A
Lepeska, Bohumir
Rainville, David Paul
Reich, Hans Jurgen
Runquist, Alfonse William
Treichel, Paul Morgan, Jr
West, Robert C

WYOMING
Branthaver, Jan Franklin
Sullivan, Brian Patrick

ALBERTA
Boere, Rene Theodoor
Chivers, Tristram
Cowie, Martin
Graham, William Arthur Grover
Sanger, Alan Rodney

BRITISH COLUMBIA
James, Brian Robert
McAuley, Alexander
Mitchell, Reginald Harry
Orvig, Chris
Pomeroy, Roland Kenneth

MANITOBA
Janzen, Alexander Frank

NEWFOUNDLAND
Lucas, Colin Robert

NOVA SCOTIA
Bridgeo, William Alphonsus
Burford, Neil
Piorko, Adam M

ONTARIO
Alper, Howard
Baines, Kim Marie
Barrett, Peter Fowler
Brook, Adrian Gibbs
Buncel, Erwin
Carty, Arthur John
Drake, John Edward
Eliades, Theo I
Hemmings, Raymond Thomas
Holloway, Clive Edward
Lautens, Mark
Makhija, Ramesh C
Miller, Jack Martin
Payne, Nicholas Charles
Rempel, Garry Llewellyn
Walker, Alan

QUEBEC
Butler, Ian Sydney
Harrod, John Frank
Onyszchuk, Mario
Shaver, Alan Garnet
Theophanides, Theophile
Wuest, James D

OTHER COUNTRIES
Chan, Albert Sun Chi
Housecroft, Catherine Elizabeth
Irgolic, Kurt Johann
Irvine, Stuart James Curzon

Rakita, Philip Erwin
Sarkisov, Pavel Jebraelovich
Wenclawiak, Bernd Wilhelm

Synthetic Organic & Natural Products Chemistry

ALABAMA
Andrews, Russell S, Jr
Breithaupt, Lea Joseph, Jr
Czepiel, Thomas P
Fox, Sidney Walter
Hall, David Michael
Kiely, Donald Edward
Leslie, Thomas M
Parish, Edward James
Riley, Thomas N
Secrist, John Adair, III
Struck, Robert Frederick
Whorton, Rayburn Harlen

ARIZONA
Doubek, Dennis Lee
Herald, Cherry Lou
Herald, Delbert Leon, Jr
Hoffmann, Joseph John
Howells, Thomas Alfred
Hruby, Victor J
Jungermann, Eric
Kelley, Maurice Joseph
Moore, Ana M L
Smith, Cecil Randolph, Jr
Timmermann, Barbara Nawalany

ARKANSAS
Commerford, John D
Ervin, Hollis Edward
Fu, Peter Pi-Cheng
Glendening, Norman Willard

CALIFORNIA
Acton, Edward McIntosh
Banigan, Thomas Franklin, Jr
Bird, Harold L(eslie), Jr
Boger, Dale L
Butler, Alison
Chan, David S
Chaubal, Madhukar Gajanan
Chen, Ming Sheng
Cole, Thomas Ernest
Cooper, Geoffrey Kenneth
Correia, John Sidney
Corse, Joseph Walters
Craig, John Cymerman
Crews, Phillip
Crum, James Davidson
Daub, Guido William
Dev, Vasu
Djerassi, Carl
Dougherty, Dennis A
Erdman, Timothy Robert
Flath, Robert Arthur
Fung, Steven
Gaffield, William
Gall, Walter George
Gardner, Howard Shafer
Gaston, Lyle Kenneth
Ginsberg, Theodore
Gordon, Eric Michael
Grethe, Guenter
Harris, William Merl
Heather, James Brian
Henrick, Clive Arthur
Houk, Kendall N
Jones, Todd Kevin
Jurd, Leonard
Kiely, John Steven
Koepfli, Joseph B
Kryger, Roy George
Lee, Chi-Hang
Lewis, Robert Allen
Lin, Robert I-San
Lin, Tz-Hong
Livingston, Douglas Alan
Lorance, Elmer Donald
Lucci, Robert Dominick
McGee, Lawrence Ray
Matuszak, Charles A
Micheli, Robert Angelo
Mihailovski, Alexander
Millar, Jocelyn Grenville
Miller, Russell Bryan
Molyneux, Russell John
Moos, Walter Hamilton
Morgan, Alan Raymond
Mosher, Harry Stone
Musser, John H
Nambiar, Krishnan P
Nestor, John Joseph, Jr
Nitecki, Danute Emilija
Nutt, Ruth Foelsche
Ogimachi, Naomi Neil
Ortiz de Montellano, Paul Richard
Overman, Larry Eugene
Pavlath, Attila Endre
Petryka, Zbyslaw Jan
Pierschbacher, Michael Dean
Prakash, Surya G K
Preus, Martin William
Reardon, Edward Joseph, Jr
Rivier, Jean E F
Robins, Roland Kenith

Roitman, James Nathaniel
Sanchez, Robert A
Seapy, Dave Glenn
Sheridon, Nicholas Keith
Shoolery, James Nelson
Siegel, Brock Martin
Sims, James Joseph
Singaram, Bakthan
Singleton, Vernon LeRoy
Snow, John Thomas
Spatz, David Mark
Spray, Clive Robert
Swidler, Ronald
Untch, Karl G(eorge)
Van Tamelen, Eugene Earl
Vinogradoff, Anna Patricia
Vollhardt, K Peter C
Waiss, Anthony C, Jr
Welch, Steven Charles
Wender, Paul Anthony
Williams, John Wesley
Winston, Philip Eldridge, Jr
Wong, Chi-Huey
Yavrouian, Andre
Zavarin, Eugene
Zimmerman, Mary Prislopski

COLORADO
Allen, Larry Milton
Barrett, Anthony Gerard Martin
Bartlett, William Rosebrough
Caruthers, Marvin Harry
DoMinh, Thap
Druelinger, Melvin L
Evans, David Lane
Fleming, Michael Paul
Gabel, Richard Allen
Guinn, Denise Eileen
Stermitz, Frank
Walba, David Mark
Wright, Terry L

CONNECTICUT
Arif, Shoaib
Baldwin, Ronald Martin
Brown, Keith Charles
Buchanan, Ronald Leslie
Church, Robert Fitz (Randolph)
Davis, Robert Glenn
Dirlam, John Philip
Doshan, Harold David
Dow, Robert Lee
Doyle, Terrence William
Forenza, Salvatore
Forssen, Eric Anton
Gewanter, Herman Louis
Gordon, Philip N
Holland, Gerald Fagan
Kavarnos, George James
Kluender, Harold Clinton
Kotick, Michael Paul
Kuck, Julius Anson
Luh, Yuhshi
Marfat, Anthony
Melvin, Lawrence Sherman, Jr
Perry, Clark William
Proverb, Robert Joseph
Rennhard, Hans Heinrich
Rother, Ana
Schaaf, Thomas Ken
Schile, Richard Douglas
Shapiro, Warren Barry
Sieckhaus, John Francis
Volkmann, Robert Alfred
Wasser, Richard Barkman
Witczak, Zbigniew J
Wright, Herbert Fessenden
Wuskell, Joseph P
Ziegler, Theresa Frances

DELAWARE
Chorvat, Robert John
Class, Jay Bernard
Confalone, Pat N
Crow, Edwin Lee
DeGrado, William F
Dubas, Lawrence Francis
Ehrich, F(elix) Frederick
Feltzin, Joseph
Freerksen, Robert Wayne
Fukunaga, Tadamichi
Hayek, Mason
Holyoke, Caleb William, Jr
Jenner, Edward L
Keim, Gerald Inman
Kettner, Charles Adrian
Krespan, Carl George
Loomis, Gary Lee
Monagle, Daniel J
Pratt, Herbert T
Pretka, John E
Putnam, Stearns Tyler
Reap, James John
Rogers, Donald B
Rondestvedt, Christian Scriver, Jr
Scalfarotto, Robert Emil
Schlecht, Matthew Fred
Taber, Douglass Fleming
Targett, Nancy McKeever
Thayer, Chester Arthur
Twelves, Robert Ralph
Webster, Owen Wright
Wuonola, Mark Arvid

Synthetic Organic & Natural Products Chemistry (cont)

DISTRICT OF COLUMBIA
Attaway, David Henry
Baker, Charles Wesley
Bultman, John D
Horak, Vaclav
Hudrlik, Anne Marie
Hudrlik, Paul Frederick
Klayman, Daniel Leslie
Kulawiec, Robert Joseph
Page, Samuel William
Ponaras, Anthony A
Sarmiento, Rafael Apolinar

FLORIDA
Alderson, Thomas
Battiste, Merle Andrew
Bledsoe, James O, Jr
Blossey, Erich Carl
Derfer, John Mentzer
Herz, Werner
Hull, Harry H
Krafft, Marie Elizabeth
Lee, Henry Joung
Linfield, Warner Max
Mehta, Nariman Bomanshaw
Miles, Delbert Howard
Moskowitz, Mark Lewis
Most, David S
Ohrn, Nils Yngve
Owen, Terence Cunliffe
Pop, Emil
Porter, Lee Albert
Rokach, Joshua
Romeo, John Thomas
Russell, James
Schwarting, Arthur Ernest
Shapiro, Jeffrey Paul
Sisler, Harry Hall
Spencer, John Lawrence
Teal, Peter E A
Vander Meer, Robert Kenneth
Williams, Norris Hagan

GEORGIA
Bayer, Charlene Warres
Chiu, Kirts C
Chortyk, Orestes Timothy
Chu, Chung K
Clark, Charles Kittredge
Espelie, Karl Edward
Hill, Richard Keith
Hurd, Phillip Wayne
Iacobucci, Guillermo Arturo
Knopka, W N
Kocurek, Michael Joseph
Leffingwell, John C
Malcolm, Earl Walter
Nes, William David
Pelletier, S William
Polk, Malcolm Benny
Schramm, Lee Clyde
Sweeny, James Gilbert
Tolbert, Laren Malcolm
Tyler, Jean Mary
Walker, William Comstock
Woodford, James

HAWAII
Grothaus, Paul Gerard
Patterson, Gregory Matthew Leon
Scheuer, Paul Josef
Tius, Marcus Antonius

IDAHO
Lambuth, Alan Letcher
Mosher, Michael David
Natale, Nicholas Robert

ILLINOIS
Abbott, Thomas Paul
Adlof, Richard Otto
Ager, David John
Arendsen, David Lloyd
Armstrong, P(aul) Douglas
Babler, James Harold
Bernhardt, Randal Jay
Brooks, Dee W
Buchman, Russell
Cengel, John Anthony
Chorghade, Mukund Shankar
Chu, Daniel Tim-Wo
Cole, Wayne
Cordell, Geoffrey Alan
Curtis, Veronica Anne
DeBoer, Edward Dale
Dellaria, Joseph Fred, Jr
Drengler, Keith Allan
Elliott, William John
Eskins, Kenneth
Estes, Timothy King
Field, Kurt William
Ford, Susan Heim
Friedman, Robert Bernard
Garland, Robert Bruce
Giori, Claudio
Grotefend, Alan Charles
Groziak, Michael Peter
Hofreiter, Bernard T
Hsu, Charles Fu-Jen
Jiu, James

Johnson, John Harold
Katzenellenbogen, John Albert
Kerdesky, Francis A J
Kerkman, Daniel Joseph
Klein, Larry L
Klimstra, Paul D
Ku, Yi-Yin
Lash, Timothy David
Lewis, Morton
McGorrin, Robert Joseph
Maher, George Garrison
Miller, Fredric N
Mohan, Prem
Pariza, Richard James
Perun, Thomas John
Piatak, David Michael
Powell, Richard Grant
Prasad, Ramanujam Srinivasa
Rakoff, Henry
Rankin, John Carter
Richmond, James M
Rinehart, Kenneth Lloyd
Rosenberg, Saul Howard
Rosenbrook, William, Jr
Shotwell, Odette Louise
Silverman, Richard Bruce
Tindall, Donald R
Tortorello, Anthony Joseph
Traxler, James Theodore
Tsang, Joseph Chiao-Liang
Wallen, Lowell Lawrence
Wang, Gary T
Weier, Richard Mathias
Weston, Arthur Walter
Yokelson, Howard Bruce

INDIANA
Barnett, Charles Jackson
Becker, Elizabeth Ann (White)
Bergstrom, Donald Eugene
Besch, Henry Roland, Jr
Buchanan, Russell Allen
Bucholtz, Dennis Lee
Carmack, Marvin
Chang, Ching-Jer
Cooper, Robin D G
Corey, Paul Frederick
Dreikorn, Barry Allen
Fidelle, Thomas Patrick
Fife, Wilmer Krafft
Gresham, Robert Marion
Gutowski, Gerald Edward
Hamill, Robert L
Harper, Richard Waltz
Haslanger, Martin Frederick
Helquist, Paul M
Jackson, Bill Grinnell
Kjonaas, Richard A
Kossoy, Aaron David
Long, Eric Charles
McLaughlin, Jerry Loren
Marconi, Gary G
Negishi, Ei-ichi
O'Doherty, George Oliver-Plunkett
Patwardhan, Bhalchandra H
Rapkin, Myron Colman
Robertson, Donald Edwin
Rowe, James Lincoln
Shiner, Vernon Jack, Jr
Stevenson, William Campbell
Thompson, Gerald Lee
Wildfeuer, Marvin Emanuel
Yordy, John David

IOWA
Barua, Arun B
Carew, David P
Downing, Donald Talbot
Hampton, David Clark
Kraus, George Andrew
Kraus, Kenneth Wayne
Linhardt, Robert John
Nair, Vasu
Watkins, G(ordon) Leonard
Wiemer, David F

KANSAS
Beechan, Curtis Michael
Borchardt, Ronald T
Burgstahler, Albert William
Cheng, Chia-Chung
Duncan, William Perry
Engler, Thomas A
Hua, Duy Huu
McCullough, Elizabeth Ann
Marx, Michael
Mitscher, Lester Allen
Wiley, James C, Jr

KENTUCKY
Crooks, Peter Anthony
Kennedy, John Elmo, Jr
Lauterbach, John Harvey
Luzzio, Frederick Anthony
McGinness, William George, III
Meier, Mark Stephan
Slocum, Donald Warren
Taylor, Kenneth Grant
Wong, John Lui

LOUISIANA
Bauer, Dennis Paul
Bayer, Arthur Craig
Connick, William Joseph, Jr

Ensley, Harry Eugene
Fischer, Nikolaus Hartmut
Frank, Arlen W(alker)
Gallo, August Anthony
Gonzales, Elwood John
Harper, Robert John, Jr
Klobucar, William Dirk
McGraw, Gerald Wayne
Shin, Kju Hi
Smith, Grant Warren, II
Somsen, Robert Alan
Starrett, Richmond Mullins
Stowell, John Charles
Swenson, David Harold
Theriot, Kevin Jude
Vail, Sidney Lee
Welch, Clark Moore
Willer, Rodney Lee
Zentner, Thomas Glenn

MAINE
Eames, Arnold C
Green, Brian
McKerns, Kenneth (Wilshire)
Pierce, Arleen Cecilia
Renn, Donald Walter
Spencer, Claude Franklin

MARYLAND
Aldrich, Jeffrey Richard
Brady, Robert Frederick, Jr
Burrows, Elizabeth Parker
Burton, Lester Percy Joseph
Carter, John Paul, Jr
Cavagna, Giancarlo Antonio
Chen, Hao
Chiu, Tony Man-Kuen
Christensen, Richard G
Cohen, Louis Arthur
Cragg, Gordon Mitchell
Engle, Robert Rufus
Falci, Kenneth Joseph
Fales, Henry Marshall
Feldlaufer, Mark Francis
Feyns, Liviu Valentin
Fleming, Paul Robert
Galetto, William George
Guruswamy, Vinodhini
Hackley, Brennie Elias, Jr
Henery-Logan, Kenneth Robert
Hosmane, Ramachandra Sadashiv
Inscoe, May Nilson
Jacobson, Kenneth Alan
Jacobson, Martin
James, John Cary
Klein, Michael
Lusby, William Robert
McShefferty, John
Marquez, Victor Esteban
Massie, Samuel Proctor
Miller, Stephen P F
Misra, Renuka
Mokhtari-Rejali, Nahld
Nemec, Josef
Otani, Theodore Toshiro
Paull, Kenneth Dywain
Poochikian, Guiragos K
Reynolds, Kevin A
Rice, Kenner Cralle
Roller, Peter Paul
Saunders, James Allen
Schwartz, Paul
Scovill, John Paul
Snader, Kenneth Means
Sripada, Pavanaram Kameswara
Torrence, Paul Frederick
Vandegaer, Jan Edmond
Warthen, John David, Jr
Watthey, Jeffrey William Herbert
Weser, Don Benton
Whaley, Wilson Monroe
Wheeler, James William, Jr
Yoho, Clayton W
Zeiger, William Nathaniel

MASSACHUSETTS
Ahern, David George
Ajami, Alfred Michel
Carpino, Louis A
Chase, Arleen Ruth
Clapp, Richard Crowell
Driedger, Paul Edwin
Eby, Robert Newcomer
Haber, Stephen B
Hendrickson, James Briggs
Hodge, Richard Paul
Hodgins, George Raymond
Holick, Michael Francis
Holick, Sally Ann
Hurd, Richard Nelson
Idelson, Martin
Jachimowicz, Felek
Jiang, Jack Bau-Chien
Kelley, Charles Joseph
Kishi, Yoshito
Klanfer, Karl
Krull, Ira Stanley
Leary, John Dennis
Naidus, Harold
Ocain, Timothy Donald
O'Rell, Dennis Dee
Raffauf, Robert Francis
Rosenfeld, Stuart Michael
Ruelius, Hans Winfried

Snider, Barry B
Soffer, Milton David
Stinchfield, Carleton Paul
Sullivan, Edward Augustine
Takman, Bertil Herbert
Warner, John Charles
Williams, John Russell

MICHIGAN
Aaron, Charles Sidney
Axen, Udo Friedrich
Belmont, Daniel Thomas
Berman, Ellen Myra
Blankley, Clifton John
Chen, Huai Gu
Clark, Peter David
Culbertson, Townley Payne
Darby, Nicholas
Deering, Carl F
Dewitt, Sheila Hobbs
Domagala, John Michael
Elrod, David Wayne
Elslager, Edward Faith
Frade, Peter Daniel
Greminger, George King, Jr
Hart, Harold
Herrinton, Paul Matthew
Hester, Jackson Boling, Jr
Huang, Che C
Huber, Joel E
Krueger, Robert John
Kuo, Mingshang
Laks, Peter Edward
LeBel, Norman Albert
Lipton, Michael Forrester
Logue, Marshall Woford
Marshall, Charles Wheeler
Meister, John J
Moore, Eugene Roger
Nader, Bassam Salim
Nagler, Robert Carlton
Nestrick, Terry John
Newcomb, Martin
Otto, Charlotte A
Pearlman, Bruce A
Pearson, William H
Petersen, Quentin Richard
Pfeiffer, Curtis Dudley
Putnam, Alan R
Sacks, Clifford Eugene
Schramm, John Gilbert
Skorcz, Joseph Anthony
Stille, John Robert
Suggs, William Terry
Tanis, Steven Paul
Thomas, Richard Charles
Verseput, Herman Ward
Woo, P(eter) W(ing) K(ee)
Wormser, Henry C
Wuts, Peter G M

MINNESOTA
Barany, George
Carlson, Janet Lynn
Cheng, H(wei)-H(sien)
Crooks, Stephen Lawrence
Gray, Gary Ronald
Katz, Morris Howard
Lucast, Donald Hurrell
Noland, Wayland Evan
Pletcher, Wayne Albert
Schultz, Robert John
Spessard, Gary Oliver
Throckmorton, James Rodney

MISSISSIPPI
Cibula, William Ganley
Hufford, Charles David
McChesney, James Dewey
Panetta, Charles Anthony
Sindelar, Robert D
Williamson, John S

MISSOURI
Ahmad, Moghisuddin
Asrar, Jawed
Baker, Joseph Willard
Cole, Douglas L
Coudron, Thomas A
Delaware, Dana Lewis
Durley, Richard Charles
Elliott, William H
Giarrusso, Frederick Frank
Gustafson, Mark Edward
Koenig, Karl E
Ku, Audrey Yeh
McCormick, John Pauling
Moeller, Kevin David
Plimmer, Jack Reynolds
Popp, Frank Donald
Rabjohn, Norman
Schleppnik, Alfred Adolf
Sikorski, James Alan
Slocombe, Robert Jackson
Wasson, Richard Lee

MONTANA
Richards, Geoffrey Norman

NEBRASKA
Gold, Barry Irwin
Hulce, Martin R
Takemura, Kaz H(orace)
Vennerstrom, Jonathan Lee

Wheeler, Desmond Michael Sherlock

NEVADA
Lightner, David A

NEW HAMPSHIRE
Bohm, Howard A
Danforth, Raymond Hewes
Gribble, Gordon W
Neil, Thomas C
Woodbury, Richard Paul

NEW JERSEY
Adams, Richard Ernest
Addor, Roger Williams
Allen, Richard Charles
Andrade, John Robert
Armbruster, David Charles
Augustine, Robert Leo
Bamberger, Curt
Batcho, Andrew David
Bathala, Mohinder S
Bernholz, William Francis
Bertz, Steven Howard
Bodanszky, Miklos
Borris, Robert P
Bose, Ajay Kumar
Brand, William Wayne
Burbaum, Beverly Wolgast
Carter, Richard John
Cavender, Patricia Lee
Chan, Ka-Kong
Christensen, Burton Grant
Christenson, Philip A
Cohen, Noal
Crawley, Lantz Stephen
Dahill, Robert T, Jr
Delaney, Edward Joseph
Draper, Richard William
Durette, Philippe Lionel
Evans, Ralph H, Jr
Evers, William John
Fahey, John Leonard
Finkelstein, Jacob
Fobare, William Floyd
Focella, Antonino
Fong, Jones W
Fost, Dennis L
Gall, Martin
Garfinkel, Harmon Mark
Grabowski, Edward Joseph John
Grosso, John A
Hagmann, William Kirk
Hardtmann, Goetz E
Hatch, Charles Eldridge, III
Heck, James Virgil
Hensens, Otto Derk
Hung, Paul Ling-Kong
Jensen, Norman P
Jirkovsky, Ivo
Joseph, John Mundancheril
Kagan, Michael Z
Kalm, Max John
Kanojia, Ramesh Maganlal
Kasparek, Stanley Vaclav
Kees, Kenneth Lewis
Klaubert, Dieter Heinz
Kolc, Jaroslav Jerry F
Kroll, Emanuel
Ladner, David William
Lam, Yiu-Kuen Tony
Lavanish, Jerome Michael
LaVoie, Edmond J
Leong, William
Liebowitz, Stephen Marc
Liesch, Jerrold Michael
Linn, Bruce Oscar
Liu, Yu-Ying
Lorey, Frank William
Luk, Kin-Chun C
MacConnell, John Griffith
MacCoss, Malcolm
Maehr, Hubert
Mallams, Alan Keith
Marmor, Robert Samuel
Martin, Lawrence Leo
Mathew, Chempolil Thomas
Mertel, Holly Edgar
Michel, Gerd Wilhelm
Miller, Thomas William
Montgomery, Richard Millar
Naipawer, Richard Edward
Olson, Gary Lee
Palmere, Raymond M
Parker, William Lawrence
Putter, Irving
Ramsden, Hugh Edwin
Ramsey, Arthur Albert
Rasmusson, Gary Henry
Reider, Paul Joseph
Reilly, Eugene Patrick
Rinehart, Jay Kent
Roland, Dennis Michael
Rosegay, Avery
Rossi, Robert Daniel
Schaffner, Carl Paul
Schwartz, Herbert
Schwarz, John Samuel Paul
Sekula, Bernard Charles
Shih, Jenn-Shyong
Sieh, David Henry
Skotnicki, Jerauld S
Slade, Joel S
Slusarchyk, William Allen
Sodano, Charles Stanley
Stein, Reinhardt P
Sugathan, Kanneth Kochappan
Tilley, Jefferson Wright
Vlattas, Isidoros
Wanat, Stanley Frank
Wildman, George Thomas
Wilkinson, Raymond George
Williams, Thomas Henry
Zalay, Andrew W(illiam)
Zask, Arie
Ziegler, John Benjamin

NEW MEXICO
Cheavens, Thomas Henry
Guziec, Lynn Erin
Ott, Donald George
Smith, Wayne Howard
Zeigler, John Martin

NEW YORK
Abate, Kenneth
Agosta, William Carleton
Alks, Vitauts
Anderson, Wayne Keith
Ashen, Philip
Attygalle, Athula B
Bardos, Thomas Joseph
Baron, Arthur L
Bell, Thomas Wayne
Berdahl, Donald Richard
Bitha, Panayota
Bittman, Robert
Boeckman, Robert K, Jr
Boettger, Susan D
Borowitz, Irving Julius
Box, Vernon G S
Brust, David Philip
Buchel, Johannes A
Burrows, Cynthia Jane
Curran, William Vincent
Daves, Glenn Doyle, Jr
Dick, William Edwin, Jr
Dickinson, William Borden
Dolak, Terence Martin
Edasery, James P
Feeny, Paul Patrick
Frechet, Jean M J
Fuerst, Adolph
Ganem, Bruce
Gershon, Herman
Giguere, Raymond Joseph
Gletsos, Constantine
Goldman, Norman L
Grina, Larry Dale
Gurel, Demet
Haase, Jan Raymond
Hahn, Roger C
Hall, Frederick Keith
Happel, John
James, Daniel Shaw
Johnson, Bruce Fletcher
Jones, Clive Gareth
Juby, Peter Frederick
Kende, Andrew S
Khan, Jamil Akber
Kinnel, Robin Bryan
Kronstein, Max
Lai, Yu-Chin
LaLonde, Robert Thomas
Landon, Shayne J
Lee, Ving Jick
Leopold, Bengt
Loev, Bernard
McCormick, J Robert D
McMurry, John Edward
Mansfield, Kevin Thomas
Meinwald, Jerrold
Meinwald, Yvonne Chu
Misra, Anand Lal
Murdock, Keith Chadwick
Murphy, Daniel Barker
Nair, Ramachandran Mukundalayam Sivarama
Nakanishi, Koji
Ojima, Iwao
Oliver, Gene Leech
Pankiewicz, Krzysztof Wojciech
Petersen, Ingo Hans
Poss, Andrew Joseph
Premuzic, Eugene T
Pucknat, John Godfrey
Reich, Marvin Fred
Rich, Arthur Gilbert
Rosi, David
Schultz, Arthur George
Schulz, John Hampshire
Schwab, Linda Sue
Silverstein, Robert Milton
Siuta, Gerald Joseph
Solo, Alan Jere
Soloway, Harold
Stark, John Howard
Steinberg, David H
Svoronos, Paris D N
Tobkes, Martin
Trust, Ronald I
Van Verth, James Edward
Webster, Francis X
Weisler, Leonard
Weiss, Martin Joseph

NORTH CAROLINA
Barborak, James Carl
Bassett, Joseph Yarnall, Jr
Bell, Jimmy Holt
Beverung, Warren Neil, Jr
Bigham, Eric Cleveland
Blumenkopf, Todd Andrew
Chilton, William Scott
Clarke, Robert LaGrone
Crimmins, Michael Thomas
Culberson, Chicita Frances
Davis, Roman
Dixit, Ajit Suresh
Dufresne, Richard Frederick
Edwards, Ben E
Engle, Thomas William
Fleischer, Thomas B
Fraser-Reid, Bertram Oliver
Gains, Lawrence Howard
Hooper, Irving R
Huang, Jamin
Hudson, Albert Berry
Kelsey, John Edward
Laganis, Evan Dean
Landes, Chester Grey
Lee, Kuo-Hsiung
Long, Robert Allen
Majewski, Theodore E
Manning, David Treadway
Martin, Gary Edwin
Martin, Ned Harold
Price, Kenneth Elbert
Ranieri, Richard Leo
Richardson, Stephen Giles
Rivers, Jessie Markert
Russell, Henry Franklin
Sarjeant, Peter Thomson
Schickedantz, Paul David
Schwindeman, James Anthony
VanMiddlesworth, Frank L
Wallace, James William, Jr
Wani, Mansukhlal Chhaganlal
Waters, James Augustus
Woosley, Royce Stanley
Wyrick, Steven Dale

NORTH DAKOTA
Donnelly, Brendan James
Jasperse, Craig Peter
Khalil, Shoukry Khalil Wahba

OHIO
Awad, Albert T
Belletire, John Lewis
Bertelson, Robert Calvin
Brain, Devin King
Brillhart, Donald D
Buck, Keith Taylor
Carmichael, Wayne William
Darling, Stephen Deziel
Davis, Gerald Titus
Delvigs, Peter
Dolfini, Joseph E
Erman, William F
Fall, Harry H
Fechter, Robert Bernard
Fentiman, Allison Foulds, Jr
Gibson, Thomas William
Greenlee, Kenneth William
Holbert, Gene W(arwick)
Horne, Samuel Emmett, Jr
Johnson, David Barton
Ketcha, Daniel Michael
Kinstle, Thomas Herbert
Kurtz, David Williams
Letton, James Carey
Lollar, Robert Miller
Meinert, Walter Theodore
Merritt, Robert Edward
Ostrum, G(eorge) Kenneth
Peterson, Robert C
Portfolio, Donald Carl
Sampson, Paul
Sheets, George Henkle
Smith, Douglas Alan
Stevenson, Don R
Strobel, Rudolf G K
Taschner, Michael J
Throckmorton, Peter E
Turkel, Rickey M(artin)
Turnbull, Kenneth
Voss, Anne Coble
Wade, Peter Cawthorn
Wagner, Eugene Ross
Walker, Thomas Eugene
Ziller, Stephen A, Jr

OKLAHOMA
Bunce, Richard Alan
Eisenbraun, Edmund Julius
Nicholas, Kenneth M
Schmitz, Francis John

OREGON
Badger, Rodney Allan
Bublitz, Walter John, Jr
Dolby, Lloyd Jay
Duncan, James Alan
Gould, Steven James
Hauser, Frank Marion
Hudak, Norman John
Kelsey, Rick G
Laver, Murray Lane
McClard, Ronald Wayne
Mason, Robert Thomas
Shilling, Wilbur Leo
Szalecki, Wojciech Jozef

PENNSYLVANIA
Abou-Gharbia, Magid
Anspon, Harry Davis
Bare, Thomas M
Beavers, Ellington McHenry
Bender, Paul Elliot
Berges, David Alan
Bloomer, James L
Bock, Mark Gary
Box, Larry
Brady, Stephen Francis
Canter, Neil M
Carter, Linda G
Chang, Charles Hung
Crissman, Jack Kenneth, Jr
Curran, Dennis Patrick
Djuric, Stevan Wakefield
Erb, Dennis J
Erhan, Semih M
Evans, Ben Edward
Fairchild, Edward H
Ford, Michael Edward
Freidinger, Roger Merlin
Gibson, Raymond Edward
Giuliano, Robert Michael
Godfrey, John Carl
Gordon, Janice Taylor
Grey, Roger Allen
Gunther, Wolfgang Hans Heinrich
Gustine, David Lawrence
Hamel, Coleman Rodney
Harris, James Joseph
Heys, John Richard
Hicks, Kevin B
Hoffman, Jacob Matthew, Jr
Holden, Kenneth George
Hunt, David Allen
Hurt, Susan Schilt
Johnson, Richard William
Johnson, Wayne Orrin
Kauffman, Joel Mervin
Kowalski, Conrad John
Krow, Grant Reese
Kruse, Lawrence Ivan
Lange, Barry Clifford
Lapidus, Milton
Larson, Gerald Louis
Lavoie, Alvin Charles
Lesko, Patricia Marie
Leston, Gerd
Lopez, R C Gerald
McCurry, Patrick Matthew, Jr
Maryanoff, Cynthia Anne Milewski
Mokotoff, Michael
Monsimer, Harold Gene
Morse, Lewis David
Nelson, George Leonard
Noceti, Richard Paul
Osei-Gyimah, Peter
Prentiss, William Case
Press, Jeffery Bruce
Reingold, I(ver) David
Reitz, Allen Bernard
Robinson, Charles Albert
Ross, Stephen T
Schiff, Paul L, Jr
Shamma, Maurice
Shepard, Kenneth LeRoy
Singer, Alan G
Singh, Baldev
Smith, Amos Brittain, III
Soriano, David S
Stevens, Travis Edward
Strike, Donald Peter
Thelman, John Patrick
Thoman, Charles James
Thompson, Wayne Julius
Turner, Andrew B
Veber, Daniel Frank
Volgenau, Lewis
Wade, Peter Allen
Weinstock, Joseph
Weiss, James Allyn
Wempe, Lawrence Kyran
Werny, Frank
Widdowson, Katherine Louisa
Williams, John Roderick
Wineholt, Robert Leese
Zajac, Walter William, Jr

RHODE ISLAND
Bide, Martin John
Cane, David Earl
Elango, Varadaraj
Nugent, James F
Shimizu, Yuzuru
Slater, Schuyler G
Williard, Paul Gregory

SOUTH CAROLINA
Bailey, Carl Williams, III
Bauknight, Charles William, Jr
Dieter, Richard Karl
Farewell, John P
Haskell, Theodore Herbert, Jr
Hon, David Nyok-Sai
Liebman, Arnold Alvin
Martin, Tellis Alexander
Scruggs, Jack G
Shah, Hamish V
Smith, William Edmond
Sproul, Gordon Duane

262 / DISCIPLINE INDEX

Synthetic Organic & Natural Products Chemistry (cont)

Strauss, Roger William
Tour, James Mitchell
Ward, Benjamin F, Jr
Weinstock, Leonard M
Wise, John Thomas

SOUTH DAKOTA
Gaines, Jack Raymond
Lewis, David Edwin

TENNESSEE
Adcock, James Luther
Bennett, Word Brown, Jr
Clemens, Robert Jay
Cliffton, Michael Duane
Germroth, Ted Calvin
Ho, Patience Ching-Ru
Kashdan, David Stuart
Kuo, Chung-Ming
McConnell, Richard Leon
Olsen, John Stuart
Penner, Hellmut Philip
Rash, Fred Howard
Redfearn, Richard Daniel
Sachleben, Richard Alan
Schell, Fred Martin
Smith, Howard E
Spadafino, Leonard Peter
Waddell, Thomas Groth
White, Alan Wayne
Yau, Cheuk Chung

TEXAS
Alam, Maktoob
Aslam, Mohammad
Ball, M Isabel
Bartsch, Richard Allen
Beier, Ross Carlton
Bergbreiter, David Edward
Ciufolini, Marco A
Dahm, Karl Heinz
Davenport, Kenneth Gerald
Ellison, Robert Hardy
Gilbert, John Carl
Hillery, Herbert Vincent
Jaszberenyi, Joseph C
Lyle, Robert Edward, Jr
Malpass, Dennis B
Marchand, Alan Philip
Mulvey, Dennis Michael
Nachman, Ronald James
Parry, Ronald John
Reinecke, Manfred Gordon
Roehrig, Gerald Ralph
Sand, Ralph E
Schimelpfenig, Clarence William
Scott, Alastair Ian
Sessler, Jonathan Lawrence
Skarlos, Leonidas
Sonntag, Norman Oscar Victor
Stallings, James Cameron
Stipanovic, Robert Douglas
Teng, Jon Ie
Watson, William Harold, Jr
Wheeler, Michael Hugh
Willson, Carlton Grant
Zey, Edward G

UTAH
Balandrin, Manuel F
Broadbent, Hyrum Smith
DelMar, Eric G
Epstein, William Warren
Keeler, Richard Fairbanks
Moe, Scott Thomas
Robins, Morris Joseph
Segelman, Florence H
Van Wagenen, Bradford Carr
Walker, Edward Bell Mar

VERMONT
Hussak, Robert Edward

VIRGINIA
Beyad, Mohammed Hossain
Daylor, Francis Lawrence, Jr
Edwards, William Brundige, III
Fedrick, James Love
Gager, Forrest Lee, Jr
Harris, Paul Robert
Harvey, William Ross
Houghton, Kenneth Sinclair, Sr
Hudlicky, Milos
Hudson, Frederick Mitchell
Hussamy, Samir
Johnston, Byron E
Jones, Tappey Hughes
Kingston, David George Ian
Kinsley, Homan Benjamin, Jr
Little, Edwin Demetrius
Rodig, Oscar Rudolf
Rosenberg, Murray David
Salley, John Jones, Jr
Shipe, James R, Jr
Sprinkle, Robert Shields, III
Teng, Lina Chen
Thompson, Mark Ewell
Whitney, George Stephen

WASHINGTON
Anderson, Charles Dean
Bennett, Clifton Francis
Biskupiak, Joseph E
Duhl-Emswiler, Barbara Ann
Hansen, Michael Roy
Hopkins, Paul Brink
Ingalsbe, David Weeden
Lewis, Norman G
Malik, Sohail
Maranville, Lawrence Frank
Nelson, Randall Bruce
Tostevin, James Earle
Trotter, Patrick Casey
Wade, Leroy Grover, Jr
Wilbur, Daniel Scott
Wollwage, Paul Carl

WEST VIRGINIA
Giza, Chester Anthony
Hubbard, John Lewis
Hussey, Edward Walter
Koleske, Joseph Victor
Ode, Richard Herman

WISCONSIN
Bernardin, Leo J
Burke, Steven Douglas
Chenier, Philip John
Cook, James Minton
Dugal, Hardev Singh
Holland, Dewey G
Hollenberg, David Henry
Hornemann, Ulfert
Klopotek, David L
Lardy, Henry Arnold
Lin, Stephen Y
Manning, James Harvey
Park, Robert William
Pearl, Irwin Albert
Ratliff, Francis Tenney
Runquist, Alfonse William
Sommers, Raymond A
Wollwage, John Carl
Zinkel, Duane Forst

WYOMING
Branthaver, Jan Franklin
Seese, William Shober

PUERTO RICO
Alzerreca, Arnaldo
Carrasquillo, Arnaldo
Stephens, William Powell

ALBERTA
Bachelor, Frank William
Back, Thomas George
Liu, Hsing-Jang
Moore, Graham John

BRITISH COLUMBIA
Becker, Edward Samuel
Boehm, Robert Louis
Goodeve, Allan McCoy
Liao, Ping-huang
Manville, John Fieve
Paszner, Laszlo
Procter, Alan Robert
Shaw, A(lexander) J(ohn)
Underhill, Edward Wesley
Yunker, Mark Bernard

MANITOBA
Charlton, James Leslie
Chow, Donna Arlene
Hunter, Norman Robert
McKinnon, David M

NEW BRUNSWICK
Jankowski, Christopher K
Strunz, G(eorge) M(artin)

NEWFOUNDLAND
Georghiou, Paris Elias

NOVA SCOTIA
Chandler, Reginald Frank
Grindley, T Bruce
Piorko, Adam M
Vining, Leo Charles
White, Robert Lester

ONTARIO
Aspinall, Gerald Oliver
Baer, Hans Helmut
Baines, Kim Marie
Bannard, Robert Alexander Brock
Bersohn, Malcolm
Birnbaum, George I
Butler, Douglas Neve
Collins, Frank William, Jr
Cory, Robert Mackenzie
Court, William Arthur
Fallis, Alexander Graham
Greenhalgh, Roy
Gupta, Virendra Nath
Hay, George William
Histed, John Allan
Hopkins, Clarence Yardley
Lange, Gordon Lloyd
Lautens, Mark
Leznoff, Clifford Clark
Lyne, Leonard Murray, Sr

MacDonald, Stewart Ferguson
MacLean, David Bailey
Morand, Peter
Pare, J R Jocelyn
Ratnayake, Walisundera Mudiyanselage Nimal
Rodrigo, Russell Godfrey
Saxton, William Reginald
Singh, Bhagirath
Sowa, Walter
Starratt, Alvin Neil
Stavric, Stanislava
Still, Ian William James
Szarek, Walter Anthony
Vijay, Hari Mohan
Warnhoff, Edgar William
Weedon, Alan Charles
Wightman, Robert Harlan
Yates, Peter

QUEBEC
Alince, Bohumil
Atkinson, Joseph George
Belanger, Patrice Charles
Burnell, Robert H
Chenevert, Robert (Bernard)
Darling, Graham Davidson
Engel, Charles Robert
Escher, Emanuel
Feldman, Dorel
Ferdinandi, Eckhardt Stevan
Garneau, Francois Xavier
Goldstein, Sandu M
Hamlet, Zacharias
Kokta, Bohuslav Vaclav
Kubes, George Jiri
Perron, Yvon G
Regoli, Domenico
Scallan, Anthony Michael
Schiller, Peter Wilhelm
Van Calsteren, Marie-Rose
Wuest, James D
Yeats, Ronald Bradshaw
Young, Robert Norman

SASKATCHEWAN
Abrams, Suzanne Roberta
Gear, James Richard
Gupta, Vidya Sagar
Mikle, Janos J

OTHER COUNTRIES
Barnes, Roderick Arthur
Diederich, Francois Nico
Goldstein, Melvin Joseph
Gottlieb, Otto Richard
Graber, Robert Philip
Hall, Laurance David
Hassner, Alfred
Hummel, Hans Eckhardt
Kent, Stephen Brian Henry
Kurobane, Itsuo
Kuttab, Simon Hanna
Malherbe, Roger F
Nair, Vijay
Shapiro, Stuart
Slagel, Robert Clayton
Trofimov, Boris Alexandrovich
Wang, Chia-Lin Jeffrey

Theoretical Chemistry

ALABAMA
Baird, James Kern
McKee, Michael Leland
Visscher, Pieter Bernard

ARIZONA
Balasubramanian, Krishnan
Golden, Sidney
Lichtenberger, Dennis Lee
Nordstrom, Brian Hoyt
Salzman, William Ronald

ARKANSAS
Pulay, Peter

CALIFORNIA
Alder, Berni Julian
Allen, Thomas Lofton
Blumstein, Carl Joseph
Brewer, Leo
Brown, Nancy J
Carter, Emily Ann
Chandler, David
Chen, Joseph Cheng Yih
Cook, Robert Crossland
Creighton, John Rogers
Dawson, Kenneth Adrian
Dows, David Alan
Ewig, Carl Stephen
Fetzer, John Charles
Fink, William Henry
Fredrickson, Glenn Harold
Goddard, William Andrew, III
Goldstein, Elisheva
Haney, David N
Harris, Robert A
Hazi, Andrew Udvar
Head-Gordon, Martin Paul
Head-Gordon, Teresa Lyn
Hempel, Judith Cato
Herman, Zelek Seymour
Horsley, John Anthony
Houk, Kendall N
Hsiung, Chi-Hua Wu
Huestis, David Lee
Johnson, Oliver
Juster, Norman Joel
Keizer, Joel Edward
Kollman, Peter Andrew
Komornicki, Andrew
Konowalow, Daniel Dimitri
Lester, William Alexander, Jr
Lindenberg, Katja
Loew, Gilda Harris
McKoy, Basil Vincent
McQuarrie, Donald Allan
Marcus, Rudolph Arthur
Martens, Craig Colwell
Matlow, Sheldon Leo
Medrud, Ronald Curtis
Metiu, Horia I
Miller, James Angus
Miller, William Hughes
Morrison, Harry Lee
Oakes, John Morgan
Olafson, Barry D
Palke, William England
Payne, Philip Warren
Reiss, Howard
Ritchie, Adam Burke
Schugart, Kimberly A
Segal, Gerald A
Shuler, Kurt Egon
Shull, Harrison
Simon, John Douglas
Skolnick, Jeffrey
Stephens, Philip J
Strauss, Herbert L
Syage, Jack A
Whaley, Katharine Birgitta
Wheeler, John C
Wilemski, Gerald
Wilkins, Roger Lawrence
Winter, Nicholas Wilhelm
Wipke, W Todd
Wolf, Kathleen A
Wolfsberg, Max
Woodson, John Hodges
Wulfman, Carl E

COLORADO
Ely, James Frank
Fixman, Marshall
Ladanyi, Branka Maria
Mansfield, Roger Leo
Parson, Robert Paul
Sanderson, Robert Thomas
Stephens, Jeffrey Alan
Stewart, James Joseph Patrick
Yates, Albert Carl

CONNECTICUT
Anderson, Carol Patricia
Beveridge, David L
De Rocco, Andrew Gabriel
Jorgensen, William L
Kavarnos, George James
Keyes, Thomas Francis
Michels, H(orace) Harvey
Petersson, George A
Sanders, William Albert
Sinanoglu, Oktay

DELAWARE
Dixon, David Adams
Doren, Douglas James
Foss, Robert Paul
Morgan, John Davis, III
Shipman, Lester Lynn
Stewart, Charles Winfield, Sr
Stuper, Andrew John
Szalewicz, Krzysztof
Wasserman, Edel
Wasserman, Zelda Rakowitz
Wiley, Douglas Walker

DISTRICT OF COLUMBIA
Chiu, Lue-Yung Chow
Kaye, Jack Alan
Kertesz, Miklos
Ramaker, David Ellis
Rubin, Robert Joshua
Vertes, Akos

FLORIDA
Bartlett, Rodney Joseph
Bash, Paul Anthony
Baum, J(ames) Clayton
Boehnke, David Neal
Campbell, Hallock Cowles
Cioslowski, Jerzy
Dufty, James W
Jones, Walter H(arrison)
Linder, Bruno
Micha, David Allan
Person, Willis Bagley
Rhodes, William Clifford
Seiger, Harvey N
Zerner, Michael Charles

GEORGIA
Barbas, John Theophani
Bowman, Joel Mark
Hentschel, Hilary George
Liotta, Dennis C

Mead, Chester Alden
Peterson, Keith L
Rouvray, Dennis Henry
Schaefer, Henry Frederick, III
Uzer, Ahmet Turgay
Whetten, Robert Lloyd

ILLINOIS
Astumian, Raymond Dean
Bouman, Thomas David
Curtiss, Larry Alan
Freed, Karl F
Gislason, Eric Arni
Goodman, Gordon Louis
Josefson, Clarence Martin
Kern, Charles William
Kim, Ki Hwan
Lykos, Peter George
Makri, Nancy
Martin, Yvonne Connolly
O'Donnell, Terence J
Pariza, Richard James
Pople, John Anthony
Prais, Michael Gene
Schatz, George Chappell
Scheiner, Steve
Snyder, Gary James
Tyrrell, James
Vanderkooi, Garret
Wagner, Albert Fordyce
Wilson, Robert Steven
Wolynes, Peter Guy

INDIANA
Adelman, Steven A
Aprison, Morris Herman
Bentley, John Joseph
Boyd, Donald Bradford
Davidson, Ernest Roy
Dykstra, Clifford Elliot
Janis, F Timothy
Kosman, Warren Melvin
Langhoff, Peter Wolfgang
Lipkowitz, Kenny Barry
Malik, David Joseph
Nazaroff, George Vasily
Pimblott, Simon Martin
Schwartz, Maurice Edward
Strieder, William

IOWA
DePristo, Andrew Elliott
Kozak, John Joseph
Randic, Milan
Rider, Paul Edward, Sr
Ruedenberg, Klaus
Sando, Kenneth Martin

KANSAS
Chu, Shih I

KENTUCKY
Mueller, Sr Rita Marie

LOUISIANA
Bauer, Beverly Ann
Domelsmith, Linda Nell
Flurry, Robert Luther, Jr
Klasinc, Leo
Lee, Burnell
Politzer, Peter Andrew
Stanley, George Geoffrey
Watkins, Steven F

MARYLAND
Alexander, Millard Henry
Bunyan, Ellen Lackey Spotz
Chen, Yi-Der
Gadzuk, John William
Hunter, Lawrence Wilbert
Kleier, Daniel Anthony
Liebman, Joel Fredric
MacKerell, Alexander Donald, Jr
Marqusee, Jeffrey Alan
Silver, David Martin
Stuebing, Edward Willis
Szabo, Attila
Tupper, Kenneth Joseph
Weeks, John David

MASSACHUSETTS
Alper, Joseph Seth
Brenner, Howard
Chen, Maynard Ming-Liang
Epstein, Irving Robert
Gund, Peter Herman Lourie
Heller, Eric Johnson
Hendrickson, James Briggs
Herzfeld, Judith
Hobey, William David
Jordan, Peter C H
Kirby, Kate Page
Muthukumar, Murugappan
Oppenheim, Irwin
Pan, Yuh Kang
Phillies, George David Joseph
Phillips, Philip Wirth
Pyun, Chong Wha
Ruskai, Mary Beth
Suplinskas, Raymond Joseph
Winick, Jeremy Ross

MICHIGAN
Bartell, Lawrence Sims

Blankley, Clifton John
Blinder, Seymour Michael
Blint, Richard Joseph
Cheney, B Vernon
Crippen, Gordon Marvin
Cukier, Robert Isaac
Harrison, James Francis
Hase, William Louis
Hunt, Katharine Lois Clark
Lohr, Lawrence Luther, Jr
Riehl, James Patrick
Rohrer, Douglas C
Schnitker, Jurgen H
Turner, Almon George, Jr
Woehler, Scott Edwin

MINNESOTA
Almlof, Jan Erik
Eades, Robert Alfred
Olmsted, Richard Dale
Polnaszek, Carl Francis
Thompson, Herbert Bradford

MISSOURI
Biolsi, Louis, Jr
Duchek, John Robert
Glaser, Rainer
Hansen, Richard Lee
McGraw, Robert Leonard
Rouse, Robert Arthur
Welsh, William James

NEBRASKA
Diestler, Dennis Jon
Harris, Holly Ann

NEVADA
Shin, Hyung Kyu

NEW HAMPSHIRE
Mayne, Howard R

NEW JERSEY
Adams, William Henry
Andose, Joseph D
Chance, Ronald Richard
Cotter, Martha Ann
Elk, Seymour B
Ermler, Walter Carl
Getzin, Paula Mayer
Goodman, Lionel
Hager, Douglas Francis
Hammond, Willis Burdette
Helfand, Eugene
Paul, Edward W
Phillips, James Charles
Raghavachari, Krishnan
Raynor, Susanne
Strange, Ronald Stephen
Tully, John Charles
Upton, Thomas Hallworth
Weintraub, Herschel Jonathan R

NEW MEXICO
Bellum, John Curtis
Coppa, Nicholas V
Engelke, Raymond Pierce
Erpenbeck, Jerome John
Harary, Frank
Hay, Philip Jeffrey
Martin, Richard Lee
Ortiz, Joseph Vincent
Pack, Russell T
Peek, James Mack
Redondo, Antonio
Wadt, Willard Rogers
Walker, Robert Bridges
Werbelow, Lawrence Glen
Woodruff, Susan Beatty

NEW YORK
Berne, Bruce J
Birge, Robert Richards
Bishop, Marvin
Bommaraju, Tilak V
Box, Vernon G S
Buff, Frank Paul
Campbell, Edwin Stewart
Carpenter, Barry Keith
Chapman, Sally
Ehrenson, Stanton Jay
Ezra, Gregory Sion
Frisch, Harry Lloyd
Ginell, Robert
Hecht, Charles Edward
Hoffmann, Roald
Holmes, Curtis Frank
Kuki, Atsuo
Liao, Martha
Loring, Roger Frederic
Lounsbury, John Baldwin
Luke, Brian
McKelvey, John Murray
Marchese, Francis Thomas
Mehler, Ernest Louis
Mercier, Gustavo Alberto, Jr
Messmer, Richard Paul
Mezei, Mihaly
Miller, Kenneth John
Monaco, Regina R
Muckenfuss, Charles
Nemethy, George
Newton, Marshall Dickinson
Pechukas, Philip

Percus, Jerome K
Porter, Richard Needham
Radel, Stanley Robert
Rhodes, Yorke Edward
Rosenkrantz, Marcy Ellen
Sage, Martin Lee
Sellers, Peter Hoadley
Snyder, Lawrence Clement
Stell, George Roger
Sufrin, Janice Richman
Widom, Benjamin
Zhang, John Zeng Hui
Zielinski, Theresa Julia

NORTH CAROLINA
Anderson, Marshall W
Bernholc, Jerzy
Hegstrom, Roger Allen
Jackels, Charles Frederick
Jones(Stover), Betsy
Kohn, Michael Charles
MacPhail, Richard Allyn
Parr, Robert Ghormley
Poirier, Jacques Charles
Whangbo, Myung Hwan
Whitten, Jerry Lynn
Yang, Weitao

OHIO
Carlton, Terry Scott
Colvin, Howard Allen
Del Bene, Janet Elaine
Huff, Norman Thomas
Jaffe, Hans
McFarland, Charles Warren
Mortensen, Earl Miller
Pitzer, Russell Mosher
Seybold, Paul Grant
Shavitt, Isaiah
Singer, Sherwin
Smith, Douglas Alan
Taylor, William Johnson
Thomas, Terence Michael
Tuan, Debbie Fu-Tai
Whitman, Donald Ray
Wolken, George, Jr

OKLAHOMA
Neely, Stanley Carrell
Raff, Lionel M
Sundberg, Kenneth Randall

OREGON
Herrick, David Rawls
Horne, Frederick Herbert
McClure, David Warren
Mazo, Robert Marc

PENNSYLVANIA
Dougherty, Eugene P
Dyott, Thomas Michael
Francl, Michelle Miller
Garrison, Barbara Jane
George, Philip
Hameka, Hendrik Frederik
Irwin, Philip George
Jordan, Kenneth David
Kim, Shoon Kyung
Klein, Michael Lawrence
Kraihanzel, Charles S
Lee, Robert William
Ludwig, Oliver George
Pierce, Thomas Henry
Ranck, John Philip
Smith, David
Smith, Graham Monro
Szamosi, Janos
Thibeault, Jack Claude
Wood, Kurt Arthur
Yuan, Jian-Min
Zeroka, Daniel

RHODE ISLAND
Mason, Edward Allen
Stratt, Richard Mark

SOUTH CAROLINA
Gimarc, Benjamin M

TENNESSEE
Clarke, James Harold
Crawford, Oakley H
Gilliom, Richard D
Hefferlin, Ray (Alden)
Kovac, Jeffrey Dean
Mortimer, Robert George
Painter, Gayle Stanford
Polavarapu, Prasad Leela
Sworski, Thomas John

TEXAS
Auchmuty, Giles
Birney, David Martin
Boggs, James Ernest
Caldwell, Richard A
Cantrell, Cyrus D, III
Cotton, Frank Albert
Eaker, Charles William
Ford, George Peter
Hall, Michael Bishop
Harwood, William H
Herndon, William Cecil
Klein, Douglas J
Kouri, Donald Jack

Lippmann, David Zangwill
McCammon, James Andrew
Marshall, Paul
Matcha, Robert Louis
Parr, Christopher Alan
Pettitt, Bernard Montgomery
Rossky, Peter Jacob
Schmalz, Thomas G
Scuseria, Gustavo Enrique
Webber, Stephen Edward
White, Ronald Joseph
Wyatt, Robert Eugene

UTAH
Henderson, Douglas J
Shirts, Randall Brent

VIRGINIA
Delos, John Bernard
Farmer, Barry Louis
Greer, William Louis
Hilderbrandt, Richard L
LaBudde, Robert Arthur
Marron, Michael Thomas
Reynolds, Peter James
Richardson, Frederick S
Sabelli, Nora Hojvat
Schneider, Barry I
Shillady, Donald Douglas
Smith, Bertram Bryan, Jr
Trindle, Carl Otis

WASHINGTON
Carlson, Charles Merton
Currah, Walter E
Dunning, Thomas Harold, Jr
George, Thomas Frederick
Lybrand, Terry Paul
Moseley, W(illiam) David
Reinhardt, William Parker

WISCONSIN
Bird, R(obert) Byron
Bondeson, Stephen Ray
Curtiss, Charles Francis
England, Walter Bernard
King, Frederick Warren
Skinner, James Lauriston
Zimmerman, Howard Elliot

ALBERTA
Boere, Rene Theodoor
Clarke, Bruce Leslie
Fraga, Serafin
Grabenstetter, James Emmett
Huzinaga, Sigeru
Klobukowski, Mariusz Andrzej
Laidlaw, William George
Paul, Reginald
Thorson, Walter Rollier

BRITISH COLUMBIA
Dingle, Thomas Walter
Snider, Robert Folinsbee

MANITOBA
Gordon, Richard

NEW BRUNSWICK
Grein, Friedrich
Sichel, John Martin
Thakkar, Ajit Jamnadas

NOVA SCOTIA
Boyd, Russell Jaye
Coxon, John Anthony
Kusalik, Peter Gerard
Pacey, Philip Desmond
Wasylishen, Roderick Ernest

ONTARIO
Allnatt, Alan Richard
Bader, Richard Frederick W
Bishop, David Michael
Brumer, Paul William
Burns, George
Colpa, Johannes Pieter
Desai, Rashmi C
Henry, Bryan Roger
Kavassalis, Tom A
LeRoy, Robert James
Liu, Wing-Ki
Meath, William John
Ostlund, Neil Sinclair
Paldus, Josef
Penner, Alvin Paul
Pietro, William Joseph
Pritchard, Huw Owen
Siebrand, Willem
Snider, Neil Stanley
Valleau, John Philip
Whittington, Stuart Gordon
Wright, James Sherman

QUEBEC
Bandrauk, Andre D
Eu, Byung Chan
Sanctuary, Bryan Clifford
Whitehead, Michael Anthony

SASKATCHEWAN
Mezey, Paul G

Theoretical Chemistry (cont)

OTHER COUNTRIES
Bruns, Roy Edward
Buckingham, Amyand David
Bunge, Carlos Federico
Burnell, Louis A
Choi, Sang-il
Evleth, Earl Mansfield
Forst, Wendell
Haymet, Anthony Douglas-John
Hirshfeld, Fred Lurie
Housecroft, Catherine Elizabeth
Kay, Kenneth George
Keller, Jaime
Schleyer, Paul von Rague

Thermodynamics & Material Properties

ALABAMA
Baird, James Kern
Byrd, James Dotson
Harrison, Benjamin Keith
McKannan, Eugene Charles
Martin, James Arthur
Miller, George Paul
Pan, Chai-Fu
Thompson, Raymond G

ALASKA
Merrill, Robert Clifford, Jr

ARIZONA
Lunine, Jonathan Irving
Rupley, John Allen
Venables, John Anthony
Withers, James C

ARKANSAS
Richardson, Charles Bonner

CALIFORNIA
Alivisatos, A Paul
Altman, Robert Leon
Amis, Eric J
Athey, Robert Douglas, Jr
Balzhiser, Richard E(arl)
Baskes, Michael I
Baur, Mario Elliott
Benson, Sidney William
Bernstein, Uri
Bohlen, Steven Ralph
Breitmayer, Theodore
Brewer, Leo
Bukowinski, Mark S T
Carniglia, Stephen C(harles)
Carter, Emily Ann
Chang, Elfreda Te-Hsin
Chen, Timothy Shieh-Sheng
Condit, Ralph Howell
Cotts, David Bryan
Cotts, Patricia Metzger
Cubicciotti, Daniel David
Dawson, Kenneth Adrian
Day, Howard Wilman
Deshpande, Shirkant V
Domeniconi, Michael John
Doyle, Fiona Mary
Duff, Russell Earl
Dunn, Bruce
Even, William Roy, Jr
Feigelson, Robert Saul
Fisher, Robert Amos, Jr
Fox, Joseph M(ickle), III
Ginell, William Seaman
Gur, Turgut M
Hahn, Harold Thomas
Hem, John David
Hopper, Robert William
Hsieh, You-Lo
Hultgren, Ralph Raymond
Jing, Liu
Kay, Robert Eugene
Kidd, Richard Wayne
King, William Robert, Jr
Krikorian, Oscar Harold
Kroneberger, Gerald F
Langer, Sidney
Langerman, Neal Richard
Laub, Richard J
Lepie, Albert Helmut
Lewis, Nathan Saul
Lindner, Duane Lee
Liu, Frederick F
Ludwig, Frank Arno
McCarty, Jon Gilbert
McKenzie, William F
Martin, L(aurence) Robbin
Mason, D(avid) M(alcolm)
Meschi, David John
Miller, Donald Gabriel
Morris, John William, Jr
Munir, Zuhair A
Myronuk, Donald Joseph
Neidlinger, Hermann H
Noring, Jon Everett
Norman, John Harris
Novak, Bruce Michael
O'Loane, James Kenneth
Pask, Joseph Adam
Perry, Dale Lynn
Potter, Norman D
Rard, Joseph Antoine
Ritz-Gold, Caroline Joyce
Robin, Allen Maurice
Rock, Peter Alfred
Rothman, Albert J(oel)
Shelly, John Richard
Simpson, Howard Douglas
Skowronski, Raymund Paul
Smith, Charles Aloysius
Smith, James Hart
Smith, Louis Charles
Spiegler, Kurt Samuel
Stebbins, Jonathan F
Stern, John Hanus
Strauss, Herbert L
Szabo, Emery D
Theodorou, Doros Nicolas
Tilley, T(erry) Don
Tschoegl, Nicholas William
Urtiew, Paul Andrew
Van Hecke, Gerald Raymond
Van Thiel, Mathias
Vernon, Gregory Allen
Whatley, Thomas Alvah
Wheeler, John C
White, Jack Lee
Wilemski, Gerald
Williams, Richard Stanley
Wise, Henry
Wu, Lei
Yee, Rena

COLORADO
Arp, Vincent D
Bachman, Bonnie Jean Wilson
Blake, Daniel Melvin
Callanan, Jane Elizabeth
Chum, Helena Li
DePoorter, Gerald Leroy
Edwards, Harry Wallace
Ely, James Frank
Geller, Seymour
Hanley, Howard James Mason
Levenson, Leonard L
Montgomery, Robert L
Moore, John Jeremy
Nordstrom, Darrell Kirk
Rainwater, James Carlton
Reed, Thomas Binnington
Roshko, Alexana
Strauss, Eric L
Vaughan, John Dixon
Zoller, Paul

CONNECTICUT
Cohen, Myron Leslie
Grossman, Leonard N(athan)
Koberstein, Jeffrey Thomas
Kunz, Harold Russell
McBride, James Michael
Rie, John E
Sarada, Thyagaraja
Wellman, Russel Elmer
Wolfram, Leszek January

DELAWARE
Becker, Aaron Jay
Chiu, Jen
Chu, Sung Gun
English, Alan Dale
Mills, Patrick Leo, Sr
Ross, William D(aniel)
Sammak, Emil George

DISTRICT OF COLUMBIA
Butler, Dixon Matlock
Carter, Gesina C
Corsaro, Robert Dominic
Dragoo, Alan Lewis
Klein, Max
Klein, Philipp Hillel
Stern, Kurt Heinz

FLORIDA
Beatty, Charles Lee
Chao, Raul Edward
Dufty, James W
Hull, Harry H
Person, Willis Bagley
Seiger, Harvey N
Van Zee, Richard Jerry
Wei-Berk, Caroline
Zimmer, Martin F

GEORGIA
Anderson, Bruce Martin
Anderson, James Leroy
Anderson, Robert Lester
Clever, Henry Lawrence
Derrick, Mildred Elizabeth
Eckert, Charles Alan
Etzler, Frank M
Heric, Eugene LeRoy
Teja, Amyn Sadruddin
Walters, John Philip
Williams, Emmett Lewis

HAWAII
Ming, Li Chung

IDAHO
Christian, Jerry Dale
Jacobsen, Richard T

Miller, Richard Lloyd
Ramshaw, John David
Slansky, Cyril M
Tanner, John Eyer, Jr

ILLINOIS
Abraham, Bernard M
Ali, Naushad
Blanks, Robert F
Burns, Richard Price
Cafasso, Fred A
Chan, Sai-Kit
Curtiss, Larry Alan
Finn, Patricia Ann
Fischer, Albert Karl
Giacobbe, F W
Giori, Claudio
Gray, Linsley Shepard, Jr
Gruen, Dieter Martin
Gutfreund, Kurt
House, James Evan, Jr
House, Kathleen Ann
Huang, Leo W
Hwang, Yu-Tang
Johnson, Irving
Jonas, Jiri
Krawetz, Arthur Altshuler
Leibowitz, Leonard
Li, Yao-En
Maroni, Victor August
Nagy, Zoltan
Nordine, Paul Clemens
Perkins, Alfred J
Peters, James Empson
Poeppelmeier, Kenneth Reinhard
Primak, William L
Sachtler, Wolfgang Max Hugo
Sather, Norman F(redrick)
Savidge, Jeffrey Lee
Stout, John Willard
Wilson, Robert Steven
Zaromb, Solomon

INDIANA
Desai, Pramod D
Droege, John Walter
Fidelle, Thomas Patrick
Grimley, Robert Thomas
Henke, Mitchell C
Honig, Jurgen Michael
Liley, Peter Edward
Lin, Ho-Mu
Pielet, Howard M
Strieder, William

IOWA
Cater, Earle David
Gschneidner, Karl A(lbert), Jr
Jolls, Kenneth Robert
Miller, Gordon James
Thiel, Patricia Ann
Wurster, Dale Eric, Jr

KANSAS
Brewer, Jerome
Greene, Frank T
Lindenmann, Siegfried
Ryan, James M
Rytting, Joseph Howard
Wahlbeck, Phillip Glenn

KENTUCKY
Crim, Gary Allen
Huttenlocher, Dietrich F
Keely, William Martin
Man, Chi-Sing
Pan, Wei-Ping
Plucknett, William Kennedy
Reucroft, Philip J
Sands, Donald Edgar

LOUISIANA
Allen, Susan Davis
Gautreaux, Marcelian Francis
Gonzales, Elwood John
Raisen, Elliott
Wan, Peter J
Wendt, Richard P

MAINE
Sottery, Theodore Walter

MARYLAND
Bonnell, David William
Bonsack, James Paul
Callahan, John Joseph
Carter, John Paul
Chang, Ren-Fang
Chang, Shu-Sing
Colwell, Jack Harold
Egelhoff, William Frederick, Jr
Fifer, Robert Alan
Flynn, Joseph Henry
Fong, Jeffrey Tse-Wei
Foresti, Roy J(oseph), Jr
Freedman, Eli (Hansell)
Frey, Douglas D
Greer, Sandra Charlene
Hollenbeck, Robert Gary
Houseman, Barton L
Iwasa, Kuni H
Jovancicevic, Vladimir
Kumar, K Sharvan
Liebman, Joel Fredric

McKinney, John Edward
McQuaid, Richard William
Marinenko, George
Mopsik, Frederick Israel
Paule, Robert Charles
Powell, Edward Gordon
Price, Donna
Promisel, Nathan E
Raksis, Joseph W
Roche, Lidia Alicia
Roth, Robert S
Ruffa, Anthony Richard
Sandusky, Harold William
Sengers, Johanna M H Levelt
Tupper, Kenneth Joseph
Wang, Frederick E
Williams, Ellen D
Wong, Kin Fai
Young, Jay Alfred

MASSACHUSETTS
Agee, Carl Bernard
Berker, Ahmet Nihat
Berlowitz Tarrant, Laurence
Bonner, Francis Joseph
Buckler, Sheldon A
Burns, Roger George
Caslavsky, Jaroslav Ladislav
Chou, Kuo-Chih
Eagar, Thomas W
Evans, Michael Douglas
Foxman, Bruce Mayer
Freedman, Andrew
Gibbard, H Frank
Harman, T(heodore) C(arter)
Jankowski, Conrad M
Lapkin, Milton
Lees, Wayne Lowry
Lester, Joseph Eugene
Licht, Stuart Lawrence
Macaione, Domenic Paul
Monson, Peter A
Moulton, Russell Dana
Nelson, Keith A
Perkins, Janet Sanford
Rivin, Donald
Rose, Timothy Laurence
Rowell, Robert Lee
Sadoway, Donald Robert
Serafin, Frank G
Szafran, Zvi
Tiernan, Robert Joseph
Tuller, Harry Louis
Warner, John Charles
Weller, Paul Franklin

MICHIGAN
Atreya, Arvind
Bennett, Stephen Lawrence
Bettman, Max
Clark, Peter David
Cornilsen, Bahne Carl
Downey, Joseph Robert, Jr
Dye, James Louis
Eick, Harry Arthur
Ellis, Thomas Stephen
Essene, Eric J
Evans, B J
Filisko, Frank Edward
Harris, Kenneth
Henein, Naeim A
Hucke, Edward E
Hunt, Katharine Lois Clark
Johnson, David Alfred
Meier, Dale Joseph
Meister, John J
Miller, Robert Llewellyn
Russell, William Charles
Schlick, Shulamith
Seefurth, Randall N
Sharma, Ram Autar
Solc, Karel
Stivender, Donald Lewis
Swathirajan, S
Van Poolen, Lambert John

MINNESOTA
Anderson, George Robert
Bates, Frank S
Carpenter, John Harold
Douglas, William Hugh
Eckert, Ernst R(udolf) G(eorg)
Fletcher, Edward A(braham)
Gabor, Thomas
Haas, Larry Alfred
Larson, Eric George
Owens, Kenneth Eugene
Penney, William Harry
Pignolet, Louis H
Scriven, L E(dward), (II)
Shores, David Arthur
Suryanarayanan, Raj Gopalan

MISSOURI
Coria, Jose Conrado
Fisch, Ronald
Frederick, Raymond II
Grayson, Michael A
Holsen, James N(oble)
Holtzer, Marilyn Emerson
McGraw, Robert Leonard
Ownby, P(aul) Darrell
Painter, James Howard
Preckshot, G(eorge) W(illiam)

Robertson, David G C
Sastry, Shankara M L
Thomas, Timothy Farragut
Tria, John Joseph, Jr
Viswanath, Dabir S

MONTANA
Jennings, Paul W

NEBRASKA
Zebolsky, Donald Michael

NEVADA
Turner, Robert Harold

NEW HAMPSHIRE
Rice, Dale Wilson

NEW JERSEY
Bair, Harvey Edward
Ban, Vladimir Sinisa
Barns, Robert L
Baughman, Ray Henry
Binder, Michael
Borysko, Emil
Brazinsky, Irv(ing)
Breslauer, Kenneth John
Buck, Thomas M
Buckley, Denis Noel
Calcote, Hartwell Forrest
Choi, Byung Chang
Couchman, Peter Robert
Crerar, David Alexander
Debenedetti, Pablo Gaston
Dittman, Frank W(illard)
Dunn, William Howard
Elder, John Philip
Erenrich, Eric Howard
Fleury, Paul A
Halpern, Teodoro
Hirsch, Arthur
Irgon, Joseph
Joffe, Joseph
Koss, Valery Alexander
Li, Hsin Lang
Libowitz, George Gotthart
Loh, Roland Ru-loong
Maloy, Joseph T
Miale, Joseph Nicolas
Mraw, Stephen Charles
Newman, Stephen Alexander
Osuch, Christopher Erion
Panish, Morton B
Paul, Edward W
Pethica, Brian Anthony
Rabinovich, Eliezer M
Rollino, John
Rosen, Carol Zwick
Salomon, Mark
Schwarz, John Samuel Paul
Townsend, Palmer W
Tsonopoulos, Constantine

NEW MEXICO
Behrens, Robert George
Cady, Howard Hamilton
Coppa, Nicholas V
Haschke, John Maurice
Kerley, Gerald Irwin
Krupka, Milton Clifford
Lawson, Andrew Cowper, II
Lieberman, Morton Leonard
Loehman, Ronald Ernest
Magnani, Nicholas J
Mason, Caroline Faith Vibert
Mills, Robert Leroy
Molecke, Martin A
Mullendore, A(rthur) W(ayne)
O'Rourke, John Alvin
Peterson, Eugene James
Sherman, Robert Howard
Stampfer, Joseph Frederick
Stark, Walter Alfred, Jr
Visscher, William M
Ward, John William
Yost, Frederick Gordon

NEW YORK
Adelstein, Peter Z
Alliet, David F
Bell, Thomas Wayne
Bluhm, Terry Lee
Brun, Milivoj Konstantin
Buff, Frank Paul
Chen, Cindy Chei-Jen
Chow, Che Chung
Coffee, Robert Dodd
Conan, Robert James, Jr
Danielson, Paul Stephen
D'Heurle, Francois Max
Dowell, Flonnie
Elder, Fred A
Gancy, Alan Brian
Gardella, Joseph Augustus, Jr
Gentner, Robert F
Ginell, Robert
Haberfield, Paul
Haller, Ivan
Hillig, William Bruno
Holmes, Curtis Frank
Hurd, Jeffery L
Janik, Gerald S
Jespersen, Neil David
Jones, William Davidson

Kagansky, Larisa
Kumler, Philip L
Luthra, Krishan Lal
McCauley, James Weymann
Mendel, John Richard
Morrison, Ian Douglas
Moynihan, Cornelius Timothy
Muller, Olaf
Mullhaupt, Joseph Timothy
Myers, Clifford Earl
Neveu, Maurice C
Othmer, Donald F(rederick)
Prochazka, Svante
Questad, David Lee
Revankar, Vithal V S
Schweitzer, Donald Gerald
Silver, Herbert Graham
Stell, George Roger
Sternstein, Sanford Samuel
Streett, William Bernard
Takeuchi, Esther Sans
Taylor, Dale Frederick
Thomas, Harold Todd
Thompson, David Fred
Van Ness, Hendrick C(harles)
Von Berg, Robert L(ee)
Whang, Sung H
Widom, Benjamin
Xue, Yongpeng
Ziemniak, Stephen Eric
Zollweg, John Allman

NORTH CAROLINA
Abbey, Kathleen Mary Kyburz
Abbott, Richard Newton, Jr
Allen, Harry Clay, Jr
Goodwin, Frank Erik
Herrell, Astor Y
Hugus, Z Zimmerman, Jr
Koch, Carl Conrad
McDonald, Daniel Patrick
MacPhail, Richard Allyn
Pienta, Norbert J

OHIO
Bailey, John Clark
Baloun, Calvin H(endricks)
Bansal, Narottam Prasad
Bittker, David Arthur
Bundy, Francis P
Carbonara, Robert Stephen
Chrysochoos, John
Clark, William Alan Thomas
Datta, Ranajit K
Dembowski, Peter Vincent
Eby, Ronald Kraft
Edwards, Jimmie Garvin
Foland, Kenneth Austin
Freeberg, Fred E
Herum, Floyd L(yle)
Hirt, Andrew Michael
Jacobs, Donald Thomas
Kaufman, John Gilbert, Jr
Kaufman, Miron
Kornbrekke, Ralph Erik
Kreidler, Eric Russell
Moon, Tag Young
Nardi, John Christopher
Palffy-Muhoray, Peter
Prahl, Joseph Markel
Prusaczyk, Joseph Edward
Rogers, Charles Edwin
Sara, Raymond Vincent
Selover, Theodore Britton, Jr
Simha, Robert
Thomas, Terence Michael
Uys, Johannes Marthinus
Yarrington, Robert M

OKLAHOMA
Brady, Barry Hugh Garnet
Burchfield, Thomas Elwood
Chung, Ting-Horng
Freeman, Robert David
Frost, Jackie Gene
Gall, James William
Hays, George E(dgar)
Johnson, Timothy Walter
Lindstrom, Merlin Ray
McKeever, Stephen William Spencer
Noll, Leo Albert

OREGON
Church, Larry B
Ericksen, Jerald LaVerne
Ferrante, Michael John
Haygarth, John Charles
Horne, Frederick Herbert
Ko, Hon-Chung
McClure, David Warren
Mazo, Robert Marc
Nafziger, Ralph Hamilton
Prentice, Jack L
Schaefer, Seth Clarence
Smith, Kelly L
Tenney, Agnes

PENNSYLVANIA
Birks, Neil
Bitler, William Reynolds
Brown, Paul Edmund
Burch, Mary Kappel
Chubb, Walston
Corneliussen, Roger DuWayne

Danner, Ronald Paul
Daubert, Thomas Edward
Dunlap, Lawrence H
Enick, Robert M
Falcone, James Salvatore, Jr
Fitzgerald, Maurice E
Foster, Perry Alanson, Jr
Frank, William Benson
Fuchs, Walter
Garbuny, Max
Gardner, David Milton
Gerasimowicz, Walter Vladimir
Gittler, Franz Ludwig
Goodwin, James Gordon, Jr
Grabowski, Joseph J
Grossmann, Elihu D(avid)
Hultman, Carl A
Johnston, William V
Kanter, Ira E
Klein, Michael Lawrence
Lavin, J Gerard
MacRae, Donald Richard
Nichols, Duane Guy
Patterson, Gary David
Pierce, Thomas Henry
Plazek, Donald John
Sabol, George Paul
Shozda, Raymond John
Smyth, Donald Morgan
Steiger, Roger Arthur
Stubican, Vladimir S(tjepan)
Szamosi, Janos
Taylor, Robert Morgan
Worrell, Wayne L
Wriedt, Henry Anderson
Yang, Arthur Jing-Min

RHODE ISLAND
Giletti, Bruno John
Hess, Paul C
Stratt, Richard Mark

SOUTH CAROLINA
Haile, James Mitchell
Herdklotz, Richard James
Jaco, Charles M, Jr
Metz, Clyde Raymond
Plodinec, Matthew John
Roldan, Luis Gonzalez
Tolbert, Thomas Warren

TENNESSEE
Barber, Eugene John
Besser, John Edwin
Bleier, Alan
Bullock, Jonathan S, IV
Caflisch, George Barrett
Dudney, Nancy Johnston
Gibson, John Knight
Grugel, Richard Nelson
Habenschuss, Anton
Ho, Patience Ching-Ru
Klots, Cornelius E
Lenhert, P Galen
Li, Ying Sing
Lindemer, Terrence Bradford
Pasto, Arvid Eric
Reynolds, Jefferson Wayne
Sales, Brian Craig
Schnelle, K(arl) B(enjamin), Jr
Seaton, William Hafford
Thiessen, William Ernest
Trowbridge, Lee Douglas
Williams, Robin O('Dare)

TEXAS
Albright, John Grover
Baker, Charles Taft
Brostow, Witold Konrad
Caflisch, Robert Galen
Chao, Jing
Clingman, William Herbert, Jr
DeBerry, David Wayne
De Groot, Peter Bernard
Dufner, Douglas Carl
Eakin, Bertram E
Fredin, Leif G R
Garza, Cesar Manuel
Goodenough, John Bannister
Gooding, James Leslie
Hance, Robert Lee
Holste, James Clifton
Ignatiev, Alex
Li, Shu
Mack, Russel Travis
Margrave, John Lee
Marsh, Kenneth Neil
Munk, Petr
Pogue, Randall F
Spencer, John Edward
Tong, Youdong
Whalen, James William
Whitmire, Kenton Herbert
Wilhoit, Randolph Carroll
Wood, Scott Emerson

UTAH
Carnahan, Robert D
Eatough, Delbert J
Guillot, David G
Izatt, Reed McNeil
Miller, Akeley
Oyler, Dee Edward
Rowley, Richard L

Stringfellow, Gerald B

VERMONT
Schiffman, Robert A

VIRGINIA
Bodnar, Robert John
Erickson, Wayne Douglas
Field, Paul Eugene
Graham, Kenneth Judson
Hudson, Frederick Mitchell
Huebner, John Stephen
Jordan, Wade H(ampton), Jr
Laszlo, Tibor S
McGee, Henry A(lexander), Jr
Nuckolls, Joe Allen
Orwoll, Robert Arvid
Oyama, Shigeo Ted
Phillips, Donald Herman
Pohlmann, Juergen Lothar Wolfgang
Porzel, Francis Bernard
Schreiber, Henry Dale
Smith, Bertram Bryan, Jr
Ticknor, Leland Bruce
Toulmin, Priestley
Yates, Charlie Lee

WASHINGTON
Collins, Gary Scott
Ghiorso, Mark Stefan
Gregory, Norman Wayne
Hinman, Chester Arthur
Kruger, Albert Aaron
Leggett, Robert Dean
Lindenmeyer, Paul Henry
Rai, Dhanpat
Riehl, Jerry A
Spencer, Arthur Milton, Jr
Stuve, Eric Michael
Wilson, Charles Norman
Yunker, Wayne Harry

WISCONSIN
Atalla, Rajai Hanna
Cooper, Reid F
Draeger, Norman Arthur
Kotvis, Peter Van Dyke
Pollnow, Gilbert Frederick
Rosenberg, Robert Melvin
Sather, Glenn A(rthur)
Spencer, Brock

WYOMING
Harnsberger, Paul Michael
Hiza, Michael John, Jr
Hoffman, Edward Jack

ALBERTA
Checkel, M David
Gee, Norman
Heimann, Robert B(ertram)
Hepler, Loren George
Tschuikow-Roux, Eugene
Yeager, Howard Lane

MANITOBA
Cenkowski, Stefan
McFarlane, Joanna
Wikjord, Alfred George

NOVA SCOTIA
Kreuzer, Hans Jurgen
White, Mary Anne

ONTARIO
Conard, Bruce R
Glew, David Neville
Goodings, John Martin
Handa, Paul
Hemmings, Raymond Thomas
Jacobs, Patrick W M
Kavassalis, Tom A
Koffyberg, Francois Pierre
Kumaran, Mavinkal Kizhakkeveettil
Leaist, Derek Gordon
Lepock, James Ronald
Perlman, Maier
Tombalakian, Artin S
Weir, Ronald Douglas
Westland, Alan Duane

PRINCE EDWARD ISLAND
Madan, Mahendra Pratap

QUEBEC
Britten, Michel
Chan, Sek Kwan
Dealy, John Michael
Desnoyers, Jacques Edouard
Dorris, Gilles Marcel
Leigh, Charles Henry
Sanctuary, Bryan Clifford
Utracki, Lechoslaw Adam

SASKATCHEWAN
Kybett, Brian David

OTHER COUNTRIES
Gordon, Jeffrey Miles
Gray, Kenneth W
Haymet, Anthony Douglas-John
Irvine, Stuart James Curzon
Jacob, K Thomas
Kafri, Oded

Thermodynamics & Material Properties (cont)

Nyburg, Stanley Cecil
Steichen, Richard John

Other Chemistry

ALABAMA
Acton, Ronald Terry
Hart, Gerald Warren
Krishna, N Rama
O'Kane, Kevin Charles
Pan, Chai-Fu
Sakai, Ted Tetsuo
Sayles, David Cyril
Tamburin, Henry John

ARIZONA
Ambrose, Robert T
Denton, M Bonner
Muggli, Robert Zeno

ARKANSAS
Fu, Peter Pi-Cheng
Sears, Derek William George

CALIFORNIA
Adams, George Baker
Bartlett, Paul A
Blatt, Joel Martin
Chang, Ming-Houng (Albert)
Delaney, Margaret Lois
Domeniconi, Michael John
Gardner, Marjorie Hyer
Garfin, David Edward
Garrison, Warren Manford
George, Patricia Margaret
George-Nascimento, Carlos
Gerstl, Bruno
Harrison, Jonas P
Houle, Frances Anne
Hughes, Robert Alan
Jacob, Robert Allen
Juster, Norman Joel
Kluksdahl, Harris Eudell
Lu, Hsieng S
Ma, Joseph T
Maas, Keith Allan
Metiu, Horia I
Miyada, Don Shuso
Munger, Charles Galloway
Nixon, Alan Charles
Prakash, Surya G K
Rodriguez, Eloy
Rogers, Howard H
Scattergood, Thomas W
Stucky, Galen Dean
Swidler, Ronald
Tin-Wa, Maung
Trapani, Robert Vincent
Winterlin, Wray Laverne
Wu, Jiann-Long
Zeronian, Sarkis Haig
Zwerdling, Solomon

COLORADO
Batchelder, Arthur Roland
Bower, Nathan Wayne
Czanderna, Alvin Warren
Macalady, Donald Lee
Marsh, Gregory K
Nozik, Arthur Jack
Schroeder, Herbert A(ugust)
Seo, Eddie Tatsu

CONNECTICUT
Baker, Bernard S
Century, Bernard
Haydu, Juan B
Howard, Phillenore Drummond
Marfat, Anthony
Markham, Claire Agnes
Perry, Clark William
Thomas, Arthur L
Ziegler, Theresa Frances

DELAWARE
Bergna, Horacio Enrique
Cairncross, Allan
Ehrich, F(elix) Frederick
Gibson, Joseph W(hitton), Jr
Hodges, Charles Thomas
Pagano, Alfred Horton
Vassiliou, Eustathios
Wayrynen, Robert Ellis
Williams, Ebenezer David, Jr

DISTRICT OF COLUMBIA
Colton, Richard J
Dennis, William E
Field, Ruth Bisen
James, Alton Everette, Jr
McCormack, Mike
Morehouse, Kim Matthew
Wakelyn, Phillip Jeffrey

FLORIDA
Cohen, Martin Joseph
Cross, Timothy Albert
Dannenberg, E(li) M
Duranceau, Steven Jon

Echegoyen, Luis Alberto
Ellis, Harold Hal
Graves, Robert Joseph
Hardman, Bruce Bertolette
Hass, Michael A
Jackson, George Richard
Jungbauer, Mary Ann
Kagan, Benjamin
Lewis, Nita Aries
Mason, Kenneth M
Miles, Carl J
Moudgil, Brij Mohan
Roy, Clarence
Safron, Sanford Alan
Urone, Paul

GEORGIA
Allison, John P
Block, Toby Fran
Bowen, Paul Tyner
Brown, William E
Caughman, Henry Daniel
Derrick, Mildred Elizabeth
Goldsmith, Edward
Islam, M Safizul
Knight, James Albert, Jr
Kohl, Paul Albert
Lin, Ming Chang
Netzel, Thomas Leonard
Occelli, Mario Lorenzo
Radin, Nathan
Shapira, Raymond
Tan, Kim H
Trawick, William George
Van Assendelft, Onno Willem

HAWAII
Crane, Sheldon Cyr
Hihara-Endo, Linda Masae

ILLINOIS
Albertson, Clarence E
Boese, Robert Alan
Brown, Meyer
Doody, Marijo
Filler, Robert
Horwitz, Earl Philip
House, Kathleen Ann
Kaistha, Krishan K
McGorrin, Robert Joseph
Made-Gowda, Netkal M
Mahajan, Om Prakash
Matthews, Clifford Norman
Mertz, William J
Mohan, Prem
Moll, William Francis, Jr
Moschandreas, Demetrios J
Nuzzo, Ralph George
Plautz, Donald Melvin
Rogers, Robin Don
Ryan, Clarence E, Jr
Savit, Joseph
Schurr, Garmond Gaylord
Stafford, Fred E
Stein, Lawrence
Vissers, Donald R
Walhout, Justine I Simon
Whittington, Wesley Herbert
Wilke, Robert Nielsen
Wolff, Thomas E
Zobel, Henry Freeman

INDIANA
Chen, Victor John
Dreikorn, Barry Allen
Dubin, Paul Lee
Gantzer, Mary Lou
Hemmes, Paul Richard
Kamat, Prashant V
Reuland, Donald John

IOWA
Campion, Stephen R

KANSAS
Ramamurti, Krishnamurti
Wright, Donald C

KENTUCKY
Baumgart, Richard
Giammara, Berverly L Turner Sites
Hamilton-Kemp, Thomas Rogers
Lloyd, William Gilbert
Riley, John Thomas
Wilson, Thomas Lamont
Zembrodt, Anthony Raymond

LOUISIANA
Blanchard, Eugene Joseph
Carver, James Clark
Clarke, Margaret Alice
Greco, Edward Carl
Hankins, B(obby) E(ugene)
Hanlon, Thomas Lee
Jung, John Andrew, Jr
Kane, Gordon Philo
Marshall, Wayne Edward
Mollere, Phillip David
Morris, Cletus Eugene
Smith, Grant Warren, II
Wendt, Richard P
Zaweski, Edward F

MAINE
Tedeschi, Robert James
Yentsch, Clarice M

MARYLAND
Aaron, Herbert Samuel
Alperin, Harvey Albert
Craig, Paul N
Crawley, Jacqueline N
Darneal, Robert Lee
Fahy, Gregory Michael
Heller, Stephen Richard
Holloway, Caroline T
Kirkpatrick, Diana (Rorabaugh) M
Koch, Thomas Richard
Mackier, Bruce F
McNairy, Sidney A, Jr
Malghan, Subhaschandra Gangappa
Pattabiraman, Nagarajan
Roethel, David Albert Hill
Ryback, Ralph Simon
Saba, William George
Sharrow, Susan O'Hott
Shrake, Andrew
Sripada, Pavanaram Kameswara
Tarshis, Irvin Barry
Tower, Donald Bayley
Weser, Don Benton

MASSACHUSETTS
Barish, Leo
Caslavska, Vera Barbara
Caslavsky, Jaroslav Ladislav
Chiu, Tak-Ming
Cohen, Fredric Sumner
Cohen, Robert Edward
Fessler, William Andrew
Figueras, John
Gray, Nancy M
Gutoff, Edgar B(enjamin)
Hoffman, Richard Laird
Idelson, Martin
Liggero, Samuel Henry
Matsuda, Seigo
Moser, William Ray
Rivin, Donald
Schlaikjer, Carl Roger
Serafin, Frank G
Trotz, Samuel Isaac

MICHIGAN
Chau, Vincent
Corrigan, Dennis Arthur
Dewitt, Sheila Hobbs
Dubpernell, George
Elrod, David Wayne
Graham, John
Jaglan, Prem S
Janik, Borek
Kiechle, Frederick Leonard
Korcek, Stefan
Lee, Stephen
Lentz, Charles Wesley
Lorch, Steven Kalman
Maasberg, Albert Thomas
Nazri, Gholam-Abbas
Palaszek, Mary De Paul
Purcell, Jerry
Reising, Richard F
Riley, Bernard Jerome
Ryntz, Rose A
Sard, Richard
Schneider, John Arthur
Schubert, Jack
Schwartz, Shirley E
Seefurth, Randall N
Sharma, Ram Autar
Stevenson, Richard Marshall
Strom, Robert Michael
Swovick, Melvin Joseph
Zak, Bennie
Zanini-Fisher, Margherita

MINNESOTA
Beebe, George Warren
Douglas, David Lewis
Farber, Elliott
Grasse, Peter Brunner
Hagen, Donald Frederick
Hogle, Donald Hugh
Horner, James William, Jr
Johnson, Rodney L
Liu, Hung-Wen
Schultz, Robert John
Thompson, Phillip Gerhard

MISSISSIPPI
Hwang, Huey-Min
Little, Brenda Joyce

MISSOURI
Crutchfield, Marvin Mack
Frankoski, Stanley P
Kalota, Dennis Jerome
Knox, Walter Robert
Monzyk, Bruce Francis
Shen, Chung Yu
Slocombe, Robert Jackson
Wagenknecht, John Henry

NEBRASKA
Allington, Robert W
Kenkel, John V

NEVADA
Herzlich, Harold Joel

NEW HAMPSHIRE
Greenwald, Harold Leopold
Nadeau, Herbert Gerard
Richmond, Robert Chaffee
Zager, Ronald

NEW JERSEY
Aldrich, Haven Scott
Arbuckle, Georgia Ann
Armbruster, David Charles
Bhandarkar, Suhas D
Chandrasekhar, Prasanna
Cohen, Paul
Crawley, Lantz Stephen
Dalton, Thomas Francis
Garwood, William Everett
Gierer, Paul L
Gray, Russell Houston
Izod, Thomas Paul John
Jaffe, Sol Samson
Kline, Charles Howard
Kuehl, Guenter Hinrich
Leeds, Norma S
Lynch, Charles Andrew
Malinowski, Edmund R
Miale, Joseph Nicolas
Montgomery, Richard Millar
Mustafa, Shams
Nassau, Kurt
Orlando, Carl
Priesing, Charles Paul
Ramsden, Hugh Edwin
Shacklette, Lawrence Wayne
Sheppard, Keith George
Sugathan, Kenneth Kochappan
Vukov, Rastko
Walradt, John Pierce
Weigmann, Hans-Dietrich H
Weil, Rolf
Weston, Charles Alvin

NEW MEXICO
Hockett, John (Edwin)
Kelber, Jeffry Alan
Maggiore, Carl Jerome
Thompson, Joseph Lippard
Walton, George

NEW YORK
Acharya, Seetharama A
Beck, Curt Werner
Beer, Sylvan Zavi
Bent, Brian E
Bommaraju, Tilak V
Brill, Robert H
Cheh, Huk Yuk
Danielson, Paul Stephen
Flessner, Michael F
Giguere, Raymond Joseph
Gluck, Ronald Monroe
Goldschmidt, Eric Nathan
Gorfien, Stephen Frank
Hargitay, Bartholomew
Homan, Clarke Gilbert
Inman, Charles Gordon
James, Daniel Shaw
Joseph, Solomon
Keller, D Steven
Kohrt, Carl Fredrick
Lee, Lieng-Huang
Levitt, Morton
Lorenzo, George Albert
Luckey, George William
Markarian, Herand M
Orem, Michael William
Pochan, John Michael
Rader, Charles George
Rebel, William J
Rhodes, Yorke Edward
Rubin, Isaac D
Sage, Gloria W
Schuster, David Israel
Takeuchi, Esther Sans
Yip, Yum Keung

NORTH CAROLINA
Bereman, Robert Deane
Beverung, Warren Neil, Jr
Chalovich, Joseph M
Coury, Louis
Fabricius, Dietrich M
Freeman, Harold Stanley
Fytelson, Milton
Garrou, Philip Ernest
Hildebolt, William Morton
Irene, Eugene Arthur
Johnson, J(ames) Donald
Kurtz, A Peter
Levine, Solomon Leon
McPeters, Arnold Lawrence
Majewski, Theodore E
Noftle, Ronald Edward
Perfetti, Patricia F
Roth, Barbara
Walters, Douglas Bruce

OHIO
Bailey, John Clark
Bansal, Narottam Prasad
Barsky, Constance Kay
Bhatti, Asif

Brooman, Eric William
Didchenko, Rostislav
Dinbergs, Kornelius
Dodge, James Stanley
Eley, Richard Robert
Fischer, Mark Benjamin
Gratzl, Miklos
Halasa, Adel F
Harris, Arlo Dean
Hartsough, Robert Ray
Isenberg, Allen (Charles)
Lippert, Bruce J
Manning, Maurice
Maximovich, Michael Joseph
Merten, Helmut L
Myers, Elliot H
Ostrum, G(eorge) Kenneth
Rynbrandt, Donald Jay
Selover, Theodore Britton, Jr
Seybold, Paul Grant
Shaw, Wilfrid Garside
Sibley, Lucy Roy
Tiernan, Thomas Orville
Tinker, H(orald) Burnham
Tong, James Ying-Peh
Wagner, Eugene Ross
Wagner, Melvin Peter
Warner, Ann Marie

OKLAHOMA
Kucera, Clare H
Letcher, John Henry, III
Linder, Donald Ernst
Lyon, William Graham
Palmer, Bruce Robert
Thomson, George Herbert
Wang, Binghe

OREGON
Autrey, Robert Luis
Biermann, Christopher James
Detlefsen, William David, Jr
Jaeger, Charles Wayne
Laver, Murray Lane

PENNSYLVANIA
Bare, Thomas M
Berges, David Alan
Berkheimer, Henry Edward
Butera, Richard Anthony
Chang, Charles Hung
Cornell, John Alston
De Tommaso, Gabriel Louis
Ebert, Philip E
Feely, Wayne E
Friedman, Herbert Alter
Garbarini, Victor C
Garrison, Barbara Jane
Gerasimowicz, Walter Vladimir
Gergova, Katia M
Goodwin, James Gordon, Jr
Kayne, Fredrick Jay
Kugler, George Charles
Lai, Por-Hsiung
Louie, Ming
Luck, Stanley D
Marcelin, George
Mayer, Theodore Jack
Mehta, Prakash V
Reider, Malcolm John
Sax, Martin
Schoen, Kurt L
Shafer, Jules Alan
Sitrin, Robert David
Stryker, Lynden J
Traynor, Sean G
Urenovitch, Joseph Victor
Watterson, Kenneth Franklin
White, Malcolm Lunt
Wiehe, William Albert

RHODE ISLAND
Casey, John Edward, Jr
Ho, Kay T
Kreiser, Ralph Rank

SOUTH CAROLINA
Aelion, Claire Marjorie
Dunlap, Robert Bruce
Elzerman, Alan William
Gadsden, Richard Hamilton, Sr
Jones, Edward Stephen
Kinard, W Frank
King, Lee Curtis
Longtin, Bruce
Maier, Herbert Nathaniel
Spencer, Harold Garth

TENNESSEE
Barron, Eugene Roy
Barton, Kenneth Ray
Bleier, Alan
Gupta, Ramesh C
Gustafson, Bruce LeRoy
Harmer, David Edward
Higgins, Irwin Raymond
Horton, Charles Abell
Karve, Mohan Dattatreya
Murphy, William R
Quist, Arvin Sigvard

TEXAS
Allison, Jean Batchelor
Bullock, Kathryn Rice

Fodor, George E
Fritsche, Herbert Ahart, Jr
Hart, Paul Robert
Hausler, Rudolf H
Hendrickson, Constance McRight
Johnson, Fred Lowery, Jr
Jones, Paul Ronald
Mitchell, Roy Ernest
Prasad, Rupi
Saleh, Farida Yousry
Sullivan, Thomas Allen
Wong, Shan Shekyuk

UTAH
Aldous, Duane Leo
Dehm, Henry Christopher
Miller, Jan Dean
Pons, Stanley

VERMONT
Tremmel, Carl George

VIRGINIA
Allen, Ralph Orville, Jr
Barber, Patrick George
Boadle-Biber, Margaret Clare
Bolleter, William Theodore
Dockter, Kenneth Wylie
Gager, Forrest Lee, Jr
Hussamy, Samir
Jaisinghani, Rajan A
Maybury, Robert H
Olin, Jacqueline S
Perillo, Benjamin Anthony
Sedriks, Aristide John
Shipe, James R, Jr
Sobel, Robert Edward
Taylor, Ronald Paul
Tichenor, Robert Lauren
Turner, Anne Halligan
Walters, Allan N

WASHINGTON
Berry, Keith O
Nelson, Sidney D
Pillay, Gautam
Rohrmann, Charles A(lbert)
Styris, David Lee
Wollwage, Paul Carl

WEST VIRGINIA
Bartley, William J
Bryant, David

WISCONSIN
Berge, John Williston
Bloch, Daniel R
Fitch, Robert McLellan
Gonzalez, Hector Julian
Kotvis, Peter Van Dyke
Lee, Kathryn Adele Bunding
Mukerjee, Pasupati
Pearl, Irwin Albert
Schwab, Helmut
Sinko, John

WYOMING
Harnsberger, Paul Michael

PUERTO RICO
ALZERRECA, ARNALDO

ALBERTA
Fisher, Harold M
Gee, Norman
Taylor, Wesley Gordon
Wiebe, Leonard Irving

BRITISH COLUMBIA
Fryzuk, Michael Daniel
Strong, David F

NEW BRUNSWICK
Fairchild, Wayne Lawrence
Passmore, Jack

NEWFOUNDLAND
Shahidi, Fereidoon

NOVA SCOTIA
Coxon, John Anthony
Ellenberger, Herman Albert
Fraser, Albert Donald

ONTARIO
Bolton, James R
Burns, George
Eliades, Theo I
Kazmaier, Peter Michael
MacDougall, Daniel
Oldham, Keith Bentley
Wigfield, Yuk Yung
Wu, Tai Wing

QUEBEC
Bandrauk, Andre D
Clayton, David Walton
Lasia, Andrzej
Lessard, Jean
Serpone, Nick
Vijh, Ashok Kumar
Whitehead, Michael Anthony

OTHER COUNTRIES
Ahmed, Syed Mahmood
Grunwald, John J
Laurence, Alfred Edward
Menon, Manchery Prabhakara
Samuelsson, Bengt Ingemar

COMPUTER SCIENCES
Artificial Intelligence

ARIZONA
Findler, Nicholas Victor
Szilagyi, Miklos Nicholas

ARKANSAS
Talburt, John Randolph

CALIFORNIA
Berenji, Hamid R
Frenster, John H
Hightower, James K
Hoffman, Donald D
Jacobson, Alexander Donald
Korf, Richard E
Parkison, Roger C
White, Ray Henry

CONNECTICUT
McDermott, Drew Vincent

FLORIDA
Elsner, James Brian
Fausett, Donald Wright

GEORGIA
Covington, Michael A

ILLINOIS
Fildes, John M
Mavrovouniotis, Michael L
Raiden, Robert Lee
Ray, Sylvian Richard
Rips, Lance Jeffrey

INDIANA
Penna, Michael Anthony

MARYLAND
Dixon, John Kent
Minker, Jack
Stillman, Rona Barbara

MASSACHUSETTS
Estes, William K

MICHIGAN
Goodman, Erik David
Holland, Steven William
Reynolds, Robert Gene
Wu, Dolly Y

MISSOURI
Ali, Syed S

MONTANA
Wright, Alden Halbert

NEBRASKA
Surkan, Alvin John

NEVADA
Minor, John Threecivelous

NEW JERSEY
Bhagat, Phiroz Maneck
Sofer, William Howard
Waltz, David Leigh

NEW YORK
Allen, James Frederick
Cok, David R
Kent, Henry Johann
Leung, Yiu M
Rapaport, William Joseph
Shapiro, Stuart Charles
Weintraub, Joseph

NORTH DAKOTA
Winrich, Lonny B

OHIO
Hobart, William Chester, Jr
Waldron, Manjula Bhushan

OREGON
Douglas, Sarah A
Freiling, Michael Joseph

PENNSYLVANIA
Metaxas, Dimitri
Neveln, Bob
Rau, Lisa Fay

RHODE ISLAND
Charniak, Eugene

TEXAS
Hall, Douglas Lee
Kuipers, Benjamin Jack
Miikkulainen, Risto Pekka
Ruchelman, Maryon W

VIRGINIA
Campbell, Leo James
Deal, George Edgar
Franklin, Jude Eric

WISCONSIN
Firebaugh, Morris W

ALBERTA
Sorenson, Paul G

NOVA SCOTIA
Schuegraf, Ernst Josef

ONTARIO
Halperin, Janet R P

OTHER COUNTRIES
Vamos, Tibor

Computer Sciences, General

ALABAMA
Atkinston, Ronald A
Barnard, Anthony C L
Dean, Susan Thorpe
De Maine, Paul Alexander Desmond
Dorai-Raj, Diana Glover
Downey, James Merritt
Fullerton, Larry Wayne
Goldsmith, Donald Leon
Hazelrig, Jane
Hill, David Thomas
Irwin, John David
Leslie, Thomas M
Lowry, James Lee
McCarthy, Dennis Joseph
Pattillo, Robert Allen
Schroer, Bernard J
Seidman, Stephen Benjamin
Sides, Gary Donald
Stone, Max Wendell
White, Ronald
Zalik, Richard Albert

ALASKA
Dahlberg, Michael Lee
Lando, Barbara Ann
Larson, Frederic Roger

ARIZONA
Andrews, Gregory Richard
Barckett, Joseph Anthony
Barrett, Craig R
Bartels, Peter H
Bitter, Gary G
Carruthers, Lucy Marston
Clark, Wilburn O
Epstein, L(udwig) Ivan
Feldstein, Alan
Findler, Nicholas Victor
Fink, James Brewster
Formeister, Richard B
Gaud, William S
Holmes, William Farrar
Kakar, Rajesh Kumar
Kreidl, Tobias Joachim
Lewis, William E(rvin)
Lorents, Alden C
Lovell, Stuart Estes
Mackulak, Gerald Thomas
Peterson, James Douglas
Randall, Lawrence Kessler, Jr
Shemer, Jack Evvard
Smith, Josef Riley
Triffet, Terry
Winder, Robert Owen
Yakowitz, Sidney J

ARKANSAS
Abedi, Farrokh
Asfahl, C Ray
Berghel, Hal L
Bonner, Errol Marcus
Crisp, Robert M, Jr
Gildseth, Wayne
Mink, Lawrence Albright
Rossa, Robert Frank
Sedelow, Sally Yeates
Sims, Robert Alan
Starling, Albert Gregory
Talburt, John Randolph
Wheeler, Keith Wilson
Wilkes, Joseph Wray

CALIFORNIA
Adams, Ralph Melvin
Amdahl, Gene M
Anderson, Dennis Elmo
Anderson, Edward P(arley)
Anderson, Roger E
Aoki, Masanao
Arbib, Michael A
Arnovick, George (Norman)
Ashworth, Edwin Robert
Avizienis, Algirdas
Backus, John
Baggi, Denis Louis
Baker, James Addison
Banerjee, Utpal
Barrett, William Avon
Barsky, Brian Andrew
Beach, Sharon Sickel

Computer Sciences, General (cont)

Beaver, W(illiam) L(awrence)
Bell, Chester Gordon
Benson, Robert Franklin
Berardo, Peter Antonio
Bergquist, James William
Bernstein, Ralph
Bevc, Dimitri
Bhanu, Bir
Bharadvaj, Bala Krishnan
Bic, Lubomir
Bingham, Harry H, Jr
Blattner, Meera McCuaig
Blum, Edward K
Blum, Manuel
Bly, Sara A
Boehm, Barry William
Boltinghouse, Joseph C
Boyle, Walter Gordon, Jr
Bradley, Hugh Edward
Brown, David Randolph
Brown, Harold David
Bussell, Bertram
Butler, Jon Terry
Cantor, David Geoffrey
Carlan, Audrey M
Carlson, F Roy, Jr
Carlyle, Jack Webster
Carson, George Stephen
Case, Lloyd Allen
Cassel, Russell N
Caswell, John N(orman)
Cautis, C Victor
Chambers, Frank Warman
Chan, Tony Fan-Cheong
Chang, Y(uan)-F(eng)
Chen, Alice Tung-Hua
Cheriton, David Ross
Choudary, Prabhakara Velagapudi
Christiansen, Jerald N
Chu, Kai-Ching
Chu, Wesley W(ei-chin)
Clark, Crosman Jay
Clarke, Wilton E L
Clegg, Frederick Wingfield
Clinnick, Mansfield
Cochran, Stephen G
Cogan, Adrian Ilie
Coleman, Charles Clyde
Conant, Curtis Terry
Cooke, Ron Charles
Cottrell, Roger Leslie Anderton
Cover, Thomas Merrill
Crandall, Ira Carlton
Cuadra, Carlos A(lbert)
Cunningham, Robert Stephen
Dairiki, Ned Tsuneo
Dalrymple, Stephen Harris
Deaton, Edmund Ike
Dennis, Martha Greenberg
De Pillis, Lisette G
Despain, Alvin M(arden)
Deutsch, Laurence Peter
Dolen, Richard
Dudziak, Walter Francis
Dutt, Nikil D
Edgerley, Dennis A
Elkind, Jerome I
Elspas, B(ernard)
Engel, Robert David
Engelbart, Douglas C(arl)
English, William Kirk
Erickson, Stanley Arvid
Estabrook, Kent Gordon
Estrin, Gerald
Estrin, Thelma A
Farone, William Anthony
Fateman, Richard J
Feigenbaum, Edward A(lbert)
Feign, David
Feldman, Jerome A
Fernbach, Sidney
Fischler, Martin A(lvin)
Flamm, Daniel Lawrence
Flegal, Robert Melvin
Floyd, Robert W
Friedlander, Carl B
Friedman, Alan E
Garber, Morris Joseph
Gehrmann, John Edward
Gelinas, Robert Joseph
Gerhardt, Mark S
Geschke, Charles Matthew
Ghosh, Sakti P
Gittleman, Arthur P
Gladney, Henry M
Goddard, Charles K
Goldberg, Jacob
Gott, Euyen
Gottlieb, Peter
Gould, William E
Graham, Martin H(arold)
Greenberger, Martin
Greenwood, James Robert
Greig, Douglas Richard
Gretsky, Neil E
Grinnell, Robin Roy
Gruenberger, Fred J(oseph)
Gunckel, Thomas L, II
Gundersen, Larry Edward
Gwinn, William Dulaney
Hamlin, Griffith Askew, Jr
Hamming, Richard W
Hance, Anthony James
Hardy, John W, Jr
Harris, David R
Harrison, Michael A
Hart, Peter E(lliot)
Harvey, Everett H
Hayes, Robert Mayo
Hearn, Anthony Clem
Hecht, Herbert
Hecht, Myron J
Hellerman, Herbert
Herriot, John George
Hightower, James K
Hilker, Harry Van Der Veer, Jr
Hillam, Bruce Parks
Hinrichs, Karl
Hoffman, Howard Torrens
Holl, Manfred Matthias
Hollander, Gerhard Ludwig
Holoien, Martin O
Horowitz, Ellis
Howell, Russell W
Hsu, John Y
Hu, Te Chiang
Huang, Di-Hui (David)
Huang, Sung-Cheng
Huberman, Bernardo Abel
Hughett, Paul William
Hurwitz, Alexander
Jackson, Albert S(mith)
Jacobs, Irwin Mark
Jacobson, Raymond E
James, Kenneth Eugene
James, Philip Nickerson
Janos, William Augustus
Jefferson, Thomas Hutton, Jr
Kaelble, David Hardie
Kahan, William M
Karin, Sidney
Karplus, Walter J
Katz, Louis
Keaton, Michael John
Keddy, James Richard
Kehl, William Brunner
Kendall, Burton Nathaniel
Kenyon, Richard R(eid)
Kershaw, David Stanley
Keshavan, H R
Kimble, Gerald Wayne
Klahr, Philip
Kleinrock, Leonard
Knuth, Donald Ervin
Korchev, Dmitriy Veniaminovich
Korf, Richard E
Korpman, Ralph Andrew
Kraabel, John Stanford
Kreuzer, Lloyd Barton
Kubik, Robert N
Kull, Lorenz Anthony
Kuwabara, James S
Lamie, Edward Louis
Lamport, Leslie B
Lang, Martin T
Lange, Rolf
Larkin, K(enneth) T(rent)
Larson, Harry Thomas
Lathrop, Kaye Don
Lay, Thorne
Ledin, George, Jr
Lee, K Hung
Leedham, Clive D(ouglas)
Lelewer, Debra Ann
Levin, Roy
Levit, Robert Jules
Lewis, William Perry
Lichter, James Joseph
Lindal, Gunnar F
Linz, Peter
Lu, Adolph
Luckham, David Comptom
Luehrmann, Arthur Willett, Jr
Lum, Vincent Yu-Sun
Luqi,
Luxenberg, Harold Richard
Lynch, William C
McCann, Gilbert Donald, Jr
McCarthy, John
McCluskey, Edward Joseph
McCool, John Macalpine
McCreight, Edward M
MacDonald, Gordon James Fraser
McDonald, H(enry) S(tanton)
McMurray, Loren Robert
McQuillen, Howard Raymond
Magness, T(om) A(lan)
Maillot, Patrick Gilles
Maloy, John Owen
Mantey, Patrick E(dward)
Marder, Stanley
Maron, Melvin Earl
Marquart, Ronald Gary
Matis, Howard S
Mattson, Richard Lewis
Mead, Carver Andress
Meads, Philip Francis, Jr
Meisling, Torben (Hans)
Meissinger, Hans F
Meissner, Loren Phillip
Melvin, Jonathan David
Meyers, Gene Howard
Mihailovski, Alexander
Miller, William Frederick
Mitchell, James George
Mitchell, John Clifford
Monard, Joyce Anne
Monier, Louis Marcel
Mooney, Larry Albert
Morris, Jerry Lee, Jr
Morse, Joseph Grant
Motteler, Zane Clinton
Mulligan, Geoffrey C
Muntz, Richard Robert
Myers, George Scott, Jr
Nassi, Isaac Robert
Neher, Maynard Bruce
Nelson, Eldred (Carlyle)
Neumann, Peter G
Newsam, John M
Ng, Edward Wai-Kwok
Ng, Lawrence Chen-Yim
Niccolai, Nilo Anthony
Nickolls, John Richard
Nico, William Raymond
Niday, James Barker
Nikora, Allen P
Nilsson, Nils John
Obremski, Robert John
O'Dell, Austin Almond, Jr
Oliver, Ron
Olney, Ross David
O'Malley, Thomas Francis
Paal, Frank F
Papadimitriou, Christos
Parker, Alice Cline
Parker, Donn Blanchard
Patel, Arvindkumar Motibhai
Paulson, Boyd Colton, Jr
Pease, Marshall Carleton, III
Pennington, Ralph Hugh
Perry, Robert Nathaniel, III
Petrowski, Gary E
Phillips, Thomas Joseph
Plock, Richard James
Portnoff, Michael Rodney
Presser, Leon
Preston, Glenn Wetherby
Prokop, Jan Stuart
Ralston, Elizabeth Wall
Ramamoorthy, Chittoor V
Ramey, Daniel Bruce
Randall, Charles McWilliams
Ratcliff, Milton, Jr
Rau, Bantwal Ramakrishna
Reed, Irving Stoy
Requa, Joseph Earl
Rhodes, Edward Joseph, Jr
Richardson, John Mead
Rickard, James Joseph
Rider, Ronald Edward
Ritchie, Robert Wells
Roberts, Charles Sheldon
Robin, Allen Maurice
Rodabaugh, David Joseph
Rohm, C E Tapie, Jr
Rose, Gene Fuerst
Rosen, Irwin Gary
Rosenblatt, Daniel Bernard
Rosler, Lawrence
Rothenberg, Stephen
Rubin, Izhak
Rubin, Robert Howard
Rulifson, Johns Frederick
Russell, Gerald Frederick
Russell, Stuart Jonathan
Sangiovanni-Vincentelli, Alberto Luigi
Savitch, Walter John
Scherer, James R
Schlesinger, Stewart Irwin
Schwartz, Morton Donald
Schwartz, Sanford Bernard
Schweikert, Daniel George
Scott, Karen Christine
Seitz, S Stanley
Sengupta, Sumedha
Sequin, Carlo Heinrich
Shar, Leonard E
Shaw, Charles Alden
Shaw, George, II
Shenker, Scott Joseph
Sherman, John Edwin
Shoch, John F
Silvester, John Andrew
Simons, Roger Mayfield
Singleton, Richard Collom
Sinton, Steven Williams
Slaughter, John Brooks
Smith, Alan Jay
Smith, Douglas Wemp
Smith, George C
Smith, James Thomas
Smolarski, Dennis Chester
Solomon, Malcolm David
Sondak, Norman Edward
Sorrels, John David
Spinrad, Robert J(oseph)
Stauffer, Howard Boyer
Stone, William Ross
Sturm, Walter Allan
Suffin, Stephen Chester
Summers, Audrey Lorraine
Sutherland, Ivan Edward
Sutherland, William Robert
Talbot, Raymond James, Jr
Tang, Hwa-Tsang
Tannenbaum, Peter
Tanner, Robert Michael
Taylor, Robert William
Teague, Lavette Cox
Thacher, Henry Clarke, Jr
Thompson, Evan M
Toepfer, Richard E, Jr
Trenholme, John Burgess
Turin, George L
Ullman, Jeffrey D(avid)
Ung, Man T
Unti, Theodore Wayne Joseph
Valach, Miroslav
Vardiman, Larry
Vassiliou, Marius Simon
Vaughan, J Rodney M
Vemuri, Venkateswararao
Verona, Andrei
Vidal, Jacques J
Wagner, Raymond Lee
Waldinger, Richard J
Wallace, Graham Franklin
Walters, Richard Francis
Ware, W(illis) H(oward)
Webre, Neil Whitney
Weibell, Fred John
Wexler, Jonathan David
White, Ray Henry
Wilensky, Robert
Will, Peter Milne
Williams, Jack Rudolph
Williams, Robin
Wilmer, Michael Emory
Wilson, Edward L
Winograd, Terry Allen
Wolf, Robert Peter
Wood, Roger Charles
Wood, William Irwin
Woods, Roger David
Woolard, Henry W(aldo)
Wray, John L
Yates, Scott Raymond
Young, John W(esley), Jr
Zadeh, L(otfi) A
Zimmerman, C Duane

COLORADO

Brasch, Frederick Martin, Jr
Brown, Austin Robert, Jr
Buzbee, Billy Lewis
Clementson, Gerhardt C
Cole, Julian Wayne (Perry)
Cook, Robert Neal
Couger, J Daniel
Dandapani, Ramaswami
Dipner, Randy W
Dorn, William S
Eichelberger, W(illiam) H
Fosdick, Lloyd D(udley)
Frick, Pieter A
Frost, H(arold) Bonnell
Gentry, Willard Max, Jr
Gibson, William Loane
Hittelman, Allen M
Irons, Edgar T(owar)
Irwin, Robert Cook
Johnson, Janice Kay
Johnstone, James G(eorge)
Jordan, Harry Frederick
Kelley, Neil Davis
King, Lowell Alvin
Knight, Douglas Wayne
Lackey, Laurence
Lameiro, Gerard Francis
Lynch, Robert Michael
MacMillian, Stuart A
Malaiya, Yashwant Kumar
Marshall, Charles F
Meyer, Harvey John
Montague, Patricia Tucker
Mueller, Robert Andrew
Musick, James R
Nemeth, Evi
Paine, Richard Bradford
Pawlicki, Anthony Joseph
Platter, Sanford
Quick, James S
Rechard, Ottis William
Sargent, Howard Harrop, III
Sauer, Jon Robert
Schnabel, Robert B
Simske, Steven John
Singer, Howard Joseph
Stanger, Andrew L
Sugar, George R
Vogel, Richard E
Wackernagel, Hans Beat
Wessel, William Roy
Wiatrowski, Claude Allan
Wonsiewicz, Bud Caesar
Zeiger, H Paul

CONNECTICUT

Agresta, Joseph
Buckley, Jay Selleck, Jr
Burnett, Robert Walter
Close, Richard Thomas
Collins, Sylva Heghinian
Coulter, Paul David
Cowles, William Warren
Darcey, Terrance Michael
Dominy, Beryl W
Engel, Gerald Lawrence
Feit, Sidnie Marilyn
Frankel, Martin Richard
Harrison, Irene R

COMPUTER SCIENCES / 269

Hudak, Paul Raymond
Kozak, Gary S
Lang, George E, Jr
Levien, Roger Eli
Lof, John L(ars) C(ole)
Lovell, Bernard Wentzel
McCartin, Brian James
McCoy, Rawley D
McDermott, Drew Vincent
Mann, Richard A(rnold)
Maryanski, Fred J
Prabulos, Joseph J, Jr
Rezek, Geoffrey Robert
Ridgway, William C(ombs), III
Robbins, David Alvin
Sarfarazi, Mansoor
Scherr, Allan L
Schultz, Martin H
Sharlow, John Francis
Sholl, Howard Alfred
Simpson, Wilburn Dwain
Smith, Edgar Clarence, Jr
Stein, Alan H
Stein, David Morris
Steingiser, Samuel
Stephenson, Kenneth Edward
Stuart, James Davies
Youssef, Mary Naguib

DELAWARE
Abrahamson, Earl Arthur
Amer, Paul David
Benson, Frederic Rupert
Blaker, Robert Hockman
Bowyer, Kern M(allory)
Carberry, Judith B
Caviness, Bobby Forrester
Chester, Daniel Leon
Foster, Susan J
Golden, Kelly Paul
Graham, John Alan
Hays, John Thomas
Hirsch, Albert Edgar
Jansson, Peter Allan
Jones, Louise Hinrichsen
Milian, Alwin S, Jr
Monroe, Elizabeth McLeister
Montague, Barbara Ann
Murphy, Arthur Thomas
Pensak, David Alan
Quarry, Mary Ann
Saunders, B David
Schultz, John Lawrence
Skolnik, Herman
Taylor, Robert Burns, Jr
Ulery, Dana Lynn
Weiher, James F
Whitehouse, Bruce Alan

DISTRICT OF COLUMBIA
Andreadis, Tim Dimitri
Baker, Dennis John
Baruch, Jordan J(ay)
Bertaut, Edgard Francis
Caceres, Cesar A
Della Torre, Edward
Diemer, F(erdinand) P(eter)
Edwards, Willard
Frieder, Gideon
Hammer, Carl
Hart, Richard Cullen
Hedges, Harry G
Hodgson, James B, Jr
Hoffman, Lance J
Holland, Charles Jordan
Jones, Anita Katherine
Killion, Lawrence Eugene
Kitchens, Thomas Adren
Ledley, Robert Steven
McLean, John Dickson
Meltzer, Arnold Charles
Murphy, James John
Nelson, Larry Dean
Olmer, Jane Chasnoff
Raub, William F
Rosenblum, Lawrence Jay
Shore, John Edward
Tidball, Charles Stanley
Vandergraft, James Saul
Werdel, Judith Ann
Williams, Leland Hendry
Yamamoto, William Shigeru
Youm, Youngil

FLORIDA
Abou-Khalil, Samir
Baker, Theodore Paul
Bash, Paul Anthony
Bauer, Christian Schmid, Jr
Berk, Toby Steven
Bingham, Richard S(tephen), Jr
Bonn, T(heodore) H(ertz)
Bowers, James Clark
Carrier, Steven Theodore
Carroll, Dennis Patrick
Cengeloglu, Yilmaz
Chapman, Michael S
Chartrand, Robert Lee
Clark, Kerry Bruce
Clutterham, David Robert
Codd, Edgar Frank
Cohen, Howard Lionel
Conaway, Charles William
Cooke, William Joe

Copeland, Richard Franklin
Coulter, Neal Stanley
Couturier, Gordon W
Fine, Rana Arnold
Fishwick, Paul Anthony
Friedl, Frank Edward
Fultyn, Robert Victor
Goldwyn, Roger M(artin)
Goodman, Richard Henry
Gotterer, Malcolm Harold
Hall, David Goodsell, IV
Hand, Thomas
Harrison, Thomas J
Howard, Bernard Eufinger
Jahoda, Gerald
Jensen, Clayton Everett
Kandel, Abraham
Levow, Roy Bruce
Lewis, Brian Kreglow
Lighterman, Mark S
Lillien, Irving
Llewellyn, J(ohn) Anthony
Lofquist, George W
McMillan, Donald Ernest
Maddox, Billy Hoyte
Malthaner, W(illiam) A(mond)
Mankin, Richard Wendell
Martin, Charles John
Miler, George Gibbon, Jr
Miles, David H
Miller, Donald F
Naini, Majid M
Navlakha, Jainendra K
Rabbit, Guy
Rice, Randall Glenn
Roth, Paul Frederick
Scruggs, Frank Dell
Sheppard, Albert Parker
Sherman, Edward
Shershin, Anthony Connors
Stang, Louis George
Storm, Leo Eugene
Su, Stanley Y W
Timmer, Kathleen Mae
Tomonto, James R
Trubey, David Keith
Tulenko, James Stanley
Veinott, Cyril G
Vitagliano, Vincent J
Ward, James Audley
Wille, Luc Theo
Williams, Willie Elbert
Winton, Charles Newton
Yau, Stephen S

GEORGIA
Adams, David B
Arbogast, Richard Terrance
Arnold, Jonathan
Badre, Albert Nasib
Barnwell, Thomas Osborn, Jr
Bass, William Thomas
Blanke, Jordan Matthew
Camp, Ronnie Wayne
Caras, Gus J(ohn)
Covington, Michael A
Daniel, Leonard Rupert
Gallaher, Lawrence Joseph
Ghate, Suhas Ramkrishna
Goldman, John Abner
Graham, Roger Neill
Gupta, Rakesh Kumar
Holtum, Alfred G(erard)
Honkanen, Pentti A
Hunt, Gary W
Husson, Samir S
Jarvis, John Frederick
Koerner, T J
Krol, Joseph
Owen, Gene Scott
Pogue, Richard Ewert
Riddle, Lawrence H
Sizemore, Douglas Reece
Sparks, Arthur Godwin
Stone, David Ross
Straley, Tina
Suddath, Fred LeRoy, (Jr)
Techo, Robert
Vicory, William Anthony
Worth, Roy Eugene
Young, Raymond Hinchcliffe, Jr

HAWAII
Carlson, John Gregory
Kinariwala, Bharat K
Pager, David
Peterson, William Wesley
Walker, Terry M
Watanabe, Daniel Seishi
Weldon, Edward J, Jr

IDAHO
East, Larry Verne
Hoagland, Gordon Wood
Lawson, John D
Moore, Kenneth Virgil
Mortensen, Glen Albert
Wang, Ya-Yen Lee

ILLINOIS
Ahuja, Narendra
Arab-Ismaili, Mohammad Sharif
Ashenhurst, Robert Lovett
Ausman, Robert K

Bagley, John D(aniel)
Baldwin, Jack Timothy
Banerjee, Prithviraj
Bareiss, Erwin Hans
Belford, Geneva Grosz
Bertoncini, Peter Joseph
Bookstein, Abraham
Boyle, James Martin
Brown, Patricia Lynn
Brown, Richard Maurice
Carroll, John Terrance
Chaszeyka, Michael A(ndrew)
Cohen, Gloria
Cook, David Robert
Cook, Joseph Marion
Croke, Edward John
Divilbiss, James Leroy
Edelsbrunner, Herbert
Evens, Martha Walton
Faiman, Michael
Field, Kurt William
Fields, Thomas Henry
Filson, Don P
Finnerty, James Lawrence
Fisher, Richard Gary
Flora, Robert Henry
Fourer, Robert
Gay, Ben Douglas
Goodman, Gordon Louis
Gray, Linsley Shepard, Jr
Hammond, William Marion
Hansen, William Anthony
Hanson, Floyd Bliss
Hasdal, John Alan
Hattemer, Jimmie Ray
Hosken, William H
Jacobsen, Fred Marius
Jan, Kwan-Hwa
Johnson, Donald Elwood
Jordan, Steven Lee
Kamath, Krishna
Kaper, Hans G
Kasube, Herbert Emil
Kim, Ki Hwan
Kovacs, Eve Veronika
Kuck, David Jerome
Lamont, Patrick
Larson, Richard Gustavus
Lawrie, Duncan H
Lax, Louis Carl
Lee, Der-Tsai
Levenberg, Milton Irwin
Levy, Allan Henry
Liu, Chung Laung
Loui, Michael Conrad
McKnight, Richard D
Mittman, Benjamin
Murata, Tadao
Muroga, Saburo
Newhart, M(ary) Joan
Nieman, George Carroll
O'Donnell, Terence J
Palmore, Julian Ivanhoe, III
Parker, Ehi
Parr, Phyllis Graham
Polychronopoulos, Constantine Dimitrius
Powers, Michael Jerome
Prais, Michael Gene
Price, David Thomas
Pyle, K Roger
Rafelson, Max Emanuel, Jr
Raffenetti, Richard Charles
Reed, Daniel A
Reetz, Harold Frank, Jr
Renneke, David Richard
Rickert, Neil William
Ring, James George
Robertson, James E(vans)
Sawyer, S Prentiss
Scanlon, Jack M
Schipma, Peter B
Schwab, Gary Michael
Sellers, Donald Roscoe
Smarr, Larry Lee
Soare, Robert I
Steele, Ian McKay
Stenberg, Charles Gustave
Swanson, Don R
Swoyer, Vincent Harry
Tomkins, Marion Louise
Ts'o, Timothy On-To
Van Ness, James E(dward)
Wagner, Gerald C
Wells, Jane Frances
Wheeler, Bruce Christopher
Wilcox, Lee Roy
Williams, Martha E
Yoh, John K
Yu, Clement Tak
Yule, Herbert Phillip

INDIANA
Anderson-Mauser, Linda Marie
Atallah, Mikhail Jibrayil
Brown, Buck F(erguson)
Carlson, James C
Carlson, Lee Arnold
Carroll, John T, III
Costello, Donald F
Criss, Darrell E
Dalphin, John Francis
Davis, Grayson Steven
DeLong, Allyn F
DeMillo, Richard A

Drufenbrock, Diane
Easton, Richard J
Elharrar, Victor
Fuelling, Clinton Paul
Gersting, John Marshall, Jr
Gersting, Judith Lee
Guckel, Gunter
Gutowski, Gerald Edward
Hamblen, John Wesley
Hanes, Harold
Henry, Eugene W
Hofstadter, Douglas Richard
Houstis, Elias N
Huffman, John Curtis
Kallman, Ralph Arthur
Kent, Earle Lewis
Kuzel, Norbert R
Laxer, Cary
LoSecco, John M
Ma, Cynthia Sanman
Mansfield, Maynard Joseph
Marconi, Gary G
Marshall, Francis J
Moore, Emily Allyn
Moser, John William, Jr
Pollock, G Donald
Prosser, Franklin Pierce
Putnam, Thomas Milton
Rego, Vernon J
Rice, John Rischard
Robertson, Edward L
Rosen, Saul
Smith, Peter David
Smucker, Arthur Allan
Spafford, Eugene Howard
Springer, George
Stanton, Charles Madison
Thomas, Robert Jay
Thompson, Howard Doyle
Thuente, David Joseph
Weber, Janet Crosby
Wise, David Stephen
Young, Frank Hood
Yovits, Marshall Clinton

IOWA
Alton, Donald Alvin
Brearley, Harrington C(ooper), Jr
Garfield, Alan J
Hoffman, Eric Alfred
Jacob, Richard L
Kawai, Masataka
Lambert, Robert J
Lutz, Robert William
Meints, Clifford Leroy
Rila, Charles Clinton
Small, Arnold McCollum, Jr
Yohe, James Michael
Zettel, Larry Joseph

KANSAS
Bulgren, William Gerald
Carpenter, Kenneth Halsey
Conrow, Kenneth
Crowther, Robert Hamblett
Grzymala-Busse, Jerzy Witold
Hagen, Lawrence J
Hooker, Mark L
Legler, Warren Karl
Schweppe, Earl Justin
Stein, Marvin L
Tran, Loc Binh
Unger, Elizabeth Ann
Van Swaay, Maarten
Wallace, Victor Lew
Wright, Donald C

KENTUCKY
Alter, Ronald
Carpenter, Dwight William
Davidson, Jeffrey Neal
Davis, Chester L
Davis, Thomas Haydn
Finkel, Raphael Ari
Franke, Charles H
Lanska, Douglas John
Lindauer, George Conrad
Livingood, Marvin D(uane)
Looney, Stephen Warwick
McCord, Michael Campbell
Pearson, Earl F
Schnare, Paul Stewart

LOUISIANA
Agarwal, Arun Kumar
Bauer, Beverly Ann
Bedell, Louis Robert
Benard, Mark
Bonnette, Della T
Chen, Peter Pin-Shan
Dominick, Wayne Dennis
Gajendar, Nandigam
Henry, Herman Luther, Jr
Jones, Daniel Elven
Kak, Subhash Chandra
Keys, L Ken
Lefton, Lew Edward
Ovunc, Bulent Ahmet
Roquemore, Leroy

MAINE
Carter, William Caswell
Kepron, Michael Raymond

Computer Sciences, General (cont)

MARYLAND
Adler, Sanford Charles
Agrawala, Ashok Kumar
Aloimonos, Yiannis John
Ashman, Michael Nathan
Atchison, William Franklin
Barbe, David Franklin
Basili, Victor Robert
Bateman, Barry Lynn
Behforooz, Ali
Belliveau, Louis J
Berger, Robert Lewis
Berger, William J
Blanc, Robert Peter
Blevins, Gilbert Sanders
Blue, James Lawrence
Blum, Joseph
Blum, Robert Allan
Boisvert, Ronald Fernand
Braatz, James Anthony
Bukowski, Richard William
Burrows, James H
Cantor, David S
Carlstead, Edward Meredith
Chapin, Edward William, Jr
Chellappa, Ramalingam
Chen, Ching-Nien
Chen, Lily Ching-Chun
Cherniavsky, Ellen Abelson
Chi, Donald Nan-Hua
Chi, L K
Clark, Bill Pat
Cohen, Gerson H
Conklin, James Byron, Jr
Coriell, Kathleen Patricia
Cornyn, John Joseph
Cotton, Ira Walter
Criss, Thomas Benjamin
Das, Prasanta
Dasler, Adolph Richard
Deaven, Dennis George
Dhyse, Frederick George
Dubois, Andre T
Eaton, Barbra L
Elliot, Joe Oliver
Feldman, Alfred Philip
Fleming, James Joseph
Forman, Richard Allan
Franks, David A
Fried, Jerrold
Gabriele, Thomas L
Garvin, James Brian
Geckle, William Jude
Geduldig, Donald
Gelberg, Alan
Ginevan, Michael Edward
Glaser, Edmund M
Gleissner, Gene Heiden
Glennan, T(homas) Keith
Goldfinger, Andrew David
Goldstein, Charles M
Goldstein, Gordon D(avid)
Gordon, Daniel Israel
Graham, Dale Elliott
Graham, Edward Underwood
Grosman, Louis Hirsch
Gross, James Harrison
Gull, Cloyd Dake
Hendrickson, Adolph C(arl)
Hevner, Alan Raymond
Hodes, Louis
Hollis, Jan Michael
Horna, Otakar Anthony
Howard, Sethanne
Hsu, Chen C
Hybl, Albert
Ingram, Glenn R
Jefferson, David Kenoss
Jefferson, Donald Earl
Jenkins, Robert Edward
Kahn, Arthur B
Kanal, Laveen Nanik
Keenan, Thomas Aquinas
King, Joseph Herbert
Kippenberger, Donald Justin
Kissman, Henry Marcel
Knox, Robert Gaylord
Kosaraju, S Rao
Kroll, Bernard Hilton
Kurzhals, Peter R(alph)
Kyle, Herbert Lee
Landsburg, Alexander Charles
Lawson, Mildred Wiker
Lepson, Benjamin
Lidtke, Doris Keefe
Ligomenides, Panos Aristides
Lindberg, Donald Allan Bror
Little, John Llewellyn
Little, Joyce Currie
Lozier, Daniel William
Lyon, Gordon Edward
McRae, Vincent Vernon
Maisel, Herbert
Marimont, Rosalind Brownstone
Martin, John A
May, Everette Lee, Jr
Messina, Carla Gretchen
Miller, Raymond Edward
Miller, Robert Gerard
Minker, Jack
Mohler, William C
Moshman, Jack
Nau, Dana S
Nusbickel, Edward M, Jr
Oliver, Paul Alfred
Ortmeyer, Heidi Karen
Oza, Dipak H
Penn, Howard Lewis
Pickle, Linda Williams
Prewitt, Judith Martha Shimansky
Quinn, Frank Russell
Raff, Samuel J
Reeker, Larry Henry
Rich, Robert Peter
Rivkin, Maxcy
Robock, Alan
Rombach, Hans Dieter
Sachs, Lester Marvin
Saltman, Roy G
Sammet, Jean E
Sandusky, Harold William
Sayer, John Samuel
Schlesinger, Judith Diane
Schmidt, Edward Matthews
Schneider, John H
Schneider, William C
Schrum, Mary Irene Knoller
Sewell, Winifred
Shah, Shirish
Sher, Alvin Harvey
Shotland, Lawrence M
Sidhu, Deepinder Pal
Siegel, Elliot Robert
Silverton, James Vincent
Smith, Mark Stephen
Snow, Milton Leonard
Solomon, Jay Murrie
Spornick, Lynna
Srivastava, Sudhir
Stadter, James Thomas
Steven, Alasdair C
Stickler, Mitchell Gene
Stolz, Walter S
Stone, Philip M
Sullivan, Francis E
Taragin, Morton Frank
Thakor, Nitish Vyomesh
Thoma, George Ranjan
Tippett, James T
Tripathi, Satish K
Trotter, Gordon Trumbull
Valley, Sharon Louise
Weigman, Bernard J
White, Eugene L
Wooster, Harold Abbott
Yedinak, Peter Demerton
Zamora, Antonio
Zimmermann, Mark Edward
Zusman, Fred Selwyn

MASSACHUSETTS
Adrion, William Richards
Akers, Sheldon Buckingham, Jr
Ash, Michael Edward
Bailey, Duane W
Ball, John Allen
Barngrover, Debra Anne
Beaton, Albert E
Belady, Laszlo Antal
Bhandarkar, Dileep Pandurang
Bolker, Ethan D
Branscomb, Lewis McAdory
Brayer, Kenneth
Briney, Robert Edward
Bruce, Kim Barry
Buono, John Arthur
Buzen, Jeffrey Peter
Cameron, Alastair Graham Walter
Cappallo, Roger James
Celmaster, William Noah
Chakrabarti, Supriya
Champine, George Allen
Cheatham, Thomas Edward, Jr
Chen, Chi-Hau
Chen, Ching-chih
Chen, Robert Chia-Hua
Chu, J Chuan
Clarke, Lori A
Cohen, Howard David
Connors, Philip Irving
Cooke, Richard A
Corbato, Fernando Jose
Croft, Bruce
Dechene, Lucy Irene
Dennis, Jack Bonnell
Dertouzos, Michael L
Donovan, John Joseph
Dowd, John P
Dudgeon, Dan Ed
Eckhouse, Richard Henry
Eckstein, Jonathan
Evans, Thomas George
Fano, Robert M(ario)
Fariss, Thomas Lee
Floyd, William Beckwith
Friedman, Joyce Barbara
Fuller, Samuel Henry
Giuliano, Vincent E
Glorioso, Robert M
Gold, Bernard
Gordon, Kurtiss Jay
Gorgone, John T
Graham, Robert Montrose
Griffin, William G
Grodzinsky, Alan J
Grosz, Barbara Jean
Gruener, William B
Hafen, Elizabeth Susan Scott
Hahn, Robert S(impson)
Harrison, Ralph Joseph
Hayden, Thomas Day
Heitman, Richard Edgar
Hershberg, Philip I
Holst, William Frederick
Hope, Lawrence Latimer
Hyatt, Raymond R, Jr
Ingham, Kenneth R
Johnson, Douglas William
Kaman, Charles Henry
Klensin, John
Klinedinst, Keith Allen
Kocher, Bryan S
Krause, Irvin
Kung, Hsiang-Tsung
Kupferberg, Lenn C
Lampson, Butler Wright
Laning, J Halcombe
Lechner, Robert Joseph
Levesque, Allen Henry
Levine, Randolph Herbert
Liskov, Barbara H
Logcher, Robert Daniel
Lynch, Nancy Ann
McKnight, Lee Warren
Maeks, Joel
Makhoul, John Ibrahim
Martin, Richard Hadley, Jr
Martz, Eric
Menzin, Margaret Schoenberg
Meyer, Albert Ronald
Miffitt, Donald Charles
Milder, Fredric Lloyd
Moler, Cleve B
Morton, Donald John
Moses, Joel
Mountain, David Charles, Jr
Nitecki, Zbigniew
Pample, Roland D
Paquette, Gerard Arthur
Prerau, David Stewart
Press, William Henry
Rabin, Michael O
Raffel, Jack I
Rao, Ramgopal P
Riseman, Edward M
Rivest, Ronald L
Robbat, Albert, Jr
Roberts, Louis W
Ross, Douglas Taylor
Rowell, Robert Lee
Saltzer, Jerome H(oward)
Salveter, Sharon Caroline
Scheff, Benson H(offman)
Serr, Frederick E
Shaffer, Harry Leonard
Shambroom, Wiliam David
Shawcross, William Edgerton
Shum, Annie Waiching
Slack, Warner Vincent
Sprott, George F
Sproull, Robert Fletcher
Stewart, Lawrence Colm
Stiffler, Jack Justin
Stillwell, Richard Newhall
Storer, James E(dward)
Szonyi, Geza
Teng, Chojan
Theobald, Charles Edwin, Jr
Thornton, Richard D(ouglas)
Tompkins, Howard E(dward)
Townley, Judy Ann
Tweed, David George
Van Deusen, Paul Cook
Van Meter, David
Vetterling, William Thomas
Walter, Charlton M
Wand, Mitchell
Waters, Richard C(abot)
Weinberg, I Jack
Winett, Joel M
Wood, David Belden
Woods, William A
Zachary, Norman

MICHIGAN
Anderson, Robert Hunt
Atreya, Arvind
Baker, Robert Henry, Jr
Barnard, Robert D(ane), Jr
Bollinger, Robert Otto
Briggs, Darinka Zigic
Caron, E(dgar) Louis
Cheydleur, Benjamin Frederic
Conrad, Michael
Conway, Lynn Ann
Coyle, Peter
Cullati, Arthur G
Denman, Harry Harroun
Dershem, Herbert L
Dubes, Richard C
Eichhorn, J(acob)
Elshoff, James L(ester)
Farah, Badie Naiem
Finerman, A(aron)
Flanders, Harley
Fry, James Palmer
Furlong, Robert B
Galler, Bernard Aaron
Getty, Ward Douglas
Gjostein, Norman A
Hee, Christopher Edward
Hicks, Darrell Lee
Hillig, Kurt Walter, II
Holland, John Henry
Holland, Steven William
Howe, William Jeffrey
Islam, Mohammed N
Johnson, Whitney Larsen
Kallander, John William
Kolopajlo, Lawrence Hugh
Laurance, Neal L
Lowther, John Lincoln
McLaughlin, Renate
Marshall, Dale Earnest
Meteer, James William
Nelson, James Donald
Patt, Yale Nance
Petzold, Edgar
Phillips, Richard Lang
Pynnonen, Bruce W
Ramamurthy, Amurthur C
Reynolds, Robert Gene
Rollinger, Charles N(icholas)
Rosen, Jeffrey Kenneth
Rosenbaum, Manuel
Rovner, David Richard
Rundensteiner, Elke Angelika
Schriber, Thomas J
Schuetzle, Dennis
Simpson, William Albert
Singh, Lal Pratap S
Sokol, Robert James
Stout, Quentin Fielden
Tihansky, Diane Rice
Tsao, Nai-Kuan
Tuchinsky, Philip Martin
Vishnubhotla, Sarma Raghunadha
Watts, Jeffrey Lynn
Weinshank, Donald Jerome
Westervelt, Franklin Herbert
Williams, John Albert
Windeknecht, Thomas George
Wojcik, Anthony Stephen
Wolfson, Seymour J
Wu, Hai
Wu, Hofu

MINNESOTA
Ackerman, Eugene
Anderson, Ronald E
Austin, Donald Murray
Cohen, Arnold A
Crooks, Stephen Lawrence
Drew, Bruce Arthur
Du, David Hung-Chang
Dueltgen, Ronald Rex
Ek, Alan Ryan
Ellis, Lynda Betty
Emeagwali, Philip Chukwurah
Gatewood, Lael Cranmer
Geokezas, Meletios
Glewwe, Carl W(illiam)
Gupta, Satish Chander
Henderson, Donald Lee
Jacobs, David R, Jr
Krueger, Eugene Rex
Kumar, Vipin
Lin, Benjamin Ming-Ren
McQuarrie, Donald G
Marshall, John Clifford
Meeder, Jeanne Elizabeth
Oelberg, Thomas Jonathon
Pong, Ting-Chuen
Robb, Richard A
Schneider, G Michael
Schuldt, Marcus Dale
Seebach, J Arthur, Jr
Slagle, James R
Slavin, Joanne Louise
Smith, Robert Elijah
Stebbings, William Lee
Steen, Lynn Arthur
Wetmore, Clifford Major
Wolfe, Barbara Blair
Yasko, Richard N

MISSISSIPPI
Coleman, Thomas George
Colonias, John S
Miller, James Edward
Taylor-Cade, Ruth Ann

MISSOURI
Agarwal, Ramesh K
Ali, Syed S
Arvidson, Raymond Ernst
Banakar, Umesh Virupaksh
Benson, Robert John
Blackwell, Paul K, II
Bodine, Richard Shearon
Bosanquet, Louis Percival
Bradburn, Gregory Russell
Cohick, A Doyle, Jr
Cornelius, Steven Gregory
Cottrell, Roy
Cox, Jerome R(ockhold), Jr
Franklin, Mark A
Giarrusso, Frederick Frank
Griffith, Virgil Vernon
Groenweghe, Leo Carl Denis
Heard, John Thibaut, Jr
Ho, Chung You (Peter)

Jones, Ronald Dale
Keller, Roy Fred
Krone, Lester H(erman), Jr
Lee, Ralph Edward
Lembke, Roger Roy
MacDonald, Carolyn Trott
Moore, James Thomas
Morgan, Nancy H
Morley, Robert Emmett, Jr
Neuman, Rosalind Joyce
Peng, Yongren Benjamin
Ramsey, Robert Bruce
Schmidt, Bruno (Francis)
Stalling, David Laurence
Stuth, Charles James
Talbott, Ted Delwyn
Taylor, Ralph Dale
Thomas, Lewis Jones, Jr
Tyrer, Harry Wakeley
Wagner, Joseph Edward
Walsh, Robert Jerome

MONTANA
Amend, John Robert
Banaugh, Robert Peter
Combie, Joan D
Manhart, Robert (Audley)
Pierre, Donald Arthur
Weaver, Donald K(essler), Jr

NEBRASKA
Clements, Gregory Leland
Downing, John Scott
Fairchild, Robert Wayne
Gale, Douglas Shannon, II
Kolade, Alabi E
Leung, Joseph Yuk-Tong
Longley, William Warren, Jr
Scudder, Jeffrey Eric
Seth, Sharad Chandra
Sharp, Edward A
Zebolsky, Donald Michael

NEVADA
Brady, Allen H
Haun, Robert Dee, Jr
Juliussen, J Egil
Marsh, David Paul
Minor, John Threecivelous
Nassersharif, Bahram
Telford, James Wardrop

NEW HAMPSHIRE
Abbott, Robert Classie
Epstein, Harvey Irwin
Gagliardi, Ugo Oscar
Klotz, Louis Herman
Kuo, Shan Sun
Kurtz, Thomas Eugene
McKeeman, William Marshall
Taffe, William John
Tourgee, Ronald Alan

NEW JERSEY
Afshar, Siroos K
Agalloco, James Paul
Amarel, S(aul)
Amron, Irving
Anderson, Milton Merrill
Anderson, Terry Lee
Andrews, Ronald Allen
Baker, Brenda Sue
Barnes, Derek A
Barsky, James
Becker, Sheldon Theodore
Beckman, Frank Samuel
Bergeron, Robert F(rancis) (Terry), Jr
Bobeck, Andrew H
Boylan, Edward S
Brandmaier, Harold Edmund
Burkhardt, Kenneth J
Cai, Jin-yi
Chambers, John McKinley
Cheo, Li-hsiang Aria S
Chovan, James Peter
Coffman, Edward G, Jr
Copenhaver, Thomas Wesley
Cornell, W(arren) A(lvan)
Coughran, William Marvin, Jr
Crowley, Thomas Henry
Cutler, Frank Allen, Jr
DeBaun, Robert Matthew
Desjardins, Raoul
Dickerson, L(oren) L(ester), Jr
DiMasi, Gabriel Joseph
Dwyer, Harry, III
Eastwood, Abraham Bagot
Eisner, Mark Joseph
Fischer, Ronald Howard
Fix, Kathleen A
Flores, Ivan
Forys, Leonard J
Fox, Phyllis
Freeman, Herbert
Freund, Roland Wilhelm
Fu, Hui-Hsing
Gabriel, Edwin Z
Gargaro, Anthony
Gear, Charles William
George, Kenneth Dudley
Glass, John Richard
Goffman, Martin
Goldstein, Philip
Golin, Stuart

Gossmann, Hans Joachim
Gourary, Barry Sholom
Goyal, Suresh
Grandolfo, Marian Carmela
Green, Edwin James
Gross, Arthur Gerald
Grosso, Anthony J
Hackman, Martin Robert
Halaby, Sami Assad
Hall, Gene Stephen
Hall, Homer James
Harris, James Ridout
Hill, David G
Holzmann, Gerard J
Huang, Jennming Stephen
Huber, Richard V
Iorns, Terry Vern
Jackson, Thomas A J
Johannes, Virgil Ivancich
Johnson, Stephen Curtis
Joseph, John Mundancheril
Jurkat, Martin Peter
Kaita, Robert
Kangovi, Sach
Karney, Charles Fielding Finch
Karol, Mark J
Kasparek, Stanley Vaclav
Kaufman, Linda
Kellgren, John
Khalifa, Ramzi A
Kirch, Murray R
Kowalski, Ludwik
Krzyzanowski, Paul
Kulikowski, Casimir A
LaMastro, Robert Anthony
Larson, Eric Heath
Levine, Richard S
Levy, Leon Sholom
Lichtenberger, Gerald Burton
Lohse, David John
Love, L J Cline
McCaslin, Darrell
McGuinness, Deborah Louise
McIlroy, M Douglas
Maclennan, Carol G
Mahoney, Michael S
Malbrock, Jane C
Mamelak, Joseph Simon
Manickam, Janardhan
Marlin, Robert Lewis
Marmor, Robert Samuel
Marsh, John L
Matey, James Regis
Mauzey, Peter T
Mehta, Atul Mansukhbhai
Miller, Robert Alan
Moroni, Antonio
Most, Joseph Morris
Myers, Jeffrey
Newman, Stephen Alexander
Newton, Paul Edward
Nicholls, Gerald P
Nonnenmann, Uwe
O'Gorman, James
Ott, Teunis Jan
Philipp, Ronald E
Pinson, Elliot N
Polonsky, Ivan Paul
Popper, Robert David
Raychaudhuri, Kamal Kumar
Riggs, Richard
Robins, Jack
Roland, Dennis Michael
Rollino, John
Rosenblum, Martin Jacob
Rubin, Arthur I(srael)
Sannuti, Peddapullaiah
Schmidt, Barnet Michael
Shaer, Norman Robert
Shine, Robert John
Shipley, Edward Nicholas
Shober, Robert Anthony
Shulman, Herbert Byron
Siegel, Jeffry A
Silberschatz, Abraham
Smagorinsky, Joseph
Steben, John D
Steiglitz, Kenneth
Stevens, John G
Stevenson, Robert Louis
Stoddard, James H
Strange, Ronald Stephen
Sublette, Ivan H(ugh)
Tarjan, Robert Endre
Turoff, Murray
Uhrig, Jerome Lee
Verma, Pramode Kumar
Wagner, Richard Carl
Walsh, Teresa Marie
Weinberger, Peter Jay
Weiss, C Dennis
Weiss, Stanley H
Whigan, Daisy B
White, Myron Edward
Woo, Nam-Sung
Wright, Margaret Hagen
Zwass, Vladimir

NEW MEXICO
Adams, J Mack
Ahmed, Nasir
Barsis, Edwin Howard
Bell, Stoughton
Bleyl, Robert Lingren

Brainerd, Walter Scott
Campbell, Katherine Smith
Clauser, Milton John
Curtis, Wesley E
Davey, William Robert
Dempster, William Fred
Divett, Robert Thomas
Engelhardt, Albert George
Erdal, Bruce Robert
Erickson, Kenneth Lynn
Faber, Vance
Fronek, Donald Karel
Henderson, Dale Barlow
Johnston, John B(everley)
Jones, Merrill C(alvin)
Kensek, Ronald P
Korman, N(athaniel) I(rving)
Lawrence, Raymond Jeffery
Luger, George F
McDonald, Timothy Scott
Marker, Thomas F(ranklin)
Miller, Edmund K(enneth)
Morrison, Donald Ross
Northrup, Clyde John Marshall, Jr
Phister, Montgomery, Jr
Reinfelds, Juris
Snell, Charles Murrell
Sprinkle, James Kent, Jr
Stanbro, William David
Stark, Richard Harlan
Sung, Andrew Hsi-Lin
Tapp, Charles W(illiam)
Terrell, C(harles) W(illiam)
Thorne, Billy Joe
Trowbridge, George Cecil
Wells, Mark Brimhall
Wienke, Bruce Ray
Wilkins, Ronald Wayne
Willbanks, Emily West

NEW YORK
Allen, Frances Elizabeth
Allen, James Frederick
Alt, Franz L
Anastassiou, Dimitris
Anderson, Albert Edward
Andrews, Mark Allen
Anshel, Michael
Apte, Chidanand
Archibald, Julius A, Jr
Arden, Bruce Wesley
Asprey, Winifred Alice
Augenstein, Moshe
Bahl, Lalit R(ai)
Ball, George William
Bashkow, Theodore R(obert)
Batter, John F(rederic), Jr
Bednowitz, Allan Lloyd
Berger, Jay Manton
Berger, Toby
Bienstock, Daniel
Birnbaum, David
Bishop, Marvin
Blau, Lawrence Martin
Bloch, Alan
Borden, Edward B
Bourbakis, Niklaos G
Brennan, John Paul
Brown, David T
Brown, Theodore D
Bunch, Phillip Carter
Burnham, Dwight Comber
Cabeen, Samuel Kirkland
Campbell, Graham Hays
Capwell, Robert J
Carey, Bernard Joseph
Chan, Tat-Hung
Chang, John H(si-Teh)
Chi, Benjamin E
Christensen, Robert Lee
Cocke, John
Cohen, Daniel Isaac Aryeh
Cohen, Irving Allan
Cohen, Irwin A
Cohen, Martin O
Cohn, Harvey
Collins, Carol Desormeau
Cornacchio, Joseph V(incent)
Coutchie, Pamela Ann
Curtis, Ronald S
D'Auria, Thomas A
De Lillo, Nicholas Joseph
DeSieno, Robert P
Diamond, Fred Irwin
Dimmler, D(ietrich) Gerd
Douglas, Craig Carl
Drossman, Melvyn Miles
Dumoulin, Charles Lucian
Eberlein, Patricia James
Eckert, Richard Raymond
Eydgahi, Ali Mohammadzadeh
Falk, Catherine T
Fama, Donald Francis
Farber, Andrew R
Federighi, Francis D
Feeser, Larry James
Feingold, Alex Jay
Finlay, Thomas Hiram
Flannery, John B, Jr
Fleisher, Harold
Foley, Kenneth John
Foreman, Bruce Milburn, Jr
Fox, Geoffrey Charles
Frank, Thomas Stolley

Frantz, Andrew Gibson
Frazer, W(illiam) Donald
Freiman, Charles
Friedberg, Carl E
Gabay, Jonathan Glenn
Gabelman, Irving J(acob)
Galatianos, Anthony Athanassios
Gause, Donald C
Gelernter, Herbert Leo
Ghozati, Seyed-Ali
Gilchrist, Bruce
Gilman, Sr John Frances
Glimm, James Gilbert
Goertzel, Gerald
Goldberg, Conrad Stewart
Golden, Robert K
Goldstein, Lawrence Howard
Gonzalez, Robert Anthony
Goodman, Seymour
Gordon, Irving
Greenberg, Donald P
Gries, David
Gustavson, Fred Gehrung
Hainline, Louise
Hammer, Richard Benjamin
Hannay, David G
Hanson, Jonathan C
Harrison, Malcolm Charles
Hartmanis, Juris
Heidelberger, Philip
Heller, Jack
Heller, William R
Heppa, Douglas Van
Herrington, Lee Pierce
Hershenov, Joseph
Hoenig, Alan
Hong, Se June
Hsiao, Mu-Yue
Hubbard, Richard Alexander, II
Hudak, Michael J
Huddleston, John Vincent
Impagliazzo, John
Jabbur, Ramzi Jibrail
Jain, Duli Chandra
Jesaitis, Raymond G
Johnson, Wallace E
Jones, Leonidas John
Joyner, William Henry, Jr
Kane, John Vincent, Jr
Kaplan, Ehud
Karron, Daniel B
Kim, Carl Stephen
Kinzly, Robert Edward
Kirchmayer, Leon K
Klerer, Melvin
Komiak, James Joseph
Kotchoubey, Andrew
Kwok, Thomas Yu-Kiu
Kyburg, Henry
Laderman, Julian David
Lamster, Hal B
Lazareth, Otto William, Jr
Lerner, Rita Guggenheim
Leung, Yiu M
Lewis, Philip M
Lidofsky, Leon Julian
Lubowsky, Jack
Luk, Franklin T
Macaluso, Pat
McGloin, Paul Arthur
McLaughlin, Harry Wright
McNaughton, Robert
Maguire, Gerald Quentin, Jr
Maissel, Leon I
Mansfield, Victor Neil
Marcuvitz, Nathan
Marr, Robert B
Matthews, Dwight Earl
Meenan, Peter Michael
Mercer, Robert Leroy
Merritt, Susan Mary
Merten, Alan Gilbert
Miranker, Willard Lee
Mohlke, Byron Henry
Moniot, Robert Keith
Monz, Pauline
Morgenstern, Matthew
Morrison, John B
Moyne, John Abel
Muench, Donald Leo
Muller, Otto Helmuth
Musser, David Rea
Nagel, Harry Leslie
Nagy, George
Nevison, Christopher H
Notaro, Anthony
Ortel, William C(harles) G(ormley)
Padalino, Stephen John
Pardee, Otway O'Meara
Parikh, Hemant Bhupendra
Parker, Ronald W
Peled, Abraham
Pence, Harry Edmond
Pifko, Allan Bert
Polivka, Raymond Peter
Postman, Robert Derek
Preiser, Stanley
Priore, Roger L
Ralston, Anthony
Ramaley, James Francis
Rauscher, Tomlinson Gene
Reckhow, Warren Addison
Reilly, Edwin David, Jr
Richter, Donald

Computer Sciences, General (cont)

Rogers, Edwin Henry
Rosberg, Zvi
Rosenfeld, Jack Lee
Rosenkrantz, Daniel J
Roth, John Paul
Rudolph, Luther Day
Rutledge, Joseph Dela
Sachs, Martin William
Sakoda, William John
Sargent, Robert George
Schmitt, Erich
Schneider, Fred Barry
Schoenfeld, Robert Louis
Schwartz, Jacob Theodore
Schwartz, Mischa
Selman, Alan L
Shamash, Yacov A
Shapiro, Donald M
Shapiro, Stephen D
Shapiro, Stuart Charles
Shaw, David Elliot
Sherman, Philip Martin
Shilepsky, Arnold Charles
Shooman, Martin L
Shriver, Bruce Douglas
Shuey, R(ichard) L(yman)
Sibert, Elbert Ernest
Sklarew, Robert J
Smith, Bernard
Sorter, Peter F
Spiegel, Robert
Srihari, Sargur N
Star, Martin Leon
Stearns, Richard Edwin
Stevens, William Y(eaton)
Storm, Edward Francis
Su, Stephen Y H
Summers, Phillip Dale
Sweeney, Thomas Francis
Swinehart, James Stephen
Szymanski, Boleslaw K
Tassinari, Silvio John
Tewarson, Reginald P
Tobin, Michael
Todd, Aaron Rodwell
Tracz, Will
Traub, Joseph Frederick
Trefethen, Lloyd Nicholas
Tycko, Daniel H
Uretsky, Myron
Vollmer, Frederick Wolfer
Voss, Richard Frederick
Wadell, Lyle H
Wald, Alvin Stanley
Walker, Edward John
Waltzer, Wayne C
Wang, Chu Ping
Wang, Hsin-Pang
Weinberger, Arnold
Weinstein, David Alan
Whitby, Owen
White, Gerald M(ilton)
White, William Wallace
Wilkinson, John Peter Darrell
Williams, George Harry
Winkler, Leonard P
Wolf, Carol Euwema
Wong, Chak-Kuen
Worthing, Jurgen
Yu, Kai-Bor

NORTH CAROLINA

Agrawal, Dharma Prakash
Allen, Charles Michael
Barr, Roger Coke
Biermann, Alan Wales
Brooks, Frederick P(hillips), Jr
Calingaert, Peter
Carlson, William Theodore
Chen, Michael Yu-Men
Chen, Su-shing
Chou, Wushow
Christian, Wolfgang C
Dinse, Gregg Ernest
Dotson, Marilyn Knight
Epstein, George
Ferentz, Melvin
Foley, Gary J
Fuchs, Henry
Gallie, Thomas Muir
Gelenbe, Erol
Gemperline, Margaret Mary Cetera
Grau, Albert A
Greim, Barbara Ann
Groves, William Ernest
Gustafson, Carl Gustaf, Jr
Hadeen, Kenneth Doyle
Halton, John Henry
Hammond, William Edward
Hartzema, Abraham Gijsbert
Hildebrandt, Theodore Ware
Karr, Alan Francis
Kerr, Sandria Neidus
Kurtz, A Peter
Lapicki, Gregory
Long, Andrew Fleming, Jr
Lucas, Carol N
McAllister, David Franklin
McDaniel, Roger Lanier, Jr
Mago, Gyula Antal

Malindzak, George S, Jr
Marco, Gino Joseph
Marinos, Pete Nick
Meyer, Carl Dean, Jr
Murray, Francis Joseph
Neidinger, Richard Dean
Norgaard, Nicholas J
Norris, Fletcher R
Nutter, James I(rving)
Pizer, Stephen M
Plemmons, Robert James
Portier, Christopher Jude
Ramm, Dietolf
Riemann, James Michael
Roberts, Stephen D
Sawyer, John Wesley
Schell, Joseph Francis
Shelly, James H
Snodgrass, Rex Jackson
Snyder, Wesley Edwin
Stanat, Donald Ford
Stead, William Wallace
Struve, William George
Sussenguth, Edward H
Teague, David Boyce
Tlalka, Jacek
Trivedi, Kishor Shridharbhai
Trussell, Henry Joel
Wagner, Robert Alan
Weiss, Stephen Fredrick
Wetmore, David Eugene
Whipkey, Kenneth Lee
Whitehurst, Garnett Brooks
Wright, William V(aughn)
Wright, William Vale

NORTH DAKOTA

Lieberman, Milton Eugene
Magel, Kenneth I
Tareski, Val Gerard
Vasey, Edfred H
Winrich, Lonny B

OHIO

Adeli, Hojjat
Beverly, Robert Edward, III
Blake, James Elwood
Buoni, John J
Burden, Richard L
Canfield, James Howard
Chamis, Alice Yanosko
Coe, Kenneth Loren
Cooke, Charles C
Cruce, William L R
Damian, Carol G
Darby, Ralph Lewis
Davis, Henry Werner
Davis, Michael E
Ernst, George W
Feld, William Adam
Fleming, Paul Daniel, III
Foulk, Clinton Ross
Fraser, Alex Stewart
Friedlander, Ira Ray
Goodson, Alan Leslie
Graham, Paul Whitener Link
Gregory, Thomas Bradford
Hartrum, Thomas Charles
Heines, Thomas Samuel
Hess, Evelyn V
Hill, Ann Gertrude
Hsiao, David Kai-Mei
Hulley, Clair Montrose
James, Thomas Ray
Jehn, Lawrence Andrew
Jekeli, Christopher
Jendrek, Eugene Francis, Jr
Kerr, Douglas S
Keyes, Marion Alvah, IV
Kwatra, Subhash Chander
Lamont, Gary Byron
LaRue, Robert D(ean)
Lechner, Joseph H
Little, Richard Allen
Liu, Ming-Tsan
Lutton, John Kazuo
McDonough, James Francis
Massey, L(ester) G(eorge)
Mathis, Robert Fletcher
Minchak, Robert John
Mochel, Virgil Dale
Mulhausen, Hedy Ann
Muller, Mervin Edgar
Murdoch, Arthur
Nardi, John Christopher
Neaderhouser Purdy, Carla Cecilia
Neff, Raymond Kenneth
Nicholson, Victor Alvin
Niemann, Theodore Frank
Nikolai, Paul John
Ogden, William Frederick
O'Neill, Edward True
Orr, Charles Henry
Ouimet, Alfred J, Jr
Papachristou, Christos A
Patrick, Edward Alfred
Petrarca, Anthony Edward
Platau, Gerard Oscar
Potoczny, Henry Basil
Purvis, John Thomas
Quam, David Lawrence
Rajsuman, Rochit
Randels, James Bennett
Ratliff, Priscilla N

Rattan, Kuldip Singh
Reilly, Charles Bernard
Restemeyer, William Edward
Rich, Ronald Lee
Ricord, Louis Chester
Ross, Charles Burton
Roth, Mark A
Rubin, Stanley G(erald)
Ryan, Anne Webster
Santos, Eugene (Sy)
Schaefer, Donald John
Sellers, James Allen
Shields, George Seamon
Slotterbeck-Baker, Oberta Ann
Snider, John William
Solow, Daniel
Spialter, Leonard
Sterling, Leon Samuel
Sterrett, Andrew
Stickney, Alan Craig
Stobaugh, Robert Earl
Stoner, Ronald Edward
Suter, Bruce Wilsey
Way, Frederick, III
White, Lee James
Wigington, Ronald L
Wolfe, Robert Kenneth
Zaye, David F

OKLAHOMA

Anderson, Paul Dean
Bednar, Jonnie Bee
Brady, Barry Hugh Garnet
Calvert, Jon Channing
Cheung, John Yan-Poon
Crafton, Paul A(rthur)
Eby, Harold Hildenbrandt
Fisher, Donald D
Folk, Earl Donald
Hale, William Henry, Jr
Hedrick, George Ellwood, III
Lauffer, Donald Eugene
McClain, Gerald Ray
McGurk, Donald J
Minor, John Threecivelous
Monn, Donald Edgar
Naymik, Daniel Allan
Noll, Leo Albert
Page, Rector Lee
Phillips, John Richard
Reimer, Dennis D
Rutledge, Carl Thomas
Summers, Gregory Lawson
Testerman, Jack Duane
Tomaja, David Louis
Usher, W(illia)m Mack
Vasicek, Daniel J
Vinatieri, James Edward
Walker, Billy Kenneth
Winter, William Kenneth

OREGON

Alin, John Suemper
Beyer, Terry
Coop, Leonard Bryan
Currie, James Orr, Jr
Davenport, Wilbur B(ayley), Jr
Demarest, Harold Hunt, Jr
Douglas, Sarah A
Ecklund, Earl Frank, Jr
Edwards, James Mark
Feldman, Milton H
Freiling, Michael Joseph
Gilbert, Barrie
Holmes, Zoe Ann
Jones, Donlan F(rancis)
Kieburtz, Richard B(ruce)
Moursund, David G
Pursglove, Laurence Albert
Rudd, Walter Greyson
Skelton, John Edward
Struble, George W
Terwilliger, Don William
Tinney, William Frank
Udovic, Daniel
Wagner, Orvin Edson

PENNSYLVANIA

Abend, Kenneth
Aburdene, Maurice Felix
Adams, William Sanders
Aiken, Robert McLean
Alvino, William Michael
Anderson, Jay Martin
Angstadt, Carol Newborg
Auerbach, Isaac L
Aviles, Rafael G
Badler, Norman Ira
Bajcsy, Ruzena K
Bamberger, Judy
Banerji, Ranan Bihari
Barnhart, Barry B
Bass, Leonard Joel
Baumann, Dwight Maylon Billy
Berard, Anthony D, Jr
Bowlden, Henry James
Box, Larry
Buriok, Gerald Michael
Burness, James Hubert
Burnett, Roger Macdonald
Cain, James Thomas
Caldwell, Christopher Sterling
Carbonell, Jaime Guillermo
Carr, John Weber, III

Chang, C Hsiung
Chaplin, Norman John
Clemons, Eric K
Coleman, Anna M
Crawley, James Winston, Jr
Cupper, Robert
Director, Stephen William
Dwyer, Thomas A
Dyott, Thomas Michael
Eckert, J Presper
Eddy, William Frost
Eisenstein, Bruce A
Emmons, Larrimore Browneller
Ertekin, Turgay
Fabrey, James Douglas
Farley, Belmont Greenlee
Farren, Ann Louise
Favret, A(ndrew) G(illigan)
Feng, Tse-yun
Frazier, James Lewis
Frederick, David Eugene
Friedman, Frank L
Fugger, Joseph
Furst, Merrick Lee
Garfield, Eugene
Gerasimowicz, Walter Vladimir
Gibbs, Norman Edgar
Goldwasser, Samuel M
Golumbic, Martin Charles
Gorn, Saul
Gottlieb, Frederick Jay
Gould, William Allen
Grabner, Elise M
Grabowski, Thomas J
Gross, Michael R
Habermann, Arie Nicolaas
Hansen, John C
Hermans, Hans J
Herzfeld, Valerius E
Hockswender, Thomas Richard
Hoover, L Ronald
Hurwitz, Jan Krosst
Huthnance, Edward Dennis, Jr
Irwin, Mary Jane
Isett, Robert David
Janicki, Casimir A
Joshi, Aravind Krishna
Kahler, Albert Comstock, III
Kashkoush, Ismail I
Kazahaya, Masahiro Matt
Kelley, Jay Hilary
Kent, Allen
Kirk, David Blackburn
Kleban, Morton H
Klotz, Eugene Arthur
Koffman, Elliot B
Korsh, James F
Kozik, Eugene
Kusic, George Larry, Jr
Laird, Donald T(homas)
Leas, J(ohn) W(esley)
Leibholz, Stephen W
Licini, Jerome Carl
Loscher, Robert A
McAllister, Marialuisa N
McArthur, William George
McLaughlin, Francis X(avier)
Mahaffy, John Harlan
Mason, Thomas Joseph
Mayer, Joerg Werner Peter
Meyer, John Sigmund
Miller, G(erson) H(arry)
Mitchell, Tom M
Moravec, Hans Peter
Murgie, Samuel A
Nair, K Aiyappan
Newell, Allen
O'Neill, John Cornelius
Olsen, Glenn W
Patton, Peter C(lyde)
Paulish, Daniel John
Penman, Paul D
Plonsky, Andrew Walter
Pool, James C T
Pruett, John Robert
Prywes, Noah S(hmarya)
Rau, Lisa Fay
Reddy, Raj
Reigel, Earl William
Reynolds, John C
Ridener, Fred Louis, Jr
Roemer, Richard Arthur
Rogers, Ralph Loucks
Roskies, Ralph Zvi
Ross, Arthur Leonard
Rush, James E
Schwerer, Frederick Carl
Seetharam, Ramnath (Ram)
Sher, Irving Harold
Shuck, John Winfield
Sieber, James Leo
Smith, Chester Martin, Jr
Sullivan, George Allen
Tobias, Russell Lawrence
Tulenko, Thomas Norman
Vladutz, George E
Wagh, Meghanad D
Wagner, Clifford Henry
Walchli, Harold E(dward)
Walker, David Kenneth
Wei, Susanna
Weinberger, Edward Bertram
Wiley, Samuel J
Wilkins, Raymond Leslie

Wismer, Robert Kingsley
Wolstenholme, Wayne W
Wu, John Naichi
Zaphyr, Peter Anthony

RHODE ISLAND
Carney, Edward J
Dewhurst, Peter
Gabriel, Richard Francis
Hemmerle, William J
Krikorian, John Sarkis, Jr
Preparata, Franco Paolo
Savage, John Edmund
Strauss, Charles Michael
Vitter, Jeffrey Scott
Wegner, Peter

SOUTH CAROLINA
Akpan, Edward
Brons, Kenneth Allyn
Cannon, Robert L
Dobbins, James Gregory Hall
Eastman, Caroline Merriam
Foster, Kent Ellsworth
Gamble, Robert Oscar
Gillette, Paul Crawford
Hammond, Joseph Langhorne, Jr
Hopkins, Laurie Boyle
Klemm, James L
Lam, Chan F
Leathrum, James Frederick
Luedeman, John Keith
Moore, Alexander Mazyck
Nanney, Thomas Ray
Porter, Hayden Samuel, Jr
Rowlett, Russell Johnston, Jr
Suich, John Edward
Turner, Albert Joseph, Jr
Wheat, Joseph Allen
Wise, William Curtis

SOUTH DAKOTA
Koepsell, Paul L(oel)

TENNESSEE
Bailes, Gordon Lee
Ball, Raiford Mill
Blass, William Errol
Braunstein, Helen Mentcher
Burton, John Williams
Campbell, Warren Elwood
Carter, Harvey Pate
Child, Harry Ray
Chitwood, Howard
Ferguson, Robert Lynn
Fischer, Charlotte Froese
Fischer, Patrick Carl
Garzon, Max
Gomberg, Joan Susan
Gray, William Harvey
Holdeman, Jonas Tillman, Jr
Hopkins, Richard Allen
Hutcheson, Paul Henry
Kiech, Earl Lockett
Kirchner, Frederick Karl
Lane, Charles A
Lea, James Wesley, Jr
LeBlanc, Larry Joseph
Murphy, Brian Donal
Otis, Marshall Voigt
Payne, DeWitt Allen
Petersen, Harold, Jr
Pfuderer, Helen A
Pleasant, James Carroll
Poore, Jesse H, Jr
Ross, Clay Campbell, Jr
Roussin, Robert Warren
Rowan, William Hamilton, Jr
Sherman, Gordon R
Springer, John Mervin
Umholtz, Clyde Allan
Van Rij, Willem Idaniel
Voorhees, Larry Donald
Wignall, George Denis

TEXAS
Albertson, Harold D
Ali, Moonis
Alo, Richard Anthony
Atkinson, Edward Neely
Aucoin, Paschal Joseph, Jr
Benedict, George Frederick
Bowdon, Edward K(night), Sr
Brown, Barry W
Brown, Jack Harold Upton
Browne, James Clayton
Carter, Elmer Buzby
Cecil, David Rolf
Childs, S(elma) Bart
Crawford, Richard Haygood, Jr
Curtin, Richard B
Daniel, James Wilson
Darwin, James T, Jr
DeGroot, Doug
Dijkstra, Edsger Wybe
Dobrott, Robert D
Dodson, David Scott
Dolling, David Stanley
Drew, Dan Dale
Eckles, Wesley Webster, Jr
Ellwood, Brooks B
Feagin, Terry
Ford, Charles Erwin
Frailey, Dennis J(ohn)
Franco, Zachary Martin
Freeman, Maynard Lloyd
Frick, John P
Gillette, Kevin Keith
Gladwell, Ian
Gorry, G Anthony
Hall, Douglas Lee
Hamill, Dennis W
Hanne, John R
Hardcastle, Donald Lee
Herbert, Morley Allen
Hernandez, Norma
Hixon, Sumner B
Huang, Jung-Chang
Jennings, Alfred Roy, Jr
Johnson, Lloyd N(ewhall)
Kaenel, Reginald A(lfred)
Keller, Thomas W
Kim, Won
King, Willis Kwongtsu
Konkel, David Anthony
Konstam, Aaron Harry
Koplyay, Janos Bernath
Kozmetsky, George
Ku, Bernard Siu-Man
Lacewell, Ronald Dale
Lam, Simon Shin-Sing
Lee, Ellen Szeto
Linn, John Charles
McCoy, John Harold
McGovern, Terence Joseph
Mack, Russel Travis
McLeod, Raymond, Jr
McSherry, Diana Hartridge
Malek, Miroslaw
Mao, Shing
Marcotte, Ronald Edward
Markenscoff, Pauline
Matula, David William
Maxwell, Donald A
Monk, Clayborne Morris
Moore, J Strother
Moore, Kris
Newhouse, Albert
Nickel, James Alvin
Novak, Gordon Shaw, Jr
Olness, Fredrick Iver
Parberry, Ian
Parker, Arthur L (Pete)
Parr, Christopher Alan
Pervin, William Joseph
Peterson, Lynn Louise Meister
Pitts, Gerald Nelson
Pooch, Udo Walter
Poucher, William B
Poulsen, Lawrence Leroy
Pyle, Leonard Duane
Rambow, Frederick H K
Randlett, Herbert Eldridge, Jr
Ransom, C J
Richardson, Arthur Jerold
Rinehart, Wilbur Allan
Ring, Dennis Randall
Rodriguez, Carlos Eduardo
Rudin, Bernard D
Schlessinger, Bernard S
Schuster, Eugene F
Schutz, Bob Ewald
Self, Glendon Danna
Shieh, Leang-San
Shooter, Jack Allen
Simmons, Dick Bedford
Smith, Philip Wesley
Smith, Rolf C, Jr
Srinivasan, Ramachandra Srini
Stein, William Edward
Stuart, Joe Don
Sudborough, Ivan Hal
Szygenda, Stephen Anthony
Volz, Richard A
Wang, Yuan R
Warlick, Charles Henry
Weiser, Alan
Whorton, Elbert Benjamin
Wiggins, James Wendell
Wilcox, Roberta Arlene
Wilkov, Robert Spencer
Williams, Glen Nordyke
Wolter, Jan D(ithmar)
Wood, Craig Adams
Wright, James Foley
Wrotenbery, Paul Taylor
Yang, Chao-Chih
Yeh, Raymond T
Yorio, Thomas
Zagata, Michael DeForest

UTAH
Bowman, Carlos Morales
Butterfield, Veloy Hansen, Jr
Carlson, Gary
Chabries, Douglas M
Cox, Benjamin Vincent
Gates, Henry Stillman
Hurst, Rex LeRoy
Jones, Merrell Robert
Lewis, Lawrence Guy
Park, Robert Lynn
Pope, Wendell LaVon
Rasmussen, V Philip, Jr
Riesenfeld, Richard F
Stockham, Thomas Greenway, Jr
Stokes, Gordon Ellis
Walker, LeRoy Harold

VERMONT
Ballou, Donald Henry
Ellis-Monaghan, John J
Gray, Kenneth Stewart
Rankin, Joanna Marie
Wiitala, Stephen Allen

VIRGINIA
Abdali, Syed Kamal
Abrams, Marshall D
Alter, Ralph
Anderson, Walter L(eonard)
Aqualino, Alan
Artna-Cohen, Agda
Balint, Francis Joseph
Barnes, Bruce Herbert
Batson, Alan Percy
Burns, William, Jr
Carterette, Edward Calvin Hayes
Catlin, Avery
Cerf, Vinton Gray
Cheng, George Chiwo
Chimenti, Frank A
Chou, Tsai-Chia Peter
Claus, Alfons Jozef
Cook, Robert Patterson
Cornforth, Clarence Michael
Daly, Robert Francis
Davis, Ruth Margaret
Dengel, Ottmar H
Denning, Peter James
Eldridge, Charles A
Elkins, Judith Molinar
Epstein, Samuel David
Farrell, Robert Michael
Feustel, Edward Alvin
Feyock, Stefan
Fisher, Gordon McCrea
Fox, Edward A
Freitag, Harlow
Friedman, Fred Jay
Gander, George William
Garcia, Oscar Nicolas
Goranson, H T
Heath, Lenwood S
Heffron, W(alter) Gordon
Heterick, Robert Cary, Jr
Hodge, Bartow
Hoffman, Karla Leigh
Hopper, Grace Murray
Huddle, Benjamin Paul, Jr
Hudgins, Aubrey C, Jr
Irick, Paul Eugene
Ivanetich, Richard John
Jacob, Robert J(oseph) K(assel)
Kahn, Robert Elliot
King, George, III
Koch, Carl Fred
Kripke, Bernard Robert
Lanzano, Bernadine Clare
Leczynski, Barbara Ann
Lee, John A N
Lefebvre, Yvon
Lehmann, John R(ichard)
Lewis, Jesse C
Lipner, Steven Barnett
Liss, Ivan Barry
Lobb, Barry Lee
McCutchen, Samuel P(roctor)
McFarlane, Kenneth Walter
MacKinney, Arland Lee
Mast, Joseph Willis
Mata-Toledo, Ramon Alberto
Meyrowitz, Alan Lester
Michalski, Ryszard Spencer
Moraff, Howard
Moritz, Barry Kyler
Mullen, Joseph Matthew
Murray, Jeanne Morris
Nance, Richard E
Neher, Dean Royce
Norris, Eugene Michael
Perry, Dennis Gordon
Perry, John Stephen
Pratt, Terrence Wendall
Pyke, Thomas Nicholas, Jr
Reisler, Donald Laurence
Rine, David C
Rosenberg, Murray David
Rosenfeld, Robert L
Roussos, Constantine
Sabelli, Nora Hojvat
Schneck, Paul Bennett
Schneider, Philip Allen David
Seiner, John Milton
Shetler, Antoinette (Toni)
Shoosmith, John Norman
Shrier, Stefan
Sime, David Gilbert
Small, Timothy Michael
Sorrell, Gary Lee
Stevens, Donald Meade
Taggart, Keith Anthony
Talbot, Richard Burritt
Thiel, Thomas J
Tole, John Roy
Torio, Joyce Clarke
Tripp, John Stephen
Van Tilborg, André Marcel
Voigt, Robert Gary
Warfield, J(ohn) N(elson)
Weaver, Alfred Charles
Weinstein, Stanley Edwin
Whitaker, William Armstrong
Wilcox, Lyle C(hester)
Wist, Abund Ottokar
Wolf, Eric W
Wulf, William Allan
Yarbrough, Lynn Douglas

WASHINGTON
Benda, Miroslav
Bolme, Mark W
Brink, James E
Britt, Patricia Marie
Cole, Charles Ralph
Cutlip, William Frederick
Drevdahl, Elmer R(andolph)
Edison, Larry Alvin
Engleman, Christian L
Farnum, Peter
Golde, Hellmut
Gravitz, Sidney I
Jenner, David Charles
Kehl, Theodore H
Klepper, John Richard
Kowalik, Janusz Szczesny
Morrill, Ralph Alfred
Newman, Paul Harold
Noe, Jerre D(onald)
Oster, Clarence Alfred
Parker, Richard Alan
Richardson, Michael Lewellyn
Shirley, Frank Connard
Siegel, Andrew Francis
Smilen, Lowell I
Somani, Arun Kumar
Stokes, Gerald Madison
Thomas, James
Thompson, Robert Harry
Tinker, John Frank
Underwood, Douglas Haines

WEST VIRGINIA
Atkins, John Marshall
Becker, Stanley Leonard
Brown, Paul B
Chahryar, Hamid N
Jefimenko, Oleg D
Larson, Gary Eugene
Leech, H(arry) William
Muth, Wayne Allen

WISCONSIN
Chen, Steve S
Cohen, Bernard Allan
Cook, David Marsden
Davida, George I
Davidson, Charles H(enry)
Desautels, Edouard Joseph
Divjak, August A
Dyer, Charles Robert
Eggert, Arthur Arnold
Evans, James Stuart
Firebaugh, Morris W
Fitzwater, Donald (Robert)
Fossum, Timothy V
Heinen, James Albin
Hutchinson, George Keating
Koenig, Eldo C(lyde)
Landweber, Lawrence H
Levine, Leonard P
Littlewood, Roland Kay
Malkus, David Starr
Miller, Paul Dean
Moses, Gregory Allen
Northouse, Richard A
Pollnow, Gilbert Frederick
Ramakrishnan, Raghu
Rang, Edward Roy
Rhyner, Charles R
Schultz, David Harold
Stahl, Neil
Uhr, Leonard Merrick

WYOMING
Bauer, Henry Raymond, III
Gastl, George Clifford
Magee, Michael Jack
Rowland, John H
Smith, William Kirby
Zhang, Renduo

PUERTO RICO
Lewis, Brian Murray
Montes, Maria Eugenia

ALBERTA
Bennion, Douglas Wilford
Chang, Chi
Fu, Cheng-Tze
George, Ronald Edison
Harvey, Ross Buschlen
Honsaker, John Leonard
Plambeck, James Alan
Prepas, Ellie E

BRITISH COLUMBIA
Clark, Stanley Ross
Davis, Wayne Alton
Davison, Allan John
Dill, John C
Dower, Gordon Ewbank
Ehle, Byron Leonard
Gilmore, Paul Carl
Hafer, Louis James
Halabisky, Lorne Stanley
Harrop, Ronald

Computer Sciences, General (cont)

Kennedy, James M
MacDonald, John Lauchlin
Manning, Eric G(eorge)
Odeh, Robert Eugene
Sheehan, Bernard Stephen
Sterling, Theodor David
Van Emden, Maarten Herman
Wade, Adrian Paul

MANITOBA
King, Peter Ramsay
Tait, John Charles
Williams, Hugh Cowie

NEW BRUNSWICK
Wasson, W(alter) Dana

NEWFOUNDLAND
Longerich, Henry Perry
Shieh, John Shunen
Wroblewski, Joseph S

NOVA SCOTIA
Kruse, Robert Leroy
McEwan, Alan Thomas
Matthews, Frederick White

ONTARIO
Argyropoulos, Stavros Andreas
Ball, Gordon Charles
Bartels, Richard Harold
Batra, Tilak Raj
Bauer, Michael Anthony
Beatty, John C
Beylerian, Nurel
Bioleau, Luc J R
Boorn, Andrew William
Borodin, Allan Bertram
Boulton, Peter Irwin Paul
Bush, George Clark
Calder, Peter N
Chu, Chun-Lung (George)
Clarke, James Newton
Cowan, Donald D
Fabian, Robert John
Fairman, Frederick Walker
Gaherty, Geoffrey George
Gentleman, Jane Forer
Gentleman, William Morven
Gotlieb, C(alvin) C(arl)
Graham, James W
Guevremont, Roger M
Haynes, Robert Brian
Hull, Thomas Edward
Ionescu, Dan
Jardine, D(onald) A(ndrew)
Jorch, Harald Heinrich
Lasker, George Eric
Lawson, John Douglas
Linders, James Gus
Meads, Jon A
Molder, S(annu)
Morris, Lawrence Robert
Munro, J Ian
Raaphorst, G Peter
Read, Ronald Cedric
Redish, Kenneth Adair
Reiter, Raymond
Riordon, J(ohn) Spruce
Rowe, Ronald Kerry
Slonim, Jacob
Steiner, George
Tavares, Stafford Emanuel
Thierrin, Gabriel
Thomas, Paul A V
Vanstone, Scott Alexander
Waddington, Raymond
Watson, Jeffrey
West, Eric Neil
White, George Michael
Wilson, John Cleland
Wong, Johnny Wai-Nang
Wood, Derick

QUEBEC
Agarwal, Vinod Kumar
Bui, Tien Dai
Davies, Roger
Duong, Quang P
Ferguson, Michael John
Jordan, Byron Dale
Lecours, Michel
Newborn, Monroe M
Opatrny, Jaroslav
Poussart, Denis
Shizgal, Harry M
Suen, Ching Yee
Tao, Lixiu
Vaucher, Jean G
Zucker, Steven Warren

SASKATCHEWAN
Cercone, Nicholas Joseph
Gupta, Madan Mohan
Symes, Lawrence Richard

OTHER COUNTRIES
Arimoto, Suguru
Bellanger, Maurice G
Blackburn, Jacob Floyd

Brown, Richard Harland
Burkhardt, Walter H
Camiz, Sergio
Cartwright, Hugh Manning
Chanson, Samuel T
Chen, Tien Chi
Cryer, Colin Walker
De Renobales, Mertxe
Fiesler, Emile
Fujiwara, Hideo
Glover, Francis Nicholas
Gray, Kenneth W
Gritzmann, Peter
Halkias, Christos
Hall, Peter
Humphrey, Albert S
Hura, Gurdeep Singh
Ilten, David Frederick
Ku, Victor Chia-Tai
Levialdi, Stefano
Lieth, Helmut Heinrich Friedrich
Low, Chow-Eng
Muskat, Joseph Baruch
Nestvold, Elwood Olaf
Park, Chan Mo
Pau, Louis F
Querinjean, Pierre Joseph
Raviv, Josef
Sargent, David Fisher
Scheiter, B Joseph Paul
Sekine, Yasuji
Shafi, Mohammad
Shamir, Adi
Shen, Vincent Y
Siklossy, Laurent
Siegel, Herbert
Smoliar, Stephen William
Vamos, Tibor
Wilkes, Maurice V

Computer Engineering

CALIFORNIA
Agostini, Romain Camille
Ferrari, Domenico
Patterson, David Andrew
Sharma, Nita
Smith, Steven Eslie

COLORADO
Mandics, Peter Alexander

CONNECTICUT
Smith, Donald Arthur

FLORIDA
Balaban, Murat Omer
Fishwick, Paul Anthony
Perrotta, Alessandro
Yau, Stephen S

GEORGIA
Bridges, Alan L

ILLINOIS
Emmel, Robert Shafer
Fuchs, W Kent
Sekar, Raj

INDIANA
Jamieson, Leah H

IOWA
Rudolphi, Thomas Joseph

LOUISIANA
Fuchs, Richard E(arl)

MASSACHUSETTS
Menon, Premachandran R(ama)
Moss, J Eliot B

MICHIGAN
Smith, Joseph D

NEW JERSEY
Cazes, Jack
Kahrs, Mark William
Mersten, Gerald Stuart

NEW YORK
Leung, Yiu M
Luk, Franklin T
Notaro, Anthony
Padegs, Andris
Rosenfeld, Jack Lee
Stevens, William Y(eaton)
Tracz, Will
Wong, Edward Chor-Cheung

NORTH CAROLINA
Gelenbe, Erol
Yamamoto, Toshiaki

OHIO
Gratzl, Miklos
Hobart, William Chester, Jr
Moon, Donald Lee

PENNSYLVANIA
Irwin, Mary Jane
Krutz, Ronald L

TEXAS
Giese, Robert Paul
Jump, J Robert
Keller, Thomas W
Ruchelman, Maryon W
Swartzlander, Earl Eugene, Jr

VIRGINIA
Dengel, Ottmar H

WEST VIRGINIA
Chahryar, Hamid N

ALBERTA
Ulagaraj, Munivandy Seydunganallur

ONTARIO
Smith, Kenneth Carless
Venetsanopoulos, Anastasios Nicolaos

QUEBEC
Tao, Lixiu

Hardware Systems

ALABAMA
Jolley, Homer Richard
Stone, Max Wendell

ALASKA
Watkins, Brenton John

ARIZONA
Benfield, Charles W(illiam)
Dybvig, Paul Henry
Haynes, Munro K
Preston, Kendall, Jr
Shoemaker, Richard Lee
Stafford, John William
Woodfill, Marvin Carl

CALIFORNIA
Albert, Paul A(ndre)
Allen, Charles A
Bardin, Russell Keith
Bertin, Michael C
Bic, Lubomir
Birecki, Henryk
Bush, George Edward
Butler, Jon Terry
Carter, J Lawrence (Larry)
Chen, Tsai Hwa
Crosbie, Roy Edward
Davis, Alan Lynn
Dolen, Richard
Dutt, Nikil D
Eaton, John Kelly
Fasang, Patrick Pad
Flynn, Michael J
Friedlander, Carl B
Gesner, Bruce D
Harker, J M
Hennessy, John LeRoy
Herzog, Gerald B(ernard)
Hu, Sung Chiao
Kalbach, John Frederick
Kirk, Donald Evan
Kodres, Uno Robert
Kreuzer, Lloyd Barton
Langdon, Glen George, Jr
Latta, Gordon
Levine, Howard Bernard
Lieber, Richard L
Lin, Wen-C(hun)
Loomis, Herschel Hare, Jr
Meagher, Donald Joseph
Melvin, Jonathan David
Meyer, Lhary
Michaelson, Jerry Dean
Mitra, Sanjit K
Monier, Louis Marcel
Motteler, Zane Clinton
Napolitano, Leonard Michael, Jr
Nikora, Allen P
Parker, Alice Cline
Patterson, David Andrew
Phipps, Peter Beverley Powell
Prasanna Kumar, V K
Requa, Joseph Earl
Roberts, Charles Sheldon
Schechtman, Barry H
Schwartz, Morton Donald
Shaka, Athan James
Shar, Leonard E
Sherden, David J
Silvester, John Andrew
Smith, Steven Eslie
Sterling, Warren Martin
Strohman, Rollin Dean
Sung, Chia-Hsiang
Thomas, Hubert Jon
Thompson, David A
Willens, Ronald H
Winzenread, Marvin Russell

COLORADO
Allen, James Lamar
Bennett, W Scott
Brady, Douglas MacPherson
Cathey, Wade Thomas, Jr
Clark, Jon D
Etter, Delores Maria
Frick, Pieter A

Furcinitti, Paul Stephen
Johnson, Gearold Robert
Lee, Yung-Cheng
Malaiya, Yashwant Kumar
Mueller, Robert Andrew
Wiatrowski, Claude Allan

CONNECTICUT
Goldman, Ernest Harold
Hudak, Paul Raymond
McDonald, John Charles

DELAWARE
Amer, Paul David
Christie, Phillip
Holob, Gary M
Moore, Charles B(ernard)
Scofield, Dillon Foster

DISTRICT OF COLUMBIA
Bloch, Erich
Curtin, Frank Michael
Friedman, Arthur Daniel
Meltzer, Arnold Charles

FLORIDA
Bonn, T(heodore) H(ertz)
Cengeloglu, Yilmaz
Coates, Clarence L(eroy), Jr
Coulter, Neal Stanley
Ege, Raimund K
Hoffman, Thomas R(ipton)
Marcovitz, Alan Bernard
Moreno, Wilfrido A
Sahni, Sartaj Kumar

GEORGIA
Alford, Cecil Orie
Bomar, Lucien Clay
Cohn, Charles Erwin
Croft, George Thomas
Currie, Nicholas Charles
Husson, Samir S
Subramanian, Mani M
Techo, Robert
Tummala, Rao Ramamohana

HAWAII
Kinariwala, Bharat K

ILLINOIS
Chien, Andrew Andai
Coplien, James O
DeFanti, Thomas A
Domanik, Richard Anthony
Flora, Robert Henry
Fuchs, W Kent
Hajj, Ibrahim Nasri
Kaplan, Daniel Moshe
Karaali, Orhan
Kubitz, William John
Merkelo, Henri
Murphy, Gordon J
Padua, David A
Pardo, Richard Claude
Polychronopoulos, Constantine Dimitrius
Ray, Sylvian Richard
Reed, Daniel A

INDIANA
Aprahamian, Ani
Brown, Buck F(erguson)
Esch, Harald Erich
Harbron, Thomas Richard
Hinkle, Charles N(elson)
Place, Ralph L
Siegel, Howard Jay
Stanley, Gerald R
Western, Arthur Boyd
Wise, David Stephen

IOWA
Stewart, Robert Murray, Jr

KENTUCKY
Villareal, Ramiro

LOUISIANA
Burke, Robert Wayne

MAINE
Carter, William Caswell

MARYLAND
Alperin, Harvey Albert
Berson, Alan
Buell, Duncan Alan
Hammond, Charles E
Meyer, Gerard G L
Papian, William Nathaniel
South, Hugh Miles

MASSACHUSETTS
Adrion, William Richards
Chang, Robin
Cotter, Douglas Adrian
Dennison, Byron Lee
Eckhouse, Richard Henry
Foster, Caxton Croxford
Guenther, R(ichard)
Hofstetter, Edward
Hopkins, Albert Lafayette, Jr
Kung, Hsiang-Tsung
Lis, Steven Andrew

Menon, Premachandran R(ama)
Moss, J Eliot B
Raffel, Jack I
Rosenberg, Arnold Leonard
Simovici, Dan
Speliotis, Dennis Elias
Stewart, Lawrence Colm
Stiffler, Jack Justin
Strong, Robert Michael

MICHIGAN
Becher, William D(on)
Be Ment, Spencer L
Bickart, Theodore Albert
Brumm, Douglas B(ruce)
Horvath, Ralph S(teve)
Larrowe, Boyd T
Nevius, Timothy Alfred
Patt, Yale Nance
Scott, Norman R(oss)
Sloan, Martha Ann
Westervelt, Franklin Herbert
Wojcik, Anthony Stephen

MINNESOTA
Du, David Hung-Chang
Franta, William Roy
Gilbert, Barry Kent
Kain, Richard Yerkes
Lo, David S(hih-Fang)
MacKellar, William John
Thomborson, Clark D(avid)

MISSISSIPPI
Elsherbeni, Atef Zakaria

MISSOURI
Atwood, Charles LeRoy
Baer, Robert W
Franklin, Mark A
McFarland, William D
Morgan, Nancy H
Morley, Robert Emmett, Jr
Moss, Randy Hays
Rouse, Robert Arthur
Taylor, Ralph Dale

NEBRASKA
Gale, Douglas Shannon, II

NEVADA
Anderson, Arthur George

NEW HAMPSHIRE
Brender, Ronald Franklin
Crowell, Merton Howard
Gagliardi, Ugo Oscar

NEW JERSEY
Burkhardt, Kenneth J
Carr, William N
Crochiere, Ronald E
Duttweiler, Donald Lars
Dwyer, Harry, III
French, Larry J
Gittleson, Stephen Mark
Goldstein, Philip
Graf, Hans Peter
Haas, Zygmunt
Horn, David Nicholas
Horng, Shi-Jinn
Kahrs, Mark William
Katz, Sheldon Lane
Kobrin, Robert Jay
Krzyzanowski, Paul
Kuo, Ying L
Lo, Arthur W(unien)
Mersten, Gerald Stuart
Michelson, Leslie Paul
Mondal, Kalyan
Murthy, Srinivasa K R
Nelson, Gregory Victor
Ninke, William Herbert
Ourmazd, Abbas
Schlam, Elliott
Schmidt, Barnet Michael
Sears, Raymond Warrick, Jr
Shahbender, R(abah) A(bd-El-Rahman)
Shieh, Ching-Chyuan
Silberschatz, Abraham
Taylor, Harold Evans
Thompson, Kenneth Lane
Tsaliovich, Anatoly
Woo, Nam-Sung
Zwass, Vladimir

NEW MEXICO
Butler, Harold S
Cress, Daniel Hugg
Giesler, Gregg Carl
Hardin, James T
Phister, Montgomery, Jr
Smith, William Conrad

NEW YORK
Albicki, Alexander
Balabanian, Norman
Bashkow, Theodore R(obert)
Bernstein, Herbert Jacob
Bourbakis, Niklaos G
Brantley, William Cain, Jr
Brennemann, Andrew E(rnest), Jr
Capwell, Robert J
Carleton, Herbert Ruck

Chrien, Robert Edward
Clarke, Kenneth Kingsley
Cole, James A
Costello, Richard Gray
Csermely, Thomas J(ohn)
Debany, Warren Harding, Jr
Delgado-Frias, Jose
Eichelberger, Edward B
Fiedler, Harold Joseph
Fordon, Wilfred Aaron
Frank, Thomas Stolley
Gabay, Jonathan Glenn
Gottlieb, Allan
Heller, William R
Huang, Ming-Hui
Keyes, Robert William
Kotchoubey, Andrew
Landauer, Rolf William
Leung, Yiu M
Li, Chung-Sheng
Lorenzen, Jerry Alan
Maguire, Gerald Quentin, Jr
Maissel, Leon I
Morgenstern, Matthew
Myers, Robert Anthony
Padegs, Andris
Peled, Abraham
Pingali, Keshav Kumar
Pomerene, James Herbert
Puttlitz, Karl Joseph
Rauscher, Tomlinson Gene
Rogers, Edwin Henry
Rosenkrantz, Daniel J
Schamberger, Robert Dean
Shooman, Martin L
Spencer, David R
Stern, Miklos
Su, Stephen Y H
Syed, Ashfaquzzaman
Tatarczuk, Joseph Richard
Torng, Hwa-Chung
Trexler, Frederick David
Wittie, Larry Dawson
Wong, Edward Chor-Cheung
Zukowski, Charles Albert

NORTH CAROLINA
Adelberger, Rexford E
Chen, Su-shing
D'Arruda, Jose Joaquim
Franzon, Paul Damian
Gray, James P
Hutchins, William R(eagh)
Javel, Eric
Johnson, G Allan
Kanopoulos, Nick
Nagle, H Troy
Pearlstein, Robert David
Wright, William V(aughn)
Zahed, Hyder Ali

NORTH DAKOTA
Tallman, Dennis Earl

OHIO
Dehne, George Clark
Gilfert, James C(lare)
Hollingsworth, Ralph George
Hubin, Wilbert N
Ismail, Amin Rashid
James, Thomas Ray
Jendrek, Eugene Francis, Jr
Mathis, Robert Fletcher
Moon, Tag Young
Papachristou, Christos A
Rajsuman, Rochit
Ramamoorthy, Panapakkam A
Rose, Charles William
Ross, Charles Burton
Taylor, Barney Edsel
Webster, Allen E
Zheng, Yuan Fang

OKLAHOMA
Grigsby, Ronald Davis
Gunter, Deborah Ann
VanDerwiele, James Milton

OREGON
Kocher, Carl A
Schwartz, James William
Stewart, Bradley Clayton

PENNSYLVANIA
Aburdene, Maurice Felix
Arrington, Wendell S
Barbacci, Mario R
Druffel, Larry Edward
Fisher, Tom Lyons
Goutmann, Michel Marcel
Grassi, Vincent G
Hardesty, Patrick Thomas
Harris, Lowell Dee
Humphrey, Watts S
Irwin, Mary Jane
Klafter, Richard D(avid)
Larky, Arthur I(rving)
Mitchell, Tom M
Muir, Patrick Fred
Murphy, Bernard T
Pinkerton, John Edward
Reigel, Earl William
Rhodes, Donald Frederick
Vastola, Francis J

Vogt, William G(eorge)
Wagh, Meghanad D
Walker, David Kenneth
Warme, Paul Kenneth

RHODE ISLAND
Gabriel, Richard Francis

SOUTH CAROLINA
Bennet, Archie Wayne
Bennett, Archie Wayne
Goode, Scott Roy
Lambert, Jerry Roy

SOUTH DAKOTA
Riemenschneider, Albert Louis

TENNESSEE
Bishop, Asa Orin, Jr
Campbell, Warren Elwood
Rajan, Periasamy Karivaratha
Skinner, George T
Trivedi, Mohan Manubhai

TEXAS
Abraham, Jacob A
Bossen, Douglas C
Brakefield, James Charles
Brock, James Harvey
Franke, Ernest A
Garner, Harvey L(ouis)
Gonzalez, Mario J
Grobman, Warren David
Hayman, Alan Conrad
Jump, J Robert
Knapp, Roger Dale
Lewis, Donald Richard
Lipovski, Gerald John (Jack)
Malek, Miroslaw
Markenscoff, Pauline
Martin, Norman Marshall
Nather, Roy Edward
Nute, C Thomas
Perez, Ricardo
Pritchard, John Paul, Jr
Smith, Reid Garfield
Swartzlander, Earl Eugene, Jr
Vines, Darrell Lee
Watt, Joseph T(ee), Jr

UTAH
Hollaar, Lee Allen
Smith, Kent Farrell

VIRGINIA
Alter, Ralph
Anderson, Walter L(eonard)
Beale, Guy Otis
Brenner, Alfred Ephraim
Cook, Robert Patterson
Crawford, Daniel J
Eldridge, Charles A
Fisher, Arthur Douglas
Garcia, Albert B
Garcia, Oscar Nicolas
Guerber, Howard P(aul)
Hilger, James Eugene
Larson, Arvid Gunnar
Lehmann, John R(ichard)
McGrath, Arthur Kevin
Rony, Peter R(oland)
Saltz, Joel Haskin
Toida, Shunichi
Waxman, Ronald
Webber, John Clinton

WASHINGTON
Preikschat, F(ritz) K(arl)
Seamans, David A(lvin)
Smith, Burton Jordan
Somani, Arun Kumar
Spillman, Richard Jay

WEST VIRGINIA
Tewksbury, Stuart K

WISCONSIN
Brookshear, James Glenn
Dietmeyer, Donald L
Dobson, David A
Harris, J Douglas
Hind, Joseph Edward
Jacobi, George (Thomas)
Levine, Leonard P
Moore, Edward Forrest
Rosenthal, Jeffrey
Viswanathan, Ramaswami

WYOMING
Howell, Robert Richard

BRITISH COLUMBIA
Dimopoulos, Nikitas J
Hafer, Louis James

NEWFOUNDLAND
Rahman, Md Azizur

ONTARIO
Emraghy, Hoda Abdel-Kader
Jullien, Graham Arnold
Majithia, Jayanti
Mouftah, Hussein T
Thomas, Paul A V

Wilson, John D(ouglas)

QUEBEC
Agarwal, Vinod Kumar
Atwood, John William
De Mori, Renato
Desai, Bipin C
Eddy, Nelson Wallace
Marin, Miguel Angel

OTHER COUNTRIES
Burkhardt, Walter H
Chi, Chao Shu
Fujiwara, Hideo
Goguen, Joseph A
Halkias, Christos
Kim, Myunghwan
Kuroyanagi, Noriyoshi
Lewis, Arnold Leroy, II
Maas, Peter

Information Science & Systems

ALABAMA
Coleman, Tommy Lee

ARIZONA
Lorents, Alden C

ARKANSAS
Alguire, Mary Slanina (Hirsch)
Lloyd, Harris Horton
Sedelow, Sally Yeates
Sedelow, Walter Alfred, Jr

CALIFORNIA
Agostini, Romain Camille
Alper, Marshall Edward
Birnbaum, Joel S
Block, Richard B
Bradley, Hugh Edward
Butz, Robert Frederick
Chock, Margaret Irvine
Clark, Peter O(sgoode)
Cooper, William S
Corwin, Dennis Lee
Engelbart, Douglas C(arl)
Fa'arman, Alfred
Firdman, Henry Eric
Freeman, Linton Clarke
Glaser, Myron B(arnard)
Grethe, Guenter
Groner, Gabriel F(rederick)
Guner, Osman Fatih
Hankins, Hesterly G, III
Hellerman, Herbert
Hsu, Charles Teh-Ching
Jackson, Durward P
Jacobson, Allan Stanley (Bud)
Kerfoot, Branch Price
Kimme, Ernest Godfrey
Lientz, Bennet Price
Lim, Hong Seh
Lyman, John (Henry)
McCarthy, John Lockhart
Maillot, Patrick Gilles
Mathis, Ronald Floyd
Mayer, William Joseph
Minckler, Tate Muldown
Morris, Cecil Arthur, Sr
Mulligan, Geoffrey C
Papadimitriou, Christos
Rohm, C E Tapie, Jr
Rulifson, Johns Frederick
Rusch, Peter Frederick
Schmidt, Robert Milton
Selinger, Patricia Griffiths
Siddiqee, Muhammad Waheeduddin
Star, Jeffrey L
Stoms, David Michael
Strand, Timothy Carl
Teague, Lavette Cox
Williams, Donald Spencer
Willner, Alan Eli
Wong, Kwan Y
Zebroski, Edwin L

COLORADO
Asherin, Duane Arthur
Call, Patrick Joseph
Horowitz, Isaac
Reiter, Elmar Rudolf

CONNECTICUT
Maizell, Robert Edward
Martin, R Keith
Nair, Sreedhar
Newman, Robert Weidenthal
Rezek, Geoffrey Robert
Stuck, Barton W

DELAWARE
Bartkus, Edward Peter
Lehr, Gary Fulton
Skolnik, Herman
Talley, John Herbert

DISTRICT OF COLUMBIA
Garavelli, John Stephen
Orthner, Helmuth F
Panar, Manuel
Poon, Bing Toy
Stein, Marjorie Leiter

Information Science & Systems (cont)

FLORIDA
Couturier, Gordon W
Froehlich, Fritz Edgar
Kinnie, Irvin Gray
Korwek, Alexander Donald
Lurie, Arnold Paul
Remmel, Randall James
Starke, Albert Carl, Jr
Swart, William W
Trubey, David Keith
Zanakis, Stelios (Steve) H

GEORGIA
Aronson, Jay E
Blanke, Jordan Matthew
Hoogenboom, Gerrit
Kollig, Heinz Philipp
Remillard, Marguerite Madden
Slamecka, Vladimir
Zunde, Pranas

HAWAII
Zisk, Stanley Harris

ILLINOIS
Beecher, Christopher W W
Bookstein, Abraham
Buntrock, Robert Edward
Eggan, Lawrence Carl
Emmel, Robert Shafer
Evens, Martha Walton
Foster, George A, Jr
Haddad, Abraham H(erzl)
Hoffman, Gerald M
Joshi, Umeshwar Prasad
Paddock, Robert Alton
Raiden, Robert Lee
Wells, Jane Frances
Whisler, Kenneth Eugene
Wolff, Thomas E

INDIANA
Eberts, Ray Edward
Koch, Kay Frances
Moore, Emily Allyn
Robertson, Edward L
Thompson, Wilmer Leigh
Wynne, Bayard Edmund

KANSAS
Meier, Robert J
Roberts, James Arnold
Unger, Elizabeth Ann

KENTUCKY
Cooke, Samuel Leonard
Tran, Long Trieu

LOUISIANA
Cline, Robert William
Kraft, Donald Harris
Sharma, Bhudev

MAINE
Guidi, John Neil

MARYLAND
Arsham, Hossein
Bagg, Thomas Campbell
Blum, Bruce I
Buell, Duncan Alan
Campbell, William J
Clark, Joseph E(dward)
Cotton, Ira Walter
Henderson, Madeline M Berry
Hevner, Alan Raymond
Johnson, Frederick Carroll
Kornman, Brent D
Leese, John Albert
Little, Joyce Currie
Munro, Ronald Gordon
Randolph, Lynwood Parker
Reeker, Larry Henry
Robbins, Robert John
Saloman, Edward Barry
Shotland, Lawrence M
Stillman, Rona Barbara
Thoma, George Ranjan

MASSACHUSETTS
Barlas, Julie S
Brayer, Kenneth
Champine, George Allen
Dyer, Charles Austen
Ernst, Martin L
Fortmann, Thomas Edward
Friedenstein, Hanna
Gilfix, Edward Leon
Giuliano, Vincent E
Gorgone, John T
Griffin, William G
Johnson, Corinne Lessig
McKnight, Lee Warren
Morris, Robert
Moss, J Eliot B
O'Neil, Patrick Eugene
Prerau, David Stewart
Ross, Douglas Taylor
Royston, Richard John
Stonier, Tom Ted

Wasserman, Jerry

MICHIGAN
Aupperle, Eric Max
Coburn, Joel Thomas
Farah, Badie Naiem
Kulier, Charles Peter
Rundensteiner, Elke Angelika
Sawatari, Takeo
Shahabuddin, Syed
Westland, Roger D(ean)

MINNESOTA
Beeler, George W, Jr
Dybvig, Douglas Howard
Frank, William Charles
Hoffmann, Thomas Russell
March, Salvatore T
Pong, Ting-Chuen
Rust, Lawrence Wayne, Jr

MISSISSIPPI
Jennings, David Phipps

MISSOURI
Ball, William E(rnest)
Kochtanek, Thomas Richard
Morgan, Nancy H

NEBRASKA
Grisso, Robert Dwight
Hubbard, Kenneth Gene
Nelson, Don Jerome

NEVADA
Miel, George J

NEW HAMPSHIRE
Egan, John Frederick

NEW JERSEY
Abeles, Francine
Allen, Robert B
Andrews, Ronald Allen
Brown, William Stanley
Carney, Richard William James
Garber, H(irsh) Newton
Grinberg-Funes, Ricardo A
Hang, Hsueh-Ming
Horn, David Nicholas
Howell, Mary Gertrude
Huber, Melvin Lefever
Johnson, Layne Mark
Kaback, Stuart Mark
Kantor, Paul B
Luk, Kin-Chun C
McKenna, James
Marcus, Phillip Ronald
Mersten, Gerald Stuart
Popper, Robert David
Poucher, John Scott
Rosin, Robert Fisher
Shieh, Ching-Chyuan
Sinclair, Brett Jason
Verdu, Sergio
Waltz, David Leigh
Woodruff, Hugh Boyd
Zwass, Vladimir

NEW MEXICO
Caves, Carlton Morris

NEW YORK
Anderson, John Bailey
Ballou, Donald Pollard
Bernstein, Herbert Jacob
Chang, Ifay F
Chi, Benjamin E
Coden, Michael H
Doganaksoy, Necip
Farb, Edith H
Fine, Terrence Leon
Gabay, Jonathan Glenn
Galatianos, Anthony Athanassios
Goldschmidt, Eric Nathan
Golibersuch, David Clarence
Griffith, Daniel Alva
Hannay, David G
Harris, Alex L
Hsu, Donald K
Hunt, David Michael
Koplowitz, Jack
Lerman, Steven I
Li, Chung-Sheng
Lommel, J(ames) M(yles)
McDonald, Robert Skillings
McGee, James Patrick
McGee-Russell, Samuel M
Mowshowitz, Abbe
O'Brien, Anne T
Pechacek, Terry Frank
Quaile, James Patrick
Salton, Gerard
Salwen, Martin J
Sherman, Philip Martin
Shuey, R(ichard) L(yman)
Szymanski, Boleslaw K
Windsor, Donald Arthur

NORTH CAROLINA
Godschalk, David Robinson
Kurtz, A Peter
Shackelford, Ernest Dabney
Smith, William Adams, Jr

Starmer, C Frank
Williams, Myra Nicol

NORTH DAKOTA
Magel, Kenneth I

OHIO
Chamis, Alice Yanosko
Dunn, Horton, Jr
Hollis, William Frederick
Kaufman, John Gilbert, Jr
Marble, Duane Francis
Weisgerber, David Wendelin

OKLAHOMA
McDevitt, Daniel Bernard
Urban, Timothy I

OREGON
Beaulieu, John David
Keltner, Llew

PENNSYLVANIA
Anandalingam, G
Bhagavatula, VijayaKumar
Blair, Grant Clark
Cardarelli, Joseph S
Confer, Gregory Lee
Di Cuollo, C John
Farlee, Rodney Dale
Franz, Edmund C(larence)
Garfield, Eugene
Gledhill, Ronald James
Goutmann, Michel Marcel
Grover, Carole Lee
Ho, Thomas Inn Min
Kam, Moshe
Kennedy, Harvey Edward
Korfhage, Robert R
Krishnan, Ramayya
Li, C(hing) C(hung)
Palmer, Jon (Carl)
Rau, Lisa Fay
Skoner, Peter Raymond
Sonnemann, George
Treu, Siegfried
Wei, Susanna

SOUTH CAROLINA
Eastman, Caroline Merriam

SOUTH DAKOTA
Leslie, Jerome Russell
Paterson, Colin James

TENNESSEE
Gonzalez, Rafael C
Kanciruk, Paul
LeBlanc, Larry Joseph
Sittig, Dean Forrest
Upadhyaya, Belle Raghavendra
Voorhees, Larry Donald

TEXAS
Ball, Millicent (Penny)
Bartley, William Call
Bishop, Robert H
Blakley, George Robert
Brock, James Harvey
Cameron, Bruce Francis
Danburg, Jerome Samuel
Fix, Ronald Edward
Gorry, G Anthony
Hall, Douglas Lee
Harmon, Glynn
Hedstrom, John Richard
Hennessey, Audrey Kathleen
Jackson, Eugene Bernard
McLeod, Raymond, Jr
Mendez, Victor Manuel
Mikiten, Terry Michael
Randolph, Paul Herbert
Sobol, Marion Gross
Wilhoit, Randolph Carroll
Wood, Kristin Lee
Wyllys, Ronald Eugene
Yao, James T-P

UTAH
Bowman, Carlos Morales
Stokes, Gordon Ellis

VIRGINIA
Awad, Elias M
Campbell, Leo James
Carr, Roger Byington
Cerf, Vinton Gray
Detmer, Don Eugene
Dockter, Kenneth Wylie
Fox, Edward A
Franklin, Jude Eric
Gregory, Arthur Robert
Hodge, Donald Ray
Hunt, V Daniel
Ivanetich, Richard John
Johnson, Charles Minor
Kaminski, Paul G
Neeper, Ralph Arnold
Rine, David C
Roberts, David Craig
Shenolikar, Ashok Kumar
Shetler, Antoinette (Toni)
Zilcer, Janet Ann

WASHINGTON
Mendelson, Martin
Morrill, Ralph Alfred
Shin, Suk-han

WISCONSIN
McClenahan, William St Clair
Ramakrishnan, Raghu

ALBERTA
Gong, Peng

BRITISH COLUMBIA
Agathoklis, Panajotis

NEW BRUNSWICK
Faig, Wolfgang
Lee, Y C

NEWFOUNDLAND
Thomeier, Siegfried

NOVA SCOTIA
Schuegraf, Ernst Josef

ONTARIO
Bauer, Michael Anthony
Famili, Abolfazl
Mouftah, Hussein T
Sablatash, Mike
Slonim, Jacob
Vincett, Paul Stamford

QUEBEC
Davidson, Colin Henry
De Mori, Renato
Li, Zi-Cai
Morgera, Salvatore Domenic
Yalovsky, Morty A

OTHER COUNTRIES
Argos, Patrick
Arimoto, Suguru
Garcia-Santesmases, Jose Miguel
Inose, Hiroshi
Kafri, Oded
Kaiser, Wolfgang A
Levialdi, Stefano
Malah, David
Ros, Herman H

Intelligent Systems

ALABAMA
Carlo, Waldemar Alberto
Habtemariam, Tsegaye
Harding, Thomas Hague
Reilly, Kevin Denis
Werkheiser, Arthur H, Jr

ARIZONA
Cochran, Jeffery Keith
Findler, Nicholas Victor
Iverson, A Evan
Meieran, Eugene Stuart
Ryan, Thomas Wilton
Schowengerdt, Robert Alan
Young, Kenneth Christie

ARKANSAS
Asfahl, C Ray
Sedelow, Sally Yeates
Sedelow, Walter Alfred, Jr
Wrobel, Joseph Stephen

CALIFORNIA
Abbott, Dean William
Abbott, Seth R
Agoston, Max Karl
Ahumada, Albert Jil, Jr
Akonteh, Benny Ambrose
Altes, Richard Alan
Antonsson, Erik Karl
Arens, Yigal
Beach, Sharon Sickel
Bekey, George A(lbert)
Bendat, Julius Samuel
Berenji, Hamid R
Bhanu, Bir
Bic, Lubomir
Birnir, Björn
Bobrow, Daniel G
Brutlag, Douglas Lee
Calabrese, Philip G
Chang, Tien-Lin
Cooper, William S
Curry, Bo(stick) U
Curtis, Earl Clifton, Jr
Dev, Parvati
Dornfeld, David Alan
Duda, Richard Oswald
Dym, Clive L
Eckhardt, Wilfried Otto
Elspas, B(ernard)
Erman, Lee Daniel
Firdman, Henry Eric
Fischler, Martin A(lvin)
Freedy, Amos
Friedlander, Carl B
Gesner, Bruce D
Green, Claude Cordell
Groce, David Eiben
Gutfinger, Dan Eli

COMPUTER SCIENCES / 277

Hackwood, Susan
Hamlin, Griffith Askew, Jr
Healey, Anthony J
Herbst, Noel Martin
Hightower, James K
Holeman, Dennis Leigh
Holtzman, Samuel
Hopfield, John Joseph
Hudson, Cecil Ivan, Jr
Jacob, George Korathu
Jacobson, Alexander Donald
Kagiwada, Harriet Hatsune Natsuyama
Kaplan, Ronald M
Klahr, Philip
Kreutz-Delgado, Kenneth Keith
Leavy, Paul Matthew
Limb, John Ormond
Lin, Wen-C(hun)
McCarthy, John Michael
Manna, Zohar
Martin, Jim Frank
Matsumura, Kenneth N
Mendel, Jerry M
Moran, Thomas Patrick
Mueller, Thomas Joseph
Murphy, Roy Emerson
Nassi, Isaac Robert
Nelson, David A
Nesbit, Richard Allison
Nielsen, Norman Russell
Painter, Jeffrey Farrar
Parker, Donn Blanchard
Parkison, Roger C
Patterson, Tim J
Pearl, Judea
Perkins, Walton A, III
Phillips, Veril LeRoy
Ralston, Elizabeth Wall
Requicha, Aristides A G
Rowe, Lawrence A
Rulifson, Johns Frederick
Russell, Stuart Jonathan
Sadler, Charles Robinson, Jr
Sakaguchi, Dianne Koster
Salisbury, Stanley R
Schoenfeld, Alan Henry
Shepanski, John Francis
Skrzypek, Josef
Surko, Pamela Toni
Teague, Tommy Kay
Thomas, Graham Havens
Van Klaveren, Nico
Wagner, Carl E
Wagner, William Gerard
Wexler, Jonathan David
Williams, Donald Spencer
Wipke, W Todd
Woodriff, Roger L
Zadeh, L(otfi) A

COLORADO
Augusteijn, Marijke Francina
Clark, Jon D
Glover, Fred William
MacMillian, Stuart A
Miller, Betty M (Tinklepaugh)
Mueller, Robert Andrew
Phillips, Keith L
Reiter, Elmar Rudolf
Troxell, Wade Oakes
Veirs, Val Rhodes

CONNECTICUT
Dreyfus, Marc George
Hanson, Trevor Russell

DELAWARE
Chester, Daniel Leon
Dhurjati, Prasad S
Lehr, Gary Fulton
Lund, John Turner
Owens, Aaron James
Shipman, Lester Lynn

DISTRICT OF COLUMBIA
Hammer, Charles F
Schafrik, Robert Edward
Tangney, John Francis

FLORIDA
Biegel, John E
Blatt, Joel Herman
Bressler, Steven L
Buoni, Frederick Buell
Cengeloglu, Yilmaz
Clarke, Thomas Lowe
Detweiler, Steven Lawrence
Duersch, Ralph R
Fishwick, Paul Anthony
Hand, Thomas
Hoffman, Frederick
Kandel, Abraham
Kirmse, Dale William
Marquis, David Alan
Martsolf, J David
Navlakha, Jainendra K
Nevill, Gale E(rwin), Jr
Peart, Robert McDermand
Seireg, Ali A
Tebbe, Dennis Lee
Tulenko, James Stanley
Ulug, Esin M
Winton, Charles Newton
Young, Tzay Y

GEORGIA
Alford, Cecil Orie
Aronson, Jay E
Foley, James David
Gibson, John Michael
Lee, Kok-Meng
Nevins, Arthur James
Prince, M(orris) David

HAWAII
Itoga, Stephen Yukio

IDAHO
Johnson, John Alan
Stock, Molly Wilford

ILLINOIS
Agrawal, Arun Kumar
Ahuja, Narendra
Boyle, James Martin
Catrambone, Joseph Anthony, Sr
Chang, Shi Kuo
Coley, Ronald Frank
Crocker, Diane Winston
DeFanti, Thomas A
Evens, Martha Walton
Gabriel, John R
Harrington, Joseph Anthony
Heller, Barbara Ruth
Henschen, Lawrence Joseph
Krug, Samuel Edward
Lamont, Patrick
McAlpin, John Harris
Mount, Bertha Lauritzen
Polychronopoulos, Constantine Dimitrius
Wallis, Walter Denis
Wilson, Howell Kenneth

INDIANA
Ersoy, Okan Kadri
Hanson, Andrew Jorgen
Kak, Avinash Carl
Lehto, Mark R
Nof, Shimon Y
Patterson, Larry K
Salvendy, Gavriel
Siegel, Howard Jay
Stampfli, Joseph
Uhran, John Joseph, Jr

IOWA
Jeyapalan, Kandiah
Myers, Glenn Alexander
Quinn, Loyd Yost

KANSAS
Farokhi, Saeed
Grzymala-Busse, Jerzy Witold
Roddis, Winifred Mary Kim
Shanmugan, K Sam

KENTUCKY
Jain, Ravinder Kumar
McGinness, William George, III

LOUISIANA
Burke, Robert Wayne
Gajendar, Nandigam
Kraft, Donald Harris

MAINE
Guidi, John Neil
Tucker, Allen B

MARYLAND
Albus, James S
Benokraitis, Vitalius
Blum, Bruce I
Blum, Harry
Bonavita, Nino Louis
Campbell, William J
Chang, Alfred Tieh-Chun
Chapin, Edward William, Jr
Chellappa, Ramalingam
Corliss, John Burt
Darden, Lindley
Eisenman, Richard Leo
Finkelstein, Robert
Hartman, Patrick James
Hevner, Alan Raymond
Hodge, David Charles
Horst, John Albert
Jenkins, Robert Edward
Jones, George R
Keenan, Thomas Aquinas
Kornman, Brent D
Ligomenides, Panos Aristides
Minker, Jack
Nanzetta, Philip Newcomb
Nau, Dana S
Powell, Edward Gordon
Reeker, Larry Henry
Rosenfeld, Azriel
Schlesinger, Judith Diane
Sidhu, Deepinder Pal
Sigillito, Vincent George
Tripathi, Satish K
Vogl, Thomas Paul
Weaver, Christopher Scot

MASSACHUSETTS
Becker, David Stewart
Blidberg, D Richard
Bovopoulos, Andreas D
Carpenter, Gail Alexandra
Collins, Allan Meakin
Doyle, Jon
Dudgeon, Dan Ed
Evans, Thomas George
Frawley, William James
Futrelle, Robert Peel
Hirschman, Lynette
Kolodzy, Paul John
Lacoss, Richard Thaddee
McConnell, Robert Kendall
Madden, Stephen James, Jr
Nesbeda, Paul
Pauker, Stephen Gary
Prerau, David Stewart
Riseman, Edward M
Roberts, Louis W
Royston, Richard John
Salveter, Sharon Caroline
Seibel, Frederick Truman
Smith, Raoul Normand
Strong, Robert Michael
Waters, Richard C(abot)
Woods, William A

MICHIGAN
Conrad, Michael
Elrod, David Wayne
Farah, Badie Naiem
Hassoun, Mohamad Hussein
Holland, John Henry
Kochen, Manfred
Krawetz, Stephen Andrew
Ku, Albert B
Lindsay, Robert Kendall
Marko, Kenneth Andrew
Page, Carl Victor
Ramamurthy, Amurthur C
Reynolds, Robert Gene
Rillings, James H
Rosenbaum, Manuel
Rundensteiner, Elke Angelika
Sawatari, Takeo
Sethi, Ishwar Krishan
Stivender, Donald Lewis
Stockman, George C
Teichroew, Daniel
Tuchinsky, Philip Martin
Westervelt, Franklin Herbert

MINNESOTA
Bahn, Robert Carlton
Connelly, Donald Patrick
Gini, Maria Luigia
Joseph, Earl Clark, II
Knutson, Charles Dwaine
Kumar, Vipin
La Bonte, Anton Edward
Pierce, Keith Robert
Pong, Ting-Chuen
Sadjadi, Firooz Ahmadi
Spurrell, Francis Arthur

MISSISSIPPI
Shim, Jung P

MISSOURI
Atwood, Charles LeRoy
Ball, William E(rnest)
Franz, John Matthias
Magill, Robert Earle
Peterson, Gerald E
Rodin, Ervin Y
Sabharwal, Chaman Lal
Wagner, Robert G
Yohe, Cleon Russell

MONTANA
Latham, Don Jay
Wright, Alden Halbert

NEBRASKA
Surkan, Alvin John

NEVADA
Minor, John Threecivelous

NEW HAMPSHIRE
Abbott, Robert Classie
Frantz, Daniel Raymond
Misra, Alok C

NEW JERSEY
Afshar, Siroos K
Amarel, S(aul)
Bernstein, Lawrence
Bosacchi, Bruno
Burkhardt, Kenneth J
Cooper, Paul W
Dwyer, Harry, III
Gabbe, John Daniel
Gale, William Arthur
Graf, Hans Peter
Grek, Boris
Hong, Allan Jixian
Hubbard, William Marshall
Jackel, Lawrence David
Jurkat, Martin Peter
Kalley, Gordon S
Kirch, Murray R
Kuhl, Frank Peter, Jr
McGuinness, Deborah Louise
Most, Joseph Morris
Nonnenmann, Uwe
Schlam, Elliott
Shieh, Ching-Chyuan
Stengel, Robert Frank
Tishby, Naftali Z
Verma, Pramode Kumar
Waltz, David Leigh
Weissman, Paul Morton
Woodruff, Hugh Boyd

NEW MEXICO
Castle, John Granville, Jr
Enke, Christie George
Frank, Robert Morris
Gilbert, Alton Lee
Luger, George F
Niemczyk, Thomas M
Osbourn, Gordon Cecil
Ross, Timothy Jack
Salzman, Gary Clyde
Sung, Andrew Hsi-Lin

NEW YORK
Allen, James Frederick
Anderson, Peter Gordon
Apte, Chidanand
Bourbakis, Niklaos G
Campbell, Gregory August
Comly, James B
Damerau, Frederick Jacob
Dimmler, D(ietrich) Gerd
Fenrich, Richard Karl
Ferguson, David Lawrence
Gabay, Jonathan Glenn
Galatianos, Anthony Athanassios
Genova, James John
Golibersuch, David Clarence
Griesmer, James Hugo
Hajela, Prabhat
Hastings, Harold Morris
Herrington, Lee Pierce
Hong, Se June
Huang, Ming-Hui
Hudak, Michael J
Kalogeropoulos, Theodore E
Karnaugh, Maurice
Kaufman, Sol
Klerer, Melvin
Klir, George Jiri
Kornberg, Fred
Kyburg, Henry
Layzer, Arthur James
Li, Chou H(siung)
McGee, James Patrick
Morgenstern, Matthew
Moyne, John Abel
Nerode, Anil
Pavlidis, Theo
Rauscher, Tomlinson Gene
Sakoda, William John
Sanderson, Arthur Clark
Schwartz, Jacob Theodore
Seeley, Thomas Dyer
Shapiro, Stuart Charles
Shasha, Dennis E
Shore, Richard A
Strand, Richard Carl
Summers, Phillip Dale
Taylor, James Hugh
Torng, Hwa-Chung
Walters, Deborah K W
Wang, Shu Lung
Westbrook, J(ack) H(all)
Wittie, Larry Dawson
Wolf, Walter Alan
Zdan, William
Zic, Eric A

NORTH CAROLINA
Bell, Norman R(obert)
Chang, Jeffrey C F
Chen, Su-shing
Dixon, N(orman) Rex
Loveland, Donald William
Ras, Zbigniew Wieslaw
Scott, Donald Ray
Utku, Senol

NORTH DAKOTA
Winrich, Lonny B

OHIO
Adeli, Hojjat
Burghart, James H(enry)
Chandrasekaran, Balakrishnan
Chawla, Mangal Dass
Genaidy, Ashraf Mohamed
Hall, Ernest Lenard
Hollingsworth, Ralph George
Law, Eric W
Ramamoorthy, Panapakkam A
Santos, Eugene (Sy)
Schlipf, John Stewart
Servais, Ronald Albert
Slotterbeck-Baker, Oberta Ann
Slutzky, Gale David
Speicher, Carl Eugene
Sterling, Leon Samuel
Suter, Bruce Wilsey
Tamburino, Louis A
Waldron, Manjula Bhushan
Zheng, Yuan Fang

OKLAHOMA
Hochhaus, Larry

Intelligent Systems (cont)

Koller, Glenn R
Naymik, Daniel Allan
Paxson, John Ralph

OREGON
Bregar, William S
Brown, Cynthia Ann
Coop, Leonard Bryan
Freiling, Michael Joseph
Lendaris, George G(regory)
Schmisseur, Wilson Edward
Stewart, Bradley Clayton
Udovic, Daniel

PENNSYLVANIA
Abremski, Kenneth Edward
Aiken, Robert McLean
Altschuler, Martin David
Andrews, Peter Bruce
Badler, Norman Ira
Banerji, Ranan Bihari
Berliner, Hans Jack
Bowlden, Henry James
Buchanan, Bruce G
Carbonell, Jaime Guillermo
Carr, John Weber, III
Chaplin, Norman John
Dax, Frank Robert
Elder, James Franklin, Jr
Farlee, Rodney Dale
Finin, Timothy Wilking
Forgy, Charles Lanny
Fuchs, Walter
Glaser, Robert
Goldwasser, Samuel M
Hahn, Peter Mathias
Herman, Martin
Hildebrandt, Thomas Henry
Kam, Moshe
Klafter, Richard D(avid)
Koffman, Elliot B
Krishnan, Ramayya
Li, C(hing) C(hung)
Meystel, Alexander Michael
Mitchell, Tom M
Moser, Gene Wendell
Mowrey, Gary Lee
Rau, Lisa Fay
Reigel, Earl William
Siegel, Melvin Walter
Simon, Herbert A
Stern, Richard Martin, Jr
Stover, James Anderson, Jr
Treu, Siegfried
Verhanovitz, Richard Frank
Vladutz, George E
Vogt, William G(eorge)
Werner, Gerhard

RHODE ISLAND
Fasching, James Le Roy
Kirschenbaum, Susan S
McClure, Donald Ernest
Streit, Roy Leon

SOUTH CAROLINA
Cannon, Robert L
Hammond, Joseph Langhorne, Jr
Lambert, Jerry Roy

SOUTH DAKOTA
Hodges, Neal Howard, II

TENNESSEE
Bailes, Gordon Lee
Bartell, Steven Michael
Bourne, John Ross
Campbell, Warren Elwood
Franklin, Stanley Phillip
Gonzalez, Rafael C
Kawamura, Kazuhiko
Rodgers, Billy Russell
Schell, Fred Martin
Siirola, Jeffrey John
Trivedi, Mohan Manubhai
Uhrig, Robert Eugene
Umholtz, Clyde Allan

TEXAS
Aldridge, Jack Paxton, III
Ali, Moonis
Alo, Richard Anthony
Barstow, David Robbins
Beck, John Robert
Bendapudi, Kasi Visweswararao
DeGroot, Doug
Deming, Stanley Norris
Ewell, James John, Jr
Franke, Ernest A
Frazer, Marshall Everett
Hall, Douglas Lee
Hennessey, Audrey Kathleen
Konstam, Aaron Harry
Kuipers, Benjamin Jack
Kyle, Thomas Gail
Levinson, Stuart Alan
Miikkulainen, Risto Pekka
Mikiten, Terry Michael
Mitchell, Owen Robert
Novak, Gordon Shaw, Jr
Peterson, Lynn Louise Meister
Pitts, Gerald Nelson

Self, Glendon Danna
Smith, Reid Garfield
Smith, Rolf C, Jr
Sobey, Arthur Edward, Jr
Stiller, Peter Frederick
Stuart, Joe Don
Styblinski, Maciej A
Walkup, John Frank
Wolter, Jan D(ithmar)
Yao, James T-P

VIRGINIA
Awad, Elias M
Butler, Charles Thomas
Bynum, William Lee
Chou, Tsai-Chia Peter
Deal, George Edgar
Farrell, Robert Michael
Fink, Lester Harold
Fisher, Arthur Douglas
Fox, Edward A
Franklin, Jude Eric
Garcia, Oscar Nicolas
Holzmann, Ernest G(unther)
Jones, Terry Lee
Kelly, Michael David
Kenyon, Stephen C
Levis, Alexander Henry
Loatman, Robert Bruce
McGrath, Arthur Kevin
Mandelberg, Martin
Masterson, Kleber Sanlin, Jr
Mata-Toledo, Ramon Alberto
Meyrowitz, Alan Lester
Moraff, Howard
Orcutt, Bruce Call
Sebastian, Richard Lee
Settle, Frank Alexander, Jr
Toida, Shunichi

WASHINGTON
Fetter, William Allan
Haralick, Robert M
Kowalik, Janusz Szczesny
Liu, Chen-Ching
MacVicar-Whelan, Patrick James
Morrill, Ralph Alfred
Savol, Andrej Martin
Waltar, Alan Edward

WEST VIRGINIA
Chahryar, Hamid N

WISCONSIN
Dyer, Charles Robert
Eggert, Arthur Arnold
Firebaugh, Morris W
Levine, Leonard P
Moore, Edward Forrest
Ramakrishnan, Raghu

WYOMING
Cowles, John Richard
Magee, Michael Jack

ALBERTA
Gourishankar, V (Gouri)
Marsland, T(homas) Anthony
Martin, John Scott

BRITISH COLUMBIA
Calvert, Thomas W(illiam)
Dimopoulos, Nikitas J
Dumont, Guy Albert
Goodenough, David George
Gruver, William A
Meech, John Athol
Rosenberg, Richard Stuart
Thomson, Alan John
Wade, Adrian Paul

NEWFOUNDLAND
Shieh, John Shunen

NOVA SCOTIA
Oliver, Leslie Howard

ONTARIO
Argyropoulos, Stavros Andreas
Bauer, Michael Anthony
Chou, Jordan Quan Ban
Cohn-Sfetcu, Sorin
Dastur, Ardeshir Rustom
Emraghy, Hoda Abdel-Kader
Famili, Abolfazl
Geddes, Keith Oliver
Giles, Robin
Henry, Roger P
Ionescu, Dan
MacRae, Andrew Richard
Ratz, H(erbert) C(harles)
Slonim, Jacob
Stillman, Martin John
Wilson, John D(ouglas)

QUEBEC
De Mori, Renato
Desai, Bipin C
Guttmann, Ronald David
Hay, Donald Robert
Levine, Martin David
Newborn, Monroe M
Patel, Rajnikant V
Suen, Ching Yee

Zucker, Steven Warren

OTHER COUNTRIES
Arimoto, Suguru
Chi, Chao Shu
Fiesler, Emile
Garcia-Santesmases, Jose Miguel
Goguen, Joseph A
Levialdi, Stefano
Pau, Louis F
Singh, Madan Gopal
Siu, Tsunpui Oswald
Tzafestas, Spyros G
Vamos, Tibor
Woo, Kwang Bang
Zabriskie, Franklin Robert

Software Systems

ALABAMA
Alexander, David Michael
Biggs, Albert Wayne
Copeland, David Anthony
Davis, Carl George
Jolley, Homer Richard
Klip, Dorothea A
O'Kane, Kevin Charles
Roe, James Maurice, Jr
Seidman, Stephen Benjamin
Sides, Gary Donald
Stone, Max Wendell
Yeh, Pu-Sen

ALASKA
Hulsey, J Leroy
Rogers, James Joseph
Watkins, Brenton John

ARIZONA
Andrews, Gregory Richard
Arabyan, Ara
Comba, Paul Gustavo
Contractor, Dinshaw N
Daniel, Sam Mordochai
Dybvig, Paul Henry
Foltz, Craig Billig
Frieden, Bernard Roy
Gall, Donald Alan
Hayes, Donald Scott
Kieffer, Hugh Hartman
Kreidl, Tobias Joachim
Lorents, Alden C
Palusinski, Olgierd Aleksander
Slaughter, Charles D
Sridharan, Natesa S
Wymore, Albert Wayne
Zielen, Albin John

ARKANSAS
Talburt, John Randolph

CALIFORNIA
Adinoff, Bernard
Agostini, Romain Camille
Agoston, Max Karl
Anderson, Dennis Elmo
Armstrong, John William
Backus, John
Bahn, Gilbert S(chuyler)
Bailey, Michael John
Banerjee, Utpal
Barany, Ronald
Bar-Cohen, Yoseph
Barry, James Dale
Barsky, Brian Andrew
Basinger, Richard Craig
Benson, Donald Charles
Berardo, Peter Antonio
Bertin, Michael C
Bic, Lubomir
Blattner, Meera McCuaig
Blinn, James Frederick
Blumstein, Carl Joseph
Bobrow, Daniel G
Bowles, Kenneth Ludlam
Boyle, Walter Gordon, Jr
Brutlag, Douglas Lee
Burg, John Parker
Burton, Donald Eugene
Cantin, Gilles
Carter, J Lawrence (Larry)
Chen, Alice Tung-Hua
Cheriton, David Ross
Cherlin, George Yale
Chern, Ming-Fen Myra
Creighton, John Rogers
Crosbie, Roy Edward
Cuadra, Carlos A(lbert)
Cunningham, Mary Elizabeth
Cunningham, Robert Stephen
Dalrymple, Stephen Harris
Davis, Alan Lynn
Deutsch, Laurence Peter
Dixon, Wilfrid Joseph
Doolittle, Robert Frederick, II
Dutt, Nikil D
Dwarakanath, Manchagondanahalli H
Edelman, Jay Barry
Eisberg, Robert Martin
Elliott, Denis Anthony
Elspas, B(ernard)
Emerson, Thomas James
Erman, Lee Daniel

Ewig, Carl Stephen
Faulkner, Thomas Richard
Ferrari, Domenico
Firdman, Henry Eric
Fletcher, John George
Forster, Julian
Forsythe, Alan Barry
Freiling, Edward Clawson
Friedlander, Carl B
Gallegos, Emilio Juan
Gaposchkin, Peter John Arthur
Garren, Alper A
Giannetti, Ronald A
Gladney, Henry M
Goheen, Lola Coleman
Gott, Euyen
Gottlieb, Peter
Graham, Susan Lois
Gray, Paul
Green, Claude Cordell
Greenberger, Martin
Grethe, Guenter
Groner, Gabriel F(rederick)
Hankins, Hesterly G, III
Hanscom, Roger H
Harrison, Michael A
Hecht, Herbert
Hegenauer, Jack C
Hennessy, John LeRoy
Herbst, Noel Martin
Hershey, Allen Vincent
Hightower, James K
Hilliker, David Lee
Hodson, William Myron
Hopponen, Jerry Dale
Huebsch, Ian O
Hughett, Paul William
Innes, William Beveridge
Irvine, Cynthia Emberson
Jacobson, Allan Stanley (Bud)
Jewett, Don L
Johnson, Mark Scott
Judd, Stanley H
Kaelble, David Hardie
Kaliski, Martin Edward
Kevorkian, Aram K
Kissinger, David George
Klahr, Philip
Koenig, Daniel Rene
Korchev, Dmitriy Veniaminovich
Korpman, Ralph Andrew
Kreuzer, Lloyd Barton
Kritz-Silverstein, Donna
Kuan, Teh S
Kuhlmann, Karl Frederick
Laiken, Nora Dawn
Landgraf, William Charles
Lang, Martin T
Latta, Gordon
Ledin, George, Jr
Leif, Robert Cary
Lelewer, Debra Ann
Levine, Howard Bernard
Lewis, Nina Alissa
Licht, Paul
Lim, Hong Seh
Lipeles, Martin
Lodato, Michael W
Lupash, Lawrence O
Luqi,
McCarthy, John Lockhart
McCarthy, Mary Anne
McKay, Dale Robert
Magness, T(om) A(lan)
Maker, Paul Donne
Marcus, Rudolph Julius
Marquart, Ronald Gary
Martin, Gordon Eugene
Melkanoff, Michel Allan
Melvin, Jonathan David
Meyers, Gene Howard
Mitchell, James George
Mitchell, John Clifford
Mitra, Sanjit K
Moran, Thomas Patrick
Morris, Cecil Arthur, Sr
Motteler, Zane Clinton
Muchmore, Robert B(oyer)
Mulligan, Geoffrey C
Murphy, Roy Emerson
Nassi, Isaac Robert
Nelson, Eldred (Carlyle)
Neumann, Peter G
Niccolai, Nilo Anthony
Nico, William Raymond
Nielsen, Norman Russell
Nikora, Allen P
Oddson, John Keith
Olsen, Edward Tait
Painter, Jeffrey Farrar
Pamidi, Prabhakar Ramarao
Parkison, Roger C
Payne, Philip Warren
Perloff, David Steven
Persky, George
Phillips, Veril LeRoy
Prag, Arthur Barry
Pratt, Vaughan Ronald
Pridmore-Brown, David Clifford
Pye, Earl Louis
Ratcliff, Milton, Jr
Rawal, Kanti M
Requa, Joseph Earl
Requicha, Aristides A G

Roberts, Charles Sheldon
Rogers, Howard H
Rowe, Lawrence A
Rudge, William Edwin
Rulifson, Johns Frederick
Runge, Richard John
Saffren, Melvin Michael
Sakaguchi, Dianne Koster
Saunders, Robert M(allough)
Schlafly, Roger
Schreiner, Robert Nicholas, Jr
Schwall, Richard Joseph
Schwartz, Morton Donald
Selinger, Patricia Griffiths
Shaka, Athan James
Shar, Leonard E
Sharma, Nita
Silvester, John Andrew
Simons, Barbara Bluestein
Smith, Frederick T(ucker)
Smith, Steven Eslie
Solomon, Malcolm David
Sovers, Ojars Juris
Strickland, Erasmus Hardin
Strohman, Rollin Dean
Sung, Chia-Hsiaing
Suri, Ashok
Surko, Pamela Toni
Szentirmai, George
Taylor, Richard N
Teague, Tommy Kay
Teeter, Richard Malcolm
Tilson, Bret Ransom
Tobey, Arthur Robert
VanAntwerp, Craig Lewis
Wagner, William Gerard
Walker, Kelsey, Jr
Wallace, Graham Franklin
Walton, James Stephen
Ward, David Gene
Weaver, William Bruce
Wilbarger, Edward Stanley, Jr
Williams, Morris Evan
Winkelmann, Frederick Charles
Worden, Paul Wellman, Jr
Zuppero, Anthony Charles

COLORADO
Abshier, Curtis Brent
Allen, James Lamar
Boggs, George Johnson
Brasch, Frederick Martin, Jr
Call, Patrick Joseph
Chugh, Ashok Kumar
Clark, Jon D
Eichelberger, W(illiam) H
Etter, Delores Maria
Gibson, William Loane
Glover, Fred William
Hittle, Douglas Carl
Johnson, Gearold Robert
MacMillian, Stuart A
Miesch, Alfred Thomas
Moore, Carla Jean
Morrison, John Stuart
Mueller, Robert Andrew
Munro, Richard Harding
Pearson, John Richard
Robertson, Jerold C
Schnabel, Robert B
Sloss, Peter William
Smith, William Bridges
Stewart, James Joseph Patrick
Waite, William McCastline
Walden, Jack M
Wiatrowski, Claude Allan

CONNECTICUT
Baumgarten, Alexander
Cheng, David H S
Collins, John Barrett
Darcey, Terrance Michael
Engel, Gerald Lawrence
Greenwood, Ivan Anderson
Hanson, Trevor Russell
Harrison, Irene R
Henkel, James Gregory
Hudak, Paul Raymond
McDonald, John Charles
Martinez, Robert Manuel
Relyea, Douglas Irving
Rusling, James Francis
Savitzky, Abraham
Scherr, Allan L
Seitelman, Leon Harold
Zubal, I George

DELAWARE
Amer, Paul David
Edwards, David Owen
Lehr, Gary Fulton
Nielsen, Paul Herron
Ray, Thomas Shelby
Scofield, Dillon Foster
Ulery, Dana Lynn
Van Dyk, John William

DISTRICT OF COLUMBIA
Apple, Martin Allen
Baker, Dennis John
Casten, Richard G
Ching, Chauncey T K
Curtin, Frank Michael
Denning, Dorothy Elizabeth Robling

Dosier, Larry Waddell
Filliben, James John
Finn, Edward J
Kaplan, David Jeremy
Kaplan, George Harry
McLean, John Dickson
Olmer, Jane Chasnoff
Spohr, Daniel Arthur

FLORIDA
Ancker-Johnson, Betsy
Aubel, Joseph Lee
Bellenot, Steven F
Bonn, T(heodore) H(ertz)
Callan, Edwin Joseph
Cengeloglu, Yilmaz
Clark, Kerry Bruce
Coulter, Neal Stanley
Ege, Raimund K
Garrett, James Richard
Hand, Thomas
Hosni, Yasser Ali
Hughes, Charles Edward
Johnson, Harold Hunt
Kelley, Myron Truman
Kolhoff, M(arvin) J(oseph)
Levow, Roy Bruce
Lloyd, Laurance H(enry)
Makowski, Lee
Millor, Steven Ross
Navlakha, Jainendra K
Selfridge, Ralph Gordon
Storm, Leo Eugene
Travis, Russell Burton
Veinott, Cyril G
Winton, Charles Newton
Yau, Stephen S
Young, John William

GEORGIA
Blanke, Jordan Matthew
Bridges, Alan L
Brubaker, Leonard Hathaway
Cohn, Charles Erwin
De Sa, Richard John
Fletcher, James Erving
Foley, James David
Gallaher, Lawrence Joseph
Hudson, Sigmund Nyrop
McClure, Donald Allan
Mersereau, Russell Manning
Owen, Gene Scott
Subramanian, Mani M
Tolbert, Laren Malcolm
Tooke, William Raymond, Jr

HAWAII
Johnson, Carl Edward
Kinariwala, Bharat K
Lindsay, Kenneth Lawson
Walker, Terry M

IDAHO
Anderegg, Doyle Edward
Mortensen, Glen Albert
Potter, David Eric

ILLINOIS
Aagaard, James S(tuart)
Al-Khafaji, Amir Wadi Nasif
Bagley, John D(aniel)
Bartlett, J Frederick
Bombeck, Charles Thomas, III
Bowles, Joseph Edward
Boyle, James Martin
Chien, Andrew Andai
Cohen, Stanley
Coplien, James O
Cornwell, Larry Wilmer
Cowell, Wayne Russell
Davis, A Douglas
DeFanti, Thomas A
Doerner, Robert Carl
Flora, Robert Henry
Gabriel, John R
Gay, Ben Douglas
Grace, Thom P
Guenther, Peter T
Gupta, Udaiprakash I
Hagstrom, Ray Theodore
Heinicke, Peter Hart
Horve, Leslie A
Johnson, Donald Elwood
Johnson, Paul Lorentz
Kerkman, Daniel Joseph
Land, Robert H
McConnell, H(oward) M(arion)
Marr, William Wei-Yi
Miller, Robert Carl
Mount, Bertha Lauritzen
Murata, Tadao
Nezrick, Frank Albert
Orthwein, W(illiam) C(oe)
Padua, David A
Phelan, James Joseph
Polychronopoulos, Constantine Dimitrius
Powers, Michael Jerome
Raffenetti, Richard Charles
Raiden, Robert Lee
Reed, Daniel A
Rickert, Neil William
Singer, Richard Alan
Stark, Henry
Stutte, Linda Gail

Thomas, Gerald H
Toppel, Bert Jack
Udler, Dmitry
Ward, Charles Eugene Willoughby
Wells, Jane Frances
Yu, Clement Tak
Yule, Herbert Phillip

INDIANA
Boyd, Donald Bradford
Bunker, Bruce Alan
Dalphin, John Francis
Harbron, Thomas Richard
Hellenthal, Ronald Allen
Helman, William Phillip
Jamieson, Leah H
Kuzel, Norbert R
Moore, Emily Allyn
Purdom, Paul W, Jr
Rego, Vernon J
Reklaitis, Gintaras Victor
Robertson, Edward L
Siegel, Howard Jay
Spafford, Eugene Howard
Uhran, John Joseph, Jr
Vigdor, Steven Elliot
Weber, Janet Crosby
Wright, Jeffery Regan
Young, Frank Hood

IOWA
Alexander, Roger Keith
Christian, Lauren L
Fleck, Arthur C
Garfield, Alan J
Mischke, Charles R(ussell)
Tannehill, John Charles

KANSAS
Culvahouse, Jack Wayne
Giri, Jagannath
Kurt, Carl Edward
Wallace, Victor Lew

KENTUCKY
Villareal, Ramiro

LOUISIANA
Andrus, Jan Frederick
Brans, Carl Henry
Chen, Peter Pin-Shan
Denny, William F
Gajendar, Nandigam
Klyce, Stephen Downing
Roquemore, Leroy
Young, Myron H(wai-Hsi)

MAINE
Gordon, Geoffrey Arthur
Guidi, John Neil
Snyder, Arnold Lee, Jr
Tucker, Allen B

MARYLAND
Abramson, Fredric David
Agresti, William W
Babcock, Anita Kathleen
Basili, Victor Robert
Behforooz, Ali
Benokraitis, Vitalius
Blum, Bruce I
Brant, Larry James
Bush, David E
Butler, Louis Peter
Chapin, Edward William, Jr
Chi, Richard Nan-Hua
Clark, Gary Edwin
Cornyn, John Joseph
Deprit, Andre A(lbert) M(aurice)
Douglas, Lloyd Evans
Evans, David L
Fortner, Brand I
Franks, David A
Goldstein, Charles M
Goldstein, Larry Joel
Graves, Harvey W(ilbur), Jr
Hammond, Charles E
Hevner, Alan Raymond
Hiatt, James Lee
Huffington, Norris J(ackson), Jr
Huneycutt, James Ernest, Jr
Jenkins, Robert Edward
Johnson, David Lee
Kahaner, David Kenneth
Kahn, Arthur B
Keenan, Thomas Aquinas
Kornman, Brent D
Kowalski, Richard
Lakein, Richard Bruce
Leach, Ronald
Lidtke, Doris Keefe
McCarn, Davis Barton
MacDonald, William
McNaughton, James Larry
May, Everette Lee, Jr
Paull, Kenneth Dywain
Penhollow, John O
Perry, Peter M
Pisacane, Vincent L
Poltorak, Andrew Stephen
Reeker, Larry Henry
Remondini, David Joseph
Rombach, Hans Dieter
Salwin, Arthur Elliott

Sammet, Jean E
Schlesinger, Judith Diane
Sidhu, Deepinder Pal
Singhal, Jaya Asthana
Smith, Richard Lloyd
South, Hugh Miles
Stickler, Mitchell Gene
Stillman, Rona Barbara
Tissue, Eric Bruce
Tompkins, Robert Charles
Tripathi, Satish K
Weir, Edward Earl, II
Wende, Charles David
Whitmore, Bradley Charles
Zelkowitz, Marvin Victor

MASSACHUSETTS
Abrahams, Paul W
Adrion, William Richards
Aronson, James Ries
Barlas, Julie S
Belady, Laszlo Antal
Bertera, James H
Bloomfield, David Peter
Bruce, Kim Barry
Celmaster, William Noah
Champine, George Allen
Chang, Robin
Cheatham, Thomas Edward, Jr
Chew, Frances Sze-Ling
Clarke, Lori A
Cohen, Howard David
Cooperman, Gene David
Cullinane, Thomas Paul
Dhar, Sachidulal
Dyer, Charles Austen
Eckhouse, Richard Henry
Evans, Thomas George
Figueras, John
Fitzgerald, Marie Anton
Floyd, William Beckwith
Foster, Caxton Croxford
Giuliano, Vincent E
Graham, Robert Montrose
Griffin, William G
Hahn, Robert S(impson)
Halperin-Maya, Miriam Patricia
Heitman, Richard Edgar
Hofstetter, Edward
Hollister, Charlotte Ann
Janowitz, Melvin Fiva
King, William Connor
Knighten, Robert Lee
Lees, David Eric Berman
Lin, Alice Lee Lan
Long, Alan K
Lukas, Joan Donaldson
McConnell, Robert Kendall
MacDougall, Edward Bruce
Meal, Janet Hawkins
Mooers, Calvin Northrup
Moss, J Eliot B
Newman, Kenneth Wilfred
O'Neil, Elizabeth Jean
O'Neil, Patrick Eugene
Pope, Mary E
Ransil, Bernard J(erome)
Reinschmidt, Kenneth F(rank)
Ross, Douglas Taylor
Ross, Edward William, Jr
Royston, Richard John
Rubin, Allen Gershon
Salveter, Sharon Caroline
Salzberg, Betty
Scheff, Benson H(offman)
Schomer, Donald Lee
Sciore, Edward
Seibel, Frederick Truman
Shaffer, Harry Leonard
Shorb, Alan McKean
Simovici, Dan
Smith, Raoul Normand
Sneddon, Leigh
Snow, Beatrice Lee
Stewart, Lawrence Colm
Strecker, William D
Theobald, Charles Edwin, Jr
Tompkins, Howard E(dward)
Toscano, William Michael
Vidale, Richard F(rancis)
Wand, Mitchell
Waters, Richard C(abot)
Waud, Douglas Russell
Whitney, Cynthia Kolb
Williams, James G
Yonda, Alfred William

MICHIGAN
Bergmann, Dietrich R(udolf)
Bickart, Theodore Albert
Bookstein, Fred Leon
Campbell, Wilbur Harold
Cantor, David Milton
Caron, E(dgar) Louis
Chen, Min-Shih
Clauer, C Robert, Jr
Duchamp, David James
Eagen, Charles Frederick
Estabrook, George Frederick
Freedman, M(orris) David
Grosky, William Irvin
Herman, John Edward
Jacoby, Ronald Lee
Killgoar, Paul Charles, Jr

Software Systems (cont)

Knoll, Alan Howard
Kootsey, Joseph Mailen
Ku, Albert B
Lenker, Susan Stamm
Lustgarten, Ronald Krisses
Metzler, Carl Maust
Rajlich, Vaclav Thomas
Reynolds, Robert Gene
Rosen, Jeffrey Kenneth
Rundensteiner, Elke Angelika
Skaff, Michael Samuel
Teichroew, Daniel
Tihansky, Diane Rice
Tsao, Nai-Kuan
Tuchinsky, Philip Martin
Turley, June Williams
Wasielewski, Paul Francis
Wims, Andrew Montgomery

MINNESOTA
Conlin, Bernard Joseph
Douglas, William Hugh
Du, David Hung-Chang
Fan, David P
Franta, William Roy
Gini, Maria Luigia
Kain, Richard Yerkes
Magnuson, Vincent Richard
Marshall, John Clifford
Polnaszek, Carl Francis
Pong, Ting-Chuen
Riley, Donald Ray
Schneider, G Michael
Spurrell, Francis Arthur
Thomborson, Clark D(avid)
Zitney, Stephen Edward

MISSISSIPPI
Elsherbeni, Atef Zakaria
Pearce, David Harry

MISSOURI
Ali, Syed S
Atwood, Charles LeRoy
Baer, Robert W
Ball, William E(rnest)
Cannon, John Francis
Koederitz, Leonard Frederick
McFarland, William D
Magill, Robert Earle
Moffatt, David John
Morgan, Nancy H
Rouse, Robert Arthur
Smith, William R
Styron, Clarence Edward, Jr
Zobrist, George W(inston)

NEBRASKA
Jenkins, Thomas Gordon
Scudder, Jeffrey Eric

NEW HAMPSHIRE
Brender, Ronald Franklin
Farag, Ihab Hanna
Frantz, Daniel Raymond
Gagliardi, Ugo Oscar
Linnell, Richard D(ean)
MacGillivray, Jeffrey Charles
McGinness, James E
Misra, Alok C
Tourgee, Ronald Alan

NEW JERSEY
Afshar, Siroos K
Andose, Joseph D
Appel, Andrew Wilson
Barnes, Derek A
Becker, Richard Alan
Belanger, David Gerald
Bergeron, Robert F(rancis) (Terry), Jr
Bernhardt, Ernest C(arl)
Bernstein, Lawrence
Bhagat, Phiroz Maneck
Bouboulis, Constantine Joseph
Burkhardt, Kenneth J
Catanese, Carmen Anthony
Chambers, John McKinley
Crane, Roger L
Delaney, Edward Joseph
Dobkin, David Paul
Duttweiler, Donald Lars
Elder, John Philip
Feiner, Alexander
Fischell, David R
Foregger, Thomas H
French, Larry J
Gabbe, John Daniel
Gargaro, Anthony
Gewirtz, Allan
Ginsberg, Barry Howard
Gittleson, Stephen Mark
Golin, Stuart
Halemane, Thirumala Raya
Henrich, Christopher John
Horng, Shi-Jinn
Huber, Richard V
Hulse, Russell Alan
Jeck, Richard Kahr
Kahrs, Mark William
Kauffman, Ellwood
Kelsey, Edward Joseph
Kennedy, Anthony John

Kim, Uing W
Kobrin, Robert Jay
Krzyzanowski, Paul
Lechleider, J W
Lee, Keun Myung
McGuinness, Deborah Louise
Maleeny, Robert Timothy
Manganaro, James Lawrence
Marcus, Robert Troy
Marscher, William Donnelly
Michelson, Leslie Paul
Mondal, Kalyan
Musa, John D
Nelson, Gregory Victor
Nonnenmann, Uwe
Ordille, Carol Maria
Polhemus, Neil W
Popper, Robert David
Prekopa, Andras
Reiman, Allan H
Robins, Jack
Robrock, Richard Barker, II
Rosin, Robert Fisher
Roy, Herbert C
Russell, Frederick A(rthur)
Sams, Burnett Henry, III
Saniee, Iraj
Schindler, Max J
Schivell, John Francis
Scholz, Lawrence Charles
Sears, Raymond Warrick, Jr
Shaer, Norman Robert
Shipley, Edward Nicholas
Silberschatz, Abraham
Sinclair, Brett Jason
Strother, J(ohn) A(lan)
Thompson, Kenneth Lane
Tyson, J Anthony
Van Wyk, Christopher John
Weinstein, Stephen B
Weintraub, Herschel Jonathan R
Weiss, C Dennis
Woo, Nam-Sung
Wright, Margaret Hagen
Zabusky, Norman J
Zwass, Vladimir

NEW MEXICO
Bergeron, Kenneth Donald
Boland, W Robert
Brainerd, Walter Scott
Butler, Harold S
Campbell, John Raymond
Cobb, Loren
Diel, Joseph Henry
Emigh, Charles Robert
Forslund, David Wallace
Giesler, Gregg Carl
Godfrey, Thomas Nigel King
Hartman, James Keith
Howell, Jo Ann Shaw
Hurd, Jon Rickey
Jones, Eric Daniel
Lambert, Howard W
Lilley, John Richard
Marks, Thomas, Jr
Mason, Rodney Jackson
Miller, Alan R(obert)
Nevitt, Thomas D
Phister, Montgomery, Jr
Poore, Emery Ray Vaughn
Reinfelds, Juris
Rypka, Eugene Weston
Sicilian, James Michael
Smith, Brian Thomas
Sung, Andrew Hsi-Lin
Williams, Robert Allen
Wood, John Herbert

NEW YORK
Agarwal, Ramesh Chandra
Anastassiou, Dimitris
Archibald, Julius A, Jr
Arin, Kemal
Augenstein, Moshe
Ball, George William
Barnett, Michael Peter
Bell, Duncan Hadley
Bernstein, Herbert Jacob
Boeck, William Louis
Bourbakis, Niklaos G
Bourne, Philip Eric
Brantley, William Cain, Jr
Brinch-Hansen, Per
Brunelle, Eugene John, Jr
Bunch, Phillip Carter
Burstein, Samuel Z
Chan, Tat-Hung
Cline, Harvey Ellis
Cole, James A
Conway, Richard Walter
Cornacchio, Joseph V(incent)
Coutchie, Pamela Ann
Curtis, Ronald S
Dasgupta, Gautam
Davenport, Lesley
De Lillo, Nicholas Joseph
Diebold, John Brock
Douglas, Craig Carl
Duek, Eduardo Enrique
Esch, Louis James
Fenrich, Richard Karl
Flaherty, Joseph E
Folts, Dwight David

Galatianos, Anthony Athanassios
Goldberg, Conrad Stewart
Gottlieb, Allan
Greco, William Robert
Gustavson, Fred Gehrung
Hannay, David G
Harrington, Steven Jay
Hartwig, Curtis P
Heppa, Douglas Van
Herbert, Marc L
Huddleston, John Vincent
Jaffe, Morry
Jain, Duli Chandra
Janak, James Francis
Jones, Leonidas John
Kaltofen, Erich L
Karnaugh, Maurice
Kiphart, Kerry
Klerer, Melvin
Koplowitz, Jack
Kotchoubey, Andrew
Lamster, Hal B
La Tourrette, James Thomas
Leith, ArDean
Levine, Stephen Alan
Lichtig, Leo Kenneth
Litke, John David
Liu, Chamond
Loach, Kenneth William
Macaluso, Pat
Maguire, Gerald Quentin, Jr
Marschke, Stephen Frank
Merritt, Susan Mary
Merten, Alan Gilbert
Monmonier, Mark
Morgenstern, Matthew
Morse, William M
Mosteller, Henry Walter
Pifko, Allan Bert
Pingali, Keshav Kumar
Ramaley, James Francis
Rauscher, Tomlinson Gene
Rich, Kenneth Eugene
Robinson, Harry
Rohlf, F James
Rosenkrantz, Daniel J
Ruston, Henry
Saunders, Burt A
Sayre, David
Schwartz, Jacob Theodore
Sevian, Walter Andrew
Sherman, Philip Martin
Shooman, Martin L
Sinclair, Douglas C
Spang, H Austin, III
Spencer, David R
Srihari, Sargur N
Stenzel, Wolfram G
Strand, Richard Carl
Szymanski, Boleslaw K
Tatarczuk, Joseph Richard
Thomas, Telfer Lawson
Tracz, Will
Tuli, Jagdish Kumar
Tycko, Daniel H
Wilson, Jack Martin
Wittie, Larry Dawson
Wong, Edward Chor-Cheung
Worthing, Jurgen

NORTH CAROLINA
Chen, Su-shing
Conners, Carmen Keith
Ferentz, Melvin
Gould, Christopher Robert
Gray, James P
Higgins, Robert H
Javel, Eric
Johnson, G Allan
Latch, Dana May
Macon, Nathaniel
Reif, John H
Simon, Sheridan Alan
Smith-Thomas, Barbara
Sussenguth, Edward H
Sutcliffe, William Humphrey, Jr
Wallingford, John Stuart
Weiner, Daniel Lee
Wright, William V(aughn)
Yarger, William E

NORTH DAKOTA
Magel, Kenneth I
Winrich, Lonny B

OHIO
Adeli, Hojjat
Brown, John Boyer
Buoni, John J
Chamis, Alice Yanosko
Chapple, Paul James
Chiu, Victor
Christensen, James Henry
Craig, Rachael Grace
Dehne, George Clark
Giamati, Charles C, Jr
Greene, David C
Hall, Franklin Robert
Hartrum, Thomas Charles
Hollingsworth, Ralph George
Holtman, Mark Steven
Huff, Norman Thomas
Hulley, Clair Montrose
Ismail, Amin Rashid

Johnson, Robert Oscar
Juang, Ling Ling
Kerr, Douglas S
Klopman, Gilles
Lake, Robin Benjamin
Lipinsky, Edward Solomon
Mamrak, Sandra Ann
Mathis, Robert Fletcher
Mills, Wendell Holmes, Jr
Neff, Raymond Kenneth
Oliver, Joel Day
Peck, Lyman Colt
Petrarca, Anthony Edward
Potoczny, Henry Basil
Rose, Charles William
Roth, Mark A
Rounsley, Robert R(ichard)
Santos, Eugene (Sy)
Shick, Philip E(dwin)
Slotterbeck-Baker, Oberta Ann
Stansloski, Donald Wayne
Sterling, Leon Samuel
Waller, Michael Holton
Wang, Paul Shyh-Horng
Waren, Allan D(avid)
Winslow, Leon E
Zweben, Stuart Harvey

OKLAHOMA
Albright, James Curtice
Anderson, Paul Dean
Fisher, Donald D
Frame, Harlan D
Graves, Toby Robert
Grigsby, Ronald Davis
Hamming, Mynard C
Jennemann, Vincent Francis
Walker, Billy Kenneth
Zetik, Donald Frank

OREGON
Anderson, Tera Lougenia
Douglas, Sarah A
Ecklund, Earl Frank, Jr
Freiling, Michael Joseph
Hamlet, Richard Graham
Schmisseur, Wilson Edward
Schwartz, James William
Shapiro, Leonard David
Simpson, James Edward
Stewart, Bradley Clayton
Urquhart, N Scott
Wickes, William Castles

PENNSYLVANIA
Aburdene, Maurice Felix
Altschuler, Martin David
Ambs, William Joseph
Arrington, Wendell S
Bamberger, Judy
Barbacci, Mario R
Beck, Robert Edward
Beizer, Boris
Bowlden, Henry James
Bradley, James Henry Stobart
Carr, John Weber, III
Cendes, Zoltan Joseph
Chang, C Hsiung
Confer, Gregory Lee
Cupper, Robert
Detwiler, John Stephen
Druffel, Larry Edward
Elder, James Franklin, Jr
Ellison, Frank Oscar
Ermentrout, George Bard
Ertekin, Turgay
Fan, Ningping
Fenves, Steven J(oseph)
Finin, Timothy Wilking
Frederick, David Eugene
Gottlieb, Frederick Jay
Greenfield, Roy Jay
Haghighat, Alireza
Harbison, S P
Ho, Thomas Inn Min
Humphrey, Watts S
Hurwitz, Jan Krosst
Hyatt, Robert Monroe
Kaslow, David Edward
Kelemen, Charles F
Kleban, Morton H
Kline, Richard William
Korsh, James F
Larky, Arthur I(rving)
Lee, Insup
Lustbader, Edward David
Martin, Aaron J
Monsimer, Harold Gene
Murgie, Samuel A
Nash, David Henry George
Pierson, Ellery Merwin
Pirnot, Thomas Leonard
Reigel, Earl William
Ross, Bradley Alfred
Ryan, Thomas Arthur, Jr
Satyanarayanan, Mahadev
Scandura, Joseph M
Scranton, Bruce Edward
Shaw, Mary M
Sommer, Holger Thomas
Szepesi, Zoltan Paul John
Tackie, Michael N
Treu, Siegfried
Vastola, Francis J

Warme, Paul Kenneth
Watson, William Martin, Jr
Wiley, Samuel J
Wilkins, Raymond Leslie
Young, Frederick J(ohn)

RHODE ISLAND
Doeppner, Thomas Walter, Jr
Gabriel, Richard Francis
Kawaters, Woody H
Rossi, Joseph Stephen
Vitter, Jeffrey Scott

SOUTH CAROLINA
Cannon, Robert L
Dworjanyn, Lee O(leh)
Goode, Scott Roy
Gregory, Michael Vladimir
Lambert, Jerry Roy
Schwartz, Arnold Edward

SOUTH DAKOTA
Reuter, William L(ee)

TENNESSEE
Bailes, Gordon Lee
Bronzini, Michael Stephen
Campbell, Warren Elwood
Dobosy, Ronald Joseph
Gray, William Harvey
Harrison, Robert Edwin
Hayter, John Bingley
Iskander, Shafik Kamel
Lane, Eric Trent
Lenhert, P Galen
Martin, Harry Lee
Ordman, Edward Thorne
Patterson, Malcolm Robert
Petersen, Harold, Jr
Pribor, Hugo C
Raridon, Richard Jay
Reister, David B(ryan)
Rivas, Marian Lucy
Rome, James Alan
Sayer, Royce Orlando
Schach, Stephen Ronald
Sincovec, Richard Frank
Sworski, Thomas John
Trivedi, Mohan Manubhai
Trowbridge, Lee Douglas
Wheeler, Orville Eugene
Wohlfort, Sam Willis

TEXAS
Aberth, Oliver George
Abraham, Jacob A
Bare, Charles L
Barstow, David Robbins
Beck, John Robert
Bessman, Joel David
Brakefield, James Charles
Cameron, Bruce Francis
Campbell, Charles Edgar
Chakravarty, Indranil
Chen, Peter
Clingman, William Herbert, Jr
Czerwinski, Edmund William
Dolling, David Stanley
Ewell, James John, Jr
French, Robert Leonard
Frenger, Paul F
Garner, Harvey L(ouis)
Gentle, James Eddie
Girou, Michael L
Glaspie, Donald Lee
Gonzalez, Mario J
Haenni, David Richard
Hanne, John R
Hasling, Jill Freeman
Hedstrom, John Richard
Heinze, William Daniel
Hennessey, Audrey Kathleen
Katsnelson, Lev Z
Kennedy, Ken
Kingsley, Henry A(delbert)
Knapp, Roger Dale
Kochhar, Rajindar Kumar
Konstam, Aaron Harry
Leiss, Ernst L
Leventhal, Stephen Henry
Levinson, Stuart Alan
Lurix, Paul Leslie, Jr
Markenscoff, Pauline
Massell, Wulf F
Midgley, James Eardley
Mistree, Farrokh
Moore, Thomas Matthew
Mortada, Mohamed
Nachlinger, R Ray
Nather, Roy Edward
Novak, Gordon Shaw, Jr
Palmeira, Ricardo Antonio Ribeiro
Parberry, Ian
Peterson, Lynn Louise Meister
Potts, Mark John
Ransom, C J
Reeder, Charles Edgar
Rubin, Richard Mark
Ruchelman, Maryon W
Seth, Mohan S
Shen, Chin-Wen
Shilstone, James Maxwell, Jr
Smith, Michael Kavanagh
Smith, Reid Garfield

Tibbals, Harry Fred, III
Vinson, David Berwick
Volz, Richard A
Ward, Phillip Wayne
Woodward, Joe William

UTAH
Embley, David Wayne
Herrin, Charles Selby
Hoggan, Daniel Hunter
Hollaar, Lee Allen
Karren, Kenneth W
Stokes, Gordon Ellis
Stonehocker, Garth Hill

VIRGINIA
Bingham, Billy Elias
Brenner, Alfred Ephraim
Calm, James M
Cherniavsky, John Charles
Cook, Robert Patterson
Dendrou, Stergios
Fitts, Richard Earl
Friedman, Fred Jay
Garcia, Albert B
Gardenier, John Stark
Gowdy, Robert Henry
Hall, Otis F
Hartung, Homer Arthur
Hartzler, Alfred James
Henry, Neil Wylie
Jacob, Robert J(oseph) K(assel)
Kelly, Michael David
Liceaga, Carlos Arturo
Lipner, Steven Barnett
McGean, Thomas J
McGrath, Arthur Kevin
Mata-Toledo, Ramon Alberto
Montague, Stephen
Nance, Richard E
Orcutt, Bruce Call
Pitts, Thomas Griffin
Pratt, Terrence Wendall
Reierson, James (Dutton)
Richardson, Henry Russell
Rine, David C
Sage, Andrew Patrick
Saltz, Joel Haskin
Shetler, Antoinette (Toni)
Sjogren, Robert W, Jr
Van Tilborg, André Marcel
Walker, Robert Paul
Yang, Ta-Lun

WASHINGTON
Benson, David Bernard
Bolme, Mark W
Britt, Patricia Marie
Carter, John Lemuel, Jr
Clark, William Greer
Golde, Hellmut
Henry, Robert R
Hinthorne, James Roscoe
Kerlick, George David
Korn, Granino A(rthur)
Kouzes, Richard Thomas
Larrabee, Allan Roger
Lazowska, Edward Delano
MacKichan, Barry Bruce
MacLaren, Malcolm Donald
Malone, Stephen D
Mobley, Curtis Dale
Nickerson, Robert Fletcher
Padilla, Andrew, Jr
Person, James Carl
Rehfield, David Michael
Sargent, Murray, III
Savol, Andrej Martin
Sheridan, John Roger
Smith, Burton Jordan
Spillman, Richard Jay
Tripp, Leonard L
Walker, Sharon Leslie
Wither, Ross Plummer

WEST VIRGINIA
Fisher, John F
Ghigo, Frank Dunnington

WISCONSIN
Bates, Douglas Martin
Chi, Che
Clark, Harlan Eugene
Dalrymple, David Lawrence
Davida, George I
Fossum, Timothy V
Gelatt, Charles Daniel, Jr
Harris, J Douglas
Heinen, James Albin
Hinze, Harry Clifford
Jacobi, George (Thomas)
Kao, Wen-Hong
Ramakrishnan, Raghu
Rosenthal, Jeffrey
Solomon, Marvin H

WYOMING
Bauer, Henry Raymond, III
Eddy, David Maxon
Tauer, Jane E

PUERTO RICO
Peinado, Rolando E

ALBERTA
Klobukowski, Mariusz Andrzej
Sorenson, Paul G

BRITISH COLUMBIA
Booth, Andrew Donald
Dill, John C
Hafer, Louis James
Marshall, Peter Lawrence
Morbey, Christopher Leon
Wade, Adrian Paul
Zhang, Ziyang

MANITOBA
Perkins, Harold Jackson

NEW BRUNSWICK
Lee, Y C
Vanicek, Petr

NEWFOUNDLAND
Rahman, Md Azizur
Shieh, John Shunen

NOVA SCOTIA
El-Hawary, Mohamed El-Aref
Kruse, Robert Leroy
Oliver, Leslie Howard
Rautaharju, Pentti M

ONTARIO
Argyropoulos, Stavros Andreas
Bauer, Michael Anthony
Blackwell, Alan Trevor
Dastur, Ardeshir Rustom
Emraghy, Hoda Abdel-Kader
Farrar, John Keith
Gentleman, William Morven
Jarvis, Roger George
Jeejeebhoy, Khursheed Nowrojee
Nash, John Christopher
Newnham, Robert Montague
Ogata, Hisashi
Okashimo, Katsumi
Roe, Peter Hugh O'Neil
Salvadori, Antonio
Slonim, Jacob
Taylor, David James
Tompa, Frank William
Waddington, Raymond

QUEBEC
Alagar, Vangalur S
Atwood, John William
Beron, Patrick
Chaubey, Yogendra Prasad
De Mori, Renato
Desai, Bipin C
Fabrikant, Valery Isaak
Opatrny, Jaroslav
Seldin, Jonathan Paul
Tao, Lixiu
Tomas, Francisco

SASKATCHEWAN
Lapp, Martin Stanley
Symes, Lawrence Richard

OTHER COUNTRIES
Bywater, Anthony Colin
Chanson, Samuel T
Collins, George Edwin
Czepyha, Chester George Reinhold
Davis, Edward Alex
Goguen, Joseph A
Humphrey, Albert S
Maas, Peter
Nievergelt, Jurg
Revesz, Zsolt
Shafi, Mohammad
Shen, Vincent Y
Siu, Tsunpui Oswald
Thorson, John Wells

Theory

ALABAMA
Dean, Susan Thorpe
Seidman, Stephen Benjamin

ALASKA
Lando, Barbara Ann

ARIZONA
Maier, Robert S

ARKANSAS
Talburt, John Randolph

CALIFORNIA
Backus, John
Bull, Everett L, Jr
Butler, Jon Terry
Carter, J Lawrence (Larry)
Drobot, Vladimir
Emerson, Thomas James
Enderton, Herbert Bruce
Fagin, Ronald
Fraser, Grant Adam
Fredman, Michael Lawrence
Ginsburg, Seymour
Greibach, Sheila Adele
Hankins, Hesterly G, III
Karp, Richard M
Korchev, Dmitriy Veniaminovich
Lawler, Eugene L(eighton)
Lelewer, Debra Ann
Lomax, Harvard
Manna, Zohar
Meagher, Donald Joseph
Melkanoff, Michel Allan
Mitchell, John Clifford
Monier, Louis Marcel
Moschovakis, Yiannis N
Motteler, Zane Clinton
Murphy, Roy Emerson
Nassi, Isaac Robert
Nico, William Raymond
Pearl, Judea
Pease, Marshall Carleton, III
Pixley, Alden F
Pratt, Vaughan Ronald
Simons, Barbara Bluestein
Spanier, Edwin Henry
Stockmeyer, Larry Joseph
Tilson, Bret Ransom
Weeks, Dennis Alan

COLORADO
Edmundson, Harold Parkins
Manvel, Bennet

CONNECTICUT
Engel, Gerald Lawrence
Hajela, Dan
Hudak, Paul Raymond
Nilson, Edwin Norman
Rao, Valluru Bhavanarayana

DELAWARE
Case, John William
Caviness, Bobby Forrester

DISTRICT OF COLUMBIA
Della Torre, Edward
Friedman, Arthur Daniel

FLORIDA
Cenzer, Douglas
Hughes, Charles Edward
Levitz, Hilbert
Navlakha, Jainendra K
Pedersen, John F
Roy, Dev Kumar
Sahni, Sartaj Kumar
Stark, William Richard

GEORGIA
Belinfante, Johan G F
Pomerance, Carl

ILLINOIS
Chang, Shi Kuo
Grace, Thom P
Gray, John Walker
Gupta, Udaiprakash I
Kalai, Ehud
Larson, Richard Gustavus
Lee, Der-Tsai
Livingston, Marilyn Laurene
Loui, Michael Conrad
Packel, Edward Wesler
Reingold, Edward Martin
Tier, Charles
Wallis, Walter Denis
Wells, Jane Frances
West, Douglas Brent

INDIANA
Atallah, Mikhail Jibrayil
Drufenbrock, Diane
Frederickson, Greg N
Fuelling, Clinton Paul
Gersting, Judith Lee
Purdom, Paul W, Jr
Robertson, Edward L
Szpankowski, Wojciech
Wise, David Stephen
Young, Frank Hood

IOWA
Alton, Donald Alvin

KANSAS
Strecker, George Edison

KENTUCKY
Lewis, Forbes Downer

LOUISIANA
Burke, Robert Wayne
Denny, William F
Kak, Subhash Chandra
Roquemore, Leroy

MAINE
Tucker, Allen B

MARYLAND
Atanasoff, John Vincent
Behforooz, Ali
Gasarch, William Jan
Jablonski, Daniel Gary
Kosaraju, S Rao
Miller, Raymond Edward
Nau, Dana S

Theory (cont)

Rombach, Hans Dieter
Sadowsky, John

MASSACHUSETTS
Adrion, William Richards
Albertson, Michael Owen
Bruce, Kim Barry
Chang, Robin
Cooperman, Gene David
D'Alarcao, Hugo T
Doyle, Jon
Dyer, Charles Austen
Elias, Peter
Faber, Richard Leon
Gacs, Peter
Gessel, Ira Martin
Homer, Steven Elliott
Joni, Saj-nicole A
Levesque, Allen Henry
Levin, Leonid A
O'Rourke, Joseph
Ostrovsky, Rafail M
Rogers, Hartley, Jr
Rosenberg, Arnold Leonard
Ross, Douglas Taylor
Sciore, Edward
Simovici, Dan
Stonier, Tom Ted
Wand, Mitchell

MICHIGAN
Bachelis, Gregory Frank
Hinman, Peter Greayer
Malm, Donald E G
Rajlich, Vaclav Thomas
Stout, Quentin Fielden
Tihansky, Diane Rice
Windeknecht, Thomas George

MINNESOTA
Pierce, Keith Robert
Pour-El, Marian Boykan
Seebach, J Arthur, Jr
Thomborson, Clark D(avid)

MISSOURI
Ali, Syed S
Bruening, James Theodore
Nichols, Robert Ted
Peterson, Gerald E

MONTANA
Wright, Alden Halbert

NEBRASKA
Leung, Joseph Yuk-Tong
Magliveras, Spyros Simos

NEVADA
Minor, John Threecivelous

NEW HAMPSHIRE
Bent, Samuel W
Bogart, Kenneth Paul

NEW JERSEY
Abeles, Francine
Afshar, Siroos K
Allender, Eric Warren
Amarel, S(aul)
Boesch, Francis Theodore
Brown, William Stanley
Dobkin, David Paul
Garey, Michael Randolph
Holzmann, Gerard J
Kirch, Murray R
Lagarias, Jeffrey Clark
Sontag, Eduardo Daniel
Steiger, William Lee
Tindell, Ralph S
Van Wyk, Christopher John
Winkler, Peter Mann

NEW MEXICO
Harary, Frank
Sung, Andrew Hsi-Lin

NEW YORK
Anderson, Peter Gordon
Ball, George William
Bencsath, Katalin A
Chan, Tat-Hung
Cohen, Daniel Isaac Aryeh
Constable, Robert L
Coppersmith, Don
De Lillo, Nicholas Joseph
Di Paola, Robert Arnold
Goldfarb, Donald
Goodman, Jacob Eli
Gross, Jonathan Light
Hannay, David G
Hopcroft, John E(dward)
Inselberg, Alfred
Kaltofen, Erich L
Kershenbaum, Aaron
Klir, George Jiri
McConnell, Jeffrey Joseph
Merritt, Susan Mary
Mitchell, Joseph Shannon Baird
Nerode, Anil
Pan, Victor
Parikh, Rohit Jivanlal

Rosenkrantz, Daniel J
Selman, Alan L
Shasha, Dennis E
Shore, Richard A
Shub, Michael I
Sobczak, Thomas Victor
Stearns, Richard Edwin
Van Slyke, Richard M
Vasquez, Alphonse Thomas
Wagner, Eric G

NORTH CAROLINA
Blanchet-Sadri, Francine
Latch, Dana May
Loveland, Donald William
McAllister, David Franklin
Ras, Zbigniew Wieslaw
Reif, John H
Smith-Thomas, Barbara

OHIO
Friedman, Harvey Martin
Hartrum, Thomas Charles
Johnson, Robert Oscar
Potoczny, Henry Basil
Roth, Mark A
Rothstein, Jerome
Schlipf, John Stewart
Wells, Charles Frederick

OKLAHOMA
Letcher, John Henry, III
Walker, Billy Kenneth

OREGON
Brown, Cynthia Ann
Hamlet, Richard Graham

PENNSYLVANIA
Becker, Stephen Fraley
Cupper, Robert
Furst, Merrick Lee
Kelemen, Charles F
Korsh, James F
Meystel, Alexander Michael
Shamos, Michael Ian
Sieber, James Leo
Simpson, Stephen G
Wagh, Meghanad D

RHODE ISLAND
Preparata, Franco Paolo
Savage, John Edmund
Simons, Roger Alan
Vitter, Jeffrey Scott

SOUTH CAROLINA
Comer, Stephen Daniel
Hammond, Joseph Langhorne, Jr
Jamison, Robert Edward

TENNESSEE
Franklin, Stanley Phillip
Ordman, Edward Thorne

TEXAS
Abraham, Jacob A
Borm, Alfred Ervin
Fix, George Joseph
Leiss, Ernst L
Malek, Miroslaw
Matula, David William
Parberry, Ian
Ruchelman, Maryon W
Wang, Yuan R
Wolter, Jan D(ithmar)

UTAH
Campbell, Douglas Michael
Higgins, John Clayborn

VERMONT
Weed, Lawrence Leonard

VIRGINIA
Abdali, Syed Kamal
Cherniavsky, John Charles
Friedman, Fred Jay
Heath, Lenwood S
Lipner, Steven Barnett
Mata-Toledo, Ramon Alberto
Pratt, Terrence Wendall
Rine, David C
Spresser, Diane Mar
Stockmeyer, Paul Kelly

WASHINGTON
Benson, David Bernard
Kowalik, Janusz Szczesny
Somani, Arun Kumar
Storwick, Robert Martin
Young, Paul Ruel

WEST VIRGINIA
Chahryar, Hamid N

WISCONSIN
Bach, Eric
Kinsinger, Richard Estyn
Moore, Edward Forrest
Solomon, Marvin H

WYOMING
Cowles, John Richard

PUERTO RICO
Peinado, Rolando E

ONTARIO
Brzozowski, Janusz Antoni
Colbourn, Charles Joseph
Cook, Stephen Arthur
Davison, J Leslie
Dixon, John Douglas
Kent, Clement F

QUEBEC
Leroux, Pierre
Newborn, Monroe M

OTHER COUNTRIES
Andrej, Ladislav
Collins, George Edwin
Fiesler, Emile
Fujiwara, Hideo
Goguen, Joseph A
Myerson, Gerald
Nievergelt, Jurg

Other Computer Sciences

ALABAMA
Klymenko, Victor

ARIZONA
Batson, Raymond Milner
Daniel, Sam Mordochai
Dickinson, Frank N
El-Ghazaly, Samir M
Justice, Keith Evans
Lorents, Alden C
Soderblom, Laurence Albert
Zeigler, Bernard Philip

CALIFORNIA
Baba, Paul David
Bailey, Michael John
Barsky, Brian Andrew
Bernstein, Ralph
Croft, Paul Douglas
Crowell, John Marshall
Cumberland, William Glen
Dorfman, Leslie Joseph
Engel, Jan Marcin
Gailar, Owen H
Henderson, Sheri Dawn
Hotz, Henry Palmer
Keshavan, H R
Korchev, Dmitriy Veniaminovich
Langdon, Allan Bruce
Lathrop, Richard C(harles)
Ma, Joseph T
Mathews, M(ax) V(ernon)
O'Keefe, Michael Adrian
Prasanna Kumar, V K
Requa, Joseph Earl
Sangiovanni-Vincentelli, Alberto Luigi
Shevel, Wilbert Lee
Smith, Bradley Richard
Spencer, James Eugene
Thacker, Charles
Thompson, Timothy J
Willens, Ronald H

COLORADO
Burkley, Richard M
Horton, Clifford E(dward)
Lett, Gregory Scott
Olson, George Gilbert
Ralston, Margarete A
Sloss, Peter William
Stanger, Andrew L

CONNECTICUT
Engel, Gerald Lawrence
Frankel, Martin Richard
Keyes, David Elliot
Mandelbrot, Benoit B
Nelson, Roger Peter
Ortner, Mary Joanne

DELAWARE
Klein, Michael Tully
Ray, Thomas Shelby

DISTRICT OF COLUMBIA
April, Robert Wayne
Denning, Dorothy Elizabeth Robling
Hawkins, Morris, Jr
James, Alton Everette, Jr
Olmer, Jane Chasnoff

FLORIDA
Chartrand, Robert Lee
Fischler, Drake Anthony
Hand, Thomas
Jungbauer, Mary Ann
Levow, Roy Bruce
Mount, Gary Arthur
Naini, Majid M
Peterson, Ernest A
Rikvold, Per Arne
Sigmon, Kermit Neal
Starke, Albert Carl, Jr
Wu, Jin Zhong

GEORGIA
Cohn, Charles Erwin

Foley, James David
Hill, David W(illiam)
Husson, Samir S
Prince, M(orris) David
Zunde, Pranas

ILLINOIS
Chang, Shi Kuo
Grace, Thom P
Koenig, Michael Edward Davison
Kubitz, William John
O'Morchoe, Patricia Jean
Saied, Faisal
Saylor, Paul Edward
Skeel, Robert David
Wilke, Robert Nielsen

INDIANA
Harbron, Thomas Richard
Markee, Katherine Madigan
Rego, Vernon J
Spicer, Donald Z
Wasserman, Gerald Steward
Wise, David Stephen
Young, Frank Hood

IOWA
Burrill, Claude Wesley
Coy, Daniel Charles
Garfield, Alan J

KANSAS
Gordon, Michael Andrew
Lenhert, Donald H
Unger, Elizabeth Ann

MAINE
Carter, William Caswell

MARYLAND
Adams, James Alan
Belliveau, Louis J
Boggs, Paul Thomas
Boisvert, Ronald Fernand
Chapin, Douglas Scott
Craig, Paul N
Demmerle, Alan Michael
Draper, Richard Noel
Eisner, Howard
Ephremides, Anthony
Fowler, Howland Auchincloss
Gary, Robert
Goldin, Edwin
Heilprin, Laurence Bedford
Hobbs, Robert Wesley
Lepley, Arthur Ray
Levine, Elissa Robin
Liverman, James Leslie
Marotta, Charles Rocco
Marx, Egon
Newman, David Bruce, Jr
Pallett, David Stephen
Pattabiraman, Nagarajan
Saltman, Roy G
Schlesinger, Judith Diane
Smith, James Alan
Stuelpnagel, John Clay
Sullivan, Francis E
Turnrose, Barry Edmund
Valley, Sharon Louise
White, Eugene L

MASSACHUSETTS
Berera, Geetha Poonacha
Berkovits, Shimshon
Brenner, John Francis
Chang, Ching Shung
Cooperman, Gene David
Crossman, David W
Flowers, Woodie C
Graham, Robert Montrose
Hartline, Peter Haldan
Hershberg, Philip I
Heyerdahl, Eugene Gerhardt
Hoffman, David Allen
Marinaccio, Paul J
Mehra, Raman Kumar
Mesirov, Jill Portner
Murch, Laurence Everett
Olivo, Richard Francis
Pample, Roland D
Read, Philip Lloyd
Richter, Stephen L(awrence)
Royston, Richard John
St Mary, Donald Frank
Sherman, Thomas Oakley
Winfrey, Richard Cameron

MICHIGAN
Cullati, Arthur G
Field, David Anthony
LePage, Raoul
Levine, Simon P
Marhsall, Hal G
Roc, Philip Lawrence
Tsao, Nai-Kuan
Verhey, Roger Frank
Zamorano, Lucia Jopehina

MINNESOTA
Franta, William Roy
Rosen, Judah Ben

MISSISSIPPI
Andrews, Gordon Louis

MISSOURI
Honnell, Pierre M(arcel)
Hoover, Loretta White
Rogic, Milorad Mihailo
Wagner, Robert G

NEBRASKA
Walters, Randall Keith

NEW HAMPSHIRE
Black, Sydney D
Gagliardi, Ugo Oscar
Kurtz, Thomas Eugene

NEW JERSEY
Cohen, George Lester
Dobkin, David Paul
Franks, Richard Lee
Gear, Charles William
Goffman, Martin
Halpern, Donald F
Hut, Piet
Jacquin, Arnaud Eric
Kalley, Gordon S
Letcher, David Wayne
Lucas, Henry C, Jr
McGuinness, Deborah Louise
Quinlan, Daniel A
Randall, Karen T
Schrenk, George L
Sinclair, Brett Jason
Stanton, Robert E
Weiss, Alan

NEW MEXICO
Alldredge, Gerald Palmer
Amann, James Francis
Burks, Christian
Finkner, Morris Dale
Fries, Ralph Jay
Gula, William Peter
Hanson, Kenneth Merrill
Herman, George
Reinfelds, Juris
Williams, Robert Allen

NEW YORK
Ames, Stanley Richard
Archibald, Julius A, Jr
Augenstein, Moshe
Blumenson, Leslie Eli
Boettger, Susan D
Borowitz, Irving Julius
Bourne, Philip Eric
Braun, Ludwig
Camazine, Scott
Cok, David R
Conant, Francis Paine
Dudewicz, Edward John
Duggin, Michael J
Eydgahi, Ali Mohammadzadeh
Jedruch, Jacek
Kaplan, Martin Charles
Karron, Daniel B
Kenett, Ron
Knowles, Richard James Robert
Kriss, Michael Allen
Leith, ArDean
McConnell, Jeffrey Joseph
Maguire, Gerald Quentin, Jr
Merritt, Susan Mary
Miller, Russ
Mitchell, Joan LaVerne
Pan, Victor
Pennington, Keith Samuel
Schwartz, Mischa
Sobczak, Thomas Victor
Thomas, Richard Alan
Tuel, William Gole, Jr
Wadell, Lyle H
Walters, Deborah K W
Weinberger, Arnold
Yoffa, Ellen June
Zic, Eric A

NORTH CAROLINA
Beeler, Joe R, Jr
Fuchs, Henry
Holcomb, Charles Edward
Land, Ming Huey
Lutz, Michael W
Noe, Frances Elsie
Oblinger, Diana Gelene
Ras, Zbigniew Wieslaw
Shackelford, Walter McDonald
Stead, William Wallace
Sussenguth, Edward H
Trussell, Henry Joel
Vaughan, Douglas Stanwood
Woodbury, Max Atkin

NORTH DAKOTA
Kemper, Gene Allen
Magel, Kenneth I
Uherka, David Jerome

OHIO
Becker, Lawrence Charles
Easterday, Jack L(eroy)
Fitzmaurice, Nessan
Ghandakly, Adel Ahmad

Griffith, Walter M, Jr
Hassell, John Allen
Hern, Thomas Albert
Hollingsworth, Ralph George
Kertz, George J
Knox, Francis Stratton, III
Levin, Jerome Allen
Ockerman, Herbert W
Potoczny, Henry Basil
Siervogel, Roger M

OKLAHOMA
Argyros, Ioannis Konstantinos
Cheung, John Yan-Poon
Fisher, Donald D
McClain, Gerald Ray
Nofziger, David Lynn
Robinson, Enders Anthony

OREGON
Jaeger, Charles Wayne
Matthes, Steven Allen

PENNSYLVANIA
Bergmann, Ernest Eisenhardt
Blair, Grant Clark
Chase, Gene Barry
Green, Michael H(enry)
Grim, Larry B
Grinberg, Eric L
Harris, Lowell Dee
Hinton, Raymond Price
Jeruchim, Michel Claude
Kresh, J Yasha
Lambert, Joseph Michael
Metaxas, Dimitri
Meystel, Alexander Michael
Moss, William Wayne
Paulish, Daniel John
Santora, Norman Julian
Shar, Albert O
Szyld, Daniel Benjamin
Treu, Siegfried
Wei, Susanna

RHODE ISLAND
McConeghy, Matthew H
Savage, John Edmund

SOUTH CAROLINA
Balintfy, Joseph L
Davis, Joseph B

SOUTH DAKOTA
Helsdon, John H, Jr

TENNESSEE
Baillargeon, Victor Paul
Dongarra, Jack Joseph
Evans, Joe Smith
Garzon, Max
Sittig, Dean Forrest
Sorrells, Frank Douglas
Watson, Evelyn E

TEXAS
Crossley, Peter Anthony
Cusachs, Louis Chopin
Dukatz, Ervin L, Jr
Dyke, Bennett
Gemignani, Michael C
Gonzalez, Mario J
Penz, P Andrew
Potter, Robert Joseph
Renka, Robert Joseph
Treat, Charles Herbert
Watt, Joseph T(ee), Jr
Whittlesey, John R B

UTAH
Bowman, Carlos Morales
Cohen, Elaine
Gardner, Reed McArthur
Hoggan, Daniel Hunter
Tadjeran, Hamid

VIRGINIA
Anderson, Walter L(eonard)
Ell, William M
Field, Paul Eugene
Hamilton, Thomas Charles
Jacob, Robert J(oseph) K(assel)
Kilpatrick, S James, Jr
Lai, Chintu (Vincent C)
Latta, John Neal
Moritz, Barry Kyler
Patrick, Merrell Lee
Rossotti, Charles Ossola
Roussos, Constantine
Shetler, Antoinette (Toni)
Welt, Isaac Davidson
Wolf, Eric W

WASHINGTON
Bax, Nicholas John
Edison, Larry Alvin
Millham, Charles Blanchard

WISCONSIN
Borkovitz, Henry S
Kresch, Alan J
Rolley, Robert Ewell
Strikwerda, John Charles

BRITISH COLUMBIA
Benston, Margaret Lowe
Brantingham, Patricia Louise
Ward, Lawrence McCue

NOVA SCOTIA
Barzilai, Jonathan
Moriarty, Kevin Joseph
Schuegraf, Ernst Josef

ONTARIO
Baltacioglu, Mehmet Necip
Geddes, Keith Oliver
Gough, Sidney Roger
Jackson, Kenneth Ronald
Law, Cecil E
Waddington, Raymond
Wright, Joseph D

QUEBEC
Desai, Bipin C
Newborn, Monroe M
Opatrny, Jaroslav
Poussart, Denis

SASKATCHEWAN
Law, Alan Greenwell

OTHER COUNTRIES
Hura, Gurdeep Singh
Lehman, Meir M
Massey, James L

ENGINEERING

Aeronautical & Astronautical Engineering

ALABAMA
Anderson, Bernard Jeffrey
Bailey, J Earl
Bailey, Wayne Lewis
Blackmon, James B
Brainerd, Jerome J(ames)
Brandon, Walter Wiley, Jr
Cochran, John Euell, Jr
Cost, Thomas Lee
Cutchins, Malcolm Armstrong
Dannenberg, Konrad K
De Grandpre, Jean Louis
Doughty, Julian O
Dunn, Anne Roberts
French, Kenneth Edward
Geissler, Ernst D(ietrich)
Haeussermann, Walter
Hermann, R(udolf)
Hollub, Raymond M(athew)
Hung, Ru J
Krause, Helmut G L
Lee, Thomas J
McAuley, Van Alfon
Martin, James Arthur
Moore, Ronald Lee
Morgan, Bernard S(tanley), Jr
Moses, Ray Napoleon, Jr
Nola, Frank Joseph
Ramachandran, Narayanan
Rees, Eberhard F M
Reese, Bruce Alan
Rey, William K(enneth)
Ritter, A(lfred)
Roe, James Maurice, Jr
Sforzini, Richard Henry
Smith, Steven Patrick Decland
Turner, Charlie Daniel, Jr
Weeks, George Eliot
Willenberg, Harvey Jack
Williams, James C(lifford), III
Wu, Shi Tsan

ARIZONA
Ahmed, Saad Attia
Bluestein, Theodore
Butler, Blaine R(aymond), Jr
Chen, Chuan Fang
Dittemore, David H
Fung, K Y
Glick, Robert L
Hirleman, Edwin Daniel, Jr
Makepeace, Gershom Reynolds
Mardian, James K W
Matthews, James B
Metzger, Darryl E
Miller, Harry
O'Leary, Brian Todd
Parks, E(dwin) K(etchum)
Ramohalli, Kumar Nanjunda Rao
Rummel, Robert Wiland
Saric, William Samuel
Sears, William R(ees)
Singhal, Avinash Chandra
Swan, Peter A
Vincent, Thomas Lange
Wallace, C(harles) E(dward)
Witten, Mark Lee
Wygnanski, Israel Jerzy

ARKANSAS
Burchard, John Kenneth
Carr, Gerald Paul

CALIFORNIA
Abbott, Dean William
Abzug, M(alcolm) J
Acheson, Louis Kruzan, Jr
Albers, James Arthur
Aldridge, Edward C, Jr
Araki, Minoru S
Ardema, Mark D
Ashley, Holt
Atwood, John Leland
Bailey, Don Matthew
Baldwin, Barrett S, Jr
Ballhaus, William F(rancis)
Barlow, Edward J(oseph)
Batdorf, Samuel Burbridge
Becker, Randolph Armin
Bell, M(arion) W(etherbee) Jack
Bharadvaj, Bala Krishnan
Biblarz, Oscar
Bittner, Harlan Fletcher
Blackwelder, Ron F
Blasingame, Benjamin P(aul)
Bleviss, Zegmund O(scar)
Blumenthal, Irwin S(imeon)
Bono, Philip
Bright, Peter Bowman
Broadwell, James E(ugene)
Bromberg, R(obert)
Brown, Alan Charlton
Burden, Harvey Worth
Caflisch, Russel Edward
Cannon, Robert H, Jr
Caren, Robert Poston
Carlson, Richard M
Carman, Robert Lincoln, Jr
Casani, John R
Castle, Karen G
Chamberlain, Robert Glenn
Chambers, Robert J
Chapman, Dean R(oden)
Chapman, Gary Theodore
Charwat, Andrew F(ranciszek)
Chen, Tsai Hwa
Chen, Wen H
Chin, Jin H
Cho, Alfred Chih-Fang
Clauser, Francis Hettinger
Cohen, Clarence B(udd)
Cohen, Lawrence Mark
Coles, Donald (Earl)
Compton, Dale L(eonard)
Cooper, George Emery
Copper, John A(lan)
Currie, Malcolm R
Cutforth, Howard Glen
Damonte, John Batista
Davis, Robert A(rthur)
Daybell, Melvin Drew
Deets, Dwain Aaron
Donovan, Allen F(rancis)
Eggers, A(lfred) J(ohn), Jr
Elachi, Charles
Elliott, David Duncan
Elverum, Gerard William, Jr
Farber, Joseph
Feign, David
Feinstein, Charles David
Feldman, Nathaniel E
Field, Lester
Fishburne, Edward Stokes, III
Fleming, Alan Wayne
Forbrich, Carl A, Jr
Forsberg, Kevin
Frazier, Edward Nelson
French, Edward P(erry)
Friedmann, Peretz Peter
Frye, William Emerson
Fuhrman, Robert Alexander
Fuhs, Allen E(ugene)
Gaskell, Robert Weyand
Gat, Nahum
Gedeon, Geza S(cholcz)
Gevarter, William Bradley
Green, Allen T
Grobecker, Alan J
Gunkel, Robert James
Hackett, Colin Edwin
Haines, Richard Foster
Hall, Charles Frederick
Haloulakos, Vassilios E
Hansen, Grant Lewis
Hansmann, Douglas R
Harter, George A
Hawkins, Willis M(oore)
Hedgepeth, John M(ills)
Heinemann, Edward H
Heppe, R Richard
Hermsen, Robert W
Herrmann, George
Hess, R(onald) A(ndrew)
Hickman, Roy Scott
Hoff, N(icholas) J(ohn)
Hoffmann, Jon Arnold
Horner, Richard E(lmer)
Hsu, Chieh-Su
Hu, Steve Seng-Chiu
Hult, John Luther
Hyatt, Abraham
Iorillo, Anthony J
Jackson, Durward P
Jacobs, Michael Moises
Jeffs, George W
Jones, Robert Thomas
Judge, Roger John Richard

Aeronautical & Astronautical Engineering (cont)

Kane, Thomas R(eif)
Kaplan, Richard E
Karamcheti, K(rishnamurty)
Kayton, Myron
Kelly, Robert Edward
Kennel, John Maurice
Kim, J John
King, Hartley H(ughes)
Knuth, Eldon L(uverne)
Kosmatka, John Benedict
Kraabel, John Stanford
Kraus, Samuel
Kropp, John Leo
Kuan, Teh S
Kunc, Joseph Anthony
Kwok, Munson Arthur
Laderman, A(rnold) J(oseph)
Lai, Kai Sun
Lampert, Seymour
Landecker, Peter Bruce
Landsbaum, Ellis M(erle)
Larson, Edward William, Jr
Lathrop, Richard C(harles)
Layton, Thomas William
Lederer, Jerome F
Leibowitz, Lewis Phillip
Leite, Richard Joseph
Leitmann, G(eorge)
Libbrecht, Kenneth
Libby, Paul A(ndrews)
Liepman, H(ans) P(eter)
Lillie, Charles Frederick
Lin, Shao-Chi
Lindley, Charles A(lexander)
Lindsey, Gerald Herbert
Ling, Rung Tai
Lissaman, Peter Barry Stuart
Liu, Chung-Yen
Lomax, Harvard
Lopina, Robert F(erguson)
Lovelace, Alan Mathieson
Lowi, Alvin, Jr
Lynch, David Dexter
MacCready, Paul Beattie, Jr
McDonald, William True
McKenzie, Robert Lawrence
McLaughlin, William Irving
Macomber, Thomas Wesson
McRuer, Duane Torrance
Madan, Ram Chand
Mager, Artur
Magness, T(om) A(lan)
Manganiello, Eugene J(oseph)
Marcus, Bruce David
Martin, Elmer Dale
Maslach, George James
Maxworthy, Tony
Meecham, William Coryell
Meissinger, Hans F
Melese, Gilbert B(ernard)
Melkanoff, Michel Allan
Miatech, Gerald James
Mingori, Diamond Lewis
Mirels, Harold
Moe, Osborne Kenneth
Monkewitz, Peter Alexis
Morgan, Lucian L(loyd)
Morris, Brooks T(heron)
Munson, Albert G
Myers, Dale Dehaven
Nachtsheim, Philip Robert
Netzer, David Willis
Neu, John Ternay
Ng, Lawrence Chen-Yim
Nguyen, Caroline Phuongdung
Nicolaides, John Dudley
Nielan, Paul E
Nielsen, Helmer L(ouis)
Niles, Philip William Benjamin
Pamidi, Prabhakar Ramarao
Park, Chul
Parker, Donn Blanchard
Parkin, Blaine R(aphael)
Peeters, Randall Louis
Peterson, Victor Lowell
Pierucci, Mauro
Pinkel, B(enjamin)
Platzer, Maximilian Franz
Plotkin, Allen
Pottsepp, L(embit)
Preston, Robert Arthur
Puckett, Allen Emerson
Purcell, Everett Wayne
Putt, John Ward
Rauch, Herbert Emil
Rauch, Lawrence L(ee)
Rauch, Richard Travis
Raymond, Arthur E(mmons)
Reardon, Frederick H(enry)
Rechtin, Eberhardt
Redekopp, Larry G
Refhield, Lawrence Wilmer
Revell, James D(ewey)
Reynolds, Harry Lincoln
Rich, Ben R
Ring, Robert E
Riparbelli, Carlo
Rivers, William J(ones)
Roberts, Leonard
Rockwell, Robert Lawrence

Rodabaugh, David Joseph
Rosen, Alan
Rossen, Joel N(orman)
Rott, Nicholas
Rubin, Sheldon
Rudy, Thomas Philip
Safonov, Michael G
Savas, Omer
Sawyer, Robert Fennell
Schairer, Robert S(org)
Schalla, Clarence August
Schamberg, Richard
Schaufele, Roger Donald
Schjelderup, Hassel Charles
Schmit, Lucien A(ndre), Jr
Schneider, Alan M(ichael)
Schurmeier, Harris McIntosh
Scott, Paul Brunson
Seide, Paul
Seiff, Alvin
Sellars, John R(andolph)
Seltzer, Leon Z(ee)
Shaffar, Scott William
Shepherd, Joseph Emmett
Shiffman, Max
Shu, Frank H
Shvartz, Esar
Sirignano, William Alfonso
Smith, A(pollo) M(ilton) O(lin)
Smith, George C
Snyder, Charles Thomas
Solomon, George E
Sparling, Rebecca Hall
Spier, Edward Ellis
Spiro, Irving J
Spokoyny, Felix E
Spreiter, John R(obert)
Springer, George S
Stella, Paul M
Stevenson, Robin
Stewart, H(omer) J(oseph)
Stone, Howard N(ordas)
Stoolman, Leo
Synolakis, Costas Emmanuel
Syvertson, Clarence A
Tang, Homer H(o)
Tang, Stephen Shien-Pu
Thelen, Charles John
Thompson, Milton Orville
Thompson, W P(aul)
Throner, Guy Charles
Trubert, Marc
Tubbs, Eldred Frank
Turchan, Otto Charles
Unal, Aynur
Vajk, J(oseph) Peter
Vanderplaats, Garret Niel
Victor, Andrew C
Vincenti, Walter G(uido)
Vinokur, Marcel
Vlay, George John
Waaland, Irving T
Wagner, Raymond Lee
Walker, Kelsey, Jr
Wang, Ji Ching
Warren, Walter R(aymond), Jr
Wazzan, A R Frank
Weissman, Paul Robert
Welmers, Everett Thomas
Wertz, James Richard
White, Moreno J
Whittier, James S(pencer)
Williams, E(dgar)
Williams, Forman A(rthur)
Withee, Wallace Walter
Wittry, John P(eter)
Wood, Carlos C
Wood, Robert M(cLane)
Woolard, Henry W(aldo)
Yaggy, Paul Francis
Yakura, James K
Yang, H(sun) T(iao)
Yeager, Charles Elwood (Chuck)
Yeh, Paul Pao
Yildiz, Alaettin

COLORADO
Barber, Robert Edwin
Born, George Henry
Brown, Alison Kay
Bunting, Jackie Ondra
Chow, Chuen-Yen
Cochran, Leighton Scott
Colvis, John Paris
Culp, Robert D(udley)
Daley, Daniel H
Donovan, Terrence John
Fester, Dale A(rthur)
Freymuth, Peter
Gibson, William Loane
Goldburg, Arnold
Jansson, David Guild
Kantha, Lakshmi
Kemper, William Alexander
Klenknecht, Kenneth S(amuel)
Kohlman, David L(eslie)
Loehrke, Richard Irwin
Mansfield, Roger Leo
Marshall, Charles F
Mitchell, Charles Elliott
Morgenthaler, George William
Murphy, John Michael
Paynter, Howard L
Peterson, Harry C(larence)

Regenbrecht, D(ouglas) E(dward)
Sandborn, Virgil A
Seebass, Alfred Richard, III
Siuru, William D, Jr
Uberoi, M(ahinder) S(ingh)
Wall, Edward Thomas
Westfall, Richard Merrill
Wilbur, Paul James
Winn, C Byron

CONNECTICUT
Barnett, Mark
Campbell, George S(tuart)
Chu, Boa-Teh
Cohen, Myron Leslie
Davis, Roger L
Duvivier, Jean Fernand
Fink, Martin Ronald
Ftaclas, Christ
Hajek, Thomas J
Jenney, David S
Kaman, Charles Huron
Keyes, David Elliot
Lambrakis, Konstantine Christos
McCoy, Rawley D
Nilson, Edwin Norman
O'Brien, Robert L
Ojalvo, Irving U
Parker, Jack Steele
Paterson, Robert W
Pitkin, Edward Thaddeus
Porter, Richard W(illiam)
Robinson, Donald W(allace), Jr
Shainin, Dorian
Sreenivasan, Katepalli Raju
Verdon, Joseph Michael

DELAWARE
Danberg, James E(dward)
Schwartz, Leonard William

DISTRICT OF COLUMBIA
Brueckner, Guenter Erich
Callaham, Michael Burks
Chen, Davidson Tah-Chuen
Duncan, Robert C
Durocher, Cort Louis
Fan, Dah-Nien
Flax, Alexander H(enry)
French, Francis William
Gould, Phillip
Green, Richard James
Harris, Wesley Leroy
Holloway, Harry
Holt, Alan Craig
Korkegi, Robert Hani
Liebowitz, Harold
Oran, Elaine Surick
Pearson, Jeremiah W, III
Randolph, James E
Rose, Raymond Edward
Schriever, Bernard Adolf
Stever, H Guyford
Thompson, James Robert, Jr
Townsend, Marjorie Rhodes
Whang, Yun Chow
Widnall, Sheila Evans
Zlotnick, Martin

FLORIDA
Anderson, Roland Carl
Aronson, M(oses)
Balaguer, John P
Bigley, Harry Andrew, Jr
Cates, Harold Thomas
Chamberlain, John
Chow, Wen Lung
Claridge, Richard Allen
Clark, John F
Cloutier, James Robert
Coar, Richard J
Cortright, Edgar Maurice
Dalehite, Thomas H
Davis, Duane M
Diaz, Nils Juan
Dreves, Robert G(eorge)
Eisenberg, Martin A(llan)
Fearn, Richard L(ee)
Fleddermann, Richard G(rayson)
Gillette, Frank C, Jr
Hammond, Charles Eugene
Harmon, G Lamar
Haviland, Robert P(aul)
Hawkins, Robert C
Hoover, John W(esley)
Kerr, Robert Lowell
Krishnamurthy, Lakshminarayanan
Leadon, Bernard M(atthew)
Lijewski, Lawrence Edward
Lund, Frederick H(enry)
McLafferty, George H(oagland), Jr
Mulholland, John Derral
Peterson, James Robert
Proctor, Charles Lafayette, II
Russell, John Masters
Scaringe, Robert P
Scheuing, Richard A(lbert)
Schimmel, Walter Paul
Smith, Jerome Allan
Suciu, S(piridon) N
Sun, Chang-Tsan
Thomas, Garland Leon
Von Ohain, Hans Joachim
Wimberly, C Ray

Yong, Yan

GEORGIA
Adomian, George
Atluri, Satya N
Cahill, Jones F(rancis)
Carlson, Robert L
Cassanova, Robert Anthony
Craig, James I
Cremens, Walter Samuel
Gray, Robin B(ryant)
Hackett, James E
Hubbart, James E
Lewis, David S(loan), Jr
Loewy, Robert G(ustav)
Pierce, G(eorge) Alvin
Price, Edward Warren
Raj, Pradeep
Strahle, Warren C(harles)
Zinn, Ben T

HAWAII
Cheng, Ping
Stuiver, W(illem)

IDAHO
Elias, Thomas Ittan
Gutzman, Philip Charles
Law, John
North, Paul

ILLINOIS
Barthel, Harold O(scar)
Benjamin, Roland John
Brewster, Marcus Quinn
Bullard, Clark W
Coverstone-Carroll, Victoria L
D'Souza, Anthony Frank
Dutton, Jonathan Craig
Errede, Steven Michael
George, John Angelos
Hofer, Kenneth Emil
Holmes, L(awrence) B(ruce)
Hugelman, Rodney D(ale)
Kentzer, Czeslaw P(awel)
Malloy, Donald Jon
Matalon, Moshe
Morel, Thomas
Morkovin, Mark V(ladimir)
Ormsbee, Allen I(ves)
Palmore, Julian Ivanhoe, III
Prussing, John E(dward)
Sentman, Lee H(anley), III
Sivier, Kenneth R(obert)
Tipei, Nicolae
Uherka, Kenneth Leroy
White, Robert Allan
Wolf, Ludwig, Jr

INDIANA
Alspaugh, Dale W(illiam)
Gad-el-Hak, Mohamed
Gustafson, Winthrop A(dolph)
Herrick, Thomas J(efferson)
Hoffman, Joe Douglas
Jumper, Eric J
Kareem, Ahsan
Lykoudis, Paul S
Marshall, Francis J
Mueller, Thomas J
Osborn, J(ohn) R(obert)
Razdan, Mohan Kishen
Schmidt, David Kelso
Sen, Mihir
Skelton, Robert Eugene
Snare, Leroy Earl
Thompson, Howard Doyle
Wallace, F Blake
Yang, Henry T Y

IOWA
Cook, William John
Hall, Jerry Lee
Hsu, Cheng-Ting
Iversen, James D(elano)
Jischke, Martin C(harles)
Ma, Benjamin Mingli
Miller, Richard Keith
Northup, Larry L(ee)
Seversike, Leverne K
Sturges, Leroy D
Tannehill, John Charles

KANSAS
Cogley, Allen C
Cook, Everett L
Downing, David Royal
Ellis, David R
Farokhi, Saeed
Giri, Jagannath
Hosni, Mohammad Hosein
Lan, Chuan-Tau Edward
Muirhead, Vincent Uriel
Razak, Charles Kenneth
Roskam, Jan
Tiffany, Charles F
Wattson, Robert K(ean), Jr
Wentz, William Henry, (Jr)
Zumwalt, Glen W(allace)

LOUISIANA
Callens, Earl Eugene, Jr
Chieri, P(ericle) A(driano)
Courter, R(obert) W(ayne)

Kythe, Prem Kishore
Miller, Percy Hugh
Tewell, Joseph Robert
Thornton, William Edgar
Wetzel, Albert John

MARYLAND
Abed, Eyad Husni
Antman, Stuart S
Augustine, Norman R
Bainum, Peter Montgomery
Ballhaus, William Francis, Jr
Bastress, E(rnest) Karl
Beattie, Donald A
Berger, William J
Bernard, Peter Simon
Bersch, Charles Frank
Billig, Frederick S(tucky)
Burns, Bruce Peter
Butler, Louis Peter
Caveny, Leonard Hugh
Cipriano, Leonard Francis
Cleveland, William Grover, Jr
Cronvich, Lester Louis
Day, LeRoy E(dward)
Deprit, Andre A(lbert) M(aurice)
Diamante, John Matthew
Dunham, David Waring
Eades, James B(everly), Jr
Eaton, Alvin Ralph
Edelson, Burton Irving
Edwards, Alan M
Ely, Raymond Lloyd
Epstein, Gabriel Leo
Evans, Larry Gerald
Fabunmi, James Ayinde
Fansler, Kevin Spain
Fifer, Robert Alan
Fischell, Robert E
Fleig, Albert J, Jr
Floyd, J F R(abardy)
Gartrell, Charles Frederick
Gessow, Alfred
Goldstein, Charles M
Gordon, Gary Donald
Gull, Theodore Raymond
Gupta, Ashwani Kumar
Hammersmith, John L(eo)
Hazen, David Comstock
Hobbs, Robert Wesley
Hornbuckle, Franklin L
Huffington, Norris J(ackson), Jr
Johnson, David Simonds
Jones, Everett
Kalil, Ford
Katzoff, Samuel
Kemelhor, Robert Elias
Kiebler, John W(illiam)
Kitchens, Clarence Wesley, Jr
Korobkin, Irving
Ku, Jentung
Kurzhals, Peter R(alph)
Lankford, J(ohn) L(lewellyn)
Levine, Robert S(idney)
Liu, Han-Shou
Lull, David B
McCourt, A(ndrew) W(ahlert)
McNaughton, James Larry
Mahle, Christoph E
Markley, Francis Landis
Marsten, Richard B(arry)
Martin, John J(oseph)
Mathieu, Richard D(etwiler)
Mayers, Jean
Mecherikunnel, Ann Pottanat
Morgan, Walter L(eroy)
Mumford, Willard R
Murphy, C(harles) H(enry), Jr
Obremski, Henry J(ohn)
Oza, Dipak H
Paddack, Stephen J(oseph)
Paik, Ho Jung
Phinney, Ralph E(dward)
Pieper, George Francis
Powers, John Orin
Presser, Cary
Promisel, Nathan E
Raabe, Herbert P(aul)
Reeves, Richard Allen
Reilly, James Patrick
Reupke, William Albert
Riley, Claude Frank, Jr
Rivello, Robert Matthew
Rogers, David Freeman
Rosen, Milton W(illiam)
Rueger, Lauren J(ohn)
Saarlas, Maido
Schneider, William C
Schuman, William John, Jr
Scialdone, John Joseph
Scipio, L(ouis) Albert, II
Sevik, Maurice
Smith, Richard Lloyd
Stadter, James Thomas
Street, William G(eorge)
Sturek, Walter Beynon
Tai, Tsze Cheng
Thompson, James Laurence, Jr
Trainor, James H
Tschunko, Hubert F A
Urbach, Herman B
Utgoff, Vadym V
Weber, Richard Rand
Whittle, Frank

Yeh, Tsyh Tyan
Yin, Frank Chi-Pong
Zien, Tse-Fou

MASSACHUSETTS
Ash, Michael Edward
Baron, Judson R(ichard)
Battin, Richard H(orace)
Bender, Welcome W(illiam)
Budiansky, Bernard
Burke, Shawn Edmund
Cordero, Julio
Counselman, Charles Claude, III
Covert, Eugene Edzards
Crawley, Edward Francis
Delvaille, John Paul
Detra, Ralph W(illiam)
Dow, Paul C(rowther), Jr
Duffy, Robert A
Dugundji, John
Eschenroeder, Alan Quade
Fay, James A(lan)
Fraser, Donald C
Frey, Elmer Jacob
Garvey, R(obert) Michael
Gavin, Joseph Gleason, Jr
Gionfriddo, Maurice Paul
Gold, Harris
Gregers-Hansen, Vilhelm
Greitzer, Edward Marc
Hawthorne, William (Rede)
Hoag, David Garratt
Hollister, Walter M(ark)
Humi, Mayer
Hursh, John W(oodworth)
Johnson, Aldie E(dwin), Jr
Kautz, Frederick Alton, II
Kelley, Albert J(oseph)
Kerrebrock, Jack Leo
Krebs, James N
Kreifeldt, John Gene
Kumar, Kaplesh
Landahl, Marten T
Lees, Sidney
Lemnios, A(ndrew) Z
Leung, Woon Fong (Wallace)
Lien, Hwachii
Lightfoot, Ralph B(utterworth)
Lurie, Harold
Mastronardi, Richard
Morse, Francis
Murman, Earll Morton
Neumann, Gerhard
Oman, Charles McMaster
Pallone, Adrian Joseph
Pan, Coda H T
Petschek, Harry E
Pian, Carlson Chao-Ping
Pian, Theodore H(sueh) H(uang)
Reeves, Barry L(ucas)
Reinecke, William Gerald
Schmidt, George Thomas
Shea, Joseph F(rancis)
Stickler, David Bruce
Sutton, Emmett Albert
Tong, Pin
Topping, Richard Francis
Trilling, Leon
Udelson, Daniel G(erald)
Vander Velde, W(allace) E(arl)
Ventres, Charles Samuel
Wachman, Harold Yehuda
Waldman, George D(ewey)
Wilk, Leonard Stephen
Witmer, Emmett A(tlee)
Witt, August Ferdinand
Wong, Po Kee

MICHIGAN
Adamson, Thomas C(harles), Jr
Alexandridis, Alexander A
Amick, James L(ewis)
Bitondo, Domenic
Blaser, Dwight A
Chen, Francis Hap-Kwong
Eisley, Joe G(riffin)
Faeth, Gerard Michael
Galan, Louis
Greenwood, D(onald) T(heodore)
Howell, Larry James
Ibrahim, Raouf A
Liu, Vi-Cheng
Lund, Charles Edward
Narain, Amitabh
Nicholls, J(ames) A(rthur)
Phillips, Richard Lang
Powell, Kenneth Grant
Roe, Philip Lawrence
Rumpf, R(obert) J(ohn)
Segel, L(eonard)
Sichel, Martin
Spreitzer, William Matthew
Vinh, Nguyen Xuan
Whicker, Donald
Willmarth, William W(alter)

MINNESOTA
Abraham, John
Beavers, Gordon Stanley
Cunningham, Thomas B
Greene, Christopher Storm
Harvey, Charles Arthur
Holdhusen, James S(tafford)
Moran, John P

Reilly, Richard J
Scott, Charles James
Stolarik, Eugene
Stone, Charles Richard
Sutton, Matthew Albert
Wilson, Theodore A(lexander)

MISSISSIPPI
Bennett, Albert George, Jr
Cliett, Charles Buren
Hackett, Robert M(oore)
Smith, Allie Maitland
Thompson, Joe Floyd

MISSOURI
Agarwal, Ramesh K
Bogar, Thomas John
Bower, William Walter
Branahl, Erwin Fred
Dharani, Lokeswarappa R
Flowers, Harold L(ee)
Graff, George Stephen
Hakkinen, Raimo Jaakko
Hohenemser, Kurt Heinrich
Holsen, James N(oble)
Kibens, Valdis
Kotansky, D(onald) R(ichard)
Koval, Leslie R(obert)
Kozlowski, Don Robert
Leutzinger, Rudolph L(eslie)
Major, Schwab Samuel, Jr
Mirowitz, L(eo) I(saak)
Nelson, Harlan Frederick
Oetting, Robert B(enfield)
Painter, James Howard
Perisho, Clarence H(oward)
Peterson, Gerald E
Poynton, Joseph Patrick
Rinehart, Walter Arley
Roos, Frederick William
Spaid, Frank William
Tsoulfanidis, Nicholas
Ulrich, Benjamin H(arrison), Jr
Warder, Richard C, Jr
Weissenburger, Jason T
Yardley, John Finley

NEBRASKA
Elias, Samy E G
Lu, Pau-Chang
Martin, Charles Wayne

NEVADA
Miel, George J
Pepper, Darrell Weldon
Wells, William Raymond

NEW HAMPSHIRE
Kantrowitz, Arthur (Robert)
Lopez, Guido Wilfred
Morduchow, Morris
Peline, Val P

NEW JERSEY
Benaroya, Haym
Berman, Julian
Blackmore, Denis Louis
Bogdonoff, Seymour (Moses)
Brill, Yvonne Claeys
Clymer, Arthur Benjamin
Curtiss, Howard C(rosby), Jr
Davis, William F
DePalma, Anthony Michael
Durbin, Enoch Job
Florio, Pasquale J, Jr
Graham, (Frank) Dunstan
Hazelrigg, George Arthur, Jr
Kam, Gar Lai
Kornhauser, Alain Lucien
Lam, Sau-Hai
Lewellen, William Stephen
Loman, James Mark
Mikolajczak, Alojzy Antoni
Minter, Jerry Burnett
Orszag, Steven Alan
Schmidlin, Albertus Ernest
Schnapf, Abraham
Scholz, Lawrence Charles
Sernas, Valentinas A
Shefer, Joshua
Sisto, Fernando
Stengel, Robert Frank
Strother, J(ohn) A(lan)
Swern, Frederic Lee
Yu, Yi-Yuan

NEW MEXICO
Baty, Richard Samuel
Berrie, David William
Bertin, John Joseph
Chen, Er-Ping
Curry, Warren H(enry)
Durham, Franklin P(atton)
El-Genk, Mohamed Shafik
Fradkin, David Barry
Garcia, Carlos E(rnesto)
Hartley, Danny L
Maydew, Randall C
Mayer, Harris Louis
Moulds, William Joseph
Savage, Charles Francis
Scharn, Herman Otto Friedrich
Schmitt, Harrison Hagan
Touryan, Kenell James

Williamson, Walton E, Jr
Woods, Robert Octavius

NEW YORK
Agosta, Vito
Ahmadi, Goodarz
Anderson, Roy E
Armen, Harry, Jr
Auer, Peter Louis
Austin, Fred
Banerjee, Prasanta Kumar
Barrows, John Frederick
Bennett, Leon
Boley, Bruno Adrian
Bonan, Eugene J
Boyer, Donald Wayne
Breuhaus, W(aldemar) O(tto)
Brower, William B, Jr
Burns, Joseph A
Busnaina, Ahmed A
Caporalli, Benso L
Caughey, David Alan
Dosanjh, Darshan S(ingh)
Duffy, Robert E(dward)
Ellis, William Rufus
Falk, Theodore J(ohn)
Flaherty, Joseph E
Fleisig, Ross
Foreman, Kenneth M
Fried, Erwin
George, A(lbert) R(ichard)
George, William Kenneth, Jr
Gilmore, Arthur W
Givi, Peyman
Goldstein, Stanley P(hilip)
Gran, Richard J
Grey, Jerry
Grossman, Norman
Guman, William J(ohn)
Hagerup, Henrik J(ohan)
Hajela, Prabhat
Hedrick, Ira Grant
Higuchi, Hiroshi
Hussain, Moayyed A
Isom, Morris P
Jedruch, Jacek
Kapila, Ashwani Kumar
Kellogg, Spencer, II
Kelly, Thomas Joseph
Kempner, Joseph
Klosner, Jerome M
LaGraff, John Erwin
Lane, Frank
Lessen, Martin
Libove, Charles
Litchford, George B
Loeffler, Albert L, Jr
Longman, Richard Winston
Longobardo, Anna Kazanjian
Lumley, John L(eask)
Mack, Charles Edward, Jr
Marrone, Paul Vincent
Mead, Lawrence Myers
Metzger, Ernest Hugh
Milliken, W(illiam) F(ranklin), Jr
Nardo, Sebastian V(incent)
Pifko, Allan Bert
Rae, William J
Raj, Rishi S
Reismann, Herbert
Resler, E(dwin) L(ouis), Jr
Retelle, John Powers, Jr
Scala, Sinclaire M(aximilian)
Schriro, George R
Sforza, Pasquale M
Shaffer, Bernard W(illiam)
Shamash, Yacov A
Shen, S(han) F(u)
Shepherd, D(ennis) G(ranville)
Sherman, Zachary
Shube, Eugene E
Smedfjeld, John B
Sobel, Kenneth Mark
Soures, John Michael
Talapatra, Dipak Chandra
Taulbee, Dale B(ruce)
Torrance, Kenneth E(ric)
Treanor, Charles Edward
Vaicaitis, Rimas
Valentine, Daniel T
Visich, Marian, Jr
Watkins, Charles B
Wehle, Louis Brandeis, Jr
Weingarten, Norman C
Whiteside, James Brooks
Wilson, James William Alexander
Wolf, Daniel Star
Yu, Chia-Ping

NORTH CAROLINA
DeJarnette, Fred Roark
Dowell, Earl Hugh
Dunn, Joseph Charles
Hale, Francis Joseph
Hassan, Hassan Ahmad
Neal, C Leon
Torquato, Salvatore

OHIO
Armstrong, Neil A
Ault, George Mervin
Badertscher, Robert F(rederick)
Bagby, Frederick L(air)
Bahr, Donald Walter

Aeronautical & Astronautical Engineering (cont)

Bailey, Cecil Dewitt
Banda, Siva S
Barranger, John P
Bluford, Guion Stewart, Jr
Bodonyi, Richard James
Breuer, Delmar W
Brilliant, Howard Michael
Burggraf, Odus R
Butz, Donald Josef
Chuang, Henry Ning
Donovan, Leo F(rancis)
Douglas, Donald Wills, Jr
Drake, Michael L
Dunford, Edsel D
Dunn, Robert Garvin
Edse, Rudolph
Fiore, Anthony William
Fitzmaurice, Nessan
Foster, Michael Ralph
Franke, Milton Eugene
Gatewood, Buford Echols
Glasgow, John Charles
Golovin, Michael N
Gorla, Rama S R
Gorland, Sol H
Graham, Robert William
Greber, Isaac
Grigger, David John
Groetsch, Charles William
Guderley, Karl Gottfried
Gunasekera, Jay Sarath
Halford, Gary Ross
Hamed, Awatef A
Hankey, Wilbur Leason, Jr
Hart, David Charles
Hearth, Donald Payne
Herbert, Thorwald
Hsia, John S
Keith, Theo Gordon, Jr
Keshock, Edward G
Kroll, Robert J
Krysiak, Joseph Edward
Larsen, Harold Cecil
Lee, John D(avid)
Lee, Jon H(yunkoo)
Leissa, A(rthur) W(illiam)
Lenhart, Jack G
Lieblein, Seymour
Lundin, Bruce T(heodore)
McBride, J(ames) W(allace)
Macke, H(arry) Jerry
Mall, Shankar
Marek, Cecil John
Moeckel, W(olfgang) E(rnst)
Mohn, Walter Rosing
Mularz, Edward Julius
Olson, Walter T
Pachter, Meir
Pearson, Jerome
Peterson, George P
Povinelli, Louis A
Prahl, Joseph Markel
Quam, David Lawrence
Ramalingam, Mysore Loganathan
Reshotko, Eli
Richardson, David Louis
Rowe, Brian H
Rubin, Stanley G(erald)
Sajben, Miklos
Salkind, Michael Jay
Schultz, Edwin Robert
Schweiger, Marvin I
Serafini, John S
Servais, Ronald Albert
Sheskin, Theodore Jerome
Simitses, George John
Stetson, Kenneth F(rancis)
Tabakoff, Widen
Turchi, Peter John
Van Fossen, Don B
Voisard, Walter Bryan
Von Eschen, Garvin L(eonard)
Wilson, Jack
Wisler, David Charles
Wolaver, Lynn E(llsworth)

OKLAHOMA
Bert, Charles Wesley
Blick, Edward F(orrest)
Earls, James Roe
Emanuel, George
Ketcham, Bruce V(alentine)
Rasmussen, Maurice L
Striz, Alfred Gerhard
Swaim, Robert Lee
Vasicek, Daniel J
Weston, Kenneth Clayton
Zigrang, Denis Joseph

OREGON
Plummer, James Walter

PENNSYLVANIA
Amon Parisi, Cristina Hortensia
Blythe, Philip Anthony
Brunner, Mathias J
Cernansky, Nicholas P
Charp, Solomon
Chigier, Norman
Dash, Sanford Mark

Daywitt, James Edward
Fabish, Thomas John
Griffin, Jerry Howard
Henderson, Robert E
Hnatiuk, Bohdan T
Hull, John Laurence
Karras, Thomas William
Kashkoush, Ismail I
Lakshminarayana, B
Lancaster, Otis Ewing
Lipsky, Stephen E
Long, Lyle Norman
McCormick, Barnes W(arnock), Jr
McLaughlin, Philip V(an Doren), Jr
Mazzitelli, Frederick R(occo)
Merz, Richard A
Nolan, Edward J
Odrey, Nicholas Gerald
Paolino, Michael A
Patton, Peter C(lyde)
Piasecki, Frank Nicholas
Pinckney, Robert L
Ramos, Juan Ignacio
Rodini, Benjamin Thomas, Jr
Rosen, Gerald Harris
Ross, Arthur Leonard
Rumbarger, John H
Schulman, Marvin
Sharma, Mangalore Gokulanand
Shuch, H Paul
Steg, L(eo)
Walker, James David Allan
Wang, Albert Show-Dwo
Whirlow, Donald Kent
Williams, Max L(ea), Jr
Zhou, Qian
Zweben, Carl Henry

RHODE ISLAND
Hayward, John T(ucker)
Liu, J(oseph) T(su) C(hieh)
McEligot, Donald M(arinus)
Mellberg, Leonard Evert

SOUTH CAROLINA
Tull, William J

SOUTH DAKOTA
Winker, James A(nthony)

TENNESSEE
Burton, Robert Lee
Collins, Frank Gibson
Daniel, Donald Clifton
Garrison, George Walker
Harwell, Kenneth Edwin
Keefer, Dennis Ralph
Kraft, Edward Michael
McGregor, Wheeler Kesey, Jr
Potter, John Leith
Roth, J(ohn) Reece
Shahrokhi, Firouz
Sissom, Leighton E(sten)
Skinner, George T
Toline, Francis Raymond
Wheeler, Orville Eugene
Whitfield, Jack D
Wu, Ying-Chu Lin (Susan)
Young, Robert L(yle)

TEXAS
Adams, Richard E
Anderson, Dale Arden
Anderson, Edward Everett
Ball, Kenneth Steven
Bishop, Robert H
Bland, William M, Jr
Bond, Aleck C
Brock, James Harvey
Carey, Graham Francis
Chevalier, Howard L
Cohen, Aaron
Cronk, Alfred E(dward)
Dalley, Joseph W(inthrop)
Dalton, Charles
Dietz, William C
Dolling, David Stanley
Epp, Chirold Delain
Faget, Maxime A(llan)
Fairchild, Jack
Feagin, Terry
Fenter, Felix West
Fowler, Wallace T(homas)
Gourdine, Meredith Charles
Haisler, Walter Ervin
Henize, Karl Gordon
Hesse, Walter J(ohn)
Hinrichs, Paul Rutland
Hull, David G(eorge)
Hunt, Louis Roberts
Hussain, A K M Fazle
Jackson, Eugene Bernard
Johnson, Johnny Albert
Jones, Jerold W
Junkins, John Lee
Juricic, Davor
Kerwin, Joseph Peter
Koplyay, Janos Bernath
Kraft, Christopher Columbus, Jr
Kranz, Eugene Francis
Lamb, J(amie) Parker, Jr
Levine, Joseph H
Loftus, Joseph P, Jr
Lowy, Stanley H(oward)

Lu, Frank Kerping
McBride, Jon Andrew
Miele, Angelo
Miksad, Richard Walter
Miller, Gerald E
Modisette, Jerry L
Morris, Owen G
Musa, Samuel A
Nicks, Oran Wesley
Norton, David Jerry
Page, Robert Henry
Payne, Fred R(ay)
Pray, Donald George
Purser, Paul Emil
Rand, James Leland
Ransleben, Guido E(rnst), Jr
Ray, Robert Landon
Rodenberger, Charles Alvard
Sadler, Stanley Gene
Schapery, Richard Allan
Schutz, Bob Ewald
Secrest, Bruce Gill
Spindler, Max
Stanovsky, Joseph Jerry
Stearman, Ronald Oran
Sullivan, Kathryn D
Sullivan, Thomas Allen
Szebehely, Victor
Talent, David Leroy
Tapley, Byron D(ean)
Thomas, Richard Eugene
Vance, John Milton
Ward, Donald Thomas
Webb, Theodore Stratton, Jr
Westkaemper, John C(onrad)
Widmer, Robert H
Young, John W(atts)

UTAH
Clyde, Calvin G(eary)
Dulock, Victor A, Jr
Hoeppner, David William
Krejci, Robert Henry
Long, David G
Polve, James Herschal
Strozier, James Kinard
Zeamer, Richard Jere

VERMONT
Hermance, C(larke) E(dson)

VIRGINIA
Anders, William A
Beach, Harry Lee, Jr
Bingham, Billy Elias
Bishara, Michael Nageeb
Bolender, Carroll H
Bressler, Barry Lee
Brooks, Thomas Furman
Campbell, James Franklin
Carlson, Harry William
Chang, George Chunyi
Cliff, Eugene M
Cooley, William C
Cooper, Earl Dana
Crawford, Daniel J
Cross, Ernest James, Jr
Crossfield, A Scott
Curtin, Robert H
Donaldson, Coleman DuPont
Donivan, Frank Forbes, Jr
Duberg, John E(dward)
Elias, Antonio L
Erickson, Wayne Douglas
Everett, Warren S
Fabian, John M
Finke, Reinald Guy
Finkelstein, Jay Laurence
Flack, Ronald Dumont, Jr
Freck, Peter G
Freitag, Robert Frederick
Gilbert, Arthur Charles
Gilruth, Robert R(owe)
Goodwin, Francis E
Graham, Kenneth Judson
Greenberg, Arthur Bernard
Gunzburger, Max Donald
Haviland, John Kenneth
Henderson, Charles B(rooke)
Hidalgo, Henry
Holloway, Paul Fayette
Hou, Gene Jean-Win
Houbolt, John C(ornelius)
Huston, Robert James
Joshi, Suresh Meghashyam
Juang, Jer-Nan
Kaminski, Paul G
Kapania, Rakesh Kumar
Kaplan, Marshall Harvey
Keever, David Bruce
Keigler, John Edward
King, Merrill Kenneth
Lauver, Dean C
Lea, George Koo
Lee, Fred C
Lichtenberg, Byron Kurt
Lobb, R(odmond) Kenneth
Lovell, Robert R(oland)
Ludwig, George H
Lutze, Frederick Henry, Jr
McDivitt, James Alton
Manfredi, Arthur Frank, Jr
Marchman, James F(ranklin), III
Margrave, Thomas Ewing, Jr

Michael, William Herbert, Jr
Myers, Michael Kenneth
Ng, Wing-Fai
Perkins, Courtland D(avis)
Petersen, Richard Herman
Pruett, Charles David
Puster, Richard Lee
Pyke, Thomas Nicholas, Jr
Raju, Ivatury Sanyasi
Roberts, William Woodruff, Jr
Rokni, Mohammad Ali
Roy, Gabriel D
Sabadell, Alberto Jose
Savage, William F(rederick)
Schetz, Joseph A
Schy, Albert Abe
Scott, John E(dward), Jr
Seelig, Jakob Williams
Seiner, John Milton
Simmonds, James G
Singh, Mahendra Pal
Sobieszczanski-Sobieski, Jaroslaw
Summers, George Donald
Telionis, Demetri Pyrros
Theon, John Speridon
Thompson, David Walker
Tiwari, Surendra Nath
Tole, John Roy
Tolson, Robert Heath
Tsang, Wai Lin
Waesche, R(ichard) H(enley) Woodward
Weinstein, Leonard Murrey
Welch, Jasper Arthur, Jr
Whitesides, John Lindsey, Jr
Whitlock, Charles Henry
Wilson, Herbert Alexander, Jr
Wolfhard, Hans Georg
Wood, Albert D(ouglas)
Yates, Edward Carson, Jr
Zakrzewski, Thomas Michael

WASHINGTON
Ahlstrom, Harlow G(arth)
Arenz, Robert James
Bollard, R(ichard) John H
Bruckner, Adam Peter
Buonamici, Rino
Christiansen, Walter Henry
Clark, Robert Newhall
Clarkson, Mark H(all)
Cosgrove, Benjamin A
Delisi, Donald Paul
Dixon, Robert Jerome, Jr
Dvorak, Frank Arthur
Fetter, William Allan
Fyfe, I(an) Millar
Gravitz, Sidney I
Hamilton, William Thorne
Hebeler, Henry K
Heilenday, Frank W (Tod)
Hertzberg, Abraham
Holtby, Kenneth F
Kennet, Haim
Kurosaka, Mitsuru
Lin, H(ua)
McCarthy, Douglas Robert
McMurtrey, Lawrence J
Martin, George C(oleman)
Pennell, Maynard L
Pilet, Stanford Christian
Quist, William Edward
Rae, William H, Jr
Robel, Gregory Frank
Roetman, Ernest Levane
Rubbert, Paul Edward
Russell, David A
Schairer, G(eorge) S(wift)
Schmidt, Eckart W
Shank, Maurice E(dwin)
Speer, Fridtjof Alfred
Steiner, John Edward
Street, Robert Elliott
Sutter, Joseph F
Tam, Patrick Yui-Chiu
Taya, Minoru
Varanasi, Suryanarayana Rao
Walsh, John Breffni
Walton, Vincent Michael
Wilson, Thornton Arnold

WEST VIRGINIA
Fanucci, Jerome B(en)
Kuhlman, John Michael
Loth, John Lodewyk
Peters, Penn A
Smith, James Earl

WISCONSIN
Bonazza, Riccardo
Lovell, Edward George
Moore, Gary T
Viets, Hermann
Wnuk, Michael Peter
Zelazo, Nathaniel Kachorek

WYOMING
Adams, Donald F
Pell, Kynric M(artin)
Werner, Frank D(avid)
Wheasler, Robert

ALBERTA
Marsden, D(avid) J(ohn)

BRITISH COLUMBIA
Miura, Robert Mitsuru
Modi, V J
Muggeridge, Derek Brian

NOVA SCOTIA
Cochkanoff, O(rest)
Eames, M(ichael) C(urtis)

ONTARIO
Altman, Samuel Pinover
Barron, Ronald Michael
Chan, Yat Yung
Chu, Wing Tin
Cockshutt, E(ric) P(hilip)
Cowper, George Richard
DeLeeuw, J H
Dunn, D(onald) W(illiam)
Elfstrom, Gary Macdonald
Etkin, Bernard
French, J(ohn) Barry
Gladwell, Graham M L
Glass, I(rvine) I(srael)
Hughes, Peter C(arlisle)
Jacobson, Stuart Lee
Kind, Richard John
Koul, Ashok Kumar
Krishnappa, Govindappa
Lapp, P(hilip) A(lexander)
Lawson, John Douglas
Lukasiewicz, Julius
Matar, Said E
Moffatt, William Craig
Molder, S(annu)
Orlik-Ruckemann, Kazimierz Jerzy
Pindera, Jerzy Tadeusz
Plett, Edelbert Gregory
Prince, Robert Harry
Reid, Lloyd Duff
Ribner, Herbert Spencer
Richarz, Werner Gunter
Rimrott, F(riedrich) P(aul) J(ohannes)
Stevinson, Harry Thompson
Street, Kenneth Norman
Sullivan, Philip Albert
Tennyson, Roderick C
Tyler, R(onald) A(nthony)
Wallace, William
Weick, Richard Fred

QUEBEC
Bhat, Rama B
Galanis, Nicolas
Habashi, Wagdi George
Luckert, H(ans) J(oachim)
Mavriplis, F
Newman, B(arry) G(eorge)
Ryan, Norman Daniel
Saber, Aaron Jaan
Vatistas, Georgios H

SASKATCHEWAN
Fuller, Gerald Arthur
Walerian, Szyszkowski

OTHER COUNTRIES
Al-Zubaidy, Sarim Naji
Chiu, Huei-Huang
Idelsohn, Sergio Rodolfo
Ince, A Nejat
Miller, Rene H(arcourt)
Morino, Luigi
Nomura, Yasumasa
Ramachandran, Venkataraman
Sato, Gentei
Van Hoften, James Dougal Adrianus
Voutsas, Alexander Matthew (Voutsadakis)
Vrebalovich, Thomas
Willeke, Klaus

Agricultural Engineering

ALABAMA
Corley, Tom Edward
Flood, Clifford Arrington, Jr
Hill, David Thomas
Johnson, Loyd
Renoll, Elmo Smith
Rochester, Eugene Wallace
Schafer, Robert Louis
Taylor, James H(obert)
Turnquist, Paul Kenneth

ARIZONA
Becker, Clarence F(rederick)
Bessey, Paul Mack
Buras, Nathan
Cannon, M(oody) Dale
Clemmens, Albert J
Fangmeier, Delmar Dean
Foster, Kennith Earl
Jordan, Kenneth A(llan)
Rasmussen, William Otto
Replogle, John A(sher)
Smerdon, Ernest Thomas
Woodruff, Neil Parker
Woolhiser, David A(rthur)

ARKANSAS
Berry, Ivan Leroy
Bryan, Billy Bird
Rokeby, Thomas R(upert) C(ollinson)

Sutherland, G Russell

CALIFORNIA
Adams, William John, Jr
Akesson, Norman B(erndt)
Bery, Mahendera K
Chang, Andrew C
Fridley, Robert B
Goss, John R(ay)
Grinnell, Robin Roy
Hermsmeier, Lee F
Hickok, Robert Baker
Kepner, Robert Allen
Loper, Willard H(ewitt)
Matthews, Floyd V(ernon), Jr
Merriam, John L(afayette)
Mohsenin, Nuri N
Mueller, Charles Carsten
Neubauer, L(oren) W(enzel)
O'Brien, Michael
Pitts, Donald James
Robertson, George Harcourt
Saltveit, Mikal Endre, Jr
Strohman, Rollin Dean
Wallender, Wesley William

COLORADO
Corey, Arthur Thomas
DeCoursey, Donn G(ene)
Doering, Eugene J(ohnson)
Evans, Norman A(llen)
Frasier, Gary Wayne
Harper, Judson M(orse)
Heermann, Dale F(rank)
Jensen, Marvin E(li)
Mickelson, Rome H
Thierstein, Gerald E

CONNECTICUT
Aldrich, Robert A(dams)
Bibeau, Thomas Clifford

DELAWARE
Collins, Norman Edward, Jr
Dwivedy, Ramesh C
Ritter, William Frederick
Scarborough, Ernest N

DISTRICT OF COLUMBIA
Endahl, Lowell Jerome
Finney, Essex Eugene, Jr

FLORIDA
Bagnall, Larry Owen
Clayton, Joe Edward
Coppock, Glenn E(dgar)
Felton, Kenneth E(ugene)
Fluck, Richard Conard
Freeman, George R(oland)
Harrison, Dalton S(idney)
Holtan, Heggie Nordahl
Isaacs, Gerald W
Nordstedt, Roger Arlo
Overman, Allen Ray
Peart, Robert McDermand
Sager, John Clutton
Shaw, Lawrance Neil
Stephens, John C(arnes)
Teixeira, Arthur Alves
Wheeler, William Crawford
Whitney, Jodie Doyle
Zachariah, Gerald L

GEORGIA
Arkin, Gerald Franklin
Burgoa, Benali
Butler, James Lee
Clark, Rex L
Drury, Liston Nathaniel
Ghate, Suhas Ramkrishna
Henderson, George Edwin
Hoogenboom, Gerrit
Lawrence, Kurt C
Nelson, S(tuart) O(wen)
Sheppard, David Craig
Smith, Ralph E(dward)
Thomas, Adrian Wesley
Threadgill, Ernest Dale
White, Harold D(ouglas)

HAWAII
Bartosik, Alexander Michael
Gitlin, Harris Martlin
Liang, Tung
Liu, Wei
Wang, Jaw-Kai

IDAHO
Bloomsburg, George L
Bondurant, James A(llison)
Brakensiek, Donald Lloyd
Dixon, John E(lvin)
Fitzsimmons, Delbert Wayne
Halterman, Jerry J
Humpherys, Allan S(tratford)
Johnson, Clifton W
Kincaid, Dennis Campbell
Neibling, William Howard
Pair, Claude H(erman)
Peterson, Charles Loren
Schreiber, David Laurence
Trout, Thomas James
Williams, Larry G

ILLINOIS
Cheryan, Munir
Curtis, James O(wen)
Day, Donald Lee
Fonken, David W(alter)
Goering, Carroll Eugene
Hansen, Ralph W(aldo)
Hay, Ralph C
Hirschi, Michael Carl
Hummel, John William
Hunt, Donnell Ray
Jones, Benjamin A(ngus), Jr
Lloyd, Monte
Muehling, Arthur J
Olver, Elwood Forrest
Paulsen, Marvin Russell
Pickett, Leroy Kenneth
Puckett, Hoyle Brooks
Rodda, Errol David
Shove, Gene C(lere)
Siemens, John Cornelius
Steinberg, Marvin Phillip
Wakefield, Ernest Henry
Walter, Gordon H
Wang, Pie-Yi
Yoerger, Roger R

INDIANA
Dale, Alvin C
Gibson, Harry Gene
Hinkle, Charles N(elson)
Huggins, Larry Francis
Johns, Dennis Michael
Monke, Edwin J
Richey, Clarence B(entley)
Wai, Wing-Kin

IOWA
Baker, James LeRoy
Beer, Craig
Buchele, W(esley) F(isher)
Colvin, Thomas Stuart
Curry, Norval H
Hull, Dale O
Ives, Norton C(onrad)
Johnson, Howard P
Johnson, Lawrence Alan
Lane, Orris John, Jr
Marley, Stephen J
Melvin, Stewart W
Meyer, Vernon M(ilo)

KANSAS
Clark, Stanley Joe
Hagen, Lawrence J
Johnson, William Howard
Larson, George H(erbert)
Lyles, Leon
Reh, John W
Schaper, Laurence Teis
Spillman, Charles Kennard

KENTUCKY
Barfield, Billy Joe
Frank, John L
Henson, W(iley) H(ix), Jr
Parker, B(laine) F(rank)
Ross, Ira Joseph
Smith, Edward M(anson)
Walker, John Neal
Wells, Larry Gene
Yoder, Elmon Eugene

LOUISIANA
Braud, Harry J
Brown, William Henry
Cochran, Billy Juan
Fouss, James L(awrence)
Parish, Richard Lee
Robbins, Jackie Wayne Darmon
Rogers, James Samuel
Singh, Vijay P

MAINE
Hunter, James H
Klinge, Albert Frederick
Rowe, Richard J(ay)

MARYLAND
Ahrens, M(iles) Conner
Altman, Landy B, Jr
Amerman, Carroll R(ichard)
Green, Robert Lamar
Harris, Wesley Lamar
Johnson, Arthur Thomas
Langley, Maurice N(athan)
Norris, Karl H(oward)
Ross, David Stanley
Salomonson, Vincent Victor
Taylor, Leonard S
Van Schilfgaarde, Jan
Yaramanoglu, Melih
Yeck, Robert Gilbert

MASSACHUSETTS
Rosenau, John (Rudolph)

MICHIGAN
Anderson, James Henry
Bakker-Arkema, Frederik Wilte
Bickert, William George
Butchbaker, Allen F
Esmay, Merle L(inden)
Hahn, Russell H

McColly, Howard F(ranklin)
Mackson, Chester John
Maley, Wayne A
Marshall, Dale Earnest
Merva, George E

MINNESOTA
Allred, E(van) R(ich)
Bates, Donald W(esley)
Filkke, Arnold M(aurice)
Gupta, Satish Chander
Larson, Curtis L(uverne)
Morey, Robert Vance
Schertz, Cletus E

MISSISSIPPI
Colwick, Rex Floyd
Foster, George Rainey
Fox, William R(obert)
Hendricks, Donovan Edward
Matthes, Ralph Kenneth, Jr
Meyer, Lawrence Donald
Mutchler, Calvin Kendal
Welch, George Burns

MISSOURI
Brooker, Donald Brown
Day, Cecil LeRoy
Frisby, James Curtis
Hewitt, Andrew
Hjelmfelt, Allen T, Jr
Kishore, Ganesh M
McFate, Kenneth L(everne)
Meador, Neil Franklin
Saran, Chitaranjan
Wilson, Clyde Livingston

NEBRASKA
Edwards, Donald M(ervin)
Grisso, Robert Dwight
Hahn, George LeRoy
Hanna, Milford A
Hoffman, Glenn Jerrald
Kleis, Robert W(illiam)
Leviticus, Louis I
Olson, Emanuel A(ole)
Splinter, William Eldon
Stetson, LaVerne Ellis
Thompson, Thomas Leigh
Vanderholm, Dale Henry
Von Bargen, Kenneth Louis
Wittmuss, Howard D(ale)

NEVADA
Guitjens, Johannes C

NEW JERSEY
Johnson, Curtis Allen
Lund, Daryl B
Mears, David R
Ponessa, Joseph Thomas
Singley, Mark E(ldridge)

NEW MEXICO
Abernathy, George Henry
Morin, George Cardinal Albert
Rebuck, Ernest C(harles)
Yonas, Gerold

NEW YORK
Capps, Susan Glass
Cooke, James Robert
Furry, Ronald B(ay)
Guest, Richard W(illiam)
Gunkel, Wesley W
Huntington, David Hans
Longhouse, Alfred Delbert
Millier, William F(rederick)
Rehkugler, Gerald E(dwin)
Scott, Norman Roy
Virk, Kashmir Singh

NORTH CAROLINA
Abrams, Charlie Frank, Jr
Barker, James Cathey
Bowen, Henry Dittimus
Cleland, John Gregory
Hassan, Awatif E
Hassler, F(rancis) J(efferson)
Huang, Barney K(uo-Yen)
Humphries, Ervin G(rigg)
Johnson, William Hugh
Kriz, George James
Lalor, William Francis
Overcash, Michael Ray
Rohrbach, Roger P(hillip)
Safley, Lawson McKinney, Jr
Sanoff, Henry
Skaggs, Richard Wayne
Suggs, Charles Wilson
Whitaker, Thomas Burton
Wiser, Edward H(empstead)
Young, James H

NORTH DAKOTA
Bauer, Armand
French, Ernest W(ebster)
Pratt, George L(ewis)
Promersberger, William J
Witz, Richard L

OHIO
Bondurant, Byron L(ee)
Brazee, Ross D

Agricultural Engineering (cont)

Curry, R(obert) Bruce
Fox, Robert Dean
Hall, Glenn Eugene
Hamdy, Mohamed Yousry
Herum, Floyd L(yle)
Huber, Samuel G(eorge)
Keener, Harold Marion
Lal, Rattan
Lamp, Benson J
Nelson, Gordon L(eon)
Rausch, David Leon
Reeve, Ronald C(ropper)
Roller, Warren L(eon)
Schwab, Glenn O(rville)
Sharma, Shri C
Short, Ted H
Taiganides, E Paul

OKLAHOMA
Batchelder, David G(eorge)
Brusewitz, Gerald Henry
Clary, Bobby Leland
Crow, Franklin Romig
Haan, Charles Thomas
Noyes, Ronald T
Porterfield, Jay G
Roth, Lawrence O(rval)
Simpson, Ocleris C
Thompson, David Russell
Whitney, Richard Wilbur

OREGON
Booster, Dean Emerson
Hansen, Hugh J
Hashimoto, Andrew G
Hellickson, Martin Leon
Kirk, Dale E(arl)
Miner, J Ronald
Miner, John Ronald
Moore, James Allan

PENNSYLVANIA
Anderson, Paul Milton
Buffington, Dennis Elvin
Daum, Donald Richard
Kjelgaard, William L
Manbeck, Harvey B
Morrow, Charles T(erry)
Myers, Earl A(braham)
Persson, Sverker
Skromme, Lawrence H
Walton, Harold V(incent)
White, John W

SOUTH CAROLINA
Bunn, Joe M(illard)
Drew, Leland Overbey
Fischer, James Roland
Lambert, Jerry Roy
Ligon, James T(eddie)
Snell, A(bsalom) W(est)
Webb, Byron Kenneth
White, Richard Kenneth
Williamson, Robert Elmore

SOUTH DAKOTA
Hellickson, Mylo A
Moe, Dennis L
Myers, Victor (Ira)

TENNESSEE
Baxter, Denver O(len)
Helweg, Otto Jennings
Henry, Zachary Adolphus
Jones, Larry Warner
McDow, John J(ett)
Sewell, John I

TEXAS
Allison, Robert Dean
Bockhop, Clarence William
Boyer, Robert Elston
Dvoracek, Marvin John
Engler, Cady Roy
Fryrear, Donald W
Gavande, Sampat A
Hauser, Victor La Vern
Hiler, Edward Allan
Hollingsworth, Joseph Pettus
Howell, Terry Allen
Kirk, Ivan Wayne
Krishna, J Hari
Kunze, Otto Robert
Morrison, John Eddy, Jr
Musick, Jack T(hompson)
Nixon, Paul R(obert)
Onstad, Charles Arnold
Reddell, Donald Lee
Stermer, Raymond A
Ulich, Willie Lee

UTAH
Allen, Richard Glen
Anderson, Bruce Holmes
Hargreaves, George H(enry)
James, David Winston
Keller, Jack
Kidder, Ernest H(igley)
Peralta, Richard Carl
Rasmussen, V Philip, Jr
Silver, Barnard Stewart
Skogerboe, Gaylord Vincent

Stringham, Glen Evan
Willardson, Lyman S(essions)

VERMONT
Bornstein, Joseph

VIRGINIA
Corey, Gilbert
Hall, Carl W(illiam)
Haugh, C(larence) Gene
Hurst, Homer T(heodore)
Mason, J(ohn) Philip Hanson, Jr
Miller, Adolphus James
Schlegelmilch, Reuben Orville
Shanholtz, Vernon Odell
Wright, Farrin Scott

WASHINGTON
Campbell, John C(arl)
Cykler, John Freuler
Davis, Denny Cecil
Hermanson, Ronald Eldon
James, Larry George
Kevorkian, Jirair
King, Larry Gene
McCool, D(onald) K
Powell, Albert E
Saxton, Keith E

WEST VIRGINIA
Diener, Robert G
Peters, Penn A

WISCONSIN
Amundson, Clyde Howard
Barrington, Gordon P
Bruhn, H(jalmar) D(iehl)
Bubenzer, Gary Dean
Buelow, Frederick H(enry)
Converse, James Clarence
Cramer, Calvin O
Evans, James Ornette
Kinch, Donald M(iles)
Northouse, Richard A
Schuler, Ronald Theodore
Thompson, Paul DeVries

WYOMING
Pochop, Larry Otto
Smith, James Lee

PUERTO RICO
Goyal, Megh R
Rodriguez-Arias, Jorge H

ALBERTA
Domier, Kenneth Walter
Sommerfeldt, Theron G

BRITISH COLUMBIA
Liao, Ping-huang
Staley, L(eonard) M(aurice)

MANITOBA
Cenkowski, Stefan
Jayas, Digvir Singh
Laliberte, Garland E
Lapp, H(erbert) M(elbourne)
Muir, William Ernest
Townsend, James Skeoch

ONTARIO
Dickinson, William Trevor
Mittal, Gauri S
Thansandote, Artnarong
Wong, Jo Yung

QUEBEC
Alward, Ron E
Broughton, Robert Stephen
McKyes, Edward
St-Yves, Angele

SASKATCHEWAN
Norum, Donald Iver

OTHER COUNTRIES
Finkel, Herman J(acob)
Preston, Thomas Alexander
Singh, Gajendra
Stout, Bill A(lvin)
Ward, Gerald T(empleton)

Artificial Intelligence

ARIZONA
Sridharan, Natesa S

CALIFORNIA
Adams, John Andrew
Berenji, Hamid R
Przystupa, Marek Antoni
Schatzki, Thomas Ferdinant
Wipke, W Todd

CONNECTICUT
Smith, Donald Arthur

DELAWARE
Chitra, Surya P

FLORIDA
Doi, Shinobu

Ulug, Esin M

GEORGIA
Covington, Michael A
Peterson, Keith L

ILLINOIS
Karaali, Orhan

INDIANA
Uhran, John Joseph, Jr
Wai, Wing-Kin

MARYLAND
Minker, Jack

MASSACHUSETTS
Dyer, Charles Austen
Johnson, Walter Hudson
Lacoss, Richard Thaddee

MISSOURI
Ball, William E(rnest)

NEW YORK
Bandyopadhyay, Amitabha
Chang, Ching Ming
Jabbour, Kamal
Leisman, Gerald
Pellegrino, Francesco

NORTH CAROLINA
Woodland, Joseph

PENNSYLVANIA
Kolar, Michael Joseph

RHODE ISLAND
Raykhman, Aleksandr

TENNESSEE
Malasri, Siripong
Uhrig, Robert Eugene

TEXAS
Liu, Kai
Philippe, Edouard Antoine

VIRGINIA
Chen, Jiande
Werbos, Paul John

ONTARIO
Famili, Abolfazl
Halperin, Janet R P

Bioengineering & Biomedical Engineering

ALABAMA
Goldman, Jay
Harding, Thomas Hague
Lucas, Linda C
McCutcheon, Martin J
Makhijani, Vinod B
Melville, Joel George
Rigney, Earnest Douglas, Jr
Ryder, Robert M
Sheffield, L Thomas
Stokely, Ernest Mitchell
Twieg, Donald Baker
Vacik, James P
Wu, Xizeng

ARIZONA
Ahmed, Saad Attia
Arabyan, Ara
Bahill, A(ndrew) Terry
Barr, Ronald Edward
Dorson, William John
Gall, Donald Alan
Gross, Joseph F
Guilbeau, Eric J
Holloway, G Allen, Jr
Kauffman, John W
Koeneman, James Bryant
McRae, Lorin Post
Mylrea, Kenneth C
Preston, Kendall, Jr
Sauer, Barry W
Smith, Josef Riley
Stiles, Philip Glenn
Yamaguchi, Gary T

ARKANSAS
Clausen, Edgar Clemens
Schmitt, Neil Martin
Wear, James Otto

CALIFORNIA
Addy, Tralance Obuama
Akonteh, Benny Ambrose
Allen, Charles William
Altes, Richard Alan
Alvi, Zahoor M
Amstutz, Harlan Cabot
Angell, James Browne
Anger, Hal Oscar
Astrahan, Melvin Alan
Bacskai, Brian James
Beljan, John Richard
Berger, Stanley A(llan)
Bery, Mahendera K

Black, Kirby Samuel
Blanch, Harvey Warren
Bliss, James C(harles)
Boulton, Roger Brett
Bradley, A Freeman
Brant, Albert Wade
Britt, Edward Joseph
Butterworth, Thomas Austin
Canawati, Hanna N
Cardullo, Richard Anthony
Carew, Thomas Edward
Chan, David S
Chenoweth, Dennis Edwin
Chou, Chung-Kwang
Chow, John Lap Hong
Codrington, Robert Smith
Cohn, Theodore E
Cook, Albert Moore
Coombs, William, Jr
Covell, James Wachob
Cromwell, Leslie
D'Argenio, David Z
Daughters, George T, II
Davey, Trevor B(lakely)
Dev, Parvati
DiStefano, Joseph John, III
Eckenhoff, James Benjamin
Edmonds, Peter Derek
El-Refai, Mahmoud F
Estridge, Trudy Donette
Estrin, Thelma A
Fung, Yuan-Cheng B(ertaam)
Ghazanshahi, Shahin D
Glantz, Stanton Arnold
Goldstein, Walter Elliott
Gough, David Arthur
Gray, Joe William
Green, Philip S
Gutfinger, Dan Eli
Haas, Gustav Frederick
Haines, Richard Foster
Hansmann, Douglas R
Harder, James Albert
Hayes, Thomas L
Hemmingsen, Edvard A(lfred)
Hill, Bruce Colman
Horres, Charles Russell, Jr
Huang, Sung-Cheng
Hughes, Everett C
Huszczuk, Andrew Huszcza
Iberall, Arthur Saul
Ingels, Neil Barton, Jr
Intaglietta, Marcos
Jacob, George Korathu
Jenkins, Steven J
Jewett, Don L
Jones, Jerry Latham
Jones, Joie Pierce
Karuza, Sarunas Kazys
Keller, Edward Lowell
Kelly, Donald Horton
Khoo, Michael D
Kim, Kwang-Jin
Knight, Patricia Marie
Kohler, George Oscar
Leake, Donald L
Leifer, Larry J
Lele, Padmakar Pratap
Levy, Donald M(arc)
Lewis, Edwin Reynolds
Lewis, Steven M
Likuski, Robert Keith
Lim, H(enry) C(hol)
Llaurado, Josep G
Logsdon, Donald Francis, Jr
Lundblad, Roger Lauren
Lyman, John (Henry)
McCulloch, Andrew Douglas
McRuer, Duane Torrance
Madan, Ram Chand
Marmarelis, Vasilis Z
Matsumura, Kenneth N
Matthys, Eric Francois
Mazzoni, Coleen Michelle
Meerbaum, Samuel
Midura, Thaddeus
Miller, Sol
Mirhashemi, Soheila
Montgomery, Leslie D
Morse, Michael S
Mote, C(layton) D(aniel), Jr
Nimni, Marcel Efraim
Nunan, Craig S(pencer)
Pacela, Allan F
Pangborn, Rose Marie Valdes
Paulsen, A William
Pedersen, Leo Damborg
Pfost, R Fred
Pinto, John Gilbert
Pittman, Ray Calvin
Powell, Frank Ludwig, Jr
Ramey, Melvin Richard
Randle, Robert James
Rasor, Ned S(haurer)
Robertson, George Harcourt
Rockwell, Robert Lawrence
Rugge, Henry F
Ryan, Wen
Sadler, Charles Robinson, Jr
Saha, Subrata
Saunders, Frank Austin
Schmid-Schoenbein, Geert W
Schneider, Alan M(ichael)
Shackelford, James Floyd

Singer, Jerome Ralph
Siposs, George G
Skalak, Richard
Skrzypek, Josef
Small, James Graydon
Smith, Robert Elphin
Smith, Warren Drew
Specht, Donald Francis
Spencer, Cherrill Melanie
Stark, Lawrence
Steele, Charles Richard
Sternberg, Moshe
Stroeve, Pieter
Stuhmiller, James Hamilton
Susskind, Charles
Szeto, Andrew Y J
Tanenbaum, Basil Samuel
Terdiman, Joseph Franklin
Thompson, Noel Page
Throner, Guy Charles
Unal, Aynur
Vurek, Gerald G
Waldman, Lewis K
Walton, James Stephen
Ward, David Gene
Waring, Worden
Waterland, Larry R
Wayland, J(ames) Harold
Webbon, Bruce Warren
Weibell, Fred John
Weinberg, Daniel I
Werblin, Frank Simon
White, Moreno J
Wiberg, Donald M
Wieland, Bruce Wendell
Wirta, Roy W(illiam)
Yamashiro, Stanley Motohiro
Yeung, King-Wah Walter
Yildiz, Alaettin
Zweifach, Benjamin William

COLORADO
Aunon, Jorge Ignacio
Dunn, Richard Lee
Ellis, Donald Griffith
Gabridge, Michael Gregory
Gamow, Rustem Igor
Hale, John
Hinman, Norman Dean
Histand, Michael B(enjamin)
Markert, Clement Lawrence
Morgan, R John
Morrison, Charles Freeman, Jr
Newkirk, John Burt
Painter, Kent
Pan, Huo-Ping
Pauley, James Donald
Rakow, Allen Leslie
Scherer, Ronald Callaway
Simske, Steven John
Sodal, Ingvar E
Taylor, Kenneth Doyle
Tengerdy, Robert Paul

CONNECTICUT
Berglund, Larry Glenn
Bronzino, Joseph Daniel
Cohen, Myron Leslie
Crawford, John Okerson
Darcey, Terrance Michael
Dreyfus, Marc George
Hlavacek, Robert Allen
Horváth, Csaba Gyula
Hou, Kenneth C
Krummel, William Marion
Lambrakis, Konstantine Christos
Marchand, Nathan
Nehorai, Arye
Ranalli, Anthony William
Reisman, Harold Bernard
Salafia, W(illiam) Ronald
Schetky, Laurence McDonald
Solonche, David Joshua
Stadnicki, Stanley Walter, Jr
Tuteur, Franz Benjamin
Wernau, William Charles

DELAWARE
Allen, Robert Harry
Bischoff, Kenneth Bruce
Dhurjati, Prasad S
Dwivedy, Ramesh C
Jefferies, Steven
Kilkson, Henn
Kingsbury, Herbert B
Messinger-Rapport, Barbara J
Pressman, Norman Jules
Read, Robert E
Santare, Michael Harold
Willis, Frank Marsden

DISTRICT OF COLUMBIA
Douple, Evan Barr
Jones, Janice Lorraine
Katona, Peter Geza
Ledley, Robert Steven
Ligler, Frances Smith
Parasuraman, Raja
Salu, Yehuda
Schnur, Joel Martin
Tozeren, Aydin
Youm, Youngil

FLORIDA
Aronson, M(oses)
Binford, Jesse Stone, Jr
Blatt, Joel Herman
Block, Seymour Stanton
Bramante, Pietro Ottavio
Bressler, Steven L
Callaghan, Frank J
Coulter, Wallace H
Deutsch, S(id)
Enger, Carl Christian
Finney, Roy Pelham
Freeman, Neil Julian
Gaudy, Anthony F, Jr
Gibbs, Charles Howard
Gittelman, Donald Henry
Goldberg, Eugene P
Goldstein, Mark Kane
Heller, Zindel Herbert
Henson, Carl P
Huckaba, Charles Edwin
Kato, Yasushi
Kline, Jacob
Lodwick, Gwilym Savage
McMillan, Donald Ernest
Martin, Charles John
Mayrovitz, Harvey N
Milne, Edward Lawrence
Nagel, Joachim Hans
Nunez, German
Phillips, Winfred M(arshall)
Piotrowski, George
Price, Joel McClendon
Purdy, Alan Harris
Ramsey, Maynard, III
Seireg, Ali A
Spurlock, Jack Marion
Tiederman, William Gregg, Jr
Walburn, Frederick J
Whipple, Royson Newton
Williamson, Donald Elwin

GEORGIA
Bhatia, Darshan Singh
Busch, Kenneth Louis
Cassanova, Robert Anthony
English, Arthur William
Gibson, John Michael
Hamada, Spencer Hiroshi
Ku, David Nelson
May, Sheldon William
Morman, Michael T
Nerem, Robert Michael
Searle, John Randolph
Threadgill, Ernest Dale
Toler, J C
Wang, Johnson Jenn-Hwa
Warner, Harold
Wolf, Steven L
Yoganathan, Ajit Prithiviraj
Zhu, Cheng
Ziffer, Jack

HAWAII
Gitlin, Harris Martlin
Koide, Frank T
Moser, Roy Edgar
Pruder, Gary David

IDAHO
Riemke, Richard Allan

ILLINOIS
Agarwal, Gyan C
Bacus, James William
Barenberg, Sumner
Bettice, John Allen
Birnholz, Jason Cordell
Borso, Charles S
Breillatt, Julian Paul, Jr
Brown, Sherman Daniel
Childress, Dudley Stephen
Cork, Douglas J
Cullum, Malford Eugene
Dallos, Peter John
Ducoff, Howard S
Dunn, Floyd
Eckmann, David M
Elble, Rodger Jacob
Feinberg, Barry N
Flinn, James Edwin
Francis, Howard Thomas
Frizzell, Leon Albert
Goldstick, Thomas Karl
Gottlieb, Gerald Lane
Gupta, Ramesh
Holmes, Kenneth Robert
Houk, James Charles
Hua, Ping
Hughes, John Russell
Jacobs, John Edward
Jadvar, Hossein
Joung, John Jongin
Kaganov, Alan Lawrence
Kertesz, Andrew (Endre)
Lauffenburger, Douglas Alan
Layton, Terry North
Lin, James Chih-I
Mavrovouniotis, Michael L
Miller, Irving F(ranklin)
Mockros, Lyle F(red)
Moore, John Fitzallen
O'Brien, William Daniel, Jr
Papoutsakis, Eleftherios Terry
Phillips, James Woodward
Punwani, Dharam Vir
Rawlings, Charles Adrian
Reynolds, Larry Owen
Rosen, Arthur Leonard
Rymer, William Zev
Sahakian, Alan Varteres
Sather, Norman F(redrick)
Schnell, Gene Wheeler
Sellers, Donald Roscoe
Steinberg, Marvin Phillip
Trimble, John Leonard
Weil, Max Harry
Wheeler, Bruce Christopher
Wolf, Ludwig, Jr
Yeates, Donovan B
Zuber, B(ert) L

INDIANA
Abel, Larry Allen
Amlaner, Charles Joseph, Jr
Brannon-Peppas, Lisa
Brooks, Austin Edward
Burr, David Bentley
DeWitt, David P
Fearnot, Neal Edward
Geddes, Leslie Alexander
Hinds, Marvin Harold
Hinkle, Charles N(elson)
Hulbert, Samuel Foster
Kessler, David Phillip
Koivo, Antti J
Laxer, Cary
Liska, Bernard Joseph
Maass-Moreno, Roberto
Marks, Jay Stewart
Mastrototaro, John Joseph
Peppas, Nikolaos Athanassiou
Potvin, Alfred Raoul
Queener, Stephen Wyatt
Rothe, Carl Frederick
Stiver, James Frederick
Turner, Charles Hall
Voelz, Michael H
Vogelhut, Paul Otto
Wasserman, Gerald Steward

IOWA
Buck, James R
Carlson, David L
Collins, Steve Michael
Crosby, Lon Owen
Engen, Richard Lee
Goel, Vijay Kumar
Huston, Jeffrey Charles
Lakes, Roderic Stephen
Liu, Young King
Myers, Glenn Alexander
Park, Joon B
Rethwisch, David Gerard
Siebes, Maria
Smith, Norman B
Titze, Ingo Roland
Young, Donald F(redrick)

KANSAS
Cooke, Francis W
Erickson, Howard Hugh
Erickson, Larry Eugene
Farrell, Eugene Patrick
Smith, Raymond V(irgil)
Zehr, John E

KENTUCKY
Bruce, Eugene N
Edwards, Richard Glenn
Fuller, Peter McAfee
Hanley, Thomas Richard
Harris, Patrick Donald
Knapp, Charles Francis
Lafferty, James Francis
Lai-Fook, Stephen J
McCook, Robert Devon
Patwardham, Abhijit R
Randall, David Clark
Squires, Robert Wright

LOUISIANA
Bellina, Joseph Henry
Bundy, Kirk Jon
Chen, Isaac I H
Conrad, Steven A
Cook, Stephen D
Day, Donal Forest
Ewing, Channing Lester
Gaver, Donald P, III
Hale, Paul Nolen, Jr
Huckabay, Houston Keller
Klyce, Stephen Downing
Moulder, Peter Vincent, Jr
Peattie, Robert Addison
Sarkar, Nikhil Kumar
Schubert, Roy W
Seaman, Ronald L
Solomonow, Moshe
Sterling, Arthur MacLean
Thomas, Kevin Anthony
Walker, Cedric Frank

MAINE
Bush, George F(ranklin)
Hodgkin, Brian Charles

MARYLAND
Abbrecht, Peter H
Ashman, Michael Nathan
Basser, Peter J
Baust, John G
Berson, Alan
Bhagat, Hitesh Rameshchandra
Bowen, Rafael Lee
Brody, William R
Bruley, Duane Frederick
Brunner, Martha J
Bungay, Peter M
Carlson, Drew E
Carmines, David Vaughn
Chadwick, Richard Simeon
Chang, Zhaohua
Charles, Harry Krewson, Jr
Chen, Ching-Nien
Clark, Carl Cyrus
Cohen, Gerald Stanley
Covell, David Gene
Daniels, William Fowler, Sr
DeLoatch, Eugene M
Douglas, Andrew Sholto
Eden, Murray
Fischell, Robert E
Fishman, Elliot Keith
Flanigan, William Francis, Jr
Fleshman, James Wilson, Jr
Frommer, Peter Leslie
Gabelnick, Henry Lewis
Geckle, William Jude
Geduldig, Donald
Giddens, Don P(eyton)
Goldstein, Moise Herbert, Jr
Goldstein, Seth Richard
Gordon, Stephen L
Graves, Willard L
Gupta, Vaikunth N
Hambrecht, Frederick Terry
Hamilton, Bruce King
Hamilton, Leroy Leslie
Harrison, George H
Haudenschild, Christian C
Heetderks, William John
Heinz, John Michael
Johnson, Arthur Thomas
Jones, Everett
Khachaturian, Zaven Setrak
Langlykke, Asger Funder
Leong, Kam W
LeRoy, André François
Mackie, Linda H
Massey, Joe Thomas
Massof, Robert W
Maughan, W Lowell
Meyer, Richard Arthur
Michelsen, Arve
Moreira, Antonio R
Neugroschl, Daniel
Newcomb, Robert Wayne
Popel, Aleksander S
Roth, Bradley John
Ruttimann, Urs E
Salcman, Michael
Schmidt, Edward Matthews
Schramm, Lawrence Peter
Severns, Matthew Lincoln
Sheppard, Norman F, Jr
Stromberg, Robert Remson
Swift, David Leslie
Thakor, Nitish Vyomesh
Tsitlik, Joshua E
Ulbrecht, Jaromir Josef
Vargas, Fernando Figueroa
Waterstrat, Richard Milton
Watson, John Thomas
Webb, George N
Weisfeldt, Myron Lee
Yang, Xiaowei
Yin, Frank Chi-Pong

MASSACHUSETTS
Aronow, Saul
Atala, Anthony
Barbino, Gilda A
Barngrover, Debra Anne
Breslau, Barry Richard
Bustead, Ronald Lorima, Jr
Chang, Kuo Wei
Charm, Stanley E
Chesler, David Alan
Clayton, J(oe) T(odd)
Collins, James Joseph
Colton, Clark Kenneth
Cook, Nathan Henry
Cooney, Charles Leland
Coury, Arthur Joseph
Cravalho, Ernest G
Cuffin, B(enjamin) Neil
Dale, William
De Luca, Carlo J
Dennison, Byron Lee
Dewey, C(larence) Forbes, Jr
Doane, Marshall Gordon
Downer, Nancy Wuerth
Drinker, Philip A
Eckhouse, Richard Henry
Feldman, Charles Lawrence
Fine, Samuel
Fishman, Philip M
Gaw, C Vernon
Gordy, Edwin
Grodzinsky, Alan J

Bioengineering & Biomedical Engineering (cont)

Gross, David John
Harris, Wayne G
Hatch, Randolph Thomas
He, Bin
Herzlinger, George Arthur
Hoffman, Allen Herbert
Horowitz, Arie
Jackson, Andrew C
Jain, Rakesh Kumar
Johnson, Richard Noring
Kamen, Gary P
Kamm, Roger Dale
Klibanov, Alexander M
Krasner, Jerome L
Kreifeldt, John Gene
Lampi, Rauno Andrew
Langer, Robert Samuel
Lederman, David Mordechai
Lees, Sidney
Lefebvre, Xavier P
Lutchen, Kenneth R
Maloney, John F
Mann, Robert W(ellesley)
Maran, Janice Wengerd
Mark, Roger G
Mastenbrook, S Martin, Jr
Meldon, Jerry Harris
Merfeld, Daniel Michael
Mikic, Bora
Mix, Thomas W
Mountain, David Charles, Jr
Norton, Robert Leo
Olderman, Jerry
Oman, Charles McMaster
Pan, Coda H T
Paul, Igor
Pedersen, Peder Christian
Petschek, Harry E
Peura, Robert Allan
Poirier, Victor L
Poon, Chi-Sang
Pope, Mary E
Rose, Timothy Laurence
Sahatjian, Ronald Alexander
Schoen, Frederick J
Shapiro, Ascher H(erman)
Shapiro, Howard Maurice
Sheridan, Thomas Brown
Singh, Param Indar
Sodickson, Lester A
Suki, Bela
Tuomy, Justin M(atthew)
Voigt, Herbert Frederick
Wall, Conrad, III
Wallace, David H
Wang, Lawrence K
Weinberg, Crispin Bernard
White, Augustus Aaron, III
Whitney, Lester F(rank)
Wiegner, Allen W
Young, Laurence Retman
Yuan, Fan
Zamenhof, Robert G A

MICHIGAN
Anderson, David J
Anderson, Thomas Edward
Asgar, Kamal
Be Ment, Spencer L
Cain, Charles Alan
Canale, Raymond Patrick
Chaffin, Don B
Chen, Michael Ming
Cox, Mary E
Cunningham, Fay Lavere
Francis, Ray Llewellyn
Fyhrie, David Paul
Goldstein, Steven Alan
Hillegas, William Joseph
Horvath, Ralph S(teve)
Kim, Changhyun
King, Albert Ignatius
Kohn, David H
Kootsey, Joseph Mailen
Levine, Simon P
McGrath, John Joseph
Meyerhoff, Mark Elliot
Midgley, A Rees, Jr
Miller, Thomas Lee
Nuttall, Alfred L
Nyquist, Gerald Warren
O'Donnell, Matthew
Radin, Eric Leon
Repa, Brian Stephen
Reynolds, Herbert McGaughey
Robertson, John Harvey
Schramm, John Gilbert
Schultz, Albert Barry
Soutas-Little, Robert William
Stein, Paul David
Viano, David Charles
Wang, Henry Y
Wiitanen, Wayne Alfred
Williams, William James
Wolterink, Lester Floyd
Yang, Victor Chi-Min
Yang, Wen Jei

MINNESOTA
Bakken, Earl E

Buchwald, Henry
Carim, Hatim Mohamed
Clapper, David Lee
Cornelissen Guillaume, Germaine G
Daniel, John Mahendra Kumar
Erdman, Arthur Guy
Finkelstein, Stanley Michael
Fredrickson, Arnold G(erhard)
Gibbons, Donald Frank
Greenleaf, James Fowler
Hu, Wei-Shou
Kallok, Michael John
Keshaviah, Prakash Ramnathpur
Khraibi, Ali A
Kvalseth, Tarald Oddvar
Mikhail, Adel Ayad
Miller, Nicholas Carl
Mueller, Rolf Karl
Neft, Nivard
Nicoloff, Demetre M
Olson, Walter Harold
Ray, Charles Dean
Schmitt, Otto Herbert
Shu, Mark Chong-Sheng
Soechting, John F
Standing, Charles Nicholas
Starkebaum, Warren L
Stephanedes, Yorgos Jordan
Thornton, Arnold William
Timm, Gerald Wayne

MISSISSIPPI
Matthes, Ralph Kenneth, Jr
Montani, Jean-Pierre
Pearce, David Harry
Sadana, Ajit

MISSOURI
Allen, William Corwin
Bajpai, Rakesh Kumar
Graham, Charles
Guccione, Julius Matteo
Hahn, Allen W
Landiss, Daniel Jay
Larson, Kenneth Blaine
McFarland, William D
Mavis, James Osbert
Morley, Robert Emmett, Jr
Peng, Yongren Benjamin
Rubin, Bruce Kalman
Saran, Chitaranjan
Schuder, John Claude
Thomas, Lewis Jones, Jr
Titus, Dudley Seymour
Twardowski, Zbylut Jozef
Tyrer, Harry Wakeley
Yasuda, Hirotsugu

MONTANA
Banaugh, Robert Peter
Jurist, John Michael

NEBRASKA
Chakkalakal, Dennis Abraham
Ellingson, Robert James
Haack, Donald C(arl)
Leviticus, Louis I

NEW HAMPSHIRE
Baumann, Hans D
Brous, Don W
Converse, Alvin O
Daubenspeck, John Andrew
Hayes, Kirby Maxwell
Kantrowitz, Arthur (Robert)
Strohbehn, John Walter

NEW JERSEY
Aberman, Harold Mark
Akerboom, Jack
Amory, David William
Atwater, Beauford W
Bernstein, Alan D
Cahn, Frederick
Campfield, L Arthur
Castellana, Frank Sebastian
Chen, Nai Y
Chu, Horn Dean
Clements, Wayne Irwin
Connor, John Arthur
Crump, Jesse Franklin
Davis, Thomas Arthur
Drzewiecki, Gary
Edelman, Norman H
Fischell, David R
Flam, Eric
Franse, Renard
Goggins, Jean A
Goyal, Suresh
Greenstein, Teddy
Hall, Joseph L
Halpern, Teodoro
Hong, Allan Jixian
Houston, Vern Lynn
Huang, Ching-Rong
Jacobs, Carl Henry
Kangovi, Sach
Kaufman, Arnold
Keller, Kenneth H(arrison)
Konikoff, John Jacob
Kristol, David Sol
Langrana, Noshir A
Li, John Kong-Jiann
Liebig, William John

Litman, Irving Ira
Lunn, Anthony Crowther
Maclin, Ernest
McNicholas, James J
Malkin, Robert Allen
Mallison, George Franklin
Masurekar, Prakash Sharatchandra
Mears, David R
Meyer, Andrew U
Micheli-Tzanakou, Evangelia
Pilla, Arthur Anthony
Reisman, Stanley S
Ronel, Samuel Hanan
Salkind, Alvin J
Semmlow, John Leonard
Shumate, Starling Everett, II
Silver, Frederick Howard
Sofer, Samir Salim
Stanton, Robert E
Stickle, Gene P
Szarka, Laszlo Joseph
Tojo, Kakuji
Torregrossa, Robert Emile
Trivedi, Nayan B
Tzanakou, M Evangelia
Weiss, Lawrence H(eisler)
Welkowitz, Walter
West, John M(aurice)
Williams, Maryon Johnston, Jr
Yarmush, Martin Leon
Ziskin, Marvin Carl

NEW MEXICO
Cho, Michael Yongkook
Doss, James Daniel
Fukushima, Eiichi
Gray, Edwin R
Johnston, Roger Glenn
Novak, James Lawrence
Roubicek, Rudolf V
Sinha, Dipen N
Welford, Norman Traviss
Wilkins, Ebtisam A M Seoudi

NEW YORK
Akers, Charles Kenton
Alexander, Harold
Baier, Robert Edward
Belfort, Georges
Bell, Duncan Hadley
Bennett, Leon
Berndt, Christopher Charles
Bizios, Rena
Blesser, William B
Borden, Edward B
Bottomley, Paul Arthur
Branch, Garland Marion, Jr
Brennan, John Paul
Brenner, Mortimer W
Brunski, John Beyer
Bungay, Henry Robert, III
Capps, Susan Glass
Carstensen, Edwin L(orenz)
Cheney, Margaret
Clark, Alfred, Jr
Cluxton, David H
Cohen, Irving Allan
Cokelet, Giles R(oy)
Cowin, Stephen Corteen
Csermely, Thomas J(ohn)
Dahl, John Robert
Dobelle, William Harvey
Doremus, Robert Heward
Duffey, Michael Eugene
Engbretson, Gustav Alan
Falk, Theodore J(ohn)
Fewkes, Robert Charles Joseph
Fordon, Wilfred Aaron
Fordyce, Wayne Edgar
Frankel, Victor H
Gilmore, Robert Snee
Goodhue, Charles Thomas
Greatbatch, Wilson
Greene, Peter Richard
Harris, Jack Kenyon
Hart, Howard Roscoe, Jr
Hovnanian, H(rair) Philip
Hrushesky, William John Michael
Kim, Young Joo
Kinnen, Erwin
Kletsky, Earl J(ustin)
Knowles, Richard James Robert
Kohn, Michael
Konnerth, Karl Louis
Lai, W(ei) Michael
Leary, James Francis
Lee, George C
Lee, Richard Shao-Lin
Leisman, Gerald
Lenchner, Nathaniel Herbert
Levine, Sumner Norton
Lieber, Baruch B
Litz, Lawrence Marvin
Liu, Mao-Zu
LoGerfo, John J
Lubowsky, Jack
Luther, Herbert George
Mann, Kenneth A
Mates, Robert Edward
Maurice, David Myer
Mellins, Robert B
Meltzer, Richard Stuart
Meth, Irving Marvin
Miller, Christine E

Moghadam, Omid A
Morris, Thomas Wilde
Moss, Gerald
Moss, Melvin Lionel
Mow, Van C
Mozley, James Marshall, Jr
Murphy, Eugene F(rancis)
Newell, Jonathan Clark
Oldshue, J(ames) Y(oung)
Ostrander, Lee E
Otter, Mark William
Palmer, Harvey John
Pierson, Richard Norris, Jr
Quan, Weilun
Rabbany, Sina Y
Ranu, Harcharan Singh
Redington, Rowland Wells
Riffenburgh, Robert Harry
Roy, Rob
Savic, Michael I
Schoenfeld, Robert Louis
Scott, Peter Douglas
Shapiro, Paul Jonathon
Shuler, Michael Louis
Slezak, Jane Ann
Smith, Lowell Scott
Smith, Robert L
Sotirchos, Stratis V
Susskind, Herbert
Tichauer, Erwin Rudolph
Tobin, Michael
Torre, Douglas Paul
Tsai, Gow-Jen
Tycko, Daniel H
Vance, Miles Elliott
Viernstein, Lawrence J
Vogelman, Joseph H(erbert)
Wald, Alvin Stanley
Weinbaum, Sheldon
Weinstein, Herbert
Weissman, Michael Herbert
Wertman, Louis
Wheeless, Leon Lum, Jr
Wiegert, Philip E
Xu, Xuemin
Yellin, Edward L
Yoon, Hyo Sub
Zablow, Leonard
Zelman, Allen
Zwislocki, Jozef John

NORTH CAROLINA
Abrams, Charlie Frank, Jr
Archie, Joseph Patrick, Jr
Barnes, Ralph W
Barr, Roger Coke
Blanchard, Susan M
Clark, Howard Garmany
Coulter, Norman Arthur, Jr
Defoggi, Ernest
DeGuzman, Allan Francis
Eberhardt, Allen Craig
Floyd, Carey E, Jr
Gerhardt, Don John
Glower, Donald D, Jr
Hammond, William Edward
Haynes, John Lennis
Hochmuth, Robert Milo
Hooper, Irving R
Hsiao, Henry Shih-Chan
Huang, Barney K(uo-Yen)
Huang, Eng-Shang (Clark)
Hutchins, Phillip Michael
Jaszczak, Ronald Jack
Kelley, Thomas F
Klitzman, Bruce
Knisley, Stephen
Krassowska, Wanda
Lawson, Dewey Tull
Leatherman, Nelson E(arle)
Lucas, Carol N
Malindzak, George S, Jr
Mercer, Robert R
Mosberg, Arnold T
Nagle, H Troy
Owens, Boone Bailey
Pasipoularides, Ares D
Pearlstein, Robert David
Periasamy, Ammasi
Plonsey, Robert
Sandok, Paul Louis
Smith, Charles Eugene
Starmer, C Frank
Thubrikar, Mano J
Truskey, George A
Tsui, Benjamin Ming Wah
Weiner, Richard D
Wiley, Albert Lee, Jr
Wolbarsht, Myron Lee
Wolcott, Thomas Gordon
Zucker, Robert Martin

NORTH DAKOTA
Bares, William Anthony
Mathsen, Don Verden

OHIO
Andonian, Arsavir Takfor
Bahniuk, Eugene
Bennett, G(ary) F
Berliner, Lawrence J
Brown, Stanley Alfred
Brunner, Gordon Francis
Chovan, John David

Chu, Mamerto Loarca
Clark, David Lee
Dell'Osso, Louis Frank
Durand, Dominique M
Engin, Ali Erkan
Fatemi, Ray S
Fink, David Jordan
Frank, Thomas Paul
Friedman, Morton Harold
Genaidy, Ashraf Mohamed
Glaser, Roger Michael
Greber, Isaac
Greenberg, David B(ernard)
Grood, Edward S
Hall, Ernest Lenard
Hangartner, Thomas Niklaus
Hartrum, Thomas Charles
Hinman, Channing L
Hughes, Kenneth E(ugene)
Kahn, Alan Richard
Katz, J Lawrence
Katzen, Raphael
Kinzel, Gary Lee
Knox, Francis Stratton, III
Ko, Wen Hsiung
Kwatra, Subhash Chander
LaManna, Joseph Charles
Lee, Sunggyu
Leininger, Robert Irvin
Levy, Matthew Nathan
McComis, William T(homas)
McGoron, Anthony Joseph
Macklin, Martin
McMillin, Carl Richard
Marchant, Roger E
Marras, William Steven
Martin, Paul Joseph
Miraldi, Floro D
Moore, Thomas Joseph
Mortimer, J(ohn) Thomas
Mudry, Karen Michele
Neuman, Michael R
Peckham, P Hunter
Phillips, Chandler Allen
Primiano, Frank Paul, Jr
Prohaska, Oho J
Raftopoulos, Demetrios D
Reddy, Narender Pabbathi
Repperger, Daniel William
Reynolds, David B
Ricord, Louis Chester
Rittgers, Stanley Earle
Roberts, Cynthia J
Rudy, Yoram
Saidel, Gerald Maxwell
Slonim, Arnold Robert
Taylor, Bruce Cahill
Thomas, Cecil Wayne
Tuovinen, Olli Heikki
Verstraete, Mary Clare
Von Gierke, Henning Edgar
Waldron, Manjula Bhushan
Wasserman, Donald Eugene
Webb, Paul
Weed, Herman Roscoe
Wolaver, Lynn E(llsworth)
Wong, Anthony Sai-Hung
Yoon, Jong Sik

OKLAHOMA
Berbari, Edward J
Campbell, John Alexander
Collins, William Edward
Dormer, Kenneth John
Foutch, Gary Lynn
Neathery, Raymond Franklin
Snow, Clyde Collins
Stover, Enos Loy
Troelstra, Arne

OREGON
Ehrmantraut, Harry Charles
Fields, R(ance) Wayne
Skiens, William Eugene
Smith, Kelly L
Yang, Hoya Y

PENNSYLVANIA
Alteveer, Robert Jan George
Altschuler, Martin David
Angstadt, Carol Newborg
Baish, James William
Batterman, Steven C(harles)
Berkowitz, David Andrew
Bilgutay, Nihat Mustafa
Borovetz, Harvey S
Boston, John Robert
Brighton, John Austin
Briller, Stanley A
Buchsbaum, Gershon
Buerk, Donald Gene
Cape, Edward
Cavanagh, Peter R
Chapman, John Donald
Ching, Stephen Wing-Fook
Cox, Robert Harold
Detwiler, John Stephen
Dinges, David Francis
Eisenstein, Bruce A
Fan, Ningping
Fromm, Eli
Gee, William
George, David T
Geselowitz, David B(eryl)

Giannovario, Joseph Anthony
Grossmann, Elihu D(avid)
Hargens, Charles William, III
Harris, Lowell Dee
Johnson, Ogden Carl
Kameneva, Marina Vitaly
Klafter, Richard D(avid)
Kline, Donald Edgar
Kornfield, A(lfred) T(heodore)
Kozak, Wlodzimierz M
Kresh, J Yasha
Kwatny, Eugene Michael
Lankford, Edward B
Li, C(hing) C(hung)
Litt, Mitchell
Longini, Richard Leon
Loscher, Robert A
Metaxas, Dimitri
Natoli, John
Newhouse, Vernon Leopold
Nobel, Joel J
Noordergraaf, Abraham
Onik, Gary M
Parker, Jennifer Ware
Pelleg, Amir
Pennock, Bernard Eugene
Pollack, Solomon R
Purdy, David Lawrence
Rajagopal, K R
Reid, John Mitchell
Robinson, Charles J
Russell, Alan James
Samuelson, H Vaughn
Scherer, Peter William
Schulman, Marvin
Schwan, Herman Paul
Sciamanda, Robert Joseph
Sharma, Mangalore Gokulanand
Sharpless, Thomas Kite
Sheers, William Sadler
Shung, K Kirk
Sitrin, Robert David
Spicher, John L
Spooner, Robert Bruce
Stone, Douglas Roy
Sutter, Philip Henry
Tarbell, John M
Thomas, Donald H(arvey)
Ultman, James Stuart
Villafana, Theodore
Voloshin, Arkady S
Wampler, D Eugene
Weisz, Paul Burg
Whinnery, James Elliott
Williams, Albert J, Jr
Wolken, Jerome Jay

RHODE ISLAND
Barnett, Stanley M(arvin)
Christenson, Lisa
Galletti, Pierre Marie
Hoffman, Joseph Ellsworth, Jr
Marsh, Donald Jay
Richardson, Peter Damian

SOUTH CAROLINA
Akpan, Edward
Friedman, Richard Joel
Hargest, Thomas Sewell
Lam, Chan F
McCutcheon, Ernest P
McNamee, James Emerson
Moyle, David Douglas
Sias, Fred R, Jr
Turner, John Lindsey
Voit, Eberhard Otto
Von Recum, Andreas F
Watson, Philip Donald

TENNESSEE
Bahner, Carl Tabb
Barach, John Paul
Campbell, William B(uford), Jr
Collins, Jerry C
Davison, Brian Henry
Deivanayagan, Subramanian
Donaldson, Terrence Lee
Eckstein, Eugene Charles
Galloway, Robert L
Gan, Rong Zhu
Genung, Richard Keith
Goulding, Charles Edwin, Jr
Harris, Thomas R(aymond)
King, Paul Harvey
McLeod, William D
Ogg, Robert James
Overholser, Knowles Arthur
Partridge, Lloyd Donald
Proctor, Kenneth Gordon
Roselli, Robert J
Schreuders, Paul D
Sittig, Dean Forrest
Slack, Steven M
Terry, Fred Herbert
Turitto, Vincent Thomas
Wasserman, Jack F
Wikswo, John Peter, Jr
Yen, Michael R T

TEXAS
Aggarwal, Shanti J
Alexander, William Carter
Alfrey, Clarence P, Jr
Armeniades, Constantine D

Athanasiou, Kyriacos A
Baker, Lee Edward
Boriek, Aladin Mohammed
Brown, Jack Harold Upton
David, Yadin B
Diller, Kenneth Ray
Dunbar, Bonnie J
Dunham, James George
Dykstra, Jerald Paul
Eberhart, Robert Clyde
Eberle, Jon William
Eggers, Fred M
Fleenor, Marvin Bension
Fletcher, John Lynn
Frenger, Paul F
Georgiou, George
Gilmour-Stallsworth, Lisa K
Gutierrez, Guillermo
Hartley, Craig Jay
Hellums, Jesse David
Homsy, Charles Albert
Howard, Lorn Lambier
Johnston, Melvin Roscoe
Jones, William B
Journeay, Glen Eugene
Kolesar, Edward S
Koppa, Rodger J
Krock, Larry Paul
Krouskop, Thomas Alan
Lioi, Anthony Pasquale
McFadyen, Gary M
McIntire, Larry V(ern)
McIntyre, John Armin
McKay, Colin B
Marshall, Robert P(aul)
Mikos, Antonios G
Miller, Gerald E
Ong, Poen Sing
Pandy, Marcus G
Patel, Anil S
Philippe, Edouard Antoine
Popovich, Robert Peter
Ramsey, Jerry Dwain
Rollwitz, William Lloyd
Schermerhorn, John W
Senatore, Ford Fortunato
Sheppard, Louis Clarke
Skolnick, Malcolm Harris
Slotta, Larry Stewart
Soltes, Edward John
Srebro, Richard
Srinivasan, Ramachandra Srini
Summers, Richard Lee
Sweat, Vincent Eugene
Throckmorton, Gaylord Scott
Thurston, George Butte
Triano, John Joseph
Von Maltzahn, Wolf W
White, Ronald Joseph
Zhong, Pei

UTAH
Andrade, Joseph D
Berggren, Michael J
Billings, R Gail
Daniels, Alma U(riah)
Gardner, Reed McArthur
Germane, Geoffrey James
Horch, Kenneth William
Kim, Sung Wan
Kopecek, Jindrich
Kratochvil, Jiri
Lee, James Norman
Normann, Richard A
Olsen, Don B
Peterson, Stephen Craig
Powers, Linda Sue
Solea, Kenneth A
Strozier, James Kinard
Trujillo, Edward Michael
Tuckett, Robert P
Warner, Homer R

VERMONT
Dean, Robert Charles, Jr
Ellis, David M
Keller, Tony S
Lipson, Richard L
Sachs, Thomas Dudley

VIRGINIA
Ackerman, Roy Alan
Adams, James Milton
Alter, Ralph
Anne, Antharvedi
Arp, Leon Joseph
Attinger, Ernst Otto
Barrett, James R
Beran, Robert Lynn
Boadle-Biber, Margaret Clare
Chen, Jiande
Cheng, George Chiwo
Chi, Michael
Clarke, Alexander Mallory
Diller, Thomas Eugene
Edlich, Richard French
Eppink, Richard Theodore
Evans, Francis Gaynor
Gabriel, Barbra L
Gillies, George Thomas
Grant, John Wallace
Heineken, Frederick George
Hutchinson, Thomas Eugene
Jaron, Dov

Kim, Yong Il
Lee, Jen-Shih
Lemp, John Frederick, Jr
Lichtenberg, Byron Kurt
Llewellyn, Gerald Cecil
Mikulecky, Donald C
Moon, Peter Clayton
Moraff, Howard
Myers, Donald Albin
Reswick, James Bigelow
Rich, George F
Sudarshan, T S
Summers, George Donald
Theodoridis, George Constantin
Tole, John Roy
Updike, Otis L(ee), Jr
Vaughan, Christopher Leonard
Wayne, Jennifer Susan
Wilkins, Judd Rice
Wist, Abund Ottokar

WASHINGTON
Andersen, Jonny
Auth, David C
Babb, Albert L(eslie)
Bassingthwaighte, James B
Bhansali, Praful V
Binder, Marc David
Canfield, Robert Charles
Cholvin, Neal R
Daly, Colin Henry
Forster, Fred Kurt
Fuchs, Albert Frederick
Halbert, Sheridan A
Heideger, William J(oseph)
Hoffman, Allan Sachs
Hsu, Chin Shung
Huntsman, Lee L
Jacobson, John Obert
Janata, Jiri
Kaune, William Tyler
Klepper, John Richard
Lago, James
Lee, James Moon
Luft, John Herman
MacGinitie, Laura A
Miller, Douglas Lawrence
Newman, Paul Harold
Pollack, Gerald H
Ratner, Buddy Dennis
Reeves, Jerry John
Riederer-Henderson, Mary Ann
Ringo, John Alan
Robkin, Maurice
Rushmer, Robert Frazer
Sanders, Joan Elizabeth
Staiff, Donald C
Stear, Edwin Byron
Tam, Patrick Yui-Chiu
Taylor, Eugene M
Van der Werff, Terry Jay
Yee, Sinclair Shee-Sing

WEST VIRGINIA
Albertson, John Newman, Jr
Crum, Edward Hibbert
Kumar, Alok
Mucino, Victor H
Shaeiwitz, Joseph Alan
Shuck, Lowell Zane
Zeng, Shengke

WISCONSIN
Balmer, Robert Theodore
Barnicki, Steven L
Cataldi, Horace A(nthony)
Cohen, Bernard Allan
Divjak, August A
Flax, Stephen Wayne
Geisler, C(hris) D(aniel)
Harrington, Sandra Serena
Hinkes, Thomas Michael
Jeutter, Dean Curtis
Linehan, John Henry
Prieto, Thomas E
Sances, Anthony, Jr
Schopler, Harry A
Sfat, Michael R(udolph)
Sheehy, Michael Joseph
Silver-Thorn, M Barbara
Skatrud, Thomas Joseph
Smith, Michael James
Thompson, Paul DeVries
Tompkins, Willis Judson
Trautman, Jack Carl
Vanderheiden, Gregg
Vonderby, Ray
Warner, H Jack
Webster, John Goodwin
Whiffen, James Douglass
Widera, Georg Ernst Otto
Yoganandan, Narayan
Zuperku, Edward John

WYOMING
Ferris, Clifford D
Steadman, John William

PUERTO RICO
Patra, Amit Lal

ALBERTA
Jack, Thomas Richard
Rajotte, Ray V

Bioengineering & Biomedical Engineering (cont)

Shysh, Alec
Wayman, Morris
Williams, Michael C(harles)

BRITISH COLUMBIA
Dower, Gordon Ewbank
Hoffer, J(oaquin) A(ndres)
Mantle, J(ohn) B(ertram)
Milsum, John H
Morrison, James Barbour
Shaw, A(lexander) J(ohn)

MANITOBA
Delecki, Andrew
Gordon, Richard

NEW BRUNSWICK
Scott, Robert Nelson

NOVA SCOTIA
Rautaharju, Pentti M
Wong, Alan Yau Kuen

ONTARIO
Albisser, Anthony Michael
Bewtra, Jatinder Kumar
Brach, Eugene Jenö
Brash, John Law
Cairns, William Louis
Cobbold, R S C
Diosady, Levente Laszlo
Fenster, Aaron
Finlay, Joseph Bryan
Ghista, Dhanjoo Noshir
Hill, Martha Adele
Hopps, John Alexander
Jacobson, Stuart Lee
James, David F
Joy, Michael Lawrence Grahame
Kennedy, James Cecil
Kennedy, Kevin Joseph
Kim, Sukyoung
Krull, Ulrich Jorg
Kunov, Hans
Landolt, Jack Peter
Leung, Lai-Wo Stan
Loeb, Gerald Eli
McNeice, Gregory Malcolm
Norwich, Kenneth Howard
Roach, Margot Ruth
Robinson, Campbell William
Sherebrin, Marvin Harold
Slutsky, Arthur
Smith, Kenneth Carless
Sun, Anthony Mein-Fang
Thansandote, Artnarong
White, Denis Naldrett
Winter, David Arthur
Wu, Tai Wing
Zingg, Walter

PRINCE EDWARD ISLAND
Amend, James Frederick
Lodge, Malcolm A

QUEBEC
Bates, Jason H T
Billette, Jacques
Boyarsky, Abraham Joseph
Chang, Thomas Ming Swi
Cooper, David Gordon
Davis, John F(rederick)
Drouin, Gilbert
Frojmovic, Maurice Mony
Gulrajani, Ramesh Mulchand
Kearney, Robert Edward
Leduy, Anh
Roberge, Fernand Adrien
Roy, Guy
Volesky, Bohumil
Zucker, Steven Warren

SASKATCHEWAN
Cullimore, Denis Roy
Gupta, Madan Mohan
Macdonald, Douglas Gordon
Pollak, Viktor A

OTHER COUNTRIES
Agathos, Spiros Nicholas
Barbenel, Joseph Cyril
Bar-Yishay, Ephraim
Chang, H K
Cooke, T Derek V
De Bruyne, Pieter
Dinnar, Uri
Dunn, Irving John
Eik-Nes, Kristen Borger
Elad, David
Fujiwara, Keigi
Hazeyama, Yuji
Lee, Choong Woong
Liepsch, Dieter W
Monos, Emil
Moore, James Edward, Jr
O'Neill, Patricia Lynn
Orphanoudakis, Stelios Constantine
Rodriguez, Federico Angel
Shapiro, Stuart
Suga, Hiroyuki

Tipans, Igors O
Westerhof, Nicolaas
Woo, Kwang Bang
Yang, Ho Seung
Yoshida, Fumitake

Biological Engineering

ALABAMA
Taylor, Kenneth Boivin

CALIFORNIA
Downing, Willis G, Jr
Friedman, Milton Joe
Marmarelis, Vasilis Z
Martinez-Uriegas, Eugenio
Mishra, Nirmal K
Pitcher, Wayne Harold, Jr
Rabinowitz, Israel Nathan
Reilly, Dorothea Eleanor
Ryan, Wen
Sung, Kuo-Li Paul
Zamost, Bruce Lee

COLORADO
Palmer, Bradley M
Seely, Robert J
Wyman, Charles Ely

DELAWARE
Anton, David L
Work, Dennis M

DISTRICT OF COLUMBIA
Hassan, Aftab Syed

FLORIDA
Dewanjee, Mrinal K
Teixeira, Arthur Alves

GEORGIA
Brockbank, Kelvin Gordon Mashader

IDAHO
DeShazer, James Arthur
Korus, Roger Alan
Riemke, Richard Allan

ILLINOIS
Kim, Byung J
Papatheofanis, Frank John

IOWA
Dordick, Jonathan Seth

KENTUCKY
Wang, Yi-Tin

MARYLAND
Kasianowicz, John James
Sung, Cynthia

MASSACHUSETTS
Margolin, Alexey L
Remmel, Ronald Sylvester
Toner, Mehmet

MINNESOTA
Iaizzo, Paul Anthony

MISSISSIPPI
Gilbert, Jerome A

MISSOURI
Heldman, Dennis Ray

NEBRASKA
Hoffman, Glenn Jerrald

NEW JERSEY
Lewandowski, Gordon A
Yu, Stephen Kin-Cheun
Zoltan, Bart Joseph

NEW MEXICO
Cho, Michael Yongkook
Khandan, Nirmala N
Roubicek, Rudolf V

NEW YORK
Capps, Susan Glass
Chu, Samuel Sheung-Tak
Elander, Richard Paul
Tsai, Gow-Jen

OREGON
Hashimoto, Andrew G

PENNSYLVANIA
Dong, Cheng
Drake, B(illy) Blandin
Garwin, Jeffrey Lloyd

SOUTH CAROLINA
Dille, John Emmanuel

TENNESSEE
Koh, Sung-Cheol
Prokop, Ales

TEXAS
Baldwin, John Timothy
Holtzapple, Mark Thomas

McIntire, Larry V(ern)
Menon, Venugopal Balakrishna
Takatani, Setsuo
Wu, Hsin-i

UTAH
Olsen, Edwin Carl, III

VIRGINIA
Kirpekar, Abhay C

WISCONSIN
Lelkes, Peter Istvan

PUERTO RICO
Patra, Amit Lal

ONTARIO
Kosaric, Naim
Legge, Raymond Louis
Mittal, Gauri S

QUEBEC
Couillard, Denis

Ceramics Engineering

ALABAMA
Gates, D(aniel) W(illiam)
Heystek, Hendrik
Pattillo, Robert Allen
Yee, Tin Boo

ARIZONA
Carpenter, Ray Warren
Hansen, Kent W(endrich)
Kingery, W David
Savrun, Ender
Smyth, Jay Russell
Sundahl, Robert Charles, Jr
Uhlmann, Donald Robert
Willson, Donald Bruce
Withers, James C

CALIFORNIA
Baba, Paul David
Baldwin, Chandler Milnes
Bieler, Barrie Hill
Bragg, Robert H(enry)
Brown, William E(ric)
Bunshah, Rointan F(ramroze)
Carniglia, Stephen C(harles)
Clarke, David Richard
Cline, Carl F
Dejonghe, Lutgard C
Dunn, Bruce
Evans, Anthony Glyn
Feigelson, Robert Saul
Ferreira, Laurence E
Fletcher, Peter C
Gulden, Terry Dale
Gur, Turgut M
Halden, Frank
Harvey, A(lexander)
Hoenig, Clarence L
Hopper, Robert William
Johnson, Sylvia Marian
Kim, Ki Hong
Kuan, Teh S
Kurtz, Peter, Jr
Langdon, Terence George
Leipold, Martin H(enry)
Lin, Charlie Yeongching
McCreight, Louis R(alph)
Macha, Milo
McKee, W(illiam) Dean, Jr
Mecartney, Martha Lynn
Meiser, Michael David
Munir, Zuhair A
Nacamu, Robert Larry
Nieh, Tai-Gang
Nutt, Steven R
Paquette, David George
Pask, Joseph Adam
Prindle, William Roscoe
Ram, Michael Jay
Risbud, Subhash Hanamant
Rothman, Albert J(oel)
Sachdev, John
Savitz, Maxine Lazarus
Scott, Garland Elmo, Jr
Seiling, Alfred William
Shackelford, James Floyd
Sines, George, Jr
Spera, Frank John
Van Dreser, Merton Lawrence
Watson, James Frederic
Weinland, Stuart Louis
White, Jack Lee
Yang, Jenn-Ming
Yeh, Harry Chiang
Yoon, Rick K

COLORADO
DePoorter, Gerald Leroy
Dinneen, Gerald Uel
Koenig, John Henry
Moore, John Jeremy
Roshko, Alexana
Speil, Sidney
Thurnauer, Hans

CONNECTICUT
Abraham, Thomas
Eppler, Richard A
Grossman, Leonard N(athan)
Hulse, Charles O
Kilbourn, Barry T
Lemkey, Frankin David
Myers, Mark B
Nath, Dilip K
Newman, Robert Weidenthal
Popplewell, James Malcolm
Prasad, Arun
Smith, Donald Arthur
Strife, James Richard

DELAWARE
Arots, Joseph B(artholomew)
Burn, Ian
Chowdhry, Uma
Hartmann, Hans S(iegfried)
Hinton, Jonathan Wayne
Scherer, George Walter
Schiroky, Gerhard H
Slack, Lyle Howard
Tomic, Ernst Alois
Urquhart, Andrew Willard
Williams, Richard Anderson

DISTRICT OF COLUMBIA
Aggarwal, Ishwar D
Gottschall, Robert James
Kramer, Bruce Michael
Rao, Jaganmohan Boppana Lakshmi
Schioler, Liselotte Jensen
Tokar, Michael
Wolf, Stanley Myron

FLORIDA
Beck, Warren R(andall)
Berger, Richard Lee
Birch, Raymond E(mbree)
Clark, David Edward
Crandall, William B
Hench, Larry Leroy
McCracken, Walter John
Mecholsky, John Joseph, Jr
Mitoff, S(tephan) P(aul)
Moudgil, Brij Mohan
Owens, James Samuel
Ricker, Richard W(ilson)
Stokes, Charles Anderson
Ucci, Pompelio Angelo
Whitney, Ellsworth Dow
Wygant, J(ames) F(rederic)

GEORGIA
Chandan, Harish Chandra
Chapman, Alan T
Lackey, Walter Jackson
Logan, Kathryn Vance
Moody, Willis E, Jr
Pentecost, Joseph L(uther)
Proctor, W(illiam) J(efferson), Jr
Sanders, T H, Jr
Schulz, David Arthur
Soora, Siva Shunmugam
Starr, Thomas Louis

IDAHO
Tallman, Richard Louis
Weyand, John David

ILLINOIS
Albertson, Clarence E
Berg, M(orris)
Bergeron, Clifton George
Berkelhamer, Louis H(arry)
Bickelhaupt, R(oy) E(dward)
Bratschun, William R(udolph)
Brown, Sherman Daniel
Bunting, Bruce Gordon
Cosper, David Russell
Crowley, Michael Summers
Fenske, George R
Fessler, Raymond R
Johnson, D(avid) Lynn
Mason, Thomas Oliver
Myles, Kevin Michael
Poeppel, Roger Brian
Richardson, Joel Albert
Rothman, Steven J
Teng, Mao-Hua

INDIANA
Byers, Stanley A
Kennel, William E(lmer)
Kuo, Charles C Y
Lyon, K(enneth) C(assingham)
Tucker, Robert C, Jr
Wallace, F Blake

IOWA
Berard, Michael F
McGee, Thomas Donald
Martin, S W
Park, Joon B
Patterson, John W(illiam)
Wilder, David Randolph

KANSAS
Cooke, Francis W
Frazier, A Joel
Grisafe, David Anthony

KENTUCKY
Brock, Louis Milton
Currie, Thomas Eswin
Patrick, Robert F(ranklin)
Sarma, Atul C

LOUISIANA
Eaton, Harvill Carlton

MAINE
Hopkins, George Robert

MARYLAND
Adams, Edward Franklin
Bhalla, Sushil K
Block, Stanley
Chuang, Tze-jer
Cooperstein, Raymond
Coyle, Thomas Davidson
Diness, Arthur M(ichael)
Dolhert, Leonard Edward
Economos, Geo(rge)
Fishman, Steven Gerald
Freiman, Stephen Weil
Gillich, William John
Guruswamy, Vinodhini
Haller, Wolfgang Karl
Hovmand, Svend
Kacker, Raghu N
Kerkar, Awdhoot Vasant
Kerper, Matthew J(ulius)
Lundsager, C(hristian) Bent
Machlin, Irving
McIlvain, Jess Hall
Malghan, Subhaschandra Gangappa
Merz, Kenneth M(alcolm), Jr
Munro, Ronald Gordon
Murray, Peter
Nagle, Dennis Charles
O'Keefe, John Aloysius
Parker, Frederick John
Quinn, George David
Ricker, Richard Edmond
Simiu, Emil
Van Echo, Andrew
Wiederhorn, Sheldon M
Willis, James Byron
Winzer, Stephen Randolph

MASSACHUSETTS
Aldrich, Ralph Edward
Alliegro, Richard Alan
Ault, N(eil) N(orman)
Bates, Carl H
Blum, John Bennett
Blum, Seymour L
Chang, Frank C
Coble, R(obert) L(ouis)
Elliott, John Frank
Haggerty, John S
Keat, Paul Powell
Khattak, Chandra Prakash
Kim, Han Joong
Kumar, Kaplesh
Lee, D(on) William
Li, Rounan
Lu, Grant
Mahlo, Edwin K(urt)
Messier, Donald Royal
Pal, Uday Bhanu
Passmore, Edmund M
Peters, Edward Tehle
Poirier, Victor L
Quackenbush, Carr Lane W
Rhodes, William Holman
Schmidt, Werner H(ans)
Schroter, Stanislaw Gustaw
Schwartz, Thomas Alan
Seyferth, Dietmar
Suresh, Subra
Thieme, Cornelis Leo Hans
Trostel, Louis J(acob), Jr
Tuller, Harry Louis
Tustison, Randal Wayne
Vasilos, Thomas
Viechnicki, Dennis J
Wuensch, Bernhardt J(ohn)

MICHIGAN
Allan, George B
Baney, Ronald Howard
Brubaker, Burton D(ale)
Chandra, Grish
Cornilsen, Bahne Carl
Fawcett, Timothy Goss
Hauth, Willard Ellsworth, III
Hucke, Edward E
Humenik, Michael, Jr
Jain, Kailash Chandra
Lipowitz, Jonathan
Platts, Dennis Robert
Pynnonen, Bruce W
Russell, William Charles
Subramanian, K N
Tien, Tseng-Ying
Van Vlack, Lawrence H(all)
Whalen, Thomas J(ohn)
Willson, Philip James

MINNESOTA
Bailey, Joseph T(homas)
Fleming, Peter B
Huffine, Coy L(ee)
McHenry, Kelly David

Owens, Kenneth Eugene
Sowman, Harold G

MISSISSIPPI
Wehr, Allan Gordon

MISSOURI
Anderson, Harlan U(rie)
Crookston, J(ames) A(damson)
Day, D(elbert) E(dwin)
French, James Edwin
Grant, Sheldon Kerry
Harmon, Robert Wayne
Heimann, Robert L
Hunter, Orville, Jr
Mattox, Douglas M
Ownby, P(aul) Darrell
Peng, Yongren Benjamin
Ramey, Roy Richard
Sparlin, Don Merle
Viswanath, Dabir S

NEVADA
Bradt, Richard Carl
Eichbaum, Barlane Ronald

NEW HAMPSHIRE
Beasley, Wayne M(achon)
Marion, Robert Howard
Rice, Dale Wilson

NEW JERSEY
Abendroth, Reinhard P(aul)
Allegretti, John E
Bachman, George S(trickler)
Berkman, Samuel
Bhandarkar, Suhas D
Blank, Stuart Lawrence
Charvat, F(edia) R(udolf)
Comeforo, Jay E(ugene)
Dentai, Andrew G
Johnson, David W, Jr
Jones, John Taylor
Khan, Saad Akhtar
Kramer, Carolyn Margaret
Kurkjian, Charles R(obert)
LaMastro, Robert Anthony
Lemaire, Paul J
Ling, Hung Chi
Loh, Roland Ru-loong
McLaren, Malcolm G(rant)
Newman, Stephen Alexander
Rabinovich, Eliezer M
Rorabaugh, Donald T
Rosen, Carol Zwick
Shanefield, Daniel J
Speronello, Barry Keven
Tenzer, Rudolf Kurt
Tischler, Oscar
Voss, Kenneth Edwin
Wachtman, John Bryan, Jr
Wenzel, John Thompson
Yan, Man Fei
Yoon, Euijoon

NEW MEXICO
Beauchamp, Edwin Knight
Diver, Richard Boyer, Jr
Eagan, Robert John
Hall, Charles A(insley)
Hockett, John (Edwin)
Kreidl, Norbert J(oachim)
Land, Cecil E(lvin)
Loehman, Ronald Ernest
Magnani, Nicholas J
Matthews, R(obert) B(ruce)
Michalske, Terry Arthur
Milewski, John Vincent
Mitchell, Terence Edward
Sheinberg, Haskell
Stoddard, Stephen D(avidson)
Wilcox, Paul Denton
Wilder, James Andrew, Jr

NEW YORK
Adams, P B
Beall, George Halsey
Borom, Marcus P(reston)
Boyd, David Charles
Britton, Marvin Gale
Brownell, Wayne E(rnest)
Brun, Milivoj Konstantin
Buhsmer, Charles P
Chyung, Kenneth
Danielson, Paul Stephen
Doremus, Robert Heward
Duke, David Allen
Fehette, H(owells Achille) Van Derck
Frechette, Van Derck
Giannelis, Emmanuel P
Greskovich, Charles David
Grossman, David G
Guile, Donald Lloyd
Gupta, Krishna Murari
Hammer, Richard Benjamin
Herman, Herbert
Hillig, William Bruno
Jacoby, William R(ichard)
Johnson, Curtis Alan
King, Alexander Harvey
Lachman, Irwin Morris
Lakatos, Andras Imre
Lane, Richard L
Lanford, William Armistead

Luthra, Krishan Lal
McCauley, James Weymann
MacDowell, John Fraser
McMurtry, Carl Hewes
Meiling, Gerald Stewart
Meloon, David Rand
Minnear, William Paul
Morfopoulos, Vassilis C(onstantinos) P
Mueller, Edward E(ugene)
Nowick, A(rthur) S(tanley)
Ott, Walter Richard
Pasco, Robert William
Prochazka, Svante
Psioda, Joseph Adam
Pye, Lenwood D(avid)
Reed, James Stalford
Reich, Ismar M(eyer)
Revankar, Vithal V S
Rossington, David Ralph
Rothermel, Joseph Jackson
Schreiber, Edward
Shaw, Robert Reeves
Shelby, James Elbert
Smith, Gail Preston
Smith, Russell D
Snyder, Robert Lyman
Spriggs, Richard Moore
Srinivasan, Makuteswaran
Thompson, David Allen
Thompson, David Fred
Thompson, Robert Alan
Wang, Xingwu
Ward, Thomas J(ulian)
Westbrook, J(ack) H(all)
Wexell, Dale Richard
Wilson, James M
Zimar, Frank

NORTH CAROLINA
Davis, Robert F(oster)
Gillooly, George Rice
Graff, William (Arthur)
Hamme, John Valentine
Hurt, John Calvin
Hutchins, John R(ichard), III
Landes, Chester Grey
Manning, Charles Richard, Jr
Palmour, Hayne, III
Stoops, R(obert) F(ranklin)

OHIO
Alam, M Khairul
Baker, Robert J(ethro), Jr
Bansal, Narottam Prasad
Beals, Robert J(ennings)
Bradley, Ronald W
Buchanan, Relva Chester
Choudhary, Manoj Kumar
Chuck, Leon
Cooper, Alfred R, Jr
Copp, Albert Nils
DiCarlo, James Anthony
Drummond, Charles Henry, III
Duckworth, Winston H(oward)
Duderstadt, Edward C(harles)
Dutta, Sunil
Flynn, Ronald Thomas
Graham, Henry Collins
Graham, John W
Graham, Paul Whitener Link
Harmer, Richard Sharpless
Heuer, Arthur Harold
Hill, Brian
Huff, Norman Thomas
Johnson, Howard B(eattie)
Kampe, Dennis James
Kawasaki, Edwin Pope
Koenig, Charles Jacob
Kreidler, Eric Russell
Krysiak, Joseph Edward
La Mers, Thomas Herbert
Land, Peter L
Lee, Haynes A
Lennon, John W(illiam)
McCoy, Robert Allyn
Marchant, David Dennis
Mohn, Walter Rosing
Moore, Arthur William
Parthasarathy, Triplicane Asuri
Petersen, Fredrick Adolph
Piper, Ervin L
Pirooz, Perry Parviz
Rai, Amarendra Kumar
Readey, Dennis W(illiam)
Rosa, Casimir Joseph
Rosenfield, Alan R(obert)
Ruh, Robert
Ryder, Robert J
St Pierre, P(hilippe) D(ouglas) S
Sara, Raymond Vincent
Semler, Charles Edward
Shonebarger, F(rancis) J(oseph)
Shook, William Beattie
Stambaugh, Edgel Pryce
Stewart, Daniel Robert
Stover, E(dward) R(oy)
Swift, Howard R(aymond)
Tallan, Norman M
Tooley, F(ay) V(aNisle)
Venkatu, Doulatabad A
Wurst, John Charles

OKLAHOMA
Harris, Jesse Ray

PENNSYLVANIA
Baran, George Roman
Belitskus, David
Bhalla, Amar S
Biggers, James Virgil
Blachere, Jean R
Chitale, Sanjay Madhav
Cook, Charles S
Cox, John E(dward)
Davis, R(obert) E(lliot)
Emlemdi, Hasan Bashir
Esmen, Nurtan A(lan)
Garg, Diwakar
Gebhardt, Joseph John
Goldman, Kenneth M(arvin)
Goodwin, Charles Arthur
Greenberg, Charles Bernard
Gruver, Robert Martin
Gupta, Tapan Kumar
Hall, Gary R
Harmer, Martin Paul
Harrison, Don Edward
Harvey, Frances J, II
Hogg, Richard
Hummel, F(loyd) A(llen)
Jain, Himanshu
Jang, Sei Joo
Kilp, Gerald R
Kirchner, H(enry) P(aul)
Koczak, Michael Julius
Kotyk, Michael
Kumta, Prashant Nagesh
Lidman, William G
Lo, W(ing) C(heuk)
MacZura, George
Messier, Russell
Mistler, Richard Edward
Nair, K Manikantan
Orr, Leighton
Osborn, Elburt Franklin
Perrotta, Anthony Joseph
Proske, Joseph Walter
Ray, Siba Prasad
Rindone, Guy E(dward)
Ruh, Edwin
Shapiro, Zalman Mordecai
Sieger, John S(ylvester)
Smid, Robert John
Smothers, William Joseph
Smyth, Donald Morgan
Staut, Ronald
Steiger, Roger Arthur
Stetson, Harold W(ilbur)
Stubican, Vladimir S(tjepan)
Tressler, Richard Ernest
Waldman, Jeffrey
Wang, Ke-Chin
Wasylyk, John Stanley
Weber, Frank L
Williams, David Bernard
Williamson, William O(wen)
Wood, Susan
Worrell, Wayne L
Yoldas, Bulent Erturk

RHODE ISLAND
Reynolds, Charles C
Rockett, Thomas John

SOUTH CAROLINA
Coffeen, W(illiam) W(eber)
Haertling, Gene Henry
Insley, Robert H(iteshew)
Lee, Burtrand Insung
Lefort, Henry G(erard)
Lewis, Gordon
Piper, John
Quadir, Tariq
Robinson, Gilbert C(hase)

TENNESSEE
Bloom, Everett E
Boatner, Lynn Allen
Campbell, William B(uford), Jr
Dahotre, Narendra Bapurao
Dudney, Nancy Johnston
Eyerly, George B(rown)
Harms, William Otto
Holcombe, Cressie Earl, Jr
Klueh, Ronald Lloyd
Lotts, Adolphus Lloyd
Pasto, Arvid Eric
Scott, J(ames) L(ouis)
Stradley, James Grant

TEXAS
Arrowood, Roy Mitchell, Jr
Brown, Richard Martin
Cao, Hengchu
Chen, In-Gann
Cleek, Given Wood
Diller, Kenneth Ray
Dunbar, Bonnie J
Hillery, Herbert Vincent
Johnson, Elwin L Pete
McNamara, Edward P(aul)
Marcus, Harris L
Parikh, N(iranjan) M
Rulon, Richard M
Stradley, Norman H(enry)
Thornton, Hubert Richard
Warner, Bert Joseph

Ceramics Engineering (cont)

UTAH
Carnahan, Robert D
Horton, Ralph M
Hyatt, Edmond Preston
Rasmussen, Jewell J
Rigby, E(ugene) B(ertrand)
Weed, Grant B(arg)

VIRGINIA
Brown, Jesse J, Jr
Clarke, Alan R
Desu, Seshu Babu
Hasselman, Didericus Petrus Hermannus
Johnson, David Harley
Kahn, Manfred
Kerr, John (Jack) M(artin)
Lynch, Eugene Darrel
Ormsby, W(alter) Clayton
Puster, Richard Lee
Rice, Roy Warren
Schreiber, Henry Dale
Spitzig, William Andrew
Vojnovich, Theodore
Wilcox, Benjamin A

WASHINGTON
Anderson, Harlan John
Bates, J(unior) Lambert
Burns, Robert Ward
Chikalla, Thomas D(avid)
Ding, Jow-Lian
Einziger, Robert E
Fischbach, David Bibb
Hart, Patrick E(ugene)
Hinman, Chester Arthur
Kalonji, Gretchen
Kruger, Owen L
Miller, Alan Dale
Roberts, J T Adrian
Scott, William D(oane)
Stang, Robert George
Weber, William J
Whittemore, O(sgood) J(ames)
Wilson, Charles Norman
Woods, Keith Newell

WEST VIRGINIA
Olcott, Eugene L
Zhang, Xiao

WISCONSIN
Dunn, Stanley Austin
Fournelle, Raymond Albert
Johnson, J(ames) R(obert)
Lenling, William James
Rohatgi, Pradeep Kumar
Szpot, Bruce F

ALBERTA
Heimann, Robert B(ertram)

BRITISH COLUMBIA
Chaklader, Asoke Chandra Das

NOVA SCOTIA
Gow, K V

ONTARIO
Kim, Sukyoung
King, Hubert Wylam
Kuriakose, Areekattuthazhayil
Nicholson, Patrick Stephen
Prasad, S E
Runnalls, O(liver) John C(lyve)
Sadler, Arthur Graham

QUEBEC
Angers, Roch
Saint-Jacques, Robert G

OTHER COUNTRIES
Jacob, K Thomas
Kim, Young-Gil
Kishi, Keiji
Montierth, Max Romney
Yu, Chyang John

Chemical Engineering

ALABAMA
Achorn, Frank P
April, Gary Charles
Arnold, David Walker
Barber, J(ames) C(orbett)
Barrier, J Wayne
Blouin, Glenn M(organ)
Curry, James Eugene
Dahlin, Robert S
Galil, Fahmy
Harrison, Benjamin Keith
Hart, David R
Hatcher, William Julian, Jr
Hignett, Travis P(orter)
Hill, David Thomas
Hsu, Andrew C T
Hubbuch, Theodore N(orbert)
Lanewala, Mohammed A
McKinley, Marvin Dyal
McMinn, Curtis J
Nieberlein, Vernon Adolph
Pelt, Roland J

Phillips, Alvin B(urt)
Renard, Jean Joseph
Rhoades, Richard G
Rodriguez, Harold Vernon
Roos, C(harles) William
Schrodt, Verle N(ewton)
Stark, John, Jr
Stonecypher, Thomas E(dward)
Veazey, Thomas Mabry
Walsh, William K
White, Niles C
Yett, Fowler Redford

ALASKA
Merrill, Robert Clifford, Jr
Ostermann, Russell Dean
Williams, Frank Lynn

ARIZONA
Anthes, John Allen
Berman, Neil Sheldon
Calcaterra, Robert John
Catterall, William E(dward)
Clark, Ezekail Louis
Dorson, William John
Edwards, Richard M(odlin)
Gealer, Roy L(ee)
Henderson, James Monroe
Kaufman, C(harles) W(esley)
Kleinschmidt, Eric Walker
Langdon, William Mondeng
Liaw, Hang Ming
Mohan, J(oseph) C(harles), Jr
Post, Roy G
Randolph, Alan Dean
Reed, R(obert) M(arion LaFollette)
Rehm, Thomas R(oger)
Reiser, Castle O
Rosler, Richard S(tephen)
Sater, Vernon E(ugene)
Scott, Ralph Asa, Jr
Talbert, Norwood K(eith)
Utagikar, Ajit Purushottam
Weber, James H(arold)
Wendt, Jost O L

ARKANSAS
Bocquet, Philip E(dmund)
Bonner, Errol Marcus
Burchard, John Kenneth
Clausen, Edgar Clemens
Couper, James R(iley)
Day, James Meikle
Fortner, Wendell Lee
Gaddy, James Leoma
Havens, Jerry Arnold
Juhl, William G
Oxford, C(harles) W(illiam)
Ransford, George Henry
Shelley, William J
Welker, J(ohn) Reed

CALIFORNIA
Adelman, Barnet Reuben
Alexander, Earl L(ogan), Jr
Allen, David Thomas
Anderson, Robert Neil
Appleman, Gabriel
Arias, Jose Manuel
Armstrong, Don L
Aroyan, H(arry) J(ames)
Arslancan, Ahmet N
Augood, Derek Raymond
Aziz, Khalid
Bacastow, Robert Bruce
Bae, Jae Ho
Baerg, William
Baldauf, Gunther H(erman)
Balzhiser, Richard E(arl)
Bareis, David W(illard)
Barr, Frank T(homas)
Barry, Michael Lee
Beek, John
Bell, Alexis T
Bengtson, Kermit (Bernard)
Benson, Dale B(ulen)
Berg, Clyde H O
Bernath, L(ouis)
Berriman, Lester P
Bery, Mahendera K
Blanch, Harvey Warren
Blue, E(manuel) M(orse)
Bogart, Marcel J(ean) P(aul)
Boulton, Roger Brett
Brace, Robert Allen
Braithwaite, Charles Henry, Jr
Breitmayer, Theodore
Brice, Donat B(ennes)
Bridge, Alan G
Brigham, William Everett
Bromley, Le Roy Alton
Brown, Mark William
Buck, F(rank) A(lan) Mackinnon
Busch, Joseph Sherman
Butler, G(erard)
Caenepeel, Christopher Leon
Canning, T(homas) F
Carley, James F(rench)
Cashin, Kenneth D(elbert)
Chandrasekaran, Santosh Kumar
Charlesworth, Robert K(oridon)
Che, Stanley Chia-Lin
Chin, Jin H
Chin, Yu-Ren

Chittenden, David H
Chompff, Alfred J(ohan)
Chrysikopoulos, Constantinos V
Chu, Ju Chin
Clazie, Ronald N(orris)
Cleland, Jeffrey Lynn
Cocks, George Gosson
Colmenares, Carlos Adolfo
Constant, Clinton
Cook, Glenn Melvin
Cook, Paul M
Coons, Fred F(leming)
Cooper, Wilson Wayne
Crandall, Edward D
Culler, F(loyd) L(eRoy), Jr
Current, Jerry Hall
Curry, Fitz-Roy Edward
Dance, Eldred Leroy
Davis, Bruce W
Deleray, Arthur Loyd
Denn, Morton M(ace)
Detz, Clifford M
Di Zio, Steven F(rank)
Dobbins, John Potter
Dougherty, Elmer Lloyd, Jr
Duffin, John H
Durai-swamy, Kandaswamy
Durland, John R(oyden)
Eagle, Sam
Edwards, Arthur L
Elton, Edward Francis
Engleman, Victor Solomon
Epperly, W Robert
Evans, James William
Favorite, John R
Feiler, William A, Jr
Feinleib, Morris
Ferm, Richard L
Ferreira, Lawrence E
Flagan, Richard Charles
Flamm, Daniel Lawrence
Fok, Samuel S(hiu) M(ing)
Foss, Alan Stuart
Foster, E(lton) Gordon
Fox, Joseph M(ickle), III
Frattini, Paul L
Fredrickson, Glenn Harold
Friedlander, Sheldon K
Friend, William L
Frischmuth, Robert Wellington
Furukawa, David Hiroshi
Gardner, Howard Shafer
Garrett, Donald E(verett)
Garrett, L(uther) W(eaver), Jr
Gast, Alice Petry
Gates, Bruce C(lark)
Gavalas, George R(ousetos)
Gay, Richard Leslie
Gilkeson, M(urray) Mack
Goddard, Joe Dean
Goldfrank, Max
Golding, Brage
Goldstein, Walter Elliott
Goren, Simon L
Gouw, T(an) H(ok)
Green, Stanley J(oseph)
Grimaldi, John Vincent
Grossberg, Arnold Lewis
Gunness, R(obert) C(harles)
Hahn, Harold Thomas
Haley, Kenneth William
Hamai, James Y
Hamilton, J(ames) Hugh
Hammond, R Philip
Hanna, Owen Titus
Hanson, D(onald) N(orman)
Haritatos, Nicholas John
Harrison, Jonas P
Hass, Robert Henry
Hausmann, Werner Karl
Heibel, John T(homas)
Heil, John F, Jr
Hennig, Harvey
Hermsen, Robert W
Hertwig, Waldemar R
Hester, John Nelson
Hidy, George Martel
Higgins, Brian Gavin
Hong, Ki C(hoong)
Hovorka, Robert Bartlett
Hsieh, Paul Yao Tong
Hubred, Gale L
Huff, James Eli
Iglesia, Enrique
Irving, James P
Isenberg, Lionel
Ishihara, Kohei
Jacobs, Joseph John
Jacobson, Robert Leroy
Jaros, Stanley E(dward)
Jones, Jerry Latham
Judson, Burton Frederick
Kaellis, Joseph
Kafesjian, R(alph)
Kalvinskas, John J(oseph)
Kane, Daniel E(dwin)
Kaneko, Thomas Motomi
Katz, Herbert M(arvin)
Keaton, Michael John
Kendall, Robert McCutcheon
Kennedy, James Vern
Kent, Clifford Eugene
Kertamus, Norbert John
Kieschnick, W(illiam) F(rederick)

King, C(ary) Judson, III
Klipstein, David Hampton
Knight, Patricia Marie
Knowlton, Floyd M(arion)
Knuth, Eldon L(uverne)
Kohl, A(rthur) L(ionel)
Kornfield, Julia Ann
Kouzel, Bernard
Kroneberger, Gerald F
Kuehne, Donald Leroy
Lai, San-Cheng
Laity, David Sanford
Lambert, Walter Paul
Lapple, Charles E
Lavendel, Henry W
Leal, L Gary
Lehmann, A(ldo) Spencer
Leslie, James C
Leventhal, Leon
Levine, Charles (Arthur)
Levy, Ricardo Benjamin
Li, Pei-Ching
Lichtblau, Irwin Milton
Lieberman, Alvin
Lim, H(enry) C(hol)
Lind, Wilton H(oward)
Lindner, Elek
Lindsay, W(esley) N(ewton)
Liu, Ming-Biann
Lockhart, F(rank) J(ones)
Longwell, P(aul) A(lan)
Lovell, Robert Edmund
Lowi, Alvin, Jr
Lund, J Kenneth
Lunde, Kenneth E(van)
Lynn, Scott
McBride, Lyle E(rwin), Jr
McCarty, Jon Gilbert
McCoy, Benjamin J(oe)
McCune, Conwell Clayton
McKay, Richard A(lan)
McKisson, R(aleigh) L(lewellyn)
McKoy, Vincent
Madix, Robert James
Maimoni, Arturo
Mansfeld, Florian Berthold
Margolis, Stephen Barry
Mason, D(avid) M(alcolm)
Mason, Harold Frederick
Mason, John L(atimer)
Mathews, Larry Arthur
Mavity, Victor T(homas), Jr
Mayer, Jerome F
Mears, David Elliott
Mecartney, Martha Lynn
Meiners, Henry C(ito)
Meldau, R(obert) F(rederick)
Metzler, Charles Virgil
Meyer, Brad Anthony
Michels, Lloyd R
Mikolaj, Paul G(eorge)
Milligan, Robert T(homas)
Minet, Ronald G(eorge)
Mitchell, Reginald Eugene
Moll, Albert James
Morari, Manfred
Morgal, Paul Walter
Morgan, Lucian L(loyd)
Morrison, Malcolm Cameron
Muller, Rolf Hugo
Munger, Charles Galloway
Myers, John E(arle)
Naworski, Joseph Sylvester, Jr
Nazaroff, William W
Nelson, David A
Newman, John Scott
Newsom, Herbert Charles
Nguyen, Caroline Phuongdung
Nicholson, W(illiam) J(oseph)
Nixon, Alan Charles
Nobe, Ken
Noring, Jon Everett
Oldenburg, C(harles) C(lifford)
Oldenkamp, Richard D(ouglas)
Oliver, Earl Davis
Olson, Austin C(arlen)
Opfell, John Burton
Ore, Fernando
O'Rear, Dennis John
Orr, Franklin M, Jr
Ouano, Augustus Ceniza
Palen, Joseph W
Pan, Bingham Y(ing) K(uei)
Pastell, Daniel L(ouis)
Pearce, Frank G
Pearl, W(esley) L(loyd)
Pearson, Dale Sheldon
Petersen, Eugene E(dward)
Phillips, John Richard
Piasecki, Leonard R(ichard)
Pigford, Thomas H(arrington)
Pitcher, Wayne Harold, Jr
Porter, Marcellus Clay
Prausnitz, John Michael
Quentin, George Heinz
Raben, Irwin A(bram)
Rabussay, Dietmar Paul
Ramaswami, Devabhaktuni
Ramey, H(enry) J(ackson), Jr
Ravicz, Arthur Eugene
Remedios, E(dward) C(harles)
Remer, Donald Sherwood
Richter, George Neal
Rinker, Robert G(ene)

Ritter, R(obert) Brown
Robertson, George Harcourt
Robertson, Glenn D(avid), Jr
Robillard, Geoffrey
Robin, Allen Maurice
Robinson, Richard C(lark)
Romig, Robert P
Ross, Philip Norman, Jr
Rossen, Joel N(orman)
Rothman, Albert J(oel)
Rubin, Barney
Ruskin, Arnold M(ilton)
Russell, Charles Roberts
Russell, Grant E(dwin)
Rutz, Lenard O(tto)
Ryan, Wen
Sanborn, Charles E(van)
Sandall, Orville Cecil
Sanders, Charles F(ranklin), Jr
Sarem, Amir M Sam
Satre, Rodrick Iverson
Sawyer, Frederick George
Schlatter, James Cameron
Schlinger, W(arren) G(leason)
Schneider, George Ronald
Schneider, Kenneth John
Schumacher, William John
Scott, John W(alter)
Seborg, Dale Edward
Seinfeld, John H
Semrau, Konrad (Troxel)
Serbia, George William
Sesonske, Alexander
Shah, Manesh J(agmohan)
Shair, Fredrick H
Sharma, Bhavender Paul
Sherwood, Albert E(dward)
Shrontz, John William
Shuler, Patrick James
Siegfried, William
Simkin, Donald Jules
Simon, Robert H
Simpson, Howard Douglas
Singh, Hanwant B
Slivinsky, Sandra Harriet
Smith, A(llen) N(athan)
Smith, Joe M(auk)
Smith, Ralph Carlisle
Snyder, Nathan W(illiam)
Spiegler, Kurt Samuel
Stein, Richard Ballin
Stephens, Douglas Robert
Stillman, Richard Ernest
Stroeve, Pieter
Sun, Yun-Chung
Sundberg, John Edwin
Swidler, Ronald
Switzer, Robert L
Theodorou, Doros Nicolas
Tobias, Charles W
Todd, Eric E(dward)
Trilling, Charles A(lexander)
Valle-Riestra, J(oseph) Frank
Van Klaveren, Nico
Van Vorst, William D
Varteressian, K(egham) A(rshavir)
Vause, Edwin H(amilton)
Vermeulen, Theodore (Cole)
Vilker, Vincent Lee
Walker, Jimmy Newton
Walsh, William J
Wang, Chiu-Sen
Waterland, Larry R
Wazzan, A R Frank
Weinberg, William Henry
Weir, Alexander, Jr
Westerman, Edwin J(ames)
Whitehead, Kenneth E
Whitney, Gina Marie
Whittam, James Henry
Wilde, D(ouglass) J(ames)
Wilemski, Gerald
Wilke, Charles R
Williams, Alan K
Winter, Olaf Hermann
Witham, Clyde Lester
Wolf, Irving W
Wong, James B(ok)
Wong, Morton Min
Wright, Roger M
Yang, Meiling T
Yasui, George
Yen, I-Kuen
Yen, Teh Fu
Yuan, Shao-yuen
Yuan, Sidney Wei Kwun
Yum, Su Il
Ziegenhagen, Allyn James

COLORADO
Baldwin, Lionel V
Baldwin, Robert Milton
Barrick, P(aul) L(atrell)
Barry, Henry F
Beck, Steven R
Chase, Curtis Alden, Jr
Christiansen, Robert M(ilton)
Clough, David Edwards
Colver, C(harles) Phillip
Danzberger, Alexander Harris
Davis, John Albert, Jr
Davison, J(oseph) W(ade)
Drayer, Dennis Eugene
Ely, James Frank

Falconer, John Lucien
Flynn, Thomas M(urray)
Gary, James H(ubert)
Giller, E(dward) B(onfoy)
Gilliland, John L(awrence), Jr
Gogarty, W(illiam) B(arney)
Golden, John O(rville)
Graue, Dennis Jerome
Grobner, Paul Josef
Harper, Judson M(orse)
Hauser, Ray Louis
Herring, Richard Norman
Hinman, Norman Dean
Jargon, Jerry Robert
Jha, Mahesh Chandra
Jones, Stanley C(ulver)
Kelchner, Burton L(ewis)
Keller, Frederick Albert, Jr
Kidnay, Arthur J
Klein, James H(enry)
Knight, Bruce L
Kompala, Dhinakar S
Koons, David Swarner
Krantz, William Bernard
Kridel, Donald Joseph
Lauer, B(yron) E(lmer)
Lewis, Clifford Jackson
Linden, James Carl
Mahajan, Roop Lal
Navratil, James Dale
Noble, Richard Daniel
Olien, Neil Arnold
Parsons, Robert W(estwood)
Peters, Max S(tone)
Plummer, Mark Alan
Poettmann, Fred H(einz)
Rakow, Allen Leslie
Ramirez, W Fred
Rothfeld, Leonard B(enjamin)
Sani, Robert L(e Roy)
Savory, Leonard E(rwin)
Schmidt, Alan Frederick
Severson, Donald E(verett)
Streib, W(illiam) C(harles)
Timblin, Lloyd O, Jr
Timmerhaus, K(laus) D(ieter)
VanderLinden, Carl R
Vestal, Charles Russell
West, Ronald E(mmett)
Wyman, Charles Ely
York, J(esse) Louis
Zahradnik, Raymond Louis
Zimmerman, Carle Clark, Jr

CONNECTICUT
Baker, Bernard S
Bell, James Paul
Bennett, Carroll O(sborn)
Bilhorn, John Merlyn
Bollyky, L(aszlo) Joseph
Borden, George Wayne
Bozzuto, Carl Richard
Bretton, R(andolph) H(enry)
Bunney, Benjamin Stephenson
Butensky, Martin Samuel
Cadogan, W(illiam) P(atrick)
Carpenter, Kent Heisley
Casberg, John Martin
Chan, Wing Cheng Raymond
Cheatham, Robert Gary
Cole, Stephen H(ervey)
Corwin, H(arold) E(arl)
Coughlin, Robert William
Cutlip, Michael B
Denton, William Irwin
DiBenedetto, Anthony T
Dobay, Donald Gene
Dobry, Reuven
Eppler, Richard A
Forman, J(oseph) Charles
Geitz, R(obert) C(harles)
George, Edward Thomas
Glenn, Roland Douglas
Goldstein, Marvin Sherwood
Gupta, Dharam Vir
Haller, Gary Lee
Harding, R(onald) H(ugh)
Hersh, Charles K(arrer)
Horváth, Csaba Gyula
Hou, Kenneth C
Howard, G(eorge) Michael
Hsu, Nelson Nae-Ching
Hyde, John Welford
I, Ting-Po
Kellner, Henry L(ouis)
Kesten, Arthur S(idney)
Klei, Herbert Edward, Jr
Koberstein, Jeffrey Thomas
Korin, Amos
Kroll, Charles L(ouis)
Kunz, Harold Russell
Kurose, George
Leinroth, Jean Paul, Jr
Lobo, Walter E(der)
Lunde, Peter J
McCardell, W(illiam) M(arkham)
McKibbins, Samuel Wayne
McLain, William Harvey
Marsh, Bertrand Duane
Mitchell, Robert L(ynne)
Murai, Kotaro
Pai, Venkatrao K
Prabulos, Joseph J, Jr
Ramakrishnan, Terizhandur S

Reed, Charles E(li)
Rie, John E
Rosner, Daniel E(dwin)
Sabnis, Suman T
Schiessl, H(enry) W(illiam)
Schoen, Herbert M
Schoenbrunn, Erwin F(rederick)
Schrage, F Eugene
Schwier, Chris Edward
Shaw, Montgomery Throop
Simerl, L(inton) E(arl)
Sinha, Vinod T(arkeshwar)
Skrivan, J(oseph) F(rancis)
Slotnick, Herbert
Stutzman, Leroy F
Suen, T(zeng) J(iueq)
Taylor, Roy Jasper
Thomas, Arthur L
Traskos, Richard Thomas
Walker, Charles A(llen)
Weidenbaum, Sherman S
Weiss, Robert Alan
Welch, John F, Jr
Wernau, William Charles

DELAWARE
Akell, Robert B(erry)
Albers, Robert Edward
Allan, A J Gordon
Anderson, Howard W(ayne)
Arots, Joseph B(artholomew)
Babcock, Byron D(ale)
Babcock, Dale F
Balder, Jay Royal
Bannister, Robert Grimshaw
Barteau, Mark Alan
Bedingfield, Charles H(osmer)
Benmark, Leslie Ann
Bischoff, Kenneth Bruce
Borden, James B
Bours, William A(lsop), III
Brandreth, Dale Alden
Brill, Thomas Barton
Buckley, Page Scott
Busche, Robert M(arion)
Bydal, Bruce A
Carter, Richard P(ence)
Chitra, Surya P
Cichelli, Mario T(homas)
Clark, H(arold) B(lack)
Cohen, Edward David
Comings, Edward Walter
Cooper, Stuart L
Corty, Claude
Cunning, Joe David
Davies, Reg
Dentel, Steven Keith
Dhurjati, Prasad S
Dietz, John W
Dodge, Donald W(illiam)
Doerner, William A(llen)
Eidman, Richard August Louis
Evans, Franklin James, Jr
Foley, Henry Charles
Foster, Henry D(orroh)
Gibson, Joseph W(hitton), Jr
Glaeser, Hans Hellmut
Goffinet, Edward P(eter), Jr
Gonzalez, Raul A(lberto)
Gossage, Thomas Layton
Gray, Joseph B(urnham)
Greenewalt, Crawford Hallock
Habibi, Kamran
Hackman, Elmer Ellsworth, III
Hecht, J(ames) L(ee)
Hochberg, Jerome
Holob, Gary M
Hoopes, John W(alker), Jr
Hughes, John I(ngram)
Hull, Donald R(obert)
Immediata, Tony Michael
Jefferies, Steven
Kaler, Eric William
Kallal, R(obert) J(ohn)
Kamack, H(arry) J(oseph)
Kemp, Harold Steen
Kilkson, Henn
King, Charles O(rrin)
Klein, Michael Tully
Kouba, Delore Loren
Lamb, David E(rnest)
Lamb, William Bolitho
Leffew, Kenneth W
Littleton, H(arold) T(homas) J(ackson)
Lombardo, R(osario) J(oseph)
Long, John Reed
Luckring, R(ichard) M(ichael)
McCullough, Roy Lynn
McCune, Leroy K(iley)
McGeorge, A(rthur), Jr
McNeely, James Braden
Manogue, William H(enry)
Mehra, Vinodkumar S
Metzner, Arthur B(erthold)
Miller, Donald Nelson
Mills, Patrick Leo, Sr
Moch, Irving W
Mongan, Edwin Lawrence, Jr
Muendel, Carl H(einrich)
Ogunnaike, Babatunde Ayodeji
Olson, Jon H
Pecorini, Hector A(ndrew)
Pell, Mel
Pierce, Robert Henry Horace, Jr

Pontius, E(ugene) C(ameron)
Rabe, Allen E
Rawling, Frank L(eslie), Jr
Rea, David Richard
Reis, Paul G(eorge)
Rick, Christian E(dward)
Riggle, J(ohn) W(ebster)
Roberts, John Burnham
Rodriguez-Parada, Jose Manuel
Ross, William D(aniel)
Royer, Dennis Jack
Rush, Frank E(dward), Jr
Russell, T W Fraser
Ryder, D(avid) F(rank)
St John, Daniel Shelton
Sandler, Stanley I
Sashihara, Thomas F(ujio)
Sauerbrunn, Robert Dewey
Scherger, Dale Albert
Schiewetz, D(on) B(oyd)
Schotte, William
Sears, Leo A
Secor, Robert M(iller)
Senecal, Vance E(van)
Shedrick, Carl F(ranklin)
Shine, Annette Dudek
Smith, Donald W(anamaker)
Smith, Ronald W
Stine, William H, Jr
Tallman, J(ohn) C(ornwell)
Ten Eyck, Edward H(anlon), Jr
Thayer, Chester Arthur
Thomas, Lee W(ilson)
Tomkowit, Thaddeus W(alter)
Uy, William Cheng
Vick Roy, Thomas Rogers
Wagner, Martin Gerald
Walmsley, Peter N(ewton)
Wilkens, George A(lbert)
Winer, Richard

DISTRICT OF COLUMBIA
Aggarwal, Ishwar D
Bowen, David Hywel Michael
Buzzelli, Donald Edward
Cosgarea, Andrew, Jr
Goodman, Eli I
Hill, Christopher Thomas
Kahn, Walter K(urt)
Kirkbride, Chalmer Gatlin
Kordoski, Edward William
Lang, Peter Michael
Pings, C(ornelius) J(ohn)
Ramsey, Jerry Warren
Rosen, Howard Neal
Schmalzer, David Keith
Shrier, Adam Louis
Stevenson, F Dee
Topper, Leonard
Trice, Virgil Garnett, Jr
Wallace, Carl J

FLORIDA
Albert, R(obert) E(yer)
Ariet, Mario
Balaban, Murat Omer
Benedict, Manson
Bergelin, Olaf P(reysz)
Beuther, Harold
Blase, Edwin W(illiam)
Boehme, Werner Richard
Bonnell, James Monroe
Brooks, John A(lbert)
Brown, David
Browning, Joe Leon
Chamberlain, Donald F(rank)
Chao, Raul Edward
Cline, Kenneth Charles
Cornelius, E(dward) B(ernard)
Curry, Thomas Harvey
Dailey, Charles E(lmer), III
Danly, Donald Ernest
Davidson, Mark Rogers
Dubravcic, Milan Frane
Edelson, David
Elzinga, D(onald) Jack
Ermenc, Eugene D
Fahien, Ray W
Feagin, Roy C(hester)
Feinman, J(erome)
Fricke, Arthur Lee
Friedland, Daniel
Friedman, Raymond
Gander, Frederick W(illiam)
Garwood, Maurice F
Gillin, James
Giraudi, Carlo
Gittelman, Donald Henry
Goll, Robert John
Harrow, Lee Salem
Henderson, James B(rooke)
Henry, John Patrick, Jr
Hill, Frank B(ruce)
Hirsch, Donald Earl
Horrigan, Robert V(incent)
Hosler, F(earl) Ramon
Hsu, Tsong-Han
Huckaba, Charles Edwin
Hyman, Seymour C
Jaeger, Marc Jules
Johns, Lewis E(dward), Jr
Jurgensen, Delbert F(rederick), Jr
Kaghan, Walter S(eidel)
Kapner, Robert S(idney)

Chemical Engineering (cont)

Kircher, Morton S(ummer)
Kirmse, Dale William
Klausner, James Frederick
Kowalczyk, Leon S(tanislaw)
Kraybill, Richard R(eist)
Kumar, Ganesh N
Leibson, Irving
List, Harvey L(awrence)
Lynn, R(alph) Emerson
McAfee, Jerry
McFedries, Robert, Jr
Maleady, N(oel) R(ichard)
Martin, Ralph H(arding)
Mason, D(onald) R(omagne)
May, Frank Pierce
Mickley, Harold S(omers)
Miles, Harry V(ictor)
Milton, Robert Mitchell
Moudgil, Brij Mohan
Normile, Hubert Clarence
Nusim, Stanley Herbert
Osborne, Franklin Talmage
Packard, Richard Darrell
Parker, Howard Ashley, Jr
Polejes, J(acob) D
Roe, William P(rice)
Ross, Fred Michael
Schimmel, Walter Paul
Sellers, John William
Selund, Robert B(ernard)
Seugling, Earl William, Jr
Shair, Robert C
Sherwin, Martin Barry
Spurlock, Jack Marion
Stahl, Joel S
Stana, Regis Richard
Stern, Milton
Summers, Hugh B(loomer), Jr
Szonntagh, Eugene L(eslie)
Teller, Aaron Joseph
Topcik, Barry
Travnicek, Edward Adolph
Tyner, Mack
Walker, Robert D(ixon), Jr
Waller, Richard Conrad
Wang, Tsen Chen
Watkins, Charles H(enry)
Wiles, Robert Allan
Wilson, Robert E(lwood)
Wornick, Robert C(harles)
Zegel, William Case
Zimmer, Martin F

GEORGIA
Bebbington, W(illiam) P(earson)
Caras, Gus J(ohn)
Cornell, Stephen Watson
Dooley, William Paul
Eckert, Charles Alan
Ernst, William Robert
Forney, Larry J
Grubb, H(omer) V(ernon)
Gupta, Rakesh Kumar
Harbin, William T(homas)
Hart, John Robert
Hicks, Harold E(ugene)
Hicks, William B(ruce)
Holley, Richard Howard
Holt, David Lowell
Huang, Denis K
Jones, Alan Richard
Knight, James Albert, Jr
Lewis, H(erbert) Clay
Lyons, Anthony Vincent
McDonough, Thomas Joseph
Marshall, Clair Addison
Matteson, Michael Jude
Morman, Michael T
Nerem, Robert Michael
Nichols, G(eorge) Starr
Orr, Clyde, Jr
Peake, Thaddeus Andrew, III
Poehlein, Gary Wayne
Price, Charles R(onald)
Robbins, Wayne Brian
Roberts, Ronnie Spencer
Rousseau, Ronald William
Schneider, Alfred
Schwartz, Robert John
Sommerfeld, Jude T
Soora, Siva Shunmugam
Stelson, Arthur Wesley
Szostak, Rosemarie
Techo, Robert
Teja, Amyn Sadruddin
Toledo, Romeo Trance
Tooke, William Raymond, Jr
Wethern, James Douglas
White, Mark Gilmore
Wick, Timothy M
Winnick, Jack
Wiseman, William H(oward)
Yirak, Jack J(unior)
Yoganathan, Ajit Prithiviraj

HAWAII
Antal, Michael Jerry, Jr
Brantley, Lee Reed
Krock, Hans-Jürgen
Moy, James Hee
Takahashi, Patrick Kenji

IDAHO
Bower, J(ohn) R(oy), Jr
Edwards, Louis Laird, Jr
Felt, Rowland Earl
Hanson, George H(enry)
Hyde, William W
Iyer, Ravi
Jackson, M(elbourne) L(eslie)
Jobe, Lowell A(rthur)
Korus, Roger Alan
Miller, Reid C
Obenchain, Carl F(ranklin)
Ortiz, Marcos German
Paige, David M(arsh)
Scheldorf, Jay J(ohn)
Slansky, Cyril M
Warner, R(ichard) E(lmore)

ILLINOIS
Agarwal, Ashok Kumar
Alkire, Richard Collin
Alsberg, Henry
Alwitt, Robert S(amuel)
Arastoopour, Hamid
Babcock, Lyndon Ross, (Jr)
Babu, Suresh Pandurangam
Baldwin, Richard H(arold)
Bankoff, S(eymour) George
Bengtson, Harlan Holger
Berdichevsky, Alexander
Bezella, Winfred August
Biljetina, Richard
Blanks, Robert F
Boehler, Robert A
Bowman, Walker H(ill)
Brazelton, William T(homas)
Brockmeier, Norman Frederick
Brown, Charles S(aville)
Brown, George Martin
Brown, Sherman Daniel
Brusenback, Robert A(llen)
Bryan, William L
Bukacek, Richard F
Burris, Leslie
Butt, John B(aecher)
Carr, Stephen Howard
Cengel, John Anthony
Chao, Tai Siang
Chen, Juh Wah
Cheryan, Munir
Chilenskas, Albert Andrew
Clark, John Peter, III
Clauson, W(arren) W(illiam)
Cohen, William C(harles)
Coleman, Lester F
Conn, Arthur L(eonard)
Crawford, Raymond Maxwell, Jr
Cronauer, Donald (Charles)
Davis, Edward Nathan
Devgun, Jas S
Didwania, Hanuman Prasad
Dranoff, Joshua S(imon)
Ekman, Frank
Falkenstein, Gary Lee
Fiedelman, Howard W(illiam)
Fildes, John M
Fink, Joanne Krupey
Finn, Patricia Ann
Fisher, Robert Earl
Fitzpatrick, Joseph A
Flinn, James Edwin
Forgac, John Michael
From, Charles A(ugustus), Jr
Gabor, John Dewain
Gavlin, Gilbert
Gembicki, Stanley Arthur
Giacobbe, F W
Goldman, Arthur Joseph
Goldstick, Thomas Karl
Gordon, Gerald Arthur
Griffin, Edward L(awrence), Jr
Groves, Michael John
Gupta, Ramesh
Hacker, David S(olomon)
Hagenbach, W(illiam) P(aul)
Hanratty, Thomas J(oseph)
Harris, Ronald David
Heaston, Robert Joseph
Heil, Richard Wendell
Hesketh, Howard E
Honea, Franklin Ivan
Huang, Shu-Jen Wu
Hundley, John Gower
Hwang, Yu-Tang
Im, Un Kyung
Irvin, Howard H
Ivins, Richard O(rville)
Jacobsen, Fred Marius
Johnson, Terry R(obert)
Jostlein, Hans
Joung, John Jongin
Junk, William A(rthur), Jr
Kamm, Gilbert G(eorge)
Khan, Amanullah Rashid
Kiefer, John Harold
King, Edward P(eter)
Klein, Robert L
Koenig, James J(acob)
Konopinski, Virgil J
Kramer, John Michael
Kramer, Sheldon J
Kruse, Carl William
Kumar, Romesh
Kung, Harold Hing Chuen
Kyle, Martin Lawrence
Lambert, John B(oyd)
Lauffenburger, Douglas Alan
Lawroski, Stephen
Lee, Bernard S
Lester, George Ronald
Lewandowski, Melvin A
Li, Norman N(ian-Tze)
Li, Yao-En
Linden, Henry R(obert)
Long, George
McHenry, K(eith) W(elles), Jr
McKnight, Richard D
Mah, Richard S(ze) H(ao)
Mannella, Gene Gordon
Mansoori, G Ali
Margolis, Asher J(acob)
Marianowski, Leonard George
Masel, Richard Isaac
Mason, David McArthur
Mateles, Richard I
Mavrovouniotis, Michael L
May, Walter Grant
Meter, Donald M(ervyn)
Miller, David
Miller, Irving F(ranklin)
Miller, Shelby A(lexander)
Mosby, James Francis
Mulcahey, Thomas P
Murad, Sohail
Myles, Kevin Michael
Nagy, Zoltan
Nordine, Paul Clemens
Onischak, Michael
Osri, Stanley M(aurice)
Ottino, Julio Mario
Papoutsakis, Eleftherios Terry
Parent, Joseph D(ominic)
Patel, Mayur
Pearlstein, Arne Jacob
Pecina, Richard W
Peterson, Kenneth C(arl)
Pierce, R(obert) Dean
Poulikakos, Dimos
Pritzlaff, August H(enry), Jr
Punwani, Dharam Vir
Pyrcioch, Eugene Joseph
Rakowski, Robert F
Rasche, John Frederick
Reeve, Aubrey C
Rest, David
Rittmann, Bruce Edward
Robins, Norman Alan
Robinson, Ralph M(yer)
Ryskin, Gregory
Sather, Norman F(redrick)
Savidge, Jeffrey Lee
Schoepfer, Arthur E(ric)
Schowalter, William Raymond
Sebree, Bruce Randall
Seebauer, Edmund Gerard
Seefeldt, Waldemar Bernhard
Sherrill, J(oseph) C(yril)
Shimotake, Hiroshi
Singh, Shyam N
Snow, Richard Huntley
Sohr, Robert Trueman
Solvik, R S(ven)
Staats, William R
Stadtherr, Mark Allen
Steinberg, Marvin Phillip
Stevens, William F(oster)
Strong, E(rwin) R(ayford)
Swanson, Bernet S(teven)
Swearingen, John Eldred
Szepe, Stephen
Thiele, Ernest
Thodos, George
Thompson, John
Travelli, Armando
Troscinski, Edwin S
Tsai, Boh Chang
Tsaros, C(onstantine) L(ouis)
Udani, Kanakkumar Harilal
Veeravalli, Madhavan
Vora, Manu Kishandas
Wadden, Richard Albert
Wang, Hwa-Chi
Wang, Pie-Yi
Wasan, Darsh T
Westwater, J(ames) W(illiam)
Wilson, Gerald Gene
Wilson, Thomas Edward
Witter, Lloyd David
Wong, Wang Mo
Yao, Neng-Ping
Zaromb, Solomon
Zheng, Shuming

INDIANA
Abegg, Carl F(rank)
Albright, Lyle F(rederick)
Andres, Ronald Paul
Banchero, J(ulius) T(homas)
Booe, J(ames) M(arvin)
Bowden, Warren W(illiam)
Brannon-Peppas, Lisa
Carberry, James John
Cartmell, Robert Root
Caskey, Jerry Allan
Chao, Kwang Chu
Cosway, Harry F(rancis)
Delgass, W Nicholas
Eckert, Roger E(arl)
Egly, Richard S(amuel)
Emery, Alden H(ayes), Jr
Fagan, J(ohn) R(obert)
Flock, John William
Greenkorn, Robert A(lbert)
Hite, S(amuel) C(harles)
Houze, Richard Neal
Hubble, Billy Ray
Hurley, Robert Edward
Jeffries, Quentin Ray
Johns, Dennis Michael
Kesler, G(eorge) H(enry)
Kessler, David Phillip
Kohn, James P(aul)
Kuo, Charles C Y
Ladisch, Michael R
Liley, Peter Edward
Lin, Ho-Mu
Ling, Ting H(ung)
Liu, Tung
Machtinger, Lawrence Arnold
Moore, Noel E(dward)
Nirschl, Joseph Peter
Nolting, Henry Frederick
Okos, Martin Robert
Park, Kinam
Peppas, Nikolaos Athanassiou
Ranade, Madhukar G
Razdan, Mohan Kishen
Reklaitis, Gintaras Victor
Schmitz, Roger A(nthony)
Squires, Robert George
Strieder, William
Tsao, George T
Varma, Arvind
Wang, Jin-Liang
Wankat, Phillip Charles
Williams, Theodore J(oseph)

IOWA
Abraham, William H
Berry, Clyde Marvin
Boylan, D(avid) R(ay)
Brown, Melvin Henry
Burnet, George
Carmichael, Gregory Richard
Coy, Daniel Charles
Dordick, Jonathan Seth
Egger, Carl Thomas
Gaertner, Richard F(rancis)
Johnson, Donald Lee
Jolls, Kenneth Robert
Kammermeyer, Karl
Larson, Maurice A(llen)
Peterson, Idelle M(arietta)
Reilly, Peter John
Rethwisch, David Gerard
Schustek, George W(illiam), Jr
Seagrave, Richard C(harles)
Sheeler, John B(riggs)
Tate, R(oger) W(allace)
Ulrichson, Dean LeRoy
Vetter, Arthur Frederick
Wheelock, Thomas David

KANSAS
Akins, Richard G(lenn)
Bates, Herbert T(empleton)
Brewer, Jerome
Chung, Do Sup
Crowther, Robert Hamblett
Cupit, Charles R(ichard)
Erickson, Larry Eugene
Fan, Liang-Tseng
Faw, Richard E
Fujimoto, Gordon Takeo
Green, Don Wesley
Honstead, William Henry
Jacobs, Louis John
Kyle, Benjamin G(ayle)
McElroy, Albert Dean
Maloney, J(ames) O(Hara)
Mesler, R(ussell) B(ernard)
Rosson, H(arold) F(rank)
Ryan, James M
Soodsma, James Franklin
Sutherland, John B(ennett)
Swift, George W(illiam)
Willhite, Glen Paul
Young, David A(nthony)

KENTUCKY
Flegenheimer, Harold H(ansleo)
Fleischman, Marvin
Funk, James Ellis
Gerhard, Earl R(obert)
Hamrin, Charles E(dward), Jr
Hanley, Thomas Richard
Harper, Dean Owen
Hoertz, Charles David, Jr
Jensen, Randolph A(ugust)
Kohnhorst, Earl Eugene
Laukhuf, Walden Louis Shelburne
Livingood, Marvin D(uane)
Lockwood, Frances Ellen
Mapes, William Henry
Mayland, Bertrand Jesse
Mellinger, Gary Andreas
Moore, Howard Francis
Plank, Charles Andrews
Ray, Asit Kumar
Sanford, Robert Alois
Schrodt, James Thomas
Wells, William Lochridge

LOUISIANA

Arthur, Jett Clinton, Jr
Ashby, B(illy) B(ob)
Bailey, Raymond Victor
Barona, Narses
Bertolette, W(illiam) deB(enneville)
Biggs, R Dale
Bouloucon, Peter
Bunch, David William
Callihan, Clayton D
Carpenter, Frank G(ilbert)
Carr, James W(oodford), Jr
Causey, Ardee
Chen, Hoffman Hor-Fu
Chew, Woodrow W(ilson)
Clavenna, LeRoy Russell
Coates, Jesse
Collier, John Robert
Constant, William David
Cordiner, James B(eattie), Jr
Corripio, Armando Benito
Davidson, Donald H(oward)
Davis, P(hilip) C
Dearth, James Dean
Decossas, Kenneth Miles
Ellison, Bart T
Fang, Cheng-Shen
Gautreaux, Marcelian Francis
Goerner, Joseph Kofahl
Griffin, Gregory Lee
Hansen, Robert M(arius)
Harrison, Douglas P
Hill, Archibold G
Ho, Chong Cheong
Hu, John Nan-Hai
Huckabay, Houston Keller
Johnson, Adrian Earl, Jr
Karkalits, Olin Carroll, Jr
Killgore, Charles A
Kingrea, C(harles) L(eo)
Knoepfler, Nestor B(eyer)
Koltun, Stanley Phelps
Lippman, Alfred
Loechelt, Cecil P(aul)
McCarthy, Danny W
McLaughlin, Edward
Malone, James W(illiam)
Martinez, John L(uis)
Mayer, Francis X(avier)
Mukherjee, Debi Prasad
Murrill, Paul W(hitfield)
Pike, Ralph W
Polack, Joseph A(lbert)
Ponter, Anthony Barrie
Pressburg, Bernard S(amuel)
Reneau, Daniel Dugan, Jr
Riley, Kenneth Lloyd
Robson, Harry Edwin
Samuels, Martin E(lmer)
Schucker, Robert Charles
Shuck, Frank O
Sterling, Arthur MacLean
Sullivan, Samuel Lane, Jr
Thibodeaux, Louis J
Valsaraj, Kalliat Thazhathuveetil
Voorhies, Alexis, Jr
Wan, Peter J
Weaver, R(obert) E(dgar) C(oleman)
Wilkins, Bert, Jr

MAINE

Chase, Andrew J(ackson)
Genco, Joseph Michael
Gorham, John Francis
Hill, Marquita K
Meftah, Bachir
Mumme, Kenneth Irving
Orcutt, John C
Rather, James B, Jr
Shipman, C(harles) William
Thompson, Edward Valentine
Zieminski, S(tefan) A(ntoni)

MARYLAND

Adams, Edward Franklin
Adams, James Miller
Anderson, W(endell) L
Barclay, James A(lexander)
Barney, Duane Lowell
Beckmann, Robert B(ader)
Bournia, Anthony
Braude, George Leon
Brodsky, Allen
Bruley, Duane Frederick
Bungay, Peter M
Cadman, Theodore W
Camp, Albert T(alcott)
Carlon, Hugh Robert
Chamberlain, David Leroy, Jr
Chiu, Ching Ching
Choi, Kyu-Yong
Cleveland, William Grover, Jr
Collins, William Henry
Cooperstein, Raymond
Dang, Vi Duong
Dart, Jack Calhoun
Dashiell, Thomas Ronald
Donohue, Marc David
Farmer, William S(ilas), Jr
Fishman, Steven Gerald
Foresti, Roy J(oseph), Jr
Frey, Douglas D
Gabelnick, Henry Lewis
Gamson, Bernard W(illiam)

Gessner, Adolf Wilhelm
Gomezplata, Albert
Graebert, Eric W
Greenfeld, Sidney H(oward)
Gupta, Ashwani Kumar
Hader, Rodney N(eal)
Haller, Elden D
Hamilton, Bruce King
Harper, C(harles) A(rthur)
Harris, B(enjamin) L(ouis)
Heath, George A(ugustine)
Hegedus, L Louis
Herold, Keith Evan
Herrick, Elbert Charles
Hovmand, Svend
Hsu, Stephen M
Jashnani, Indru
Jovancicevic, Vladimir
Katz, Joseph L
Kayser, Richard Francis
Kerkar, Awdhoot Vasant
Khatib-Rahbar, Mohsen
Kirby, Ralph C(loudsberry)
Knoedler, Elmer L
Ku, Chia-Soon
Langlykke, Asger Funder
Lard, Edwin Webster
Leong, Kam W
Levine, Robert S(idney)
Lieberman, Richard Barry
Lowe, William Webb
Lull, David B
McAvoy, Thomas John
McBride, Gordon Williams
McElroy, Wilbur Renfrew
McMullen, Bryce H
Moreira, Antonio R
Moses, Saul
Muller, Robert E
Parczewski, Krzysztof I(gnacy)
Perry, Charles William
Plumlee, Karl Warren
Prentice, Geoffrey Allan
Presser, Cary
Preusch, Peter Charles
Quandt, Earl Raymond, Jr
Ranade, Madhav (Arun) Bhaskar
Rankin, Sidney
Regan, Thomas M(ichael)
Reilly, Hugh Thomas
Rice, James K
Richman, David M(artin)
Rockwell, Theodore
Rzeszotarski, Waclaw Janusz
Saltzman, W Mark
Schlain, David
Semerjian, Hratch G
Sengers, Jan V
Sengers, Johanna M H Levelt
Shane, Presson S
Sheppard, Norman F, Jr
Smith, Elmer Robert
Smith, Theodore G
Smutz, Morton
Steinberg, Ronald T
Swift, David Leslie
Szego, George C(harles)
Tippens, Dorr E(ugene) F(elt)
Ulanowicz, Robert Edward
Ulbrecht, Jaromir Josef
Villet, Ruxton Herrer
Walton, Ray Daniel, Jr
Wang, Francis Wei-Yu
Weigand, William Adam
White, Edmund W(illiam)
Wing, James
Winters, C(harles) E(rnest)
Wong, Kin Fai
Yates, Harold W(illiam)
Zafiriou, Evanghelos
Zdravkovich, Vera
Zehner, Lee Randall

MASSACHUSETTS

Agarwal, Jagdish Chandra
Altieri, A(ngelo) M(ichael)
Ammlung, Richard Lee
Andersen, L(aird) Bryce
Apelian, Diran
Baddour, Raymond F(rederick)
Barbino, Gilda A
Blake, Thomas R
Blankshtein, Daniel
Bloomfield, David Peter
Bodman, Samuel Wright, III
Bonner, Francis Joseph
Botsaris, Gregory D(ionysios)
Brenner, Howard
Breslau, Barry Richard
Brown, Robert A
Brown, Robert Lee
Buonopane, Ralph A(nthony)
Chang, Tai Ming
Chen, Chuan Ju
Chen, Ning Hsing
Chen, Shiou-Shan
Chen, Sung Jen
Chu, Nori Yaw-Chyuan
Clopper, Herschel
Cohen, Robert Edward
Cole, Harvey M
Cooney, Charles Leland
Cooper, William Wailes
Dale, William

Dalzell, William Howard
Dauphiné, T(honet) C(harles)
Dave, Raju S
DeCosta, Peter F(rancis)
Deen, William Murray
De Filippi, R(ichard) P(aul)
De Los Reyes, Gaston
De Noto, Thomas Gerald
Donovan, James
Douglas, J(ames) M(errill)
Drake, Elisabeth Mertz
Eby, Robert Newcomer
Ehrenfeld, John R(oos)
Eldridge, John W(illiam)
Evans, Lawrence B(oyd)
Fagerson, Irving Seymour
Fariss, Robert Hardy
Feakes, Frank
Ferris, Theodore Vincent
Fine, David H
Fisher, Harold Wallace
George, James Henry Bryn
Goettler, Lloyd Arnold
Gold, Harris
Gopikanth, M L
Gutoff, Edgar B(enjamin)
Hausslein, Robert William
Heitman, Richard Edgar
Hoffman, Paul Roger
Hoffman, Richard Laird
Holmes, Douglas Burnham
Hottel, Hoyt C(larke)
Howard, Jack Benny
Huang, Jan-Chan
Jacobson, Murray M
Jain, Rakesh Kumar
Jalan, Vinod Motilal
Jensen, Klavs Fleming
Johnson, Stephen Allen
Kaiser, Robert
Kingston, George C
Kirk, R(obert) S(tewart)
Kittrell, James Raymond
Klanfer, Karl
Kogos, L(aurence)
Kolodzy, Paul John
Kranich, Wilmer Leroy
Kreiling, William H(erman)
Kruse, Robert Louis
Kulwicki, Bernard Michael
Kusik, Charles Lembit
La Mantia, Charles R
Latanision, Ronald Michael
Laurence, Robert L(ionel)
Lee, Eric Kin-Lam
Leung, Woon Fong (Wallace)
Lindroos, Arthur E(dward)
Longwell, John Ploeger
Luft, Ludwig
Lurie, Robert M(andel)
Ma, Yi Hua
McFarland, Charles Manter
Magarian, Charles Aram
Magnanti, Thomas L
Malloy, John B
Mason, Edward Archibald
Meldon, Jerry Harris
Michaels, Alan Sherman
Miekka, Richard G(eorge)
Mir, Leon
Mix, Thomas W
Monson, Peter A
Morbey, Graham Kenneth
Mueller, Robert Kirk
Myers, Robert Frederick
Neue, Uwe Dieter
Novack, Joseph
Peng, Fred Ming-Sheng
Petrovic, Louis John
Prodany, Nicholas V
Ragone, David Vincent
Reichard, H(arold) F(orrest)
Reid, Robert C(lark)
Sarofim, Adel Fares
Satterfield, Charles N(elson)
Schoenberg, Theodore
Schutte, A(ugust) H(enry)
Schwartzberg, Henry G
Selke, William A
Short, John Lawson
Simon, Robert H(erbert) M(elvin)
Sioui, Richard Henry
Smith, Kenneth A
Snedeker, Robert A(udley)
Sonnichsen, Harold Marvin
Spangler, Lora Lee
Stephanopoulos, Gregory
Stephens, Richard Harry
Stevens, J(ames) I(rwin)
Stone, H Nathan
Sussman, M(artin) V(ictor)
Szekely, Julian
Tester, Jefferson William
Thieme, Cornelis Leo Hans
Thomas, Martha Jane Bergin
Thorstensen, Thomas Clayton
Towns, Donald Lionel
Van Wormer, Kenneth A(ugustus), Jr
Vartanian, Leo
Wagner, Robert E(arl)
Wallace, David H
Wang, Daniel I-Chyau
Wang, Lawrence K
Wei, James

Weiss, Alvin H(arvey)
Westmoreland, Phillip R
Williams, Glenn C(arber)
Winter, Horst Henning
Zotos, John

MICHIGAN

Amin, Sanjay Indubhai
Anderson, Donald K(eith)
Atreya, Arvind
Atwell, William Henry
Ben, Manuel
Bens, Frederick Peter
Breuer, Max Everett
Briggs, Dale Edward
Buss, David R(ichard)
Camp, David Thomas
Caplan, John D(avid)
Carmouche, L(ouis) N(orman)
Carnahan, Brice
Cha, Dae Yang
Chiu, Thomas T
Cho, Byong Kwon
Clarke, Richard Penfield
Cooper, Carl (Major)
Crittenden, John Charles
Cunningham, Fay Lavere
Curl, Rane L(ocke)
Daniels, Stacy Leroy
Donahue, Francis M(artin)
Ebach, Earl A
Eichhorn, J(acob)
Eldred, Norman Orville
Fauver, Vernon A(rthur)
Fenn, H(oward) N(athan)
Finch, C(harles) R(ichard)
Flynn, John M(athew)
Fogler, Hugh Scott
Foister, Robert Thomas
Forsblad, Ingemar Bjorn
Francis, Ray Llewellyn
Gagnon, Eugene Gerald
Gandhi, Harendra Sakarlal
Ginn, Robert Ford
Graves, Harold E(dward)
Gulari, Esin
Hall, Richard Harold
Hand, James Henry
Hawley, Martin C
Heiss, John F
Hill, Robert F
Hinsch, James E
Hubbard, Carolyn Parks
Hutchinson, Kenneth A
Hyun, Kun Sup
Jariwala, Sharad Lallubhai
Joshi, Mukund Shankar
Kadlec, Robert Henry
Kempe, Lloyd L(ute)
Kenaga, Duane Leroy
Kim, Byung Ro
Klimisch, Richard L
Klimpel, Richard Robert
Korcek, Stefan
Kummer, Joseph T
Kummler, Ralph H
Larson, John Grant
Lee, Do Ik
Leng, Douglas E
Lenton, Philip A(lfred)
Leopold, Wilbur Richard, III
Lewis, John G(alen)
Li, Chi
Li, Shin-Hwa
Maasberg, Albert Thomas
McKinstry, Karl Alexander
McMicking, James H(arvey)
McMillan, Michael Lathrop
Macriss, Robert A
Martz, Lyle E(rwin)
Merritt, R(obert) W(alter)
Mickelson, Richard W
Moll, Harold Wesbrook
Moore, Eugene Roger
Murchison, Craig Brian
Narain, Amitabh
Narayan, Ramani
Nelson, John Arthur
Nesbitt, Carl C
Nevius, Timothy Alfred
Ng, Ka Yuen Simon
Otrhalek, Joseph V
Revzin, Arnold
Robbins, Lanny Arnold
Rossi, Giuseppe
Rothe, Erhard William
Rounds, Fred G
Russell, William Charles
Salinger, Rudolf Michael
Savage, Albert B
Savage, Phillip E
Schwing, Richard C
Smith, Frederick W(ilson)
Smith, Joseph D
Soh, Sung Kuk
Stevenson, Richard Marshall
Stynes, Stanley K
Tison, Richard Perry
Tsang, Peter H S
Voelker, C(larence) E(lmer)
Voorhees, Howard R(obert)
Vukasovich, Mark Samuel
Wang, Henry Y
Wegst, W(alter) F(rederick)

Chemical Engineering (cont)

Weissman, Eugene Y(ehuda)
Wilkes, James Oscroft
Wilkinson, Bruce W(endell)
Williams, Eugene H(ughes)
Williams, G(eorge) Brymer
Ying, Ramona Yun-Ching
Yost, John R(obarts), Jr
Young, David Caldwell
Young, Edwin H(arold)

MINNESOTA

Anderson, Harvey L(awrence)
Aris, Rutherford
Baker, Michael Harry
Baria, Dorab Nacroze
Bates, Frank S
Benner, Blair Richard
Boening, Paul Henrik
Bohon, Robert Lynn
Britz, Galen C
Brown, Robert Ordway
Carr, Robert Wilson, Jr
David, Moses M
Dreier, William Matthews, Jr
Dreshfield, Arthur C(harles), Jr
Drew, Bruce Arthur
Eades, Robert Alfred
Erickson, Eugene E
Erwin, James V
Fletcher, Edward A(braham)
Fredrickson, Arnold G(erhard)
Geankoplis, Christie J(ohn)
Hanby, John Estes, Jr
Hann, G(eorge) C(harles)
Haun, J(ames) W(illiam)
Hu, Wei-Shou
Huffine, Coy L(ee)
Isbin, Herbert S(tanford)
Jensen, Timothy B(erg)
Johnson, Lennart Ingemar
Johnson, W(illiam) C
Kwong, Joseph N(eng) S(hun)
Lai, Juey Hong
Macosko, Christopher Ward
Manske, Wendell J(ames)
Miller, Nicholas Carl
Moison, Robert Leon
Nam, Sehyun
Patel, Kalyanji U
Pearson, John W(illiam)
Penney, William Harry
Podas, William M(orris)
Ranz, William E(dwin)
Reinhart, Richard D
Sandvig, Robert L(eRoy)
Sapakie, Sidney Freidin
Schmidt, Lanny D
Schoenherr, Roman Uhrich
Schoon, David Jacob
Shor, Steven Michael
Smuk, John Michael
Standing, Charles Nicholas
Strauss, H(oward) J(erome)
Thangaraj, Sandy
Tirrell, Matthew Vincent
Werth, Richard George
Yang, Jih Hsin (Jason)
Zitney, Stephen Edward

MISSISSIPPI

Anderson, Frank A(bel)
Aven, Russell E(dward)
Cade, Ruth Ann
Cornell, David
George, Clifford Eugene
Kuo, Chiang-Hai
Leybourne, A(llen) E(dward), III
Norman, Roger Atkinson, Jr
Persell, Ralph M(ountjoy)
Sadana, Ajit
Sukanek, Peter Charles
Wehr, Allan Gordon

MISSOURI

Adams, C(harles) Howard
Anderson, Fletcher N
Bajpai, Rakesh Kumar
Barker, George Edward
Bosanquet, Louis Percival
Brahmbhatt, Sudhir R
Carter, Don E
Carter, Lee
Chae, Young C
Cheng, William J(en) P(u)
Chien, Henry H(ung-Yeh)
Cova, Dario R
Cowherd, Chatten, Jr
Crocker, Burton B(lair)
Crosser, Orrin Kingsbery
Crum, Glen F(rancis)
Deam, James Richard
De Chazal, L(ouis) E(dmond) Marc
Dmytryszyn, Myron
Donovan, John Richard
Dudukovic, Milorad
Fellinger, L(owell) L(ee)
Ferchaud, John B(artholomew)
Findley, Marshall E(wing)
Fowler, Frank Cavan
Gadberry, Howard M(ilton)
Geisman, Raymond August, Sr
Getty, Robert J(ohn)

Gibbs, Marvin E
Grice, Harvey H(oward)
Harakas, N(icholas) Konstantinos
Harland, Ronald Scott
Harris, J(ohn) S(terling)
Heitsch, Charles Weyand
Hines, Anthony Loring
Holsen, James N(oble)
Howard, Walter B(urke)
Hudson, Robert B
Johnson, James W(inston)
Jones, Otha Clyde
Kardos, John Louis
Kessler, Nathan
Kim, Keun Young
Koederitz, Leonard Frederick
Kramer, Richard Melvyn
Lannert, Kent Philip
Lapple, Walter C(hristian)
Lee, Roberto
Luebbers, Ralph H(enry)
Luecke, Richard H
Marc de Chazal, L E
McKelvey, James M(organ)
Marrero, Thomas Raphael
Mason, Norbert
Motard, R(odolphe) L(eo)
Nellums, Robert (Overman)
Null, Harold R
Otto, Robert Emil
Patterson, Gary Kent
Peng, Yongren Benjamin
Preckshot, G(eorge) W(illiam)
Privott, Wilbur Joseph, Jr
Proctor, Stanley Irving, Jr
Retzloff, David George
Robertson, David G C
Rosen, Edward M(arshall)
Rosen, Stephen L(ouis)
Scallet, Barrett Lerner
Schott, Jeffrey Howard
Schroy, Jerry M
Schuler, Rudolph William
Sisler, Charles Carleton
Snyder, Joseph Quincy
Sparks, Robert Edward
Stone, Bobbie Dean
Storvick, Truman S(ophus)
Strunk, Mailand Rainey
Sutterby, John Lloyd
Teng, James
Thompson, Dudley
Throdahl, Monte C(orden)
Vandegrift, Alfred Eugene
Viswanath, Dabir S
Volz, William K(urt)
Waggoner, Raymond C
Walsh, Robert Jerome
Wasson, Richard Lee
Weddell, David S(tover)
Wohl, Martin H
Wulfman, David Swinton
Young, Henry H(ans)

MONTANA

Berg, Lloyd
Characklis, William Gregory
Deibert, Max Curtis
McCandless, Frank Philip
Nickelson, Robert L(eland)
Sears, John T
Twidwell, Larry G

NEBRASKA

Eastin, John A
Gilbert, Richard E(arle)
Scheller, W(illiam) A(lfred)
Tao, L(uh) C(heng)
Timm, Delmar C

NEVADA

Cox, Neil D
Dobo, Emerick Joseph
Eichbaum, Barlane Ronald
Hendrix, James Lauris
Herzlich, Harold Joel
MacDonald, David J
Miller, E(ugene)
Miller, Wayne L(eroy)
Prater, C(harles) D(wight)

NEW HAMPSHIRE

Bertrand, Rene Robert
Converse, Alvin O
Dewey, Bradley, Jr
Fan, Stephen S(hu-Tu)
Farag, Ihab Hanna
Fricke, Edwin Francis
Mathur, Virendra Kumar
Secord, Robert N
Shinskey, Francis Gregway
Smith, Frank W(illiam)
Sundberg, Donald Charles
Sze, Morgan Chuan-Yuan
Ulrich, Gael Dennis

NEW JERSEY

Adams, James Mills
Adler, Irwin L
Agalloco, James Paul
Ahlert, Robert Christian
Albohn, Arthur R(aymond)
Andreatch, Anthony J
Armamento, Eduardo T

Armbrecht, Frank Maurice, Jr
Bajars, Laimonis
Bakal, Abraham I
Bartok, William
Beinfest, Sidney
Berkman, Samuel
Bevans, Rowland S(cott)
Bhagat, Phiroz Maneck
Bhandarkar, Suhas D
Bieber, Harold H
Bieber, Herman
Biesenberger, Joseph A
Bisio, Attilio L
Blank, Zvi
Blyler, Lee Landis, Jr
Bockelmann, John B(urggraf)
Bohrer, Thomas Lee
Bonacci, John C
Bowen, J Hartley, Jr
Bragg, Leslie B(artlett)
Brazinsky, Irv(ing)
Bryant, Howard Sewall
Bull, Daniel Newell
Burke, Robert F
Carluccio, Frank
Cart, Eldred Nolen, Jr
Cecchi, Joseph Leonard
Cerkanowicz, Anthony Edward
Cevasco, Albert Anthony
Chappelear, David C(onrad)
Chen, Nai Y
Cheng, Shang I
Cheremisinoff, Paul N
Chien, Luther C
Chimes, Daniel
Choi, Byung Chang
Chu, Horn Dean
Cipriani, Cipriano
Ciprios, George
Cruice, William James
Czeropski, Robert S(tephen)
Dautzenberg, Frits Mathia
Davis, Thomas Arthur
Debenedetti, Pablo Gaston
Deem, William Brady
Deen, Harold E(ugene)
De Lancey, George Byers
Detrick, John K(ent)
DiMasi, Gabriel Joseph
DiSalvo, Walter A
Dittman, Fred W(illard)
Drew, Stephen W
Dunning, Ranald G(ardner)
Eck, John Clifford
Eddinger, Ralph Tracy
Engel, Lawrence J
Farrauto, Robert Joseph
Fass, Stephen M
Feder, David O
Feder, Raymond L
Fischer, Ronald Howard
Fogg, Edward T(hompson)
Foroulis, Z Andrew
Fowles, Patrick Ernest
Franse, Renard
Frohlich, Gerhard J
Fung, Shun Cheng
Gagliardi, George N(icholas)
Gans, Manfred
Gartside, Robert N(ifong)
Garwood, William Everett
Giddings, Sydney Arthur
Glass, Werner
Gogos, Costas G
Goksel, Mehmet Adnan
Goldstein, David
Graessley, William W(alter)
Gray, Charles A(ugustus)
Gray, Ralph Donald, Jr
Greenstein, Teddy
Gregory, Richard Alan, Jr
Grelecki, Chester
Griskey, R(ichard) G(eorge)
Grubman, Wallace Karl
Grumer, Eugene Lawrence
Guernsey, Edwin O(wens)
Haag, Werner O
Halik, Raymond R(ichard)
Han, Byung Joon
Hanesian, Deran
Heath, Carl E(rnest), Jr
Heath, D(onald) P
Heck, Ronald Marshall
Heney, Lysle Joseph, Jr
Higginson, George W, Jr
Hnatow, Miquel Alexander
Hoffman, Robert Frank
Hong, Allan Jixian
Hopkins, Charles B(everley), Jr
Horzepa, John Philip
Hsu, Kenneth Hsuehchia
Huang, C C
Huang, Ching-Rong
Huang, Jennming Stephen
Hundert, Irwin
Huppert, Irwin Neil
Jackson, Roy
Jacolev, Leon
Jagel, Kenneth I(rwin), Jr
Joffe, Joseph
Johnson, Douglas L
Johnson, Ernest F(rederick), (Jr)
Johnson, James M(elton)
Johnston, Christian William

Kandiner, Harold J(ack)
Kane, Ronald S(teven)
Kaplan, Joel Howard
Karoly, Gabriel
Kaufman, Arnold
Kaufmann, Thomas G(erald)
Kaup, Edgar George
Kavesh, Sheldon
Keller, Kenneth H(arrison)
Khan, Saad Akhtar
Kheshgi, Haroon S
Kibbel, William H, Jr
Klenke, Edward Frederick, Jr
Kmak, Walter S(teven)
Kydd, Paul Harriman
Lai, Kuo-Yann
Larson, Eric Heath
Lederman, Peter B
Lee, Wei-Kuo
Lee, Wooyoung
Lessard, Richard R
Levinson, Sidney Bernard
Lewandowski, Gordon A
Lewis, Theodore
Lifson, William E(ugene)
Liu, David H(o-Feng)
Lobo, Paul A(llan)
Lovinger, Andrew Joseph
Lowenstein, J(ack) G(ert)
Lund, Daryl B
McCaffrey, David Saxer, Jr
McCormick, John E
McGarvey, Francis X(avier)
McNicholas, James J
Manganaro, James Lawrence
Mantell, Charles L(etnam)
Mathis, James Forrest
Matsen, John M(orris)
Matsuoka, Shiro
Maxwell, Bryce
Metzger, Charles O
Midler, Michael, Jr
Miller, Robert H(enry)
Miner, Robert Scott, Jr
Mraw, Stephen Charles
Munger, Stanley H(iram)
Murphy, Thomas Daniel
Murthy, Andiappan Kumaresa Sundara
Nace, Donald Miller
Newberger, Mark
Nguyen, Hien Vu
O'Brien, Gerald Joseph
Park, Young D
Patel, Rutton Dinshaw
Peltzman, Alan
Perlmutter, Arthur
Pfeffer, Robert
Prapas, Aristotle G
Prizer, Charles J(ohn)
Prud'Homme, Robert Krafft
Quan, Xina Shu-Wen
Register, Richard Alan
Reilly, James William
Rhee, Aaron Seung-Joon
Robertson, Jerry L(ewis)
Robinson, Jerome David
Robinson, Robert George
Rosensweig, Ronald E(llis)
Rupp, Walter H(oward)
Russel, William Bailey
Sacks, Martin Edward
Salkind, Alvin J
Salzarulo, Leonard Michael
Sartor, Anthony
Saville, D(udley) A(lbert)
Sawin, Steven P
Schmidle, Claude Joseph
Schmidt, John P
Seltzer, Edward
Sermolins, Maris Andris
Sestanj, Kazimir
Shah, Hasmukh N
Sharp, Hugh T
Shaw, Henry
Short, W(illiam) Leigh
Shrensel, J(ulius)
Shumate, Starling Everett, II
Silla, Harry
Silvestri, Anthony John
Singleton, Bert
Skaperdas, George T(heodore)
Sofer, Samir Salim
Song, Jung Hyun
Sonn, George Frank
Spence, Sydney P(ayton)
Speronello, Barry Keven
Sprow, Frank Barker
Steele, Lawrence Russell
Sturzenegger, August
Suciu, George Dan
Sundaresan, Sankaran
Swabb, Lawrence E(dward), Jr
Swanson, David Bernard
Sweed, Norman Harris
Sywe, Bei-Shen
Tamarelli, Alan Wayne
Tan, Victor
Taylor, William Francis
Tegge, B(ruce) R(obert)
Tio, Cesario O
Tischler, Oscar
Todd, David Burton
Tolsma, Jacob
Toner, Richard K(enneth)

Tsao, Utah
Tsonopoulos, Constantine
Tunkel, Steven Joseph
Udani, Lalit Kumar Harilal
Vardi, Joseph
Vieth, Wolf R(andolph)
Vriens, Gerard N(ichols)
Weekman, Vern W(illiam), Jr
Weil, Benjamin Henry
Weinstein, Norman J(acob)
Weiss, Lawrence H(eisler)
Wenis, Edward
West, John M(aurice)
Westerdahl, Raymond P
Wildman, George Thomas
Williams, Larry McClease
Wolynic, Edward Thomas
Wright, Stuart R(edmond)
Yan, Tsoung-Yuan
Yarmush, Martin Leon
Yourtee, John Boteler
Yu, Stephen Kin-Cheun
Yurchak, Sergei
Zonis, Irwin S(amuel)
Zoss, Abraham Oscar
Zudkevitch, David

NEW MEXICO
Amos, Donald E
Benziger, Theodore Michell
Bhada, Rohinton(Ron) K
Bopp, Gordon R(onald)
Brabson, George Dana, Jr
Brown, Lee F(rancis)
Burick, Richard Joseph
Cheng, Yung-Sung
Cho, Michael Yongkook
Christiansen, David Ernest
Edeskuty, F(rederick) J(ames)
El-Genk, Mohamed Shafik
Erickson, Kenneth Lynn
Guevara, Francisco A(ntonio)
Gurley, Lawrence Ray
Holmes, John Thomas
Kiley, Leo Austin
Kropschot, Richard H
Lam, Kin Leung
Lehman, Hugh Roberts
Lynch, Richard Wallace
MacFarlane, Donald Robert
Marcy, Willard
Mauzy, Michael P
Mead, Richard Wilson
Mokler, Brian Victor
Mullins, Lawrence A
Nathan, Charles C(arb)
Nevitt, Thomas D
Nowak, Edwin James
Nuttall, Herbert Ericksen, Jr
Patton, John Tinsman
Raseman, Chad J(oseph)
Sheinberg, Haskell
Short, Wallace W(alter)
Snyder, Albert W
Stephans, Richard A
Tapscott, Robert Edwin
Thode, E(dward) F(rederick)
Whan, Glenn A(lan)
Williams, James Marvin
Wilson, Donald Bruce

NEW YORK
Ackerberg, Robert C(yril)
Allerton, Joseph
Altwicker, Elmar Robert
Anderson, J(ohn) E(rling)
Anthony, Thomas Richard
Atkinson, Joseph Ferris
Aust, Richard Bert
Baginski, F(rank) C(harles)
Baier, Robert Edward
Balassa, Leslie Ladislaus
Banholzer, William Frank
Barduhn, Allen J(ohn)
Baron, Seymour
Barone, Louis J(oseph)
Batey, Robert William
Begell, William
Belfort, Georges
Benenati, R(obert) F(rancis)
Bequette, B Wayne
Berry, E Janet
Bitha, Panayota
Bizios, Rena
Bloomer, Oscar T(heodore)
Booser, E(arl) R(ichard)
Brown, Pamela Ann McElroy
Brunjes, A(ustin) S
Brutvan, Donald Richard
Burris, Conrad Timothy
Campbell, Gregory August
Chadwick, George F(redrick)
Chandler, Horace W
Cheh, Huk Yuk
Chen, B(enjamin) T(eh-Kung)
Chen, Shaw-Horng
Chertow, Bernard
Cheung, Harry
Child, Edward T(aylor)
Chin, Der-Tau
Chou, Chung-Chi
Chueh, Chun Fei
Clark, Alfred
Coffee, Robert Dodd

Cohen, Stephen Robert
Cokelet, Giles R(oy)
Cole, Robert
Coler, Myron Abraham
Conti, James J(oseph)
Coppersmith, Frederick Martin
Dahneke, Barton Eugene
Dastin, Samuel J
Davidson, H(arold)
DeCrosta, Edward Francis, Jr
Dessauer, John Hans
Dexter, Theodore Henry
Dietz, Russell Noel
Dugliss, Charles H(osea)
Duncan, Thomas Michael
Eaton, George T(homas)
Eberts, Robert Eugene
Emmert, R(ichard) E(ugene)
Fayon, A(bram) M(iko)
Feinberg, Martin Robert
Ferron, John R(oyal)
Fewkes, Robert Charles Joseph
Finn, R(obert) K(aul)
Follows, Alan Greaves
Fontijn, Arthur
Formanek, R(obert) J(oseph)
Foster, Richard N(orman)
Friedly, John C
Gale, Stephen Bruce
Ghezzo, Mario
Gill, William N(elson)
Gilmont, Roger
Givi, Peyman
Good, Robert James
Gossett, James Michael
Gould, Robert Kinkade
Graff, Robert A
Grayson, Herbert G
Green, Harry J(ames), Jr
Gregory, Clarence Leslie, Jr
Grohse, Edward William
Gubbins, Keith E(dmund)
Haas, Frederick Carl
Hall, Kimball Parker
Han, Chang Dae
Handman, Stanley E
Happel, John
Harriott, Peter
Harwood, Colin Frederick
Heeks, R(obert) E(ugene)
Heisig, Charles G(ladstone)
Hirtzel, Cynthia S
Hlavacek, Vladimir
Hollein, Helen Conway Faris
House, Gary Lawrence
Huebschmann, John W
Ing, Samuel W(ei-Hsing), Jr
Isaacs, Leslie Laszlo
Janik, Gerald S
Jelinek, Robert V(incent)
Jenekhe, Samson A
Jorne, Jacob
Kagansky, Larisa
Kagetsu, T(adashi) J(ack)
Kant, Fred H(ugo)
Keeney, Norwood Henry, Jr
Kerr, Donald L(aurens)
Kim, Young Joo
Kiphart, Kerry
Kiser, Kenneth M(aynard)
Klotzbach, Robert J(ames)
Knauer, Bruce Richard
Kolb, Frederick J(ohn), Jr
Kosky, Philip George
Kraft, William Gerald
Kramer, Franklin
Krinsky, Herman Y
Krutchen, Charles M(arion)
Kuo, Yue
Kwentus, Gerald K(enneth)
Landau, Ralph
Lapin, Abraham
Lashmet, Peter K(erns)
Laster, Richard
Leavitt, Fred W
Lee, Tzoong-Chyh
Leonard, Edward (Francis)
Lhila, Ramesh Chand
Li, Chung-Hsiung
Lichtenwalner, Hart K
Littman, Howard
Litz, Lawrence Marvin
Loeffler, Albert L, Jr
Luce, James Edward
McCamy, Calvin S
McConnell, C(harles) W(illiam)
Makarewicz, Peter James
Manowitz, Bernard
Marion, C(harles) P(arker)
Mathad, Gangadhara S
Matienzo, Luis J
Mead, William J(asper)
Meng, Karl H(all)
Merrill, Robert P
Meyer, Walter
Mika, Thomas Stephen
Miller, Melvin J
Mizma, Edward John
Morgan, Frank W
Morgan, Morris Herbert
Mount, Eldridge Milford, III
Mundel, August B(aer)
Mustacchi, Henry
Myerson, Allan Stuart

Mytelka, Alan Ira
Naphtali, Leonard Mathias
Nauman, Edward Bruce
Nazem, Faramarz Franz
Nissan, Alfred H(eskel)
Obey, James H(oward)
Oldshue, J(ames) Y(oung)
Othmer, Donald F(rederick)
Ott, Henry C(arl)
Palladino, William Joseph
Palmer, Harvey John
Panagiotopoulos, Athanassios Z
Panos, Peter S
Parikh, Hemant Bhupendra
Park, Paul Heechung
Pearson, Glen Hamilton
Peet, Robert G(uthrie)
Peters, Theodore, Jr
Poirier, Robert Victor
Powers, Dale Robert
Preisler, Joseph J(ohn)
Rao, Mentreddi Anandha
Reber, Raymond Andrew
Reeves, Robert R
Reid, Allen Francis
Revankar, Vithal V S
Reynolds, Joseph
Rice, Philip A
Richards, Jack Lester
Robb, Walter L(ee)
Rodriguez, Ferdinand
Rosen, Milton Jacques
Rubin, Bruce Joel
Ruckenstein, Eli
Ruschak, Kenneth John
Salemme, Robert Michael
Salzano, Francis J(ohn)
Schachter, Rozalie
Scheele, George F(rederick)
Scheibel, Edward G(eorge)
Schulz, Helmut Wilhelm
Schwan, Judith A
Seltzer, Stanley
Shade, Ray W(alton)
Shah, Ramesh Keshavlal
Shaw, John A
Sheeran, Stanley Robert
Shen, Thomas T
Shiao, Daniel Da-Fong
Shinnar, Reuel
Shuler, Michael Louis
Shulman, Herman L
Slater, C Stewart
Smith, Julian Cleveland
Smura, Bronislaw Bernard
Somekh, George S
Sotirchos, Stratis V
Spevack, Jerome
Springer, Dwight Sylvan
Srinivasaraghavan, Rengachari
Staples, Basil George
Staub, Fred W
Staudenmayer, William J(oseph)
Steinberg, Meyer
Stenuf, Theodore Joseph
Stern, Silviu Alexander
Sturges, Stuart
Subramanian, Gopal
Subramanian, Ram Shankar
Susskind, Herbert
Tang, Ignatius Ning-Bang
Tavlarides, Lawrence Lasky
Taylor, Ross
Tecklenburg, Harry
Terwilliger, James Paul
Themelis, Nickolas John
Thomas, Leo John
Thornton, Roy Fred
Tien, Chi
Tsai, Gow-Jen
Van Ness, Hendrick C(harles)
Von Berg, Robert L(ee)
Wang, Muhao S
Wang, Shu Lung
Ward, Thomas J(ulian)
Ward, William J, III
Wayner, Peter C, Jr
Weber, Arthur Phineas
Weber, Thomas W(illiam)
Weinstein, Herbert
Weller, S(ol) W(illiam)
Wey, Jong-Shinn
Whittingham, M(ichael) Stanley
Wiegandt, Herbert F(rederick)
Wilcox, W(illiam) R(oss)
Woodmansee, Donald Ernest
Wotzak, Gregory Paul
Wu, Wen Pao
Wu, William Chi-Liang
Yang, Ralph Tzu-Bow
Yu, Wen-Shi
Zenz, Frederick A(nton)
Ziegler, Edward N
Ziemniak, Stephen Eric

NORTH CAROLINA
Aneja, Viney P
Apperson, Charles Hamilton
Armstrong, A(rthur) A(lexander), Jr
Beatty, K(enneth) O(rion), Jr
Bobalek, Edward G(eorge)
Bowen, Joshua Shelton, Jr
Carpenter, Benjamin H(arrison)
Carr, Dodd S(tewart)

Chandra, Suresh
Chang, John Chung-Shih
Chang, Richard C(hi-Cheng)
Chantry, William Amdor
Cipau, Gabriel R
Cleland, John Gregory
Coli, G(uido) J(ohn), Jr
Farrar, R(ichard) E(dward)
Felder, Richard Mark
Ferrell, James K(io)
Foley, Gary W
Fulton, James W(illiam)
Fytelson, Milton
Gangwal, Santosh Kumar
Gray, Walter C(larke)
Hall, Carol Klein
Hemphill, Adley W(alton)
Hochmuth, Robert Milo
Holcomb, Charles Edward
Holland, Lyman Lyle
Hopfenberg, Harold Bruce
Humphreys, Kenneth K
Johnston, Peter Ramsey
Jones, Samuel O'Brien
King, Franklin G, Jr
Knowlton, Robert Charles
Lao, Yan-Jeong
Larson, Richard I
McCracken, Philip Glen
Manson, Allison Ray
Marsland, David B(oyd)
Martin, Donald Crowell
Moelter, Gregory Martin
Nutter, James I(rving)
Ollis, David F(rederick)
Overcash, Michael Ray
Owens, Boone Bailey
Richardson, Frances Marian
Roberts, George W(illard)
Roblin, John M
Roop, Robert Kenneth
Saeman, W(alter) C(arl)
Samfield, Max
Scheffe, Richard Donald
Serad, George A
Sherer, James Pressly
Stahel, Edward P(aul)
Stenger, William J(ames), Sr
Sullivan, F(rederick) W(illiam), III
Timmins, Robert Stone
Torquato, Salvatore
Trenholm, Andrew Rutledge
Whitehurst, Brooks M
Wilkes, William Roy
Winston, Hubert
Wood, Rodney David
Zevnik, Francis C(lair)
Zumwalt, Lloyd Robert

NORTH DAKOTA
Bierwagen, Gordon Paul
Hasan, Abu Rashid
Ludlow, Douglas Kent
Owens, Thomas Charles
Watt, David Milne, Jr

OHIO
Adler, Robert J
Ahmad, Shamim
Alam, M Khairul
Alcorn, William R(obert)
Angus, John Cotton
Atwood, Glenn A
Ayer, Howard Earle
Baasel, William David
Bahr, Donald Walter
Baid, Kushalkumar Moolchand
Baker, D(ale) B(urdette)
Baloun, Calvin H(endricks)
Bardasz, Ewa Alice
Bares, William G
Beckham, Robert R(ound)
Bennett, G(ary) F
Bierlein, James A(llison)
Binning, Robert Christie
Biswas, Pratim
Blaisdell, John Lewis
Bollinger, Edward H(arry)
Bonner, David Calhoun
Bostian, Harry E(dward)
Boyne, William Joseph
Bradbury, Elmer J(oseph)
Brillhart, Donald D
Brodkey, R(obert) S(tanley)
Brosilow, Coleman B
Brown, Glenn R(obbins)
Brunner, Carl Alan
Brunner, Gordon Francis
Bruno, David Joseph
Burte, Harris M(erl)
Chaloud, J(ohn) Hoyt
Chari, Nallan C
Cheng, Stephen Zheng Di
Choudhary, Manoj Kumar
Christensen, Craig Mitchell
Christensen, James Henry
Chung, Benjamin T F
Cinadr, Bernard F(rank)
Coaker, A(nthony) William
Colby, Edward Eugene
Conklin, R(oger) N(orton)
Cornelius, Billy Dean
Coulman, G(eorge) A(lbert)
Darby, Ralph Lewis

Chemical Engineering (cont)

Dickey, David S
DiLiddo, Bart A(nthony)
Dinos, Nicholas
Dodge, James Stanley
Donovan, Leo F(rancis)
Drake, George M(arshall), Jr
Dunn, Robert Garvin
Duval, Leonard A
Edwards, Robert V(alentino)
Ewing, R(obert) A(rno)
Fan, Liang-Shih
Farrell, J(oseph) B(rendan)
Faust, Charles L(awson)
Fenn, Robert William, III
Fisch, Herbert A(lbert)
Fleming, Paul Daniel, III
Flitcraft, R(ichard) K(irby), II
Forsyth, Thomas Henry
Frank, Sylvan Gerald
Freeberg, Fred E
Fried, Joel Robert
Friedman, Morton Harold
Frye, C(lifton) G(eorge)
Gieseke, James Arnold
Ginning, P(aul) R(oll)
Goldberger, W(illiam) M(organ)
Grasser, Bruce Howard
Greenberg, David B(ernard)
Greene, Howard Lyman
Grigger, David John
Grisaffe, Salvatore J
Gurklis, John A(nthony)
Haering, Edwin Raymond
Hedley, William H(enby)
Heines, Thomas Samuel
Heinold, Robert H
Henderson, Courtland M
Hershey, Daniel
Hershey, Harry Chenault
Huddleston, George Richmond, Jr
Hwang, Sun-Tak
Ishida, Hatsuo
Jenkins, Robert George
Jones, Millard Lawrence, Jr
Kao, Yuen-Koh
Katzen, Raphael
Kawasaki, Edwin Pope
Kay, Webster Bice
Kell, Robert M
Kenat, Thomas Arthur
Kendall, H(arold) B(enne)
Khang, Soon-Jai
Kim, Byung Cho
Knaebel, Kent Schofield
Knowlton, David A
Knox, Kenneth L
Koegle, John S(tuart)
Krieble, James G(erhard)
Lacksonen, James W(alter)
Lahti, Leslie Erwin
Lee, Jon H(yunkoo)
Lee, Sunggyu
Lemlich, Robert
Lewis, George R(obert)
Licht, W(illiam), Jr
Lindstedt, P(aul) M
Liu, Chung-Chiun
Lyons, Carl J(ohn)
Martin, John B(ruce)
Massey, L(ester) G(eorge)
Measamer, S(chubert) G(ernt)
Meinecke, Eberhard A
Metanomski, Wladyslaw Val
Michalakos, Peter Michael
Minges, Merrill Loren
Moon, George D(onald), Jr
Moore, Arthur William
Mosser, John Snavely
Noda, Isao
Oster, Eugene Arthur
Oxley, Joseph H(ubbard)
Ozkan, Umit Sivrioglu
Poling, Bruce Earl
Price, John Avner
Purdon, James Ralph, Jr
Rains, Roger Keranen
Rau, Allen H
Roberts, Robert William
Roberts, Timothy R
Romesberg, Floyd Eugene
Rosa, Casimir Joseph
Rounsley, Robert R(ichard)
Rutherford, William M(organ)
Scaccia, Carl
Schlaudecker, George F(rederick)
Schmitt, George Frederick, Jr
Schooley, Arthur Thomas
Schultz, Thomas J
Schumm, Brooke, Jr
Selden, George
Selover, Theodore Britton, Jr
Servais, Ronald Albert
Sharma, Shri C
Sheets, George Henkle
Shick, Philip E(dwin)
Sinner, Donald H
Skidmore, Duane R(ichard)
Snavely, Cloyd A(rten)
Sommer, John G
Springer, Allan Matthew
Stevenson, James Francis
Stockman, Charles H(enry)
Stoops, Charles E(mmet), Jr
Strop, Hans R
Stubblebine, Warren
Sweeney, Thomas L(eonard)
Sylvester, Nicholas Dominic
Szirmay, Leslie v
Throckmorton, Peter E
Turner, Andrew
Updegrove, Louis B
Vahldiek, Fred W(illiam)
Wallace, Michael Dwight
Waller, Michael Holton
Warshay, Marvin
Weber, Lester George
Webster, Allen E
Weisman, Joel
Wiederhold, Edward W(illiam)
Winn, Hugh
Wolfe, Robert Kenneth
Yang, Philip Yung-Chin
Yarrington, Robert M
Yu, Thomas Huei-Chung
Zager, Stanley E(dward)
Zakin, Jacques L(ouis)

OKLAHOMA

Adams, Don
Albright, Melvin A
Aldag, Arthur William, Jr
Arnold, D(onald) S(mith)
Becraft, Lloyd G(rainger)
Bell, Kenneth J(ohn)
Bray, Bruce G(lenn)
Casad, Burton M
Cerro, Ramon Luis
Cheng, Paul J(ih) T(ien)
Chung, Ting-Horng
Clark, George C(harles)
Crynes, Billy Lee
Dale, Glenn H(ilburn)
Daniels, Raymond D(eWitt)
Dew, John N(orman)
Doane, Elliott P
Donaldson, Erle C
Ellington, Rex T(ruesdale), Jr
Foutch, Gary Lynn
Gall, James William
Gilliland, Joe E(dward)
Goddin, C(lifton) S(ylvanus)
Graves, Toby Robert
Green, Ray Charles
Harrison, R(oland) H(enry)
Harwell, Jeffrey Harry
Hays, George E(dgar)
Herbolsheimer, Glenn
Hughes, Kenneth James
Irvin, Howard Brownlee
Johnson, Marvin M
Johnson, Paul H(ilton)
Joshi, Sadanand D
Koerner, E(rnest) L(ee)
Kriegel, Monroe W(erner)
Leder, Frederic
Lilley, David Grantham
Logan, R(ichard) S(utton)
Madden, Michael Preston
Maddox, R(obert) N(ott)
Manning, Francis S(cott)
Marwil, S(tanley) J(ackson)
Mihm, John Clifford
O'Rear, Edgar Allen, III
Philoon, Wallace C, Jr
Pollock, L(yle) W(illiam)
Rein, Robert G, Jr
Richardson, Raymond C(harles)
Robinson, Robert L(ouis), Jr
Sattler, Robert E(dward)
Skinner, Joseph L
Sliepcevich, Cedomir M
Starling, Kenneth Earl
Stover, Dennis Eugene
Tham, Min Kwan
Thompson, Richard E(ugene)
Thomson, George Herbert
Vanderveen, John Warren
Van De Steeg, Garet Edward
Vidaurri, Fernando C, Jr
Warzel, L(awrence) A(lfred)
Weis, Robert E(dward)
West, John B(ernard)
Williamson, William Burton
Wisecarver, Keith Douglas
Wood, Harold Sinclair
Wride, W(illiam) James
Yarborough, Lyman
Zetik, Donald Frank

OREGON

Barkelew, Chandler H(arrison)
Cleland, Franklin Andrew
Haygarth, John Charles
Jansen, George, Jr
Knudsen, J(ames) G(eorge)
Landsberg, Arne
Levenspiel, Octave
McDuffie, Norton G(raham), Jr
Maze, Robert Craig
Meredith, Robert E(ugene)
Miner, J Ronald
Mrazek, Robert Vernon
Olsen, Richard Standal
Smith, Kelly L
Tucker, W(illiam) Henry
Wicks, Charles E(dward)
Woods, W(allace) Kelly

PENNSYLVANIA

Adams, Roy Melville
Alexander, Stuart David
Aly, Abdel Fattah
Amero, R(obert) C
Anderson, James B
Anderson, John Leonard
Archer, David Horace
Austin, Leonard G(eorge)
Aviles, Rafael G
Baird, Michael Jefferson
Baker, Frank William
Baker, Rees Terence Keith
Banks, Robert R(ae)
Barton, Paul
Bauer, William, Jr
Bay, Theodosios (Ted)
Beck, William F(rank)
Berty, Jozsef M
Biegler, Lorenz Theodor
Bielecki, Edwin J(oseph)
Bienstock, D(aniel)
Bogash, R(ichard)
Brainard, Alan J
Braun, Walter G(ustav)
Breston, Joseph N(orbert)
Brey, R(obert) N(ewton)
Brian, P(ierre) L(eonce) Thibaut
Buerk, Donald Gene
Byers, R Lee
Camp, Frederick William
Caram, Hugo Simon
Carr, Norman L(oren)
Castellano, Salvatore, Mario
Champagne, Paul Ernest
Chappelow, Cecil Clendis, III
Chaudhury, Manoj Kumar
Chelemer, Harold
Chen, John Chun-Chien
Chen, Michael S K
Chen-Tsai, Charlotte Hsiao-yu
Chiang, S(hiao) H(ung)
Chitale, Sanjay Madhav
Chun, Sun Woong
Churchill, Stuart W(inston)
Clump, Curtis William
Cobb, James Temple, Jr
Coghlan, David B(uell)
Condo, Albert Carman, Jr
Corbo, Vincent James
Cost, J(oe) L(ewis)
Cotabish, Harry N(elson)
Coughanowr, D(onald) R(ay)
Craig, James Clifford, Jr
Crits, George J(ohn)
Dalton, Augustine Ivanhoe, Jr
Danner, Ronald Paul
Daubert, Thomas Edward
Davis, Burl Edward
Dean, Sheldon Williams, Jr
Dell, M(anuel) Benjamin
Dent, Anthony L
DeVries, Frederick William
Dicciani, Nance Katherine
Doelp, Louis C(onrad), Jr
Dohany, Julius Eugene
Donald, Dennis Scott
Dromgold, Luther D
Duda, J(ohn) L(arry)
Dunckhorst, F(austino) T
Dwyer, Francis Gerard
Ebert, Philip E
El-Aasser, Mohamed S
Engel, Alfred J
Enick, Robert M
Erskine, Donald B
Fear, J(ames) Van Dyck
Feathers, William D
Fisher, Sallie Ann
Foltz, Donald Richard
Forney, Albert J
Forney, R(obert) C(lyde)
Forsman, William C(omstock)
Foster, Perry Alanson, Jr
Franco, Nicholas Benjamin
Frenklach, Michael Y
Frumerman, Robert
Garber, Charles A
Garg, Diwakar
Gee, Edwin Austin
Georgakis, Christos
Giffen, Robert H(enry)
Glandt, Eduardo Daniel
Goldthwait, R(ichard) G(raham)
Goodwin, James Gordon, Jr
Gordon, Lyle J
Gottschlich, Chad F
Gottscho, Alfred M(orton)
Grace, Harold P(adget)
Grassi, Vincent G
Graves, David J(ames)
Griffith, Michael Grey
Grossmann, Elihu D(avid)
Grossmann, Ignacio E
Guger, Charles Edmund, Jr
Gupta, Vijai Prakash
Gurol, Mirat D
Hager, Wayne R
Halpern, Benjamin David
Hammack, William S
Hanauer, David
Hartzog, David G(eorge)
Hawkins, A(lbert) W(illiam)
Haynes, William P
Hein, R(owland) F(rank)
Helfferich, Friedrich G
Henly, Robert Stuart
Herman, Frederick Louis
Herman, Richard Gerald
Hess, Dennis William
Hoerner, George M, Jr
Holder, Gerald D
Hopson, Kevin Matthew
Horn, Lyle William
Horst, Ralph L, Jr
Humphrey, Arthur E(arl)
Iezzi, Robert Aldo
Isakoff, Sheldon Erwin
Jarrett, Noel
Jester, William A
Jhon, Myung S
Jones, Roger Franklin
Jordan, Robert Kenneth
Jost, Donald E
Kabel, Robert L(ynn)
Kakar, Anand Swaroop
Kanter, Ira E
Keairns, Dale Lee
Keith, Frederick W(alter), Jr
Keller, Rudolf
Kelly, Conrad Michael
King, William Emmett, Jr
Kitzes, Arnold S(tanley)
Kivnick, Arnold
Klaus, E(lmer) Erwin
Klaus, Ronald Louis
Kline, Richard William
Klinzing, George Engelbert
Klugherz, Peter D(avid)
Ko, Edmond Inq-Ming
Korchak, Ernest I(an)
Ladenheim, Harry
Lancet, Michael Savage
Lantos, P(eter) R(ichard)
Larson, Kenneth Curtis
Lavin, J Gerard
Lawrence, Sigmund J(oseph)
Lawson, Neal D(evere)
Lee, Young Hie
Lerner, B(ernard) J
Levy, Edward Kenneth
Li, Kun
Lindt, Jan Thomas
Litt, Mitchell
Lloyd, Robert
Lloyd, Wallis A(llen)
Lokay, Joseph Donald
Loscher, Robert A
Lugar, Richard Charles
Luthy, Richard Godfrey
Luyben, William Landes
McCormick, Robert H(enry)
McDonnell, Leo F(rancis)
McKinley, Clyde
McNeil, Kenneth Martin
MacRae, Donald Richard
Manka, Dan P
Manning, R(obert) E(dward)
Mao, Chung-Ling
Marcelin, George
Martin, Jay Ronald
Matulevicius, Edward S(tephen)
Mendelsohn, Morris A
Meyer, Bernard Henry
Misra, Sudhan Sekher
Montgomery, Ronald Eugene
Moore, Robert Byron
Moore, Walter Calvin
Morgan, Arthur I, Jr
Morris, Stanley M
Mount, Lloyd Gordon
Mutharasan, Rajakkannu
Myers, Alan Louis
Myers, John Adams
Myron, Thomas L(eo)
Natoli, John
Nedwick, John Joseph
Nemeth, Edward Joseph
Neufeld, Ronald David
Nichols, Duane Guy
Nolan, Edward J
Novak, Darwin Albert, Jr
Orphanides, Gus George
Pandis, Spyros N
Parchen, Frank Raymond, Jr
Pardee, William A(ugustus)
Parker, Jennifer Ware
Parker, Robert Orion
Pasceri, Ralph Edward
Paxton, R(alph) R(obert)
Peel, James Edwin
Penn, William B
Perlmutter, Daniel D
Pessen, Helmut
Petrich, Robert Paul
Petronio, Marco
Petura, John C
Pinckney, Robert L
Pinkston, John Turner
Pommersheim, James Martin
Powers, Gary James
Prane, Joseph W(illiam)
Prieve, Dennis Charles
Pulsifer, Allen Huntington
Qazi, Mahmood A
Quinn, John A(lbert)

Racunas, Bernard J
Raimondi, Pietro
Ramezan, Massood
Redmount, Melvin B(ernard)
Reeder, Clyde
Regan, Raymond Wesley
Reiff, Harry Elmer
Retallick, William Bennett
Rice, William James
Rosenthal, Howard
Ross, Bradley Alfred
Rothfus, Robert R(andle)
Rudershausen, Charles Gerald
Ruppel, Thomas Conrad
Russell, Alan James
Sampson, Ronald N
Sanabor, Louis John
Sattler, Frank A(nton)
Sawicki, John Edward
Sayre, Clifford M(orrill), Jr
Scattergood, Edgar Morris
Schell, George W(ashington)
Schiesser, W(illiam) E(dward)
Schiffman, Louis F
Schlesinger, Martin D(avid)
Schnaible, H(arold) W(illiam)
Schultz, J(erome) S(amson)
Schultz, Jane Schwartz
Schwartz, Albert B
Scigliano, J Michael
Seider, Warren David
Seiner, Jerome Allan
Shellenberger, Donald J(ames)
Sheridan, John Joseph, III
Siddiqui, Habib
Sides, Paul Joseph
Sieger, John S(ylvester)
Singleton, Alan Herbert
Sinnett, Carl E(arl)
Sittenfield, Marcus
Smith, George C(unningham)
Snyder, William James
Solt, Paul E
Soung, Wen Y
Sprowles, Donald O(tte)
Stein, Fred P(aul)
Stephenson, Robert L
Stone, Douglas Roy
Struck, Robert T(heodore)
Sweeny, Robert F(rancis)
Sykes, James Aubrey, Jr
Tackie, Michael N
Tallmadge, J(ohn) A(llen), Jr
Tang, Y(u) S(un)
Thompson, Ralph Newell
Thompson, Sheldon Lee
Thygeson, John R(obert), Jr
Tierney, John W(illiam)
Tilton, Robert Daymond
Tinkler, Jack D(onald)
Tobias, George S
Toor, H(erbert) L(awrence)
Trunzo, Floyd F(rank)
Turner, Howard S(inclair)
Ultman, James Stuart
Vance, William Harrison
Vannice, Merlin Albert
Vidt, Edward James
Vrentas, Christine Mary
Vrentas, James Spiro
Wahnsiedler, Walter Edward
Waldman, L(ouis) A(braham)
Walpert, George W
Wamsley, W(elcome) W(illard)
Warburton, Charles E, Jr
Weddell, George G(ray)
Weimer, Robert Fredrick
Weinberger, Charles Brian
Weisser, Eugene P
Weisz, Paul Burg
Wenzel, Leonard A(ndrew)
Werner, John Ellis
West, A(rnold) Sumner
Westerberg, Arthur William
Wettach, William
Wetzel, Roland H(erman)
White, Robert E(dward)
Wilder, Harry D(ouglas)
Wilkens, John Albert
Wilkens, Lucile Shanes
Wilson, Harry W(alton), Jr
Wilson, Lawrence Albert, Jr
Wu, Dao-Tsing
Yang, Wen-Ching
Yavorsky, Paul M(ichael)
Yerazunis, Stephen
Zamanzadeh, Mehrooz
Zeigler, A(lfred) G(eyer)
Zhou, Qian
Ziering, Lance K
Zolotorofe, Donald Lee

RHODE ISLAND
Barnett, Stanley M(arvin)
Calo, Joseph Manuel
Graham, Ronald A(rthur)
Hutzler, Leroy, III
Knickle, Harold Norman
Liu, J(oseph) T(su) C(hieh)
Madsen, N(iels)
Mason, Edward Allen
Ricklin, Saul
Rose, Vincent C(elmer)
Satas, Donatas

Sprowles, Jolyon Charles
Suuberg, Eric Michael
Thompson, A(lexander) Ralph
Votta, Ferdinand, Jr

SOUTH CAROLINA
Alley, Forrest C
Ayen, Richard J(ohn)
Barron, Charles
Beckwith, William Frederick
Boelter, Edwin D, Jr
Bradley, Robert Foster
Bryan, James Clarence
Buckham, James A(ndrew)
Clarke, James
Cogan, Jerry Albert, Jr
Corbitt, Maurice R(ay)
Darnell, W(illiam) H(eaden)
Davis, Milton W(ickers), Jr
Dworjanyn, Lee O(leh)
Gibbons, Joseph H(arrison)
Grady, Cecil Paul Leslie, Jr
Greene, George C, III
Groce, William Henry (Bill)
Haile, James Mitchell
Hale, Raymond Joseph
Harshman, Richard C(alvert)
Herdklotz, Richard James
Hon, David Nyok-Sai
Hootman, Harry Edward
Hubbell, Douglas Osborne
Jennings, Alfred S(tonebraker)
Johnson, Ben S(lemmons), Jr
Krumrei, W(illiam) C(larence)
Lee, Craig Chun-Kuo
Longtin, Bruce
Melsheimer, Stephen Samuel
Morris, J(ames) William
Mullins, Joseph Chester
O'Neill, John H(enry), Jr
O'Rear, Steward William
Otto, Wolfgang Karl Ferdinand
Proctor, J(ohn) F(rancis)
Robinson, William Courtney, Jr
Schilson, Robert E(arl)
Squires, Paul Herman
Swank, Robert Roy, Jr
Terry, Stuart Lee
Webster, D(onald) S(teele)
White, Ralph E

SOUTH DAKOTA
Han, Kenneth N

TENNESSEE
Anderson, Roger W(hiting)
Bagrodia, Shriram
Beall, S(amuel) E, Jr
Bigelow, John E(aly)
Bloom, Sanford Gilbert
Bogue, Donald Chapman
Bond, Walter D
Bopp, C(harles) Dan
Boyle, William Robert
Buck, George Sumner, Jr
Bullock, Jonathan S, IV
Burwell, Calvin C
Byers, Charles Harry
Cain, Carl, Jr
Claiborne, H(arry) C(lyde)
Cochran, Henry Douglas, Jr
Coleman, Charles F(ranklin)
Crawford, Lloyd W(illiam)
Crouse, David J, Jr
Culberson, Oran L(ouis)
Cummings, Peter Thomas
Davison, Brian Henry
DeVan, Jackson H
Donaldson, Terrence Lee
Duckworth, William C(apell)
Finch, Gaylord Kirkwood
Fort, Tomlinson
Frazier, George Clark, Jr
Fulkerson, William
Gano, Richard W
Garber, Harold Jerome
Genung, Richard Keith
Godbee, H(erschel) W(illcox)
Gregory, Dale R(ogers)
Haas, Paul Arnold
Harris, Thomas R(aymond)
Henry, Jonathan Flake
Hightower, Jesse Robert
Hoffman, H(erbert) W(illiam)
Hsu, Hsien-Wen
James, V(irgil) Eugene
Janna, William Sied
Jasny, George R
Jensen, J(ohn) H(enry), Jr
Judkins, Roddie Reagan
Kasten, Paul R(udolph)
Keller, Charles A(lbert)
Klein, Jerry Alan
Leuze, Rex Ernest
Levin, Seymour A(rthur)
McClung, Robert W
McGill, Robert Mayo
McNeese, Leonard Eugene
Mahlman, Harvey Arthur
Mailen, James Clifford
Malling, Gerald F
Murray, Lawrence P(atterson), Jr
North, Edward D(avid)
Ornitz, Barry Louis

Overholser, Knowles Arthur
Parish, Trueman Davis
Parsly, Lewis F(uller), Jr
Perona, Joseph James
Phillips, Bobby Mal
Poese, Lester E
Polahar, Andrew Francis
Prados, John W(illiam)
Rodgers, Billy Russell
Rosenthal, Murray Wilford
Roth, John Austin
Russell, Ross F
Ryon, Allen Dale
Sanders, John P(aul), Sr
Schnelle, K(arl) B(enjamin), Jr
Scott, Charles D(avid)
Sears, Mildred Bonham
Seaton, William Hafford
Sheth, Atul C
Siirola, Jeffrey John
Silverman, Meyer David
Singh, Suman Priyadarshi Narain
Sissom, Leighton E(sten)
Sittel, Chester Nachand
Spruiell, Joseph E(arl)
Tanner, Robert Dennis
Thomas, David Glen
Threadgill, W(alter) D(ennis)
Turitto, Vincent Thomas
Umholtz, Clyde Allan
Van Horn, Wendell Earl
Vaughen, Victor C(ornelius) A(dolph)
Wang, Chien Bang
Watson, Jack Samuel
Woodard, Kenneth Eugene, Jr
Yarbrough, David Wylie
Yee, William C
Zachariasen, K(arsten) A(ndreas)

TEXAS
Acciarri, Jerry A(nthony)
Akgerman, Aydin
Ali, Yusuf
Al-Saadoon, Faleh T
Amundson, Neal R
Anthony, Rayford Gaines
Armeniades, Constantine D
Atkins, George T(yng)
Baeder, Donald L(ee)
Bannon, Robert Patrick
Barlow, Joel William
Barrere, Clem A, Jr
Bawa, Mohendra S
Beck, Curt B(uxton)
Bethea, Robert Morrison
Biles, W(illiam) R(oy)
Blum, Harold A(rthur)
Blytas, George Constantin
Bodnar, Stephen J
Bradford, John R(oss)
Brennecke, Henry Martin
Brennecke, Llewellyn F(rancis)
Brock, James Rush
Broodo, Archie
Brooke, M(axey)
Brown, Robert Griffith
Brunsting, Elmer H(enry)
Bush, Warren Van Ness
Camero, Arthur Anthony
Campbell, William M
Cardner, David V
Cares, William Ronald
Carradine, William Radell, Jr
Chen, Anthony Hing
Chen, Edward Chuck-Ming
Chew, Ju-Nam
Chu, Chieh
Chyu, Ming-Chien
Cier, H(arry) E(vans)
Claassen, E(dwin) J(ack), Jr
Claridge, E(lmond) L(owell)
Clark, Stanley Preston
Clunie, Thomas John
Coldren, Clarke L(incoln)
Colten, Oscar A(aron)
Cook, Evin Lee
Crookston, Reid B
Crump, John Joseph
Cruse, Carl Max
Cummings, George H(erbert)
Cupples, Barrett L(eMoyne)
Darby, Ronald
Davis, Samuel Henry, Jr
Davison, Richard Read
Davitt, Harry James, Jr
Dillard, Robert Garing, Jr
Domask, W(illiam) G(erhard)
Dotterweich, Frank H(enry)
Doumas, A(rthur) C(onstantinos)
Dukler, A(braham) E(manuel)
Duncan, Dennis Andrew
Durbin, Leonel Damien
Dye, Robert F(ulton)
Dyson, Derek C(harlesworth)
Eakin, Bertram E
Economou, Demetre J
Edgar, Thomas Flynn
Edwards, Victor Henry
Eggers, Fred M
Ehlig-Economides, Christine Anna
Ekerdt, John Gilbert
Engler, Cady Roy
Eubank, Philip Toby
Eubanks, L(loyd) Stanley

Ewing, Richard Edward
Fair, James R(utherford)
Fewel, Kenneth Jack, Jr
Flumerfelt, Raymond W
Fontaine, Marc F(rancis)
Fox, George Edward
Frantz, Joseph Foster
Furgason, Robert Roy
Fyfe, Richard Ross
Gidley, J(ohn) L(ynn)
Gilliland, Harold Eugene
Gilmour-Stallsworth, Lisa K
Goldblum, David Kiva
Goring, Geoffrey E(dward)
Graham, Harold L(aVerne)
Gregory, A(lvin) R(ay)
Grieves, Robert Belanger
Griffin, John R(obert)
Grimsby, F(rank) Norman
Gully, Arnold J(arvis)
Gwyn, J(ohn) E(dward)
Haase, Donald J(ames)
Hales, Hugh B(radley)
Hall, Kenneth Richard
Hammond, James W
Harris, William Birch
Hart, Paul Robert
Heenan, William A
Heichelheim, Hubert Reed
Hellums, Jesse David
Hellwig, L(angley) R(oberts)
Henley, Ernest J(ustus)
Hightower, Joe W(alter)
Hill, Arthur S
Himmelblau, David M(autner)
Hirasaki, George J
Hirsch, Robert Louis
Hite, J Roger
Hocott, C(laude) R(ichard)
Hoffman, T(errence) W(illiam)
Holland, Charles D(onald)
Holmes, Larry A
Holste, James Clifton
Holtzapple, Mark Thomas
Honeywell, Wallace I(rving)
Hopper, Jack R
Hougen, Joel O(liver)
Huang, Wann Sheng
Humphrey, Jimmy Luther
Hurlburt, H(arvey) Zeh
Ivey, E(dwin) H(arry), Jr
Jelen, Frederic Charles
Jiles, Charles William
Johnson, Lloyd N(ewhall)
Kalfoglou, George
Kemp, L(ebbeus) C(ourtright), Jr
Kingsley, Henry A(delbert)
Klecka, Miroslav Ezidor
Klein, V(ernon) A(lfred)
Knipp, Ernest A, (Jr)
Kobayashi, Riki
Koch, Howard A(lexander)
Koemtzopoulos, C Robert
Koepf, Ernest Henry
Koppel, Lowell B
Koros, William John
Krautz, Fred Gerhard
Landis, E K
Latour, Pierre Richard
Lawler, James Henry Lawrence
Lescarboura, Jaime Aquiles
Levy, Robert Edward
Lewis, James Pettis
Lloyd, Douglas Roy
Long, R(obert) B(yron)
Lowell, Philip S(iverly)
Ludwig, Allen Clarence, Sr
Luss, Dan
Lyon, John B(ennett)
McCutchan, Roy T(homas)
McIntire, Larry V(ern)
McIntire, Louis V(ictor)
McKee, Herbert C(harles)
McKetta, John J, Jr
McNiel, James S(amuel), Jr
Maddox, Larry A(llen)
Mai, Klaus L(udwig)
Marberry, James E(dward)
Marple, Stanley, Jr
Marsh, Kenneth Neil
Martin, Thomas L(yle), Jr
Massey, Michael John
Mattax, Calvin Coolidge
Messenger, Joseph Umlah
Miller, Clarence A(lphonso)
Miller, Richard Linn
Mitchell, James Emmett
Moltzan, Herbert John
Morris, Herbert Comstock
Morrison, Milton Edward
Morse, N(orman) L(ester)
Moser, James Howard
Mosher, Donald Raymond
Mullikin, Richard V(ickers)
Nicolaisen, B(ernard) H(enry)
Ochiai, Shinya
O'Connell, Harry E(dward)
Okamoto, K Keith
Olson, John S
Padgett, Algie Ross
Park, Vernon Kee
Parker, Arthur L (Pete)
Parker, Harry W(illiam)

Chemical Engineering (cont)

Patil, Kashinath Z(iparu)
Paul, Donald Ross
Pennington, Robert Elija
Perkins, Thomas K(eeble)
Pitt, Woodrow Wilson, Jr
Plunkett, Roy J
Poddar, Syamal K
Popovich, Robert Peter
Powell, Richard James
Prengle, H(erman) William, Jr
Pritchett, P(hilip) W(alter)
Rai, Charanjit
Rajagopalan, Raj
Randlett, Herbert Eldridge, Jr
Rase, Howard F
Rhinehart, Robert Russell, II
Richardson, James T(homas)
Richardson, Joseph Gerald
Robbins, Roger Alan
Roberts, Louis Reed
Roeger, Anton, III
Russell, James N(elson), Jr
Schechter, Robert Samuel
Schrader, R(obert) J
Schrage, Robert W
Scinta, James
Senatore, Ford Fortunato
Serth, Robert William
Shanfield, Henry
Sheppard, Louis Clarke
Silberberg, I(rwin) Harold
Simon, Eric
Slattery, John C
Slusser, M(arion) L(iles)
Somerville, George R
Souby, Armand Max
Speed, Raymond A(ndrew)
Stewart, J(ames) R(ush), Jr
Stone, Herbert L(osson)
Styring, Ralph E
Sullivan, Thomas Allen
Swearingen, Judson Sterling
Tao, Frank F
Teague, Abner F
Tetlow, Norman Jay
Thatcher, C(harles) M(anson)
Thompson, William Horn
Tiedemann, Herman Henry
Tiller, F(rank) M(onterey)
Tock, Richard William
Trachtenberg, Isaac
Tully, Philip C(ochran)
Van Horn, Lloyd Dixon
Volpe, P(eter) J, Jr
Von Rosenberg, Dale Ursini
von Rosenberg, H(ermann) E(ugene)
Wagner, John Philip
Walker, Richard E
Walls, Hugh A(lan)
Warren, Francis A(lbert)
Warren, Kenneth Wayne
Weatherford, W(illiam) D(ewey), Jr
Wheeler, Joe Darr
White, Bernard Henry
Wiener, L(udwig) D(avid)
Wilde, Kenneth Alfred
Williams, Curtis Chandler, III
Willoughby, Sarah Margaret C(laypool)
Wilson, J(ames) W(oodrow)
Wissler, Eugene H(arley)
Wong, S(oon) Y(uck)
Woods, J(ohn) M(elville)
Woodward, Joe William
Worley, Frank L, Jr
Wurth, Thomas Joseph
Yaws, Carl Lloyd
Yeung, Reginald Sze-Chit

UTAH
Adams, Michael Curtis
Barker, Dee H(eaton)
Bartholomew, Calvin Henry
Bennion, Douglas Noel
Budge, Wallace Don
Chesworth, Robert Hadden
Coates, Ralph L
Dahlstrom, Donald A(lbert)
De Nevers, Noel Howard
Fay, John Edward, II
Fletcher, Thomas Harvey
Hanks, Richard W(ylie)
Horton, M Duane
Kelsey, Stephen Jorgensen
Larsen, James Victor
Lighty, JoAnn Slama
Miller, Jan Dean
Oblad, Alex Golden
Othmer, Hans George
Pope, Bill Jordan
Powers, John E(dward)
Rowley, Richard L
Ryan, Norman W(allace)
Salt, Dale L(ambourne)
Seader, J(unior) D(eVere)
Silver, Barnard Stewart
Smoot, Leon Douglas
Sohn, Hong Yong
Solea, Kenneth A
Thirkill, John D
Trujillo, Edward Michael
Tyler, Austin Lamont
Zeamer, Richard Jere

VERMONT
Bhatt, Girish M
Dean, Robert Charles, Jr
Dritschilo, William
Jones, John F(rederick)
Vivian, J(ohnson) Edward

VIRGINIA
Bartlett, John W(esley)
Beach, Robert L
Beyer, Gerhard H(arold)
Bickling, Charles Robert
Blum, Edward H(oward)
Brown, Winton
Burka, Maria Karpati
Bushey, Gordon Lake
Chapin, Douglas McCall
Chiou, Minshon Jebb
Conger, William Lawrence
Cox, Edwin, III
Dardoufas, Kimon C
Davis, Bernard Eric
Dedrick, Robert L(yle)
DePoy, Phil Eugene
Diller, Thomas Eugene
Dorsey, Clark L(awler), Jr
Durfee, Robert Lewis
Erickson, Wayne Douglas
Fein, Harvey L(ester)
Ferguson, Thomas Lee
Fisher, Farley
Foerster, Edward L(eRoy), Sr
Friel, Daniel Denwood
Gaden, Elmer L(ewis), Jr
Gainer, John Lloyd
Gallini, John B(attista)
Glasser, Wolfgang Gerhard
Gordon, John S(tevens)
Gross, George C(onrad)
Halsey, Brenton S
Harris, Paul Robert
Hauxwell, Gerald Dean
Hemm, Robert Virgil
Herbst, Mark Joseph
Hill, Roger W(arren)
Hinkle, Barton L(eslie)
Hollingsworth, David S
Hudson, J(ohn) L
Hur, J James
Hussamy, Samir
Jaisinghani, Rajan A
Jimeson, Robert M(acKay), Jr
Joebstl, Johann Anton
Jonnard, Aimison
Keeler, Roger Norris
King, Merrill Kenneth
Kirkendall, Ernest Oliver
Kirpekar, Abhay C
Kirtley, Thomas L(loyd)
Knap, James E(li)
Lacy, W(illiam) J(ohn)
Lasser, Howard Gilbert
LeVan, Martin Douglas, Jr
Lih, Marshall Min-Shing
Lilleleht, L(embit) U(no)
Ling, James Gi-Ming
MacClaren, Robert H
McGee, Henry A(lexander), Jr
McHale, Edward Thomas
McKenna, John Dennis
Malloy, Alfred Marcus
Marchello, Joseph M(aurice)
Markels, Michael, Jr
Mayfield, Lewis G
Mertes, Frank Peter, Jr
Mischke, Roland A(lan)
Mukherjee, Tapan Kumar
O'Connell, John P
Oliver, Robert C(arl)
Oyama, Shigeo Ted
Patterson, Earl E(dgar)
Perry, John E(dward)
Puyear, Donald E(mpson)
Raymond, Dale Rodney
Roberts, Irving
Roco, Mihail Constantin
Rony, Peter R(oland)
Sabadell, Alberto Jose
Schaffner, Robert M(ichael)
Setterstrom, Carl A(lbert)
Shockley, Gilbert R
Sikri, Atam P
Southworth, Raymond W(illiam)
Squires, Arthur Morton
Stoddard, C(arl) Kerby
Tinsley, Richard Sterling
Uhl, V(incent) W(illiam)
Walther, James Eugene
Washington, James M(acknight)
Wills, George B(ailey)
Winslow, Charles Ellis, Jr
Young, Earle F(rancis), Jr
Zaborsky, Oskar Rudolf

WASHINGTON
Ackerman, C(arl) D(avid)
Amberg, Herman R(obert)
Austin, George T(homas)
Babb, Albert L(eslie)
Barker, James J(oseph)
Beaton, Roy Howard
Beck, Theodore R(ichard)
Berg, John Calvin
Berger, Albert Jeffrey
Bierman, Sidney Roy
Bolme, Donald W(eston)
Bolme, Mark W
Bowdish, Frank W(illiam)
Bowen, J(ewell) Ray
Brehm, William Frederick
Burns, Robert Ward
Clark, Donald Eldon
Cook, John Carey
Damon, Robert A(rthur)
Davis, E(arl) James
Dickinson, Dean Richard
Divine, James R(obert)
Dobratz, Carroll J
Duncan, James Byron
Edde, Howard Jasper
Evans, Thomas F(rederick)
Falco, James William
Faletti, Duane W
Finlayson, Bruce Alan
Finnigan, J(erome) W(oodruff)
Fosberg, Theodore M(ichael)
Foster, Norman Charles
Fullam, Harold Thomas
Garlid, Kermit L(eroy)
Hales, Jeremy M(organ)
Hanthorn, Howard E(ugene)
Heideger, William J(oseph)
Hendrickson, Waldemar Forrsel
Hirsch, Horst Eberhard
Johanson, L(ennart) N(oble)
Johnson, B(enjamin) M(artineau)
Kaser, J(ohn) D(onald)
Kerbecek, Arthur J(oseph), Jr
Koehmstedt, Paul Leon
Krieger, Barbara Brockett
Krier, Carol Alnoth
LaFontaine, Thomas E
Lago, James
Lee, James Moon
Lindstrom, Duaine Gerald
McCarthy, Joseph L(ePage)
McKean, William Thomas, Jr
Mahalingam, R
Moore, Raymond H
Orth, John C(arl)
Padilla, Andrew, Jr
Pearson, Donald A
Pilat, Michael Joseph
Pillay, Gautam
Putnam, G(arth) L(ouis)
Richardson, Gerald Laverne
Ricker, Neil Lawrence
Rieck, H(enry) G(eorge)
Roake, William Earl
Rohrmann, Charles A(lbert)
Romero, Jacob B
Rosenwald, Gary W
Sehmel, George Albert
Sleicher, Charles A
Stuve, Eric Michael
Thomson, William Joseph
Wilburn, Norman Patrick
Woodfield, F(rank) W(illiam), Jr
Worthington, Ralph Eric
Zollars, Richard Lee

WEST VIRGINIA
Augstkalns, Valdis Ansis
Bailie, Richard Colsten
Beaton, Daniel H(arper)
Beeson, Justin Leo
Breneman, William C
Chaty, John Culver
Clark, Nigel Norman
Cope, Charles S(amuel)
Crum, Edward Hibbert
De Hoff, George R(oland)
Gillmore, Donald W(ood)
Green, R(alph) V(ernon)
Guthrie, Hugh D
Hazelton, Russell Frank
Howell, John H(ancock)
Hwu, Mark Chung-Kong
Keller, George E, II
Larsen, Howland Aikens
Mark, Christopher
Mellow, Ernest W(esley)
Powell, G(eorge) M(atthews), III
Shaeiwitz, Joseph Alan
Tatomer, Harry Nicholas
Verma, Surendra Kumar
Wales, Charles E
Ware, Charles Harvey, Jr
Weiser, Robert B(ruce)
Wertman, William Thomas
Wilson, John Sheridan

WISCONSIN
Amundson, Clyde Howard
Atalla, Rajai Hanna
Bajikar, Sateesh S
Bird, R(obert) Byron
Bringer, Robert Paul
Carstensen, Jens T(huroe)
Chapman, Thomas Woodring
Christiansen, Alfred W
Ciriacks, John A(lfred)
Coberly, Camden Arthur
Doshi, Mahendra R
Duffie, John A(twater)
DuTemple, Octave J
Gehrke, Willard H
Gluckstein, Martin E(dwin)
Gottschalk, Robert Neal
Grace, Thomas Michael
Grindrod, Paul (Edward)
Han, Shu-Tang
Hansen, Paul B(ernard)
Hanway, John E(dgar), Jr
Hill, Charles Graham, Jr
Holland, Dewey G
Hurley, James R(obert)
Katz, William J(acob)
Kim, Sangtae
Kotvis, Peter Van Dyke
Koutsky, James A
Kurath, Sheldon Frank
Lightfoot, E(dwin) N(iblock), Jr
Makela, Lloyd Edward
Manning, James Harvey
Megahed, Sid A
Parker, Peter Eddy
Powell, Hugh N
Ray, W(illis) Harmon
Reid, Robert Lelon
Rudd, D(ale) F(rederick)
Sather, Glenn A(rthur)
Schaffer, Erwin Lambert
Seale, Dianne B
Sell, Nancy Jean
Sfat, Michael R(udolph)
Smith, Buford Don
Springer, Edward L(ester)
Stewart, W(arren) E(arl)
Tiedemann, William Harold
Wegner, Theodore H
Whitney, Roy P(owell)
Williams, Duane Alwin
Wingert, Louis Eugene

WYOMING
Carpenter, Harry C(lifford)
Cooney, David Ogden
Deans, Harry A
Gunn, Robert Dewey
Haynes, Henry William, Jr
Hiza, Michael John, Jr
Hoffman, Edward Jack
Lawroski, Harry
Mingle, John O(rville)
Morrow, Norman Robert
Odasz, F(rancis) B(ernard), Jr
Raynes, Bertram C(hester)
Silver, Howard Findlay
Stinson, Donald Leo

PUERTO RICO
Baus, Bernard V(illars)
Bonnet, Juan A, Jr
Carter, Leo F
Mandavilli, Satya N
Rodriguez, Luis F
Souto Bachiller, Fernando Alberto

ALBERTA
Brown, R(onald) A(nderson) S(teven)
Butler, R(oger) M(oore)
Dalla Lana, I(vo) G(iovanni)
Drum, I(an) M(ondelet)
Fu, Cheng-Tze
Govier, G(eorge) W(heeler)
Gregory, Garry Allen
Mohtadi, Farhang
Nandakumar, Krishnaswamy
Otto, Fred Douglas
Pan, Chuen Yong
Rhodes, E(dward)
Tollefson, Eric Lars
Wanke, Sieghard Ernst
Wayman, Morris
Williams, Michael C(harles)
Wood, Reginald Kenneth

BRITISH COLUMBIA
Brimacombe, James Keith
Dumont, Guy Albert
Epstein, Norman
Godard, Hugh P(hillips)
Grace, John Ross
Kerekes, Richard Joseph
Levelton, B(ruce) Harding
Lielmezs, Janis
Meech, John Athol
Meisen, Axel
Pinder, Kenneth Lyle
Shaw, A(lexander) J(ohn)
Thompson, D(onald) W(illiam)

MANITOBA
Bassim, Mohamad Nabil
Rosinger, Eva L J

NEW BRUNSWICK
Kristmanson, Daniel D
LeBel, Roland Guy
Morris, David Rowland
Picot, Jules Jean Charles
Ruthven, Douglas M
Tory, Elmer Melvin

NOVA SCOTIA
Chen, B(ih) H(wa)
McMillan, Alan F

ONTARIO
Adamek, Eduard Georg
AL-Hashimi, Ahmad

Bacon, David W
Baird, Malcolm Henry Inglis
Basmadjian, Diran
Becker, Henry A
Beeckmans, Jan Maria
Benedek, Andrew
Bergougnou, Maurice A(medee)
Bodnar, Louis Eugene
Bowman, C(lement) W(illis)
Brash, John Law
Bulani, Walter
Burns, C(harles) M(ichael)
Cale, William Robert
Campbell, W(illiam) M(unro)
Carr, Jack A(lbert)
Chaffey, Charles Elswood
Charles, Michael Edward
Chase, John Donald
Chen, Erh-Chun
Cho, Sang Ha
Chrones, James
Clark, Reginald Harold
Conard, Bruce R
Cranford, William B(rett)
Crowe, Cameron Macmillan
De Marco, F(rank) A(nthony)
Diaz, Carlos Manuel
Diosady, Levente Laszlo
Douglas, Peter Lewis
Downie, John
Duever, Thomas Albert
Dullien, Francis A L
Ettel, Victor Alexander
Fahidy, Thomas Z(oltan)
Feuerstein, Irwin
Furter, W(illiam) F(rederick)
Gall, Carl Evert
Hamielec, Alvin Edward
Handa, Paul
Hatcher, S(tanley) Ronald
Hayduk, Walter
Holtslander, William John
Hudgins, Robert R(oss)
Hummel, Richard Line
Jardine, John McNair
Kennedy, Kevin Joseph
Kosaric, Naim
Laughlin, R(obert) G(ardiner) W(illis)
Leaist, Derek Gordon
Leask, R(aymond) A(lexander)
Lu, Benjamin C(hih) Y(eu)
Luus, R(ein)
Macdonald, Ian Francis
Mann, Ranveer S
Mann, Ronald Francis
Matsuura, Takeshi
Miller, Alistair Ian
Missen, R(onald) W(illiam)
Mitsoulis, Evan
Mittal, Gauri S
Moo-Young, Murray
Munns, W(illiam) O
Mutton, Donald Barrett
Napier, Douglas Herbert
Nicolle, Francois Marcel Andre
Nirdosh, Inderjit
Osberg, G(ustav) L(awrence)
Parker, Wayne Jeffery
Pei, David Chung-Tze
Penlidis, Alexander
Piggott, Michael R(antell)
Pogorski, Louis August
Reilly, Park McKnight
Rempel, Garry Llewellyn
Robinson, Campbell William
Rosehart, Robert George
Rowzee, E(dwin) R(alph)
Rubin, Leon Julius
Sandler, Samuel
Schlesinger, Mordechay
Scott, Donald S(trong)
Shemilt, L(eslie) W(ebster)
Silveston, Peter Lewis
Singer, Eugen
Smith, James W(ilmer)
Souhrada, Frank
Southam, Frederick William
Sowa, Walter
Spink, D(onald) R(ichard)
Spinner, Irving Herbert
Sullivan, John Leslie
Tennankore, Kannan Nagarajan
Tombalakian, Artin S
Trass, O(lev)
Tzoganakis, Costas
Van Wagner, Charles Edward
Vijayan, Sivaraman
Vlachopoulos, John A(postolos)
Weir, Ronald Douglas
Williams, Norman S W
Wojciechowski, Bohdan Wieslaw
Woods, Donald Robert
Wright, Joseph D
Wynnyckyj, John Rostyslav

QUEBEC
Beron, Patrick
Carreau, Pierre
Charette, Andre
Cholette, A(lbert)
Cloutier, Leonce
Cooke, Norman E(dward)
Croctogino, Reinhold Hermann
Dealy, John Michael

Douglas, W J Murray
Feldman, Dorel
Gauvin, William H
Kamal, Musa Rasim
Klassen, J(ohn)
Kubes, George Jiri
Lavalle, H Claude
Leduy, Anh
Lee, John Hak Shan
Monsaroff, Adolph
Mujumdar, Arun Sadashiv
Roy, Paul-H(enri)
Thirion, Jean Paul Joseph
Waid, Ted Henry

SASKATCHEWAN
Bakhshi, Narendra Nath
Catania, Peter J
DeCoursey, W(illiam) J(ames)
Esmail, Mohamed Nabil
Macdonald, Douglas Gordon
Mikle, Janos J
Shook, Clifton Arnold

OTHER COUNTRIES
Ahmed, Syed Mahmood
Astarita, Gianni
Bailey, James E(dwin)
Chang, H K
Chen, Fred Fen Chuan
Cullinan, Harry T(homas), Jr
Dunn, Irving John
Economides, Michael John
Esterson, Gerald L(ee)
Fey, George Ting-Kuo
Fok, Siu Yuen
Fu, Yuan C(hin)
Halemane, Keshava Prasad
Hamada, Mokhtar M
Hochman, Jack M(artin)
Howell, John Anthony
Kocatas, Babur M(ehmet)
Lee, Sidney
Liepsch, Dieter W
Liu, Ta-Jo
McRae, Wayne A
Mathews, Joseph F(ranklin)
Miyano, Kenjiro
Mou, Duen-Gang
Park, Chan Mo
Spottiswood, David James
Stam, Jos
Van Weert, Gezinus
Vasilakos, Nicholas Petrou
Williams, F(ord) Campbell
Yoshida, Fumitake
Zaczepinski, Sioma

Civil Engineering

ALABAMA
Appleton, Joseph Hayne
Blakney, William G G
Carlton, Thomas A, Jr
Cost, Thomas Lee
Costes, Nicholas Constantine
Gilbert, John Andrew
Hardin, Edwin M(ilton)
Jackson, John Elwin, Jr
Judkins, Joseph Faulcon, Jr
Kallsen, Henry Alvin
Kandhal, Prithvi Singh
Keith, Warren Gray
Killingsworth, R(oy) W(illiam)
Marek, Charles R(obert)
Melville, Joel George
Omar, Husam Anwar
Segner, Edmund Peter, Jr
Turner, Daniel Shelton
Walters, James Vernon
Zehrt, William H(arold)

ALASKA
Alter, Amos Joseph
Bennett, F Lawrence
Hulsey, J Leroy
Johansen, Nils Ivar
Sweet, Larry Ross
Tilsworth, Timothy
Zarling, John P

ARIZONA
Betz, Mathew J(oseph)
Blackburn, Jack Bailey
Brinker, Russell Charles
Burbank, Nathan C, Jr
Carder, David Ross
Clemmens, Albert J
Contractor, Dinshaw N
Govil, Sanjay
Havers, John Alan
Hill, Louis A, Jr
Ince, Simon
Klock, John W
Krajcinovic, Dusan
Kriegh, James Douglas
Laursen, E(mmett) M(orton)
Lundgren, Harry Richard
Maddock, Thomas, Jr
Matthias, Judson S
Mays, Larry Wesley
Morris, Gene Ray
Newlin, Charles W(illiam)

Newlin, Philip Blaine
Nordby, Gene M
Reich, Brian M
Renard, Kenneth G
Replogle, John A(sher)
Richard, Ralph Michael
Rouse, Hunter
Shupe, John W(allace)
Singhal, Avinash Chandra
Smerdon, Ernest Thomas
Sorooshian, Soroosh
Sultan, Hassan Ahmed
Upchurch, Jonathan Everett
Woolhiser, David A(rthur)

ARKANSAS
Alguire, Mary Slanina (Hirsch)
Andrews, John Frank
Bissett, J(ames) R(obert)
Boyd, Donald Edward
Fletcher, Alan G(ordon)
Ford, Miller Clell, Jr
Heiple, Loren Ray
Jellinger, Thomas Christian
Jong, Ing-Chang

CALIFORNIA
Abdel-Ghaffar, Ahmed Mansour
Agardy, Franklin J
Alexander, Ira H(enris)
Alexander, Robert L
Ang, Alfredo H(ua)-S(ing)
Aroni, Samuel
Atwood, John Leland
Baker, Warren J
Banks, Harvey Oren
Bea, Robert G
Birkimer, Donald Leo
Blume, John A(ugust)
Brahma, Chandra Sekhar
Breckenridge, Robert A(rthur)
Bresler, B(oris)
Brock, Richard R
Brooks, Norman H(errick)
Bugg, Sterling L(owe)
Cassidy, John J(oseph)
Castellan, Norman J
Chadwick, Wallace Lacy
Chang, Howard
Chelapati, Chunduri V(enkata)
Chen, Carl W(an-Cheng)
Cheney, James A
Chi, Cheng-Ching
Chiang, George C(hihming)
Chopra, Anil K
Chou, Larry I-Hui
Chu, Kuang-Han
Chuang, Kuen-Puo (Ken)
Cooke, James Barry
Cornell, C(arl) Allin
Crawford, Norman Holmes
Cross, Ralph Herbert, III
Daganzo, Carlos Francisco
Daily, James W(allace)
Davis, Harmer E
Dong, Richard Gene
Dong, Stanley B
Douglas, James
Dowell, Douglas C
Dracup, John Albert
Eberhart, H(oward) D(avis)
Elder, Rex Alfred
Eliassen, Rolf
Evans, T(homas) H(ayhurst)
Everts, Craig Hamilton
Felton, Lewis P(eter)
Ferrara, Thomas Ciro
Filippou, Filip C
Fletcher, David Quentin
Fondahl, John W(alker)
Forrest, James Benjamin
Gabriel, Lester H
Gabrielsen, Bernard L
Garbarini, Edgar Joseph
Georgakakos, Konstantine P
Gere, James Monroe
Gerwick, Ben Clifford, Jr
Glenn, Richard A(llen)
Goodman, Richard E
Gordon, Ruth Vida
Grant, Eugene Lodewick
Guymon, Gary L
Hackel, Lloyd Anthony
Hahne, Henry V
Ham, Lee Edward
Hamilton, Gordon Wayne
Hammond, David G
Hanna, George P, Jr
Harder, James Albert
Hayes, Thomas Jay, III
Henderson, D(elbert) W
Herrmann, George
Herrmann, Leonard R(alph)
Heuze, Francois E
Hickok, Robert Baker
Housner, George W(illiam)
Hung, You-Tsai Joseph
Hwang, Li-San
Idriss, Izzat M
Ingram, Gerald E(ugene)
Iselin, Donald G
Jacobs, Joseph Donovan
Jeng, Raymond Ing-Song
Jenkins, David Isaac

Jennings, Paul C(hristian)
Johnston, Roy G
Karn, Richard Wendall
Kennedy, Robert P
Kiely, John Roche
Kim, Young C
Kiremidjian, Anne Setian
Kostyrko, George Jurij
Krishnamoorthy, Govindarajalu
Krone, Ray B
Kruger, Paul
Lam, Tenny N(icolas)
Larock, Bruce E
Lau, John H
Leidersdorf, Craig B
Lemcoe, M M(arshall)
Leonhard, William E
Leps, Thomas MacMaster
Lin, Tung Yen
Love, Joseph E(ugene), Jr
Lubliner, J(acob)
Luk, King Sing
Luscher, Ulrich
Lysmer, John
McCammon, Lewis B(rown), Jr
McCarty, Perry L(ee)
May, Adolf D(arlington), Jr
Medwadowski, Stefan J
Merriam, John L(afayette)
Merritt, J(oshua) L(evering), Jr
Mitchell, James K(enneth)
Moehle, Jack P
Monismith, Carl L(eroy)
Moore, William W
Morel-Seytoux, Hubert Jean
Morgali, James R
Morgan, James John
Morris, Brooks T(heron)
Morris, Henry Madison, Jr
Mow, Maurice
Mueller, Charles Carsten
Naar, Jacques
Napolitano, Leonard Michael, Jr
Narasimhan, Thiruppudaimarudhur N
Nelson, Richard Bartel
Nemat-Nasser, Siavouche
Nordell, William James
Nowatzki, Edward Alexander
Oglesby, Clarkson Hill
Ongerth, Henry J
Orlob, Gerald T
Oswald, William J
Parker, Henry Whipple
Paulson, Boyd Colton, Jr
Penzien, Joseph
Perloff, David Steven
Pietrzak, Lawrence Michael
Pincus, Howard Jonah
Popov, E(gor) P(aul)
Post, J(ames) L(ewis)
Raichlen, Fredric
Ramey, Melvin Richard
Ratner, Robert (Stephen)
Riggs, Louis William
Ritchie, Stephen G
Rogers, Gifford Eugene
Rosen, Alan
Rowland, Walter Francis
Rubin, Sheldon
Rubinstein, Moshe Fajwel
Rudavsky, Alexander Bohdan
Saha, Subrata
Sarkaria, Gurmukh S
Schiff, Anshel J
Schmit, Lucien A(ndre), Jr
Scordelis, Alexander Costicas
Scott, Ronald F(raser)
Scott, Verne H(arry)
Seaman, Lynn
Seide, Paul
Shanteau, Robert Marshall
Sharpe, Roland Leonard
Shaw, Warren A(rthur)
Shen, Hsieh Wen
Shepherd, Robin
Sicular, George M
Simunek, Jiri
Singh, Rameshwar
Spicher, Robert G
Statt, Terry G
Steinbrugge, Karl V
Stenstrom, Michael Knudson
Stolzenbach, Keith Densmore
Stratton, Frank E(dward)
Sundaram, Panchanatham N
Supersad, Jankie Nanan
Sutherland, Louis Carr
Sve, Charles
Synolakis, Costas Emmanuel
Taylor, Robert Leroy
Teague, Lavette Cox
Teicholz, Paul M
Todd, David Keith
Vanoni, Vito A(ugust)
Venuti, William J(oseph)
Wallender, Wesley William
Watters, Gary Z
Westmann, Russell A
Wiegel, Robert L
Wiggins, John H(enry), Jr
Williams, John A(rthur)
Wilson, Basil W(rigley)
Wilson, Edward L
Yeh, William Wen-Gong

Civil Engineering (cont)

Yen, Bing Cheng
Ying, William H
Yu, David U L
Zwoyer, Eugene

COLORADO
Bartlett, Paul Eugene
Charlie, Wayne Alexander
Chugh, Ashok Kumar
Cochran, Leighton Scott
Criswell, Marvin Eugene
Dahl, Arthur Richard
Day, David Allen
DeCoursey, Donn G(ene)
Evans, Norman A(llen)
Faddick, Robert Raymond
Fead, John William Norman
Feldman, Arthur
Feng, Chuan C(hung) D(avid)
Flack, J(ohn) E(rnest)
Frangopol, Dan Mircea
Frasier, Gary Wayne
Gerstle, Kurt H
Gessler, Johannes
Goble, George G
Hendricks, David Warren
Holtz, Wesley G
Hughes, William Carroll
Illangasekare, Tissa H
Jibson, Randall W
Johnson, Arnold I(van)
Johnstone, James G(eorge)
Ketchum, Milo S
Ko, Hon-Yim
Koelzer, Victor A
Linstedt, Kermit Daniel
McLean, Francis Glen
Mays, John Rushing
Medearis, Kenneth Gordon
Moody, Martin L(uther)
Osterberg, J(orj) O(scar)
Pak, Ronald Y S
Rautenstraus, R(oland) C(urt)
Richardson, Everett V
Romig, William D(avis)
Savage, William Zuger
Schirmer, Howard August, Jr
Schuster, Robert Lee
Simons, Daryl B
Smith, Roger Elton
Strom, Oren Grant
Summers, Luis Henry
Sunada, Daniel K(atsuto)
Suprenant, Bruce A
Tabler, Ronald Dwight
Thompson, Erik G(rinde)
Tulin, Leonard George
Van der Heijde, Paul Karel Maria
Wu, Jonathan Tzong
Yang, Chih Ted

CONNECTICUT
Alcantara, Victor Franco
DeWolf, John T
Epstein, Howard I
Flannelly, William G
Frantz, Gregory Clayton
Hemond, Conrad J(oseph), Jr
Johnston, E(lwood) Russell, Jr
Lai, Jai-Lue
Levinson, Herbert S
Lin, Jia Ding
Long, Richard Paul
Nadel, Norman A
Olster, Elliot Frederick
Panuzio, Frank L
Pfrang, Edward Oscar
Posey, C(hesley) J(ohnston)
Reyna, Luis Guillermo
Stephens, Jack E(dward)

DELAWARE
Cheng, Alexander H-D
Chesson, Eugene, Jr
Dalrymple, Robert Anthony
Dentel, Steven Keith
Goel, Kailash C(handra)
Hsiao, George Chia-Chu
Huang, Chin Pao
Hume, Harold Frederick
Jones, Russel C(ameron)
Kaliakin, Victor Nicholas
Kerr, Arnold D
Kikuchi, Shinya
Kobayashi, Nobuhisa
Koch, Carl Mark
Nicholls, Robert Lee
Svendsen, Ib Arne
Wu, Jin
Yang, C(heng) Y(i)

DISTRICT OF COLUMBIA
Adams, Francis L(ee)
Basdekas, Nicholas Leonidas
Bell, Bruce Arnold
Blanchard, Bruce
Caywood, James Alexander, III
Chung, Riley M
Culver, Charles George
Deason, Jonathan P
Deen, Thomas B
Eggenberger, Andrew Jon

Hampton, Delon
Jones, Irving Wendell
Kao, Timothy Wu
McGinnis, David Franklin, Jr
Mermel, Thaddeus Walter
Ochs, Walter J
Odar, Fuat
Parks, Vincent Joseph
Roberts, Paul Osborne, Jr
Smith, Waldo E(dward)
Soteriades, Michael C(osmas)
Toridis, Theodore George
Yachnis, Michael

FLORIDA
Anderson, Melvin W(illiam)
Berger, Richard Lee
Boucher, Raymond
Brotherson, Donald E
Brumer, Milton
Carrier, W(illiam) David, III
Chalabi, A Fattah
Chiu, Tsao Yi
Christensen, Bent Aksel
Chryssafopoulos, Hanka Wanda Sobczak
Chryssafopoulos, Nicholas
Claridge, Richard Allen
Covault, Donald O
Dallman, Paul Jerald
Dantin, Elvin J, Sr
Dean, Donald L(ee)
Dean, Robert George
Dietz, Jess Clay
Drucker, Daniel Charles
Edson, Charles Grant
Ege, Raimund K
Fogarty, William Joseph
Givens, Paul Edward
Goodson, James Brown, Jr
Grinter, Linton E(lias)
Harrenstien, Howard P(aul)
Harris, Lee Errol
Hartman, John Paul
Heaney, James Patrick
Huang, Y(en) T(i)
Jenkins, David R(ichard)
Kersten, Robert D(onavon)
Kuzmanovic, B(ogdan) O(gnjan)
Lambe, T(homas) William
Langfelder, Leonard Jay
Lin, Y(u) K(weng)
Mantell, M(urray) I(rwin)
Mecholsky, John Joseph, Jr
Michejda, Oskar
Miller, Charles Leslie
Nichols, James Carlile
Nunnally, Stephens Watson
Partheniades, Emmanuel
Patterson, Archibald Oscar
Richards, A(lvin) M(aurer)
Ruth, Byron E
Scarlatos, Panagiotis Dimitrios
Schaub, James H(amilton)
Schmertmann, John H(enry)
Sirkin, Alan N
Stevens, John A(lexander)
Surti, Vasant H
Viessman, Warren, Jr
Vitagliano, Vincent J
Wekezer, Jerzy Wladyslaw
Yong, Yan
Zollo, Ronald Francis

GEORGIA
Amirtharajah, Appiah
Barksdale, Richard Dillon
Barnwell, Thomas Osborn, Jr
Bowen, Paul Tyner
Brumund, William Frank
Caseman, A(ustin) Bert
Chang, Chin Hao
Chiu, Kirts C
Circeo, Louis Joseph, Jr
Craig, James I
Fitzgerald, J(ohn) Edmund
Fragaszy, Richard J
Fulton, Robert E(arle)
Galeano, Sergio F(rancis)
Goodno, Barry John
Kahn, Lawrence F
Lnenicka, William J(oseph)
Rouhani, Shahrokh
Sangster, William M(cCoy)
Saunders, Fred Michael
Sowers, George F(rederick)
Spetnagel, Theodore John
Stanley, Luticious Bryan, Jr
Stelson, T(homas) E(ugene)
Thomas, Adrian Wesley
Thornton, William Aloysius
Tooles, Calvin W(arren)
Traina, Paul J(oseph)

HAWAII
Chiu, Arthur Nang Lick
Cox, Richard Horton
Flachsbart, Peter George
Fok, Yu-Si
Fuka, Louis Richard
Gerritsen, Franciscus
Gitlin, Harris Martlin
Go, Mateo Lian Poa
Grace, Robert Archibald
Hamada, Harold Seichi

Hihara-Endo, Linda Masae
Hufschmidt, Maynard Michael
Krock, Hans-Jürgen
Matsuda, Fujio
Nielsen, N Norby
Ogata, Akio
Papacostas, Constantinos Symeon
Prevedouros, Panos D
Robertson, Ian Nicol
Saxena, Narendra K
Singh, Amarjit
Taoka, George Takashi
Teng, Michelle Hsiao Tsing
Wu, I-Pai
Young, Reginald H F

IDAHO
Humpherys, Allan S(tratford)
Johnson, Clifton W
Kincaid, Dennis Campbell
Sargent, Charles
Schreiber, David Laurence
Swiger, William F
Trout, Thomas James
Wilbur, Lyman D

ILLINOIS
Adams, John Rodger
Al-Khafaji, Amir Wadi Nasif
Ardis, Colby V, Jr
Babcock, Lyndon Ross, (Jr)
Barenberg, Ernest J(ohn)
Bazant, Zdenek P(avel)
Bell, Charles Eugene, Jr
Berry, Donald S(tilwell)
Bowles, Joseph Edward
Boyer, LeRoy T
Briscoe, John William
Calderon, Alberto Pedro
Chakrabarti, Subrata K
Chugh, Yoginder Paul
Cording, Edward J
Corley, William Gene
Daniel, Isaac M
Davisson, M T
Dempsey, Barry J
Ditmars, John David
Dobrovolny, Jerry S(tanley)
Fites, Donald Vester
Fonken, David W(alter)
Fraenkel, Stephen Joseph
Fucik, Edward Montford
Gamble, William Leo
Garcia, Marcelo Horacio
Gemmell, Robert S(tinson)
Gerstner, Robert W(illiam)
Ghosh, Satyendra Kumar
Gnaedinger, John P(hillip)
Guralnich, Sidney Aaron
Gurfinkel, German R
Hall, W(illiam) J(oel)
Haltiwanger, John D(avid)
Hanna, Steven J(ohn)
Hawkins, Neil Middleton
Hay, William Walter
Heil, Richard Wendell
Hendron, Alfred J, Jr
Herrin, Moreland
Hofer, Kenneth Emil
Hognestad, Eivind
Holland, Eugene Paul
Holsen, Thomas Michael
Ireland, Herbert O(rin)
Keer, Leon M
Kesler, Clyde E(rvin)
Khachaturian, Narbey
Kim, Byung J
Klieger, Paul
Korn, Alfred
Krizek, Raymond John
Lawrence, Frederick Van Buren, Jr
Liu, Wing Kam
Lopez, Leonard Anthony
Lue-Hing, Cecil
McKee, Keith Earl
Maxwell, William Hall Christie
Maynard, Theodore Roberts
Merkel, Frederick Karl
Mockros, Lyle F(red)
Mosborg, Robert J(ohn)
Muehling, Arthur J
Munse, William H(erman)
Murtha, Joseph P
Muvdi, Bichara B
O'Connor, John Thomas
Paintal, Amreek Singh
Pfeffer, John T
Prakash, Anand
Regunathan, Perialwar
Rittmann, Bruce Edward
Robinson, Arthur R(ichard)
Rodda, Errol David
Roy, Dipak
Russell, Henry George
Sami, Sedat
Saxena, Surendra K
Schlesinger, Lee
Shaffer, Louis Richard
Shah, Surendra P
Siess, Chester P(aul)
Silver, Marshall Lawrence
Smedskjaer, Lars Christian
Sozen, M(ete) A(vni)
Stallmeyer, J(ames) E(dward)

Tang, Wilson H
Thompson, Marshall Ray
Valocchi, Albert Joseph
Walker, William Hamilton
Wenzel, Harry G, Jr
Wojnowski, Daniel Allen
Wong, Kam Wu
Woods, Kenneth R
Yen, Ben Chie
Zenz, David R

INDIANA
Bell, J(ohn) M
Chen, Wai-Fah
Delleur, Jacques W(illiam)
Drnevich, Vincent Paul
Eckelman, Carl A
Fredrich, Augustine Joseph
Gaunt, John Thixton
Goetz, William H(arner)
Graves, Leroy D
Gray, William Guerin
Halpin, Daniel William
Huang, Nai-Chien
Kareem, Ahsan
Leonards, G(erald) A(llen)
Lewis, A(lbert) D(ale) M(ilton)
Lobo, Cecil T(homas)
Lovell, Charles W(illiam), Jr
McEntyre, John G(erald)
McLaughlin, J(ohn) F(rancis)
Meyers, Vernon J
Michael, Harold Louis
Miles, Robert D(ouglas)
Rao, R(amachandra) A
Satterly, Gilbert T(hompson)
Scholer, Charles Frey
Scott, Marion B(oardman)
Sinha, Kumares C
Skibniewski, Miroslaw Jan
Taylor, Richard L
Tenney, Mark W
Waling, J(oseph) L(ee)
Warder, David Lee
Winslow, Douglas Nathaniel
Wood, Leonard E(ugene)
Wright, Jeffery Regan
Yang, Henry T Y
Yeh, Pai-T(ao)

IOWA
Austin, Tom Al
Baumann, E(dward) Robert
Branson, Dan E(arle)
Brewer, Kenneth Alvin
Cleasby, John Leroy
Dague, Richard R(ay)
Ekberg, Carl E(dwin), Jr
Fung, Honpong
Greimann, Lowell F
Handy, Richard L(incoln)
Hoover, James M(yron)
Hubbard, Philip G(amaliel)
Jeyapalan, Kandiah
Kane, Harrison
Kennedy, John Fisher
Klaiber, Fred Wayne
Lane, Orris John, Jr
Lee, Dah-Yinn
McCauley, Howard W
Maze, Thomas Harold
Miller, Richard Keith
Morgan, Paul E(merson)
Nakato, Tatsuaki
Nixon, Wilfrid Austin
Odgaard, A Jacob
Oulman, Charles S
Parkin, Gene F
Patel, Virendra C
Sanders, W(allace) W(olfred), Jr
Sheeler, John B(riggs)

KANSAS
Cook, Everett L
Cooper, Peter B(ruce)
Darwin, David
Ellis, Harold Bernard
Hoffman, Jerry C
Huang, Chi-Lung Dominic
Kahn, Charles Howard
Kurt, Carl Edward
Lee, Joe
McBean, Robert Parker
McCabe, Steven Lee
McKinney, Ross E(rwin)
Moore, Raymond Knox
Roddis, Winifred Mary Kim
Rolfe, Stanley Theodore
Schaper, Laurence Teis
Smith, Bob L(ee)
Smith, Robert Lee
Snell, Robert Ross
Surampalli, Rao Yadagiri
Swartz, Stuart Endsley
Wilhelm, William Jean
Willems, Nicholas
Yu, Yun-Sheng

KENTUCKY
Blythe, David K(nox)
Crabtree, Koby Takayashi
Fardo, Stephen W
Gesund, Hans
Hardin, Bobby Ott

Hutchinson, John W(endle)
Kao, David Teh-Yu
Kinman, Riley Nelson
Lauderdale, Robert A(mis), Jr
Paz, Mario Meir
Sweigard, Richard Joseph
Vaziri, Menouchehr
Wang, Shien Tsun
Warner, Richard Charles
Yoder, Elmon Eugene

LOUISIANA
Aguilar, Rodolfo J
Arman, Ara
Benedict, Barry Arden
Bruce, Robert Nolan, Jr
Courtney, John Charles
Dalia, Frank J
Grimwood, Charles
Johnson, Lee H(arnie)
Kazmann, Raphael Gabriel
Lee, Griff C
Lemke, Calvin A(ubrey)
McGhee, Terence Joseph
Mayer, J(ohn) K(ing)
Ovunc, Bulent Ahmet
Painter, Jack T(imberlake)
Price, Bobby Earl
Roberts, Freddy Lee
Singh, Vijay P
Stopher, Peter Robert
Szabo, A(ugust) J(ohn)
Tumay, Mehmet Taner
Wilson, Joe Robert
Wintz, William A, Jr

MAINE
Gorrill, William R
Greenwood, George W(atkins)
Kleinschmidt, R Stevens
Nilson, Arthur H(umphrey)

MARYLAND
Asrar, Ghassam R
Austin, John H(enry)
Babrauskas, Vytenis
Blevins, R(alph) W(allace)
Browzin, Boris S(ergeevich)
Brush, Lucien M(unson), Jr
Bryan, Edward H
Cattaneo, L(ouis) E(mile)
Chen, Benjamin Yun-Hai
Chuang, Tze-jer
Corn, Morton
Corotis, Ross Barry
Davidson, Bruce M
Douglas, Andrew Sholto
Duane, David Bierlein
Ellingwood, Bruce Russell
Eny, Desire M(arc)
Frank, Carolyn
Gaum, Carl H
Geyer, John Charles
Green, Richard Stedman
Greenfeld, Sidney H(oward)
Harris, Leonard Andrew
Jabbour, Kahtan Nicolas
Linaweaver, Frank Pierce
Mallory, Charles William
Meyers, Bernard Leonard
Michalowski, Radoslaw Lucas
Moran, David Dunstan
Morris, Alan
Ostrom, Thomas Ross
Papet, Louis M
Rosenberg, Arnold Morry
Sabnis, Gajanan Mahadeo
Saville, Thorndike, Jr
Scanlan, Robert Harris
Schruben, John H
Shalowitz, Erwin Emmanuel
Simiu, Emil
Steiner, Henry M
Theuer, Paul John
Tholen, Albert David
Way, George H
Wright, Richard N(ewport)
Wright, Thomas Wilson
Yaramanoglu, Melih
Yokel, Felix Y

MASSACHUSETTS
Aldrich, Harl P, Jr
Barnhart, Cynthia
Ben-Akiva, Moshe E
Berger, Bernard Ben
Blanc, Frederic C
Bras, Rafael Luis
Brebbia, Carlos Alberto
Breuning, Siegfried M
Brocard, Dominique Nicolas
Brown, Linfield Cutter
Carver, Charles E(llsworth), Jr
Chang, Ching Shung
Chen, Ming M
Collura, John
De Neufville, Richard Lawrence
Dietz, Albert (George Henry)
Entekhabi, Dara
Fiering, Myron B
Firnkas, Sepp
Fitzgerald, Robert William
Garrelick, Joel Marc
Gartner, Nathan Hart

Gumpertz, Werner H(erbert)
Hansen, R(obert) J(oseph)
Hecker, George Ernst
Hirschfeld, Ronald Colman
Hope, Elizabeth Greeley
Kachinsky, Robert Joseph
Kelsey, Ronald A(lbert)
Keshavan, Krishnaswamiengar
Ladd, Charles Cushing, III
LeMessurier, William James
Liepins, Atis Aivars
Logcher, Robert Daniel
Madsen, Ole Secher
Male, James William
Marks, David Hunter
Marston, George Andrews
Martland, Carl Douglas
Mason, William C
Mei, Chiang C(hung)
Miller, Melton M, Jr
Moavenzadeh, Fred
Morel, Francois M M
Murphy, Peter John
Ostendorf, David William
Platt, Milton M
Reinschmidt, Kenneth F(rank)
Schwartz, Thomas Alan
Selig, Ernest Theodore
Sheffi, Yosef
Shuldiner, Paul W(illiam)
Sutcliffe, Samuel
Tarkoy, Peter J
Thomas, Harold A(llen), Jr
Wang, Lawrence K
Webster, Lee Alan
White, Merit P(enniman)
Whitman, Robert V(an Duyne)
Wilson, Nigel Henry Moir

MICHIGAN
Andersland, Orlando Baldwin
Armstrong, John Morrison
Baillod, Charles Robert
Beaufait, Frederick W
Bergmann, Dietrich R(udolf)
Bulkley, Jonathan William
Canale, Raymond Patrick
Cleveland, Donald Edward
Cutts, Charles E(ugene)
Datta, Tapan K
Hanson, Norman Walter
Hanson, Robert D(uane)
Huang, Eugene Yuching
Johnston, Clair C
Kaldjian, Movses J(eremy)
Khasnabis, Snehamay
Kim, Byung Ro
Knoll, Alan Howard
Ku, Albert B
Lubkin, James Leigh
McCauley, Robert F(orrestelle)
Morman, Kenneth N
Nowak, Andrzej S
Pao, Richard H(sien) F(eng)
Paulson, James M(arvin)
Prasuhn, Alan Lee
Quinn, Frank Hugh
Richart, F(rank) E(dwin), Jr
Saul, William Edward
Shkolnikov, Moisey B
Smith, Hadley J(ames)
Snyder, Virgil W(ard)
Watwood, Vernon Bell, Jr
Woods, Richard David
Wright, Steven Jay
Wu, Kuang Ming
Wylie, Evan Benjamin

MINNESOTA
Allred, E(van) R(ich)
Arndt, Roger Edward Anthony
Blaisdell, Fred W(illiam)
Boening, Paul Henrik
Bowers, C(harles) Edward
Crouch, Steven L
Emeagwali, Philip Chukwurah
Galambos, Theodore V
Goodman, L(awrence) E(ugene)
Gulliver, John Stephen
Johnson, Gerald Winford
Kersten, Miles S(tokes)
Khraibi, Ali A
Ling, Joseph Tso-Ti
Reynolds, James Harold
Silberman, Edward
Stephanedes, Yorgos Jordan
Sterling, Raymond Leslie
Straub, Conrad P(aul)

MISSISSIPPI
Abdulrahman, Mustafa Salih
Bowie, Andrew J(ackson)
Brown, Fred(erick) R(aymond)
Corey, Marion Willson
Cunny, Robert William
DeLeeuw, Samuel Leonard
George, Kalankamary Pily
Hackett, Robert M(oore)
Hudson, Robert Y(oung)
Johnson, L(awrence) D(avid)
McCrae, John Leonidas
Mather, Bryant
Perry, Edward Belk
Priest, Melville S(tanton)

Shindala, Adnan
Shockley, W(oodland) G(ray)
Wang, Sam S(hu) Y(i)
Woodburn, Russell

MISSOURI
Andrews, William Allen
Baldwin, James W(arren), Jr
Banerji, Shankha K
Barr, David John
Cheng, Franklin Yih
Cronin, James Lawrence, Jr
Douty, Richard T
Edgerley, Edward, Jr
Emanuel, Jack Howard
Fowler, Timothy John
Gevecker, Vernon A(rthur) C(harles)
Goran, Robert Charles
Gould, Phillip L
Guccione, Julius Matteo
Guell, David Lee
Hatheway, Allen Wayne
Hauck, George F(rederick) W(olfgang)
Heagler, John B(ay), Jr
Hemphill, Louis
Hjelmfelt, Allen T, Jr
Jawad, Maan Hamid
Jester, Guy Earlscort
Liu, Henry
McCarthy, John F(rancis)
Mains, Robert M(arvin)
Mansur, Charles I(saiah)
Munger, Paul R
Prakash, Shamsher
Robinson, Thomas B
Rockaway, John D, Jr
Rosenkoetter, Gerald Edwin
Salmons, John Robert
Schmidt, Norbert Otto
Schwartz, Henry Gerard, Jr
Senne, Joseph Harold, Jr
Szabo, Barna Aladar
Yu, Wei Wen

MONTANA
Friel, Leroy Lawrence
Hunt, William A(lfred)
Peavy, Howard Sidney
Scheer, Alfred C(arl)
Videon, Fred F(rancis)
Walker, Leland J

NEBRASKA
Andersen, Dewey Richard
Hammer, Mark J(ohn)
Marlette, Ralph R(oy)
Martin, Charles Wayne
Phelps, George Clayton
Riveland, A(rvin) R(oy)
Swihart, G(erald) R(obert)
Tunnicliff, David George

NEVADA
Brandstetter, Albin
Daemen, Jaak J K
Epps, Jon Albert
Ghosh, Amitava
Guitjens, Johannes C
Helm, Donald Cairney
Merdinger, Charles J(ohn)
Nelson, R William
Orcutt, Richard G(atton)
Salmon, Charles G(erald)
Thornburn, Thomas H(ampton)

NEW HAMPSHIRE
Batchelder, Gerald M(yles)
Klotz, Louis Herman
Long, Carl F(erdinand)
Stearns, S(tephen) Russell
Wilson, Donald Alfred

NEW JERSEY
Balaguru, Perumalsamy N
Becht, Charles, IV
Billington, David Perkins
Blumberg, Alan Fred
Borg, Sidney Fred
Bruno, Michael Stephen
Caudill, Reggie Jackson
Chae, Yong Suk
Chan, Paul C
Cheng, David H(ong)
Derucher, Kenneth Noel
Esrig, Melvin I
Garrelts, Jewell Milan
Genetelli, Emil J
Granstrom, Marvin L(e Roy)
Haefeli, Robert J(ames)
Horn, Harry Moore
Iyer, Ram R
Kagan, Harvey Alexander
Killam, Everett Herbert
Kim, Uing W
Kraft, Walter H
Mark, Robert
Monahan, Edward James
Nathan, Kurt
Nawy, Edward George
Pfafflin, James Reid
Pignataro, Louis J(ames)
Pincus, George
Plummer, Dirk Arnold
Prevost, Jean Herve

Raamot, Tonis
Schmid, Werner E(duard)
Severud, Fred N
Spillers, William R
Thomann, Robert V
Uchrin, Christopher George
Vanmarcke, Erik Hector
Wiesenfeld, Joel

NEW MEXICO
Alzheimer, William Edmund
Baerwald, John E(dward)
Bleyl, Robert Lingren
Christensen, N(ephi) A(lbert)
Clough, Richard H(udson)
Colp, John Lewis
Dillon, Robert Morton
Hall, Jerome William
Hernandez, John W(hitlock)
Howell, Gregory A
Hulme, Bernie Lee
Hulsbos, C(ornie) (Leonard)
Khandan, Nirmala N
Lunsford, Jesse V(ernon)
Maggard, Samuel P
Martinez, Jose E(leazar)
Nirmalakhandan, Nagamany
Peck, Ralph B(razelton)
Perret, William Riker
Ross, Timothy Jack
Thomson, Bruce M
Triandafilidis, George Emmanuel
Varan, Cyrus O
Von Riesemann, Walter Arthur
Yamada, Tetsuji
Zimmerman, Roger M

NEW YORK
Abel, John Fredrick
Ahmad, Jameel
Ahmadi, Goodarz
Alben, Katherine Turner
Alpern, Milton
Alvarez, Ronald Julian
Armen, Harry, Jr
Armenakas, Anthony Emanuel
Bandyopadhyay, Amitabha
Banerjee, Prasanta Kumar
Baron, Melvin L(eon)
Barry, B(enjamin) Austin
Batson, Gordon B
Betti, Raimondo
Bieniek, Maciej P
Binger, Wilson Valentine
Birnstiel, Charles
Brandt, G(eorge) Donald
Brennan, Paul Joseph
Brown, William Augustin
Cantilli, Edmund Joseph
Cataldo, Joseph C
Clemence, Samuel Patton
Cohen, Edward
Collins, Anthony Gordon
Constantinou, Michalakis
Dasgupta, Gautam
Deleanu, Aristide Alexandru-Ion
Dempsey, John Patrick
Deresiewicz, Herbert
Dicker, Daniel
Di Maggio, Frank Louis
Dobry, Ricardo
Ezra, Arthur Abraham
Feeser, Larry James
Fisher, Gordon P(age)
Florman, Samuel C
Fogel, Charles M(orton)
Fox, George A
Gallagher, Richard Hugo
Gennaro, Joseph J(ohn)
Gergely, Peter
Gossett, James Michael
Grigoriu, Mircea Dan
Haines, Daniel Webster
Harlow, H(enry) Gilbert
Helander, Martin Erik Gustav
Huddleston, John Vincent
Irwin, Lynne Howard
Isada, Nelson M
Jirka, Gerhard Hermann
Jordan, Mark H(enry)
Kaarsberg, Ernest Andersen
Kahn, Elliott H
Kulhawy, Fred Howard
Lai, W(ei) Michael
Lawler, John Patrick
Lee, George C
Lee, Robert Bumjung
Libove, Charles
Liggett, James Alexander
Lowe, John, III
Lynn, Walter R(oyal)
McDonald, Donald
McGuire, William
Mahtab, M Ashraf
Mandel, James A
Mead, Lawrence Myers
Mendoza-Cabrales, Cesar
Meredith, Dale Dean
Meyburg, Arnim Hans
Meyer, Christian
Mijovic, Jovan
Miller, George Maurice
Morabito, Bruno P
Neal, John Alva

Civil Engineering (cont)

O'Connor, Donald J
O'Rourke, Thomas Denis
Paaswell, Robert E(mil)
Palevsky, Gerald
Pao, Yih-Hsing
Pekoz, Teoman
Prawel, Sherwood Peter, Jr
Rihm, Alexander, Jr
Robertson, Leslie Earl
Rumer, Ralph R
Salvadori, M(ario) G(iorgio)
Shaw, Richard P(aul)
Shen, Hung Tao
Sherman, Zachary
Slate, Floyd Owen
Soifer, Herman
Soong, Tsu-Teh
Spencer, James W(endell)
Stahl, Frank Ludwig
Swanson, Robert Lawrence
Tatlow, Richard H(enry), III
Testa, Rene B(iaggio)
Triantafyllou, George S
Turnquist, Mark Alan
Velzy, Charles O
Vemula, Subba Rao
Wang, Muhao S
Wang, Ping Chun
Weidlinger, Paul
White, Richard Norman
Yapa, Poojitha Dahanayake
Zimmie, Thomas Frank

NORTH CAROLINA
Amein, Michael
Brown, Earl Ivan, II
Burton, Ralph Gaines
Chanlett, Emil T(heodore)
Clark, Charles Edward
Cribbins, Paul Day
Evett, Jack B(urnie)
Fadum, Ralph Eigil
Fisher, C(harles) Page, Jr
Galler, William Sylvan
Gupta, Ajaya Kumar
Hanson, John M
Horn, J(ohn) W(illiam)
Humenik, Frank James
Iverson, F Kenneth
Kashef, A(bdel-Aziz) I(smail)
King, L(ee) Ellis
Matzen, Vernon Charles
Medina, Miguel Angel, Jr
Miranda, Constancio F
Pas, Eric Ivan
Peirce, James Jeffrey
Petroski, Henry J
Priory, Richard
Sharkoff, Eugene Gibb
Smallwood, Charles, Jr
Smith, J C
Tung, Chi Chao
Utku, Bisulay Bereket
Utku, Senol
Wahls, Harvey E(dward)
Zia, Paul Z

NORTH DAKOTA
Fossum, Guilford O
Mason, Earl Sewell
Phillips, Monte Leroy

OHIO
Adeli, Hojjat
Bakos, Jack David, Jr
Bishop, Paul Leslie
Bondurant, Byron L(ee)
Cernica, John N
Chan, Yupo
Chawla, Mangal Dass
Chen, Chao-Hsing Stanley
Chen, T(ien) Y(i)
Cook, John P(hilip)
Cosens, Kenneth W
Crum, Ralph G
Fertis, Demeter G(eorge)
Fleck, William G(eorge)
Fok, Thomas Dso Yun
Fu, Kuan-Chen
Gupta, Jiwan Das
Hobbs, Benjamin F
Howe, Robert T(heodore)
Hyland, John R(oth)
Kaneshige, Harry Masato
Kellerstrass, Ernst Junior
Koo, Benjamin
Korda, Peter E
Laushey, Louis M(cNeal)
Leis, Brian Norman
Lestingi, Joseph Francis
Lyon, John Grimson
McDonough, James Francis
Minich, Marlin
Nemeth, Zoltan Anthony
Ojalvo, Morris
Olesen, Douglas Eugene
Palazotto, Anthony Nicholas
Papadakis, Constantine N
Preul, Herbert C
Rai, Iqbal Singh
Ricca, Vincent Thomas
Saada, Adel Selim
Sandhu, Ranbir Singh
Shah, Kanti L
Simon, Andrew L
Stiefel, Robert Carl
Treiterer, Joseph
Venkayya, Vipperla
Wandmacher, Cornelius
Ward, Roscoe Fredrick
Willeke, Gene E
Wu, Tien Hsing

OKLAHOMA
Abdel-Hady, M(ohamed) A(hmed)
Brady, Barry Hugh Garnet
Canter, Larry Wayne
Cox, Gordon F N
Dawkins, William Paul
Dunn, Clark A(llan)
Earls, James Roe
Forney, Bill E
Harp, Jimmy Frank
Manke, Phillip Gordon
Parcher, James V(ernon)
Ree, William O(scar)
Reid, George W(illard)
Sack, Ronald Leslie
Stover, Enos Loy
Streebin, Leale E
Summers, Gregory Lawson

OREGON
Burgess, Fred J
Clough, Ray William
Erzurumlu, H Chik
Hashimoto, Andrew G
Huber, Wayne Charles
Klingeman, Peter C
Kocaoglu, Dundar F
Lall, B Kent
Mueller, Wendelin Henry, III
Rad, Franz N
Sangrey, Dwight A
Schroeder, Warren Lee
Sollitt, Charles Kevin
Tewinkel, G Carper

PENNSYLVANIA
Abrams, Joel Ivan
Aron, Gert
Au, Tung
Barnoff, Robert Mark
Batterman, Steven C(harles)
Becker, Stanley J
Beedle, Lynn Simpson
Bielak, Jacobo
Bieniawski, Zdzislaw Tadeusz
Bjorhovde, Reidar
Brungraber, Robert J
Bullen, Allan Graham Robert
Cady, Philip Dale
Christiano, Paul P
Clark, John W(ood)
Condo, Albert Carman, Jr
Daniels, John Hartley
D'Appolonia, Elio
Davinroy, Thomas Bernard
Dejaiffe, Ernest
Driscoll, George C
Durkee, Jackson L
Dzombak, David Adam
Elsworth, Derek
Errera, Samuel J(oseph)
Fenves, Steven J(oseph)
Fisher, John William
Fleming, John F
Gidley, James Scott
Goel, Ram Parkash
Gotolski, William H(enry)
Hamel, James V(ictor)
Hartmann, Alois J(oseph)
Haythornthwaite, Robert M(orphet)
Hendrickson, Chris Thompson
Higgins, Frederick B(enjamin), Jr
Hribar, John Anthony
Kim, Jai Bin
Knaster, Tatyana
Koliner, Ralph
LaGrega, Michael Denny
Lennon, Gerard Patrick
Lepore, John A(nthony)
Lin, Larry Y H
Loigman, Harold
Lorsch, Harold G
Lu, Le-Wu
Luthy, Richard Godfrey
McClure, Richard Mark
McDonnell, Archie Joseph
McLaughlin, Philip V(an Doren), Jr
McMichael, Francis Clay
McNamee, Bernard M
Mangelsdorf, Clark P
Meyer, Wolfgang E(berhard)
Myers, Earl A(braham)
Nesbitt, John B
Ostapenko, A(lexis)
Pangborn, Robert Northrup
Plonsky, Andrew Walter
Popovics, Sandor
Quimpo, Rafael Gonzales
Rachford, Thomas Milton
Ramanathan, M
Reed, Joseph Raymond
Regan, Raymond Wesley
Rolf, Richard L(awrence)
Roll, Frederic
Romualdi, James P
Rumpf, John L
Saigal, Sunil
Samples, William R(ead)
Schlimm, Gerard Henry
Schuster, James J(ohn)
Sooky, Attila A(rpad)
Stallard, Michael Lee
Swedlow, Jerold Lindsay
Thompson, Ansel Frederick, Jr
Untrauer, Raymond E(rnest)
Van Horn, D(avid) A(lan)
Viest, Ivan M
Viscomi, B(runo) Vincent
Voight, Barry
Vuchic, Vukan R
Weggel, John Richard
White, Elizabeth Loczi
Yen, Ben-Tseng
Younkin, Larry Myrle
Zandi, Iraj

RHODE ISLAND
Clifton, Rodney James
Dafermos, Stella
McEwen, Everett E(dwin)
Poon, Calvin P C
Pretzer, C Andrew
Silva, Armand Joseph

SOUTH CAROLINA
Alexander, William Davidson, III
Anand, Subhash Chandra
Anderson, Thomas L(eonard)
Awadalla, Nabil G
Bridges, Donald Norris
Clark, J(ames) Edwin
Dysart, Benjamin Clay, III
Glenn, George R(embert), Sr
Grady, Cecil Paul Leslie, Jr
Henry, Harold Robert
Jennett, Joseph Charles
Ledbetter, W(illiam) B(url)
McCormac, Jack Clark
Saxe, Harry Charles
Schwartz, Arnold Edward
Sparks, Peter Robert

SOUTH DAKOTA
Goodman, James R
Koepsell, Paul L(oel)
Ramakrishnan, Venkataswamy
Wegman, Steven M
Zogorski, John Steward

TENNESSEE
Armentrout, Daryl Ralph
Brewington, Percy, Jr
Bronzini, Michael Stephen
Brown, Bevan W, Jr
Cooper, Alfred J(oseph)
Foster, Edwin Powell
Freitag, Dean R(ichard)
Gray, William Harvey
Grecco, William L
Green, Robert S(mith)
Harrawood, Paul
Helweg, Otto Jennings
Hensley, Marble John, Sr
Humphreys, Jack Bishop
Madhavan, Kunchithapatham
Malasri, Siripong
Morrison, Thomas Golden
Perry, Randy L
Rowan, William Hamilton, Jr
Rowe, Robert S(eaman)
Rozenberg, J(uda) E(ber)
Sapirie, S(amuel) R(alph)
Schilling, Charles H(enry)
Smith, Dallas Glen, Jr
Tschantz, Bruce A
Wheeler, Orville Eugene
Whitehurst, Eldridge Augustus

TEXAS
Armstrong, James Clyde
Austin, Walter J
Bartel, Herbert H(erman), Jr
Beale, Luther A(lton)
Beard, Leo Roy
Bendapudi, Kasi Visweswararao
Benson, Fred J(acob)
Betts, Austin Wortham
Bigham, Robert Eric
Boyer, Robert Elston
Breen, John E(dward)
Brown, Daniel Mason
Brown, Robert Wade
Burns, Ned Hamilton
Buth, Carl Eugene
Caffey, James Enoch
Charbeneau, Randall Jay
Chiang, Chen Y
Clark, Robert Alfred
Collins, Michael Albert
Collipp, Bruce Garfield
Coyle, Harry Michael
DeHart, Robert C(harles)
Delflache, Andre P
Everard, Noel James
Focht, John Arnold, Jr
Furlong, Richard W
Gallaway, Bob Mitchel
George, James Francis
Geyling, F(ranz) Th(omas)
Gloyna, Earnest F(rederick)
Graff, William J(ohn), Jr
Grieves, Robert Belanger
Hadley, William Owen
Halff, Albert Henry
Hann, Roy William, Jr
Haynes, John J
Herbich, John Bronislaw
Hirsch, Teddy James
Holcomb, Robert M(arion)
Holley, Edward R
Holliday, George Hayes
Holt, Edward C(hester), Jr
Huang, Tseng
Hudson, William Ronald
Ivey, Don Louis
Jirsa, James O
Junkins, Jerry R
Kennedy, Thomas William
Kiesling, Ernst W(illie)
Koehn, Enno
Kohl, John C(layton)
Lazar, Benjamin Edward
Li, Wen-Hsiung
Loehr, Raymond Charles
Lowery, Lee Leon, Jr
Lu, Frank Kerping
Ludwig, Allen Clarence, Sr
Lytton, Robert Leonard
McClelland, Bramlette
McCullough, Benjamin Franklin
Machemehl, Jerry Lee
Mak, King Kuen
Manning, Sherrell Dane
Mathewson, Christopher Colville
Matlock, Hudson
Maxwell, Donald A
Michie, Jarvis D
Mohraz, Bijan
Monsees, James E
Moore, Walter L(eon)
Morgan, Carl William
Morgan, James Richard
Naismith, James Pomeroy
Nicastro, David Harlan
Olson, Roy Edwin
O'Neill, Michael Wayne
Parate, Natthu Sonbaji
Pavlovich, Raymond Doran
Phillips, Joseph D
Pinnell, Charles
Price, William Charles
Qasim, Syed Reazul
Rao, Shankaranarayana Ramohallinanjunda
Reddell, Donald Lee
Reese, Lymon C(lifton)
Richardson, Clarence Wade
Roesset, Jose M
Rogers, Bruce G(eorge)
Rooke, Allen Driscoll, Jr
Rozendal, David Bernard
Samson, Charles Harold
Shah, Haresh C
Simonis, John Charles
Sims, James R(edding)
Slotta, Larry Stewart
Spindler, Max
Stanovsky, Joseph Jerry
Streltsova, Tatiana D
Stubbs, Norris
Tarquin, Anthony Joseph
Thompson, Louis Jean
Truitt, Marcus M(cCafferty)
Tucker, Richard Lee
Vallabhan, C V Girija
Vann, W(illiam) Pennington
Veletsos, A(nestis)
Wah, Thein
Walton, Charles Michael
White, N(ikolas) F(rederick)
Williams, Glen Nordyke
Wolf, Harold William
Wright, Stephen Gailord
Yao, James T-P
Yura, Joseph Andrew

UTAH
Allen, Richard Glen
Anderson, Bruce Holmes
Anderson, Douglas I
Bagley, Jay M(errill)
Christiansen, J(erald) E(mmett)
Clyde, Calvin G(eary)
Enke, Glenn L
Firmage, D(avid) Allan
Flammer, Gordon H(ans)
Fuhriman, D(ean) K(enneth)
Ghosh, Sambhunath
Grenney, William James
Hansen, Vaughn Ernest
Hargreaves, George H(enry)
Hoggan, Daniel Hunter
Israelsen, C Earl
Jeppson, Roland W
Karren, Kenneth W
Keller, Jack
Merritt, LaVere Barrus
Nordquist, Edwin C(lyde)
Olsen, Edwin Carl, III
Peralta, Richard Carl
Rich, Elliot

Riley, John Paul
Sidle, Roy Carl
Tullis, J Paul
Waddell, Kidd M
Wilson, Arnold
Yu, Jason C

VERMONT
Butler, Donald J(oseph)
Cassell, Eugene Alan
Olson, James Paul
Oppenlander, Joseph Clarence
Pinder, George Francis
Pyper, Gordon R(ichardson)

VIRGINIA
Blakey, Lewis Horrigan
Boardman, Gregory Dale
Buchanan, Thomas Joseph
Byrd, Lloyd G
Chang, George Chunyi
Chi, Michael
Chong, Ken Pin
Clarke, Frederick James
Cole, Ralph I(ttleson)
Cox, William Edward
Cragwall, J(oseph) S(amuel), Jr
Curtin, Robert H
Dendrou, Stergios
Diercks, Frederick O(tto)
Dobyns, Samuel Witten
Douma, Jacob H
Doyle, Frederick Joseph
Drewry, William Alton
Duncan, James M
Echols, Charles E(rnest)
Eppink, Richard Theodore
Estes, Edward Richard
Everett, Warren S
Fearnsides, John Joseph
Ferguson, George E(rnest)
Galvin, Cyril Jerome, Jr
Goins, Truman
Gray, George A(lexander)
Hakala, William Walter
Heller, Robert A
Heterick, Robert Cary, Jr
Higgins, Thomas Ernest
Hobeika, Antoine George
Hoel, Lester A
Holzer, Siegfried Mathias
Houbolt, John C(ornelius)
Houck, Mark Hedrich
Hudson, Charles Michael
Jennings, Richard Louis
Johnson, George Patrick
Knapp, John Williams
Kreipke, Merrill Vincent
Kuesel, Thomas Robert
Kuhlthau, A(lden) R(obert)
Lai, Chintu (Vincent C)
Lamb, James C(hristian), III
Larew, H(iram) Gordon
Lee, John A N
Lewis, Russell M(acLean)
Lipner, Steven Barnett
Liu, Tony Chen-Yeh
McCormick, Fred C(ampbell)
McEwen, Robert B
Moore, Joseph Herbert
Morris, David
Morris, John Woodland
Ni, Chen-Chou
Noor, Ahmed Khairy
Nordlund, R(aymond) L(ouis)
Pasko, Thomas Joseph, Jr
Pilkey, Walter David
Pletta, Dan Henry
Reilly, Thomas E
Rojiani, Kamal B
Scalzi, John Baptist
Schad, Theodore M(acNeeve)
Schaefer, Francis T
Seelig, Jakob Williams
Senich, Donald
Simpson, James R(ussell)
Singh, Mahendra Pal
Snyder, Franklin F
Tsai, Frank Y
Wagner, John Edward
Walker, Richard David
Walker, William R
Wang, Leon Ru-Liang
Weichel, Hugo
Weyler, Michael E
Whitlock, Charles Henry
Wieczorek, Gerald Francis
Woo, Dah-Cheng
Worrall, Richard D
Yang, Ta-Lun
Zuk, William

WASHINGTON
Amberg, Herman R(obert)
Anderson, Arthur R(oland)
Anderson, Richard Gregory
Bacon, Vinton Walker
Bender, Donald Lee
Berg, Richard Harold
Bhagat, Surinder Kumar
Bogan, Richard Herbert
Brown, Colin Bertram
Burges, Stephen John
Carlson, Dale Arvid

Carpenter, James E(dwin)
Chace, Alden Buffington, Jr
Christiansen, John V
Colcord, J E
Cole, Jon A(rthur)
De Goes, Louis
Edde, Howard Jasper
Evans, Roger James
Glenne, Bard
Hennes, Robert G(raham)
Jansen, Robert Bruce
Kent, Joseph C(han)
King, Larry Gene
Law, Albert G(iles)
Lottman, Robert P(owell)
Luck, Leon D(an)
Mast, Robert F
Mattock, Alan H
Mylroie, Willa W
Nece, Ronald Elliott
Pilat, Michael Joseph
Robbins, Richard J
Roberson, John A(rthur)
Roeder, Charles William
Rossano, August Thomas
Sawhill, Roy Bond
Skilling, John Bower
Skrinde, Rolf T
Sorensen, Harold C(harles)
Stockdale, William K
Sylvester, Robert O(hrum)
Tedesko, Anton
Terrel, Ronald Lee
Tichy, Robert J
Vasarhelyi, Desi D
Wenk, Edward, Jr

WEST VIRGINIA
Conway, Richard A
Jenkins, Charles Robert
Kemp, Emory Leland
Thornton, Stafford E

WISCONSIN
Appel, David W(oodhull)
Bartel, Fred F(rank)
Bauer, Kurt W
Berthouex, Paul Mac
Boyle, William C(harles)
Busby, Edward Oliver
Christensen, Erik Regnar
Crandall, Lee W(alter)
Day, Harold J
Edil, Tuncer Berat
Faherty, Keith F
Freas, Alan D('Yarmett)
Hoopes, John A
Johnson, John E(dwin)
Karadi, Gabor
Katz, William J(acob)
Kiefer, Ralph W
Kipp, Raymond J
Klus, John P
Larkin, Lawrence A(lbert)
Lillesand, Thomas Martin
Murphy, William G(rove)
Naik, Tarun Ratilal
Polkowski, Lawrence B(enjamin)
Roderick, Gilbert Leroy
Schaffer, Erwin Lambert
Scherz, James Phillip
Shaikh, A(bdul) Fattah
Sherman, D(onald) R
Wolf, Paul R
Zahn, John J
Zanoni, Alphonse E(ligius)

WYOMING
Bellamy, John C
Dolan, Charles W
Humenick, Michael John, II
Kim, Sang Soo
Lamb, Donald R(oy)
Rechard, Paul A(lbert)
Smith, James Lee
Tung, Yeou-Koung
Wilson, Eugene Madison

PUERTO RICO
Lluch, Jose Francisco
Santiago-Melendez, Miguel

ALBERTA
Adams, Peter Frederick Gordon
Bakker, Jaap Jelle
De Paiva, Henry Albert Rawdon
Ellyin, Fernand
Epstein, Marcelo
Glockner, Peter G
Hunt, Robert Nelson
Johnston, Colin Deane
Joshi, Ramesh Chandra
Kivisild, Hans R(obert)
Kulak, Geoffrey Luther
Loov, Robert Edmund
MacGregor, James Grierson
Morgenstern, N(orbert) R
Murray, David William
Nasseri, Touraj
Rajaratnam, N(allamuthu)
Scott, J(ohn) D(onald)
Simmonds, Sidney Herbert
Smith, Daniel Walter
Thomson, Stanley

BRITISH COLUMBIA
Bell, H(arry) R(ich)
Buckland, Peter Graham
Cherry, S(heldon)
Isaacson, Michael
Johnson, Joe W
Lind, Niels Christian
Mavinic, Donald Stephen
Mindess, Sidney
Ruus, E(ugen)
Sexsmith, Robert G

MANITOBA
Baracos, Andrew
Lansdown, A(llen) M(aurice)
Rizkalla, Sami H
Soliman, Afifi Hassan
Sparling, Arthur Bambridge

NEW BRUNSWICK
Davar, K(ersi) S
Faig, Wolfgang
Lin, Kwan-Chow
Wells, David Ernest

NEWFOUNDLAND
Sharp, James Jack

NOVA SCOTIA
Mufti, Aftab A
Sastry, Vankamamidi VRN

ONTARIO
Baltacioglu, Mehmet Necip
Batchelor, B(arrington) D(eVere)
Bewtra, Jatinder Kumar
Beylerian, Nurel
Bozozuk, Michael
Byer, Philip Howard
Clark, Ferrers Robert Scougall
Coakley, John Phillip
Davenport, Alan Garnett
Davis, Merritt McGregor
Ellis, J S
Gracie, G(ordon)
Handa, V(irender) K(umar)
Harris, P(hilip) J(ohn)
Hipel, Keith William
Hope, Brian Bradshaw
Humar, Jagmohan Lal
Huseyin, Koncay
Irwin, Peter Anthony
Jones, L(lewellyn) E(dward)
Kamphuis, J(ohn) William
Kennedy, John B
Kennedy, Russell Jordan
Kenney, T Cameron
Khan, Ata M
Kirk, Donald Wayne
Krishnappa, Govindappa
Krishnappan, Bommanna Gounder
Leutheusser, H(ans) J(oachim)
Litvan, Gerard Gabriel
McCorquodale, John Alexander
MacInnis, Cameron
McLaughlin, Wallace Alvin
McNeice, Gregory Malcolm
Matyas, E(lmer) Leslie
Milligan, Victor
Monforton, Gerard Roland
Ng, Simon S F
Novak, Milos
Parker, Wayne Jeffery
Poucher, Mellor Proctor
Quigley, Robert Murvin
Raymond, Gerald Patrick
Roorda, John
Rowe, Ronald Kerry
Selvadurai, A P S
Sharan, Shailendra Kishore
Sherbourne, Archibald Norbert
Thompson, John Carl
Timusk, John
Topper, T(imothy) H(amilton)
Turkstra, Carl J
Wilson, Kenneth Charles
Wright, Douglas Tyndall
Wright, Peter Murrell
Zytner, Richard G

QUEBEC
Bourque, Paul N(orbert)
Brandenberger, Arthur J
Broughton, Robert Stephen
Brunelle, Paul-Edouard
Douglass, Matthew McCartney
Gallez, Bernard
Giroux, Yves M(arie)
Hanna, Adel
Johns, Kenneth Charles
Joly, George W(ilfred)
Ladanyi, Branko
McKyes, Edward
Marsh, Cedric
Paultre, Patrick
Pekau, Oscar A
Poorooshasb, Hormozd Bahman
Redwood, R(ichard) G(eorge)
Stathopoulos, Theodore
Troitsky, Michael S(erge)

SASKATCHEWAN
Curtis, Fred Allen
Fuller, Gerald Arthur

Hosain, Mahbub Ul
Mollard, John D
Nasser, Karim Wade
Norum, Donald Iver
Viraraghavan, Thiruvenkatachari

OTHER COUNTRIES
Bond, Marvin Thomas
Capodaglio, Andrea G
Ersoy, Ugur
Finkel, Herman J(acob)
Hansen, Torben Christen
Jones, P(hilip) H(arrhy)
Laura, Patricio Adolfo Antonio
Lohtia, Rajinder Paul
McNown, John S
Naudascher, Eduard
Thurlimann, Bruno
Van Hoften, James Dougal Adrianus

Computer Engineering

ALABAMA
Copeland, David Anthony

CALIFORNIA
Atherton, Robert W
Bogy, David B
Brayton, Robert K
Crosbie, Roy Edward
Eaton, J(ames) H(oward)
Harker, J M
Harrison, Michael A
Kolsky, Harwood George
Lee, Kenneth
McCulloch, Andrew Douglas
Volpe, Richard Alan
Wong, Kwan Y
Yao, Jason

COLORADO
Andrews, Michael

CONNECTICUT
Ewing, Martin Sipple
Shirer, Donald Leroy

DISTRICT OF COLUMBIA
Meltzer, Arnold Charles

FLORIDA
Doi, Shinobu
Yuan, Jiann-Shiun

GEORGIA
Bridges, Alan L

ILLINOIS
Emmel, Robert Shafer
Fuchs, W Kent
Hess, George J
Jan, Kwan-Hwa
Karaali, Orhan

INDIANA
Jamieson, Leah H
Wai, Wing-Kin

LOUISIANA
Yu, Fang Xin

MARYLAND
Burke, Edward Raymond
Grabow, Barry Edward
Shelton, Robert Duane

MASSACHUSETTS
Fortmann, Thomas Edward
Lanzkron, Rolf W
Wasserbauer, John Gilmary

MICHIGAN
Chen, Yudong
Ganesan, Subramaniam
Sloan, Martha Ann

MISSOURI
Hahn, James H
Tyrer, Harry Wakeley

NEBRASKA
Leung, Joseph Yuk-Tong
Nelson, Don Jerome

NEVADA
Juliussen, J Egil

NEW JERSEY
Kahrs, Mark William
McPherson, Ross
Musa, John D
Waltz, David Leigh

NEW MEXICO
DeVries, Ronald Clifford

NEW YORK
Chen, Alek Chi-Heng
Delgado-Frias, Jose
Fenrich, Richard Karl
Jabbour, Kamal
Li, Chung-Sheng
Notaro, Anthony

Computer Engineering (cont)

Sasserath, Jay N
Sri-Jayanthal, Sri Muthythamby
Varshney, Pramod Kumar
Wittie, Larry Dawson
Zhou, Zhen-Hong

NORTH CAROLINA
Nie, Dalin

OHIO
Hobart, William Chester, Jr
Kapoor, Vikram J
Rajsuman, Rochit

OREGON
Chrzanowska-Jeske, Malgorzata Ewa

PENNSYLVANIA
Goodman, Alan Lawrence
Hildebrandt, Thomas Henry
Tzeng, Kenneth Kai-Ming

TEXAS
Jump, J Robert
Otis, John Noel
Ott, Granville E
Rhyne, V(ernon) Thomas
Volz, Richard A

UTAH
Bunch, Kyle

VIRGINIA
Gray, F(estus) Gail
Lea, George Koo
Liceaga, Carlos Arturo
Weaver, Alfred Charles

WISCONSIN
Uicker, John Joseph, Jr

ALBERTA
Ulagaraj, Munivandy Seydunganallur

BRITISH COLUMBIA
Agathoklis, Panajotis
Dimopoulos, Nikitas J

MANITOBA
Delecki, Andrew

ONTARIO
Coll, David C
Holford, Richard Moore

Electrical Engineering

ALABAMA
Aldridge, Melvin D(ayne)
Biggs, Albert Wayne
Deck, Howard Joseph
Dezenberg, George John
Dodd, Curtis Wilson
Dowdle, Joseph C(lyde)
Gilbert, Stephen Marc
Graf, E(dward) R(aymond)
Grigsby, Leonard Lee
Haeussermann, Walter
Halijak, Charles A(ugust)
Honnell, Martial A(lfred)
Irwin, John David
Jaeger, Richard Charles
Jarem, John
Johnson, David Edsel
Johnson, Johnny R(ay)
Kowel, Stephen Thomas
Lovingood, Judson Allison
Lowry, James Lee
Lueg, Russell E
McAuley, Van Alfon
McDuff, Odis P(elham)
MacIntyre, John R(ichard)
Moore, Fletcher Brooks
Morley, Lloyd Albert
Mott, Harold
Nalley, Donald Woodrow
Oglesby, Sabert, Jr
Polge, Robert J
Porter, William A
Poularikas, Alexander D
Randall, Joseph Lindsay
Rekoff, M(ichael) G(eorge), Jr
Rouse, John Wilson, Jr
Thurstone, Robert Leon

ALASKA
Bates, Howard Francis
Kokjer, Kenneth Jordan
Sweet, Larry Ross

ARIZONA
Akers, Lex Alan
Andeen, Richard E
Bahill, A(ndrew) Terry
Balanis, Constantine A
Barnes, Howard Clarence
Carlile, Robert Nichols
Choi, Kwan-Yiu Calvin
Clark, Wilburn O
Collins, F(red) R(obert)
DeMassa, Thomas A
Dereniak, Eustace Leonard
Dickieson, Alton C
Duncan, James W
Eberhard, Everett
El-Ghazaly, Samir M
Fahey, Walter John
Ferry, David K
Fliegel, Frederick Martin
Fordemwalt, James Newton
Galloway, Kenneth Franklin
Gerhard, Glen Carl
Gilkeson, Robert Fairbairn
Glenner, E J
Graham, Le Roy Cullen
Hamilton, Douglas J(ames)
Handy, Robert M(axwell)
Haynes, Munro K
Hessemer, Robert A(ndrew), Jr
Hill, Fredrick J
Hoehn, A(lfred) J(oseph)
Hoenig, Stuart Alfred
Howard, William Gates, Jr
Huelsman, Lawrence Paul
Huffman, Tommie Ray
Hunt, Bobby Ray
Johnson, C(harles) Bruce
Johnson, Vern Ray
Karady, George Gyorgy
Kelly, Richard W(alter)
Kerwin, William J(ames)
Krasnow, Marvin Ellman
Kreutel, Randall William, Jr
Lonsdale, Edward Middlebrook
McRae, Lorin Post
Metz, Donald C(harles)
Mulvey, James Patrick
Myers, Ronald G
Mylrea, Kenneth C
Nabours, Robert Eugene
Palais, Joseph C(yrus)
Palusinski, Olgierd Aleksander
Peterson, Harold A(lbert)
Porcello, Leonard J(oseph)
Prince, John Luther, III
Ray, Howard Eugene
Reagan, John Albert
Ren, Shang Yuan
Russell, Paul E(dgar)
Ryan, Thomas Wilton
Schroder, Dieter K
Schultz, Donald Gene
Shemer, Jack Evvard
Stott, Brian
Szilagyi, Miklos Nicholas
Thompson, Truet B(radford)
Tice, Thomas E(arl)
Ulich, Bobby Lee
Wait, John V
Welch, H(omer) William
White, Nathaniel Miller
Witulski, Arthur Frank
Wolfe, William Louis, Jr
Zeigler, Bernard Philip
Zimmer, Carl R(ichard)
Ziolkowski, Richard Walter

ARKANSAS
Cardwell, David Michael
Kaupp, Verne H
McCloud, Hal Emerson, Jr
Mix, Dwight Franklin
Muly, Emil Christopher, Jr
Setian, Leo
Stephenson, Stanley E(lbert)
Yaz, Edwin Engin
Yeargan, Jerry Reese

CALIFORNIA
Abbott, Dean William
Abbott, Wilton R(obert)
Altes, Richard Alan
Amazeen, Paul Gerard
Ames, John Wendell
Ammann, E(ugene) O(tto)
Anderson, Edward P(arley)
Anderson, George William
Anderson, Paul Maurice
Anderson, Warren R(onald)
Anderson, William W
Andrews, Austin Michael, II
Angelakos, Diogenes James
Angell, James Browne
Armantrout, Guy Alan
Arnold, James S(loan)
Baer, Walter S
Baggi, Denis Louis
Ballantyne, Catherine Cox
Barnes, C(asper) W(illiam), Jr
Barnum, James Robert
Barrett, William Avon
Barrick, Donald Edward
Basin, M(ichael) A(bram)
Beaver, W(illiam) L(awrence)
Beckwith, Sterling
Bedrosian, E(dward)
Belcher, Melvin B
Bell, Charles Vester
Bell, Chester Gordon
Bentley, Kenton Earl
Berg, Myles Renver
Berk, Aristid D
Berlekamp, Elwyn R(alph)
Bernstein, Ralph
Bernstein, Robert Lee
Bershad, Neil Jeremy
Bevc, Vladislav
Bhanu, Bir
Biedenbender, Michael David
Birdsall, Charles Kennedy
Blair, William Emanuel
Blauschild, Robert Alan
Bloxham, Laurence Hastings
Blumenthal, Irwin S(imeon)
Boileau, Oliver C
Boltinghouse, Joseph C
Bond, Frederick E
Borst, David W(ellington)
Bouldry, John M(iller)
Bracewell, Ronald Newbold
Brandewie, Richard A(nthony)
Braverman, David J(ohn)
Brayton, Robert K
Breuer, Melvin A
Bridges, William Bruce
Briggs, Faye Alaye
Brodersen, Robert W
Brown, David Randolph
Burg, John Parker
Burgin, George Hans
Bush, Gary Graham
Butler, Jack F
Byram, George Wayne
Caird, John Allyn
Castro, Peter S(alvatore)
Caswell, John N(orman)
Chan, Bertram Kim Cheong
Chan, Shu-Gar
Chan, Shu-Park
Chan, Vincent Sikhung
Chang, Juang-Chi (Joseph)
Chang, Tien-Lin
Chang, William S C
Chen, Kao
Chen, Wen H
Cheng, Tsen-Chung
Childs, William Henry
Chin, Maw-Rong
Chodos, Steven Leslie
Choma, John, Jr
Chou, Chung-Kwang
Chu, Kai-Ching
Chu, Wesley W(ei-chin)
Chua, Leon O(ng)
Clark, Lloyd Douglas
Clark, Peter O(sgoode)
Clark, William Merle, Jr
Clegg, Frederick Wingfield
Clemens, Jon K(aufmann)
Cline, J(ack) F(ribley)
Cockrum, Richard Henry
Cohen, Robert Lewis
Cohn, George I(rving)
Cone, Donald R(oy)
Cook, Charles J
Cooper, Franklin Seaney
Cooper, Martin
Corry, Thomas M
Cox, Donald Clyde
Crandall, Ira Carlton
Crane, Hewitt David
Croft, Thomas A(rthur)
Crombie, Douglass D
Crowell, Clarence Robert
Crowley, Joseph Michael
Curry, George Richard
Cutrona, Louis J
Dairiki, Setsuo
Damonte, John Batista
D'Argenio, David Z
Datta, Samir Kumar
Davis, Alan Lynn
Dawson, Charles H
Day, Walter R, Jr
De Figueiredo, Rui J P
De Meo, Edgar Anthony
De Micheli, Giovanni
Dempsey, Martin E(wald)
Depp, Joseph George
Derenzo, Stephen Edward
Deri, Robert Joseph
Despain, Alvin M(arden)
Dethlefsen, Rolf
Dixon, Harry S(terling)
Dixon, Paul King
Dolby, Ray Milton
Donn, Cheng
Dorf, Richard C
Dost, Martin Hans-Ulrich
Doty, Robert L
Dowell, Jerry Tray
Duda, Richard Oswald
Durbeck, Robert C(harles)
Dutton, Robert W
Duyan, Peter
Dwarakanath, Manchagondanahalli H
Eargle, John Morgan
Eaton, John Kelly
Edwards, Byron N
Eldon, Charles A
Elfant, Robert F
El-Kareh, Auguste Badih
Ellion, M Edmund
Elliott, Douglas Floyd
Elliott, Robert S(tratman)
Elsey, John C(harles)
Enenstein, Norman H(arry)
Engelbart, Douglas C(arl)
Eshleman, Von R(ussel)
Estrin, Thelma A
Evans, Todd Edwin
Everest, F(rederick) Alton
Everhart, Thomas E(ugene)
Evtuhov, Viktor
Fa'arman, Alfred
Falcone, Roger Wirth
Fasang, Patrick Pad
Feinstein, Joseph
Feldman, David
Feldman, Nathaniel E
Fenner, Wayne Robert
Fenwick, Robert B
Ferber, Robert R
Ferrari, Domenico
Ferrari, Leonard
Fetterman, Harold Ralph
Fialer, Philip A
Finn, Harold M
Fire, Philip
Firestone, William L(ouis)
Firle, Tomas E(rasmus)
Flynn, Michael J
Ford, John Philip
Forster, Julian
Foy, Wade H(ampton)
Franklin, Gene F(arthing)
Fraser-Smith, Antony Charles
Freeman, James J
Frerking, Margaret Ann
Fried, Walter Rudolf
Frisco, L(ouis) J(oseph)
Gagliardi, Robert M
Gardner, Floyd M
Garwood, Roland William, Jr
Gatts, Thomas F
Geiges, K S
Geminder, Robert
Gentile, Ralph G
Gesner, Bruce D
Ghausi, Mohammed Shuaib
Ghazanshahi, Shahin D
Ghose, Rabindra Nath
Gibbons, James F
Gibson, Earl Doyle
Gillis, James Thompson
Gillott, Donald H
Giroux, Vincent A(rthur)
Glazer, Judith
Gleghorn, G(eorge) J(ay), (Jr)
Goldberg, Jacob
Goldman, Stanford
Golomb, Solomon Wolf
Goodman, Joseph Wilfred
Gossard, Arthur Charles
Gould, Roy W(alter)
Grant, Eugene F(redrick)
Gray, Paul R
Gray, Robert Molten
Graybill, Howard W
Greene, Frank Sullivan, Jr
Greenfield, Eugene W(illis)
Griffiths, Lloyd Joseph
Grigsby, John Lynn
Grimes, Craig Alan
Grinberg, Jan
Grinnell, Robin Roy
Gunckel, Thomas L, II
Gutfinger, Dan Eli
Hageman, Donald Henry
Hai, Francis
Hakimi, S(eifollah) L(ouis)
Halsted, A(bel) Stevens
Hamilton, J(ames) Hugh
Hankins, George Thomas
Hansen, Grant Lewis
Hansen, Robert C(linton)
Harris, Jay H
Harris, Stephen Ernest
Hawryluk, Andrew Michael
Hawthorne, Edward I
Helfman, Howard N
Helliwell, R(obert) A(rthur)
Hellman, Martin Edward
Helstrom, Carl Wilhelm
Hempstead, Robert Douglas
Hench, David Le Roy
Hendricks, Charles D(urrell), Jr
Hendry, George Orr
Herb, John A
Herbst, Noel Martin
Herget, Charles John
Heritage, Jonathan Paul
Herman, Ray Michael
Hermsen, Richard J
Herzog, Gerald B(ernard)
Heyborne, Robert L(inford)
Hibbs, Andrew D
Higa, Walter Hiroichi
Hingorani, Narain G
Hoagland, A(lbert) S(miley)
Hodges, David A(lbert)
Hoff, Marcian Edward, Jr
Holmes, Dale Arthur
Hooper, Catherine Evelyn
Hoover, William G(eorge)
Hopkin, A(rthur) M(cMurrin)
Hord, William Eugene
Howard, H(enry) Taylor
Hsia, Yukun
Hsu, Jang-Yu
Hu, Chenming
Hu, Sung Chiao

Huang, Chaofu
Huang, Di-Hui (David)
Huang, Tiao-yuan
Hubbs, Robert A(llen)
Huber, Edward Allen
Huffman, David A(lbert)
Hughett, Paul William
Hunt, Eugene B
Huntley, David
Hurrell, John Patrick
Huskey, Harry D(ouglas)
Hwang, Kai
Isberg, Clifford A
Itoh, Tatsuo
Jackson, Gary Leslie
Jacob, George Korathu
Jacobs, Irwin Mark
Jacobson, Alexander Donald
James, Jack N
Jansen, Frank
Jayadev, T S
Jedynak, Leo
Johannessen, Jack
Johnson, Horace Richard
Johnson, Jerome H
Jones, E(dward) M(cClung) T(hompson)
Jones, Lincoln D
Jory, Howard Roberts
Joseph, Peter D(aniel)
Justice, Raymond
Jutamulia, Suganda
Kagiwada, Reynold Shigeru
Kailath, Thomas
Kalbach, John Frederick
Kaliski, Martin Edward
Karp, Arthur
Karunasiri, Gamani
Kastenholz, Claude E(dward)
Kaufman, Alvin B(eryl)
Kavanagh, Karen L
Kearney, Joseph W(illiam)
Kennedy, W Keith, Jr
Kenyon, Richard R(eid)
Keshavan, H R
Khalona, Ramon Antonio
Kilmer, Louis Charles
Kim, John K
King, Harry J
King, Howard E
King, Ray J(ohn)
Kino, Gordon Stanley
Kirchner, Ernst Karl
Kirk, Donald Evan
Klestadt, Bernard
Klinger, Allen
Koetsch, Philip
Koford, James Shingle
Konrad, Gerhard T(hies)
Koopman, Ronald P
Kopp, Eugene H(oward)
Kosai, Kenneth
Kotzebue, Kenneth Lee
Kramer, Noah Herbert
Krause, Lloyd O(scar)
Krener, Arthur James
Kreutz-Delgado, Kenneth Keith
Krishnan, Venkatanama
Krumm, Charles Ferdinand
Kuehl, Hans H(enry)
Kuh, Ernest Shiu-Jen
Kulke, Bernhard
Kummer, W(olfgang) H(elmut)
Kuo, Franklin F(a-Kun)
Lafranchi, Edward Alvin
Lally, Philip M(arshall)
Lane, Richard Neil
Lapin, A I E
Larkin, K(enneth) T(rent)
Lathrop, Richard C(harles)
Latorre, V(ictor) R(obert)
Lee, C(hia) H(uan)
Lee, Ray H(ui-choung)
Lee, Wai-Hon
Lee, William Chien-Yeh
Leedham, Clive D(ouglas)
Leger, Robert M(arsh)
Lehan, Frank W(elborn)
Leichner, Gene H(oward)
Lemke, James Underwood
Lender, Adam
Leonhard, William E
Lerche, Richard Allan
Levan, Nhan
Levine, Arnold Milton
Levine, Daniel
Levy, Donald M(arc)
Lewis, James Laban, III
Lichtenberg, Allan J
Lidow, Eric
Lieberman, Michael A
Likuski, Robert Keith
Lim, Hong Seh
Lim, Teong Cheng
Limb, John Ormond
Lin, Anthony T
Lin, Mao-Shiu
Lin, Wen-C(hun)
Lindal, Gunnar F
Linvill, John G(rimes)
Liu, Jia-ming
Lock, Kenneth
Lodge, Chester Ray
Long, H(ugh) M(ontgomery)
Long, Stephen Ingalls

Loomis, Herschel Hare, Jr
Lord, Harold Wilbur
Lovell, Robert Edmund
Lubcke, Harry Raymond
Lum, Vincent Yu-Sun
Lupash, Lawrence O
Lusignan, Bruce Burr
McBride, Lyle E(rwin), Jr
McCarthy, Francis Desmond
McCluskey, Edward Joseph
McCool, John Macalpine
McCurdy, Alan Hugh
McDonald, H(enry) S(tanton)
McGhee, Robert B
McGill, Thomas Conley, Jr
McLennan, Miles A
Macmillan, Robert S(mith)
McMurty, Burton J
McNamee, Lawrence Paul
Macovski, Albert
McRuer, Duane Torrance
Magill, David Thomas
Maldonado, C(lifford) Daniel
Maloney, Timothy James
Maneatis, George A
Manning, Laurence A(lbert)
Manohar, Aneesh
Mansfield, Ralph
Mantey, Patrick E(dward)
Marmarelis, Vasilis Z
Marquart, Ronald Gary
Martel, Hardy C(ross)
Martin, Steven Roden
Marx, Kenneth Donald
Marxheimer, Rene B
Mason, V Bradford
Mathews, M(ax) V(ernon)
Matthaei, George L(awrence)
Matthews, E(dgar) W(esley), Jr
Mattson, Richard Lewis
Meagher, Donald Joseph
Meeker, Donald J
Mehdi, Imran
Mei, Kenneth K
Mendel, Jerry M
Meng, Shien-Yi
Mertz, Robert Leroy
Messerschmitt, David G
Mettler, Ruben Frederick
Milstein, Laurence
Mireshghi, Ali
Misheloff, Michael Norman
Mitchell, Richard Lee
Mitzner, Kenneth Martin
Mo, Charles Tse Chin
Moe, Maynard L
Mohr, Milton Ernst
Molinder, John Irving
Moll, John L
Money, Lloyd J(ean)
Monson, James Edward
Montgomery, Richard A(lan)
Moore, Ernest J(ulius)
Morgan, Michael Allen
Morris, Fred(erick) W(illiam)
Mortensen, Richard E
Morton, G A
Motley, David Malcolm
Mueller, George E(dwin)
Muller, Richard Stephen
Mulligan, James H(enry), Jr
Munushian, Jack
Murphy, Roy Emerson
Nanevicz, Joseph E
Nathanson, Weston Irwin
Nazaroff, William W
Nease, Robert F
Nelson, Charles G(arthe)
Neumann, Peter G
Newman, John Joseph
Newman, M(orris) M
Ng, Lawrence Chen-Yim
Nguyen, Caroline Phuongdung
Nguyen, Thuan Van
Nickolls, John Richard
Nikora, Allen R
Nilsson, Nils John
Nishi, Yoshio
Nunan, Craig S(pencer)
Ohlson, John E
Okada, R(obert) H(arry)
Oldham, William G
Orchard, Henry John
Ordung, Philip F(ranklin)
Ostwald, Peter Frederic
Owyang, Gilbert Hsiaopin
Oza, Kandarp Govindlal
Pacela, Allan F
Packard, David
Page, D(errick) J(ohn)
Pantell, Richard
Parker, Don
Parker, Sydney R(ichard)
Parks, Robert J
Partanen, Jouni Pekka
Patterson, Tim J
Pawula, Robert Francis
Pease, Roger Fabian Wedgwood
Pederson, Donald O(scar)
Peterson, Allen Montgomery
Pettit, Ray Howard
Pfost, R Fred
Pianetta, Piero Antonio
Pickering, W(illiam) H(ayward)

Pierce, John Robinson
Pinson, John C(arver)
Plummer, James D
Poggio, Andrew John
Pogorzelski, Ronald James
Polak, Elijah
Posner, Edward Charles
Potts, Byron C
Poulter, Howard C
Powell, Ronald Allan
Powers, John Patrick
Prabhakar, Jagdish Chandra
Purl, O(liver) Thomas
Quate, Calvin F(orrest)
Rahmat-Samii, Yahya
Ramamoorthy, Chittoor V
Ramo, Simon
Rauch, Donald J(ohn)
Rauch, Herbert Emil
Reddy, Parvathareddy Balarami
Redfield, David
Redinbo, G Robert
Reed, Eugene D
Reed, Irving Stoy
Rhodes, Ian Burton
Ricardi, Leon J
Rice, Dennis Keith
Riese, Russell L(loyd)
Roberts, Lawrence G
Robinson, Guner Suzek
Rode, Jonathan Pace
Roney, Robert K(enneth)
Rosen, C(harles) A(braham)
Rosenheim, D(onald) E(dwin)
Ross, Hugh Courtney
Rubin, Izhak
Sack, E(dgar) A(lbert), Jr
Safonov, Michael G
Salihi, Jalal T(awfiq)
Salzer, John M(ichael)
Sangiovanni-Vincentelli, Alberto Luigi
Saunders, Robert M(allough)
Sawchuk, Alexander Andrew
Schaffner, Gerald
Scharfetter, D(onald) L
Scheibe, Paul Otto
Scheuch, Don Ralph
Schinzinger, Roland
Schlesinger, Robert Jackson
Schmars, William Thomas
Scholtz, Robert A
Schorr, Herbert
Schott, Frederick W(illiam)
Schwartz, Morton Donald
Schwarz, Steven E
Sclar, Nathan
Sebald, Anthony Vincent
Segal, Alexander
Seib, David Henry
Seitz, S Stanley
Seling, Theodore Victor
Senge, George H
Sensiper, S(amuel)
Sepmeyer, L(udwig) W(illiam)
Shaffner, Richard Owen
Shaw, Elden K
Shaw, George, II
Shevel, Wilbert Lee
Shimada, Katsunori
Shin, Hyungcheol
Shoch, John F
Sie, Charles H
Siegman, A(nthony) E(dward)
Silverman, Leonard
Simon, Marvin Kenneth
Simonen, Thomas Charles
Simpson, Richard Allan
Skalnik, J(ohn) G(ordon)
Skilling, Hugh
Sklansky, J(ack)
Skomal, Edward N
Skrzypek, Josef
Slaton, Jack H(amilton)
Slocum, Richard William
Smith, Otto J(oseph) M(itchell)
Smith, Ralph J(udson)
Smith, Steven Eslie
Sognefest, Peter William
Soohoo, Ronald Franklin
Specht, Donald Francis
Srour, Joseph Ralph
Stahley, William
Staprans, Armand
Steenson, Bernard O(wen)
Steier, William H(enry)
Steinberg, Richard
Steiner, James W(esley)
Stenzel, Reiner Ludwig
Stinson, Donald Cline
Stoll, Paul James
Stolte, Charles
Stout, Thomas Melville
Strand, Timothy Carl
Stubberud, Allen Roger
Sun, Cheng
Sung, Chia-Hsiaing
Suran, Jerome J
Sutherland, Ivan Edward
Sutherland, William Robert
Sweeney, Lawrence Earl, Jr
Szentirmai, George
Tabak, Mark David
Tahiliani, Vasu H
Tang, Denny Duan-Lee

Tanner, Robert Michael
Tatic-Lucic, Svetlana
Thaler, G(eorge) J(ulius)
Thomas, David Tipton
Thomas, Hubert Jon
Thomas, John B(owman)
Thomasian, Aram John
Thompson, David A
Tilles, Abe
Tilley, Brian John
Tomasetta, Louis Ralph
Tortonese, Marco
Travis, J(ohn) C(harles)
Trujillo, Stephen Michael
Tu, Charles Wuching
Tunis, C(yril) J(ames)
Turin, George L
Tuttle, David F(ears)
Tyler, George Leonard
Uebbing, John Julian
Vallerga, Bernard A
Van Atta, Lester Clare
Vane, Arthur B(ayard)
Varaiya, Pravin Pratap
Vartanian, Perry H(atch), Jr
Veigele, William John
Vemuri, Venkateswararao
Vickers, Roger Spencer
Viglione, Sam S
Villard, Oswald G(arrison), Jr
Villeneuve, A(lfred) T(homas)
Viterbi, Andrew J
Vitols, Visvaldis Alberts
Vlay, George John
Volpe, Richard Alan
Vorhaus, James Louis
Wada, George
Wade, Glen
Walden, Robert Henry
Walter, Carlton H(arry)
Wang, Carl C T
Wang, Ji Ching
Wang, Jon Y
Wang, Paul Keng Chieh
Wang, Thomas Nie-Chin
Ward, John Robert
Washburn, Harold W(illiams)
Waterman, Alan T(ower), Jr
Waters, Joe William
Watkins, Dean Allen
Webber, Stanley Eugene
Weber, Charles L
Weckler, Gene Peter
Weinberg, Daniel I
Weissman, Robert Henry
Welch, Lloyd Richard
Welsh, David Edward
Wertz, Harvey J
Wheeler, Harold A(lden)
Whinnery, John R(oy)
White, George Matthews
White, Robert Lee
White, Stanley A
Whitmer, Robert Morehouse
Will, Peter Milne
Williams, Jack Rudolph
Willner, Alan Eli
Willson, Alan Neil, Jr
Wilson, Barbara Ann
Wilson, Perry Baker
Wilson, William John
Wilt, Michael
Wolf, Jack Keil
Wolff, Milo Mitchell
Wood, F(rederick) B(ernard)
Woodbury, Eric John
Wozencraft, John McReynolds
Wu, Felix F
Wu, Te-Kao
Wuebbles, Donald J
Yaggy, Paul Francis
Yamakawa, Kazuo Alan
Yao, Jason
Yao, Kung
Yeh, Cavour W
Yeh, Hen-Geul
Yeh, Paul Pao
Yeung, King-Wah Walter
Young, James R(alph)
Young, Konrad Kwang-Leei
Yousif, Salah Mohammad
Zarem, Abe Mordecai
Zeidler, James Robert
Zeoli, G(ene) W(esley)
Zucker, Oved Shlomo Frank

COLORADO
Allen, James Lamar
Avery, Susan Kathryn
Baird, Jack R
Baird, Ramon Condie
Barnes, Frank Stephenson
Bennett, W Scott
Bloom, L(ouis) R(ichard)
Booton, Richard C(rittenden), Jr
Bradford, Phillips Verner
Brady, Douglas MacPherson
Brandauer, Carl M(artin)
Britton, Charles Cooper
Broome, Paul W(allace)
Brown, Alison Kay
Brubaker, Thomas Allen
Calvert, James Bowles
Cathey, Wade Thomas, Jr

310 / DISCIPLINE INDEX

Electrical Engineering (cont)

Chen, Di
Chernow, Fred
Clifford, Steven Francis
Dichtl, Rudolph John
Eichelberger, W(illiam) H
Elkins, Lincoln F
Engen, Glenn Forrest
Enloe, Louis Henry
Etter, Delores Maria
Faucett, Robert E
Flavin, Michael Austin
Flock, Warren L(incoln)
Frost, H(arold) Bonnell
Fry, Francis J(ohn)
Fuchs, Ewald Franz
Fuller, Jackson Franklin
Gage, Donald S(hepard)
Gless, George E
Goldfarb, Ronald B
Hanna, William J(ohnson)
Haupt, Randy Larry
Hayes, Russell E
Henkel, Richard Luther
Hermann, Allen Max
Hess, Howard M
Hill, David Allan
Hjelme, Dag Roar
Horowitz, Isaac
Horton, Clifford E(dward)
Hull, Joseph A
Hume, Wayne C
Johnk, Carl T(heodore) A(dolf)
Jordan, Harry Frederick
Kanda, Motohisa
Kazmerski, Lawrence L
Kemmerly, Jack E(llsworth)
Kremer, Russell Eugene
Krenz, Jerrold H(enry)
Lewin, Leonard
Lubell, Jerry Ira
Ma, Mark T
Majerfeld, Arnoldo
Malaiya, Yashwant Kumar
Maler, George J(oseph)
Maley, S(amuel) W(ayne)
Marks, Roger Bradley
Mathys, Peter
Mattson, Roy Henry
Maxwell, Lee M(edill)
May, William G(ambrill)
Mengel, J T
Minnick, Robert C
Morgan, Alvin H(anson)
Morrison, John Stuart
Nahman, Norris S(tanley)
Nesenbergs, Martin
Oleszek, Gerald Michael
Ondrejka, Arthur Richard
O'Neill, John Francis
Ralston, Margarete A
Randa, James P
Schiffmacher, E(dward) R(obert)
Schleif, Ferber Robert
Serafin, Robert Joseph
Sisson, Ray L
Smith, Ernest Ketcham
Smith, William Bridges
Staebler, David Lloyd
Stone, J(ack) L(ee)
Summers, Claude M
Taylor, Kenneth Doyle
Thomas, Joe Ed
Twombly, John C
Utlaut, William Frederick
Van Pelt, Richard W(arren)
Wainwright, Ray M
Waite, William McCastline
Walden, Jack M
Wall, Edward Thomas
Webb, James R
Webb, Richard C(larence)
Wiatrowski, Claude Allan
Wilmsen, Carl William
Winn, C Byron
Wu, Min-Yen
Ziemer, Rodger Edmund

CONNECTICUT

Akkapeddi, Prasad Rao
Bennett, William Ralph
Bird, Leslie V(aughn)
Calabrese, Carmelo
Cheng, David H S
Cowles, William Warren
Crawford, John Okerson
Cunningham, W(alter) J(ack)
Davenport, Lee Losee
Deneberg, Jeffrey N
Dubin, Fred S
Ellis, Lynn W
Ewing, Martin Sipple
Fitchen, Franklin Charles
Flynn, T(homas) F(rancis)
Fox, Martin Dale
Gaidis, Michael Christopher
Galbiati, Louis J
Glomb, Walter L
Hajela, Dan
Izenour, George C(harles)
Jain, Faquir C
Javidi, Bahram
Johnson, Loering M
Kerby, Hoyle Ray
Kirwin, Gerald James
Knapp, Charles H
Krummel, William Marion
Lewis, T(homas) Skipwith
Liu, Qing-Huo
Lof, John L(ars) C(ole)
Lovell, Bernard Wentzel
McDonald, John Charles
Mack, Donald R(oy)
Marchand, Nathan
Melehy, Mahmoud Ahmed
Nehorai, Arye
Northrop, Robert Burr
O'Brien, Brian
Quazi, Azizul H(aque)
Reed, Joseph
Rich, Leonard G
Ruckebusch, Guy Bernard
Sapega, A(ugust) E(dward)
Schultheiss, Peter M(ax)
Schwartz, Michael H
Schwarz, Frank
Seely, Samuel
Songster, Gerard F(rancis)
Sparacino, Robert R
Stoker, Warren C
Troutman, Ronald R
Tuteur, Franz Benjamin
Volpe, Gerald T
Wearly, William L
Zweig, Felix

DELAWARE

Barnett, Allen M
Berger, Paul Raymond
Bigliano, Robert P(aul)
Boer, Karl Wolfgang
Bolgiano, Louis Paul, Jr
Bowyer, Kern M(allory)
Dell, Curtis G(eorge)
Hardy, Henry Benjamin, Jr
Ih, Charles Chung-Sen
McNeely, James Braden
Maher, John Philip
Messinger-Rapport, Barbara J
Moore, Charles B(ernard)
Murphy, Arthur Thomas
Vance, Paul A(ndrew), Jr
Walsh, Robert R(eddington)

DISTRICT OF COLUMBIA

Adams, Francis L(ee)
Aein, Joseph Morris
Berman, Gerald Adrian
Campbell, Arthur B
Campillo, Anthony Joseph
Cassidy, Esther Christmas
Della Torre, Edward
Diemer, F(erdinand) P(eter)
Du, Li-Jen
English, William Joseph
Friebele, Edward Joseph
Friedman, Arthur Daniel
Gallagher, John M(ichael), Jr
Grayson, Lawrence P(eter)
Harmuth, Henning Friedolf
Heacock, E(arl) Larry
Huband, Frank L
Kalmus, Henry P(aul)
Koo, Kee P
Kyriakopoulos, Nicholas
Lee, Alfred M
Meister, Robert
Meltzer, Arnold Charles
Mermel, Thaddeus Walter
Montgomery, G(eorge) Franklin
Nelson, David Brian
Nguyen, Charles Cuong
Niedenfuhr, Francis W(illiam)
Parasuraman, Raja
Pei, Richard Yu-Sien
Pickholtz, Raymond L
Rao, Jaganmohan Boppana Lakshmi
Schildcrout, Michael
Schriever, Richard L
Sjogren, Jon Arne
Stephanakis, Stavros John
Susman, Leon
Thomas, Leonard William, Sr
Tolstoy, Alexandra
Trunk, Gerard Vernon
Wong, E(ugene)
Yen, Nai-Chyuan

FLORIDA

Aaron, M Robert
Adair, James Edwin
Alexiou, Arthur George
Amey, William G(reenville)
Ashley, J(ames) Robert
Bernhardt, Richard Charles
Bixby, W(illiam) Herbert
Blecher, F(ranklin) H(ugh)
Boll, H(arry) J(oseph)
Bullock, Thomas Edward, Jr
Burdick, Glenn Arthur
Caldwell, Melba Carstarphen
Callaghan, Frank J
Callahan, Thomas William
Carroll, Dennis Patrick
Chen, Tsong-Ming
Chen, Wayne H(wa-Wei)
Childers, Donald Gene
Cloutier, James Robert
Concordia, Charles
Cookson, Albert Ernest
Couch, Leon Worthington, II
Couturier, Gordon W
Dacey, George Clement
Davis, Carl F
DeLoach, Bernard Collins, Jr
Doi, Shinobu
Dolin, Richard
Donaldson, Merle Richard
Dougherty, John William
Duersch, Ralph R
Easton, Ivan G(eorge)
Einspruch, Norman G(erald)
Elgerd, O(lle) I(ngemar)
Field, George S
Finlon, Francis P(aul)
Flick, Carl
Forsman, M(arion) E(dwin)
Fossum, Jerry G
Gonzalez, Guillermo
Goss, Charles Rapp, Jr
Gould, G(erald) G(eza)
Gruen, H(arold)
Hammer, Jacob
Hancock, John C(oulter)
Harrison, Thomas J
Haviland, Robert P(aul)
Hebb, Maurice F, Jr
Henning, Rudolf E
Hoeppner, Conrad Henry
Hoffman, Thomas R(ipton)
Holley, Charles H
Hollis, J(ohn) Searcy
Houts, Ronald C(arl)
Hoyer-Ellefsen, Sigurd
Huber, William Richard, III
Jain, Vijay Kumar
Johnson, E(well) Calvin
Johnson, J(oseph) Stuart
Jury, Eliahu I(braham)
Kennedy, David P
Kessler, William J(oseph)
Kline, Jacob
Kolhoff, M(arvin) J(oseph)
Kornblith, Lester
Kostopoulos, George
Krausche, Dolores Smoleny
Lade, Robert Walter
Lai, David Chin
Laschever, Norman Lewis
LeVine, D(onald) J(ay)
Li, Sheng-San
Lister, Charles Allan
Malocha, Donald C
Malthaner, W(illiam) A(mond)
Mandil, I Harry
Mathews, Bruce Eugene
Melvin, E(ugene) A(very)
Messenger, Roger Alan
Metzger, Sidney
Migliaro, Marco William
Miller, Hillard Craig
Moreno, Wilfrido A
Moskowitz, Ronald
Naini, Majid M
Nuese, Charles J
O'Malley, John Richard
Packer, Lewis C
Palmer, Ralph Lee
Peebles, Peyton Z, Jr
Perini, Jose
Perrotta, Alessandro
Puckett, Russell Elwood
Ramond, Pierre Michel
Revay, Andrew W, Jr
Riblet, Henry B
Ross, Gerald Fred
Roth, Paul Frederick
Scruggs, Frank Dell
Sells, Jackson S(tuart)
Shaffer, Charles V(ernon)
Sias, Frederick Ralph
Smith, Jack R(eginald)
Smythe, Robert C
Snyder, R L
Spooner, M(orton) G(ailend)
Staudhammer, John
Stavis, Gus
Stodola, Edwin King
Swartz, William Edward, Jr
Tapia, M(oiez) A(hmedale)
Taylor, Fredrick James
Tebbe, Dennis Lee
Timoshenko, Gregory Stephen
Tou, Julius T(su) L(ieh)
Towner, O W
Traexler, John F
Tusting, Robert Frederick
Ulm, Lester, Jr
Ulug, Esin M
Uman, Martin A(llan)
Ungvichian, Vichate
Van Ligten, Raoul Fredrik
Veinott, Cyril G
Vemuri, Suryanarayana (Suri)
Wade, Thomas Edward
Wallace, James D
Weaver, Lynn E(dward)
Weber, Harry P(itt)
Weller, Richard Irwin
Whitman, Lawrence C
Wiseman, Robert S
Yacoub, Kamal
Yau, Stephen S
Yii, Roland
Yon, E(ugene) T
Young, Tzay Y
Yuan, Jiann-Shiun

GEORGIA

Adomian, George
Alford, Cecil Orie
Baxter, Samuel G
Boland, Joseph S(amuel), III
Bourne, Henry C(lark), Jr
Bruder, Jospeh Albert
Callahan, Leslie G, Jr
Cohen, Leonard George
Connelly, J(oseph) Alvin
Cook, James H, Jr
Currie, Charles H
Dees, Julian Worth
Derrick, Robert P
Dutton, John C
Ecker, Harry Allen
Esogbue, Augustine O
Evans, Thomas P
Evans, W E
Feeney, Robert K
Fey, Willard Russell
Fielder, D(aniel) C(urtis)
Finn, D(avid) L(ester)
Fisher, Reed Edward
Galeano, Sergio F(rancis)
Gaylord, Thomas Keith
Gray, James S
Griffin, Clayton Houstoun
Gudehus, Donald Henry
Harrison, Gordon R
Holtum, Alfred G(erard)
Honkanen, Pentti A
Hunt, William Daniel
Jokerst, Nan Marie
Jones, William B(enjamin), Jr
Joy, Edward Bennett
Kalb, John W
Landgren, John Jeffrey
Lee, Kok-Meng
McClellan, James Harold
Martin, Roy A
Mersereau, Russell Manning
Moad, M(ohamed) F(ares)
Moss, Richard Wallace
Owens, William Richard
Paris, Demetrius T
Peatman, John B(urling)
Ray, Dale C(arney)
Rhodes, William Terrill
Rodrigue, George Pierre
Rupf, John Albert, Jr
Russell, John Lynn, Jr
Ryan, Charles Edward, Jr
Schafer, Ronald W
Su, Kendall L(ing-Chiao)
Subramanian, Mani M
Thumann, Albert
Vail, Charles R(owe)
Vann, William L(onnie)
Wallace, John M, Jr
Wang, Johnson Jenn-Hwa
Webb, Roger P(aul)
Yokelson, Bernard J(ulius)

HAWAII

Abramson, N(orman)
Gaarder, Newell Thomas
Granborg, Bertil Svante Mikael
Hwang, Hu Hsien
Kadota, T Theodore
Kinariwala, Bharat K
Lin, Shu
McFee, Richard
Peterson, William Wesley
Roelofs, Thomas Harwood
Weaver, Paul Franklin
Yuen, Paul C(han)

IDAHO

Baily, Everett M
DeBow, W Brad
Demuth, Howard B
Gray, Earl E
Hall, J(ames) A(lexander)
Hyde, William W
Iyer, Ravi
Law, John
Mann, Paul
Meyer, Orville R
Parish, William R
Rigas, Anthony L
Stratford, R P
Stuffle, Roy Eugene
Trinko, Joseph Richard, Jr

ILLINOIS

Aagaard, James S(tuart)
Agarwal, Gyan C
Ahuja, Narendra
Armington, Ralph Elmer
Arzbaecher, Robert
Basar, Ronald John
Blahut, Richard E
Briley, Bruce Edwin
Brodwin, Morris E(llis)
Brown, David P(aul)
Brown, Julius

Burnham, Robert Danner
Burtness, Roger William
Chang, Herbert Yu-Pang
Chen, W(ai) K(ai)
Chew, Weng Cho
Christianson, Clinton Curtis
Conant, Roger C
Cooper, Duane H(erbert)
Coplien, James O
Demaeyer, Bruce R
Deschamps, Georges Armand
DeTemple, Thomas Albert
Dunn, Floyd
Dyson, John Douglas
Eden, James Gary
Eilers, Carl G
Emery, W(illis) L(aurens)
Engel, Joel S
Epstein, Max
Erwin, David B(ishop), Sr
Friedlander, Alan L
Frizzell, Leon Albert
Fuchs, W Kent
Gaddy, Oscar
Gallagher, David Alden
Gardner, Chester Stone
Gillette, Richard F
Godhwani, Arjun
Gottlieb, Gerald Lane
Haddad, Abraham H(erzl)
Hajj, Ibrahim Nasri
Hammond, William Marion
Hess, Karl
Himmelstein, Sydney
Holmes, Kenneth Robert
Hrbek, George W(illiam)
Hua, Ping
Huang, Thomas Shi-Tao
Hunsinger, Bill Jo
Jan, Kwan-Hwa
Jenkins, William Kenneth
Johnson, Milton R(aymond), Jr
Jones, J(ohn) L(loyd), Jr
Jordan, Edward C(onrad)
Kang, Sung-Mo Steve
Karaali, Orhan
Kim, Kyekyoon K(evin)
Knop, Charles M(ilton)
Krein, Philip T
Kumar, Panganamala Ramana
Kuo, Benjamin Chung-I
Lawrie, Duncan H
Laxpati, Sharad R
Lipinski, Walter C(harles)
Lo, Y(uen) T(ze)
Luplow, Wayne Charles
McConnell, H(oward) M(arion)
Maclay, G Jordan
Mast, P(lessa) Edward
Mayes, Paul E(ugene)
Messinger, Henry Peter
Metze, Gernot
Miller, W(endell) E(arl)
Murata, Tadao
Murphy, Gordon J
Naylor, David L
Neuhalfen, Andrew J
Newell, Darrell E
Newton, John S
Noonan, John Robert
O'Brien, William Daniel, Jr
Overbye, Thomas Jeffrey
Pai, Anantha M
Pai, Mangalore Anantha
Parker, Ehi
Perkins, William Randolph
Phillips, Thomas James
Plonus, Martin
Poppelbaum, Wolfgang Johannes
Priemer, Roland
Ransom, Preston Lee
Reichard, Grant Wesley
Rekasius, Zenonas V
Reynolds, Larry Owen
Robbins, R(oger) W(ellington)
Rodriguez, Jacinto
Rutledge, Robert B
Saeks, Richard E
Sahakian, Alan Varteres
Saltzberg, Theodore
Sanathanan, C(hathilingath) K
Sarwate, Dilip Vishwanath
Sauer, Peter William
Savic, Stanley D
Scanlon, Jack M
Siegel, Jonathan Howard
Smith, James G(ilbert)
Smith, Leslie Garrett
Stein, Herbert Joseph
Stinaff, Russell Dalton
Studtmann, George H
Swenson, G(eorge) W(arner), Jr
Taflove, Allen
Toppeto, Alphonse A
Trick, Timothy Noel
Trimble, John Leonard
Turnbull, Robert James
Ulrich, Werner
Uslenghi, Piergiorgio L
Van Ness, James E(dward)
Van Valkenburg, M(ac) E(lwyn)
Venema, Harry J(ames)
Verdeyen, Joseph T
Wakefield, Ernest Henry

Wax, Nelson
Webb, Harold Donivan
Weber, Erwin Wilbur
Weinberg, Philip
Wey, Albert Chin-Tang
Wheeler, Bruce Christopher
Zell, Blair Paul

INDIANA
Acker, Frank Earl
Atallah, Mikhail Jibrayil
Bass, Steven Craig
Berry, William B(ernard)
Brown, Buck F(erguson)
Chen, Chin-Lin
Chittick, K(enneth) A
Cohn, David L(eslie)
Coyle, Edward John
Criss, Darrell E
Detraz, Orville R
Dunipace, Kenneth Robert
Ersoy, Okan Kadri
Fearnot, Neal Edward
Friedlaender, Fritz J(osef)
Fukunaga, Keinosuke
Gabriel, Garabet J(acob)
Gallagher, Neal Charles
Gelopulos, Demosthenes Peter
Gunshor, Robert Lewis
Hammann, John William
Hayt, William H(art), Jr
Henke, Mitchell C
Henry, Eugene W
Heydt, Gerald Thomas
Jamieson, Leah H
Kashyap, Rangasami Laksminarayana
Kendall, Perry E(ugene)
Kim, Young Duc
Koivo, Antti J
Kramer, Robert Allen
Krause, Paul Carl, Jr
Lafuse, Harry G
Landgrebe, David Allen
Lin, P(en) M(in)
Lindenlaub, John Charles
Liu, Ruey-Wen
Lundstrom, Mark Steven
McGillem, C(lare) D(uane)
Mahmoud, Aly Ahmed
Melloch, Michael Raymond
Michel, Anthony Nikolaus
Mowle, Frederic J
Neudeck, Gerold W(alter)
Ogborn, Lawrence L
Pierret, Robert Francis
Pierson, Edward S
Rikoski, Richard Anthony
Rogers, Charles C
Ross, Irvine E
Sabbagh, Harold A(braham)
Schalliol, Willis Lee
Schwartz, Richard John
Shelley, Austin L(inn)
Shewan, William
Siegel, Howard Jay
Sinha, Akhouri Suresh Chandra
Skelton, Robert Eugene
Snow, John Thomas
Stanley, Gerald R
Szpankowski, Wojciech
Thompson, Hannis Woodson, Jr
Uhran, John Joseph, Jr
Vocke, Merlyn C
Wai, Wing-Kin
Williams, Walter Jackson, Jr
Winton, Henry J
Wintz, P(aul) A

IOWA
Andersland, Mark Steven
Anderson, R(obert) M(orris), Jr
Basart, John Philip
Brearley, Harrington C(ooper), Jr
Carlson, David L
Chyung, Dong Hak
Coady, Larry B
Collins, Steve Michael
Dalal, Vikram
Fanslow, Glenn E
Fouad, Abdel Aziz
Geiger, Randall L
Getting, Peter Alexander
Hale, Harry W(illiam)
Horton, Richard E
Hsieh, Hsung-Cheng
Hubbard, Philip G(amaliel)
Jones, Edwin C, Jr
Kopplin, J(ulius) O(tto)
Lamont, John W(illiam)
Lonngren, Karl E(rik)
Lunde, Barbara Kegerreis, (BK)
Malik, Norbert Richard
Mericle, Morris H
Myers, Glenn Alexander
Pierce, Roger J
Pohm, A(rthur) V(incent)
Potter, Allan G
Reddy, Sudhakar M
Robinson, John Paul
Smay, Terry A
Stephenson, David Town
Titze, Ingo Roland
Wilson, Robert D(owning)

KANSAS
Carpenter, Kenneth Halsey
Cottom, Melvin C(lyde)
Daugherty, Don G(ene)
Dellwig, Louis Field
Grzymala-Busse, Jerzy Witold
Holtzman, Julian Charles
Hummels, Donald Ray
Jakowatz, Charles V
Johnson, Gary Lee
Legler, Warren Karl
Lenhert, Donald H
Moore, Richard K(err)
Norris, Roy Howard
Roberts, James Arnold
Rowland, James Richard
Sawan, Mahmoud Edwin
Schrag, Robert L(eroy)
Shanmugan, K Sam
Smith, W(illiam) P(ayne)
Trank, John W
Weller, Lawrence Allenby

KENTUCKY
Bordoloi, Kiron
Bradley, Eugene Bradford
Cathey, Jimmie Joe
Chenoweth, Darrel L
Cox, Hollace Lawton, Jr
Distler, Raymond Jewel
Fardo, Stephen W
Jackson, John Sterling
Nasar, Syed Abu
Pierce, William H
Scheer, Donald Jordan
Wagner, William S

LOUISIANA
Adams, Leonard C(aldwell)
Azzam, Rasheed M A
Bishop, Richard Ray
Burke, Robert Wayne
Cline, Robert William
Colgrove, Steven Gray
Cospolich, James D
Council, Marion Earl
Cronvich, James A(nthony)
Draayer, Jerry Paul
Drake, Robert L
Harlow, Charles Alton
Kak, Subhash Chandra
Kinney, Ralph A
Klos, William Anton
Leon, B(enjamin) J(oseph)
Marshak, Alan Howard
Rice, David A
Roemer, Louis Edward
Sperry, Claude J, Jr
Tan, Owen T
Tewell, Joseph Robert
Tims, Eugene F(rancis)
Trahan, Russell Edward, Jr
Vliet, Daniel H(endricks)
Voss, Charles Henry, Jr
Ward, Ronald Wayne
Williamson, Edward P
Young, Robert Lee

MAINE
Hodgkin, Brian Charles
McMillan, Brockway
Otto, Fred Bishop
Soria, Rodolfo M(aximiliano)
Turner, Walter W(eeks)
Vetelino, John Frank
Wilms, Hugo John, Jr
Wuorinen, John H, Jr

MARYLAND
Abed, Eyad Husni
Alers, George A
Atia, Ali Ezz Eldin
Babcock, Anita Kathleen
Baker, Francis Edward, Jr
Barbe, David Franklin
Bargellini, P(ier) L(uigi)
Barker, John L, Jr
Barrack, Carroll Marlin
Basham, Ray S(cott)
Bauer, Peter W
Beach, Eugene Huff
Beaumariage, D(onald) C(urtis)
Belanger, Brian Charles
Bellmore, Mandell
Best, George E(dward)
Blake, Lamont Vincent
Blewett, John P
Blum, Robert Allan
Bluzer, Neal
Brancato, Emanuel L
Brodsky, Marc Herbert
Brody, William R
Bukowski, Richard William
Burdge, Geoffrey Lynn
Campanella, Samuel Joseph
Charles, Harry Krewson, Jr
Cheng, David Keun
Chi, L K
Cohen, Stanley Alvin
Conklin, James Byron, Jr
Cooper, R(obert) S(hanklin)
Cottony, Herman Vladimir
Cotts, Arthur C(lement)
Davidson, Frederic M

Davisson, Lee David
Dean, Charles E(arle)
DeClaris, N(icholas)
Degenford, James Edward
Del Grosso, Vincent Alfred
DeLoatch, Eugene M
Destler, William W
Dickens, Lawrence Edward
Dixon, John Kent
Durling, Allen E(dgar)
Durrani, Sajjad H(aidar)
Dzimianski, John W(illiam)
Edelson, Burton Irving
Eden, Murray
Ephremides, Anthony
Epstein, Henry
Evans, Alan G
Farnsworth, Richard Kent
Fitzgerald, Duane Glenn
Frankel, Herbert
Freeman, Ernest Robert
Frey, Jeffrey
Fthenakis, Emanuel John
Gedulig, Donald
Geist, J(ohn) C(harles)
Geneaux, Nancy Lynne
Getsinger, William J
Glennan, T(homas) Keith
Goldhirsh, Julius
Golding, Leonard S
Goldstein, Seth Richard
Gore, Willis C(arroll)
Gorozdos, Richard E(mmerich)
Gottfried, Paul
Grabow, Barry Edward
Graves, Willard L
Gupta, Ramesh K
Gupta, Vaikunth N
Habib, Edmund J
Hall, Arthur David, III
Hanson, William A
Harger, Robert Owens
Harris, Forest K
Harris, Jack R
Henderson, William Boyd
Henry, James H(erbert)
Hermach, Francis L
Hochuli, Urs Erwin
Hofmann, Charles Bartholomew
Hornbuckle, Franklin L
Horst, John Albert
Huggins, W(illiam) H(erbert)
Jablonski, Daniel Gary
Jackson, William David
Jamieson, J(ohn) A(nthony)
Jenkins, Robert Edward
Johnson, David Lee
Jones, T(homas) Benjamin
Kanal, Laveen Nanik
Kantor, Gideon
Karulkar, Pramod C
Kopanski, Joseph J
Kotter, F(red) Ralph
Legeckis, Richard Vytautas
Lettieri, Thomas Robert
Levine, William Silver
Lowney, Jeremiah Ralph
Lupton, William Hamilton
McCourt, A(ndrew) W(ahlert)
McDonough, Robert Newton
McDuffie, George E(addy), Jr
Manley, Oscar Paul
Marcus, Steven Irl
Marple, Stanley Lawrence, Jr
Marsten, Richard B(arry)
Melngailis, John
Metze, George M
Meyer, Gerard G L
Meyer, Richard Arthur
Michalowicz, Joseph C(asimir)
Miller, Raymond Edward
Moore, Robert Avery
Morris, Alan
Moulton, Arthur B(ertram)
Munasinghe, Mohan P
Nathanson, Fred E(liot)
Newcomb, Robert Wayne
Oettinger, Frank Frederic
Ohman, Gunnar P(eter)
Ostaff, William A(llen)
Ott, Edward
Page, Chester H
Palmer, Charles Harvey
Palmer, Laurence Clive
Papadopoulos, Konstantinos Dennis
Papian, William Nathaniel
Pinkston, John Turner, III
Pollack, Louis
Potyraj, Paul Anthony
Prabhakar, Arati
Preisman, Albert
Price, Michael Glendon
Pritchard, Wilbur L(ouis)
Pugsley, James H(arwood)
Pullen, Keats A(bbott), Jr
Quinn, Thomas Patrick
Raabe, Herbert P(aul)
Rhee, Moon-Jhong
Richardson, John Marshall
Ricker, Richard Edmond
Riganati, John Philip
Riley, Patrick Eugene
Robertson, T(homas) M(ills)
Rosenblum, Howard Edwin

Electrical Engineering (cont)

Roth, Michael William
Rugh, Wilson J(ohn), II
Sadowsky, John
Schleif, Robert Ferber
Schutzman, Elias
Schwee, Leonard Joseph
Shelton, Robert Duane
Sheppard, Norman F, Jr
Skillman, William A
Skinner, Dale Dean
Sommer, Helmut
Sommerman, George
Stevens, Howard Odell, (Jr)
Strohbehn, Kim
Strull, Gene
Talaat, Mostafa E(zzat)
Tampico, Joseph
Tang, Cha-Mei
Taylor, Leonard S
Thakor, Nitish Vyomesh
Thoma, George Ranjan
Tisdale, Glenn E(van)
Tretter, Steven Alan
Tsitlik, Joshua E
Tudbury, Chester A
Valenzuela, Gaspar Rodolfo
Wagner, Thomas Charles Gordon
Walleigh, Robert S(huler)
Walter, John Fitler
Wang, Ting-I
Wang, Yen Chu
Waynant, Ronald William
Webb, George N
Weinberger, Lester
Wolf, A(lfred) A(braham)
Wolff, Edward A
Yang, Xiaowei
Yu, Anthony Woonchiu
Zafiriou, Evanghelos
Zarwyn, B(erthold)
Zelkowitz, Marvin Victor

MASSACHUSETTS
Abraham, L(eonard) G(ladstone), Jr
Adlerstein, Michael Gene
Alexander, Henry R(ichard)
Allen, Jonathan
Allen, Ryne C
Anderson, John G(aston)
Aprille, Thomas Joseph, Jr
Athans, Michael
Avila, Charles Francis
Bangert, John T(heodore)
Barbour, William E, Jr
Barton, David Knox
Bello, P(hillip)
Bennett, C Leonard
Beranek, Leo Leroy
Berger, Robert
Bers, Abraham
Beusch, John U
Bicknell, William Edmund
Blair, John
Bloom, Jerome H(ershel)
Bobrow, Leonard S(aul)
Bogert, Bruce Plympton
Bovopoulos, Andreas D
Bozler, Carl O
Brierley, Steven K
Brown, Robert Lee
Bruce, James Donald
Bugnolo, Dimitri Spartaco
Burrows, Michael L(eonard)
Carlson, Herbert Christian, Jr
Caron, Paul R(onald)
Carpenter, Richard M
Carver, Keith Ross
Carver, O(liver) T(homas)
Castro, Alfred A
Chen, Chi-Hau
Chu, J Chuan
Clifton, Brian John
Cogbill, Bell A
Colby, George Vincent, Jr
Connors, Philip Irving
Cook, Charles Emerson
Corry, Andrew F
Cronin-Golomb, Mark
Cuffin, B(enjamin) Neil
Denholm, Alec Stuart
Dennison, Byron Lee
Deno, Don W
Dertouzos, Michael L
Dettinger, David
Dixit, Sudhir S
Dorschner, Terry Anthony
Douglas, Edward Curtis
Drane, Charles Joseph, Jr
Dudgeon, Dan Ed
Eaves, Reuben Elco
Edwards, Frederick H(orton)
Eggimann, Wilhelm Hans
Ehrman, Leonard
Elias, Peter
Elmer, William B
Emshwiller, Maclellan
Enzmann, Robert D
Essigmann, Martin W(hite)
Everett, Robert R(ivers)
Faflick, Carl E(dward)
Fan, John C C
Fano, Robert M(ario)
Farnham, Sherman B
Feldman, James Michael
Figwer, J(ozef) Jacek
Fisher, Franklin A
Fishman, Philip M
Fortmann, Thomas Edward
Franks, Lewis Embree
Frazier, Richard H
Freed, Charles
Freedman, Jerome
Fujimoto, James G
Fuller, Samuel Henry
Gallager, Robert G
Garodnick, Joseph
Garvey, R(obert) Michael
Gaw, C Vernon
Gentile, Thomas Joseph
Giger, Adolf J
Gilfix, Edward Leon
Gilinson, Philip J(ulius), Jr
Giordano, Arthur Anthony
Glorioso, Robert M
Gold, Bernard
Goldberg, Aaron Joseph
Goldberg, Harold Seymour
Goldberg, Seymour
Goldner, Ronald B
Goldstein, Edward
Gonsalves, Lenine M
Gonsalves, Robert Arthur
Gould, Leonard A(braham)
Graneau, Peter
Gray, Paul Edward
Gray, Truman S(tretcher)
Gregory, Bob Lee
Grogan, William R
Gross, Fritz A
Hall, W(illiam) M(ott)
Harrington, Edmund Aloysius, Jr
Haus, Hermann A(nton)
Hedlund, Donald A
Heuchling, Theodore P
Hofstetter, Edward
Holmes, D Brainerd
Holmstrom, Frank Ross
Holst, William Frederick
Holt, Frederick Sheppard
Horowitz, Larry Lowell
Howell, Alvin H(arold)
Hurwitz, Charles E(lliot)
Ilic, Marija
Inada, Hitoshi
Ince, William J(ohn)
Jakes, W(illiam) C(hester)
Johnson, Walter Hudson
Jones, Nolan T(homas)
Kassakian, John Gabriel
Keast, Craig Lewis
Keicher, William Eugene
Kennedy, Robert Spayde
King, Ronold (Wyeth Percival)
Kolodziejski, Leslie Ann
Kong, Jin Au
Kostishack, Daniel F(rank)
Kramer, J David R, Jr
Kramer, Robert
Kudlich, Robert A
Kuo, Yen-Long
Kusko, Alexander
Kyhl, Robert Louis
Lamensdorf, David
Lanzkron, Rolf W
Larson, Richard Charles
Lee, Thomas Henry
Lehr, Carlton G(orney)
Lele, Shreedhar G
Lemnios, William Zachary
Lemons, Thomas M
Levesque, Allen Henry
Levin, Robert Edmond
Levison, William H(enry)
Liberman, Allen Harvey
Linden, Kurt Joseph
Litzenberger, Leonard Nelson
Lyons, William Gregory
McAulay, Robert J
McBee, W(arren) D(ale)
McCord, Harold Kenneth
McIntosh, Robert Edward
McVicar, Kenneth E(arl)
McWhorter, Alan L
Maeks, Joel
Makhoul, John Ibrahim
Malme, Charles I(rving)
Manasse, Fred Kurt
Margolin, Jerome
Martin, Frederick Wight
Mavretic, Anton
Melngailis, Ivars
Mitter, Sanjoy
Monopoli, Richard V(ito)
Montgomery, Donald Bruce
Moran, James Michael, Jr
Morgenthaler, Frederic R(ichard)
Morrow, Walter E, Jr
Mountain, David Charles, Jr
Mulpur, Arun K
Mulukutla, Sarma Sreerama
Murphy, Daniel John
Norris, Richard C
Olsen, Kenneth
Oppenheim, Alan Victor
Osepchuk, John M
Paik, S(ungik) F(rancis)
Parker, Ronald R
Payne, Richard Steven
Peake, William Tower
Pedersen, Peder Christian
Penchina, Claude Michel
Penfield, Paul, Jr
Pennington, Anthony James
Perreault, David Alfred
Peura, Robert Allan
Plumb, John Laverne
Poon, Chi-Sang
Pozar, David Michael
Prerau, David Stewart
Proakis, John George
Pucel, Robert A(lbin)
Raemer, Harold R
Raffel, Jack I
Rafuse, Robert P(endleton)
Reich, Herbert Joseph
Reif, L Rafael
Reiffen, Barney
Reinisch, Bodo Walter
Reintjes, J Francis
Rheinstein, John
Riblet, Gordon Potter
Rios, Pedro Agustin
Riseberg, Leslie Allen
Riseman, Edward M
Roadstrum, William H(enry)
Roberge, James Kerr
Robinson, D(enis) M(orrell)
Rochefort, John S
Rodbell, Donald S
Rona, Mehmet
Rosenkranz, Philip William
Ross, Myron Jay
Rotman, Walter
Rubin, Milton D(avid)
Rudko, Robert I
Ruina, J(ack) P(hilip)
Sakshaug, Eugene C
Salah, Joseph E
Saltzer, Jerome H(oward)
Sarles, F(rederick) Williams
Schattenburg, Mark Lee
Schaubert, Daniel Harold
Scherer, Harold Nicholas, Jr
Schetzen, Martin
Schmidt, George Thomas
Schoendorf, William H(arris)
Schonhoff, Thomas Arthur
Seifert, William W(alther)
Sethares, James C(ostas)
Shaffer, Harry Leonard
Shapiro, Jeffrey Howard
Shrader, William Whitney
Siebert, W(illiam) M(cConway)
Silevich, Michael B
Singer, Joshua J
Smith, Henry I
Smith, Paul E, Jr
Smullin, Louis Dijour
Sneddon, Leigh
Spears, David Lewis
Spencer, Gordon Reed
Spergel, Philip
Staelin, David Hudson
Stevens, Kenneth N(oble)
Stevenson, David Michael
Strong, Robert Michael
Stuart, John W(arren)
Sun, Fang-Kuo
Swift, Calvin Thomas
Tancrell, Roger Henry
Tang, Tin-Wei
Teager, Herbert Martin
Temme, Donald H(enry)
Teng, Chojan
Theobald, Charles Edwin, Jr
Thornton, Richard D(ouglas)
Troxel, Donald Eugene
Tsandoulas, Gerasimos Nicholas
Tsang, Dean Zensh
Tsaur, Bor-Yeu
Tweed, David George
Uhlir, Arthur, Jr
Unlu, M Selim
Van Meter, David
Voigt, Herbert Frederick
Wagner, David Kendall
Wang, Christine A
Ward, Harold Richard
Ward, James
Wasserburger, John Gilmary
Wedlock, Bruce D(aniels)
Weng, Lih-Jyh
Wetzstein, H(anns) J(uergen)
Wheeler, Ned Brent
White, David Calvin
Whitehouse, David R(empfer)
Wiederhold, Pieter Rijk
Wiegner, Allen W
Wiesner, J(erome) B
Wilk, Leonard Stephen
Williamson, Richard Cardinal
Wilson, Gerald Loomis
Wogrin, Conrad A(nthony)
Woll, Harry Jean
Wu, Pei-Rin
Yaghjian, Arthur David
Yngvesson, K Sigfrid
Zheng, Xiao Lu
Zimmermann, H(enry) J(oseph)
Zornig, John Grant
Zraket, Charles Anthony

MICHIGAN
Ackenhusen, John Goodyear
Aleksoff, Carl Chris
Alton, Everett Donald
Anderson, Walter T(heodore)
Asmussen, Jes
Atwood, Donald J
Aupperle, Eric Max
Barr, Robert Ortha, Jr
Beglau, David Alan
Be Ment, Spencer L
Bickart, Theodore Albert
Bird, Michael Wesley
Bockemuehl, R(obert) R(ussell)
Borcherts, Robert H
Brammer, Forest E(vert)
Brown, Richard K(emp)
Brown, Verne R
Brumm, Douglas B(ruce)
Bryant, John H(arold)
Calahan, Donald A
Carey, John Joseph
Caron, E(dgar) Louis
Chau, Hin Fai
Chen, Kan
Chen, Kun-Mu
Chen, Yudong
Chuang, Kuei
Clark, John R(ay)
Conway, Lynn Ann
Davidson, Edward S(teinberg)
Dow, W(illiam) G(ould)
Dubes, Richard C
Eesley, Gary
Enns, Mark K
Fienup, James R
Frank, Max
Freedman, M(orris) David
Gaerttner, Martin R
Gagnon, Eugene Gerald
Getty, Ward Douglas
Ghoneim, Youssef Ahmed
Haddad, George I
Hanysz, Eugene Arthur
Hassoun, Mohamad Hussein
Hemdal, John Frederick
Hiatt, Ralph Ellis
Hillig, Kurt Walter, II
Ho, Bong
Holland, Steven William
Islam, Mohammed N
Jacovides, Linos J
Jain, Anil Kumar
Jones, R(ichard) James
Kerr, William
Khargonekar, Pramod P
Klingler, Eugene H(erman)
Knoll, Alan Howard
Koenig, Herman E
Kreer, John B(elshaw)
Kulkarni, Anand K
LaHaie, Ivan Joseph
LaRocca, Anthony Joseph
Larrowe, Boyd T
Larrowe, Vernon L
Lewis, Robert Warren
Lomax, Ronald J(ames)
Luxon, James Thomas
MacGregor, Donald Max
McMahon, E(dward) Lawrence
Macnee, Alan B(reck)
Malaczynski, Gerard W
Meisel, Jerome
Miller, M(urray) H(enri)
Minck, Robert W
Nagy, Andrew F
Nagy, Louis Leonard
Nevius, Timothy Alfred
Nyquist, Dennis Paul
Ohlsson, Robert Louis
Park, Gerald L(eslie)
Peck, D Stewart
Pendleton, Wesley William
Piatkowski, Thomas Frank
Polis, Michael Philip
Reid, Richard J(ames)
Reitan, Daniel Kinseth
Ribbens, William B(ennett)
Rigas, Harriett B
Rohde, Steve Mark
Roll, James Robert
Salam, Fathi M A
Sattinger, Irvin J(ack)
Schubring, Norman W(illiam)
Sebastian, Franklin W
Sellers, Ernest E(dwin)
Sharpe, Charles Bruce
Sloan, Martha Ann
Smith, Robert W
Soper, Jon Allen
Stocker, Donald V(ernon)
Tai, Chen-To
Turner, Robert
Ulaby, Fawwaz Tayssir
Upatnieks, Juris
Vander Kooi, Lambert Ray
Vaz, Nuno A
Von Tersch, Lawrence W
Wang, Charles C(hen-Ding)
Weil, Herschel
Weng, Tung Hsiang
Wennerberg, A(llan) L(orens)

Williams, William James
Witt, Howard Russell
Zanini-Fisher, Margherita

MINNESOTA
Albertson, Vernon D(uane)
Bailey, Fredric Nelson
Champlin, Keith S(chaffner)
Cohen, Arnold A
Davidson, Donald
Dinneen, Gerald Paul
Dixon, John D(ouglas)
Du, David Hung-Chang
DuBois, John R(oger)
Geddes, John Joseph
Geokezas, Meletios
Georgiou, Tryphon T
Gilbert, Barry Kent
Graves, Wayne H(aigh)
Greene, Christopher Storm
Gruber, Carl L(awrence)
Harvey, Charles Arthur
Haxby, B(ernard) V(an Loan)
Higgins, Robert Arthur
Hocker, George Benjamin
Holte, James Edward
Kain, Richard Yerkes
Kaveh, Mostafa
Keitel, Glenn H(oward)
Killpatrick, Joseph E
Kinney, Larry Lee
Krueger, Jack N
Kumar, K S P
La Bonte, Anton Edward
Lambert, Robert F
Lea, Wayne Adair
Lee, Gordon M(elvin)
Lee, T(homas) S(hao-Chung)
Lindemann, Wallace W(aldo)
Lo, David S(hih-Fang)
McLane, Robert Clayton
Mohan, Narendra
Mooers, Howard T(heodore)
Nelson, Gerald Duane
Olyphant, Murray, Jr
Pinckaers, B(althasar) Hubert
Riaz, M(ahmoud)
Robbins, William Perry
Sackett, W(illiam) T(ecumseh), Jr
Sadjadi, Firooz Ahmadi
Schmitz, Harold Gregory
Shepherd, W(illiam) G(erald)
Smith, David Philip
Stephanedes, Yorgos Jordan
Thiede, Edwin Carl
Timm, Gerald Wayne
Tran, Nang Tri
Tufte, Obert Norman
Wollenberg, Bruce Frederick

MISSISSIPPI
Ball, Billie Joe
Elsherbeni, Atef Zakaria
Ferguson, Mary Hobson
Fry, Thomas R
Gassaway, James D
Glisson, Allen Wilburn, Jr
Hanson, Donald Farness
Herring, John Wesley, Jr
Ingels, Franklin M(uranyi)
Jacob, Paul B(ernard), Jr
Kajfez, Darko
Kishk, Ahmed A
McDaniel, Willie L(ee), Jr
Miller, David Burke
Pearce, David Harry
Ramsdale, Dan Jerry
Simrall, Harry C(harles) F(leming)
Smith, Charles Edward
Wier, David Dewey

MISSOURI
Adams, Gayle E(ldredge)
Betten, J Robert
Bialecke, Edward P
Brown, Lloyd Robert
Carson, Ralph S(t Clair)
Carter, Robert L(eroy)
Combs, Robert Glade
Constant, Paul C, Jr
Cox, Jerome R(ockhold), Jr
Denman, Eugene D(ale)
Dreifke, Gerald E(dmond)
DuBroff, Richard Edward
Everhart, James G
Farmer, John William
Glaenzer, Richard H
Goldstein, Julius L
Graham, Robert Reavis
Gruber, B(ernard) A
Hahn, James H
Harbourt, Cyrus Oscar
Harmon, Robert Wayne
Hicks, Patricia Fain
Hoft, Richard Gibson
Honnell, Pierre M(arcel)
Huddleston, Philip Lee
Indeck, Ronald S
Kimball, Charles Newton
Kline, Raymond Milton
Koopman, Richard J(ohn) W(alter)
Landiss, Daniel Jay
Lemon, Leslie Roy
Lindsay, James Edward, Jr

McFarland, William D
McLaren, Robert Wayne
McMillan, Gregory Keith
McPherson, George, Jr
Mak, Sioe Tho
Marshall, Stanley V(ernon)
Martin, Edwin J, Jr
Morley, Robert Emmett, Jr
Moss, Randy Hays
Muller, Marcel Wettstein
Nau, Robert H(enry)
Rao, Vittal Srirangam
Richards, Earl Frederick
Rosenbaum, Fred J(erome)
Ross, Donald K(enneth)
Ross, Monte
Ryan, Carl Ray
Sherman, Byron Wesley
Shrauner, Barbara Abraham
Slivinsky, Charles R
Sparlin, Don Merle
Spielman, Barry
Stanek, Eldon Keith
Tudor, James R
Tyrer, Harry Wakeley
Waid, Rex A(dney)
Waidelich, D(onald) L(ong)
Winter, David F(erdinand)
Wolfe, Charles Morgan
Wu, Cheng-Hsiao
Zobrist, George W(inston)

MONTANA
Bennett, Byron J(irden)
Durnford, Robert F(red)
Gerez, Victor
Hanton, John Patrick
Knox, James L(ester)
Manhart, Robert (Audley)
Pierre, Donald Arthur

NEBRASKA
Bahar, Ezekiel
Bashara, N(icolas) M
Chung, Hui-Ying
Dillon, Rodney O
Edison, Allen Ray
Hultquist, Paul F(redrick)
Lagerstrom, John E(mil)
Myers, Grant G
Nelson, Don Jerome
Olive, David Wolph
Robison, Wendall C(loyd)
Seth, Sharad Chandra
Soukup, Rodney Joseph
Stetson, LaVerne Ellis
Voelker, Robert Heth
Woods, Joseph
Woollam, John Arthur

NEVADA
Betts, Attie L(ester)
Brogan, William L
Messenger, George Clement
Miel, George J
Rawat, Banmali Singh
Tryon, John G(riggs)

NEW HAMPSHIRE
Adjemian, Haroutioon
Blesser, Barry Allen
Brous, Don W
Buffler, Charles Rogers
Carroll, Lee Francis
Clark, Ronald Rogers
Crane, Robert Kendall
Crowell, Merton Howard
Glanz, Filson H
Hough, Richard R(alston)
Hraba, John Burnett
Hutchinson, Charles E(dgar)
Juhola, Carl
Klarman, Karl J(oseph)
Laaspere, Thomas
Melvin, Donald Walter
Murdoch, Joseph B(ert)
Pokoski, John Leonard
Scott, Ronald E
Seifer, Arnold David
Simmons, Alan J(ay)
Tillinghast, John Avery
Von-Recklinghausen, Daniel R

NEW JERSEY
Amitay, Noach
Anderson, G(eorge) M(ullen)
Anderson, Robert E(dwin)
Anderson, Stanley William
Andrekson, Peter A(vo)
Andrews, Frederick T, Jr
Ashcroft, Dale Leroy
Assadourian, Fred
Atal, Bishnu S
Audet, Sarah Anne
Ball, William Henry Warren
Bean, John Condon
Beard, Richard B
Becken, Eugene D
Bernstein, Alan D
Bernstein, Lawrence
Bhagat, Pramode Kumar
Bharj, Sarjit Singh
Bitzer, Richard Allen
Blanck, Andrew R(ichard)

Boesch, Francis Theodore
Bond, Robert Harold
Bondy, Michael F
Boyd, Gary Delane
Brandinger, Jay J
Bridges, Thomas James
Brilliant, Martin Barry
Brown, Irmel Nelson
Brownell, R(ichard) M(iller)
Bucher, T(homas) T(albot) Nelson
Buchner, Morgan M(allory), Jr
Butler, Herbert I
Cammarata, John
Capasso, Federico
Carter, Ashley Hale
Castenschiold, Rene
Chen, Fang Shang
Cheo, Li-hsiang Aria S
Chernak, Jess
Cheung, Shiu Ming
Chirlian, Paul M(ichael)
Cho, Alfred Y
Chu, Ta-Shing
Cimini, Leonard Joseph, Jr
Clements, Wayne Irwin
Cohen, Barry George
Cohen, Edwin
Coleman, Norman P, Jr
Cooper, Paul W
Cornell, W(arren) A(lvan)
Coughran, William Marvin, Jr
Crane, Roger L
Crochiere, Ronald E
Curtice, Walter R
Curtis, Thomas Hasbrook
Damen, Theo C
Davey, James R
David, Edward E(mil), Jr
Davis, William F
DeMarinis, Bernard Daniel
Denes, Peter B
Denlinger, Edgar J
Denno, Khalil I
Denton, Richard T
Detig, Robert Henry
Dickerson, L(oren) L(ester), Jr
Dickinson, Bradley William
Dorros, Irwin
Drucker, Harris
Duttweiler, Donald Lars
Eager, George S, Jr
Easton, Elmer C(harles)
Eldumiati, Ismail Ibrahim
Elmendorf, Charles Halsey, III
Estrada, Herbert
Feiner, Alexander
Fich, Sylvan
Field, George Robert
Fischer, Russell Jon
Fishbein, William
Flanagan, James L(oton)
Flores, Ivan
Flory, Leslie E(arl)
Forys, Leonard J
Franks, Richard Lee
Friedland, Bernard
Fukui, H(atsuaki)
Gabriel, Edwin Z
Gelnovatch, V
Giuffrida, Thomas Salvatore
Gloge, Detlef Christoph
Goodfriend, Lewis S(tone)
Gummel, Hermann K
Haas, Zygmunt
Hall, Joseph L
Hang, Hsueh-Ming
Hartung, John
Haskell, Barry G
Hazelrigg, George Arthur, Jr
Heffes, Harry
Heirman, Donald N
Hershenov, B(ernard)
Herskowitz, Gerald Joseph
Hilibrand, J(ack)
Himmel, Leon
Holman, Wayne J(ames), Jr
Hong, Won-Pyo
House, Arthur Stephen
Houston, Vern Lynn
Hsu, H(wei) P(iao)
Hsuan, Hulbert C S
Jacquin, Arnaud Eric
Jagerman, David Lewis
Jain, Sushil C
Jayant, Nuggehally S
Joel, Amos Edward, Jr
Johannes, Virgil Ivancich
Johnston, Wilbur Dexter, Jr
Kaiser, James F(rederick)
Kaiser, Peter
Karol, Mark J
Katz, Sheldon Lane
Kessler, Frederick Melvyn
Kessler, Irving Jack
Kieburtz, R(obert) Bruce
King, B(ernard) G(eorge)
Klapper, Jacob
Kleckner, Willard R
Kluver, J(ohan) W(ilhelm)
Kobayashi, Hisashi
Koch, Thomas L
Kovats, Andre
Krevsky, Seymour
Krzyzanowski, Paul

Kuhl, Frank Peter, Jr
Kunhardt, Erich Enrique
Kuo, Ying L
Labianca, Frank Michael
Langseth, Rollin Edward
Lasky, Jack Samuel
Lechner, Bernard J
Lee, Keun Myung
Lee, T(ien) P(ei)
Lemberg, Howard Lee
Leu, Ming C
Levine, Nathan
Li, John Kong-Jiann
Li, Tingye
Lichtenberger, Gerald Burton
Lin, Chinlon
Lin, Dong Liang
Lindberg, Craig Robert
Liu, Bede
Liu, S(hing) G(ong)
Lo, Arthur W(unien)
Lopresti, Philip V(incent)
Lucky, Robert W
Maa, Jer-Shen
McDermott, Kevin J
McLane, George Francis
McNair, Irving M, Jr
Malkin, Robert Allen
Manz, August Frederick
Marcatili, Enrique A J
Marcus, Robert Boris
Mattes, Hans George
Mauzey, Peter T
Mayhew, Thomas R
Mayo, John S
Meola, Robert Ralph
Mersten, Gerald Stuart
Meyer, Andrew U
Minter, Jerry Burnett
Misra, Raj Pratap
Mitchell, Olga Mary Mracek
Mondal, Kalyan
Myers, George Henry
Nelson, W(inston) L(owell)
Netravali, Arun Narayan
Newhouse, Russell C(onwell)
Nowogrodzki, M(arkus)
Ohm, E(dward) A(llen)
Olive, Joseph P
Olken, Melvin I
Olmstead, John Aaron
O'Neill, Eugene F
Osifchin, Nicholas
Padalino, Joseph John
Parkins, Bowen Edward
Partovi, Afshin
Patton, W(illard) T(homas)
Personick, Stewart David
Pfafflin, James Reid
Pinson, Elliot N
Poor, Harold Vincent
Powers, K(erns) H(arrington)
Quinlan, Daniel A
Rabiner, Lawrence Richard
Rainal, Attilio Joseph
Rastani, Kasra
Raybon, Gregory
Reisman, Stanley S
Riggs, Victoria G
Rips, E(rvine) M(ilton)
Rosenberg, Aaron E(dward)
Rosenstark, Sol(omon)
Rosenthal, Lee
Rosenthal, Louis Aaron
Ross, Ian M(unro)
Rothschild, Gilbert Robert
Rowe, H(arrison) E(dward)
Rubinstein, Charles B(enjamin)
Russell, Frederick A(rthur)
Sagal, Matthew Warren
Saleh, Adel Abdel Moneim
Schaefer, Jacob W
Schiff, Leonard Norman
Schink, F E
Schmidt, Barnet Michael
Scholz, Lawrence Charles
Schulke, Herbert Ardis, Jr
Schwartz, Stuart Carl
Seman, George William
Shaer, Norman Robert
Shamis, Sidney S
Shea, Timothy Edward
Shefer, Joshua
Shen, Tek-Ming
Sherman, Ronald
Sherrick, Carl Edwin
Shulman, Herbert Byron
Silver, Howard I(ra)
Slepian, David
Sloane, Neil James Alexander
Smith, Peter William E
Sohn, Kenneth S (Kyu Suk)
Sontag, Eduardo Daniel
Stach, Joseph
Stampfl, Rudolf A
Standley, Robert Dean
Stevens, James Everell
Strano, Joseph J
Strauss, Walter
Sublette, Ivan H(ugh)
Suozzi, Joseph John
Sywe, Bei-Shen
Taylor, George William
Thelin, Lowell Charles

314 / DISCIPLINE INDEX

Electrical Engineering (cont)

Thomas, David Gilbert
Thomas, Gary Lee
Thouret, Wolfgang E(mery)
Tien, P(ing) K(ing)
Tow, James
Tsang, Won
Uhrig, Jerome Lee
Urkowitz, Harry
Vahaviolos, Sotirios J
Valenzuela, Reinaldo A
Vallese, Lucio M(ario)
Verdu, Sergio
Verma, Pramode Kumar
Von Aulock, Wilhelm Heinrich
Walsh, Teresa Marie
Weinberg, Louis
Weisbecker, Henry B
Weissenburger, Don William
Whitman, Gerald Martin
Wier, Joseph M(arion)
Wilkinson, W(illiam) C(layton)
Williams, Maryon Johnston, Jr
Williams, Thomas R(ay)
Winthrop, Joel Albert
Wood, Obert Reeves, II
Wyndrum, Ralph William, Jr
Wyner, Aaron D
Xie, Ya-Hong
Yadvish, Robert D
Yi-Yan, Alfredo
Young, William Rae
Yue, On-Ching

NEW MEXICO

Adler, Richard John
Anderson, Richard Ernest
Baty, Richard Samuel
Baum, Carl E(dward)
Bolie, Victor Wayne
Bradshaw, Martin Daniel
Davis, L(loyd) Wayne
DeVries, Ronald Clifford
Diver, Richard Boyer, Jr
Dorato, Peter
Doss, James Daniel
Duncan, Richard H(enry)
Elliott, Dana Edgar
Engelhardt, Albert George
Erteza, Ahmed
Feiertag, Thomas Harold
Freeman, Bruce L, Jr
Fronek, Donald Karel
Giles, Michael Kent
Gover, James E
Grannemann, W(ayne) W(illis)
Griffin, Patrick J
Gwyn, Charles William
Hardin, James T
Harrison, Charles Wagner, Jr
Humphries, Stanley, Jr
Jameson, Robert A
Jamshidi, Mohammad Mo
Joppa, Richard M
Karni, Shlomo
Kazda, Louis F(rank)
Kelley, Gregory M
Knudsen, Harold Knud
Linford, Rulon Kesler
Lucky, George W(illiam)
Maez, Albert R
Milton, Osborne
Morgan, Johnie Derald
Myers, David Richard
Newby, Neal D(ow)
Noel, Bruce William
Novak, James Lawrence
Owen, Thomas E(dwin)
Paul, Robert Hugh
Rachele, Henry
Sandmeier, Henry Armin
Schueler, Donald G(eorge)
Short, Wallace W(alter)
Stevens, Roger Templeton
Tallerico, Paul Joseph
Tapp, Charles Millard
Taylor, Javin Morse
Thompson, Wiley Ernest
Tsao, Jeffrey Yeenien
VanDevender, John Pace
Weber, Harry A
Wecksung, George William
Welber, Irwin
Whitman, Rollin Lawrence
Wiggins, Carl M
Williams, James D
Wing, Janet E (Sweedyk) Bendt
Yonas, Gerold

NEW YORK

Abetti, Pier Antonio
Adams, Arlon Taylor
Adler, Michael Stuart
Agarwal, Ramesh Chandra
Agrawal, Govind P(rasad)
Ahmed-Zaid, Said
Ahn, K(ie) Y(eung)
Albicki, Alexander
Alfano, Robert R
Almasi, George Stanley
Anagnostopoulos, Constantine N
Anantha, Narasipur Gundappa
Anastassiou, Dimitris
Anderson, John Bailey
Anderson, John M(elvin)
Angevine, Oliver Lawrence
Ankrum, Paul Denzel
Apai, Gustav Richard, II
Apte, Chidanand
Arden, Bruce Wesley
Bachman, Henry L(ee)
Baertsch, Richard D
Bahl, Lalit R(ai)
Balabanian, Norman
Ballantyne, Joseph M(errill)
Barthold, Lionel O
Baule, Gerhard M
Baum, Eleanor Kushel
Berger, Toby
Bewley, Loyal V
Blanchard, Fletcher A(ugustus), Jr
Blazey, Richard N
Blesser, William B
Bocko, Mark Frederick
Bolgiano, Ralph, Jr
Bolle, Donald Martin
Bond, Norman T(uttle)
Boorstyn, Robert Roy
Borrego, Jose M
Brennemann, Andrew E(rnest), Jr
Brenner, Egon
Bresler, Aaron D
Brock, Robert H, Jr
Brower, Allen S(pencer)
Brown, Burton Primrose
Brown, Dale Marius
Buchta, John C(harles)
Bulusu, Dutt V
Butler, Walter John
Caprio, James R
Carlson, A Bruce
Cassedy, Edward S(pencer), Jr
Chan, Jack-Kang
Chang, Ifay F
Chang, Sheldon S L
Chass, Jacob
Chen, Arthur Chih-Mei
Chen, Cheng-Lin
Cheo, Bernard Ru-Shao
Chestnut, H(arold)
Chow, Chao K
Chow, Tat-Sing Paul
Christiansen, Donald David
Chubb, Charles F(risbie), Jr
Clarke, Kenneth Kingsley
Close, Charles M(ollison)
Cohen, Gerald H(oward)
Cole, James A
Colin, Lawrence
Constable, James Harris
Cornacchio, Joseph V(incent)
Costello, Richard Gray
Cotellessa, Robert F(rancis)
Cottingham, James Garry
Craft, Harold Dumont, Jr
Craig, Edward J(oseph)
Dalman, G(isli) Conrad
Daskam, Edward, Jr
Debany, Warren Harding, Jr
De France, Joseph J
Dehn, Rudolph A(lbert)
Del Toro, Vincent
De Mello, F Paul
Demerdash, Nabeel A O
Dennard, Robert H
DeRusso, Paul M(adden)
Desrochers, Alan Alfred
Diament, Paul
Diamond, Fred Irwin
Dickey, Frank R(amsey), Jr
Difranco, Julius V
Dill, Frederick H(ayes), Jr
Duggin, Michael J
Duncan, Charles Clifford
Eastman, Dean Eric
Eastman, Lester F(uess)
Eichelberger, Edward B
Eichmann, George
Eisenberg, Lawrence
Elbaum, Marek
Ellert, Frederick J
Ellis, William Rufus
Erickson, William Harry
Eveleigh, Virgil W(illiam)
Everett, Woodrow W, III
Farley, Donald T, Jr
Feiker, George E(dward), Jr
Feisel, Lyle Dean
Felsen, Leopold Benno
Feth, George C(larence)
Fiedler, Harold Joseph
Fink, Donald G(len)
Florman, Monte
Fordon, Wilfred Aaron
Forster, William H(all)
Forsyth, Eric Boyland
Fowler, James A
Frederick, Dean Kimball
Frisch, I(van) T
Fuller, Lynn Fenton
Geiss, Gunther R(ichard)
George, Nicholas
Gerace, Paul Louis
Gerhardt, Lester A
Ghozati, Seyed-Ali
Gildersleeve, Richard E
Giordano, Anthony B(runo)
Givone, Donald Daniel
Glasford, Glenn M(ilton)
Glinkowski, Mietek T
Goldsmith, Paul Felix
Goldstein, Lawrence Howard
Gonzalez, Edgar
Goodheart, Clarence F(rancis)
Gran, Richard J
Grant, Ian S
Gratian, J(oseph) Warren
Greenwood, Allan Nunns
Groner, Carl Fred
Gronner, A(lfred) D(ouglas)
Grover, Paul L, Jr
Gruenberg, Harry
Gutmann, Ronald Jay
Haddad, Richard A
Hall, Richard Travis
Hammam, M Shawky
Happ, Harvey Heinz
Harley, Naomi Hallden
Harrington, Dean Butler
Harrington, Roger F(uller)
Harris, Bernard
Harris, Lawson P(arks)
Harvey, Alexander Louis
Hauspurg, Arthur
Hayter, Walter R, Jr
Hedman, Dale E
Hesse, M(ax) H(arry)
Hessel, Alexander
Honsinger, Vernon Bertram
Hood, Jerry A
Hopcroft, John E(dward)
Houde, Robert A
Hsiang, Thomas Y
Hsu, Charles Ching-Hsiang
Hsu, Donald K
Hu, Ming-Kuei
Huening, Walter C, Jr
Hunt, Donald F(ulper)
Hyman, Abraham
Indusi, Joseph Paul
Jabbour, Kamal
Jacobsen, William E
Jatlow, J(acob) L(awrence)
Javid, Mansour
Jefferies, Michael John
Jelinek, Frederick
Jocoy, Edward Henry
Johnson, Charles Richard, Jr
Johnson, I Birger
Jones, Leonidas John
Kadin, Alan Mitchell
Kaneko, Hisashi
Karatzas, Ioannis
Karnaugh, Maurice
Kear, Edward B, Jr
Keller, Arthur Charles
Kelley, Michael C
Kent, Gordon
Ketchen, Mark B
Kinnen, Edwin
Kirchgessner, Joseph L
Kirchmayer, Leon K
Kiszenick, Walter
Kletsky, Earl J(ustin)
Kliman, Gerald Burt
Koehler, Richard Frederick, Jr
Komiak, James Joseph
Koplowitz, Jack
Kopp, Richard E
Korwin-Pawlowski, Michael Lech
Kranc, George M(aximilian)
Krembs, G(eorge) M(ichael)
Kwok, Thomas Yu-Kiu
Laghari, Javaid Rasoolbux
La Tourrette, James Thomas
Lauber, Thornton Stuart
Lee, Robert E
Leland, Harold R(obert)
Le Page, Wilbur R(eed)
Levi, Enrico
Levine, Alfred Martin
Ley, B James
Li, Zheng
Linder, Clarence H
Lindley, Kenneth Eugene
Linke, Simpson
Litman, Richard
Litynski, Daniel Mitchell
Liu, Mao-Zu
Lounsbury, John Baldwin
Low, Paul R
Lurch, E(dward) Norman
Lurkis, Alexander
Ma, William Hsioh-Lien
McCary, Richard O
McDonald, John F(rancis)
MacDonald, Noel C
McGaughan, Henry S(tockwell)
McGrath, Eugene R
McIsaac, Paul R(owley)
McMurry, William
Madhu, Swaminathan
Mahrous, Haroun
Maling, George Croswell, Jr
Malone, Dennis P(hilip)
Maltz, Martin Sidney
Margolis, Stephen G(oodfriend)
Marsocci, Velio Arthur
Mathes, Kenneth Natt
Matick, Richard Edward
Mayer, James W(alter)
Meadows, Henry E(merson), Jr
Merriam, Charles Wolcott, III
Meth, Irving Marvin
Mgrdechian, Raffee
Middleton, David
Mihran, Theodore Gregory
Modestino, James William
Moghadam, Omid A
Morris, George Michael
Morris, James Eliot
Morris, Thomas Wilde
Mosteller, Henry Walter
Mundel, August B(aer)
Nagy, George
Nation, John
Nawrocky, Roman Jaroslaw
Nelson, John Keith
Nichols, Benjamin
Nisteruk, Chester Joseph
Oden, Peter Howland
Oh, Se Jeung
O'Kean, Herman C
Padegs, Andris
Pardee, Otway O'Meara
Parker, Kevin James
Parks, Thomas William
Patel, Sharad A
Paul, George, Jr
Pavlidis, Theo
Paxton, K Bradley
Peng, Song-Tsuen
Perry, Michael Paul
Peskin, Arnold Michael
Peters, Philip H
Pflug, Donald Ralph
Pivnichny, John R
Plotkin, Martin
Possin, George Edward
Pottle, Christopher
Powell, Noble R
Quan, Weilun
Raghuveer, Mysore R
Rappaport, Stephen S
Richardson, Robert Lloyd
Richman, Donald
Ringlee, Robert J
Robinson, Charles C(anfield)
Rodems, James D
Rootenberg, Jacob
Rose, Kenneth
Rosenfeld, Jack Lee
Rosenthal, Saul W
Ross, John B(ye)
Rothschild, David (Seymour)
Rubin, Bruce Joel
Rubin, Joel E(dward)
Ruehli, Albert Emil
Rushing, Allen Joseph
Russo, Roy Lawrence
Ryerson, Joseph L
Sanderson, Arthur Clark
Sandler, Melvin
Sanford, Richard Selden
Sarachik, Philip E(ugene)
Sarjeant, Walter James
Schanker, Jacob Z
Schlesinger, S Perry
Schultz, William C(arl)
Schwartz, Mischa
Schwartz, Richard F(rederick)
Schwartzman, Leon
Schwarz, Ralph J
Sen, Amiya K
Sen, Ujjal
Shamash, Yacov A
Shapiro, Sidney
Sharbaugh, Amandus Harry
Shaw, David T
Shaw, Leonard G
Shen, Wu-Mian
Shinners, Stanley Marvin
Shmoys, Jerry
Shulman, Carl
Silverman, Bernard
Simpson, Murray
Slade, Bernard Newton
Slaymaker, Frank Harris
Smilowitz, Bernard
Smith, Albert A, Jr
Smith, David R(ichard)
Smith, Edward J(oseph)
Smith, Joseph H, Jr
Smith, Robert L
Snelling, Christopher
Sobel, Kenneth Mark
Spang, H Austin, III
Sri-Jayantha, Sri Muthythamby
Stancampiano, Charles Vincent
Staples, Basil George
Star, Joseph
Stern, Miklos
Strait, Bradley Justus
Stratton, Roy Franklin, Jr
Strauss, Leonard
Strohm, Warren B(ruce)
Stutt, Charles A(dolphus)
Su, Stephen Y H
Suenaga, Masaki
Tamir, Theodor
Tang, Donald T(ao-Nan)
Tanguay, A(rmand) R(ene)
Taub, Herbert
Taub, Jesse J
Taylor, James Hugh

Tehon, Stephen Whittier
Teich, Malvin Carl
Thau, Frederick E
Thiel, Frank L(ouis)
Thomann, Gary C
Thorp, James Shelby
Timms, Robert J
Titlebaum, Edward Lawrence
Truxal, John G(roff)
Tschang, Pin-Seng
Tseng, Fung-I
Tsividis, Yannis P
Tuan, Hang-Sheng
Tuel, William Gole, Jr
Unger, S(tephen) H(erbert)
Van Der Voorn, Peter C
Van Wagner, Edward M
Varshney, Pramod Kumar
Viertl, John Ruediger Mader
Vrana, Norman M
Waag, Robert Charles
Wang, Wen I
Wang, Xingwu
Watson, P(ercy) Keith
Weissman, David E(verett)
Welsh, James P
Weng, Wu Tsung
Westover, Thomas A(rchie)
Whalen, James Joseph
Wheeless, Leon Lum, Jr
Wiley, Richard G
Wilson, Delano D
Wilson, James William Alexander
Winkler, Leonard P
Winsor, Lauriston P(earce)
Winter, Herbert
Witt, Christopher John
Wobschall, Darold C
Wolf, Edward D
Wong, Edward Chor-Cheung
Wood, Allen John
Woodard, David W
Woods, John William
Worthing, Jurgen
Wouk, Victor
Wysocki, Joseph J(ohn)
Yao, David D
Yasar, Tugrul
York, Raymond A
Youla, Dante C
Yu, Hwa Nien
Yu, Kai-Bor
Zeidenbergs, Girts
Zemanian, Armen Humpartsoum
Zhou, Zhen-Hong
Zic, Eric A
Zimmerman, Stanley William
Zukowski, Charles Albert

NORTH CAROLINA
Agrawal, Dharma Prakash
Allen, Charles Michael
Barclay, W(illiam) J(ohn)
Bedair, Salah M
Bell, Norman R(obert)
Bellack, Jack H
Bitzer, Donald L
Burton, Ralph Gaines
Casey, Horace Craig, Jr
Chou, Wushow
Coleman, Robert J(oseph)
Dickson, LeRoy David
Easter, William Taylor
Eckels, Arthur R(aymond)
Enquist, Paul Michael
Fair, Richard Barton
Flood, Walter A(loysius)
Franzon, Paul Damian
Gelenbe, Erol
Gordy, Thomas D(aniel)
Grady, Perry Linwood
Hacker, Herbert, Jr
Harris, Roy H
Hauser, John Reid
Hester, Donald L
Hoadley, George B(urnham)
Jones, Creighton Clinton
Kauffman, James Frank
Kim, Chang-Sik
Kirschbaum, H(erbert) S(pencer)
Lamorte, Michael Francis
Leech, Robert Leland
Lovick, Robert Clyde
McAlpine, George Albert
McCumber, Dean Everett
Masnari, Nino A
Massoud, Hisham Z
Matthews, N(eely) F(orsyth) J(ones)
Mink, James Walter
Monteith, Larry King
Moore, Edward T(owson)
Nagle, H Troy
Nolte, Loren W
O'Neal, John Benjamin, Jr
Owen, Harry A(shton), Jr
Peterson, David West
Pilkington, Theo C(lyde)
Redmon, John King
Reenstra, Arthur Leonard
Reinhart, Stanley E, Jr
Rhodes, Donald R(obert)
Rubloff, Gary W
Samuels, Robert Lynn
Sander, William August, III

Sandok, Paul Louis
Simons, Mayrant, Jr
Smith, Charles Eugene
Snyder, Wesley Edwin
Stafford, Bruce H(ollen)
Steer, Michael Bernard
Struve, William George
Suttle, Jimmie Ray
Tischer, Frederick Joseph
Tommerdahl, James B
Trussell, Henry Joel
Vander Lugt, Anthony
Wayne, Burton Howard
Weber, Ernst
Wilson, Thomas G(eorge)
Wright, William V(aughn)
Wylie, C J

NORTH DAKOTA
Anderson, Edwin Myron
Bares, William Anthony
Bertnolli, Edward Clarence
Kuruganty, Sastry P
Tareski, Val Gerard
Thomforde, C(lifford) J(ohn)
Wang, Jun
Yuvarajan, Subbaraya

OHIO
Adams, William Lawreson
Banda, Siva S
Barranger, John P
Batcher, Kenneth Edward
Bibyk, Steven B
Brandeberry, James E
Brown, Frank Markham
Buckingham, William Thomas
Burghart, James H(enry)
Burhans, Ralph W(ellman)
Campbell, Richard M
Ceasar, Gerald P
Chen, Chiou Shiun
Chen, Hollis C(hing)
Chipman, R(obert) A(very)
Chovan, John David
Colbert, Charles
Collin, Robert E(manuel)
Compton, Ralph Theodore, Jr
Cornetet, Wendell Hillis, Jr
Cowan, John D, Jr
Cruz, Jose B(ejar), Jr
Dakin, James Thomas
Damon, Edward K(ennan)
Dayton, James Anthony, Jr
D'Azzo, John Joachim
Dell'Osso, Louis Frank
Dorsey, Robert T
El-Naggar, Mohammed Ismail
Eltimsahy, Adel H
Emmerling, Edward J
Engelmann, Richard H(enry)
Eppers, William C
Ernst, George W
Essman, Joseph Edward
Ewing, Donald J(ames), Jr
Faiman, Robert N(eil)
Falk, Harold Charles
Farison, James Blair
Fenton, Robert E
Forestieri, Americo F
Galm, James Michael
Garbacz, Robert J
Ghandakly, Adel Ahmad
Golovin, Michael N
Goradia, Chandra P
Grant, Michael P(eter)
Gruber, Sheldon
Grumbach, Robert S(tephen)
Hall, Ernest Lenard
Hancock, Harold E(dwin)
Harpster, Joseph
Hazen, Richard R(ay)
Hemami, Hooshang
Hill, Jack Douglas
Hobart, William Chester, Jr
Hodge, Daniel B
Hogan, James J(oseph)
Holmes, Roger Arnold
Horing, Sheldon
Horton, Billy Mitchusson
Hsu, Hsiung
Ismail, Amin Rashid
Jackson, Warren, Jr
Johnson, Ray O
Jordan, Howard Emerson
Kapoor, Vikram J
Kashkari, Chaman Nath
Kazimierczuk, Marian K
Keister, James E
Klock, Harold F(rancis)
Ko, H(sien) C(hing)
Koozekanai, Said H
Kramer, Raymond Edward
Kramerich, George L
Kraus, John Daniel
Ksienski, A(haron)
Kwatra, Subhash Chander
Lamont, Gary Byron
Lawler, Martin Timothy
Lawton, John G
Lee, Kai-Fong
Lemke, Ronald Dennis
Levis, C(urt) A(lbert)
Lewis, Donald Edward

Long, Ronald K(illworth)
Lott, James Anthony
McFarland, Richard Herbert
McGoron, Anthony Joseph
McNair, Robert J(ohn), Jr
Manning, Robert M
Mayhan, Robert J(oseph)
Mehalic, Mark Andrew
Menq, Chia-Hsiang
Mergler, H(arry) W(inston)
Middendorf, William H
Moffatt, David Lloyd
Moon, Donald Lee
Mudry, Karen Michele
Munk, Benedikt Aage
Neuman, Michael R
Olson, Karl William
Osterbrock, Carl H
Pachter, Meir
Pao, Yoh-Han
Papachristou, Christos A
Pathak, Prabhakar H
Peters, Leon, Jr
Puglielli, Vincent George
Qureshi, A H
Raible, Robert H(enry)
Rajsuman, Rochit
Ramamoorthy, Panapakkam A
Rattan, Kuldip Singh
Repperger, Daniel William
Restemeyer, William Edward
Richmond, J(ack) H(ubert)
Ringel, Steven Adam
Rose, Harold Wayne
Rose, Mitchell
Ross, Lawrence John
Rowe, Joseph E(verett)
Schell, Allan Carter
Schultz, Ronald G(len)
Sebo, Stephen Andrew
Shenoi, Belle Anantha
Shirk, B(rian) Thomas
Sibila, Kenneth Francis
Skinner, John Paul
Soska, Geary Victor
Springer, Robert Harold
Steckl, Andrew Jules
Stewart, John
Strnat, Karl J
Stuart, Thomas Andrew
Thiele, Gary Alan
Thomas, Cecil Wayne
Thorbjornsen, Arthur Robert
Thurston, M(arlin) O(akes)
Trautman, DeForest L Woody
Tsui, James Bao-Yen
Vassell, Gregory S
Waldron, Manjula Bhushan
Ware, Brendan J
Warren, Claude Earl
Webster, Allen E
Wee, William Go
Weed, Herman Roscoe
Weimer, F(rank) Carlin
White, Lee James
Wigington, Ronald L
Yeh, Hsi-Han
Zavodney, Lawrence Dennis
Zheng, Yuan Fang

OKLAHOMA
Baker, James E
Basore, B(ennett) L(ee)
Batchman, Theodore E
Berbari, Edward J
Bilger, Hans Rudolf
Breipohl, Arthur M
Buffum, C(harles) Emery
Cheung, John Yan-Poon
Cole, Charles F(rederick), Jr
Cronenwett, William Treadwell
Cruz, J R
Cummins, Richard L
El-Ibiary, Mohamed Yousif
Engle, Jessie Ann
Kuriger, William Louis
Lal, Manohar
Lee, Samuel C
Lingelbach, D(aniel) D(ee)
McCollom, Kenneth A(llen)
Mulholland, Robert J(oseph)
Ramakumar, Ramachandra Gupta
Richardson, J Mark
Schwenker, J(ohn) E(dwin)
Smith, Gerald Lynn
Strattan, Robert Dean
Tuma, Gerald
VanDerwiele, James Milton
Walker, Gene B(ert)
Williams, Thomas Henry Lee
Yarlagadda, Radha Krishna Rao
Zelby, Leon W

OREGON
Alexander, Gerald Corwin
Annestrand, Stig A
Arthur, John Read, Jr
Berglund, Carl Neil
Blumhagen, Vern Allen
Casperson, Lee Wendel
Chartier, Vernon Lee
Chrzanowska-Jeske, Malgorzata Ewa
Collins, John A(ddison)
Davenport, Wilbur B(ayley), Jr

Ede, Alan Winthrop
Edwards, James Mark
Engelking, Paul Craig
Engle, John Franklin
Gens, Ralph S
Ginsburg, Charles P
Hansen, Hugh J
Hayes, Thomas B
Herzog, James Herman
Holmes, James Frederick
Isaacson, Dennis Lee
Johnston, George I
Jones, Donlan F(rancis)
Lendaris, George G(regory)
Liu-Ger, Tsu-Huei
Magnusson, Philip C(ooper)
Matson, Leslie Emmet, Jr
Mohler, Ronald Rutt
Morgan, Merle L(oren)
Schaumann, Rolf
Schrumpf, Barry James
Short, Robert Allen
Skinner, Richard Emery
Stewart, Bradley Clayton
Stone, Solon Allen
Taylor, Carson William
Temes, Gabor Charles
Tinney, William Frank
Tripathi, Vijai Kumar
Wager, John Fisher
Yamaguchi, Tadanori
Yau, Leopoldo D

PENNSYLVANIA
Abend, Kenneth
Aburdene, Maurice Felix
Adams, Raymond F
Adams, William Sanders
Allen, Charles C(orletta)
Anderson, A Eugene
Arora, Vijay Kumar
Ashok, S
Aston, Richard
Batchelor, John W
Baxter, Ronald Dale
Bedrosian, Samuel D
Berger, Daniel S
Berkowitz, Raymond S
Bilgutay, Nihat Mustafa
Bloor, W(illiam) Spencer
Bockosh, George R
Bose, Nirmal K
Boston, John Robert
Brastins, Auseklis
Brown, Homer E
Brown, John Lawrence, Jr
Buchsbaum, Gershon
Cain, James Thomas
Caplan, Aubrey E
Carpenter, Lynn Allen
Casasent, David P(aul)
Cendes, Zoltan Joseph
Charap, Stanley H
Charp, Solomon
Cohen, Jeffrey M
Colclaser, Robert Gerald
Cookson, Alan Howard
Coren, Richard L
Cross, Leslie Eric
Dahlke, Walter Emil
Das, Mukunda B
Davis, Paul Cooper
Delansky, James F
Denshaw, Joseph Moreau
Detwiler, John Stephen
Dick, George W
Director, Stephen William
Dodds, Wellesley Jamison
Dordick, Herbert S(halom)
Dorny, C Nelson
Douglas, Joseph Francis
Drumwright, Thomas Franklin, Jr
Dumbri, Austin C
Dunn, Charles Nord
Eisenstein, Bruce A
Embree, M(ilton) L(uther)
Engheta, Nader
Etzweiler, George Arthur
Faber, Donald Stuart
Falconer, Thomas Hugh
Fan, Ningping
Feero, William E
Fegley, Kenneth A(llen)
Felker, Jean Howard
Feng, Tse-yun
Finin, Timothy Wilking
Fischer, John Edward
Frost, L(eslie) S(wift)
Furfari, Frank A
Goell, James E(manuel)
Goff, Kenneth W(ade)
Goldwasser, Samuel M
Goodman, Robert M(endel)
Gorham, R C
Gorog, Istvan
Gray, H(arry) J(oshua)
Grimes, Dale M(ills)
Guyker, William C, Jr
Haber, Fred
Hamilton, Howard Britton
Hamilton, William Howard
Harder, Edwin L
Hargens, Charles William, III
Harney, Robert Charles

Electrical Engineering (cont)

Harrold, Ronald Thomas
Herczfeld, Peter Robert
Hielscher, Frank Henning
Hildebrandt, Thomas Henry
Hileman, Andrew R
Hoburg, James Frederick
Hockenberry, Terry Oliver
Hoelzeman, Ronald G
Hollister, Floyd Hill
Hoover, L Ronald
Hounshell, David A
Hutcheson, J(ohn) A(lister)
Hutzler, John R
Hyatt, Robert Monroe
Jaggard, Dwight Lincoln
Jang, Sei Joo
Jeruchim, Michel Claude
Johnson, Alfred Theodore, Jr
Jones, A(ndrew) R(oss)
Jones, Clifford Kenneth
Jordan, Angel G
Kam, Moshe
Karakash, John J
Kassam, Saleem Abdulali
Kaufman, William Morris
Keizer, Eugene O(rville)
Kilgore, Lee A
Kimblin, Clive William
Kingrea, James I, Jr
Klafter, Richard D(avid)
Klein, Gerald I(rwin)
Koepfinger, J L
Koffman, Elliot B
Kritikos, Haralambos N
Kryder, Mark Howard
Ku, Y(u) H(siu)
Kurfess, Thomas Roland
Lakhtakia, Akhlesh
Larky, Arthur I(rving)
Lavi, Abraham
Leas, J(ohn) W(esley)
Leibholz, Stephen W
Lex, R(owland) G(arber), Jr
Li, C(hing) C(hung)
Liberman, Irving
Lory, Henry James
Lowry, Lewis Roy, Jr
McAdam, Will
McCormick, Paul R(ichard)
McCracken, Leslie G(uy), Jr
McCrumm, J(ohn) D(oench)
McGarity, Arthur Edwin
McMurtry, George James
MacRae, Donald Richard
Magison, Ernest Carroll
Maltby, Frederick L(athrop)
Mastascusa, Edward John
Matcovich, Thomas J(ames)
Mathews, C A
Mathews, John David
Mathias, Robert A(ddison)
Meiksin, Zvi H(ans)
Metzner, John J(acob)
Mickle, Marlin Homer
Miller, David L
Milnes, Arthur G(eorge)
Mitchell, John Douglas
Mitchell, Tom M
Molter, Lynne Ann
Mouly, Raymond J
Muir, Patrick Fred
Murphy, John N
Nelson, Richard L(loyd)
Neuman, Charles P(aul)
Nisbet, John S(tirling)
Nygren, Stephen Fredrick
Olson, H(ilding) M
Padulo, Louis
Paulish, Daniel John
Pearce, Charles Walter
Penn, William B
Perks, Norman William
Peterson, Wilbur Carroll
Phares, Alain Joseph
Pittman, G(eorge) F(rank), Jr
Przybysz, John Xavier
Putman, Thomas Harold
Rabii, Sohrab
Riemersma, H(enry)
Robinson, Charles J
Robinson, James William
Rohrer, Ronald A
Ross, William J(ohn)
Rowlands, R(ichard) O(wen)
Roy, Pradip Kumar
Rubin, Herbert
Rubinoff, Morris
Salati, Octavio M(ario)
Schatz, Edward R(alph)
Schroeder, Alfred C(hristian)
Schwan, Herman Paul
Shen, D(avid) W(ei) C(hi)
Shibib, M Ayman
Shobert, Erle Irwin, II
Showers, Ralph M(orris)
Shung, K Kirk
Siewiorek, Daniel Paul
Simaan, Marwan
Smith, David Beach
Smith, Richard G(rant)
Smith, Warren LaVerne
Smock, Dale Owen

Smyth, Michael P(aul)
Spandorfer, Lester M
Spielvogel, Lawrence George
Stein, Jack J(oseph)
Steinberg, Bernard D
Stern, Richard Martin, Jr
Stone, Robert P(orter)
Sun, Hun H
Sutter, Philip Henry
Sverdrup, Edward F
Sze, Tsung Wei
Szedon, John R(obert)
Talvacchio, John
Taylor, Harry Elmer
Thompson, Francis Tracy
Thompson, William, Jr
Tomiyasu, Kiyo
Trainer, Michael Norman
Travis, Irven
Tretiak, Oleh John
Tzeng, Kenneth Kai-Ming
Uher, Richard Anthony
Van Der Spiegel, Jan
Van Gelder, Arthur
Varadan, Vasundara Venkatraman
Varadan, Vijay K
Vogt, William G(eorge)
Voshall, Roy Edward
Wagh, Meghanad D
Wahl, A(rthur) J
Warren, S Reid, Jr
Warshawsky, Jay
Weaver, L(elland) A(ustin) C(harles)
Weeks, L(oraine) H(ubert)
Wemple, Stuart H(arry)
Whitehead, Daniel L(ee)
Whitney, Eugene C
Williams, Albert J, Jr
Wolfe, Reuben Edward
Young, Frederick J(ohn)
Young, Raymond H(ykes)
Yu, Francis T S
Zebrowitz, S(tanley)
Zocholl, Stanley E

RHODE ISLAND
Bartram, James F(ranklin)
Cooper, David B
Daly, James C(affrey)
Davis, John Sheldon
Hazeltine, Barrett
Hoffman, Joseph Ellsworth, Jr
Jackson, Leland Brooks
Krikorian, John Sarkis, Jr
Kushner, Harold J(oseph)
Lengyel, G(abriel)
Logan, Joseph Skinner
Middleton, Foster H(ugh)
Park, Robert H
Petrou, Panayiotis Polydorou
Polk, C(harles)
Prince, Mack J(ames)
Savage, John Edmund
Spence, John Edwin
Streit, Roy Leon
Sullivan, Edmund Joseph
Tufts, Donald Winston
Wolovich, William Anthony

SOUTH CAROLINA
Askins, Harold Williams, Jr
Behnke, Wallace B, Jr
Bennet, Archie Wayne
Bennett, Archie Wayne
Bourkoff, Etan
Dumin, David Joseph
Eccles, William J
Ernst, Edward W
Fellers, Rufus Gustavus
Garzon, Ruben Dario
Geary, Leo Charles
Gilliland, Bobby Eugene
Gowdy, John Norman
Halford, Jake Hallie
Hammond, Joseph Langhorne, Jr
Hogan, Joseph C(harles)
Lam, Chan F
Lathrop, Jay Wallace
Long, Jim T(homas)
Luh, Johnson Yang-Seng
McCarter, James Thomas
Maricq, John
Martin, John Campbell
Pearson, Lonnie Wilson
Peeples, Johnston William
Pursley, Michael Bader
Sias, Fred R, Jr
Simpson, John W(istar)
Snelsire, Robert W
Sudarshan, Tangali S
Thurston, James N(orton)
Yang, Tsute

SOUTH DAKOTA
Cannon, Patrick Joseph
Cox, Cyrus W(illiam)
Gowen, Richard J
Hanson, Jerry Lee
Hughes, William (Lewis)
Kurtenbach, Aelred J(oseph)
Manning, M(elvin) L(ane)
Reuter, William L(ee)
Riemenschneider, Albert Louis
Sander, Duane E

Smith, Paul Letton, Jr
Storry, Junis O(liver)

TENNESSEE
Adams, Raymond Kenneth
Bell, Zane W
Benson, Robert Wilmer
Bishop, Asa Orin, Jr
Blalock, Theron Vaughn
Bose, Bimal K
Brodersen, Arthur James
Cadzow, James A(rchie)
Chowdhuri, Pritindra
Cook, George Edward
Dicker, Paul Edward
Essler, Warren O(rvel)
Gonzalez, Rafael C
Green, Walter L(uther)
Harter, James A(ndrew)
Hoffman, G(raham) W(alter)
Hung, James Chen
Jermann, William Howard
Johnson, L(ee) Ensign
Joseph, Roy D
Kawamura, Kazuhiko
Kennedy, Eldredge Johnson
Leffell, W(ill) O(tis)
Lyon, James F
Malkani, Mohan J
Mosko, Sigmund W
Murakami, Masanori
Nash, Robert T
Neff, Herbert Preston, Jr
Oakes, Lester C
Ogg, Robert James
Pace, Marshall Osteen
Parrish, Edward Alton, Jr
Pearsall, S(amuel) H(aff)
Pierce, John Frank
Rajan, Periasamy Karivaratha
Rochelle, Robert W(hite)
Roth, J(ohn) Reece
Shockley, Thomas D(ewey), Jr
Terry, Fred Herbert
Tillman, J(ames) D(avid), Jr
Waller, John Wayne
Weaver, Charles Hadley
White, Edward John

TEXAS
Aggarwal, Jagdishkumar Keshoram
Aiken, James G
Allen, John Burton
Anderson, Dale Arden
Anderson, Wallace L(ee)
Bailey, James Stephen
Baker, Wilfred E(dmund)
Balasinski, Artur
Barton, James Brockman
Bean, Wendell Clebern
Becker, Michael Franklin
Benning, Carl J, Jr
Bhattacharyya, Shankar P
Biard, James R
Bishop, Robert H
Bossen, Douglas C
Bowdon, Edward K(night), Sr
Bryant, Michael David
Burrus, Charles Sidney
Cantrell, Cyrus D, III
Cash, Floyd Lee
Caswell, Gregory K
Chakravarty, Indranil
Chang, Christopher Teh-Min
Chen, Mo-Shing
Chenoweth, R(obert) D(ean)
Cherrington, Blake Edward
Choate, William Clay
Collins, Francis Allen
Cooke, James Louis
Cory, William Eugene
Cox, Dennis Dean
Crossley, Peter Anthony
Crum, Floyd M(axilas)
David, Yadin B
Denison, John Scott
Dodd, Edward Elliott
Donaldson, W(illis) Lyle
Dougal, Arwin A(delbert)
Drazenovic, Perunicic
Duesterhoeft, William Charles, Jr
Dunham, James George
Dupuis, Russell Dean
Dutcher, Clinton Harvey, Jr
Ehsani, Mehrdad
Eknoyan, Ohannes
Ekstrom, Michael P
Esquivel, Agerico Liwag
Fannin, Bob M(eredith)
Fink, Lyman R
Fite, Lloyd Emery
Fitzer, Jack
Franke, Ernest A
Friberg, Emil Edwards
Friedrich, Otto Martin, Jr
Gordon, William E(dwin)
Green, Terry C
Griffith, M S
Gully, John Houston
Gupta, Someshwar C
Haden, Clovis Roland
Hagler, Marion O(tho)
Hanne, John R

Hatsell, Charles Proctor
Haupt, L(ewis) M(cDowell), Jr
Hayden, Edgar C(lay)
Hayre, Harbhajan Singh
Heizer, Kenneth W
Hixson, Elmer LaVerne
Hoehn, G(ustave) L(eo)
Hoffmann-Pinther, Peter Hugo
Hopkins, George H, Jr
Howard, Lorn Lambier
Huang, Jung-Chang
Huebner, George L(ee), Jr
Hunt, Louis Roberts
Jackson, Edwin L
Johnson, Richard Leon
Kaenel, Reginald A(lfred)
Katsnelson, Lev Z
Kesler, Oren Byrl
Kilby, Jack St Clair
King, Willis Kwongtsu
Koepsel, Wellington Wesley
Korges, Emerson
Kostelnicek, Richard J
Krezdorn, Roy R
Krishen, Kumar
Kristiansen, Magne
Ku, Bernard Siu-Man
Kunze, Otto Robert
Leeds, J Venn, Jr
Lengel, Robert Charles
Leonard, William F
Lewis, Frank Leroy
Liang, Charles Shih-Tung
Ling, Hao
Liu, Kai
Long, Stuart A
MacFarlane, Duncan Leo
Malek, Miroslaw
Markenscoff, Pauline
Matzen, Walter T(heodore)
Maxum, Bernard J
Mays, Robert, Jr
Mitchell, Owen Robert
Musa, Samuel A
Neikirk, Dean P
Nevels, Robert Dudley
Newton, Richard Wayne
Novak, Gordon Shaw, Jr
Ober, Raimund Johannes
Ong, Poen Sing
Opiela, Alexander David, Jr
Ott, Granville E
Overzet, Lawrence John
Paskusz, Gerhard F
Paterson, James Lenander
Patton, Alton DeWitt
Pearson, James Boyd, Jr
Peikari, Behrouz
Perez, Ricardo
Portnoy, William M
Powers, Edward Joseph, Jr
Powers, Robert S(inclair), Jr
Pritchett, William Carr
Provencio, Jesus Roberto
Raju, Satyanarayana G V
Renner, Darwin S(prathard)
Reynolds, Jack
Rhyne, V(ernon) Thomas
Ribe, M(arshall) L(ouis)
Robbins, Roger Alan
Rogers, Phil H
Russell, B Don
Salis, Andrew E
Sandberg, I(rwin) W(alter)
Satterwhite, Ramon S(tewart)
Schell, Robert Ray
Schoenberger, Michael
Schulz, Richard Burkart
Shah, Pradeep L
Shen, Liang Chi
Shepherd, Mark, Jr
Sheppard, Louis Clarke
Shieh, Leang-San
Shrime, George P
Simpson, Richard S
Singh, Vijay Pal
Smith, Harold W(ood)
Smith, Jack
Sobol, Harold
Srinath, Mandyam Dhati
Srinivasan, Ramachandra Srini
Stockton, James Evan
Stone, Orville L
Straiton, Archie Waugh
Streetman, Ben Garland
Szygenda, Stephen Anthony
Tasch, Al Felix, Jr
Taylor, Herbert Lyndon
Thorn, Donald Childress
Tittel, Frank K(laus)
Trench, William Frederick
Vance, Edward F(lavus)
Victor, Joe Mayer
Vines, Darrell Lee
Volz, Richard A
Von Maltzahn, Wolf W
Wakeland, William Richard
Walkup, John Frank
Wang, Yuan R
Watt, Joseph T(ee), Jr
Welch, Ashley James
Weldon, William Forrest
Wharton, Russell Perry
Widerquist, V(ernon) R(oberts)

Willman, Joseph F(rank)
Wilton, Donald Robert
Winje, Russell A
Wischmeyer, Carl R(iehle)
Woodson, Herbert H(orace)
Young, James Forrest
Zrudsky, Donald Richard

UTAH
Atwood, Kenneth W
Baker, Kay Dayne
Barlow, Mark Owens
Berrett, Paul O(rin)
Bowman, Lawrence Sieman
Bunch, Kyle
Chabries, Douglas M
Chadwick, Duane G(eorge)
Christensen, Douglas Allen
Clark, Clayton
Clegg, John C(ardwell)
Cole, Larry S
Comer, David J
Durney, Carl H(odson)
Embry, Bertis L(loyd)
Evans, David C(annon)
Feucht, Donald Lee
Gandhi, Om P
Harris, Ronney D
Hollaar, Lee Allen
Kieda, David Basil
Long, David G
Losee, Ferril A
Rushforth, Craig Knewel
Skinner, Robert L
Smith, Kent Farrell
Stephenson, Robert E(ldon)
Stringfellow, Gerald B
Woodbury, Richard C

VERMONT
Anderson, R(ichard) L(ouis)
Clark, Donald Lyndon
Critchlow, D(ale)
Ellis, David M
Ellis-Monaghan, John J
Evering, Frederick Christian, Jr
Golden, Kenneth Ivan
Gray, Kenneth Stewart
Higgins, John J
Quinn, Robert M(ichael)
Rush, Stanley
Stapper, Charles Henri
Wallis, Clifford Merrill
Williams, Ronald Wendell

VIRGINIA
Abrams, Marshall D
Aller, James Curwood
Aylor, James Hiram
Baeumler, Howard William
Bahl, Inder Jit
Batte, William Granville
Beale, Guy Otis
Beran, Robert Lynn
Blackwell, William A(llen)
Blasbalg, Herman
Bordogna, Joseph
Bostian, Charles William
Brown, Gary S
Bruno, Ronald C
Bunting, Christopher David
Campbell, Francis James
Caperley, Peter H
Casazza, John Andrew
Chapin, Douglas McCall
Chen, Jiande
Chen, Pi-Fuay
Chien, Yi-Tzuu
Choi, Junho
Cibulka, Frank
Clark, Ralph Leigh
Claus, Richard Otto
Coffey, Timothy
Cook, Gerald
Curry, Thomas F(ortson)
Daly, Robert Francis
Davis, William Arthur
Dean, William C(orner)
Delnore, Victor Eli
Denning, Peter James
De Wolf, David Alter
Dharamsi, Amin N
Dickman, Robert Laurence
Doles, John Henry, III
Duff, William G
Dwyer, Samuel J, III
Edyvane, John
Egan, Raymond D(avis)
Ellersick, Fred W(illiam)
Entzminger, John N
Farrell, Robert Michael
Farrukh, Usamah Omar
Fath, George R
Feustel, Edward Alvin
Fink, Lester Harold
Fitts, Richard Earl
Frank, Howard
Franklin, Jude Eric
Fripp, Archibald Linley
Gabriel, William Francis
Garcia, Albert B
Garcia, Oscar Nicolas
Gee, Sherman
Gibson, John E(gan)

Ginsberg, Murry B(enjamin)
Godbole, Sadashiva Shankar
Goodman, A(lvin) M(alcolm)
Gordon, John P(etersen)
Grafton, Robert Bruce
Gray, F(estus) Gail
Guerber, Howard P(aul)
Hall, Albert C(arruthers)
Hammond, Marvin H, Jr
Harrington, Robert Joseph
Hasenfus, Harold J(oseph)
Heebner, David Richard
Heffron, W(alter) Gordon
Heidbreder, Glenn R
Herbst, Mark Joseph
Holt, Charles A(sbury)
Holzmann, Ernest G(unther)
Hopkins, Mansell Herbert, Jr
Huber, Paul W(illiam)
Humphris, Robert R
Hwang, William Gaong
Inigo, Rafael Madrigal
Jacobs, Ira
Jaron, Dov
Johnson, Preston Benton
Jordan, Kenneth L(ouis), Jr
Joshi, Ravindra Prabhakar
Joshi, Suresh Meghashyam
Juang, Jer-Nan
Jud, Henry G
Kahn, Manfred
Kahn, Robert Elliot
Kahne, Stephen James
Kaminski, Paul G
Keblawi, Feisal Said
Keigler, John Edward
Keiser, Bernhard E(dward)
Kelly, Peter Michael
Kelsey, Eugene Lloyd
Kenyon, Stephen C
Kim, Yong Il
Kruger, Richard Paul
Landes, Hugh S(tevenson)
Larson, Arvid Gunnar
Latta, John Neal
Lee, Fred C
Lehmann, John R(ichard)
Lenoir, William Benjamin
Lerner, Norman Conrad
McCutchen, Samuel P(roctor)
McGregor, Dennis Nicholas
McVey, Eugene Steven
McWright, Glen Martin
Madan, Rabinder Nath
Manriquez, Rolando Paredes
Mattauch, Robert Joseph
Mazzoni, Omar S
Miller, Robert H
Moseley, Sherrard Thomas
Nashman, Alvin E
Nichols, Lee L(ochhead), Jr
Nunnally, Huey Neal
Palmer, James D(aniel)
Park, John Howard, Jr
Phadke, Arun G
Phelan, James Frederick
Poon, Ting-Chung
Powley, George R(einhold)
Pryor, C(abell) Nicholas, Jr
Pulvari, Charles F
Raab, Harry Frederick, Jr
Rader, Louis T(elemacus)
Russell, James Madison, III
Sage, Andrew Patrick
Sanders, John D
Schlegelmilch, Reuben Orville
Schneider, Ralph Jacob
Scoggin, James F, Jr
Shenolikar, Ashok Kumar
Shur, Michael
Siegel, Clifford M(yron)
Singleton, James L
Smith, B(lanchard) D(rake), Jr
Smits, Talivaldis I(vars)
Srikanth, Sivasankaran
Stanley, William Daniel
Stephenson, Frederick William
Stuart, Charles Edward
Stutzman, Warren Lee
Tabak, Daniel
Temes, Clifford Lawrence
Thompson, Anthony Richard
Toida, Shunichi
Tsang, Wai Lin
Van Landingham, Hugh F(och)
Violette, Joseph Lawrence Norman
Walker, Loren Haines
Walsh, Edward Joseph
Waxman, Ronald
Weik, Martin H, Jr
Wolla, Maurice L(eRoy)
Wu, Yung-Kuang

WASHINGTON
Ackerlind, E(rik)
Andersen, Jonny
Auth, David C
Baker, Richard A
Bernard, Gary Dale
Bhansali, Praful V
Blackburn, John Lewis
Bose, Anjan
Brownlee, Donald E

Clark, Robert Newhall
Craine, Lloyd Bernard
Damborg, Mark J(ohannes)
Daniels, Patricia D
Dow, Daniel G(ould)
Dowis, W J
Eckersley, Alfred
El-Sharkawi, Mohamed A
Flechsig, Alfred J, Jr
Golde, Hellmut
Hanna, Jeff
Haralick, Robert M
Heindsmann, T(heodore) E(dward)
Heisler, Rodney
Helms, Ward Julian
Holden, Alistair David Craig
Hower, Glen L
Hsu, C(hih) C(hi)
Hsu, Chin Shung
Isaak, Robert D(eets)
Ishimaru, Akira
Kim, Jae Hoon
Koerber, George G(regory)
Kofoid, Melvin J(ulius)
Lauritzen, Peter O
Liu, Chen-Ching
Lytle, Dean Winton
Masden, Glenn W(illiam)
Meditch, James S
Metz, Peter Robert
Mosher, Clifford Coleman, III
Noe, Jerre D(onald)
Noges, Endrik
Nutley, Hugh
Olsen, Robert Gerner
Oman, Henry
Peden, Irene C(arswell)
Preikschat, Ekkehard
Preikschat, F(ritz) K(arl)
Reynolds, Donald Kelly
Richardson, Richard Laurel
Ringo, John Alan
Robel, Gregory Frank
Rogers, Joseph Wood
Savol, Andrej Martin
Schrader, David Hawley
Schweitzer, Edmund Oscar, III
Seamans, David A(lvin)
Seliga, Thomas A
Somani, Arun Kumar
Spillman, Richard Jay
Stevens, Richard F
Stewart, John L(awrence)
Szablya, John F(rancis)
Tsang, Leung
Venkata, Subrahmanyam Saraswati
Walsh, John Breffni
Waltar, Alan Edward
Walton, Vincent Michael
West, Herschel J
Wood, Francis Patrick
Yore, Eugene Elliott
Young, James Arthur, Jr

WEST VIRGINIA
Blackwell, Lyle Marvin
Breece, Harry T, III
Brown, Paul B
Cannon, Walton Wayne
Cooley, Wils LaHugh
Corum, James Frederic
Fischer, Robert Lee
Jones, Edwin C
Keener, E(verett) L(ee)
Nutter, Roy Sterling, Jr
Parker, David Hiram
Russell, James A(lvin), Jr
Smith, Nelson S(tuart), Jr
Spaniol, Craig
Whittington, Bernard W
Zeng, Shengke

WISCONSIN
Battocletti, Joseph H(enry)
Bernstein, Theodore
Beyer, James B
Birkemeier, William P
Boettcher, Harold P(aul)
Bollinger, John G(ustave)
Botez, Dan
Brauer, John Robert
Burrage, Lawrence Minott
Cerrina, Francesco
Christensen, Erik Regnar
Davidson, Charles H(enry)
Denton, Denice Dee
Dietmeyer, Donald L
Eriksson, Larry John
Farrow, John H
Gray, Paul Eugene
Greiner, Richard A(nton)
Heinen, James Albin
Higgins, Thomas James
Horgan, James D(onald)
Hummert, George Thomas
Hyzer, William Gordon
Ishii, T(homas) Koryu
Jacobi, George (Thomas)
Kerkman, Russel John
Klein, Carl Frederick
Lind, Robert Wayne
Lipo, Thomas Anthony
Long, Willis Franklin
Martinson, Edwin O(scar)

Moeller, Arthur Charles
Morin, Dorris Clinton
Nash, William Hart
Niederjohn, Russell James
Nondahl, Thomas Arthur
Northouse, Richard A
Novotny, Donald Wayne
Prager, Stewart Charles
Rideout, Vincent C(harles)
Rosenthal, Jeffrey
Saleh, Bahaa E A
Scharer, John Edward
Schlicke, Heinz M
Schmitz, Norbert Lewis
Schotz, Larry
Scidmore, Allan K
Seshadri, Sengadu Rangaswamy
Shohet, Juda Leon
Siedband, Melvin Paul
Skiles, James J(ean)
Smith, Luther Michael
Staats, Gustav W(illiam)
Steber, George Rudolph
Stremler, Ferrel G
Thompson, Paul DeVries
Tompkins, Willis Judson
Vairavan, Kasivisvanathan
Vernon, Ronald J
Walter, Gilbert G
Webster, John Goodwin
Wu, Sherman H
Zheng, Xiaoci

WYOMING
Beach, R(upert) K(enneth)
Constantinides, Christos T(heodorou)
Hakes, Samuel D(uncan)
Jacquot, Raymond C
Long, F(rancis) M(ark)
Rhodine, Charles Norman
Spenner, Frank J(ohn)
Tauer, Jane E
Weeks, Richard W(illiam)

PUERTO RICO
Hussain, Malek Gholoum Malek
Mercado-Jimenez, Teodoro

ALBERTA
Berg, Gunnar Johannes
Cormack, George Douglas
Fedosejevs, Robert
Goud, Paul A
Gourishankar, V (Gouri)
Malik, Om Parkash
Nutakki, Dharma Rao
Stein, Richard Adolph
Streets, Rubert Burley, Jr
Trofimenkoff, Frederick N(icholas)

BRITISH COLUMBIA
Agathoklis, Panajotis
Antoniou, A(ndreas)
Beddoes, M(ichael) P(eter)
Booth, Andrew Donald
Davies, Michael Shapland
Dumont, Guy Albert
Hoefer, Wolfgang Johannes Reinhard
Jull, Edward Vincent
Kerr, J(ames) S(anford) Stephenson
Lu, Wu Sheng
Moore, A(rthur) Donald
Shulman, David-Dima
Soudack, Avrum Chaim
Stuchly, Maria Anna
Stuchly, Stanislaw S
Wedepohl, Leonhard M
Young, Lawrence
Yu, Yao Nan
Zielinski, Adam

MANITOBA
Bridges, E(rnest)
Hamid, Michael
Johnson, R(ichard) A(llan)
Kao, Kwan Chi
Kim, Hyong Kap
Shafai, Lotfollah
Swift, Glenn W(illiam)

NEW BRUNSWICK
Lewis, J(ohn) E(ugene)
Scott, Robert Nelson
Wasson, W(alter) Dana

NEWFOUNDLAND
Rahman, Md Azizur
Vetter, William J

NOVA SCOTIA
Baird, Charles Robert
Bhartia, Prakash
El-Hawary, Mohamed El-Aref
El-Masry, Ezz Ismail

ONTARIO
Abdelmalek, Nabin N
Ahmed, Nasir Uddin
Aitken, G(eorge) J(ohn) M(urray)
Balmain, Keith George
Bandler, John William
Beal, Jim C(ampbell)
Bhatnagar, Yashraj Kishore
Biringer, Paul P(eter)

DISCIPLINE INDEX

Electrical Engineering (cont)

Blake, Ian Fraser
Broughton, M(ervyn) B(lythe)
Campbell, C(olin) K(ydd)
Campling, C(harles) H(ugh) R(amsay)
Carr, Jan
Castle, George Samuel Peter
Celinski, Olgierd J(erzy) Z(dzislaw)
Chow, Yung Leonard
Coll, David C
Copeland, Miles Alexander
Costache, George Ioan
Davidson, Grant E(dward)
Densley, John R
Dewan, Shashi B
Di Cenzo, Colin D
Dohoo, Roy McGregor
Ellis, Jack Barry
Endrenyi, Janos
Fahmy, Moustafa Mahmoud
Fairman, Frederick Walker
Falconer, David Duncan
Findlay, Raymond D(avid)
Georganas, Nicolas D
Goldenberg, Andrew Avi
Hackam, Reuben
Ham, James M(ilton)
Hamacher, V(incent) Carl
Hanff, Ernest Salo
Heasell, E(dwin) L(ovell)
Ionescu, Dan
Janischewskyj, W
Ji, Guangda Winston
Jullien, Graham Arnold
Kalra, S(urindra) N(ath)
Kavanagh, Robert John
Kitai, Reuven
Kruus, Jaan
Kundur, Prabha Shankar
Kunov, Hans
Landolt, Jack Peter
Leach, Barrie William
Lemyre, C(lement)
Ling, Daniel
McCausland, Ian
McGee, William F
McLane, Peter John
Majithia, Jayanti
Mark, Jon Wei
Mathur, R(adhey) M(ohan)
Morris, Lawrence Robert
Mouftah, Hussein T
Mungall, Allan George
Penstone, S(idney) Robert
Perlman, Maier
Plant, B J
Raju, Govinda
Raney, Russell Keith
Ratz, H(erbert) C(harles)
Reeve, John
Reza, Fazlollah M
Riordon, J(ohn) Spruce
Ristic, Velimir Mihailo
Roe, Peter Hugh O'Neil
Salama, Clement Andre Tewfik
Selvakumar, Chettypalayam Ramanathan
Semlyen, Adam
Sinclair, George
Sinha, Naresh Kumar
Slemon, Gordon R(ichard)
Stevinson, Harry Thompson
Tavares, Stafford Emanuel
Thansandote, Artnarong
Thomas, Raye Edward
Tsao, Sai Hoi
Van Driel, Henry Martin
Venetsanopoulos, Anastasios Nicolaos
Vervoort, Gerardus
Vlach, Jiri
Wallingford, Errol E(lwood)
Watt, Lynn A(lexander) K(eeling)
Wei, L(ing) Y(un)
Wilson, John D(ouglas)
Wittke, Paul H
Wong, George Shoung-Koon

PRINCE EDWARD ISLAND
Lodge, Malcolm A

QUEBEC
Adler, Eric L
Agarwal, Vinod Kumar
Aktik, Cetin
Bartnikas, Ray
Belanger, Pierre Rolland
Bhattacharyya, Bibhuti Bhusan
Boulet, J Lionel
Boyarsky, Abraham Joseph
Crine, Jean-Pierre C
Dumas, Jean
Farnell, G(erald) W(illiam)
Giguere, Joseph Charles
Hayes, Jeremiah Francis
Jumarie, Guy Michael
Kahrizi, Mojtaba
Lecours, Michel
Lefebvre, Mario
Levine, Martin David
Malowany, Alfred Stephen
Maruvada, P Sarma
Morgera, Salvatore Domenic
Mukhedkar, Dinkar
Paquet, Jean Guy

Patel, Rajnikant V
Pavlasek, Tomas J(an) F(rantisek)
Plotkin, Eugene Isaak
Ramachandran, Venkatanarayana D
Roberge, Fernand Adrien
Saint-Arnaud, Raymond
Silvester, Peter Peet
St-Jean, Guy
Swamy, Mayasandra Nanjundiah Srikanta
Zames, George
Zucker, Steven Warren

SASKATCHEWAN
Billinton, Roy
Blachford, Cameron W
Boyle, A(rchibald) R(aymond)
Gupta, Madan Mohan
Hirose, Akira
Kasap, Safa O
Pollak, Viktor A

OTHER COUNTRIES
Aleksandrov, Georgij Nikolaevich
Angelini, Arnaldo M
Balk, Pieter
Beckhoff, Gerhard Franz
Benson, Frank Atkinson
Boxman, Raymond Leon
Broers, Alec N
Burckhardt, Christoph B
Callier, Frank Maria
Chang, Chun-Yen
Chu, Kwo Ray
Conforti, Evandro
Czepyha, Chester George Reinhold
Fettweis, Alfred Leo Maria
Flurscheim, Cedric Harold
Forrer, Max P(aul)
Gauster, Wilhelm Friedrich
Giarola, Attilio Jose
Jespersen, Nils Vidar
Kagawa, Yukio
Kaiser, Wolfgang A
Kao, Charles K
Khatib, Hisham M
Kim, Myunghwan
Kimura, Hidenori
Kirby, Richard C(yril)
Kishi, Keiji
Ku, Victor Chia-Tai
Kurata, Mamoru
Kurokawa, Kaneyuki
Kwok, Hoi S
Lee, Choong Woong
Lindell, Ismo Veikko
Liu, Chao-Han
Luckey, Paul David, Jr
Lueder, Ernst H
Luo, Peilin
Malah, David
Mansour, Mohamed
Maqusi, Mohammad
Mark, James Wai-Kee
Marom, Emanuel
Massey, James L
Mayeda, Wataru
Narasimhan, Mandayam A
Orphanoudakis, Stelios Constantine
Poloujadoff, Michel E
Renfrew, Robert Morrison
Saba, Shoichi
Sakai, Hiroshi
Sakurai, Yoshifumi
Sato, Gentei
Sekimoto, Tadahiro
Sekine, Yasuji
Sessler, Gerhard Martin
Shtrikman, Shmuel
Steen, Robert Frederick
Strintzis, Michael Gerassimos
Tarui, Yasuo
Tzafestas, Spyros G
Udo, Tatsuo
Un, Chong Kwan
Ungerboeck, Gottfried
Van Raalte, John A
Vidyasagar, Mathukumalli
Watanabe, Akira
Wing, Omar
Winn, Edward Barriere
Woo, Kwang Bang
Yamamoto, Mitsuyoshi
Yanabu, Satoru
Yeh, K(ung) C(hie)
Young, Ian Theodore
Yu, Chyang John
Zeheb, Ezra

Electronics Engineering

ALABAMA
Brennan, Lawrence Edward
Dixon, Julian Thomas
Edlin, George Robert
Fullerton, Larry Wayne
Gilbert, Stephen Marc
Honnell, Martial A(lfred)
Jaeger, Richard Charles
Nola, Frank Joseph
Post, M(aurice) Dean
Thomas, Albert Lee, Jr
Watson, Raymond Coke, Jr

White, Ronald
Wolin, Samuel

ALASKA
Hallinan, Thomas James

ARIZONA
Bahill, A(ndrew) Terry
Balanis, Constantine A
Bao, Qingcheng
Benfield, Charles W(illiam)
Bluestein, Theodore
Cabaret, Joseph Ronald
Chandra, Abhijit
Craib, James F(rederick)
De La Moneda, Francisco Homero
Dougherty, John Joseph
Dybvig, Paul Henry
El-Ghazaly, Samir M
Everett, Paul Marvin
Falk, Thomas
Fordemwalt, James Newton
Formeister, Richard B
Galloway, Kenneth Franklin
Garrett, Lane Sayre
Gerhard, Glen Carl
Hickernell, Fred Slocum
Hoehn, A(lfred) J(oseph)
Hohl, Jakob Hans
Holmgren, Paul
Jackson, Kenneth Arthur
Jones, Roger C(lyde)
Karady, George Gyorgy
Kaufman, Irving
Kerwin, William J(ames)
Liaw, Hang Ming
McRae, Lorin Post
Mylrea, Kenneth C
Palmer, James McLean
Perper, Lloyd
Pokorny, Gerold E(rwin)
Rutledge, James Luther
Schroder, Dieter K
Scott, Alwyn C
Selvidge, Harner
Silvern, Leonard Charles
Stafford, John William
Wasson, James Walter
Weinstein, Ronald S
Weller, Edward F(rank), Jr
Witulski, Arthur Frank
Zeigler, Bernard Philip
Zelenka, Jerry Stephen

ARKANSAS
Cardwell, David Michael

CALIFORNIA
Abraham, W(ayne) G(ordon)
Allen, Charles A
Alves, Ronald V
Anderson, Warren R(onald)
Andres, John Milton
Angell, James Browne
Anger, Hal Oscar
Antoniak, Charles Edward
Atchison, F(red) Stanley
Aukland, Jerry C
Bacon, Lyle C(holwell), Jr
Baker, Richard H(owell)
Bardin, Russell Keith
Barhydt, Hamilton
Begovich, Nicholas A(nthony)
Bennett, Alan Jerome
Berg, Myles Renver
Berger, Josef
Bernstein, Uri
Bertram, Sidney
Bharat, Ramasesha
Bhaumik, Mani Lal
Bishop, Harold (Oswald)
Bixler, Otto E
Blasingame, Benjamin P(aul)
Blumenthal, Irwin S(imeon)
Borst, David W(ellington)
Breitwieser, Charles J(ohn)
Brewer, George R
Brodie, Ivor
Brothers, James Townsend
Bullis, William Murray
Burnett, Lowell Jay
Caldwell, J(oseph) J(efferson), Jr
Carlan, Alan J
Carlyle, Jack Webster
Cass, Thomas Robert
Chan, Bertram Kim Cheong
Chang, Juang-Chi (Joseph)
Chang, Tien-Lin
Chen, Tsai Hwa
Chiang, Anne
Chin, Maw-Rong
Chodorow, Marvin
Choi, Won Kil
Christie, John McDougall
Clarke, Lucien Gill
Cogan, Adrian Ilie
Conant, Curtis Terry
Cone, Donald R(oy)
Cook, Albert Moore
Cormier, Reginald Albert
Cox, Donald Clyde
Crain, Cullen Malone
Crandall, Ira Carlton
Crapuchettes, Paul W(ythe)

Creighton, John Rogers
Crosbie, Roy Edward
Culler, Donald Merrill
Cutler, Leonard Samuel
Daehler, Max, Jr
Damonte, John Batista
Davis, William R
De Gaston, Alexis Neal
DeGrasse, Robert W(oodman)
Demetrescu, M
De Micheli, Giovanni
Dempsey, Martin E(wald)
Deshpande, Shivajirao M
Dev, Parvati
Dines, Eugene L
Dorfan, David Elliot
Dyce, Rolf Buchanan
Eargle, John Morgan
Eden, Richard Carl
Edson, William A(lden)
Eldon, Charles A
Ellingboe, J(ules) K
Erdley, Harold F(rederick)
Evans, Bob Overton
Everhart, Thomas E(ugene)
Ewing, Gerald Dean
Fasang, Patrick Pad
Feldman, Nathaniel E
Feng, Joseph Shao-Ying
Ferrari, Domenico
Fialer, Philip A
Fleming, Alan Wayne
Fonck, Eugene J
Frankel, Sidney
Franklin, Gene F(arthing)
Frescura, Bert Louis
Fried, Walter Rudolf
Gamo, Hideya
Gardner, Floyd M
Gibson, Atholl Allen Vear
Glazer, Judith
Grebene, Alan B
Green, Allen T
Green, Philip S
Grinnell, Robin Roy
Grobecker, Alan J
Groner, Gabriel F(rederick)
Groner, Paul Stephen
Gunckel, Thomas L, II
Gupta, Madhu Sudan
Haas, Gustav Frederick
Hackett, Le Roy Huntington, Jr
Haeff, Andrew V(asily)
Hagstrom, Stig Bernt
Haller, Eugene Ernest
Hankins, George Thomas
Hansen, Grant Lewis
Hardy, Vernon E
Harker, J M
Harper, Francis Edward
Harris, James Stewart, Jr
Hartwick, Thomas Stanley
Hearn, Robert Henderson
Hellerman, Herbert
Helmer, John
Henning, Harley Barry
Herman, E(lvin) E(ugene)
Hewlett, W(illiam) R(edington)
Hingorani, Narain G
Hinrichs, Karl
Hoch, Orion L
Hodges, Dean T, Jr
Hoffman, James Tracy
Hooper, Catherine Evelyn
Hornak, Thomas
Hovanessian, Shahen Alexander
Hrubesh, Lawrence W(ayne)
Hsia, Yukun
Hu, Chenming
Huang, Chaofu
Huang, Tiao-yuan
Hult, John Luther
Hummel, Steven G
Huszczuk, Andrew Huszcza
Ikezi, Hiroyuki
Isaman, Francis
Jacob, George Korathu
Jacobson, Raymond E
Jansen, Michael
Jobs, Steven P
Johannessen, Jack
Johnson, David Leroy
Johnson, Robert A
Jones, E(dward) M(cClung) T(hompson)
Jones, Earle Douglas
Jordan, Gary Blake
Josias, Conrad S(eymour)
Kafka, Robert W(illiam)
Kagiwada, Reynold Shigeru
Kahn, Frederic Jay
Kaisel, S(tanley) F(rancis)
Kalafus, Rudolph M
Kamins, Theodore I
Kamphoefner, Fred J(ohn)
Kanter, Helmut
Kao, Tai-Wu
Karp, Arthur
Karuza, Sarunas Kazys
Kaufman, Alvin B(eryl)
Kayton, Myron
Kazan, Benjamin
Kearney, Joseph W(illiam)
Kelly, Kenneth C
Kerfoot, Branch Price

Khalona, Ramon Antonio
King, Harry J
King, Howard E
Kinnison, Gerald Lee
Kjar, Raymond Arthur
Knighten, James Leo
Koetsch, Philip
Konrad, Gerhard T(hies)
Kosai, Kenneth
Kozlowski, Lester Joseph
Krause, Lloyd O(scar)
Kroemer, Herbert
Lacy, Peter D(empsey)
Lafranchi, Edward Alvin
Lally, Philip M(arshall)
Lampert, Carl Matthew
Lao, Binnee Yanbing
Larson, Harry Thomas
Lau, John H
Lau, S S
Law, Hsiang-Yi David
Leadabrand, Ray Laurence
Leavy, Paul Matthew
Leedham, Clive D(ouglas)
Lehovec, Kurt
Leipold, Martin H(enry)
Lender, Adam
Lepoff, Jack H
Leung, Charles Cheung-Wan
Levenson, Marc David
Levy, Ralph
Lewis, John E
Li, Chia-Chuan
Lieberman, Michael A
Limb, John Ormond
Lin, Anthony T
Lin, Wen-C(hun)
Lisman, Perry Hall
Liu, Chi-Sheng
Lonky, Martin Leonard
Lopina, Robert F(erguson)
Lord, Harold Wilbur
Love, Sydney Francis
Lubman, David
Lynch, Frank W
McCaldin, J(ames) O(eland)
McColl, Malcolm
McCurdy, Alan Hugh
McDonald, William True
Macha, Milo
Mack, Dick A
McKay, Dale Robert
Mackay, Ralph Stuart
McNutt, Michael John
McQuillen, Howard Raymond
McWhorter, Malcolm M(yers)
Main, William Francis
Maloney, Timothy James
Maloy, John Owen
Manes, Kennneth Rene
Margerum, Donald L(ee)
Marshall, J Howard, III
Martin, Steven Roden
Matare, Herbert Franz
Mathews, W(arren) E(dward)
Mayper, V(ictor), Jr
Meagher, Donald Joseph
Meggers, William F(rederick), Jr
Meng, Shien-Yi
Meyer, Lhary
Michaelson, Jerry Dean
Middlebrook, R(obert) D(avid)
Miller, Arnold
Mitra, Sanjit K
Molnar, Imre
Morris, Fred(erick) W(illiam)
Morrow, Charles Tabor
Muchmore, Robert B(oyer)
Muller, Richard Stephen
Mykkanen, Donald L
Naqvi, Iqbal Mehdi
Nelson, Richard Burton
Nelson, Richard David
Ohlson, John E
Oleesky, Samuel S(imon)
Ozguz, Volkan Husnu
Page, D(errick) J(ohn)
Palmer, John Parker
Paoli, Thomas Lee
Parker, Alice Cline
Parker, Don
Parode, L(owell) C(arr)
Patel, Arvindkumar Motibhai
Persky, George
Pfost, R Fred
Pierce, John Robinson
Porter, John E(dward)
Portnoff, Michael Rodney
Presnell, Ronald I
Preston, Glenn Wetherby
Ranftl, Robert M(atthew)
Rao, R(angaiya) A(swathanarayana)
Rauch, Lawrence L(ee)
Reiche, Ludwig P(ercy)
Requicha, Aristides A G
Rickard, John Terrell
Robinson, Lloyd Burdette
Rogers, Howard H
Ross, Hugh Courtney
Rubenson, J(oseph) G(eorge)
Rude, Paul A
Rynn, Nathan
Sachdev, Suresh
Sadler, Charles Robinson, Jr

Safonov, Michael G
Sample, Steven Browning
Samulon, Henry A
Sangiovanni-Vincentelli, Alberto Luigi
Sawyer, David Erickson
Schechter, Joel Ernest
Schweikert, Daniel George
Scott, Paul Brunson
Sensiper, S(amuel)
Sequin, Carlo Heinrich
Sewell, Curtis, Jr
Shank, Charles Vernon
Shar, Leonard E
Shaw, Charles Alden
Sheingold, Abraham
Shevel, Wilbert Lee
Shin, Hyungcheol
Shin, Soo H
Silverstein, Elliot Morton
Sinclair, Robert
Slaton, Jack H(amilton)
Smith, George Foster
Smith, Warren Drew
Sokolich, William Gary
Spieler, Helmuth
Spinrad, Robert J(oseph)
Spitzer, William George
Stack (Stachiewicz), B(ogdan) R(oman)
Stapelbroek, Maryn G
Staudhammer, Peter
Steenson, Bernard O(wen)
Sterling, Warren Martin
Stern, Arthur Paul
Stewart, James A
Stewart, Richard William
Sturm, Walter Allan
Szentirmai, George
Sziklai, George C(lifford)
Tauber, Richard Norman
Thiene, Paul G(eorge)
Thomas, Hubert Jon
Tilles, Abe
Torgow, Eugene N
Trenholme, John Burgess
Trubert, Marc
Tuszynski, Alfons Alfred
Vane, Arthur B(ayard)
Vaughan, J Rodney M
Victor, Walter K
Viswanathan, Chand R
Vlay, George John
Voreades, Demetrios
Waldhauer, F D
Wang, Kang-Lung
Wang, Shyh
Ware, W(illis) H(oward)
Washburn, Jack
Webster, Emilia
White, Richard Manning
White, Stanley A
Wieder, Harry H
Wiedow, Carl Paul
Willner, Alan Eli
Wilson, Robert Gray
Wilson, William John
Wilts, Charles H(arold)
Wintroub, Herbert Jack
Wistreich, George A
Wright, John Marlin
Yang, Chung Ching
Yao, Jason
Yeh, Hen-Geul
Yeh, Paul Pao
Yeh, Yea-Chuan Milton
Yeh, Yin
Yeung, King-Wah Walter
Young, John A
Young, Konrad Kwang-Leei
Young, Richard D
Yue, A(lfred) S(hui-Choh)
Zamboni, Luciano
Zhou, Simon Zheng
Zuleeg, Rainer

COLORADO
Abshier, Curtis Brent
Adams, James Russell
Allen, James Lamar
Altschuler, Helmut Martin
Barnes, James Allen
Bauer, Charles Edward
Becker, F(loyd) K(enneth)
Boyd, Malcolm R(obert)
Brady, Douglas MacPherson
Brasch, Frederick Martin, Jr
Bushnell, Robert Hempstead
Crawford, Myron Lloyd
Dick, Donald Edward
Dixon, Robert Clyde
Eichelberger, W(illiam) H
Ellis, Donald Griffith
Elmore, Kimberly Laurence
Emery, Keith Allen
Engstrom, Herbert Leonard
Estin, Arthur John
Frick, Pieter A
Gardner, Edward Eugene
Gray, James Edward
Gupta, Kuldip Chand
Hanson, Donald Wayne
Haydon, George William
Hill, David Allan
Kanda, Motohisa
Lally, Vincent Edward

Lebiedzik, Jozef
Lee, Yung-Cheng
Lewin, Leonard
Lubell, Jerry Ira
McAllister, Robert Wallace
McManamon, Peter Michael
Mahajan, Roop Lal
Mathys, Peter
May, William G(ambrill)
Moddel, Garret R
Nahman, Norris S(tanley)
Oleszek, Gerald Michael
Ralston, Margarete A
Reiff, Glenn Austin
Sargent, Howard Harrop, III
Schiffmacher, E(dward) R(obert)
She, Chiao-Yao
Staehelin, Lucas Andrew
Strauch, Richard G
Sugar, George R
Swanson, Lawrence Ray
Tsao-Wu, Nelson Tsin
Van Pelt, Richard W(arren)
Wellman, Dennis Lee
Wertz, Ronald Duane
Zhang, Bing-Rong

CONNECTICUT
Barker, Richard Clark
Best, Stanley Gordon
Brenholdt, Irving R
Brienza, Michael Joseph
Burkhard, Mahlon Daniel
Calabrese, Carmelo
Carter, G Clifford
Castonguay, Richard Norman
Chang, Richard Kounai
Foyt, Arthur George, Jr
Gaertner, Wolfgang Wilhelm
Glomb, Walter L
Leonard, Robert F
Leonberger, Frederick John
Levenstein, Harold
Marchand, Nathan
Nehorai, Arye
Newman, Robert Weidenthal
Popplewell, James Malcolm
Rau, Richard Raymond
Reed, Joseph
Reed, Mark Arthur
Rydz, John S
Schwartz, Michael H
Schwarz, Frank
Shankar, Ramamurti
Stein, Richard Jay
Taylor, Geoff W
Troutman, Ronald R
Tuteur, Franz Benjamin
Zubal, I George

DELAWARE
Bowyer, Kern M(allory)
Bragagnolo, Julio Alfredo
Burgess, John S(tanley)
Hegedus, Steven Scott
Hunsperger, Robert G(eorge)
Ih, Charles Chung-Sen
Jansson, Peter Allan
Nielsen, Paul Herron
Silzars, Aris
Young, M(ilton) G(abriel)

DISTRICT OF COLUMBIA
Anderson, Gordon Wood
Bloch, Erich
Borgiotti, Giorgio V
Borsuk, Gerald M
Donahue, D Joseph
Galane, Irma B(ereston)
Gordon, William Bernard
Henry, Warren Elliott
Jones, Howard St Claire, Jr
Jordan, Arthur Kent
Killion, Lawrence Eugene
Knudson, Alvin Richard
Kohl, Walter H(einrich)
Kyriakopoulos, Nicholas
Leak, Lee Virn
Lebow, Irwin L(eon)
McKay, Jack Alexander
McMahon, John Michael
Marcus, Michael Jay
Ngai, Kia Ling
Nguyen, Charles Cuong
Norton, Mahlon H
Pickholtz, Raymond L
Roitman, Peter
Rojas, Richard Raimond
Roosevelt, C(ornelius) V(an) S(chaak)
Skolnik, Merrill I
Tangney, John Francis
Taylor, Archer S
Thomas, Leonard William, Sr
Townsend, Marjorie Rhodes
Webb, Denis Conrad
Weinberg, Donald Lewis
Wing, William Hinshaw
Young, Leo

FLORIDA
Adhav, Ratnakar Shankar
Ancker-Johnson, Betsy
Bowers, James Clark
Cates, Harold Thomas

Chen, Tsong-Ming
Clark, John F
Cohen, Martin Joseph
Cookson, Albert Ernest
Coulter, Wallace H
Couturier, Gordon W
Davidson, Mark Rogers
Doyon, Leonard Roger
Enger, Carl Christian
Everett, Woodrow Wilson, Jr
Fischlschweiger, Werner
Fossum, Jerry G
Froehlich, Fritz Edgar
Gottesman, Stephen T
Gruen, H(arold)
Hamman, Donald Jay
Henning, Rudolf E
Herwald, Seymour W(illis)
Hoeppner, Conrad Henry
Hoffman, Thomas R(ipton)
Holst, Gerald Carl
Houts, Ronald C(arl)
Hoyer-Ellefsen, Sigurd
Huber, William Richard, III
Huebner, Jay Stanley
Johnson, Raymond C, Jr
Keenan, Robert Kenneth
Kelso, John Morris
Kiewit, David Arnold
Kimble, Allan Wayne
Kornblith, Lester
Kretzmer, Ernest R(udolf)
Lachs, Gerard
LeVine, D(onald) J(ay)
Li Kam Wa, Patrick
Lindholm, Fredrik Arthur
Lister, Charles Allan
Ludwig, Matthias Heinz
Lund, Frederick H(enry)
Mertens, Lawrence E(dwin)
Miler, George Gibbon, Jr
Mulson, Joseph F
Newman, Roger
Oman, Robert Milton
Perrotta, Alessandro
Revay, Andrew W, Jr
Rosenblum, Harold
Ryder, John Douglass
Schmidt, Klaus H
Shea, Richard Franklin
Sheppard, Albert Parker
Singh, Rama Shankar
Suski, Henry M(ieczyslaw)
Tusting, Robert Frederick
Wade, Thomas Edward
Wang, Ru-Tsang
Watson, J(ames) Kenneth
Wu, Jin Zhong
Young, C(harles) Gilbert
Yuan, Jiann-Shiun
Ziernicki, Robert S

GEORGIA
Bodnar, Donald George
Bomar, Lucien Clay
Bruder, Jospeh Albert
Camp, Ronnie Wayne
Currie, Nicholas Charles
Fisher, Reed Edward
Gibson, John Michael
Grace, Donald J
Holtum, Alfred G(erard)
Hooper, John William
Jones, Betty Ruth
Jones, William B(enjamin), Jr
Joy, Edward Bennett
Long, Maurice W(ayne)
McMillan, Robert Walker
Meindl, James D
Milkovic, Miran
Moss, Richard Wallace
Paulin, Jerome John
Pence, Ira Wilson, Jr
Pidgeon, Rezin E, Jr
Pippin, John Eldon
Prince, M(orris) David
Robertson, Douglas Welby
Roper, Robert George
Sayle, William, II
Singletary, Thomas Alexander
Stoneburner, Daniel Lee
Tummala, Rao Ramamohana
Vail, Charles R(owe)
Vann, William L(onnie)
Wiltse, James Cornelius
Witt, Samuel N(ewton), Jr
Yokelson, Bernard J(ulius)

HAWAII
Au, Whitlow W L
Crane, Sheldon Cyr
Holm-Kennedy, James William
Koide, Frank T
Zisk, Stanley Harris

IDAHO
DeBow, W Brad
Stuffle, Roy Eugene

ILLINOIS
Adler, Robert
Agrawal, Arun Kumar
Ahuja, Narendra
Baier, William H

Electronics Engineering (cont)

Banerjee, Prithviraj
Borso, Charles S
Camras, Marvin
Chien, Andrew Andai
Coleman, James J
Coleman, P(aul) D(are)
DeVault, Don Charles
Domanik, Richard Anthony
Dyson, John Douglas
Eilers, Carl G
Errede, Steven Michael
Feinberg, Barry N
Gabriel, John R
Gupta, Nand K
Gupta, Udaiprakash I
Hajj, Ibrahim Nasri
Holonyak, N(ick), Jr
Hua, Ping
Hupert, Julius Jan Marian
Jan, Kwan-Hwa
Jostlein, Hans
Kang, Sung-Mo Steve
Kaplan, Daniel Moshe
Kaplan, Sam H
Kessler, Lawrence W
Kim, Kyekyoon K(evin)
Kitchen, W J
Kumar, Panganamala Ramana
Kustom, Robert L
Laxpati, Sharad R
Lin, James Chih-I
Loos, James Stavert
Luplow, Wayne Charles
Macrander, Albert Tiemen
Merkelo, Henri
Mikulski, James J(oseph)
Millhouse, Edward W, Jr
Murphy, Gordon J
Neuhalfen, Andrew J
Offner, Franklin Faller
O'Neil, John J
Parker, Ehi
Parzen, Philip
Raccah, Paul M(ordecai)
Rudnick, Stanley J
Schwab, Gary Michael
Smedskjaer, Lars Christian
Sobel, Alan
Staunton, John Joseph Jameson
Studtmann, George H
Taflove, Allen
Toppeto, Alphonse A
Venema, Harry J(ames)
Windhorn, Thomas H
Zell, Blair Paul
Zitter, Robert Nathan

INDIANA
Almquist, Donald John
Brown, Buck F(erguson)
Detraz, Orville R
Fearnot, Neal Edward
Hammann, John William
Johnson, Charles F
Martin, William L, Jr
Neudeck, Gerold W(alter)
Ogborn, Lawrence L
Richter, Ward Robert
Roth, Lawrence Max
Schwartz, Richard John
Stanley, Gerald R
Stein, Frank S
Thomas, Lucius Ponder
Vogelhut, Paul Otto
White, Samuel Grandford, Jr

IOWA
Haendel, Richard Stone
Henrici, Carl Ressler
Jolls, Kenneth Robert
Pierce, Roger J

KANSAS
Moore, Richard K(err)
Norris, Roy Howard
Richard, Patrick
Shanmugan, K Sam
Wright, Donald C

KENTUCKY
Cathey, Jimmie Joe
Fardo, Stephen W
Frank, John L
Kadaba, Prasad Krishna
Steele, Earl L(arsen)

LOUISIANA
Bedell, Louis Robert
Cullen, John Knox, Jr
Erath, Louis W
Grimwood, Charles
Jones, Daniel Elven
Keys, L Ken
Marshak, Alan Howard
Seaman, Ronald L
Trahan, Russell Edward, Jr

MAINE
Holshouser, Don F
Snyder, Arnold Lee, Jr

MARYLAND
Adams, Robert John
Allen, John L(oyd)
Atanasoff, John Vincent
Atia, Ali Ezz Eldin
Barbe, David Franklin
Barrack, Carroll Marlin
Beal, Robert Carl
Bennighof, R(aymond) H(oward)
Berman, Michael Roy
Bishop, Walton B
Bluzer, Nathan
Brodsky, Marc H
Bromberger-Barnea, Baruch (Berthold)
Brown, Harry Benjamin
Bukowski, Richard William
Cacciamani, Eugene Richard, Jr
Campanella, Samuel Joseph
Carp, Gerald
Coates, R(obert) J(ay)
Cohen, Stanley Alvin
Corak, William Sydney
Cornett, Richard Orin
Cresswell, Michael William
Daly, John Anthony
Daniel, Charles Dwelle, Jr
DeLamater, Edward Doane
Demmerle, Alan Michael
Drzewiecki, Tadeusz Maria
Eisner, Howard
Evans, Alan G
Fink, Don Roger
Flanigan, William Francis, Jr
Flory, Thomas Reherd
Frommer, Peter Leslie
Gabriele, Thomas L
Gammell, Paul M
Garver, Robert Vernon
Ghoshtagore, Rathindra Nath
Goepel, Charles Albert
Golding, Leonard S
Goldstein, Gordon D(avid)
Granatstein, Victor Lawrence
Greenhouse, Harold Mitchell
Gupta, Ramesh K
Gupta, Vaikunth N
Halpin, Joseph John
Hammond, Charles E
Hellwig, Helmut Wilhelm
Henderson, Beauford Earl
Ho, Henry Sokore
Horna, Otakar Anthony
Hornbuckle, Franklin L
Hyde, Geoffrey
Jackson, William David
Kanal, Laveen Nanik
Kaplan, Alexander E
Kiebler, John W(illiam)
Kravitz, Lawrence C
Kundu, Mukul Ranjan
Larrabee, R(obert) D(ean)
Leydorf, Glenn E(dwin)
Ligomenides, Panos Aristides
Lin, Hung Chang
Lowney, Jeremiah Ralph
Maccabee, Bruce Sargent
McConoughey, Samuel R
McLean, Flynn Barry
Mahle, Christoph E
Marsten, Richard B(arry)
Masters, Robert Wayne
Mayo, Santos
Michaelis, Michael
Michelsen, Arve
Miller, Kenneth M, Sr
Mills, Thomas K
Minneman, Milton J(ay)
Moore, Robert Avery
Munson, J(ohn) C(hristian)
Neale, Elaine Anne
Newcomb, Robert Wayne
Newman, David Bruce, Jr
Nusbickel, Edward M, Jr
Nylen, Marie Ussing
Oettinger, Frank Frederic
Oscar, Irving S
Palmer, Charles Harvey
Pande, Krishna P
Parakkal, Paul Fab
Paul, Dilip Kumar
Pirraglia, Joseph A
Pollack, Louis
Potzick, James Edward
Preisman, Albert
Pullen, Keats A(bbott), Jr
Raabe, Herbert P(aul)
Raines, Jeremy Keith
Rathbun, Edwin Roy, Jr
Riley, Patrick Eugene
Rosenblum, Howard Edwin
Saba, William George
Sann, Klaus Heinrich
Schafft, Harry Arthur
Schmidt, Edward Matthews
Severns, Matthew Lincoln
Shakur, Asif Mohammed
Shelton, Emma
Sheridan, Michael N
Sicotte, Raymond L
Singh, Amarjit
South, Hugh Miles
Stickler, Mitchell Gene
Sugai, Iwao
Sztankay, Zoltan Geza
Thakor, Nitish Vyomesh
Thoma, George Ranjan
Timm, Raymond Stanley
Tippett, James T
Vergara, William Charles
Wagner, James W
Watters, Edward C(harles), Jr
Weaver, Christopher Scot
Weber, Richard Rand
Weinreb, Sander
Woolston, Daniel D
Yang, Xiaowei
Young, C(harles), Jr

MASSACHUSETTS
Abraham, L(eonard) G(ladstone), Jr
Adlerstein, Michael Gene
Alexander, Michael Norman
Allen, Thomas John
Altshuler, Edward E
Aprille, Thomas Joseph, Jr
Ball, John Allen
Band, Hans Eduard
Banderet, Louis Eugene
Barrington, A(lfred) E(ric)
Barton, David Knox
Bell, Richard Oman
Benton, Stephen Anthony
Blake, Carl
Blidberg, D Richard
Blum, John Bennett
Bose, Amar G(opal)
Bowman, David F(rancis)
Brayer, Kenneth
Breton, J Raymond
Brierley, Steven K
Brookner, Eli
Bugnolo, Dimitri Spartaco
Bussgang, J(ulian) J(akob)
Cardiasmenos, Apostle George
Castro, Alfred A
Chakrabarti, Supriya
Chang, Robin
Chase, Charles Elroy, Jr
Chave, Alan Dana
Clark, Melville, Jr
Clifton, Brian John
Colby, George Vincent, Jr
Comer, Joseph John
Corey, Brian K
Counselman, Charles Claude, III
Cronson, Harry Marvin
Dale, Brian
Dallos, Andras
Davis, Charles Freeman, Jr
Decker, C(harles) David
Dixit, Sudhir S
Drouilhet, Paul R
Dyer, James Arthur
Eaves, Reuben Elco
Fano, Robert M(ario)
Feist, Wolfgang Martin
Figwer, J(ozef) Jacek
Forney, G(eorge) David, Jr
Forrester, Jay W
Foster, Caxton Croxford
Fowler, Charles A(lbert)
Fraser, Donald C
Galvin, Aaron A
Garvey, R(obert) Michael
Gaw, C Vernon
Geary, John Charles
Giordano, Arthur Anthony
Grabel, Arvin
Grass, Albert M(elvin)
Gray, Truman S(tretcher)
Gregers-Hansen, Vilhelm
Groginsky, Herbert Leonard
Gross, Fritz A
Guidice, Donald Anthony
Hart, Harold M(artin)
Haugsjaa, Paul O
Hempstead, Charles Francis
Hershberg, Philip I
Higgins, W(illiam) F(rancis)
Hockney, Richard L
Holmstrom, Frank Ross
Holway, Lowell Hoyt, Jr
Hope, Lawrence Latimer
Horowitz, Barry Martin
Hurlburt, Douglas Herendeen
Hursh, John W(oodworth)
Ilic, Marija
Ippen, Erich Peter
James, George Ellert
Kanter, Irving
Keicher, William Eugene
Kincaid, Thomas Gardiner
Kosowsky, David I
Kovaly, John J
Kovatch, George
Krasner, Jerome L
Kyhl, Robert Louis
Lanyon, Hubert Peter David
Levesque, Allen Henry
Loretz, Thomas J
McCue, John Joseph Gerald
Mack, Richard Bruce
McKnight, Lee Warren
Manley, Harold J(ames)
Mavretic, Anton
Miffitt, Donald Charles
Milburn, Nancy Stafford
Montazer, G Hosein
Moore, Robert Edmund
Morrow, Walter E, Jr
Mullen, James A
Mulukutla, Sarma Sreerama
Naka, F(umio) Robert
Navon, David H
Nesbeda, Paul
Nesline, Frederick William, Jr
Oliner, Arthur A(aron)
Palubinskas, Felix Stanley
Pearlman, Michael R
Penndorf, Rudolf
Peura, Robert Allan
Poon, Chi-Sang
Price, Robert
Raffel, Jack I
Rediker, Robert Harmon
Reich, Herbert Joseph
Reif, L Rafael
Reintjes, J Francis
Remmel, Ronald Sylvester
Renner, Gerard W
Roberge, James Kerr
Rochefort, John S
Rudenberg, H(ermann) Gunther
Ruze, John
Safford, Richard Whiley
Sage, Jay Peter
Schloemann, Ernst
Schreiber, William F(rancis)
Schwartz, Jack
Searle, Campbell L(each)
Senturia, Stephen David
Shen, Hao-Ming
Shepherd, Freeman Daniel, Jr
Singleton, John Byrne
Skolnikoff, Eugene B
Smith, Alan Bradford
Sommer, Alfred Hermann
Statz, Hermann
Stevenson, David Michael
Stewart, Lawrence Colm
Stiffler, Jack Justin
Stiglitz, Irvin G
Stockman, Harry E
Strelzoff, Alan G
Strimling, Walter Eugene
Teng, Chojan
Thun, Rudolf Eduard
Tsandoulas, Gerasimos Nicholas
Tsaur, Bor-Yeu
Tuller, Harry Louis
Ward, James
Wasserman, Jerry
Wilensky, Samuel
Yannoni, Nicholas

MICHIGAN
Baldwin, Keith Malcolm
Becher, William D(on)
Bhattacharya, Pallab Kumar
Birdsall, Theodore G
Brown, William M(ilton)
Brumm, Douglas B(ruce)
Bryant, John H(arold)
Cheydleur, Benjamin Frederic
Crary, Selden Bronson
Dow, W(illiam) G(ould)
Eagen, Charles Frederick
Eldred, Norman Orville
Essner, Edward Stanley
Frank, Max
Gaerttner, Martin R
Getty, Ward Douglas
Ghoneim, Youssef Ahmed
Giacoletto, L(awrence) J(oseph)
Gustafson, Herold Richard
Haddox, Mark
Heremans, Joseph P
Horvath, Ralph S(teve)
Irani, Keki B
Jacovides, Linos J
Jain, Kailash Chandra
Johansen, Elmer L
Johnson, Wayne Jon
Jones, R(ichard) James
Khargonekar, Pramod P
Kim, Changhyun
Kozma, Adam
Kulkarni, Anand K
Larrowe, Vernon L
Li, Shin-Hwa
Lomax, Ronald J(ames)
MacGregor, Donald Max
Malaczynski, Gerard W
Meitzler, Allen Henry
Miller, M(urray) H(enri)
Nagy, Louis Leonard
Nazerian, Keyvan
Pavlidis, Dimitris
Pryor, Roger Welton
Ristenbatt, Marlin P
Salam, Fathi M A
Schlax, T(imothy) Roth
Schubring, Norman W(illiam)
Sengupta, Dipak L(al)
Sethi, Ishwar Krishan
Sloan, Martha Ann
Wennerberg, A(llan) L(orens)
Woodyard, James Robert
Yeh, Chai

MINNESOTA
Bailey, Fredric Nelson

Champlin, Keith S(chaffner)
Dubbe, Richard F
Follingstad, Henry George
Fritze, Curtis W(illiam)
Gallo, Charles Francis
George, Peter Kurt
Glewwe, Carl W(illiam)
Gruber, Carl L(awrence)
Haxby, B(ernard) V(an Loan)
Lin, Benjamin Ming-Ren
Lutes, Olin S
Mooers, Howard T(heodore)
Olyphant, Murray, Jr
Premanand, Visvanatha
Sadjadi, Firooz Ahmadi
Schumer, Douglas Brian
Stebbings, William Lee
Sutton, Matthew Albert
Thiede, Edwin Carl
Tran, Nang Tri
Van Doeren, Richard Edgerly
Werth, Richard George
Wesenberg, Clarence L
Yasko, Richard N

MISSISSIPPI
Smith, Charles Edward

MISSOURI
Brackmann, Richard Theodore
Carson, Ralph S(t Clair)
Ellison, John Vogelsanger
Flowers, Harold L(ee)
Franklin, Mark A
Gordon, Albert Raye
Griffith, Virgil Vernon
Hahn, James H
Hill, Dale Eugene
Kozlowski, Don Robert
Landiss, Daniel Jay
Leopold, Daniel J
Manson, Donald Joseph
Moss, Randy Hays
Pickard, William Freeman
Richards, Earl Frederick
Rosenbaum, Fred J(erome)
Ross, Monte
Ryan, Carl Ray
Taylor, Ralph Dale
Wagner, Robert G
White, Warren D
Zobrist, George W(inston)

MONTANA
Gerez, Victor
Latham, Don Jay
Pierre, Donald Arthur
Robinson, Clark Shove, Jr
Weaver, Donald K(essler), Jr

NEBRASKA
Bahar, Ezekiel
Lenz, Charles Eldon

NEVADA
Haun, Robert Dee, Jr
Johnson, Bruce Paul
Rawat, Banmali Singh
Telford, James Wardrop

NEW HAMPSHIRE
Corum, Kenneth Lawrence
Egan, John Frederick
Frost, Albert D(enver)
Hanson, Per Roland
Jeffery, Lawrence R
Misra, Alok C
Simmons, Alan J(ay)
Stephenson, J(ohn) Gregg
Von-Recklinghausen, Daniel R
Weiler, Margaret Horton
Winn, Alden L(ewis)
Wolff, Nikolaus Emanuel

NEW JERSEY
Alig, Roger Casanova
Allen, Jonathan
Amitay, Noach
Anandan, Munisamy
Anderson, Milton Merrill
Anderson, Robert E(dwin)
Avins, Jack
Ballato, Arthur
Bechtle, Daniel Wayne
Berthold, Joseph Ernest
Bhagat, Pramode Kumar
Bharj, Sarjit Singh
Blanck, Andrew R(ichard)
Blicher, Adolph
Bloom, Stanley
Borkan, Harold
Bosacchi, Bruno
Brown, Alfred Bruce, Jr
Brown, Irmel Nelson
Brudner, Harvey Jerome
Bucher, T(homas) T(albot) Nelson
Buckley, R Russ
Candy, J(ames) C(harles)
Carnes, James Edward
Carr, William N
Castenschiold, Rene
Cazes, Jack
Celler, George K
Chand, Naresh

Chang, Tao-Yuan
Cho, Alfred Y
Chu, Sung Nee George
Chung, Yun C
Chynoweth, Alan Gerald
Clements, Wayne Irwin
Coffman, Edward G, Jr
Cohen, Abraham Bernard
Crochiere, Ronald E
Daly, Daniel Francis, Jr
D'Asaro, L Arthur
Denlinger, Edgar J
Derkits, Gustav
Detig, Robert Henry
Dickerson, L(oren) L(ester), Jr
Dickey, E(dward) T(hompson)
Dingle, Raymond
Dixon, Samuel, Jr
Dodabalapur, Ananth
Dodington, Sven Henry Marriott
Dolny, Gary Mark
Dwyer, Harry, III
Enstrom, Ronald Edward
Farber, Herman
Feuer, Mark David
Finkel, Leonard
Fishbein, William
French, Larry J
Freund, Robert Stanley
Gabriel, Edwin Z
Gibson, James John
Giordmaine, Joseph Anthony
Goldberg, Edwin A(llen)
Goldstein, Robert Lawrence
Golin, Stuart
Goodman, David Joel
Gossmann, Hans Joachim
Grimmell, William C
Grosso, Anthony J
Gualtieri, Devlin Michael
Guenzer, Charles S P
Hakim, Edward Bernard
Hang, Hsueh-Ming
Harris, James Ridout
Hasegawa, Ryusuke
Hernqvist, Karl Gerhard
Herold, E(dward) W(illiam)
Hilibrand, J(ack)
Hittinger, William C(harles)
Hollywood, John M(atthew)
Hong, Won-Pyo
Hoover, Charles Wilson, Jr
Horn, David Nicholas
Hubbard, William Marshall
Hughes, James Sinclair
Hwang, Cherng-Jia
Johannes, Virgil Ivancich
Johnson, Walter C(urtis)
Kaminow, Ivan Paul
Katzmann, Fred L
Kesselman, Warren Arthur
Kieburtz, R(obert) Bruce
Kihn, Harry
Klapper, Jacob
Knausenberger, Wulf H
Kniazuk, Michael
Kogelnik, H W
Krevsky, Seymour
Kuo, Ying L
Lawson, James Robert
Lechner, Bernard J
Levi, Anthony Frederic John
Li, John Kong-Jiann
Lin, Chinlon
Link, Fred M
Lubowe, Anthony G(arner)
Lunsford, Ralph D
Luryi, Serge
Maa, Jer-Shen
McLane, George Francis
McNair, Irving M, Jr
Mac Rae, Alfred Urquhart
MacRae, Alfred Urquhart
Mangulis, Visvaldis
Martin, Thomas A(ddenbrook)
Mauzey, Peter T
Meixler, Lewis Donald
Mesner, Max H(utchinson)
Millar, J(ulian) Z(immerman)
Millman, Sidney
Minter, Jerry Burnett
Mondal, Kalyan
Morgan, Dennis Raymond
Mueller, Charles W(illiam)
Murthy, Srinivasa K R
Natapoff, Marshall
Ninke, William Herbert
Onyshkevych, Lubomyr S
Orlando, Carl
Ourmazd, Abbas
Panayotatos, Paul
Partovi, Afshin
Pei, Shin-Shem
Perlman, Barry Stuart
Rainal, Attilio Joseph
Ravindra, Nuggehalli Muthanna
Reingold, Irving
Riddle, George Herbert Needham
Robrock, Richard Barker, II
Rulf, Benjamin
Russell, Frederick A(rthur)
Saltzberg, Burton R
Schlam, Elliott
Schmidt, Barnet Michael

Schneider, Sol
Schulke, Herbert Ardis, Jr
Schulte, Harry John, Jr
Schwering, Felix
Sears, Raymond Warrick, Jr
Segelken, John Maurice
Semmlow, John Leonard
Shahbender, R(abah) A(bd-El-Rahman)
Shumate, Paul William, Jr
Sipress, Jack M
Sterzer, Fred
Strother, J(ohn) A(lan)
Suhir, Ephraim
Taylor, Harold Evans
Thornton, C G
Tishby, Naftali Z
Tsaliovich, Anatoly
Uhrig, Jerome Lee
Ullrich, Felix Thomas
Van Etten, James P(aul)
Verma, Pramode Kumar
Wagner, Sigurd
Wang, Chao Chen
Wang, Chen-Show
Warters, William Dennis
White, Lawrence Keith
Wilkinson, W(illiam) C(layton)
Wittenberg, Albert M
Woo, Nam-Sung
Wullert, John R, II
Yi-Yan, Alfredo
Yoon, Euijoon
Zaininger, Karl Heinz
Ziegler, Hans K(onrad)

NEW MEXICO
Anderson, Richard Ernest
Berrie, David William
Brock, Ernest George
Brueck, Steven Roy Julien
Burr, Alexander Fuller
Castle, John Granville, Jr
Cole, Edward Issac, Jr
Dawes, William Redin, Jr
Doss, James Daniel
England, Talmadge Ray
Enke, Christie George
Gilbert, Alton Lee
Gourley, Paul Lee
Gover, James E
Gurbaxani, Shyam Hassomal
Haberstich, Albert
Hankins, Timothy Hamilton
Hardin, James T
Hines, William Curtis
Jones, James Jordan
Korman, N(athaniel) I(rving)
Land, Cecil E(lvin)
Linford, Rulon Kesler
Maez, Albert R
Marker, Thomas F(ranklin)
Metzger, Daniel Schaffer
Miller, Edmund K(enneth)
Norris, Carroll Boyd, Jr
Northrup, Clyde John Marshall, Jr
Overhage, Carl F J
Parker, Joseph R(ichard)
Potter, James Martin
Rach, Randolph Carl
Rigrod, William W
Roberts, Peter Morse
Schueler, Donald G(eorge)
Shafer, Robert E
Shipley, James Parish, Jr
Sinha, Dipen N
Smith, William Conrad
Stein, William Earl
Swingle, Donald Morgan
Tapp, Charles Millard

NEW YORK
Abetti, Pier Antonio
Ahmed-Zaid, Said
Albicki, Alexander
Anderson, Roy E
Anderson, Wayne Arthur
Apte, Chidanand
Arams, F(rank) R(obert)
Bachman, Henry L(ee)
Baertsch, Richard D
Ballantyne, Joseph M(errill)
Basavaiah, Suryadevara
Bertoni, Henry Louis
Bilderback, Donald Heywood
Bingham, J Peter
Bloch, Alan
Borrego, Jose M
Bradshaw, John Alden
Brantley, William Cain, Jr
Brennemann, Andrew E(rnest), Jr
Buchta, John C(harles)
Busnaina, Ahmed A
Carlin, Herbert J
Chen, Alek Chi-Heng
Chiang, Yuen-Sheng
Chow, Tat-Sing Paul
Christiansen, Donald David
Clarke, Kenneth Kingsley
Coden, Michael H
Comly, James B
Coppola, Patrick Paul
Cottingham, James Garry
Daijavad, Shahrokh
Damouth, David Earl

Das, Pankaj K
Debany, Warren Harding, Jr
De France, Joseph J
Delgado-Frias, Jose
Derman, Samuel
Dermit, George
Diamond, Fred Irwin
Dimmler, D(ietrich) Gerd
DiStefano, Thomas Herman
Drossman, Melvyn Miles
Eisenberg, Lawrence
Engeler, William E
Fang, Frank F
Feerst, Irwin
Feth, George C(larence)
Fey, Curt F
Fisher, George M C
Forster, William H(all)
Freundlich, Martin M
Fried, Erwin
Gabelman, Irving J(acob)
Genova, James John
Ghezzo, Mario
Glinkowski, Mietek T
Golden, Robert K
Good, William E
Gran, Richard J
Gratian, J(oseph) Warren
Green, Paul E(liot), Jr
Grosewald, Peter
Grumet, Alex
Hahnel, Alwin
Hall, Robert Noel
Harnden, John D, Jr
Harris, Bernard
Hartwig, Curtis P
Hawkins, Gilbert Allan
Herman, Lawrence
Higinbotham, William Alfred
Hill, Cliff Otis
Hsiang, Thomas Y
Hunts, Barney Dean
Hyman, Abraham
Isaacson, Michael Saul
Jatlow, J(acob) L(awrence)
Jelinek, Frederick
Jorne, Jacob
Kane, John Vincent, Jr
Keonjian, Edward
Ketchen, Mark B
Keyes, Robert William
Kiang, Ying Chao
Kiehl, Richard Arthur
King, Marvin
Kinnen, Edwin
Kornberg, Fred
Kornreich, Philipp G
Kritz, J(acob)
Kuo, Yue
Kuper, J B Horner
Kwok, Thomas Yu-Kiu
Lafferty, James M(artin)
LaMuth, Henry Lewis
Laponsky, Alfred Baer
La Russa, Joseph Anthony
Lebenbaum, Matthew T(obriner)
Lee, Young-Hoon
Litchford, George B
Littauer, Raphael Max
Logue, J(oseph) C(arl)
Lubowsky, Jack
Ludwig, Gerald W
Lustig, Howard E(ric)
McGee-Russell, Samuel M
McMurry, William
Maldonado, Juan Ramon
Martens, Alexander E(ugene)
Mercer, Kermit R
Meth, Irving Marvin
Metzger, Ernest Hugh
Mgrdechian, Raffee
Mihran, Theodore Gregory
Moe, George Wylbur
Moghadam, Omid A
Mogro-Campero, Antonio
Morris, James Eliot
Morrison, John B
Morse, William M
Mortenson, Kenneth Ernest
Moy, Dan
Nelson, David Elmer
Ning, Tak Hung
Oh, Byungdu
Okwit, Seymour
Ortel, William C(harles) G(ormley)
Packard, Karle Sanborn, Jr
Palocz, Istvan
Paraszczak, Jurij Rostyslan
Parks, Harold George
Peled, Abraham
Platner, Edward D
Plotkin, Martin
Polychronakos, Venetios Alexander
Powell, Noble R
Quan, Weilun
Queen, Daniel
Raghuveer, Mysore R
Redington, Rowland Wells
Rosen, Paul
Rosenthal, Saul W
Ruddick, James John
Sackman, George Lawrence
Sai-Halasz, George Anthony
Salzberg, Bernard

Electronics Engineering (cont)

Sante, Daniel P(aul)
Sasserath, Jay N
Satya, Akella V S
Savic, Michael I
Schachter, Rozalie
Schamberger, Robert Dean
Schanker, Jacob Z
Schindler, Joe Paul
Schoenfeld, Robert Louis
Schuster, Frederick Lee
Schwartz, Mischa
Schwartzman, Leon
Schwarz, Ralph J
Sewell, Frank Anderson, Jr
Shamash, Yacov A
Shaw, John A
Sheryll, Richard Perry
Shevack, Hilda N
Silverman, Gordon
Simpson, Murray
Singer, Barry M
Sorokin, Peter
Spencer, David R
Spiegel, Robert
Spielman, Harold S
Spiro, Julius
Srinivasan, G(urumakonda) R
Star, Joseph
Stenger, Richard J
Stringer, L(oren) F(rank)
Su, Stephen Y H
Szymanski, Boleslaw K
Temple, Victor Albert Keith
Terman, Lewis Madison
Thiel, Frank L(ouis)
Trevoy, Donald James
Trexler, Frederick David
Varshney, Pramod Kumar
Vogelman, Joseph H(erbert)
Wald, Alvin Stanley
Walter, William Trump
Webster, Harold Frank
Wei, Ching-Yeu
Whalen, James Joseph
Wheatley, W(illiam) A(rthur)
Wie, Chu Ryang
Wiesner, Leo
Wilson, James William Alexander
Witt, Christopher John
Woods, John William
Worthing, Jurgen
Wouk, Victor
Yang, Edward S
Zdan, William
Zukowski, Charles Albert

NORTH CAROLINA
Appleby, Robert H
Bailey, Kincheon Hubert, Jr
Bailey, Robert Brian
Baliga, B Jayant
Brinn, Jack Elliott, Jr
Buiting, Francis P
Burger, Robert M
Casey, Horace Craig, Jr
Defoggi, Ernest
Dotson, Marilyn Knight
Easter, William Taylor
Franzon, Paul Damian
Gray, James P
Hackenbrock, Charles Robert
Hauser, John Reid
Haynes, John Lenneis
Hutchins, William R(eagh)
Kanopoulos, Nick
Kelly, John Henry
Kuhn, Matthew
Leamy, Harry John
Lovick, Robert Clyde
McEnally, Terence Ernest, Jr
Massoud, Hisham Z
Moses, Montrose James
Nagle, H Troy
Russell, Lewis Keith
Sander, William August, III
Steer, Michael Bernard
Stroscio, Michael Anthony
Tischer, Frederick Joseph
Van Horn, Diane Lillian
Wakeman, Charles B
Walters, Mark David
Wooten, Frank Thomas
Wortman, Jimmie J(ack)

NORTH DAKOTA
Kuruganty, Sastry P
Yuvarajan, Subbaraya

OHIO
Antler, Morton
Arnett, Jerry Butler
Barranger, John P
Berthold, John William, III
Blessinger, Michael Anthony
Connolly, Denis Joseph
Cummings, Charles Arnold
Cummins, Stewart Edward
Dell'Osso, Louis Frank
De Lucia, Frank Charles
Deubner, Russell L(eigh)
Duff, J(ack) E(rrol)
Duff, Robert Hodge

Easterday, Jack L(eroy)
El-Naggar, Mohammed Ismail
Falkenbach, George J(oseph)
Fritsch, Klaus
Gilfert, James C(lare)
Hazen, Richard R(ay)
Hill, James Stewart
Hostetler, Jeptha Ray
Hunter, William Winslow, Jr
Ingram, David Christopher
Ismail, Amin Rashid
Kazimierczuk, Marian K
Keim, John Eugene
Keyes, Marion Alvah, IV
Ko, Wen Hsiung
Kosel, Peter Bohdan
Kosmahl, Henry G
Laskowski, Edward L
Lawson, Kenneth Dare
Madden, Thomas Lee
Middleton, Arthur Everts
Miller, Carl Henry, Jr
Moon, Donald Lee
Morrill, Charles D(uncker)
Neuman, Michael R
Papachristou, Christos A
Pinchak, Alfred Cyril
Ramamoorthy, Panapakkam A
Redlich, Robert Walter
Rooney, Victor Martin
Rose, Harold Wayne
Rothstein, Jerome
Schneider, John Matthew
Schutta, James Thomas
Snider, John William
Suter, Bruce Wilsey
Thorbjornsen, Arthur Robert
Vlcek, Donald Henry
Wickersham, Charles Edward, Jr
Wigington, Ronald L
Williams, Richard Alvin
Wong, Anthony Sai-Hung

OKLAHOMA
Baker, James E
Berbari, Edward J
Brown, Graydon L
Buck, Richard F
Cook, Charles F(oster), Jr
Cronenwett, William Treadwell
Grischkowsky, Daniel Richard
Hadley, Charles F(ranklin)
Hairfield, Harrell D, Jr
Hawley, Paul F(rederick)
McDevitt, Daniel Bernard
Robinson, Enders Anthony
Schwenker, J(ohn) E(dwin)
Taylor, William L
VanDerwiele, James Milton

OREGON
Chrzanowska-Jeske, Malgorzata Ewa
Clark, James Orie, II
Collins, John A(ddison)
Ede, Alan Winthrop
Engelbrecht, Rudolf S
Gilbert, Barrie
Graham, Beardsley
Hackleman, David E
Morgan, Merle L(oren)
Owen, Sydney John Thomas
Paarsons, James Delbert
Plummer, James Walter
Schaumann, Rolf
Schwartz, James William
Skinner, Richard Emery
Stringer, Gene Arthur
Wagner, Orvin Edson
Yamaguchi, Tadanori
Yau, Leopoldo D

PENNSYLVANIA
Abend, Kenneth
Adams, Raymond F
Advani, Sunder
Akruk, Samir Rizk
Alexander, Frank Creighton, Jr
Allen, Charles C(orletta)
Anouchi, Abraham Y
Arora, Vijay Kumar
Ashok, S
Barber, M(ark) R
Beier, Eugene William
Belohoubek, Erwin F
Bhagavatula, VijayaKumar
Bilgutay, Nihat Mustafa
Bowers, Klaus Dieter
Bowler, David Livingstone
Bradley, W(illiam) E(arle)
Brastins, Auseklis
Burton, Larry C(lark)
Cendes, Zoltan Joseph
Chaplin, Norman John
Clapp, Richard Gardner
Davis, Jon Preston
Davis, Paul Cooper
Davis, R(obert) E(lliot)
Dick, George W
Director, Stephen William
Druffel, Larry Edward
Dumbri, Austin C
Eberhardt, Nikolai
Elder, H E
Emtage, Peter Roesch

Fan, Ningping
Farr, Kenneth E(dward)
Favret, A(ndrew) G(illigan)
Fisher, Tom Lyons
Flattery, David Kevin
Fonash, Stephan J(oseph)
Gewartowski, J(ames) W(alter)
Goldman, Robert Barnett
Goldwasser, Samuel M
Goutmann, Michel Marcel
Grace, Harold P(adget)
Grim, Larry B
Hahn, Peter Mathias
Harrold, Ronald Thomas
Hess, Dennis William
Hinton, Raymond Price
Jaffe, Donald
Jang, Sei Joo
Jenny, Hans K
Jordan, Angel G
Kaufman, William Morris
Kazahaya, Masahiro Matt
Keizer, Eugene O(rville)
Kryder, Mark Howard
Langer, Dietrich Wilhelm
Laughlin, David Eugene
Leas, J(ohn) W(esley)
Lee, Chin-Chiu
Lempert, Joseph
Licini, Jerome Carl
Lipsky, Stephen E
Long, Alton Los, Jr
Lory, Henry James
Lu, Chih Yuan
Martin, Aaron J
Meiksin, Zvi H(ans)
Mitchell, John Douglas
Mowrey, Gary Lee
Murphy, Bernard T
Nagel, G(eorge) W(ood)
Niewenhuis, Robert James
Oneal, Glen, Jr
O'Neill, John Cornelius
Ontell, Marcia
Paulish, Daniel John
Peterson, James Oliver
Pinkerton, John Edward
Post, Irving Gilbert
Putman, Thomas Harold
Revesz, George
Riebman, Leon
Rodrigues, Anil Noel
Roy, Pradip Kumar
Sabnis, Anant Govind
Schnable, George Luther
Schroeder, Alfred C(hristian)
Shackelford, Charles L(ewis)
Shibib, M Ayman
Showers, Ralph M(orris)
Shuch, H Paul
Shung, K Kirk
Sibul, Leon Henry
Simaan, Marwan
Smits, Friedolf M
Stehle, Philip McLellan
Stinger, Henry J(oseph)
Stone, Douglas Roy
Stone, Robert P(orter)
Szepesi, Zoltan Paul John
Thompson, Eric Douglas
Thompson, Fred C
Tomiyasu, Kiyo
Trotter, Nancy Louisa
Van Der Spiegel, Jan
Wagner, David Loren
White, George Rowland
Zebrowitz, S(tanley)

RHODE ISLAND
Baker, Walter L(ouis)
Grossi, Mario Dario
Krikorian, John Sarkis, Jr
Kroenert, John Theodore
Petrou, Panayiotis Polydorou

SOUTH CAROLINA
Askins, Harold Williams, Jr
Blevins, Maurice Everett
Bourkoff, Etan
Doherty, William Humphrey
Dumin, David Joseph
Faust, John William, Jr
Garzon, Ruben Dario
Harrison, James William, Jr
Hogan, Joseph C(harles)
Miller, Lee Stephen
Pritchard, Dalton H
Pursley, Michael Bader
Sawers, James Richard, Jr
Sheppard, Emory Lamar

SOUTH DAKOTA
Brown, Michael Lee
Nelson, Vernon Ronald
Reuter, William L(ee)
Ross, Keith Alan
Swiden, LaDell Ray

TENNESSEE
Ball, Frances Louise
Barr, Dennis Brannon
Bose, Bimal K
Bundy, Robert W(endel)
Davidson, Robert C

Douglass, Terry Dean
Eads, B G
George, Ted Mason
Joy, David Charles
Lyon, James F
Mosko, Sigmund W
Ornitz, Barry Louis
Pearsall, S(amuel) H(aff)
Rajan, Periasamy Karivaratha
Rome, James Alan
Trivedi, Mohan Manubhai
Watson, Robert Lowrie
Williams, Charles Wesley

TEXAS
Abraham, Jacob A
Balasinski, Artur
Banerjee, Sanjay Kumar
Barton, James Brockman
Basham, Jerald F
Bate, Robert Thomas
Bhuva, Rohit L
Biard, James R
Brakefield, James Charles
Bronaugh, Edwin Lee
Brown, Glenn Lamar
Callas, Gerald
Carvajal, Fernando David
Chamberlain, Nugent Francis
Chang, Christopher Teh-Min
Chu, Shirley Shan-Chi
Chu, Ting Li
Chyu, Ming-Chien
Cole, David F
Collins, Dean Robert
Cory, William Eugene
Crossley, Peter Anthony
Curtin, Richard B
Davidson, David Lee
Dougal, Arwin A(delbert)
Dykstra, Jerald Paul
Eknoyan, Ohannes
Esquivel, Agerico Liwag
Feit, Eugene David
Frazer, Marshall Everett
Frensley, William Robert
Garza, Cesar Manuel
Ghowsi, Kiumars
Godbey, John Kirby
Gordon, Millard F(reeman)
Grissom, Charles Franklin
Grobman, Warren David
Hahn, Larry Alan
Harper, James George
Hartley, Craig Jay
Hasty, Turner Elilah
Hazen, Gary Alan
Hinrichs, Paul Rutland
Hopkins, George H, Jr
Iyer, Ramakrishnan S
Kirk, Wiley Price
Krishen, Kumar
Kristiansen, Magne
Lawrence, Joseph D, Jr
Levine, Jules David
Linder, John Scott
Ling, Hao
McDavid, James Michael
Mao, Shing
Mitchell, John Charles
O'Keefe, Dennis Robert
Overzet, Lawrence John
Parker, Cleofus Varren, Jr
Penn, Thomas Clifton
Philippe, Edouard Antoine
Prabhu, Vasant K
Pritchard, John Paul, Jr
Pritchett, William Carr
Rabson, Thomas A(velyn)
Rhodes, John Rathbone
Riter, Stephen
Schroen, Walter
Schulz, Richard Burkart
Schwetman, Herbert Dewitt
Singh, Vijay Pal
Sloan, Tod Burns
Smith, Reid Garfield
Sobol, Harold
Stenback, Wayne Albert
Stone, Orville L
Strieter, Frederick John
Styblinski, Maciej A
Summers, Richard Lee
Swartzlander, Earl Eugene, Jr
Taylor, Morris Chapman
Tong, Youdong
Trachtenberg, Isaac
Truchard, James Joseph
Von Maltzahn, Wolf W
White, Charles F(loyd)
Yanagisawa, Samuel T

UTAH
Atwood, Kenneth W
Barlow, Mark Owens
Butterfield, Veloy Hansen, Jr
Clegg, John C(ardwell)
Cox, Benjamin Vincent
Eernisse, Errol Peter
Grow, Richard W
Hollaar, Lee Allen
Huber, Robert John
Jeffery, Rondo Nelden
Jolley, David Kent

Lawton, Robert Arthur
Lee, James Norman
Smith, Kent Farrell
Stonehocker, Garth Hill

VERMONT
Adler, Eric
Gray, Kenneth Stewart
Halpern, William
Pires, Renato Guedes
Pricer, Wilbur David
Raab, Frederick Herbert
Roth, Wilfred
Wallis, Clifford Merrill

VIRGINIA
Alcaraz, Ernest Charles
Alter, Ralph
Anderson, Walter L(eonard)
Bahl, Inder Jit
Bailey, Marion Crawford
Beale, Guy Otis
Blume, Hans-Juergen Christian
Bordogna, Joseph
Burns, William, Jr
Cabral, Guy Antony
Choi, Junho
Chun, Myung K(i)
Cibulka, Frank
Cooper, Larry Russell
Cosby, Lynwood Anthony
Crawford, Daniel J
Curry, Thomas F(ortson)
Daniel, Kenneth Wayne
Davis, William Arthur
Dessouky, Dessouky Ahmad
Desu, Seshu Babu
Duff, William G
Entzminger, John N
Fei, Ding-Yu
Florman, Edwin F(rank)
Ganguly, Suman
Gee, Sherman
Gerry, Edward T
Ginsberg, Murry B(enjamin)
Goodman, A(lvin) M(alcolm)
Guerber, Howard P(aul)
Haar, Jack Luther
Hammond, Marvin H, Jr
Hartzler, Alfred James
Hasegawa, Ichiro
Heebner, David Richard
Hilger, James Eugene
Howard, Dean Denton
Hvatum, Hein
Inigo, Rafael Madrigal
James, W(ilbur) Gerald
Jaron, Dov
Joyner, Weyland Thomas, Jr
Kalab, Bruno Marie
Keblawi, Feisal Said
Keigler, John Edward
Kellett, Claud Marvin
Kennedy, Andrew John
Kerr, Anthony Robert
Koch, Carl Fred
Kooij, Theo
Larson, Arvid Gunnar
Lee, Fred C
Lehmann, John R(ichard)
Liceaga, Carlos Arturo
McCutchen, Samuel P(roctor)
McLucas, John L(uther)
McWright, Glen Martin
Mandelberg, Martin
Mayer, Cornell Henry
Mazzoni, Omar S
Moraff, Howard
Moseley, Sherrard Thomas
Neil, George Randall
Nunnally, Huey Neal
Nyman, Thomas Harry
Pan, Shing-Kuo
Patterson, James Douglas
Peake, Harold J(ackson)
Plait, Alan Oscar
Pollard, John Henry
Potter, Richard R(alph)
Pry, Robert Henry
Raab, Harry Frederick, Jr
Rakestraw, James William
Richard, Robert H(enry)
Robinson, Arthur S
Rogers, Thomas F
Rudmin, Joseph Webster
Schlegelmilch, Reuben Orville
Schneck, Paul Bennett
Schneider, Ralph Jacob
Schutt, Dale W
Seibel, Hugo Rudolf
Seiler, Steven Wing
Siegel, Clifford M(yron)
Smith, Carey Daniel
Smith, Sidney Taylor
Spitzer, Cary Redford
Srikanth, Sivasankaran
Stephenson, Frederick William
Stermer, Robert L, Jr
Stone, Harris B(obby)
Tole, John Roy
Tripp, John Stephen
Tsang, Wai Lin
VanDerlaske, Dennis P
Weinstein, Iram J

Wolicki, Eligius Anthony
Woods, Roy Alexander
Worthington, John Wilbur
Yang, Ta-Lun

WASHINGTON
Boatman, Edwin S
Craine, Lloyd Bernard
Daniel, J(ack) Leland
Darling, Robert Bruce
Dunham, Glen Curtis
Gillingham, Robert J
Gunderson, Leslie Charles
Heilenday, Frank W (Tod)
Heindsmann, T(heodore) E(dward)
Henry, Robert R
Hirsch, Horst Eberhard
Hoisington, David B(oysen)
Hueter, Theodor Friedrich
Kim, Jae Hoon
Lorenzen, Howard O(tto)
Morton, Randall Eugene
Nomura, Kaworu Carl
Peterson, Arnold (Per Gustaf)
Pichal, Henri Thomas, II
Preikschat, F(ritz) K(arl)
Ringo, John Alan
Russell, James Torrance
Smilen, Lowell I
Stanley, Hugh P
Thorne, Charles M(orris)
Tsang, Leung

WEST VIRGINIA
Gould, Henry Wadsworth
Osborn, Claiborn Lee

WISCONSIN
Anderson, James Gerard
Anway, Allen R
Bajikar, Sateesh S
Bomba, Steven James
Brauer, John Robert
Coggins, James Ray
Crawmer, Daryl E
Davis, Thomas William
Divjak, August A
Dunn, Stanley Austin
Flax, Stephen Wayne
Hind, Joseph Edward
Jeutter, Dean Curtis
Kerkman, Russel John
Klein, Carl Frederick
Nordman, James Emery
Schlicke, Heinz M
Thompson, Paul DeVries
Vanderheiden, Gregg
Webster, John Goodwin
Yin, Jun-Jie
Zelazo, Nathaniel Kachorek

WYOMING
Constantinides, Christos T(heodorou)

ALBERTA
Cormack, George Douglas
Haslett, James William
Hunt, Robert Nelson
Johnston, Ronald Harvey
Liu, Dennis Dong

BRITISH COLUMBIA
Deen, Mohamed Jamal
Hafer, Louis James
Jull, Edward Vincent
Menon, Thuppalay K
Pomeroy, Richard James
Sallos, Joseph
Zielinski, Adam

MANITOBA
Bridges, E(rnest)
Delecki, Andrew
Shafai, Lotfollah

NEWFOUNDLAND
Rahman, Md Azizur

NOVA SCOTIA
Baird, Charles Robert
Bhartia, Prakash
El-Masry, Ezz Ismail
Roger, William Alexander
Wood, Eunice Marjorie

ONTARIO
Bahadur, Birendra
Bandler, John William
Barclay, Alexander Primrose Hutcheson
Bhatnagar, Yashraj Kishore
Bigham, Clifford Bruce
Brach, Eugene Jenö
Brannen, Eric
Carr, Jan
Carroll, John Millar
Celinski, Olgierd J(erzy) Z(dzislaw)
Chamberlain, S(avvas) G(eorgiou)
Chow, Yung Leonard
Cohn-Sfetcu, Sorin
Costain, Cecil Clifford
Earnshaw, John W
Fahmy, Moustafa Mahmoud
Fewer, Darrell R(aymond)
Ganza, Kresimir Peter

Greenaway, Keith R(ogers)
Hayes, J Edmund
Hobson, John Peter
Jull, George W(alter)
Jullien, Graham Arnold
Jurkus, Algirdas Petras
Keeler, John S(cott)
Kuiper-Goodman, Tine
Langille, Robert C(arden)
Lemyre, C(lement)
Lindsey, George Roy
McIntosh, Bruce Andrew
Mark, Jon Wei
Mouftah, Hussein T
Nevitt, H J Barrington
Northwood, Derek Owen
Ratz, H(erbert) C(harles)
Rauch, Sol
Ryan, Dave
Sablatash, Mike
Salama, Clement Andre Tewfik
Sedra, Adel S
Selvakumar, Chettypalayam Ramanathan
Shewchun, John
Siddall, Ernest
Slater, Keith
Smith, Kenneth Carless
Tavares, Stafford Emanuel
Thansandote, Artnarong
Tsao, Sai Hoi
Zukotynski, Stefan

QUEBEC
Agarwal, Vinod Kumar
Archibald, Frederick S
Atwood, John William
Blostein, Maier Lionel
Champness, Clifford Harry
Cheng, Richard M H
Ferguson, Michael John
Ghosh, Asoke Kumar
Gregory, Brian Charles
Gulrajani, Ramesh Mulchand
Lafontaine, Jean-Gabriel
L'Archeveque, Real Viateur
Lecours, Michel
Lombos, Bela Anthony
Maciejko, Roman
Patel, Rajnikant V
Schwelb, Otto
Shkarofsky, Issie Peter
Van Vliet, Carolyne Marina

OTHER COUNTRIES
Bellanger, Maurice G
Benson, Frank Atkinson
Boxman, Raymond Leon
Carassa, Francesco
Chang, Chun-Yen
Chang, L(eroy) L(i-Gong)
Chang, Morris
Chase, Robert L(loyd)
Chi, Chao Shu
Chu, Kwo Ray
Conforti, Evandro
Cullen, Alexander Lamb
De Bruyne, Pieter
Esposito, Raffaele
Eykhoff, Pieter
Fettweis, Alfred Leo Maria
Fujiwara, Hideo
Gindsberg, Joseph
Granger, John Van Nuys
Haggis, Geoffrey Harvey
Halkias, Christos
Heiblum, Mordehai
Hugon, Jean S
Ince, A Nejat
Inose, Hiroshi
Jespersen, Nils Vidar
Kaiser, Wolfgang A
Kao, Charles K
Kim, Dae Mann
Kishi, Keiji
Kodali, V Prasad
Kurokawa, Kaneyuki
Kuroyanagi, Noriyoshi
Lahiri, Syamal Kumar
Lee, Choong Woong
Lehman, Meir M
Lewis, Arnold Leroy, II
Lindell, Ismo Veikko
Luo, Peilin
McInerney, John Gerard
Malah, David
Maqusi, Mohammad
Marom, Emanuel
Mengali, Umberto
Misugi, Takahiko
Okamura, Sogo
Onoe, Morio
Pau, Louis F
Purbo, Onno Widodo
Ros, Herman H
Rozzi, Tullio
Sakurai, Yoshifumi
Sato, Gentei
Struzak, Ryszard G
Stumpers, Frans Louis H M
Suematsu, Yasuharu
Sugaya, Hiroshi
Tarui, Yasuo
Toshio, Makimoto
Un, Chong Kwan

Unger, Hans-Georg
Ushioda, Sukekatsu
Vander Vorst, Andre
Van Overstraeten, Roger Joseph
Wallmark, J(ohn) Torkel
Wilkes, Maurice V
Zeheb, Ezra

Engineering, General

ALABAMA
Carden, Arnold Eugene
Dannenberg, Konrad K
Edlin, George Robert
Flinn, David R
Gilbert, Stephen Marc
Johnson, Clarence Eugene
Nola, Frank Joseph
Tidwell, Eugene Delbert
Vance, Ollie Lawrence

ALASKA
Sweet, Larry Ross

ARIZONA
Brown, Gordon S
Brown, Paul Wayne
Corcoran, William P
Edlin, Frank E
Garrett, Lane Sayre
Govil, Sanjay
McGuirk, William Joseph
McKenney, Dean Brinton
Miklofsky, Haaren A(lbert)
Osborn, Donald Earl
Randall, Lawrence Kessler, Jr
Rice, Warren
Rowand, Will H
Thomson, Quentin Robert
Yamaguchi, Gary T

ARKANSAS
Crisp, Robert M, Jr
Sutherland, G Russell
Yaz, Edwin Engin

CALIFORNIA
Abbott, Seth R
Adelman, Barnet Reuben
Agrawal, Suphal Prakash
Albers, James Arthur
Arnold, Frank R(obert)
Arnold, Walter Frank
Arslancan, Ahmet N
Attwood, David Thomas, Jr
Atwood, John Leland
Ayars, James Earl
Bahn, Gilbert S(chuyler)
Balakrishnan, A V
Bhandarkar, Mangalore Dilip
Blankenship, Victor D(ale)
Blink, James Allen
Bourke, Robert Hathaway
Brahma, Chandra Sekhar
Brand, Donald A
Brennen, Christopher E
Bridges, William Bruce
Buchberg, Harry
Burnett, James R
Cannon, Robert H, Jr
Cardwell, William Thomas, Jr
Carroll, William J
Chambre, Paul L
Chivers, Hugh John
Chu, Chung-Yu Chester
Clarke, Lucien Gill
Cleland, Laurence Lynn
Cohn, Seymour B(ernard)
Conly, John F
Crooke, Robert C
Dahlberg, Richard Craig
Davies, John Tudor
Davis, Frank W
Davis, W Kenneth
Denavit, Jacques
Desoer, Charles A(uguste)
Dhir, Vijay Kumar
Dixon, Harry S(terling)
Donnan, William W
Drezner, Stephen M
Durbeck, Robert C(harles)
Dyer, James Lee
Fan, Chien
Feller, Robert S, Sr
Fergin, Richard Kenneth
Fitzpatrick, Gary Owen
Foster, John Stuart, Jr
Freedy, Amos
Friend, William L
Futernick, Kenneth
Garrick, B(ernell) John
Gasich, Welko E
Gay, Charles Francis
Ghose, Rabindra Nath
Gilmartin, Thomas Joseph
Gleghorn, G(eorge) J(ay), (Jr)
Grant, Eugene Lodewick
Gunderson, Norman O
Hadley, Jeffery A
Haloulakos, Vassilios E
Hamilton, J(ames) Hugh
Hammond, R Philip
Hanson, Merle Edwin

Engineering, General (cont)

Harmen, Raymond A
Hesse, Christian August
Hingorani, Narain G
Hoffman, Myron A(rnold)
Homer, Paul Bruce
Howard, Ronald A(rthur)
Hubka, William Frank
Hughes, Thomas Joseph
Ingram, Gerald E(ugene)
Ishihara, Kohei
Johannessen, Jack
Johnson, Conor Deane
Johnson, Paul L
Johnson, Robert L
Jones, Clarence S
Kesselring, John Paul
Kim, Kwang-Jin
King, William Stanely
Kittredge, Clifford Proctor
Krause, Ralph A(lvin)
Kulgein, Norman Gerald
Kvaas, T(horvald) Arthur
Larson, Harry Thomas
Lee, K Hung
Leonard, A(nthony)
Li, Seung P(ing)
Lloyd, Edward C(harles)
Lynch, Frank W
McDonald, Henry C
McManigal, Paul Gabriel Moulin
Maillot, Patrick Gilles
Marxheimer, Rene B
Meeker, Donald J
Milstein, Frederick
Mohamed, Farghalli Abdelrahman
Molinder, John Irving
Moorhouse, Douglas C
Muskat, Morris
Mykkanen, Donald L
Nash, James Richard
Nelson, Arthur L(ee)
Olfe, D(aniel)
Oliver, Bernard M(ore)
O'Neill, R(ussell) R(ichard)
Ormsby, Robert B, Jr
Oshman, M Kenneth
Palter, N(orman) H(oward)
Parden, Robert James
Parker, N(orman) F(rancis)
Parme, Alfred L
Patel, Chandra Kumar Naranbhai
Platus, Daniel Herschel
Reyhner, Theodore O
Richardson, Neal A(llen)
Roe, Arnold
Rosenstein, A(llen) B
Sackman, Jerome L(eo)
Sadeh, Willy Zeev
Salinas, David
Sarkaria, Gurmukh S
Schmitendorf, William E
Schonberg, Russell George
Sommers, William P(aul)
Spear, Robert Clinton
Sperling, Jacob L
Spindt, Charles A (Capp)
Spitzer, Irwin Asher
Standing, Marshall B
Staudhammer, Peter
Stearns, H(orace) Myrl
Stimmell, K G
Stockwell, Norman D
Stone, Henry E
Sworder, David D
Tang, Stephen Shien-Pu
Thomas, Frank J(oseph)
Thomas, Mitchell
Thomson, William Tyrrell
Thorpe, Howard A(lan)
Tribus, Myron
Trorey, A(lan) W(ilson)
Vassiliou, Marius Simon
Waldhauer, F D
Welsh, David Edward
Wheaton, Elmer Paul
Whipple, Christopher George
Wilkinson, Eugene P Dennis
Wood, Carlos C
Wood, Edward C(halmers)
Woods, Ralph Arthur
Wooldridge, Dean E
Wyllie, Loring A

COLORADO
Albert, Harrison Bernard
Beck, Betty Anne
Beckstead, Leo William
Brown, Alison Kay
Chamberlain, A(drian) R(amond)
Eyman, Earl Duane
Jansson, David Guild
Krill, Arthur Melvin
MacGregor, Ronald John
McManamon, Peter Michael
Mercer, Robert Allen
Perez, Jean-Yves
Poirot, James Wesley
Roberts, Howard C(reighton)
Stauffer, Jack B
Timblin, Lloyd O, Jr
Verschoor, J(ack) D(ahlstrom)
Wainwright, William Lloyd

CONNECTICUT
Brody, Harold David
Burkhard, Mahlon Daniel
Dubin, Fred S
Elrod, Harold G(lenn), Jr
Frey, Carl
Gerber, H Joseph
Gordon, Robert Boyd
Hood, Edward E, Jr
Jordan, Donald J
Ketchman, Jeffrey
Shichman, D(aniel)
Shuey, Merlin Arthur
Sternberg, Robert Langley
Strough, Robert I(rving)
Suprynowicz, Vincent A

DELAWARE
Scherger, Dale Albert
Uy, William Cheng
Yiannos, Peter N

DISTRICT OF COLUMBIA
Frair, Karen Lee
Green, Richard James
Mermel, Thaddeus Walter
Nichols, Kenneth David
Oran, Elaine Surick
Paige, Hilliard W
Sexton, Ken
Stanley, Thomas P
Thomas, Leonard William, Sr
Townsend, Marjorie Rhodes
Weiss, Leonard
Willenbrock, Frederick Karl

FLORIDA
Amon, Max
Berkey, Donald C
Concordia, Charles
Cox, Parker Graham
Crane, Lee Stanley
Davis, Duane M
Davison, Beaumont
Dean, Donald L(ee)
Dilpare, Armand Leon
Dowdell, Rodger B(irtwell)
Ewald, William Philip
Florea, Harold R(obert)
Gaffney, F(rancis) J(oseph)
Gold, Edward
Goldwyn, Roger M(artin)
Groves, Donald George
Hamming, Kenneth W
Henning, Rudolf E
Hirsch, Donald Earl
Hollis, Mark Dexter
Kalman, Rudolf Emil
Kolhoff, M(arvin) J(oseph)
McCune, Francis K(imber)
Maeder, P(aul) F(ritz)
Milton, James E(dmund)
Mumma, Albert G
Penney, Gaylord W
Price, Donald Ray
Sirkin, Alan N
Squires, Lombard
Stefanakos, Elias Kyriakos
Swain, Geoffrey W
Walsh, Edward Kyran
Wiebusch, Charles Fred
Wimberly, C Ray
Zimmer, Martin F

GEORGIA
Antolovich, Stephen D
Cassanova, Robert Anthony
Cullison, William Lester
Dickert, Herman A(lonzo)
Fisher, Reed Edward
Ghate, Suhas Ramkrishna
Green, G W
Gupton, Guy Winfred, Jr
Kamen, Edward Walter
Stanley, Luticious Bryan, Jr

HAWAII
Gersch, Will
Yee, Alfred A

IDAHO
Bondurant, James A(llison)
Byers, Roland O
Froes, Francis Herbert (Sam)
Martin, J(ames) W(illiam)
Williams, George Arthur

ILLINOIS
Ban, Stephen Dennis
Basar, Ronald John
Burks, G Edwin
Christianson, Clinton Curtis
Conry, Thomas Francis
Drickamer, Harry George
Firstman, Sidney I(rving)
Giacobbe, F W
Gose, Earl E(ugene)
Hull, John R
Johari, Om
Kliphardt, Raymond A(dolph)
Lischer, Ludwig F
Morin, Charles Raymond
Patel, Mayur
Ryskin, Gregory
Safdari, Yahya Bhai
Sanathanan, C(hathilingath) K
Schallert, William Francis
Schlesinger, Lee
Shabana, Ahmed Abdelraouf
Shipley, Roch Joseph
Sohr, Robert Trueman
Staunton, John Joseph Jameson
Stukel, James Joseph
Tao, Liang Neng
Weber, J K Richard
Woods, Kenneth R
Wool, Richard P
Wright, Maurice Arthur

INDIANA
Acker, Frank Earl
Atallah, Mikhail Jibrayil
Bogdanoff, John Lee
Burden, Stanley Lee, Jr
Farris, Thomas N
Kennedy, J(ohn) R(obert)
Kronmiller, C W
Mercer, Walter Ronald
Nagle, Edward John
Odor, David Lee
Pierson, Edward S
Plants, Helen Lester
Sadeghi, Farshid
Saucedo, Ismael G
Sen, Mihir
Skibniewski, Miroslaw Jan
Vanderbilt, Vern C, Jr
Voelz, Michael H

IOWA
Boylan, D(avid) R(ay)
Braaten, Melvin Ole
Firstenberger, B(urnett) G(eorge)
Greer, Raymond T(homas)
Kortanek, Kenneth O
Mack, Michael J
Mather, Roger Frederick
Northup, Larry L(ee)
Siebes, Maria

KANSAS
Frohmberg, Richard P
Gowdy, Kenneth King
Hosni, Mohammad Hosein
Joyner, H(oward) Sajon
Kirmser, P(hilip) G(eorge)
Lindholm, John C
Razak, Charles Kenneth
Robinson, M John
Smith, Howard Wesley
Spears, Sholto Marion
Williams, Wayne Watson

KENTUCKY
Beattie, Horace S
Cathey, Jimmie Joe
Gold, Harold Eugene
Johnson, Alan Arthur
Viswanadham, Ramamurthy K

LOUISIANA
Abdelhamied, Kadry A
Dareing, Donald W
Decossas, Kenneth Miles
Latorre, Robert George
Robert, Kearny Quinn, Jr
Sabbaghian, Mehdy
Snyder, Harold Jack
Thomas, Kevin Anthony
Thurman, Henry L, Jr

MAINE
Kinnaird, Richard Farrell

MARYLAND
Adams, Edward Franklin
Adams, Laurence J
Bascunana, Jose Luis
Bukowski, Richard William
Burns, Bruce Peter
Burns, John Joseph, Jr
Castella, Frank Robert
Chiogioji, Melvin Hiroaki
Christensen, Richard G
Cohen, Stanley Alvin
Coulter, James B
Coutinho, John
Dame, Richard Edward
Darr, J(ack) E(dwin)
Field, Herbert Cyre
Gottfried, Paul
Graves, Harvey W(ilbur), Jr
Harman, George Gibson, Jr
Heider, Shirley (Scott) A(mborn)
Herold, Keith Evan
Johnston, Thomas M(atkins)
Katsanis, D(avid) J(ohn)
Kemelhor, Robert Elias
Knazek, Richard Allan
Ku, Harry Hsien Hsiang
Lundsager, C(hristian) Bent
Makofski, Robert Anthony
Quinn, George David
Rabinow, Jacob
Rostenbach, Royal E(dwin)
Sallet, Dirse Wilkis
Sanders, Robert Charles
Schubauer, Galen B
Schwoerer, F(rank)
Smith, Joseph Collins
Talbott, Edwin M
Theuer, Paul John
Troutman, James Scott
Walsh, James Paul
Walter, Donald K
Wimenitz, Francis Nathaniel
Zuber, Novak

MASSACHUSETTS
Apelian, Diran
Bowen, H Kent
Bradner, Mead
Drew, Philip Garfield
Emanuel, Alexander Eigeles
Fariss, Thomas Lee
Forrester, Jay W
Frosch, Robert Alan
Glaser, Peter E(dward)
Hockney, Richard L
Hubbard, M(alcolm) M(acGregor)
Keil, Alfred Adolf Heinrich
Kroner, Klaus E(rlendur)
Lacoss, Richard Thaddee
Latham, Allen, Jr
Lees, David Eric Berman
McCune, William James, Jr
McDonald, Donald C
Mack, Michael E
Markey, Winston Roscoe
Mastronardi, Richard
Merrill, E(dward) W(ilson)
Mulpur, Arun K
Osgood, Elmer Clayton
Phillips, Thomas Leonard
Pickering, Frank E
Pirri, Anthony Nicholas
Pulling, Nathaniel H(osler)
Richardson, Allyn (St Clair)
Roadstrum, William H(enry)
Robinson, D(enis) M(orrell)
Rogowski, A(ugustus) R(udolph)
Scherer, Harold Nicholas, Jr
Schmergel, Gabriel
Schwartz, Thomas Alan
Shepherd, James E
Sikina, Thomas
Sosnowski, Thomas Patrick
Sprague, Robert C
Webster, Lee Alan

MICHIGAN
Alexandridis, Alexander A
Batrin, George Leslie
Chou, Clifford Chi Fong
Eltinge, Lamont
Fancher, Paul S(trimple)
Fisher, Edward Richard
Gilbert, E(dward) O(tis)
Gilbert, Elmer G(rant)
Goldsberry, Ronald E
Greene, Bruce Edgar
Hrovat, Davorin
Jaicks, Frederick G
Luxon, James Thomas
Luzzi, Theodore E, Jr
Musinski, Donald Louis
Nagy, Louis Leonard
Olte, Andrejs
Powers, William Francis
Prostak, Arnold S
Ressler, Neil William
Riley, Bernard Jerome
Ryant, Charles J(oseph), Jr
Stansel, John Charles
Thomson, Robert Francis
Wagner, Harvey Arthur
Weber, Walter J, Jr
Westervelt, Franklin Herbert

MINNESOTA
Bateson, Robert Neil
Mahmoodi, Parviz
Rao, Prakash
Scriven, L E(dward), (II)
Shulman, Yechiel
Stadtherr, Leon

MISSISSIPPI
Eubanks, William Hunter
Foster, George Rainey
Happ, Stafford Coleman
Rosenhan, A Kirk
Taylor-Cade, Ruth Ann

MISSOURI
Byrnes, Christopher Ian
Deam, James Richard
Glauz, William Donald
McDonnell, Sanford
Major, Schwab Samuel, Jr
Mathiprakasam, Balakrishnan
Oglesby, David Berger
Peltier, Eugene J
Roberts, J(asper) Kent
Schweiker, Jerry W

MONTANA
Friel, Leroy Lawrence
Taylor, William Robert

NEBRASKA
Lenz, Charles Eldon

Pao, Y(en)-C(hing)
Smith, Durward A

NEVADA
Snaper, Alvin Allyn
Tryon, John G(riggs)

NEW HAMPSHIRE
Bahr, Karl Edward
Fletcher, William L
Klarman, Karl J(oseph)
Laaspere, Thomas
Mellor, Malcolm
Noble, Charles Carmin
Spehrley, Charles W, Jr
Wallis, Graham B

NEW JERSEY
Biskeborn, Merle Chester
Brill, Yvonne Claeys
Clements, Wayne Irwin
Coleman, Norman P, Jr
Courtney-Pratt, Jeofry Stuart
DiMasi, Gabriel Joseph
Domeshek, S(ol)
Epstein, Seymour
Feinblum, David Alan
Fox, A(rthur) Gardner
Grunes, Robert Lewis
Jain, Sushil C
Kandoian, A(rmig) G(hevont)
Lechner, Bernard J
Le Mee, Jean M
Lepselter, Martin P
Maclin, Ernest
Marcus, Phillip Ronald
Mikolajczak, Alojzy Antoni
Negin, Michael
Newman, Joseph Herbert
Peskin, Richard Leonard
Phadke, Madhav Shridhar
Puri, Narindra Nath
Rigassio, James Louis
St Onge, G H
Sharobeam, Monir Hanna
Shashidhara, Nagalapur Sastry
Shinozuka, Masanobu
Steigelmann, William Henry
Stevens, William D
Weiss, C Dennis

NEW MEXICO
Broyles, Carter D
Cronenberg, August William
Dempster, William Fred
Diver, Richard Boyer, Jr
Finch, Thomas Wesley
Howard, William Jack
Icerman, Larry
Jankowski, Francis James
Kammerdiener, John Luther
Perkins, Roger Bruce
Russell, John Henry
Savage, Charles Francis
Vortman, L(uke) J(erome)
Welber, Irwin
Wylie, Kyral Francis

NEW YORK
Amero, Bernard Alan
Ausubel, Jesse Huntley
Bandyopadhyay, Amitabha
Baron, Seymour
Baum, Eleanor Kushel
Bigelow, John Edward
Bradburd, Ervin M
Broadwin, Alan
Cheng, Yean Fu
Chu, Richard Chao-Fan
Cornacchio, Joseph V(incent)
Curtis, Huntington W(oodman)
Eissenberg, David M(artin)
Fama, Donald Francis
Fitzroy, Nancy Deloye
Flannery, John B, Jr
Friedman, Morton (Benjamin)
Glinkowski, Mietek T
Goertzel, Gerald
Greene, Peter Richard
Haag, Fred George
Haddad, Jerrier Abdo
Hall, Kimball Parker
Jefferies, Michael John
Jensen, Richard Alan
Krishna, C R
Lammie, James L
Lee, Charles Northam
Levy, Robert S(amuel)
Logan, H(enry) L(eon)
McMurry, William
Madey, Robert W
Matkovich, Vlado Ivan
Morabito, Bruno P
Parsegian, V(ozcan) Lawrence
Penwell, Richard Carlton
Rosenberg, Paul
Samuels, Reuben
Schanker, Jacob Z
Schultz, Andrew, Jr
Shaffer, Bernard W(illiam)
Shames, Irving H
Shube, Eugene E
Simons, Gene R
Snyder, Robert L(eon)

Soifer, Herman
Udolf, Roy
Ullmann, John E
Velzy, Charles O
Vosburgh, Kirby Gannett
Weber, Arthur Phineas
Wendt, Robert L(ouis)
Wilcock, Donald F(rederick)
Witt, Christopher John
Wood, Wallace D(ean)
Zic, Eric A
Ziegler, Robert C(harles)

NORTH CAROLINA
Bailey, Kincheon Hubert, Jr
Baker, Charles Ray
Chin, Robert Allen
Clark, Charles Edward
Gerhardt, Don John
Hess, Daniel Nicholas
Land, Ming Huey
Lee, William States
Owen, Warren H
Wollin, Goesta
Woodland, Joseph
Wright, William Vale

NORTH DAKOTA
Rieder, William G(ary)

OHIO
Arnett, Jerry Butler
Beale, William Taylor
Bruno, David Joseph
Butz, Donald Josef
Ciricillo, Samuel F
Cook, James Robert
Cooke, Charles C
Cosgrove, Stanley Leonard
Cummings, Charles Arnold
Damianov, Vladimir B
Dunn, Robert Garvin
Eckert, John S
El-Naggar, Mohammed Ismail
Erbacher, John Kornel
Grigger, David John
Henderson, Courtland M
Hulley, Clair Montrose
Jones, Trevor O
Klineberg, John Michael
LaRue, Robert D(ean)
Lownie, H(arold) W(illiam), Jr
Lynch, T(homas) E(lwin)
McCauley, Roy B, Jr
McNichols, Roger J(effrey)
Manson, S(amuel) S(tanford)
Messick, Roger E
Middendorf, William H
Newby, John R
Pugh, John W(illiam)
Putnam, Abbott (Allen)
Ramalingam, Mysore Loganathan
Robe, Thurlow Richard
Rubin, Saul H
Seagle, Stan R
Sevin, E(ugene)
Soska, Geary Victor
Sutton, J(ames) L(owell)
Voigt, Charles Frederick
Voisard, Walter Bryan
White, Willis S, Jr
Wolfe, Robert Kenneth
Zavodney, Lawrence Dennis

OKLAHOMA
Baker, James E
Basore, B(ennett) L(ee)
DeLacerda, Fred G
Donahue, Hayden Hackney
Forney, Bill E
Hays, George E(dgar)
Norton, Joseph R(andolph)
Rice, Charles Edward

OREGON
Fields, R(ance) Wayne
Gray, A(ugustine) H(eard), Jr
Skinner, Richard Emery

PENNSYLVANIA
Advani, Sunder
Anouchi, Abraham Y
Bandel, Hannskarl
Bissinger, Barnard Hinkle
Brown, Chester Harvey, Jr
Burtis, Theodore A
Carfagno, Salvatore P
Champagne, Paul Ernest
Charp, Solomon
Chitale, Sanjay Madhav
Chou, Pei Chi (Peter)
Davis, R(obert) E(lliot)
DebRoy, Tarasankar
Dorne, Arthur
Falconer, Thomas Hugh
Feero, William E
Forscher, Frederick
Gebhart, Benjamin
Geldmacher, R(obert) C(arl)
Hager, Wayne R
Hayes, Miles V(an Valzah)
Haythornthwaite, Robert M(orphet)
Hoburg, James Frederick
Howland, Frank L

Knaster, Tatyana
Koch, Ronald N
Kreh, E(dward) J(oseph), Jr
Langston, Charles Adam
Lanphier, Robert C
Larson, Thomas D
Lauchle, Gerald Clyde
Lazaridis, Anastas
Loxley, Thomas Edward
Marks, Peter J
Meredith, David Bruce
Merz, Richard A
Michael, Norman
Mouly, Raymond J
O'Donnell, William James
Palmer, Rufus N(elson)
Pan, Yuan-Siang
Paolino, Michael A
Peterson, Richard Walter
Prywes, Noah S(hmarya)
Schenk, H(arold) L(ouis), Jr
Schulman, Marvin
Shobert, Erle Irwin, II
Silverstein, Calvin C(arlton)
Soung, Wen Y
Spencer, Arthur Coe, II
Stone, M(orris) D
Takeuchi, Kenji
Tobias, Philip E
Tuffey, Thomas J
Viest, Ivan M
Walchli, Harold E(dward)
Williams, David Bernard
Wilson, Geoffrey Leonard
Wolgemuth, Carl Hess
Wray, Porter R
Wronski, Christopher R

RHODE ISLAND
Kowalski, Tadeusz
Lerner, Samuel
Morse, Theodore Frederick
Pearson, Allan Einar
Spero, Caesar A(nthony), Jr
Symonds, Paul S(outhworth)

SOUTH CAROLINA
Bridges, Donald Norris
Chao, Yuh J (Bill)
Harrison, James William, Jr
Hogan, Joseph C(harles)
Jaco, Charles M, Jr
Kellerman, Karl F(rederic)
Russell, Allen Stevenson
Sandrapaty, Ramachandra Rao
Williamson, Robert Elmore

SOUTH DAKOTA
Schleusener, Richard A

TENNESSEE
Buchanan, George R(ichard)
Burwell, Calvin C
Dahotre, Narendra Bapurao
Foster, Edwin Powell
Frounfelker, Robert E
Gat, Uri
Janna, William Sied
Kimmons, George H
Kinser, Donald LeRoy
Lim, Alexander Te
Moulden, Trevor Holmes
Pandeya, Prakash N
Reister, David B(ryan)
Riewald, Paul Gordon
Ross, Harley Harris
Sheth, Atul C
Simhan, Raj
Spruiell, Joseph E(arl)
Trauger, Donald Byron
Waters, Dean Allison
Wichner, Robert Paul

TEXAS
Abramson, H(yman) Norman
Albertson, Harold D
Austin, T Louis, Jr
Biard, James R
Bohacek, Peter Karl
Bovay, Harry Elmo, Jr
Boyer, Robert Elston
Bravenec, Edward V
Burkey, Roland Steven
Childs, S(elma) Bart
Chyu, Ming-Chien
Cragon, Harvey George
Daigh, John D(avid)
Davis, Alfred V
Dhudshia, Vallabh H
French, Robert Leonard
Friberg, Emil Edwards
Frick, John P
Glaspie, Donald Lee
Godwin, James Basil, Jr
Griffin, Richard B
Harper, James George
Hillaker, Harry J
Hyman, William A
Kiel, Otis Gerald
Lawrence, Joseph D, Jr
Liu, Kai
MacFarlane, Duncan Leo
Masten, Michael K
Meyer, W(illiam) Keith

Musgrave, Albert Wayne
Odeh, A(ziz) S(alim)
Raju, Satyanarayana G V
Rodenberger, Charles Alvard
Sawyer, Ralph Stanley
Schuhmann, Robert Ewald
Schweppe, Joseph L(ouis)
Sjoberg, Sigurd A
Stucker, Harry T
Vance, John Milton
Ward, Donald Thomas
Widmer, Robert H
Yuan, Robert L

UTAH
Allen, Dell K
Chesworth, Robert Hadden
Jacobsen, Stephen C
Johnson, Robert R(oyce)
MacGregor, Douglas
Nelson, Mark Adams

VERMONT
Foote, Vernon Stuart, Jr
Howard, Robert T(urner)
Kintner, Edwin E

VIRGINIA
Balwanz, William Walter
Bishara, Michael Nageeb
Chapin, Douglas McCall
Devens, W(illiam) George
Duscha, Lloyd A
Ell, William M
Foster, Richard B(ergeron)
Fubini, Eugene G(hiron)
Garcia, Albert B
Goins, Truman
Gordon, Ronald Stanton
Hall, Carl W(illiam)
Hallanger, Lawrence William
Hammond, Marvin H, Jr
Hauxwell, Gerald Dean
Heller, Agnes S
Krueger, Peter George
Leon, H(erman) I
Lerner, Norman Conrad
Levine, Martin
Lowry, Ralph A(ddison)
McNichols, Gerald Robert
Marschall, Albert Rhoades
Mazzoni, Omar S
Mehta, Gurmukh D
Meirovitch, L(eonard)
Nuckolls, Joe Allen
Preusch, Charles D
Raju, Ivatury Sanyasi
Rall, Lloyd L(ouis)
Shrier, Stefan
Sikri, Atam P
Summers, George Donald
Thorp, Benjamin A
Webb, George Randolph
Wilkinson, Thomas Lloyd, Jr

WASHINGTON
Bell, Milo C
Boone, James Lightholder
Clevenger, William A
Clough, G Wayne
Condit, Philip M
Glosten, Lawrence R
Gunderson, Leslie Charles
Hirschfelder, John Joseph
Morgan, Jeff
Pearson, Carl E
Stanton, K Neil
Thorne, Charles M(orris)
Withington, Holden W

WEST VIRGINIA
Clark, Nigel Norman
Fischer, Robert Lee
Mark, Christopher
Muth, Wayne Allen
Peters, Penn A
Strickland, Larry Dean

WISCONSIN
Berthouex, Paul Mac
Bishop, Charles Joseph
Bomba, Steven James
Crawmer, Daryl E
Hodges, Lawrence H
Kenny, Andrew Augustine
Klein, Carl Frederick
Lenz, Arno T(homas)
McMurray, David Claude
Ranganathan, Brahmanpalli
  Narasimhamurthy
Richardson, Bobbie L
Schaffer, Erwin Lambert
Zanoni, Alphonse E(ligius)

WYOMING
Canfield, Carl Rex, Jr
Dolan, Charles W

ALBERTA
Karim, Ghazi A
Micko, Michael M
Nasseri, Touraj
Verschuren, Jacobus Petrus

Engineering, General (cont)

BRITISH COLUMBIA
Buckland, Peter Graham
Newbury, Robert W(illiam)
Srivastava, Krishan

MANITOBA
Cahoon, John Raymond

NEW BRUNSWICK
Chrzanowski, Adam

NOVA SCOTIA
Nugent, Sherwin Thomas
Sastry, Vankamamidi VRN

ONTARIO
Balakrishnan, Narayanaswamy
Baronet, Clifford Nelson
Biggs, Ronald C(larke)
Carr, Jan
Fortier, Y O
Jones, L(lewellyn) E(dward)
Lapp, P(hilip) A(lexander)
Matar, Said E
Novak, Milos
Vijay, Mohan Madan

PRINCE EDWARD ISLAND
Lodge, Malcolm A

QUEBEC
Bakhiet, Atef
Bhat, Rama B
Charette, Andre
Cherna, John C(harles)
Hanna, Adel
Marsh, Cedric
Pekau, Oscar A
Sankar, Seshadri
Shaw, Robert Fletcher
Strasser, John Albert
Thompson, Allan Lloyd
Verschingel, Roger H C

SASKATCHEWAN
Riemer, Paul

OTHER COUNTRIES
Ayres, Robert U
Eichenberger, Hans P
Jacob, K Thomas
Mylonas, Constantine
Saba, Shoichi
Von Hoerner, Sebastian

Engineering Mechanics

ALABAMA
Beckett, Royce E(lliott)
Boykin, William H(enry), Jr
Cochran, John Euell, Jr
Costes, Nicholas Constantine
Cutchins, Malcolm Armstrong
Davis, Anthony Michael John
Davis, Carl George
Doughty, Julian O
Evces, Charles Richard
Gambrell, Samuel C, Jr
Gilbert, John Andrew
Hill, James Lafe
Jackson, John Elwin, Jr
Jones, Stanley E
Jordan, William D(itmer)
Kattus, J Robert
Killingsworth, R(oy) W(illiam)
Melville, Joel George
Schutzenhofer, Luke A
Turner, Charlie Daniel, Jr
Vance, Ollie Lawrence

ALASKA
Johansen, Nils Ivar
Skudrzyk, Frank J

ARIZONA
Arabyan, Ara
Chandra, Abhijit
Chen, Stanley Shiao-Hsiung
Chenea, Paul F(ranklin)
Govil, Sanjay
Krajcinovic, Dusan
Malvick, Allan J(ames)
Morrow, JoDean
Neff, Richmond C(lark)
Saric, William Samuel
Savrun, Ender
Shaw, Milton C(layton)
Singhal, Avinash Chandra
Stafford, John William
Ulich, Bobby Lee
Wallace, C(harles) E(dward)

ARKANSAS
Boyd, Donald Edward
Jong, Ing-Chang
Ma, Er-Chieh
Smith, E(astman)

CALIFORNIA
Abdel-Ghaffar, Ahmed Mansour
Adler, William Fred
Aggarwal, H(ans) R(aj)
Ardema, Mark D
Arnold, Frank R(obert)
Asaro, Robert John
Atchley, Bill Lee
Batdorf, Samuel Burbridge
Beal, Thomas R
Bendisz, Kazimierz
Blythe, William Richard
Boltinghouse, Joseph C
Bonora, Anthony Charles
Caligiuri, Robert Domenic
Cannon, Robert H, Jr
Cantin, Gilles
Carleone, Joseph
Casey, James
Chelapati, Chunduri V(enkata)
Chi, Cheng-Ching
Chiang, George C(hihming)
Chou, Larry I-Hui
Christensen, Richard Monson
Chuang, Kuen-Puo (Ken)
Collins, Jon David
Colton, James Dale
Das, Mihir Kumar
Dashner, Peter Alan
DeBra, Daniel Brown
Dong, Richard Gene
Dost, Martin Hans-Ulrich
Dowell, Douglas C
Dym, Clive L
Eggers, A(lfred) J(ohn), Jr
Fisher, Franklin E(ugene)
Fleming, Alan Wayne
Flora, Edward B(enjamin)
Forrest, James Benjamin
Forsberg, Kevin
Geminder, Robert
Gere, James Monroe
Gleghorn, G(eorge) J(ay), (Jr)
Goldsmith, Werner
Hackel, Lloyd Anthony
Hackett, Colin Edwin
Hahn, Hong Thomas
Herrmann, George
Hodge, Philip G(ibson), Jr
Hoff, N(icholas) J(ohn)
Horne, Roland Nicholas
Hsu, Chieh-Su
Hutchinson, James R(ichard)
Ingram, Gerald E(ugene)
Ito, Y(asuo) Marvin
Jacobson, Marcus
Johnson, Conor Deane
Kaul, Maharaj Krishen
Keller, Joseph Bishop
Kelly, Robert Edward
Kosmatka, John Benedict
Kraus, Samuel
Kwok, Munson Arthur
Lampert, Seymour
Larson, Edward William, Jr
Lau, John H
Leal, L Gary
Leckie, Frederick Alexander
Logan, James Columbus
Lysmer, John
Ma, Fai
McCulloch, Andrew Douglas
Madan, Ram Chand
Maewal, Akhilesh
Mal, Ajit Kumar
Meissinger, Hans F
Meriam, James Lathrop
Merritt, J(oshua) L(evering), Jr
Mikes, Peter
Monkewitz, Peter Alexis
Mote, C(layton) D(aniel), Jr
Mow, C(hao) C(how)
Muhlbauer, Karlheinz Christoph
Murch, S(tanley) Allan
Napolitano, Leonard Michael, Jr
Nelson, Richard Bartel
Nemat-Nasser, Siavouche
Nordell, William James
Pamidi, Prabhakar Ramarao
Pfost, R Fred
Piersol, Allan Gerald
Pierucci, Mauro
Pincus, Howard Jonah
Pinto, John Gilbert
Plotkin, Allen
Rau, Charles Alfred, Jr
Rubin, Sheldon
Schmit, Lucien A(ndre), Jr
Seaman, Lynn
Seide, Paul
Skalak, Richard
Spier, Edward Ellis
Spreiter, John R(obert)
Stevenson, Merlon Lynn
Stockel, Ivar H(oward)
Stout, Ray Bernard
Strickland, Gordon Edward, Jr
Suer, H(erbert) S
Szego, Peter A
Trubert, Marc
Twiss, Robert John
Unal, Aynur
Vanblarigan, Peter
Vanderplaats, Garret Niel
Vanyo, James Patrick
Wallerstein, David Vandermere
Walton, James Stephen
Wang, Paul Keng Chieh
Wehausen, John Vrooman
Weingarten, Victor I
Whittier, James S(pencer)
Wilbarger, Edward Stanley, Jr
Yeung, Ronald Wai-Chun
Zickel, John

COLORADO
Bacon, Merle D
Burnham, Marvin William
Cermak, J(ack) E(dward)
Chugh, Ashok Kumar
Criswell, Marvin Eugene
Datta, Subhendu Kumar
Feldman, Arthur
Frangopol, Dan Mircea
Geers, Thomas L
Gerdeen, James C
Medearis, Kenneth Gordon
Peterson, Harry C(larence)
Plows, William Herbert
Thompson, Erik G(rinde)
Thurston, Gaylen Aubrey
Weidman, Patrick Dan

CONNECTICUT
Cheatham, Robert Gary
Chin, Charles L(ee) D(ong)
Crossley, F(rancis) R(endel) Erskine
DeWolf, John T
Fink, Martin Ronald
Ketchman, Jeffrey
Mann, Richard A(rnold)
Nilson, Edwin Norman
Schetky, Laurence McDonald
Schile, Richard Douglas
Solecki, Roman
Verdon, Joseph Michael
Work, Clyde E(verette)
Wuthrich, Paul

DELAWARE
Cheng, Alexander H-D
Faupel, Joseph H(erman)
Hirsch, Albert Edgar
Kaliakin, Victor Nicholas
Kerr, Arnold D
Kingsbury, Herbert B
Lucas, James M
Santare, Michael Harold
Szeri, Andras Zoltan

DISTRICT OF COLUMBIA
Abrahamson, George R(aymond)
Basdekas, Nicholas Leonidas
Belsheim, Robert Oscar
Jones, Douglas Linwood
Moll, Magnus
Nachman, Arje
Niedenfuhr, Francis W(illiam)
Parks, Vincent Joseph
Youm, Youngil

FLORIDA
Beil, Robert J(unior)
Block, David L
Catz, Jerome
Chiu, Tsao Yi
DeHart, Arnold O'Dell
Dilpare, Armand Leon
Ebrahimi, Fereshteh
Eisenberg, Martin A(llan)
Griffith, John E(dward)
Halperin, Don A(kiba)
Hsieh, Chung Kuo
Jenkins, David R(ichard)
Kurzweg, Ulrich H(ermann)
Lijewski, Lawrence Edward
Lin, Y(u) K(weng)
Martin, Charles John
Neff, Thomas O'Neil
Nevill, Gale E(rwin), Jr
Oline, Larry Ward
Perrotta, Alessandro
Plass, Harold J(ohn), Jr
Ranov, Theodor
Reichard, Ronnal Paul
Sidebottom, Omar M(arion)
Spears, Richard Kent
Stevens, Karl Kent
Sun, Chang-Tsan
Villanueva, Jose
Von Ohain, Hans Joachim
Ward, Leonard George
Wimberly, C Ray
Wiseman, H(arry) A(lexander) B(enjamin)
Yong, Yan

GEORGIA
Chang, Chin Hao
Ginsberg, Jerry Hal
Gupta, Rakesh Kumar
King, Wilton W(ayt)
McGill, David John
Neitzel, George Paul
Orloff, David Ira
Raville, Milton E(dward)
Sangster, William M(cCoy)
Spetnagel, Theodore John
Thornton, William Aloysius
Wang, James Ting-Shun
Wempner, Gerald Arthur

HAWAII
Burgess, John C(argill)
Fuka, Louis Richard
Stuiver, W(illem)
Taoka, George Takashi
Teng, Michelle Hsiao Tsing

ILLINOIS
Achenbach, Jan Drewes
Adrian, Ronald John
Ban, Stephen Dennis
Bazant, Zdenek P(avel)
Benjamin, Roland John
Berry, Gregory Franklin
Chakrabarti, Subrata K
Cheng, Herbert S
Chugh, Yoginder Paul
Conry, Thomas Francis
Cusano, Cristino
Daniel, Isaac M
Dantzig, Jonathan A
Davis, Philip K
Dutton, Jonathan Craig
Edelstein, Warren Stanley
Erwin, Lewis
Francis, Philip Hamilton
Hofer, Kenneth Emil
Hsui, Albert Tong-Kwan
Keer, Leon M
Kesler, Clyde E(rvin)
Koves, William John
Kramer, John Michael
Kumar, Sudhir
Liu, Wing Kam
Mataga, Peter Andrew
Miller, Robert Earl
Mockros, Lyle F(red)
Moran, Thomas J
Napadensky, Hyla S
Orthwein, W(illiam) C(oe)
Phillips, James Woodward
Pickett, Leroy Kenneth
Reichard, Grant Wesley
Riahi, Daniel Nourollah
Robinson, Arthur R(ichard)
Ruhl, Roland Luther
Schnobrich, William Courtney
Schoeberle, Daniel F
Sciammarella, Caesar August
Shabana, Ahmed Abdelraouf
Shack, William John
Socie, Darrell Frederick
Tang, Wilson H
Ting, Thomas C(hi) T(sai)
Tucker, Charles L
Walker, John Scott
Weeks, Richard William
White, Robert Allan
Wiley, Jack Cleveland
Wilson, William Robert Dunwoody
Wolosewicz, Ronald Mitchell
Worley, Will J
Wright, Maurice Arthur
Yen, Ben Chie
Youngdahl, Carl Kerr

INDIANA
Bernhard, Robert J
Chen, Wai-Fah
Gad-el-Hak, Mohamed
Hartsaw, William O
Huang, Nai-Chien
Jones, James Darren
Kareem, Ahsan
Kobayashi, F(rancis) M(asao)
Lee, L(awrence) H(wa) N(i)
Lyons, Jerry L
Mortimer, Kenneth
Sen, Mihir
Snow, John Thomas
Strandhagen, Adolf G(ustav)
Sun, Chin-Teh
Sweet, Arnold Lawrence

IOWA
Akers, Arthur
Arora, Jasbir Singh
Choi, Kyung Kook
Cook, William John
Goel, Vijay Kumar
Hekker, Roeland M T
Huston, Jeffrey Charles
Lakes, Roderic Stephen
Landweber, Louis
McConnell, Kenneth G
Miller, Richard Keith
Nariboli, Gundo A
Nixon, Wilfrid Austin
Riley, William F(ranklin)
Rim, Kwan
Rogge, Thomas Ray
Rudolphi, Thomas Joseph
Sturges, Leroy D
Thompson, Robert Bruce
Tsai, Y(u)-M(in)
Zachary, Loren William

KANSAS
Huang, Chi-Lung Dominic
Lindholm, John C
McBean, Robert Parker
McCabe, Steven Lee
Yu, Yun-Sheng

KENTUCKY
Bakanowski, Stephen Michael
Brock, Louis Milton
Dillon, Oscar Wendell, Jr
Hantman, Robert Gary
Huttenlocher, Dietrich F
Leigh, Donald C
Lu, Wei-yang
Man, Chi-Sing
Miller, C Eugene
Shippy, David James
Wang, Shien Tsun

LOUISIANA
Hewitt, Hudy C, Jr
Hibbeler, Russell Charles
Rubinstein, Asher A
Yu, Fang Xin

MAINE
Nilson, Arthur H(umphrey)

MARYLAND
Alwan, Abdul-Mehsin
Antman, Stuart S
Arsenault, Richard Joseph
Bainum, Peter Montgomery
Basser, Peter J
Belliveau, Louis J
Berger, Bruce S
Bernard, Peter Simon
Burns, Bruce Peter
Butler, Thomas W(esley)
Camponeschi, Eugene Thomas, Jr
Chadwick, Richard Simeon
Chuang, Tze-jer
Cunniff, Patrick F
Dhir, Surendra Kumar
Eades, James B(everly), Jr
Eitzen, Donald Gene
Ellingwood, Bruce Russell
Everstine, Gordon Carl
Fabunmi, James Ayinde
Flynn, Paul D(avid)
Fong, Jeffrey Tse-Wei
Haberman, William L(awrence)
Huffington, Norris J(ackson), Jr
Jabbour, Kahtan Nicolas
Jones, Thomas Scott
Kitchens, Clarence Wesley, Jr
Koh, Severino Legarda
Lange, Eugene Albert
Menton, Robert Thomas
Michalowski, Radoslaw Lucas
Mordfin, Leonard
Peppin, Richard J
Poulose, Pathickal K
Rich, Harry Louis
Sallet, Dirse Wilkis
Schuman, William John, Jr
Sharpe, William Norman, Jr
Simiu, Emil
Smith, Russell Aubrey
Stein, Robert Alfred
Tai, Tsze Cheng
Tsai, Lung-Wen
Walsh, James Paul
Wang, Shou-Ling

MASSACHUSETTS
Adams, George G
Archer, Robert Raymond
Budiansky, Bernard
Castro, Alfred A
Chang, Ching Shung
Cook, Nathan Henry
Cooper, W(illiam) E(ugene)
Covert, Eugene Edzards
De Fazio, Thomas Luca
Hoffman, Allen Herbert
Kachanov, Mark L
Lardner, Thomas Joseph
Lemnios, A(ndrew) Z
Levasseur, Kenneth M
MacGregor, C(harles) W(inters)
Motherway, Joseph E
Pan, Coda H T
Pang, Yuan
Raju, Palanichamy Pillai
Rapperport, Eugene J
Reed, F(lood) Everett
Rice, James R
Rossettos, John N(icholas)
Sanders, J(ohn) Lyell, Jr
Sharp, A(rnold) G(ideon)
Suresh, Subra
Thompson, Charles
Tong, Pin
Tugal, Halil
Wong, Po Kee

MICHIGAN
Alexandridis, Alexander A
Altiero, Nicholas James, Jr
Barber, James Richard
Bolander, Richard
Browne, Alan Lampe
Caulk, David Allen
Chen, Francis Hap-Kwong
Chou, Clifford Chi Fong
Cloud, Gary Lee
Comninou, Maria
Dawson, D(onald) E(merson)
DeSilva, Carl Nevin

Evaldson, Rune L
Ezzat, Hazem Ahmed
Ghoneim, Youssef Ahmed
Greenwood, D(onald) T(heodore)
Herzog, Bertram
Hess, Robert L(awrence)
Hovanesian, Joseph Der
Howell, Larry James
Iacocca, Lee A
Ibrahim, Raouf A
Im, Jang Hi
Kamal, Mounir Mark
Kammash, Terry
Karnopp, Bruce Harvey
Kline, Kenneth A(lan)
Ku, Albert B
Kubis, Joseph J(ohn)
Kullgren, Thomas Edward
Kurajian, George Masrob
Li, Chin-Hsiu
Liu, Yi
Lord, Harold Wesley
Lubkin, James Leigh
Lund, Charles Edward
Medick, Matthew A
Morman, Kenneth N
Narain, Amitabh
Narayanaswamy, Onbathiveli S
Nefske, Donald Joseph
Nyquist, Gerald Warren
Passerello, Chris Edward
Pavlidis, Dimitris
Pence, Thomas James
Robbins, D(elmar) Hurley
Rosinski, Michael A
Sachs, Herbert K(onrad)
Saul, William Edward
Schmidt, Robert
Schultz, Albert Barry
Shkolnikov, Moisey B
Sikarskie, David L(awrence)
Smith, Hadley J(ames)
Snyder, Virgil W(ard)
Soutas-Little, Robert William
Subramanian, K N
Sung, Shung H
Tuchinsky, Philip Martin
Whicker, Donald
Wolf, Joseph A(llen), Jr
Wolf, Louis W
Wu, Hai

MINNESOTA
Beavers, Gordon Stanley
Garrard, William Lash, Jr
Gulliver, John Stephen
Hsiao, Chih Chun
Kallok, Michael John
Luskin, Mitchell B
Roska, Fred James
Sarkar, Kamalaksha
Shu, Mark Chong-Sheng
Sterling, Raymond Leslie
Warner, William Hamer

MISSISSIPPI
Butler, Dwain Kent
Cade, Ruth Ann
Carnes, Walter Rosamond
DeLeeuw, Samuel Leonard
Hackett, Robert M(oore)
Houston, James Robert
Scott, Charley

MISSOURI
Baldwin, James W(arren), Jr
Batra, Romesh Chander
Blundell, James Kenneth
Cheng, Franklin Yih
Cunningham, Floyd Mitchell
Davis, Robert Lane
Dharani, Lokeswarappa R
Fowler, Timothy John
Guccione, Julius Matteo
Haas, Charles John
Hansen, Peter Gardner
Hornsey, Edward Eugene
Jawad, Maan Hamid
Koval, Leslie R(obert)
Leutzinger, Rudolph L(eslie)
Liu, Henry
Mains, Robert M(arvin)
Mathiprakasam, Balakrishnan
Mirowitz, L(eo) I(saak)
Nikolai, Robert Joseph
Sastry, Shankara M L
Weissenburger, Jason T

MONTANA
Banaugh, Robert Peter
Friel, Leroy Lawrence

NEBRASKA
Alberts, Russell
Beatty, Millard F
Haack, Donald C(arl)
James, Merlin Lehn
Martin, Charles Wayne
Pierce, Donald N(orman)
Smith, Gerald M(ax)
Young, Lyle E(ugene)

NEVADA
Ghosh, Amitava

NEW HAMPSHIRE
Hibler, William David, III
Klotz, Louis Herman
Long, Carl F(erdinand)
Morduchow, Morris
Spehrley, Charles W, Jr

NEW JERSEY
Balaguru, Perumalsamy N
Becht, Charles, IV
Benaroya, Haym
Caudill, Reggie Jackson
Chae, Yong Suk
Clymer, Arthur Benjamin
Denno, Khalil I
Dill, Ellis Harold
Frauenthal, James Clay
Goyal, Suresh
Karcher, Guido George
Killam, Everett Herbert
Langrana, Noshir A
Lee, Peter Chung-Yi
Lee, Wei-Kuo
Lubowe, Anthony G(arner)
Mark, Robert
Marscher, William Donnelly
Miller, Edward
Pae, K(ook) D(ong)
Pai, David H(sien)-C(hung)
Pan, Ko Chang
Philipp, Ronald E
Pope, Daniel Loring
Prevost, Jean Herve
Schmidlin, Albertus Ernest
Sharobeam, Monir Hanna
Shinozuka, Masanobu
Sisto, Fernando
Spillers, William R
Thomas, Ralph Henry, Sr
Tsakonas, Stavros
Weil, Rolf
Wilson, Charles Elmer
Yadvish, Robert D
Yu, Yi-Yuan
Zabusky, Norman J

NEW MEXICO
Albrecht, Bohumil
Alzheimer, William Edmund
Chen, Er-Ping
Clouser, William Sands
Davison, Lee Walker
Drumheller, Douglas Schaeffer
Ebrahimi, Nader Dabir
Herrmann, Walter
Ju, Frederick D
Lowe, Terry Curtis
McTigue, David Francis
Moulds, William Joseph
Reuter, Robert Carl, Jr
Sutherland, Herbert James
Von Riesemann, Walter Arthur

NEW YORK
Ahmad, Jameel
Ahmadi, Goodarz
Armen, Harry, Jr
Austin, Fred
Banerjee, Prasanta Kumar
Begell, William
Bennett, Leon
Benson, Richard Carter
Betti, Raimondo
Birnstiel, Charles
Boley, Bruno Adrian
Cheng, Yean Fu
Chiang, Fu-Pen
Constantinou, Michalakis
Cozzarelli, Francis A(nthony)
Dempsey, John Patrick
Deresiewicz, Herbert
Haines, Daniel Webster
Hajela, Prabhat
Herron, Isom H
Higuchi, Hiroshi
Hoenig, Alan
Huddleston, John Vincent
Hussain, Moayyed A
Hutchings, William Frank
Hwang, S Steve
Irwin, Arthur S(amuel)
Jwo, Chin-Hung
Kapila, Ashwani Kumar
Kayani, Joseph Thomas
Klosner, Jerome M
Krempl, Erhard
Kuchar, Norman Russell
Lai, W(ei) Michael
Lee, Erastus Henry
Lee, Richard Shao-Lin
Lessen, Martin
Levy, Alan Joseph
Libove, Charles
Lin, Sung P
Mendel, John Richard
Menkes, Sherwood Bradford
Moon, Francis C
Morfopoulos, Vassilis C(onstantinos) P
Mowbray, Donald F
Panlilio, Filadelfo
Pifko, Allan Bert
Questad, David Lee
Sachse, Wolfgang H
Sancaktar, Erol

Shaffer, Bernard W(illiam)
Shaw, Richard P(aul)
Srinivasan, Vijay
Talapatra, Dipak Chandra
Testa, Rene B(iaggio)
Thomas, John Howard
Vafakos, William P(aul)
Vemula, Subba Rao
Vidosic, J(oseph) P(aul)
Wagner, John George
Ward, Lawrence W(aterman)
Weidlinger, Paul
White, Richard Norman
Whiteside, James Brooks
Wilkinson, John Peter Darrell

NORTH CAROLINA
Anderson, John P
Batra, Subhash Kumar
DeHoff, Paul Henry, Jr
Dow, Thomas Alva
Gupta, Ajaya Kumar
Hamann, Donald Dale
Hassan, Awatif E
Havner, Kerry S(huford)
Humphries, Ervin G(rigg)
McDonald, P(atrick) H(ill), Jr
Maday, Clarence Joseph
Matzen, Vernon Charles
Neal, C Leon
Pasipoularides, Ares D
Petroski, Henry J
Sandor, George N(ason)
Schmiedeshoff, Frederick William
Sharkoff, Eugene Gibb
Shield, Richard Thorpe
Snyder, Robert Douglas
Sorrell, Furman Y(ates), Jr
Utku, Senol

NORTH DAKOTA
Maurer, Karl Gustav
Rieder, William G(ary)

OHIO
Andonian, Arsavir Takfor
Bogner, Fred K
Campbell, James Edward
Chakko, Mathew K(anjhirathinkal)
Chamis, Christos Constantinos
Chen, Chao-Hsing Stanley
Chu, Mamerto Loarca
Chuck, Leon
Clausen, William E(arle)
Cummings, Charles Arnold
Damianov, Vladimir B
Dickey, David S
Drake, Michael L
Fu, Li-Sheng William
Gabb, Timothy Paul
Gent, Alan Neville
Glasgow, John Charles
Gorla, Rama S R
Halford, Gary Ross
Ho, Fanghuai H(ubert)
Holden, Frank C(harles)
Horton, Billy Mitchusson
Hulbert, Lewis E(ugene)
Jain, Vinod Kumar
Kicher, Thomas Patrick
Korda, Peter E
Kroll, Robert J
Lee, Jon H(yunkoo)
Lee, June Key
Leis, Brian Norman
Leissa, A(rthur) W(illiam)
Lemke, Ronald Dennis
Lestingi, Joseph Francis
Lewandowski, John Joseph
Macke, H(arry) Jerry
Maier, Leo Robert, Jr
Minich, Marlin
Nara, Harry R(aymond)
Ojalvo, Morris
Popelar, Carl H(arry)
Raftopoulos, Demetrios D
Rai, Iqbal Singh
Ridha, R(aouf) A
Robe, Thurlow Richard
Robinson, David Nelson
Saada, Adel Selim
Sierakowski, Robert L
Smith, Robert Emery
Stouffer, Donald Carl
Torvik, Peter J
Tsao, Bang-Hung
Vafai, Kambiz
Verstraete, Mary Clare
Wagoner, Robert H
Walter, Joseph David
Whitney, James Martin
Zavodney, Lawrence Dennis

OKLAHOMA
Appl, Franklin John
Bert, Charles Wesley
Blenkarn, Kenneth Ardley
Earls, James Roe
Neathery, Raymond Franklin
Nolte, Kenneth George
Norton, Joseph R(andolph)
Striz, Alfred Gerhard
Weston, Kenneth Clayton

Engineering Mechanics (cont)

OREGON
Calder, Clarence Andrew
Ericksen, Jerald LaVerne
Ethington, Robert Loren

PENNSYLVANIA
Barsom, John M
Batterman, Steven C(harles)
Bielak, Jacobo
Bjorhovde, Reidar
Bretz, Philip Eric
Bucci, Robert James
Carelli, Mario Domenico
Chan, Siu-Kee
Confer, Gregory Lee
Conway, Joseph C, Jr
Dash, Sanford Mark
DeSanto, Daniel Frank
DiTaranto, Rocco A
Duncombe, E(liot)
Elsworth, Derek
Goel, Ram Parkash
Griffin, Jerry Howard
Hardy, H(enry) Reginald, Jr
Hayek, Sabih I
Haythornthwaite, Robert M(orphet)
Hettche, Leroy Raymond
Hu, L(ing) W(en)
Jolles, Mitchell Ira
Kinnavy, M(artin) G(erald)
Ko, Frank K
Koch, Ronald N
Lakhtakia, Akhlesh
Leinbach, Ralph C, Jr
Likins, Peter William
Lorsch, Harold G
McCoy, John J
McLaughlin, Philip V(an Doren), Jr
McNitt, Richard Paul
Manjoine, Michael J(oseph)
Meyer, Paul A
Morehouse, Chauncey Anderson
Neubert, Vernon H
Newman, John B(ullen)
Novak, Stephen Robert
Ochs, Stefan A(lbert)
Odrey, Nicholas Gerald
Ostapenko, A(lexis)
Pangborn, Robert Northrup
Patton, Peter C(lyde)
Paul, B(urton)
Pfennigwerth, Paul Leroy
Plonsky, Andrew Walter
Raimondi, Albert Anthony
Rajagopal, K R
Rodini, Benjamin Thomas, Jr
Rogers, H(arry) C(arton), Jr
Ross, Arthur Leonard
Rumbarger, John H
Ruud, Clayton Olaf
Saigal, Sunil
Scheetz, Howard A(nsel)
Sharma, Mangalore Gokulanand
Sinclair, Glenn Bruce
Soler, Alan I(srael)
Sonnemann, George
Steg, L(eo)
Tuba, I Stephen
Varadan, Vasundara Venkatraman
Varadan, Vijay K
Vierck, Robert K
Voloshin, Arkady S
Williams, Max L(ea), Jr
Wu, John Naichi
Yen, Ben-Tseng
Zamrik, Sam Yusuf
Zhu, Dong
Zweben, Carl Henry

RHODE ISLAND
Clifton, Rodney James
Duffy, Jacques Wayne
Ferrante, W(illiam) R(obert)
Koch, Robert Michael
Symonds, Paul S(outhworth)

SOUTH CAROLINA
Anand, Subhash Chandra
Awadalla, Nabil G
Bauld, Nelson Robert, Jr
Castro, Walter Ernest
Chao, Yuh J (Bill)
Goree, James Gleason
Sparks, Peter Robert
Turner, John Lindsey
Uldrick, John Paul
Waugh, John David
Yau, Wen-Foo

TENNESSEE
Baker, Allen Jerome
Blass, Joseph J(ohn)
Buchanan, George R(ichard)
Bundy, Robert W(endel)
Carley, Thomas Gerald
Foster, Edwin Powell
Iskander, Shafik Kamel
Janna, William Sied
Keedy, Hugh F(orrest)
Landes, John D
Lee, Ching-Wen
Lee, Donald William

Pandeya, Prakash N
Pawel, Janet Elizabeth
Pih, Hui
Pugh, Claud Ervin
Shobe, L(ouis) Raymon
Slade, Edward Colin
Smith, Dallas Glen, Jr
Snyder, William Thomas
Stoneking, Jerry Edward
Wasserman, Jack F
Wheeler, Orville Eugene
Yen, Michael R T

TEXAS
Anderson, Charles E, Jr
Athanasiou, Kyriacos A
Bendapudi, Kasi Visweswararao
Brock, James Harvey
Burger, Christian P
Cao, Hengchu
Carey, Graham Francis
Carroll, Michael M
Chevalier, Howard L
Dalley, Joseph W(inthrop)
Diller, Kenneth Ray
Ebner, Stanley Gadd
Eichberger, Le Roy Carl
Fischer, Ferdinand Joseph
Gaines, J H
Geyling, F(ranz) Th(omas)
Gladwell, Ian
Hussain, A K M Fazle
Johnson, Brann
Jones, William B
Jordan, Neal F(rancis)
Juricic, Davor
Kana, Daniel D(avid)
Kanninen, Melvin Fred
Kinra, Vikram Kumar
Lawrence, Kent L(ee)
Lutes, Loren Daniel
Lytton, Robert Leonard
Mack, Lawrence R(iedling)
Marder, Michael P
Marshek, Kurt M
Monsees, James E
Moon, Tessie Jo
Morgan, James Richard
Muster, Douglas Frederick
Nachlinger, R Ray
Nordgren, Ronald Paul
Oden, John Tinsley
Patrick, Wesley Clare
Pray, Donald George
Ripperger, Eugene Arman
Schneider, William Charles
Schruben, Johanna Stenzel
Simonis, John Charles
Solberg, Ruell Floyd, Jr
Stanovsky, Joseph Jerry
Stern, Morris
Stubbs, Norris
Tien, John Kai
Vallabhan, C V Girija
Von Maltzahn, Wolf W
Waid, Margaret Cowsar
Walton, Jay R
Wang, Chao-Cheng
Ward, Donald Thomas
Warren, William Ernest
Watson, Hal, Jr
Wenzel, Alexander B
Wilcox, Marion Walter

UTAH
Adams, Brent Larsen
Barton, Cliff S
Hoeppner, David William
Karren, Kenneth W
Nairn, John Arthur
Rotz, Christopher Alan
Silver, Barnard Stewart

VIRGINIA
Barton, Furman W(yche)
Beran, Robert Lynn
Berry, James G(ilbert)
Chang, George Chunyi
Chi, Michael
Chisholm, Douglas Blanchard
Chong, Ken Pin
Cooper, Henry Franklyn, Jr
Davidson, John Richard
Davis, John Grady, Jr
Duke, John Christian, Jr
Edlich, Richard French
Eppink, Richard Theodore
Fisher, Samuel Sturm
Frederick, Daniel
Gunter, Edgar Jackson, Jr
Hasenfus, Harold J(oseph)
Heller, Robert A
Herakovich, Carl Thomas
Horgan, Cornelius Oliver
Hou, Gene Jean-Win
Hudson, Charles Michael
Jones, Robert Millard
Juang, Jer-Nan
Kapania, Rakesh Kumar
Landgraf, Ronald William
Lea, George Koo
Maderspach, Victor
Maher, F(rancis) J(oseph)
Mook, Dean Tritschler

Moon, Peter Clayton
Newman, James Charles, Jr
Pletta, Dan Henry
Post, Daniel
Raju, Ivatury Sanyasi
Rosenfeld, Robert L
Rosenthal, F(elix)
Senseny, Paul Edward
Singh, Mahendra Pal
Smith, C(harles) William
Spitzig, William Andrew
Stinchcomb, Wayne Webster
Sword, James Howard
Tompkins, Stephen Stern
Tsai, Frank Y
Wayne, Jennifer Susan
Wolko, Howard Stephen
Wood, Albert D(ouglas)
Yang, Ta-Lun
Yates, Edward Carson, Jr

WASHINGTON
Chalupnik, James Dvorak
Ding, Jow-Lian
Dusto, Arthur Ronald
Duvall, George Evered
Goodstein, Robert
Hamilton, C Howard
Hartz, Billy J
Johnson, Jay Allan
Levinson, Mark
Madden, James H(oward)
Merchant, Howard Carl
Merckx, Kenneth R(ing)
Reyhner, Theodore Alison
Sherrer, Robert E(ugene)
Sorensen, Harold C(harles)
Sutherland, Earl C
Varanasi, Suryanarayana Rao
Wolak, Jan

WEST VIRGINIA
Beeson, Justin Leo
Mucino, Victor H
Shuck, Lowell Zane
Venable, Wallace Starr

WISCONSIN
Bodine, Robert Y(oung)
Burck, Larry Harold
Cheng, Shun
Cook, Robert D(avis)
Crandall, Lee W(alter)
Crawmer, Daryl E
Cutler, Verne Clifton
Daane, Robert A
Edil, Tuncer Berat
Engelstad, Roxann Louise
Huang, T(zu) C(huen)
Hung, James Y(un-Yann)
Johnson, John E(dwin)
Johnson, Millard Wallace, Jr
Kenny, Andrew Augustine
Larkin, Lawrence A(lbert)
Lovell, Edward George
Malkus, David Starr
Naik, Tarun Ratilal
Pillai, Thankappan A K
Richard, Terry Gordon
Rowlands, Robert Edward
Saemann, Jesse C(harles), Jr
Schaffer, Erwin Lambert
Schmieg, Glenn Melwood
Widera, Georg Ernst Otto
Wnuk, Michael Peter
Yoganandan, Narayan
Zahn, John J

WYOMING
Adams, Donald F
Dolan, Charles W

PUERTO RICO
Goyal, Megh R
Khan, Winston
Patra, Amit Lal

ALBERTA
Epstein, Marcelo
Singh, Mansa C

BRITISH COLUMBIA
Mantle, J(ohn) B(ertram)
Pomeroy, Richard James

MANITOBA
Thomas, Robert Spencer David
Wilms, Ernest Victor

NOVA SCOTIA
Cochkanoff, O(rest)
Sastry, Vankamamidi VRN

ONTARIO
Barron, Ronald Michael
Cowper, George Richard
Huseyin, Koncay
Pindera, Jerzy Tadeusz
Rimrott, F(riedrich) P(aul) J(ohannes)
Schwaighofer, Joseph
Selvadurai, A P S
Sharan, Shailendra Kishore
Shelson, W(illiam)
Sherbourne, Archibald Norbert

Siddell, Derreck
Thompson, John Carl
Wong, Jo Yung

QUEBEC
Bhat, Rama B
Dealy, John Michael
Fabrikant, Valery Isaak
Hoa, Suong Van
Johns, Kenneth Charles
McKyes, Edward
Marsh, Cedric
Pekau, Oscar A
Stathopoulos, Theodore
Vatistas, Georgios H

OTHER COUNTRIES
Al-Zubaidy, Sarim Naji
Bodner, Sol R(ubin)
Burnett, Jerrold J
Chwang, Allen Tse-Yung
Ersoy, Ugur
Laura, Patricio Adolfo Antonio
Lee, Choung Mook
Revesz, Zsolt
Sunahara, Yoshifumi
Thurlimann, Bruno
Tipans, Igors O

Engineering Physics

ALABAMA
Askew, Raymond Fike
Blackmon, James B
Cushing, Kenneth Mayhew
Mookherji, Tripty Kumar
Regner, John LaVerne
Wilhold, Gilbert A

ALASKA
Beebee, John Christopher
Hunsucker, Robert Dudley
Wentink, Tunis, Jr

ARIZONA
El-Ghazaly, Samir M
Falk, Thomas
Fung, K Y
Hight, Ralph Dale
Mardian, James K W
Osborn, Donald Earl
Perper, Lloyd
Peyghambarian, Nasser
Rasmussen, William Otto
Schowengerdt, Robert Alan
Silvern, Leonard Charles
Szilagyi, Miklos Nicholas
Utagikar, Ajit Purushottam
Weddell, James Blount
Welch, H(omer) William
Ziolkowski, Richard Walter

CALIFORNIA
Adams, John Andrew
Adler, William Fred
Ames, Lawrence Lowell
Amirkhanian, Varouj David
Andersen, Wilford Hoyt
Ballantyne, Catherine Cox
Bardsley, James Norman
Barhydt, Hamilton
Batdorf, Samuel Burbridge
Bharat, Ramasesha
Billman, Kenneth William
Blankenship, Victor D(ale)
Block, Richard B
Bradner, Hugh
Chesnut, Walter G
Clarke, David Richard
Clemens, Bruce Montgomery
Cohen, Lawrence Mark
Collins, Carter Compton
Cooper, Alfred William Madison
Coward, David Hand
Crandall, Michael Grain
Deri, Robert Joseph
Deshpande, Shivajirao M
Dethlefsen, Rolf
Duneer, Arthur Gustav, Jr
Eby, Frank Shilling
Edmonds, Peter Derek
Edwall, Dennis Dean
Egermeier, R(obert) P(aul)
Elverum, Gerard William, Jr
Eukel, Warren W(enzl)
Everhart, Thomas E(ugene)
Fillius, Walker
Forbes, Judith L
Frank, Alan M
Freedman, Stuart Jay
Freis, Robert P
Fultz, Brent T
Garrett, Steven Lurie
Gin, W(inston)
Ginsburgh, Irwin
Glick, Harold Alan
Gould, Harvey Allen
Grant, Eugene F(redrick)
Grove, Andrew S
Heer, Ewald
Hendricks, Charles D(urrell), Jr
Hendry, George Orr
Herb, John A

ENGINEERING / 329

Hesselink, Lambertus
Hoover, Carol Griswold
Hsia, Yukun
Hult, John Luther
Jaklevic, Joseph Michael
Jansen, Frank
Jory, Howard Roberts
Judge, Roger John Richard
Karlsson, Eric Allan
Karp, Arthur
Kasper, Gerhard
Kavanagh, Karen L
Kennel, John Maurice
Khan, Mahbub R
Kilmer, Louis Charles
Kittel, Peter
Kupfer, John Carlton
Lagarias, John S(amuel)
Lai, Kai Sun
Lam, Leo Kongsui
Lapp, M(arshall)
Lee, Kenneth
Leibowitz, Lewis Phillip
Lemke, James Underwood
Leung, Charles Cheung-Wan
Leung, Ka-Ngo
Likuski, Robert Keith
Liu, Frederick F
Liu, Jia-ming
Longley-Cook, Mark T
Loo, Billy Wei-Yu
Lowi, Alvin, Jr
Marcus, John Stanley
Martin, Gordon Eugene
Maserjian, Joseph
Matthews, Stephen M
Miller, Joseph
Miller, Roger Heering
Mireshghi, Ali
Moir, Ralph Wayne
Moore, Edgar Tilden, Jr
Muller, Richard Stephen
Naqvi, Iqbal Mehdi
Nelson, Raymond Adolph
Nooker, Eugene L(eRoy)
Novotny, Vlad Joseph
Nunan, Craig S(pencer)
O'Donnell, C(edric) F(inton)
Palmer, John Parker
Parks, Lewis Arthur
Perkins, Kenneth L(ee)
Pomraning, Gerald C
Rasor, Ned S(haurer)
Rathmann, Carl Erich
Reinheimer, Julian
Remley, Marlin Eugene
Reynolds, Harry Lincoln
Rosenblum, Stephen Saul
Saito, Theodore T
Salisbury, Stanley R
Salsig, William Winter, Jr
Sarwinski, Raymond Edmund
Scott, Franklin Robert
Scott, Paul Brunson
Shaffner, Richard Owen
Sher, Arden
Silverstein, Elliot Morton
Simonen, Thomas Charles
Sinclair, Kenneth F(rancis)
Spector, Clarence J(acob)
Spencer, Cherrill Melanie
Spindt, Charles A (Capp)
Stalder, Kenneth R
Starr, Chauncey
Stewart, Richard William
Stirn, Richard J
Stockel, Ivar H(oward)
Suchard, Steven Norman
Tatic-Lucic, Svetlana
Terhune, Robert William
Thompson, David A
Thomson, James Alex L
Trapp, Robert F(rank)
Tricoles, Gus P
Urtiew, Paul Andrew
Vanblarigan, Peter
Vaughan, J Rodney M
Veigele, William John
Victor, Andrew C
Vinokur, Marcel
Viswanathan, R
Wagner, Carl E
Waldner, Michael
Wang, Jon Y
Warburton, William Kurtz
Warren, Mashuri Laird
Watt, Robert Douglas
Webber, Donald Salyer
Webster, Emilia
Weiss, Jeffrey Martin
Wessel, Frank J
Winkelmann, Frederick Charles
Yap, Fung Yen
Yeh, Edward H Y
Young, Louise Gray
Zucca, Ricardo
Zuleeg, Rainer
Zuppero, Anthony Charles

COLORADO
Augusteijn, Marijke Francina
Brown, Edmund Hosmer
Bushnell, Robert Hempstead
Cathey, Wade Thomas, Jr

Chisholm, James Joseph
Gardner, Edward Eugene
Hanley, Howard James Mason
Harrington, Robert D(ean)
Hilchie, Douglas Walter
Hoffman, John Raleigh
Jaye, Seymour
Kremer, Russell Eugene
McAllister, Robert Wallace
Marks, Roger Bradley
Morse, J(erome) G(ilbert)
Radebaugh, Ray
Schwiesow, Ronald Lee
Verschoor, J(ack) D(ahlstrom)
Wertz, Ronald Duane

CONNECTICUT
Antkiw, Stephen
Brown, Arthur A(ustin)
Burwell, Wayne Gregory
Dreyfus, Marc George
Erlandson, Paul M(cKillop)
Greenwood, Ivan Anderson
Holzman, George Robert
Hufnagel, Robert Ernest
Paolini, Francis Rudolph
Rau, Richard Raymond
Reed, Joseph
Similon, Philippe Louis
Sreenivasan, Katepalli Raju
Taylor, Geoff W
Tittman, Jay
Troutman, Ronald R
Wormser, Eric M
Wuthrich, Paul

DELAWARE
Bierlein, John David
Christie, Phillip
Hardy, Henry Benjamin, Jr
Hirsch, Albert Edgar
Jansson, Peter Allan
Moore, Charles B(ernard)
Owens, Aaron James
Scofield, Dillon Foster

DISTRICT OF COLUMBIA
Beckner, Everet Hess
Borgiotti, Giorgio V
Boris, Jay Paul
Cambel, Ali B(ulent)
Forester, Donald Wayne
Huband, Frank L
Nachman, Arje
Needels, Theodore S
Pikus, Irwin Mark
Schriever, Richard L
Spohr, Daniel Arthur

FLORIDA
Adhav, Ratnakar Shankar
Anghaie, Samim
Baird, Alfred Michael
Cherepanov, Genady Petrovich
Clarke, Thomas Lowe
Cox, John David
Fry, James N
Hansel, Paul G(eorge)
Harmon, G Lamar
Kinnmark, Ingemar Per Erland
Klausner, James Frederick
Krishnamurthy, Lakshminarayanan
Langford, David
Melich, Michael Edward
Min, Kwang-Shik
Pendleton, Winston Kent, III
Penney, Gaylord W
Sah, Chih-Tang
Schimmel, Walter Paul
Tomonto, James R
Tusting, Robert Frederick
Vollmer, James
Young, C(harles) Gilbert

GEORGIA
Batten, George L, Jr
Croft, George Thomas
Currie, Nicholas Charles
Fisher, Reed Edward
Jokerst, Nan Marie
Lau, Jark Chong
Lorber, Herbert William
McMillan, Robert Walker

HAWAII
Craven, John P
Curtis, George Darwin
Frazer, L Neil
Fuka, Louis Richard

IDAHO
Carpenter, Stuart Gordon
Mott, Jack Edward
Spencer, Paul Roger
Woodall, David Monroe

ILLINOIS
Bjorkland, John A(lexander)
Cohn, Gerald Edward
Coleman, P(aul) D(are)
Cooper, William Edward
Fast, Ronald Walter
Flora, Robert Henry
Gallagher, David Alden

Herzenberg, Caroline Stuart Littlejohn
Horve, Leslie A
Jostlein, Hans
Kang, Sung-Mo Steve
Kim, Kyekyoon K(evin)
Kuchnir, Moyses
Kustom, Robert L
Macrander, Albert Tiemen
Mielke, Robert L
Mokadam, Raghunath G(anpatrao)
Morkoc, Hadis
Pardo, Richard Claude
Saxena, Satish Chandra
Shabana, Ahmed Abdelraouf
Shea, Michael Francis
Sloma, Leonard Vincent
Spradley, Joseph Leonard
Sproul, William Dallas
Till, Charles Edgar
Uherka, Kenneth Leroy
Venema, Harry J(ames)
Wolsky, Alan Martin
Yamada, Ryuji
Zelac, Ronald Edward

INDIANA
Ersoy, Okan Kadri
Henke, Mitchell C
Ho, Cho-Yen
Kramer, Robert Allen
Schwartz, Richard John
Snare, Leroy Earl
Voelz, Frederick

IOWA
Schaefer, Joseph Albert
Thompson, Robert Bruce

KANSAS
Gray, Tom J
Stockli, Martin P
Unz, Hillel

KENTUCKY
Bakanowski, Stephen Michael
Frank, John L

LOUISIANA
Holden, William R(obert)
Witriol, Norman Martin

MAINE
Nicol, James

MARYLAND
Alers, George A
Birnbaum, George
Blessing, Gerald Vincent
Blum, Norman Allen
Bur, Anthony J
Castruccio, Peter Adalbert
Chern, Engmin James
Connerney, John E P
Connor, Joseph Gerard, Jr
Cook, Richard Kaufman
Dauber, Edwin George
Dhir, Surendra Kumar
Doiron, Theodore Danos
Ethridge, Noel Harold
Fabunmi, James Ayinde
Fong, Jeffrey Tse-Wei
French, Joseph Cull
Fthenakis, Emanuel John
Gartrell, Charles Frederick
Geckle, William Jude
Gibson, Harold F(loyd)
Gillich, William John
Gonano, John Roland
Gordon, Gary Donald
Hauser, Michael George
Henry, James H(erbert)
Herzfeld, Charles Maria
Huang, Thomas Tsung-Tse
Hubbell, John Howard
Huttlin, George Anthony
Hyde, Geoffrey
Jensen, Arthur Seigfried
Kalil, Ford
Larrabee, R(obert) D(ean)
Lull, David B
Lundholm, J(oseph) G(ideon), Jr
McCoubrey, Arthur Orlando
Malarkey, Edward Cornelius
Melngailis, John
Merkel, George
Metze, George M
Moulton, James Frank, Jr
Ostaff, William A(llen)
Pearl, John Christopher
Pierce, Harry Frederick
Pisacane, Vincent L
Preisman, Albert
Reader, Wayne Truman
Reilly, James Patrick
Rueger, Lauren J(ohn)
Sanders, Robert Charles
Sank, Victor L
Scialdone, John Joseph
Skiff, Frederick Norman
Stone, Albert Mordecai
Turner, Robert Davison
Waldo, George Van Pelt, Jr
Warner, Brent A
Weigman, Bernard J

Zien, Tse-Fou
Zirkind, Ralph

MASSACHUSETTS
Allen, Steven
Antal, John Joseph
Arganbright, Donald G
Bechis, Kenneth Paul
Brown, Walter Redvers John
Castro, Alfred A
Chandler, Charles H(orace)
Cordero, Julio
Dakss, Mark Ludmer
Donnelly, Joseph P(eter)
Dyer, James Arthur
Esch, Robin E
Fan, John C C
Feinleib, Julius
Garvey, R(obert) Michael
Guenther, R(ichard)
Holt, Frederick Sheppard
Huguenin, George Richard
Jones, Robert Allan
Jones, William J
Kamentsky, Louis A
Kanter, Irving
Kniazzeh, Alfredo G(iovanni) F(rancesco)
Kolodziejski, Leslie Ann
Kovaly, John J
Lederman, David Mordechai
Lien, Hwachii
Mack, Michael E
Mulpur, Arun K
Nowak, Welville B(erenson)
Oettinger, Peter Ernest
Olderman, Jerry
Orloff, Lawrence
Rapperport, Eugene J
Robinson, D(enis) M(orrell)
Schneider, Robert Julius
Schwartz, Jack
Sethares, James C(ostas)
Shapiro, Edward K Edik
Singleton, John Byrne
Smith, Alan Bradford
Sosnowski, Thomas Patrick
Speliotis, Dennis Elias
Tancrell, Roger Henry
Tugal, Halil
Wagner, David Kendall
Weggel, Robert John
Weiss, Jerald Aubrey
Winer, Bette Marcia Tarmey
Woskov, Paul Peter
Yngvesson, K Sigfrid

MICHIGAN
Bolander, Richard
Brown, Steven Michael
Dow, W(illiam) G(ould)
Gilgenbach, Ronald Matthew
Johnson, Wayne Jon
Joseph, Bernard William
Kim, Changhyun
Liu, Yi
Logothetis, Eleftherios Miltiadis
Marko, Kenneth Andrew
Miller, Carl Elmer
Morgan, Roger John
Peterson, Lauren Michael
Piccirelli, Robert Anthony
Schumacher, Berthold Walter
Smith, Hadley J(ames)
Thomson, Robert Francis

MINNESOTA
Agarwal, Vijendra Kumar
DiBona, Peter James
George, Peter Kurt
Hocker, George Benjamin
Keister, Jamieson Charles
Schumer, Douglas Brian
Shu, Mark Chong-Sheng
Yasko, Richard N

MISSOURI
Carter, Robert L(eroy)
Kibens, Valdis
Kunze, Jay Frederick
Lind, Arthur Charles
Lowe, Forrest Gilbert
McMillan, Gregory Keith
Northrip, John Willard

NEBRASKA
Dowben, Peter Arnold
Rohde, Suzanne Louise
Soukup, Rodney Joseph

NEW HAMPSHIRE
Buffler, Charles Rogers
Crowell, Merton Howard
Linnell, Richard D(ean)
Lopez, Guido Wilfred
Spehrley, Charles W, Jr
Temple, Peter Lawrence
Yen, Yin-Chao

NEW JERSEY
Berkman, Samuel
Bharj, Sarjit Singh
Bjorkholm, John Ernst
Browne, Robert Glenn

Engineering Physics (cont)

Brucker, Edward Byerly
Catanese, Carmen Anthony
Dutta, Mitra
Jahn, Robert G(eorge)
Jassby, Daniel Lewis
Kaita, Robert
Kam, Gar Lai
Kapron, Felix Paul
Kheshgi, Haroon S
Kugel, Henry W
Kunhardt, Erich Enrique
Kunzler, John Eugene
Lanzerotti, Louis John
Lechner, Bernard J
Leheny, Robert Francis
Liao, Paul Foo-Hung
Michaelis, Paul Charles
Miles, Richard Bryant
Pargellis, Andrew Nason
Rainal, Attilio Joseph
Rothleder, Stephen David
Schivell, John Francis
Schneider, Sol
Schrenk, George L
Schulte, Harry John, Jr
Shahbender, R(abah) A(bd-El-Rahman)
Shen, Tek-Ming
Wachtell, George Peter
Weissenburger, Don William
Wiesenfeld, Jay Martin
Williams, Neal Thomas
Yadvish, Robert D
Zweig, Gilbert

NEW MEXICO
Barsis, Edwin Howard
Benjamin, Robert Fredric
Bergeron, Kenneth Donald
Bowman, Allen Lee
Chamberlin, Edwin Phillip
Clark, Wallace Thomas, III
Davis, L(loyd) Wayne
Dempster, William Fred
Dingus, Ronald Shane
Drumheller, Douglas Schaeffer
Engelhardt, Albert George
Farmer, William Michael
Fradkin, David Barry
Gorham, Elaine Deborah
Gurbaxani, Shyam Hassomal
Haberstich, Albert
Hartman, James Keith
Henderson, Dale Barlow
Holm, Dale M
Jameson, Robert A
Janney, Donald Herbert
McNally, James Henry
Metzger, Daniel Schaffer
Oyer, Alden Tremaine
Palmer, Byron Allen
Parker, Joseph R(ichard)
Perret, William Riker
Schneider, Jacob David
Sze, Robert Chia-Ting
Touryan, Kenell James

NEW YORK
Baker, Alan Gardner
Benumof, Reuben
Bickmore, John Tarry
Bilderback, Donald Heywood
Birnbaum, David
Blumenthal, Ralph Herbert
Brehm, Lawrence Paul
Brunelle, Eugene John, Jr
Buhrman, Robert Alan
Cady, K Bingham
Chappell, Richard Lee
Chen, Alek Chi-Heng
Cheng, Yean Fu
Cheo, Bernard Ru-Shao
Cohen, Mitchell S(immons)
Cole, Robert
Constable, James Harris
Cottingham, James Garry
Flom, Donald Gordon
Friedman, Jack P
Genin, Dennis Joseph
George, Nicholas
Glinkowski, Mietek T
Greene, Peter Richard
Grunwald, Hubert Peter
Harris, Donald R, Jr
Hartmann, George Cole
Hoisie, Adolfy
Hoover, Thomas Earl
Jwo, Chin-Hung
Kasprzak, Lucian A
Kaup, David James
Komiak, James Joseph
Kuchar, Norman Russell
Lakatos, Andras Imre
LaMuth, Henry Lewis
Li, Ching-Hsiung
Llimatainen, T(oivo) M(atthew)
Likharev, Konstantin K
Lindstrom, Gary J
Lowther, Frank Eugene
MacDonald, Noel C
McGuire, Stephen Craig
Mahler, David S
Malaviya, Bimal K

Maling, George Croswell, Jr
Marcus, Michael Alan
Margulies, Marcel
Moe, George Wylbur
Morris, George Michael
Mortenson, Kenneth Ernest
Nation, John
Nicolosi, Joseph Anthony
Osgood, Richard Magee, Jr
Parks, Harold George
Pernick, Benjamin J
Podowski, Michael Zbigniew
Rudinger, George
Sai-Halasz, George Anthony
Salzberg, Bernard
Siegel, Benjamin Morton
Simpson, Murray
Slade, Paul Graham
Spoelhof, Charles Peter
Star, Joseph
Stenzel, Wolfram G
Stephenson, Thomas E(dgar)
Stevens, William Y(eaton)
Stover, Raymond Webster
Talley, Robert Lee
Temple, Victor Albert Keith
Thomas, J Earl
Vahey, David William
Vosburgh, Kirby Gannett
Walker, Michael Stephen
Weber, Arthur Phineas
Weeks, William Thomas
Wei, Ching-Yeu
Wertheimer, Alan Lee
Wysocki, Joseph J(ohn)

NORTH CAROLINA
Dayton, Benjamin Bonney
Dow, Thomas Alva
Grainger, John Joseph
Hinesley, Carl Phillip
Irene, Eugene Arthur
Jones, Creighton Clinton
Lawless, Philip Austin
Nader, John S(haheen)
Quisenberry, Karl Spangler, Jr
Winkler, Linda Irene

OHIO
Andrews, Merrill Leroy
Bennett, Foster Clyde
Bhattacharya, Rabi Sankar
Brown, Robert William
Curtis, Lorenzo Jan
Dakin, James Thomas
Fritsch, Klaus
Golovin, Michael N
Griffing, David Francis
Horton, Billy Mitchusson
Hunter, William Winslow, Jr
Keim, John Eugene
Landstrom, D(onald) Karl
Mayers, Richard Ralph
Reiling, Gilbert Henry
Rich, Joseph William
Seldin, Emanuel Judah
Smith, James Edward, Jr
Smith, Robert Emery
Springer, Robert Harold
Thompson, Hugh Ansley
Turchi, Peter John

OKLAHOMA
Blais, Roger Nathaniel
Kuenhold, Kenneth Alan
Lal, Manohar
Ryan, Stewart Richard

OREGON
Gerlach, Robert Louis
Gillette, Philip Roger
Matson, Leslie Emmet, Jr
Paarsons, James Delbert
Wiley, William Charles
Yau, Leopoldo D

PENNSYLVANIA
Arnold, James Norman
Artman, Joseph Oscar
Ashok, S
Bilaniuk, Oleksa-Myron
Boer, Karl Wolfgang
Brunner, Mathias J
Carelli, Mario Domenico
Chaiken, Robert Francis
Cook, Donald Bowker
Cookson, Alan Howard
Deis, Daniel Wayne
Dick, George W
Emtage, Peter Roesch
Fonash, Stephan J(oseph)
Frederick, David Eugene
Garbuny, Max
Grimes, Dale M(ills)
Hansen, Robert Jack
Harrold, Ronald Thomas
Hoburg, James Frederick
Hurwitz, Jan Krosst
Jones, Clifford Kenneth
Kahn, David
Kanter, Ira E
Kimblin, Clive William
Kryder, Mark Howard
Lempert, Joseph

Mahajan, Subhash
Messier, Russell
Meyer, Paul A
Ochs, Stefan A(lbert)
Oder, Robin Roy
Pinckney, Robert L
Pinkerton, John Edward
Roy, Pradip Kumar
Schenk, H(arold) L(ouis), Jr
Spira, Joel Solon
Steinbruegge, Kenneth Brian
Stinger, Henry J(oseph)
Tingle, William Herbert
Trainer, Michael Norman
Uher, Richard Anthony
Van Roggen, Arend
Varadan, Vijay K
Wilson, Walter R
Zemel, Jay N(orman)

RHODE ISLAND
Avery, Donald Hills
Nurmikko, Arto V

SOUTH CAROLINA
Avegeropoulos, G
Harrison, James William, Jr

SOUTH DAKOTA
Cannon, Patrick Joseph

TENNESSEE
Allison, Stephen William
Baker, Charles Clayton
Blankenship, James Lynn
Burton, Robert Lee
Foster, Edwin Powell
Kiech, Earl Lockett
Lee, Donald William
LeVert, Francis E
Longmire, Martin Shelling
McGregor, Wheeler Kesey, Jr
Maienschein, Fred (Conrad)
Morgan, Ora Billy, Jr
Morrison, Thomas Golden
Pearsall, S(amuel) H(aff)
Roth, J(ohn) Reece
Schwenterly, Stanley William, III
Slade, Edward Colin
Walter, F John
Watson, Robert Lowrie
Wilcox, William Jenkins, Jr

TEXAS
Anderson, Charles E, Jr
Baker, Wilfred E(dmund)
Birchak, James Robert
Boers, Jack E
Boyer, Lester Leroy
Brannon, H Raymond, Jr
Cao, Hengchu
Chivian, Jay Simon
Dhudshia, Vallabh H
Dykstra, Jerald Paul
Eberhart, Robert Clyde
Griffin, Richard B
Hebert, Joel J
Hewett, Lionel Donnell
Ignatiev, Alex
Jordan, Neal F(rancis)
Kristiansen, Magne
Lacy, Lewis L
Marlow, William Henry
Matzkanin, George Andrew
Moore, Thomas Matthew
Nimitz, Walter W von
Parker, Jerald Vawer
Ransom, C J
Richard, Richard Ray
Skinner, G(eorge) M(acGillivray)
Stone, Orville L
Tao, Frank F
Waggoner, James Arthur
Wehring, Bernard William
Wenzel, Alexander B
Young, James Forrest

UTAH
Adams, Brent Larsen
Bunch, Kyle
Clegg, John C(ardwell)
Cohen, Richard M
Jeffery, Rondo Nelden
Jolley, David Kent
Lund, Mark Wylie
Ward, Roger Wilson
Zeamer, Richard Jere

VERMONT
Ellis-Monaghan, John J

VIRGINIA
Bahl, Inder Jit
Bailey, Marion Crawford
Benesch, Jay F
Burley, Carlton Edwin
Cardman, Lawrence Santo
Church, Charles Henry
De Wolf, David Alter
Hafele, Joseph Carl
Hammer, Jacob Meyer
Hasenfus, Harold J(oseph)
Herbst, Mark Joseph
Hwang, William Gaong

James, W(ilbur) Gerald
Johnson, Robert Edward
Krassner, Jerry
Kuhlmann-Wilsdorf, Doris
Moon, Peter Clayton
Neubauer, Werner George
Ni, Chen-Chou
Pettus, William Gower
Porzel, Francis Bernard
Ramey, Robert Lee
Reiss, Keith Westcott
Sager, Earl Vincent
Schremp, Edward Jay
Shenolikar, Ashok Kumar
Siebentritt, Carl R, Jr
Smith, Sidney Taylor
Weichel, Hugo
Wolicki, Eligius Anthony

WASHINGTON
Bruckner, Adam Peter
Center, Robert E
Christiansen, Walter Henry
Das, K Bhagwan
Gupta, Yogendra M(ohan)
Kim, Jae Hoon
Nichols, Davis Betz
Pichal, Henri Thomas, II
Russell, David A
Russell, James Torrance
Safai, Morteza
Soreide, David Christien
Stitch, Malcolm Lane
Stoebe, Thomas Gaines

WEST VIRGINIA
Corum, James Frederic
Littleton, John Edward
Parker, David Hiram
Spaniol, Craig

WISCONSIN
Bomba, Steven James
Boom, Roger Wright
Hyzer, William Gordon
Kinsinger, Richard Estyn
Klein, Carl Frederick
Nordman, James Emery
Schlicke, Heinz M
Schmieg, Glenn Melwood
Tonner, Brian P

PUERTO RICO
Hussain, Malek Gholoum Malek

ALBERTA
Cormack, George Douglas
Fedosejevs, Robert
Offenberger, Allan Anthony

BRITISH COLUMBIA
Bittner, John William
Deen, Mohamed Jamal
Jull, Edward Vincent
Mackenzie, George Henry
Menon, Thuppalay K
Palmer, James Frederick
Tiedje, J Thomas

MANITOBA
Johnson, R(ichard) A(llan)
Loly, Peter Douglas
Mathur, Maya Swarup
Wilkins, Brian John Samuel

NEW BRUNSWICK
Boorne, Ronald Albert
Wells, David Ernest

NOVA SCOTIA
Elliott, James Arthur
El-Masry, Ezz Ismail
Roger, William Alexander

ONTARIO
Berezin, Alexander A
Brach, Eugene Jenö
Brimacombe, Robert Kenneth
Chang, Jen-Shih
Chou, Jordan Quan Ban
Cohn-Sfetcu, Sorin
Critoph, E(ugene)
Davies, John Arthur
Dunlop, David John
Gold, Lorne W
Grodski, Juliusz Jan
Harms, Archie A
Jackson, David Phillip
Johnston, G(ordon) W(illiam)
King, Hubert Wylam
Kumaran, Mavinkal Kizhakkeveettil
Lipshitz, Stanley Paul
Lynch, Gerard Francis
McGee, William F
McNamara, Allen Garnet
Napier, Douglas Herbert
Orlik-Ruckemann, Kazimierz Jerzy
Ottensmeyer, Frank Peter
Pindera, Jerzy Tadeusz
Raju, Govinda
Rauch, Sol
Rochon, Paul Leo
Rosen, Marc Allen
Selvakumar, Chettypalayam Ramanathan

Siddell, Derreck
Sonnenberg, Hardy
Thompson, David Allan
Uffen, Robert James
Yevick, David Owen

QUEBEC
Aktik, Cetin
Bose, Tapan Kumar
Champness, Clifford Harry
Chan, Sek Kwan
Gregory, Brian Charles
Hetherington, Donald Wordsworth
Izquierdo, Ricardo
Maciejko, Roman
Meunier, Michel
Power, Joan F
Schwelb, Otto
Terreault, Bernard J E J
Vatistas, Georgios H
Wertheimer, Michael Robert

SASKATCHEWAN
Nikiforuk, P(eter) N

OTHER COUNTRIES
Beckwith, Steven Van Walter
Benson, Frank Atkinson
Bryngdahl, Olof
Gordon, Jeffrey Miles
Mengali, Umberto
Purbo, Onno Widodo
Sato, Gentei
Shah, Mirza Mohammed
Steffen, Juerg
Sugaya, Hiroshi
Unger, Hans-Georg

Fuel Technology & Petroleum Engineering

ALABAMA
Barrier, J Wayne
Dahlin, Robert S

ALASKA
Merrill, Robert Clifford, Jr
Ostermann, Russell Dean
Sharma, Ghanshyam D

ARIZONA
Ramohalli, Kumar Nanjunda Rao

CALIFORNIA
Adicoff, Arnold
Bauer, Robert Forest
Blunt, Martin Julian
Bridge, Alan G
Buck, F(rank) A(lan) Mackinnon
Burnham, Alan Kent
Campbell, Thomas Cooper
Chesnut, Dwayne A(llen)
Chilingarian, George V(aros)
Compton, Leslie Ellwyn
Coppel, Claude Peter
Culler, F(loyd) L(eRoy), Jr
Davidson, Lynn Blair
Davis, Bruce W
Day, Robert J(ames)
Dickinson, Wade
Dolbear, Geoffrey Emerson
Durai-swamy, Kandaswamy
Engleman, Victor Solomon
Fox, Joseph M(ickle), III
Frost, Kenneth Almeron, Jr
Goss, John R(ay)
Greene, Charles Richard
Handy, Lyman Lee
Hass, Robert Henry
Heinemann, Heinz
Hennig, Harvey
Hess, Patrick Henry
Horne, Roland Nicholas
Irving, James P
Johnson, Carl Emil, Jr
Keaton, Michael John
Keller, James Lloyd
Kertamus, Norbert John
Klotz, James Allen
Kohl, A(rthur) L(ionel)
Kroneberger, Gerald F
Krueger, Roland Frederick
Kuehne, Donald Leroy
Lewis, Arthur Edward
Low, Hans
Ludwig, Harvey F
McKinney, Ted Meredith
Mallett, William Robert
Maloney, Kenneth Long
Marsden, Sullivan S(amuel), Jr
Mason, Harold Frederick
Meiners, Henry C(ito)
Meldau, R(obert) F(rederick)
Meyer, Jarold Alan
Meyers, Robert Allen
Muskat, Morris
Nelson, David A
Nguyen, Caroline Phuongdung
Nunn, Roland Cicero
Offen, George Richard
Orr, Franklin M, Jr
Penner, S(tanford) S(olomon)
Petersen, Eugene E(dward)

Ratcliffe, Charles Thomas
Richter, George Neal
Robinson, Richard C(lark)
Rosenberg, Sanders David
Rothman, Albert J(oel)
Sanders, Charles F(ranklin), Jr
Schwartz, Daniel Evan
Shrontz, John William
Shuler, Patrick James
Sifferman, Thomas Raymond
Silcox, William Henry
Steffgen, Frederick Williams
Stephens, Douglas Robert
Sullivan, Richard Frederick
Sundberg, John Edwin
Turchan, Otto Charles
Uhl, Arthur E(dward)
Vermeulen, Theodore (Cole)
Williams, Alan K
Witherspoon, Paul A(dams), Jr

COLORADO
Allred, V(ictor) Dean
Baldwin, Robert Milton
Breitenbach, E A
Burdge, David Newman
Collins, Royal Eugene
Gibbons, David Louis
Goldburg, Arnold
Graue, Dennis Jerome
Jargon, Jerry Robert
Jha, Mahesh Chandra
Kazemi, Hossein
Keder, Wilbert Eugene
Mast, Richard F(redrick)
Milne, Thomas Anderson
Osterhoudt, Walter Jabez
Tosch, William Conrad

CONNECTICUT
Boggs, Steven Allan
Bozzuto, Carl Richard
Galstaun, Lionel Samuel
Goldstein, Marvin Sherwood
Gray, Thomas James
Hyde, John Welford
Ramakrishnan, Terizhandur S
Rosner, Daniel E(dwin)
Ruckebusch, Guy Bernard
Seery, Daniel J
Solomon, Peter R
Stephenson, Kenneth Edward
Thomas, Arthur L

DELAWARE
Cantwell, Edward N(orton), Jr
Cheng, Alexander H-D
Klein, Michael Tully
Mills, George Alexander

DISTRICT OF COLUMBIA
Arnold, William Howard, Jr
Boron, David J
Hershey, Robert Lewis

FLORIDA
Chao, Raul Edward
Coats, Keith Hal
Francis, Stanley Arthur
Givens, Paul Edward
Hirt, Thomas J(ames)
Normile, Hubert Clarence
Parker, Howard Ashley, Jr
Schneider, Richard Theodore
Seireg, Ali A
Stokes, Charles Anderson

GEORGIA
Hughes, Richard V(an Voorhees)
Winnick, Jack

HAWAII
Rollins, Ronald Roy

IDAHO
Hyde, William W
Reno, Harley W

ILLINOIS
Agarwal, Ashok Kumar
Arastoopour, Hamid
Assanis, Dennis N
Bunting, Bruce Gordon
Clark, Stephen Darrough
Conn, Arthur L(eonard)
Gembicki, Stanley Arthur
Gupta, Ramesh
Hutchings, L(eRoi) E(arl)
Kamath, Krishna
Klass, Donald Leroy
Krawetz, Arthur Altshuler
Kruse, Carl William
Kukes, Simon G
Lester, George Ronald
Linden, Henry R(obert)
Markuszewski, Richard
Moyer, H(allard) C(harles)
Murdoch, Bruce Thomas
Myles, Kevin Michael
Nandi, Satyendra Prosad
Punwani, Dharam Vir
Rose, Walter Deane
Rusinko, Frank, Jr
Schwartz, Michael Muni

Staats, William R
Wang, Michael Quanlu
Wert, Charles Allen

INDIANA
Cartmell, Robert Root
Droege, John Walter
Shankland, Rodney Veeder
Stehouwer, David Mark
Valia, Hardarshan S
Voelz, Frederick

IOWA
Chen, Lea-Der
Egger, Carl Thomas
Fisher, Ray W
Hall, Jerry Lee

KANSAS
Crowther, Robert Hamblett
DeForest, Elbert M
Preston, Floyd W
Ryan, James M
Smith, Raymond V(irgil)
Underwood, James Ross, Jr

KENTUCKY
Givens, Edwin Neil
Lloyd, William Gilbert
Mitchell, Maurice McClellan, Jr
Pan, Wei-Ping
Von Schonfeldt, Hilmar Armin
Wells, William Lochridge

LOUISIANA
Bourgoyne, Adam T, Jr
Brown, Leo Dale
Huckabay, Houston Keller
Ledford, Thomas Howard
McCarthy, Danny W
Reichle, Alfred Douglas

MAINE
Meftah, Bachir

MARYLAND
Argauer, Robert John
Bascunana, Jose Luis
Coates, Arthur Donwell
Foresti, Roy J(oseph), Jr
Gupta, Ashwani Kumar
Hornbuckle, Franklin L
Kolobielski, Marjan
Krynitsky, John Alexander
McQuaid, Richard William
Mangen, Lawrence Raymond
Plumlee, Karl Warren
Presser, Cary
Tuttle, Kenneth Lewis
White, Edmund W(illiam)

MASSACHUSETTS
Armington, Alton
Hals, Finn
Hottel, Hoyt C(larke)
Jalan, Vinod Motilal
Johnson, Stephen Allen
Kincaid, Thomas Gardiner
Kranich, Wilmer Leroy
Leung, Woon Fong (Wallace)
Petrovic, Louis John
Rowell, Robert Lee
Tester, Jefferson William

MICHIGAN
Eldred, Norman Orville
Eltinge, Lamont
Hemingway, William Wayne, Sr
Henein, Naeim A
Hubbard, Carolyn Parks
Kabel, Richard Harvey
Klimpel, Richard Robert
Korcek, Stefan
Piccirelli, Robert Anthony
Savage, Phillip E
Schramm, John Gilbert
Shelef, Mordecai
Smith, Joseph D
Stansel, John Charles
Willermet, Pierre Andre

MINNESOTA
Ayeni, Babatunde J

MISSISSIPPI
Cotton, Frank Ethridge, Jr

MISSOURI
Carter, Lee
Fowler, Frank Cavan
Koederitz, Leonard Frederick
Preckshot, G(eorge) W(illiam)

MONTANA
Bradley, Daniel Joseph

NEW HAMPSHIRE
Hunt, Everett Clair

NEW JERSEY
Bachman, Kenneth Charles
Bhagat, Phiroz Maneck
Bieber, Herman
Calcote, Hartwell Forrest

Chen, Nai Y
Dautzenberg, Frits Mathia
Denno, Khalil I
Eddinger, Ralph Tracy
Fischer, Ronald Howard
Fung, Shun Chong
Gethner, Jon Steven
Glassman, Irvin
Gorbaty, Martin Leo
Graham, Margaret Helen
Grandolfo, Marian Carmela
Griffith, Martin G
Grumer, Eugene Lawrence
Haag, Werner O
Halik, Raymond R(ichard)
Heath, Carl E(rnest), Jr
Jacolev, Leon
Kang, Chia-Chen Chu
Karcher, Guido George
Koehl, William John, Jr
Lee, Wooyoung
Lessard, Richard R
Mraw, Stephen Charles
Munsell, Monroe Wallwork
Murthy, Andiappan Kumaresa Sundara
Panzer, Jerome
Rankel, Lillian Ann
Rebick, Charles
Schlosberg, Richard Henry
Smyers, William Hays
Suciu, George Dan
Tiethof, Jack Alan
Upton, Thomas Hallworth
Van Hook, James Paul
Weinstein, Norman J(acob)
Yan, Tsoung-Yuan

NEW MEXICO
Castle, John Granville, Jr
Drumheller, Douglas Schaeffer
Finch, Thomas Wesley
Granoff, Barry
Lee, David Oi
Mead, Richard Wilson
Taber, Joseph John
Vanderborgh, Nicholas Ernest

NEW YORK
Agosta, Vito
Child, Edward T(aylor)
Corbeels, Roger
Davis, Marshall Earl
Fein, R(ichard) S(aul)
Givi, Peyman
Grohse, Edward William
Handman, Stanley E
Harris, Colin C(yril)
Hileman, Robert E
Kaarsberg, Ernest Andersen
Keller, D Steven
Kim, Young Joo
Kolaian, Jack H
Krishna, C R
Lemanski, Michael Francis
Neveu, Maurice C
Othmer, Donald F(rederick)
Pollart, Dale Flavian
Reber, Raymond Andrew
Virk, Kashmir Singh
Wang, Lin-Shu
Weinstein, Herbert
Wiese, Warren M(elvin)

NORTH CAROLINA
Arnold, Allen Parker
Eckerlin, Herbert Martin
Gangwal, Santosh Kumar
Winston, Hubert

NORTH DAKOTA
Hasan, Abu Rashid
Ludlow, Douglas Kent
Mathsen, Don Verden
Willson, Warrack Grant

OHIO
Dorer, Casper John
Dunn, Robert Garvin
Engdahl, Richard Bott
Fan, Liang-Shih
Harmon, David Elmer, Jr
Hazard, Herbert Ray
Jenkins, Robert George
Kampe, Dennis James
Kim, Byung Cho
Lee, Sunggyu
Massey, L(ester) G(eorge)
Metcalfe, Joseph Edward, III
Mourad, A George
Murphy, Michael John
Putnam, Abbott (Allen)
Reid, William T(homas)
Shick, Philip E(dwin)
Slider, H C (Slip)
Sprafka, Robert J
Standish, Norman Weston
Webb, Thomas Howard
Yarrington, Robert M

OKLAHOMA
Allen, Thomas Oscar
Azar, Jamal J
Baldwin, Roger Allan
Beckett, James Reid

Fuel Technology & Petroleum Engineering (cont)

Blenkarn, Kenneth Ardley
Blick, Edward F(orrest)
Brown, Kermit Earl
Brown, Larry Patrick
Bryant, Rebecca Smith
Campbell, John Morgan, Sr
Carlile, Robert Elliot
Chatenever, Alfred
Chung, Ting-Horng
Coffman, Harold H
Dauben, Dwight Lewis
Davis, Robert Elliott
Eastman, Alan D
Elkins, L(loyd) E(dwin)
Fast, C(larence) R(obert)
Ferrell, Howard H
Forgotson, James Morris, Jr
Gall, James William
Geffen, T(heodore) M(orton)
Growcock, Frederick Bruce
Guerrero, E T
Hall, Colby D(ixon), Jr
Hays, George E(dgar)
Heath, Larman Jefferson
Helander, Donald P(eter)
Howard, Robert Adrian
Hurn, R(ichard) W(ilson)
Johnston, Harlin Dee
Kleinschmidt, Roger Frederick
Knapp, Roy M
Lal, Manohar
Lilley, David Grantham
Menzie, Donald E
Pollock, L(yle) W(illiam)
Raghavan, Rajagopal
Root, Paul John
Shell, Francis Joseph
Sheth, Ketankumar K
Sloan, Norman Grady
Thompson, Richard E(ugene)
Tiab, Djebbar
Vinatieri, James Edward
West, John B(ernard)
Wimmer, Donn Braden
Yeh, Nai-Shyong
Zetik, Donald Frank

PENNSYLVANIA
Acheson, Willard Phillips
Amero, R(obert) C
Archer, David Horace
Baird, Michael Jefferson
Barmby, David Stanley
Bissey, Luther Trauger
Blaustein, Bernard Daniel
Boodman, Norman S
Camp, Frederick William
Catchen, Gary Lee
Cernansky, Nicholas P
Champagne, Paul Ernest
Charmbury, H(erbert) Beecher
Chigier, Norman
Churchill, Stuart W(inston)
Cobb, James Temple, Jr
Conroy, James Strickler
Davis, Burl Edward
Del Bel, Elsio
Dell, M(anuel) Benjamin
Enick, Robert M
Ertekin, Turgay
Finseth, Dennis Henry
Fulton, Paul F(ranklin)
Grumer, Joseph
Hofer, Lawrence John Edward
Holder, Gerald D
Kanter, Ira E
Levy, Edward Kenneth
Lovell, Harold Lemuel
McCormick, Robert H(enry)
Meredith, David Bruce
Meyer, Wolfgang E(berhard)
Muir, Patrick Fred
Nichols, Duane Guy
Nishioka, Masaharu
Oder, Robin Roy
Palmer, Howard Benedict
Petura, John C
Podgers, Alexander Robert
Romovacek, George R
Schiffman, Louis F
Siddiqui, Habib
Sinnett, Carl E(arl)
Soung, Wen Y
Stahl, C(harles) D(rew)
Szabo, Miklos Tamas
Vastola, Francis J
Venuto, Paul B
Vick, Gerald Kieth
Vidt, Edward James
Warren, J(oseph) E(mmet)
Wender, Irving
Whyte, Thaddeus E, Jr
Wilkens, John Albert
Zauderer, Bert

RHODE ISLAND
Suuberg, Eric Michael

SOUTH DAKOTA
Wegman, Steven M

TENNESSEE
Hodgson, J(effrey) William
Phung, Doan Lien
Sheth, Atul C
Singh, Suman Priyadarshi Narain

TEXAS
Abernathy, Bobby F
Abouhalkah, T A
Ali, Yusuf
Al-Saadoon, Faleh T
Anderson, Gordon MacKenzie
Arganbright, Robert Philip
Arnold, C(harles) W(illiam)
Baber, Burl B, Jr
Bare, Charles L
Barnes, Allen Lawrence
Bradley, H(oward) B(ishop)
Brunsting, Elmer H(enry)
Bush, Warren Van Ness
Bussian, Alfred Erich
Calhoun, John C, Jr
Carroll, Michael M
Caudle, Ben Hall
Chew, Ju-Nam
Chien, Sze-Foo
Claridge, E(lmond) L(owell)
Clark, James Donald
Clementz, David Michael
Closmann, Philip Joseph
Cocanower, R(obert) D(unlavy)
Collipp, Bruce Garfield
Crandall, John R
Crawford, Duane Austin
Crawford, Paul B(erlowitz)
Crenshaw, Paul L
Crouse, Philip Charles
Crump, John Joseph
Curtis, Lawrence B
Dellinger, Thomas Baynes
Dorfman, Myron Herbert
Dunlap, Henry Francis
Eakin, Bertram E
Eckles, Wesley Webster, Jr
Ewing, Richard Edward
Fewel, Kenneth Jack, Jr
Fontaine, Marc F(rancis)
Gilchrist, Ralph E(dward)
Gilliland, Harold Eugene
Gilmore, Robert Beattie
Godwin, James Basil, Jr
Gray, Kenneth Eugene
Gregory, A(lvin) R(ay)
Gruy, Henry Jones
Haberman, John Phillip
Harris, Martin H
Heinrich, R(aymond) L(awrence)
Holliday, George Hayes
Holste, James Clifton
Hrkel, Edward James
Huang, Wann Sheng
Ivey, E(dwin) H(arry), Jr
Jahns, Hans O(tto)
Jennings, Alfred Roy, Jr
Johnson, Lloyd N(ewhall)
Johnson, Mary Lynn Miller
Jones, Marvin Richard
Jorden, James Roy
Kalfoglou, George
Kellerhals, Glen E
Kelton, Frank Caleb
Lake, Larry Wayne
Lee, William John
Lestz, Sidney J
Levine, J(oseph) S(amuel)
Levy, Robert Edward
Little, Jack Edward
Lurix, Paul Leslie, Jr
McLeod, Harry O'Neal, Jr
Massey, Michael John
Mattax, Calvin Coolidge
Matthews, Charles Sedwick
Melrose, James C
Monaghan, Patrick Henry
Morgan, Bryan Edward
Morse, R(ichard) A(rden)
Mortada, Mohamed
Nabor, George W(illiam)
Nancarrow, Warren George
Neavel, Richard Charles
Nemeth, Laszlo K
Nordgren, Ronald Paul
Olson, Danford Harold
Parker, Harry W(illiam)
Patton, Charles C(lifford)
Peaceman, Donald W(illiam)
Perkins, Thomas K(eeble)
Poddar, Syamal K
Podio, Augusto L
Posey, Daniel Earl
Rambow, Frederick H K
Richardson, Jasper E
Richardson, Joseph Gerald
Roebuck, Isaac Field
Roeger, Anton, III
Rowe, Allen McGhee, Jr
Russell, Donald Glenn
Russell, Thomas Webb
Seth, Mohan S
Shen, Chin-Wen
Sinclair, A Richard
Smith, Joseph Patrick
Souby, Armand Max
Stacy, T(homas) D(onnie)
Stalkup, Fred I(rving), Jr
Strange, Lloyd K(eith)
Streltsova, Tatiana D
Tao, Frank F
Thompson, William Horn
Von Gonten, (William) Douglas
Von Rosenberg, Dale Ursini
Wagner, John Philip
Weatherford, W(illiam) D(ewey), Jr
White, N(ikolas) F(rederick)
Whiting, R(obert) L(ouis)
Wieland, Denton R
Wilhoit, Randolph Carroll
Winegartner, Edgar Carl

UTAH
Bishop, Jay Lyman
Bunger, James Walter
Hill, George Richard
Langheinrich, Armin P(aul)
Sohn, Hong Yong
Trujillo, Edward Michael
Tuddenham, W(illiam) Marvin
Wiser, Wendell H(aslam)

VERMONT
Lang, William Harry

VIRGINIA
Bayliss, John Temple
Gordon, John S(tevens)
Hamrick, Joseph Thomas
Harney, Brian Michael
Hazlett, Robert Neil
Henderson, Charles B(rooke)
Jimeson, Robert M(acKay), Jr
Leonard, Joseph Thomas
Pitts, Thomas Griffin
Wayland, Russell Gibson

WASHINGTON
Holzman, George
Johnson, A(lfred) Burtron, Jr
Newman, Darrell Francis
Rosenwald, Gary W
Spencer, Arthur Milton, Jr

WEST VIRGINIA
Abel, William T
Chisholm, William Preston
Clark, Nigel Norman
Das, Kamalendu
Gillmore, Donald W(ood)
Gwinn, James E
Rieke, Herman Henry, III
Shuck, Lowell Zane
Spaniol, Craig
Thakur, Pramod Chandra
Wasson, James A(llen)
Wertman, William Thomas
Wilson, John Sheridan

WISCONSIN
Gluckstein, Martin E(dwin)

WYOMING
Biggs, Paul
Dorrence, Samuel Michael
Hoffman, Edward Jack
Huff, Kenneth O
Kim, Sang Soo
Marchant, Leland Condo
Odasz, F(rancis) B(ernard), Jr
Stinson, Donald Leo
Weinbrandt, Richard M

ALBERTA
Bennion, Douglas Wilford
Bird, Gordon Winslow
Brown, R(onald) A(nderson) S(teven)
Butler, R(oger) M(oore)
Chakravorty, S(ailendra) K(umer)
Checkel, M David
Dranchuk, Peter Michael
Flock, Donald Louis
Hepler, Loren George
Kisman, Kenneth Edwin
Mungan, Necmettin
Scott, J(ohn) D(onald)

BRITISH COLUMBIA
Grace, John Ross

MANITOBA
Ruth, Douglas Warren

ONTARIO
Bergougnou, Maurice A(medee)
Eliades, Theo I
Fingas, Merv Floyd
Macdonald, Ian Francis
Napier, Douglas Herbert
Plett, Edelbert Gregory
Rosen, Marc Allen
Shelson, W(illiam)
Walsh, John Heritage
Wojciechowski, Bohdan Wieslaw
Wynnyckyj, John Rostyslav

SASKATCHEWAN
Faber, Albert John

OTHER COUNTRIES
Boustany, Kamel
Economides, Michael John
Fu, Yuan C(hin)
Lingane, Peter James
Vasilakos, Nicholas Petrou
Zaczepinski, Sioma

Industrial & Manufacturing Engineering

ALABAMA
Brown, Robert Alan
Clark, Charles Edward
Cox, Julius Grady
Goldman, Jay
Hool, James N
Kallsen, Henry Alvin
Kassner, James Lyle, Jr
Smith, Leo Anthony
Unger, Vernon Edwin, Jr
Webster, Dennis Burton
Workman, Gary Lee
Wyskida, Richard Martin

ALASKA
Bennett, F Lawrence

ARIZONA
Askin, Ronald Gene
Bailey, James Edward
Beaumariage, Terrence Gilbert
Bedworth, David D
Chandra, Abhijit
Cochran, Jeffery Keith
Cowles, Harold Andrews
Ferrell, William Russell
Fitch, W(illiam) Chester
Hoyt, Charles D, Jr
Keats, John Bert
Kleinschmidt, Eric Walker
Mackulak, Gerald Thomas
Macleod, Hugh Angus
Metz, Donald C(harles)
Montgomery, Douglas C(arter)
Moor, William C(hattle)
Seifert, Deborah Roeckner
Settles, F Stan
Shaw, Milton C(layton)
Slaughter, Charles D
Thompson, Truet B(radford)
Van Arsdel, John Hedde
Wymore, Albert Wayne
Young, Hewitt H

ARKANSAS
Asfahl, C Ray
Crisp, Robert M, Jr
Imhoff, John Leonard
McBryde, Vernon E(ugene)
Thomas, Walter E
Wilkes, Joseph Wray

CALIFORNIA
Anjard, Ronald P, Sr
Antonsson, Erik Karl
Arnwine, William Carrol
Arslancan, Ahmet N
Bernard, Douglas Alan
Bertin, Michael C
Block, Richard B
Bonora, Anthony Charles
Brandeau, Margaret Louise
Campbell, Bonita Jean
Chu, Chung-Yu Chester
Cook, Charles J
Cooper, George Emery
Czuha, Michael, Jr
Das, Mihir Kumar
Deriso, Richard Bruce
Dornfeld, David Alan
Edwards, Ward Dennis
Eldon, Charles A
Fleischer, Gerald A
Grant, Eugene Lodewick
Gray, Paul
Greene, Charles Richard
Grimaldi, John Vincent
Hahn, Hong Thomas
Hahne, Henry V
Hammond, Martin L
Harrison, George Conrad, Jr
Hausman, Warren H
Hecht, Myron J
Herrmann, George
Inoue, Michael Shigeru
Ireson, W(illiam) Grant
Jacobson, Albert H(erman), Jr
Jillie, Don W
Kaliski, Martin Edward
Laitin, Howard
Lalehzarian, Hamo
Lave, Roy E(llis), Jr
Lichter, James Joseph
Lindsay, Glenn Frank
Luther, Lester Charles
Martin, George
Mohamed, Farghalli Abdelrahman
Morgan, Donald E(arle)
Nadler, Gerald
Nguyen, Luu Thanh
Oakford, Robert Vernon
Philipson, Lloyd Lewis

Remer, Donald Sherwood
Robinson, Gordon Heath
Rowe, Lawrence A
Satre, Rodrick Iverson
Schwartz, Daniel M(ax)
Schwartzbart, Harry
Shanthikumar, Jeyaveerasingam George
Singleton, Henry E
Thompson, David A(lfred)
Throner, Guy Charles
Vomhof, Daniel William
Wallack, Paul Mark
Wong, Kwan Y
Wright, Edward Kenneth
Wymer, Joseph Peter
Yamakawa, Kazuo Alan

COLORADO
Bauer, Charles Edward
Ferguson, William Sidney
Giffin, Walter C(harles)
Lazarus, Steven S
Litwhiler, Daniel W
McLean, Francis Glen
Ostwald, Phillip F
Sandstrom, Donald James
Tischendorf, John Allen
Troxell, Wade Oakes
Vejvoda, Edward
Wall, Edward Thomas
Wertz, Ronald Duane
Williams, Robert J(ames)

CONNECTICUT
Corwin, H(arold) E(arl)
De Groot, Sybil Gramlich
Goldstein, Marvin Sherwood
Gupta, Dharam Vir
Haag, Robert Edwin
Kleinfeld, Ira H
Martin, R Keith
Rezek, Geoffrey Robert
Schwartz, Michael H
Shainin, Dorian
Stein, Richard Jay
Wingfield, Edward Christian
Wuthrich, Paul

DELAWARE
Gay, Frank P
Kikuchi, Shinya
Kiviat, Fred E
Long, John Reed
Stewart, Edward William

DISTRICT OF COLUMBIA
Donahue, D Joseph
Esterling, Donald M
Kramer, Bruce Michael
Kyriakopoulos, Nicholas
Trice, Virgil Garnett, Jr
Waters, Robert Charles

FLORIDA
Blatt, Joel Herman
Brown, John Lott
Carpenter, James L(inwood), Jr
Crane, Lee Stanley
Doering, Robert Distler
Doyon, Leonard Roger
Elzinga, D(onald) Jack
Francis, Richard Lane
Givens, Paul Edward
Gold, Edward
Haug, Arthur John
Hayes, Edward J(ames)
Hosni, Yasser Ali
Hoyer-Ellefsen, Sigurd
Hsu, Tsong-Han
Kruesi, William R
Kunce, Henry Warren
Leavenworth, Richard S
Muth, Eginhard Joerg
Nagy, Steven
Nattress, John Andrew
Nunez, German
Proctor, Charles Lafayette
Schrader, George Frederick
Shaw, Lawrance Neil
Swart, William W
Van Ligten, Raoul Fredrik
Ward, Leonard George
Whitehouse, Gary E
Wiener, Earl Louis
Wilcox, Donald Brooks
Zanakis, Stelios (Steve) H

GEORGIA
Aronson, Jay E
Banks, Jerry
Callahan, Leslie G, Jr
Feldman, Edwin B(arry)
Fey, Willard Russell
Fyffe, David Eugene
Gambrell, Carroll B(lake), Jr
Hines, William W
Iannicelli, Joseph
Jarvis, John Frederick
Jarvis, John J
Johnson, Cecil Gray
Johnson, Lynwood Albert
Krol, Joseph
Lehrer, Robert N(athaniel)
Malcolm, Earl Walter

Pence, Ira Wilson, Jr
Porter, Alan Leslie
Ratliff, Hugh Donald
Rogers, Harvey Wilbur
Rogers, Nelson K
Sadosky, Thomas Lee
Thuesen, Gerald Jorgen
Vicory, William Anthony
Wadsworth, Harrison M(orton)
White, John Austin, Jr
Wood, Robert E
Zunde, Pranas

HAWAII
Bartosik, Alexander Michael

IDAHO
Lambuth, Alan Letcher

ILLINOIS
Bayne, James Wilmer
Bell, Charles Eugene, Jr
Berkelhamer, Louis H(arry)
Block, Stanley M(arlin)
Brauer, Roger L
Cannon, Howard S(uckling)
Cook, Harry Edgar
Dessouky, Mohamed Ibrahim
Doyle, Lawrence Edward
Francis, Philip Hamilton
Frey, Donald N(elson)
Glod, Edward Francis
Haugen, Robert Kenneth
Hurter, Arthur P
McFee, Donald Ray
McKee, Keith Earl
Miller, Floyd Glenn
Morrow, Richard M
O'Connor, John Thomas
Parker, Ehi
Phillips, Rohan Hilary
Pigage, Leo C(harles)
Poulikakos, Dimos
Pritzker, Robert A
Rubenstein, Albert Harold
Ruhl, Roland Luther
Shipley, Roch Joseph
Siegal, Burton L
Spector, Leo Francis
Thompson, Charles William Nelson
Wilson, William Robert Dunwoody

INDIANA
Amrine, Harold Thomas
Barany, James W
Compton, Walter Dale
Eaton, Paul Bernard
Eberts, Ray Edward
English, Robert E
Gorman, Eugene Francis
Greene, James H
Gresham, Robert Marion
Hemmes, Paul Richard
Klein, Howard Joseph
Lehto, Mark R
Leimkuhler, Ferdinand F
Nagle, Edward John
Nof, Shimon Y
Ovens, William George
Pielet, Howard M
Pritsker, A(drian) Alan B(eryl)
Salvendy, Gavriel
Saucedo, Ismael G
Solberg, James J
Sweet, Arnold Lawrence
Thomas, Marlin Uluess
Wynne, Bayard Edmund

IOWA
Adams, S Keith
Buck, James R
Eldin, Hamed Kamal
Hempstead, J(ean) C(harles)
Hillyard, Lawrence R(obertson)
Kusiak, Andrew
Mack, Michael J
McRoberts, Keith L
Maze, Thomas Harold
Papadakis, Emmanuel Philippos
Patterson, John W(illiam)
Rocklin, Isadore J
Smith, Gerald Wavern
Vardeman, Stephen Bruce
Vaughn, Richard Clements

KANSAS
Connor, Sidney G
Eckhoff, Norman Dean
Graham, A Richard
Harnett, R Michael
Konz, Stephan A
Kramer, Bradley Alan
Lambert, Brian Kerry
Malzahn, Don Edwin
Tiffany, Charles F
Tillman, Frank A

KENTUCKY
Cathey, Jimmie Joe
Evans, Gerald William
Kohnhorst, Earl Eugene
Leep, Herman Ross
Lu, Wei-yang
Tran, Long Trieu

LOUISIANA
Cospolich, James D
Fontenot, Martin Mayance, Jr
Hale, Paul Nolen, Jr
Henry, Herman Luther, Jr
Keys, L Ken
Kraft, Donald Harris
Mann, Lawrence, Jr
Ristroph, John Heard
Rubinstein, Asher A
Siebel, M(athias) P(aul)

MAINE
Webster, Karl Smith

MARYLAND
Albus, James S
Berger, Harold
Carver, Gary Paul
Doiron, Theodore Danos
Haffner, Richard William
Hovmand, Svend
Jackson, Richard H F
Kanarowski, S(tanley) M(artin)
Katsanis, D(avid) J(ohn)
Kemelhor, Robert Elias
McLaughlin, William Lowndes
Nusbickel, Edward M, Jr
Steiner, Henry M
Tarrants, William Eugene
Van Cott, Harold Porter
Weiss, Roland George

MASSACHUSETTS
Addiss, Richard Robert, Jr
Adler, Ralph Peter Isaac
Bailey, Milton
Brandler, Philip
Brown, Robert Lee
Chandler, Charles H(orace)
Chryssostomidis, Chryssostomos
Cook, Nathan Henry
Cullinane, Thomas Paul
De Fazio, Thomas Luca
De Los Reyes, Gaston
Dyer, James Arthur
Edelman, Julian
Emmons, Howard W(ilson)
Feinstein, Leonard Gordon
Gilfix, Edward Leon
Graves, Robert John
Gupta, Surendra Mohan
Hahn, Robert S(impson)
Hulbert, Thomas Eugene
Krause, Irvin
Kroner, Klaus E(rlendur)
Leamon, Tom B
Magnanti, Thomas L
Meal, Harlan C
Nelson, J(ohn) Byron
Norton, Robert Leo
Papageorgiou, John Constantine
Rising, Edward James
Scudder, Henry J, III
Story, Anne Winthrop
Subramanian, Krishnamoorthy
Van der Burg, Sjirk
Vartanian, Leo
Wagner, David Kendall

MICHIGAN
Aswad, A Adnan
Atkin, Rupert Lloyd
Batrin, George Leslie
Carson, Gordon B(loom)
Chaffin, Don B
Chen, Kan
Chen, Yudong
Dilley, David Ross
Eaton, Robert James
Farah, Badie Naiem
Flinn, Richard A(loysius)
Goodman, Erik David
Hauth, Willard Ellsworth, III
Kachhal, Swatantra Kumar
Knappenberger, Herbert Allan
Lamberson, Leonard Roy
Lin, Paul Kuang-Hsien
Lorenz, John Douglas
Lumsdaine, Edward
Maley, Wayne A
Marks, Craig
Murty, Katta Gopalakrishna
Pandit, Sudhakar Madhavrao
Parsons, John Thoren
Pollock, Stephen M
Rollinger, Charles N(icholas)
Rosinski, Michael A
Sahney, Vinod K
Schriber, Thomas J
Schumacher, Berthold Walter
Shkolnikov, Moisey B
Spurgeon, William Marion
Stevenson, Robin
Taraman, Khalil Showky
Thomson, Robert Francis
Tompkins, Curtis Johnston
Weiner, Steven Allan
Weinstein, Jeremy Saul
Williams, Gwynne
Wolf, Franklin Kreamer
Wu, Hai

MINNESOTA
Hoffmann, Thomas Russell
Jacobson, Allen F
Kvalseth, Tarald Oddvar
McElrath, Gayle W(illiam)
Robertsen, John Alan
Smith, Thomas Jay
Starr, Patrick Joseph
Wade, Michael George

MISSISSIPPI
Cotton, Frank Ethridge, Jr
Johnson, Larry Ray
Oswalt, Jesse Harrell
Parker, Murl Wayne

MISSOURI
Blundell, James Kenneth
Bockserman, Robert Julian
Brahmbhatt, Sudhir R
Brumbaugh, Philip
David, Larry Gene
Donovan, John Richard
Eastman, Robert M(erriam)
Gillespie, LaRoux King
Klein, Cerry M
Miller, Owen Winston
Oliver, Larry Ray
Vandegrift, Alfred Eugene

MONTANA
Gibson, David F(rederic)

NEBRASKA
Elias, Samy E G
Hoffman, Richard Otto
Riley, Michael Waltermier
Schneider, Morris Henry
Schniederjans, Marc James
Von Bargen, Kenneth Louis

NEW HAMPSHIRE
Bahr, Karl Edward
Black, Sydney D
Carreker, R(oland) P(olk), Jr
Hunt, Everett Clair
Wang, Der-Shi

NEW JERSEY
Albin, Susan Lee
Anandan, Munisamy
Armamento, Eduardo T
Beinfest, Sidney
Boucher, Thomas Owen
Browne, Robert Glenn
Davis, William F
Donovan, Richard C
Elsayed, Elsayed Abdelrazik
Enell, John Warren
Franse, Renard
Geskin, Ernest S
Hoover, Charles Wilson, Jr
Huang, Jennming Stephen
Hunter, John Stuart
Jain, Sushil C
Kalley, Gordon S
Kellgren, John
Khalifa, Ramzi A
Landwehr, James M
Leu, Ming C
Liebig, William John
Luxhoj, James Thomas
McDermott, Kevin J
Magliola-Zoch, Doris
Marcus, Phillip Ronald
Mihalisin, John Raymond
Pargellis, Andrew Nason
Pasteelnick, Louis Andre
Perrotta, James
Polhemus, Neil W
Riley, David Waegar
Schoenfeld, Theodore Mark
Sermolins, Maris Andris
Snowman, Alfred
Thomas, Stanislaus S(tephen)
Vyas, Brijesh
Walsh, Teresa Marie
Whitt, Ward
Wildman, George Thomas
Yoon, Kwangsun Paul
Young, Li

NEW MEXICO
Alzheimer, William Edmund
Benziger, Theodore Michell
George, Raymond S
Jamshidi, Mohammad Mo
Lee, David Oi
Maez, Albert R
Martin, Richard Alan
Reuter, Robert Carl, Jr
Thode, E(dward) F(rederick)
Woods, Robert Octavius

NEW YORK
Aly, Adel Ahmed
Axelrod, Norman Nathan
Balassa, Leslie Ladislaus
Barrows, John Frederick
Beer, Sylvan Zavi
Benson, Richard Carter
Benzinger, James Robert
Berg, Daniel
Bialas, Wayne Francis

Industrial & Manufacturing Engineering (cont)

Bienstock, Daniel
Bommaraju, Tilak V
Broadwin, Alan
Crooke, James Stratton
Desrochers, Alan Alfred
Dizer, John Thomas, Jr
Doganaksoy, Necip
Drury, Colin Gordon
Dudewicz, Edward John
Fasullo, Eugene J
Frazer, J(ohn) Ronald
Gellman, Charles
Goldfarb, Donald
Gottesman, Elihu
Greenberg, Jacob
Haddock, Jorge
Happel, John
Hauser, Norbert
Heath, David Clay
Helander, Martin Erik Gustav
Jaffe, William J(ulian)
Karwan, Mark Henry
Kenett, Ron
Lindstrom, Gary J
McGee, James Patrick
Maxwell, William L
Meenan, Peter Michael
Menkes, Sherwood Bradford
Morfopoulos, Vassilis C(onstantinos) P
Mundel, August B(aer)
Mustacchi, Henry
Nanda, Ravinder
Penney, Carl Murray
Rosenberg, Paul
Sargent, Robert George
Schultz, Andrew, Jr
Simons, Gene R
Solomon, Jack
Srinivasan, Vijay
Stankiewicz, Raymond
Summers, Phillip Dale
Swalm, Ralph Oehrle
Thomas, Warren H(afford)
Thompson, Robert Alan
Ullmann, John E
Voelcker, Herbert B(ernhard)
Wang, Hsin-Pang
Wilczynski, Janusz S
Wright, Roger Neal
Yasar, Tugrul
Young, William Donald, Jr

NORTH CAROLINA
Appleby, Robert H
Ayoub, Mahmoud A
Bernhard, Richard Harold
Bockstahler, Theodore Edwin
Dyrkacz, W William
Hurt, Alfred B(urman), Jr
Kilpatrick, Kerry Edwards
King, L(ee) Ellis
Lalor, William Francis
Mallik, Arup Kumar
Meier, Wilbur L(eroy), Jr
Morgan, Donald Lee
Nutter, James I(rving)
Roberts, Stephen D
Smith, William Adams, Jr
Stidham, Shaler, Jr
Tartt, Thomas Edward

NORTH DAKOTA
Meldrum, Alan Hayward
Stanislao, Joseph
Wang, Jun

OHIO
Altan, Taylan
Beckham, Robert R(ound)
Bishop, Albert B
Carmichael, Donald C(harles)
Chrissis, James W
Clark, Gordon Meredith
Daschbach, James McCloskey
Duttweiler, Russell E
Erbacher, John Kornel
Fraker, John Richard
Fritsch, Charles A(nthony)
Genaidy, Ashraf Mohamed
Gilby, Stephen Warner
Gunasekera, Jay Sarath
Hall, Ernest Lenard
Hulley, Clair Montrose
Hyndman, John Robert
Jain, Vinod Kumar
Kaye, Christopher J
Keim, John Eugene
Kishel, Chester Joseph
Klosterman, Albert Leonard
Lahoti, Goverdhan Das
Lee, Peter Wankyoon
McNichols, Roger J(effrey)
Marras, William Steven
Menq, Chia-Hsiang
Merchant, Mylon Eugene
Miller, Richard Allen
Pace, Henry Alexander
Preszler, Alan Melvin
Roberts, Timothy R
Rockwell, Thomas H
Schutta, James Thomas
Sheskin, Theodore Jerome
Shetterly, Donivan Max
Slutzky, Gale David
Smith, George Leonard, Jr
Soska, Geary Victor
Srivatsan, Tirumalai Srinivas
Ventresca, Carol
Williams, Robert Lloyd
Wolfe, Robert Kenneth

OKLAHOMA
Card, Roger John
Case, Kenneth E(ugene)
Ferguson, Earl J
Foote, Bobbie Leon
Greene, Timothy J
Mize, Joe H(enry)
Schuermann, Allen Clark, Jr
Shamblin, James E
Shapiro, Robert Allen
Terrell, Marvin Palmer
Turner, Wayne Connelly
Urban, Timothy I
VanDerwiele, James Milton

OREGON
Kocaoglu, Dundar F

PENNSYLVANIA
Barsom, John M
Bise, Christopher John
Bockosh, George R
Bukowski, Julia Victoria
Burnham, Donald C
Dalal, Harish Maneklal
Dorny, C Nelson
Furfari, Frank A
Gardiner, Keith M
Garg, Diwakar
Georgakis, Christos
Goodman, Alan Lawrence
Gottfried, Byron S(tuart)
Groover, Mikell
Ham, Inyong
Johnson, Littleton Wales
Johnston-Feller, Ruth M
Kashkoush, Ismail I
Keyt, Donald E
Kingrea, James I, Jr
Kipp, Egbert Mason
Kottcamp, Edward H, Jr
Kuhn, Howard A
Kurfess, Thomas Roland
Lidman, William G
McCune, Duncan Chalmers
McDonnell, Leo F(rancis)
Mayer, George Emil
Mouly, Raymond J
Niebel, B(enjamin) W(illard)
Odrey, Nicholas Gerald
Penman, Paul D
Podgers, Alexander Robert
Procknow, Donald Eugene
Rogers, H(arry) C(arton), Jr
Ruud, Clayton Olaf
Saluja, Jagdish Kumar
Scigliano, J Michael
Thuering, George Lewis
Ventura, Jose Antonio
Wagner, David Loren
Weindling, Joachim I(gnace)
Whyte, Thaddeus E, Jr
Wolfe, Harvey
Woodburn, Wilton A
Zhu, Dong

RHODE ISLAND
Needleman, Alan
Olson, David Gardels

SOUTH CAROLINA
Chisman, James Allen
Graulty, Robert Thomas
Holme, Thomas T(imings)
Kimbler, Delbert Lee
Leonard, Michael Steven
Sandrapaty, Ramachandra Rao
Zumbrunnen, David Arnold

SOUTH DAKOTA
Johnson, Marvin Melrose
Swiden, LaDell Ray

TENNESSEE
Bontadelli, James Albert
Cox, Ronald Baker
Deivanayagan, Subramanian
Gano, Richard W
Gilbreath, Sidney Gordon, III
Isley, James Don
Lundy, Ted Sadler
Lyle, Benjamin Franklin
Pearsall, S(amuel) H(aff)
Prairie, Michael L
Sissom, Leighton E(sten)
Slaughter, Gerald M
Smith, James Ross
Sundaram, R Meenakshi
Yen, Michael R T

TEXAS
Ayoub, Mohamed Mohamed
Barna, Gabriel George
Beightler, Charles Sprague
Bennett, George Kemble
Blank, Leland T
Bravenec, Edward V
Bryant, Michael David
Caswell, Gregory K
CoVan, Jack Phillip
Dudek, R(ichard) A(lbert)
Edgar, Thomas Flynn
Fletcher, John Lynn
Gates, David G(ordon)
Geyling, F(ranz) Th(omas)
Gupton, Paul Stephen
Hoffman, Herbert I(rving)
Hogg, Gary Lynn
Koehn, Enno
Koppa, Rodger J
Lewis, Frank Leroy
Lutz, Raymond
McNees, Roger Wayne
Marshek, Kurt M
Meier, France Arnett
Miller, Gerald E
Mills, John James
Moon, Tessie Jo
Ozan, M Turgut
Phillips, Don T
Poage, Scott T(abor)
Ramaswamy, Kizhanatham V
Ramsey, Jerry Dwain
Rhodes, Allen Franklin
Smith, Milton Louis
Stevens, Gladstone Taylor, Jr
Street, Robert Lewis
Tims, George B(arton), Jr
Triano, John Joseph
Vernon, Ralph Jackson
Wilhelm, Wilbert Edward
Wolter, Jan D(ithmar)
Wood, Kristin Lee
Yakin, Mustafa Zafer

UTAH
Baum, Sanford
Nelson, Mark Adams
Rotz, Christopher Alan
Turley, Richard Eyring

VIRGINIA
Agee, Marvin H
Arp, Leon Joseph
Betti, John A
Cetron, Marvin J
Chiou, Minshon Jebb
Ell, William M
Fabrycky, Wolter J
Healy, John Joseph
Hunt, V Daniel
Jorstad, John Leonard
Kinsley, Homan Benjamin, Jr
Ling, James Gi-Ming
Maloney, Kenneth Morgan
Moore, James Mendon
Owczarski, William A(nthony)
Preusch, Charles D
Siebentritt, Carl R, Jr
Sink, David Scott
Smith, Michael Claude
Wierwille, Walter W(erner)
Williges, Robert Carl

WASHINGTON
Anderson, Richard Gregory
Barbosa-Canovas, Gustavo Victor
Bernard, Gary Dale
Bethel, James Samuel
Campbell, John C(arl)
Drui, Albert Burnell
Drumheller, Kirk
Johnson, Stephen Thomas
Jorgensen, Jens Erik
Kantowitz, Barry H
Kapur, Kailash C
Madden, James H(oward)
Maranville, Lawrence Frank
Newman, Paul Harold
Ramulu, Mamidala
Savol, Andrej Martin
Tenenbaum, Michael
Verson, Scott
Wolak, Jan
Woodfield, F(rank) W(illiam), Jr

WEST VIRGINIA
Breneman, William C
Mucino, Victor H
Stoll, Richard E

WISCONSIN
Bisgaard, Søren
Bollinger, John G(ustave)
Brisley, Chester L(avoyen)
Chang, Tsong-How
DeVries, Marvin Frank
Farrow, John H
Gonzalez, Hector Julian
Gustafson, David Harold
Jacobi, George (Thomas)
James, Charles Franklin, Jr
Jared, Alva Harden
Lentz, Mark Steven
Moy, William A(rthur)
Robinson, Stephen Michael
Samuel, Jay Morris
Saxena, Umesh
Schotz, Larry
Smith, Michael James
Steudel, Harold Jude
Suri, Rajan
Szpot, Bruce F
Zelazo, Nathaniel Kachorek

WYOMING
Odasz, F(rancis) B(ernard), Jr

PUERTO RICO
Rodriguez, Luis F

ALBERTA
Silver, Edward Allan
Sprague, James Clyde

BRITISH COLUMBIA
Franz, Norman Charles
Mitten, Loring G(oodwin)

NEW BRUNSWICK
Pham-Gia, Thu

NOVA SCOTIA
Kujath, Marek Ryszard
Wilson, George Peter

ONTARIO
Beevis, David
Buzacott, J(ohn) A(lan)
Carroll, John Millar
Cohn, Stanley Howard
Dhillon, Balbir Singh
Fahmy, Moustafa Mahmoud
Fenton, Robert George
Jeswiet, Jacob
Schey, John Anthony
Smith, Kenneth Carless
Steiner, George
Wong, George Shoung-Koon

QUEBEC
Bakhiet, Atef
Cherna, John C(harles)
Osman, M(ohamed) O M
Saber, Aaron Jaan
Sankar, Seshadri

OTHER COUNTRIES
De Vedia, Luis Alberto
Hamacher, Horst W
Irvine, Stuart James Curzon
Jost, Hans Peter
Kodali, V Prasad
Malik, Mazhar Ali Khan
Pau, Louis F
Raouf, Abdul
Sakai, Hiroshi
Sekine, Yasuji
Singh, Madan Gopal
Steffen, Juerg
Stephens, Kenneth S
Van Weert, Gezinus

Marine Engineering

CALIFORNIA
Bushnell, James Judson
Ortolano, Ralph J
Paulling, J(ohn) R(andolph), Jr

FLORIDA
Mooney, John Bradford, Jr
Scarlatos, Panagiotis Dimitrios
Yuan, Jiann-Shiun

IOWA
Pierce, Roger J

MARYLAND
Noblesse, Francis
Tuttle, Kenneth Lewis
Urbach, Herman B

MASSACHUSETTS
Malme, Charles I(rving)
Silvers, J(ohn) P(hillip)

MINNESOTA
Emeagwali, Philip Chukwurah

NEW YORK
Berndt, Christopher Charles
Dastin, Samuel J
Femenia, Jose
Pierson, Willard James, Jr
Stephens, Olin James, II

OHIO
Spokane, Robert Bruce

RHODE ISLAND
Koch, Robert Michael
Nadolink, Richard Hughes

TEXAS
Fewel, Kenneth Jack, Jr
Miksad, Richard Walter
Nordgren, Ronald Paul

VIRGINIA
Spitzig, William Andrew

BRITISH COLUMBIA
Muggeridge, Derek Brian

Materials Science Engineering

ALABAMA
Agresti, David George
Andrews, Russell S, Jr
Beindorff, Arthur Baker
Biblis, Evangelos J
Cataldo, Charles Eugene
Dean, David Lee
Hall, David Michael
Hardin, Ian Russell
Haygreen, John G
Hubbell, Wayne Charles
Janowski, Gregg Michael
Kattus, J Robert
Lemons, Jack Eugene
McKannan, Eugene Charles
Marek, Charles R(obert)
Naumann, Robert Jordan
Nikles, David Eugene
Pattillo, Robert Allen
Penn, Benjamin Grant
Richards, Nolan Earle
Rigney, Earnest Douglas, Jr
Stefanescu, Doru Michael
Thompson, Raymond G
Vohra, Yogesh K
Warfield, Carol Larson
Willenberg, Harvey Jack
Workman, Gary Lee

ALASKA
Skudrzyk, Frank J

ARIZONA
Agnihotri, Ram K
Albrecht, Edward Daniel
Berg, Howard Martin
Burke, William James
Carpenter, Ray Warren
Cowley, John Maxwell
Dodd, Charles Gardner
Eyring, LeRoy
Fairbanks, Harold V
Falco, Charles Maurice
George, Thomas D
Hight, Ralph Dale
Hiskey, J Brent
Hunter, William Leslie
Jackson, Kenneth Arthur
Jacobson, Michael Ray
Kingery, W David
Kleinschmidt, Eric Walker
Krajcinovic, Dusan
Krasnow, Marvin Ellman
Leeser, David O(scar)
Liaw, Hang Ming
Neilson, George Francis, Jr
Orton, George W
Osborn, Donald Earl
Ramohalli, Kumar Nanjunda Rao
Ren, Shang Yuan
Savrun, Ender
Schroder, Dieter K
Shafer, M(errill) W(ilbert)
Shaw, Milton C(layton)
Simmons, Paul C
Singhal, Avinash Chandra
Smyth, Jay Russell
Sundahl, Robert Charles, Jr
Swalin, Richard Arthur
Swink, Laurence N
Tsong, Ignatius Siu Tung
Uhlmann, Donald Robert
Wagner, James Bruce, Jr
Willson, Donald Bruce
Withers, James C

ARKANSAS
Keplinger, Orin Clawson

CALIFORNIA
Adicoff, Arnold
Adinoff, Bernard
Adler, William Fred
Agrawal, Suphal Prakash
Akhtar, Masyood
Ang, Tjoan-Liem
Anjard, Ronald P, Sr
Ardell, Alan Jay
Arnold, Walter Frank
Asaro, Robert John
Atkinson, Russell H
Baba, Paul David
Baglin, John Edward Eric
Bajaj, Jagmohan
Baldwin, Chandler Milnes
Ballantyne, Catherine Cox
Barasch, Werner
Bar-Cohen, Yoseph
Barnett, David M
Baskes, Michael I
Bateman, John Hugh
Bean, Ross Coleman
Belmares, Hector
Berdahl, Paul Hilland
Biedenbender, Michael David

Bienenstock, Arthur Irwin
Bilow, Norman
Bowman, Robert Clark, Jr
Bragg, Robert H(enry)
Brundle, Christopher Richard
Bube, Richard Howard
Bullis, William Murray
Bullock, Ronald Elvin
Burland, Donald Maxwell
Burmeister, Robert Alfred
Bush, Gary Graham
Caligiuri, Robert Domenic
Carley, James F(rench)
Carniglia, Stephen C(harles)
Carruthers, John Robert
Cass, Thomas Robert
Chang, Kaung-Jain
Chen, C(hih) W(en)
Chen, Tu
Chiang, Anne
Chidsey, Christopher E
Chiou, C(harles)
Chompff, Alfred J(ohan)
Chou, Yungnien John
Christensen, Richard Monson
Chung, Dae Hyun
Clark, Lloyd Douglas
Clarke, David Richard
Clemens, Bruce Montgomery
Cline, Carl F
Coburn, John Wyllie
Cockrell, Robert Alexander
Coleman, Charles Clyde
Colodny, Paul Charles
Comley, Peter Nigel
Condit, Ralph Howell
Cope, Oswald James
Dapkus, Paul Daniel
Dejonghe, Lutgard C
Dempsey, Martin E(wald)
Denn, Morton M(ace)
Docks, Edward Leon
Dorward, Ralph C(larence)
Dowell, Armstrong Manly
Doyle, Fiona Mary
Dunn, Bruce
Durham, William Bryan
Dyball, Christopher John
Eden, Richard Carl
Edwall, Dennis Dean
Edwards, Eugene H
Eichinger, Bruce Edward
Epstein, George
Erickson, Jon W
Estill, Wesley Boyd
Evans, Anthony Glyn
Even, William Roy, Jr
Farrar, John
Feather, David Hoover
Feay, Darrell Charles
Fedors, Robert Francis
Feigelson, Robert Saul
Feng, Joseph Shao-Ying
Fishman, Norman
Flamm, Daniel Lawrence
Flinn, Paul Anthony
Foiles, Stephen Martin
Forbes, James Franklin
Forbes, Judith L
Frattini, Paul L
Freise, Earl J
Fultz, Brent T
Gay, Richard Leslie
Geballe, Theodore Henry
Gehman, Bruce Lawrence
Geiss, Roy Howard
Gentile, Anthony L
Gershenzon, M(urray)
Gill, William D(elahaye)
Gilman, John Joseph
Glazer, Judith
Goetzel, Claus G(uenter)
Goo, Edward Kwock Wai
Gossard, Arthur Charles
Graper, Edward Bowen
Green, Allen T
Green, Harry Western, II
Gronsky, Ronald
Gulden, Terry Dale
Gur, Turgut M
Hackett, Le Roy Huntington, Jr
Haegel, Nancy M
Hahn, Hong Thomas
Hai, Francis
Hall, James Timothy
Haller, Eugene Ernest
Hammond, Martin L
Hanafee, James Eugene
Harkins, Carl Girvin
Harris, Daniel Charles
Hartsough, Larry Dowd
Hauber, Janet Elaine
Haycock, Ernest William
Hempstead, Robert Douglas
Herb, John A
Hertzler, Barry Lee
Hiel, Clement
Hopper, Robert William
Hrubesh, Lawrence W(ayne)
Hsieh, You-Lo
Huang, Tiao-yuan
Hummel, Steven G
Hunt, Arlon Jason
Iglesia, Enrique

Ivanov, Igor C
Jaffee, Robert I(saac)
Jaklevic, Joseph Michael
James, Edward, Jr
James, Michael Royston
Jansen, Frank
Johanson, William Richard
Johnson, Sylvia Marian
Johnson, William Lewis
Johnson, William Robert
Jones, Brian Herbert
Jones, Robin L(eslie)
Jorgensen, Paul J
Kaae, James Lewis
Kalfayan, Sarkis Hagop
Kamins, Theodore I
Kamp, David Allen
Karlsson, Eric Allan
Kashar, Lawrence Joseph
Kavanagh, Karen L
Kay, Robert Eugene
Kendig, Martin William
Khan, Mahbub R
Kilmer, Louis Charles
King, Ray J(ohn)
Knight, Patricia Marie
Koenig, Nathan Hart
Kolek, Robert Louis
Kolyer, John M
Kosmatka, John Benedict
Krikorian, Esther
Krivanek, Ondrej Ladislav
Kroemer, Herbert
Kuan, Teh S
Kupfer, John Carlton
LaChance, Murdock Henry
Lamb, Walter Robert
Lampert, Carl Matthew
Landel, Robert Franklin
Langdon, Terence George
Langer, Sidney
Lau, S S
Lauer, James Lothar
Lavernia, Enrique Jose
Lee, Kenneth
Lee, Stuart M(ilton)
Leeg, Kenton J(ames)
Leeper, Harold Murray
Lehovec, Kurt
Leipold, Martin H(enry)
Lemke, James Underwood
Leslie, James C
Leverton, Walter Frederick
Lewis, Nathan Saul
Li, Chia-Chuan
Lin, Charlie Yeongching
Logan, James Columbus
Long, Stephen Ingalls
Lovelace, Alan Mathieson
Lucas, Glenn E
Lugassy, Armand Amram
McCreight, Louis R(alph)
McHugh, Stuart Lawrence
McKaveney, James P
Mackenzie, John D(ouglas)
McNutt, Michael John
Maloney, Thomas James
Mansfeld, Florian Berthold
Mar, Raymond W
Marcus, John Stanley
Margerum, John David
Marshall, Grayson William, Jr
Marshall, Sally J
Martin, George
Martin, Gordon Eugene
Martin, Jim Frank
Maserjian, Joseph
Masters, Burton Joseph
Matthys, Eric Francois
Mecartney, Martha Lynn
Mehta, Povindar Kumar
Meike, Annemarie
Meiser, Michael David
Merriam, Marshal F(redric)
Meyer, Stephen Frederick
Mikami, Harry M
Mikes, Peter
Miller, Edward
Milstein, Frederick
Mitzner, Kenneth Martin
Mohamed, Farghalli Abdelrahman
Molau, Gunther Erich
Moon, Donald W(ayne)
Mooney, John Bernard
Morris, John William, Jr
Mukherjee, Amiya K
Muller, Richard Stephen
Munir, Zuhair A
Needles, Howard Lee
Nellis, William J
Nelson, Howard Gustave
Nelson, Norvell John
Nguyen, Luu Thanh
Nieh, Tai-Gang
Nishi, Yoshio
Nix, William Dale
Nooker, Eugene L(eRoy)
Norman, John Harris
Novotny, Vlad Joseph
Nutt, Steven R
Odette, G(eorge) Robert
O'Keefe, Michael Adrian
O'Meara, David Lillis
Ono, Kanji

Paciorek, Kazimiera J L
Packer, Charles M
Paquette, David George
Parrish, William
Pask, Joseph Adam
Pathania, Rajeshwar S
Pearson, Dale Sheldon
Peeters, Randall Louis
Perkins, Kenneth L(ee)
Petroff, Pierre Marc
Pianetta, Piero Antonio
Pollack, Louis Rubin
Prindle, William Roscoe
Przystupa, Marek Antoni
Pye, Earl Louis
Rabolt, John Francis
Rau, Charles Alfred, Jr
Redfield, David
Regulski, Thomas Walter
Reynolds, Richard Alan
Richman, Roger H
Riley, Robert Lee
Risbud, Subhash Hanamant
Ritchie, Robert Oliver
Romanowski, Christopher Andrew
Rosenberg, Sanders David
Roy, Prodyot
Rubin, Barney
Rubin, Louis
Rudee, Mervyn Lea
Ruskin, Arnold M(ilton)
Sachdev, Suresh
Sahbari, Javad Jabbari
Salovey, Ronald
Saperstein, David Dorn
Sashital, Sanat Ramanath
Schalla, Clarence August
Schjelderup, Hassel Charles
Schwab, Michael
Schwartzbart, Harry
Sclar, Nathan
Seaman, Lynn
Searcy, A(lan) W(inn)
Sedgwick, Robert T
Seiling, Alfred William
Shackelford, James Floyd
Shapiro, Isadore
Shelton, Robert Neal
Sher, Arden
Sherby, Oleg D(imitri)
Sherman, Arthur
Shin, Soo H
Shine, Carl
Shinohara, Makoto
Shockey, Donald Albert
Sie, Charles H
Simnad, Massoud T
Sinclair, Robert
Skowronski, Raymund Paul
Slivinsky, Sandra Harriet
Smith, Charles Aloysius
Spier, Edward Ellis
Spitzer, William George
Sree Harsha, Karnamadakala S
Stapelbroek, Maryn G
Steinberg, Daniel J
Stephens, Douglas Robert
Stetson, Alvin Rae
Stetson, Robert F
Stevenson, David Austin
Stirn, Richard J
Stolte, Charles
Stout, Ray Bernard
Stringer, John
Stroeve, Pieter
Suits, James Carr
Swanson, David G, Jr
Tauber, Richard Norman
Tennant, William Emerson
Thomas, Gareth
Thompson, Larry Dean
Tiller, William Arthur
Tombrello, Thomas Anthony, Jr
Treves, David
Tsai, Ming-Jong
Tu, Charles Wuching
Tu, King-Ning
Tung, Lu Ho
Twiss, Robert John
Vanblarigan, Peter
Viswanathan, R
Volksen, Willi
Vreeland, Thad, Jr
Vroom, David Archie
Wagner, C(hristian) N(ikolaus) J(ohann)
Wang, Kang-Lung
Wang, Wei-E
Warburton, William Kurtz
Washburn, Jack
Weber, Carl Joseph
Weber, Eicke Richard
Weinland, Stuart Louis
Westmacott, Kenneth Harry
White, Jack Lee
White, Moreno J
Whitney, Gina Marie
Wiener, Sidney
Williams, Morris Evan
Williams, Richard Stanley
Williamson, Robert Brady
Winkler, DeLoss Emmet
Wood, David S(hotwell)
Wood, David Wells
Wright, Charles Cathbert

Materials Science Engineering (cont)

Wudl, Fred
Yang, Jenn-Ming
Yasui, George
Yoon, Rick J
Young, Konrad Kwang-Leei
Yu, Karl Ka-Chung
Zebroski, Edwin L
Zuleeg, Rainer

COLORADO
Adams, James Russell
Arp, Vincent D
Bachman, Bonnie Jean Wilson
Bauer, Charles Edward
Beck, Betty Anne
Blake, Daniel Melvin
Bodig, Jozsef
Brady, Brian T
Ciszek, Ted F
Czanderna, Alvin Warren
Datta, Subhendu Kumar
Dunn, Richard Lee
Ellis, Donald Griffith
Feldman, Arthur
Gardner, Edward Eugene
Goldfarb, Ronald B
Grobner, Paul Josef
Hauser, Ray Louis
Kazmerski, Lawrence L
Krantz, William Bernard
Kremer, Russell Eugene
Lebiedzik, Jozef
Lefever, Robert Allen
Levenson, Leonard L
Mackinney, Herbert William
Mahajan, Roop Lal
Moddel, Garret R
Moore, John Jeremy
Noufi, Rommel
Olson, David LeRoy
Owen, Robert Barry
Pankove, Jacques I
Pitts, John Roland
Predecki, Paul K
Quick, Nathaniel Richard
Rauch, Gary Clark
Richards, Harold Rex
Roshko, Alexana
Sandstrom, Donald James
Schmidt, Alan Frederick
Schramm, Raymond Eugene
Shuler, Craig Edward
Simon, Nancy Jane
Strauss, Eric L
Timblin, Lloyd O, Jr
VanderLinden, Carl R
Webster, Larry Dale
Yarar, Baki
Zoller, Paul

CONNECTICUT
Abraham, Thomas
Agulian, Samuel Kevork
Arnold, John Richard
Berry, Richard C(hisholm)
Blackwood, Andrew W
Bradley, Elihu F(rancis)
Brody, Harold David
Castonguay, Richard Norman
Cheatham, Robert Gary
Cheng, Francis Sheng-Hsiung
Chin, Charles L(ee) D(ong)
Cooke, Theodore Frederic
Easterbrook, Eliot Knights
Emond, George T
Eppler, Richard A
Gardner, Fred Marvin
Giamei, Anthony Francis
Goldberg, A Jon
Gray, Thomas James
Grossman, Leonard N(athan)
Gutierrez-Miravete, Ernesto
Haacke, Gottfried
Haag, Robert Edwin
Hauser, Martin
Horhota, Stephen Thomas
Hsu, Nelson Nae-Ching
Johnson, Julian Frank
Kenig, Marvin Jerry
Kessel, Quentin Cattell
Koberstein, Jeffrey Thomas
Landsman, Douglas Anderson
Langeland, Kaare
Layden, George Kavanaugh
Lemkey, Frankin David
Levin, S Benedict
Levine, Leon
Maizell, Robert Edward
Myers, Mark B
Nath, Dilip K
Nielsen, John Merle
Olster, Elliot Frederick
Parthasarathi, Manavasi Narasimhan
Popplewell, James Malcolm
Potter, Donald Irwin
Powers, Joseph
Reed, Mark Arthur
Reffner, John A
Rie, John E
Roberts, George P
Rosner, Daniel E(dwin)
Schwier, Chris Edward
Scola, Daniel Anthony
Shalaby, Shalaby W
Shaw, Montgomery Throop
Shichman, D(aniel)
Sprague, G(eorge) Sidney
Strife, James Richard
Sutton, W(illard) H(olmes)
Thompson, Earl Ryan
Traskos, Richard Thomas
Wheeler, Edward Stubbs

DELAWARE
Akeley, David Francis
Bair, Thomas Irvin
Becker, Aaron Jay
Bierlein, John David
Bindloss, William
Birkenhauer, Robert Joseph
Boer, Karl Wolfgang
Brown, Stewart Cliff
Cessna, Lawrence C, Jr
Chesson, Eugene, Jr
Chou, Tsu-Wei
Chowdhry, Uma
Cooper, Stuart L
Cooper, Terence Alfred
Dawson, Robert Louis
Dexter, Stephen C
English, Alan Dale
Faller, James George
Ferriss, Donald P
Firment, Lawrence Edward
Freed, Robert Lloyd
Funck, Dennis Light
Gibbs, Thomas W(atson)
Goodman, Albert
Gorrafa, Adly Abdel-Moniem
Grant, David Evans
Greenfield, Irwin G
Hall, Ian Wavell
Hall, Robert B
Hardie, William George
Hardy, Henry Benjamin, Jr
Harrell, Jerald Rice
Hartmann, Hans S(iegfried)
Hershkowitz, Robert L
Holob, Gary M
Hsiao, Benjamin S
Hurford, Thomas Rowland
Immediata, Tony Michael
Jolley, John Eric
Katz, Manfred
Keidel, Frederick Andrew
Kramer, John J(acob)
Kwolek, Stephanie Louise
Landoll, Leo Michael
Lewis, John Raymond
Lin, Kuang-Farn
Logothetis, Anestis Leonidas
Loken, Halvar Young
McCullough, Roy Lynn
McNeely, James Braden
McTigue, Frank Henry
Markiewitz, Kenneth Helmut
Meakin, John David
Nelb, Gary William
Nicholls, Robert Lee
Onn, David Goodwin
Ostmann, Bernard George
Paulson, Charles Maxwell, Jr
Pontrelli, Gene J
Pretka, John E
Ransom, J(ohn) T(hompson)
Sands, Seymour
Schaefgen, John Raymond
Schenker, Henry Hans
Scherer, George Walter
Schiroky, Gerhard H
Schodt, Kathleen Patricia
Schultz, Jerold M
Scofield, Dillon Foster
Shealy, Otis Lester
Shoaf, Charles Jefferson
Staley, Ralph Horton
Tabb, David Leo
Tomic, Ernst Alois
Tuites, Donald Edgar
Urquhart, Andrew Willard
Van Trump, James Edmond
Vassallo, Donald Arthur
Vriesen, Calvin W
Wagner, Richard Lloyd
Weeks, Gregory Paul
Whitehouse, Bruce Alan
Williams, Richard Anderson

DISTRICT OF COLUMBIA
Aggarwal, Ishwar D
Alic, John A
Ayers, Jack Duane
Belsheim, Robert Oscar
Cosgarea, Andrew, Jr
Crouch, Roger Keith
Davis, Lance A(lan)
Dragoo, Alan Lewis
Gottschall, Robert James
Grabowski, Kenneth S
Henry, Richard Lynn
Jawed, Inam
Jordan, Albert Gustav
Klein, Max
Klein, Philipp Hillel
Kramer, Bruce Michael
Lashmore, David S
Lawson, Charles Alden
McNeil, Michael Brewer
Maxwell, Paul Charles
Moon, Deug Woon
O'Grady, William Edward
Pande, Chandra Shekhar
Radcliffe, S Victor
Rath, Bhakta Bhusan
Schafrik, Robert Edward
Schioler, Liselotte Jensen
Schnur, Joel Martin
Smith, Leslie E
Sprague, James Alan
Weinstock, Harold
Wilsey, Neal David
Wolf, Stanley Myron

FLORIDA
Abbaschian, Reza
Ahmann, Donald H(enry)
Ambrose, John Russell
Anghaie, Samim
Anusavice, Kenneth John
Batich, Christopher David
Beatty, Charles Lee
Biver, Carl John, Jr
Bonifazi, Stephen
Burns, Michael J
Callan, Edwin Joseph
Catz, Jerome
Cherepanov, Genady Petrovich
Clark, David Edward
Coke, C(hauncey) Eugene
Crandall, William B
Dannenberg, E(li) M
Davidson, Mark Rogers
Drucker, Daniel Charles
Dunbar, Richard Alan
Ebrahimi, Fereshteh
Fogg, Peter John
Fullman, R(obert) L(ouis)
Goldberg, David C
Goldberg, Eugene P
Gould, Robert William
Groves, Donald George
Gutbezahl, Boris
Hartley, Craig Sheridan
Hascicek, Yusuf Suat
Hayes, Edward J(ames)
Hecht, Ralph J
Herczog, Andrew
Holloway, Paul Howard
Hopkins, George C
Huffman, Jacob Brainard
Hummel, Rolf Erich
Huston, Ernest Lee
Jones, Kevin Scott
Kiewit, David Arnold
Kumar, Ganesh N
Lind, David Melvin
Linz, Arthur
McAlpine, Kenneth Donald
McCracken, Walter John
McKoy, James Benjamin, Jr
McSwain, Richard Horace
Maleady, N(oel) R(ichard)
Mandelkern, Leo
Mason, D(onald) R(omagne)
Masters, Larry William
Mecholsky, John Joseph, Jr
Milne, Edward Lawrence
Mizell, Louis Richard
Moore, Joseph B
Moudgil, Brij Mohan
Nemeth, Joseph
Pepinsky, Raymond
Postelnek, William
Potts, James Edward
Reichard, Ronnal Paul
Reynolds, Samuel D, Jr
Rice, Stephen Landon
Ridgway, James Stratman
Rikvold, Per Arne
Russell, Charles Addison
Russell, James
Saunders, James Henry
Seelbach, Charles William
Seth, Brij B
Shane, Robert S
Shottafer, James Edward
Silver, Frank Morris
Smith, William Fortune
Spears, Richard Kent
Springborn, Robert Carl
Steadman, Thomas Ree
Stickley, C(arlisle) Martin
Sun, Chang-Tsan
Sundaram, Swaminatha
Swain, Geoffrey W
Swartz, William Edward, Jr
Ting, Robert Yen-Ying
Vassamillet, Lawrence Francois
Weddleton, Richard Francis
Wei-Berk, Caroline
Whitney, Ellsworth Dow
Wille, Luc Theo
Willwerth, Lawrence James
Wiseman, Robert S
Yavorsky, John Michael
Yu, A Tobey
Zollo, Ronald Francis

GEORGIA
Antolovich, Stephen D
Atluri, Satya N
Bacon, Roger
Chandan, Harish Chandra
Colton, Jonathan Stuart
Coyne, Edward James, Jr
Cremens, Walter Samuel
D'Annessa, A(nthony) T(homas)
Danyluk, Steven
Dickert, Herman A(lonzo)
Fitzgerald, J(ohn) Edmund
Hawkins, Walter Lincoln
Kalish, David
Kaufman, Stephen
Kocurek, Michael Joseph
Korda, Edward J(ohn)
Krochmal, Jerome J(acob)
Logan, Kathryn Vance
Lundberg, John L(auren)
McCarthy, Neil Justin, Jr
Meyers, Carolyn Winstead
Morman, Michael T
Rees, William Smith, Jr
Rice, James Thomas
Rodrigue, George Pierre
Rodriguez, Augusto
Roobol, Norman R
Samoilov, Sergey Michael
Samuels, Robert Joel
Sanders, T H, Jr
Schulz, David Arthur
Simmons, George Allen
Soora, Siva Shunmugam
Spauschus, Hans O
Spetnagel, Theodore John
Starr, Thomas Louis
Stratton, Robert Alan
Tooke, William Raymond, Jr
Tummala, Rao Ramamohana
Weber, Robert Emil
Wellons, Jesse Davis, III
Yen, William Maoshung
Yeske, Ronald A

HAWAII
Brantley, Lee Reed
Daniel, Thomas Henry
Hihara, Lloyd Hiromi

IDAHO
Bobeck, Gene E
Brown, Harry Lester
Buescher, Brent J
Froes, Francis Herbert (Sam)
Korth, Gary E
Miller, Richard Lloyd
Place, Thomas Alan
Switendick, Alfred Carl
Telschow, Kenneth Louis
Walters, Leon C
Weyand, John David

ILLINOIS
Albertson, Clarence E
Aldred, Anthony T
Allen, Charles W(illard)
Alwitt, Robert S(amuel)
Baitinger, William F, Jr
Barnett, Scott A
Bazant, Zdenek P(avel)
Berkelhamer, Louis H(arry)
Bhattacharya, Debanshu
Birnbaum, H(oward) K(ent)
Bradley, Steven Arthur
Bratschun, William R(udolph)
Bridgeford, Douglas Joseph
Broutman, Lawrence Jay
Burnham, Robert Danner
Carr, Stephen Howard
Chan, Sai-Kit
Chang, Franklin Shih Chuan
Chang, R P H
Chang, Shu-Pei
Chen, Haydn H D
Chiou, Wen-An
Chung, Yip-Wah
Cohen, J(erome) B(ernard)
Connolly, James D
Cook, Harry Edgar
Crawford, Roy Kent
Crowley, Michael Summers
Damusis, Adolfas
Daniel, Isaac M
Dantzig, Jonathan A
De Rijk, Waldemar G
Dutta, Pulak
Eades, John Alwyn
Edelstein, Warren Stanley
Erck, Robert Alan
Erwin, Lewis
Falk, John Carl
Falkenstein, Gary Lee
Fenske, George R
Fine, Morris Eugene
Firestone, Ross Francis
Fraenkel, Stephen Joseph
Frey, Donald N(elson)
Gaylord, Richard J
Geil, Phillip H
Gembicki, Stanley Arthur
Gibson, J Murray
Girolami, Gregory Scott
Granick, Steve

Greener, Evan H
Griffith, James H
Gruber, Eugene E, Jr
Guggenheim, Stephen
Hardwidge, Edward Albert
Hawkins, Neil Middleton
Hersh, Herbert N
Hofer, Kenneth Emil
Ika, Prasad Venkata
Irvin, Howard H
Isbister, Roger John
Johnson, D(avid) Lynn
Joung, John Jongin
Kallend, John Scott
Kamm, Gilbert G(eorge)
Kaufmann, Elton Neil
Kelman, L(eRoy) R
Kim, Kyekyoon K(evin)
Klemm, Waldemar Arthur, Jr
Knox, Jack Rowles
Kramer, John Michael
Lam, Nghi Quoc
Lautenschlager, Eugene Paul
Lauterbach, Richard Thomas
Luplow, Wayne Charles
McKee, Keith Earl
Macrander, Albert Tiemen
Marks, Laurence D
Mason, Thomas Oliver
Mataga, Peter Andrew
Meshii, Masahiro
Metz, Florence Irene
Mirabella, Francis Michael, Jr
Mura, Toshio
Muzyczko, Thaddeus Marion
Nash, Philip Graham
Neuhalfen, Andrew J
Noonan, John Robert
Nordine, Paul Clemens
Olson, Gregory Bruce
Paschke, Edward Ernest
Pincus, Irving
Poeppel, Roger Brian
Raccah, Paul M(ordecai)
Raheel, Mastura
Rastogi, Prabhat Kumar
Rehn, Lynn Eduard
Reichard, Grant Wesley
Rey, Charles Albert
Rostoker, William
Rothman, Steven J
Routbort, Jules Lazar
Rowe, Charles David
Rowland, Theodore Justin
Rusinko, Frank, Jr
Sandrik, James Leslie
Seebauer, Edmund Gerard
Seidman, David N(athaniel)
Shaneyfelt, Duane L
Shenoy, Gopal K
Shipley, Roch Joseph
Siegel, Richard W(hite)
Sinha, Shome Nath
Socie, Darrell Frederick
Sproul, William Dallas
Stavrolakis, J(ames) A(lexander)
Stowell, James Kent
Stubbins, James Frederick
Stupp, Samuel Isaac
Su, Cheh-Jen
Suslick, Kenneth Sanders
Teng, Mao-Hua
Tsai, Boh Chang
Tucker, Charles L
Tuomi, Donald
Voorhees, Peter Willis
Weber, J K Richard
Weeks, Richard William
Weertman, Julia Randall
Wert, Charles Allen
Wessels, Bruce Warren
Wiedersich, H(artmut)
Wimber, R(ay) Ted
Wirtz, Gerald Paul
Wojnowski, Daniel Allen
Wolf, Dieter
Wright, Maurice Arthur

INDIANA
Bellina, Joseph James, Jr
Bement, A(rden) L(ee), Jr
Benford, Arthur E
Booe, J(ames) M(arvin)
Bradway, Keith E
Brannon-Peppas, Lisa
Chernoff, Donald Alan
Dayananda, Mysore Ananthamurthy
Diamond, Sidney
Dolch, William Lee
Eckelman, Carl A
Fritzlen, Glenn A
Gallucci, Robert Russell
Grace, Richard E(dward)
Gresham, Robert Marion
Jacobs, Martin Irwin
Kazem, Sayyed M
Koller, Charles Richard
Kuczynski, George Czeslaw
Kuo, Charles C Y
Liberti, Frank Nunzio
Liedl, Gerald L(eRoy)
McAleece, Donald John
McGinness, James Donald
Mindlin, Harold

Peppas, Nikolaos Athanassiou
Phillips, Ralph W
Ranade, Madhukar G
Sato, Hiroshi
Saucedo, Ismael G
Schalliol, Willis Lee
Solomon, Alvin Arnold
Taylor, Raymond Ellory
Valia, Hardarshan S
Vest, Robert W(ilson)
Winslow, Douglas Nathaniel

IOWA
Beaudry, Bernard Joseph
Buck, Otto
Eaton, David Leo
Gaertner, Richard F(rancis)
Greer, Raymond T(homas)
Hauenstein, Jack David
Lakes, Roderic Stephen
Lane, Orris John, Jr
Larsen, William L(awrence)
McGee, Thomas Donald
Moriarty, John Lawrence, Jr
Nariboli, Gundo A
Nixon, Wilfrid Austin
Park, Joon B
Rethwisch, David Gerard
Rocklin, Isadore J
Schaefer, Joseph Albert
Schmidt, Frederick Allen
Thompson, Robert Bruce
Trivedi, Rohit K
Wilder, David Randolph

KANSAS
Best, Cecil H(amilton)
Cooke, Francis W
Darwin, David
Frohmberg, Richard P
Grisafe, David Anthony
Grosskreutz, Joseph Charles
Huang, Chi-Lung Dominic
Rose, Kenneth E(ugene)
Scherrer, Joseph Henry
Sherwood, Peter Miles Anson

KENTUCKY
Coe, Gordon Randolph
DeLong, Lance Eric
Giammara, Berverly L Turner Sites
Gillis, Peter Paul
Goren, Alan Charles
Huffman, Gerald P
Johnson, Alan Arthur
Kuhn, William E(rik)
Lipscomb, Nathan Thornton
Lu, Wei-yang
Mellinger, Gary Andreas
Mihelich, John L
Puleo, David A
Reucroft, Philip J
Sarma, Atul C
Viswanadham, Ramamurthy K

LOUISIANA
Allen, Susan Davis
Beckman, Joseph Alfred
Bedell, Louis Robert
Bourdillon, Antony John
Bundy, Kirk Jon
Calamari, Timothy A, Jr
Choong, Elvin T
Diwan, Ravinder Mohan
Eaton, Harvill Carlton
Haefner, A(lbert) J(ohn)
Laurent, Sebastian Marc
Li, Hsueh Ming
McMillin, Charles W
Naidu, Seetala V
Penton, Harold Roy, Jr
Reid, John David
Rubinstein, Asher A
Snyder, Harold Jack
Somsen, Roger Alan
Thomas, Kevin Anthony
Vashishta, Priya Darshan
Vigo, Tyrone Lawrence
Wickson, Edward James
Zentner, Thomas Glenn

MAINE
Hopkins, George Robert
Lepoutre, Pierre
Meftah, Bachir
Thompson, Edward Valentine

MARYLAND
Ahearn, John Stephen
Armstrong, R(onald) W(illiam)
Arsenault, Richard Joseph
Basser, Peter J
Bennett, Lawrence Herman
Benson, Richard C
Berger, Harold
Bersch, Charles Frank
Block, Ira
Bowen, Rafael Lee
Brauer, Gerhard Max
Brodsky, Marc Herbert
Butler, Thomas W(esley)
Camponeschi, Eugene Thomas, Jr
Carter, John Paul
Cezairliyan, Ared

Chang, Shu-Sing
Charles, Harry Krewson, Jr
Chern, Engmin James
Chien, Chia-Ling
Chimenti, Dale Everett
Chuang, Tze-jer
Clark, Alan Fred
Corak, William Sydney
Coriell, Sam Ray
Davis, George Thomas
Davis, Guy Donald
Diness, Arthur M(ichael)
Dolhert, Leonard Edward
Downing, Robert Gregory
Economos, Geo(rge)
Eiss, Abraham L(ouis)
Ernest, Michael Vance, Sr
Fanconi, Bruno Mario
Feldman, Albert
Ferington, Thomas Edwin
Fishman, Steven Gerald
Franke, Gene Louis
Frazer, Benjamin Chalmers
Frederick, William George DeMott
Frederikse, Hans Pieter Roetert
Frohnsdorff, Geoffrey James Carl
Fulmer, Glenn Elton
Gammell, Paul M
Ghoshtagore, Rathindra Nath
Gillich, William John
Goktepe, Omer Faruk
Green, John Arthur Savage
Green, Robert E(dward), Jr
Greenfeld, Sidney H(oward)
Greenhouse, Harold Mitchell
Haffner, Richard William
Han, Charles Chih-Chao
Harper, C(harles) A(rthur)
Hasson, Dennis Francis
Hastie, John William
Hoffman, John D
Hoguet, Robert Gerard
Horowitz, Emanuel
Hovmand, Svend
Hsu, Stephen M
Hua, Susan Zonglu
Huffington, Norris J(ackson), Jr
Ibrahim, A Mahammad
Jones, Thomas Scott
Kaiser, Debra Lee
Kanarowski, S(tanley) M(artin)
Karulkar, Pramod C
Kirkendall, Thomas Dodge
Koh, Severino Legarda
Kopanski, Joseph J
Kreider, Kenneth Gruber
Kruger, Jerome
Krumbein, Simeon Joseph
Kumar, K Sharvan
Kuriyama, Masao
Levin, Roger L(ee)
Liang, Shoudeng
Loss, Frank J
Mabie, Curtis Parsons
McElroy, Wilbur Renfrew
McFadden, Geoffrey Bey
Machlin, Irving
Malghan, Subhaschandra Gangappa
Marcinkowski, M(arion) J(ohn)
Merz, Kenneth M(alcolm), Jr
Metze, George M
Meyers, Bernard Leonard
Miller, Gerald R
Moorjani, Kishin
Mordfin, Leonard
Murphey, Wayne K
Nagle, Dennis Charles
Nelson, Elton Glen
Neugroschl, Daniel
Newbury, Dale Elwood
Ostrofsky, Bernard
Parker, Frederick John
Paul, Dilip Kumar
Peters, Alan Winthrop
Poehler, Theodore O
Poulose, Pathickal K
Promisel, Nathan E
Rasberry, Stanley Dexter
Raskin, Betty Lou
Reno, Robert Charles
Revesz, Akos George
Rice, James K
Ricker, Richard Edmond
Ritter, Joseph John
Rosenberg, Arnold Morry
Rosenstein, Alan Herbert
Rozner, Alexander
Ruby, Stanley
Schaefer, Robert J
Schruben, John H
Schwartz, Lyle H
Schwartz, M(urray) A(rthur)
Scialdone, John Joseph
Sharp, William Broom Alexander
Skalny, Jan Peter
Skrabek, Emanuel Andrew
Smith, Jack Carlton
Smith, John Henry
Staker, Michael Ray
Steiner, Bruce
Stromberg, Robert Remson
Thomson, Robb M(ilton)
Thurber, Willis Robert
Van Echo, Andrew

Van Houten, Robert
Venkatesan, Thirumalai
Wacker, George Adolf
Wagner, Herman Leon
Wagner, James W
Waltrup, Paul John
Wang, Francis Wei-Yu
Warshaw, Israel
Warshaw, Stanley I(rving)
Waterstrat, Richard Milton
Winzer, Stephen Randolph
Wolock, Irvin
Yeh, Kwan-Nan
Yolken, Howard Thomas
Zwilsky, Klaus M(ax)

MASSACHUSETTS
Abkowitz, Stanley
Ackerman, Jerome Leonard
Adler, Ralph Peter Isaac
Alexander, Michael Norman
Allen, Lisa Parechanian
Allen, Samuel Miller
Ammlung, Richard Lee
Apelian, Diran
Arganbright, Donald G
Argon, Ali Suphi
Aronin, Lewis Richard
Argy, Dimitri
Averbach, B(enjamin) L(ewis)
Baboian, Robert
Bates, Carl H
Baublitz, Millard, Jr
Berera, Geetha Poonacha
Bernal G, Enrique
Bever, Michael B(erliner)
Biederman, Ronald R
Bliss, David Francis
Blum, Seymour L
Blumstein, Rita Blattberg
Bonner, Francis Joseph
Bougas, James Andrew
Brack, Karl
Buono, John Arthur
Bush, John Burchard, Jr
Butterworth, George A M
Chang, Frank C
Chen, Chuan Ju
Chin, David
Chu, Nori Yaw-Chyuan
Chung, Frank H
Churchill, Geoffrey Barker
Clopper, Herschel
Cohen, Merrill
Cohen, Morris
Collins, Aliki Karipidou
Cooke, Richard A
Cooper, W(illiam) E(ugene)
Cosgrove, James Francis
Cotten, George Richard
Dale, William
Darby, Joseph B(ranch), Jr
Dave, Raju S
David, Donald J
Deckert, Cheryl A
Delagi, Richard Gregory
Desper, Clyde Richard
Dietz, Albert (George Henry)
Dionne, Gerald Francis
Ditchek, Brian Michael
Druy, Mark Arnold
Eagar, Thomas W
Eschenroeder, Alan Quade
Evans, Michael Douglas
Fan, John C C
Feinstein, Leonard Gordon
Fettes, Edward Mackay
Flemings, Merton Corson
Folweiler, Robert Cooper
French, David N(ichols)
Garrett, Paul Daniel
Gatos, H(arry) C(onstantine)
Giessen, Bill C(ormann)
Godrick, Joseph Adam
Goela, Jitendra Singh
Goettler, Lloyd Arnold
Goldstein, Joseph I
Golub, Samuel J(oseph)
Gopikanth, M L
Grant, Nicholas J(ohn)
Haas, Howard Clyde
Hagnauer, Gary Lee
Handy, Carleton Thomas
Harman, T(heodore) C(arter)
Harrison, Ralph Joseph
Haugsjaa, Paul O
Hiatt, Norman Arthur
Hill, Loren Wallace
Hoadley, Robert Bruce
Horn, Mark William
Illinger, Joyce Lefever
Iseler, Gerald William
Israel, Stanley C
Isserow, Saul
Jalan, Vinod Motilal
Jensen, Klavs Fleming
Johnson, Walter Roland
Jones, Alan A
Kachanov, Mark L
Kantor, Simon William
Karasz, Frank Erwin
Kelsey, Ronald A(lbert)
Kennedy, Edward Francis
Khattak, Chandra Prakash

Materials Science Engineering (cont)

Kimerling, Lionel Cooper
Klinedinst, Keith Allen
Kolodziejski, Leslie Ann
Krumhansl, James Arthur
Kula, Eric Bertil
Kumar, Kaplesh
Latanision, Ronald Michael
Lauer, Robert B
Lavin, Edward
Lee, D(on) William
Lee, Kang In
Lees, Wayne Lowry
Lement, Bernard S
Levitt, Albert P
Li, Rounan
Liau, Zong-Long
Lin, Alice Lee Lan
Linden, Kurt Joseph
Livingston, James Duane
Lu, Grant
McGarry, Frederick J(erome)
MacGregor, C(harles) W(inters)
Maher, Galeb Hamid
Malozemoff, Alexis P
Marinaccio, Paul J
Martin, Richard Hadley, Jr
Masi, James Vincent
Matsuda, Seigo
Meyer, Robert Bruce
Milgrom, Jack
Minot, Michael Jay
Mix, Thomas W
Mont, George Edward
Moore, Robert Edmund
Morbey, Graham Kenneth
Moustakas, Theodore D
Natansohn, Samuel
Novich, Bruce Eric
Nowak, Welville B(erenson)
O'Connell, Richard John
Ogilvie, Robert Edward
O'Rell, Dennis Dee
Owen, Walter S
Pal, Uday Bhanu
Park, B J
Penchina, Claude Michel
Peters, Edward Tehle
Platt, Milton M
Pope, Mary E
Powers, Donald Howard, Jr
Quackenbush, Carr Lane W
Quynn, Richard Grayson
Reif, L Rafael
Rice, James R
Ritter, John Earl, Jr
Rose, Robert M(ichael)
Russell, Kenneth Calvin
Sadoway, Donald Robert
Sagalyn, Paul Leon
Sahatjian, Ronald Alexander
Sastri, Suri A
Sawan, Samuel Paul
Schloemann, Ernst
Schoen, Frederick J
Schroter, Stanislaw Gustaw
Schuler, Alan Norman
Schwartz, Thomas Alan
Schwensfeir, Robert James, Jr
Servi, I(talo) S(olomon)
Shapiro, Edward K Edik
Sharp, A(rnold) G(ideon)
Shaughnessy, Thomas Patrick
Shu, Larry Steven
Sioui, Richard Henry
Smith, Donald Ross
Sonnichsen, Harold Marvin
Spaepen, Frans
Spangler, Lora Lee
Strauss, Alan Jay
Sun, Shuwei
Suplinskas, Raymond Joseph
Suresh, Subra
Thieme, Cornelis Leo Hans
Thomas, Edwin Lorimer
Thun, Rudolf Eduard
Tsaur, Bor-Yeu
Tuler, Floyd Robert
Tuller, Harry Louis
Tustison, Randal Wayne
Tzeng, Wen-Shian Vincent
Udipi, Kishore
Uhlig, Herbert H(enry)
Vander Sande, John Bruce
Viechnicki, Dennis J
Wald, Fritz Veit
Wang, Christine A
Wasserbauer, John Gilmary
Weiner, Louis I
Wentworth, Stanley Earl
Williams, David John
Williams, John Russell
Wineman, Robert Judson
Winter, Horst Henning
Yau, Chiou Ching
Zheng, Xiao Lu
Zotos, John

MICHIGAN

Aaron, Howard Berton
Arends, Charles Bradford
Baxter, William John
Bhattacharya, Pallab Kumar
Bierlein, John Carl
Bigelow, Wilbur Charles
Bilello, John Charles
Blurton, Keith F
Brown, Steven Michael
Camp, David Thomas
Clark, Peter David
Clarke, Roy
Corey, Clark L(awrence)
Courtney, Thomas Hugh
Cox, Mary E
Craig, Robert George
Dasgupta, Rathindra
Davidovits, Joseph
Dearlove, Thomas John
Decker, R(aymond) F(rank)
Dreyfuss, Patricia
Eldis, George Thomas
Ellis, Thomas Stephen
Filisko, Frank Edward
Fisher, Galen Bruce
Flaim, Thomas Alfred
Foister, Robert Thomas
Fortuna, Edward Michael, Jr
Francis, Ray Llewellyn
Fuerst, Carlton Dwight
Gallagher, James A
Gardon, John Leslie
Garrett, David L, Jr
Ghosh, Amit Kumar
Gjostein, Norman A
Greene, Bruce Edgar
Harwood, Julius J
Hauth, Willard Ellsworth, III
Heckel, Richard W(ayne)
Heminghous, William Wayne, Sr
Heremans, Joseph P
Heybey, Otfried Willibald Georg
Hillegas, William Joseph
Hohnke, Dieter Karl
Hosford, William Fuller, Jr
Hucke, Edward E
Im, Jang Hi
Jacko, Michael George
Jain, Kailash Chandra
Jarvis, Lactance Aubrey
Jones, Frank Norton
Jones, Frederick Goodwin
Keinath, Steven Ernest
Kohn, David H
Kresta, Jiri Erik
Kulkarni, Anand K
Kullgren, Thomas Edward
Langley, Neal Roger
Le Beau, Stephen Edward
Li, Shin-Hwa
Lipowitz, Jonathan
Little, Robert E(ugene)
Logothetis, Eleftherios Miltiadis
Lund, Anders Edward
McKeever, L Dennis
Mansfield, Marc L
Meier, Dale Joseph
Miller, Carl Elmer
Miller, Robert Llewellyn
Montgomery, Donald Joseph
Morgan, Roger John
Morman, Kenneth N
Mukherjee, Kalinath
Narayanaswamy, Onbathiveli S
Newman, Seymour
O'Brien, William Joseph
Paton, Neil (Eric)
Pehlke, Robert Donald
Pence, Thomas James
Pronko, Peter Paul
Pryor, Roger Welton
Reynard, Kennard Anthony
Rhee, Seong Kwan
Robertson, Richard Earl
Rol, Pieter Klaas
Rosenberg, Richard Carl
Rossi, Giuseppe
Rowland, Sattley Clark
Russell, William Charles
Rutter, Edward Walter, Jr
Schrenk, Walter John
Schubring, Norman W(illiam)
Sell, Jeffrey Alan
Sliker, Alan
Smith, Darrell Wayne
Smith, John Robert
Soh, Sung Kuk
Srolovitz, David Joseph
Subramanian, K N
Suchsland, Otto
Suits, Bryan Halyburton
Summitt, W(illiam) Robert
Sun, Bernard Ching-Huey
Tillitson, Edward Walter
Van Vlack, Lawrence H(all)
Vukasovich, Mark Samuel
Wang, Yar-Ming
Warpehoski, Martha Anna
Whalen, Thomas J(ohn)
Williams, Fredrick David
Williams, Gwynne
Wyzgoski, Michael Gary
Yee, Albert Fan
Yeh, Gregory Soh-Yu

MINNESOTA

Bryan, Thomas T
Burkstrand, James Michael
Chamberlain, Craig Stanley
Chelikowsky, James Robert
David, Moses M
Diedrich, James Loren
Esmay, Donald Levern
Fick, Herbert John
George, Peter Kurt
Gerberich, William Warren
Goken, Garold Lee
Goldenberg, Barbara Lou
Goldman, Allen Marshall
Gruber, Carl L(awrence)
Hendricks, James Owen
Hill, Brian Kellogg
Huffine, Coy L(ee)
Johnson, Lennart Ingemar
Johnson, Robert F
Katz, William
Koepke, Barry George
Kuder, Robert Clarence
Langager, Bruce Allen
Luskin, Mitchell B
McHenry, Kelly David
Macosko, Christopher Ward
Miller, Wilmer Glenn
Mularie, William Mack
Nam, Sehyun
Olyphant, Murray, Jr
Oriani, Richard Anthony
Punderson, John Oliver
Rao, Prakash
Ross, Stuart Thom
Sandberg, Carl Lorens
Sarkar, Kamalaksha
Shores, David Arthur
Thornton, Arnold William
Tirrell, Matthew Vincent
Tran, Nang Tri
Tsai, Ching-Long
Wright, Charles Dean

MISSISSIPPI

Barnes, Hoyt Michael
Johnson, L(awrence) D(avid)
Mather, Bryant
Mather, Katharine Kniskern
Taylor, Fred William

MISSOURI

Baldwin, James W(arren), Jr
Burke, James Joseph
Carlsson, Anders Einar
Carpenter, James Franklin
Carter, Robert L(eroy)
Creighton, Donald L
Darby, Joseph Raymond
Day, D(elbert) E(dwin)
Dharani, Lokeswarappa R
Farmer, John William
Frushour, Bruce George
Gadberry, Howard M(ilton)
Gibbons, Patrick C(handler)
Goran, Robert Charles
Greenley, Robert Z
Hale, Edward Boyd
Harmon, Robert Wayne
Hill, Dale Eugene
Indeck, Ronald S
Kenyon, Allen Stewart
Kurz, James Eckhardt
Lamelas, Francisco Javier
Leopold, Daniel J
Lind, Arthur Charles
McGinnes, Edgar Allan, Jr
Mattox, Douglas M
Millich, Frank
Monfore, Gervaise Edwin
Mori, Erik Jun
Niesse, John Edgar
Ownby, P(aul) Darrell
Prelas, Mark Antonio
Ross, Frederick Keith
Sastry, Shankara M L
Sears, J Kern
Sonnino, Carlo Benvenuto
Sparlin, Don Merle
Switzer, Jay Alan
Walsh, Robert Jerome
Wehrmann, Ralph F(rederick)
Winholtz, Robert Andrew
Wohl, Martin H
Woodbrey, James C
Yasuda, Hirotsugu

MONTANA

Bradley, Daniel Joseph
Koch, Peter

NEBRASKA

Dowben, Peter Arnold
Liou, Sy-Hwang
Phelps, George Clayton
Pierce, Donald N(orman)
Rohde, Suzanne Louise
Sellmyer, David Julian
Tunnicliff, David George

NEVADA

Chandra, Dhanesh
Clements, Linda L
Epps, Jon Albert

Skaggs, Robert L
Snaper, Alvin Allyn

NEW HAMPSHIRE

Backofen, Walter A
Comerford, Matthias F(rancis)
Creagh-Dexter, Linda T
Ge, Weikun
Hill, John Ledyard
Hynes, Thomas Vincent
Light, Thomas Burwell
Marion, Robert Howard
Queneau, Paul E(tienne)
Rice, Dale Wilson
Schulson, Erland Maxwell
Varnerin, Lawrence J(ohn)
Wang, Victor S F
Webb, Richard Lansing

NEW JERSEY

Abrahams, M(arvin) S(idney)
Anandan, Munisamy
Anthony, Philip John
Arendt, Volker Dietrich
Ariemma, Sidney
Balaguru, Perumalsamy N
Ballato, Arthur
Bardoliwalla, Dinshaw Framroze
Bassett, Alton Herman
Batlogg, Bertram
Battisti, Angelo James
Baughman, Ray Henry
Baum, Gerald A(llan)
Bentz, Bryan Lloyd
Berardinelli, Frank Michael
Berenbaum, Morris Benjamin
Berkman, Samuel
Besso, Michael M
Bevk, Joze
Bhandarkar, Suhas D
Biskeborn, Merle Chester
Biswas, Dipak R
Blanck, Andrew R(ichard)
Blank, Stuart Lawrence
Borysko, Emil
Bovey, Frank Alden
Bowden, Murrae John Stanley
Bowen, J Hartley, Jr
Brenner, Douglas
Broer, Matthijs Meno
Brown, Walter Lyons
Buckley, Denis Noel
Buhks, Ephraim
Burrus, Charles Andrew, Jr
Buteau, L(eon) J
Cadet, Gardy
Cais, Rudolf Edmund
Canter, Nathan H
Cassola, Charles A(lfred)
Cecchi, Joseph Leonard
Celler, George K
Chan, Maureen Gillen
Chand, Naresh
Chang, Chuan Chung
Chang, Tao-Yuan
Chatterjee, Pronoy Kumar
Chen, Catherine S H
Chen, Shuhchung Steve
Chen, William Kwo-Wei
Chenicek, Albert George
Chlanda, Frederick P
Cho, Alfred Y
Chu, Sung Nee George
Chynoweth, Alan Gerald
Cohen, Abraham Bernard
Conger, Robert Perrigo
Connolly, John Charles
Couchman, Peter Robert
Cullen, Glenn Wherry
Das, Santosh Kumar
Davidson, Theodore
DeBona, Bruce Todd
Derkits, Gustav
Derucher, Kenneth Noel
Dittman, Frank W(illard)
Dodabalapur, Ananth
Donovan, Richard C
Drelich, Arthur (Herbert)
Dubey, Madan
Dyer, John
Eachus, Spencer William
Ebert, Lawrence Burton
Ettenberg, Michael
Fahrenholtz, Susan Roseno
Farrauto, Robert Joseph
Feldman, Leonard Cecil
Finch, Rogers B(urton)
Fischer, Traugott Erwin
Forster, Eric Otto
Frankenthal, Robert Peter
Gaylord, Norman Grant
Gillham, John K
Glass, Alastair Malcolm
Green, Martin Laurence
Griskey, R(ichard) G(eorge)
Gross, Leonard
Grunes, Robert Lewis
Gualtieri, Devlin Michael
Hale, Warren Frederick
Hannon, Martin J
Hanson, James Edward
Harpell, Gary Allan
Hobstetter, John Norman
Hong, Minghwie

Hudson, Steven David
Hughes, O Richard
Isaacson, Robert B
Jacobs, Carl Henry
Jaffe, Michael
Jin, Sungho
Jirgensons, Arnold
Johnson, David W, Jr
Johnston, John Eric
Johnston, Wilbur Dexter, Jr
Jordan, Andrew Stephen
Kalnin, Ilmar L
Karol, Frederick J
Kear, Bernard Henry
Keramidas, Vassilis George
Khan, Saad Akhtar
Kirshenbaum, Gerald Steven
Klein, Imrich
Knausenberger, Wulf H
Koul, Maharaj Kishen
Kramer, Carolyn Margaret
Kresge, Edward Nathan
Kunzler, John Eugene
Kveglis, Albert Andrew
Landrock, Arthur Harold
Larkin, William Albert
Lemaire, Paul J
Leupold, Herbert August
Levi, David Winterton
Levy, Alan
Li, Hong
Liang, Keng-San
Liebermann, Howard Horst
Ling, Hung Chi
Loh, Roland Ru-loong
Louis, Kwok Toy
Lowell, A(rthur) I(rwin)
Lundberg, Robert Dean
Lunn, Anthony Crowther
Lunt, Harry Edward
Maa, Jer-Shen
MacChesney, John Burnette
McGregor, Walter
MacRae, Alfred Urquhart
Mandel, Andrea Sue
Manders, Peter William
Marcus, Robert Boris
Mihalisin, John Raymond
Miller, Bernard
Miskel, John Joseph, Jr
Mitchell, Olga Mary Mracek
Morris, Robert Craig
Muldrow, Charles Norment, Jr
Murarka, Shyam Prasad
Murthy, N(arasimhaiah) Sanjeeva
Nassau, Kurt
Ohring, Milton
Ong, Chung-Jian Jerry
Pacansky, Thomas John
Pae, K(ook) D(ong)
Panish, Morton B
Patel, Gordhanbhai Nathalal
Pebly, Harry E
Phillips, Julia M
Pinch, Harry Louis
Plymale, Charles E
Poate, John Milo
Polhemus, Neil W
Polk, Conrad Joseph
Prevorsek, Dusan Ciril
Quarles, Richard Wingfield
Rabinovich, Eliezer M
Racette, George William
Ramharack, Roopram
Ravindra, Nuggehalli Muthanna
Raybon, Gregory
Ray-Chaudhuri, Dilip K
Register, Richard Alan
Reich, Leo
Reich, Murray H
Rorabaugh, Donald T
Royce, Barrie Saunders Hart
Rutherford, John L(oftus)
Saferstein, Lowell G
Saxon, Robert
Scharfstein, Lawrence Robert
Schlabach, T(om) D(aniel)
Schoenfeld, Theodore Mark
Schonfeld, Edward
Schwab, Frederick Charles
Segelken, John Maurice
Shallcross, Frank V(an Loon)
Shelton, James Churchill
Sheppard, Keith George
Sinclair, James Douglas
Sivco, Deborah L
Smith, Carl Hofland
Smith, Donald Eugene
Snowman, Alfred
Song, Won-Ryul
Sorenson, Wayne Richard
Sprague, Basil Sheldon
Subramanyam, Dilip Kumar
Suchow, Lawrence
Suhir, Ephraim
Sywe, Bei-Shen
Takahashi, Akio
Tan, Victor
Tatyrek, Alfred Frank
Tauber, Arthur
Tenzer, Rudolf Kurt
Terenzi, Joseph F
Thomas, Stanislaus S(tephen)
Thompson, Darrell Robert

Thompson, Larry Flack
Tischler, Oscar
Townsend, Palmer W
Van Uitert, LeGrand G(erard)
Ver Strate, Gary William
Von Neida, Allyn Robert
Wagner, Richard S(iegfried)
Wang, Chih Chun
Wang, Tsuey Tang
Wasserman, David
Weil, Rolf
Weiss, Lawrence H(eisler)
Weissmann, Gerd Friedrich Horst
Wernick, Jack H(arry)
Westerdahl, Carolyn Ann Lovejoy
White, Lawrence Keith
Williams, Bernard Leo
Wong, Yiu-Huen
Woodward, Ted K
Xie, Ya-Hong
Yadvish, Robert D
Yoon, Euijoon
Zipfel, Christie Lewis

NEW MEXICO
Adler, Richard John
Alam, Mansoor
Arnold, Charles, Jr
Assink, Roger Alyn
Beauchamp, Edwin Knight
Benicewicz, Brian Chester
Binder, Irwin
Chason, Eric
Christensen, N(ephi) A(lbert)
Claytor, Thomas Nelson
Clinard, Frank Welch, Jr
Curro, John Gillette
Dick, Jerry Joel
Diegle, Ronald Bruce
Doss, James Daniel
Ericksen, Richard Harold
Farnum, Eugene Howard
Follstaedt, David Martin
Fries, Ralph Jay
Frost, Harold Maurice, III
Gerstle, Francis Peter, Jr
Gillen, Kenneth Todd
Gourley, Paul Lee
Kenna, Bernard Thomas
Kocks, U(lrich) Fred
Kreidl, Norbert J(oachim)
Kurtz, Steven Ross
Laquer, Henry L
Liepins, Raimond
Lowe, Terry Curtis
McGuire, Joseph Clive
Matthews, R(obert) B(ruce)
Miller, Alan R(obert)
Milton, Osborne
Mitchell, Terence Edward
Moir, David Chandler
Morosin, Bruno
Mullendore, A(rthur) W(ayne)
Northrop, David A(mos)
Northrup, Clyde John Marshall, Jr
O'Rourke, John Alvin
Parkin, Don Merrill
Peterson, Charles Leslie
Picraux, Samuel Thomas
Pike, Gordon E
Pope, Larry Elmer
Rohde, Richard Whitney
Romig, Alton Dale, Jr
Salzbrenner, Richard John
Saxton, Harry James
Schaefer, Dale Wesley
Seeger, Philip Anthony
Seiffert, Stephen Lockhart
Skaggs, Samuel Robert
Staudhammer, Karl P
Sutherland, Herbert James
Taylor, Gene Warren
Travis, James Roland
Tsao, Jeffrey Yeenien
Wallace, Terry Charles, Sr
Wawersik, Wolfgang R
Westwood, Albert Ronald Clifton
Wilder, James Andrew, Jr
Wu, Xin Di
Yost, Frederick Gordon

NEW YORK
Abate, Kenneth
Adelstein, Peter Z
Adler, Philip N(athan)
Albers, Francis C
Albrecht, William Melvin
Altemose, Vincent O
Altscher, Siegfried
Anderson, Wayne Arthur
Anthony, Thomas Richard
Aven, Manuel
Banholzer, William Frank
Barr, Donald Eugene
Bayer, Raymond George
Benzinger, James Robert
Berkes, John Stephan
Berndt, Christopher Charles
Beshers, Daniel N(ewson)
Bilderback, Donald Heywood
Blakely, J(ohn) M
Boesch, William J
Bolon, Roger B
Borom, Marcus P(reston)

Borrelli, Nicholas Francis
Briant, Clyde Leonard
Brun, Milivoj Konstantin
Budinski, Kenneth Gerard
Bulusu, Dutt V
Burke, J(oseph) E(ldrid)
Burns, Stephen James
Capwell, Robert J
Cargill, George Slade, III
Carrier, Gerald Burton
Chan, Siu-Wai
Chang, Chin-An
Cheh, Huk Yuk
Chen, Inan
Chen, Tsang Jan
Chen, Yu-Pei
Chiang, Yuen-Sheng
Chow, Tat-Sing Paul
Chung, Deborah Duen Ling
Chyung, Kenneth
Coden, Michael H
Cozzarelli, Francis A(nthony)
Culver, Richard S
Cunningham, Michael Paul
Czajkowski, Carl Joseph
Dastin, Samuel J
Davidson, Robert W
Dermit, George
De Zeeuw, Carl Henri
D'Heurle, Francois Max
Doremus, Robert Heward
Dowell, Flonnie
Duclos, Steven J
Dudley, Michael
Duquette, David J(oseph)
Dvorak, George J
Edelglass, Stephen Mark
Erhardt, Peter Franklin
Fay, Homer
Feger, Claudius
Feige, Norman G
Felty, Evan J
Fields, Alfred E
Fleischer, Robert Louis
Flom, Donald Gordon
Fox, Adrian Samuel
Fritz, K(enneth) E(arl)
Fuerniss, Stephen Joseph
Gambino, Richard Joseph
George, Philip Donald
Ghandhi, Sorab Khushro
Ghosh, Arup Kumar
Giannelis, Emmanuel P
Gigliotti, Michael Francis Xavier
Gilmore, Robert Snee
Gluck, Ronald Monroe
Goland, Allen Nathan
Good, Robert James
Gorbatsevich, Serge N
Gould, Robert Kinkade
Greenbaum, Steven Garry
Guile, Donald Lloyd
Gupta, Devendra
Gupta, Krishna Murari
Hall, Ernest Leroy
Hart, Edward Walter
Hartwig, Curtis P
Headrick, Randall L
Herley, Patrick James
Herman, Herbert
Hillig, William Bruno
Hirsh, Merle Norman
Hobbs, Stanley Young
Holland, Hans J
Homan, Clarke Gilbert
Horowitz, Carl
Hovel, Harold John
Hudson, John B(alch)
Hurd, Jeffery L
Interrante, Leonard V
Isaacs, Hugh Solomon
Isaacs, Leslie Laszlo
Jackson, Melvin Robert
Jenekhe, Samson A
Jensen, Richard Alan
Johansson, Sune
Johnson, Curtis Alan
Julinao, Peter C
Kadin, Alan Mitchell
Kamath, Vasanth Rathnakar
Kambour, Roger Peabody
Kamdar, Madhusudan H
Kammer, Paul A
Karz, Robert Stephen
Kasprzak, Lucian A
King, Alexander Harvey
Klein, Gerald Wayne
Kohrt, Carl Fredrick
Kosky, Philip George
Kramer, Edward J(ohn)
Krempl, Erhard
Kuan, Tung-Sheng
Kuo, Yue
Kwok, Thomas Yu-Kiu
Laghari, Javaid Rasoolbux
Lane, Richard L
Lanford, William Armistead
Lavine, James Philip
Lay, Kenneth W(ilbur)
Lee, Daeyong
Lee, Minyoung
Leigh, Richard Woodward
Lever, Reginald Frank
Li, Che-Yu

Li, J(ames) C(hen) M(in)
Lindberg, Vern Wilton
Liu, Hao-Wen
Lorenzen, Jerry Alan
Luborsky, Fred Everett
Ludeke, Rudolf
Luthra, Krishan Lal
Lynn, Kelvin G
Lynn, Merrill
McCauley, James Weymann
MacCrone, Robert K
MacDonald, Noel C
MacDowell, John Fraser
McElligott, Peter Edward
Machlin, E(ugene) S(olomon)
Mackay, Raymond Arthur
MacKenzie, Donald Robertson
McMurtry, Carl Hewes
Mahler, David S
Marwick, Alan David
Mason, John Hugh
Mathes, Kenneth Natt
Matienzo, Luis J
Matkovich, Vlado Ivan
Maurer, Gernant E
Mayer, James W(alter)
Meloon, David Rand
Meyerson, Bernard Steele
Mijovic, Jovan
Mogro-Campero, Antonio
Mondolfo, L(ucio) F(austo)
Moore, Robert Stephens
Mort, Joseph
Mount, Eldridge Milford, III
Moy, Dan
Moynihan, Cornelius Timothy
Mueller, Edward E(ugene)
Mulford, Robert Alan
Muller, Olaf
Murphy, Eugene F(rancis)
Muzyka, Donald Richard
Naar, Raymond Zacharias
Nauman, Edward Bruce
Neal, John Alva
Nelson, John Keith
Nowick, A(rthur) S(tanley)
Ober, Christopher Kemper
Oh, Byungdu
O'Reilly, James Michael
Otocka, Edward Paul
Ozimek, Edward Joseph
Pai, Damodar Mangalore
Palladino, William Joseph
Partch, Richard Earl
Passoja, Dann E
Pearce, Eli M
Pedroza, Gregorio Cruz
Pemrick, Raymond Edward
Penn, Lynn Sharon
Penwell, Richard Carlton
Persans, Peter D
Petrie, Sarah Elaine
Pillar, Walter Oscar
Pliskin, William Aaron
Prager, Martin
Psioda, Joseph Adam
Puttlitz, Karl Joseph
Quesnel, David John
Questad, David Lee
Rasmussen, Don Henry
Reardon, Joseph Daniel
Revankar, Vithal V S
Robinson, Charles C(anfield)
Roedel, George Frederick
Romankiw, Lubomyr Taras
Ruckman, Mark Warren
Rungta, Ravi
Sachse, Wolfgang H
Sadagopan, Varadachari
Saha, Bijay S
Sancaktar, Erol
Sanger, Gregory Marshall
Sass, Stephen L
Sasserath, Jay N
Satya, Akella V S
Schiff, Eric Allan
Schroder, Klaus
Schulze, Walter Arthur
Schwartz, Leon Joseph
Seigle, L(eslie) L(ouis)
Seltzer, Raymond
Setchell, John Stanford, Jr
Shaw, Robert Reeves
Shelby, James Elbert
Sheryll, Richard Perry
Shih, Kwang Kuo
Shoup, Robert D
Silcox, John
Skelly, David W
Slate, Floyd Owen
Smith, Russell D
Snyder, Robert L(eon)
Snyder, Robert Lyman
Solomon, Harvey Donald
Spriggs, Richard Moore
Srinivasan, G(urumukonda) R
Srinivasan, Makuteswaran
Srinivasan, Rangaswamy
Sternstein, Sanford Samuel
Stookey, Stanley Donald
Strasser, Alfred Anthony
Sullivan, Peter Kevin
Sutherland, Judith Elliott

Materials Science Engineering (cont)

Swanson, Alan Wayne
Sweet, Richard Clark
Talapatra, Dipak Chandra
Thompson, David Fred
Tonelli, John P, Jr
Totta, Paul Anthony
Trevoy, Donald James
Valyi, Emery I
Vanier, Peter Eugene
Vaughan, Michael Thomas
Vidali, Gian Franco
Vitus, Carissima Marie
Von Gutfeld, Robert J
Vook, Richard Werner
Wachtell, Richard L(loyd)
Wagner, John George
Waltcher, Irving
Walter, William Trump
Wang, Franklin Fu-Yen
Ward, Samuel Abner
Webb, Watt Wetmore
Wei, Ching-Yeu
Weiss, Volker
Welch, David O(tis)
Westbrook, J(ack) H(all)
Wetherhold, Robert Campbell
Wexell, Dale Richard
Whang, Sung H
Whiteside, James Brooks
Whittingham, M(ichael) Stanley
Wie, Chu Ryang
Wilcox, W(illiam) R(oss)
Wilson, James M
Wittmer, Marc F
Woodard, David W
Wosinski, John Francis
Wright, Roger Neal
Wu, Wen Pao
Yudelson, Joseph Samuel
Yuh, Huoy-Jen
Zakraysek, Louis
Zeldin, Arkady N
Zhou, Zhen-Hong

NORTH CAROLINA

Andrady, Anthony Lakshman
Austin, William W(yatt)
Bailey, Robert Brian
Barnhardt, Robert Alexander
Batra, Subhash Kumar
Beck, Keith Russell
Beeler, Joe R, Jr
Bernholc, Jerzy
Berry, B(rian) S(hepherd)
Brittain, John O(liver)
Buchanan, David Royal
Burton, Ralph Gaines
Cates, David Marshall
Clarke, John Ross
Clator, Irvin Garrett
Cocks, Franklin H
Cole, Jerome Foster
Conn, Paul Kohler
Conrad, Hans
Cuculo, John Anthony
Cuomo, Jerome John
Daumit, Gene Philip
Davis, Robert F(oster)
Douglas, Robert Alden
Drechsel, Paul David
DuBroff, William
Dyrkacz, W William
El-Shiekh, Aly H
Enquist, Paul Michael
Fornes, Raymond Earl
Fox, Bradley Alan
Gilooly, George Rice
Goldstein, Irving Solomon
Goodwin, Frank Erik
Grady, Perry Linwood
Graham, Louis Atkins
Gupta, Bhupender Singh
Hall, Peter M
Hall, Seymour Gerald
Hamby, Dame Scott
Hamner, William Frederick
Haseley, Edward Albert
Havner, Kerry S(huford)
Haynie, Fred Hollis
Hersh, Solomon Philip
Hren, John J(oseph)
Hsieh, Henry Lien
Humphreys, Kenneth K
Hurt, John Calvin
Irene, Eugene Arthur
Kirk, Wilber Wolfe
Koch, Carl Conrad
Kusy, Robert Peter
Leamy, Harry John
Linton, Richard William
Lord, Peter Reeves
McNeil, Laurie Elizabeth
Masnari, Nino A
Massoud, Hisham Z
Mohanty, Ganesh Prasad
Narayan, Jagdish
Nash, James Lewis, Jr
Olf, Heinz Gunther
Patil, Arvind Shankar
Pearsall, George W(ilbur)

Powers, Edward James
Prater, John Thomas
Price, Howard Charles
Ramanan, V R V
Redmond, John Peter
Reeber, Robert Richard
Reisman, Arnold
Rozgonyi, George A
Shepard, Marion L(averne)
Stoops, R(obert) F(ranklin)
Tanquary, Albert Charles
Taylor, Duane Francis
Tonelli, Alan Edward
Walsh, Edward John
Walters, Mark David
Wechsler, Monroe S(tanley)
Williams, Joel Lawson
Wooten, Willis Carl, Jr

NORTH DAKOTA

Adams, David George
Bierwagen, Gordon Paul

OHIO

Adams, Harold Elwood
Aggarwal, Sundar Lal
Antler, Morton
Baer, Eric
Bagley, Brian G
Baid, Kushalkumar Moolchand
Ball, Lawrence Ernest
Baloun, Calvin H(endricks)
Bansal, Narottam Prasad
Barkley, John R
Bethea, Tristram Walker, III
Bhattacharya, Rabi Sankar
Birle, John David
Blackwell, John
Bohm, Georg G A
Bonner, David Calhoun
Bouton, Thomas Chester
Bovenkerk, Harold P(aul)
Brandon, Clement Edwin
Brantley, William Arthur
Brooman, Eric Williams
Brown, Stanley Alfred
Buchanan, Reiva Chester
Bufkin, Billy George
Burte, Harris M(erl)
Buttner, F(rederick) H(oward)
Calderon, Nissim
Campbell, James Edward
Campbell, Robert Wayne
Carbonara, Robert Stephen
Carmichael, Donald C(harles)
Causa, Alfredo G
Chakko, Mathew K(anjhirathinkal)
Choudhary, Manoj Kumar
Christian, John B
Chuck, Leon
Clark, William Alan Thomas
Conant, Floyd Sanford
Copp, Albert Nils
Covitch, Michael J
D'Ianni, James Donato
DiCarlo, James Anthony
Divis, Roy Richard
Drake, Michael L
Duckworth, Winston H(oward)
Dudek, Thomas Joseph
Duderstadt, Edward C(harles)
Durand, Edward Allen
Duttweiler, Russell E
Eckstein, Bernard Hans
Ells, Frederick Richard
England, Richard Jay
Falkenbach, George J(oseph)
Fisher, George A, Jr
Gabb, Timothy Paul
Gates, John E(dward)
Gates, Raymond Dee
Gebelein, Charles G
Gerace, Michael Joseph
Germann, Richard P(aul)
Ginning, P(aul) R(oll)
Golovin, Michael N
Graham, Paul Whitener Link
Green, Robert Patrick
Griffin, John Leander
Griffin, Richard Norman
Griffith, Walter M, Jr
Gromelski, Stanley John, Jr
Gunasekera, Jay Sarath
Halford, Gary Ross
Hall, Judd Lewis
Hamed, Gary Ray
Haque, Azeez C
Harmer, Richard Sharpless
Harrington, Roy Victor
Harris, Frank Wayne
Hassler, Craig Reinhold
Henderson, Courtland M
Henry, Leonard Francis, III
Hickok, Robert Lee
Hoch, Michael
Hough, Ralph L
Hsu, William Yang-Hsing
Hughes, Kenneth E(ugene)
Ingram, David Christopher
Ishida, Hatsuo
Jain, Vinod Kumar
Jayaraman, Narayanan
Jayne, Theodore D
Kampe, Dennis James

Kanakkanatt, Antony
Kanakkanatt, Sebastian Varghese
Katz, J Lawrence
Kaufman, John Gilbert, Jr
Kaye, Christopher J
Kennedy, Joseph Paul
Kinstle, James Francis
Koch, Ronald Joseph
Koenig, Jack L
Kollen, Wendell James
Koros, Peter J
Kot, Richard Anthony
Krause, Horatio Henry
Kreidler, Eric Russell
Kroenke, William Joseph
Krysiak, Joseph Edward
Lad, Robert Augustin
Lando, Jerome B
Landstrom, D(onald) Karl
Lee, Min-Shiu
Lee, Peter Wankyoon
Lehn, William Lee
Leis, Brian Norman
Lewandowski, John Joseph
Li, George Su-Hsiang
Livingston, Daniel Isadore
Lohr, Delmar Frederick, Jr
McCoy, Robert Allyn
McDonel, Everett Timothy
McGinniss, Vincent Daniel
McIntyre, Donald
McMillin, Carl Richard
Mall, Shankar
Mark, James Edward
Markworth, Alan John
Meinecke, Eberhard A
Mikesell, Sharell Lee
Miller, William Reynolds, Jr
Minges, Merrill Loren
Miyoshi, Kazuhisa
Mobley, Carroll Edward
Mohn, Walter Rosing
Moore, Arthur William
Morin, Brian Gerald
Morral, F(acundo) R(olf)
Nakajima, Nobuyuki
O'Leary, Kevin Joseph
Payer, Joe Howard
Perrin, James Stuart
Peterson, Robert C
Peterson, Timothy Lee
Piirma, Irja
Porter, Leo Earle
Proctor, David George
Prusas, Zenon C
Pugh, John W(illiam)
Purdon, James Ralph, Jr
Pyle, James Johnston
Rai, Amarendra Kumar
Ramalingam, Mysore Loganathan
Reardon, Joseph Patrick
Reuter, Robert A
Rexer, Joachim
Rigney, David Arthur
Ringel, Steven Adam
Roe, Ryong-Joon
Roehrig, Frederick Karl
Rosenfield, Alan R(obert)
Rupert, John Paul
Sahai, Yogeshwar
St John, Douglas Francis
Salkind, Michael Jay
Sara, Raymond Vincent
Saraceno, Anthony Joseph
Sargent, Gordon Alfred
Sawyer, Baldwin
Schmidt, Donald L
Schmitt, George Frederick, Jr
Seldin, Emanuel Judah
Semler, Charles Edward
Servais, Ronald Albert
Shroff, Ramesh N
Sicka, Richard Walter
Sinclair, Richard Glenn, II
Sommer, John L
Srivatsan, Tirumalai Srinivas
Stover, E(dward) R(oy)
Strnat, Karl J
Summers, James William
Tallan, Norman M
Thomas, Joseph Francis, Jr
Tsao, Bang-Hung
Uebele, Curtis Eugene
Uscheek, Dave Petrovich
Uys, Johannes Marthinus
Venkatu, Doulatabad A
Versic, Ronald James
Wagoner, Robert H
Warner, Walter Charles
Welsch, Gerhard Egon
Westermann, Fred Ernst
White, James L(indsay)
Wickersham, Charles Edward, Jr
Wilde, Bryan Edmund
Williams, James Case
Williams, Wendell Sterling
Wittebort, Jules I
Wlodek, Stanley T
Wolff, Gunther Arthur
Wong, Wai-Mai Tsang
Yang, Philip Yung-Chin
Yu, Thomas Huei-Chung

OKLAHOMA

Cook, Charles F(oster), Jr
Cox, Gordon F N
Daniels, Raymond D(eWitt)
Growcock, Frederick Bruce
Highsmith, Ronald Earl
Jones, Faber Benjamin
Kline, Ronald Alan
Lou, Alex Yih-Chung
Shue, Robert Sidney
Striz, Alfred Gerhard
Thomason, William Hugh
Zelinski, Robert Paul

OREGON

Arthur, John Read, Jr
Daellenbach, Charles Byron
Devletian, Jack H
Dooley, George Joseph, III
Ethington, Robert Loren
Ford, Wayne Keith
Gerlach, Robert Louis
Huang, Zhijian
Lomas, Charles Gardner
Merz, Paul Louis
Nielsen, Lawrence Ernie
Owen, Sydney John Thomas
Paarsons, James Delbert
Roberts, C Sheldon
Seaman, Geoffrey Vincent F
Sleight, Arthur William
Wager, John Fisher
Wood, William Edwin
Yau, Leopoldo D

PENNSYLVANIA

Agarwala, Vinod Shanker
Albert, Robert Lee
Alley, Richard Blaine
Alper, Allen Myron
Arkles, Barry Charles
Armor, John N
Ashok, S
Baker, Rees Terence Keith
Balaba, Willy Mukama
Baran, George Roman
Baratta, Anthony J
Barnes, Mary Westergaard
Barsom, John M
Bartovics, Albert
Bauer, C(harles) L(loyd)
Bauman, Bernard D
Behrens, Ernst Wilhelm
Belitskus, David
Bhalla, Amar S
Blankenhorn, Paul Richard
Bluhm, Harold Frederick
Boer, Karl Wolfgang
Bolstad, Luther
Boltax, Alvin
Bramfitt, Bruce Livingston
Bretz, Philip Eric
Brown, Norman
Brungraber, Robert J
Bucci, Robert James
Bucher, John Henry
Buck, Jean Coberg
Chan, Siu-Kee
Chaudhury, Manoj Kumar
Chen-Tsai, Charlotte Hsiao-yu
Cheung, Peter Pak Lun
Chitale, Sanjay Madhav
Chou, Y(e) T(sang)
Claiborne, C Clair
Clark, J(ohn) B(everley)
Coleman, Michael Murray
Conyne, Richard Francis
Cookson, Alan Howard
Corneliussen, Roger DuWayne
Dalal, Harish Maneklal
Das, Suryya Kumar
Dax, Frank Robert
DeBroy, Tarasankar
Deeg, Emil W(olfgang)
DeGraef, Marc Julie
De Luccia, John Jerry
Doak, Kenneth Worley
Doty, W(illiam) D'Orville
Egami, Takeshi
Ehrhart, Wendell A
Emlemdi, Hasan Bashir
Erhan, Semih M
Eror, Nicholas George, Jr
Field, Nathan David
Fitzgerald, Maurice E
Flattery, David Kevin
Fong, James T(se-Ming)
Forscher, Frederick
Frost, Lawrence William
Garbarini, Victor C
Garber, Charles A
Garg, Diwakar
Gebhardt, Joseph John
German, Randall Michael
Giannovario, Joseph Anthony
Gillis, Marina N
Gittler, Franz Ludwig
Goddu, Robert Fenno
Goel, Ram Parkash
Goldman, Kenneth M(arvin)
Goodyear, William Frederick, Jr
Graham, Charles D(anne), Jr
Greenberg, Charles Bernard
Gruenwald, Geza

ENGINEERING / 341

Hall, Gary R
Harkins, Thomas Regis
Harmer, Martin Paul
Harrison, Ian Roland
Hartman, Marvis Edgar
Harvey, Frances J, II
Herzog, Leonard Frederick, II
Hess, Dennis William
Hettche, Leroy Raymond
Hildeman, Gregory John
Hillner, Edward
Hofferth, Burt Frederick
Hopkins, Richard H(enry)
Houlihan, John Frank
Howell, Paul Raymond
Hu, L(ing) W(en)
Hu, William H(sun)
Hunsicker, Harold Yundt
Iezzi, Robert Aldo
Ikeda, Richard Masayoshi
Irwin, William Edward
Isakoff, Sheldon Erwin
Jaffe, Donald
Jain, Himanshu
Jang, Sei Joo
Johnson, Gerald Glenn, Jr
Johnston, William V
Jones, Roger Franklin
Jordan, Robert Kenneth
Kakar, Anand Swaroop
Kang, Joohee
Katz, Lewis E
Khan, Parwaiz Ashraf Ali
Khare, Ashok Kumar
Kim, Hong C
Kitazawa, George
Klapproth, William Jacob, Jr
Kline, Donald Edgar
Klingsberg, Cyrus
Knaster, Tatyana
Knox, Bruce E
Ko, Frank K
Koczak, Michael Julius
Kolar, Michael Joseph
Komarneni, Sridhar
Korostoff, Edward
Kottcamp, Edward H, Jr
Koziar, Joseph Cleveland
Kraft, R(alph) Wayne
Kraitchman, Jerome
Krishnaswamy, S V
Kuhn, Howard A
Kumta, Prashant Nagesh
Kun, Zoltan Kokai
Lake, Robert D
Lakhtakia, Akhlesh
Langsam, Michael
Lannin, Jeffrey S
Lawley, Alan
Leonard, Laurence
Libsch, Joseph F(rancis)
Lindt, Jan Thomas
Long, Alton Los, Jr
Lord, Arthur E, Jr
Loria, Edward Albert
Louie, Ming
Love, Gordon Ross
Lu, Chih Yuan
Luck, Russell M
McMahon, Charles J, Jr
Macmillan, Norman Hillas
Magill, Joseph Henry
Mahajan, Subhash
Mahlman, Bert H
Manganello, S(amuel) J(ohn)
Manjoine, Michael J(oseph)
Marton, Joseph
Mayer, George
Mayer, George Emil
Mehta, Sudhir
Meibohm, Edgar Paul Hubert
Meier, Joseph Francis
Messier, Russell
Michael, Norman
Moss, Herbert Irwin
Muldawer, Leonard
Nair, K Manikantan
Nakahara, Shohei
Niebel, B(enjamin) W(illard)
Novak, Stephen Robert
Nuessle, Albert Christian
Pangborn, Robert Northrup
Paul, Anand Justin
Peffer, John Roscoe
Perzak, Frank John
Peterson, Richard Walter
Pfennigwerth, Paul Leroy
Pike, Ralph Edwin
Pinckney, Robert L
Plazek, Donald John
Pollack, Solomon R
Pope, David Peter
Popovics, Sandor
Prane, Joseph W(illiam)
Prout, James Harold
Ray, Siba Prasad
Reynolds, Claude Lewis, Jr
Rogers, H(arry) C(arton), Jr
Romovacek, George R
Rowe, Anne Prine
Roy, Della M(artin)
Roy, Rustum
Ryba, Earle Richard
Sabol, George Paul

Sander, Louis Frank
Scheetz, Howard A(nsel)
Schimmel, Karl Francis
Schwerer, Frederick Carl
Seiner, Jerome Allan
Shabel, Barrie Steven
Shaler, Amos J(ohnson)
Sharma, Mangalore Gokulanand
Siddiqui, Habib
Sieger, John S(ylvester)
Simkovich, George
Simmons, Richard Paul
Singh, Narsingh B
Smith, Amos Brittain, III
Smith, Deane Kingsley, Jr
Smith, James David Blackhall
Smith, W Novis, Jr
Smyth, Donald Morgan
Snider, Albert Monroe, Jr
Spitznagel, John A
Stefanou, Harry
Steiger, Roger Arthur
Stengle, William Bernard
Stinger, Henry J(oseph)
Sutter, Philip Henry
Swartz, John Croucher
Sykes, James Aubrey, Jr
Talvacchio, John
Thomas, David Alden
Thomas, Donald E(arl)
Thompson, Anthony W
Thrower, Peter Albert
Treadwell, Kenneth Myron
Tressler, Richard Ernest
Uz, Mehmet
Van Der Spiegel, Jan
Vander Voort, George Frederic
Varadan, Vasundara Venkatraman
Varadan, Vijay K
Varrese, Francis Raymond
Vedam, Kuppuswamy
Venable, Emerson
Verleur, Hans Willem
Verma, Deepak Kumar
Wagner, J Robert
Waldman, Jeffrey
Waldman, L(ouis) A(braham)
Walker, Augustus Chapman
Walker, Philip L(eroy), Jr
Wallace, William Edward
Wei, Robert Peh-Ying
Wei, Yen
Weiner, Robert Allen
Wells, Ralph Gordon
Werner, F(red) E(ugene)
Werny, Frank
White, William Blaine
Whyte, Thaddeus E, Jr
Wilder, Harry D(ouglas)
Williams, David Bernard
Wood, Susan
Work, William James
Worrell, Wayne L
Wynblatt, Paul P
Yang, Arthur Jing-Min
Zamanzadeh, Mehrooz
Zhang, Qiming
Zhu, Dong
Zweben, Carl Henry

RHODE ISLAND

Avery, Donald Hills
Caroselli, Remus Francis
Clifton, Rodney James
Findley, William N(ichols)
Freund, Lambert Ben
Gurland, Joseph
Logan, Joseph Skinner
Richman, Marc H(erbert)
Rockett, Thomas John

SOUTH CAROLINA

Akpan, Edward
Aspland, John Richard
Avegeropoulos, G
Awadalla, Nabil G
Browne, Colin Lanfear
Caskey, George R, Jr
Chase, Vernon Lindsay
Diefendorf, Russell Judd
Drews, Michael James
Dumin, David Joseph
Faust, John William, Jr
Fowler, John Rayford
Goldstein, Herman Bernard
Goswami, Bhuvenesh C
Hay, Ian Leslie
Hendrix, James Easton
Hopkins, Allen John
Jarvis, Christine Woodruff
Kimmel, Robert Michael
Krantz, Karl Walter
LaFleur, Kermit Stillman
Lee, Burtrand Insung
McDonell, William Robert
Minford, James Dean
Mosley, Wilbur Clanton, Jr
Moyle, David Douglas
Quadir, Tariq
Rack, Henry Johann
Randall, Michael Steven
Rootare, Hillar Muidar
Schwartz, Elmer G(eorge)
Stevens, James Levon

Sturcken, Edward Francis
Taras, Michael Andrew
Taylor, Peter Anthony
Terry, Stuart Lee
Tolbert, Thomas Warren
Wicks, George Gary
Young, Franklin Alden, Jr
Zumbrunnen, David Arnold

SOUTH DAKOTA

Cannon, Patrick Joseph
Han, Kenneth N
Ross, Keith Alan

TENNESSEE

Armentrout, Daryl Ralph
Bagrodia, Shriram
Bardos, Denes I(stvan)
Behr, Eldon August
Besmann, Theodore Martin
Besser, John Edwin
Billington, Douglas S(heldon)
Bleier, Alan
Bryan, Robert H(owell)
Budai, John David
Burton, Robert Lee
Canonico, Domenic Andrew
Chambers, Ralph Arnold
Cunningham, John Edward
Dahotre, Narendra Bapurao
Davis, Burns
Dorsey, George Francis
Ford, James Arthur
Fort, Tomlinson
Gilkey, Russell
Goodwin, Gene M
Googin, John M
Gorbatkin, Steven M
Gray, Theodore Flint, Jr
Grossbeck, Martin Lester
Grugel, Richard Nelson
Harms, William Otto
Holcombe, Cressie Earl, Jr
Horak, James Albert
Horton, Joseph Arno, Jr
Horton, Linda Louise Schiestle
Ice, Gene Emery
Iskander, Shafik Kamel
Joy, David Charles
Judkins, Roddie Reagan
Kinser, Donald LeRoy
Klueh, Ronald Lloyd
Langley, Robert Archie
Lichter, Barry D(avid)
Liu, Chain T
Lowndes, Douglas H, Jr
Lundy, Ted Sadler
McHargue, Carl J(ack)
Mansur, Louis Kenneth
Mayberry, Thomas Carlyle
Newland, Gordon Clay
Pasto, Arvid Eric
Pawel, Janet Elizabeth
Pearson, R(ay) L(eon)
Pennycook, Stephen John
Phillips, Paul J
Pugh, Claud Ervin
Ramey, Harmon Hobson, Jr
Riewald, Paul Gordon
Roberto, James Blair
Sales, Brian Craig
Scherpereel, Donald E
Scott, Herbert Andrew
Scott, J(ames) L(ouis)
Simhan, Raj
Slaughter, Gerald M
Smith, Michael James
Sparks, Cullie J(ames), Jr
Spooner, Stephen
Spruiell, Joseph E(arl)
Stiegler, James O
Stroud, Robert Wayne
Swindeman, Robert W
Wachs, Alan Leonard
Waters, Dean Allison
Weeks, Robert A
Wert, James J
Wildman, Gary Cecil
Williams, Robin O('Dare)
Wunderlich, Bernhard
Yoo, Man Hyong

TEXAS

Allen, Roland Emery
Armeniades, Constantine D
Arrowood, Roy Mitchell, Jr
Arthur, Marion Abrahams
Aufdermarsh, Carl Albert, Jr
Balasinski, Artur
Banerjee, Sanjay Kumar
Barton, John R
Bokros, J(ack) C(hester)
Bourell, David Lee
Bovay, Harry Elmo, Jr
Bradley, Walter L
Bravenec, Edward V
Brostow, Witold Konrad
Brotzen, Franz R(ichard)
Bruins, Paul F(astenau)
Burger, Christian P
Burkart, Leonard F
Cao, Hengchu
Chen, In-Gann
Chen, Michael Chia-Chao

Chopra, Dev Raj
Chu, Ting Li
Cole, David F
Daues, Gregory W, Jr
Daugherty, Kenneth E
Davidson, David Lee
Davison, Sol
Dhudshia, Vallabh H
Dufner, Douglas Carl
Dukatz, Ervin L, Jr
Eberhart, Robert Clyde
Economou, Demetre J
Eyrick, Theodore B
Feit, Eugene David
Fish, John G
Forest, Edward
Frick, John P
Geyling, F(ranz) Th(omas)
Ghowsi, Kiumars
Glosser, Robert
Golden, David E
Goodwyn, Jack Ray
Griffin, John R(obert)
Griffin, Richard B
Gruber, George J
Gully, John Houston
Gupton, Paul Stephen
Harper, James George
Hasty, Turner Elilah
Hausler, Rudolf H
Henderson, Gregory Wayne
Hill, Robert William
Holliday, George Hayes
Holmes, Larry A
Huege, Fred Robert
Ignatiev, Alex
Johnson, Elwin L Pete
Jones, William B
Kanninen, Melvin Fred
Kaye, Howard
Kirk, Wiley Price
Koemtzopoulos, C Robert
Kolesar, Edward S
Koros, William John
Lacy, Lewis L
Levine, Jules David
Lewis, James Pettis
Lian, Shawn
Lindholm, Ulric S
Lloyd, Douglas Roy
Lytton, Robert Leonard
McConnell, Duncan
McDaniel, Floyd Delbert, Sr
McDavid, James Michael
Machacek, Oldrich
Marcus, Harris L
Matzkanin, George Andrew
Mills, John James
Moon, Tessie Jo
Moore, Thomas Matthew
Moss, Simon Charles
Murr, Lawrence Eugene
Nemphos, Speros Peter
Olstowski, Franciszek
Packman, Paul Frederick
Park, Vernon Kee
Patriarca, Peter
Paul, Donald Ross
Pavlovich, Raymond Doran
Perez, Ricardo
Petersen, Donald H
Porter, Vernon Ray
Porter, Wilbur Arthur
Powers, John Michael
Raba, Carl Franz, Jr
Rabson, Thomas A(velyn)
Ralls, Kenneth M(ichael)
Rao, Shankaranarayana Ramohallinanjunda
Reed, Thomas Freeman
Rosenmayer, Charles Thomas
Salama, Kamel
Sanchez, Isaac Cornelius
Schapery, Richard Allan
Sharma, Suresh C
Shaw, Don W
Sheshtawy, Adel A
Shilstone, James Maxwell, Jr
Spencer, Gregory Fielder
Stahl, Glenn Allan
Stehling, Ferdinand Christian
Steinfink, Hugo
Teller, Cecil Martin, II
Thornton, Joseph Scott
Tien, John Kai
Trachtenberg, Isaac
Turner, William Danny
Uralil, Francis Stephen
Wang, Paul Weily
Warren, William Ernest
Williamson, Luther Howard
Winegartner, Edgar Carl
Yao, Joe
Yuan, Robert L

UTAH

Adams, Brent Larsen
Anderson, Douglas I
Andrade, Joseph D
Benner, Robert E
Boyd, Richard Hays
Carnahan, Robert D
Cohen, Richard M
Cook, Melvin Alonzo

Materials Science Engineering (cont)

Daniels, Alma U(riah)
Hoeppner, David William
Horton, Ralph M
Kim, Sung Wan
Nairn, John Arthur
Nelson, Mark Adams
Orava, R(aimo) Norman
Rotz, Christopher Alan
Sohn, Hong Yong
Stringfellow, Gerald B
Sudweeks, Walter Bentley
Talbot, Eugene L(eroy)
Taylor, Philip Craig
Wilson, Arnold
Woodbury, Richard C

VERMONT
Anderson, R(ichard) L(ouis)
Berens, Alan Robert
Bhatt, Girish M
Furukawa, Toshiharu
Howard, Robert T(urner)
Hutchins, Gudrun A
Pires, Renato Guedes
Schiffman, Robert A
Von Turkovich, Branimir F(rancis)

VIRGINIA
Almeter, Frank M(urray)
Armstrong, Robert G
Barker, Robert Edward, Jr
Bikales, Norbert M
Birnbaum, Leon S
Buckley, John Dennis
Burley, Carlton Edwin
Campbell, Francis James
Cantrell, John H(arris)
Chiou, Minshon Jebb
Cook, Desmond C
Cooper, Khershed Pessie
Duesbery, Michael Serge
Duke, John Christian, Jr
Farago, John
Farmer, Barry Louis
Fisher, Farley
Frederick, Daniel
Fripp, Archibald Linley
Gangloff, Richard Paul
Goodman, A(lvin) M(alcolm)
Goodson, Louie Aubrey, Jr
Gordon, Ronald Stanton
Graham, Kenneth Judson
Hahn, Henry
Hanneman, Rodney E
Hasselman, Didericus Petrus Hermannus
Haworth, W(illiam) Lancelot
Hendricks, Robert Wayne
Henneke, Edmund George, II
Hibbard, Walter Rollo, Jr
Hodge, James Dwight
Holt, William Henry
Houska, Charles Robert
Hove, John Edward
Hudson, Charles Michael
Hutchinson, Thomas Eugene
Ijaz, Lubna Razia
Jesser, William Augustus
Johnson, Robert Alan
Johnson, William Randolph, Jr
Johnston, Norman Joseph
Kaznoff, Alexis I(van)
Kinsley, Homan Benjamin, Jr
Kirkendall, Ernest Oliver
Kranbuehl, David Edwin
Kuhlmann-Wilsdorf, Doris
Landgraf, Ronald William
Lane, Joseph Robert
Larsen-Basse, Jorn
Lawless, Kenneth Robert
Lawrence, David Joseph
Lodoen, Gary Arthur
Long, Edward Richardson, Jr
Lowe, A(rthur) L(ee), Jr
Lytton, Jack L(ester)
Maahs, Howard Gordon
Maloney, Kenneth Morgan
Milford, George Noel, Jr
Moon, Peter Clayton
Newman, James Charles, Jr
Outlaw, Ronald Allen
Patterson, James Douglas
Plamondon, Joseph Edward
Pletta, Dan Henry
Pohlmann, Juergen Lothar Wolfgang
Preusch, Charles D
Puster, Richard Lee
Reifsnider, Kenneth Leonard
Reynik, Robert John
Rice, Roy Warren
Rijke, Arie Marie
Robertson, W(illiam) D(onald)
Rothwarf, Frederick
St Clair, Anne King
St Clair, Terry Lee
Schuurmans, Hendrik J L
Sedriks, Aristide John
Senseny, Paul Edward
Sherbeck, L Adair
Shur, Michael
Siau, John Finn

Smidt, Fred August, Jr
Squire, David R
Steele, Lendell Eugene
Stein, Bland Allen
Stinchcomb, Wayne Webster
Sudarshan, T S
Sumner, Barbara Elaine
Taggart, G(eorge) Bruce
Tompkins, Stephen Stern
Unnam, Jalaiah
Van Ness, Kenneth E
Van Reuth, Edward C
Wawner, Franklin Edward, Jr
Wayland, Rosser Lee, Jr
Wilcox, Benjamin A
Wilkes, Garth L
Wilsdorf, Heinz G(erhard) F(riedrich)
Wood, George Marshall
Wynne, Kenneth Joseph

WASHINGTON
Baer, Donald Ray
Bates, J(unior) Lambert
Boyer, Rodney Raymond
Brimhall, J(ohn) L
Bryant, Ben S
Burns, Robert Ward
Clark, Donald Eldon
Collins, Gary Scott
Cordingly, Richard Henry
Daniel, J(ack) Leland
Danko, Joseph Christopher
Das, K Bhagwan
Demirel, T(urgut)
Ding, Jow-Lian
Doran, Donald George
Dunham, Glen Curtis
Duran, Servet A(hmet)
Einziger, Robert E
Ellis, Everett Lincoln
Eustis, William Henry
Evans, Thomas Walter
Fischbach, David Bibb
Gelles, David Stephen
Guion, Thomas Hyman
Hamilton, C Howard
Hannay, Norman Bruce
Hansen, Michael Roy
Hinman, Chester Arthur
Huang, Fan-Hsiung Frank
Johns, William E
Johnson, A(lfred) Burton, Jr
Johnson, Jay Allan
Jones, Russell Howard
Kalonji, Gretchen
Karagianes, Manuel Tom
Keating, John Joseph
Kent, Ronald Allan
Klement, William, Jr
Knotek, Michael Louis
Leney, Lawrence
Lindenmeyer, Paul Henry
Mahalingam, R
Maloney, Thomas M
Matlock, John Hudson
Megraw, Robert Arthur
Polonis, Douglas Hugh
Quist, William Edward
Ramulu, Mamidala
Roake, William Earl
Roberts, J T Adrian
Smith, Brad Keller
Stang, Robert George
Stoebe, Thomas Gaines
Subramanian, Ravanasamudram Venkatachalam
Taya, Minoru
Taylor, Murray East
Tichy, Robert J
Tinder, Richard F(ranchere)
Tingey, Garth Leroy
Trotter, Patrick Casey
Weber, William J
Wilson, Charles Norman
Woodfield, F(rank) W(illiam), Jr
Woods, Keith Newell

WEST VIRGINIA
Bryant, George Macon
Crist, John Benjamin
De Barbadillo, John Joseph
Hamilton, John Robert
Knight, Alan Campbell
Koleske, Joseph Victor
Seehra, Mohindar Singh
Smith, Walton Ramsay
Sperati, Carleton Angelo
Winston, Anthony
Zhang, Xiao

WISCONSIN
Aita, Carolyn Rubin
Bajikar, Sateesh S
Baker, George Severt
Barr, Tery Lynn
Blanchard, James Page
Booth, Roger Wright
Botez, Dan
Burck, Larry Harold
Casper, Lawrence Allen
Cataldi, Horace A(nthony)
Caulfield, Daniel Francis
Christiansen, Alfred W
Cooper, Reid F

Crawmer, Daryl E
Day, Anthony R
Denton, Denice Dee
Draeger, Norman Arthur
Dunn, Stanley Austin
Ferry, John Douglass
Fischer, Richard Martin, Jr
Fournelle, Raymond Albert
Fripiat, Jose J
Gross, James Richard
Hauser, Edward Russell
Hill, Charles Graham, Jr
Hinkes, Thomas Michael
Isebrands, Judson G
Isenberg, Irving Harry
Johnson, J(ames) R(obert)
Johnson, John E(dwin)
Kehres, Paul W(illiam)
Lagally, Max Gunter
Larbalestier, David C
Lemm, Arthur Warren
Lenling, William James
Loper, Carl R(ichard), Jr
Mangat, Pawitterjit Singh
Morris, Marion Clyde
Neumann, Joachim Peter
Nordman, James Emery
Pearl, Irwin Albert
Pillai, Thankappan A K
Randall, Francis James
Rechtin, Michael David
Reichenbacher, Paul H
Richard, Terry Gordon
Rohatgi, Pradeep Kumar
Rowlands, Robert Edward
Sanyer, Necmi
Schwarz, Eckhard C A
Svoboda, Glenn Richard
Tonner, Brian P
Van den Akker, Johannes Archibald
Verbrugge, Calvin James
Wnuk, Michael Peter
Young, Raymond A
Zerbe, John Irwin
Zheng, Xiaoci

WYOMING
Adams, Donald F
Dolan, Charles W

PUERTO RICO
Schwartz, Abraham

ALBERTA
Egerton, Raymond Frank
Micko, Michael M
Vroom, Alan Heard

BRITISH COLUMBIA
Brimacombe, James Keith
Davis, Gerald Gordon
Deen, Mohamed Jamal
Franz, Norman Charles
Garner, Andrew
Hardwicke, Norman Lawson
Hatton, John Victor
Mindess, Sidney
Paszner, Laszlo
Tromans, D(esmond)
Vitovec, Franz H
Wiles, David M
Young, Lawrence

MANITOBA
Bassim, Mohamad Nabil
Cahoon, John Raymond
Chaturvedi, Mahesh Chandra
Dutton, Roger
Kao, Kwan Chi
Simpson, Leonard Angus
Wilkins, Brian John Samuel
Woo, Chung-Ho

NEW BRUNSWICK
Schneider, Marc H
Sebastian, Leslie Paul

NEWFOUNDLAND
Molgaard, Johannes

NOVA SCOTIA
Jones, Derek William
Steinitz, Michael Otto

ONTARIO
Alexandru, Lupu
Aust, Karl T(homas)
Bahadur, Birendra
Basinski, Zbigniew S
Berezin, Alexander A
Bhatnagar, Yashraj Kishore
Biggs, Ronald C(larke)
Bratina, Woymir John
Cameron, Irvine R
Carlsson, David James
Cocivera, Michael
Convey, John
Cox, B(rian)
Dawes, David Haddon
Densley, John R
Geach, George Alwyn
Golemba, Frank John
Hackam, Reuben
Hope, Brian Bradshaw

Howard-Lock, Helen Elaine
Huang, Robert Y M
Hunt, Charles Edmund Laurence
Hurd, Colin Michael
Irons, Gordon Alexander
Ives, Michael Brian
Jackman, Thomas Edward
Ji, Guangda Winston
Jorch, Harald Heinrich
Kavassalis, Tom A
Kim, Sukyoung
King, Hubert Wylam
Koul, Ashok Kumar
Krausz, Alexander Stephen
Litvan, Gerard Gabriel
Logan, Charles Donald
McGeer, James Peter
Miller, W Alfred
Morris, Larry Arthur
Nicholson, Patrick Stephen
Northwood, Derek Owen
O'Driscoll, Kenneth F(rancis)
Pascual, Roberto
Piggott, Michael R(antell)
Pilliar, Robert Mathews
Pindera, Jerzy Tadeusz
Plumtree, A(lan)
Pundsack, Arnold L
Purdy, Gary Rush
Ramachandran, Vangipuram S
Ruda, Harry Eugen
Russell, Kenneth Edwin
Sahoo, Mahi
Saxton, William Reginald
Siddell, Derreck
Slater, Keith
Smeltzer, Walter William
Smith, Dennis Clifford
Street, Kenneth Norman
Sundararajan, Pudupadi Ranganathan
Thompson, David Allan
Tyson, William Russell
Varin, Robert Andrzej
Vincett, Paul Stamford
Wallace, William
Weir, Ronald Douglas
Williams, Harry Leverne
Witzman, Sorin
Yan, Maxwell Menuhin
Zukotynski, Stefan

PRINCE EDWARD ISLAND
Burch, George Nelson Blair

QUEBEC
Ajersch, Frank
Angers, Roch
Bartnikas, Ray
Champness, Clifford Harry
Crine, Jean-Pierre C
Dealy, John Michael
Edington, Jeffrey William
Feldman, Dorel
Hanna, Adel
Hay, Donald Robert
Hoa, Suong Van
Izquierdo, Ricardo
Jonas, John Joseph
Jordan, Byron Dale
Kahrizi, Mojtaba
Kokta, Bohuslav Vaclav
Koran, Zoltan
Lombos, Bela Anthony
Manley, Rockliffe St John
Marsh, Cedric
Masut, Remo Antonio
Meunier, Michel
Osman, M(ohamed) O M
Page, Derek Howard
Patterson, Donald Duke
Prud'homme, Robert Emery
Rigaud, Michel Jean
Ryan, Norman Daniel
Saint-Jacques, Robert G
St Pierre, Leon Edward
Strasser, John Albert
Tardif, Henri Pierre
Utracki, Lechoslaw Adam
Van Neste, Andre
Wertheimer, Michael Robert

SASKATCHEWAN
Kasap, Safa O

OTHER COUNTRIES
Anantharaman, Tanjore Ramachandra
Balk, Pieter
Barbenel, Joseph Cyril
Brown, Peter
Chang, Chun-Yen
Chang, L(eroy) L(i-Gong)
Davidson, Daniel Lee
De Vedia, Luis Alberto
French, William George
Hansen, Torben Christen
Hashin, Zvi
Hornbogen, Erhard
Isaacs, Philip Klein
Jacob, K Thomas
Katz, Gerald
Kim, Young-Gil
Kishi, Keiji
Lahiri, Syamal Kumar
Lin, Otto Chui Chau

Markovitz, Hershel
Meier, James Archibald
Neuse, Eberhard Wilhelm
Nowotny, Hans
Otsuka, Kazuhiro
Peterlin, Anton
Purbo, Onno Widodo
Ramachandran, Venkataraman
Reid, William John
Rodriguez, Federico Angel
Sakurai, Yoshifumi
Sessler, Gerhard Martin
Shimizu, Ken'ichi
Solari, Mario Jose Adolfo
Van Raalte, John A
Wasa, Kiyotaka
Yu, Chyang John

Mechanical Engineering

ALABAMA
Barfield, Robert F(redrick) (Bob)
Bussell, William Harrison
Crawford, Martin
Crocker, Malcolm John
Davis, Carl George
Doughty, Julian O
Dybczak, Z(bigniew) W(ladyslaw)
French, Kenneth Edward
Gilbert, John Andrew
Hung, Ru J
Jackson, John Elwin, Jr
Jones, Edward O(scar), Jr
Jones, Jess Harold
Jordan, William D(itmer)
Liu, C(hang) K(eng)
Liu, Frank C
Makhijani, Vinod B
Maples, Glennon
Morris, James Allen
Pears, Coultas D
Ramachandran, Narayanan
Schaetzle, Walter J(acob)
Schroer, Bernard J
Talbot, T(homas) F
Thompson, Byrd Thomas, Jr
Thompson, Kenneth O(rval)
Walker, William F(red)
Yeh, Pu-Sen
Zalik, Richard Albert

ALASKA
Zarling, John P

ARIZONA
Ahmed, Saad Attia
Arnell, Walter James
Backus, Charles E
Baroczy, Charles J(ohn)
Barr, Lawrence Dale
Beakley, George Carroll, Jr
Bean, James J(oseph)
Chandra, Abhijit
Chen, Chuan Fang
Cunningham, Richard G(reenlaw)
Davidson, Joseph Killworth
Evans, Donovan Lee
Fernando, Harindra Joseph
Florschuetz, Leon W(alter)
Fung, K Y
Gatley, William Stuart
Glick, Robert L
Govil, Sanjay
Hepworth, H(arry) Kent
Hilliard, Ronnie Lewis
Hirleman, Edwin Daniel, Jr
Jahsman, William Edward
Jordan, Richard Charles
Koeneman, James Bryant
Krajcinovic, Dusan
Limbert, Douglas A(lan)
Lundstrom, Louis C
McGuirk, William Joseph
Matsch, L(ee) A(llan)
Matthews, James B
Metzger, Darryl E
Miller, Harry
Osborn, Donald Earl
Peck, Robert E
Perkins, Henry Crawford, Jr
Ragsdell, Kenneth Martin
Ratkowski, Donald J
Russell, Paul E(dgar)
Saric, William Samuel
Shaw, Milton C(layton)
Slaughter, Charles D
So, Ronald Ming Cho
Stafford, John William
Suriano, F(rancis) J(oseph)
Swan, Peter A
Thomas, R E
Van Sant, James Hurley, Jr
Wood, Bruce
Wood, Byard Dean
Wygnanski, Israel Jerzy
Yamaguchi, Gary T

ARKANSAS
Akin, Jim Howard
Deaver, Franklin Kennedy
Gilbrech, Donald Albert
Gleason, James Gordon
Jong, Ing-Chang

Kedzie, Donald P
Krohn, John Leslie
Manos, William P
Reis, Irvin L

CALIFORNIA
Abdel-Ghaffar, Ahmed Mansour
Acosta, Allan James
Adams, William John, Jr
Addy, Tralance Obuama
Akin, Lee Stanley
Allen, Charles William
Alvares, Norman J
Antonsson, Erik Karl
Ardema, Mark D
Arnold, Frank R(obert)
Arthur, Paul D(avid)
Ashcroft, Frederick H
Auksmann, Boris
Austin, Arthur Leroy
Bailey, Michael John
Barkan, P(hilip)
Battenburg, Joseph R
Beadle, Charles Wilson
Beighley, Clair M(yron)
Bell, Robert Alan
Bendisz, Kazimierz
Berger, Stanley A(llan)
Bernsen, Sidney A
Bevill, Vincent (Darell)
Biblarz, Oscar
Blackketter, Dennis O
Blackwelder, Ron F
Blake, Alexander
Blasingame, Benjamin P(aul)
Blink, James Allen
Bogy, David B
Bolt, Robert O'Connor
Boltinghouse, Joseph C
Bonin, John H(enry)
Bonora, Anthony Charles
Bowman, Craig T
Brandt, H(arry)
Bray, A Philip
Brennen, Christopher E
Bright, Peter Bowman
Brock, John E(dison)
Broido, Jeffrey Hale
Brown, Tony Ray
Bruch, John C(larence), Jr
Burden, Harvey Worth
Bushnell, James Judson
Caligiuri, Robert Domenic
Campbell, Edward Michael
Cantin, Gilles
Carr, Robert Charles
Cass, Glen R
Castelli, Vittorio
Caton, Jerald A
Chang, Daniel P Y
Charley, Philip J(ames)
Chen, Charles Shin-Yang
Cheng, Edward Teh-Chang
Chenoweth, James Merl
Chi, Cheng-Ching
Chou, Larry I-Hui
Choudhury, P Roy
Christensen, Richard Monson
Chu, Chung-Yu Chester
Clauser, Francis Hettinger
Clothier, Robert Frederic
Cochran, David L(eo)
Colton, James Dale
Comley, Peter Nigel
Conant, Curtis Terry
Conlon, William Martin
Conn, Robert William
Cornet, I(srael) I(saac)
Culick, Fred E(llsworth) C(low)
Damonte, John Batista
D'Ardenne, Walter H
Das, Mihir Kumar
Dau, Gary John
Davey, Trevor B(lakely)
Davidson, Ernest
Davis, Charles Packard
Dean, Richard A
Deckert, Curtis Kenneth
Dergarabedian, Paul
Derr, Ronald Louis
Dignon, Jane
Dipprey, Duane F(loyd)
Dong, Richard Gene
Dornfeld, David Alan
Dowell, Douglas C
Dudley, Darle W
Duffield, Jack Jay
Durbeck, Robert C(harles)
Dutt, Nikil D
Eaton, John Kelly
Edelman, Walter E(ugene), Jr
Edwards, Donald K
Egermeier, R(obert) P(aul)
Ellion, M Edmund
Eustis, Robert H(enry)
Evans, Gregory Herbert
Falcon, Joseph A
Falcone, Patricia Kuntz
Fan, Chien
Faulders, Charles R(aymond)
Feinstein, Charles David
Feltz, Charles Henderson
Ferziger, Joel H(enry)
Finnie, I(ain)

Fischer, George K
Fisher, Franklin E(ugene)
Flagan, Richard Charles
Flora, Edward B(enjamin)
Folsom, Richard G(ilman)
Freberg, C(arl) Roger
Frey, Chris(tian) M(iller)
Friedmann, Peretz Peter
Frisch, Joseph
Fuhs, Allen E(ugene)
Fung, Sui-an
Garbaccio, Donald Howard
Garry, Frederick W
Gay, Richard Leslie
Gerpheide, John H
Giedt, W(arren) H(arding)
Glenn, Lewis Alan
Gojny, Frank
Goluba, Raymond William
Goodwine, James K, Jr
Gordon, Hayden S(amuel)
Gould, William Richard
Grassi, Raymond Charles
Greif, Ralph
Haberman, Charles Morris
Haener, Juan
Hahn, Hong Thomas
Han, L(it) S(ien)
Hansmann, Douglas R
Harder, James Albert
Harker, J M
Harvey, A(lexander)
Hauber, Janet Elaine
Haughton, Kenneth E
Healey, Anthony J
Helfman, Howard N
Henning, Carl Douglas
Hickman, Roy Scott
Hiel, Clement
Hirasuna, Alan Ryo
Hodge, Philip G(ibson), Jr
Hoffmann, Jon Arnold
Holeman, Dennis Leigh
Horne, Roland Nicholas
Hovingh, Jack
Hoyt, Jack W(allace)
Hsu, Chieh-Su
Huang, Francis F
Hudson, Donald E(llis)
Hussain, Nihad A
Ishihara, Kohei
Jakubowski, Gerald S
Johanson, Jerry Ray
Johnson, Conor Deane
Johnson, David Leroy
Johnston, James P(aul)
Jory, Howard Roberts
Kane, E(neas) D(illon)
Karnopp, Dean Charles
Katz, Robert
Kaufman, Boris
Kays, William Morrow
Kayton, Myron
Kelleher, Matthew D(ennis)
Keller, Joseph Edward, Jr
Kelly, Robert Edward
Kemper, John D(ustin)
Kennedy, Ian Manning
Kesselring, John Paul
Kim, J John
King, Hartley H(ughes)
Klestadt, Bernard
Kline, Stephen Jay
Klipstein, David Hampton
Kobayashi, Shiro
Kosmatka, John Benedict
Kraabel, John Stanford
Kruger, Charles Herman, Jr
Laderman, A(rnold) J(oseph)
La Fleur, James Kemble
Lang, Thomas G(lenn)
Lavernia, Enrique Jose
Lavine, Adrienne Gail
Lay, Thorne
Leifer, Larry J
Leitmann, G(eorge)
Levy, Salomon
Lewis, Francis Hotchkiss, Jr
Lick, Wilbert James
Lindsey, Gerald Herbert
Liu, Chen Ya
London, A(lexander) L(ouis)
Lorenzen, Coby
Low, Lawrence J(acob)
Lowi, Alvin, Jr
Lucas, John W
Luongo, Cesar Augusto
Ma, Fai
McCarthy, John Michael
McClure, Eldon Ray
McKillop, Allan A
McLeod, John Hugh, Jr
McMillan, Oden J
Macomber, Thomas Wesson
Madan, Ram Chand
Madsen, Richard Alfred
Maewal, Akhilesh
Majumdar, Arunava
Marcus, Bruce David
Margolis, Stephen Barry
Marner, Wilbur Joseph
Martin, George
Marto, Paul James
Masri, Sami F(aiz)

Massier, Paul Ferdinand
Matheny, James Donald
Matthys, Eric Francois
Maulbetsch, John Stewart
Maxworthy, Tony
Melese, Gilbert B(ernard)
Meriam, James Lathrop
Merilo, Mati
Meyer, Brad Anthony
Meyers, Marc Andre
Mikesell, Walter R, Jr
Miklowitz, Julius
Mills, Anthony Francis
Mitchner, Morton
Moffat, Robert J
Moon, Donald W(ayne)
Moore, John R(obert)
Mow, C(hao) C(how)
Myers, Blake
Myronuk, Donald Joseph
Nathenson, Manuel
Netzer, David Willis
Nguyen, Luu Thanh
Nieh, Tai-Gang
Nielan, Paul E
Nielsen, Helmer L(ouis)
Niles, Philip William Benjamin
Nooker, Eugene L(eRoy)
Noring, Jon Everett
Nunn, Robert Harry
Nypan, Lester Jens
Oehlberg, Richard N
Offen, George Richard
O'Hern, Eugene A
O'Meara, David Lillis
Oppenheim, A(ntoni) K(azimierz)
Orcutt, John Arthur
Owens, William Leo
Pagni, Patrick John
Pamidi, Prabhakar Ramarao
Patzer, Eric John
Pearson, John
Pedersen, Knud B(orge)
Pefley, Richard K
Pellinen, Donald Gary
Petrone, Rocco A
Pinkel, B(enjamin)
Pinto, John Gilbert
Potter, Richard C(arter)
Pressman, Ada Irene
Przystupa, Marek Antoni
Pucci, Paul F(rancis)
Rathmann, Carl Erich
Rau, Charles Alfred, Jr
Reardon, Frederick H(enry)
Reynolds, William Craig
Ritchie, Robert Oliver
Rivers, William J(ones)
Robison, D(elbert) E(arl)
Ross, Bernard
Roth, Bernard
Rubin, Sheldon
Russell, T(homas) L(ee)
Russell, William T(reloar)
Saha, Subrata
Salerno, Louis Joseph
Salmassy, Omar K
Salsig, William Winter, Jr
Samson, Sten
Sanders, Charles F(ranklin), Jr
Sarpkaya, Turgut
Savas, Omer
Sawyer, Robert Fennell
Schalla, Clarence August
Scharton, Terry Don
Schefer, Robert Wilfred
Schmid-Schoenbein, Geert W
Schreiner, Robert Nicholas, Jr
Schrock, Virgil E(dwin)
Schurman, Glenn August
Schurmeier, Harris McIntosh
Schwartz, Daniel M(ax)
Schwartzbart, Harry
Seban, Ralph A
Seide, Paul
Shaffar, Scott William
Shah, Ramesh Trikamlal
Shanthikumar, Jeyaveerasingam George
Shelly, John Richard
Shepherd, Joseph Emmett
Shoup, Terry Emerson
Sifferman, Thomas Raymond
Silcox, William Henry
Sirignano, William Alfonso
Smith, Donald Stanley
Snyder, Nathan W(illiam)
Spier, Edward Ellis
Spinks, John Lee
Spitzer, Irwin Asher
Spokoyny, Felix E
Springett, David Roy
Statt, Terry G
Steidel, Robert F(rancis), Jr
Stockel, Ivar H(oward)
Stone, Robert K(emper)
Stout, Ray Bernard
Street, Robert L(ynnwood)
Stuhmiller, James Hamilton
Sutherland, Ivan Edward
Swearengen, Jack Clayton
Sweeney, Donald Wesley
Takahashi, Yasundo
Talbot, Lawrence
Tellep, Daniel M

Mechanical Engineering (cont)

Thomas, Floyd W, Jr
Thomas, Graham Havens
Thrasher, L(awrence) W(illiam)
Throner, Guy Charles
Trezek, George J
Trubert, Marc
Tsao, Ching H
Unt, Hillar
Uyehara, Otto A(rthur)
Vanblarigan, Peter
Vanderplaats, Garret Niel
Velkoff, Henry Rene
Walton, James Stephen
Wang, Ji Ching
Wang, Wei-E
Wasley, Richard J(unior)
Welsh, David Edward
Whittier, James S(pencer)
Wieland, Bruce Wendell
Wilde, D(ouglass) J(ames)
Wildmann, Manfred
Williams, Harry Edwin
Wood, Allen D(oane)
Wray, John L
Yang, Chun Chuan
Yeung, King-Wah Walter
Yeung, Ronald Wai-Chun
Yez, Martin S(imon)
Yildiz, Alaettin
Youngdahl, Paul F
Yuan, Sidney Wei Kwun
Zeren, Richard William
Zickel, John
Zivi, Samuel M(eisner)

COLORADO
Barber, Robert Edwin
Burnham, Marvin William
Carlson, Lawrence Evan
Chelton, Dudley B(oyd)
Crawford, Richard H
Daily, John W
Datta, Subhendu Kumar
Durham, Michael Dean
Ellis, Donald Griffith
Frangopol, Dan Mircea
Garvey, Daniel Cyril
Geers, Thomas L
Gosink, Joan P
Haberstroh, Robert D
Hittle, Douglas Carl
Jansson, David Guild
Johnson, Gearold Robert
Kassoy, David R
Kaufman, Harold Richard
Kerr, Robert McDougall
Kober, C(arl) L(eopold)
Kreith, Frank
Krill, Arthur Melvin
Ladd, Conrad Mervyn
Lee, Yung-Cheng
Loehrke, Richard Irwin
Mahajan, Roop Lal
Marshall, Charles F
Melsheimer, Frank Murphy
Meroney, Robert N
Miller, Paul Leroy, Jr
Mitchell, Charles Elliott
Mote, Jimmy Dale
Murphy, John Michael
Norwood, Richard E(llis)
Pak, Ronald Y S
Paynter, Howard L
Peterson, Harry C(larence)
Radebaugh, Ray
Regenbrecht, D(ouglas) E(dward)
Rennat, Harry O(laf)
Schmidt, Alan Frederick
Siuru, William D, Jr
Smith, Frederick Willis
Stauffer, Jack B
Steward, W(illis) G(ene)
Suh, Chung-Ha
Summers, Luis Henry
Swanson, Lawrence Ray
Thompson, Erik G(rinde)
Troxell, Wade Oakes
Tuttle, Elizabeth R
Van Pelt, Richard W(arren)
Verschoor, J(ack) D(ahlstrom)
Wilbur, Paul James
Zoller, Paul

CONNECTICUT
Arnoldi, Walter Edwin
Berglund, Larry Glenn
Best, Stanley Gordon
Bowley, Wallace William
Bozzuto, Carl Richard
Brancato, Leo J(ohn)
Brand, Ronald S(cott)
Burridge, Robert
Burwell, Wayne Gregory
Chin, Charles L(ee) D(ong)
Chiu, Yih-Ping
Cohen, Myron Leslie
Crawford, John Okerson
Crossley, F(rancis) R(endel) Erskine
Dabora, Eli K
Davis, Roger L
Dubin, Fred S
Fink, Martin Ronald
Franz, Anselm
Garrett, Richard E
Hobbs, David E
Johnson, Bruce Virgil
Kazerounian, Kazem
Ketchman, Jeffrey
Keyes, David Elliot
Kimberley, John A
Kunz, Harold Russell
Lemkey, Frankin David
McFadden, Peter W(illiam)
Mack, Donald R(oy)
Ojalvo, Irving U
Paterson, Robert W
Peracchio, Aldo Anthony
Pitkin, Edward Thaddeus
Robinson, Donald W(allace), Jr
Rosner, Daniel E(dwin)
Sheets, Herman E(rnest)
Shichman, D(aniel)
Shuey, Merlin Arthur
Smith, Donald Arthur
Smith, Melvin I
Smooke, Mitchell D
Sreenivasan, Katepalli Raju
Williams, John Ernest
Wuthrich, Paul

DELAWARE
Advani, Suresh Gopaldas
Bydal, Bruce A
Cantwell, Edward N(orton), Jr
Chou, Tsu-Wei
Dentel, Steven Keith
Dexter, Stephen C
Kaliakin, Victor Nicholas
Kingsbury, Herbert B
McKee, David Edward
Moore, Ralph Leslie
Murphy, Arthur Thomas
Santare, Michael Harold
Schwartz, Leonard William
Szeri, Andras Zoltan
Washburn, Robert Latham
Willis, Frank Marsden
Work, Dennis M
Zimmerman, John R(ichard)

DISTRICT OF COLUMBIA
Alic, John A
Baer, Robert Lloyd
Baz, Amr Mahmoud Sabry
Boehler, Gabriel D(ominique)
Bosnak, Robert J
Cooper, Thomas Edward
Frair, Karen Lee
Gallagher, William J(oseph)
Gould, Phillip
Huber, Peter William
Jones, Douglas Linwood
Kelly, George Eugene
Kelnhofer, William Joseph
Korkegi, Robert Hani
Kramer, Bruce Michael
Larson, Charles Fred
Ojalvo, Morris S(olomon)
Parks, Vincent Joseph
Pei, Richard Yu-Sien
Reis, Victor H
Robertson, A(lexander) F(rancis)
Rockett, John A
Salmon, William Cooper
Walker, M Lucius, Jr
Warnick, Walter Lee
Wilmotte, Raymond M
Youm, Youngil
Zucchetto, James John

FLORIDA
Adt, Robert (Roy)
Anghaie, Samim
Anusavice, Kenneth John
Bigley, Harry Andrew, Jr
Billings, Charles Edgar
Bober, William
Bowman, Thomas Eugene
Buzyna, George
Case, Robert Oliver
Catz, Jerome
Chen, Ching-Jen
Chow, Wen Lung
Claridge, Richard Allen
Coar, Richard J
Concordia, Charles
Cross, Ralph Emerson
DeHart, Arnold O'Dell
Diaz, Nils Juan
Dilpare, Armand Leon
Doering, Robert Distler
Drucker, Daniel Charles
Edwards, Thomas Claude
Etherington, Harold
Farber, Erich A(lexander)
Foster, J Earl
Gaither, Robert Barker
Gencsoy, Hasan Tahsin
Hahn, C(harles) Archie, Jr
Harrisberger, Edgar Lee
Hartley, Craig Sheridan
Hartman, John Paul
Hayes, Edward J(ames)
Hoeppner, Conrad Henry
Hosler, E(arl) Ramon
Hosni, Yasser Ali
Howell, Ronald Hunter
Hoyer-Ellefsen, Sigurd
Hrones, John Anthony
Hsieh, Chung Kuo
Juvinall, Robert C
Kerr, Robert Lowell
King, Blake
Kinney, Robert Bruce
Klausner, James Frederick
Kranc, Stanley Charles
Krishnamurthy, Lakshminarayanan
Langford, David
Lijewski, Lawrence Edward
Lundgren, Dale A(llen)
McCracken, Walter John
Mandil, I Harry
Miller, Robert Gerry
Mittleman, John
Nimmo, Bruce Glen
Oliver, Calvin C(leek)
Phillips, Winfred M(arshall)
Piotrowski, George
Proctor, Charles Lafayette, II
Psarouthakis, John
Reichard, Ronnal Paul
Reisbig, Ronald Luther
Roan, Vernon P
Scaringe, Robert P
Schimmel, Walter Paul
Schmidtke, R(ichard) A(llen)
Scott, Kenneth Elsner
Scott, Linus Albert
Seireg, Ali A
Shaw, Lawrance Neil
Sheppard, Donald M(ax)
Sias, Frederick Ralph
Silverstein, Abe
Silvestri, George J, Jr
Suciu, S(piridon) N
Sun, Chang-Tsan
Sundaram, Swaminatha
Teixeira, Arthur Alves
Tiederman, William Gregg, Jr
Traexler, John F
Ward, Leonard George
Warner, Thomas Clark, Jr
Wimberly, C Ray
Witzell, O(tto) W(illiam)
Woll, Edward
Zimmerman, Norman H(erbert)
Zubko, L(eonard) M(artin)

GEORGIA
Abdel-Khalik, Said Ibrahim
Antolovich, Stephen D
Atluri, Satya N
Barnett, Samuel C(larence)
Black, William Z(achary)
Colton, Jonathan Stuart
Craig, James I
Desai, Prateen V
Dickerson, Stephen L(ang)
Durbetaki, Pandeli
Freeston, W Denney, Jr
Ghate, Suhas Ramkrishna
Ginsberg, Jerry Hal
Goglia, M(ario) J(oseph)
Hodgdon, F(rank) E(llis)
Kadaba, Prasanna V
Knight, Lee H, Jr
Komkov, Vadim
Krol, Joseph
Ku, David Nelson
Lau, Jark Chong
Lee, Kok-Meng
Loewy, Robert G(ustav)
Matula, Richard A
May, Gerald Ware
Meyers, Carolyn Winstead
Neitzel, George Paul
Nerem, Robert Michael
Orloff, David Ira
Rouse, William Bradford
Salant, Richard Frank
Shakun, Wallace
Stalford, Harold Lenn
Stanley, Luticious Bryan, Jr
Staton, Rocker Theodore, Jr
Strahle, Warren C(harles)
Vachon, Reginald Irenee
Vicory, William Anthony
Wempner, Gerald Arthur
Wepfer, William J
Winer, Ward Otis
Wu, James Chen-Yuan
Zinn, Ben T

HAWAII
Antal, Michael Jerry, Jr
Bartosik, Alexander Michael
Cheng, Ping
Chou, James C S
Fand, Richard Meyer
Fox, Joel S
Fuka, Louis Richard
Hihara, Lloyd Hiromi

IDAHO
Elias, Thomas Ittan
Jacobsen, Richard T
Korth, Gary E
Larson, Jay Reinhold
Ortiz, Marcos German
Riemke, Richard Allan

Sohal, Manohar Singh
Stewart, Richard Byron
Trout, Thomas James

ILLINOIS
Addy, Alva LeRoy
Adrian, Ronald John
Anderson, Thomas Patrick
Aref, Hassan
Assanis, Dennis N
Balcerzak, Marion John
Bayne, James Wilmer
Bell, Charles Eugene, Jr
Benjamin, Roland John
Berdichevsky, Alexander
Berry, Gregory Franklin
Block, Stanley M(arlin)
Brauer, Roger L
Brewster, Marcus Quinn
Bullard, Clark W
Bunting, Bruce Gordon
Chakrabarti, Subrata K
Chao, B(ei) T(se)
Chaszeyka, Michael A(ndrew)
Chato, John C(lark)
Choi, Stephen U S
Chung, Paul M(yungha)
Clausing, A(rthur) M(arvin)
Conry, Thomas Francis
Cusano, Cristino
Daniel, Isaac M
Dantzig, Jonathan A
Deen, James Robert
Deitrich, L(awrence) Walter
Dix, Rollin Cumming
Dolan, Thomas J(ames)
D'Souza, Anthony Frank
Dunning, Ernest Leon
Dutton, Jonathan Craig
Erwin, Lewis
Eshleman, Ronald L
Farhadieh, Rouyentan
Francis, John Elbert
Freedman, Steven I(rwin)
Frey, Donald N(elson)
Georgiadis, John G
Graae, Johan E A
Gupta, Krishna Chandra
Halleen, Robert M(arvin)
Hartenberg, Richard S(cheunemann)
Hartnett, James P(atrick)
Holmes, L(awrence) B(ruce)
Horve, Leslie A
Hull, William L(avaldin)
Jefferson, Thomas Bradley
Jones, Barclay G(eorge)
Kaganov, Alan Lawrence
Kalpakjian, Serope
Kasuba, Romualdas
Kern, Roy Fredrick
Konzo, Seichi
Korst, Helmut Hans
Kottas, Harry
Koves, William John
Kovitz, Arthur A(braham)
Leach, James L(indsay)
Lemke, Donald G(eorge)
Levine, Michael W
Litvin, Faydor L
Liu, Wing Kam
Lottes, P(aul) A(lbert)
McKee, Keith Earl
Mapother, Dillon Edward
Marr, William Wei-Yi
Martinec, Emil Louis
Matalon, Moshe
Mehta, N(avnit) C(haganlal)
Miller, Floyd Glenn
Minkowycz, W J
Mokadam, Raghunath G(anpatrao)
Moran, Thomas J
Morel, Thomas
Morkovin, Mark V(ladimir)
Nachtman, Elliot Simon
O'Bryant, David Claude
Okamura, Kiyohisa
Olmstead, William Edward
Orthwein, W(illiam) C(oe)
Pearlstein, Arne Jacob
Peters, James Empson
Phillips, Rohan Hilary
Pickett, Leroy Kenneth
Porter, Robert William
Poulikakos, Dimos
Reichard, Grant Wesley
Rettaliata, John Theodore
Riahi, Daniel Nourollah
Ruhl, Roland Luther
Salzenstein, Marvin A(braham)
Sather, Norman F(redrick)
Sekar, Raj
Shabana, Ahmed Abdelraouf
Shack, William John
Siegal, Burton L
Singh, Shyam N
Soo, Shao-Lee
Spector, Leo Francis
Spotts, M(erhyle) F(ranklin)
Thomas, Anthony
Ting, Thomas C(hi) T(sai)
Tipei, Nicolae
Trigger, Kenneth James
Tucker, Charles L
Tuzson, John J(anos)

Uherka, Kenneth Leroy
Walker, John Scott
Wang, Pie-Yi
Warinner, Douglas Keith
Weber, Norman
Weeks, Richard William
Wessler, Max Alden
White, Robert Allan
Wiley, Jack Cleveland
Wilson, William Robert Dunwoody
Wirtz, Gerald Paul
Wolf, Ludwig, Jr
Wolosewicz, Ronald Mitchell
Woods, Kenneth R
Worek, William Martin
Worley, Will J
Wurm, Jaroslav
Yoerger, Roger R
Youngdahl, Carl Kerr
Zell, Blair Paul

INDIANA
Barash, Moshe M
Bergdolt, Vollmar Edgar
Bernhard, Robert J
Brach, Raymond M
Brown, Charles L(eonard)
Carey, Alfred W(illiam), Jr
Carroll, John T, III
Citron, Stephen J
Cohen, Raymond
Dalphin, John Francis
DeWitt, David P
Fox, Robert William
Gad-el-Hak, Mohamed
Gibson, Harry Gene
Gibson, James Darrell
Goldschmidt, Victor W
Gorman, Eugene Francis
Hall, A(llen) S(trickland), Jr
Hamilton, James F(rancis)
Hansen, Arthur G(ene)
Hartsaw, William O
Hawks, Keith Harold
Ho, Cho-Yen
Hoffman, Joe Douglas
Holowenko, A(lfred) R(ichard)
Hooper, Irvin P(latt)
Howland, George Russell
Huang, Nai-Chien
Huffman, John Curtis
Incropera, Frank P
Jerger, E(dward) W
Jones, James Darren
Kareem, Ahsan
Kazem, Sayyed M
Kruger, Fred W
Laurendeau, Normand Maurice
L'Ecuyer, Mel R
Lehmann, Gilbert Mark
Liley, Peter Edward
Lyons, Jerry L
McAleece, Donald John
McComas, Stuart T
McDonald, Alan T(aylor)
Mercer, Walter Ronald
Messal, Edward Emil
Modrey, Joseph
Osborn, J(ohn) R(obert)
Ovens, William George
Pearson, Joseph T(atem)
Phillips, James W
Pierson, Edward S
Quinn, C Jack
Ramadhyani, Satish
Raven, Francis Harvey
Razdan, Mohan Kishen
Schoenhals, Robert James
Sen, Mihir
Shahidi, Freydoon
Sieloff, Ronald F
Smith, Charles O(liver)
Soedel, Werner
Stevenson, Warren H
Szewczyk, Albin A
Taylor, Raymond Ellory
Thompson, Howard Doyle
Trachman, Edward G
Tree, David R
Turner, Charles Hall
Wallace, F Blake
Wark, Kenneth, Jr
Warner, Cecil F(rancis)
Widener, Edward Ladd, Sr
Yang, Henry T Y
Yang, Kwang-Tzu
Zoss, Leslie M(ilton)

IOWA
Akers, Arthur
Baumgarten, Joseph Russell
Beckermann, Christoph
Chen, Lea-Der
Choi, Kyung Kook
Cook, William John
Fellinger, Robert Cecil
Goel, Vijay Kumar
Gurll, Nelson
Hall, Jerry Lee
Haug, Edward J, Jr
Junkhan, George H
Kusiak, Andrew
Lance, George M(ilward)
Landweber, Louis

McConnell, Kenneth G
Mack, Michael J
Madsen, Donald H(oward)
Miller, Richard Keith
Mischke, Charles R(ussell)
Myers, Thomas Wilmer
Northup, Larry L(ee)
Okiishi, Theodore Hisao
Peters, Leo Charles
Pletcher, Richard H
Rocklin, Isadore J
Scholz, Paul Drummond
Serovy, George K(aspar)
Stephens, Ralph Ivan
Tannehill, John Charles
Trummel, J(ohn) Merle

KANSAS
Annis, Jason Carl
Appl, Fredric Carl
Barr, B(illie) Griffith
Burmeister, Louis C
Cheng, Lester (Le-Chung)
Forman, George W
Giri, Jagannath
Gorton, Robert Lester
Gosman, Albert Louis
Gowdy, Kenneth King
Graham, A Richard
Hosni, Mohammad Hosein
Huang, Chi-Lung Dominic
Hwang, Ching-Lai
Johnson, Richard T(errell)
Jovanovic, M(ilan) K(osta)
Kipp, Harold Lyman
Lindholm, John C
Lyles, Leon
McCabe, Steven Lee
McNall, Preston E(ssex), Jr
Razak, Charles Kenneth
Robinson, M John
Smith, Raymond V(irgil)
Swenson, Donald Otis
Thompson, Joseph Garth
Turnquist, Ralph Otto
Walker, Hugh S(anders)
Wentz, William Henry, (Jr)

KENTUCKY
Conley, Weld E
Cremers, Clifford J
Drake, Robert M, Jr
Eaton, Thomas Eldon
Edwards, Richard Glenn
Frank, John L
Funk, James Ellis
Gard, O(liver) W(illiam)
Gold, Harold Eugene
Hahn, Ottfried J
Hantman, Robert Gary
Leep, Herman Ross
Leigh, Donald C
Lu, Wei-yang
Marshall, Maurice K(eith)
Mason, Harry Louis
Shupe, Dean Stanley
Tran, Long Trieu
Wilhoit, James Cammack, Jr

LOUISIANA
Adams, Kenneth H
Barron, Randall F(ranklin)
Chieri, P(ericle) A(driano)
Cospolich, James D
Dareing, Donald W
Eaton, Harvill Carlton
Hamilton, D(ewitt) C(linton), Jr
Hewitt, Hudy C, Jr
Lowther, James David
Munchmeyer, Frederick Clarke
Parish, Richard Lee
Peyronnin, Chester A(rthur), Jr
Rotty, Ralph M(cGee)
Russo, Edwin Price
Siebel, M(athias) P(aul)
Sogin, H(arold) H
Thigpen, J(oseph) J(ackson)
Thompson, Hugh Allison
Trammell, Grover J(ackson), Jr
Warrington, Robert O'Neil, Jr
Whitehouse, Gerald D(ean)
Whitehurst, Charles A(ugustus)

MAINE
Grosh, Richard J(oseph)
Hill, Richard C(onrad)
Meftah, Bachir
Rohsenow, Warren M(ax)
Sucec, James
Webster, Karl Smith
Weinstein, Alvin Seymour
Wilbur, L(eslie) C(lifford)

MARYLAND
Adams, James Alan
Anand, Davinder K
Armstrong, R(onald) W(illiam)
Baker, Kenneth L(eroy)
Barr, William A(lexander)
Bascunana, Jose Luis
Baum, Howard Richard
Berger, Bruce S
Bernard, Peter Simon
Brown, Paul Joseph

Burns, Bruce Peter
Burns, Timothy John
Camponeschi, Eugene Thomas, Jr
Chang, Zhaohua
Chen, Benjamin Yun-Hai
Cleveland, William Grover, Jr
Dailey, George
Dally, James William
Dieter, George E(llwood), Jr
Douglas, Andrew Sholto
Drzewiecki, Tadeusz Maria
Dugoff, Howard
Duncan, James Playford
Fabic, Stanislav
Fabunmi, James Ayinde
Flynn, Paul D(avid)
Foa, J(oseph) V(ictor)
Fowell, Andrew John
Franke, Gene Louis
Goldstein, Seth Richard
Gupta, Ashwani Kumar
Hardis, Leonard
Hartman, Patrick James
Hasson, Dennis Francis
Hawkins, William M(adison), Jr
Haynes, Emanuel
Heller, Edward Lincoln
Herold, Keith Evan
Hill, James Edward
Hopenfield, Joram
Hsu, Stephen M
Huang, Thomas Tsung-Tse
Hunter, Lawrence Wilbert
Irwin, George Rankin
Jabbour, Kahtan Nicolas
Jackson, William David
Jones, Everett
Khan, Akhtar Salamat
Khatri, Hiralal C
Kiang, Robert L
Kitchens, Clarence Wesley, Jr
Koh, Severino Legarda
Kramer, David Buckley
Ku, Jentung
Kusuda, Tamami
Landsburg, Alexander Charles
Lange, Eugene Albert
Laurenson, Robert Mark
Lopardo, Vincent Joseph
Loss, Frank J
Lu, Yeh-Pei
Lundsager, C(hristian) Bent
Marks, Colin H
Marlowe, Donald E(dward)
Martin, John J(oseph)
Menton, Robert Thomas
Mintz, Fred
Mitler, Henri Emmanuel
Mordfin, Leonard
Morris, Alan
Nash, Jonathon Michael
Obremski, Henry J(ohn)
Peppin, Richard J
Perrone, Nicholas
Popel, Aleksander S
Poulose, Pathickal K
Presser, Cary
Quintiere, James G
Radermacher, Reinhard
Rich, Harry Louis
Riley, Claude Frank, Jr
Ritter, Joseph John
Rostamian, Rouben
Rozner, Alexander
Rushing, Frank C
Saarlas, Maido
Sallet, Dirse Wilkis
Sanders, Robert Charles
Sandusky, Harold William
Schleiter, Thomas Gerard
Schwartz, John T
Scialdone, John Joseph
Seigel, Arnold E(lliott)
Semerjian, Hratch G
Sevik, Maurice
Sharpe, William Norman, Jr
Smedley, William Michael
Smith, John Henry
Smith, Russell Aubrey
Smylie, Robert Edwin
Staker, Michael Ray
Strombotne, Richard L(amar)
Sturek, Walter Beynon
Tai, Tsze Cheng
Thiruvengadam, Alagu Pillai
Tsai, Lung-Wen
Tuttle, Kenneth Lewis
Urbach, Herman B
Wallace, James M
Walston, William H(oward), Jr
Way, George H
Weckesser, Louis Benjamin
Yeh, Tsyh Tyan
Young, C(harles), Jr
Zmola, Paul C(arl)

MASSACHUSETTS
Adams, George G
Allen, Steven
Alpert, Ronald L
Ambs, Lawrence Lacy
Anderson, John Edward
Argon, Ali Suphi
Astill, Kenneth Norman

Avila, Charles Francis
Bailey, Bruce M(onroe)
Barclay, Stanton Dewitt
Blake, Thomas R
Bloomfield, David Peter
Bolz, R(ay) E(mil)
Borden, Roger R(ichmond)
Brosens, Pierre Joseph
Brown, Robert Lee
Burke, Shawn Edmund
Carson, John William
Chryssostomidis, Chryssostomos
Chu, Chauncey C
Cipolla, John William, Jr
Clark, Llewellyn Evans
Cook, Nathan Henry
Cooper, W(illiam) E(ugene)
Copley, Lawrence Gordon
Corey, Harold Scott
Crandall, Stephen H
Darkazalli, Ghazi
Daskin, W(alter)
De Fazio, Thomas Luca
Delagi, Richard Gregory
De Los Reyes, Gaston
DiPippo, Ronald
Dittfach, John Harland
Dixon, John R
Dubowsky, Steven
Dunn, John Frederick, Jr
Ehrich, Fredric F(ranklin)
Fay, James A(lan)
Feldman, Charles Lawrence
Foster, Arthur R(owe)
Glicksman, Leon Robert
Godrick, Joseph Adam
Goela, Jitendra Singh
Gold, Harris
Goss, William Paul
Greif, Robert
Greitzer, Edward Marc
Griffith, Peter
Hahn, Robert S(impson)
Hals, Finn
Haworth, D(onald) R(obert)
Hawthorne, William (Rede)
Heywood, John Benjamin
Hill, Albert G
Hoffman, Allen Herbert
Hurt, William C, Jr
Iwasa, Yukikazu
Jakus, Karl
Johnson, Aldie E(dwin), Jr
Johnson, Walter Roland
Kachanov, Mark L
Kamien, C Zelman
Katz, Israel
Kazimi, Mujid S
Kerney, Peter Joseph
Kirkpatrick, E(dward) T(homson)
Krause, Irvin
Kronauer, Richard Ernest
Lee, Harvey Shui-Hong
Lee, Shih-Ying
Lemnios, A(ndrew) Z
Leung, Woon Fong (Wallace)
Levitt, Albert P
Lindberg, Edward E
Loretz, Thomas J
McClintock, Frank A(mbrose)
McGowan, Jon Gerald
MacGregor, C(harles) W(inters)
Malkin, Stephen
Mann, Robert W(ellesley)
Manning, Jerome Edward
Mark, Melvin
Mastronardi, Richard
Miaoulis, Ioannis N
Mikic, Bora
Modak, Ashok Trimbak
Moore, Robert Edmund
Motherway, Joseph E
Murch, Laurence Everett
Nelson, Frederick Carl
Norton, Robert Leo
Nye, Edwin (Packard)
Pan, Coda H T
Park, B J
Paul, Igor
Picha, Kenneth G(eorge)
Poirier, Victor L
Poli, Corrado
Pope, Joseph
Pope, Mary E
Probstein, Ronald F(ilmore)
Pulling, Nathaniel H(osler)
Rabinowicz, Ernest
Rapperport, Eugene J
Reed, F(lood) Everett
Reeves, Barry L(ucas)
Reichenbach, George Sheridan
Rios, Pedro Agustin
Rosenthal, Richard Alan
Rossettos, John N(icholas)
Schenck, Hilbert Van Nydeck, Jr
Shapiro, Ascher H(erman)
Sharp, A(rnold) G(ideon)
Sheridan, Thomas Brown
Silvers, J(ohn) P(hillip)
Sioui, Richard Henry
Smith, Joseph LeConte, Jr
Sonin, Ain A(nts)
Staszesky, Francis M
Stekly, Z J John

Mechanical Engineering (cont)

Stevens, Herbert H(owe), Jr
Subramanian, Krishnamoorthy
Suh, Nam Pyo
Sunderland, James Edward
Thompson, Charles
Toner, Mehmet
Tong, Pin
Toong, Tau-Yi
Topping, Richard Francis
Toscano, William Michael
Udelson, Daniel G(erald)
Ungar, Eric E(dward)
Vest, Charles Marstiller
Wadleigh, Kenneth R(obert)
Weinberg, Marc Steven
Wheildon, W(illiam) M(axwell), Jr
Whitney, Daniel Eugene
Williams, James Henry, Jr
Wilson, David Gordon
Winfrey, Richard Cameron
Yener, Yaman
Yuan, Fan
Zalosh, Robert Geoffrey
Zinsmeister, George Emil
Zotos, John
Zwiep, Donald N

MICHIGAN
Adams, William Eugene
Agnew, William G(eorge)
Alexandridis, Alexander A
Amann, Charles A(lbert)
Atkin, Rupert Lloyd
Atreya, Arvind
Barber, James Richard
Barnes, Gerald Joseph
Batrin, George Leslie
Beck, James V(ere)
Beltz, Charles R(obert)
Blaser, Dwight A
Boddy, David Edwin
Bolt, Jay A(rthur)
Brehob, W(ayne) M
Caplan, John D(avid)
Chace, Milton A
Chen, Francis Hap-Kwong
Chen, Michael Ming
Chen, Yudong
Chesebrough, Harry E
Cisler, Walker L(ee)
Clark, John A(lden)
Clark, Samuel Kelly
Cloud, Gary Lee
Cole, David Edward
Colucci, Joseph M(ichael)
Datsko, Joseph
Despres, Thomas A
Dhanak, Amritlal M(aganlal)
Eaton, Robert James
Edgerton, Robert Howard
Ellis, Robert William
Eltinge, Lamont
Ezzat, Hazem Ahmed
Felbeck, David K(niseley)
Fleischer, Henry
Foss, John F
Frutiger, Robert Lester
Fyhrie, David Paul
Galan, Louis
Gibson, Harold J(ames)
Gillespie, Thomas David
Gratch, Serge
Greene, Bruce Edgar
Habib, Izzeddin Salim
Haviland, Merrill L
Hays, Donald F(rank)
Henein, Naeim A
Hrovat, Davorin
Hunstad, Norman A(llen)
Ibrahim, Raouf A
Im, Jang Hi
Ishihara, Teruo (Terry)
John, James Edward Albert
Johnson, John Harris
Jones, R(ichard) James
Kabel, Richard Harvey
Kauppila, Raymond William
Kehrl, Howard H
Keller, Robert B
Kerber, Ronald Lee
King, Albert Ignatius
Klomp, Edward
Krieger, Roger B
Kullgren, Thomas Edward
Kurajian, George Masrob
Lamberson, Leonard Roy
Lett, Philip W(ood), Jr
Li, Chin-Hsiu
Libertiny, George Zoltan
Little, Robert E(ugene)
Lord, Harold Wesley
Lubkin, James Leigh
Ludema, Kenneth C
Lumsdaine, Edward
Lund, Charles Edward
McGrath, John Joseph
Martin, George H(enry)
Merte, Herman, Jr
Mirsky, W(illiam)
Misch, Herbert Louis
Narain, Amitabh
Nefske, Donald Joseph

Nyquist, Gerald Warren
Oh, Hilario Lim
Pandit, Sudhakar Madhavrao
Passerello, Chris Edward
Pearson, J(ohn) Raymond
Pence, Thomas James
Petersen, Donald E
Peterson, Richard Carl
Petrick, Ernest N(icholas)
Petrof, Robert Charles
Piccirelli, Robert Anthony
Potter, Merle C(larence)
Prasad, Biren
Quader, Ather Abdul
Rhee, Seong Kwan
Rivard, Jerome G
Rodgers, John James
Rohde, Steve Mark
Roll, James Robert
Rollinger, Charles N(icholas)
Rosenberg, Richard Carl
Rosenberg, Ronald C(arl)
Rosinski, Michael A
Sagi, Charles J(oseph)
Schrenk, Walter John
Schultz, Albert Barry
Schwerzler, Dennis David
Segel, L(eonard)
Shapton, William Robert
Shigley, Joseph E(dward)
Shipinski, John
Shkolnikov, Moisey B
Singh, Trilochan (Hardeep)
Smith, Gene E
Smith, Hadley J(ames)
Sonntag, Richard E
Soutas-Little, Robert William
Stivender, Donald Lewis
Suryanarayana, Narasipur Venkataram
Taplin, Lael Brent
Taraman, Khalil Showky
Taulbee, Carl D(onald)
Thompson, James Lowry
Tison, Richard Perry
Van Poolen, Lambert John
Van Wylen, Gordon J(ohn)
Wang, Dazong
Wedekind, Gilbert Leroy
Whicker, Donald
Willermet, Pierre Andre
Williams, Gwynne
Wineman, Alan Stuart
Wolf, Joseph A(llen), Jr
Wu, Hai
Wu, Hofu
Wu, Shien-Ming
Yang, Wen Jei
Yu, Mason K

MINNESOTA
Abraham, John
Bailey, Fredric Nelson
Beavers, Gordon Stanley
Blackshear, Perry L(ynnfield), Jr
Chase, Thomas Richard
Daniel, John Mahendra Kumar
Davidson, Donald
Erdman, Arthur Guy
Fingerson, Leroy Malvin
Frohrib, Darrell A
Goldstein, R(ichard) J(ay)
Heberlein, Joachim Viktor Rudolf
Ibele, Warren Edward
Johnson, Robert L(awrence)
Joseph, Daniel D
Joseph, Earl Clark, II
Kallok, Michael John
Keinath, Gerald E
Keshaviah, Prakash Ramnathpur
Kittelson, David Burnelle
Knappe, Laverne F
Kulacki, Francis Alfred
Larson, Vincent H(ennix)
Li, Kam W(u)
Liu, Benjamin Y H
McElrath, Gayle W(illiam)
McMurry, Peter Howard
Marple, Virgil Alan
Ogata, Katsuhiko
Olson, Reuben Magnus
Patankar, Suhas V
Pfender, Emil
Ramalingam, Subbiah
Riley, Donald Ray
Sarkar, Kamalaksha
Shu, Mark Chong-Sheng
Simon, Terrence William
Smith, David Philip
Sparrow, E(phraim) M(aurice)
Strait, John
Uban, Stephen A

MISSISSIPPI
Altenkirch, Robert Ames
Bouchillon, Charles W
Brady, Eugene F
Brenkert, Karl, Jr
Brown, Joseph M
Carley, Charles Team, Jr
Fox, John Arthur
Jasper, Martin Theophilus
Powe, Ralph Elward
Rosenhan, A Kirk
Scott, Charley

MISSOURI
Armaly, Bassem Farid
Batra, Romesh Chander
Blundell, James Kenneth
Brooker, Donald Brown
Carter, Don E
Chen, Ta-Shen
Creighton, Donald L
Crosbie, Alfred Linden
Culp, Archie W
Dharani, Lokeswarappa R
Dreifke, Gerald E(dmond)
Duffield, Roger C
Faucett, T(homas) R(ichard)
Flanigan, Virgil James
Fowler, Timothy John
Francis, Lyman L(eslie)
Gardner, Richard A
Gillespie, LaRoux King
Guccione, Julius Matteo
Heimann, Robert L
Hines, Anthony Loring
Husar, Rudolf Bertalan
Kimel, William R(obert)
Kovacs, Sandor J, Jr
Koval, Leslie R(obert)
Lehnhoff, Terry Franklin
Leutzinger, Rudolph L(eslie)
Look, Dwight Chester, Jr
Lowe, Forrest Gilbert
Loyalka, Sudarshan Kumar
Lysen, John C
Mathiprakasam, Balakrishnan
Miles, John B(ruce)
Oetting, Robert B(enfield)
Oliver, Larry Ray
Painter, James Howard
Perisho, Clarence H(oward)
Pringle, Oran A(llan)
Remington, C(harles) R(oy), Jr
Sauer, Harry John, Jr
Schott, Jeffrey Howard
Smith, William R
Sutera, Salvatore P
Warder, Richard C, Jr
Weissenburger, Jason T
Winholtz, Robert Andrew
Young, Ross Darell

MONTANA
Albini, Frank Addison
Gerez, Victor
Herndon, Charles L(ee)
Townes, Harry W(arren)

NEBRASKA
Beatty, Millard F
Lauck, Francis W
Leviticus, Louis I
Lou, David Yeong-Suei
Lu, Pau-Chang
Ma, Wen
McDougal, Robert Nelson
Martin, Charles Wayne
Newhouse, Keith N
Peters, Alexander Robert
Rohde, Suzanne Louise
Wittke, Dayton D
Wolford, James C
Woods, Joseph

NEVADA
Anderson, James T(reat)
Boehm, Robert Foty
Brogan, William L
Chandra, Dhanesh
McKee, Robert B(ruce), Jr
Nassersharif, Bahram
Pepper, Darrell Weldon
Turner, Robert Harold
Wirtz, Richard Anthony

NEW HAMPSHIRE
Allmendinger, E Eugene
Baumann, Hans D
Black, Sydney D
Block, James A
Browne, W(illiam) H(ooper)
Colon, Frazier Paige
Feely, Frank Joseph, Jr
Fricke, Edwin Francis
Hill, Percy Holmes
Hoagland, Lawrence Clay
Jenike, Andrew W(itold)
Limbert, David Edwin
Liston, Ronald Argyle
Lopez, Guido Wilfred
Lunardini, Virgil J(oseph), Jr
Savage, Godfrey H
Spehrley, Charles W, Jr
Temple, Peter Lawrence

NEW JERSEY
Becht, Charles, IV
Ben-Israel, Adi
Berman, Arthur
Bhagat, Pramode Kumar
Brandmaier, Harold Edmund
Briggs, David Griffith
Brunell, Karl
Bush, Charles Edward
Caudill, Reggie Jackson
Cerkanowicz, Anthony Edward
Chen, Rong Yaw
Cho, Soung Moo
Connolly, John Charles
Daman, Ernest L
Davis, William F
DePalma, Anthony Michael
Domeshek, S(ol)
Donovan, Richard C
Duerr, J Stephen
Durbin, Enoch Job
Ehrlich, I(ra) Robert
Enell, John Warren
Fenster, Saul K
File, Joseph
Flaherty, Franklin Trimby, Jr
Florio, Pasquale J, Jr
Frenkiel, Richard H
Gabriel, Edwin Z
Geskin, Ernest S
Glassman, Irvin
Goyal, Suresh
Griskey, R(ichard) G(eorge)
Griswold, Edward Mansfield
Groeger, Theodore Oskar
Grunes, Robert Lewis
Harford, James J
Hazelrigg, George Arthur, Jr
Henrikson, Frank W
Hollenbach, Edwin
Hoover, Charles Wilson, Jr
Hrycak, Peter
Hsieh, Jui Sheng
Hung, Tonney H M
Imming, Harry S(tanley)
Iyer, Ram R
Jackson, Andrew
Jacobs, Carl Henry
Jaluria, Yogesh
Johnson, Kurt P
Kane, Ronald S(teven)
Kangovi, Sach
Karcher, Guido George
Khalifa, Ramzi A
Kirchner, Robert P
Kovats, Andre
Langrana, Noshir A
Leu, Ming C
Li, Hsin Lang
Lubowe, Anthony G(arner)
Luxhoj, James Thomas
McCullough, Robert William
McGregor, Walter
Magee, Richard Stephen
Mandel, Andrea Sue
Manson, Simon V(erril)
Marscher, William Donnelly
Matthews, E(dward) K
Maxwell, Bryce
Mellor, George Lincoln, Jr
Michaelis, Paul Charles
Miller, Edward
Moshey, Edward A
Mutter, William Hugh
Orszag, Steven Alan
Pae, K(ook) D(ong)
Percarpio, Edward P(eter)
Perry, Erik David
Philipp, Ronald E
Polaner, Jerome L(ester)
Prasad, Marehalli Gopalan
Pschunder, Ralph J(osef)
Raichel, Daniel R(ichter)
Rhee, Aaron Seung-Joon
Sachs, Harvey Maurice
Schlink, F(rederick) J(ohn)
Schmidlin, Albertus Ernest
Schnabel, George Joseph
Schnapf, Abraham
Segelken, John Maurice
Sernas, Valentinas A
Shah, Hasmukh N
Sharobeam, Monir Hanna
Sisto, Fernando
Socolow, Robert H(arry)
Stamper, Eugene
Stengel, Robert Frank
Suhir, Ephraim
Temkin, Samuel
Thangam, Siva
Thomas, Stanislaus S(tephen)
Thurston, Robert Norton
Weinzimmer, Fred
Wells, Herbert Arthur
Werner, Daniel Paul
Wesner, John William
Williams, Maryon Johnston, Jr
Wilson, Charles Elmer
Young, Li
Yu, Yi-Yuan
Zebib, Abdelfattah M G

NEW MEXICO
Bertin, John Joseph
Blackwell, Bennie Francis
Blejwas, Thomas Edward
Bryant, Lawrence E, Jr
Burchett, O Neill J
Burick, Richard Joseph
Clouser, William Sands
Cobble, Milan Houston
Diver, Richard Boyer, Jr
Dunn, Richard B
Ebbesen, Lynn Royce
Ebrahimi, Nader Dabir
El-Genk, Mohamed Shafik

Ellett, D Maxwell
Ford, C(larence) Quentin
Garcia, Carlos E(rnesto)
Gibbs, Terry Ralph
Gilbert, Joel Sterling
Harary, Frank
Hardin, James T
Hsu, Y(oun) C(hang)
Jones, Orval Elmer
Kashiwa, Bryan Andrew
Kelley, Gregory M
Kimbrell, Jack T(heodore)
Lam, Kin Leung
Lawrence, Raymond Jeffery
Lee, David Oi
Lowe, Terry Curtis
Matthews, Larryl Kent
Metzger, Daniel Schaffer
Moulds, William Joseph
Mulholland, George
O'Rourke, Peter John
Philbin, Jeffrey Stephen
Powell, R(ay) B(edenkapp)
Reynolds, Stephen Edward
Sheffield, Stephen Anderson
Sheinberg, Haskell
Shouman, A(hmad) R(aafat)
Stephans, Richard A
Straight, James William
Sullivan, J Al
Thurston, Rodney Sundbye
Touryan, Kenell James
Travis, John Richard
Von Riesemann, Walter Arthur
Wawersik, Wolfgang R
Wildin, Maurice W(ilbert)
Worstell, Hairston G(eorge)
Yeh, Hsu-Chi

NEW YORK
Adams, Donald E
Agosta, Vito
Ahmadi, Goodarz
Aronson, Herbert
Auer, Peter Louis
Banerjee, Prasanta Kumar
Barrows, John Frederick
Bartel, Donald L
Batter, John F(rederic), Jr
Baum, Richard T
Bayer, Raymond George
Begell, William
Bennett, Gerald William
Bennett, Leon
Benson, Richard Carter
Bergles, Arthur E(dward)
Birnstiel, Charles
Blesser, William B
Boley, Bruno Adrian
Bolton, Joseph Aloysius
Booker, J(ohn) F
Brown, Dale H(enry)
Brownlow, Roger W
Burns, Stephen James
Busnaina, Ahmed A
Cess, Robert D
Chandler, Jasper S(chell)
Chang, Ching Ming
Chang, Shih-Yung
Chen, Alek Chi-Heng
Cheng, Yean Fu
Chiang, Fu-Pen
Chinitz, Wallace
Clark, Alfred, Jr
Coffin, Louis F(ussell), Jr
Conta, Bart(holomew) J(oseph)
Crooke, James Stratton
Csermely, Thomas J(ohn)
Dagan, Zeev
Dahneke, Barton Eugene
Dastin, Samuel J
De Boer, P(ieter) C(ornelis) Tobias
Deresiewicz, Herbert
Deshpande, Narayan V(aman)
Dibelius, Norman Richard
Dizer, John Thomas, Jr
Drucker, E(ugene) E(lias)
Eisenstadt, Raymond
Ellson, Robert A
Endsley, L(ouis) E(ugene), Jr
Engel, Peter Andras
Ezra, Arthur Abraham
Femenia, Jose
Freudenstein, Ferdinand
Fried, Erwin
Fritz, Robert J(acob)
Fuller, Dudley D(ean)
Gans, Roger Frederick
Garmezy, R(obert) H(arper)
Gazda, I(rving) W(illiam)
George, A(lbert) R(ichard)
George, William Kenneth, Jr
Gibbon, Norman Charles
Ginsberg, Theodore
Givi, Peyman
Goldmann, Kurt
Goldspiel, Solomon
Graebel, William P(aul)
Granet, Irving
Greene, Peter Richard
Grisoli, John Joseph
Grossman, Perry L
Gupta, Pradeep Kumar
Haag, Fred George

Haid, D(avid) A(ugustus)
Haines, Daniel Webster
Hall, Richard Travis
Handman, Stanley E
Harper, Kennard W(atson)
Harvey, Douglass Coate
Heimburg, Richard W
Hendrie, Joseph Mallam
Hetnarski, Richard Bozyslaw
Hoenig, Alan
Hollander, Milton B(ernard)
Hollenberg, Joel Warren
Holmes, Philip John
Hussain, Moayyed A
Hutchings, William Frank
Hwang, S Steve
Hwang, Shy-Shung
Imber, Murray
Irvine, Thomas Francis
Jedruch, Jacek
Jefferies, Michael John
Jensen, Richard Alan
Jones, Charles
Jwo, Chin-Hung
Kapila, Ashwani Kumar
Karlekar, Bhalchandra Vasudeo
Kayani, Joseph Thomas
Kear, Edward B, Jr
Kelly, Thomas Joseph
Kempner, Joseph
Kenyon, Richard A(lbert)
Kessler, Thomas J
Ketchum, Gardner M(ason)
Kramer, Aaron R
Krauter, Allan Irving
Krishna, C R
Kroeger, Peter G
Kuchar, Norman Russell
Kyanka, George Harry
Lahey, Richard Thomas, Jr
Landzberg, Abraham H(arold)
La Russa, Joseph Anthony
Laskaris, Trifon Evangelos
Lauber, Thornton Stuart
Lee, H(o) C(hong)
Lee, Richard Shao-Lin
Lessen, Martin
Levitsky, Myron
Li, Chung-Hsiung
Li, J(ames) C(hen) M(in)
Libove, Charles
Loeber, John Frederick
Loewen, Erwin G
Longman, Richard Winston
Longobardo, Anna Kazanjian
Longobardo, Guy S
Lowen, Gerard G
Lowen, W(alter)
Lumley, John L(eask)
Lyman, Frederic A
McCalley, Robert B(ruce), Jr
Maclean, Walter M
Martens, Hinrich R
Mates, Robert Edward
Menkes, Sherwood Bradford
Miller, Christine E
Mochel, Myron George
Montag, Mordechai
Moore, Franklin K(ingston)
Mosher, Melville Calvin
Murphy, Eugene F(rancis)
Naar, Raymond Zacharias
Namer, Izak
Osborn, Henry H(ooper)
Othmer, Donald F(rederick)
Pan, Huo-Hsi
Perkins, Kenneth Roy
Perkins, Richard W, Jr
Phelan, R(ichard) M(agruder)
Pipes, Robert Byron
Pita, Edward Gerald
Podowski, Michael Zbigniew
Purdy, D(avid) C(arl)
Quesnel, David John
Rae, William J
Raj, Rishi S
Reber, Raymond Andrew
Rice, John T(homas)
Rohatgi, Upendra Singh
Ross, Donald Edward
Rungta, Ravi
Sancaktar, Erol
Sanders, W(illiam) Thomas
Scher, Robert Sander
Schroeder, John
Schwartzman, Leon
Sessler, John George
Sforza, Pasquale M
Shaffer, Bernard W(illiam)
Shah, Ramesh Keshavlal
Shanebrook, J(ohn) Richard
Shapiro, Paul Jonathon
Shepherd, D(ennis) G(ranville)
Sherman, Zachary
Shube, Eugene E
Simon, Albert
Sneck, Henry James, (Jr)
Somerscales, Euan Francis Cuthbert
Somerset, James H
Sri-Jayanthal, Sri Muthythamby
Srinivasan, Vijay
Staub, Fred W
Sternlicht, B(eno)
Sumner, Eric E

Tabi, Rifat
Talapatra, Dipak Chandra
Tan, Chor-Weng
Thompson, Robert Alan
Thorsen, Richard Stanley
Torrance, Kenneth E(ric)
Tourin, Richard Harold
Triantafyllou, George S
Vafakos, William P(aul)
Valentine, Daniel T
Vassallo, Franklin A(llen)
Velzy, Charles O
Vidosic, J(oseph) P(aul)
Virk, Kashmir Singh
Voelcker, Herbert B(ernhard)
Vohr, John H
Vopat, William A
Wagner, John George
Wallenfels, Miklos
Wang, Hsin-Pang
Wang, Kuo King
Wang, Lin-Shu
Watkins, Charles B
Wehe, Robert L(ouis)
Weinbaum, Sheldon
Wenig, Harold G(eorge)
Wetherhold, Robert Campbell
Whiteside, James Brooks
Wiese, Warren M(elvin)
Wiggins, Edwin George
Willmert, Kenneth Dale
Wilson, Merle R(obert)
Wolkoff, Harold
Wright, Roger Neal
Wulff, Wolfgang
Zimmer, G(eorge) A(rthur)
Zubaly, Robert B(arnes)
Zweig, John E

NORTH CAROLINA
Barrett, Rolin F(arrar)
Bejan, Adrian
Block, M Sabel
Boyers, Albert Sage
Brown, Theodore Cecil
Burton, Ralph A(shby)
Chaddock, Jack B
Chaplin, Frank S(prague)
Childers, Robert Warren
Clark, Charles Edward
Cleland, John Gregory
Doster, Joseph Michael
Dow, Thomas Alva
Eberhardt, Allen Craig
Eckerlin, Herbert Martin
Elsevier, Ernest
El-Shiekh, Aly H
Garg, Devendra Prakash
Gupta, Ajaya Kumar
Hale, Francis Joseph
Hall, Peter M
Harman, Charles M(organ)
Hill, Herbert Hewett
Hirshburg, Robert Ivan
Hochmuth, Robert Milo
Hurt, Alfred B(urman), Jr
Jennings, Burgess H(ill)
Jordan, Edward Daniel
Keltie, Richard Francis
Kim, Rhyn H(yun)
Krassowska, Wanda
Kusy, Robert Peter
Lalor, William Francis
McDonald, P(atrick) H(ill), Jr
McElhaney, James Harry
Martin, David E(dwin)
Min, Tony C(harles)
Neal, C Leon
Odman, Mehmet Talat
Ozisik, M Necati
Pearsall, George W(ilbur)
Petroski, Henry J
Reece, Joe Wilson
Sandor, George N(ason)
Smetana, Frederick Otto
Sorrell, Furman Y(ates), Jr
Stephenson, Harold Patty
Sud, Ish
Torquato, Salvatore
Yamamoto, Toshiaki
Zorowski, Carl F(rank)

NORTH DAKOTA
Bibel, George
Mathsen, Don Verden
Pfister, Philip Carl
Rieder, William G(ary)

OHIO
Adams, Otto Eugene, Jr
Alam, M Khairul
Altan, Taylan
Anderson, William J
Andonian, Arsavir Takfor
Bagby, Frederick L(air)
Bahnfleth, Donald R
Bahniuk, Eugene
Bahr, Donald Walter
Barchfeld, Francis John
Beans, Elroy William
Bechtel, Stephen E
Berthold, John William, III
Bhattacharya, Amit
Bhushan, Bharat

Bisson, Edmond E(mile)
Biswas, Pratim
Bobo, Melvin
Bolz, Harold A(ugust)
Boulger, Francis W(illiam)
Bovenkerk, Harold P(aul)
Brown, Max Louis
Brush, John (Burke)
Buttner, F(rederick) H(oward)
Chakko, Mathew K(anjhirathinkal)
Chawla, Mangal Dass
Chu, Mamerto Loarca
Chuang, Henry Ning
Chuck, Leon
Chung, Benjamin T F
Ciricillo, Samuel F
Collins, Jack A(dam)
Damianov, Vladimir B
D'isa, Frank Angelo
Domholdt, Lowell Curtis
Eibling, James A(lexander)
Engdahl, Richard Bott
Faulkner, Lynn L
Field, Michael
Franke, Milton Eugene
Freche, John C(harles)
Friar, Billy W(ade)
Frink, Donald W
Fritsch, Charles A(nthony)
Gabb, Timothy Paul
Gerhardt, Jon Stuart
Glaeser, William A(lfred)
Glasgow, John Charles
Glower, Donald D(uane)
Gorla, Rama S R
Gorland, Sol H
Graham, Robert William
Greber, Isaac
Greene, Neil E(dward)
Gregory, R(obert) Lee
Gunasekera, Jay Sarath
Hall, Glenn Eugene
Hamed, Awatef A
Hardy, Richard Allen
Hazard, Herbert Ray
Heath, George L
Hemsworth, Martin C
Herbert, Thorwald
Ho, Fanghuai H(ubert)
Horton, Billy Mitchusson
Houser, Donald Russell
Hursh, Robert W(illiam)
Huss, Harry O(tto)
Huston, Ronald L
Irey, Richard Kenneth
Jaikrishnan, Kadambi Rajgopal
Jain, Vinod Kumar
Janzow, Edward F(rank)
Jeffries, Neal Powell
Jehn, Lawrence Andrew
Jeng, Duen-Ren
Johnson, Ralph Sterling, Jr
Kalivoda, Frank E, Jr
Keim, John Eugene
Keith, Theo Gordon, Jr
Kepler, Harold B(enton)
Kerka, William (Frank)
Keshock, Edward G
Kinzel, Gary Lee
Klosterman, Albert Leonard
Krysiak, Joseph Edward
Lahoti, Goverdhan Das
Lai, Ying-San
La Mers, Thomas Herbert
Landstrom, D(onald) Karl
Lawler, Martin Timothy
Lee, Peter Wankyoon
Leissa, A(rthur) W(illiam)
Lenhart, Jack G
Lestingi, Joseph Francis
Lewandowski, John Joseph
Lundin, Bruce T(heodore)
Macke, H(arry) Jerry
Maier, Leo Robert, Jr
Mall, Shankar
Marshall, Kenneth D
Mayers, Richard Ralph
Menq, Chia-Hsiang
Merchant, Mylon Eugene
Messinger, Richard C
Miller, William B
Miller, William Ralph
Morgan, William R(ichard)
Morse, Ivan E, Jr
Mularz, Edward Julius
Murphy, Michael John
Orndorff, Roy Lee, Jr
Ostrach, Simon
Peckham, P Hunter
Pinchak, Alfred Cyril
Powell, Gordon W
Prahl, Joseph Markel
Putnam, Abbott (Allen)
Ramalingam, Mysore Loganathan
Reddy, Narender Pabbathi
Reshotko, Eli
Rich, Joseph William
Richley, E(dward) A(nthony)
Rittgers, Stanley Earle
Robinson, Richard Alan
Rockstroh, Todd Jay
Rodman, Charles William
Sanders, John Claytor
Savage, Michael

Mechanical Engineering (cont)

Scavuzzo, Rudolph J, Jr
Schauer, John Joseph
Schleef, Daniel J
Schoch, Daniel Anthony
Sepsy, Charles Frank
Serafini, John S
Shine, Andrew J(oseph)
Siegel, Robert
Singh, Rajendra
Smith, Howard Edwin
Smith, L(eroy) H(arrington), Jr
Smith, Marion L(eroy)
Soni, Atmaram Harilal
Southam, Donald Lee
Srivatsan, Tirumalai Srinivas
Starkey, Walter L(eroy)
Sutton, George E
Tarantine, Frank J(ames)
Thompson, Hugh Ansley
Thorpe, James F(ranklin)
Tsao, Bang-Hung
Tse, Francis S
Voisard, Walter Bryan
Voorhees, John E
Waldron, Kenneth John
Waller, Michael Holton
Webster, Allen E
Whitacre, Gale R(obert)
White, John Joseph, III
Wilkinson, William H(adley)
Winiarz, Marek Leon
Zaman, Khairul B M Q
Zavodney, Lawrence Dennis

OKLAHOMA
Appl, Franklin John
Bennett, A(rthur) D(avid)
Bert, Charles Wesley
Blenkarn, Kenneth Ardley
Boggs, James H(arlow)
Chapel, Raymond Eugene
Cornelius, Archie J(unior)
Earls, James Roe
Egle, Davis Max
Emanuel, George
Fitch, Ernest Chester, Jr
Forney, Bill E
Gitzendanner, L G
Gollahalli, Subramanyam Ramappa
Guerrero, E T
Heath, Larman Jefferson
Hurn, R(ichard) W(ilson)
Joshi, Sadanand D
Lilley, David Grantham
Love, Tom Jay, Jr
Lowery, R(ichard) L
McQuiston, Faye C(lem)
Moretti, Peter M
Neathery, Raymond Franklin
Parker, Jerald D
Reid, Karl Nevelle, Jr
Riley, Richard King
Rybicki, Edmund Frank
Sheth, Ketankumar K
Striz, Alfred Gerhard
Upthegrove, W(illiam) R(eid)
Weston, Kenneth Clayton
Zigrang, Denis Joseph
Zirkle, Larry Don

OREGON
Blickensderfer, Robert
Boubel, Richard W
Calder, Clarence Andrew
Devletian, Jack H
Fanger, Carleton G(eorge)
Hansen, Hugh J
Larson, M(ilton) B(yrd)
Light, Kenneth Freeman
Migliore, Herman James
Perry, Robert W(illiam)
Popovich, M(ilosh)
Reistad, Gordon Mackenze
Reiter, William Frederick, Jr
Savery, Clyde William
Scanlan, J(ack) A(ddison), Jr
Thornburgh, George E(arl)
Welty, James Richard
Zaworski, R(obert) J(oseph)

PENNSYLVANIA
Advani, Sunder
Akay, Adnan
Amon Parisi, Cristina Hortensia
Antes, Harry W
Avitzur, Betzalel
Ayyaswamy, Portonovo S
Baish, James William
Bau, Haim Heinrich
Baumann, Dwight Maylon Billy
Benner, Russell Edward
Berkof, Richard Stanley
Bridenbaugh, Peter R
Brighton, John Austin
Brown, Forbes Taylor
Brunner, Mathias J
Bucci, Robert James
Burnham, Donald C
Caporali, Ronald Van
Carlson, H(enning) Maurice
Carmi, Shlomo
Catanach, Wallace M, Jr

Cernansky, Nicholas P
Chan, Siu-Kee
Chang, C Hsiung
Chigier, Norman
Chou, Pei Chi (Peter)
Chuang, Strong Chieu-Hsiung
Cohen, Ira M
Confer, Gregory Lee
Cook, Charles S
Curlee, Neil J(ames), Jr
Daum, Donald Richard
Devine, John Edward
Dillio, Charles Carmen
Dougall, Richard S(tephen)
Dunn, J(ohn) Howard
Everett, James Legrand, III
Falconer, Thomas Hugh
Farouk, Bakhtier
Fischer, Edward G(eorge)
Fong, James T(se-Ming)
Fukushima, Toshiyuki
Fullerton, H(erbert) P(almer)
Gaylord, Eber William
Geiger, Gene E(dward)
Goel, Ram Parkash
Goldberg, Joseph
Griffin, Jerry Howard
Gwiazda, S(tanley) J(ohn)
Hansen, Robert Jack
Harvey, Frances J, II
Heinsohn, Robert J(ennings)
Henderson, Robert E
Herrick, Joseph Raymond
Holl, J(ohn) William
Hughes, Arthur D(ouglas)
Hughes, William F(rank)
Hull, John Laurence
Hyatt, Robert Monroe
Ibrahim, Baky Badie
Jacobs, Harold Robert
Juster, Allan
Kameneva, Marina Vitaly
Karlovitz, Bela
Kashkoush, Ismail I
Kaufman, Howard Norman
Keyser, David Richard
Keyt, Donald E
Khonsari, Michael M
Kim, Hong S
Kingrea, James I, Jr
Kinnavy, M(artin) G(erald)
Klaus, Ronald Louis
Koch, Ronald N
Koopmann, Gary Hugo
Kowal, George M
Kroon, R(einout) P(ieter)
Kuhn, Howard A
Kurfess, Thomas Roland
Lakshminarayana, B
Lavanchy, Andre C(hristian)
Lavin, J Gerard
Lawrence, Kenneth Bridge
Lazaridis, Anastas
Leidy, Blaine I(rvin)
Levy, Bernard, Jr
Levy, Edward Kenneth
Lior, Noam
Long, Lyle Norman
Lorsch, Harold G
Loxley, Thomas Edward
Lucas, Robert Alan
McAssey, Edward V, Jr
McLaughlin, Philip V(an Doren), Jr
Macpherson, Alistair Kenneth
Mahaffy, John Harlan
Manjoine, Michael J(oseph)
Marston, Charles H
Matulevicius, Edward S(tephen)
May, Harold E(dward)
Mehta, Kishor Singh
Mendler, Oliver J(ohn)
Mercer, Samuel, Jr
Meredith, David Bruce
Merz, Richard A
Metaxas, Dimitri
Meyer, Wolfgang E(berhard)
Molyneux, John Ecob
Morrill, Bernard
Novak, Stephen Robert
Ochs, Stefan A(lbert)
Olson, Donald Richard
Orr, Leighton
Osterle, J(ohn) F(letcher)
Owczarek, J(erzy) A(ntoni)
Paolino, Michael A
Park, William H
Paul, B(urton)
Pfennigwerth, Paul Leroy
Podgers, Alexander Robert
Progelhof, Richard Carl
Purdy, David Lawrence
Putman, Thomas Harold
Raimondi, Albert Anthony
Rajagopal, K R
Ramezan, Massood
Ramos, Juan Ignacio
Reethof, Gerhard
Remick, Forrest J(erome)
Roberts, Richard
Robertson, Donald
Rodini, Benjamin Thomas, Jr
Rodrigues, Anil Noel
Rohrer, Wesley M, Jr
Rose, Joseph Lawrence

Ross, Arthur Leonard
Rouleau, Wilfred T(homas)
Rubin, Edward S
Rumbarger, John H
Satz, Ronald Wayne
Schmidt, Alfred Otto
Schmidt, Frank W(illiam)
Schneider, Robert W(illiam)
Settles, Gary Stuart
Shearer, J(esse) Lowen
Silverstein, Calvin C(arlton)
Sinclair, Glenn Bruce
Solt, Paul E
Sommer, Holger Thomas
Spielvogel, Lawrence George
Stokey, W(illiam) F(armer)
Stradling, Lester J(ames), Jr
Tallian, Tibor E(ugene)
Thomas, Donald H(arvey)
Treadwell, Kenneth Myron
Tsu, T(sung) C(hi)
Tuba, I Stephen
Walker, James David Allan
Wallace, Paul William
Wambold, James Charles
Webb, Ralph L
Weindling, Joachim I(gnace)
Weinstock, Manuel
Whitman, Alan M
Wiebe, Donald
Williams, Albert J, Jr
Williams, Max L(ea), Jr
Wolgemuth, Carl Hess
Woo, Savio Lau-Yuen
Woodburn, Wilton A
Wright, Dexter V(ail)
Wu, John Naichi
Yeh, Hsuan
Young, Frederick J(ohn)
Zaiser, James Norman
Zauderer, Bert
Zhou, Qian
Zhu, Dong
Zimmermann, F(rancis) J(ohn)

RHODE ISLAND
Boothroyd, Geoffrey
Bradbury, Donald
Brady, John F
Butler, Howard W(allace)
Dahl, Norman C(hristian)
Dewhurst, Peter
Dobbins, Richard Andrew
Dunlap, Richard M(orris)
Fernandes, John Henry
Kestin, J(oseph)
Kim, Thomas Joon-Mock
Koch, Robert Michael
Lessmann, Richard Carl
Liu, J(oseph) T(su) C(hieh)
McEligot, Donald M(arinus)
Nadolink, Richard Hughes
Nash, Charles Dudley, Jr
Ng, Kam Wing
Palm, William John, III
Richardson, Peter Damian
Sibulkin, Merwin
Spero, Caesar A(nthony), Jr
Test, Frederick L(aurent)
Thielsch, Helmut
White, Frank M

SOUTH CAROLINA
Awadalla, Nabil G
Bishop, Eugene H(arlan)
Byars, Edward F(ord)
Chao, Yuh J (Bill)
Christy, William O(liver)
Drane, John Wanzer
Elrod, Alvon Creighton
Flipse, John E
Garzon, Ruben Dario
Graulty, Robert Thomas
Herdklotz, Richard James
Hester, Jarrett Charles
Irwin, Lafayette K(ey)
Jens, Wayne H(enry)
Joseph, J Walter, Jr
McCarter, James Thomas
McMillan, Harry King
Overcamp, Thomas Joseph
Paul, Frank W(aters)
Przirembel, Christian E G
Rudisill, Carl Sidney
Sandrapaty, Ramachandra Rao
Townes, George Anderson
Turner, John Lindsey
Uldrick, John Paul
Ward, David Aloysius
Westerman, William Joseph, II
Zumbrunnen, David Arnold

SOUTH DAKOTA
Chiang, Chao-Wang
Scofield, Gordon L(loyd)

TENNESSEE
Attaway, Cecil Reid
Baker, Merl
Barnett, Paul Edward
Boudreau, William F(rancis)
Brown, Bevan W, Jr
Bundy, Robert W(endel)
Burton, Robert Lee

Cox, Ronald Baker
Deivanayagan, Subramanian
Eckstein, Eugene Charles
Edmondson, Andrew Joseph
Flandro, Gary A
Fraas, Arthur P(aul)
Frost, Walter
Gan, Rong Zhu
Garrison, George Walker
Gat, Uri
Gilbert, R(obert) J(ames)
Gray, William Harvey
Greenstreet, William (B) LaVon
Hodgson, J(effrey) William
Holland, Ray W(alter)
Houghton, James Richard
Iskander, Shafik Kamel
Isley, James Don
Janna, William Sied
Johnson, William S(tanley)
Kelso, Richard Miles
Kerr, Hugh Barkley
Kidd, George Joseph, Jr
Kinyon, Brice W(hitman)
Kress, Thomas Sylvester
Lim, Alexander Te
Martin, Harry Lee
Maxwell, Robert L(ouis)
Mellor, Arthur M(cLeod)
Milligan, Mancil W(ood)
Mullikin, H(arwood) F(ranklin)
Pandeya, Prakash N
Patterson, Malcolm Robert
Pih, Hui
Pitts, Donald Ross
Potter, John Leith
Pugh, Claud Ervin
Ray, John Delbert
Scherpereel, Donald E
Sissom, Leighton E(sten)
Slade, Edward Colin
Snyder, William Thomas
Sorrells, Frank Douglas
Stair, William K(enneth)
Sun, Pu-Ning
Sundaram, R Meenakshi
Swim, William B(axter)
Swindeman, Robert W
Swisher, George Monroe
Tunnell, William C(lotworthy)
Ventrice, Marie Busck
Wasserman, Jack F
Williamson, John W
Witten, Alan Joel
Wright, William F(red)
Young, Robert L(yle)
Zuhr, Raymond Arthur

TEXAS
Anderson, Dale Arden
Anderson, Edward Everett
Athanasiou, Kyriacos A
Baber, Burl B, Jr
Baker, Wilfred E(dmund)
Ball, Kenneth Steven
Barbin, Allen R
Barker, C(alvin) L R
Barnes, Allen Lawrence
Bergman, Theodore L
Bland, William M, Jr
Bourell, David Lee
Bradley, Walter L
Brown, Howard (Earl)
Brown, Otto G
Bryant, Michael David
Burger, Christian P
Chapman, Alan J(esse)
Cheatham, John B(ane), Jr
Chien, Sze-Foo
Chyu, Ming-Chien
Clark, L(yle) G(erald)
Cooper, John C, Jr
Crawford, Richard Haygood, Jr
Cunningham, Robert Ashley
Dalton, Charles
Deily, Fredric H(arry)
Diller, Kenneth Ray
Dodge, Franklin Tiffany
Eberhart, Robert Clyde
Eyrick, Theodore R
Feltz, Donald Everett
Fewel, Kenneth Jack, Jr
Finch, Robert D
Fletcher, Leroy S(tevenson)
Flowers, Daniel F(ort)
Friberg, Emil Edwards
Glaspie, Donald Lee
Godwin, James Basil, Jr
Goodzeit, Carl Leonard
Graff, William J(ohn), Jr
Graiff, Leonard B(aldine)
Gully, John Houston
Gupton, Paul Stephen
Haji-Sheikh, Abdolhossein
Harkey, Jack W
Hebert, Joel J
Helmers, Donald Jacob
Henderson, Gregory Wayne
Hesse, Walter J(ohn)
Hoffman, Herbert I(rving)
Holman, Jack Philip
Hrkel, Edward James
Hussain, A K M Fazle
Jenkins, Lawrence E

Johnson, David B
Jones, Jerold W
Jones, Marvin Richard
Jonsson, John Erik
Jordan, Duane Paul
Juricic, Davor
Kanninen, Melvin Fred
Kearns, Robert William
Kreisle, Leonardt F(erdinand)
Kunze, Otto Robert
Lamb, J(amie) Parker, Jr
Lawrence, James Harold, Jr
Lawrence, Kent L(ee)
Lengel, Robert Charles
Lewis, James Pettis
Lienhard, John H(enry)
Ling, F(rederick) F(ongsun)
Lioi, Anthony Pasquale
Liu, Kai
Lu, Frank Kerping
McBee, Frank W(ilkins), Jr
Mack, Russel Travis
Marshek, Kurt M
Martinez, Jose E(dwardo)
Mistree, Farrokh
Moon, Tessie Jo
Mouritsen, T(horvald) Edgar
Muster, Douglas Frederick
Ochiai, Shinya
Packman, Paul Frederick
Page, Robert Henry
Panton, R(onald) L(ee)
Perry, John Vivian, Jr
Peterman, David A
Plapp, John E(lmer)
Pope, Larry Debbs
Powell, Alan
Powers, John Michael
Powers, Louis John
Pray, Donald George
Purvis, Merton Brown
Rabins, Michael J
Rand, James Leland
Reddy, Junuthula N
Reynolds, Elbert Brunner, Jr
Rhodes, Allen Franklin
Richardson, Herbert Heath
Rodenberger, Charles Alvard
Rylander, Henry Grady, Jr
Sadler, Stanley Gene
Schmidt, P(hilip) S(tephen)
Short, Byron Elliott
Simmang, C(lifford) M(ax)
Simonis, John Charles
Sinclair, A Richard
Slotta, Larry Stewart
Solberg, Ruell Floyd, Jr
Spanos, Pol Dimitrios
Springer, Karl Joseph
Staph, Horace E(ugene)
Strange, Lloyd K(eith)
Swope, Richard Dale
Teller, Cecil Martin, II
Tesar, Delbert
Tien, John Kai
Traver, Alfred Ellis
Treat, Charles Herbert
Turner, William Danny
Vance, John Milton
Vitkovits, John A(ndrew)
Vliet, Gary Clark
Waid, Margaret Cowsar
Wang, Chao-Cheng
Weldon, William Forrest
Wenzel, Alexander B
Westkaemper, John C(onrad)
Weynand, Edmund E
Wick, Robert S(enters)
Widmer, Robert H
Wilhelm, Wilbert Edward
Wilson, Claude Leonard
Winnie, Dayle David
Witte, Larry C(laude)
Wood, Charles D
Wood, Kristin Lee
Woolf, J(ack) R(oyce)
Young, Dana
Young, H(enry) Ben, Jr

UTAH
Andersen, Blaine Wright
Cannon, John N(elson)
DeVries, K Lawrence
Fletcher, Thomas Harvey
Free, Joseph Carl
Germane, Geoffrey James
Glenn, Alan Holton
Green, Sidney J
Heaton, Howard S(pring)
Hoeppner, David William
Holdredge, Russell M
Isaacson, Lavar King
Johnson, Arlo F
Limpert, Rudolf
MacGregor, Douglas
Moser, Alma P
Nairn, John Arthur
Rotz, Christopher Alan
Sandquist, Gary Marlin
Silver, Barnard Stewart
Snow, Donald Ray
Strozier, James Kinard
Turley, Richard Eyring
Warner, Charles Y

Zeamer, Richard Jere

VERMONT
Dean, Robert Charles, Jr
Francis, Gerald Peter
Hundal, Mahendra S(ingh)
Japikse, David
Keller, Tony S
Outwater, John O(gden)
Pope, Malcolm H
Tartaglia, Paul Edward
Tuthill, Arthur F(rederick)
Von Turkovich, Branimir F(rancis)
Wallace, Donald Macpherson, Jr

VIRGINIA
Allaire, Paul Eugene
Anderson, Willard Woodbury
Ash, Robert Lafayette
Bauer, Douglas Clifford
Beach, Harry Lee, Jr
Beach, Robert L
Beard, James Taylor
Beran, Robert Lynn
Betti, John A
Bishara, Michael Nageeb
Brooks, Thomas Furman
Bushnell, Dennis Meyer
Calm, James M
Carpenter, James E(ugene)
Chappelle, Thomas W
Charvonia, David Alan
Chen, Leslie H(ung)
Comparin, Robert A(nton)
Cooley, William C
Cooper, Henry Franklyn, Jr
Cox, J E
Davidson, John Richard
Deddens, J(ames) C(arroll)
Diller, Thomas Eugene
Dutt, Gautam Shankar
Edyvane, John
Eiss, Norman Smith, Jr
Ell, William M
Fei, Ding-Yu
Flack, Ronald Dumont, Jr
Foster, Eugene L(ewis)
Fuller, Christopher Robert
Gorges, Heinz A(ugust)
Gouse, S William, Jr
Grant, John Wallace
Grunder, Hermann August
Gunter, Edgar Jackson, Jr
Hall, Carl W(illiam)
Hallanger, Lawrence William
Hamrick, Joseph Thomas
Hasenfus, Harold J(oseph)
Hill, Floyd Isom
Hoffman, Kenneth Charles
Hou, Gene Jean-Win
Inman, Daniel John
Jacobs, Theodore Alan
Johnson, David Harley
Johnson, Glen Eric
Jones, J(ames) B(everly)
Kapania, Rakesh Kumar
Kauzlarich, James J(oseph)
Larsen-Basse, Jorn
Layson, William M(cIntyre)
Leonard, Robert Gresham
Levis, Alexander Henry
Mabie, Hamilton Horth
McGean, Thomas J
Manfredi, Arthur Frank, Jr
Mehta, Gurmukh D
Mindak, Robert Joseph
Moen, Walter B(oniface)
Moore, James W(allace) I
Moore, Robert E(dward)
Morgan, Charles D(avid)
Moses, Hal Lynwood
Ng, Wing-Fai
Ramberg, Steven Eric
Raney, J(ohn) P(hilip)
Reddy, J Narasimh
Reese, Ronald Malcolm
Roberts, A(lbert) S(idney), Jr
Roco, Mihail Constantin
Rokni, Mohammad Ali
Rosenthal, F(elix)
Roy, Gabriel D
Russell, John Alvin
Savage, William F(rederick)
Schonstedt, Erick O(scar)
Sensenyi, Paul Edward
Silveira, Milton Anthony
Smith, Carey Daniel
Snyder, Glenn J(acob)
Stein, Bland Allen
Stiegler, T(heodore) Donald
Strawn, Oliver P(erry), Jr
Szeless, Adorjan Gyula
Tompkins, Stephen Stern
Townsend, Miles Averill
Wang, C(hiao) J(en)
Weyler, Michael E
Whitelaw, R(obert) L(eslie)
Wilkinson, Thomas Lloyd, Jr
Wolko, Howard Stephen
Wood, Albert D(ouglas)
Wood, Houston Gilleylen, III
Yates, Charlie Lee
Yates, Edward Carson, Jr
Young, Maurice Isaac

WASHINGTON
Arenz, Robert James
Barbosa-Canovas, Gustavo Victor
Beaubien, Stewart James
Birkebak, Richard C(larence)
Bruckner, Adam Peter
Buonamici, Rino
Bush, S(pencer) H(arrison)
Childs, Morris E
Claudson, T(homas) T(ucker)
Crain, Richard Willson, Jr
Cross, Edward F
Currah, Walter E
Cykler, John Freuler
Day, Emmett E(lbert)
Depew, Creighton A
Ding, Jow-Lian
Drumheller, Kirk
Emery, Ashley F
Firey, Joseph Carl
Forster, Fred Kurt
Gajewski, W(alter) M(ichael)
Galle, Kurt R(obert)
Gessner, Frederick B(enedict)
Hale, Leonard Allen
Hamilton, C Howard
Jacobson, John Obert
Johnson, B(enjamin) M(artineau)
Johnson, Stephen Thomas
Jorgensen, Jens Erik
Kemper, Robert Schooley, Jr
Kippenhan, C(harles) J(acob)
Kobayashi, Albert S(atoshi)
LaFontaine, Thomas E
Leavenworth, Howard W, Jr
Levinson, Mark
McCormick, Norman Joseph
McFeron, D(ean) E(arl)
McMurtrey, Lawrence J
Madden, James H(oward)
Masden, Glenn W(illiam)
Matte, Joseph, III
Merchant, Howard Carl
Mills, B(lake) D(avid), Jr
Paynter, Gerald C(lyde)
Pratt, David Terry
Ramulu, Mamidala
Renkey, Edmund Joseph, Jr
Richmond, William D
Rieck, H(enry) G(eorge)
Robbins, Richard J
Robel, Stephen B(ernard)
Sanders, Joan Elizabeth
Sandwith, Colin John
Sengupta, Gautam
Shank, Maurice E(dwin)
Spencer, Max M(arlin)
Stock, David Earl
Taggart, Raymond
Tam, Patrick Yui-Chiu
Taya, Minoru
Trent, Donald Stephen
Viggers, Robert F
Wolak, Jan
Yore, Eugene Elliott

WEST VIRGINIA
Chiang, Han-Shing
Clark, Nigel Norman
Kuhlman, John Michael
Kumar, Alok
Lantz, Thomas Lee
Mucino, Victor H
Nelson, Leonard C(arl)
Shuck, Lowell Zane
Smith, James Earl
Sneckenberger, John Edward
Steinhardt, Emil J
Vucich, M(ichael) G(eorge)
Wilson, John Sheridan

WISCONSIN
Balmer, Robert Theodore
Barrington, Gordon P
Beachley, Norman Henry
Beckman, William A
Blanchard, James Page
Bollinger, John G(ustave)
Bonazza, Riccardo
Borman, Gary Lee
Bradish, John Patrick
Bruhn, H(jalmar) D(iehl)
Burck, Larry Harold
Burstein, Sol
Chan, Shih Hung
DeVries, Marvin Frank
Doshi, Mahendra R
Dries, William Charles
Duffie, Neil Arthur
El-Wakil, M(ohamed)
Engelstad, Roxann Louise
Fournelle, Raymond Albert
Frank, Arne
Gaggioli, Richard A
Garg, Arun
Gopal, Raj
Grogan, Paul J(oseph)
Henke, Russell W
Horntvedt, Earl W
Kenny, Andrew Augustine
Koeller, Ralph Carl
Kohli, Dilip
Landis, Fred
Lentz, Mark Steven

Linehan, John Henry
McMurray, David Claude
Mitchell, John Wright
Myers, Glen E(verett)
Myers, Phillip S(amuel)
Nigro, Nicholas J
Obert, Edward Fredric
Pavelic, Vjekoslav
Reid, Robert Lelon
Rohatgi, Pradeep Kumar
Rowlands, Robert Edward
Snyder, Warren Edward
Steudel, Harold Jude
Stumpe, Warren Robert
Szpot, Bruce F
Tsao, Keh Cheng
Uicker, John Joseph, Jr
Viets, Hermann
Weiss, Stanley
Widera, Georg Ernst Otto
Wittnebert, Fred R
Zahn, John J
Zelazo, Nathaniel Kachorek

WYOMING
Adams, Donald F
Chittenden, William A
Jacquot, Raymond G
Pell, Kynric M(artin)
Sutherland, Robert L(ouis)
Wheasler, Robert

PUERTO RICO
Patra, Amit Lal
Sasscer, Donald S(tuart)
Soderstrom, Kenneth G(unnar)

ALBERTA
Bellow, Donald Grant
Checkel, M David
Dale, J(ames) D(ouglas)
Feingold, Adolph
Kentfield, John Alan Charles
Lock, G(erald) S(eymour) H(unter)
Mikulcik, E(dwin) C(harles)
Norrie, D(ouglas) H
Rodkiewicz, Czeslaw Mateusz
Sadler, G(erald) W(esley)

BRITISH COLUMBIA
Brimacombe, James Keith
Hill, Philip G(raham)
Kerekes, Richard Joseph
Ko, Pak Lim
Leith, William Cumming
Mantle, J(ohn) B(ertram)
Milsum, John H
Morrison, James Barbour
Pomeroy, Richard James
Salcudean, Martha Eva
Tabarrok, B(ehrouz)
Wighton, John L(atta)

MANITOBA
Cenkowski, Stefan
Crosthwaite, John Leslie
Ruth, Douglas Warren

NEW BRUNSWICK
Neill, Robert D

NEWFOUNDLAND
Molgaard, Johannes

NOVA SCOTIA
Cochkanoff, O(rest)
Kujath, Marek Ryszard
Wilson, George Peter

ONTARIO
Baines, William Douglas
Baronet, Clifford Nelson
Brzustowski, Thomas Anthony
Chandrashekar, Muthu
Chou, Jordan Quan Ban
Cockshutt, E(ric) P(hilip)
Colborne, William George
Corneil, Ernest Ray
Currie, Iain George
Dalrymple, Desmond Grant
Dubey, Rajendra Narain
Emraghy, Hoda Abdel-Kader
Famili, Abolfazl
Fenton, Robert George
Feraday, Melville Albert
Finlay, Joseph Bryan
Foreman, John E(dward) K(endall)
Foster, John S
Gilbert, W(illiam) D(ouglas)
Gladwell, Graham M L
Goldenberg, Andrew Avi
Hageniers, Omer Leon
Hancox, William Thomas
Henry, Roger P
Hooper, F(rank) C(lements)
Howard, Joseph H(erman) G(regg)
Jeswiet, Jacob
Jones, L(lewellyn) E(dward)
Judd, Ross Leonard
Kardos, Geza
Keffer, James F
Kind, Richard John
Koul, Ashok Kumar
Krausz, Alexander Stephen

Mechanical Engineering (cont)

Krishnappa, Govindappa
Ledwell, Thomas Austin
Lee, Benedict Huk Kun
Lee, Yung
Leutheusser, H(ans) J(oachim)
McCammond, David
McDonald, T(homas) W(illiam)
Makomaski, Andrzej Henryk
Marsters, Gerald F
Matar, Said E
Mercer, Alexander McDowell
Moroz, William James
Munro, Michael Brian
Nelles, John Sumner
North, Henry E(rick) T(uisku)
North, W(alter) Paul Tuisku
Orlik-Ruckemann, Kazimierz Jerzy
Perlman, Maier
Pick, Roy James
Pike, J(ohn) G(ibson)
Plett, Edelbert Gregory
Plumtree, A(lan)
Rice, W(illiam) B(othwell)
Rimrott, F(riedrich) P(aul) J(ohannes)
Rogers, James Terence, (Jr)
Rosen, Marc Allen
Round, G(eorge) F(rederick)
Rush, Charles Kenneth
Saravanamuttoo, Herbert Ian H
Schey, John Anthony
Shelson, W(illiam)
Southwell, P(eter) H(enry)
Tarasuk, John David
Tavoularis, Stavros
Varin, Robert Andrzej
Venter, Ronald Daniel
Vijay, Mohan Madan
Ward, Charles Albert
Wong, Jo Yung

PRINCE EDWARD ISLAND
Lodge, Malcolm A

QUEBEC
Alward, Ron E
Berlyn, Robin Wilfrid
Bhat, Rama B
Bui, Tien Rung
Charette, Andre
Cheng, Richard M H
Croctogino, Reinhold Hermann
Dore, Roland
Galanis, Nicolas
Habashi, Wagdi George
Hoa, Suong Van
Kiss, Laszlo Istvan
Kwok, Clyde Chi Kai
Lee, John Hak Shan
Lin, Sui
Massoud, Monir Fouad
Mavriplis, F
Osman, M(ohamed) O M
Ryan, Norman Daniel
Sankar, Seshadri
Sankar, Thiagas Sriram
Vatistas, Georgios H

SASKATCHEWAN
Esmail, Mohamed Nabil
Hedlin, Charles P(eter)
Koh, Eusebio Legarda
Nikiforuk, P(eter) N
Ukrainetz, Paul Ruvim
Walerian, Szyszkowski

OTHER COUNTRIES
Al-Zubaidy, Sarim Naji
Bligh, Thomas Percival
Bodner, Sol R(ubin)
Chiu, Huei-Huang
Chwang, Allen Tse-Yung
De Vedia, Luis Alberto
Duncan, John L(eask)
Elad, David
Flurscheim, Cedric Harold
Gordon, Jeffrey Miles
Hiroyasu, Hiroyuji
Idelsohn, Sergio Rodolfo
Jespersen, Nils Vidar
Jost, Hans Peter
Lalas, Demetrius P
Lee, Choung Mook
Lee, Peter H Y
Liepsch, Dieter W
Nomura, Yasumasa
Otsuka, Kazuhiro
Pessen, David W
Revesz, Zsolt
Romero, Alejandro F
Sarmiento, Gustavo Sanchez
Shah, Mirza Mohammed
Solari, Mario Jose Adolfo
Tanner, Roger Ian
Taplin, David Michael Robert
Voutsas, Alexander Matthew (Voutsadakis)
Ward, Gerald T(empleton)
Wendland, Wolfgang Leopold
Wolf, Helmut
Yamamoto, Mitsuyoshi

Metallurgy & Physical Metallurgical Engineering

ALABAMA
Cataldo, Charles Eugene
Chia, E Henry
Frye, John H, Jr
Geppert, Gerard Allen
Holland, James Read
Janowski, Gregg Michael
Jemian, Wartan A(rmin)
Kattus, J Robert
Lucas, William R(ay)
Piwonka, Thomas Squire
Rampacek, Carl
Scheiner, Bernard James
Stanczyk, Martin Henry
Stefanescu, Doru Michael
Stein, Dale Franklin
Thompson, Raymond G
Vohra, Yogesh K
Wilcox, Roy Carl

ARIZONA
Albrecht, Edward Daniel
Arbiter, Nathaniel
Bhappu, Roshan B
Blickwede, D(onald) J(ohnson)
Carpenter, Ray Warren
Das, Subodh Kumar
Fairbanks, Harold V
Frame, John W
Hendrickson, Lester Ellsworth
Hiskey, J Brent
Keating, Kenneth L(ee)
Kebler, Richard William
Kessler, Harold D
Kuhn, Martin Clifford
Leeser, David O(scar)
Levinson, David W
McPherson, Donald J(ames)
Meieran, Eugene Stuart
Murphy, Daniel John
Peck, John F
Quadt, R(aymond) A(dolph)
Savrun, Ender
Simmons, Paul C
Swalin, Richard Arthur
Taub, James M
Walker, Walter W(yrick)
Withers, James C
Zukas, Eugene G

CALIFORNIA
Agrawal, Suphal Prakash
Albert, Paul A(ndre)
Anjard, Ronald P, Sr
Armijo, Joseph Sam
Asaro, Robert John
Baldwin, Chandler Milnes
Baldwin, David Hale
Bar-Cohen, Yoseph
Baskes, Michael I
Beebe, Robert Ray
Bertossa, Robert C
Bhandarkar, Mangalore Dilip
Bonora, Anthony Charles
Bragg, Robert H(enry)
Bray, Robert S
Brewer, Leo
Brimm, Eugene Oskar
Buhl, Robert Frank
Bunshah, Rointan F(ramroze)
Burke, Edmund C(harles)
Caligiuri, Robert Domenic
Cape, Arthur Tregoning (Tommy)
Chandler, Willis Thomas
Charley, Philip J(ames)
Chen, C(hih) W(en)
Chen, Tu
Childs, Wylie J
Chiou, C(harles)
Clemens, Bruce Montgomery
Cline, Carl F
Comley, Peter Nigel
Crossley, Frank Alphonso
Dasher, John
Dejonghe, Lutgard C
Denney, Joseph M(yers)
Dorward, Ralph C(larence)
Douglass, David Leslie
Doyle, Fiona Mary
Economy, George
El Guindy, Mahmoud Ismail
Elsner, Norbert Bernard
Evans, Donald B
Evans, James William
Feuerstein, Seymour
Finnegan, Walter Daniel
Freise, Earl J
French, Edward P(erry)
Fuerstenau, D(ouglas) W(inston)
Fultz, Brent T
Gehman, Bruce Lawrence
Gerlach, John Norman
Glazer, Judith
Goetzel, Claus G(uenter)
Goldberg, Alfred
Gordon, Barry Monroe
Gordon, Gerald M
Goubau, Wolfgang M
Gronsky, Ronald
Guttenplan, Jack David
Hackett, Le Roy Huntington, Jr
Harrison, J(ohn) D(avid)
Hauber, Janet Elaine
Helfrich, Wayne J(on)
Hill, Russell John
Hubred, Gale L
Hultgren, Ralph Raymond
Hunnicutt, Richard P(earce)
Jaffe, Leonard David
Jaffee, Robert I(saac)
Johnson, William Robert
Jones, Brian Herbert
Jones, Robin L(eslie)
Kaneko, Thomas Motomi
Kashar, Lawrence Joseph
Kendig, Martin William
Kikuchi, Ryoichi
Kim, Ki Hong
Kneip, G(eorge) D(ewey), Jr
Koppenaal, Theodore J
LaChance, Murdock Henry
Lampson, Francis Keith
Langdon, Terence George
Lavernia, Enrique Jose
Lewis, J(ack) R(ockley)
Li, Chia-Chuan
Lin, Charlie Yeongching
Littauer, Ernest Lucius
Loring, Blake M(arshall)
Lucas, Glenn E
Marshall, Sally J
Martin, George
Mecartney, Martha Lynn
Meyer, Brad Anthony
Meyers, Marc Andre
Mills, G(eorge) J(acob)
Mitchell, Michael Roger
Moazed, K L
Mohamed, Farghalli Abdelrahman
Mooz, William Ernst
Mukherjee, Amiya K
Murray, George T(homas)
Nachman, Joseph F(rank)
Nieh, Tai-Gang
Nutt, Steven R
Paine, T(homas) O(tten)
Parker, Earl Randall
Pask, Joseph Adam
Pathania, Rajeshwar S
Prindle, William Roscoe
Pritchett, Thomas Ronald
Przystupa, Marek Antoni
Rau, Charles Alfred, Jr
Raymond, Louis
Risbud, Subhash Hanamant
Ritchie, Robert Oliver
Roberts, George A(dam)
Romanowski, Christopher Andrew
Rosenbaum, H(erman) S(olomon)
Rossin, A David
Schjelderup, Hassel Charles
Schuyten, John
Schwartzbart, Harry
Shackelford, James Floyd
Sherby, Oleg D(imitri)
Shine, Carl
Shyne, J(ohn) C(ornelius)
Sinclair, Robert
Sines, George, Jr
Skinner, Loren Courtland, II
Slivinsky, Sandra Harriet
Sparling, Rebecca Hall
Stein, Richard Ballin
Steinberg, M(orris) A(lbert)
Stetson, Alvin Rae
Stewart, Robert Daniel
Strahl, Erwin Otto
Stringer, John
Sumsion, H(enry) T(heodore)
Syrett, Barry Christopher
Tanner, Lee Elliot
Tauber, Richard Norman
Teitel, Robert J(errell)
Thiers, Eugene Andres
Thomas, Gareth
Thompson, Larry Dean
Tietz, Thomas E(dwin)
Tiller, William Arthur
Tilles, Abe
Toensing, C(larence) H(erman)
Tomlinson, John Lashier
Vreeland, Thad, Jr
Wadsworth, Jeffrey
Wagner, C(hristian) N(ikolaus) J(ohann)
Waisman, Joseph L
Wang, John Ling-Fai
Watson, James Frederic
Westerman, Edwin J(ames)
Westmacott, Kenneth Harry
White, Jack Lee
Wong, Morton Min
Woods, Ralph Arthur
Yang, Charles Chin-Tze
Yang, Jenn-Ming
Yeh, Harry Chiang
Yoon, Rick J

COLORADO
Averill, William Allen
Beckstead, Leo William
Bull, W(illiam) Rex
Carpenter, Steve Haycock
Copeland, William D
Edwards, Glen Robert
Fields, Davis S(tuart), Jr
Fischer, R(oland) B(arton)
Fraikor, Frederick John
Goth, John W
Grobner, Paul Josef
Haas, Frank C
Hager, John P(atrick)
Heiple, Clinton R
Holzwarth, James C
Jha, Mahesh Chandra
Krauss, George
Lewis, Clifford Jackson
Lewis, Homer Dick
McKinnell, W(illiam) P(arks), Jr
Michal, Eugene J(oseph)
Moore, John Jeremy
Mosher, Don R(aymond)
Mote, Jimmy Dale
Mueller, William M(artin)
Newkirk, John Burt
Olson, David LeRoy
Peretti, Ettore A(lex)
Quick, Nathaniel Richard
Rairden, John Ruel, III
Rauch, Gary Clark
Read, David Thomas
Reed, Richard P
Roshko, Alexana
Sandstrom, Donald James
Selle, James Edward
Simons, Sanford L(awrence)
Vogt, Thomas Clarence, Jr
Ward, Milton Hawkins
Wonsiewicz, Bud Caesar
Yarar, Baki
Yukawa, S(umio)

CONNECTICUT
Abraham, Thomas
Arnold, John Richard
Blackwood, Andrew W
Bradley, Elihu F(rancis)
Burghoff, Henry L
Cheatham, Robert Gary
Chernock, Warren Philip
Clark, C(harles) C(anfield)
Cole, Stephen H(ervey)
Devereux, Owen Francis
Donachie, Matthew J(ohn), Jr
Duhl, David N
Galligan, James M
Gell, Maurice L
Giamei, Anthony Francis
Goldberg, A Jon
Gutierrez-Miravete, Ernesto
Haag, Robert Edwin
Hamjian, Harry J
Howes, Trevor Denis
Joesten, Raymond
Kattamis, Theodoulos Zenon
Lemkey, Frankin David
McEvily, Arthur Joseph, Jr
Mannweiler, Gordon B(annatyne)
Morral, John Eric
Parthasarathi, Manavasi Narasimhan
Pellier, Laurence (Delisle)
Perkins, Roger A(llan)
Popplewell, James Malcolm
Potter, Donald Irwin
Prasad, Arun
Pryor, Michael J
Schetky, Laurence McDonald
Snow, David Baker
Sperry, Philip Roger
Strife, James Richard
Sullivan, Cornelius Patrick, Jr
Swan, D(avid)
Tangel, O(scar) F(rank)
Thomas, Alvin David, Jr
Williams, John Ernest
Winter, Joseph
Zackay, Victor Francis

DELAWARE
Birchenall, Charles Ernest
Cook, Leslie G(ladstone)
Dexter, Stephen C
Ferriss, Donald P
Gemmell, Gordon D(ouglas)
Gibbs, Thomas W(atson)
Greenfield, Irwin G
Hall, Ian Wavell
Hentschel, Robert A(dolf) A(ndrew)
Kramer, John J(acob)
Lin, Kuang-Farn
Ransom, J(ohn) T(hompson)
Streicher, Michael A(lfred)
Urquhart, Andrew Willard

DISTRICT OF COLUMBIA
Aaronson, Hubert Irving
Ayers, Jack Duane
Cosgarea, Andrew, Jr
Cullen, Thomas M
Dennis, William E
Gottschall, Robert James
Horton, Kenneth Edwin
Jordan, Albert Gustav
Kaplan, Robert S
Lashmore, David S
Meussner, R(ussell) A(llen)
Michel, David John
Pande, Chandra Shekhar
Radcliffe, S Victor

Rath, Bhakta Bhusan
Ritter, Donald Lawrence
Schafrik, Robert Edward
Sullivan, Anna Mannevillette
Tokar, Michael
Vardiman, Ronald G
Vold, Carl Leroy
Wolf, Stanley Myron

FLORIDA
Abbaschian, Reza
Anusavice, Kenneth John
Campbell, Hallock Cowles
Carney, D(ennis) J(oseph)
Caserio, Martin J
DeHoff, Robert Thomas
Dudley, Richard H(arrison)
Ebrahimi, Fereshteh
Etherington, Harold
Fullman, R(obert) L(ouis)
Garwood, Maurice F
Gensamer, Maxwell
Gold, Edward
Goldberg, David C
Hartley, Craig Sheridan
Hartt, William Handy
Hascicek, Yusuf Suat
Haseman, Joseph Fish
Hosni, Yasser Ali
Hummel, Rolf Erich
Huston, Ernest Lee
Kennelley, James A
Kudman, Irwin
Lubker, Robert A(lfred)
MacDonnell, Wilfred Donald
McSwain, Richard Horace
Reed-Hill, Robert E(llis)
Roe, William P(rice)
Seth, Brij B
Smith, Edward S
Smith, William Fortune
Stern, Milton
Vassamillet, Lawrence Francois
Verink, Ellis D(aniel), Jr
Wang, K P
Ward, Leonard George
Wood, John D(udley)
Wyatt, James L(uther)

GEORGIA
Aseff, George V
Chandan, Harish Chandra
Cofer, Daniel Baxter
Coyne, Edward James, Jr
Cremens, Walter Samuel
D'Annessa, A(nthony) T(homas)
Danyluk, Steven
Engel, Niels N(ikolaj)
Grenga, Helen E(va)
Hart, Raymond Kenneth
Hochman, Robert F(rancis)
Iannicelli, Joseph
Kalish, David
Korda, Edward J(ohn)
Logan, Kathryn Vance
Meyers, Carolyn Winstead
Sanders, T H, Jr
Underwood, Ervin E(dgar)
Williams, Emmett Lewis
Yeske, Ronald A
Young, C(larence) B(ernard) F(ehrler)

HAWAII
Hamaker, John C, Jr
Hihara, Lloyd Hiromi

IDAHO
Bartlett, R(obert) W(atkins)
Bobeck, Gene E
Clifton, Donald F(rederic)
Emigh, G Donald
Griffith, W(illiam) A(lexander)
Keefer, Donald Walker
Korth, Gary E
Miller, Richard Lloyd
Place, Thomas Alan
Prisbrey, Keith A
Richardson, Lee S(pencer)
Tallman, Richard Louis
Walters, Leon C

ILLINOIS
Altstetter, Carl J
Askill, John
Barnett, Scott A
Battles, James E(verett)
Beck, Paul Adams
Bhattacharya, Debanshu
Bohl, Robert W
Bradley, Steven Arthur
Breen, Dale H
Breyer, Norman Nathan
Cannon, Howard S(uckling)
Chen, Haydn H D
Chiswik, Haim H
Cohen, J(erome) B(ernard)
Cork, Douglas J
Erck, Robert Alan
Fenske, George R
Fessler, Raymond R
Fradin, Frank Yale
Frost, Brian R T
Gordon, P(aul)
Gruber, Eugene E, Jr

Jacobi, Anthony Mark
Kamm, Gilbert G(eorge)
Kelman, L(eRoy) R
Kern, Roy Fredrick
Keyser, N(aaman) H(enry)
Kittel, J Howard
Lam, Daniel J
Langenberg, F(rederick) C(harles)
Lawrence, Frederick Van Buren, Jr
Liu, Yung Yuan
Mueller, Melvin H(enry)
Myles, Kevin Michael
Nachtman, Elliot Simon
Nash, Philip Graham
Olson, Gregory Bruce
Packer, Kenneth Frederick
Rastogi, Prabhat Kumar
Ripling, E(dward) J
Rostoker, William
Rothman, Steven J
Rowland, Theodore Justin
Sandoz, George
Scott, Tom E
Shack, William John
Shapiro, Eugene
Shapiro, Stanley
Shenoy, Gopal K
Shipley, Roch Joseph
Siegel, Richard W(hite)
Sinha, Shome Nath
Smedskjaer, Lars Christian
Socie, Darrell Frederick
Sproul, William Dallas
Sticha, Ernest August
Teng, Mao-Hua
Tuomi, Donald
Van Pelt, Richard H
Voorhees, Peter Willis
Walter, Gordon H
Wayman, C(larence) Marvin
Weeks, Richard William
Weertman, Julia Randall
Wensch, Glen W(illiam)
Westlake, Donald G(ilbert)
Wiedersich, H(artmut)
Wimber, R(ay) Ted
Wojnowski, Daniel Allen
Wright, Maurice Arthur
Zambrow, J(ohn) L(ucian)

INDIANA
Balajee, Shankverm R
Bellina, Joseph James, Jr
Chaubal, Pinakin Chintamani
Croat, John Joseph
Dayananda, Mysore Ananthamurthy
Drake, Justin R
Dukelow, Donald Allen
Eaton, Paul Bernard
Fosnacht, Donald Ralph
Fritzlen, Glenn A
Gaskell, David R
Gorman, Eugene Francis
Grace, Richard E(dward)
Hass, Kenneth Philip
Herman, Marvin
Hruska, Samuel Joseph
Hughes, Ian Frank
Kaye, Albert L(ouis)
Kelley, Thomas Neil
Klein, Howard Joseph
Lazaridis, Nassos A(thanasius)
Lee, Harvie Ho
Liedl, Gerald L(eRoy)
Luerssen, Frank W
Miller, Albert Eugene
Ovens, William George
Pearson, Donald A
Pielet, Howard M
Pops, Horace
Ranade, Madhukar G
Read, Robert H
Saucedo, Ismael G
Schuhmann, R(einhardt), Jr
Sherry, John M
Smith, Charles O(liver)
Tucker, Robert C, Jr
Young, William Ben

IOWA
Beaudry, Bernard Joseph
Buck, Otto
Carlson, Oscar Norman
Gschneidner, Karl A(lbert), Jr
Johnson, Donald L(ee)
Kayser, Francis X
Larsen, William L(awrence)
Mather, Roger Frederick
Moriarty, John Lawrence, Jr
Patterson, John W(illiam)
Peterson, David T
Rocklin, Isadore J
Schmidt, Frederick Allen
Smith, John F(rancis)
Trivedi, Rohit K
Verhoeven, John Daniel

KANSAS
Cooke, Francis W
Rose, Kenneth E(ugene)
Roth, Thomas Allan

KENTUCKY
Kuhn, William E(rik)

Mateer, Richard S(helby)
Mihelich, John L
Morris, James G
Stine, Philip Andrew
Viswanadham, Ramamurthy K

LOUISIANA
Bundy, Kirk Jon
Diwan, Ravinder Mohan
Kinchen, David G
Oyler, Glenn W(alker)
Raman, Aravamudhan
Rubinstein, Asher A
Snyder, Harold Jack

MAINE
Taylor, Anthony Boswell

MARYLAND
Alers, George A
Armstrong, R(onald) W(illiam)
Arsenault, Richard Joseph
Beachem, Cedric D
Boettinger, William James
Cahn, John W(erner)
Carter, John Paul
Coriell, Sam Ray
Cuthill, John R(obert)
Das, Badri N(arayan)
DeCraene, Denis Fredrick
Dieter, George E(llwood), Jr
Doan, Arthur Sumner, Jr
Early, James G(arland)
Economos, Geo(rge)
Fields, Richard Joel
Fishman, Steven Gerald
Fletcher, Stewart G(ailey)
Fraker, Anna Clyde
Franke, Gene Louis
Goode, Robert J
Hahn, W(alter) C(harles), Jr
Hayes, Earl T
Hirschhorn, Joel S(tephen)
Karulkar, Pramod C
Kirby, Ralph C(loudsberry)
Kramer, Irvin Raymond
Kreider, Kenneth Gruber
Kumar, K Sharvan
Lange, Eugene Albert
Lauriente, Michael
Loss, Frank J
McCawley, Frank X(avier)
Machlin, Irving
Manning, John Randolph
Moore, George A(ndrew)
Murray, Peter
Nash, Ralph Robert
Newbury, Dale Elwood
Nusbickel, Edward M, Jr
Poulose, Pathickal K
Pugh, E(dison) Neville
Reuther, Theodore Carl, Jr
Ricker, Richard Edmond
Schaefer, Robert J
Sharp, William Broom Alexander
Smith, John Henry
Solow, Max
Staker, Michael Ray
Van Echo, Andrew
Wacker, George Adolf
Waterstrat, Richard Milton
Weber, Clifford E
Yakowitz, Harvey
Yolken, Howard Thomas
Zapffe, Carl Andrew
Zwilsky, Klaus M(ax)

MASSACHUSETTS
Abkowitz, Stanley
Adler, Ralph Peter Isaac
Allen, Ryne C
Allen, Samuel Miller
Allen, Steven
Apelian, Diran
Argy, Dimitri
Aronin, Lewis Richard
Bachner, Frank Joseph
Balluffi, R(obert) W(eierter)
Bever, Michael B(erliner)
Biederman, Ronald R
Bolsaitis, Pedro
Bornemann, Alfred
Bruggeman, Gordon Arthur
Buffum, Donald C
Chang, Frank C
Chen, Sung Jen
Chou, Kuo-Chih
Cohen, Morris
Coleman, George W(illiam)
Cooke, Richard A
Cornie, James Allen
Coyne, James E
David, Donald J
Delagi, Richard Gregory
Ditchek, Brian Michael
Eagar, Thomas W
Elliott, John Frank
Evans, Michael Douglas
Feinstein, Leonard Gordon
Flemings, Merton Corson
Freedman, George
French, David N(ichols)
French, Robert Dexter
Gangulee, Amitava

ENGINEERING / 351

Garrett, Paul Daniel
Godrick, Joseph Adam
Goldstein, Joseph I
Grant, Nicholas J(ohn)
Gregory, Eric
Haggerty, John S
Harvey, Walter William
Horwedel, Charles Richard
Isserow, Saul
Johnson, Walter Roland
Khattak, Chandra Prakash
Kim, Han Joong
Kleis, John Dieffenbach
Krashes, David
Kula, Eric Bertil
Lander, Horace N(orman)
Latanision, Ronald Michael
Lement, Bernard S
Livingston, James Duane
McFarland, Charles Manter
MacGregor, C(harles) W(inters)
Matsuda, Seigo
Nowak, Welville B(erenson)
Old, Bruce S(cott)
Owen, Walter S
Pal, Uday Bhanu
Parkinson, R(obert) E(dward)
Passmore, Edmund M
Pelloux, Regis M N
Petrovic, Louis John
Ragone, David Vincent
Rapperport, Eugene J
Rhodes, William Holman
Sadoway, Donald Robert
Sastri, Suri A
Schoen, Frederick J
Schwensfeir, Robert James, Jr
Shen, Yuan-Shou
Smith, Cyril Stanley
Spaepen, Frans
Strauss, Alan Jay
Subramanian, Krishnamoorthy
Sung, Changmo
Suresh, Subra
Szekely, Julian
Thieme, Cornelis Leo Hans
Tustison, Randal Wayne
Tzeng, Wen-Shian Vincent
Wang, Chih-Chung
Zabielski, Chester V
Zotos, John

MICHIGAN
Aaron, Howard Berton
Albright, Darryl Louis
Beaman, Donald Robert
Beardmore, Peter
Ben, Manuel
Bens, Frederick Peter
Bilello, John Charles
Birkle, A(dolph) John
Bolling, Gustaf Fredric
Cole, Gerald Sidney
Corey, Clark L(awrence)
Courtney, Thomas Hugh
Davies, Richard Glyn
Doane, Douglas V
Eldis, George Thomas
Flinn, Richard A(loysius)
Foreman, Robert Walter
Ghosh, Amit Kumar
Gibala, Ronald
Giszczak, Thaddeus
Gjostein, Norman A
Hagel, William C(arl)
Hallada, Calvin James
Harris, Kenneth
Harvey, Douglas J
Harwood, Julius J
Heldt, Lloyd A
Heminghous, William Wayne, Sr
Hendrickson, Alfred A
Hill, John Campbell
Hohnke, Dieter Karl
Hucke, Edward E
Hughel, Thomas J(osiah)
Humenik, Michael, Jr
Ivanso, Eugene V
Jain, Kailash Chandra
Jones, Frederick Goodwin
Keeler, Stuart P
Kovacs, Bela Victor
Lahr, Gilbert M
Le Beau, Stephen Edward
Mikkola, Donald E(mil)
Mould, Peter Roy
Mukherjee, Kalinath
Nagler, Charles Arthur
Nesbitt, Carl C
Paton, Neil (Eric)
Pehlke, Robert Donald
Quigg, Richard J
Rausch, Doyle W
Reynolds, E(dward) E(vans)
Riegger, Otto K
Robinson, George H(enry)
Rohrig, Ignatius A
Sard, Richard
Sharma, Ram Autar
Shoemaker, Robert Harold
Sinnott, M(aurice) J(oseph)
Smith, Darrell Wayne
Spencer, Andrew R
Sponseller, D(avid) L(ester)

352 / DISCIPLINE INDEX

Metallurgy & Physical Metallurgical Engineering (cont)

Srolovitz, David Joseph
Stevenson, Robin
Stickels, Charles A
Subramanian, K N
Summitt, W(illiam) Robert
Thayer, Duane M
Thomson, Robert Francis
Tischer, Ragnar P(ascal)
Trojan, Paul K
Turcotte, William Arthur
Van Vlack, Lawrence H(all)
Wang, Yar-Ming
Whalen, Thomas J(ohn)
Williams, Gwynne
Winterbottom, W L
Young, Edwin H(arold)

MINNESOTA
Benner, Blair Richard
Bitsianes, Gust
Bleifuss, Rodney L
Dorenfeld, Adrian C
Fisher, Edward S
Geiger, Gordon Harold
Gerberich, William Warren
Huffine, Coy L(ee)
Hultgren, Frank Alexander
Iwasaki, Iwao
Khalafalla, Sanaa E
Larson, Andrew Hessler
McHenry, Kelly David
Maciolek, Ralph Bartholomew
Nam, Sehyun
Nicholson, Morris E(mmons), Jr
Oriani, Richard Anthony
Premanand, Visvanatha
Ross, Stuart Thom
Sartell, Jack A(lbert)
Schmit, Joseph Lawrence
Shah, Ishwarlal D
Shores, David Arthur
Staehle, Roger Washburne

MISSISSIPPI
Wehr, Allan Gordon

MISSOURI
Askeland, Donald Raymond
Bolon, Albert Eugene
Brasunas, Anton Des
Fine, Morris M(ilton)
Hansen, John Paul
Kisslinger, Fred
Larson, Kenneth Blaine
Leighly, Hollis Philip, Jr
Morris, Arthur Edward
Neumeier, Leander Anthony
O'Keefe, Thomas Joseph
Robertson, David G C
Sastry, Shankara M L
Sonnino, Carlo Benvenuto
Watson, John Leslie
Weart, Harry W(aldron)
Winholtz, Robert Andrew
Wolf, Robert V(alentin)

MONTANA
Griffiths, Vernon
Twidwell, Larry G
Zucker, Gordon L(eslie)

NEBRASKA
Nelson, Russell C
Rohde, Suzanne Louise

NEVADA
Chandra, Dhanesh
Clements, Linda L
Eichbaum, Barlane Ronald
Ferrell, Edward F(rancis)
Fuerstenau, M(aurice) C(lark)
Jones, Denny Alan
Skaggs, Robert L

NEW HAMPSHIRE
Backofen, Walter A
Carreker, R(oland) P(olk), Jr
Colligan, George Austin
Dancy, Terence E(rnest)
Epremian, E(dward)
Gardner, Frank S(treeter)
Marion, Robert Howard
Queneau, Paul E(tienne)
Schulson, Erland Maxwell

NEW JERSEY
Abendroth, Reinhard P(aul)
Albrethsen, A(drian) E(dysel)
Avery, Howard S
Bakish, Robert
Bevk, Joze
Boni, Robert Eugene
Brytczuk, Walter L(ucas)
Chapin, Henry J(acob)
Chatterji, Debajyoti
Connolly, John Charles
Covert, Roger A(llen)
Das, Santosh Kumar
DeLuca, Robert D(avid)

Dittman, Frank W(illard)
Duerr, J Stephen
Eash, John T(rimble)
Feder, David O
Finke, Guenther Bruno
Foerster, George Stephen
Gagnebin, Albert Paul
Geskin, Ernest S
Green, Martin Laurence
Grunes, Robert Lewis
Grupen, William Brightman
Jin, Sungho
Kalish, Herbert S(aul)
Karcher, Guido George
Koch, Frederick Bayard
Koul, Maharaj Kishen
Kurtz, Anthony David
Lanam, Richard Delbert, Jr
Lemaire, Paul J
Levine, Richard S
Li, Hsin Lang
Liebermann, Howard Horst
Lunsford, Ralph D
Lunt, Harry Edward
Manders, Peter William
Mathias, Joseph Simon
Mihalisin, John Raymond
Mitchell, Olga Mary Mracek
Morris, Robert Craig
Moyer, Kenneth Harold
Murakami, Masanori
Murarka, Shyam Prasad
Ohring, Milton
Opie, William R(obert)
Patel, Jamshed R(uttonshaw)
Peters, Dale Thompson
Polk, Conrad Joseph
Rorabaugh, Donald T
Rosen, Carol Zwick
Rutherford, John L(oftus)
Scharfstein, Lawrence Robert
Schlabach, T(om) D(aniel)
Sheppard, Keith George
Snowman, Alfred
Subramanyam, Dilip Kumar
Tanenbaum, Morris
Taylor, Jean Ellen
Von Neida, Allyn Robert
Vyas, Brijesh
Weil, Rolf
Wernick, Jack H(arry)
Winsor, Frederick James
Yan, Man Fei
Yokelson, M(arshall) V(ictor)
Yoon, Euijoon

NEW MEXICO
Alam, Mansoor
Armstrong, Philip E(dward)
Baker, Don H(obart), Jr
Behrens, Robert George
Catlett, Duane Stewart
Clouser, William Sands
Cost, James R(ichard)
Dickinson, James M(illard)
Diegle, Ronald Bruce
Eckelmeyer, Kenneth Hall
Ericksen, Richard Harold
Fritschy, J(ohn) Melvin
George, Timothy Gordon
Hecker, Siegfried Stephen
Hockett, John (Edwin)
Lawson, Andrew Cowper, II
Lowe, Terry Curtis
Magnani, Nicholas J
Mead, Richard Wilson
Miller, Alan R(obert)
Milton, Osborne
Mitchell, David Wesley
Mitchell, Terence Edward
Mulford, Robert Neal Ramsay
Munson, Darrell E(ugene)
O'Rourke, John Alvin
Pope, Larry Elmer
Reiswig, Robert D(avid)
Romig, Alton Dale, Jr
Salzbrenner, Richard John
Sheinberg, Haskell
Tonks, Davis Loel
Van Den Avyle, James Albert
Yost, Frederick Gordon

NEW YORK
Adler, Philip N(athan)
Albers, Francis C
Albrecht, William Melvin
Aldrich, Robert George
Begley, James Andrew
Benz, M G
Berndt, Christopher Charles
Beshers, Daniel N(ewson)
Born, Allen
Bradley, William W(hitney)
Brauer, Joseph B(ertram)
Brock, Geoffrey E(dgar)
Bronnes, Robert L(ewis)
Brophy, Jere H(all)
Budinski, Kenneth Gerard
Burr, Arthur Albert
Carapella, S(am) C(harles), Jr
Cargill, George Slade, III
Castleman, L(ouis) S(amuel)
Cech, Robert E(dward)
Chan, Siu-Wai

Charles, R(ichard) J(oseph)
Chaudhari, Praveen
Chung, Deborah Duen Ling
Chyung, Kenneth
Clayton, Clive Robert
Clum, James Avery
Cocca, M(ichael) A(nthony)
Cordovi, Marcel A(very)
Curran, Robert M
Czajkowski, Carl Joseph
Death, Frank Stuart
Decker, L(ouis) H
D'Heurle, Francois Max
Duby, Paul F(rancois)
Duquette, David J(oseph)
Edelglass, Stephen Mark
Estes, John H
Fiedler, H(oward) C(harles)
Fischer, George J
Flom, Donald Gordon
Floreen, Stephen
Gigliotti, Michael Francis Xavier
Glicksman, Martin E
Goldspiel, Solomon
Grosewald, Peter
Gupta, Devendra
Gupta, Krishna Murari
Gurinsky, David H(arris)
Hall, Ernest Leroy
Harris, Colin C(yril)
Hausner, Henry H(erman)
Herman, Herbert
Homan, Clarke Gilbert
Jackson, Melvin Robert
Judd, Gary
Kammer, Paul A
Kellogg, Herbert H(umphrey)
Kim, Jonathan Jang-Ho
King, Alexander Harvey
Kotval, Pesho Sohrab
Kuan, Tung-Sheng
Le Maistre, Christopher William
Lenel, Fritz (Victor)
Li, Chou H(siung)
Lommel, J(ames) M(yles)
Lord, Arthur N(elson)
Luborsky, Fred Everett
Luthra, Krishan Lal
Lyman, W(ilkes) Stuart
McCarthy, John Joseph
Margolin, Harold
Marwick, Alan David
Maurer, Gernant E
Mayadas, A Frank
Minnear, William Paul
Mondolfo, L(ucio) F(austo)
Morfopoulos, Vassilis C(onstantinos) P
Morris, William Guy
Mulford, Robert Alan
Muzyka, Donald Richard
Nippes, Ernest F(rederick)
Pasco, Robert William
Pollock, D(aniel) D(avid)
Prager, Martin
Prater, T(homas) A(llen)
Preisler, Joseph J(ohn)
Psioda, Joseph Adam
Puttlitz, Karl Joseph
Pye, Lenwood D(avid)
Quesnel, David John
Reardon, Joseph Daniel
Rowe, Raymond Grant
Rungta, Ravi
Saha, Bijay S
Sasserath, Jay N
Satya, Akella V S
Scala, E(raldus) (Pete)
Schadler, Harvey W(alter)
Schoefer, Ernest A(lexander)
Schulte, Robert Lawrence
Seigle, L(eslie) L(ouis)
Sims, Chester Thomas
Singh, Mrityunjay
Smith, G(eorge) V
Solomon, Harvey Donald
Srinivasan, Makuteswaran
Stoloff, Norman Stanley
Strasser, Alfred Anthony
Suenaga, Masaki
Sweet, Richard Clark
Szekely, Andrew Geza
Tavlarides, Lawrence Lasky
Taylor, Dale Frederick
Thompson, Charles Denison
Tobin, Albert George
Totta, Paul Anthony
Urban, Joseph
Van den Sype, Jaak Stefaan
Vermilyea, D(avid) A(ugustus)
Wachtell, Richard L(loyd)
Weeks, John R(andel), IV
Weinig, Sheldon
Whang, Sung H
Woodford, David A(ubrey)
Wright, Roger Neal
Zakraysek, Louis

NORTH CAROLINA
Austin, William W(yatt)
Bailey, John Albert
Beiser, Carl A(dolph)
Cocks, Franklin H
Cox, John Jay, Jr
Dyrkacz, W William

Fahmy, Abdel Aziz
Fox, Bradley Alan
Goodwin, Frank Erik
Haissig, Manfred
Haynie, Fred Hollis
Hinesley, Carl Phillip
Holshouser, W(illiam) L(uther)
Kirk, Wilber Wolfe
Koch, Carl Conrad
Kozlik, Roland A
Kusy, Robert Peter
Pearsall, George W(ilbur)
Prater, John Thomas
Reeber, Robert Richard
Scattergood, Ronald O
Shepard, Marion L(averne)
Stadelmaier, H(ans) H(einrich)

OHIO
Adams, Robert R(oyston)
Ahmed, Shaffiq Uddin
Alam, M Khairul
Altan, Taylan
Arnold, Lynn Ellis
Aukrust, Egil
Ault, George Mervin
Babcock, Donald Eric
Baloun, Calvin H(endricks)
Bartlett, Edwin S(outhworth)
Bauer, James H(arry)
Beck, Franklin H(orne)
Bidwell, Lawrence Romaine
Bloom, Joseph Morris
Bomberger, H(oward) B(rubaker), Jr
Boulger, Francis W(illiam)
Burte, Harris M(erl)
Butler, John F(rancis)
Buttner, F(rederick) H(oward)
Campbell, James Edward
Carbonara, Robert Stephen
Carlson, Robert G(ustav)
Carmichael, Donald C(harles)
Cary, Howard Bradford
Choudhary, Manoj Kumar
Clark, William Alan Thomas
Clauer, Allan Henry
Cooper, Thomas D(avid)
Copp, Albert Nils
Cordea, James Nicholas
Dawson, Steven Michael
Dembowski, Peter Vincent
Dickerson, R(onald) F(rank)
Dreshfield, Robert Lewis
Duttweiler, Russell E
Ebert, Lynn J
Focke, Arthur E(ldridge)
Foster, Ellis L(ouis)
Freche, John C(harles)
Gabb, Timothy Paul
Gegel, Harold L(ouis)
Gelles, S(tanley) H(arold)
Gilby, Stephen Warner
Glaeser, William A(lfred)
Gonser, Bruce Winfred
Graham, John W
Griffith, Cecil Baker
Griffith, Walter M, Jr
Grisaffe, Salvatore J
Gurklis, John A(nthony)
Hall, Albert M(angold)
Harrison, Robert W
Heckard, David Custer
Henry, Donald J
Holden, Frank C(harles)
Hook, Rollin Earl
Howden, David Gordon
Ingram, David Christopher
Jackson, Curtis M(aitland)
Jacobs, J(ames) H(arrison)
Jayaraman, Narayanan
Johnson, Ralph Sterling, Jr
Kaczynski, Don
Kahles, John Frank
Kampe, Dennis James
Kawasaki, Edwin Pope
Klems, George J
Koros, Peter J
Koster, W(illiam) P(feiffer)
Kot, Richard Anthony
Krashin, Bernard R(obert)
Leber, Sam
Lee, Peter Wankyoon
Leontis, T(homas) E(rnest)
Lewandowski, John Joseph
Lipsitt, Harry A(llan)
Littmann, Martin F(rederick)
Lownie, H(arold) W(illiam), Jr
McCall, James Lodge
McCoy, Robert Allyn
Marschall, Charles W(alter)
Maykuth, D(aniel) J(ohn)
Mezey, Eugene Julius
Mobley, Carroll Edward
Mohn, Walter Rosing
Morral, F(acundo) R(olf)
Myers, James R(ussell)
Newby, John R
Nikkel, Henry
Paine, Robert Madison
Parthasarathy, Triplicane Asuri
Payer, Joe Howard
Perlmutter, Isaac
Pierce, Cyril Marvin
Pool, Monte J

Powell, Gordon W
Pugh, John W(illiam)
Rai, Amarendra Kumar
Rapp, Robert Anthony
Ray, Alden E(arl)
Readey, Dennis W(illiam)
Rengstorff, George W(illard) P(epper)
Ricksecker, Ralph E
Rigney, David Arthur
Roehrig, Frederick Karl
Rosa, Casimir Joseph
Rosenfield, Alan R(obert)
St Pierre, George R(oland)
St Pierre, P(hilippe) D(ouglas) S
Saxer, Richard Karl
Seagle, Stan R
Seltzer, Martin S
Shah, Bhupendra Umedchand
Shemenski, Robert Martin
Shewmon, Paul G(riffith)
Stover, E(dward) R(oy)
Thellmann, Edward L
Thomas, Joseph Francis, Jr
Thomas, Seth Richard
Troiano, A(lexander) R(obert)
Tsao, Bang-Hung
Uebele, Curtis Eugene
Uys, Johannes Marthinus
Venkatu, Doulatabad A
Villar, James Walter
Wagoner, Robert H
Wakelin, David Herbert
Wallace, John F(rancis)
Welsch, Gerhard Egon
Westerman, Arthur B(aer)
Westermann, Fred Ernst
Wickersham, Charles Edward, Jr
Wilde, Bryan Edmund
Williams, James Case
Wlodek, Stanley T

OKLAHOMA
Block, Robert Jay
Daniels, Raymond D(eWitt)
Growcock, Frederick Bruce
Radd, F(rederick) J(ohn)
Richards, Kenneth Julian
Riggs, Olen Lonnie, Jr
Thomason, William Hugh

OREGON
Banning, Lloyd H(arold)
Beall, Robert Allan
Blickensderfer, Robert
Czyzewski, Harry
Daellenbach, Charles Byron
Dash, John
Devletian, Jack H
Dooley, George Joseph, III
Grange, Raymond A
Humphrey, J Richard
Nafziger, Ralph Hamilton
Olleman, Roger D(ean)
Olsen, Richard Standal
Paarsons, James Delbert
Roberts, C Sheldon
Schaefer, Seth Clarence
Siemens, Richard Ernest

PENNSYLVANIA
Agarwala, Vinod Shanker
Aggen, George
Albert, Robert Lee
Antes, Harry W
Aplan, Frank F(ulton)
Aronson, Arthur H
Aspden, Robert George
Bajaj, Ram
Baran, George Roman
Barnartt, Sidney
Bauer, C(harles) L(loyd)
Bell, Gordon M(oreton)
Berman, David Alvin
Bhat, Gopal Krishna
Billman, Fred Richard
Birks, Neil
Bitler, William Reynolds
Bjorhovde, Reidar
Blayden, Lee Chandler
Boltax, Alvin
Bonewitz, Robert Allen
Bowman, Kenneth Aaron
Bramfitt, Bruce Livingston
Brenner, Sidney S(iegfried)
Bretz, Philip Eric
Bridenbaugh, Peter R
Brondyke, Kenneth J
Bryner, Joseph S(anson)
Bucci, Robert James
Bucher, John Henry
Burkart, Milton W(alter)
Bush, Glenn W
Caffrey, Robert E
Caton, Robert Luther
Chou, Y(e) T(sang)
Chubb, Walston
Clark, J(ohn) B(everley)
Conard, George P(owell)
Corbett, Robert B(arnhart)
Daga, Raman Lall
Dax, Frank Robert
DebRoy, Tarasankar
DeGraef, Marc Julie
Dell, M(anuel) Benjamin
DeLong, William T
De Luccia, John Jerry
Denny, J(ohn) P(almer)
Derge, G(erhard) (Julius)
Deurbrouck, Albert William
Dorschu, Karl E(dward)
Doty, W(illiam) D'Orville
Dulis, Edward J(ohn)
Emerick, Harold B(urton)
Emlemdi, Hasan Bashir
Englehart, Edwin Thomas, Jr
Enrietto, Joseph Francis
Ferrari, Harry M
Fiore, Nicholas F
Fischer, Robert B
Fitterer, G(eorge) R(aymond)
Foltz, Donald Richard
Fonseca, Anthony Gutierre
Franz, Edmund C(larence)
Fricke, William G(eorge), Jr
Fruehan, Richard J
Gardiner, Keith M
Gill, C(harles) Burroughs
Girifalco, Louis A(nthony)
Goldman, Kenneth M(arvin)
Grabowski, Thomas J
Graham, Arthur H(ughes)
Graham, Charles D(anne), Jr
Harkness, Samuel Dacke
Harvey, Frances J, II
Heger, James J
Hertzberg, Richard Warren
Hildeman, Gregory John
Hillner, Edward
Hogg, Richard
Hoke, John Henry
Hopkins, Richard H(enry)
Horne, G(erald) T(erence)
Howell, Paul Raymond
Hu, William H(sun)
Hull, Frederick Charles
Hultman, Carl A
Hunsicker, Harold Yundt
Iezzi, Robert Aldo
Jaffe, Donald
Jain, Himanshu
Jarrett, Noel
Jensen, Craig Leebens
Johnson, C Walter
Johnston, William V
Jordan, Robert Kenneth
Katz, Lewis E
Katz, Owen M
Keller, Rudolf
Khan, Parwaiz Ashraf Ali
Khare, Ashok Kumar
Kilp, Gerald R
Kim, Hong C
Koczak, Michael Julius
Korchynsky, M(ichael)
Kosco, John C(arroll)
Koss, Donald A
Kotyk, Michael
Kraft, R(alph) Wayne
Kumta, Prashant Nagesh
Kun, Zoltan Kokai
Lai, Ralph Wei-Meen
Laird, Campbell
Langford, George
Laughlin, David Eugene
Lawley, Alan
Leinbach, Ralph C, Jr
Lena, Adolph J
Leonard, Laurence
Libsch, Joseph F(rancis)
Lidman, William G
Lifka, Bernard William
London, Gilbert J(ulius)
Loria, Edward Albert
Lustman, Benjamin
McCaughey, Joseph M
McGeady, Leon Joseph
McMahon, Charles J, Jr
MacRae, Donald Richard
Mahajan, Subhash
Maloney, James L
Manganello, S(amuel) J(ohn)
Manjoine, Michael J(oseph)
Marder, A R
Massalski, T(adeusz) B(ronislaw)
Mayer, George
Mears, Dana Christopher
Mehrabian, Robert
Mehrkam, Quentin D
Meier, Gerald Herbert
Michalak, Joseph T(homas)
Misra, Sudhan Sekher
Mullendore, James Alan
Murphy, William J(ames)
Nilan, Thomas George
Novak, Stephen Robert
Orehotsky, John Lewis
Ostrowski, Edward Joseph
Pangborn, Robert Northrup
Paul, Anand Justin
Pavlovic, Dusan M(ilos)
Paxton, H(arold) W(illiam)
Peiffer, Howard R
Pense, Alan Wiggins
Pettit, Frederick S
Pickering, Howard W
Pinnow, Kenneth Elmer
Pope, David Peter
Porter, Lew F(orster)
Proske, Joseph Walter
Radford, Kenneth Charles
Ray, Siba Prasad
Reid, H(arry) F(rancis), Jr
Roberts, Malcolm John
Rogers, H(arry) C(arton), Jr
Rossin, P(eter) C(harles)
Rowe, Anne Prine
Roy, Pradip Kumar
Ruud, Clayton Olaf
Ryba, Earle Richard
Sabol, George Paul
Sander, Louis Frank
Schaeffer, Gene Thomas
Schmidt, Richard
Shabel, Barrie Steven
Shellenberger, Donald J(ames)
Sheridan, John Joseph, III
Shields, Bruce Maclean
Simkovich, George
Smith, James S(terrett)
Spaeder, Carl Edward, Jr
Spitznagel, John A
Sprowls, Donald O(tte)
Staley, James T
Starr, C Dean
Stephenson, Edward T
Stephenson, Robert L
Stoehr, Robert Allen
Stout, Robert Daniel
Stumpf, H(arry) C(linch)
Tankins, Edwin S
Tarby, Stephen Kenneth
Taylor, A
Tepper, Frederick
Thompson, Anthony W
Thornburg, Donald Richard
Torok, Theodore Elwyn
Towner, R(aymond) J(ay)
Townsend, Herbert Earl, Jr
Turkdogan, Ethem Tugrul
Turnbull, G(ordon) Keith
Uz, Mehmet
Vander Voort, George Frederic
Varrese, Francis Raymond
Waldman, Jeffrey
Weber, Frank L
Werner, F(red) E(ugene)
Werner, John Ellis
Whiteley, Roger L
Whitlow, Graham Anthony
Wiehe, William Albert
Williams, David Bernard
Wilson, Alexander D
Wood, Susan
Woodburn, Wilton A
Wriedt, Henry Anderson
Wynblatt, Paul P
Yeniscavich, William
Zamanzadeh, Mehrooz
Zuspan, G William

RHODE ISLAND
Avery, Donald Hills
Gurland, Joseph
Reynolds, Charles C
Richman, Marc H(erbert)
Rockett, Thomas John

SOUTH CAROLINA
Akpan, Edward
Carter, Giles Frederick
Lee, Craig Chun-Kuo
Linnert, George Edwin
McDonell, William Robert
Nevitt, Michael Vogt
Piper, John
Quadir, Tariq
Rack, Henry Johann
Rideout, Sheldon P
Robinson, William Courtney, Jr
Schilling, William Frederick
Spicer, Clifford W
Sturcken, Edward Francis
Weiss, Michael Karl
Wolf, James S

SOUTH DAKOTA
Han, Kenneth N

TENNESSEE
Bayuzick, Robert J(ohn)
Borie, Bernard Simon
Brinkman, Charles R
Brooks, Charlie R(ay)
Buchheit, Richard D(ale)
Bullock, Jonathan S, IV
Canonico, Domenic Andrew
Coleman, Charles F(ranklin)
Dahotre, Narendra Bapurao
David, Stanislaus Antony
DeVan, Jackson H
Flanagan, William F(rancis)
Ford, James Arthur
Glasser, Julian
Goodwin, Gene M
Gorbatkin, Steven M
Gray, Allen G(ibbs)
Gray, Robert J
Griess, John Christian, Jr
Grossbeck, Martin Lester
Harms, William Otto
Holcombe, Cressie Earl, Jr
Horak, James Albert
Jessen, Nicholas C, Jr
Klueh, Ronald Lloyd
Koger, John W
Lotts, Adolphus Lloyd
Lundin, Carl D
McElroy, David L(ouis)
Manly, William D
Miller, Ronald Eldon
Oliver, Bennie F(rank)
Pardue, William M
Pawel, Richard E
Slade, Edward Colin
Spooner, Stephen
Spruiell, Joseph E(arl)
Stansbury, E(le) E(ugene)
Stiegler, James O
Sutherland, Charles William
Swindeman, Robert W
Weir, James Robert, Jr
Wert, James J
Williams, Robin O('Dare)
Yoakum, Anna Margaret
Yoo, Man Hyong
Yust, Charles S(imon)

TEXAS
Anderson, Robert Clark
Arrowood, Roy Mitchell, Jr
Bokros, J(ack) C(hester)
Bourell, David Lee
Bradley, Walter L
Bravenec, Edward V
Brotzen, Franz R(ichard)
Caudle, Danny Dearl
Chen, In-Gann
Dobrott, Robert D
Eliezer, Zwy
Frick, John P
Gray, John Malcolm
Griffin, Richard B
Gupton, Paul Stephen
Harper, James George
Harris, William J
Henderson, Gregory Wayne
Jerner, R Craig
Johnson, Robert M
Jones, William B
Kennelley, Kevin James
Koh, P(un) K(ien)
Lautzenheiser, Clarence Eric
Leverant, Gerald Robert
McLellan, Rex B
Marcus, Harris L
Marshall, Robert P(aul)
Mertens, Frederick Paul
Moore, Thomas Matthew
Murr, Lawrence Eugene
Parikh, N(iranjan) M
Patriarca, Peter
Patton, Charles C(lifford)
Prill, Arnold L
Ralls, Kenneth M(ichael)
Robinson, J(ames) Michael
Rosenmayer, Charles Thomas
Runke, Sidney Morris
Salama, Kamel
Scales, Stanley R
Schlitt, William Joseph, III
Spencer, Thomas H
Stark, J(ohn) P(aul), Jr
Staudenmayer, Ralph
Tems, Robin Douglas
Thomas, P(aul) D(aniel)
Tien, John Kai
Walsh, Kenneth Albert
Watkins, Maurice
Winegartner, Edgar Carl
Wiseman, Carl D

UTAH
Adams, Brent Larsen
Alexander, Guy B
Bishop, Jay Lyman
Bradford, Harold R(awsel)
Byrne, J(oseph) Gerald
Carnahan, Robert D
Dahlstrom, Donald A(lbert)
Daniels, Alma U(riah)
Healy, George W(illiam)
Henrie, Thomas A
Horton, Ralph M
McKinney, William Alan
Malouf, Emil Edward
Michaelson, S(tanley) D(ay)
Miller, Jan Dean
Olson, Ferron Allred
Orava, R(aimo) Norman
Pitt, Charles H
Prater, John D
Sohn, Hong Yong
Spear, Carl D(avid)
Talbot, Eugene L(eroy)
Tuddenham, W(illiam) Marvin
Wadsworth, Milton E(lliot)

VIRGINIA
Almeter, Frank M(urray)
Cook, Desmond C
Cooper, Khershed Pessie
Davis, Bernard Eric
Gangloff, Richard Paul
Gillmor, R(obert) N(iles)
Hahn, Henry
Haworth, W(illiam) Lancelot

Metallurgy & Physical Metallurgical Engineering (cont)

Healy, John Joseph
Hendricks, Robert Wayne
Joebstl, Johann Anton
Jones, Thomas S
Jorstad, John Leonard
Kirkendall, Ernest Oliver
Kuhlmann-Wilsdorf, Doris
Landgraf, Ronald William
Lane, Joseph Robert
Larsen-Basse, Jorn
Leslie, William C(airns)
Levy, Sander Alvin
Lowe, A(rthur) L(ee), Jr
Lynd, Langtry Emmett
Lytton, Jack L(ester)
Owczarski, William A(nthony)
Preusch, Charles D
Sansonetti, S John
Schwartz, Ira A(rthur)
Sedriks, Aristide John
Spangler, Grant Edward
Spencer, Chester W(allace)
Spitzig, William Andrew
Starke, Edgar Arlin, Jr
Stein, Bland Allen
Sudarshan, T S
Vandermeer, R(oy) A
Wilcox, Benjamin A
Zinkham, Robert Edward

WASHINGTON
Archbold, Thomas Frank
Basmajian, John Aram
Bates, J(unior) Lambert
Bowdish, Frank W(illiam)
Boyer, Rodney Raymond
Brehm, William Frederick
Brimhall, J(ohn) L
Bush, S(pencer) H(arrison)
Chikalla, Thomas D(avid)
Claudson, T(homas) T(ucker)
Danko, Joseph Christopher
Das, K Bhagwan
DeMoney, Fred William
Duran, Servet A(hmet)
Ekenes, J Martin
Evans, Ersel Arthur
Finlay, Walter L(eonard)
Fischbach, David Bibb
Gelles, David Stephen
Hamilton, C Howard
Hirsch, Horst Eberhard
Hirth, John P(rice)
Holt, Richard E(dwin)
Huang, Fan-Hsiung Frank
Jones, Russell Howard
Kalonji, Gretchen
Kissin, G(erald) H(arvey)
Kruger, Owen L
Leavenworth, Howard W, Jr
Leggett, Robert Dean
Masson, D(ouglas) Bruce
Matlock, John Hudson
Minor, James E(rnest)
Peterson, Warren Stanley
Polonis, Douglas Hugh
Quist, William Edward
Rao, Yalamanchili Krishna
Roberts, J T Adrian
Sheely, W(allace) F(ranklyn)
Stang, Robert George
Sutherland, Earl C
Sweet, John W
Tenenbaum, Michael
Westerman, Richard Earl
Wick, O(swald) J
Woods, Keith Newell

WEST VIRGINIA
Berry, Vinod K
De Barbadillo, John Joseph
Eiselstein, Herbert Louis
Lemke, Thomas Franklin
Meloy, Thomas Phillips
Olcott, Eugene L
Stoll, Richard E
Zhang, Xiao

WISCONSIN
Baker, George Severt
Blumenthal, Robert N
Burck, Larry Harold
Chang, Y Austin
Curtis, Ralph Wendell
Daykin, Robert P
DeVries, Marvin Frank
Dodd, Richard Arthur
Dries, William Charles
Fournelle, Raymond Albert
Heine, Richard W
Lenling, William James
Loper, Carl R(ichard), Jr
Mangat, Pawitterjit Singh
Megahed, Sid A
Michael, Arthur B
Moll, Richard A
Neumann, Joachim Peter
Ramsey, Paul W
Ranganathan, Brahmanpalli Narasimhamurthy
Richard, Terry Gordon
Rohatgi, Pradeep Kumar
Samuel, Jay Morris
Schultz, Jay Ward
Wasson, L(oerwood) C(harles)
Weiss, Stanley
Wnuk, Michael Peter
Worzala, F(rank) John

WYOMING
Laman, Jerry Thomas

ALBERTA
Bradford, Samuel Arthur
Kuntz, Garland Parke Paul
Wayman, Michael Lash

BRITISH COLUMBIA
Alden, Thomas H(yde)
Barer, Ralph David
Brimacombe, James Keith
Brown, L(aurence) C(laydon)
Garner, Andrew
Hames, F(rederick) A(rthur)
Leja, J(an)
Meech, John Athol
Mular, A(ndrew) L(ouis)
Peters, B(runo) Frank
Peters, E(rnest)
Poling, George Wesley
Tromans, D(esmond)
Vitovec, Franz H
Wynne-Edwards, Hugh Robert

MANITOBA
Chaturvedi, Mahesh Chandra
Dutton, Roger
Rosinger, Herbert Eugene
Wilkins, Brian John Samuel
Wright, Michael George

NOVA SCOTIA
Gow, K V

ONTARIO
Agar, G(ordon) E(dward)
Alfred, Louis Charles Roland
Argyropoulos, Stavros Andreas
Ashbrook, Allan William
Aust, Karl T(homas)
Awadalla, Farouk Tawfik
Bennett, W Donald
Bibby, Malcolm
Bratina, Woymir John
Buhr, Robert K
Cameron, John
Caplan, Donald
Carpenter, Graham John Charles
Conard, Bruce R
Craig, George Black
Cupp, Calvin R
Diaz, Carlos Manuel
Ells, Charles Edward
Embury, John David
Ettel, Victor Alexander
Franklin, Ursula Martius
Gow, William Alexander
Hamilton, Bruce M
Hansson, Carolyn M
Heffernan, Gerald R
Holt, Richard Thomas
Hunt, Charles Edmund Laurence
Illis, Alexander
Irons, Gordon Alexander
Ives, Michael Brian
Keys, John David
King, Hubert Wylam
Kirkaldy, J(ohn) S(amuel)
Koul, Ashok Kumar
Lakshmanan, Vaikuntam Iyer
Lu, Wei-Kao
McGeer, James Peter
Mackay, W(illiam) B(rydon) F(raser)
Miller, W Alfred
Morris, Larry Arthur
Northwood, Derek Owen
Orton, John Paul
Pascual, Roberto
Piercy, George Robert
Plumtree, A(lan)
Purdy, Gary Rush
Roberts, W(illiam) Neil
Runnalls, O(liver) John C(lyve)
Sahoo, Mahi
Schey, John Anthony
Sekerka, Ivan
Siddell, Derreck
Smeltzer, Walter William
Spink, D(onald) R(ichard)
Street, Kenneth Norman
Thornhill, Philip G
Toguri, James M
Varin, Robert Andrzej
Wallace, William
Walsh, John Heritage
Watt, Daniel Frank
Winegard, William Charles
Wynnyckyj, John Rostyslav
Youdelis, W(illiam) V(incent)

QUEBEC
Claisse, Fernand
Galibois, Andre
Habashi, Fathi
Hoa, Suong Van
Jonas, John Joseph
Kiss, Laszlo Istvan
Lavigne, Maurice J
McQueen, Hugh J
Ryan, Norman Daniel
Saint-Jacques, Robert G
Van Neste, Andre
Zepfel, William F

SASKATCHEWAN
Barton, Richard J

OTHER COUNTRIES
Anantharaman, Tanjore Ramachandra
Belton, Geoffrey Richard
Boxman, Raymond Leon
Daniel, John Sagar
De Vedia, Luis Alberto
Hornbogen, Erhard
Idelsohn, Sergio Rodolfo
Kim, Young-Gil
Ossipyan Yuriy, Andrew
Otsuka, Kazuhiro
Ramachandran, Venkataraman
Revesz, Zsolt
Rivier, Nicolas Yves
Sarmiento, Gustavo Sanchez
Shimizu, Ken'ichi
Solari, Mario Jose Adolfo
Spacil, Henry Stephen
Spottiswood, David James
Taplin, David Michael Robert
Tedmon, Craig Seward, Jr
Van Weert, Gezinus

Mining Engineering

ALABAMA
Ahrenholz, H(erman) William
Bakker, Martin Gerard
Dahl, Hilbert Douglas
Morley, Lloyd Albert
Park, Duk-Won

ALASKA
Cook, Donald J
Johansen, Nils Ivar
Rao, Pemmasani Dharma
Skudrzyk, Frank J
Wolff, Ernest N

ARIZONA
Hiskey, J Brent

CALIFORNIA
Beebe, Robert Ray
Boynton, W(illiam) W(entworth)
Cahn, David Stephen
Conger, Harry M
Gabelman, John Warren
Hartman, Howard L(evi)
Havard, John Francis
Hesse, Christian August
Heuze, Francois E
Hong, Ki C(hoong)
Kam, James Ting-Kong
Post, J(ames) L(ewis)
Pray, Ralph Emerson
Spokes, Ernest M(elvern)
Wallerstein, David Vandermere

COLORADO
Agapito, J F T
Ayler, Maynard Franklin
Chugh, Ashok Kumar
Gentry, Donald William
Grosvenor, Niles E(arl)
Howard, Thomas E(dward)
Hustrulid, William A
Lewis, Clifford Jackson
Michal, Eugene J(oseph)
Panek, Louis A(nthony)
Ponder, Herman
Porter, Darrell Dean
Ward, Milton Hawkins
Westfall, Richard Merrill

CONNECTICUT
Milliken, Frank R

DISTRICT OF COLUMBIA
Roosevelt, C(ornelius) V(an) S(chaak)
Yancik, Joseph J

FLORIDA
Abbaschian, Reza
Garnar, Thomas E(dward), Jr
Goudarzi, Gus (Hossein)
Poundstone, William N
Wang, K P

GEORGIA
Moody, Willis E, Jr

HAWAII
Cheng, Ping
Cruickshank, Michael James
Rollins, Ronald Roy

IDAHO
Austin, Carl Fulton
Bartlett, R(obert) W(atkins)
Blake, Wilson
Froes, Francis Herbert (Sam)

ILLINOIS
Chugh, Yoginder Paul
Eadie, George Robert
Heil, Richard Wendell

INDIANA
Lazaridis, Nassos A(thanasius)

KANSAS
Annis, Jason Carl

KENTUCKY
Leonard, Joseph William
Parekh, Bhupendra Kumar
Sweigard, Richard Joseph

LOUISIANA
Kazmann, Raphael Gabriel
Ponter, Anthony Barrie

MARYLAND
Horst, John Albert
Malghan, Subhaschandra Gangappa
Morgan, John D(avis)

MASSACHUSETTS
Tarkoy, Peter J

MICHIGAN
Freyberger, Wilfred L(awson)
Greuer, Rudolf E A
Klimpel, Richard Robert
Sarkar, Nitis
Snyder, Virgil W(ard)

MINNESOTA
Atchison, Thomas Calvin, Jr
Benner, Blair Richard
Crouch, Steven L
Dorenfeld, Adrian C
Fairhurst, C(harles)
Johnson, Thys B(rentwood)
Yardley, Donald H

MISSISSIPPI
Mather, Katharine Kniskern

MISSOURI
Barr, David John
Carmichael, Ronald L(ad)
Haas, Charles John
Hatheway, Allen Wayne
Saperstein, Lee W(aldo)
Scott, James J
Summers, David Archibold
Watson, John Leslie

MONTANA
Stout, Koehler

NEVADA
Daemen, Jaak J K
Ghosh, Amitava
Schilling, John H(arold)
Smith, Ross W

NEW JERSEY
Albrethsen, A(drian) E(dysel)
Avery, Howard S
Hassialis, Menelaos D(imitri)
Jin, Sungho
Maxim, Leslie Daniel
Yan, Tsoung-Yuan

NEW MEXICO
Baker, Don H(obart), Jr
Chen, Er-Ping
Coursen, David Linn
Griswold, George B
Sutherland, Herbert James
Wawersik, Wolfgang R

NEW YORK
Born, Allen
Boshkov, Stefan
Gambs, Gerard Charles
Harris, Colin C(yril)
Litz, Lawrence Marvin
Mahtab, M Ashraf
Malozemoff, Plato
Mika, Thomas Stephen
Wane, Malcolm T(rafford)
Yegulalp, Tuncel Mustafa

NORTH CAROLINA
Hamme, John Valentine
Humphreys, Kenneth K

OHIO
Babcock, Donald Eric
Copp, Albert Nils
Konya, Calvin Joseph

OKLAHOMA
Brady, Barry Hugh Garnet
Bray, Bruce G(lenn)

OREGON
Daellenbach, Charles Byron

PENNSYLVANIA
Aplan, Frank F(ulton)
Bell, Gordon M(oreton)
Bieniawski, Zdzislaw Tadeusz
Bise, Christopher John
Bockosh, George R
Core, Jesse F
Deurbrouck, Albert William
du Breuil, Felix L(emaigre)
Falkie, Thomas Victor
Fonseca, Anthony Gutierre
Hertzberg, Martin
Kelley, Jay Hilary
Kleinman, Robert L P
Lovell, Harold Lemuel
Mangelsdorf, Clark P
Mitchell, Donald W(illiam)
Mowrey, Gary Lee
Murphy, John N
Phelps, Lee Barry
Ramani, Raja Venkat
Rumbarger, John H
Skoner, Peter Raymond

RHODE ISLAND
Raykhman, Aleksandr

SOUTH DAKOTA
Han, Kenneth N

TEXAS
Bigham, Robert Eric
Camp, Frank A, III
Hoskins, Earl R, Jr
Mathewson, Christopher Colville
Monsees, James E
Parate, Natthu Sonbaji
Patrick, Wesley Clare
Russell, James E(dward)
Sahinen, Winston Martin
Schlitt, William Joseph, III
Shen, Chin-Wen
Winegartner, Edgar Carl

UTAH
Bhaskar, Ragula
Hucka, Vladimir Joseph
Michaelson, S(tanley) D(ay)
Miller, Jan Dean
Spedden, H Rush
Zavodni, Zavis Marian

VIRGINIA
Ary, T S
Cooley, William C
Faulkner, Gavin John
Foreman, William Edwin
Karfakis, Mario George
Karmis, Michael E
Lucas, J B
Topuz, Ertugrul S
Torgersen, Paul E
Wayland, Russell Gibson
Wright, Fred(erick) D(unstan)

WASHINGTON
Bowdish, Frank W(illiam)
Drevdahl, Elmer R(andolph)
Finlay, Walter L(eonard)
Hoskins, John Richard
Robbins, Richard J

WEST VIRGINIA
Adler, Lawrence
Chiang, Han-Shing
Khair, Abdul Wahab
Peng, Syd Syh-Deng
Thakur, Pramod Chandra

WISCONSIN
Dinter Brown, Ludmila

WYOMING
Laman, Jerry Thomas

ALBERTA
Patching, Thomas
Scott, J(ohn) D(onald)

BRITISH COLUMBIA
Hollister, Victor F
Meech, John Athol
Mular, A(ndrew) L(ouis)

ONTARIO
Baird, Malcolm Henry Inglis
Calder, Peter N
Diaz, Carlos Manuel
Graham, A(lbert) Ronald
Simpson, Frank
Udd, John Eaman

QUEBEC
Carbonneau, Come
Poorooshasb, Hormozd Bahman

OTHER COUNTRIES
Heerema, Ruurd Herre
Spottiswood, David James
Wolfe, John A(llen)

Nuclear Engineering

ALABAMA
Hollis, Daniel Lester, Jr
Morris, James Allen
Perry, Nelson Allen
Willenberg, Harvey Jack
Williams, Louis Gressett

ARIZONA
Albrecht, Edward Daniel
Ganapol, Barry Douglas
Hetrick, David LeRoy
Kirk, William Leroy
Koch, Leonard John
Koeneman, James Bryant
McKlveen, John William
Nelson, George William
Post, Roy G
Seale, Robert L(ewis)

ARKANSAS
Krohn, John Leslie

CALIFORNIA
Ahlfeld, Charles Edward
Auerbach, Jerome Martin
Balzhiser, Richard E(arl)
Bareis, David W(illard)
Benjamin, Arlin James
Bernath, L(ouis)
Bernsen, Sidney A
Blink, James Allen
Bolt, Robert O'Connor
Breitmayer, Theodore
Broido, Jeffrey Hale
Budnitz, Robert Jay
Bullock, Ronald Elvin
Busch, Joseph Sherman
Bush, Gary Graham
Carver, John Guill
Chan, Bertram Kim Cheong
Cheng, Edward Teh-Chang
Childs, Wylie J
Cohen, Karl (Paley)
Conlon, William Martin
Conn, Robert William
Culler, F(loyd) L(eRoy), Jr
Dahlberg, Richard Craig
D'Ardenne, Walter H
Dau, Gary John
Dean, Richard A
Deleray, Arthur Loyd
De Micheli, Giovanni
Dowdy, William Louis
Dunning, John Ray, Jr
Eggen, Donald T(ripp)
Ehrlich, Richard
Erdmann, Robert Charles
Esselman, Walter H
Fenech, Henri J
Forsen, Harold K(ay)
Freis, Robert P
Fullmer, George Clinton
Goodjohn, Albert J
Gritton, Eugene Charles
Grossman, Lawrence M(orton)
Gyorey, Geza Leslie
Haake, Eugene Vincent
Hallet, Raymon William, Jr
Hammond, R Philip
Harris, Sigmund Paul
Hecht, Myron J
Heckman, Richard Ainsworth
Hendry, George Orr
Hill, Ernest E(lwood)
Holdren, John Paul
Hovingh, Jack
Howe, John P(erry)
Howerton, Robert James
Judge, Frank D
Kaplan, Selig N(eil)
Karin, Sidney
Kastenberg, William Edward
Keller, Joseph Edward, Jr
Kendrick, Hugh
Kim, Jinchoon
Koenig, Daniel Rene
Koopman, Ronald P
Kull, Lorenz Anthony
Kupfer, John Carlton
Landsbaum, Ellis M(erle)
Langer, Sidney
Lerche, Richard Allan
Leung, Ka-Ngo
Levenson, Milton
Leverett, M(iles) C(orrington)
Lewis, Kenneth D
Loewenstein, Walter B
Long, Alexander B
Lucas, Glenn E
Lurie, Norman A(lan)
McCarthy, Walter J, Jr
McCreless, Thomas Griswold
Marto, Paul James
Melese, Gilbert B(ernard)
Merilo, Mati
Merker, Milton
Miatech, Gerald James
Michels, Lloyd R
Mireshghi, Ali
Moir, Ralph Wayne
Monard, Joyce Anne
Montague, L David
Morewitz, Harry Alan
Musket, Ronald George
Naymark, Sherman
O'Dell, Austin Almond, Jr
Odette, G(eorge) Robert
Oehlberg, Richard N
Olander, D(onald) R(aymond)
Orcutt, John Arthur
Orphan, Victor John
Papay, Lawrence T
Parks, Lewis Arthur
Partanen, Jouni Pekka
Passell, Thomas Oliver
Pathania, Rajeshwar S
Pedersen, Knud B(orge)
Perkins, Sterrett Theodore
Plebuch, Richard Karl
Pollack, Louis Rubin
Pomraning, Gerald C
Porter, Gary Dean
Profio, A(medeus) Edward
Ramaswami, Devabhaktuni
Rao, Atambir
Remley, Marlin Eugene
Reynolds, Harry Lincoln
Rickard, Corwin Lloyd
Risbud, Subhash Hanamant
Rossin, A David
Rumble, Edmund Taylor, III
Russell, Charles Roberts
Sampson, Henry T
Schalla, Clarence August
Schlaug, Robert Noel
Schrock, Virgil E(dwin)
Scott, Franklin Robert
Sesonske, Alexander
Sher, Rudolph
Siegel, Sidney
Simnad, Massoud T
Skavdahl, R(ichard) E(arl)
Slocum, Richard William
Smathers, James Burton
Snyder, Nathan W(illiam)
Stuhmiller, James Hamilton
Swanson, David G, Jr
Talley, Wilson K(inter)
Taylor, John Joseph
Theofanous, Theofanis George
Trippe, Anthony Philip
Turchan, Otto Charles
Vagelatos, Nicholas
Wagner, Carl E
Wang, Wei-E
Wazzan, A R Frank
Weber, Hans Josef
Williams, Alan K
Wolfe, Bertram
Wray, John L
Zebroski, Edwin L
Zimmerman, Elmer Leroy

COLORADO
Graham, John
Harlan, Ronald A
Jha, Mahesh Chandra
Ladd, Conrad Mervyn
Lewis, Homer Dick
Lubell, Jerry Ira
Meroney, Robert N
Reichardt, John William
Smith, Richard Cecil
Stauffer, Jack B
Wilbur, Paul James

CONNECTICUT
Blackshaw, George Lansing
Brenholdt, Irving R
Chernock, Warren Philip
Cobern, Martin E
Hood, Edward E, Jr
Johnson, Loering M
Jones, Richard Bradley
Korin, Amos
Krisst, Raymond John
Lichtenberger, Harold V
Prabulos, Joseph J, Jr
Rohan, Paul E(dward)
Storrs, Charles Lysander
Wachspress, Eugene Leon

DELAWARE
Kamack, H(arry) J(oseph)
Stewart, John Woods
Vick Roy, Thomas Rogers

DISTRICT OF COLUMBIA
Andreadis, Tim Dimitri
Bachkosky, John M (Jack)
Baer, Robert Lloyd
Bosnak, Robert J
Dowdy, Edward Joseph
Ferguson, George Alonzo
Florig, Henry Keith
Goodman, Eli I
Grabowski, Kenneth S
Griffith, Jerry Dice
Killion, Lawrence Eugene
Lang, Peter Michael
Marcus, Gail Halpern
Michel, David John
Odar, Fuat
Oktay, Erol
Rosen, Sol
Shon, Frederick John
Thomas, Cecil Owen, Jr
Tokar, Michael
Trice, Virgil Garnett, Jr

FLORIDA
Anghaie, Samim
Becker, Martin
Benedict, Manson
Buoni, Frederick Buell
Carroll, Edward Elmer
Cox, John David
Dalton, G(eorge) Ronald
Davis, Duane M
Diaz, Nils Juan
Dunford, James Marshall
Etherington, Harold
Hill, Frank B(ruce)
Humphries, Jack Thomas
Hungerford, Herbert Eugene
Jacobs, Alan M(artin)
Kerlin, Thomas W
Kornblith, Lester
Langford, David
Mandil, I Harry
Pendleton, Winston Kent, III
Rosenthal, Henry Bernard
Shea, Richard Franklin
Suciu, S(piridon) N
Tomonto, James R
Trubey, David Keith
Tulenko, James Stanley
Van Sciver, Steven W
Veziroglu, T Nejat
West, John M
Wethington, John A(bner), Jr

GEORGIA
Abdel-Khalik, Said Ibrahim
Circeo, Louis Joseph, Jr
Clement, Joseph D(ale)
Davis, Monte V
Eichholz, Geoffrey G(unther)
Graham, Walter Waverly, III
Griffin, Clayton Houstoun
Holt, David Lowell
Jessen, Nicholas C, Sr
Kallfelz, John Michael
Karam, Ratib A(braham)
Oakes, Thomas Wyatt
Russell, John Lynn, Jr
Rust, James Harold
Schneider, Alfred
Schutt, Paul Frederick
Shonka, Joseph John
Stacey, Weston Monroe, Jr
Visner, Sidney
Wilkins, J Ernest, Jr

HAWAII
Russo, Anthony R

IDAHO
Bickel, John Henry
Bills, Charles Wayne
Buescher, Brent J
Carpenter, Stuart Gordon
Chapman, Ray LaVar
DeBow, W Brad
Gasidlo, Joseph Michael
Gehrke, Robert James
Gutzman, Philip Charles
Hanson, George H(enry)
Harper, Henry Amos, Jr
Hyde, William W
Jobe, Lowell A(rthur)
Johnson, Stanley O(wen)
Kaiser, Richard Edward
Larson, Jay Reinhold
Leyse, Carl F(erdinand)
Long, John Kelley
McFarlane, Harold Finley
Majumdar, Debaprasad (Debu)
Meyer, Orville R
Moore, Kenneth Virgil
Mortensen, Glen Albert
Obenchain, Carl F(ranklin)
Olson, Willard Orvin
Ortiz, Marcos German
Reno, Harley W
Rice, Charles Merton
Riemke, Richard Allan
Sohal, Manohar Singh
Tanner, John Eyer, Jr
Trinko, Joseph Richard, Jr
Wilson, Albert E
Wood, Richard Ellet
Woodall, David Monroe

ILLINOIS
Axford, Roy Arthur
Bezella, Winfred August
Bhattacharyya, Samit Kumar
Braid, Thomas Hamilton
Braun, Joseph Carl
Carpenter, John Marland
Cember, Herman
Chang, Yoon Il
Crawford, Raymond Maxwell, Jr
Deen, James Robert
Deitrich, L(awrence) Walter
De Volpi, Alexander
Dickerman, Charles Edward
Doerner, Robert Carl
Errede, Steven Michael

Nuclear Engineering (cont)

Fenske, George R
Fink, Charles Lloyd
Flynn, Kevin Francis
Goldman, Arthur Joseph
Gruber, Eugene E, Jr
Guenther, Peter T
Hang, Daniel F
Haugen, Robert Kenneth
Holtzman, Richard Beves
Horve, Leslie A
Ivins, Richard O(rville)
Jones, Barclay G(eorge)
Kelman, L(eRoy) R
Kittel, J Howard
Lawroski, Stephen
Lellouche, Gerald S
LeSage, Leo G
Liaw, Jye Ren
Lipinski, Walter C(harles)
Liu, Yung Yuan
Macherey, Robert E
McKnight, Richard D
Malloy, Donald Jon
Marchaterre, John Frederick
Marr, William Wei-Yi
Massel, Gary Alan
Meneghetti, David
Miley, G(eorge) H(unter)
Monson, Harry O
Moran, Thomas J
Murdoch, Bruce Thomas
Pizzica, Philip Andrew
Redman, William Charles
Rose, David
Seefeldt, Waldemar Bernhard
Sha, William T
Shaftman, David Harry
Snelgrove, James Lewis
Steindler, Martin Joseph
Stubbins, James Frederick
Toppel, Bert Jack
Travelli, Armando
Uherka, Kenneth Leroy
Wensch, Glen W(illiam)
Woodruff, William Lee

INDIANA
Bement, A(rden) L(ee), Jr
Clikeman, Franklyn Miles
Kramer, Robert Allen
Lucey, John William
Lykoudis, Paul S
Ott, Karl O
Tucker, Robert C, Jr
Wellman, Henry Nelson

IOWA
Edelson, Martin Charles
Hendrickson, Richard A(llan)
Rohach, Alfred F(ranklin)

KANSAS
Eckhoff, Norman Dean
Faw, Richard E
Robinson, M John
Shultis, J Kenneth
Simons, Gale Gene
Weller, Lawrence Allenby

KENTUCKY
Davis, Thomas Haydn
Eaton, Thomas Eldon
Lindauer, George Conrad
Ulrich, Aaron Jack

LOUISIANA
Courtney, John Charles
Goodman, Alan Leonard
Hibbeler, Russell Charles
Lambremont, Edward Nelson
Snyder, Harold Jack
Young, Myron H(wai-Hsi)

MAINE
Hopkins, George Robert
Wilbur, L(eslie) C(lifford)

MARYLAND
Adamantiades, Achilles G
Almenas, Kazys Kestutis
Arsenault, Richard Joseph
Barrett, Richard John
Beckjord, Eric Stephen
Blush, Steven Michael
Bournia, Anthony
Buchanan, John Donald
Buck, John Henry
Bustard, Thomas Stratton
Carter, Robert Emerson
Congel, Frank Joseph
Coursey, Bert Marcel
Dean, Stephen Odell
Duffey, Dick
Eisenhauer, Charles Martin
Eiss, Abraham L(ouis)
Englehart, Richard W(ilson)
Fabic, Stanislav
Farmer, William S(ilas), Jr
Fitzgerald, Duane Glenn
Goktepe, Omer Faruk
Graves, Harvey W(ilbur), Jr
Grotenhuis, Marshall

Hawkins, William M(adison), Jr
Huddleston, Charles Martin
Jabbour, Kahtan Nicolas
Kazi, Abdul Halim
Khatib-Rahbar, Mohsen
Ku, Jentung
Loftness, Robert L(eland)
Loss, Frank J
McLeod, Norman Barrie
Mallory, Charles William
Marotta, Charles Rocco
Miller, Hugh Hunt
Munno, Frank J
Myers, Peter Briggs
Papadopoulos, Konstantinos Dennis
Reupke, William Albert
Reuther, Theodore Carl, Jr
Riel, Gordon Kienzle
Rockwell, Theodore
Rotariu, George Julian
Roush, Marvin Leroy
Sallet, Dirse Wilkis
Sanders, Robert Charles
Schleiter, Thomas Gerard
Schroeder, Frank, Jr
Schwoerer, F(rank)
Shalowitz, Erwin Emmanuel
Song, Yo Taik
Staker, Michael Ray
Stone, Philip M
Trombka, Jacob Israel
Tropf, Cheryl Griffiths
Van Echo, Andrew
Van Houten, Robert
Walton, Ray Daniel, Jr
Weber, Clifford E
Wiggins, Peter F
Willis, Charles Algert
Zmola, Paul C(arl)

MASSACHUSETTS
Abernathy, Frederick Henry
Bowman, H(arry) Frederick
Brown, Gilbert J
Carnesale, Albert
Connors, Philip Irving
Cooper, W(illiam) E(ugene)
Delagi, Richard Gregory
Driscoll, Michael John
Eoll, John Gordon
Gyftopoulos, Elias P(anayiotis)
Hansen, Kent F(orrest)
Harling, Otto Karl
Huffman, Fred Norman
Hutchinson, Ian Horner
Kautz, Frederick Alton, II
Kazimi, Mujid S
Kelsey, Ronald A(lbert)
Landis, John W
Lanning, David D(ayton)
Latanision, Ronald Michael
Lurie, Harold
Rasmussen, Norman Carl
Russell, Kenneth Calvin
Sigmar, Dieter Joseph
Silvers, J(ohn) P(hillip)
Stevens, J(ames) I(rwin)
Yener, Yaman
Yip, Sidney

MICHIGAN
Akcasu, Ahmet Ziyaeddin
Dow, W(illiam) G(ould)
Duderstadt, James
Eldred, Norman Orville
Gilgenbach, Ronald Matthew
Gomberg, Henry J(acob)
Haidler, William B(ernard)
Hill, Robert F
Kammash, Terry
Kerr, William
Knoll, Glenn F
Lee, John Chaeseung
Mayer, Frederick Joseph
Miller, Herman Lunden
Shure, Fred C(harles)
Summerfield, George Clark
Vincent, Dietrich H(ermann)

MINNESOTA
Fisher, Edward S
Isbin, Herbert S(tanford)
Keister, Jamieson Charles
Mohan, Narendra

MISSISSIPPI
Paulk, John Irvine

MISSOURI
Bolon, Albert Eugene
Carter, Robert L(eroy)
Donovan, John Richard
Edwards, Doyle Ray
Glascock, Michael Dean
Lowe, Forrest Gilbert
Loyalka, Sudarshan Kumar
Richards, Earl Frederick
Tsoulfanidis, Nicholas
Werner, Samuel Alfred

MONTANA
Cohn, Paul Daniel

NEBRASKA
Blotcky, Alan Jay
Borchert, Harold R
Wittke, Dayton D

NEVADA
Brandstetter, Albin
Miller, Wayne L(eroy)
Nassersharif, Bahram
Thamer, B(urton) J(ohn)

NEW HAMPSHIRE
Fricke, Edwin Francis
Price, Glenn Albert

NEW JERSEY
Adams, James Mills
Audet, Sarah Anne
Bartnoff, Shepard
Bush, Charles Edward
Cho, Soung Moo
Clark, Philip Raymond
Denno, Khalil I
Fu, Hui-Hsing
George, Kenneth Dudley
Gordon, R(obert)
Halik, Raymond R(ichard)
Hendrickson, Tom A(llen)
Kane, Ronald S(teven)
Kovats, Andre
Levine, Jerry David
Loman, James Mark
Long, Robert Leroy
McNicholas, James J
Mark, Robert
Reisman, Otto
Rothleder, Stephen David
Shaw, Henry
Shober, Robert Anthony
Towner, Harry H
Vann, Joseph M
Wachtell, George Peter
Weinzimmer, Fred
Welt, Martin A
Yoshikawa, Shoichi

NEW MEXICO
Baldwin, George C
Bergeron, Kenneth Donald
Booth, Lawrence A(shby)
Booth, Thomas Edward
Boudreau, Jay Edmond
Bryant, Lawrence E, Jr
Burick, Richard Joseph
Burman, Robert L
Caldwell, John Thomas
Chambers, William Hyland
Chen, Er-Ping
Drumm, Clifton Russell
El-Genk, Mohamed Shafik
England, Talmadge Ray
Fan, Wesley C
Gibbs, Terry Ralph
Gover, James E
Griffin, Patrick J
Guevara, Francisco A(ntonio)
Haight, Robert Cameron
Hoover, Mark Douglas
Humphries, Stanley, Jr
Jackson, James Fredrick
Jankowski, Francis James
Kelley, Gregory M
Kensek, Ronald P
Knief, Ronald Allen
Lam, Kin Leung
Lanter, Robert Jackson
Lee, David Mallin
Leonard, Ellen Marie
MacFarlane, Donald Robert
McGuire, Joseph Clive
McLeod, John
Marlow, Keith Winton
Matthews, R(obert) B(ruce)
Mazarakis, Michael Gerassimos
Menlove, Howard Olsen
Miller, Warren Fletcher, Jr
Milton, Osborne
Muir, Douglas William
Nickel, George H(erman)
Nuttall, Herbert Ericksen, Jr
O'Dell, Ralph Douglas
Philbin, Jeffrey Stephen
Powers, Dana Auburn
Pryor, Richard J
Rahal, Leo James
Renken, James Howard
Sandmeier, Henry Armin
Sicilian, James Michael
Snyder, Albert W
Sprinkle, James Kent, Jr
Stephans, Richard A
Stevenson, Michael Gail
Tanaka, Nobuyuki
Terrell, C(harles) W(illiam)
Thomas, Charles Carlisle, Jr
Travis, John Richard
Valentine, James K
Vigil, John Carlos
Von Riesemann, Walter Arthur
Watson, Clayton Wilbur
Williams, James Marvin
Williams, Michael D
Yarnell, John Leonard

NEW YORK
Anthony, Donald Joseph
Bari, Robert Allan
Baron, Seymour
Baum, John W
Block, Robert Charles
Carew, John Francis
Chan, Shu Fun
Chappell, Richard Lee
Chen, Inan
Clark, David Delano
Cokinos, Dimitrios
Coppersmith, Frederick Martin
Diamond, David J(oseph)
Feinberg, Robert Jacob
Fishbone, Leslie Gary
Fried, Erwin
Fullwood, Ralph Roy
Goldmann, Kurt
Gralnick, Samuel Louis
Hammer, David Andrew
Hanson, Albert L
Harris, Donald R, Jr
Hartmann, Francis Xavier
Hendrie, Joseph Mallam
Higinbotham, William Alfred
Hockenbury, Robert Wesley
Hofer, Gregory George
Hoisie, Adolfy
Ishida, Takanobu
Jedruch, Jacek
Jensen, Betty Klainminc
Kane, John Vincent, Jr
Kato, Walter Yoneo
Kayani, Joseph Thomas
King, Alexander Harvey
Kitchen, Sumner Wendell
Knight, Bruce Winton, (Jr)
Krieger, Gary Lawrence
Kroeger, Peter G
Lahey, Richard Thomas, Jr
Lazareth, Otto William, Jr
Lee, Richard Shao-Lin
Levine, Melvin Mordecai
Lidofsky, Leon Julian
Lin, Mow Shiah
Lubitz, Cecil Robert
McGuire, Stephen Craig
Marschke, Stephen Frank
Marwick, Alan David
Meyer, Walter
Ott, Henry C(arl)
Perkins, Kenneth Roy
Podowski, Michael Zbigniew
Powell, James R, Jr
Simon, Albert
Stephenson, Thomas E(dgar)
Strasser, Alfred Anthony
Sturges, Stuart
Susskind, Herbert
Takahashi, Hiroshi
Von Berg, Robert L(ee)
Vortuba, Jan
Ward, Thomas J(ulian)
Weber, Arthur Phineas
Weng, Wu Tsung
White, Frederick Andrew
Wulff, Wolfgang

NORTH CAROLINA
Ahearne, John Francis
Burton, Ralph Gaines
Callihan, Dixon
Doster, Joseph Michael
Dudziak, Donald John
Fahmy, Abdel Aziz
Gardner, Robin P(ierce)
Jordan, Edward Daniel
MacMillan, J(ohn) H(enry)
Murray, Raymond L(eRoy)
Siewert, Charles Edward
Squire, Alexander
Turinsky, Paul Josef
Verghese, Kuruvilla
Watson, James E, Jr
Wiley, Albert Lee, Jr
Zaalouk, Mohamed Gamal
Zumwalt, Lloyd Robert

OHIO
Anno, James Nelson
Bailey, Robert E
Basham, Samuel Jerome, Jr
Beverly, Robert Edward, III
Bhattacharya, Ashok Kumar
Bridgman, Charles James
Carbiener, Wayne Alan
Denning, Richard Smith
Janzow, Edward F(rank)
John, George
Kaplan, Arthur Lewis
Kelly, Kevin Anthony
Keshock, Edward G
Lundin, Bruce T(heodore)
Miller, Don Wilson
Nakamura, Shoichiro
Robinson, Richard Alan
Rudy, Clifford R
Sanford, Edward Richard
Shapiro, Alvin
Stockman, Charles H(enry)
Sund, Raymond Earl
Vanderburg, Vance Dilks
Weisman, Joel

Woodard, Ralph Emerson

OKLAHOMA
Bray, Bruce G(lenn)
Howard, Robert Adrian
Zigrang, Denis Joseph

OREGON
Chezem, Curtis Gordon
Kolar, Oscar Clinton
Ringle, John Clayton
Ruby, Lawrence
Woods, W(allace) Kelly

PENNSYLVANIA
Baratta, Anthony J
Boltax, Alvin
Boyer, Vincent S
Brey, R(obert) N(ewton)
Carelli, Mario Domenico
Carfagno, Salvatore P
Catchen, Gary Lee
Chapin, David Lambert
Chelemer, Harold
Chi, John Wen Hua
Chubb, Walston
Cohen, Paul
Curlee, Neil J(ames), Jr
Diethorn, Ward Samuel
Doub, William Blake
Fischer, John Edward
Foderaro, Anthony Harolde
Forscher, Frederick
Freeman, Louis Barton
Giffen, Robert H(enry)
Green, Lawrence
Haghighat, Alireza
Hardy, Judson, Jr
Harkness, Samuel Dacke
Hogan, Aloysius Joseph, Jr
Houlihan, John Frank
Impink, Albert J(oseph), Jr
Jacobi, W(illiam) M(allett)
Jester, William A
Johnston, William V
Kitzes, Arnold S(tanley)
Klevans, Edward Harris
Kline, Donald Edgar
Kolar, Michael Joseph
Kowal, George M
Levine, Samuel Harold
Lippincott, Ezra Parvin
Lochstet, William A
Loscher, Robert A
Mahaffy, John Harlan
Mezger, Fritz Walter William
Michael, Norman
Palladino, Nunzio J(oseph)
Palusamy, Sam Subbu
Pfennigwerth, Paul Leroy
Redfield, John A(lden)
Remick, Forrest J(erome)
Saluja, Jagdish Kumar
Sarram, Mehdi
Shure, Kalman
Siddiqui, Habib
Silverstein, Calvin C(arlton)
Smid, Robert John
Steinbruegge, Kenneth Brian
Stern, Theodore
Tang, Y(u) S(un)
Treadwell, Kenneth Myron
Vance, William Harrison
Waldman, L(ouis) A(braham)
Weddell, George G(ray)
Witzig, Warren Frank
Yao, Shi Chune
Yeniscavich, William

RHODE ISLAND
Rose, Vincent C(elmer)

SOUTH CAROLINA
Baumann, Norman Paul
Benjamin, Richard Walter
Bradley, Robert Foster
Bridges, Donald Norris
Chao, Yuh J (Bill)
Church, John Phillips
Clark, Hugh Kidder
Crandall, John Lou
Dworjanyn, Lee O(leh)
Fjeld, Robert Alan
Gaines, Albert L(owery)
Godfrey, W Lynn
Graulty, Robert Thomas
Graves, William Ewing
Gregory, Michael Vladimir
Groh, Harold John
Hahn, Walter I
Hennelly, Edward Joseph
Hootman, Harry Edward
Inhaber, Herbert
Johnson, Ben S(lemmons), Jr
Macaulay, Andrew James
McCrosson, F Joseph
Parkinson, Thomas Franklin
Schwartz, Elmer G(eorge)
Swank, Robert Roy, Jr
Townes, George Anderson
Williamson, Thomas Garnett
Zumbrunnen, David Arnold

TENNESSEE
Ackermann, Norbert Joseph, Jr
Armentrout, Daryl Ralph
Auxier, John A
Baer, Thomas Strickland
Baker, Charles Clayton
Besmann, Theodore Martin
Bigelow, John E(aly)
Booth, Ray S
Brewington, Percy, Jr
Bryan, Robert H(owell)
Burwell, Calvin C
Claiborne, H(arry) C(lyde)
Cole, Thomas Earle
Cope, David Franklin
Crume, Enyeart Charles, Jr
DeVan, Jackson H
Fontana, Mario H
Garber, Harold Jerome
Gat, Uri
Godbee, H(erschel) W(illcox)
Gray, Allen G(ibbs)
Grossbeck, Martin Lester
Haas, Paul Arnold
Harmer, David Edward
Horak, James Albert
Johnson, Elizabeth Briggs
Judkins, Roddie Reagan
Kasten, Paul R(udolph)
Kress, Thomas Sylvester
Larkin, William (Joseph)
LeVert, Francis E
Lotts, Adolphus Lloyd
MacPherson, Herbert Grenfell
Maerker, Richard Erwin
Maienschein, Fred (Conrad)
Mansur, Louis Kenneth
Martin, Harry Lee
Mihalczo, John Thomas
Morgan, Karl Ziegler
Morgan, Marvin Thomas
Morgan, Ora Billy, Jr
Mott, Julian Edward
Mullikin, H(arwood) F(ranklin)
Murakami, Masanori
Nestor, C William, Jr
Osborne, Morris Floyd
Pardue, William M
Partain, Clarence Leon
Pasqua, Pietro F
Phung, Doan Lien
Roussin, Robert Warren
Row, Thomas Henry
Sanders, John P(aul), Sr
Shor, Arthur Joseph
Stone, Robert Sidney
Stradley, James Grant
Toline, Francis Raymond
Trauger, Donald Byron
Uhrig, Robert Eugene
Upadhyaya, Belle Raghavendra
Webster, Curtis Cleveland
Weinberg, Alvin Martin

TEXAS
Betts, Austin Wortham
Bland, William M, Jr
Cochran, Robert Glenn
Dodson, David Scott
Feltz, Donald Everett
Hirsch, Robert Louis
Iddings, Frank Allen
Klein, Dale Edward
Koen, Billy Vaughn
Levine, Joseph H
Marshall, Robert P(aul)
Overmyer, Robert Franklin
Peddicord, Kenneth Lee
Poston, John Ware
Preeg, William Edward
Rubin, Richard Mark
Shieh, Paulinus Shee-Shan
Simonis, John Charles
Wainerdi, Richard E(lliott)
Watson, Jerry M
Wehring, Bernard William
Wells, Michael Byron

UTAH
Chesworth, Robert Hadden
Lloyd, Ray Dix
Rogers, Vern Child
Sandquist, Gary Marlin
Turley, Richard Eyring

VIRGINIA
Almeter, Frank M(urray)
Ball, Russell Martin
Bartlett, John W(esley)
Bauer, Douglas Clifford
Bingham, Billy Elias
Campbell, Francis James
Chapin, Douglas McCall
Deddens, J(ames) C(arroll)
DePoy, Phil Eugene
Deuster, Ralph W(illiam)
Dorning, John Joseph
Finke, Reinald Guy
Finkelstein, Jay Laurence
Fitzgibbons, J(ohn) D(avid)
Foy, C Allan
Hanauer, Stephen H(enry)
Hauxwell, Gerald Dean
Johnson, W Reed

Kelly, James L(eslie)
Kinder, Thomas Hartley
Kurstedt, Harold Albert, Jr
Meem, J(ames) Lawrence, Jr
Mertes, Frank Peter, Jr
Mock, John E(dwin)
Muehlhause, Carl Oliver
Mulder, Robert Udo
Neil, George Randall
Pettus, William Gower
Raab, Harry Frederick, Jr
Roberts, A(lbert) S(idney), Jr
Savage, William F(rederick)
Senich, Donald
Siebentritt, Carl R, Jr
Steele, Lendell Eugene
Townsend, Lawrence Willard
Whitelaw, R(obert) L(eslie)
Zweifel, Paul Frederick

WASHINGTON
Albrecht, Robert William
Anderson, Harlan John
Astley, Eugene Roy
Babb, Albert L(eslie)
Barker, James J(oseph)
Beaton, Roy Howard
Bowdish, Frank W(illiam)
Bunch, Wilbur Lyle
Buonamici, Rino
Bush, S(pencer) H(arrison)
Campbell, Milton Hugh
Carter, John Lemuel, Jr
Carter, Leland LaVelle
Clayton, Eugene Duane
Danko, Joseph Christopher
Davies, Craig Edward
Evans, Ersel Arthur
Evans, Thomas Walter
Fisher, Darrell R
Gajewski, W(alter) M(ichael)
Garlid, Kermit L(eroy)
Gold, Raymond
Gore, Bryan Frank
Hanna, Jeff
Hendrickson, Waldemar Forrsel
Hill, Orville Farrow
Hofmann, Peter L(udwig)
Hopkins, Horace H, Jr
Huang, Fan-Hsiung Frank
Johnson, A(lfred) Burtron, Jr
Keating, John Joseph
Leggett, Robert Dean
Lewis, Milton
Lindstrom, Duaine Gerald
Little, Winston Woodard, Jr
Lloyd, Raymond Clare
McCormick, Norman Joseph
Matte, Joseph, III
Morton, Randall Eugene
Newman, Darrell Francis
Nicholson, Richard Benjamin
Reynolds, Roger Smith
Rieck, H(enry) G(eorge)
Roake, William Earl
Roberts, J T Adrian
Robkin, Maurice
Rohrmann, Charles A(lbert)
Shen, Peter Ko-Chun
Spinrad, Bernard Israel
Timmons, Darrol Holt
Trent, Donald Stephen
Waltar, Alan Edward
Weber, William J
Wedlick, Harold Lee
Wood, Donald Eugene
Woodfield, F(rank) W(illiam), Jr
Woodruff, Gene L(owry)
Woods, Keith Newell
Yoshikawa, Herbert Hiroshi

WEST VIRGINIA
Spaniol, Craig
Wilson, John Sheridan

WISCONSIN
Barschall, Henry Herman
Bevelacqua, Joseph John
Blanchard, James Page
Bonazza, Riccardo
Boom, Roger Wright
Burstein, Sol
Carbon, Max W(illiam)
Chan, Shih Hung
Christensen, Erik Regnar
DuTemple, Octave J
Emmert, Gilbert A
Falk, Edward D
Kulcinski, Gerald La Vern
Moses, Gregory Allen

WYOMING
Lawroski, Harry
Mingle, John O(rville)

PUERTO RICO
Bonnet, Juan A, Jr
Rodriguez, Luis F
Sasscer, Donald S(tuart)

BRITISH COLUMBIA
Palmer, James Frederick

MANITOBA
Barclay, Francis Walter
Crosthwaite, John Leslie
Jovanovich, Jovan Vojislav
McDonnell, Francis Nicholas
Rosinger, Herbert Eugene
Sargent, Frederick Peter
Smith, Roger M
Woo, Chung-Ho

ONTARIO
Andrews, Douglas Guy
Biggs, Ronald C(larke)
Bigham, Clifford Bruce
Cox, B(rian)
Craig, Donald Spence
Dastur, Ardeshir Rustom
Dickson, Lawrence William
Furter, W(illiam) F(rederick)
Garland, William James
Gow, William Alexander
Hancox, William Thomas
Harms, Archie A
Hastings, Ian James
Holtslander, William John
Jackson, David Phillip
Nelles, John Sumner
Nirdosh, Inderjit
Robertson, J(ohn) A(rchibald) L(aw)
Rogers, James Terence, (Jr)
Rosen, Marc Allen
Shemilt, L(eslie) W(ebster)
Siddall, Ernest
Vijay, Mohan Madan

QUEBEC
L'Archeveque, Real Viateur
Paskievici, Wladimir
Pla, Conrado

SASKATCHEWAN
Amundrud, Donald Lorne

OTHER COUNTRIES
Angelini, Arnaldo M
Burnett, Jerrold J
Gauster, Wilhelm Belrupt
Gauster, Wilhelm Friedrich
Mathews, Donald R(ichard)
Mulas, Pablo Marcelo
Rodriguez, Federico Angel
Solari, Mario Jose Adolfo
Williams, Michael Maurice Rudolph

Operations Research

ALABAMA
Brown, Robert Alan
Cox, Julius Grady
Griffin, Marvin A
Haak, Frederik Albertus
Regner, John LaVerne
Smith, Ernest Lee, Jr
Unger, Vernon Edwin, Jr
White, Charles Raymond
Wyskida, Richard Martin

ALASKA
Bennett, F Lawrence

ARIZONA
Askin, Ronald Gene
Bergman, Ray E(ldon)
Buras, Nathan
Clemmens, Albert J
Cochran, Jeffery Keith
Duckstein, Lucien
Harris, DeVerle Porter
Lewis, William E(rvin)
Lord, William B
Mackulak, Gerald Thomas
Neuts, Marcel Fernand
Sorooshian, Soroosh
Szidarovszky, Ferenc

ARKANSAS
Carr, Gerald Paul
Taha, Hamdy Abdelaziz

CALIFORNIA
Ancker, Clinton J(ames), Jr
Anderson, George William
Arnwine, William Carrol
Arrow, Kenneth J
Arthur, William Brian
Aster, Robert Wesley
Baer, Adrian Donald
Basinger, Richard Craig
Beckwith, Richard Edward
Benjamin, Arlin James
Bhandarkar, Mangalore Dilip
Blumberg, Mark Stuart
Borsting, Jack Raymond
Bradley, Hugh Edward
Brandeau, Margaret Louise
Bratt, Bengt Erik
Bryson, Marion Ritchie
Capp, Walter B(ernard)
Chamberlain, Robert Glenn
Channel, Lawrence Edwin
Coile, Russell Cleven
Cottle, Richard W
Cretin, Shan

Operations Research (cont)

Daganzo, Carlos Francisco
Davidson, Lynn Blair
Dean, Burton Victor
Dresch, Francis William
Dreyfus, Stuart Ernest
Earhart, Richard Wilmot
Eaves, Burchet Curtis
Erickson, Stanley Arvid
Farquhar, Peter Henry
Feinstein, Charles David
Forrest, Robert Neagle
Freiling, Edward Clawson
Gafarian, Antranig Vaughn
Gray, Paul
Greenberg, Bernard
Groce, David Eiben
Harrison, Don Edward, Jr
Holtzman, Samuel
Hong, Ki C(hoong)
Horne, Roland Nicholas
Hudson, Cecil Ivan, Jr
Ingram, Gerald E(ugene)
Jensen, Paul Edward T
Jewell, William S(ylvester)
Keeney, Ralph Lyons
Kelley, Charles Thomas, Jr
Kendall, Burton Nathaniel
Koenigsberg, Ernest
Kreutz-Delgado, Kenneth Keith
Laitin, Howard
Lalchandani, Atam Prakash
Lam, Tenny N(icolas)
Lambert, Walter Paul
Lave, Roy E(llis), Jr
Lawler, Eugene L(eighton)
Lieberman, Gerald J
Lim, Hong Seh
Lindsay, Glenn Frank
Lucas, William Franklin
McMasters, Alan Wayne
Manne, Alan S
Mendelssohn, Roy
Miles, Ralph Fraley, Jr
Mood, Alexander McFarlane
Morris, Peter Alan
Nahmias, Steven
Oliver, Robert Marquam
Oren, Shmuel Shimon
Ott, Wayne R
Paulson, Boyd Colton, Jr
Philipson, Lloyd Lewis
Pietrzak, Lawrence Michael
Richards, Francis Russell
Rockower, Edward Brandt
Rowntree, Robert Fredric
Schamberg, Richard
Schlesinger, Robert Jackson
Scholtz, Robert A
Schrady, David Alan
Schweikert, Daniel George
Shanteau, Robert Marshall
Shanthikumar, Jeyaveerasingam George
Shedler, Gerald Stuart
Silvester, John Andrew
Stauffer, Howard Boyer
Steingold, Harold
Stinson, Perri June
Sweeney, James Lee
Torbett, Emerson Arlin
Trauring, Mitchell
Vanderplaats, Garret Niel
Veinott, Arthur Fales, Jr
Wallerstein, David Vandermere
Whipple, Christopher George
Wilde, D(ouglass) J(ames)
Wollmer, Richard Dietrich
Wong, Derek
Wright, Edward Kenneth
Wu, Felix F
Yu, Oliver Shukiang
Zimmerman, Elmer Leroy

COLORADO
Clementson, Gerhardt C
Lameiro, Gerard Francis
Lazzarini, Albert John
Litwhiler, Daniel W
Morrison, John Stuart
Osborn, Ronald George
Tischendorf, John Allen
Underwood, Robert Gordon
Wackernagel, Hans Beat

CONNECTICUT
Abraham, Thomas
DeDecker, Hendrik Kamiel Johannes
Duvivier, Jean Fernand
Friedman, Don Gene
Hajela, Dan
Kleinfeld, Ira H
Newman, Robert Weidenthal
Powell, Bruce Allan
Stein, David Morris
Terry, Herbert
Williams, David G(erald)

DELAWARE
Elterich, G Joachim
Kikuchi, Shinya
Kiviat, Fred E
Stark, Robert M

DISTRICT OF COLUMBIA
Baz, Amr Mahmoud Sabry
Kaplan, David Jeremy
Klemm, Rebecca Jane
Knight, John C (Ian)
Machol, Robert E
Needels, Theodore S
Niedenfuhr, Francis W(illiam)
Seglie, Ernest Augustus
Singpurwalla, Nozer Drabsha
Soland, Richard Martin
Waters, Robert Charles
Willard, Daniel

FLORIDA
Buoni, Frederick Buell
Doyon, Leonard Roger
Duersch, Ralph R
Elzinga, D(onald) Jack
Francis, Richard Lane
Hanson, Morgan A
Hertz, David Bendel
Johnson, Mark Edward
Leavenworth, Richard S
Lloyd, Laurance H(enry)
Lund, Frederick H(enry)
Moder, Joseph J(ohn)
Muth, Eginhard Joerg
Olmstead, Paul Smith
Peart, Robert McDermand
Roth, Paul Frederick
Sahni, Sartaj Kumar
Sivazlian, Boghos D
Stana, Regis Richard
Swart, William W
Whitehouse, Gary E

GEORGIA
Aronson, Jay E
Banks, Jerry
Buckalew, Louis Walter
Callahan, Leslie G, Jr
Cohen, Alonzo Clifford, Jr
Dickerson, Stephen L(ang)
Esogbue, Augustine O
Fyffe, David Eugene
Grum, Allen Frederick
Jarvis, John J
Johnson, Ellis Lane
Johnson, Lynwood Albert
Longini, Ira Mann, Jr
Nemhauser, George L
Shakun, Wallace
Thomas, Michael E(dward)
Ware, Glenn Oren
White, John Austin, Jr
Wood, Robert E

HAWAII
Anderson, Gerald M
Lawson, Joel S(mith)
Liang, Tung
Papacostas, Constantinos Symeon
Wylly, Alexander

IDAHO
Bickel, John Henry
Ortiz, Marcos German
Stuffle, Roy Eugene

ILLINOIS
Caywood, Thomas E
Chitgopekar, Sharad Shankarrao
Chugh, Yoginder Paul
Conry, Thomas Francis
Cooper, Mary Weis
Cornwell, Larry Wilmer
Dessouky, Mohamed Ibrahim
Firstman, Sidney I(rving)
Fourer, Robert
Gillette, Richard F
Hassan, Mohammad Zia
Hoffman, Gerald M
Hurter, Arthur P
Kalai, Ehud
Koenig, Michael Edward Davison
Kumar, Panganamala Ramana
Liebman, Jon C(harles)
Liebman, Judith Stenzel
McAlpin, John Harris
Maltz, Michael D
Miller, Robert Carl
Orden, Alex
Phelan, James Joseph
Phillips, Rohan Hilary
Smith, Spencer B
Steinberg, David Israel
Thomopoulos, Nick Ted
West, Douglas Brent

INDIANA
Cerimele, Benito Joseph
Coyle, Edward John
Hockstra, Dale Jon
Leimkuhler, Ferdinand F
Lyons, Jerry L
Morin, Thomas Lee
Nof, Shimon Y
Reklaitis, Gintaras Victor
Skibniewski, Miroslaw Jan
Solberg, James J
Sweet, Arnold Lawrence
Szpankowski, Wojciech
Thomas, Marlin Uluess

Thuente, David Joseph
Wright, Jeffery Regan

IOWA
Andersland, Mark Steven
Kortanek, Kenneth O
Kusiak, Andrew
Liittschwager, John M(ilton)
McRoberts, Keith L
Sposito, Vincent Anthony

KANSAS
Grosh, Doris Lloyd
Harnett, R Michael
Hearne, Horace Clark, Jr
Kramer, Bradley Alan
Lee, E(ugene) Stanley
Lee, Joe
Yu, Yun-Sheng

KENTUCKY
Evans, Gerald William
Feibes, Walter
Seeger, Charles Ronald
Vaziri, Menouchehr
Warner, Richard Charles

LOUISIANA
Kraft, Donald Harris
McCarthy, Danny W
Naidu, Seetala V
Ristroph, John Heard
Spaht, Carlos G, II

MAINE
Wilms, Hugo John, Jr

MARYLAND
Adler, Sanford Charles
Arsham, Hossein
Barrack, Carroll Marlin
Barrett, Richard John
Bateman, Barry Lynn
Belliveau, Louis J
Burkhalter, Barton R
Chacko, George Kutticka
Chiogioji, Melvin Hiroaki
Costanza, Robert
Cushen, Walter Edward
Daly, John Anthony
Das, Prasanta
Egner, Donald Otto
Eisner, Howard
Ellingwood, Bruce Russell
Ellwein, Leon Burnell
Englund, John Arthur
Flagle, Charles D(enhard)
Flum, Robert S(amuel), Sr
Frank, Carolyn
Fromovitz, Stan
Gardner, Sherwin
Garver, Robert Vernon
Gottfried, Paul
Grubbs, Frank Ephraim
Harris, Carl Matthew
Hartman, Patrick James
Henry, James H(erbert)
Honig, John Gerhart
Hsieh, Richard Kuochi
Johnson, Russell Dee, Jr
Katcher, David Abraham
Korobkin, Irving
Levine, Eugene
More, Kenneth Riddell
Moshman, Jack
Offutt, William Franklin
Poltorak, Andrew Stephen
ReVelle, Charles S
Samuel, Aryeh Hermann
Steiner, Henry M
Temperley, Judith Kantack
Troutman, James Scott
Voss, Paul Joseph
Walton, Ray Daniel, Jr
Wilson, John Phillips

MASSACHUSETTS
Barnhart, Cynthia
Bovopoulos, Andreas D
Brandler, Philip
Cline, James Edward
Cullinane, Thomas Paul
DeCosta, Peter F(rancis)
De Neufville, Richard Lawrence
Drake, Alvin William
Eckstein, Jonathan
Ernst, Martin L
Fariss, Thomas Lee
Frey, Elmer Jacob
Gartner, Nathan Hart
Giglio, Richard John
Graves, Robert John
Gupta, Surendra Mohan
Heitman, Richard Edgar
Hertweck, Gerald
Ilic, Marija
Kreifeldt, John Gene
Larson, Richard Charles
Little, John Dutton Conant
Magnanti, Thomas L
Martland, Carl Douglas
Meal, Harlan C
Meal, Janet Hawkins
Papageorgiou, John Constantine

Reinschmidt, Kenneth F(rank)
Russo, Richard F
Safford, Richard Whiley
Sheffi, Yosef
Sun, Fang-Kuo
Thorpe, Rodney Warren
Tobin, Roger Lee
Wolfe, Harry Bernard
Zachary, Norman

MICHIGAN
Bergmann, Dietrich R(udolf)
Bonder, Seth
Chakravarthy, Srinivasaraghavan
Dittmann, John Paul
Flaim, Thomas Alfred
Goodman, Erik David
Horvath, William John
Kachhal, Swatantra Kumar
Khasnabis, Snehamay
Klimpel, Richard Robert
Kochen, Manfred
Kshirsagar, Anant Madhav
Lamberson, Leonard Roy
Li, Chi
Magerlein, James Michael
Pandit, Sudhakar Madhavrao
Polis, Michael Philip
Pollock, Stephen M
Roll, James Robert
Sahney, Vinod K
Schriber, Thomas J
Taraman, Khalil Showky
Tompkins, Curtis Johnston
Wang, Dazong
Weiner, Steven Allan
Weinstein, Jeremy Saul
Wolf, Franklin Kreamer
Wolman, Eric

MINNESOTA
Ayeni, Babatunde J
Carstens, Allan Matlock
Hoffmann, Thomas Russell
Johnson, Thys B(rentwood)
Joseph, Earl Clark, II
Knutson, Charles Dwaine
Richards, James L
Starr, Patrick Joseph
Wollenberg, Bruce Frederick

MISSISSIPPI
Rosenhan, A Kirk

MISSOURI
Glauz, William Donald
Klein, Cerry M
Krone, Lester H(erman), Jr
Murden, W(illiam) P(aul), Jr
Taibleson, Mitchell H

MONTANA
Jameson, William J, Jr

NEBRASKA
Leung, Joseph Yuk-Tong

NEW HAMPSHIRE
Bahr, Karl Edward
Jeffery, Lawrence R
Jolly, Stuart Martin
Stancl, Donald L

NEW JERSEY
Ahsanullah, Mohammad
Albin, Susan Lee
Ben-Israel, Adi
Boucher, Thomas Owen
Chasalow, Ivan G
Cornell, W(arren) A(lvan)
Cullen, Daniel Edward
Eckler, Albert Ross
Elsayed, Elsayed Abdelrazik
Fask, Alan S
Forys, Leonard J
Garber, H(irsh) Newton
Gourary, Barry Sholom
Grumer, Eugene Lawrence
Haas, Zygmunt
Hunter, John Stuart
Iyer, Ram R
Kantor, Paul B
Kaplan, Joel Howard
Kobayashi, Hisashi
Luxhoj, James Thomas
McCallum, Charles John, Jr
Maxim, Leslie Daniel
Ott, Teunis Jan
Polhemus, Neil W
Ramaswami, Vaidyanathan
Rothkopf, Michael H
Saniee, Iraj
Sarakwash, Michael
Schneider, Alfred Marcel
Segal, Moshe
Segers, Richard George
Sengupta, Bhaskar
Sherman, Ronald
Thomas, Ronald Emerson
Thomas, Stanislaus S(tephen)
Tsaliovich, Anatoly
Weir, William Thomas
Whitt, Ward
Wood, Eric F

Yoon, Kwangsun Paul

NEW MEXICO
Berrie, David William
Flanagan, Robert Joseph
Jamshidi, Mohammad Mo
Martz, Harry Franklin, Jr
Walvekar, Arun Govind

NEW YORK
Aly, Adel Ahmed
Balachandran, Kashi Ramamurthi
Barnett, Michael Peter
Benson, Richard Carter
Bialas, Wayne Francis
Brown, Theodore D
Burke, Paul J
Burros, Raymond Herbert
Conway, Richard Walter
Derman, Cyrus
Doganaksoy, Necip
Dudewicz, Edward John
Fishbone, Leslie Gary
Goel, Amrit Lal
Harris, Colin C(yril)
Heidelberger, Philip
Helly, Walter S
Kabak, Irwin William
Karwan, Mark Henry
Kaufman, Sol
Kershenbaum, Aaron
Kingston, Paul L
Klein, Morton
Leibowitz, Martin Albert
Lyons, Joseph Paul
Mitchell, Joseph Shannon Baird
Mosteller, Henry Walter
Owen, Joel
Richter, Donald
Sargent, Robert George
Schweitzer, Paul Jerome
Shaw, Leonard G
Shoemaker, Christine Annette
Stedinger, Jery Russell
Swann, Dale William
Thomas, Warren H(afford)
Todd, Michael Jeremy
Turnbull, Bruce William
Turnquist, Mark Alan
Ullmann, John E
Van Slyke, Richard M
Vidosic, J(oseph) P(aul)
White, William Wallace
Wilks, Daniel S
Wood, Paul Mix
Yablon, Marvin
Yao, David D
Yegulalp, Tuncel Mustafa

NORTH CAROLINA
Ball, Michael Owen
Bernhard, Richard Harold
Elmaghraby, Salah Eldin
Galler, William Sylvan
Gold, Harvey Joseph
Hodgson, Thom Joel
Honeycutt, Thomas Lynn
Karr, Alan Francis
Kilpatrick, Kerry Edwards
Martin, LeRoy Brown, Jr
Meier, Wilbur L(eroy), Jr
Papadopoulos, Alex Spero
Peterson, Elmor Lee
Roth, Walter John
Stidham, Shaler, Jr
Tolle, Jon Wright

NORTH DAKOTA
Hare, Robert Ritzinger, Jr
Kuruganty, Sastry P
Wang, Jun

OHIO
Bishop, Albert B
Chan, Yupo
Chrissis, James W
Clark, Gordon Meredith
Cruz, Jose B(ejar), Jr
Daschbach, James McCloskey
Emmons, Hamilton
Flowers, Allan Dale
Gleim, Clyde Edgar
Hurt, James Joseph
Kowal, Norman Edward
McNair, Robert J(ohn), Jr
Martino, Joseph Paul
O'Neill, Edward True
Replogle, Clyde R
Sheskin, Theodore Jerome
Shock, Robert Charles
Solow, Daniel
Theusch, Colleen Joan
Ventresca, Carol
Waren, Allan D(avid)
Weisman, Joel
Williams, Robert Lloyd
Yu, Greta

OKLAHOMA
Foote, Bobbie Leon
Johnston, Harlin Dee
Mize, Joe H(enry)
Schuermann, Allen Clark, Jr
Terrell, Marvin Palmer

Turner, Wayne Connelly
Urban, Timothy I
Yeh, Nai-Shyong

OREGON
Boyd, Dean Weldon
Gillette, Philip Roger
Kocaoglu, Dundar F

PENNSYLVANIA
Anandalingam, G
Arora, Vijay Kumar
Blumstein, Alfred
Bolmarcich, Joseph John
Clapp, Richard Gardner
Dorny, C Nelson
Gidley, James Scott
Gottfried, Byron S(tuart)
Grossmann, Ignacio E
Hildebrand, David Kent
Kaslow, David Edward
Korsh, James F
Kozik, Eugene
Krishnan, Ramayya
Kushner, Harvey D(avid)
Leibholz, Stephen W
Levine, Samuel Harold
Longini, Richard Leon
Love, Carl G(eorge)
McCune, Duncan Chalmers
McGarity, Arthur Edwin
McLaughlin, Francis X(avier)
Mazumdar, Mainak
Mazzitelli, Frederick R(occo)
Odrey, Nicholas Gerald
Pierskalla, William P
Popovics, Sandor
Rosenshine, Matthew
Roush, William Burdette
Satz, Ronald Wayne
Stephenson, Robert L
Stover, James Anderson, Jr
Ventura, Jose Antonio
Weindling, Joachim I(gnace)
Wolfe, Harvey
Zaphyr, Peter Anthony
Zindler, Richard Eugene

RHODE ISLAND
Olson, David Gardels
Robertshaw, Joseph Earl

SOUTH CAROLINA
Chisman, James Allen
Drane, John Wanzer
Dysart, Benjamin Clay, III
Inhaber, Herbert
Kimbler, Delbert Lee
Shier, Douglas Robert

TENNESSEE
Bartell, Steven Michael
Bronzini, Michael Stephen
Cummings, Peter Thomas
LeBlanc, Larry Joseph
Lyle, Benjamin Franklin
Reister, David B(ryan)
Rowan, William Hamilton, Jr
Sherman, Gordon R
Siirola, Jeffrey John
Smith, James Ross
Sundaram, R Meenakshi

TEXAS
Bare, Charles L
Beightler, Charles Sprague
Bennet, George Kemble, Jr
Bennett, George Kemble
Disney, Ralph L(ynde)
Edgar, Thomas Flynn
Feldman, Richard Martin
Giese, Robert Paul
Greaney, William A
Harbaugh, Allan Wilson
Herman, Robert
Hogg, Gary Lynn
Innis, George Seth
Jensen, Paul Allen
Klingman, Darwin Dee
Lesso, William George
Lytton, Robert Leonard
Mak, King Kuen
Meier, France Arnett
Mistree, Farrokh
Modisette, Jerry L
Moore, Bill C
Ozan, M Turgut
Page, Thornton Leigh
Patton, Alton DeWitt
Poage, Scott T(abor)
Rhinehart, Robert Russell, II
Schrage, Robert W
Self, Glendon Danna
Sielken, Robert Lewis, Jr
Walton, Charles Michael
Wang, Yuan R
Wilhelm, Wilbert Edward
Woods, J(ohn) M(elville)
Yakin, Mustafa Zafer
Zey, Edward G

UTAH
Anderson, Douglas I
Baum, Sanford

Davey, Gerald Leland
Lilieholm, Robert John
Peralta, Richard Carl

VERMONT
Brown, Robert Goodell

VIRGINIA
Barrois, Bertrand C
Bothwell, Frank Edgar
Brown, Gerald Leonard
Campbell, Leo James
Cornforth, Clarence Michael
DePoy, Phil Eugene
Dockery, John T
Dorsey, Clark L(awler), Jr
Eldridge, Charles F
Epstein, Samuel David
Farrell, Robert Michael
Finkelstein, Jay Laurence
Freck, Peter G
Gale, Harold Walter
Gardenier, John Stark
Garrett, Benjamin Caywood
Golub, Abraham
Greenberg, Irwin
Gross, Donald
Grotte, Jeffrey Harlow
Haering, George
Ho, S(hui)
Hobeika, Antoine George
Hodge, Ronald Ray
Hoffman, Karla Leigh
Hoffman, Kenneth Charles
Houck, Mark Hedrich
Hurley, William Jordan
Ivanetich, Richard John
Jimeson, Robert M(acKay), Jr
Johnson, Glen Eric
Johnson, Robert Ward
Kaplan, Robert Lewis
Kapos, Ervin
Karns, Charles W(esley)
Keblawi, Feisal Said
Keever, David Bruce
Kneece, Roland Royce, Jr
Kydes, Andy Steve
Lerner, Norman Conrad
Levis, Alexander Henry
Ling, James Gi-Ming
Lowry, Philip Holt
Lundegard, Robert James
McNichols, Gerald Robert
Malcolm, Janet May
Marlow, William Henry
Masterson, Kleber Sanlin, Jr
May, Donald Curtis, Jr
Montague, Stephen
Narula, Subhash Chander
Neuendorffer, Joseph Alfred
Overholt, John Lough
Pokrant, Marvin Arthur
Ratches, James Arthur
Reynolds, Marion Rudolph, Jr
Rosenbaum, David Mark
Rowe, William David
Sage, Andrew Patrick
Seelig, Jakob Williams
Smith, Michael Claude
Spivack, Harvey Marvin
Tabak, Daniel
Weinstein, Iram J
Wilkinson, William Lyle

WASHINGTON
Chaney, Robin W
Lloyd, Raymond Clare
Pilet, Stanford Christian
Pina, Eduardo Isidorio
Ruby, Michael Gordon
Storwick, Robert Martin
Swartzman, Gordon Leni
Verson, Scott
Walsh, John Breffni

WEST VIRGINIA
Elmes, Gregory Arthur
Smith, Walton Ramsay
Stoll, Richard E

WISCONSIN
Chang, Tsong-How
Moy, William A(rthur)
Myers, Vernon W
Robinson, Stephen Michael
Schmidt, John Richard
Schotz, Larry
Suri, Rajan

WYOMING
Eddy, David Maxon
Tung, Yeou-Koung

ALBERTA
Chan, Wah Chun
Elliott, Robert James
Fu, Cheng-Tze
Hamza, Mohamed Hamed
Silver, Edward Allan

BRITISH COLUMBIA
Dumont, Guy Albert
Lambe, Thomas Anthony
Thomson, Alan John

NOVA SCOTIA
Eames, M(ichael) C(urtis)
El-Hawary, Mohamed El-Aref

ONTARIO
Bandler, John William
Buzacott, J(ohn) A(lan)
Colbourn, Charles Joseph
Handa, V(irender) K(umar)
Hilldrup, David J
Hipel, Keith William
Hopkins, Nigel John
Landolt, Jack Peter
Law, Cecil E
Magazine, Michael Jay
Mark, Jon Wei
Penner, Alvin Paul
Posner, Morton Jacob
Shelson, W(illiam)
Simard, Albert Joseph

QUEBEC
Ferguson, Michael John
Loulou, Richard Jacques

OTHER COUNTRIES
Chanson, Samuel T
Davis, Edward Alex
Dunn, Irving John
Halemane, Keshava Prasad
Lueder, Ernst H
Malik, Mazhar Ali Khan
Park, Chan Mo
Thissen, Wil A
Yazicigil, Hasan

Optoelectronics

ARIZONA
Gaskill, Jack Donald
Jackson, Kenneth Arthur
Krasnow, Marvin Ellman

CALIFORNIA
Ballantyne, Catherine Cox
Biblarz, Oscar
Bridges, William Bruce
Bube, Richard Howard
Dapkus, Paul Daniel
Ellion, M Edmund
Fergason, James L
Gamo, Hideya
Gat, Nahum
Gershenzon, M(urray)
Hackett, Le Roy Huntington, Jr
Harris, James Stewart, Jr
Jutamulia, Suganda
Knize, Randall James
Levine, Bruce Martin
Lieberman, Robert Arthur
Massey, Gail Austin
Nelson, Richard David
Ozguz, Volkan Husnu
Paoli, Thomas Lee
Richter, Thomas A
Seib, David Henry
Wang, Michael Renxun
Washburn, Jack
Willner, Alan Eli

CONNECTICUT
Brenholdt, Irving R
Javidi, Bahram

DELAWARE
Golden, Kelly Paul

FLORIDA
Doi, Shinobu
Holloway, Paul Howard
Li Kam Wa, Patrick

GEORGIA
Cohen, Leonard George

ILLINOIS
Wong, Kam Wu

KENTUCKY
Cohn, Robert Warren

MARYLAND
Dauber, Edwin George
Grabow, Barry Edward
Hua, Susan Zonglu
Rall, Jonathan Andrew Reiley
Ravitsky, Charles
Simonis, George Jerome

MASSACHUSETTS
Donnelly, Joseph P(eter)
Lees, David Eric Berman
Linden, Kurt Joseph
Tsang, Dean Zensh
Unlu, M Selim
Williamson, Richard Cardinal
Zheng, Xiao Lu

NEW JERSEY
Buckley, Denis Noel
Burrus, Charles Andrew, Jr
Glass, Alastair Malcolm

Optoelectronics (cont)

Hershenov, B(ernard)
Hong, Won-Pyo
Kaminow, Ivan Paul
Keramidas, Vassilis George
Leheny, Robert Francis
Ling, Hung Chi
Mollenauer, Linn F
Sipress, Jack M
Sywe, Bei-Shen
Wiesenfeld, Jay Martin
Wood, Thomas H

NEW MEXICO
Osinski, Marek Andrzej

NEW YORK
Axelrod, Norman Nathan
Brennemann, Andrew E(rnest), Jr
Bulusu, Dutt V
Chen, Yu-Pei
Das, Pankaj K
Jenekhe, Samson A
Kaplan, Martin Charles
Li, Chung-Sheng
Liu, Yung Sheng
Myers, Robert Anthony
Sen, Ujjal
Stern, Miklos
Wie, Chu Ryang
Woodard, David W

NORTH CAROLINA
Enquist, Paul Michael

OHIO
Johnson, Ray O
Kosel, Peter Bohdan

OREGON
Solanki, Raj

PENNSYLVANIA
Kun, Zoltan Kokai
Shuch, H Paul
Steinbruegge, Kenneth Brian
Tingle, William Herbert
Voloshin, Arkady S

TEXAS
Bartley, William Call
Caswell, Gregory K

VIRGINIA
Jacobs, Ira
Joshi, Ravindra Prabhakar
Lawrence, David Joseph

ONTARIO
Bhatnagar, Yashraj Kishore
Brimacombe, Robert Kenneth

QUEBEC
Ghosh, Asoke Kumar

OTHER COUNTRIES
Benson, Frank Atkinson
Friesem, Albert Asher
Lueder, Ernst H
Snyder, Allan Whitnack

Polymer Engineering

ALABAMA
McKannan, Eugene Charles
Tice, Thomas Robert

ARIZONA
Uhlmann, Donald Robert

CALIFORNIA
Carley, James F(rench)
Chatfield, Dale Alton
Chompff, Alfred J(ohan)
Fishman, Norman
Fok, Samuel S(hiu) M(ing)
Frattini, Paul L
Giants, Thomas W
Lin, Charlie Yeongching
Matthys, Eric Francois
Nguyen, Luu Thanh
Russell, Thomas Paul
Saltman, William Mose
Seiling, Alfred William
Smith, Thor Lowe
Stroeve, Pieter
Szabo, Emery D
Tschoegl, Nicholas William
Witham, Clyde Lester

COLORADO
Bachman, Bonnie Jean Wilson
Pitts, Malcolm John
Quick, Nathaniel Richard

CONNECTICUT
Casberg, John Martin
Gupta, Dharam Vir
Hsu, Nelson Nae-Ching
Kontos, Emanuel G
Prasad, Arun

Schwier, Chris Edward
Scola, Daniel Anthony

DELAWARE
Cooper, Stuart L
Fleming, Richard Allan
Hardie, William George
Hsiao, Benjamin S
Kassal, Robert James
Kohan, Melvin Ira
Leffew, Kenneth W
Norton, Lilburn Lafayette
Ogunnaike, Babatunde Ayodeji
Rodriguez-Parada, Jose Manuel
Sample, Paul E(dward)
Van Trump, James Edmond
Vick Roy, Thomas Rogers
Wagner, Martin Gerald

DISTRICT OF COLUMBIA
Stiehler, Robert D(aniel)

FLORIDA
Owen, Gwilym Emyr, Jr
Ridgway, James Stratman

GEORGIA
Colton, Jonathan Stuart

ILLINOIS
Arastoopour, Hamid
Bartimes, George F
Carr, Stephen Howard
Crist, Buckley, Jr
Fessler, Raymond R
Hass, Alvin A
Lustig, Stanley
Ottino, Julio Mario
Richardson, Joel Albert
Sternberg, Shmuel Mookie
Tucker, Charles L
Veeravalli, Madhavan

INDIANA
Benford, Arthur E
Brannon-Peppas, Lisa
Eckert, Roger E(arl)
Koller, Charles Richard

IOWA
Park, Joon B

KANSAS
Macke, Gerald Fred
Young, David A(nthony)

LOUISIANA
Huckabay, Houston Keller

MARYLAND
Choi, Kyu-Yong
Ibrahim, A Mahammad
Kanarowski, S(tanley) M(artin)
Kerkar, Awdhoot Vasant
Lieberman, Richard Barry
Lundsager, C(hristian) Bent
Neugroschl, Daniel

MASSACHUSETTS
Argon, Ali Suphi
Chen, Chuan Ju
Churchill, Geoffrey Barker
Deanin, Rudolph D
Doshi, Anil G
Goettler, Lloyd Arnold
Huang, Jan-Chan
Kingston, George C
Knapczyk, Jerome Walter
Manning, Monis Joseph
Mehta, Mahesh J
Park, B J
Spangler, Lora Lee
Van der Burg, Sjirk
Wilde, Anthony Flory
Williams, John Russell

MICHIGAN
Dickie, Ray Alexander
Filisko, Frank Edward
Foister, Robert Thomas
Gulari, Esin
Heminghous, William Wayne, Sr
Hyun, Kun Sup
Klempner, Daniel
Laverty, John Joseph
Li, Chi
Moore, Eugene Roger
Narayan, Ramani
Newman, Seymour
Soh, Sung Kuk
Wineman, Alan Stuart

MINNESOTA
Bates, Frank S
Daniel, John Mahendra Kumar
Nam, Sehyun
Thalacker, Victor Paul
Thornton, Arnold William

MISSISSIPPI
Hackett, Robert M(oore)
Seymour, Raymond B(enedict)

MISSOURI
Blum, Frank D
Burke, James Joseph
Harland, Ronald Scott
Lind, Arthur Charles

NEVADA
Clements, Linda L
Herzlich, Harold Joel

NEW HAMPSHIRE
Wang, Victor S F

NEW JERSEY
Bair, Harvey Edward
Blyler, Lee Landis, Jr
Filas, Robert William
Fishman, David H
Giddings, Sydney Arthur
Gregory, Richard Alan, Jr
Han, Byung Joon
Hansen, Ralph Holm
Hudson, Steven David
Hughes, O Richard
Khan, Saad Akhtar
Kirshenbaum, Gerald Steven
Landrock, Arthur Harold
Lee, Wei-Kuo
Lee, Wooyoung
Lem, Kwok Wai
Murthy, N(arasimhaiah) Sanjeeva
Register, Richard Alan
Rhee, Aaron Seung-Joon
Sacks, William
Segelken, John Maurice
Shah, Hasmukh N
Silver, Frederick Howard
Song, Jung Hyun
Tamarelli, Alan Wayne
Todd, David Burton
Weiss, Lawrence H(eisler)
Yourtee, John Boteler

NEW MEXICO
Liepins, Raimond
Stampfer, Joseph Frederick

NEW YORK
Campbell, Gregory August
Feger, Claudius
Horowitz, Carl
Immergut, Edmund H(einz)
Jensen, Richard Alan
Kosky, Philip George
Lhila, Ramesh Chand
Mount, Eldridge Milford, III
Nauman, Edward Bruce
Parikh, Hemant Bhupendra
Pasco, Robert William
Pearson, Glen Hamilton
Petrie, Sarah Elaine
Sancaktar, Erol
Shen, Wu-Mian
Sternstein, Sanford Samuel
Wu, Wen Pao

NORTH CAROLINA
Aneja, Viney P
Chantry, William Amdor
Clark, Howard Garmany
Hurt, Alfred B(urman), Jr
Roop, Robert Kenneth

OHIO
Ahmad, Shamim
Baid, Kushalkumar Moolchand
Bohm, Georg G A
Bradley, Ronald W
Cheng, Stephen Zheng Di
Covitch, Michael J
Drake, Michael L
Eby, Ronald Kraft
Gent, Alan Neville
Gratzl, Miklos
Grimm, Robert Arthur
Jamieson, Alexander MacRae
Kaye, Christopher J
Kinstle, James Francis
Lee, Sunggyu
Leis, Brian Norman
McCoy, Robert Allyn
Nelson, Wayne Franklin
Orndorff, Roy Lee, Jr
Roberts, Robert William
Roberts, Timothy R
Schmitt, George Frederick, Jr
Standish, Norman Weston
Stevenson, James Francis
Thompson, Hugh Ansley
Throne, James Louis
Tsai, Chung-Chieh
Weber, Lester George
White, James L(indsay)
Wideman, Lawson Gibson
Young, Patrick Henry

OKLAHOMA
Efner, Howard F
Rotenberg, Don Harris

OREGON
Skiens, William Eugene

PENNSYLVANIA
Balaba, Willy Mukama
Chen-Tsai, Charlotte Hsiao-yu
Claiborne, C Clair
Dohany, Julius Eugene
Ehrig, Raymond John
Ellis, Jeffrey Raymond
Georgakis, Christos
Grace, Harold P(adget)
Grassi, Vincent G
Gruenwald, Geza
Hull, John Laurence
Kim, Hong C
Lindsey, Roland Gray
Lynch, Thomas John
Parchen, Frank Raymond, Jr
Prane, Joseph W(illiam)
Tackie, Michael N
Vrentas, Christine Mary
Vrentas, James Spiro

RHODE ISLAND
Avery, Donald Hills
Barnett, Stanley M(arvin)
Rockett, Thomas John

SOUTH CAROLINA
Barron, Charles
Wishman, Marvin

TENNESSEE
Bagrodia, Shriram
Bundy, Robert W(endel)
Griswold, Phillip Dwight
Phillips, Bobby Mal
Phillips, Paul J
Wang, Chien Bang

TEXAS
Armeniades, Constantine D
Deviney, Marvin Lee, Jr
Kennelley, Kevin James
Krautz, Fred Gerhard
Mendelson, Robert Allen
Paul, Donald Ross
Pritchett, P(hilip) W(alter)
Scinta, James

UTAH
DeVries, K Lawrence
Nairn, John Arthur

VERMONT
Bhatt, Girish M

VIRGINIA
Chiou, Minshon Jebb
Kinsley, Homan Benjamin, Jr
Wilkinson, Thomas Lloyd, Jr

WASHINGTON
Johns, William E
Lyman, Donald Joseph

WEST VIRGINIA
De Hoff, George R(oland)

WISCONSIN
Bajikar, Sateesh S
Cleary, James William
Denton, Denice Dee
Lepeska, Bohumir
Parnell, Donald Ray

ALBERTA
Williams, Michael C(harles)

ONTARIO
Duever, Thomas Albert
Mitsoulis, Evan
Penlidis, Alexander
Slater, Gary W
Slater, Keith
Tzoganakis, Costas

OTHER COUNTRIES
Tanner, Roger Ian

Sanitary & Environmental Engineering

ALABAMA
Anderson, Bernard Jeffrey
Barrier, J Wayne
Cushing, Kenneth Mayhew
Dahlin, Robert S
Hill, David Thomas
Leonard, Kathleen Mary
Marchant, Guillaume Henri (Wim), Jr
Perry, Nelson Allen
Walters, James Vernon
Wehner, Donald C
Whittle, George Patterson

ALASKA
Brown, Edward James
Tilsworth, Timothy

ARIZONA
Brusseau, Mark Lewis
Burke, James L(ee)
Gerba, Charles Peter

Gutierrez, Leonard Villapando, Jr
Householder, Michael K
Klock, John W
McKlveen, John William
Orton, George W
Poole, H K
Sears, Donald Richard

ARKANSAS
Andrews, John Frank
Fortner, Wendell Lee
Parker, David Garland
Snow, Donald L(oesch)

CALIFORNIA
Agardy, Franklin J
Akesson, Norman B(erndt)
Allen, George Herbert
Asante-Duah, D Kofi
Bery, Mahendera K
Bourke, Robert Hathaway
Brooks, Norman H(errick)
Cass, Glen R
Chang, Andrew C
Chang, Daniel P Y
Chen, Carl W(an-Cheng)
Chen, Kenneth Yat-Yi
Chrysikopoulos, Constantinos V
Chu, Chung-Yu Chester
Church, Richard Lee
Conlon, William Martin
Cooper, Robert Chauncey
Cota, Harold Maurice
Cummings, John Patrick
Diaz, Luis Florentino
Elimelech, Menachem
Everett, Lorne Gordon
Flagan, Richard Charles
Foxworthy, James E(rnest)
Furtado, Victor Cunha
Ghassemi, Masood
Goltz, Mark Neil
Guymon, Gary L
Hanna, George P, Jr
Heckman, Richard Ainsworth
Johansson, Karl Richard
Jones, Jerry Latham
Jones, Rebecca Anne
Kam, James Ting-Kong
Karlsson, Eric Allan
Kerri, Kenneth D
Kroneberger, Gerald F
Kuwabara, James S
Lagarias, John S(amuel)
Lambert, Walter Paul
Lim, H(enry) C(hol)
Longley-Cook, Mark T
Ludwig, Harvey F
Luscher, Ulrich
Lysyj, Ihor
Nazaroff, William W
Olson, Austin C(arlen)
Olson, Betty H
Orlob, Gerald T
Ott, Wayne R
Paustenbach, Dennis J
Perrine, R(ichard) L(eroy)
Porcella, Donald Burke
Robeck, Gordon G
Rogers, Gifford Eugene
Rossiter, Bryant William
Rudavsky, Alexander Bohdan
Satre, Rodrick Iverson
Scherfig, Jan W
Sharma, Prasanta
Slote, Lawrence
Smith, Robert Gordon
Spicher, Robert G
Spokoyny, Felix E
Statt, Terry G
Stenstrom, Michael Knudson
Sutton, Donald Dunsmore
Tchobanoglous, George
Thomas, Jerome Francis
Wade, Richard Lincoln
Waterland, Larry R
Wayne, Lowell Grant
Wilbarger, Edward Stanley, Jr
Wilson, Katherine Woods
Yen, I-Kuen
Yen, Teh Fu

COLORADO
Arnott, Robert A
Bradshaw, William Newman
Church, Brooks Davis
Durham, Michael Dean
Hittle, Douglas Carl
Illangasekare, Tissa H
Jorden, Roger M
Masse, Arthur N
Navratil, James Dale
Rakow, Allen Leslie
Rothfeld, Leonard B(enjamin)
Sabo, Julius Jay
Sherrill, Bette Cecile Benham
Vaseen, V(esper) Albert
Ward, Robert Carl
West, Ronald E(mmett)
Yarar, Baki

CONNECTICUT
Alcantara, Victor Franco
Bollyky, L(aszlo) Joseph

Bozzuto, Carl Richard
Dobay, Donald Gene
Fitzpatrick, Michele
Widmer, Wilbur James

DELAWARE
Allen, Herbert E
Carberry, Judith B
Dam, Rudy Johan
Dentel, Steven Keith
Dwivedy, Ramesh C
Falk, Lloyd L(eopold)
Goel, Kailash C(handra)
Habibi, Kamran
Huang, Chin Pao
Hume, Harold Frederick
Jacobs, Emmett S
Koch, Carl Mark
Scherger, Dale Albert
Smith, Donald W(anamaker)

DISTRICT OF COLUMBIA
Bell, Bruce Arnold
Briscoe, John
Brown, Norman Louis
Cotruvo, Joseph Alfred
Deason, Jonathan P
Mermel, Thaddeus Walter
Middleton, John T(ylor)
Nebiker, John Herbert
Shuster, Kenneth Ashton
Silver, Francis
Walker, John Martin
Wolf, Stanley Myron

FLORIDA
Allen, Eric Raymond
Bevis, Herbert A(nderson)
Billings, Charles Edgar
Bolch, William Emmett, Jr
Dallman, Paul Jerald
Delfino, Joseph John
Dietz, Jess Clay
Duranceau, Steven Jon
Echelberger, Wayne F, Jr
Hiser, Homer Wendell
Jackson, Daniel Francis
Lundgren, Dale A(llen)
Madhanagopal, T
Nemerow, Nelson L(eonard)
Nichols, James Carlile
Nordstedt, Roger Arlo
Proctor, Charles Lafayette, II
Scarlatos, Panagiotis Dimitrios
Shair, Robert C
Sirkin, Alan N
Stephens, N(olan) Thomas
Stokes, Charles Anderson
Sullivan, James Haddon, Jr
Symons, George E(dgar)
Teller, Aaron Joseph
Wang, Tsen Chen
Wanielista, Martin Paul
Zegel, William Case

GEORGIA
Amirtharajah, Appiah
Bowen, Paul Tyner
Chiu, Kirts C
Circeo, Louis Joseph, Jr
Eckert, Charles Alan
Galeano, Sergio F(rancis)
Hedden, Kenneth Forsythe
Hill, David W(illiam)
Hooper, Richard Paul
Jones, Alan Richard
Oakes, Thomas Wyatt
Rogers, Harvey Wilbur
Sanders, Walter MacDonald, III
Saunders, Fred Michael
Schliessmann, D(onald) J(oseph)
Shonka, Joseph John
Stanley, Luticious Bryan, Jr
Stelson, Arthur Wesley

HAWAII
Hihara-Endo, Linda Masae
Krock, Hans-Jürgen
Lau, L(eung Ku) Stephen
Papacostas, Constantinos Symeon
Teng, Michelle Hsiao Tsing
Young, Reginald H F

IDAHO
Engen, Ivar A
Korus, Roger Alan
Schreiber, David Laurence
Wallace, Alfred Thomas

ILLINOIS
Babcock, Lyndon Ross, (Jr)
Bengtson, Harlan Holger
Bickelhaupt, R(oy) E(dward)
Clark, John Peter, III
Ditmars, John David
Ekman, Frank
Engelbrecht, R(ichard) S(tevens)
Hagenbach, W(illiam) P(aul)
Heil, Richard Wendell
Hermann, Edward Robert
Herricks, Edwin E
Hirschi, Michael Carl
Holsen, Thomas Michael
Hsu, Deh Yuan

Kanter, Manuel Allen
Kim, Byung J
Klass, Donald Leroy
Liebman, Jon C(harles)
Lue-Hing, Cecil
McFee, Donald Ray
Minear, Roger Allan
Muehling, Arthur J
Nesselson, Eugene J(oseph)
Noll, Kenneth E(ugene)
Patterson, James W(illiam)
Prakasam, Tata B S
Reeve, Aubrey C
Regunathan, Perialwar
Rittmann, Bruce Edward
Roy, Dipak
Schaller, Robin Edward
Snoeyink, Vernon Leroy
Snow, Richard Huntley
Thomsen, Kurt O
Wadden, Richard Albert
Wang, Hwa-Chi
Wang, Michael Quanlu
Wilson, Thomas Edward
Yen, Ben Chie
Zenz, David R

INDIANA
Etzel, James Edward
Lee, Linda Shahrabani
Lin, Ping-Wha
Pichtel, John Robert
Shultz, Clifford Glen
Taylor, Richard L
Tenney, Mark W

IOWA
Austin, Tom Al
Baumann, E(dward) Robert
Cleasby, John Leroy
Dague, Richard R(ay)
Egger, Carl Thomas
Parkin, Gene F
Paulson, Wayne Lee
Schnoor, Jerald L

KANSAS
Cohen, Jules Bernard
Erickson, Larry Eugene
Hosni, Mohammad Hosein
Lane, Dennis Del
Metzler, Dwight F
Surampalli, Rao Yadagiri

KENTUCKY
Edwards, Richard Glenn
Fleischman, Marvin
Foree, Edward G(olden)
Jain, Ravinder Kumar
Jensen, Randolph A(ugust)
Tucker, Irwin William
Wang, Yi-Tin
Warner, Richard Charles

LOUISIANA
Fontenot, Martin Mayance, Jr
Grimwood, Charles
Jones, Daniel Elven
Reimers, Robert T
Singh, Vijay P
Sterling, Arthur MacLean
Szabo, A(ugust) J(ohn)
Valsaraj, Kalliat Thazhathuveetil
Wrobel, William Eugene

MAINE
Rock, Chet A

MARYLAND
Anderson, William Carl
Bascunana, Jose Luis
Baummer, J Charles, Jr
Bryan, Edward H
Burrows, William Dickinson
Cleave, Mary L
Cookson, John Thomas, Jr
Fought, Sheryl Kristine
Gessner, Adolf Wilhelm
Gilmore, Thomas Meyer
Goldberg, Irving David
Gupta, Gian Chand
Hao, Oliver J
Herold, Keith Evan
Hirschhorn, Joel S(tephen)
Kawata, Kazuyoshi
LeRoy, André François
Linaweaver, Frank Pierce
Linder, Seymour Martin
Morhardt, Josef Emil, IV
O'Melia, Charles R(ichard)
Pritchard, Donald William
ReVelle, Charles S
Rice, James K
Seagle, Edgar Franklin
Shipman, Harold R
Tuttle, Kenneth Lewis
Varma, Man Mohan
Walker, William J, Jr
Walter, Donald K
Weinberger, Leon Walter
Wong, Kin Fai

MASSACHUSETTS
Blanc, Frederic C

Egan, Bruce A
Ehrenfeld, John R(oos)
Eklund, Karl E
Fay, James A(lan)
Fennelly, Paul Francis
Golay, Michael W
Harleman, Donald R(obert) F(ergusson)
Hart, Frederick Lewis
Holmes, Douglas Burnham
Kachinsky, Robert Joseph
Keshavan, Krishnaswamiengar
Male, James William
Marks, David Hunter
Morel, Francois M M
Murphy, Brian Logan
Newman, Bernard
Olsen, Robert Thorvald
Ostendorf, David William
Pojasek, Robert B
Ram, Neil Marshall
Reckhow, David Alan
Reimold, Robert J
Teal, John Moline
Tester, Jefferson William
Thorstensen, Thomas Clayton
Wang, Lawrence K

MICHIGAN
Abriola, Linda Marie
Batterman, Stuart A
Crittenden, John Charles
Daniels, Stacy Leroy
Domke, Charles J
Hubbard, Carolyn Parks
Huebner, George J
Kim, Byung Ro
Klempner, Daniel
Morman, Kenneth N
Ryant, Charles J(oseph), Jr
Savage, Phillip E
Scavia, Donald
Siddiqui, Waheed Hasan
Smith, Joseph D
Teichroew, Daniel
Walles, Wilhelm Egbert
Wright, Steven Jay

MINNESOTA
Baria, Dorab Naoroze
Boening, Paul Henrik
Dobbins, Durell Cecil
Eisenreich, Steven John
Gulliver, John Stephen
Haun, J(ames) W(illiam)
Hepworth, Malcolm T(homas)
Johnson, Walter K
Ling, Joseph Tso-Ti
McMurry, Peter Howard
Reynolds, James Harold
Shen, Lester Sheng-wei
Stefan, Heinz G
Susag, Russell H(arry)
Uban, Stephen A

MISSISSIPPI
Sherrard, Joseph Holmes
Wolverton, Billy Charles

MISSOURI
Bajpai, Rakesh Kumar
Banerji, Shankha K
Hatheway, Allen Wayne
Hoogheem, Thomas John
Koederitz, Leonard Frederick
Modesitt, Donald Ernest
Mossman, Deborah Jean
Rawlings, Gary Don
Robinson, Thomas B
Ryckman, DeVere Wellington
Schroy, Jerry M
Schwartz, Henry Gerard, Jr
Sievers, Dennis Morline

MONTANA
Deibert, Max Curtis
Peavy, Howard Sidney

NEBRASKA
Eldred, Kenneth M
Hoffman, Glenn Jerrald

NEVADA
Middlebrooks, Eddie Joe

NEW HAMPSHIRE
Ballestero, Thomas P
Frost, Terrence Parker
Jackson, Herbert William
Lynch, Daniel Roger
Sproul, Otis J

NEW JERSEY
Blank, Zvi
Blumberg, Alan Fred
Bruno, Michael Stephen
Brytczuk, Walter L(ucas)
Cohen, Irving David
Czeropski, Robert S(tephen)
Dobi, John Steven
Graedel, Thomas Eldon
Haefeli, Robert J(ames)
Halasi-Kun, George Joseph
Iyer, Ram R
Kleckner, Willard R

Sanitary & Environmental Engineering (cont)

Krishnamurthy, Subrahmanya
Lederman, Peter B
Lessard, Richard R
Lewandowski, Gordon A
Mallison, George Franklin
Manganaro, James Lawrence
O'Brien, Gerald Joseph
Pfafflin, James Reid
Sartor, Anthony
Skovronek, Herbert Samuel
Toro, Richard Frank
Uchrin, Christopher George
Weinstein, Norman J(acob)

NEW MEXICO
Alam, Mansoor
Goodgame, Thomas H
Holmes, John Thomas
Khandan, Nirmala N
Lee, William Wai-Lim
Mauzy, Michael P
Nirmalakhandan, Nagamany
Nuttall, Herbert Ericksen, Jr
Peil, Kelly M
Roubicek, Rudolf V
Ryti, Randall Todd
Sivinski, Jacek Stefan
Thomson, Bruce M
Yeh, Hsu-Chi

NEW YORK
Agosta, Vito
Ahmad, Jameel
Atkinson, Joseph Ferris
Aulenbach, Donald Bruce
Belfort, Georges
Bove, John L
Cataldo, Joseph C
Chadwick, George F(redrick)
Clesceri, Nicholas Louis
Collins, Anthony Gordon
Dick, Richard Irwin
Gale, Stephen Bruce
Gellman, Isaiah
Gossett, James Michael
Grohse, Edward William
Hirtzel, Cynthia S
Hovey, Harry Henry, Jr
Jeris, John S(tratis)
Jirka, Gerhard Hermann
Kiphart, Kerry
Koppelman, Lee Edward
Lin, Wei
Loucks, Daniel Peter
Lynn, Walter R(oyal)
Miller, George Maurice
Molof, Alan H(artley)
Murphy, Kenneth Robert
Oldshue, J(ames) Y(oung)
Palevsky, Gerald
Robinson, Myron
Shen, Hung Tao
Shen, Thomas T
Shoemaker, Christine Annette
Soifer, Herman
Srinivasaraghavan, Rengachari
Star, Joseph
Stedinger, Jery Russell
Tavlarides, Lawrence Lasky
Valentine, Daniel T
Velzy, Charles O
Wagner, Peter Ewing
Wang, Muhao S
Yapa, Poojitha Dahanayake
Ziegler, Edward N
Ziegler, Robert C(harles)
Zimmie, Thomas Frank

NORTH CAROLINA
Ballard, Lewis Franklin
Barker, James Cathey
Bell, Carlos G(oodwin), Jr
Bobalek, Edward G(eorge)
Chang, John Chung-Shih
Clark, Charles Edward
Daumit, Gene Philip
Ensor, David Samuel
Fox, Donald Lee
Hamner, William Frederick
Hanes, N Bruce
Kohl, Jerome
Larsen, Ralph Irving
Lawless, Philip Austin
Medina, Miguel Angel, Jr
Moeller, D(ade) W(illiam)
Okun, D(aniel) A(lexander)
Peirce, James Jeffrey
Reckhow, Kenneth Howland
Reist, Parker Cramer
Roop, Robert Kenneth
Saeman, W(alter) C(arl)
Scheffe, Richard Donald
Senzel, Alan Joseph
Singer, Philip C
Stern, A(rthur) C(ecil)
Trenholm, Andrew Rutledge
Winston, Hubert
Yamamoto, Toshiaki

NORTH DAKOTA
Ludlow, Douglas Kent

OHIO
Armstrong, James Anthony
Ayer, Howard Earle
Bennett, G(ary) F
Bishop, Paul Leslie
Biswas, Pratim
Bondurant, Byron L(ee)
Bostian, Harry (Edward)
Brooman, Eric William
Clark, C Scott
Engdahl, Richard Bott
Fried, Joel Robert
Geldreich, Edwin E(mery)
Grube, Walter E, Jr
Gruber, Charles W
Gurklis, John A(nthony)
Hannah, Sidney Allison
Herdendorf, Charles Edward, III
Hyland, John R(oth)
Kellerstrass, Ernst Junior
Kim, Byung Cho
Michalakos, Peter Michael
Morand, James M(cHuron)
Moulton, Edward Q(uentin)
Murdoch, Lawrence Corlies, III
Noss, Richard Robert
Olesen, Douglas Eugene
Papadakis, Constantine N
Preul, Herbert C
Prober, Richard
Saltzman, Bernard Edwin
Scarpino, Pasquale Valentine
Smith, James Edward, Jr
Springer, Allan Matthew
Stephan, David George
Stiefel, Robert Carl
Sykes, Robert Martin
Taiganides, E Paul
Turner, Andrew
Wideman, Lawson Gibson
Willeke, Gene E

OKLAHOMA
De Vries, Richard N
Foutch, Gary Lynn
Graves, Toby Robert
Harwell, Jeffrey Harry
Kincannon, Donny Frank
Koerner, E(rnest) L(ee)
Madden, Michael Preston
Stover, Enos Loy
Turner, Wayne Connelly

OREGON
Boubel, Richard W
Dunnette, David Arthur
Hashimoto, Andrew G
Hsiung, Andrew K
Huber, Wayne Charles
Huntzicker, James John
Miner, J Ronald
Moore, James Allan
Schaumburg, Frank David
Tucker, W(illiam) Henry
Wohlers, Henry Carl

PENNSYLVANIA
Anandalingam, G
Atkins, Patrick Riley
Berty, Jozsef M
Bhatla, Manmohan N (Bart)
Blayden, Lee Chandler
Cernansky, Nicholas P
Chang, C Hsiung
Cobb, James Temple, Jr
Cooper, Joseph E
Corbin, Michael H
Davidson, Cliff Ian
Deb, Arun K
Dicciani, Nance Katherine
Doelp, Louis C(onrad), Jr
Dzombak, David Adam
Esmen, Nurtan A(lan)
Gidley, James Scott
Gilardi, Edward Francis
Gill, C(harles) Burroughs
Grossmann, Elihu D(avid)
Gurol, Mirat D
Haas, Charles Nathan
Hofer, Lawrence John Edward
Johnson, Glenn M
Keenan, John Douglas
Kugelman, Irwin Jay
LaGrega, Michael Denny
Lawrence, Alonzo William
Loigman, Harold
Loxley, Thomas Edward
Luthy, Richard Godfrey
McGarity, Arthur Edwin
McMichael, Francis Clay
Manka, Dan P
Meyer, Ellen McGhee
Neufeld, Ronald David
Pandis, Spyros N
Pena, Jorge Augusto
Penman, Paul D
Petura, John C
Pipes, Wesley O
Pohland, Frederick G
Qazi, Mahmood A

Ramani, Raja Venkat
Reed, Joseph Raymond
Regan, Raymond Wesley
Rosenstock, Paul Daniel
Sagik, Bernard Phillip
Samples, William R(ead)
Sawicki, John Edward
Schoenberger, Robert J(oseph)
Shaler, Amos J(ohnson)
Shapiro, Maurice A
Sooky, Attila A(rpad)
Spear, Jo-Walter
Sproull, Wayne Treber
Stallard, Michael Lee
Stephenson, Robert L
Thompson, Ansel Frederick, Jr
Touhill, Charles Joseph
Unz, Richard F(rederick)
Weston, Roy Francis
Wilkens, John Albert
Woodruff, Kenneth Lee
Yohe, Thomas Lester
Zhou, Qian

RHODE ISLAND
Dobbins, Richard Andrew
Hutzler, Leroy, III
Poon, Calvin P C

SOUTH CAROLINA
Aelion, Claire Marjorie
Carnes, Richard Albert
Dysart, Benjamin Clay, III
Elzerman, Alan William
Grady, Cecil Paul Leslie, Jr
Jennett, Joseph Charles
Keinath, Thomas M
Overcamp, Thomas Joseph
Parkinson, Thomas Franklin
Patterson, C(leo) Maurice
Rich, Linvil G(ene)
Townes, George Anderson
Wang, Wun-Cheng W(oodrow)
White, Richard Kenneth
Wixson, Bobby Guinn

SOUTH DAKOTA
Hodges, Neal Howard, II
Zogorski, John Steward

TENNESSEE
Bonner, William Paul
Cheng, Meng-Dawn
Clarke, Ann Neistadt
Cowser, Kenneth Emery
Eckenfelder, William Wesley, Jr
Farrington, Joseph Kirby
Gallagher, Michael Terrance
Gano, Richard W
Genung, Richard Keith
Ghosh, Mriganka M(ouli)
Godbee, H(erschel) W(illcox)
Hensley, Marble John, Sr
Kelso, Richard Miles
Parker, Frank L
Perry, Lloyd Holden
Polahar, Andrew Francis
Roesler, Joseph Frank
Schnelle, K(arl) B(enjamin), Jr
Singh, Suman Priyadarshi Narain
Smith, John W(arren)
Smith, Michael James
Thackston, Edward Lee

TEXAS
Akgerman, Aydin
Armstrong, Neal Earl
Azad, Hardam Singh
Batchelor, Bill
Bodre, Robert Joseph
Busch, Arthur Winston
Caffey, James Enoch
Camp, Frank A, III
Charbeneau, Randall Jay
Chiang, Chen Y
Collins, Michael Albert
Crump, John Joseph
Davis, Ernst Michael
Eaton, David J
Fix, Ronald Edward
Gavande, Sampat A
Giammona, Charles P, Jr
Gilmour-Stallsworth, Lisa K
Gloyna, Earnest F(rederick)
Goldblum, David Kiva
Halff, Albert Henry
Harbordt, C(harles) Michael
Henderson, Gregory Wayne
Hoage, Terrell Rudolph
Howe, Richard Samuel
McKee, Herbert C(harles)
Malina, Joseph F(rancis), Jr
Morrow, Norman Louis
Myrick, Henry Nugent
Naismith, James Pomeroy
Parate, Natthu Sonbaji
Qasim, Syed Reazul
Reynolds, Tom Davidson
Sorber, Charles Arthur
Stanford, Geoffrey
Stubbeman, Robert Frank
Symons, James M(artin)

Tarquin, Anthony Joseph
Tiller, F(rank) M(onterey)
Wagner, John Philip
Wolf, Harold William

UTAH
Ghosh, Sambhunath
Lee, Jeffrey Stephen
Merritt, LaVere Barrus
Rogers, Vern Child
Ryan, Norman W(allace)
Sigler, John William

VERMONT
Hussak, Robert Edward
Pyper, Gordon R(ichardson)

VIRGINIA
Beach, Robert L
Boardman, Gregory Dale
Cawley, William Arthur
Chi, Michael
Drewry, William Alton
Echols, Charles E(rnest)
Fang, Ching Seng
Higgins, Thomas Ernest
Houck, Mark Hedrich
Jaisinghani, Rajan A
Kornicker, William Alan
Lamb, James C(hristian), III
Lassiter, William Stone
McKenna, John Dennis
Myers, Charles Edwin
Olem, Harvey
Randall, Clifford W(endell)
Reilly, Thomas E
Shea, Timothy Guy
Simpson, James R(ussell)
Spofford, Walter O, Jr
Trout, Dennis Alan
Westfield, James D
Young, Earle F(rancis), Jr

WASHINGTON
Berg, Richard Harold
Bhagat, Surinder Kumar
Bogan, Richard Herbert
Carlson, Dale Arvid
Cormack, James Frederick
Darley, Ellis Fleck
Duncan, James Byron
Edde, Howard Jasper
Emery, Richard Meyer
Ewing, Ben B
Glenne, Bard
Jones, Kay H
LaFontaine, Thomas E
Mahalingam, R
Mar, Brian W(ayne)
Peterson, Warren Stanley
Rosenwald, Gary W
Ruby, Michael Gordon
Skrinde, Rolf T
Sylvester, Robert O(hrum)

WEST VIRGINIA
Conway, Richard A
McCall, Robert G
Mark, Christopher

WISCONSIN
Arthur, Robert M
Boyle, William C(harles)
Christensen, Erik Regnar
Converse, James Clarence
Hanway, John E(dgar), Jr
Jayne, Jack Edgar
Kipp, Raymond J
Lentz, Mark Steven
Zanoni, Alphonse E(ligius)

WYOMING
Champlin, Robert L

PUERTO RICO
Bonnet, Juan A, Jr

ALBERTA
Smith, Daniel Walter

BRITISH COLUMBIA
Grace, John Ross
Liao, Ping-huang
Walden, C(ecil) Craig

NEW BRUNSWICK
Lin, Kwan-Chow

ONTARIO
AL-Hashimi, Ahmad
Bewtra, Jatinder Kumar
Boadway, John Douglas
Byer, Philip Howard
Gorman, William Alan
Heinke, Gerhard William
Kennedy, Kevin Joseph
Kosaric, Naim
Legge, Raymond Louis
McBean, Edward A
Murphy, Keith Lawson
Nicolle, Francois Marcel Andre
Parker, Wayne Jeffery
Robinson, Campbell William

Shemilt, L(eslie) W(ebster)
Slawson, Peter (Robert)
VandenHazel, Bessel J
Warith, Mostafa Abdel
Zytner, Richard G

QUEBEC
Beron, Patrick
Couillard, Denis
De la Noue, Joel Jean-Louis
Leduy, Anh
Tyagi, Rajeshwar Dayal
Volesky, Bohumil

SASKATCHEWAN
Cullimore, Denis Roy
Curtis, Fred Allen
Viraraghavan, Thiruvenkatachari

OTHER COUNTRIES
Aoyama, Isao
Bond, Marvin Thomas
Capodaglio, Andrea G
Grigoropoulos, Sotirios G(regory)
Hoadley, Alfred Warner
Huang, Ju-Chang
Jokl, Miloslav Vladimir
Kimura, Ken-ichi
Liepsch, Dieter W
Rodriguez, Federico Angel
Sakai, Hiroshi
Shuval, Hillel Isaiah
Stumm, Werner
Yao, Kuan Mu

Systems Design & Systems Science

ALABAMA
Blackmon, James B
Boykin, William H(enry), Jr
Brown, Robert Alan
Fullerton, Larry Wayne
Haeussermann, Walter
McAuley, Van Alfon
Martin, James Arthur
Rekoff, M(ichael) G(eorge), Jr
Twieg, Donald Baker

ALASKA
Rogers, James Joseph
Skudrzyk, Frank J

ARIZONA
Askin, Ronald Gene
Bahill, A(ndrew) Terry
Bartocha, Bodo
Bluestein, Theodore
Burke, James L(ee)
Clark, Wilburn O
Cochran, Jeffery Keith
Dahl, Randy Lynn
Falk, Thomas
Ferrell, William Russell
Formeister, Richard B
Mackulak, Gerald Thomas
Miller, Harry
Preston, Kendall, Jr
Ramler, W(arren) J(oseph)
Ryan, Thomas Wilton
Seifert, Deborah Roeckner
Silvern, Leonard Charles
Slaughter, Charles F
Sorooshian, Soroosh
Swan, Peter A
Szidarovszky, Ferenc
Wasson, James Walter
Welch, H(omer) William
Wymore, Albert Wayne
Zeigler, Bernard Philip
Zelenka, Jerry Stephen

ARKANSAS
Carr, Gerald Paul
Sedelow, Walter Alfred, Jr

CALIFORNIA
Abbott, Dean William
Acheson, Louis Kruzan, Jr
Akonteh, Benny Ambrose
Alspach, Daniel Lee
Amara, Roy C
Ames, John Wendell
Anderson, George William
Antonsson, Erik Karl
Austin, Arthur Leroy
Baggerly, Leo L
Barr, Kevin Patrick
Berenji, Hamid R
Berg, Myles Renver
Bertram, Sidney
Beyster, J Robert
Blackwell, Charles C, Jr
Bleviss, Zegmund O(scar)
Bloxham, Laurence Hastings
Bradley, Hugh Edward
Bradner, Hugh
Bryson, Arthur E(arl), Jr
Bush, George Edward
Caligiuri, Joseph Frank
Cannon, Robert H, Jr
Carlyle, Jack Webster
Carson, George Stephen
Carver, John Guill
Cashen, John F
Caswell, John N(orman)
Chamberlain, Robert Glenn
Channel, Lawrence Edwin
Chen, Kao
Chen, Tsai Hwa
Chester, Arthur Noble
Chi, Cheng-Ching
Christiansen, Jerald N
Clark, Crosman Jay
Clemens, Jon K(aufmann)
Coleman, Paul Jerome, Jr
Culler, Donald Merrill
Davis, William R
Deckert, Curtis Kenneth
Depp, Joseph George
Dornfeld, David Alan
Dost, Martin Hans-Ulrich
Doyle, Walter M
Drugan, James Richard
Duffield, Jack Jay
Dunn, Donald A(llen)
Eden, Richard Carl
Egermeier, R(obert) P(aul)
Eggers, A(lfred) J(ohn), Jr
Elliott, David Duncan
Elverum, Gerard William, Jr
Faulkner, Thomas Richard
Feign, David
Firdman, Henry Eric
Fleischer, Gerald A
Fleming, Alan Wayne
Forbes, Judith L
Forsberg, Kevin
Fox, Joseph M(ickle), III
Franklin, Gene F(arthing)
Friedman, George J(erry)
Furst, Ulrich Richard
Gat, Nahum
Gillis, James Thompson
Gilmartin, Thomas Joseph
Gleghorn, G(eorge) J(ay), (Jr)
Goddard, Earl G(ascoigne)
Goheen, Lola Coleman
Grace, O Donn
Grant, Eugene F(redrick)
Gray, Paul
Green, Allen T
Gritton, Eugene Charles
Grossman, Jack Joseph
Gustavson, Marvin Ronald
Haake, Eugene Vincent
Hackwood, Susan
Hallet, Raymon William, Jr
Hamilton, Carole Lois
Hanson, Joe A
Hartwick, Thomas Stanley
Healey, Anthony J
Hecht, Herbert
Hecht, Myron J
Heer, Ewald
Hendrix, C(harles) E(dmund)
Hewer, Gary Arthur
Hilker, Harry Van Der Veer, Jr
Hoffman, Howard Torrens
Hoffman, James Tracy
Holder, Harold Douglas
Holeman, Dennis Leigh
Hollander, Gerhard Ludwig
Holsztynski, Wlodzimierz
Holtzman, Samuel
Horn, Kenneth Porter
Hsu, Chieh-Su
Huberman, Marshall Norman
Hult, John Luther
Hwang, Kai
Iberall, Arthur Saul
Ibrahim, Medhat Ahmed Helmy
Isenberg, Lionel
Jackson, Albert S(mith)
Jackson, Durward P
Jacobs, Michael Moises
Jacobson, Albert H(erman), Jr
Jaffe, Leonard David
Jenkins, Steven J
Johannessen, Jack
Johnson, Reynold B
Kagiwada, Harriet Hatsune Natsuyama
Kaliski, Martin Edward
Kalvinskas, John J(oseph)
Kayton, Myron
Kelly, Kenneth C
Kendall, Burton Nathaniel
Kennel, John Maurice
Kerfoot, Branch Price
Kinnison, Gerald Lee
Klestadt, Bernard
Krause, Lloyd O(scar)
Kreutz-Delgado, Kenneth Keith
Kubik, Robert N
Laitin, Howard
Lampert, Seymour
Landecker, Peter Bruce
Langdon, Glen George, Jr
Larson, Harry Thomas
Laub, Alan John
Layton, Thomas William
Leavy, Paul Matthew
Leibowitz, Lewis Phillip
Leite, Richard Joseph
Leitmann, G(eorge)
Lientz, Bennet Price
Lillie, Charles Frederick
Lindley, Charles A(lexander)
Lipeles, Martin
Lisman, Perry Hall
Liu, Frederick F
Longmuir, Alan Gordon
Lovell, Robert Edmund
Luenberger, David Gilbert
Lyman, John (Henry)
McBride, Lyle E(rwin), Jr
McDonald, William True
McQuillen, Howard Raymond
McRuer, Duane Torrance
Maewal, Akhilesh
Magness, T(om) A(lan)
Maloy, John Owen
Mantey, Patrick E(dward)
Marmarelis, Vasilis Z
Mathews, W(arren) E(dward)
Meissinger, Hans F
Meyer, Lhary
Michaelson, Jerry Dean
Mikes, Peter
Miles, Ralph Fraley, Jr
Mingori, Diamond Lewis
Mitra, Sanjit K
Mitsutomi, T(akashi)
Molinder, John Irving
Molnar, Imre
Moran, Thomas Patrick
Morrow, Charles Tabor
Morse, Joseph Grant
Mulligan, Geoffrey C
Muntz, Richard Robert
Mykkanen, Donald L
Nadler, Gerald
Nesbit, Richard Allison
Newton, Arthur R
Nickolls, John Richard
Nielsen, Norman Russell
Nunan, Craig S(pencer)
Nunn, Robert Harry
Omura, Jimmy Kazuhiro
Onton, Aare
Orlob, Gerald T
Owen, Paul Howard
Park, Edward C(ahill), Jr
Parker, Alice Cline
Paulson, Boyd Colton, Jr
Pearl, Judea
Pedersen, Leo Damborg
Pelka, David Gerard
Philipson, Lloyd Lewis
Pietrzak, Lawrence Michael
Pinson, John C(arver)
Pinto, John Gilbert
Pister, Karl S(tark)
Poggio, Andrew John
Pressman, Ada Irene
Pye, Earl Louis
Ratner, Robert (Stephen)
Rau, Bantwal Ramakrishna
Rauch, Herbert Emil
Rechnitzer, Andreas Buchwald
Rechtin, Eberhardt
Reddy, Parvathareddy Balarami
Rickard, John Terrell
Roney, Robert K(enneth)
Rubin, Itzhak
Ruskin, Arnold M(ilton)
Safonov, Michael G
Sakaguchi, Dianne Koster
Samson, Sten
Schlesinger, Robert Jackson
Schneider, Alan M(ichael)
Schreiner, Robert Nicholas, Jr
Schweikert, Daniel George
Scudder, Harvey Israel
Seitz, S Stanley
Senge, George H
Sensiper, S(amuel)
Serdengecti, Sedat
Shepherd, Robin
Shoch, John F
Siljak, Dragoslav (D)
Silverstein, Elliot Morton
Slaton, Jack H(amilton)
Smith, Frederick T(ucker)
Smith, Otto J(oseph) M(itchell)
Smith, Stephen Allen
Sondak, Norman Edward
Speyer, Jason L
Spinrad, Robert J(oseph)
Spitzer, Irwin Asher
Steingold, Harold
Stenstrom, Michael Knudson
Stephanou, Stephen Emmanuel
Stern, Albert Victor
Stevenson, Robin
Stewart, Richard William
Sturm, Walter Allan
Sung, Chia-Hsiaing
Surko, Pamela Toni
Swanson, Eric Richmond
Talbot, Raymond James, Jr
Teague, Lavette Cox
Titus, Harold
Toepfer, Richard E, Jr
Tooley, Richard Douglas
Tyler, Christopher William
Wagner, Raymond Lee
Walker, Warren Elliott
Wallerstein, David Vandermere
Wang, H E Frank
Wang, Ji Ching
Ware, W(illis) H(oward)
Warren, Mashuri Laird
Weber, Charles L
Weinberg, Daniel I
Weiss, Herbert Klemm
Weissenberger, Stein
Welmers, Everett Thomas
Wertz, James Richard
White, Moreno J
Wiberg, Donald M
Wilde, D(ouglass) J(ames)
Williams, Donald Spencer
Wong, Kwan Y
Wray, John L
Wright, Edward Kenneth
Wu, Felix F
Yao, Jason
Yeh, Hen-Geul
Young, Richard D
Zander, Andrew Thomas
Zebroski, Edwin L
Zuppero, Anthony Charles

COLORADO
Barber, Robert Edwin
Brown, Alison Kay
Frangopol, Dan Mircea
Frick, Pieter A
Gupta, Kuldip Chand
Jansson, David Guild
Johnson, Gearold Robert
Kelley, Neil Davis
Koldewyn, William A
Lazzarini, Albert John
Lund, Donald S
Mandics, Peter Alexander
Marshall, Charles F
Morrison, John Stuart
Munro, Richard Harding
Smith, William Bridges
Wackernagel, Hans Beat
Wall, Edward Thomas
Wertz, Ronald Duane

CONNECTICUT
Cooper, James William
Dubin, Fred S
Goldman, Ernest Harold
Halpern, Howard S
Johnson, Loering M
Ketchman, Jeffrey
Krummel, William Marion
Levenstein, Harold
Levien, Roger Eli
Mack, Donald R(oy)
Martin, R Keith
Narendra, Kumpati S
O'Brien, Robert L
Porter, Richard W(illiam)
Poultney, Sherman King
Rezek, Geoffrey Robert
Simpson, Wilburn Dwain
Stein, David Morris
Stuck, Barton W
Telfair, William Boys

DELAWARE
Arots, Joseph B(artholomew)
Bartkus, Edward Peter
Kikuchi, Shinya
Leffew, Kenneth W
Murphy, Arthur Thomas
Nicholls, Robert Lee
Vick Roy, Thomas Rogers

DISTRICT OF COLUMBIA
Baker, Dennis John
Baz, Amr Mahmoud Sabry
Bloch, Erich
Callaham, Michael Burks
Huband, Frank L
James, Alton Everette, Jr
Kyriakopoulos, Nicholas
Marcus, Michael Jay
Moll, Magnus
Randolph, James E
Zucchetto, James John

FLORIDA
Anderson, Roland Carl
Ariet, Mario
Bonn, T(heodore) H(ertz)
Braswell, Robert Neil
Calhoun, Ralph Vernon, Jr
Capehart, Barney Lee
Chartrand, Robert Lee
Coates, Clarence L(eroy), Jr
Concordia, Charles
Cox, John David
Dilpare, Armand Leon
Dougherty, John William
Duersch, Ralph R
Everett, Woodrow Wilson, Jr
Flick, Carl
Hammer, Jacob
Hampp, Edward Gottlieb
Harmon, G Lamar
Haviland, Robert P(aul)
Herwald, Seymour W(illis)
Howard, Bernard Eufinger
Kinnie, Irvin Gray

Systems Design & Systems Science (cont)

Kirmse, Dale William
Klock, Benny LeRoy
Korwek, Alexander Donald
Kunce, Henry Warren
Lund, Frederick H(enry)
Mahig, Joseph
Melich, Michael Edward
Mertens, Lawrence E(dwin)
Miler, George Gibbon, Jr
Mitoff, S(tephan) P(aul)
Nevill, Gale E(rwin), Jr
Peterson, James Robert
Proctor, Charles Lafayette
Roth, Paul Frederick
Sager, John Clutton
Scaringe, Robert P
Scruggs, Frank Dell
Smerage, Glen H
Spurlock, Jack Marion
Stahl, Joel S
Storm, Leo Eugene
Suciu, S(piridon) N
Swain, Ralph Warner
Travis, Russell Burton
Tulenko, James Stanley
Tusting, Robert Frederick
Ulug, Esin M
Whitehouse, Gary E

GEORGIA
Braden, Charles Hosea
Callahan, Leslie G, Jr
Dickerson, Stephen L(ang)
Dutton, John C
Esogbue, Augustine O
Fey, Willard Russell
Fletcher, James Erving
Hines, William W
Johnson, Cecil Gray
Komkoy, Vadim
Krol, Joseph
Landgren, John Jeffrey
Lee, Kok-Meng
Lehrer, Robert N(athaniel)
Lorber, Herbert William
Nessmith, Josh T(homas), Jr
Pence, Ira Wilson, Jr
Rouse, William Bradford
Stalford, Harold Lenn
Subramanian, Mani M
Tummala, Rao Ramamohana
Vachon, Reginald Irenee
White, Harold D(ouglas)
Zunde, Pranas

HAWAII
Johnson, Carl Edward
Sagle, Arthur A

IDAHO
DeBow, W Brad
Jobe, Lowell A(rthur)
Potter, David Eric

ILLINOIS
Agrawal, Arun Kumar
Berry, Gregory Franklin
Coplien, James O
Dessouky, Mohamed Ibrahim
Fildes, John M
Finn, Patricia Ann
Frey, Donald N(elson)
Graupe, Daniel
Guenther, Peter T
Gupta, Nand K
Haddad, Abraham H(erzl)
Huck, Morris Glen
Hurter, Arthur P
Jacobsen, Fred Marius
Kang, Sung-Mo Steve
Koenig, Michael Edward Davison
Kuchnir, Moyses
Kumar, Panganamala Ramana
Luplow, Wayne Charles
Mavrovouniotis, Michael L
Melsa, James Louis
Mikulski, James J(oseph)
Miller, Floyd Glenn
Moore, John Fitzallen
Murphy, Gordon J
Okamura, Kiyohisa
O'Neill, William D(ennis)
Paddock, Robert Alton
Patel, Mayur
Phelan, James Joseph
Ruhl, Roland Luther
Siegal, Burton L
Thompson, Charles William Nelson
Toy, William W
Worley, Will J
Wurm, Jaroslav
Zell, Blair Paul

INDIANA
Carlson, James C
Coyle, Edward John
Fearnot, Neal Edward
Kendall, Perry E(ugene)
Lyons, Jerry L
Nof, Shimon Y
Reklaitis, Gintaras Victor
Sain, Michael K(ent)
Schmidt, David Kelso
Sinha, Kumares C
Thomas, Marlin Uluess
Williams, Theodore J(oseph)
Wolber, William George
Wood, Frederick Starr
Wright, Jeffery Regan
Wynne, Bayard Edmund

IOWA
Akers, Arthur
Chyung, Dong Hak
Haendel, Richard Stone
Hekker, Roeland M T
Kortanek, Kenneth O
Kusiak, Andrew
Lardner, James F
Mischke, Charles R(ussell)
Ong, John Nathan, Jr
Robinson, John Paul

KANSAS
Gray, Tom J
Harnett, R Michael
Kramer, Bradley Alan
Lee, E(ugene) Stanley
Robinson, M John
Sawan, Mahmoud Edwin
Thompson, Joseph Garth
Welch, Stephen Melwood

KENTUCKY
Cooke, Samuel Leonard
Evans, Gerald William
Gold, Harold Eugene
Livingood, Marvin D(uane)
Merker, Stephen Louis
Squires, Robert Wright
Vaziri, Menouchehr

LOUISIANA
Cospolich, James D
Latorre, Robert George
McCarthy, Danny W
Ristroph, John Heard
Tewell, Joseph Robert

MAINE
Senders, John W

MARYLAND
Abed, Eyad Husni
Anand, Davinder K
Atia, Ali Ezz Eldin
Baker, Francis Edward, Jr
Battin, William James
Blevins, Gilbert Sanders
Bugenhagen, Thomas Gordon
Burkhalter, Barton R
Butler, Louis Peter
Campbell, William J
Chi, Donald Nan-Hua
Choi, Kyu-Yong
Cohen, Stanley Alvin
Cotton, Ira Walter
Das, Prasanta
Drzewiecki, Tadeusz Maria
Eaton, Alvin Ralph
Eckerman, Jerome
Eisner, Howard
Ellwein, Leon Burnell
Ely, Raymond Lloyd
Evans, Alan G
Fischel, David
Fisher, Dale John
Fishman, Elliot Keith
Fleig, Albert J, Jr
Frank, Carolyn
Freeman, Ernest Robert
Fthenakis, Emanuel John
Gabriele, Thomas L
Gasparovic, Richard Francis
Gleissner, Gene Heiden
Goepel, Charles Albert
Goldfinger, Andrew David
Gottfried, Paul
Graham, Edward Underwood
Grant, Conrad Joseph
Hall, Arthur David, III
Hammond, Charles E
Hanson, William A
Johnson, David Simonds
Kanal, Laveen Nanik
Kiebler, John W(illiam)
Kossiakoff, Alexander
Kroeger, Richard Alan
Ligomenides, Panos Aristides
Mahle, Christoph E
Makofski, Robert Anthony
Malarkey, Edward Cornelius
Martin, John A
Messing, Fredric
Meyer, Gerard G L
Michelsen, Arve
Minneman, Milton J(ay)
Morgan, Walter L(eroy)
Mumford, Willard R
Paul, Dilip Kumar
Penhollow, John O
Pisacane, Vincent L
Powell, Edward Gordon
Resnick, Joel B
Reupke, William Albert
Riley, Patrick Eugene
Rivkin, Maxcy
Rueger, Lauren J(ohn)
Saltman, Roy G
Sayer, John Samuel
Schneider, William C
Schruben, John H
Sharma, Dinesh C
Sleight, Thomas Perry
Thompson, Robert John, Jr
Tropf, William Jacob
Turner, Robert Davison
Vogelberger, Peter John, Jr
Voss, Paul Joseph
Weber, Richard Rand
Weiss, Roland George
Yaramanoglu, Melih
Zafiriou, Evanghelos

MASSACHUSETTS
Aprille, Thomas Joseph, Jr
Bennett, C Leonard
Blidberg, D Richard
Bloomfield, David Peter
Boodman, David Morris
Bovopoulos, Andreas D
Brayer, Kenneth
Brookner, Eli
Bussgang, J(ulian) J(akob)
Caron, Paul R(onald)
Clarke, Lori A
Colby, George Vincent, Jr
Cotter, Douglas Adrian
Cullinane, Thomas Paul
Dakss, Mark Ludmer
De Fazio, Thomas Luca
De Los Reyes, Gaston
De Neufville, Richard Lawrence
Douglas, J(ames) M(errill)
Dow, Paul C(rowther), Jr
Dubowsky, Steven
Duffy, Robert A
Eaves, Reuben Elco
Faflick, Carl E(dward)
Finn, John Thomas
Forrester, Jay W
Frey, Elmer Jacob
Gartner, Nathan Hart
Gelb, Arthur
Gilfix, Edward Leon
Gupta, Surendra Mohan
Hals, Finn
Hearn, David Russell
Hempstead, Charles Francis
Ho, Yu-Chi
Holst, William Frederick
Hopkins, Albert Lafayette, Jr
Horowitz, Larry Lowell
Hursh, John W(oodworth)
Ilic, Marija
Keicher, William Eugene
Kelley, Albert J(oseph)
King, William Connor
Kovaly, John J
Kovatch, George
Kreifeldt, John Gene
Lacoss, Richard Thaddee
Lees, Sidney
Lerner, Moisey
Levison, William H(enry)
Luft, Ludwig
Luther, Holger Martin
McBee, W(arren) D(ale)
Mastronardi, Richard
Mavretic, Anton
Merfeld, Daniel Michael
Miranda, Henry A, Jr
Mulukutla, Sarma Sreerama
Nelson, J(ohn) Byron
Nesbeda, Paul
Ogar, George W(illiam)
Paynter, Henry M(artyn)
Reinschmidt, Kenneth F(rank)
Richter, Stephen L(awrence)
Rubin, Milton D(avid)
Russell, George Albert
Safford, Richard Whiley
Sasiela, Richard
Schetzen, Martin
Schmidt, George Thomas
Schwartz, Jack
Seamans, R(obert) C(hanning), Jr
Seibel, Frederick Truman
Shea, Joseph F(rancis)
Sheridan, Thomas Brown
Shorb, Alan McKean
Sorenson, Harold Wayne
Story, Anne Winthrop
Strelzoff, Alan G
Sun, Fang-Kuo
Thun, Rudolf Eduard
Tweed, David George
Walter, Charlton M
Wetzstein, H(anns) J(uergen)
Wheeler, Ned Brent
Winer, Bette Marcia Tarmey

MICHIGAN
Asik, Joseph R
Aswad, A Adnan
Ayres, Robert Allen
Barnard, Robert D(ane), Jr
Barr, Robert Ortha, Jr
Bergmann, Dietrich R(udolf)
Beutler, Frederick J(oseph)
Borcherts, Robert H
Camp, David Thomas
Chen, Kan
Flaim, Thomas Alfred
Fleischer, Henry
Freedman, M(orris) David
Ghoneim, Youssef Ahmed
Gjostein, Norman A
Goodman, Erik David
Hamilton, William Eugene, Jr
Haynes, Dean L
Hrovat, Davorin
Jones, R(ichard) James
Khargonekar, Pramod P
Khasnabis, Snehamay
Kozma, Adam
Larrowe, Boyd T
Lett, Philip W(ood), Jr
McClamroch, N Harris
Mordukhovich, Boris S
O'Donnell, Matthew
Petrick, Ernest N(icholas)
Polis, Michael Philip
Powers, William Francis
Rillings, James H
Rohde, Steve Mark
Roll, James Robert
Root, William L(ucas)
Rosenberg, Ronald C(arl)
Sahney, Vinod K
Salam, Fathi M A
Savageau, Michael Antonio
Schlax, T(imothy) Roth
Schriber, Thomas J
Scott, Norman R(oss)
Sethi, Ishwar Krishan
Shapton, William Robert
Stivender, Donald Lewis
Teichroew, Daniel
Thompson, James Lowry
Turner, Robert
Windeknecht, Thomas George

MINNESOTA
Anderson, Willard Eugene
Cunningham, Thomas B
Erdman, Arthur Guy
Glewwe, Carl W(illiam)
Greene, Christopher Storm
Johnson, Howard Arthur, Sr
Knutson, Charles Dwaine
Kumar, K S P
Kvalseth, Tarald Oddvar
Lee, E(rnest) Bruce
Starr, Patrick Joseph
Stephanedes, Yorgos Jordan
Thiede, Edwin Carl
Wollenberg, Bruce Frederick

MISSISSIPPI
Miller, James Edward

MISSOURI
Atwood, Charles LeRoy
Babcock, Daniel Lawrence
Blundell, James Kenneth
Brackmann, Richard Theodore
Creighton, Donald L
Deam, James Richard
Flowers, Harold L(ee)
Franklin, Mark A
Glauz, William Donald
Griffith, Virgil Vernon
Katz, Israel Norman
Lemon, Leslie Roy
Linder, Solomon Leon
McMillan, Gregory Keith
Richards, Earl Frederick
Taylor, Ralph Dale
Zaborszky, John
Zobrist, George W(inston)

MONTANA
Gerez, Victor
Pierre, Donald Arthur
Stanislao, Bettie Chloe Carter

NEBRASKA
Lenz, Charles Eldon
Von Bargen, Kenneth Louis

NEW HAMPSHIRE
Jeffery, Lawrence R
Jolly, Stuart Martin
Klarman, Karl J(oseph)
Limbert, David Edwin
Linnell, Richard D(ean)
Seifer, Arnold David
Taft, Charles Kirkland

NEW JERSEY
Amarel, S(aul)
Amitay, Noach
Anderson, Milton Merrill
Andrews, Ronald Allen
Belanger, David Gerald
Bergman, Robert
Bhagat, Pramode Kumar
Boucher, Thomas Owen
Brenner, Douglas
Brown, Irmel Nelson

Browne, Robert Glenn
Canzonier, Walter J
Castenschiold, Rene
Cheung, Shiu Ming
Clymer, Arthur Benjamin
Curtis, Thomas Hasbrook
DeMarinis, Bernard Daniel
Dobkin, David Paul
Dorros, Irwin
Falconer, Warren Edgar
Feiner, Alexander
Franse, Renard
Fretwell, Lyman Jefferson, Jr
Gabriel, Oscar V
Grauman, Joseph Uri
Haas, Zygmunt
Harris, James Ridout
Hsu, Jay C
Huang, Jennming Stephen
Huber, Richard V
Hung, Tonney H M
Johannes, Virgil Ivancich
Kalley, Gordon S
Kantor, Paul B
Kaplan, Joel Howard
Kesselman, Warren Arthur
Kobayashi, Hisashi
Lawson, James Robert
Leu, Ming C
Lucas, Henry C, Jr
Maclin, Ernest
McNicholas, James J
Mangulis, Visvaldis
Mauzey, Peter T
Meyer, Andrew U
Murthy, Srinivasa K R
Poor, Harold Vincent
Poucher, John Scott
Prekopa, Andras
Raychaudhuri, Kamal Kumar
Robrock, Richard Barker, II
Rosenblum, Martin Jacob
Rosenthal, Lee
Sachs, Harvey Maurice
Saniee, Iraj
Sannuti, Peddapullaiah
Scales, William Webb
Schaefer, Jacob W
Schnapf, Abraham
Scholz, Lawrence Charles
Sears, Raymond Warrick, Jr
Semmlow, John Leonard
Shefer, Joshua
Shelton, James Churchill
Sherman, Ronald
Shieh, Ching-Chyuan
Shipley, Edward Nicholas
Shulman, Herbert Byron
Sipress, Jack M
Sontag, Eduardo Daniel
Steben, John D
Stengel, Robert Frank
Strother, J(ohn) A(lan)
Turoff, Murray
Tzanakou, M Evangelia
Uhrig, Jerome Lee
Weir, William Thomas
Weiss, C Dennis
Westerman, Howard Robert
Young, Li

NEW MEXICO
Baty, Richard Samuel
Clark, Wallace Thomas, III
Coulter, Claude Alton
Curtis, Wesley E
Dempster, William Fred
Gilbert, Alton Lee
Hakkila, Eero Arnold
Lee, David Mallin
Lundergan, Charles Donald
Mayer, Harris Louis
Novak, James Lawrence
Shipley, James Parish, Jr
Sitney, Lawrence Raymond
Snider, Donald Edward
Swingle, Donald Morgan
Welber, Irwin
Woods, Robert Octavius

NEW YORK
Ahmed-Zaid, Said
Aly, Adel Ahmed
Anastassiou, Dimitris
Anderson, Roy E
Axelrod, Norman Nathan
Balabanian, Norman
Bonan, Eugene J
Brownlow, Roger W
Burros, Raymond Herbert
Carlson, A Bruce
Carruthers, Raymond Ingalls
Chang, John H(si-Teh)
Chen, Chi-Tsong
Christiansen, Donald David
Clarke, Kenneth Kingsley
Cohen, Edwin
Costello, Richard Gray
Dahm, Douglas Barrett
Damouth, David Earl
Desrochers, Alan Alfred
Difranco, Julius V
Dimmler, D(ietrich) Gerd
Fine, Terrence Leon

Geiss, Gunther R(ichard)
Goel, Narendra Swarup
Gran, Richard J
Haas, David Jean
Hajela, Prabhat
Hauser, Norbert
Hetzel, Richard Ernest
Higinbotham, William Alfred
Howland, Howard Chase
Hutchings, William Frank
Jones, Leonidas John
Karnaugh, Maurice
Kershenbaum, Aaron
Kinzly, Robert Edward
Kirchmayer, Leon K
Klir, George Jiri
Kopp, Richard E
La Russa, Joseph Anthony
Leigh, Richard Woodward
Levine, Stephen Alan
Ling, Huei
Litchford, George B
Logue, J(oseph) C(arl)
Longman, Richard Winston
Longobardo, Anna Kazanjian
Lowen, W(alter)
Lustig, Howard E(ric)
Maltz, Martin Sidney
Mead, Lawrence Myers
Meenan, Peter Michael
Menkes, Sherwood Bradford
Metzger, Ernest Hugh
Monmonier, Mark
Morton, Roger Roy Adams
Mosteller, Henry Walter
Notaro, Anthony
Nyrop, Jan Peter
Ortel, William C(harles) G(ormley)
Palmedo, Philip F
Papoulis, Athanasios
Pomerene, James Herbert
Raghuveer, Mysore R
Rich, Kenneth Eugene
Rogers, Edwin Henry
Rushing, Allen Joseph
Sarachik, Philip E(ugene)
Sargent, Robert George
Saridis, George N
Sevian, Walter Andrew
Shen, Chi-Neng
Sherman, Philip Martin
Sheryll, Richard Perry
Shevack, Hilda N
Shooman, Martin L
Shuey, R(ichard) L(yman)
Sinclair, Douglas C
Smith, Bernard
Sobczak, Thomas Victor
Sobel, Kenneth Mark
Spang, H Austin, III
Spencer, David R
Sri-Jayanthal, Sri Muthythamby
Sternlicht, B(eno)
Stone, Harold S
Summers, Phillip Dale
Swartz, James Lawrence
Tatarczuk, Joseph Richard
Taylor, James Hugh
Thompson, Robert Alan
Thorndike, Alan Moulton
Tiemann, Jerome J
Torng, Hwa-Chung
Van Slyke, Richard M
Varshney, Pramod Kumar
Victor, Jonathan David
Wang, Shu Lung
Weingarten, Norman C
White, William Wallace
Willmert, Kenneth Dale
Winter, Herbert
Witt, Christopher John
Yablon, Marvin
Yao, David D
Yip, Kwok Leung
Youla, Dante C
Zdan, William

NORTH CAROLINA
Allen, Charles Michael
Bernhard, Richard Harold
Coulter, Norman Arthur, Jr
Defoggi, Ernest
Dudziak, Donald John
Eckerlin, Herbert Martin
Franzon, Paul Damian
Gangwal, Santosh Kumar
Garg, Devendra Prakash
Gerhardt, Don John
Hutchins, William R(eagh)
Kanopoulos, Nick
McCumber, Dean Everett
Marinos, Pete Nick
Meier, Wilbur L(eroy), Jr
Nutter, James I(rving)
Smith, William Adams, Jr
Struve, William George
Zaalouk, Mohamed Gamal

NORTH DAKOTA
Wang, Jun
Yuvarajan, Subbaraya

OHIO
Ash, Raymond H(ouston)

Bernard, John Wilfrid
Bishop, Albert B
Burghart, James H(enry)
Chen, Chiou Shiun
Chovan, John David
Christensen, James Henry
Ciricillo, Samuel F
Cruz, Jose B(ejar), Jr
Cummings, Charles Arnold
Damianov, Vladimir B
D'Azzo, John Joachim
Easterday, Jack L(eroy)
El-Naggar, Mohammed Ismail
Faymon, Karl A(lois)
Fritsch, Charles A(nthony)
Ghandakly, Adel Ahmad
Giamati, Charles C, Jr
Gorland, Sol H
Grant, Michael P(eter)
Hajek, Otomar
Ho, Fanghuai H(ubert)
Hobbs, Benjamin F
Hoy, Casey William
Jayne, Theodore D
Kaya, Azmi
Khalimsky, Efim D
Klosterman, Albert Leonard
Kramerich, George L
Kwatra, Subhash Chander
Lamont, Gary Byron
Laskowski, Edward L
Lefkowitz, Irving
Lenhart, Jack G
Liu, Ming-Tsan
McGoron, Anthony Joseph
McNichols, Roger J(effrey)
Marble, Duane Francis
Menq, Chia-Hsiang
Quam, David Lawrence
Rose, Charles William
Schultz, Edwin Robert
Shetterly, Donivan Max
Shick, Philip E(dwin)
Sprafka, Robert J
Steele, Jack Ellwood
Vassell, Gregory S
Wallace, Michael Dwight
Wyman, Bostwick Frampton

OKLAHOMA
Baker, James E
Ellington, Rex T(ruesdale), Jr
Fagan, John Edward
Howard, Robert Eugene
McDevitt, Daniel Bernard
Martner, Samuel (Theodore)
Mize, Joe H(enry)
Richardson, J Mark
Schuermann, Allen Clark, Jr
Terrell, Marvin Palmer

OREGON
Boyd, Dean Weldon
Gillette, Philip Roger
Linstone, Harold A

PENNSYLVANIA
Abend, Kenneth
Allen, Charles C(orletta)
Anandalingam, G
Anouchi, Abraham Y
Baumann, Dwight Maylon Billy
Bedrosian, Samuel D
Bhagavatula, VijayaKumar
Biegler, Lorenz Theodor
Bise, Christopher John
Bradley, James Henry Stobart
Brunner, Mathias J
Bukowski, Julia Victoria
Butkiewicz, Edward Thomas
Carbonell, Jaime Guillermo
Cardarelli, Joseph S
Charp, Solomon
Clapp, Richard Gardner
Dax, Frank Robert
Dorny, C Nelson
Eisenberg, Lawrence
Eisenstein, Bruce A
Feng, Tse-yun
Frumerman, Robert
Fuchs, Walter
Georgakis, Christos
Gilstein, Jacob Burrill
Goldman, Robert Barnett
Goutmann, Michel Marcel
Griffin, Jerry Howard
Grossmann, Ignacio E
Harney, Robert Charles
Hendrickson, Chris Thompson
Johnson, Stanley Harris
Kam, Moshe
Kaslow, David Edward
Kazahaya, Masahiro Matt
Kraft, R(alph) Wayne
Krendel, Ezra Simon
Ku, Y(u) H(siu)
Kurfess, Thomas Roland
Kusic, George Larry, Jr
Leibholz, Stephen W
Likins, Peter William
Longini, Richard Leon
Loxley, Thomas Edward
McGarity, Arthur Edwin
Murgie, Samuel A

Nichols, Duane Guy
Popovics, Sandor
Putman, Thomas Harold
Satz, Ronald Wayne
Scigliano, J Michael
Scranton, Bruce Edward
Sherwood, Bruce Arne
Shirk, Richard Jay
Sibul, Leon Henry
Siegel, Melvin Walter
Smith, David Beach
Stover, James Anderson, Jr
Thomas, Donald H(arvey)
Tierney, John W(illiam)
Tobias, Philip E
Uher, Richard Anthony
Verhanovitz, Richard Frank
Vogt, William G(eorge)
Walker, David Kenneth
Weinstock, W
Zygmont, Anthony J

RHODE ISLAND
Bartram, James F(ranklin)
Ehrlich, Stanley L(eonard)
Hazeltine, Barrett
Palm, William John, III
Petrou, Panayiotis Polydorou
Raykhman, Aleksandr
Rerick, Mark Newton
Robertshaw, Joseph Earl

SOUTH CAROLINA
Balintfy, Joseph L
Blevins, Maurice Everett
Gaines, Albert L(owery)
Klemm, James L
Pursley, Michael Bader

SOUTH DAKOTA
Hodges, Neal Howard, II
Oliver, Thomas Kilbury
Reuter, William L(ee)
Riemenschneider, Albert Louis

TENNESSEE
Arlinghaus, Heinrich Franz
Bartell, Steven Michael
Bontadelli, James Albert
Cox, Ronald Baker
Dabbs, John Wilson Thomas
Emanuel, William Robert
Haaland, Carsten Meyer
Hill, Hubert Mack
House, Robert W(illiam)
Lane, Eric Trent
Martin, Harry Lee
Siirola, Jeffrey John
Upadhyaya, Belle Raghavendra

TEXAS
Bennett, George Kemble
Bhuva, Rohit L
Bishop, Robert H
Blank, Leland T
Chakravarty, Indranil
Collins, Francis Allen
Cory, William Eugene
Crump, John Joseph
Curtin, Richard B
Dhudshia, Vallabh H
Dodd, Edward Elliott
Edgar, Thomas Flynn
Ewell, James John, Jr
Fletcher, John Lynn
Fontana, Robert E
French, Robert Leonard
Garner, Harvey L(ouis)
Hamill, Dennis W
Hayman, Alan Conrad
Hennessey, Audrey Kathleen
Kesler, Oren Byrl
Koehn, Enno
Ladde, Gangaram Shivlingappa
Lawrence, Joseph D, Jr
Levine, Jules David
Linder, John Scott
Lipovski, Gerald John (Jack)
Marshek, Kurt M
Martin, Norman Marshall
Matlock, Gibb B
Maxwell, Donald A
Miller, Maurice Max
Mistree, Farrokh
Moser, James Howard
Mouritsen, T(horvald) Edgar
Nute, C Thomas
Ober, Raimund Johannes
Ochiai, Shinya
Ostrofsky, Benjamin
Paddison, Fletcher C
Parker, Arthur L (Pete)
Patton, Alton DeWitt
Phillips, Don T
Potter, Robert Joseph
Randolph, Paul Herbert
Rodenberger, Charles Alvard
Samson, Charles Harold
Secrest, Everett Leigh
Summers, Richard Lee
Teague, Abner F
Tibbals, Harry Fred, III
VantHull, LORIN L
Wakeland, William Richard
Walkup, John Frank

Systems Design & Systems Science (cont)

Ward, Phillip Wayne
White, Charles F(loyd)
Widmer, Robert H
Wilhelm, Wilbert Edward
Wood, Kristin Lee
Yakin, Mustafa Zafer

UTAH
Baum, Sanford
Butterfield, Veloy Hansen, Jr
Clegg, John C(ardwell)
Cox, Benjamin Vincent
Dulock, Victor A, Jr
Hoggan, Daniel Hunter
Jolley, David Kent
Krejci, Robert Henry
MacGregor, Douglas
Sandquist, Gary Marlin
Turley, Richard Eyring
Yorks, Terence Preston

VERMONT
Ferris-Prabhu, Albert Victor Michael

VIRGINIA
Agee, Marvin H
Attinger, Ernst Otto
Ball, Joseph Anthony
Barmby, John G(lennon)
Beam, Walter R(aleigh)
Bingham, Billy Elias
Bordogna, Joseph
Browell, Edward Vern
Burka, Maria Karpati
Cerf, Vinton Gray
Charvonia, David Alan
Cheng, George Chiwo
Childress, Otis Steele, Jr
Choi, Junho
Cook, Charles William
Coram, Donald Sidney
Crawford, Daniel J
Daniel, Kenneth Wayne
Doles, John Henry, III
Donivan, Frank Forbes, Jr
Edyvane, John
Entzminger, John N
Epstein, Samuel David
Fabrycky, Wolter J
Fearnsides, John Joseph
Fink, Lester Harold
Finkelstein, Jay Laurence
Fletcher, James Chipman
Gamble, Thomas Dean
Gerlach, A(lbert) A(ugust)
Gibson, John E(gan)
Godbole, Sadashiva Shankar
Gordon, John S(tevens)
Gouse, S William, Jr
Greinke, Everett D
Guerber, Howard P(aul)
Hammond, Marvin H, Jr
Heffron, W(alter) Gordon
Herm, Ronald Richard
Hobeika, Antoine George
Hodge, Donald Ray
Holzmann, Ernest G(unther)
Hou, Gene Jean-Win
Houck, Mark Hedrich
Hunt, V Daniel
Hwang, John Dzen
James, W(ilbur) Gerald
Johnson, Charles Minor
Johnson, George Patrick
Joshi, Suresh Meghashyam
Kahn, Robert Elliot
Kahne, Stephen James
Kaplan, Robert Lewis
Keblawi, Feisal Said
Keever, David Bruce
Kelly, Henry Charles
Kissell, Kenneth Eugene
Kuhlthau, A(lden) R(obert)
Larson, Arvid Gunnar
Leon, H(erman) I
Levis, Alexander Henry
Ludwig, George H
McGean, Thomas J
McGregor, Dennis Nicholas
McKenna, John Dennis
McRee, G(riffith) J(ohn), Jr
Madan, Rabinder Nath
Manfredi, Arthur Frank, Jr
Marsh, Howard Stephen
Mehta, Gurmukh D
Murray, Jeanne Morris
Myers, Donald Albin
Nyman, Thomas Harry
Perry, Dennis Gordon
Pyke, Thomas Nicholas, Jr
Reierson, James (Dutton)
Richards, Paul Bland
Rowe, William David
Sage, Andrew Patrick
Sawyer, James W
Schlegelmilch, Reuben Orville
Schwartz, Jay W(illiam)
Schy, Albert Abe
Seward, William Davis
Shenolikar, Ashok Kumar

Smith, Michael Claude
Sobieszczanski-Sobieski, Jaroslaw
Sorrell, Gary Lee
Thompson, Anthony Richard
Tsang, Wai Lin
Walker, Robert Paul
Warfield, J(ohn) N(elson)
Waxman, Ronald
Webber, John Clinton
Weinstein, Iram J
Wells, John Morgan, Jr
Whitelaw, R(obert) L(eslie)
Wilcox, Lyle C(hester)
Woeste, Frank Edward
Wolf, Eric W
Worthington, John Wilbur

WASHINGTON
Blackburn, John Lewis
Driver, Garth Edward
Gravitz, Sidney I
Hueter, Theodor Friedrich
MacVicar-Whelan, Patrick James
Madden, James H(oward)
Mar, Brian W(ayne)
Millham, Charles Blanchard
Preikschat, F(ritz) K(arl)
Ricker, Neil Lawrence
Stear, Edwin Byron
Tripp, Leonard L
Yore, Eugene Elliott

WEST VIRGINIA
Tewksbury, Stuart K

WISCONSIN
Bishop, Charles Joseph
Bomba, Steven James
Crandall, Lee W(alter)
Dobson, David A
Eriksson, Larry John
Goodson, Raymond Eugene
Hanway, John E(dgar), Jr
Heinen, James Albin
Henke, Russell W
Northouse, Richard A
Sanborn, I B
Schultz, Hilbert Kenneth
Suri, Rajan
Szpot, Bruce F

WYOMING
Constantinides, Christos T(heodorou)
Mingle, John O(rville)

PUERTO RICO
Hussain, Malek Gholoum Malek
Soderstrom, Kenneth G(unnar)
Turner, Kenneth Clyde

ALBERTA
Chan, Wah Chun
Hamza, Mohamed Hamed

BRITISH COLUMBIA
Agathoklis, Panajotis
Gruver, William A
Pomeroy, Richard James

MANITOBA
Johnson, R(ichard) A(llan)

NEW BRUNSWICK
Lee, Y C

NEWFOUNDLAND
Shieh, John Shunen

NOVA SCOTIA
Baird, Charles Robert
Eames, M(ichael) C(urtis)
Kujath, Marek Ryszard

ONTARIO
Ahmed, Nasir Uddin
Altman, Samuel Pinover
Bandler, John William
Beevis, David
Byer, Philip Howard
Chou, Jordan Quan Ban
Cohn, Stanley Howard
Corneil, Ernest Ray
Davison, E(dward) J(oseph)
Fairman, Frederick Walker
Falconer, David Duncan
Georganas, Nicolas D
Goldenberg, Andrew Avi
Grodski, Juliusz Jan
Henry, Roger P
Hipel, Keith William
Husejin, Koncay
Law, Cecil E
McClymont, Kenneth
Mittal, Gauri S
Nevitt, H J Barrington
Ratz, H(erbert) C(harles)
Rauch, Sol
Reza, Fazlollah M
Riordon, J(ohn) Spruce
Roe, Peter Hugh O'Neil
Ryan, Dave
Siddall, Ernest
Simard, Albert Joseph
Slocombe, Donald Scott

Smith, Harold William
Wilson, John D(ouglas)
Witzman, Sorin
Wong, Jo Yung
Wonham, W Murray
Wright, Douglas Tyndall
Wright, Joseph D

QUEBEC
Bui, Tien Rung
Cheng, Richard M H
Davidson, Colin Henry
Ferguson, Michael John
Jumarie, Guy Michael
Lecours, Michel
Marin, Miguel Angel
Patel, Rajnikant V
Plotkin, Eugene Isaak
Sankar, Seshadri

SASKATCHEWAN
Curtis, Fred Allen
Fuller, Gerald Arthur
Gupta, Madan Mohan
Koh, Eusebio Legarda
Nikiforuk, P(eter) N
Walerian, Szyszkowski

OTHER COUNTRIES
Arimoto, Suguru
Bryngdahl, Olof
Chanson, Samuel T
Czepyha, Chester George Reinhold
Esterson, Gerald L(ee)
Eykhoff, Pieter
Fettweis, Alfred Leo Maria
Fitzhugh, Henry Allen
Halemane, Keshava Prasad
Ince, A Nejat
Inose, Hiroshi
Jespersen, Nils Vidar
Kimura, Hidenori
Kuroyanagi, Noriyoshi
Lehman, Meir M
Lindquist, Anders Gunnar
Mansour, Mohamed
Renfrew, Robert Morrison
Ros, Herman H
Sekine, Yasuji
Shah, Mirza Mohammed
Singh, Madan Gopal
Sunahara, Yoshifumi
Thissen, Wil A
Tzafestas, Spyros G
Willems, Jan C
Yazicigil, Hasan
Zeheb, Ezra

Other Engineering

ALABAMA
Clark, Charles Edward
Hart, David R
Johnson, Evert William
Park, Duk-Won
Smoot, George Fitzgerald

ALASKA
Gedney, Larry Daniel

ARIZONA
Barrett, Harrison Hooker
Batson, Raymond Milner
Benfield, Charles W(illiam)
Freyermuth, Clifford L
Helbert, John N
Kurz, Richard J
Makepeace, Gershom Reynolds
Mulvey, James Patrick
Reid, Ralph R
Young, Hewitt H

ARKANSAS
Bond, Robert Levi
Cardwell, David Michael
Carr, Gerald Paul
Glendening, Norman Willard
Sims, Robert Alan

CALIFORNIA
Agbabian, Mihran S
Agrawal, Suphal Prakash
Anger, Hal Oscar
Ashcroft, Frederick H
Aster, Robert Wesley
Auksmann, Boris
Aziz, Khalid
Baba, Paul David
Bontrager, Ernest L
Booth, Newell Ormond
Braithwaite, Charles Henry, Jr
Byer, Robert L
Byram, George Wayne
Cook, Charles J
Cook, Neville G W
Crutchfield, Charlie
Damos, Diane Lynn
Davis, Harmer E
Downing, Willis G, Jr
Elings, Virgil Bruce
Ewing, Gerald Dean
Fischer, George K
Forster, Julian
Fung, Yuan-Cheng B(ertaam)

Gearhart, Roger A
Graham, Martin H(arold)
Grier, Herbert E
Hackwood, Susan
Hayes, Thomas Jay, III
Hess, Cecil F
Hingorani, Narain G
Hurd, Walter Leroy, Jr
John, Joseph
Kennedy, Robert P
Kivenson, Gilbert
Laub, Alan John
Lee, Hua
Leidersdorf, Craig B
Longley-Cook, Mark T
Ma, Joseph T
McIntyre, Marie Claire
Maxfield, Bruce Wright
Merritt, J(oshua) L(evering), Jr
Meyer, Lhary
Munger, Charles Galloway
Nooker, Eugene L(eRoy)
Nowatzki, Edward Alexander
O'Donnell, Ashton Jay
Paquette, Robert George
Rechnitzer, Andreas Buchwald
Remer, Donald Sherwood
Robinson, Lloyd Burdette
Robinson, Merton Arnold
Schade, Henry A(drian)
Schmidt, Raymond LeRoy
Sengupta, Sailes Kumar
Sharpe, Roland Leonard
Shelly, John Richard
Shemdin, Omar H
Siljak, Dragoslav (D)
Smith, Levering
Subramanya, Shiva
Suer, H(erbert) S
Swanson, Eric Richmond
Sweeney, James Lee
Swerling, Peter
Thiers, Eugene Andres
Thomas, Graham Havens
Van Leer, John Cloud
Volpe, Richard Alan
Vomhof, Daniel William
Waldhauer, F D
Wallerstein, Edward Perry
Wenzel, James Gottlieb
Williamson, Robert Brady

COLORADO
Cochran, Leighton Scott
Cohen, Robert
Holtz, Wesley G
Hotchkiss, William Rouse
Huisjen, Martin Albert
Keller, Frederick Albert, Jr
Kelley, Neil Davis
Knollenberg, Robert George
Kompala, Dhinakar S
Lightsey, Paul Alden
Nesenbergs, Martin
Peacock, Roy Norman
Plows, William Herbert
Puckorius, Paul Ronald
Schmidt, Alan Frederick
Strom, Oren Grant
Ward, Milton Hawkins
Weber, James R
Zimmerer, Robert W

CONNECTICUT
Crossley, F(rancis) R(endel) Erskine
Drozdowicz, Zbigniew Maria
Engelberger, Joseph Frederick
Hill, David Lawrence
Levinson, Herbert S
Long, Richard Paul
Mancheski, Frederick J
Pooley, Alan Setzler
Powers, Joseph
Sheets, Herman E(rnest)
Shuey, Merlin Arthur
Tepas, Donald Irving
Weinstein, Curt David

DELAWARE
Arots, Joseph B(artholomew)
Benmark, Leslie Ann
Dalrymple, Robert Anthony
Hanzel, Robert Stephen
Kobayashi, Nobuhisa
Levy, Paul F
Moore, Ralph Leslie
Sheer, M Lana

DISTRICT OF COLUMBIA
Baruch, Jordan J(ay)
Cambel, Ali B(ulent)
Caywood, James Alexander, III
Coulter, Henry W
Lee, Alfred M
O'Fallon, Nancy McCumber
Parasuraman, Raja

FLORIDA
Alexiou, Arthur George
Arnold, Henry Albert
Beatty, Charles Lee
Braman, Robert Steven
Brandt, Bruce Losure
Campbell, Hallock Cowles
Cassidy, William F

Dallman, Paul Jerald
Daubin, Scott C
Davidson, James Blaine
Dunford, James Marshall
Freeman, Neil Julian
Friedman, Raymond
Froehlich, Fritz Edgar
Koff, Bernard Louis
McAllister, Raymond Francis
Rice, Stephen Landon
Rona, Donna C
Rosenthal, Henry Bernard
Smith, Malcolm (Kinmonth)
Williamson, Donald Elwin
Wylde, Ronald James
Zborowski, Andrew

GEORGIA
Buckalew, Louis Walter
Cremens, Walter Samuel
Milkovic, Miran
Rogers, Nelson K
Woods, Wendell David

HAWAII
Bretschneider, Charles Leroy
Gerritsen, Franciscus
Kaye, Wilbur (Irving)
Spielvogel, Lester Q

IDAHO
Budwig, Ralph S
Engen, Ivar A
Spencer, Paul Roger

ILLINOIS
Brauer, Roger L
Cheng, Herbert S
Dolgoff, Abraham
Layton, Terry North
Mielke, Robert L
Savic, Stanley D
Saxena, Satish Chandra
Sebree, Bruce Randall
Toy, William W
Wise, Evan Michael
Worley, Will J
Wurm, Jaroslav

INDIANA
Dando, William Arthur
Ladisch, Michael R
McAleece, Donald John
Nolting, Henry Frederick
Skibniewski, Miroslaw Jan

IOWA
Brown, Robert Grover
Nixon, Wilfrid Austin
Porter, Max L

KANSAS
Dean, Thomas Scott
Norris, Roy Howard
Razak, Charles Kenneth
Willhite, Glen Paul

KENTUCKY
Parekh, Bhupendra Kumar
Uhl, Robert H

LOUISIANA
Bourdillon, Antony John
Greco, Edward Carl
Kinchen, David G
Lee, Griff C
Rice, David A

MAINE
Gibson, Richard C(ushing)
Norton, Karl Kenneth

MARYLAND
Acheson, Donald Theodore
Amde, Amde M
Babrauskas, Vytenis
Basser, Peter J
Belanger, Brian Charles
Berger, Harold
Brenner, Fivel Cecil
Dauber, Edwin George
Doctor, Norman J(oseph)
Fowell, Andrew John
Franke, Gene Louis
Frey, Henry Richard
Gheorghiu, Paul
Gilford, Leon
Hartman, Patrick James
Hyde, Geoffrey
Johnson, Bruce
Kacker, Raghu N
Kalil, Ford
Lange, Eugene Albert
Levine, Robert S(idney)
McConoughey, Samuel R
McIlvain, Jess Hall
Morgan, William Bruce
Navarro, Joseph Anthony
Palmer, Timothy Trow
Papian, William Nathaniel
Potzick, James Edward
Quintiere, James G
Rao, Gopalakrishna M
Rochlin, Phillip
Schulkin, Morris
Sikora, Jerome Paul

Spivak, Steven Mark
Stein, Robert Alfred
Sutter, David Franklin
Ventre, Francis Thomas
Walker, James Harris
Weaver, Christopher Scot
Weinschel, Bruno Oscar
Wing, John Faxon

MASSACHUSETTS
Allen, Steven
Chryssostomidis, Chryssostomos
Clarke, Edward Nielsen
Coor, Thomas
Copley, Lawrence Gordon
DiNardi, Salvatore Robert
Ehrlich, Daniel Jacob
Field, Hermann Haviland
Fitzgerald, Robert William
Hertweck, Gerald
Hutchinson, John Woodside
Krasner, Jerome L
Krofta, Milos
LeMessurier, William James
Lemons, Thomas M
Levin, Robert Edmond
MacMillan, Douglas Clark
Masubuchi, Koichi
Newman, J Nicholas
Park, B J
Pearlman, Michael R
Pope, Joseph
Read, Philip Lloyd
Reznicek, Bernard William
Shrader, William Whitney
Spergel, Philip
Wilkinson, Robert Haydn
Williams, Albert James, III
Zalosh, Robert Geoffrey

MICHIGAN
Batrin, George Leslie
Beck, Robert Frederick
Benford, Harry (Bell)
Chapman, Robert Earl, Jr
Chou, Clifford Chi Fong
Datta, Tapan K
Evans, Leonard
Fancher, Paul S(trimple)
Fortuna, Edward Michael, Jr
Greene, Bruce Edgar
Levine, Simon P
Prasad, Biren
Reising, Richard F
Risdon, Thomas Joseph
Rodgers, John James
Saul, William Edward
Taraman, Khalil Showky
Turcotte, William Arthur
Yagle, Raymond A(rthur)

MINNESOTA
Harvey, Charles Arthur
Johnson, Robert L(awrence)
Mooers, Howard T(heodore)
Sterling, Raymond Leslie

MISSISSIPPI
Aughenbaugh, Nolan B(laine)
Bryant, Glenn D(onald)
Cade, Ruth Ann
Rosenhan, A Kirk

MISSOURI
Bockserman, Robert Julian
Kurtz, Vincent E
Monzyk, Bruce Francis
Morgan, Robert P
Natani, Kirmach
Poynton, Joseph Patrick
Saran, Chitaranjan
Schaefer, Joseph Thomas
Wulfman, David Swinton

NEW HAMPSHIRE
Allmendinger, E Eugene
Crowell, Merton Howard
Kasper, Joseph F, Jr
Wang, Der-Shi
Wilson, Donald Alfred

NEW JERSEY
Abrahams, Elihu
Biagetti, Richard Victor
Chan, Chun Kin
DeMarinis, Bernard Daniel
Dodington, Sven Henry Marriott
Endo, Youichi
Heirman, Donald N
Holt, Vernon Emerson
Huang, C C
Hwang, Helen H L
Kiessling, George Anthony
Killam, Everett Herbert
Kleckner, Willard R
Kluver, J(ohan) W(ilhelm)
Liebe, Donald Charles
Lund, Daryl B
McGregor, Walter
Mandel, Andrea Sue
Manz, August Frederick
Miner, Robert Scott, Jr
Newman, Joseph Herbert
Ponessa, Joseph Thomas
Quinlan, Daniel A
Rhee, Aaron Seung-Joon

Rothschild, Gilbert Robert
Sams, Burnett Henry, III
Savitsky, Daniel
Schoenberg, Leonard Norman
Schofield, Robert Edwin
Sharp, Hugh T
Stiles, Lynn F, Jr
Wadlow, David
Warren, Clifford A
Wentworth, John W

NEW MEXICO
Edeskuty, F(rederick) J(ames)
Elliott, Dana Edgar
Finch, Thomas Wesley
Keepin, George Robert, Jr
Moulds, William Joseph
Neeper, Donald Andrew
Savage, Charles Francis
Stephans, Richard A
Strong, Ronald Dean
Sugerman, Leonard Richard
Valentine, James K

NEW YORK
Abkowitz, Martin A(aron)
Bandyopadhyay, Amitabha
Barone, Louis J(oseph)
Bayer, Raymond George
Bechhofer, Robert E(ric)
Belden, David Leigh
Fein, R(ichard) S(aul)
Fisch, Oscar
Goodwin, Arthur VanKleek
Gould, James P
Greenspan, Joseph
Hall, Kimball Parker
Herman, Herbert
Hetzel, Richard Ernest
Hoover, Thomas Earl
Kahn, Elliott H
Kammer, Paul A
Klein, Harvey Gerald
Lakatos, Andras Imre
Lang, Martin
Le Maistre, Christopher William
Leone, James A
Lindstrom, Gary J
Lurkis, Alexander
McShane, William R
Malaviya, Bimal K
Mathes, Kenneth Natt
Meth, Irving Marvin
Meyburg, Arnim Hans
Meyer, Robert Walter
Pajari, George Edward
Palladino, William Joseph
Pegram, John B
Pennington, Keith Samuel
Pernick, Benjamin J
Phillips, Anthony
Radeka, Veljko
Raghuveer, Mysore R
Rothman, Herbert B
Schlee, Frank Herman
Sinclair, Douglas C
Staples, Basil George
Thomas, Richard Alan
Tobin, Michael
Vemula, Subba Rao
Vogel, Alfred Morris
Welsh, James P
Wiese, Warren M(elvin)
Wiggins, Edwin George
Wotherspoon, Neil
Wouk, Victor
Youla, Dante C
Zubaly, Robert B(arnes)

NORTH CAROLINA
Ayoub, Mahmoud A
Ballard, Lewis Franklin
Buchanan, David Royal
Chin, Robert Allen
Ciccolella, Joseph A
Downs, Robert Jack
Fisher, C(harles) Page, Jr
Grainger, John Joseph
Humphreys, Kenneth K
Kirkbride, L(ouis) D(ale)
Nader, John S(haheen)
Sherer, James Pressly
Utku, Bisulay Bereket

OHIO
Bisson, Edmond E(mile)
Butz, Donald Josef
Ghosh, Sanjib Kumar
Gratzl, Miklos
Gregory, Thomas Bradford
Grimm, Robert Arthur
Houpis, Constantine H
Keyes, Marion Alvah, IV
Nash, David Byer
Sahai, Yogeshwar
Singh, Rajendra
Voisard, Walter Bryan
Waller, Michael Holton

OKLAHOMA
Forney, Bill E
Hendrickson, John Robert
Jones, Jack Raymond
Summers, Jerry C
Winter, William Kenneth
Zigrang, Denis Joseph

OREGON
Davis, Alvin Herbert
Devletian, Jack H
DuFresne, Ann
Duncan, Walter E(dwin)
Landers, John Herbert, Jr
Walsh, Don

PENNSYLVANIA
Bergmann, Ernest Eisenhardt
Bjorhovde, Reidar
Brown, Neil Harry
Cavanagh, Peter R
Cohen, Anna (Foner)
Drumwright, Thomas Franklin, Jr
Ebert, Philip E
Hager, Wayne R
Hertzberg, Martin
Howe, Richard Hildreth
Ibrahim, Baky Badie
Kaufman, William Morris
Ko, Frank K
Kraus, C Raymond
Krishnaswamy, S V
Levin, Michael H(oward)
Magaziner, Henry Jonas
Magison, Ernest Carroll
Mehta, Kishor Singh
Moore, Walter Calvin
Mouly, Raymond J
Oliver, Morris Albert
Ostapenko, A(lexis)
Rohrer, Ronald A
Sheppard, Walter Lee, Jr
Staley, James T
Sterk, Andrew A
Stroud, Jackson Swavely
Van Roggen, Arend
Wagner, J Robert
Werkman, Joyce
Westine, Peter Sven
White, Eugene Wilbert
Wiehe, William Albert
Zimmermann, F(rancis) J(ohn)

RHODE ISLAND
Ellison, William Theodore
Kirschenbaum, Susan S
Silva, Armand Joseph
Silverman, Harvey Fox
Spaulding, Malcolm Lindhurst
Sternling, Charles V
Tyce, Robert Charles

SOUTH CAROLINA
Sawers, James Richard, Jr
Singleton, Chloe Joi
Sudarshan, Tangali S

SOUTH DAKOTA
Oliver, Thomas Kilbury

TENNESSEE
Ackermann, Norbert Joseph, Jr
Baer, Thomas Strickland
Devgan, Satinderpaul Singh
Goans, Ronald Earl
Godbee, H(erschel) W(illcox)
Hensley, Marble John, Sr
Isley, James Don
Kopp, Manfred Kurt
Kress, Thomas Sylvester
Lubell, Martin S
Mueller, Theodore Rolf
Ornitz, Barry Louis
Pate, Samuel Ralph
Pitcher, DePuyster Gilbert
Prairie, Michael L
Raudorf, Thomas Walter
Rayside, John Stuart
Slaughter, Gerald M

TEXAS
Abrams, Albert
Bigham, Robert Eric
Briggs, Edward M
Collipp, Bruce Garfield
Crenshaw, Paul L
Eggers, Fred M
Ehlig-Economides, Christine Anna
Fowler, David Wayne
Geer, Ronald L
Gourdine, Meredith Charles
Halff, Albert Henry
Hanson, Bernold M
Hinterberger, Henry
Hopkins, Robert Charles
Hyman, William A
Johnstone, C(harles) Wilkin
Levine, Jules David
Machemehl, Jerry Lee
McNiel, James S(amuel), Jr
Messenger, Joseph Umlah
Moore, Walter P, Jr
Naismith, James Pomeroy
Packman, Paul Frederick
Page, Robert Henry
Powell, Alan
Scinta, James
Senatore, Ford Fortunato
Smith, Richard Thomas
Spencer, Thomas H
Wagner, John Philip
Wenzel, Alexander B
Whiting, Allen R

UTAH
Bishop, Jay Lyman
Israelsen, C Earl

Other Engineering (cont)

MacGregor, Douglas

VERMONT
Arns, Robert George
Linnell, Robert Hartley
Smith, Mark K
Warfield, George

VIRGINIA
Arthur, Richard J(ardine)
Claus, Alfons Jozef
Dalton, Roger Wayne
Gooding, Robert C
Haimes, Yacov Y
Hallanger, Lawrence William
Harrington, Robert Joseph
King, Merrill Kenneth
Lai, Chintu (Vincent C)
Leopold, Reuven
Levy, Sander Alvin
Moore, Joseph Herbert
Neeper, Ralph Arnold
Newman, James Charles, Jr
Ng, Wing-Fai
Phillips, William H
Reese, Ronald Malcolm
Rojiani, Kamal B
Schetz, Joseph A
Searle, Willard F, Jr
Whitcomb, Richard T
Whitelaw, R(obert) L(eslie)
Woeste, Frank Edward
Yeager, Paul Ray

WASHINGTON
Adams, William Mansfield
Barbosa-Canovas, Gustavo Victor
Bolme, Donald W(eston)
Butler, Don
Das, K Bhagwan
Hebeler, Henry K
Katz, Yale H
Pigott, George M
Pillay, Gautam
Rasco, Barbara A
Sandstrom, Wayne Mark
Sutherland, Earl C

WEST VIRGINIA
Lantz, Thomas Lee
Wales, Charles E

WISCONSIN
Campbell, Edward J
Falk, Edward D
Moore, Gary T
Smith, Luther Michael
Vanderheiden, Gregg

PUERTO RICO
Mandavilli, Satya N
Soderstrom, Kenneth G(unnar)

ALBERTA
Fisher, Harold M
Merta de Velehrad, Jan
Nigg, Benno M
Vroom, Alan Heard

BRITISH COLUMBIA
Bell, H(arry) R(ich)
Buckland, Peter Graham
Thyer, Norman Harold

MANITOBA
Anderson, Christian Donald
Jayas, Digvir Singh

NEW BRUNSWICK
Faig, Wolfgang
Vanicek, Petr

NEWFOUNDLAND
Bajzak, Denes

NOVA SCOTIA
Eames, M(ichael) C(urtis)

ONTARIO
Carr, Jan
Etkin, Bernard
Goldenberg, Andrew Avi
Gracie, G(ordon)
Jones, Alun Richard
Kamphuis, J(ohn) William
Keeler, John S(cott)
Stirling, Andrew John

QUEBEC
Devereaux, William A

SASKATCHEWAN
Seshadri, Rangaswamy

OTHER COUNTRIES
Al-Zubaidy, Sarim Naji
Bayev, Alexander A
Farquhar, Gale Burton
Lohtia, Rajinder Paul
Naudascher, Eduard
Richards, Adrian F
Toyoda, Shoichiro
Tzafestas, Spyros G

ENVIRONMENTAL, EARTH & MARINE SCIENCES

Atmospheric Dynamics

ALABAMA
Fowlis, William Webster

ARIZONA
Bougher, Stephen Wesley
Dickinson, Robert Earl
Fernando, Harindra Joseph

CALIFORNIA
Chan, Kwoklong Roland
Chang, Chih-Pei
Chew, Frank
Chiu, Yam-Tsi
Chu, Peter Cheng
Covey, Curtis Charles
Haltiner, George Joseph
Ingersoll, Andrew Perry
Kilb, Ralph Wolfgang
Maas, Stephen Joseph
Marcus, Philip Stephen
Molenkamp, Charles Richard
Schubert, Gerald
Seiff, Alvin
Simon, Richard L
Street, Robert L(ynnwood)
Tag, Paul Mark
Wagner, Kit Kern
Weare, Bryan C
Williams, Roger Terry
Wurtele, Morton Gaither

COLORADO
Albritton, Daniel L
Avery, Susan Kathryn
Balsley, Ben Burton
Baumhefner, David Paul
Diaz, Henry Frank
Dirks, Richard Allen
Elmore, Kimberly Laurence
Gage, Kenneth Seaver
Gall, Robert Lawrence
Gille, John Charles
Gilman, Peter A
Hanson, Howard P(aul)
Johnson, Richard Harlan
Kantha, Lakshmi
Kasahara, Akira
Kennedy, Patrick James
LeMone, Margaret Anne
Lenschow, Donald Henry
McWilliams, James Cyrus
Massman, William Joseph
Richmond, Arthur Dean
Rotunno, Richard
Snyder, Howard Arthur
Thaler, Eric Ronald
Trenberth, Kevin Edward
Van Zandt, Thomas Edward
Wessel, William Roy

CONNECTICUT
Brooks, Douglas Lee
Saltzman, Barry

DELAWARE
St John, Daniel Shelton

DISTRICT OF COLUMBIA
Deen, Thomas B
Handel, Mark David
Kalnay, Eugenia
Kousky, Vernon E
Rao, Desiraju Bhavanarayana
Wagner, Andrew James

FLORIDA
Barcilon, Albert I
Bleck, Rainer
Clarke, Allan James
Mankin, Richard Wendell
Minzner, Raymond Arthur
Mooers, Christopher Northrup Kennard
Olson, Donald B
Pfeffer, Richard Lawrence
Ruscher, Paul H

GEORGIA
Alyea, Fred Nelson
Justus, Carl Gerald
Roper, Robert George

HAWAII
Wang, Bin

ILLINOIS
Adrian, Ronald John
Coulter, Richard Lincoln
Fultz, Dave
Kuo, Hsiao-Lan
Man-Kin, Mak
Moschandreas, Demetrios J
Platzman, George William

INDIANA
Agee, Ernest M(ason)
Haines, Patrick A
Smith, Phillip J
Snow, John Thomas

Sun, Wen-Yin

IOWA
Hall, Jerry Lee

KANSAS
Braaten, David A

LOUISIANA
Rouse, Lawrence James, Jr

MAINE
Gordon, Geoffrey Arthur

MARYLAND
Adler, Robert Frederick
Baer, Ferdinand
Benton, George Stock
Busalacchi, Antonio James, Jr
Conrath, Barney Jay
Diamante, John Matthew
Einaudi, Franco
Flasar, F Michael
Holloway, John Leith, Jr
Pirraglia, Joseph A
Reeves, Robert William
Rice, Clifford Paul
Robock, Alan
Rodenhuis, David Roy
Scala, John Richard
Schemm, Charles Edward
Strobel, Darrell Fred

MASSACHUSETTS
Champion, Kenneth Stanley Warner
Denig, William Francis
Lindzen, Richard Siegmund
Nehrkorn, Thomas
Pedlosky, Joseph
Rosen, Richard David
Salah, Joseph E
Salstein, David A
Sanders, Frederick
Shapiro, Ralph
Spiegel, Stanley Lawrence
Stone, Peter Hunter

MICHIGAN
Banks, Peter Morgan
Boyd, John Philip
Heilman, Warren Emanuel

MINNESOTA
Pitcher, Eric John

MISSISSIPPI
Hurlburt, Harley Ernest

MISSOURI
Lemon, Leslie Roy
Moore, James Thomas

NEBRASKA
Verma, Shashi Bhushan
Wehrbein, William Mead

NEVADA
Telford, James Wardrop

NEW HAMPSHIRE
Clark, Ronald Rogers

NEW JERSEY
Cheremisinoff, Paul N
Hamilton, Kevin
Hayashi, Yoshikazu
Held, Isaac Meyer
Lau, Ngar-Cheung
Lewellen, William Stephen
Lipps, Frank B
Mahlman, Jerry David
Manabe, Syukuro
Mellor, George Lincoln, Jr
Oort, Abraham H
Ross, Bruce Brian
Smagorinsky, Joseph
Wetherald, Richard Tryon
Williams, Gareth Pierce

NEW MEXICO
Argo, Paul Emmett
Bleiweiss, Max Phillip
Campbell, John Raymond
Malone, Robert Charles
Orgill, Montie M
Rachele, Henry
Rokop, Donald J
Yamada, Tetsuji

NEW YORK
Bosart, Lance Frank
Cane, Mark A
Gross, Stanley H
Halitsky, James
Hameed, Sultan
Rao, Samohineeveesu Trivikrama
Rind, David Harold
Tenenbaum, Joel
Wang, Lin-Shu

NORTH CAROLINA
Bach, Walther Debele, Jr
Saucier, Walter Joseph
Vukovich, Fred Matthew

NORTH DAKOTA
Grainger, Cedric Anthony

OHIO
Burggraf, Odus R
Rayner, John Norman
Tuan, Tai-Fu

OKLAHOMA
Coffman, Harold H
Davies-Jones, Robert Peter
Hane, Carl Edward
Lilly, Douglas Keith
Sasaki, Yoshi Kazu

OREGON
Dews, Edmund

PENNSYLVANIA
Albrecht, Bruce Allen
Dutton, John Altnow
Wyngaard, John C

SOUTH CAROLINA
Larsen, Miguel Folkmar

TENNESSEE
Dobosy, Ronald Joseph
Schnelle, K(arl) B(enjamin), Jr
Skinner, George T

TEXAS
Brundidge, Kenneth Cloud
Chang, Ping
Das, Phanindramohan
Few, Arthur Allen
Gillette, Kevin Keith
Johnson, Francis Severin
Markowski, Gregory Ray
Peterson, Richard Emil

UTAH
Geisler, John Edmund
Liou, Kuo-Nan
Long, David G
Paegle, Julia Nogues
Stonehocker, Garth Hill

VIRGINIA
Curtin, Brian Thomas
Grose, William Lyman

WASHINGTON
Baldwin, Mark Phillip
Brown, Robert Alan
Clark, Kenneth Courtright
Delisi, Donald Paul
Harrison, Don Edmunds
Hernandez, Gonzalo J
Holton, James R
Overland, James Edward
Rhines, Peter Broomell
Sarachik, Edward S
Stock, David Earl
Tung, Ka-Kit

WISCONSIN
Horn, Lyle Henry
Houghton, David Drew

WYOMING
Marwitz, John D

PUERTO RICO
Soderstrom, Kenneth G(unnar)

ALBERTA
Leahey, Douglas McAdam

ONTARIO
Barrie, Leonard Arthur
Cho, Han-Ru
Gultepe, Ismail
Hedley, Mark Allen
Khandekar, Madhav Laxman
Peltier, William Richard
Shepherd, Gordon Greeley

QUEBEC
Derome, Jacques Florian
Mysak, Lawrence Alexander

OTHER COUNTRIES
Kitaigorodskii, Sergei Alexander
Lalas, Demetrius P
Liu, Chao-Han
Spence, Thomas Wayne
Subrahmanyam, D
Tolstoy, Ivan

Atmospheric Chemistry & Physics

ALABAMA
Abbas, Mian Mohammad
Anderson, Bernard Jeffrey
Davis, John Moulton
Helminger, Paul Andrew
Kassner, James Lyle, Jr
Meagher, James Francis
Stephens, James Briscoe
Stephens, Timothy Lee
Williamson, Ashley Deas

ALASKA
Shaw, Glenn Edmond
Wentink, Tunis, Jr

ARIZONA
Bougher, Stephen Wesley
Heaps, Melvin George
Herbert, Floyd Leigh
Hood, Lon(nie) Lamar
Jackson, Ray Dean
Krider, Edmund Philip
Layton, Richard Gary
Lunine, Jonathan Irving
Nordstrom, Brian Hoyt
O'Brien, Keran
Smith, Mark Alan
Twomey, Sean Andrew
Young, Kenneth Christie

ARKANSAS
Berger, Jerry Eugene
Richardson, Charles Bonner
Steele, Kenneth F

CALIFORNIA
Allen, Charles Freeman
Anderson, Kinsey Amor
Andre, Michael Paul
Ashburn, Edward V
Atkinson, Roger
Baker, Neal Kenton
Beer, Reinhard
Bonner, Norman Andrew
Borucki, William Joseph
Byers, Horace Robert
Cahill, Thomas A
Cain, William F
Cass, Glen R
Chahine, Moustafa Toufic
Chan, Kwoklong Roland
Chiu, Yam-Tsi
Christensen, Andrew Brent
Cicerone, Ralph John
Collis, Ronald Thomas
Colovos, George
Cooper, Alfred William Madison
Coulson, Kinsell Leroy
Cruikshank, Dale Paul
Daisey, Joan M
Dignon, Jane
Dowling, Jerome M
Eatough, Norman L
Ellis, Elizabeth Carol
Farber, Robert James
Farmer, Crofton Bernard
Feynman, Joan
Friedlander, Sheldon K
Gibbons, Mathew Gerald
Gilmore, Forrest Richard
Gordon, Robert Julian
Grossman, Jack Joseph
Hall, Freeman Franklin, Jr
Hamlin, Daniel Allen
Hansen, Anthony David Anders
Hardy, Kenneth Reginald
Hill, Robert Dickson
Hoffmann, Michael Robert
Innes, William Beveridge
Jacobs, Michael Moises
Jaffe, Sigmund
Jaklevic, Joseph Michael
Jayaweera, Kolf
Judge, Roger John Richard
Kalathil, James Sakaria
Katen, Paul C
Kilb, Ralph Wolfgang
Landman, Donald Alan
Lang, Valerie Ilona
Lange, Rolf
Larmore, Lewis
Lauer, George
Leite, Richard Joseph
Lieberman, Alvin
Lipeles, Martin
Loos, Hendricus G
Maas, Stephen Joseph
Mahadevan, Parameswar
Manalis, Melvyn S
Martin, L(aurence) Robbin
Mathews, Larry Arthur
Menzies, Robert Thomas
Moe, Mildred Minasian
Moe, Osborne Kenneth
Molenkamp, Charles Richard
Myers, Benjamin Franklin, Jr
Nero, Anthony V, Jr
Okumura, Mitchio
Old, Thomas Eugene
Paige, David A
Pang, Kevin
Park, Chul
Park, Chung Gun
Paulson, Suzanne Elizabeth
Penner, Joyce Elaine
Pitts, James Ninde, Jr
Poppoff, Ilia George
Prag, Arthur Barry
Prather, Michael John
Pueschel, Rudolf Franz
Randall, Charles McWilliams
Richards, L(orenzo) Willard
Rognlien, Thomas Dale
Rosen, Leonard Craig
Rowland, F Sherwood
Rundel, Robert Dean
Salanave, Leon Edward
Schwall, Richard Joseph
Sentman, Davis Daniel
Sextro, Richard George
Sharp, Richard Dana
Shemansky, Donald Eugene
Simon, Richard L
Singh, Hanwant B
Smith, Charles Francis, Jr
Straus, Joe Melvin
Swenson, Gary Russell
Tag, Paul Mark
Thorne, Richard Mansergh
Toon, Owen Brian
Topol, Leo Eli
Turco, Richard Peter
Underwood, James Henry
Vedder, James Forrest
Vernazza, Jorge Enrique
Voss, Henry David
Walton, John Joseph
Waters, Joe William
Whitten, Robert Craig, Jr
Winer, Arthur Melvyn
Wuebbles, Donald J
Young, Louise Gray

COLORADO
Albritton, Daniel L
Amme, Robert Clyde
Anderson, Larry Gene
Balsley, Ben Burton
Birks, John William
Boatman, Joe Francis
Brault, James William
Bushnell, Robert Hempstead
Caracena, Fernando
Carbone, Richard Edward
Chappell, Charles Franklin
Clark, Wallace Lee
Cotton, William Reuben
Cox, Stephen Kent
Curtis, George William
Dana, Robert Watson
Davies, Kenneth
Derr, Vernon Ellsworth
Durham, Michael Dean
Dye, James Eugene
Eddy, John Allen
Edwards, Harry Wallace
Fehsenfeld, Fredrick C
Foote, Garvin Brant
Frisch, Alfred Shelby
Fritz, Richard Blair
Gage, Kenneth Seaver
Gille, John Charles
Goldan, Paul David
Hofmann, David John
Holzer, Thomas Edward
Hord, Charles W
Howard, Carleton James
Johnson, Richard Harlan
Joselyn, Jo Ann Cram
Kennedy, Patrick James
Kiehl, Jeffrey Theodore
Kleis, William Delong
Knight, Charles Alfred
Lenschow, Donald Henry
Lincoln, Jeannette Virginia
Little, Charles Gordon
Lodge, James Piatt, Jr
Mankin, William Gray
Martell, Edward A
Middleton, Paulette Bauer
Morgan, William Lowell
Murcray, David Guy
Murcray, Frank James
Neff, William David
Ohtake, Takeshi
Peterson, James T
Peterson, Vern Leroy
Pike, Julian M
Post, Madison John
Querfeld, Charles William
Radke, Lawrence Frederick
Richmond, Arthur Dean
Rosinski, Jan
Rufenach, Clifford L
Rush, Charles Merle
Ruttenberg, Stanley
Schonbeck, Niels Daniel
Schwiesow, Ronald Lee
Sinclair, Peter C
Solomon, Susan
Stedman, Donald Hugh
Thomas, Gary E
Van Valin, Charles Carroll
Webster, Peter John
Wellman, Dennis Lee
Westwater, Edgeworth Rupert

CONNECTICUT
Castillo, Raymond A
Poultney, Sherman King
Robinson, George David

DELAWARE
Glasgow, Louis Charles
Pierrard, John Martin

DISTRICT OF COLUMBIA
Appleby, John Frederick
Apruzese, John Patrick
Auclair, Allan Nelson Douglas
Carrigan, Richard Alfred
Fitzgerald, James W
Hawkins, Gerald Stanley
Kaye, Jack Alan
McPherson, Ronald P
Ohring, George
Picone, J Michael
Summers, Michael Earl
Tarpley, Jerald Dan
Uman, Myron F
Weinreb, Michael Philip
Withbroe, George Lund

FLORIDA
Allen, Eric Raymond
Block, David L
Hill, Frank B(ruce)
Killinger, Dennis K
Llewellyn, Ralph A
Minzner, Raymond Arthur
Myerson, Albert Leon
Ostlund, H Gote
Reynolds, George William, Jr
Rogers, Lewis Henry
Ruscher, Paul H
Winchester, John W
Windsor, John Golay, Jr
Zuo, Yuegang

GEORGIA
Chameides, William Lloyd
Justus, Carl Gerald
Patterson, Edward Matthew
Stelson, Arthur Wesley
Wine, Paul Harris

HAWAII
Canfield, Richard Charles
Fullerton, Charles Michael
Huebert, Barry Joe

IDAHO
Farwell, Sherry Owen
Mills, James Ignatius

ILLINOIS
Beard, Kenneth Van Kirke
Birchfield, Gene Edward
Cook, David Robert
Coulter, Richard Lincoln
Frederick, John Eugene
Gaffney, Jeffrey Steven
Kraakevik, James Henry
Marley, Nancy Alice
Murphy, Thomas Joseph
Reck, Ruth Annette
Semonin, Richard Gerard
Sheft, Irving
Wang, Hwa-Chi
Wesely, Marvin Larry

INDIANA
Landreth, Ronald Ray
Pribush, Robert A

IOWA
Carmichael, Gregory Richard
Yarger, Douglas Neal

KANSAS
Lambert, Jack Leeper

KENTUCKY
Ray, Asit Kumar

LOUISIANA
Broeg, Charles Burton
Klasinc, Leo

MAINE
Blau, Henry Hess, Jr
Cronn, Dagmar Rais

MARYLAND
Aikin, Arthur Coldren
Allen, John Edward, Jr
Bass, Arnold Marvin
Bauer, Ernest
Benesch, William Milton
Brace, Larry Harold
Brasunas, John Charles
Chandra, Sushil
Ellingson, Robert George
Epstein, Gabriel Leo
Feldman, Paul Donald
Fifer, Robert Alan
Goldberg, Richard Aran
Goldenbaum, George Charles
Gordon, Glen Everett
Hanel, Rudolf A
Hedin, Alan Edgar
Herrero, Federico Antonio
Hicks, Bruce Boundy
Hudson, Robert Douglas
Joiner, R(eginald) Gracen
Julienne, Paul Sebastian
Kostiuk, Theodor
Krueger, Arlin James
Kurylo, Michael John, III
Kyle, Herbert Lee
Larson, Reginald Einar
Lufkin, Daniel Harlow
McPeters, Richard Douglas
Martinez, Richard Isaac
Neupert, Werner Martin
Omidvar, Kazem
Piotrowicz, Stephen R
Potemra, Thomas Andrew
Prabhakara, Cuddapah
Quinn, Thomas Patrick
Rosenberg, Theodore Jay
Rosenfield, Joan E
Schatten, Kenneth Howard
Shettle, Eric Payson
Silver, David Martin
Stewart, Richard Willis
Stief, Louis J
Strobel, Darrell Fred
Thompson, Owen Edward
Tilford, Shelby G
Weller, Charles Stagg, Jr
Wiscombe, Warren Jackman

MASSACHUSETTS
Aarons, Jules
Anderson, James Gilbert
Birstein, Seymour J
Carlson, Herbert Christian, Jr
Chakrabarti, Supriya
Champion, Kenneth Stanley Warner
Engelman, Arthur
Eoll, John Gordon
Fitzgerald, Donald Ray
Gallagher, Charles Clifton
Gardner, James A
Glover, Kenneth Merle
Hedlund, Donald A
Huffman, Robert Eugene
Kaplan, Lewis David
Kolb, Charles Eugene, Jr
Levine, Randolph Herbert
McClatchey, Robert Alan
Markson, Ralph Joseph
Molina, Mario Jose
Morris, Robert Alan
Nehrkorn, Thomas
Newell, Reginald Edward
Paulsen, Duane E
Penndorf, Rudolf
Picard, Richard Henry
Ratner, Michael Ira
Robertson, David C
Rothman, Laurence Sidney
Sagalyn, Rita C
Sprengnether, Michele M
Stakutis, Vincent John
Swider, William, Jr
Toman, Kurt
Tomljanovich, Nicholas Matthew
Traub, Wesley Arthur
Victor, George A
Viggiano, Albert
Volz, Frederic Ernst
Whitney, Cynthia Kolb
Winick, Jeremy Ross
Wofsy, Steven Charles
Yamartino, Robert J

MICHIGAN
Atreya, Sushil Kumar
Burek, Anthony John
Chang, Tai Yup
Dingle, Albert Nelson
Donahue, Thomas Michael
Drayson, Sydney Roland
Francisco, Joseph Salvadore, Jr
Gorse, Robert August, Jr
Heilman, Warren Emanuel
Japar, Steven Martin
Kelly, Nelson Allen
Klimisch, Richard L
Kuhn, William R
Nagy, Andrew F
Rogers, Jerry Dale
Sharp, William Edward, III
Wallington, Timothy John
Wolff, George Thomas

MINNESOTA
Freier, George David
Mauersberger, Konrad

MISSOURI
Bragg, Susan Lynn
Carstens, John C
DuBroff, Richard Edward
Hagen, Donald E
Macias, Edward S
Nelson, George Driver
Pallmann, Albert J
Plummer, Patricia Lynne Moore
Podzimek, Josef
Vaughan, William Mace
Whitefield, Philip Douglas

MONTANA
Latham, Don Jay
Peavy, Howard Sidney

NEBRASKA
Billesbach, David P
Verma, Shashi Bhushan

NEVADA
Hales, J Vern
Hallett, John
Hudson, James Gary

Atmospheric Chemistry & Physics (cont)

Johnson, Brian James
Lowenthal, Douglas H
McElroy, James Lee
Pierson, William R
Pitter, Richard Leon
Scott, William Taussig
Tanner, Roger Lee
Taylor, George Evans, Jr
Telford, James Wardrop

NEW HAMPSHIRE
Taffe, William John

NEW JERSEY
Feit, Julius
Graedel, Thomas Eldon
Jeck, Richard Kahr
Lau, Ngar-Cheung
Lioy, Paul James
McAfee, Kenneth Bailey, Jr
Mahlman, Jerry David
Martin, John David
Ramaswamy, Venkatachalam
Weschler, Charles John

NEW MEXICO
Bleiweiss, Max Phillip
Brook, Marx
Chylek, Petr
Cunningham, Paul Thomas
Czuchlewski, Stephen John
Fowler, Malcolm McFarland
Goldman, S Robert
Guthals, Paul Robert
Jones, James Jordan
Karl, Robert Raymond, Jr
Maier, William Bryan, II
Mason, Allen Smith
Miller, August
Morse, Fred A
Orgill, Montie M
Peek, H(arry) Milton
Popp, Carl John
Porch, William Morgan
Sedlacek, William Adam
Shively, Frank Thomas
Snider, Donald Edward
Streit, Gerald Edward
Tisone, Gary C
Wilkening, Marvin H
Winn, William Paul
Yamada, Tetsuji
Zinn, John

NEW YORK
Akers, Charles Kenton
Blanchard, Duncan Cromwell
Boeck, William Louis
Cess, Robert D
Chu, Liang Tung
Cipriano, Ramon John
De Zafra, Robert Lee
Farley, Donald T, Jr
Greenebaum, Michael
Hameed, Sultan
Hindman, Edward Evans
Hirsh, Merle Norman
Hopke, Philip Karl
Janik, Gerald S
Jordan, David M
Keesee, Robert George
Kim, Jai Soo
Kumai, Motoi
Lovett, Gary Martin
McLaren, Eugene Herbert
Michalsky, Joseph Jan, Jr
Millman, George Harold
Mohnen, Volker A
Newman, Leonard
Oppenheimer, Michael
Penney, Carl Murray
Ram, Michael
Rao, Samohineeveesu Trivikrama
Schwartz, Stephen Eugene
Senum, Gunnar Ivar
Takacs, Gerald Alan
Tang, Ignatius Ning-Bang
Varanasi, Prasad
Wiesenfeld, John Richard

NORTH CAROLINA
Aneja, Viney P
Braham, Roscoe Riley, Jr
Clements, John B(elton)
Fox, Donald Lee
Gay, Bruce Wallace, Jr
Jackels, Charles Frederick
Linton, Richard William
Odman, Mehmet Talat
Saucier, Walter Joseph
Saxena, Vinod Kumar
Scheffe, Richard Donald
Schmeltekopf, A
Shreffler, Jack Henry
Swank, Wayne T
Weaver, Ervin Eugene
Wiener, Russell Warren
Wilson, William Enoch, Jr

NORTH DAKOTA
Alkezweeny, Abdul Jabbar
Grainger, Cedric Anthony
Stith, Jeffrey Len

OHIO
Armstrong, James Anthony
Bhatti, Asif
Biswas, Pratim
Cantrell, Joseph Sires
Ferguson, Dale Curtis
Fordyce, James Stuart
Fox, Robert Dean
Gibson, Edward George
Hwang, Soon Muk
Levy, Arthur
Murphy, Michael John
Saltzman, Bernard Edwin
Sticksel, Philip Rice
Sverdrup, George Michael

OKLAHOMA
Beasley, William Harold
Blais, Roger Nathaniel
Buck, Richard F
Hane, Carl Edward
Rust, Walter David

OREGON
Hard, Thomas Michael
Khalil, M Aslam Khan

PENNSYLVANIA
Albrecht, Bruce Allen
Ambs, William Joseph
Bandy, Alan Ray
Bohren, Craig Frederick
Cohen, Leonard David
Cooney, John Anthony
de Pena, Rosa G
Fraser, Alistair Bisson
Frenklach, Michael Y
Friend, James Philip
Heicklen, Julian Phillip
Kasting, James Fraser
Kay, Jack Garvin
Lamb, Dennis
Maroulis, Peter James
Mathews, John David
Mitchell, John Douglas
Pandis, Spyros N
Philbrick, Charles Russell
Sheffield, Ann Elizabeth
Thomson, Dennis Walter
Thornton, Donald Carlton
Witkowski, Robert Edward

RHODE ISLAND
Calo, Joseph Manuel
Duce, Robert Arthur
Fasching, James Le Roy
Merrill, John T
Rahn, Kenneth A

SOUTH CAROLINA
Hahn, Walter I
Jones, Ulysses Simpson, Jr
Porter, Hayden Samuel, Jr

SOUTH DAKOTA
Helsdon, John H, Jr
Smith, Paul Letton, Jr
Welch, Ronald Maurice

TENNESSEE
Blass, William Errol
Cheng, Meng-Dawn
Hochanadel, Clarence Joseph
Lorenz, Philip Jack, Jr
MacQueen, Robert Moffat
Worsham, Lesa Marie Spacek

TEXAS
Ballard, Harold Noble
Bering, Edgar Andrew, III
Borst, Walter Ludwig
Breig, Edward Louis
Das, Phanindramohan
Few, Arthur Allen
Heelis, Roderick Antony
Hoffman, John Harold
Johnson, Francis Severin
Jurica, Gerald Michael
Marlow, William Henry
Orville, Richard Edmonds
Roeder, Robert Charles
Sheldon, William Robert
Stuart, Joe Don
Wilheit, Thomas Turner

UTAH
Ballif, Jae R
Eatough, Delbert J
Fukuta, Norihiko
Graff, Darrell Jay
Hansen, Lee Duane
Heaney, Robert John
Liou, Kuo-Nan
Mangelson, Nolan Farrin

VIRGINIA
Barasch, Guy Errol
Brewer, Dana Alice
Brockman, Philip
Cutter, Gregory Allan
Dolezalek, Hans
Galloway, James Neville
Garstang, Michael
Grant, William B
Grose, William Lyman
Herm, Ronald Richard
Hooke, William Hines
Howard, Russell Alfred
Levine, Joel S
McCormick, Michael Patrick
Phillips, Donald Herman
Remsberg, Ellis Edward
Ruhnke, Lothar Hasso
Ruskin, Robert Edward
Szuszczewicz, Edward Paul
Todd, Edward Payson
Trout, Dennis Alan
Wood, George Marshall
Worden, Simon Peter

WASHINGTON
Baldwin, Mark Phillip
Barchet, William Richard
Baughcum, Steven Lee
Cole, Robert Stephen
Dana, M Terry
Doran, J Christopher
Giddings, William Paul
Harrison, Halstead
Hernandez, Gonzalo J
Hill, Herbert Henderson, Jr
Hobbs, Peter Victor
Katsaros, Kristina B
Laulainen, Nels Stephen
Lee, Richard Norman
Maki, Arthur George, Jr
Nutley, Hugh
Person, James Carl
Riehl, Jerry A
Ruby, Michael Gordon
Sloane, Christine Scheid
Stokes, Gerald Madison
Wukelic, George Edward

WISCONSIN
Kuhn, Peter Mouat
Peace, George Earl, Jr

WYOMING
Marwitz, John D
Vali, Gabor

ALBERTA
Anger, Clifford D
Checkel, M David
Grabenstetter, James Emmett
Honsaker, John Leonard
Leahey, Douglas McAdam
Strausz, Otto Peter

ONTARIO
Anlauf, Kurt Guenther
Bunce, Nigel James
Cho, Han-Ru
Dickson, Lawrence William
Fingas, Merv Floyd
Gattinger, Richard Larry
Gultepe, Ismail
Hedley, Mark Allen
Iribarne, Julio Victor
List, Roland
Lowe, Robert Peter
McConnell, John Charles
Mackay, Gervase Ian
Martin, Hans Carl
Megaw, William James
Muldrew, Donald Boyd
Niki, Hiromi
Sanderson, H Preston
Schemenauer, Robert Stuart
Schiff, Harold Irvin
Schroeder, William Henry
Shepherd, Gordon Greeley
Singer, Eugen
Singleton, Donald Lee
St-Maurice, Jean-Pierre
Tang, You-Zhi
Wallis, Donald Douglas James H
Whelpdale, Douglas Murray

QUEBEC
Chisholm, Alexander James
Davies, Roger
Leighton, Henry George

SASKATCHEWAN
Maybank, John

OTHER COUNTRIES
Andreae, Meinrat Otto
Box, Michael Allister
Chen, Chen-Tung Arthur
Czepyha, Chester George Reinhold
Levin, Zev
Long, Alexis Boris
Mølhave, Lars
Papee, Henry Michael
Taylor, Fredric William
Willeke, Klaus

Atmospheric Sciences, General

ALABAMA
Essenwanger, Oskar M
Hung, Ru J
Pruitt, Kenneth M
Reynolds, George Warren
Smith, Robert Earl
Stewart, Dorothy Anne
Vaughan, William Walton

ALASKA
Bigler, Stuart Grazier
Bowling, Sue Ann
Brown, Neal B
Diemer, Edward Devlin
Fahl, Charles Byron
Kelley, John Joseph, II
Watkins, Brenton John
Weller, Gunter Ernst
Wendler, Gerd Dierk

ARIZONA
Baron, William Robert
Battan, Louis Joseph
Bergen, James David
Brazel, Anthony James
Burgoyne, Edward Eynon
Chapman, Clark Russell
Comrie, Andrew Charles
Dickinson, Robert Earl
Greenberg, Milton
Heaps, Melvin George
Herman, Benjamin Morris
Hood, Lon(nie) Lamar
Hughes, Malcolm Kenneth
Idso, Sherwood B
Jones, David Lloyd
Krider, Edmund Philip
Palmer, James McLean
Parrish, Judith Totman
Pinter, Paul James, Jr
Reagan, John Albert
Schotland, Richard Morton
Sprague, Ann Louise
Vogelmann, Andrew Mark
Young, Kenneth Christie

ARKANSAS
McCarty, Clark William

CALIFORNIA
Adam, David Peter
Akesson, Norman B(erndt)
Berry, Edwin X
Bornstein, Robert D
Brand, Samson
Byers, Horace Robert
Chang, Chih-Pei
Chivers, Hugh John
Clark, Nathan Edward
Coulson, Kinsell Leroy
Court, Arnold
Covey, Curtis Charles
Daniels, Jeffrey Irwin
Davidson, Kenneth LaVern
Deuble, John Lewis, Jr
Dignon, Jane
Dorman, Clive Edgar
Ellis, Elizabeth Carol
Elsberry, Russell Leonard
Frank, Sidney Raymond
Freis, Robert P
Fullmer, George Clinton
Fymat, Alain L
Gates, William Lawrence
Georgakakos, Konstantine P
Gottlieb, Peter
Grobecker, Alan J
Guy, George Anderson
Hall, Freeman Franklin, Jr
Hallanger, Norman Lawrence
Halpern, David
Haltiner, George Joseph
Hamilton, Harry Lemuel, Jr
Haney, Robert Lee
Hardy, Kenneth Reginald
Harmen, Raymond A
Hauser, Rolland Keith
Hazelwood, R(obert) Nichols
Hewson, Edgar Wendell
Hill, Robert Dickson
Holl, Manfred Matthias
Hovermale, John Bruce
Howell, Wallace Egbert
Huebsch, Ian O
Huschke, Ralph Ernest
Ingersoll, Andrew Perry
Janssen, Michael Allen
Jung, Glenn Harold
Kagiwada, Harriet Hatsune Natsuyama
Katz, Richard Whitmore
Kern, Clifford Dalton
Kliore, Arvydas J(oseph)
Koenig, Lloyd Randall
Krick, Irving Parkhurst
Lang, Valerie Ilona
La Rue, Jerrold A
Lauer, George
Lee, Hsi-Nan
Leith, Cecil Eldon, Jr
Levine, Bruce Martin
Lier, John
Lindal, Gunnar F

Liu, Wingyuen Timothy
McIntyre, Adelbert
MacKay, Kenneth Pierce, Jr
Martin, Terry Zachry
Mendis, Devamitta Asoka
Miatech, Gerald James
Mooney, Margaret L
Namias, Jerome
Pitts, James Ninde, Jr
Pollack, James Barney
Potter, Gerald Lee
Ramanathan, Veerabhadran
Read, Robert G
Renard, Robert Joseph
Roads, John Owen
Robinson, Lewis Howe
Russell, Philip Boyd
Ryan, Bill Chatten
St Amand, Pierre
Schneider, Stephen Henry
Seiff, Alvin
Shemansky, Donald Eugene
Shinn, Joseph Hancock
Simon, Richard L
Singh, Hanwant B
Siquig, Richard Anthony
Smith, Theodore Beaton
Somerville, Richard Chapin James
Spinks, John Lee
Stidd, Charles Ketchum
Stinson, James Robert
Swanson, Robert Nels
Tag, Paul Mark
Terrile, Richard John
Tess, Roy William Henry
Toon, Owen Brian
Trogler, William C
Turco, Richard Peter
Van der Bijl, William
Vardiman, Larry
Wang, Jen Yu
Wang, Jon Y
Waterman, Alan T(ower), Jr
Waters, Joe William
Weare, Bryan C
West, Robert A
Wood, Fergus James
Wuebbles, Donald J
Yap, Fung Yen

COLORADO
Anthes, Richard Allen
Balsley, Ben Burton
Barry, Roger Graham
Baynton, Harold Wilbert
Beran, Donald Wilmer
Blumen, William
Boatman, Joe Francis
Boudreau, Robert Donald
Carbone, Richard Edward
Cerni, Todd Andrew
Chamberlain, A(drian) R(amond)
Chappell, Charles Franklin
Cotton, William Reuben
Cox, Stephen Kent
Curry, Judith Ann
Dabberdt, Walter F
Diaz, Henry Frank
Donovan, Terrence John
Eberhard, Wynn Lowell
Elmore, Kimberly Laurence
Esposito, Larry Wayne
Firor, John William
Foote, Garvin Brant
Fox, Douglas Gary
Frisinger, H Howard, II
Gage, Kenneth Seaver
Gille, John Charles
Girz, Cecilia Griffith
Gossard, Earl Everett
Gray, William Mason
Hanson, Howard P(aul)
Heymsfield, Andrew Joel
Hughes, William Carroll
Hutchinson, Gordon Lee
Johnson, Richard Harlan
Kahan, Archie M
Kelley, Neil Davis
Kellogg, William Welch
Kennedy, Patrick James
Kraus, Eric Bradshaw
Lally, Vincent Edward
LeMone, Margaret Anne
Lenschow, Donald Henry
London, Julius
McCabe, Gregory James, Jr
McCarthy, John
McKee, Thomas Benjamin
Mahler, Robert John
Mandics, Peter Alexander
Marlatt, William Edgar
Martinelli, Mario, Jr
Meitin, Jose Garcia, Jr
Meroney, Robert N
Mielke, Paul W, Jr
Newton, Chester Whittier
Ohtake, Takeshi
Parton, William Julian, Jr
Peterson, James T
Peterson, Vern Leroy
Pielke, Roger Alvin
Purdom, James Francis Whitehurst
Radke, Lawrence Frederick
Reiter, Elmar Rudolf

Renne, David Smith
Riehl, Herbert
Rottman, Gary James
Rotunno, Richard
Schwiesow, Ronald Lee
Serafin, Robert Joseph
Thaler, Eric Ronald
Thompson, Starley Lee
Trenberth, Kevin Edward
Veal, Donald L
Vonder Haar, Thomas Henry
Washington, Warren Morton
Webster, Peter John
Wellman, Dennis Lee
Williams, Merlin Charles
Wooldridge, Gene Lysle

CONNECTICUT
Aylor, Donald Earl
Castillo, Raymond A
Friedman, Don Gene
Herron, Michael Myrl
Jackson, C(harles) Ian
Monahan, Edward Charles
Newman, Steven Barry
Pandolfo, Joseph P
Paskausky, David Frank
Russo, John A, Jr
Turekian, Karl Karekin
Waggoner, Paul Edward

DELAWARE
Kalkstein, Laurence Saul
Mather, John Russell

DISTRICT OF COLUMBIA
Appleby, John Frederick
Bergman, Kenneth Harris
Bierly, Eugene Wendell
Brueckner, Guenter Erich
Crosby, David S
Gutman, George Garik
Handel, Mark David
Hecht, Alan David
Hornstein, John Stanley
Kalnay, Eugenia
Lehman, Richard Lawrence
Lindstrom, Eric Jon
Long, Paul Eastwood, Jr
McGinnis, David Franklin, Jr
Ohring, George
Quiroz, Roderick S
Summers, Michael Earl
Tarpley, Jerald Dan
Wagner, Andrew James
Ward, John Henry
White, Robert M

FLORIDA
Baum, Werner A
Brandli, Henry William
Clark, John F
Dubach, Leland L
Elsner, James Brian
Emiliani, Cesare
Fuelberg, Henry Ernest
Gerber, John Francis
Gleeson, Thomas Alexander
Haines, Donald Arthur
Harris, D Lee
Henning, Rudolf E
Hiser, Homer Wendell
Jamrich, John Xavier
Jensen, Clayton Everett
Jordan, Charles Lemuel
Kelso, John Morris
Krishnamurti, Tiruvalam N
La Seur, Noel Edwin
Legler, David M
Lhermitte, Roger M
Long, Robert Radcliffe
Lundgren, Dale A(llen)
McFadden, James Douglas
Martsolf, J David
Mecklenburg, Roy Albert
Minzner, Raymond Arthur
Neuberger, Hans Hermann
Newstein, Herman
O'Brien, James J
Olivero, John Joseph, Jr
Ray, Peter Sawin
Rosenthal, Stanley Lawrence
Ruscher, Paul H
Sheets, Robert Chester
Sinclair, Thomas Russell
Stephens, Jesse Jerald
Stuart, David W
Weinberg, Jerry L
Zhu, Xiaorong

GEORGIA
Arbogast, Richard Terrance
Blanton, Jackson Orin
Cooper, Charles Dewey
Evans, William Buell
Grams, Gerald William
Justus, Carl Gerald
Kiang, Chia Szu
Meentemeyer, Vernon George
Parker, Albert John
Peake, Thaddeus Andrew, III
Roper, Robert George
Steiner, Jean Louise
Taylor, Howard Edward

HAWAII
Chiu, Wan-Cheng
Chuan, Raymond Lu-Po
Murakami, Takio
Pyle, Robert Lawrence
Sadler, James C
Wang, Bin
Wyrtki, Klaus

IDAHO
Wright, James Louis

ILLINOIS
Ackerman, Bernice
Ardis, Colby V, Jr
Beard, Kenneth Van Kirke
Changnon, Stanley Alcide, Jr
Cook, David Robert
Coulter, Richard Lincoln
Dailey, Paul William, Jr
Dordick, Isadore
Fisher, Perry Wright
Frederick, John Eugene
Fujita, Tetsuya T
Handler, Paul
Muehling, Arthur J
Muller, Dietrich
Sasamori, Takashi
Semonin, Richard Gerard
Shannon, Jack Dee
Smith, Leslie Garrett
Staver, Allen Ernest
Stout, Glenn Emanuel
Webb, Harold Donivan
Wendland, Wayne Marcel

INDIANA
Gommel, William Raymond
Newman, James Edward
Oliver, John Edward
Smith, Phillip J
Vincent, Dayton George

IOWA
Shaw, Robert Harold
Waite, Paul J

KANSAS
Bark, Laurence Dean
Braaten, David A
Eagleman, Joe R
Hagen, Lawrence J
McNulty, Richard Paul
Welch, Stephen Melwood

KENTUCKY
Crawford, Nicholas Charles
Stewart, Robert Earl

LOUISIANA
Dessauer, Herbert Clay
Glenn, Alfred Hill
Hsu, Shih-Ang
Muller, Robert Albert

MAINE
Cronn, Dagmar Rais
Dethier, Bernard Emile
Gordon, Geoffrey Arthur
Snow, Joseph William
Snyder, Arnold Lee, Jr

MARYLAND
Abbey, Robert Fred, Jr
Acheson, Donald Theodore
Adler, Robert Frederick
Anderson, David Leslie
Asrar, Ghassam R
Atlas, David
Baer, Ferdinand
Baer, Ledolph
Bandeen, William Reid
Barrientos, Celso Saquitan
Bauer, Ernest
Benton, George Stock
Berning, Warren Walt
Carlon, Hugh Robert
Carlstead, Edward Meredith
Cavalieri, Donald Joseph
Chang, Alfred Tieh-Chun
Conrath, Barney Jay
Cressman, George Parmley
Deaven, Dennis George
Ellingson, Robert George
Elliott, William Paul
Endal, Andrew Samson
Epstein, Edward Selig
Fraser, Robert Stuart
Fritz, Sigmund
Gaffen, Dian Judith
Gilliland, Ronald Lynn
Glahn, Harry Ross
Goldberg, Richard Aran
Greenstone, Reynold
Gruber, Arnold
Hallgren, Richard E
Hart, Michael H
Hasler, Arthur Frederick
Heffter, Jerome L
Heppner, James P
Heymsfield, Gerald M
Holland, Joshua Zalman
Huang, Joseph Chi Kan
Hudlow, Michael Dale

Johnson, David Simonds
Johnson, Harry McClure
Kaiser, Jack Allen Charles
King, Michael Dumont
Kyle, Herbert Lee
Lavoie, Ronald Leonard
Leese, John Albert
Leight, Walter Gilbert
McBryde, F(elix) Webster
McGovern, Wayne Ernest
Maier, Eugene Jacob Rudolph
Means, Lynn L
Mecherikunnel, Ann Pottanat
Meyer, James Henry
Mirabito, John A
Mogil, H Michael
Parkinson, Claire Lucille
Parkinson, Claire Lucille
Price, John Charles
Rall, Jonathan Andrew Reiley
Rango, Albert
Rao, P Krishna
Rasmusson, Eugene Martin
Reeves, Robert William
Robock, Alan
Rosenfield, Joan E
Rubin, Morton Joseph
Ruff, Irwin S
Salomonson, Vincent Victor
Schmugge, Thomas Joseph
Schneider, Edwin Kahn
Seguin, Ward Raymond
Shettle, Eric Payson
Sibeck, David G
Simpson, Joanne
Stackpole, John Duke
Strong, Alan Earl
Tepper, Morris
Thompson, Owen Edward
Tracton, Martin Steven
Uhart, Michael Scott
Vernekar, Anandu Devarao
Vestal, Claude Kendrick
Wang, Ting-I
Young, George Anthony
Zankel, Kenneth L

MASSACHUSETTS
Anstey, Robert L
Austin, James Murdoch
Austin, Pauline Morrow
Barnes, Arnold Appleton, Jr
Bass, Arthur
Beckerle, John C
Bell, Barbara
Bradley, Raymond Stuart
Brooks, Edward Morgan
Carlson, Herbert Christian, Jr
Chisholm, Donald Alexander
Clough, Shepard Anthony
Colinvaux, Paul Alfred
Conover, John Hoagland
Cunningham, Robert M
Darling, Eugene Merrill, Jr
Delvaille, John Paul
Emanual, Kerry Andrew
Entekhabi, Dara
Faller, Alan Judson
Fougere, Paul Francis
Gray, Douglas Carmon
Hanna, Steven Rogers
Harris, Miles Fitzgerald
Howard, John Nelson
Humi, Mayer
Hummel, John Richard
Johnson, Douglas William
Kaplan, Lewis David
Lai, Shu Tim
Lorenz, Edward Norton
McClatchey, Robert Alan
McMenamin, Mark Allan
Myers, Robert Frederick
Nehrkorn, Thomas
Newell, Reginald Edward
Norment, Hillyer Gavin
Olmez, Ilhan
Penndorf, Rudolf
Ramakrishna, Kilaparti
Rork, Eugene Wallace
Salah, Joseph E
Scuderi, Louis Anthony
Sellers, Francis Bachman
Spengler, Kenneth C
Staelin, David Hudson
Stergis, Christos George
Stump, Robert
Twitchell, Paul F
Wheeler, Ned Brent
Wilson, F Wesley, Jr

MICHIGAN
Aron, Robert
Atreya, Sushil Kumar
Batterman, Stuart A
Dingle, Albert Nelson
Gates, David Murray
Heilman, Warren Emanuel
Kuhn, William R
Kummler, Ralph H
Mason, Conrad Jerome
Menard, Albert Robert, III
Portman, Donald James
Ruffner, James Alan
Schwab, David John

Atmospheric Sciences, General (cont)

Wolff, George Thomas
Wu, Hofu
Yerg, Donald G

MINNESOTA
Anderson, Frances Jean
Baker, Donald Gardner
Dahlberg, Duane Arlen
Lawrence, Donald Buermann
McClure, Benjamin Thompson
McMurry, Peter Howard

MISSISSIPPI
Cross, Ralph Donald
Hurlburt, Harley Ernest

MISSOURI
Cowherd, Chatten, Jr
Darkow, Grant Lyle
Decker, Wayne Leroy
Kung, Ernest Chen-Tsun
Lemon, Leslie Roy
Lin, Yeong-Jer
Moore, James Thomas
Ostby, Frederick Paul, Jr
Podzimek, Josef
Rao, Gandikota V
Schaefer, Joseph Thomas

MONTANA
Caprio, Joseph Michael

NEBRASKA
Blad, Blaine L
Hahn, George LeRoy
Hubbard, Kenneth Gene
Lawson, Merlin Paul
Wehrbein, William Mead

NEVADA
Flueck, John A
Gilbert, Dewayne Everett
Hoffer, Thomas Edward
Hudson, James Gary
Klieforth, Harold Ernest
Krenkel, Peter Ashton
Leipper, Dale F
Murino, Clifford John
Pepper, Darrell Weldon
Pitter, Richard Leon
Randerson, Darryl

NEW HAMPSHIRE
Bernstein, Abram Bernard
Bilello, Michael Anthony
Crane, Robert Kendall
Federer, C Anthony
Jastrow, Robert
Perovich, Donald Kole
Snively, Leslie O

NEW JERSEY
Avissar, Roni
Becken, Eugene D
Broccoli, Anthony Joseph
Bryan, Kirk, (Jr)
Hamilton, Kevin
Havens, Abram Vaughn
House, Edward Holcombe
Hsu, Chin-Fei
Jeck, Richard Kahr
Krey, Philip W
Kurihara, Yoshio
Lau, Ngar-Cheung
Lipps, Frank B
Manabe, Syukuro
Miyakoda, Kikuro
Ramaswamy, Venkatachalam
Robinson, David Alton
Spelman, Michael John
Stouffer, Ronald Jay
Wetherald, Richard Tryon

NEW MEXICO
Barr, Sumner
Breiland, John Gustavson
Chylek, Petr
Cionco, Ronald Martin
Clements, William Earl
Davis, Cecil Gilbert
Gerstl, Siegfried Adolf Wilhelm
Hoard, Donald Ellsworth
Kelly, Robert Emmett
Mason, Allen Smith
Mokler, Brian Victor
Novlan, David John
Orgill, Montie M
Raymond, David James
Reed, Jack Wilson
Reifsnyder, William Edward
Schery, Stephen Dale
Yamada, Tetsuji

NEW YORK
Ausubel, Jesse Huntley
Biscaye, Pierre Eginton
Blum, Samuel Emil
Cess, Robert D
Chadwick, George F(redrick)
Chermack, Eugene E A
Cipriano, Ramon John
Czapski, Ulrich Hans
Demerjian, Kenneth Leo
Dubins, Mortimer Ira
Egan, Walter George
Ellis, William Rufus
Geer, Ira W
Geller, Marvin Alan
Gross, Stanley H
Halitsky, James
Hameed, Sultan
Havens, James Meryle
Herrington, Lee Pierce
Hindman, Edward Evans
Hubbard, John Edward
Kaimal, Jagadish Chandran
Kornfield, Jack I
Laghari, Javaid Rasoolbux
Michael, Paul Andrew
Millman, George Harold
Murphy, Kenneth Robert
Myer, Glenn Evans
Neumann, Paul Gerhard
Oglesby, Ray Thurmond
Pack, Albert Boyd
Pierson, Willard James, Jr
Ram, Michael
Rampino, Michael Robert
Rao, Samohineeveesu Trivikrama
Rind, David Harold
Sato, Makiko
Stolov, Harold L
Travis, Larry Dean
Varanasi, Prasad
Vickers, William W
Wilks, Daniel S

NORTH CAROLINA
Alpert, Leo
Arya, Satya Pal
Binkowski, Francis Stanley
Brotak, Edward Allen
Ching, Jason Kwock Sung
Crutcher, Harold L
Decker, Clifford Earl, Jr
Dickerson, Willard Addison
Droessler, Earl George
Dubach, Harold William
Foley, Gary J
Fox, Donald Lee
Hadeen, Kenneth Doyle
Haggard, William Henry
Haynie, Fred Hollis
Heck, Walter Webb
Jayanty, R K M
Karl, Thomas Richard
Kirk, Wilber Wolfe
Nie, Dalin
Odman, Mehmet Talat
Pooler, Francis, Jr
Ramage, Colin Stokes
Robinson, Peter John
Rodney, Earnest Abram
Saucier, Walter Joseph
Saxena, Vinod Kumar
Sethuraman, S
Shreffler, Jack Henry
Steila, Donald
Swift, Lloyd Wesley, Jr
Wiener, Russell Warren
Wollin, Goesta

NORTH DAKOTA
Enz, John Walter
Grainger, Cedric Anthony

OHIO
Arnfield, Anthony John
Brazee, Ross D
Cholak, Jacob
Gordon, Sydney Michael
James, Philip Benjamin
Rayner, John Norman
Sticksel, Philip Rice
Weiss, Harold Samuel

OKLAHOMA
Bluestein, Howard Bruce
Brock, Fred Vincent
Brown, Rodger Alan
Burgess, Donald Wayne
Carr, Meg Brady
Doswell, Charles Arthur, III
Doviak, Richard J
Duncan, Lewis Mannan
Eddy, George Amos
Hane, Carl Edward
Kessler, Edwin, 3rd
Kimpel, James Froome
Lamb, Peter James
Maddox, Robert Alan
Murray, Frederick Nelson
Richman, Michael B
Taylor, William L

OREGON
Coakley, James Alexander, Jr
Decker, Fred William
Murphy, Allan Hunt
Neilson, Ronald Price
Niem, Wendy Adams
Paulson, Clayton Arvid
Peterson, Ernest W
Quinn, William Hewes
Wolf, Marvin Abraham

PENNSYLVANIA
Albrecht, Bruce Allen
Barron, Eric James
Baurer, Theodore
Blackadar, Alfred Kimball
Cahir, John Joseph
Carlson, Toby Nahum
DeWalle, David Russell
Guinan, Edward F
Hosler, Charles Luther, Jr
Illick, J(ohn) Rowland
Jaggard, Dwight Lincoln
Justham, Stephen Alton
Kreitzberg, Carl William
Laws, Kenneth Lee
Lee, John Denis
McFarland, Mack
Mitchell, John Douglas
Russo, Joseph Martin
Sion, Edward Michael
Spalding, George Robert
Tang, Chung-Muh

RHODE ISLAND
Godshall, Fredric Allen
Kawaters, Woody H
Merrill, John T
Webb, Thompson, III

SOUTH CAROLINA
Crawford, Todd V
Davis, Francis Kaye
Gentry, Robert Cecil
Hofstetter, Kenneth John
Ulbrich, Carlton Wilbur
Weber, Allen Howard

SOUTH DAKOTA
Helsdon, John H, Jr
Orville, Harold Duvall
Schleusener, Richard A
Winker, James A(nthony)

TENNESSEE
Amundsen, Clifford C
Baldocchi, Dennis D
Blasing, Terence Jack
Gifford, Franklin Andrew, Jr
Holdeman, Jonas Tillman, Jr
Hosker, Rayford Peter, Jr
Kohland, William Francis
Murphy, Brian Donal

TEXAS
Brooks, David Arthur
Brundidge, Kenneth Cloud
Chamberlain, Joseph Wyan
Clark, Robert Alfred
Djuric, Dusan
Driscoll, Dennis Michael
Few, Arthur Allen
Franceschini, Guy Arthur
Frank, Neil LaVerne
Freeman, John Clinton, Jr
Goldman, Joseph L
Greene, Donald Miller
Griffiths, John Frederick
Haragan, Donald Robert
Harrison, Wilks Douglas
Hasling, Jill Freeman
Hill, Thomas Westfall
Kyle, Thomas Gail
Ledley, Tamara Shapiro
McDaniel, Willard Rich
Miksad, Richard Walter
Monahan, Hubert Harvey
Moser, James Howard
Moyer, Vance Edwards
Pitts, David Eugene
Runnels, Robert Clayton
Sass, Ronald L
Scoggins, James R
Tinsley, Brian Alfred
Wagner, Norman Keith
Webb, Willis Lee
Woodward, Wayne Anthony

UTAH
Allen, Richard Glen
Astling, Elford George
Baker, Kay Dayne
Cramer, Harrison Emery
Dickson, Don Robert
Liou, Kuo-Nan
Malek, Esmaiel
Snellman, Leonard W

VERMONT
Minsinger, William Elliot
Spar, Jerome

VIRGINIA
Beal, Stuart Kirkham
Benton, Duane Marshall
Bishop, Joseph Michael
Browell, Edward Vern
Chang, Chieh Chien
Cooley, Duane Stuart
Deepak, Adarsh
Farrukh, Usamah Omar
Fein, Jay Sheldon
Friday, Elbert W, Jr
Ganguly, Suman
Garstang, Michael
Gilman, Donald Lawrence
Greenfield, Richard Sherman
Harding, Duane Douglass
Hicks, Steacy Dopp
Jess, Edward Orland
McCormick, Michael Patrick
Melfi, Leonard Theodore, Jr
Miller, John Frederick
Murino, Vincent S
Parry, Hubert Dean
Peck, Eugene Lincoln
Perry, John Stephen
Rigby, Malcolm
Russell, James Madison, III
Siegel, Clifford M(yron)
Spilhaus, Athelstan Frederick
Swissler, Thomas James
Theon, John Speridon
Tolson, Robert Heath
Trizna, Dennis Benedict
Trout, Dennis Alan
White, Fred D(onald)

WASHINGTON
Asher, William Edward
Badgley, Franklin Ilsley
Baldwin, Mark Phillip
Barchet, William Richard
Businger, Joost Alois
Campbell, Gaylon Sanford
Craine, Lloyd Bernard
Droppo, James Garnet, Jr
Elderkin, Charles Edwin
Fleagle, Robert Guthrie
Fremouw, Edward Joseph
Frenzen, Paul
Fritschen, Leo J
Fuquay, James Jenkins
Hadlock, Ronald K
Hartmann, Dennis Lee
Hilst, Glenn Rudolph
Jones, Kay H
Katz, Yale H
Laevastu, Taivo
Lansinger, John Marcus
Leovy, Conway B
Mobley, Curtis Dale
Muench, Robin Davie
Reed, Richard John
Roden, Gunnar Ivo
Sehmel, George Albert
Seliga, Thomas A
Sloane, Christine Scheid
Thompson, Aylmer Henry
Tung, Ka-Kit
Wallace, John Michael
Weiss, Richard Raymond

WEST VIRGINIA
Covey, Winton Guy, Jr

WISCONSIN
Anderson, Charles Edward
Bretherton, Francis P
Bryson, Reid Allen
Haig, Thomas O
Hastenrath, Stefan Ludwig
Hedden, Gregory Dexter
Horn, Lyle Henry
Houghton, David Drew
Johnson, Donald R
Kahl, Jonathan D W
Kuhn, Peter Mouat
Martin, David William
Miller, David Hewitt
Mitchell, Val Leonard
Moran, Joseph Michael
Nealson, Kenneth Henry
Ragotzkie, Robert Austin
Reinsel, Gregory Charles
Sharkey, Thomas D
Sikdar, Dhirendra N
Stearns, Charles R
Suchman, David
Suomi, Verner Edward
Wang, Pao-Kuan
Young, John A

WYOMING
Marwitz, John D

PUERTO RICO
McDowell, Dawson Clayborn

ALBERTA
Fletcher, Roy Jackson
Giovinetto, Mario Bartolome
Hage, Keith Donald
Leahey, Douglas McAdam
Major, David John

BRITISH COLUMBIA
Davis, Roderick Leigh
Humphries, Robert Gordon
Marko, John Robert
Mathewes, Rolf Walter
Oke, Timothy Richard
Thyer, Norman Harold
Walker, Edward Robert
Wilson, Richard Garth

ENVIRONMENTAL, EARTH & MARINE SCIENCES / 373

MANITOBA
Amiro, Brian Douglas

NEW BRUNSWICK
Hawkes, Robert Lewis

NOVA SCOTIA
Elliott, James Arthur

ONTARIO
Anlauf, Kurt Guenther
Baier, Wolfgang
Barrie, Leonard Arthur
Bhartendu, Srivastava
Blevis, Bertram Charles
Boville, Byron Walter
Cho, Han-Ru
Donelan, Mark Anthony
Ganza, Kresimir Peter
Gillespie, Terry James
Godson, Warren Lehman
Gultepe, Ismail
Hedley, Mark Allen
Iribarne, Julio Victor
Khandekar, Madhav Laxman
McBean, Edward A
McConnell, John Charles
Mackay, Gervase Ian
McTaggart-Cowan, Patrick Duncan
Martin, Hans Carl
Moorcroft, Donald Ross
Moroz, William James
Munn, Robert Edward
Sanderson, H Preston
Stinson, Michael Roy
Strachan, William Michael John
Thomas, Morley Keith

QUEBEC
Barthakur, Nayana N
Langleben, Manuel Phillip
Leighton, Henry George
Paulin, Gaston (Ludger)
Rogers, Roddy R

SASKATCHEWAN
Ripley, Earle Allison

OTHER COUNTRIES
Blackburn, Jacob Floyd
Box, Michael Allister
Bray, John Roger
Bridgman, Howard Allen
Hagfors, Tor
Levin, Zev
Lloyd, Christopher Raymond
Sakamoto, Clarence M
Stoeber, Werner
Willeke, Klaus
Williams, Michael Maurice Rudolph
Wilson, Cynthia
Zdunkowski, Wilford G

Earth Sciences, General

ALABAMA
Canis, Wayne F
Costes, Nicholas Constantine

ALASKA
Brown, Neal B
Miller, Lance Davison

ARIZONA
Baker, Victor Richard
Baron, William Robert
Bates, Charles Carpenter
Brazel, Anthony James
Brusseau, Mark Lewis
Chapman, Clark Russell
Comrie, Andrew Charles
Damon, Paul Edward
Dean, Jeffrey Stewart
Dodge, Theodore A
Dorn, Ronald I
Fellows, Larry Dean
Hartmann, William K
Hutchinson, Charles F
Jones, David Lloyd
Kamilli, Robert Joseph
Loomis, Frederick B
Lucchitta, Baerbel Koesters
Mead, James Irving
Moke, Charles Burdette
Parrish, Judith Totman
Sonett, Charles Philip
Wallace, Terry Charles, Jr
Wierenga, Peter J

ARKANSAS
Sears, Derek William George

CALIFORNIA
Austin, Steven Arthur
Bailey, Roy Alden
Bakun, William Henry
Berg, Henry Clay
Bernstein, Ralph
Bertoldi, Gilbert LeRoy
Bohlen, Steven Ralph
Brahma, Chandra Sekhar
Campbell, Kenneth Eugene, Jr
Carlson, Carl E
Carney, Heath Joseph
Carr, Michael H
Carrigan, Charles Roger
Carsola, Alfred James
Castle, Robert O
Chesnut, Dwayne A(llen)
Ciancanelli, Eugene Vincent
Clague, David A
Coleman, Robert Griffin
Covey, Curtis Charles
Dalrymple, Gary Brent
Denton, Joan E
Drake, Robert E
Fischer, Alfred George
Galehouse, Jon Scott
Gottlieb, Peter
Grantz, Arthur
Hanan, Barry Benton
Heusinkveld, Myron Ellis
Hirschfeld, Sue Ellen
Hodges, Carroll Ann
Holzer, Alfred
Huber, Norman King
Iberall, Arthur Saul
Isherwood, William Frank
Iyer, Hariharaiyer Mahadeva
Karinen, Arthur Eli
Kasameyer, Paul William
Kennedy, Burton Mack
Koenig, James Bennett
Koestel, Mark Alfred
Krauskopf, Konrad Bates
Ku, Teh-Lung
Lay, Thorne
Leps, Thomas MacMaster
Lier, John
Lippitt, Louis
Lopes-Gautier, Rosaly
Luu, Jane
McHugh, Stuart Lawrence
Mack, Seymour
Mahood, Gail Ann
Majmundar, Hasmukhrai Hiralal
Mooney, Walter D
Moore, James Gregory
Morris, John David
Morton, Douglas M
Muffler, Leroy John Patrick
Nelson, Robert M
Nemat-Nasser, Siavouche
Nielson, Jane Ellen
Nur, Amos M
Orwig, Eugene R, Jr
Paige, David A
Phoenix, David A
Piel, Kenneth Martin
Pierce, William G
Reynolds, Michael David
Rossbacher, Lisa Ann
Servos, Kurt
Shreve, Ronald Lee
Silver, Eli Alfred
Smith, Richard Avery
Spence, Harlan Ernest
Sternberg, Hilgard O'Reilly
Stolper, Edward Manin
Stout, Ray Bernard
Sundaram, Panchanatham N
Switzer, Paul
Sylvester, Arthur Gibbs
Tewes, Howard Allan
Tilling, Robert Ingersoll
Tombrello, Thomas Anthony, Jr
Wadsworth, William Bingham
Welch, Robin Ivor
Wells, Ronald Allen
Wells, Stephen Gene
Werth, Glenn Conrad
Williams, Alan Evan
Zumberge, James Herbert

COLORADO
Armbrustmacher, Theodore J
Ayler, Maynard Franklin
Barber, George Arthur
Blair, Robert G
Bryant, Bruce Hazelton
Bryant, Donald G
Burroughs, Richard Lee
Butler, Arthur P
Charlie, Wayne Alexander
Coates, Donald Allen
Cole, James Channing
Crockett, Allen Bruce
Crone, Anthony Joseph
Diment, William Horace
Downs, William Fredrick
Driscoll, Richard Stark
Ellingson, Jack A
Goetz, Alexander Franklin Hermann
Hay, William Winn
Henrickson, Eiler Leonard
Herrmann, Raymond
Hills, Francis Allan
Hittelman, Allen M
Holcombe, Troy Leon
Hotchkiss, William Rouse
Illangasekare, Tissa H
Kenny, Ray
Lockley, Martin Gaudin
McCabe, Gregory James, Jr
MacCarthy, Patrick
Machette, Michael N
Malde, Harold Edwin
Menzer, Fred J
Morrison, Roger Barron
Myers, Donald Arthur
Palmer, Allison Ralph
Pickett, George R
Pierce, Kenneth Lee
Porter, Kenneth Raymond
Raup, Robert Bruce, Jr
Rold, John W
Rowley, Peter DeWitt
Schleicher, David Lawrence
Sheridan, Douglas Maynard
Sidder, Gary Brian
Tatsumoto, Mitsunobu
Toy, Terrence J
Tsvankin, Ilya Daniel
Walker, Theodore Roscoe
Warme, John Edward

CONNECTICUT
Anderson, Carol Patricia
Buss, Leo William
Castillo, Raymond A
Chriss, Terry Michael
Clebnik, Sherman Michael
Friedman, Don Gene
Jackson, C(harles) Ian
Patton, Peter C
Tolderlund, Douglas Stanley
Wescott, Roger Williams

DELAWARE
Keidel, Frederick Andrew
Temple, Stanley

DISTRICT OF COLUMBIA
Adey, Walter Hamilton
Baker, D(onald) James
Carlson, Richard Walter
Chung, Riley M
Davis, Robert E
Doumani, George Alexander
Geelhoed, Glenn William
Gutman, George Garik
Lindstrom, Eric Jon
MacCracken, Michael Calvin
Matson, Michael
Melickian, Gary Edward
Prewitt, Charles Thompson
Price, Jonathan Greenway
Rattien, Stephen
Rogers, Thomas Hardin
Simkin, Thomas Edward
Towe, Kenneth McCarn
Wilson, Frederick Albert

FLORIDA
Andregg, Charles Harold
Beck, Warren R(andall)
Clarke, Allan James
Compton, John S
Lovejoy, Donald Walker
Maurrasse, Florentin Jean-Marie Robert
Pirkle, Fredric Lee
Sanford, Malcolm Thomas
Schmidt, Walter
Tanner, William Francis, Jr
Tull, James Franklin
Winters, Stephen Samuel

GEORGIA
Chatelain, Edward Ellis
Dod, Bruce Douglas
Hayes, Willis B
Tan, Kim H
Whitney, James Arthur

HAWAII
Garcia, Michael Omar
Hey, Richard N
Khan, Mohammad Asad
Meech, Karen Jean
Moberly, Ralph M
Peterson, Frank Lynn
Self, Stephen
Zisk, Stanley Harris

IDAHO
Knutson, Carroll Field
Ulliman, Joseph James

ILLINOIS
Altpeter, Lawrence L, Jr
Anderson, Richard Charles
Berry, Gregory Franklin
Chao, Sherman S
Chew, Weng Cho
Chou, Chen-Lin
Flynn, John Joseph
Fraunfelter, George H
Gealy, William James
Leary, Richard Lee
Leighton, Morris Wellman
Lesht, Barry Mark
Mann, Christian John
Masters, John Michael
Maynard, Theodore Roberts
Peck, Theodore Richard
Poulikakos, Dimos
Rodolfo, Kelvin S
Sepkoski, J(oseph) J(ohn), Jr
Trask, Charles Brian

INDIANA
Biggs, Maurice Earl
Hamburger, Michael Wile
Kim, Yeong Ell
Martin, Charles Wellington
Park, Richard Avery, IV
Rosenberg, Gary David
Votaw, Robert Barnett

KANSAS
Clark, George Richmond, II
Moore, Richard K(err)
Welch, Jerome E
White, Stephen Edward

KENTUCKY
Blackburn, John Robert
Ettensohn, Francis Robert
Hoffman, Wayne Larry
Singh, Raman J

LOUISIANA
DeLaune, Ronald D
Rogers, James Edwin
Walker, Harley Jesse

MAINE
Nelson, Robert Edward

MARYLAND
Asrar, Ghassam R
Barker, John L, Jr
Brown, Michael
Chang, Alfred Tieh-Chun
Chery, Donald Luke, Jr
Cohen, Steven Charles
Comiso, Josefino Cacas
Crampton, Janet Wert
French, Bevan Meredith
Gadsby, Dwight Maxon
Garvin, James Brian
Hartle, Richard Eastham
Koblinsky, Chester John
Kolstad, George Andrew
Leatherman, Stephen Parker
Leedy, Daniel Loney
McCammon, Helen Mary
Mecherikunnel, Ann Pottanat
Meredith, Leslie Hugh
Mielke, James Edward
Mogil, H Michael
Nazarova, Katherine A
Ostenso, Ned Allen
Parkinson, Claire Lucille
Parkinson, Claire Lucille
Phair, George
Price, John Charles
Raff, Samuel J
Rall, Jonathan Andrew Reiley
Salomonson, Vincent Victor
Schulkin, Morris
Shotland, Lawrence M
Stifel, Peter Beekman
Wells, Eddie N
Wolman, Markley Gordon

MASSACHUSETTS
Ballard, Robert D
Bradley, Raymond Stuart
Greenstein, Benjamin Joel
Grove, Timothy Lynn
Hager, Bradford Hoadley
Hepburn, John Christopher
Hunt, John Meacham
Knebel, Harley John
Lin, Jian
McNutt, Marcia Kemper
Marvin, Ursula Bailey
Molnar, Peter Hale
Nunes, Paul Donald
Oldale, Robert Nicholas
Ridge, John Charles
Romey, William Dowden
Shimizu, Nobumichi
Takahashi, Kozo
Tarkoy, Peter J
Terzaghi, Ruth Doggett
Williamson, Peter George

MICHIGAN
Atreya, Sushil Kumar
Batterman, Stuart A
Blinn, Lorena Virginia
Lefebvre, Richard Harold
MacMahan, Horace Arthur, Jr
Menninga, Clarence
Ronca, Luciano Bruno
Rusch, Wilbert H, Sr

MINNESOTA
Edwards, Richard Lawrence
Hudleston, Peter John
Parker, Ronald Bruce
Rapp, George Robert, Jr

MISSISSIPPI
Bruce, John Goodall, Jr
Meylan, Maurice Andre
Perry, Edward Belk
Saucier, Roger Thomas
Seglund, James Arnold
Strom, Eric William

Earth Sciences, General (cont)

MISSOURI
Arvidson, Raymond Ernst
Ashworth, William Bruce, Jr
Brown, Donald M
Snowden, Jesse O
Unklesbay, Athel Glyde

MONTANA
Hyndman, Donald William

NEBRASKA
Doran, John Walsh
Goss, David
Grew, Priscilla Croswell Perkins

NEVADA
Gifford, Gerald F
Helm, Donald Cairney
Henry, Christopher D
Krenkel, Peter Ashton
Raines, Gary L
Savage, Martha Kane
Trexler, Dennis Thomas

NEW HAMPSHIRE
Adams, Samuel S
Oreskes, Naomi
Richter, Dorothy Anne
Springsteen, Kathryn Rose Mooney

NEW JERSEY
Agron, Sam Lazrus
Hanson, James Edward
Hordon, Robert M
Kennish, Michael Joseph
Lindberg, Craig Robert
Manabe, Syukuro
Monahan, Edward James
Robinson, David Alton
Sarmiento, Jorge Louis
Strausberg, Sanford I
Weisberg, Joseph Simpson

NEW MEXICO
Arion, Douglas
Bish, David Lee
Broxton, David Edward
Colp, John Lewis
Gancarz, Alexander John
Gerstl, Siegfried Adolf Wilhelm
Grambling, Jeffrey A
Kuellmer, Frederick John
Love, David Waxham
McTigue, David Francis
Manley, Kim
Shomaker, John Wayne
Trauger, Frederick Dale
Wawersik, Wolfgang R

NEW YORK
Bhattacharji, Somdev
Bond, Gerard C
Brett, Carlton Elliot
De Laubenfels, David John
Donnelly, Thomas Wallace
Eaton, Gordon Pryor
Egan, Walter George
Fleischer, Robert Louis
Hayes, Dennis E
Henderson, Floyd M
Henderson, Thomas E
Jacob, Klaus H
Kent, Dennis V
Kumai, Motoi
Lindsley, Donald Hale
Mullins, Henry Thomas
Rampino, Michael Robert
Rokitka, Mary Anne
Savage, E Lynn
Schoonen, Martin A A
Stanley, Edward Alexander
Tavel, Judith Fibkins
Trexler, Frederick David
Van Burkalow, Anastasia

NORTH CAROLINA
Brockhaus, John Albert
Cannon, Ralph S, Jr
Corliss, Bruce Hayward
Overcash, Michael Ray
Parsons, Willard H
Read, Floyd M
Ross, Thomas Edward
Stuart, Alfred Wright
Towell, William Earnest

OHIO
Bladh, Katherine Laing
Bladh, Kenneth W
Bundy, Francis P
Friberg, LaVerne Marvin
Fuller, James Osborn
Harris-Noel, Ann (F H) Graetsch
Kovach, Jack
Marble, Duane Francis
Mayer, Victor James
Nash, David Byer
Noble, Allen G
Savin, Samuel Marvin
Unrug, Raphael
Wojtal, Steven Francis

OKLAHOMA
Cardott, Brian Joseph
Henderson, Frederick Bradley, III
Johnson, Kenneth Sutherland
Martner, Samuel (Theodore)
Root, Paul John
Thomsen, Leon
Visher, Glenn S
Williams, Thomas Henry Lee

OREGON
Addicott, Warren O
Drake, Ellen Tan
Edwards, James Mark
Lawrence, Robert D
Lupton, John Edward
McKinney, William Mark
Quinn, William Hewes
Young, J Lowell

PENNSYLVANIA
Barron, Eric James
Beutner, Edward C
Brantley, Susan Louise
Briggs, Reginald Peter
Cassidy, William Arthur
Crawford, Maria Luisa Buse
Cuff, David J
Elsworth, Derek
Fritz, Lawrence William
Justham, Stephen Alton
Kasting, James Fraser
Kodama, Kenneth Philip
Krishtalka, Leonard
Luce, Robert James
Meisler, Harold
Pauline, Ronald Frank
Pfefferkorn, Hermann Wilhelm
Rizza, Paul Frederick
Rohl, Arthur N
Schultz, Lane D
Shultz, Charles H
Simonds, John Ormsbee
Taylor, Alan H
Tuffey, Thomas J
Wegweiser, Arthur E

RHODE ISLAND
Cain, J(ames) Allan
Pieters, Carle M

SOUTH CAROLINA
Crawford, Todd V
Gardner, Leonard Robert
Hawkins, Richard Horace
James, L(aurence) Allan

TENNESSEE
Bell, Persa Raymond
Fullerton, Ralph O
Gentry, Robert Vance
Jumper, Sidney Roberts
McSween, Harry Y, Jr
Miller, Calvin F
Siesser, William Gary
Tamura, Tsuneo
Yau, Cheuk Chung

TEXAS
Berkhout, Aart W J
Brown, Glenn Lamar
Burton, Robert Clyde
Chronic, John
DeFord, Ronald K
Dorfman, Myron Herbert
Eckelmann, Walter R
Eifler, Gus Kearney, Jr
Elsik, William Clinton
Gore, Dorothy J
Greene, Donald Miller
Halpern, Martin
Harrison, Wilks Douglas
Hudson, Robert Frank
Jahns, Hans O(tto)
Jamieson-Magathan, Esther R
Jennings, Alfred Roy, Jr
Johnston, David Hervey
Jones, Ernest Austin, Jr
Judd, Frank Wayne
Kaufman, Sidney
Kerans, Charles
LaBree, Theodore Robert
Lafon, Guy Michel
Lambert, Paul Wayne
Ledley, Tamara Shapiro
Lofgren, Gary Ernest
Murphy, Daniel L
Murray, Grover Elmer
Nestell, Merlynd Keith
Nicks, Oran Wesley
Nussmann, David George
Pearson, Daniel Bester, III
Phinney, William Charles
Reaser, Donald Frederick
Reso, Anthony
Robertson, James Douglas
Rowe, Marvin W
Schink, David R
Sprunt, Eve Silver
Tyler, Noel
Whitford-Stark, James Leslie
Wilson, Clark Roland
Woodruff, Charles Marsh, Jr
Woods, Raymond D

Zhou, Hua Wei

UTAH
Case, James B(oyce)
Cerling, Thure E
Christiansen, Eric H
Freethey, Geoffrey Ward
Gates, Joseph Spencer
Harris, John Michael
Mosher, Loren Cameron
Ross, Howard Persing
Vaughn, Danny Mack

VERMONT
Coish, Raymond Alpheaus
Hussak, Robert Edward
Ratté, Charles A

VIRGINIA
Bowker, David Edwin
Burstyn, Harold Lewis
Cameron, Maryellen
Campbell, Russell Harper
Corell, Robert Walden
Corwin, Gilbert
Doyle, Frederick Joseph
Finkelman, Robert Barry
Gaines, Alan McCulloch
Gushee, Beatrice Eleanor
Haq, Bilal U
Hearn, Bernard Carter, Jr
James, Odette Bricmont
Johnson, Robert Ward
Leo, Gerhard William
Lipin, Bruce Reed
Macko, Stephen Alexander
Martens, James Hart Curry
Miller, William Lawrence
Milton, Nancy Melissa
Molnia, Bruce Franklin
Mulder, Robert Udo
Naeser, Charles Wilbur
Naeser, Nancy Dearien
Newton, Elisabeth G
Odum, William Eugene
Outerbridge, William Fulwood
Padovani, Elaine Reeves
Said, Rushdi
Sato, Motoaki
Sayala, Dasharatham (Dash)
Spitzer, Cary Redford
Stebbins, Robert H
Suiter, Marilyn J
Van der Vink, Gregory E
Wayland, Russell Gibson
Welch, Jasper Arthur, Jr
Wolff, Roger Glen

WASHINGTON
Ames, Lloyd Leroy, Jr
Armstrong, Frank C(larkson Felix)
Barnes, Ross Owen
Cowan, Darrel Sidney
Dzurisin, Daniel
Heilman, Paul E
Hodges, Floyd Norman
Holcomb, Robin Terry
Keller, C Kent
Malone, Stephen D
Owens, Edward Henry
Riley, Robert Gene
Roberts, George E(dward)
Sanford, Thomas Bayes
Shin, Suk-han
Symonds, Robert B

WEST VIRGINIA
Khair, Abdul Wahab

WISCONSIN
Borowiecki, Barbara Zakrzewska
Bryson, Reid Allen
Fowler, Gerald Allan
Gernant, Robert Everett
Horn, Lyle Henry
Moran, Joseph Michael
Paull, Rachel Krebs
Read, William F
Valley, John Williams

WYOMING
Frost, Carol Denison
Huff, Kenneth O
Mateer, Niall John

ALBERTA
Charlesworth, Henry A K
Coen, Gerald Marvin
Goodarzi, Fariborz
Hunt, C Warren
McMillan, Neil John
Macqueen, Roger Webb
Mossop, Grant Dilworth
Nesbitt, Bruce Edward
Rogerson, Robert James

BRITISH COLUMBIA
Bevier, Mary Lou
Davis, Roderick Leigh
Wheeler, John Oliver

MANITOBA
Sheppard, Marsha Isabell
Smil, Vaclav

Welsted, John Edward
Whitaker, Sidney Hopkins
Wilkins, Brian John Samuel

NEW BRUNSWICK
Allen, Clifford Marsden
Ferguson, Laing

NOVA SCOTIA
Zentilli, Marcos

ONTARIO
Arkani-Hamed, Jafar
Bell, Richard Thomas
Brown, Richard L
Cherry, John A
Cowan, W R
Csillag, Ferenc
Dredge, Lynda Ann
Fawcett, James Jeffrey
Flint, Jean-Jacques
Fox, Joseph S
Gussow, W(illia)m C(arruthers)
Jackson, Togwell Alexander
Long, Darrel Graham Francis
MacCrimmon, Hugh Ross
Niki, Hiromi
Percival, John A
Rowe, Ronald Kerry
Terasmae, Jaan
Veizer, Ján
Wahl, William G

QUEBEC
Angers, Denis Arthur
Brandenberger, Arthur J
Chagnon, Jean Yves
David, Peter P
Dubois, Jean-Marie M
Ladanyi, Branko
Margolis, Hank A
Yong, Raymond N

SASKATCHEWAN
St Arnaud, Roland Joseph
Slater, George P
Strathdee, Graeme Gilroy

OTHER COUNTRIES
Alonso, Ricardo N
Dengo, Gabriel
Economides, Michael John
Kubik, Peter W
Lloyd, Christopher Raymond
Rich, Thomas Hewitt
Ugolini, Fiorenzo Cesare

Environmental Sciences, General

ALABAMA
Baker, June Marshall
Campbell, Robert Terry
Cushing, Kenneth Mayhew
Dahlin, Robert S
Fowler, Charles Sidney
Huluka, Gobena
Hung, Ru J
McCarl, Henry Newton
McDonald, Jack Raymond
Marano, Gerald Alfred
Marchant, Guillaume Henri (Wim), Jr
Meagher, James Francis
Nelson, David Herman
Olson, Willard Paul
Reynolds, George Warren
Rivers, Douglas Bernard
Rogers, Hugo H, Jr
Smith, Wallace Britton
Vacik, James P
Wallace, Dennis D
Wehner, Donald C
Wetzel, Robert George
Workman, William Edward

ALASKA
Fahl, Charles Byron
Fink, Thomas Robert
Morrison, John Albert
Mowatt, Thomas C
Rice, Stanley Donald
Rockwell, Julius, Jr
Schindler, John Frederick

ARIZONA
Avery, Charles Carrington
Bradley, Michael Douglas
Brathovde, James Robert
Brazel, Anthony James
Brusseau, Mark Lewis
Comrie, Andrew Charles
Day, Arden Dexter
Diehn, Bodo
Dorn, Ronald I
Dworkin, Judith Marcia
Fernando, Harindra Joseph
Gerba, Charles Peter
Haase, Edward Francis
Hodges, Carl Norris
Hutchinson, Charles F
Idso, Sherwood B
Kay, Fenton Ray
Klopatek, Jeffrey Matthew

McKlveen, John William
Nordstrom, Brian Hoyt
Poole, H K
Ramsey, John Charles
Robertson, Frederick Noel
Roth, Eldon Sherwood
Sears, Donald Richard
Shields, Jimmie Lee
Twomey, Sean Andrew
Wait, James Richard
Wierenga, Peter J
Zwolinski, Malcolm John

ARKANSAS
Anderson, Robbin Colyer
Andrews, Luther David
Bonner, Errol Marcus
Burchard, John Kenneth
Chowdhury, Parimal
Day, James Meikle
Ervin, Hollis Edward
Fortner, Wendell Lee
Meyer, Richard Lee
Wolf, Duane Carl

CALIFORNIA
Akmal, Naim
Anspaugh, Lynn Richard
Asante-Duah, D Kofi
Augood, Derek Raymond
Baez, Albert Vinicio
Bakus, Gerald Joseph
Barbour, Michael G
Baston, Janet Evelyn
Baugh, Ann Lawrence
Baur, Mario Elliott
Beck, Albert J
Becking, Rudolf (Willem)
Bencala, Kenneth Edward
Bendix, Selina (Weinbaum)
Bennett, Charles Franklin
Benson, Andrew Alm
Bils, Robert F
Bishop, Keith C, III
Bissing, Donald Ray
Blume, Frederick Duane
Blunt, Martin Julian
Brown, Stephen L(awrence)
Bubeck, Robert Clayton
Buck, Alan Charles
Cahn, David Stephen
Callahan, Joan Rea
Carney, Heath Joseph
Carsola, Alfred James
Chamberlain, Dilworth Woolley
Chang, Andrew C
Chang, David Bing Jue
Chang, Shih-Ger
Chatigny, Mark A
Chen, Yang-Jen
Cliath, Mark Marshall
Cluff, Lloyd Sterling
Cole, Richard
Cooper, Wilson Wayne
Cosgrove, Gerald Edward
Crawford, Richard Whittier
Crosby, Donald Gibson
Daniels, Jeffrey Irwin
Davis, Jay C
Davis, William Ellsmore, Jr
Davis, William Jackson
Dedrick, Kent Gentry
DeJongh, Don C
de Latour, Christopher
Dessel, Norman F
Deuble, John Lewis, Jr
Doner, Harvey Ervin
Dowdy, William Louis
Doyle, Fiona Mary
Dzieciuch, Matthew Andrew
Eastmond, David Albert
Edwards, J Gordon
Ehrlich, Anne Howland
Elimelech, Menachem
Ellstrand, Norman Carl
Elsberry, Russell Leonard
Farmer, Walter Joseph
Ferguson, William E
Fischback, Bryant C
Frank, Sidney Raymond
Freiling, Edward Clawson
Froebe, Larry R
Gaal, Robert A P
Gee, Allen
Gehri, Dennis Clark
Goh, Kean S
Goltz, Mark Neil
Gordon, Robert Julian
Grantz, David Arthur
Greenfield, Stanley Marshall
Grubbs, Charles Leslie
Hamersma, J Warren
Hanna, George P, Jr
Hardebeck, Ellen Jean
Harrach, Robert James
Harris, Sigmund Paul
Harte, John
Hass, Robert Henry
Hazelwood, R(obert) Nichols
Hemmingsen, Barbara Bruff
Hester, Norman Eric
Hodson, William Myron
Holdren, John Paul
Houpis, James Louis Joseph

Hovanec, B(ernard) Michael
Howe, Robert Kenneth
Hrubesh, Lawrence W(ayne)
Hu, Steve Seng-Chiu
Hughes, Robert Alan
Hygh, Earl Hampton
Innes, William Beveridge
Isherwood, William Frank
Ives, John (Jack) David
John, Walter
Jones, Rebecca Anne
Kallman, Burton Jay
Karuza, Sarunas Kazys
Kashar, Lawrence Joseph
Kasper, Gerhard
Kassakhian, Garabet Haroutioun
Katen, Paul C
Kirch, Patrick Vinton
Kland, Mathilde June
Klein, David Henry
Knight, Allen Warner
Koenig, James Bennett
Koestel, Mark Alfred
Kothny, Evaldo Luis
Ku, Teh-Lung
Kuehne, Donald Leroy
Kuwabara, James S
Lagarias, John S(amuel)
Laitin, Howard
Lamb, Sandra Ina
Landolph, Joseph Richard, Jr
Langerman, Neal Richard
Lapple, Charles E
Lauer, George
Layton, David Warren
Leon, Henry A
Letey, John, Jr
Lick, Wilbert James
Lier, John
Lindberg, R(obert) G(ene)
Lissaman, Peter Barry Stuart
Longley-Cook, Mark T
Ludwig, Claus Berthold
Lustig, Max
MacCready, Paul Beattie, Jr
MacKay, Kenneth Pierce, Jr
McKenzie, Donald Edward
McKone, Thomas Edward
Mahadevan, Parameswar
Malouf, George M
Martin, Michael
Mathews, Larry Arthur
Matthews, Stephen M
Melis, Anastasios
Mertes, Leal Anne Kerry
Mikel, Thomas Kelly, Jr
Mikkelsen, Duane Soren
Mode, V(incent) Alan
Morse, John Thomas
Nagy, Kenneth Alex
Nero, Anthony V, Jr
Newsom, Bernard Dean
Oechel, Walter C
Offen, George Richard
Olson, Betty H
Orloff, Neil
Ott, Wayne R
Owen, Paul Howard
Paige, David A
Patterson, Clair Cameron
Paulson, Suzanne Elizabeth
Perlmutter, Milton Manuel
Perrine, R(ichard) L(eroy)
Pestrong, Raymond
Phalan, Robert F
Pierce, Matthew Lee
Piersol, Allan Gerald
Pipkin, Bernard Wallace
Pitts, James Ninde, Jr
Porcella, Donald Burke
Quan, William
Que Hee, Shane Stephen
Ragaini, Richard Charles
Ramspott, Lawrence Dewey
Rateaver, Bargyla
Remson, Irwin
Resh, Vincent Harry
Riffer, Richard
Riley, Robert Lee
Roberts, Stephen Winston
Robilliard, Gordon Allan
Robinson, Lewis Howe
Robinson, Paul Ronald
Robison, William Lewis
Rogers, Gifford Eugene
Rosen, Alan
Rosen, Leonard Craig
Rossbacher, Lisa Ann
Rossiter, Bryant William
Rowland, F Sherwood
Russell, Philip Boyd
Sage, Orrin Grant, Jr
Salmon, Eli J
Sanders, Brenda Marie
Satre, Rodrick Iverson
Schaleger, Larry L
Schneider, Meier
Schneider, Stephen Henry
Scow, Kate Marie
Segal, Alexander
Sextro, Richard George
Shaffar, Scott William
Sheppard, Asher R
Shinn, Joseph Hancock

Shuh, David Kelly
Simon, Richard L
Singh, Hanwant B
Smith, Charles Francis, Jr
Smith, Leverett Ralph
Spencer, Herbert W, III
Spicher, Robert G
Sprague, Robert W
Star, Jeffrey L
Starr, Mortimer Paul
Stenstrom, Michael Knudson
Straughan, Isdale (Dale) Margaret
Suffet, I H (Mel)
Swanson, Robert Nels
Swearengen, Jack Clayton
Switzer, Paul
Tewes, Howard Allan
Thomas, Jerome Francis
Thomas, Walter Dill, Jr
Thompson, Chester Ray
Thompson, Rosemary Ann
Thornton, John Irvin
Tilling, Robert Ingersoll
Tin, Hla Ngwe
Toor, Arthur
VanBruggen, Ariena H C
Vilkitis, James Richard
Voeks, Robert Allen
Wade, Richard Lincoln
Walter, Hartmut
Warren, Mashuri Laird
Waterland, Larry R
Wauchope, Robert Donald
Wayne, Lowell Grant
Weathers, Wesley Wayne
Weiss, Fred Toby
Wentworth, Carl M, Jr
Westman, Walter Emil
Wilcox, Bruce Alexander
Wilcox, Howard Albert
Wilson, Katherine Woods
Winer, Arthur Melvyn
Witham, Clyde Lester
Wuebbles, Donald J
Wyzga, Ronald Edward
Yamamoto, Sachio
Ziegler, Carole L
Ziemer, Robert Ruhl

COLORADO
Arnott, Robert A
Boatman, Joe Francis
Boyd, Josephine Watson
Bradshaw, William Newman
Bush, Alfred Lerner
Chappell, Willard Ray
Chiou, Cary T(sair)
Christian, Wayne Gillespie
Cohen, Ronald R H
Colton, Roger Burnham
Craig, Roy Phillip
Crockett, Allen Bruce
Dabberdt, Walter F
Danzberger, Alexander Harris
Dewey, Fred McAlpin
Diaz, Henry Frank
Diment, William Horace
Duray, John R
Eberhard, Wynn Lowell
Edwards, Harry Wallace
Evans, David Lane
Fischer, William Henry
Fox, Douglas Gary
Hanson, Howard P(aul)
Havlick, Spenser Woodworth
Herrmann, Raymond
Hutchinson, Gordon Lee
Ivory, Thomas Martin, III
Johnson, Ross Byron
Johnson, Verner Carl
Kalabokidis, Kostas D
Kearney, Philip Daniel
Kemper, William Alexander
Kenny, Ray
Knight, William Glenn
Koizumi, Carl Jan
Krausz, Stephen
Langworthy, William Clayton
Legal, Casimer Claudius, Jr
McAuliffe, Clayton Doyle
McCabe, Gregory James, Jr
McKown, Cora F
Martin, Stephen George
Meroney, Robert N
Mosier, Arvin Ray
Navratil, James Dale
Nordstrom, Darrell Kirk
Olhoeft, Gary Roy
Parker, H Dennison
Peterson, James T
Porter, Kenneth Raymond
Ratte, James C
Reynolds, Richard Truman
Schmidt, S K
Smith, Dwight Raymond
Stauffer, Jack B
Steele, Timothy Doak
Thompson, Milton Avery
Thompson, Starley Lee
Timblin, Lloyd O, Jr
Toy, Terrence J
Tucker, Alan
Waldren, Charles Allen
Webster, Peter John

Weist, William Godfrey, Jr
Wellman, Dennis Lee
White, Gilbert Fowler
Wildeman, Thomas Raymond
York, J(esse) Louis

CONNECTICUT
Baillie, Priscilla Woods
Bentz, Alan P(aul)
Bollyky, L(aszlo) Joseph
Constant, Frank Woodbridge
Cornell, Alan
Damman, Antoni Willem Hermanus
DeSanto, Robert Spilka
Dobay, Donald Gene
Dobbs, Gregory Melville
Egler, Frank Edwin
Friedlander, Henry Z
Friedman, Howard Stephen
Gauthier, George James
Geballe, Gordon Theodore
Groff, Donald William
Groth, Richard Henry
Haakonsen, Harry Olav
Haas, Thomas J
Hauser, Martin
Jackson, C(harles) Ian
Koch, Richard Carl
Kozak, Gary S
Krause, Leonard Anthony
Mattina, Mary Jane Incorvia
Mellett, James Silvan
Remington, Charles Lee
Renfro, William Charles
Rho, Jinnque
Rusling, James Francis
Seaman, Gregory G
Shaw, Brenda Roberts
Spiegel, Zane
Stuart, James Davies
Voorhies, John Davidson

DELAWARE
Allen, Herbert E
Crippen, Raymond Charles
Ehrlich, Robert Stark
Ellis, Patricia Mench
Geer, Richard P
Goel, Kailash C(handra)
Harvey, John, Jr
Hatchard, William Reginald
Huang, Chin Pao
Jacobs, Emmett S
Kalkstein, Laurence Saul
Krone, Lawrence James
McGee, Charles E
Martin, Wayne Holderness
Mathre, Owen Bertwell
Milian, Alwin S, Jr
Moses, Francis Guy
Narvaez, Richard
Pagano, Alfred Horton
Pierrard, John Martin
Potrafke, Earl Mark
Scherger, Dale Albert
Sherman, John Walter
Sparks, Donald
Spoljaric, Nenad
Temple, Stanley
Tolman, Chadwick Alma
Vassiliou, Eustathios
Willmott, Cort James

DISTRICT OF COLUMBIA
April, Robert Wayne
Argus, Mary Frances
Ballantine, David Stephen
Bannerman, Douglas George
Beard, Charles Irvin
Beehler, Bruce McPherson
Bell, Bruce Arnold
Breen, Joseph John
Bushman, John Branson
Carrigan, Richard Alfred
Cash, Gordon Graham
Cuatrecasas, Jose
DeCicco, Benedict Thomas
Doumani, George Alexander
El-Ashry, Mohamed T
Feeney, Gloria Comulada
Goodland, Robert James A
Guimond, Richard Joseph
Hershey, Robert Lewis
Hirzy, John William
Jhirad, David John
Keitt, George Wannamaker, Jr
Krinsley, Daniel B
Lee, Robert E, Jr
Mac, Michael John
Middleton, John T(ylor)
Needels, Theodore S
Neihof, Rex A
Paris, Oscar Hall
Phillips, Gary Wilson
Plost, Charles
Powell, Justin Christopher
Preer, James Randolph
Rattien, Stephen
Richardson, Allan Charles Barbour
Rogers, Senta S(tephanie)
Rosenberg, Norman J
Rubinoff, Roberta Wolff
Schneider, Bernard Arnold
Sexton, Ken

Environmental Sciences, General (cont)

Sterrett, Kay Fife
Sudia, Theodore William
Terrell, Charles R
Uman, Myron F
Varga, Gideon Michael, Jr
Wakelyn, Phillip Jeffrey
Walker, John Martin
Warnick, Walter Lee
Wilkinson, John Edwin
Zeeman, Maurice George
Zucchetto, James John

FLORIDA
Allen, Eric Raymond
Barile, Diane Dunmire
Belanger, Thomas V
Besch, Emerson Louis
Billings, Charles Edgar
Brooker, Hampton Ralph
Burrill, Robert Meredith
Carter, James Harrison, II
Chryssafopoulos, Hanka Wanda Sobczak
Clegern, Robert Wayne
Cooper, William James
Coppoc, William Joseph
Delfino, Joseph John
De Lorge, John Oldham
Dietz, Jess Clay
Donnalley, James R, Jr
Donoghue, Joseph F
Ekberg, Donald Roy
Fitzpatrick, George
Fourqurean, James Warren
Frazier, Stephen Earl
Frei, Sr John Karen
Fuller, Harold Wayne
Gnaedinger, Richard H
Goldberg, Walter M
Goldstein, Harold William
Goss, Gary Jack
Hanley, James Richard, Jr
Harrold, Gordon Coleson
Haviland, Robert P(aul)
Herndon, Roy C
Jackson, Daniel Francis
Jenkins, Dale Wilson
Jones, John A(rthur)
Klacsmann, John Anthony
Langford, David
Lim, Daniel V
Llewellyn, Ralph A
Lundgren, Dale A(llen)
McCowen, Max Creager
Maskal, John
Mason, D(onald) R(omagne)
Menzer, Robert Everett
Meryman, Charles Dale
Miles, Carl J
Miskimen, George William
Mushinsky, Henry Richard
Myers, Richard Lee
Myerson, Albert Leon
Nelson, Gordon Leigh
Nelson, John William
Nichols, James Carlile
Parsons, Lawrence Reed
Posner, Gerald Seymour
Radomski, Jack London
Rao, Palakurthi Suresh Chandra
Rechcigl, John E
Rice, Stanley Alan
Rona, Donna C
Ruscher, Paul H
Shair, Robert C
Silverman, David Norman
Sirkin, Alan N
Smith, William Mayo
Snow, Eleanour Anne
Solon, Leonard Raymond
Stamper, James Harris
Suman, Daniel Oscar
Teller, Aaron Joseph
Tiffany, William James, III
Virnstein, Robert W
Walker, Charles R
Wander, Joseph Day
Wang, Tsen Chen
White, Arlynn Quinton, Jr
Windsor, John Golay, Jr
Yost, Richard A
Zuo, Yuegang

GEORGIA
Alberts, James Joseph
Amirtharajah, Appiah
Baarda, David Gene
Bailey, George William
Baughman, George Larkins
Bayer, Charlene Warres
Bozeman, John Russell
Brown, William E
Burns, Lawrence Anthony
Carver, Robert E
Chameides, William Lloyd
Chandler, James Harry, III
Cofer, Harland E, Jr
Ellington, James Jackson
Fineman, Manuel Nathan
Gaffney, Peter Edward
Garrison, Arthur Wayne
Ghuman, Gian Singh
Glooschenko, Walter Arthur
Greear, Philip French-Carson
Hargrove, Robert John
Hedden, Kenneth Forsythe
Hill, David W(illiam)
Hooper, Richard Paul
Hoover, Thomas Burdett
Justus, Carl Gerald
Kollig, Heinz Philipp
Lee, Richard Fayao
Milham, Robert Carr
Nichols, Michael Charles
Oakes, Thomas Wyatt
Odum, Eugene Pleasants
Peake, Thaddeus Andrew, III
Quisenberry, Dan Ray
Remillard, Marguerite Madden
Richardson, Susan D
Robbins, Wayne Brian
Rodriguez, Augusto
Rogers, Harvey Wilbur
Schultz, Donald Paul
Shure, Donald Joseph
Spetnagel, Theodore John
Sumartojo, Jojok
Sutton, William Wallace
Vander Velde, George
Wiegel, Juergen K
Williams, Daniel James
Windom, Herbert Lynn
Woodall, William Robert, Jr
Ziffer, Jack

HAWAII
Coles, Stephen Lee
Cox, Doak Carey
Curtis, George Darwin
Flachsbart, Peter George
Frystak, Ronald Wayne
Fujioka, Roger Sadao
Halbig, Joseph Benjamin
Hodgson, Lynn Morrison
Hufschmidt, Maynard Michael
Mackenzie, Frederick Theodore
Morin, Marie Patricia
Segal, Earl
Siegel, Sanford Marvin
Teramura, Alan Hiroshi
Winget, Robert Newell

IDAHO
Clark, William Hilton
Ferguson, James Homer
Gehrke, Robert James
Gesell, Thomas Frederick
Hollenbaugh, Kenneth Malcolm
Knutson, Carroll Field
Lehrsch, Gary Allen
Li, Guang Chao
Reno, Harley W
Welhan, John Andrew
Wright, Kenneth James

ILLINOIS
Akin, Cavit
Allen, John Kay
Babcock, Lyndon Ross, (Jr)
Banwart, Wayne Lee
Bath, Donald Alan
Baumgarten, Ronald J
Benioff, Paul
Bernath, Tibor
Bhatti, Neeloo
Bjorklund, Richard Guy
Brand, Raymond Howard
Carlson, Gerald Eugene
Casella, Alexander Joseph
Clark, Stephen Darrough
Doehler, Robert William
Erickson, Mitchell Drake
Flynn, Kevin Francis
Flynn, Robert James
Foster, John Webster
Frank, James Richard
Garvin, Paul Joseph, Jr
Gibbons, Larry V
Gilbert, Thomas Lewis
Girard, G Tanner
Glod, Edward Francis
Hagenbach, W(illiam) P(aul)
Hall, Stephen Kenneth
Hallenbeck, William Hackett
Harrison, Paul C
Harrison, Wyman
Herendeen, Robert Albert
Herricks, Edwin E
Hirschi, Michael Carl
Hockman, Deborah C
Holsen, Thomas Michael
Hsui, Albert Tong-Kwan
Hutchcroft, Alan Charles
Jarke, Frank Henry
Johnson, John Harold
Khan, Mohammed Abdul Quddus
Kim, Byung J
Klass, Donald Leroy
Knake, Ellery Louis
Konopinski, Virgil J
LaFevers, James Ronald
Lerman, Abraham
Lesht, Barry Mark
Loehle, Craig S
Markuszewski, Richard
Marmer, Gary James
Mendelsohn, Marshall H
Minear, Roger Allan
Murphy, Thomas Joseph
Paddock, Robert Alton
Pal, Dhiraj
Paulson, Glenn
Penrose, William Roy
Piwoni, Marvin Dennis
Prakasam, Tata B S
Reilly, Christopher Aloysius, Jr
Rodolfo, Kelvin S
Roy, Dipak
Schaeffer, David Joseph
Schaller, Robin Edward
Seitz, Wesley Donald
Shen, Sin-Yan
Shen, Sinyan
Singh, Shyam N
Smiciklas, Kenneth Donald
Sohr, Robert Trueman
Southern, William Edward
Stearner, Sigrid Phyllis
Stetter, Joseph Robert
Stratton, James Forrest
Streets, David George
Stull, Elisabeth Ann
Surgi, Marion Rene
Sytsma, Louis Frederick
Thomsen, Kurt O
Wadden, Richard Albert
Wahlgren, Morris A
Wang, Michael Quanlu
Wesolowski, Wayne Edward
Wetegrove, Robert Lloyd
Wingender, Ronald John
Wolff, Robert John
Woods, Kenneth R
Zar, Jerrold Howard
Zelac, Ronald Edward
Zenz, David R

INDIANA
Alliston, Charles Walter
Becker, Elizabeth Ann (White)
Born, Gordon Stuart
Bumpus, John Arthur
Carlson, Merle Winslow
Cassidy, Harold Gomes
Chowdhury, Dipak Kumar
Crawford, Thomas Michael
Dyer, Rolla McIntyre, Jr
Eberly, William Robert
Elmore, David
Hagen, Richard Eugene
Hites, Ronald Atlee
Hubble, Billy Ray
Hurst, Robert Nelson
Kulpa, Charles Frank, Jr
Landreth, Ronald Ray
Leap, Darrell Ivan
Lee, Linda Shahrabani
Leonard, Jack E
McFee, William Warren
Mengel, David Bruce
Michaud, Howard H
Miller, Richard William
Mirsky, Arthur
Nelson, Craig Eugene
Parkhurst, David Frank
Pichtel, John Robert
Pyle, James L
Racke, Kenneth David
Schaible, Robert Hilton
Speece, Susan Phillips
Squiers, John Richard
Voelz, Frederick
Widener, Edward Ladd, Sr
Winkler, Erhard Mario
Wolt, Jeffrey Duaine

IOWA
Bremner, John McColl
Carmichael, Robert Stewart
Drake, Lon David
Graham, Benjamin Franklin
Karlen, Douglas Lawrence
Kniseley, Richard Newman
Matteson, Patricia Claire
Nakato, Tatsuaki
Parkin, Gene F
Schnoor, Jerald L
Walters, James Carter
Watkins, Stanley Read

KANSAS
Braaten, David A
Grant, Stanley Cameron
Hoffman, Jerry C
McCullough, Elizabeth Ann
Pierson, David W
Platt, Dwight Rich
Schaper, Laurence Teis
Soodsma, James Franklin
Steeples, Donald Wallace
Surampalli, Rao Yadagiri
Welch, Jerome E
Wright, Donald C

KENTUCKY
Birge, Wesley Joe
Gauri, Kharaiti Lal
Hoffman, Wayne Larry
Jensen, Randolph A(ugust)
Kiefer, John David
Livingood, Marvin D(uane)
Musacchia, X J
Pearson, William Dean
Sarma, Atul C
Schneiderman, Steven S
Seay, Thomas Nash
Stanonis, Francis Leo
Stevenson, Robert Jan
Volp, Robert Francis
Wang, Yi-Tin

LOUISIANA
Bahr, Leonard M, Jr
Bundy, Kirk Jon
Cartledge, Frank
Curry, Mary Grace
DeLaune, Ronald D
DeLeon, Ildefonso R
Ferrell, Ray Edward, Jr
Flournoy, Robert Wilson
Fontenot, Martin Mayance, Jr
Loden, Michael Simpson
Mills, Earl Ronald
Poirrier, Michael Anthony
Robinson, James William
Sartin, Austin Albert
Seigel, Richard Allyn
Thibodeaux, Louis J
Thompson, James Joseph
Valsaraj, Kalliat Thazhathuveetil
Wiewiorowski, Tadeusz Karol
Wilson, John Thomas
Wrobel, William Eugene

MAINE
Chute, Robert Maurice
Cronan, Christopher Shaw
Cronn, Dagmar Rais
Eastler, Thomas Edward
Fitch, John Henry
Howd, Frank Hawver
Rock, Chet A
Snow, Joseph William
Taylor, Anthony Boswell
Tewhey, John David
Whitten, Maurice Mason

MARYLAND
Alter, Harvey
Asrar, Ghassam R
Barbour, Michael Thomas
Bauer, Ernest
Baummer, J Charles, Jr
Bausum, Howard Thomas
Berkson, Harold
Beroza, Morton
Bregman, Jacob Israel
Caret, Robert Laurent
Carton, Robert John
Chamberlain, David Leroy, Jr
Coble, Anna Jane
Colle, Ronald
Costanza, Robert
Cothern, Charles Richard
Creasia, Donald Anthony
Dashiell, Thomas Ronald
Dasler, Adolph Richard
Deitzer, Gerald Francis
De Planque, Gail
Englehart, Richard W(ilson)
Fanning, Delvin Seymour
Fink, Don Roger
Fisher, Thomas Richard, Jr
Foresti, Roy J(oseph), Jr
Fowler, Bruce Andrew
Friedman, Fred K
Gaffen, Dian Judith
Galler, Sidney Roland
Gavis, Jerome
Gerritsen, Jeroen
Gessner, Adolf Wilhelm
Gevantman, Lewis Herman
Gift, James J
Goktepe, Omer Faruk
Gould, Jack Richard
Graves, Willard L
Grotenhuis, Marshall
Gupta, Gian Chand
Guruswamy, Vinodhini
Heath, George A(ugustine)
Heath, Robert Gardner
Helz, George Rudolph
Herrett, Richard Allison
Hirschhorn, Joel S(tephen)
Hodge, Frederick Allen
Hopkins, Homer Thawley
Hubbard, Donald
Hunt, Charles Maxwell
Johnson, Charles C, Jr
Joiner, R(eginald) Gracen
Josephs, Melvin Jay
Katcher, David Abraham
Kelsey, Morris Irwin
Khare, Mohan
Kirkland, James T
Kline, Jerry Robert
Kraybill, Herman Fink
Krizek, Donald Thomas
Krumbein, Simeon Joseph
Kurylo, Michael John, III
Kushner, Lawrence Maurice
Larson, Reginald Einar
Lee, Edward Hsien-Chi

Levine, Elissa Robin
Liggett, Walter Stewart, Jr
Linder, Seymour Martin
McBryde, F(elix) Webster
McCammon, Helen Mary
McVey, James Paul
Matta, Joseph Edward
May, Ira Philip
Melancon, Mark J
Menez, Ernani Guingona
Mihursky, Joseph Anthony
Miller, Raymond Earl
Mogil, H Michael
Mountford, Kent
Mulchi, Charles Lee
Munasinghe, Mohan P
Murray, Laura
Muser, Marc
Myers, Lawrence Stanley, Jr
Nanzetta, Philip Newcomb
Newbury, Dale Elwood
Ostenso, Ned Allen
Panayappan, Ramanathan
Plumlee, Karl Warren
Pomerantz, Irwin Herman
Rankin, Sidney
Reichel, William Louis
Rosenblatt, David Hirsch
Ross, Philip
Row, Clark
Rubin, Robert Jay
Sansone, Eric Brandfon
Sarver, Emory William
Seifried, Harold Edwin
Sellner, Kevin Gregory
Shah, Shirish
Shykind, Edwin B
Smith, Elmer Robert
Sonawane, Babasaheb R
Spangler, Glenn Edward
Swift, David Leslie
Talbot, Donald R(oy)
Topping, Joseph John
Traystman, Richard J
Uhart, Michael Scott
Weir, Edward Earl, II
Weisberg, Stephen Barry
White, Harris Herman
Yakowitz, Harvey
Young, Alvin L
Young, George Anthony
Young, Jay Alfred
Zankel, Kenneth L

MASSACHUSETTS
Bade, Maria Leipelt
Barske, Philip
Bass, Arthur
Berlandi, Francis Joseph
Bertin, Robert Ian
Bowley, Donovan Robin
Busby, William Fisher, Jr
Butler, James Newton
Castro, Gonzalo
Chernosky, Edwin Jasper
Chisholm, Sallie Watson
Clark, William Cummin
Cohen, Merrill
Crandlemere, Robert Wayne
Dahm, Donald J
DeCosta, Peter F(rancis)
Dixon, Brian Gilbert
Ehrenfeld, John R(oos)
Epstein, Eliot
Eschenroeder, Alan Quade
Farrington, John William
Fay, James A(lan)
Feder, William Adolph
Fennelly, Paul Francis
Fessler, William Andrew
Field, Hermann Haviland
Fix, Richard Conrad
Foster, Charles Henry Wheelwright
Genes, Andrew Nicholas
Golay, Michael W
Gold, Harris
Graham, John D
Gregory, Constantine J
Hadlock, Charles Robert
Harvey, Walter William
Hoffman, Paul Felix
Horne, Ralph Albert
Horzempa, Lewis Michael
Hummel, John Richard
Jankowski, Conrad M
Jop, Krzysztof M
Kaplan, David Lee
Kazmaier, Harold Eugene
Keast, David N(orris)
Lajtha, Kate
Lee, Kai Nien
Licht, Stuart Lawrence
Long, Nancy Carol
Marini, Robert C
Marquis, Judith Kathleen
Moir, Ronald Brown, Jr
Murphy, Brian Logan
Nickerson, Norton Hart
Olmez, Ilhan
Ozonoff, David Michael
Pandolf, Kent Barry
Platt, John Rader
Pojasek, Robert B
Primack, Richard Bart

Prodany, Nicholas W
Ram, Neil Marshall
Ramakrishna, Kilaparti
Reckhow, David Alan
Redington, Charles Bahr
Reimold, Robert J
Rhoads, Donald Cave
Richards, F Paul
Schempp, Ellory
Stevens, J(ames) I(rwin)
Stinchfield, Carleton Paul
Trotz, Samuel Isaac
Turner, Jefferson Taylor
Wallace, Robert William
Wright, Richard T
Wun, Chun Kwun

MICHIGAN
Ben, Manuel
Bhatnagar, Lakshmi
Blanchard, Fred Ayres
Bone, Larry Irvin
Brewer, George E(ugene) F(rancis)
Buis, Patricia Frances
Burlington, Roy Frederick
Cadle, Steven Howard
Carpenter, Michael Kevin
Chang, Tai Yup
Cho, Byong Kwon
Chock, David Poileng
Crummett, Warren B
Daniels, Stacy Leroy
D'Itri, Frank M
Domke, Charles J
Farber, Hugh Arthur
Ficsor, Gyula
France, W(alter) DeWayne, Jr
Gardner, Wayne Stanley
Gates, David Murray
Gelderloos, Orin Glenn
Hahne, Rolf Mathieu August
Halberstadt, Marcel Leon
Hoerger, Fred Donald
Hough, Richard Anton
Jensen, David James
Johnson, Ray Leland
Kelly, Nelson Allen
Kerfoot, Wilson Charles
Kevern, Niles Russell
Kim, Byung Ro
Klempner, Daniel
Klimisch, Richard L
Kullgren, Thomas Edward
Kummler, Ralph H
Kunkle, George Robert
Landrum, Peter Franklin
McClelland, Nina Irene
Mallinson, George Greisen
Medlin, Julie Anne Jones
Miller, Robert Rush
Moolenaar, Robert John
Moore, Thomas Edwin
Nestrick, Terry John
Niese, Jeffrey Neal
Northup, Melvin Lee
Nriagu, Jerome Okon
O'Shea, Timothy Allan
Otto, Klaus
Powitz, Robert W
Prostak, Arnold S
Reeves, Andrew Louis
Reynolds, John Z
Robbins, John Alan
Rogers, Jerry Dale
Ronca, Luciano Bruno
Schuetzle, Dennis
Schwab, David John
Solomon, Allen M
Stevens, Violete L
Stoermer, Eugene F
Stynes, Stanley K
Tomboulian, Paul
Tuesday, Charles Sheffield
Walles, Wilhelm Egbert
Whitten, Bertwell Kneeland
Williams, Ronald Lloyde
Wolff, George Thomas
Wood, Jack Sheehan
Young, David Caldwell

MINNESOTA
Abeles, Tom P
Arthur, John W
Barber, Donald E
Berube, Robert
Bohon, Robert Lynn
Brezonik, Patrick Lee
Buchwald, Caryl Edward
Cheng, H(wei)-H(sien)
Clapp, C(harles) Edward
Dobbins, Durell Cecil
Eisenreich, Steven John
Gorham, Eville
Hagen, Donald Frederick
Henry, Mary Gerard
Krug, Edward Charles
Lande, Sheldon Sidney
McMurry, Peter Howard
McNaught, Donald Curtis
Olness, Alan
Oxborrow, Gordon Squires
Peck, John Hubert
Rapp, George Robert, Jr
Reynolds, James Harold

Richason, Benjamin Franklin, Jr
Roessler, Charles Ervin
Sautter, Chester A
Stefan, Heinz G
Swain, Edward Balcom
Thompson, Fay Morgen
Ward, Wallace Dixon
Wetmore, Clifford Major

MISSISSIPPI
Davis, Robert Gene
Greever, Joe Carroll
Hendricks, Donovan Edward
Kushlan, James A
Minyard, James Patrick
Overstreet, Robin Miles
Paulson, Oscar Lawrence
Saucier, Roger Thomas
Shields, Fletcher Douglas, Jr
Strom, Eric William
Wolverton, Billy Charles

MISSOURI
Armstrong, Daniel Wayne
Brahmbhatt, Sudhir R
Burst, John Frederick
Callis, Clayton Fowler
Carpenter, Will Dockery
Churchill, Ralph John
Cowherd, Chatten, Jr
Crocker, Burton B(lair)
De Buhr, Larry Eugene
Dietrich, Martin Walter
Eigner, Joseph
Frye, Charles Isaac
Gledhill, William Emerson
Godt, Henry Charles, Jr
Hallas, Laurence Edward
Hasan, Syed Eqbal
Henderson, Gray Stirling
Hewitt, Andrew
Holloway, Thomas Thornton
Holroyd, Louis Vincent
Hoogheem, Thomas John
Jungclaus, Gregory Alan
Kaley, Robert George, II
Klein, Andrew John
Kramer, Richard Melvyn
Kupchella, Charles E
Lawless, Edward William
Long, Lawrence William
Malik, Joseph Martin
Mieure, James Philip
Mori, Erik Jun
Murray, Robert Wallace
Raven, Peter Hamilton
Rawlings, Gary Don
Rueppel, Melvin Leslie
Saeger, Victor William
Schroy, Jerry M
Sheets, Ralph Waldo
Snowden, Jesse O
Soundararajan, Rengarajan
Stern, Daniel Henry
Stern, Michele Suchard
Swisher, Robert Donald
Vandegrift, Alfred Eugene
Vaughan, William Mace
Whitefield, Philip Douglas
Wilcox, Harold Kendall
Worley, Jimmy Weldon

MONTANA
Erickson, Ronald E
Knox, James L(ester)
Montagne, John M
Peavy, Howard Sidney
Temple, Kenneth Loren

NEBRASKA
Carson, Steven Douglas
Fairchild, Robert Wayne
Grew, Priscilla Croswell Perkins
Louda, Svata Mary
Ma, Wen
Maier, Charles Robert
Pekas, Jerome Charles
Phelps, George Clayton
Raun, Earle Spangler
Walsh, Gary Lynn

NEVADA
Barth, Delbert Sylvester
Bishop, William P
Deacon, James Everett
Dickson, Frank Wilson
Flueck, John A
Jobst, Joel Edward
Kinnison, Robert Ray
Krenkel, Peter Ashton
McElroy, James Lee
Meier, Eugene Paul
Miles, Maurice Jarvis
Miller, Watkins Wilford
Nauman, Charles Hartley
Pitter, Richard Leon
Poziomek, Edward John
Taranik, James Vladimir
Trexler, Dennis Thomas
Wiemeyer, Stanley Norton
Yousef, Mohamed Khalil

NEW HAMPSHIRE
Bernstein, Abram Bernard

Bilello, Michael Anthony
Blum, Joel David
Chamberlain, C Page
Danforth, Raymond Hewes
Farag, Ihab Hanna
Fogleman, Wavell Wainwright
Friedland, Andrew J
Frost, Terrence Parker
Jackson, Herbert William
Lynch, Daniel Roger
Petrasek, Emil John
Stepenuck, Stephen Joseph, Jr
Strohbehn, John Walter
Weber, James Harold

NEW JERSEY
Agron, Sam Lazrus
Barnes, Robert Lee
Booman, Keith Albert
Borowsky, Harry Herbert
Cheremisinoff, Paul N
Ciaccio, Leonard Louis
Colby, Richard H
Cooper, Keith Raymond
Crerar, David Alexander
Dalton, Thomas Francis
Dayan, Jason Edward
Delaney, Bernard T
Dobi, John Steven
Ekstrom, Lincoln
Falk, Charles David
Farrauto, Robert Joseph
Faust, Samuel Denton
Finstein, Melvin S
Fishman, David H
Frederick, James R
Greenberg, Arthur
Greenberg, Michael Richard
Hall, Herbert Joseph
Hall, Homer James
Holt, Vernon Emerson
Hrycak, Peter
Hundert, Irwin
Iglewicz, Raja
Jackson, Thomas A J
Kahn, Donald Jay
Keating, Kathleen Irwin
Kheshgi, Haroon S
Koehl, William John, Jr
Krey, Philip W
Krishnamurthy, Subrahmanya
Kukin, Ira
Leone, Ida Alba
Levine, Richard S
Lewellen, William Stephen
Lewis, Thomas Brinley
Ling, Hubert
Lioy, Paul James
London, Mark David
McGuire, David Kelty
Makofske, William Joseph
Manabe, Syukuro
Martin, John David
Menczel, Jehuda H
Morgan, Mark Douglas
Nicholls, Gerald P
Priesing, Charles Paul
Prince, Roger Charles
Ramaswamy, Venkatachalam
Rau, Eric
Rigler, Neil Edward
Rothstein, Edwin C(arl)
Rupp, Walter H(oward)
Salvesen, Robert H
Sarmiento, Jorge Louis
Schubert, Rudolf
Shaw, Henry
Skovronek, Herbert Samuel
Socolow, Robert H(arry)
Sugam, Richard Jay
Tamarelli, Alan Wayne
Tatyrek, Alfred Frank
Uchrin, Christopher George
Vallese, Frank M
Vanden Heuvel, William John Adrian, III
Van Pelt, Wesley Richard
Watts, Daniel Jay
Weinstein, Norman J(acob)
Weisberg, Joseph Simpson
Wenzel, John Thompson
Williams, Robert H
Wilson, Charles Elmer
Witkowski, Mark Robert
Wolfe, Peter E
Wood, Eric F
Yan, Tsoung-Yuan

NEW MEXICO
Barton, Larry Lumir
Bhada, Rohinton(Ron) K
Binder, Irwin
Bingham, Felton Wells
Brown, James Hemphill
Broxton, David Edward
Cheng, Yung-Sung
Clark, Ronald Duane
Close, Donald Alan
Darnall, Dennis W
Deal, Dwight Edward
Essington, Edward Herbert
Fries, James Andrew
Fritz, Georgia T(homas)
Guthrie, George D, Jr

Environmental Sciences, General (cont)

Hadley, William Melvin
Hansen, Wayne Richard
Henderson, Thomas Richard
Holloway, Richard George
Holmes, John Thomas
Hoover, Mark Douglas
Jones, Kirkland Lee
Kenna, Bernard Thomas
Khandan, Nirmala N
Kosiewicz, Stanley Timothy
Kuellmer, Frederick John
Laughlin, Alexander William
Lewis, James Vernon
McLemore, Virginia Teresa Hill
Mason, Caroline Faith Vibert
Merritt, Melvin Leroy
Mokler, Brian Victor
Niemczyk, Thomas M
Nirmalakhandan, Nagamany
Palmer, Byron Allen
Phillips, Fred Melville
Polzer, Wilfred L
Popp, Carl John
Reimann, Bernhard Erwin Ferdinand
Rohde, Richard Whitney
Sayre, William Olaf
Schery, Stephen Dale
Shore, Fred L
Smith, James Lewis
Stanbro, William David
Tapscott, Robert Edwin
Thompson, Joseph Lippard
Tisue, George Thomas
Wayland, James Robert, Jr
Williams, Mary Carol
Wilson, Lee
Wolberg, Donald Lester
Yeh, Hsu-Chi

NEW YORK
Abraham, Rajender
Alben, Katherine Turner
Altwicker, Elmar Robert
Andrus, Richard Edward
Babich, Harvey
Baer, Norbert Sebastian
Barber, Eugene Douglas
Bedford, Barbara Lynn
Berger, Selman A
Beyea, Jan Edgar
Boger, Phillip David
Bokuniewicz, Henry Joseph
Boylen, Charles William
Brandwein, Paul Franz
Bulloff, Jack John
Bush, David Graves
Cappel, C Robert
Carbone, Gabriel
Cardenas, Raúl R, Jr
Ceprini, Mario Q
Chadwick, George F(redrick)
Clarke, Raymond Dennis
Coch, Nicholas Kyros
Cohen, Beverly Singer
Cohen, Norman
Cole, Jonathan Jay
Collins, Anthony Gordon
Conant, Francis Paine
Coppersmith, Frederick Martin
Dahneke, Barton Eugene
Dietz, Russell Noel
DiGaudio, Mary Rose
Dille, Kenneth Leroy
Dondero, Norman Carl
Dooley, James Keith
Durst, Richard Allen
Edelstein, William Alan
Elder, Fred A
Fakundiny, Robert Harry
Feinberg, Robert Jacob
Fischman, Harlow Kenneth
Freudenthal, Hugo David
Garrell, Martin Henry
Gellman, Isaiah
Ginzburg, Lev R
Goldfarb, Theodore D
Gossett, James Michael
Gray, D Anthony
Grayson, Herbert G
Hairston, Nelson George, Jr
Hang, Yong Deng
Harley, Naomi Hallden
Haynes, James Mitchell
Hecht, Stephen Samuel
Henderson, Floyd M
Hennigan, Robert Dwyer
Henshaw, Robert Eugene
Hewitt, Philip Cooper
Hoffmann, Dietrich
Horvat, Robert Emil
Howarth, Robert W
Hunt, David Michael
Jackson, Kenneth William
Jacobson, Jay Stanley
Jenkins, Philip Winder
Jensen, Betty Klainminc
Jirka, Gerhard Hermann
Kennedy, Edwin Russell
Klanderman, Bruce Holmes
Klingele, Harold Otto
Kohut, Robert John
Konigsberg, Alvin Stuart
Koppelman, Lee Edward
Krieger, Gary Lawrence
Kumai, Motoi
Lasday, Albert Henry
Levandowsky, Michael
Lieberman, Arthur Stuart
Locke, David Creighton
Lockhart, Haines Boots, Jr
Loggins, Donald Anthony
Lovett, Gary Martin
Ludlum, Kenneth Hills
Lundgren, Lawrence William, Jr
Ma, Maw-Suen
McCune, Delbert Charles
McNeil, Richard Jerome
Makarewicz, Joseph Chester
Marschke, Stephen Frank
Millstein, Jeffrey Alan
Molloy, Andrew A
Muller, Ernest Hathaway
Nathanson, Benjamin
Nealy, Carson Louis
O'Keefe, Patrick William
Owens, Patrick M
Padnos, Norman
Pajari, George Edward
Palmer, James F
Pereira, Martin Rodrigues
Putman, George Wendell
Rae, Stephen
Rana, Mohammad A
Rifkind, Arleen B
Ring, James Walter
Robertson, William, IV
Robinson, Joseph Edward
Robinson, Myron
Sage, Gloria W
Salotto, Anthony W
Sambrotto, Raymond Nicholas
Sass, Daniel B
Savage, E Lynn
Scala, Sinclaire M(aximilian)
Schulz, Helmut Wilhelm
Serafy, D Keith
Sheeran, Stanley Robert
Shen, Thomas T
Siegart, William Raymond
Siegfried, Clifford Anton
Simons, Edward Louis
Smardon, Richard Clay
Smethie, William Massie, Jr
Smith, Thomas Harry Francis
Sternberg, Stephen Stanley
Sterrett, Frances Susan
Stotzky, Guenther
Toribara, Taft Yutaka
Weiss, Dennis
Weiss, Jonas
White, James Carrick
Wolff, Manfred Paul (Fred)
Wolfson, Leonard Louis
Yang, John Yun-Wen

NORTH CAROLINA
Andrady, Anthony Lakshman
Andrews, Richard Nigel Lyon
Baranski, Michael Joseph
Booth, Donald Campbell
Bruce, Richard Conrad
Bruck, Robert Ian
Bursey, Joan Tesarek
Carlson, William Theodore
Carpenter, Benjamin H(arrison)
Carson, Bonnie L Bachert
Chang, John Chung-Shih
Collins, Jeffrey Jay
Decker, Clifford Earl, Jr
Dement, John M
Elias, Robert William
Ellison, Alfred Harris
Ensor, David Samuel
Fischer, Janet Jordan
Flint, Elizabeth Parker
Fouts, James Ralph
Gardner, Donald Eugene
Gillespie, Arthur Samuel, Jr
Glaze, William H
Godschalk, David Robinson
Haile, Clarence Lee
Hamner, William Frederick
Harris, Robert L, Jr
Hartford, Winslow H
Haseman, Joseph Kyd
Heck, Walter Webb
Hess, Daniel Nicholas
Holland, Marjorie Miriam
Jayanty, R K M
Karl, Thomas Richard
Knapp, Kenneth T
Knight, Clifford Burnham
Kodavanti, Prasada Rao S
Kohl, Jerome
Kolenbrander, Lawrence Gene
Kuenzler, Edward Julian
Lao, Yan-Jeong
Lapp, Thomas William
Lawless, Philip Austin
LeBaron, Homer McKay
Leiserson, Lee
McKinney, James David
Maier, Robert Hawthorne
Malling, Heinrich Valdemar
Marcus, Allan H
Neher, Deborah A
Odman, Mehmet Talat
O'Neal, Thomas Denny
Patterson, David Thomas
Peirce, James Jeffrey
Pellizzari, Edo Domenico
Pfaender, Frederic Karl
Phelps, Allen Warner
Purchase, Earl Ralph
Qualls, Robert Gerald
Ross, Richard Henry, Jr
Ross, Thomas Edward
Sanoff, Henry
Scott, Donald Ray
Shafer, Steven Ray
Shendrikar, Arun D
Shreffler, Jack Henry
Shuman, Mark S
Singer, Philip C
Sneade, Barbara Herbert
Sobsey, Mark David
Sonnenfeld, Gerald
Stenger, William J(ames), Sr
Stern, A(rthur) C(ecil)
Stillwell, Harold Daniel
Sumner, Darrell Dean
Swift, Lloyd Wesley, Jr
Textoris, Daniel Andrew
Trenholm, Andrew Rutledge
Turner, Alvis Greely
Volpe, Rosalind Ann
Walsh, Edward John
Walters, Douglas Bruce
Weaver, Ervin Eugene
Webster, William David
Wilson, Lauren R

NORTH DAKOTA
Alkezweeny, Abdul Jabbar
Bickel, Edwin David
Ludlow, Douglas Kent

OHIO
Ayer, Howard Earle
Baasel, William David
Baumann, Paul C
Bendure, Robert J
Bimber, Russell Morrow
Bishop, Paul Leslie
Bleckmann, Charles Allen
Boerner, Ralph E J
Bohrman, Jeffrey Stephen
Burrows, Kerilyn Christine
Carfagno, Daniel Gaetano
Compton, Ell Dee
Disinger, John Franklin
Eckert, Donald James
Fan, Liang-Shih
Feairheller, William Russell, Jr
Fisher, Gerald Lionel
Forsyth, Jane Louise
Fortner, Rosanne White
Frank, Glenn William
Gieseke, James Arnold
Gordon, Gilbert
Grube, Walter E, Jr
Gruber, Charles W
Hannah, Sidney Allison
Harris, Richard Lee
Hess, George G
Hill, James Stewart
Innes, John Edwin
Kagen, Herbert Paul
Kallenberger, Waldo Elbert
Kawahara, Fred Katsumi
Keiser, Terry Dean
Kim, Byung Cho
Ku, Han San
Lollar, Robert Miller
Lustick, Sheldon Irving
Lyon, John Grimson
McClaugherty, Charles Anson
McKenzie, Garry Donald
McQuigg, Robert Duncan
Mao, Hsiang-Kuen Mark
Marble, Duane Francis
Marks, Alfred Finlay
Matisoff, Gerald
Miller, Harvey Alfred
Mitsch, William Joseph
Mukhtar, Hasan
Nobis, John Francis
Purvis, John Thomas
Ray, John Robert
Rubin, Alan J
Ruedisili, Lon Chester
Saraceno, Anthony Joseph
Scarborough, Vernon Lee
Scarpino, Pasquale Valentine
Shaw, Wilfrid Garside
Smith, Geoffrey W
Springer, Allan Matthew
Sticksel, Philip Rice
Stoldt, Stephen Howard
Sydnor, Thomas Davis
Taylor, Michael Lee
Thomas, Craig Eugene
Toeniskoetter, Richard Henry
Tuovinen, Olli Heikki
Utgard, Russell Oliver
Voigt, Charles Frederick
Weeks, Thomas Joseph, Jr
Whiting, Peter John
Yanko, William Harry
Zimmerman, Ernest Frederick

OKLAHOMA
Bryant, Rebecca Smith
Coleman, Nancy Pees
Crockett, Jerry J
Dorris, Troy Clyde
Eby, Harold Hildenbrandt
Enfield, Carl George
Fish, Wayne William
Fletcher, John Samuel
Hounslow, Arthur William
Johnson, Kenneth Sutherland
Koerner, E(rnest) L(ee)
Miller, Helen Carter
Miller, John Walcott
Monn, Donald Edgar
Mulholland, Robert J(oseph)
Nalley, Elizabeth Ann
Pollock, L(yle) W(illiam)
Puls, Robert W
Runnels, John Hugh
Shmaefsky, Brian Robert
Smith, Samuel Joseph
Summers, Gregory Lawson
Thurman, Lloy Duane
Van De Steeg, Garet Edward

OREGON
Amundson, Robert Gale
Buscemi, Philip Augustus
Chapman, Gary Adair
Church, Marshall Robbins
Dunnette, David Arthur
Ferraro, Steven Peter
Fish, William
Goodney, David Edgar
Higgins, Paul Daniel
Huntzicker, James John
Jarrell, Wesley Michael
Lubchenco, Jane
Neilson, Ronald Price
Pizzimenti, John Joseph
Reporter, Minocher C
Seyb, Leslie Philip
Siemens, Richard Ernest
Stones, Robert C
Thorson, Thomas Bertel
Tingey, David Thomas
Warkentin, Benno Peter
Williams, Evan Thomas
Wohlers, Henry Carl
Wolf, Marvin Abraham
Worrest, Robert Charles

PENNSYLVANIA
Andelman, Julian Barry
Aplan, Frank F(ulton)
Arnold, Dean Edward
Baurer, Theodore
Birchem, Regina
Bott, Thomas Lee
Brady, Keith Bryan Craig
Browning, Daniel Dwight
Bushnell, Kent O
Caruso, Sebastian Charles
Casey, Adria Catala
Chen, Audrey Weng-Jang
Cheng, Cheng-Yin
Claiborne, C Clair
Clark, Roger Alan
Cohen, Anna (Foner)
Cohen, Bernard Leonard
Commito, John Angelo
Crits, George J(ohn)
Cunningham, Virginia
Daniels-Severs, Anne Elizabeth
Denoncourt, Robert Francis
Doelp, Louis C(onrad), Jr
Dzombak, David Adam
Einhorn, Philip A
Emrich, Grover Harry
Fischhoff, Baruch
Fletcher, Frank William
Fletcher, Ronald D
Franz, Craig Joseph
Gardner, Thomas William
Gertz, Steven Michael
Giffen, Robert H(enry)
Glazier, Douglas Stewart
Goodman, Donald
Goodspeed, Robert Marshall
Graybill, Donald Lee
Gurol, Mirat D
Haas, Charles Nathan
Hogan, Aloysius Joseph, Jr
Hutchinson, Peter J
Keenan, John Douglas
Keller, Joseph Herbert
Khosah, Robinson Panganai
Kleinman, Robert L P
Komarneni, Sridhar
Ku, Peh Sun
Kurtz, David Allan
LaGrega, Michael Denny
Larkin, Robert Hayden
Levin, Michael H(oward)
Lochstet, William A
Lynch, Thomas John
McCarthy, Raymond Lawrence
MacDonald, Hubert C, Jr
McDonnell, Archie Joseph
McDonnell, Leo F(rancis)

McFarland, Mack
McGovern, John Joseph
Mann, Keith Olin
Maroulis, Peter James
Melamed, Sidney
Mellinger, Michael Vance
Meredith, David Bruce
Morgan, M(illett) Granger
Myron, Thomas L(eo)
Neufeld, Ronald David
Patil, Ganapati P
Pees, Samuel T(homas)
Pena, Jorge Augusto
Peterson, Jack Kenneth
Poss, Stanley M
Press, Linda Seghers
Puglia, Charles David
Rizza, Paul Frederick
Rohl, Arthur N
Rose, Arthur William
Rosenstock, Paul Daniel
Ruppel, Thomas Conrad
Saluja, Jagdish Kumar
Sawicki, John Edward
Schultz, Lane D
Seidell, Barbara Castens
Sharer, Cyrus J
Simonds, John Ormsbee
Sittenfield, Marcus
Smith, James Stanley
Snyder, Donald Benjamin
Sproull, Wayne Treber
Stanley, Edward Livingston
Stehly, David Norvin
Taylor, David Cobb
Unz, Richard F(rederick)
Venable, Emerson
Villafana, Theodore
Ward, Laird Gordon Lindsay
Wicker, Robert Kirk
Williams, Frederick McGee
Woodruff, Kenneth Lee
Yohe, Thomas Lester

RHODE ISLAND
Brungs, William Aloysius
Burroughs, Richard
Cain, J(ames) Allan
Donnelly, Grace Marie
Hermance, John Francis
Ho, Kay T
Kawaters, Woody H
Lazell, James Draper
McConeghy, Matthew H
Miller, Don Curtis
Poon, Calvin P C
Tarzwell, Clarence Matthew
Ward, Harold Roy

SOUTH CAROLINA
Abernathy, A(twell) Ray
Aelion, Claire Marjorie
Benjamin, Richard Walter
Blood, Elizabeth Reid
Camberato, James John
Carnes, Richard Albert
Caruccio, Frank Thomas
Conner, William Henry
Coulston, Mary Lou
Crawford, Todd V
Dame, Richard Franklin
Davis, Raymond F
Doig, Marion Tilton, III
Fowler, John Rayford
Franklin, Ralph E
Geidel, Gwendelyn
Goodroad, Lewis Leonard
Greene, George C, III
Groce, William Henry (Bill)
Hofstetter, Kenneth John
Holcomb, Herman Perry
Inhaber, Herbert
James, L(aurence) Allan
Jones, Ulysses Simpson, Jr
Lacher, Thomas Edward, Jr
Lee, Craig Chun-Kuo
McKellar, Henry Northington, Jr
Morris, J(ames) William
Morris, James T
Muska, Carl Frank
Newman, Michael Charles
Nichols, Chester Encell
Overcamp, Thomas Joseph
Swank, Robert Roy, Jr
Teska, William Reinhold
Vernberg, Winona B
Wang, Wun-Cheng W(oodrow)
White, Richard Kenneth
Wilhite, Elmer Lee
Wixson, Bobby Guinn

SOUTH DAKOTA
Berry, Charles Richard, Jr
Draeger, William Charles
Zogorski, John Steward

TENNESSEE
Auerbach, Stanley Irving
Baldocchi, Dennis D
Ball, Mary Uhrich
Bonner, William Paul
Boston, Charles Ray
Chung, King-Thom
Churnet, Habte Giorgis

Clarke, James Harold
Clauberg, Martin
Coburn, Corbett Benjamin, Jr
Cowser, Kenneth Emery
Eddlemon, Gerald Kirk
Elmore, James Lewis
Fulkerson, William
Garber, Robert William
Garten, Charles Thomas, Jr
Graveel, John Gerard
Haywood, Frederick F
Helweg, Otto Jennings
Jolley, Robert Louis
Jones, Larry Warner
Kathman, R Deedee
Koh, Sung-Cheol
Lee, Suk Young
Lindberg, Steven Eric
Loar, James M
McCarthy, John F
McDuffie, Bruce
McLaughlin, Samuel Brown
Marland, Gregg (Hinton)
Mullen, Michael David
Nyquist, Jonathan Eugene
Parker, Frank L
Pearson, Donald Emanual
Pfuderer, Helen A
Richmond, Chester Robert
Rodgers, Billy Russell
Roop, Robert Dickinson
Rose, Kenneth Alan
Row, Thomas Henry
Scroggie, Lucy E
Sharples, Frances Ellen
Sheth, Atul C
Smith, Michael James
Stow, Stephen Harrington
Thoma, Roy E
Tolbert, Virginia Rose
Turner, Robert Spilman
Van Hook, Robert Irving, Jr
Van Winkle, Webster, Jr
Vo-Dinh, Tuan
Webb, J(ohn) Warren
White, David Cleaveland
Witten, Alan Joel
Wohlfort, Sam Willis
Yarbro, Claude Lee, Jr

TEXAS
Abernathy, Bobby F
Albach, Roger Fred
Allen, Peter Martin
Applegate, Howard George
Armstrong, Neal Earl
Beran, Jo Allan
Bethea, Robert Morrison
Boyer, Lester Leroy
Briske, David D
Brooks, James Mark
Busch, Marianna Anderson
Camp, Frank A, III
Campbell, Michael David
Cares, William Ronald
Carlton, Donald Morrill
Chaney, Allan Harold
Chiang, Chen Y
Chrzanowski, Thomas Henry
Clark, William Jesse
Cocke, David Leath
Cole, Larry Lee
Convertino, Victor Anthony
Corcoran, Marjorie
Crawford, George Wolf
Cronkright, Walter Allyn, Jr
Daugherty, Kenneth E
Davis, Ernst Michael
Dillard, Robert Garing, Jr
Doyle, Frank Lawrence
Elder, Vincent Allen
Espey, Lawrence Lee
Estes, Frances Lorraine
Evans, John Edward
Finley, Robert James
Forest, Edward
French, William Edwin
Giammona, Charles P, Jr
Giese, Robert Paul
Gilmore, Robert Beattie
Girardot, Peter Raymond
Goodwin, John Thomas, Jr
Gore, Dorothy J
Grant, Clarence Lewis
Heinze, John Edward
Herbst, Richard Peter
Holliday, George Hayes
Howatson, John
Howe, Richard Samuel
Ickes, William K
Iyer, Ramakrishnan S
Johnson, Mary Lynn Miller
Keith, Lawrence H
Key, Morris Dale
Ledley, Tamara Shapiro
Lind, Owen Thomas
Loeppert, Richard Henry, Jr
Longley, Glenn, Jr
Louden, L Richard
McKee, Herbert C(harles)
Maki, Alan Walter
Massey, Michael John
Melia, Michael Brendan
Merchant, Michael Edward

Miller, James Richard
Morrow, Norman Louis
Mullineaux, Richard Denison
O'Connor, Rod
Parate, Natthu Sonbaji
Patel, Bhagwandas Mavjibhai
Pearson, Daniel Bester, III
Pirie, Robert Gordon
Pitts, David Eugene
Powell, Michael Robert
Powell, Richard James
Price, William Charles
Purser, Paul Emil
Ray, James P
Reaser, Donald Frederick
Reese, Weldon Harold
Rennie, Thomas Howard
Rhodes, John Rathbone
Rodriguez, Charles F
Rosenthal, Michael R
Rozas, Lawrence Paul
Rudenberg, Frank Hermann
Smith, Joseph Patrick
Smith, Michael Alexis
Soloyanis, Susan Constance
Standford, Geoffery Brian
Stanford, Geoffery
Steele, James Harlan
Sterner, Robert Warner
Sullivan, Arthur Lyon
Thompson, Richard John
Trieff, Norman Martin
Urdy, Charles Eugene
Warren, Kenneth Wayne
Watterston, Kenneth Gordon
Whiteside, Charles Hugh
Williams, Calvit Herndon, Jr
Wilson, Bobby L
Wright, James Foley
Zlatkis, Albert

UTAH
Bishop, Jay Lyman
Bollinger, William Hugh
Brierley, Corale Louise
Carey, John Hugh
Cerling, Thure E
Dorward-King, Elaine Jay
Dueser, Raymond D
Gates, Henry Stillman
Ghosh, Sambhunath
Hill, Archie Clyde
Israelsen, C Earl
Jewell, Paul William
Jurinak, Jerome Joseph
Muir, Melvin K
Rogers, Vern Child
Rothenberg, Mortimer Abraham
Sidle, Roy Carl
Sigler, John William
Wood, O Lew

VERMONT
Abajian, Paul G
Douglass, Irwin Bruce
Dritschilo, William
Gregory, Robert Aaron
Hall, Ross Hume
Keenan, Robert Gregory
Kneip, Theodore Joseph
Linnell, Robert Hartley
Pool, Edwin Lewis
Provost, Ronald Harold
Trombulak, Stephen Christopher
VanderMeer, Canute
Westing, Arthur H

VIRGINIA
Alden, Raymond W, III
Baedecker, Mary Jo
Bailey, Susan Goodman
Bass, Michael Lawrence
Batra, Karam Vir
Bender, Michael E
Bennett, Gordon Daniel
Benton, Duane Marshall
Brewer, Dana Alice
Briggs, Jeffrey L
Britton, Maxwell Edwin
Brown, Richard Dee
Carpenter, Richard A
Carstea, Dumitru
Christian, Jack G
Cochran, Thomas B
Conn, William David
Cutter, Gregory Allan
Davis, William Spencer
English, Bruce Vaughan
Exton, Reginald John
Fisher, Farley
Gager, Forrest Lee, Jr
Gaines, Alan McCulloch
Galloway, James Neville
Goodell, Horace Grant
Gould, David Huntington
Grunder, Fred Irwin
Haque, Rizwanul
Heath, Alan Gard
Hemingway, Bruce Sherman
Henderson, Ulysses Virgil, Jr
Holtzman, Golde Ivan
Hufford, Terry Lee
Jennings, Allen Lee
Johnston, Pauline Kay

Jordan, Wade H(ampton), Jr
Jurinski, Neil B(ernard)
Lamb, James C(hristian), III
Landa, Edward Robert
Langford, Russell Hal
Lee, James A
Le Febvre, Edward Ellsworth
Leung, Wing Hai
Lewis, Cornelius Crawford
Lipnick, Robert Louis
Ludwig, George H
McCormick, Michael Patrick
McCormick-Ray, M Geraldine
McEwen, Robert B
McKenna, John Dennis
Mahoney, Bernard Launcelot, Jr
Martens, William Stephen
Miller, William Lawrence
Munday, John Clingman, Jr
Myers, Thomas DeWitt
Neiheisel, James
Ney, Ronald E, Jr
Olem, Harvey
O'Neill, Eileen Jane
Parrish, David Joe
Raymond, Dale Rodney
Roberts, David Craig
Ross, Malcolm
Rowe, William David
Rule, Joseph Houston
Sahli, Brenda Payne
Sayala, Dasharatham (Dash)
Schneider, Robert
Seipel, John Howard
Sellers, Cletus Miller, Jr
Shelkin, Barry David
Sikri, Atam P
Spofford, Walter O, Jr
Talbot, Lee Merriam
Trout, Dennis Alan
Van der Vink, Gregory E
Wallace, Lance Arthur
Willis, Lloyd L, II
Wong, George Tin Fuk
Zelazny, Lucian Walter
Zieman, Joseph Crowe, Jr

WASHINGTON
Adams, William Mansfield
Asher, William Edward
Beug, Michael William
Booth, Pieter Nathaniel
Brewer, William Augustus
Brown, George Willard, Jr
Brown, Robert Alan
Cannon, William Charles
Cellarius, Richard Andrew
Cormack, James Frederick
Curl, Herbert (Charles), Jr
Darley, Ellis Fleck
Droppo, James Garnet, Jr
Drumheller, Kirk
Duncan, Leonard Clinton
Easterbrook, Don J
Friedrich, Donald Martin
Fruchter, Jonathan S
Fulmer, Charles V
Goheen, Steven Charles
Gray, Robert H
Hanson, Wayne Carlyle
Hardy, John Thomas
Hill, Orville Farrow
Hinman, George Wheeler
Karr, James Richard
Keller, C Kent
King, Larry Gene
Knotek, Michael Louis
Kolb, James A
Lansinger, John Marcus
Malik, Sohail
Mayer, Julian Richard
Moore, Emmett Burris, Jr
Mopper, Kenneth
Palmquist, Robert Clarence
Pessl, Fred, Jr
Phillips, Richard Dean
Pool, Karl Hallman
Price, Keith Robinson
Proctor, Charles Mahan
Rosengreen, Theodore E
Scheffer, Victor B
Schermer, Eugene DeWayne
Sehmel, George Albert
Serne, Roger Jeffrey
Shin, Suk-han
Sklarew, Deborah S
Sloane, Christine Scheid
Stevens, Todd Owen
Stewart, Donald Borden
Stottlemyre, James Arthur
Stout, Virginia Falk
Strand, John A, III
Thompson, James Arthur
VanBlaricom, Glenn R
Vaughan, Burton Eugene
Wood, Donald Eugene

WEST VIRGINIA
Conway, Richard A
Elmes, Gregory Arthur
Sandridge, Robert Lee

WISCONSIN
Becker, Edward Brooks

Environmental Sciences, General (cont)

Bennett, James Peter
Bisgaard, Søren
Borowiecki, Barbara Zakrzewska
Bringer, Robert Paul
Brooks, Arthur S
Ciriacks, John A(lfred)
Curtin, Charles Gilbert
Doshi, Mahendra R
Dutton, Herbert Jasper
Gleiter, Melvin Earl
Grunewald, Ralph
Hedden, Gregory Dexter
Heggestad, Howard Edwin
Horton, Joseph William
Hynek, Robert James
Jayne, Jack Edgar
Johnson, Barry Lee
Johnson, Wendel J
Jordan, William R, III
Minnich, John Edwin
Moore, Leonard Oro
Morgan, Michael Dean
Morton, Stephen Dana
Muto, Peter
Parnell, Donald Ray
Peace, George Earl, Jr
Reinsel, Gregory Charles
Remsen, Charles C, III
Saryan, Leon Aram
Trainer, Daniel Olney
Trischan, Glenn M
Willis, Harold Lester

WYOMING
Gloss, Steven Paul
Hoffman, Edward Jack
Katta, Jayaram Reddy
Poulson, Richard Edwin
Vance, George Floyd

PUERTO RICO
Carrasquillo, Arnaldo
Infante, Gabriel A
Lopez, Jose Manuel
Rodriguez, Luis F
Sasscer, Donald S(tuart)
Toranzos, Gary Antonio

ALBERTA
Agbeti, Michael Domingo
Apps, Michael John
Barendregt, Rene William
Fedorak, Phillip Michael
Feng, Joseph C
Francis, Mike McD
Gong, Peng
Janzen, Helmut Henry
Kariel, Herbert G
Lane, Robert Kenneth
Mackay, William Charles
Major, David John
Seghers, Benoni Hendrik
Smith, Daniel Walter
Tollefson, Eric Lars
Vitt, Dale Hadley
Weale, Gareth Pryce

BRITISH COLUMBIA
Arnold, Ralph Gunther
Bunce, Hubert William
Clark, Malcolm John Roy
Davis, Roderick Leigh
Hayward, John S
Hocking, Martin Blake
Lewis, Alan Graham
Macdonald, Robie Wilton
Mitchell, Reginald Harry
Oliver, Barry Gordon
Peterman, Randall Martin
Runeckles, Victor Charles
Shaw, A(lexander) J(ohn)
Vermeer, Kees
Walker, Ian Richard
Weaver, Ralph Sherman

MANITOBA
Amiro, Brian Douglas
Jovanovich, Jovan Vojislav
Rosenberg, David Michael
Rosinger, Eva L J
Saluja, Preet Pal Singh
Sheppard, Marsha Isabell
Sheppard, Stephen Charles
Smil, Vaclav
Sze, Yu-Keung
Tait, John Charles
Vandergraaf, Tjalle T
Wikjord, Alfred George

NEW BRUNSWICK
Fairchild, Wayne Lawrence
Lakshminarayana, J S S
Lin, Kwan-Chow
Paim, Uno

NEWFOUNDLAND
Georghiou, Paris Elias
Hellou, Jocelyne

NOVA SCOTIA
Bridgeo, William Alphonsus
Brown, Richard George Bolney
Buckley, Dale Eliot
Chatt, Amares
Freeman, Harry Cleveland
Stiles, David A
Vandermeulen, John Henri
Warman, Philip Robert

ONTARIO
Adamek, Eduard Georg
Afghan, Baderuddin Khan
AL-Hashimi, Ahmad
Allen, Gordon Ainslie
Belanger, Jacqueline M R
Bidleman, Terry Frank
Birmingham, Brendan Charles
Biswas, Asit Kumar
Bolton, James R
Brown, Robert Melbourne
Buckner, Charles Henry
Burton, Ian
Carlisle, David Brez
Cedar, Frank James
Chua, Kian Eng
Coakley, John Phillip
Csillag, Ferenc
Dillon, Peter J
Dredge, Lynda Ann
Effer, W R
Ewins, Peter J
Ford, Derek Clifford
Fortescue, John A C
Fyles, John Gladstone
Gillham, Robert Winston
Grant, Douglas Roderick
Green, Roger Harrison
Greenhalgh, Roy
Hare, Frederick Kenneth
Hill, Patrick Arthur
Hopton, Frederick James
Houston, Arthur Hillier
Ingraham, Thomas Robert
Jackson, Togwell Alexander
Jones, Philip Arthur
Jorgensen, Erik
Keddy, Paul Anthony
Kelso, John Richard Murray
Kesik, Andrzej B
Krantzberg, Gail
Lawrence, John
Lee, David Robert
Leppard, Gary Grant
Lesage, Suzanne
Liu, Dickson Lee Shen
McAdie, Henry George
McBean, Edward A
Mack, Alexander Ross
McKenney, Donald Joseph
Maguire, Robert James
Makhija, Ramesh C
Matheson, De Loss H(ealy)
Meresz, Otto
Morley, Harold Victor
Moroz, William James
Morris, Larry Arthur
Nevitt, H J Barrington
Nicholls, Kenneth Howard
Nirdosh, Inderjit
Osborne, Richard Vincent
Patel, Girishchandra Babubhai
Philogene, Bernard J R
Quigley, Robert Murvin
Roots, Ernest Frederick
Ruel, Maurice M J
Schroeder, William Henry
Singer, Eugen
Singh, Surinder Shah
Slawson, Peter (Robert)
Slocombe, Donald Scott
Smol, John Paul
Stairs, Robert Ardagh
Stebelsky, Ihor
Strachan, William Michael John
Sturgeon, Ralph Edward
Sundaram, K M Somu
Svoboda, Josef
Tang, You-Zhi
Topp, Edward
VandenHazel, Bessel J
Warith, Mostafa Abdel
Warner, Barry Gregory
Whelpdale, Douglas Murray
Wilkinson, Thomas Preston
Williams, David Dudley
Williams, David Trevor
Williams, George Ronald
Wolkoff, Aaron Wilfred
Wonnacott, Thomas Herbert
Yang, Paul Wang
Young, David Matheson

QUEBEC
Angers, Denis Arthur
Barthakur, Nayana N
Beland, Pierre
Bonn, Ferdinand J
Bouchard, Michel Andre
Campbell, Peter G C
Chagnon, Jean Yves
Cooper, David Gordon
Couillard, Denis
Desjardins, Claude W
Dubois, Jean-Marie M
Gangloff, Pierre
Kalff, Jacob
Kingsley, Michael Charles Stephen
Lalancette, Jean-Marc
Paulin, Gaston (Ludger)
Prescott, Jacques
Shaw, Robert Fletcher
Tessier, Andre
Tyagi, Rajeshwar Dayal
Yong, Raymond N

SASKATCHEWAN
Cassidy, Richard Murray
Chew, Hemming
Forsyth, Douglas John
Huang, P M
Tanino, Karen Kikumi
Viraraghavan, Thiruvenkatachari
Waddington, John

OTHER COUNTRIES
Aoyama, Isao
Bloom, Arthur David
Brydon, James Emerson
Carney, Gordon C
Chen, Chen-Tung Arthur
Dahl, Arthur Lyon
Davis, Edward Alex
Dettmann, Mary Elizabeth
Grigoropoulos, Sotirios G(regory)
Hutzinger, Otto
Jokl, Miloslav Vladimir
Lalas, Demetrius P
Lieth, Helmut Heinrich Friedrich
Mou, Duen-Gang
Rabl, Ari
Raveendran, Ekarath
Richmond, Robert H
Stam, Jos
Stoeber, Werner
Tsuda, Roy Toshio
Wenclawiak, Bernd Wilhelm

Fuel Technology & Petroleum Engineering

ALABAMA
Mancini, Ernest Anthony

ALASKA
Ostermann, Russell Dean

CALIFORNIA
Abdel-Baset, Mahmoud B
Bahn, Gilbert S(chuyler)
Baugh, Ann Lawrence
Blumstein, Carl Joseph
Brown, Glenn A
Brownscombe, Eugene Russell
Compton, Leslie Ellwyn
Epperly, W Robert
Hill, Richard William
Lewis, Robert Allen
Long, H(ugh) M(ontgomery)
Nixon, Alan Charles
Nur, Amos M
Rainis, Andrew
Schmidt, Raymond LeRoy
Shrontz, John William
Sifferman, Thomas Raymond
Silcox, William Henry
Sinton, Steven Williams
Sparling, Rebecca Hall
Uhl, Arthur E(dward)
Wright, Charles Cathbert
Yen, Teh Fu

COLORADO
Fischer, William Henry
Giulianelli, James Louis
Gogarty, W(illiam) B(arney)
Montgomery, Robert L
Osmond, John C, Jr
Pickett, George R
Pitts, Malcolm John
Smith, Glenn S
Vogt, Thomas Clarence, Jr

CONNECTICUT
Bilhorn, John Merlyn
Cobern, Martin E
Gupta, Dharam Vir

DELAWARE
Cantwell, Edward N(orton), Jr

DISTRICT OF COLUMBIA
Brown, Norman Louis
Melickian, Gary Edward
Neihof, Rex A
Ramsey, Jerry Warren
Shrier, Adam Louis
Varga, Gideon Michael, Jr

FLORIDA
Nersasian, Arthur

GEORGIA
Stelson, Arthur Wesley
Woodall, William Robert, Jr

IDAHO
Maloof, Giles Wilson

ILLINOIS
Bickelhaupt, R(oy) E(dward)
Buchanan, David Hamilton
Chou, Chen-Lin
Gorody, Anthony Wagner
Harrington, Joseph Anthony
Kanter, Manuel Allen
Kruse, Carl William
Kukes, Simon G
Marr, William Wei-Yi
Masel, Richard Isaac
Simon, Jack Aaron
Singh, Shyam N
Wilson, Gerald Gene

INDIANA
Gibson, Harry Gene
Shankland, Rodney Veeder
Stehouwer, David Mark

KENTUCKY
Cobb, James C
Huffman, Gerald P

LOUISIANA
Bourgoyne, Adam T, Jr
Forman, McLain J
Jones, Paul Hastings
Ledford, Thomas Howard
Neel, Thomas H
Sartin, Austin Albert
Schucker, Robert Charles

MARYLAND
Becker, Donald Arthur
Gessner, Adolf Wilhelm

MASSACHUSETTS
Frey, Elmer Jacob
Johnson, Stephen Allen
Robbat, Albert, Jr
Tester, Jefferson William

MICHIGAN
Furey, Robert Lawrence
Henein, Naeim A
Thompson, Paul Woodard
Tuesday, Charles Sheffield

MISSISSIPPI
George, Clifford Eugene
Knight, Wilbur Hall

MISSOURI
Goebel, Edwin DeWayne
Soundararajan, Rengarajan

NEVADA
Kukal, Gerald Courtney

NEW JERSEY
Blackburn, Gary Ray
Brenner, Douglas
Chen, Nai Y
Farrington, Thomas Allan
Hundert, Irwin
Jacolev, Leon
Mitchell, Thomas Owen
Salvesen, Robert H
Schutz, Donald Frank

NEW MEXICO
Bowsher, Arthur LeRoy
Hill, Mary Rae
Northrop, David A(mos)
Smith, James Lewis
Taber, Joseph John
Williams, Joel Mann, Jr

NEW YORK
Douglas, Craig Carl
Forero, Enrique
Happel, John
Nealy, Carson Louis
Pannella, Giorgio
Sotirchos, Stratis V
Virk, Kashmir Singh

NORTH CAROLINA
O'Connor, Lila Hunt
Quisenberry, Karl Spangler, Jr
Smith, Norman Cutler

NORTH DAKOTA
Harris, Steven H

OHIO
Bittker, David Arthur
Burow, Duane Frueh
Hall, Ronald Henry
Lewis, Irwin C
Mezey, Eugene Julius
Murphy, Michael John
Putnam, Abbott (Allen)
Stoldt, Stephen Howard

OKLAHOMA
Albright, James Curtice
Blais, Roger Nathaniel
Boade, Rodney Russett
Bryant, Rebecca Smith

Ellington, Rex T(ruesdale), Jr
Hurn, R(ichard) W(ilson)
Hyne, Norman John
Joshi, Sadanand D
Lorenz, Philip Boalt
Madden, Michael Preston
Menzie, Donald E
Pober, Kenneth William
Sloan, Norman Grady
Yeh, Nai-Shyong

PENNSYLVANIA
Aplan, Frank F(ulton)
Chigier, Norman
Durante, Vincent Anthony
Friedman, Sidney
Holder, Gerald D
Kenney, William Lawrence
Lovell, Harold Lemuel
Natoli, John
Nishioka, Masaharu
Noceti, Richard Paul
Nolan, Edward J
Retcofsky, Herbert L
Soung, Wen Y
Wicker, Robert Kirk

TENNESSEE
Guerin, Michael Richard
Rodgers, Billy Russell
Singh, Suman Priyadarshi Narain

TEXAS
Baker, Donald Roy
Barrow, Thomas D
Brown, James Michael
Campbell, William M
Closmann, Philip Joseph
Crawford, Duane Austin
Crouse, Philip Charles
Desai, Kantilal Panachand
DiFoggio, Rocco
Dorfman, Myron Herbert
Earlougher, Robert Charles, Jr
Font, Robert G
James, Bela Michael
Lacy, Lewis L
Leventhal, Stephen Henry
Levien, Louise
Lewis, James Pettis
Massell, Wulf F
Maute, Robert Edgar
Meyer, W(illiam) Keith
Milliken, Spencer Rankin
Mortada, Mohamed
Murphy, Daniel L
Nabor, George W(illiam)
Nussmann, David George
Prats, Michael
Randolph, Philip L
Rodriguez, Charles F
Roebuck, Isaac Field
Siegfried, Robert Wayne, II
Souby, Armand Max
Sprunt, Eve Silver
Stead, Frederick L
Thomas, Estes Centennial, III
Van Rensburg, Willem Cornelius Janse
Witterholt, Edward John

UTAH
Bartholomew, Calvin Henry
Comeford, John J(ack)
Ryan, Norman W(allace)

VIRGINIA
Bartis, James Thomas
Everett, Warren S
Gordon, John S(tevens)
Hudson, Frederick Mitchell
Sikri, Atam P

WASHINGTON
LaFontaine, Thomas E
Spencer, Arthur Milton, Jr

WEST VIRGINIA
Lamey, Steven Charles
Strickland, Larry Dean

WISCONSIN
Steinhart, John Shannon

WYOMING
Branthaver, Jan Franklin
Meyer, Edmond Gerald
Miknis, Francis Paul
Poulson, Richard Edwin
Robinson, Wilbur Eugene

ALBERTA
Campbell, John Duncan
Collins, David Albert Charles
Fedorak, Phillip Michael
Feingold, Adolph
Goodarzi, Fariborz
Overfield, Robert Edward
Tollefson, Eric Lars
Wiggins, Ernest James

BRITISH COLUMBIA
Bustin, Robert Marc

NEWFOUNDLAND
King, Michael Stuart

NOVA SCOTIA
Lynch, Brian Maurice

ONTARIO
Gussow, W(illia)m C(arruthers)
Walsh, John Heritage
Williams, Peter J

QUEBEC
Leigh, Charles Henry

OTHER COUNTRIES
Raveendran, Ekarath

Geochemistry

ALABAMA
Chalokwu, Christopher Iloba
Cook, Robert Bigham, Jr
Griffin, Robert Alfred
Isphording, Wayne Carter
Lyons, William Berry

ALASKA
Barsdate, Robert John
Hawkins, Daniel Ballou
Karl, Susan Margaret
Mowatt, Thomas C
Swanson, Samuel Edward

ARIZONA
Boynton, William Vandegrift
Burt, Donald McLain
Buseck, Peter R
Chase, Clement Grasham
Damon, Paul Edward
Drake, Michael Julian
El Wardani, Sayed Aly
Fisher, Frederick Stephen
Ganguly, Jibamitra
Graf, Donald Lee
Guilbert, John M
Haynes, Caleb Vance, Jr
Holloway, John Requa
Jull, Anthony John Timothy
Kamilli, Diana Chapman
Kamilli, Robert Joseph
Knowlton, Gregory Dean
Larimer, John William
Learned, Robert Eugene
Lewis, John Simpson
Long, Austin
McMillan, Paul Francis
Moore, Carleton Bryant
Nagy, Bartholomew Stephen
Norton, Denis Locklin
Robertson, Frederick Noel
Swindle, Timothy Dale
Titley, Spencer Rowe

ARKANSAS
Nix, Joe Franklin
Steele, Kenneth F

CALIFORNIA
Albee, Arden Leroy
Allan, John Ridgway
Arth, Joseph George, Jr
Austin, Steven Arthur
Bacastow, Robert Bruce
Bada, Jeffrey L
Bailey, Roy Alden
Bass, Manuel N
Beaty, David Wayne
Bernstein, Lawrence Richard
Bischoff, James Louden
Boettcher, Arthur Lee
Bohlen, Steven Ralph
Brimhall, George H
Brown, Gordon Edgar, Jr
Bubeck, Robert Clayton
Bukowinski, Mark S T
Burnett, Donald Stacy
Burnham, Alan Kent
Burtner, Roger Lee
Cain, William F
Carmichael, Ian Stuart
Carpenter, Alden B
Carr, Michael H
Chapman, Robert Mills
Chodos, Arthur A
Chow, Tsaihwa James
Christiansen, Robert Lorenz
Coe, Robert Stephen
Cohen, Lewis H
Conomos, Tasso John
Craig, Harmon
Cruikshank, Dale Paul
Czamanske, Gerald Kent
Dalrymple, Gary Brent
Davidson, Jon Paul
Day, Howard Wilman
DePaolo, Donald J
Des Marais, David John
Dignon, Jane
Einaudi, Marco Tullio
Epstein, Samuel
Ernst, Wallace Gary
Fedder, Steven Lee
Fournier, Robert Orville

Garlick, George Donald
Gerlach, John Norman
Glassley, William Edward
Goldberg, Edward D
Hanan, Barry Benton
Hein, James R
Helgeson, Harold Charles
Hem, John David
Housley, Robert Melvin
Hurst, Richard William
Irwin, James Joseph
Isherwood, Dana Joan
Jacobson, Stephen Richard
Jeanloz, Raymond
Kaplan, Issac R
Kastner, Miriam
Keeling, Charles David
Keith, Terry Eugene Clark
Kennedy, Burton Mack
Kennedy, Vance Clifford
Kharaka, Yousif Khoshu
Kothny, Evaldo Luis
Kramer, John Howard
Krauskopf, Konrad Bates
Ku, Teh-Lung
Kvenvolden, Keith Arthur
Lal, Devendra
Lanphere, Marvin Alder
Lind, Carol Johnson
Liou, Juhn G
Lugmair, Guenter Wilhelm
Macdougall, John Douglas
McKenzie, William F
Magoon, Leslie Blake, III
Mahood, Gail Ann
Majmundar, Hasmukhrai Hiralal
Manning, Craig Edward
Meike, Annemarie
Merrin, Seymour
Metzger, Albert E
Michalowski, Joseph Thomas
Moiseyev, Alexis N
Moore, James Gregory
Nelson, Robert M
Nielson, Jane Ellen
Niemeyer, Sidney
Papanastassiou, Dimitri A
Parks, George A(lbert)
Patterson, Clair Cameron
Pierce, Matthew Lee
Piper, David Zink
Pray, Ralph Emerson
Rau, Gregory Hudson
Redfield, William David
Reynolds, John Hamilton
Rosenfeld, John L
Russ, Guston Price, III
Saleeby, Jason Brian
Sarna-Wojcicki, Andrei M
Silver, Leon Theodore
Silverman, Sol Robert
Smalley, Robert Gordon
Smith, Robert Leland
Snetsinger, Kenneth George
Spera, Frank John
Sposito, Garrison
Stebbins, Jonathan F
Stensrud, Howard Lewis
Stolper, Edward Manin
Stonehouse, Harold Bertram
Taylor, Hugh P, Jr
Thompson, John Michael
Tilling, Robert Ingersoll
Tilton, George Robert
Turco, Richard Peter
van de Kamp, Peter Cornelis
Warnke, Detlef Andreas
Warren, Paul Horton
Wasson, John Taylor
Webster, Clyde Leroy, Jr
White, Donald Edward
Williams, Alan Evan
Williams, Peter M
Williams, Quentin Christopher
Wyllie, Peter John
Ziegler, Carole L

COLORADO
Benson, Larry V
Berger, Byron Roland
Bisque, Ramon Edward
Byers, Frank Milton, Jr
Cadigan, Robert Allen
Chaffee, Maurice A(hlborn)
Chao, Tsun Tien
Chiou, Cary T(sair)
Christensen, Odin Dale
Christie, Joseph Herman
Church, Stanley Eugene
Clark, Benton C
Cowgill, Ursula Moser
Dean, Walter E, Jr
Dembicki, Harry, Jr
Downs, William Fredrick
Dubiel, Russell F
Edwards, Kenneth Ward
Erslev, Eric Allan
Ferguson, William Sidney
Ferris, Bernard Joe
Friedman, Irving
Griffitts, Wallace Rush
Heyl, Allen Van, Jr
Hill, Walter Edward, Jr
Hills, Francis Allan

Hinners, Noel W
Hughes, Travis Hubert
Keighin, Charles William
Kenny, Ray
Klusman, Ronald William
Knight, William Glenn
Langmuir, Donald
McAuliffe, Clayton Doyle
McCallum, Malcolm E
Mason, Malcolm
Miesch, Alfred Thomas
Miller, William Robert
Modreski, Peter John
Nash, J(ohn) Thomas
Nordstrom, Darrell Kirk
Olhoeft, Gary Roy
Palacas, James George
Patton, James Winton
Rall, Raymond Wallace
Raup, Omer Beaver
Reitsema, Robert Harold
Romberger, Samuel B
Rouse, George Elverton
Runnells, Donald DeMar
Rye, Robert O
Sanford, Richard Frederick
Selverstone, Jane Elizabeth
Severson, Ronald Charles
Sidder, Gary Brian
Sommerfeld, Richard Arthur
Szabo, Barney Julius
Tatsumoto, Mitsunobu
Vine, James David
Wahl, Floyd Michael
Wershaw, Robert Lawrence
Wildeman, Thomas Raymond
Yang, In Che
Young, Edward Joseph
Zartman, Robert Eugene

CONNECTICUT
Berner, Robert A
Herron, Michael Myrl
Lasaga, Antonio C
Megrue, George Henry
Rye, Danny Michael
Skinner, Brian John
Turekian, Karl Karekin

DELAWARE
Allen, Herbert E
Culberson, Charles Henry
Leavens, Peter Backus
Luther, George William, III
Sharp, Jonathan Hawley

DISTRICT OF COLUMBIA
Carlson, Richard Walter
Clarke, Roy Slayton, Jr
Dasch, Ernest Julius
Fleischer, Michael
Fredriksson, Kurt A
Fudali, Robert F
Hare, Peter Edgar
Hoering, Thomas Carl
Lawson, Charles Alden
Mao, Ho-Kwang
Mason, Brian Harold
Mysen, Bjorn Olav
Price, Jonathan Greenway
Rumble, Douglas, III
Siegel, Frederic Richard
Usselman, Thomas Michael
Wright, Thomas Oscar

FLORIDA
Braids, Olin Capron
Burnett, William Craig
Compton, John S
Delfino, Joseph John
Fanning, Kent Abram
Fourqurean, James Warren
Gale, Paula M
Minkin, Jean Albert
Mueller, Paul Allen
Ragland, Paul C
Sackett, William Malcolm
Sokoloff, Vladimir P
Suman, Daniel Oscar
Thrailkill, John
Warburton, David Lewis
Zhu, Xiaorong

GEORGIA
Alberts, James Joseph
Bailey, George William
Brown, David Smith
Dod, Bruce Douglas
Hanson, Hiram Stanley
Holzer, Gunther Ulrich
Howard, James Hatten, III
Noakes, John Edward
Pollard, Charles Oscar, Jr
Railsback, Loren Bruce
Walters, John Philip
Wampler, Jesse Marion
Weaver, Charles Edward
Wenner, David Bruce
Whitney, James Arthur

HAWAII
Clark, Allen LeRoy
Erdman, John Gordon
Garcia, Michael Omar

Geochemistry (cont)

Halbig, Joseph Benjamin
Mackenzie, Frederick Theodore
Manghnani, Murli Hukumal
Ming, Li Chung
Muenow, David W
Sansone, Francis Joseph
Self, Stephen
Taylor, G Jeffrey
Thornber, Carl Richard

IDAHO
Farwell, Sherry Owen
Harrison, William Earl
Knobel, LeRoy Lyle
Welch, Jane Marie
Welhan, John Andrew

ILLINOIS
Anders, Edward
Anderson, Thomas Frank
Bennett, Glenn Allen
Berg, Jonathan H
Bishop, Finley Charles
Chou, Chen-Lin
Clayton, Robert Norman
Davis, Andrew Morgan
Goldsmith, Julian Royce
Gorody, Anthony Wagner
Grossman, Lawrence
Guggenheim, Stephen
Hayatsu, Ryoichi
Holt, Ben Dance
Kirkpatrick, Robert James
Koster Van Groos, August Ferdinand
Lerman, Abraham
Lewin, Seymour Z
Liu, Mian
Marley, Nancy Alice
Minear, Roger Allan
Moll, William Francis, Jr
Montgomery, Carla Westlund
Moore, Duane Milton
Perry, Eugene Carleton, Jr
Scouten, Charles George
Sood, Manmohan K
Stueber, Alan Michael
Teng, Mao-Hua
Winans, Randall Edward

INDIANA
Dyer, Rolla McIntyre, Jr
Elmore, David
Hayes, John Michael
Heydegger, Helmut Roland
Krothe, Noel C
Kullerud, Gunnar
Lipschutz, Michael Elazar
Meyer, Henry Oostenwald Albertijn
Pichtel, John Robert
Shaffer, Nelson Ross
Shieh, Yuch-Ning
Towell, David Garrett
Wintsch, Robert P

IOWA
Kracher, Alfred
Lemish, John
Nordlie, Bert Edward
Seifert, Karl E

KANSAS
Angino, Ernest Edward
Cullers, Robert Lee
Jones, Lois Marilyn
Van Schmus, William Randall
Welch, Jerome E
Whittemore, Donald Osgood
Zeller, Edward Jacob

KENTUCKY
Cobb, James C
Ehmann, William Donald

LOUISIANA
Allen, Gary Curtiss
Byerly, Gary Ray
Ferrell, Ray Edward, Jr
Ham, Russell Allen
Overton, Edward Beardslee
Thibodeaux, Louis J
Wood, Elwyn Devere

MAINE
Cronan, Christopher Shaw
Garside, Christopher
Howd, Frank Hawver
Mayer, Lawrence M
Murphy, Mary Ellen
Tewhey, John David

MARYLAND
Arem, Joel Edward
Bricker, Owen P, III
Brown, Michael
Corliss, John Burt
Cornwell, Jeffrey C
Dunning, Herbert Neal
Everett, Ardell Gordon
Ferry, John Mott
Fisher, Thomas Richard, Jr
French, Bevan Meredith
Helz, George Rudolph

Jones, Blair Francis
Kelly, William Robert
Krueger, Arlin James
Luth, William Clair
MacGregor, Ian Duncan
Marsh, Bruce David
May, Ira Philip
May, Irving
Mielke, James Edward
O'Keefe, John Aloysius
Piotrowicz, Stephen R
Riedel, Gerhardt Frederick
Shakur, Asif Mohammed
Thomas, Herman H
Trombka, Jacob Israel
Walter, Louis S
Weidner, Jerry R
Wells, Eddie N

MASSACHUSETTS
Agee, Carl Bernard
Bates, Carl H
Bell, Peter M
Boehm, Paul David
Brownlow, Arthur Hume
Burns, Roger George
Dacey, John W H
Dennen, William Henry
Deuser, Werner Georg
Donohue, John J
Eby, G Nelson
Farrington, John William
Frey, Frederick August
Grove, Timothy Lynn
Hager, Jutta Lore
Hart, Stanley Robert
Hepburn, John Christopher
Holland, Heinrich Dieter
Horzempa, Lewis Michael
Hunt, John Meacham
Jacobsen, Stein Bjornar
King, Kenneth, Jr
Leahy, Richard Gordon
Manheim, Frank T
Marshall, Donald James
Morse, Stearns Anthony
Naylor, Richard Stevens
Nunes, Paul Donald
Roedder, Edwin Woods
Sayles, Frederick Livermore
Shimizu, Nobumichi
Siever, Raymond
Teal, John Moline
Thompson, Geoffrey
Thompson, James Burleigh, Jr
Thompson, Mary Eleanor
Whelan, Jean King
Yuretich, Richard Francis

MICHIGAN
Barcelona, Michael Joseph
Buis, Patricia Frances
Essene, Eric J
Furlong, Robert B
Gardner, Wayne Stanley
Leenheer, Mary Janeth
Meyers, Philip Alan
Nriagu, Jerome Okon
O'Neil, James R
Vogel, Thomas A

MINNESOTA
Brezonik, Patrick Lee
Edwards, Richard Lawrence
Gorham, Eville
Krug, Edward Charles
Murthy, Varanasi Rama
Parker, Ronald Bruce

MISSISSIPPI
Bishop, Allen David, Jr
Sundeen, Daniel Alvin

MISSOURI
Bolter, Ernst A
Crozaz, Ghislaine M
Haskin, Larry A
Johns, William Davis
Jungclaus, Gregory Alan
Korotev, Randall Lee
Mantei, Erwin Joseph
Manuel, Oliver K
Podosek, Frank A

MONTANA
Alt, David D
Engel, Albert Edward John
Lange, Ian M
Volborth, Alexis

NEVADA
Cloke, Paul LeRoy
Dickson, Frank Wilson
Emerson, Donald Orville
Garside, Larry Joe
Henry, Christopher D
Hess, John Warren
Hsu, Liang-Chi
Kukal, Gerald Courtney

NEW HAMPSHIRE
Blum, Joel David
Chamberlain, C Page
Friedland, Andrew J

Gaudette, Henri Eugene
Hines, Mark Edward
McDowell, William H
Reynolds, Robert Coltart, Jr
Weber, James Harold

NEW JERSEY
Catanzaro, Edward John
Crerar, David Alexander
Feely, Herbert William
Ganapathy, Ramachandran
Gavini, Muralidhara B
McVey, Jeffrey King
Mitchell, Thomas Owen
Puffer, John H
Sarmiento, Jorge Louis
Schutz, Donald Frank
Sugam, Richard Jay
Takahashi, Taro

NEW MEXICO
Attrep, Moses, Jr
Baldridge, Warren Scott
Bild, Richard Wayne
Bish, David Lee
Broadhead, Ronald Frigon
Brookins, Douglas Gridley
Broxton, David Edward
Bryant, Ernest Atherton
Buden, Rosemary V
Cannon, Helen Leighton
Charles, Robert Wilson
Condie, Kent Carl
Erdal, Bruce Robert
Ewing, Rodney Charles
Floran, Robert John
Gancarz, Alexander John
Giordano, Thomas Henry
Grambling, Jeffrey A
Kuellmer, Frederick John
Laughlin, Alexander William
McLemore, Virginia Teresa Hill
Meijer, Arend
Morris, David Albert
Norris, Andrew Edward
Northrop, David A(mos)
Ogard, Allen E
Papike, James Joseph
Phillips, Fred Melville
Polzer, Wilfred L
Popp, Carl John
Renault, Jacques Roland
Ristvet, Byron Leo
Rokop, Donald J
Shannon, Spencer Sweet, Jr
Trujillo, Patricio Eduardo
Wolfsberg, Kurt

NEW YORK
Aller, Robert Curwood
Barnard, Walther M
Beall, George Halsey
Berkley, John Lee
Bickford, Marion Eugene
Biscaye, Pierre Eginton
Boger, Phillip David
Broecker, Wallace
Brueckner, Hannes Kurt
Carl, James Dudley
Charola, Asuncion Elena
Clemency, Charles V
DeLong, Stephen Edwin
Fisher, Nicholas Seth
Fountain, John Crothers
Froelich, Philip Nissen
Gaffey, Michael James
Hanson, Gilbert N
Howarth, Robert W
Kay, Robert Woodbury
Kelly, John Russell
Lattimer, James Michael
Lindsley, Donald Hale
Longhi, John
Ma, Maw-Suen
McLaren, Eugene Herbert
McLennan, Scott Mellin
Miller, Donald Spencer
Moniot, Robert Keith
Pajari, George Edward
Premuzic, Eugene T
Putman, George Wendell
Reitan, Paul Hartman
Rutstein, Martin S
Schoonen, Martin A A
Scranton, Mary Isabelle
Smethie, William Massie, Jr
Speidel, David H
Thomas, John Jenks
Thurber, David Lawrence
Vairavamurthy, Murthy Appathurai
Wallace, Douglas William Roy
White, William Michael
Whitney, Philip Roy
Woodmansee, Donald Ernest

NORTH CAROLINA
Bray, John Thomas
Feiss, Paul Geoffrey
Fullagar, Paul David
Grove, Thurman Lee
Harris, William Burleigh
Livingstone, Daniel Archibald
Qualls, Robert Gerald
Rogers, John James William

Willey, Joan DeWitt

NORTH DAKOTA
McCarthy, Gregory Joseph
Sugihara, James Masanobu

OHIO
Banks, Philip Oren
Bladh, Katherine Laing
Bladh, Kenneth W
Carlson, Ernest Howard
Dahl, Peter Steffen
Dollimore, David
Faure, Gunter
Flynn, Ronald Thomas
Foland, Kenneth Austin
Gregor, Clunie Bryan
Grube, Walter E, Jr
Kilinc, Attila Ishak
Koucky, Frank Louis, Jr
Kovach, Jack
Law, Eric W
Lo, Howard H
Matisoff, Gerald
Maynard, James Barry
Means, Jeffrey Lynn
Moody, Judith Barbara
Pushkar, Paul
Savin, Samuel Marvin
Stout, Barbara Elizabeth

OKLAHOMA
Al-Shaieb, Zuhair Fouad
Ames, Roger Lyman
Barker, Colin G
Cardott, Brian Joseph
Collins, Arlee Gene
Enfield, Carl George
Fisher, John Berton
Gilbert, Murray Charles
Henderson, Frederick Bradley, III
Hitzman, Donald Oliver
Ho, Thomas Tong-Yun
Hounslow, Arthur William
Kokesh, Fritz Carl
London, David
Philp, Richard Paul
Powell, Benjamin Neff
Puls, Robert W
Smith, James Edward
Thompson, Robert Richard
Williams, Jack A

OREGON
Bennington, Kenneth Oliver
Field, Cyrus West
Fish, William
Goles, Gordon George
Goodney, David Edgar
Holser, William Thomas
Huh, Chih-An
Lotspeich, Frederick Benjamin
Lupton, John Edward
McBirney, Alexander Robert
Nafziger, Ralph Hamilton
Powell, James Lawrence

PENNSYLVANIA
Allen, Jack C, Jr
Allen, John Christopher
Arthur, Michael Allan
Barnes, Hubert Lloyd
Brady, Keith Bryan Craig
Brantley, Susan Louise
Cassidy, William Arthur
Clavan, Walter
Coe, Charles Gardner
Cohen, Alvin Jerome
Crawford, William Arthur
Deines, Peter
Eggler, David Hewitt
Goldstein, Theodore Philip
Grubbs, Donald Keeble
Heald, Emerson Francis
Herman, Richard Gerald
Herzog, Leonard Frederick, II
Hodgson, Robert Arnold
Hutchinson, Peter J
Keith, MacKenzie Lawrence
Kerrick, Derrill M
Kohman, Truman Paul
Kump, Lee Robert
Lukert, Michael T
Mangelsdorf, Paul Christoph, Jr
Maroulis, Peter James
Niemitz, Jeffrey William
Ohmoto, Hiroshi
Osborn, Elburt Franklin
Rohl, Arthur N
Rose, Arthur William
Roy, Della M(artin)
Schmalz, Robert Fowler
Sclar, Charles Bertram
Seidell, Barbara Castens
Spear, Jo-Walter
Suhr, Norman Henry
Ulmer, Gene Carleton
White, William Blaine
Witkowski, Robert Edward
Yeh, Gour-Tsyh

RHODE ISLAND
Abell, Paul Irving
Duce, Robert Arthur

Gearing, Juanita Newman
Giletti, Bruno John
Hermes, O Don
Leinen, Margaret Sandra
Pilson, Michael Edward Quinton
Quinn, James Gerard

SOUTH CAROLINA
Clayton, Donald Delbert
Gardner, Leonard Robert
Geidel, Gwendelyn
Moore, Willard S
Nichols, Chester Encell
Shervais, John Walter
Stakes, Debra Sue
Strom, Richard Nelsen
Warner, Richard Dudley

SOUTH DAKOTA
Jonte, John Haworth
Paterson, Colin James

TENNESSEE
Ault, Wayne Urban
Bohlmann, Edward Gustav
Churnet, Habte Giorgis
Emanuel, William Robert
Kopp, Otto Charles
Lindberg, Steven Eric
McSween, Harry Y, Jr
Marland, Gregg (Hinton)
Naney, Michael Terrance
Stow, Stephen Harrington
Taylor, Lawrence August
Turner, Robert Spilman

TEXAS
Adams, John Allan Stewart, Sr
Anthony, Elizabeth Youngblood
Arp, Gerald Kench
Baker, Donald Roy
Bence, Alfred Edward
Bernard, Bernie Boyd
Bishop, Richard Stearns
Bogard, Donald Dale
Botto, Robert Irving
Boutton, Thomas William
Brown, Robert Wade
Burkart, Burke
Casagrande, Daniel Joseph
Claypool, George Edwin
Clementz, David Michael
Clendening, John Albert
Cooper, James Erwin
Crick, Rex Edward
Dahl, Harry Martin
Eastwood, Raymond L
Eckelmann, Walter R
Elthon, Donald L
Ewing, Thomas Edward
Fang, Jiasong
Gibson, Everett Kay, Jr
Golden, Dadigamuwage Chandrasiri
Gooding, James Leslie
Haas, Herbert
Halpern, Martin
Ho, Clara Lin
Holdaway, Michael Jon
Howatson, John
Ingerson, Fred Earl
Kaiser, William Richard
Koepnick, Richard Borland
Kulla, Jean B
Lafon, Guy Michel
Land, Lynton S
Levien, Louise
Lewis, Donald Richard
Lindstrom, David John
Lindstrom, Marilyn Martin
Loeppert, Richard Henry, Jr
Lofgren, Gary Ernest
Long, Leon Eugene
Louden, L Richard
McDowell, Fred Wallace
McGuire, Anne Vaughan
Matthews, Martin David
Mitterer, Richard Max
Monaghan, Patrick Henry
Morse, John Wilbur
Nussmann, David George
Nyquist, Laurence Elwood
Orr, Wilson Lee
Pannell, Keith Howard
Parker, Patrick LeGrand
Parker, Robert Hallett
Pearson, Daniel Bester, III
Pearson, Frederick Joseph
Phinney, William Charles
Pogue, Randall F
Potts, Mark John
Presnall, Dean C
Riese, Walter Charles Rusty
Roe, Glenn Dana
Rogers, Marion Alan
Schink, David R
Schumacher, Dietmar
Schwartz, Robert Donald
Schwarzer, Theresa Flynn
Scott, Martha Richter
Smith, Michael Alexis
Sommer, Sheldon E
Stormer, John Charles, Jr
Strom, E(dwin) Thomas
Von Rosenberg, Dale Ursini

Walters, Lester James, Jr
Warshauer, Steven Michael
Whelan, Thomas, III
Young, William Allen
Zlatkis, Albert

UTAH
Adams, Michael Curtis
Cerling, Thure E
Christiansen, Eric H
Jewell, Paul William
Jurinak, Jerome Joseph
Kolesar, Peter Thomas
Nash, William Purcell
Parry, William Thomas
Phillips, Lee Revell
Wolgemuth, Kenneth Mark

VERMONT
Coish, Raymond Alpheaus
Furukawa, Toshiharu
Westerman, David Scott

VIRGINIA
Allen, Ralph Orville, Jr
Baedecker, Mary Jo
Baedecker, Philip A
Bethke, Philip Martin
Bodnar, Robert John
Busenberg, Eurybiades
Craig, James Roland
Cunningham, Charles G
Dickinson, Stanley Key, Jr
Fabbi, Brent Peter
Finkelman, Robert Barry
Fisher, James Russell
Frye, Keith
Galloway, James Neville
Grossman, Jeffrey N
Hanna, William Jefferson
Hanshaw, Bruce Busser
Hearn, Bernard Carter, Jr
Hemingway, Bruce Sherman
Herr, Frank Leaman, Jr
Huang, Wen Hsing
Huebner, John Stephen
Huggett, Robert James
Kornicker, William Alan
Landa, Edward Robert
Leo, Gerhard William
Levine, Joel S
Lipin, Bruce Reed
Macko, Stephen Alexander
Moore, Raymond Kenworthy
Morgan, John Walter
Mose, Douglas George
Padovani, Elaine Reeves
Palmer, Curtis Allyn
Pickering, Ranard Jackson
Rankin, Douglas Whiting
Ridge, John Drew
Roseboom, Eugene Holloway, Jr
Ross, Malcolm
Rubin, Meyer
Rule, Joseph Houston
Sato, Motoaki
Sayala, Dasharatham (Dash)
Schoen, Robert
Schreiber, Henry Dale
Sherwood, William Cullen
Simon, Frederick Otto
Smith, Richard Elbridge
Snyder, John L
Stein, Holly Jayne
Stewart, David Benjamin
Tanner, Allan Bain
Toulmin, Priestley
Walker, Alta Sharon
Weill, Daniel Francis
Wong, George Tin Fuk
Wood, Warren Wilbur

WASHINGTON
Ames, Lloyd Leroy, Jr
Bamford, Robert Wendell
Barnes, Ross Owen
Brownlee, Donald E
Chang, Yew Chun
Crecelius, Eric A
Gerlach, Terrence Melvin
Ghiorso, Mark Stefan
Hedges, John Ivan
Hodges, Floyd Norman
Houck, John Candee
Jenne, Everett A
Keller, C Kent
Laul, Jagdish Chander
Mopper, Kenneth
Olson, Edwin Andrew
Quay, Paul Douglas
Riley, Robert Gene
Serne, Roger Jeffrey
Sherman, David Michael
Sklarew, Deborah S
Stuiver, Minze
Symonds, Robert B
Walker, Sharon Leslie
Weis, Paul Lester
Wright, Judith

WEST VIRGINIA
Rauch, Henry William
Renton, John Johnston

WISCONSIN
Bowser, Carl
Brown, Philip Edward
Fripiat, Jose J
Hoffman, James Irvie
Krezoski, John R
Mudrey, Michael George, Jr
Myers, Charles R
Rosson, Reinhardt Arthur
Stenstrom, Richard Charles
Valley, John Williams
West, Walter Scott

WYOMING
Branthaver, Jan Franklin
Felbeck, George Theodore, Jr
Frost, Carol Denison
Katta, Jayaram Reddy
Robinson, Wilbur Eugene

ALBERTA
Baadsgaard, Halfdan
Bird, Gordon Winslow
Cumming, George Leslie
Ghent, Edward Dale
Goodarzi, Fariborz
Hitchon, Brian
Hunt, C Warren
Lambert, Richard St John
Macqueen, Roger Webb
Nesbitt, Bruce Edward
Schmidt, Volkmar
Strausz, Otto Peter
Vitt, Dale Hadley

BRITISH COLUMBIA
Armstrong, Richard Lee
Arnold, Ralph Gunther
Bevier, Mary Lou
Clark, Lloyd Allen
Kieffer, Susan Werner
Macdonald, Robie Wilton
Russell, Richard Doncaster
Sinclair, Alastair James

MANITOBA
Brunskill, Gregg John
Cerny, Petr
Hecky, Robert Eugene
Smil, Vaclav
Vandergraaf, Tjalle T
Vilks, Peter

NEW BRUNSWICK
Allen, Clifford Marsden

NEWFOUNDLAND
Haines, Robert Ivor
Longerich, Henry Perry

NOVA SCOTIA
Buckley, Dale Eliot
Cooke, Robert Clark
Loring, Douglas Howard

ONTARIO
AL-Hashimi, Ahmad
Anderson, Gregor Munro
Appleyard, Edward Clair
Baragar, William Robert Arthur
Brown, Seward Ralph
Cabri, Louis J
Carlisle, David Brez
Crocket, James Harvie
Dayal, Ramesh
De Kimpe, Christian Robert
Dillon, Peter J
Duke, John Murray
Edgar, Alan D
Emslie, Ronald Frank
Fortescue, John A C
Fox, Joseph S
Fyfe, William Sefton
George, Albert El Deeb
Grieve, Richard Andrew
Haynes, Simon John
Jackson, Togwell Alexander
James, Richard Stephen
Kramer, James R(ichard)
Kretz, Ralph
Krogh, Thomas Edvard
Lesage, Suzanne
McNutt, Robert Harold
MacRae, Neil D
Maxwell, John Alfred
Mudroch, Alena
Pearce, Thomas Hulme
Percival, Jeanne Barbara
Roeder, Peter Ludwig
Schwarcz, Henry Philip
Scott, Steven Donald
Shaw, Denis Martin
Smith, Terence E
Thode, Henry George
Turek, Andrew
Veizer, Ján
Weiler, Roland R

QUEBEC
Brown, Alex Cyril
Campbell, Peter G C
Hesse, Reinhard
Hillaire-Marcel, Claude
Khalil, Michel

MacLean, Wallace H
Ouellet, Marcel
Power, Joan F
Tessier, Andre
Webber, George Roger

SASKATCHEWAN
Chew, Hemming
Davies, John Clifford
Huang, P M

OTHER COUNTRIES
Andreae, Meinrat Otto
Finlayson, James Bruce
Fritz, Peter
Hofmann, Albrecht Werner
Loeliger, David A
Ugolini, Fiorenzo Cesare
Walton, Wayne J A, Jr

Geology, Applied, Economic & Engineering

ALABAMA
Bearce, Denny N
Carrington, Thomas Jack
Chalokwu, Christopher Iloba
Cook, Robert Bigham, Jr
Groshong, Richard Hughes, Jr
Hastings, Earl L
Jones, Douglas Epps
LaMoreaux, Philip Elmer
McCarl, Henry Newton
Mancini, Ernest Anthony
Marchant, Guillaume Henri (Wim), Jr
Moser, Paul H
Neathery, Thornton Lee
Simpson, Thomas A
Workman, William Edward

ALASKA
Brewer, Max C(lifton)
Calderwood, Keith Wright
Ferrians, Oscar John, Jr
Johansen, Nils Ivar
Luttrell, Eric Martin
Mangus, Marvin D
Miller, Lance Davison
Mowatt, Thomas C
Reed, Bruce Loring
Schmidt, Ruth A M
Triplehorn, Don Murray
Turner, Donald Lloyd
Weber, Florence Robinson
Wolff, Ernest N

ARIZONA
Agenbroad, Larry Delmar
Breed, Carol Sameth
Burt, Donald McLain
Childs, Orlo E
Cotera, Augustus S, Jr
Cree, Allan
Dapples, Edward Charles
DeCook, Kenneth James
Dietz, Robert Sinclair
Elston, Donald (Parker)
Emery, Philip Anthony
Fellows, Larry Dean
Fink, James Brewster
Fisher, Frederick Stephen
Ganguly, Jibamitra
Gillis, James E, Jr
Greeley, Ronald
Guilbert, John M
Harris, DeVerle Porter
Hawkes, Herbert Edwin, Jr
Haynes, Caleb Vance, Jr
Heede, Burchard Heinrich
Johnsen, John Herbert
Kamilli, Robert Joseph
Kiersch, George Alfred
Kremp, Gerhard Otto Wilhelm
Lacy, W(illard) C(arleton)
Lance, John Franklin
Learned, Robert Eugene
Loring, William Bacheller
Lundin, Robert Folke
McCauley, John Francis
Newman, William L
Norton, Denis Locklin
Ohle, Ernest Linwood
Page, Norman J
Peters, William Callier
Pewe, Troy Lewis
Richardson, Randall Miller
Roddy, David John
Rush, Richard William
Schaber, Gerald Gene
Schreiber, Joseph Frederick, Jr
Shenkel, Claude W, Jr
Stump, Edmund
Sumner, John Stewart
Swanberg, Chandler A
Swann, Gordon Alfred
Thurston, William R
Titley, Spencer Rowe
Wait, James Richard
Willis, Clifford Leon
Wilson, Richard Fairfield

Geology, Applied, Economic & Engineering (cont)

ARKANSAS
Bergendahl, Maximilian Hilmar
Jackson, Kern Chandler
Konig, Ronald H
MacDonald, Harold Carleton
Mack, Leslie Eugene
Manger, Walter Leroy
Oglesby, Gayle Arden
Spivey, Robert Charles
Steele, Kenneth F
Wagner, George Hoyt

CALIFORNIA
Abrams, Michael J
Adent, William A
Albee, Arden Leroy
Alfors, John Theodore
Allan, John Ridgway
Allen, Clarence Roderic
Arnal, Robert Emile
Ashley, Roger Parkmand
Avent, Jon C
Balthaser, Lawrence Harold
Barron, John Arthur
Bass, Manuel N
Beaty, David Wayne
Berg, Henry Clay
Bevc, Dimitri
Birman, Joseph Harold
Blackerby, Bruce Alfred
Blake, Milton Clark, Jr
Blom, Christian James
Bohannon, Robert Gary
Bonham, Lawrence Cook
Bonilla, Manuel George
Borax, Eugene
Borg, Iris Y P
Brabb, Earl Edward
Brahma, Chandra Sekhar
Bredehoeft, John Dallas
Brimhall, George H
Brogan, George Edward
Brown, Glenn A
Bunch, Theodore Eugene
Carlisle, Donald
Case, James Edward
Castle, Robert O
Chapman, Robert Mills
Christensen, Mark Newell
Churkin, Michael, Jr
Ciancanelli, Eugene Vincent
Clark, Bruce R
Clements, Thomas
Clifton, Hugh Edward
Cluff, Lloyd Sterling
Coash, John Russell
Coats, Robert Roy
Cogen, William Maurice
Coleman, Robert Griffin
Conant, Louis Cowles
Conrey, Bert L
Cook, Harry E, III
Cox, Dennis Purver
Creasey, Savill Cyrus
Curtis, Garniss Hearfield
Dewart, Gilbert
Donath, Fred Arthur
Douglas, Hugh
Dreyer, Robert Marx
Drugg, Warren Sowle
Easton, William Heyden
Eberlein, George Donald
Edmund, Rudolph William
Ehlig, Perry Lawrence
Ehrreich, Albert LeRoy
Einaudi, Marco Tullio
Elders, Wilfred Allan
Elliott, William J
Evans, James R
Evitt, William Robert
Filippone, Walter R
Finney, Stanley Charles
Fisher, Richard Virgil
Ford, Arthur B
Foster, Robert John
Frautschy, Jeffery Dean
Frederickson, Arman Frederick
Friberg, James Frederick
Gabelman, John Warren
Garrison, Robert Edward
Gastil, R Gordon
Gealy, Elizabeth Lee
Goodman, Richard E
Graham, Stephan Alan
Graves, Roy William, Jr
Gray, Clifton Herschel, Jr
Greene, Robert Carl
Greenman, Norman
Griggs, Gary B
Gromme, Charles Sherman
Gryc, George
Gurnis, Michael
Gutstadt, Allan Morton
Guyton, James W
Hanscom, Roger H
Harbaugh, John Warvelle
Harpster, Robert E
Hart, Earl W
Harwood, David Smith
Havard, John Francis
Hawkins, James Wilbur, Jr
Hein, James R
Helin, Eleanor Kay
Henshaw, Paul Carrington
Herber, Lawrence Justin
Hesse, Christian August
Hietanen-Makela, Anna (Martta)
Higgins, James Woodrow
Hill, Donald Gardner
Hill, Mason Lowell
Hill, Merton Earle, III
Hirschfeld, Sue Ellen
Hodges, Carroll Ann
Hodges, Lance Thomas
Holwerda, James G
Holzer, Thomas Lequear
Hoskins, Cortez William
Hotz, Preston Enslow
Humphrey, Thomas Milton, Jr
Hurst, Richard William
Inman, Douglas Lamar
Isherwood, William Frank
Jacobson, Stephen Richard
James, Laurence Beresford
Janke, Norman Charles
Jizba, Zdenek Vaclav
Johnson, Vard Hayes
Jones, David Lawrence
Kam, James Ting-Kong
Kaska, Harold Victor
Keith, Terry Eugene Clark
Kirby, James M
Kistler, Ronald Wayne
Klinger, Allen
Koestel, Mark Alfred
Kopf, Rudolph William
Kramer, John Howard
Kruger, Fredrick Christian
Krummenacher, Daniel
LaFountain, Lester James, Jr
Lanphere, Marvin Alder
Leighton, Freeman Beach
Leith, Carlton James
Leps, Thomas MacMaster
Lewis, Arthur Edward
Lipman, Peter Waldman
Lippmann, Marcelo Julio
Lipps, Jere Henry
Loomis, Alden Albert
Lumsden, William Watt, Jr
Lyon, Ronald James Pearson
McCrone, Alistair William
Macdonald, Kenneth Craig
McGeary, David F R
McKague, Herbert Lawrence
McLean, Hugh
McNitt, James R
Maddock, Marshall
Magoon, Leslie Blake, III
Majmundar, Hasmukhrai Hiralal
Mandra, York T
Martin, Roger Charles
Matthews, Robert A
Merifield, Paul M
Miatech, Gerald James
Moiseyev, Alexis N
Moore, Henry J, II
Moran, William Rodes
Morris, Hal Tryon
Morris, John David
Morris, William Joseph
Muessig, Siegfried Joseph
Muffler, Leroy John Patrick
Murphy, Michael A
Murray, Bruce C
Nash, Douglas B
Nelson, Carlton Hans
Nygreen, Paul W
Oakeshott, Gordon B
Olmsted, Franklin Howard
Olson, Jerry Chipman
Otte, Carel
Pampeyan, Earl H
Parker, Calvin Alfred
Patton, William Wallace, Jr
Peak, Wilferd Warner
Pestrong, Raymond
Peterson, Donald William
Phillips, Richard P
Phoenix, David A
Pike, Richard Joseph
Pincus, Howard Jonah
Pipkin, Bernard Wallace
Plafker, George
Pomeroy, John S
Pray, Ralph Emerson
Proctor, Richard James
Ptacek, Anton D
Ramspott, Lawrence Dewey
Rex, Robert Walter
Richter, Raymond C
Rigsby, George Pierce
Rodda, Peter Ulisse
Rose, Robert Leon
Sabins, Floyd F
Sage, Orrin Grant, Jr
St Amand, Pierre
Schneiderman, Jill Stephanie
Schwartz, Daniel Evan
Seiden, Hy
Seiders, Victor Mann
Sieh, Kerry Edward
Singer, Donald Allen
Slosson, James E
Smalley, Robert Gordon
Smedes, Harry Wynn
Smith, Raymond James
Smith, Robert Leland
Smith-Evernden, Roberta Katherine
Snavely, Parke Detweiler, Jr
Stager, Harold Keith
Stanley, George M
Stensrud, Howard Lewis
Stewart, John Harris
Stout, Martin Lindy
Stuart-Alexander, Desiree Elizabeth
Sullivan, Raymond
Sullwold, Harold H
Sundaram, Panchanatham N
Swinney, Chauncey Melvin
Tabor, Rowland Whitney
Terry, Richard D
Testa, Stephen M
Theodore, Ted George
Thompson, George Albert
Tooker, Edwin Wilson
Towse, Donald Frederick
Travers, William Brailsford
Troxel, Bennie Wyatt
Truesdell, Alfred Hemingway
Turk, Leland Jan
Tyler, John Howard
Valentine, James William
van de Kamp, Peter Cornelis
Volbrecht, Stanley Gordon
Waggoner, Eugene B
Wahrhaftig, Clyde (Adolph)
Walawender, Michael John
Wallace, Robert Earl
Wasserburg, Gerald Joseph
Watson, Kenneth De Pencier
Weed, Elaine Greening Ames
Weiss, Lionel Edward
Wentworth, Carl M, Jr
White, Donald Edward
Williams, Alan Evan
Williams, Floyd James
Williams, John Wharton
Willis, David Edwin
Wilson, L Kenneth
Wilt, Michael
Witherspoon, Paul A(dams), Jr
Wrucke, Chester Theodore, Jr
Wyllie, Peter John
Yeend, Warren Ernest
Zenger, Donald Henry
Ziony, Joseph Israel
Zumberge, James Herbert

COLORADO
Adams, John Wagstaff
Ager, Thomas Alan
Ahlbrandt, Thomas Stuart
Aleinikoff, John Nicholas
Alpha, Andrew G
Applegate, James Keith
Arbenz, Johann Kaspar
Ayler, Maynard Franklin
Bailey, R V
Ball, Mahlon M
Barber, George Arthur
Barker, Fred
Berger, Byron Roland
Berry, George Willard
Birkeland, Peter Wessel
Blair, Robert William
Bohor, Bruce Forbes
Braun, Lewis Timothy
Broin, Thayne Leo
Bromfield, Calvin Stanton
Bucknam, Robert Campbell
Bush, Alfred Lerner
Byers, Frank Milton, Jr
Cadigan, Robert Allen
Cady, Wallace Martin
Callender, Jonathan Ferris
Canney, Frank Cogswell
Cathcart, James B
Chaffee, Maurice A(hlborn)
Chamberlain, Theodore Klock
Chenoweth, William Lyman
Chico, Raymundo Jose
Choquette, Philip Wheeler
Christensen, Odin Dale
Clift, William Orrin
Cluff, Robert Murri
Coates, Donald Allen
Cobban, William Aubrey
Colton, Roger Burnham
Connor, Jon James
Cowart, Vicki Jane
Craig, Dexter Hildreth
Crandell, Dwight Raymond
Cummings, David
Curry, James Kenneth
Curtis, Bruce Franklin
Cygan, Norbert Everett
Dahl, Arthur Richard
Daviess, Steven Norman
Desborough, George A
Dobrovolny, Ernest
Doehring, Donald O
Dolly, Edward Dawson
Downs, William Fredrick
Dubiel, Russell F
Eckel, Edwin Butt
Edwards, John D
Eicher, Don Lauren
Elder, Curtis Harold
Erskine, Christopher Forbes
Everhart, Donald Lough
Fails, Thomas Glenn
Ferris, Bernard Joe
Ferris, Clinton S, Jr
Finch, Warren Irvin
Fleming, Robert William
Flores, Romeo M
Foster, Norman Holland
Frezon, Sherwood Earl
Friedman, Jules Daniel
Goldich, Samuel Stephen
Graham, Charles Edward
Griffitts, Wallace Rush
Grose, Thomas Lucius Trowbridge
Grutt, Eugene Wadsworth, Jr
Hardaway, John E(vans)
Harms, John Conrad
Harrison, Jack Edward
Harshman, Elbert Nelson
Haun, John Daniel
Hayes, John Bernard
Hays, William Henry
Henrickson, Eiler Leonard
Heyl, Allen Van, Jr
Hittelman, Allen M
Hobbs, Samuel Warren
Howard, Thomas E(dward)
Hughes, Travis Hubert
Hull, Joseph Poyer Deyo, Jr
Hutchinson, Richard William
Hutchinson, Robert Maskiell
Jacobson, Jimmy Joe
Jibson, Randall W
Johnson, Frederick Arthur, Jr
Johnson, Robert Britten
Johnson, Ross Byron
Johnstone, James G(eorge)
Jordan, James N
Kanizay, Stephen Peter
Keefer, William Richard
Keighin, Charles William
Kent, Harry Christison
Knight, Fred G
Kohls, Donald W
Kupfer, Donald Harry
Lackey, Laurence
Landon, Robert E
Langmuir, Donald
Larson, Edwin E
Lee, Keenan
LeMasurier, Wesley Ernest
Leonard, B(enjamin) F(ranklin), (III)
Lewis, John Hubbard
Lindsey, David Allen
Lohman, Stanley William
Longley, W(illiam) Warren
Lowell, James Diller
Luft, Stanley Jeremie
McCallum, Malcolm E
McCubbin, Donald Gene
MacKenzie, David Brindley
MacLachlan, James Crawford
Mallory, William Wyman
Marvin, Richard F(rederick)
Mason, Malcolm
Mast, Richard F(redrick)
Maughan, Edwin Kelly
Maxwell, Charles Henry
Meade, Robert Heber, Jr
Meyer, Harvey John
Miller, Betty M (Tinklepaugh)
Moench, Robert Hadley
Moore, David A
Moore, David Warren
Mullineaux, Donal Ray
Murphy, John Francis
Mytton, James W
Nash, J(ohn) Thomas
Newman, Karl Robert
Nichols, Donald Ray
Olson, Richard Hubbell
Osmond, John C, Jr
Osterhoudt, Walter Jabez
Oxley, Philip
Panek, Louis A(nthony)
Parker, John Marchbank
Pennington, Wayne David
Peterman, Zell Edwin
Peterson, Fred
Pillmore, Charles Lee
Ponder, Herman
Powell, James Daniel
Power, Walter Robert
Prather, Thomas Leigh
Pratt, Walden Penfield
Prichard, George Edwards
Quirke, Terence Thomas, Jr
Rall, Raymond Wallace
Raup, Omer Beaver
Reed, John Calvin, Jr
Robinson, Peter
Rold, John W
Romberger, Samuel B
Rooney, Lawrence Frederick
Rosenblum, Sam
Rouse, George Elverton
Rowley, Peter DeWitt
Sable, Edward George
Sanborn, Albert Francis
Sanford, Richard Frederick
Savage, William Zuger
Schmidt, Dwight Lyman

Schuster, Robert Lee
Scott, James Henry
Scott, Robert Blackburn
Seeland, David Arthur
Segerstrom, Kenneth
Shawe, Daniel Reeves
Sheridan, Douglas Maynard
Sidder, Gary Brian
Silberman, Miles Louis
Simpson, Howard Edwin
Sims, Paul Kibler
Skipp, Betty Ann
Sloss, Peter William
Smyth, Joseph Richard
Sommerfeld, Richard Arthur
Spencer, Charles Winthrop
Staatz, Mortimer Hay
Stark, Philip Herald
Strobell, John Dixon, Jr
Stucky, Richard K(eith)
Taylor, Michael E
Terriere, Robert T
Thomasson, M Ray
Thompson, Thomas Luman
Thompson, Thomas Luther
Thompson, Tommy Burt
Thomsen, Harry Ludwig
Thorman, Charles Hadley
Tschanz, Charles McFarland
Turner, Daniel Stoughton
Turner, Mortimer Darling
Twenhofel, William Stephens
Tweto, Ogden
Van Zant, Kent Lee
Varnes, David Joseph
Vine, James David
Visher, Frank N
Wallace, Stewart Raynor
Warner, Lawrence Allen
Weimer, Robert J
Weist, William Godfrey, Jr
Weitz, Joseph Leonard
Wenrich, Karen Jane
Westphal, Warren Henry
Wilson, Elizabeth Allen
Wilson, John Coe
Wingate, Frederick Huston
Witkind, Irving Jerome
Wray, John Lee

CONNECTICUT
Baskerville, Charles Alexander
Fox, G(eorge) Sidney
Frasche, Dean F
Groff, Donald William
Hyde, John Welford
Levin, S Benedict
MacFadyen, John Archibald, Jr
Miller, Franklin Stuart
Orville, Philip Moore
Skinner, Brian John
Spiegel, Zane
Waage, Karl Mensch
Winchell, Horace

DELAWARE
Jordan, Robert R
Kraft, John Christian
Nicholls, Robert Lee
Pickett, Thomas Ernest
Stiles, A(lvin) B(arber)
Talley, John Herbert

DISTRICT OF COLUMBIA
Appleman, Daniel Everett
Beckner, Everet Hess
Benson, William Edward Barnes
Boyd, Francis R
Carroll, Gerald V
Duguid, James Otto
Dutro, John Thomas, Jr
Fiske, Richard Sewell
Justus, Philip Stanley
Knox, Arthur Stewart
Melickian, Gary Edward
O'Donnell, Edward
Parham, Walter Edward
Pierce, Jack Warren
Price, Jonathan Greenway
Rogers, Thomas Hardin
Saunders, Ronald Stephen
Sheldon, Richard P
Zeizel, A(rthur) J(ohn)

FLORIDA
Arden, Daniel Douglas
Buie, Bennett Frank
Chauvin, Robert S
Chawner, William Donald
Chryssafopoulos, Hanka Wanda Sobczak
Chryssafopoulos, Nicholas
Craig, Alan Knowlton
Deere, Don U(el)
DeVore, George Warren
Freeland, George Lockwood
Garnar, Thomas E(dward), Jr
Geraghty, James Joseph
Goudarzi, Gus (Hossein)
Guild, Philip White
Hartman, John Paul
Hoagland, Alan D
Joensuu, Oiva I
Kadey, Frederic L, Jr
Kirkemo, Harold

Kisvarsanyi, Geza
Langfelder, Leonard Jay
Lemon, Roy Richard Henry
Nichols, Robert Leslie
O'Hara, Norbert Wilhelm
Opdyke, Neil
Osmond, John Kenneth
Parker, Garald G, Sr
Patton, Thomas Hudson
Pirkle, Fredric Lee
Pullen, Milton William, Jr
Randazzo, Anthony Frank
Rona, Peter Arnold
Schmidt, Walter
Smith, Douglas Lee
Snyder, Walter Stanley
Spangler, Daniel Patrick
Stehli, Francis Greenough
Thrailkill, John
Travis, Russell Burton
Upchurch, Sam Bayliss
Werner, Harry Emil
Wise, Sherwood Willing, Jr

GEORGIA
Allard, Gilles Olivier
Allen, Arthur T, Jr
Barksdale, Richard Dillon
Carver, Robert E
Cofer, Harland E, Jr
Darrell, James Harris, II
Ghuman, Gian Singh
Gonzales, Serge
Grant, Willard H
Hanson, Hiram Stanley
Henry, Vernon James, Jr
Herz, Norman
Howard, James Dolan
Hurst, Vernon James
Koch, George Schneider, Jr
Little, Robert Lewis
Sanders, Richard Pat
Shrum, John W
Size, William Bachtrup
Smith, John M
Sowers, George F(rederick)
Sumartojo, Jojok
Weaver, Charles Edward

HAWAII
Arkle, Thomas, Jr
Chave, Keith Ernest
Clark, Allen LeRoy
Cox, Doak Carey
Cruickshank, Michael James
Divis, Allan Francis
Erdman, John Gordon
Fan, Pow-Foong
Helsley, Charles Everett
Kemp, Paul James
Lockwood, John Paul
Malahoff, Alexander
Pankiwskyj, Kost Andrij
Robertson, Forbes
Sinton, John Maynard

IDAHO
Austin, Carl Fulton
Bonnichsen, Bill
Emigh, G Donald
Hall, William Bartlett
Hoggan, Roger D
Hollenbaugh, Kenneth Malcolm
Isaacson, Peter Edwin
Jones, Robert William
Knobel, LeRoy Lyle
Knutson, Carroll Field
Laval, William Norris
Miller, Maynard Malcolm
Reid, Rolland Ramsay
Siems, Peter Laurence
Waag, Charles Joseph
Warner, Mont Marcellus
Williams, George Arthur
Wilson, Monte Dale
Wood, Spencer Hoffman

ILLINOIS
Archbold, Norbert L
Baxter, James Watson
Berg, Richard Carl
Bergstrom, Robert Edward
Block, Douglas Alfred
Bradbury, James Clifford
Buschbach, Thomas Charles
Carozzi, Albert Victor
Carpenter, Philip John
Casella, Clarence J
Chou, Chen-Lin
Collinson, Charles William
Corbett, Robert G
Crelling, John Crawford
Damberger, Heinz H
Dolgoff, Abraham
DuMontelle, Paul Bertrand
Dutcher, Russell Richardson
Flemal, Ronald Charles
Foster, John Webster
Fryxell, Fritiof Melvin
Gealy, William James
Gorody, Anthony Wagner
Gross, David Lee
Haddock, Gerald Hugh
Harrison, Wyman

Hart, Richard Royce
Hess, David Filbert
Hoffer, Abraham
Ireland, Herbert O(rin)
Kent, Lois Schoonover
Langenheim, Ralph Louis, Jr
Leighton, Morris Wellman
McKay, Edward Donald, III
Mann, Christian John
Masters, John Michael
Maynard, Theodore Roberts
Mikulic, Donald George
Norby, Roger Dale
Okal, Emile Andre
Olsen, Edward John
Paige, Russell Alston
Peppers, Russel A
Qutub, Musa Y
Rue, Edward Evans
Sandvik, Peter Olaf
Sasman, Robert T
Saxena, Surendra K
Schlanger, Seymour Oscar
Schmidt, Richard Arthur
Searight, Thomas Kay
Shabica, Charles Wright
Simon, Jack Aaron
Speed, Robert Clarke
Sundelius, Harold Wesley
Trask, Charles Brian
Utgaard, John Edward
Weiss, Malcolm Pickett
Wills, Donald L

INDIANA
Biggs, Maurice Earl
Boneham, Roger Frederick
Carr, Donald Dean
Chowdhury, Dipak Kumar
Cleveland, John H
Davis, Michael Edward
Fairley, William Merle
Gray, Henry Hamilton
Horowitz, Alan Stanley
Judd, William Robert
Kane, Henry Edward
Levandowski, Donald William
Madison, James Allen
Mirsky, Arthur
Murray, Haydn Herbert
Owen, Donald Eugene
Ruhe, Robert Victory
Scholer, Charles Frey
Shaffer, Nelson Ross
Suttner, Lee Joseph
Totten, Stanley Martin
Valia, Hardarshan S
Vitaliano, Charles Joseph
Vitaliano, Dorothy Brauneck
Votaw, Robert Barnett
West, Terry Ronald
Wier, Charles Eugene
Winkler, Erhard Mario

IOWA
Anderson, Wayne I
Carmichael, Robert Stewart
De Nault, Kenneth J
Dorheim, Fredrick Houge
Drake, Lon David
Foster, Charles Thomas, Jr
Hallberg, George Robert
Handy, Richard L(incoln)
Hansman, Robert H
Hendriks, Herbert Edward
Hershey, H Garland
Hoff, Darrel Barton
Lemish, John
Lohnes, Robert Alan
Nordlie, Bert Edward
O'Brien, Dennis Craig
Vondra, Carl Frank

KANSAS
Avcin, Matthew John, Jr
Baars, Donald Lee
Berg, J(ohn) Robert
Brady, Lawrence Lee
Bridge, Thomas E
Davis, John Clements
Enos, Paul
Frazier, A Joel
Grant, Stanley Cameron
Gundersen, James Novotny
Hambleton, William Weldon
Jones, Lois Marilyn
Leonard, Roy J
Merriam, Daniel F(rancis)
Merrill, William Meredith
Oros, Margaret Olava (Erickson)
Parizek, Eldon Joseph
Robison, Richard Ashby
Steeples, Donald Wallace
Tasch, Paul
Walters, Robert F
Williams, Wayne Watson

KENTUCKY
Brant, Russell Alan
Brown, William Randall
Campbell, Lois Jeannette
Chesnut, Donald R, Jr
Clark, Armin Lee
Cobb, James C

Conkin, James E
Drahovzal, James Alan
Ferm, John Charles
Gauri, Kharaiti Lal
Gildersleeve, Benjamin
Hagan, Wallace Woodrow
Haney, Donald C
Hunt, Graham Hugh
Kiefer, John David
MacQuown, William Charles, Jr
Philley, John Calvin
Seeger, Charles Ronald
Sendlein, Lyle V A
Smith, Gilbert Edwin
Stanonis, Francis Leo
Sweigard, Richard Joseph
Whaley, Peter Walter

LOUISIANA
Eisenstatt, Phillip
Fairchild, Joseph Virgil, Jr
Forman, McLain J
Franklin, George Joseph
Franks, Stephen Guest
Frey, Maurice G
Gagliano, Sherwood Moneer
Goldthwaite, Duncan
Groat, Charles George
Hartman, James Austin
Herrmann, Leo Anthony
Jones, Paul Hastings
Kumar, Madhurendu B
Leutze, Willard Parker
MacDougall, Robert Douglas
McGehee, Richard Vernon
Moore, Clyde H, Jr
Roberts, Harry Heil
Shaw, Nolan Gail
Skinner, Hubert Clayton
Stephens, Raymond Weathers, Jr
Tabbert, Robert L
Thomson, Alan
Utterback, Donald D
Van Lopik, Jack Richard
Von Almen, William Frederick I
Weidie, Alfred Edward
Wilcox, Ronald Erwin
Young, Leonard M

MAINE
Eastler, Thomas Edward
Farnsworth, Roy Lothrop
Fisher, Irving Sanborn
Gates, Olcott
Guidotti, Charles V
Howd, Frank Hawver
Norton, Stephen Allen
Osberg, Philip Henry
Splettstoesser, John Frederick
Tewhey, John David
Trefethen, Joseph Muzzy

MARYLAND
Arem, Joel Edward
Baker, Arthur A
Barker, John L, Jr
Beattie, Donald A
Brush, Lucien M(unson), Jr
Cleaves, Emery Taylor
Cohee, George Vincent
Doan, David Bentley
Dorr, John Van Nostrand, II
Everett, Ardell Gordon
Felsher, Murray
Gair, Jacob Eugene
Gaum, Carl H
Hamburger, Richard
Honkala, Fred Saul
Irvine, T Neil
Kappel, Ellen Sue
Karklins, Olgerts Longins
Kinney, Douglas Merrill
Lesure, Frank Gardner
Luth, William Clair
Mangen, Lawrence Raymond
May, Ira Philip
Mielke, James Edward
Miller, Ralph LeRoy
Nininger, Robert D
Ozol, Michael Arvid
Patterson, Charles Meade
Pavlides, Louis
Petrie, William Leo
Pettijohn, Francis John
Phair, George
Rozanski, George
Salisbury, John William, Jr
Segovia, Antonio
Wargotz, Eric S
Warren, Charles Reynolds
Weaver, Kenneth Newcomer
Yokel, Felix Y
Zall, Linda S
Zen, E-an

MASSACHUSETTS
Berggren, William Alfred
Blackey, Edwin Arthur, Jr
Bromery, Randolph Wilson
Brophy, Gerald Patrick
Brown, George D, Jr
Brownlow, Arthur Hume
Bryan, Wilfred Bottrill
Burchfiel, Burrell Clark

Geology, Applied, Economic & Engineering (cont)

Burns, Daniel Robert
Caldwell, Dabney Withers
Carr, Jerome Brian
Dennen, William Henry
Donohue, David Arthur Timothy
Donohue, John J
El-Baz, Farouk
Enzmann, Robert D
Farquhar, Oswald Cornell
Foose, Richard Martin
Foote, Freeman
Fox, William Templeton
Frimpter, Michael Howard
Furlong, Ira E
Genes, Andrew Nicholas
Gheith, Mohamed A
Hager, Jutta Lore
Hart, Stanley Robert
Hope, Elizabeth Greeley
Jacobsen, Stein Bjornar
Jenks, William Furness
McConnell, Robert Kendall
Motts, Ward Sundt
Patterson, Joseph Gilbert
Petersen, Ulrich
Pinson, William Hamet, Jr
Reuss, Robert L
Robinson, Peter
Roy, David C
Schalk, Marshall
Schultz, John Russell
Skehan, SJ James William
Socolow, Arthur A
Stalcup, Marvel C
Tarkoy, Peter J
Terzaghi, Ruth Doggett
Uchupi, Elazar
Walsh, Joseph Broughton
Wise, Donald U

MICHIGAN
Andersland, Orlando Baldwin
Buis, Patricia Frances
Catacosinos, Paul Anthony
Coyle, Peter
Fisher, James Harold
Hendrix, Thomas Eugene
Kalliokoski, Jorma Osmo Kalervo
Kehew, Alan Everett
Kelly, William Crowley
Kesler, Stephen Edward
Knowles, David M
Kunkle, George Robert
McKee, Edith Merritt
MacTavish, John N
Mills, Madolia Massey
Moore, Wayne Elden
Ogden, Lawrence
Peterson, Dallas Odell
Ronca, Luciano Bruno
Straw, William Thomas
Tharin, James Cotter
Trow, James
Vogel, Thomas A
Ward, Richard Floyd
Whitney, Marion Isabelle
Wicander, Edwin Reed

MINNESOTA
Bayer, Thomas Norton
Blake, Rolland Laws
Bleifuss, Rodney L
Bright, Robert C
Buchwald, Caryl Edward
Donovan, John Francis
Farnham, Paul Rex
Hoagberg, Rudolph Karl
Lepp, Henry
Matsch, Charles Leo
Meyers, James Harlan
Morey, Glenn Bernhardt
Richason, Benjamin Franklin, Jr
Southwick, David Leroy
Sparling, Dale R
Swain, Frederick Morrill, Jr
Tuthill, Samuel James
Walton, Matt Savage
Weiblen, Paul Willard
Willard, Robert Jackson

MISSISSIPPI
Brunton, George Delbert
Butler, Dwain Kent
Coleman, Neil Lloyd
Egloff, Julius, III
Happ, Stafford Coleman
Henderson, Arnold Richard
Johnson, Henry Stanley, Jr
Knight, Wilbur Hall
Krinitzsky, E(llis) L(ouis)
Laswell, Troy James
Lowrie, Allen
Malone, Philip Garcin
Mather, Bryant
Mather, Katharine Kniskern
Meylan, Maurice Andre
Paulson, Oscar Lawrence
Perry, Edward Belk
Reynolds, William Roger
Riggs, Karl A, Jr

Rucker, James Bivin
Russell, Ernest Everett
Seglund, James Arnold
Shamburger, John Herbert
Strom, Eric William

MISSOURI
Arvidson, Raymond Ernst
Barr, David John
Belt, Charles Banks, Jr
Bolter, Ernst A
Coveney, Raymond Martin, Jr
Danser, James W
Echols, Dorothy Jung
Freeman, Thomas J
Gadberry, Howard M(ilton)
Hagni, Richard D
Hasan, Syed Eqbal
Hatheway, Allen Wayne
Hilpman, Paul Lorenz
Hopkins, M E
Houseknecht, David Wayne
Howe, Wallace Brady
Howell, Dillon Lee
Jester, Guy Earlscort
Keller, Walter David
Kisvarsanyi, Eva Bognar
Knox, Burnal Ray
Levin, Harold Leonard
Prakash, Shamsher
Rockaway, John D, Jr
Ryan, Edward McNeill
Stitt, James Harry
Thomson, Kenneth Clair
Tibbs, Nicholas H
Tourtelot, Harry Allison
Viele, George Washington

MONTANA
Alt, David D
Balster, Clifford Arthur
Berg, Richard Blake
Dresser, Hugh W
Earll, Fred Nelson
Engel, Albert Edward John
Fanshawe, John Richardson, II
Groff, Sidney Lavern
Lange, Ian M
Peterson, James Algert
Ruppel, Edward Thompson
Stout, Koehler
Temple, Kenneth Loren
Thompson, Gary Gene
Walker, Leland J
Weidman, Robert McMaster
Winston, Donald

NEBRASKA
Bentall, Ray
Carlson, Marvin Paul
Dreeszen, Vincent Harold
Grew, Priscilla Croswell Perkins
Meade, Grayson Eichelberger
Nettleton, Wiley Dennis
Shroder, John Ford, Jr
Stout, Thompson Mylan
Tanner, Lloyd George
Underwood, Lloyd B
Wayne, William John

NEVADA
Bonham, Harold F, Jr
Carr, James Russell
Daemen, Jaak J K
Dickson, Frank Wilson
Ellinwood, Howard Lyman
Emerson, Donald Orville
Felts, Wayne Moore
Fiero, George William, Jr
Flint, Delos Edward
Fuerstenau, M(aurice) C(lark)
Garside, Larry Joe
Ghosh, Amitava
Gustafson, Lewis Brigham
Helm, Donald Cairney
Henry, Christopher D
Hibbard, Malcolm Jackman
Horton, Robert Carlton
Jones, Michael Baxter
Kaufmann, Robert Frank
Kukal, Gerald Courtney
Larson, Lawrence T
Lintz, Joseph, Jr
Littleton, Robert T
Lucas, James Robert
Manning, John Craige
Papke, Keith George
Payne, Anthony
Peck, John H
Ritter, Dale Franklin
Robinson, Raymond Francis
Schilling, John H(arold)
Slemmons, David Burton
Stone, John Grover, II
Taranik, James Vladimir
Trexler, Dennis Thomas
Watters, Robert James
Wyman, Richard Vaughn

NEW HAMPSHIRE
Birnie, Richard Williams
Boudette, Eugene L
Denny, Charles Storrow
Gaudette, Henri Eugene

Hoag, Roland Boyden, Jr
Holton, Adolphus
Hume, James David
Johnson, Gary Dean
Klotz, Louis Herman
Lyons, John Bartholomew
McKim, Harlan L
Page, Lincoln Ridler
Richter, Dorothy Anne

NEW JERSEY
Agron, Sam Lazrus
Bonini, William E
Brons, Cornelius Hendrick
Dike, Paul Alexander
Faust, George Tobias
Gilliland, William Nathan
Grow, George Copernicus, Jr
Hargraves, Robert Bero
Harrison, Jack Lamar
Manspeizer, Warren
Metsger, Robert William
Meyer, Richard Fastabend
Monahan, Edward James
Navarra, John Gabriel
Suppe, John
Widmer, Kemble
Yasso, Warren E

NEW MEXICO
Anderson, Roger Yates
Austin, George Stephen
Bejnar, Waldemere
Bieberman, Robert Arthur
Bish, David Lee
Broadhead, Ronald Frigon
Buden, Rosemary V
Chapin, Charles Edward
Clemons, Russell Edward
Cunningham, John E
Deal, Dwight Edward
Duffy, Clarence John
Elston, Wolfgang Eugene
Erikson, Mary Jane
Finney, Joseph J
Floran, Robert John
Flower, Rousseau Hayner
Geissman, John William
Griswold, George B
Havenor, Kay Charles
Hawley, John William
Heiken, Grant Harvey
Jacobsen, Lynn C
Kottlowski, Frank Edward
Lattman, Laurence Harold
Laughlin, Alexander William
McLemore, Virginia Teresa Hill
Meijer, Arend
Morin, George Cardinal Albert
Morris, David Albert
Neal, James Thomas
Rapaport, Irving
Renault, Jacques Roland
Riecker, Robert E
Shannon, Spencer Sweet, Jr
Shomaker, John Wayne
Smith, Clay Taylor
Trauger, Frederick Dale
Weber, Robert Harrison
Wengerd, Sherman Alexander
Whetten, John T
Wilson, Lee

NEW YORK
Adams, Robert W
Bhattacharji, Somdev
Bird, John Malcolm
Boger, Phillip David
Boshkov, Stefan
Cazeau, Charles J
Clemence, Samuel Patton
Coates, Donald Robert
Coch, Nicholas Kyros
Dubins, Mortimer Ira
Dunn, James Robert
Elberty, William Turner, Jr
Fairbridge, Rhodes Whitmore
Fakundiny, Robert Harry
Fickies, Robert H
Foster, Helen Laura
Gillett, Lawrence B
Gilman, Richard Atwood
Hatheway, Richard Brackett
Hawkins, William Max
Hay, Robert E
Hays, James D
Hewitt, Philip Cooper
Heyl, George Richard
Hutchison, David M
Isachsen, Yngvar William
Jacoby, Russell Stephen
Jicha, Henry Louis, Jr
Kulhawy, Fred Howard
Langwy, Chester Charles, Jr
Liebe, Richard Milton
Loring, Arthur Paul
Lowe, Kurt Emil
Ludman, Allan
Lundgren, Lawrence William, Jr
Mahtab, M Ashraf
Manos, Constantine T
Mattson, Peter Humphrey
Maurer, Robert Eugene
Mercer, Kermit R

Morisawa, Marie
Muller, Ernest Hathaway
Muller, Otto Helmuth
Multer, H Gray
Nair, Ramachandran Mukundalayam Sivarama
Newell, Norman Dennis
O'Rourke, Thomas Denis
Philbrick, Shailer Shaw
Potter, Donald B
Prucha, John James
Reitan, Paul Hartman
Rhodes, Frank Harold Trevor
Roberson, Herman Ellis
Robinson, Joseph Edward
Rodriguez, Joaquin
Ryan, William B F
Sanders, John Essington
Sargent, Kenneth Albert
Sirkin, Leslie A
Sorauf, James E
Speidel, David H
Stewart, John Conyngham
Sturm, Edward
Sutton, Robert George
Teichert, Curt
Vollmer, Frederick Wolfer
Wagner, Gerald Roy
Wang, Kia K
Wohlford, Duane Dennis
Wolff, Manfred Paul (Fred)
Young, Richard A

NORTH CAROLINA
Brown, Charles Quentin
Brown, Henry Seawell
Butler, James Robert
Callahan, John Edward
Carpenter, Robert Halstead
Chapman, John Judson
Feiss, Paul Geoffrey
Ferguson, Herman White
Forsythe, Randall D
Heroy, William Bayard, Jr
Horstman, Arden William
Karson, Jeffrey Alan
Kashef, A(bdel-Aziz) I(smail)
Marlowe, James Irvin
Mauger, Richard L
Mock, Steven James
Neal, Donald Wade
Parker, John Mason, III
Riggs, Stanley R
Smith, Norman Cutler
Teague, Kefton Harding
Textoris, Daniel Andrew
Tyrrell, Willis W, Jr
White, William Alexander

NORTH DAKOTA
Bickel, Edwin David
Brophy, John Allen
Karner, Frank Richard
Phillips, Monte Leroy
Reid, John Reynolds, Jr

OHIO
Anderson, John Jerome
Babcock, Loren Edward
Banks, Philip Oren
Bates, Robert Latimer
Bergstrom, Stig Magnus
Bernhagen, Ralph John
Bever, James Edward
Bladh, Kenneth W
Bonnett, Richard Brian
Carlson, Ernest Howard
Carlton, Richard Walter
Caster, Kenneth Edward
Collins, Horace Rutter
Colombini, Victor Domenic
Coogan, Alan H
Cropp, Frederick William, III
Eckelmann, Frank Donald
Elliot, David Hawksley
Emerson, John Wilford
Everett, K R
Fay, Philip S
Fenner, Peter
Frank, Glenn William
Fuller, James Osborn
Gerrard, Thomas Aquinas
Harmon, David Elmer, Jr
Harris, C Earl, Jr
Harris-Noel, Ann (F H) Graetsch
Heimlich, Richard Allen
Hyland, John R(oth)
Kahle, Charles F
Khawaja, Ikram Ullah
Kneller, William Arthur
Kulander, Byron Rodney
Laughon, Robert Bush
McKenzie, Garry Donald
Mancuso, Joseph J
Maynard, James Barry
Means, Jeffrey Lynn
Moody, Judith Barbara
Morris, Robert William
Murdoch, Lawrence Corlies, III
Nance, Richard Damian
Nash, David Byer
Noel, James A
Pope, John Keyler
Potter, Paul Edwin

Pride, Douglas Elbridge
Pryor, Wayne Arthur
Pushkar, Paul
Reinhart, Roy Herbert
Root, Samuel I
Schoonmaker, George Russell
Schumm, Brooke, Jr
Skinner, Walter Swart
Springer, George Henry
Steel, Warren G
Summerson, Charles Henry
Unrug, Raphael
Utgard, Russell Oliver
Villar, James Walter
Webb, Peter Noel
Weitz, John Hills
White, John Francis
White, Sidney Edward
Wiese, Robert George, Jr
Wu, Tien Hsing

OKLAHOMA
Barczak, Virgil J
Bennison, Allan P
Bradley, John Samuel
Busch, Daniel Adolph
Cardott, Brian Joseph
Chenoweth, Philip Andrew
Dickey, Parke Atherton
Erickson, John William
Espach, Ralph H, Jr
Forgotson, James Morris, Jr
Friedman, Samuel Arthur
Goldstein, August, Jr
Green, Thom Henning
Hedlund, Richard Warren
Helander, Donald P(eter)
Henderson, Frederick Bradley, III
Holmes, Clifford Newton
Horn, Myron K
Hyne, Norman John
Johnson, Kenneth Sutherland
Joshi, Sadanand D
Kent, Douglas Charles
Kleen, Harold J
Koller, Glenn R
Konkel, Philip M
London, David
Mankin, Charles John
Masters, Bruce Allen
Menzie, Donald E
Meyers, William C
Moritz, Carl Albert
Mumma, Martin Dale
Naff, John Davis
Oliphant, Charles Winfield
Powell, Benjamin Neff
Root, Paul John
Sanderson, George Albert
Scott, Robert W
Shelton, John Wayne
Stewart, Gary Franklin
Stone, John Elmer
Stover, Dennis Eugene
Tillman, Roderick W
Wilson, Leonard Richard
Zimmer, Louis George

OREGON
Agnew, Allen Francis
Allen, John Eliot
Baldwin, Ewart Merlin
Ball, Douglas
Barnett, Stockton Gordon, III
Beaulieu, John David
Benson, Gilbert Thomas
Boucot, Arthur James
Field, Cyrus West
Griggs, Allan Bingham
Griswold, Daniel H(alsey)
Hook, John W
Hull, Donald Albert
Kays, M Allan
Lotspeich, Frederick Benjamin
Orr, William N
Palmer, Leonard A
Purdom, William Berlin
Rice, Jack Morris
Sangrey, Dwight A
Schroeder, Warren Lee
Staples, Lloyd William
Swanston, Douglas Neil
Taylor, Edward Morgan
Yeats, Robert Sheppard
Youngquist, Walter

PENNSYLVANIA
Adams, John Kendal
Allen, Jack C, Jr
Andrews, George William
Ballmann, Donald Lawrence
Barnes, Hubert Lloyd
Barron, Eric James
Biederman, Edwin Williams, Jr
Bopp, Frederick, III
Brady, Keith Bryan Craig
Briggs, Reginald Peter
Bushnell, Kent O
Clark, Roger Alan
Davis, Alan
Eggler, David Hewitt
Faas, Richard William
Fergusson, William Blake
Fletcher, Frank William
Frantz, Wendelin R
Garber, Murray S
Gold, David Percy
Goodspeed, Robert Marshall
Gray, Ralph J
Greenberg, Seymour Samuel
Grossman, I G
Grubbs, Donald Keeble
Haefner, Richard Charles
Hamel, James V(ictor)
Harker, Robert Ian
Heck, Edward Timmel
Hoskins, Donald Martin
Howe, Richard Hildreth
Illick, J(ohn) Rowland
Inners, Jon David
Jacobs, Alan Martin
Jenkins, George Robert
Kunasz, Ihor Andrew
Lidiak, Edward George
Lukert, Michael T
Mehta, Sudhir
Myer, George Henry
Nagell, Raymond H
Parizek, Richard Rudolph
Pees, Samuel T(homas)
Rose, Arthur William
Ryan, John Donald
Schneider, Michael Charles
Schultz, Lane D
Sclar, Charles Bertram
Sims, Samuel John
Skolnick, Herbert
Slingerland, Rudy Lynn
Smith, Arthur R
Stephenson, Robert Charles
Stingelin, Ronald Werner
Tomikel, John
Trexler, John Peter
Voight, Barry
Williams, Eugene G
Wright, Lauren Albert
Young, John Albion, Jr

RHODE ISLAND
Cain, J(ames) Allan
Moore, George Emerson, Jr
Parmentier, Edgar M(arc)

SOUTH CAROLINA
Bartholomew, Mervin Jerome
Cohen, Arthur David
Colquhoun, Donald John
Ehrlich, Robert
Eyer, Jerome Arlan
Fisher, Stanley Parkins
Geidel, Gwendolyn
Gipson, Mack, Jr
Glenn, George R(embert), Sr
Griffin, Villard Stuart, Jr
Haselton, George Montgomery
Hayes, Miles O
Lawrence, David Reed
Nairn, Alan Eben Mackenzie
Nichols, Chester Encell
Pirkle, Earl C
Pirkle, William Arthur
Snipes, David Strange
Visvanathan, T R

SOUTH DAKOTA
Davis, Briant LeRoy
Gries, John Paul
Lisenbee, Alvis Lee
McGregor, Duncan J
Mickelson, John Chester
Norton, James Jennings
Parrish, David Keith
Paterson, Colin James
Redden, Jack A
Stevenson, Robert Evans
Tipton, Merlin J

TENNESSEE
Bergenback, Richard Edward
Brown, Bevan W, Jr
Byerly, Don Wayne
Churnet, Habte Giorgis
Deboo, Phili B
Deininger, Robert W
Delcourt, Paul Allen
Dodson, Chester Lee
Grant, Leland F(auntleroy)
Helton, Walter Lee
Hill, William T
Hopkins, Richard Allen
Klepser, Harry John
Kopp, Otto Charles
Lumsden, David Norman
Luther, Edward Turner
Maher, Stuart Wilder
Mason, John Frederick
Naney, Michael Terrance
Reesman, Arthur Lee
Stearns, Richard Gordon
Wilson, Robert Lake

TEXAS
Abernathy, Bobby F
Agatston, Robert Stephen
Ahr, Wayne Merrill
Allen, Peter Martin
Al-Saadoon, Faleh T
Amsbury, David Leonard
Anderson, Duwayne Marlo
Anthony, Elizabeth Youngblood
Anthony, John Williams
Aronow, Saul
Arp, Gerald Kench
Baie, Lyle Fredrick
Ball, Stanton Mock
Bally, Albert Walter
Barnes, Virgil Everett
Barrow, Thomas D
Beaver, H H
Bence, Alfred Edward
Berg, Robert R
Berkhout, Aart W J
Bingler, Edward Charles
Bishop, Margaret S
Bishop, Richard Stearns
Blackwelder, Blake Winfield
Bloch, Salman
Boon, John Daniel, Jr
Boyer, Robert Ernst
Bromberger, Samuel H
Brooks, James Elwood
Brown, Leonard Franklin, Jr
Buis, Otto J
Bullard, Fred Mason
Burgess, Jack D
Burk, Creighton
Burkart, Burke
Burke, Kevin Charles
Burton, Robert Clyde
Butler, John C
Caffey, James Enoch
Campbell, Michael David
Castano, John Roman
Chakravarty, Indranil
Chuber, Stewart
Clabaugh, Stephen Edmund
Clark, Howard Charles, Jr
Clark, Kenneth Frederick
Claypool, George Edwin
Clendening, John Albert
Cochrum, Arthur L
Coffman, James Bruce
Cook, Theodore Davis
Cordell, Robert James
Dahl, Harry Martin
Dally, Jesse LeRoy
Daugherty, Franklin W
Davies, David K
Denison, Rodger Espy
Dodge, Charles Fremont
Dodge, Rebecca L
Dorfman, Myron Herbert
Doyle, Frank Lawrence
Duke, Michael B
Dunlap, Henry Francis
Durham, Clarence Orson, Jr
Eckles, Wesley Webster, Jr
Edwards, Marc Benjamin
Eifler, Gus Kearney, Jr
Elam, Jack Gordon
Ellison, Samuel Porter, Jr
Eveland, Harmon Edwin
Feazel, Charles Tibbals
Fett, John D
Finley, Robert James
Fisher, William Lawrence
Flawn, Peter Tyrrell
Folk, Stewart H
Font, Robert G
Fountain, Lewis Spencer
French, William Edwin
Gallagher, John Joseph, Jr
Gealy, John Robert
Ghaly, Tharwat Shahata
Gibson, George R
Gould, Howard Ross
Granath, James Wilton
Gregory, A(lvin) R(ay)
Gruy, Henry Jones
Gucwa, Paul Ramon
Gustavson, Thomas Carl
Guven, Necip
Haas, Merrill Wilber
Hadley, Kate Hill
Haeberle, Frederick Roland
Halbouty, Michel Thomas
Haller, Charles Regis
Hardin, George C, Jr
Harris, Rae Lawrence, Jr
Hayward, Oliver Thomas
Helwig, James A
Hewitt, Charles Hayden
Heyman, Louis
Heymann, Dieter
High, Lee Rawdon, Jr
Hills, John Moore
Hirsch, John Michele
Hoffer, Jerry M
Holden, Frederick Thompson
Hood, William Calvin
Horz, Friedrich
Howard, John Hall
Huffington, Roy Michael
Hurley, Neal Lilburn
Ingerson, Fred Earl
Jackson, Martin Patrick Arden
Jeffords, Russell MacGregor
Jennings, Albert Ray
Jodry, Richard L
Johnson, Carlton Robert
Johnson, Lloyd N(ewhall)
Jones, Ernest Austin, Jr
Jones, Eugene Laverne
Jones, Theodore Sidney
Kaiser, William Richard
Keenmon, Kendall Andrews
Kerans, Charles
Kidwell, Albert Laws
King, Elbert Aubrey, Jr
Knoll, Kenneth Mark
Koenig, Karl Joseph
Koepnick, Richard Borland
Kulla, Jean B
Laird, Wilson Morrow
Land, Lynton S
Lane, Donald Wilson
Lawrence, Philip Linwood
LeBlanc, Arthur Edgar
Lehman, David Hershey
Lehmann, Elroy Paul
Levinson, Stuart Alan
Longacre, Susan Ann Burton
Louden, L Richard
McCulloh, Thane H
McDaniel, Willard Rich
McFarlan, Edward, Jr
McGookey, Donald Paul
McIver, Norman L
Maddocks, Rosalie Frances
Massell, Wulf F
Mathewson, Christopher Colville
Matthews, Martin David
Mattis, Allen Francis
Mattox, Richard Benjamin
Matuszak, David Robert
Maxwell, John Crawford
Mazzullo, James Michael
Merrill, Glen Kenton
Merrill, Robert Kimball
Meyer, Franz O
Milling, Marcus Eugene
Monsees, James E
Morrison, Donald Allen
Morton, Robert Alex
Mundt, Philip A
Murany, Ernest Elmer
Murphy, Daniel L
Murray, Grover Elmer
Murray, Joseph Buford
Mutis-Duplat, Emilio
Nanz, Robert Hamilton, Jr
Nugent, Robert Charles
Owen, Donald Edward
Pampe, William R
Parker, Travis Jay
Patrick, Wesley Clare
Peterson, Hazel Agnes
Phillips, Joseph D
Pierce, Robert Wesley
Pirie, Robert Gordon
Prescott, Basil Osborne
Price, William Charles
Pritchett, William Carr
Raba, Carl Franz, Jr
Rambow, Frederick H K
Reaser, Donald Frederick
Reeves, C C, Jr
Reeves, Robert Grier (Lefevre)
Reso, Anthony
Richardson, Joseph Gerald
Ries, Edward Richard
Riese, Walter Charles Rusty
Riley, Charles Marshall
Roebuck, Isaac Field
Roehl, Perry Owen
Rogers, Marion Alan
Roper, Paul James
Rose, Peter R
Rose, William Dake
Rosen, Norman Charles
Sahinen, Winston Martin
St John, Bill
Saunders, Donald Frederick
Schramm, Martin William, Jr
Schroeder, Melvin Carroll
Schumacher, Dietmar
Schwarzer, Theresa Flynn
Scobey, Ellis Hurlbut
Shipman, Ross Lovelace
Slodowski, Thomas R
Smith, Charles Isaac
Smith, Jan G
Smith, Michael Alexis
Sneider, Robert Morton
Snelson, Sigmund
Soderman, J William
Spencer, Alexander Burke
Stanton, Robert James, Jr
Stead, Frederick L
Strickland, John Willis
Sullivan, Herbert J
Sullivan, Kathryn D
Swanson, Donald Charles
Swanson, Eric Rice
Thompson, Samuel, III
Tonking, William Harry
Trace, Robert Denny
Tyler, Noel
Van Rensburg, Willem Cornelius Janse
Van Siclen, DeWitt Clinton
Walper, Jack Louis
Warshauer, Steven Michael
Watkins, Jackie Lloyd
Wegner, Robert Carl
Wermund, Edmund Gerald, Jr
White, David Archer

Geology, Applied, Economic & Engineering (cont)

White, N(ikolas) F(rederick)
Wiese, John Herbert
Wilde, Garner Lee
Woodruff, Charles Marsh, Jr
Young, William Allen

UTAH
Albee, Howard F
Allison, Merle Lee
Alvord, Donald C
Bissell, Harold Joseph
Bullock, Kenneth C
Bushman, Jess Richard
Christiansen, Eric H
Christiansen, Francis Wyman
Doelling, Hellmut Hans
Gates, Joseph Spencer
Godfrey, Andrew Elliott
Hamblin, William Kenneth
Hintze, Lehi Ferdinand
Hood, James Warren
Jensen, Mead LeRoy
Lohrengel, Carl Frederick, II
McCalpin, James P
McCleary, Jefferson Rand
McCoy, Roger Michael
Moore, Joseph Neal
Mosher, Loren Cameron
Moyle, Richard W
Nielson, Dennis Lon
Pashley, Emil Frederick, Jr
Proctor, Paul Dean
Rawson, Richard Ray
Ross, Howard Persing
Smith, Robert Baer
Southard, Alvin Reid
Spotts, John Hugh
Vaughn, Danny Mack
Welsh, John Elliott, Sr
Whelan, James Arthur
Willden, Charles Ronald
Zavodni, Zavis Marian

VERMONT
Jaffe, Howard William
Klemme, Hugh Douglas
Ratté, Charles A
Stanley, Rolfe S
Stoiber, Richard Edwin
Warren, John Stanley

VIRGINIA
Andrews, Robert Sanborn
Bartlett, Charles Samuel, Jr
Barton, Paul Booth, Jr
Bennett, Gordon Daniel
Bergin, Marion Joseph
Bethke, Philip Martin
Bick, Kenneth F
Bowen, Zeddie Paul
Brobst, Donald Albert
Brown, Glen Francis
Carter, William Douglas
Chao, Edward Ching-Te
Clark, Sandra Helen Becker
Conley, James Franklin
Craig, James Roland
Cronin, Thomas Mark
Cunningham, Charles G
Davidson, David Francis
Doe, Bruce R
Dragonetti, John Joseph
Drew, Lawrence James
Eggleston, Jane R
Ellison, Robert L
Epstein, Jack Burton
Ericksen, George Edward
Eriksson, Kenneth Andrew
Ern, Ernest Henry
Froelich, Albert Joseph
Fryklund, Verne Charles, Jr
Gawarecki, Stephen Jerome
Gluskoter, Harold Jay
Goldsmith, Richard
Goodell, Horace Grant
Greenwood, William R
Hackett, Orwoll Milton
Harbour, Jerry
Hart, Dabney Gardner
Heald, Pamela
Hearn, Bernard Carter, Jr
Hemley, John Julian
Herd, Darrell Gilbert
Hobbs, Charles Roderick Bruce, Jr
Horton, James Wright, Jr
Hunt, Lee McCaa
Husted, John E
Inderbitzen, Anton Louis
Karfakis, Mario George
Kauffman, Marvin Earl
Kearns, Lance Edward
Konikow, Leonard Franklin
Kozak, Samuel J
Kreipke, Merrill Vincent
Krushensky, Richard D
Kydes, Andy Steve
Lohman, Kenneth Elmo
Lowry, Wallace Dean
Lynd, Langtry Emmett
McCammon, Richard B
Maccini, John Andrew
McGuire, Odell
Masters, Charles Day
Meissner, Charles Roebling, Jr
Milici, Robert Calvin
Milliman, John D
Milton, Daniel Jeremy
Monroe, Watson Hiner
Naeser, Charles Wilbur
Neiheisel, James
Newton, Elisabeth G
Noble, Edwin Austin
Outerbridge, William Fulwood
Ovenshine, A Thomas
Pasko, Thomas Joseph, Jr
Patterson, Sam H
Peck, Dallas Lynn
Pickering, Ranard Jackson
Ratcliffe, Nicholas Morley
Reilly, Thomas E
Reinemund, John Adam
Reinhardt, Juergen
Ridge, John Drew
Rioux, Robert Lester
Riva, Joseph Peter, Jr
Robbins, Eleanora Iberall
Robertson, Eugene Corley
Robinson, Edwin S
Roseboom, Eugene Holloway, Jr
Ryder, Robert Thomas
Said, Rushdi
Schmidt, Robert Gordon
Scolaro, Reginald Joseph
Sears, Charles Edward
Shelkin, Barry David
Sinha, Akhuary Krishna
Smith, Richard Elbridge
Smith, William Lee
Spencer, Edgar Winston
Stebbins, Robert H
Stein, Holly Jayne
Sutter, John Frederick
Toulmin, Priestley
Wagner, John Edward
Wayland, Russell Gibson
Weiss, Martin
Wieczorek, Gerald Francis
Wiesnet, Donald Richard
Wolff, Roger Glen
Wood, Leonard Alton
Wood, Leonard E(ugene)
Wright, Thomas L
Zadnik, Valentine Edward

WASHINGTON
Anderson, Norman Roderick
Armstrong, Frank C(larkson Felix)
Bamford, Robert Wendell
Banks, Norman Guy
Brewer, William Augustus
Brown, Donald Jerould
Brown, Randall Emory
Caggiano, Joseph Anthony, Jr
Carson, Robert James, III
Cheney, Eric Swenson
Christman, Robert Adam
Costa, John Emil
De Goes, Louis
Demirel, T(urgut)
Easterbrook, Don J
Ellis, Ross Courtland
Finlay, Walter L(eonard)
Foit, Franklin Frederick, Jr
Foley, Michael Glen
Fulmer, Charles V
Galster, Richard W
Gerlach, Terrence Melvin
Hallet, Bernard
Henricksen, Thomas Alva
Hoblitt, Richard Patrick
Holmes, Mark Lawrence
Hopkins, William Stephen
James, Harold Lloyd
Karlstrom, Thor Nels Vincent
Kesler, Thomas L
Kiilsgaard, Thor H
McHuron, Clark Ernest
Mallory, Virgil Standish
Michelsen, Ari Montgomery
Miller, C Dan
Oles, Keith Floyd
Olson, Edwin Andrew
Peterson, Richard Charles
Pierson, Thomas Charles
Pollack, Jerome Marvin
Rau, Weldon Willis
Roberts, Ralph Jackson
Rosengreen, Theodore E
Ross, Charles Alexander
Scott, Kevin M
Scott, William Edward
Snook, James Ronald
Sorem, Ronald Keith
Steele, William Kenneth
Stewart, Richard John
Terrel, Ronald Lee
Threet, Richard Lowell
Vance, Joseph Alan
Waldron, Howard Hamilton (Hank)
Webster, Gary Dean
Weis, Paul Lester
Wodzicki, Antoni
Wolfe, Edward W

WEST VIRGINIA
Behling, Robert Edward
Bird, Samuel Oscar, II
Chen, Ping-Fan
Donaldson, Alan C
Erwin, Robert Bruce
Gwinn, James E
Lessing, Peter
Ludlum, John Charles
Murphy, Allen Emerson
Rieke, Herman Henry, III
Ting, Francis Ta-Chuan

WISCONSIN
Broughton, William Albert
Brown, Bruce Elliot
Brown, Philip Edward
Byers, Charles Wesley
Cameron, Eugene Nathan
Dickas, Albert Binkley
Edil, Tuncer Berat
Emmons, Richard Conrad
Gates, Robert Maynard
Govin, Charles Thomas, Jr
LaBerge, Gene L
Lasca, Norman P, Jr
Meyer, Robert Paul
Mudrey, Michael George, Jr
Mursky, Gregory
Ostrom, Meredith Eggers
Palmquist, John Charles
Paull, Richard Allen
Robertson, James Magruder
Ross, Theodore William
Shea, James H
Stenstrom, Richard Charles
Stieglitz, Ronald Dennis
Tychsen, Paul C
Weege, Randall James
Weis, Leonard Walter
West, Walter Scott
Willis, Ronald Porter

WYOMING
Borgman, Leon Emry
Curry, William Hirst, III
Glass, Gary Bertram
Houston, Robert S
Huff, Kenneth O
Love, John David
Metzger, William John
Snoke, Arthur Wilmot

PUERTO RICO
Brooks, Harold Kelly

ALBERTA
Andrichuk, John Michael
Burwash, Ronald Allan
Campbell, Finley Alexander
Campbell, John Duncan
Cecile, Michael Peter
Cruden, David Milne
Folinsbee, Robert Edward
Furnival, G M
Halferdahl, Laurence Bowes
Henderson, Gerald Gordon Lewis
Hills, Leonard Vincent
Hriskevich, Michael Edward
Hunt, C Warren
Klassen, Rudolph Waldemar
Levinson, Alfred Abraham
McMillan, Neil John
Masters, John Alan
Mellon, George Barry
Morgenstern, N(orbert) R
Neale, Ernest Richard Ward
Nesbitt, Bruce Edward
Norford, Brian Seeley
Nowlan, Gofrey S
Okulitch, Andrew Vladimir
Okulitch, Vladimir Joseph
Oliver, Thomas Albert
Ollerenshaw, Neil Campbell
Peirce, John Wentworth
Raasch, Gilbert O
Scafe, Donald William
Schmidt, Volkmar
Smith, Dorian Glen Whitney
Stelck, Charles Richard
Tankard, Anthony James
Wardlaw, Norman Claude
Young, Frederick Griffin

BRITISH COLUMBIA
Arnold, Ralph Gunther
Barnes, William Charles
Bustin, Robert Marc
Byrne, Peter M
Chase, Richard L
Clague, John Joseph
Clark, Lloyd Allen
Cooke, Herbert Basil Sutton
Danner, Wilbert Roosevelt
Gabrielse, Hubert
Gazin, Charles Lewis
Greenwood, Hugh J
Jackson, Lionel Eric
Kieffer, Susan Werner
Mathews, William Henry
Mothersill, John Sydney
Murray, James W
Payne, Myron William
Pocock, Stanley Albert John
Robinson, Gershon Duvall
Roddick, James Archibald
Sinclair, Alastair James
Soregaroli, A(rthur) E(arl)
Sutherland-Brown, Atholl
Tozer, E T
Warren, Harry Verney
Wheeler, John Oliver
Wynne-Edwards, Hugh Robert

MANITOBA
Anderson, Donald T
Baracos, Andrew
Cerny, Petr
Lajtai, Emery Zoltan
McAllister, Arnold Lloyd
Teller, James Tobias
Turnock, A C
Whitaker, Sidney Hopkins
Wilson, Harry David Bruce

NEW BRUNSWICK
Moore, John Carman Gailey
Mossman, David John

NEWFOUNDLAND
King, Arthur Francis
King, Michael Stuart
McKillop, J(ohn) H
Williams, Harold

NOVA SCOTIA
Cormier, Randal
Greenblatt, Jayson Herschel
Hacquebard, Peter Albertus
Milligan, George Clinton
Riddell, John Evans
Shaw, William S
Williamson, Douglas Harris
Zentilli, Marcos

ONTARIO
Anderson, Francis David
Appleyard, Edward Clair
Armstrong, Herbert Stoker
Baragar, William Robert Arthur
Beales, Francis William
Bell, Richard Thomas
Blackadar, Robert Gordon
Brummer, Johannes J
Burk, Cornelius Franklin, Jr
Calder, Peter N
Cameron, Robert Alan
Card, Kenneth D
Chandler, Frederick William
Chua, Kian Eng
Church, William Richard
Coakley, John Phillip
Cowan, W R
Crocket, James Harvie
Cumming, Leslie Merrill
Currie, John Bickell
Davies, James Frederick
Dence, Michael Robert
Donaldson, John Allan
Duke, John Murray
Evans, James Eric Lloyd
Franklin, James McWillie
Frarey, Murray James
Fulton, Robert John
Fyles, John Gladstone
Gibb, Richard A
Goranson, Edwin Alexander
Grant, Douglas Roderick
Gussow, W(illia)m C(arruthers)
Haynes, Simon John
Hill, Patrick Arthur
Hodder, Robert William
Hoffman, Douglas Weir
Jenness, Stuart Edward
Joubin, Franc Renault
Karrow, Paul Frederick
Kenney, T Cameron
Kissin, Stephen Alexander
Kramer, James R(ichard)
Lee, Hulbert Austin
Leech, Geoffrey Bosdin
Lenz, Alfred C
Lumbers, Sydney Blake
McLaren, Digby Johns
MacRae, Neil D
Menzies, John
Michener, Charles Edward
Moore, John Marshall, Jr
Naldrett, Anthony James
Norris, Geoffrey
Palmer, H Currie
Pearce, George William
Petruk, William
Prince, Alan Theodore
Pye, Edgar George
Reesor, John Elgin
Roots, Ernest Frederick
Scott, John Stanley
Scott, Steven Donald
Smith, Charles Haddon
Sonnenfeld, Peter
Spooner, Ed Thornton Casswell
Stesky, Robert Michael
Symons, David Thorburn Arthur
Terasmae, Jaan
Thorpe, Ralph Irving
Tovell, Walter Massey
Van Loon, Jon Clement

Westermann, Gerd Ernst Gerold
Yole, Raymond William

QUEBEC
Blais, Roger A
Brown, Alex Cyril
Chagnon, Jean Yves
Chown, Edward Holton
Dionne, Jean-Claude
Doig, Ronald
Eakins, Peter Russell
Globensky, Yvon Raoul
Ladanyi, Branko
MacLean, Wallace H
Mamet, Bernard Leon
Martignole, Jacques
Martin, Robert Francois Churchill
Mountjoy, Eric W
Mukherji, Kalyan Kumar
Perrault, Guy
Poorooshasb, Hormozd Bahman
Sabourn, Robert Joseph Edmond
Seguin, Maurice Krisholm
Webber, George Roger

SASKATCHEWAN
Christiansen, E A
Coleman, Leslie Charles
Davies, John Clifford
Kent, Donald Martin
Kupsch, Walter Oscar
Langford, Fred F
Mollard, John D
Vigrass, Laurence William

OTHER COUNTRIES
Alonso, Ricardo N
Arce, Jose Edgar
Charlier, Roger Henri
Conolly, John R
Fritz, Peter
Govett, Gerald James
Hoke, John Humphreys
Hsu, Kenneth Jinghwa
Jolly, Janice Laurene Willard
Kennedy, Michael John
Libby, Willard Gurnea
Lynts, George Willard
Van Den Bold, Willem Aaldert
Von Huene, Roland
Walton, Wayne J A, Jr
Whitten, Eric Harold Timothy
Winkler, Virgil Dean
Wolfe, John A(llen)
Yazicigil, Hasan

Geology, Structural

ALABAMA
Groshong, Richard Hughes, Jr

ALASKA
Karl, Susan Margaret
Miller, Lance Davison
Weber, Florence Robinson

ARIZONA
Barnes, Charles Winfred
Chase, Clement Grasham
Coney, Peter James
Cree, Allan
Davis, George Herbert
Duffield, Wendell Arthur
Fink, Jonathan Harry
Lucchitta, Baerbel Koesters
Lucchitta, Ivo
McCullough, Edgar Joseph, Jr
Ragan, Donal Mackenzie
Sellin, H A
Shoemaker, Eugene Merle
Titley, Spencer Rowe
Wallace, Terry Charles, Jr
Whorton, Chester D

ARKANSAS
Konig, Ronald H

CALIFORNIA
Alvarez, Walter
Bailey, Roy Alden
Beard, Charles Noble
Beaty, David Wayne
Berkland, James Omer
Bird, Peter
Brew, David Alan
Carr, Michael H
Chang, Kaung-Jain
Chapman, Robert Mills
Christiansen, Robert Lorenz
Christie, John McDougall
Ciancanelli, Eugene Vincent
Clark, Malcolm Mallory
Colburn, Ivan Paul
Compton, Robert Ross
Creely, Robert Scott
Crowell, John Chambers
Cruikshank, Dale Paul
Csejtey, Bela, Jr
Davis, Gregory Arlen
Day, Howard Wilman
DePaolo, Donald J
Donath, Fred Arthur
Durham, William Bryan
Farrington, William Benford
Gabelman, John Warren
Galehouse, Jon Scott
Grantz, Arthur
Green, Harry Western, II
Hall, Clarence Albert, Jr
Hanan, Barry Benton
Harpster, Robert E
Hopson, Clifford Andrae
Hose, Richard K
Howard, Keith Arthur
Ingersoll, Raymond Vail
Irwin, James Joseph
Janke, Norman Charles
Kirby, James M
Kopf, Rudolph William
Loney, Robert Ahlberg
Lyon, Ronald James Pearson
McKague, Herbert Lawrence
Mackevett, Edward M
Moores, Eldridge Morton
Morris, John David
Nielson, Jane Ellen
Nur, Amos M
Oertel, Gerhard Friedrich
Olson, Jerry Chipman
Page, Benjamin Markham
Page, Robert Alan, Jr
Pampeyan, Earl H
Pierce, William G
Pincus, Howard Jonah
Polit, Andres Cassard
Rosenfeld, John L
Rudnyk, Marian E
Saleeby, Jason Brian
Sarna-Wojcicki, Andrei M
Seff, Philip
Sieh, Kerry Edward
Silver, Eli Alfred
Smith, Robert Leland
Sylvester, Arthur Gibbs
Tabor, Rowland Whitney
Thompson, George Albert
Tobisch, Othmar Tardin
Twiss, Robert John
Vasco, Donald Wyman
Wadsworth, William Bingham
Wenk, Hans-Rudolf
Wentworth, Carl M, Jr
Wright, William Herbert, III
Zoback, Mary Lou Chetlain

COLORADO
Anderson, Ernest R
Ayler, Maynard Franklin
Braddock, William A
Bryant, Bruce Hazelton
Cole, James Channing
Cowart, Vicki Jane
Drewes, Harald D
Elder, Curtis Harold
Erslev, Eric Allan
Fails, Thomas Glenn
Friedman, Jules Daniel
Fuchs, Robert L
Grose, Thomas Lucius Trowbridge
Hamilton, Warren Bell
Henrickson, Eiler Leonard
Heyl, Allen Van, Jr
Hustrulid, William A
Keefer, William Richard
Ketner, Keith B
Kupfer, Donald Harry
Link, Peter K
Lowell, James Diller
McCallum, Malcolm E
Moore, David A
Moore, David Warren
Oriel, Steven S
Parker, John Marchbank
Pierce, Kenneth Lee
Pillmore, Charles Lee
Raup, Robert Bruce, Jr
Rold, John W
Rowley, Peter DeWitt
Sable, Edward George
Sanborn, Albert Francis
Sass, Louis Carl
Selverstone, Jane Elizabeth
Shanks, Wayne C, III
Sheridan, Douglas Maynard
Silberling, Norman John
Skipp, Betty Ann
Swanson, Vernon E
Trimble, Donald E
Verbeek, Earl Raymond
Warner, Lawrence Allen
Wheeler, Russell Leonard
Zartman, Robert Eugene

CONNECTICUT
DeBoer, Jelle
Rodgers, John

DISTRICT OF COLUMBIA
Brocoum, Stephan John
Erwin, Douglas Hamilton
Knox, Arthur Stewart
Nolan, Thomas Brennan
Rogers, Thomas Hardin
Rumble, Douglas, III
Watters, Thomas Robert
Wilson, Frederick Albert

FLORIDA
Draper, Grenville
Rosendahl, Bruce Ray
Snow, Eleanour Anne
Tull, James Franklin

GEORGIA
Goodwin, Sidney S

HAWAII
Moore, Gregory Frank
Thornber, Carl Richard

IDAHO
Hall, William Bartlett

ILLINOIS
Buschbach, Thomas Charles
Klasner, John Samuel
Liu, Mian
Rue, Edward Evans
Woodland, Bertram George

INDIANA
Chen, Wai-Fah
Chowdhury, Dipak Kumar
Dunning, Jeremy David
Hamburger, Michael Wile
Martin, Charles Wellington
Rigert, James Aloysius
Wintsch, Robert P

IOWA
Hoppin, Richard Arthur

KANSAS
Aber, James Sandusky
Blythe, Jack Gordon
Gries, John Charles
Underwood, James Ross, Jr

KENTUCKY
Brown, William Randall
Ewers, Ralph O
Hagan, Wallace Woodrow
Rast, Nicholas
Seeger, Charles Ronald
Von Schonfeldt, Hilmar Armin

LOUISIANA
Forman, McLain J
Frey, Maurice G
Kinsland, Gary Lynn
Kumar, Madhurendu B
Wilcox, Ronald Erwin

MAINE
Hussey, Arthur M, II

MARYLAND
Allenby, Richard John, Jr
Baker, Arthur A
Beal, Robert Carl
Brown, Michael
Edwards, Jonathan, Jr
Fisher, George Wescott
Pavlides, Louis

MASSACHUSETTS
Ballard, Robert D
Bonnichsen, Martha Miller
Brace, William Francis
Burger, Henry Robert, III
Hepburn, John Christopher
Herda, Hans-Heinrich Wolfgang
Hoffman, Paul Felix
Kamen-Kaye, Maurice
Kleinrock, Martin Charles
Lin, Jian
McGill, George Emmert
Takahashi, Kozo
Thompson, Margaret Douglas
Tucholke, Brian Edward

MICHIGAN
Frankforter, Weldon D
Huntoon, Jacqueline E
McKee, Edith Merritt
Van der Voo, Rob

MINNESOTA
Holst, Timothy Bailey
Hudleston, Peter John
Parker, Ronald Bruce
Southwick, David Leroy
Walton, Matt Savage

MISSISSIPPI
Egloff, Julius, III
Knight, Wilbur Hall
Morris, Gerald Brooks
Seglund, James Arnold

MISSOURI
Arvidson, Raymond Ernst
Goebel, Edwin DeWayne
McKinnon, William Beall
Summers, David Archibold

MONTANA
Baker, David Warren
Hyndman, Donald William
Montagne, John M

NEBRASKA
Kaplan, Sanford Sandy
Nelson, Robert B

NEVADA
Felts, Wayne Moore
Garside, Larry Joe
Henry, Christopher D
Mifflin, Martin David
Peck, John H
Schweickert, Richard Allan
Slemmons, David Burton
Wilbanks, John Randall

NEW HAMPSHIRE
Billings, Marland Pratt
Bothner, Wallace Arthur
Richter, Dorothy Anne
Spink, Walter John

NEW JERSEY
Agron, Sam Lazrus
Hamil, Martha M
Metsger, Robert William

NEW MEXICO
Baldridge, Warren Scott
Budding, Antonius Jacob
Cserna, Eugene George
Geissman, John William
Grambling, Jeffrey A
Love, David Waxham
Schmitt, Harrison Hagan
Shomaker, John Wayne
Woodward, Lee Albert

NEW YORK
Bayly, M Brian
Bhattacharji, Somdev
Bird, John Malcolm
Brock, Patrick W G
Brueckner, Hannes Kurt
Bursnall, John Treharne
Cassie, Robert MacGregor
Diebold, John Brock
Fakundiny, Robert Harry
Foster, Helen Laura
Hayes, Dennis E
Heyl, George Richard
Isachsen, Yngvar William
Jacob, Klaus H
Karig, Daniel Edmund
Kidd, William Spencer Francis
King, John Stuart
Liebling, Richard Stephen
MacDonald, William David
Mattson, Peter Humphrey
Means, Winthrop D
Muller, Otto Helmuth
Prucha, John James
Seyfert, Carl K, Jr
Squyres, Steven Weldon
Teichert, Curt
Thomas, John Jenks
Vandiver, Bradford B
Vollmer, Frederick Wolfer

NORTH CAROLINA
Butler, James Robert
Dennison, John Manley
Forsythe, Randall D
Karson, Jeffrey Alan
Lowry, Jean
Marlowe, James Irvin
Miller, Daniel Newton, Jr
Raymond, Loren Arthur
Smith, Norman Cutler

OHIO
Baldwin, Arthur Dwight, Jr
Bever, James Edward
Burford, Arthur Edgar
Harmon, David Elmer, Jr
Kulander, Byron Rodney
Law, Eric W
Mahard, Richard H
Murdoch, Lawrence Corlies, III
Nance, Richard Damian
Richard, Benjamin H
Scotford, David Matteson
Skinner, William Robert
Unrug, Raphael
Wojtal, Steven Francis

OKLAHOMA
Ahern, Judson Lewis
Espach, Ralph H, Jr
Meyerhoff, Arthur Augustus
Murray, Frederick Nelson
Sondergeld, Carl Henderson
Stearns, David Winrod
Williams, Lyman O

OREGON
Lawrence, Robert D
Moore, George William
Yeats, Robert Sheppard

PENNSYLVANIA
Alley, Richard Blaine
Clark, Roger Alan
Eggler, David Hewitt
Faill, Rodger Tanner
Haefner, Richard Charles

Geology, Structural (cont)

Hodgson, Robert Arnold
Langston, Charles Adam
Nickelsen, Richard Peter
Pees, Samuel T(homas)
Platt, Lucian B
Voight, Barry
Washburn, Robert Henry

RHODE ISLAND
McMaster, Robert Luscher
Moore, George Emerson, Jr
Tullis, Julia Ann
Tullis, Terry Edson

SOUTH CAROLINA
Secor, Donald Terry, Jr

SOUTH DAKOTA
Lisenbee, Alvis Lee
Parrish, David Keith

TENNESSEE
Hatcher, Robert Dean, Jr

TEXAS
Arrowood, Roy Mitchell, Jr
Ave Lallemant, Hans Gerhard
Berkhout, Aart W J
Bishop, Richard Stearns
Blackwelder, Blake Winfield
Buis, Otto J
Carter, Neville Louis
Cebull, Stanley Edward
Chatterjee, Sankar
Crouse, Philip Charles
Dalziel, Ian William Drummond
Dodge, Rebecca L
Dunn, David Evan
Eifler, Gus Kearney, Jr
Ewing, Thomas Edward
Folk, Stewart H
Font, Robert G
Friedman, Melvin
Gallagher, John Joseph, Jr
Granath, James Wilton
Green, Arthur R
Gruy, Henry Jones
Helwig, James A
Hilde, Thomas Wayne Clark
Hoge, Harry Porter
Huffington, Roy Michael
Ingerson, Fred Earl
Jackson, Martin Patrick Arden
Johnson, Brann
Lawrence, Philip Linwood
Logan, John Merle
Mosher, Sharon
Muehlberger, William Rudolf
Mutis-Duplat, Emilio
Nielsen, Kent Christopher
Norman, Carl Edgar
Pauken, Robert John
Reaser, Donald Frederick
Roper, Paul James
Russell, James E(dward)
St John, Bill
Stead, Frederick L
Swanson, Eric Rice
Van Siclen, DeWitt Clinton
Whitford-Stark, James Leslie
Wiltschko, David Vilander
Zhou, Hua Wei

UTAH
Allison, Merle Lee
Fidlar, Marion M
Hardy, Clyde Thomas
McCleary, Jefferson Rand
Nielson, Dennis Lon
Proctor, Paul Dean

VERMONT
Westerman, David Scott

VIRGINIA
Brent, William B
Drake, Avery Ala, Jr
Eggleston, Jane R
Faust, Charles R
Gawarecki, Stephen Jerome
Goodwin, Bruce K
Harbour, Jerry
Hatch, Norman Lowrie, Jr
Herd, Darrell Gilbert
Horton, James Wright, Jr
McDowell, Robert Carter
Milici, Robert Calvin
Nelson, Arthur Edward
Offield, Terry Watson
Rankin, Douglas Whiting
Robertson, Eugene Corley
Russ, David Perry
Stewart, David Benjamin
Weems, Robert Edwin

WASHINGTON
Armstrong, Frank C(larkson Felix)
Beck, Myrl Emil, Jr
Cowan, Darrel Sidney
Kesler, Thomas L
Lowes, Brian Edward
Smith, Brad Keller

Swanson, Donald Alan
Talbot, James Lawrence
Wahlstrom, Ernest E

WEST VIRGINIA
Gwinn, James E
Khair, Abdul Wahab
Lessing, Peter
Peng, Syd Syh-Deng
Shumaker, Robert C

WISCONSIN
Craddock, J(ohn) Campbell
Dickas, Albert Binkley
Dutch, Steven Ian
Stenstrom, Richard Charles

WYOMING
Oman, Carl Henry
Snoke, Arthur Wilmot

ALBERTA
Cook, Frederick Ahrens
Cruden, David Milne
Cuny, Robert Michael
Hunt, C Warren
Okulitch, Andrew Vladimir
Tankard, Anthony James

BRITISH COLUMBIA
Armstrong, Richard Lee
Bustin, Robert Marc
Chase, Richard L
Monger, James William Heron
Wheeler, John Oliver

NEW BRUNSWICK
Allen, Clifford Marsden

NOVA SCOTIA
Keppie, John D
Milligan, George Clinton
Stevens, George Richard
Zentilli, Marcos

ONTARIO
Babcock, Elkanah Andrew
Bell, Richard Thomas
Fox, Joseph S
Grieve, Richard Andrew
Gussow, W(illia)m C(arruthers)
Haynes, Simon John
Kehlenbeck, Manfred Max
Percival, John A
Poole, William Hope
Price, Raymond Alex
Rousell, Don Herbert
Starkey, John
Udd, John Eaman

QUEBEC
Auger, Paul Emile
Chown, Edward Holton
Laurin, Andre Frederic
Mountjoy, Eric W

SASKATCHEWAN
Stauffer, Mel R

OTHER COUNTRIES
Dengo, Gabriel
Dewey, John Frederick
Kennedy, Michael John
McIntyre, Donald B
Sales, John Keith
Trumpy, D Rudolf
Wolfe, John A(llen)

Geomorphology & Glaciology

ALASKA
Benson, Carl Sidney
Hamilton, Thomas Dudley
Hopkins, David Moody
Long, William Ellis
Osterkamp, Thomas Eugene
Ragle, Richard Harrison
Reger, Richard David
Weber, Florence Robinson
Weeks, Wilford Frank

ARIZONA
Baker, Victor Richard
Bergen, James David
Breed, Carol Sameth
Breed, William Joseph
Bull, William Benham
Chase, Clement Grasham
Childs, Orlo E
Dorn, Ronald I
Graf, William L
Greeley, Ronald
Haynes, Caleb Vance, Jr
Heede, Burchard Heinrich
Lucchitta, Baerbel Koesters
Lucchitta, Ivo
Pewe, Troy Lewis
Schoenwetter, James

CALIFORNIA
Armstrong, Baxter Hardin
Brice, James Coble
Campbell, Catherine Chase

Carr, Michael H
Clark, Malcolm Mallory
Curtis, Garniss Hearfield
Dewart, Gilbert
Forrest, James Benjamin
Hart, Earl W
Higgins, Charles Graham
Huber, Norman King
Ives, John (Jack) David
Kamb, Walter Barclay
Keller, Edward Anthony
Kirch, Patrick Vinton
Leopold, Luna Bergere
Malin, Michael Charles
Mertes, Leal Anne Kerry
Nelson, Robert M
Nobles, Laurence Hewit
Norris, Robert Matheson
Page, Robert Alan, Jr
Rhodes, Dallas D
Rossbacher, Lisa Ann
Rudnyk, Marian E
Sharp, Robert Phillip
Shreve, Ronald Lee
Sieh, Kerry Edward
Sternberg, Hilgard O'Reilly
Thompson, Warren Charles
Thorne, Robert Folger
Warnke, Detlef Andreas
Wells, Stephen Gene
Zumberge, James Herbert

COLORADO
Andrews, John Thomas
Behrendt, John Charles
Birkeland, Peter Wessel
Bradley, William Crane
Caine, T Nelson
Carrara, Paul Edward
Coates, Donald Allen
Cooley, Richard Lewis
Crone, Anthony Joseph
Doehring, Donald O
Emmett, William W
Fernald, Arthur Thomas
Friedman, Jules Daniel
Fullerton, David Stanley
Hadley, Richard Frederick
Holcombe, Troy Leon
Jibson, Randall W
Krantz, William Bernard
Lackey, Laurence
Luft, Stanley Jeremie
Madole, Richard
Malde, Harold Edwin
Maughan, Edwin Kelly
Meier, Mark Frederick
Moore, David Warren
Morrison, Roger Barron
Nichols, Donald Ray
Pierce, Kenneth Lee
Schumm, Stanley Alfred
Smith, J Dungan
Strobel, John Dixon, Jr
Tabler, Ronald Dwight
Toy, Terrence J
Turner, Mortimer Darling

CONNECTICUT
Baskerville, Charles Alexander
Clebnik, Sherman Michael
Hack, John Tilton
Patton, Peter C
Spiegel, Zane
Tolderlund, Douglas Stanley

DELAWARE
Kraft, John Christian

DISTRICT OF COLUMBIA
Melickian, Gary Edward
Vogt, Peter Richard
Zimbelman, James Ray

FLORIDA
Chryssafopoulos, Hanka Wanda Sobczak
Collins, Mary Elizabeth
Draper, Grenville
Finkl, Charles William, II
Goldthwait, Richard Parker
Lovejoy, Donald Walker
Parker, Garald G, Sr

GEORGIA
Shonka, Joseph John

IDAHO
Hall, William Bartlett

ILLINOIS
Anderson, Richard Charles
Berg, Richard Carl
Harris, Stanley Edwards, Jr
Johnson, William Hilton
McKay, Edward Donald, III
Masters, John Michael
Paige, Russell Alston
Reams, Max Warren
Wallace, Ronald Gary

INDIANA
Bleuer, Ned Kermit
Melhorn, Wilton Newton
Reshkin, Mark

IOWA
Handy, Richard L(incoln)
Hussey, Keith Morgan
Prior, Jean Cutler
Walters, James Carter

KANSAS
Aber, James Sandusky
Blythe, Jack Gordon
Dort, Wakefield, Jr

KENTUCKY
Ettensohn, Francis Robert
Tuttle, Sherwood Dodge

LOUISIANA
Coleman, James Malcolm
Walker, Harley Jesse

MAINE
Kellogg, Thomas B
Koons, Donaldson
Novak, Irwin Daniel
Splettstoesser, John Frederick

MARYLAND
Cleaves, Emery Taylor
Garvin, James Brian
Kearney, Michael Sean
Kravitz, Joseph Henry
Leatherman, Stephen Parker
Prestegaard, Karen Leah
Rango, Albert
Warren, Charles Reynolds
Wolman, Markley Gordon

MASSACHUSETTS
Bradley, Raymond Stuart
Colman, Steven Michael
Genes, Andrew Nicholas
Hartshorn, Joseph Harold
Kleinrock, Martin Charles
Lewis, Laurence A
Newman, William Alexander
Newton, Robert Morgan
Ridge, John Charles
Southard, John Brelsford
Williams, Richard Sugden, Jr

MICHIGAN
Clark, James Alan
Eschman, Donald Frazier
Farrand, William Richard
Fekety, F Robert, Jr
Kehew, Alan Everett
Schmaltz, Lloyd John
Taylor, Lawrence Dow
Tenbrink, Norman Wayne

MINNESOTA
Cushing, Edward John
Hooke, Roger LeBaron
Hudleston, Peter John
Lawrence, Donald Buermann
Matsch, Charles Leo
Richason, Benjamin Franklin, Jr
Van Alstine, James Bruce

MISSISSIPPI
Happ, Stafford Coleman
Mylroie, John Eglinton
Otvos, Ervin George
Saucier, Roger Thomas

MISSOURI
Knox, Burnal Ray
McKinnon, William Beall
Miles, Randall Jay
Rockaway, John D, Jr
Schroeder, Walter Albert
Watson, Richard Allan

MONTANA
Montagne, John M

NEBRASKA
Diffendal, Robert Francis, Jr
Fairchild, Robert Wayne
Shroder, John Ford, Jr
Tanner, Lloyd George
Wayne, William John

NEVADA
Ritter, Dale Franklin
Slemmons, David Burton

NEW HAMPSHIRE
Ackley, Stephen Fred
Hibler, William David, III
Mayewski, Paul Andrew

NEW JERSEY
Judson, Sheldon
Psuty, Norbert Phillip
Robinson, David Alton
Wolfe, Peter E
Yasso, Warren E

NEW MEXICO
Hawley, John William
Hill, Mary Rae
Love, David Waxham
Manley, Kim
Stearns, Charles Edward

ENVIRONMENTAL, EARTH & MARINE SCIENCES / 391

NEW YORK
Abrahams, Athol Denis
Bloom, Arthur Leroy
Calkin, Parker E
Christman, Edward Arthur
Coch, Nicholas Kyros
Fleisher, Penrod Jay
Foster, Helen Laura
Jacobs, Stanley S
Johnson, Kenneth George
Kidd, William Spencer Francis
Liebling, Richard Stephen
Lougeay, Ray Leonard
Marschke, Stephen Frank
Morisawa, Marie
Muller, Ernest Hathaway
Squyres, Steven Weldon
Street, James Stewart
Van Burkalow, Anastasia

NORTH CAROLINA
Dennison, John Manley
Mock, Steven James
Stephenson, Richard Allen
Stuart, Alfred Wright

NORTH DAKOTA
Clausen, Eric Neil
Reid, John Reynolds, Jr

OHIO
Andreas, Barbara Kloha
Bonnett, Richard Brian
Calhoun, Frank Gilbert
Craig, Rachael Grace
Forsyth, Jane Louise
Laughon, Robert Bush
McKenzie, Garry Donald
Meinecke, Eberhard A
Nash, David Byer
Nave, Floyd Roger
Rich, Charles Clayton
Smith, Geoffrey W
Stewart, David Perry
Utgard, Russell Oliver
Whillans, Ian Morley
Whiting, Peter John

OKLAHOMA
Cannon, Philip Jan
Cox, Gordon F N
Myers, Arthur John
Salisbury, Neil Elliot

OREGON
Clark, Peter Underwood
McKinney, William Mark
Niem, Wendy Adams

PENNSYLVANIA
Adovasio, James Michael
Alley, Richard Blaine
Chapman, William Frank
Evenson, Edward B
Gardner, Thomas William
Hodgson, Robert Arnold
Potter, Noel, Jr
Sevon, William David, III

SOUTH CAROLINA
Gardner, Leonard Robert
Haselton, George Montgomery
James, L(aurence) Allan
Pirkle, William Arthur

SOUTH DAKOTA
Rahn, Perry H

TENNESSEE
Gomberg, Joan Susan
Lietzke, David Albert

TEXAS
Allen, Peter Martin
Bohacs, Kevin Michael
Doyle, Frank Lawrence
Finley, Robert James
Gustavson, Thomas Carl
Hall, Stephen Austin
Harrison, Wilks Douglas
Johnson, Brann
Knoll, Kenneth Mark
McPherson, John G(ordon)
Melia, Michael Brendan
Perez, Francisco Luis
Reeves, C C, Jr
Robertson, James Douglas
Rutford, Robert Hoxie
Soloyanis, Susan Constance
Woodruff, Charles Marsh, Jr

UTAH
Cerling, Thure E
Godfrey, Andrew Elliott
McCalpin, James P
Stokes, William Lee
Vaughn, Danny Mack

VERMONT
Larsen, Frederick Duane

VIRGINIA
Deye, James A
Galvin, Cyril Jerome, Jr

Garner, Hessle Filmore
Gawarecki, Stephen Jerome
Goldsmith, Richard
Herd, Darrell Gilbert
Molnia, Bruce Franklin
Outerbridge, William Fulwood
Russ, David Perry
Weems, Robert Edwin
Wieczorek, Gerald Francis
Wright, Lynn Donelson

WASHINGTON
Anderson, Norman Roderick
Caggiano, Joseph Anthony, Jr
Carson, Robert James, III
Costa, John Emil
De Goes, Louis
Dunne, Thomas
Easterbrook, Don J
Foley, Michael Glen
Grootes, Pieter Meiert
Hallet, Bernard
Hodge, Steven McNiven
Holcomb, Robin Terry
Kaatz, Martin Richard
Kiver, Eugene P
Megahan, Walter Franklin
Miller, C Dan
Owens, Edward Henry
Palmquist, Robert Clarence
Pessl, Fred, Jr
Pierson, Thomas Charles
Rosengreen, Theodore E
Scott, William Edward
Threet, Richard Lowell
Washburn, Albert Lincoln

WISCONSIN
Bentley, Charles Raymond
Borowiecki, Barbara Zakrzewska
Hendee, Wiliam R
Hole, Francis Doan
Knox, James Clarence
Lasca, Norman P, Jr
Schneider, Allan Frank
Stieglitz, Ronald Dennis

WYOMING
Marston, Richard Alan
Mears, Brainerd, Jr

ALBERTA
Barendregt, Rene William
Cruden, David Milne
Giovinetto, Mario Bartolome
Honsaker, John Leonard
Klassen, Rudolph Waldemar
Rogerson, Robert James

BRITISH COLUMBIA
Armstrong, John Edward
Clague, John Joseph
Clarke, Garry K C
Foster, Harold Douglas
Jackson, Lionel Eric
Slaymaker, Herbert Olav

MANITOBA
Teller, James Tobias
Welsted, John Edward

ONTARIO
Adams, William Alfred
Blake, Weston, Jr
Bunting, Brian Talbot
Davidson-Arnott, Robin G D
Dredge, Lynda Ann
Dreimanis, Aleksis
Flint, Jean-Jacques
Ford, Derek Clifford
Greenwood, Brian
Johnson, Peter Graham
Karrow, Paul Frederick
Kesik, Andrzej B
LaValle, Placido Dominick
Lee, Hulbert Austin
Menzies, John
Prest, Victor Kent
Stalker, Archibald MacSween
St-Onge, Denis Alderic
Terasmae, Jaan
Wilkinson, Thomas Preston

QUEBEC
Bonn, Ferdinand J
Bouchard, Michel Andre
Dubois, Jean-Marie M
Elson, John Albert
Gangloff, Pierre
Ladanyi, Branko
Langleben, Manuel Phillip

SASKATCHEWAN
Chew, Hemming
Langford, Fred F
Mollard, John D

OTHER COUNTRIES
Dury, George H
Hattersley-Smith, Geoffrey Francis
Jopling, Alan Victor
Van Andel, Tjeerd Hendrik

Geophysics

ALABAMA
Dalins, Ilmars
Kenyon, Patricia May
Smith, Robert Earl

ALASKA
Akasofu, Syun-Ichi
Belon, Albert Edward
Brewer, Max C(lifton)
Brown, Neal B
Davis, Thomas Neil
Degen, Vladimir
Gedney, Larry Daniel
Hallinan, Thomas James
Morack, John Ludwig
Stenbaek-Nielsen, Hans C
Stone, David B
Watkins, Brenton John
Weeks, Wilford Frank
Wescott, Eugene Michael
Wilson, Charles R
Wyss, Max

ARIZONA
Bates, Charles Carpenter
Best, David Malcolm
Butler, Robert Franklin
Chase, Clement Grasham
Elston, Donald (Parker)
Fink, James Brewster
Fink, Jonathan Harry
Herbert, Floyd Leigh
Kieffer, Hugh Hartman
Levy, Eugene Howard
Lunine, Jonathan Irving
Melosh, Henry Jay, IV
O'Leary, Brian Todd
Richardson, Randall Miller
Roddy, David John
Sass, John Harvey
Soderblom, Laurence Albert
Sonett, Charles Philip
Sumner, John Stewart
Swanberg, Chandler A
Wallace, Terry Charles, Jr
West, Robert Elmer
Wildey, Robert Leroy

ARKANSAS
Leming, Charles William

CALIFORNIA
Abdel-Ghaffar, Ahmed Mansour
Ahrens, Thomas J
Aki, Keiiti
Alexander, Ira H(enris)
Allen, Clarence Roderic
Anderson, Don Lynn
Anderson, Orson Lamar
Atwater, Tanya Maria
Bachman, Richard Thomas
Backus, George Edward
Bakun, William Henry
Barnes, David Fitz
Becker, Alex
Berryman, James Garland
Bevc, Dimitri
Biehler, Shawn
Bird, Peter
Bolt, Bruce A
Borchardt, Glenn (Arnold)
Borucki, William Joseph
Bradner, Hugh
Braginsky, Stanislav Iosifovich
Breiner, Sheldon
Brooke, John Percival
Brown, Maurice Vertner
Bukowinski, Mark S T
Burg, John Parker
Burnett, Lowell Jay
Busse, Friedrich Hermann
Carrigan, Charles Roger
Case, James Edward
Cherry, Jesse Theodore
Christiansen, Robert Lorenz
Chung, Dae Hyun
Cicerone, Ralph John
Claerbout, Jon F
Coe, Robert Stephen
Coleman, Paul Jerome, Jr
Costantino, Marc Shaw
Cross, Ralph Herbert, III
Curtis, Garniss Hearfield
Decker, Robert Wayne
Dewart, Gilbert
Dey, Abhijit
Doell, Richard Rayman
Donath, Fred Arthur
Donoho, David Leigh
Dorman, LeRoy Myron
Dyal, Palmer
Eaton, Jerry Paul
Eittreim, Stephen L
Esposito, Pasquale Bernard
Ewing, Clair Eugene
Fillippone, Walter R
Flatte, Stanley Martin
Fliegel, Henry Frederick
Frank, Sidney Raymond
Fraser-Smith, Antony Charles
Fuller, Michael D
Gaal, Robert A P

Galehouse, Jon Scott
Gaskell, Robert Weyand
Gilbert, James Freeman
Gimlett, James I
Goins, Neal Rodney
Green, Harry Western, II
Grine, Donald Reaville
Griscom, Andrew
Grobecker, Alan J
Hall, Robert Earl
Hanks, Thomas Colgrove
Hannon, Willard James, Jr
Hardy, Rolland L(ee)
Harkrider, David Garrison
Harrison, John Christopher
Healy, John H
Hearst, Joseph R
Heusinkveld, Myron Ellis
Hill, David Paul
Hill, Donald Gardner
Housley, Robert Melvin
Huebsch, Ian O
Inman, Douglas Lamar
Irwin, James Joseph
Isherwood, William Frank
Iyer, Hariharaiyer Mahadeva
Jachens, Robert C
Jackson, Bernard Vernon
Jackson, David Diether
Jeanloz, Raymond
Jones, Stanley Bennett
Joyner, William B
Kahle, Anne B
Kanamori, Hiroo
Kasameyer, Paul William
Kaula, William Mason
Kennedy, Burton Mack
Kerr, Donald M, Jr
Khurana, Krishan Kumar
King, Chi-Yu
Kirk, John Gallatin
Kliore, Arvydas J(oseph)
Knopoff, Leon
Knudsen, William Claire
Lachenbruch, Arthur Herold
Lay, Thorne
Le, Guan
Lee, Tien-Chang
Lee, William Hung Kan
Leonard, Robert Stuart
Lin, Robert Peichung
Lin, Wunan
Lindh, Allan Goddard
Lippitt, Louis
Lipson, Joseph Issac
Lopes-Gautier, Rosaly
Luyendyk, Bruce Peter
MacDonald, Gordon James Fraser
Macdonald, Kenneth Craig
MacDoran, Peter Frank
McEvilly, Thomas V
McGarr, Arthur
McHugh, Stuart Lawrence
McIntyre, Marie Claire
McKenzie, William F
McNally, Karen Cook
McPherron, Robert Lloyd
Mal, Ajit Kumar
Martin, Terry Zachry
Meadows, Mark Allan
Moe, Osborne Kenneth
Mooney, Walter D
Morrison, Huntly Frank
Nakanishi, Keith Koji
Nason, Robert Dohrmann
Nathenson, Manuel
Nellis, William J
Northrop, John
Nur, Amos M
Nurmia, Matti Juhani
O'Keefe, John Dugan
Orcutt, John Arthur
Page, Robert Alan, Jr
Palluconi, Frank Don
Pang, Kevin
Petersen, Carl Frank
Pfluke, John H
Phillips, Richard P
Poppoff, Ilia George
Porter, Lawrence Delpino
Portnoff, Michael Rodney
Pray, Ralph Emerson
Reynolds, John Hamilton
Reynolds, Ray Thomas
Romanowicz, Barbara Anna
Roy, Robert Francis
Runge, Richard John
Russell, Christopher Thomas
Russell, Philip Boyd
Ryan, Jack A
Ryu, Jisoo Vinsky
St Amand, Pierre
Schock, Robert Norman
Scholl, James Francis
Schubert, Gerald
Schwing, Franklin Burton
Sclater, John George
Sentman, Davis Daniel
Shaw, Herbert Richard
Shor, George G, Jr
Shreve, Ronald Lee
Silver, Eli Alfred
Sovers, Ojars Juris
Spence, Harlan Ernest

Geophysics (cont)

Springer, Donald Lee
Stacey, John Sydney
Stephenson, Lee Palmer
Stevenson, David John
Stocker, Richard Louis
Sundaram, Panchanatham N
Swift, Charles Moore, Jr
Teng, Ta-Liang
Terrile, Richard John
Thatcher, Wayne Raymond
Thompson, George Albert
Thompson, Timothy J
Tooley, Richard Douglas
Tuman, Vladimir Shlimon
Tyler, George Leonard
Vacquier, Victor
Vanyo, James Patrick
Vasco, Donald Wyman
Vassiliou, Marius Simon
Verhoogen, John
Verosub, Kenneth Lee
Vickers, Roger Spencer
Vidale, John Emilio
Walt, Martin
Ward, Peter Langdon
Warren, David Henry
Warshaw, Stephen I
Wasserburg, Gerald Joseph
Weiss, Jeffrey Martin
Wells, Ronald Allen
Widess, Moses B
Wiggins, John H(enry), Jr
Williams, James Gerard
Williams, Quentin Christopher
Willis, David Edwin
Wilt, Michael
Wood, Fergus James
Wright, Frederick Hamilton
Yoder, Charles Finney
Zoback, Mark D
Zoback, Mary Lou Chetlain
Zumberge, James Frederick

COLORADO
Abshier, Curtis Brent
Albritton, Daniel L
Alldredge, Leroy Romney
Allen, Joe Haskell
Applegate, James Keith
Archambeau, Charles Bruce
Bailey, Dana Kavanagh
Balch, Alfred Hudson
Ball, Mahlon M
Behrendt, John Charles
Benton, Edward Rowell
Born, George Henry
Bufe, Charles Glenn
Campbell, Wallace Hall
Chang, Bunwoo Bertram
Chinnery, Michael Alistair
Cummings, David
Dana, Robert Watson
Diment, William Horace
Donovan, Terrence John
Engdahl, Eric Robert
Faller, James E
Friedman, Jules Daniel
Goetz, Alexander Franklin Hermann
Gosink, Joan P
Grose, Thomas Lucius Trowbridge
Hansen, Richard Olaf
Hilt, Richard Leighton
Hittelman, Allen M
Holmer, Ralph Carrol
Hoover, Donald Brunton
Jacobson, Jimmy Joe
Johnson, Verner Carl
Kane, Martin Francis
Kisslinger, Carl
Kleinkopf, Merlin Dean
Koizumi, Carl Jan
Lackey, Laurence
Lander, James French
Larner, Kenneth Lee
Larson, Edwin E
Leinbach, F Harold
Lincoln, Jeannette Virginia
Mason, Malcolm
Meyers, Herbert
Nabighian, Misac N
Obradovich, John Dinko
Olhoeft, Gary Roy
Osterhoudt, Walter Jabez
Pakiser, Louis Charles, Jr
Parker, John Marchbank
Pawlowicz, Edmund F
Pennington, Wayne David
Peterson, Vern Leroy
Pickett, George R
Plows, William Herbert
Richmond, Arthur Dean
Romig, Phillip Richardson
Sanborn, Albert Francis
Sargent, Howard Harrop, III
Savage, William Zuger
Scales, John Alan
Schiffmacher, E(dward) R(obert)
Scott, James Henry
Shedlock, Kaye M
Singer, Howard Joseph
Smith, J Dungan
Snyder, Howard Arthur

Speiser, Theodore Wesley
Spence, William J
Spencer, Joseph Walter
Tatarskii, Valerian I
Thomsen, Harry Ludwig
Tsvankin, Ilya Daniel
Urban, Thomas Charles
Viksne, Andy
Watson, Kenneth
White, James Edward
Williams, David Lee
Wold, Richard John
Wunderman, Richard Lloyd

CONNECTICUT
Aitken, Janet M
Burridge, Robert
Clark, Sydney P, Jr
DeBoer, Jelle
Dowling, John J
Friedman, Don Gene
Gordon, Robert Boyd
Kleinberg, Robert Leonard
Liu, Qing-Huo
O'Brien, Brian
Plona, Thomas Joseph
Schweitzer, Jeffrey Stewart
Tittman, Jay
Wiggins, Ralphe

DELAWARE
Hanson, Roy Eugene
Ness, Norman Frederick

DISTRICT OF COLUMBIA
Aldrich, Lyman Thomas
Berkson, Jonathan Milton
Boss, Alan Paul
Chung, Riley M
Crary, Albert Paddock
Greenewalt, David
Gutman, George Garik
Hart, Pembroke J
Huba, Joseph D
James, David Evan
Lawson, Charles Alden
Linde, Alan Trevor
Mao, Ho-Kwang
Mead, Gilbert Dunbar
Meier, Robert R
Press, Frank
Reilly, Michael Hunt
Riemer, Robert Lee
Senftle, Frank Edward
Smith, Philip Meek
Smith, Waldo E(dward)
Tolstoy, Alexandra
Usselman, Thomas Michael
Vogt, Peter Richard
Watters, Thomas Robert
Wetherill, George West
Zimbelman, James Ray

FLORIDA
Cain, Joseph Carter, III
Draper, Grenville
Harrison, Christopher George Alick
Helava, U(uno) V(ilho)
Jin, Rong-Sheng
Long, James Alvin
McLeroy, Edward Glenn
O'Hara, Norbert Wilhelm
Olivero, John Joseph, Jr
Olson, Donald B
Opdyke, Neil
Orlin, Hyman
Parker, Garald G, Sr
Rona, Peter Arnold
Rosendahl, Bruce Ray
Sivjee, Gulamabas Gulamhusen

GEORGIA
Long, Leland Timothy
Lowell, Robert Paul
Whitney, James Arthur

HAWAII
Berg, Eduard
Daniel, Thomas Henry
Frazer, L Neil
Furumoto, Augustine S
Helsley, Charles Everett
Hey, Richard N
Johnson, Carl Edward
Keating, Barbara Helen
Khan, Mohammad Asad
Malahoff, Alexander
Manghnani, Murli Hukumal
Ming, Li Chung
Moore, Gregory Frank
Raleigh, Cecil Baring
Rose, John Creighton
Saxena, Narendra K
Self, Stephen
Vitousek, Martin J
Walker, Daniel Alvin

IDAHO
Maloof, Giles Wilson

ILLINOIS
Carpenter, Philip John
Ervin, C Patrick
Fish, Ferol F, Jr

Foster, John Webster
Hoffer, Abraham
Hsui, Albert Tong-Kwan
Klasner, John Samuel
Liu, Mian
McGinnis, Lyle David
Okal, Emile Andre
Riahi, Daniel Nourollah
Stein, Seth Avram
White, W(illiam) Arthur

INDIANA
Biggs, Maurice Earl
Blakely, Robert Fraser
Chowdhury, Dipak Kumar
Christensen, Nikolas Ivan
Hamburger, Michael Wile
Hinze, William James
Kim, Yeong Ell
Kivioja, Lassi A
Madison, James Allen
Mead, Judson
Rudman, Albert J

IOWA
Carmichael, Robert Stewart
Hartung, Jack Burdair, Sr
Jeyapalan, Kandiah

KANSAS
Hambleton, William Weldon
Steeples, Donald Wallace
Surampalli, Rao Yadagiri

KENTUCKY
Brock, Louis Milton
Seeger, Charles Ronald
Sendlein, Lyle V A

LOUISIANA
Bradley, Marshall Rice
Breeding, J Ernest, Jr
French, William Stanley
Higgs, Robert Hughes
Kinsland, Gary Lynn
Meriwether, John R
Ward, Ronald Wayne

MAINE
Wall, Robert Ecki

MARYLAND
Alers, Perry Baldwin
Alexander, Joseph Kunkle
Allenby, Richard John, Jr
Behannon, Kenneth Wayne
Bostrom, Carl Otto
Buchanan, John Donald
Carter, William Eugene
Chovitz, Bernard H
Clark, Thomas Arvid
Coates, R(obert) J(ay)
Cohen, Steven Charles
Comiso, Josefino Cacas
Connerney, John E P
Degnan, John James, III
Dick, Richard Dean
Elsasser, Walter M
Fink, Don Roger
Fischer, Irene Kaminka
Flinn, Edward Ambrose
Garvin, James Brian
Hammond, Allen Lee
Heirtzler, James Ransom
Heppner, James P
Herrero, Federico Antonio
Kappel, Ellen Sue
King, Joseph Herbert
Kohler, Max A
Krueger, Arlin James
Langel, Robert Allan
Liu, Han-Shou
Maier, Eugene Jacob Rudolph
Marsh, Bruce David
Martin, James Milton
Meredith, Leslie Hugh
Nazarova, Katherine V
O'Keefe, John Aloysius
Olson, Peter Lee
Paik, Ho Jung
Penhollow, John O
Pisacane, Vincent L
Robertson, Douglas Scott
Romick, Gerald J
Salisbury, John William, Jr
Shotland, Lawrence M
Sibeck, David G
Smith, David Edmund
Spilhaus, Athelstan Frederick, Jr
Stone, Albert Mordecai
Svendsen, Kendall Lorraine
Taylor, Patrick Timothy
Thomas, Herman H
Trainor, James H
Wells, Eddie N
Williams, Donald J

MASSACHUSETTS
Agee, Carl Bernard
Beckerle, John C
Bell, Peter M
Birch, Albert Francis
Bogert, Bruce Plympton
Bowhill, Sidney Allan

Bowin, Carl Otto
Brecher, Aviva
Bromery, Randolph Wilson
Brown, Laurie Lizbeth
Brynjolfsson, Ari
Bunce, Elizabeth Thompson
Burger, Henry Robert, III
Burns, Daniel Robert
Carr, Jerome Brian
Chave, Alan Dana
Cheng, Chuen Hon
Corey, Brian E
Counselman, Charles Claude, III
Dainty, Anton Michael
Dandekar, Balkrishna S
Denig, William Francis
Donohue, John J
Dziewonski, Adam Marian
Eckhardt, Donald Henry
Ewing, John I
Frisk, George Vladimir
Goody, Richard (Mead)
Guertin, Ralph Francis
Guidice, Donald Anthony
Hager, Bradford Hoadley
Hager, Jutta Lore
Hope, Elizabeth Greeley
Hoskins, Hartley
Itzkan, Irving
Jacobsen, Stein Bjornar
Jordan, Thomas Hillman
Kachanov, Mark L
King, Robert Wilson, Jr
Kleinrock, Martin Charles
Knecht, David Jordan
Leblanc, Gabriel
Lin, Jian
Little, Sarah Alden
McConnell, Robert Kendall
McNutt, Marcia Kemper
Madden, Stephen James, Jr
Madden, Theodore Richard
Mahoney, William C
Marone, Chris James
Michael, Irving
Molnar, Peter Hale
O'Connell, Richard John
Petschek, Harry E
Pike, Charles P
Rice, James R
Ridge, John Charles
Rooney, Thomas Peter
Rosen, Richard David
Rothwell, Paul L
Shaw, Peter Robert
Shuman, Bertram Marvin
Silverman, Sam M
Slowey, Jack William
Solomon, Sean Carl
Stephen, Ralph A
Toksoz, Mehmet Nafi
Tucholke, Brian Edward
Van Tassel, Roger A
Von Herzen, Richard P
Walsh, Joseph Broughton
Watt, James Peter
Whitehead, John Andrews

MICHIGAN
Clark, James Alan
Clauer, C Robert, Jr
Cloud, Gary Lee
England, Anthony W
Frantti, Gordon Earl
Huntoon, Jacqueline E
Jacobs, Stanley J
Nagy, Andrew F
Pollack, Henry Nathan
Smith, Thomas Jefferson
Trow, James
Van der Voo, Rob
Walker, James Callan Gray

MINNESOTA
Atchison, Thomas Calvin, Jr
Banerjee, Subir Kumar
Farnham, Paul Rex
Fogelson, David Eugene
Mauersberger, Konrad
Yuen, David Alexander

MISSISSIPPI
Ballard, James Alan
Butler, Dwain Kent
Chang, Frank Keng
Davis, Thomas Mooney
Geddes, Wilburt Hale
Lowrie, Allen
Malone, Philip Garcin
Morris, Gerald Brooks
Sprague, Vance Glover, Jr

MISSOURI
DuBroff, Richard Edward
Heinrich, Ross Raymond
Herrmann, Robert Bernard
Hunt, Mahlon Seymour
McFarland, Robert Harold
McKinnon, William Beall
Mitchell, Brian James
Rechtien, Richard Douglas
Rupert, Gerald Bruce
Stauder, William
Vincenz, Stanislaw Aleksander

ENVIRONMENTAL, EARTH & MARINE SCIENCES

White, Jon M

MONTANA
Banaugh, Robert Peter
Bird, Harvey Harold
Sill, William Robert
Wideman, Charles James

NEVADA
Anderson, John Gregg
Brune, James N
Hess, John Warren
Peck, John H
Savage, Martha Kane
Slemmons, David Burton

NEW HAMPSHIRE
Ackley, Stephen Fred
Bothner, Wallace Arthur
Chamberlain, C Page
Colbeck, Samuel C
Hibler, William David, III
Hoag, Roland Boyden, Jr
Perovich, Donald Kole
Simmons, Gene

NEW JERSEY
Alsop, Leonard E
Bonini, William E
Cox, Michael
Dahlen, Francis Anthony, Jr
Jeter, Hewitt Webb
Lanzerotti, Louis John
Maclennan, Carol G
Metsger, Robert William
Miyakoda, Kikuro
Morgan, William Jason
Nolet, Guust M
Orlanski, Isidoro
Phinney, Robert A
Sheridan, Robert E
Suppe, John
Takahashi, Taro
Thiruvathukal, John Varkey
Thompson, William Baldwin
Tuleya, Robert E
Weissenburger, Don William

NEW MEXICO
Bullard, Edwin Roscoe, Jr
Close, Donald Alan
Cress, Daniel Hugg
Derr, John Sebring
Deupree, Robert G
Fehler, Michael C
Fugelso, Leif Erik
Geissman, John William
Gross, Gerardo Wolfgang
Hansen, Wayne Richard
Holmes, Charles Robert
Huestis, Stephen Porter
Kuiper, Logan Keith
Lee, David Oi
Lysne, Peter C
McNaughton, Michael Walford
McTigue, David Francis
Marker, Thomas F(ranklin)
Perret, William Riker
Phillips, John Lynch
Reiter, Marshall Allan
Riecker, Robert E
Roberts, Peter Morse
Sanders, Walter L
Sanford, Allan Robert
Sayre, William Olaf
Searls, Craig Allen
Snell, Charles Murrell
Sun, James Ming-Shan
Tripp, R(ussell) Maurice
Weart, Wendell D

NEW YORK
Bassett, William Akers
Bokuniewicz, Henry Joseph
Bolgiano, Ralph, Jr
Bollinger, Gilbert A
Cathles, Lawrence MacLagan, III
Cisne, John Luther
Cozzarelli, Francis A(nthony)
Czapski, Ulrich Hans
Dickman, Steven Richard
Diebold, John Brock
Egan, Walter George
Fitzpatrick, Robert Charles
Fleischer, Robert Louis
Flood, Roger Donald
Gilmore, Robert Snee
Hayes, Dennis E
Heyl, George Richard
Hodge, Dennis
Isacks, Bryan L
Jacob, Klaus H
Kaarsberg, Ernest Andersen
Kastens, Kim Anne
Katz, Samuel
Kent, Dennis V
Kuo, John Tsung-fen
Langseth, Marcus G
Liebermann, Robert C
Liu, Mao-Zu
MacDonald, William David
Magorian, Thomas R
Meisel, David Dering
Meyer, Robert Jay

Millman, George Harold
Moniot, Robert Keith
Muller, Otto Helmuth
Oliver, Jack Ertle
Parks, Thomas William
Prucha, John James
Reitan, Paul Hartman
Remo, John Lucien
Revetta, Frank Alexander
Richards, Paul Granston
Rimai, Donald Saul
Rind, David Harold
Robinson, Joseph Edward
Ruddick, James John
Scholz, Christopher Henry
Schreiber, Edward
Senus, Walter Joseph
Sutton, George Harry
Sykes, Lynn Ray
Turcotte, Donald Lawson
Vaughan, Michael Thomas
Weidner, Donald J
Woodmansee, Donald Ernest
Wu, Francis Taming

NORTH CAROLINA
Almy, Charles C, Jr
Heroy, William Bayard, Jr
Johnson, Thomas Charles
Karson, Jeffrey Alan
Stavn, Robert Hans
Straub, Karl David
Worzel, John Lamar

OHIO
Ahmad, Moid Uddin
Bossler, John David
Goad, Clyde Clarenton
Janzow, Edward F(rank)
Kellerstrass, Ernst Junior
Kulander, Byron Rodney
Kunze, A(dolf) W(ilhelm) Gerhard
Mourad, A George
Mueller, Ivan I
Noltimier, Hallan Costello
Rapp, Richard Henry
Richard, Benjamin H
Stierman, Donald John
Walter, Edward Joseph
Whillans, Ian Morley
Wolfe, Paul Jay

OKLAHOMA
Ahern, Judson Lewis
Brown, Graydon L
Byerly, Perry Edward
Chambers, Richard Lee
Crowe, Christopher
Du Bois, Robert Lee
Duncan, Lewis Mannan
Elliott, Sheldon Ellwood
Frisillo, Albert Lawrence
Gutowski, Paul Ramsden
Hauge, Paul Stephen
Hill, Stephen James
Jennemann, Vincent Francis
Koller, Glenn R
Martner, Samuel (Theodore)
Meyerhoff, Arthur Augustus
Richman, Michael B
Robinson, Enders Anthony
Sigal, Richard Frederick
Smith, James Edward
Sondergeld, Carl Henderson
Souder, Wallace William
Stoeckley, Thomas Robert
Stone, John Floyd
Thapar, Mangat Rai
Thomsen, Leon
Treitel, Sven

OREGON
Couch, Richard W
Demarest, Harold Hunt, Jr
Mather, Keith Benson
Trehu, Anne Martine
Waymire, Edward Charles

PENNSYLVANIA
Bell, Maurice Evan
Bennett, Lee Cotton, Jr
Biondi, Manfred Anthony
Greenfield, Roy Jay
Hodgson, Robert Arnold
Howell, Benjamin F, Jr
Kodama, Kenneth Philip
Langston, Charles Adam
Lavin, Peter Masland
Mowrey, Gary Lee
Pilant, Walter L
Schmidt, Victor A
Simaan, Marwan
Sittel, Karl
Strick, Ellis
Wuenschel, Paul Clarence

RHODE ISLAND
Hermance, John Francis
McMaster, Robert Luscher
Polk, C(harles)
Schultz, Peter Hewlett
Tullis, Terry Edson

SOUTH CAROLINA
Eyer, Jerome Arlan
Meriwether, John Williams, Jr
Secor, Donald Terry, Jr

SOUTH DAKOTA
Davis, Briant LeRoy

TENNESSEE
Churnet, Habte Giorgis
Dorman, Henry James
Gentry, Robert Vance
Gomberg, Joan Susan
Haywood, Frederick F
Hopkins, Richard Allen
Nyquist, Jonathan Eugene
Thomson, Kerr Clive

TEXAS
Anderson, Edmund Hughes
Angona, Frank Anthony
Aucoin, Paschal Joseph, Jr
Backus, Milo M
Barrow, Thomas D
Baskir, Emanuel
Bean, Robert Jay
Behrens, Earl William
Berkhout, Aart W J
Bernard, Bernie Boyd
Birchak, James Robert
Blackwelder, Blake Winfield
Brown, Leonard Franklin, Jr
Bucy, J Fred
Burke, William Henry
Bussian, Alfred Erich
Carter, Neville Louis
Cheney, Monroe G
Chiburis, Edward Frank
Clark, Howard Charles, Jr
Clayton, Neal
Clough, John Wendell
Cook, Clarence Sharp
Cook, Ernest Ewart
Cutler, Roger T
Dahm, Cornelius George
Damuth, John Erwin
Danburg, Jerome Samuel
Deaton, Bobby Charles
De Bremaecker, Jean-Claude
Deeming, Terence James
Delflache, Andre P
Dent, Brian Edward
Doser, Diane Irene
Dunlap, Henry Francis
Eastwood, Raymond L
Eaton, Jerome F
Eisner, Elmer
Ellwood, Brooks B
Fahlquist, Davis A
Feagin, Frank J
Fertl, Walter Hans
Fett, John D
Fountain, Lewis Spencer
Galbraith, James Nelson, Jr
Gallagher, John Joseph, Jr
Gangi, Anthony Frank
Gardner, Gerald Henry Fraser
Ginsburg, Merrill Stuart
Goforth, Thomas Tucker
Granath, James Wilton
Gregory, A(lvin) R(ay)
Hall, David Joseph
Handin, John Walter
Hanss, Robert Edward
Hardage, Bob Adrian
Hasbrook, Arthur F(erdinand)
Heinze, William Daniel
Herrin, Eugene Thornton, Jr
Hilde, Thomas Wayne Clark
Hilterman, Fred John
Hopkins, George H, Jr
Hopkins, John Raymond
Hoskins, Earl R, Jr
Huffington, Roy Michael
Hurley, Neal Lilburn
Iddings, Frank Allen
Johnson, Brann
Johnson, William W
Johnston, David Hervey
Jordan, Neal F(rancis)
Justice, James Horace
Kaufman, Sidney
Keller, George Randy, Jr
Kern, John W
Kibler, Kenneth G
Kulla, Jean B
La Fehr, Thomas Robert
Landrum, Ralph Avery, Jr
Lawrence, Philip Linwood
Leiss, Ernst L
Levien, Louise
Levin, Franklyn Kussel
McCormack, Harold Robert
Massell, Wulf F
Mateker, Emil Joseph, Jr
Matumoto, Tosimatu
Miller, Max K
Moorhead, William Dean
Musgrave, Albert Wayne
Mut, Stuart Creighton
Neidell, Norman Samson
Nielsen, Kent Christopher
Nosal, Eugene Adam
Nugent, Robert Charles

Nussmann, David George
Palmeira, Ricardo Antonio Ribeiro
Pan, Poh-Hsi
Peeples, Wayne Jacobson
Peters, Jack Warren
Peterson, Donald Neil
Phillips, Joseph D
Pilger, Rex Herbert, Jr
Potts, Mark John
Price, William Charles
Pritchett, William Carr
Ranganayaki, Rambabu Pothireddy
Reed, Dale Hardy
Reeves, Robert Grier (Lefevre)
Rezak, Richard
Rinehart, Wilbur Allan
Robertson, James Douglas
Rosenbaum, Joseph Hans
Savit, Carl Hertz
Sbar, Marc Lewis
Schoenberger, Michael
Schramm, Martin William, Jr
Schutz, Bob Ewald
Seriff, Aaron Jay
Sheriff, Robert Edward
Shurbet, Deskin Hunt, Jr
Siegfried, Robert Wayne, II
Soloyanis, Susan Constance
Sorrells, Gordon Guthrey
Spencer, Terry Warren
Sprunt, Eve Silver
Srnka, Leonard James
Stoffa, Paul L
Sumner, John Randolph
Tajima, Toshiki
Talwani, Manik
Tixier, Maurice Pierre
Truxillo, Stanton George
Ward-McLemore, Ethel
Watkins, Joel Smith, Jr
Wegner, Robert Carl
Wells, Frederick Joseph
Whittlesey, John R B
Wiggins, James Wendell
Wilson, Clark Roland
Wilson, John Human
Wiltschko, David Vilander
Winbow, Graham Arthur
Witterholt, Edward John
Wu, Changsheng
Yarger, Harold Lee
Zemanek, Joseph, Jr
Zhou, Hua Wei
Zimmerman, Carol Jean

UTAH
Allison, Merle Lee
Benson, Alvin K
Cook, Kenneth Lorimer
Foster, Robert H
Greene, Gordon William
Ross, Howard Persing
Smith, Robert Baer

VERMONT
Drake, Charles Lum

VIRGINIA
Algermissen, Sylvester Theodore
Bennett, Gordon Daniel
Berg, Joseph Wilbur, Jr
Blandford, Robert Roy
Bowker, David Edwin
Cameron, Maryellen
Chubb, Talbot Albert
Costain, John Kendall
Davis, Charles Mitchell, Jr
Delnore, Victor Eli
DeNoyer, John M
Dickinson, Stanley Key, Jr
Faust, Charles R
Gamble, Thomas Dean
Gawarecki, Stephen Jerome
Hamilton, Robert Morrison
Hanna, William F
Hartline, Beverly Karplus
Hays, James Fred
Hemingway, Bruce Sherman
Hersey, John B
Hosterman, John W
Huebner, John Stephen
Johnson, Leonard Evans
Johnson, Robert Edward
King, Elizabeth Raymond
Koch, Carl Fred
Lanzano, Paolo
Lenoir, William Benjamin
Levine, Joel S
Michael, William Herbert, Jr
Milton, Nancy Melissa
Morrow, Richard Joseph
Orr, Marshall H
Radoski, Henry Robert
Robertson, Eugene Corley
Robinson, Edwin S
Romney, Carl Fredrick
Russ, David Perry
Ryall, Alan S, Jr
Sax, Robert Louis
Senseny, Paul Edward
Sims, John David
Singer, S(iegfried) Fred
Snoke, J Arthur
Spall, Henry Roger

Geophysics (cont)

**Spinrad, Richard William
Stewart, David Benjamin
Tanner, Allan Bain
Ugincius, Peter
Unger, John Duey
Van der Vink, Gregory E
Wallace, Raymond Howard, Jr
Wesson, Robert Laughlin
Whitcomb, James Hall

WASHINGTON
Adams, William Mansfield
Beck, Myrl Emil, Jr
Booker, John Ratcliffe
Bostrom, Robert Christian
Brown, Robert Alan
Bull, Colin Bruce Bradley
Caggiano, Joseph Anthony, Jr
Crosson, Robert Scott
Danes, Zdenko Frankenberger
Dehlinger, Peter
Dzurisin, Daniel
Heacock, Richard Ralph
Henyey, Frank S
Hoblitt, Richard Patrick
Hoch, Richmond Joel
Holcomb, Robin Terry
Holmes, Mark Lawrence
Hussong, Donald MacGregor
Johnson, Harlan Paul
Kenney, James Francis
Lansinger, John Marcus
Lister, Clive R B
Malone, Stephen D
Merrill, Ronald Thomas
Olsen, Kenneth Harold
Parks, George Kung
Quigley, Robert James
Raymond, Charles Forest
Schwarz, Sigmund D
Smith, Brad Keller
Smith, Stewart W
Steele, William Kenneth
Stottlemyre, James Arthur
Veress, Sandor A

WISCONSIN
Bentley, Charles Raymond
Clay, Clarence Samuel
Cooper, Reid F
Laudon, Thomas S
Meyer, Richard Ernst
Meyer, Robert Paul
Mortimer, Clifford Hiley
Mudrey, Michael George, Jr
Thurber, Clifford Hawes
Wang, Herbert Fan

WYOMING
Shive, Peter Northrop
Smithson, Scott Busby
Tauer, Jane E

ALBERTA
Argenal-Arauz, Roger
Cook, Frederick Ahrens
Cruden, David Milne
Cumming, George Leslie
Deegan, Ross Alfred
Georgi, Daniel Taylan
Gough, Denis Ian
Jessop, Alan Michael
Kanasewich, Ernest Raymond
Lerbekmo, John Franklin
Peirce, John Wentworth

BRITISH COLUMBIA
Barker, Alfred Stanley, Jr
Best, Melvyn Edward
Clarke, Garry K C
Clowes, Ronald Martin
Dosso, Harry William
Ellis, Robert Malcolm
Galt, John (Alexander)
Horita, Robert Eiji
Hyndman, Roy D
Irving, Edward
Kieffer, Susan Werner
Lewis, Trevor John
Mothersill, John Sydney
Nafe, John Elliott
Oldenburg, Douglas William
Russell, Richard Doncaster
Strangway, David W
Thomson, David James
Ulrych, Tadeusz Jan
Watanabe, Tomiya
Weaver, John Trevor
Weichert, Dieter Horst

MANITOBA
Anderson, Christian Donald
Hall, Donald Herbert

NEW BRUNSWICK
Burke, Kenneth B S
Hamilton, Angus Cameron
Vanicek, Petr

NEWFOUNDLAND
Deutsch, Ernst Robert
King, Michael Stuart

Rochester, Michael Grant
Wright, James Arthur

NOVA SCOTIA
Blanchard, Jonathan Ewart
Keen, Charlotte Elizabeth
Ravindra, Ravi
Reynolds, Peter Herbert
Salisbury, Matthew Harold
Srivastava, Surat Prasad
Stevens, George Richard

ONTARIO
Arkani-Hamed, Jafar
Berry, Michael John
Chettle, David Robert
Dence, Michael Robert
Drury, Malcolm John
Dunlop, David John
Garland, George David
Gibb, Richard A
Gladwell, Graham M L
Grieve, Richard Andrew
Hare, Frederick Kenneth
Haworth, Richard Thomas
Hobson, George Donald
Hodgson, John Humphrey
Jones, Alister Vallance
McNamara, Allen Garnet
Manchee, Eric Best
Mansinha, Lalatendu
Mereu, Robert Frank
Morley, Lawrence Whitaker
Palmer, H Currie
Pearce, George William
Peltier, William Richard
Smylie, Douglas Edwin
Stesky, Robert Michael
Symons, David Thorburn Arthur
Thomas, Michael David
Uffen, Robert James
Watson, Michael Douglas
Weber, Jean Robert
Wesson, Paul Stephen
West, Gordon Fox
Wilson, John Tuzo
York, Derek H

QUEBEC
Crossley, David John
Doig, Ronald
Mukherji, Kalyan Kumar
Poorooshasb, Hormozd Bahman
Seguin, Maurice Krisholm

OTHER COUNTRIES
Arce, Jose Edgar
Clement, William Glenn
Geller, Robert James
Hoke, John Humphreys
Horai, Ki-iti
Kozai, Yoshihide
Lomnitz, Cinna
Mason, Ronald George
Nestvold, Elwood Olaf
Parsignault, Daniel Raymond
Pekeris, Chaim Leib
Rees, Manfred Hugh
Sobolev, Nikolai V
Tanis, James Iran
Taylor, Howard Lawrence
Tolstoy, Ivan
Vozoff, Keeva
Woodside, John Moffatt

Hydrology & Water Resources

ALABAMA
Bayne, David Roberge
Boyd, Claude Elson
Cook, Robert Bigham, Jr
Fowler, Charles Sidney
LaMoreaux, Philip Elmer
Leonard, Kathleen Mary
Molz, Fred John, III
Moser, Paul H
Smoot, George Fitzgerald
Warman, James Clark
Wetzel, Robert George

ALASKA
Brown, Edward James
Fahl, Charles Byron
Hawkins, Daniel Ballou
LaPerriere, Jacqueline Doyle
Long, William Ellis
Rogers, James Joseph
Slaughter, Charles Wesley

ARIZONA
Agenbroad, Larry Delmar
Avery, Charles Carrington
Baker, Malchus Brooks, Jr
Baker, Victor Richard
Bergen, James David
Bouwer, Herman
Bradley, Michael Douglas
Brusseau, Mark Lewis
Buras, Nathan
Callahan, Joseph Thomas
Clemmens, Albert J
Cluff, Carwin Brent
Contractor, Dinshaw N

Davis, Edwin Alden
Davis, Stanley Nelson
DeCook, Kenneth James
Dickinson, Robert Earl
Dworkin, Judith Marcia
Emery, Philip Anthony
Evans, Daniel Donald
Fink, James Brewster
Gay, Lloyd Wesley
Graf, William L
Hawkins, Richard Holmes
Heede, Burchard Heinrich
Idso, Sherwood B
Ince, Simon
Lane, Leonard James
Limpert, Frederick Arthur
Lord, William B
Manera, Paul Allen
Mays, Larry Wesley
Montgomery, Errol Lee
Neuman, Shlomo Peter
Osborn, Herbert B
Rasmussen, William Otto
Reich, Brian M
Renard, Kenneth G
Resnick, Sol Donald
Simpson, Eugene Sidney
Sorooshian, Soroosh
Sumner, John Stewart
Swenson, Frank Albert
Thames, John Long
Woolhiser, David A(rthur)
Zwolinski, Malcolm John

ARKANSAS
Mack, Leslie Eugene
Steele, Kenneth F

CALIFORNIA
Albert, Jerry David
Anderson, Henry Walter
Asante-Duah, D Kofi
Bean, Robert Taylor
Bencala, Kenneth Edward
Benes, Norman Stanley
Bertoldi, Gilbert LeRoy
Bevc, Dimitri
Blunt, Martin Julian
Bredehoeft, John Dallas
Brice, James Coble
Brooks, Norman H(errick)
Bubeck, Robert Clayton
Carnahan, Chalon Lucius
Carney, Heath Joseph
Cassidy, John J(oseph)
Chen, Carl W(an-Cheng)
Chen, Cheng-lung
Chesnut, Dwayne A(llen)
Chrysikopoulos, Constantinos V
Coats, Robert N
Cooper, Charles F
Corwin, Dennis Lee
Court, Arnold
Crawford, Norman Holmes
Dracup, John Albert
Everett, Lorne Gordon
Flashman, Stuart Milton
Georgakakos, Konstantine P
Getzen, Rufus Thomas
Goltz, Mark Neil
Griggs, Gary B
Guymon, Gary L
Hagan, Robert M(ower)
Ham, Lee Edward
Hanna, George P, Jr
Hassler, Thomas J
Humphrey, Thomas Milton, Jr
Huntley, David
Hwang, Li-San
Janke, Norman Charles
Jeng, Raymond Ing-Song
Kam, James Ting-Kong
Karn, Richard Wendall
Kennedy, Vance Clifford
Kharaka, Yousif Khoshu
Kramer, John Howard
Kuwabara, James S
Lambert, Walter Paul
Lee, Tien-Chang
Leps, Thomas MacMaster
Letey, John, Jr
Lippmann, Marcelo Julio
Lucas, Joe Nathan
Lysyj, Ihor
McKone, Thomas Edward
Manalis, Melvyn S
Mertes, Leal Anne Kerry
Morel-Seytoux, Hubert Jean
Morris, Henry Madison, Jr
Narasimhan, Thiruppudaimarudhur N
Orlob, Gerald T
Oster, James Donald
Phoenix, David A
Remson, Irwin
Rogers, Gifford Eugene
Rudavsky, Alexander Bohdan
Saltonstall, Clarence William, Jr
Sarkaria, Gurmukh S
Scott, Verne H(arry)
Simunek, Jiri
Slack, Keith Vollmer
Snyder, Charles Theodore
Spieker, Andrew Maute
Sposito, Garrison

Sutton, Donald Dunsmore
Testa, Stephen M
Todd, David Keith
Turk, Leland Jan
Volz, Michael George
Wells, Stephen Gene
White, Donald Edward
Wilt, Michael
Yates, Scott Raymond
Ziemer, Robert Ruhl

COLORADO
Benson, Larry V
Blackburn, Wilbert Howard
Branson, Farrel Allen
Caine, T Nelson
Clebsch, Alfred, Jr
Cohen, Ronald R H
Cooley, Richard Lewis
Curtis, Bruce Franklin
DeCoursey, Donn G(ene)
Diment, William Horace
Dunn, Darrel Eugene
Eisenlohr, W(illiam) S(tewart), Jr
Emmett, William W
Erskine, Christopher Forbes
Frasier, Gary Wayne
Girz, Cecilia Griffith
Gosink, Joan P
Hadley, Richard Frederick
Hardaway, John E(vans)
Herrmann, Raymond
Hoffer, Roger M(ilton)
Hotchkiss, William Rouse
Hughes, William Carroll
Illangasekare, Tissa H
Johnson, Arnold I(van)
Johnson, Verner Carl
Jones, Everett Bruce
Koelzer, Victor A
Langmuir, Donald
McCabe, Gregory James, Jr
McLaughlin, Thad G
Meade, Robert Heber, Jr
Meiman, James R
Mickelson, Rome H
Miller, Glen A
Moore, John Ezra
Morrison, Roger Barron
Nordstrom, Darrell Kirk
Olhoeft, Gary Roy
Peckham, Alan Embree
Richardson, Everett V
Runnells, Donald DeMar
Sanford, Richard Frederick
Schumm, Stanley Alfred
Smith, Ralph Emerson
Smith, Roger Elton
Sommerfeld, Richard Arthur
Spence, William J
Steele, Timothy Doak
Striffler, William D
Tabler, Ronald Dwight
Toy, Terrence J
Urban, Thomas Charles
Van der Heijde, Paul Karel Maria
Vaseen, V(esper) Albert
Visher, Frank N
Walker, Theodore Roscoe
Ward, Robert Carl
Weist, William Godfrey, Jr
Wershaw, Robert Lawrence
White, Gilbert Fowler
Yang, Chih Ted
Yang, In Che

CONNECTICUT
Bock, Paul
Fox, G(eorge) Sidney
Patton, Peter C
Spiegel, Zane

DELAWARE
Cheng, Alexander H-D
Ellis, Patricia Mench
Habibi, Kamran
Mather, John Russell
Talley, John Herbert
Willmott, Cort James

DISTRICT OF COLUMBIA
Bell, Bruce Arnold
Blanchard, Bruce
Briscoe, John
Chung, Riley M
Duguid, James Otto
Jones, Alice J
Knopman, Debra Sara
McGinnis, David Franklin, Jr
Mercer, James Wayne
Rattien, Stephen
Smith, Waldo E(dward)
Stringfield, Victor Timothy
Terrell, Charles R
Walrafen, George Edouard
Warnick, Walter Lee
Watt, Hame Mamadov
Zeizel, A(rthur) J(ohn)

FLORIDA
Anderson, Melvin W(illiam)
Ballerini, Rocco
Barile, Diane Dunmire
Belanger, Thomas V**

Bermes, Boris John
Chryssafopoulos, Nicholas
Conover, Clyde S(tuart)
Davis, John Armstrong
Dietz, Jess Clay
Gale, Paula M
Hammond, Luther Carlisle
Jackson, Daniel Francis
Kantrowitz, Irwin H
Kersten, Robert D(onavon)
Martsolf, J David
Nichols, James Carlile
O'Hara, Norbert Wilhelm
Parker, Garald G, Sr
Patterson, Archibald Oscar
Randazzo, Anthony Frank
Rao, Palakurthi Suresh Chandra
Scarlatos, Panagiotis Dimitrios
Steinman, Alan David
Stephens, John C(arnes)
Stewart, Mark Thurston
Tanner, William Francis, Jr
Thrailkill, John
Upchurch, Sam Bayliss
Viessman, Warren, Jr
Warburton, David Lewis
Wozab, David Hyrum

GEORGIA
Arkin, Gerald Franklin
Barnwell, Thomas Osborn, Jr
Bowen, Paul Tyner
Burgoa, Benali
Carver, Robert E
Cofer, Harland E, Jr
Hewlett, John David
Hill, David W(illiam)
Hooper, Richard Paul
Johnston, Richard H
Langdale, George Wilfred
Lineback, Jerry A(lvin)
Mace, Arnett C, Jr
Meyer, Judy Lynn
Rogers, Harvey Wilbur
Snyder, Willard Monroe
Steiner, Jean Louise
Thomas, Adrian Wesley
Vorhis, Robert C
Wallace, James Robert
Wenner, David Bruce

HAWAII
Cox, Doak Carey
Ekern, Paul Chester
Fast, Arlo Wade
Hihara-Endo, Linda Masae
Hufschmidt, Maynard Michael
Lau, L(eung Ku) Stephen
Merriam, Robert Arnold
Peterson, Frank Lynn
Wu, I-Pai

IDAHO
Brakensiek, Donald Lloyd
Clark, William Hilton
Crawford, Ronald Lyle
Kincaid, Dennis Campbell
Knobel, LeRoy Lyle
Knutson, Carroll Field
Lehrsch, Gary Allen
Neibling, William Howard
Schreiber, David Laurence
Welhan, John Andrew
Williams, Roy Edward

ILLINOIS
Adams, John Rodger
Ardis, Colby V, Jr
Aubertin, Gerald Martin
Boast, Charles Warren
Carpenter, Philip John
Cartwright, Keros
Corbett, Robert G
Ditmars, John David
Foster, John Webster
Garcia, Marcelo Horacio
Georgiadis, John G
Gorody, Anthony Wagner
Heigold, Paul C
Herzog, Beverly Leah
Hirschi, Michael Carl
Kempton, John P(aul)
Lesht, Barry Mark
McKay, Edward Donald, III
Marmer, Gary James
Maxwell, William Hall Christie
Paintal, Amreek Singh
Piwoni, Marvin Dennis
Prakash, Anand
Sasman, Robert T
Schwab, Gary Michael
Semonin, Richard Gerard
Stout, Glenn Emanuel
Valocchi, Albert Joseph
Wendland, Wayne Marcel
Yen, Ben Chie

INDIANA
Fredrich, Augustine Joseph
Gray, William Guerin
Hellenthal, Ronald Allen
Krothe, Noel C
Leap, Darrell Ivan
McCafferty, William Patrick

Rao, R(amachandra) A
Shaffer, Nelson Ross
Spacie, Anne

IOWA
Austin, Tom Al
Hallberg, George Robert
Hatfield, Jerry Lee
Hershey, H Garland
Kirkham, Don
Koch, Donald Leroy
Nakato, Tatsuaki
Odgaard, A Jacob
Prior, Jean Cutler
Schnoor, Jerald L

KANSAS
Aber, James Sandusky
Beck, Henry V
Eklund, Darrel Lee
Gries, John Charles
Kromm, David Elwyn
Metzler, Dwight F
Skidmore, Edward Lyman
Steeples, Donald Wallace
White, Stephen Edward
Whittemore, Donald Osgood

KENTUCKY
Barfield, Billy Joe
Budde, Mary Laurence
Coltharp, George B
Crawford, Nicholas Charles
Kiefer, John David
Warner, Richard Charles

LOUISIANA
Ferrell, Ray Edward, Jr
Grimwood, Charles
Jones, Paul Hastings
Kazmann, Raphael Gabriel
Latorre, Robert George
Loden, Michael Simpson
Murray, Stephen Patrick
Nyman, Dale James
Singh, Vijay P
Wilson, John Thomas
Wrobel, William Eugene
Yu, Fang Xin

MAINE
Kleinschmidt, R Stevens
Prescott, Glenn Carleton, Jr
Snow, Joseph William
Tewhey, John David
Yentsch, Clarice M

MARYLAND
Baummer, J Charles, Jr
Browzin, Boris S(ergeevich)
Chery, Donald Luke, Jr
Clarke, Frank Eldridge
Davis, George H
Everett, Ardell Gordon
Fanning, Delvin Seymour
Farnsworth, Richard Kent
Gaum, Carl H
Hudlow, Michael Dale
Jones, Blair Francis
Kohler, Max A
Linaweaver, Frank Pierce
Mack, Frederick K
May, Ira Philip
Munasinghe, Mohan P
Obremski, Henry J(ohn)
Papadopulos, Stavros Stefanu
Prestegaard, Karen Leah
Ragan, Robert Malcolm
Randall, John Douglas
Rango, Albert
ReVelle, Charles S
Salomonson, Vincent Victor
Schmugge, Thomas Joseph
Sternberg, Yaron Moshe
Wolman, Markley Gordon
Yaramanoglu, Melih

MASSACHUSETTS
Bowley, Donovan Robin
Bras, Rafael Luis
Carr, Jerome Brian
Deegan, Linda Ann
Eagleson, Peter Sturges
Entekhabi, Dara
Frimpter, Michael Howard
Hillel, Daniel
Male, James William
Ostendorf, David William

MICHIGAN
Abriola, Linda Marie
Burton, Thomas Maxie
Clark, James Alan
D'Itri, Frank M
Holland-Beeton, Ruth Elizabeth
Kehew, Alan Everett
Kunkle, George Robert
Nriagu, Jerome Okon
Quinn, Frank Hugh
Reynolds, John Z
Ronca, Luciano Bruno
Scavia, Donald
Tomboulian, Paul
Wright, Steven Jay

Wylie, Evan Benjamin

MINNESOTA
Alexander, Emmit Calvin, Jr
Boening, Paul Henrik
Downing, William Lawrence
Gulliver, John Stephen
Gupta, Satish Chander
Klemer, Andrew Robert
Krug, Edward Charles
Latterell, Joseph J
Linden, Dennis Robert
Lonergan, Dennis Arthur
McConville, David Raymond
McNaught, Donald Curtis
Perry, James Alfred
Pfannkuch, Hans Olaf
Silberman, Edward
Song, Charles Chieh-Shyang
Stefan, Heinz G
Tuthill, Samuel James
Walton, Matt Savage

MISSISSIPPI
Baker, Robert Andrew
Cooper, Charles Morris
Cross, Ralph Donald
Foster, George Rainey
Mutchler, Calvin Kendal
Priest, Melville S(tanton)
Shields, Fletcher Douglas, Jr
Strom, Eric William
Ursic, Stanley John

MISSOURI
Axthelm, Deon D
Barr, David John
Frye, Charles Isaac
Goebel, Edwin DeWayne
Henson, Bob Londes
Hjelmfelt, Allen T, Jr
Hoogheem, Thomas John
Jester, Guy Earlscort
Schroeder, Walter Albert
Settergren, Carl David
Stern, Michele Suchard
Unklesbay, Athel Glyde
Warner, Don Lee

NEBRASKA
Dragoun, Frank J
Dreesen, Vincent Harold
Hubbard, Kenneth Gene
Pederson, Darryll Thoralf
Power, James Francis
Skopp, Joseph Michael
Swartzendruber, Dale

NEVADA
Brandstetter, Albin
Case, Clinton Meredith
Fenske, Paul Roderick
Gifford, Gerald F
Helm, Donald Cairney
Hess, John Warren
Kaufmann, Robert Frank
McMillion, Leslie Glen, Sr
Mifflin, Martin David
Nelson, R William
Nork, William Edward
Peck, John H
Pepper, Darrell Weldon
Ritter, Dale Franklin
Worts, George Frank, Jr

NEW HAMPSHIRE
Ballestero, Thomas P
Bilello, Michael Anthony
Dingman, Stanley Lawrence
Federer, C Anthony
Hall, Francis Ramey
Hoag, Roland Boyden, Jr
Long, Carl F(erdinand)
Lynch, Daniel Roger

NEW JERSEY
Avissar, Roni
Dobi, John Steven
Halasi-Kun, George Joseph
Hordon, Robert M
Hsu, Chin-Fei
Kasabach, Haig F
Kim, Uing W
Klenke, Edward Frederick, Jr
Metsger, Robert William
Morgan, Mark Douglas
Nathan, Kurt
Stinson, Mary Krystyna
Strausberg, Sanford I
Uchrin, Christopher George
Wetherald, Richard Tryon
Widmer, Kemble
Wolfe, Peter E
Wood, Eric F

NEW MEXICO
Aldon, Earl F
Gutjahr, Allan L
Hawley, John William
Kuiper, Logan Keith
Lee, William Wai-Lim
Mauzy, Michael P
Morin, George Cardinal Albert
Phillips, Fred Melville

Rebuck, Ernest C(harles)
Sayre, William Olaf
Shomaker, John Wayne
Trainer, Frank W
Trauger, Frederick Dale
Tromble, John M
Wilson, Lee

NEW YORK
Abrahams, Athol Denis
Atkinson, Joseph Ferris
Black, Peter Elliott
Breed, Helen Illick
Brutsaert, Wilfried
Dicker, Daniel
Eschner, Arthur Richard
Forest, Herman Silva
Hennigan, Robert Dwyer
Hubbard, John Edward
Jarvis, Richard S
Jirka, Gerhard Hermann
LaFleur, Robert George
Loucks, Daniel Peter
Lougeay, Ray Leonard
Makarewicz, Joseph Chester
Mendoza-Cabrales, Cesar
Molloy, Andrew A
Morisawa, Marie
Muessig, Paul Henry
Naidu, Janakiram Ramaswamy
Nathanson, Benjamin
Palmer, Arthur N
Putman, George Wendell
Robinson, Joseph Edward
Schoonen, Martin A A
Sevian, Walter Andrew
Shen, Hung Tao
Shoemaker, Christine Annette
Siegfried, Clifford Anton
Stedinger, Jery Russell
Wallis, James Richard
Wang, Hsin-Pang
Wang, Muhao S
Yapa, Poojitha Dahanayake
Zimmie, Thomas Frank

NORTH CAROLINA
Barker, James Cathey
Blackwell, Floyd Oris
Forsythe, Randall D
Heath, Ralph Carr
Hellmers, Henry
Johnson, Thomas Charles
Karl, Thomas Richard
Kashef, A(bdel-Aziz) I(smail)
Kolenbrander, Lawrence Gene
LeGrand, Harry E
Okun, D(aniel) A(lexander)
Reckhow, Kenneth Howland
Reice, Seth Robert
Richardson, Curtis John
Robinson, Peter John
Ross, Thomas Edward
Scheffe, Richard Donald
Sneade, Barbara Herbert
Swank, Wayne T
Swift, Lloyd Wesley, Jr
Van Eck, Willem Adolph
Welby, Charles William

NORTH DAKOTA
Mason, Earl Sewell

OHIO
Ahmad, Moid Uddin
Bondurant, Byron L(ee)
Carmichael, Wayne William
Chawla, Mangal Dass
Cooke, George Dennis
Craig, Rachael Grace
Eckstein, Yoram
Farvolden, Robert Norman
Fuller, James Osborn
Hobbs, Benjamin F
Hyland, John R(oth)
Lal, Rattan
La Mers, Thomas Herbert
Lehr, Jay H
Lyon, John Grimson
Matisoff, Gerald
Moulton, Edward Q(uentin)
Murdoch, Lawrence Corlies, III
Noss, Richard Robert
Papadakis, Constantine N
Rausch, David Leon
Ricca, Vincent Thomas
Ruedisili, Lon Chester
Scarborough, Vernon Lee
Schmidt, Ronald Grover
Stiefel, Robert Carl
Whiting, Peter John
Wilmoth, Benton M
Wu, Tien Hsing

OKLAHOMA
Dunlap, William Joe
Haan, Charles Thomas
Hounslow, Arthur William
Johnson, Kenneth Sutherland
Knapp, Roy M
Pettyjohn, Wayne A
Salisbury, Neil Elliot
Schmelling, Stephen Gordon
Stone, John Floyd

Hydrology & Water Resources (cont)

Stover, Dennis Eugene

OREGON
Brown, George Wallace
Church, Marshall Robbins
Griswold, Daniel H(alsey)
Huber, Wayne Charles
Klingeman, Peter C
Linn, DeVon Wayne
Neilson, Ronald Price
Norris, Logan Allen
Sollitt, Charles Kevin
Waymire, Edward Charles
Wheeler, Richard Hunting

PENNSYLVANIA
Alley, Richard Blaine
Aron, Gert
Brady, Keith Bryan Craig
Brantley, Susan Louise
Carlson, Toby Nahum
Carson, Bobb
Chiu, Chao-Lin
Clark, Roger Alan
DeWalle, David Russell
Doelp, Louis C(onrad), Jr
Dzombak, David Adam
Elsworth, Derek
Emrich, Grover Harry
Fletcher, Frank William
Garber, Murray S
Gardner, Thomas William
Harrison, Samuel S
Hutchinson, Peter J
Keenan, John Douglas
Kleinman, Robert L P
Lowright, Richard Henry
McDonnell, Archie Joseph
Meisler, Harold
Parizek, Richard Rudolph
Petura, John C
Quimpo, Rafael Gonzales
Reed, Joseph Raymond
Rosenstock, Paul Daniel
Schmalz, Robert Fowler
Schultz, Lane D
Sooky, Attila A(rpad)
Stehly, David Norvin
Yeh, Gour-Tsyh
Younkin, Larry Myrle
Zeigler, A(lfred) G(eyer)

RHODE ISLAND
Golet, Francis Charles
Hermance, John Francis
Tarzwell, Clarence Matthew

SOUTH CAROLINA
Caruccio, Frank Thomas
Dysart, Benjamin Clay, III
Gardner, Leonard Robert
Geidel, Gwendolyn
Greene, George C, III
Henry, Harold Robert
James, L(aurence) Allan
McKellar, Henry Northington, Jr
Newman, Michael Charles
Nichols, Chester Encell
Strom, Richard Nelsen

SOUTH DAKOTA
LeRoux, Edmund Frank
Rahn, Perry H
Zogorski, John Steward

TENNESSEE
Brown, Bevan W, Jr
Dodson, Chester Lee
Elmore, James Lewis
Gamliel, Amir
Huff, Dale Duane
Lee, Donald William
Nyquist, Jonathan Eugene
Parker, Frank L
Patric, James Holton
Roop, Robert Dickinson
Turner, Robert Spilman

TEXAS
Allen, Peter Martin
Armstrong, Neal Earl
Beard, Leo Roy
Berkebile, Charles Alan
Caffey, James Enoch
Campbell, Michael David
Chang, Mingteh
Charbeneau, Randall Jay
Chew, Ju-Nam
Chiang, Chen Y
Collins, Michael Albert
Cotner, James Bryan
Daugherty, Franklin W
Dodge, Rebecca L
Doyle, Frank Lawrence
Ewing, Richard Edward
Fix, Ronald Edward
French, William Edwin
Giammona, Charles P, Jr
Gordon, Ellis Davis
Guyton, William F

Halff, Albert Henry
Haragan, Donald Robert
Holt, Charles Lee Roy, Jr
Johnson, Carlton Robert
Jones, Ernest Austin, Jr
Jones, Ordie Reginal
Kaiser, William Richard
Krishna, J Hari
Lind, Owen Thomas
McNaughton, Duncan A
Mathewson, Christopher Colville
Moore, Walter L(eon)
Mortada, Mohamed
Pearson, Frederick Joseph
Phillips, Joseph D
Rast, Walter, Jr
Reddell, Donald Lee
Richardson, Clarence Wade
Russell, Thomas Webb
Sharp, John Malcolm, Jr
Slotta, Larry Stewart
Streltsova, Tatiana D
Wiegand, Craig Loren
Wiese, John Herbert
Woodruff, Charles Marsh, Jr

UTAH
Adams, Michael Curtis
Allen, Richard Glen
Arnow, Theodore
Clyde, Calvin G(eary)
DeByle, Norbert V
Dobrowolski, James Phillip
Freethey, Geoffrey Ward
Gates, Joseph Spencer
Godfrey, Andrew Elliott
Goode, Harry Donald
Hansen, Vaughn Ernest
Hargreaves, George H(enry)
Hart, George Emerson, Jr
Hood, James Warren
Israelsen, C Earl
Jewell, Paul William
Keller, Jack
Kolesar, Peter Thomas
Malek, Esmaiel
Pashley, Emil Frederick, Jr
Peralta, Richard Carl
Sidle, Roy Carl
Waddell, Kidd M
Wood, Timothy E

VERMONT
Cushman, Robert Vittum
McIntosh, Alan William
Pinder, George Francis

VIRGINIA
Back, William
Baedecker, Mary Jo
Bennett, Gordon Daniel
Blakey, Lewis Horrigan
Buchanan, Thomas Joseph
Carstea, Dumitru
Cohen, Philip
Cohn, Timothy A
Cox, William Edward
Cragwall, J(oseph) S(amuel), Jr
Dendrou, Stergios
Echols, Charles E(rnest)
Hackett, Orwoll Milton
Holtzman, Golde Ivan
Hornberger, George Milton
Kelly, Mahlon George, Jr
Konikow, Leonard Franklin
Kornicker, William Alan
Langford, Russell Hal
Miller, John Frederick
Moody, David Wright
Myers, Charles Edwin
Olem, Harvey
O'Neill, Eileen Jane
Peck, Eugene Lincoln
Pickering, Ranard Jackson
Reilly, Thomas E
Rickert, David A
Roco, Mihail Constantin
Rosenshein, Joseph Samuel
Sayala, Dasharatham (Dash)
Schad, Theodore M(acNeeve)
Schaefer, Francis T
Schneider, Robert
Spofford, Walter O, Jr
Tsai, Frank Y
Wallace, Raymond Howard, Jr
Wieczorek, Gerald Francis
Wiesnet, Donald Richard
Wolff, Roger Glen
Woo, Dah-Cheng
Wood, Leonard Alton
Wood, Warren Wilbur
Zubkoff, Paul L(eon)

WASHINGTON
Adams, William Mansfield
Bender, Donald Lee
Bethlahmy, Nedavia
Brewer, William Augustus
Brown, Randall Emory
Burges, Stephen John
Cole, Charles Ralph
Costa, John Emil
Dunne, Thomas
Foley, Michael Glen

Foxworthy, Bruce L
Harr, Robert Dennis
Hodges, Floyd Norman
Karr, James Richard
Keller, C Kent
King, Larry Gene
Law, Albert G(iles)
Megahan, Walter Franklin
Michelsen, Ari Montgomery
Millham, Charles Blanchard
Orsborn, John F
Pierson, Thomas Charles
Rosengreen, Theodore E
Wooldridge, David Dilley

WEST VIRGINIA
Gwinn, James E
Kotcon, James Bernard
Kuhlman, John Michael
Rauch, Henry William

WISCONSIN
Anderson, Mary Pikul
Bubenzer, Gary Dean
Cain, John Manford
Drescher, William James
Dukenschein, Jeanne Therese
Fetter, Charles Willard, Jr
Govin, Charles Thomas, Jr
Green, Theodore, III
Horntvedt, Earl W
Knox, James Clarence
Lindorff, David Everett
Miller, David Hewitt
Orth, Paul Gerhardt
Stenstrom, Richard Charles
Stieglitz, Ronald Dennis
Waller, Roger Milton

WYOMING
Gloss, Steven Paul
Marston, Richard Alan
Rechard, Paul A(lbert)
Sturges, David L
Tung, Yeou-Koung
Zhang, Renduo

ALBERTA
Barendregt, Rene William
Chang, Chi
Feng, Joseph C
Lane, Robert Kenneth
Prepas, Ellie E
Rogerson, Robert James
Singh, Teja
Smith, Daniel Walter
Sommerfeldt, Theron G
Swanson, Robert Harold
Toth, Jozsef
Verschuren, Jacobus Petrus
Wallis, Peter Malcolm

BRITISH COLUMBIA
Clark, Robert H
Clarke, Garry K C
Foster, Harold Douglas
Freeze, Roy Allan
Golding, Douglas Lawrence
Jackson, Lionel Eric
Newbury, Robert W(illiam)
Perrin, Christopher John
Ruus, E(ugen)
Slaymaker, Herbert Olav

MANITOBA
Welsted, John Edward

NEW BRUNSWICK
Chow, Thien Lien
Davar, K(ersi) S

ONTARIO
Biswas, Asit Kumar
Burton, Ian
Cherry, John A
Dickinson, William Trevor
Dillon, Peter J
Elrick, David Emerson
Flint, Jean-Jacques
Ford, Derek Clifford
Gillham, Robert Winston
Gultepe, Ismail
Hughes, George Muggah
Jones, L(lewellyn) E(dward)
Kruus, Jaan
LaValle, Placido Dominick
Lee, David Robert
Lennox, Donald Haughton
McBean, Edward A
Meyboom, Peter
Moltyaner, Grigory
Poulton, Donald John
Schemenauer, Robert Stuart
Simpson, Frank
Tan, Chin Sheng
Wilkinson, Thomas Preston

QUEBEC
Beron, Patrick
Bouchard, Michel Andre
Ouellet, Marcel
Tessier, Andre

SASKATCHEWAN
Chew, Hemming
Cullimore, Denis Roy
Fuller, Gerald Arthur
Mollard, John D
Norum, Donald Iver
Tsang, Gee
Viraraghavan, Thiruvenkatachari

OTHER COUNTRIES
Briand, Frederic Jean-Paul
Capodaglio, Andrea G
Raveendran, Ekarath
Ward, Gerald T(empleton)
Williams, Michael Maurice Rudolph
Yazicigil, Hasan

Marine Sciences, General

ALABAMA
Brady, Yolanda J
Cowan, James Howard, Jr
Kenyon, Patricia May
Modlin, Richard Frank
Williams, Louis Gressett

ALASKA
Ahlgren, Molly O
Dahlberg, Michael Lee
Forest, Kathryn Jo
Hameedi, Mohammad Jawed
Hatch, Scott Alexander
Highsmith, Raymond C
Mathisen, Ole Alfred
Olsen, James Calvin
Severin, Kenneth Paul
Shirley, Thomas Clifton
Stekoll, Michael Steven
Thomas, Gary Lee
Weeks, Wilford Frank

ARIZONA
Benfield, Charles W(illiam)
Glenn, Edward Perry
Parrish, Judith Totman

CALIFORNIA
Benson, Andrew Alm
Berry, William Benjamin Newell
Boehlert, George Walter
Bottjer, David John
Brand, Samson
Bray, Richard Newton
Brumbaugh, Joe H
Cheng, Lanna
Clague, David A
Cohen, Anne Carolyn Constant
Cohen, Daniel Morris
Cowles, David Lyle
Coyer, James A
Cummings, William Charles
Davis, Gary Everett
Davis, William Jackson
Dedrick, Kent Gentry
Delaney, Margaret Lois
Dietz, Thomas John
Di Girolamo, Rudolph Gerard
Dill, Robert Floyd
Douglas, Robert G
Eittreim, Stephen L
Fiedler, Paul Charles
Field, A J
Fusaro, Craig Allen
Halpern, David
Ham, Lee Edward
Hanggi, Evelyn Betty
Hanson, Joe A
Haygood, Margo Genevieve
Hemmingsen, Barbara Bruff
Hill, Merton Earle, III
Horn, Michael Hastings
Hunt, George Lester, Jr
Karl, Herman Adolf
Kitting, Christopher Lee
Kremer, Patricia McCarthy
Lafferty, Kevin D
Lange, Carina Beatriz
Lapota, David
Leidersdorf, Craig B
Lindner, Elek
Macdonald, Kenneth Craig
Mackenzie, Kenneth Victor
Martin, Joel William
Matthews, Kathleen Ryan
McLeod, Samuel Albert
Morse, Daniel E
Mullen, Ashley John
Mulligan, Timothy James
Murray, Steven Nelsen
Muscatine, Leonard
Nybakken, James W
Oglesby, Larry Calmer
Orcutt, Harold George
Owen, Robert W, Jr
Parrish, Richard Henry
Pearse, Vicki Buchsbaum
Phleger, Charles Frederick
Potts, Donald Cameron
Powers, Dennis A
Prasad, Kota S
Rau, Gregory Hudson
Ripley, William Ellis
Robilliard, Gordon Allan

Roth, Ariel A
Sakagawa, Gary Toshio
Saleeby, Jason Brian
Sanders, Brenda Marie
Schmieder, Robert W
Schwing, Franklin Burton
Sharp, Gary Duane
Smith, Kenneth Lawrence, Jr
Soule, Dorothy (Fisher)
Squires, Dale Edward
Stewart, Brent Scott
Stowe, Keith S
Stramski, Dariusz
Straughan, Isdale (Dale) Margaret
Tegner, Mia Jean
Thompson, Rosemary Ann
Yeung, Ronald Wai-Chun

COLORADO
Brown, Lewis Marvin
Hay, William Winn
Kent, Harry Christison
Mykles, Donald Lee
Osterhoudt, Walter Jabez
Shanks, Wayne C, III
Shopp, George Milton, Jr

CONNECTICUT
Baillie, Priscilla Woods
Booth, Charles E
Brooks, Douglas Lee
Dawson, Margaret Ann
Farmer, Mary Winifred
Fell, Paul Erven
Haakonsen, Harry Olav
Katz, Max
Lee, Douglas Scott
Monahan, Edward Charles
Tolderlund, Douglas Stanley
Turekian, Karl Karekin

DELAWARE
Karlson, Ronald Henry
Kobayashi, Nobuhisa
Luther, George William, III
Maurmeyer, Evelyn Mary
Schuler, Victor Joseph
Swann, Charles Paul
Targett, Nancy McKeever
Targett, Timothy Erwin
Taylor, Malcolm Herbert
Thoroughgood, Carolyn A

DISTRICT OF COLUMBIA
Adey, Walter Hamilton
Barber, Mary Combs
Cairns, Stephen Douglas
Diemer, F(erdinand) P(eter)
Feeney, Gloria Comulada
Gaffney, Paul G, II
Lindstrom, Eric Jon
Littler, Mark Masterton
Rubinoff, Roberta Wolff
Sze, Philip
Talbot, Frank Hamilton
Terrell, Charles R
Urban, Edward Robert, Jr
Vogt, Peter Richard
Williams, Austin Beatty

FLORIDA
Alevizon, William
Barkalow, Derek Talbot
Betzer, Peter Robin
Bloom, Stephen Allen
Courtenay, Walter Rowe, Jr
Craig, Alan Knowlton
Dodge, Richard E
Donoghue, Joseph F
Feddern, Henry A
Finkl, Charles William, II
Fourqurean, James Warren
Gifford, John A
Gleeson, Richard Alan
Hallock-Muller, Pamela
Joyce, Edwin A, Jr
Kerr, John Polk
Legler, David M
Lutz, Peter Louis
McAllister, Raymond Francis
Marcus, Nancy Helen
Mariscal, Richard North
Maurrasse, Florentin Jean-Marie Robert
Miller, James Woodell
Muller-Karger, Frank Edgar
Nelson, Walter Garnet
Phlips, Edward Julien
Prince, Eric D
Reynolds, John Elliott, III
Rice, Stanley Alan
Richards, William Joseph
Smith, Sharon Louise
Stafford, Robert Oppen
Tamplin, Mark Lewis
Tappert, Frederick Drach
Taylor, Barrie Frederick
Teaf, Christopher Morris
Tiffany, William James, III
Tucker, Gail Susan
Virnstein, Robert W
White, Arlynn Quinton, Jr
Wiebusch, Charles Fred
Wise, Sherwood Willing, Jr
Wyneken, Jeanette

Young, Craig Marden
Zhu, Xiaorong
Zikakis, John Philip

GEORGIA
Adkison, Daniel Lee
Alberts, James Joseph
Blanton, Jackson Orin
Burns, Lawrence Anthony
Goldstein, Susan T
Kneib, Ronald Thomas
Newell, Steven Young
Pomeroy, Lawrence Richards
Rogers, Peter H
Walls, Nancy Williams
Weissburg, Marc Joel
Wiebe, William John

HAWAII
Ako, Harry Mu Kwong Ching
Bailey-Brock, Julie Helen
Coles, Stephen Lee
Curtis, George Darwin
Eldredge, Lucius G
Garcia, Michael Omar
Helfrich, Philip
Hey, Richard N
Hirota, Jed
Hodgson, Lynn Morrison
Hunter, Cynthia L
Krock, Hans-Jürgen
Krupp, David Alan
Landry, Michael Raymond
McCoy, Floyd W, Jr
Moberly, Ralph M
Moore, Gregory Frank
Polovina, Jeffrey Joseph
Radtke, Richard Lynn
Uchida, Richard Noboru
Wyban, James A
Wyrtki, Klaus
Zisk, Stanley Harris

ILLINOIS
Assanis, Dennis N
Bieler, Rudiger
Braker, William Paul
Dyer, William Gerald
Lima, Gail M
Tindall, Donald R
Towle, David Walter
Voight, Janet Ruth

INDIANA
Hulbert, Matthew H
Spacie, Anne

IOWA
Budd, Ann Freeman
Landweber, Louis

KENTUCKY
Chesnut, Donald R, Jr

LOUISIANA
Bahr, Leonard M, Jr
Bauer, Raymond Thomas
Chesney, Edward Joseph, Jr
Deaton, Lewis Edward
Defenbaugh, Richard Eugene
DeLaune, Ronald D
Herke, William Herbert
Mills, Earl Ronald
Morris, Halcyon Ellen McNeil
Sammarco, Paul William
Shaw, Richard Francis
Turner, Robert Eugene
Van Lopik, Jack Richard
Walker, Harley Jesse
Weidner, Earl

MAINE
Borei, Hans Georg
Brawley, Susan Howard
Garside, Christopher
Hawes, Robert Oscar
King, Gary Michael
Larsen, Peter Foster
Marcotte, Brian Michael
Mazurkiewicz, Michael
Ridgway, George Junior
Welch, Walter Raynes

MARYLAND
Anderson, Robert Simpers
Baer, Ledolph
Berkson, Harold
Busalacchi, Antonio James, Jr
Cammen, Leon Matthew
Capone, Douglas George
Cronin, Thomas Wells
Ebert, James David
Fink, Don Roger
Fisher, Thomas Richard, Jr
Gloersen, Per
Helz, George Rudolph
Hines, Anson Hemingway
Kappel, Ellen Sue
Kayar, Susan Rennie
Kearney, Michael Sean
Koblinsky, Chester John
Kravitz, Joseph Henry
Leatherman, Stephen Parker
Leese, John Albert

McCammon, Helen Mary
McVey, James Paul
Menez, Ernani Guingona
Mielke, James Edward
Mihursky, Joseph Anthony
Moran, David Dunstan
Murray, Laura
Nazarova, Katherine A
Nelson, Jay Arlen
Osman, Richard William
Phelps, Harriette Longacre
Pritchard, Donald William
Reader, Wayne Truman
Rich, Harry Louis
Riedel, Gerhardt Frederick
Robertson, Andrew
Sanders, James Grady
Shykind, Edwin B
Stubblefield, William Lynn
Szego, George C(harles)
Taylor, Richard Timothy
Veitch, Fletcher Pearre, Jr
Weisberg, Stephen Barry
Wolfe, Douglas Arthur

MASSACHUSETTS
Atema, Jelle
Aubrey, David Glenn
Ballard, Robert D
Broadus, James Matthew
Bryan, Wilfred Bottrill
Burris, John Edward
Butler, James Newton
Dacey, John W H
Dixon, Brian Gilbert
Hillman, Robert Edward
Jearld, Ambrose, Jr
Kleinrock, Martin Charles
Knebel, Harley John
Lin, Jian
Little, Sarah Alden
McLaren, J Philip
McNutt, Marcia Kemper
Marshall, Nelson
Mayo, Barbara Shuler
Parker, Henry Seabury, III
Pearce, John Bodell
Pechenik, Jan A
Poag, Claude Wylie
Prescott, John Hernage
Reed, F(lood) Everett
Reimold, Robert J
Richards, F Paul
Robinson, William Edward
Scheltema, Amelie Hains
Stephen, Ralph A
Turner, Jefferson Taylor
Wallace, Gordon Thomas
West, Arthur James, II
White, Alan Whitcomb
Wigley, Roland L
Wirsen, Carl O, Jr

MICHIGAN
Boyd, John Philip
Bratt, Albertus Dirk
Gardner, Wayne Stanley
Hawley, Nathan
Lehman, John Theodore
Moll, Russell Addison
Nyman, Melvin Andrew
Schwab, David John

MINNESOTA
Baker, Shirley Marie
Edwards, Richard Lawrence
Murdock, Gordon Robert

MISSISSIPPI
Anderson, Gary
Geddes, Wilburt Hale
Hurlburt, Harley Ernest
Lawler, Adrian Russell
Meylan, Maurice Andre
Morris, Gerald Brooks
Ramsdale, Dan Jerry
Sprague, Vance Glover, Jr

MISSOURI
Kuhns, John Farrell
Sharp, John Roland
Snowden, Jesse O

MONTANA
Stanley, George Dabney, Jr

NEVADA
Deacon, James Everett

NEW HAMPSHIRE
Hines, Mark Edward
Lesser, Michael Patrick
McDowell, William H
Weber, James Harold

NEW JERSEY
Brine, Charles James
Bruno, Michael Stephen
Canzonier, Walter J
Cooper, Keith Raymond
Farmanfarmaian, Allahverdi
Glaser, Keith Brian
Grassle, Judith Payne
Kennish, Michael Joseph

London, Mark David
Pacheco, Anthony Louis
Psuty, Norbert Phillip
Sugam, Richard Jay

NEW MEXICO
Stearns, Charles Edward
Stricker, Stephen Alexander

NEW YORK
Biscaye, Pierre Eginton
Bokuniewicz, Henry Joseph
Boynton, John E
Brandt, Stephen Bernard
Cheung, Paul James
Clarke, Raymond Dennis
Coffroth, Mary Alice
Conover, David Olmstead
Cousteau, Jacques-Yves
Diebold, John Brock
Dugolinsky, Brent Kerns
Finks, Robert Melvin
Flood, Roger Donald
Gerard, Valrie Ann
Hayes, Dennis E
Jacobs, Stanley S
Kaplan, Eugene Herbert
Karig, Daniel Edmund
Kastens, Kim Anne
Kent, Dennis V
Lasker, Howard Robert
Levinton, Jeffrey Sheldon
Marra, John Frank
Multer, H Gray
Nuzzi, Robert
Quigley, James P
Rivest, Brian Roger
Scranton, Mary Isabelle
Serafy, D Keith
Shumway, Sandra Elisabeth
Smardon, Richard Clay
Smethie, William Massie, Jr
Sutton, George Harry
Triantafyllou, George S
Vairavamurthy, Murthy Appathurai
Webber, Edgar Ernest
Weyl, Peter K

NORTH CAROLINA
Bird, Kimon T
Chester, Alexander Jeffrey
Eggleston, David Bryan
Forward, Richard Blair, Jr
Govoni, John Jeffrey
Horton, Carl Frederick
Janowitz, Gerald S(aul)
Johnson, Thomas Charles
Karson, Jeffrey Alan
Kuenzler, Edward Julian
Lindquist, David Gregory
Link, Garnett William, Jr
Marlowe, James Irvin
Oliver, James David
Ramus, Joseph S
Rittschof, Daniel
Rulifson, Roger Allen
Schwartz, Frank Joseph
Searles, Richard Brownlee
Sizemore, Ronald Kelly
Stoskopf, Michael Kerry
Sunda, William George
Thayer, Gordon Wallace
Webster, William David
Williams, Ann Houston

NORTH DAKOTA
Lieberman, Milton Eugene

OHIO
Boczar, Barbara Ann
Fortner, Rosanne White
Hamlett, William Cornelius
Hummon, William Dale
Hutchinson, Frederick Edward
Keiser, Terry Dean
Miller, Arnold I
Sabourin, Thomas Donald
Saksena, Vishnu P
Teeter, James Wallis

OKLAHOMA
Tillman, Roderick W
Visher, Glenn S

OREGON
Cimberg, Robert Lawrence
Drake, Ellen Tan
Duffield, Deborah Ann
Lubchenco, Jane
Malouf, Robert Edward
Sherr, Barry Frederick
Sherr, Evelyn Brown
Terwilliger, Nora Barclay
Trehu, Anne Martine
Wood, Anne Michelle

PENNSYLVANIA
Barron, Eric James
Bowen, Vaughan Tabor
Bursey, Charles Robert
Carson, Bobb
Commito, John Angelo
Franz, Craig Joseph
Guida, Vincent George

Marine Sciences, General (cont)

Haase, Bruce Lee
Kump, Lee Robert
Lambertsen, Richard H
Miller, Richard Lee
Niemitz, Jeffrey William
Sayre, Clifford M(orrill), Jr
Steele, Craig William
Taylor, Rhoda E
Williams, Frederick McGee

RHODE ISLAND
Benoit, Richard J
Bertness, Mark David
Brungs, William Aloysius
Gould, Mark D
Leinen, Margaret Sandra
Miller, Don Curtis
Prell, Warren Lee
Rice, Michael Alan

SOUTH CAROLINA
Browdy, Craig Lawrence
Conner, William Henry
Coulston, Mary Lou
Dame, Richard Franklin
Dean, John Mark
Feller, Robert Jarman
Higerd, Thomas Braden
Lundstrom, Ronald Charles
McKellar, Henry Northington, Jr
Morris, James T
Smith, Theodore Isaac Jogues
Stakes, Debra Sue

SOUTH DAKOTA
Wiersma, Daniel

TENNESSEE
Bartell, Steven Michael
Brinkhurst, Ralph O
Cushing, Bruce S
Simco, Bill Al

TEXAS
Aldrich, David Virgil
Baughn, Robert Elroy
Bohacs, Kevin Michael
Caillouet, Charles W, Jr
Chrzanowski, Thomas Henry
Damuth, John Erwin
DiMichele, Leonard Vincent
Dunton, Kenneth Harlow
Foster, John Robert
Gittings, Stephen Reed
Hanlon, Roger Thomas
Hasling, Jill Freeman
Hellier, Thomas Robert, Jr
La Claire, John Willard, II
McCain, John Charles
McCloy, James Murl
Maki, Alan Walter
Mazzullo, James Michael
Montagna, Paul A
Neill, William Harold
Parker, Robert Hallett
Pirie, Robert Gordon
Ray, James P
Rennie, Thomas Howard
Roberts, Susan Jean
Rozas, Lawrence Paul
Smith, Malcolm Crawford, Jr
Suttle, Curtis
Wohlschlag, Donald Eugene
Zimmerman, Carol Jean

UTAH
Jensen, David
Waddell, Kidd M

VIRGINIA
Alden, Raymond W, III
Burreson, Eugene M
Chang, William Y B
Curtin, Brian Thomas
Diaz, Robert James
Diehl, Fred A
Goodell, Horace Grant
Huggett, Robert James
Hurdle, Burton Garrison
Keinath, John Allen
Leung, Wing Hai
Loesch, Joseph G
Luckenbach, Mark Wayne
Lynch, Maurice Patrick
McCormick-Ray, M Geraldine
Munday, John Clingman, Jr
Musick, John A
Myers, Thomas DeWitt
Neves, Richard Joseph
Odum, William Eugene
Patterson, Mark Robert
Perkins, Frank Overton
Peters, Esther Caroline
Rippen, Thomas Edward
Siapno, William David
Spinrad, Richard William
Weinstein, Alan Ira
Wingard, Christopher Jon
Wood, Robertson Harris Langley
Zieman, Joseph Crowe, Jr

WASHINGTON
Benson, Keith Rodney
Bernard, Eddie Nolan
Chace, Alden Buffington, Jr
Crabtree, David Melvin
Damkaer, David Martin
Hardy, John Thomas
Harry, George Yost
Hirschfelder, John Joseph
Kolb, James A
Long, Edward R
Mearns, Alan John
Miles, Edward Lancelot
Mills, Claudia Eileen
Muench, Robin Davie
Muller-Parker, Gisele Therese
Pearson, Walter Howard
Pietsch, Theodore Wells
Ross, June Rosa Pitt
Simenstad, Charles Arthur
Sorem, Ronald Keith
Sulkin, Stephen David
Swartzman, Gordon Leni
Templeton, William Lees
Thom, Ronald Mark

WEST VIRGINIA
Anderson, Douglas Poole
Henderson-Arzapalo, Anne Patricia

WISCONSIN
Amundson, Clyde Howard
Besch, Gordon Otto Carl
Brooks, Arthur S
Fowler, Gerald Allan
Gernant, Robert Everett
Hedden, Gregory Dexter
Horntvedt, Earl W
Krezoski, John R
Magnuson, John Joseph
Remsen, Charles C, III
Rosson, Reinhardt Arthur
Wimpee, Charles F

PUERTO RICO
Voltzow, Janice

ALBERTA
Peirce, John Wentworth

BRITISH COLUMBIA
Chase, Richard L
Domenici, Paolo
Fankboner, Peter Vaughn
Hartwick, Earl Brian
Kelly, Michael Thomas
Mackie, George Owen
Muggeridge, Derek Brian
Perry, Richard Ian
Tunnicliffe, Verena Julia
Vermeer, Kees
Yunker, Mark Bernard

NEW BRUNSWICK
Chopin, Thierry Bernard Raymond
Frantsi, Christopher
Hanson, John Mark
Needler, Alfred Walker Holinshead
Pohle, Gerhard Werner
Taylor, Andrew Ronald Argo
Wells, David Ernest

NEWFOUNDLAND
Anderson, John Truman
Gogan, Niall Joseph
Haedrich, Richard L
Hiscott, Richard Nicholas
Khan, Rasul Azim
Steele, Vladislava Julie

NOVA SCOTIA
Cook, Robert Harry
Daborn, Graham Richard
Dadswell, Michael John
Jamieson, William David
Odense, Paul Holger
Sinclair, Michael Mackay
Stobo, Wayne Thomas
Vandermeulen, John Henri

ONTARIO
Arkani-Hamed, Jafar
Bidleman, Terry Frank
Geurts, Marie Anne H L
Green, Roger Harrison
LaValle, Placido Dominick
Leppard, Gary Grant
Ruel, Maurice M J
Sale, Peter Francis
Smith, Lorraine Catherine
Williams, David Dudley
Winterbottom, Richard

QUEBEC
Beland, Pierre
Browman, Howard Irving
Digby, Peter Saki Bassett
Filteau, Gabriel
Percy, Jonathan Arthur

SASKATCHEWAN
Hobson, Keith Alan

OTHER COUNTRIES
Briand, Frederic Jean-Paul
Chadwick, Nanette Elizabeth
Dahl, Arthur Lyon
Fowler, Scott Wellington
Marsh, James Alexander, Jr
Ni, I-Hsun
Paulay, Gustav
Richards, Adrian F
Rodriguez, Gilberto
South, Graham Robin
Takeuchi, Kiyoshi Hiro
Thierstein, Hans Rudolf

Mineralogy-Petrology

ALABAMA
Chalokwu, Christopher Iloba
Cook, Robert Bigham, Jr
Ehlers, Ernest George
Fang, Jen-Ho
Isphording, Wayne Carter
Kenyon, Patricia May
Prescott, Paul Ithel

ALASKA
Karl, Susan Margaret
Mowatt, Thomas C
Swanson, Samuel Edward

ARIZONA
Boyd, George Addison
Burt, Donald McLain
Buseck, Peter R
Duffield, Wendell Arthur
Fink, Jonathan Harry
Garske, David Herman
Gasparrini, Claudia
Guilbert, John M
Holloway, John Requa
Kamilli, Diana Chapman
Kamilli, Robert Joseph
Kepper, John C
Loomis, Timothy Patrick
McMillan, Paul Francis
Page, Norman J
Titley, Spencer Rowe
Wallace, Terry Charles, Jr
Williams, Sidney Arthur

ARKANSAS
Bergendahl, Maximilian Hilmar
Oglesby, Gayle Arden

CALIFORNIA
Akella, Jagannadham
Albee, Arden Leroy
Allan, John Ridgway
Anderson, Orson Lamar
Austin, James Arthur
Bailey, Roy Alden
Beaty, David Wayne
Bernstein, Lawrence Richard
Bieler, Barrie Hill
Boettcher, Arthur Lee
Bohlen, Steven Ralph
Boles, James Richard
Borg, Iris Y P
Brimhall, George H
Brooks, Elwood Ralph
Brown, Gordon Edgar, Jr
Bukowinski, Mark S T
Burnett, John L
Burtner, Roger Lee
Carmichael, Ian Stuart
Chmura-Meyer, Carol A
Christiansen, Robert Lorenz
Clague, David A
Coleman, Robert Griffin
Collins, Lorence Gene
Compton, Robert Ross
Cook, Harry E, III
Csejtey, Bela, Jr
Davidson, Jon Paul
Day, Howard Wilman
Dodge, Franklin C W
Doner, Harvey Ervin
Dusel-Bacon, Cynthia
Ehlig, Perry Lawrence
Einaudi, Marco Tullio
Ernst, Wallace Gary
Fernandez, Louis Anthony
Ford, Arthur B
Fryer, Charles W
Gaal, Robert A P
Gabelman, John Warren
Garlick, George Donald
Green, Harry Western, II
Greene, Robert Carl
Hanan, Barry Benton
Harpster, Robert E
Harwood, David Smith
Hein, James R
Herber, Lawrence Justin
Hill, Hamilton Stanton
Hodges, Lance Thomas
Hopson, Clifford Andrae
Janke, Norman Charles
Jeanloz, Raymond
Kamb, Walter Barclay
Keith, Terry Eugene Clark
Liddicoat, Richard Thomas, Jr
Lind, Carol Johnson
Liou, Juhn G
Loney, Robert Ahlberg
Longshore, John David
McFadden, Lucy-Ann Adams
McKague, Herbert Lawrence
McKenzie, William F
McLean, Hugh
Mahood, Gail Ann
Majmundar, Hasmukhrai Hiralal
Manning, Craig Edward
Mattinson, James Meikle
Meike, Annemarie
Mikami, Harry M
Moore, James Gregory
Moores, Eldridge Morton
Olson, Jerry Chipman
Parrish, William
Peterson, Donald William
Plummer, Charles Carlton
Ramspott, Lawrence Dewey
Rhoades, James David
Rosenfeld, John L
Ross, Donald Clarence
Rossman, George Robert
Saleeby, Jason Brian
Sarna-Wojcicki, Andrei M
Schneiderman, Jill Stephanie
Silver, Leon Theodore
Smith, Robert Leland
Snetsinger, Kenneth George
Spera, Frank John
Stebbins, Jonathan F
Stolper, Edward Manin
Strahl, Erwin Otto
Swinney, Chauncey Melvin
Sylvester, Arthur Gibbs
Tabor, Rowland Whitney
Taylor, Hugh P, Jr
Testa, Stephen M
Tilling, Robert Ingersoll
Vallier, Tracy L
Wadsworth, William Bingham
Walawender, Michael John
Warren, Paul Horton
Wenk, Hans-Rudolf
Williams, Alan Evan
Williams, Quentin Christopher
Wilshire, Howard Gordon
Winchell, Robert E
Wise, William Stewart
Woyski, Margaret Skillman

COLORADO
Agapito, J F T
Bohor, Bruce Forbes
Bryant, Bruce Hazelton
Cadigan, Robert Allen
Campbell, John Arthur
Cathcart, James B
Christensen, Odin Dale
Cole, James Channing
Desborough, George A
Dixon, Helen Roberta
Elder, Curtis Harold
Erslev, Eric Allan
Griffitts, Wallace Rush
Grose, Thomas Lucius Trowbridge
Gude, Arthur James, 3rd
Hayes, John Bernard
Henrickson, Eiler Leonard
Hill, Walter Edward, Jr
Hills, Francis Allan
Jansen, George James
Keighin, Charles William
Kuntz, Mel Anton
McCallum, Malcolm E
Miesch, Alfred Thomas
Modreski, Peter John
Munoz, James Loomis
Ponder, Herman
Sanford, Richard Frederick
Selverstone, Jane Elizabeth
Sheppard, Richard A
Sheridan, Douglas Maynard
Sidder, Gary Brian
Smyth, Joseph Richard
Thompson, Tommy Burt
Verbeek, Earl Raymond
Wahl, Floyd Michael
Walker, Theodore Roscoe
Wallace, Chester Alan
Warner, Lawrence Allen
Wilcox, Ray Everett
Wunderman, Richard Lloyd
Young, Edward Joseph
Zartman, Robert Eugene

CONNECTICUT
Frueh, Alfred Joseph, Jr
Joesten, Raymond
Liese, Homer C
Philpotts, Anthony Robert
Piotrowski, Joseph Martin
Skinner, H Catherine W
Winchell, Horace

DELAWARE
Leavens, Peter Backus
Spoljaric, Nenad
Stiles, A(lvin) B(arber)
Thompson, Allan M

DISTRICT OF COLUMBIA
Finger, Larry W

Fleischer, Michael
Fudali, Robert F
Hazen, Robert Miller
Lawson, Charles Alden
Mao, Ho-Kwang
Mysen, Bjorn Olav
Parham, Walter Edward
Prewitt, Charles Thompson
Price, Jonathan Greenway
Rumble, Douglas, III
Simkin, Thomas Edward
Usselman, Thomas Michael
Wilson, Frederick Albert

FLORIDA
Blanchard, Frank Nelson
Compton, John S
Councill, Richard J
Draper, Grenville
Eades, James L
Gielisse, Peter Jacob Maria
Kadey, Frederic L, Jr
Lee, Harley Clyde
Minkin, Jean Albert
Mueller, Paul Allen
Nagle, Frederick, Jr
Ragland, Paul C
Rosendahl, Bruce Ray
Snow, Eleanour Anne
Travis, Russell Burton
Tull, James Franklin
Young, Craig Marden

GEORGIA
Chatelain, Edward Ellis
Cofer, Harland E, Jr
Dod, Bruce Douglas
Giardini, Armando Alfonzo
Gore, Pamela J(eanne) W(heeless)
Herz, Norman
Logan, Kathryn Vance
Pollard, Charles Oscar, Jr
Railsback, Loren Bruce
Smith, John M
Sumartojo, Jojok
Weaver, Charles Edward
Weaver, Robert Michael
Whitney, James Arthur

HAWAII
Garcia, Michael Omar
Keil, Klaus
McCoy, Floyd W, Jr
Ming, Li Chung
Scott, Edward Robert Dalton
Sinton, John Maynard
Taylor, G Jeffrey
Thornber, Carl Richard

IDAHO
Bonnichsen, Bill
Welch, Jane Marie

ILLINOIS
Ahmed, Wase U
Amos, Dewey Harold
Anderson, Alfred Titus, Jr
Baxter, James Watson
Berg, Jonathan H
Bishop, Finley Charles
Chapman, Carleton Abramson
Chiou, Wen-An
Doehler, Robert William
Glass, Herbert David
Goldsmith, Julian Royce
Guggenheim, Stephen
Haddock, Gerald Hugh
Harvey, Richard David
Hay, Richard Le Roy
Henderson, Donald Munro
Hess, David Filbert
Jones, Robert L
Kirchner, James Gary
Kirkpatrick, Robert James
Liu, Mian
Masters, John Michael
Moll, William Francis, Jr
Montgomery, Carla Westlund
Moore, Duane Milton
Moore, Paul Brian
Newton, Robert Chaffer
Odom, Ira Edgar
Reams, Max Warren
Sandberg, Philip A
Smith, Joseph Victor
Sood, Manmohan K
Speed, Robert Clarke
Steele, Ian McKay
Woodland, Bertram George

INDIANA
Brock, Kenneth Jack
Christensen, Nikolas Ivan
Madison, James Allen
Martin, Charles Wellington
Meyer, Henry Oostenwald Albertijn
Murray, Haydn Herbert
Roepke, Harlan Hugh
Shaffer, Nelson Ross
Shieh, Yuch-Ning
White, Joe Lloyd
Wintsch, Robert P

IOWA
De Nault, Kenneth J
Foster, Charles Thomas, Jr
Garvin, Paul Lawrence
McCormick, George R

KANSAS
Twiss, Page Charles
Van Schmus, William Randall

KENTUCKY
Barnhisel, Richard I
Clark, Armin Lee
Stanonis, Francis Leo

LOUISIANA
Allen, Gary Curtiss
Byerly, Gary Ray
Curry, Mary Grace
Ferrell, Ray Edward, Jr
Franks, Stephen Guest
Kinsland, Gary Lynn
McDowell, John Parmelee
Sartin, Austin Albert
Simmons, William Bruce, Jr
Wilcox, Ronald Erwin

MAINE
Grew, Edward Sturgis
Guidotti, Charles V
Morris, Horton Harold

MARYLAND
Arem, Joel Edward
Brown, Michael
Chang, Luke Li-Yu
Darneal, Robert Lee
Fanning, Delvin Seymour
Ferry, John Mott
Fisher, George Wescott
French, Bevan Meredith
Jones, Blair Francis
Kappel, Ellen Sue
MacGregor, Ian Duncan
Marsh, Bruce David
Mathew, Mathai
Parker, Frederick John
Pavlides, Louis
Phair, George
Roth, Robert S
Thomas, Herman H
Veblen, David Rodli
Winzer, Stephen Randolph
Wittels, Mark C
Yoder, Hatten Schuyler, Jr
Zen, E-an

MASSACHUSETTS
Agee, Carl Bernard
Aronson, James Ries
Bonnichsen, Martha Miller
Boutilier, Robert Francis
Brophy, Gerald Patrick
Brownlow, Arthur Hume
Burnham, Charles Wilson
Burns, Roger George
Dick, Henry Jonathan Biddle
Donohue, John J
Fairbairn, Harold Williams
Frondel, Clifford
Gheith, Mohamed A
Grove, Timothy Lynn
Hepburn, John Christopher
Hurlbut, Cornelius Searle, Jr
Jacobsen, Stein Bjornar
Marvin, Ursula Bailey
Morse, Stearns Anthony
Naylor, Richard Stevens
Nunes, Paul Donald
Peters, Edward Tehle
Robinson, Peter
Shimizu, Nobumichi
Siever, Raymond
Thompson, James Burleigh, Jr
Watt, James Peter
Wobus, Reinhard Arthur

MICHIGAN
Buis, Patricia Frances
Dietrich, Richard Vincent
Essene, Eric J
Furlong, Robert B
Heinrich, Eberhardt William
Peacor, Donald Ralph
Ruotsala, Albert P
Sibley, Duncan Fawcett
Van Vlack, Lawrence H(all)
Vogel, Thomas A
Young, Davis Alan

MINNESOTA
Boardman, Shelby Jett
Grant, James Alexander
Green, John Chandler
Hatch, Robert Alchin
Rapp, George Robert, Jr
Walton, Matt Savage
Weiblen, Paul Willard
Zoltai, Tibor

MISSISSIPPI
Brunton, George Delbert
Collins, Johnnie B
Mather, Katharine Kniskern

Reynolds, William Roger
Sundeen, Daniel Alvin

MISSOURI
Frye, Charles Isaac
Hagni, Richard D
Himmelberg, Glen Ray
Keller, Walter David
Kisvarsanyi, Eva Bognar
Lowell, Gary Richard
Maxwell, Dwight Thomas
Snowden, Jesse O
Thomson, Kenneth Clair

MONTANA
Hyndman, Donald William
Wehrenberg, John P

NEBRASKA
Treves, Samuel Blain

NEVADA
Emerson, Donald Orville
Felts, Wayne Moore
Garside, Larry Joe
Hsu, Liang-Chi
Kukal, Gerald Courtney
Larson, Lawrence T
Monroe, Eugene Alan
Pough, Frederick Harvey
Smith, Eugene I(rwin)
Wilbanks, John Randall

NEW HAMPSHIRE
Blum, Joel David
Chamberlain, C Page
Reynolds, Robert Coltart, Jr
Richter, Dorothy Anne
Schneer, Cecil J

NEW JERSEY
Bundy, Wayne Miley
Carr, Michael John
Douglas, Lowell Arthur
Faust, George Tobias
Hamil, Martha M
Hamilton, Charles Leroy
Hewins, Roger Herbert
Hollister, Lincoln Steffens
Puffer, John H
Vassiliou, Andreas H

NEW MEXICO
Baldridge, Warren Scott
Bish, David Lee
Broadhead, Ronald Frigon
Broxton, David Edward
Budding, Antonius Jacob
Buden, Rosemary V
Charles, Robert Wilson
Condie, Kent Carl
Eichelberger, John Charles
Ewing, Rodney Charles
Finney, Joseph J
Floran, Robert John
Geissman, John William
Grambling, Jeffrey A
Guthrie, George D, Jr
Klein, Cornelis
Kudo, Albert Masakiyo
Kuellmer, Frederick John
McLemore, Virginia Teresa Hill
Meijer, Arend
Papike, James Joseph
Renault, Jacques Roland
Sun, James Ming-Shan

NEW YORK
Baird, Gordon Cardwell
Barnard, Walther M
Bassett, William Akers
Berkley, John Lee
Bickford, Marion Eugene
Bird, John Malcolm
Boone, Gary M
Britton, Marvin Gale
Brock, Patrick W G
Bursnall, John Trehearne
Clemency, Charles V
DeLong, Stephen Edwin
DeVries, Robert Charles
Dodd, Robert Taylor
Foster, Helen Laura
Friedman, Gerald Manfred
Giese, Rossman Frederick, Jr
Haas, Werner E L
Hanson, Gilbert N
Hatheway, Richard Brackett
Hawkins, William Max
Heyl, George Richard
Hodge, Dennis
Holland, Harry D
Jacoby, Russell Stephen
Kay, Suzanne Mahlburg
Langer, Arthur M
Leung, Irene Sheung-Ying
Liebling, Richard Stephen
Lindsley, Donald Hale
Longhi, John
Lowe, Kurt Emil
Ludman, Allan
Mattson, Peter Humphrey
Miyashiro, Akiho
Mumpton, Frederick Albert

Pajari, George Edward
Prinz, Martin
Prucha, John James
Reitan, Paul Hartman
Roberson, Herman Ellis
Rutstein, Martin S
Scharf, Walter
Schreiber, B Charlotte
Seyfert, Carl K, Jr
Sheridan, Michael Francis
Sturm, Edward
Teichert, Curt
Thomas, John Jenks
Vandiver, Bradford B
Vaughan, Michael Thomas
Vollmer, Frederick Wolfer
Walker, David
Weber, Julius
Whitney, Philip Roy
Williams, Grahame John Bramald
Wohlford, Duane Dennis
Wosinski, John Francis

NORTH CAROLINA
Abbott, Richard Newton, Jr
Butler, James Robert
Harris, William Burleigh
Howard, Clarence Edward
Neal, Donald Wade
Perkins, Ronald Dee
Raymond, Loren Arthur
Reeber, Robert Richard
Rogers, John James William
Textoris, Daniel Andrew

NORTH DAKOTA
McCarthy, Gregory Joseph
Martin, DeWayne

OHIO
Bain, Roger J
Beard, William Clarence
Bever, James Edward
Bladh, Katherine Laing
Bladh, Kenneth W
Calhoun, Frank Gilbert
Carlson, Ernest Howard
Carlton, Richard Walter
Cook, William R, Jr
Dahl, Peter Steffen
Elliot, David Hawksley
Foland, Kenneth Austin
Foreman, Dennis Walden, Jr
Friberg, LaVerne Marvin
Heimlich, Richard Allen
Hoekstra, Karl Egmond
Huff, Warren D
Koucky, Frank Louis, Jr
Laughon, Robert Bush
Law, Eric W
Lo, Howard H
Malcuit, Robert Joseph
Moody, Judith Barbara
Nance, Richard Damian
Schmidt, Mark Thomas
Scotford, David Matteson
Semler, Charles Edward
Skinner, William Robert
Tettenhorst, Rodney Tampa
Versic, Ronald James
Weitz, John Hills

OKLAHOMA
Al-Shaieb, Zuhair Fouad
Barczak, Virgil J
Blatt, Harvey
Gilbert, Murray Charles
Hounslow, Arthur William
London, David
Pittman, Edward D
Powell, Benjamin Neff
Tillman, Roderick W

OREGON
Borst, Roger Lee
Hammond, Paul Ellsworth
Kays, M Allan
McBirney, Alexander Robert
Powell, James Lawrence
Staples, Lloyd William
Taubeneck, William Harris
Taylor, Edward Morgan

PENNSYLVANIA
Allen, Jack C, Jr
Allen, John Christopher
Bates, Thomas Fulcher
Biederman, Edwin Williams, Jr
Clavan, Walter
Cohen, Alvin Jerome
Crawford, Maria Luisa Buse
Crawford, William Arthur
Davis, Alan
Dunlap, Lawrence H
Eggler, David Hewitt
Erickson, Edwin Sylvester, Jr
Faill, Rodger Tanner
Gold, David Percy
Gray, Ralph J
Griffiths, John Cedric
Haefner, Richard Charles
Johnson, Leon Joseph
Kerrick, Derrill M
Komarneni, Sridhar

Mineralogy-Petrology (cont)

Myer, George Henry
Osborn, Elburt Franklin
Perrotta, Anthony Joseph
Rohl, Arthur N
Sclar, Charles Bertram
Short, Nicholas Martin
Shultz, Charles H
Simpson, Dale R
Smith, Deane Kingsley, Jr
Thornton, Charles Perkins
Vernon, William Wallace
Walther, Frank H
White, Eugene Wilbert
Wiebe, Robert A
Williams, Eugene G
Williamson, William O(wen)

RHODE ISLAND
Giletti, Bruno John
Hermes, O Don
Hess, Paul C
Moore, George Emerson, Jr
Rutherford, Malcolm John
Sigurdsson, Haraldur
Yund, Richard Allen

SOUTH CAROLINA
Carew, James L
Shervais, John Walter
Stakes, Debra Sue
Strom, Richard Nelsen
Warner, Richard Dudley

SOUTH DAKOTA
Norton, James Jennings
Paterson, Colin James

TENNESSEE
Deininger, Robert W
Gavasci, Anna Teresa
Kohland, William Francis
Kopp, Otto Charles
McSween, Harry Y, Jr
Naney, Michael Terrance
Pasto, Arvid Eric
Stow, Stephen Harrington
Tamura, Tsuneo
Taylor, Lawrence August

TEXAS
Ahr, Wayne Merrill
Anthony, Elizabeth Youngblood
Anthony, John Williams
Asquith, George Benjamin
Baker, Donald Roy
Barker, Daniel Stephen
Bence, Alfred Edward
Berkebile, Charles Alan
Butler, John C
Carman, Max Fleming, Jr
Cepeda, Joseph Cherubini
Clabaugh, Stephen Edmund
Clementz, David Michael
Clendening, John Albert
Dahl, Harry Martin
Dickey, John Sloan, Jr
Eastwood, Raymond L
Ehlmann, Arthur J
Elthon, Donald L
Ewing, Thomas Edward
Folk, Robert Louis
Ghaly, Tharwat Shahata
Golden, Dadigamuwage Chandrasiri
Gore, Dorothy J
Guven, Necip
Hewitt, Charles Hayden
Heyman, Louis
Hixon, Sumner B
Holdaway, Michael Jon
Horz, Friedrich
Huege, Fred Robert
Ingerson, Fred Earl
Jonas, Edward Charles
Jones, Ernest Austin, Jr
Koepnick, Richard Borland
Kunze, George William
Levien, Louise
Lofgren, Gary Ernest
Longacre, Susan Ann Burton
McBride, Earle Francis
McCaleb, Stanley B
McConnell, Duncan
McGuire, Anne Vaughan
McPherson, John G(ordon)
Milford, Murray Hudson
Morrison, Donald Allen
Mosher, Sharon
Mutis-Duplat, Emilio
Namy, Jerome Nicholas
Ohashi, Yoshikazu
Phinney, William Charles
Presnall, Dean C
Smith, Douglas
Stevenson, Ralph Girard, Jr
Stormer, John Charles, Jr
Swanson, Eric Rice
Tieh, Thomas Ta-Pin
Whitford-Stark, James Leslie

UTAH
Christiansen, Eric H
Fiesinger, Donald William

Griffen, Dana Thomas
Morris, Elliot Cobia
Nash, William Purcell
Parry, William Thomas
Phillips, William Revell
Picard, M Dane

VERMONT
Ratté, Charles A
Westerman, David Scott

VIRGINIA
Bethke, Philip Martin
Bloss, Fred Donald
Bodnar, Robert John
Brett, Robin
Calver, James Lewis
Corwin, Gilbert
Craig, James Roland
Cunningham, Charles G
Ericksen, George Edward
Ern, Ernest Henry
Evans, Howard Tasker, Jr
Finkelman, Robert Barry
Frye, Keith
Gibbs, Gerald V
Hays, James Fred
Heald, Pamela
Hearn, Bernard Carter, Jr
Helz, Rosalind Tuthill
Hemingway, Bruce Sherman
Horton, James Wright, Jr
Huang, Wen Hsing
Huebner, John Stephen
James, Odette Bricmont
Kearns, Lance Edward
Leo, Gerhard William
Lipin, Bruce Reed
Lyons, Paul Christopher
McCartan, Lucy
Martens, James Hart Curry
Mitchell, Richard Scott
Moore, Raymond Kenworthy
Naeser, Nancy Dearien
Nelson, Bruce Warren
Nord, Gordon Ludwig, Jr
Padovani, Elaine Reeves
Rankin, Douglas Whiting
Ribbe, Paul Hubert
Robie, Richard Allen
Roseboom, Eugene Holloway, Jr
Ross, Malcolm
Stewart, David Benjamin
Sukow, Wayne William
Toulmin, Priestley
Weill, Daniel Francis
Williams, Richard John
Wright, Thomas L
Zelazny, Lucian Walter
Zilczer, Janet Ann

WASHINGTON
Ames, Lloyd Leroy, Jr
Evans, Bernard William
Foit, Franklin Frederick, Jr
Gerlach, Terrence Melvin
Ghiorso, Mark Stefan
Ghose, Subrata
Hinthorne, James Roscoe
Hodges, Floyd Norman
Huestis, Laurence Dean
Kesler, Thomas L
Kittrick, James Allen
Macklin, John Welton
Sherman, David Michael
Smith, Brad Keller
Sorem, Ronald Keith
Swanson, Donald Alan
Symonds, Robert B
Weis, Paul Lester

WEST VIRGINIA
Heald, Milton Tidd
Kalb, G William
Shumaker, Robert C

WISCONSIN
Bailey, Sturges Williams
Brown, Philip Edward
Cameron, Eugene Nathan
Cooper, Reid F
Medaris, L Gordon, Jr
Mudrey, Michael George, Jr
Valley, John Williams
Woodard, Henry Herman, Jr

WYOMING
Frost, Carol Denison
Houston, Robert S
Smithson, Scott Busby

ALBERTA
Burwash, Ronald Allan
Cook, Frederick Ahrens
Foscolos, Anthony E
Ghent, Edward Dale
Heimann, Robert B(ertram)
Lambert, Richard St John
Lerbekmo, John Franklin
Morton, Roger David
Nesbitt, Bruce Edward
Schmidt, Volkmar
Smith, Dorian Glen Whitney

BRITISH COLUMBIA
Armstrong, Richard Lee
Bevier, Mary Lou
Chase, Richard L
Greenwood, Hugh J
Kieffer, Susan Werner
McTaggart, Kenneth C
Nuffield, Edward Wilfrid
Soregaroli, A(rthur) E(arl)
Warren, Harry Verney

MANITOBA
Cerny, Petr
Ferguson, Robert Bury
Hawthorne, Frank Christopher

NEW BRUNSWICK
Allen, Clifford Marsden
Mossman, David John

NEWFOUNDLAND
Hiscott, Richard Nicholas

NOVA SCOTIA
Schenk, Paul Edward
Zentilli, Marcos

ONTARIO
Appleyard, Edward Clair
Baragar, William Robert Arthur
Blackburn, William Howard
Cabri, Louis J
Corlett, Mabel Isobel
Davies, James Frederick
De Kimpe, Christian Robert
Duke, John Murray
Edgar, Alan D
Emslie, Ronald Frank
Gait, Robert Irwin
Gittins, John
Graham, A(lbert) Ronald
Grieve, Richard Andrew
Harris, Donald C
James, Richard Stephen
Jolly, Wayne Travis
Kissin, Stephen Alexander
Kodama, Hideomi
Lumbers, Sydney Blake
Mandarino, Joseph Anthony
Moore, John Marshall, Jr
Pearce, Thomas Hulme
Percival, Jeanne Barbara
Percival, John A
Quon, David Shi Haung
Roeder, Peter Ludwig
Rucklidge, John Christopher
Smith, Terence E
Starkey, John
Veizer, Ján
Wahl, William G
Wicks, Frederick John

QUEBEC
Chown, Edward Holton
Donnay, Joseph Desire Hubert
Feininger, Tomas
Laurent, Roger
Laurin, Andre Frederic
Laverdiere, Marc Richard
Ledoux, Robert Louis
Stevenson, Louise Stevens
Trzcienski, Walter Edward, Jr

SASKATCHEWAN
Davies, John Clifford
St Arnaud, Roland Joseph

OTHER COUNTRIES
Baur, Werner Heinz
Bayliss, Peter
Brydon, James Emerson
Dengo, Gabriel
Jolly, Janice Laurene Willard
Kennedy, Michael John
Walton, Wayne J A, Jr

Oceanography

ALABAMA
Cowan, James Howard, Jr
Davis, Anthony Michael John
Fowlis, William Webster
Lyons, William Berry
Reynolds, George Warren

ALASKA
Alexander, Vera
Burrell, David Colin
Elsner, Robert
Feder, Howard Mitchell
Goering, John James
Hameedi, Mohammad Jawed
Kelley, John Joseph, II
McRoy, C Peter
Mathisen, Ole Alfred
Naidu, Angi Satyanarayan
Nayudu, Y Rammohanroy
Paul, Augustus John, III
Royer, Thomas Clark
Sharma, Ghanshyam D
Shaw, David George
Smoker, William Williams
Wing, Bruce Larry

Wright, Frederick Fenning

ARIZONA
Agenbroad, Larry Delmar
Bates, Charles Carpenter
Dietz, Robert Sinclair
El Wardani, Sayed Aly
Fernando, Harindra Joseph
Osterberg, Charles Lamar

CALIFORNIA
Alldredge, Alice Louise
Arp, Alissa Jan
Arrhenius, Gustaf Olof Svante
Arthur, Robert Siple
Bacastow, Robert Bruce
Barnett, Tim P
Bascom, Willard
Beers, John R
Berger, Wolfgang Helmut
Bernstein, Robert Lee
Berry, Richard Warren
Boehlert, George Walter
Bourke, Robert Hathaway
Bradshaw, John Stratlii
Brinton, Edward
Bubeck, Robert Clayton
Buffington, Edwin Conger
Burnett, Bryan Reeder
Carlson, Paul Roland
Carsola, Alfred James
Chamberlain, Dilworth Woolley
Chew, Frank
Childress, James J
Chow, Tsaihwa James
Chu, Peter Cheng
Clark, Nathan Edward
Collins, Curtis Allan
Conomos, Tasso John
Cox, Charles Shipley
Coyer, James A
Craig, Harmon
Crandell, George Frank
Cross, Ralph Herbert, III
Cummings, William Charles
Curray, Joseph Ross
Davis, Curtiss Owen
Davis, Gary Everett
Davis, Russ E
Dayton, Paul K
Delong, Edward F
Demond, Joan
Di Girolamo, Rudolph Gerard
Dorman, Clive Edgar
Dorman, LeRoy Myron
Dugdale, Richard Cooper
Eittreim, Stephen L
Elsberry, Russell Leonard
Eppley, Richard Wayne
Everts, Craig Hamilton
Fast, Thomas Normand
Fiedler, Paul Charles
Field, Michael Ehrenhart
Filloux, Jean H
Fisher, Robert Lloyd
Flatte, Stanley Martin
Ford, Richard Fiske
Foster, Michael Simmler
Foster, Theodore Dean
Fu, Lee Lueng
Fuhrman, Jed Alan
Fusaro, Craig Allen
Galt, Charles Parker, Jr
Gardner, James Vincent
Garwood, Roland William, Jr
Gast, James Avery
Geminder, Robert
Gerrodette, Timothy
Gibson, Carl H
Gordon, Malcolm Stephen
Gorsline, Donn Sherrin
Griggs, Gary B
Groves, Gordon William
Haderlie, Eugene Clinton
Hall, Robert Earl
Halpern, David
Hamilton, Edwin Lee
Hammond, Douglas Ellenwood
Haney, Robert Lee
Harrison, Florence Louise
Haury, Loren Richard
Hein, James R
Holl, Manfred Matthias
Holliday, Dale Vance
Holmes, Robert W
Honey, Richard Churchill
Hubred, Gale L
Hunt, Arlon Jason
Hunter, John Roe
Hurley, Robert Joseph
Hwang, Li-San
Inman, Douglas Lamar
Iyer, Hariharaiyer Mahadeva
Joseph, James
Joy, Joseph Wayne
Jung, Glenn Harold
Kaye, George Thomas
Kelley, James Charles
Kenyon, Kern Ellsworth
Knauss, John Atkinson
Knox, Robert Arthur
Kremer, James Nevin
Kremer, Patricia McCarthy
Ku, Teh-Lung

ENVIRONMENTAL, EARTH & MARINE SCIENCES / 401

La Fond, Eugene Cecil
Lange, Carina Beatriz
Lapota, David
Leiderdsorf, Craig B
Liu, Wingyuen Timothy
Loomis, Alden Albert
Lynch, Richard Vance, III
McBlair, William
McCarthy, Francis Davey
Macdonald, Kenneth Craig
McGeary, David F R
McGowan, John Arthur
McIntyre, Marie Claire
Mackenzie, Kenneth Victor
McLain, Douglas Robert
Martin, John Holland
Mathewson, James H
Medwin, Herman
Mikel, Thomas Kelly, Jr
Mohr, John Luther
Mulligan, Timothy James
Mullin, Michael Mahlon
Munk, Walter Heinrich
Newman, William Anderson
Nichols, Frederic Hone
Noble, Paula Jane
Normark, William Raymond
North, Wheeler James
Nygreen, Paul W
Owen, Robert W, Jr
Paquette, Robert George
Park, Taisooe
Parrish, Richard Henry
Pattabhiraman, Tammanur R
Patton, John Stuart
Penry, Deborah L
Pequegnat, Linda Franklin Haithcock
Pequegnat, Willis Eugene
Phillips, David William
Phillips, Richard P
Phleger, Fred B
Pickwell, George Vincent
Pieper, Richard Edward
Pinkel, Robert
Potter, David Samuel
Potts, Donald Cameron
Powell, Thomas Mabrey
Prasad, Kota S
Rau, Gregory Hudson
Read, Robert G
Rechnitzer, Andreas Buchwald
Reeburgh, William Scott
Reid, Joseph Lee
Richards, Thomas L
Riedel, William Rex
Roach, David Michael
Rogers, David Peter
Roth, Ariel A
Ruby, Edward George
Saffo, Mary Beth
Schwing, Franklin Burton
Sclater, John George
Seckel, Gunter Rudolf
Seibel, Erwin
Semtner, Albert Julius, Jr
Seymour, Richard Jones
Sharp, Gary Duane
Show, Ivan Tristan
Silver, Eli Alfred
Silver, Mary Wilcox
Slack, Keith Vollmer
Smith, Kenneth Lawrence, Jr
Smith, Raymond Calvin
Smith-Evernden, Roberta Katherine
Soule, Dorothy (Fisher)
Spies, Robert Bernard
Spiess, Fred Noel
Stephens, John Stewart, Jr
Stevenson, Robert Everett
Stramski, Dariusz
Street, Robert L(ynnwood)
Sund, Paul N
Tegner, Mia Jean
Terry, Richard D
Thomas, William Hewitt
Thompson, Warren Charles
Trench, Robert Kent
Tsuchiya, Mizuki
Van Leer, John Cloud
Wang, Frank Feng Hui
Ward, Bess B
Warnke, Detlef Andreas
Weare, Bryan C
Wilde, Pat
Wilkie, Donald W
Williams, Peter M
Wilson, Basil W(rigley)
Wimberley, Stanley
Wood, Fergus James
Yayanos, A Aristides
Zetler, Bernard David

COLORADO
Behrendt, John Charles
Born, George Henry
Chamberlain, Theodore Klock
Emery, William Jackson
Frisch, Alfred Shelby
Gossard, Earl Everett
Hanson, Howard P(aul)
Hay, William Winn
Holcombe, Troy Leon
Holland, William Robert
Kantha, Lakshmi

Kraus, Eric Bradshaw
Loughridge, Michael Samuel
McWilliams, James Cyrus
Moore, Carla Jean
Rufenach, Clifford L
Sawyer, Constance B
Shideler, Gerald Lee
Sloss, Peter William
Smith, J Dungan
Weihaupt, John George
Williams, David Lee

CONNECTICUT
Bohlen, Walter Franklin
Carlton, James Theodore
Chriss, Terry Michael
Cooper, Richard Arthur
Farmer, Mary Winifred
Fitzgerald, William Francis
Haffner, Rudolph Eric
Hanks, James Elden
Horne, Gregory Stuart
McGill, David A
Monahan, Edward Charles
Natarajan, Kottayam Viswanathan
Pandolfo, Joseph P
Paskausky, David Frank
Renfro, William Charles
Squires, Donald Fleming
Thurberg, Frederick Peter
Tolderlund, Douglas Stanley
Veronis, George
Welsh, Barbara Lathrop

DELAWARE
Culberson, Charles Henry
Daiber, Franklin Carl
Epifanio, Charles Edward
Garvine, Richard William
Glass, Billy Price
Karlson, Ronald Henry
Kobayashi, Nobuhisa
Luther, George William, III
Maurmeyer, Evelyn Mary
Miller, Douglas Charles
Price, Kent Sparks, Jr
Sharp, Jonathan Hawley
Svendsen, Ib Arne
Targett, Nancy McKeever
Targett, Timothy Erwin
Webster, Ferris
Wu, Jin

DISTRICT OF COLUMBIA
Baker, D(onald) James
Benson, Richard Hall
Berkson, Jonathan Milton
Bernor, Raymond Louis
Christofferson, Eric
Handel, Mark David
Henry, Warren Elliott
Knowlton, Robert Earle
Lindstrom, Eric Jon
Manning, Raymond B
Maynard, Nancy Gray
Patterson, Steven Leroy
Pellenbarg, Robert Ernest
Pyle, Thomas Edward
Rao, Desiraju Bhavanarayana
Roper, Clyde Forrest Eugene
Rosenblum, Lawrence Jay
Shen, Colin Yunkang
Stanley, Daniel Jean
Wilkerson, John Christopher

FLORIDA
Alam, Mohammed Khurshid
Atwood, Donald Keith
Betzer, Peter Robin
Bezdek, Hugo Frank
Bullis, Harvey Raymond, Jr
Burnett, William Craig
Burney, Curtis Michael
Butler, Philip Alan
Byrd, Isaac Burlin
Byrne, Robert Howard
Carder, Kendall L
Clarke, Allan James
Compton, John S
Cooper, William James
Corcoran, Eugene Francis
Dean, Robert George
De Sylva, Donald Perrin
Dobkin, Sheldon
Doyle, Larry James
Dubbelday, Pieter Steven
Duedall, Iver Warren
Emiliani, Cesare
Fanning, Kent Abram
Fine, Rana Arnold
Freeland, George Lockwood
Gilbert, Perry Webster
Gordon, Howard R
Griffin, George Melvin, Jr
Hallock-Muller, Pamela
Hansen, Donald Vernon
Harris, D Lee
Harris, Lee Errol
Herrnkind, William Frank
Hopkins, Thomas Lee
Hsueh, Ya
Kruczynski, William Leonard
Legler, David M
Lovejoy, Donald Walker

Ludeke, Carl Arthur
McAllister, Raymond Francis
McFadden, James Douglas
McLeroy, Edward Glenn
Mahadevan, Selvakumaran
Marcus, Nancy Helen
Mariscal, Richard North
Maul, George A
Mertens, Lawrence E(dwin)
Mitsui, Akira
Mooers, Christopher Northrup Kennard
Muller-Karger, Frank Edgar
Myrberg, Arthur August, Jr
Nakamura, Eugene Leroy
Nelson, Walter Garnet
Norris, Dean Rayburn
O'Brien, James J
Odum, Howard Thomas
Olson, Donald B
Ostlund, H Gote
Prager, Michael Haskell
Proni, John Roberto
Reichard, Ronnal Paul
Rona, Donna C
Rona, Peter Arnold
Smith, Ned Philip
Smith, Sharon Louise
Southam, John Ralph
Stewart, Harris Bates, Jr
Sturges, Wilton, III
Suman, Daniel Oscar
Swain, Geoffrey W
Taylor, Barrie Frederick
Thistle, David
Virnstein, Robert W
Walsh, John Joseph
Wanless, Harold Rogers
Weatherly, Georges Lloyd
Weisberg, Robert H
Wiebusch, Charles Fred
Wiebush, Joseph Roy
Windsor, John Golay, Jr
Zuo, Yuegang

GEORGIA
Blanton, Jackson Orin
Glooschenko, Walter Arthur
Hanson, Roger Brian
Harding, James Lombard
Hayes, Willis B
Henry, Vernon James, Jr
Menzel, David Washington
Meyer, Judy Lynn
Noakes, John Edward
Pomeroy, Lawrence Richards
Railsback, Loren Bruce
Windom, Herbert Lynn

HAWAII
Abbott, Isabella Aiona
Bailey-Brock, Julie Helen
Bathen, Karl Hans
Berg, Carl John, Jr
Bienfang, Paul Kenneth
Bretschneider, Charles Leroy
Clarke, Thomas Arthur
Cruickshank, Michael James
Daniel, Thomas Henry
Forster, William Owen
Gallagher, Brent S
Gopalakrishnan, Kakkala
Grigg, Richard Wyman
Hacker, Peter Wolfgang
Hey, Richard N
Keating, Barbara Helen
Kroenke, Loren William
Landry, Michael Raymond
Laurs, Robert Michael
Laws, Edward Allen
Loomis, Harold George
McCoy, Floyd W, Jr
Mackenzie, Frederick Theodore
Mader, Charles Lavern
Magaard, Lorenz Carl
Peterson, Melvin Norman Adolph
Polovina, Jeffrey Joseph
Randall, John Ernest
Russo, Anthony R
Sansone, Francis Joseph
Spielvogel, Lester Q
Wang, Bin
Woodcock, Alfred Herbert
Wyrtki, Klaus
Young, Richard Edward

IDAHO
Welhan, John Andrew

ILLINOIS
Chiou, Wen-An
Foster, Merrill W
Lesht, Barry Mark
Michelson, Irving
Platzman, George William
Rodolfo, Kelvin S
Shabica, Charles Wright
Thomsen, Kurt O
Walton, William Ralph

INDIANA
Christensen, Nikolas Ivan

IOWA
Hoffmann, Richard John

Walters, James Carter

LOUISIANA
Avent, Robert M
Bouma, Arnold Heiko
Breeding, J Ernest, Jr
Carney, Robert Spencer
Chesney, Edward Joseph, Jr
Defenbaugh, Richard Eugene
Drummond, Kenneth Herbert
Harris, Alva H
Hastings, Robert Wayne
Hoese, Hinton Dickson
Huh, Oscar Karl
Korgen, Benjamin Jeffry
Murray, Stephen Patrick
Patrick, William H, Jr
Rouse, Lawrence James, Jr
Sammarco, Paul William
Sen Gupta, Barun Kumar
Shaw, Richard Francis
Stiffey, Arthur V
Suhayda, Joseph Nicholas
Turner, Robert Eugene
Wiseman, William Joseph, Jr

MAINE
Andersen, Robert Arthur
Dean, David
De Witt, Hugh Hamilton
Fink, Loyd Kenneth, Jr
Garside, Christopher
Gilbert, James Robert
Gilmartin, Malvern
McAlice, Bernard John
Mayer, Lawrence M
Puri, Kewal Krishan
Revelante, Noelia
Schnitker, Detmar
Townsend, David Warren
Watling, Les
Welch, Walter Raynes
Yentsch, Charles Samuel
Yentsch, Clarice M

MARYLAND
Andersen, Neil Richard
Apel, John Ralph
Baer, Ledolph
Barrientos, Celso Saquitan
Bjerkaas, Allan Wayne
Busalacchi, Antonio James, Jr
Calder, John Archer
Cammen, Leon Matthew
Capone, Douglas George
Cavalieri, Donald Joseph
Chen, Benjamin Yun-Hai
Comiso, Josefino Cacas
Corliss, John Burt
Cornwell, Jeffrey C
Dantzler, Herman Lee, Jr
Del Grosso, Vincent Alfred
Diamante, John Matthew
Duane, David Bierlein
Elliott, William Paul
Etter, Paul Courtney
Evans, David L
Felsher, Murray
Fischer, Eugene Charles
Fisher, Thomas Richard, Jr
Frey, Henry Richard
Fry, John Craig
Gereben, Istvan B
Gerritsen, Jeroen
Gloersen, Per
Hakkinen, Sirpa Marja Anneli
Hanssen, George Lyle
Houde, Edward Donald
Huang, Joseph Chi Kan
Jefferson, Donald Earl
Johnson, Harry McClure
Kaiser, Jack Allen Charles
Koblinsky, Chester John
Kravitz, Joseph Henry
Lambert, Richard Bowles, Jr
Larson, Reginald Einar
Leese, John Albert
Legeckis, Richard Vytautas
Malone, Thomas C
Menez, Ernani Guingona
Mied, Richard Paul
Monaldo, Francis Michael
Mountford, Kent
Park, Paul Kilho
Parkinson, Claire Lucille
Parkinson, Claire Lucille
Phillips, Owen M
Piotrowicz, Stephen R
Pritchard, Donald William
Purcell, Jennifer Estelle
Rao, P Krishna
Rebach, Steve
Reeves, Robert William
Riedel, Gerhardt Frederick
Robertson, Andrew
Sanders, James Grady
Saville, Thorndike, Jr
Schemm, Charles Edward
Schneider, Eric Davis
Schulkin, Morris
Sellner, Kevin Gregory
Shykind, Edwin B
Sindermann, Carl James
Sissenwine, Michael P

Oceanography (cont)

Small, Eugene Beach
Snow, Milton Leonard
Spilhaus, Athelstan Frederick, Jr
Stoecker, Diane Kastelowitz
Strong, Alan Earl
Taylor, Walter Rowland
Tenore, Kenneth Robert
Valenzuela, Gaspar Rodolfo
Warsh, Catherine Evelyn
Weller, Charles Stagg, Jr
White, Harris Herman
Williams, Jerome
Williams, Robert Glenn
Wilson, William Stanley
Wolfe, Douglas Arthur

MASSACHUSETTS
Aubrey, David Glenn
Backus, Richard Haven
Baird, Ronald C
Ballard, Robert D
Beardsley, Robert Cruce
Beckerle, John C
Beckwitt, Richard David
Blidberg, D Richard
Boehm, Paul David
Bower, Amy Sue
Brink, Kenneth Harold
Bryden, Harry Leonard
Butman, Cheryl Ann
Capuzzo, Judith M
Carr, Jerome Brian
Chave, Alan Dana
Chisholm, Sallie Watson
Dillon, William Patrick
Emery, Kenneth Orris
Faller, Alan Judson
Farrington, John William
Fofonoff, Nicholas Paul
Folger, David W
Frisk, George Vladimir
Gibb, Thomas Robinson Pirie
Giese, Graham Sherwood
Gordon, Bernard Ludwig
Greenstein, Benjamin Joel
Grice, George Daniel, Jr
Guillard, Robert Russell Louis
Harbison, G Richard
Hollister, Charles Davis
Irish, James David
Johnson, David Ashby
Kelts, Larry Jim
Knebel, Harley John
Krause, Daniel, Jr
Lippson, Robert Lloyd
Livingston, Hugh Duncan
Luyten, James Reindert
McCarthy, James Joseph
McLaren, J Philip
Marshall, Nelson
Mayo, Charles Atkins, III
Miller, Arthur R
Mollo-Christensen, Erik Leonard
Montgomery, Raymond Braislin
Newell, Reginald Edward
Noiseux, Claude Francois
O'Brien, Francis Xavier
Payne, Richard Earl
Pedlosky, Joseph
Poag, Claude Wylie
Price, James Franklin
Prindle, Bryce
Richardson, Philip Livermore
Rizzoli-Malanotte, Paola
Robinson, Allan Richard
Ross, David A
Sayles, Frederick Livermore
Scheltema, Rudolf S
Schmitt, Raymond W, Jr
Schmitz, William Joseph, Jr
Sharp, A(rnold) G(ideon)
Shaw, Peter Robert
Smith, Frederick Edward
Southard, John Brelsford
Spencer, Derek W
Spiegel, Stanley Lawrence
Stalcup, Marvel C
Stommel, Henry Melson
Taft, Jay Leslie
Takahashi, Kozo
Thompson, Geoffrey
Tucholke, Brian Edward
Turner, Jefferson Taylor
Vine, Allyn Collins
Wallace, Gordon Thomas
Warren, Bruce Alfred
Weller, Robert Andrew
Whitehead, John Andrews
Wiebe, Peter Howard
Williams, Albert James, III
Wilson, Philo Calhoun
Wirsen, Carl O, Jr
Wofsy, Steven Charles
Wright, William Redwood
Wunsch, Carl Isaac

MICHIGAN
Andresen, Norman A
Birdsall, Theodore G
Boojum, Basil John
Gardner, Wayne Stanley
Hawley, Nathan

Leenheer, Mary Janeth
Liu, Paul Chi
Meadows, Guy Allen
Meyers, Philip Alan
Moore, Theodore Carlton, Jr
Owen, Robert Michael
Passino, Dora R May
Rea, David Kenerson
Schwab, David John
Smith, Vann Elliott
Wynne, Michael James

MINNESOTA
Bedford, William Brian
Edwards, Richard Lawrence

MISSISSIPPI
Asper, Vernon Lowell
Ballard, James Alan
Bennett, Richard Harold
Biesiot, Patricia Marie
Bruce, John Goodall, Jr
Brunner, Charlotte
Dunn, Dean A(lan)
Egloff, Julius, III
Einwich, Anna Maria
Farwell, Robert William
Fay, Temple Harold
Ferer, Kenneth Michael
Fleischer, Peter
Geddes, Wilburt Hale
Green, Albert Wise
Hallock, Zachariah R
Hurlburt, Harley Ernest
Keen, Dorothy Jean
La Violette, Paul Estronza
Lins, Thomas Wesley
Lowrie, Allen
Overstreet, Robin Miles
Priest, Melville S(tanton)
Rucker, James Bivin
Saunders, Kim David
Sprague, Vance Glover, Jr
Woodmansee, Robert Asbury

MISSOURI
Armstrong, Daniel Wayne
Jones, Everet Clyde
Otto, David Arthur

NEVADA
Leipper, Dale F
Lowenthal, Douglas H

NEW HAMPSHIRE
Anderson, Franz Elmer
Brown, Wendell Stimpson
Hibler, William David, III
Ingham, Merton Charles
Jones, Galen Everts
Lesser, Michael Patrick
Lynch, Daniel Roger
Perovich, Donald Kole

NEW JERSEY
Blumberg, Alan Fred
Brine, Charles James
Bruno, Michael Stephen
Deffeyes, Kenneth Stover
Dorfman, Donald
Grassle, John Frederick
Hamby, Robert Jay
Haskin, Harold H
Herring, H(ugh) James
Hires, Richard Ives
Kennedy, Anthony John
Kennish, Michael Joseph
Kraeuter, John Norman
Lo Pinto, Richard William
Lutz, Richard Arthur
Mellor, George Lincoln, Jr
Meyerson, Arthur Lee
Sarmiento, Jorge Louis
Stouffer, Ronald Jay
Thiruvathukal, John Varkey
Thomann, Robert V

NEW MEXICO
Campbell, John Raymond
Degenhardt, William George
Huestis, Stephen Porter
Malone, Robert Charles
Sayre, William Olaf
Simmons, William Frederick
Wade, Richard Archer
Wengerd, Sherman Alexander
Whetten, John T

NEW YORK
Anderson, O(rvil) Roger
Biscaye, Pierre Eginton
Blanchard, Duncan Cromwell
Bokuniewicz, Henry Joseph
Bowman, James Floyd, II
Bowman, Malcolm James
Brandt, Stephen Bernard
Brothers, Edward Bruce
Cane, Mark A
Carpenter, Edward J
Coles, James Stacy
Cousteau, Jacques-Yves
Cutler, Edward Bayler
Dickman, Steven Richard
Dugolinsky, Brent Kerns

Falkowski, Paul Gordon
Fisher, Nicholas Seth
Flood, Roger Donald
Froelich, Philip Nissen
Gallagher, Jane Chispa
Geller, Marvin Alan
Gordon, Arnold L
Haefner, Paul Aloysius, Jr
Houghton, Robert W
Howarth, Robert W
Hunkins, Kenneth
Jacobs, Stanley S
Jacobson, Melvin Joseph
Kastens, Kim Anne
Katz, Eli Joel
Kelly, John Russell
Langseth, Marcus G
McIntyre, Andrew
Marra, John Frank
Moeller, Henry William
Mullins, Henry Thomas
Myer, Glenn Evans
Naidu, Janakiram Ramaswamy
Neumann, Paul Gerhard
Nittrouer, Charles A
Nuzzi, Robert
O'Connor, Joel Sturges
Okubo, Akira
Pasby, Brian
Pierson, Willard James, Jr
Rhee, G-Yull
Rind, David Harold
Ryan, William B F
Sambrotto, Raymond Nicholas
Sancetta, Constance Antonina
Schubel, Jerry Robert
Scranton, Mary Isabelle
Shaw, Richard P(aul)
Sheryll, Richard Perry
Siegel, Alvin
Smethie, William Massie, Jr
Stevens, Richard S
Swanson, Robert Lawrence
Tietjen, John H
Vairavamurthy, Murthy Appathurai
Wallace, Douglas William Roy
Yalkovsky, Rafael

NORTH CAROLINA
Bane, John McGuire, Jr
Carter, Harry Hart
Chester, Alexander Jeffrey
Cleary, William James
Corliss, Bruce Hayward
Dubach, Harold William
Frankenberg, Dirk
Govoni, John Jeffrey
Gutknecht, John William
Humm, Harold Judson
Janowitz, Gerald S(aul)
Kamykowski, Daniel
Karl, Thomas Richard
Knowles, Charles Ernest
McCrary, Anne Bowden
Martens, Christopher Sargent
Neumann, A Conrad
Perkins, Ronald Dee
Peterson, Charles Henry
Pilkey, Orrin H, Jr
Stavn, Robert Hans
Sunda, William George
Sutcliffe, William Humphrey, Jr
Thayer, Paul Arthur
Vaughan, Douglas Stanwood
Vukovich, Fred Matthew
Willey, Joan DeWitt
Wolcott, Thomas Gordon

OHIO
Bieri, Robert
Boardman, Mark R
Carlton, Richard Walter
Herdendorf, Charles Edward, III
Hillis, Llewellya
Matisoff, Gerald
Savin, Samuel Marvin
Stockman, Charles H(enry)
Unrug, Raphael
Wilson, Mark Allan
Worsley, Thomas R

OKLAHOMA
Cox, Gordon F N

OREGON
Burt, Wayne Vincent
Byrne, John Vincent
Caldwell, Douglas Ray
Carey, Andrew Galbraith, Jr
Davis, Michael William
Drake, Ellen Tan
Fish, William
Frey, Bruce Edward
Frolander, Herbert Farley
Gonor, Jefferson John
Gordon, Louis Irwin
Holton, Robert Lawrence
Huh, Chih-An
Huyer, Adriana (Jane)
Keller, George H
Komar, Paul Dwight
Kulm, LaVerne Duane
Lupton, John Edward
McConnaughey, Bayard Harlow

Mesecar, Roderick Smit
Miller, Charles Benedict
Morita, Richard Yukio
Neal, Victor Thomas
Nelson, David Michael
Neshyba, Steve
Paulson, Clayton Arvid
Pillsbury, Dale Ronald
Quinn, William Hewes
Rasmussen, Lois E Little
Sherr, Barry Frederick
Sherr, Evelyn Brown
Small, Lawrence Frederick
Smith, Robert Lloyd
Sollitt, Charles Kevin
Walsh, Don
Wood, Anne Michelle
Zaneveld, Jacques Ronald Victor

PENNSYLVANIA
Arthur, Michael Allan
Bennett, Lee Cotton, Jr
Biggs, R B
Bona, Jerry Lloyd
Carson, Bobb
Commito, John Angelo
Dudzinski, Diane Marie
Herman, Sidney Samuel
Markham, James J
Oostdam, Bernard Lodewijk
Pees, Samuel T(homas)
Sanders, Robert W
Schmalz, Robert Fowler
Shaler, Amos J(ohnson)
Soong, Yin Shang
Strick, Ellis
Wegweiser, Arthur E

RHODE ISLAND
Bhatt, Jagdish J
Browning, David Gunter
Burroughs, Richard
Callahan, Jeffrey Edwin
Duce, Robert Arthur
Fox, Mary Frances
Fox, Paul Jeffrey
Gearing, Juanita Newman
Hammen, Susan Lum
Hargraves, Paul E
Jossi, Jack William
Kester, Dana R
Leinen, Margaret Sandra
McMaster, Robert Luscher
Mayer, Garry Franklin
Mellberg, Leonard Evert
Napora, Theodore Alexander
Nixon, Scott West
Olson, David Gardels
Oviatt, Candace Ann
Pilson, Michael Edward Quinton
Prell, Warren Lee
Rossby, Hans Thomas
Sherman, Kenneth
Sieburth, John McNeill
Smayda, Theodore John
Swift, Dorothy Garrison
Swift, Elijah, V
Wahle, Richard Andreas
Watts, Dennis Randolph
Wimbush, Mark
Winn, Howard Elliott

SOUTH CAROLINA
Burrell, Victor Gregory, Jr
Feller, Robert Jarman
Hayes, David Wayne
Hayes, Miles O
Moore, Willard S
Vernberg, Frank John
Williams, Douglas Francis

TENNESSEE
Cutshall, Norman Hollis
McLean, Richard Bea
Siesser, William Gary
Smith, Walker O, Jr
Witten, Alan Joel

TEXAS
Abbott, William Harold
Ahr, Wayne Merrill
Anderson, Aubrey Lee
Anderson, John B
Baie, Lyle Fredrick
Behrens, Earl William
Berner, Leo Dewitte, Jr
Biggs, Douglas Craig
Blanton, William George
Bright, Thomas J
Brooks, David Arthur
Bryant, William Richards
Buffler, Richard Thurman
Chang, Ping
Clayton, William Howard
Cotner, James Bryan
Damuth, John Erwin
Darnell, Rezneat Milton
Dunton, Kenneth Harlow
Ellwood, Brooks B
Elthon, Donald L
Fahlquist, Davis A
Feazel, Charles Tibbals
Franceschini, Guy Arthur
French, William Edwin

ENVIRONMENTAL, EARTH & MARINE SCIENCES / 403

Fryxell, Greta Albrecht
Gartner, Stefan, Jr
Geyer, Richard Adam
Giammona, Charles P, Jr
Gittings, Stephen Reed
Goldman, Joseph L
Guinasso, Norman Louis, Jr
Herbich, John Bronislaw
Hilde, Thomas Wayne Clark
Hildebrand, Henry H
James, Bela Michael
Lawrence, Philip Linwood
Loeblich, Alfred Richard, III
Maxwell, Arthur Eugene
Miksad, Richard Walter
Montagna, Paul A
Nowlin, Worth D, Jr
Oetking, Philip
Oppenheimer, Carl Henry, Jr
Presley, Bobby Joe
Ray, James P
Ray, Sammy Mehedy
Reid, Robert Osborne
Rezak, Richard
Roels, Oswald A
Rowe, Gilbert Thomas
Schink, David R
Schlemmer, Frederick Charles, II
Schutz, Bob Ewald
Stewart, Robert Henry
Suttle, Curtis
Vastano, Andrew Charles
West, Bruce Joseph
Williams, Glen Nordyke
Wong, William Wai-Lun
Wormuth, John Hazen
Wursig, Bernd Gerhard
Zimmerman, Carol Jean

UTAH
Chesworth, Robert Hadden
Jewell, Paul William
Long, David G
Roth, Peter Hans

VIRGINIA
Andrews, James Einar
Atkinson, Larry P
Bailey, James Stuart
Berman, Alan
Bishop, Joseph Michael
Blandford, Robert Roy
Briggs, Jeffrey L
Briscoe, Melbourne George
Brown, Gary S
Byrne, Robert John
Castagna, Michael
Codispoti, Louis Anthony
Csanady, Gabriel Tibor
Curtin, Brian Thomas
Cutter, Gregory Allan
Daniel, Kenneth Wayne
Delnore, Victor Eli
Dendrou, Stergios
Dolezalek, Hans
Dorman, Craig Emery
Dugan, John Philip
Dunstan, William Morgan
Edgar, Norman Terence
Fine, Michael Lawrence
Gaines, Gregory
Galvin, Cyril Jerome, Jr
Grabowski, Walter John
Grant, George C
Green, Edward Jewett
Gross, M Grant, Jr
Hargis, William Jennings, Jr
Heinrichs, Donald Frederick
Hicks, Steacy Dopp
Howard, William Eager, III
Hunt, Lee McCaa
Hurdle, Burton Garrison
Inderbitzen, Anton Louis
Johnson, George Leonard
Johnson, Ronald Ernest
Kelly, Mahlon George, Jr
Kinder, Thomas Hartley
Kirwan, Albert Dennis, Jr
Kornicker, William Alan
Kroll, John Ernest
Kuo, Albert Yi-Shuong
Lepple, Frederick Karl
Leung, Wing Hai
Lill, Gordon Grigsby
Ludwick, John Calvin, Jr
McGregor, Bonnie A
Macko, Stephen Alexander
Malfait, Bruce Terry
Marshall, Harold George
Milliman, John D
Orth, Robert Joseph
Osborn, Kenneth Wakeman
Parvulescu, Antares
Patterson, Mark Robert
Penhale, Polly Ann
Ramberg, Steven Eric
Raymond-Savage, Anne
Schwiderski, Ernst Walter
Snyder, J Edward, Jr
Spilhaus, Athelstan Frederick
Spinrad, Richard William
Stuart, Charles Edward
Swift, Donald J P
Sylvester, Joseph Robert

Tatro, Peter Richard
Tipper, Ronald Charles
Trizna, Dennis Benedict
Webb, Kenneth L
Wells, John Morgan, Jr
Whitcomb, James Hall
Witting, James M
Wong, George Tin Fuk
Wright, Lynn Donelson
Yates, Charlie Lee
Zieman, Joseph Crowe, Jr

WASHINGTON
Aagaard, Knut
Ahmed, Saiyed I
Alverson, Dayton L
Anderson, Donald Hervin
Anderson, James Jay
Aron, William Irwin
Asher, William Edward
Baker, Edward Thomas
Banse, Karl
Barker, Morris Wayne
Barnes, Clifford Adrian
Barnes, Ross Owen
Bernard, Eddie Nolan
Best, Edgar Allan
Booth, Beatrice Crosby
Burns, Robert Earle
Buske, Norman L
Cannon, Glenn Albert
Carpenter, Roy
Chace, Alden Buffington, Jr
Chew, Kenneth Kendall
Coachman, Lawrence Keyes
Cokelet, Edward Davis
Collias, Eugene Evans
Crabtree, David Melvin
Creager, Joe Scott
Crecelius, Eric A
Curl, Herbert (Charles), Jr
Damkaer, David Martin
Dehlinger, Peter
Delisi, Donald Paul
Deming, Jody W
Duxbury, Alyn Crandall
Ebbesmeyer, Curtis Charles
Eickstaedt, Lawrence Lee
Favorite, Felix
Feely, Richard Alan
Frost, Bruce Wesley
Gregg, Michael Charles
Gunderson, Donald Raymond
Hardy, John Thomas
Harlett, John Charles
Harrison, Don Edmunds
Hawkes, Joyce W
Healy, Michael L
Heath, G Ross
Hedges, John Ivan
Hickey, Barbara Mary
Holcomb, Robin Terry
Holmes, Mark Lawrence
Hood, Donald Wilbur
Ichiye, Takashi
Jennings, Charles David
Johnson, Gregory Conrad
Jumars, Peter Alfred
Kunze, Eric
Laevastu, Taivo
Lavelle, John William
Lingren, Wesley Earl
Long, Edward R
McCormick, Norman Joseph
McManus, Dean Alvis
Martin, Seelye
Mills, Claudia Eileen
Mobley, Curtis Dale
Mopper, Kenneth
Muench, Robin Davie
Muller-Parker, Gisele Therese
Napp, Jeffrey M
Neve, Richard Anthony
Olsen, Kenneth Harold
Overland, James Edward
Pearson, Walter Howard
Perry, Mary Jane
Plant, William James
Rattray, Maurice, Jr
Reed, Ronald Keith
Rhines, Peter Broomell
Rieck, H(enry) G(eorge)
Roden, Gunnar Ivo
Salo, Ernest Olavi
Sanford, Thomas Bayes
Sarachik, Edward S
Schwartz, Maurice Leo
Serne, Roger Jeffrey
Skov, Niels A
Smith, David Clement, IV
Sorem, Ronald Keith
Sparks, Albert Kirk
Sternberg, Richard Walter
Strathmann, Richard Ray
Taft, Bruce A
Taylor, Peter Berkley
Thom, Ronald Mark
Trent, Donald Stephen
Untersteiner, Norbert
VanBlaricom, Glenn R
Walker, Sharon Leslie
Wooster, Warren Scriver
Wright, Judith

WEST VIRGINIA
Calhoon, Donald Alan

WISCONSIN
Bentley, Charles Raymond
Difford, Winthrop Cecil
Edgington, David Norman
Meyer, Robert Paul
Mortimer, Clifford Hiley
Ragotzkie, Robert Austin

PUERTO RICO
Hernandez Avila, Manuel Luis
Lopez, Jose Manuel
Morelock, Jack

ALBERTA
Georgi, Daniel Taylan
Giovinetto, Mario Bartolome
Lane, Robert Kenneth
Scafe, Donald William

BRITISH COLUMBIA
Booth, Ian Jeremy
Burling, Ronald William
Calvert, Stephen Edward
Clements, Reginald Montgomery
Hughes, Blyth Alvin
Hyndman, Roy D
Krauel, David Paul
Lalli, Carol Marie
LeBlond, Paul Henri
Levings, Colin David
Lewis, Alan Graham
Lewis, Edward Lyn
Littlepage, Jack Leroy
Macdonald, Robie Wilton
Mackas, David Lloyd
Marko, John Robert
Parsons, Timothy Richard
Perry, Richard Ian
Pickard, George Lawson
Pond, George Stephen
Skud, Bernard Einar
Stockner, John G
Tabata, Susumu
Taylor, Frank John Rupert (Max)
Thomson, Keith A
Thomson, Richard Edward
Tunnicliffe, Verena Julia
Waldichuk, Michael
Wong, Chi-Shing
Yunker, Mark Bernard
Zhang, Ziyang

NEW BRUNSWICK
Kohler, Carl
Vanicek, Petr

NEWFOUNDLAND
Anderson, John Truman
Haedrich, Richard L
Hellou, Jocelyne
Pepin, P
Rivkin, Richard Bob
Wroblewski, Joseph S

NOVA SCOTIA
Boyd, Carl M
Bridgeo, William Alphonsus
Brown, Richard George Bolney
Buckley, Dale Eliot
Cameron, Barry Winston
Castell, John Daniel
Cooke, Robert Clark
Cullen, John Joseph
Dickie, Lloyd Merlin
Elliott, James Arthur
Gradstein, Felix Marcel
King, Lewis H
Longhurst, Alan R
Mann, Kenneth H
Mills, Eric Leonard
Needler, George Treglohan
Platt, Trevor Charles
Scarratt, David Johnson
Silvert, William Lawrence
Sinclair, Michael Mackay
Smith, Stuart D
Tan, Francis C
Tremblay, Michael John
Trites, Ronald Wilmot

ONTARIO
Barica, Jan M
Campbell, Neil John
Chau, Yiu-Kee
Donelan, Mark Anthony
Drury, Malcolm John
Forrester, Warren David
Fyfe, William Sefton
Gaskin, David Edward
Greenwood, Brian
Khandekar, Madhav Laxman
Scott, Steven Donald
Shih, Chang-Tai
Trick, Charles Gordon

QUEBEC
Brunel, Pierre
Castonguay, Martin
Desjardins, Claude W
Dunbar, Maxwell John
Hesse, Reinhard

Ingram, Richard Grant
Khalil, Michel
Legendre, Louis
Leggett, William C
Levasseur, Maurice Edgar
Lewis, John Bradley
Mansfield, Arthur Walter
Mysak, Lawrence Alexander
Packard, Theodore Train
Pounder, Elton Roy
Reiswig, Henry Michael

SASKATCHEWAN
Evans, Marlene Sandra

OTHER COUNTRIES
Andreae, Meinrat Otto
Bakun, Andrew
Caperon, John
Charlier, Roger Henri
Chen, Chen-Tung Arthur
Fowler, Scott Wellington
Frakes, Lawrence Austin
Jones, Robert Sidney
Kitaigorodskii, Sergei Alexander
Koslow, Julian Anthony
Krause, Dale Curtiss
Mandelbaum, Hugo
Murty, Tadepalli Satyanarayana
Rodriguez, Gilberto
Sleeter, Thomas David
Spence, Thomas Wayne
Thierstein, Hans Rudolf
Thompson, Peter Allan
Van Andel, Tjeerd Hendrik
Williams, Francis

Paleontology

ALABAMA
Beals, Harold Oliver
Copeland, Charles Wesley, Jr
Hoover, Richard Brice
Huff, William J
Jones, Douglas Epps
Lamb, George Marion
Mancini, Ernest Anthony
Stock, Carl William

ALASKA
Guthrie, Russell Dale
Severin, Kenneth Paul

ARIZONA
Agenbroad, Larry Delmar
Batten, Roger Lyman
Beus, Stanley S
Canright, James Edward
Cohen, Andrew Scott
Colbert, Edwin Harris
Davis, Owen Kent
Flessa, Karl Walter
Jackson, Stephen Thomas
Kremp, Gerhard Otto Wilhelm
Lindsay, Everett Harold, Jr
Lundin, Robert Folke
Mead, James Irving
Nations, Jack Dale
Parrish, Judith Totman
White, John Anderson

ARKANSAS
Edson, James Edward, Jr
Manger, Walter Leroy
Thurmond, John Tydings

CALIFORNIA
Adam, David Peter
Allison, Carol Wagner
Allison, Richard C
Archibald, James David
Awramik, Stanley Michael
Axelrod, Daniel Isaac
Barnes, Lawrence Gayle
Barron, John Arthur
Berger, Wolfgang Helmut
Black, Craig C
Bottjer, David John
Brand, Leonard Roy
Buchheim, H Paul
Bukry, John David
Callison, George
Campbell, Catherine Chase
Campbell, Kenneth Eugene, Jr
Chmura-Meyer, Carol A
Clemens, William Alvin
Cohen, Anne Carolyn Constant
Cowen, Richard
Demond, Joan
Downs, Theodore
Drugg, Warren Sowle
Durham, John Wyatt
Easton, William Heyden
Estes, Richard
Etheridge, Richard Emmett
Evitt, William Robert
Finney, Stanley Charles
Fischer, Alfred George
Fry, Wayne Lyle
Gregory, Joseph Tracy
Hall, Donald D
Hickman, Carole Stentz
Hill, Merton Earle, III

Paleontology (cont)

Howard, Hildegarde
Hutchison, John Howard
Ingle, James Chesney, Jr
Jacobson, Stephen Richard
Jakway, George Elmer
Jenkins, Floyd Albert
Jones, David Lawrence
Kaska, Harold Victor
Kennett, James Peter
Kern, John Philip
Kitts, David Burlingame
Kopf, Rudolph William
Krejsa, Richard Joseph
Lane, Bernard Owen
Laporte, Leo Fredric
Lindberg, David Robert
Lipps, Jere Henry
Loeblich, Alfred Richard, Jr
Loeblich, Helen Nina (Tappan)
Lumsden, William Watt, Jr
McCrone, Alistair William
McHenry, Henry Malcolm
McLeod, Samuel Albert
Marshall, Charles Richard
Mintz, Leigh Wayne
Myrick, Albert Charles, Jr
Noble, Paula Jane
Olson, Everett Claire
Padian, Kevin
Parker, Frances Lawrence
Piel, Kenneth Martin
Prothero, Donald Ross
Riedel, William Rex
Rudnyk, Marian E
Saul, Lou Ella Rankin
Savage, Donald Elvin
Schopf, James William
Smith, Judith Terry
Smith-Evernden, Roberta Katherine
Snyder, Charles Theodore
Stevens, Calvin H
Swift, Camm Churchill
Tiffney, Bruce Haynes
Vaughn, Peter Paul
Warter, Janet Kirchner
Warter, Stuart L
Welles, Samuel Paul
Whistler, David Paul
Wiggins, Virgil Dale
Wilson, Edward Carl
Woodburne, Michael O

COLORADO
Ager, Thomas Alan
Bown, Thomas Michael
Cygan, Norbert Everett
Dubiel, Russell F
Elias, Scott Armstrong
Fagerstrom, John Alfred
Forester, Richard Monroe
Guennel, Gottfried Kurt
Harris, Judith Anne Van Couvering
Kauffman, Erle Galen
Kent, Bion H
Kent, Harry Christison
Kissling, Don Lester
Kosanke, Robert Max
Lewis, George Edward
Lockley, Martin Gaudin
Nichols, Douglas James
Nichols, Kathryn Marion
Palmer, Allison Ralph
Rall, Raymond Wallace
Robinson, Peter
Ross, Reuben James, Jr
Silberling, Norman John
Skipp, Betty Ann
Stucky, Richard K(eith)
Taylor, Michael E
Van Zant, Kent Lee

CONNECTICUT
Frankel, Larry
Galton, Peter Malcolm
Hickey, Leo Joseph
Hill, Andrew
MacClintock, Copeland
Ostrom, John Harold

DELAWARE
Kraft, John Christian

DISTRICT OF COLUMBIA
Benson, Richard Hall
Berdan, Jean Milton
Cheetham, Alan Herbert
Coates, Anthony George
De Queiroz, Kevin
DiMichele, William Anthony
Domning, Daryl Paul
Emry, Robert John
Erwin, Douglas Hamilton
Grant, Richard Evans
Hecht, Alan David
Hussain, Syed Taseer
Neuman, Robert Ballin
Nowicke, Joan Weiland
Oliver, William Albert, Jr
Ray, Clayton Edward
Sando, William Jasper
Sohn, Israel Gregory
Towe, Kenneth McCarn

Waller, Thomas Richard
West, Robert MacLellan
Whitmore, Frank Clifford, Jr
Yochelson, Ellis Leon

FLORIDA
Bock, Wayne Dean
Brack-Hanes, Sheila Delfeld
Curren, Caleb
Dilcher, David L
Dodge, Richard E
Edmund, Alexander Gordon
Gu, Binhe
Hallock-Muller, Pamela
Maurrasse, Florentin Jean-Marie Robert
Nicol, David
Patton, Thomas Hudson
Shaak, Graig Dennis
Webb, Sawney David
Winters, Stephen Samuel
Wise, Sherwood Willing, Jr

GEORGIA
Bishop, Gale Arden
Chatelain, Edward Ellis
Cramer, Howard Ross
Darrell, James Harris, II
Fritz, William J
Goldstein, Susan T
Gore, Pamela J(eanne) W(heeless)
Laerm, Joshua
Lipps, Emma Lewis
Rich, Fredrick James

HAWAII
Keating, Barbara Helen
McCoy, Scott, Jr

IDAHO
Isaacson, Peter Edwin
Spinosa, Claude

ILLINOIS
Abler, William Lewis
Bardack, David
Barghusen, Herbert Richard
Baxter, James Watson
Blake, Daniel Bryan
Bolt, John Ryan
DeMar, Robert E
Eggert, Donald A
Flynn, John Joseph
Foster, Merrill W
Frankenberg, Julian Myron
Fraunfelter, George H
Greaves, Walter Stalker
Griffiths, Thomas Alan
Haddock, Gerald Hugh
Hopson, James A
Jablonski, David
Kent, Lois Schoonover
Kethley, John Bryan
Kluessendorf, Joanne
Kolata, Dennis Robert
Langenheim, Ralph Louis, Jr
Leary, Richard Lee
Ling, Hsin Yi
Matten, Lawrence Charles
Mikulic, Donald George
Nitecki, Matthew H
Norby, Rodney Dale
Peppers, Russel A
Phillips, Tom Lee
Raup, David Malcolm
Reed, Charles Allen
Sandberg, Philip A
Saunders, Jeffrey John
Sepkoski, J(oseph) J(ohn), Jr
Sereno, Paul C
Singer, Rolf
Stidd, Benton Maurice
Stratton, James Forrest
Utgaard, John Edward
Van Valen, Leigh Maiorana
Walton, William Ralph
Ziegler, Alfred M

INDIANA
Dodd, James Robert
Dolph, Gary Edward
Gutschick, Raymond Charles
Hattin, Donald Edward
Horowitz, Alan Stanley
Howe, Robert C
Lane, Norman Gary
Orr, Robert William
Pachut, Joseph F(rancis), Jr
Park, Richard Avery, IV
Rexroad, Carl Buckner
Rosenberg, Gary David
Shaver, Robert Harold
Votaw, Robert Barnett
Zinsmeister, William John

IOWA
Baker, Richard Graves
Budd, Ann Freeman
Ciochon, Russell Lynn
Glenister, Brian Frederick
Hansman, Robert H
Hinman, Eugene Edward
Klapper, Gilbert
Semken, Holmes Alford, Jr

KANSAS
Baxter, Robert Wilson
Blythe, Jack Gordon
Clark, George Richmond, II
Green, Morton
Kaesler, Roger LeRoy
Maples, Christopher Grant
Martin, Larry Dean
Potter, Frank Walter, Jr
Robison, Richard Ashby
Rowell, Albert John
Schultze, Hans-Peter
Thomasson, Joseph R
West, Ronald Robert
Wilson, Robert Warren
Zakrzewski, Richard Jerome

KENTUCKY
Chesnut, Donald R, Jr
Clark, Armin Lee
Conkin, James E
Hamon, J Hill
Helfrich, Charles Thomas
Roberts, Thomas Glasdir
Singh, Raman J

LOUISIANA
Craig, William Warren
Everett, Robert W, Jr
Glawe, Lloyd Neil
Hazel, Joseph Ernest
Horodyski, Robert Joseph
Kessinger, Walter Paul, Jr
Lieux, Meredith Hoag
Parsley, Ronald Lee
Schiebout, Judith Ann
Sen Gupta, Barun Kumar
Vokes, Emily Hoskins
Vokes, Harold Ernest
Von Almen, William Frederick I

MAINE
Nelson, Robert Edward
Pestana, Harold Richard
Schnitker, Detmar

MARYLAND
Buzas, Martin A
Hotton, Nicholas, III
McCammon, Helen Mary
Moran, David Dunstan
Osman, Richard William
Rose, Kenneth David
Scala, John Richard
Stanley, Steven Mitchell
Stifel, Peter Beekman

MASSACHUSETTS
Bailey, Richard Hendricks
Berggren, William Alfred
Brown, George D, Jr
Coombs, Margery Chalifoux
Crompton, Alfred W
Curran, Harold Allen
Davis, William Edwin, Jr
Giffin, Emily Buchholtz
Gould, Stephen Jay
Greenstein, Benjamin Joel
Jenkins, Farish Alston, Jr
Knoll, Andrew Herbert
McMenamin, Mark Allan
Meszoely, Charles Aladar Maria
Pitrat, Charles William
Poag, Claude Wylie
Roth, John L, Jr
Shrock, Robert Rakes
Williamson, Peter George

MICHIGAN
Burnham, Robyn Jeanette
Collins, Laurel Smith
Frankforter, Weldon D
Gingerich, Philip Derstine
Hayward, James Lloyd
Holman, J Alan
Kerfoot, Wilson Charles
Kesling, Robert Vernon
Miller, Robert Rush
Ritland, Richard Martin
Seltin, Richard James
Smith, Bennett Holly
Stoermer, Eugene F
Wicander, Edwin Reed

MINNESOTA
Bayer, Thomas Norton
Carlson, Keith J
Darby, David G
Heller, Robert Leo
Lewis, Standley Eugene
Melchior, Robert Charles
Sloan, Robert Evan
Sparling, Dale R
Van Alstine, James Bruce
Webers, Gerald F
Wright, Herbert Edgar, Jr

MISSISSIPPI
Brunner, Charlotte
Dunn, Dean A(lan)

MISSOURI
Ashworth, William Bruce, Jr
Echols, Dorothy Jung

Ethinton, Raymond Lindsay
Finger, Terry Richard
Hawksley, Oscar
Koenig, John Waldo
Kurtz, Vincent E
Levin, Harold Leonard
Miller, James Frederick
Oboh, Francisca Emiede
Rasmussen, David Tab
Stitt, James Harry
Unklesbay, Athel Glyde
Wood, Joseph M

MONTANA
Dresser, Hugh W
Fields, Robert William
Melton, William Grover, Jr
Stanley, George Dabney, Jr
Thompson, Gary Gene

NEBRASKA
Diffendal, Robert Francis, Jr
Hunt, Robert M, Jr
Stout, Thompson Mylan
Voorhies, Michael Reginald

NEVADA
Firby, James R

NEW HAMPSHIRE
Andrews, Henry Nathaniel, Jr
Tischler, Herbert

NEW JERSEY
Alexander, Richard Raymond
Lutz, Richard Arthur
McGhee, George Rufus, Jr
Theokritoff, George
Wood, Albert E(lmer)

NEW MEXICO
Bowsher, Arthur LeRoy
Flower, Rousseau Hayner
Holloway, Richard George
Johnston, Roger Glenn
Lochman-Balk, Christina
McComas, David John
Pierce, Robert William
Wolberg, Donald Lester

NEW YORK
Banks, Harlan Parker
Beerbower, James Richard
Bell, Michael Allen
Brenner, Gilbert J
Bretsky, Peter William
Brett, Carlton Elliot
Brower, James Clinton
Cisne, John Luther
Crepet, William Louis
Delson, Eric
Eldredge, Niles
Emerson, William Keith
Erickson, John Mark
Finks, Robert Melvin
Fisher, Donald William
Gillette, Norman John
Grierson, James Douglas
Habib, Daniel
Heaslip, William Graham
Hewitt, Philip Cooper
Hoover, Peter Redfield
Kennedy, Kenneth Adrian Raine
Krause, David Wilfred
Landing, Ed
Lasker, Howard Robert
Laub, Richard Steven
Levinton, Jeffrey Sheldon
Linsley, Robert Martin
Lund, Richard
McCune, Amy Reed
MacIntyre, Giles T
McKenna, Malcolm Carnegie
Macomber, Richard Wiltz
Mendel, Frank C
Nair, Ramachandran Mukundalayam Sivarama
Newell, Norman Dennis
Newton, Cathryn Ruth
Novacek, Michael John
Pannella, Giorgio
Rhodes, Frank Harold Trevor
Rickard, Lawrence Vroman
Rightmire, George Philip
Sancetta, Constance Antonina
Sass, Daniel B
Scatterday, James Ware
Schaeffer, Bobb
Shaw, Frederick Carleton
Sirkin, Leslie A
Steineck, Paul Lewis
Tattersall, Ian
Tedford, Richard Hall
Teichert, Curt
Tesmer, Irving Howard
Titus, Robert Charles
Wahlert, John Howard
Walker, Philip Caleb
Weiss, Dennis
Wells, John West

NORTH CAROLINA
Carter, Joseph Gaylord
Cooper, Gustav Arthur

ENVIRONMENTAL, EARTH & MARINE SCIENCES / 405

Corliss, Bruce Hayward
Dennison, John Manley
Gensel, Patricia Gabbey
McKinney, Frank Kenneth
Neal, Donald Wade
St Jean, Joseph, Jr
Simons, Elwyn L
Welby, Charles William
Zullo, Victor August

NORTH DAKOTA
Holland, F D, Jr
Kelley, Patricia Hagelin

OHIO
Ausich, William Irl
Babcock, Loren Edward
Bergstrom, Stig Magnus
Bork, Kennard Baker
Camp, Mark Jeffrey
Caster, Kenneth Edward
Chitaley, Shyamala D
Coogan, Alan H
Cropp, Frederick William, III
Davis, Richard Arnold
Feldmann, Rodney Mansfield
Hannibal, Joseph Timothy
Hoare, Richard David
Howe, John A
Kovach, Jack
Latimer, Bruce Millikin
McCall, Peter Law
Mapes, Gene Kathleen
Mapes, Royal Herbert
Meyer, David Lachlan
Miller, Arnold I
Morris, Robert William
Nave, Floyd Roger
Norton, Charles Warren
Pope, John Keyler
Reinhart, Roy Herbert
Steinker, Don Cooper
Sturgeon, Myron Thomas
Taylor, Edith L
Taylor, Thomas Norwood
Teeter, James Wallis
Webb, Peter Noel
Williams, Michael Eugene
Wilson, Mark Allan

OKLAHOMA
Amsden, Thomas William
Cardott, Brian Joseph
Grayson, John Francis
Harper, Charles Woods, Jr
Hedlund, Richard Warren
Masters, Bruce Allen
Meyers, William C
Sanderson, George Albert
Scott, Robert W
Sutherland, Patrick Kennedy
Wilson, Leonard Richard

OREGON
Addicott, Warren O
Boucot, Arthur James
Coffin, Harold Glen
Drake, Ellen Tan
Fisk, Lanny Herbert
Gray, Jane
Lamb, James L
Retallack, Gregory John
Roberts, Michael Foster
Roellig, Harold Frederick
Savage, Norman Michael
Sherwood, Martha Allen
Thoms, Richard Edwin
Youngquist, Walter

PENNSYLVANIA
Andrews, George William
Baird, Donald
Berman, David S
Carter, John Lyman
Cuffey, Roger James
Dawson, Mary (Ruth)
Dodson, Peter
Faas, Richard William
Hill, Frederick Conrad
Inners, Jon David
King, James Edward
Krishtalka, Leonard
Mann, Keith Olin
Miklausen, Anthony J
Penny, John Sloyan
Pfefferkorn, Hermann Wilhelm
Rackoff, Jerome S
Ramsdell, Robert Cole
Rollins, Harold Bert
Rosenberg, Gary
Saunders, William Bruce
Spackman, William, Jr
Thayer, Charles Walter
Thomas, Roger David Keen
Thomson, Keith Stewart
Traverse, Alfred
Wegweiser, Arthur E

RHODE ISLAND
Imbrie, John
Janis, Christine Marie
Prell, Warren Lee
Webb, Thompson, III
Zavada, Michael Stephan

SOUTH CAROLINA
Carew, James L
Cohen, Arthur David
Engelhardt, Donald Wayne
Lawrence, David Reed
Williams, Douglas Francis

SOUTH DAKOTA
Bjork, Philip R
Macdonald, James Reid
Martin, James Edward

TENNESSEE
Bordine, Burton W
Deboo, Phili B
Knox, Larry William
McLaughlin, Robert Everett
Siesser, William Gary

TEXAS
Abbott, William Harold
Beyer, Arthur Frederick
Bonem, Rena Mae
Chadwick, Arthur Vorce
Chatterjee, Sankar
Clarke, Robert Travis
Cornell, William Crowninshield
Crick, Rex Edward
Delevoryas, Theodore
DuBar, Jules R
Durden, Christopher John
Echols, Joan
Elsik, William Clinton
Fang, Jiasong
Feazel, Charles Tibbals
Fisher, William Lawrence
Gartner, Stefan, Jr
Gibson, Lee B
Gillette, David Duane
Gombos, Andrew Michael, Jr
Hall, Stephen Austin
Harris, Arthur Horne
James, Gideon T
Jamieson-Magathan, Esther R
Kocurko, Michael John
Lane, Harold Richard
Langston, Wann, Jr
LeBlanc, Arthur Edgar
LeMone, David V
Levinson, Stuart Alan
Lundelius, Ernest Luther, Jr
McAlester, Arcie Lee, Jr
McNulty, Charles Lee, Jr
Macurda, Donald Bradford, Jr
Maddocks, Rosalie Frances
Melia, Michael Brendan
Merrill, Glen Kenton
Pampe, William R
Perkins, Bobby Frank
Pierce, Robert Wesley
Pruitt, Basil Arthur, Jr
Reading, John Frank
Rohr, David M
Schultz, Gerald Edward
Schumacher, Dietmar
Shane, John Denis
Shaw, Alan Bosworth
Sprinkle, James (Thomas)
Stanton, Robert James, Jr
Steele, David Gentry
Stover, Lewis Eugene
Sullivan, Herbert J
Vincent, Jerry William
Weart, Richard Claude
Wilde, Garner Lee
Yancey, Thomas Erwin
Young, Keith Preston
Zingula, Richard Paul

UTAH
Ash, Sidney Roy
Baer, James L
Ball, Terry Briggs
Ekdale, Allan Anton
Harris, John Michael
Hintze, Lehi Ferdinand
Jensen, David
Liddell, William David
Madsen, James Henry, Jr
Miller, Wade Elliott, II
Mosher, Loren Cameron
Moyle, Richard W
Rigby, J Keith
Roth, Peter Hans
Stokes, William Lee

VERMONT
Trombulak, Stephen Christopher

VIRGINIA
Bowen, Zeddie Paul
Cronin, Thomas Mark
Douglass, Raymond Charles
Edwards, Lucy Elaine
Eggleston, Jane R
Ellison, Robert L
Gibson, Thomas George
Haq, Bilal U
Koch, Carl Fred
Lohman, Kenneth Elmo
Nelson, Clifford M(elvin)
Pojeta, John, Jr
Robbins, Eleanora Iberall
Said, Rushdi

Scheckler, Stephen Edward
Skog, Judith Ellen
Sohl, Norman Frederick
Spencer, Randall Scott
Weems, Robert Edwin
Wilson, John William, III
Winston, Judith Ellen

WASHINGTON
Eck, Gerald Gilbert
Gilmour, Ernest Henry
Hopkins, William Stephen
Howe, Herbert James
Kohn, Alan Jacobs
Lane, Wallace
Lowther, John Stewart
Mallory, Virgil Standish
Mehringer, Peter Joseph, Jr
Rau, Weldon Willis
Ross, Charles Alexander
Ross, June Rosa Pitt
Simmonds, Robert T
Tsukada, Matsuo
Webster, Gary Dean
Wright, Judith

WEST VIRGINIA
Bird, Samuel Oscar, II

WISCONSIN
Clark, David Leigh
Fowler, Gerald Allan
Gernant, Robert Everett
Kitchell, Jennifer Ann
Long, Charles Alan
Maher, Louis James, Jr
Mendelson, Carl Victor
Paull, Rachel Krebs
Westphal, Klaus Wilhem

WYOMING
Boyd, Donald Wilkin
Cvancara, Alan Milton
Frerichs, William Edward
Lillegraven, Jason Arthur
Mateer, Niall John
Oman, Carl Henry

ALBERTA
Campbell, John Duncan
Chatterton, Brian Douglas Eyre
Currie, Philip John
Fox, Richard Carr
Hunt, Robert Nelson
Naylor, Bruce Gordon
Norford, Brian Seeley
Nowlan, Gofrey S
Okulitch, Vladimir Joseph
Osborn, Jeffrey Whitaker
Russell, Anthony Patrick
Stelck, Charles Richard
Wall, John Hallett
Wilson, Mark Vincent Hardman

BRITISH COLUMBIA
Hebda, Richard Joseph
Mathewes, Rolf Walter
Monger, James William Heron
Nelson, Samuel James
Pocock, Stanley Albert John
Rouse, Glenn Everett
Smith, Paul L
Walker, Ian Richard

MANITOBA
Hecky, Robert Eugene

NEW BRUNSWICK
Ferguson, Laing
Logan, Alan

NEWFOUNDLAND
May, Sherry Jan

NOVA SCOTIA
Cameron, Barry Winston
Gradstein, Felix Marcel
Moore, Reginald George
Vilks, Gustavs
Zodrow, Erwin Lorenz

ONTARIO
Begun, David Rene
Brose, David Stephen
Caldwell, William Glen Elliot
Churcher, Charles Stephen
Collins, Desmond H
Copeland, Murray John
Copper, Paul
Fritz, William Harold
Harington, Charles Richard
Jarzen, David MacArthur
Kobluk, David Ronald
McLaren, Digby Johns
Morgan, Alan Vivian
Norris, Geoffrey
Radforth, Norman William
Risk, Michael John
Russell, Dale A
Russell, Loris Shano
Terasmae, Jaan

QUEBEC
Beland, Pierre

Carroll, Robert Lynn
Clark, Thomas Henry
Hillaire-Marcel, Claude
Lesperance, Pierre J
Richard, Pierre Joseph Herve
Riva, John F
Stearn, Colin William

SASKATCHEWAN
Basinger, James Frederick
Braun, Willi Karl
Sarjeant, William Antony Swithin

OTHER COUNTRIES
Alonso, Ricardo N
Archangelsky, Sergio
Dettmann, Mary Elizabeth
Erdtmann, Bernd Dietrich
Hageman, Steven James
Lynts, George Willard
Paulay, Gustav
Rich, Thomas Hewitt
Schram, Frederick R
Thierstein, Hans Rudolf
Trumpy, D Rudolf
Van Den Bold, Willem Aaldert
Vickers-Rich, Patricia

Physical Geography

ALABAMA
Quattrochi, Dale Anthony
Williams, Aaron, Jr

ARIZONA
Baron, William Robert
Berlin, Graydon Lennis
Brazel, Anthony James
Breed, Carol Sameth
Childs, Orlo E
Comrie, Andrew Charles
Crosswhite, Frank Samuel
Gillis, James E, Jr
Hutchinson, Charles F
Kremp, Gerhard Otto Wilhelm
Linehan, Urban Joseph
Manera, Paul Allen
Marcus, Melvin Gerald

CALIFORNIA
Beard, Charles Noble
Gordon, Burton LeRoy
Ives, John (Jack) David
Lier, John
McArthur, David Samuel
McElhoe, Forrest Lester, Jr
Pease, Robert Wright
Pike, Richard Joseph
Rees, John David
Sauer, Jonathan Deininger
Star, Jeffrey L
Sternberg, Hilgard O'Reilly
Stevens, Calvin H
Stoms, David Michael
Thorne, Robert Folger
Vermeij, Geerat Jacobus
Voeks, Robert Allen

COLORADO
Carrara, Paul Edward
Diaz, Henry Frank
Lehrer, Paul Lindner
Mason, Malcolm
Redente, Edward Francis
Steen-McIntyre, Virginia Carol

CONNECTICUT
Sakalowsky, Peter Paul, Jr

DELAWARE
Kalkstein, Laurence Saul
Mather, John Russell
Willmott, Cort James

DISTRICT OF COLUMBIA
Abler, Ronald Francis
Matson, Michael
Strommen, Norton Duane

FLORIDA
Berry, Leonard
Burrill, Robert Meredith
Chauvin, Robert S
Craig, Alan Knowlton
Cross, Clark Irwin
Marcus, Robert Brown

GEORGIA
Meentemeyer, Vernon George
Parker, Albert John

ILLINOIS
Anderson, Richard Charles
Changnon, Stanley Alcide, Jr
Espenshade, Edward Bowman, Jr
Gabler, Robert Earl
McMillan, R Bruce
Sharpe, David McCurry
Wendland, Wayne Marcel

INDIANA
Dando, William Arthur
Mausel, Paul Warner

Physical Geography (cont)

Oliver, John Edward

IOWA
Akin, Wallace Elmus
Malanson, George Patrick
Prior, Jean Cutler

KANSAS
Kromm, David Elwyn
White, Stephen Edward

KENTUCKY
Schwendeman, Joseph Raymond, Jr

LOUISIANA
Gagliano, Sherwood Moneer
Muller, Robert Albert
Walker, Harley Jesse
Webb, Robert MacHardy

MARYLAND
Campbell, William J
Chovitz, Bernard H
Fonaroff, Leonard Schuyler
Kearney, Michael Sean
McBryde, F(elix) Webster
Pritchard, Donald William
Scala, John Richard

MASSACHUSETTS
Bradley, Raymond Stuart
MacDougall, Edward Bruce

MICHIGAN
Aron, Robert
Aufdemberge, Theodore Paul
Erhart, Rainer R
Outcalt, Samuel Irvine
Sommers, Lawrence M

MINNESOTA
Richason, Benjamin Franklin, Jr

MISSISSIPPI
Cross, Ralph Donald
Saucier, Roger Thomas

MISSOURI
Schroeder, Walter Albert
Stauffer, Truman Parker, Sr

NEBRASKA
Lawson, Merlin Paul

NEVADA
Mifflin, Martin David

NEW JERSEY
Hordon, Robert M
Psuty, Norbert Phillip
Robinson, David Alton
Wetherald, Richard Tryon

NEW MEXICO
Hill, Mary Rae
Holloway, Richard George

NEW YORK
Abrahams, Athol Denis
Batty, J Michael
De Laubenfels, David John
Ebert, Charles H V
Henderson, Floyd M
Jarvis, Richard S
Lougeay, Ray Leonard
Metzger, Ernest Hugh
Monmonier, Mark
Muller, Ernest Hathaway
Palmer, James F
Van Burkalow, Anastasia
Vuilleumier, Francois

NORTH CAROLINA
Hidore, John J
Nunnally, Nelson Rudolph
Rees, John
Robinson, Peter John
Ross, Thomas Edward
Stillwell, Harold Daniel

OHIO
Arnfield, Anthony John
Basile, Robert Manlius
Crawford, Paul Vincent
Mahard, Richard H
Noble, Allen G
Ray, John Robert
Rayner, John Norman

OREGON
Johannessen, Carl L
McKinney, William Mark
Niem, Wendy Adams
Ripple, William John

PENNSYLVANIA
Justham, Stephen Alton
Sharer, Cyrus J
Taylor, Alan H
Wilhelm, Eugene J, Jr

TENNESSEE
Fullerton, Ralph O
Jumper, Sidney Roberts

TEXAS
Butzer, Karl W(ilhelm)
Fix, Ronald Edward
Gore, Dorothy J
Humphries, James Edward, Jr
Kimber, Clarissa Therese
Lambert, Paul Wayne
McCloy, James Murl
Perez, Francisco Luis
Whitford-Stark, James Leslie

UTAH
Foster, Robert H
Godfrey, Andrew Elliott
Grey, Alan Hopwood
McCoy, Roger Michael
Stevens, Dale John
Vaughn, Danny Mack

VERMONT
Gade, Daniel W
VanderMeer, Canute

VIRGINIA
Dorman, Craig Emery
Pontius, Steven Kent
Wright, Lynn Donelson

WASHINGTON
Kaatz, Martin Richard

WEST VIRGINIA
Calzonetti, Frank J
Elmes, Gregory Arthur
Murphy, Allen Emerson

WISCONSIN
Bhatia, Shyam Sunder
Knox, James Clarence

WYOMING
Marston, Richard Alan
Reider, Richard Gary

ALBERTA
Klassen, Rudolph Waldemar
McPherson, Harold James
Rogerson, Robert James

BRITISH COLUMBIA
Barnes, Christopher Richard
Neilsen, Gerald Henry
Oke, Timothy Richard
Slaymaker, Herbert Olav

MANITOBA
Rounds, Richard Clifford
Welsted, John Edward

ONTARIO
Bunting, Brian Talbot
Forbes, Bruce Cameron
Ford, Derek Clifford
Gracie, G(ordon)
Greenwood, Brian
Hewitt, Kenneth
Kesik, Andrzej B
King, Roger Hatton
Menzies, John
Morgan, Alan Vivian
Phillips, Brian Antony Morley
Prest, Victor Kent
Reeds, Lloyd George
Sanderson, Marie Elizabeth
Whyte, Anne V T
Wilkinson, Thomas Preston

QUEBEC
Bonn, Ferdinand J
Dionne, Jean-Claude
Gangloff, Pierre

OTHER COUNTRIES
Bridgman, Howard Allen
Charlier, Roger Henri
Francke, Oscar F
Rich, Thomas Hewitt

Stratigraphy-Sedimentation

ALABAMA
Copeland, Charles Wesley, Jr
Hooks, William Gary
Huff, William J
Lamb, George Marion
Mancini, Ernest Anthony
Stock, Carl William

ALASKA
Karl, Susan Margaret
Naidu, Angi Satyanarayan

ARIZONA
Breed, Carol Sameth
Cohen, Andrew Scott
Davis, Owen Kent
Dickinson, William Richard
Kepper, John C
Kremp, Gerhard Otto Wilhelm
Lucchitta, Ivo
Nations, Jack Dale
Schreiber, Joseph Frederick, Jr

ARKANSAS
Edson, James Edward, Jr
Kehler, Philip Leroy
Vosburg, David Lee

CALIFORNIA
Abbott, Patrick Leon
Allan, John Ridgway
Allison, Richard C
Alvarez, Walter
Austin, Steven Arthur
Bachman, Richard Thomas
Bassett, Henry Gordon
Bottjer, David John
Buchheim, H Paul
Burtner, Roger Lee
Campbell, Kenneth Eugene, Jr
Chapman, Robert Mills
Chmura-Meyer, Carol A
Ciancanelli, Eugene Vincent
Colburn, Ivan Paul
Cook, Harry E, III
Fischer, Alfred George
Ford, John Philip
Friberg, James Frederick
Graham, Stephan Alan
Grantz, Arthur
Gutstadt, Allan Morton
Hill, Merton Earle, III
Hodges, Lance Thomas
Hose, Richard K
Hunter, Ralph Eugene
Ingersoll, Raymond Vail
Ingle, James Chesney, Jr
Kaska, Harold Victor
Kirby, James M
Kopf, Rudolph William
Lange, Carina Beatriz
Laporte, Leo Fredric
Lowe, Donald Ray
McCrone, Alistair William
McLean, Hugh
Mann, John Allen
Mertes, Leal Anne Kerry
Mintz, Leigh Wayne
Murphy, Michael A
Noble, Paula Jane
Osborne, Robert Howard
Pampeyan, Earl H
Parker, Frances Lawrence
Peterson, Gary Lee
Prothero, Donald Ross
Russell, Richard Dana
Sarna-Wojcicki, Andrei M
Schwartz, Daniel Evan
Sieh, Kerry Edward
Smith-Evernden, Roberta Katherine
Stevens, Calvin H
Tabor, Rowland Whitney
Warnke, Detlef Andreas
Wells, Stephen Gene
Wentworth, Carl M, Jr

COLORADO
Ahlbrandt, Thomas Stuart
Bryant, Bruce Hazelton
Cadigan, Robert Allen
Campbell, John Arthur
Cathcart, James B
Clift, William Orrin
Crone, Anthony Joseph
Cross, Timothy Aureal
Dean, Walter E, Jr
Dubiel, Russell F
Elder, Curtis Harold
Ethridge, Frank Gulde
Fails, Thomas Glenn
Fassett, James Ernest
Flores, Romeo M
Hay, William Winn
Hayes, John Bernard
Holcombe, Troy Leon
Keefer, William Richard
Keighin, Charles William
Kent, Harry Christison
Ketner, Keith B
Lewis, George Edward
Link, Peter K
Lockley, Martin Gaudin
Luft, Stanley Jeremie
McCubbin, Donald Gene
MacKenzie, David Brindley
Maughan, Edwin Kelly
Miller, Betty M (Tinklepaugh)
Moore, David A
Morrison, Robert Barron
Nichols, Kathryn Marion
Palmer, Allison Ralph
Parker, John Marchbank
Peterson, Fred
Pillmore, Charles Lee
Rall, Raymond Wallace
Raup, Robert Bruce, Jr
Ross, Reuben James, Jr
Sable, Edward George
Sanborn, Albert Francis
Shideler, Gerald Lee
Siemers-Blay, Charles T
Silberling, Norman John
Smith, J Dungan

Steen-McIntyre, Virginia Carol
Taylor, Michael E
Vine, James David
Walker, Theodore Roscoe
Wallace, Chester Alan
Warme, John Edward

CONNECTICUT
Chriss, Terry Michael
Drobnyk, John Wendel
Hickey, Leo Joseph
Hill, Andrew
Horne, Gregory Stuart
Nadeau, Betty Kellett
Rodgers, John

DELAWARE
Benson, Richard Norman
Jordan, Robert R
Kraft, John Christian
Maurmeyer, Evelyn Mary
Talley, John Herbert
Thompson, Allan M

DISTRICT OF COLUMBIA
Boosman, Jaap Wim
Coates, Anthony George
Emry, Robert John
Hussain, Syed Taseer
Lindholm, Roy Charles
Rogers, Thomas Hardin
Rumble, Douglas, III
Sando, William Jasper
Tracey, Joshua Irving, Jr
Wright, Thomas Oscar

FLORIDA
Council, Richard J
Davis, Richard Albert, Jr
Donoghue, Joseph F
Finkl, Charles William, II
Freeland, George Lockwood
Gifford, John A
Ginsburg, Robert Nathan
Halley, Robert Bruce
Hallock-Muller, Pamela
Lovejoy, Donald Walker
McLeroy, Edward Glenn
Maurrasse, Florentin Jean-Marie Robert
Pirkle, Fredric Lee
Randazzo, Anthony Frank
Rosendahl, Bruce Ray
Schmidt, Walter
Tanner, William Francis, Jr
Wanless, Harold Rogers
Winters, Stephen Samuel

GEORGIA
Carver, Robert E
Chatelain, Edward Ellis
Cramer, Howard Ross
Fritz, William J
Gonzales, Serge
Gore, Pamela J(eanne) W(heeless)
Lineback, Jerry A(lvin)
Railsback, Loren Bruce
Rich, Fredrick James
Rich, Mark
Weaver, Charles Edward

HAWAII
Keating, Barbara Helen
McCoy, Floyd W, Jr
Mackenzie, Frederick Theodore
Moberly, Ralph M
Moore, Gregory Frank
Self, Stephen

IDAHO
Isaacson, Peter Edwin
Ore, Henry Thomas
Spinosa, Claude

ILLINOIS
Baxter, James Watson
Berg, Richard Carl
Block, Douglas Alfred
Buschbach, Thomas Charles
Doehler, Robert William
Flynn, John Joseph
Foster, Merrill W
Fraunfelter, George H
Gealy, William James
Hart, Richard Royce
Jablonski, David
Johnson, William Hilton
Kluessendorf, Joanne
Kolata, Dennis Robert
Langenheim, Ralph Louis, Jr
Leary, Richard Lee
Leighton, Morris Wellman
McKay, Edward Donald, III
Mann, Christian John
Mikulic, Donald George
Norby, Rodney Dale
Reams, Max Warren
Rue, Edward Evans
Simon, Jack Aaron
Smith, Norman Dwight
Trask, Charles Brian
Walton, William Ralph
Ziegler, Alfred M

INDIANA
Dodd, James Robert
Droste, John Brown
Fraser, Gordon Simon
Gutschick, Raymond Charles
Hattin, Donald Edward
Mirsky, Arthur
Pachut, Joseph F(rancis), Jr
Roepke, Harlan Hugh
Valia, Hardarshan S
Votaw, Robert Barnett

IOWA
Heckel, Philip Henry
Hinman, Eugene Edward
Kramer, Theodore Tivadar
Swett, Keene

KANSAS
Aber, James Sandusky
Avcin, Matthew John, Jr
Baars, Donald Lee
Blythe, Jack Gordon
Gerhard, Lee C
Jackson, Roscoe George, II
Maples, Christopher Grant
Twiss, Page Charles
Walton, Anthony Warrick

KENTUCKY
Chesnut, Donald R, Jr
Clark, Armin Lee
Helfrich, Charles Thomas
Kiefer, John David
Roberts, Thomas Glasdir
Singh, Raman J
Thomas, William Andrew

LOUISIANA
Braunstein, Jules
Craig, William Warren
Forman, McLain J
Hazel, Joseph Ernest
Horodyski, Robert Joseph
Huh, Oscar Karl
Kumar, Madhurendu B
Lock, Brian Edward
Meriwether, John R
Sartin, Austin Albert
Vokes, Harold Ernest
Ward, William Cruse

MAINE
Kellogg, Thomas B
Rosenfeld, Melvin Arthur

MARYLAND
Jakab, George Joseph
Kravitz, Joseph Henry
Miller, Ralph LeRoy
Pavlides, Louis
Stifel, Peter Beekman

MASSACHUSETTS
Belt, Edward Scudder
Binford, Michael W
Greenstein, Benjamin Joel
Hager, Jutta Lore
Hoffman, Paul Felix
Hubert, John Frederick
Knoll, Andrew Herbert
McMenamin, Mark Allan
Pitrat, Charles William
Poag, Claude Wylie
Rhoads, Donald Cave
Ridge, John Charles
Roth, John L, Jr
Schlee, John Stevens
Shrock, Robert Rakes
Southard, John Brelsford
Takahashi, Kozo
Tucholke, Brian Edward
White, Brian
Williamson, Peter George
Wilson, Philo Calhoun
Yuretich, Richard Francis

MICHIGAN
Briggs, Darinka Zigic
Catacosinos, Paul Anthony
Hawley, Nathan
Huntoon, Jacqueline E
McKee, Edith Merritt
Neal, William Joseph
Nordeng, Stephen C
Prouty, Chilton E
Rea, David Kenerson
Wilkinson, Bruce H

MINNESOTA
Crews, Anita L
Davis, Margaret Bryan
Heller, Robert Leo
Morey, Glenn Bernhardt
Sloan, Robert Evan
Sparling, Dale R
Van Alstine, James Bruce

MISSISSIPPI
Brunner, Charlotte
Dunn, Dean A(lan)
Egloff, Julius, III
Fleischer, Peter
Happ, Stafford Coleman

Laswell, Troy James
Lowrie, Allen
Otvos, Ervin George
Reynolds, William Roger
Seglund, James Arnold

MISSOURI
Brill, Kenneth Gray, Jr
Echols, Dorothy Jung
Frye, Charles Isaac
Gentile, Richard J
Goebel, Edwin DeWayne
Knox, Burnal Ray
Kurtz, Vincent E
Miller, James Frederick
Oboh, Francisca Emiede
Spreng, Alfred Carl
Unklesbay, Athel Glyde

MONTANA
Balster, Clifford Arthur
Fields, Robert William
Moore, Johnnie Nathan
Murray, Raymond Carl

NEBRASKA
Carlson, Marvin Paul
Diffendal, Robert Francis, Jr
Kaplan, Sanford Sandy

NEVADA
Felts, Wayne Moore

NEW HAMPSHIRE
Spink, Walter John
Tischler, Herbert

NEW JERSEY
Klein, George D
Metz, Robert
Olsson, Richard Keith
Van Houten, Franklyn Bosworth

NEW MEXICO
Bowsher, Arthur LeRoy
Broadhead, Ronald Frigon
Floran, Robert John
Kottlowski, Frank Edward
Love, David Waxham
McTigue, David Francis
Manley, Kim
Pierce, Robert William
Wengerd, Sherman Alexander
Wolberg, Donald Lester

NEW YORK
Abrahams, Athol Denis
Bowman, James Floyd, II
Brett, Carlton Elliot
Cisne, John Luther
Coch, Nicholas Kyros
Dawson, James Clifford
Dugolinsky, Brent Kerns
Finks, Robert Melvin
Flood, Roger Donald
Friedman, Gerald Manfred
Hand, Bryce Moyer
Johnson, Kenneth George
Kent, Dennis V
Kidd, William Spencer Francis
Krause, David Wilfred
Landing, Ed
Laub, Richard Steven
Lund, Richard
McLennan, Scott Mellin
Macomber, Richard Wiltz
Manos, Constantine T
Muskatt, Herman S
Newell, Norman Dennis
Newton, Cathryn Ruth
O'Brien, Neal Ray
Rampino, Michael Robert
Rickard, Lawrence Vroman
Sancetta, Constance Antonina
Savage, E Lynn
Scatterday, James Ware
Schreiber, B Charlotte
Tesmer, Irving Howard
Wang, Kia K
Wolff, Manfred Paul (Fred)
Woodrow, Donald L

NORTH CAROLINA
Banerjee, Umesh Chandra
Briggs, Garrett
Cleary, William James
Dennison, John Manley
Harris, William Burleigh
Heron, S Duncan, Jr
Ingram, Roy Lee
Johnson, Thomas Charles
Marlowe, James Irvin
Miller, Daniel Newton, Jr
Neal, Donald Wade
Textoris, Daniel Andrew
Webb, Fred, Jr
Welby, Charles William
Zullo, Victor August

NORTH DAKOTA
Kelley, Patricia Hagelin

OHIO
Babcock, Loren Edward

Bain, Roger J
Boardman, Mark R
Bork, Kennard Baker
Burford, Arthur Edgar
Carlton, Richard Walter
Collinson, James W
Fuller, James Osborn
Harmon, David Elmer, Jr
Hatfield, Craig
Henniger, Bernard Robert
Herdendorf, Charles Edward, III
Kovach, Jack
McWilliams, Robert Gene
Mapes, Gene Kathleen
Martin, Wayne Dudley
Miller, Arnold I
Morris, Robert William
Sturgeon, Myron Thomas
Wilson, Mark Allan

OKLAHOMA
Boellstorff, John David
Chambers, Richard Lee
Chenoweth, Philip Andrew
Espach, Ralph H, Jr
Friedman, Samuel Arthur
Masters, Bruce Allen
Murray, Frederick Nelson
Pittman, Edward D
Powell, Benjamin Neff
Sanderson, George Albert
Smit, David Ernst
Suhm, Raymond Walter
Tillman, Roderick W
Visher, Glenn S

OREGON
Beaulieu, John David
Boggs, Sam, Jr
Clark, Peter Underwood
Fisk, Lanny Herbert
Krinsley, David
Niem, Alan Randolph
Niem, Wendy Adams
Retallack, Gregory John
Savage, Norman Michael

PENNSYLVANIA
Adovasio, James Michael
Anderson, Edwin J
Carson, Bobb
Cotter, Edward
Donahue, Jack David
Faill, Rodger Tanner
Frantz, Wendelin R
Gardner, Thomas William
Goodwin, Peter Warren
Guber, Albert Lee
Hanson, Henry W A, III
Hutchinson, Peter J
Inners, Jon David
Jordan, William Malcolm
Lowright, Richard Henry
Mann, Keith Olin
Pfefferkorn, Hermann Wilhelm
Ramsdell, Robert Cole
Saunders, William Bruce
Schmalz, Robert Fowler
Seidell, Barbara Castens
Slingerland, Rudy Lynn
Trexler, John Peter
Washburn, Robert Henry
Wegweiser, Arthur E

RHODE ISLAND
Burroughs, Richard
Head, James William, III
Leinen, Margaret Sandra
McMaster, Robert Luscher
Matthews, Robley Knight
Prell, Warren Lee

SOUTH CAROLINA
Carew, James L
Cohen, Arthur David
Ehrlich, Robert
Gipson, Mack, Jr
Kanes, William H
Pirkle, William Arthur

SOUTH DAKOTA
Martin, James Edward

TENNESSEE
Brode, William Edward
Helton, Walter Lee
Klepser, Harry John
Siesser, William Gary
Walker, Kenneth Russell

TEXAS
Abbott, William Harold
Ahr, Wayne Merrill
Amsbury, David Leonard
Baker, Donald Roy
Bebout, Don Gray
Behrens, Earl William
Blackwelder, Blake Winfield
Bohacs, Kevin Michael
Brennan, Daniel Joseph
Brooks, James Elwood
Brown, Leonard Franklin, Jr
Buis, Otto J
Burton, Robert Clyde

Chadwick, Arthur Vorce
Chatterjee, Sankar
Crick, Rex Edward
Dahl, Harry Martin
Damuth, John Erwin
DuBar, Jules R
Edwards, Marc Benjamin
Eifler, Gus Kearney, Jr
Elsik, William Clinton
Ewing, Thomas Edward
Feazel, Charles Tibbals
Finley, Robert James
Folk, Stewart H
Galloway, William Edmond
Gombos, Andrew Michael, Jr
Granath, James Wilton
Gruy, Henry Jones
Gustavson, Thomas Carl
Halpern, Martin
Heyman, Louis
Hixon, Sumner B
Huffington, Roy Michael
Jackson, Martin Patrick Arden
Jones, James Ogden
Kaiser, William Richard
Kerans, Charles
Kocurko, Michael John
Koepnick, Richard Borland
Lafon, Guy Michel
Lane, Harold Richard
Leblanc, Rufus Joseph
Longacre, Susan Ann Burton
McFarlan, Edward, Jr
McPherson, John G(ordon)
Macurda, Donald Bradford, Jr
Matthews, Martin David
Mattis, Allen Francis
Mazzullo, James Michael
Melia, Michael Brendan
Merrill, Glen Kenton
Meyer, Franz O
Milling, Marcus Eugene
Moiola, Richard James
Morton, Robert Alex
Murphy, Daniel L
Murray, Grover Elmer
Namy, Jerome Nicholas
North, William Gordon
Nugent, Robert Charles
Owen, Donald Edward
Pierce, Robert Wesley
Pirie, Robert Gordon
Reso, Anthony
Rezak, Richard
Riese, Walter Charles Rusty
Rohr, David M
Roper, Paul James
Rose, Peter R
Rosen, Norman Charles
St John, Bill
Schumacher, Dietmar
Shane, John Denis
Shaw, Alan Bosworth
Smith, Michael Alexis
Soloyanis, Susan Constance
Spencer, Alexander Burke
Stead, Frederick L
Stoudt, Emily Laws
Swanson, Donald Charles
Swanson, Eric Rice
Thompson, Samuel, III
Tyler, Noel
Warshauer, Steven Michael
Weart, Richard Claude
Wilde, Garner Lee
Wilson, James Lee
Wright, Ramil Carter
Yancey, Thomas Erwin
Young, Keith Preston

UTAH
Ash, Sidney Roy
Ekdale, Allan Anton
Fidlar, Marion M
Freethey, Geoffrey Ward
Hintze, Lehi Ferdinand
Kolesar, Peter Thomas
McCleary, Jefferson Rand
Mosher, Loren Cameron
Oaks, Robert Quincy, Jr
Petersen, Morris Smith
Stokes, William Lee

VERMONT
Baldwin, Brewster

VIRGINIA
Brent, William B
Darby, Dennis Arnold
Edwards, Lucy Elaine
Eggleston, Jane R
Eriksson, Kenneth Andrew
Galvin, Cyril Jerome, Jr
Hadley, Donald G
Haq, Bilal U
Harbour, Jerry
Lohman, Kenneth Elmo
McCartan, Lucy
Martens, James Hart Curry
Masters, Charles Day
Meissner, Charles Roebling, Jr
Milici, Robert Calvin
Molnia, Bruce Franklin
Neiheisel, James

Stratigraphy-Sedimentation (cont)

Nelson, Bruce Warren
Nelson, Clifford M(elvin)
Outerbridge, William Fulwood
Reinhardt, Juergen
Russ, David Perry
Ryder, Robert Thomas
Said, Rushdi
Schwab, Frederick Lyon
Sims, John David
Spencer, Randall Scott
Swift, Donald J P
Weems, Robert Edwin
Whisonant, Robert Clyde
Wright, Lynn Donelson

WASHINGTON
Caggiano, Joseph Anthony, Jr
Kesler, Thomas L
Miller, C Dan
Oles, Keith Floyd
Owens, Edward Henry
Pierson, Thomas Charles
Rau, Weldon Willis
Ross, Charles Alexander
Scott, William Edward
Suczek, Christopher Anne
Webster, Gary Dean
Wright, Judith

WEST VIRGINIA
Khair, Abdul Wahab
Smosna, Richard Allan

WISCONSIN
Dott, Robert Henry, Jr
Fowler, Gerald Allan
Gernant, Robert Everett
Laudon, Thomas S
McKee, James W
Mendelson, Carl Victor
Paull, Rachel Krebs
Pray, Lloyd Charles
Schneider, Allan Frank
Stieglitz, Ronald Dennis
West, Walter Scott

WYOMING
Boyd, Donald Wilkin
Cvancara, Alan Milton
Huff, Kenneth O
Lillegraven, Jason Arthur
Oman, Carl Henry

ALBERTA
Andrichuk, John Michael
Baillie, Andrew Dollar
Cecile, Michael Peter
Currie, Philip John
Lerbekmo, John Franklin
Macqueen, Roger Webb
Mossop, Grant Dilworth
Nowlan, Gofrey S
Schmidt, Volkmar
Stelck, Charles Richard
Tankard, Anthony James
Wilson, Mark Vincent Hardman
Young, Frederick Griffin

BRITISH COLUMBIA
Barnes, Christopher Richard
Clague, John Joseph
Hebda, Richard Joseph
Jackson, Lionel Eric
Macdonald, Robie Wilton
Monger, James William Heron
Nelson, Samuel James
Payne, Myron William
Smith, Paul L
Stott, Donald Franklin

MANITOBA
Teller, James Tobias

NEW BRUNSWICK
Ferguson, Laing

NEWFOUNDLAND
Harper, John David
Hiscott, Richard Nicholas
King, Arthur Francis

NOVA SCOTIA
Cameron, Barry Winston
Schenk, Paul Edward
Zodrow, Erwin Lorenz

ONTARIO
Bell, Richard Thomas
Caldwell, William Glen Elliot
Chandler, Frederick William
Coakley, John Phillip
Dredge, Lynda Ann
Franklin, James McWillie
Geurts, Marie Anne H L
Greenwood, Brian
Jarzen, David MacArthur
Karrow, Paul Frederick
Lawson, David Edward
Long, Darrel Graham Francis
McLaren, Digby Johns

Martini, Ireneo Peter
Menzies, John
Miall, Andrew D
Middleton, Gerard Viner
Poole, William Hope
Prest, Victor Kent
Rukavina, Norman Andrew
Simpson, Frank
St-Onge, Denis Alderic
Walker, Roger Geoffrey
Yole, Raymond William
Young, Grant McAdam

QUEBEC
Bouchard, Michel Andre
Clark, Thomas Henry
Hesse, Reinhard
Lesperance, Pierre J
Mountjoy, Eric W
Stearn, Colin William

SASKATCHEWAN
Basinger, James Frederick
Braun, Willi Karl
Hendry, Hugh Edward
Kent, Donald Martin

OTHER COUNTRIES
Alonso, Ricardo N
Dettmann, Mary Elizabeth
Frakes, Lawrence Austin
Hageman, Steven James
Kennedy, Michael John
Lloyd, Christopher Raymond
Thierstein, Hans Rudolf
Trumpy, D Rudolf
Van Andel, Tjeerd Hendrik
Vickers-Rich, Patricia

Other Environmental, Earth & Marine Sciences

ALABAMA
Daniel, Thomas W, Jr
Guarino, Anthony Michael
Hazlegrove, Leven S
Quattrochi, Dale Anthony

ALASKA
Hamilton, Thomas Dudley
Hawkins, Daniel Ballou
Hunsucker, Robert Dudley
Ragle, Richard Harrison
Reger, Richard David
Rockwell, Julius, Jr
Severin, Kenneth Paul
Turner, Donald Lloyd

ARIZONA
Bannister, Bryant
Batson, Raymond Milner
Burke, James L(ee)
Glenn, Edward Perry
Harris, DeVerle Porter
Jones, David Lloyd
Jull, Anthony John Timothy
Lewis, John Simpson
Moore, Carleton Bryant
Murphy, Michael Joseph
Schoenwetter, James
Smiley, Terah Leroy
Soren, David

CALIFORNIA
Alvarino, Angeles
Arthur, Susan Peterson
Bada, Jeffrey L
Barnett, Tim P
Birnbaum, Milton
Blum, Fred A
Brownell, Robert Leo, Jr
Cain, William F
Chang, Ming-Houng (Albert)
Clark, J Desmond
Cluff, Lloyd Sterling
Coats, Robert Roy
Collins, Lorence Gene
Corwin, Dennis Lee
Costa, Daniel Paul
Deuble, John Lewis, Jr
Feller, Robert S, Sr
Gaal, Robert A P
Grivetti, Louis Evan
Harte, John
Hodges, Carroll Ann
Hoffmann, Michael Robert
Jaffe, Leonard David
Kalvinskas, John J(oseph)
Kavanaugh, David Henry
Kirch, Patrick Vinton
Koenig, James Bennett
Lyon, Waldo (Kampmeier)
Mabey, William Ray
Macdougall, John Douglas
Maciolek, John A
Manalis, Melvyn S
Matthews, Robert A
Mattinson, James Meikle
Morrow, Charles Tabor
Nickel, Phillip Arnold
Niemeyer, Sidney
Oberlander, Theodore M
Osborne, Michael Andrew

Owen, Paul Howard
Phillips, Richard P
Phillips, Thomas Joseph
Prasad, Kota S
Ramey, Daniel Bruce
Rogers, David Peter
Schwing, Franklin Burton
Seibel, Erwin
Serat, William Felkner
Sextro, Richard George
Silcox, William Henry
Skinner, G(eorge) William
Sleep, Norman H
Smith, George Irving
Spokes, Ernest M(elvern)
Swanson, Eric Richmond
Sweeney, James Lee
Thuillier, Richard Howard
Trapani, Robert Vincent
Udvardy, Miklos Dezso Ferenc
Ward, Peter Langdon
Weaver, Ellen Cleminshaw
White, Donald Edward
Wilkening, Laurel Lynn

COLORADO
Barry, Roger Graham
Born, George Henry
Bufe, Charles Glenn
Dewey, James W
Dunahay, Terri Goodman
Fullerton, David Stanley
Huss, Glenn I
Kenny, Ray
Landon, Susan Melinda
Lewis, George Edward
Lightsey, Paul Alden
Lind, David Arthur
Litsey, Linus R
Lockley, Martin Gaudin
Luckey, Thomas Donnell
Malde, Harold Edwin
Radke, Lawrence Frederick
Richmond, Gerald Martin
Simons, Frank
Steen-McIntyre, Virginia Carol
Strobell, John Dixon, Jr
Verdeal, Kathey Marie
Weihaupt, John George

CONNECTICUT
Antkiw, Stephen
Farmer, Mary Winifred
Haas, Thomas J
Houston, Walter Scott
Pooley, Alan Setzler

DELAWARE
Frankenburg, Peter Edgar
Gibbs, Ronald John

DISTRICT OF COLUMBIA
Barber, Mary Combs
Blanchard, Bruce
Brown, Norman Louis
Doumani, George Alexander
Gutman, George Garik
Holloway, John Thomas
Meserve, Richard A
Parham, Walter Edward
Rattien, Stephen
Stern, Thomas Whital
Warnick, Walter Lee
Watt, Hame Mamadov
Whitmore, Frank Clifford, Jr
Zimbelman, James Ray

FLORIDA
Alexiou, Arthur George
Belanger, Thomas V
Burrill, Robert Meredith
Craig, Alan Knowlton
Curren, Caleb
Fisher, David E
Freeland, George Lockwood
Genthner, Barbara Robyn Sharak
Gifford, John A
Hartley, Arnold Manchester
Harwell, Mark Alan
Heinen, Joel Thomas
Hurley, Patrick Mason
Ivey, Marvin L(ee)
Kitchens, Wiley M
McAllister, Raymond Francis
Maturo, Frank J S, Jr
Muller-Karger, Frank Edgar
Norris, Dean Rayburn
O'Hara, Norbert Wilhelm
Prince, Eric D
Rathbun, Carlisle Baxter, Jr
Rechcigl, John E
Rosenthal, Henry Bernard
Schmidt, Walter
Swain, Geoffrey W
Tanner, William Francis, Jr
Tomonto, James R
Windsor, John Golay, Jr

GEORGIA
Coleman, David Cowan
Dallmeyer, R David
Gore, Pamela J(eanne) W(heeless)
Howard, James Hatten, III
McClure, Donald Allan

Porter, James W
Remillard, Marguerite Madden
Shotts, Emmett Booker, Jr
Weber, William Mark
Zepp, Richard Gardner

HAWAII
Hubbard, Harold Mead
Keil, Klaus

IDAHO
Harrison, William Earl

ILLINOIS
Anderson, Alfred Titus, Jr
Anderson, Richard Vernon
Berg, Richard Carl
Casella, Alexander Joseph
Chiou, Wen-An
Espenshade, Edward Bowman, Jr
Johnson, William Hilton
Lewis, Roy Stephen
Mikulic, Donald George
Montgomery, Carla Westlund
Rouffa, Albert Stanley
Savic, Stanley D
Thomsen, Kurt O
Yopp, John Herman

INDIANA
Funk, Emerson Gornflow, Jr
Howe, Robert C
Melhorn, Wilton Newton

IOWA
Baker, Richard Graves
Christiansen, Kenneth Allen
Cressie, Noel A C
Hammond, Earl Gullette
Hartung, Jack Burdair, Sr
Semken, Holmes Alford, Jr

KANSAS
Beck, Henry V
Dellwig, Louis Field
Dort, Wakefield, Jr
Hoffman, Jerry C
Underwood, James Ross, Jr

KENTUCKY
Flegenheimer, Harold H(ansleo)

LOUISIANA
Chesney, Edward Joseph, Jr
Hazel, Joseph Ernest
Huh, Oscar Karl
Thibodeaux, Louis J
Wilcox, Ronald Erwin

MAINE
Borns, Harold William, Jr
Kates, Robert
Kellogg, Thomas B
King, Gary Michael
Page, David Sanborn
Splettstoesser, John Frederick
Yentsch, Clarice M

MARYLAND
Busalacchi, Antonio James, Jr
Crampton, Janet Wert
Diamante, John Matthew
Downing, Mary Brigetta
Evans, Alan G
Foster, Nancy Marie
French, Bevan Meredith
Gary, Robert
Gloersen, Per
Gould, Jack Richard
Hollinger, James Pippert
Leatherman, Stephen Parker
Lowman, Paul Daniel, Jr
Nelson, Judd Owen
Neuschel, Sherman K
Riedel, Gerhardt Frederick
Riel, Gordon Kienzle
Sellner, Kevin Gregory
Young, George Anthony

MASSACHUSETTS
Baird, Ronald C
Broadus, James Matthew
Burns, Daniel Robert
Chang, Ching Shung
Colinvaux, Paul Alfred
Hope, Elizabeth Greeley
Lamberg-Karlovsky, Clifford Charles
Lewis, Laurence A
McGill, George Emmett
McMenamin, Mark Allan
Nunes, Paul Donald
Pojasek, Robert B
Prescott, John Hernage
Teal, John Moline
Washburn, H Bradford, Jr
Wood, John Armstead

MICHIGAN
Arlinghaus, Sandra Judith Lach
Clark, James Alan
Corey, Kenneth Edward
Curl, Rane L(ocke)
Farber, Hugh Arthur
Farrand, William Richard

Frankforter, Weldon D
McCarville, Michael Edward
McKee, Edith Merritt
Marhsall, Hal G
Mills, Madolia Massey
Peterson, Lauren Michael
Rose, William Ingersoll, Jr
Sicko-Goad, Linda May
Sommers, Lawrence M
Tomboulian, Paul

MINNESOTA
Abrahamson, Dean Edwin
Davis, Margaret Bryan
Eisenreich, Steven John
Johnson, Robert Glenn
Murdock, Gordon Robert
Steck, Daniel John
Thompson, Phillip Gerhard
Van Alstine, James Bruce

MISSISSIPPI
Ballard, James Alan
Bowen, Richard Lee
Cooper, Charles Morris
Hwang, Huey-Min
Kushlan, James A
Lee, Charles Richard
Morgan, Eddie Mack, Jr

MISSOURI
Podosek, Frank A
Schroy, Jerry M
Soundararajan, Rengarajan
Stalling, David Laurence

MONTANA
Montagne, John M

NEBRASKA
Pritchard, Mary (Louise) Hanson

NEVADA
Evans, David L
Taranik, James Vladimir
Weide, David L

NEW HAMPSHIRE
Gaudette, Henri Eugene
Hines, Mark Edward
Phillips, Norman

NEW JERSEY
Borg, Sidney Fred
Brons, Cornelius Hendrick
Buckley, Glenn R
Dalton, Thomas Francis
Flannery, Brian Paul
Hordon, Robert M
Klein, George D
Pfafflin, James Reid
Pierce, George Edward
Stinson, Mary Krystyna
Strausberg, Sanford I
Weis, Peddrick

NEW MEXICO
Bachman, George O
Bild, Richard Wayne
Brainard, James R
Elston, Wolfgang Eugene
Grant, Philip R, Jr
Greiner, Norman Roy
Hillsman, Matthew Jerome
Kudo, Albert Masakiyo
Kyle, Philip R

NEW YORK
Beim, Howard
Boynton, John E
Breed, Helen Illick
Carter, Timothy Howard
Cole, Jonathan Jay
Conant, Francis Paine
Dodge, Cleveland John
Duggin, Michael J
Elahi, Nasik
Fox, William
Freudenthal, Hugo David
Graham, Jack Bennett
Gray, D Anthony
Heusser, Linda Olga
Jacobs, Stanley S
Jacobson, Melvin Joseph
Karig, Daniel Edmund
Kelly, John Russell
Koppelman, Lee Edward
Liguori, Vincent Robert
Loring, Arthur Paul
Monmonier, Mark
Muller, William A
Nottebohm, Fernando
Premuzic, Eugene T
Ram, Michael
Rampino, Michael Robert
Rao, Samohineeveesu Trivikrama
Scatterday, James Ware
Schnell, George Adam
Senus, Walter Joseph
Sheridan, Michael Francis
Siemankowski, Francis Theodore
Squyres, Steven Weldon
Swanson, Robert Lawrence
Tobach, Ethel

Tonelli, John P, Jr
Van Burkalow, Anastasia
Vandiver, Bradford B
Wallis, James Richard
Ward, Lawrence W(aterman)
Winter, John Henry
Wolff, Manfred Paul (Fred)
Wurster, Charles F

NORTH CAROLINA
Almy, Charles C, Jr
Aneja, Viney P
Bonaventura, Celia Jean
Campbell, Peter Hallock
Cheng, Pi-Wan
Eckerlin, Herbert Martin
Fisher, C(harles) Page, Jr
Forsythe, Randall D
Lindquist, David Gregory
Livingstone, Daniel Archibald
Nader, John S(haheen)
Rittschof, Daniel
Schaaf, William Edward
Watson, James E, Jr
Welby, Charles William

NORTH DAKOTA
Grainger, Cedric Anthony
Martin, DeWayne
Schwert, Donald Peters

OHIO
Amjad, Zahid
Bishop, Paul Leslie
Burford, Arthur Edgar
Carver, Eugene Arthur
Duval, Leonard A
Faure, Gunter
Ghosh, Sanjib Kumar
Gilloteaux, Jacques Jean-Marie A
Gonzalez, Paula
Hannibal, Joseph Timothy
Harris-Noel, Ann (F H) Graetsch
Jekeli, Christopher
Rice, William Abbott
Staker, Robert D
Taylor, Edith L
Uotila, Urho A(ntti Kalevi)
Wilmoth, Benton M
Wingard, Paul Sidney

OKLAHOMA
Boellstorff, John David
Dunlap, William Joe
Meyerhoff, Arthur Augustus

OREGON
Benedict, Ellen Maring
Fish, William
Huh, Chih-An
Kay, Michael Aaron
Lotspeich, Frederick Benjamin
Weber, Lavern J

PENNSYLVANIA
Arthur, Michael Allan
Bikerman, Michael
Booth, Bruce L
Carlson, Toby Nahum
Fritz, Lawrence William
Griffiths, John Cedric
Guinan, Edward F
Kline, Richard William
Kugelman, Irwin Jay
Kump, Lee Robert
Kurtz, David Allan
Long, Kenneth Maynard
Mutmansky, Jan Martin
Neste, Sherman Lester
Peck, Robert McCracken
Sabloff, Jeremy Arac
Simonson, John C
Smith, James Donaldson

RHODE ISLAND
Ho, Kay T
Larson, Roger
Oviatt, Candace Ann
Quinn, James Gerard
Sigurdsson, Haraldur
Silva, Armand Joseph
Webb, Thompson, III

SOUTH CAROLINA
Browdy, Craig Lawrence
Coull, Bruce Charles
Davis, Raymond F
Elzerman, Alan William
Hawkins, Richard Horace
Parker, Julian E, III
Zingmark, Richard G

TENNESSEE
Baer, Thomas Strickland
Berven, Barry Allen
Dreier, RaNaye Beth
McNutt, Charles Harrison
Sharp, Aaron John

TEXAS
Amsbury, David Leonard
Arnold, J Barto, III
Bass, George F
Black, David Charles

Bullard, Fred Mason
Campbell, Charles Edgar
Cepeda, Joseph Cherubini
Clark, Donald Ray, Jr
Colaco, Joseph P
Denley, David R
Dodge, Rebecca L
Haas, Herbert
Halpern, Martin
Hanson, Bernold M
Hay-Roe, Hugh
Kaufman, Sidney
Kimber, Clarissa Therese
Leblanc, Rufus Joseph
Ledley, Tamara Shapiro
Murray, Grover Elmer
Pearson, Daniel Bester, III
Potter, Andrew Elwin, Jr
Roper, Paul James
Schmandt, Jurgen
Shelby, T H
Standford, Geoffery Brian
Stotler, Raymond T
Taylor, Robert Dalton
Thill, Ronald E
Vail, Peter R

UTAH
Cramer, Harrison Emery
Freethey, Geoffrey Ward
Ghosh, Sambhunath
Haslem, William Joshua
Malek, Esmaiel
Sigler, John William

VERMONT
Stoiber, Richard Edwin
VanderMeer, Canute

VIRGINIA
Beal, Stuart Kirkham
Benton, Duane Marshall
Bernthal, Frederick Michael
Blum, Linda Kay
Calm, James M
Calver, James Lewis
Delnore, Victor Eli
Glover, Lynn, III
Haq, Bilal U
Huggett, Robert James
Krushensky, Richard D
Lacy, W(illiam) J(ohn)
Masters, Charles Day
Mose, Douglas George
Murden, William Roland, Jr
Myers, Charles Kern
Noble, Vincent Edward
Pontius, Steven Kent
Robbins, Eleanora Iberall
Spangenberg, Dorothy Breslin
Squibb, Robert E
Sutter, John Frederick
Theon, John Speridon
Tipper, Ronald Charles
Todd, Edward Payson
Tsai, Frank Y
Walker, Grayson Watkins
Weiss, Martin
Wright, Thomas L
Zimmerman, Peter David

WASHINGTON
Anderson, George Cameron
Fluharty, David Lincoln
Grootes, Pieter Meiert
Holmes, Mark Lawrence
Jenne, Everett A
Lipscomb, David M
Miller, C Dan
Porter, Stephen Cummings
Robkin, Maurice
Ross, Charles Alexander
Simenstad, Charles Arthur
Templeton, William Lees
Varanasi, Usha
Washburn, Albert Lincoln

WEST VIRGINIA
Henderson-Arzapalo, Anne Patricia
Torres, Adolfo M

WISCONSIN
Lasca, Norman P, Jr
Lillesand, Thomas Martin
Maher, Louis James, Jr
Moore, Gary T
Mortimer, Clifford Hiley
Rhyner, Charles R

WYOMING
Barlow, James A, Jr
Doerges, John E
Hayden-Wing, Larry Dean
Howell, Robert Richard
Steidtmann, James R

ALBERTA
Goodarzi, Fariborz
Johnson, Daniel Lloyd
Nasseri, Touraj
Rutter, Nathaniel Westlund

BRITISH COLUMBIA
Armstrong, John Edward

Marko, John Robert
Sutherland-Brown, Atholl

NEW BRUNSWICK
Lakshminarayana, J S S

NOVA SCOTIA
Bidwell, Roger Grafton Shelford
Buckley, Dale Eliot
Daborn, Graham Richard
King, Lewis H
Reynolds, Peter Herbert
Stevens, George Richard

ONTARIO
Bend, John Richard
Brose, David Stephen
Cale, William Robert
Csillag, Ferenc
Czuba, Margaret
Dawes, David Haddon
Dayal, Ramesh
Dence, Michael Robert
Dillon, Peter J
Donaldson, John Allan
Eade, Kenneth Edgar
Farquhar, Ronald McCunn
Gorman, William Alan
Hackett, Peter Andrew
Halliday, Ian
Hare, Frederick Kenneth
Jackson, Togwell Alexander
Jolly, Wayne Travis
Kerr, Peter Donald
Krantzberg, Gail
Kudo, Akira
McNutt, Robert Harold
Morgan, Alan Vivian
Percival, Jeanne Barbara
Poulton, Donald John
Ruel, Maurice M J
Selvadurai, A P S
Simpson, Frank
Sly, Peter G
Vollenweider, Richard A
Whelpdale, Douglas Murray
Williams, Peter J
Woo, Ming-Ko

QUEBEC
Belanger, Pierre Andre
Denis, Paul-Yves
Dubois, Jean-Marie M
Elson, John Albert
Hillaire-Marcel, Claude
Khalil, Michel
Prescott, Jacques
Sainte-Marie, Bernard

OTHER COUNTRIES
Fowler, Scott Wellington
Hofmann, Albrecht Werner
Hughes, Charles James

MATHEMATICS

Algebra

ALABAMA
Brackin, Eddy Joe
Gibson, Robert Wilder
Hou, Roger Hsiang-Dah
Kim, Ki Hang

ALASKA
Lando, Barbara Ann

ARIZONA
Clay, James Ray
Jacobowitz, Ronald
Moore, Nadine Hanson
Pierce, Richard Scott
Toubassi, Elias Hanna
Velez, William Yslas

ARKANSAS
Madison, Bernard L
Rossa, Robert Frank
Schein, Boris M

CALIFORNIA
Akasaki, Takeo
Baker, Kirby Alan
Bergman, George Mark
Blasius, Don Malcolm
Blattner, Robert J(ames)
Block, Richard Earl
Cannonito, Frank Benjamin
Carlson, David Hilding
Celik, Hasan Ali
Cutler, Doyle O
Eklof, Paul Christian
Estes, Dennis Ray
Fisher, William Bruce
Fraser, Grant Adam
Friedberg, Solomon
Gaposchkin, Peter John Arthur
Gaskell, Robert Weyand
Gerstein, Larry J
Givant, Steven Roger
Goodearl, Kenneth Ralph
Grigori, Artur
Guralnick, Robert Michael

Algebra (cont)

Hales, Alfred Washington
Henkin, Leon (Albert)
Hopponen, Jerry Dale
Horn, Alfred
Johnsen, Eugene Carlyle
Lam, Tsit-Yuen
Lanski, Charles Philip
Lawson, Charles L
Lelewer, Debra Ann
Luce, R(obert) Duncan
McKenzie, Ralph Nelson
Marcus, Marvin
Mena, Roberto Abraham
Merris, Russell Lloyd
Mignone, Robert Joseph
Minc, Henryk
Montgomery, M Susan
Oppenheim, Joseph Harold
Patenaude, Robert Alan
Phillips, Veril LeRoy
Pixley, Alden F
Pratt, Vaughan Ronald
Ralston, Elizabeth Wall
Ratliff, Louis Jackson, Jr
Rhodes, John Lewis
Rieffel, Marc A
Rodabaugh, David Joseph
Rosenberg, Alex
Small, Lance W
Smith, Velma Merriline
Smoke, William Henry
Sun, Hugo Sui-Hwan
Tang, Alfred Sho-Yu
Tang, Hwa-Tsang
Taussky, Olga
Teague, Tommy Kay
Tilson, Bret Ransom
Yaqub, Adil Mohamed
Zelmanowitz, Julius Martin

COLORADO
Briggs, William Egbert
Brown, Gordon Elliott
Derr, James Bernard
Lundgren, J Richard
Roth, Richard Lewis
Struik, Ruth Rebekka

CONNECTICUT
Gürsey, Feza
Hager, Anthony Wood
Howe, Roger
Hurley, James Frederick
Mostov, George Daniel
Nilson, Edwin Norman
Reid, James Dolan
Spiegel, Eugene

DELAWARE
Caviness, Bobby Forrester
Ebert, Gary Lee
Federighi, Enrico Thomas
Saunders, B David

DISTRICT OF COLUMBIA
Braunfeld, Peter George
Rosenblatt, David
Sjogren, Jon Arne
Teller, John Roger

FLORIDA
Brewer, James W
Clark, William Edwin
Gilmer, Robert
Hoffman, Frederick
Kreimer, Herbert Frederick, Jr
Magarian, Elizabeth Ann
Miller, Don Dalzell
Mott, Joe Leonard
Pedersen, John F
Timmer, Kathleen Mae
Wells, Jacqueline Gaye
Williams, Gareth
Yellen, Jay
Zerla, Fredric James

GEORGIA
Aust, Catherine Cowan
Belinfante, Johan G F
Bevis, Jean Harwell
Bompart, Billy Earl
Carlson, Jon Frederick
Falconer, Etta Zuber
Nail, Billy Ray
Neff, Mary Muskoff
Norwood, Gerald
Robinson, Daniel Alfred
Sharp, Thomas Joseph
Stone, David Ross
Straley, Tina
Watts, Sherrill Glenn

HAWAII
Nobusawa, Nobuo
Sagle, Arthur A

IDAHO
Higdem, Roger Leon

ILLINOIS
Bateman, Felice Davidson
Blair, William David
Blau, Harvey Isaac
Boyle, James Martin
Camburn, Marvin E
Clover, William John, Jr
Comerford, Leo P, Jr
Cutler, Janice Zemanek
Earnest, Andrew George
Evans, E Graham, Jr
Fossum, Robert Merle
Friedlander, Eric Mark
Glauberman, George Isaac
Haboush, William Joseph
Hsu, Nai-Chao
Janusz, Gerald Joseph
Kasube, Herbert Emil
Laible, Jon Morse
Lamont, Patrick
Larson, Richard Gustavus
Lazerson, Earl Edwin
Leibhardt, Edward
MacLane, Saunders
Otto, Albert Dean
Paley, Hiram
Parr, James Theodore
Price, David Thomas
Priddy, Stewart Beauregard
Reznick, Bruce Arie
Robinson, Derek John Scott
Stein, Michael Roger
Ullom, Stephen Virgil
Wagreich, Philip Donald
Wallis, Walter Denis
Walter, John Harris
Waterman, Peter Lewis
Weichsel, Paul M
Weiner, Louis Max
Wilcox, Lee Roy
Yau, Stephen Shing-Toung
Zelinsky, Daniel

INDIANA
Azumaya, Goro
Beehler, Jerry R
Drazin, Michael Peter
Drufenbrock, Diane
Hahn, Alexander J
Johns, Dennis Michael
Kronstein, Karl Martin
Kuczkowski, Joseph Edward
Machtinger, Lawrence Arnold
MacKenzie, Robert Earl
Murphy, Catherine Mary
Perlis, Sam
Phan, Kok-Wee
Pohl, Victoria Mary
Pollak, Barth
Shahidi, Freydoon
Sommese, Andrew John
Wong, Warren James

IOWA
Adelberg, Arnold M
Anderson, Daniel D
Barnes, Wilfred E
Dickson, Spencer E
Ferguson, Pamela A
Fuller, Kent Ralph
Hogben, Leslie
Johnson, Eugene W
Kleinfeld, Erwin
Peake, Edmund James, Jr
Pigozzi, Don

KANSAS
Davis, Elwyn H
Emerson, Marion Preston
McCarthy, Paul Joseph
Shult, Ernest E

KENTUCKY
Beidleman, James C
Eaves, James Clifton
Elder, Harvey Lynn
Enochs, Edgar Earle
Franke, Charles H
Gird, Steven Richard
Kearns, Thomas J
LeVan, Marijo O'Connor
Wallace, Kyle David

LOUISIANA
Benard, Mark
Cordes, Craig M
Fuchs, Laszlo
Heatherly, Henry Edward
Hildebrant, John A
McLean, R T
Ohm, Jack Elton
Spaht, Carlos G, II

MAINE
Johnson, Robert Wells
Lange, Gail Laura

MARYLAND
Adams, William Wells
Andre, Peter P
Cohen, Joel M
Douglas, Lloyd Evans
Draper, Richard Noel
Dribin, Daniel Maccabaeus
Ehrlich, Gertrude
Evans, David L
Goldhaber, Jacob Kopel
Good, Richard Albert
Helzer, Gerry A
Hutchinson, George Allen
Katz, Victor Joseph

MASSACHUSETTS
Altman, Allen Burchard
Auslander, Maurice
Bennett, Mary Katherine
Berkovits, Shimshon
Bernstein, Joseph N
Blackett, Donald Watson
Cattani, Eduardo
Cox, David Archibald
D'Alarcao, Hugo T
Dechene, Lucy Irene
Eisenbud, David
Fogarty, John Charde
Guterman, Martin Mayr
Hill, Victor Ernst, IV
Holland, Samuel S, Jr
Janowitz, Melvin Fiva
Joni, Saj-nicole A
Kass, Seymour
Kenney, Margaret June
Knighten, Robert Lee
Koshy, Thomas
Kundert, Esayas G
Levasseur, Kenneth M
Lusztig, George
McQuarrie, Bruce Cale
Malone, Joseph James
Martindale, Wallace S
Mattuck, Arthur Paul
Mauldon, James Grenfell
Morris, Robert
Newman, Kenneth Wilfred
Pollatsek, Harriet Suzanne
Reynolds, William Francis
Rudvalis, Arunas
Schafer, Richard Donald
Stanley, Richard Peter
Svanes, Torgny
Weiss, Edwin

MICHIGAN
Althoen, Steven Clark
Arlinghaus, William Charles
Feldman, Chester
Fink, John Bergeman
Griess, Robert Louis, Jr
Heuvers, Konrad John
Hochster, Melvin
Howard, Paul Edward
McLaughlin, Jack Enloe
Morash, Ronald
O'Neill, John Dacey
Petro, John William
Phillips, Richard E, Jr
Simpson, William Albert
Slaby, Harold Theodore
Tomber, Marvin L
Tsai, Chester E

MINNESOTA
Avelsgaard, Roger A
Dyer-Bennet, John
Fenrick, Maureen Helen
Friedman, Avner
Gaal, Ilse Lisl Novak
Gallian, Joseph A
Harris, Morton E
Jonsson, Bjarni
Pierce, Keith Robert
Roberts, Joel Laurence
Schue, John R
Sizer, Walter Scott
Wheaton, Burdette Carl

MISSISSIPPI
Fay, Temple Harold

MISSOURI
Bruening, James Theodore
Daly, John Francis
Huckaba, James Albert
Neuman, Rosalind Joyce
Riles, James Byrum
Spitznagel, Edward Lawrence, Jr
Stuth, Charles James
Tikoo, Mohan L
Yohe, Cleon Russell
Zemmer, Joseph Lawrence, Jr

NEBRASKA
Chouinard, Leo George, II
Larsen, Leland Malvern
Larsen, Max Dean
Leavitt, William Grenfell
Magliveras, Spyros Simos
Meakin, John C
Walters, Randall Keith
Wiegand, Sylvia Margaret

NEVADA
Shiue, Peter Jay-Shyong
Verma, Sadanand

NEW HAMPSHIRE
Bickel, Thomas Fulcher
Snapper, Ernst

NEW JERSEY
Cohn, Richard Moses
Faith, Carl Clifton
Faltings, Gerd
Foregger, Thomas H
Gaer, Marvin Charles
Gelfand, Israel M
Gewirtz, Allan
Gilman, Robert Hugh
Goldberg, Vladislav V
Gunning, Robert Clifford
Herlands, Charles William
Hughes, James Sinclair
Infante, Ronald Peter
Keigher, William Francis
Lepowsky, James Ivan
Malbrock, Jane C
O'Nan, Michael Ernest
Osofsky, Barbara Langer
Papantonopoulou, Aigli Helen
Plank, Donald Leroy
San Soucie, Robert Louis
Taft, Earl J
Wagner, Richard Carl
Weibel, Charles Alexander
Wilson, Robert Lee

NEW MEXICO
Drumm, Clifton Russell
Faber, Vance
Propes, Ernest (A)
Walker, Carol L
Wisner, Robert Joel

NEW YORK
Alex, Leo James
Barshay, Jacob
Baruch, Marjory Jean
Bencsath, Katalin A
Billera, Louis Joseph
Bing, Kurt
Blanton, John David
Catefloris, Vasily C
Childs, Lindsay Nathan
Clark, David M
Deleanu, Aristide Alexandru-Ion
Demerdash, Nabeel A O
Di Paola, Robert Arnold
Edwards, Harold M
Feingold, Alex Jay
Feuer, Richard Dennis
Geoghegan, Ross
Goodman, Jacob Eli
Grabois, Neil
Harrison, Nancy Evelyn King
Hoisie, Adolfy
Hoobler, Raymond Taylor
Kaltofen, Erich L
Kleiman, Howard
Knapp, Anthony William
Knee, David Isaac
Kohls, Carl William
Kolchin, Ellis Robert
Lawvere, Francis William
Leary, Francis Christian
Lewis, L Gaunce, Jr
McCarthy, Donald John
Pinkham, Henry Charles
Saracino, Daniel Harrison
Seppala-Holtzman, David N
Sit, William Yu
Spiro, Claudia Alison
Tamari, Dov
Ventriglia, Anthony E
Wagner, Eric G
Watts, Charles Edward

NORTH CAROLINA
Bernhardt, Robert L, III
Blanchet-Sadri, Francine
Byrd, Kenneth Alfred
Geissinger, Ladnor Dale
Graves, William Howard
Greim, Barbara Ann
Hartwig, Robert Eduard
Jambor, Paul Emil
Kumar, Shrawan
Latch, Dana May
Luh, Jiang
Mewborn, Ancel Clyde
Putcha, Mohan S
Schlessinger, Michael
Smith, James Reaves
Smith, William Walker
Sowell, Katye Marie Oliver
Stitzinger, Ernest Lester
Vaughan, Theresa Phillips
Warner, Seth L
Wilson, Jack Charles

OHIO
Buchthal, David C
Cox, Milton D
Ferrar, Joseph C
Gagola, Stephen Michael, Jr
Glass, Andrew Martin William
Gregory, Thomas Bradford
Holland, Wilbur Charles
Holyoke, Thomas Campbell
Hsia, John S
Hungerford, Thomas W
Jain, Surender K
Khalimsky, Efim D
Lee, Dong Hoon
Lindstrom, Wendell Don
McCleary, Stephen Hill

Rutter, Edgar A, Jr
Satyanarayana, Motupalli
Sehgal, Surinder K
Shock, Robert Charles
Sholander, Marlow
Silverman, Robert
Stander, Joseph W
Steinberg, Stuart Alvin
Sterling, Leon Samuel
Stickney, Alan Craig
Vancko, Robert Michael
Wang, Paul Shyh-Horng
Wells, Charles Frederick
Wyman, Bostwick Frampton
Zassenhaus, Hans J

OKLAHOMA
Gunter, Deborah Ann
McCullough, Darryl J

OREGON
Alin, John Suemper
Byrne, John Richard
Enneking, Marjorie
Fein, Burton Ira
Wright, Charles R B
Yuzvinsky, Sergey

PENNSYLVANIA
Arrington, Elayne
Beck, Robert Edward
Brooks, James O
Cable, Charles Allen
Cassel, David Wayne
Chein, Orin Nathaniel
Coughlin, Raymond Francis
Cunningham, Joel L
Deskins, Wilbur Eugene
Formanek, Edward William
Furst, Merrick Lee
Gerstenhaber, Murray
Harbater, David
James, Donald Gordon
Klotz, Eugene Arthur
Kushner, Harvey
Pirnot, Thomas Leonard
Schattschneider, Doris Jean
Shuck, John Winfield
Tzeng, Kenneth Kai-Ming
Vaserstein, Leonid, IV
Ware, Roger Perry
Waterhouse, William Charles
Wong, Roman Woon-Ching

RHODE ISLAND
Beauregard, Raymond A
Casey, John Edward, Jr
Harris, Bruno

SOUTH CAROLINA
Brawley, Joel Vincent, Jr
Comer, Stephen Daniel
Edwards, Constance Carver
Gilbert, Jimmie D
LaTorre, Donald Rutledge
Luedeman, John Keith
McNulty, George Frank
Markham, Thomas Lowell

TENNESSEE
Dixon, Edmond Dale
Dobbs, David Earl
Lea, James Wesley, Jr
Mazeres, Reginald Merle
Myers, Carrol Bruce
Suh, Tae-il

TEXAS
Armendariz, Efraim Pacillas
Berberian, Sterling Khazag
Borm, Alfred Ervin
Brooks, Dorothy Lynn
Brown, Dennison Robert
Coppage, William Eugene
Cude, Joe E
Doran, Robert Stuart
Durbin, John Riley
Fincher, Bobby Lee
Friesen, Donald Kent
Geller, Susan Carol
Gerth, Frank E, III
Gustafson, William Howard
Hall, Carl Eldridge
Hart, R Neal
Johnson, Johnny Albert
McCown, Malcolm G
Maxson, Carlton J
Moore, Marion E
O'Malley, Matthew Joseph
Piziak, Robert
Salingaros, Nikos Angelos
Saltman, David J
Shieh, Leang-San
Smith, Martha Kathleen
Stiller, Peter Frederick
Tarwater, Jan Dalton
Vaughan, Nick Hampton
Vick, George R
Weaver, Milo Wesley

UTAH
Carlson, James Andrew
Gersten, Stephen M
Gross, Fletcher

Milicic, Dragan
Moore, Hal G
Robinson, Donald Wilford
Scott, William Raymond

VERMONT
Emerson, John David
Foote, Richard Martin
Wiitala, Stephen Allen

VIRGINIA
Arnberg, Robert Lewis
Arnold, Jimmy Thomas
Burian, Richard M
Crittenden, Rebecca Slover
Deveney, James Kevin
Drew, John H
Ethington, Donald Travis
Green, Edward Lewis
Greville, Thomas Nall Eden
Johnson, Charles Royal
Jones, Eleanor Green Dawley
Kuhn, Nicholas J
Lipman, Marc Joseph
McNeil, Phillip Eugene
Meyrowitz, Alan Lester
Montague, Stephen
Neeper, Ralph Arnold
Paige, Eugene
Parry, Charles J
Parshall, Brian J
Schafer, Alice Turner
Scott, Leonard Lewy, Jr
Smith, Emma Breedlove
Thaler, Alvin Isaac
Ward, Harold Nathaniel
Wei, Diana Yun Dee (Fan)

WASHINGTON
Cunnea, William M
Dalla, Ronald Harold
Dimitroff, George Ernest
Dubois, Donald Ward
Kallaher, Michael Joseph
MacLaren, Malcolm Donald
Mills, Janet E
Sullivan, John Brendan
Underwood, Douglas Haines

WEST VIRGINIA
Cavalier, John F
Hogan, John Wesley
Kim, Jin Bai

WISCONSIN
Bell, James Henry
Fossum, Timothy V
Fournelle, Thomas Albert
Isaacs, I(rving) Martin
Levy, Lawrence S
Osborn, J Marshall, Jr
Passman, Donald Steven
Teply, Mark Lawrence
Terwilliger, Paul M
Weston, Kenneth W

WYOMING
Bridges, William G
Gastl, George Clifford
Porter, A Duane

PUERTO RICO
Moreno, Oscar
Peinado, Rolando E

ALBERTA
Bercov, Ronald David
Hoo, Cheong Seng
Mollin, Richard Anthony
Rhemtulla, Akbar Hussein

BRITISH COLUMBIA
Alspach, Brian Roger
MacDonald, John Lauchlin
Mekler, Alan
Moyls, Benjamin Nelson
Page, Stanley Stephen
Reilly, Norman Raymund
Westwick, Roy

MANITOBA
Gratzer, George
Losey, Gerald Otis
Mendelsohn, Nathan Saul

NOVA SCOTIA
Grant, Douglass Lloyd
Kruse, Robert Leroy

ONTARIO
Arthur, James Greig
Bell, Howard E
Choi, Man-Duen
Coxeter, Harold Scott MacDonald
Cummings, Larry Jean
Davis, Chandler
Davis, Harry Floyd, II
Dixon, John Douglas
Djokovic, Dragomir Z
Dlab, Vlastimil
Elliott, George Arthur
Fleischer, Isidore
Mueller, Bruno J W
Nel, Louis Daniel

Orzech, Morris
Poland, John C
Racine, Michel Louis
Ribenboim, Paulo
Ribes, Luis
Riehm, Carl Richard
Robinson, Gilbert de Beauregard
Shklov, Nathan
Taylor, David James

QUEBEC
Barr, Michael
Kilambi, Srinvasacharyulu
Lin, Paul C S
Sabidussi, Gert Otto
Schlomiuk, Dana

OTHER COUNTRIES
Collins, George Edwin
Horwitz, Lawrence Paul
Hughes, Daniel Richard
Kambayashi, Tatsuji
Myerson, Gerald
Smith, Harry Francis
Thomason, Robert Wayne

Analysis & Functional Analysis

ALABAMA
Govil, Narendra Kumar
Howell, James Levert
Klip, Dorothea A
Kurtz, Lawrence Alfred
Moore, Robert Laurens
Plunkett, Robert Lee
Wang, James Li-Ming

ARIZONA
Barckett, Joseph Anthony
Barnhill, Robert E
DeVito, Carl Louis
Faris, William Guignard
Feldstein, Alan
Garder, Arthur
Hughes, Eugene Morgan
Laetsch, Theodore Willis
Leonard, John Lander
Reagan, John Albert
Smith, Harvey Alvin

ARKANSAS
Annulis, John T
Cochran, Allan Chester
Feldman, William A
Mullins, Charles Warren
Summers, William Hunley

CALIFORNIA
Amelin, Charles Francis
Anderson, Clifford Harold
Antosiewicz, Henry Albert
Arens, Richard Friederich
Banerjee, Utpal
Berg, Paul Walter
Blattner, Robert J(ames)
Bruckner, Andrew M
Caflisch, Russel Edward
Carson, George Stephen
Chang, Sun-Yung Alice
Cherlin, George Yale
Chernoff, Paul Robert
Coleman, Courtney (Stafford)
Crandall, Michael Grain
Curtis, Philip Chadsey, Jr
DePrima, Charles Raymond
Dickson, Lawrence John
Di Franco, Roland B
Dixon, Michael John
Donoghue, William F, Jr
Donoho, David Leigh
Drobot, Vladimir
Elderkin, Richard Howard
Fattorini, Hector Osvaldo
Feinstein, Charles David
Feldman, Jacob
Ferguson, Le Baron O
Finn, Robert
Gamelin, Theodore W
Genensky, Samuel Milton
Gilbert, Richard Carl
Goodearl, Kenneth Ralph
Grabiner, Sandy
Greene, Robert Everist
Greenhall, Charles August
Gretsky, Neil E
Gruenberger, Fred J(oseph)
Halmos, Paul Richard
Helson, Henry
Helton, J William
Hench, David Le Roy
Herriot, John George
Hirsch, Morris William
Huddleston, Robert E
Hunt, Robert Weldon
James, Ralph L
James, Robert Clarke
Jewell, Nicholas Patrick
Juberg, Richard Kent
Keller, Herbert Bishop
Krabbe, Gregers Louis
Kurowski, Gary John
Lai, Kai Sun
Langford, Eric Siddon

Lapidus, Michel Laurent
Lashley, Gerald Ernest
Lesley, Frank David
Li, Hung Chiang
Li, Peter Wai-Kwong
Lippman, Gary Edwin
Liu, Tai-Ping
Lu, Kau U
Luxemburg, Wilhelmus Anthonius Josephus
McCarthy, Mary Anne
Mignone, Robert Joseph
Miles, Frank Belsley
Nur, Hussain Sayid
Oliger, Joseph Emmert
Osserman, Robert
Pinney, Edmund Joy
Radbill, John R(ussell)
Radnitz, Alan
Rao, Malempati Madhusudana
Reich, Simeon
Rieffel, Marc A
Riggle, Everett C
Rodabaugh, David Joseph
Rosen, Irwin Gary
Royden, Halsey Lawrence
Satyanarayana, Upadhyayula V
Schechter, Martin
Schiffer, Menahem Max
Shapiro, Victor Lenard
Shiffman, Max
Simon, Barry Martin
Simons, Stephen
Smith, Marianne (Ruth) Freundlich
Thacher, Henry Clarke, Jr
Todd, John
Vasco, Donald Wyman
Verona, Andrei
Voiculescu, Dan-Virgil
Ware, Buck
Watson, Velvin Richard
Weiss, Max Leslie
Widom, Harold
Williams, Ruth Jeannette
Wulbert, Daniel Eliot

COLORADO
Bleistein, Norman
Brown, Edmund Hosmer
Buzbee, Billy Lewis
Daly, James Edward
Darst, Richard B
Gibson, William Loane
Gill, John Paul, Jr
Hagin, Frank G
Hagler, James Neil
Hundhausen, Joan Rohrer
Jones, William B
Manteuffel, Thomas Albert
Phillips, Keith L
Ramsay, Arlan (Bruce)
Shank, George Deane
Thaler, Eric Ronald
Thron, Wolfgang Joseph
Walter, Martin Edward

CONNECTICUT
Abikoff, William
Ahlberg, John Harold
Beals, Richard William
Blei, Ron Charles
Gine-Masdeu, Evarist
Gosselin, Richard Pettengill
Howe, Roger
Leibowitz, Gerald Martin
Robbins, David Alvin
Schlesinger, Ernest Carl
Sidney, Stuart Jay

DELAWARE
Libera, Richard Joseph
Livingston, Albert Edward
Weinacht, Richard Jay

DISTRICT OF COLUMBIA
Chang, I-Lok
Long, Paul Eastwood, Jr
Robinson, E Arthur, Jr

FLORIDA
Bellenot, Steven F
Cenzer, Douglas
Goodman, Richard Henry
Goodner, Dwight Benjamin
Isaacs, Godfrey Leonard
Ismail, Mourad E H
McArthur, Charles Wilson
McWilliams, Ralph David
Minzner, Raymond Arthur
Mukherjea, Arunava
Nagle, Richard Kent
Oberlin, Daniel Malcolm
Palmer, Kenneth
Ratti, Jogindar Singh
Rubin, Richard Lee
Saff, Edward Barry
Squier, Donald Platte
Young, Eutiquio Chua

GEORGIA
Azoff, Edward Arthur
Batterson, Steven L
Bouldin, Richard H
Cain, George Lee, Jr

412 / DISCIPLINE INDEX

Analysis & Functional Analysis (cont)

Cash, Dewey Byron
Clark, Douglas Napier
Green, William Lohr
Harrell, Evans Malott, II
Herod, James Victor
Komkov, Vadim
Landgren, John Jeffrey
Meyer, Gunter Hubert
Miller, Paul George
Osborn, James Maxwell
Riddle, Lawrence H
Ripy, Sara Louise
Shabazz, Abdulalim A
Waltman, Paul Elvis
Worth, Roy Eugene
Youse, Bevan K

HAWAII
Bear, Herbert S, Jr
Iha, Franklin Takashi

IDAHO
Bobisud, Larry Eugene
Hausrath, Alan Richard
McClure, John Arthur

ILLINOIS
Albrecht, Felix Robert
Bank, Steven Barry
Barron, Emmanuel Nicholas
Bellow, Alexandra
Berkson, Earl Robert
Berndt, Bruce Carl
Buntinas, Martin George
Burkholder, Donald Lyman
Calderon, Calixto Pedro
Carroll, Robert Wayne
Day, Mahlon Marsh
Diamond, Harold George
Driscoll, Richard James
Feinstein, Irwin K
Friedberg, Stephen Howard
Gasper, George, Jr
Ho, Chung-Wu
Ionescu Tulcea, Cassius
Jerome, Joseph Walter
Jones, Roger L
Kammler, David W
Lubin, Arthur Richard
McAlpin, John Harris
Packel, Edward Wesler
Peck, Emily Mann
Peck, Newton Tenney
Peressini, Anthony L
Philipp, Walter V
Redmond, Donald Michael
Reznick, Bruce Arie
Rickert, Neil William
Robinson, Clark
Sherbert, Donald R
Sons, Linda Ruth
Waterman, Peter Lewis
Wenger, John C
Wetzel, John Edwin
Wheeler, Robert Francis
Yau, Stephen Shing-Toung
Zettl, Anton

INDIANA
Aliprantis, Charalambos Dionisios
Brothers, John Edwin
Chihara, Theodore Seio
Conte, Samuel D
Cowen, Carl Claudius, Jr
Drasin, David
Gautschi, Walter
Golomb, Michael
Jerison, Meyer
Kallman, Ralph Arthur
Kaminker, Jerome Alvin
Legg, David Alan
Lyons, Russell David
McKinney, Earl H
Penney, Richard Cole
Rhoades, Billy Eugene
Rothman, Neal Jules
Schober, Glenn E
Shahidi, Freydoon
Sommese, Andrew John
Specht, Edward John
Stampfli, Joseph
Stanton, Charles Madison
Stoll, Wilhelm
Swift, William Clement
Synowiec, John A
Torchinsky, Alberto
Zachmanoglou, Eleftherios Charalambos
Zink, Robert Edwin

IOWA
Alexander, Roger Keith
Carlson, Bille Chandler
Dahiya, Rajbir Singh
Jorgensen, Palle E T
Peters, Justin
Ton-That, Tuong

KANSAS
Acker, Andrew French, III
Bajaj, Prem Nath

Brown, Robert Dillon
Burckel, Robert Bruce
Calys, Emanuel G
Eltze, Ervin Marvin
Himmelberg, Charles John
Linscheid, Harold Wilbert
Nicholls, Peter John
Paschke, William Lindall
Ramm, Alexander G
Van Vleck, Fred Scott

KENTUCKY
Buckholtz, James Donnell
Howard, Henry Cobourn
King, Amy P
Perry, Peter Anton
Porter, James Franklin
Van Winter, Clasine

LOUISIANA
Agarwal, Arun Kumar
Anders, Edward B
Chan, Chiu Yeung
Conway, Edward Daire, III
Denny, William F
Dorroh, James Robert
Kalka, Morris
Knill, Ronald John
Kuo, Hui-Hsiung
Kythe, Prem Kishore
Lefton, Lew Edward
Liukkonen, John Robie
McGehee, Oscar Carruth
McKinney, Alfred Lee
Richardson, Leonard Frederick
Roques, Alban Joseph
Swetharanyam, Lalitha

MAINE
Balakrishnan, V K
Brace, John Wells
Elliott, H(elen) Margaret
Todaro, Michael P

MARYLAND
Alexander, James Crew
Antman, Stuart S
Babuska, Ivo Milan
Benedetto, John Joseph
Berger, Alan Eric
Calinger, Ronald Steve
Cawley, Robert
Cohen, Edgar A, Jr
Cohen, Joel M
Cook, Clarence Harlan
Diesen, Carl Edwin
Douglas, Lloyd Evans
Douglis, Avron
Ellis, Robert L
Goldberg, Seymour
Heins, Maurice Haskell
Horvath, John Michael
Hubbard, Bertie Earl
Hummel, James Alexander
Huneycutt, James Ernest, Jr
Johnson, David Lee
Johnson, Raymond Lewis
Kirwan, William English
Kleppner, Adam
Krantz, Steven George
Lay, David Clark
Lepson, Benjamin
Lozier, Daniel William
Mahar, Thomas J
Manley, Oscar Paul
May, Everette Lee, Jr
Neri, Umberto
Newcomb, Robert Wayne
Raphael, Louise Arakelian
Rosenberg, Jonathan Micah
Saworotnow, Parfeny Pavolich
Shiffman, Bernard
Stephens, Arthur Brooke
Sullivan, Francis E
Wallace, Alton Smith
Warner, Charles Robert
Witzgall, Christoph Johann
Zabronsky, Herman
Zame, William Robin
Zedek, Mishael

MASSACHUSETTS
Bailey, Duane W
Ben-Akiva, Moshe E
Berkey, Dennis Dale
Berkovits, Shimshon
Bernstein, Joseph N
Bilodeau, Gerald Gustave
Breton, J Raymond
Cole, Nancy
Cook, Thurlow Adrean
Cormack, Allan MacLeod
Dudley, Richard Mansfield
Dyer, James Arthur
Filgo, Holland Cleveland, Jr
Fountain, Leonard Du Bois
Giger, Adolf J
Grossberg, Stephen
Helgason, Sigurdur
Holland, Samuel S, Jr
Holmes, Richard Bruce
Horowitz, Larry Lowell
Kamowitz, Herbert M
Kon, Mark A

Lam, Kui Chuen
Melrose, Richard B
Parrott, Stephen Kinsley
Quinto, Eric Todd
Ruskai, Mary Beth
St Mary, Donald Frank
Seeley, Robert T
Sherman, Thomas Oakley
Shuchat, Alan Howard
Siu, Yum-Tong
Stone, Dorothy Maharam
Warga, Jack

MICHIGAN
Anderson, Glen Douglas
Babbitt, Donald George
Bachelis, Gregory Frank
Barnard, Robert D(ane), Jr
Bartle, Robert Gardner
Cesari, Lamberto
Dickson, Douglas Grassel
Duren, Peter Larkin
Fast, Henryk
Feldman, Chester
Fleming, Richard J
Fornaess, John Erik
Hansen, Lowell John
Herzog, Fritz
Howard, Paul Edward
Johnson, George Philip
Lick, Don R
Loebl, Richard Ira
McLaughlin, Renate
Marik, Jan
Miklavcic, Milan
Mordukhovich, Boris S
Nelson, James Donald
Nyman, Melvin Andrew
Piranian, George
Reade, Maxwell Ossian
Salehi, Habib
Simpson, William Albert
Skaff, Michael Samuel
Stoline, Michael Ross
Stout, Quentin Fielden
Vinh, Nguyen Xuan
Weil, Clifford Edward

MINNESOTA
Fabes, Eugene Barry
Gil de Lamadrid, Jesus
Gulliver, Robert David, II
Hejhal, Dennis Arnold
Hilding, Stephen R
Humke, Paul Daniel
Jodeit, Max A, Jr
Keynes, Harvey Bayard
Littman, Walter
McCarthy, Charles Alan
Mericle, R Bruce
Munro, William Delmar
Nitsche, Johannes Carl Christian
Olver, Peter John
Pour-El, Marian Boykan
Vessey, Theodore Alan

MISSISSIPPI
Causey, William McLain
Doblin, Stephen Alan
Mastin, Charles Wayne
Solomon, Jimmy Lloyd
Spikes, Paul Wenton

MISSOURI
Andalafte, Edward Ziegler
Baernstein, Albert, II
Connett, William C
Crownover, Richard McCranie
Gillilan, James Horace
Haimo, Deborah Tepper
Hall, Leon Morris, Jr
Harris, Beverly Howard
Nussbaum, Edward
Ornstein, Wilhelm
Peterson, George E
Pursell, Lyle Eugene
Rigler, A Kellam
Rochberg, Richard Howard
Spielman, Barry
Taibleson, Mitchell H
Tikoo, Mohan L
Utz, Winfield Roy, Jr
Wilson, Edward Nathan

MONTANA
Barrett, Louis Carl
Derrick, William Richard
Jameson, William J, Jr
Wend, David Van Vranken
Yale, Irl Keith

NEBRASKA
Fuller, Derek Joseph Haggard
Jorgensen, T, Jr
Maloney, John P
Skoug, David L

NEVADA
Gupta, Chaitan Prakash
Tompson, Robert Norman

NEW HAMPSHIRE
Lahr, Charles Dwight
Lamperti, John Williams

NEW JERSEY
Andrushkiw, Roman Ihor
Badalamenti, Anthony Francis
Beckenstein, Edward
Bidwell, Leonard Nathan
Blackmore, Denis Louis
Browder, Felix Earl
Byrne, George D
Caffarelli, Luis Angel
Chai, Winchung A
Cheung, Shiu Ming
Christodoulou, Demetrios
Coleman, Bernard David
Corwin, Lawrence Jay
Fefferman, Charles Louis
Gaer, Marvin Charles
Goldberg, Vladislav V
Goodman, Roe William
Grimmell, William C
Gunning, Robert Clifford
Johnson, Robert Shepard
Kim, Moon W
Lieb, Elliott Hershel
Majda, Andrew J
Makar, Boshra Halim
Malbrock, Jane C
Muckenhoupt, Benjamin
Nussbaum, Roger David
Poiani, Eileen Louise
Saccoman, John Joseph
Schmidlin, Albertus Ernest
Steiner, Gilbert
Sverdlove, Ronald
Taylor, Jean Ellen
Walsh, Bertram (John)
Weiss, Alan
Wheeden, Richard Lee

NEW MEXICO
Arterburn, David Roe
Boland, W Robert
Campbell, Katherine Smith
Dendy, Joel Eugene, Jr
Gilfeather, Frank L
Hersh, Reuben
Hyman, James Macklin
Kist, Joseph Edmund
McGehee, Ralph Marshall
Sanderson, James George
Smith, Brian Thomas
Swartz, Blair Kinch
Zemach, Charles

NEW YORK
Adhout, Shahla Marvizi
Barth, Karl Frederick
Berresford, Geoffrey Case
Blanton, John David
Boyce, William Edward
Bramble, James H
Cargo, Gerald Thomas
Chan, Jack-Kang
Chou, Ching
Church, Philip Throop
Coburn, Lewis Alan
Dodziuk, Jozef
Earle, Clifford John, Jr
Eberlein, Patricia James
Ginsberg, Jonathan I
Goodman, Nicolas Daniels
Gordon, Hugh
Greenleaf, Frederick P
Hartnett, William Edward
Hilbert, Stephen Russell
Hochstadt, Harry
Isaacson, David
Iwaniec, Tadeusz
Jenkins, Joe Wiley
Johnson, Guy, Jr
Kent, James Ronald Fraser
Kim, Woo Jong
Knapp, Anthony William
Kranzer, Herbert C
Kraus, Jon Eric
Lardy, Lawrence James
Larsen, Ronald John
Lavine, Richard Bengt
Lindberg, John Albert, Jr
MacGregor, Thomas Harold
Machover, Maurice
McLaughlin, Harry Wright
McLaughlin, Joyce Rogers
Miller, Sanford Stuart
Mitchell, Josephine Margaret
Moritz, Roger Homer
Mosak, Richard David
Narici, Lawrence Robert
Nirenberg, Louis
Park, Samuel
Pasciak, Joseph Edward
Phong, Duong Hong
Piech, Margaret Ann
Raimi, Ralph Alexis
Range, R Michael
Rosenfeld, Norman Samuel
Roytburd, Victor
Sherman, Malcolm J
Smith, James F
Strichartz, Robert Stephen
Todd, Aaron Rodwell
Trasher, Donald Watson
Waterman, Daniel
Weill, Georges Gustave
Weiss, Norman Jay

MATHEMATICS / 413

Zielezny, Zbigniew Henryk
Zukowski, Charles Albert

NORTH CAROLINA
Campbell, Stephen La Vern
Carmichael, Richard Dudley
Cato, Benjamin Ralph, Jr
Dankel, Thaddeus George, Jr
Dettman, John Warren
Dunn, Joseph Charles
Embry-Wardrop, Mary Rodriguez
Faulkner, Gary Doyle
Graves, William Howard
Kerzman, Norberto Luis Maria
Klein, Benjamin Garrett
Lambert, Alan L
Neidinger, Richard Dean
Petersen, Karl Endel
Pfaltzgraff, John Andrew
Sagan, Hans
Singh, Kanhaya Lal
Weinstock, Barnet Mordecai
Wogen, Warren Ronald

NORTH DAKOTA
Gregory, Michael Baird

OHIO
Andrews, George Harold
Baker, John Warren
Bojanic, Ranko
Buoni, John J
Carroll, Francis W
Carson, Robert Cleland
Cavaretta, Alfred S
Chalkley, Roger
Chao, Jia-Arng
Dombrowski, Joanne Marie
Dumas, H Scott
Faires, John Douglas
Fobes, Melcher Prince
Foreman, Matthew D
Fridy, John Albert
Groetsch, Charles William
Hart, David Charles
Horner, James M
Juang, Ling Ling
Laatsch, Richard G
Laha, Radha Govinda
Lee, Dong Hoon
McLeod, Robert Melvin
Merkes, Edward Peter
Mills, Wendell Holmes, Jr
Minda, Carl David
Mityagin, Boris Samuel
Nevai, Paul
Parson, Louise Alayne
Piotrowski, Zbigniew
Prentice, Wilbert Neil
Shah, Swarupchand Mohanlal
Smith, Mark Andrew
Steinlage, Ralph Cletus
Ward, Douglas Eric
Zimering, Shimshon

OKLAHOMA
Cassens, Patrick
Choike, James Richard
Eliason, Stanley B
Engle, Jessie Ann
Goodey, Paul Ronald
Gunter, Deborah Ann
Huff, William Nathan
Keener, Marvin Stanford
Owens, G(lenda) Kay
Rakestraw, Roy Martin

OREGON
Anselone, Philip Marshall
Bhattacharyya, Ramendra Kumar
Dell, Roger Marcus
Gilkey, Peter Belden
Parks, Harold Raymond
Petersen, Bent Edvard
Phillips, N Christopher
Solmon, Donald Clyde
Tam, Kwok-Wai

PENNSYLVANIA
Abu-Shumays, Ibrahim Khalil
Bollinger, Richard Coleman
Bona, Jerry Lloyd
Bressoud, David Marius
Caldwell, Christopher Sterling
Conrad, Bruce
Deutsch, Frank Richard
Erdelyi, Ivan Nicholas
Faires, Barbara Trader
Goldstine, Herman Heine
Grinberg, Eric L
Grosswald, Emil
Haber, Seymour
Hahn, Kyong Taik
Hartmann, Frederick W
Holder, Leonard Irvin
Kadison, Richard Vincent
Kanwal, Ram Prakash
Kazdan, Jerry Lawrence
King, Jerry Porter
Knopp, Marvin Isadore
Kolodner, Ignace I(zaak)
Lambert, Joseph Michael
Merrill, Samuel, III
Mitchell, Theodore

Navarro-Bermudez, Francisco Jose
Nicolaides, R A
Passow, Eli (Aaron)
Patton, Peter C(lyde)
Rung, Donald Charles, Jr
Sardinas, August A
Schaffer, Juan Jorge
Schild, Albert
Schweinsberg, Allen Ross
Scranton, Bruce Edward
Sours, Richard Eugene
Sturdevant, Eugene J
Sussman, Myron Maurice
Transue, William Reagle
Yood, Bertram
Yukich, Joseph E

RHODE ISLAND
Cole, Brian J
Davis, Philip J
Driver, Rodney D
Federer, Herbert
Ladas, Gerasimos
Sine, Robert C
Strauss, Walter A
Woolf, William Blauvelt

SOUTH CAROLINA
Bennett, Colin
Cleaver, Charles E
Jamison, Robert Edward
Kostreva, Michael Martin
Reneke, James Allen
Ruckle, William Henry
Sharpley, Robert Caldwell
Stoll, Manfred
Warner, Daniel Douglas

TENNESSEE
Austin, Billy Ray
Britton, Otha Leon
Chitwood, Howard
Conway, John Bligh
Cook, James Marion
Hinton, Don Barker
Jain, Mahendra Kumar
Jordan, George Samuel
Kiech, Earl Lockett
Puckette, Stephen Elliott
Rozema, Edward Ralph
Schaefer, Philip William
Soni, Kusum
Ward, Robert Cleveland
Webb, Glenn Francis

TEXAS
Alo, Richard Anthony
Anderson, Larry A
Atchison, Thomas Andrew
Bakelman, Ilya J(acob)
Barnard, Roger W
Berberian, Sterling Khazag
Bernau, Simon J
Bilyeu, Russell Gene
Brown, Dennison Robert
Cheney, Elliott Ward, Jr
Clark, William Dean
Coppin, Charles Arthur
Corduneanu, Constantin C
Cox, Dennis Dean
Daniel, James Wilson
Darwin, James T, Jr
Dawson, David Fleming
Deeter, Charles Raymond
Dodson, David Scott
Doran, Robert Stuart
Franco, Zachary Martin
Fulks, Watson
Fulling, Stephen Albert
Heath, Larry Francis
Hickey, Jimmy Ray
Huggins, Frank Norris
Johnson, William Buhmann
Justice, James Horace
Kay, Alvin John
Kimes, Thomas Fredric
Klipple, Edmund Chester
Lorentz, George G
Mauldin, Richard Daniel
Meyer, Harold David
Narcowich, Francis Joseph
Naugle, Norman Wakefield
Ober, Raimund Johannes
Parrish, Herbert Charles
Pearcy, Carl Mark, Jr
Perry, William Leon
Polking, John C
Renka, Robert Joseph
Roach, Francis Aubra
Robinson, Charles Dee
Sanders, Bobby Lee
Smith, Philip Wesley
Srnka, Leonard James
Strauss, Monty Joseph
Taliaferro, Steven Douglas
Tapia, Richard
Tidmore, Eugene F
Vaaler, Jeffrey David
Vastano, Andrew Charles
Walton, Jay R
Ward, Joseph J
Wheeler, Mary Fanett

UTAH
Bates, Peter William
Campbell, Douglas Michael
Carlson, James Andrew
Gustafson, Grant Bernard
Lewis, Lawrence Guy
McDonald, Keith Leon
Milicic, Dragan
Miller, Richard Roy
Skarda, R Vencil, Jr
Snow, Donald Ray
Taylor, Joseph Lawrence
Wilcox, Calvin Hayden

VERMONT
Gross, Kenneth Irwin
Martin, Robert Paul
Oughstun, Kurt Edmund

VIRGINIA
Arnberg, Robert Lewis
Ball, Joseph Anthony
Bartelt, Martin William
Bolstein, Arnold Richard
Chambers, Charles MacKay
Chartres, Bruce A
Collier, Manning Gary
Deal, Albert Leonard, III
Dozier, Lewis Bryant
Dunkl, Charles Francis
Edyvane, John
Elkins, Judith Molinar
Ethington, Donald Travis
Hannsgen, Kenneth Bruce
Howland, James Secord
Jones, Eleanor Green Dawley
Kripke, Bernard Robert
Lasiecka, Irena
Lucey, Juliana Margaret
Lutzer, David John
Martin, Nathaniel Frizzel Grafton
Olin, Robert
Ortega, James M
Rehm, Allan Stanley
Renardy, Michael
Rosenblum, Marvin
Ryff, John V
Scanlon, Charles Harris
Smith, James Breedlove
Summerville, Richard Marion
Swetits, John Joseph
Wells, Benjamin B, Jr
Wheeler, Robert Lee
Wood, James Alan

WASHINGTON
Boas, Ralph Philip, Jr
Burkhart, Richard Henry
Chaney, Robin W
DeTemple, Duane William
Firkins, John Frederick
Folland, Gerald Budge
Hewitt, Edwin
Horner, Donald Ray
Hsu, Chin Shung
Johnson, Roy Andrew
Lowell, Sherman Cabot
MacKichan, Barry Bruce
Marshall, Donald E
Newton, Tyre Alexander
Nievergelt, Yves
Pyke, Ronald
Ragozin, David Lawrence
Read, Thomas Thornton
Robel, Gregory Frank
Schaerf, Henry Maximilian
Shin, Suk-han
Tripp, Leonard L
Warfield, Virginia McShane
Williams, Lynn Dolores
Zafran, Misha

WEST VIRGINIA
Cavalier, John F
Hiergeist, Franz Xavier

WISCONSIN
Askey, Richard Allen
Buck, Robert Creighton
Engert, Martin
Forelli, Frank John
Hall, Robert Lester
Kurtz, Thomas Gordon
Levin, Jacob Joseph
Marden, Morris
Mullins, Robert Emmet
Nagel, Alexander
O'Malley, Richard John
Rabinowitz, Paul H
Rall, Louis Baker
Rudin, Walter
Shah, Ghulam M
Shea, Daniel Francis
Solomon, Donald W
Terwilliger, Paul M
Walter, Gilbert G
Ziegler, Michael Robert

WYOMING
Roth, Ben G

PUERTO RICO
Bobonis, Augusto

ALBERTA
Andersen, Kenneth F
Johnson, Dudley Paul
Riemenschneider, Sherman Delbert
Rod, David Lawrence

BRITISH COLUMBIA
Anderson, R F V
Boyd, David William
Bures, Donald John (Charles)
Casselman, William Allen
Cayford, Afton Herbert
Ehle, Byron Leonard
Granirer, Edmond
Leeming, David John
Pfaffenberger, William Elmer
Sion, Maurice
Srivastava, Hari Mohan
Srivastava, Rekha
Swanson, Charles Andrew
Ton, Bui An
Wray, Stephen Donald

MANITOBA
Finlayson, Henry C
Usmani, Riaz Ahmad

NEW BRUNSWICK
Pham-Gia, Thu

NEWFOUNDLAND
Burry, John Henry William
Rees, Rolf Stephen
Singh, Sankatha Prasad
Thomeier, Siegfried

NOVA SCOTIA
Dunn, Kenneth Arthur
El-Hawary, Mohamed El-Aref
Fillmore, Peter Arthur
Grant, Douglass Lloyd
Tan, Kok-Keong

ONTARIO
Abdelmalek, Nabin N
Aczel, Janos D
Akhtar, Mohammad Humayoun
Arthur, James Greig
Atkinson, Harold Russell
Baker, John Alexander
Barbeau, Edward Joseph, Jr
Billigheimer, Claude Elias
Borwein, David
Bryan, Robert Neff
Caradus, Selwyn Ross
Chakravartty, Iswar C
Chang, Shao-chien
Choi, Man-Duen
Cross, George Elliot
Davis, Chandler
Davis, Harry Floyd, II
Duff, George Francis Denton
Eames, William
Elliott, George Arthur
Fleischer, Isidore
Giles, Robin
Graham, Colin C
Greiner, Peter Charles
Headley, Velmer Bentley
Heinig, Hans Paul
Henniger, James Perry
Howland, James Lucien
Husain, Taqdir
Kerman, R A
Langford, William Finlay
Macphail, Moray St John
Mercer, Alexander McDowell
Muldoon, Martin E
Nel, Louis Daniel
Prugovecki, Eduard
Rahman, Mizanur
Rosenthal, Peter (Michael)
Russell, Dennis C
Sablatash, Mike
Schubert, Cedric F
Stewart, James Drewry
Whitfield, John Howard Mervyn
Wigley, Neil Marchand

QUEBEC
Boyarsky, Abraham Joseph
Choksi, Jal R
Gora, Pawel
Herschorn, Michael
Herz, Carl Samuel
Koosis, Paul
Lin, Paul C S
Schlomiuk, Dana
Taylor, John Christopher
Zaidman, Samuel

SASKATCHEWAN
Kaul, S K
Koh, Eusebio Legarda
Law, Alan Greenwell
Sato, Daihachiro

OTHER COUNTRIES
Artemiadis, Nicholas
Chesson, Peter Leith
Horwitz, Lawrence Paul
Jurkat, Wolfgang Bernhard
Kantorovitz, Shmuel
Korevaar, Jacob

414 / DISCIPLINE INDEX

Analysis & Functional Analysis (cont)

Maltese, George J
Roseman, Joseph Jacob
Ryan, Robert Dean
Sato, Mikio
Wong, Roderick Sue-Cheun
Yang, Chung-Chun

Applied Mathematics

ALABAMA
Cochran, John Euell, Jr
Cook, Frederick Lee
Davis, Anthony Michael John
Fitzpatrick, Philip Matthew
Gilbert, Stephen Marc
Hatfield, John Dempsey
Holt, William Robert
Howell, James Levert
Jackson, John Elwin, Jr
Johnson, Johnny R(ay)
Jones, Stanley E
Kenyon, Patricia May
McAuley, Van Alfon
McDonald, Jack Raymond
Yett, Fowler Redford
Zalik, Richard Albert

ALASKA
Beebee, John Christopher

ARIZONA
Ahmed, Saad Attia
Bergman, Ray E(ldon)
Cushing, Jim Michael
Daniel, Sam Mordochai
Faris, William Guignard
Findler, Nicholas Victor
Fung, K Y
Hetrick, David LeRoy
Hoffman, William Charles
Iverson, A Evan
Kyrala, Ali
Lamb, George Lawrence, Jr
Lomen, David Orlando
Maier, Robert S
Melia, Fulvio
Myers, Donald Earl
Perko, Lawrence Marion
Saric, William Samuel
Schowengerdt, Robert Alan
Secomb, Timothy W
Smith, Hal Leslie
Szidarovszky, Ferenc
Walling, Derald Dee
Wildey, Robert Leroy
Ziolkowski, Richard Walter

ARKANSAS
Dunn, James Eldon
Perryman, John Keith
Yaz, Edwin Engin

CALIFORNIA
Ablow, Clarence Maurice
Amster, Harvey Jerome
Ancker, Clinton J(ames), Jr
Anderson, Charles Hammond
Anderson, Paul Maurice
Anderson, William Niles, Jr
Ardema, Mark D
Arnold, Frank R(obert)
Asaro, Robert John
Aster, Robert Wesley
Banerjee, Utpal
Banks, Dallas O
Batchelder, William Howard
Bate, George Lee
Beaver, W(illiam) L(awrence)
Bendat, Julius Samuel
Berg, Paul Walter
Berger, Stanley A(llan)
Bergquist, James William
Berryman, James Garland
Blackwell, Charles C, Jr
Blum, Lenore Carol
Blum, Marvin
Bohachevsky, Ihor O
Bolt, Bruce A
Borrelli, Robert L
Brandeau, Margaret Louise
Brown, Harold David
Bruch, John C(larence), Jr
Bucy, Richard Snowden
Bunch, James R
Burgin, George Hans
Burke, James Edward
Busenberg, Stavros Nicholas
Caflisch, Russel Edward
Cantin, Gilles
Carmona, Rene A
Chamberlain, Robert Glenn
Chambre, Paul L
Chan, Bertram Kim Cheong
Chang, Tien-Lin
Chase, David Marion
Chen, Yang-Jen
Cheng, Edward Teh-Chang
Cheng, Ralph T(a-Shun)
Cherlin, George Yale

Chou, Larry I-Hui
Chrysikopoulos, Constantinos V
Chu, Kai-Ching
Chu, Peter Cheng
Clark, Crosman Jay
Cohen, Donald Sussman
Cohen, Moses E
Concus, Paul
Cooke, Kenneth Lloyd
Corngold, Noel Robert David
Cruise, Donald Richard
Cumberbatch, Ellis
Dalrymple, Stephen Harris
Dashner, Peter Alan
Debreu, Gerard
De Figueiredo, Rui J P
DeFranco, Ronald James
De Micheli, Giovanni
De Pillis, Lisette G
Derenzo, Stephen Edward
Deriso, Richard Bruce
Dickson, Lawrence John
Di Franco, Roland B
Dost, Martin Hans-Ulrich
Dym, Clive L
Edgerley, Dennis A
Eisemann, Kurt
Elderkin, Richard Howard
Eng, Genghmun
Fan, Hsin Ya
Fattorini, Hector Osvaldo
Faulkner, Frank David
Fernandez, Alberto Antonio
Filippenko, Vladimir I
Finn, Robert
Finney, Ross Lee, III
Fisk, Robert Spencer
Franklin, Joel Nicholas
Frenzen, Christopher Lee
Fried, Walter Rudolf
Friedman, George J(erry)
Fu, Lee Lueng
Garfin, Louis
Garwood, Roland William, Jr
Gaskell, Robert Eugene
Gibson, James (Benjamin)
Gilbarg, David
Gill, Stephen Paschall
Gillis, James Thompson
Glaser de Lugo, Frank
Glauz, Robert Doran
Goheen, Lola Coleman
Gold, Richard Robert
Golub, Gene H
Grace, O Donn
Grant, Eugene F(redrick)
Gretsky, Neil E
Gupta, Madhu Sudan
Guymon, Gary L
Haloulakos, Vassilios E
Hardy, Rolland L(ee)
Hastings, Alan Matthew
Hausman, Arthur Herbert
Hedgepeth, John M(ills)
Heflinger, Lee Opert
Helton, F Joanne
Helton, J William
Hewer, Gary Arthur
Hirsch, Morris William
Ho, Hung-Ta
Holoien, Martin O
Holt, Maurice
Hoover, Carol Griswold
Hovanessian, Shahen Alexander
Hu, Steve Seng-Chiu
Hughett, Paul William
Hunt, Robert Weldon
Jacobson, Alexander Donald
Jefferson, Thomas Hutton, Jr
Johnsen, Eugene Carlyle
Jordan, James A, Jr
Juncosa, Mario Leon
Junge, Douglas
Kaellis, Joseph
Kalensher, Bernard Earl
Kaul, Maharaj Krishen
Keller, Edward Lee
Keller, Herbert Bishop
Keller, Joseph Bishop
Kevorkian, Aram K
Khalona, Ramon Antonio
Khurana, Krishan Kumar
Kidder, Ray Edward
Kimble, Gerald Wayne
Klotz, James Allen
Knobloch, Edgar
Knowles, James Kenyon
Koh, Robert Cy
Kraabel, John Stanford
Krener, Arthur James
Lamberson, Roland H
Lang, Martin T
Lange, Charles Gene
Lapidus, Michel Laurent
Lathrop, Richard C(harles)
Lau, John H
Laub, Alan John
Lawson, Charles L
Lax, Melvin David
Layton, Thomas William
Levine, Harold
Lick, Wilbert James
Lindal, Gunnar F
Little, Thomas Morton

Lomax, Harvard
Lu, Kau U
Luce, R(obert) Duncan
Lupash, Lawrence O
Luxenberg, Harold Richard
Ma, Fai
McMurray, Loren Robert
Maillot, Patrick Gilles
Malmuth, Norman David
Marcus, Philip Stephen
Margolis, Stephen Barry
Maria, Narendra Lal
Martin, Elmer Dale
Meadows, Mark Allan
Meecham, William Coryell
Meissner, Loren Phillip
Metcalf, Frederic Thomas
Miles, John Wilder
Mitchell, Reginald Eugene
Moore, Douglas Houston
Moore, Edgar Tilden, Jr
Morgan, Donald E(arle)
Morizumi, S James
Mow, C(hao) C(how)
Muchmore, Robert B(oyer)
Nachbar, William
Nathanson, Weston Irwin
Nemat-Nasser, Siavouche
Neustadter, Siegfried Friedrich
Newell, Gordon Frank
Ng, Edward Wai-Kwok
Ng, Lawrence Chen-Yim
Oddson, John Keith
O'Dell, Austin Almond, Jr
Oliger, Joseph Emmert
Osher, Stanley Joel
Painter, Jeffrey Farrar
Parker, Don
Parker, Donn Blanchard
Parlett, Beresford
Patel, Vithalbhai Ambalal
Phillips, Ralph Saul
Pierce, John Gregory
Pinkel, B(enjamin)
Plock, Richard James
Plotkin, Allen
Poggio, Andrew John
Polit, Andres Cassard
Pomraning, Gerald C
Porter, Lawrence Delpino
Portnoff, Michael Rodney
Pridmore-Brown, David Clifford
Prim, Robert Clay, III
Purcell, Everett Wayne
Rahmat-Samii, Yahya
Ramey, Daniel Bruce
Rao, Jammalamadaka S
Reissner, Eric
Richards, Francis Russell
Roberts, Leonard
Robinson, Lewis Howe
Rocke, David M
Rockwell, Robert Lawrence
Rosen, Irwin Gary
Rosenblatt, Daniel Bernard
Rosenblatt, Murray
Runge, Richard John
Saffman, Philip Geoffrey
Salinas, David
Sangren, Ward Conrad
Schechter, Martin
Scheibe, Paul Otto
Schlafly, Roger
Schneider, Stanley
Schoenstadt, Arthur Loring
Scholl, James Francis
Serat, William Felkner
Shen, Chung Yi
Shi, Yun Yuan
Shiffman, Max
Shnider, Ruth Wolkow
Simons, Roger Mayfield
Sirignano, William Alfonso
Smith, Frederick T(ucker)
Spanier, Jerome
Stauffer, Howard Boyer
Synolakis, Costas Emmanuel
Szego, Peter A
Talley, Wilson K(inter)
Taub, Abraham Haskel
Taussky, Olga
Thorne, Charles Joseph
Tracy, Craig Arnold
Trenholme, John Burgess
Troesch, Beat Andreas
Tu, Yih-O
Vajk, J(oseph) Peter
Van Bibber, Karl Albert
Vardiman, Larry
Vaughan, J Rodney M
Walker, Kelsey, Jr
Wang, Paul Keng Chieh
Warshaw, Stephen I
Weber, Charles L
Weeks, Dennis Alan
Wehausen, John Vrooman
Wells, Willard H
Whitham, Gerald Beresford
Williams, Ruth Jeannette
Willman, Warren Walton
Winter, Donald F
Wollmer, Richard Dietrich
Wood, Roger Charles
Woolard, Henry W(aldo)

Wulfman, Carl E
Yates, Scott Raymond
Yee, Kane Shee-Gong
Yeh, Yea-Chuan Milton
Yeomans, Donald Keith
Yeung, Ronald Wai-Chun
Ziegenhagen, Allyn James

COLORADO
Abshier, Curtis Brent
Bebernes, Jerrold William
Benton, Edward Rowell
Bleistein, Norman
Bosch, William W
Brady, Brian T
Coffey, Mark William
Colvis, John Paris
Cook, Robert Neal
Crawford, Myron Lloyd
Datta, Subhendu Kumar
Elmore, Kimberly Laurence
Esposito, Larry Wayne
Geers, Thomas L
Greenberg, Herbert Julius
Hector, David Lawrence
Heine, George Winfield, III
Hermes, Henry
Hittle, Douglas Carl
Hoots, Felix R
Hundhausen, Joan Rohrer
Irwin, Robert Cook
Jones, Richard Hunn
Kerr, Robert McDougall
Lett, Gregory Scott
Lundgren, J Richard
Lynch, Robert Michael
McCormick, Stephen Fahrney
McKnight, Randy Sherwood
Massman, William Joseph
Miller, Arnold Reed
Miller, James Gilbert
Morgenthaler, George William
Mueller, Raymond Karl
Nahman, Norris S(tanley)
Nesenbergs, Martin
Noble, Richard Daniel
Oneil, Stephen Vincent
Pak, Ronald Y S
Phillipson, Paul Edgar
Poore, Aubrey Bonner
Pruess, Steven Arthur
Purdom, James Francis Whitehurst
Ralston, Margarete A
Sani, Robert L(e Roy)
Savage, William Zuger
Scales, John Alan
Schmid, John Carolus
Seebass, Alfred Richard, III
Shedlock, Kaye M
Slutz, Ralph Jeffery
Swanson, Lawrence Ray
Thaler, Eric Ronald
Underwood, Robert Gordon
Urban, Thomas Charles
Walden, Jack M
Walter, Martin Edward
Wessel, William Roy
West, Anita
Windholz, Walter M

CONNECTICUT
Agresta, Joseph
Ahlberg, John Harold
Barnett, Mark
Burridge, Robert
Chan, Wing Cheng Raymond
Coulter, Paul David
Crofts, Geoffrey
Goldstein, Rubin
Gutierrez-Miravete, Ernesto
Hajela, Dan
Harding, R(onald) H(ugh)
Harrison, Irene R
Keyes, David Elliot
Levien, Roger Eli
Liu, Qing-Huo
McCartin, Brian James
Mandelbrot, Benoit B
Masaitis, Ceslovas
Nehorai, Arye
Ojalvo, Irving U
Phillips, Peter Charles B
Ramakrishnan, Terizhandur S
Reyna, Luis Guillermo
Ruckebusch, Guy Bernard
Sachdeva, Baldev Krishan
Seitelman, Leon Harold
Sharlow, John Francis
Smooke, Mitchell D
Stuck, Barton W
Verdon, Joseph Michael
Wachman, Murray
Wachspress, Eugene Leon

DELAWARE
Angell, Thomas Strong
Bydal, Bruce A
Caviness, Bobby Forrester
Chitra, Surya P
Cook-Ioannidis, Leslie Pamela
Hsiao, George Chia-Chu
Jones, Louise Hinrichsen
Kerr, Arnold D
Kleinman, Ralph Ellis

MacDonald, James
Mehl, James Bernard
Mills, Patrick Leo, Sr
Ogunnaike, Babatunde Ayodeji
Owens, Aaron James
St John, Daniel Shelton
Schwartz, Leonard William
Stakgold, Ivar
Taylor, Howard Milton, III
Trabant, Edward Arthur
Ulery, Dana Lynn
Weinacht, Richard Jay

DISTRICT OF COLUMBIA
Baer, Ralph Norman
Borgiotti, Giorgio V
Casten, Richard G
Cohen, Michael Paul
English, William Joseph
Filliben, James John
Grossman, John Mark
Hall, William Spencer
Harris, Wesley Leroy
Holland, Charles Jordan
Kalnay, Eugenia
Kirwan, Donald Frazier
Kupperman, Robert Harris
Lehman, Richard Lawrence
Nachman, Arje
Nelson, David Brian
Nelson, Larry Dean
Nguyen, Charles Cuong
Palmadesso, Peter Joseph
Rockett, John A
Rojas, Richard Raimond
Sáenz, Albert William
Schweizer, Francois
Stokes, Arnold Paul
Tolstoy, Alexandra
Tucker, John Richard
Vandergraft, James Saul
Watt, Hame Mamadov
Weiss, Leonard
Yen, Nai-Chyuan

FLORIDA
Asner, Bernarad A, Jr
Beil, Robert J(unior)
Bezdek, James Christian
Bober, William
Chang, Lena
Chen, Ching-Jen
Cherepanov, Genady Petrovich
Chow, Wen Lung
Christiano, John G
Clark, Mary Elizabeth
Clarke, Allan James
Clarke, Thomas Lowe
Cloutier, James Robert
Coats, Keith Hal
Dean, Donald L(ee)
Debnath, Lokenath
Elsner, James Brian
Fausett, Donald Wright
Fausett, Laurene van Camp
Fossum, Jerry G
Freeman, Neil Julian
Garrett, James Richard
George, John Harold
Goldwyn, Roger M(artin)
Goodman, Richard Henry
Hager, William Ward
Hammer, Jacob
Howard, Bernard Eufinger
Howard, Louis Norberg
Hsieh, Chung Kuo
Hunter, Christopher
Ismail, Mourad E H
Jacobs, Alan M(artin)
Jacobs, Elliott Warren
Johnson, Harold Hunt
Keesling, James Edgar
Kinnmark, Ingemar Per Erland
Klausner, James Frederick
Krishnamurthy, Lakshminarayanan
Kurzweg, Ulrich H(ermann)
Lacher, Robert Christopher
Lade, Robert Walter
Leitner, Alfred
Lick, Dale W
Lindholm, Fredrik Arthur
Loper, David Eric
McCormick, Clyde Truman
Manougian, Manoug N
Martin, Charles John
Meacham, Robert Colegrove
Merrill, John Ellsworth
Mesterton-Gibbons, Michael Patrick
Nagle, Richard Kent
Palmer, Kenneth
Piquette, Jean C
Rautenstrauch, Carl Peter
Sigmon, Kermit Neal
Tam, Christopher K W
Todd, Hollis N
Tsokos, Chris Peter
Walsh, Edward Kyran
Williams, Carol Ann
Williams, Gareth
Young, John William

GEORGIA
Adomian, George
Atluri, Satya N

Boal, Jan List
Bodnar, Donald George
Dixon, Robert Malcolm
Ginsberg, Jerry Hal
Ho, Dar-Veig
Hooper, Richard Paul
Huberty, Carl J
Hutcherson, Joseph William
Joyner, Ronald Wayne
Komkov, Vadim
Landgren, John Jeffrey
Longini, Ira Mann, Jr
Massey, Fredrick Alan
Meyer, Gunter Hubert
Mickens, Ronald Elbert
Neitzel, George Paul
Rice, Peter Milton
Shonkwiler, Ronald Wesley
Sizemore, Douglas Reece
Sloan, Alan David
Smith, William Allen
Stalford, Harold Lenn
Wang, Johnson Jenn-Hwa
Wilkins, J Ernest, Jr
Wiltse, James Cornelius

HAWAII
Antal, Michael Jerry, Jr
Coburn, Richard Karl
Frazer, L Neil
Loomis, Harold George
Sagle, Arthur A
Tuan, San Fu
Watanabe, Daniel Seishi

IDAHO
Bobisud, Larry Eugene
Hausrath, Alan Richard
Hoagland, Gordon Wood
Johnson, John Alan
Knobel, LeRoy Lyle
Maloof, Giles Wilson
Moore, Kenneth Virgil
Mortensen, Glen Albert
Shaw, Charles Bergman, Jr

ILLINOIS
Albrecht, Felix Robert
Al-Khafaji, Amir Wadi Nasif
Allgaier, Darrell Emory
Aref, Hassan
Ashenhurst, Robert Lovett
Barcilon, Victor
Bareiss, Erwin Hans
Barron, Emmanuel Nicholas
Barston, Eugene Myron
Bart, George Raymond
Berger, Neil Everett
Berger, Steven Barry
Bernstein, Barry
Birnholz, Jason Cordell
Bookstein, Abraham
Buckmaster, John David
Campbell, David Kelly
Carroll, Robert Wayne
Chew, Weng Cho
Cook, Joseph Marion
Cowan, Jack David
Davis, Stephen H(oward)
Dupont, Todd F
Edelstein, Warren Stanley
Evans, William Paul
Fink, Charles Lloyd
Friedlander, Susan Jean
Georgiadis, John G
Golomski, William Arthur
Gruber, Eugene E, Jr
Hakala, Reino William
Hanson, Floyd Bliss
Hearn, Dwight D
Hsui, Albert Tong-Kwan
Hwang, Yu-Tang
Jerome, Joseph Walter
Jerrard, Richard Patterson
Kammler, David W
Kaper, Hans G
Kath, William Lawrence
Kentzer, Czeslaw P(awel)
Kramer, John Michael
Lebovitz, Norman Ronald
Liu, Wing Kam
McLinden, Lynn
Mann, Christian John
Mann, James Edward, Jr
Matalon, Moshe
Matkowsky, Bernard J
Maxwell, William Hall Christie
Moran, Thomas J
Muller, David Eugene
Mura, Toshio
Nagylaki, Thomas Andrew
Olmstead, William Edward
Ottino, Julio Mario
Palmore, Julian Ivanhoe, III
Papoutsakis, Eleftherios Terry
Pericak-Spector, Kathleen Anne
Prussing, John E(dward)
Ramanathan, Ganapathiagraharam V
Reynolds, Larry Owen
Riahi, Daniel Nourollah
Ritt, Robert King
Saari, Donald Gene
Saied, Faisal
Scott, Meckinley

Secrest, Donald H
Shaftman, David Harry
Skeel, Robert David
Tier, Charles
Tipei, Nicolae
Toy, William W
Travelli, Armando
Udler, Dmitry
Wacker, William Dennis
Wilson, Howell Kenneth
Zettl, Anton

INDIANA
Berkovitz, Leonard David
Carroll, John T, III
Chancey, Charles Clifton
Chen, Wai-Fah
Citron, Stephen J
Criss, Darrell E
Ersoy, Okan Kadri
Fink, James Paul
Fuelling, Clinton Paul
Gersting, John Marshall, Jr
Golomb, Michael
Haaser, Norman Bray
Hansen, Arthur G(ene)
Houstis, Elias N
Lee, Norman K
Luke, Jon Christian
Lynch, Robert Emmett
Madan, Ved P
Murphy, Catherine Mary
Ng, Bartholomew Sung-Hong
Reid, William Hill
Rice, John Rischard
Sweet, Arnold Lawrence
Synowiec, John A
Szpankowski, Wojciech
Thieme, Melvin T
Thompson, Maynard
Thuente, David Joseph
Thurber, James Kent
Weston, Vaughan Hatherley

IOWA
Alexander, Roger Keith
Atkinson, Kendall E
Buck, James R
Carlson, Bille Chandler
Choi, Kyung Kook
Dahiya, Rajbir Singh
Evans, David Hunden
Haug, Edward J, Jr
Hethcote, Herbert Wayne
Horton, Robert, Jr
Jeyapalan, Kandiah
Jorgensen, Palle E T
Kawai, Masataka
Lambert, Robert J
Landweber, Louis
Luban, Marshall
Luecke, Glenn Richard
Miller, Richard Keith
Nariboli, Gundo A
Reckase, Mark Daniel
Rudolphi, Thomas Joseph
Seifert, George
Sheeler, John B(riggs)
Weiss, Harry Joseph

KANSAS
Acker, Andrew French, III
Davis, Elwyn H
Elcrat, Alan Ross
Kirmser, P(hilip) G(eorge)
Lerner, David Evan
Miller, Forest R
Ramm, Alexander G
Shultis, J Kenneth
Van Vleck, Fred Scott

KENTUCKY
Brock, Louis Milton
Davis, Chester L
Evans, Gerald William
Fairweather, Graeme
Hantman, Robert Gary
Howard, Henry Cobourn
Man, Chi-Sing
Perry, Peter Anton
Ray, Asit Kumar
Russell, Marvin W

LOUISIANA
Andrus, Jan Frederick
Bradley, Marshall Rice
Chan, Chiu Yeung
Chen, Isaac I H
Draayer, Jerry Paul
Foote, Joe Reeder
Gajendar, Nandigam
Heatherly, Henry Edward
Kythe, Prem Kishore
Lefton, Lew Edward
Maxfield, John Edward
Puri, Pratap
Roquemore, Leroy
Sharma, Bhudev
Temple, Austin Limiel, (Jr)
Tipler, Frank Jennings, III
Ward, Ronald Wayne
Warner, Isiah Manuel
Wenzel, Alan Richard
Yannitell, Daniel W

Yu, Fang Xin

MAINE
Bowie, Oscar L
Farlow, Stanley Jerome
McMillan, Brockway

MARYLAND
Abed, Eyad Husni
Alexander, James Crew
Antman, Stuart S
Atanasoff, John Vincent
Babcock, Anita Kathleen
Babuska, Ivo Milan
Baer, Ferdinand
Bainum, Peter Montgomery
Baum, Howard Richard
Benedetto, John Joseph
Benokraitis, Vitalius
Berger, Alan Eric
Berger, Bruce S
Bernard, Peter Simon
Bishop, Walton B
Blue, James Lawrence
Boggs, Paul Thomas
Bromberg, Eleazer
Bruley, Duane Frederick
Burns, Timothy John
Cantor, David S
Cawley, Robert
Celmins, Aivars Karlis Richards
Chadwick, Richard Simeon
Chi, Donald Nan-Hua
Chi, L K
Ciment, Melvyn
Cohen, Edgar A, Jr
Colvin, Burton Houston
Covell, David Gene
Criss, John W
Criss, Thomas Benjamin
Davis, Frederic I
Eades, James B(everly), Jr
Ehrlich, Louis William
Elliott, David LeRoy
Ely, Raymond Lloyd
Everstine, Gordon Carl
Fletcher, John Edward
Fong, Jeffrey Tse-Wei
Fthenakis, Emanuel John
Gates, Sylvester J, Jr
Goldman, Alan Joseph
Goldstein, Charles M
Grebogi, Celso
Hammersmith, John L(eo)
Hammond, Allen Lee
Herczynski, Andrzej
Herrmann, Robert Arthur
Hitchcock, Daniel Augustus
Hollenbeck, Robert Gary
Horak, Martin George
Horst, John Albert
Hubbell, John Howard
Hummel, James Alexander
Jackson, Richard H F
Johnson, Frederick Carroll
Kahaner, David Kenneth
Kahn, Arthur B
Kattakuzhy, George Chacko
Kellogg, Royal Bruce
Kitchens, Clarence Wesley, Jr
Kuttler, James Robert
Lee, David Allan
Lozier, Daniel William
Lynn, Yen-Mow
McFadden, Geoffrey Bey
Mahar, Thomas J
Marimont, Rosalind Brownstone
Martin, Monroe Harnish
Maslen, Stephen Harold
Massell, Paul Barry
Menton, Robert Thomas
Miller, Raymond Edward
Nau, Dana S
Noblesse, Francis
Odle, John William
Olver, Frank William John
Osborn, John Edward
Oser, Hans Joerg
Pai, S(hih) I
Peredo, Mauricio
Rand, Robert Collom
Raphael, Louise Arakelian
Reader, Wayne Truman
Rehm, Ronald George
Rinzel, John Matthew
Roberts, Richard Calvin
Rosenstock, Herbert Bernhard
Rostamian, Rouben
Sadowsky, John
Schmid, Lawrence Alfred
Seidman, Thomas I(srael)
Serfling, Robert Joseph
Shaw, Harry, Jr
Simmons, John Arthur
Singhal, Jaya Asthana
Skiff, Frederick Norman
Smith, James Alan
Solomon, Jay Murrie
Stadter, James Thomas
Sullivan, Francis E
Tissue, Eric Bruce
Tropf, Cheryl Griffiths
Turner, Robert Davison
Valenzuela, Gaspar Rodolfo

Applied Mathematics (cont)

Vangel, Mark Geoffrey
Weiss, George Herbert
Weiss, Michael David
Wells, William T
Wright, Thomas Wilson
Wu, Ching-Sheng
Yang, Xiaowei
Yorke, James Alan
Zabronsky, Herman
Zien, Tse-Fou

MASSACHUSETTS

Abbott, Douglas E(ugene)
Anderson, Donald Gordon Marcus
Archer, Robert Raymond
Ash, Michael Edward
Berman, Robert Hiram
Bernstein, Joseph N
Blake, Thomas R
Bower, Amy Sue
Brockett, Roger Ware
Burke, Shawn Edmund
Burns, Daniel Robert
Butler, James Preston
Calabi, Lorenzo
Cappallo, Roger James
Carpenter, Richard M
Carrier, George Francis
Cipolla, John William, Jr
Clough, Shepard Anthony
Cooperman, Gene David
Crowell, Julian
Cuffin, B(enjamin) Neil
Davis, Paul William
Delvaille, John Paul
Drane, Charles Joseph, Jr
Esch, Robin E
Estes, William K
Fiering, Myron B
Fishman, Philip M
Fougere, Paul Francis
Freund, Robert M
Granoff, Barry
Greenspan, Harvey Philip
Hadlock, Charles Robert
Herda, Hans-Heinrich Wolfgang
Ho, Yu-Chi
Holst, William Frederick
Holt, Frederick Sheppard
Holway, Lowell Hoyt, Jr
Hovorka, John
Kanter, Irving
Kitchin, John Francis
Klein, William
Kon, Mark A
Lechner, Robert Joseph
Lemnios, A(ndrew) Z
Lin, Chia Chiao
Lindzen, Richard Siegmund
Lo, Andrew W
McElroy, Michael Brendan
Maskin, Eric S
Meldon, Jerry Harris
Miller, William Brunner
Moler, Cleve B
Mulpur, Arun K
Nehrkorn, Thomas
Nesbeda, Paul
Noiseux, Claude Francois
O'Connell, Richard John
Oettinger, Anthony Gervin
Ogilvie, T(homas) Francis
O'Neil, Elizabeth Jean
Ostrovsky, Rafail M
Palubinskas, Felix Stanley
Pang, Yuan
Quinto, Eric Todd
Richter, Stephen L(awrence)
Ross, Edward William, Jr
Rossettos, John N(icholas)
Roth, Robert Steele
St Mary, Donald Frank
Salstein, David A
Schetzen, Martin
Senechal, Marjorie Lee
Shannon, Claude Elwood
Shapiro, Ralph
Sherman, Thomas Oakley
Shu, Larry Steven
Spiegel, Stanley Lawrence
Stiffler, Jack Justin
Stone, Peter Hunter
Strang, Gilbert
Sun, Fang-Kuo
Sun, Shuwei
Tait, Kevin S
Teng, Chojan
Thompson, Charles
Thorpe, Rodney Warren
Toomre, Alar
Toscano, William Michael
Warga, Jack
Weinberg, I Jack
Wong, Po Kee
Woods, William A
Wunsch, Abraham David
Yaghjian, Arthur David
Yamartino, Robert J

MICHIGAN

Abriola, Linda Marie
Alavi, Yousef
Anderson, Howard Benjamin
Birdsall, Theodore G
Bolander, Richard
Boojum, Basil John
Boyd, John Philip
Brown, James Ward
Chang, Jhy-Jiun
Cho, Byong Kwon
Chock, David Poileng
Chow, Pao Liu
Crittenden, John Charles
Curl, Rane L(ocke)
Dawson, D(onald) E(merson)
Denman, Harry Harroun
Dolph, Charles Laurie
Field, David Anthony
Flanders, Harley
Habib, Izzeddin Salim
Hee, Christopher Edward
Hicks, Darrell Lee
Hill, Bruce M
Holland, John Henry
Huntoon, Jacqueline E
Ibrahim, Raouf A
Kincaid, Wilfred Macdonald
Kochen, Manfred
Kubis, Joseph J(ohn)
LaHaie, Ivan Joseph
Larsen, Edward William
Lubman, David Mitchell
Lumsdaine, Edward
Lund, Charles Edward
Miklavcic, Milan
Mordukhovich, Boris S
Murphy, Brian Boru
Nesbitt, Cecil James
Pence, Thomas James
Petersen, Donald Ralph
Polis, Michael Philip
Powell, Kenneth Grant
Ramamurthy, Amurthur C
Root, William L(ucas)
Salam, Fathi M A
Schmidt, Robert
Schneider, Eric West
Scott, Richard Anthony
Simon, Carl Paul
Soutas-Little, Robert William
Spahn, Robert Joseph
Thompson, James Lowry
Ullman, Nelly Szabo
Wang, Chang-Yi
Wasserman, Arthur Gabriel
Wasserman, Robert H
Weil, Herschel
Wineman, Alan Stuart
Wong, Pui Kei
Yang, Wei-Hsuin
Yen, David Hsien-Yao

MINNESOTA

Abraham, John
Anderson, Willard Eugene
Aris, Rutherford
Austin, Donald Murray
Cassola, Robert Louis
Castore, Glen M
Cornelissen Guillaume, Germaine G
Drew, Bruce Arthur
Emeagwali, Philip Chukwurah
Follingstad, Henry George
Gaal, Steven Alexander
Harvey, Charles Arthur
Infante, Ettore F
Keister, Jamieson Charles
Kemper, John Thomas
Lodge, Timothy Patrick
Lundberg, Gustave Harold
Luskin, Mitchell B
Miller, Willard, Jr
Nitsche, Johannes Carl Christian
Olver, Peter John
Poling, Craig
Robb, Richard A
Rosen, Judah Ben
Schuldt, Spencer Burt
Scriven, L E(dward), (II)
Warner, William Hamer
Weinberger, Hans Felix
Wolfe, Barbara Blair
Woodward, Paul Ralph
Zitney, Stephen Edward

MISSISSIPPI

Cade, Ruth Ann
Davis, Thomas Mooney
Fay, Temple Harold
George, Clifford Eugene
Green, Albert Wise
Johnston, Russell Shayne
Kishk, Ahmed A
Koshel, Richard Donald
Rohde, Florence Virginia
Solomon, Jimmy Lloyd
Warsi, Zahir U A

MISSOURI

Agarwal, Ramesh K
Ahlbrandt, Calvin Dale
Beem, John Kelly
Bruening, James Theodore
Cowherd, Chatten, Jr
Edwards, Carol Abe
Edwards, Terry Winslow
Henson, Bob Londes
Ho, Chung You (Peter)
Huddleston, Philip Lee
Johnson, Darell James
Katz, Israel Norman
Krone, Lester H(erman), Jr
MacDonald, Carolyn Trott
Ornstein, Wilhelm
Pettey, Dix Hayes
Petty, Clinton Myers
Plummer, Otho Raymond
Polowy, Henry
Retzloff, David George
Rodin, Ervin Y
Sabharwal, Chaman Lal
Shrauner, Barbara Abraham
Spielman, Barry
Tarn, Tzyh-Jong

MONTANA

Barrett, Louis Carl
Derrick, William Richard
Grossman, Stanley I
Hess, Lindsay LaRoy
Wright, Alden Halbert

NEBRASKA

Beatty, Millard F
Conway, James Joseph
Haeder, Paul Albert
Maloney, John P
Schniederjans, Marc James
Scudder, Jeffrey Eric
Surkan, Alvin John

NEVADA

Lindstrom, Fredrick Thomas
Miel, George J
Nassersharif, Bahram
Rawat, Banmali Singh
Tompson, Robert Norman
Turner, Robert Harold
Wells, William Raymond

NEW HAMPSHIRE

Albert, Arthur Edward
Black, Sydney D
Buffler, Charles Rogers
Darlington, Sidney
Kuo, Shan Sun
Meeker, Loren David
Morduchow, Morris
Radlow, James
Seifer, Arnold David

NEW JERSEY

Ahluwalia, Daljit Singh
Aizenman, Michael
Andrushkiw, Roman Ihor
Ball, William Henry Warren
Ben-Israel, Adi
Bergeron, Robert F(rancis) (Terry), Jr
Blackmore, Denis Louis
Brandmaier, Harold Edmund
Caffarelli, Luis Angel
Castor, William Stuart, Jr
Chai, Winchung A
Cheo, Li-hsiang Aria S
Chien, Victor
Cho, Soung Moo
Cohen, Edwin
Coleman, Norman P, Jr
Coughran, William Marvin, Jr
Crane, Roger L
Crochiere, Ronald E
Elk, Seymour B
Flannery, Brian Paul
Fornberg, Bengt
Foschini, Gerard Joseph
Frauenthal, James Clay
Freund, Roland Wilhelm
Gaer, Marvin Charles
Gear, Charles William
Gelfand, Israel M
Goldberg, Vladislav V
Gossmann, Hans Joachim
Gross, Arthur Gerald
Holford, Richard L
Holland, Paul William
Houston, Vern Lynn
Hughes, James Sinclair
Kam, Gar Lai
Kaufman, Linda
Kheshgi, Haroon S
Kim, Uing W
Kirch, Murray R
Koss, Valery Alexander
Kriegsmann, Gregory A
Labianca, Frank Michael
Larson, Eric Heath
Levin, Simon Asher
Levine, Lawrence Elliott
Lieb, Elliott Hershel
Lindberg, Craig Robert
Lioy, Paul James
Lucantoni, David Michael
Majda, Andrew J
Mamelak, Joseph Simon
Mangulis, Visvaldis
Mazo, James Emery
Morgan, Dennis Raymond
Morgan, Samuel Pope
Morrison, John Allan
Neal, Scotty Ray
Nguyen, Hien Vu
Nolet, Guust M
Orszag, Steven Alan
Plummer, Dirk Arnold
Rebhuhn, Deborah
Reiman, Allan H
Roberts, Fred Stephen
Rosenblum, Martin Jacob
Saltzberg, Burton R
Schryer, Norman Loren
Shober, Robert Anthony
Sinclair, J Cameron
Sisto, Fernando
Slepian, David
Snygg, John Morrow
Sontag, Eduardo Daniel
Stehney, Ann Kathryn
Steiger, William Lee
Stepleman, Robert Saul
Stevens, John G
Stickler, David Collier
Strauss, Walter
Suhir, Ephraim
Sundaresan, Sankaran
Sverdlove, Ronald
Thomas, Larry Emerson
Vann, Joseph M
Van Roosbroeck, Willy Werner
Weiss, Alan
White, Benjamin Steven
Whitman, Gerald Martin
Williams, Gareth Pierce
Wilson, Lynn O
Winthrop, Joel Albert
Witsenhausen, Hans S
Yu, Yi-Yuan
Zabusky, Norman J
Zatzkis, Henry
Zebib, Abdelfattah M G
Zitron, Norman Ralph

NEW MEXICO

Allen, Richard Crenshaw, Jr
Amos, Donald E
Auer, Lawrence H
Barnes, Daniel C
Barr, George E
Baum, Carl E(dward)
Bedell, Kevin Shawn
Beyer, William A
Boland, W Robert
Brackbill, Jeremiah U
Bradshaw, Martin Daniel
Cogburn, Robert Francis
Dietz, David
Erickson, Kenneth Lynn
Faber, Vance
Fugelso, Leif Erik
Godfrey, Brendan Berry
Hanson, Kenneth Merrill
Harlow, Francis Harvey, Jr
Hulme, Bernie Lee
Julian, William H
Kelly, Robert Emmett
Kuiper, Logan Keith
Luehr, Charles Poling
McGehee, Ralph Marshall
McKay, Michael Darrell
Metropolis, Nicholas Constantine
Mueller, Marvin Martin
Norwood, Frederick Reyes
Perelson, Alan Stuart
Rach, Randolph Carl
Rachele, Henry
Reuter, Robert Carl, Jr
Scharn, Herman Otto Friedrich
Schultz, Rodney Brian
Searls, Craig Allen
Shively, Frank Thomas
Steinberg, Stanly
Stone, Alexander Paul
Summers, Donald Lee
Swartz, Blair Kinch
Thompson, Robert James
Thorne, Billy Joe
Wallace, Terry Charles, Sr
Walvekar, Arun Govind
Wecksung, George William
Weigle, Robert Edward
White, George Nichols, Jr
Wiggins, Carl M
Wing, G(eorge) Milton
Zemach, Charles

NEW YORK

Ackerberg, Robert C(yril)
Acrivos, Andreas
Agarwal, Ramesh Chandra
Ahmed-Zaid, Said
Alt, Franz L
Arin, Kemal
Ascher, Marcia
Austin, Fred
Ballou, Donald Pollard
Barnett, Michael Peter
Bauer, Frances Brand
Bernstein, Herbert Jacob
Berresford, Geoffrey Case
Bewley, Loyal V
Bienstock, Daniel
Bland, Robert Gary
Blank, Albert Abraham
Bonan, Eugene J
Boyce, William Edward

Brady, William Gordon
Bramble, James H
Braun, Martin
Brunelle, Eugene John, Jr
Burstein, Samuel Z
Callender, Robert
Cane, Mark A
Chen, Yung Ming
Cheney, Margaret
Chesshire, Geoffrey S
Chester, Clive Ronald
Childress, William Stephen
Chu, Chia-Kun
Clark, Alfred, Jr
Coden, Michael H
Cohen, Hirsh G
Cohn, Harvey
Cole, Julian D(avid)
Cooley, James William
Cozzarelli, Francis A(nthony)
Craxton, Robert Stephen
Cullum, Jane Kehoe
Curtis, Ronald S
Dasgupta, Gautam
Demerdash, Nabeel A O
Dempsey, John Patrick
DeNoyer, Linda Kay
Deresiewicz, Herbert
Diament, Paul
Dicker, Daniel
Doering, Charles Rogers
Dolezal, Vaclav J
Douglas, Craig Carl
Drenick, Rudolf F
Ecker, Joseph George
Fagot, Wilfred Clark
Fama, Donald Francis
Fialkow, Aaron David
Finkelstein, Abraham Bernard
Flaherty, Joseph E
Fleishman, Bernard Abraham
Foster, Kenneth William
Friedman, Jack P
Gazis, Denos Constantinos
Geer, James Francis
Geiss, Gunther R(ichard)
Genova, James John
Gerst, Irving
Glasser, M Lawrence
Glimm, James Gilbert
Goldfarb, Donald
Goldstein, Charles Irwin
Gomory, Ralph E
Goodman, Theodore R(obert)
Gordon, Sheldon P
Graff, Samuel M
Griffith, Daniel Alva
Gulati, Suresh T
Gustavson, Fred Gehrung
Haines, Charles Wills
Handelman, George Herman
Heller, William R
Heppa, Douglas Van
Herron, Isom H
Hershenov, Joseph
Hetnarski, Richard Bozyslaw
Hirtzel, Cynthia S
Hochstadt, Harry
Hoenig, Alan
Holmes, Philip John
Hussain, Moayyed A
Imber, Murray
Indusi, Joseph Paul
Isaacson, David
Isom, Morris P
Jacobson, Melvin Joseph
Jerri, Abdul J
Kapila, Ashwani Kumar
Kaplan, Paul
Karatzas, Ioannis
Karp, Samuel Noah
Kaufman, Sol
Kaup, David James
Kayani, Joseph Thomas
Kenett, Ron
King, Alan Jonathan
Klein, John Sharpless
Knight, Bruce Winton, (Jr)
Kohls, Carl William
Kotchoubey, Andrew
Kranzer, Herbert C
Lauber, Thornton Stuart
Leibovich, Sidney
Leibowitz, Martin Albert
Lessen, Martin
Levy, Alan Joseph
Levy, Paul
Lew, John S
Liban, Eric
Liboff, Richard L
Lindquist, William Brent
Loughlin, Timothy Arthur
Lovass-Nagy, Victor
McCamy, Calvin S
MacGillivray, Archibald Dean
McKinstrie, Colin J(ohn)
McLaughlin, Joyce Rogers
McMurry, William
Maltz, Martin Sidney
Marr, Robert B
Marshall, Clifford Wallace
Meisel, David Dering
Michiels, Leo Paul
Miller, Kenneth Sielke

Morawetz, Cathleen Synge
Moretti, G(ino)
Moritz, Roger Homer
Mosher, Melville Calvin
Nerode, Anil
Newburg, Edward A
Newman, Charles Michael
Niman, John
Nyrop, Jan Peter
Odeh, Farouk M
Papoulis, Athanasios
Parker, Francis Dunbar
Pasciak, Joseph Edward
Paxton, K Bradley
Payne, Lawrence Edward
Payton, Robert Gilbert
Peierls, Ronald F
Pernick, Benjamin J
Pflug, Donald Ralph
Podowski, Michael Zbigniew
Pohle, Frederick V
Powers, David Leusch
Ritterman, Murray B
Rogers, Edwin Henry
Roskes, Gerald J
Rothenberg, Ronald Isaac
Roytburd, Victor
Ruschak, Kenneth John
Sayre, David
Schriro, George R
Sevian, Walter Andrew
Seyler, Charles Eugene
Shapiro, Donald M
Siegmann, William Lewis
Siggia, Eric Dean
Singer, Barry M
Singh, Chanchal
Sobel, Kenneth Mark
Soong, Tsu-Teh
Sotirchos, Stratis V
Srivastav, Ram Prasad
Stanionis, Victor Adam
Steinberg, Herbert Aaron
Steinmetz, William John
Swann, Dale William
Talham, Robert J
Tang, Ignatius Ning-Bang
Tewarson, Reginald P
Thomas, John Howard
Ting, Lu
Trefethen, Lloyd Nicholas
Triantafyllou, George S
Ventriglia, Anthony E
Wagner, John George
Weill, Georges Gustave
Weitzner, Harold
Winograd, Shmuel
Wolf, Henry
Wolfe, Philip
Wong, Chak-Kuen
Wu, Lilian Shiao-Yen
Yanowitch, Michael
Yarmush, David Leon
Zablow, Leonard
Zemanian, Armen Humpartsoum

NORTH CAROLINA
Baker, Charles Ray
Batra, Subhash Kumar
Bishir, John William
Burniston, Ernest Edmund
Campbell, Stephen La Vern
Cato, Benjamin Ralph, Jr
Clark, James Samuel
Davis, Charles Alfred
Doster, Joseph Michael
Dunn, Joseph Charles
Faulkner, Gary Doyle
Gardner, Robert B
Hall, George Lincoln
Hall, Peter M
Hartwig, Robert Eduard
Hassan, Awatif E
Havner, Kerry S(huford)
Jaszczak, Ronald Jack
Jawa, Manjit S
Keltie, Richard Francis
Lapicki, Gregory
Lucas, Thomas Ramsey
McAllister, David Franklin
Macon, Nathaniel
Mann, William Robert
Marlin, Joe Alton
Mullikin, Thomas Wilson
Pao, Chia-Ven
Peterson, David West
Pizer, Stephen M
Reckhow, Kenneth Howland
Roberts, Jerry Allan
Rose, Milton Edward
Saeman, W(alter) C(arl)
Sandor, George N(ason)
Spickerman, William Reed
Sussenguth, Edward H
Swallow, William Hutchinson
Teague, David Boyce
Tlalka, Jacek
Turinsky, Paul Josef
Utku, Senol
Van der Vaart, Hubertus Robert
Wallingford, John Stuart
Wilson, James Blake
Wu, Julian Juh-Ren

NORTH DAKOTA
Hare, Robert Ritzinger, Jr
Kemper, Gene Allen
Kurtze, Douglas Alan
Rieder, William G(ary)
Schwalm, William A
Uherka, David Jerome

OHIO
Bailey, James L
Black, Richard H
Bodonyi, Richard James
Brazee, Ross D
Brown, Robert William
Bullock, Robert M, III
Buoni, John J
Chen, Chao-Hsing Stanley
Chen, Hollis C(hing)
Chung, Benjamin T F
Clark, Robert Arthur
Compton, Ralph Theodore, Jr
Craig, Rachael Grace
Davis, Bruce Wilson
Dumas, H Scott
Durand, Dominique M
Fitzmaurice, Nessan
Fleming, Paul Daniel, III
Gatewood, Buford Echols
Goldberg, Samuel
Goldstein, Marvin E
Gorla, Rama S R
Groetsch, Charles William
Guo, Shumei
Gustafson, Steven Carl
Hajek, Otomar
Hart, David Charles
Herbert, Thorwald
Horner, James M
Hull, David Lee
Jain, Surender K
Jehn, Lawrence Andrew
Jeng, Duen-Ren
Johnson, Robert Oscar
Khamis, Harry Joseph
Klingbeil, Werner Walter
Kummer, Martin
Lake, Robin Benjamin
Leitman, Marshall J
Long, Clifford A
Mathis, Robert Fletcher
Messick, Roger E
Mills, Wendell Holmes, Jr
Mityagin, Boris Samuel
Mugler, Dale H
Ogg, Frank Chappell, Jr
Quam, David Lawrence
Quesada, Antonio F
Quinn, Dennis Wayne
Ray-Chaudhuri, Dwijendra Kumar
Restemeyer, William Edward
Reynolds, David Stephen
Richardson, David Louis
Rounsley, Robert R(ichard)
Rubin, Stanley G(erald)
Shankar, Hari
Silver, Gary Lee
Smith, Robert Emery
Steinlage, Ralph Cletus
Suter, Bruce Wilsey
Tamburino, Louis A
Van Horn, David Downing
Vayo, Harris Westcott
Wang, Paul Shyh-Horng
Ward, Douglas Eric
Wen, Shih-Liang
Wolaver, Lynn E(llsworth)
Wyman, Bostwick Frampton

OKLAHOMA
Carr, Meg Brady
Cerro, Ramon Luis
Choike, James Richard
Elliott, Sheldon Ellwood
Fisher, Donald D
Jones, Jack Raymond
Keener, Marvin Stanford
Kemp, Louis Franklin, Jr
Mulholland, Robert J(oseph)
Nofziger, John Lynn
Phillips, John Richard
Richardson, J Mark
Root, Paul John
Vasicek, Daniel J
Walker, Billy Kenneth
Williams, Thomas Henry Lee
Winter, William Kenneth

OREGON
Horne, Frederick Herbert
McClure, David Warren
McLeod, Robin James Young
Murphy, Lea Frances
Narasimhan, Mysore N L
Solmon, Donald Clyde
Tauber, Selmo
Taylor, Carson William
Tewinkel, G Carper
Tinney, William Frank

PENNSYLVANIA
Abu-Shumays, Ibrahim Khalil
Acheson, Willard Phillips
Akay, Adnan
Allen, Charles C(orletta)

MATHEMATICS / 417

Amon Parisi, Cristina Hortensia
Balas, Egon
Bau, Haim Heinrich
Becker, Stanley J
Beggs, William John
Bielak, Jacobo
Blaisdell, Ernest Atwell, Jr
Blythe, Philip Anthony
Bona, Jerry Lloyd
Bose, Nirmal K
Brown, John Lawrence, Jr
Cendes, Zoltan Joseph
Chan, Siu-Kee
Chelemer, Harold
Chuang, Strong Chieu-Hsiung
Conrad, Bruce
Crawford, John David
Dash, Sanford Mark
Davidon, William Cooper
Davies, Kenneth Thomas Reed
Daywitt, James Edward
DeSanto, Daniel Frank
Dwyer, Thomas A
Edelen, Dominic Gardiner Bowling
Engheta, Nader
Ermentrout, George Bard
Ertekin, Turgay
Fabrey, James Douglas
Federowicz, Alexander John
Fisher, David George
Ford, William Frank
Freeman, Louis Barton
Gordon, William John
Gorman, Arthur Daniel
Grassi, Vincent G
Grimes, Dale M(ills)
Grover, Carole Lee
Gurtin, Morton Edward
Hageman, Louis Alfred
Haghighat, Alireza
Hall, Charles Allan
Hickman, Warren David
Hoburg, James Frederick
Holland, Burt S
Howland, Frank L
Jech, Thomas J
Kazakia, Jacob Yakovos
Kolar, Michael Joseph
Kolodner, Ignace I(zaak)
Lakhtakia, Akhlesh
Lepore, John A(nthony)
Lokay, Joseph Donald
Lynn, Roger Yen Shen
McLain, David Kenneth
Mallett, Russell Lloyd
Merrill, Samuel, III
Michalik, Edmund Richard
Molyneux, John Ecob
Moore, Walter Calvin
Muller, Karl Frederick
Murgie, Samuel A
Nash, David Henry George
Neuman, Charles P(aul)
Nolan, Edward J
Novak, Darwin Albert, Jr
Owen, David R
Pierce, Allan Dale
Pool, James C T
Porsching, Thomas August
Rajagopal, K R
Ramos, Juan Ignacio
Rapp, Paul Ernest
Rheinboldt, Werner Carl
Rhodes, Donald Frederick
Rivlin, Ronald Samuel
Rodrigues, Anil Noel
Rorres, Chris
Sawyers, Kenneth Norman
Sibul, Leon Henry
Simaan, Marwan
Sinclair, Glenn Bruce
Smith, Gerald Francis
Smith, Warren LaVerne
Spencer, Arthur Coe, II
Starling, James Lyne
Szyld, Daniel Benjamin
Thompson, Gerald Luther
Thompson, William, Jr
Tozier, John E
Troy, William Christopher
Tuba, I Stephen
Vrentas, Christine Mary
Vrentas, James Spiro
Wagner, Clifford Henry
Walker, James David Allan
Wayne, Clarence Eugene
Weiner, Robert Allen
Williams, Frederick McGee
Woo, Tse-Chien
Woodburn, Wilton A
Wu, John Naichi
Yang, Arthur Jing-Min
Yukich, Joseph E

RHODE ISLAND
Bisshopp, Frederic Edward
Dafermos, Constantine M
Driver, Rodney D
Falb, Peter L
Fleming, Wendell Helms
Freiberger, Walter Frederick
Gottschalk, Walter Helbig
Grenander, Ulf
Krikorian, John Sarkis, Jr

Applied Mathematics (cont)

Ladas, Gerasimos
Liu, Pan-Tai
McClure, Donald Ernest
Pipkin, Allen Compere
Raykhman, Aleksandr
Roxin, Emilio O
Sirovich, Lawrence
Strauss, Charles Michael
Streit, Roy Leon
Verma, Ghasi Ram
Weiner, Jerome Harris

SOUTH CAROLINA
Felling, William E(dward)
Fennell, Robert E
Honeck, Henry Charles
Hunt, Hurshell Harvey
Klemm, James L
Kostreva, Michael Martin
Proctor, Thomas Gilmer
Sheppard, Emory Lamar
Uldrick, John Paul
Voit, Eberhard Otto

SOUTH DAKOTA
Gaalswyk, Arie
Raab, Wallace Albert

TENNESSEE
Andrew, Merle M
Baker, Allen Jerome
Beauchamp, John J
Bigelow, John E(aly)
Bownds, John Marvin
Cline, Randall Eugene
Crooke, Philip Schuyler
Cross, James Thomas
Daniel, Donald Clifton
Demetriou, Charles Arthur
Dobosy, Ronald Joseph
Dongarra, Jack Joseph
Emanuel, William Robert
Fischer, Charlotte Froese
Gamliel, Amir
Gan, Rong Zhu
Gomberg, Joan Susan
Hallam, Thomas Guy
Henry, Jonathan Flake
Holdeman, Jonas Tillman, Jr
Hopkins, Richard Allen
Hutcheson, Paul Henry
Jordan, George Samuel
Kraft, Edward Michael
Nowlin, Charles Henry
Pugh, Claud Ervin
Rajan, Periasamy Karivaratha
Reddy, Kapuluru Chandrasekhara
Rowan, William Hamilton, Jr
Rozema, Edward Ralph
Schaefer, Philip William
Schumaker, Larry L
Sorrells, Frank Douglas
Stone, Robert Sidney
Strauss, Alvin Manosh
Turitto, Vincent Thomas
Whittle, Charles Edward, Jr

TEXAS
Anderson, Charles E, Jr
Anderson, David H
Anderson, Ronald M
Aronofsky, Julius S
Atchison, Thomas Andrew
Auchmuty, Giles
Bakelman, Ilya J(acob)
Ball, Kenneth Steven
Barton, James Brockman
Bennett, George Kemble
Bhattacharyya, Shankar P
Biedenharn, Lawrence C, Jr
Bigham, Robert Eric
Bowen, Ray M
Boyer, Lester Leroy
Breig, Edward Louis
Brockett, Patrick Lee
Brown, Dennison Robert
Brown, Glenn Lamar
Brown, Robert Wade
Bryant, Michael David
Caflisch, Robert Galen
Carey, Graham Francis
Carroll, Michael M
Chang, Ping
Charbeneau, Randall Jay
Chen, Goong
Chen, Peter
Cheney, Elliott Ward, Jr
Corduneanu, Constantin C
Cowan, Russell (Walter)
Deeming, Terence James
Deeter, Charles Raymond
Dodson, David Scott
Doser, Diane Irene
Eargle, George Marvin
Ehlig-Economides, Christine Anna
Eubank, Randall Lester
Ewell, James John, Jr
Ewing, Richard Edward
Fischer, Ferdinand Joseph
Fix, George Joseph
Fulks, Watson
Gardner, Clifford S

Giese, Robert Paul
Gilmour-Stallsworth, Lisa K
Gladwell, Ian
Goldman, Joseph L
Golubitsky, Martin Aaron
Haberman, Richard
Hartfiel, Darald Joe
Hasling, Jill Freeman
Hazen, Gary Alan
Heelis, Roderick Antony
Hunt, Louis Roberts
Hurlburt, H(arvey) Zeh
Johnson, Johnny Albert
Junkins, John Lee
Justice, James Horace
Karal, Frank Charles, Jr
Katsnelson, Lev Z
Keown, Ernest Ray
Kesler, Oren Byrl
Knudsen, John R
Koplyay, Janos Bernath
Leventhal, Stephen Henry
Li, Wen-Hsiung
Liemohn, Harold Benjamin
Long, Stuart A
Lurix, Paul Leslie, Jr
MacFarlane, Duncan Leo
McIntyre, Robert Gerald
Markowski, Gregory Ray
Mauldin, Richard Daniel
Midgley, James Eardley
Mitchell, A Richard
Moorhead, William Dean
Morris, William Lewis
Nabor, George W(illiam)
Nachlinger, R Ray
Narcowich, Francis Joseph
Naugle, Norman Wakefield
Nickel, James Alvin
Ober, Raimund Johannes
Ochiai, Shinya
Olson, Danford Harold
Pearcy, Carl Mark, Jr
Peterson, Lynn Louise Meister
Prats, Michael
Rajagopalan, Raj
Renka, Robert Joseph
Robbins, Kay Ann
Roberts, Thomas M
Rosenbaum, Joseph Hans
Rudin, Bernard D
Sadler, Stanley Gene
Sandberg, I(rwin) W(alter)
Sanders, Bobby Lee
Schruben, Johanna Stenzel
Schulz, Richard Burkart
Seth, Mohan S
Shieh, Leang-San
Smith, Philip Wesley
Srnka, Leonard James
Stone, Herbert L(osson)
Stubbs, Norris
Styblinski, Maciej A
Tao, Frank F
Trench, William Frederick
Trost, Hans-Jochen
Turner, Danny William
Von Rosenberg, Dale Ursini
Waid, Margaret Cowsar
Walton, Jay R
Warren, William Ernest
Weiser, Alan
White, Ronald Joseph
Wiggins, James Wendell
Williams, Glen Nordyke
Wiltschko, David Vilander
Witterholt, Edward John
Yakin, Mustafa Zafer
Zhou, Hua Wei

UTAH
Alfeld, Peter
Bates, Peter William
Coles, William Jeffrey
Davey, Gerald Leland
Fletcher, Harvey Junior
Fletcher, Thomas Harvey
Horn, Roger Alan
Kelsey, Stephen Jorgensen
Larsen, Kenneth Martin
Othmer, Hans George
Paegle, Julia Nogues
Sandquist, Gary Marlin
Skarda, R Vencil, Jr
Snow, Donald Ray
Stenger, Frank
Tadjeran, Hamid
Walker, Homer Franklin
Wilcox, Calvin Hayden
Windham, Michael Parks

VERMONT
Olinick, Michael
Oughstun, Kurt Edmund
Pinder, George Francis

VIRGINIA
Adam, John Anthony
Andersen, Carl Marius
Arnberg, Robert Lewis
Bailey, Marion Crawford
Bakhshi, Vidya Sagar
Bartelt, Martin William
Buchal, Robert Norman

Burns, John Allen
Collier, Manning Gary
Coram, Donald Sidney
Davidson, John Richard
Davis, William Arthur
Debney, George Charles, Jr
Dorning, John Joseph
Dozier, Lewis Bryant
Elkins, Judith Molinar
Evans, David Arthur
Gale, Harold Walter
Gerlach, A(lbert) A(ugust)
Gray, F(estus) Gail
Grotte, Jeffrey Harlow
Gunter, Edgar Jackson, Jr
Gunzburger, Max Donald
Hamilton, Thomas Charles
Hannsgen, Kenneth Bruce
Henry, Myron S
Herdman, Terry Lee
Hoffman, Karla Leigh
Horgan, Cornelius Oliver
Houbolt, John C(ornelius)
Hunt, Leon Gibson
Hurley, William Jordan
Hwang, John Dzen
Hwang, William Gaong
Johnson, Charles Royal
Joshi, Suresh Meghashyam
Juang, Jer-Nan
Kapania, Rakesh Kumar
Kay, Irvin (William)
Kroll, John Ernest
Kydes, Andy Steve
LaBudde, Robert Arthur
Lanzano, Paolo
Lasiecka, Irena
LeVan, Martin Douglas, Jr
Low, Emmet Francis, Jr
McNeil, Phillip Eugene
Mandelberg, Martin
Mansfield, Lois E
Marlow, William Henry
Michalowicz, Joseph Victor
Monacella, Vincent Joseph
Morse, Burt Jules
Myers, Michael Kenneth
Nayfeh, Ali Hasan
Perry, Charles Rufus, Jr
Peterson, Janet Sylvia
Pettus, William Gower
Poon, Ting-Chung
Pruett, Charles David
Raychowdhury, Pratip Nath
Rehm, Allan Stanley
Renardy, Michael
Renardy, Yuriko
Roberts, William Woodruff, Jr
Rokni, Mohammad Ali
Saltz, Joel Haskin
Schmeelk, John Frank
Schwiderski, Ernst Walter
Sebastian, Richard Lee
Shoosmith, John Norman
Shrier, Stefan
Simmonds, James G
Smith, Sidney Taylor
Snyder, Herbert Howard
Sobieszczanski-Sobieski, Jaroslaw
Stone, Lawrence David
Tompkins, Stephen Stern
Triggiani, Roberto
Tweed, John
Tyson, John Jeanes
Vaughan, Christopher Leonard
Wayne, Jennifer Susan
Werbos, Paul John
Wohl, Philip R

WASHINGTON
Berger, Albert Jeffrey
Burkhart, Richard Henry
Cochran, James Alan
Cokelet, Edward Davis
Criminale, William Oliver, Jr
Falco, James William
Galas, David John
Gibbs, Alan Gregory
Goldstein, Allen A
Harrison, Don Edmunds
Kevorkian, Jirair
Lessor, Delbert Leroy
LeVeque, Randall J
Levinson, Mark
Marshall, Donald E
Moody, Michael Eugene
Nievergelt, Yves
O'Malley, Robert Edmund, Jr
Rehfield, David Michael
Reinhardt, William Parker
Richardson, Robert Laurel
Rimbey, Peter Raymond
Robel, Gregory Frank
Roetman, Ernest Levane
Storwick, Robert Martin
Street, Robert Elliott
Trent, Donald Stephen
Tripp, Leonard L
Tung, Ka-Kit
Wan, Frederic Yui-Ming

WEST VIRGINIA
Brady, Dennis Patrick
Bryant, Robert William

Cavalier, John F
Mark, Christopher
Squire, William
Thakur, Pramod Chandra
Trapp, George E, Jr

WISCONSIN
Balmer, Robert Theodore
Dickey, Ronald Wayne
Divjak, August A
Hickman, James Charles
Johnson, Millard Wallace, Jr
Karreman, Herman Felix
Malkus, David Starr
Mangasarian, Olvi Leon
Moore, Robert H
Morin, Dornis Clinton
Pollnow, Gilbert Frederick
Rang, Edward Roy
Schultz, David Harold
Shah, Ghulam M
Shen, Mei-Chang
Skinner, Lindsay A
Stahl, Neil
Strikwerda, John Charles
Tiedemann, William Harold
Uhlenbrock, Dietrich A
Wang, Pao-Kuan

WYOMING
Zhang, Renduo

PUERTO RICO
Khan, Winston

ALBERTA
Aggarwala, Bhagwan D
Blais, J A Rodrigue
Collins, David Albert Charles
Dhaliwal, Ranjit S
Epstein, Marcelo
Grabenstetter, James Emmett
Ludwig, Garry (Gerhard Adolf)
Majumdar, Samir Ranjan
Nandakumar, Krishnaswamy
Westbrook, David Rex

BRITISH COLUMBIA
Clark, Colin Whitcomb
Fahlman, Gregory Gaylord
Graham, George Alfred Cecil
Gruver, William A
Halabisky, Lorne Stanley
Haussmann, Ulrich Gunther
Hughes, Blyth Alvin
Lardner, Robin Willmott
Leimanis, Eugene
Lu, Wu Sheng
Ludwig, Donald A
Millar, Robert Fyfe
Miura, Robert Mitsuru
Reed, William J
Seymour, Brian Richard
Shoemaker, Edward Milton
Singh, Manohar
Srivastava, Hari Mohan
Srivastava, Rekha

MANITOBA
Barclay, Francis Walter
Basilevsky, Alexander
Jayas, Digvir Singh
Thomas, Robert Spencer David
Usmani, Riaz Ahmad
Zeiler, Frederick

NEW BRUNSWICK
Tory, Elmer Melvin

NEWFOUNDLAND
May, Sherry Jan
Rochester, Michael Grant

NOVA SCOTIA
Barzilai, Jonathan
Bhartia, Prakash
Kujath, Marek Ryszard
Mohammed, Auyuab
Tan, Kok-Keong

ONTARIO
Aczel, Janos D
Ahmed, Nasir Uddin
Baillie, Donald Chesley
Barrie, Leonard Arthur
Barron, Ronald Michael
Barton, Richard Donald
Billigheimer, Claude Elias
Blackwell, John Henry
Campbell, Louis Lorne
Campeau, Radu Ioan
Carver, Michael Bruce
Chan, Yat Yung
Cowper, George Richard
Davison, Sydney George
Derzko, Nicholas Anthony
Duever, Thomas Albert
Duff, George Francis Denton
Dunn, D(onald) W(illiam)
Ehrman, Joachim Benedict
Fairman, Frederick Walker
Fraser, Peter Arthur
Georganas, Nicolas D
Gladwell, Graham M L

Hancox, William Thomas
Hogarth, Jacke Edwin
Horbatsch, Marko M
Howland, James Lucien
Huschilt, John
Iverson, Kenneth Eugene
Jackson, Kenneth Ronald
Jarvis, Roger George
Kesarwani, Roop Narain
Kirby, Bruce John
Langford, William Finlay
Lawson, John Douglas
Leach, Barrie William
Lehma, Alfred Baker
Lipshitz, Stanley Paul
McGee, William F
Mandelis, Andreas
Mandl, Paul
Mercer, Alexander McDowell
Molder, S(annu)
Mufti, Izhar H
Nash, John Christopher
Naylor, Derek
Nerenberg, Morton Abraham
Norminton, Edward Joseph
Oldham, Keith Bentley
Pfalzner, Paul Michael
Ponzo, Peter James
Rahman, Mizanur
Ranger, Keith Brian
Ray, Ajit Kumar
Reza, Fazlollah M
Rimrott, F(riedrich) P(aul) J(ohannes)
Ross, Roderick Alexander
Rowe, Ronald Kerry
Sablatash, Mike
Selvadurai, A P S
Sherbourne, Archibald Norbert
Sida, Derek William
Stauffer, Allan Daniel
Sullivan, Paul Joseph
Tzoganakis, Costas
Wainwright, John
Wesson, Paul Stephen
Wonham, W Murray
Woodside, William

QUEBEC
Charette, Andre
Clarke, Francis
Davies, Roger
Fabrikant, Valery Isaak
Jumarie, Guy Michael
Lefebvre, Mario
Li, Zi-Cai
Luckert, H(ans) J(oachim)
Morgera, Salvatore Domenic
Mysak, Lawrence Alexander
Roth, Charles
Sankoff, David
Schlomiuk, Dana
Tam, Kwok Kuen
Zlobec, Sanjo

SASKATCHEWAN
Esmail, Mohamed Nabil
Koh, Eusebio Legarda
Law, Alan Greenwell
Symes, Lawrence Richard
Walerian, Szyszkowski

OTHER COUNTRIES
Aumann, Robert John
Balinski, Michel Louis
Blackburn, Jacob Floyd
Callier, Frank Maria
Chwang, Allen Tse-Yung
Dinnar, Uri
Dougalis, Vassilios
Greenblatt, Seth Alan
Hamacher, Horst W
Hawking, Stephen W
Ho, Yew Kam
Hochberg, Kenneth J
Hsieh, Din-Yu
Idelsohn, Sergio Rodolfo
Kimura, Hidenori
Kydoniefs, Anastasios D
Lapostolle, Pierre Marcel
Laura, Patricio Adolfo Antonio
Lindquist, Anders Gunnar
Liu, Ta-Jo
Luo, Peilin
Malah, David
Maqusi, Mohammad
Mark, James Wai-Kee
May, Robert McCredie
Nievergelt, Jurg
Orphanoudakis, Stelios Constantine
Raviv, Josef
Roseman, Joseph Jacob
Sarmiento, Gustavo Sanchez
Segel, Lee Aaron
Stoneham, Richard George
Taylor, Howard Lawrence
Tolstoy, Ivan
Wendland, Wolfgang Leopold
White, Simon David Manton
Willems, Jan C
Wong, Roderick Sue-Cheun
Wu, Lancelot T L
Yang, Chung-Chun
Zeheb, Ezra

Biomathematics

ALABAMA
Cook, Frederick Lee
Katholi, Charles Robinson
Siler, William MacDowell
Turner, Malcolm Elijah

ARIZONA
Aickin, Mikel G
Colburn, Wayne Alan
Dickinson, Frank N
Dorson, William John
Jacobowitz, Ronald
Kessler, John Otto
Smith, Hal Leslie

CALIFORNIA
Altes, Richard Alan
Biles, Charles Morgan
Bremermann, Hans J
Busenberg, Stavros Nicholas
Carew, Thomas Edward
Cavalli-Sforza, Luigi Luca
Cooke, Kenneth Lloyd
Cooper, William S
Elderkin, Richard Howard
Evans, John W
Feldman, Marcus William
Greever, John
Hastings, Alan Matthew
Hirsch, Morris William
Hood, J Myron
Huang, Sung-Cheng
Keller, Joseph Bishop
Kope, Robert Glenn
Lamberson, Roland H
Landahl, Herbert Daniel
Lein, Allen
Licko, Vojtech
Mitchell, John Alexander
Neustadter, Siegfried Friedrich
Oster, George F
Purdue, Peter
Richardson, Irvin Whaley
Scheibe, Paul Otto
Sinsheimer, Janet Suzanne
Slatkin, Montgomery (Wilson)
Whittemore, Alice S
Wiggins, Alvin Dennie
Williams, William Arnold

COLORADO
Briggs, William Egbert
Miller, Arnold Reed
Reiner, John Maximilian
Warren, Peter

CONNECTICUT
Collins, Sylva Heghinian
Darcey, Terrance Michael
McCartin, Brian James
Wachman, Murray

DISTRICT OF COLUMBIA
Jernigan, Robert Wayne

FLORIDA
Jacobson, Jerry I
Jones, Albert Cleveland
Keesling, James Edgar
Lyman, Gary Herbert
Stark, William Richard
White, Timothy Lee

GEORGIA
Arnold, Jonathan
Mickens, Ronald Elbert
Shonkwiler, Ronald Wesley
Waltman, Paul Elvis

HAWAII
Polovina, Jeffrey Joseph

IDAHO
Bobisud, Larry Eugene
Hausrath, Alan Richard

ILLINOIS
Calderon, Calixto Pedro
Cowan, Jack David
Deysach, Lawrence George
Elble, Rodger Jacob
Feinstein, Irwin K
Lauffenburger, Douglas Alan
Rosenblatt, Karin Ann
Skeel, Robert David
Smeach, Stephen Charles

INDIANA
Cerimele, Benito Joseph
Madan, Ved P
Thompson, Maynard

IOWA
Cornette, James L
Dexter, Franklin, III
Hethcote, Herbert Wayne
Kirkham, Don

KANSAS
Mainster, Martin Aron
Stewart, William Henry

KENTUCKY
Hayden, Thomas Lee
Hirsch, Henry Richard
McMorris, Fred Raymond

LOUISIANA
Heatherly, Henry Edward

MARYLAND
Alexander, James Crew
Altschul, Stephen Frank
Chen, Tar Timothy
Eden, Murray
Feldman, Jacob J
Hiatt, Caspar Wistar, III
Jessup, Gordon L, Jr
Lachin, John Marion, III
Langenberg, Patricia Warrington
Marimont, Rosalind Brownstone
Massell, Paul Barry
Mazur, Joel
Serfling, Robert Joseph
Verter, Joel I
Yorke, James Alan

MASSACHUSETTS
Butler, James Preston
Carpenter, Gail Alexandra
Collins, James Joseph
Fishman, Philip M
Hattis, Dale B
Janowitz, Melvin Fiva
Keller, Evelyn Fox
Lagakos, Stephen William
Stanley, Kenneth Earl

MICHIGAN
Conrad, Michael
Harrison, Michael Jay
Metzler, Carl Maust
Nyman, Melvin Andrew
Simon, Carl Paul

MINNESOTA
Ek, Alan Ryan
Kemper, John Thomas
Rescigno, Aldo

MISSOURI
DeFacio, W Brian
Green, Philip Palmer
Larson, Kenneth Blaine
Luxon, Bruce Arlie
Taibleson, Mitchell H

MONTANA
Derrick, William Richard

NEBRASKA
Jenkins, Thomas Gordon
Schutz, Wilfred M
Ueda, Clarence Tad

NEW JERSEY
Ahsanullah, Mohammad
Andrushkiw, Roman Ihor
Cronin, Jane Smiley
Frauenthal, James Clay
Levin, Simon Asher
Pilla, Arthur Anthony
Sverdlove, Ronald
Tank, David W

NEW MEXICO
Cobb, Loren
Diel, Joseph Henry
Gibbs, William Royal
Ortiz, Melchor D
Perelson, Alan Stuart

NEW YORK
Altshuler, Bernard
Anderson, Allan George
Bell, Jonathan George
Benham, Craig John
Berresford, Geoffrey Case
Blessing, John A
Blumenson, Leslie Eli
Brennan, Murray F
Chandran, V Ravi
Hartnett, William Edward
Hood, Donald C
Isaacson, David
Leibovic, K Nicholas
Lubowsky, Jack
Nemethy, George
Norton, Larry
Priore, Roger L
Rabbany, Sina Y
Ruppert, David
Schimmel, Herbert
Shalloway, David Irwin
Sherman, Malcolm J
Simon, William
Tamari, Dov
Teich, Malvin Carl
Yarmush, David Leon

NORTH CAROLINA
Bishir, John William
Faulkner, Gary Doyle
Floyd, George E, Jr
Franke, John Erwin
Johnson, Thomas
Lobaugh, Bruce
Portier, Christopher Jude
Swallow, William Hutchinson
Van der Vaart, Hubertus Robert
Woodbury, Max Atkin

NORTH DAKOTA
Penland, James Granville

OHIO
Bhargava, Triloki Nath
Friedlander, Ira Ray

OKLAHOMA
Mulholland, Robert J(oseph)
Weeks, David Lee

OREGON
Mullooly, John P
Murphy, Lea Frances

PENNSYLVANIA
Ermentrout, George Bard
Hodgson, Jonathan Peter Edward
Karreman, George
Merrill, Samuel, III
Rorres, Chris
Scherer, Peter William
Tallarida, Ronald Joseph

SOUTH CAROLINA
Bohning, Daryl Eugene
Luedeman, John Keith

TENNESSEE
Bernard, Selden Robert
Dale, Virginia House
DeAngelis, Donald Lee
Demetriou, Charles Arthur
Dixon, Edmond Dale
Hallam, Thomas Guy
Partain, Clarence Leon
Partridge, Lloyd Donald
Rose, Kenneth Alan
Tan, Wai-Yuan
Zeighami, Elaine Ann

TEXAS
Atkinson, Edward Neely
Blakley, George Robert
Brakefield, James Charles
Brown, Barry W
Eisenfeld, Jerome
Gutierrez, Guillermo
Jansson, Birger
Saltzberg, Bernard
Sterner, Robert Warner
Thames, Howard Davis, Jr
White, Robert Allen
White, Ronald Joseph
Zimmerman, Stuart O

UTAH
Coles, William Jeffrey
Othmer, Hans George

VIRGINIA
Grant, John Wallace
Heineken, Frederick George
Hohenboken, William Daniel
Holtzman, Golde Ivan
Hutchinson, Thomas Eugene
Leczynski, Barbara Ann
Orcutt, Bruce Call
Wohl, Philip R

WASHINGTON
Britt, Patricia Marie
Conquest, Loveday Loyce
Francis, Robert Colgate
Gunderson, Donald Raymond
Moody, Michael Eugene
Moolgavkar, Suresh Hiraji
Siegel, Andrew Francis
Swartzman, Gordon Leni
Van der Werff, Terry Jay
Wijsman, Ellen Marie

WISCONSIN
Chover, Joshua
Klein, John Peter
Kurtz, Thomas Gordon
Schultz, David Harold
Walter, Gilbert G

PUERTO RICO
Peinado, Rolando E

ALBERTA
Freedman, Herbert I
Voorhees, Burton Hamilton

BRITISH COLUMBIA
Anderson, R F V
Hewgill, Denton Elwood
Ludwig, Donald A
Miura, Robert Mitsuru
Seymour, Brian Richard

MANITOBA
Somorjai, Rajmund Lewis

NOVA SCOTIA
Rosen, Robert

Biomathematics (cont)

ONTARIO
Bertell, Rosalie
Ferrier, Jack Moreland
Halfon, Efraim
Miller, Donald Richard
Rapoport, Anatol
Zuker, Michael

QUEBEC
Boyarsky, Abraham Joseph
Jolicoeur, Pierre
Lacroix, Norbert Hector Joseph
Sankoff, David

SASKATCHEWAN
Spurr, David Tupper

OTHER COUNTRIES
Bartlett, Maurice S
Hochberg, Kenneth J
Phelps, Frederick Martin, IV
Segel, Lee Aaron

Combinatorics & Finite Mathematics

ALABAMA
Seidman, Stephen Benjamin
Slater, Peter John

ALASKA
Beebee, John Christopher
Hulsey, J Leroy

ARIZONA
Clay, James Ray
Leonard, John Lander
Palusinski, Olgierd Aleksander

ARKANSAS
Schein, Boris M

CALIFORNIA
Alexanderson, Gerald Lee
Block, Richard Earl
Brandeau, Margaret Louise
Butler, Jon Terry
Carter, J Lawrence (Larry)
Dean, Richard Albert
Ford, Lester Randolph, Jr
Fraser, Grant Adam
Fredman, Michael Lawrence
Ghosh, Subir
Hakimi, S(eifollah) L(ouis)
Hales, Alfred Washington
Harper, Lawrence Hueston
Harrison, Michael A
Johnsen, Eugene Carlyle
Kaellis, Joseph
Kevorkian, Aram K
Lawler, Eugene L(eighton)
McKenzie, Ralph Nelson
Merris, Russell Lloyd
Minc, Henryk
Nathanson, Weston Irwin
Pease, Marshall Carleton, III
Rebman, Kenneth Ralph
Reid, Kenneth Brooks
Rhodes, John Lewis
Ringel, Gerhard
Sallee, G Thomas
Senge, George H
Simons, Barbara Bluestein
Stockmeyer, Larry Joseph
Sun, Hugo Sui-Hwan
Tanner, Robert Michael
Weeks, Dennis Alan
Welch, Lloyd Richard
Williamson, Stanley Gill

COLORADO
Clements, George Francis
Irwin, Robert Cook
Lundgren, J Richard
Manvel, Bennet
Maybee, John Stanley
Nemeth, Evi
Phillips, Keith L

CONNECTICUT
Blei, Ron Charles
Fine, Benjamin
Howe, Roger
Reid, James Dolan
Spiegel, Eugene
Stein, Alan H
Wolk, Elliot Samuel

DELAWARE
Ebert, Gary Lee

DISTRICT OF COLUMBIA
Garavelli, John Stephen
Sjogren, Jon Arne
Stein, Marjorie Leiter

FLORIDA
Butson, Alton Thomas
Cenzer, Douglas
Clark, William Edwin
Drake, David Allyn
Hoffman, Frederick
Ismail, Mourad E H
Olmstead, Paul Smith
Pedersen, John F
Pettofrezzo, Anthony J
Roy, Dev Kumar
Shershin, Anthony Connors

GEORGIA
Canfield, Earl Rodney
Duke, Richard Alter
Neff, Mary Muskoff
Rouvray, Dennis Henry
Straley, Tina

ILLINOIS
Cormier, Romae Joseph
Eggan, Lawrence Carl
Grace, Thom P
Gupta, Udaiprakash I
Kasube, Herbert Emil
Lee, Der-Tsai
Livingston, Marilyn Laurene
Otto, Albert Dean
Reingold, Edward Martin
Reznick, Bruce Arie
Vanden Eynden, Charles Lawrence
Wallis, Walter Denis
Weichsel, Paul M
West, Douglas Brent

INDIANA
Beineke, Lowell Wayne
Kallman, Ralph Arthur
Murphy, Catherine Mary
Peltier, Charles Francis
Penna, Michael Anthony
Pippert, Raymond Elmer
Swift, William Clement

IOWA
Ferguson, Pamela A
Robinson, John Paul

KANSAS
Davis, Elwyn H
Galvin, Fred
Shult, Ernest E

KENTUCKY
McMorris, Fred Raymond

LOUISIANA
Berman, David Michael
Maxfield, Margaret Waugh
Sharma, Bhudev

MAINE
Balakrishnan, V K

MARYLAND
Brant, Larry James
Franks, David A
Rukhin, Andrew Leo
Siegel, Martha J
Singhal, Jaya Asthana
Wierman, John C

MASSACHUSETTS
Albertson, Michael Owen
Bolker, Ethan D
Dechene, Lucy Irene
Freund, Robert M
Gessel, Ira Martin
Hoffer, Alan R
Hyatt, Raymond R, Jr
Janowitz, Melvin Fiva
Joni, Saj-nicole A
Kenney, Margaret June
Koshy, Thomas
Levasseur, Kenneth M
Levin, Leonid A
O'Neil, Patrick Eugene
O'Rourke, Joseph
Ostrovsky, Rafail M
Rosenberg, Arnold Leonard
Simovici, Dan
Smith, Raoul Normand
Stanley, Richard Peter
Woods, William A

MICHIGAN
Alavi, Yousef
Arlinghaus, William Charles
Bachelis, Gregory Frank
Heuvers, Konrad John
Howard, Paul Edward
Lick, Don R
Stout, Quentin Fielden
VanderJagt, Donald W
White, Arthur Thomas, II

MINNESOTA
Anderson, Sabra Sullivan
Gallian, Joseph A
Hutchinson, Joan Prince

MISSOURI
Klein, Cerry M

MONTANA
Thompson, Edward Otis

NEBRASKA
Chouinard, Leo George, II
Kramer, Earl Sidney
Magliveras, Spyros Simos
Walters, Randall Keith

NEVADA
Shiue, Peter Jay-Shyong

NEW HAMPSHIRE
Appel, Kenneth I
Bogart, Kenneth Paul

NEW JERSEY
Allender, Eric Warren
Blickstein, Stuart I
Boesch, Francis Theodore
Dekker, Jacob Christoph Edmond
Faith, Carl Clifton
Foregger, Thomas H
Garey, Michael Randolph
Gewirtz, Allan
Holzmann, Gerard J
Kessler, Irving Jack
Lagarias, Jeffrey Clark
Lepowsky, James Ivan
Lesniak, Linda
Mills, William Harold
Naus, Joseph Irwin
Roberts, Fred Stephen
Saniee, Iraj
Tindell, Ralph S
Tucker, Albert William
Winkler, Peter Mann
Zierler, Neal

NEW MEXICO
Bradshaw, Martin Daniel
Entringer, Roger Charles
Faber, Vance
Harary, Frank
Hulme, Bernie Lee
Simmons, Gustavus James

NEW YORK
Anderson, Allan George
Baruch, Marjory Jean
Bencsath, Katalin A
Billera, Louis Joseph
Bing, Kurt
Buckley, Fred Thomas
Cohen, Daniel Isaac Aryeh
Coppersmith, Don
Cusick, Thomas William
Ecker, Joseph George
Feuer, Richard Dennis
Gerst, Irving
Goodman, Jacob Eli
Graver, Jack Edward
Green, Alwin Clark
Griesmer, James Hugo
Gross, Jonathan Light
Hausner, Melvin
Johnson, Sandra Lee
Kingston, Paul L
Knee, David Isaac
Kwong, Yui-Hoi Harris
Leary, Francis Christian
Martin, George Edward
Mattson, Harold F, Jr
Melter, Robert Alan
Mitchell, Joseph Shannon Baird
Mowshowitz, Abbe
Nathanson, Melvyn Bernard
Percus, Jerome K
Powers, David Leusch
Ralston, Anthony
Rispoli, Fred Joseph
Rosenthal, John William
Sellers, Peter Hoadley
Tamari, Dov
Tucker, Thomas William
Vasquez, Alphonse Thomas
Watkins, Mark E
Zaslavsky, Thomas
Zemanian, Armen Humpartsoum

NORTH CAROLINA
Blanchet-Sadri, Francine
Geissinger, Ladnor Dale
Hartwig, Robert Eduard
Hodel, Margaret Jones
Kay, David Clifford
Klerlein, Joseph Ballard
Reif, John H
Reiter, Harold Braun
Sagan, Hans

OHIO
Barger, Samuel Floyd
Bhargava, Triloki Nath
Gagola, Stephen Michael, Jr
Hahn, Samuel Wilfred
Hales, Raleigh Stanton, Jr
Hsia, John S
Khalimsky, Efim D
Maneri, Carl C
Nicholson, Victor Alvin
Ray-Chaudhuri, Dwijendra Kumar
Silverman, Robert
Solow, Daniel

OREGON
Stalley, Robert Delmer

PENNSYLVANIA
Andrews, George Eyre
Assmus, Edward F(erdinand), Jr
Balas, Egon
Bressoud, David Marius
Korfhage, Robert R
Mullen, Gary Lee
Schattschneider, Doris Jean
Simpson, Stephen G
Thompson, Gerald Luther
Tzeng, Kenneth Kai-Ming
Ventura, Jose Antonio
Wolfe, Dorothy Wexler

RHODE ISLAND
Simons, Roger Alan

SOUTH CAROLINA
Brawley, Joel Vincent, Jr
Hopkins, Laurie Boyle
Jamison, Robert Edward
Kostreva, Michael Martin
Luedeman, John Keith
McNulty, George Frank
Saxena, Subhash Chandra
Shier, Douglas Robert

TENNESSEE
Faudree, Ralph Jasper, Jr
Ordman, Edward Thorne
Plummer, Michael David
Schelp, Richard Herbert

TEXAS
Blakley, George Robert
Cecil, David Rolf
Girou, Michael L
Hartfiel, Darald Joe
Matula, David William
Parberry, Ian
Tarwater, Jan Dalton

UTAH
Gross, Fletcher
Skarda, R Vencil, Jr
Snow, Donald Ray

VERMONT
Foote, Richard Martin

VIRGINIA
Collier, Manning Gary
Gray, F(estus) Gail
Heath, Lenwood S
Hoffman, Karla Leigh
Johnson, Charles Royal
Lipman, Marc Joseph
Norris, Eugene Michael
Rehm, Allan Stanley
Ringeisen, Richard Delose
Rosenbaum, David Mark
Roussos, Constantine
Spresser, Diane Mar
Stockmeyer, Paul Kelly
Ward, Harold Nathaniel

WASHINGTON
DeTemple, Duane William
Grunbaum, Branko
Hansen, Rodney Thor
Kallaher, Michael Joseph
Long, Calvin Thomas
Nijenhuis, Albert
Webb, William Albert

WEST VIRGINIA
Gould, Henry Wadsworth

WISCONSIN
Bach, Eric
Barkauskas, Anthony Edward
Crowe, Donald Warren
Fournelle, Thomas Albert
Harris, Bernard
Kaye, Norman Joseph
Moore, Edward Forrest
Pemantle, Robin A
Solomon, Marvin H
Terwilliger, Paul M

WYOMING
Bridges, William G
Di Paola, Jane Walsh
Porter, A Duane
Shank, Herbert S

PUERTO RICO
Moreno, Oscar

ALBERTA
Moon, John Wesley
Rhemtulla, Akbar Hussein

BRITISH COLUMBIA
Alspach, Brian Roger
Boyd, David William
Casselman, William Allen
Matheson, David Stewart
Moyls, Benjamin Nelson
Riddell, James
Totten, James Edward

MANITOBA
Batten, Lynn Margaret

Dowling, Diane Mary
Mendelsohn, Nathan Saul

NEWFOUNDLAND
Rees, Rolf Stephen

NOVA SCOTIA
Kruse, Robert Leroy

ONTARIO
Balakrishnan, Narayanaswany
Colbourn, Charles Joseph
Coxeter, Harold Scott MacDonald
Davison, J Leslie
Djokovic, Dragomir Z
Garner, Cyril Wilbur Luther
Lehma, Alfred Baker
Lehman, Alfred Baker
Pullman, Norman J
Roe, Peter Hugh O'Neil
Schellenberg, Paul Jacob
Steiner, George
Vanstone, Scott Alexander
Whittington, Stuart Gordon

QUEBEC
Allen, Harold Don
Brown, William G
Leroux, Pierre
Moser, William O J
Sabidussi, Gert Otto

SASKATCHEWAN
Sato, Daihachiro

OTHER COUNTRIES
Balinski, Michel Louis
Bannai, Eichi
Brown, Richard Harland
Gritzmann, Peter
Hamacher, Horst W
Myerson, Gerald
Olive, Gloria

Geometry

ARIZONA
Clay, James Ray
Groemer, Helmut (Johann)

CALIFORNIA
Agoston, Max Karl
Alexanderson, Gerald Lee
Amemiya, Frances (Louise) Campbell
Barnett, David
Barsky, Brian Andrew
Blattner, Robert J(ames)
Blinn, James Frederick
Casson, Andrew J
Chakerian, Gulbank Donald
Chan, Jean B
Chern, Shiing-Shen
Collier, James Bryan
Feldman, Louis A
Finn, Robert
Green, Mark Lee
Greene, Robert Everist
Krener, Arthur James
Labarre, Anthony E, Jr
Lewis, George McCormick
Li, Peter Wai-Kwong
Millman, Richard Steven
Moore, John Douglas
Olwell, Kevin D
Osserman, Robert
Rabin, Jeffrey Mark
Richman, David Paul
Ringel, Gerhard
Sallee, G Thomas
Schlafly, Roger
Tuller, Annita
Wavrik, John J
Weinstein, Alan David
Williamson, Robert Emmett
Wolf, Joseph Albert
Wu, Hung-Hsi

COLORADO
Derr, James Bernard

CONNECTICUT
Abikoff, William
Howe, Roger
Mostov, George Daniel
Seitelman, Leon Harold
Wachspress, Eugene Leon

DELAWARE
Ebert, Gary Lee

FLORIDA
Ehrlich, Paul Ewing
Johnson, Harold Hunt
Magarian, Elizabeth Ann

GEORGIA
Batterson, Steven L
Sparks, Arthur Godwin

ILLINOIS
Abler, William Lewis
Albrecht, Felix Robert
Bishop, Richard Lawrence

Earnest, Andrew George
Eichten, Estia Joseph
Fossum, Robert Merle
Haboush, William Joseph
Heitsch, James Lawrence
Ho, Chung-Wu
Jordan, Steven Lee
Leibhardt, Edward
Litvin, Faydor L
Portnoy, Esther
Wagreich, Philip Donald
Waterman, Peter Lewis
Wetzel, John Edwin
Yau, Stephen Shing-Toung

INDIANA
Ludwig, Hubert Joseph
Mast, Cecil B
Peltier, Charles Francis
Penna, Michael Anthony
Penney, Richard Cole
Pohl, Victoria Mary
Sommese, Andrew John
Specht, Edward John
Stanton, Charles Madison

IOWA
Adelberg, Arnold M
Johnson, Norman L
Jorgensen, Palle E T
Kleinfeld, Erwin
Randell, Richard
Ton-That, Tuong

KANSAS
Lerner, David Evan
Shult, Ernest E

KENTUCKY
Lowman, Bertha Pauline

MAINE
Murphy, Grattan Patrick

MARYLAND
Blum, Harry
Fischer, Irene Kaminka
Gray, Alfred
Maloney, Clifford Joseph
Meyerson, Mark Daniel
Rosenberg, Jonathan Micah
Shiffman, Bernard
Witzgall, Christoph Johann

MASSACHUSETTS
Altman, Allen Burchard
Bennett, Mary Katherine
Bolker, Ethan D
Cattani, Eduardo
Cecil, Thomas E
Cox, David Archibald
Halperin-Maya, Miriam Patricia
Helgason, Sigurdur
Hoffer, Alan R
Hoffman, David Allen
Mann, Larry N
Matsusaka, Teruhisa
Mattuck, Arthur Paul
Mauldon, James Grenfell
Melrose, Richard B
O'Rourke, Joseph
Sacks, Jonathan
Senechal, Marjorie Lee
Struik, Dirk Jan
Tanimoto, Taffee Tadashi
Whitman, Andrew Peter

MICHIGAN
Albaugh, A Henry
Arlinghaus, Sandra Judith Lach
Bookstein, Fred Leon
Chen, Bang-Yen
Houh, Chorng Shi
Jones, Harold Trainer
Whitmore, Edward Hugh

MINNESOTA
Gulliver, Robert David, II
Nitsche, Johannes Carl Christian
Sperber, Steven Irwin

MISSOURI
Andalafte, Edward Ziegler
Beem, John Kelly
Cantwell, John Christopher
Freese, Raymond William
Jenkins, James Allister
Jensen, Gary Richard
Mashhoon, Bahram
Petty, Clinton Myers
Valentine, Joseph Earl
Wilson, Edward Nathan

MONTANA
Thompson, Edward Otis

NEBRASKA
Gross, Mildred Lucile
Magliveras, Spyros Simos

NEW HAMPSHIRE
Arkowitz, Martin Arthur
Gordon, Carolyn Sue
Snapper, Ernst

NEW JERSEY
Elk, Seymour B
Faith, Carl Clifton
Faltings, Gerd
Goldberg, Vladislav V
Papantonopoulou, Aigli Helen
Posamentier, Alfred S
Taylor, Jean Ellen
Van Wyk, Christopher John
Weinstein, Tilla
Young, Li

NEW MEXICO
Zund, Joseph David

NEW YORK
Abbott, John S
Adhout, Shahla Marvizi
Anderson, Peter Gordon
Barber, Sherburne Frederick
Barshay, Jacob
Billera, Louis Joseph
Blanton, John David
Chavel, Isaac
Dodziuk, Jozef
Dubisch, Russell John
Feldman, Edgar A
Fialkow, Aaron David
Gerber, Leon E
Goodman, Jacob Eli
Green, Alwin Clark
Hausner, Melvin
Henderson, David Wilson
Iwaniec, Tadeusz
Johnson, Sandra Lee
Martin, George Edward
Mitchell, Joseph Shannon Baird
Nirenberg, Louis
Pinkham, Henry Charles
Range, R Michael
Seppala-Holtzman, David N
Srinivasan, Vijay
Sweeney, Thomas Francis
Trasher, Donald Watson
Vasquez, Alphonse Thomas
Warner, Frederic Cooper
Zaslavsky, Thomas
Zaustinsky, Eugene Michael

NORTH CAROLINA
Eberlein, Patrick Barry
Fulp, Ronald Owen
Gardner, Robert B
Jambor, Paul Emil
Kay, David Clifford
King, Lunsford Richardson
Kumar, Shrawan
Schell, Joseph Francis
Sowell, Katye Marie Oliver
Stitzinger, Ernest Lester
Stroud, Junius Brutus
York, James Wesley, Jr

OHIO
Berton, John Andrew
Cummins, Kenneth Burdette
Ferrar, Joseph C
Houston, William Bernard, Jr
Klosterman, Albert Leonard
Kullman, David Elmer
Lee, Dong Hoon
Maneri, Carl C
Sholander, Marlow
Silverman, Robert
Weber, Waldemar Carl
Yaqub, Jill Courtaney Donaldson Spencer

OKLAHOMA
Breen, Marilyn
Goodey, Paul Ronald
McCullough, Darryl J

OREGON
Bhattacharyya, Ramendra Kumar
Flaherty, Francis Joseph
Gilkey, Peter Belden
Isenberg, James Allen
Koch, Richard Moncrief
Parks, Harold Raymond
Tewinkel, G Carper
Todd, Philip Hamish

PENNSYLVANIA
Baildon, John David
Buyske, Steve George
Ehrmann, Rita Mae
Grinberg, Eric L
Harbater, David
James, Donald Gordon
Kazdan, Jerry Lawrence
Schaffer, Juan Jorge
Schattschneider, Doris Jean
Warner, Frank Wilson, III
Waterhouse, William Charles

RHODE ISLAND
Datta, Dilip Kumar
Rudolph, Lee

SOUTH CAROLINA
Jamison, Robert Edward
Layton, William Isaac
Saxena, Subhash Chandra

Wylie, Clarence Raymond, Jr

TENNESSEE
Dobbs, David Earl

TEXAS
Aberth, Oliver George
Bakelman, Ilya J(acob)
Golubitsky, Martin Aaron
McCown, Malcolm G
Stiller, Peter Frederick
Vaaler, Jeffrey David
Vick, George R
Weaver, Milo Wesley
Whittlesey, John R B

UTAH
Carlson, James Andrew
Clemens, Charles Herbert
Milicic, Dragan
Toledo, Domingo

VIRGINIA
Gold, Sydell Perlmutter
Gowdy, Robert Henry
Parshall, Brian J

WASHINGTON
Blumenthal, Robert Allan
DeTemple, Duane William
Dubois, Donald Ward
Duchamp, Thomas Eugene
Firkins, John Frederick
Grunbaum, Branko
Kallaher, Michael Joseph
Nievergelt, Yves
Nijenhuis, Albert

WEST VIRGINIA
Corum, James Frederic

WISCONSIN
Bleicher, Michael Nathaniel
Crowe, Donald Warren

WYOMING
Di Paola, Jane Walsh
Roth, Ben G

ALBERTA
Antonelli, Peter Louis
Honsaker, John Leonard

BRITISH COLUMBIA
Alspach, Brian Roger
Boyd, David William
Lancaster, George Maurice
Totten, James Edward

MANITOBA
Batten, Lynn Margaret
Thomas, Robert Spencer David

NEW BRUNSWICK
Dekster, Boris Veniamin

ONTARIO
Coxeter, Harold Scott MacDonald
Djokovic, Dragomir Z
Garner, Cyril Wilbur Luther
Mann, Robert Bruce
Robinson, Gilbert de Beauregard
Scherk, Peter
Sherk, Frank Arthur
Whitfield, John Howard Mervyn

QUEBEC
Schlomiuk, Dana

SASKATCHEWAN
Sato, Daihachiro

OTHER COUNTRIES
Crapo, Henry Howland
Gritzmann, Peter
Hughes, Daniel Richard
Kambayashi, Tatsuji
Lay, Steven R
Thomason, Robert Wayne

Logic

ARIZONA
Smarandache, Florentin

ARKANSAS
Berghel, Hal L
Schein, Boris M

CALIFORNIA
Addison, John West, Jr
Arens, Yigal
Bischof, Nobert S
Blum, Lenore Carol
Bull, Everett L, Jr
Calabrese, Philip G
Cannonito, Frank Benjamin
Chang, Chen Chung
Church, Alonzo
Cooper, William S
Deutsch, Laurence Peter
Eklof, Paul Christian
Enderton, Herbert Bruce

Logic (cont)

Fagin, Ronald
Feferman, Solomon
Fraser, Grant Adam
Givant, Steven Roger
Halmos, Paul Richard
Halpern, James Daniel
Henkin, Leon (Albert)
Lender, Adam
McKenzie, Ralph Nelson
McLaughlin, William Irving
Manaster, Alfred B
Mignone, Robert Joseph
Moschovakis, Joan Rand
Moschovakis, Yiannis N
Pixley, Alden F
Pratt, Vaughan Ronald
Russell, Stuart Jonathan
Vaught, Robert L
Wiebe, Richard Penner
Wolf, Robert Stanley

COLORADO
Reinhardt, William Nelson

CONNECTICUT
Goldman, Ernest Harold
Lerman, Manuel
Sharlow, John Francis
Wood, Carol Saunders

DELAWARE
Case, John William
Chester, Daniel Leon

DISTRICT OF COLUMBIA
Curtin, Frank Michael
McLean, John Dickson
Nelson, David
Rosenblatt, David

FLORIDA
Cenzer, Douglas
Hughes, Charles Edward
Levitz, Hilbert
Miller, Don Dalzell
Pedersen, John F
Roy, Dev Kumar
Stark, William Richard

GEORGIA
Covington, Michael A

ILLINOIS
Baldwin, John Theodore
Clover, William John, Jr
Cormier, Romae Joseph
Henson, C Ward
Jockusch, Carl Groos, Jr
MacLane, Saunders
Raiden, Robert Lee
Rips, Lance Jeffrey
Soare, Robert I
Takeuti, Gaisi
Wilson, Howell Kenneth

INDIANA
Cunningham, Ellen M
Peltier, Charles Francis
Rubin, Jean E
Wheeler, William Hollis

IOWA
Alton, Donald Alvin
Nelson, George C
Pigozzi, Don

KANSAS
Galvin, Fred
Roitman, Judy

KENTUCKY
Crawford, Robert R

MAINE
Pogorzelski, Henry Andrew

MARYLAND
Bernstein, Allen Richard
Chapin, Edward William, Jr
Feldman, Robert H L
Gasarch, William Jan
Green, Judy
Herrmann, Robert Arthur
Hodes, Louis
Horna, Otakar Anthony
Kueker, David William
Owings, James Claggett, Jr
Rynasiewicz, Robert Alan
Smith, Mark Stephen
Suppe, Frederick (Roy)

MASSACHUSETTS
Barnes, Arnold Appleton, Jr
Bruce, Kim Barry
Gacs, Peter
Henle, James Marston
Hirschman, Lynette
Homer, Steven Elliott
Levin, Leonid A
Lukas, Joan Donaldson
Marsh, William Ernest
Meyer, Albert Ronald

Quine, Willard V
Rogers, Hartley, Jr
Sacks, Gerald Enoch
Shorb, Alan McKean
Wand, Mitchell

MICHIGAN
Hinman, Peter Greayer
Howard, Paul Edward
Lenker, Susan Stamm
Stob, Michael Jay
Wojcik, Anthony Stephen

MINNESOTA
Fuhrken, Gebhard
Gaal, Ilse Lisl Novak
Pierce, Keith Robert
Pour-El, Marian Boykan

MISSOURI
Andalafte, Edward Ziegler

MONTANA
Barrett, Louis Carl

NEW JERSEY
Abeles, Francine
Dekker, Jacob Christoph Edmond
Nonnenmann, Uwe
Osofsky, Barbara Langer
Rosenstein, Joseph Geoffrey
Van Wyk, Christopher John
Winkler, Peter Mann

NEW MEXICO
Diel, Joseph Henry
Rypka, Eugene Weston

NEW YORK
Allen, James Frederick
Bing, Kurt
Bowen, Kenneth Alan
Chan, Tat-Hung
Cogan, Edward J
Constable, Robert L
Davis, Martin (David)
De Lillo, Nicholas Joseph
Di Paola, Robert Arnold
Goodman, Nicolas Daniels
Hartnett, William Edward
Hechler, Stephen Herman
Hughes, Harold K
Kopperman, Ralph David
Lercher, Bruce L
Mendelson, Elliott
Morley, Michael Darwin
Nerode, Anil
Parikh, Rohit Jivanlal
Platek, Richard Alan
Rapaport, William Joseph
Rosenthal, John William
Rutledge, Joseph Dela
Saracino, Daniel Harrison
Shapiro, Stuart Charles
Shore, Richard A
Sibert, Elbert Ernest
Sweeney, Thomas Francis
Wang, Hao
Williams, Scott Warner
Zuckerman, Martin Michael

NORTH CAROLINA
Blanchet-Sadri, Francine
Hodel, Richard Earl
Loveland, Donald William
Mitchell, William John
Ras, Zbigniew Wieslaw

OHIO
Applebaum, Charles H
Foreman, Matthew D
Friedman, Harvey Martin
Glass, Andrew Martin William
McCloskey, Teresemarie
McLarty, Colin Slator
Mendenhall, Robert Vernon
Schlipf, John Stewart
Swardson, Mary Anne
Wells, Charles Frederick

OREGON
Parks, Harold Raymond
Penk, Anna Michaelides

PENNSYLVANIA
Andrews, Peter Bruce
Asenjo, Florencio Gonzalez
Chase, Gene Barry
Dawson, John William, Jr
Finin, Timothy Wilking
Fisher, Marshall L
Furst, Merrick Lee
Jech, Thomas J
Nevoln, Bob
Paulos, John Allen
Rescher, Nicholas
Simpson, Stephen G

RHODE ISLAND
Simons, Roger Alan

SOUTH CAROLINA
Comer, Stephen Daniel
McNulty, George Frank

TENNESSEE
Dixon, Edmond Dale
Purisch, Steven Donald

TEXAS
DeBakey, Lois
DeBakey, Selma
Hart, R Neal
Martin, Norman Marshall
Vobach, Arnold R

VERMONT
Kreider, Donald Lester

VIRGINIA
Cherniavsky, John Charles
Dubin, Henry Charles
Loatman, Robert Bruce
Orcutt, Bruce Call
Schneider, Philip Allen David

WASHINGTON
Young, Paul Ruel

WISCONSIN
Fournelle, Thomas Albert
Keisler, Howard Jerome
Kleene, Stephen Cole
Solomon, Marvin H

WYOMING
Cowles, John Richard

BRITISH COLUMBIA
Mekler, Alan
Thomason, Steven Karl

NEWFOUNDLAND
May, Sherry Jan

ONTARIO
Chang, Shao-chien
Cook, Stephen Arthur
Fleischer, Isidore
Giles, Robin
Kent, Clement F
Olin, Philip
Tall, Franklin David

QUEBEC
Daigneault, Aubert
Hatcher, William S
Seldin, Jonathan Paul

OTHER COUNTRIES
Collins, George Edwin
Janos, Ludvik
Vamos, Tibor

Mathematical Statistics

ALABAMA
Bradley, Edwin Luther, Jr
Bradley, Laurence A
Chow, Bryant
Drake, Albert Estern
Essenwanger, Oskar M
Gibbons, Jean Dickinson
Holt, William Robert
Hurst, David Charles
Mishra, Satya Narayan
Trawinski, Benon John
Trawinski, Irene Patricia Monahan
Welker, Everett Linus
Wheeler, Ruric E

ALASKA
Van Veldhuizen, Philip Androcles

ARIZONA
Anderson, Mary Ruth
Barckett, Joseph Anthony
Denny, John Leighton
Falk, Thomas
Freund, John Ernst
Heinle, Preston Joseph
Jacobowitz, Ronald
Kakar, Rajesh Kumar
Keats, John Bert
Kececioglu, D(imitri) B(asil)
Mittal, Yashaswini Deval
Myers, Donald Earl
Walling, Derald Dee
Yakowitz, Sidney J
Young, Dennis Lee

ARKANSAS
Alguire, Mary Slanina (Hirsch)
Brown, A Hayden, Jr
Dunn, James Eldon
McBryde, Vernon E(ugene)
Senseman, Scott Allen
Smith, Robert Paul
Springer, Melvin Dale
Strub, Mike Robert
Walls, Robert Clarence

CALIFORNIA
Alder, Henry Ludwig
Aldous, David J
Anderson, Theodore Wilbur
Antoniak, Charles Edward
Apostol, Tom M

Arnwine, William Carrol
Arthur, Susan Peterson
Astrachan, Max
Baker, George Allen
Barlow, Richard Eugene
Beaver, Robert John
Beckwith, Richard Edward
Bell, Charles Bernard, Jr
Bendisz, Kazimierz
Benson, Robert Franklin
Bentley, Donald Lyon
Bickel, Peter J
Birnbaum, Michael Henry
Blackwell, David (Harold)
Blischke, Wallace Robert
Book, Stephen Alan
Brillinger, David Ross
Burg, John Parker
Cahn, David Stephen
Carlyle, Jack Webster
Chen, Chung Wei
Chern, Ming-Fen Myra
Chernick, Michael Ross
Cover, Thomas Merrill
Cumberland, William Glen
Curry, Bo(stick) U
David, Florence N
DeArmon, Ira Alexander, Jr
Donoho, David Leigh
Dresch, Francis William
Drugan, James Richard
Efron, Bradley
Elashoff, Janet Dixon
Esary, James Daniel
Fearn, Dean Henry
Ferguson, Thomas S
Forsythe, Alan Barry
Franti, Charles Elmer
Freedman, David A
Gavin, Lloyd Alvin
Geng, Shu
Ghosh, Sakti P
Ghosh, Subir
Gill, Dhanwant Singh
Gokhale, Dattaprabhakar V
Gordon, Louis
Gordon, Milton Andrew
Halmos, Paul Richard
Heitkamp, Norman Denis
Hewer, Gary Arthur
Hodges, Joseph Lawson, Jr
Hoel, Paul Gerhard
Hsu, Yu-Sheng
Iglehart, Donald Lee
Ivanov, Igor C
Jacobs, Patricia Anne
Jayachandran, Toke
Jennrich, Robert I
Jewell, Nicholas Patrick
Johns, Milton Vernon, Jr
Johnson, Carl William
Kalensher, Bernard Earl
Kane, Daniel E(dwin)
Karlin, Samuel
Katz, Richard Whitmore
Kimme, Ernest Godfrey
Krieger, Henry Alan
Kritz-Silverstein, Donna
Lai, Tze Leung
Lange, Kenneth L
Larson, Harold Joseph
Le Cam, Lucien Marie
Lehmann, Erich Leo
Levine, Bruce Martin
Lewis, Peter Adrian Walter
Li, Hung Chiang
Lieberman, Gerald J
McCall, Chester Hayden, Jr
MacDonald, Gordon James Fraser
Maksoudian, Y Leon
Mann, Nancy Robbins
Massey, Frank Jones, Jr
Meecham, William Coryell
Mickey, Max Ray, Jr
Milch, Paul R
Molnar, Imre
Mood, Alexander McFarlane
Moser, Joseph M
Myers, Verne Steele
Myhre-Hollerman, Janet M
Newcomb, Robert Lewis
Olkin, Ingram
Olshen, Richard Allen
Park, Chong Jin
Park, Heebok
Payne, Holland I
Piersol, Allan Gerald
Preisler, Haiganoush Krikorian
Press, S James
Ramey, Daniel Bruce
Randle, Robert James
Rao, Jammalamadaka S
Ray, Rose Marie
Read, Robert Richard
Resnikoff, George Joseph
Richards, Francis Russell
Rocke, David M
Romano, Albert
Rosenblatt, Murray
Scheuer, Ernest Martin
Schwall, Richard Joseph
Sengupta, Sumedha
Sevacherian, Vahram
Shen, Nien-Tsu

Show, Ivan Tristan
Siegmund, David O
Singleton, Richard Collom
Smith, Robert William
Smith, Wayne Earl
Smoke, Mary E
Solomon, Herbert
Sprowls, Riley Clay
Stauffer, Howard Boyer
Stewart, Leland Taylor
Stone, Charles Joel
Suich, Ronald Charles
Suppes, Patrick
Swan, Shanna Helen
Switzer, Paul
Thacker, John Charles
Trumbo, Bruce Edward
Tysver, Joseph Bryce
Ury, Hans Konrad
Wang, Peter Cheng-Chao
Waterman, Michael S
Watkins, William
Wiggins, Alvin Dennie
Wu, Sing-Chou

COLORADO
Bardwell, George
Barnes, James Allen
Bosch, William W
Bowden, David Clark
Chase, Gerald Roy
Crow, Edwin Louis
Davis, Wilford Lavern
Eberhart, Steve A
Gentry, Willard Max, Jr
Graybill, Franklin A
Healy, William Carleton, Jr
Heine, George Winfield, III
Heiny, Robert Lowell
Houston, Samuel Robert
Hume, Merril Wayne
Kafadar, Karen
Mielke, Paul W, Jr
Miesch, Alfred Thomas
Morgenthaler, George William
Peterson, Gary A
Remmenga, Elmer Edwin
Schmid, John Carolus
Shedlock, Kaye M
Srivastava, Jagdish Narain
Tischendorf, John Allen
Tweedie, Richard Lewis
Weiss, Irving

CONNECTICUT
Anscombe, Francis John
Fischer, Diana Bradbury
Frankel, Martin Richard
Gelfand, Alan Enoch
Gine-Masdeu, Evarist
Goldstein, Rubin
Hartigan, John A
Kleinfeld, Ira H
Levenstein, Harold
Moore, Melita Holly
Mukhopadhyay, Nitis
Page, Edgar J(oseph)
Peduzzi, Peter N
Phillips, Peter Charles B
Posten, Harry Owen
Rao, Valluru Bhavanarayana
Savage, I Richard
Shorr, Bernard
Singer, Burton Herbert
Tuteur, Franz Benjamin
Vitale, Richard Albert
Woods, Jimmie Dale

DELAWARE
Chitra, Surya P
Eissner, Robert M
Fellner, William Henry
Lucas, James M
Marquardt, Donald Wesley
Ogunnaike, Babatunde Ayodeji
Snee, Ronald D
Taylor, Howard Milton, III

DISTRICT OF COLUMBIA
Abramson, Lee Richard
Bishop, Yvonne M M
Bowker, Albert Hosmer
Chase, Gary Andrew
Cohen, Michael Paul
Crosby, David S
Deming, W(illiam) Edwards
Filliben, James John
Gastwirth, Joseph L
Goldfield, Edwin David
Hammer, Carl
Hayek, Lee-Ann Collins
Hedlund, James Howard
Jabine, Thomas Boyd
Jernigan, Robert Wayne
Klemm, Rebecca Jane
Knopman, Debra Sara
Kordoski, Edward William
Korin, Basil Peter
Kullback, Solomon
Lilliefors, Hubert W
Mandel, Benjamin J
Mann, Charles Roy
Melnick, Daniel
Nayak, Tapan Kumar

Olmer, Jane Chasnoff
Robinette, Charles Dennis
Rosenblatt, David
Schroeder, Anita Gayle
Spruill, Nancy Lyon
Straf, Miron L
Tortora, Robert D
Wallis, W Allen
Wiener, Howard Lawrence

FLORIDA
Ballerini, Rocco
Bingham, Richard S(tephen), Jr
Brain, Carlos W
Byrkit, Donald Raymond
Carrier, Steven Theodore
Chang, Lena
Chew, Victor
Coldwell, Robert Lynn
Cornell, John Andrew
Dutton, Arthur Morlan
Eichhorn, Heinrich Karl
Elsner, James Brian
Frishman, Fred
George, John Harold
Haase, Richard Henry
Hall, David Goodsell, IV
Hollander, Myles
Huerta, Manuel Andres
Jamrich, John Xavier
Johnson, Mark Edward
Kirmse, Dale William
Lin, Pi-Erh
Littell, Ramon Clarence
Magarian, Elizabeth Ann
Martin, Frank Garland
Meeter, Duane Anthony
Mendenhall, William, III
Millor, Steven Ross
Moder, Joseph J(ohn)
Nguyen, Ru
Olmstead, Paul Smith
Ostle, Bernard
Pirkle, Fredric Lee
Proschan, Frank
Randles, Ronald Herman
Rao, Pejaver Vishwamber
Roth, Raymond Edward
Sartain, Jerry Burton
Scheaffer, Richard Lewis
Schott, James Robert
Sethuraman, Jayaram
Shapiro, Samuel S
Somerville, Paul Noble
Stamper, James Harris
Storm, Leo Eugene
Tebbe, Dennis Lee
Tsokos, Chris Peter
Ungvichian, Vichate
Williams, Willie Elbert
Yang, Mark Chao-Kuen

GEORGIA
Arnold, Jonathan
Bargmann, Rolf Erwin
Bhagia, Gobind Shewakram
Bradley, Ralph Allan
Caras, Gus J(ohn)
Carter, Melvin Winsor
Chen, Hubert Jan-Peing
Cohen, Alonzo Clifford, Jr
Coyne, Edward James, Jr
Koch, George Schneider, Jr
Kutner, Michael Henry
Lavender, DeWitt Earl
Lyons, Nancy I
Mabry, John William
Neter, John
Pepper, William Donald
Riddle, Lawrence H
Taylor, Robert Lee
Thompson, Shirley Williams
Thompson, William Oxley, II
Tong, Yung Liang
Wadsworth, Harrison M(orton)
Ware, Glenn Oren
Zarnoch, Stanley Joseph

HAWAII
Flachsbart, Peter George
Miura, Carole K Masutani

IDAHO
Butler, Calvin Charles
Everson, Dale O
Higdem, Roger Leon
Juola, Robert C
Laughlin, James Stanley
Li, Guang Chao
Stuffle, Roy Eugene

ILLINOIS
Abrams, Israel Jacob
Agnew, Robert A
Bahadur, Raghu Raj
Berk, Kenneth N
Bohrer, Robert Edward
Burkholder, Donald Lyman
Chakrabarti, Subrata K
Chitgopekar, Sharad Shankarrao
DeBoer, Edward Dale
De Rijk, Waldemar G
Dwass, Meyer
Dyson, John Douglas

Eaton, Paul Wentland
Edge, Orlyn P
Fox, James David
Friedberg, Stephen Howard
Georgakis, Constantine
Golomski, William Arthur
Harter, H(arman) Leon
Hedayat, A Samad
Heller, Barbara Ruth
Ionescu Tulcea, Cassius
Jacobsen, Fred Marius
Kalai, Ehud
Knight, Frank B
Krug, Samuel Edward
Kruskal, William Henry
Laud, Purushottam Waman
Li, Jane Chiao
Madansky, Albert
Marden, John Iglehart
Martinsek, Adam Thomas
Miller, Floyd Glenn
Monrad, Ditlev
Niles, Walter Dulany, II
Pendergrass, Robert Nixon
Philipp, Walter V
Portnoy, Stephen Lane
Roberts, Harry Vivian
Rodgers, James Foster
Roth, Arthur Jason
Sacks, Jerome
Sclove, Stanley Louis
Scott, Meckinley
Seefeldt, Waldemar Bernhard
Stigler, Stephen Mack
Stout, William F
Tamhane, Ajit C
Taneja, Vidya Sagar
Tarver, Mae-Goodwin
Thisted, Ronald Aaron
Tiao, George Ching-Hwuan
Wacker, William Dennis
Wallace, David Lee
Wasserman, Stanley
Wijsman, Robert Arthur
Wolter, Kirk Marcus

INDIANA
Ali, Mir Masoom
Anderson, Virgil Lee
Bhattacharya, Rabindra Nath
Bradley, Richard Crane, Jr
Costello, Donald F
Cote, Louis J
Coyle, Edward John
Eberts, Ray Edward
Eckert, Roger E(arl)
Hamblen, John Wesley
Harrington, Rodney B
Hicks, Charles Robert
King, Edgar Pearce
Kinney, John James
Klein, Howard Joseph
Ma, Cynthia Sanman
McCabe, George Paul, Jr
Moore, David Sheldon
Murphy, John Riffe
Norton, James Augustus, Jr
Obenchain, Robert Lincoln
Rubin, Herman
Sampson, Charles Berlin
Samuels, Myra Lee
Studden, William John
Tan, James Chien-Hua
Townsend, Douglas Wayne

IOWA
Alberda, Willis John
Athreya, Krishna Balasundaram
Berger, Philip Jeffrey
Braaten, Melvin Ole
Bright, Harold Frederick
Canfield, Earle Lloyd
Christian, Lauren L
Cressie, Noel A C
Cryer, Jonathan D
David, Herbert Aron
Dykstra, Richard Lynn
Feldt, Leonard Samuel
Fuller, Wayne Arthur
Geiger, Randall L
Harville, David Arthur
Hickman, Roy D
Hinz, Paul Norman
Hogg, Robert Vincent, Jr
Huntsberger, David Vernon
Isaacson, Stanley Leonard
Jebe, Emil H
Kempthorne, Oscar
Lott, Fred Wilbur, Jr
Meeker, William Quackenbush, Jr
Pollak, Edward
Reckase, Mark Daniel
Robertson, Tim
Russo, Ralph P
Snider, Bill Carl F
Sukhatme, Balkrishna Vasudeo
Sukhatme, Shashikala Balkrishna
Vardeman, Stephen Bruce

KANSAS
Arteaga, Lucio
Bulgren, William Gerald
Chopra, Dharam-Vir
Feyerherm, Arlin Martin

Fryer, Holly Claire
Grosh, Doris Lloyd
Higgins, James Jacob
Johnson, Dallas Eugene
Kemp, Kenneth E
Paulie, M Catherine Therese
Platt, Ronald Dean
Price, Griffith Baley
Shanmugan, K Sam
Stewart, William Henry
Thomas, Harold Lee
White, Stephen Edward

KENTUCKY
Allen, David Mitchell
Anderson, Richard L
Bhapkar, Vasant Prabhakar
Feibes, Walter
Govindarajulu, Zakkula
Griffith, William Schuler
O'Connor, Carol Alf
Smith, Timothy Andre
Vaziri, Menouchehr

LOUISIANA
Baldwin, Virgil Clark, Jr
Bennett, Lonnie Truman
Boullion, Thomas L
Bradley, Marshall Rice
Burford, Roger Lewis
Crump, Kenny Sherman
Dickinson, Peter Charles
Escobar, Luis A
Guzman Foresti, Miguel Angel
Kinchen, David G
Koonce, Kenneth Lowell
Kuo, Hui-Hsiung
Montalvo, Joseph G, Jr
Sharma, Bhudev
Timon, William Edward, Jr
Warner, Isiah Manuel
Watkins, Terry Anderson
Young, John Coleman

MAINE
McMillan, Brockway

MARYLAND
Abbey, Robert Fred, Jr
Alling, David Wheelock
Arsham, Hossein
Austin, Homer Wellington
Babcock, Anita Kathleen
Bartko, John Jaroslav
Behforooz, Ali
Blanche, Ernest Evred
Brant, Larry James
Bryant, Edward Clark
Cameron, Joseph Marion
Celmins, Aivars Karlis Richards
Chacko, George Kuttickal
Chiacchierini, Richard Philip
Coakley, Kevin Jerome
Correa Villasenor, Adolfo
Der Simonian, Rebecca
Diamond, Earl Louis
Douglas, Lloyd Evans
Dubey, Satya D(eva)
Duncan, Acheson Johnston
Eberhardt, Keith Randall
Eisenhart, Churchill
Engel, Joseph H(enry)
Fairweather, William Ross
Fisher, Gail Feimster
Gart, John Jacob
Geller, Harvey
Geller, Nancy L
Gilford, Leon
Goktepe, Omer Faruk
Gordon, Chester Murray
Gordon, Tavia
Greenhouse, Samuel William
Grubbs, Frank Ephraim
Hanson, Robert Harold
Hogben, David
Kacker, Raghu N
Kaplan, David L
Kattakuzhy, George Chacko
Kligman, Ronald Lee
Kotz, Samuel
Ku, Harry Hsien Hsiang
Lachin, John Marion, III
Land, Charles Even
Langenberg, Patricia Warrington
Lao, Chang Sheng
Lechner, James Albert
Lee, Young Jack
Leone, Fred Charles
Lepson, Benjamin
Levine, Eugene
Liggett, Walter Stewart, Jr
Linder, Forrest Edward
Locke, Ben Zion
Luther, Lonnie W
Maar, James Richard
Maisel, Herbert
Maloney, Clifford Joseph
Mandel, John
Marzetti, Lawrence Arthur
Merrill, Warner Jay, Jr
Mikulski, Piotr W
Miller, Robert Gerard
Milton, Roy Charles
Moshman, Jack

Mathematical Statistics (cont)

Murphy, Thomas James
Navarro, Joseph Anthony
Odle, John William
Paule, Robert Charles
Pickle, Linda Williams
Ponnapalli, Ramachandramurty
Raff, Morton Spencer
Rastogi, Suresh Chandra
Robinson, Richard Carleton, Jr
Robock, Alan
Rosenblatt, Joan Raup
Royall, Richard Miles
Rubinstein, Lawrence Victor
Rukhin, Andrew Leo
Serfling, Robert Joseph
Shapiro, Sam
Sirken, Monroe Gilbert
Smith, Paul John
Stalling, David L
Steinberg, Joseph
Steinberg, Seth Michael
Tarone, Robert Ernest
Tepping, Benjamin Joseph
Thiebaux, H Jean
Tissue, Eric Bruce
Tripathi, Satish K
Vangel, Mark Geoffrey
Waksberg, Joseph
Weir, Edward Earl, II
Weiss, Michael David
Wells, William T
Wierman, John C
Williford, William Olin
Yang, Grace L
Zoellner, John Arthur

MASSACHUSETTS
Arnold, Kenneth James
Ash, Arlene Sandra
Beaton, Albert E
Ben-Akiva, Moshe E
Bert, Mark Henry
Bussgang, J(ulian) J(akob)
Cardello, Armand Vincent
Chernoff, Herman
Chisholm, Donald Alexander
Cox, Lawrence Henry
D'Agostino, Ralph B
Diaconis, Persi
Doerfler, Thomas Eugene
Dudley, Richard Mansfield
Esch, Robin E
Foulis, David James
Freeman, Harold Adolph
Giordano, Arthur Anthony
Gutoff, Edgar B(enjamin)
Hoaglin, David Caster
Horowitz, Joseph
Hsieh, Hui-Kuang
Hyatt, Raymond R, Jr
Kitchin, John Francis
Kolakowski, Donald Louis
Lagakos, Stephen William
Lavin, Philip Todd
Lemeshow, Stanley Alan
Levin, Leonid A
Liu, Jun S
Lo, Andrew W
Marino, Richard Matthew
Mehta, Mahendra
Mosteller, C Frederick
Ross, Edward William, Jr
Rubin, Donald Bruce
Siegel, Armand
Skibinsky, Morris
Stanley, Kenneth Earl
Thomas, Abdelnour Simon
Tsiatis, Anastasios A
Van Deusen, Paul Cook
Ware, James H
Waud, Douglas Russell
Wei, Lee-Jen
Welsch, Roy Elmer
Wolock, Fred Walter
Zelen, Marvin

MICHIGAN
Bryant, Barbara Everitt
Chakravarthy, Srinivasaraghavan
Crary, Selden Bronson
Cress, Charles Edwin
Dijkers, Marcellinus P
Dubes, Richard C
Ericson, William Arnold
Fox, Martin
Gilliland, Dennis Crippen
Hannan, James Francis
Hill, Bruce M
Johnson, Whitney Larsen
Kincaid, Wilfred Macdonald
Knappenberger, Herbert Allan
Koo, Delia Wei
Koul, Hira Lal
Kowalski, Charles Joseph
Kshirsagar, Anant Madhav
Lamberson, Leonard Roy
Leabo, Dick Albert
Lenker, Susan Stamm
LePage, Raoul
Lin, Paul Kuang-Hsien
Marchand, Margaret O
Muirhead, Robb John

Pandit, Sudhakar Madhavrao
Powell, James Henry
Root, William L(ucas)
Shantaram, Rajagopal
Sievers, Gerald Lester
Spivey, Walter Allen
Stapleton, James H
Starr, Norman
Stoline, Michael Ross
Tanis, Elliot Alan
Tsao, Chia Kuei
Whalen, Thomas J(ohn)
Zemach, Rita

MINNESOTA
Alders, C Dean
Ayeni, Babatunde J
Balcziak, Louis William
Bingham, Christopher
Britz, Galen C
Buehler, Robert Joseph
Busch, Robert Henry
Eaton, Morris Leroy
Franta, William Roy
Geisser, Seymour
Giese, David Lyle
Hansen, Leslie Bennett
Hicks, Dale R
Hurwicz, Leonid
Larntz, Kinley
Le, Chap Than
McHenry, Kelly David
Martin, Frank Burke
Perry, Robert Leonard
Rust, Lawrence Wayne, Jr
Sentz, James Curtis
Singh, Rajinder
Smith, John Henry
Smith, Robert Elijah
Sudderth, William David
Weisberg, Sanford
Wolf, Frank Louis
Zelterman, Daniel

MISSISSIPPI
Butler, Charles Morgan
Gerard, Patrick Dale
Sharpe, Thomas R

MISSOURI
Bain, Lee J
Baltz, Howard Burl
Barnett, William Arnold
Basu, Asit Prakas
Carlson, Roger
Davis, James Wendell
Flora, Jairus Dale, Jr
Franck, Wallace Edmundt
Hewett, John Earl
Katti, Shriniwas Keshav
King, Terry Lee
Moore, James Thomas
Nemeth, Margaret Ann
Ott, R(ay) Lyman, Jr
Oviatt, Charles Dixon
Spitznagel, Edward Lawrence, Jr
Stalling, David Laurence
Templeton, Alan Robert
Tsutakawa, Robert K
Wright, Farroll Tim

MONTANA
Loftsgaarden, Don Owen
Reinhardt, Howard Earl
Taylor, William Robert

NEBRASKA
Ahlschwede, William T
Gessaman, Margaret Palmer
Schutz, Wilfred M

NEVADA
Carr, James Russell
Kinnison, Robert Ray
Lowenthal, Douglas H
Miller, Forest Leonard, Jr
Starks, Thomas Harold

NEW HAMPSHIRE
Abbott, Robert Classie
Albert, Arthur Edward
Lamperti, John Williams
Tourgee, Ronald Alan

NEW JERSEY
Ahsanullah, Mohammad
Aizenman, Michael
Becker, Richard Alan
Brandinger, Jay J
Chambers, John McKinley
Chassan, Jacob Bernard
Cheung, Shiu Ming
Chien, Victor
Church, Eugene Lent
Cleveland, William S
Coale, Ansley J
Dalal, Siddhartha Ramanlal
Denby, Lorraine
Eckler, Albert Ross
Fask, Alan S
Frey, William Carl
Gabbe, John Daniel
Gale, William Arthur
Gethner, Jon Steven

Gilmer, George Hudson
Gnanadesikan, Ramanathan
Gold, Daniel Howard
Green, Edwin James
Holland, Paul William
Hsu, Chin-Fei
Hunter, John Stuart
Hwang, Frank K
Hwang, Helen H L
Jain, Aridaman Kumar
Johnston, Christian William
Jurkat, Martin Peter
Kruskal, Joseph Bernard
Labianca, Frank Michael
LaMastro, Robert Anthony
Lambert, Diane
Landwehr, James M
Loader, Clive Roland
Lucantoni, David Michael
Magliola-Zoch, Doris
Mallows, Colin Lingwood
Mihram, George Arthur
Miller, Irwin
Murphy, Ray Bradford
Murphy, Thomas Daniel
Naus, Joseph Irwin
Ordille, Carol Maria
Pasteelnick, Louis Andre
Phadke, Madhav Shridhar
Poor, Harold Vincent
Radner, Roy
Rainal, Attilio Joseph
Ramaswami, Vaidyanathan
Riggs, Richard
Robbins, Herbert Ellis
Robbins, Naomi Bograd
Sak, Joseph
Sarakwash, Michael
Schneider, Alfred Marcel
Smouse, Peter Edgar
Steiger, William Lee
Strawderman, William E
Sykes, Donald Joseph
Teicher, Henry
Thomas, Ronald Emerson
Tukey, John Wilder
Turner, Edwin Lewis
Vardi, Yehuda
Verdu, Sergio
Watson, Geoffrey Stuart
Wood, Eric F
Yoon, Kwangsun Paul

NEW MEXICO
Bruckner, Lawrence Adam
Campbell, Katherine Smith
Cardenas, Manuel
Cobb, Loren
Cogburn, Robert Francis
Faulkenberry, Gerald David
Finkner, Morris Dale
George, Timothy Gordon
Goldman, Aaron Sampson
Gutjahr, Allan L
Julian, William H
Koopmans, Lambert Herman
Kraichnan, Robert Harry
Lohrding, Ronald Keith
McKay, Michael Darrell
Martz, Harry Franklin, Jr
Ortiz, Melchor, Jr
Qualls, Clifford Ray
Rogers, Gerald Stanley
Ryti, Randall Todd
Waller, Ray Albert
Zeigler, Royal Keith

NEW YORK
Altland, Henry Wolf
Anderson, Allan George
Aroian, Leo Avedis
Arvesen, James Norman
Banerjee, Kali Shankar
Barr, Donald R
Bechhofer, Robert E(ric)
Beil, Gary Milton
Bialas, Wayne Francis
Blackman, Samuel William
Blessing, John A
Bross, Irwin Dudley Jackson
Brown, Lawrence D
Capobianco, Michael F
Cargo, Gerald Thomas
Chatterjee, Samprit
Clatworthy, Willard Hubert
Cok, David R
Collins, James Joseph
Crooke, James Stratton
Cunia, Tiberius
Danziger, Lawrence
Derman, Cyrus
Desu, Manavala Mahamunulu
Doganaksoy, Necip
Ehrenfeld, Sylvain
Emsley, James Alan Burns
Farrell, Roger Hamlin
Ferguson, David Lawrence
Finch, Stephen Joseph
Fine, Terrence Leon
Goel, Amrit Lal
Goldfarb, Nathan
Gordon, Florence S
Gordon, Sheldon P
Griffith, Daniel Alva

Hachigian, Jack
Hahn, Gerald John
Hall, William Jackson
Hanson, David Lee
Harris, Stewart
Herbach, Leon Howard
Hunte, Beryl Eleanor
Jackson, James Edward
Kao, Samuel Chung-Siung
Karatzas, Ioannis
Kaufman, Sol
Keilson, Julian
King, Alan Jonathan
Kolesar, Peter John
Kurnow, Ernest
Laska, Eugene
Lawton, William Harvey
Layzer, Arthur James
Lerman, Steven I
Levene, Howard
Levin, Bruce
Li, Chou H(siung)
Lieberman, Burton Barnet
Lininger, Lloyd Lesley
Lorenzen, Jerry Alan
McCarthy, Philip John
Mahtab, M Ashraf
Marshall, Clifford Wallace
Mehrotra, Kishan Gopal
Meier, Paul
Melnick, Edward Lawrence
Mike, Valerie
Morrison, Nathan
Mudholkar, Govind S
Mundel, August B(aer)
Nagel, Harry Leslie
Nelson, Wayne Bryce
Norton, Larry
Oakes, David
Owen, Joel
Paulson, Edward
Prabhu, Narahari Umanath
Rao, Poduri S R S
Richter, Donald
Riffenburgh, Robert Harry
Rosberg, Zvi
Roshwalb, Irving
Rothenberg, Ronald Isaac
Ruppert, David
Schilling, Edward George
Schmee, Josef
Schoefer, Ernest A(lexander)
Searle, Shayle Robert
Sedransk, Joseph Henry
Shah, Bhupendra K
Shane, Harold D
Sherman, Malcolm J
Sievers, Sally Riedel
Simon, Gary Albert
Singh, Chanchal
Slezak, Jane Ann
Sommer, Charles John
Star, Martin Leon
Stedinger, Jery Russell
Stephany, Edward O
Stiteler, William Merle, III
Straight, H Joseph
Tanner, Martin Abba
Terzuoli, Andrew Joseph
Turnbull, Bruce William
Weiss, Lionel Ira
Welch, Peter D
Whitby, Owen
Wilkinson, John Wesley
Yablon, Marvin
Yakir, Benjamin
Zacks, Shelemyahu

NORTH CAROLINA
Arnold, Harvey James
Bayless, David Lee
Benrud, Charles Harris
Bergsten, Jane Williams
Bhattacharyya, Bibhuti Bhushan
Boos, Dennis Dale
Burdick, Donald Smiley
Carpenter, Benjamin H(arrison)
Dinse, Gregg Ernest
Dotson, Marilyn Knight
Eisen, Eugene J
Farthing, Barton Roby
Gerig, Thomas Michael
Gordy, Thomas D(aniel)
Gruber, Helen Elizabeth
Hader, Robert John
Hafley, William LeRoy
Hankins, Gerald Robert
Haseman, Joseph Kyd
Haynie, Fred Hollis
Herr, David Guy
Holbert, Donald
Hooke, Robert
Horner, Theodore Wright
Horvitz, Daniel Goodman
Johnson, Norman Lloyd
Johnson, Thomas
Karr, Alan Francis
Keepler, Manuel
Koehler, Truman L, Jr
Koop, John C
Kozelka, Robert M
LeDuc, Sharon Kay
Lessler, Judith Thomasson
Lewis, Paul Ollin

Linnerud, Ardell Chester
Manson, Allison Ray
Margolin, Barry Herbert
Mason, David Dickenson
Mason, Robert Edward
Monroe, Robert James
Nelson, A Carl, Jr
Nie, Dalin
Norgaard, Nicholas J
Papadopoulos, Alex Spero
Peterson, David West
Poole, William Kenneth
Portier, Christopher Jude
Potthoff, Richard Frederick
Powers, William Allen, III
Quade, Dana Edward Anthony
Quesenberry, Charles P
Reinfurt, Donald William
Rhyne, A Leonard
Sen, Pranab Kumar
Shah, Babubhai Vadilal
Singh, Kanhaya Lal
Smith, Walter Laws
Steel, Robert George Douglas
Van der Vaart, Hubertus Robert
Weiner, Daniel Lee
Wesler, Oscar
Whipkey, Kenneth Lee
Whitmore, Roy Walter
Winkler, Robert Lewis
Zeng, Zhao-Bang

NORTH DAKOTA
Hertsgaard, Doris M Fisher
Landry, Richard Georges
Williams, John Delane
Winger, Milton Eugene

OHIO
Barr, David Ross
Berens, Alan Paul
Beyer, William Hyman
Bhargava, Triloki Nath
Blumenthal, Saul
Caldito, Gloria C
Chern, Wen Shyong
Chrissis, James W
Cooper, Tommye
Daschbach, James McCloskey
Deddens, James Albert
Disko, Mildred Anne
Goel, Prem Kumar
Guo, Shumei
Gupta, Arjun Kumar
Gustafson, Steven Carl
Hull, David Lee
Khamis, Harry Joseph
Low, Leone Yarborough
McCloskey, John W
McLemore, Benjamin Henry, Jr
McNichols, Roger J(effrey)
Moeschberger, Melvin Lee
Muller, Mervin Edgar
Notz, William Irwin
Orban, John Edward
Park, Won Joon
Patton, Jon Michael
Powers, Jean D
Ray-Chaudhuri, Dwijendra Kumar
Reilly, Charles Bernard
Reynolds, David Stephen
Rimm, Alfred A
Rohatgi, Vijay
Ross, Louis
Santner, Joseph Frank
Santner, Thomas Joseph
Schutta, James Thomas
Seabaugh, Pyrtle W
Silverman, Robert
Sterrett, Andrew
Sweet, Leonard
Whitney, Donald Ransom
Willke, Thomas Aloys
Yiamouyiannis, John Andrew

OKLAHOMA
Anderson, Paul Dean
Carr, Meg Brady
Chambers, Richard Lee
Folk, Earl Donald
Folks, John Leroy
Koller, Glenn R
Lee, Lloyd Lieh-Shen
Parker, Don Earl
Pope, Paul Terrell
Testerman, Jack Duane
Vasicek, Daniel J
Weeks, David Lee

OREGON
Ahuja, Jagdish C
Andrews, Fred Charles
Birkes, David Spencer
Brunk, Hugh Daniel
Calvin, Lyle D
Enneking, Eugene A
Jones, Donald Akers
McKean, Joseph Walter, Jr
Magwire, Craig A
Murphy, Allan Hunt
Nelsen, Roger Bain
Peterson, Reider Sverre
Ramsey, Fred Lawrence
Rowe, Kenneth Eugene

Seely, Justus Frandsen
Tate, Robert Flemming
Truax, Donald R
Urquhart, N Scott

PENNSYLVANIA
Adams, John William
Antle, Charles Edward
Bai, Zhi Dong
Beggs, William John
Blaisdell, Ernest Atwell, Jr
Block, Henry William
Callen, Herbert Bernard
Chelemer, Harold
Clelland, Richard Cook
De Cani, John Stapley
DeGroot, Morris H
Duncan, George Thomas
Eddy, William Frost
Eisenberg, Bennett
Emlemdi, Hasan Bashir
Esmen, Nurtan A(lan)
Feigelson, Eric Dennis
Fienberg, Stephen Elliott
Fryling, Robert Howard
Galambos, Janos
Ghosh, Bhaskar Kumar
Ginsburg, Herbert
Gleser, Leon Jay
Grandage, Arnold Herbert Edward
Hargrove, George Lynn
Harkness, William Leonard
Hettmansperger, Thomas Philip
Hildebrand, David Kent
Holland, Burt S
Hultquist, Robert Allan
Iglewicz, Boris
Isett, Robert David
Iversen, Gudmund R
Jeruchim, Michel Claude
Kadane, Joseph Born
Kassam, Saleem Abdulali
Ko, Frank K
Koller, Robert Dene
Kushner, Harvey
Kuzmak, Joseph Milton
Lautenberger, William J
Lehoczky, John Paul
Lindsay, Bruce George
McCune, Duncan Chalmers
McLaughlin, Francis X(avier)
McLennan, James Alan, Jr
Mehta, Jatinder S
Merrill, Samuel, III
Meyer, John Sigmund
Michalik, Edmund Richard
Morrison, Donald Franklin
Murray, Donald Shipley
Nair, K Aiyappan
Parnes, Milton N
Patil, Ganapati P
Pickands, James, III
Pierson, Ellery Merwin
Pinski, Gabriel
Richards, Winston Ashton
Rigby, Paul Herbert
Ryan, Thomas Arthur, Jr
Sanathanan, Lalitha P
Schervish, Mark John
Seltzer, James Edward
Shaman, Paul
Singh, Jagbir
Sokolowski, Henry Alfred
Sorensen, Frederick Allen
Sullivan, George Allen
Treadwell, Kenneth Myron
Vaux, James Edward, Jr
Ventura, Jose Antonio
Wagner, Clifford Henry
Wei, Susanna
Wei, William Wu-Shyong

RHODE ISLAND
Carney, Edward J
Grenander, Ulf
Hanumara, Ramachandra Choudary
Hemmerle, William J
Jarrett, Jeffrey E
Lawing, William Dennis
McClure, Donald Ernest
Rossi, Joseph Stephen
Streit, Roy Leon

SOUTH CAROLINA
Byrd, Wilbert Preston
Dobbins, James Gregory Hall
Drane, John Wanzer
Felling, William E(dward)
Gross, Alan John
Holmes, Paul Thayer
Hubbell, Douglas Osborne
Hunt, Hurshell Harvey
Ling, Robert Francis
Nelson, Peter Reid
Padgett, William Jowayne
Russell, Charles Bradley
Surak, John Godfrey
Visvanathan, T R
Walker, James Wilson
Yu, Kai Fun

SOUTH DAKOTA
Brown, Michael Lee

TENNESSEE
Bayne, Charles Kenneth
Beauchamp, John J
Bowman, Kimiko Osada
Cheng, Meng-Dawn
Cross, James Thomas
Finch, Gaylord Kirkwood
Gardiner, Donald Andrew
Ginnings, Gerald Keith
Gosslee, David Gilbert
Lasater, Herbert Alan
Leon, Ramon V
Lowe, James Urban, II
McHenry, Hugh Lansden
McNutt, Charles Harrison
Parr, William Chris
Redman, Charles Edwin
Samejima, Fumiko
Sherman, Gordon R
Somes, Grant William
Tan, Wai-Yuan
Upadhyaya, Belle Raghavendra
Uppuluri, V R Rao
Walker, Grayson Howard
Williams, Robin O('Dare)
Younger, Mary Sue

TEXAS
Adams, Jasper Emmett, Jr
Atkinson, Edward Neely
Bedgood, Dale Ray
Bhat, Uggappakodi Narayan
Bland, Richard P
Bratcher, Thomas Lester
Brockett, Patrick Lee
Broemeling, Lyle David
Brooks, Dorothy Lynn
Browne, Richard Harold
Carroll, Raymond James
Cecil, David Rolf
Chen, Peter
Conover, William Jay
Cooke, William Peyton, Jr
Cox, Dennis Dean
Cruse, Carl Max
Duran, Benjamin S
Eubank, Randall Lester
Ford, Charles Erwin
Freund, Rudolf Jakob
Gehan, Edmund A
Gentle, James Eddie
Hallum, Cecil Ralph
Han, Chien-Pai
Harrist, Ronald Baxter
Hatch, John Phillip
Hinkley, David Victor
Hocking, Ronald Raymond
John, Peter William Meredith
Johnson, Richard Leon
Johnston, Walter Edward
Kimeldorf, George S
Kirk, Roger E
Kussmaul, Keith
Larsen, Russell D
Lee, Eun Sul
Matis, James Henry
Maute, Robert Edgar
Maxwell, Donald A
Michalek, Joel Edmund
Moore, Kris
Newton, Howard Joseph
Nickel, James Alvin
Odell, Patrick L
Owen, Donald Bruce
Parzen, Emanuel
Prabhu, Vasant K
Ringer, Larry Joel
Rosenblatt, Judah Isser
Sager, Thomas William
Schucany, William Roger
Schuster, Eugene F
Sielken, Robert Lewis, Jr
Smith, James Douglas
Smith, Patricia Lee
Smith, William Boyce
Sobol, Marion Gross
Stein, William Edward
Styblinski, Maciej A
Thall, Peter Francis
Thompson, James Robert
Turner, Danny William
Turner, John Charles
Vance, Joseph Francis
Van Ness, John Winslow
Webster, John Thomas
Wehrly, Thomas Edward
Wichern, Dean William
Wiorkowski, John James
Woodward, Wayne Anthony
Wright, Henry Albert
Wyllys, Ronald Eugene

UTAH
Bryce, Gale Rex
Hilton, Horace Gill
Hurst, Rex LeRoy
Nielson, Howard Curtis
Oyler, Dee Edward
Paegle, Julia Nogues
Reading, James Cardon
Rencher, Alvin C
Richards, Dale Owen
Turner, David L(ee)
Walker, LeRoy Harold

VERMONT
Costanza, Michael Charles
Emerson, John David
Haugh, Larry Douglas
Hill, David Byrne
Wooding, William Minor

VIRGINIA
Arnold, Jesse Charles
Bailar, Barbara Ann
Bauer, David Francis
Berger, Beverly Jane
Bolstein, Arnold Richard
Brignoli Gable, Carol
Bush, Norman
Corwin, Thomas Lewis
Davidson, John Richard
Dick, Ronald Stewart
Drew, Lawrence James
Evans, David Arthur
Fink, Lester Harold
Gardenier, John Stark
Good, Irving John
Harshbarger, Boyd
Healy, John Joseph
Heller, Agnes S
Henry, Neil Wylie
Hinkelmann, Klaus Heinrich
Irick, Paul Eugene
Jensen, Donald Ray
Krutchkoff, Richard Gerald
Kullback, Joseph Henry
Kupperman, Morton
Leczynski, Barbara Ann
Loatman, Robert Bruce
Lundegard, Robert James
McLaughlin, Gerald Wayne
Mensing, Richard Walter
Mikhail, Nabih N
Minton, Paul Dixon
Moyer, Leroy
Myers, Raymond Harold
Narula, Subhash Chander
O'Neill, Brian
Perry, Charles Rufus, Jr
Pirie, Walter Ronald
Ramirez, Donald Edward
Reynolds, Marion Rudolph, Jr
Richardson, Henry Russell
Rowe, William David
Schneider, Philip Allen David
Solomon, Vasanth Balan
Stroup, Cynthia Roxane
Tang, Victor Kuang-Tao
Taylor, Samuel James
Walpole, Ronald Edgar
Wegman, Edward Joseph
Wine, Russell Lowell

WASHINGTON
Bennett, Carl Allen
Bogyo, Thomas P
Bolme, Mark W
Chapman, Roger Charles
Chiu, John Shih-Yao
Conquest, Loveday Loyce
Farnum, Peter
Ghurye, Sudhish G
Hansen, Rodney Thor
Kappenman, Russell Francis
Nicholson, Wesley Lathrop
Perlman, Michael David
Pina, Eduardo Isidorio
Prentice, Ross L
Pyke, Ronald
Riedel, Eberhard Karl
Roberts, Norman Hailstone
Russell, Thomas Solon
Saunders, Sam Cundiff
Siegel, Andrew Francis
Thompson, Elizabeth Alison
Van Belle, Gerald

WEST VIRGINIA
Brady, Dennis Patrick
Chaty, John Culver
Wearden, Stanley

WISCONSIN
Bates, Douglas Martin
Berthouex, Paul Mac
Bhattacharyya, Gouri Kanta
Bisgaard, Søren
Box, George Edward Pelham
Chang, Tsong-How
Draper, Norman Richard
Farrow, John H
Girard, Dennis Michael
Gopal, Raj
Gurland, John
Harris, Bernard
Hickman, James Charles
Hill, William Joseph
Johnson, Charles Henry
Jowett, David
Kaye, Norman Joseph
Klein, John Peter
Lawrence, Willard Earl
Lochner, Robert Herman
Miller, Paul Dean
Miller, Robert Burnham
Ogden, Robert Verl
Pemantle, Robin A
Reinsel, Gregory Charles

Mathematical Statistics (cont)

Silverman, Franklin Harold
Stoneman, David McNeel
Wahba, Grace
Walter, Gilbert G

WYOMING
Borgman, Leon Emry
Guenther, William Charles

PUERTO RICO
Bangdiwala, Ishver Surchand
Sahai, Hardeo

ALBERTA
Blais, J A Rodrigue
Consul, Prem Chandra
Enns, Ernest Gerhard
Kelker, Douglas
Scheinberg, Eliyahu
Wani, Jagannath K

BRITISH COLUMBIA
Booth, Andrew Donald
Eaves, David Magill
Hall, John Wilfred
Kaller, Cecil Louis
Marshall, Peter Lawrence
Odeh, Robert Eugene
Peterson, Raymond Glen
Petkau, Albert John
Promnitz, Lawrence Charles
Routledge, Richard Donovan
Srivastava, Rekha
Stephens, Michael A
Villegas, Cesareo
Zidek, James Victor

MANITOBA
Basilevsky, Alexander
Chan, Lai Kow
Kocherlakota, Kathleen
Kocherlakota, Subrahmaniam
Paul, Gilbert Ivan
Sinha, Snehesh Kumar
Zeiler, Frederick

NEW BRUNSWICK
Mureika, Roman A
Pham-Gia, Thu

NEWFOUNDLAND
Hiscott, Richard Nicholas

NOVA SCOTIA
Grant, Douglass Lloyd
Kabe, Dattatraya G
Wentzell, Peter Dale

ONTARIO
Anderson, Christine Michaela
Andrews, David F
Bacon, David W
Balakrishnan, Narayanaswamy
Batra, Tilak Raj
Behara, Minaketan
Blyth, Colin Ross
Brainerd, Barron
Dale, Douglas Keith
Dawson, Donald Andrew
DeLury, Daniel Bertrand
Denzel, George Eugene
Fraser, Donald Alexander Stuart
Gentleman, Jane Forer
Glew, David Neville
Godson, Warren Lehman
Graham, John Elwood
Haq, M Safiul
Hipel, Keith William
Jamieson, Derek Maitland
Jardine, John McNair
Kalbfleisch, James G
Kalbfleisch, John David
MacNeill, Ian B
Mohanty, Sri Gopal
Mullen, Kenneth
Nash, John Christopher
Paull, Allan E
Rao, Jonnagadda Nalini Kanth
Reilly, Park McKnight
Rust, Velma Irene
Schaufele, Ronald A
Shklov, Nathan
Sprott, David Arthur
Srivastava, Muni Shanker
Stroud, Thomas William Felix
Thompson, Mary Elinore
Tracy, Derrick Shannon
Wasan, Madanlal T
Watts, Donald George
West, Eric Neil
Wolynetz, Mark Stanley
Wong, Chi Song
Wonnacott, Thomas Herbert

QUEBEC
Allen, Harold Don
Chaubey, Yogendra Prasad
Duong, Quang P
Giri, Narayan C
Jumarie, Guy Michael
Lefebvre, Mario
Lehman, Eugene H

Maag, Urs Richard
Mathai, Arak M
Morgera, Salvatore Domenic
Murty, M Ram
Plotkin, Eugene Isaak
Sankoff, David
Seshadri, Vanamamalai
Srivastava, Triloki N
St-Pierre, Jacques
Yalovsky, Morty A

SASKATCHEWAN
O'Shaughnessy, Charles Dennis

OTHER COUNTRIES
Allen, Robin Leslie
Basu, Debabrata
Brockwell, Peter John
Camiz, Sergio
Chesson, Peter Leith
Das Gupta, Somesh
Deely, John Joseph
Foglio, Susana
Francis, Ivor Stuart
Garcia-Santesmases, Jose Miguel
Greenblatt, Seth Alan
Hochberg, Kenneth J
Konijn, Hendrik Salomon
McCue, Edmund Bradley
Malik, Mazhar Ali Khan
Mara, Michael Kelly
Mengali, Umberto
Philippou, Andreas Nicolaou
Samuel-Cahn, Ester
Siddiqui, Mohammed Moinuddin
Snyder, Mitchell
Stephens, Kenneth S
Sunahara, Yoshifumi
Van Eeden, Constance
Wu, Lancelot T L

Mathematics, General

ALABAMA
Ainsworth, Oscar Richard
Ball, Richard William
Benington, Frederick
Blake, John Archibald
Bond, Albert F
Burton, Leonard Pattillo
Cappas, C
Chow, Bryant
Cook, Frederick Lee
Crittenden, Richard James
Easterday, Kenneth E
Gibson, Peter Murray
Goldsmith, Donald Leon
Hill, Paul Daniel
Hobby, Charles R
Howell, James Levert
Hutchison, Jeanne S
Kiser, Lola Frances
Locker, John L
Lovingood, Judson Allison
McGill, Suzanne
Mishra, Satya Narayan
Moore, Robert Laurens
Neggers, Joseph
Peeples, William Dewey, Jr
Rice, Barbara Slyder
Rodgers, Aubrey
Roush, Fred William
Segal, Arthur Cherny
Shipman, Jerry
Stocks, Douglas Roscoe, Jr
Stone, Max Wendell
Vaughan, Loy Ottis, Jr
Vinson, Richard G
Wilson, Walter Lucien, Jr

ALASKA
Van Veldhuizen, Philip Androcles

ARIZONA
Bedient, Jack DeWitt
Cheema, Mohindar Singh
Clay, James Ray
Dougherty, John Joseph
Franz, Otto Gustav
Goldstein, Myron
Gray, Allan, Jr
Griego, Richard Jerome
Grove, Larry Charles
Hodges, Carl Norris
Johnson, Lee Murphy
Kelly, John Beckwith
Larney, Violet Hachmeister
Leonard, John Lander
Little, Charles Edward
Lovelock, David
McDonald, John N
Meserve, Bruce Elwyn
Moore, Charles Godat
Moore, John Douglas
Nering, Evar Dare
Nicolaenko, Basil
Pollin, Jack Murph
Rieke, Carol Anger
Rund, Hanno
Sansone, Fred J
Savage, Nevin William
Sherman, Thomas Lawrence
Smarandache, Florentin

Smith, Harvey Alvin
Smith, Lehi Tingen
Steiner, Eugene Francis
Stewart, Donald George
Strebe, David Diedrich
Thompson, Richard Bruce
Walter, Everett L
Wood, Bruce

ARKANSAS
Fuller, Roy Joseph
Hudson, Frank M
Kimura, Naoki
Linnstaedter, Jerry Leroy
Meek, James Latham
Moon, Wilchor David
Moore, Marvin G
Oldham, Bill Wayne
Orton, William R, Jr
Ottinger, Carol Blanche
Pilgrim, Donald
Scroggs, James Edward
Smith, Robert Paul
Starling, Albert Gregory
Walls, Robert Clarence
Wild, Wayne Grant

CALIFORNIA
Abraham, Ralph Herman
Adams, Ralph Melvin
Aissen, Michael Israel
Albert, Eugene
Alexander, Elaine E
Ali, Mir Kursheed
Allen, Arnold Oral
Amemiya, Frances (Louise) Campbell
Anderson, A Keith
Anderson, Donald Werner
Apostol, Tom M
Armacost, William L
Arnold, Hubert Andrew
Arnold, Robert Fairbanks
Arveson, William Barnes
Astrachan, Max
Astromoff, Andrew
Austin, Charles Ward
Bade, William George
Baldwin, Harry L, Jr
Barker, William Hamblin, II
Bars, Itzhak
Basinger, Richard Craig
Basor, Estelle
Baxendale, Peter Herbert
Becker, Gerald Anthony
Beechler, Barbara Jean
Bell, Charles Bernard, Jr
Bender, Edward Anton
Benedicty, Mario
Benson, Donald Charles
Berlekamp, Elwyn R(alph)
Biles, Charles Morgan
Biriuk, George
Birnbaum, Sidney
Birnir, Björn
Bischof, Nobert S
Bishop, Errett A
Blackwell, David (Harold)
Block, Richard Earl
Blum, Edward K
Bourne, Samuel G
Bowman, Eugene W
Bredon, Glen E
Bressler, David Wilson
Brooks, Robert W
Brown, George William
Calvert, Ralph Lowell
Cantor, David Geoffrey
Casey, James
Ceder, Jack G
Chan, Tony Fan-Cheong
Chang, I-Dee
Chang, Y(uan)-F(eng)
Chatland, Harold
Chernick, Michael Ross
Chicks, Charles Hampton
Chinn, Phyllis Zweig
Chorin, Alexandre J
Chow, Tseng Yeh
Christianson, Deann
Christopher, John
Chung, Kai Lai
Church, Alonzo
Clark, Charles Lester
Clarke, Wilton E L
Coddington, Earl Alexander
Cohen, Paul Joseph
Coleman, Robert Frederick
Cottle, Richard W
Councilman, Samuel G
Cullen, Theodore John
Cummings, Frederick W
Cusick, Larry William
Cutler, Leonard Samuel
David, Florence N
Davison, Walter Francis
Dean, Richard Albert
Deaton, Edmund Ike
DeLand, Edward Charles
Demmel, James W
De Pillis, John
Dewar, Jacqueline M
Dimsdale, Bernard
Douglas, Robert James
Drobnies, Saul Isaac

Dubins, Lester Eli
Duncan, Donald Gordon
Dye, Henry Abel
Edelson, Allan L
Edelson, Sidney
Effros, Edward George
Ellis, Wade
Emerson, Thomas James
Ernest, John Arthur
Fan, Ky
Farrell, Edward Joseph
Fateman, Richard J
Feldman, Jerome A
Feldman, Leonard
Feldman, Raisa Epstein
Fendel, Daniel
Fisher, Newman
Fitzgerald, Carl
Ford, Lester Randolph, Jr
Foster, Lorraine L
Frankel, Theodore Thomas
Fuller, F(rancis) Brock
Gale, David
Garsia, Adriano Mario
Gaver, Donald Paul
Gavin, Lloyd Alvin
Getoor, Ronald Kay
Ghose, Rabindra Nath
Gibson, James (Benjamin)
Gieseker, David
Gill, Dhanwant Singh
Gillman, David
Gittleman, Arthur P
Goberstein, Simon M
Golomb, Solomon Wolf
Gordon, Samuel Robert
Goss, Robert Nichols
Gould, William E
Grady, Michael D
Gragg, William Bryant, Jr
Grannan, Ralph T
Graves, Glenn William
Green, John Willie
Greenberg, Marvin Jay
Greenspan, Bernard
Grossman, Nathaniel
Gruenberger, Fred J(oseph)
Halberg, Charles John August, Jr
Hales, Alfred Washington
Halkin, Hubert
Haltiner, George Joseph
Hardy, John W, Jr
Hart, Garry Dewaine
Hartshorne, Robert (Robin) Cope
Harvey, Albert Raymond
Hayes, Charles Amos, Jr
Hayes, Robert Mayo
Healy, Paul William
Hedstrom, Gerald Walter
Henriksen, Melvin
Heynick, Louis Norman
Hirsch, Peter M
Hobbs, Billy Frankel
Hochschild, Gerhard P
Holl, Manfred Matthias
Holmes, Calvin Virgil
Holsztynski, Wlodzimierz
Hood, J Myron
Hopponen, Jerry Dale
Householder, Alston Scott
Householder, James Earl
Howell, Russell W
Hsu, Yu-Sheng
Hudson, George Elbert
Hughart, Stanley Parlett
Hughes, Thomas Rogers, Jr
Hurd, Cuthbert Corwin
Huskey, Harry D(ouglas)
Ingram, John Michael
Jacobs, Judith E
Jaffa, Robert E
James, Ralph L
Javaher, James N
Jayachandran, Toke
Juncosa, Mario Leon
Kahan, William M
Kalisch, Gerhard Karl
Kanarik, Rosella
Kaplansky, Irving
Karush, William
Kato, Tosio
Kelley, Allen Frederick, Jr
Kelley, John Le Roy
Kelly, Paul J
Kevorkian, Aram K
Kipps, Thomas Charles
Knuth, Donald Ervin
Kobayashi, Shoshichi
Konheim, Alan G
Koval, Daniel
Krakowski, Fred
Kreith, Kurt
Krom, Melven R
Lai, Tze Leung
Landesman, Edward Milton
Lane, Richard Neil
Lange, Lester Henry
Lanski, Charles Philip
Larmore, Lawrence Louis
Lashof, Richard Kenneth
Latta, Gordon
Lawson, Charles L
Lax, Melvin David
Lee, K Hung

MATHEMATICS / 427

Lehman, R Sherman
Lehmer, Derrick Henry
Leipnik, Roy Bergh
Lender, Adam
Lengyel, Bela Adalbert
Lenstra, Hendrik W
Lin, Tung-Po
Liu, Chi-Sheng
Lockhart, Brooks Javins
Lonsdale, Carol Jean
Lovaglia, Anthony Richard
Luttmann, Frederick William
Lutz, Donald Alexander
Lynch, William C
McCready, Thomas Arthur
Mach, George Robert
McKenzie, Ralph Nelson
McLeod, Edward Blake
Maksoudian, Y Leon
Manjarrez, Victor M
Marble, Frank E(arl)
Marcus, Brian Harry
Mardellis, Anthony
Margulies, William George
Marley, Gerald C
Marsden, Jerrold Eldon
Martinez-Uriegas, Eugenio
Matlow, Sheldon Leo
Mendis, Devamitta Asoka
Meredith, David B
Miech, Ronald Joseph
Milgram, Richard James
Mitchem, John Alan
Monier, Louis Marcel
Moore, Calvin C
Morgan, Christopher L
Morgan, John Clifford, II
Morgan, Kathryn A
Morris, George William
Moser, Louise Elizabeth
Mumme, Judith E
Murphy, James Lee
Myers, William Howard
Najar, Rudolph Michael
Neustadter, Siegfried Friedrich
Newcomb, Robert Lewis
Newman, Morris
Niccolai, Nilo Anthony
Norton, Donald Alan
Nower, Leon
Olshen, Richard Allen
Orchard, Henry John
Orland, George H
Ornstein, Donald Samuel
Osher, Stanley Joel
Pacholka, Kenneth
Pease, Marshall Carleton, III
Peglar, George W
Pfeffer, Washek F
Pflugfelder, Hala
Phillips, Ralph Saul
Pignataro, Augustus
Port, Sidney C
Potts, Donald Harry
Price, Charles Morton
Proskurowski, Wlodzimierz
Protter, Murray Harold
Purcell, Everett Wayne
Purvis, Colbert Thaxton
Radnitz, Alan
Ralston, James Vickroy, Jr
Ramakrishnan, Dinakar
Ratner, Marina
Rebman, Kenneth Ralph
Redheffer, Raymond Moos
Reed, Irving Stoy
Rhodes, John Lewis
Riley, James Daniel
Rissanen, Jorma Johannes
Ritchie, Robert Wells
Robertson, James Byron
Robertson, Thomas N
Robinson, Raphael Mitchel
Rodin, Burton
Rohrl, Helmut
Romanowicz, Barbara Anna
Rose, Gene Fuerst
Rosenfeld, Melvin
Rosenlicht, Maxwell
Rothschild, Bruce Lee
Rozsnyai, Balazs
Ruggles, Ivan Dale
Rush, David Eugene
Sabharwal, Ranjit Singh
Saltz, Daniel
Samelson, Hans
Samuelson, Donald James
Sanderson, Judson
Sarason, Donald Erik
Satyanarayana, Upadhyayula V
Savitch, Walter John
Scharlemann, Martin G
Scheinok, Perry Aaron
Schindler, Guenter Martin
Schlesinger, Stewart Irwin
Schneider, Harold O
Schoen, Richard M(elvin)
Schoenfeld, Alan Henry
Seppi, Edward Joseph
Shapley, Lloyd Stowell
Sharpe, Michael John
Shen, Nien-Tsu
Shepherd, William Lloyd
Sherry, Edwin J

Shiffman, Max
Shiflett, Ray Calvin
Short, Donald Ray, Jr
Shreve, David Carr
Shubert, Bruno Otto
Simon, Arthur Bernard
Sloss, James M
Smale, Stephen
Smith, Donald Ray
Smith, James Thomas
Smith, Larry
Smith, Marianne (Ruth) Freundlich
Smith, Marion Bush, Jr
Solovay, Robert M
Sowder, Larry K
Specht, Robert Dickerson
Sprecher, David A
Stanek, Peter
Stannard, William A
Stein, Charles M
Stein, James D, Jr
Stein, Sherman Kopald
Steinberg, Robert
Stiel, Edsel Ford
Stillman, Richard Ernest
Stone, Alexander Glattstein
Stone, Charles Joel
Stralka, Albert R
Strauch, Ralph Eugene
Summers, Audrey Lorraine
Sutherland, Ivan Edward
Swann, Howard Story Gray
Swenson, Donald Adolph
Takesaki, Masamichi
Tang, Hwa-Tsang
Tannenbaum, Peter
Taylor, Angus Ellis
Terry, Raymond Douglas
Thompson, Robert Charles
Thorp, Edward O
Thurston, William P
Tolsted, Elmer Beaumont
Topp, William Robert
Trahan, Donald Herbert
Tromba, Anthony Joseph
Tropp, Henry S
Tucker, Roy Wilbur
Tuller, Annita
Tully, Edward Joseph, Jr
Ung, Man T
Van de Wetering, Richard Lee
Vaught, Robert L
Veigele, William John
Venit, Stewart Mark
Verdina, Joseph
Vojta, Paul Alan
Wagoner, Ronald Lewis
Waiter, Serge-Albert
Wales, David Bertram
Wallen, Clarence Joseph
Waterman, Michael S
Watkins, William
Weidlich, John Edward, Jr
Weinbaum, Carl Martin
Welch, Lloyd Richard
Welmers, Everett Thomas
Wernick, Robert J
Wets, Roger J B
White, Alvin Murray
White, Paul A
Whiteman, Albert Leon
Whitmore, William Francis
Wilde, Carroll Orville
Willoughby, Ralph Arthur
Winzenread, Marvin Russell
Wong, K(wee) C
Woo, Norman Tzu Teh
Wooldridge, Kent Ernest
Wrede, Robert C, Jr
Wright, Elisabeth Muriel Jane
Wright, Russell E
Wulff, John Leland
Yale, Paul B
Yeh, James Jui-Tin
Zeitlin, Joel Loeb

COLORADO
Allgower, Eugene L
Baggett, Lawrence W
Bailey, Charles W
Bardwell, George
Bechtel, Robert D
Bennett, W Scott
Beylkin, Gregory
Boes, Ardel J
Bosch, William W
Brown, Austin Robert, Jr
Butler, Walter Cassius
Buzbee, Billy Lewis
Cavanagh, Timothy D
Cicerone, Carol Mitsuko
Cohen, Joel Seymour
Colvis, John Paris
Crow, Edwin Louis
Darst, Richard B
Deal, Ervin R
DeMeyer, Frank R
Dorn, William S
Easton, Robert Walter
Eberhard, Wynn Lowell
Ellis, Homer Godsey
Eyman, Earl Duane
Farnell, Albert Bennett
Fisch, Forest Norland

Fischer, Irwin
Fox, Bennett L
Frisinger, H Howard, II
Gaines, Robert Earl
Gilbert, Gordon R
Gliner, Gail S
Goodrich, Robert Kent
Grefsrud, Gary W
Gudder, Stanley Phillip
Gustafson, Karl Edwin
Hansman, Margaret Mary
Hardy, Darel W
Heerema, Nickolas
Hodges, John Herbert
Holley, Frieda Koster
Holley, Richard Andrew
Hoots, Felix R
Hufford, George (Allen)
Hundhausen, Joan Rohrer
Kafadar, Karen
Kelman, Robert Bernard
Klopfenstein, Kenneth F
LeVeque, William Judson
Levinger, Bernard Werner
Litwhiler, Daniel W
Lundell, Albert Thomas
McNerney, Chuck
MacRae, Robert E
Magnus, Arne
Meyer, Burnett Chandler
Mielke, Paul W, Jr
Milligan, Merle Wallace
Monk, James Donald
Montague, Patricia Tucker
Mycielski, Jan
Niemann, Ralph Henry
Painter, Richard J
Payne, Stanley E
Phillips, David Lowell
Popejoy, William Dean
Raab, Joseph A
Rechard, Ottis William
Reeves, Roy Franklin
Roeder, David Herbert
Roth, Richard Lewis
Sather, Duane Paul
Schmid, John Carolus
Schmidt, Wolfgang M
Simmons, George Finlay
Spencer, Donald Clayton
Srivastava, Jagdish Narain
Stein, Frederick Max
Taylor, Walter Fuller
Thomas, James William
Vilms, Jaak
Warren, Peter
Windholz, Walter M
Wouk, Arthur
Wyss, Walter
Zirakzadeh, Aboulghassem

CONNECTICUT
Aaboe, Asger (Hartvig)
Anderson, George Albert
Bannister, Peter Robert
Brunell, Gloria Florette
Comfort, William Wistar
De Gray, Ronald Willoughby
Eby, Edward Stuart, Sr
Eisenberg, Sheldon Merven
Feit, Walter
Garland, Howard
Hedlund, Gustav Arnold
Hirsch, Warren Maurice
Ho, Grace Ping-Poo
Jacobson, Florence Dorfman
Jacobson, Nathan
Kagan, Joel (David)
Kakutani, Shizuo
Klimczak, Walter John
Lang, Serge
Linton, Fred E J
Locke, Stanley
MacDonnell, Joseph Francis
Markham, Elizabeth Mary
Mostow, George Daniel
Neuwirth, Jerome H
Nigam, Lakshmi Narayan
Poliferno, Mario Joseph
Raney, George Neal
Rickart, Charles Earl
Rosenbaum, Robert Abraham
Roulier, John Arthur
Schmerl, James H
Schrader, Dorothy Virginia
Seligman, George Benham
Sells, Jean Thurber
Shaffer, Dorothy Browne
Singleton, Robert Richmond
Spencer, Domina Eberle (Mrs Parry Moon)
Sternberg, Robert Langley
Stewart, Robert Clarence
Tollefson, Jeffrey L
Vinsonhaler, Charles I
Walde, Ralph Eldon
Welna, Cecilia
Whitley, William Thurmon
Whittlesey, Emmet Finlay
Wolk, Elliot Samuel
Woods, Jimmie Dale

DELAWARE
Baxter, Willard Ellis

Case, John William
Colton, David L
Doerner, William A(llen)
Gilbert, Robert Pertsch
Graham, John Alan
Kamack, H(arry) J(oseph)
Marquardt, Donald Wesley
Nashed, Mohammed Zuhair Zaki
Pellicciaro, Edward Joseph
Roselle, David Paul
Thickstun, William Russell, Jr
Wenger, Ronald Harold

DISTRICT OF COLUMBIA
Blankfield, Alan
Bobo, Edwin Ray
Bogdan, Victor Michael
Bradley, John Spurgeon
Braunfeld, Peter George
Donaldson, James A
Egan, Howard L
Gordon, William Bernard
Gray, Mary Wheat
Hayek, Lee-Ann Collins
Hensel, Gustav
Hoffman, Kenneth Myron
Jacobs, Marc Quillen
Katz, Irving
Kelber, Charles Norman
Kenyon, Hewitt
Lagnese, John Edward
Levy, Mark B
Lukacs, Eugene
Matson, Michael
Moller, Raymond William
Osgood, Charles Freeman
Rose, Raymond Edward
Rosier, Ronald Crosby
Siry, Joseph William
Sokoloski, Martin Michael
Stein, Marjorie Leiter
Stokes, Arnold Paul
Tucker, John Richard
Voytuk, James A
Williams, Leland Hendry

FLORIDA
Adney, Joseph Elliott, Jr
Agins, Barnett Robert
Andregg, Charles Harold
Bagley, Robert Waller
Bednarek, Alexander R
Blake, Robert George
Brooks, James Keith
Bryant, John Logan
Bullis, George LeRoy
Butson, Alton Thomas
Byrkit, Donald Raymond
Chae, Soo Bong
Chen, Wayne H(wa-Wei)
Cowling, Vincent Frederick
Debnath, Lokenath
DeSua, Frank Crispin
Eachus, Joseph Jackson
Fagin, Karin
Fass, Arnold Lionel
Fine, Rana Arnold
Fitzgerald, Lawrence Terrell
Garman, Brian Lee
Garrett, James Richard
Gleyzal, Andre
Goldberg, Estelle Maxine
Goldberg, Lee Dresden
Goodman, Adolph Winkler
Grams, Anne P
Heil, Wolfgang Heinrich
Hertzig, David
Hoffman, Ruth I
Horton, Thomas Roscoe
Jones, Roy Carl, Jr
Kalman, Rudolf Emil
Keedy, Mervin Laverne
Kelley, John Ernest
Kerr, Donald R
Lakshmikantham, Vangipuram
Leja, Stanislaw
Levitz, Hilbert
Liang, Joseph Jen-Yin
Lick, Dale W
Lighterman, Mark S
Livingood, John N B
Lofquist, George W
McArthur, Charles Wilson
McClure, Clair Wylie
McDougle, Paul E
Maddox, Billy Hoyte
Martin, Kenneth Edward
Meacham, Robert Colegrove
Medlin, Gene Woodard
Miles, Ernest Percy, Jr
Moore, Theral Orvis
Nichols, Eugene Douglas
Nielsen, Kaj Leo
Norman, Edward
Novosad, Robert S
Palmer, Kenneth
Pop-Stojanovic, Zoran Rista
Rose, Donald Clayton
Russell, John Masters
Ryan, Peter Michael
Sanders, Oliver Paul
Selfridge, Ralph Gordon
Sigmon, Kermit Neal
Sinclair, Annette

Mathematics, General (cont)

Smith, William K
Smythe, Robert C
Snider, Arthur David
Snover, James Edward
Squier, Donald Platte
Stiles, Wilbur J
Taylor, Michael Dee
Turner, Ralph Waldo
Ungvichian, Vichate
Van Zee, Richard Jerry
Varma, Arun Kumar
Villemure, M Paul James
Wahab, James Hatton
Ward, James Audley
Winton, Charles Newton
Wolfson, Kenneth Graham
Zame, Alan

GEORGIA
Aust, Catherine Cowan
Barnes, Earl Russell
Bompart, Billy Earl
Brahana, Thomas Roy
Buccino, Alphonse
Clemmons, John B
Cliett, Otis Jay
Croft, George Thomas
Devries, David J
Duquette, Alfred L
Edwards, Charles Henry, Jr
Evans, Trevor
Evans, William Buell
Ford, David A
Goslin, Roy Nelson
Hahn, Hwa Suk
Hale, Jack Kenneth
Hamrick, Anna Katherine Barr
Hardy, Flournoy Lane
Hollingsworth, John Gressett
Holtum, Alfred G(erard)
Horne, James Grady, Jr
Hudson, Anne Lester
Jackson, Prince A, Jr
Johnson, Roger D, Jr
Kamen, Edward Walter
Kammerer, William John
Karam, Ratib A(braham)
Kasriel, Robert H
Kilpatrick, Jeremy
Kunze, Ray A
Lavender, DeWitt Earl
Line, John Paul
McKinney, Max Terral
Mahavier, William S
Neff, John David
Oliker, Vladimir
Pittman, Chatty Roger
Pomerance, Carl
Robinson, Robert W
Schleusner, John William
Shonkwiler, Ronald Wesley
Skypek, Dora Helen
Smith, William Allen
Sparks, Arthur Godwin
Taylor, Howard Edward
Taylor, Robert Lee
Tong, Yung Liang
Tonne, Philip Charles
Wall, Donald Dines
Walton, Harriett J
Wilson, James William
Worth, Roy Eugene
Zander, Vernon Emil

HAWAII
Allday, Christopher J
Freese, Ralph
Heacox, William Dale
Luther, Norman Y
Oliphant, Malcolm William
Pager, David
Pitcher, Tom Stephen

IDAHO
Benson, Dean Clifton
Ferguson, David John
Henry, Boyd (Herbert)
Higdem, Roger Leon
Juola, Robert C
Mech, William Paul
Richardson, Lee S(pencer)
Wang, Ya-Yen Lee
Ward, Frederick Roger

ILLINOIS
Allard, Nona Mary
Allen, Frank B
Alperin, Jonathan L
Amick, Charles James
Angotti, Rodney
Ash, J Marshall
Ash, Robert B
Atkins, Ferrel
Austin, Donald Guy
Baily, Walter Lewis, Jr
Bank, Steven Barry
Barratt, Michael George
Bateman, Felice Davidson
Bateman, Paul Trevier
Becker, Jerry Page
Berg, Ira David
Berman, Joel David
Billingsley, Patrick Paul
Bohrer, Robert Edward
Brigham, Nelson Allen
Brown, Joseph Ross
Burkholder, Donald Lyman
Burton, Theodore Allen
Butler, Margaret K
Calderon, Alberto Pedro
Catrambone, Joseph Anthony, Sr
Clough, Robert Ragan
Coon, Lewis Hulbert
Cormier, Romae Joseph
Cramton, Thomas James
Curtis, Herbert John
D'Angelo, John Philip
Darsow, William Frank
Deliyannis, Platon Constantine
Devinatz, Allen
Doob, Joseph Leo
Dornhoff, Larry Lee
Dwinger, Philip
Ecker, Edwin D
Edelsbrunner, Herbert
Evens, Leonard
Ferguson, William Allen
Fisher, Stephen D
Foland, Neal Eugene
Foulser, David A
Francis, George Konrad
Frank, Maurice Jerome
Franz, Edgar Arthur
Gates, Leslie Dean, Jr
Georgakis, Constantine
Goldberg, Samuel I
Graves, Robert Lawrence
Gray, John Walker
Greenstein, David Snellenburg
Griffith, Phillip A
Grimmer, Ronald Calvin
Gundersen, Roy Melvin
Haken, Wolfgang
Hamstrom, Mary-Elizabeth
Hasdal, John Allan
Hattemer, Jimmie Ray
Hawkins, Neil Middleton
Hedayat, A Samad
Helms, Lester LaVerne
Hsu, Nai-Chao
Insel, Arnold J
Johnson, Donald Elwood
Johnson, Robert Leroy
Jones, R E Douglas
Jordan, Steven Lee
Kahn, Daniel Stephen
Kuller, Robert G
Kwong, Man Kam
Langebartel, Ray Gartner
Larson, Richard Gustavus
Leonard, Henry Siggins, Jr
Lin, Sue Chin
Loeb, Peter Albert
McAlpin, John Harris
MacLane, Saunders
McLinden, Lynn
Marcus, Philip Selmar
Matlis, Eben
Maxwell, Charles Neville
Miles, Joseph Belsley
Mock, Gordon Duane
Monrad, Ditlev
Moore, Robert Alonzo
Morris, Herbert Allen
Moschandreas, Demetrios J
Mount, Kenneth R
Nanda, Jagdish L
Nelson, Harry Ernest
O'Malley, Mary Therese
Osborn, Howard Ashley
Oursler, Clellie Curtis
Parker, E T
Parr, Phyllis Graham
Pearson, Hans Lennart
Peck, Emily Mann
Pinsky, Mark A
Pless, Vera S
Porta, Horacio A
Pranger, Walter
Putnam, Alfred Lunt
Radford, David Eugene
Reiner, Irma Moses
Reingold, Haim
Reisel, Robert Benedict
Rickert, Neil William
Rodine, Robert Henry
Rotman, Joseph Jonah
Rubel, Lee Albert
Rutledge, Robert B
Saari, Donald Gene
Sawinski, Vincent John
Schumaker, John Abraham
Schupp, Paul Eugene
Scott, Edward Joseph
Sekar, Raj
Shelton, Ronald M
Shively, Ralph Leland
Shryock, A Jerry
Soare, Robert I
Sons, Linda Ruth
Srinivasan, Bhama
Stein, Michael Roger
Stipanowich, Joseph Jean
Stolarsky, Kenneth B
Sturley, Eric Avern
Sullivan, Michael Joseph, Jr
Suzuki, Michio
Swan, Richard Gordon
Szeto, George
Ting, Tsuan Wu
Tondeur, Philippe
Troy, Daniel Joseph
Troyer, Robert James
Tuzar, Jaroslav
Uhl, John Jerry, Jr
Vaughan, Herbert Edward
Wagreich, Philip Donald
Wantland, Evelyn Kendrick
Weichsel, Paul M
Weinberg, Elliot Carl
Weiner, Louis Max
Weinzweig, Avrum Israel
Welland, Robert Roy
Weller, Glenn Peter
Wendt, Arnold
Williams, Eddie Robert
Wirszup, Izaak
Wong, Yuen-Fat
Yntema, Mary Katherine
Zaring, Wilson Miles
Zygmund, Antoni

INDIANA
Abhyankar, Shreeram
Anderson, John R
Anderson, Virgil Lee
Bailey, Herbert R
Balka, Don Stephen
Beineke, Lowell Wayne
Berkovitz, Leonard David
Bittinger, Marvin Lowell
Branges, Louis de
Brown, Arlen
Brown, Lawrence G
Buesking, Clarence W
Chess, Karin V T
Cooley, Robert Lee
Cooney, Miriam Patrick
Cunningham, Ellen M
Dalphin, John Francis
DeMillo, Richard A
Diekhans, Herbert Henry
Drufenbrock, Diane
Finco, Arthur A
Fishback, William Thompson
Forbes, Jack Edwin
Fry, Cleota Gage
Fuller, William Richard
Gambill, Robert Arnold
Gass, Clinton Burke
Goetz, Abraham
Golomb, Michael
Goodman, Victor Wayne
Gottlieb, Daniel Henry
Gupta, Shanti Swarup
Haas, Felix
Hahn, Alexander J
Hamblen, John Wesley
Hanes, Harold
Hill, Robert Joe
Hood, Rodney Taber
Humberd, Jesse David
Hutton, Lucreda Ann
Kallman, Ralph Arthur
Kane, Robert B
Kaplan, Samuel
Keller, Marion Wiles
Kinney, John James
Kinsey, David Webster
Klinger, William Russell
Lipman, Joseph
Lowengrub, Morton
Ludwig, Hubert Joseph
McCormick, Roy L
Machtinger, Lawrence Arnold
Mansfield, Maynard Joseph
Martin, William L, Jr
Mastrototaro, John Joseph
Mielke, Paul Theodore
Miller, John Grier
Minassian, Donald Paul
Morrel, Bernard Baldwin
Morrill, John Elliott
Morrison, Clarence C
Neugebauer, Christoph Johannes
Neuhouser, David Lee
Novodvorsky, Mark Benjamin
O'Meara, O Timothy
Otter, Richard Robert
Peterson, Harold LeRoy
Putnam, Calvin Richard
Reed, Ellen Elizabeth
Regan, Francis
Rhoades, Billy Eugene
Robold, Alice Ilene
Rubin, Herman
Rueve, Charles Richard
Schober, Glenn E
Schraut, Kenneth Charles
Schultz, Reinhard Edward
Seifert, Ralph Louis
Sherman, Gary Joseph
Silverman, Edward
Smith, Peter David
Smith, Shelby Dean
Springer, George
Stanton, Nancy Kahn
Stephens, Stanley LaVerne
Svoboda, Rudy George
Swift, William Clement
Synowiec, John A
Townsend, Douglas Wayne
Tucker, Robert C, Jr
Vuckovic, Vladeta
Webster, Merritt Samuel
Weitsman, Allen William
Williams, Lynn Roy
Wilson, David E
Yackel, James W
Ziemer, William P
Zimmerman, Lester J
Zwick, Earl J

IOWA
Abian, Alexander
Athreya, Krishna Balasundaram
Atkinson, Kendall E
Barnes, Wilfred E
Burrill, Claude Wesley
Camillo, Victor Peter
Canfield, Earle Lloyd
Dahiya, Rajbir Singh
Fink, Arlington M
Friedell, John C
Fuller, Kent Ralph
Geraghty, Michael A
Gillam, Basil Early
Graber, Leland D
Hart, Lawrence Alan
Hentzel, Irvin Roy
Herman, Eugene Alexander
Hill, Edward T
Hinrichsen, John James Luett
Homer, Roger Harry
Jenkins, Terry Lloyd
Khurana, Surjit Singh
Kirk, William Arthur
Kleiner, Alexander F, Jr
Kleinfeld, Margaret Humm
Kosier, Frank J
Lambert, Robert J
Lediaev, John P
Levine, Howard Allen
Lin, Bor-Luh
Lindsay, Charles McCown
Lott, Fred Wilbur, Jr
Mathews, Jerold Chase
Oehmke, Robert H
Oxley, Theron D, Jr
Rothlisberger, Hazel Marie
Rudolph, William Brown
Russo, Ralph P
Schurrer, Augusta
Sprague, Richard Howard
Tondra, Richard John
Trytten, George Norman
Waltmann, William Lee
Wilkinson, Jack Dale
Wright, Fred Marion
Yohe, James Michael
Zettel, Larry Joseph

KANSAS
Adams, Robert D
Arteaga, Lucio
Beougher, Elton Earl
Brady, Stephen W
Brown, Robert Dillon
Buser, Mary Paul
Chaney, George L
Chopra, Dharam-Vir
Conrad, Paul
Creese, Thomas Morton
Dixon, Lyle Junior
Flusser, Peter R
Fuller, Leonard Eugene
Galvin, Fred
Haggard, J D
Hanna, Martin Slafter
Hearne, Horace Clark, Jr
Hight, Donald Wayne
Johnson, Dallas Eugene
Johnston, Andrea
Kriegsman, Helen
Laws, Leonard Stewart
McClendon, James Fred
Marr, John Maurice
Meier, Robert J
Parker, Willard Albert
Paulie, M Catherine Therese
Price, Griffith Baley
Schweppe, Earl Justin
Stahl, Saul
Stamey, William Lee
Stein, Marvin L
Strecker, George Edison
Stromberg, Karl Robert
Szeptycki, Pawel
Thomas, Harold Lee
Upmeier, Harald
Van Vleck, Fred Scott
Votaw, Charles Isac
Wedel, Arnold Marion
Young, Paul McClure

KENTUCKY
Alter, Ronald
Barksdale, James Bryan, Jr
Boyce, Stephen Scott
Cantrell, Grady Leon
Coleman, Donald Brooks
Cox, Raymond H
Davitt, Richard Michael
Elder, Harvey Lynn

Elsey, Margaret Grace
Enochs, Edgar Earle
Fugate, Joseph B
Gariepy, Ronald F
Geeslin, Roger Harold
Govindarajulu, Zakkula
Harris, Henson
Hayden, Thomas Lee
Howard, Aughtum Smith
King, Amy P
Lane, Bennie Ray
Looney, Stephen Warwick
Lowman, Bertha Pauline
McFadden, Robert B
McGlasson, Alvin Garnett
Moldoveanu, C Serban
Mostert, Paul Stallings
Newbery, A Chris
Oppelt, John Andrew
Rishel, Raymond Warren
Royster, Wimberly Calvin
Sabel, Clara Ann
Scorsone, Francesco G
Sitaraman, Yegnaseshan
Smith, Joe K
Stokes, Joseph Franklin
Wallace, Kyle David
Watson, Martha F
Wells, James Howard

LOUISIANA
Agarwal, Arun Kumar
Anderson, Richard Davis
Andrew, David Robert
Andrews, Bethlehem Kottes
Attebery, Billy Joe
Authement, Ray Paul
Bennett, Lonnie Truman
Birtel, Frank T
Boullion, Thomas L
Butts, Hubert S
Chadick, Stan R
Collins, Heron Sherwood
Conner, Pierre Euclide, Jr
Corley, Glyn Jackson
Crump, Kenny Sherman
Dearth, James Dean
Dupree, Daniel Edward
Ellis, James Watson
Ford, Patrick Lang
Fritsche, Richard T
Gans, David
Goldstein, Jerome Arthur
Greechie, Richard Joseph
Griffin, Ernest Lyle
Keisler, James Edwin
Kelly, Edgar Preston, Jr
Koch, Robert Jacob
Lawson, Jimmie Don
Lefton, Lew Edward
Ligh, Steve
Maxfield, John Edward
Mislove, Michael William
Morris, Halcyon Ellen McNeil
Mulcrone, Thomas Francis
Newman, Rogers J
Ohm, Jack Elton
Ohme, Paul Adolph
Ohmer, Merlin Maurice
Quigley, Frank Douglas
Rees, Charles Sparks
Rees, Paul Klein
Retherford, James Ronald
Rogers, James Ted, Jr
Rosencrans, Steven I
Ryan, Donald Edwin
Scholz, Dan Robert
Simmons, David Rae
Smith, Charles R
Spaht, Carlos G, II
Stoltzfus, Neal W

MAINE
Carter, William Caswell
Haines, David Clark
Hsu, Yu Kao
Johnson, Robert Wells
Leonard, William Wilson
Locke, Philip M
Mairhuber, John Carl
Merriell, David McCray
Munroe, Marshall Evans
Northam, Edward Stafford
Norton, Karl Kenneth
Page, Robert Leroy
Raisbeck, Gordon
Semon, Mark David
Ward, James Edward, III

MARYLAND
Atchison, William Franklin
Auslander, Joseph
Austin, Homer Wellington
Berger, William J
Bernstein, Allen Richard
Betz, Ebon Elbert
Bhatia, Nam Parshad
Blanche, Ernest Evred
Blum, Joseph
Boisvert, Ronald Fernand
Cacciamani, Eugene Richard, Jr
Carasso, Alfred Samuel
Chang, Elizabeth B
Chellappa, Ramalingam

Chiacchierini, Richard Philip
Chu, Sherwood Cheng-Wu
Cohen, Edgar A, Jr
Colvin, Burton Houston
Crampton, Theodore Henry Miller
Cuthill, Elizabeth
Dame, Richard Edward
Di Rienzi, Joseph
Dixon, John Kent
Dowling, Marie Augustine
Dubey, Satya D(eva)
Eastham, James Norman
Ehrlich, Gertrude
Elder, Alexander Stowell
Embree, Earl Owen
Engel, Joseph H(enry)
Epstein, Bernard
Ernest, Michael Vance, Sr
Faust, William R
Fitzpatrick, Patrick Michael
Franks, David A
Fusaro, Bernard A
Gabriele, Thomas L
Giese, John H
Gleissner, Gene Heiden
Goldman, Alan Joseph
Gonzalez-Fernandez, Jose Maria
Gordon, Chester Murray
Gorman, John Richard
Gramick, Jeannine
Greenberg, James M
Gross, Fred
Gulick, Sidney L (Denny), III
Hanson, Robert Harold
Henney, Dagmar Renate
Herrmann, Robert Arthur
Horelick, Brindell
Hummel, James Alexander
Igusa, Jun-Ichi
Jablonski, Felix Joseph
Jones, John Paul
Kalme, John S
Komm, Horace
Koppelman, Elaine
Kowalski, Richard
Lawson, Mildred Wiker
Lepson, Benjamin
Li, Ming Chiang
Lightner, James Edward
Lomonaco, Samuel James, Jr
Lopez-Escobar, Edgar George Kenneth
Marotta, Charles Rocco
May, Everette Lee, Jr
Miller, A Eugene
Miller, Robert Gerard
Moulis, Edward Jean, Jr
Nanzetta, Philip Newcomb
O'Brien, Robert
Odle, John William
Oliver, Paul Alfred
Parker, Rodger D
Penn, Howard Lewis
Pierce, Harry Frederick
Pittenger, Arthur O
Ponnapalli, Ramachandramurty
Prewitt, Judith Martha Shimansky
Reed, Coke S
Rheinstein, Peter Howard
Rich, Robert Peter
Rudolph, Ray Ronald
Rukhin, Andrew Leo
Sagan, Leon Francis
Sampson, Joseph Harold
Saworotnow, Parfeny Pavolich
Schafer, James A
Schleiter, Thomas Gerard
Siegel, Martha J
Slepian, Paul
Smith, Paul John
Spohn, William Gideon, Jr
Strohl, George Ralph, Jr
Stuelpnagel, John Clay
Syski, Ryszard
Tepper, Morris
Theilheimer, Feodor
Toller, Gary Neil
Truesdell, Clifford Ambrose, III
Van Tuyl, Andrew Heuer
Von Bun, Friedrich Otto
Wardlaw, William Patterson
Washington, Lawrence C
Watters, Edward C(harles), Jr
Weissberg, Alfred
Wierman, John C
Willcox, Alfred Burton
Wilson, John Phillips
Wolfe, Carvel Stewart
Young, James Howard
Zelkowitz, Marvin Victor
Zucker, Steven Mark

MASSACHUSETTS
Ahlfors, Lars Valerian
Allen, Stephen Ives
Armacost, David Lee
Artin, Michael
Auslander, Bernice Liberman
Azpeitia, Alfonso Gil
Baglivo, Jenny Antoinette
Bandes, Dean
Bennett, Mary Katherine
Berger, Melvyn Stuart
Birkhoff, Garrett
Blackett, Donald Watson

Blackett, Shirley Allart
Borrego, Joseph Thomas
Bott, Raoul H
Buchsbaum, David Alvin
Catlin, Donald E
Chaiken, Jan Michael
Chen, Yu Why
Cohen, David Warren
Cohen, Haskell
Cole, Nancy
Cox, David Archibald
Crabtree, Douglas Everett
Cullen, Helen Frances
Darling, Eugene Merrill, Jr
Dickinson, David (James)
Doyle, Jon
Durfee, William Hetherington
Eckstein, Jonathan
Evans, Thomas George
Fishman, Robert Sumner
Foulis, David James
Freedman, Marvin I
Freier, Jerome Bernard
Freyre, Raoul Manuel
Galmarino, Alberto Raul
Gleason, Andrew Mattei
Guillemin, Victor W
Hajian, Arshag B
Hardell, William John
Hawkins, Thomas William, Jr
Hayes, Dallas T
Helgason, Sigurdur
Hildebrand, Francis Begnaud
Hoffer, Alan R
Hogan, Guy T
Hurlburt, Douglas Herendeen
Isles, David Frederick
Jacob, Henry George, Jr
Jensen, Barbara Lynne
Kamowitz, Herbert M
Kazhdan, David
Kearns, Donald Allen
Kennison, John Frederick
Killam, Eleanor
Kleis, John Dieffenbach
Kleitman, Daniel J
Kostant, Bertram
Kung, Hsiang-Tsung
Lanzkron, Rolf W
Levine, Harold
Luther, Holger Martin
Lutts, John A
McBrien, Vincent Owen
McCabe, Robert Lyden
MacDonnell, John Joseph
McGuigan, Robert Alister, Jr
Mackey, George W
Magee, John Francis
Magnanti, Thomas L
Manes, Ernest Gene
Martindale, Wallace S
Mauldon, James Grenfell
Mazur, Barry
Menzin, Margaret Schoenberg
Miller, William Brunner
Minsky, Marvin Lee
Mumford, David Bryant
Nitecki, Zbigniew
Ozimkoski, Raymond Edward
Palais, Richard Sheldon
Papert, Seymour A
Parrott, Stephen Kinsley
Perkins, Peter
Peters, Stefan
Peterson, Franklin Paul
Porter, Richard D
Ramras, Mark Bernard
Resnikoff, Howard L
Rogers, Hartley, Jr
Rota, Gian-Carlo
Schafer, Richard Donald
Scheid, Francis
Schlesinger, James William
Schmid, Wilfried
Schoen, Kenneth
Schweizer, Berthold
Segal, Irving Ezra
Senechal, Lester John
Sethares, George C
Shahin, Jamal Khalil
Shanahan, Patrick
Singer, Isadore Manual
Smith, John Howard
Smith, Luther W
Spencer, Guilford Lawson, II
Staknis, Victor Richard
Starr, Norton
Sternberg, Vita Shlomo
Stockton, Doris S
Stone, Arthur Harold
Strang, W(illiam) Gilbert
Strimling, Walter Eugene
Stubbe, John Sunapee
Sullivan, Joseph Arthur
Sulski, Leonard C
Sutcliffe, Samuel
Tanimoto, Taffee Tadashi
Thomas, Abdelnour Simon
Vogan, David A, Jr
Whitehead, George William
Whitman, Philip Martin
Wilcox, Howard Joseph
Williamson, Susan
Wilson, F Wesley, Jr

Winston, Arthur William
Yau, Shing-Tung

MICHIGAN
Albaugh, A Henry
Allan, George B
Arlinghaus, Sandra Judith Lach
Arlinghaus, William Charles
Bartels, Robert Christian Frank
Bouwsma, Ward D
Bragg, Louis Richard
Brown, Morton
Buckeye, Donald Andrew
Buckley, Joseph Thaddeus
Caldwell, William V
Chartrand, Gary
Clarke, Allen Bruce
Eisenstadt, Bertram Joseph
Feeman, George Franklin
Feldman, Chester
Flanders, Harley
Folkert, Jay Ernest
Frame, James Sutherland
Froemke, Jon
Fryxell, Ronald C
Fuerst, Carlton Dwight
Funkenbusch, Walter W
Fyhrie, David Paul
Galler, Bernard Aaron
Gehring, Frederick William
Gerlach, Eberhard
Gilliland, Dennis Crippen
Gilpin, Michael James
Hay, George Edward
Heikkinen, Donald D
Heins, Albert Edward
Heuvers, Konrad John
Higman, Donald Gordon
Hocking, John Gilbert
Holter, Marvin Rosenkrantz
Hoppensteadt, Frank Charles
Hsieh, Po-Fang (Philip)
Irwin, John McCormick
Jones, Phillip Sanford
Kaplan, Wilfred
Kapoor, S F
Kelly, Leroy Milton
Kincaid, Wilfred Macdonald
Kolopajlo, Lawrence Hugh
Koo, Delia Wei
Kosanovich, Robert Joseph
Krause, Eugene Franklin
Kugler, Lawrence Dean
Kuipers, Jack
Le Beau, Stephen Edward
LePage, Raoul
Lewis, Donald John
Li, Tien-Yien
Lindsay, Robert Kendall
Longyear, Judith Querida
Ludden, Gerald D
Lyjak, Robert Fred
McCoy, Thomas LaRue
McCully, Joseph C
McKay, James Harold
McNeill, Robert Bradley
Maffett, Andrew L
Marchand, Margaret O
Montgomery, Hugh Lowell
Moore, Warren Keith
Morash, Ronald
Murphy, Brian Boru
Nelson, James Donald
Nemeth, Abraham
Nemitz, William Charles
Nielson, George Marius
Nyman, Melvin Andrew
Nyquist, Gerald Warren
Palmer, Edgar M
Papp, F(rancis) J(oseph)
Pixley, Emily Chandler
Plotkin, Jacob Manuel
Rajnak, Stanley L
Schochet, Claude Lewis
Schreiner, Erik Andrew
Schuur, Jerry D
Simon, Carl Paul
Singh, Lal Pratap S
Sinha, Indranand
Sinke, Carl
Skaff, Michael Samuel
Sledd, William T
Smith, Thomas Jefferson
Smoller, Joel A
Sonneborn, Lee Meyers
Stevenson, Richard Marshall
Stewart, Bonnie Madison
Storer, Thomas
Stortz, Clarence B
Swank, Rolland Laverne
Swart, William Lee
Tanis, Elliot Alan
Tihansky, Diane Rice
Titus, Charles Joseph
Ullman, Joseph Leonard
Van Zwalenberg, George
Verhey, Roger Frank
Vichich, Thomas E
Waggoner, Wilbur J
Wasserman, Arthur Gabriel
Winter, David John
Zwier, Paul J

Mathematics, General (cont)

MINNESOTA
Aagard, Roger L
Aeppli, Alfred
Anderson, Sabra Sullivan
Aronson, Donald Gary
Barany, George
Braden, Charles McMurray
Brauer, George Ulrich
Bridgman, George Henry
Burgstahler, Sylvan
Calehuff, Girard Lester
Carlson, Philip R
Carr, Ralph W
Coonce, Harry B
Eagon, John Alonzo
Ellefson, Ralph Donald
Fleming, Walter
Fosdick, Roger L(ee)
Fox, David William
Friedman, Avner
Fristedt, Bert
George, Melvin Douglas
Gershenson, Hillel Halkin
Goblirsch, Richard Paul
Green, Leon William
Harper, Laurence Raymond, Jr
Heuer, Charles Vernon
Heuer, Gerald Arthur
Hunzeker, Hubert LaVon
Jarvinen, Richard Dalvin
Jenkins, Howard Bryner
Kahn, Donald W
Konhauser, Joseph Daniel Edward
Krueger, Eugene Rex
Lindgren, Bernard William
Lindgren, Gordon Edward
Loud, Warren Simms
McGehee, Richard Paul
McLean, Jeffery Thomas
Markus, Lawrence
Meyers, Norman George
Miracle, Chester Lee
Mordue, Dale Lewis
Nau, Richard William
Nitsche, Johannes Carl Christian
Orey, Steven
Pedoe, Daniel
Perisho, Clarence R
Prikry, Karel Libor
Reich, Edgar
Richards, Jonathan Ian
Richter, Wayne H
Roberts, A Wayne
Roberts, Joel Laurence
Roosenraad, Cris Thomas
Rust, Lawrence Wayne, Jr
Sattinger, David H
Schuster, Seymour
Seebach, J Arthur, Jr
Sell, George Roger
Serrin, James B
Sethna, Patarasp R(ustomji)
Sibuya, Yasutaka
Slagle, James R
Steen, Lynn Arthur
Storvick, David A
Sudderth, William David
Thomsen, Warren Jessen
Varberg, Dale Elthon
Walczak, Hubert R

MISSISSIPPI
Colonias, John S
Garner, Jackie Bass
Heller, Robert, Jr
King, Robert William
McCall, Jerry C
Riggs, Schultz
Sheffield, Roy Dexter
Shive, Robert Allen, Jr
Smith, Gaston
Solomon, Jimmy Lloyd
Stokes, Russell Aubrey
Tilley, John Leonard
Truax, Robert Lloyd
Webster, Porter Grigsby

MISSOURI
Ahlbrandt, Calvin Dale
Balbes, Raymond
Blackwell, Paul K, II
Bodine, Richard Shearon
Boothby, William Munger
Burcham, Paul Baker
Campbell, Larry N
Erkiletian, Dickran Hagop, Jr
Flowers, Joe D
Foran, James Michael
Francis, Richard L
Fulton, John David
Goldberg, Merrill B
Green, Philip Palmer
Grimm, Louis John
Hall, Leon Morris, Jr
Hatfield, Charles, Jr
Hicks, Troy L
Hirschman, Isidore Isaac, Jr
Jenkins, James Allister
Johnson, Charles Andrew
Johnson, Larry K
Keller, Roy Fred
Kenner, Morton Roy

Kezlan, Thomas Phillip
King, Terry Lee
Kubicek, John D
Lange, Leo Jerome
Levine, Jeffrey
Liebnitz, Paul W
McDowell, Robert Hull
McIntosh, William David
Nicholson, Eugene Haines
Padberg, Harriet A
Pearson, Bennie Jake
Rant, William Howard
Reeder, John Hamilton
Schrader, Keith William
Schwartz, Alan Lee
Sentilles, F Dennis, Jr
Shiflett, Lilburn Thomas
Smith, William R
Stanojevic, Caslav V
Stilwell, Kenneth James
Stumpff, Howard Keith
Thro, Mary Patricia
Weiss, Guido Leopold
Welland, Grant Vincent
White, Warren D
Wilke, Frederick Walter
Wright, David Lee
Wright, Farroll Tim

MONTANA
Barrett, Louis Carl
Baussus von Luetzow, Hans Gerhard
Daniel, Victor Wayne
Goebel, Jack Bruce
Jamison, William H
Lott, Johnny Warren
McAllister, Byron Leon
McKelvey, Robert William
McRae, Daniel George
Manis, Merle E
Swartz, William John
Taylor, Donald Curtis
Thompson, Edward Otis

NEBRASKA
Chivukula, Ramamohana Rao
Cox, Henry Miot
Downing, John Scott
Gross, Mildred Lucile
Heisler, Joseph Patrick
Jackson, Lloyd K
Jaswal, Sitaram Singh
Kramer, Earl Sidney
Leitzel, James Robert C
Lenz, Charles Eldon
Leonhardt, Earl A
Lewis, William James
McDougal, Robert Nelson
Masters, Christopher Fanstone
Meakin, John C
Meisters, Gary Hosler
Mientka, Walter Eugene
Miller, Donald Wright
Mordeson, John N
Nelson, Theodora S
Pao, Y(en)-C(hing)
Pickens, Charles Glenn
Scheerer, Anne Elizabeth
Sharp, Edward A
Shores, Thomas Stephen
Thornton, Melvin Chandler
Wallen, Stanley Eugene
Wampler, Joe Forrest
Zechmann, Albert W

NEVADA
Aizley, Paul
Beesley, Edward Maurice
Blackadar, Bruce Evan
Brady, Allen H
Davis, Robert Dabney
Fitting-Gifford, Marjorie Ann Premo
Graham, Malcolm
McMinn, Trevor James
Pfaff, Donald Chesley

NEW HAMPSHIRE
Baumgartner, James Earl
Brown, Edward Martin
Copeland, Arthur Herbert, Jr
Jacoby, Alexander Robb
Jeffery, Lawrence R
Meeker, Loren David
Moore, Berrien, III
Nelson, Lloyd Steadman
Norman, Robert Zane
Prosser, Reese Trego
Ross, Shepley Littlefield
Ryan, James Michael
Shore, Samuel David
Slesnick, William Ellis
Snell, James Laurie
Stancl, Donald L
Wixson, Eldwin A, Jr

NEW JERSEY
Almgren, Frederick Justin, Jr
Anderson, Stanley William
Anton, Howard
Bean, Ralph J
Beckman, Frank Samuel
Bein, Donald
Benes, Vaclav Edvard
Bidwell, Leonard Nathan

Borel, Armand
Bourgain, Jean
Browder, William
Bush, Charles Edward
Cai, Jin-yi
Canavan, Robert I
Clark, Frank Eugene
Crabtree, James Bruce
Daubechies, Ingrid
Davis, Morton David
Davis, Robert Benjamin
Dekker, Jacob Christoph Edmond
Deligne, Pierre R
Denby, Lorraine
Elliott, Joanne
Faith, Carl Clifton
Flatto, Leopold
Fox, Phyllis
Freund, Roland Wilhelm
Gelfand, Israel M
Geshner, Robert Andrew
Gilbert, Edgar Nelson
Gonshor, Harry
Gorenstein, Daniel
Graham, Ronald Lewis
Griffiths, Phillip A
Guilfoyle, Richard Howard
Gurk, Herbert Morton
Haenisch, Siegfried
Hausdoerffer, William H
Hazard, Katharine Elizabeth
Henrich, Christopher John
Houle, Joseph E
Hoyt, William Lind
Hwang, Helen H L
Jagerman, David Lewis
Johnson, Donald Glen
Jurkat, Martin Peter
Kalmanson, Kenneth
Karel, Martin Lewis
Karol, Mark J
Kauffman, Ellwood
Keen, Linda
Kees, Kenneth Lewis
Kemperman, Johannes Henricus
  Bernardus
Kessler, Irving Jack
Kochen, Simon Bernard
Kohn, Joseph John
Kovach, Ladis Daniel
Kruskal, Joseph Bernard
Kruskal, Martin David
Kuhn, Harold William
Kuntz, Richard A
Lagarias, Jeffrey Clark
Langlands, Robert P
Lapidus, Arnold
Leader, Solomon
Lechleider, J W
Lewis, Charles Joseph
Magliola-Zoch, Doris
Majda, Andrew J
Makar, Boshra Halim
Maletsky, Evan M
Nelson, Joseph Edward
Ness, Linda Ann
Ortiz-Martinez, Aury
Petryshyn, Walter Volodymyr
Poiani, Eileen Louise
Pollak, Henry Otto
Polonsky, Ivan Paul
Posamentier, Alfred S
Puri, Narindra Nath
Ramanujan, Melapalayam Srinivasan
Riggs, Richard
Rinaldi, Leonard Daniel
Rosenberg, Herman
Roth, Rodney J
Sams, Burnett Henry, III
Segers, Richard George
Selberg, Atle
Shepp, Lawrence Alan
Shimura, John
Sloane, Neil James Alexander
Sloyan, Mary Stephanie
Spencer, Thomas
Stein, Elias M
Stevenson, Robert Louis
Stewart, Ruth Carol
Stoddard, James H
Sweeney, William John
Tong, Mary Powderly
Treves, Jean Francois
Trotter, Hale Freeman
Vasconcelos, Wolner V
Wagner, Richard Carl
Waldinger, Hermann V
Weil, Andre
Weinberger, Peter Jay
White, Myron Edward
Wiles, Andrew J
Williams, Gareth Pierce
Williams, Vernon
Winkler, Peter Mann
Zierler, Neal
Zimmerberg, Hyman Joseph

NEW MEXICO
Bell, Stoughton
Beyer, William A
Brenton, June Grimm
Christensen, N(ephi) A(lbert)
Davis, Jeffrey Robert
DeMarr, Ralph Elgin

DePree, John Deryck
Giever, John Bertram
Ginsparg, Paul H
Guinn, Theodore
Hahn, Liang-Shin
Harris, Reece Thomas
Hillman, Abraham P
Johnson, Ralph T, Jr
Kist, Joseph Edmund
Kruse, Arthur Herman
Loustaunau, Joaquin
Mark, J Carson
Metzler, Richard Clyde
Morrison, Donald Ross
Novlan, David John
O'Rourke, Peter John
Peterson, Donald Palmer
Pimbley, George Herbert, Jr
Scharn, Herman Otto Friedrich
Schlosser, Jon A
Searcy, Charles Jackson
Swartz, Charles W
Thompson, Robert James
Trowbridge, George Cecil
Wackerle, Jerry Donald
Walker, Elbert Abner
Wells, Mark Brimhall
Wendroff, Burton
Willbanks, Emily West
Wilson, Carroll Klepper
Winkler, Max Albert
Zeigler, Royal Keith

NEW YORK
Abbott, John S
Adhout, Shahla Marvizi
Adler, Alfred
Adler, Roy Lee
Akst, Geoffrey R
Albicki, Alexander
Alex, Leo James
Alling, Norman Larrabee
Alt, Franz L
Anastasio, Salvatore
Anshel, Michael
Archibald, Julius A, Jr
Archibald, Ralph George
Arkin, Joseph
Asprey, Winifred Alice
Auslander, Louis
Bachman, George
Bailey, William T
Barback, Joseph
Barcus, William Dickson, Jr
Bass, Hyman
Baumslag, Gilbert
Bazer, Jack
Bell, Jonathan George
Beltrami, Edward J
Berkowitz, Jerome
Bernkopf, Michael
Bernstein, Irwin S
Bers, Lipman
Bick, Theodore A
Bing, Kurt
Birman, Joan Sylvia
Blackman, Jerome
Bloch, Alan
Brown, David T
Buckley, Fred Thomas
Bumcrot, Robert J
Burling, James P
Butler, Lewis Clark
Capobianco, Michael F
Carpenter, James Edward
Chako, Nicholas
Chen, Yuh-Ching
Chilton, Bruce L
Chow, Yuan Shih
Chuckrow, Vicki G
Cohen, Daniel Isaac Aryeh
Cohn, Harvey
Craft, George Arthur
Danese, Arthur E
Dauben, Joseph W
Davis, Edward Dewey
De Carlo, Charles R
Deming, Robert W
DePalma, James John
Desrochers, Alan Alfred
Di Paola, Robert Arnold
Douglas, Ronald George
Dowds, Richard E
Dristy, Forrest E
Durkovic, Russell George
Dutka, Jacques
Ebin, David G
Egan, Francis P
Eilenberg, Samuel
Eisele, Carolyn
Fadell, Albert George
Fairchild, William Warren
Farrell, F Thomas
Feroe, John Albert
Foisy, Hector B
Forman, William
Forray, Marvin Julian
Fox, William Cassidy
Freilich, Gerald
Freimer, Marshall Leonard
Fu, Lorraine Shao-Yen
Fuchs, Wolfgang Heinrich
Gallagher, Patrick Ximenes
Garabedian, Paul Roesel

Garcia, Mariano
Garnet, Hyman R
Gelbart, Abe
Gelbaum, Bernard Russell
Gilbert, Paul Wilner
Gilman, Sr John Frances
Godino, Charles F
Goldberg, Kenneth Philip
Goldman, Malcolm
Goldstein, Max
Goldstein, Richard
Grad, Arthur
Griesmer, James Hugo
Grosof, Miriam Schapiro
Gross, Leonard
Grossman, George
Grossman, Norman
Gucker, Gail
Guggenheimer, Heinrich Walter
Gunderson, Norman Gustav
Hachigian, Jack
Hailpern, Raoul
Hall, Dick Wick
Hanson, David Lee
Harrison, Malcolm Charles
Hartmanis, Juris
Head, Thomas James
Heller, Alex
Hemmingsen, Erik
Herbach, Leon Howard
Hershenov, Joseph
Herwitz, Paul Stanley
Hilbert, Stephen Russell
Hilton, Peter John
Hirshon, Ronald
Hochberg, Murray
Hoffman, Alan Jerome
Hollingsworth, Jack W
Huie, Carmen Wah-Kit
Hunte, Beryl Eleanor
Hurwitz, Solomon
Isaac, Richard Eugene
Isaacson, Eugene
Isbell, John Rolfe
Iwaniec, Tadeusz
Jenks, Richard D
John, Fritz
Johnson, Sandra Lee
Kahn, Peter Jack
Kao, Samuel Chung-Siung
Kazarinoff, Nicholas D
Kelisky, Richard Paul
Kesten, Harry
Kim, Woo Jong
Klimko, Eugene M
Knee, David Isaac
Kolchin, Ellis Robert
Koranyi, Adam
Kra, Irwin
Kronk, Hudson V
Kurss, Herbert
Laderman, Julian David
Larsen, Charles McLoud
Laska, Eugene
Lawler, John Patrick
Lawson, Herbert Blaine, Jr
Lax, Anneli
Lax, Peter David
Leela, Srinivasa (G)
Lefkowitz, Ruth Samson
Lefton, Phyllis
Lemke, Carlton Edward
Levenson, Morris E
Levin, Gerson
Levine, Leo Meyer
Lieberman, Burton Barnet
Lisman, Henry
Lissner, David
Liu, Chamond
Lorch, Edgar Raymond
Lubell, David
Luchins, Edith Hirsch
Lutwak, Erwin
McAuley, Louis Floyd
McCabe, John Patrick
McGloin, Paul Arthur
McKean, Henry P
McNaughton, Robert
Magill, Kenneth Derwood, Jr
Malkevitch, Joseph
Mansfield, Larry Everett
Marcus, Michael Barry
Maskit, Bernard
Melnick, Edward Lawrence
Meyer, Paul Richard
Meyer, Walter Joseph
Miller, Frederick
Miller, Kenneth Sielke
Milnor, John Willard
Miranker, Willard Lee
Moise, Edwin Evariste
Montague, Harriet Frances
Moore, John Coleman
Moreno, Carlos Julio
Mott, Thomas
Muench, Donald Leo
Mylroie, Victor L
Nachbin, Leopoldo
Neisendorfer, Joseph
Nevison, Christopher H
Newberger, Edward
Nirenberg, Louis
O'Brien, Redmond R
O'Brien, Ronald J

Ogawa, Hajimu
Olson, Frank R
Padalino, Stephen John
Pardee, Otway O'Meara
Peluso, Ada
Phillips, Esther Rodlitz
Piech, Margaret Ann
Polimeni, Albert D
Pollack, Richard
Pomeranz, Janet Bellcourt
Postman, Robert Derek
Pownall, Malcolm Wilmor
Ramaley, James Francis
Randol, Burton
Rao, Poduri S R S
Ravenel, Douglas Conner
Rees, Mina S
Resnick, Sidney I
Rhee, Haewun
Rivlin, Theodore J
Robusto, C Carl
Rockhill, Theron D
Rockoff, Maxine Lieberman
Rose, Israel Harold
Rosenbloom, Paul Charles
Rosenthal, John William
Roth, John Paul
Rothaus, Oscar Seymour
Rothenberg, Ronald Isaac
Sacksteder, Richard Carl
Sah, Chih-Han
Sandorfi, Andrew M J
Scalora, Frank Salvatore
Schaefer, Paul Theodore
Schanuel, Stephen Hoel
Schaumberger, Norman
Schindler, Susan
Schlissel, Arthur
Seckler, Bernard David
Segal, Sanford Leonard
Seiken, Arnold
Sellers, Peter Hoadley
Shabanowitz, Harry
Shapiro, Jack Sol
Shilepsky, Arnold Charles
Shimamoto, Yoshio
Shub, Michael I
Shulman, Harold
Silberger, Donald Morison
Sloan, Robert W
Small, Donald Bridgham
Small, William Andrew
Snow, Wolfe
Sohmer, Bernard
Spencer, Armond E
Spitzer, Frank L
Spoelhof, Charles Peter
Standish, Charles Junior
Steinberg, David H
Stemple, Joel G
Stephany, Edward O
Stephens, Clarence Francis
Sterling, Nicholas J
Stern, Samuel T
Sternstein, Martin
Stevenson, John Crabtree
Stoker, James Johnston
Stolfo, Salvatore Joseph
Stoner, George Green
Storm, Edward Francis
Straight, H Joseph
Strait, Peggy
Strand, Richard Carl
Strasser, Elvira Rapaport
Strichartz, Robert Stephen
Sullivan, Dennis P
Swick, Kenneth Eugene
Szusz, Peter
Tamari, Dov
Taylor, Alan D
Taylor, Francis B
Texter, John
Thomas, Edward Sandusky, Jr
Thorpe, John A(lden)
Tian, Gang
Tucciarone, John Peter
Tucker, Alan Curtiss
Turner, Edward C
Ullman, Arthur William James
Uschold, Richard L
Vasquez, Alphonse Thomas
Victor, Jonathan David
Vogeli, Bruce R
Wagner, Eric G
Wallach, Sylvan
Waltcher, Azelle Brown
Waterman, Daniel
Watts, Charles Edward
Wernick, William
White, Albert George, Jr
Wilken, Donald Rayl
Williams, Scott Warner
Wilson, Paul Robert
Wohlgelernter, Devora Kasachkoff
Wolf, Ira Kenneth
Wolsson, Kenneth
Wormser, Gary Paul
Wunderlich, Marvin C
Yu, Kai-Bor
Ziebur, Allen Douglas
Zlot, William Leonard
Zuckerberg, Hyam L
Zuckerman, Israel

NORTH CAROLINA
Allard, William Kenneth
Bernard, Richard Ryerson
Blackburn, Thomas Henry
Bright, George Walter
Bryant, Robert Leamon
Burch, Benjamin Clay
Cameron, Edward Alexander
Campbell, Stephen La Vern
Carlitz, Leonard
Cato, Benjamin Ralph, Jr
Chandler, Richard Edward
Chandra, Jagdish
Charlton, Harvey Johnson
Church, Charles Alexander, Jr
Cleland, John Gregory
Cooke, Henry Charles
Davis, Kenneth Joseph
Davis, Myrtis
Davis, Robert Lloyd
Dressel, Francis George
Eakin, Richard R
Epstein, George
Fletcher, William Thomas
Franke, John Erwin
Frazier, Robert Carl
Gentry, Ivey Clenton
Gentry, Karl Ray
Gilman, Albert F, III
Goodman, Sue Ellen
Gordh, George Rudolph, Jr
Gordy, Thomas D(aniel)
Graham, Ray Logan
Grau, Albert A
Gurganus, Kenneth Rufus
Halton, John Henry
Hansen, Donald Joseph
Harrington, Walter Joel
Herbst, Robert Taylor
Herr, David Guy
Hildebrandt, Theodore Ware
Hooke, Robert
Houston, Johnny Lee
Jackson, Robert Bruce, Jr
Johnson, Phillip Eugene
Johnson, Wendell Gilbert
Kallianpur, Gopinath
Karlof, John Knox
Kerr, Sandria Neidus
Kimbell, Julia S
King, Lunsford Richardson
Kneale, Samuel George
Koh, Kwangil
Levine, Jack
Macey, Wade Thomas
Manduley, Ilma Morell
Martin, LeRoy Brown, Jr
Meyer, Carl Dean, Jr
Mobley, Jean Bellingrath
Murray, Francis Joseph
Neidinger, Richard Dean
Nelson, A Carl, Jr
Norris, Fletcher R
Page, Nelson Franklin
Paur, Sandra Orley
Peterson, Elmor Lee
Plemmons, Robert James
Posey, Eldon Eugene
Ramanan, V R V
Roberts, John Henderson
Rodney, Earnest Abram
Rose, Nicholas John
Rutledge, Dorothy Stallworth
Saunders, Frank Wendell
Schaeffer, David George
Scoville, Richard Arthur
Selgrade, James Francis
Shell, Donald Lewis
Shelly, James H
Shoenfield, Joseph Robert
Silber, Robert
Simms, Nathan Frank, Jr
Singer, Michael
Smith, Douglas D
Smith, Fred R, Jr
Sonner, Johann
Sowell, Katye Marie Oliver
Speece, Herbert E
Stone, Erika Mares
Struble, Raimond Aldrich
Sutton, Louise Nixon
Taylor, Jerry Duncan
Thomas, John Pelham
Tolle, Jon Wright
Ullrich, David Frederick
Van Alstyne, John Pruyn
Vaughan, Jerry Eugene
Wahl, Jonathan Michael
Wesler, Oscar
Whipkey, Kenneth Lee
Willett, Richard Michael
Wright, Fred Boyer

NORTH DAKOTA
Bzoch, Ronald Charles
Koch, Charles Frederick
McBride, Woodrow H
Mathsen, Ronald M
Robinson, Thomas John
Rue, James Sandvik
Uherka, David Jerome
Winger, Milton Eugene

OHIO
Allen, Harry Prince
Andrews, George Harold
Arnoff, E Leonard
Atalla, Robert E
Atkins, Henry Pearce
Beane, Donald Gene
Bell, Harold
Bernstein, Barbara Elaine
Beyer, William Hyman
Bohn, Sherman Elwood
Bolger, Edward M
Bonar, Daniel Donald
Brockman, Harold W
Brown, Dean Raymond
Brown, Richard Kettel
Brown, Robert Bruce
Buchthal, David C
Buck, Charles (Carpenter)
Burden, Richard L
Capel, Charles Edward
Cassim, Joseph Yusuf Khan
Chalkley, Roger
Chidambaraswamy, Jayanthi
Clemens, Stanley Ray
Cummins, Kenneth Burdette
Deever, David Livingstone
Denbow, Carl (Herbert)
Diestel, Joseph
Disko, Mildred Anne
Drobot, Stefan
Duval, Leonard A
Egar, Joseph Michael
Eldridge, Klaus Emil
Ericksen, Wilhelm Skjetstad
Ferguson, Harry
Flaspohier, David
Foreman, Matthew D
Foulk, Clinton Ross
Fricke, Gerd
Garrow, Robert Joseph
Gass, Frederick Stuart
Gilligan, Lawrence G
Goldberg, Samuel
Graue, Louis Charles
Gruber, H Thomas
Guderley, Karl Gottfried
Haber, Robert Morton
Hahn, Samuel Wilfred
Hales, Raleigh Stanton, Jr
Hampton, Charles Robert
Hern, Thomas Albert
Huneke, John Philip
Hungerford, Thomas W
Isenecker, Lawrence Elmer
James, Thomas Ray
Jasper, Samuel Jacob
Jehn, Lawrence Andrew
Jekeli, Christopher
Jones, John, Jr
Kaufmann, Alvern Walter
Knight, Lyman Coleman
Kullman, David Elmer
Laha, Radha Govinda
Lair, Alan Van
Leach, Ernest Bronson
Leake, Lowell, Jr
Leetch, James Frederick
Levine, Maita Faye
Lipsich, H David
Little, Richard Allen
Long, Clifford A
Low, Marc E
MacDowell, Robert W
Markley, William A, Jr
Marty, Roger Henry
Maxwell, Glenn
Mehr, Cyrus B
Meyers, Leroy Frederick
Miller, James Roland
Mityagin, Boris Samuel
Moscovici, Henri
Nelson, Raymond John
Neuzil, John Paul
Nikolai, Paul John
Notz, William Irwin
Oestreicher, Hans Laurenz
Ogden, William Frederick
Osterburg, James
Park, Chull
Park, Won Joon
Peck, Lyman Colt
Pepper, Paul Milton
Pinzka, Charles Frederick
Piper, Ervin L
Powell, Robert Ellis
Rabson, Gustave
Ralley, Thomas G
Rattan, Kuldip Singh
Ray-Chaudhuri, Dwijendra Kumar
Riedl, John Orth
Riggle, Timothy A
Riner, John William
Robinson, Stewart Marshall
Rolwing, Raymond H
Ross, Louis
Sachs, David
Saltzer, Charles
Santos, Eugene (Sy)
Schaefer, Donald John
Schlegel, Donald Louis
Schultz, James Edward
Schuurmann, Frederick James
Semon, Warren Lloyd

Mathematics, General (cont)

Shields, Paul Calvin
Shoemaker, Richard W
Sholander, Marlow
Silberger, Allan Joseph
Smith, Frank A
Smith, James L
Smith, Mark Andrew
Smith, Robert Sefton
Spielberg, Stephen E
Spring, Ray Frederick
Staley, David H
Stanton, Robert Joseph
Steele, William F
Stein, Ivie, Jr
Steinlage, Ralph Cletus
Sterrett, Andrew
Stickney, Alan Craig
Tikson, Michael
Townsend, Ralph N
Tung, John Shih-Hsiung
Ungar, Gerald S
Vance, Elbridge Putnam
Vancko, Robert Michael
Varga, Richard S
Waits, Bert Kerr
Walum, Herbert
Wen, Shih-Liang
Wente, Henry Christian
Wilson, Eric Leroy
Wilson, Richard Michael
Yantis, Richard P
Yozwiak, Bernard James
Zilber, Joseph Abraham

OKLAHOMA

Agnew, Jeanne Le Caine
Albright, James Curtice
Bednar, Jonnie Bee
Bertholf, Dennis E
Boyce, Donald Joe
Briggs, Phillip D
Burchard, Hermann Georg
Cairns, Thomas W
Davies-Jones, Robert Peter
Ewing, George McNaught
George, Dick Leon
Goff, Gerald K
Gray, Samuel Hutchison
Huneke, Harold Vernon
Kotlarski, Ignacy Icchak
Krattiger, John Trubert
Loman, M Laverne
McKellips, Terral Lane
Magid, Andy Roy
Naymik, Daniel Allan
Page, Rector Lee
Perel, William Morris
Phelps, Jack
Pope, Paul Terrell
Reimer, Dennis D
Springer, Charles Eugene
Switzer, Laura Mae
Uehara, Hiroshi
Wagner, Harry Mahlon

OREGON

Ahuja, Jagdish C
Anderson, Frank Wylie
Anderson, Tera Lougenia
Arnold, Bradford Henry
Ballantine, C S
Balogh, Charles B
Barrar, Richard Blaine
Bhattacharyya, Ramendra Kumar
Brandon, Robert L
Brunk, Hugh Daniel
Burdg, Donald Eugene
Burton, Robert M, Jr
Butler, John Ben, Jr
Carter, David Southard
Cater, Frank Sydney
Chrestenson, Hubert Edwin
Civin, Paul
Crittenden, Richard B
Curtis, Charles Whittlesey
Enneking, Marjorie
Firey, William James
Freeman, Robert
Ghent, Kenneth Smith
Graham, Beardsley
Gray, A(ugustine) H(eard), Jr
Hall, Richard S
Harrison, David Kent
Haskell, Charles Thomson
Hutzenlaub, John F
Hyers, Donald Holmes
Iltis, Donald Richard
Jensen, Bruce A
Kaplan, Edward Lynn
Khalil, M Aslam Khan
Kunkle, Donald Edward
Maier, Eugene Alfred
Montgomery, Richard Glee
Nelsen, Roger Bain
Newberger, Stuart Marshall
Owens, Robert W
Palmer, Theodore W
Rempfer, Robert Weir
Rio, Sheldon T
Roberts, Joseph Buffington
Ross, Kenneth Allen
Schmidt, Harvey John, Jr

Seely, Justus Frandsen
Simons, William Haddock
Simpson, James Edward
Smith, John Wolfgang
Smith, Kennan Taylor
Stanley, Robert Lauren
Swanson, Leonard George
Whitesitt, John D
Whitesitt, John Eldon
Wilson, Howard Le Roy
Wyse, Frank Oliver

PENNSYLVANIA

Adams, William Sanders
Anderle, Richard
Andrews, George Eyre
Archer, David Horace
Argabright, Loren N
Arnold, Leslie K
Arrington, Elayne
Ayoub, Christine Williams
Ayoub, Raymond G Dimitri
Baric, Lee Wilmer
Bartley, Edward Francis
Bartlow, Thomas L
Bartoo, James Breese
Bates, Grace Elizabeth
Baum, Paul Frank
Bedient, Phillip E
Bickel, Robert John
Bissinger, Barnard Hinkle
Block, I(saac) Edward
Buriok, Gerald Michael
Busby, Robert Clark
Calabi, Eugenio
Castellano, Salvatore, Mario
Chao, Chong-Yun
Chow, Shin-Kien
Coffman, Charles Vernon
Crawley, James Winston, Jr
Cullen, Charles G
Cunningham, Frederic, Jr
Dank, Milton
Dice, Stanley Frost
Duffin, Richard James
Dunham, William Wade
Earl, Boyd L
Ehrenpreis, Leon
England, James Walton
Evans, Edward William
Fell, James Michael Gardner
Filano, Albert E
Freyd, Peter John
Glasner, Moses
Golumbic, Martin Charles
Gould, William Allen
Gowdy, Spenser O
Greene, Curtis
Hall, James Emerson
Harrell, Ronald Earl
Hartmann, Frederick W
Hastings, Stuart
Hearsey, Bryan Vandiver
Heath, Robert Winship
Heckscher, Stevens
Herpel, Coleman
Hickman, Warren David
Hinton, Raymond Price
Holder, Leonard Irvin
Holzinger, Joseph Rose
Homann, Frederick Anthony
Hostetler, Robert Paul
Hsiung, Chuan Chih
Hunter, Robert P
Jacobson, Bernard
Kadison, Richard Vincent
Kamel, Hyman
Kelemen, Charles F
Kerr, Carl E
Khabbaz, Samir Anton
Khalil, Mohamed Thanaa
Kinderlehrer, David (Samuel)
Kirk, David Blackburn
Kolman, Bernard
Kolodner, Ignace I(zaak)
Korfhage, Robert R
Krall, Allan M
Krall, Harry Levern
Kresh, J Yasha
Lancaster, Otis Ewing
Larson, Roland Edwin
Laush, George
Lee, Hua-Tsun
Levitan, Michael Leonard
Light, John Henry
Lindgren, William Frederick
McAllister, Gregory Thomas, Jr
McAllister, Marialuisa N
McCammon, Mary
MacCamy, Richard C
Machusko, Andrew Joseph, Jr
Maclay, Charles Wylie
Masani, Pesi Rustom
Maserick, Peter H
Mayer, Joerg Werner Peter
Mehta, Jatinder S
Metzger, Thomas Andrew
Meyer, John Sigmund
Miehle, William
Miller, G(erson) H(arry)
Misra, Dhirendra N
Mode, Charles J
Moore, Richard Allan
Nee, M Coleman

Nichols, John C
Noll, Walter
Ohl, Donald Gordon
Oler, Norman
Olsen, Glenn W
Ord, John Allyn
Ossesia, Michel Germain
Padulo, Louis
Paine, Dwight Milton
Pederson, Roger Noel
Phelps, Dean G
Pickands, James, III
Pinter, Charles Claude
Pitcher, Arthur Everett
Porter, Gerald Joseph
Porter, John Robert
Portmann, Walter Oddo
Rabenstein, Albert Louis
Ray, David Scott
Raymon, Louis
Reed, Joel
Reich, Daniel
Riggle, John H
Rosen, David
Rosenstein, George Morris, Jr
Rowlands, R(ichard) O(wen)
Saaty, Thomas L
Schechter, Murray
Schiller, John Joseph
Schultz, Blanche Beatrice
Sebesta, Charles Frederick
Shaffer, Douglas Howerth
Shale, David
Shatz, Stephen S
Shuck, John Winfield
Sigler, Laurence Edward
Silver, Edward A
Skeath, J Edward
Slaughter, Frank Gill, Jr
Smith, David Alexander
Snyder, Andrew Kagey
Strauser, Wilbur Alexander
Taylor, Floyd Heckman
Thomas, George Brinton, Jr
Thomas, Joseph Charles
Tillman, Stephen Joel
Tozier, John E
Wagner, Daniel Hobson
Warren, Richard Hawks
Wei, William Wu-Shyong
Weiss, Sol
Western, Donald Ward
Wilansky, Albert
Williams, William Orville
Wolf, Charles Trostle
Wolfe, Dorothy Wexler
Wong, Bing Kuen
Wong, Roman Woon-Ching
Wyler, Oswald
Yang, Chung-Tao
Yukich, Joseph E
Zaphyr, Peter Anthony
Zindler, Richard Eugene
Zitarelli, David Earl

RHODE ISLAND

Davis, Philip J
Gabriel, Richard Francis
Geman, Stuart Alan
McKillop, Lucille Mary
Rosen, Michael Ira
Sine, Robert C
Steward, Robert F
Stewart, Frank Moore
Vitter, Jeffrey Scott
Woolf, William Blauvelt

SOUTH CAROLINA

Abernathy, Robert O
Allen, Roger W, Jr
Ashy, Peter Jawad
Bell, Curtis Porter
Bose, Anil Kumar
Clark, Hugh Kidder
Cohn, Leslie
Dobbins, James Gregory Hall
Dodson, Norman Elmer
Fray, Robert Dutton
Hammett, Michael E
Hare, William Ray, Jr
Harley, Peter W, III
Hedberg, Marguerite Zeigel
Hind, Alfred Thomas, Jr
Hodges, Billy Gene
Hopkins, Laurie Boyle
Huff, Charles William
Kenelly, John Willis, Jr
Kirkwood, Charles Edward, Jr
Laskar, Renu Chakravarti
LaTorre, Donald Rutledge
Lytle, Raymond Alfred
Nicol, Charles Albert
Padgett, William Jowayne
Phillips, Robert Gibson
Poole, John Terry
Russell, Charles Bradley
Saxena, Subhash Chandra
Stephenson, Robert Moffatt, Jr
Sutton, Charles Samuel
Ulmer, Millard B
Waldman, Alan S
Womble, Eugene Wilson
Yu, Kai Fun

SOUTH DAKOTA

Carlson, John W
Cook, Cleland V
Fors, Elton W
Grimm, Carl Albert
Haigh, William E
Kranzler, Albert William
Miller, Bruce Linn
Pedersen, Franklin D
Rognlie, Dale Murray

TENNESSEE

Alsmiller, Rufard G, Jr
Alvarez, Laurence Richards
Arenstorf, Richard F
Austin, Billy Ray
Baumeyer, Joel Bernard
Beauchamp, John J
Bowling, Floyd E
Brooks, Sam Raymond
Bryant, Billy Finney
Carruth, James Harvey
Celauro, Francis L
Cheatham, Thomas J
Childress, Denver Ray
Croom, Frederick Hailey
Dauer, Jerald Paul
Dent, William Hunter, Jr
Dixon, Edmond Dale
Evans, Joe Smith
Forrest, Thomas Douglas
Frandsen, Henry
Ginnings, Gerald Keith
Goldberg, Richard Robinson
Haddock, John R
Horton, Linda Louise Schiestle
Jamison, King W, II
Jordan, George Samuel
Kalin, Robert
Kaltenborn, Howard Scholl
Kerce, Robert H
King, Calvin Elijah
Kreiling, Daryl
Lu, Mary Kwang-Ruey Chao
McConnel, Robert Merriman
McCowan, Otis Blakely
McHenry, Hugh Lansden
Mathews, Harry T
Megibben, Charles Kimbrough
Nymann, DeWayne Stanley
Oliver, Carl Edward
Parr, William Chris
Pleasant, James Carroll
Plummer, Michael David
Poole, George Douglas
Potter, Thomas Franklin
Ratner, Lawrence Theodore
Ross, Clay Campbell, Jr
Rousseau, Cecil Clyde
Sakhare, Vishwa M
Shobe, L(ouis) Raymon
Smith, James H
Spraker, Harold Stephen
Stallmann, Friedemann Wilhelm
Stephens, Harold W
Stevenson, Everett E
Textor, Robin Edward
Tirman, Alvin
Turner, Lincoln Hulley
Vanaman, Sherman Benton
Wade, William Raymond, II
Wagner, Carl George
Walker, David Tutherly
Ware, James Gareth
Webster, Curtis Cleveland
Wesson, James Robert
Williams, George Kenneth
Wright, Harvel Amos
Yanushka, Arthur

TEXAS

Aberth, Oliver George
Abramson, H(yman) Norman
Adams, Jasper Emmett, Jr
Ahmad, Shair
Alexander, Forrest Doyle
Allen, John Ed
Amir-Moez, Ali R
Appling, William David Love
Ashton, Joseph Benjamin
Bailey, Donald Forest
Bakelman, Ilya J(acob)
Ball, Billy Joe
Batten, George Washington, Jr
Bean, William Clifton
Bedgood, Dale Ray
Bethel, Edward Lee
Bichteler, Klaus Richard
Bledsoe, Horace Willie Lee
Bledsoe, Woodrow Wilson
Boone, James Robert
Borosh, Itshak
Bourgin, David Gordon
Boyer, Delmar Lee
Bradford, James C
Brooks, Dorothy Lynn
Byrd, Richard Dowell
Cannon, John Rozier
Cannon, Raymond Joseph, Jr
Carter, Frederick J
Charnes, Abraham
Chatfield, John A
Christy, John Harlan
Chui, Charles Kam-Tai

MATHEMATICS / 433

Colquitt, Landon Augustus
Connor, Frank Field
Cooke, William Peyton, Jr
Cooley, Adrian B, Jr
Cranford, Robert Henry
Crim, Sterling Cromwell
Crowson, Henry L
Daigh, John D(avid)
Darwin, James T, Jr
Davis, Robert Clay
De Korvin, Andre
Dijkstra, Edsger Wybe
Dollard, John D
Drew, Dan Dale
Dubay, George Henry
Duke, John Walter
Duran, Benjamin S
Eaton, William Thomas
Edmonson, Don Elton
Etgen, Garret Jay
Faucett, William Munroe
Faust, Sr Claude Marie
Fitzgibbon, William Edward, III
Fitzpatrick, Ben, Jr
Ford, Wayne
Freeman, John Clinton, Jr
Friedberg, Michael
Friedman, Charles Nathaniel
Fulks, Watson
Garner, Meridon Vestal
Gillette, Kevin Keith
Gillman, Leonard
Gilmore, Earl Howard
Gladman, Charles Herman
Goldblum, David Kiva
Greenspan, Donald
Greenwood, Robert Ewing
Guseman, Lawrence Frank, Jr
Guy, William Thomas, Jr
Hagan, Melvin Roy
Hall, Carl Eldridge
Hanne, John R
Hardt, Robert Miller
Harrist, Ronald Baxter
Harvey, F Reese
Hausen, Jutta
Hedstrom, John Richard
Hernandez, Norma
Hocking, Ronald Raymond
Hodge, James Edgar
Huffman, Louie Clarence
Huggins, Frank Norris
Hughes, David Knox
Innis, George Seth
Johnson, Gordon Gustav
Johnston, Ernest Raymond
Jones, Benjamin Franklin, Jr
Kellogg, Charles Nathaniel
Kiel, Otis Gerald
Kirk, Joe Eckley, Jr
Konen, Harry P
Kowalik, Virgil C
Lacey, Howard Elton
Lau, Yiu-Wa August
Lewis, Daniel Ralph
Lewis, Paul Weldon
Lipps, Frederick Wiessner
McCann, Roger C
McCoy, Dorothy
McLaughlin, Thomas G
Martin, Norman Marshall
Matlock, Gibb B
Mattingly, Glen E
Melvin, Cruse Douglas
Meux, John Wesley
Mezzino, Michael Joseph, Jr
Miller, Max K
Mitchell, Roger W
Mohat, John Theodore
Molloy, Marilyn
Moore, Bill C
Morgan, Raymond Victor, Jr
Muecke, Herbert Oscar
Murray, Christopher Brock
Nestell, Merlynd Keith
Neuberger, John William
Newhouse, Albert
Northcutt, Robert Allan
Nymann, James Eugene
Osborn, Roger (Cook)
Oxford, Edwin
Palas, Frank Joseph
Parrish, Herbert Charles
Pervin, William Joseph
Pfeiffer, Paul Edwin
Pitts, Jon T
Piziak, Robert
Polking, John C
Price, Kenneth Hugh
Proctor, Clarke Wayne
Pyle, Leonard Duane
Rachford, Henry Herbert, Jr
Rahe, Maurice Hampton
Reaves, Harry Lee
Richardson, Arthur Jerold
Rigby, Fred Durnford
Riggs, Charles Lathan
Robinson, Charles Dee
Robinson, Ivor
Rodriguez, Argelia Velez
Rodriguez, Dennis Milton
Rolf, Howard Leroy
Samn, Sherwood
Sanchez, David A
Schatz, Joseph Arthur
Schruben, Johanna Stenzel
Schumaker, Robert Louis
Schuster, Eugene F
Secrest, Bruce Gill
Senechalle, David Albert
Skarlos, Leonidas
Slye, John Marshall
Smith, Kirby Campbell
Smith, Rose Marie
Spellman, John W
Stark, Jeremiah Milton
Starr, David Wright
Stecher, Michael
Stenger, William
Stokes, William Glenn
Strauss, Frederick Bodo
Strauss, Monty Joseph
Tate, John T
Thompson, James Robert
Thrall, Robert McDowell
Trafton, Laurence Munro
Treisman, Philip Uri
Treybig, Leon Bruce
Tucker, Charles Thomas
Uhlenbeck, Karen K
Vance, Joseph Francis
Veech, William Austin
Vick, James Whitfield
Vobach, Arnold R
Walston, Dale Edouard
Warlick, Charles Henry
Wells, Raymond O'Neil, Jr
White, John Thomas
Whitmore, Ralph M
Wilkerson, Robert C
Williams, Bennie B
Williams, Richard Kelso
Wood, Craig Adams
Worrell, John Mays, Jr
Young, David Monaghan, Jr
Younglove, James Newton

UTAH
Allan, David Wayne
Brooks, Robert M
Burgess, Cecil Edmund
Cannon, Lawrence Orson
Chamberlin, Richard Eliot
Davis, Edward Allan
Elich, Joe
Fife, Paul Chase
Foster, James Henderson
Garner, Lynn E
Gill, Gurcharan S
Gross, Fletcher
Hammond, Robert Grenfell
Higgins, John Clayborn
Hillam, Kenneth L
Hyde, Kendell Heman
Jamison, Ronald D
Mason, Jesse David
Miller, Richard Roy
Nielson, Howard Curtis
Peterson, John M
Robinson, Donald Wilford
Rossi, Hugo
Schmitt, Klaus
Speiser, Robert David
Suprunowicz, Konrad
Tucker, Don Harrell
Vinograde, Bernard
Wattenberg, Franklin Arvey
Wickes, Harry E
Willett, Douglas W
Windham, Michael Parks
Wolfe, James H
Yearout, Paul Harmon, Jr

VERMONT
Arzt, Sholom
Ballou, Donald Henry
Chamberlain, Erling William
Cooke, Roger Lee
Heed, Joseph James
Hill, David Byrne
Izzo, Joseph Anthony, Jr
Moser, Donald Eugene
Olinick, Michael
Peterson, Bruce Bigelow
Van Der Linde, Reinhoud H
White, Christopher Clarke
Williamson, Richard Edmund

VIRGINIA
Aull, Charles Edward
Baeumler, Howard William
Barnes, Bruce Herbert
Barrett, Lida Kittrell
Botts, Truman Arthur
Chambers, Barbara Mae Fromm
Collier, Manning Gary
Culbertson, George Edward
Dobyns, Roy A
Drew, John H
Dunkl, Charles Francis
Ethington, Donald Travis
Fang, Joong J
Farley, Reuben William
Fisher, Gordon McCrea
Foy, C Allan
Friedman, Fred Jay
Gaskill, Irving E
Glynn, William Allen
Good, Irving John
Grandy, Charles Creed
Greenwood, Joseph Albert
Greinke, Everett D
Greville, Thomas Nall Eden
Hancock, V(ernon) Ray
Heath, Lenwood S
Hendricks, Walter James
Hershner, Ivan Raymond, Jr
Holub, James Robert
Humphreys, Mabel Gweneth
Johnson, Lee W
Johnson, Robert S
Johnston, Robert Howard
Jones, Eleanor Green Dawley
Kneece, Roland Royce, Jr
Kottman, Clifford Alfons
Krause, Ralph M
Lanzano, Bernadine Clare
Lawrence, James Franklin
Leitzel, Joan Phillips
Lewis, Jesse C
Liverman, Thomas Phillip George
McDonald, Bernard Robert
Malbon, Wendell Endicott
Mast, Joseph Willis
Mayo, Thomas Tabb, IV
Mikhail, Nabih N
Minton, Paul Dixon
Natrella, Joseph
Ottoson, Harold
Owens, Robert Hunter
Pace, Wesley Emory
Perry, John Murray
Perry, William James
Phelan, James Frederick
Phillips, Robert Bass, Jr
Pitt, Loren Dallas
Quinn, Frank S
Ramirez, Donald Edward
Reid, Lois Jean
Riess, Ronald Dean
Ringeisen, Richard Delose
Roberts, Robert Abram
Root, David Harley
Rovnyak, James L
Russell, David L
Rutland, Leon W
Sanders, William Mack
Schneck, Paul Bennett
Scott, Leland Latham
Selden, Dudley Byrd
Sharp, Henry, Jr
Shlesinger, Michael
Shockley, James Edgar
Smith, Emma Breedlove
Smith, James Clarence, Jr
Stolberg, Harold Josef
Summerville, Richard Marion
Tang, Victor Kuang-Tao
Taylor, Elizabeth Beaman Hesch
Trampus, Anthony
Van Tilborg, André Marcel
Wagner, Richard Lorraine, Jr
Walpole, Ronald Edgar
Ward, Harold Nathaniel
Wei, Diana Yun Dee (Fan)
Weinstein, Stanley Edwin
Weiss, Robert John
Whitaker, Mack Lee
Wilson, Robert Lee
Yang, Chao-Hui
Zilczer, Janet Ann

WASHINGTON
Arsove, Maynard Goodwin
Ball, Ralph Wayne
Beaumont, Ross Allen
Benda, Miroslav
Blumenthal, Robert McCallum
Brink, James E
Brooks, David C
Brownell, Frank Herbert, III
Bushaw, Donald (Wayne)
Comstock, Dale Robert
Corson, Harry Herbert
Craswell, Keith J
Cunnea, William M
Curjel, Caspar Robert
Cutlip, William Frederick
Davis, Thomas Austin
Dean, Robert Yost
Dekker, David Bliss
DeTemple, Duane William
Dubisch, Roy
Dunn, Samuel L
Edison, Larry Alvin
Firkins, John Frederick
Goodrick, Richard Edward
Graham, C Robin
Greenberg, Ralph
Hall, Wayne Hawkins
Hamel, Ray O
Hansen, Rodney Thor
Herzog, John Orlando
Higgins, Theodore Parker
Hirschfelder, John Joseph
Jans, James Patrick
Jordan, James Henry
Kas, Arnold
Kingston, John Maurice
Klee, Victor La Rue, Jr
LeVeque, Randall J
Lind, Douglas A
Lister, Frederick Monie
Lloyd, Raymond Clare
Luke, Stanley D
Martin, Bernard Loyal
Montzingo, Lloyd J, Jr
Morrow, James Allen, Jr
Namioka, Isaac
Nunke, Ronald John
Oster, Clarence Alfred
Ostrom, Theodore Gleason
Perlman, Michael David
Phelps, Robert Ralph
Pyke, Ronald
Reay, John R
Reudink, Douglas O
Robertson, Jack M
Rockafellar, Ralph Tyrrell
Sarason, Leonard
Schaerf, Henry Maximilian
Smith, David Clement, IV
Smith, Stewart W
Snell, Robert Isaac
Stout, Edgar Lee
Sullivan, Hugh D
Sullivan, John Brendan
Thompson, Robert Harry
Toskey, Burnett Roland
Tull, Jack Phillip
Underwood, Douglas Haines
Van Enkevort, Ronald Lee
Wiser, Horace Clare
Woll, John William, Jr
Yagi, Fumio

WEST VIRGINIA
Gould, Henry Wadsworth
Hogan, John Wesley
LaRue, James Arthur
Morris, Peter Craig
Simons, William Harris

WISCONSIN
Barnett, Jeffrey Charles
Beck, Anatole
Bender, Phillip R
Bleicher, Michael Nathaniel
Borman, Gary Lee
Brauer, Fred
Braunschweiger, Christian Carl
Brookshear, James Glenn
Brualdi, Richard Anthony
Bruck, Richard Hubert
Carlson, Stanley L
Chover, Joshua
Chow, Yutze
Conner, Howard Emmett
De Boor, Carl (Wilhelm Reinhold)
Flanagan, Carroll Edward
Forelli, Frank John
Gade, Edward Herman Henry, III
Girard, Dennis Michael
Gollmar, Dorothy May
Gough, Lillian
Gunji, Hiroshi
Hanneken, Clemens
Harris, J Douglas
Harvey, John Grover
Kaye, Norman Joseph
Klatt, Gary Brandt
Kuenzi, Norbert James
Liu, Matthew J P
Matchett, Andrew James
Meitler, Carolyn Louise
Meyer, Richard Ernst
Miller, Gordon Lee
Moore, Robert H
Moyer, John Clarence
Ney, Peter E
Orlik, Peter Paul
Parter, Seymour Victor
Poss, Richard Leon
Prielipp, Robert Walter
Robbin, Joel W
Rudin, Mary Ellen
Rudin, Walter
Solomon, Louis
Spitzbart, Abraham
Turner, Robert E L
Van Ryzin, Martina
Voichick, Michael
Wahlstrom, Lawrence F
Wainger, Stephen
Wasow, Wolfgang Richard
Weibel, Armella
Wilson, Robert Lee, Jr

WYOMING
Collins, James R
Husain, Syed Alamdar
Rowland, John H
Shader, Leslie Elwin
Varineau, Verne John

PUERTO RICO
Francis, Eugene A
Laplaza, Miguel Luis
Montes, Maria Eugenia

ALBERTA
Christian, Robert Roland
Consul, Prem Chandra
Garg, Krishna Murari
Jones, J P
Klamkin, Murray S

Mathematics, General (cont)

Lowig, Henry Francis Joseph
Macki, Jack W
McKinney, Richard Leroy
Milner, Eric Charles
Mollin, Richard Anthony
Muldowney, James
Mungan, Necmettin
Rhemtulla, Akbar Hussein
Schaer, Jonathan
Stone, Michael Gates
Timourian, James Gregory
Varadarajan, Kalathoor
Wong, James Chin-Sze
Zvengrowski, Peter Daniel

BRITISH COLUMBIA
Belluce, Lawrence P
Brantingham, Patricia Louise
Bullen, Peter Southcott
Bures, Donald John (Charles)
Chacon, Rafael Van Severen
Chang, Bomshik
Divinsky, Nathan Joseph
Fournier, John J F
Freedman, Allen Roy
Gargett, Ann E
Hewgill, Denton Elwood
Hinrichs, Lowell A
Hurd, Albert Emerson
MacDonald, John Lauchlin
Murdoch, David Carruthers
Restrepo, Rodrigo Alvaro
Rieckhoff, Klaus E
Robertson, Malcolm Slingsby
Srivastava, Hari Mohan
Srivastava, Rekha
Thyer, Norman Harold
Whittaker, James Victor

MANITOBA
Armstrong, Kenneth William
Basilevsky, Alexander
Gupta, Narain Datt
Mendelsohn, Nathan Saul
Sichler, Jiri Jan
Williams, Hugh Cowie

NEW BRUNSWICK
Crawford, William Stanley Hayes
Dekster, Boris Veniamin
Sullivan, Donald
Thompson, Jon H

NEWFOUNDLAND
Burry, John Henry William
Fekete, Antal E
Lal, Mohan
Shawyer, Bruce L R

NOVA SCOTIA
Asadulla, Syed
Oliver, Leslie Howard
Pearson, Terrance Laverne
Snow, Douglas Oscar
Swaminathan, Srinivasa
Tingley, Arnold Jackson

ONTARIO
Abrham, Jaromir Vaclav
Auer, Jan Willem
Banaschewski, Bernhard
Barbeau, Edward Joseph, Jr
Behara, Minaketan
Berman, Gerald
Bierstone, Edward
Bower, Oliver Kenneth
Brainerd, Barron
Bush, George Clark
Choi, Man-Duen
Coleman, Albert John
Csorgo, Miklos
Cvetanovic, R J
Davidson, Kenneth R
Davison, Thomas Matthew Kerr
Dawson, Donald Andrew
DeLury, Daniel Bertrand
Drury, Malcolm John
Dunham, Charles Burton
Ellers, Erich Werner
Gadamer, Ernst Oscar
Gentleman, William Morven
Graham, James W
Graham, John Elwood
Griffith, John Sidney
Haruki, Hiroshi
Headley, Velmer Bentley
Heinig, Hans Paul
Hilldrup, David J
Jamieson, Derek Maitland
Kannappan, Palaniappan
Kaufman, Hyman
Kent, Clement F
Laframboise, Marc Alexander
LeBel, Jean Eugene
Lehma, Alfred Baker
Linders, James Gus
Lorch, Lee (Alexander)
McCallion, William James
McKiernan, Michel Amedee
MacNeill, Ian B
Macphail, Moray St John
Mueller, Bruno J W
Mullen, Kenneth
Mullin, Ronald Cleveland
Murasugi, Kunio
Nesheim, Michael Ernest
Norman, Robert Daniel
Ostry, Sylvia
Pandey, Jagdish Narayan
Pelletier, Joan Wick
Pope, Noel Kynaston
Pundsack, Arnold L
Puttaswamaiah, Bannikuppe M
Read, Ronald Cedric
Rice, Norman Molesworth
Roberts, Leslie Gordon
Robinson, Gilbert de Beauregard
Rooney, Paul George
Rosenthal, Peter (Michael)
Schubert, Cedric F
Shenitzer, Abe
Thierrin, Gabriel
Tutte, William Thomas
Ursell, John Henry
Vaillancourt, Remi Etienne
Vanstone, J R
Verner, James Hamilton
Woodside, William
Younger, Daniel H

QUEBEC
Allen, Harold Don
Barr, Michael
Chaubey, Yogendra Prasad
Driscoll, John G
Dubuc, Serge
Duncan, Richard Dale
Goodspeed, Frederick Maynard (Cogswell)
Herschorn, Michael
Joffe, Anatole
Jonsson, Wilbur Jacob
Klemola, Tapio
Labute, John Paul
Lacroix, Norbert Hector Joseph
Lambek, Joachim
Lederman, Frank L
Leroux, Pierre
Lin, Paul C S
Rattray, Basil Andrew
Rosenberg, Ivo George
Sayeki, Hidemitsu
Schlomiuk, Norbert
Srivastava, Triloki N
Turgeon, Jean

SASKATCHEWAN
Amundrud, Donald Lorne
Blum, Richard
Conlan, James
Saini, Girdhari Lal

OTHER COUNTRIES
Adams, Helen Elizabeth
Andrea, Stephen Alfred
Brady, Wray Grayson
Brown, John Wesley
Camiz, Sergio
Chalk, John
Chan, Albert Sun Chi
Cherenack, Paul Francis
D'Ambrosio, Ubiratan
Eells, James
Fiesler, Emile
Foglio, Susana
Furstenberg, Hillel
Hofmann, Karl Heinrich
Janos, Ludvik
Klaasen, Gene Allen
Kwun, Kyung Whan
Lambert, William M, Jr
Lay, Steven R
Lee, Chung N
McCue, Edmund Bradley
Maltese, George J
Miller, Charles Frederick, III
Mond, Bertram
Moser, Jurgen (Kurt)
Nohel, John Adolph
Olive, Gloria
Olum, Paul
Pekeris, Chaim Leib
Quillen, Daniel G
Rabinowitz, Philip
Rich, Michael
Roseman, Joseph Jacob
Rosen, William G
Ryan, Robert Dean
Sakai, Shoichiro
Schurle, Arlo Willard
Settles, Ronald Dean
Stavroudis, Orestes Nicholas
Stoneham, Richard George
Thompson, J G
Wallace, Andrew Hugh
Warne, Ronson Joseph
Wendland, Wolfgang Leopold
Woods, Edward James
Yang, Chung-Chun

Number Theory

ALABAMA
Mattics, Leon Eugene

ARIZONA
Smarandache, Florentin
Velez, William Yslas

CALIFORNIA
Alder, Henry Ludwig
Alexanderson, Gerald Lee
Aschbacher, Michael
Blasius, Don Malcolm
Drobot, Vladimir
Estes, Dennis Ray
Fried, Michael D
Friedberg, Solomon
Garrison, Betty Bernhardt
Gerstein, Larry J
Gruenberger, Fred J(oseph)
Guralnick, Robert Michael
Hales, Alfred Washington
Harris, Vincent Crockett
Hilliker, David Lee
Lenstra, Hendrik W
Levit, Robert Jules
Lippa, Erik Alexander
Lu, Kau U
Ringel, Gerhard
Stark, Harold Mead
Sun, Hugo Sui-Hwan
Sweet, Melvin Millard
Taussky, Olga
Terras, Audrey Anne
Thomas, Paul Emery

COLORADO
Derr, James Bernard
Hightower, Collin James
Marsh, Donald Charles Burr
Mathys, Peter
Rearick, David F
Roeder, David William

CONNECTICUT
Eigel, Edwin George, Jr
Reid, James Dolan
Stein, Alan H

DELAWARE
Federighi, Enrico Thomas

DISTRICT OF COLUMBIA
Vegh, Emanuel

FLORIDA
DeLeon, Morris Jack
Goodman, Adolph Winkler
Sakmar, Ismail Aydin
Zame, Alan

GEORGIA
Bompart, Billy Earl
Pomerance, Carl

ILLINOIS
Ballew, David Wayne
Bateman, Paul Trevier
Berndt, Bruce Carl
Bloch, Spencer J
Cormier, Romae Joseph
Diamond, Harold George
Earnest, Andrew George
Eggan, Lawrence Carl
Halberstam, Heini
Hildebrand, Adolf J
Janusz, Gerald Joseph
Kasube, Herbert Emil
Lamont, Patrick
Lazerson, Earl Edwin
Philipp, Walter V
Redmond, Donald Michael
Reznick, Bruce Arie
Ullom, Stephen Virgil
Vanden Eynden, Charles Lawrence

INDIANA
Dudley, Underwood
Hahn, Alexander J
Murphy, Catherine Mary
Pollak, Henry Otto
Shahidi, Freydoon
Wagstaff, Samuel Standfield, Jr
Yates, Richard Lee

IOWA
Kutzko, Philip C

KANSAS
Dressler, Robert Eugene
McCarthy, Paul Joseph

KENTUCKY
LeVan, Marijo O'Connor
Lowman, Bertha Pauline

LOUISIANA
Maxfield, Margaret Waugh

MAINE
Johnson, Robert Wells
Norton, Karl Kenneth
Pogorzelski, Henry Andrew

MARYLAND
Adams, William Wells
Buell, Duncan Alan
Ferguson, Helaman Rolfe Pratt
Goldstein, Larry Joel
Katz, Victor Joseph
Lakein, Richard Bruce
Massell, Paul Barry
Sadowsky, John
Spohn, William Gideon, Jr
Washington, Lawrence C

MASSACHUSETTS
Ankeny, Nesmith Cornett
Cox, David Archibald
Gessel, Ira Martin
Hayes, David R
Kenney, Margaret June
Koshy, Thomas
Mazur, Barry
Orr, Richard Clayton
Ostrovsky, Rafail M
Senechal, Marjorie Lee

MICHIGAN
Arlinghaus, William Charles
Calloway, Jean Mitchener
Gioia, Anthony Alfred
Herzog, Fritz
Malm, Donald E G
Papp, F(rancis) J(oseph)

MINNESOTA
Gaal, Ilse Lisl Novak
Hejhal, Dennis Arnold

NEBRASKA
Leitzel, James Robert C

NEVADA
Shiue, Peter Jay-Shyong

NEW JERSEY
Bumby, Richard Thomas
Charlap, Leonard Stanton
Endo, Youichi
Faltings, Gerd
Lagarias, Jeffrey Clark
Mills, William Harold
Sinai, Yakov G

NEW YORK
Alex, Leo James
Arpaia, Pasquale J
Barshay, Jacob
Brzenk, Ronald Michael
Childs, Lindsay Nathan
Chudnovsky, Gregory V
Cohen, Herman Jacob
Cohn, Harvey
Cusick, Thomas William
Edwards, Harold M
Goldfeld, Dorian
Grabois, Neil
Grossman, George
Hoobler, Raymond Taylor
Jacquet, Herve
Kaltofen, Erich L
Kleiman, Howard
Kwong, Yui-Hoi Harris
Lefton, Phyllis
Moritz, Roger Homer
Nathanson, Melvyn Bernard
Segal, Sanford Leonard
Sheingorn, Mark
Spiro, Claudia Alison
Wunderlich, Marvin C

NORTH CAROLINA
Haggard, Paul Wintzel
Hall, George Lincoln
Hodel, Margaret Jones
Howard, Fredric Timothy
Vaughan, Theresa Phillips

OHIO
Flicker, Yuval Zvi
Glass, Andrew Martin William
Gold, Robert
Hsia, John S
Parson, Louise Alayne
Paul, Jerome L
Ponomarev, Paul
Rubin, Karl C
Sellers, James Allen
Steiner, Ray Phillip
Theusch, Colleen Joan
Wang, Paul Shyh-Horng
Zassenhaus, Hans J

OREGON
Ecklund, Earl Frank, Jr
Flahibe, Mary Elizabeth
Stalley, Robert Delmer

PENNSYLVANIA
Andrews, George Eyre
Beck, Stephen D
Bressoud, David Marius
Brownawell, Woodrow Dale
Cable, Charles Allen
Galambos, Janos
Grosswald, Emil
Hagis, Peter, Jr

Harbater, David
James, Donald Gordon
Knopp, Marvin Isadore
Li, Wen-Ch'ing Winnie
Mullen, Gary Lee
Rosen, David
Shuck, John Winfield
Vaserstein, Leonid, IV
Waterhouse, William Charles

RHODE ISLAND
Durst, Lincoln Kearney
Lichtenbaum, Stephen
Lubin, Jonathan Darby

SOUTH CAROLINA
Green, Harold Rugby

TENNESSEE
Cross, James Thomas
Dobbs, David Earl
Wood, Bobby Eugene

TEXAS
Arnold, David M
Blakley, George Robert
Borosh, Itshak
Chen, Goong
Franco, Zachary Martin
Gerth, Frank E, III
Hardy, John Thomas
Hart, R Neal
Hazlewood, Donald Gene
Hensley, Douglas Austin
Stiller, Peter Frederick
Vaaler, Jeffrey David
Weaver, Milo Wesley

UTAH
Skarda, R Vencil, Jr

VERMONT
Foote, Richard Martin

VIRGINIA
Dozier, Lewis Bryant
Parry, Charles J
Rabung, John Russell
Sunley, Judith S
Thaler, Alvin Isaac

WASHINGTON
Hansen, Rodney Thor
Long, Calvin Thomas
Webb, William Albert

WISCONSIN
Bach, Eric
Bange, David W
Bleicher, Michael Nathaniel
Smart, John Roderick
Terwilliger, Paul M

WYOMING
Cowles, John Richard
Porter, A Duane

ALBERTA
Mollin, Richard Anthony

BRITISH COLUMBIA
Boyd, David William
Cayford, Afton Herbert
Riddell, James

MANITOBA
Mendelsohn, Nathan Saul

NEWFOUNDLAND
Rideout, Donald Eric

ONTARIO
Arthur, James Greig
Barbeau, Edward Joseph, Jr
Chakravartty, Iswar C
Davison, J Leslie
Dixon, John Douglas
Friedlander, John Benjamin
Handelman, David E
Ribenboim, Paulo
Riehm, Carl Richard
Stewart, Cameron Leigh
Williams, Kenneth Stuart

QUEBEC
Murty, M Ram

SASKATCHEWAN
Sato, Daihachiro

OTHER COUNTRIES
Gelbart, Stephen Samuel
Iwasawa, Kenkichi
Muskat, Joseph Baruch
Myerson, Gerald
Philippou, Andreas Nicolaou
Sato, Mikio
Stoneham, Richard George
Thomason, Robert Wayne
Zagier, Don Bernard

Operations Research

ALABAMA
Fowler, Bruce Wayne
Griffin, Marvin A
Schroer, Bernard J
Smith, Ernest Lee, Jr
White, Charles Raymond

ALASKA
Larson, Frederic Roger

ARIZONA
Duckstein, Lucien
Seifert, Deborah Roeckner
Smith, Harvey Alvin
Walling, Derald Dee
Wymore, Albert Wayne

ARKANSAS
Taha, Hamdy Abdelaziz

CALIFORNIA
Antoniak, Charles Edward
Arthur, William Brian
Barker, William Hamblin, II
Book, Stephen Alan
Bratt, Bengt Erik
Bryson, Marion Ritchie
Capp, Walter B(ernard)
Chernick, Michael Ross
Christman, Arthur Castner, Jr
Coile, Russell Cleven
Dantzig, George Bernard
Dean, Burton Victor
Debreu, Gerard
Dreyfus, Stuart Ernest
Drobnies, Saul Isaac
Eaves, Burchet Curtis
Eisemann, Kurt
Emerson, Thomas James
Eriksen, Stuart P
Farquhar, Peter Henry
Forrest, Robert Neagle
Gafarian, Antranig Vaughn
Goheen, Lola Coleman
Gretsky, Neil E
Hakimi, S(eifollah) L(ouis)
Heitkamp, Norman Denis
Hillier, Frederick Stanton
Holder, Harold Douglas
Howard, Gilbert Thoreau
Hu, Te Chiang
Iglehart, Donald Lee
Isenberg, Lionel
Jacobs, Patricia Anne
Jensen, Paul Edward T
Jewell, William S(ylvester)
Keeler, Emmett Brown
Keeney, Ralph Lyons
Kimme, Ernest Godfrey
Klinger, Allen
Koenigsberg, Ernest
Krieger, Henry Alan
Lieberman, Gerald J
Lodato, Michael W
Lucas, William Franklin
McCall, Chester Hayden, Jr
McMasters, Alan Wayne
Marshall, Kneale Thomas
Mendelssohn, Roy
Milch, Paul R
Miles, Ralph Fraley, Jr
Morris, Peter Alan
Nahmias, Steven
Nielsen, Norman Russell
O'Dell, Jean Marland
Oliver, Robert Marquam
Opfell, John Burton
Oren, Shmuel Shimon
Prim, Robert Clay, III
Purdue, Peter
Richards, Francis Russell
Rosenblatt, Daniel Bernard
Scheuer, Ernest Martin
Schrady, David Alan
Shanthikumar, Jeyaveerasingam George
Shapiro, Edwin Seymour
Shedler, Gerald Stuart
Simons, Barbara Bluestein
Smallwood, Richard Dale
Smith, Stephen Allen
Smith, Wayne Earl
Stinson, Perri June
Torbett, Emerson Arlin
Trauring, Mitchell
Tysver, Joseph Bryce
Veinott, Arthur Fales, Jr
Walker, Warren Elliott
Weaver, William Bruce
Wollmer, Richard Dietrich
Wong, Derek
Yu, Oliver Shukiang

COLORADO
Darst, Richard B
DeCoursey, Donn G(ene)
Heine, George Winfield, III
Hoots, Felix R
Lett, Gregory Scott
Litwhiler, Daniel W
Morgenthaler, George William
Schnabel, Robert B
White, Franklin Estabrook

CONNECTICUT
Ghen, David C
Mukhopadhyay, Nitis
Powell, Bruce Allan
Rao, Valluru Bhavanarayana
Shorr, Bernard
Stein, David Morris
Terry, Herbert
Youssef, Mary Naguib

DELAWARE
Eissner, Robert M
Elterich, G Joachim
Stark, Robert M

DISTRICT OF COLUMBIA
Ching, Chauncey T K
Gould, Phillip
Houck, Frank Scanland
Kaplan, David Jeremy
Knopman, Debra Sara
Kupperman, Robert Harris
Marcus, Michael Jay
Offutt, Susan Elizabeth
Singpurwalla, Nozer Drabsha
Soland, Richard Martin
Spruill, Nancy Lyon
Stein, Marjorie Leiter
Wiener, Howard Lawrence
Willard, Daniel

FLORIDA
Buoni, Frederick Buell
Fausett, Donald Wright
Haase, Richard Henry
Halsey, John Joseph
Hanson, Morgan A
Heaney, James Patrick
Hertz, David Bendel
Karp, Abraham E
Kunce, Henry Warren
Lloyd, Laurance H(enry)
Melich, Michael Edward
Powers, Joseph Edward
Rubin, Richard Lee
Shershin, Anthony Connors
Sivazlian, Boghos D
Tsokos, Chris Peter
Zanakis, Stelios (Steve) H

GEORGIA
Cohen, Alonzo Clifford, Jr
Grum, Allen Frederick
Johnson, Ellis Lane
Lorber, Herbert William
Pepper, William Donald
Serfozo, Richard Frank
Shakun, Wallace
Thomas, Michael E(dward)

HAWAII
Hiller, Robert Ellis
Wylly, Alexander

IDAHO
Laughlin, James Stanley

ILLINOIS
Agnew, Robert A
Arenson, Donald L
Barron, Emmanuel Nicholas
Blair, Charles Eugene
Caywood, Thomas E
Clover, William John, Jr
Cook, Joseph Marion
Cooper, Mary Weis
Cornwell, Larry Wilmer
Eggan, Lawrence Carl
Fourer, Robert
Kalai, Ehud
Lee, Der-Tsai
McLinden, Lynn
Maltz, Michael D
Massel, Gary Alan
Orden, Alex
Raghavan, Thirukkannamangai E S
Reingold, Edward Martin
Smith, Spencer B
Steinberg, David Israel
Thomopoulos, Nick Ted
Tier, Charles
Wacker, William Dennis

INDIANA
Fuelling, Clinton Paul
Hockstra, Dale Jon
Morin, Thomas Lee
Obenchain, Robert Lincoln
Stampfli, Joseph
Thuente, David Joseph
Wynne, Bayard Edmund

IOWA
Buck, James R
Kortanek, Kenneth O
Liittschwager, John M(ilton)
Sposito, Vincent Anthony
Vardeman, Stephen Bruce

KANSAS
Meier, Robert J
Price, Griffith Baley
Wallace, Victor Lew

KENTUCKY
Griffith, William Schuler

LOUISIANA
Andrus, Jan Frederick
Denny, William F

MAINE
Allen, Roger Baker
Balakrishnan, V K

MARYLAND
Adler, Sanford Charles
Arsham, Hossein
Benokraitis, Vitalius
Boggs, Paul Thomas
Bugenhagen, Thomas Gordon
Cherniavsky, Ellen Abelson
Cushen, Walter Edward
Eisenman, Richard Leo
Engel, Joseph H(enry)
Englund, John Arthur
Finkelstein, Robert
Flagle, Charles D(enhard)
Fromovitz, Stan
Gass, Saul Irving
Gilford, Leon
Goldman, Alan Joseph
Graves, Willard L
Harris, Carl Matthew
Heinberg, Milton
Honig, John Gerhart
Jackson, Richard H F
Johnson, Russell Dee, Jr
Kattakuzhy, George Chacko
Leight, Walter Gilbert
McCarn, Davis Barton
Marimont, Rosalind Brownstone
Meyer, Gerard G L
Moses, Edward Joel
Odle, John William
Offutt, William Franklin
Raff, Samuel J
Robinson, Richard Carleton, Jr
Singhal, Jaya Asthana
Steinwachs, Donald Michael
Voss, Paul Joseph
Wallace, Alton Smith
Witzgall, Christoph Johann
Young, Hobart Peyton
Zabronsky, Herman

MASSACHUSETTS
Ben-Akiva, Moshe E
Blackett, Donald Watson
Bras, Rafael Luis
Buzen, Jeffrey Peter
Chaiken, Jan Michael
Cline, James Edward
Cox, Lawrence Henry
Drake, Alvin William
Eckstein, Jonathan
Eklund, Karl E
Esch, Robin E
Freund, Robert M
Gangulee, Amitava
Giglio, Richard John
Hertweck, Gerald
Herz, Matthew Lawrence
Holmes, Richard Bruce
Larson, Richard Charles
Lewinson, Victor A
Little, John Dutton Conant
O'Neil, Patrick Eugene
Russo, Richard F
Shorb, Alan McKean
Shuchat, Alan Howard
Tobin, Roger Lee
Whitman, Philip Martin
Wolfe, Harry Bernard

MICHIGAN
Bonder, Seth
Chakravarthy, Srinivasaraghavan
Dittmann, John Paul
Hamilton, William Eugene, Jr
Kochen, Manfred
Lenker, Susan Stamm
Magerlein, James Michael
Mordukhovich, Boris S
Murty, Katta Gopalakrishna
Patterson, Richard L
Pollock, Stephen M
Saperstein, Alvin Martin
Simpson, William Albert

MINNESOTA
Anderson, Sabra Sullivan
Carstens, Allan Matlock
Heuer, Gerald Arthur
Johnson, Howard Arthur, Sr
Richards, James L
Rosen, Judah Ben

MISSISSIPPI
Butler, Charles Morgan

MISSOURI
Frisby, James Curtis

MONTANA
Hess, Lindsay LaRoy
McRae, Daniel George

Operations Research (cont)

NEBRASKA
Castater, Robert Dewitt
Schniederjans, Marc James
Walters, Randall Keith

NEVADA
Fain, William Wharton
Flueck, John A

NEW HAMPSHIRE
Bogart, Kenneth Paul
Bossard, David Charles
Hunt, Everett Clair

NEW JERSEY
Blickstein, Stuart I
Boesch, Francis Theodore
Boylan, Edward S
Chasalow, Ivan G
Chien, Victor
Cinlar, Erhan
Cullen, Daniel Edward
DeBaun, Robert Matthew
Eisner, Mark Joseph
Garber, H(irsh) Newton
Green, Edwin James
Greenberg, Michael Richard
Lucantoni, David Michael
McCallum, Charles John, Jr
Magliola-Zoch, Doris
Maxim, Leslie Daniel
Mihram, George Arthur
Naus, Joseph Irwin
Prekopa, Andras
Roberts, Fred Stephen
Robertson, Jerry L(ewis)
Rothkopf, Michael H
Sarakwash, Michael
Segal, Moshe
Sengupta, Bhaskar
Tucker, Albert William
Vardi, Yehuda
Weibel, Charles Alexander
Whitt, Ward
Wright, Margaret Hagen
Yoon, Kwangsun Paul

NEW MEXICO
Baty, Richard Samuel
Bell, Stoughton
Boland, W Robert
Farmer, William Michael
Gutjahr, Allan L
Hulme, Bernie Lee
Ross, Timothy Jack
Rypka, Eugene Weston
Selden, Robert Wentworth
Summers, Donald Lee

NEW YORK
Balachandran, Kashi Ramamurthi
Ballou, Donald Pollard
Bencsath, Katalin A
Bienstock, Daniel
Billera, Louis Joseph
Bland, Robert Gary
Burke, Paul J
Chatterjee, Samprit
Conway, Richard Walter
Ecker, Joseph George
Goldfarb, Donald
Gomory, Ralph E
Green, Alwin Clark
Heath, David Clay
Helly, Walter S
Jabbur, Ramzi Jibrail
Kabak, Irwin William
Keilson, Julian
Kershenbaum, Aaron
King, Alan Jonathan
Kleiman, Howard
Klein, Morton
Kolesar, Peter John
Lyons, Joseph Paul
Nagel, Harry Leslie
Nevison, Christopher H
Pan, Victor
Richter, Donald
Riffenburgh, Robert Harry
Rispoli, Fred Joseph
Rosberg, Zvi
Rothenberg, Ronald Isaac
Schweitzer, Paul Jerome
Singh, Chanchal
Smith, Bernard
Stein, Arthur
Todd, Michael Jeremy
Van Slyke, Richard M
White, William Wallace
Wolfe, Philip
Yao, David D

NORTH CAROLINA
Aydlett, Louise Clate
Ball, Michael Owen
Elmaghraby, Salah Eldin
Fishman, George Samuel
Gelenbe, Erol
Hodgson, Thom Joel
Honeycutt, Thomas Lynn
Jambor, Paul Emil
McAllister, David Franklin

Nelson, A Carl, Jr
Peterson, David West
Potthoff, Richard Frederick
Roth, Walter John
Stidham, Shaler, Jr
Wallingford, John Stuart

OHIO
Arnoff, E Leonard
Chrissis, James W
Emmons, Hamilton
Flowers, Allan Dale
Hales, Raleigh Stanton, Jr
Hurt, James Joseph
Martino, Joseph Paul
Mills, Wendell Holmes, Jr
Nielsen, Philip Edward
Patton, Jon Michael
Piotrowski, Zbigniew
Reisman, Arnold
Schultz, Edwin Robert
Solow, Daniel
Staker, Robert D
Ventresca, Carol
Ward, Douglas Eric
Waren, Allan D(avid)
Wicke, Howard Henry
Yantis, Richard P

OKLAHOMA
Schuermann, Allen Clark, Jr

OREGON
Linstone, Harold A
Murphy, Allan Hunt
Tam, Kwok-Wai
Tinney, William Frank

PENNSYLVANIA
Abramson, Stanley L
Antle, Charles Edward
Arnold, Leslie K
Balas, Egon
Beck, Robert Edward
Belkin, Barry
Biegler, Lorenz Theodor
Block, Henry William
Bolmarcich, Joseph John
Kushner, Harvey D(avid)
Lehoczky, John Paul
Mazumdar, Mainak
Mitchell, Theodore
Mutmansky, Jan Martin
Pierskalla, William P
Rosenshine, Matthew
Scranton, Bruce Edward
Szyld, Daniel Benjamin
Thompson, Gerald Luther
Vaserstein, Leonid, IV
Weindling, Joachim I(gnace)
Wolfe, Harvey
Wolfe, Reuben Edward

RHODE ISLAND
Dafermos, Stella
Jarrett, Jeffrey E
Olson, David Gardels

SOUTH CAROLINA
Balintfy, Joseph L
Dobbins, James Gregory Hall
Kostreva, Michael Martin
Ruckle, William Henry

TENNESSEE
Cummings, Peter Thomas
Demetriou, Charles Arthur
Leon, Ramon V
Umholtz, Clyde Allan

TEXAS
Beck, John Robert
Bennet, George Kemble, Jr
Bhat, Uggappakodi Narayan
Blank, Leland T
Brockett, Patrick Lee
Feldman, Richard Martin
Gillette, Kevin Keith
Greaney, William A
Harbaugh, Allan Wilson
Ignizio, James Paul
Jensen, Paul Allen
Johnston, Walter Edward
Kimeldorf, George S
Klingman, Darwin Dee
Lea, James Dighton
Lesso, William George
Matula, David William
Nickel, James Alvin
Noonan, James Waring
Patrick, Wesley Clare
Randolph, Paul Herbert
Self, Glendon Danna
Smith, Philip Wesley
Sobol, Marion Gross
Stein, William Edward
Weiser, Dan
Wu, Hsin-i

UTAH
Haslem, William Joshua
Richards, Dale Owen
Stearman, Roebert L(yle)
Walker, LeRoy Harold

Windham, Michael Parks

VERMONT
Abbott, Stephen Dudley
Brown, Robert Goodell
Olinick, Michael

VIRGINIA
Arnberg, Robert Lewis
Barmby, John G(lennon)
Berghoefer, Fred G
Bolstein, Arnold Richard
Bothwell, Frank Edgar
Cooper, Henry Franklyn, Jr
Coram, Donald Sidney
Cornforth, Clarence Michael
Desiderio, Anthony Michael
Drew, John H
Dubin, Henry Charles
Emlet, Harry Elsworth, Jr
Golub, Abraham
Greenberg, Irwin
Greer, William Louis
Gross, Donald
Haering, George
Hall, Otis F
Hartzler, Alfred James
Ho, S(hui)
Hudgins, Aubrey C, Jr
Hwang, William Gaong
Kapos, Ervin
Karns, Charles W(esley)
Kydes, Andy Steve
LaBudde, Robert Arthur
Lawwill, Stanley Joseph
Lowry, Philip Holt
McNichols, Gerald Robert
Malcolm, Janet May
Marlow, William Henry
May, Donald Curtis, Jr
Michalowicz, Joseph Victor
Mikhail, Nabih N
Montague, Stephen
Overholt, John Lough
Pokrant, Marvin Arthur
Rehm, Allan Stanley
Reierson, James (Dutton)
Reynolds, Marion Rudolph, Jr
Richards, Paul Bland
Richardson, Henry Russell
Ringeisen, Richard Delose
Rudmin, Joseph Webster
Schwartz, Benjamin L
Sessler, John Charles
Shrier, Stefan
Sorrell, Gary Lee
Stone, Lawrence David
Walker, Robert Paul
Wilkinson, William Lyle
Woisard, Edwin Lewis

WASHINGTON
Bare, Barry Bruce
Edison, Larry Alvin
Heilenday, Frank W (Tod)
Liu, Chen-Ching
Millham, Charles Blanchard
Moore, Richard Lee
Rockafellar, Ralph Tyrrell
Rustagi, Krishna Prasad

WISCONSIN
Robinson, Stephen Michael
Schultz, Hilbert Kenneth
Stahl, Neil

ALBERTA
Enns, Ernest Gerhard
Silver, Edward Allan

BRITISH COLUMBIA
Gruver, William A
Lambe, Thomas Anthony
Welch, David Warren

MANITOBA
Zeiler, Frederick

NEWFOUNDLAND
May, Sherry Jan

NOVA SCOTIA
Barzilai, Jonathan

ONTARIO
Caron, Richard John
Colbourn, Charles Joseph
Davis, Ronald Stuart
Hopkins, Nigel John
Lindsey, George Roy
Magazine, Michael Jay
Mayberry, John Patterson
Mohanty, Sri Gopal
Nash, John Christopher
North, Henry E(rick) T(uisku)
Ostry, Sylvia
Paull, Allan E
Posner, Morton Jacob
Pressman, Irwin Samuel
Stauffer, Allan Daniel
Steiner, George

QUEBEC
Clarke, Francis

Duong, Quang P
Galibois, Andre
Loulou, Richard Jacques
Sankoff, David
Tao, Lixiu

OTHER COUNTRIES
Balinski, Michel Louis
Davis, Edward Alex
Garcia-Santesmases, Jose Miguel
Greenblatt, Seth Alan
Gritzmann, Peter
Halemane, Keshava Prasad
Hamacher, Horst W
Lindquist, Anders Gunnar
Philippou, Andreas Nicolaou

Physical Mathematics

ALABAMA
Chang, Mou-Hsiung
Cook, Frederick Lee
Fowler, Bruce Wayne
Lee, Ching Tsung

ARIZONA
Faris, William Guignard
Hoehn, A(lfred) J(oseph)
Iverson, A Evan
Lomont, John S
Rund, Hanno
Triffet, Terry
Twomey, Sean Andrew

ARKANSAS
Lieber, Michael

CALIFORNIA
Antolak, Arlyn Joe
Aurilia, Antonio
Berdjis, Fazlollah
Bernard, Douglas Alan
Birnir, Björn
Bleick, Willard Evan
Burgoyne, Peter Nicholas
Burke, James Edward
Cohen, E Richard
Dedrick, Kent Gentry
Derenzo, Stephen Edward
Finn, Robert
Freis, Robert P
Garrison, John Carson
Gillespie, Daniel Thomas
Hershey, Allen Vincent
Hovingh, Jack
Hudson, George Elbert
Hutchin, Richard Ariel
Klapman, Solomon Joel
Klein, Abel
Krause, Lloyd O(scar)
Landshoff, Rolf
Langdon, Allan Bruce
Lapidus, Michel Laurent
Lu, Kau U
Mayer, Meinhard Edwin
Moore, Mortimer Norman
Morris, George William
Nachamkin, Jack
Neustadter, Siegfried Friedrich
Newman, William I
O'Malley, Thomas Francis
Park, Chong Jin
Saffren, Melvin Michael
Scholl, James Francis
Shepanski, John Francis
Taub, Abraham Haskel
Twersky, Victor
Unal, Aynur
Villeneuve, A(lfred) T(homas)
Wagner, Richard John
Wilson, Robert Norton
Yeh, Cavour W

COLORADO
Augusteijn, Marijke Francina
Coffey, Mark William
Ekin, Jack W
Gall, Robert Lawrence
Graham, John
Hundhausen, Joan Rohrer
Kassoy, David R
Marchand, Jean-Paul
Poore, Aubrey Bonner
Walter, Martin Edward
Whitten, Barbara L

CONNECTICUT
Kohlmayr, Gerhard Franz
McCartin, Brian James
Sachdeva, Baldev Krishan
Wong, Maurice King Fan

DELAWARE
Angell, Thomas Strong
Hirsch, Albert Edgar
Morgan, John Davis, III

DISTRICT OF COLUMBIA
Grossman, John Mark
Langworthy, James Brian
Meijer, Paul Herman Ernst
Nachman, Arje
Yen, Nai-Chyuan

FLORIDA
Albright, John Rupp
Curtright, Thomas Lynn
Debnath, Lokenath
Dubbelday, Pieter Steven
Emch, Gerard G
Kelley, Robert Lee
Neuringer, Joseph Louis
Nunn, Walter M(elrose), Jr
Seiger, Harvey N
Toupin, Richard A

GEORGIA
Harrell, Evans Malott, II
Mickens, Ronald Elbert
Warsi, Nazir Ahmed

ILLINOIS
Carhart, Richard Alan
Carroll, Robert Wayne
Ceperley, David Matthew
Cook, Joseph Marion
Dykla, John J
Ho, Chung-Wu
Jerome, Joseph Walter
Kaper, Hans G
Kath, William Lawrence
Matkowsky, Bernard J
Oono, Yoshitsugu
Prais, Michael Gene
Redmount, Ian H
Siegert, Arnold John Frederick
Smarr, Larry Lee
Tao, Rongjia

INDIANA
Challifour, John Lee
Lenard, Andrew
Madan, Ved P
Penna, Michael Anthony

IOWA
Alexander, Roger Keith
Carlson, Bille Chandler
Hammer, Charles Lawrence
Jorgensen, Palle E T
Struck-Marcell, Curtis John
Ton-That, Tuong

KANSAS
Acker, Andrew French, III
Lerner, David Evan
Miller, Forest R
Ramm, Alexander G
Unz, Hillel

KENTUCKY
Man, Chi-Sing
Perry, Peter Anton
Van Winter, Clasine

LOUISIANA
Brans, Carl Henry
Goldstein, Jerome Arthur
Heatherly, Henry Edward
Knill, Ronald John
Kuo, Hui-Hsiung
Kyame, Joseph John
Runnels, Kelli

MARYLAND
Berger, Martin Jacob
Blue, James Lawrence
Criss, John W
Dehn, James Theodore
Fischel, David
Fisher, Michael Ellis
Gates, Sylvester J, Jr
Granger, Robert A, II
Herman, Richard Howard
Hu, Bei-Lok Bernard
Kuttler, James Robert
Massell, Paul Barry
Rehm, Ronald George
Rostamian, Rouben
Wallace, Alton Smith

MASSACHUSETTS
Baublitz, Millard, Jr
Bernstein, Joseph N
Cronin-Golomb, Mark
Humi, Mayer
Jaffe, Arthur Michael
Kon, Mark A
Melrose, Richard B
Parker, Lee Ward
Parrott, Stephen Kinsley
Rizzoli-Malanotte, Paola
Ruskai, Mary Beth
Schay, Geza

MICHIGAN
Fairchild, Edward Elwood, Jr
Khargonekar, Pramod P
Lomax, Ronald J(ames)
Marriott, Richard
Miklavcic, Milan
Murphy, Brian Boru
Rauch, Jeffrey Baron
Williams, David Noel

MINNESOTA
Bonne, Ulrich
Follingstad, Henry George

Luskin, Mitchell B
Miller, Willard, Jr
Olver, Peter John
Pour-El, Marian Boykan

MISSISSIPPI
Rayborn, Grayson Hanks

MISSOURI
Beem, John Kelly
Bender, Carl Martin
Katz, Israel Norman
Ornstein, Wilhelm
Penico, Anthony Joseph
Pettey, Dix Hayes
Plummer, Otho Raymond
Retzloff, David George

NEW HAMPSHIRE
MacGillivray, Jeffrey Charles
Shepard, Harvey Kenneth

NEW JERSEY
Andrushkiw, Roman Ihor
Benaroya, Haym
Budny, Robert Vierling
Carter, Ashley Hale
Chien, Victor
Christodoulou, Demetrios
Dyson, Freeman John
Goldin, Gerald Alan
Koss, Valery Alexander
Lepowsky, James Ivan
McKenna, James
Nachtergaele, Bruno Leo Zulma
Rulf, Benjamin
Schrenk, George L
Seidl, Frederick Gabriel Paul
Sverdlove, Ronald
Taylor, Jean Ellen
White, Benjamin Steven
Wightman, Arthur Strong

NEW MEXICO
Baker, George Allen, Jr
Becker, Wilhelm
Biggs, Frank
Bolie, Victor Wayne
Campbell, John Raymond
Critchfield, Charles Louis
Devaney, Joseph James
Dietz, David
Griffin, Patrick J
Hurd, Jon Rickey
Hyman, James Macklin
Lee, Clarence Edgar
Louck, James Donald
Mark, J Carson
Norton, John Leslie
O'Dell, Ralph Douglas
Ross, Timothy Jack
Sandford, Maxwell Tenbrook, II
Wing, Janet E (Sweedyk) Bendt

NEW YORK
Blank, Albert Abraham
Bowick, Mark John
Boyce, William Edward
Chudnovsky, Gregory V
Dasgupta, Gautam
Deady, Matthew William
Deift, Percy Alec
Dicker, Daniel
Dickman, Steven Richard
Doering, Charles Rogers
Fagot, Wilfred Clark
Feigenbaum, Mitchell Jay
Feingold, Alex Jay
Fisher, Edward B
Fleishman, Bernard Abraham
Geer, James Francis
Glimm, James Gilbert
Guillen, Michael Arthur
Habetler, George Joseph
Harrington, Steven Jay
Herron, Isom H
Isaacson, David
Laks, David Bejnesh
Mane, Sateesh Ramchandra
Marcus, Paul Malcolm
Marcuvitz, Nathan
Margulies, Marcel
Newman, Charles Michael
Nirenberg, Louis
Phong, Duong Hong
Roskes, Gerald J
Roytburd, Victor
Rubenfeld, Lester A
Schlitt, Dan Webb
Siegmann, William Lewis
Stell, George Roger
Wang, Xingwu
Weng, Wu Tsung
Wolf, Emil
Yarmush, David Leon
Zipfel, Warren Roger

NORTH CAROLINA
Brown, J(ohn) David
Cotanch, Stephen Robert
Dankel, Thaddeus George, Jr
Dillard, Margaret Bleick
Dolan, Louise Ann
Fulp, Ronald Owen

Reed, Michael Charles
Teague, David Boyce
York, James Wesley, Jr

OHIO
Burggraf, Odus R
Butz, Donald Josef
Dumas, H Scott
Goad, Clyde Clarenton
Kaplan, Bernard
Kummer, Martin
Liska, John W
Long, Clifford A
Macklin, Philip Alan
Plybon, Benjamin Francis
Tamburino, Louis A
Weinstock, Robert
Wen, Shih-Liang
Wicke, Howard Henry

OKLAHOMA
Fowler, Richard Gildart
Lal, Manohar
Robinson, Enders Anthony

OREGON
Ericksen, Jerald LaVerne
Guenther, Ronald Bernard
Isenberg, James Allen
Kocher, Carl A
Murphy, Lea Frances
Warner, David Charles
Wheeler, Nicholas Allan

PENNSYLVANIA
Ambs, William Joseph
Bradley, James Henry Stobart
Diaz, Mario Claudio
Farouk, Bakhtier
Gorman, Arthur Daniel
Kadison, Richard Vincent
Muir, Patrick Fred
Porter, John Robert
Roman, Paul
Rorres, Chris
Rosen, Gerald Harris
Sarram, Mehdi
Wayne, Clarence Eugene
Weiner, Brian Lewis

RHODE ISLAND
Childs, Donald Ray

SOUTH CAROLINA
Stevens, James Levon

TENNESSEE
Dory, Robert Allan

TEXAS
Auchmuty, Giles
Aucoin, Paschal Joseph, Jr
Bronaugh, Edwin Lee
Fix, George Joseph
Haberman, Richard
Hartfiel, Darald Joe
Heelis, Roderick Antony
Katsnelson, Lev Z
Lewis, Ira Wayne
McCoy, Jimmy Jewell
Narcowich, Francis Joseph
Nordlander, Peter Jan Arne
Olenick, Richard Peter
Olness, Fredrick Iver
Ozsvath, Istvan
Prabhu, Vasant K
Radin, Charles Lewis
Roberts, Thomas M
Salingaros, Nikos Angelos
Voigt, Gerd-Hannes
Walton, Jay R
Wang, Chao-Cheng
Wu, Hsin-i

UTAH
Bates, Peter William
Wilcox, Calvin Hayden

VIRGINIA
Adam, John Anthony
Bowden, Robert Lee, Jr
Bragg, Lincoln Ellsworth
Dorning, John Joseph
Farrukh, Usamah Omar
Greenberg, William
Hafele, Joseph Carl
Hagedorn, George Allan
Horgan, Cornelius Oliver
Howland, James Secord
Imbrie, John Z
Lasiecka, Irena
Parvulescu, Antares
Raychowdhury, Pratip Nath
Renardy, Yuriko
Richards, Paul Bland
Simmonds, James G
Thomas, Lawrence E
Wheeler, Robert Lee
Wohl, Philip R
Zweifel, Paul Frederick

WASHINGTON
Burkhart, Richard Henry
Cochran, James Alan

Gangolli, Ramesh A
Moseley, W(illiam) David
Nievergelt, Yves

WISCONSIN
Cook, David Marsden
Durand, Loyal
Stewart, W(arren) E(arl)
Strikwerda, John Charles
Uhlenbrock, Dietrich A

PUERTO RICO
Khan, Winston

ALBERTA
Blais, J A Rodrigue
Capri, Anton Zizi
Epstein, Marcelo
Kunzle, Hans Peter
Ludwig, Garry (Gerhard Adolf)

BRITISH COLUMBIA
Millar, Robert Fyfe
Srivastava, Hari Mohan
Weaver, Ralph Sherman

MANITOBA
Cohen, Harley

ONTARIO
Duff, George Francis Denton
Elliott, George Arthur
Huschilt, John
Jackson, David Phillip
Langford, William Finlay
Lee-Whiting, Graham Edward
Lipshitz, Stanley Paul
Moltyaner, Grigory
Nerenberg, Morton Abraham
Paldus, Josef
Prugovecki, Eduard
Sen, Dipak Kumar
Tross, Ralph G

QUEBEC
Boivin, Alberic
Clarke, Francis
Fabrikant, Valery Isaak
Grmela, Miroslav
Kroger, Helmut Karl

SASKATCHEWAN
Law, Alan Greenwell

OTHER COUNTRIES
Andrej, Ladislav
Lanford, Oscar E, III
Liu, Ta-Jo
Nussenzveig, Herch Moyses
Rivier, Nicolas Yves
Roseman, Joseph Jacob
Rosenbaum, Marcos
Sato, Mikio
Schutz, Bernard Frederick
Williams, Michael Maurice Rudolph
Wu, Yong-Shi

Probability

ALABAMA
Chang, Mou-Hsiung
Hudson, William Nathaniel
Schroer, Bernard J

ARIZONA
Barckett, Joseph Anthony
Daniel, Sam Mordochai
Faris, William Guignard
Frieden, Bernard Roy
Griego, Richard Jerome
Maier, Robert S
Mittal, Yashaswini Deval
Neuts, Marcel Fernand

CALIFORNIA
Aldous, David J
Anderson, Paul Maurice
Blischke, Wallace Robert
Book, Stephen Alan
Calabrese, Philip G
Chernick, Michael Ross
Fearn, Dean Henry
Feldman, Jacob
Froebe, Larry R
Georgakakos, Konstantine P
Gillis, James Thompson
Gray, Robert Molten
Haight, Frank Avery
Harmen, Raymond A
Harris, Theodore Edward
Helstrom, Carl Wilhelm
Henning, Harley Barry
Hsu, Yu-Sheng
Iglehart, Donald Lee
Jacobs, Patricia Anne
Jewell, Nicholas Patrick
Khalona, Ramon Antonio
Kimme, Ernest Godfrey
Krener, Arthur James
Krieger, Henry Alan
Kuhlmann, Karl Frederick
Langdon, Glen George, Jr
Lange, Kenneth L

Probability (cont)

Le Cam, Lucien Marie
Liggett, Thomas Milton
Milch, Paul R
Mo, Charles Tse Chin
Molnar, Imre
Mood, Alexander McFarlane
Ott, Wayne R
Papanicolaou, George Constantine
Patterson, Tim J
Rao, Jammalamadaka S
Rao, Malempati Madhusudana
Rosenblatt, Murray
Sharpe, Michael John
Tucker, Howard Gregory
Weaver, William Bruce
Weber, Charles L
Weiner, Howard Jacob
Wendel, James G
Wiggins, Alvin Dennie
Williams, Jack Rudolph
Williams, Ruth Jeannette

COLORADO
Crow, Edwin Louis
Darst, Richard B
Etter, Delores Maria
Heine, George Winfield, III
Kafadar, Karen
Marchand, Jean-Paul
Mathys, Peter
Nesenbergs, Martin
Schmid, John Carolus
Tweedie, Richard Lewis

CONNECTICUT
Blei, Ron Charles
Cordery, Robert Arthur
Gine-Masdeu, Evarist
Kleinfeld, Ira H
Mukhopadhyay, Nitis
Phillips, Peter Charles B
Rao, Valluru Bhavanarayana
Ruckebusch, Guy Bernard
Stuck, Barton W
Vitale, Richard Albert

DISTRICT OF COLUMBIA
Cohen, Michael Paul
Filliben, James John
Hammer, Carl
Nayak, Tapan Kumar
Robinson, E Arthur, Jr
Rosenblatt, David
Wagner, Andrew James

FLORIDA
Ballerini, Rocco
Haase, Richard Henry
Keesling, James Edgar
Leysieffer, Frederick Walter
Lin, Y(u) K(weng)
Mukherjea, Arunava
Muth, Eginhard Joerg
Schott, James Robert
Sethuraman, Jayaram
Zame, Alan

GEORGIA
Kutner, Michael Henry
Lavender, DeWitt Earl
Shonkwiler, Ronald Wesley

HAWAII
Kadota, T Theodore

IDAHO
Bickel, John Henry

ILLINOIS
Barron, Emmanuel Nicholas
Bellow, Alexandra
Burkholder, Donald Lyman
Calderon, Calixto Pedro
Clover, William John, Jr
Heller, Barbara Ruth
Ionescu Tulcea, Cassius
Jones, Roger L
Knight, Frank B
Laud, Purushottam Waman
Philipp, Walter V
Tang, Wilson H
Tier, Charles
Wacker, William Dennis
Wolter, Kirk Marcus

INDIANA
Beekman, John Alfred
Bhattacharya, Rabindra Nath
Bradley, Richard Crane, Jr
Carlson, Lee Arnold
Kinney, John James
Lyons, Russell David
Pimblott, Simon Martin
Protter, Philip Elliott
Rego, Vernon J
Rhoades, Billy Eugene

IOWA
Andersland, Mark Steven
Athreya, Krishna Balasundaram
Cressie, Noel A C
Reckase, Mark Daniel

Russo, Ralph P

KANSAS
Burckel, Robert Bruce
Ramm, Alexander G
Stewart, William Henry
Stockli, Martin P

KENTUCKY
Griffith, William Schuler

LOUISIANA
Abbott, James H
Kuo, Hui-Hsiung
Sundar, P

MAINE
McMillan, Brockway
Todaro, Michael P

MARYLAND
Abbey, Robert Fred, Jr
Austin, Homer Wellington
Cacciamani, Eugene Richard, Jr
Cherniavsky, Ellen Abelson
Cohen, Edgar A, Jr
Ephremides, Anthony
Huddleston, Charles Martin
Huneycutt, James Ernest, Jr
Kattakuzhy, George Chacko
Kotz, Samuel
Lao, Chang Sheng
Lechner, James Albert
Robinson, Richard Carleton, Jr
Rukhin, Andrew Leo
Sawortonow, Parfeny Pavolich
Serfling, Robert Joseph
Siegel, Martha J
Smith, Paul John
Sponsler, George C
Syski, Ryszard
Weiss, Michael David
Wierman, John C
Zabronsky, Herman

MASSACHUSETTS
Bras, Rafael Luis
Dudley, Richard Mansfield
Elias, Peter
Ellis, Richard Steven
Entekhabi, Dara
Fougere, Paul Francis
Gacs, Peter
Giordano, Arthur Anthony
Graham, John D
Hahn, Marjorie G
Horowitz, Joseph
Horowitz, Larry Lowell
Kitchin, John Francis
Kon, Mark A
Liu, Jun S
Lo, Andrew W
Mauldon, James Grenfell
Rogers, Hartley, Jr
Schonhoff, Thomas Arthur
Servi, Leslie D
Skibinsky, Morris
Stevens, Herbert H(owe), Jr

MICHIGAN
Albaugh, A Henry
Chakravarthy, Srinivasaraghavan
Chow, Pao Liu
Clarke, Allen Bruce
Hill, Bruce M
LePage, Raoul
Roe, Byron Paul
Tanis, Elliot Alan

MINNESOTA
Ayeni, Babatunde J
Gray, Lawrence Firman
Pruitt, William Edwin
Sudderth, William David

MISSOURI
King, Terry Lee
Sawyer, Stanley Arthur

NEBRASKA
Conway, James Joseph
Heaney, Robert Proulx
Skoug, David L

NEW HAMPSHIRE
Jolly, Stuart Martin
Lamperti, John Williams

NEW JERSEY
Albin, Susan Lee
Benaroya, Haym
Boylan, Edward S
Cimini, Leonard Joseph, Jr
Cinlar, Erhan
Landwehr, James M
Loader, Clive Roland
Lucantoni, David Michael
McKenna, James
Naus, Joseph Irwin
Ott, Teunis Jan
Prekopa, Andras
Riggs, Richard
Saltzberg, Burton R
Steiger, William Lee

Vardi, Yehuda
Verdu, Sergio
Weiss, Alan
White, Benjamin Steven
Whitt, Ward

NEW MEXICO
Beyer, William A
Bruckner, Lawrence Adam
Cobb, Loren
Cogburn, Robert Francis
Davis, James Avery
Gutjahr, Allan L
Silver, Richard N

NEW YORK
Anderson, Allan George
Balachandran, Kashi Ramamurthi
Berman, Simeon Moses
Berresford, Geoffrey Case
Blessing, John A
Boyce, William Edward
Chan, Jack-Kang
Cohen, Joel Ephraim
Debany, Warren Harding, Jr
Dynkin, Eugene B
Falk, Harold
Fine, Terrence Leon
Haag, Fred George
Hanson, David Lee
Heath, David Clay
Heidelberger, Philip
Hirtzel, Cynthia S
Karatzas, Ioannis
Kyburg, Henry
Lawton, William Harvey
Lax, Melvin
Moritz, Roger Homer
Newman, Charles Michael
Oakes, David
Prabhu, Narahari Umanath
Rosberg, Zvi
Ruppert, David
Scalora, Frank Salvatore
Singh, Chanchal
Sweeney, Thomas Francis
Tucker, Thomas William
Wallis, James Richard
Wilks, Daniel S
Yablon, Marvin
Yakir, Benjamin

NORTH CAROLINA
Bergsten, Jane Williams
Bernhard, Richard Harold
Bishir, John William
Dankel, Thaddeus George, Jr
Guess, Harry Adelbert
Jordan, Edward Daniel
Karr, Alan Francis
Keepler, Manuel
Lawler, Gregory Francis
Lewis, Paul Ollin
Menius, Arthur Clayton, Jr
Nelson, A Carl, Jr
Petersen, Karl Endel
Stidham, Shaler, Jr

OHIO
Bhargava, Triloki Nath
Breitenberger, Ernst
Caldito, Gloria C
Cavaretta, Alfred S
De Acosta, Alejandro
Fan, Liang-Shih
Goel, Prem Kumar
Hern, Thomas Albert
Neaderhouser Purdy, Carla Cecilia
Rayner, John Norman
Stackelberg, Olaf Patrick
Sterrett, Andrew
Ward, Douglas Eric
Woyczynski, Wojbor Andrzej

OKLAHOMA
Chambers, Richard Lee
Hochhaus, Larry
Richardson, J Mark

OREGON
Chartier, Vernon Lee
Demarest, Harold Hunt, Jr
Engelbrecht, Rudolf S
Nelsen, Roger Bain
Waymire, Edward Charles

PENNSYLVANIA
Arnold, Leslie K
Beggs, William John
Belkin, Barry
Blaisdell, Ernest Atwell, Jr
Block, Henry William
Eisenberg, Bennett
Fryling, Robert Howard
Galambos, Janos
Gleser, Leon Jay
Grinberg, Eric L
Jech, Thomas J
Jeruchim, Michel Claude
Lehoczky, John Paul
Mehta, Jatinder S
Mitra, Shashanka S
Schervish, Mark John
Vaserstein, Leonid, IV

Wagner, Clifford Henry
Yukich, Joseph E

RHODE ISLAND
Bartram, James F(ranklin)
Kushner, Harold J(oseph)

SOUTH CAROLINA
Pursley, Michael Bader
Rust, Philip Frederick

TENNESSEE
Dobosy, Ronald Joseph
Groer, Peter Gerold
Leon, Ramon V
Perey, Francis George
Tan, Wai-Yuan

TEXAS
Bhat, Uggappakodi Narayan
Brockett, Patrick Lee
Cecil, David Rolf
Cox, Dennis Dean
Eubank, Randall Lester
Hensley, Douglas Austin
Johnson, Johnny Albert
Kimeldorf, George S
Marder, Michael P
Mauldin, Richard Daniel
Pfeiffer, Paul Edwin
Randolph, Paul Herbert
Schuster, Eugene F
Stein, William Edward
Thall, Peter Francis
Turner, Danny William
Vaaler, Jeffrey David
Ward-McLemore, Ethel
West, Bruce Joseph
Wheeler, Joe Darr
Whitmore, Ralph M
Wilson, Thomas Leon
Wright, James Foley
Yao, James T-P

UTAH
Walker, LeRoy Harold

VERMONT
Emerson, John David

VIRGINIA
Bolstein, Arnold Richard
Foy, C Allan
Hendricks, Walter James
Henry, Neil Wylie
Imbrie, John Z
Kaminski, Paul G
Kratzke, Thomas Martin
Lundegard, Robert James
Martin, Nathaniel Frizzel Grafton
Mensing, Richard Walter
Michalowicz, Joseph Victor
Minton, Paul Dixon
Richardson, Henry Russell
Singh, Mahendra Pal
Stone, Lawrence David

WASHINGTON
Anderson, Richard Gregory
Band, William
Burkhart, Richard Henry
Conquest, Loveday Loyce
Perlman, Michael David
Pyke, Ronald
Saunders, Sam Cundiff
Schaerf, Henry Maximilian

WEST VIRGINIA
Brady, Dennis Patrick

WISCONSIN
Gurland, John
Harris, Bernard
Kaye, Norman Joseph
Keisler, Howard Jerome
Kurtz, Thomas Gordon
Pemantle, Robin A
Reinsel, Gregory Charles

WYOMING
Tung, Yeou-Koung

PUERTO RICO
Hussain, Malek Gholoum Malek

ALBERTA
Elliott, Robert James
Johnson, Dudley Paul

BRITISH COLUMBIA
Anderson, R F V
Haussmann, Ulrich Gunther
Reed, William J

MANITOBA
Basilevsky, Alexander

NEW BRUNSWICK
Mureika, Roman A
Pham-Gia, Thu

NEWFOUNDLAND
Burry, John Henry William

ONTARIO
Brainerd, Barron
Campbell, Louis Lorne
Denzel, George Eugene
Mark, Jon Wei
Mayberry, John Patterson
Schaufele, Ronald A

QUEBEC
Chaubey, Yogendra Prasad
Choksi, Jal R
Duong, Quang P
Herz, Carl Samuel
Lefebvre, Mario
Stathopoulos, Theodore

SASKATCHEWAN
Tomkins, Robert James

OTHER COUNTRIES
Brockwell, Peter John
Brown, Richard Harland
Chesson, Peter Leith
Greenblatt, Seth Alan
Hochberg, Kenneth J
Mengali, Umberto
Philippou, Andreas Nicolaou
Shapiro, Jesse Marshall
Sunahara, Yoshifumi

Topology

ALABAMA
Plunkett, Robert Lee
Rogers, Jack Wyndall, Jr
Stocks, Douglas Roscoe, Jr

ARIZONA
Grace, Edward Everett
Thompson, Richard Bruce

ARKANSAS
Cochran, Allan Chester
Feldman, William A
Madison, Bernard L
Schein, Boris M

CALIFORNIA
Agoston, Max Karl
Arnold, Hubert Andrew
Borges, Carlos Rego
Brown, Robert Freeman
Clarke, Wilton E L
Davis, Alan Lynn
Day, Jane Maxwell
De Sapio, Rodolfo Vittorio
Elderkin, Richard Howard
Feldman, Louis A
Flegal, Robert Melvin
Freedman, Michael Hartley
Greever, John
Hagopian, Charles Lemuel
Hirsch, Morris William
Jones, Floyd Burton
Kirby, Robion C
Lin, James Peicheng
Mignone, Robert Joseph
Millett, Kenneth Cary
Morgan, John Clifford, II
Morris, Cecil Arthur, Sr
Newcomb, Robert Lewis
Pines, Alexander
Rabin, Jeffrey Mark
Satyanarayana, Upadhyayula V
Smith, Alton Hutchison
Spanier, Edwin Henry
Stallings, John Robert, Jr
Stern, Ronald John
Thomas, Paul Emery
Verona, Andrei
Vought, Eldon Jon
Williamson, Robert Emmett

COLORADO
Holley, Frieda Koster
Lundell, Albert Thomas
Mansfield, Roger Leo
Thron, Wolfgang Joseph

CONNECTICUT
Comfort, William Wistar
Hager, Anthony Wood
Kim, Soon-Kyu
Lang, George E, Jr
Lee, Ronnie
Massey, William S
Mostov, George Daniel
O'Neill, Edward John
Reddy, William L
Szczarba, Robert Henry
Troyer, Stephanie Fantl
Whitley, William Thurmon
Zahler, Raphael

DELAWARE
Bellamy, David P
Sloyer, Clifford V, Jr

DISTRICT OF COLUMBIA
Robinson, E Arthur, Jr
Sjogren, Jon Arne

FLORIDA
Brechner, Beverly Lorraine
Duda, Edwin
Heil, Wolfgang Heinrich
Keesling, James Edgar
Lacher, Robert Christopher
Lin, Shwu-Yeng Tzen
Lin, You-Feng
Mielke, Marvin V
Sumners, DeWitt L
Wright, Thomas Perrin, Jr

GEORGIA
Batterson, Steven L
Cain, George Lee, Jr
Cantrell, James Cecil
Howell, Leonard Rudolph, Jr
Hudson, Sigmund Nyrop
Penney, David Emory
Wheeler, Ed R
Worth, Roy Eugene

HAWAII
Dovermann, Karl Heinz
Papacostas, Constantinos Symeon

IDAHO
Christenson, Charles O
Cobb, John Iverson
Voxman, William L

ILLINOIS
Albrecht, Felix Robert
Bousfield, Aldridge Knight
Clough, Robert Ragan
Craggs, Robert F
Friedlander, Eric Mark
Gray, Brayton
Gugenheim, Victor Kurt Alfred Morris
Hansen, William Anthony
Heitsch, James Lawrence
Ho, Chung-Wu
Jerrard, Richard Patterson
Jungck, Gerald Frederick
MacLane, Saunders
Mahowald, Mark Edward
May, J Peter
Mount, Bertha Lauritzen
Osborn, Howard Ashley
Powers, Michael Jerome
Priddy, Stewart Beauregard
Rosenholtz, Ira N
Stein, Michael Roger
Tangora, Martin Charles
Waterman, Peter Lewis
Wheeler, Robert Francis
Wood, John William
Yau, Stephen Shing-Toung

INDIANA
Jaworowski, Jan W
Kaminker, Jerome Alvin
Peltier, Charles Francis
Rigdon, Robert David
Schultz, Reinhard Edward
Wilkerson, Clarence Wendell, Jr

IOWA
Randell, Richard
Sanderson, Donald Eugene
Steiner, Anne Kercheval

KANSAS
Bajaj, Prem Nath
Himmelberg, Charles John
Muenzenberger, Thomas Bourque
Porter, Jack R
Roitman, Judy
Strecker, George Edison

LOUISIANA
Hildebrant, John A
Knill, Ronald John
Lawson, Jimmie Don
Swetharanyam, Lalitha

MAINE
Murphy, Grattan Patrick

MARYLAND
Alexander, James Crew
Boardman, John Michael
Cohen, Joel M
Lehner, Guydo R
Meyer, Jean-Pierre
Meyerson, Mark Daniel
Rosenberg, Jonathan Micah
Saworotnow, Parfeny Pavolich

MASSACHUSETTS
Brown, Edgar Henry, Jr
Cenkl, Bohumil
Eisenberg, Murray
Gilmore, Maurice Eugene
Ku, Hsu-Tung
Ku, Mei-Chin Hsiao
Lavallee, Lorraine Doris
Levine, Jerome Paul
Lusztig, George
MacPherson, Robert Duncan
Mann, Larry N
Munkres, James Raymond
O'Rourke, Joseph
Peterson, Franklin Paul

Sacks, Jonathan
Su, Jin-Chen

MICHIGAN
Arlinghaus, Sandra Judith Lach
Blefko, Robert L
Davis, Harvey Samuel
Handel, David
Hee, Christopher Edward
King, Larry Michael
Kister, James Milton
Moran, Daniel Austin
O'Neill, Ronald C
Rhee, Choon Jai
Stubblefield, Beauregard
Venema, Gerard Alan
Wasserman, Arthur Gabriel
Wechsler, Martin T

MINNESOTA
Frank, David Lewis
Gropen, Arthur Louis
Humke, Paul Daniel
Seebach, J Arthur, Jr

MISSISSIPPI
Cook, David Edwin
Fay, Temple Harold

MISSOURI
Cantwell, John Christopher
Freiwald, Ronald Charles
Haddock, Aubura Glen
Ingram, William Thomas
Jenkins, James Allister
McIntosh, William David
Pettey, Dix Hayes
Tikoo, Mohan L
Wenner, Bruce Richard

NEBRASKA
Downing, John Scott
Halfar, Edwin

NEVADA
Verma, Sadanand

NEW HAMPSHIRE
Arkowitz, Martin Arthur
Crowell, Richard Henry
Stancl, Mildred Luzader

NEW JERSEY
Bean, Ralph J
Bidwell, Leonard Nathan
Blackmore, Denis Louis
Foregger, Thomas H
Gilman, Robert Hugh
Kosinski, Antoni A
Landweber, Peter Steven
Plank, Donald Leroy
Tucker, Albert William
Weibel, Charles Alexander
Woodruff, Edythe Parker

NEW MEXICO
Lambert, Howard W
Mandelkern, Mark
Mann, Benjamin Michael
Pengelley, David John

NEW YORK
Anderson, Peter Gordon
Bendersky, Martin
Benham, Craig John
Berstein, Israel
Blanton, John David
Cappell, Sylvain Edward
Chan, Jack-Kang
Church, Philip Throop
Cohen, Herman Jacob
Deleanu, Aristide Alexandru-Ion
Ferry, Steven C
Geoghegan, Ross
Gottlieb, Allan
Gurel, Okan
Hartnett, William Edward
Hastings, Harold Morris
Hechler, Stephen Herman
Henderson, David Wilson
Houghton, Charles Joseph
Itzkowitz, Gerald Lee
Iwaniec, Tadeusz
Johnson, Sandra Lee
Karron, Daniel B
Kopperman, Ralph David
Kumpel, Paul Gremminger
Lewis, L Gaunce, Jr
Livesay, George Roger
McAuley, Patricia Tulley
McDuff, Dusa Margaret
Meyer, Paul Richard
Neisendorfer, Joseph
Seppala-Holtzman, David N
Singer, William Merrill
Todd, Aaron Rodwell
Wan, Yieh-Hei
Watts, Charles Edward
West, James Edward
Williams, Scott Warner

NORTH CAROLINA
Damon, James Norman
Durham, Harvey Ralph

Duvall, Paul Frazier, Jr
Faulkner, Gary Doyle
Franke, John Erwin
Gordh, George Rudolph, Jr
Hodel, Richard Earl
King, Lunsford Richardson
Kraines, David Paul
Kumar, Shrawan
Lane, Ernest Paul
Latch, Dana May
Montgomery, Deane
Rowlett, Russell Johnston, III
Sagan, Hans
Sher, Richard B
Smith-Thomas, Barbara
Vaughan, Jerry Eugene

OHIO
Baker, John Warren
Bentley, Herschel Lamar
Burke, Dennis Keith
Cameron, Douglas Ewan
Davis, Michael Walter
Khalimsky, Efim D
Klein, Albert Jonathan
Lee, Dong Hoon
Marty, Roger Henry
Nicholson, Victor Alvin
O'Brien, Thomas V
Piotrowski, Zbigniew
Schneider, Harold William
Swardson, Mary Anne
Wicke, Howard Henry

OKLAHOMA
Jobe, John M
McCullough, Darryl J
Rubin, Leonard Roy

OREGON
Bradshaw, Gay
Harris, John Kenneth
Phillips, N Christopher
Schori, Richard M
Ward, Lewis Edes, Jr

PENNSYLVANIA
Armentrout, Steve
Baildon, John David
Berard, Anthony D, Jr
Christoph, Francis Theodore, Jr
Conrad, Bruce
Davis, Donald Miller
Grabner, Elise M
Harbater, David
Hodgson, Jonathan Peter Edward
Jech, Thomas J
Sieber, James Leo
Stasheff, James Dillon
Warner, Frank Wilson, III

RHODE ISLAND
Gottschalk, Walter Helbig
Harris, Bruno
Jaco, William Howard
Rudolph, Lee

SOUTH CAROLINA
Nyikos, Peter Joseph
Yang, Jeong Sheng

SOUTH DAKOTA
Pedersen, Katherine L

TENNESSEE
Austin, Billy Ray
Bryant, Billy Finney
Chitwood, Howard
Daverman, Robert Jay
Husch, Lawrence S
Lea, James Wesley, Jr
Purisch, Steven Donald

TEXAS
Anderson, Edmund Hughes
Born, Alfred Ervin
Brown, Dennison Robert
Cook, Howard
Cude, Joe E
Doran, Robert Stuart
Girou, Michael L
Hall, Carl Eldridge
Hart, R Neal
Hempel, John Paul
Huggins, Frank Norris
Lewis, Ira Wayne
Mauldin, Richard Daniel
Pixley, Carl Preston
Singh, Sukhjit
Smith, Spurgeon Eugene
Starbird, Michael Peter
Vobach, Arnold R
Wagner, Neal Richard
Williams, Robert Fones

UTAH
Burgess, Cecil Edmund
Case, James Hughson
Fearnley, Lawrence
Gersten, Stephen M
Glaser, Leslie
Rushing, Thomas Benny
Toledo, Domingo
Wright, David Grant

Topology (cont)

VIRGINIA
Barrett, Lida Kittrell
Chang, Lay Nam
Coram, Donald Sidney
Dickman, Raymond F, Jr
Fletcher, Peter
Floyd, Edwin Earl
Haver, William Emery
Kent, Joseph Francis
Kuhn, Nicholas J
Levy, Ronald Fred
Loatman, Robert Bruce
Lobb, Barry Lee
Lutzer, David John
McCoy, Robert A
Morris, Joseph Richard
Patty, Clarence Wayne
Sawyer, Jane Orrock
Smith, James Clarence, Jr
Walker, Robert Paul

WASHINGTON
Blumenthal, Robert Allan
Firkins, John Frederick
Michael, Ernest Arthur
Ryeburn, David
Schaerf, Henry Maximilian
Segal, Jack
Sims, Benjamin Turner
Yandl, Andre

WEST VIRGINIA
Hogan, John Wesley

WISCONSIN
Ancel, Fredric D
Davis, Thomas William
Fadell, Edward Richard
Fournelle, Thomas Albert
Harris, J Douglas
McMillan, Daniel Russell, Jr

WYOMING
Gastl, George Clifford

ALBERTA
Antonelli, Peter Louis
Hoo, Cheong Seng
Murdeshwar, Mangesh Ganesh
Varadarajan, Kalathoor
Zvengrowski, Peter Daniel

BRITISH COLUMBIA
Douglas, Roy Rene
Lancaster, George Maurice
MacDonald, John Lauchlin

MANITOBA
Rayburn, Marlon Cecil, Jr

NEWFOUNDLAND
Thomeier, Siegfried

NOVA SCOTIA
Grant, Douglass Lloyd
Tan, Kok-Keong

ONTARIO
Coxeter, Harold Scott MacDonald
Fleischer, Isidore
Halperin, Stephen
Kochman, Stanley Oscar
Naimpally, Somashekhar Amrith
Nel, Louis Daniel
Schirmer, Helga H
Tall, Franklin David

QUEBEC
Alagar, Vangalur S

SASKATCHEWAN
Kaul, S K
Mezey, Paul G
Tymchatyn, Edward Dmytro

OTHER COUNTRIES
Kinoshita, Shin'ichi
Thomason, Robert Wayne
Zagier, Don Bernard

Givant, Steven Roger
Grabiner, Judith Victor
Lang, Martin T
Lee, K Hung
Levine, Howard Bernard
Liu, David Shiao-Kung
McCarthy, Mary Anne
Marder, Stanley
Olwell, Kevin D
Painter, Jeffrey Farrar
Patel, Arvindkumar Motibhai
Pearson, Scott Roberts
Rosenberg, Alex
Sansom, Richard E
Shestakov, Aleksei Ilyich
Simon, Horst D
Smolarski, Dennis Chester
Thiessen, Richard E
Thompson, Patrick W
Urquidi-MacDonald, Mirna
Woodriff, Roger L

COLORADO
Ablowitz, Mark Jay
Glover, Fred William
Holley, Frieda Koster
Lee, Kotik Kai
Walter, Martin Edward

CONNECTICUT
Fischer, Michael
Frankel, Martin Richard
Goldstein, Rubin
Sharlow, John Francis

DELAWARE
Kittlitz, Rudolf Gottlieb, Jr

DISTRICT OF COLUMBIA
Bohi, Douglas Ray
Graber, James Stanley

FLORIDA
Allen, Jon Charles
Clutterham, David Robert
Cooke, William Joe
Halpern, Leopold (Ernst)
Johnson, Harold Hunt
Martinez, Jorge
Ritter, Gerhard X
Shershin, Anthony Connors
Taylor, Michael Dee

GEORGIA
Bevis, Jean Harwell
Smith, William Allen

HAWAII
Anderson, Gerald M

IDAHO
Hoagland, Gordon Wood
Maloof, Giles Wilson

ILLINOIS
Becker, Gary Stanley
Hannon, Bruce Michael
Heller, Barbara Ruth
Litvin, Faydor L
Reiter, Stanley
Rodgers, James Foster
Saylor, Paul Edward
Smith, Stephen D
Usiskin, Zalman P
Wood, John William

INDIANA
Hood, Rodney Taber
Madan, Ved P
Minassian, Donald Paul
Taylor, Gladys Gillman

IOWA
Maxey, E James
Ton-That, Tuong

KANSAS
Strecker, George Edison

KENTUCKY
Looney, Stephen Warwick

LOUISIANA
Davis, Lawrence H
Goldstein, Jerome Arthur
Kythe, Prem Kishore
Sundar, P

MARYLAND
Austin, Homer Wellington
Boisvert, Ronald Fernand
Calinger, Ronald Steve
Draper, Richard Noel
Elliott, David LeRoy
Engel, Joseph H(enry)
Grebogi, Celso
Kowalski, Richard
Krishnaprasad, Perinkulam S
Lopez-Escobar, Edgar George Kenneth
Lozier, Daniel William
Preisman, Albert
Shiffman, Bernard

MASSACHUSETTS
Ankeny, Nesmith Cornett
Berkovits, Shimshon
Bezuszka, Stanley John
Cole, Nancy
Gacs, Peter
Grossberg, Stephen
Hsiao, William C
Kenney, Margaret June
Lam, Kui Chuen
McOwen, Robert C
Shum, Annie Waiching

MICHIGAN
Bickart, Theodore Albert
Field, David Anthony
McLaughlin, Renate
Murphy, Brian Boru
Papp, F(rancis) J(oseph)
Parks, Terry Everett
Roe, Philip Lawrence
Verhey, Roger Frank

MINNESOTA
Chipman, John Somerset
Kemper, John Thomas
Keynes, Harvey Bayard

MISSISSIPPI
Doblin, Stephen Alan

MISSOURI
Kubicek, John D
Seydel, Robert E
Silverstein, Martin L
Tikoo, Mohan L

MONTANA
Lott, Johnny Warren
Thompson, Edward Otis

NEBRASKA
Maloney, John P
Meakin, John C

NEW HAMPSHIRE
Meeker, Loren David

NEW JERSEY
Ahsanullah, Mohammad
Beals, R Michael
Belanger, David Gerald
Brand, William Wayne
Endo, Youichi
Gilman, Jane P
Goldstein, Sheldon
Hunter, John Stuart
Jacquin, Arnaud Eric
Leef, Audrey V
Mather, John Norman
Poiani, Eileen Louise
Rosenstein, Joseph Geoffrey
Sen, Tapas K

NEW MEXICO
Ahmed, Nasir
Booth, Thomas Edward

NEW YORK
Ascher, Marcia
Ascher, Robert
Braun, Ludwig
Chen, Chi-Tsong
Cok, David R
Cox, John Theodore
Daly, John T
Edwards, Harold M
Feuer, Richard Dennis
Guckenheimer, John
Gustavson, Fred Gehrung
Haddock, Jorge
Hastings, Harold Morris
Kline, Morris
LeVine, Micheal Joseph
Moskowitz, Martin A
Pan, Victor
Pattee, Howard Hunt, Jr
Shevack, Hilda N
Stearns, Richard Edwin
Stone, William C

NORTH CAROLINA
Burbeck, Christina Anderson
Katzin, Gerald Howard
Lambert, Alan L
Lucas, Carol N
Sowell, Katye Marie Oliver
Tyndall, John Raymond
Wu, Julian Juh-Ren

OHIO
Barger, Samuel Floyd
Burden, Richard L
Chiu, Victor
Davis, Sheldon W
Kertz, George J
St John, Ralph C
Schneider, Harold William
Steinlage, Ralph Cletus

OKLAHOMA
Aichele, Douglas B
Argyros, Ioannis Konstantinos
Briggs, Phillip D

Carr, Meg Brady
Jones, Jack Raymond
Richman, Michael B
Weeks, David Lee

OREGON
Ballantine, C S
Seitz, Gary M

PENNSYLVANIA
Baker, Ronald G
Chase, Gene Barry
Frieze, Alan Michael
Harris, Zellig S
Klotz, Eugene Arthur
McWhirter, James Herman
Mehta, Jatinder S
Pommersheim, James Martin
Scott, Dana S
Szyld, Daniel Benjamin
Tozier, John E
Williams, Albert J, Jr

RHODE ISLAND
Laurence, Geoffrey Cameron
Rudolph, Lee

SOUTH CAROLINA
Henry, Harold Robert
Saxena, Subhash Chandra

TENNESSEE
Adams, Dennis Ray
Beauchamp, John J
Davidson, Robert C
Hill, Hubert Mack

TEXAS
Carey, Graham Francis
Dennis, John Emory, Jr
Gladwell, Ian
Hernandez, Norma
Ladde, Gangaram Shivlingappa
Linn, John Charles
Renka, Robert Joseph
Semmes, Stephen William
Trench, William Frederick

UTAH
Alfeld, Peter
Cohen, Elaine
Horn, Roger Alan

VIRGINIA
Asterita, Mary Frances
Burns, William, Jr
Crowley, James M
Kratzke, Thomas Martin
Lasiecka, Irena
Smith, Emma Breedlove
Southworth, Raymond W(illiam)
Thomas, Lawrence E

WASHINGTON
Curtis, Wendell D
Gangolli, Ramesh A
Lind, Douglas A

WISCONSIN
Meitler, Carolyn Louise
Parter, Seymour Victor
Schneider, Hans
Strikwerda, John Charles

ALBERTA
Andersen, Kenneth F
Blais, J A Rodrigue
Muldowney, James
Tait, Robert James

BRITISH COLUMBIA
Das, Anadijiban
Miura, Robert Mitsuru

NEWFOUNDLAND
Goodaire, Edgar George

ONTARIO
Aczel, Janos D
Carlisle, David Brez
Choi, Man-Duen
Davison, E(dward) J(oseph)
Denzel, George Eugene
Duever, Thomas Albert
Lehma, Alfred Baker
Muller, Eric Rene
Poulton, Donald John

QUEBEC
Alagar, Vangalur S
Hall, Richard L
Lin, Paul C S

OTHER COUNTRIES
Dougalis, Vassilios
Lambert, William M, Jr

Other Mathematics

ALABAMA
Shipman, Jerry

ARIZONA
Smarandache, Florentin

ARKANSAS
Thornton, Kent W

CALIFORNIA
Benesch, Samuel Eli
Cates, Marshall L
Cherlin, George Yale
Chu, Kai-Ching
Dalrymple, Stephen Harris
Desoer, Charles A(uguste)
Fisher, William Bruce
Fried, Michael D
Garber, Morris Joseph

MEDICAL & HEALTH SCIENCES

Audiology & Speech Pathology

ALABAMA
Cooper, Eugene Bruce
Sheeley, Eugene C
Wester, Derin C

ARIZONA
Nation, James Edward

CALIFORNIA
Cleary, James W
Simmons, Dwayne Deangelo

COLORADO
Blager, Florence Berman
Scherer, Ronald Callaway

DISTRICT OF COLUMBIA
Weiss, Ira Paul

FLORIDA
Brown, William Samuel, Jr
Rothman, Howard Barry

HAWAII
Yates, James T

ILLINOIS
Hoshiko, Michael S
Schultz, Martin C

INDIANA
Hoops, Richard Allen
Ritter, E Gene

IOWA
Moll, Kenneth Leon
Schwartz, Ralph Jerome
Small, Arnold McCollum, Jr

LOUISIANA
Berlin, Charles I
Collins, Mary Jane
Cullen, John Knox, Jr

MARYLAND
Brownell, William Edward
Causey, G(eorge) Donald
Moscicki, Eve Karin

MASSACHUSETTS
Goldberg, Sheldon Sumner

MICHIGAN
Dolan, David Francis

NEBRASKA
Cochran, John Rodney

NEW JERSEY
Prasad, Marehalli Gopalan

NEW MEXICO
Stewart, Joseph Letie

NEW YORK
Castle, William Eugene
Dickson, Stanley
Katz, Jack

NORTH CAROLINA
Royster, Julia Doswell

OHIO
Hyman, Melvin
Lesner, Sharon A
Nixon, Charles William

TENNESSEE
Mendel, Maurice I
Nabelek, Anna K
Stewart, James Monroe

WASHINGTON
Lipscomb, David M
Pichal, Henri Thomas, II
Yantis, Phillip Alexander

WEST VIRGINIA
Lass, Norman J(ay)
Ruscello, Dennis Michael

WISCONSIN
Ritchie, Betty Caraway

ONTARIO
Harrison, Robert Victor
Kunov, Hans

OTHER COUNTRIES
Bergman, Moe

Dentistry

ALABAMA
Baker, John Rowland
Feagin, Frederick F
Graves, Carl N

Jeffcoat, Marjorie K
Keller, Stanley E
Kiyono, Hiroshi
Koulourides, Theodore I
Manson-Hing, Lincoln Roy
Menaker, Lewis
Navia, Juan Marcelo
Retief, Daniel Hugo
Ringsdorf, Warren Marshall, Jr
Speed, Edwin Maurice
Waite, Peter Daniel
Weatherford, Thomas Waller, III

ARIZONA
Sarid, Dror

CALIFORNIA
Abrams, Albert Maurice
Allerton, Samuel E
Ast, David Bernard
Barber, Thomas King
Bernard, George W
Bertolami, Charles Nicholas
Chierici, George J
Dixon, Andrew Derart
Dummett, Clifton Orrin
Duperon, Donald Francis
Evans, Caswell A, Jr
Feller, Ralph Paul
Greenspan, John Simon
Hansen, Louis Stephen
Ingle, John Ide
Jarvis, William Tyler
Jendresen, Malcolm Dan
Jones, Edward George
Kaplan, Herman
Kapur, Krishan Kishore
Krol, Arthur J
Leake, Donald L
Lugassy, Armand Amram
Lundblad, Roger Lauren
Marshall, Grayson William, Jr
Miller, Arthur Joseph
Newbrun, Ernest
Robinson, Hamilton Burrows Greaves
Schoen, Max H
Schonfeld, Steven Emanuel
Simmons, Norman Stanley
Slavkin, Harold Charles
Stark, Marvin Michael
Steinman, Ralph R
Weinstock, Alfred
Wescott, William B
Westmoreland, Winfred William
White, Stuart Cossitt
Wirthlin, Milton Robert, Jr
Woolley, LeGrand H
Wycoff, Samuel John
Yagiela, John Allen
Zernik, Joseph

COLORADO
Eames, Wilmer B
Greene, David Lee
Krikos, George Alexander
Meskin, Lawrence Henry
Sharma, Brahma Dutta

CONNECTICUT
Burstone, Charles Justin
Celesk, Roger A
Coykendall, Alan Littlefield
Goldberg, A Jon
Hand, Arthur Ralph
Katz, Ralph Verne
Krutchkoff, David James
Macko, Douglas John
Nalbandian, John
Nanda, Ravindra
Norton, Louis Arthur
Nuki, Klaus
Poole, Andrew E
Prasad, Arun
Rossomando, Edward Frederick
Tinanoff, Norman
Venham, Larry Lee
Weinstein, Sam
Zubal, I George

DELAWARE
Jefferies, Steven
Schweiger, James W

DISTRICT OF COLUMBIA
Calhoun, Noah Robert
Ferrigno, Peter D
Gupta, Om Prakash
Littleton, Preston A, Jr
Sinkford, Jeanne C

FLORIDA
Anusavice, Kenneth John
Bennett, Carroll G
Clark, William Burton, IV
Corn, Herman
Duany, Luis F, Jr
Gibbs, Charles Howard
Grodberg, Marcus Gordon
Hassell, Thomas Michael
Kerr, I Lawrence
Lobene, Ralph Rufino
Mackenzie, Richard Stanley
Medina, Jose Enrique
Nizel, Abraham Edward

Simring, Marvin
Sweeney, Thomas Patrick
Vesely, David Lynn
Wisotzky, Joel
Zamikoff, Irving Ira

GEORGIA
Benoit, Peter Wells
Bustos-Valdes, Sergio Enrique
Ciarlone, Alfred Edward
Clark, James William
Dirksen, Thomas Reed
Fritz, Michael E
Gangarosa, Louis Paul, Sr
Hawkins, Isaac Kinney
Pashley, David Henry
Sharawy, Mohamed
Urbanek, Vincent Edward
Weatherred, Jackie G
Whitford, Gary M
Zwemer, Thomas J

ILLINOIS
Aduss, Howard
Cleall, John Frederick
Cohen, Gloria
Corruccini, Robert Spencer
Coy, Richard Eugene
Dahlberg, Albert A
De Rijk, Waldemar G
Douglas, Bruce L
Engel, Milton Baer
Flores, Samson Sol
Gaik, Geraldine Catherine
Goepp, Robert August
Gowgiel, Joseph Michael
Graber, T(ouro) M(or)
Grandel, Eugene Robert
Jacobson, Marvin
Martinez-Lopez, Norman Petronio
Murphy, Richard Allan
Norman, Richard Daviess
Rao, Gopal Subba
Rosenberg, Henry Mark
Rosenstein, Sheldon William
Scapino, Robert Peter
Steffek, Anthony J
Teuscher, George William
Turner, Donald W
Waterhouse, John P
Yale, Seymour Hershel

INDIANA
Bixler, David
Bogan, Robert L
Garner, LaForrest D
Gregory, Richard Lee
Henderson, Hala Zawawi
Hennon, David Kent
Hine, Maynard Kiplinger
Lund, Melvin Robert
McDonald, Ralph Earl
Patterson, Samuel S
Potter, Rosario H Yap
Standish, Samuel Miles
Starkey, Paul Edward
Stookey, George K
Swenson, Henry Maurice
Tomich, Charles Edward
Wagner, Morris

IOWA
Bishara, Samir Edward
Bjorndal, Arne Magne
Chan, Kai Chiu
Grigsby, William Redman
Hammond, Harold Logan
Jacobs, Richard M
Johnson, Wallace W
Kremenak, Charles Robert
Laffoon, John
McLeran, James Herbert
Nowak, Arthur John
Olin, William (Harold), Sr
Sabiston, Charles Barker, Jr
Staley, Robert Newton
Thayer, Keith Evans

KANSAS
Stewart, Jack Lauren

KENTUCKY
Boyer, Harold Edwin
Crim, Gary Allen
Elwood, William K
Farman, Allan George
Mann, Wallace Vernon, Jr
Nash, David Allen
Parkins, Frederick Milton
Podshadley, Arlon George
Saxe, Stanley Richard
Smith, Timothy Andre
Spedding, Robert H
Staat, Robert Henry
Stewart, Arthur Van
Wesley, Robert Cook
Wittwer, John William

LOUISIANA
Alam, Syed Qamar
Nakamoto, Tetsuo
Petri, William Henry, III
Waldrep, Alfred Carson, Jr

MARYLAND
Baum, Bruce J
Bowen, Rafael Lee
Brauer, Gerhard Max
Chow, Laurence Chung-Lung
Cohen, Lois K
Cotton, William Robert
Davidson, William Martin
Dionne, Raymond A
Driscoll, William
Gartner, Leslie Paul
Gobel, Stephen
Grewe, John Mitchell
Joly, Olga G
Kendrick, Francis Joseph
Kinnard, Matthew Anderson
Kirkpatrick, Diana (Rorabaugh) M
Larson, Rachel Harris (Mrs John Watson Henry)
Loe, Harald
Loebenstein, William Vaille
McCauley, H(enry) Berton
Means, Craig Ray
Minah, Glenn Ernest
Myslinski, Norbert Raymond
Nylen, Marie Ussing
Ogden, Ingram Wesley
Patel, Prafull Raojibhai
Pinkus, Lawrence Mark
Ranney, Richard Raymond
Redman, Robert Shelton
Rizzo, Anthony Augustine
Rovelstad, Gordon Henry
Seibel, Werner
Takagi, Shozo
Tesk, John A
Valega, Thomas Michael
Whitehurst, Virgil Edwards

MASSACHUSETTS
Allukian, Myron
Auskaps, Aina Marija
Baratz, Robert Sears
Caslavska, Vera Barbara
Gianelly, Anthony Alfred
Goldhaber, Paul
Gron, Poul
Harris, Melvyn H
Hein, John William
Henry, Joseph L
Howell, Thomas Howard
Johansen, Erling
Keith, David Alexander
Kramer, Gerald M
Kronman, Joseph Henry
Moorrees, Coenraad Frans August
Niederman, Richard
Roberts, Francis Donald
Ruben, Morris P
Slavin, Ovid
Smith, Daniel James
Smulow, Jerome B
Sonis, Stephen Thomas
Susi, Frank Robert
Tenca, Joseph Ignatius
Turesky, Samuel Saul
Wells, Herbert
Williams, David Lloyd
Yen, Peter Kai Jen

MICHIGAN
Ash, Major McKinley, Jr
Avery, James Knuckey
Brooks, Sharon Lynn
Burdi, Alphonse R
Burt, Brian Aubrey
Carlson, David Sten
Charbeneau, Gerald T
Christiansen, Richard Louis
Craig, Robert George
Harris, James Edward
Holmstedt, Jan Olle Valter
Lillie, John Howard
Loesche, Walter J
McNamara, James Alyn, Jr
Millard, Herbert Dean
Morawa, Arnold Peter
Ramfjord, Sigurd
Richards, Albert Gustav
Rowe, Nathaniel H
Steiman, Henry Robert
Strachan, Donald Stewart
Striffler, David Frank

MINNESOTA
Cervenka, Jaroslav
Douglas, William Hugh
Elzay, Richard Paul
Gibilisco, Joseph
Goodkind, Richard Jerry
Gorlin, Robert James
Martens, Leslie Vernon
Meyer, Maurice Wesley
Oliver, Richard Charles
Schachtele, Charles Francis
Schaffer, Erwin Michael
Speidel, T(homas) Michael
Till, Michael John
Witkop, Carl Jacob, Jr
Yamane, George M

MISSISSIPPI
Alexander, William Nebel

Dentistry (cont)

MISSOURI
Bensinger, David August
Cobb, Charles Madison
Nikolai, Robert Joseph
Zoeller, Gilbert Norbert

MONTANA
Ludwig, Theodore Frederick

NEBRASKA
Bradley, Richard E
Kramer, William S
Kutler, Benton
Rinne, Vernon Wilmer
Sullivan, Robert Emmett

NEW HAMPSHIRE
Hill, Percy Holmes
Zumbrunnen, Charles Edward

NEW JERSEY
Alfano, Michael Charles
Baron, Hazen Jay
Bolden, Theodore Edward
Chance, Kenneth Bernard
Chasens, Abram I
Cinotti, William Ralph
Curro, Frederick A
Di Paolo, Rocco John
Evans, Ralph H, Jr
Gershon, Sol D
Lucca, John J
Mellberg, James Richard
Najjar, Talib A
Parker, LeRoy A, Jr
Pianotti, Roland Salvatore
Reussner, George Henry
Rivetti, Henry Conrad
Shovlin, Francis Edward

NEW MEXICO
Bleiweiss, Max Phillip
Yudkowsky, Elias B

NEW YORK
Archard, Howell Osborne
Baer, Paul Nathan
Blechman, Harry
Boucher, Louis Jack
Ciancio, Sebastian Gene
Di Salvo, Nicholas Armand
Featherstone, John Douglas Bernard
Fischman, Stuart L
Gilmour, Marion Nyholm H
Goldberg, Louis J
Golub, Lorne Malcolm
Goswami, Meeta J
Gottsegen, Robert
Green, Larry J
Hanley, Kevin Joseph
Hoffman, Robert
Horowitz, Sidney Lester
Jones, Melvin D
Kahn, Norman
Kaslick, Ralph Sidney
Kaufman, Edward Godfrey
Kleinberg, Israel
Lamster, Ira Barry
Lenchner, Nathaniel Herbert
Leverett, Dennis Hugh
Linkow, Leonard I
McHugh, William Dennis
Mandel, Irwin D
Morris, Melvin Lewis
Moss-Salentijn, Letty
Mundorff-Shrestha, Sheila Ann
Oaks, J Howard
Ortman, Harold R
Powell, Richard Anthony
Quartararo, Ignatius Nicholas
Reyes-Guerra, Antonio
Rifkin, Barry Richard
Roistacher, Seymour Lester
Schaaf, Norman George
Scopp, Irwin Walter
Singh, Inder Jit
Sirianni, Joyce E
Sreebny, Leo Morris
Stahl, S Sigmund
Staple, Peter Hugh
Zegarelli, Edward Victor
Zero, Domenick Thomas

NORTH CAROLINA
Adisman, I Kenneth
Barker, Ben D
Bawden, James Wyatt
Ennever, John Joseph
Hylander, William Leroy
Johnston, Malcolm Campbell
Lindahl, Roy Lawrence
Marble, Howard Bennett, Jr
Nelson, Robert Mellinger
Oldenburg, Theodore Richard
Proffit, William R
Quinn, Galen Warren
Shankle, Robert Jack
Via, William Fredrick, Jr
Warren, Donald W
Webster, William Phillip
Williams, Ray Clayton
Woolridge, Edward Daniel

NORTH DAKOTA
Vogel, Gerald Lee

OHIO
Alley, Keith Edward
Amjad, Zahid
Blozis, George G
Chen, Wen Yuan
Cooley, William Edward
Dew, William Calland
Ferencz, Nicholas
Goorey, Nancy Reynolds
Horton, John Edward
Israel, Harry, III
Jordan, Freddie L
Lydy, David Lee
Rossi, Edward P
Sciulli, Paul William
Temple, Robert Dwight
Williams, Benjamin Hayden
Woelfel, Julian Bradford

OKLAHOMA
Glass, Richard Thomas
Howes, Robert Ingersoll, Jr
Miranda, Frank Joseph
Shapiro, Stewart

OREGON
Bruckner, Robert Joseph
Buck, Douglas L
Hedtke, James Lee
Howard, William Weaver
Jastak, J Theodore
Kinersly, Thorn
Riviere, George Robert
Saffir, Arthur Joel
Sorenson, Fred M
Tracy, William E
Van Hassel, Henry John
Walters, Roland Dick
Zingeser, Maurice Roy

PENNSYLVANIA
Ackerman, James L
Bassiouny, Mohamed Ali
Brendlinger, Darwin
Brightman, Vernon
Cohen, David Walter
Cooper, Stephen Allen
Cornell, John Alston
Farber, Paul Alan
Lindemeyer, Rochelle G
Listgarten, Max
Mazaheri, Mohammad
Milligan, Wilbert Harvey, III
Oka, Seishi William
Oliet, Seymour
Rapp, Robert
Rosan, Burton
Segal, Alan H
Seltzer, Samuel
Smudski, James W
Stevens, Roy Harris
Stiff, Robert H
Taichman, Norton Stanley
Vergona, Kathleen Anne Dobrosielski
Winkler, Sheldon
Yankell, Samuel L

SOUTH CAROLINA
Corcoran, John W
Cordova-Salinas, Maria Asuncion
Edwards, James Burrows
Fingar, Walter Wiggs
Lawson, Benjamin F
Morris, Alvin Leonard
Nuckles, Douglas Boyd
White, Edward

TENNESSEE
Behrents, Rolf Gordon
Gibbs, Samuel Julian
Gotcher, Jack Everett
Harris, Edward Frederick
Jurand, Jerry George
Keim, Robert Gerald
Lynch, Denis Patrick
McKnight, James Pope
Nunez, Loys Joseph
Richardson, Elisha Roscoe
Siskin, Milton
Smith, Roy Martin
Sticht, Frank Davis
Weber, Faustin N

TEXAS
Allen, Don Lee
Allen, Edward Patrick
Biggerstaff, Robert Huggins
Binnie, William Hugh
Blanton, Patricia Louise
Byrd, David Lamar
Cochran, David Lee
Cottone, James Anthony
Del Rio, Carlos Eduardo
DePaola, Dominick Philip
Englander, Harold Robert
Fultz, R Paul
Gage, Tommy Wilton
Gerding, Thomas G
Goaz, Paul William
Guentherman, Robert Henry
Harris, Norman Oliver
Hatch, John Phillip
Henry, Clay Allen
Hoskins, Sam Whitworth, Jr
Hurt, William Clarence
Kalkwarf, Kenneth Lee
Lambert, Joseph Parker
Langland, Olaf Elmer
McConnell, Duncan
Madden, Richard M
Miller, Edward Godfrey, Jr
Nelson, John Franklin
Ohlenbusch, Robert Eugene
Olson, John Victor
Powers, John Michael
Rawls, Henry Ralph
Seliger, William George
Shillitoe, Edward John
Storey, Arthur Thomas
Suddick, Richard Phillips
Sweeney, Edward Arthur
Taylor, Paul Peak
Throckmorton, Gaylord Scott
Vogel, John
Williams, Fred Eugene

UTAH
James, Garth A

VIRGINIA
Abbott, David Michael
Butler, James Hansel
Gallardo-Carpentier, Adriana
Isaacson, Robert John
Wiebusch, F B
Wittemann, Joseph Klaus

WASHINGTON
Bolender, Charles L
Canfield, Robert Charles
Hall, Stanton Harris
Hodson, Jean Turnbaugh
Johnson, Robert H
Moffett, Benjamin Charles, Jr
Moore, Alton Wallace
Page, Roy Christopher
Riedel, Richard Anthony
Stern, Irving B
Stibbs, Gerald Denike
Swoope, Charles C
Way, Jon Leong
Wehner, Alfred Peter

WEST VIRGINIA
Biddington, William Robert
Bouquot, Jerry Elmer
Calhoun, Donald Alan
Overberger, James Edwin
Overman, Dennis Orton

WISCONSIN
Goggins, John Francis

PUERTO RICO
Ortiz, Araceli

ALBERTA
Eggert, Frank Michael
Haryett, Rowland D
Holland, Graham Rex
MacRae, Patrick Daniel
Osborn, Jeffrey Whitaker
Sperber, Geoffrey Hilliard
Thomas, Norman Randall
Wood, Norman Kenyon

BRITISH COLUMBIA
Beagrie, George Simpson
Bennett, Ian Cecil
Kraintz, Leon
Leung, So Wah
Neilson, John Warrington

NOVA SCOTIA
McLean, James Douglas
Sykora, Oskar P

ONTARIO
Brooke, Ralph Ian
Galil, Khadry Ahmed
Hunter, W(illiam) Stuart
Kim, Sukyoung
Mamandras, Antonios H
Mayhall, John Tarkington
Melcher, Antony Henry
Nikiforuk, Gordon
Popovich, Frank
Sandham, Herbert James
Saunders, Shelley Rae
Sessle, Barry John
Speck, John Edward
Ten Cate, Arnold Richard
Wu, Alan SeeMing

QUEBEC
Chan, Eddie Chin Sun

SASKATCHEWAN
Ambrose, Ernest R

OTHER COUNTRIES
Barbenel, Joseph Cyril
Stallard, Richard E
Wei, Stephen Hon Yin

Environmental Health

ALABAMA
Bailey, William C
Coker, Samuel Terry
Dalvi, Ramesh R
Hood, Ronald David
Mundy, Roy Lee
Rahn, Ronald Otto
Rajanna, Bettaiya
Singh, Jarnail
Wallace, Dennis D

ALASKA
Mills, William J, Jr
Tilsworth, Timothy

ARIZONA
Barnes, Howard Clarence
Beck, John R(oland)
Colburn, Wayne Alan
Lebowitz, Michael David
Witten, Mark Lee

ARKANSAS
Brewster, Marjorie Ann
Chang, Louis Wai-Wah
Chowdhury, Parimal
Howell, Robert T
Kadlubar, Fred F
Smith, Carroll Ward
Snow, Donald L(oesch)
Townsend, James Willis

CALIFORNIA
Amoore, John Ernest
Anspaugh, Lynn Richard
Asante-Duah, D Kofi
Barnes, Paul Richard
Beard, Rodney Rau
Bernstein, Leslie
Bhatnagar, Rajendra Sahai
Bils, Robert F
Brantingham, Charles Ross
Brown, Harold Victor
Brown, Stephen L(awrence)
Buck, Alan Charles
Campbell, Kirby I
Cass, Glen R
Chang, Daniel P Y
Chatigny, Mark A
Collins, James Francis
Cooper, Robert Chauncey
Coughlin, James Robert
Daisey, Joan M
Danse, Ilene H Raisfeld
DePass, Linval R
Detels, Roger
Dobson, R Lowry
Duhl, Leonard J
Eastmond, David Albert
Eckert, Curtis Dale
Embree, James Willard, Jr
Esposito, Michael Salvatore
Fabrikant, Irene Berger
Fallon, Joseph Greenleaf
Froebe, Larry R
Garland, Cedric Frank
Goldman, Marvin
Hackett, Nora Reed
Hallesy, Duane Wesley
Hubbard, Eric R
Hubert, Helen Betty
Hughes, James Paul
Jensen, Ronald Harry
Johansson, Karl Richard
Judd, Stanley H
Kilburn, Kaye Hatch
Kleinman, Michael Thomas
Kohn, Henry Irving
Koschier, Francis Joseph
Larkin, Edward Charles
Last, Jerold Alan
Leo, Albert Joseph
Low, Hans
McGaughey, Charles Gilbert
McKone, Thomas Edward
Mayron, Lewis Walter
Newell, Gordon Wilfred
Niklowitz, Werner Johannes
Owen, Robert L
Palmer, Alan
Pelletier, Charles A
Pierce, James Otto, II
Porcella, Donald Burke
Rice, Robert Hafling
Richters, Arnis
Sagan, Leonard A
Salmon, Eli J
Schienle, Jan Hoops
Sheneman, Jack Marshall
Shvartz, Esar
Slote, Lawrence
Sparling, Rebecca Hall
Springer, Wayne Richard
Sterner, James Hervi
Stonebraker, Peter Michael
Swan, Shanna Helen
Ury, Hans Konrad
Vanderlaan, Martin
Victery, Winona Whitwell
Volz, Michael George
Wagner, Jeames Arthur
Whipple, Christopher George

Whittenberger, James Laverre
Wiley, Lynn M
Wood, Corinne Shear
Wyzga, Ronald Edward
Yager, Janice L Winter
Zinkl, Joseph Grandjean

COLORADO
Aldrich, Franklin Dalton
Anderson, Robert
Andrews, Ken J
Audesirk, Gerald Joseph
Barnes, Allan Marion
Batchelder, Arthur Roland
Best, Jay Boyd
Boatman, Joe Francis
Brooks, Bradford O
Irons, Richard Davis
Ivory, Thomas Martin, III
Marsh, Gregory K
Shopp, George Milton, Jr
Stallones, Lorann
Ulvedal, Frode
Wilson, Vincent L

CONNECTICUT
Borgia, Julian Frank
Cohen, Steven Donald
Hinkle, Lawrence Earl, Jr
Lieberman, James
Matheson, Dale Whitney
Milzoff, Joel Robert
Monro, Alastair Macleod
Nadel, Ethan Richard
Rho, Jinnque
Stitt, John Thomas
Stolwijk, Jan Adrianus Jozef

DELAWARE
Frawley, John Paul
Garland, Charles E
Krone, Lawrence James
Patterson, Gordon Derby, Jr
Pell, Sidney
Potrafke, Earl Mark
Staples, Robert Edward

DISTRICT OF COLUMBIA
Arcos, Joseph (Charles)
Breen, Joseph John
Byrd, Daniel Madison, III
Chiazze, Leonard, Jr
Cotruvo, Joseph Alfred
Florig, Henry Keith
Goeringer, Gerald Conrad
Hill, Richard Norman
Kimbrough, Renate Dora
Kimmel, Carole Anne
Lee, Cheng-Chun
Lunchick, Curt
McEwen, Gerald Noah, Jr
Moore, John Arthur
Nadolney, Carlton H
Prival, Michael Joseph
Rall, David Platt
Rao, Mamidanna S
Rhoden, Richard Allan
Richardson, Allan Charles Barbour
Rogers, Senta S(tephanie)
Schwartz, Sorell Lee
Sexton, Ken
Smith, David Allen
Todhunter, John Anthony
Tompkins, George Jonathan
Weinberg, Myron Simon
Wolff, Arthur Harold
Zeeman, Maurice George
Zook, Bernard Charles

FLORIDA
Abraham, William Michael
Billings, Charles Edgar
Brenner, Richard Joseph
Clark, Mary Elizabeth
Crabtree, Ross Edward
Dougherty, Ralph C
Echelberger, Wayne F, Jr
Freudenthal, Ralph Ira
Gittelman, Donald Henry
Grabowski, Casimer Thaddeus
Harrold, Gordon Coleson
Lowe, Ronald Edsel
Lu, Frank Chao
Myers, Richard Lee
Teaf, Christopher Morris
Thoma, Richard William
Tiffany, William James, III
Walker, Charles R
Wolsky, Sumner Paul

GEORGIA
Batra, Gopal Krishan
Berg, George G
Bernstein, Robert Steven
Bryan, Frank Leon
Curley, Winifred H
Falk, Henry
Fineman, Manuel Nathan
Galeano, Sergio F(rancis)
Hargrove, Robert John
Hedden, Kenneth Forsythe
Inge, Walter Herndon, Jr
Kilbourne, Edwin Michael
Lisella, Frank Scott

McGrath, William Robert
Mitchell, Frank L
Mumtaz, Mohammad Moizuddin
Oakes, Thomas Wyatt
Oakley, Godfrey Porter, Jr
Peake, Thaddeus Andrew, III
Pratt, Harry Davis
Reinhardt, Donald Joseph
Schliessmann, D(onald) J(oseph)
Singh, Harpal P
Stevenson, Dennis A
Stone, Connie J
Thacker, Stephen Brady
Thun, Michael John
Urso, Paul
Wolven-Garrett, Anne M

HAWAII
Baker, Paul Thornell
Crane, Sheldon Cyr
Hertlein, Fred, III
Kodama, Arthur Masayoshi

IDAHO
Gesell, Thomas Frederick
Hillyard, Ira William
Parker, Robert Davis Rickard

ILLINOIS
Benkendorf, Carol Ann
Bernath, Tibor
Boulos, Badi Mansour
Brenniman, Gary Russell
Carnow, Bertram Warren
Conibear, Shirley Ann
Cossairt, Jack Donald
Doege, Theodore Charles
Ehrlich, Richard
Fenters, James Dean
Francis, Bettina Magnus
Garvin, Paul Joseph, Jr
Golchert, Norbert William
Hermann, Edward Robert
Hubbard, Lincoln Beals
Huberman, Eliezer
Jacobson, Marvin
Konopinski, Virgil J
Lasley, Stephen Michael
Leikin, Jerrold Blair
McCormick, David Loyd
McFee, Donald Ray
Oleckno, William Anton
Savic, Stanley D
Stoffer, Robert Llewellyn
Wadden, Richard Albert
Zelac, Ronald Edward

INDIANA
Dando, William Arthur
Evans, Michael Allen
Kessler, Wayne Vincent
Klaunig, James E
Landolt, Robert Raymond
McMasters, Donald L
Slavik, Nelson Sigman
Vodicnik, Mary Jo
Weinrich, Alan Jeffrey
Wolff, Ronald Keith

IOWA
Chung, Ronald Aloysius
Hausler, William John, Jr
Kniseley, Richard Newman
Long, Keith Royce
Schnoor, Jerald L

KANSAS
Cohen, Jules Bernard
Ferraro, John Anthony
Hinshaw, Charles Theron, Jr
Neuberger, John Stephen
Oehme, Frederick Wolfgang
Pierce, John Thomas
Urban, James Edward

KENTUCKY
Chen, Theresa S
Christensen, Ralph C(hresten)
Hoye, Robert Earl
Myers, Steven Richard
Tseng, Michael Tsung

LOUISIANA
Brody, Arnold R
Flournoy, Robert Wilson
Franklin, William Elwood
Makielski, Sarah Kimball
Mangham, Jesse Roger
Mehendale, Harihara Mahadeva
O'Neil, Carol Elliot
Pfeifer, Gerard David

MAINE
Paigen, Beverly Joyce
Stoudt, Howard Webster

MARYLAND
Alavanja, Michael Charles Robert
Baker, Frank
Baker, Susan Pardee
Becker, Donald Arthur
Beebe, Gilbert Wheeler
Benbrook, Charles M
Biersner, Robert John

Bricker, Jerome Gough
Brodsky, Allen
Bromberger-Barnea, Baruch (Berthold)
Cantor, Kenneth P
Carton, Robert John
Cheh, Albert Mei-Chu
Chisolm, James Julian, Jr
Colucci, Anthony Vito
Cookson, John Thomas, Jr
Correa Villasenor, Adolfo
Cosmides, George James
Creasia, Donald Anthony
Dacre, Jack Craven
De Luca, Luigi Maria
Duggar, Benjamin Charles
Dyer, Robert Francis, Jr
Fechter, Laurence David
Feinman, Susan (Birnbaum)
Fowler, Bruce Andrew
Frazier, John Melvin
Gelboin, Harry Victor
Glowa, John Robert
Gori, Gio Batta
Gould, Jack Richard
Gray, Allan P
Groopman, John Davis
Guilarte, Tomas R
Hamill, Peter Van Vechten
Heath, Robert Gardner
Hennings, Henry
Hodge, Frederick Allen
Hsia, Mong Tseng Stephen
Jakab, George Joseph
Johnson, Carl Boone
Johnson, Charles C, Jr
Johnson, David Norseen
Jones, LeeRoy G(eorge)
Kawata, Kazuyoshi
Keefer, Larry Kay
Lee, I P
Levin, Barbara Chernov
Levin, Gilbert Victor
Levine, Richard Joseph
Lewis, Stephen B
Miller, Raymond Earl
Mills, William Andy
Minthorn, Martin Lloyd, Jr
Morton, Joseph James Pandozzi
Munn, John Irvin
Palmer, Winifred G
Phillips, Grace Briggs
Proctor, Donald Frederick
Rodgers, Imogene Sevin
Rosenberg, Saul H
Saffiotti, Umberto
Schaub, Stephen Alexander
Seagle, Edgar Franklin
Silbergeld, Ellen K
Sonawane, Babasaheb R
Swift, David Leslie
Traystman, Richard J
Valdes, James John
Wachs, Melvin Walter
Yodaiken, Ralph Emile

MASSACHUSETTS
Albertini, David Fred
Brain, Joseph David
Breckenridge, John Robert
Burse, Richard Luck
Camougis, George
Chivian, Eric Seth
Chou, Iih-Nan (George)
Clapp, Richard W
Crusberg, Theodore Clifford
DiNardi, Salvatore Robert
Epstein, Eliot
Ferris, Benjamin Greeley, Jr
Gaudette, Leo Eward
Goldberg, Sheldon Sumner
Goldman, Ralph Frederick
Gottlieb, Marise Suss
Gray, Douglas Carmon
Hammond, Sally Katharine
Hattis, Dale B
Hubbard, Roger W
Landowne, Milton
Modest, Edward Julian
Nauss, Kathleen Minihan
Ofner, Peter
Ozonoff, David Michael
Peters, Howard August
Riley, Robert Lord
Shapiro, Jacob
Sivak, Andrew
Speizer, Frank Erwin
Strauss, Harlee S
Szlyk-Modrow, Patricia Carol
Thomas, Clayton Lay
Thomas, Gail B
Trichopoulos, Dimitrios V
Wegman, David Howe
Weiss, Scott T
Wenger, Christian Bruce

MICHIGAN
Adrounie, V Harry
Anderson, Marvin David
Batterman, Stuart A
Dauphinais, Raymond Joseph
Deininger, Rolf A
Dragun, James
Evans, David Arnold
Farber, Hugh Arthur

Ficsor, Gyula
Garg, Bhagwan D
Gray, Robert Howard
Hart, Donald John
Hartung, Rolf
Hoerger, Fred Donald
Humphrey, Harold Edward Burton, Jr
Krebs, William H
Krishna, Gopala
Leng, Marguerite Lambert
Meeks, Robert G
Moolenaar, Robert John
Musch, David C(harles)
Nriagu, Jerome Okon
Palmer, Kenneth Charles
Perrin, Eugene Victor
Powitz, Robert W
Richardson, Rudy James
Roslinski, Lawrence Michael
Sikarskie, James Gerard
Toth, Paul Eugene
Trosko, James Edward
Wu, Hofu

MINNESOTA
Arthur, John W
Barber, Donald E
Cornelissen Guillaume, Germaine G
Dahlberg, Duane Arlen
Levitan, Alexander Allen
Messing, Rita Bailey
Oxborrow, Gordon Squires
Roessler, Charles Ervin
Roy, Robert Russell
Ruschmeyer, Orlando R
Thompson, Fay Morgen
Thompson, Phillip Gerhard
Vesley, Donald
Ward, Wallace Dixon
Willard, Paul W

MISSISSIPPI
Kopfler, Frederick Charles

MISSOURI
Badr, Mostafa
Beck, William J
Black, Wayne Edward
Blenden, Donald C
Brahmbhatt, Sudhir R
Brown, Olen Ray
Craddock, John Harvey
Gay, Don Douglas
Hallas, Laurence Edward
Kupchella, Charles E
Lawless, Edward William
Lower, William Russell
Perry, Horace Mitchell, Jr
Saran, Chitaranjan
Schwartz, Henry Gerard, Jr
Spiers, Donald Ellis
Weeks, Robert Joe
Wells, Frank Edward
Wilkinson, Ralph Russell

NEBRASKA
Berndt, William O
Schneider, Norman Richard

NEVADA
Kinnison, Robert Ray
Nauman, Charles Hartley
Scholler, Jean
Wegst, Walter F, Jr

NEW HAMPSHIRE
Hatch, Frederick Tasker

NEW JERSEY
Abdel-Rahman, Mohamed Shawky
Beispiel, Myron
Bogden, John Dennis
Borowsky, Harry Herbert
Churg, Jacob
Feasley, Charles Frederick
Ford, Daniel Morgan
Garro, Anthony Joseph
George, Kenneth Dudley
Gochfeld, Michael
Goldstein, Bernard David
Hutcheon, Duncan Elliot
Iglewicz, Raja
Johnson, Bobby Ray
Klenke, Edward Frederick, Jr
Lewis, Neil Jeffrey
Lioy, Paul James
Lyman, Frank Lewis
Mallison, George Franklin
Mehlman, Myron A
Moe, Robert Anthony
Newton, Paul Edward
Rolle, F Robert
Schutz, Donald Frank
Snyder, Robert
Stillman, John Edgar
Taylor, Bernard Franklin
Van Pelt, Wesley Richard
Wilkening, George Martin
Witz, Gisela

NEW MEXICO
Bachrach, Arthur Julian
Chen, David J
Cheng, Yung-Sung

Environmental Health (cont)

Erikson, Mary Jane
Finch, Gregory Lee
Griffin, Travis Barton
Hobbs, Charles Henry
Johnson, Neil Francis
Mauderly, Joe L
Mewhinney, James Albert
Mokler, Brian Victor
Norris, Andrew Edward
Schulte, Harry Frank
Seiler, Fritz A
Thrasher, Jack D
Tillery, Marvin Ishmael
Wilkening, Marvin H
Williams, Mary Carol
Yeh, Hsu-Chi

NEW YORK
Abraham, Jerrold L
Altshuler, Bernard
Amborski, Leonard Edward
Aschner, Michael
Baum, John W
Becker, Robert O
Blank, Robert H
Brophy, Mary O'Reilly
Burns, Fredric Jay
Bush, David Graves
Carsten, Arland L
Chishti, Muhammad Azhar
Christman, Edward Arthur
Cohen, Beverly Singer
Cohen, Norman
Collins, James Joseph
Conway, John Bell
Cory-Slechta, Deborah Ann
David, Oliver Joseph
Dicello, John Francis, Jr
Eadon, George Albert
Eckert, Richard Raymond
Elahi, Nasik
Ely, Thomas Sharpless
Ferber, Kelvin Halket
Ferin, Juraj
Fernandez, Jose Martin
Frenkel, Krystyna
Freudenthal, Peter
Garte, Seymour Jay
Gates, Allen H(azen), Jr
Gellman, Isaiah
Gunnison, Albert Farrington
Guttenplan, Joseph B
Haley, Nancy Jean
Harley, Naomi Hallden
Hecht, Stephen Samuel
Holtzman, Seymour
Hull, Andrew P
Hyman, Abraham
Jenkins, Philip Winder
Katz, Gary Victor
Landrigan, Philip J
Lawrence, David A
Leach, Leonard Joseph
Lengemann, Frederick William
Lippmann, Morton
Lockhart, Haines Boots, Jr
McCann, Michael Francis
Mellins, Robert B
Michaelson, Solomon M
Miller, Albert
Mitacek, Eugene Jaroslav
Morris, Samuel Cary, III
Naidu, Janakiram Ramaswamy
Nakatsugawa, Tsutomu
Narahara, Hiromichi Tsuda
Neal, Michael William
Nicholson, William Jamieson
Nuzzi, Robert
Oberdorster, Gunter
Palevsky, Gerald
Parsons, Patrick Jeremy
Pasternack, Bernard Samuel
Pence, Harry Edmond
Pereira, Martin Rodrigues
Poe, Robert Hilleary
Rhodes, Robert Shaw
Rifkind, Arleen B
Robinson, Myron
Rossman, Toby Gale
Savage, E Lynn
Schachter, E Neil
Schachter, Michael B
Schecter, Arnold Joel
Schlesinger, Richard B
Selikoff, Irving John
Shapiro, Raymond E
Shen, Thomas T
Shore, Roy E
Smith, Frank Ackroyd
Solomon, Jerome Jay
Spiegel, Allen David
Squires, Richard Felt
Sternberg, Stephen Stanley
Stotzky, Guenther
Straus, David Bradley
Thompson, John C, Jr
Wilson, Robert Hallowell
Zaki, Mahfou H

NORTH CAROLINA
Andersen, Melvin Ernest
Anderson, Marshall W
Axtell, Richard Charles
Ayoub, Mahmoud A
Barnes, Donald Wesley
Barrett, J(ames) Carl
Birnbaum, Linda Silber
Blackwell, Floyd Oris
Boorman, Gary Alexis
Bunn, William Bernice, III
Carson, Bonnie L Bachert
Claxton, Larry Davis
Coffin, David L
Cole, Jerome Foster
Crawford-Brown, Douglas John
De Serres, Frederick Joseph
Dieter, Michael Phillip
Dunnick, June K
Eisenbud, Merril
Elder, Joe Allen
Fjellstedt, Thorsten A
Folinsbee, Lawrence John
Fouts, James Ralph
Fraser, David Allison
Gardner, Donald Eugene
Garner, Reuben John
Ghanayem, Burhan I
Goyer, Robert Andrew
Hamilton, Pat Brooks
Hart, Larry Glen
Haseman, Joseph Kyd
Hass, James Ronald
Hatch, Gary Ephraim
Hersh, Solomon Philip
Hook, Gary Edward Raumati
Jameson, Charles William
Johnson, Franklin M
Kaplan, Berton Harris
Kodavanti, Urmila P
Korach, Kenneth Steven
McClellan, Roger Orville
McLachlan, John Alan
McRee, Donald Ikerd
Malindzak, George S, Jr
Marcus, Allan H
Melnick, Ronald L
Mennear, John Hartley
Nesnow, Stephen Charles
O'Neil, John J
Popp, James Alan
Pritchard, John B
Rabinowitz, James Robert
Rolofson, George Lawrence
Ross, Jeffrey Alan
Sannes, Philip Loren
Sassaman, Anne Phillips
Savitz, David Alan
Schonwalder, Christopher O
Senzel, Alan Joseph
Shiffman, Morris A
Sobsey, Mark David
Stenger, William J(ames), Sr
Stopford, Woodhall
Suk, William Alfred
Sumner, Darrell Dean
Waters, Michael Dee
Wiener, Russell Warren
Wright, James Francis

NORTH DAKOTA
Hein, David William

OHIO
Albert, Roy Ernest
Ayer, Howard Earle
Bingham, Eula
Bogar, Louis Charles
Buncher, Charles Ralph
Caplan, Paul E
Cody, Terence Edward
Cooper, Gary Pettus
Couri, Daniel
Crable, John Vincent
D'Ambrosio, Steven M
Dodd, Darol Ennis
Elmets, Craig Allan
Farrier, Noel John
Haxhiu, Musa A
Henschel, Austin
Hopps, Howard Carl
Knox, Francis Stratton, III
Loper, John C
Lovett, Joseph
Mukhtar, Hasan
Murthy, Raman Chittaram
Nebert, Daniel Walter
Niemeier, Richard William
Nixon, Charles William
Nygaard, Oddvar Frithjof
Oleinick, Nancy Landy
Pereira, Michael Alan
Rice, Eugene Ward
Robbins, Norman
Sabourin, Thomas Donald
Saltzman, Bernard Edwin
Schanbacher, Floyd Leon
Skelly, Michael Francis
Slonim, Arnold Robert
Smith, Carl Clinton
Stoner, Gary David
Suskind, Raymond Robert
Tabor, Marvin Wilson
Von Gierke, Henning Edgar
Weiss, Harold Samuel
Wong, Laurence Cheng Kong
Yeary, Roger A

Yoon, Jong Sik

OKLAHOMA
Branch, John Curtis
Coleman, Nancy Pees
Coleman, Ronald Leon
Melton, Carlton E, Jr
Miranda, Frank Joseph

OREGON
Dunnette, David Arthur
Elia, Victor John
Hemphill, Delbert Dean
Matarazzo, Joseph Dominic
Parrott, Marshall Ward
Spencer, Peter Simner

PENNSYLVANIA
Anderson, David Martin
Atkins, Patrick Riley
Chan, Ping-Kwong
Danchik, Richard S
Deckert, Fred W
Dinman, Bertram David
Dixit, Rakesh
Duryea, William R
Esmen, Nurtan A(lan)
Fletcher, Ronald D
Fogleman, Ralph William
Foxx, Richard Michael
Gottlieb, Karen Ann
Hager, Nathaniel Ellmaker, Jr
Hajjar, Nicolas Philippe
Kamon, Eliezer
Karol, Meryl Helene
Kennedy, Ann Randtke
Kleban, Morton H
Levine, Geoffrey
Mason, Thomas Joseph
Mattison, Donald Roger
Metry, Amir Alfi
Mount, Lloyd Gordon
Neufeld, Ronald David
Nobel, Joel J
Poss, Stanley M
Rothman, Alan Michael
Sagik, Bernard Phillip
Shank, Kenneth Eugene
Shapiro, Maurice A
Shoub, Earle Phelps
Sooky, Attila A(rpad)
Urbach, Frederick
Weiss, William

RHODE ISLAND
Donnelly, Grace Marie
Hutzler, Leroy, III
Rossi, Joseph Stephen
Shaikh, Zahir Ahmad
Worthen, Leonard Robert

SOUTH CAROLINA
Darnell, W(illiam) H(eaden)
Muska, Carl Frank
Oswald, Edward Odell
Patterson, C(leo) Maurice

SOUTH DAKOTA
Schlenker, Evelyn Heymann

TENNESSEE
Barton, Charles Julian, Sr
Close, David Matzen
Dudney, Charles Sherman
Dumont, James Nicholas
Gehrs, Carl William
Guerin, Michael Richard
Iglar, Albert Francis, Jr
McDuffie, Bruce
Marchok, Ann Catherine
Morgan, Monroe Talton, Sr
Schultz, Terry Wayne
Uziel, Mayo
Vo-Dinh, Tuan
Watson, Annetta Paule

TEXAS
Albrecht, Thomas Blair
Bailey, Everett Murl, Jr
Barnes, Charles M
Biles, Robert Wayne
Campbell, Carlos Boyd Godfrey
Campbell, Dan Norvell
Davis, Ernst Michael
Dodson, Ronald Franklin
Domask, W(illiam) G(erhard)
Ellis, Kenneth Joseph
Fox, Donald A
Garcia, Hector D
Gothelf, Bernard
Gruber, George J
Guentzel, M Neal
Harbordt, C(harles) Michael
Heffley, James D
Hill, Reba Michels
Holguin, Alfonso Hudson
Jauchem, James Robert
Journeay, Glen Eugene
Kusnetz, Howard L
Kutzman, Raymond Stanley
Ledbetter, Joe O(verton)
Lengel, Robert Charles
Loftin, Karin Christiane
Lowry, William Thomas

Nachtwey, David Stuart
Nechay, Bohdan Roman
Pier, Stanley Morton
Powell, Michael Robert
Randerath, Kurt
Smolensky, Michael Hale
Taylor, Robert Dalton
Thompson, Richard John
Ulrich, Roger Steffen
Ward, Jonathan Bishop, Jr
Williams, Calvit Herndon, Jr
Wolf, Harold William
Yaden, Senka Long

UTAH
Archer, Victor Eugene
Bennett, Jesse Harland
Gortatowski, Melvin Jerome
Melton, Arthur Richard
Moser, Royce, Jr
Schiager, Keith Jerome
Wrenn, McDonald Edward

VERMONT
Falk, Leslie Alan
Hussak, Robert Edward
Keenan, Robert Gregory
Kneip, Theodore Joseph
Linnell, Robert Hartley
McIntosh, Alan William

VIRGINIA
Atrakchi, Aisar Hasan
Blanke, Robert Vernon
Boardman, Gregory Dale
Brown, Elise Ann Brandenburger
Donohue, Joyce Morrissey
Furr, Aaron Keith
Gould, David Huntington
Gregory, Arthur Robert
Grunder, Fred Irwin
Hill, Jim T
Jennings, Allen Lee
Lee, James A
McCormick-Ray, M Geraldine
McNerney, James Murtha
Mahboubi, Ezzat
Martin, James Henry, III
Melton, Thomas Mason
Meyer, Alvin F, Jr
Morton, John Dudley
Rodricks, Joseph Victor
Rosenbaum, David Mark
Ross, Malcolm
Sahli, Brenda Payne
Stroup, Cynthia Roxane
Tardiff, Robert G
Wands, Ralph Clinton
Wasti, Khizar
Woo, Yin-tak

WASHINGTON
Ammann, Harriet Maria
Benditt, Earl Philip
Bhagat, Surinder Kumar
Boatman, Edwin S
Johnstone, Donald Lee
Koenig, Jane Quinn
Lovely, Richard Herbert
Luchtel, Daniel Lee
Reid, Donald House
Sanders, Charles Leonard, Jr
Smith, Victor Herbert
Toohey, Richard Edward
Valanis, Barbara Mayleas
Wedlick, Harold Lee
Wilson, John Thomas, Jr
Young, Francis Allan

WEST VIRGINIA
Bouquot, Jerry Elmer
McCall, Robert G
Olenchock, Stephen Anthony
Ong, Tong-man
Sheikh, Kazim

WISCONSIN
Arthur, Robert M
Gasiorkiewicz, Eugene Constantine
Hake, Carl (Louis)
Johnson, Barry Lee
Kitzke, Eugene David
Krezoski, John R
Marx, James John, Jr
Raff, Hershel
Saryan, Leon Aram
Siegesmund, Kenneth A
Smith, Michael James
Trainer, Daniel Olney

ALBERTA
Coppock, Robert Walter
Man, Shu-Fan Paul

BRITISH COLUMBIA
Banister, Eric Wilton
Bates, David Vincent
Davison, Allan John
Law, Francis C P
Petkau, Albert John

MANITOBA
Pip, Eva
Sheppard, Marsha Isabell

Weeks, John Leonard

NOVA SCOTIA
Chatt, Amares

ONTARIO
Basu, Prasanta Kumar
Bend, John Richard
Bertell, Rosalie
Bioleau, Luc J R
Burton, Ian
Chettle, David Robert
Cummins, Joseph E
Hickman, John Roy
Hopton, Frederick James
Lees, Ronald Edward
Liu, Dickson Lee Shen
Morris, Larry Arthur
Robertson, James McDonald
Roy, Marie Lessard
Sanderson, H Preston
Somers, Emmanuel
Subramanian, Kunnath Sundara
Uchida, Irene Ayako
Warith, Mostafa Abdel
Warren, Douglas Robson
Williams, Norman S W

PRINCE EDWARD ISLAND
Singh, Amreek

QUEBEC
Becklake, Margaret Rigsby
Eddy, Nelson Wallace
Fournier, Michel
Maruvada, P Sarma
Page, Michel
Siemiatycki, Jack
Theriault, Gilles P

SASKATCHEWAN
Blakley, Barry Raymond

OTHER COUNTRIES
Aoyama, Isao
Back, Kenneth Charles
Bar-Yishay, Ephraim
Bernstein, David Maier
Brydon, James Emerson
Emmett, Edward Anthony
Goldsmith, John Rothchild
Greene, Velvl William
Haschke, Ferdinand
Hilbert, Morton S
Hoadley, Alfred Warner
Jokl, Miloslav Vladimir
Lowrance, William Wilson, Jr
Radford, Edward Parish
Shiraki, Keizo
Shuval, Hillel Isaiah
Stam, Jos
Strehlow, Clifford David
Thiessen, Jacob Willem
Valencia, Mauro Eduardo
Van Loveren, Henk
Willeke, Klaus

Health Physics

CALIFORNIA
Brooks, George Austin
Brown, Stephen L(awrence)
McCall, Richard C
Shrivastava, Prakash Narayan

GEORGIA
Shonka, Joseph John

ILLINOIS
Holtzman, Richard Beves

IOWA
Breuer, George Michael

KANSAS
Neuberger, John Stephen

LOUISIANA
Datta, Ratna

MARYLAND
Buchanan, John Donald
Hodge, Frederick Allen
Walker, William J, Jr
Yaniv, Shlomo Stefan

MASSACHUSETTS
Baratta, Edmond John

MICHIGAN
Zamorano, Lucia Jopehina

MINNESOTA
Roessler, Charles Ervin

MISSOURI
Styron, Clarence Edward, Jr

NEW MEXICO
Guilmette, Raymond Alfred

NEW YORK
Christman, Edward Arthur

Cohen, Norman
Meinhold, Charles Boyd

OHIO
Olson, Carl Thomas

OREGON
Parrott, Marshall Ward

PENNSYLVANIA
Laws, Priscilla Watson
Wald, Niel

SOUTH CAROLINA
Mitchell, Hugh Bertron
Parkinson, Thomas Franklin

TENNESSEE
Berven, Barry Allen
Darden, Edgar Bascomb
Gibbs, Samuel Julian

TEXAS
Murphy, Paul Henry
Perry, Robert Riley

VIRGINIA
Landa, Edward Robert

WASHINGTON
Kathren, Ronald Laurence
Toohey, Richard Edward

WISCONSIN
Piacsek, Bela Emery

ONTARIO
Rogers, David William Oliver

QUEBEC
Podgorsak, Ervin B

Hospital Administration

ALABAMA
Clark, Charles Edward
Roe, James Maurice, Jr
Smith, John Stephen

ARIZONA
Burkholder, Peter M
Schneller, Eugene S

ARKANSAS
Bates, Stephen Roger
Lloyd, Harris Horton
Towbin, Eugene Jonas

CALIFORNIA
Affeldt, John E
Bixler, Otto C
Bodfish, Ralph E
Cretin, Shan
Hansen, Howard Edward
Haughton, James Gray
Hudgins, Arthur J
Makowka, Leonard
Merchant, Roland Samuel, Sr
Mills, Gary Kenith
Nolan, Janiece Simmons
Overstreet, James Wilkins
Sams, Bruce Jones, Jr
Scitovsky, Anne A
Urbach, Karl Frederic
Weinstein, Irwin M

COLORADO
Bonn, Ethel May
Jafek, Bruce William

CONNECTICUT
Bondy, Philip K
Foster, James Henry
Klimek, Joseph John
Wetstone, Howard J

DISTRICT OF COLUMBIA
Shine, Kenneth I
Stark, Nathan Julius

FLORIDA
Brightwell, Dennis Richard
Kessler, Richard Howard
Rosenkrantz, Jacob Alvin
Terenzio, Joseph V

GEORGIA
Clark, Winston Craig

ILLINOIS
Best, William Robert
Hand, Roger
Kinsinger, Jack Burl
Korn, Roy Joseph
Paynter, Camen Russell
Pine, Michael B
Prabhudesai, Mukund M
Schyve, Paul Milton
Singer, Irwin
Tourlentes, Thomas Theodore

INDIANA
Evans, Michael Allen

KANSAS
Coles, Anna Bailey
Simpson, William Stewart

LOUISIANA
Grace, Marcellus
Threefoot, Sam Abraham

MARYLAND
Abramson, Fredric David
Chernow, Bart
Goldman, Stephen Shepard
Grimes, Carolyn E
Hausch, H George
Klarman, Herbert E
Wilson, Maureen O

MASSACHUSETTS
Buchanan, J Robert
Czeisler, Charles Andrew
Gentry, John Tilmon
Grossman, Jerome H
Lederman, David Mordechai
Lipinski, Boguslaw
Litsky, Bertha Yanis
Winkelman, James W

MICHIGAN
Dauphinais, Raymond Joseph
Griffith, John Randall
Kachhal, Swatantra Kumar
Rosenberg, Jerry C
Sahney, Vinod K
Sorscher, Alan J

MISSOURI
Hoover, Loretta White
Whitten, Elmer Hammond

NEW HAMPSHIRE
Kaiser, C William

NEW JERSEY
Brubaker, Paul Eugene, II
Dunikoski, Leonard Karol, Jr
Light, Donald W
McDonnell, William Vincent
Royce, Paul C

NEW MEXICO
Omer, George Elbert, Jr

NEW YORK
Cohen, Harris L
Davis, Carolyne K
English, Joseph T
Famighetti, Louise Orto
Fischel, Edward Elliot
Frosch, William Arthur
Gaintner, John Richard
Gerstman, Hubert Louis
Gibofsky, Allan
Joshi, Sharad Gopal
Kaslick, Ralph Sidney
Lichtig, Leo Kenneth
Mattson, Marlin Roy Albin
Reichgott, Michael Joel
Riehle, Robert Arthur, Jr
Rosner, Fred
Sidel, Victor William
Spaulding, Stephen Waasa
Spear, Paul William
Wolin, Harold Leonard
Yolles, Stanley Fausst

NORTH CAROLINA
McMahon, John Alexander
Meyer, George Wilbur
Weil, Thomas P
Wilson, Ruby Leila

OHIO
Altose, Murray D
Anderson, Ruth Mapes
Chen, Wen Yuan
Crespo, Jorge H
Morrow, Grant, III
Neuhauser, Duncan B
Rogers, Richard C
Schubert, William K
Vogel, Thomas Timothy

OKLAHOMA
Donahue, Hayden Hackney
Lynn, Thomas Neil, Jr
Thomas, Ellidee Dotson
Vanhoutte, Jean Jacques
Whitcomb, Walter Henry

OREGON
Clark, William Melvin, Jr

PENNSYLVANIA
Arce, A Anthony
Bondi, Amedeo
Deitrick, John E
Gershon, Samuel
Gray, Frieda Gersh
Jungkind, Donald Lee
Sovie, Margaret D
Yu, Lucy Cha

SOUTH CAROLINA
Johnson, James Allen, Jr

TENNESSEE
Christman, Luther P
Ross, Joseph C
Skoutakis, Vasilios A

TEXAS
Aunc, Janet
Burdine, John Alton
Mendel, Julius Louis
Mitra, Sunanda
Myers, Paul Walter

UTAH
Lee, Jeffrey Stephen

VIRGINIA
Adams, Nancy R
Chaudhuri, Tapan K
Levine, Jules Ivan
Maturi, Vincent Francis

WASHINGTON
Smith, Victor Herbert
Verson, Scott

WEST VIRGINIA
Victor, Leonard Baker

WISCONSIN
Treffert, Darold Allen

BRITISH COLUMBIA
Garg, Arun K

ONTARIO
Finlay, Joseph Bryan
Hansebout, Robert Roger
Rosenberg, Gilbert Mortimer

QUEBEC
Cantor, Ena D
Contandriopoulos, Andre-Pierre
Joly, Louis Philippe
Palayew, Max Jacob

OTHER COUNTRIES
Ichikawa, Shuichi

Medical Devices & Medical Diagnostics

ARIZONA
Denny, John Leighton
Dorson, William John

CALIFORNIA
Dafforn, Geoffrey Alan
Delemos, Robert A
Gray, Joe William
Hopf, Harriet Dudey Williams
Jewett, Don L
Jutamulia, Suganda
Kuwahara, Steven Sadao
Martz, Harry Edward, Jr
Phelps, Michael Edward
Singh, Jai Prakash
Sokolich, William Gary
Taft, David Dakin
Trippe, Anthony Philip
Vreman, Hendrik Jan
Wunderman, Irwin

CONNECTICUT
Rafferty, Terence Damien
Schetky, Laurence McDonald
Wackers, Frans J Th

DELAWARE
Craig, Alan Daniel

FLORIDA
Callaghan, Frank J
Coulter, Wallace H
Nagel, Joachim Hans

GEORGIA
Brockbank, Kelvin Gordon Mashader

IDAHO
Beck, Sidney M

ILLINOIS
Bender, James Gregg
Breillatt, Julian Paul, Jr
Kapsalis, Andreas A
Layton, Terry North
Martin, Ronald Lavern
Northup, Sharon Joan
Sternberg, Shmuel Mookie
Stromberg, LaWayne Roland
Wilber, David James
Young, Jay Maitland

INDIANA
Bonderman, Dean P
Free, Alfred Henry
Gaur, Pramod Kumar
Gunter, Claude Ray
Pepper, Daniel Allen

IOWA
Naides, Stanley J
Watkins, G(ordon) Leonard

Medical Devices & Medical Diagnostics (cont)

KANSAS
Sobczynski, Dorota
Sobczynski, Radek

LOUISIANA
Parent, Richard Alfred

MARYLAND
Bruck, Stephen Desiderius
Calton, Gary Jim
Colburn, Theodore R
Dillon, James Greenslade
Elespuru, Rosalie K
Estrin, Norman Frederick
Fisher, Dale John
Horowitz, Emanuel
Nossal, Ralph J
Severns, Matthew Lincoln
Sloan, Michael Allan
Tsitlik, Joshua E
Van De Merwe, Willem Pieter
Waxdal, M J

MASSACHUSETTS
Ackerman, Jerome Leonard
Atala, Anthony
Baratz, Robert Sears
Dewey, C(larence) Forbes, Jr
He, Bin
Krasner, Jerome L
Merfeld, Daniel Michael
Metz, Kenneth
Mix, Thomas W
Olderman, Jerry
Webster, Edward William

MICHIGAN
Kruse, James Alexander

MINNESOTA
Chesney, Charles Frederic
Daniel, John Mahendra Kumar
Tihon, Claude

MISSOURI
Das, Manik Lal
Shoemaker, James Daniel
Siegel, Barry Alan

NEW JERSEY
Aberman, Harold Mark
Li, John Kong-Jiann
Lunn, Anthony Crowther
Nichols, Joseph
Rosenstraus, Maurice Jay
Williams, Bernard Leo

NEW MEXICO
Tripp, R(ussell) Maurice

NEW YORK
Brennan, John Paul
Broadwin, Alan
Case, Robert B
Cohen, Harris L
Jayme, David Woodward
Lane, Lewis Behr
Lepp, Cyrus Andrew
McCormack, Patricia Marie
Quan, Weilun
Smith, Lowell Scott
Taub, Aaron M
Zonereich, Samuel

NORTH CAROLINA
Caruolo, Edward Vitangelo
Childers, Robert Warren
Eberhardt, Allen Craig
Haynes, John Lenneis
Kern, Frank Howard
Krassowska, Wanda
Price, Howard Charles

OHIO
Anderson, James Morley
Friedlander, Miriam Alice
Hangartner, Thomas Niklaus
Hughes, Kenneth E(ugene)
Levy, Richard Philip
Phillips, Chandler Allen
Spokane, Robert Bruce

OKLAHOMA
Altmiller, Dale Henry
Campbell, Gregory Alan
Johnson, Gerald, III
Veltri, Robert William

OREGON
Drummond, James Edgar
Keltner, Llew
Skiens, William Eugene

PENNSYLVANIA
Liu, Andrew Tze-chiu

RHODE ISLAND
Hopkins, Robert West

SOUTH CAROLINA
Leman, Robert B
Stewart, William C
Von Recum, Andreas F

TENNESSEE
Fordham, J(ames) Lynn
Kabalka, George Walter
Ogg, Robert James
Steen, R Grant
Vnencak-Jones, Cindy Lenore

TEXAS
Aspelin, Gary B
Bettinger, George E
Ferguson, James Joseph, IV
Gerding, Thomas G
Hyman, William A
Loubser, Paul Gerhard
Patel, Anil S

VIRGINIA
Fox, Kenneth Richard
Vaughan, Christopher Leonard

BRITISH COLUMBIA
Eaves, Allen Charles Edward
Harrop, Ronald
Kubik, Philip Roman

OTHER COUNTRIES
Dinnar, Uri
Elad, David
Ling, Chung-Mei
Ros, Herman H

Medical Physics

ARIZONA
Baldwin, Ann Linda
Begovac, Paul C

CALIFORNIA
Berry, Michael James
Forte, John Gaetano
Haymond, Herman Ralph
Litt, Lawrence
Shrivastava, Prakash Narayan

COLORADO
Adams, Gail Dayton

FLORIDA
Milne, Edward Lawrence
Nagel, Joachim Hans

ILLINOIS
Beck, Robert Nason
Doi, Kunio
Metz, Charles Edgar
Spokas, John J
Stinchcomb, Thomas Glenn

KENTUCKY
Christensen, Ralph C(hresten)
Fingar, Victor H
Simmons, Guy Held, Jr

LOUISIANA
Datta, Ratna

MARYLAND
Tsitlik, Joshua E
Walker, William J, Jr
Zink, Sandra

MASSACHUSETTS
Bjarngard, Bengt E
Itzkan, Irving
Rabin, Monroe Stephen Zane
Webster, Edward William

MINNESOTA
Gray, Joel Edward

NEW YORK
Anbar, Michael
Goodwin, Paul Newcomb
Koutcher, Jason Arthur
Leith, ArDean
Yip, Kwok Leung

NORTH CAROLINA
DeGuzman, Allan Francis
Tsui, Benjamin Ming Wah

OHIO
Brown, Robert William
Iwamoto, Harriet S
MacIntyre, William James

OKLAHOMA
Anderson, David Walter

PENNSYLVANIA
Sashin, Donald
Waterman, Frank Melvin

SOUTH CAROLINA
Mitchell, Hugh Bertron

TENNESSEE
Gibbs, Samuel Julian

Gronemeyer, Suzanne Alsop

TEXAS
Bencomo, Jose A, Jr
Murphy, Paul Henry
Perry, Robert Riley

VIRGINIA
Agarwal, Suresh Kumar
Wist, Abund Ottokar

WASHINGTON
Bichsel, Hans
Sikov, Melvin Richard

WEST VIRGINIA
Douglass, Kenneth Harmon

WISCONSIN
DeWerd, Larry Albert
Mackie, Thomas Rockwell
Siedband, Melvin Paul
Sorenson, James Alfred
Weindruch, Richard

BRITISH COLUMBIA
Lam, Gabriel Kit Ying

ONTARIO
Ji, Guangda Winston
Rainbow, Andrew James
Rogers, David William Oliver

QUEBEC
Pla, Conrado
Podgorsak, Ervin B
Schreiner, Lawrence John

Medical Sciences, General

ALABAMA
Anderson, Peter Glennie
Borton, Thomas Ernest
Dacheux, Ramon F, II
Elson, Charles O, III
Fuller, Gerald M
Gay, Renate Erika
Gay, Steffen
Jeffcoat, Marjorie K
Lloyd, L Keith
McCombs, Candace Cragen
Montgomery, John R
Roozen, Kenneth James
Schafer, James Arthur
Taylor, Kenneth Boivin
Turco, Jenifer
Wilborn, Walter Harrison

ALASKA
DeLapp, Tina Davis

ARIZONA
Birkby, Walter H
Blouin, Leonard Thomas
Bowen, Theodore
Dereniak, Eustace Leonard
Hendrix, Mary J C
Kazal, Louis Anthony
Koldovsky, Otakar
Kurth, Janice H
Lebowitz, Michael David
Manciet, Lorraine Hanna
Morgan, Wayne Joseph
Peric-Knowlton, Wlatka

ARKANSAS
Chang, Louis Wai-Wah
Epstein, Joshua
Pasley, James Neville
Schwarz, Anton

CALIFORNIA
Adair, Suzanne Frank
Adams, Thomas Edwards
Aly, Raza
Andrews, Brian S
Antoine, John Eugene
Arcadi, John Albert
Ascher, Michael S
Bailey, David Nelson
Bainton, Cedric R
Berk, Arnold J
Bertolami, Charles Nicholas
Bibel, Robin Jan
Bissell, Dwight Montgomery
Bloomfield, Harold H
Bolaffi, Janice Lerner
Braemer, Allen C
Brown, Warren Shelburne, Jr
Brunell, Philip Alfred
Budny, John Arnold
Buehring, Gertrude Case
Butz, Robert Frederick
Carlson, Don Marvin
Chen, Benjamin P P
Choudary, Prabhakara Velagapudi
Chuong, Cheng-Ming
Clewe, Thomas Hailey
Cohen, Morris
Cooper, Howard K
Cosman, Bard Clifford
Dasgupta, Asim
David, Gary Samuel

Delemos, Robert A
Demetrescu, M
Dobson, R Lowry
Donnelly, Timothy C
Dorfman, Leslie Joseph
Downing, Michael Richard
Epstein, John Howard
Erickson, Jeanne Marie
Fallon, James Harry
Felch, William C
Fischell, Tim Alexander
Flacke, Werner Ernst
Francke, Uta
Fuhrman, Frederick Alexander
Fukuda, Minoru
Garratty, George
Gehlsen, Kurt Ronald
Geiselman, Paula J
Geokas, Michael C
Giorgi, Janis V(incent)
Gong, William C
Goodpasture, Jessie Carrol
Gottlieb, Michael Stuart
Gray, Joe William
Grody, Wayne William
Gupta, Rishab Kumar
Harrison, Michael R
Hayes, Thomas L
Hayflick, Leonard
Haynes, Alfred
Hecht, Elizabeth Anne
Heiner, Douglas C
Heuser, Gunnar
Hoth, Daniel F
Huang, Sung-Cheng
Hughes, Walter William, III
Ingram, Marylou
Insel, Paul Anthony
Jackson, Craig Merton
Kang, Chang-Yuil
Kempter, Charles Prentiss
Kiprov, Dobri D
Kloner, Robert A
Knapp, Theodore Martin
Krauss, Ronald
Laiken, Nora Dawn
Largman, Corey
Lee, Paul L
Leffert, Hyam Lerner
Levine, Jon David
Lin, Ming-Fong
Lin, Robert I-San
Linker-Israeli, Mariana
Lipsick, Joseph Steven
Lopo, Alina C
Lu, John Kuew-Hsiung
MacFarlane, Malcolm David
Mack, Thomas McCulloch
Mann, Lewis Theodore, Jr
Mason, Dean Towle
Mestril, Ruben
Miyada, Don Shuso
Mooney, Larry Albert
Moos, Walter Hamilton
Moxley, John H, III
Najafi, Ahmad
Nelson, Gary Joe
Neville, James Ryan
Nicholson, Arnold Eugene
Ning, Shoucheng
Novack, Gary Dean
Pallavicini, Maria Georgina
Patton, John Stuart
Pearl, Ronald G
Peroutka, Stephen Joseph
Philpott, Delbert E
Pierschbacher, Michael Dean
Pitts, Robert Gary
Povzhitkov, Moysey Michael
Powanda, Michael Christopher
Powell, Frank Ludwig, Jr
Prusiner, Stanley Ben
Quismorio, Francisco P, Jr
Ram, Michael Jay
Reeve, Peter
Remington, Jack Samuel
Rho, Joon H
Rinderknecht, Heinrich
Rosenfeld, George
Rosenfeld, Michael G
Rousell, Ralph Henry
Rubenstein, Howard S
Ruggeri, Zaverio Marcello
Saha, Subrata
Sams, Bruce Jones, Jr
Schienle, Jan Hoops
Schiller, Neal Leander
Schmid, Peter
Scott, W(illiam) Richard
Seegmiller, Jarvis Edwin
Seppi, Edward Joseph
Shortliffe, Edward Hance
Shvartz, Esar
Slater, Grant Gay
Spiller, Gene Alan
Strautz, Robert Lee
Sweet, Benjamin Hersh
Taylor, Anna Newman
Tewari, Sujata Lahiri
Thompson, John Frederick
Vehar, Gordon Allen
Victery, Winona Whitwell
Vreman, Hendrik Jan
Wagner, Jeames Arthur

Wangler, Roger Dean
Weinreb, Robert Neal
Weinstein, Irwin M
Westfahl, Pamela Kay
Williams, Julian Carroll
Winet, Howard
Withers, Hubert Rodney
Wood, Corinne Shear
Yamini, Sohrab
Yildiz, Alaettin
Yu, David Tak Yan
Zuckerkandl, Emile

COLORADO
Allen, Larry Milton
Dobersen, Michael J
Ewing, Dean Edgar
Fan, Leland Lane
Hager, Jean Carol
Nelson, Bernard W
Powell, Robert Lee
Tosch, William Conrad
Urban, Richard William

CONNECTICUT
Agulian, Samuel Kevork
Ballantyne, Bryan
Baltimore, Robert Samuel
Bondy, Philip K
Borgia, Julian Frank
Boron, Walter Frank
Cohen, Donald J
Conrad, Eugene Anthony
Dooley, Joseph Francis
Heyd, Allen
Hlavacek, Robert Allen
Khan, Abdul Jamil
Kidd, Kenneth Kay
Letts, Lindsay Gordon
Loose, Leland David
McIlhenny, Hugh M
Manner, Richard John
Matovcik, Lisa M
Milne, George McLean, Jr
Nadel, Ethan Richard
O'Looney, Patricia Anne
O'Neill, James F
Pinson, Ellis Rex, Jr
Rafferty, Terence Damien
Rockwell, Sara Campbell
Ryan, Richard Patrick
Schaaf, Thomas Ken
Schafer, David Edward
Schmeer, Arline Catherine
Stadnicki, Stanley Walter, Jr
Stuart, James Davies
Taylor, Kenneth J W
Waxman, Stephen George
Zahler, Raphael
Zaret, Barry L

DELAWARE
Dautlick, Joseph X
Eleuterio, Marianne Kingsbury
Ginnard, Charles Raymond

DISTRICT OF COLUMBIA
Allen, Robert Erwin
Archer, Juanita Almetta
Diggs, John W
Harmon, John W
Hassan, Aftab Syed
Ison-Franklin, Eleanor Lutia
Jose, Pedro A
Kant, Gloria Jean
Kettel, Louis John
MacDonald, Mhairi Graham
Miller, Linda Jean
Rao, Mamidanna S
Rapisardi, Salvatore C
Read, Merrill Stafford
Sarin, Prem S

FLORIDA
Abitbol, Carolyn Larkins
Baker, Carleton Harold
Ben, Max
Berlin, Nathaniel Isaac
Bliznakov, Emile George
Callaghan, Frank J
Cameron, Don Frank
Carter, James Harrison, II
DeRousseau, C(arol) Jean
Drummond, Willa H
Emerson, Geraldine Mariellen
Fackler, Martin L
Farah, Alfred Emil
Fisher, Waldo Reynolds
Goihman-Yahr, Mauricio
Hampp, Edward Gottlieb
Hurst, Josephine M
Joftes, David Lion
Kline, Jacob
Lyman, Gary Herbert
McMillan, Donald Ernest
Maue-Dickson, Wilma
Murchison, Thomas Edgar
Noble, Nancy Lee
Nonoyama, Meihan
Perez, Guido O
Phillips, Michael Ian
Pratt, Melanie M
Ramsey, James Marvin
Reininger, Edward Joseph

Ringel, Samuel Morris
Roberts, Thomas L
Salas, Pedro Jose I
Simring, Marvin
Soliman, Karam Farag Attia
Spielberger, Charles Donald
Stevens, Bruce Russell
Storrs, Eleanor Emerett
Stroud, Robert Malone
Sypert, George Walter
Thornthwaite, Jerry T
Tsibris, John-Constantine Michael
Wilcox, Christopher Stuart
Williams, Joseph Francis
Wissow, Lennard Jay

GEORGIA
Adkison, Linda Russell
Binnicker, Pamela Caroline
Brown, Hugh Keith
Colborn, Gene Louis
Dailey, Harry A
Herman, Chester Joseph
Hicks, Heraline Elaine
Innes, David Lyn
Jones, Robert Leroy
Kaiser, Robert L
Klein, Luella
Kolbeck, Ralph Carl
Krauss, Jonathan Seth
Satcher, David
Thacker, Stephen Brady
Warren, McWilson
Warren, Stephen Theodore
Woodford, James

HAWAII
Marchette, Nyven John
Yang, Hong-Yi

IDAHO
Hillyard, Ira William

ILLINOIS
Albrecht, Ronald Frank
Anderson, Kenning M
Aramini, Jan Marie
Belford, R Linn
Biller, Jose
Bone, Roger C
Brody, Jacob A
Buckman, Robert W
Cabana, Veneracion Garganta
Chan, Yun Lai
Christensen, Mary Lucas
Cohen, Gloria
Doege, Theodore Charles
Ecanow, Bernard
Ernest, J Terry
Fang, Victor Shengkuen
Ferguson, James L
Friedman, Yochanan
Fuchs, Elaine V
Gaik, Geraldine Catherine
Glisson, Silas Nease
Goebel, Maristella
Gross, Thomas Lester
Hast, Malcolm Howard
Hjelle, Joseph Thomas
Hua, Ping
Johari, Om
Jones, Peter Hadley
Kang, David Soosang
Kass, Leon Richard
Kim, Yoon Berm
Kim, Yung Dai
Kirschner, Robert H
Leikin, Jerrold Blair
Lin, James Chih-I
Loeb, Jerod M
Lopez-Majano, Vincent
McCorquodale, Maureen Marie
Manyam, Bala Venktesha
Moticka, Edward James
Needleman, Saul Ben
Nevalainen, David Eric
Opgenorth, Terry John
Ostrow, David Henry
Plate, Janet Margaret
Pope, Richard M
Potempa, Lawrence Albert
Rao, Mrinalini Chatta
Reft, Chester Stanley
Rosen, Arthur Leonard
Smith, Douglas Calvin
Snyder, Ann Knabb
Stevens, Joseph Alfred
Tobin, Martin John
Tomita, Joseph Tsuneki
Yelich, Michael Ralph

INDIANA
Biggs, Homer Gates
Broxmeyer, Hal Edward
Donner, David Bruce
Eisenstein, Barry I
Goh, Edward Hua Seng
Gray, Peter Norman
Haak, Richard Arlen
Macchia, Donald Dean
McNeill, Kenny Earl
Paladino, Frank Vincent
Park, Kinam
Svoboda, Gordon H

Tacker, Willis Arnold, Jr
Vasko, Michael Richard
Voelz, Michael H
Yoder, John Menly

IOWA
Bell, Edward Frank
Conn, P Michael
Fellows, Robert Ellis, Jr
Hoffman, Eric Alfred
Koontz, Frank P
Maruyama, George Masao
Montgomery, Rex
Schrott, Helmut Gunther
Siebes, Maria
Stratton, Donald Brendan
Thompson, Robert Gary
Widness, John Andrew

KANSAS
Davis, John Stewart
Gattone, Vincent H, II
Goetz, Kenneth Lee
Norris, Patricia Ann
Oehme, Frederick Wolfgang
Singhal, Ram P
Stevens, David Robert
Tomita, Tatsuo

KENTUCKY
Cohn, David Valor
Harris, Patrick Donald
Kasarskis, Edward Joseph
Porter, William Hudson
Rees, Earl Douglas
Straus, Robert
Tseng, Michael Tsung
Urano, Muneyasu
Williams, Walter Michael

LOUISIANA
Bagby, Gregory John
Bazan, Nicolas Guillermo
Chen, Hoffman Hor-Fu
Dessauer, Herbert Clay
Frohlich, Edward David
Messerli, Franz Hannes
Newsome, David Anthony
Peacock, Milton O
Samuels, Arthur Seymour
Shiau, Yih-Fu
Spitzer, Judy A
Xie, Jianming

MAINE
Kent, Barbara
Norton, James Michael
Weiss, John Jay

MARYLAND
Abramson, I Jerome
Adams, Junius Greene, III
Adelstein, Robert Simon
Baldwin, Wendy Harmer
Balinsky, Doris
Balow, James E
Baquet, Claudia
Baxter, James Hubert
Benevento, Louis Anthony
Bergey, Gregory Kent
Berman, Howard Mitchell
Berson, Alan
Biswas, Robin Michael
Blackman, Marc Roy
Blaustein, Mordecai P
Boyd, Virginia Ann Lewis
Bricker, Jerome Gough
Bunger, Rolf
Calabi, Ornella
Carlson, Drew E
Carski, Theodore Robert
Catt, Kevin John
Chang, Henry
Chaudhari, Anshumali
Chen, Joseph Ke-Chou
Chernow, Bart
Chirigos, Michael Anthony
Cushman, Samuel Wright
Darby, Eleanor Muriel Kapp
Daza, Carlos Hernan
Deuster, Patricia Anne
Dewys, William Dale
Estrin, Norman Frederick
Fahy, Gregory Michael
Fakunding, John Leonard
Feigal, Ellen G
Feinstone, Stephen Mark
Fertziger, Allen Philip
Fex, Jorgen
Fischell, Robert E
Fitch, Kenneth Leonard
Flaherty, John Thomas
Foster, Willis Roy
Fredrickson, Donald Sharp
Freeman, Julia Berg
Froehlich, Luz
Glenn, John Frazier
Goldstein, David Stanley
Goldwater, William Henry
Gonda, Matthew Allen
Gorden, Phillip
Gordon, Stephen L
Green, Jerome George
Greenough, William B, III

Gruber, Kenneth Allen
Handwerger, Barry S
Haseltine, Florence Pat
Hearn, Henry James, Jr
Heyse, Stephen P
Holmes, Kathryn Voelker
Horan, Michael
Irwin, David
Johnston, Gerald Samuel
Kakefuda, Tsusyoshi
Kendrick, Francis Joseph
Kety, Seymour S
Kinnard, Matthew Anderson
Klombers, Norman
Kneller, Robert William
Koch, Thomas Richard
Kramer, Barnett Sheldon
Kresina, Thomas Francis
Lawrence, Dale Nolan
Lee-Ham, Doo Young
Lenfant, Claude J M
LeRoith, Derek
Levine, Alan Stewart
Levine, Richard Joseph
Linzer, Melvin
Lipicky, Raymond John
Litterst, Charles Lawrence
Lobel, Steven A
Longo, Dan L
Lymangrover, John R
MacVittie, Thomas Joseph
Magrath, Ian
Manganiello, Vincent Charles
Marban, Eduardo
Marsden, Halsey M
Maslow, David E
Massof, Robert W
Matheson, Nina W
Maumenee, Irene Hussels
Meryman, Harold Thayer
Miller, Charles A
Miller, Frederick William
Minners, Howard Alyn
Mishler, John Milton
Moshell, Alan N
Mouradian, M Maral
Murphy, Donald G
Nelson, James Harold
Parkman, Paul Douglas
Paskins-Hurlburt, Andrea Jeanne
Pearson, Jay Dee
Porter, Roger J
Price, Thomas R
Rawlings, Samuel Craig
Rhim, Johng Sik
Robbins, Jacob
Robbins, Jay Howard
Rovelstad, Gordon Henry
Sack, George H(enry), Jr
Salcman, Michael
Savage, Peter
Schechter, Alan Neil
Shevach, Ethan Menahem
Shih, Thomas Y
Simpson, Ian Alexander
Slife, Charles W
Sommer, Alfred
Spooner, Peter Michael
Sztein, Marcelo Benjamin
Talbot, Bernard
Talbot, John Mayo
Tankersley, Donald
Taube, Sheila Efron
Traub, Richard Kimberley
Valega, Thomas Michael
Wallace, Craig Kesting
Wolff, David A
Wortman, Bernard
Yanagihara, Richard

MASSACHUSETTS
Alston, Theodore A
Anderson, Rosalind Coogan
Aroesty, Julian Max
Axelrod, Lloyd
Beggs, Alan H, III
Castronovo, Frank Paul, Jr
Chiu, Tak-Ming
Davis, Michael Allan
Dayal, Yogeshwar
DeSanctis, Roman William
Driedger, Paul Edwin
Eastwood, Gregory Lindsay
Federman, Daniel D
Feld, Michael S
Fox, Irving Harvey
Fox, Maurice Sanford
Fox, Thomas Oren
Furie, Bruce
Gange, Richard William
Ganley, Oswald Harold
George, Harvey
Goldman, Harvey
Goldstein, Richard Neal
Grace, Norman David
Grossman, Jerome H
Haimes, Howard B
Hallock, Gilbert Vinton
Hawiger, Jack Jacek
Hilgar, Arthur Gilbert
Holick, Michael Francis
Hoop, Bernard, Jr
Jain, Rakesh Kumar
Jung, Lawrence Kwok Leung

MEDICAL & HEALTH SCIENCES / 447

Medical Sciences, General (cont)

Kaminskas, Edvardas
Khaw, Ban-An
Klempner, Mark Steven
Koul, Omanand
Lees, Robert S
Libby, Peter
Lloyd, Weldon S
Loscalzo, Joseph
Modest, Edward Julian
Moskowitz, Michael Arthur
Murray, Joseph E
Olsen, Bjorn Reino
Palubinskas, Felix Stanley
Pandian, Natesa G
Pfeffer, Janice Marie
Pinkus, Jack Leon
Pino, Richard M
Rigney, David Roth
Rosen, Fred Saul
Roy, Sayon
Schaub, Robert George
Schorr, Lisbeth Bamberger
Spragg, Jocelyn
Szlyk-Modrow, Patricia Carol
Thorp, James Wilson
Tomera, John F
Trinkaus-Randall, Vickery E
Vanderburg, Charles R
Villee, Dorothy Balzer
Waud, Douglas Russell
Weinberg, Crispin Bernard
Weinstein, Robert
Wurtman, Richard Jay
Zetter, Bruce Robert

MICHIGAN
Baumiller, Robert C
Berent, Stanley
Bouma, Hessel, III
Briggs, Josephine P
Burnett, Jean Bullard
Cerny, Joseph Charles
Cohen, Bennett J
Dore-Duffy, Paula
Ellis, C N
Glover, Roy Andrew
Grant, Rhoda
Gray, Robert Howard
Hashimoto, Ken
Heinrikson, Robert L
Kagan, Fred
Klurfeld, David Michael
Kruse, James Alexander
McCann, Daisy S
Mahajan, Sudesh K
Marcelo, Cynthia Luz
Medlin, Julie Anne Jones
Meites, Joseph
Mizukami, Hiroshi
Mohrland, J Scott
Parfitt, A(lwyn) M(ichael)
Poulik, Miroslav Dave
Sorscher, Alan J
Swovick, Melvin Joseph
Viano, David Charles
Voorhees, John James
Wei, Wei-Zen
Zak, Bennie

MINNESOTA
Adams, Ernest Clarence
Bach, Marilyn Lee
Brown, David Mitchell
Dousa, Thomas Patrick
Iaizzo, Paul Anthony
Johnson, Susan Bissen
Lambert, Edward Howard
Leon, Arthur Sol
Leung, Benjamin Shuet-Kin
Mikhail, Adel Ayad
O'Connor, Michael Kieran
Robb, Richard A
Sawchuk, Ronald John
Schwartz, Samuel
Severson, Arlen Raynold
Spilker, Bert
Tamsky, Morgan Jerome
Thompson, Edward William
Wood, Earl Howard

MISSISSIPPI
Rockhold, Robin William
Shirley, Aaron
Steinberg, Martin H

MISSOURI
Baumann, Thiema Wolf
Blim, Richard Don
Bretton, Randolph Henry, Jr
Covey, Douglas Floyd
Coyer, Philip Exton
Emami, Bahman
Gardner, Jerry David
Green, Eric Douglas
Karl, Michael M
Lin, Hsiu-san
Morley, John E
Munns, Theodore Willard
Quay, Wilbur Brooks
Rubin, Bruce Kalman

Sadler, J(asper) Evan
Schuck, James Michael
Sharma, Krishna
Shoemaker, James Daniel
Stephens, Robert Eric
Strauss, Arnold Wilbur
Taylor, John Joseph
Thomas, Lewis Jones, Jr
Tyagi, Suresh C
Weber, Morton M
Welply, Joseph Kevin

MONTANA
Dorward, David William

NEBRASKA
Dalley, Arthur Frederick, II
Gendelman, Howard Eliot
Johnson, John Arnold
Knoop, Floyd C
Stratta, Robert Joseph

NEVADA
Newmark, Stephen

NEW HAMPSHIRE
Brous, Don W
Edwards, Brian Ronald
Musiek, Frank Edward
Stibitz, George Robert
Swartz, Harold M

NEW JERSEY
Arcese, Paul Salvatore
Audet, Sarah Anne
Baker, Herman
Blumenthal, Rosalyn D
Bonner, R Alan
Borg, Donald Cecil
Boyle, Joseph, III
Chao, Yu-Sheng
Chimes, Daniel
Clayton, John Mark
Controulis, John
Curro, Frederick A
Desjardins, Raoul
Duran, Walter Nunez
Emele, Jane Frances
Firschein, Hilliard E
Flam, Eric
Ford, Daniel Morgan
Gately, Maurice Kent
Gorodetzky, Charles W
Green, Erika Ana
Grinberg-Funes, Ricardo A
Gullo, Vincent Philip
Jabbar, Gina Marie
Ji, Sungchul
Kamiyama, Mikio
Kapp, Robert Wesley, Jr
Kronenthal, Richard Leonard
Kumbaraci-Jones, Nuran Melek
Laemle, Lois K
Leeds, Morton W
Leung, Albert Yuk-Sing
Leventhal, Howard
Levy, Alan
Levy, David Edward
Levy, Martin J Linden
Littlewood, Peter B
McDowell, Wilbur Benedict
McGregor, Walter
McKinstry, Doris Naomi
Margolskee, Robert F
Massa, Tobias
Matthijssen, Charles
Moreland, Suzanne
Nakhla, Atif Mounir
Nichols, Joseph
Orlando, Carl
OSullivan, Joseph
Parsonnet, Victor
Peets, Edwin Arnold
Perhach, James Lawrence
Pestka, Sidney
Plotkin, Diane Joyce
Pobiner, Bonnie Fay
Ponzio, Nicholas Michael
Reuben, Roberta C
Ritchie, David Malcolm
Roberts, Ronald C
Rosenblum, Irwin Yale
Rushton, Alan R
Sheikh, Maqsood A
Shin, Seung-il
Sterbenz, Francis Joseph
Swislocki, Norbert Ira
Wasserman, Martin Allan
Watkins, Tom R
Weir, James Henry, III
Wilson, Alan C

NEW MEXICO
Cole, Edward Issac, Jr
Crawford, Michael Howard
Elias, Laurence
Reyes, Philip
Welford, Norman Traviss
Wild, Gaynor (Clarke)
Woodfin, Beulah Marie

NEW YORK
Althuis, Thomas Henry
Altman, Kurt Ison

Antzelevitch, Charles
April, Ernest W
Armstrong, Donald
Aronson, Ronald Stephen
Ascensao, Joao L
Aull, Felice
Badalamente, Marie Ann
Beal, Myron Clarence
Berger, Frank Milan
Bhattacharya, Jahar
Blau, Lawrence Martin
Borer, Jeffrey Stephen
Bottomley, Paul Arthur
Boutjdir, Mohamed
Brown, W Virgil
Burchfiel, James Lee
Bystryn, Jean-Claude
Case, Robert B
Cassell, Eric J
Chanana, Arjun Dev
Chauhan, Abha
Chess, Leonard
Chou, Ting-Chao
Coderre, Jeffrey Albert
Cohen, Harris L
Conard, Robert Allen
Cort, Susannah
Counts, David Francis
David, Oliver Joseph
Delmonte, Lilian
Devereux, Richard Blyton
DiMauro, Salvatore
Dole, Vincent Paul
Doty, Stephen Bruce
Edelheit, Lewis S
Eder, Howard Abram
Edwards, Adrian L
Egan, Edmund Alfred
Eisinger, Josef
Elsbach, Peter
Erickson, Wayne Francis
Factor, Stephen M
Famighetti, Louise Orto
Field, Michael
Finlay, Thomas Hiram
Finster, Mieczyslaw
Gambert, Steven Ross
Gardner, Daniel
Gelbard, Alan Stewart
Gerolimatos, Barbara
Gintautas, Jonas
Goldman, James Eliot
Golub, Lorne Malcolm
Gonzalez, Eulogio Raphael
Goodman, Jerome
Gordon, Harry William
Gotschlich, Emil C
Grimwood, Brian Gene
Ham, Richard John
Hatcher, Victor Bernard
Heller, Milton David
Hershey, Linda Ann
Isenberg, Henry David
Jaanus, Siret Desiree
Kaplan, John Ervin
Kappas, Attallah
Karr, James Presby
Kashin, Philip
Kaufman, Albert Irving
Kohler, Constance Anne
Kostreva, David Robert
Kream, Jacob
Lamster, Ira Barry
Lang, Charles H
Lebwohl, Mark Gabriel
Lee, Ching-Tse
Lieberman, Abraham N
Lin, Jane Huey-Chai
Lutton, John D
Maitra, Subir R
Matthews, Dwight Earl
Meyer, Harry Martin, Jr
Moor-Jankowski, J
Morin, John Edward
Morris, John Emory
Natta, Clayton Lyle
Oakes, David
O'Brien, Anne T
Ornt, Daniel Burrows
Ottenbacher, Kenneth John
Padawer, Jacques
Papsidero, Lawrence D
Pentney, Roberta Pierson
Pesce, Michael A
Pittman, Kenneth Arthur
Pogo, A Oscar
Pollack, Robert Elliot
Rattazzi, Mario Cristiano
Reichberg, Samuel Bringeissen
Reichert, Leo E, Jr
Reit, Barry
Rich, Marvin R
Rivlin, Richard S
Roboz, John
Rothman, Francoise
Ruddell, Alanna
Ruggiero, David A
Sachdev, Om Prakash
Schenck, John Frederic
Schwartz, Ernest
Sehgal, Pravinkumar B
Seon, Ben K
Siemann, Dietmar W
Sitarz, Anneliese Lotte

Spangler, Robert Alan
Spaulding, Stephen Waasa
Srivastava, Suresh Chandra
Stanley, Evan Richard
Stark, John Howard
Stolfi, Robert Louis
Stryker, Martin H
Szabo, Piroska Ludwig
Thornborough, John Randle
Toth, Eugene J
Triner, Lubos
Trust, Ronald I
Vacirca, Salvatore John
Vladutiu, Adrian O
Vladutiu, Georgirene Dietrich
Volkman, David J
Waksman, Byron Halstead
Wasserheit, Judith Nina
Watson, James Dewey
Weber, Julius
Weiner, Matei
Weitkamp, Lowell R
Wellner, Daniel
Wineburg, Elliot N
Wolfson, Leonard Louis
Wolin, Harold Leonard
Wollman, Leo
Yao, Alice C
Yeh, Samuel D J
Zonereich, Samuel
Zumoff, Barnett

NORTH CAROLINA
Anderegg, Robert James
Bentzel, Carl Johan
Bode, Arthur Palfrey
Caterson, Bruce
Chen, Michael Yu-Men
Dunnick, June K
Frisell, Wilhelm Richard
Hershfield, Michael Steven
Hicks, T Philip
Kucera, Louis Stephen
Laupus, William E
Lawson, Edward Earle
Lee, Martin Jerome
Loomis, Carson Robert
Louie, Dexter Stephen
McManus, Thomas (Joseph)
Misek, Bernard
Murdock, Harold Russell
Mustafa, Syed Jamal
Niedel, James Edward
Noe, Frances Elsie
Olsen, Elise Arline
Putman, Charles E
Reimer, Keith A
Reinhard, John Frederick, Jr
Sadler, Thomas William
Sassaman, Anne Phillips
Schiffman, Susan S
Swift, Michael
Thubrikar, Mano J
Tlsty, Thea Dorothy
Toole, James Francis

OHIO
Adragna, Norma C
Atkins, Charles Gilmore
Barnes, H Verdain
Blanton, Ronald Edward
Carothers, Donna June
Centner, Rosemary Louise
Clanton, Thomas L
Clifford, Donald H
Crafts, Roger Conant
Curry, John Joseph, III
Davis, Pamela Bowes
Dell'Osso, Louis Frank
Dirckx, John H
Dobelbower, Ralph Riddall, Jr
Ely, Daniel Lee
Fenoglio, Cecilia M
Francis, Marion David
Friedlander, Miriam Alice
Geha, Alexander Salim
Green, Ralph
Haas, Erwin
Harris, Richard Lee
Hinkle, George Henry
Holtkamp, Dorsey Emil
Homer, George Mohn
Jamasbi, Roudabeh J
Johnson, J David
Klein, LeRoy
Latimer, Bruce Millikin
Levy, Richard Philip
Lindower, John Oliver
Mannino, Joseph Robert
Nahata, Milap Chand
Preszler, Alan Melvin
Reddy, Narender Pabbathi
Resnick, Martin I
Riker, Donald Kay
Rodman, Harvey Meyer
Rosenberg, Howard C
Stuhlman, Robert August
Sutherland, James McKenzie
Watanakunakorn, Chatrchai
Westerman, Philip William

OKLAHOMA
Calvert, Jon Channing
Collins, William Edward

Corder, Clinton Nicholas
Cunningham, Madeleine White
Dubowski, Kurt M(ax)
Epstein, Robert B
Fine, Douglas P
King, Mary Margaret
Lee, Lela A
McKee, Patrick Allen
Melton, Carlton E, Jr
Muneer, Razia Sultana
Snow, Clyde Collins
Watson, Gary Hunter

OREGON
Andresen, Michael Christian
Bussman, John W
Fields, R(ance) Wayne
Keltner, Llew
Mela-Riker, Leena Marja
Prescott, Gerald H
Robinson, Arthur B
Severson, Herbert H
Swank, Roy Laver
Urba, Walter John
Zimmerman, Earl Abram

PENNSYLVANIA
Agarwal, Jai Bhagwah
Alhadeff, Jack Abraham
Alteveer, Robert Jan George
Baechler, Charles Albert
Baker, Lester
Bhaduri, Saumya
Biebuyck, Julien Francois
Bly, Chauncey Goodrich
Botelho, Stella Yates
Boxenbaum, Harold George
Cavanagh, Peter R
Chapman, John Donald
Cheng, Hung-Yuan
Cooper, Stephen Allen
Davis, Richard A
DeRubertis, Frederick R
Diamond, Leila
Duryea, William R
Ellis, Demetrius
Fabrizio, Angelina M
Flynn, John Thomas
Freeman, Joseph Theodore
Garwin, Jeffrey Lloyd
Glick, John H
Godfrey, John Carl
Goldberg, Martin
Henderson, George Richard
Hobby, Gladys Lounsbury
Iwatsuki, Shunzaburo
Johnson, Ernest Walter
Kissick, William Lee
Klein, Herbert A
Koelle, Winifred Angenent
Kowlessar, O Dhodanand
Ladda, Roger Louis
Laufer, Igor
Lentini, Eugene Alfred Anthony
Levine, Geoffrey
Liberti, Paul A
Lloyd, Thomas A
McKenna, William Gillies
Magargal, Larry Elliot
Meek, Edward Stanley
Miller, Barry
Mulvihill, John Joseph
Murer, Erik Homann
Okunewick, James Philip
Oldham, James Warren
Orenstein, David M
Perchonock, Carl David
Pirone, Louis Anthony
Poupard, James Arthur
Raikow, Radmila Boruvka
Revesz, George
Rottenberg, Hagai
Ryan, James Patrick
Simmons, Daryl Michael
Six, Howard R
Smith, John Bryan
Smith, Roger Bruce
Strauss, Jerome Frank, III
Strom, Brian Leslie
Vickers, Florence Foster
Weinbaum, George
Whinnery, James Elliott
Woo, Savio Lau-Yuen
Zagon, Ian Stuart

RHODE ISLAND
Carpenter, Charles C J
Christenson, Lisa
Galletti, Pierre Marie
Jackson, Ivor Michael David
O'Leary, Gerard Paul, Jr
Shank, Peter R
Vezeridis, Michael Panagiotis

SOUTH CAROLINA
Ainsworth, Sterling K
Beck, Ronald Richard
Gee, Adrian Philip
Goust, Jean Michel
Horger, Edgar Olin, III
Kaiser, Charles Frederick
Pittman, Fred Estes
Rathbun, Ted Allan
Yard, Allan Stanley

SOUTH DAKOTA
Martin, Douglas Stuart

TENNESSEE
Beaty, Orren, III
Bhattacharya, Syamal Kanti
Chandler, Robert Walter
Crawford, Lloyd V
Crofford, Oscar Bledsoe
Gillespie, Elizabeth
Glassman, Armand Barry
Goans, Ronald Earl
Harris, Thomas R(aymond)
Huggins, James Anthony
Hunt, James Calvin
Lynch, Denis Patrick
Nanney, Lillian Bradley
Proctor, Kenneth Gordon
Pui, Ching-Hon
Richmond, Chester Robert
Robertson, David
Skoutakis, Vasilios A
Stone, William Lawrence
Sundell, Hakan W
Wimalasena, Jay
Zemel, Michael Barry

TEXAS
Bessman, Joel David
Brown, Jack Harold Upton
Burton, Karen Poliner
Butler, Bruce David
Costanzi, John J
Cottone, James Anthony
Crass, Maurice Frederick, III
DeBakey, Lois
DeBakey, Selma
Deloach, John Rooker
Doody, Rachelle Smith
Dorn, Gordon Lee
Dudrick, Stanley John
Entman, Mark Lawrence
Franklin, Thomas Doyal, Jr
Fujitani, Roy Masami
Goldberg, Leonard H
Gregerman, Robert Isaac
Hecht, Gerald
Hewett-Emmett, David
Hudspeth, Albert James
Hudspeth, Emmett Leroy
Hussian, Richard A
Johnson, Raleigh Francis, Jr
Kerwin, Joseph Peter
Kit, Saul
Klein, Peter Douglas
Kuo, Lih
LaBree, Theodore Robert
Land, Geoffrey Allison
Lees, George Edward
Lehmkuhl, L Don
Levin, Harvey Steven
Liepa, George Uldis
Loftin, Karin Christiane
Mark, Lloyd K
Marshall, Gailen D, Jr
Marshburn, Paul Bartow
Mastromarino, Anthony John
Mundy, Gregory Robert
Nishioka, Kenji
Quarles, John Monroe
Riser, Mary Elizabeth
Rolfe, Rial Dewitt
Rosenfeld, Charles Richard
Sampson, Herschel Wayne
Sant'Ambrogio, Giuseppe
Shannon, Wilburn Allen, Jr
Shotwell, Thomas Knight
Skelley, Dean Sutherland
Sloan, Tod Burns
Taegtmeyer, Heinrich
Tilbury, Roy Sidney
Trentin, John Joseph
Waldbillig, Ronald Charles
Walker, Cheryl Lyn
Wilcox, Roberta Arlene
Wilson, L Britt
Zwaan, Johan Thomas

UTAH
Albertine, Kurt H
Border, Wayne Allen
Galinsky, Raymond Ethan
Olson, Randall J
Shelby, Nancy Jane
Westenfelder, Christof

VERMONT
Ellis, David M
Webb, George Dayton

VIRGINIA
Asterita, Mary Frances
Barrett, Richard John
Bempong, Maxwell Alexander
Chaudhuri, Tapan K
Cooper, John Allen Dicks
DeLorenzo, Robert John
DeSesso, John Michael
Diwan, Bhalchandra Apparao
Elford, Howard Lee
Gargus, James L
Grider, John Raymond
Lepper, Mark H
Loria, Roger Moshe

Malik, Vedpal Singh
Mandell, Gerald Lee
Martensen, Todd Martin
Nygard, Neal Richard
Seipel, John Howard
Stetka, Daniel George
Swain, David P
Thorner, Michael Oliver
Turner, Terry Tomo
Wei, Enoch Ping
Weiss, Daniel Leigh
Williams, Gerald Albert
Williams, Patricia Bell

WASHINGTON
Ammann, Harriet Maria
Artru, Alan Arthur
Bhansali, Praful V
Canfield, Robert Charles
Couser, William Griffith
Haviland, James West
Hildebrandt, Jacob
Ho, Lydia Su-yong
Huntsman, Lee L
Kenney, Gerald
Lakshminarayan, S
McCabe, Donald Lee
Maguire, Yu Ping
Mannik, Mart
Mendelson, Martin
Mitchell, Michael Ernst
Rosenblatt, Roger Alan
Salk, Darrell John
Schiffrin, Milton Julius
Spackman, Darrel H
Wehner, Alfred Peter
Whatmore, George Bernard

WEST VIRGINIA
Chang, William Wei-Lien
Pore, Robert Scott
Rhoten, William Blocher

WISCONSIN
Allred, Lawrence Ervin (Joe)
Bortin, Mortimer M
Buchanan-Davidson, Dorothy Jean
Crowe, Dennis Timothy, Jr
Daniels, Farrington, Jr
Engerman, Ronald Lester
Hinkes, Thomas Michael
Ladinsky, Judith L
Manning, James Harvey
Mazess, Richard B
Ndon, John A
Perry, Billy Wayne
Pruss, Thaddeus P
Raff, Hershel
Sieber, Fritz
Sothmann, Mark Steven
Stowe, David F
Sufit, Robert Louis
Uno, Hideo
Walker, Duard Lee
Warltier, David Charles
Woodson, Robert D

PUERTO RICO
Alcala, Jose Ramon
Banerjee, Dipak Kumar
Bangdiwala, Ishver Surchand
De La Sierra, Angell O
Ramirez-Ronda, Carlos Hector
Rodriguez, Jose Raul

ALBERTA
Hollenberg, Morley Donald
Hutchison, Kenneth James
Jones, Richard Lee
Koopmans, Henry Sjoerd
Leung, Alexander Kwok-Chu
Lorscheider, Fritz Louis
Roth, Sheldon H
Stell, William Kenyon
Yoon, Ji-Won

BRITISH COLUMBIA
Applegarth, Derek A
Bates, David Vincent
Friedman, Jan Marshall
Frohlich, Jiri J
Kraintz, Leon
Lederis, Karolis (Karl)
Milsom, William Kenneth
Olive, Peggy Louise
Perry, Thomas Lockwood
Skala, Josef Petr
Wong, Norman L M

MANITOBA
Bertalanffy, Felix D
Bowman, John Maxwell
Choy, Patrick C
Gerrard, Jonathan M
Glavin, Gary Bertrun
Polimeni, Philip Iniziato
Rosenmann, Edward A
Schacter, Brent Allan

NEWFOUNDLAND
Michalak, Thomas Ireneusz

NOVA SCOTIA
Armour, John Andrew

Carruthers, Samuel George
Dickson, Douglas Howard
Fraser, Albert Donald
Murray, T J
Tan, Meng Hee

ONTARIO
Baker, Michael Allen
Barr, Ronald Duncan
Basu, Prasanta Kumar
Biro, George P
Buick, Fred, Jr
Campbell, James Fulton
Capaldi, Dante James
Cheng, Hazel Pei-Ling
Doane, Frances Whitman
Evans, John R
Forstner, Janet Ferguson
Froese, Alison Barbara
Ghista, Dhanjoo Noshir
Girvin, John Patterson
Goldenberg, Gerald J
Hagen, Paul Beo
Harwood-Nash, Derek Clive
Inman, Robert Davies
Jeejeebhoy, Khursheed Nowrojee
Jones, Douglas L
Kako, Kyohei Joe
Kerbel, Robert Stephen
Kinlough-Rathbone, Roelene L
Kooh, Sang Whay
MacRae, Andrew Richard
Mazurkiewicz-Kwilecki, Irena Maria
Merskey, Harold
Mickle, Donald Alexander
Mishra, Ram K
Montemurro, Donald Gilbert
Osmond, Daniel Harcourt
Papp, Kim Alexander
Phillips, Robert Allan
Pross, Hugh Frederick
Renaud, Leo P
Rosenthal, Kenneth Lee
Schimmer, Bernard Paul
Somers, Emmanuel
Steiner, George
Szewczuk, Myron Ross
Tepperman, Barry Lorne
Tobin, Sidney Morris
Uchida, Irene Ayako
Vadas, Peter
Worton, Ronald Gilbert

QUEBEC
Aranda, Jacob Velasco
Belanger, Luc
Braun, Peter Eric
Chang, Thomas Ming Swi
Freeman, Carolyn Ruth
Frojmovic, Maurice Mony
Gagnon, Claude
Goltzman, David
Heisler, Seymour
Kearney, Robert Edward
Lussier, Andre (Joseph Alfred)
Nair, Vasavan N P
Outerbridge, John Stuart
Palayew, Max Jacob
Ponka, Premysl
Regoli, Domenico
Roy, Claude Charles
Sandor, Thomas
Scriver, Charles Robert
Skamene, Emil
Stephens, Heather R
Thomson, David M P
Zidulka, Arnold

SASKATCHEWAN
Kalra, Jawahar

OTHER COUNTRIES
Bar-Yishay, Ephraim
Binder, Bernd R
Bodmer, Walter Fred
Bojalil, Luis Felipe
Dinnar, Uri
Fitzgerald, John Desmond
Goldberg, Burton David
Gomez, Arturo
Gorski, Andrzej
Ha, Tai-You
Hashimoto, Paulo Hitonari
Hirokawa, Katsuiku
Hounsfield, G Newbold
Huang, Kun-Yen
Ikehara, Yukio
Izui, Shozo
Johnsen, Dennis O
Jonzon, Anders
Kawamura, Hiroshi
Kreider, Eunice S
Lu, Guo-Wei
Merlini, Giampaolo
Miyasaka, Kyoko
Modabber, Farrokh Z
Monos, Emil
Mori, Wataru
Prywes, Moshe
Querinjean, Pierre Joseph
Sasaki, Hidetada
Sheng, Hwai-Ping
Spiess, Joachim
Strous, Ger J

Medical Sciences, General (cont)

Takai, Yasuyuki
Takeuchi, Kiyoshi Hiro
Tanphaichitr, Vichai
Tung, Che-Se
Vane, John Robert
Van Furth, Ralph
Vogel, Carl-Wilhelm E
Wu, Albert M
Yamada, Eichi

Medicine

ALABAMA
Alford, Charles Aaron, Jr
Allen, Lois Brenda
Allison, Ronald C
Bailey, William C
Ball, Gene V
Bargeron, Lionel Malcolm, Jr
Batson, Oscar Randolph
Baxley, William Allison
Benton, John William, Jr
Bishop, Sanford Parsons
Blackburn, Will R
Boshell, Buris Raye
Bradley, Laurence A
Brascho, Donn Joseph
Brogdon, Byron Gilliam
Bryan, Thornton Embry
Butterworth, Charles E, Jr
Carlo, Waldemar Alberto
Cassady, George
Cassell, Gail Houston
Caulfield, James Benjamin
Ceballos, Ricardo
Cheraskin, Emanuel
Conrad, Marcel E
Cooper, Max Dale
Culpepper, Roy M
Daniel, William A, Jr
Diasio, Robert Bart
Durant, John Ridgway
Dyken, Paul Richard
Eddleman, Elvia Etheridge, Jr
Evanochko, William Thomas
Finklea, John F
Flowers, Charles E, Jr
Fraser, Robert Gordon
Furner, Raymond Lynn
Gardner, William Albert, Jr
Garrett, J Marshall
Gay, Renate Erika
Gay, Steffen
Geer, Jack Charles
Gleysteen, John Jacob
Griffin, Frank M, Jr
Hageman, Gilbert Robert
Halsey, James H, Jr
Hefner, Lloyd Lee
Hill, Samuel Richardson, Jr
Hirschowitz, Basil Isaac
Holley, Howard Lamar
Hughes, Edwin R
Lecklitner, Myron Lynn
Lee, Daisy Si
Levene, Ralph Zalman
Lobuglio, Albert Francis
Logic, Joseph Richard
Lorincz, Andrew Endre
Lupton, Charles Hamilton, Jr
McCann, William Peter
McMahon, John Martin
Meredith, Ruby Frances
Michalski, Joseph Potter
Montgomery, John R
Moreno, Hernan
Myers, Ira Lee
Oberman, Albert
Oh, Shin Joong
Oparil, Suzanne
Osment, Lamar Sutton
Palmisano, Paul Anthony
Pass, Robert Floyd
Peterson, Raymond Dale August
Philips, Joseph Bond, III
Pittman, James Allen, Jr
Reque, Paul Gerhard
Roth, Robert Earl
Russell, Richard Olney, Jr
Sams, Wiley Mitchell, Jr
Schnaper, Harold Warren
Segrest, Jere Palmer
Sheehy, Thomas W
Sheffield, L Thomas
Shin, Myung Soo
Shingleton, Hugh Maurice
Siegal, Gene Philip
Simpson-Herren, Linda
Skalka, Harold Walter
Stagno, Sergio Bruno
Stover, Samuel Landis
Tamura, Tsunenobu
Tiller, Ralph Earl
Waite, Peter Daniel
Winternitz, William Welch

ALASKA
Mills, William J, Jr

ARIZONA
Adler, Charles H
Aikawa, Jerry Kazuo
Bernhard, Victor Montwid
Bowden, George Timothy
Boyse, Edward A
Brandenburg, Robert O
Brosin, Henry Walter
Burrows, Benjamin
Butler, Byron C
Capp, Michael Paul
Cherny, Walter B
Daly, David DeRouen
Denny, William F
Evans, Tommy Nicholas
Fagan, Timothy Charles
Fischer, Harry William
Fisher, William Gary
Garewal, Harinder Singh
Gordon, Richard Seymour
Heine, Melvin Wayne
Hersh, Evan Manuel
Holloway, G Allen, Jr
Hudgings, Daniel W
Huestis, Douglas William
Hull, Hugh Boden
Iserson, Kenneth Victor
Johnson, Sture Archie Mansfield
Joy, Vincent Anthony
Katz, Murray Alan
Kay, Marguerite M B
Kelly, John V
Klotz, Arthur Paul
Koelsche, Giles Alexander
Lewis, Alan Ervin
Marcus, Frank I
Morgan, Wayne Joseph
Nugent, Charles Arter, Jr
Ogden, David Anderson
Ordway, Nelson Kneeland
Orient, Jane Michel
Palumbo, Pasquale John
Parsons, William Belle, Jr
Patton, Dennis David
Pettit, George Robert
Pinnas, Jacob Louis
Quan, Stuart F
Quinlivan, William Leslie G
Rifkind, David
Sandberg, Avery Aba
Sibley, William Austin
Smith, Josef Riley
Snyder, Richard Gerald
Stephens, Charles Arthur Lloyd, Jr
Tatelman, Maurice
Weil, Andrew Thomas
Werner, Sidney Charles
Yakaitis, Ronald William
Zauder, Howard L

ARKANSAS
Abernathy, Robert Shields
Bard, David S
Barnhard, Howard Jerome
Bates, Joseph H
Bates, Stephen Roger
Beard, Owen Wayne
Berry, Daisilee H
Borden, Craig W
Diner, Wilma Canada
Doherty, James Edward, III
Dungan, William Thompson
Ebert, Richard Vincent
Flanigan, Stevenson
Flanigan, William J
Haut, Arthur
Hawks, Byron Lovejoy
Hiller, Frederick Charles
Jasin, Hugo E
Lester, Roger
Martin, Richard Harvey
Sanders, Louis Lee
Stead, William White
Texter, E Clinton, Jr
Towbin, Eugene Jonas

CALIFORNIA
Abildgaard, Charles Frederick
Adams, Forrest Hood
Adams, Robert Monroe
Agras, William Stewart
Agre, Karl
Agress, Clarence M
Ainsworth, Cameron
Allen, Joseph Garrott
Alpert, Louis Katz
Amador, Elias
Ament, Marvin Earl
Amstutz, Harlan Cabot
Anger, Hal Oscar
Archambeau, John Orin
Archibald, Kenneth C
Arieff, Allen I
Arnold, Harry L(oren), Jr
Ash, Lawrence Robert
Ashburn, William Lee
Austin, Donald Franklin
Babior, Bernard M
Bacchus, Habeeb
Bagshaw, Malcolm A
Bainton, Dorothy Ford
Balchum, Oscar Joseph
Barnard, R James
Barnes, Asa, Jr
Barnett, Eugene Victor
Barrett, Peter Van Doren
Barron, Charles Irwin
Basham, Teresa
Baum, David
Baxter, John Darling
Beall, Gildon Noel
Beard, Rodney Rau
Beck, John Christian
Behrens, Herbert Charles
Behrman, Richard Elliot
Beigelman, Paul Maurice
Beljan, John Richard
Bennett, Cleaves M
Bennett, Leslie R
Berek, Jonathan S
Berk, Jack Edward
Berk, Robert Norton
Berkowitz, Carol Diane
Bernstein, Gerald Sanford
Bernstein, Leslie
Bernstein, Sol
Berryman, George Hugh
Bessman, Alice Neuman
Bessman, Samuel Paul
Bettman, Jerome Wolf
Beutler, Ernest
Bierman, Howard Richard
Biglieri, Edward George
Bikle, Daniel David
Billingham, John
Billings, Charles Edgar, Jr
Bing, Richard John
Bissell, Dwight Montgomery
Black, Robert L
Blahd, William Henry
Blankenhorn, David Henry
Blantz, Roland C
Block, Jerome Bernard
Bluestein, Harry Gilbert
Bodfish, Ralph E
Bolt, Robert James
Bookstein, Joseph Jacob
Bordin, Gerald M
Bozdech, Marek Jiri
Bragonier, John Robert
Brandt, Ira Kive
Brasel, Jo Anne
Brecher, George
Breslow, Lester
Bricker, Neal S
Bristow, Lonnie Robert
Brook, Robert H
Brown, David Edward
Brown, Ellen
Brown, J Martin
Brown, James K
Brown, Stuart Irwin
Brunell, Philip Alfred
Buchsbaum, Monte Stuart
Burnham, Thomas K
Byfield, John Eric
Cady, Lee (De), Jr
Canham, John Edward
Carbone, John Vito
Carlsson, Erik
Carmel, Ralph
Carregal, Enrique Jose Alvarez
Carson, Dennis A
Carter, Kenneth
Carter, Paul Richard
Castles, James Joseph, Jr
Catlin, Don Hardt
Catz, Boris
Cavalieri, Ralph R
Chapman, Sharon K
Charles, M Arthur
Chase, Robert A
Cheitlin, Melvin Donald
Cherry, James Donald
Chisari, Francis Vincent
Chopra, Inder Jit
Chow, John Lap Hong
Church, Joseph August
Cleaver, James Edward
Clewe, Thomas Hailey
Coburn, Jack Wesley
Cohen, Ellis N
Cohn, Major Lloyd
Collen, Morris F
Collins, Robert C
Colman, Martin
Connor, James D
Coodley, Eugene Leon
Cooper, Howard K
Cooper, Maurice Zealot
Cooper, William Clark
Costin, Anatol
Couperus, Molleurus
Covel, Mitchel Dale
Cox, John William
Crandall, Edward D
Crandall, Paul Herbert
Crawford, Donald W
Crede, Robert H
Crocker, Thomas Timothy
Cropp, Gerd J A
Cunningham, Mary Elizabeth
Curnutte, John Tolliver, III
Curry, Cynthia Jane
Curry, Francis J
Cutler, Robert W P
Dallman, Peter R
Davidson, Ezra C, Jr
Dawson, Chandler R
Deedwania, Prakash Chandra
Delemos, Robert A
Deller, John Joseph
Detels, Roger
Diamond, Ivan
Diamond, Louis Klein
Dietrich, Shelby Lee
Dignam, William Joseph
Dobbins, John Potter
Dobbs, Leland George
Doe, Richard P
Donnell, George Nino
Dorfman, Leslie Joseph
Doroshow, James Halpern
Ducsay, Charles Andrew
Dunham, Wolcott Balestier
Durfee, Raphael B
Eakins, Kenneth E
Ebbin, Allan J
Edwards, Charles C
Eichenholz, Alfred
Eilber, Frederick Richard
Eisenberg, Ronald Lee
Ellestad, Myrvin H
Eltherington, Lorne
Emmanouilides, George Christos
Engel, William King
Engleman, Ephraim Philip
Epstein, Charles Joseph
Epstein, Ervin Harold
Epstein, John Howard
Epstein, Lois Barth
Epstein, Wallace Victor
Epstein, William L
Espana, Carlos
Estrada, Norma Ruth
Evans, Harrison Silas
Ezrin, Calvin
Fainstat, Theodore
Fairley, Henry Barrie Fleming
Falcone, A B
Falkner, Frank Tardrew
Falsetti, Herman Leo
Fanestil, Darrell Dean
Farber, Eugene M
Farber, Seymour Morgan
Farquhar, John William
Farr, Richard Studley
Favour, Cutting Broad
Feeney-Burns, Mary Lynette
Felton, Jean Spencer
Figueroa, William Gutierrez
Finegold, Sydney Martin
Fischell, Tim Alexander
Fisher, Delbert A
Fisher, June Marion
Fishman, Robert Allen
Fishman, William Harold
Flacke, Joan Wareham
Foos, Robert Young
Forsham, Peter Hugh
Fortes, George (Peter Alexander)
Fowler, William Mayo, Jr
Foye, Laurance V, Jr
Frasier, S Douglas
Freeman, Gustave
Freeman, Leon David
Frenster, John H
Frick, Oscar L
Friedenberg, Richard M
Friedman, Gary David
Friedman, Paul J
Friedman, William Foster
Friedmann, Theodore
Friou, George Jacob
Fujimoto, Atsuko Ono
Fukuyama, Kimie
Furthmayr, Heinz
Gabor, Andrew John
Gardin, Julius M
Gardner, Murray Briggs
Garoutte, Bill Charles
Gentile, Dominick E
George, Kattunilathu Oommen
Giammona, Samuel T
Gibbs, Gordon Everett
Gibson, Count Dillon, Jr
Gigli, Irma
Gildenhorn, Hyman L
Gill, Gordon Nelson
Giorgi, Elsie A
Giorgio, Anthony Joseph
Gitnick, Gary L
Glaser, Robert Joy
Glass, Laurel Ellen
Glorig, Aram
Glovsky, M Michael
Gluck, Louis
Glushien, Arthur Samuel
Gofman, John William
Golbus, Mitchell S
Goldberg, Mark Arthur
Golde, David William
Goldfien, Alan
Goldsmith, Ralph Samuel
Goldstein, Abraham M B
Goltz, Robert W
Gonick, Harvey C
Goodman, Michael Gordon
Gordan, Gilbert Saul
Gottfried, Eugene Leslie
Gottlieb, Abraham Mitchell
Goulian, Mehran

MEDICAL & HEALTH SCIENCES / 451

Govan, Duncan Eben
Gray, Gary M
Gross, Ruth T
Grover, Robert Frederic
Grumbach, Melvin Malcolm
Gubersky, Victor R
Guenther, Donna Marie
Gulyassy, Paul Francis
Guth, Paul Henry
Gwinup, Grant
Hackney, Jack Dean
Haddad, Zack H
Hahn, Bevra H
Haig, Pierre Vahe
Hall, Michael Oakley
Hall, Thomas Livingston
Hallett, Wilbur Y
Halsted, Charles H
Hamburger, Robert Newfield
Hamilton, William Kennon
Hammond, George Denman
Hanafee, William Norman
Hansen, Howard Edward
Hansen, James E
Hanson, Virgil
Harris, Edward Day, Jr
Harvey, Birt
Hasterlik, Robert Joseph
Havel, Richard Joseph
Hays, Daniel Mauger
Hays, Esther Fincher
Haywood, L Julian
Heiner, Douglas C
Heinrichs, W LeRoy
Henry, James Paget
Henzl, Milan Rastislav
Hepler, Opal Elsie
Hepler, Robert S
Herber, Raymond
Herrmann, Christian, Jr
Hershman, Jerome Marshall
Heuser, Eva T
Hewitt, William Lane
Heyman, Melvin Bernard
Heymann, Michael Alexander
Hill, Edward C
Hirst, Albert Edmund
Hodgman, Joan Elizabeth
Hoeprich, Paul Daniel
Hoffman, Julien Ivor Ellis
Hofmann, Alan Frederick
Hohn, Arno R
Holaday, William J
Holland, Paul Vincent
Hollenberg, Milton
Holliday, Malcolm A
Hollingsworth, James W
Holman, Halsted Reid
Hon, Edward Harry Gee
Honrubia, Vicente
Hopkins, Gordon Bruce
Horton, Richard
Horwitz, David Larry
Horwitz, Marcus Aaron
Hostetler, Karl Yoder
House, Leland Richmond
Howell, Doris Ahlee
Hoyt, William F
Hryniuk, William
Hubbard, Richard W
Hughes, Everett C
Hughes, James Paul
Hultgren, Herbert Nils
Irie, Reiko Furuse
Irwin, Charles Edwin, Jr
Iseri, Lloyd T
Jaffe, Robert B
Jain, Naresh C
Jampolsky, Arthur
Jarmakani, Jay M
Jawetz, Ernest
Jelinek, Bohdan
Jelliffe, Derrick Brian
Jelliffe, Roger Woodham
Jensen, Gordon D
Johns, Varner Jay, Jr
Johnson, B Lamar, Jr
Johnson, Cage Saul
Jones, Malcolm David
Jones, Oliver William
Jordan, Stanley Clark
Kagan, Benjamin M
Kambara, George Kiyoshi
Kan, Yuet Wai
Kane, John Power
Kaplan, Samuel
Kaplan, Selna L
Kaplan, Solomon Alexander
Kashyap, Moti Lal
Katz, Barrett
Katz, David Harvey
Katz, Ronald Lewis
Katzman, Robert
Kaufman, Herbert S
Kaufman, Joseph J
Keane, J R
Kedes, Laurence H
Keens, Thomas George
Keller, Margaret Anne
Kelln, Elmer
Khwaja, Tasneem Afzal
Kilburn, Kaye Hatch
Kim, Young Shik
Kirschbaum, Thomas H

Klaustermeyer, William Berner
Klement, Vaclav
Klinenberg, James Robert
Koch, Richard
Kohlstaedt, Kenneth George
Kolb, Felix Oscar
Korc, Murray
Korenman, Stanley G
Kornbluth, Richard Syd
Kornreich, Helen Kass
Korpman, Ralph Andrew
Korsch, Barbara M
Kramer, Henry Herman
Krasno, Louis Richard
Kretchmer, Norman
Krevans, Julius Richard
Kriss, Joseph P
Krupp, Marcus Abraham
Kumagai, Lindy Fumio
Kurnick, John Edmund
Kurnick, Nathaniel Bertrand
Kurohara, Samuel S
Landau, Joseph White
Langdon, Edward Allen
Langer, Glenn A
Larkin, Edward Charles
Lashof, Joyce Cohen
Lasser, Elliott Charles
Lau, Francis You King
Lawrence, David M
Lazarus, Gerald Sylvan
Leake, Donald L
Lee, Garrett
Lee, Peter Van Arsdale
Leedom, John Milton
Leighton, Joseph
Lele, Padmakar Pratap
Leone, Lucile P
Lepeschkin, Eugene
Levin, Jack
Levin, Seymour R
Levine, Jon David
Levy, Ronald
Lewis, Charles E
Lewis, Jerry Parker
Lewis, Richard John
Lewiston, Norman James
Lieberman, Jack
Lillington, Glen Alan
Lin, Tz-Hong
Linde, Leonard M
Linfoot, John Ardis
Lingappa, Jaisri Rao
Linna, Timo Juhani
Lippa, Erik Alexander
Lippa, Linda Susan Mottow
Llaurado, Josep G
Longhurst, John Charles
Lowe, Harry J
Lowenstein, Jerold Marvin
Lusted, Lee Browning
McAllister, Robert Milton
McCarron, Margaret Mary
McDonnel, Gerald M
Mace, John Weldon
McGerity, Joseph Loehr
Mack, Thomas McCulloch
MacKenzie, Malcolm R
Maffly, Roy Herrick
Magnuson, Harold Joseph
Maibach, Howard I
Makker, Sudesh Paul
Mann, Leslie Bernard
Manning, Phil Richard
Margulis, Alexander Rafailo
Markell, Edward Kingsmill
Markham, Charles Henry
Maronde, Robert Francis
Marshall, John Romney
Mason, Dean Towle
Masouredis, Serafeim Panogiotis
Massie, Barry Michael
Mathews, Kenneth Pine
Mathies, Allen Wray, Jr
Matsumura, Kenneth N
Mellinkoff, Sherman Mussoff
Menkes, John H
Merigan, Thomas Charles, Jr
Merrill, James Allen
Metherell, Alexander Franz
Meyer, Paul
Meyer, Richard David
Michaels, David D
Milby, Thomas Hutchinson
Miles, James S
Miller, James Nathaniel
Miller, John Johnston, III
Miller, Stephen Herschel
Mills, Don Harper
Milne, Eric Campbell
Mitchell, Harold Hugh
Mitchell, Malcolm Stuart
Mitchell, Robert Dalton
Mock, David Clinton, Jr
Mohler, John George
Montgomery, Theodore Ashton
Morgan, Beverly Carver
Morton, Donald Lee
Moser, Kenneth Miles
Mosier, H David, Jr
Mosley, James W
Muller, George Heinz
Murray, John Frederic
Mustacchi, Piero Omar de Zamalek

Myers, Bryan D
Nagy, Stephen Mears, Jr
Nathwani, Bharat N
Nayak, Debi Prosad
Nelson, Thomas Lothian
Nerlich, William Edward
Newton, Thomas Hans
Niden, Albert H
Niebauer, John J
Noonan, Charles D
Norman, David
Norris, Forbes Holten, Jr
Nyhan, William Leo
O'Brien, John S
O'Connor, Daniel Thomas
O'Connor, George Richard
Oldendorf, William Henry
Olsen, Clarence Wilmott
Oppenheimer, Steven Bernard
Orr, Beatrice Yewer
Ovenfors, Carl-Olof Nils Sten
Owen, Robert L
Packman, Seymour
Palubinskas, Alphonse J
Parer, Julian Thomas
Parker, Donal C
Parker, Robert G
Parmley, William W
Paul, Jerome Thomas
Payne, Rose Marise
Pearce, Morton Lee
Pearl, Ronald G
Pedersen, Roger Arnold
Perkins, Herbert Asa
Peterlin, Boris Matija
Peterson, Charles Marquis
Peterson, Donald I
Peterson, John Eric
Petrakis, Nicholas Louis
Pettit, Thomas Henry
Pevehouse, Byron C
Phibbs, Roderic H
Philippart, Michel Paul
Phillips, Theodore Locke
Piel, Carolyn F
Pilgrim, Hyman Ira
Pischel, Ken Donald
Pitesky, Isadore
Pitkin, Roy Macbeth
Place, Virgil Alan
Platt, Lawrence David
Polley, Margaret J
Powanda, Michael Christopher
Powell, Clinton Cobb
Power, Gordon G
Price, David C
Prince, David Allan
Quismorio, Francisco P, Jr
Rahimtoola, Shahbudin Hooseinally
Ramachandran, Vilayanur Subramanian
Ranney, Helen M
Rapaport, Samuel I
Raskin, Neil Hugh
Raventos, Antolin, IV
Rector, Floyd Clinton, Jr
Redeker, Allan Grant
Rees, Rees Bynon
Reilly, Emmett B
Reisner, Ronald M
Reitz, Richard Elmer
Remington, Jack Samuel
Resch, Joseph Anthony
Reynolds, Telfer Barkley
Richman, Douglas Daniel
Rider, Joseph Alfred
Riemenschneider, Paul Arthur
Robbins, Dick Lamson
Robinson, William Sidney
Roemer, Milton Irwin
Rosen, Peter
Rosenberg, Saul Allen
Rosenman, Ray Harold
Ross, Gordon
Ross, John, Jr
Rubanyi, Gabor Michael
Rubenstein, Edward
Rubenstein, Howard S
Rudolph, Abraham Morris
Ruggeri, Zaverio Marcello
Rytand, David A
Sagan, Leonard A
Saha, Pamela S
Salk, Jonas Edward
Salkin, David
Samloff, I Michael
Sandberg, Eugene Carl
Sapico, Francisco L
Sartoris, David John
Schacher, John Fredrick
Schaefer, George
Scheie, Harold Glendon
Scherger, Joseph E
Schlegel, Robert John
Schmid, Rudi
Schmidt, Robert Milton
Schneider, Edward Lewis
Schneiderman, Lawrence J
Schoenfield, Leslie Jack
Schrier, Stanley Leonard
Schroeder, John Speer
Schulman, Irving
Schulz, Jeanette
Schwabe, Arthur David
Schwartz, Herbert C

Schwartz, William Benjamin
Seaman, William E
Sebastian, Anthony
Seeger, Robert Charles
Seidel, James Stephen
Selzer, Arthur
Severinghaus, John Wendell
Shames, David Marshall
Sharma, Arjun D
Shearn, Martin Alvin
Sheline, Glenn Elmer
Shinefield, Henry R
Shipp, Joseph Calvin
Shnider, Sol M
Shore, Nomie Abraham
Shortliffe, Edward Hance
Shulman, Ira Andrew
Siegel, Lawrence Sheldon
Siegel, Michael Elliot
Siemsen, Jan Karl
Simmons, Daniel Harold
Simon, Allan Lester
Simon, Harold J
Siperstein, Marvin David
Skoog, William Arthur
Slater, James Munro
Sleisenger, Marvin Herbert
Smith, Lloyd Hollingworth, Jr
Smith, Norman Ty
Smithwick, Elizabeth Mary
Snape, William J, Jr
Soeldner, John Stuart
Sokolow, Maurice
Soll, Andrew H
Solomon, David Harris
Sparkes, Robert Stanley
Spector, Samuel
Spector, Sheldon Laurence
Spencer, E Martin
Spitler, Lynn E
Stamey, Thomas Alexander
Staple, Tom Weinberg
Steen, Stephen N
Stein, Myron
Steinfeld, Jesse Leonard
Stern, Robert
Sterner, James Hervi
Stiehm, E Richard
Stockdale, Frank Edward
Stone, Daniel Boxall
Stoughton, Richard Baker
Straatsma, Bradley Ralph
Strauss, Bernard
Stringfellow, Dale Alan
Strober, Samuel
Swan, Harold James Charles
Sweetman, Lawrence
Swerdloff, Ronald S
Swezey, Robert Leonard
Tager, Ira Bruce
Takasugi, Mitsuo
Talley, Robert Boyd
Tanaka, Kouichi Robert
Tarver, Harold
Tatter, Dorothy
Thaler, M Michael
Thomas, Heriberto Victor
Thompson, William Benbow, Jr
Thrupp, Lauri David
Tierney, Donald Frank
Tobis, Jerome Sanford
Tooley, William Henry
Topping, Norman Hawkins
Torrance, Daniel J
Tourtellotte, Wallace William
Towers, Bernard
Tranquada, Robert Ernest
Trelford, John D
Triche, Timothy J
Troy, Frederic Arthur
Tubis, Manuel
Tune, Bruce Malcolm
Tupper, Charles John
Turner, John Dean
Tyan, Marvin L
Udall, John Alfred
Urbach, Karl Frederic
Valentine, William Newton
Van Den Noort, Stanley
Van Der Meulen, Joseph Pierre
Van Dop, Cornelis
Varki, Ajit Pothan
Varon, Myron Izak
Vaughan, John Heath
Vaughan, Victor Clarence, III
Vermund, Halvor
Vohs, James A
Von Leden, Hans Victor
Vredevoe, Lawrence A
Wallerstein, Ralph Oliver, Jr
Walter, Richard D
Waltz, Arthur G
Ward, Paul H
Warner, Willis L
Waskell, Lucy A
Wasserman, Karlman
Wasserman, Stephen I
Wat, Bo Ying
Watts, Malcolm S M
Waxman, Alan David
Way, Walter
Webber, Milo M
Wehrle, Paul F
Weichsel, Morton E, Jr

Medicine (cont)

Weiner, Herbert
Weinreb, Robert Neal
Weinstein, Irwin M
Weisman, Harold
Werbach, Melvyn Roy
Werner, Sanford Benson
West, John B
Weston, Raymond E
Wheeler, Henry Orson
Whittenberger, James Laverre
Whorton, M Donald
Wiederholt, Wigbert C
Wilbur, David Wesley
Wilbur, Dwight Locke
Williams, James Hutchison
Williams, Lawrence Ernest
Wilson, Charles B
Wilson, Miriam Geisendorfer
Winget, Charles M
Winkelstein, Warren, Jr
Wishner, Kathleen L
Witschi, Hanspeter R
Wolfe, Allan Marvin
Wong-Staal, Flossie
Woodruff, John H, Jr
Wright, Harry Tucker, Jr
Wu, Sing-Yung
Yamini, Sohrab
Young, Lionel Wesley
Young, Robert, Jr
Young, Robert Rice
Yow, Martha Dukes
Zatz, Leslie M
Zboralske, F Frank
Zeldis, Jerome B
Zeleznick, Lowell D
Zerez, Charles Raymond
Ziegler, Michael Gregory
Zimmerman, Emery Gilroy
Zvaifler, Nathan J

COLORADO

Aldrich, Franklin Dalton
Allen, Robert H
Anderson, Paul Nathaniel
Angleton, George M
Arend, William Phelps
Austin, James Henry
Baker, Robert Norton
Balkany, Thomas Jay
Ballinger, Carter M
Battaglia, Frederick Camillo
Berthrong, Morgan
Betz, George
Bock, S Allan
Bowling, Franklin Lee
Brettell, Herbert R
Bunn, Paul A, Jr
Chai, Hyman
Chan, Laurence Kwong-fai
Cherniack, Reuben Mitchell
Claman, Henry Neumann
Corby, Donald G
Daves, Marvin Lewis
Eickhoff, Theodore C
Eisele, Charles Wesley
Eliot, Robert S
Ellis, Philip Paul
Falliers, Constantine J
Fan, Leland Lane
Gelfand, Erwin William
Gerschenson, Lazaro E
Gersten, Jerome William
Gilden, Donald Harvey
Githens, John Horace, Jr
Glode, Leonard Michael
Gorson, Robert O
Green, Larry A
Gross, George Alvin
Guzelian, Philip Samuel
Hall, Alan H
Hammermeister, Karl E
Hofeldt, Fred Dan
Hollister, Alan Scudder
Horwitz, Kathryn Bloch
Horwitz, Lawrence D
Hudson, Bruce William
Hutt, Martin P
Jacobson, Eugene Donald
Kappler, John W
Katz, Fred H
Kern, Fred, Jr
King, Talmadge Everett, Jr
Kinzie, Jeannie Jones
Kirkpatrick, Charles Harvey
Lenke, Roger Rand
Liechty, Richard Dale
McDowell, Marion Edward
Makowski, Edgar Leonard
Marine, William Murphy
Martin, Richard Jay
Moore, George Eugene
Mueller, John Frederick
Naughton, Michael A
Nelson, Albert Wendell
Nelson, Harold Stanley
Nora, Audrey Hart
Nora, James Jackson
O'Brien, Donough
Peters, Bruce Harry
Petty, Thomas Lee
Poland, Jack Dean

Prasad, Kedar N
Reeves, John T
Repine, John E
Robinson, Arthur
Rosenwasser, Lanny Jeffery
Rudnick, Michael Dennis
Schooley, Robert T
Schrier, Robert William
Shepard, Richard Hance
Sokol, Ronald Jay
Solomons, Clive (Charles)
Spaulding, Harry Samuel, Jr
Sussman, Karl Edgar
Talmage, David Wilson
Taylor, Edward Stewart
Tofe, Andrew John
Tyler, Kenneth Laurence
Weil, John Victor
Williams, Ronald Lee

CONNECTICUT

Allaudeen, Hameedsulthan Sheik
Andriole, Vincent T
Ariyan, Stephan
Aronson, Peter S
Ballantyne, Bryan
Barash, Paul G
Baumgarten, Alexander
Berliner, Robert William
Bertino, Joseph R
Binder, Henry J
Blechner, Jack Norman
Boas, Norman Francis
Bondy, Philip K
Bonner, Daniel Patrick
Boron, Walter Frank
Boulpaep, Emile L
Bove, Joseph Richard
Boyer, James Lorenzen
Brandt, J Leonard
Braverman, Irwin Merton
Breg, William Roy
Bunney, Benjamin Stephenson
Burrow, Gerard N
Cadman, Edwin Clarence
Cheney, Charles Brooker
Cohen, Lawrence Sorel
Conn, Harold O
Cook, Charles Davenport
Curnen, Edward Charles, Jr
DeGraff, Arthur C, Jr
DeVita, Vincent T, Jr
Dillard, Morris, Jr
Donaldson, Robert M, Jr
Donta, Sam Theodore
Dorsky, David Isaac
Duff, Raymond Stanley
Ebbert, Arthur, Jr
Eisenfeld, Arnold Joel
Ezekowitz, Michael David
Feinstein, Alvan Richard
Fischer, James Joseph
Fleeson, William
Fong, Jack Sun-Chik
Freston, James W
Freymann, John Gordon
Gee, J Bernard L
Gefter, William Irvin
Genel, Myron
Glaser, Gilbert Herbert
Goodyer, Allan Victor
Greene, Nicholas Misplee
Greenspan, Richard H
Gross, Ian
Groszmann, Roberto Jose
Heller, John Herbert
Hendler, Ernesto Danilo
Hinkle, Lawrence Earl, Jr
Hinz, Carl Frederick, Jr
Hobbins, John Clark
Hoffer, Paul B
Horwitz, Ralph Irving
Hosain, Fazle
Howard, Rufus Oliver
Janeway, Charles Alderson, Jr
Jensen, Keith Edwin
Katz, Arnold Martin
Khan, Abdul Jamil
Kirchner, John Albert
Klimek, Joseph John
Korn, Joseph Howard
Krause, Peter James
Lange, Robert Carl
Lathem, Willoughby
Leonard, Martha Frances
Lerner, Aaron Bunsen
Levine, Robert John
Liao, Sung Jui
Lowman, Robert Morris
Lurie, Alan Gordon
Lytton, Bernard
McCarthy, Paul Louis
McGuire, Joseph Smith, Jr
Macko, Douglas John
Malawista, Stephen E
Markowltz, Milton
Marks, Paul A
Martin, R(ufus) Russell
Massey, Robert Unruh
Miller, Herbert Chauncey
Miller, I George, Jr
Morris, John McLean
Morse, Edward Everett
Murray, Francis Joseph

Naftolin, Frederick
Nair, Sreedhar
Nanda, Ravindra
Niederman, James Corson
O'Rourke, James
Papac, Rose
Patterson, John Ward
Pearson, Howard Allen
Perillie, Pasquale E
Rafferty, Terence Damien
Raisz, Lawrence Gideon
Randt, Clark Thorp
Reiskin, Allan B
Richards, Frank Frederick
Robinson, Donald Stetson
Rosenfield, Arthur Ted
Rothfield, Naomi Fox
Sachs, Frederick Lee
Sasaki, Clarence Takashi
Sears, Marvin Lloyd
Shehadi, William Henry
Sikand, Rajinder S
Silver, George Albert
Smith, Brian Richard
Sobol, Bruce J
Solnit, Albert J
Soufer, Robert
Spencer, Richard Paul
Spiro, Howard Marget
Sulavik, Stephen B
Sulzer-Azaroff, Beth
Talner, Norman Stanley
Tamborlane, William Valentine
Taub, Arthur
Taylor, Kenneth J W
Thompson, Hartwell Greene, Jr
Thornton, George Fred
Van Hoff, Jack
Wackers, Frans J Th
Walker, James Elliot Cabot
Weiss, Robert Martin
Welch, Annemarie S
Wetstone, Howard J
Whittemore, Ruth
Xu, Zhi Xin
Zaret, Barry L
Ziegra, Sumner Root

DELAWARE

Graham, David Tredway
Harley, Robison Dooling
Messinger-Rapport, Barbara J
Reinhardt, Charles Francis
Sarrif, Awni M
Zajac, Barbara Ann

DISTRICT OF COLUMBIA

Abramson, David C
Alving, Carl Richard
Archer, Juanita Almetta
Avery, Gordon B
Barnet, Ann B
Becker, Kenneth Louis
Bellanti, Joseph A
Birx, Deborah L
Bourne, Peter
Bulger, Roger James
Burg, Maurice B
Burger, Edward James, Jr
Burman, Kenneth Dale
Burris, James F
Caceres, Cesar A
Calcagno, Philip Louis
Callaway, Clifford Wayne
Canary, John J(oseph)
Carpentier, Robert George
Cheng, Tsung O
Chopra, Joginder Gurbux
Chung, Ed Baik
Coakley, Charles Seymour
Cummings, Nancy Boucot
Deutsch, Stanley
Dillard, Martin Gregory
Egeberg, Roger O
Ein, Daniel
Elliott, Larry Paul
Fabro, Sergio
Farrar, John Thruston
Felts, William Robert
Filipescu, Nicolae
Finkelstein, James David
Finnerty, Frank Ambrose, Jr
Fishbein, William Nichols
Fox, Samuel Mickle, III
Freis, Edward David
Fromm, Hans
Gelmann, Edward P
Giannini, Margaret Joan
Gootenberg, Joseph Eric
Gordon, James Samuel
Grigsby, Margaret Elizabeth
Grizzard, Michael B
Harvey, John Collins
Hellman, Louis M
Heman-Ackah, Samuel Monie
Henry, Walter Lester, Jr
Hicks, Jocelyn Muriel
Himmelsbach, Clifton Keck
Holzman, Gerald Bruce
Hone, Jennifer
Hugh, Rudolph
Jackson, Dudley Pennington
Jackson, Marvin Alexander
James, Alton Everette, Jr

Kaminetzky, Harold Alexander
Katz, Paul K
Katz, Sol
Kimmel, Carole Anne
Kimmel, Gary Lewis
Kobrine, Arthur
Kulczycki, Lucas Luke
Law, David H
Lee, Philip Randolph
Lessin, Lawrence Stephen
Lilienfield, Lawrence Spencer
Loo, Ti Li
Luessenhop, Alfred John
McAfee, John Gilmour
MacDonald, Mhairi Graham
McLean, Ian William
Macnamara, Thomas E
Manley, Audrey Forbes
Martin, Malcolm Mencer
Massaro, Donald John
Mullan, Fitzhugh
Nelson, Alan R
Novello, Antonia Coello
O'Doherty, Desmond Sylvester
Ozer, Mark Norman
Palestine, Alan G
Parker, Richard H
Parrish, Alvin Edward
Patel, Dali Jehangir
Pellegrino, Edmund Daniel
Perry, Seymour Monroe
Pincus, Jonathan Henry
Pipberger, Hubert V
Potter, John F
Queenan, John T
Ramsey, Elizabeth Mapelsden
Recant, Lillian
Rennert, Owen M
Rockoff, Seymour David
Rose, John Charles
Sabin, Albert B(ruce)
Sampson, Calvin Coolidge
Schechter, Geraldine Poppa
Scheele, Leonard Andrew
Scott, Roland Boyd
Sever, John Louis
Silva, Omega Logan
Simopoulos, Artemis Panageotis
Sinks, Lucius Frederick
Sly, Ridge Michael
Sphar, Raymond Leslie
Stapleton, John F
Stark, Nathan Julius
Stemmler, Edward J
Swanson, August George
Tidball, Charles Stanley
Twigg, Homer Lee
Utz, John Philip
Vick, James
Wartofsky, Leonard
Watkin, Donald M
Weingold, Allan Byrne
Weintraub, Herbert D
Werkman, Sidney Lee
Werner, Mario
Williams, James Thomas
Winchester, James Frank
Wray, H Linton
Wright, Daniel Godwin

FLORIDA

Abell, Murray Richardson
Aldrete, Jorge Antonio
Alexander, Ralph William
Allinson, Morris Jonathan Carl
Alpert, Seymour
Andersen, Thorkild Waino
Anderson, Douglas Richard
Ansbacher, Stefan
Awad, William Michel, Jr
Ayoub, Elia Moussa
Baiardi, John Charles
Balducci, Lodovico
Barness, Lewis Abraham
Befeler, Benjamin
Behnke, Roy Herbert
Behrman, Samuel J
Ben, Max
Berman, Irwin
Bernstein, Stanley H
Berry, Maxwell (Rufus)
Biersdorf, William Richard
Bisno, Alan Lester
Blank, Harvey
Block, Edward R
Blum, Leon Leib
Bourgoignie, Jacques J
Boyce, Henry Worth, Jr
Bramante, Pietro Ottavio
Brooks, Stuart Merrill
Bruce, David Lionel
Bucklin, Robert van Zandt
Bukantz, Samuel Charles
Cade, James Robert
Cavanagh, Denis
Challoner, David Reynolds
Chandler, J Ryan
Chervenick, Paul A
Cintron, Guillermo B
Cluff, Leighton Eggertsen
Cody, D Thane
Conn, Jerome W
Connolly, John Francis
Cooke, Robert E

Craig, Stanley Harold
Craythorne, N W Brian
Culver, James F
Curran, John S
Cutler, Paul
Davis, Joseph Harrison
Dawber, Thomas Royle
Del Regato, Juan A
Eitzman, Donald V
Elfenbein, Gerald Jay
Enneking, William Fisher
Epstein, Murray
Evans, Arthur T
Fagin, Karin
Fayos, Juan Vallvey
Feingold, Alfred
Flipse, Martin Eugene
Fogel, Bernard J
Freund, Gerhard
Friedland, Fritz
Friedman, Ben I
Galindo, Anibal H
Gelband, Henry
Gessner, Ira Harold
Ginsberg, Myron David
Gold, Mark Stephen
Gold, Martin I
Goldberg, Lee Dresden
Goldberg, Stephen
Goldman, Allan Larry
Good, Robert Alan
Graven, Stanley N
Gravenstein, Joachim Stefan
Graybiel, Ashton
Greenfield, George B
Griffitts, James John
Groel, John Trueman
Groover, Marshall Eugene, Jr
Grunberg, Emanuel
Gubner, Richard S
Hadden, John Winthrop
Halprin, Kenneth M
Hartmann, Robert Carl
Hebert, Clarence Louis
Hecht, Frederick
Hinkley, Robert Edwin, Jr
Hodes, Philip J
Hogan, Daniel James
Hornicek, Francis John
Hornick, Richard B
Howard, Guy Allen
Howell, David Sanders
Howell, Ralph Rodney
Hulet, William Henry
Inana, George
Jensen, Wallace Norup
Kalser, Martin
Keller, Robert H
Kessler, Richard Howard
Kinney, Michael J
Kiplinger, Glenn Francis
Klemperer, Martin R
Lasseter, Kenneth Carlyle
Lawton, Alfred Henry
Lemberg, Louis
Lenes, Bruce Allan
Lessner, Howard E
Levy, Norman Stuart
Lian, Eric Chun-Yet
Little, William Asa
Lockey, Richard Funk
Lodwick, Gwilym Savage
Loeser, Eugene William
Lombardi, Max H
MacDonald, Richard Annis
McGuigan, James E
McIntosh, Henry Deane
McKenzie, John Maxwell
McMillan, Donald Ernest
Mallams, John Thomas
Malone, John Irvin
Marks, Meyer Benjamin
Marriott, Henry Joseph Llewellyn
Marston, Robert Quarles
Martinez, Luis Osvaldo
Michael, Max, Jr
Miller, Billie Lynn
Million, Rodney Reiff
Mindlin, Rowland L
Mitus, Wladyslaw J
Modell, Jerome Herbert
Moreland, Alvin Franklin
Muench, Karl Hugo
Murphy, William Parry, Jr
Murtagh, Frederick Reed
Nagel, Eugene L
Neims, Allen Howard
Nelson, Erland
Nelson, Russell Andrew
Norton, Edward W D
Noyes, Ward David
Okulski, Thomas Alexander
Olson, Kenneth B
Palmer, Roger
Papper, Emanuel Martin
Peltier, Hubert Conrad
Pepine, Carl John
Persky, Lester
Pollara, Bernard
Prather, Elbert Charlton
Prockop, Leon D
Reich, George Arthur
Rhamy, Robert Keith
Rhodin, Johannes A G
Riser, Wayne H
Roberts, Norbert Joseph
Root, Allen William
Rosenbloom, Arlan Lee
Rosenkrantz, Jacob Alvin
Rosomoff, Hubert Lawrence
Rothschild, Marcus Adolphus
Rovelstad, Randolph Andrew
Rubin, Melvin Lynne
Ryan, James Walter
Samet, Philip
Scheinberg, Peritz
Schiebler, Gerold Ludwig
Schiff, Leon
Schmidt, Paul Joseph
Segal, Alvin
Sharp, John Turner
Sheehan, David Vincent
Shipley, Thorne
Shively, John Adrian
Singleton, George Terrell
Sinkovics, Joseph G
Smith, Donn Leroy
Smith, Richard Thomas
So, Antero Go
Sommer, Leonard Samuel
Spellacy, William Nelson
Storaasli, John Phillip
Straumfjord, Jon Vidalin, Jr
Streiff, Richard Reinhart
Streitfeld, Murray Mark
Szentivanyi, Andor
Taylor, William Jape
Thomas, William Clark, Jr
Turner, Robert Alexander
Vail, Edwin George
Valdes-Dapena, Marie A
VanItallie, Theodore Bertus
Van Mierop, Lodewyk H S
Vasey, Frank Barnett
Vesely, David Lynn
Victorica, Benjamin (Eduardo)
Waife, Sholom Omi
Wallach, Stanley
Wanner, Adam
Warren, Joel
Waterhouse, Keith R
Weidler, Donald John
Weiss, Charles Frederick
Westlake, Robert Elmer, Sr
Williams, Ralph C, Jr
Witte, John Jacob
Yeh, Billy Kuo-Jiun
Young, Lawrence Eugene
Yunis, Adel A

GEORGIA
Alonso, Kenneth B
Baisden, Charles Robert
Ballard, Marguerite Candler
Bernstein, Robert Steven
Blumberg, Richard Winston
Bransome, Edwin D, Jr
Broome, Claire V
Brown, Herbert Eugene
Brown, Walter John
Brubaker, Leonard Hathaway
Bruner, Barbara Stephenson
Bryans, Charles Iverson, Jr
Butcher, Brian T
Byrd, J Rogers
Chew, William Hubert, Jr
Clark, Winston Craig
Collipp, Platon Jack
Connell-Tatum, Elizabeth Bishop
Cooper, Gerald Rice
Copeland, Robert B
Corley, Charles Calhoun, Jr
Cox, Frederick Eugene
Criswell, Bennie Sue
Curran, James W
Davey, Winthrop Newbury
Davidson, John Keay, III
Desai, Rajendra C
Dowda, F William
Edelhauser, Henry F
Egan, Robert L
Ellison, Lois Taylor
Engler, Harold S
Evans, Edwin Curtis
Evatt, Bruce Lee
Faguet, Guy B
Falk, Henry
Feldman, Daniel S
Feldman, Elaine Bossak
Feringa, Earl Robert
Flowers, Nancy Carolyn
Foege, William Herbert
Franch, Robert H
Frank, Martin J
Frederickson, Evan Lloyd
Galambos, John Thomas
Gallagher, Brian Boru
Garrettson, Lorne Keith
Gerrity, Ross Gordon
Glass, Roger I M
Golbey, Robert (Bruce)
Goldman, John Abner
Grayzel, Arthur I
Green, Robert Castleman
Groth, Donald Paul
Guinan, Mary Elizabeth
Gunter, Bobby J
Gusdon, John Paul
Haskins, Arthur L, Jr
Herman, Chester Joseph
Herrmann, Kenneth L
Hoffman, William Hubert
Hollowell, Joseph Gurney, Jr
Hudson, James Bloomer
Huff, Thomas Allen
Hug, Carl Casimir, Jr
Hughes, James Mitchell
Huguley, Charles Mason, Jr
Hunt, Andrew Dickson
Jarvis, Jack Reynolds
Jarvis, William R
Jeffery, Geoffrey Marron
Jones, Jimmy Barthel
Joyner, Ronald Wayne
Kahn, Henry Slater
Karp, Herbert Rubin
Keene, Willis Riggs
Kilbourne, Edwin Michael
Klein, Luella
Knobloch, Hilda
LeVeau, Barney Francis
Little, Robert Colby
Long, Wilmer Newton, Jr
McDuffie, Frederic Clement
McPherson, James C, Jr
Martinez-Maldonado, Manuel
Merin, Robert Gillespie
Middleton, Henry Moore, III
Millar, John Donald
Mitch, William Evans
Moores, Russell R
Nahmias, Andre Joseph
Nemeroff, Charles Barnet
Oliver, John Eoff, Jr
Orenstein, Walter Albert
Parks, John S
Per-Lee, John H
Pool, Winford H, Jr
Raiford, Morgan B
Ramos, Harold Smith
Rasmussen, Howard
Rickles, Frederick R
Rimland, David
Robertson, Alex F
Rodey, Glenn Eugene
Rogers, James Virgil, Jr
Roper, Maryann
Roper, William L
Savage, John Edward
Schlant, Robert C
Schroder, Jack Spalding
Schultz, Everett Hoyle, Jr
Schwartz, James F
Sellers, Thomas F, Jr
Sharp, John T(homas)
Shlevin, Harold H
Shulman, Jones A(lvin)
Smith, Jesse Graham, Jr
Snider, Dixie Edward, Jr
Spitznagel, John Keith
Steinhaus, John Edward
Steinschneider, Alfred
Stevens, John Joseph
Talledo, Oscar Eduardo
Tedesco, Francis J
Thun, Michael John
Tuttle, Elbert P, Jr
Walton, Kenneth Nelson
Ward, Richard S
Waring, George O, III
Warren, McWilson
Webster, Paul Daniel, III
Wege, William Richard
Wenger, Nanette Kass
Whitsett, Carolyn F
Wigh, Russell
Willis, Isaac
Wilson, Colon Hayes, Jr
Yancey, Asa G, Sr

HAWAII
Bass, James W
Batkin, Stanley
Blaisdell, Richard Kekuni
Char, Donald F B
Furusawa, Eiichi
Gilbert, Fred Ivan, Jr
Hammar, Sherrel L
Lance, Eugene Mitchell
Linman, James William
Massey, Douglas Gordon
Mills, George Hiilani
Oishi, Noboru
Person, Donald Ames
Quisenberry, Walter Brown
Schatz, Irwin Jacob
Sharma, Santosh Devraj
Siddiqui, Wasim A
Smith, Richard A
Ueland, Kent
Whang, Robert
Yang, Hong-Yi

IDAHO
Schwartz, Theodore Benoni
Vestal, Robert Elden

ILLINOIS
Abraira, Carlos
Abramson, David Irvin
Adams, Evelyn Eleanor
Adams, John R
Adelson, Bernard Henry
Adler, Solomon Stanley
Al-Bazzaz, Faiq J
Ambre, John Joseph
Andersen, Burton
Anderson, Kenning M
Archer, John Dale
Arnason, Barry Gilbert Wyatt
Bailie, Michael David
Balagot, Reuben Castillo
Barclay, William R
Barghusen, Herbert Richard
Barton, Evan Mansfield
Baumeister, Carl Frederick
Becker, Michael Allen
Beem, Marc O
Beer, Alan E
Benson, Donald Warren
Bergen, Donna Catherine
Berger, Sheldon
Berkson, David M
Bernstein, Joel Edward
Bernstein, Lionel M
Berry, Leonidas Harris
Best, William Robert
Betts, Henry Brognard
Bharati, Saroja
Biller, Jose
Birnholz, Jason Cordell
Bitran, Jacob David
Bloomfield, Daniel Kermit
Bluefarb, Samuel M
Boehm, John Joseph
Bond, Howard Edward
Borkon, Eli Leroy
Boulos, Badi Mansour
Brandfonbrener, Martin
Brasitus, Thomas
Bruetman, Martin Edgardo
Brunner, Edward A
Buckman, Robert W
Bulkley, George
Burd, Laurence Ira
Cassel, Christine Karen
Celesia, Gastone G
Cera, Lee M
Childers, Roderick W
Christian, Joseph Ralph
Chudwin, David S
Cibils, Luis Angel
Clark, John Whitcomb
Coe, Fredric Lawrence
Cohen, Edward P
Cohen, Lawrence
Cohen, Sidney
Colbert, Marvin J
Cole, Madison Brooks, Jr
Collins, Vincent J
Conibear, Shirley Ann
Cotsonas, Nicholas John, Jr
Cugell, David Wolf
Dau, Peter Caine
Davis, Floyd Asher
Davis, Joseph Richard
Day, Emerson
Deangelis, Catherine D
DeCosta, Edwin J
De Forest, Ralph Edwin
DeGroot, Leslie Jacob
Del Greco, Francesco
DePeyster, Frederick A
Diamond, Seymour
Dmowski, W Paul
Dray, Sheldon
Duda, Eugene Edward
Earle, David Prince, Jr
Economou, Steven George
Egan, Richard L
Elble, Rodger Jacob
Elliott, William John
Erickson, James C, III
Farber, Marilyn Diane
Fellner, Susan K
Fennessy, John James
Firlit, Casimir Francis
Fisher, Ben
Fisher, Leonard V
Fozzard, Harry A
Freedman, Philip
Freese, Uwe Ernst
Freinkel, Ruth Kimmelstiel
Fried, Walter
Friederici, Hartmann H R
Frischer, Henri
Gartner, Lawrence Mitchel
Gerbie, Albert B
Gerding, Dale Nicholas
Gewertz, Bruce Labe
Giovacchini, Peter L
Glick, Gerald
Glisson, Silas Nease
Gold, Jay Joseph
Golomb, Harvey Morris
Goodheart, Clyde Raymond
Grayhack, John Thomas
Green, David
Green, Orville
Griem, Melvin Luther
Griem, Sylvia F
Grossman, Burton Jay
Gu, Yan
Gunnar, Rolf McMillan
Gurney, Clifford W
Hammond, James B

Medicine (cont)

Hanauer, Stephen B
Hand, Roger
Haring, Olga M
Harris, Jules Eli
Hasegawa, Junji
Hastreiter, Alois Rudolf
Heck, Robert Skinrood
Heller, Paul
Hellman, Samuel
Hendrickson, Frank R
Herting, Robert Leslie
Hicks, James Thomas
Hirsch, Lawrence Leonard
Holmes, Albert William, Jr
Huggins, Charles Brenton
Iber, Frank Lynn
Jackson, George G
Jacobson, Leon Orris
Joseph, Donald J
Kaizer, Herbert
Kalyan-Raman, Krishna
Kang, David Soosang
Kanofsky, Jeffrey Ronald
Kaplan, Ervin
Karinattu, Joseph J
Katz, Adrian I
Kernis, Marten Murray
Khalili, Ali A
Kim, Yoon Berm
Kimball, Chase Patterson
Kingdon, Henry Shannon
Kirsner, Joseph Barnett
Knospe, William H
Koenig, Harold
Kohrman, Arthur Fisher
Kraft, Sumner Charles
Kwaan, Hau Cheong
Langman, Craig Bradford
Lathrop, Katherine Austin
Lax, Louis Carl
Leff, Alan R
Leikin, Jerrold Blair
Lesinski, John Silvester
Levin, Murray Laurence
Levitan, Ruven
Levitsky, Lynne Lipton
Lewis, Victor L
Lichter, Edward A
Liebner, Edwin J
Limarzi, Louis Robert
Lindheimer, Marshall D
Littman, Armand
Lorincz, Allan Levente
Louis, John
Lounsbury, Franklin
Love, Leon
Lurain, John Robert, III
McCormick, Kenneth James
McGuire, Christine H
Mafee, Mahmood Forootan
Malkinson, Frederick David
Marr, James Joseph
Mayor, Gilbert Harold
Meadows, W Robert
Mendel, Gerald Alan
Meredith, Stephen Charles
Metzger, Boyd Ernest
Mewissen, Dieudonne Jean
Meyer, Richard Charles
Millichap, J(oseph) Gordon
Mirkin, Bernard Leo
Moawad, Atef H
Molitch, Mark E
Morrell, Frank
Morris, Charles Elliot
Moser, Robert Harlan
Moy, Richard Henry
Muehrcke, Robert C
Nadler, Charles Fenger
Nadler, Henry Louis
Nahrwold, David Lange
Nayyar, Rajinder
Neill, William Alexander
Newell, Frank William
Newman, Louis Benjamin
O'Conor, Vincent John, Jr
Olson, Stanley William
Onal, Ergun
Ostrow, Jay Donald
Owen, Oliver Elon
Pachman, Daniel James
Pachman, Lauren M
Page, Malcolm I
Papatheofanis, Frank John
Paterson, Philip Y
Patterson, Roy
Paul, Milton Holiday
Paynter, Camen Russell
Peraino, Carl
Perlman, Robert
Petersen, Edward S
Phair, John P
Pilz, Clifford G
Plotke, Frederick
Potter, Elizabeth Vaughan
Poznanski, Andrew K
Prinz, Richard Allen
Pullman, Theodore Neil
Rabinovich, Sergio Rospigliosi
Radwanska, Ewa
Reba, Richard Charney
Reisberg, Boris Elliott

Resnekov, Leon
Rhodes, Mitchell Lee
Roddick, John William, Jr
Roguska-Kyts, Jadwiga
Romack, Frank Eldon
Roos, Raymond Philip
Rosenberg, Henry Mark
Rosenfield, Robert Lee
Rosenthal, Ira Maurice
Rossi, Ennio Claudio
Rowley, Janet D
Rubenstein, Arthur Harold
Sabesin, Seymour Marshall
Sammons, James Harris
Samter, Max
Scanu, Angelo M
Schmid, Frank Richard
Schmitz, Robert L
Schneider, Arthur Sanford
Schoenberger, James A
Schulman, Sidney
Schumacher, Gebhard Friederich B
Schwartz, Steven Otto
Schweppe, John S
Sciarra, John J
Shambaugh, George E, III
Shoch, David Eugene
Shulman, Morton
Shulman, Stanford Taylor
Siegel, George Jacob
Silverstein, Edward Allen
Simon, Norman M
Singer, Donald H
Singer, Ira
Singer, Irwin
Singh, Sant Parkash
Sisson, George Allen
Skosey, John Lyle
Smith, Theodore Craig
Soffer, Alfred
Solomon, Lawrence Marvin
Sorensen, Leif Boge
Spaeth, Ralph
Spivey, Bruce Eldon
Stamler, Jeremiah
Stanley, Malcolm McClain
Stefani, Stefano
Steigmann, Frederick
Stepto, Robert Charles
Stumpf, David Allen
Suker, Jacob Robert
Swerdlow, Martin A
Swinnen, Lode J
Swiryn, Steven
Taylor, D Dax
Taylor, Samuel G, III
Thilenius, Otto G
Thompson, Phebe Kirsten
Thorp, Frank Kedzie
Toback, F(rederick) Gary
Tobin, John Robert, Jr
Tobin, Martin John
Torok, Nicholas
Traisman, Howard Sevin
Truong, Xuan Thoai
Tso, Mark On-Man
Ultmann, John Ernest
Valvassori, Galdino E
Van Fossan, Donald Duane
Vesselinovitch, Stan Dushan
Visotsky, Harold M
Vye, Malcolm Vincent
Waldstein, Sheldon Saul
Waterhouse, John P
Weil, Max Harry
Wessel, Hans U
Wied, George Ludwig
Wiener, Stanley L
Wilber, David James
Wilbur, Richard Sloan
Wilensky, Jacob T
Winter, Robert John
Wolf, James Stuart
Wolter, Janet
Wong, Paul Wing-Kon
Wu, Anna Fang
Yachnin, Stanley
Yang, Sen-Lian
Yunus, Muhammad Basharat
Zatuchni, Gerald Irving
Zeiss, Chester Raymond
Zsigmond, Elemer K

INDIANA

Aldo-Benson, Marlene Ann
Babbs, Charles Frederick
Bang, Nils Ulrik
Baxter, Neal Edward
Beering, Steven Claus
Bergstein, Jerry Michael
Bhatti, Waqar Hamid
Blix, Susanne
Bond, William H
Botero, J M
Carter, James Edward
Clark, Charles Malcolm, Jr
Daly, Walter J
Dansby, Doris
De Myer, William Erl
Dexter, Richard Newman
Dobben, Glen D
Dyke, Richard Warren
Dyken, Mark Lewis
Eckenhoff, James Edward

Feigenbaum, Harvey
Fisch, Charles
Fischer, A(lbert) Alan
FitzGerald, Joseph Arthur
Franson, Timothy Raymond
Gailani, Salman
Garrett, Robert Austin
Goldstein, David Joel
Gordee, Robert Stouffer
Grayson, Merrill
Henry, David P, II
Hicks, Edward James
Hodes, Marion Edward
Irwin, Glenn Ward, Jr
Johnson, Charles F
Johnston, Cyrus Conrad, Jr
Judson, Walter Emery
Kim, Kil Chol
Klatte, Eugene
Knoebel, Suzanne Buckner
Leser, Ralph Ulrich
Li, Ting Kai
Lumeng, Lawrence
McKinney, Thurman Dwight
Manders, Karl Lee
Martz, Bill L
Merritt, Doris Honig
Meyers, Michael Clinton
Mocharla, Raman
Mulcahy, John Joseph
Munsick, Robert Alliot
Nelson, Robert Leon
Norins, Arthur Leonard
Oppenheim, Bernard Edward
Owen, George Murdock
Powell, Richard Cinclair
Ridolfo, Anthony Sylvester
Rohn, Robert Jones
Siddiqui, Aslam Rasheed
Stonehill, Robert Berrell
Surawicz, Borys
Test, Charles Edward
Thompson, Wilmer Leigh
Walther, Joseph Edward
Wappner, Rebecca Sue
Weaver, David Dawson
Weber, George
Weinberger, Myron Hilmar
Wellman, Henry Nelson
White, Arthur C
Wright, Joseph William, Jr
Yune, Heun Yung
Zipes, Douglas Peter

IOWA

Abboud, Francois Mitry
Armstrong, Mark I
Bedell, George Noble
Bell, William E
Beller, Fritz K
Blodi, Frederick Christopher
Caplan, Richard Melvin
Christensen, James
Clark, Robert A
Clifton, James Albert
Crump, Malcolm Hart
DeGowin, Richard Louis
Dexter, Franklin, III
Eckhardt, Richard Dale
Eckstein, John William
Filer, Lloyd Jackson, Jr
Fomon, Samuel Joseph
Funk, David Crozier
Goplerud, Clifford P
Gurll, Nelson
Heistad, Donald Dean
Hodges, Robert Edgar
Hom, Ben Lin
Hubel, Kenneth Andrew
January, Lewis Edward
Kaelber, William Walbridge
Kasik, John Edward
Kerber, Richard E
Kieso, Robert Alfred
Koehler, P Ruben
Kolder, Hansjoerg E
Kos, Clair Michael
Kunesh, Jerry Paul
Latourette, Howard Bennett
Levy, Leonard Alvin
McCabe, Brian Francis
Morgan, Donald Pryse
Morriss, Frank Howard, Jr
Moyers, Jack
Naides, Stanley J
Nauseef, William Michael
Niebyl, Jennifer Robinson
Peterson, Richard Elsworth
Ponseti, Ignacio Vives
Read, Charles H
Richerson, Hal Bates
Schedl, Harold Paul
Schrott, Helmut Gunther
Scott, William Edwin
Seebohm, Paul Minor
Smith, Ian Maclean
Solomons, Gerald
Soper, Robert Tunnicliff
Spector, Arthur Abraham
Strauss, John S
Strauss, Ronald George
Summers, Robert Wendell
Theilen, Ernest Otto
Thomas, John Martin

Thompson, Herbert Stanley
Thompson, Robert Gary
Tranel, Daniel T
Van Leeuwen, Gerard
Weinberger, Miles
Weingeist, Thomas Alan
Welsh, Michael James
Widness, John Andrew
Zellweger, Hans Ulrich
Ziegler, Ekhard E

KANSAS

Allen, Max Scott
Bolinger, Robert E
Brown, Robert Wayne
Burket, George Edward
Calkins, William Graham
Chang, Chae Han Joseph
Cho, Cheng T
Consigli, Richard Albert
Cook, James Dennis
Cowley, Benjamin D, Jr
Diederich, Dennis A
Dujovne, Carlos A
Dunn, Marvin I
Foster, David Bernard
Goldstein, Elliot
Grantham, Jared James
Greenberger, Norton Jerald
Kennedy, James A
Kimler, Bruce Franklin
Kimmel, Joe Robert
Krantz, Kermit Edward
Lemoine, Albert N, Jr
Liu, Chien
Mainster, Martin Aron
Manning, Robert Thomas
Meek, Joseph Chester, Jr
Pingleton, Susan Kasper
Plapp, Frederick Vaughn
Price, James Gordon
Pugh, David Milton
Redford, John W B
Rhodes, James B
Roy, William R
Shaad, Dorothy Jean
Strafuss, Albert Charles
Templeton, Arch W
Tosh, Fred Eugene
Walaszek, Edward Joseph
Waxman, David
Wilson, Sloan Jacob
Ziegler, Dewey Kiper

KENTUCKY

Andersen, Roger Allen
Andrews, Billy Franklin
Aronoff, George R
Barton, John Robert
Baumann, Robert Jay
Belker, Arnold M
Bosomworth, Peter Palliser
Burki, Nausherwan Khan
Carrasquer, Gaspar
Christopherson, William Martin
Clark, David Barrett
Cummings, Norman Allen
Digenis, George A
Espinosa, Enrique
Friedell, Gilbert H
Glenn, James Francis
Greene, John W, Jr
Haynes, Douglas Martin
Holland, Nancy H
James, Mary Frances
Keeney, Arthur Hail
Kraman, Steve Seth
Lanska, Douglas John
McLeish, Kenneth R
McQuillen, Michael Paul
Mandelstam, Paul
Marciniak, Ewa J
Markesbery, William Ray
Mazzoleni, Alberto
Miller, Ralph English
Noonan, Jacqueline Anne
Parker, Joseph Corbin, Jr
Raff, Martin Jay
Redinger, Richard Norman
Rees, Earl Douglas
Rigor, Benjamin Morales, Sr
Rollo, Frank David
Rosenbaum, Harold Dennis
Schwab, John J
Scott, Ralph Mason
Shih, Wei Jen
Slevin, John Thomas
Steele, James Patrick
Steiner, Marion Rothberg
Tamburro, Carlo Horace
Thind, Gurdarshan S
Thompson, John S
Urbscheit, Nancy Lee
Weisskopf, Bernard
Whayne, Tom French
Williams, Walter Michael
Wolfe, Walter McIlhaney
Yam, Lung Tsiong
Zimmerman, Thom J

LOUISIANA

Andes, W(illard) Abe
Bellina, Joseph Henry
Berenson, Gerald Sanders

MEDICAL & HEALTH SCIENCES / 455

Burch, Robert Emmett
Burch, Robert Ray
Chin, Hong Woo
Clemetson, Charles Alan Blake
Conrad, Steven A
Corrigan, James John, Jr
D'Alessandro-Bacigalupo, Antonio
Daniels, Robert (Sanford)
Davis, William Duncan, Jr
Dickey, Richard Palmer
Duncan, Margaret Caroline
Epstein, Arthur William
Ferriss, Gregory Stark
Frohlich, Edward David
Fruthaler, George James, Jr
Fulginiti, Vincent Anthony
Gagliano, Nicholas Charles
Gallant, Donald
Ganley, James Powell
George, Ronald Baylis
Giles, Thomas Davis
Glancy, David L
Gonzalez, Francisco Manuel
Gordon, Douglas Littleton
Gould, Harry J, III
Gum, Oren Berkley
Hackett, Earl R
Haik, George Michel
Hamrick, Joseph Thomas
Hastings, Robert Clyde
Heath, Robert Galbraith
Hejtmancik, Milton R
Herbst, John J
Hollis, Walter Jesse
Hyman, Edward Sidney
Hyslop, Newton Everett, Jr
Ivens, Mary Sue
Jung, Rodney Clifton
Kadan, Savitri Singh
Kastl, Peter Robert
Kaufman, Herbert Edward
Krahenbuhl, James Lee
Krogstad, Donald John
Krupp, Iris M
Lang, Erich Karl
Lehrer, Samuel Bruce
Lindsey, Edward Stormont
Locke, William
Lopez-Santolino, Alfredo
Lyons, George D
McGrath, Hugh, Jr
McHardy, George Gordon
McLaurin, James Walter
McMahon, Francis Gilbert
Meyers, Philip Henry
Mickal, Abe
Millikan, Larry Edward
Mizell, Merle
Mogabgab, William Joseph
Morgan, Lee Roy, Jr
Myers, Melvil Bertrand, Jr
Nice, Charles Monroe, Jr
Ochsner, Seymour Fiske
Osofsky, Howard J
Paddison, Richard Milton
Pankey, George Atkinson
Peacock, Milton O
Pearson, James Eldon
Phillips, John Hunter, Jr
Pou, Jack Wendell
Puyau, Francis A
Ratz, John Louis
Ray, Clarence Thorpe
Re, Richard N
Reed, Richard Jay
Rigby, Perry G
Roberts, James Allen
Roheim, Paul Samuel
Rosenthal, J William
Rothschild, Henry
Rutledge, Lewis James
Salvaggio, John Edmond
Sander, Gary Edward
Sappenfield, Robert W
Scher, Charles D
Schimek, Robert Alfred
Schlegel, Jorgen Ulrik
Seltzer, Benjamin
Seto, Jane Mei-Chun
Shushan, Morris
Stewart, William Huffman
Stuckey, Walter Jackson, Jr
Summer, Warren R
Thornton, William Edgar
Threefoot, Sam Abraham
Trufant, Samuel Adams
Udall, John Nicholas, Jr
Vanselow, Neal A
Vial, Lester Joseph, Jr
Wallin, John David
Walsh, John Joseph
Waring, William Winburn
Webster, Douglas B
Weed, John Conant
Weill, Hans
Weiss, Thomas E
White, Charles A, Jr
Wilson, John T
Wolf, Robert E
Xie, Jianming

MAINE
Aronson, Frederick Rupp
Barker, Jane Ellen

Collins, H Douglas
Cronkhite, Leonard Woolsey, Jr
Cross, Harold Dick
Doby, Tibor
Fellers, Francis Xavier
Gezon, Horace Martin
Hayes, Guy Scull
Korst, Donald Richardson
LaMarche, Paul H
Lamdin, Ezra
McEwen, Currier
Pannill, Fitzhugh Carter, Jr
Phelps, Paulding
Rabe, Edward Frederick
Rand, Peter W
Spock, Benjamin McLane

MARYLAND
Abbrecht, Peter H
Adelstein, Robert Simon
Adkinson, Newton Franklin, Jr
Alexander, Duane Frederick
Allen, Willard M
Alling, David Wheelock
Alter, Harvey James
Amato, R Stephen S
Ambudkar, Indu S
Anderson, James Howard
Anderson, Lucy Macdonald
Andres, Reubin
Anhalt, Grant James
Asher, David Michael
Ashman, Michael Nathan
Asper, Samuel Phillips
Aszalos, Adorjan
Aurbach, Gerald Donald
Baer, Harold
Baker, Timothy D
Barker, Lewellys Franklin
Barry, Kevin Gerard
Beisel, William R
Bell, William Robert, Jr
Bennett, John E
Berendes, Heinz Werner
Bereston, Eugene Sydney
Bering, Edgar Andrew, Jr
Berkower, Ira
Berne, Bernard H
Berzofsky, Jay Arthur
Blackman, Marc Roy
Blaese, R Michael
Blake, David Andrew
Blattner, William Albert
Bleecker, Eugene R
Bohr, Vilhelm Alfred
Bono, Vincent Horace, Jr
Borsos, Tibor
Bosma, James F
Brewer, H Bryan, Jr
Brown, Paul Wheeler
Brubaker, Merlin L
Brusilow, Saul W
Burnett, Joseph W
Burton, Benjamin Theodore
Bush, David E
Calvert, Richard John
Canfield, Craig Jennings
Carski, Theodore Robert
Catz, Charlotte Schifra
Chanock, Robert Merritt
Charache, Patricia
Charache, Samuel
Charnas, Lawrence Richard
Chase, Thomas Newell
Chen, Hao
Chernoff, Amoz Immanuel
Chernow, Bart
Cheson, Bruce David
Childs, Barton
Chisolm, James Julian, Jr
Choppin, Purnell Whittington
Clemmens, Raymond Leopold
Cogan, David Glendenning
Cohen, Sheldon Gilbert
Cohn, Robert
Connor, Thomas Byrne
Corfman, Philip Albert
Cornblath, Marvin
Correa Villasenor, Adolfo
Cotter, Edward F
Crout, J Richard
Cummings, Martin Marc
Cutler, Gordon Butler, Jr
Cutler, Jeffrey Alan
Daniels, Worth B, Jr
Davidson, Jack Dougan
Davis, Leroy Thomas
De Carlo, John, Jr
Decker, John Laws
DeLaCruz, Felix F
De Monasterio, Francisco M
Dennis, John Murray
Deslattes, Richard D, Jr
Dewys, William Dale
Dhindsa, Dharam Singh
Didisheim, Paul
Diggs, Carter Lee
Dipple, Anthony
Di Sant'Agnese, Paul Emilio Artom
Donner, Martin W
Dornette, William Henry Lueders
Dorst, John Phillips
Drachman, Daniel Bruce
Drucker, William Richard

Dubois, Andre T
Ducker, Thomas Barbee
Duncan, Katherine
Durkan, James P
Dyer, Robert Francis, Jr
Edelman, Robert
Elders, Minnie Joycelyn
Elin, Ronald John
Elisberg, Bennett La Dolce
Epstein, Stephen Edward
Evans, Charles Hawes
Evans, Virginia John
Fast, Patricia E
Fauci, Anthony S
Feigal, Ellen G
Felsenthal, Gerald
Ferencz, Charlotte
Ferguson, Earl Wilson
Ferguson, James Joseph, Jr
Ferrans, Victor Joaquin
Fine, Ben Sion
Finley, Joanne Elizabeth
Fischinger, Peter John
Fishman, Elliot Keith
Fleisher, Thomas A
Foley, H Thomas
Forbes, Allan Louis
Foster, Willis Roy
Fratantoni, Joseph Charles
Fraumeni, Joseph F, Jr
Fredrickson, Donald Sharp
Freeman, John Mark
Friedman, Lawrence
Frommer, Peter Leslie
Fuller, Richard Kenneth
Gajdusek, Daniel Carleton
Gallo-Torres, Hugo E
Genecin, Abraham
Gerber, Naomi Lynn Hurwitz
Gibson, Sam Thompson
Gill, John Russell, Jr
Gillum, Richard Frank
Giovacchini, Rubert Peter
Glick, Samuel Shipley
Gohagan, John Kenneth
Gold, Philip William
Goldberg, Morton Falk
Goldberger, Michael Eric
Golden, Archie Sidney
Goldfine, Lewis John
Goldstein, David Stanley
Goldstein, Robert
Goldstein, Robert Arnold
Golumbic, Norma
Gordis, Leon
Graff, Thomas D
Graham, George G
Granger, Donald Lee
Green, Jerome George
Greenough, William B, III
Greenwald, Peter
Greisman, Sheldon Edward
Grimley, Philip M
Groves, Eric Stedman
Guthrie, Eugene Harding
Haddy, Francis John
Hahn, Richard David
Hallett, Mark
Hamburger, Anne W
Hamill, Peter Van Vechten
Handler, Joseph S
Hardegree, Mary Carolyn
Harlan, William R, Jr
Harrison, Harold Edward
Harter, Donald Harry
Harvey, Abner McGehee
Haudenschild, Christian C
Heald, Felix Pierpont, Jr
Heetderks, William John
Hegyeli, Ruth I E J
Heineman, Frederick W
Hejtmancik, James Fielding
Helmsen, Ralph John
Helrich, Martin
Henkin, Robert I
Heptinstall, Robert Hodgson
Heyssel, Robert M
Higgins, Millicent Williams Payne
Hill, Charles Earl
Hinman, Edward John
Hirshman, Carol A
Hoeg, Jeffrey Michael
Hoffman, Paul Ned
Holmes, Randall Kent
Holtzman, Neil Anton
Horner, William Harry
Howie, Donald Lavern
Hoyer, Leon William
Hume, John Chandler
Hunter, Oscar Benwood, Jr
Hutchins, Grover MacGregor
Ihde, Daniel Carlyle
Iype, Pullolickal Thomas
Jacox, Ada K
Jaffe, Elaine Sarkin
Jewett, Hugh Judge
Johns, Richard James
Johnson, Richard T
Johnston, Gerald Samuel
Jordan, William Stone, Jr
Joy, Robert John Thomas
Kaplan, Richard Stephen
Kaslow, Richard Alan
Keiser, Harry Robert

Kennedy, Charles
Kennedy, Thomas James, Jr
Kessler, Irving Isar
Kickler, Thomas Steven
Klee, Claude Blenc
Klee, Gerald D'Arcy
Kline, Ira
Knazek, Richard Allan
Knox, David Lalonde
Knox, Gaylord Shearer
Kopin, Irwin J
Kowalewski, Edward Joseph
Kowarski, A Avinoam
Kramer, Barnett Sheldon
Kramer, Norman Clifford
Krause, Richard Michael
Kuff, Edward Louis
Kupfer, Carl
Kwiterovich, Peter O, Jr
Lake, Charles Raymond
Landon, John Campbell
Latham, Patricia Suzanne
Law, Lloyd William
Lawrence, Dale Nolan
Lederer, William Jonathan
Lederman, Howard Mark
Legters, Llewellyn J
Leonard, Edward Joseph
Leonard, James Joseph
LeRoith, Derek
Leventhal, Brigid Gray
Leventhal, Carl M
Levine, Arthur Samuel
Levine, David Morris
Levine, Paul Howard
Liang, Isabella Y S
Lichtenstein, Lawrence M
Lim, David J
Lipicky, Raymond John
Littlefield, John Walley
Litvak, Austin S
Liu, Pu
Longo, Dan L
Lorber, Mortimer
Loriaux, D Lynn
Lowe, Charles Upton
Lutwak, Leo
McCune, Susan K
McHugh, Paul Rodney
McIndoe, Darrell W
McIntyre, Patricia Ann
McKenna, Thomas Michael
McKhann, Guy Mead
McKusick, Victor Almon
Magrath, Ian
Maher, John Francis
Malech, Harry Lewis
Marban, Eduardo
Mardiney, Michael Ralph, Jr
Margolis, Simeon
Marzella, Louis
Matanoski, Genevieve M
Matheson, Nina W
Mayer, Richard F
Meredith, Orsell Montgomery
Merril, Carl R
Metcalfe, Dean Darrel
Meyers, Wayne Marvin
Mickel, Hubert Sheldon
Migeon, Claude Jean
Miller, Laurence Herbert
Miller, Louis Howard
Miller, Neil Richard
Miller, Robert Warwick
Mills, James Louis
Mitchell, Thomas George
Mohler, William C
Monath, Thomas P
Mond, James Jacob
Monroe, Russell Ronald
Moss, Joel
Murphy, Patrick Aidan
Murphy, Robert Patrick
Mushinski, J Frederic
Myers, Charles
Nager, George Theodore
Naunton, Ralph Frederick
Nelson, Karin Becker
Neumann, Ronald D
Neva, Franklin Allen
Niedermeyer, Ernst F
Nightingale, Stuart L
Nisula, Bruce Carl
Obnibene, Frederick Peter
Obrams, Gunta Iris
Oldfield, Edward Hudson
Order, Stanley Elias
Oski, Frank
Ottesen, Eric Albert
Owens, Albert Henry, Jr
Packard, Barbara B K
Paige, David M
Pamnani, Motilal Bhagwandas
Park, Lee Crandall
Parker, Robert Tarbert
Patz, Arnall
Pearse, Warren Harland
Pearson, John William
Peck, Carl Curtis
Permutt, Solbert
Petty, William Clayton
Pierce, Nathaniel Field
Pilkerton, A Raymond
Pinter, Gabriel George

Medicine (cont)

Pitcairn, Donald M
Platt, William Rady
Plotnick, Gary David
Plotz, Paul Hunter
Poirier, Miriam Christine Mohrhoff
Pollard, Thomas Dean
Pollycove, Myron
Post, Robert M
Proctor, Donald Frederick
Prout, George Russell, Jr
Prout, Thaddeus Edmund
Quinnan, Gerald Vincent, Jr
Rapoport, Stanley I
Raskin, Joan
Reichelderfer, Thomas Elmer
Reid, Clarice D
Reinig, James William
Reuber, Melvin D
Rheinstein, Peter Howard
Richards, Richard Davison
Richardson, Edward Henderson, Jr
Robbins, Jay Howard
Robertson, James Sydnor
Rodbard, David
Rodrigues, Merlyn M
Rogawski, Michael Andrew
Romansky, Monroe James
Rosen, Saul W
Ross, Richard Starr
Roth, Harold Philmore
Russell, Philip King
Ruzicka, Francis Frederick, Jr
Saba, George Peter, II
Sabol, Steven Layne
Sack, George H(enry), Jr
Sack, Richard Bradley
Salik, Julian Oswald
Samelson, Lawrence Elliot
Santos, George Wesley
Sapir, Daniel Gustave
Saudek, Christopher D
Scheibel, Leonard William
Schnittman, Steven Marc
Schoenberg, Bruce Stuart
Schoolman, Harold M
Schwartz, John T
Seidel, Henry Murray
Seltser, Raymond
Sharma, Gopal Chandra
Sheridan, Philip Henry
Shimizu, Hiroshi
Shnider, Bruce I
Shulman, Lawrence Edward
Sidbury, James Buren, Jr
Sidell, Frederick R
Siegel, Jay Philip
Simpson, David Gordon
Singewald, Martin Louis
Sloan, Michael Allan
Smith, Bruce H
Smith, Mark Stephen
Smith, Phillip Doyle
Snow, James Byron, Jr
Sommer, Alfred
Song, Byoung-Joon
Sporn, Michael Benjamin
Spragins, Melchijah
Stanton, Mearl Fredrick
Starfield, Barbara Helen
Sternberger, Ludwig Amadeus
Stevens, Harold
Stolley, Paul David
Straus, Stephen Ezra
Strickland, George Thomas
Stromberg, Kurt
Summers, Richard James
Swergold, Gary David
Tabor, Edward
Talbot, John Mayo
Taube, Sheila Efron
Taylor, Carl Ernest
Taylor, David Neely
Temple, Robert Jay
Thomas, George Howard
Triantaphyllopoulos, Demetrios
Triantaphyllopoulos, Eugenie
Tschudy, Donald P
Turkeltaub, Paul Charles
Vaitukaitis, Judith L
Valentine, Martin Douglas
Van Echo, David Andrew
Van Metre, Thomas Earle, Jr
Veech, Richard L
Vermund, Sten Halvor
Vogelstein, Bert
Vorosmarti, James, Jr
Vydelingum, Nadarajen Ameerdanaden
Wagley, Philip Franklin
Wagner, Henry George
Wagner, Henry N, Jr
Waldmann, Thomas A
Waldrop, Francis N
Walker, Wilbur Gordon
Wallace, Craig Kesting
Wallach, Edward Eliot
Walser, Mackenzie
Wargotz, Eric S
Watkins, Paul Allan
Weight, Forrest F
Weintraub, Bruce Dale
Weisfeldt, Myron Lee
Welsch, Federico
Westphal, Heiner J
Whelton, Andrew
Whidden, Stanley John
White, John David
Whitehorn, William Victor
Whitley, Joseph Efird
Whitley, Nancy O'Neil
Williams, Roger Lea
Williamson, Charles Elvin
Wilson, Donald Edward
Wilson, Wyndham Hopkins
Wolff, Frederick William
Woodruff, James Donald
Woodward, Theodore Englar
Work, Henry Harcus
Yaes, Robert Joel
Yaffe, Sumner J
Yarchoan, Robert
Yerg, Raymond A
Yin, Frank Chi-Pong
Young, Eric D
Yuspa, Stuart Howard
Zee, David Samuel
Zeligman, Israel
Zierler, Kenneth
Zimmerman, Hyman Joseph
Zinkham, William Howard
Zubrod, Charles Gordon
Zukel, William John

MASSACHUSETTS

Abelmann, Walter H
Agnello, Vincent
Ahmed, A Razzaque
Aisenberg, Alan Clifford
Alper, Chester Allan
Alper, Milton H
Altschule, Mark David
Anderson, Thomas Page
Andres, Giuseppe A
Arias, Irwin Monroe
Arnaout, M Amin
Aroesty, Julian Max
Arvan, Peter
Atkins, Elisha
Ausiello, Dennis Arthur
Austen, K(arl) Frank
Avery, Mary Ellen
Avruch, Joseph
Axelrod, Lloyd
Baden, Howard Philip
Baratz, Robert Sears
Barger, A(braham) Clifford
Barlow, Charles F
Barnett, Guy Octo
Barry, Joan
Baum, Jules Leonard
Beck, William Samson
Bennison, Bertrand Earl
Benson, Herbert
Berenberg, William
Berger, Harvey J
Bernfield, Merton Ronald
Bernhard, Jeffrey David
Bistrian, Bruce Ryan
Black, Paul H
Blau, Monte
Bloomberg, Wilfred
Bonkovsky, Herbert Lloyd
Boyer, Markley Holmes
Braunwald, Eugene
Brenner, Barry Morton
Brill, A Bertrand
Broder, Martin Ivan
Brody, Jerome Saul
Brown, Robert Stephen
Bucher, Nancy L R
Bunn, Howard Franklin
Burwell, E Langdon
Canellos, George P
Caplan, Louis Robert
Carey, Martin Conrad
Carvalho, Angelina C A
Cassidy, Carl Eugene
Caviness, Verne Strudwick, Jr
Chang, Te Wen
Charney, Evan
Chasin, Werner David
Chin, William W
Chivian, Eric Seth
Chobanian, Aram V
Chodosh, Sanford
Christlieb, Albert Richard
Clark, Sam Lillard, Jr
Coffman, Jay D
Cohen, Alan Seymour
Cone, Thomas E, Jr
Costanza, Mary E
Cotran, Ramzi S
Craven, Donald Edward
Crawford, John Douglas
Crigler, John F, Jr
Crocker, Allen Carrol
Crowley, William Francis, Jr
Dammin, Gustave John
David, John R
Davidson, Charles Sprecher
Dealy, James Bond, Jr
De Cherney, Alan Hersh
Delabarre, Everett Merrill, Jr
Dershwitz, Mark
Desforges, Jane Fay
Dickson, James Francis, III
Dohlman, Claes Henrik
Douglas, Pamela Susan
Drachman, David A
Eastwood, Gregory Lindsay
Ebert, Robert H
Ehlers, Mario Ralf Werner
Eisenberg, Sidney Edwin
Elkinton, J(oseph) Russell
Ellis, Elliot F
Epstein, Franklin Harold
Federman, Daniel D
Field, James Bernard
Fine, Samuel
Fischbach, Gerald David
Fitzpatrick, Thomas Bernard
Fiumara, Nicholas J
Fox, Irving Harvey
Fox, James Gahan
Frazier, Howard Stanley
Freed, Murray Monroe
Freedberg, Abraham Stone
Frei, Emil, III
Friedman, Ephraim
Fulmer, Hugh Scott
Galaburda, Albert Mark
Garrard, Sterling Davis
Geha, Raif S
Gellis, Sydney Saul
Gerald, Park S
Gerety, Robert John
Godine, John Elliott
Goldberg, Irvin H(yman)
Goldberg, Sheldon Sumner
Gorbach, Sherwood Leslie
Gorelick, Kenneth J
Gottlieb, Marise Suss
Grace, Norman David
Grand, Richard Joseph
Grant, Walter Morton
Greenblatt, David J
Griffith, Robert W
Gross, Jerome
Grossman, William
Guralnick, Walter
Haber, Edgar
Hales, Charles A
Hallock, Gilbert Vinton
Halpern, Daniel
Handin, Robert I
Hanshaw, James Barry
Hawiger, Jack Jacek
Hedley-Whyte, John
Heinrich, Gerhard
Herrmann, John Bellows
Heyda, Donald William
Hiatt, Howard Haym
Hiregowdara, Dananagoud
Hnatowich, Donald John
Hollander, William
Holman, B Leonard
Homburger, Freddy
Horton, Edward S
Ingram, Roland Harrison
Inui, Thomas S
Irwin, Richard Stephen
Ishikawa, Sadamu
Isselbacher, Kurt Julius
Jameson, James Larry
Jandl, James Harriman
Johnson, Elsie Ernest
Kaminskas, Edvardas
Kannel, William B
Kaplan, Melvin Hyman
Kassirer, Jerome Paul
Kazemi, Homayoun
Kelly, Dorothy Helen
Kelton, Diane Elizabeth
Keusch, Gerald Tilden
Kieff, Elliott Dan
Kim, James Kyung-Hee
Kinsbourne, Marcel
Kitz, Richard J
Klempner, Mark Steven
Kliman, Allan
Knapp, Peter Hobart
Koch-Weser, Dieter
Koch-Weser, Jan
Koff, Raymond Steven
Kolodny, Gerald Mordecai
Kosasky, Harold Jack
Kramer, Philip
Krane, Stephen Martin
Kressel, Herbert Yehude
Kritzman, Julius
Kruskal, Benjamin A
Kupchik, Herbert Z
Kupferman, Allan
Kwan, Paul Wing-Ling
Kyriakis, John M
Lamb, George Alexander
Lamont, John Thomas
Landowne, Milton
Laster, Leonard
Laurenzi, Gustave
Leach, Robert Ellis
Leaf, Alexander
Lebowitz, Elliot
Leder, Philip
Lees, Robert S
Lessell, Simmons
Levine, Herbert Jerome
Levinsky, Norman George
Levinson, Gilbert E
Leviton, Alan
Levy, Harvey Louis
Li, James CC
Lin, Chi-Wei
Lipsitt, Don Richard
Livingston, David M
Loewenstein, Matthew Samuel
London, Irving M
Loscalzo, Joseph
Lou, Peter Louis
Lowenstein, Edward
Lown, Bernard
Maas, Richard Louis
McCabe, William R
McCaughan, Donald
McNally, Elizabeth Mary
McNeil, Barbara Joyce
Madias, Nicolaos E
Madoc-Jones, Hywel
Madoff, Morton A
Mann, James
Marble, Alexander
Marchant, Douglas J
Marcus, Elliot M
Marks, Leon Joseph
Mathews-Roth, Micheline Mary
Medearis, Donald N, Jr
Melby, James Christian
Mellins, Harry Zachary
Mendelson, Jack H
Mihm, Martin C, Jr
Moloney, William Curry
Monette, Francis C
Montgomery, William Wayne
Moolten, Frederick London
Moore-Ede, Martin C
Mordes, John Peter
Morgan, James Philip
Nadas, Alexander Sandor
Nadol, Bronislaw Joseph, Jr
Nandy, Kalidas
Nanji, Amin Akbarali
Navab, Farhad
Neer, Robert M
Neumeyer, David Alexander
Newcombe, David S
Pasternack, Mark Steven
Pathak, Madhukar
Patterson, William Bradford
Pauker, Stephen Gary
Paul, Oglesby
Paul, Robert E, Jr
Pechet, Liberto
Pfister, Richard Charles
Pino, Richard M
Plaut, Andrew George
Pontoppidan, Henning
Potts, John Thomas, Jr
Prout, Curtis
Rabkin, Mitchell T
Rankin, Joel Sender
Ransil, Bernard J(erome)
Raymond, Samuel
Reichlin, Seymour
Relman, Arnold Seymour
Rencricca, Nicholas John
Reynolds, Robert N
Richardson, George S
Richman, Justin Lewis
Richmond, Julius Benjamin
Rock, Paul Bernard
Rosenberg, Irwin Harold
Rosenberg, Isadore Nathan
Rosenberg, Robert D
Roubenoff, Ronenn
Ruprecht, Ruth Margrit
Russell, Robert M
Ryan, John F
Ryan, Una Scully
Sabin, Thomas Daniel
Sacks, David B
Salhanick, Hilton Aaron
Sanazaro, Paul Joseph
Sandson, John Ivan
Saper, Clifford B
Sawin, Clark Timothy
Schimmel, Elihu Myron
Schlossman, Stuart Franklin
Schmidt, Kurt F
Schuknecht, Harold Frederick
Schur, Peter Henry
Schwartz, Bernard
Schwartz, Edith Richmond
Schwartz, Robert Stewart
Segal, Rosalind A
Sencer, David Judson
Senior, Boris
Shapiro, Howard Maurice
Shear, Leroy
Sheffer, Albert L
Shih, Vivian Ean
Sifneos, Peter E
Silva, Patricio
Slack, Warner Vincent
Small, Donald MacFarland
Smith, Thomas W
Snider, Gordon Lloyd
Snyder, Louis Michael
Sobel, Edna H
Spivak, Jerry Lepow
Spodick, David Howard
Stanbury, John Bruton
Steele, Glenn Daniel, Jr
Stoeckle, John Duane
Stolbach, Leo Lucien
Stollerman, Gene Howard

Stone, Peter H
Strom, Terry Barton
Strong, Mervyn Stuart
Suit, Herman Day
Swartz, Morton N
Talbot, Nathan Bill
Tarlov, Alvin Richard
Tashjian, Armen H, Jr
Tauber, Alfred Imre
Taveras, Juan M
Taylor, Isaac Montrose
Thier, Samuel Osiah
Thomas, Clayton Lay
Thorn, George W
Tishler, Peter Verveer
Toomey, James Michael
Travis, David M
Treves, S T
Trier, Jerry Steven
Twitchell, Thomas Evans
Tyler, H Richard
Van Dam, Jacques
Vandam, Leroy David
VanPraagh, Richard
Von Essen, Carl Francois
Wacker, Warren Ernest Clyde
Wang, Chiu-Chen
Watkins, Elton, Jr
Waud, Barbara E
Wegman, David Howe
Weinberger, Steven Elliott
Weinstein, Louis
Weinstein, Robert
Weintraub, Lewis Robert
Weiss, Scott T
Weller, Peter Fahey
Weller, Thomas Huckle
Wilgram, George Friederich
Witherell, Egilda DeAmicis
Wodinsky, Isidore
Wolf, Gerald Lee
Wolff, Peter Hartwig
Wolff, Sheldon Malcolm
Wyler, David J
Wyman, Stanley M
Yunis, Edmond J
Zamecnik, Paul Charles
Zurier, Robert B

MICHIGAN
Al-Sarraf, Muhyi
Anderson, David G
Anderson, Marvin David
Ansbacher, Rudi
Axelrod, Arnold Raymond
Bacon, George Edgar
Bagchi, Mihir
Barr, Mason, Jr
Bauer, Jere Marklee
Beierwaites, William Henry
Beitins, Inese Zinta
Bender, Leonard Franklin
Biermann, Janet Sybil
Bissell, Grosvenor Willse
Block, Duane Llewellyn
Bole, Giles G
Bone, Henry G
Brennan, Michael James
Buchanan, Robert Alexander
Burdi, Alphonse R
Caldwell, John R
Castor, Cecil William
Cerny, Joseph Charles
Chou, Ching-Chung
Christenson, Paul John
Clapper, Muir
Cohen, Flossie
Corbett, Thomas Hughes
Cuatrecasas, Pedro
Curtis, George Clifton
Davenport, Horace Willard
Dekornfeld, Thomas John
DeMuth, George Richard
Diaz, Fernando G
Dick, Macdonald, II
Dilts, Preston Vine, Jr
Diokno, Ananias Cornejo
Duff, Ivan Francis
Ellis, C N
Enzer, Norbert Beverley
Fajans, Stefan Stanislaus
Fekety, F Robert, Jr
Fernandez-Madrid, Felix
Fitzgerald, Robert Hannon, Jr
Floyd, John Claiborne, Jr
Frank, Donald Joseph
Frank, Robert Neil
Freimanis, Atis K
Gabrielsen, Trygve O
Gelehrter, Thomas David
Gilman, Sid
Gilroy, John
Goldstein, Sidney
Goodman, Jay Irwin
Gossain, Ved Vyas
Gottschalk, Alexander
Green, Robert A
Grossman, Herbert Jules
Hashimoto, Ken
Heifetz, Carl Louis
Henderson, John Warren
Henderson, John Woodworth
Henley, Keith Stuart
Holt, John Floyd

Horwitz, Jerome Philip
Houts, Larry Lee
Howatt, William Frederick
Hsu, Chen-Hsing
Hubbard, William Neill, Jr
Jacobson, Arnold P
Jampel, Robert Steven
Johnson, Tom Milroy
Joseph, Ramon R
Julius, Stevo
Kaplan, Joseph
Kauffman, Carol A
King, Charles Miller
Kithier, Karel
Knopf, Ralph Fred
Kruse, James Alexander
Kuhl, David Edmund
Lapides, Jack
Lentz, Paul Jackson, Jr
Lerner, Albert Martin
Levine, Steven Richard
Lisak, Robert Philip
Livingood, Clarence Swinehart
Lockwood, Dean H
Lovell, Robert Gibson
Lusher, Jeanne Marie
McCormick, J Justin
McGrath, Charles Morris
Mack, Robert Emmet
McLean, James Amos
Mader, Ivan John
Magen, Myron S
Mahajan, Sudesh K
Maher, Veronica Mary
Martel, William
Maruyama, Yosh
Matovinovic, Josip
Mattson, Joan C
Michelakis, Andrew M
Mikkelsen, William Mitchell
Miller, Orlando Jack
Mitchell, Jerry R
Moghissi, Kamran S
Mohrland, J Scott
Morley, George W
Murray, Dennis Lynn
Murray, Raymond Harold
Nyboer, Jan
Oberman, Harold A
Oliver, William J
Olson, Nels Robert
Osborn, June Elaine
Ownby, Dennis Randall
Palmer, Kenneth Charles
Patterson, Maria Jevitz
Perrin, Eugene Victor
Perry, Harold
Pinkus, Hermann (Karl Benno)
Pitt, Bertram
Pomerleau, Ovide F
Potchen, E James
Poulik, Miroslav Dave
Powsner, Edward R
Prasad, Ananda S
Puro, Donald George
Pysh, Joseph John
Rapp, Robert
Rebuck, John Walter
Redding, Foster Kinyon
Reed, Melvin LeRoy
Riley, Michael Verity
Rogers, William Leslie
Rovner, David Richard
Sander, C Maureen
Schmaier, Alvin Harold
Schteingart, David E
Senagore, Anthony J
Shafer, A William
Shanberge, Jacob N
Sherman, Alfred Isaac
Simon, Michael Richard
Smith, Edwin Mark
Sokol, Robert James
Sorscher, Alan J
Soulen, Renate Leroi
Sparks, Robert D
Spink, Gordon Clayton
Spitz, Werner Uri
Stein, Paul David
Stern, Aaron Milton
Strang, Ruth Hancock
Suhrland, Leif George
Sullivan, Donita B
Swisher, Scott Neil
Tannen, Richard L
Thoene, Jess Gilbert
Thompson, George Richard
Thornbury, John R
Tishkoff, Garson Harold
Triebwasser, John
Vaitkevicius, Vainutis K
Vaughn, Clarence Benjamin
Vavra, James Joseph
Voorhees, John James
Waggoner, Raymond Walter
Ward, Robert C
Weber, Wendell W
Weg, John Gerard
Wegman, Myron Ezra
Weil, William B, Jr
Weiss, Joseph Jacob
Welsch, Clifford William, Jr
Weng, (Frank) Tzong-Ruey
Wicha, Max S

Wiggins, Roger C
Wiley, John W
Wilhelm, Rudolf Ernst
Willis, Park Weed, III
Wollschlaeger, Gertraud
Wolter, J Reimer
Yang, Gene Ching-Hua
Zarafonetis, Chris John Dimiter
Zweifler, Andrew J

MINNESOTA
Amplatz, Kurt
Andersen, Howard Arne
Ansari, Azam U
Awad, Essam A
Bacaner, Marvin Bernard
Baldus, William Phillip
Balfour, Henry H, Jr
Bartholomew, Lloyd Gibson
Bayrd, Edwin Dorrance
Berge, Kenneth G
Berman, Reuben
Bisel, Harry Ferree
Bowie, Edward John Walter
Brunning, Richard Dale
Buckley, Joseph J
Burchell, Howard Bertram
Burke, Edmund C
Cain, James Clarence
Carryer, Haddon McCutchen
Cervenka, Jaroslav
Ciriacy, Edward W
Cohn, Jay Norman
Connelly, Donald Patrick
Corbin, Kendall Brooks
Crowley, Leonard Vincent
Dai, Xue Zheng (Charlie)
Dalmasso, Agustin Pascual
Daugherty, Guy Wilson
Dewald, Gordon Wayne
Dousa, Thomas Patrick
Duane, William Charles, Jr
Du Shane, James William
Duvall, Arndt John, III
Eaton, John Wallace
Ebner, Timothy John
Ettinger, Milton G
Etzwiler, Donnell D
Fairbairn, John F, II
Fairbanks, Virgil
Ferris, Thomas Francis
Fisch, Robert O
Fraley, Elwin E
Francis, Gary Stuart
Fuller, Benjamin Franklin, Jr
Gastineau, Clifford Felix
Gault, N(eal) L, Jr
Gebhard, Roger Lee
Geraci, Joseph E
Ghazali, Masood Raheem
Gilbertsen, Victor Adolph
Gleich, Gerald J
Gobel, Frederick L
Goetz, Frederick Charles
Goldstein, Norman Philip
Gorman, Colum A
Gross, John Burgess
Haase, Ashley Thomson
Hagedorn, Albert Berner
Hall, Wendell Howard
Harris, Jean Louise
Hebbel, Robert P
Hermans, Paul E
Hill, John Roger
Hollenhorst, Robert William, Sr
Holtzman, Jordan L
Housmans, Philippe Robert H P
Hyatt, Robert Eliot
Jacob, Harry S
Jenkins, Robert Brian
Jepson, William W
Jessen, Carl Roger
Johnson, Joseph Richard
Juergens, John Louis
Kaplan, Manuel E
Keane, William Francis
Kearns, Thomas P
Kennedy, Byrl James
Kennedy, William Robert
Kiang, David Teh-Ming
King, Richard Allen
Kottke, Bruce Allen
Kottke, Frederic James
Krivit, William
Kurland, Leonard T
Kyle, Robert Arthur
Lakin, James D
LeBien, Tucker W
Levitan, Alexander Allen
Levitt, Michael D
Levitt, Seymour H
Logan, George Bryan
Loken, Merle Kenneth
Lucas, Alexander Ralph
Lucas, Russell Vail, Jr
Luthra, Harvinder Singh
Lynch, Peter John
McCollister, Robert John
McConahey, William McConnell, Jr
McCullough, John Jeffrey
McGill, Douglas B
McMahon, M Molly
McPherson, Thomas C(oatsworth)
Malkasian, George D, Jr

Martin, Gordon Mather
Michael, Alfred Frederick, Jr
Michels, Lester David
Millhouse, Oliver Eugene
Milliner, Eric Killmon
Moertel, Charles George
Morlock, Carl G
Mulder, Donald William
Mulhausen, Robert Oscar
Murray, M(urray) John
Nelson, John Daniel
Nichols, Donald Richardson
Nuttall, Frank Q
Obrien, Peter Charles
O'Connell, Edward J
Olsen, Arthur Martin
Oppenheimer, Jack Hans
Page, Arthur R
Paller, Mark Stephen
Perry, Harold Otto
Peterson, Harold Oscar
Peterson, Phillip Keith
Phillips, Sidney Frederick
Pierre, Robert V
Polzin, David J
Popkin, Michael Kenneth
Prem, Konald Arthur
Quie, Paul Gerhardt
Rahman, Yueh Erh
Randall, Raymond Victor
Reed, Charles E
Rehder, Kai
Reitemeier, Richard Joseph
Riggs, Byron Lawrence
Rogers, Roy Steele, III
Rohlfing, Stephen Roy
Romero, Juan Carlos
Rosenberg, Murray David
Sabath, Leon David
Schultz, Alvin Leroy
Schuman, Leonard Michael
Scudamore, Harold Hunter
Shorter, Roy Gerrard
Siekert, Robert George
Simon, Geza
Smith, Lucian Anderson
Spilker, Bert
Spittell, John A, Jr
Sprague, Randall George
Stickler, Gunnar B
Stillwell, George Keith
Strong, Cameron Gordon
Theologides, Athanasios
Tobian, Louis
Torres, Fernando
Tuna, Naip
Ulstrom, Robert A
Van Bergen, Frederick Hall
Vennes, Jack A
Verby, John E
Vernier, Robert L
Wang, Yang
Warwick, Warren J
Webster, David Dyer
Weinshilboum, Richard Merle
Weir, Edward Kenneth
Weissler, Arnold M
Westmoreland, Barbara Fenn
Whisnant, Jack Page
Wild, John Julian
Winchell, C Paul
Winkelmann, Richard Knisely
Wirtschafter, Jonathan Dine
Wood, Michael Bruce
Zelterman, Daniel
Zieve, Leslie

MISSISSIPPI
Batson, Blair Everett
Batson, Margaret Bailly
Bell, Warren Napier
Blake, Thomas Mathews
Brooks, Thomas Joseph, Jr
Corbett, James John
Currier, Robert David
Evers, Carl Gustav
Grenfell, Raymond Frederic
Haerer, Armin Friedrich
Hellems, Harper Keith
Holder, Thomas M
Hutchison, William Forrest
Jackson, John Fenwick
Johnson, Ben Butler
Johnson, Samuel Britton
LeBlanc, Michael H
Lockwood, William Rutledge
Long, Billy Wayne
Montani, Jean-Pierre
Morrison, John Coulter
Shenefelt, Ray Eldon
Steinberg, Martin H
Tavassoli, Mehdi
Uzodinma, John E
Watson, David Goulding
Wiser, Winfred Lavern

MISSOURI
Abdou, Nabih I
Alpers, David Hershel
Anderson, Philip Carr
Atkinson, John Patterson
Avioli, Louis
Bajaj, S Paul
Barbero, Giulio J

Medicine (cont)

Bauer, John Harry
Bergmann, Steven R
Bier, Dennis Martin
Boyarsky, Saul
Brodeur, Armand Edward
Brown, Elmer Burrell
Bryant, Lester Richard
Burke, William Joseph
Burns, Thomas Wade
Cassidy, James T
Chaplin, Hugh, Jr
Chapman, Ramona Marie
Chu, Jen-Yih
Cole, Barbara Ruth
Colten, Harvey Radin
Colwill, Jack M
Crosby, William Holmes, Sr
Cryer, Philip Eugene
Daly, James William
Danforth, William H
Danis, Peter Godfrey
Daughaday, William Hamilton
Davis, Bernard Booth
Diehl, Antoni Mills
Dimond, Edmunds Grey
Dodge, Philip Rogers
Donati, Robert M
Dueker, David Kenneth
Eggers, George W Nordholtz, Jr
Eliasson, Sven Gustav
Ericson, Avis J
Fabian, Leonard William
Fernandez-Pol, Jose Alberto
Ferrendelli, James Anthony
Fitch, Coy Dean
Fletcher, Anthony Phillips
Fletcher, James W
Forker, E Lee
Frawley, Thomas Francis
Fredrickson, John Murray
Freeman, Arnold I
Gallagher, Neil Ignatius
Gantner, George E, Jr
Garner, Harold E
Gilula, Louis Arnold
Givler, Robert L
Graham, Robert
Griffin, William Thomas
Griggs, Douglas M, Jr
Grunt, Jerome Alvin
Guze, Samuel Barry
Hagan, John Charles, III
Hall, David Goodsell, III
Harford, Carl Gayler
Harvey, Joseph Eldon
Hellerstein, Stanley
Herzig, Geoffrey Peter
Hillman, Richard Ephraim
Hodges, Glenn R(oss)
Horenstein, Simon
Horwitz, Edwin M
Hyers, Thomas Morgan
Ide, Carl Heinz
Karl, Michael M
Keenan, William Jerome
King, Morris Kenton
Kinsella, Ralph A, Jr
Kipnis, David Morris
Kissane, John M
Klahr, Saulo
Knight, William Allen, Jr
Kornfeld, Stuart Arthur
Kovacs, Sandor J, Jr
Kulczycki, Anthony, Jr
Lake, Lorraine Frances
Landau, William M
Lennartz, Michelle R
Loeb, Virgil, Jr
Lonigro, Andrew Joseph
Luxon, Bruce Arlie
Majerus, Philip W
Masters, William Howell
Middelkamp, John Neal
Mooradian, Arshag Dertad
Mudd, J Gerard
Nahm, Moon H
Noback, Richardson K
Nuetzel, John Arlington
Oliver, G Charles
Olson, Lloyd Clarence
Ortwerth, Beryl John
Parker, Brent M
Parker, Charles W
Parker, Mary Langston
Pearlman, Alan L
Perez, Carlos A
Perkoff, Gerald Thomas
Perry, Horace Mitchell, Jr
Pestronk, Alan
Pierce, John Albert
Powers, William John
Purkerson, Mabel Louise
Raichle, Marcus Edward
Reinhard, Edward Humphrey
Ridings, Gus Ray
Ritter, Hubert August
Rolf, Clyde Norman
Royal, Henry Duval
Santiago, Julio Victor
Schonfeld, Gustav
Schultz, Irwin
Schweiss, John Francis
Shank, Robert Ely
Shoemaker, James Daniel
Shuter, Eli Ronald
Siegel, Barry Alan
Sirridge, Marjorie Spurrier
Slatopolsky, Eduardo
Slavin, Raymond Granam
Stephen, Charles Ronald
Stoneman, William, III
Stroud, Malcolm Herbert
Thoma, George Edward
Twardowski, Zbylut Jozef
Vietti, Teresa Jane
Welch, Michael John
Weldon, Virginia V
Wenner, Herbert Allan
Whyte, Michael Peter
Wilkinson, Charles Brock
Woolsey, Robert M
Worthington, Ronald Edward
Yarbro, John Williamson

MONTANA

Boggs, Dane Ruffner
Opitz, John Marius
Pincus, Seth Henry
Priest, Jean Lane Hirsch
Swanson, John L
Williams, Roger Stewart

NEBRASKA

Aita, John Andrew
Angle, William Dodge
Beattie, Craig W
Booth, Richard W
Brody, Alfred Walter
Clifford, George O
Colombo, John Louis
Cromwell, Norman Henry
Dalrymple, Glenn Vogt
Davis, Richard Bradley
Duckworth, William Clifford
Engel, Toby Ross
Foley, John F
Fuenning, Samuel Isaiah
Fusaro, Ramon Michael
Gendelman, Howard Eliot
Gifford, Harold
Grissom, Robert Leslie
Heaney, Robert Proulx
Hunt, Howard Beeman
Keim, Lon William
Kessinger, Margaret Anne
Klassen, Lynell W
Kobayashi, Roger Hideo
Luby, Robert James
Lynch, Henry T
McIntire, Matilda S
Malashock, Edward Marvin
Millikan, Clark Harold
Mirvish, Sidney Solomon
Mohiuddin, Syed M
Mooring, Paul K
Nagel, Donald Lewis
Nair, Chandra Kunju
O'Brien, Richard Lee
Paustian, Frederick Franz
Pearson, Paul (Hammond)
Ruddon, Raymond Walter, Jr
Sanders, W Eugene, Jr
Skultety, Francis Miles
Sorrell, Michael Floyd
Stratta, Robert Joseph
Tempero, Margaret Ann
Tobin, Richard Bruce
Toth, Bela
Waggener, Ronald E
Ware, Frederick
Yonkers, Anthony J

NEVADA

Buxton, Iain Laurie Offord
DeFelice, Eugene Anthony
Dehne, Edward James
Ferguson, Roger K
Manalo, Pacita
Pool, Peter Edward
Whipple, Gerald Howard
Zanjani, Esmail Dabaghchian

NEW HAMPSHIRE

Almy, Thomas Pattison
Brooks, John Graham
Chalmers, Thomas Clark
Chapman, Carleton Burke
Clendenning, William Edmund
Guinane, James Edward
Kelley, Maurice Leslie, Jr
Kessner, David Morton
McIntyre, Oswald Ross
Mudge, Gilbert Horton
Ritzman, Thomas A
Rolett, Ellis Lawrence
Rous, Stephen N
Sox, Harold C
Strauss, Bella S
Valtin, Heinz
Zubkoff, Michael

NEW JERSEY

Ahmed, S Sultan
Alger, Elizabeth A
Amory, David William
Auerbach, Oscar
Aviv, Abraham
Ayvazian, L Fred
Banerjee, Tapas Kumar
Behrle, Franklin C
Bergen, Stanley S, Jr
Bierenbaum, Marvin L
Block, A Jay
Boksay, Istvan Janos Endre
Boyle, Joseph, III
Breen, James Langhorne
Burack, Walter Richard
Carter, Stephen Keith
Carver, David Harold
Chilton, Neal Warwick
Chinard, Francis Pierre
Cinotti, Alfonse A
Ciosek, Carl Peter, Jr
Conn, Hadley Lewis, Jr
Cook, Stuart D
Coutinho, Claude Bernard
Cross, Richard James
Crump, Jesse Franklin
Daly, John F
Das, Kiron Moy
Davies, Richard O
Day-Salvatore, Debra-Lynn
De Salva, Salvatore Joseph
Douglas, Robert Gordon, Jr
Dunn, Jonathan C
Duvoisin, Roger C
Edelman, Norman H
Eisinger, Robert Peter
Ellison, Rose Ruth
Eng, Robert H K
Ertel, Norman H
Evans, Hugh E
Finch, Stuart Cecil
Ford, Neville Finch
Franciosa, Joseph Anthony
Gaudino, Mario
Ginsberg, Barry Howard
Gochfeld, Michael
Goger, Pauline Rohm
Goldenberg, David Milton
Goldsmith, Michael Allen
Goldstein, Bernard David
Goldstein, Gideon
Goodhart, Robert Stanley
Goodkind, Morton Jay
Gorodetzky, Charles W
Gross, Peter A
Guo, Jian Zhong
Haft, Jacob I
Hall, Thomas Christopher
Hill, George James, II
Holler, Jacob William
Hutcheon, Duncan Elliot
Hutter, Robert V P
Imaeda, Tamotsu
Jacobus, David Penman
Jaffe, Ernst Richard
Johanson, Waldemar Gustave, Jr
Katsampes, Chris Peter
Khachadurian, Avedis K
Kirschner, Marvin Abraham
Krey, Phoebe Regina
Lane, Alexander Z
Lanzoni, Vincent
Lasfargues, Etienne Yves
Lattes, Raffaele
Lawrason, F Douglas
Layman, William Arthur
Leevy, Carroll M
Lehrer, Harold Z
Leibowitz, Michael Jonathan
Leitz, Victoria Mary
LeSher, Dean Allen
Levine, Robert
Levy, David Edward
Lipkin, George
Lourenco, Ruy Valentim
Louria, Donald Bruce
Marshall, Carter Lee
Medway, William
Merriam, George Rennell, Jr
Mezey, Kalman C
Moolten, Sylvan E
Morse, Bernard S
Motz, Robin Owen
Ngai, Shih Hsun
Niederland, William G
Nies, Alan Sheffer
Oleske, James Matthew
Panush, Richard Sheldon
Perr, Irwin Norman
Pinsky, Carl Muni
Poliak, Aaron
Polinsky, Ronald John
Poorvin, David Walter
Prineas, John William
Quinones, Mark A
Raetz, Christian Rudolf Hubert
Raskova, Jana D
Regan, Timothy Joseph
Royce, Paul C
Salans, Lester Barry
Scaglione, Peter Robert
Schleifer, Steven Jay
Scolnick, Edward M
Seneca, Harry
Singer, Robert Meril
Slater, Eve Elizabeth
Spector, Reynold
Stambaugh, John Edgar, Jr
Stern, Elizabeth Kay
Udem, Stephen Alexander
Vaun, William Stratin
Vukovich, Robert Anthony
Wachs, Gerald N
Wallach, Jacques Burton
Walters, Thomas Richard
Wedeen, Richard P
Weiss, Stanley H
Weisse, Allen B
Wildnauer, Richard Harry
Winters, Robert Wayne
Woske, Harry Max
Xiong, Yimin
Zuckerman, Leo

NEW MEXICO

Abrams, Jonathan
Appenzeller, Otto
Brown, Harold
Carter, Robert Eldred
Cobb, John Candler
Cochran, Paul Terry
Crawford, Michael Howard
Davis, Larry Ernest
Douglas, William Kennedy
Florman, Alfred Leonard
Frenkel, Jacob Karl
Gardner, Kenneth Drake, Jr
Leach, John Kline
Levy, Leo
Lindeman, Robert D
Mason, William van Horn
Morris, Don Melvin
Overturf, Gary D
Palmer, Darwin L
Pena, Hugo Gabriel
Raju, Mudundi Ramakrishna
Rhodes, Buck Austin
Schottstaedt, William Walter
Sloan, Robert Dye
Tempest, Bruce Dean
Tobey, Robert Allen
Voelz, George Leo
Vorherr, Helmuth Wilhelm
Walker, A Earl
Wax, Joan
Zamora, Paul O

NEW YORK

Abolhassni, Mohsen
Abramson, Allan Lewis
Ackerman, Bruce David
Ackerman, Norman Bernard
Adamson, John William
Adelson, Mark David
Adler, Alexandra
Ahrens, Edward Hamblin, Jr
Aisen, Philip
Al-Askari, Salah
Aldrich, Thomas K
Alexander, Leslie Luther
Allison, David B
Altman, Lawrence Kimball
Altner, Peter Christian
Ambrus, Clara Maria
Ambrus, Julian Lawrence
Anderson, Albert Douglas
Appelbaum, Emanuel
Archibald, Reginald MacGregor
Aronow, Wilbert Solomon
Aronson, Ronald Stephen
Artusio, Joseph F, Jr
Ascensao, Joao L
Ashutosh, Kumar
Askanazi, Jeffrey
Atkins, Harold Lewis
Atlas, Steven Alan
Atwater, Edward Congdon
Auchincloss, Joseph Howland, Jr
Auld, Peter A McF
Bachvaroff, Radoslav J
Baer, Rudolf L
Baker, David A
Baker, David H
Baldini, James Thomas
Balint, John Alexander
Ballentine, Rudolph Miller
Balthazar, Emil Jacque
Bank, Arthur
Bank, Norman
Bannon, Robert Edward
Barka, Tibor
Barnett, Henry Lewis
Barondess, Jeremiah A
Barron, Bruce Albrecht
Barron, Kevin D
Bartal, Arie H
Baruch, Sulamita B
Barzel, Uriel S
Bateman, John Laurens
Baum, John
Baum, Stephen Graham
Bauman, Norman
Baxter, Donald Henry
Beal, Myron Clarence
Beam, Thomas Roger
Bearn, Alexander Gordon
Becker, David Victor
Becker, Joshua A
Beckerman, Barry Lee
Beebe, Richard Townsend
Beinfield, William Harvey
Bell, Alfred Lee Loomis, Jr

MEDICAL & HEALTH SCIENCES / 459

Bell, James Milton
Bennett, James Anthony
Bennett, John M
Beranbaum, Samuel Louis
Berger, Lawrence
Bergstrom, William H
Berkman, James Israel
Berlyne, Geoffrey Merton
Bertles, John F
Bhattacharya, Jahar
Biedler, June Lee
Biempica, Luis
Bigger, J Thomas, Jr
Bito, Laszlo Z
Blaufox, Morton D
Bleicher, Sheldon Joseph
Bloomfield, Dennis Alexander
Blumenstock, David A
Bockman, Richard Steven
Bogdonoff, Morton David
Boley, Scott Jason
Bollet, Alfred Jay
Bond, Victor Potter
Borer, Jeffrey Stephen
Borkowsky, William
Bosniak, Morton A
Bradford, James Carrow
Bradner, William Turnbull
Brand, Leonard
Brandaleone, Harold
Brandriss, Michael W
Braunstein, Joseph David
Brennan, John Paul
Brennan, Murray F
Brentjens, Jan R
Brick, Irving B
Briehl, Robin Walt
Briggs, Donald K
Brightman, I Jay
Brody, Bernard B
Brown, Audrey Kathleen
Brown, David Frederick
Brown, Lawrence S, Jr
Bruck, Erika
Buhac, Ivo
Burchenal, Joseph Holland
Burde, Ronald Marshall
Burgener, Francis Andre
Buschke, Herman
Bushinsky, David Allen
Butler, John Joseph
Butler, Vincent Paul, Jr
Buxbaum, Joel N
Byles, Peter Henry
Cabrera, Edelberto Jose
Cahill, Kevin M
Camazine, Scott
Canfield, Robert E
Canizares, Orlando
Cannon, Paul Jude
Carlson, Harold Ernest
Carr, Edward Albert, Jr
Carr, Ronald E
Carruthers, Christopher
Carter, Anne Cohen
Carter, David Martin
Case, Robert B
Chandra, Pradeep
Chang, Chu Huai
Chase, Norman E
Chase, Randolph Montieth, Jr
Chasis, Herbert
Chess, Leonard
Chey, William Y
Chilcote, Max Eli
Chodos, Robert Bruno
Choi, Tai-Soon
Christy, Nicholas Pierson
Chu, Florence Chien-Hwa
Chutkow, Jerry Grant
Clark, Duncan William
Clark, Julian Joseph
Clarkson, Bayard D
Cochran, George Van Brunt
Cockett, Abraham Timothy K
Cohen, Alfred M
Cohen, Bernard
Cohen, Burton D
Cohen, Harris L
Cohen, Irving Allan
Cohen, Jordan J
Cohen, Jules
Cohen, Noel Lee
Cohlan, Sidney Quex
Cohn, Peter Frank
Cohn, Zanvil A
Cole, Harold S
Coleman, Morton
Coller, Barry Spencer
Collins, George H
Colman, Neville
Cooper, Louis Zucker
Copley, Alfred Lewin
Cote, Lucien Joseph
Cowger, Marilyn L
Craig, Albert Burchfield, Jr
Craig, John Philip
Cronkite, Eugene Pitcher
Crystal, Ronald George
Cucci, Cesare Eleuterio
Cunningham, Nicholas C
Cunningham-Rundles, Charlotte
Dack, Simon
Dancis, Joseph

Danilowicz, Delores Ann
Dao, Thomas Ling Yuan
Davis, James N(orman)
Davis, Paul Joseph
Day, Stacey Biswas
Dean, David Campbell
Deane, Norman
DeCosse, Jerome J
Deibel, Rudolf
Delagi, Edward F
DeLand, Frank H
Dell, Ralph Bishop
Demeter, Steven
Deming, Quentin Burritt
Demis, Dermot Joseph
Deuschle, Kurt W
DiMauro, Salvatore
Di Paola, Mario
DJang, Arthur H K
Dolgin, Martin
Dornfest, Burton S
Dorsey, Thomas Edward
Douglas, Gordon Watkins
Dow, Bruce MacGregor
Downey, John A
Doyle, Eugenie F
Doyle, Joseph Theobald
Dreizen, Paul
Drusin, Lewis Martin
Dubroff, Lewis Michael
Duffy, Philip
Duncalf, Deryck
Durr, Friedrich (E)
Dutton, Cynthia Baldwin
Dutton, Robert Edward, Jr
Dworetzky, Murray
Eagle, Harry
Edelmann, Chester M, Jr
Edelson, Paul J
Egan, Edmund Alfred
Eich, Robert
Eichna, Ludwig Waldemar
Eisenberg, M Michael
Elizan, Teresita S
Elkin, Milton
Engel, George Libman
Engle, Mary Allen English
Engle, Ralph Landis, Jr
Escher, Doris Jane Wolf
Eschner, Edward George
Fabry, Thomas Lester
Fahn, Stanley
Faloon, William Wassell
Farber, Saul Joseph
Feld, Leonard Gary
Feldman, B Robert
Felig, Philip
Felman, Yehudi M
Ferris, Philip
Ferrone, Soldano
Field, Michael
Finberg, Laurence
Fink, Austin Ira
Finkbeiner, John A
Finlay, Jonathan Lester
Fischel, Edward Elliot
Fischman, Donald A
Fish, Irving
Fleischmajer, Raul
Fleischman, Alan R
Foldes, Francis Ferenc
Foley, Kathleen M
Foley, William Thomas
Fontana, Vincent J
Forbes, Gilbert Burnett
Fox, Arthur Charles
Frame, Paul S
Francis, Charles K
Frank, Lawrence
Frantz, Andrew Gibson
Frazer, John P
Freedberg, Irwin Mark
Freedman, Aaron David
Freeman, Richard B
Freitag, Julia Louise
French, Joseph H
Friedewald, William Thomas
Friedman, Alan Herbert
Friedman, Eli A
Friedman, Emanuel A
Friedman, Irwin
Frishman, William Howard
Fritts, Harry Washington, Jr
Froelich, Ernest
Frost, E A M
Fuchs, Fritz Friedrich
Fulop, Milford
Galdston, Morton
Galin, Miles A
Gambert, Steven Ross
Gardner, Lytt Irvine
Gary, Nancy E
Garza, Cutberto
Geffen, Abraham
Geiger, H Jack
Gellhorn, Alfred
Gerber, Donald Albert
German, James Lafayette, III
Gershberg, Herbert
Gershengorn, Marvin Carl
Gershman, Lewis C
Gershon, Anne A
Gertler, Menard M
Ghosh, Nimai Kumar

Gibbs, David Lee
Giblin, Denis Richard
Gibofsky, Allan
Gidari, Anthony Salvatore
Gilbert, Harriet S
Gilbert, Robert
Gillies, Alastair J
Ginsberg-Fellner, Fredda Vita
Glabman, Sheldon
Glaser, Warren
Glass, George B Jerzy
Glomski, Chester Anthony
Goldensohn, Eli Samuel
Goldring, Roberta M
Goldschmidt, Bernard Morton
Goldsmith, Lowell Alan
Gollub, Seymour
Goodgold, Joseph
Goodman, DeWitt Stetten
Goodman, Richard S
Gootman, Norman Lerner
Gorlin, Richard
Gottlieb, Arlan J
Gould, Lawrence A
Gramiak, Raymond
Granger, Carl V
Greene, David Gorham
Greene, William Allan
Greenwald, Edward S
Gregory, Daniel Hayes
Griffiths, Raymond Bert
Griner, Paul F
Grob, David
Grollman, Arthur Patrick
Gross, Ludwik
Grossman, Jacob
Grynbaum, Bruce B
Gulotta, Stephen Joseph
Gupta, Sanjeev
Gurney, Ramsdell
Gusberg, Saul Bernard
Haas, Albert B
Habicht, Jean-Pierre
Hadley, Susan Jane
Haggerty, Robert Johns
Haines, Thomas Henry
Halevy, Simon
Hall, Charles A
Ham, Richard John
Hambrick, George Walter, Jr
Hamerman, David Jay
Hamill, Robert W
Hamilton, Byron Bruce
Hamilton, Leonard Derwent
Han, Jaok
Han, Tin
Hanks, Edgar C
Harber, Leonard C
Harris, Henry William
Harris, Matthew N
Harrison, Robert Walker, III
Hass, William K
Hauser, Willard Allen
Hawrylko, Eugenia Anna
Haymovits, Asher
Heagarty, Margaret Caroline
Helson, Lawrence
Henderson, Edward S
Henkind, Paul
Henley, Walter L
Henry, John Bernard
Henshaw, Edgar Cummings
Herbert, Victor
Herr, Harry Wallace
Hershey, Solomon George
Higgins, James Thomas, Jr
Hilal, Sadek K
Hinterbuchner, Catherine Nicolaides
Hinterbuchner, L P
Hirsch, Jacob Irwin
Hirsch, Jules
Hirschhorn, Kurt
Hirschman, Shalom Zarach
Hirshaut, Yashar
Hoberman, Henry Don
Hochwald, Gerald Martin
Holland, James Frederick
Hollander, Joshua
Hollander, Vincent Paul
Holloman, John L S, Jr
Holub, Donald Arthur
Holzman, Robert Stephen
Hood, William Boyd, Jr
Hornblass, Albert
Horton, John
Horwitz, Marshall Sydney
Houde, Raymond Wilfred
Houghton, Alan N
Hreshchyshyn, Myroslaw M
Hrushesky, William John Michael
Hutchings, Donald Edward
Hyman, Arthur Bernard
Imperato, Pascal James
Ioachim, Harry L
Jacobs, Laurence Stanton
Jacobson, Harold Gordon
Jacox, Ralph Franklin
Jaenike, John Robert
Jaffe, Eric Allen
Jaffe, Israeli Aaron
Jameson, Arthur Gregory
Janowitz, Henry David
Jarcho, Saul
Javitt, Norman B

Jean-Baptiste, Emile
Johnson, Alan J
Johnson, Horton Anton
Johnson, Kenneth Gerald
Johnston, Richard Boles, Jr
Joos, Howard Arthur
Josephson, Alan S
Joynt, Robert James
Kaiser, Irwin Herbert
Kamm, Donald E
Kanick, Virginia
Kaplan, Allen P
Kaplan, Barry Hubert
Kark, Robert Adriaan Pieter
Karmen, Arthur
Karron, Daniel B
Katz, Michael
Kayden, Herbert J
Kean, Benjamin Harrison
Keefe, Deborah Lynn
Keiser, Harold D
Kemeny, Nancy E
Kieffer, Stephen A
Kilbourne, Edwin Dennis
Kiley, John Edmund
Kim, Jae Ho
Kimberly, Robert Parker
King, Benton Davis
King, Katherine Chung-Ho
King, Robert (Bainton)
King, Thomas K C
Kissin, Benjamin
Klainer, Albert S
Klavins, Janis Viliberts
Klein, Edmund
Klein, Elena Buimovici
Klein, Richard Joseph
Kleinman, Leonard I
Klipstein, Frederick August
Klocke, Robert Albert
Kohl, Schuyler G
Kopf, Alfred Walter
Korein, Julius
Kramer, Elmer E
Kreis, Willi
Krieger, Howard Paul
Krown, Susan E
Krugman, Saul
Kuhn, Leslie A
Kumar, Vijay
Kupfer, Sherman
Kydd, David Mitchell
Lackner, Henriette
Lahita, Robert George
Lalezari, Parviz
Lamberson, Harold Vincent, Jr
Landaw, Stephen Arthur
Landrigan, Philip J
Lane, Lewis Behr
Lanzkowsky, Philip
Laragh, John Henry
Lawrence, Henry Sherwood
Lawrence, Robert Marshall
Lawrence, Robert Swan
Lebwohl, Mark Gabriel
Lee, Mathew Hung Mun
Lee, Stanley L
Lehr, David
Lehrer, Gerard Michael
Leiter, Elliot
Leleiko, Neal Simon
Lenn, Nicholas Joseph
Lepow, Martha Lipson
Lerman, Sidney
Lerner, Robert Gibbs
Leslie, Stephen Howard
Levere, Richard David
Levin, Aaron R
Levin, Morton Loeb
Levine, Bernard Benjamin
Levine, Robert Alan
Levitt, Barrie
Levitt, Marvin Frederick
Levy, Norman B
Lewis, John L, Jr
Ley, Allyn Bryson
Liang, Chang-seng
Libow, Leslie S
Lichtman, Marshall A
Lieber, Charles Saul
Lifshitz, Kenneth
Lipkin, Martin
Lippes, Jack
Lippmann, Heinz Israel
Lipsitz, Philip Joseph
Litman, Nathan
Litwin, Stephen David
Lockshin, Michael Dan
Loeb, John Nichols
Loegering, Daniel John
Lopez, Rafael
Lore, John M, Jr
Low, Niels Leo
Luhby, Adrian Leonard
Lumpkin, Lee Roy
Macario, Alberto J L
McCollester, Duncan L
McCormack, Patricia Marie
McCrory, Wallace Willard
McDowell, Fletcher Hughes
McGiff, John C
Magid, Norman Mark
Magoss, Imre V
Mahler, Richard Joseph

460 / DISCIPLINE INDEX

Medicine (cont)

Maker, Howard Smith
Makman, Maynard Harlan
Manger, William Muir
Marcus, Aaron Jacob
Mark, Herbert
Mark, Lester Charles
Marx, Gertie F
Matz, Robert
Mayberger, Harold Woodrow
Mayer, Klaus
Medici, Paul T
Melamed, Myron Roy
Mellin, Gilbert Wylie
Mellins, Robert B
Meltzer, Richard Stuart
Mendlowitz, Milton
Middleton, Elliott, Jr
Miller, Albert
Miller, Frederick
Miller, Myron
Miller, Russell Loyd, Jr
Milman, Doris H
Milmore, John Edward
Milstoc, Mayer
Mink, Irving Bernard
Mittelman, Arnold
Mohr, Jay Preston
Monheit, Alan G
Moore, Malcolm A S
Morgan, William L, Jr
Morishima, Akira
Morishima, Hisayo Oda
Morrison, John B
Moser, Marvin
Moses, Campbell, Jr
Mueller, Hiltrud S
Muller-Eberhard, Ursula
Munnell, Equinn W
Munson, Benjamin Ray
Murphy, Mary Lois
Murphy, Timothy F
Nachbar, Martin Stephen
Nachman, Ralph Louis
Nagel, Ronald Lafuente
Namba, Tatsuji
Narahara, Hiromichi Tsuda
Natta, Clayton Lyle
Nelson, Douglas A
Nelson, William Pierrepont, III
Nemerson, Yale
Nemir, Rosa Lee
Nesbitt, Robert Edward Lee, Jr
Neuwirth, Robert Samuel
New, Maria Iandolo
Nitowsky, Harold Martin
Nolan, James P
Norman, Alex
Norton, Larry
Notter, Mary Frances
Oettgen, Herbert Friedrich
Ohnuma, Takao
Old, Lloyd John
Olson, Robert Eugene
Olsson, Carl Alfred
O'Mara, Robert E
O'Reilly, Richard
Orkin, Louis R
Orlowski, Craig Charles
Ornt, Daniel Burrows
Oster, Martin William
Overweg, Norbert I A
Packman, Charles Henry
Pantuck, Eugene Joel
Parker, Margaret Maier
Parsons, Donald Frederick
Partin, John Calvin
Pasternak, Gavril William
Paton, David
Pearson, Thomas Arthur
Perlman, Ely
Phelps, Charles E
Phillips, Gerald B
Pickering, Thomas G
Pinck, Robert Lloyd
Pines, Kermit L
Piomelli, Sergio
Pi-Sunyer, F Xavier
Piver, M Steven
Plager, John Everett
Plotz, Charles M
Plum, Fred
Poe, Robert Hilleary
Poiesz, Bernard Joseph
Posner, Jerome B
Post, Joseph
Potmesil, Milam
Qazi, Qutubuddin H
Rabinowitz, Jack Grant
Rabinowitz, Ronald
Rackow, Eric C
Rackow, Herbert
Rai, Kanti R
Rao, Sreedhar P
Rao, Yalamanchili A K
Rapin, Isabelle (Mrs Harold Oaklander)
Rattazzi, Mario Cristiano
Rausen, Aaron Reuben
Reader, George Gordon
Reed, George Farrell
Reich, Nathaniel Edwin
Reichberg, Samuel Bringeissen
Reichgott, Michael Joel

Reidenberg, Marcus Milton
Reis, Donald J
Reit, Barry
Reuben, Richard N
Reynoso, Gustavo D
Rhodes, Robert Shaw
Richart, Ralph M
Richman, Robert Alan
Rieder, Ronald Frederic
Riehle, Robert Arthur, Jr
Rifkind, Richard A
Rivlin, Richard S
Rizack, Martin A
Roach, John Faunce
Rodgers, John Barclay, Jr
Roe, Daphne A
Rogers, David Elliott
Roman, Stanford A, Jr
Romney, Seymour L
Rose, Arthur L
Rose, Herbert G
Rosen, David A
Rosen, Mortimer Gilbert
Rosenfeld, Isadore
Rosenfeld, Stephen I
Rosenfield, Richard Ernest
Rosenstreich, David Leon
Rosensweig, Norton S
Rosenthal, William S
Rosner, Fred
Rosner, William
Ross, Gilbert Stuart
Rossman, Isadore
Roth, Daniel
Rothenberg, Sheldon Philip
Rothman, Francoise
Rotman, Marvin Z
Rowe, John W
Rowland, Lewis Phillip
Rowley, Peter Templeton
Ruben, Robert Joel
Rubin, Albert Louis
Rubin, Samuel H
Ruggiero, David A
Ruskin, Richard A
Russ, Gerald A
Sachs, John Richard
Sadock, Benjamin
Sager, Clifford J
Sagerman, Robert H
Said, Sami I
Sako, Kumao
Salanitre, Ernest
Sassa, Shigeru
Satran, Richard
Sawitsky, Arthur
Sawyer, William D
Scarpelli, Emile Michael
Schachter, David
Schaffner, Fenton
Schauf, Victoria
Schaumburg, Herbert Howard
Schecter, Arnold Joel
Scheig, Robert L
Scheinberg, Israel Herbert
Scherr, Lawrence
Schlaeger, Ralph
Schlagenhauff, Reinhold Eugene
Schmale, Arthur H, Jr
Schmid, Franz Anton
Schneck, Larry
Schneider, Bruce Solomon
Schreiber, Sidney S
Schulman, Harold
Schwartz, Irving Leon
Schwartz, Sheldon E
Schwartz, Stanley Allen
Schwarz, Richard Howard
Sciarra, Daniel
Scopp, Irwin Walter
Seaman, William B
Sechzer, Philip Haim
Selby, Henry M
Seligman, Stephen Jacob
Shafer, Stephen Quentin
Shafritz, David Andrew
Shalita, Alan Remi
Shamoon, Harry
Shamos, Morris Herbert
Shank, Brenda Mae Buckhold
Sherman, Samuel Murray
Shimm, Robert A
Shreeve, Walton Wallace
Sidel, Victor William
Siegal, Frederick Paul
Siegel, Morris
Silber, Robert
Silver, Jack
Silver, Lawrence
Silver, Richard Tobias
Skucas, Jovitas
Smith, Joseph Jay
Smith, Ora Kingsley
Smith, Richard Andrew
Smulyan, Harold
Snyder, Ruth Evelyn
Snyderman, Selma Eleanore
Soave, Rosemary
Solish, George Irving
Solomon, Seymour
Som, Prantika
Sonnenblick, Edmund H
Southren, A Louis
Southworth, Hamilton

Spaet, Theodore H
Spear, Paul William
Spencer, Frank Cole
Spitz, Irving Manfred
Spitzer, Adrian
Spitzer, Roger Earl
Staubitz, William Joseph
Stein, Arthur A
Stein, Ruth E K
Steiner, Morris
Steinfeld, Leonard
Steinman, Charles Robert
Stenzel, Kurt Hodgson
Sternberg, Stephen Stanley
Sternlieb, Irmin
Stevenson, Stuart Shelton
Stewart, William Andrew
Stokes, Peter E
Stone, Daniel Joseph
Stone, Martin L
Straus, Marc J
Streeten, David Henry Palmer
Stutman, Leonard Jay
Subramanian, Gopal
Susca, Louis Anthony
Susskind, Herbert
Swartz, Donald Percy
Swartz, Harry
Taggart, John Victor
Tamm, Igor
Tan, Charlotte
Tanigaki, Nobuyuki
Tapley, Donald Fraser
Taub, Robert Norman
Tennant, Bud C
Tepperman, Jay
Tessler, Arthur Ned
Texon, Meyer
Thiede, Henry A
Thomas, Lewis
Thompson, David Duvall
Tillotson, James Richard
Titus, Charles O
Torre, Douglas Paul
Tricomi, Vincent
Triner, Lubos
Troutman, Richard Charles
Trubek, Max
Tsan, Min-Fu
Turino, Gerard Michael
Ureles, Alvin L
Valentine, Fred Townsend
Vanamee, Parker
Van Poznak, Alan
Van Woert, Melvin H
Varma, Andre A O
Veenema, Ralph J
Vivona, Stefano
Volkman, David J
Voorhess, Mary Louise
Wang, Theodore Sheng-Tao
Warner, Robert
Warren, Kenneth S
Wasserman, Edward
Wasserman, Louis Robert
Watson, James Dewey
Weinstein, I Bernard
Weiss, Harvey Jerome
Weissman, Charles
Weissman, Michael Herbert
Weissmann, Gerald
Weitzman, Elliot D
Weksler, Babette Barbash
Weksler, Marc Edward
Wertheim, Arthur Robert
Wessler, Stanford
Wigger, H Joachim
Williams, Marshall Henry, Jr
Williams, Thomas Franklin
Williams, William Joseph
Wilner, George Dubar
Wilson, Michael Friend
Winawer, Sidney J
Wingate, Martin Bernard
Winick, Myron
Winkler, Norman Walter
Winnie, Glenna Barbara
Wise, Harold B
Wishnick, Marcia M
Wolf, Julius
Wolinsky, Harvey
Woodin, William Graves
Wormser, Gary Paul
Wynder, Ernst Ludwig
Yablonski, Michael Eugene
Yahr, Melvin David
Yang, Dorothy Chuan-Ying
Yang, Song-Yu
Yao, Alice C
Yarinsky, Allen
Yates, Jerome William
Yolles, Stanley Fausst
Young, Charles William
Young, John Ding-E
Young, Morris Nathan
Yu, Paul N
Yu, Ts'ai-Fan
Yuceoglu, Yusuf Ziya
Zalusky, Ralph
Zigman, Seymour
Zimmer, James Griffith
Zimmon, David Samuel
Zingesser, Lawrence H
Zitrin, Charlotte Marker

Zonereich, Samuel
Zucker, Marjorie Bass
Zugibe, Frederick T
Zumoff, Barnett

NORTH CAROLINA

Arena, Jay M
Argenzio, Robert Alan
Barnett, Thomas Buchanan
Barry, David Walter
Bass, David A
Bennett, Peter Brian
Benson, Walter Russell
Bentzel, Carl Johan
Blythe, William Brevard
Bondurant, Stuart
Bowie, James Dwight
Boyer, Cinda Marie
Bradford, William Dalton
Bream, Charles Anthony
Breitschwerdt, Edward B
Briggaman, Robert Alan
Brown, Robert Calvin
Buckalew, Vardaman M, Jr
Buckley, Charles Edward
Buckley, Rebecca Hatcher
Bunce, Paul Leslie
Callaway, Jasper Lamar
Capizzi, Robert L
Carroll, Robert Graham
Chamberlin, Harrie Rogers
Chandler, Arthur Cecil, Jr
Chen, Michael Yu-Men
Citron, David S
Clapp, James R
Clemmons, David Roberts
Clyde, Wallace Alexander, Jr
Cohen, Harvey Jay
Cohen, Philip Lawrence
Coleman, Ralph Edward
Cooper, Herbert A
Cooper, Miles Robert
Craige, Ernest
Cromartie, William James
Crounse, Robert Griffith
Dascomb, Harry Emerson
Davis, Courtland Harwell, Jr
Dees, Susan Coons
Denny, Floyd Wolfe, Jr
Dent, Sara Jamison
DiAugustine, Richard Patrick
Dillard, Margaret Bleick
Dodge, William Howard
Dunphy, Donal
Durack, David Tulloch
Dutton, Jonathan Joseph
Easterling, William Ewart, Jr
Eifrig, David Eric
Epstein, David L(ee)
Estes, Edward Harvey, Jr
Falletta, John Matthew
Farmer, Thomas Wohlsen
Felts, John Harvey
Ferree, Carolyn Ruth
Fine, Jo-David
Fischer, Janet Jordan
Fleming, William Leroy
Fordham, Christopher Columbus, III
Frothingham, Thomas Eliot
Furth, Eugene David
Gammon, Walter Ray
Gilligan, Diana Mary
Gitelman, Hillel J
Glaser, Frederick Bernard
Goldwater, Leonard John
Gottschalk, Carl William
Gray, Mary Jane
Gray, Timothy Kenney
Greiss, Frank C, Jr
Griffith, Jack Dee
Griggs, Thomas Russell
Grimson, Baird Sanford
Grisham, Joe Wheeler
Grossman, Herman
Guess, Harry Adelbert
Hadler, Nortin M
Hall, Kenneth Daland
Hallock, James A
Hammond, Charles Bessellieu
Harmel, Merel H
Harned, Herbert Spencer, Jr
Harrell, George T
Harris, Cecil Craig
Harris, Jerome Sylvan
Hayes, Donald M
Headley, Robert N
Heizer, William David
Hendricks, Charles Henning
Herion, John Carroll
Herring, William Benjamin
Heyman, Albert
Hill, Gale Bartholomew
Hinman, Alanson
Hollander, Walter, Jr
Howell, Charles Maitland
Hudson, Page
Hudson, William Rucker
Huffines, William Davis
Hulka, Jaroslav Fabian
Hunt, William B, Jr
Huntley, Robert Ross
Janeway, Richard
Johnson, George, Jr
Jorizzo, Joseph L

MEDICAL & HEALTH SCIENCES / 461

Joyner, John T, III
Kafer, Enid Rosemary
Kataria, Yash P
Katz, Samuel Lawrence
Kaufman, William
Kaufmann, John Simpson
Kempner, Walter
Kern, Frank Howard
Kiffney, Gustin Thomas, Jr
King, Theodore Matthew
Kirkman, Henry Neil, Jr
Koepke, John Arthur
Kredich, Nicholas M
Kuo, Eric Yung-Huei
Kushnick, Theodore
Kylstra, Johannes Arnold
Ladwig, Harold Allen
Lassiter, William Edmund
Laupus, William E
Lawson, Edward Earle
Lawson, Robert Barrett
Lefkowitz, Robert Joseph
Little, William C
McCall, Charles Emory
McKinney, William Markley
McLendon, William Woodard
McMillan, Campbell White
Martin, James Franklin
Maynard, Charles Douglas
Meads, Manson
Meredith, Jesse Hedgepeth
Meschan, Isadore
Meyer, George Wilbur
Miller, C Arden
Morris, James Joseph, Jr
Mueller, Robert Arthur
Nelson, James H, Jr
Newcomb, Thomas F
O'Keefe, Edward John
Oliver, Thomas K, Jr
Olsen, Elise Arline
Ott, David James
Pagano, Joseph Stephen
Palmer, Jeffress Gary
Parker, John Curtis
Pasipoularides, Ares D
Paulson, David F
Peete, Charles Henry, Jr
Pizer, Stephen M
Pryor, David Bram
Putman, Charles E
Raymond, Lawrence W
Rees, Roberts M
Reinhart, John Belvin
Renuart, Adhemar William
Richards, Frederick, II
Richter, Joel Edward
Roberts, Harold R
Rogers, Mark Charles
Rosse, Wendell Franklyn
Royster, Roger Lee
Rundles, Ralph Wayne
Saltzman, Herbert A
Sanders, Charles Addison
Sawyer, C Glenn
Scatliff, James Howard
Selkirk, James Kirkwood
Sessions, John Turner, Jr
Sheps, Cecil George
Shils, Maurice Edward
Sieker, Herbert Otto
Simon, Jimmy L
Singleton, Mary Clyde
Skinner, Newman Sheldon, Jr
Smith, Fred George, Jr
Smith, Nat E
Smith, Wirt Wilsey
Snyderman, Ralph
Spach, Madison Stockton
Sparling, Philip Frederick
Spock, Alexander
Spurr, Charles Lewis
Stead, Eugene Anson, Jr
Stead, William Wallace
Stiles, Gary L
Styron, Charles Woodrow
Sugioka, Kenneth
Takaro, Timothy
Talbert, Luther M
Taylor, B Gray
Telen, Marilyn Jo
Thomas, Francis T
Tingelstad, Jon Bunde
Toglia, Joseph U
Toole, James Francis
Tyor, Malcolm Paul
Tyroler, Herman A
Underwood, Louis Edwin
Van Wyk, Judson John
Wallace, Andrew Grover
Watts, Charles D
Waugh, William Howard
Wenger, Thomas Lee
Wheeler, Clayton Eugene, Jr
Whitaker, Clay Westerfield
Williams, Redford Brown, Jr
Wilson, Frank Crane
Wilson, Ruby Leila
Winfield, John Buckner
Wood, William Bainster
Woods, James Watson
Yarger, William E
Yount, Ernest H

NORTH DAKOTA
Bruns, Paul Donald
Keller, Reed Theodore
Severn, Charles B
Vogel, Gerald Lee

OHIO
Ackerman, Gustave Adolph, Jr
Adolph, Robert J
Allison, David Coulter
Alter, Joseph Dinsmore
Altose, Murray D
Ankeney, Jay Lloyd
Aring, Charles Dair
Bahr, Gustave Karl
Baker, Saul Phillip
Balcerzak, Stanley Paul
Banwell, John G
Barnes, H Verdain
Bashe, Winslow Jerome, Jr
Batley, Frank
Benzing, George, III
Berger, Melvin
Berman, Jerome Richard
Bernstein, I Leonard
Biehl, Joseph Park
Blaney, Donald John
Bowerfind, Edgar Sihler, Jr
Bozian, Richard C
Brown, Helen Bennett
Buchman, Elwood
Burke, John T
Butkus, Antanas
Campbell, Colin
Campbell, Earl William
Carlock, John Timothy
Carr, Richard Dean
Cassidy, Suzanne Bletterman
Chang, Jae Chan
Cherniack, Neil S
Chester, Edward Howard
Chester, Edward M
Christensen, Julien Martin
Connors, Alfred F, Jr
Conomy, John Paul
Conway, Gene Farris
Copeland, Bradley Ellsworth
Cramblett, Henry G
Crespo, Jorge H
Culbertson, William Richardson
D'Ambrosio, Steven M
Daniel, Thomas Mallon
Daoud, Georges
Davis, Pamela Bowes
Dirckx, John H
Doershuk, Carl Frederick
Donaldson, Virginia Henrietta
Dorn, Charles Richard
Duhring, John Lewis
Dworken, Harvey J
Eastwood, Douglas William
Eckel, Robert Edward
Eckstein, Richard Waldo
Eglitis, Irma
Eiben, Robert Michael
Eichner, Eduard
Elmets, Craig Allan
Farmer, Richard Gilbert
Fenoglio, Cecilia M
Fleshler, Bertram
Foley, Joseph Michael
Forster, Francis Michael
Fowler, Noble Owen
Franco-Saenz, Roberto
Freimer, Earl H
Friedell, Hymer Louis
Friedlander, Ira Ray
Friedlander, Miriam Alice
Fritz, Marc Anthony
Frohman, Lawrence Asher
Gesinski, Raymond Marion
Gifford, Ray Wallace, Jr
Glueck, Helen Iglauer
Goldstein, Gerald
Gopalakrishna, K V
Graham, Bruce Douglas
Green, Ralph
Greenwalt, Tibor Jack
Griggs, Robert C
Grupp, Gunter
Grupp, Ingrid L
Hagerman, Dwain Douglas
Hall, Philip Wells, III
Hanenson, Irwin Boris
Hangartner, Thomas Niklaus
Harris, John William
Harris, Ruth Cameron
Harrison, Donald C
Havener, William H
Haxhiu, Musa A
Haynes, Ralph Edwards
Healy, Bernadine P
Heimlich, Henry Jay
Hellerstein, Herman Kopel
Helm, Robert Albert
Hering, Thomas M
Herman, Jerome Herbert
Hess, Evelyn V
Hoekenga, Mark T
Hoobler, Sibley Worth
Horwitz, Harry
Houchens, David Paul
Howland, Willard J
Hug, George

Hutton, John James, Jr
Inkley, Scott Russell
Izant, Robert James, Jr
Jackson, Benita Marie
Jackson, Edgar B, Jr
Jacoby, Jay
Jebsen, Robert H
Kalter, Harold
Kazura, James
Kennell, John Hawks
Kerr, Andrew, Jr
Kezdi, Paul
Kimura, Kazuo Kay
Koepke, George Henry
Kogut, Maurice D
Kontras, Stella B
Kowal, Jerome
Kowal, Norman Edward
Kunin, Calvin Murry
Kurczynski, Thaddeus Walter
Kushner, Irving
Landau, Bernard Robert
Lashner, Bret Auerbach
Leier, Carl V
Lenkoski, L Douglas
Levy, Richard Philip
Liang, Tehming
Light, Irwin Joseph
Loewenstein, Joseph Edward
Loomans, Maurice Edward
Loudon, Robert G
Lowney, Edmund Dillahunty
McAdams, Arthur James
McMaster, Robert H
McNamara, Michael Joseph
Mahmoud, Adel A F
Mandal, Anil Kumar
Mannino, Joseph Robert
Mantil, Joseph Chacko
Margolin, Esar Gordon
Martin, Lester W
Medalie, Jack Harvey
Menefee, Max Gene
Mohler, Stanley Ross
Morrow, Grant, III
Mortimer, Edward Albert, Jr
Mulrow, Patrick J
Munson, Edwin Sterling
Murphy, Martin Joseph, Jr
Myers, Ronald Elwood
Nahata, Milap Chand
Nankervis, George Arthur
Neuman, Michael R
Newton, William Allen, Jr
Oestreich, Alan Emil
Ooi, Boon Seng
Pansky, Ben
Parker, Robert Frederic
Patrick, Edward Alfred
Patrick, James R
Paul, Richard Jerome
Pearson, Olof Hjalmer
Perkins, Robert Louis
Phillips, Chandler Allen
Pitner, Samuel Ellis
Pollak, Victor Eugene
Prior, John Alan
Rakita, Louis
Ratnoff, Oscar Davis
Reiner, Charles Brailove
Rejali, Abbas Mostafavi
Rittgers, Stanley Earle
Robbins, Frederick Chapman
Robinow, Meinhard
Roche, Alexander F
Rodman, Harvey Meyer
Rosenberg, Howard C
Saenger, Eugene L
Saunders, William H
Schafer, Irwin Arnold
Schiff, Gilbert Martin
Schinagl, Erich F
Schreiner, Albert William
Schubert, William K
Schwartz, Howard Julius
Scott, Ralph Carmen
Scott, William James, Jr
Selim, Mostafa Ahmed
Shelley, Walter Brown
Shields, George Seamon
Shirkey, Harry Cameron
Shumrick, Donald A
Silberstein, Edward B
Sjoersma, Albert
Skillman, Thomas G
Smith, D Edwards
Smith, Robert
Sodeman, William A
Somani, Pitambar
Sotos, Juan Fernandez
Stanberry, Lawrence Raymond
Staubus, Alfred Elsworth
Stevenson, Thomas Dickson
Stock, Charles Chester
Straffon, Ralph Atwood
Strohl, Kingman P
Suskind, Raymond Robert
Sutherland, James McKenzie
Tavill, Anthony Sydney
Teteris, Nicholas John
Troup, Stanley Burton
Tsang, Reginald C
Tubbs, Raymond R
Tucker, Harvey Michael

Turner, Edward V
Tzagournis, Manuel
Vertes, Victor
Vester, John William
Vignos, Paul Joseph, Jr
Vilter, Richard William
Wall, Robert Leroy
Warkany, Josef
Watanakunakorn, Chatrchai
Waterson, John R
Weiner, Murray
Weisman, Russell
Wentz, William Budd
Wenzel, Richard Louis
West, Clark Darwin
Wexler, Bernard Carl
White, Peter
White, Robert J
Will, John Junior
Winter, Chester Caldwell
Wolinsky, Emanuel
Wright, Francis Stuart
Wyman, Milton
Yates, Allan James
Zuspan, Frederick Paul

OKLAHOMA
Bogardus, Carl Robert, Jr
Bottomley, Richard H
Bottomley, Sylvia Stakle
Bourdeau, James Edward
Brandt, Edward Newman, Jr
Caldwell, Glyn Gordon
Calvert, Jon Channing
Campbell, John Alexander
Chrysant, Steven George
Clark, Mervin Leslie
Comp, Philip Cinnamon
Cooper, Richard Grant
Corder, Clinton Nicholas
Couch, James Russell, Jr
Crosby, Warren Melville
Czerwinski, Anthony William
Dille, John Robert
Dormer, Kenneth John
Eichner, Edward Randolph
Epstein, Robert B
Everett, Mark Allen
Farris, Bradley Kent
Gable, James Jackson, Jr
Ganesan, Devaki
Gleaton, Harriet Elizabeth
Gunn, Chesterfield Garvin, Jr
Hampton, James Wilburn
Harder, Harold Cecil
Harley, John Barker
Hope, Ronald Richmond
Ice, Rodney D
Kinasewitz, Gary Theodore
Klein, John Robert
Kosbab, Frederic Paul Gustav
Lavia, Lynn Alan
Lazzara, Ralph
Lee, Lela A
Lewis, C S, Jr
Lhotka, John Francis
Lynn, Thomas Neil, Jr
McKee, Patrick Allen
Massion, Walter Herbert
Maton, Paul Nicholas
Muchmore, Harold Gordon
Nettles, John Barnwell
Olson, Robert Leroy
Parke, David W, II
Parry, William Lockhart
Pikler, George Maurice
Reichlin, Morris
Rhoades, Everett Ronald
Seely, J Rodman
Smith, Carl Walter, Jr
Taylor, Fletcher Brandon, Jr
Thomas, Ellidee Dotson
Thurman, William Gentry
Torres-Pinedo, Ramon
Vanhoutte, Jean Jacques
Van Thiel, David H
Wenzl, James E
Whitcomb, Walter Henry
Whitsett, Thomas L
Williams, George Rainey

OREGON
Bagby, Grover Carlton
Bardana, Emil John, Jr
Bennett, Robert M
Bennett, William M
Benson, John Alexander, Jr
Bergman, Norman
Bissonnette, John Maurice
Black, Franklin Owen
Bristow, John David
Brummett, Robert E
Buchan, George Colin
Buist, Neil R M
Bussman, John W
Campbell, John Richard
Campbell, Robert A
Carter, Charles Conrad
Cereghino, James Joseph
Connor, William Elliott
De Maria, F John
DeWeese, David D
Dyson, Robert Duane
Edwards, Miles John

Medicine (cont)

Engel, Rudolf
Fraunfelder, Frederick Theodor
Frey, Bruce Edward
Grimm, Robert John
Grover, M Roberts, Jr
Herndon, Robert McCulloch
Hill, John Donald
Hu, Funan
Isom, John B
Jacob, Stanley W
Jones, Richard Theodore
Kendall, John Walker, Jr
Kohler, Peter
Lees, Martin H
Lucas, Oscar Nestor
Malinow, Manuel R
Matarazzo, Joseph Dominic
Meechan, Robert John
Menashe, Victor D
Metcalfe, James
Morris, James F
Mullins, Richard James
Norman, Douglas James
Novy, Miles Joseph
Osterud, Harold T
Patterson, James Fulton
Pirofsky, Bernard
Porter, George A
Portman, Oscar William
Reynolds, John Weston
Ritzmann, Leonard W
Schaeffer, Morris
Smith, Catherine Agnes
Stevens, Janice R
Swan, Kenneth Carl
Swank, Roy Laver
Underwood, Rex J
Walsh, John Richard
Watzke, Robert Coit
Williams, Christopher P S
Wuepper, Kirk Dean

PENNSYLVANIA

Abruzzo, John L
Adibi, Siamak A
Agarwal, Jai Bhagwah
Agus, Zalman S
Alexander, Fred
Alexander, Samuel Craighead, Jr
Alter, Milton
Asakura, Toshio
Atkins, Paul C
Austrian, Robert
Balin, Arthur Kirsner
Bannon, James Andrew
Barba, William P, II
Barker, Earl Stephens
Barranger, John Arthur
Barry, William Eugene
Bartosik, Delphine
Bartuska, Doris G
Batshaw, Mark Levitt
Baum, Stanley
Beerman, Herman
Beizer, Lawrence H
Bell, Robert Lloyd
Bennett, Hugh deEvereaux
Berlin, Cheston Milton
Berney, Steven
Bierly, Mahlon Zwingli, Jr
Bilaniuk, Larissa Tetiana
Black, Martin
Blough, Herbert Allen
Bluemle, Lewis W, Jr
Blumberg, Baruch Samuel
Bochenek, Wieslaw Janusz
Boden, Guenther
Boger, William Pierce
Boller, Francois
Bonakdarpour, Akbar
Borges, Wayne Howard
Bovee, Kenneth C
Bowden, Charles Ronald
Bowers, Paul Applegate
Bowles, Lawrence Thompson
Bradley, Matthews Ogden
Bradley, Stanley Edward
Brady, Luther Weldon, Jr
Bralow, S Philip
Brennan, James Thomas
Brest, Albert N
Bridger, Wagner H
Brightman, Vernon
Brody, Jerome Ira
Brooks, John J
Brooks, Michael Lee
Brucker, Paul Charles
Buck, Clayton Arthur
Burholt, Dennis Robert
Burkle, Joseph S
Burns, H Donald
Caddell, Joan Louise
Caliguiri, Lawrence Anthony
Cander, Leon
Carey, William Bacon
Carpenter, Gary Grant
Chait, Arnold
Chambers, Richard
Charkes, N David
Charron, Martin
Cheung, Joseph Y
Clark, James Edward

Clark, James Michael
Clark, Wallace Henderson, Jr
Cohen, Margo Nita Panush
Cohn, Robert M
Colman, Robert W
Comis, Robert Leo
Conn, Rex Boland
Cooper, David Young
Cornfeld, David
Creech, Richard Hearne
Croce, Carlo Maria
Cryer, Dennis Robert
Cutler, John Charles
Cutler, Winnifred Berg
D'Angio, Giulio J
Daniele, Ronald P
Davidoff, Frank F
Day, Harvey James
Deas, Thomas C
DeHoratius, Raphael Joseph
de Jong, Rudolph H
Demopoulos, James Thomas
DeRubertis, Frederick R
Dichter, Marc Allen
Dickman, Albert
Di George, Angelo Mario
Dinman, Bertram David
DiPalma, Joseph Rupert
Djerassi, Isaac
Doghramji, Karl
Dorwart, Bonnie Brice
Douglas, Steven Daniel
Dreifus, Leonard S
Duane, Thomas David
Earley, Laurence E
Edwards, McIver Williamson, Jr
Eggleston, Forrest Cary
Ehrlich, George Edward
Eisenberg, John Meyer
Ellis, Demetrius
Engleman, Karl
Erecinska, Maria
Erslev, Allan Jacob
Evans, Audrey Elizabeth
Faich, Gerald Alan
Farrar, George Elbert, Jr
Finestone, Albert Justin
Finkelstein, David
Fireman, Philip
Fischer, Grace Mae
Fletcher, Robert Hillman
Foley, Thomas Preston, Jr
Franke, Frederick Rahde
Frankl, William S
Freeman, James Theodore
French, Gordon Nichols
Friedman, Arnold Carl
Fromm, Gerhard Hermann
Futcher, Palmer Howard
Gabuzda, Thomas George
Gaffney, Paul Cotter
Gambescia, Joseph Marion
Garcia, Celso-Ramon
Geyer, Stanley J
Gilbert, Robert Pettibone
Glick, John H
Goldberg, Harry
Goldberg, Martin
Goldberg, Michael Ellis
Goldstein, Franz
Goldwein, Manfred Isaac
Gorin, Michael Bruce
Gosfield, Edward, Jr
Gray, Frank Davis, Jr
Greenstein, Jeffrey Ian
Grenvik, Ake N A
Gross, Peter George
Guggenheimer, James
Gurwith, Marc Joseph
Haase, Gunter R
Hann, Hie-Won L
Haskin, Marvin Edward
Haskin, Myra Ruth Singer
Haurani, Farid I
Hayashi, Teruo Terry
Hedges, Thomas Reed, Jr
Helwig, John, Jr
Herberman, Ronald Bruce
Hildreth, Eugene Augustus
Ho, Monto
Hodges, John Hendricks
Hollander, Joseph Lee
Holling, Herbert Edward
Holroyde, Christopher Peter
Holsclaw, Douglas S, Jr
Hopper, Kenneth David
Horner, George John
Horwitz, Orville
Hurley, Harry James
Huth, Edward J
Isard, Harold Joseph
Israel, Harold L
Itkin, Irving Herbert
Jackson, Laird Gray
Jacob, Leonard Steven
Jeffries, Graham Harry
Jegasothy, Brian V
Jimenez, Sergio
Johnson, Joseph Eggleston, III
Joseph, Rosaline Resnick
Joyner, Claude Reuben
Kade, Charles Frederick, Jr
Kallen, Roland Gilbert

Kaplan, Sandra Solon
Karafin, Lester
Kashatus, William C
Kauffman, Leon A
Kauffman, Stuart Alan
Kaye, Donald
Kaye, Robert
Kefalides, Nicholas Alexander
Kelley, William Nimmons
Kendall, Norman
Kerstein, Morris D
Kimbel, Philip
Kirby, Edward Paul
Kivlighn, Salah Dean
Kligman, Albert Montgomery
Kline, Irwin Kaven
Klinghoffer, June F
Klionsky, Bernard Leon
Knobler, Robert Leonard
Knudson, Alfred George, Jr
Koelle, Winifred Angenent
Krisch, Robert Earle
Krishna, Gollapudi Gopal
Kundel, Harold Louis
Ladda, Roger Louis
Laibson, Peter R
Lame, Edwin Lever
Lankford, Edward B
Laties, Alan M
Laufer, Igor
Lebenthal, Emanuel
Leberman, Paul R
Lee, Yien-Hwei
Leighton, Charles Cutler
Levey, Gerald Saul
Levinsky, Walter John
Levison, Matthew Edmund
Levison, Sandra Peltz
Levit, Edithe J
Levy, Robert I
Lewis, George Campbell, Jr
Lewis, Jessica Helen
Leymaster, Glen Ronald
Linnemann, Roger E
Lipton, Allan
London, William Thomas
Long, Joseph Pote
Longnecker, David Eugene
Luscombe, Herbert Alfred
McCloskey, Richard Vensel
Madaio, Michael P
Madow, Leo
Magargal, Larry Elliot
Maguire, Henry C, Jr
Mancall, Elliott L
Mansmann, Herbert Charles, Jr
Marinchak, Roger Alan
Martin, Christopher Michael
Martin, Donald Beckwith
Martin, John Harvey
Martin, Samuel Preston, III
Mastrangelo, Michael Joseph
Mastroianni, Luigi, Jr
Mateer, Frank Marion
Maurer, Alan H
Mayock, Robert Lee
Mazaheri, Mohammad
Meadows, Anna T
Meisler, Arnold Irwin
Melvin, John L
Michelson, Eric L
Miller, John Michael
Mills, Lewis Craig, Jr
Minn, Fredrick Louis
Montgomery, Hugh
Moore, E(arl) Neil
Moore, Mary Elizabeth
Moore, Robert Yates
Moran, John J
Morganroth, Joel
Morris, Harold Hollingsworth, Jr
Mortel, Rodrigue
Moster, Mark Leslie
Motoyama, Etsuro K
Moyer, John Henry
Murphy, John Joseph
Myers, David
Myers, Eugene Nicholas
Myers, Jack Duane
Myerson, Ralph M
Naide, Meyer
Nestico, Pasquale Francesco
Nicholas, Leslie
Nobel, Joel J
Oaks, Wilbur W
O'Brien, Charles P
Onik, Gary M
Orenstein, David M
Palmer, Robert Howard
Park, Chan H
Penneys, Raymond
Perryman, Charles Richard
Pleasure, David
Plotkin, Stanley Alan
Polgar, George
Pontarelli, Domenic Joseph
Powell, Robin Dale
Poydock, Mary Eymard
Price, Henry Locher
Prockop, Darwin J
Proctor, Julian Warrilow
Prystowsky, Harry
Pyeritz, Reed E
Raffensperger, Edward Cowell

Rattan, Satish
Reavey-Cantwell, Nelson Henry
Reinecke, Robert Dale
Reinmuth, Oscar McNaughton
Reivich, Martin
Rike, Paul Miller
Rockey, John Henry
Rodan, Gideon Alfred
Rogers, Fred Baker
Rogers, Kenneth D
Rogers, Robert M
Rohner, Thomas John
Ronis, Max Lee
Rosenblatt, Michael
Rosenkranz, Herbert S
Rosenow, Edward Carl, Jr
Roth, James Luther Aumont
Rovera, Giovanni
Rubin, Alan
Rubin, Walter
Rucinska, Ewa J
Rycheck, Russell Rule
Sachs, Marvin Leonard
Safar, Peter
Samitz, M H
Sanders, Martin E
Sazama, Kathleen
Schaedler, Russell William
Schein, Philip Samuel
Schick, Paul Kenneth
Schlezinger, Nathan Stanley
Schmickel, Roy David
Schmidt, Richard Ralph
Schnabel, Truman Gross, Jr
Schofield, Richard Alan
Schrogie, John Joseph
Schumacher, H Ralph, Jr
Schwartz, Arthur Gerald
Schwartz, Elias
Schwartz, Emanuel Elliot
Schwartz, Irving Robert
Schwartzman, Robert M
Schweighardt, Frank Kenneth
Segal, Bernard L
Segal, Stanton
Sellers, Alfred Mayer
Selzer, Michael Edgar
Senior, John Robert
Shapiro, Alvin Philip
Shapiro, Bernard
Shapiro, Sandor Solomon
Sherry, Sol
Shuman, Charles Ross
Shumway, Clare Nelson, (Jr)
Siegler, Peter Emery
Sigler, Miles Harold
Silberberg, Donald H
Silverman, Robert Eliot
Slotnick, Victor Bernard
Smith, David S
Smith, Hugo Dunlap
Smith, Jackson Bruce
Smith, Jan D
Soloff, Louis Alexander
Soma, Lawrence R
Southam, Chester Milton
Spaeth, George L
Spritzer, Albert A
Stein, George Nathan
Stoloff, Irwin Lester
Strahs, Gerald
Strauss, Jerome Frank, III
Strom, Brian Leslie
Sunderman, Frederick William
Sutnick, Alton Ivan
Szabo, Kalman Tibor
Tasman, William S
Tauxe, Welby Newlon
Taylor, Paul M
Teplick, Joseph George
Tourtellotte, Charles Dee
Tristan, Theodore A
Troen, Philip
Tse, Rose (Lou)
Tulenko, Thomas Norman
Tulsky, Emanuel Goodel
Tumen, Henry Joseph
Turchi, Joseph J
Uitto, Jouni Jorma
Urbach, Frederick
Vagnucci, Anthony Hillary
Van Scott, Eugene Joseph
Wald, Arnold
Waldman, Joseph
Walsh, Peter Newton
Wang, Allan Zu-Wu
Wang, Yen
Warren, George Harry
Waxman, Herbert Sumner
Webster, John H
Weinhouse, Sidney
Weiss, Arthur Jacobs
Weiss, William
Wesson, Laurence Goddard, Jr
Whereat, Arthur Finch
Whybrow, Peter Charles
Wilson, Frederick Sutphen
Wilson, Marjorie Price
Wing, Edward Joseph
Winkelstein, Alan
Winter, Peter Michael
Wolf, Stewart George, Jr
Wolfson, Robert Joseph
Wollman, Harry

Woloshin, Henry Jacob
Wood, Margaret Gray
Wright, Francis Howell
Young, Irving
Zatuchni, Jacob
Zelis, Robert Felix
Zimmerman, Robert Allan
Zinsser, Harry Frederick
Zweiman, Burton

RHODE ISLAND
Arnold, Mary B
Aronson, Stanley Maynard
Barnes, Frederick Walter, Jr
Calabresi, Paul
Carleton, Richard Allyn
Carpenter, Charles C J
Constantine, Herbert Patrick
Crowley, James Patrick
Cummings, Francis Joseph
Davis, Robert Paul
Dowben, Robert Morris
Fawaz Estrup, Faiza
Glicksman, Arvin Sigmund
Greer, David Steven
Hamolsky, Milton William
Jackson, Ivor Michael David
Kaplan, Stephen Robert
Lichtman, Herbert Charles
Meroney, William Hyde, III
Oh, William
Pueschel, Siegfried M
Schwartz, Robert
Silver, Alene Freudenheim
Tefft, Melvin
Thayer, Walter Raymond, Jr
Turner, Michael D
Yankee, Ronald August
Zawadzki, Zbigniew Apolinary
Zinner, Stephen Harvey

SOUTH CAROLINA
Barrett, O'Neill, Jr
Bates, G William
Best, Robert Glen
Boger, Robert Shelton
Brodoff, Bernard Noah
Brown, Arnold
Buse, John Frederick
Buse, Maria F Gordon
Carr, David Turner
Colwell, John Amory
Cooper, George, IV
Corley, John Bryson
Curry, Hiram Benjamin
Dobson, Richard Lawrence
Farrar, William Edmund
Fudenberg, H Hugh
Gillette, Paul Crawford
Goust, Jean Michel
Hester, Lawrence Lamar, Jr
Hogan, Edward L
Holper, Jacob Charles
Humphries, J O'Neal
Kane, John Joseph
Kirtley, William Raymond
Kolb, Leonard H
Lazarchick, John
Legerton, Clarence W, Jr
Leman, Robert B
LeRoy, Edward Carwile
Lin, Tu
Lopes-Virella, Maria Fernanda Leal
McConnell, Jack Baylor
McFarland, Kay Flowers
Macpherson, Roderick Ian
Malone, Michael Joseph
Maricq, Hildegard Rand
Mitchell, Hugh Bertron
Murdaugh, Herschel Victor, Jr
Nelson, George Humphry
Newberry, William Marcus
Pickett, Jackson Brittain
Pittman, Fred Estes
Ploth, David W
Postic, Bosko
Preedy, John Robert Knowlton
Putney, Floyd Johnson
Redding, Joseph Stafford
Roof, Betty Sams
Sandifer, Samuel Hope
Singh, Inderjit
Spann, James Fletcher, (Jr)
Stewart, William C
Still, Charles Neal
Stokes, David Kershaw, Jr
Swanson, Arnold Arthur
Thiers, Bruce Harris
Tripathi, Brenda Jennifer
Tripathi, Ramesh Chandra
Virella, Gabriel T
Westphal, Milton C, Jr
Wilson, Frederick Allen
Wohltmann, Hulda Justine
Woodward, Kent Thomas
Young, Gilbert Flowers, Jr

SOUTH DAKOTA
Flora, George Claude
Gregg, John Bailey
Hamm, Joseph Nicholas
Keppen, Laura Davis
Ranney, Brooks
Schramm, Mary Arthur

Wegner, Karl Heinrich
Zawada, Edward T, Jr

TENNESSEE
Adams, Robert Walker, Jr
Allen, Joseph Hunter
Allison, Fred, Jr
Altemeier, William Arthur, III
Batalden, Paul B
Beaty, Orren, III
Beauchene, Roy E
Bell, Persa Raymond
Berard, Costan William
Berman, M Lawrence
Biaggioni, Italo
Blaser, Martin Jack
Brigham, Kenneth Larry
Burk, Raymond Franklin, Jr
Burr, William Wesley, Jr
Byrd, Benjamin Franklin, Jr
Calhoun, Calvin L
Camacho, Alvro Manuel
Cantrell, William Fletcher
Carbone, Robert James
Castelnuovo-Tedesco, Pietro
Charyulu, Komanduri K N
Chesney, Russell Wallace
Christopher, Robert Paul
Churchill, John Alvord
Cooke, Charles Robert
Cox, Clair Edward, II
Crofford, Oscar Bledsoe
Cummins, Alvin J
Davis, David A
Des Prez, Roger Moister
Donald, William David
Dow, Lois Weyman
Dugdale, Marion
Elam, Lloyd C
Elliott, James H
Eyring, Edward J
Feman, Stephen Sosin
Fenichel, Gerald M
Fields, James Perry
Fletcher, Barry Davis
Flexner, John M
Foster, Henry Wendell
Franks, John Julian
Freemon, Frank Reed
Friesinger, Gottlieb Christian
Fuhr, Joseph Ernest
Ganote, Charles Edgar
Ginn, H Earl
Glassman, Armand Barry
Gompertz, Michael L
Goodwin, Robert Archer, Jr
Goswitz, Francis Andrew
Goswitz, Helen Vodopick
Granner, Daryl Kitley
Grossman, Laurence Abraham
Gupta, Ramesh C
Hak, Lawrence J
Hansen, Axel C
Hara, Saburo
Harwood, Thomas Riegel
Hayes, Raymond Leroy
Hayes, Wayland Jackson, Jr
Heimberg, Murray
Helderman, J Harold
Holaday, Duncan Asa
Hubner, Karl Franz
Hughes, James Gilliam
Hughes, Walter T
Jabbour, J T
Jacobs, Richard Lewis
Jacobson, Harry R
Jordan, Robert, Jr
Jordan, Willis Pope, Jr
Kaplan, Robert Joel
Kaplan, Stanley Baruch
Karzon, David T
King, Lloyd Elijah, Jr
Kirshner, Howard Stephen
Kitabchi, Abbas E
Kossmann, Charles Edward
Krantz, Sanford B
Kraus, Alfred Paul
Lacy, William White
Lange, Robert Dale
Lawton, Alexander R, III
Lefkowitz, Lewis Benjamin, Jr
Little, Joseph Alexander
Luton, Edgar Frank
Mauer, Alvin Marx
Montalvo, Jose Miguel
Muirhead, E Eric
Najjar, Victor Assad
Neely, Charles Lea, Jr
Netsky, Martin George
Newman, John Hughes
North, William Charles
Oates, John Alexander
Oldham, Robert Kenneth
Partain, Clarence Leon
Pate, James Wynford
Phillips, Jerry Clyde
Pincus, Theodore
Pratt, Charles Benton
Pribor, Hugo C
Price, Robert Allen
Pui, Ching-Hon
Ramsey, Lloyd Hamilton
Reddy, Churku Mohan
Rendtorff, Robert Carlisle

Reynolds, Leslie Boush, Jr
Reynolds, Vernon H
Riley, Harris D, Jr
Roberts, Lyman Jackson
Robinson, Roscoe Ross
Ross, Joseph C
Rubin, Donald Howard
Runyan, John William, Jr
Sell, Sarah H Wood
Simone, Joseph Vincent
Siskin, Milton
Smith, Bradley Edgerton
Smith, Raphael Ford
Smith, William Ogg
Snapper, James Robert
Solomon, Alan
Solomon, Solomon Sidney
Soper, Richard Graves
Stahlman, Mildred
Sullivan, Jay Michael
Summitt, Robert L
Tarleton, Gadson Jack, Jr
Terzaghi, Margaret
Teschan, Paul E
Townes, Alexander Sloan
Van Middlesworth, Lester
Wasserman, Jack F
Webster, Burnice Hoyle
Wells, Charles Edmon
Wilson, Robert John
Wood, Alastair James Johnston
Yang, Wen-Kuang
Yoo, Tai-June
Zee, Paulus

TEXAS
Abell, Creed Wills
Abrams, Steven Allen
Alexander, Charles Edward, Jr
Alexander, Drew W
Alexander, James Kermott
Alford, Bobby R
Alfrey, Clarence P, Jr
Allen, Herbert Clifton, Jr
Anderson, Karl E
Anderson, Robert E
Anuras, Sinn
Appel, Stanley Hersh
Arhelger, Roger Boyd
Armstrong, George Glaucus, Jr
Arnold, Watson Caufield
Bailey, Byron James
Baine, William Brennan
Balch, Charles Mitchell
Baron, Samuel
Bashour, Fouad A
Batsakis, John G
Beaudet, Arthur L
Becker, Frederick F
Beller, Barry M
Benjamin, Robert Stephen
Bennett, Michael
Bergstresser, Paul Richard
Bick, Rodger Lee
Bidani, Akhil
Black, Homer Selton
Blaw, Michael Ervin
Blomqvist, Carl Gunnar
Bodey, Gerald Paul, Sr
Bradley, Thomas Bernard, Jr
Bridges, Charles Hubert
Briggs, Arthur Harold
Brody, Baruch Alter
Bryan, George Thomas
Burdine, John Alton
Burzynski, Stanislaw Rajmund
Butler, William T
Calverley, John Robert
Cantrell, Elroy Taylor
Capra, J Donald
Carpenter, Robert James, Jr
Carpenter, Robert Raymond
Carr, Bruce R
Cate, Thomas Randolph
Catlin, Francis I
Catterson, Allen Duane
Chan, Lawrence Chin Bong
Chapman, John S
Chaudhuri, Tuhin
Cohen, Allen Barry
Coltman, Charles Arthur, Jr
Combes, Burton
Cooper, Melvin Wayne
Couch, Robert Barnard
Crawford, Stanley Everett
Cruz, Anatolio Benedicto, Jr
Cunningham, Glenn R
Currarino, Guido
Daeschner, Charles William, Jr
Danhof, Ivan Edward
Daniels, Jerry Claude
Dantzker, David Roy
Davis, Jefferson C
Deiss, William Paul, Jr
Desmond, Murdina MacFarquhar
Diddle, Albert W
Dietlein, Lawrence Frederick
Dietschy, John Maurice
Di Ferrante, Nicola Mario
Dobson, Harold Lawrence
Dodd, Gerald Dewey, Jr
Dodge, Warren Francis
Doody, Rachelle Smith
Dougherty, Joseph C

Dreizen, Samuel
Duflot, Leo Scott
Dunbar, Burdett Sheridan
DuPont, Herbert Lancashire
Dyck, Walter Peter
Edeiken, Jack
Edlin, John Charles
Eichenwald, Heinz Felix
Eknoyan, Garabed
Engelhardt, Hugo Tristram
Fashena, Gladys Jeannette
Feigin, Ralph David
Ferguson, Edward C, III
Ferguson, James Joseph, IV
Fernbach, Donald Joseph
Fidler, Isaiah J
Fields, William Straus
Fink, Chester Walter
Fish, Stewart Allison
Forbis, Orie Lester, Jr
Forland, Marvin
Foster, Daniel W
Franklin, Robert Ray
Freedman, David Asa
Freeman, Maynard Lloyd
Freireich, Emil J
Frenger, Paul F
Frenkel, Eugene Phillip
Frimpter, George W
Gant, Norman Ferrell, Jr
Garber, Alan J
Gaulden, Mary Esther
Gehan, Edmund A
German, Victor Frederick
Giesecke, Adolph H
Giovanella, Beppino C
Glezen, William Paul
Goldberg, Leonard H
Goldman, Armond Samuel
Grant, Arthur E
Grant, John Andrew, Jr
Green, Hubert Gordon
Green, Ronald W
Greenberg, Frank
Greenberg, Stephen B
Gregerman, Robert Isaac
Griffin, James Emmett
Gunn, Albert Edward
Gutierrez, Guillermo
Haggard, Mary Ellen
Hall, Robert Joseph
Harle, Thomas Stanley
Harris, Herbert H
Harrison, Gunyon M
Hartman, James T
Hawkins, Willard Royce
Haynie, Thomas Powell, III
Haywood, Theodore J
Heimbach, Richard Dean
Held, Berel
Herndon, James Henry, Jr
Hickey, Robert Cornelius
Hill, L(ouis) Leighton
Hill, Reba Michels
Hillis, William Daniel, Sr
Hinck, Vincent C
Holguin, Alfonso Hudson
Hollinger, F(rederick) Blaine
Holly, Frank Joseph
Holman, Gerald Hall
Holt, Charlene Poland
Hoyumpa, Anastacio Maningo
Hsu, Katharine Han Kuang
Hug, Verena
Ihler, Garret Martin
Jackson, Carmault B, Jr
James, Thomas Naum
Jenicek, John Andrew
Johnson, Larry
Johnson, William Cone
Jordon, Robert Earl
Journeay, Glen Eugene
Kallus, Frank Theodore
Kantor, Harvey Sherwin
Kaplan, Norman M
Kaufman, Raymond H
Keeler, Martin Harvey
Kirkendall, Walter Murray
Kitay, Julian I
Klein, Gordon Leslie
Knight, Vernon
Kniker, William Theodore
Kraemer, Duane Carl
Kraus, William Ludwig
Kraychy, Stephen
Kronenberg, Richard Samuel
Krusen, Edward Montgomery
Kumar, Vinay
Kuo, Lih
Kuo, Peter Te
Kurzrock, Razelle
Lamb, James Francis
Lancaster, Malcolm
Lane, Montague
Langland, Olaf Elmer
Langsjoen, Per Harald
LeBlanc, Adrian David
Ledley, Fred David
LeMaistre, Charles Aubrey
Leon, Robert Leonard
Leroy, Robert Frederick
Levin, Victor Alan
Levin, William Cohn
Lifschitz, Meyer David

464 / DISCIPLINE INDEX

Medicine (cont)

Lipshultz, Larry I
Lockhart, Lillian Hoffman
Lopez-Berestein, Gabriel
Luk, Gordon David
Lupski, James Richard
Lutherer, Lorenz O
Lynch, Edward Conover
Lynn, John R
Lynn, William Sanford
McCall, Charles B
McGanity, William James
McGovern, John Phillip
McGuire, William L
McKelvey, Eugene Mowry
McLaughlin, Peter
Maclean, Graeme Stanley
McNamara, Dan Goodrich
Marcus, Donald M
Marshall, Gailen D, Jr
Marshburn, Paul Bartow
Martinez, J Ricardo
Martinez-Lopez, Jorge Ignacio
Martt, Jack M
Medina, Daniel
Meyer, John Stirling
Miller, Jarrell E
Miller, William Franklin
Minerbo, Grace Moffat
Mintz, A Aaron
Mitchell, Jere Holloway
Mize, Charles Edward
Moldawer, Marc
Montague, Eleanor D
Moody, Eric Edward Marshall
Moreton, Robert Dulaney
Morrow, Dean Huston
Moyer, Mary Pat Sutter
Musher, Daniel Michael
Myers, Terry Lewis
Nelson, Edward Blake
Nelson, John D
Nichaman, Milton Z
Nichols, Buford Lee, Jr
Norman, Floyd (Alvin)
North, Richard Ralph
Nunneley, Sarah A
Nusynowitz, Martin Lawrence
Park, Myung Kun
Patsch, Wolfgang
Pauerstein, Carl Joseph
Paull, Barry Richard
Peake, Robert Lee
Periman, Phillip
Peters, Lester John
Peters, Paul Conrad
Petty, Charles Sutherland
Pierce, Alan Kraft
Pierce, Alexander Webster, Jr
Pinkel, Donald Paul
Piziak, Veronica Kelly
Poffenbarger, Phillip Lynn
Pollard, Richard Byrd
Powell, Don Watson
Powell, Leslie Charles
Powell, Norborne Berkeley
Pritchard, Jack Arthur
Raber, Martin Newman
Ram, C Venkata S
Reeves, T Joseph
Reuter, Stewart R
Rich, Robert Regier
Riggs, Stuart
Rodgers, Lawrence Rodney, Sr
Roffwarg, Howard Philip
Rolston, Kenneth Vijaykumar Issac
Rossen, Roger Downey
Rowen, Burt
Rudolph, Arnold Jack
Runge, Thomas Marschall
Rutledge, Felix N
Sakakini, Joseph, Jr
Sandstead, Harold Hilton
Sanford, Jay Philip
Schenker, Steven
Schneider, Rose G
Schneider, Sandra Lee
Schnur, Sidney
Schreiber, Melvyn Hirsh
Scurry, Murphy Townsend
Sears, David Alan
Seldin, Donald Wayne
Shapiro, William
Shearer, William Thomas
Shulman, Robert Jay
Simpson, Joe Leigh
Smiley, James Donald
Smith, Frank E
Smith, John Leslie, Jr
Smith, Keith Davis
Smith, Reginald Brian
Smythe, Cheves McCord
Spencer, William Albert
Spira, Melvin
Sponsel, William Eric
Sprague, Charles Cameron
Stone, Marvin J
Stout, Landon Clarke, Jr
Strong, Louise Connally
Suki, Wadi Nagib
Sullivan, Margaret P
Swischuk, Leonard Edward
Szczesniak, Raymond Albin

Taegtmeyer, Heinrich
Taylor, Fred M
Thompson, Edward Ivins Bradbridge
Thompson, Howard K, Jr
Tompsett, Ralph Raymond
Towler, Martin Lee
Travis, Luther Brisendine
Tyner, George S
Uhr, Jonathan William
Unger, Roger Harold
Vallbona, Carlos
Vanatta, John Crothers, III
Varma, Surendra K
Villacorte, Guillermo Vilar
Von Noorden, Gunter Konstantin
Wallace, Sidney
Wallace, Tracy I
Weir, Michael Ross
Weiss, Gary Bruce
Welch, Gary William
Wen, Chi-Pang
Weser, Elliot
Wigodsky, Herman S
Wildenthal, Kern
Wilkerson, James Edward
Williams, Bryan
Williams, Darryl Marlowe
Williams, John F, Jr
Williams, Robert L
Wilson, Jean Donald
Wilson, McClure
Wimer, Bruce Meade
Wolinsky, Jerry Saul
Wolma, Fred J
Worrell, John Mays, Jr
Worthen, Howard George
Wright, Stephen E
Wu, Kenneth Kun-Yu
Yamauchi, Toshio
Yatsu, Frank Michio
Yeung, Katherine Lu
Yielding, K Lemone
Young, Sue Ellen
Ziff, Morris

UTAH

Abildskov, J(unior) A
Ajax, Ernest Theodore
Archer, Victor Eugene
Athens, John William
Bailey, Richard Elmore
Baringer, J Richard
Bilbao, Marcia Kepler
Bland, Richard David
Bliss, Eugene Lawrence
Book, Linda Sue
Bragg, David Gordon
Chan, Gary Mannerstedt
Clark, Lincoln Dufton
Dethlefsen, Lyle A
Done, Alan Kimball
Ensign, Paul Roselle
Hill, Harry Raymond
Kolff, Willem Johan
Kuida, Hiroshi
Lahey, M Eugene
Lee, Glenn Richard
Matsen, John Martin
Millar, C Kay
Nelson, Don Harry
Noehren, Theodore Henry
Overall, James Carney, Jr
Parkin, James Lamar
Petajan, Jack Hougen
Prescott, Stephen M
Renzetti, Attilio D, Jr
Ruttenberg, Herbert David
Tsagaris, Theofilos John
Tyler, Frank Hill
Urry, Ronald Lee
Ward, John Robert
Williams, Roger Richard
Wilson, Dana E
Wilson, John F

VERMONT

Albertini, Richard Joseph
Bland, John (Hardesty)
Bouchard, Richard Emile
Cooper, Sheldon Mark
Dustan, Harriet Pearson
Forsyth, Ben Ralph
Gennari, F John
Gibson, Thomas Chometon
Goran, Michael I
Gump, Dieter W
Hanson, John Sherwood
Houston, Charles Snead
Howland, William Stapleton
Kelleher, Philip Conboy
Kelley, Jason
Kunin, Arthur Saul
Levy, Arthur Maurice
Lipson, Richard L
Lucey, Jerold Francis
McKay, Robert James, Jr
Martin, Herbert Lloyd
Myers, Warren Powers Laird
Page, Robert Griffith
Phillips, Carol Fenton
Reichsman, Franz Karl
Schumacher, George Adam
Sheldon, Huntington
Sims, Ethan Allen Hitchcock

Tabakin, Burton Samuel
Tampas, John Peter
Waller, Julian Arnold
Weed, Lawrence Leonard
Welsh, George W, III
Wolf, George Anthony, Jr

VIRGINIA

Abbott, Betty Jane
Ackell, Edmund Ferris
Adler, Robert Alan
Astruc, Juan
Ayers, Carlos R
Ayres, Stephen McClintock
Barranco, Sam Christopher, III
Beller, George Allan
Blackard, William Griffith
Blaylock, W Kenneth
Blizzard, Robert M
Board, John Arnold
Bornmann, Robert Clare
Boyce, William Henry
Braciale, Thomas Joseph, Jr
Bressler, Bernard
Brown, Loretta Ann Port
Burke, James Otey
Carey, Robert Munson
Cawley, Edward Philip
Chan, James C M
Chaudhuri, Tapan K
Chen, Jiande
Chevalier, Robert Louis
Chung, Choong Wha
Cole, Patricia Ellen
Cone, Clarence Donald, Jr
Connell, Alastair McCrae
Craig, James William
Crispell, Kenneth Raymond
Croft, Barbara Yoder
Davis, Charles Stewert
Davis, John Staige, IV
Dreifuss, Fritz Emanuel
Dunn, John Thornton
Edlich, Richard French
Elford, Howard Lee
Elmore, Stanley Mcdowell
Epstein, Robert Marvin
Estep, Herschel Leonard
Fariss, Bruce Lindsay
Farr, Barry Miller
Ferry, Andrew P
Fisher, Lyman McArthur
Fox, Kenneth Richard
Fu, Shu Man
Gillenwater, Jay Young
Glauser, Frederick Louis
Goldstein, Jerome Charles
Greenberg, Jerome Herbert
Greenlee, John Edward
Groschel, Dieter Hans Max
Guerrant, John Lippincott
Gwaltney, Jack Merrit, Jr
Halstead, Charles Lemuel
Hamlin, James T, III
Hendley, Joseph Owen
Hess, Michael L
Hook, Edward Watson, Jr
Hotta, Shoichi Steven
Howards, Stuart S
Hunter, Thomas Harrison
Hurt, Waverly Glenn
Irby, William Robert
James, George Watson, III
Jones, John Evan
Kamenetz, Herman Leo
Kaul, Sanjiv
Keats, Theodore Eliot
Kelly, Thaddeus Elliott
Kendig, Edwin Lawrence, Jr
Klumpp, Theodore George
Kontos, Hermes A
Koontz, Warren Woodson, Jr
Kory, Ross Conklin
Kugel, Robert Benjamin
Lee, Hyung Mo
Lee, Kyu Taik
Lester, Richard Garrison
Linden, Joel Morris
McCue, Carolyn M
McCurdy, Paul Ranney
Macdonald, Roderick, Jr
Mahboubi, Ezzat
Maio, Domenic Anthony
Makhlouf, Gabriel Michel
Mauck, Henry Page, Jr
Mayer, William Dixon
Miller, James Q
Miller, William Weaver
Moore, Edward Weldon
Moran, Thomas James
Mullinax, Perry Franklin
Nance, Walter Elmore
Neal, Marcus Pinson, Jr
Nygard, Neal Richard
Ornato, Joseph Pasquale
Owen, John Atkinson, Jr
Pariser, Harry
Park, Herbert William, III
Pastore, Peter Nicholas
Patterson, John Legerwood, Jr
Paulsen, Elsa Proehl
Pfeiffer, Carl J
Porter, Frederick Stanley, Jr
Prindle, Richard Alan

Regelson, William
Richardson, David W
Rochester, Dudley Fortescue
Rogol, Alan David
Rohn, Reuben D
Rosenberg, Stuart A
Roth, Karl Sebastian
Scheld, William Michael
Schreiner, George E
Schulman, Joseph Daniel
Scott, David Bytovetzski
Scott, Robert Bradley
Sjogren, Robert W, Jr
Smith, Maurice John Vernon
Smith, Wade Kilgore
Spencer, Frederick J
Sprinkle, Philip Martin
Stein, Richard Louis
Steinberg, Alfred David
Stevenson, Ian
Sturgill, Benjamin Caleb
Sugerman, Harvey J
Sullivan, Louis Wade
Tegtmeyer, Charles John
Telfer, Nancy
Thorner, Michael Oliver
Thorup, Oscar Andreas, Jr
Wadlington, Walter J
Wasserman, Albert J
Watlington, Charles Oscar
Watson, Robert Fletcher
Weary, Peyton Edwin
Westervelt, Frederic Ballard, Jr
Wheby, Munsey S
Wilds, Preston Lea
Williams, Gerald Albert
Wilson, Lester A, Jr
Wise, Robert Irby
Wolf, Barry
Wood, Maurice
Young, Reuben B
Zfass, Alvin Martin

WASHINGTON

Aagaard, George Nelson
Albert, Richard K
Aldrich, Robert Anderson
Altman, Leonard Charles
Anderson, Robert Authur
Avner, Ellis David
Bateman, Joseph R
Beeson, Paul Bruce
Bergman, Abraham
Bierman, Edwin Lawrence
Bigley, Robert Harry
Blackmon, John R
Bonica, John Joseph
Bornstein, Paul
Bremner, William John
Bruce, Robert Arthur
Burnell, James McIndoe
Butler, John
Chapman, Warren Howe
Chase, John David
Cheney, Frederick Wyman
Couser, William Griffith
Cullen, Bruce F
Deisher, Robert William
Deyo, Richard Alden
Dodge, Harold T
Donaldson, James Adrian
Elgee, Neil J
Eschbach, Joseph W
Fialkow, Philip Jack
Figge, David C
Figley, Melvin Morgan
Finch, Clement Alfred
Fink, Bernard Raymond
Foy, Hjordis M
Fujimoto, Wilfred Y
Geyman, John Payne
Giblett, Eloise Rosalie
Gold, Eli
Goodner, Charles Joseph
Graham, C Benjamin
Guntheroth, Warren G
Hakomori, Sen-Itiroh
Harlan, John Marshall
Haviland, James West
Hellstrom, Karl Erik Lennart
Holcenberg, John Stanley
Holmes, King Kennard
Hornbein, Thomas F
Johnson, Murray Leathers
Jones, Robert F
Kalina, Robert E
Karp, Laurence Edward
Karr, Reynold Michael, Jr
Karstens, Andres Ingver
Kelley, Vincent Charles
Kenney, Gerald
Klebanoff, Seymour J
Knopp, Robert H
Krieger, John Newton
Kuo, Cho-Chou
Lane, Wallace
Laramore, George Ernest
Lehmann, Justus Franz
Lemire, Ronald John
Loeb, Lawrence Arthur
Loeser, John David
Mackler, Bruce
Mannik, Mart
Mason, Beryl Troxell

MEDICAL & HEALTH SCIENCES / 465

Metz, Robert John Samuel
Meyer, Roger J
Moore, Daniel Charles
Morris, Lucien Ellis
Motulsky, Arno Gunther
Nelp, Wil B
Newmark, Jonathan
Ochs, Hans D
Odland, George Fisher
Paulsen, Charles Alvin
Pavlin, Edward George
Pendergrass, Thomas Wayne
Petersdorf, Robert George
Phillips, Leon A
Pious, Donald A
Pope, Charles Edward, II
Porte, Daniel, Jr
Purnell, Dallas Michael
Rasey, Janet Sue
Raskind, Wendy H
Rees, William James
Rich, Clayton
Robertson, H Thomas, II
Robertson, William O
Rosen, Henry
Rosenblatt, Roger Alan
Rosse, Cornelius
Rubin, Cyrus E
Ruvalcaba, Rogelio H A
Sandler, Harold
Sauvage, Lester Rosaire
Schwartz, Stephen Mark
Scribner, Belding Hibbard
Shepard, Thomas H
Shurtleff, David B
Simpson, David Patten
Singer, Jack W
Siscovick, David Stuart
Smith, Nathan James
Sossong, Norman D
Spadoni, Leon R
Sparkman, Donal Ross
Stamm, Stanley Jerome
Stanek, Karen Ann
Starr, Jason Leonard
Strandness, Donald Eugene, Jr
Swanson, Phillip D
Templeton, Frederic Eastland
Thomas, David Bartlett
Thomas, Edward Donnall
Thompson, Alvin Jerome
Todaro, George Joseph
Turck, Marvin
Van Arsdel, Paul Parr, Jr
Volwiler, Wade
Ward, Louis Emmerson
Weaver, W Douglas
Wedgwood, Ralph Josiah Patrick
Weisberg, Herbert
Whatmore, George Bernard
Whitney, Robert Arthur, Jr
Willson, Richard Atwood
Wood, Francis C, Jr
Wyrick, Ronald Earl
Yarington, Charles Thomas, Jr

WEST VIRGINIA
Albrink, Margaret Joralemon
Andrews, Charles Edward
Battigelli, Mario C
Bowdler, Anthony John
Einzig, Stanley
Flink, Edmund Berney
Fugo, Nicholas William
Gabriele, Orlando Frederick
Gutmann, Ludwig
Hales, Milton Reynolds
Headings, Verle Emery
Kartha, Mukund K
Klingberg, William Gene
Lapp, N LeRoy
Malin, Howard Gerald
Morgan, David Zackquill
Rath, Charles E
Sheikh, Kazim
Trotter, Robert Russell
Tryfiates, George P
Welton, William Arch

WISCONSIN
Ambuel, John Philip
Arkins, John A
Atkinson, Richard L, Jr
Azen, Edwin Allan
Bamforth, Betty Jane
Benforado, Joseph Mark
Berlow, Stanley
Blodgett, Frederic Maurice
Borden, Ernest Carleton
Brandenburg, James H
Bryan, George Terrell
Buchsbaum, Herbert Joseph
Cadmus, Robert R
Carbone, Paul P
Chun, Raymond Wai Mun
Chusid, Joseph George
Cohen, Bernard Allan
Cripps, Derek J
Crummy, Andrew B
Davis, Starkey D
Donegan, William L
Drucker, William D
Ehrlich, Edward Norman
Eichman, Peter L

Engbring, Norman H
Erwin, Chesley Para
Esterly, Nancy Burton
Farley, Eugene Shedden, Jr
Fink, Jordan Norman
Folts, John D
Gallen, William J
Goldberg, Alan Herbert
Goodfriend, Theodore L
Goodwin, James Simeon
Gould, Michael Nathan
Graziano, Frank M
Gustafson, David Harold
Hansen, Marc F
Hanson, Peter
Harb, Joseph Marshall
Harkness, Donald R
Harsch, Harold H
Hirsch, Samuel Roger
Hong, Richard
Howard, Robert Bruce
Jacobs, Lois Jean
Jefferson, James Walter
Kabler, J D
Kim, Zaezeung
Klein, Ronald
Kolesari, Gary Lee
Kubinski, Henry A
Lanphier, Edward Howell
Larson, Frank Clark
Lehman, Roger H
Lennon, Edward Joseph
Lobeck, Charles Champlin
Long, Sally Yates
McCarty, Daniel J, Jr
MacKinney, Archie Allen, Jr
Madsen, Paul O
Maki, Dennis G
Marton, Laurence Jay
Moffet, Hugh L
Morrissey, John F
Mosesson, Michael W
Mosher, Deane Fremont, Jr
Olsen, Ward Alan
Perloff, William H(arry), Jr
Peters, Henry A
Petersen, John Robert
Pisciotta, Anthony Vito
Ramirez, Guillermo
Rankin, John Horsley Grey
Rieselbach, Richard Edgar
Ritter, Mark Alfred
Rosenzweig, David Yates
Roth, Donald Alfred
Rowe, George G
Rudman, Daniel
Rusy, Ben F
Schatten, Heide
Schilling, Robert Frederick
Schlueter, Donald Paul
Schmidt, Donald Henry
Schultz, Richard Otto
Segar, William Elias
Shahidi, Nasrollah Thomas
Shrago, Earl
Silverman, Ellen-Marie
Sims, John LeRoy
Sobkowicz, Hanna Maria
Soergel, Konrad H
Sondel, Paul Mark
Spritz, Richard Andrew
Steeves, Richard Allison
Stewart, Richard Donald
Stowe, David F
Stueland, Dean T
Temin, Howard Martin
Terry, Leon Cass
Tomar, Russell H
Tormey, Douglass Cole
Treffert, Darold Allen
Turkington, Roger W
Wang, Richard I H
Ward, Richard John
Warltier, David Charles
Wen, Sung-Feng
Wolberg, William Harvey
Yale, Charles E
Youker, James Edward
Zenz, Carl
Zimmerman, Stephen William

WYOMING
Dail, Clarence Wilding
Ellwood, Paul M, Jr

PUERTO RICO
Adamsons, Karlis, Jr
Garcia-Palmieri, Mario R
Haddock, Lillian
Moreno, Esteban
Pico, Guillermo
Ramirez-Ronda, Carlos Hector
Shapiro, Douglas York
Sifontes, Jose E
Sifre, Ramon Alberto

ALBERTA
Cameron, Donald Forbes
Cochrane, William
Dawson, J W
Dewhurst, William George
Dossetor, John Beamish
Dunlop, D L
Fraser, Robert Stewart

Gilbert, James Alan Longmore
Halloran, Philip Francis
Jones, Richard Lee
Larke, R(obert) P(eter) Bryce
Leung, Alexander Kwok-Chu
McCoy, Ernest E
Man, Shu-Fan Paul
Martin, Renee Halo
Molnar, George D
Noujaim, Antoine Akl
Pearce, Keith Ian
Percy, John Smith
Read, John Hamilton
Rorstad, Otto Peder
Rossall, Richard Edward
Rudd, Noreen L
Smith, Eldon Raymond
Solez, Kim
Spence, Matthew W
Sproule, Brian J
Taylor, William Clyne
Thomson, Alan B R
Whitelaw, William Albert
Winter, Jeremy Stephen Drummond
Wyse, John Patrick Henry

BRITISH COLUMBIA
Auersperg, Nelly
Baird, Patricia A
Band, Pierre Robert
Banister, Eric Wilton
Burton, Albert Frederick
Calne, Donald Brian
Chow, Anthony W
Cleator, Iain Morrison
Dower, Gordon Ewbank
Drance, S M
Dunn, Henry George
Elliot, Alfred Johnston
Ferguson, Alexander Cunningham
Fishman, Sherold
Ford, Denys Kensington
Friedman, Jan Marshall
Friedman, Sydney Murray
Garg, Arun K
Goldie, James Hugh
Hardwick, David Francis
Jenkins, Leonard Cecil
Ledsome, John R
MacDiarmid, William Donald
Margetts, Edward Lambert
Matheson, David Stewart
Mauk, Arthur Grant
Osoba, David
Ovalle, William Keith, Jr
Price, John David Ewart
Quamme, Gary Arthur
Rotem, Chava Eve
Rowlands, Stanley
Salinas, Fernando A
Sanders, Harvey David
Silver, Hulbert Keyes Belford
Slade, H Clyde
Wada, Juhn A
Zaleski, Witold Andrew

MANITOBA
Adhikari, P K
Angel, Aubie
Beamish, Robert Earl
Birt, Arthur Robert
Bose, Deepak
Chernick, Victor
Fine, Adrian
Friesen, Rhinehart F
Greenberg, Arnold Harvey
Grewar, David
Harding, Godfrey Kynard Matthew
Haworth, James C
Hershfield, Earl S
Israels, Lyonel Garry
Kordova, Nonna
Lewis, Marion Jean
Lyons, Edward Arthur
MacEwan, Douglas W
Mathewson, Francis Alexander Lavens
Moorhouse, John A
Ronald, Allan Ross
Roulston, Thomas Mervyn
Schacter, Brent Allan
Schoemperlen, Clarence Benjamin
Thomson, Ashley Edwin
Weeks, John Leonard

NEWFOUNDLAND
Barrowman, James Adams
Bieger, Detlef
Hillman, Donald Arthur
Hillman, Elizabeth S
Pryse-Phillips, William Edward Maiben
Rees, Rolf Stephen

NOVA SCOTIA
Allen, Alexander Charles
Carr, Ronald Irving
Carruthers, Samuel George
Downie, John William
Fernandez, Louis Agnelo
Finley, John P
Goldbloom, Richard B
Grantmyre, Edward Bartlett
Hammerling, James Solomon
Hanna, Brian Dale
Hirsch, Solomon

Langley, G R
McFarlane, Ellen Sandra
Marrie, Thomas James
Moffitt, Emerson Amos
Murray, T J
Ozere, Rudolph L
Parker, William Arthur
Purkis, Ian Edward
Stewart, Chester Bryant
Tan, Meng Hee
Tupper, W R Carl
Williams, Christopher Noel
Woodbury, John F L
Wort, Arthur John

ONTARIO
Alberti, Peter W R M
Allard, Johane
Armstrong, John Briggs
Armstrong, John Buchanan
Aye, Maung Tin
Bain, Barbara
Baker, Michael Allen
Barnett, H J M
Barr, Ronald Duncan
Beck, Ivan Thomas
Bergsagel, Daniel Egil
Bienenstock, John
Bird, Charles Edward
Biro, George P
Bondy, Donald Clarence
Boone, James Edward
Boston, Robert Wesley
Brien, Francis Staples
Bryans, Alexander (McKelvey)
Carlen, Peter Louis
Chambers, Ann Franklin
Chui, David H K
Clark, David A
Clarke, W T W
Collins-Williams, Cecil
Connell, Walter Ford
Cook, Michael Arnold
Couture, Roger
Crawford, John S
Davidson, Ronald G
De Veber, Leverett L
Diosy, Andrew
Dirks, John Herbert
Dolovich, Jerry
Dorian, William D
Dube, Ian David
Dunbar, J(ohn) Scott
Egan, Thomas J
Farrar, John Keith
Freedman, Melvin Harris
Gammal, Elias Bichara
Garfinkel, Paul Earl
Goldberg, David Myer
Goldenberg, Gerald J
Goodacre, Robert Leslie
Gunton, Ramsay Willis
Haberman, Herbert Frederick
Hachinski, Vladimir
Hansebout, Robert Roger
Harwood-Nash, Derek Clive
Hawke, W A
Haynes, Robert Brian
Heagy, Fred Clark
Hegele, Robert A
Hinton, George Greenough
Hoffstein, Victor
Hollenberg, Charles H
Howe, Geoffrey Richard
Hozumi, Nobumichi
Huang, Bing Shaun
Hutton, Elaine Myrtle
Inman, Robert Davies
Jeejeebhoy, Khursheed Nowrojee
Jennings, Donald B
Kalant, Harold
Kang, C Yong
Khera, Kundan Singh
Kovacs, Bela A
Kovacs, Eve Maria
Laham, Souheil
Lala, Peeyush Kanti
Laski, Bernard
Le Riche, William Harding
Liao, Shuen-Kuei
Little, James Alexander
Logothetopoulos, J
Lott, James Stewart
Low, James Alexander
McAninch, Lloyd Nealson
McCredie, John A
McCulloch, Ernest Armstrong
McDonald, John William David
McNeill, Kenneth Gordon
McWhinney, Ian Renwick
Malkin, Aaron
Meakin, James William
Merskey, Harold
Miller, Anthony Bernard
Milliken, John Andrew
Morgan, William Keith C
Morrin, Peter Arthur Francis
Murphy, Edward G
Napke, Edward
Nisbet-Brown, Eric Robert
Ogilvie, Richard Ian
Orrego, Hector
Papp, Kim Alexander
Parker, John Orval

Medicine (cont)

Paul, Leendert Cornelis
Phillipson, Eliot Asher
Pickering, Richard Joseph
Pinkerton, Peter Harvey
Posen, Gerald
Poyton, Herbert Guy
Pross, Hugh Frederick
Pruzanski, Waldemar
Rabin, Elijah Zephania
Rapoport, Abraham
Reid, Robert Leslie
Roach, Margot Ruth
Rock, Gail Ann
Rosenberg, Gilbert Mortimer
Sackett, David Lawrence
Saleh, Wasfy Seleman
Schneiderman, Jacob Harry
Sellers, Frank Jamieson
Shane, Samuel Jacob
Shephard, Roy Jesse
Sieniewicz, David James
Simon, Jerome Barnet
Sinclair, Nicholas Roderick
Sirek, Otakar Victor
Skoryna, Stanley C
Slutsky, Arthur
Sole, Michael Joseph
Spoerel, Wolfgang Eberhart G
Steele, Robert
Stewart, Thomas Henry McKenzie
Stiller, Calvin R
Swyer, Paul Robert
Tannock, Ian Frederick
Tobin, Sidney Morris
Tugwell, Peter
Tyson, John Edward Alfred
Valberg, Leslie S
Wallace, Alexander Cameron
Warren, Douglas Robson
Watson, William Crawford
Wherrett, John Ross
White, Denis Naldrett
Whitfield, James F
Whitmore, Gordon Francis
Wigle, Ernest Douglas
Wilson, Donald Laurence
Wolfe, Bernard Martin
Woolf, C R
Zsoter, Thomas

QUEBEC
Ackman, Douglas Frederick
Aguayo, Albert J
Alpert, Elliot
Antel, Jack Perry
Aranda, Jacob Velasco
Baxter, Donald William
Becklake, Margaret Rigsby
Benard, Bernard
Benfey, Bruno Georg
Bergeron, Michel
Bolte, Edouard
Briere, Normand
Brouillette, Robert T
Burgess, John Herbert
Cameron, Douglas George
Carrière, Serge
Chicoine, Luc
Claveau, Rosario
Collin, Pierre-Paul
Cooper, Bernard A
Davignon, Jean
Delespesse, Guy Joseph
De Margerie, Jean-Marie
DeRoth, Laszlo
Ducharme, Jacques R
Dupont, Claire Hammel
Elhilali, Mostafa M
Ferguson, Ronald James
Garcia, Raul
Genest, Jacques
Gjedde, Albert
Gold, Allen
Gold, Phil
Goresky, Carl A
Guttmann, Ronald David
Guyda, Harvey John
Hamet, Pavel
Heller, Irving Henry
Joncas, Jean Harry
Jones, Geoffrey Melvill
Karpati, George
Kuchel, Otto George
Lamoureux, Gilles
Lander, Philip Howard
Lawrence, Donald Gilbert
Lefebvre, Rene
Letarte, Jacques
Little, A Brian
Lussier, Andre (Joseph Alfred)
MacDonald, R Neil
McGregor, Maurice
Macklem, Peter Tiffany
Marliss, Errol Basil
Mathieu, Roger Maurice
Maughan, George Burwell
Meyer, Francois
Mignault, Jean De L
Miller, Sandra Carol
Mortola, Jacopo Prospero
Nadeau, Reginald Antoine
Nawar, Tewfik

Nickerson, Mark
Osmond, Dennis Gordon
Palayew, Max Jacob
Pavilanis, Vytautas
Pearson Murphy, Beverly Elaine
Poirier, Louis
Polosa, Canio
Rasmussen, Theodore Brown
Rola-Pleszczynski, Marek
Rose, Bram
Rosenblatt, David Sidney
Roy, Claude Charles
Ruiz-Petrich, Elena
Sainte-Marie, Guy
Schiffrin, Ernesto Luis
Schulz, Jan Ivan
Schwab, Andreas J
Sherwin, Allan Leonard
Shuster, Joseph
Tahan, Theodore Wahba
Theriault, Gilles P
Thomson, David M P
Vinay, Patrick
Wainberg, Mark Arnold
Watters, Gordon Valentine

SASKATCHEWAN
Gerrard, John Watson
Horlick, Louis
Houston, Clarence Stuart
Kalra, Jawahar
Prasad, Kailash
Shokeir, Mohamed Hassan Kamel
Skinnider, Leo F
Smith, Colin McPherson
Wyant, Gordon Michael

OTHER COUNTRIES
Agarwal, Shyam S
Allessie, Maurits
Angelini, Corrado Italo
Bachmann, Fedor W
Baum, Gerald L
Benatar, Solomon Robert
Blanpain, Jan E
Borst-Eilers, Else
Brown, Nigel Andrew
Brunser, Oscar
Bunker, John Philip
Buzina, Ratko
Chang, H K
Chang, Kenneth Shueh-Shen
Cho, Young Won
Choi, Yong Chun
Christodoulou, John
Criss, Wayne Eldon
Dyrberg, Thomas Peter
Engel, Eric
Fernandez-Cruz, Eduardo P
Fraser, David William
Gemsa, Diethard
Gomez, Arturo
Gontier, Jean Roger
Gorski, Andrzej
Guerrero-Munoz, Frederico
Guimaraes, Armenio Costa
Hall, Peter
Hansson, Goran K
Haschke, Ferdinand
Helander, Herbert Dick Ferdinand
Herrmann, Walter L
Hillcoat, Brian Leslie
Hirano, Toshio
Hofmann, Bo
Hsu, Clement C S
Huang, Kun-Yen
Hugon, Jean S
Ichikawa, Shuichi
Izui, Shozo
Jonzon, Anders
Jordan, Robert Lawrence
Kessler, Alexander
Krakoff, Irwin Harold
Krueger, Gerhard R F
Lambrecht, Richard Merle
Lammers, Wim
Lever, Walter Frederick
Levy, David Alfred
Lewis, David Harold
Lo, Clifford W
Marquardt, Hans Wilhelm Joe
Maruyama, Koshi
Mendelsohn, Mortimer Lester
Ochoa, Severo
Partington, Michael W
Perez-Tamayo, Ruheri
Rabinovitch, Michel Pinkus
Revillard, Jean-Pierre Remy
Rossier, Alain B
Sasaki, Hidetada
Schena, Francesco Paolo
Shimaoka, Katsutaro
Shubik, Philippe
Solomons, Noel Willis
Solter, Davor
Spiess, Joachim
Steel, R Knight
Stern, Kurt
Suga, Hiroyuki
Takai, Yasuyuki
Tanphaichitr, Vichai
Thiessen, Jacob Willem
Torun, Benjamin
Van Furth, Ralph

Van Rood, Johannes J
Wakayama, Yoshihiro
Walters, Jack Henry
Wang, Soo Ray
Watanabe, Mamoru
Watanabe, Takeshi
Watanabe, Yoshio
Weissgerber, Rudolph E

Nursing

ALASKA
DeLapp, Tina Davis

ARIZONA
Peric-Knowlton, Wlatka

CALIFORNIA
Chang, Betty Lee
Dracup, Kathleen A
Estes, Carroll L
Gortner, Susan Reichert
Leone, Lucile P
Styles, Margretta M

COLORADO
Fuller, Barbara Fink Brockway

CONNECTICUT
Diers, Donna Kaye

DISTRICT OF COLUMBIA
Fishbein, Eileen Greif
Heinrich, Janet
Larson, Elaine Lucille

FLORIDA
Mauksch, Ingeborg G

GEORGIA
Binnicker, Pamela Caroline
Cotanch, Patricia Holleran

ILLINOIS
Andreoli, Kathleen Gainor
Aramini, Jan Marie
Bradford, Laura Sample
Buschmann, MaryBeth Tank
Janusek, Linda Witek
Kim, Mi Ja
Shannon, Iris R

INDIANA
Geddes, LaNelle Evelyn
McBride, Angela Barron

IOWA
Aydelotte, Myrtle K
Marshall, Leslie Brusletten

KANSAS
Coles, Anna Bailey

KENTUCKY
Fleming, Juanita W

MARYLAND
Donaldson, Sue Karen
Hinshaw, Ada Sue
Suppe, Frederick (Roy)

MASSACHUSETTS
Kibrick, Anne K

MICHIGAN
Courtney, Gladys (Atkins)
Dumas, Rhetaugh Graves
Pender, Nola J
Shea, Fredericka Palmer

MISSOURI
Rubin, Maryiln Bernice

NEW JERSEY
Iglewicz, Raja
Wasserman, Martin Allan

NEW YORK
Bullough, Vern L
Davis, Carolyne K
Ford, Loretta C
Fusco, Carmen Carrier
Golub, Sharon Bramson
Johnson, Jean Elaine
Lambertsen, Eleanor C
Lichtig, Leo Kenneth
Liebmann, Jeffrey Mitchell
Lubic, Ruth Watson
Stafford, Magdalen Marrow
Wysocki, Annette Bernadette

NORTH CAROLINA
Henning, Emilie D
Wilson, Ruby Leila

OHIO
Anderson, Ruth Mapes
Schlotfeldt, Rozella M
Stone, Kathleen Sexton

PENNSYLVANIA
Aiken, Linda H
Brooten, Dorothy

Fagin, Claire M
Gregg, Christine M
Lang, Norma M
Leddy, Susan
Lowery, Barbara J
McCorkle, Ruth
Schwartz, Doris R
Sovie, Margaret D

SOUTH CAROLINA
Darnell, W(illiam) H(eaden)
Thompson, Shirley Jean

SOUTH DAKOTA
Schramm, Mary Arthur

TENNESSEE
Christman, Luther P
Holaday, Bonnie
Norton, Virginia Marino

TEXAS
Avant, Patricia Kay Coalson
Burns, Nancy A
Chow, Rita K
Hill, Reba Michels
Pickard, Myrna Rae
Yoder Wise, Patricia S

VIRGINIA
Abdellah, Faye G
Adams, Nancy R

WASHINGTON
Barnard, Kathryn E
De Tornyay, Rheba
Gunter, Laurie M
Hegyvary, Sue Thomas
Valanis, Barbara Mayleas

WISCONSIN
Duxbury, Mitzi L
Porth, Carol Mattson

MANITOBA
Adaskin, Eleanor Jean
Sinha, Helen Luella

ONTARIO
Chapman, Jacqueline Sue
Jones, Phyllis Edith

OTHER COUNTRIES
Atinmo, Tola
Maglacas, A Mangay
Miyasaka, Kyoko

Nutrition

ALABAMA
Alexander, Herman Davis
Alvarez, Jose O
Baliga, B Surendra
Carter, Eloise
Cheraskin, Emanuel
Clark, Alfred James
Davis, Elizabeth Young
Flodin, N(estor) W(inston)
Frobish, Lowell T
Krumdieck, Carlos L
McKibbin, John Mead
Marion, James Edsel
Navia, Juan Marcelo
Parks, Paul Franklin
Prince, Charles William
Ringsdorf, Warren Marshall, Jr
Sauberlich, Howerde Edwin
Schmidt, Stephen Paul
Svacha, Anna Johnson
Tamura, Tsunenobu
Weinsier, Roland Louis

ARIZONA
Blanchard, James
Butler, Lillian Catherine
Garewal, Harinder Singh
Jeevanandam, Malayappa
Koldovsky, Otakar
Lei, David Kai Yui
Lohman, Timothy George
Moon, Thomas Edward
Palumbo, Pasquale John
Pearson, Paul Brown
Reid, Bobby Leroy
Rittenbaugh, Cheryl K
Vaughan, Linda Ann
Vavich, Mitchell George
Watson, Ronald Ross
Weber, Charles Walter
Whiting, Frank M

ARKANSAS
Bogle, Margaret L
Hopper, John Henry
Kellogg, David Wayne
Kenney, Mary Alice
Poirier, Lionel Albert
Waldroup, Park William

CALIFORNIA
Aftergood, Lilla
Allen, Lindsay Helen
Amen, Ronald Joseph

TENNESSEE
Arata, Dorothy
Bass, Mary Anna
Beauchene, Roy E
Broquist, Harry Pearson
Carruth, Betty Ruth
Chytil, Frank
Cullen, Marion Permilla
Darby, William Jefferson
Greene, Harry Lee
Hak, Lawrence J
Hansard, Samuel Leroy
Hilker, Doris M
Hunt, James Calvin
Mann, George Vernon
Pfuderer, Helen A
Reynolds, Marjorie Lavers
Sachan, Dileep Singh
Savage, Jane Ramsdell
Smith, Mary Ann Harvey
Todhunter, Elizabeth Neige
Trupin, Joel Sunrise
Zemel, Michael Barry

TEXAS
Abrams, Steven Allen
Alford, Betty Bohon
Anderson, Karl E
Arnold, Watson Caufield
Bielamowicz, Mary Claire Kinney
Brittin, Dorothy Helen Clark
Byerley, Lauri Olson
Couch, James Russell
Czajka-Narins, Dorice M
Danhof, Ivan Edward
Davis, Donald Robert
Dawson, Earl B
DePaola, Dominick Philip
Dudrick, Stanley John
Edwards, Marilyn E
Fry, Peggy Crooke
Green, Gary Miller
Heffley, James D
Helm, Raymond E
Hickson, James Forbes
Huber, Gary Louis
Keating, Robert Joseph
Kerr, George R
Klein, Gordon Leslie
Klein, Peter Douglas
Kubena, Karen Sidell
Kunkel, Harriott Orren
Kushwaha, Rampratap S
Lamb, Mina Marie Wolf
Lane, Helen W
Langford, Florence
Latimer, George Webster, Jr
Lawrence, Richard Aubrey
Liepa, George Uldis
Little, Perry L
Longenecker, John Bender
Loop, Rose-Ann
McPherson, Clara
Miller, Edward Godfrey, Jr
Miller, Sanford Arthur
Miner, James Joshua
Mize, Charles Edward
Nelson, Thomas Edward
Nichaman, Milton Z
Nichols, Buford Lee, Jr
O'Brien, Barbara Cooney
Pond, Wilson Gideon
Rassin, David Keith
Reber, Elwood Frank
Root, Elizabeth Jean
Sandstead, Harold Hilton
Shaw, Emil Gilbert
Sherrod, Lloyd B
Shulman, Robert Jay
Starcher, Barry Chapin
Stevens, Marion Bennion
Stubbs, Donald William
Sweeney, Edward Arthur
Vogel, James John
Wagner, Martin James
Williams, Darryl Marlowe
Wolinsky, Ira
Young, Eleanor Anne

UTAH
Chan, Gary Mannerstedt
Conlee, Robert Keith
Evans, Robert John
Hendricks, Deloy G
Lawson, Larry Dale
Mahoney, Arthur W
Mangelson, Farrin Leon
Oyler, Dee Edward
Page, Louise
Wilcox, Ethelwyn Bernice

VERMONT
Bartel, Lavon L
Foss, Donald C
Johnson, Robert Eugene
Kunin, Arthur Saul

VIRGINIA
Axelson, Marta Lynne
Banks, William Louis, Jr
Brignoli Gable, Carol
Brown, Arlin James
Brown, Myrtle Laurestine
Bunce, George Edwin
Christiansen, Marjorie Miner
Donohue, Joyce Morrissey
Edwards, Leslie Erroll
Evans, Joseph Liston
Flinn, Jane Margaret
Harper, Laura Jane
Hepburn, Frank
Khan, Mahmood Ahmed
Korslund, Mary Katherine
Kronfeld, David Schultz
Mahboubi, Ezzat
Morlang, Barbara Louise
Pao, Eleanor M
Phillips, Jean Allen
Ritchey, Sanford Jewell
Sassoon, Humphrey Frederick
Taper, Lillian Janette
Teague, Howard Stanley
Thye, Forrest Wallace

WASHINGTON
Anderson, Robert Authur
Bankson, Daniel Duke
Childs, Marian Tolbert
Halver, John Emil
Hard, Margaret McGregor
Harper, Alfred Edwin
Ho, Lydia Su-yong
Hoskins, Frederick Hall
McCabe, Donald Lee
McGinnis, James
Massey, Linda Kathleen Locke
Mitchell, Madeleine Enid
Smith, Sam Corry
West, Herschel J
Yamanaka, William Kiyoshi

WEST VIRGINIA
Clark, Ralph B
Jagannathan, Singanallur N
Konat, Gregory W
Lim, James Khai-Jin
Martin, William Gilbert
Sheikh, Kazim

WISCONSIN
Atkinson, Richard L, Jr
Barboriak, Joseph Jan
Benevenga, Norlin Jay
Dutton, Herbert Jasper
Elson, Charles
Ganther, Howard Edward
Greger, Janet L
Jackson, Marion Leroy
Kainski, Mercedes H
Kramer, Elizabeth
McDivitt, Maxine Estelle
Marlett, Judith Ann
Matthews, Mary Eileen
Petering, David Harold
Pringle, Dorothy Jutton
Schulz, Leslie Olmstead
Steele, Robert Darryl
Steinhart, Carol Elder

PUERTO RICO
Fernandez-Repollet, Emma D
Preston, Alan Martin
Rodriguez, Jose Raul

ALBERTA
Basu, Tapan Kumar
Clandinin, Michael Thomas
Donald, Elizabeth Ann
Koopmans, Henry Sjoerd
Renner, Ruth

BRITISH COLUMBIA
Desai, Indrajit Dayalji
Lee, Melvin
Leichter, Joseph

MANITOBA
Choy, Patrick C
Dakshinamurti, Krishnamurti
McDonald, Bruce Eugene
Vaisey-Genser, Florence Marion
Yurkowski, Michael

NEWFOUNDLAND
Brosnan, Margaret Eileen
Hulan, Howard Winston
Orr, Robin Denise Moore

ONTARIO
Anderson, G Harvey
Bayley, Henry Shaw
Beaton, George Hector
Behrens, Willy A
Bruce, W Robert
Campbell, James Alexander
Carroll, Kenneth Kitchener
Cunnane, Stephen C
Currie, Violet Evadne
Draper, Harold Hugh
Farnworth, Edward Robert
Flanagan, Peter Rutledge
Frederick, George Leonard
Gibson, Rosalind Susan
Himms-Hagen, Jean
Jeejeebhoy, Khursheed Nowrojee
Jenkins, David John Anthony
Jenkins, Kenneth James William
Lloyd, Lewis Ewan
MacRae, Andrew Richard
Milligan, Larry Patrick
Roth, René Romain
Sarwar, Ghulam
Scott, Fraser Wallace
Sibbald, Ian Ramsay
Thompson, Lilian Umale
Tryphonas, Helen
Vallerand, Andre L
Veen-Baigent, Margaret Joan
Wardlaw, Janet Melville
Yeung, David Lawrence

QUEBEC
Brisson, Germain J
Gougeon, Rejeanne
Guyda, Harvey John
Hamet, Pavel
Kuhnlein, Harriet V
Letarte, Jacques
Marliss, Errol Basil
Meyer, Francois
Roy, Claude Charles
Shizgal, Harry M

SASKATCHEWAN
Angel, Joseph Francis
Knipfel, Jerry Earl
Korsrud, Gary Olaf

OTHER COUNTRIES
Bell, Roma Raines
Brunser, Oscar
Buzina, Ratko
Casey, Clare Ellen
Dreosti, Ivor Eustace
Fairweather-Tait, Susan Jane
Gustafsson, Jan-Ake
Haschke, Ferdinand
King, Dorothy Wei (Cheng)
Leitzmann, Claus
Ling, Chung-Mei
Lo, Clifford W
Mela, David Jason
Miyasaka, Kyoko
Palmer, Sushma Mahyera
Renaud, Serge
Rio, Maria Esther
Sheng, Hwai-Ping
Shiraki, Keizo
Smith, Meredith Ford
Solomons, Noel Willis
Tanphaichitr, Vichai
Torun, Benjamin
Valencia, Mauro Eduardo

Occupational Medicine

CALIFORNIA
Adams, Robert Monroe
Cooper, William Clark
Kilburn, Kaye Hatch
Palmer, Alan
Schenker, Marc B
Spivey, Gary H

CONNECTICUT
Skinner, H Catherine W

DISTRICT OF COLUMBIA
Kimbrough, Renate Dora
Sphar, Raymond Leslie

FLORIDA
Hogan, Daniel James
Weiss, Charles Frederick

GEORGIA
Deitchman, Scott David
Mitchell, Frank L
Thun, Michael John

ILLINOIS
Carnow, Bertram Warren

MARYLAND
Dyer, Robert Francis, Jr

MASSACHUSETTS
Goldman, Ralph Frederick

MICHIGAN
Block, Duane Llewellyn
Reeves, Andrew Louis

NEW MEXICO
Tillery, Marvin Ishmael

NEW YORK
Ferin, Juraj
Miller, Albert
Zavon, Mitchell Ralph

OREGON
Morton, William Edwards

VIRGINIA
Carey, Robert Munson

WEST VIRGINIA
Thakur, Pramod Chandra

WISCONSIN
Sothmann, Mark Steven

ONTARIO
Roy, Marie Lessard

QUEBEC
Drouin, Gilbert
Dykes, Robert William
Messing, Karen

Optometry

ALABAMA
Peters, Henry Buckland
Rosenblum, William M
Wild, Bradford Williston

ARKANSAS
Smith, E(astman)

CALIFORNIA
Chang, Freddy Wilfred Lennox
Cohn, Theodore E
Dowaliby, Margaret Susanne
Enoch, Jay Martin
Gregg, James R
Haines, Richard Foster
Hawryluk, Andrew Michael
Hopping, Richard Lee
Johnson, Chris Alan
Katz, Barrett
Mandell, Robert Burton
Marg, Elwin
Martinez-Uriegas, Eugenio
Morgan, Meredith Walter
Tyler, Christopher William

CONNECTICUT
Telfair, William Boys

FLORIDA
Borish, Irvin Max
Weber, James Edward

ILLINOIS
Lit, Alfred
Rosenbloom, Alfred A, Jr

INDIANA
Allen, Merrill James
Everson, Ronald Ward
Goss, David A
Heath, Gordon Glenn
Hofstetter, Henry W
Lowther, Gerald Eugene
Soni, Prem Sarita

MARYLAND
Ferris, Frederick L
Levinson, John Z
Massof, Robert W
Michaelis, Otho Ernest, IV
Yau, King-Wai

MASSACHUSETTS
Baldwin, William Russell
Carter, John Haas
Sekuler, Robert W

MICHIGAN
Lewis, Alan Laird
Zacks, James Lee

MISSOURI
Peck, Carol King

NEW JERSEY
Roy, M S

NEW MEXICO
Wong, Siu Gum

NEW YORK
Bannon, Robert Edward
Goldrich, Stanley Gilbert
Haffner, Alden Norman
Hainline, Louise
Howland, Howard Chase
Luntz, Maurice Harold
Rosenberg, Robert

OHIO
Hebbard, Frederick Worthman
Schoessler, John Paul

OREGON
Levine, Leonard
Ludlam, William Myrton

PENNSYLVANIA
Dobson, Margaret Velma

TENNESSEE
Elberger, Andrea June
Levene, John Reuben

TEXAS
Flom, Merton Clyde
Hecht, Gerald
Pease, Paul Lorin
Pitts, Donald Graves
Stein, Jerry Michael

Optometry (cont)

VIRGINIA
Gabriel, Barbra L

WASHINGTON
Young, Francis Allan

PUERTO RICO
Afanador, Arthur Joseph

ALBERTA
Thomson, Alan B R

BRITISH COLUMBIA
Cynader, Max Sigmund

ONTARIO
Hallett, Peter Edward
Lyle, William Montgomery
Regan, David M
Remole, Arnulf
Sivak, Jacob Gershon
Woo, George Chi Shing

OTHER COUNTRIES
Hughes, Abbie Angharad

Parasitology

ALABAMA
Bailey, Wilford Sherrill
Boyles, James McGregor
Buckner, Richard Lee
Dixon, Carl Franklin
Eckman, Michael Kent
Frandsen, John Christian
Kayes, Stephen Geoffrey
Klesius, Phillip Harry

ALASKA
Esslinger, Jack Houston
Fay, Francis Hollis
Swartz, Leslie Gerard

ARIZONA
Collins, Richard Cornelius
Dewhirst, Leonard Wesley
Van Riper, Charles, III
Wilkes, Stanley Northrup

ARKANSAS
Bridgman, John Francis
Corkum, Kenneth C
Daly, James Joseph
Hinck, Lawrence Wilson
Johnson, Arthur Albin

CALIFORNIA
Amrein, Yost Ursus Lucius
Ash, Lawrence Robert
Bair, Thomas De Pinna
Basch, Paul Frederick
Beck, Albert J
Caulfield, John Philip
Chao, Jowett
Chi, Lois Wong
Cornford, Eain M
Cosgrove, Gerald Edward
Davis, Charles Edward
Demaree, Richard Spottswood, Jr
Dowell, Armstrong Manly
Furman, Deane Philip
Garcia, Richard
Golder, Thomas Keith
Hackney, Robert Ward
Harmon, Wallace Morrow
Heyneman, Donald
Ho, Ju-Shey
Horvath, Kalman
Hunderfund, Richard C
Jones, Ira
Krassner, Stuart M
Kuris, Armand Michael
Lafferty, Kevin D
MacInnis, Austin J
Mankau, Sarojam Kurudamannil
Mansour, Tag Eldin
Markell, Edward Kingsmill
Mitchell, John Alexander
Nahhas, Fuad Michael
Nichols, Barbara Ann
Nickel, Phillip Arnold
Noble, Elmer Ray
Noble, Glenn Arthur
Olson, Andrew Clarence, Jr
Owen, Robert L
Rigby, Donald W
Schacher, John Fredrick
Seidel, James Stephen
Sherman, Irwin William
Silverman, Paul Hyman
Theis, Jerold Howard
Uyeda, Charles Tsuneo
Vickrey, Herta Miller
Wagner, Edward D
Wang, Ching Chung
Weinmann, Clarence Jacob
Westervelt, Clinton Albert, Jr
Widmer, Elmer Andreas
Wong, Ming Ming
Woodruff, David Scott
Young, Robert, Jr

COLORADO
Andrews, Myron Floyd
Berens, Randolph Lee
Grieve, Robert B
Gubler, Duane J
Hathaway, Ronald Philip
Marquardt, William Charles
Schmidt, Gerald D
Vetterling, John Martin
Voth, David Richard

CONNECTICUT
Eustice, David Christopher
James, Hugo A
Krause, Peter James
Krinsky, William Lewis
Levine, Harvey Robert
Lynch, John Edward
Newborg, Michael Foxx
O'Neill, James F
Patton, Curtis LeVerne
Richards, Frank Frederick

DELAWARE
Cubberley, Virginia

DISTRICT OF COLUMBIA
Cressey, Roger F
Gravely, Sally Margaret
Hollingdale, Michael Richard
Jackson, George John
Klein, Terry Allen
Schneider, Imogene Pauline
Taylor, Diane Wallace

FLORIDA
Burridge, Michael John
CmejLa, Howard Edward
Couch, John Alexander
Courtney, Charles Hill
Crandall, Richard B
Ferguson, Malcolm Stuart
Forrester, Donald J
Fournie, John William
Friedl, Frank Edward
Lawrence, Pauline Olive
McCowen, Max Creager
Porter, James Armer, Jr
Roberts, Larry Spurgeon
Rodrick, Gary Eugene
Rossan, Richard Norman
Sengbusch, Howard George
Short, Robert Brown
Splitter, Earl John
Telford, Sam Rountree, Jr
Undeen, Albert Harold
Wood, Irwin Boyden
Young, Martin Dunaway
Zam, Stephen G, III

GEORGIA
Agosin, Moises
Ah, Hyong-Sun
Brooke, Marion Murphy
Colley, Daniel George
Damian, Raymond T
French, Frank Elwood, Jr
Hanson, William Lewis
Healy, George Richard
Husain, Ansar
Jones, Gilda Lynn
Kagan, Irving George
Kaiser, Robert L
Lammie, Patrick J
McDougal, Larry Robert
McDowell, John Willis
Maddison, Shirley Eunice
Maples, William Paul
Mathews, Henry Mabbett
Melvin, Dorothy Mae
Nguyen-Dinh, Phuc
Oliver, James Henry, Jr
Pratt, Harry Davis
Prestwood, Annie Katherine
Reid, Willard Malcolm
Roberson, Edward Lee
Saladin, Kenneth S
Schwartz, David Alan
Sulzer, Alexander Jackson
Tsang, Victor Chiu Wan
Walls, Kenneth W
Warren, McWilson

HAWAII
Desowitz, Robert
Noda, Kaoru
Siddiqui, Wasim A

IDAHO
Hibbert, Larry Eugene
Schell, Stewart Claude
Wootton, Donald Merchant

ILLINOIS
Auerbach, Earl
Chandran, Satish Raman
Chang, Kwang-Poo
Dyer, William Gerald
French, Allen Lee
Garoian, George
Huizinga, Harry William
Jaskoski, Benedict Jacob
Kniskern, Verne Burton
Larson, Ingemar W

Leathem, William Dolars
Levine, Norman Dion
Myer, Donal Gene
Nadler, Steven Anthony
Nollen, Paul Marion
Preisler, Harvey D
Ridgeway, Bill Tom
Seed, Thomas Michael
Todd, Kenneth S, Jr
Von Zellen, Bruce Walfred
Yogore, Mariano G, Jr

INDIANA
Bowen, Richard Eli
Current, William L
Denner, Melvin Walter
Diffley, Peter
Dusanic, Donald G
Gaafar, Sayed Mohammed
Jeffers, Thomas Kirk
McCracken, Richard Owen
Ott, Karen Jacobs
Saz, Howard Jay
Summers, William Allen, Sr
Vatne, Robert Dahlmeier
Weinstein, Paul P
Welker, George W
Young, Frank Nelson, Jr

IOWA
Cain, George D
Collins, Richard Francis
Greve, John Henry
Hsu, Shu Ying Li
Marty, Wayne George
Maruyama, George Masao
Mertins, James Walter
Nolf, Luther Owen
Petri, Leo Henry

KANSAS
Coil, William Herschell
Hansen, Merle Fredrick
Johnson, John Christopher, Jr
Nusser, Wilford Lee
Poeschel, Gordon Paul
Upton, Steve Jay

KENTUCKY
Drudge, J Harold
Gleason, Larry Neil
Justus, David Eldon
Lyons, Eugene T
Oetinger, David Frederick
Uglem, Gary Lee
Whittaker, Frederick Horace

LOUISIANA
Abadie, Stanley Herbert
Babero, Bert Bell
Boertje, Stanley
Brown, Kenneth Michael
Christian, Frederick Ade
D'Alessandro-Bacigalupo, Antonio
Durio, Walter O'Neal
Font, William Francis
Harris, Alva H
Jung, Rodney Clifton
Klei, Thomas Ray
Krupp, Iris M
Lanners, H Norbert
Little, Maurice Dale
Lumsden, Richard
Malek, Emile Abdel
Miller, Joseph Henry
Orihel, Thomas Charles
Sizemore, Robert Carlen
Specian, Robert David
Stewart, T Bonner
Stueben, Edmund Bruno
Swartzwelder, John Clyde
Turner, Hugh Michael
Vidrine, Malcolm Francis
Warren, Lionel Gustave
Williams, James Carl
Wiser, Mark Frederick
Yaeger, Robert George

MAINE
Chute, Robert Maurice
Gibbs, Harold Cuthbert
Meyer, Marvin Clinton
Najarian, Haig Hagop
Turner, James Henry

MARYLAND
Anderson, Norman Leigh
Ansher, Sherry Singer
Augustine, Patricia C
Bala, Shukal
Bram, Ralph A
Carney, William Patrick
Cheever, Allen
Colglazier, Merle Lee
Diamond, Louis Stanley
Diggs, Carter Lee
Djurickovic, Draginja Branko
Dubey, Jitender Prakash
Dwyer, Dennis Michael
Erickson, Duane Gordon
Fayer, Ronald
Fisher, Pearl Davidowitz
Gasbarre, Louis Charles
Gibson, Colvin Lee

Graham, Charles Lee
Gwadz, Robert Walter
Hoberg, Eric Paul
Jackson, Peter Richard
James, Stephanie Lynn
Kresina, Thomas Francis
Lawrence, Dale Nolan
Lewis, Fred A
Lichtenfels, James Ralph
Lillehoj, Hyun Soon
Lincicome, David Richard
Lunney, Joan K
Mercado, Teresa I
Miller, Louis Howard
Munson, Donald Albert
Murrell, Kenneth Darwin
O'Grady, Richard Terence
Otto, Gilbert Fred
Palmer, Timothy Trow
Pinto da Silva, Pedro Goncalves
Porter, Clarence A
Powers, Kendall Gardner
Ruff, Michael David
Scheibel, Leonard William
Sheffield, Harley George
Strickland, George Thomas
Sztein, Marcelo Benjamin
Tarshis, Irvin Barry
Taylor, David Neely
Ward, Ronald A(nthony)
Weisbroth, Steven H
Zierdt, Charles Henry

MASSACHUSETTS
Bruce, John Irvin
Chishti, Athar Husain
Coleman, Robert Marshall
Dammin, Gustave John
Duncan, Stewart
Foster, William Burnham
Honigberg, Bronislaw Mark
Humes, Arthur Grover
Hurley, Francis Joseph
Meszoely, Charles Aladar Maria
Nutting, William Brown
Pan, Steve Chia-Tung
Rivera, Ezequiel Ramirez
Spielman, Andrew
Telford, Sam Rountree, III
Weller, Peter Fahey
West, Arthur James, II
Wyler, David J

MICHIGAN
Blankespoor, Harvey Dale
Bratt, Albertus Dirk
Chobotar, Bill
Conder, George Anthony
De Giusti, Dominic Lawrence
Elslager, Edward Faith
Folz, Sylvester D
Hupe, Donald John
Jensen, James Burt
Maxey, Brian William
Peebles, Charles Robert
Rizzo, Donald Charles
Steen, Edwin Benzel
Szanto, Joseph
Twohy, Donald Wilfred
Waffle, Elizabeth Lenora
Williams, Jeffrey F
Yates, Jon Arthur

MINNESOTA
Ballard, Neil Brian
Bartel, Monroe H
Bemrick, William Joseph
Eaton, John Wallace
Gilbertson, Donald Edmund
Lange, Robert Echlin, Jr
McCue, John Francis
Meyer, Fred Paul
Paulson, Carlton
Preiss, Frederick John
Robinson, Edwin James, Jr
Schlotthauer, John Carl
Stromberg, Bert Edwin, Jr
Thompson, John Harold, Jr
Tonn, Robert James
Wagenbach, Gary Edward
Wallace, Franklin Gerhard
Wyatt, Ellis Junior

MISSISSIPPI
Hutchison, William Forrest
Lawler, Adrian Russell
Norris, Donald Earl, Jr
Overstreet, Robin Miles
White, Jesse Steven

MISSOURI
Green, Theodore James
Hall, Robert Dickinson
Miles, Donald Orval
Shelton, George Calvin
Starling, Jane Ann
Train, Carl T
Twining, Linda Carol
Wagner, Joseph Edward
Weppelman, Roger Michael

MONTANA
Mitchell, Lawrence Gustave
Speer, Clarence Arvon

Thomas, Leo Alvon
Werner, John Kirwin
Worley, David Eugene
NEBRASKA
Nickol, Brent Bonner
Phares, Cleveland Kirk
Pritchard, Mary (Louise) Hanson
NEVADA
Hunter, Kenneth W, Jr
Mead, Robert Warren
NEW HAMPSHIRE
Bullock, Wilbur Lewis
Harrises, Antonio Efthemios
Pfefferkorn, Elmer Roy, Jr
Strout, Richard Goold
NEW JERSEY
Bacha, William Joseph, Jr
Berger, Harold
Burrows, Robert Beck
D'Alesandro, Philip Anthony
Dennis, Emmet Adolphus
Dobson, Andrew Peter
Fong, Dunne
Griffo, James Vincent, Jr
Guerrero, Jorge
Hayes, Terence James
Kantor, Sidney
Katz, Frank Fred
Leaning, William Henry Dickens
Leibowitz, Michael Jonathan
Levine, Donald Martin
Ostlind, Dan A
Schenkel, Robert H
Scholl, Philip Jon
Schulman, Marvin David
Tierno, Philip M, Jr
Virkar, Raghunath Atmaram
Wang, Guang Tsan
NEW MEXICO
Davis, Larry Ernest
Duszynski, Donald Walter
Frenkel, Jacob Karl
Reyes, Philip
NEW YORK
Asch, Harold Lawrence
Barnwell, John Wesley
Bell, Robin Graham
Bettencourt, Joseph S, Jr
Clarkson, Allen Boykin, Jr
Concannon, Joseph N
Cross, George Alan Martin
Dean, David A
Despommier, Dickson
Drobeck, Hans Peter
Gold, Allen Morton
Gordon, Morris Aaron
Grimwood, Brian Gene
Habicht, Gail Sorem
Hutner, Seymour Herbert
Imperato, Pascal James
Ingalls, James Warren, Jr
Isseroff, Hadar
Jindrak, Karel
Kaplan, Eugene Herbert
Keithly, Janet Sue
Kim, Charles Wesley
Kohls, Robert E
Levin, Norman Lewis
Lo Verde, Philip Thomas
McGraw, James Carmichael
Mackiewicz, John Stanley
McNamara, Thomas Francis
Muller, Miklos
Neenan, John Patrick
Nussenzweig, Victor
Pike, Eileen Halsey
Reuter, Gerald Louis
Ritterson, Albert L
Shields, Robert James
Soave, Rosemary
Spencer, Frank
Stephenson, Lani Sue
Styles, Twitty Junius
Trager, William
Vanderberg, Jerome Philip
Vasey, Carey Edward
Weiner, Matei
Whitlock, John Hendrick
Wittner, Murray
Wood, Raymond Arthur
Yarinsky, Allen
Yolles, Stanley Fausst
Yolles, Tamarath Knigin
NORTH CAROLINA
Batte, Edward G
Behlow, Robert Frank
Coffey, James Cecil, Jr
Estevez, Enrique Gonzalo
Eure, Herman Edward
George, Charles Redgenal
Goulson, Hilton Thomas
Hampton, Carolyn Hutchins
Hendricks, James Richard
Henson, Richard Nelson
Knuckles, Joseph Lewis
LaFon, Stephen Woodrow
Lewert, Robert Murdoch

Martin, Virginia Lorelle
Noga, Edward Joseph
Pattillo, Walter Hugh, Jr
Perkins, Kenneth Warren
Pung, Oscar J
Roberts, John Fredrick
Seed, John Richard
Shepperson, Jacqueline Ruth
Sonnenfeld, Gerald
Spuller, Robert L
Weatherly, Norman F
NORTH DAKOTA
Holloway, Harry Lee, Jr
Larson, Omer R
OHIO
Aikawa, Masamichi
Arlian, Larry George
Barriga, Omar Oscar
Barrow, James Howell, Jr
Blanton, Ronald Edward
Crites, John Lee
Etges, Frank Joseph
Grassmick, Robert Alan
Hallberg, Carl William
Heck, Oscar Benjamin
Hink, Walter Fredric
Jarroll, Edward Lee, Jr
Kapral, Frank Albert
Kochanowski, Barbara Ann
Kreier, Julius Peter
Mahmoud, Adel A F
Malemud, Charles J
Myers, Betty June
Needham, Glen Ray
Pappas, Peter William
Rabalais, Francis Cleo
Rice, Eugene Ward
Schaefer, Frank William, III
Sodeman, William A
Van Zandt, Paul Doyle
OKLAHOMA
Abram, James Baker, Jr
Branch, John Curtis
Burr, John Green
Ewing, Sidney Alton
Ivey, Michael Hamilton
John, David Thomas
Jordan, Helen Elaine
Kocan, Katherine M
McKnight, Thomas John
Powders, Vernon Neil
OREGON
Anderson, John Richard
Bayne, Christopher Jeffrey
Carter, Richard Thomas
Clark, David Thurmond
McConnaughey, Bayard Harlow
Macy, Ralph William
Olson, Robert Eldon
Radovsky, Frank Jay
Rossignol, Philippe Albert
PENNSYLVANIA
AcKerman, Larry Joseph
Anderson, David Robert
Boutros, Susan Noblit
Dollahon, Norman Richard
Donnelly, John James, III
Edlind, Thomas D
Foor, W Eugene
Fremount, Henry Neil
Fried, Bernard
Harms, Clarence Eugene
Hendrix, Sherman Samuel
Humphreys, Jan Gordon
McManus, Edward Clayton
McNair, Dennis M
Mandel, John Herbert
Moss, William Wayne
Ogren, Robert Edward
Rogers, William Edwin
Ryan, James Patrick
Salerno, Ronald Anthony
Schad, Gerhard Adam
Sillman, Emmanuel I
Slifkin, Malcolm
Theodorides, Vassilios John
Vaidya, Akhil Babubhai
Walton, Bryce Calvin
Ward, William Francis
Williams, Russell Raymond
Yurchenco, John Alfonso
Zarkower, Arian
RHODE ISLAND
Flanigan, Timothy Palen
Hyland, Kerwin Ellsworth, Jr
Jones, Kenneth Wayne
Knopf, Paul M
LeBrun, Roger Arthur
Rogers, Steffen Harold
SOUTH CAROLINA
Cheng, Thomas Clement
Kowalczyk, Jeanne Stuart
Sargent, Roger Gary
Schmidt, Roger Paul
Vernberg, Winona B

SOUTH DAKOTA
Hugghins, Ernest Jay
TENNESSEE
Benz, George William
Bogitsh, Burton Jerome
Bradley, Richard E
Cypess, Raymond Harold
Elliott, Alice
Harrison, Robert Edwin
Hill, George Carver
McGavock, Walter Donald
Patton, Sharon
Prudhon, Rolland A, Jr
Risby, Edward Louis
Wilhelm, Walter Eugene
TEXAS
Box, Edith Darrow
Bristol, John Richard
Dronen, Norman Obert, Jr
Ewert, Adam
Fisher, Frank M, Jr
Fisher, William Francis
Galvin, Thomas Joseph
Hejtmancik, Kelly Erwin
Huffman, David George
Kuntz, Robert Elroy
McAllister, Chris Thomas
McGraw, John Leon, Jr
Meade, Thomas Gerald
Meola, Roger Walker
Moore, Donald Vincent
Morrison, Eston Odell
Murad, John Louis
Niederkorn, Jerry Young
Race, George Justice
Richerson, Jim Vernon
Rolston, Kenneth Vijaykumar Issac
Schlueter, Edgar Albert
Shadduck, John Allen
Slagle, Wayne Grey
Smith, Jerome H
Sogandares-Bernal, Franklin
Standifer, Lonnie Nathaniel
Stewart, George Louis
Tamsitt, James Ray
Taylor, D(orothy) Jane
Tuff, Donald Wray
Tulloch, George Sherlock
Turco, Charles Paul
Ubelaker, John E
Valdivieso, Dario
UTAH
Andersen, Ferron Lee
Fitzgerald, Paul Ray
Grundmann, Albert Wendell
Harrington, Glenn William
Healey, Mark Calvin
Heckmann, Richard Anderson
Jensen, Emron Alfred
Mason, James O
Tipton, Vernon John
Warnock, Robert G
VERMONT
Schall, Joseph Julian
VIRGINIA
Burreson, Eugene M
Cruthers, Larry Randall
Eckerlin, Ralph Peter
Fisher, Elwood
Holliman, Rhodes Burns
Mikkelsen, Ross Blake
Osborne, Paul James
Russell, Catherine Marie
Schwartzman, Joseph David
WASHINGTON
Allen, Julia Natalia
Catts, Elmer Paul
Clark, Glen W
Krieger, John Newton
Lang, Bruce Z
Rausch, Robert Lloyd
Senger, Clyde Merle
Sibley, Carol Hopkins
Stuart, Kenneth Daniel
Wescott, Richard Breslich
WEST VIRGINIA
Hall, John Edgar
Hoffman, Glenn Lyle
WISCONSIN
Burns, William Chandler
Coggins, James Ray
Guilford, Harry Garrett
Lunde, Milford Norman
McDonald, Malcolm Edwin
Nash, Reginald George
Oaks, John Adams
Peanasky, Robert Joseph
Rouse, Thomas C
Wittrock, Darwin Donald
Yoshino, Timothy Phillip
WYOMING
Bergstrom, Robert Charles
Kingston, Newton

PUERTO RICO
Hillyer, George Vanzandt
Kozek, Wieslaw Joseph
Oliver-Gonzales, Jose
ALBERTA
Arai, Hisao Philip
Baron, Robert Walter
Befus, A(lbert) Dean
Holmes, John Carl
Mahrt, Jerome L
Samuel, William Morris
Wallis, Peter Malcolm
BRITISH COLUMBIA
Bower, Susan Mae
Ching, Hilda
Margolis, Leo
MANITOBA
Novak, Marie Marta
NEW BRUNSWICK
Burt, Michael David Brunskill
Smith, Harry John
NEWFOUNDLAND
Threlfall, William
NOVA SCOTIA
Angelopoulos, Edith W
Wiles, Michael
ONTARIO
Anderson, Roy Clayton
Bourns, Thomas Kenneth Richard
Brooks, Daniel Rusk
Chadwick, June Stephens
Desser, Sherwin S
Fernando, Constantine Herbert
Lee, Eng-Hong
Mellors, Alan
Mettrick, David Francis
Slocombe, Joseph Owen Douglas
Tryphonas, Helen
QUEBEC
Meerovitch, Eugene
Prichard, Roger Kingsley
Rau, Manfred Ernst
Stevenson, Mary M
Tanner, Charles E
OTHER COUNTRIES
Bojalil, Luis Felipe
Bose, Subir Kumar
Crompton, David William Thomasson
Davidson, David Edward, Jr
Franklin, Richard Morris
Lamoyi, Edmundo
Munoz, Maria De Lourdes
Santoro, Ferrucio Fontes
Van Furth, Ralph

Pathology

ALABAMA
Abrahamson, Dale Raymond
Anderson, Peter Glennie
Bishop, Sanford Parsons
Blackburn, Will R
Caulfield, James Benjamin
Ceballos, Ricardo
Cole, George William
Cook, William Joseph
Dowling, Edmund Augustine
Edgar, Samuel Allen
Farnell, Daniel Reese
Foft, John William
Fullmer, Harold Milton
Gardner, William Albert, Jr
Gay, Renate Erika
Gay, Steffen
Geer, Jack Charles
Giambrone, Joseph James
Gillespie, George Yancey
Goldman, Ronald
Gore, Ira
Hardy, Robert W
Heath, James Eugene
Ho, Kang-Jey
Hoerr, Frederic John
Kayes, Stephen Geoffrey
Kiyono, Hiroshi
Lindsey, James Russell
Lupton, Charles Hamilton, Jr
McDonald, Jay M
Martinez, Mario Guillermo, Jr
Matukas, Victor John
Morgan, Horace C
Mowry, Robert Wilbur
Polt, Sarah Stephens
Powers, Robert D
Siegal, Gene Philip
Smith, George Thomas
Smith, Paul Clay
Wilborn, Walter Harrison
Wolfe, Lauren Gene

ALASKA
Van Pelt, Rollo Winslow, Jr

Pathology (cont)

ARIZONA
Bicknell, Edward J
Bjotvedt, George
Burkholder, Peter M
Cawley, Leo Patrick
Chin, Lincoln
Chvapil, Milos
Conroy, James D
Davis, John Robert
Golden, Alfred
Huestis, Douglas William
Layton, Jack Malcolm
McKelvie, Douglas H
Markle, Ronald A
Mulder, John Bastian
Nagle, Ray Burdell
Perper, Robert J
Reed, Raymond Edgar
Shively, James Nelson
Taylor, Richard G
Tollman, James Perry
Villanueva, Antonio R
Watrach, Adolf Michael
Weinberg, Bernd
Weinstein, Ronald S
Yohem, Karin Hummell

ARKANSAS
Beasley, Joseph Noble
Bucci, Thomas Joseph
Carter, Charleata A
Chang, Louis Wai-Wah
Fink, Louis Maier
Hsu, Su-Ming
Jones, Joe Maxey
Jones, Robin Richard
Poirier, Lionel Albert
Routh, Joseph Isaac
Sun, Chao Nien
Townsend, James Willis

CALIFORNIA
Abrams, Albert Maurice
Agnew, Leslie Robert Corbet
Allen, Raymond A
Allison, Anthony Clifford
Altera, Kenneth P
Amador, Elias
Amromin, George David
Anderson, Marilyn P
Arnold, Harry L(oren), Jr
Arquilla, Edward R
Bailey, David Nelson
Bainton, Dorothy Ford
Baluda, Marcel A
Barajas, Luciano
Barcellos-Hoff, Mary Helen
Barnes, Asa, Jr
Battifora, Hector A
Beckett, Ralph Lawrence
Beckwith, John Bruce
Benirschke, Kurt
Bennington, James Lynne
Berliner, Judith A
Bhaskar, Surindar Nath
Bidlack, Donald Eugene
Bloor, Colin Mercer
Bordin, Gerald M
Bostick, Warren Lithgow
Bozdech, Marek Jiri
Braunstein, Herbert
Bull, Brian S
Cardiff, Robert Darrell
Cardinet, George Hugh, III
Caulfield, John Philip
Chen, Stephen Shi-Hua
Chenoweth, Dennis Edwin
Cherrick, Henry M
Chisari, Francis Vincent
Choi, Byung Ho
Chuang, Hanson Yii-Kuan
Cleary, Michael
Cork, Linda K Collins
Cosgrove, Gerald Edward
Coulson, Walter F
Crawford, William Howard, Jr
DaMassa, Al John
Davis, Richard LaVerne
Deitch, Arline D
Dixon, James Francis Peter
Dorfman, Ronald F
Dowell, Armstrong Manly
Dungworth, Donald L
Dutra, Frank Robert
Earnest, Sue W
Edgington, Thomas S
Eisenson, Jon
Eng, Lawrence F
Fairchild, David George
Farquhar, Marilyn Gist
Feldman, Joseph David
Foos, Robert Young
Friedman, Nathan Baruch
Friend, Daniel S
Furthmayr, Heinz
Gardner, Murray Briggs
Garratty, George
Geller, Stephen Arthur
Gill, James Edward
Gilles, Floyd Harry
Glenner, George Geiger
Gottfried, Eugene Leslie
Gray, Constance Helen
Greenspan, John Simon
Griffin, John Henry
Grody, Wayne William
Hadley, William Keith
Hansen, Louis Stephen
Harding-Barlow, Ingeborg
Heaton, William Andrew Lambert
Henderson, Brian Edmond
Hepler, Opal Elsie
Heuser, Eva T
Hill, Rolla B, Jr
Hinton, David Earl
Holaday, William J
Holmberg, Charles Arthur
Hopkins, Gordon Bruce
Horowitz, Richard E
Hougie, Cecil
Howell, Francis V
Itabashi, Hideo Henry
Jain, Naresh C
Jasper, Donald Edward
Jensen, Hanne Margrete
Jordan, Stanley Clark
Kaneko, Jiro Jerry
Kang, Tae Wha
Kan-Mitchell, June
Kelln, Elmer
Kennedy, Peter Carleton
Kern, William H
Kikkawa, Yutaka
King, Eileen Brenneman
Kiprov, Dobri D
Kloner, Robert A
Koobs, Dick Herman
Korn, David
Kornblum, Ronald Norman
Korpman, Ralph Andrew
Kraft, Lisbeth Martha
Landing, Benjamin Harrison
Larkin, Edward Charles
Latta, Harrison
Leighton, Joseph
Lewin, Klaus J
Lewis, Alvin Edward
Lipsick, Joseph Steven
Liu, Paul Ishen
Lombard, Louise Scherger
Lubran, Myer Michael
Lucas, David Owen
MacNamee, James K
Madden, Sidney Clarence
Madin, Stewart Harvey
Makker, Sudesh Paul
Makowka, Leonard
Malamud, Nathan
Margaretten, William
Merchant, Bruce
Merrill, James Allen
Miller, Gary Arthur
Minckler, Jeff
Minckler, Tate Muldown
Miyai, Katsumi
Moyer, Dean La Roche
Moyer, Geoffrey H
Murad, Turhon A
Myhre, Byron Arnold
Nakamura, Robert Motoharu
Nathwani, Bharat N
Nichols, Barbara Ann
Nielsen, Surl L
O'Brien, John S
Oh, Jang Ok
Olander, Harvey Johan
Olson, Albert Lloyd
Osburn, Bennie Irve
Parker, John William
Peng, Shi-Kaung
Perkins, William Hughes
Peterson-Falzone, Sally Jean
Philippart, Michel Paul
Pilgeram, Laurence Oscar
Pool, Roy Ransom
Porter, David Dixon
Price, Harold M
Quismorio, Francisco P, Jr
Rasheed, Suraiya
Rearden, Carole Ann
Regezi, Joseph Alberts
Reilly, Emmett B
Rettenmier, Carl Wayne
Rice, Robert Hafling
Richters, Arnis
Roberts, James C, Jr
Rosenau, Werner
Rosenthal, Sol Roy
Roy-Burman, Pradip
Ruebner, Boris Henry
Semancik, Joseph Stephen
Sherwin, Russell P
Shott, Leonard D
Shulman, Ira Andrew
Simmons, Norman Stanley
Slager, Ursula Traugott
Sommer, Noel Frederick
Spear, Gerald Sanford
Steinberg, Bernhard
Stern, Robert
Stevens, Jack Gerald
Stevens, Lewis Axtell
Stowell, Robert Eugene
Strautz, Robert Lee
Suffin, Stephen Chester
Sunshine, Irving
Sussman, Howard H
Swanson, Virginia Lee
Tatter, Dorothy
Taylor, Clive Roy
Terry, Robert Davis
Terry, Roger
Thaler, M Michael
Thurman, Wayne Laverne
Toreson, Wilfred Earl
Triche, Timothy J
Van Lancker, Julien L
Verity, Maurice Anthony
Volk, Bruno W
Walford, Roy Lee, Jr
Wang, Nai-San
Warner, Nancy Elizabeth
Warnock, Martha L
Weihing, Robert Ralph
Weissman, Irving L
Wellington, John Sessions
Wescott, William B
Whang, Sukoo Jack
Willhite, Calvin Campbell
Withers, Hubert Rodney
Wolf, Paul Leon
Wood, David Alvra
Wood, Nancy Elizabeth
Woolley, LeGrand H
Wu, Sing-Yung
Yen, Tien-Sze Benedict
Yuen, Ted Gim Hing
Zamboni, Luciano
Zettner, Alfred
Zinkl, Joseph Grandjean

COLORADO
Alexander, Archibald Ferguson
Benjamin, Stephen Alfred
Bosley, Elizabeth Caswell
Charney, Michael
Cockerell, Gary Lee
Cox, Alvin Joseph, Jr
Davis, Robert Wilson
De Martini, James Charles
Dobersen, Michael J
Firminger, Harlan Irwin
Franklin, Wilbur A
Gardner, David Godfrey
Glock, Robert Dean
Gorthy, Willis Charles
Hoerlein, Alvin Bernard
Hollister, Alan Scudder
Irons, Richard Davis
Jensen, Rue
Kotin, Paul
Krikos, George Alexander
Norrdin, Robert W
Pierce, Gordon Barry
Repine, John E
Solomon, Gordon Charles
Speers, Wendell Carl
Swartzendruber, Douglas Edward
Takeda, Yasuhiko
Thomas, Louis Barton
Whiteman, Charles E
Wilber, Charles Grady
Young, Stuart

CONNECTICUT
Barthold, Stephen William
Baumgarten, Alexander
Boehme, Diethelm Hartmut
Booss, John
Cronin, Michael Thomas Ignatius
Cutler, Leslie Stuart
Dembitzer, Herbert
Downing, S Evans
Goldschneider, Irving
Hasson, Jack
Hukill, Peter Biggs
Jacoby, Robert Ottinger
Janeway, Charles Alderson, Jr
Kashgarian, Michael
Kerr, Kirklyn M
Krutchkoff, David James
Lee, Sin Hang
Lobdell, David Hill
Madri, Joseph Anthony
Manuelidis, Elias Emmanuel
Marchesi, Vincent T
Morrow, Jon S
Nalbandian, John
Post, John E
Schmeer, Arline Catherine
Seligson, David
Smith, Brian Richard
Snellings, William Moran
Sunderman, F William, Jr
Vidone, Romeo Albert
Yang, Tsu-Ju (Thomas)
Yesner, Raymond

DELAWARE
Burch, Ronald Martin
Cubberley, Virginia
Dohms, John Edward
Fogg, Donald Ernest
Fries, Cara Rosendale
Pegg, Philip John
Stula, Edwin Francis

DISTRICT OF COLUMBIA
Bowling, Lloyd Spencer, Sr
Bulger, Ruth Ellen
Chun, Byungkyu
Chung, Ed Baik
Fishbein, William Nichols
Hadley, Gilbert Gordon
Harshbarger, John Carl, Jr
Herman, Mary M
Hicks, Jocelyn Muriel
Higginson, John
Irey, Nelson Sumner
Jackson, Marvin Alexander
Jenson, A Bennett
Johnson, Frank Bacchus
Jones, Thomas Carlyle
Kimbrough, Renate Dora
Ligler, Frances Smith
Luke, James Lindsay
McLean, Ian William
Mann, Marion
Miller, William Robert
Mostofi, F Ksh
Mullin, Brian Robert
Nelson, James S
O'Leary, Timothy Joseph
Orenstein, Jan Marc
Pearlman, Ronald C
Ramsey, Elizabeth Mapelsden
Sampson, Calvin Coolidge
Sidransky, Herschel
Silverberg, Steven George
Sobin, Leslie Howard
Thomas, Richard Dean
Young, Elizabeth Bell
Young, Frank E
Zawisza, Julie Anne A
Zimmerman, Lorenz Eugene
Zook, Bernard Charles

FLORIDA
Abbott, Thomas B
Abell, Murray Richardson
Aleo, Joseph John
Azar, Henry A
Bensen, Jack F
Bloom, Frank
Blum, Leon Leib
Bramante, Pietro Ottavio
Bucklin, Robert van Zandt
Bzoch, Kenneth R
Cameron, Don Frank
Carpenter, Anna-Mary
Cohen, Glenn Milton
Coleman, Sylvia Ethel
Couch, John Alexander
Croker, Byron P
Dawson, Peter J
Deats, Edith Potter
Donnelly, William H
Dunning, Wilhelmina Frances
Fagin, Karin
Fournie, John William
Frank, H Lee
Fresh, James W
Gerughty, Ronald Mills
Giddens, William Ellis, Jr
Goldberg, Melvin Leonard
Greene, George W, Jr
Gross, Paul
Hackett, Raymond Lewis
Hammer, Lowell Clarke
Hampp, Edward Gottlieb
Harrison, Robert J
Hartmann, William Herman
Hassell, Thomas Michael
Hazard, John Beach
Hood, Claude Ian
Hurst, Josephine M
Inana, George
Jaques, William Everett
Jones, Russell Stine
Kahan, I Howard
Kalfayan, Bernard
Kobernick, Sidney D
Kreshover, Seymour J
Lucas, Frederick Vance, Jr
McCroskey, Robert Lee
MacDonald, Richard Annis
McEntee, Kenneth
Maclaren, Noel K
Malinin, Theodore I
Mase, Darrel Jay
Meryman, Charles Dale
Mitus, Wladyslaw J
Nicosia, Santo Valerio
Normann, Sigurd Johns
Noto, Thomas Anthony
Orthoefer, John George
Payne, Torrence P B
Pearl, Gary Steven
Pendland, Jacquelyn C
Riser, Wayne H
Robb, James Arthur
Ross, Lynne Fischer
Rowlands, David T, Jr
Schmidt, Paul Joseph
Sharf, Donald Jack
Shields, Robert Pierce
Shively, John Adrian
Siegel, Henry
Skinner, Margaret Sheppard
Smith, Albert Carl
Stanley, Harold Russell
Straumfjord, Jon Vidalin, Jr
Tompkins, Victor Norman
Valdes-Dapena, Marie A

Woodard, James Carroll

GEORGIA
Ahmed-Ansari, Aftab
Alonso, Kenneth B
Austin, Garth E
Baisden, Charles Robert
Bell, William Averill
Boone, Donald Joe
Chandler, A Bleakley
Chandler, Francis Woodrow, Jr
Chapman, Willie Lasco, Jr
Crowell, Wayne Allen
Davison, Frederick Corbet
Emerson, James L
Farhi, Diane C
Farrell, Robert Lawrence
Gerrity, Ross Gordon
Gerschenson, Mariana
Gilbert, Frederick Emerson, Jr
Godwin, John Thomas
Gravanis, Michael Basil
Halper, Jaroslava
Herman, Chester Joseph
Hobbs, Milford Leroy
Hunter, Robert L
Kapp, Judith A
Kaufman, Nathan
Kaufmann, Arnold Francis
Krauss, Jonathan Seth
Ku, David Nelson
Leitch, Gordon James
McClure, Harold Monroe
McKinney, Ralph Vincent, Jr
Miller, Paul George
Naib, Zuher M
Priest, Robert Eugene
Ragland, William Lauman, III
Sawyer, Richard Trevor
Schwartz, David Alan
Sell, Kenneth W
Sinor, Lyle Tolbot
Sisk, Dudley Byrd
Stoddard, Leland Douglas
Teabeaut, James Robert, II
Tyler, David E
Van Assendelft, Onno Willem
Vogel, Francis Stephen
Waldron, Charles A
Weathers, Dwight Ronald
Whitsett, Carolyn F

HAWAII
Hokama, Yoshitsugi
Porta, Eduardo Angel
Stemmermann, Grant N
Yang, Hong-Yi

IDAHO
Ellis, Robert William
House, Edwin W

ILLINOIS
Bailey, Orville Taylor
Bardawil, Wadi Antonio
Baron, David Alan
Bartlett, Gerald Lloyd
Beamer, Paul Donald
Bennett, Basil Taylor
Boggs, Joseph D
Bowman, James E
Buschmann, Robert J
Cabana, Veneracion Garganta
Carone, Frank
Cera, Lee M
Check, Irene J
Clasen, Raymond Adolph
Cohen, Lawrence
Crocker, Diane Winston
Dal Canto, Mauro Carlo
Davies, Peter F
Eckner, Friedrich August Otto
Emeson, Eugene Edward
Epstein, Samuel Stanley
Fierer, Joshua A
Fisher, Ben
Fisher, Cletus G
Fosslien, Egil
Friederici, Hartmann H R
Fritz, Thomas Edward
Getz, Godfrey S
Glagov, Seymour
Goepp, Robert August
Gordon, Donovan
Greenberg, Stephen Robert
Gross, Martin
Gruhn, John George
Haas, Mark
Haber, Meryl H
Hass, George Marvin
Hinojosa, Raul
Hruban, Zdenek
Huizinga, Harry William
Jacobson, Marvin
Kaminski, Edward Jozef
Kanwar, Yashpal Singh
Kent, Geoffrey
King, Donald West, Jr
King, Lester Snow
Kirschner, Robert H
Kluskens, Larry F
Kuszak, Jerome R
Landay, Alan Lee
Leestma, Jan E

Logemann, Jerilyn Ann
Lundberg, George David
McCormick, David Loyd
McDonald, Larry William
McGrew, Elizabeth Anne
Medak, Herman
Molnar, Zelma Villanyi
Molteni, Agostino
Moretz, Roger C
Muma, Nancy A
Nevalainen, David Eric
O'Morchoe, Patricia Jean
Patterson, Douglas Reid
Polet, Herman
Port, Curtis DeWitt
Prabhudesai, Mukund M
Pusch, Allen Lewis
Radosevich, James A
Rao, M S(ambasiva)
Reddy, Janardan K
Reynolds, Harry Aaron, Jr
Robertson, Abel Alfred Lazzarini, Jr
Roth, Sanford Irwin
Rowley, Donald Adams
Rubnitz, Myron Ethan
Scapino, Robert Peter
Scarpelli, Dante Giovanni
Schneider, Arthur Sanford
Schreiber, Hans
Seed, Thomas Michael
Shearer, William McCague
Sherman, Laurence A
Sherrick, Joseph C
Simon, Joseph
Sisson, Joseph A
Smith, David Waldo Edward
Sommers, Herbert M
Spargo, Benjamin H
Stein, Robert Jacob
Stepto, Robert Charles
Stevenson, George Franklin
Strano, Alfonso J
Straus, Francis Howe, II
Swerdlow, Martin A
Taswell, Howard Filmore
Taylor, D Dax
Tekeli, Sait
Toto, Patrick D
Towns, Clarence, Jr
Ts'ao, Chung-Hsin
Tso, Mark On-Man
Ts'o, Timothy On-To
Van Fossan, Donald Duane
Vye, Malcolm Vincent
Wass, John Alfred
Waterhouse, John P
Wiener, Stanley L
Wilber, Laura Ann
Willoughby, William Franklin
Wissler, Robert William
Wong, Ruth (Lau)
Wong, Ting-Wa
Zemlin, Willard R

INDIANA
Artes, Richard H
Bonderman, Dean P
Carlton, William Walter
Carter, James Edward
Cohen, Harold Karl
Culbertson, Clyde Gray
DePaoli, Alexander
Edwards, Joshua Leroy
Eisenbrandt, David Lee
Epstein, Aubrey
Gregory, Richard Lee
Griffing, William James
Harrington, Daniel Dale
Kelly, William Alva
Klaunig, James E
Lee, Shuishih Sage
Mason, Earl James
Nordschow, Carleton Deane
Past, Wallace Lyle
Pollard, Morris
Richter, Ward Robert
Robinson, Farrel Richard
Roth, Lawrence Max
Ryder, Kenneth William, Jr
Schulz, Dale Metherd
Shafer, William Gene
Shanks, James Clements, Jr
Smith, James Warren
Steer, Max David
Todd, Glen Cory
Tomich, Charles Edward
Van Sickle, David C
Van Vleet, John F
Waller, Bruce Frank
Zwicker, Gary M

IOWA
Baron, Jeffrey
Bonsib, Stephen M
Cheville, Norman F
Cook, Robert Thomas
Dailey, Morris Owen
Hammond, Harold Logan
Hardy, James C
Higa, Leslie Hideyasu
Hom, Ben Lin
Kent, Thomas Hugh
Kluge, John Paul
Koerner, Theodore Alfred William, Jr

Lynch, Richard G
Miller, Janice Margaret Lilly
Miller, Lyle Devon
Moon, Harley W
Morris, Hughlett Lewis
Osweiler, Gary D
Rose, Earl Forrest
Schelper, Robert Lawrence
Seaton, Vaughn Allen
Shires, Thomas Kay
Song, Joseph
Switzer, William Paul
Van Demark, Duane R
Williams, Dean E

KANSAS
Anderson, Harrison Clarke
Chiga, Masahiro
Coles, Embert Harvey, Jr
Cook, James Ellsworth
Cuppage, Francis Edward
Drees, David T
Dykstra, Mark Allan
Faires, Wesley Lee
Gillespie, Jerry Ray
Gray, Andrew P
Harvey, Thomas Stoltz
Hinshaw, Charles Theron, Jr
Kepes, John J
Ludvigsen, Carl W, Jr
McGregor, Douglas Hugh
Plapp, Frederick Vaughn
Reals, William Joseph
Smith, Joseph Emmitt
Sprinz, Helmuth
Stevens, David Robert
Strafuss, Albert Charles
Tasch, Paul
Tomita, Tatsuo

KENTUCKY
Blake, James Neal
Burzynski, Norbert J
Espinosa, Enrique
Farman, Allan George
Friedell, Gilbert H
Golden, Abner
Levinson, Stanley S
Markesbery, William Ray
Murray, Marvin
Parker, Joseph Corbin, Jr
Porter, William Hudson
Sabes, William Ruben
Shih, Wei Jen
Slagel, Donald E
Swerczek, Thomas Walter
Terango, Larry
Vittetoe, Marie Clare
White, Dean Kincaid
Yoneda, Kokichi

LOUISIANA
Bautista, Abraham Parana
Beeler, Myrton Freeman
Brody, Arnold R
Cannon, Donald Charles
Casey, Harold W
Constantinides, Paris
Daroca, Philip Joseph, Jr
Dhurandhar, Nina
Dunlap, Charles Edward
Eggen, Douglas Ambrose
Gerber, Michael A
Harkin, James C
Hensley, John Coleman, II
Hodgin, Ezra Clay
Holmquist, Nelson D
Ichinose, Herbert
Lehmann, Hermann Peter
McGarry, Paul Anthony
Matthews, Murray Albert
Misra, Raghunath P
Odenheimer, Kurt John Sigmund
Rampp, Donald L
Reed, Richard Jay
Roberts, Edgar D
Sarphie, Theodore G
Schor, Norberto Aaron
Scollard, David Michael
Smith, Albert Goodin
Strong, Jack Perry
Tracy, Richard E
Unterharnscheidt, Friedrich J
Vial, Lester Joseph, Jr
Welsh, Ronald
Wilson, Russell B
Xie, Jianming

MAINE
Bernstein, Seldon Edwin
Coman, Dale Rex
Goodale, Fairfield
Myers, David Daniel
Weiss, John Jay

MARYLAND
Aamodt, Roger Louis
Albrecht, Paul
Allen, Anton Markert
Anderson, Arthur O
Asofsky, Richard Marcy
Banfield, William Gethin
Bauer, Heinz
Berne, Bernard H

Bernier, Joseph Leroy
Biswas, Robin Michael
Bodammer, Joel Edward
Boone, Charles Walter
Bush, David E
Charache, Samuel
Cheever, Allen
Chu, Elizabeth Wann
Connor, Daniel Henry
Dannenberg, Arthur Milton, Jr
DeClaris, N(icholas)
Dragunsky, Eugenia M
Dubey, Jitender Prakash
Eckels, Kenneth Henry
Eddy, Hubert Allen
Eggleston, Joseph C
Elin, Ronald John
Enzinger, Franz Michael
Evarts, Ritva Poukka
Feigal, Ellen G
Ferrans, Victor Joaquin
Fornace, Albert J, Jr
Fraumeni, Joseph F, Jr
Fried, Jerrold
Friedman, Robert Morris
Gardner, Alvin Frederick
Gauhar, Aruna
Green, Marie Roder
Griffith, Linda M(ae)
Grimley, Philip M
Gross, Mircea Adrian
Hackett, Joseph Leo
Hamilton, Stanley R
Haspel, Martin Victor
Haudenschild, Christian C
Henson, Donald E
Heptinstall, Robert Hodgson
Herman, Eugene H
Hess, Helen Hope
Hilberg, Albert William
Hla, Timothy Tun
Holmes, Kathryn Voelker
Hutchins, Grover MacGregor
Iseri, Oscar Akio
Jacobs, Myron Samuel
Jaffe, Elaine Sarkin
Kendrick, Francis Joseph
Kickler, Thomas Steven
Kimball, Paul Clark
Kirschstein, Ruth Lillian
Kirsten, Werner H
Kroll, Martin Harris
Lage, Janice M
LaSalle, Bernard
Latham, Patricia Suzanne
Lerman, Michael Isaac
Liang, Isabella Y S
Liebelt, Annabel Glockler
Lindberg, Donald Allan Bror
Lindenberg, Richard
Liotta, Lance
Lipkin, Lewis Edward
Lobel, Steven A
Longo, Dan L
Lunin, Martin
McDowell, Elizabeth Mary
Malmgren, Richard Axel
Marzella, Louis
Mercado, Teresa I
Mickel, Hubert Sheldon
Micozzi, Marc S
Miller, Frank Nelson
Mishler, John Milton
O'Beirne, Andrew Jon
Olson, Jean L
Parvez, Zaheer
Petrali, John Patrick
Platt, Marvin Stanley
Platt, William Rady
Prendergast, Robert Anthony
Rabson, Alan S
Redman, Robert Shelton
Resau, James Howard
Reuber, Melvin D
Rice, Jerry Mercer
Rifkin, Erik
Roberts, David Duncan
Robison, Wilbur Gerald, Jr
Rodrigues, Merlyn M
Rosenberg, Saul H
Saffiotti, Umberto
Shamsuddin, Abulkalam Mohammad
Shin, Moon L
Smith, Bruce H
Sobel, Mark E
Squire, Robert Alfred
Stewart, Harold L
Striker, G E
Stromberg, Kurt
Strum, Judy May
Thapar, Nirwan T
Tigertt, William David
Ting, Chou-Chik
Trump, Benjamin Franklin
Varricchio, Frederick
Vickers, James Hudson
Vinores, Stanley Anthony
Von Euler, Leo Hans
Wahl, Sharon Knudson
Walton, Thomas Edward
Ward, Jerrold Michael
Wargotz, Eric S
Weinberg, Wendy C
Weisbroth, Steven H

Pathology (cont)

White, John David
Yardley, John Howard
Yodaiken, Ralph Emile
Zierdt, Charles Henry

MASSACHUSETTS
Abromson-Leeman, Sara R
Ahmed, A Razzaque
Andres, Giuseppe A
Appel, Michael Clayton
Argyris, Thomas Stephen
Badonnel, Marie-Claude H M
Balogh, Karoly
Benacerraf, Baruj
Bieber, Frederick Robert
Busch, George Jacob
Byers, Hugh Randolph
Carvalho, Angelina C A
Christensen, Thomas Gash
Christian, Howard J
Cobb, Carolus M
Collier, Robert John
Colvin, Robert B
Connolly, James L
Corson, Joseph Mackie
Cotran, Ramzi S
Dammin, Gustave John
D'Amore, Patricia Ann
Dayal, Yogeshwar
DeGirolami, Umberto
DeLellis, Ronald Albert
Diamandopolus, George Th
Dittmer, John Edward
Dvorak, Ann Marie-Tompkins
Dvorak, Harold Fisher
Fallon, John T
Federman, Micheline
Flax, Martin Howard
Fleischman, Robert Werder
Freiman, David Galland
George, Harvey
Gherardi, Gherardo Joseph
Gill, David Michael
Gimbrone, Michael Anthony, Jr
Goldman, Harvey
Gordon, Lance Kenneth
Gottlieb, Leonard Solomon
Haimes, Howard B
Handler, Alfred Harris
Hawiger, Jack Jacek
Hayes, John A
Hedley-Whyte, Elizabeth Tessa
Hertig, Arthur Tremain
Hirsch, Carl Alvin
Homburger, Freddy
Howley, Peter Maxwell
Hunt, Ronald Duncan
Jacobs, Jerome Barry
Jakobiec, Frederick Albert
Johansen, Erling
Jonas, Albert Moshe
Karnovsky, Morris John
Kim, Agnes Kyung-Hee
King, Norval William, Jr
Koch-Weser, Dieter
Kocon, Richard William
Kubin, Rosa
Kulka, Johannes Peter
Kupiec-Weglinski, Jerzy W
Kwan, Paul Wing-Ling
Laposata, Michael
Leav, Irwin
Legg, Merle Alan
Leith, John Douglas
Libby, Peter
Like, Arthur A
Locke, John Lauderdale
McCluskey, Robert Timmons
McDonagh, Jan M
MacMahon, Harold Edward
Meyer, Irving
Mihm, Martin C, Jr
Murphy, James Clair
Nanji, Amin Akbarali
Newberne, Paul M
Nielsen, Svend Woge
Olsen, Bjorn Reino
Petricciani, John C
Robbins, Stanley L
Rogers, Adrianne Ellefson
Rohovsky, Michael William
Rosen, Seymour
Russfield, Agnes Burt
Sacks, David B
Schaub, Robert George
Schneeberger, Eveline E
Schoen, Frederick J
Scully, Robert Edward
Shklar, Gerald
Sidman, Richard Leon
Siegler, Richard E
Smulow, Jerome B
Susi, Frank Robert
Szabo, Sandor
Taft, Edgar Breck
VanPraagh, Richard
Von Lichtenberg, Franz
Weinberg, David S
White, Harold J
Winkelman, James W
Wright, Thomas Carr, Jr
Yunis, Edmond J

MICHIGAN
Abrams, Gerald David
Aggarwal, Surinder K
Arnold, James Schoonover
Bedrossian, Carlos Wanes Menino
Bernstein, Jay
Bollinger, Robert Otto
Brock, Donald R
Chason, Jacob Leon
Chensue, Stephen W
Chopra, Dharam Pal
Chua, Balvin H-L
Claflin, Robert Malden
Clark, John Jefferson
Coye, Robert Dudley
Dapson, Richard W
Deegan, Michael J
De La Iglesia, Felix Alberto
Diglio, Clement Anthony
Elfont, Edna A
Elliott, George Algimon
Feenstra, Ernest Star
Floyd, Alton David
Fritts-Williams, Mary Louise Monica
Garcia, Julio H
Garg, Bhagwan D
Giacomelli, Filiberto
Gikas, Paul William
Gilbertsen, Richard B
Giraldo, Alvaro A
Goyings, Lloyd Samuel
Grant, Rhoda
Hanks, Carl Thomas
Hashimoto, Ken
Headington, John Terrence
Hicks, Samuel Pendleton
Hinerman, Dorin Lee
Hollenberg, Paul Frederick
Hunt, Charles E
Jersey, George Carl
Johnson, Herbert Gardner
Johnson, Kent J
Jones, Margaret Zee
Keahey, Kenneth Karl
Kiechle, Frederick Leonard
Kithier, Karel
Klurfeld, David Michael
Kociba, Richard Joseph
Krishan, Awtar
Landers, James Walter
Langham, Robert Fred
Leader, Robert Wardell
Lund, John Edward
McClatchey, Kenneth D
McKeever, Paul Edward
Mattson, Joan C
Moe, James Burton
Oberman, Harold A
Padgett, George Arnold
Palmer, Kenneth Charles
Perrin, Eugene Victor
Phan, Sem Hin
Piper, Richard Carl
Poulik, Miroslav Dave
Powsner, Edward R
Puro, Donald George
Raven, Clara
Rowe, Nathaniel H
Rupp, Ralph Russell
Sander, C Maureen
Schmidt, Robert W
Schnitzer, Bertram
Shanberge, Jacob N
Sleight, Stuart Duane
Sloane, Bonnie Fiedorek
Smart, James Blair
Smith, Robert James
Spitz, Werner Uri
Stowe, Howard Denison
Stromsta, Courtney Paul
Tasker, John B
Trapp, Allan Laverne
Verrill, Harland Lester
Ward, Peter A
Waxler, Glenn Lee
Webster, Harris Duane
Whitehair, Charles Kenneth
Wiener, Joseph
Wolman, Sandra R
Wolter, J Reimer
Zak, Bennie

MINNESOTA
Anderson, Robert Edwin
Anderson, W Robert
Azar, Miguel M
Baggenstoss, Archie Herbert
Bahn, Robert Carlton
Barnes, Donald McLeod
Benson, Ellis Starbranch
Brown, David Mitchell
Brunning, Richard Dale
Case, Marvin Theodore
Chesney, Charles Frederic
Connelly, Donald Patrick
Crowley, Leonard Vincent
Dahlin, David Carl
Duke, Gary Earl
Edwards, Jesse Efrem
Edwards, William Dean
Elzay, Richard Paul
Engel, Andrew G
Estensen, Richard D
Foley, William Arthur

Fraley, Elwin E
Ghazali, Masood Raheem
Gibbons, Donald Frank
Goble, Frans Cleon
Gorlin, Robert James
Hedgecock, Le Roy Darien
Henrikson, Ernest Hilmer
Homburger, Henry A
Jacobson, Joan
Johnson, Kenneth Harvey
Kammermeier, Martin A
Kersey, John H
Knabe, George W, Jr
Kottke, Bruce Allen
Kurtz, Harold John
LeBien, Tucker W
Lober, Paul Hallam
Low, Walter Cheney
McCarthy, James Benjamin
McDermott, Richard P
Manning, Patrick James
Mariani, Toni Ninetta
Mitchell, Roger Harold
Moller, Karlind Theodore
Moore, Sean Breanndan
Nelson, Robert D
Okazaki, Haruo
Osborne, Carl Andrew
Perman, Victor
Ramakrishan, S
Rao, Gundu Hirisave Rama
Romero, Juan Carlos
Sautter, Jay Howard
Sparnins, Velta L
Sung, Joo Ho
Titus, Jack L
Vallera, Daniel A
Van Kampen, Kent Rigby
Vernier, Robert L
Wallace, Larry J
Wattenberg, Lee Wolff
Witkop, Carl Jacob, Jr
Yamane, George M

MISSISSIPPI
Bloom, Sherman
Bradley, Doris P
Cruse, Julius Major, Jr
Evers, Carl Gustav
Gatling, Robert Riddick
Johnson, Warren W
Lewis, Robert Edwin, Jr
Lockwood, William Rutledge
O'Neal, Robert Munger
Overstreet, Robin Miles
Read, Virginia Hall
Shenefelt, Ray Eldon
Spann, Charles Henry
Verlangieri, Anthony Joseph

MISSOURI
Bari, Wagih A
Berrier, Harry Hilbourn
Blumenthal, Herman T
Eyestone, Willard Halsey
Gantner, George E, Jr
Givler, Robert L
Green, Eric Douglas
Hall, William Francis
Hamilton, Thomas Reid
Huang, Tim Hui-ming
Hutt-Fletcher, Lindsey Marion
Johnston, Marilyn Frances Meyers
Lacy, Paul Eston
Lagunoff, David
Lau, Brad W C
McDivitt, Robert William
Morehouse, Lawrence G
Nahm, Moon H
Olney, John William
Olson, LeRoy David
Ribelin, William Eugene
Roodman, Stanford Trent
Schmidt, Donald Arthur
Shah, Sheila
Smith, Carl Hugh
Stewart, Wellington Buel
Taylor, John Joseph
Tibbits, Donald Fay
Torack, Richard M
Townsend, John Ford
Wagner, Joseph Edward
Watanabe, Itaru S
Welch, Lin

MONTANA
Hadlow, William John
Jutila, Mark Arthur
Williams, Roger Stewart
Young, David Marshall

NEBRASKA
Berton, William Morris
Cohen, Samuel Monroe
Cox, Robert Sayre, Jr
Grace, Oliver Davies
Linder, James
Purtilo, David Theodore
Quigley, Herbert Joseph, Jr
Schenken, Jerald R
Schmitz, John Albert
Sheehan, John Francis
Todd, Gordon Livingston
Toth, Bela

Wilson, Richard Barr

NEVADA
Anderson, Bernard A
Barger, James Daniel

NEW HAMPSHIRE
Christie, Robert William
Longnecker, Daniel Sidney
Marin-Padilla, Miguel
Pape, Brian Eugene
Pettengill, Olive Standish
Rawnsley, Howard Melody
Rodgers, Frank Gerald

NEW JERSEY
Aberman, Harold Mark
Baden, Ernest
Baron, Hazen Jay
Berberian, Paul Anthony
Black, Hugh Elias
Blaiklock, Robert George
Bolden, Theodore Edward
Browning, Edward T
Chen, Thomas Shih-Nien
Cho, Eun-Sook
Churg, Jacob
Cooper, Keith Raymond
Cox, George Elton
Craig, Peter Harry
Davis, Nancy Taggart
Diener, Robert Max
Dunikoski, Leonard Karol, Jr
Durham, Stephen K
Eisenstadt, Arthur A
Ende, Norman
Eng, Robert H K
Fabry, Andras
Gambino, S(alvatore) Raymond
Girerd, Rene Jean
Goel, Mahesh C
Goldenberg, David Milton
Gupta, Godaveri Rawat
Harris, Charles
Hernandez, Juan Antonio
Hruza, Zdenek
Hutter, Robert V P
Jaffe, Ernst Richard
Jaffe, Russell M
Johnson, Allen Neill
Kauffman, Shirley Louise
Khavkin, Theodor
Kozam, George
Kung, Ted Teshih
Lev, Maurice
Lipman, Jack M
Lowndes, Herbert Edward
McCoy, John Roger
McCullough, C Bruce
McDonnell, William Vincent
McKearn, Thomas Joseph
McKenzie, Basil Everard
Madrazo, Alfonso A
Mesa-Tejada, Ricardo
Moe, Robert Anthony
Moolten, Sylvan E
Moore, Vernon Leon
Morrison, Ashton Byrom
Mulcahy, Gabriel Michael
Mushett, Charles Wilbur
Myers, Jeffrey
Ober, William B
Pensack, Joseph Michael
Ponzio, Nicholas Michael
Raska, Karel Frantisek, Jr
Raskova, Jana D
Ray, James Alton
Salgado, Ernesto D
Shea, Stephen Michael
Sheikh, Maqsood A
Shimamura, Tetsuo
Sommers, Sheldon Charles
Song, Sun Kyu
Spitznagle, Larry Allen
Stenn, Kurt S
Studzinski, George P
Swarm, Richard L(ee)
Tomeo-Dymkowski, Aline Claire
Trelstad, Robert Laurence
Vukovich, Robert Anthony
Wallach, Jacques Burton
Washko, Floyd Victor
Watson, Frank Yandle
Wolff, Marianne
Zeidman, Irving

NEW MEXICO
Bitensky, Mark Wolfe
Duszynski, Donald Walter
Frenkel, Jacob Karl
Garrett, Edgar Ray
Garry, Philip J
Hahn, Fletcher Frederick
Johnson, Neil Francis
Jordan, Scott Wilson
Key, Charles R
Kornfeld, Mario O
Saunders, George Cherdron
Smith, David Marshall
Upton, Arthur Canfield
Yudkowsky, Elias B

NEW YORK
Ablin, Richard J

Abraham, Jerrold L
Allen, Arthur Charles
Altman, Kurt Ison
Ambrus, Julian Lawrence
Anderson, Paul J
Anversa, Piero
Archard, Howell Osborne
Armstrong, Donald
Arvan, Dean Andrew
Asch, Bonnie Bradshaw
Ash, William James
Ashley, Charles Allen
Athanassiades, Thomas J
Barcos, Maurice P
Bealmear, Patricia Maria
Becker, Norwin Howard
Berkman, James Israel
Bielat, Kenneth L
Biempica, Luis
Bini, Alessandra Margherita
Bora, Sunder S
Brentjens, Jan R
Brooks, John S J
Budzilovich, Gleb Nicholas
Campbell, Wallace G, Jr
Canfield, William H
Casarett, George William
Celada, Franco
Chishti, Muhammad Azhar
Christian, John Jermyn
Chryssanthou, Chryssanthos
Collins, George H
Cooper, Norman S
Cooper, Robert Arthur, Jr
Cordon-Cardo, Carlos
Cottrell, Thomas S
Cowen, David
Cravioto, Humberto
Daoud, Assaad S
Darzynkiewicz, Zbigniew Dzierzykraj
Defendi, Vittorio
Del Cerro, Manuel
Demopoulos, Harry Byron
Denholm, Elizabeth Maria
Derby, Bennett Marsh
Deschner, Eleanor Elizabeth
DJang, Arthur H K
Doolittle, Richard L
Dorfman, Howard David
Drinnan, Alan John
Drobeck, Hans Peter
Edgcomb, John H
Ehrenreich, Theodore
Ellis, John Taylor
Fabricant, Catherine G
Factor, Stephen M
Falcone, Domenick Joseph
Feigin, Irwin Harris
Feld, Leonard Gary
Fitzgerald, Patrick James
Florentin, Rudolfo Aranda
Fritz, Katherine Elizabeth
Fromowitz, Frank B
Gil, Joan
Godfrey, Henry Philip
Godman, Gabriel C
Goldfeder, Anna
Goldfischer, Sidney L
Goldman, James Eliot
Golub, Lorne Malcolm
Gordon, Gerald Bernard
Gordon, Ronald E
Gueft, Boris
Gutstein, William H
Hagstrom, Jack Walter Carl
Haley, Nancy Jean
Hard, Gordon Charles
Hausman, Robert
Heffner, Reid Russell, Jr
Hellstrom, Harold Richard
Hirano, Asao
Hirth, Robert Stephen
Horowitz, Esther
Hubbard, John Castleman
Hurvitz, Arthur Isaac
Iatropoulos, Michael John
Ioachim, Harry L
Isenberg, Henry David
Jindrak, Karel
Joel, Darrel Dean
Johnson, Anne Bradstreet
Johnson, Horton Anton
Jones, David B
Kahn, Leonard B
Kaley, Gabor
Karmen, Arthur
Katz, Sidney
Kaufman, Mavis Anderson
Kaye, Gordon I
Kellner, Aaron
Kelly, Sally Marie
Killian, Carl Stanley
Kim, Harry Hi-Soo
Kim, Untae
King, John McKain
Klavins, Janis Viliberts
Koss, Leopold George
Krook, Lennart Per
Kuschner, Marvin
Lamberson, Harold Vincent, Jr
Lane, Bernard Paul
Lanks, Karl William
Lee, Woong Man
Lehman, John Michael

Levin, Robert Aaron
Levine, Seymour
Liu, Si-Kwang
Lumpkin, Lee Roy
Lyubsky, Sergey
McGraw, Timothy E
McHugh, William Dennis
Mankes, Russell Francis
Marboe, Charles Chostner
Mebus, Charles Albert
Mehaffey, Leathem, III
Melamed, Myron Roy
Mellors, Robert Charles
Middleton, Charles Cheavens
Miller, Frederick
Milstoc, Mayer
Minick, Charles Richard
Minkowitz, Stanley
Minor, Ronald R
Miranda, Armand F
Modafferi, Judy Hall
Mohos, Steven Charles
Mooney, Robert Arthur
Morse, Stephen S(cott)
Moss, Leo D
Moussouris, Harry
Nagler, Arnold Leon
Natta, Clayton Lyle
Neiders, Mirdza Erika
Nelson, Douglas A
Nickerson, Peter Ayers
Nussenzweig, Victor
O'Donoghue, John Lipomi
Oka, Masamichi
Panner, Bernard J
Papsidero, Lawrence D
Paronetto, Fiorenzo
Parsons, Donald Frederick
Partin, Jacqueline Surratt
Patnaik, Amiya Krishna
Pearson, Thomas Arthur
Peerschke, Ellinor Irmgard Barbara
Penney, David P
Peress, Nancy E
Peters, Theodore, Jr
Pham, Tuan Duc
Phillips, Mildred E
Pirani, Conrad Levi
Powers, James Matthew
Prince, Alfred M
Prior, John Thompson
Quimby, Fred William
Rajan, Thiruchandurai Viswanathan
Rapp, Dorothy Glaves
Reichberg, Samuel Bringeissen
Reiner, Leopold
Reynoso, Gustavo D
Richart, Ralph M
Richter, Goetz Wilfried
Rifkin, Barry Richard
Rivenson, Abraham S
Rome, Doris Spector
Roth, Daniel
Sabatini, David Domingo
Salwen, Martin J
Sands, Elaine S
Sankar, D V Siva
Santos-Buch, Charles A
Schaumburg, Herbert Howard
Schlafer, Donald H
Schlesinger, Richard B
Schroer, Richard Allen
Scott, Robert Foster
Seidman, Irving
Sharpe, William D
Shelanski, Michael L
Sher, Joanna Hollenberg
Sher, Paul Phillip
Sklarew, Robert J
Slatkin, Daniel Nathan
Socha, Wladyslaw Wojciech
Sohn, David
Sokoloff, Leon
Sparks, Charles Edward
Stahl, S Sigmund
Stark, Joel
Stein, Arthur A
Stenger, Richard J
Sternberg, Stephen Stanley
Stowens, Daniel
Taubman, Sheldon Bailey
Taylor, Charles Bruce
Tchertkoff, Victor
Terzakis, John A
Thomas, Lewis
Thomas, Wilbur Addison
Thorbecke, Geertruida Jeanette
Tomashefsky, Philip
Tulchin, Natalie
Vaage, Jan
Vladutiu, Adrian O
Vlaovic, Milan Stephen
Wagner, Bernard Meyer
Waisman, Jerry
Waksman, Byron Halstead
Weinberg, Sidney B
Weiss, Leonard
West, Norman Reed
Wigger, H Joachim
Williams, Gary Murray
Wilner, George Dubar
Wisniewski, Henry Miroslaw
Wolinsky, Harvey
Worgul, Basil Vladimir

Yu, Shiu Yeh
Zimmerman, Harry Martin

NORTH CAROLINA
Adams, Dolph Oliver
Adler, Kenneth B
Allen, James R, Jr
Bakerman, Seymour
Bell, Mary Allison
Benson, Walter Russell
Bode, Arthur Palfrey
Bolande, Robert Paul
Boorman, Gary Alexis
Borowitz, Michael J
Boyd, Jeffrey Allen
Bradford, William Dalton
Brinkhous, Kenneth Merle
Brown, Robert Calvin
Brown, Talmage Thurman, Jr
Carson, Johnny Lee
Chapman, John Franklin, Jr
Chapman, William Edward
Ciancolo, George J
Clarkson, Thomas Boston
Coffin, David L
Cooper, Herbert A
Cross, Robert Edward
Dalldorf, Frederic Gilbert
Edozien, Joseph Chike
Elchlepp, Jane G
Fetter, Bernard Frank
Fletcher, Oscar Jasper, Jr
Genter, Mary Beth
Geratz, Joachim Dieter
Gilfillan, Robert Frederick
Gilmore, Stuart Irby
Glinski, Ronald P
Goyer, Robert Andrew
Graham, Doyle Gene
Graham, John Borden
Griesemer, Richard Allen
Griggs, Thomas Russell
Grisham, Joe Wheeler
Groat, Richard Arnold
Gruber, Helen Elizabeth
Hackel, Donald Benjamin
Hamm, Thomas Edward, Jr
Hanker, Jacob S
Hartz, John William
Hoffman, Maureane Richardson
Hudson, Page
Huffines, William Davis
James, Karen K(anke)
Jennings, Robert Burgess
Jirtle, Randy L
Johnston, William Webb
Jurgelski, William, Jr
Kaufman, David Gordon
Kaufmann, William Karl
Klintworth, Gordon K
Koepke, John Arthur
Langdell, Robert Dana
Lay, John Charles
Lumb, George Dennett
Lyerly, Herbert Kim
Machemer, Robert
McConnell, Ernest Eugene
McLendon, William Woodard
McReynolds, Richard A
Marshall, Richard Blair
Massaro, Edward Joseph
Mikat, Eileen M
Morgan, Kevin Thomas
Norris, H Thomas
Nye, Sylvanus William
O'Connor, Michael L
Paluch, Edward Peter
Penick, George Dial
Popp, James Alan
Pratt, Philip Chase
Prichard, Robert Williams
Rapp, William Rodger
Reimer, Keith A
Rudel, Lawrence L
St Clair, Richard William
Shelburne, John Daniel
Smithies, Oliver
Solomon, Robert Douglas
Sommer, Joachim Rainer
Strausbauch, Paul Henry
Suzuki, Kinuko
Swenberg, James Arthur
Szczech, George Marion
Tucker, Walter Eugene, Jr
Volkman, Alvin
Widmann, Frances King
Wittels, Benjamin
Wright, James Francis

NORTH DAKOTA
Sopher, Roger Louis

OHIO
Adelson, Lester
Aikawa, Masamichi
Alden, Carl L
Anderson, James Morley
Bank, Harvey L
Barth, Rolf Frederick
Becker, Carter Miles
Bertram, Timothy Allyn
Bove, Kevin E
Brandt, John T
Capen, Charles Chabert

Carter, John Robert
Chang, Jae Chan
Chiang, John Y L
Chisolm, Guy M
Cole, Clarence Russell
Collins, William Thomas
Copeland, Bradley Ellsworth
Cornhill, John Frederick
Cross, Robert Franklin
Dahlgren, Robert R
Denine, Elliot Paul
Deodhar, Sharad Dinkar
Donnelly, Kenneth Gerald
Dun, Nae Jiuum
Fenoglio, Cecilia M
Gibson, John Phillips
Greenspan, Neil Sanford
Hamilton, Thomas Alan
Hanson, Daniel James
Harding, Clifford Vincent, III
Hartroft, Phyllis Merritt
Higgins, Jerry Mitchell
Hoff, Henry Frederick
Hopps, Howard Carl
Hutterer, Ferenc
Igel, Howard Joseph
Ingalls, William Lisle
Jennings, Frank Lamont
Kleinerman, Jerome
Koestner, Adalbert
Kwak, Yun Sik
Lamm, Michael Emanuel
Langley, Albert E
Lapham, Lowell Winship
Liebelt, Robert Arthur
Liss, Leopold
Long, John Frederick
McAdams, Arthur James
Macpherson, Colin Robertson
Mahaffey, Kathryn Rose
Malemud, Charles J
Menefee, Max Gene
Metz, David A
Mirkin, L David
Moore, Jay Winston
Moorhead, Philip Darwin
Muir, William W, III
Myers, Ronald Elwood
Nagode, Larry Allen
Newberne, James Wilson
Newton, William Ashen, Jr
Oglesbee, Michael Jerl
Opplt, Jan Jiri
Patrick, James R
Pavelic, Zlatko Paul
Perry, George
Pretlow, Thomas Garrett, II
Rabin, Erwin R
Rapp, John P
Ratliff, Norman B, Jr
Reagan, James W
Reiner, Charles Brailove
Rikihisa, Yasuko
Rodin, Alvin E
Rossi, Edward P
Rynbrandt, Donald Jay
Saif, Yehia M(ohamed)
Saul, Frank Philip
Schut, Herman A
Senhauser, Donald Albert
Shannon, Barry Thomas
Sharma, Hari M
Simmons, Harry Dady, Jr
Smith, Roger Dean
Speicher, Carl Eugene
Stoner, Gary David
Tarr, Melinda Jean
Thomas, Donald Charles
Tubbs, Raymond R
Tykocinski, Mark L
Wagner, Alan R
Washington, John A, II
Weitzner, Stanley
Yates, Allan James

OKLAHOMA
Altmiller, Dale Henry
Bell, Paul Burton, Jr
Breazile, James E
Campbell, Gregory Alan
Confer, Anthony Wayne
Fahmy, Aly
Flournoy, Dayl Jean
Gillum, Ronald Lee
Glass, Richard Thomas
Glenn, Bertis Lamon
Howard, Robert Eugene
Kocan, Katherine M
Lambird, Perry Albert
Lhotka, John Francis
McClellan, Betty Jane
Min, Kyung-Whan
Monlux, Andrew W
Nordquist, Robert Ersel
Panciera, Roger J
Roszel, Jeffie Fisher
Wickham, M Gary

OREGON
Bartley, Murray Hill, Jr
Beckstead, Jay H
Brooks, Robert E
Buchan, George Colin
Coleman, Ralph Orval, Jr

Pathology (cont)

Hutchens, Tyra Thornton
Jastak, J Theodore
Jenkins, Thomas William
Koller, Loren D
McNulty, Wilbur Palmer
Moore, Richard Donald
Niles, Nelson Robinson
Palotay, James Lajos
Prescott, Gerald H
Rickles, Norman Harold
Snyder, Stanley Paul
Spencer, Peter Simner
Stenberg, Paula E
Stroud, Richard Kim
Thompson, Charles Calvin
Young, Norton Bruce

PENNSYLVANIA

AcKerman, Larry Joseph
Allen, Henry L
Andrews, Edwin Joseph
Bagasra, Omar
Balin, Arthur Kirsner
Baserga, Renato
Benz, Edward John
Berkheiser, Samuel William
Berry, Richard G
Bibbo, Marluce
Bly, Chauncey Goodrich
Bokelman, Delwin Lee
Brooks, John J
Burgi, Ernest, Jr
Cameron, Alexander Menzies
Carlson, Arthur Stephen
Chacko, Samuel K
Chan, Ping-Kwong
Chen, Sow-Yeh
Child, Proctor Louis
Clark, Wallace Henderson, Jr
Clawson, Gary Alan
Cohen, Stanley
Colman, Robert W
Conn, Rex Boland
Damjanov, Ivan
Dekker, Andrew
Demers, Laurence Maurice
Dolphin, John Michael
Duker, Nahum Johanan
Farber, John Lewis
Farber, Phillip Andrew
Fenderson, Bruce Andrew
Fisher, Edwin Ralph
Fox, Karl Richard
Furth, John J
Garman, Robert Harvey
Gasic, Gabriel J
Gill, Thomas James, III
Gupta, Prabodh Kumar
Hall, Judy Dale (Mrs Richard Modafferi)
Hartley, Harold V, Jr
Hartman, John David
Haskins, Mark
Hubben, Klaus
Hummeler, Klaus
Jarett, Leonard
Jensen, Richard Donald
Joseph, Jeymohan
Kant, Jeffrey A
Kashatus, William C
Kelly, Alan
Kline, Irwin Kaven
Klionsky, Bernard Leon
Koffler, David
Koprowska, Irena
Kreider, John Wesley
Krieg, Arthur F
Lalley, Edward T
Leifer, Calvin
Little, Brian Woods
LiVolsi, Virginia Anne
Macartney, Lawson
McGary, Carl T
McGrath, John Thomas
Maenza, Ronald Morton
Mancall, Elliott L
Maniglia, Rosario
Martinez, A Julio
Mattison, Donald Roger
Meek, Edward Stanley
Meisler, Arnold Irwin
Michalopoulos, George
Mifflin, Theodore Edward
Miller, Arthur Simard
Ming, Si-Chun
Moossy, John
Murer, Erik Homann
Naeye, Richard L
Nass, Margit M K
Nowell, Peter Carey
Oels, Helen C
Parker, Leslie
Pietra, Giuseppe G
Poste, George Henry
Rabin, Bruce S
Ramasastry, Sai Sudarshan
Rao, Kalipatnapu Narasimha
Riley, Gene Alden
Rorke, Lucy Balian
Rothenbacher, Hansjakob
Rovera, Giovanni
Rubin, Emanuel
Rubin, Herbert
Russo, Jose
Salazar, Hernando
Saunders, Leon Z
Sazama, Kathleen
Schofield, Richard Alan
Schwartz, Heinz (Georg)
Shinozuka, Hisashi
Singh, Gurmukh
Snyder, Robert LeRoy
Steplewski, Zenon
Stewart, Gwendolyn Jane
Strayer, David S
Sunderman, Frederick William
Taichman, Norton Stanley
Tauxe, Welby Newlon
Van Zwieten, Matthew Jacobus
Warhol, Michael J
Weber, Wilfried T
Weiss, Michael Stephen
Wheeler, James English
Whiteside, Theresa L
Whitlock, Robert Henry
Winchester, Richard Albert
Witzleben, Camillus Leo
Young, Donald Stirling
Young, Irving
Yunis, Jorge J
Zeevi, Adriana
Zimmerman, Michael Raymond
Zmijewski, Chester Michael
Zopf, David Arnold

RHODE ISLAND

Bearer, Elaine L
Chang, Pei Wen
Crowley, James Patrick
Fausto, Nelson
Kane, Agnes Brezak
Kuhn, Charles, III
Lichtman, Herbert Charles
McCombs, H Louis
McCully, Kilmer Serjus
McMaster, Philip Robert Bache
Plotz, Richard Douglas
Singer, Don B
Strauss, Elliott William
Vandenburgh, Herman H
Wolke, Richard Elwood
Zacks, Sumner Irwin

SOUTH CAROLINA

Allen, Robert Carter
Balentine, J Douglas
Bryan, Charles Stone
Cannon, Albert
Cantu, Eduardo S
Fowler, Stanley D
Garvin, Abbott Julian
Hennigar, Gordon Ross, Jr
La Via, Mariano Francis
Lill, Patsy Henry
Powell, Harold
Pratt-Thomas, Harold Rawling
Rathbun, Ted Allan
Spicer, Samuel Sherman, Jr
Tripathi, Brenda Jennifer
Tripathi, Ramesh Chandra
Virella, Gabriel T
Willingham, Mark C

SOUTH DAKOTA

Bergeland, Martin E
Wegner, Karl Heinrich

TENNESSEE

Asp, Carl W
Atkinson, James Byron
Belew-Noah, Patricia W
Benz, George William
Berard, Costan William
Cheatham, William Joseph
Clapp, Neal K
Congdon, Charles C
Coogan, Philip Shields
Erickson, Cyrus Conrad
Fields, James Perry
Francisco, Jerry Thomas
Ganote, Charles Edgar
Glassman, Armand Barry
Green, Louis Douglas
Handorf, Charles Russell
Harwood, Thomas Riegel
Hoover, Richard Lee
LeQuire, Virgil Shields
Lucas, Fred Vance
Lushbaugh, Clarence Chancelum
Lynch, Denis Patrick
McCormick, William F
McGavin, Matthew Donald
Marchok, Ann Catherine
Martinez-Hernandez, Antonio
Meyrick, Barbara O
Mithcell, William Marvin
Muirhead, E Eric
Page, David L
Peterson, Harold Arthur
Pitcock, James Allison
Pribor, Hugo C
Price, Robert Allen
Scott, Robert Eugene
Smith, Roy Martin
Soper, Richard Graves
Stone, Robert Edward, Jr
Woodward, Stephen Cotter

Young, Joseph Marvin

TEXAS

Achilles, Robert F
Adams, Leslie Garry
Ahmed, Ahmed Elsayed
Allen, Robert Charles
Arhelger, Roger Boyd
Bailey, Everett Murl, Jr
Batsakis, John G
Beck, John Robert
Becker, Frederick F
Bennett, Michael
Binnie, William Hugh
Bissell, Michael G
Bost, Robert Orion
Bridges, Charles Hubert
Buffone, Gregory J
Burton, Karen Poliner
Burton, Russell Rohan
Butler, James Johnson
Campbell, Carlos Boyd Godfrey
Cimadevilla, Jose M
Coalson, Jacqueline Jones
Cooper, Ronda Fern
Cottone, James Anthony
Cowan, Daniel Francis
Dahl, Elmer Vernon
Davis, Bruce Hewat
Denko, John V
Dodson, Ronald Franklin
Dudley, Alden Woodbury, Jr
Falck, Frank James
Fidler, Isaiah J
Finegold, Milton J
Fitzgerald, James Edward
Flickinger, George Latimore, Jr
Folse, Dean Sydney
Freeman, Robert Glen
Friedberg, Errol Clive
Gazdar, Adi F
Ghidoni, John Joseph
Gibson, William Andrew
Gleiser, Chester Alexander
Graham, David Lee
Gratzner, Howard G
Greenberg, Stanly Donald
Hill, Joseph MacGlashan
Hurt, William Clarence
Ibanez, Michael Louis
Jester, James Vincent
Jones, Larry Philip
Jorgensen, James H
Keene, Harris J
Kier, Ann B
Kim, Han-Seob
Kirkpatrick, Joel Brian
Kokkinakis, Demetrius Michael
Kriesberg, Jeffrey Ira
Kurtz, Stanley Morton
Leibowitz, Julian Lazar
Lieberman, Michael Williams
Lindner, Luther Edward
McGavran, Malcolm Howard
McGill, Henry Coleman, Jr
McGregor, Douglas D
McKenna, Robert Wilson
McManus, Linda Marie
Maurer, Fred Dry
Mendel, Julius Louis
Meuten, Donald John
Migliore, Philip Joseph
Milam, John D
Minerbo, Grace Moffat
Montgomery, Philip O'Bryan, Jr
Moore, Donald Vincent
Murthy, Krishna Kesava
Ordonez, Nelson Gonzalo
Pakes, Steven P
Pinckard, Robert Neal
Race, George Justice
Rajaraman, Srinivasan
Ranney, David Francis
Reynolds, Rolland C
Schmidt, Waldemar Adrian
Schneider, Nancy Reynolds
Schwartz, Colin John
Schwartz, William Lewis
Sell, Stewart
Shadduck, John Allen
Shillitoe, Edward John
Smith, Alice Lorraine
Smith, David English
Smith, Jerome H
Smith, John Leslie, Jr
Spellman, Craig William
Spjut, Harlan Jacobson
Stacy, David Lowell
Stembridge, Vernie A(lbert)
Storts, Ralph Woodrow
Stout, Landon Clarke, Jr
Sybers, Harley D
Tessmer, Carl Frederick
Townsend, Frank Marion
Troxler, Raymond George
Turner, Robert Atwood
Ullrich, Robert Leo
Venkatachalam, Manjeri A
Villarreal, Jesse James
Walker, David Hughes
Wigodsky, Herman S
Wiig, Elisabeth Hemmersam
Wong, Shan Shekyuk
Wordinger, Robert James

Yang, Ovid Y H
Zedler, Empress Young
Zunker, Heinz Otto Hermann

UTAH

Albertine, Kurt H
Ash, Kenneth Owen
Chapman, Arthur Owen
Cordy, Donald R
Eichwald, Ernest J
Fujinami, Robert S
Hammond, Mary Elizabeth Hale
Hill, Harry Raymond
Lloyd, Ray Dix
Mecham, Merlin J
Miles, Charles P
Miner, Merthyr Leilani
Mohammad, Syed Fazal
Moody, David Edward
Olsen, Don B
Warren, Reed Parley
Wilson, John F

VERMONT

Bolton, Wesson Dudley
Clemmons, Jackson Joshua Walter
Coon, Robert William
Craighead, John Edward
Kaplow, Leonard Samuel
Korson, Roy
Luginbuhl, William Hossfeld
Mossman, Brooke T
Sheldon, Huntington
Toolan, Helene Wallace

VIRGINIA

Bobbitt, Oliver Bierne
Braciale, Thomas Joseph, Jr
Bruns, David Eugene
Burr, Helen Gunderson
Cordes, Donald Ormond
Faulconer, Robert Jamieson
Ferry, Andrew P
Fisher, Lyman McArthur
Foster, Eugene A
Gander, George William
Groschel, Dieter Hans Max
Gross, Walter Burnham
Halstead, Charles Lemuel
Hard, Richard C, Jr
Hossaini, Ali A
Hsu, Hsiu-Sheng
Kay, Saul
Koka, Mohan
Kyriazis, Andreas P
Lee, Kyu Taik
Martens, Vernon Edward
Moran, Thomas James
Normansell, David E
Peery, Thomas Martin
Peters, Esther Caroline
Rosenblum, William I
Rubenstein, Norton Michael
Russell, Catherine Marie
Russi, Simon
Salley, John Jones
Savory, John
Schwartzman, Joseph David
Shipe, James R, Jr
Sirica, Alphonse Eugene
Somlyo, Andrew Paul
Still, William James Sangster
Sturgill, Benjamin Caleb
Vennart, George Piercy
Voelker, Richard William
Weiss, Daniel Leigh

WASHINGTON

Alvord, Ellsworth Chapman, Jr
Bankson, Daniel Duke
Bean, Michael Arthur
Benditt, Earl Philip
Boatman, Edwin S
Butts, William Cunningham
Callis, James Bertram
Couser, William Griffith
Gribble, David Harold
Haggitt, Rodger C
Hall, Stanton Harris
Henson, James Bond
Hibbs, Clair Maurice
Kachmar, John Frederick
Killingsworth, Lawrence Madison
Kramer, John William
Landolt, Marsha LaMerle
Leid, R Wes
Liggitt, H Denny
Martin, George Monroe
Mottet, Norman Karle
Narayanan, A Sampath
Nelson, Karen Ann
Page, Roy Christopher
Palmer, John M
Prehn, Richmond Talbot
Prieur, David John
Purnell, Dallas Michael
Rabinovitch, Peter S
Ragan, Harvey Albert
Robinovitch, Murray R
Ross, Russell
Sanders, Charles Leonard, Jr
Schwartz, Stephen Mark
Shaw, Cheng-Mei
Smith, Dean Harley

Strandjord, Paul Edphil
Van Hoosier, Gerald L, Jr
Vogel, Arthur Mark
Wilson, Robert Burton
Wolf, Norman Sanford
Zamora, Cesario Siasoco
Zander, Donald Victor

WEST VIRGINIA
Albrink, Wilhelm Stockman
Blundell, George Phelan
Bouquot, Jerry Elmer
Chang, William Wei-Lien
Hales, Milton Reynolds
Hooper, Anne Caroline Dodge
Iammarino, Richard Michael
Morgan, Winfield Scott
Quittner, Howard
Rodman, Nathaniel Fulford, Jr
Schochet, Sydney Sigfried, Jr
Victor, Leonard Baker

WISCONSIN
Altshuler, Charles Haskell
Becker, Carl George
Beckfield, William John
Bloodworth, James Morgan Bartow, Jr
Brown, Arnold Lanehart, Jr
Collins, Richard Andrew
Eisenstein, Reuben
Erwin, Chesley Para
Fodden, John Henry
Folts, John D
Goldfarb, Stanley
Greenspan, Daniel S
Hartmann, Henrik Anton
Hinze, Harry Clifford
Hoerl, Bryan G
Hussey, Clara Veronica
Inhorn, Stanley L
Jaeschke, Walter Henry
Kuzma, Joseph Francis
Lalich, Joseph John
Larson, Frank Clark
Locke, Louis Noah
Ndon, John A
Neubecker, Robert Duane
Norback, Diane Hagemen
Oberley, Terry De Wayne
Olson, Carl
Pitot, Henry C, III
Reneau, John
Rossetti, Louis Michael
Rude, Theodore Alfred
Siegesmund, Kenneth A
Silverman, Ellen-Marie
Silverman, Franklin Harold
Uno, Hideo
Wussow, George C

WYOMING
Isaak, Dale Darwin

PUERTO RICO
Moreno, Esteban
Ortiz, Araceli

ALBERTA
Bain, Gordon Orville
Bainborough, Arthur Raymond
Bell, Harold E
Dick, Henry Marvin
Dixon, John Michael Siddons
Jimbow, Kowichi
Rewcastle, Neill Barry
Russell, James Christopher
Shnitka, Theodor Khyam
Solez, Kim
Stinson, Robert Anthony
Wilson, Frank B
Wood, Norman Kenyon
Yoon, Ji-Won

BRITISH COLUMBIA
Bower, Susan Mae
Dunn, William Lawrie
Frohlich, Jiri J
Garg, Arun K
Godolphin, William
Hardwick, David Francis
Hogg, James Cameron
Kelly, Michael Thomas
Kim, Seung U
Pearce, Richard Hugh
Salinas, Fernando A
Saunders, James Robert
Spitzer, Ralph
Thurlbeck, William Michael

MANITOBA
Adamson, Ian Young Radcliffe
Bowden, Drummond Hyde
Carr, Ian
Crowson, Charles Neville
Henderson, James Stuart
Hoogstraten, Jan
Persaud, Trivedi Vidhya Nandan
Pettigrew, Norman M

NEW BRUNSWICK
Smith, Harry John

NEWFOUNDLAND
Michalak, Thomas Ireneusz

NOVA SCOTIA
Aterman, Kurt
Fraser, Albert Donald
Ghose, Tarunendu
Hanna, Brian Dale
Tonks, Robert Stanley

ONTARIO
Asa, Sylvia L
Barr, Ronald Duncan
Baumal, Reuben
Bruce, W Robert
Campbell, James Stewart
Chander, Satish
Cutz, Ernest
De Bold, Adolfo J
Dedhar, Shoukat
De Harven, Etienne
Farber, Emmanuel
Frederick, George Leonard
Frei, Jaroslav Vaclav
Goldberg, David Myer
Gotlieb, Avrum I
Haust, M Daria
Heggtveit, Halvor Alexander
Hill, Donald P
Huang, Shao-nan
Hulland, Thomas John
Kaushik, Azad
Kisilevsky, Robert
Kovacs, Kalman T
Kuroski-De Bold, Mercedes Lina
Langille, Brian Lowell
Lee, Robert Maung Kyaw Win
Main, James Hamilton Prentice
Mills, James Herbert Lawrence
Morris, Gerald Patrick
Movat, Henry Zoltan
Mustard, James Fraser
Neufeld, Abram Herman
Nielsen, N Ole
Nisbet-Brown, Eric Robert
Orr, Frederick William
Paul, Leendert Cornelis
Phillips, Melville James
Ranadive, Narendranath Santuram
Richter, Maxwell
Ritchie, Alexander Charles
Robertson, David Murray
Rosenthal, Kenneth Lee
Rowsell, Harry Cecil
Sarma, Dittakavi S R
Silver, Malcolm David
Stanisz, Andrzej Maciej
Sturgess, Jennifer Mary
Thibert, Roger Joseph
Vadas, Peter
Wallace, Alexander Cameron
Waugh, Douglas Oliver William
Waye, John Stewart

PRINCE EDWARD ISLAND
Singh, Amreek

QUEBEC
Alavi, Misbahuddin Zafar
Bendayan, Moise
Brassard, Andre
Cantin, Marc
Cote, Roger Albert
Duguid, William Paris
Evering, Winston E N D
Jasmin, Gaetan
Karpati, George
Lee, King C
Lussier, Andre (Joseph Alfred)
Lussier, Gilles L
Meisels, Alexander
Messier, Bernard
Moore, Sean
Murthy, Mahadi Raghavandrarao Ven
Rona, George
Simard, Rene
Stejskal, Rudolf
Tremblay, Gilles
Ventura, Joaquin Calvo
Weigensberg, Bernard Irving
Yousef, Ibrahim Mohmoud

SASKATCHEWAN
Blair, Donald George Ralph
Emson, Harry Edmund
Kalra, Jawahar
Rozdilsky, Bohdan
Skinnider, Leo F
Xiang, Jim

OTHER COUNTRIES
Baldwin, Robert William
Bencosme, Sergio Arturo
Bernstein, David Maier
Blanc, William Andre
Collins, Vincent Peter
Cras, Patrick
Crompton, David William Thomasson
Czernobilsky, Bernard
Dillman, Richard Carl
Dutz, Werner
Fahimi, Hossein Dariush
Friede, Reinhard L
Fujimoto, Shigeyoshi
Gabbiani, Giulio
Gertz, Samuel David
Goldberg, Burton David

Gonzalez-Angulo, Amador
Gorski, Andrzej
Gowans, James L
Gresser, Ion
Gullino, Pietro M
Hansson, Goran K
Helander, Herbert Dick Ferdinand
Hirokawa, Katsuiku
Ide, Hiroyuki
Izui, Shozo
Jakowska, Sophie
Karasaki, Shuichi
Kerjaschki, Dontscho
Krueger, Gerhard R F
Lee, Joseph Chuen Kwun
Lever, Walter Frederick
Lewin, Lawrence M
Lin, Chin-Tarng
Ludwig, Frederic C
Maruyama, Koshi
Minowada, Jun
Mori, Wataru
Pan, In-Chang
Renaud, Serge
Rettori, Ovidio
Shubik, Philippe
Silberberg, Ruth
Stehbens, William Ellis
Stern, Kurt
Tachibana, Takehiko
Thiery, Jean Paul
Van Loveren, Henk
Wan, Abraham Tai-Hsin
Warren, Bruce Albert
Yakura, Hidetaka
Yoshida, Takeshi
Yovich, John V

Pharmacology

ALABAMA
Ayling, June E
Barker, Samuel Booth
Beaton, John McCall
Bennett, Leonard Lee, Jr
Benos, Dale John
Berecek, Kathleen Helen
Berger, Robert S
Besse, John C
Boerth, Robert Carter
Chang, Chi Hsiung
Christian, Samuel Terry
Clark, Carl Heritage
Coker, Samuel Terry
Dalvi, Ramesh R
Deneau, Gerald Antoine
Diasio, Robert Bart
Downey, James Merritt
El Dareer, Salah
Fallon, Harold Joseph
Frings, Christopher Stanton
Furner, Raymond Lynn
Guarino, Anthony Michael
Harding, Thomas Hague
Harrison, Steadman Darnell, Jr
Hill, Donald Lynch
Hocking, George Macdonald
Kaplan, Ronald S
Lecklitner, Myron Lynn
McCann, William Peter
McCarthy, Dennis Joseph
Macmillan, William Hooper
Meezan, Elias
Mundy, Roy Lee
Nair, Madhavan G
Page, John Gardner
Palmisano, Paul Anthony
Pillion, Dennis Joseph
Pruitt, Albert Wesley
Ravis, William Robert
Schneyer, Charlotte A
Sowell, John Gregory
Strada, Samuel Joseph
Struck, Robert Frederick
Susina, Stanley V
Tan, Boen Hie
Tate, Lawrence Gray
Vacik, James P
Walsh, Gerald Michael

ALASKA
DeLapp, Tina Davis

ARIZONA
Adler, Charles H
Aposhian, Hurair Vasken
Blanchard, James
Bowman, Douglas Clyde
Brands, Allen J
Brendel, Klaus
Bressler, Rubin
Buck, Stephen Henderson
Chin, Lincoln
Clayton, John Wesley, Jr
Colburn, Wayne Alan
Consroe, Paul F
Eckardt, Robert E
Fagan, Timothy Charles
Gandolfi, A Jay
Halonen, Marilyn Jean
Halpert, James Robert
Huxtable, Ryan James
Jeter, Wayburn Stewart

Krahl, Maurice Edward
Laird, Hugh Edward, II
Lindell, Thomas Jay
Lukas, Ronald John
McQueen, Charlene A
Milch, Lawrence Jacques
Palmer, John Davis
Peric-Knowlton, Wlatka
Picchioni, Albert Louis
Powis, Garth
Rowe, Verald Keith
Russell, Findlay Ewing
Schmid, Jack Robert
Sipes, Ivan Glenn
Tong, Theodore G
Weil, Andrew Thomas

ARKANSAS
Cornett, Lawrence Eugene
Davis, Virginia Eischen
Ginzel, Karl-Heinz
Greenman, David Lewis
Hanna, Calvin
Johnson, Bob Duell
Jordin, Marcus Wayne
Kadlubar, Fred F
Leakey, Julian Edwin Arundell
Light, Kim Edward
Littlefield, Neil Adair
McMillan, Donald Edgar
Paule, Merle Gale
Schieferstein, George Jacob
Seifen, Ernst
Sorenson, John R J
Stone, Joseph
Valentine, Jimmie Lloyd
Walker, Charles A
Wenger, Galen Rosenberger
Wessinger, William David
Winters, Ronald Howard
Young, John Falkner

CALIFORNIA
Agabian, Nina
Agre, Karl
Ainsworth, Earl John
Alkana, Ronald L
Alousi, Adawia A
Amer, Mohamed Samir
Anderson, Hamilton Holland
Ashe, John Herman
Azarnoff, Daniel Lester
Baldwin, Robert Charles
Barnes, Paul Richard
Batterman, Robert Coleman
Bauer, Robert Oliver
Bejar, Ezra
Benet, Leslie Z
Bennett, C Frank
Bergman, Hyman Chaim
Berman, David Albert
Berteau, Peter Edmund
Black, Kirby Samuel
Blankenship, James William
Bokoch, Gary M
Bradley, JoAnn D
Brown, Joan Heller
Brunton, Laurence
Buchmeier, Michael J
Burkhalter, Alan
Butz, Robert Frederick
Byard, James Leonard
Carson, Virginia Rosalie Gottschall
Casida, John Edward
Castles, Thomas R
Catlin, Don Hardt
Chan, Kenneth Kin-Hing
Chang, Freddy Wilfred Lennox
Chang, Yi-Han
Chapman, Sharon K
Chin, Jane Elizabeth Heng
Cho, Arthur Kenji
Chow, Samson Ah-Fu
Chuang, Ronald Yan-Li
Ciaramitaro, David A
Clark, Charles Richard
Clark, Dennis Richard
Clarke, David E
Cohen, Kenneth Samuel
Cohn, Major Lloyd
Comer, William Timmey
Corbascio, Aldo Nicola
Correia, Maria Almira
Courtney, Kenneth Randall
Cronheim, Georg Erich
Cronin, Michael John
Crooke, Stanley T
Curras, Margarita C
Dang, Peter Hung-Chen
Danse, Ilene H Raisfeld
Deedwania, Prakash Chandra
DeMet, Edward Michael
Dismukes, Robert Key
Dixon, Ross
Doroshow, James Halpern
Duckles, Sue Piper
Eakins, Kenneth E
Eddy, Lynne J
Etherington, Lorne
Embree, James Willard, Jr
Everett, Guy M
Fain, Gordon Lee
Fairchild, David George
Felton, James Steven

Pharmacology (cont)

Fields, Howard Lincoln
Flacke, Joan Wareham
Flacke, Werner Ernst
Fraser, Ian McLennan
Freeman, Leon David
Freeman, Walter Jackson, III
Friess, Seymour Louis
Fuhrman, Frederick Alexander
Furst, Arthur
Gans, Joseph Herbert
Gehrmann, John Edward
George, Robert
Geyer, Mark Allen
Giri, Shri N
Goldberg, Mark Arthur
Golder, Thomas Keith
Goldstein, Avram
Goldstein, Dora Benedict
Gong, William C
Goode, John Wolford
Gordon, Adrienne Sue
Gray, Gregory Edward
Green, Donald Eugene
Green, Sidney
Greenberg, Roland
Haley, Thomas John
Hammock, Bruce Dupree
Hance, Anthony James
Hansen, Eder Lindsay
Hecht, Elizabeth Anne
Hefti, Franz F
Henderson, Gary Lee
Hessinger, David Alwyn
Hidalgo, John
Hines, Leonard Russell
Hoener, Betty-ann
Hollinger, Mannfred Alan
Hondeghem, Luc M
Hsieh, Dennis P H
Huber, Wolfgang Karl
Ignarro, Louis Joseph
Insel, Paul Anthony
Jacobs, Robert Saul
Jain, Naresh C
Jardetzky, Oleg
Jardine, Ian
Jarvik, Murray Elias
Jelliffe, Roger Woodham
Jenden, Donald James
Jensen, Richard Arthur
Johnson, Howard (Laurence)
Johnson, Randolph Mellus
Jordan, Mary Ann
Joy, Robert McKernon
Katzung, Bertram George
Kendig, Joan Johnston
Kilgore, Wendell Warren
Killam, Eva King
Killam, Keith Fenton, Jr
Koch, Bruce D
Kodama, Jiro Kenneth
Koschier, Francis Joseph
Koths, Kirston Edward
Kuczenski, Ronald Thomas
Laiken, Nora Dawn
Lee, Chi-Ho
Lee, David Anson
Lee, Peter Van Arsdale
Lehmann, A(ldo) Spencer
Leung, Peter
Levine, Jon David
Levy, Joseph Victor
Levy, Louis
Lewis, Richard John
Lippa, Erik Alexander
Lippmann, Wilbur
Loew, Gilda Harris
Lomax, Peter
Lowinger, Paul
Lytle, Loy Denham
McCaman, Richard Eugene
MacFarlane, Malcolm David
McGaughey, Charles Gilbert
Mais, Dale E
Malone, Marvin Herbert
Mansour, Tag Eldin
Marangos, Paul Jerome
Maronde, Robert Francis
Marshall, John Foster
Mason, Dean Towle
Meerdink, Denis J
Melmon, Kenneth Lloyd
Mensah, Patricia Lucas
Menzel, Daniel B
Meyers, Frederick H
Milby, Thomas Hutchinson
Miljanich, George Paul
Miller, Jon Philip
Mitoma, Chozo
Mizuno, Nobuko S(himotori)
Morey-Holton, Emily Rene
Mule, Salvatore Joseph
Nelson, Eric Loren
Nimni, Marcel Efraim
Novack, Gary Dean
Okita, George Torao
Okun, Ronald
O'Malley, Edward Paul
Ortiz de Montellano, Paul Richard
Ottoboni, M(inna) Alice
Painter, Ruth Coburn Robbins
Patton, John Stuart
Pearl, Ronald G
Peoples, Stuart Anderson
Peroutka, Stephen Joseph
Peters, John Henry
Peters, Marvin Arthur
Peterson, Charles Marquis
Pharriss, Bruce Bailey
Phelps, Michael Edward
Place, Virgil Alan
Povzhitkov, Moysey Michael
Printz, Morton Philip
Reifenrath, William Gerald
Ridley, Peter Tone
Riedesel, Carl Clement
Riemer, Robert Kirk
Ritzmann, Ronald Fred
Rivier, Catherine L
Rivier, Jean E F
Roberts, Carmel Montgomery
Robertson, David Wayne
Rooks, Wendell Hofma, II
Rosenthale, Marvin E
Roszkowski, Adolph Peter
Rousell, Ralph Henry
Rubanyi, Gabor Michael
Sadee, Wolfgang
Saifer, Mark Gary Pierce
Sandmeyer, Esther E
Sargent, Thornton William, III
Sassenrath, Ethelda Norberg
Sattin, Albert
Saute, Robert E
Sawyer, Wilbur Henderson
Schienle, Jan Hoops
Schulman, Howard
Segal, David S
Selassie, Cynthia R
Setler, Paulette Elizabeth
Sharma, Arjun D
Shaw, Jane E
Shellenberger, Melvin Kent
Shen, Wei-Chiang
Shih, Jean Chen
Siggins, George Robert
Simmons, Dwayne Deangelo
Smith, Charles G
Smith, Martyn Thomas
Smith, Norman Ty
Spiller, Gene Alan
Stanton, Hubert Coleman
Stanton, Toni Lynn
Stark, Larry Gene
Steffey, Eugene P
Stein, Larry
Stickney, Janice Lee
Stone, Deborah Bennett
Strosberg, Arthur Martin
Strother, Allen
Sunshine, Irving
Taylor, Dermot Brownrigg
Taylor, James A
Taylor, Palmer William
Tewari, Sujata Lahiri
Thomas, Lyell Jay, Jr
Thompson, John Frederick
Thueson, David Orel
Tiedcke, Carl Heinrich Wilhelm
Torres, Martine
Tramell, Paul Richard
Trevor, Anthony John
Tsien, Richard Winyu
Tune, Bruce Malcolm
Tuttle, Ronald Ralph
Tutupalli, Lohit Venkateswara
Urquhart, John, III
Uyeno, Edward Teiso
Van Dyke, Craig
Walker, Sydney, III
Wallach, Marshall Ben
Wang, Howard Hao
Wangler, Roger Dean
Waterbury, Lowell David
Way, E Leong
Way, Walter
Wechter, William Julius
Wei, Edward T
Weinreb, Robert Neal
Weissberg, Robert Murray
Wendel, Otto Theodore, Jr
Wester, Ronald Clarence
Willhite, Calvin Campbell
Wilson, Leslie
Winters, Wallace Dudley
Woolley, Dorothy Elizabeth Schumann
Yagiela, John Allen
Yaksh, Tony Lee
Yang, Heechung
Yang, William C T
Ziegler, Michael Gregory

COLORADO

Allen, Larry Milton
Alpern, Herbert P
Appelt, Glenn David
Barrett, C Brent
Beam, Kurt George, Jr
Breault, George Omer
Carnathan, Gilbert William
Collins, Allan Clifford
Coomes, Richard Merril
Deitrich, Richard Adam
Diamond, Louis
Erwin, Virgil Gene
Fennessey, Paul V
Froede, Harry Curt
Gerber, John George
Glass, Howard George
Guzelian, Philip Samuel
Herin, Reginald Augustus
Hesterberg, Thomas William
Hilmas, Duane Eugene
Hollister, Alan Scudder
Ishii, Douglas Nobuo
Malkinson, Alvin Maynard
Murphy, Robert Carl
Musick, James R
Palmer, Michael Rule
Petersen, Dennis Roger
Richards, Edmund A
Ruth, James Allan
Sharpless, Seth Kinman
Stewart, John Morrow
Tabakoff, Boris
Thompson, John Alec
Vernadakis, Antonia
Weiner, Norman
Weliky, Irving
Williams, Ronald Lee
Wilson, Vincent L
Winstead, Jack Alan

CONNECTICUT

Aghajanian, George Kevork
Anderson, Rebecca Jane
Bacopoulos, Nicholas G
Bertino, Joseph R
Bickerton, Robert Keith
Bunney, Benjamin Stephenson
Buyniski, Joseph P
Byck, Robert
Cadman, Edwin Clarence
Century, Bernard
Chello, Paul Larson
Cheng, Yung-Chi
Conrad, Eugene Anthony
Cooper, Jack Ross
Dannies, Priscilla Shaw
Davis, Michael
DeVita, Vincent T, Jr
Devlin, Richard Gerald, Jr
Dix, Douglas Edward
Doherty, Niall Stephen
Doshan, Harold David
Douglas, James Sievers
Douglas, William Wilton
Eisenfeld, Arnold Joel
Epstein, Mary A
Escobar, Javier I
Feinstein, Maurice B
Fleming, James Stuart
Forssen, Eric Anton
Gay, Michael Howard
Gerald, Michael Charles
Gerritsen, Mary Ellen
Ghosh, Dipak K
Gillis, Charles Norman
Gunther, Jay Kenneth
Gylys, Jonas Antanas
Handschumacher, Robert Edmund
Harwood, Harold James, Jr
Haubrich, Dean Robert
Henderson, Edward George
Herbette, Leo G
Heyd, Allen
Hirsh, Eva Maria Hauptmann
Hitchcock, Margaret
Hobbs, Donald Clifford
Jahn, Reinhard
Janis, Ronald Allen
Jesmok, Gary J
Khan, Abdul Jamil
King, Theodore Oscar
Klimek, Joseph John
Lacouture, Peter George
Langner, Ronald O
Larson, Jerry King
Letts, Lindsay Gordon
Levine, Robert John
Loose, Leland David
McConnell, William Ray
McIlhenny, Hugh M
Makriyannis, Alexandros
Mayol, Robert Francis
Merker, Philip Charles
Milne, George McLean, Jr
Milzoff, Joel Robert
Moore, Peter Francis
Mycek, Mary J
Niblack, John Franklin
Nieforth, Karl Allen
Nightingale, Charles Henry
Oates, Peter Joseph
Otterness, Ivan George
Ozols, Juris
Pappano, Achilles John
Parker, Eric McFee
Pazoles, Christopher James
Pfeifer, Richard Wallace
Pinson, Ellis Rex, Jr
Possanza, Genus John
Prusoff, William Herman
Rauch, Albert Lee
Redmond, Donald Eugene, Jr
Riblet, Leslie Alfred
Ritchie, Joseph Murdoch
Robinson, Donald Stetson
Rockwell, Sara Campbell
Rosenberg, Philip
Roth, Robert Henry, Jr
Rudnick, Gary
Sartorelli, Alan Clayton
Schach von Wittenau, Manfred
Schenkman, John Boris
Schmeer, Arline Catherine
Schurig, John Eberhard
Schwartz, Pauline Mary
Scriabine, Alexander
Sheard, Michael Henry
Smilowitz, Henry Martin
Snellings, William Moran
Snider, Ray Michael
Stroebel, Charles Frederick, III
Tallman, John Francis
Taylor, Duncan Paul
Taylor, Russell James, Jr
Tyler, Tipton Ransom
Urquilla, Pedro Ramon
Weiss, Robert Martin
Weissman, Albert
Welch, Annemarie S
Yocca, Frank D

DELAWARE

Aharony, David
Aungst, Bruce J
Blum, Lee M
Brown, Barry Stephen
Buckner, Carl Kenneth
Burch, Ronald Martin
Caputo, Claudia Dr
Christoph, Greg Robert
Clark, Robert
Cook, Leonard
Galbraith, William
Giles, Ralph E
Goldberg, Morton Edward
Goldstein, Jeffrey Marc
Hartig, Paul Richard
Herblin, William Fitts
Hoffman, Howard Edgar
Kerr, Janet Spence
Kinoshita, Florence Keiko
Krell, Robert Donald
Krivanek, Neil Douglas
Lai, Chii-Ming
Lam, Gilbert Nim-Car
Li, Jack H
Loveless, Scott E
McCurdy, David Harold
Malick, Jeffrey Bevan
Malley, Linda Angevine
Maloff, Bruce L(arrie)
Nielsen, Susan Thomson
Patel, Jitendra Balkrishna
Patel, Narayan Ganesh
Price, William Alrich, Jr
Quisenberry, Richard Keith
Quon, Check Yuen
Rubin, Alan A
Salzman, Steven Kerry
Sands, Howard
Sarrif, Awni M
Schneider, Jurg Adolf
Shotzberger, Gregory Steven
Steinberg, Marshall
Stump, John M
Tam, Sang William
Turlapaty, Prasad
Vernier, Vernon George
Wheeler, Allan Gordon
Whitney, Charles Candee, Jr
Zuckerman, Joan Ellen

DISTRICT OF COLUMBIA

Abramson, Fred Paul
Alving, Carl Richard
Apple, Martin Allen
Argus, Mary Frances
Bayer, Barbara Moore
Beaver, William Thomas
Beliles, Robert Pryor
Brooker, Gary
Burris, James F
Byrd, Daniel Madison, III
Cheney, Darwin L
Chung, Eunyong
Chung, Ho
Cohn, Victor Hugo
Coomes, Marguerite Wilton
Crawford, Lester M
Deutsch, Stanley
Di Carlo, Frederick Joseph
Dickson, Robert Brent
Doctor, Bhupendra P
Dretchen, Kenneth Lewis
Ellis, Sydney
Fabro, Sergio
Fanning, George Richard
Gillis, Richard A
Hamarneh, Sami Khalaf
Heiffer, Melvin Harold
Herman, Barbara Helen
Himmelsbach, Clifton Keck
Hollinshead, Ariel Cahill
Karle, Jean Marianne
Kellar, Kenneth Jon
Kennedy, Katherine Ash
Klubes, Philip
Lee, Cheng-Chun
Loo, Ti Li
MacDonald, Mhairi Graham
Mandel, Harold George

Mazel, Paul
Milbert, Alfred Nicholas
Murphy, James John
Ohi, Seigo
Perry, David Carter
Raines, Arthur
Ramwell, Peter William
Rhoads, Allen R
Rhoden, Richard Allan
Schwartz, Sorell Lee
Sobotka, Thomas Joseph
Soller, Roger William
Straw, James Ashley
Thomas, Richard Dean
Van Arsdel, William Campbell, III
Vernikos, Joan
Vick, James
Wagstaff, David Jesse
West, William Lionel
Wise, Bradley C
Woosley, Raymond Leon
Zeeman, Maurice George

FLORIDA
Abou-Khalil, Samir
Abraham, William Michael
Baker, Stephen Phillip
Bassett, Arthur L
Battista, Sam P
Bell, John Urwin
Ben, Max
Booth, Nicholas Henry
Boxill, Gale Clark
Boyd, Eleanor H
Brazeau, Gayle A
Cintron, Guillermo B
Coffey, Ronald Gibson
Cohen, Glenn Milton
Cook, Ellsworth Barrett
Crews, Fulton T
Denber, Herman C B
Dickison, Harry Leo
Edelson, Jerome
Ehrreich, Stewart Joel
Epstein, Murray
Farah, Alfred Emil
Finger, Kenneth F
Fitzgerald, Thomas James
Frank, H Lee
Freudenthal, Ralph Ira
Freyburger, Walter Alfred
Garg, Lal Chand
Gelband, Henry
Goldstein, Burton Jack
Goodnick, Paul Joel
Greenberg, Michael John
Hackman, John Clement
Hackney, John Franklin
Hahn, Elliot F
Harbison, Raymond D
Haynes, Duncan Harold
Himes, James Albert
Hogan, Daniel James
Howes, John Francis
Hudson, Reggie Lester
Jewett, Robert Elwin
Katovich, Michael J
Kerrick, Wallace Glenn Lee
Kiplinger, Glenn Francis
Klein, Richard Lester
Korol, Bernard
Krzanowski, Joseph John, Jr
Kunisi, Venkatasubban S
Lamba, Surendar Singh
Lasseter, Kenneth Carlyle
Leibman, Kenneth Charles
Lewis, John Reed
Lindsay, Raymond H
Maren, Thomas Hartley
Martin, Frank Gene
Mayer, Richard Thomas
Menzer, Robert Everett
Merritt, Alfred M, II
Michie, David Doss
Neims, Allen Howard
Nelson, Thomas Eustis, Jr
Oberst, Fred William
Palmer, Roger
Parmar, Surendra S
Pfaffman, Madge Anna
Polson, James Bernard
Porter, Lee Albert
Rader, William Austin
Radomski, Jack London
Rennick, Barbara Ruth
Richelson, Elliott
Ringel, Samuel Morris
Rodman, Morton Joseph
Silverman, David Norman
Soliman, Karam Farag Attia
Soliman, Magdi R I
Stone, Clement A
Sturtevant, Frank Milton
Szentivanyi, Andor
Teaf, Christopher Morris
Thomas, Vera
Thureson-Klein, Asa Kristina
Tusing, Thomas William
Van Breeman, Cornelis
Vogh, Betty Pohl
Walker, Don Wesley
Wecker, Lynn
Weidler, Donald John
Weiss, Charles Frederick

Wilcox, Christopher Stuart
Williams, Joseph Francis
Wolgin, David L
Yeh, Billy Kuo-Jiun
Young, Michael David

GEORGIA
Adams, Robert Johnson
Akin, Frank Jerrel
Bain, James Arthur
Black, Asa C, Jr
Boudinot, Frank Douglas
Bowen, John Metcalf
Bransome, Edwin D, Jr
Bruckner, James Victor
Buccafusco, Jerry Joseph
Byrd, Larry Donald
Caldwell, Robert William
Carrier, Gerald O
Catravas, John D
Ciarlone, Alfred Edward
Curley, Winifred H
Denson, Donald D
Diamond, Bruce I
Dingledine, Raymond J
Ebert, Andrew Gabriel
Edwards, Gaylen Lee
Garrettson, Lorne Keith
Gatipon, Glenn Blaise
Geber, William Frederick
Glass, David Bankes
Greenbaum, Lowell Marvin
Hatch, Roger Conant
Hofman, Wendell Fey
Huber, Thomas Lee
Hug, Carl Casimir, Jr
Israili, Zafar Hasan
Iturrian, William Ben
Iuvone, Paul Michael
Karow, Armand M(onfort), Jr
Kaul, Pushkar Nath
Kirby, Margaret Loewy
Kolbeck, Ralph Carl
Kuo, Jyh-Fa
Lopez, Antonio Vincent
McGrath, William Robert
Madden, John Joseph
Merin, Robert Gillespie
Miceli, Joseph N
Minneman, Kenneth Paul
Mitch, William Evans
Mokler, Corwin Morris
Moran, Neil Clymer
Mumtaz, Mohammad Moizuddin
Nemeroff, Charles Barnet
Norred, William Preston
Potter, David Edward
Proctor, Charles Darnell
Pruett, Jack Kenneth
Schramm, Lee Clyde
Schweri, Margaret Mary
Shlevin, Harold H
Stone, Connie J
Sutherland, James Henry Richardson
Wade, Adelbert Elton
Whitford, Gary M
Wolven-Garrett, Anne M
Woodhouse, Bernard Lawrence

HAWAII
Bailey, Leslie Edgar
Blanchard, Robert Joseph
Lenney, James Francis
Lum, Bert K B
Morton, Bruce Eldine
Norton, Ted Raymond
Read, George Wesley
Shibata, Shoji

IDAHO
Cole, Franklin Ruggles
Dodson, Robin Albert
Duke, Victor Hal
Hillyard, Ira William
Mudumbi, Ramagopal Vijaya
Vestal, Robert Elden

ILLINOIS
Albrecht, Ronald Frank
Ambre, John Joseph
Anderson, David John
Anderson, Edmund George
Archer, John Dale
Atkinson, Arthur John, Jr
Baron, David Alan
Beecher, Christopher W W
Bekersky, Ihor
Bennett, Donald Raymond
Berman, Eleanor
Bernstein, Joel Edward
Berry, Charles Arthur
Bevill, Richard F, Jr
Bhargava, Hemendra Nath
Bianchi, Robert George
Bingel, Audrey Susanna
Bosmann, Harold Bruce
Boulos, Badi Mansour
Bourgault, Priscilla C
Bousquet, William F
Brioni, Jorge Daniel
Brodie, Mark S
Brown, Daniel Joseph
Browning, Ronald Anthony
Brunner, Edward A

Buck, William Boyd
Buckman, Robert W
Calandra, Joseph Carl
Caspary, Donald M
Chan, Yun Lai
Chappell, Elizabeth
Chiou, Win Loung
Chu, Sou Yie
Cline, William H, Jr
Cook, Donald Latimer
Cranston, Joseph William, Jr
Crystal, George Jeffrey
Dailey, John William
Dajani, Esam Zapher
Davis, Harvey Virgil
Davis, Joseph Richard
Davis, Oscar F
Dawson, M Joan
Dean, Richard Raymond
Dodge, Patrick William
Dubocovich, Margarita L
Ehrenpreis, Seymour
Elliott, William John
Erdos, Ervin George
Erve, Peter Raymond
Estep, Charles Blackburn
Faingold, Carl L
Farnsworth, Norman R
Feigen, Larry Philip
Feinberg, Harold
Ferguson, Hugh Carson
Fitzloff, John Frederick
Foreman, Ronald Louis
Friedman, Alexander Herbert
Garthwaite, Susan Marie
Giacobini, Ezio
Gillette, Martha Ulbrick
Glisson, Silas Nease
Green, Richard D
Hanin, Israel
Harris, Stanley Cyril
Haugen, David Allen
Heller, Michael
Henkin, Jack
Herting, Robert Leslie
Hiles, Richard Allen
Hirsch, Lawrence Leonard
Hjelle, Joseph Thomas
Ho, Andrew Kong-Sun
Hoffmann, Philip Craig
Hosey, M Marlene
Javaid, Javaid Iqbal
Jensen, Robert Alan
Jobe, Phillip Carl
Jordan, V Craig
Kabara, Jon Joseph
Kang, David Soosang
Karczmar, Alexander George
Katz, Norman L
Kesler, Darrel J
Kimura, Eugene Tatsuru
Klabunde, Richard Edwin
Knepper, Sheila M
Kochman, Ronald Lawrence
Kohli, Jai Dev
Koritz, Gary Duane
Kosman, Mary Ellen
Kulkarni, Anant Sadashiv
Kyncl, J Jaroslav
Lambert, Glenn Frederick
Lampe, Kenneth Francis
Lane, Benjamin Clay
Lasley, Stephen Michael
Le Breton, Guy C
Lee, Tony Jer-Fu
Leff, Alan R
Leikin, Jerrold Blair
Lesher, Gary Allen
Linde, Harry Wight
Loeb, Jerod M
Lorens, Stanley A
Louis, John
McCormick, David Loyd
MacKichan, Janis Jean
Manyam, Bala Venktesha
Marczynski, Thaddeus John
Martin, Yvonne Connolly
Maynert, Everett William
Mayor, Gilbert Harold
Miller, Donald Morton
Mirkin, Bernard Leo
Moawad, Atef H
Moon, Byong Hoon
Morgan, Juliet
Morris, Ralph William
Murad, Ferid
Musa, Mahmoud Nimir
Nair, Velayudhan
Nakajima, Shigehiro
Napier, T(avye) Celeste
Narahashi, Toshio
Neff-Davis, Carol Ann
Northup, Sharon Joan
Nutting, Ehard Forrest
Opgenorth, Terry John
Pai, Sadanand V
Papaioannou, Stamatios E
Perkins, William Eldredge
Peterson, Rudolph Nicholas
Plotnikoff, Nicholas Peter
Prabhu, Venkatray G
Proudfit, Carol Marie
Proudfit, Herbert K(err)
Quock, Raymond Mark

Radzialowski, Frederick M
Rao, Gopal Subba
Rasenick, Mark M
Reith, Maarten E A
Retz, Konrad Charles
Ringham, Gary Lewis
Rotenberg, Keith Saul
Sabelli, Hector C
Salafsky, Bernard P
Sanner, John Harper
Schreider, Bruce David
Schyve, Paul Milton
Seiden, Lewis S
Seidenfeld, Jerome
Sennello, Lawrence Thomas
Sherrod, Theodore Roosevelt
Siegel, Ivens Aaron
Singh, Mohinder
Smith, Steven Joel
Smith, Theodore Craig
Somani, Satu M
Spring, Bonnie Joan
Steffek, Anthony J
Steger, Richard Warren
Stein, Herman H
Stern, Paula Helene
Struthers, Barbara Joan Oft
Suarez, Kenneth Alfred
Sukowski, Ernest John
Swinnen, Lode J
Taylor, Julius David
Ten Eick, Robert Edwin
Thompson, Emmanuel Bandele
Thomson, John Ferguson
Tourlentes, Thomas Theodore
Ts'o, Timothy On-To
Van de Kar, Louis David
Vazquez, Alfredo Jorge
Wencel-Drake, June D
Wilensky, Jacob T
Williams, Michael
Wu, Chau Hsiung
Yeates, Donovan B
Yeung, Tin-Chuen
Zaroslinski, John F

INDIANA
Anderson, Robert Clarke
Babbs, Charles Frederick
Berger, James Edward
Besch, Henry Roland, Jr
Borowitz, Joseph Leo
Bosron, William F
Bowers, Larry Donald
Brater, D(onald) Craig
Bruns, Robert Frederick
Bryant, Henry Uhlman
Bumpus, John Arthur
Carlson, Gary P
Carmichael, Ralph Harry
Chio, E(ddie) Hang
Christ, Daryl Dean
Clark, William C
Cohen, Marlene Lois
Coppoc, Gordon Lloyd
Dantzig, Anne H
Diller, Erold Ray
DiMicco, Joseph Anthony
Doedens, David James
Donner, David Bruce
Dube, Gregory P
Dungan, Kendrick Webb
Echtenkamp, Stephen Frederick
Emmerson, John Lynn
Evans, Michael Allen
Fayle, Harlan Downing
Fleisch, Jerome Herbert
Forney, Robert Burns
Franklin, Ronald
Fuller, Ray W
Gehlert, Donald Richard
Gehring, Perry James
Gibson, James Edwin
Gibson, William Raymond
Goh, Edward Hua Seng
Goldstein, David Joel
Greenspan, Kalman
Griffing, William James
Hahn, Richard Allen
Harned, Ben King
Hayes, Donald Charles
Henry, David P, II
Ho, Peter Peck Koh
Hynes, Martin Dennis, III
Isom, Gary E
Kauffman, Raymond F
Kim, Kil Chol
Klaunig, James E
Laddu, Atul R
Leander, John David
Leeling, Jerry L
Lemberger, Louis
Lin, Tsung-Min
Lindstrom, Terry Donald
Macchia, Donald Dean
McKinney, Gordon R
McLaughlin, Jerry Loren
McNay, John Leeper
Maickel, Roger Philip
Mannix, Edward T
Marshall, Franklin Nick
Mason, Norman Ronald
Matsumoto, Charles
Moore, Emily Allyn

Pharmacology (cont)

Nelson, David Lloyd
Nelson, Robert Leon
Nichols, David Earl
Nickander, Rodney Carl
Odya, Charles E
Parli, C(arol) John
Pollock, G Donald
Queener, Sherry Fream
Rao, Suryanarayana K
Rebec, George Vincent
Richter, Judith Anne
Ridolfo, Anthony Sylvester
Roach, Peter John
Root, Mary Avery
Rubin, Alan
Rutledge, Charles O
Saz, Howard Jay
Schulze, Gene Edward
Seidehamel, Richard Joseph
Shea, Philip Joseph
Silbaugh, Steven A
Singhal, Radhey Lal
Sisson, George Maynard
Smith, Gerald Floyd
Spolyar, Louis William
Spratto, George R
Stark, Paul
Steinberg, Mitchell I
Sullivan, Hugh R, Jr
Sullivan, John Lawrence
Thompson, Wilmer Leigh
Thor, Karl Bruce
Van Tyle, William Kent
Vasko, Michael Richard
Vodicnik, Mary Jo
Wagle, Shreepad R
Weber, George
Weikel, John Henry, Jr
Willis, Lynn Roger
Wolen, Robert Lawrence
Wright, Walter Eugene
Yim, George Kwock Wah
Zabik, Joseph

IOWA
Ahrens, Franklin Alfred
Ascoli, Mario
Baron, Jeffrey
Benton, Byrl E
Bhatnagar, Ranbir Krishna
Bidlack, Wayne Ross
Conn, P Michael
Dutton, Gary Roger
Dyer, Donald Chester
Ghoneim, Mohamed Mansour
Granberg, Charles Boyd
Hsu, Walter Haw
Husted, Russell Forest
Johnson, Wallace W
Jones, Leslie F
Kasik, John Edward
Kunesh, Jerry Paul
Long, John Paul
Makar, Adeeb Bassili
Nair, Vasu
Niebyl, Jennifer Robinson
Orcutt, James A
Rosazza, John N
Russi, Gary Dean
Shires, Thomas Kay
Spratt, James Leo
Steele, William John
Tephly, Thomas R
Teppert, William Allan, Sr
Williamson, Harold E

KANSAS
Alper, Richard H
Browne, Ronald K
Bunag, Ruben David
Carlson, Arthur, Jr
Cheng, Chia-Chung
Davis, John Stewart
Dixon, Walter Reginald
Doull, John
Dujovne, Carlos A
Eisenbrandt, Leslie Lee
Enna, Salvatore Joseph
Faiman, Morris David
Gordon, Michael Andrew
Greene, Nathan Doyle
Kadoum, Ahmed Mohamed
Klaassen, Curtis Dean
Lanman, Robert Charles
Michaelis, Mary Louise
Nelson, Stanley Reid
Norton, Stata Elaine
Parkinson, Andrew
Pazdernik, Thomas Lowell
Poisner, Alan Mark
Ringle, David Allan
Slavik, Milan
Spohn, Herbert Emil
Takemoto, Dolores Jean
Tessel, Richard Earl
Upson, Dan W
Uyeki, Edwin M
Walaszek, Edward Joseph

KENTUCKY
Anderson, Richard L
Aronoff, George R

Benz, Frederick W
Carney, John Michael
Carr, Laurence A
Chen, Theresa S
Dagirmanjian, Rose
Diedrich, Donald Frank
Edmonds, Harvey Lee, Jr
Elkes, Joel
Flesher, James Wendell
Foster, Thomas Scott
Gordon, Helmut Albert
Gupta, Ramesh C
Heinicke, Ralph Martin
Huang, Kee-Chang
Hurst, Harrell Emerson
Jarboe, Charles Harry
Knoefel, Peter Klerner
Kulkarni, Prasad Shrikrishna
Lang, Calvin Allen
Lubawy, William Charles
Luckens, Mark Manfred
McGraw, Charles Patrick
McNicholas, Laura T
Martin, William Robert
Massik, Michael
Matheny, James Lafayette
Miller, Frederick N
Myers, Steven Richard
Nerland, Donald Eugene
Petering, Harold George
Rees, Earl Douglas
Rodriguez, Lorraine Ditzler
Rowell, Peter Putnam
Scharff, Thomas G
Schurr, Avital
Slevin, John Thomas
Tobin, Thomas
Volp, Robert Francis
Vore, Mary Edith
Waddell, William Joseph
Waite, Leonard Charles
Williams, Walter Michael
Yokel, Robert Allen
Zimmerman, Thom J

LOUISIANA
Agrawal, Krishna Chandra
Barker, Louis Allen
Beckman, Barbara Stuckey
Bobbin, Richard Peter
Boertje, Stanley
Bourn, William M
Bradley, Ronald James
Brown, Richard Don
Carter, Mary Kathleen
Chin, Hong Woo
Clarkson, Craig William
Domer, Floyd Ray
Dunn, Adrian John
Ferguson, Gary Gene
Fisher, James W
Frohlich, Edward David
Garcia, Meredith Mason
Geiger, Paul Frank
George, William Jacob
Grace, Marcellus
Greenberg, Stanley Shimen
Guth, Paul Spencer
Hastings, Robert Clyde
Hernandez, Thomas
Jenkins, William L
Judd, Robert Lee
Kadowitz, Phillip J
Komiskey, Harold Louis
Lertora, Juan J L
McNamara, Dennis B
Manno, Barbara Reynolds
Manno, Joseph Eugene
Mehendale, Harihara Mahadeva
Morgan, Lee Roy, Jr
Morrissette, Maurice Corlette
Olson, Richard David
Olubadewo, Joseph Olanrewaju
Pearson, James Eldon
Redetzki, Helmut M
Rice, David A
Robie, Norman William
Roerig, Sandra Charlene
Short, Charles Robert
Snow, Lloyd Dale
Stewart, John Joseph
Toscano, William Agostino, Jr
Venugopalan, Changaram
Wilson, John T
Wood, Charles Donald
Zavecz, James Henry

MAINE
Bird, Joseph G
Fassett, David Walter
Major, Charles Walter

MARYLAND
Adams, John George
Adamson, Richard H
Adler, Michael
Ahluwalia, Gurpreet S
Aisner, Joseph
Akiskal, Hagop S
Albuquerque, Edson Xavier
Alderdice, Marc Taylor
Alleva, Frederic Remo
Alley, Michael C
Alvares, Alvito Peter

Anderson, Arthur O
Ansher, Sherry Singer
Aronow, Lewis
Asghar, Khursheed
Axelrod, Julius
Bachur, Nicholas R, Sr
Baker, Houston Richard
Balazs, Tibor
Bareis, Donna Lynn
Barrack, Evelyn Ruth
Baskin, Steven Ivan
Batson, David Banks
Beall, James Robert
Bean, Barbara Louise
Beaven, Michael Anthony
Blake, David Andrew
Blaszkowski, Thomas P
Blomster, Ralph N
Blumberg, Peter Mitchell
Blumenthal, Herbert
Bogdanski, Donald Frank
Braude, Monique Colsenet
Bronaugh, Robert Lewis
Brookes, Neville
Bunger, Rolf
Burt, David Reed
Canfield, Craig Jennings
Caplan, Yale Howard
Carr, Charles Jelleff
Carrico, Christine Kathryn
Chaudhari, Anshumali
Chernow, Bart
Chisolm, James Julian, Jr
Chiueh, Chuang Chin
Choi, Oksoon Hong
Choudary, Jasti Bhaskararao
Chuang, De-Maw
Coffey, Donald Straley
Cohen, Jack Sidney
Cohen, Robert Martin
Colombani, Paul Michael
Commarato, Michael A
Commissiong, John Wesley
Cone, Edward Jackson
Contrera, Joseph Fabian
Cosmides, George James
Cott, Jerry Mason
Cox, Brian Martyn
Cox, George Warren
Creveling, Cyrus Robbins, Jr
Cueto, Cipriano, Jr
Dacre, Jack Craven
Davidian, Nancy McConnell
De Luca, Luigi Maria
Dingell, James V
Dionne, Raymond A
Dutta, Samarendra N
Eckardt, Michael Jon
Egorin, Merril Jon
Eldefrawi, Amira T
Eldefrawi, Mohyee E
Epstein, Stephen Edward
Fakunding, John Leonard
Farber, Theodore Myles
Finch, Robert Allen
Fitzgerald, Glenna Gibbs (Cady)
Fowler, Bruce Andrew
Friedman, Fred K
Fuchs, James Claiborne Allred
Gabay, Sabit
Gatti, Philip John
Gelboin, Harry Victor
Geller, Irving
Gillette, James Robert
Glazer, Robert Irwin
Glowa, John Robert
Gold, Philip William
Goldberg, Alan Marvin
Goldberg, Steven R
Goldstein, David Stanley
Goodwin, Frederick King
Gram, Theodore Edward
Gray, Allan P
Grieshaber, Charles K
Griffiths, Roland Redmond
Grunberg, Neil Everett
Gueriguian, John Leo
Haigler, Henry James, Sr
Hanbauer, Ingeborg
Hanig, Joseph Peter
Hardy, Lester B
Harmon, Alan Dale
Harwood, Clare Theresa
Hassett, Charles Clifford
Hattan, David Gene
Heinrich, Max Alfred, Jr
Helke, Cinda Jane
Herman, Eugene H
Hickey, Robert Joseph
Hienz, Robert Douglas
Hirshman, Carol A
Hla, Timothy Tun
Hoffmann, Conrad Edmund
Holaday, John W
Horakova, Zdenka
Hunt, Walter Andrew
Husain, Syed
Ichniowski, Casimir Thaddeus
Irving, George Washington, Jr
Irwin, Richard Leslie
Jackson, Benjamin A
Jacobowitz, David
Jacobson, Keith Hazen
Jagadeesh, Gowra G

Jasinski, Donald Robert
Jasper, Robert Lawrence
Johanson, Chris Ellyn
Johns, David Garrett
Johnson, David Norseen
Joshi, Sewa Ram
Kafka, Marian Stern
Kaiser, Joseph Anthony
Kalser, Sarah Chinn
Kandasamy, Sathasiva B
Kapetanovic, Izet Michael
Kaplan, Stanley A
Karel, Leonard
Kaufman, Joyce J
Kawalek, Joseph Casimir, Jr
Keefer, Larry Kay
Kelsey, Frances Oldham
Khan, Mushtaq Ahmad
Khazan, Naim
Kiang, Juliann G
Kincaid, Randall L
Kinnard, William J, Jr
Kinsella, James L
Kitzes, George
Kohn, Kurt William
Kohn, Leonard David
Kopin, Irwin J
Koslow, Stephen Hugh
Kramer, Barnett Sheldon
Kramer, Stanley Phillip
Kramer, William Geoffrey
Kuhar, Michael Joseph
Kunos, George
Kupferberg, Harvey J
Lake, Charles Raymond
Lathers, Claire M
Lazar-Wesley, Eliane M
Leaders, Floyd Edwin, Jr
Lee, I P
Lee, Lyndon Edmund, Jr
Lee-Ham, Doo Young
Lerner, Pauline
Levitan, Herbert
Levy, Alan C
Lietman, Paul Stanley
Lin, Shin
Lipicky, Raymond John
Litterst, Charles Lawrence
Liu, Ching-Tong
Liu, Yung-Pin
London, Edythe D
Love, Raymond Charles
Lowensohn, Howard Stanley
Ludden, Thomas Marcellus
MacCanon, Donald Moore
McCurdy, John Dennis
MacDonald, William E, Jr
Macri, Frank John
Magnani, John Louis
Mainigi, Kumar D
Majchrowicz, Edward
Manen, Carol-Ann
Marcus, Richard
Markey, Sanford Philip
Martin, George Reilly
Marwah, Joe
Marzulli, Francis Nicholas
Massari, V John
Middlebrook, John Leslie
Milman, Harry Abraham
Moreton, Julian Edward
Morton, Joseph James Pandozzi
Mouradian, M Maral
Mulinos, Michael George
Munn, John Irvin
Munson, Paul Lewis
Musselman, Nelson Page
Myers, Charles
Nemec, Josef
Nuite-Belleville, Jo Ann
O'Leary, John Francis
Oliverio, Vincent Thomas
Orahovats, Peter Dimiter
Osawa, Yoichi
Osterberg, Robert Edward
Owens, Albert Henry, Jr
Pace, Lee Crandall
Park, Lee Crandall
Parkhie, Mukund Raghunathrao
Parvez, Zaheer
Pellmar, Terry C
Pert, Candace B
Petrella, Vance John
Petrucelli, Lawrence Michael
Pickworth, Wallace Bruce
Pilch, Susan Marie
Pinkus, Lawrence Mark
Porter, Roger J
Post, Robert M
Powers, Marcelina Venus
Prasanna, Hullahalli Rangaswamy
Reed, Warren Douglas
Rehak, Matthew Joseph
Reilly, Joseph F
Resnick, Charles A
Rice, Kenner Cralle
Robinson, Cecil Howard
Robinson-White, Audrey Jean
Rogawski, Michael Andrew
Rosen, Gerald M
Rosenstein, Laurence S
Rubin, Martin Israel
Rubin, Robert Jay
Rudo, Frieda Galindo

Russell, James T
Saavedra, Juan M
Safer, Brian
Salem, Harry
Sancilio, Lawrence F
Sarvey, John Michael
Sass, Neil Leslie
Sastre, Antonio
Scheibel, Leonard William
Schleimer, Robert P
Schoenfeld, Ronald Irwin
Schuster, Charles Roberts, Jr
Schwartz, Joan Poyner
Schwartz, Paul
Scott, Charles Covert
Seamon, Kenneth Bruce
Seifried, Harold Edwin
Sharma, Gopal Chandra
Shellenberger, Thomas E
Sherman, John Foord
Shih, Tsung-Ming Anthony
Sidell, Frederick R
Sieber-Fabro, Susan M
Silbergeld, Ellen K
Sinha, Birandra Kumar
Skolnick, Phil
Smith, Douglas Lee
Smith, Jerry Morgan
Smith, Ralph Grafton
Snipes, Charles Andrew
Snyder, Solomon H
Sonawane, Babasaheb R
Song, Byoung-Joon
Steinberg, George Milton
Steranka, Larry Richard
Sung, Cynthia
Suomi, Stephen John
Szara, Stephen Istvan
Talalay, Paul
Tamminga, Carol Ann
Teske, Richard H
Thut, Paul Douglas
Titus, Elwood Owen
Tocus, Edward C
Umberger, Ernest Joy
Undem, Bradley J
Valdes, James John
Valley, Sharon Louise
Vaupel, Donald Bruce
Velletri, Paul A
Venditti, John M
Venter, J Craig
Vistica, David T
Viswanathan, C T
Vocci, Frank Joseph
Walker, Michael Dirck
Walters, Judith R
Waravdekar, Vaman Shivram
Warnick, Jordan Edward
Watkins, Clyde Andrew
Watzman, Nathan
Weeks, Maurice Harold
Weight, Forrest F
Weimer, John Thomas
Weiner, Myron
Welch, Arnold D(eMerritt)
Whitehurst, Virgil Edwards
Williams, Roger Lea
Wirth, Peter James
Wolff, Frederick William
Wolters, Robert John
Wyatt, Richard Jed
Wykes, Arthur Albert
Yaffe, Sumner J
Yang, Shen Kwei
Yeh, Shu-Yuan
Yen-Koo, Helen C
Yergey, Alfred L, III
Yu, Mei-ying Wong
Zaharko, Daniel Samuel
Zatz, Martin
Zimmerman, Hyman Joseph

MASSACHUSETTS
Adams, Jack H(erbert)
Alper, Milton H
Alston, Theodore A
Anderson, Julius Horne, Jr
Assaykeen, Tatiana Anna
Baldessarini, Ross John
Benjamin, David Marshall
Berkowitz, Barry Alan
Bird, Stephanie J(ean)
Blumberg, Harold
Boisse, Norman Robert
Brazier, Mary A B
Brown, Neal Curtis
Cameron, John Stanley
Carlson, Kristin Rowe
Castellot, John J, Jr
Castronovo, Frank Paul, Jr
Chang, Joseph Yoon
Chase, Arleen Ruth
Chodosh, Sanford
Cornman, Ivor
D'Amato, Henry Edward
Dershwitz, Mark
Dews, Peter Booth
Doolittle, Charles Herbert, III
Driedger, Paul Edwin
Ellingboe, James
Essigmann, John Martin
Fielding, Stuart
Gambill, John Douglas

Gamzu, Elkan R
Gaudette, Leo Eward
Goldberg, Irvin H(yman)
Goldmacher, Victor S
Goldman, Peter
Gorelick, Kenneth J
Gourzis, James Theophile
Greenblatt, David J
Griffith, Robert W
Gusovsky, Fabian
Hadidian, Zareh
Hadjian, Richard Albert
Hallock, Gilbert Vinton
Hartmann, Ernest Louis
Hauser, George
Hayes, Andrew Wallace, II
Hershenson, Benjamin R
Hodge, Harold Carpenter
Hollenberg, Norman Kenneth
Iuliucci, John Domenic
Jenkins, Howard Jones
Johnson, Elsie Ernest
Joyce, Nancy C
Kebabian, John Willis
Kelleher, Roger Thomson
Koch-Weser, Jan
Kornetsky, Conan
Kosersky, Donald Saadia
Kupfer, David
Kupferman, Allan
Lasagna, Louis (Cesare)
Latz, Arje
Levine, Ruth R
Levy, Deborah Louise
Lloyd, Weldon S
Lorenzo, Antonio V
Ludlum, David Blodgett
McKearney, James William
McNair, Douglas McIntosh
Madras, Bertha K
Maher, Timothy John
Maran, Janice Wengerd
Marquis, Judith Kathleen
Masek, Bruce James
Mautner, Henry George
Miczek, Klaus A
Miller, Keith Wyatt
Miller, Tracy Bertram
Mirabelli, Christopher Kevin
Modest, Edward Julian
Morgan, James Philip
Morgan, Kathleen Greive
Morse, William Herbert
Moskowitz, Michael Arthur
Muni, Indu A
Osgood, Patricia Fisher
Pelikan, Edward Warren
Pober, Zalmon
Resnick, Oscar
Riggi, Stephen Joseph
Rosenkrantz, Harris
Ruelius, Hans Winfried
Ryser, Hugues Jean-Paul
Schaub, Robert George
Schildkraut, Joseph Jacob
Schneider, Frederick Howard
Shader, Richard Irwin
Shakarjian, Michael Peter
Shuster, Louis
Signer, Ethan Royal
Smith, Emil Richard
Smith, Wendy Anne
Snyder, Benjamin Willard
Stoeckler, Johanna D
Szabo, Sandor
Tashjian, Armen H, Jr
Tomera, John F
Travis, David M
Volicer, Ladislav
Waud, Barbara E
Waud, Douglas Russell
Wells, Herbert
Wenger, Christian Bruce
Woodman, Peter William
Wurtman, Richard Jay
Yesair, David Wayne

MICHIGAN
Abel, Ernest Lawrence
Aggarwal, Surinder K
Aiken, James Wavell
Anderson, Gordon Frederick
Anderson, Thomas Edward
Aston, Roy
Bach, Michael Klaus
Bailey, Harold Edwards
Braselton, Webb Emmett, Jr
Braughler, John Mark
Braun, Werner Heinz
Briggs, Josephine P
Brody, Theodore Meyer
Buchanan, Robert Alexander
Bus, James Stanley
Cheney, B Vernon
Cochran, Kenneth William
Cohen, Sanford Ned
Collins, Robert James
Conover, James H
Cooper, Theodore
Counsell, Raymond Ernest
Dambach, George Ernest
Data, Joann L
Deuben, Roger R
Domino, Edward Felix

Drach, John Charles
DuCharme, Donald Walter
Dutta, Saradindu
Elferink, Cornelius Johan
Ellis, C N
Fischer, Lawrence J
Frade, Peter Daniel
Freedman, Robert Russell
Freeman, Arthur Scott
Galloway, Matthew Peter
Gebber, Gerard L
Gilbertson, Michael
Gilman, Martin Robert
Glazko, Anthony Joachim
Goldenthal, Edwin Ira
Goldman, Harold
Goodman, Jay Irwin
Guttendorf, Robert John
Hajratwala, Bhupendra R
Hall, Edward Dallas
Heffner, Thomas G
Herzig, David Jacob
Hollenberg, Paul Frederick
Hollingworth, Robert Michael
Hudson, Roy Davage
Hult, Richard Lee
Johnson, Garland A
Johnson, Herbert Gardner
Johnston, Raymond F
Kaplan, Harvey Robert
Kauffman, Ralph Ezra
Kramer, Sherman Francis
Krishan, Awtar
Krueger, Robert John
La Du, Bert Nichols, Jr
Lahti, Robert A
Leopold, Wilbur Richard, III
Lindblad, William John
Lucchesi, Benedict Robert
McCarty, Leslie Paul
McGovren, James Patrick
McKenna, Michael Joseph
Marcelo, Cynthia Luz
Marcoux, Frank W
Marks, Bernard Herman
Marrazzi, Mary Ann
Maxwell, Donald Robert
Medzihradsky, Fedor
Meeks, Robert G
Mehring, Jeffrey Scott
Mitchell, Jerry R
Mohrland, J Scott
Moore, Kenneth Edwin
Mostafapour, M Kazem
Neubig, Richard Robert
Njus, David Lars
Novak, Raymond Francis
Pals, Donald Theodore
Philipsen, Judith Lynne Johnson
Piercey, Montford F
Piper, Walter Nelson
Poschel, Bruno Paul Henry
Puro, Donald George
Rapundalo, Stephen T
Rech, Richard Howard
Reinke, David Albert
Reitz, Richard Henry
Roslinski, Lawrence Michael
Roth, Robert Andrew, Jr
Ruwart, Mary Jean
Schaefer, Gerald J
Schoener, Eugene Paul
Schreur, Peggy Jo Korty Dobry
Schrier, Denis J
Scicli, Alfonso Guillermo
Sebolt-Leopold, Judith S
Sethy, Vimala Hiralal
Shepard, Robert Stanley
Shipman, Charles, Jr
Siddiqui, Waheed Hasan
Slywka, Gerald William Alexander
Smith, Charles Bruce
Smith, Robert James
Smith, Thomas Charles
Spilman, Charles Hadley
Strand, James Cameron
Straw, Robert Niccolls
Swain, Henry Huntington
Tang, Andrew H
Theiss, Jeffrey Charles
Thornburg, John Elmer
Ueda, Tetsufumi
Uhde, Thomas Whitley
Uhler, Michael David
Van Dyke, Russell Austin
Vatsis, Kostas Petros
VonVoigtlander, Philip Friedrich
Vostal, Jaroslav Joseph
Vrbanac, John James
Wagner, John Garnet
Wardell, William Michael
Watanabe, Philip Glen
Webb, R Clinton
Weber, Wendell W
Weeks, James Robert
Welling, Peter Geoffrey
Wetherell, Herbert Ranson, Jr
Wilks, John William
Winbury, Martin M
Yamazaki, Russell Kazuo
Zabik, Matthew John
Zannoni, Vincent G
Zins, Gerald Raymond

MINNESOTA
Abraham, Robert Thomas
Abuzzahab, Faruk S, Sr
Ames, Matthew Martin
Beitz, Alvin James
Brimijoin, William Stephen
Broderius, Steven James
Brown, David Robert
Chesney, Charles Frederic
Cloyd, James C, III
Conard, Gordon Joseph
Cutkomp, Laurence Kremer
Dai, Xue Zheng (Charlie)
Dysken, Maurice William
Eisenberg, Richard Martin
Elde, Robert Philip
Francis, Gary Stuart
From, Arthur Harvey Leigh
Gray, Grace Warner
Hanson, Robert C
Hapke, Bern
Holtzman, Jordan L
Housmans, Philippe Robert H P
Katusic, Zvonimir
Knych, Edward Thomas
Kvam, Donald Clarence
Lee, Nancy Zee-Nee Ma
Leon, Arthur Sol
Lewis, Debra A
Loh, Horace H
Mabry, Paul Davis
Mannering, Gilbert James
Messing, Rita Bailey
Miller, Jack W
Miller, Richard Lynn
Murray, John Randolph
Nelson, Robert A
Ober, Robert Elwood
O'Dea, Robert Francis
Quebbemann, Aloysius John
Rao, Gundu Hirisave Rama
Romero, Juan Carlos
Roy, Robert Russell
Sawchuk, Ronald John
Seybold, Virginia Susan (Dick)
Shoeman, Don Walter
Sladek, Norman Elmer
Sparber, Sheldon B
Staba, Emil John
Stowe, Clarence M
Swingle, Karl F
Szurszewski, Joseph Henry
Takemori, Akira Eddie
Tedeschi, David Henry
Tempero, Kenneth Floyd
Trachte, George J
Weaver, Lawrence Clayton
Weinshilboum, Richard Merle
Wilcox, George Latimer
Zaske, Darwin Erhard
Zimmerman, Ben George
Zimmerman, Cheryl Lea

MISSISSIPPI
Alford, Geary Simmons
Chaires, Jonathan Bradford
Chambers, Howard Wayne
Davis, Wilbur Marvin
Dorough, H Wyman
ElSohly, Mahmoud Ahmed
Farley, Jerry Michael
Halaris, Angelos
Hoskins, Beth
Hume, Arthur Scott
Mercer, Henry Dwight
Pace, Henry Buford
Rockhold, Robin William
Verlangieri, Anthony Joseph
Wahba, Albert J
Walker, Larry
Waters, Irving Wade
Wilson, Marvin Cracraft
Woolverton, William L

MISSOURI
Adams, Henry Richard
Adams, Max David
Ahmed, Mahmoud S
Ahmed, Nahed K
Badr, Mostafa
Barnes, Byron Ashwood
Beinfeld, Margery Cohen
Bettonville, Paul John
Blaine, Edward Homer
Blake, Robert L
Boime, Irving
Burton, Robert Main
Byington, Keith H
Cenedella, Richard J
Chapnick, Barry M
Chiappinelli, Vincent A
Cicero, Theodore James
Coret, Irving Allen
Covey, Douglas Floyd
Ferrendelli, James Anthony
Forte, Leonard Ralph
Gold, Alvin Hirsh
Green, Theodore James
Green, Vernon Albert
Griffin, Claude Lane
Hix, Elliott Lee
Howlett, Allyn C
Hunter, Francis Edmund, Jr
Ireland, Gordon Alexander

Pharmacology (cont)

Jensen, Clyde B
Johnson, Eugene Malcolm, Jr
Johnson, Richard Dean
Kim, Yee Sik
Levinskas, George Joseph
Lonigro, Andrew Joseph
Lowry, Oliver Howe
McDougal, David Blean, Jr
Miller, Lowell D
Naeger, Leonard L
Needleman, Philip
Nickols, G Allen
Pandya, Krishnakant Hariprasad
Piepho, Robert Walter
Pike, Linda Joy
Podrebarac, Eugene George
Rolf, Clyde Norman
Russell, Robert Lee
St Omer, Vincent Victor
Shukla, Shivendra Dutt
Slatopolsky, Eduardo
Smith, Ronald Gene
Thomsen, Robert Harold
Tsai, Tsui Hsien
Turner, John T
Tuttle, Warren Wilson
Van Deripe, Donald R
Walkenbach, Ronald Joseph
Westfall, Thomas Creed
Woodhouse, Edward John
Worthington, Ronald Edward
Wosilait, Walter Daniel
Yarbro, John Williamson
Zee-Cheng, Robert Kwang-Yuen
Zenser, Terry Vernon
Zepp, Edwin Andrew

MONTANA
Bryan, Gordon Henry
Catalfomo, Philip
Johnson, Howard Ernest
Parker, Keith Krom
Van Horne, Robert Loren

NEBRASKA
Abel, Peter William
Berndt, William O
Brody, Alfred Walter
Brown, David G
Bylund, David Bruce
Corbett, Michael Dennis
Crampton, James Mylan
Deupree, Jean Durley
Dowd, Frank, Jr
Ebadi, Manuchair
Elder, John Thompson, Jr
Engel, Toby Ross
Gessert, Carl F
Grandjean, Carter Jules
Heath, Eugene Cartmill
Hexum, Terry Donald
Khan, Manzoor M
Lau, Yuen-Sum
Lockridge, Oksana Maslivec
McClurg, James Edward
Magee, Donal Francis
Murrin, Leonard Charles
Prioreschi, Plinio
Roche, Edward Browning
Ruddon, Raymond Walter, Jr
Scholar, Eric M
Scholes, Norman W
Shaw, David Harold

NEVADA
Bierkamper, George G
Buxton, Iain Laurie Offord
Ferguson, Roger K
Kelly, Raymond Crain
Pardini, Ronald Shields
Pool, Peter Edward
Potter, Gilbert David
Scholler, Jean
Westfall, David Patrick

NEW HAMPSHIRE
Dayton, Peter Gustav
Gosselin, Robert Edmond
Kensler, Charles Joseph
McCarthy, Lawrence E
Maudsley, David V
Mudge, Gilbert Horton
Pape, Brian Eugene
Smith, Roger Powell

NEW JERSEY
Abdel-Rahman, Mohamed Shawky
Abrutyn, Donald
Ahn, Ho-Sam
Amory, David William
Antonaccio, Michael John
Asaad, Magdi Mikhaeil
Aviado, Domingo Mariano
Babcock, George, Jr
Babcock, Philip Arnold
Bagdon, Robert Edward
Baker, Thomas
Balwierczak, Joseph Leonard
Banga, Harjit Singh
Barnett, Allen
Bauer, Victor John
Bautz, Gordon T
Bennett, Debra A
Beyer-Mears, Annette
Bianchi, John J
Blaiklock, Robert George
Boksay, Istvan Janos Endre
Borella, Luis Enrique
Breckenridge, Bruce (McLain)
Brezenoff, Henry Evans
Brodie, David Alan
Brogle, Richard Charles
Brostrom, Charles Otto
Brostrom, Margaret Ann
Bullock, Francis Jeremiah
Byrne, Jeffrey Edward
Cafruny, Edward Joseph
Capetola, Robert Joseph
Carlson, Richard P
Caruso, Frank San Carlo
Cayen, Mitchell Ness
Cevallos, William Hernan
Chaikin, Philip
Chand, Naresh
Chatterjee, Meeta
Chau, Thuy Thanh
Cheng, Kang
Chiu, Peter Jiunn-Shyong
Chovan, James Peter
Clay, George A
Coffin, Vicki L
Condouris, George Anthony
Conn, Hadley Lewis, Jr
Conney, Allan Howard
Conway, Paul Gary
Cotty, Val Francis
Coutinho, Claude Bernard
Crawley, Lantz Stephen
Creese, Ian N
Cressman, William Arthur
Crouthamel, William Guy
Curro, Frederick A
Cushman, David Wayne
Dain, Jeremy George
Dashman, Theodore
Davidson, Arnold B
Davies, Richard O
Davis, Harry R, Jr
Day-Salvatore, Debra-Lynn
Demarest, Keith Thomas
Dennis, Susana R K de
Desjardins, Raoul
Diamantis, William
DiFazio, Louis T
Dunikoski, Leonard Karol, Jr
Easterday, Otho Dunreath
Eckhardt, Eileen Theresa
Ellis, Daniel B(enson)
Emele, Jane Frances
Eng, Robert H K
Fawzi, Ahmad B
Fernandes, Prabhavathi Bhat
Fiedler-Nagy, Christa
Flynn, Edward Joseph
Foley, James E
Fong, Tung Ming
Frantz, Beryl May
Free, Charles Alfred
Freund, Matthew J
Frost, David
Fujimore, Tohru
Garcia, Maria Luisa
Gaut, Zane Noel
Geczik, Ronald Joseph
Geller, Herbert M
Gertner, Sheldon Bernard
Ghai, Geetha R
Gilfillan, Alasdair Mitchell
Glaser, Keith Brian
Gogerty, John Harry
Gold, Barry Ira
Goldemberg, Robert Lewis
Goldenberg, Marvin M
Goldstein, Ann L
Gomoll, Allen W
Goodman, Frank R
Gorodetzky, Charles W
Greenslade, Forrest C
Grimes, David
Grosso, John A
Grover, Gary James
Haft, Jacob I
Hageman, William E
Hagmann, William Kirk
Hahn, DoWon
Handley, Dean A
Hanna, Samir A
Harris, Don Navarro
Harrison-Johnson, Yvonne E
Harun, Joseph Stanley
Hassert, G(eorge) Lee, Jr
Heilman, William Paul
Hey, John Anthony
Hite, Mark
Hoffman, Clark Samuel, Jr
Horovitz, Zola Phillip
Howland, Richard David
Hubbard, John W
Huger, Francis P
Hutcheon, Duncan Elliot
Hyndman, Arnold Gene
Jacobson, Martin Michael
Ji, Sungchul
Jirkovsky, Ivo
Johnson, Melvin Clark
Kamm, Jerome J
Kapoor, Inder Prakash
Katz, Laurence Barry
Kauffman, Frederick C
King, John A(lbert)
Kirsten, Edward Bruce
Kissel, John Walter
Knoppers, Antonie Theodoor
Kocsis, James Joseph
Kongsamut, Sathapana
Konikoff, John Jacob
Kreutner, William
Kripalani, Kishin J
Kung, Ted Teshih
Lage, Gary Lee
Lan, Shih-Jung
Lanzoni, Vincent
Lassman, Howard B
Laurencot, Henry Jules
Leeds, Morton W
Leinweber, Franz Josef
Leitz, Frederick Henry
LeSher, Dean Allen
Letizia, Gabriel Joseph
Leung, Albert Yuk-Sing
Levenstein, Irving
Levin, Wayne
Lewis, Alan James
Lewis, Steven Craig
Lin, Chin-Chung
Lin, Reng-Lang
Lin, Yi-Jong
Ling, Alfred Soy Chou
Liu, Leroy Fong
Lowndes, Herbert Edward
Lukacsko, Alison B
Lukas, George
McGuire, John L
McIlreath, Fred J
Mackerer, Carl Robert
McKinstry, Doris Naomi
Mackway-Girardi, Ann Marie
Mandel, Lewis Richard
Martin, Lawrence Leo
Massa, Tobias
Mathur, Pershottam Prasad
Mayhew, Eric George
Mehta, Atul Mansukhbhai
Meyers, Kenneth Purcell
Mezey, Kalman C
Mezick, James Andrew
Mitala, Joseph Jerrold
Moreland, Suzanne
Morris, N Ronald
Moyer, John Allen
Murthy, Vadiraja Venkatesa
Muth, Eric Anthony
Nemecek, Georgina Marie
Nicolau, Gabriela
Nies, Alan Sheffer
Novick, William Joseph, Jr
O'Byrne, Elizabeth Milikin
O'Neill, Patrick Joseph
Orzechowski, Raymond Frank
Osborne, Melville
Oshiro, George
Pace, Daniel Gannon
Pachter, Jonathan Alan
Palmer, Keith Henry
Panagides, John
Park, Eunkyue
Paterniti, James R, Jr
Perhach, James Lawrence
Piala, Joseph J
Pitts, Barry James Roger
Pittz, Eugene P
Plummer, Albert J
Pobiner, Bonnie Fay
Pohorecky, Larissa Alexandra
Polinsky, Ronald John
Presky, David H
Prince, Herbert N
Puri, Surendra Kumar
Quinn, Gertrude Patricia
Rabii, Jamshid
Rashidbaigi, Abbas
Reiser, H Joseph
Ress, Rudyard Joseph
Ritchie, David Malcolm
Robson, Ronald D
Rodda, Bruce Edward
Rosenberg, Alberto
Rosenblum, Irwin Yale
Rothstein, Edwin C(arl)
Rubin, Bernard
Rudzik, Allan D
Saini, Ravinder Kumar
Salvador, Richard Anthony
Sanders, Marilyn Magdanz
Sapru, Hreday N
Saunders, Robert Norman
Schoepke, Hollis George
Schwartz, Morton Allen
Scott, Mary Celine
Sepinwall, Jerry
Sequeira, Joel August Louis
Shapiro, Stanley Seymour
Shellenberger, Carl H
Singer, Arnold J
Singhvi, Sampat Manakchand
Sisenwine, Samuel Fred
Snyder, Robert
Sofia, R Duane
Solomon, Thomas Allan
Soyka, Lester F
Spector, Reynold
Spoerlein, Marie Teresa
Stambaugh, John Edgar, Jr
Stefko, Paul Lowell
Steiner, Kurt Edric
Stockton, John Richard
Sugerman, Abraham Arthur
Sullivan, John William
Swenson, Christine Erica
Symchowicz, Samson
Szewczak, Mark Russell
Tabachnick, Irving I A
Thomas, Kenneth Alfred, Jr
Thompson, George Rex
Tierney, William John
Tocco, Dominick Joseph
Tolman, Edward Laurie
Tozzi, Salvatore
Traina, Vincent Michael
Triscari, Joseph
Tse, Francis Lai-Sing
Vanden Heuvel, William John Adrian, III
Vander Wende, Christina
Vane, Floie Marie
Vargas, Hugo Martin
Verebey, Karl G
Vogel, W Mark
Von Hagen, D Stanley
Vukovich, Robert Anthony
Waggoner, William Charles
Wagner, William Edward, Jr
Walker, Mary W
Wang, Hsueh-Hwa
Wang, Shih Chun
Wasserman, Martin Allan
Waugh, Margaret H
Weichman, Barry Michael
Weinstein, Stephen Henry
Weiss, George B
Weiss, Harvey Richard
Welton, Ann Frances
Whiteley, Phyllis Ellen
Wiggins, Jay Ross
Wolff, Donald John
Wu, Wen-hsien
Wulf, Ronald James

NEW MEXICO
Buss, William Charles
Coffey, John Joseph
Corcoran, George Bartlett, III
Crawford, Michael Howard
Damon, Edward G(eorge)
Garst, John Eric
Groblewski, Gerald Eugene
Hadley, William Melvin
Hobbs, Charles Henry
Hurwitz, Leon
Jones, Vernon Douglas
Leach, John Kline
Mitchell, Clifford L
North-Root, Helen May
Petersen, Donald Francis
Priola, Donald Victor
Rao, Nutakki Gouri Sankara
Rhodes, Buck Austin
Savage, Daniel Dexter
Uhlenhuth, Eberhard Henry
Vorherr, Helmuth Wilhelm
Wax, Joan

NEW YORK
Acara, Margaret A
Aiello, Edward Lawrence
Alexander, George Jay
Almon, Richard Reiling
Altar, Charles Anthony
Altszuler, Norman
Altura, Bella T
Altura, Burton Myron
Ambrus, Julian Lawrence
Anders, Marion Walter
Andersen, Kenneth J
Andrews, John Parray
Antzelevitch, Charles
Aronson, Ronald Stephen
Babish, John George
Back, Nathan
Bagchi, Sakti Prasad
Baird, Malcolm Barry
Baker, Robert George
Balis, Moses Earl
Bard, Enzo
Bardos, Thomas Joseph
Barkai, Amiram I
Barletta, Michael Anthony
Barnerjee, Shailesh Prasad
Barrett, James E
Bartus, Raymond T
Bauman, Norman
Beinfield, William Harvey
Belford, Julius
Bell, Duncan Hadley
Berger, Frank Milan
Bernacki, Ralph J
Bernheimer, Alan Weyl
Bickerman, Hylan A
Bidlack, Jean Marie
Bigger, J Thomas, Jr
Bini, Alessandra Margherita
Bito, Laszlo Z
Blanck, Thomas Joseph John
Bloch, Alexander

MEDICAL & HEALTH SCIENCES / 483

Blosser, James Carlisle
Bodnar, Richard Julius
Borch, Richard Frederic
Borgstedt, Harold Heinrich
Brooks, Robert R
Brown, Oliver Monroe
Butler, Vincent Paul, Jr
Cabrera, Edelberto Jose
Carr, Edward Albert, Jr
Castellion, Alan William
Cervoni, Peter
Chakrin, Lawrence William
Chan, Arthur Wing Kay
Chan, Peter Sinchun
Chan, W(ah) Y(ip)
Chisholm, David R
Chou, Dorthy T C T
Chou, Ting-Chao
Chryssanthou, Chryssanthos
Ciancio, Sebastian Gene
Clarkson, Allen Boykin, Jr
Clarkson, Thomas William
Claus, Thomas Harrison
Conway, Walter Donald
Cory-Slechta, Deborah Ann
Costa, Max
Costello, Christopher Hollet
Cosulich, Donna Bernice
Counts, David Francis
Coupet, Joseph
Cox, Robert H
Crain, Stanley M
Crandall, David L
Creaven, Patrick Joseph
Crew, Malcolm Charles
Crouch, Billy G
Curry, Stephen H
Darzynkiewicz, Zbigniew Dzierzykraj
Davidow, Bernard
Davis, James N(orman)
Davis, Kenneth Leon
Day, Ivana Podvalova
Demis, Dermot Joseph
DiStefano, Victor
Dole, Vincent Paul
Dolnick, Bruce J
Domantay, Antonio G
Drakontides, Anna Barbara
Drayer, Dennis Edward
Duta, Purabi
Eckhardt, Laurel Ann
Ellis, Keith Osborne
Evans, Hugh Lloyd
Fahrenbach, Marvin Jay
Ferrari, Richard Alan
Ferris, Steven Howard
Fiala, Emerich Silvio
Fink, Max
Finster, Mieczyslaw
Foley, Dennis Joseph
Foster, Kenneth William
Frishman, William Howard
Furchgott, Robert Francis
Gaver, Robert Calvin
Gessner, Peter K
Gessner, Teresa
Giering, John Edgar
Gillies, Alastair J
Gintautas, Jonas
Glassman, Jerome Martin
Glick, Stanley Dennis
Goldfarb, Joseph
Gray, William David
Greco, William Robert
Green, Jack Peter
Greenberg, Jacob
Greenberger, Lee M
Greene, Lloyd A
Greengard, Olga
Greizerstein, Hebe Beatriz
Gringauz, Alex
Grob, David
Grollman, Arthur Patrick
Guideri, Giancarlo
Gumbiner, Barry M
Guttenplan, Joseph B
Hakala, Maire Tellervo
Halevy, Simon
Halpryn, Bruce
Hamilton, Byron Bruce
Harris, Alex L
Hershey, Linda Ann
Hiller, Jacob Moses
Holland, Mary Jean Carey
Horwitz, Susan Band
Houde, Raymond Wilfred
Hough, Lindsay B
Hrushesky, William John Michael
Hudecki, Michael Stephen
Hulme, Norman Arthur
Hutner, Seymour Herbert
Iden, Charles R
Inchiosa, Mario Anthony, Jr
Ingalls, James Warren, Jr
Inturrisi, Charles E
Iodice, Arthur Alfonso
Ip, Margot Morris
Iyengar, Ravi Srinivas V
Jaanus, Siret Desiree
Jacquet, Yasuko F
Jean-Baptiste, Emile
Kahn, Norman
Kalman, Thomas Ivan
Kalsner, Stanley

Kao, Chien Yuan
Kappas, Attallah
Karl, Peter I
Kashin, Philip
Kass, Robert S
Katocs, Andrew Stephen, Jr
Keefe, Deborah Lynn
Kestenbaum, Richard Steven
Kinsolving, C Richard
Klein, Donald Franklin
Klinge, Carolyn M
Klingman, Gerda Isolde
Knaak, James Bruce
Kohler, Constance Anne
Koutcher, Jason Arthur
Kraft, Patricia Lynn
Kraus, Shirley Ruth
Kreis, Willi
Kristal, Mark Bennett
Kuntzman, Ronald Grover
Kuperman, Albert Sanford
Lai, Fong M
Lapidus, Herbert
Laties, Victor Gregory
Latimer, Clinton Narath
Lau-Cam, Cesar A
Laychock, Suzanne Gale
Leach, Leonard Joseph
Lehr, David
Lein, Pamela J
Levi, Roberto
Levine, Robert Alan
Levine, Walter (Gerald)
Levitt, Barrie
Levy, Gerhard
Liang, Chang-seng
Lifshitz, Kenneth
Loullis, Costas Christou
Lynch, Vincent De Paul
McCabe, R Tyler
McGiff, John C
McGuire, John Joseph
McIsaac, Robert James
McNamara, Thomas Francis
Madissoo, Harry
Makman, Maynard Harlan
Malbon, Craig Curtis
Malitz, Sidney
Mallov, Samuel
Margolis, Renee Kleimann
Marois, Robert Leo
Megirian, Robert
Meyers, Marian Bennett
Mihich, Enrico
Miller, Richard Kermit
Miller, Russell Loyd, Jr
Minatoya, Hiroaki
Misra, Anand Lal
Mittag, Thomas Waldemar
Modell, Walter
Moorehead, Thomas J
Morgan, John P
Morris, Marilyn Emily
Morrison, John Agnew
Morrow, Paul Edward
Munigle, Jo Anne
Nair, Ramachandran Mukundalayam Sivarama
Nakatsugawa, Tsutomu
Namba, Tatsuji
Nasjletti, Alberto
Neenan, John Patrick
Neidle, Enid Anne
Neu, Harold Conrad
Neubort, Shimon
Oberdorster, Gunter
Okamoto, Michiko
Oser, Bernard Levussove
Pagala, Murali Krishna
Paladino, Joseph Anthony
Pantuck, Eugene Joel
Parham, Margaret Payne
Pasternak, Gavril William
Pearl, William
Pentyala, Srinivas Narasimha
Pine, Martin J
Pliskin, William Aaron
Pollock, John Joseph
Pong, Schwe Fang
Pool, William Robert
Prell, George D
Reidenberg, Marcus Milton
Reilly, Margaret Anne
Reit, Barry
Reynard, Alan Mark
Rifkind, Arleen B
Riker, Walter Franklyn, Jr
Rivera-Calimlim, Leonor
Rizack, Martin A
Roboz, John
Rocci, Mario Louis, Jr
Roth, Jerome Allan
Rothman, Francoise
Rubin, Ronald Philip
Sachdev, Om Prakash
Said, Sami I
Sauro, Marie D
Schachter, E Neil
Schayer, Richard William
Schiff, Joel D
Schneider, Allan Stanford
Schwark, Wayne Stanley
Shamoian, Charles Anthony
Shargel, Leon David

Sharp, Richard Lee
Siever, Larry Joseph
Silver, Paul J
Sirotnak, Francis Michael
Skulan, Thomas William
Sloboda, Adolph Edward
Slocum, Harry Kim
Smith, Cedric Martin
Smith, Frank Ackroyd
Soderlund, David Matthew
Sohn, David
Spoor, Ryk Peter
Squires, Richard Felt
Staple, Peter Hugh
Starks, Fred
Stoewsand, Gilbert Saari
Stoll, William Russell
Sufrin, Janice Richman
Susca, Louis Anthony
Swamy, Vijay Chinnaswamy
Szeto, Hazel Hon
Terhaar, Clarence James
Titeler, Milt
Torosian, George
Traitor, Charles Eugene
Triggle, David J
Triner, Lubos
Tuttle, Richard Suneson
Van der Hoeven, Theo A
Van Poznak, Alan
Ventura, William Paul
Violante, Michael Robert
Vulliemoz, Yvonne
Wadgaonkar, Dilip Bhanudas
Wajda, Isabel
Warner, William
Weiner, Irwin M
Weinstein, Harel
Weisburger, John Hans
West, Norman Reed
Whitaker-Azmitia, Patricia Mack
Wilk, Sherwin
Williams, Curtis Alvin, Jr
Williams, David L
Winter, Jerrold Clyne, Sr
Wit, Andrew Lewis
Wolf, P S
Wolin, Michael Stuart
Wong, Lan Kan
Wong, Patrick Yui-Kwong
Yoburn, Byron Crocker
York, James Lester
Zakim, David
Zedeck, Morris Samuel
Zukin, Stephen R

NORTH CAROLINA
Abou-Donia, Mohamed Bahie
Alphin, Reevis Stancil
Aronson, A L
Barnes, Donald Wesley
Barnett, John William
Bentley, Peter John
Bernheim, Frederick
Birnbaum, Linda Silber
Bishop, Jack Belmont
Bond, James Anthony
Booth, Raymond George
Bradley, William Robinson
Brauer, Ralph Werner
Burford, Hugh Jonathan
Butterworth, Byron Edwin
Camp, Haney Bolon
Campbell, Robert Joseph
Campbell, William Howard
Capizzi, Robert L
Carr, Fred K
Carroll, Bernard James
Carroll, Robert Graham
Chhabra, Rajendra S
Chignell, Colin Francis
Childers, Steven Roger
Conners, Carmen Keith
Costa, Daniel Louis
Dar, Mohammad Saeed
Dauterman, Walter Carl
DaVanzo, John Paul
De Miranda, Paulo
DiAugustine, Richard Patrick
Dibner, Mark Douglas
Dren, Anthony Thomas
Dudley, Kenneth Harrison
Dunlap, Charles Evans, III
Dykstra, Linda A
Ehmann, Carl William
Eldridge, John Charles
Eling, Thomas Edward
Elion, Gertrude Belle
Ellinwood, Everett Hews, Jr
Ellis, Fred Wilson
Engle, Thomas William
Faber, James Edward
Fouts, James Ralph
Fyfe, James Arthur
Gaffney, Thomas Edward
Garrett, Ruby Joyce Burriss
Gatzy, John T, Jr
Ghanayem, Burhan I
Glenn, Thomas M
Goldstein, Joyce Allene
Goz, Barry
Gralla, Edward Joseph
Gray, Timothy Kenney
Harper, Curtis

Hart, Larry Glen
Hastings, Felton Leo
Hatch, Gary Ephraim
Heck, Henry d'Arcy
Heck, Jonathan Daniel
Hicks, Robert Eugene
Hicks, T Philip
Hirsch, Philip Francis
Hodgson, Ernest
Hoffman, Maureane Richardson
Hong, Jau-Shyong
Hook, Gary Edward Raumati
Howard, James Lawrence
Hueter, Francis Gordon
Ingenito, Alphonse J
Janowsky, David Steffan
Jones(Stover), Betsy
Jurgelski, William, Jr
Kaufman, William
Kaufmann, John Simpson
Kern, Frank Howard
Kirksey, Donny Frank
Kirshner, Norman
Knapp, William Arnold, Jr
Koster, Rudolf
Krasny, Harvey Charles
Lazarus, Lawrence H
Lefkowitz, Robert Joseph
Lewis, Mark Henry
Little, Patrick Joseph
Little, William C
McBay, Arthur John
McClellan, Roger Orville
McConnaughey, Mona M
McCutcheon, Rob Stewart
McLachlan, John Alan
Mailman, Richard Bernard
Matthews, Hazel Benton, Jr
Maxwell, Robert Arthur
Mennear, John Hartley
Mills, Elliott
Miya, Tom Saburo
Modrak, John Bruce
Moorefield, Herbert Hughes
Moser, Virginia Clayton
Mueller, Robert Arthur
Murdock, Harold Russell
Myers, Robert Durant
Nadler, J Victor
Namm, Donald H
Nelson, Donald J
Nichol, Charles Adam
O'Callaghan, James Patrick
Olsen, Elise Arline
Ottolenghi, Athos
Padilla, George M
Page, Rodney Lee
Paluch, Edward Peter
Parikh, Indu
Peng, Tai-Chan
Poole, Doris Theodore
Posner, Herbert S
Price, Kenneth Elbert
Pritchard, John B
Privalle, Christopher Thomas
Putney, James Wiley, Jr
Quinn, Richard Paul
Ramp, Warren Kibby
Reel, Jerry Royce
Reinhard, John Frederick, Jr
Rickert, Douglas Edward
Riviere, Jim Edmond
Rolofson, George Lawrence
Schanberg, Saul M
Schroeder, David Henry
Schwetz, Bernard Anthony
Siemens, Albert John
Slotkin, Theodore Alan
Somjen, George G
Spaulding, Theodore Charles
Spector, Thomas
Steelman, Sanford Lewis
Stern, Warren C
Stevens, James T
Stiles, Gary L
Stolk, Jon Martin
Strandhoy, Jack W
Stumpf, Walter Erich
Tadepalli, Anjaneyulu S
Thakker, Dhiren R
Thurman, Ronald Glenn
Tilson, Hugh Arval
Toverud, Svein Utheim
Van Stee, Ethard Wendel
Weiden, Mathias Herman Joseph
Welch, Richard Martin
White, Helen Lyng
Winter, Irwin Clinton
Witt, Peter Nikolaus
Wolgemuth, Richard Lee
Wooles, Wallace Ralph
Zimmerman, Thomas Paul
Zung, William Wen-Kwai

NORTH DAKOTA
Ary, Thomas Edward
Auyong, Theodore Koon-Hook
DeBoer, Benjamin
Hein, David William
Khalil, Shoukry Khalil Wahba
Messiha, Fathy S
Tanner, Noall Stevan

Pharmacology (cont)

OHIO
Abdallah, Abdulmuniem Husein
Agamanolis, Dimitris
Akbar, Huzoor
Anton, Aaron Harold
Askari, Amir
Awad, Albert T
Bachmann, Kenneth Allen
Banning, Jon Willroth
Beal, Jack Lewis
Berman, Elizabeth Faith
Bhattacharya, Amar Nath
Bianchine, Joseph Raymond
Bloomfield, Saul
Blumer, Jeffrey L
Boggs, Robert Wayne
Bohrman, Jeffrey Stephen
Breglia, Rudolph John
Bryant, Shirley Hills
Buehler, Edwin Vernon
Bunde, Carl Albert
Burkman, Allan Maurice
Chen, Chi-Po
Cheng, Hsien
Corson, Samuel Abraham
Couri, Daniel
Crutcher, Keith A
Dage, Richard Cyrus
D'Ambrosio, Steven M
Denine, Elliot Paul
Dinerstein, Robert Joseph
Dodd, Darol Ennis
Doskotch, Raymond Walter
Feller, Dennis R
Fondacaro, Joseph D
Gardier, Robert Woodward
Geha, Alexander Salim
Geho, Walter Blair
Giannini, A James
Goldstein, Sidney
Gossel, Thomas Alvin
Grupp, Gunter
Grupp, Ingrid L
Gwinn, John Frederick
Hagerman, Larry M
Hall, Madeline Molnar
Hammond, Paul B
Hayton, William Leroy
Heesch, Cheryl Miller
Hille, Kenneth R
Hinkle, George Henry
Hollander, Philip B
Holtkamp, Dorsey Emil
Hoss, Wayne Paul
Hudak, William John
Hurwitz, David Allan
Imondi, Anthony Rocco
Johnson, Carl Lynn
Johnston, John O'Neal
Khairallah, Philip Asad
Kimura, Kazuo Kay
Koerker, Robert Leland
Krimmer, Edward Charles
Krontiris-Litowitz, Johanna Kaye
Kuhn, William Lloyd
Kunin, Calvin Murry
Lang, James Frederick
Levin, Jerome Allen
Lima, John J
Lindower, John Oliver
Lutton, John Kazuo
McGoron, Anthony Joseph
Maguire, Michael Ernest
Mamrack, Mark Donovan
Mellgren, Ronald Lee
Messer, William Sherwood, Jr
Michael, William R
Michaelson, I Arthur
Mieyal, John Joseph
Millard, Ronald Wesley
Muir, William W, III
Mukhtar, Hasan
Nahata, Milap Chand
Nasrallah, Henry A
Nigrovic, Vladimir
Nolen, Granville Abraham
O'Dell, Thomas Beniah
Patil, Popat N
Pereira, Michael Alan
Powers, Thomas E
Rahwan, Ralf George
Ray, Richard Schell
Recknagel, Richard Otto
Reuning, Richard Henry
Riker, Donald Kay
Ritschel, Wolfgang Adolf
Robinson, Michael K
Rosenberg, Howard C
Saffran, Murray
Schechter, Martin David
Schinagl, Erich F
Schlender, Keith K
Schradie, Joseph
Schroeder, Friedhelm
Schwartz, Arnold
Shemano, Irving
Sherman, Gerald Philip
Shertzer, Howard Grant
Sigell, Leonard
Sjoerdsma, Albert
Skau, Kenneth Anthony
Skelly, Michael Francis
Smith, Dennison A
Somani, Pitambar
Sperelakis, Nicholas
Standaert, Frank George
Staubus, Alfred Elsworth
Stokinger, Herbert Ellsworth
Tedeschi, Ralph Earl
Tejwani, Gopi Assudomal
Tepperman, Katherine Gail
Tjioe, Sarah Archambault
Tomei, L David
Truitt, Edward Byrd, Jr
Ursillo, Richard Carmen
Van Maanen, Evert Florus
Vorhees, Charles V
Wallin, Richard Franklin
Webster, Leslie T, Jr
Wilkerson, Robert Douglas
Wilkin, Jonathan Keith
Wolfe, Seth August, Jr
Wong, Laurence Cheng Kong
Woodward, James Kenneth
Wright, George Joseph
Zigmond, Richard Eric
Zimmerman, Ernest Frederick

OKLAHOMA
Brenner, George Marvin
Bunce, Richard Alan
Burrows, George Edward
Cassidy, Brandt George
Christensen, Howard Dix
Corder, Clinton Nicholas
Crane, Charles Russell
Czerwinski, Anthony William
Ganesan, Devaki
Gass, George Hiram
Harder, Harold Cecil
Hornbrook, K Roger
Johnson, Gerald, III
Jones, Sharon Lynn
Koss, Michael Campbell
Maton, Paul Nicholas
Meier, Charles Frederick, Jr
Melton, Carlton E, Jr
Moore, Joanne Iweita
Pento, Joseph Thomas
Prabhu, Vilas Anandrao
Reinke, Lester Allen
Revzin, Alvin Morton
Rikans, Lora Elizabeth
Ringer, David P
Robinson, Casey Perry
Rolf, Lester Leo, Jr
Whitsett, Thomas L

OREGON
Andresen, Michael Christian
Bartos, Frantisek
Belknap, John Kenneth
Bennett, William M
Brummett, Robert E
Buhler, Donald Raymond
Chandler, David B
Cowan, Frederick Fletcher, Jr
Dickinson, John Otis
Fink, Gregory Burnell
Gabourel, John Dustan
Gallaher, Edward J
Gardner, Sara A
Hedtke, James Lee
Julien, Robert M
Keenan, Edward James
Kennedy, Margaret Wiener
Larson, Robert Elof
Levine, Leonard
McCawley, Elton Leeman
Murray, Thomas Francis
North, R Alan
Olsen, George Duane
Riker, William Kay
Roberts, Michael Foster
Russell, Nancy Jeanne
Ruwe, William David
Schneider, Phillip William, Jr
Smith, John Robert
Swan, Kenneth Carl
Tanz, Ralph
Weber, Lavern J

PENNSYLVANIA
Abou-Gharbia, Magid
Actor, Paul
Adler, Martin W
Agersborg, Helmer Pareli Kjerschow, Jr
Alexander, Samuel Craighead, Jr
Armstead, William M
Aronson, Carl Edward
Aston-Jones, Gary Stephen
Baer, John Elson
Bagdon, Walter Joseph
Bailey, Denis Mahlon
Baker, Alan Paul
Baker, Walter Wolf
Bannon, James Andrew
Barone, Frank Carmen
Barry, Herbert, III
Bayne, Gilbert M
Bergey, James L
Berlin, Cheston Milton
Bex, Frederick James
Beyer, Karl Henry, Jr
Bianchi, Carmine Paul
Billingsley, Melvin Lee
Block, Lawrence Howard
Boshart, Charles Ralph
Boxenbaum, Harold George
Branch, Robert Anthony
Brooks, David Patrick
Butt, Tauseef Rashid
Calesnick, Benjamin
Carrano, Richard Alfred
Cerreta, Kenneth Vincent
Chan, Ping-Kwong
Chang, Raymond S L
Chernick, Warren Sanford
Chiang, Soong Tao
Clineschmidt, Bradley Van
Comis, Robert Leo
Connamacher, Robert Henle
Connor, John D
Constantinides, Panayiotis Pericleous
Contreras, Patricia Cristina
Coon, Julius Mosher
Cooper, Stephen Allen
Cordova, Vincent Frank
Corey, Sharon Eva
Cowan, Alan
Creasey, William Alfred
Dalton, Colin
Daniels-Severs, Anne Elizabeth
Davignon, J Paul
Davis, Hugh Michael
De Groat, William C, Jr
DeHaven-Hudkins, Diane Louise
Dichter, Marc Allen
Di Gregorio, Guerino John
DiPalma, Joseph Rupert
DiPasquale, Gene
Dixit, Balwant N
Donnelly, Thomas Edward, Jr
Dubinsky, Barry
Duggan, Daniel Edward
Edwards, David J
Eichelman, Burr S, Jr
Ellison, Theodore
Engleman, Karl
Ertel, Robert James
Fanelli, George Marion, Jr
Ferko, Andrew Paul
Fletcher, Jeffrey Edward
Fogleman, Ralph William
Fong, Kei-Lai Lau
Frazer, Alan
Freeman, Alan R
Friedman, Eitan
Fromm, Gerhard Hermann
Furgiuele, Angelo Ralph
Gabriel, Karl Leonard
Gardocki, Joseph F
Gautieri, Ronald Francis
Gero, Alexander
Gershon, Samuel
Gibson, Raymond Edward
Gilman, Steven Christopher
Glauser, Elinor Mikelberg
Glauser, Stanley Charles
Goldberg, Paula Bursztyn
Goldstein, Frederick J
Gottlieb, Karen Ann
Gould, Robert James
Greene, Frank Eugene
Greenstein, Jeffrey Ian
Grego, Nicholas John
Grindel, Joseph Michael
Gross, Dennis Michael
Gussin, Robert Z
Harakal, Concetta
Harvey, John Adriance
Haugaard, Niels
Hava, Milos
Hemwall, Edwin Lyman
Henderson, George Richard
Hess, Marilyn E
Hieble, J Paul
Hitner, Henry William
Holcslaw, Terry Lee
Hook, Jerry B
Hopper, Sarah Priestly
Hucker, Howard B(enjamin)
Jackson, Edwin Kerry
Jacob, Leonard Steven
Jacoby, Henry I
Janssen, Frank Walter
Jenney, Elizabeth Holden
Jim, Kam Fook
Kade, Charles Frederick, Jr
Kaji, Hideko (Katayama)
Kallelis, Theodore S
Kaplita, Paul V
Kelliher, Gerald James
Kilpatrick, Laurie E
Kinter, Lewis Boardman
Kitzen, Jan Michael
Kivlighn, Salah Dean
Knowles, John Appleton, III
Koelle, George Brampton
Koelle, Winifred Angenent
Kopia, Gregory A
Krantz, Charles Parry
Kramer, Stanley Zachary
Krause, Stephen Myron
Krenzelok, Edward Paul
Kuhnert, Betty R
Lakoski, Joan Marie
Lambertsen, Christian James
Lazarides, Elias
Lazo, John Stephen
Lerner, Leonard Joseph
Levin, Robert Martin
Levin, Sidney Seamore
Levin, Simon Eugene
Lewis, Michael Edward
Lloyd, Thomas A
Loeb, Alex Lewis
Lotlikar, Prabhakar Dattaram
Lotti, Victor J
Lynch, Joseph J, Jr
MacDonald, James Scott
McDonald, Robert H, Jr
McElligott, James George
McKenna, Edward J
McKenna, William Gillies
Mackowiak, Elaine DeCusatis
McManus, Edward Clayton
Mann, David Edwin, Jr
Mansmann, Herbert Charles, Jr
Martin, Gregory Emmett
Matthews, Richard John, Jr
Michelson, Eric L
Milakofsky, Louis
Minsker, David Harry
Mir, Ghulam Nabi
Molinoff, Perry Brown
Mulder, Kathleen M
Murray, Rodney Brent
Myers, Howard M
Nagle, Barbara Tomassone
Nambi, Ponnal
Nass, Margit M K
Oldham, James Warren
Overton, Donald A
Packman, Albert M
Packman, Elias Wolfe
Panasevich, Robert Edward
Papacostas, Charles Arthur
Pelleg, Amir
Pendleton, Robert Grubb
Perel, James Maurice
Piperno, Elliot
Pitt, Bruce R
Poste, George Henry
Price, Henry Locher
Puglia, Charles David
Raffa, Robert B
Rattan, Satish
Reitz, Allen Bernard
Rickels, Karl
Rieders, Fredric
Riley, Gene Alden
Ritter, Carl A
Roberts, Jay
Robishaw, Janet D
Rossi, George Victor
Rucinska, Ewa J
Ruffolo, Robert Richard, Jr
Ruggieri, Michael Raymond
Saelens, David Arthur
Salganicoff, Leon
Saporito, Michael S
Sawutz, David G
Schein, Philip Samuel
Schrogie, John Joseph
Seay, Patrick H
Severs, Walter Bruce
Sevy, Roger Warren
Shank, Richard Paul
Short, John Albert
Shriver, David A
Siegman, Marion Joyce
Silver, Melvin Joel
Smith, John Bryan
Smith, Roger Bruce
Smudski, James W
Snipes, Wallace Clayton
Sprince, Herbert
Stiller, Richard L
Storella, Robert J, Jr
Strom, Brian Leslie
Summy-Long, Joan Yvette
Sweet, Charles Samuel
Taber, Robert Irving
Tepper, Lloyd Barton
Tew, Kenneth David
Tobia, Alfonso Joseph
Torchiana, Mary Louise
Torphy, Theodore John
Triolo, Anthony J
Trippodo, Nick Charles
Tulp, Orien Lee
Van Arman, Clarence Gordon
Van Kammen, Daniel Paul
Van Rossum, George Donald Victor
Venkataramanan, Raman
Vesell, Elliot S
Vickers, Florence Foster
Vickers, Stanley
Vogel, Wolfgang Hellmut
Volle, Robert Leon
Vollmer, Regis Robert
Walkenstein, Sidney S
Walz, Donald Thomas
Wang, Allan Zu-Wu
Waterhouse, Barry D
Watkins, W David
Weinryb, Ira
Weinstock, Joseph
Wendt, Robert Leo
Werner, Gerhard
Williams, Jacinta B
Wilson, Glynn
Winek, Charles L

Wingard, Lemuel Bell, Jr
Winkler, James David
Wolstenholme, Wayne W
Yellin, Tobias O
Yu, Ruey Jiin
Zaleski, Jan F
Zemaitis, Michael Alan
Zigmond, Michael Jonathan

RHODE ISLAND
Agarwal, Kailash C
Calabresi, Paul
Cha, Sungman
Crabtree, Gerald Winston
Dawicki, Doloretta Diane
DeFanti, David R
DeFeo, John Joseph
Fischer, Glenn Albert
Flanagan, Thomas Raymond
Hawrot, Edward
Heath, Dwight B(raley)
Kaplan, Stephen Robert
Miech, Ralph Patrick
Oxenkrug, Gregory Faiva
Parks, Robert Emmett, Jr
Patrick, Robert L
Shaikh, Zahir Ahmad
Shimizu, Yuzuru
Stern, Leo
Youngken, Heber Wilkinson, Jr

SOUTH CAROLINA
Allen, Donald Orrie
Baggett, Billy
Bagwell, Ervin Eugene
Benmaman, Joseph David
Cheng, Kenneth Tat-Chiu
Daniell, Herman Burch
Darby, Thomas D
Davis, Craig Wilson
Freeman, John J
Gale, Glen Roy
Gillette, Paul Crawford
Golod, William Hersh
Halushka, Perry Victor
Knapp, Daniel Roger
Kosh, Joseph William
Lin, Tu
Lindenmayer, George Earl
Lipinski, Joseph Floyd
Margolius, Harry Stephen
Newman, Walter Hayes
Privitera, Philip Joseph
Richardson, James Albert
Stewart, William C
Wagle, Gilmour Lawrence
Walle, Thomas
Walter, Wilbert George
Wang, Huei-Hsiang Lisa
Wilson, Marlene Ann
Wilson, Steven Paul
Woodard, Geoffrey Dean Leroy
Yard, Allan Stanley

SOUTH DAKOTA
Dwivedi, Chandradhar
Gross, Guilford C
Hagen, Arthur Ainsworth
Hietbrink, Bernard Edward
Kauker, Michael Lajos
Martin, Douglas Stuart
Schramm, Mary Arthur
Yutrzenka, Gerald J
Zawada, Edward T, Jr
Zeigler, David Wayne

TENNESSEE
Anjaneyulu, P S R
Ban, Thomas Arthur
Bass, Allan Delmage
Beaty, Orren, III
Berman, M Lawrence
Berney, Stuart Alan
Biaggioni, Italo
Brent, Thomas Peter
Byron, Joseph Winston
Capdevila, H Jorge
Cardoso, Sergio Steiner
Chapman, John E
Cook, George A
Crowley, William Robert
Daigneault, Ernest Albert
Farrington, Joseph Kirby
Fields, James Perry
Fridland, Arnold
Friedman, Daniel Lester
Gehrs, Carl William
Gillespie, Elizabeth
Guengerich, Frederick Peter
Hak, Lawrence J
Hancock, John Charles
Hardman, Joel G
Heimberg, Murray
Hoover, Donald Barry
Houghton, Janet Anne
Kostrzewa, Richard Michael
Lasslo, Andrew
Lawrence, William Homer
Lawson, James W
Leffler, Charles William
Lothstein, Leonard
McColl, John Duncan
MacEwen, James Douglas
Manley, Emmett S

Mayer, Steven Edward
Miyamoto, Michael Dwight
Nash, Clinton Brooks
Neal, Robert A
North, William Charles
Oates, John Alexander
Patel, Tarun B
Petrie, William Marshall
Proctor, Kenneth Gordon
Ray, Oakley S
Reed, Peter William
Robert, Timothy Albert
Roberts, DeWayne
Roberts, Lyman Jackson
Robertson, David
Sanders-Bush, Elaine
Sastry, Bhamidipaty Venkata Rama
Schmidt, Dennis Earl
Schreiber, Eric Christian
Shockley, Dolores Cooper
Skoutakis, Vasilios A
Snapper, James Robert
Spector, Sydney
Sticht, Frank Davis
Straughn, Arthur Belknap
Sulser, Fridolin
Sweetman, Brian Jack
Van Eldik, Linda Jo
Weinstein, Ira
Wells, Jack Nulk
White, Richard Paul
Wilcox, Henry G
Wiley, Ronald Gordon
Wilkinson, Grant Robert
Wimalasena, Jay
Wojciechowski, Norbert Joseph
Wood, Alastair James Johnston
Woodbury, Robert A
Woods, Maribelle

TEXAS
Acosta, Daniel, Jr
Alkadhi, Karim A
Allen, Julius Cadden
Altshuler, Harold Leon
Anderson, Karl E
Backer, Ronald Charles
Barnes, George Edgar
Benjamin, Robert Stephen
Besch, Paige Keith
Bjorndahl, Jay Mark
Blair, Murray Reid, Jr
Blum, Kenneth
Bost, Robert Orion
Briggs, Arthur Harold
Brodwick, Malcolm Stephen
Bryant, Stephen G
Buckley, Joseph Paul
Burks, Thomas F
Burton, Karen Poliner
Busch, Harris
Cantrell, Elroy Taylor
Carroll, Paul Treat
Chen, Shih-Fong
Chiou, George Chung-Yih
Clark, Wesley Gleason
Clay, Michael M
Combs, Alan B
Cooper, Cary Wayne
Crass, Maurice Frederick, III
Crawford, Isaac Lyle
Csaky, Tihamer Zoltan
Dalterio, Susan Linda
Decker, Walter Johns
Deloach, John Rooker
Dilsaver, Steven Charles
Dorris, Roy Lee
Ducis, Ilze
Edds, George Tyson
Emmett-Oglesby, Michael Wayne
Entman, Mark Lawrence
Erickson, Carlton Kuehl
Eugere, Edward Joseph
Euler, Kenneth L
Fabre, Louis Fernand, Jr
Fischer, Susan Marie
Fisher, Seymour
Foster, Donald Myers
Fox, Donald A
Frye, Gerald Dalton
Gage, Tommy Wilton
Gallagher, Joel Peter
Garriott, James Clark
Geumei, Aida M
Gilman, Alfred G
Gothelf, Bernard
Hartsfield, Sandee Morris
Heavner, James E
Heidelbaugh, Norman Dale
Heilman, Richard Dean
Henderson, George I
Hester, Richard Kelly
Hill, Reba Michels
Hillman, Gilbert R
Hilton, James Gorton
Ho, Begonia Y
Ho, Beng Thong
Ho, Dah-Hsi
Hollister, Leo E
Holly, Frank Joseph
Horning, Marjorie G
Housholder, Glenn Etta (Tholen)
Huffman, Ronald Dean
Hughes, Maysie J H

Jadhav, A L
Jandhyala, Bhagavan S
Jauchem, James Robert
Jaweed, Mazher
Johnson, Alice Ruffin
Johnson, Kenneth Maurice, Jr
Kay, David Cyril
Keats, Arthur Stanley
Keeton, Kent T
Kehrer, James Paul
Kempen, Rene Richard
Kenny, Alexander Donovan
Keplinger, Moreno Lavon
Kerley, Troy Lamar
Kimball, Aubrey Pierce
King, Thomas Scott
Knutson, Victoria P
Krivoy, William Aaron
Kroeger, Donald Charles
Lal, Harbans
Lane, Montague
Langston, Jimmy Byrd
Latimer, George Webster, Jr
Lee, Shih-Shun
Leslie, Steven Wayne
Liehr, Joachim G
Lombardini, John Barry
Lorenzetti, Ole John
Loubser, Paul Gerhard
Malin, David Herbert
Margolin, Solomon B
Marks, Gerald A
Marshburn, Paul Bartow
Martinez, J Ricardo
Medina, Miguel Angel
Mitchell, Helen C
Montgomery, Edward Harry
Moore, Erin Colleen
Mukherjee, Amal
Mundy, Gregory Robert
Nechay, Bohdan Roman
Nelson, Edward Blake
Nelson, James Arly
Newman, Robert Alwin
O'Benar, John DeMarion
Oka, Kazuhiro
Park, Myung Kun
Perkins, John Phillip
Peterson, Steven Lloyd
Pirch, James Herman
Plunkett, William Kingsbury
Poulsen, Lawrence Leroy
Ramanujam, V M Sadagopa
Ramsey, Kathleen Sommer
Randerath, Kurt
Ranney, David Francis
Riffee, William Harvey
Robison, George Alan
Ro-Choi, Tae Suk
Rose, Kathleen Mary
Ross, Elliott M
Ryan, Charles F
Sallee, Verney Lee
Saunders, Jack Palmer
Saunders, Priscilla Prince
Schonbrunn, Agnes
Schoolar, Joseph Clayton
Schultz, Fred Henry, Jr
Sherry, Clifford Joseph
Shinnick-Gallagher, Patricia L
Slaga, Thomas Joseph
Smalley, Harry Edwin
Stacy, David Lowell
Stavinoha, William Bernard
Stein, Jerry Michael
Taylor, Samuel Edwin
Tenner, Thomas Edward, Jr
Thomas, John A
Ticku, Maharaj K
Traber, Daniel Lee
Veltman, James C
Vick, Robert Lore
Walton, Charles Anthony
Wauquier, Albert
Way, James Leong
Wayner, Matthew John
Weisbrodt, Norman William
Wells, Patrick Roland
Williams, Betty L
Wilson, Carol Maggart
Wilson, L Britt
Yaden, Senka Long
Yanni, John Michael
Yeoman, Lynn Chalmers
Yeung, Katherine Lu
Yorio, Thomas

UTAH
Bennett, Jesse Harland
Done, Alan Kimball
Fingl, Edward (George)
Franz, Donald Norbert
Galinsky, Raymond Ethan
Gibb, James Woolley
Goodman, Louis Sanford
Harvey, Stewart Clyde
Karler, Ralph
Kemp, John Wilmer
Lee, Steve S
Mason, Robert C
Nichols, William Kenneth
Reed, Donal J
Segelman, Alvin Burton
Sharma, Raghubir Prasad

Simmons, Daniel L
Swinyard, Ewart Ainslie
Turkanis, Stuart Allen
Walker, Edward Bell Mar
Wardell, Joe Russell, Jr
Withrow, Clarence Dean
Wolf, Harold Herbert
Woodbury, Dixon Miles

VERMONT
Bevan, John Acton
Brown, Theodore Gates, Jr
Eckhardt, Shohreh B
Friedman, Matthew Joel
Hacker, Miles Paul
Hughes, John Russell
Jaffe, Julian Joseph
McCormack, John Joseph, Jr
Reit, Ernest Marvin I
Rosenstein, Robert
Sinclair, Peter Robert
Tritton, Thomas Richard

VIRGINIA
Aceto, Mario Domenico
Allison, Trenton B
Atrakchi, Aisar Hasan
Balster, Robert L(ouis)
Barrett, Richard John
Batra, Karam Vir
Biber, Margaret Clare Boadle
Blanke, Robert Vernon
Bleiberg, Marvin Jay
Bowe, Robert Looby
Bowman, Edward Randolph
Brand, Eugene Dew
Brase, David Arthur
Brown, Elise Ann Brandenburger
Buday, Paul Vincent
Carchman, Richard A
Castle, Manford C
Chandler, Jerry LeRoy
Ching, Melvin Chung Hing
Collins, Galen Franklin
Cone, Clarence Donald, Jr
Dannenburg, Warren Nathaniel
DeLorenzo, Robert John
Dementi, Brian Armstead
DeSesso, John Michael
Dewey, William Leo
Doherty, John Douglas
Dvorchik, Barry Howard
Egle, John Lee, Jr
Elford, Howard Lee
Eyre, Peter
Fenner-Crisp, Penelope Ann
Fitzhugh, Oscar Garth
Franko, Bernard Vincent
Freer, Richard John
Freund, Jack
Gallardo-Carpentier, Adriana
Garrison, James C
Gewirtz, David A
Glassco, William Shoemaker
Gold, Paul Ernest
Gourley, Desmond Robert Hugh
Gregory, Arthur Robert
Hackett, John Taylor
Harris, Louis Selig
Hart, Jayne Thompson
Haynes, Robert Clark, Jr
Hendricks, Albert Cates
Krop, Stephen
Kuemmerle, Nancy Benton Stevens
Larner, Joseph
Larson, Paul Stanley
Linden, Joel Morris
Lipnick, Robert Louis
Lloyd, John Willie, III
McCuen, Robert William
McNerney, James Murtha
Martin, Billy Ray
Munson, Albert Enoch
Nelson, Neal Stanley
Olson, William Arthur
Owen, Fletcher Bailey, Jr
Patrick, Graham Abner
Peach, Michael Joe
Pfeiffer, Carl J
Proakis, Anthony George
Rall, Theodore William
Reno, Frederick Edmund
Robinson, Susan Estes
Rogol, Alan David
Rosecrans, John A
Rutter, Henry Alouis, Jr
Salley, John Jones, Jr
Sando, Julianne J
Silvette, Herbert
Sjogren, Robert W, Jr
Snow, Anne E
Sperling, Frederick
Stone, John Patrick
Talbot, Richard Burritt
Tardiff, Robert G
Taylor, Jean Marie
Thorner, Michael Oliver
Tucker, Anne Nichols
Turner, Carlton Edgar
Tuttle, Jeremy Ballou
Uwaydah, Ibrahim Musa
Villar-Palasi, Carlos
Vos, Bert John
Ward, John Wesley

Pharmacology (cont)

Wasserman, Albert J
Watts, Daniel Thomas
Weir, Robert James, Jr
West, Bob
West, Fred Ralph, Jr
Williams, Patricia Bell
Woo, Yin-tak
Woods, Lauren Albert

WASHINGTON
Aagaard, George Nelson
Baillie, Thomas Allan
Baksi, Samarendra Nath
Ballou, John Edgerton
Beavo, Joseph A
Biskupiak, Joseph E
Borchard, Ronald Eugene
Bull, Richard J
Camerman, Arthur
Catterall, William A
Churchill, Lynn
Crampton, George H
Dreisbach, Robert Hastings
Dunbar, Philip Gordon
Felton, Samuel Page
Gibaldi, Milo
Gibson, Melvin Roy
Hall, Nathan Albert
Halpern, Lawrence Mayer
Harding, Joseph Warren, Jr
Hawkins, Richard L
Holcenberg, John Stanley
Horita, Akira
Johnson, William Everett
Juchau, Mont Rawlings
Klavano, Paul Arthur
Krebs, Edwin Gerhard
Levy, Rene Hanania
Loomis, Ted Albert
Lybrand, Terry Paul
Martin, Arthur Wesley
Schiffrin, Milton Julius
Schulz, John C
Share, Norman N
Su, Judy Ya-Hwa Lin
Sullivan, Maurice Francis
Taylor, Eugene M
Varanasi, Usha
Vincenzi, Frank Foster
Way, Jon Leong
West, Theodore Clinton
Whelton, Bartlett David
Woods, James Sterrett
Yarbrough, George Gibbs
Youmans, William Barton

WEST VIRGINIA
Chambers, John William
Cochrane, Robert Lowe
Colasanti, Brenda Karen
Cole, Larry King
Craig, Charles Robert
Fleming, William Wright
Hudgins, Patricia Montague
Malanga, Carl Joseph
Mawhinney, Michael G
O'Connell, Frank Dennis
Rankin, Gary O'Neal
Robinson, Robert Leo
Smith, David Joseph
Stitzel, Robert Eli
Van Dyke, Knox

WISCONSIN
Allred, Lawrence Ervin (Joe)
Barboriak, Joseph Jan
Bass, Paul
Blackwell, Barry M
Bloom, Alan S
Boyeson, Michael George
Bryan, George Terrell
Burke, David H
Campbell, William Bryson
Dodson, Vernon N
Drewett, James Gerald
Drill, Victor Alexander
Evenson, Merle Armin
Fahien, Leonard A
Fahl, William Edwin
Fischer, Paul Hamilton
Folts, John D
Fujimoto, James Masao
Gallenberg, Loretta A
Goldberg, Alan Herbert
Griffith, Owen W
Gross, Garrett John
Haavik, Coryce Ozanne
Hake, Carl (Louis)
Hardman, Harold Francis
Heath, Timothy Douglas
Hetzler, Bruce Edward
Hillard, Cecilia Jane
Hofmann, Lorenz M
Hokin-Neaverson, Mabel
Hopwood, Larry Eugene
Jefcoate, Colin R
Kochan, Robert George
Lalley, Peter Michael
Lech, John James
Meacham, Roger Hening, Jr
Murthy, Vishnubhakta Shrinivas
Osterberg, Arnold Curtis

Perloff, William H(arry), Jr
Petering, David Harold
Poland, Alan P
Pruss, Thaddeus P
Roerig, David L
Rusy, Ben F
Stewart, Richard Donald
Terry, Leon Cass
Thibodeau, Gary A
Trautman, Jack Carl
Tseng, Leon L F
Uno, Hideo
Wang, Richard I H
Warltier, David Charles
Will, James Arthur

WYOMING
Ksir, Charles Joseph
Nelson, Robert B
Seese, William Shober

PUERTO RICO
Fernandez-Repollet, Emma D
Hupka, Arthur Lee
Kaye, Sidney
Orkand, Richard K
Ramirez-Ronda, Carlos Hector
Santos-Martinez, Jesus

ALBERTA
Baker, Glen Bryan
Basu, Tapan Kumar
Biggs, David Frederick
Cook, David Alastair
Dewhurst, William George
Frank, George Barry
Henderson, Ruth McClintock
Hollenberg, Morley Donald
Huston, Mervyn James
Jaques, Louis Barker
Locock, Robert A
MacCannell, Keith Leonard
Moore, Graham John
Paton, David Murray
Pittman, Quentin J
Reiffenstein, Rhoderic John
Rorstad, Otto Peder
Roth, Sheldon H
Scott, Gerald William
Severson, David Lester
Wolowyk, Michael Walter

BRITISH COLUMBIA
Bellward, Gail Dianne
Cottle, Merva Kathryn Warren
Dost, Frank Norman
Foulks, James Grigsby
Jenkins, Leonard Cecil
Law, Francis C P
Lederis, Karolis (Karl)
McNeill, John Hugh
Perks, Anthony Manning
Salari, Hassan
Sanders, Harvey David
Sutter, Morley Carman
Tweeddale, Martin George

MANITOBA
Bihler, Ivan
Bose, Deepak
Choy, Patrick C
Dhalla, Naranjan Singh
Glavin, Gary Bertrun
Greenway, Clive Victor
Innes, Ian Rome
Kim, Ryung-Soon (Song)
Klaverkamp, John Frederick
LaBella, Frank Sebastian
Lautt, Wilfred Wayne
Pinsky, Carl
Polimeni, Philip Iniziato
Rollo, Ian McIntosh
Sitar, Daniel Samuel
Thomson, Ashley Edwin
Vitti, Trieste Guido
Weisman, Harvey

NEW BRUNSWICK
Haya, Katsuji

NEWFOUNDLAND
Bieger, Detlef
Hillman, Elizabeth S
Hoekman, Theodore Bernard
Neuman, Richard Stephen
Snedden, Walter
Virgo, Bruce Barton

NOVA SCOTIA
Armour, John Andrew
Carruthers, Samuel George
Downie, John William
Ferrier, Gregory R
Fraser, Albert Donald
Howlett, Susan Ellen
McDonald, Terence Francis
McKenzie, Gerald Malcolm
Renton, Kenneth William
Semba, Kazue
Szerb, John Conrad
White, Thomas David

ONTARIO
Appel, Warren Curtis

Ashwin, James Guy
Bend, John Richard
Beninger, Richard J
Brien, James Frederick
Burba, John Vytautas
Buttar, Harpal Singh
Camerman, Norman
Casselman, Warren Gottlieb Bruce
Ciriello, John
Cook, Michael Arnold
Daniel, Edwin Embrey
Davis, Alan
Dyer, Allan Edwin
Endrenyi, Laszlo
Epand, Richard Mayer
Ferguson, Alastair Victor
Ferguson, James Kenneth Wallace
Feuer, George
Frederick, George Leonard
Garfinkel, Paul Earl
Goldenberg, Gerald J
Gowdey, Charles Willis
Graham, Richard Charles Burwell
Grant, Donald Lloyd
Gupta, Radhey Shyam
Hagen, Paul Beo
Hirst, Maurice
Hrdina, Pavel Dusan
Huang, Bing Shaun
Israel, Yedy
Jhamandas, Khem
Kacew, Sam
Kadar, Dezso
Kalant, Harold
Kalow, Werner
Kapur, Bhushan M
Karmazyn, Morris
Khanna, Jatinder Mohan
Khera, Kundan Singh
Kovacs, Bela A
Kuiper-Goodman, Tine
Laham, Souheil
Lee, David K H
Lee, Robert Maung Kyaw Win
Lee, Tyrone Yiu-Huen
Leenen, Frans H H
Lucis, Ojars Janis
MacLeod, Stuart Maxwell
Mahon, William A
Marks, Gerald Samuel
Mayer, Joel
Mazurkiewicz-Kwilecki, Irena Maria
Mellors, Alan
Mishra, Ram K
Nakatsu, Kanji
Ogilvie, Richard Ian
Orrego, Hector
Pace Asciak, Cecil
Pang, Kim-Ching Sandy
Philp, Richard Blain
Pope, Barbara L
Rangachari, Patangi Srinivasa Kumar
Renaud, Leo P
Roschlau, Walter Hans Ernest
Schimmer, Bernard Paul
Seeman, Philip
Sellers, Edward Moncrieff
Spoerel, Wolfgang Eberhart G
Stephenson, Norman Robert
Stiller, Calvin R
Sunahara, Fred Akira
Sweeney, George Douglas
Talesnik, Jaime
Thomas, Barry Holland
Trifaró, José María
Ulpian, Carla
Waithe, William Irwin
Weaver, Lynne C

PRINCE EDWARD ISLAND
Amend, James Frederick

QUEBEC
Aranda, Jacob Velasco
Benfey, Bruno Georg
Blouin, Andre
Capek, Radan
Collier, Brian
Cuello, A Claudio
Cummings, John Rhodes
De Champlain, Jacques
Du Souich, Patrick
Ecobichon, Donald John
Escher, Emanuel
Esplin, Barbara
Ford-Hutchinson, Anthony W
Garcia, Raul
Goyer, Robert G
Hamet, Pavel
Heisler, Seymour
Jaramillo, Jorge
Kallai-Sanfacon, Mary-Ann
Lussier, Andre (Joseph Alfred)
MacIntosh, Frank Campbell
Momparler, Richard Lewis
Nair, Vasavan N P
Nickerson, Mark
Palaic, Djuro
Panisset, Jean-Claude
Pihl, Robert O
Plaa, Gabriel Leon
Powell, William St John
Regoli, Domenico
Rola-Pleszczynski, Marek

Sasyniuk, Betty Irene
Schiffrin, Ernesto Luis
Schiller, Peter Wilhelm
Sharkawi, Mahmoud
Sirois, Pierre
Srivastava, Ashok Kumar
Varma, Daya Ram
Villeneuve, Andre
Zablocka-Esplin, Barbara

SASKATCHEWAN
Gupta, Vidya Sagar
Hickie, Robert Allan
Johnson, Dennis Duane
Johnson, Gordon E
Juorio, Augusto Victor
Prasad, Kailash
Richardson, J(ohn) Steven
Sisodia, Chaturbhuj Singh
Thornhill, James Arthur

OTHER COUNTRIES
Akera, Tai
Allessie, Maurits
Back, Kenneth Charles
Bader, Hermann
Baggot, J Desmond
Barouki, Robert
Boullin, David John
Brown, Nigel Andrew
Cho, Young Won
Davidson, David Edward, Jr
De Haën, Christoph
Dunn, Christopher David Reginald
Fitzgerald, John Desmond
Fleckenstein, Albrecht
Ganten, Detlev
Gemsa, Diethard
Ghiara, Paolo
Greene, Lewis Joel
Guerrero-Munoz, Frederico
Haegele, Klaus D
Hall, Peter
Hebborn, Peter
Hedglin, Walter L
Henderson, David Andrew
Holman, Richard Bruce
Honda, Kazuo
Ide, Hiroyuki
Ishida, Yukisato
Judis, Joseph
Kang, Sungzong
Kawamura, Hiroshi
Kuriyama, Kinya
Ladinsky, Herbert
Lee, Kwang Soo
Lee, Men Hui
Lieb, William Robert
Ling, George M
Livett, Bruce G
Maragoudakis, Michael E
Markert, Claus O
Marquardt, Hans Wilhelm Joe
Martin, James Richard
Muschek, Lawrence David
Nabeshima, Toshitaka
Pina, Enrique
Reuter, Harald
Sampson, Sanford Robert
Schonbaum, Eduard
Segal, Mark
Sellin, Lawrence C
Sinha, Asru Kumar
Stebbing, Nowell
Sung, Chen-Yu
Suria, Amin
Tung, Che-Se
U'Prichard, David C
Vane, John Robert
Vanhoutte, Paul Michel
Welty, Joseph D
Wettstein, Joseph G

Pharmacy

ALABAMA
Barker, Kenneth Neil
Coker, Samuel Terry
Hocking, George Macdonald
King, William David
Ravis, William Robert
Riley, Thomas N
Susina, Stanley V
Thomasson, Claude Larry
Tice, Thomas Robert

ARIZONA
Blanchard, James
Brodie, Donald Crum
Brown, Michael
Burton, Lloyd Edward
Rowe, Thomas Dudley
Tong, Theodore G

ARKANSAS
Gloor, Walter Thomas, Jr
Lattin, Danny Lee
McCowan, James Robert
Mittelstaedt, Stanley George
Pynes, Gene Dale

CALIFORNIA
Adair, Dennis Wilton

MEDICAL & HEALTH SCIENCES / 487

Ballard, Kenneth J
Barker, Donald Young
Bartlett, Janeth Marie
Benet, Leslie Z
Blankenship, James William
Boger, Dale L
Chaubal, Madhukar Gajanan
D'Argenio, David Z
Eiler, John Joseph
Gong, William C
Havemeyer, Ruth Naomi
Hoener, Betty-ann
Hong, Keelung
Jones, Richard Elmore
Kang, Chang-Yuil
King, James C
Koda, Robert T
Lee, Chi-Ho
McCarron, Margaret Mary
Malone, Marvin Herbert
Masover, Gerald K
Matuszak, Alice Jean Boyer
Mlodozeniec, Arthur Roman
Nicholson, Arnold Eugene
Ong, John Tjoan Ho
Oppenheimer, Phillip R
Phares, Russell Eugene, Jr
Poulsen, Boyd Joseph
Rabussay, Dietmar Paul
Raff, Allan Maurice
Rowland, Ivan W
Rubas, Werner
Runkel, Richard A
Saute, Robert E
Shefter, Eli
Sorby, Donald Lloyd
Stanton, Hubert Coleman
Stimmel, Glen Lewis
Theeuwes, Felix
Thompson, John Frederick
Tozer, Thomas Nelson
Yakatan, Gerald Joseph

COLORADO
Bone, Jack Norman
Daunora, Louis George
Hammerness, Francis Carl
Williams, Ronald Lee

CONNECTICUT
Cohen, Edward Morton
Cohen, Steven Donald
Fenney, Nicholas William
Forssen, Eric Anton
Gans, Eugene Howard
Gerald, Michael Charles
Ghosh, Dipak K
Heyd, Allen
Horhota, Stephen Thomas
Kelleher, William Joseph
Kramer, Paul Alan
Lange, Winthrop Everett
Milzoff, Joel Robert
Rosenberg, Philip
Rother, Ana
Schurig, John Eberhard
Shapiro, Warren Barry
Skauen, Donald M
Tranner, Frank
Witczak, Zbigniew J
Zalucky, Theodore B

DELAWARE
Aungst, Bruce J
Lam, Gilbert Nim-Car
Rudolph, Jeffrey Stewart
Turlapaty, Prasad

DISTRICT OF COLUMBIA
Hamarneh, Sami Khalaf

FLORIDA
Araujo, Oscar Eduardo
Brazeau, Gayle A
Heller, William Mohn
Hepler, Charles Douglas
Hill, Richard A
Holstius, Elvin Albert
Hom, Foo Song
Huber, Joseph William, III
Lamba, Surendar Singh
Lindstrom, Richard Edward
Schwartz, Michael Averill
Seugling, Earl William, Jr
Soliman, Karam Farag Attia
Soliman, Magdi R I
Young, Michael David

GEORGIA
Ansel, Howard Carl
Brennan, John Joseph
Cadwallader, Donald Elton
Entrekin, Durward Neal
Feldman, Stuart
Littlejohn, Oliver Marsilius
McGrath, William Robert
Martinek, Robert George
Price, James Clarence
Schramm, Lee Clyde
Waters, Kenneth Lee
Whitworth, Clyde W

HAWAII
Patterson, Harry Robert

IDAHO
Hillyard, Ira William
Nelson, Arthur Alexander, Jr

ILLINOIS
Blake, Martin Irving
Chun, Alexander Hing Chinn
Cranston, Joseph William, Jr
Dodge, Patrick William
Farnsworth, Norman R
Fong, Harry H S
Foreman, Ronald Louis
Groves, Michael John
Heller, Michael
Hutchinson, Richard Allen
Kaistha, Krishan K
Kimura, Eugene Tatsuru
Kinghorn, Alan Douglas
Kunka, Robert Leonard
Lu, Matthias Chi-Hwa
MacKichan, Janis Jean
Mrtek, Robert George
Musa, Mahmoud Nimir
Schleif, Robert H
Singh, Mohinder
Young, James George
Yunker, Martin Henry

INDIANA
Akers, Michael James
Belcastro, Patrick Frank
Born, Gordon Stuart
Byrn, Stephen Robert
Cohen, Marlene Lois
Duvall, Ronald Nash
Evanson, Robert Verne
Gold, Gerald
Green, Mark Alan
Hamlow, Eugene Emanuel
Hem, Stanley L
Kaufman, Karl Lincoln
Kessler, Wayne Vincent
Kildsig, Dane Olin
Kuzel, Norbert R
McLaughlin, Jerry Loren
Parke, Russell Frank
Peck, Garnet E
Robbers, James Earl
Rowe, Edward John
Shaw, Margaret Ann
Shaw, Stanley Miner
Sperandio, Glen Joseph
Thakkar, Arvind Lavji
Tyler, Varro Eugene
Williams, Edward James
Wiser, Thomas Henry

IOWA
Banker, Gilbert Stephen
Baron, Jeffrey
Benton, Byrl E
Carew, David P
Granberg, Charles Boyd
Guillory, James Keith
Jones, Leslie F
Kerr, Wendle Louis
Levine, Phillip J
Makar, Adeeb Bassili
Parrott, Eugene Lee
Robinson, Robert George
Wurster, Dale E
Wurster, Dale Eric, Jr

KANSAS
Rytting, Joseph Howard
Stella, Valentino John

KENTUCKY
Crooks, Peter Anthony
Dittert, Lewis William
Kostenbauder, Harry Barr
Turco, Salvatore J
Yokel, Robert Allen

LOUISIANA
Daigle, Josephine Siragusa
Danti, August Gabriel
Grace, Marcellus

MARYLAND
Asghar, Khursheed
Augsburger, Larry Louis
Benson, Walter Roderick
Bhagat, Hitesh Rameshchandra
Calton, Gary Jim
Caplan, Yale Howard
Cone, Edward Jackson
Cosmides, George James
Gallelli, Joseph F
Hollenbeck, Robert Gary
Horakova, Zdenka
Husain, Syed
Jagadeesh, Gowra G
Kaplan, Stanley A
Knapp, David Allan
Kramer, William Geoffrey
Lamy, Peter Paul
Leaders, Floyd Edwin, Jr
Levchuk, John W
Moreton, Julian Edward
Oddis, Joseph Anthony
Osterberg, Robert Edward
Oxman, Michael Allan
Palumbo, Francis Xavier Bernard

Paul, Steven M
Poochikian, Guiragos K
Price, Thomas R
Resnick, Charles A
Rodowskas, Christopher A, Jr
Sewell, Winifred
Shangraw, Ralph F
Tousignaut, Dwight R
Valley, Sharon Louise
Viswanathan, C T
Watzman, Nathan
Weiner, Myron
Whidden, Stanley John
Young, Gerald A

MASSACHUSETTS
Bhargava, Hridaya Nath
Curtis, Robert Arthur
Gozzo, James J
Hassan, William Ephriam, Jr
Leary, John Dennis
Malakian, Artin
Mendes, Robert W
Morgan, James Philip
Pecci, Joseph
Siegal, Bernard
Silverman, Harold I
Smith, Pierre Frank
Tuckerman, Murray Moses
Woodman, Peter William

MICHIGAN
Anderson, Gordon Frederick
Barr, Martin
Bhatt, Padmamabh P
Bivins, Brack Allen
Burkholder, David Frederick
Dauphinais, Raymond Joseph
Guttendorf, Robert John
Hajratwala, Bhupendra R
Hamlin, William Earl
Hiestand, Everett Nelson
Hult, Richard Lee
Jacoby, Ronald Lee
Jensen, Erik Hugo
Kinkel, Arlyn Walter
Knoechel, Edwin Lewis
Koshy, K Thomas
Kramer, Sherman Francis
Lamb, Donald Joseph
McGovren, James Patrick
Paul, Ara Garo
Rowe, Englebert L
Shah, Shirish A
Swartz, Harry Sip
Wagner, John Garnet
Wilkinson, Paul Kenneth
Wormser, Henry C

MINNESOTA
Cloyd, James C, III
Conard, Gordon Joseph
Grant, David James William
Miller, Nicholas Carl
Rahman, Yueh Erh
Rose, Wayne Burl
Shier, Wayne Thomas
Suryanarayanan, Raj Gopalan
Tempero, Kenneth Floyd
Zimmerman, Cheryl Lea

MISSISSIPPI
Ferguson, Mary Hobson
Gafford, Lanelle Guyton
Guess, Wallace Louis
Hufford, Charles David
Sindelar, Robert D
Walker, Larry
Wilson, Marvin Cracraft

MISSOURI
Banakar, Umesh Virupaksh
Bockserman, Robert Julian
Bruns, Lester George
Ericson, Avis J
Godt, Henry Charles, Jr
Johnson, Richard Dean
Neau, Steven Henry
Ritchie, David Joseph
Slotsky, Myron Norton
Stenseth, Raymond Eugene
Tansey, Robert Paul, Sr
Zienty, Ferdinand B

MONTANA
Medora, Rustem Sohrab
Van Horne, Robert Loren
Wailes, John Leonard

NEBRASKA
Abel, Peter William
Borchert, Harold R
Brockemeyer, Eugene William
Gerraughty, Robert Joseph
Greco, Salvatore Joseph
Roche, Edward Browning
Ueda, Clarence Tad

NEVADA
Buxton, Iain Laurie Offord

NEW HAMPSHIRE
Illian, Carl Richard
Smith, Terry Douglas

NEW JERSEY
Agalloco, James Paul
Agyilirah, George Augustus
Bathala, Mohinder S
Bodin, Jerome Irwin
Bornstein, Michael
Borris, Robert P
Bowers, Roy Anderson
Brogle, Richard Charles
Caroselli, Robert Anthony
Chau, Thuy Thanh
Chertkoff, Marvin Joseph
Colaizzi, John Louis
Controulis, John
Conway, Paul Gary
Crouthamel, William Guy
Fawzi, Ahmad B
Gidwani, Ram N
Gilfillan, Alasdair Mitchell
Goldstein, Harris Sidney
Gyan, Nanik D
Henderson, Norman Leo
Hong, Wen-Hai
Huber, Raymond C
Hutchins, Hastings Harold, Sr
Infeld, Martin Howard
Jaffe, Jonah
Jain, Nemichand B
Janssen, Richard William
Katz, Irwin Alan
Kline, Berry James
Kornblum, Saul S
Leeson, Lewis Joseph
Lordi, Nicholas George
Lukas, George
Marshall, Keith
Maulding, Hawkins Valliant, Jr
Mehta, Atul Mansukhbhai
Mitala, Joseph Jerrold
Moral, Josefa Liduvina
Nashed, Wilson
Nessel, Robert J
Omar, Mostafa M
Reimann, Hans
Ress, Rudyard Joseph
Rosenberg, Howard Alan
Scarpone, A John
Sequeira, Joel August Louis
Sethi, Jitender K
Sheinaus, Harold
Simonoff, Robert
Singhvi, Sampat Manakchand
Staum, Muni M
Tierney, William John
Tojo, Kakuji
Tse, Francis Lai-Sing
Turi, Paul George
Wangemann, Robert Theodore
Wasserman, Martin Allan
Williams, Nathaniel Adeyinka
Willis, Carl Raeburn, Jr
Zatz, Joel L

NEW MEXICO
Cho, Michael Yongkook
Hadley, William Melvin
Kabat, Hugh F
Lehrman, George Philip

NEW YORK
Abdelnoor, Alexander Michael
Alks, Vitauts
Bailie, George Russell
Bartilucci, Andrew J
Baum, Martin David
Belmonte, Albert Anthony
Borisenok, Walter A
Casler, David Robert
Castellion, Alan William
Chandran, V Ravi
Chou, Dorthy T C T
Cooper, Robert Michael
Ehrke, Mary Jane
Eisen, Henry
Fields, Thomas Lynn
Fox, Chester David
Gagnon, Leo Paul
Goldberg, Arthur H
Hunke, William Allen
Jusko, William Joseph
Lachman, Leon
Lai, Fong M
Lapidus, Herbert
Larson, Allan Bennett
Levy, Gerhard
Martell, Michael Joseph, Jr
Mary, Nouri Y
Maurice, David Myer
Morris, Marilyn Emily
Overweg, Norbert I A
Paladino, Joseph Anthony
Patel, Nagin K
Redalieu, Elliot
Reuter, Gerald Louis
Rich, Arthur Gilbert
Santiago, Noemi
Sciarra, John J
Shargel, Leon David
Simonelli, Anthony Peter
Spiegel, Allen J
Stempel, Edward
Strauss, Steven
Taub, Aaron M
Tkacheff, Joseph, Jr

Pharmacy (cont)

Torosian, George
Urdang, Arnold
Violante, Michael Robert
Wadgaonkar, Dilip Bhanudas
Whitaker-Azmitia, Patricia Mack
White, Albert M

NORTH CAROLINA
Breese, George Richard
Dar, Mohammad Saeed
Dren, Anthony Thomas
Eckel, Frederick Monroe
Ghanayem, Burhan I
Hadzija, Bozena Wesley
Hall, William Earl
Hartzema, Abraham Gijsbert
Hickey, Anthony James
Hong, Donald David
McAllister, Warren Alexander
Misek, Bernard
Olsen, James Leroy
Sawardeker, Jawahar Sazro
Strandhoy, Jack W
Swarbrick, James
Taylor, William West
Wier, Jack Knight
Yeowell, David Arthur

NORTH DAKOTA
Auyong, Theodore Koon-Hook
Rosenberg, Harry
Schnell, Robert Craig
Vincent, Muriel C

OHIO
Abdallah, Abdulmuniem Husein
Billups, Norman Fredrick
Black, Curtis Doersam
Bope, Frank Willis
Breglia, Rudolph John
Brueggemeier, Robert Wayne
Chen, Chi-Po
Chen, Ker-Sang
Coran, Aubert Y
Dage, Richard Cyrus
Feller, Dennis R
Frank, Sylvan Gerald
Graef, Walter L
Hagman, Donald Eric
Hayton, William Leroy
Hinkle, George Henry
Hurwitz, David Allan
Leyda, James Perkins
Lima, John J
McVean, Duncan Edward
Messer, William Sherwood, Jr
Nahata, Milap Chand
Oesterling, Thomas O
Sherman, Gerald Philip
Shirkey, Harry Cameron
Skau, Kenneth Anthony
Sokoloski, Theodore Daniel
Stadler, Louis Benjamin
Stansloski, Donald Wayne
Theodore, Joseph M, Jr
Visconti, James Andrew
Webb, Norval Ellsworth, Jr
Zoglio, Michael Anthony

OKLAHOMA
Carubelli, Raoul
Corder, Clinton Nicholas
Ice, Rodney D
Jones, Sharon Lynn
Keller, Bernard Gerard, Jr
Nithman, Charles Joseph
Prabhu, Vilas Anandrao
Reinke, Lester Allen
Shough, Herbert Richard
Sommers, Ella Blanche
Yanchick, Victor A

OREGON
Ayres, James Walter
Block, John Harvey
Constantine, George Harmon, Jr
Sisson, Harriet E

PENNSYLVANIA
Appino, James B
Bahal, Neeta
Bahal, Surendra Mohan
Bannon, James Andrew
Boxenbaum, Harold George
Cardinal, John Robert
Cowan, Alan
Davis, Neil Monas
Goldstein, Frederick J
Herzog, Karl A
Hussar, Daniel Alexander
Kelly, Ernest L
King, Robert Edward
Krenzelok, Edward Paul
Kwan, King Chiu
Levine, Geoffrey
McGhan, William Frederick
Mackowiak, Elaine DeCusatis
Motsavage, Vincent Andrew
Nedich, Ronald Lee
ONeill, Joseph Lawrence
Osei-Gyimah, Peter
Peterson, Charles Fillmore

Restaino, Frederick A
Riley, Gene Alden
Rodell, Michael Byron
Schiff, Paul L, Jr
Schott, Hans
Smith, Roger Bruce
Sugita, Edwin T
Tice, Linwood Franklin
Venkataramanan, Raman
Wadke, Deodatt Anant
Walkling, Walter Douglas
Walsh, Arthur Campbell
Wu, Wu-Nan
Zemaitis, Michael Alan

RHODE ISLAND
Needham, Thomas E, Jr
Worthen, Leonard Robert

SOUTH CAROLINA
Cheng, Kenneth Tat-Chiu
Davis, Craig Wilson
Fincher, Julian H
Lawson, Benjamin F
Niebergall, Paul J
Orcutt, Donald Adelbert
Putney, Blake Fuqua
Sumner, Edward D

SOUTH DAKOTA
Hietbrink, Bernard Edward

TENNESSEE
Autian, John
Avis, Kenneth Edward
Berney, Stuart Alan
Caponetti, James Dante
Elowe, Louis N
Fields, James Perry
Gupta, Brij Mohan
Hak, Lawrence J
Hamner, Martin E
Lawrence, William Homer
Meyer, Marvin Chris
Nunez, Loys Joseph
Patel, Tarun R
Sheth, Bhogilal
Shukla, Atul J
Skoutakis, Vasilios A
Sticht, Frank Davis
Straughn, Arthur Belknap
Wojciechowski, Norbert Joseph

TEXAS
Bryant, Stephen G
Driever, Carl William
Eugere, Edward Joseph
Gerding, Thomas G
Gupta, Vishnu Das
Hecht, Gerald
Hickman, Eugene, Sr
Iskander, Felib Youssef
Kasi, Leela Peshkar
Kehrer, James Paul
Martin, Alfred
Monk, Clayborne Morris
Morgan, H Keith
Nishioka, Kenji
Ryan, Charles F
Schermerhorn, John W
Sheffield, William Johnson
Stavchansky, Salomon Ayzenman
Ticku, Maharaj K
Walton, Charles Anthony
Woller, William Henry

UTAH
Aldous, Duane Leo
Balandrin, Manuel F
Galinsky, Raymond Ethan
Gardner, Robert Wayne
Lee, Steve S
Moe, Scott Thomas
Segelman, Florence H
Stuart, David Marshall
Wolf, Harold Herbert
Yost, Garold Steven

VIRGINIA
Aceto, Mario Domenico
Barrett, Richard John
Batra, Karam Vir
Collins, Galen Franklin
Dvorchik, Barry Howard
Feldmann, Edward George
Griffenhagen, George Bernard
Huffman, D C, Jr
Kapadia, Abhaysingh J
Lowenthal, Werner
Penna, Richard Paul
Smith, Harold Linwood
Snow, Anne E
Stone, John Patrick
Teng, Lina Chen
Uwaydah, Ibrahim Musa
Weaver, Warren Eldred
Wood, John Henry

WASHINGTON
Dunbar, Philip Gordon
Galpin, Donald R
Hammarlund, Edwin Roy
Meadows, Gary Glenn
Plein, Elmer Michael

Plein, Joy Bickmore

WEST VIRGINIA
Lim, James Khai-Jin
Malanga, Carl Joseph
Rosenbluth, Sidney Alan

WISCONSIN
Bardell, Eunice Bonow
Bass, Paul
Blockstein, William Leonard
Krahnke, Harold C
Lalley, Peter Michael
Lemberger, August Paul
Mukerjee, Pasupati

WYOMING
Brunett, Emery W

ALBERTA
Baker, Glen Bryan
Biggs, David Frederick
Rogers, James Albert
Roth, Sheldon H
Shysh, Alec
Wiebe, Leonard Irving

BRITISH COLUMBIA
Goodeve, Allan McCoy
Miller, Robert Alon
Riedel, Bernard Edward

NOVA SCOTIA
Dudar, John Douglas
Mezei, Catherine
Parker, William Arthur
Tonks, Robert Stanley

ONTARIO
Baxter, Ross M
Endrenyi, Laszlo
Hughes, Francis Norman
Lucas, Douglas M
Nairn, John Graham
Segal, Harold Jacob
Somers, Emmanuel

QUEBEC
Braun, Peter Eric
Joly, Louis Philippe
Lamonde, Andre M
Plourde, J Rosaire

OTHER COUNTRIES
Hedglin, Walter L
Ishida, Yukisato
Nabeshima, Toshitaka
Yeon-Sook, Yun

Psychiatry

ALABAMA
Bradley, Laurence A
Froelich, Robert Earl
Garver, David L
Ryan, William George

ARIZONA
Beigel, Allan
Brands, Alvira Bernice
Cutts, Robert Irving
Felix, Robert Hanna
Finch, Stuart McIntyre
Levenson, Alan Ira
Offenkrantz, William Charles
Starr, Phillip Henry
Tapia, Fernando

ARKANSAS
Henker, Fred Oswald, III
Komoroski, Richard Andrew
McMillan, Donald Edgar
Newton, Joseph Emory O'Neal
Reese, William George

CALIFORNIA
Agras, William Stewart
Applegarth, Adrienne P
Auerback, Alfred
Barchas, Jack D
Barondes, Samuel Herbert
Barter, James T
Beisser, Arnold Ray
Bloomfield, Harold H
Brodsky, Carroll M
Brown, Joan Heller
Buchwald, Jennifer S
Bunney, William E
Burns, Marcelline
Call, Justin David
Callaway, Enoch, III
Cohn, Jay Binswanger
Courchesne, Eric
Dracup, Kathleen A
Duhl, Leonard J
Efron, Robert
Epstein, Leon J
Estrada, Norma Ruth
Evans, Harrison Silas
Feifel, Herman
Ferguson, James Mecham
Fish, Barbara
Freedman, Daniel X

Galin, David
Gericke, Otto Luke
Gislason, I(rving) Lee
Gorney, Roderic
Gottschalk, Louis August
Gray, Gregory Edward
Green, Richard
Greenblatt, Milton
Hansen, Howard Edward
Harrison, Saul I
Hecht, Elizabeth Anne
Hewitt, Robert T
Horowitz, Mardi J
Jarvik, Lissy F
Jarvik, Murray Elias
Jensen, Gordon D
Jones, Reese Tasker
Judd, Lewis Lund
Kaplan, Herman
Kaufman, Janice Norton
Keilin, Bertram
Kempler, Walter
Kline, Frank Menefee
Koran, Lorrin Michael
Kripke, Daniel Frederick
Kubler-Ross, Elisabeth
Lamb, H Richard
Leiderman, P Herbert
Liberman, Robert Paul
London, Ray William
Lord, Elizabeth Mary
Lowinger, Paul
McGinnis, James F
Marangos, Paul Jerome
Marcus, Joseph
Marshall, John Foster
Miller, Milton H
Mills, Gary Kenith
Motto, Jerome (Arthur)
Nowlis, David Peter
O'Malley, Edward Paul
Ostwald, Peter Frederic
Palmer, Beverly B
Pellegrini, Robert J
Procci, Warren R
Rappaport, Maurice
Redlich, Fredrick Carl
Riskin, Jules
Ritvo, Edward R
Ritzmann, Ronald Fred
Roessler, Robert L
Ruesch, Jurgen
Sargent, Thornton William, III
Sattin, Albert
Saunders, Virginia Fox
Schneider, Lon S
Schwartz, Donald Alan
Selzer, Melvin Lawrence
Serafetinides, Eustace A
Shapiro, David
Simmons, James Quimby, III
Simon, Alexander
Sloane, Robert Bruce
Solomon, George Freeman
Starr, Arnold
Stimmel, Glen Lewis
Storms, Lowell H
Tarjan, George
Teicher, Joseph D
Tewari, Sujata Lahiri
Thomas, Claudewell Sidney
Thomas, Robert Eugene
Tupin, Joe Paul
Turrell, Eugene Snow
Van Dyke, Craig
Walker, Sydney, III
Wallerstein, Robert Solomon
Wasserman, Martin S
Wayne, George Jerome
Weiner, Herbert
Weinrich, James D
Werbach, Melvyn Roy
West, Louis Jolyon
Woods, Sherwyn Martin
Yamamoto, Joe
Zaidel, Eran

COLORADO
Coppolillo, Henry P
Gaskill, Herbert Stockton
Kulkosky, Paul Joseph
Laudenslager, Mark LeRoy
Macdonald, John Marshall
Sander, Louis W
Shore, James H
Suinn, Richard M

CONNECTICUT
Amdur, Millard Jason
Bunney, Benjamin Stephenson
Byck, Robert
Charney, Dennis S
Cohen, Donald J
Coleman, Jules Victor
Comer, James Pierpont
Davis, Michael
Donnelly, John
Escobar, Javier I
Fleck, Stephen
Fleeson, William
Glueck, Bernard Charles
Heninger, George Robert
Henisz, Jerzy Emil
Katz, Jay

Lidz, Theodore
Mirabile, Charles Samuel, Jr
Petrakis, Ismene L
Price, Lawrence Howard
Reiser, Morton Francis
Rodin, Judith
Rounsaville, Bruce J
Ruben, Harvey L
Sernyak, Michael Joseph
Sheard, Michael Henry
Snyder, Daniel Raphael
Stroebel, Charles Frederick, III
Webb, William Logan

DELAWARE
Davis, V Terrell

DISTRICT OF COLUMBIA
Boslow, Harold Meyer
Chafetz, Morris Edward
Cowdry, Rex William
Gladue, Brian Anthony
Gordon, James Samuel
Herman, Barbara Helen
Jacobsen, Frederick Marius
Kellar, Kenneth Jon
Kleber, Herbert David
Lebensohn, Zigmond Meyer
Lieberman, Edwin James
Noshpitz, Joseph Dove
Perlin, Seymour
Rittelmeyer, Louis Frederick, Jr
Robinowitz, Carolyn Bauer
Sabshin, Melvin
Salzman, Leon
Webster, Thomas G

FLORIDA
Adams, John Evi
Apter, Nathaniel Stanley
Belmont, Herman S
Brightwell, Dennis Richard
Cohen, Sanford I
Denber, Herman C B
Denker, Martin William
Eisdorfer, Carl
Gold, Mark Stephen
Goldstein, Burton Jack
Goodnick, Paul Joel
Green, David Marvin
Kemph, John Patterson
Knopp, Walter
Levitt, LeRoy P
Mandell, Arnold J
Mantero-Atienza, Emilio
Merritt, Henry Neyron
Nagera, Humberto
Pfeiffer, Eric A
Reading, Anthony John
Richelson, Elliott
Sanberg, Paul Ronald
Sheehan, David Vincent
Sussex, James Neil
Zusman, Jack

GEORGIA
Adams, David B
Curtis, John Russell
Ermutlu, Ilhan M
Foulkes, David
Hartlage, Lawrence Clifton
Jackson, William James
Loomis, Earl Alfred, Jr
McCranie, Erasmus James
Michael, Richard Phillip
Nemeroff, Charles Barnet
Schweri, Margaret Mary
Shaffer, David
Ward, Richard S

HAWAII
Amundson, Mary Jane
Bolman, William Merton
Char, Walter F
McDermott, John Francis
Morton, Bruce Eldine

ILLINOIS
Adams, John R
Aramini, Jan Marie
Astrachan, Boris Morton
Beiser, Helen R
Boshes, Louis D
Brockman, David Dean
Brown, Meyer
Chessick, Richard D
Crawford, James Weldon
Davis, John Marcell
Davis, Oscar F
Dyrud, Jarl Edvard
Falk, Marshall Allen
Fang, Victor Shengkuen
Freedman, Lawrence Zelic
Gill, Merton
Giovacchini, Peter L
Goebel, Maristella
Goldberg, Arnold Irving
Goldsmith, Jewett
Grinker, Roy Richard, Sr
Hirsch, Jay G
Hughes, Patrick Henry
Juhasz, Stephen Eugene
Kaplan, Maurice
Kimball, Chase Patterson

Langsley, Donald Gene
McArdle, Eugene W
Masserman, Jules Homan
Metz, John Thomas
Miller, Derek Harry
Mota de Freitas, Duarte Emanuel
Musa, Mahmoud Nimir
Novick, Rudolph G
Offer, Daniel
Pollock, George Howard
Rasenick, Mark M
Rudy, Lester Howard
Sabelli, Hector C
Schulman, Jerome Lewis
Schyve, Paul Milton
Smith, Kathleen
Spring, Bonnie Joan
Taylor, Michael Alan
Thomas, Joseph Erumappettical
Tourlentes, Thomas Theodore
Trosman, Harry
Ts'o, Timothy On-To
Visotsky, Harold M

INDIANA
Altman, Joseph
Bennett, Ivan Frank
Blix, Susanne
Churchill, Don W
DeMyer, Marian Kendall
FitzGerald, Joseph Arthur
Fitzhugh-Bell, Kathleen
Greist, John Howard
Grosz, Hanus Jiri
King, Lucy Jane
Klinghammer, Erich
McBride, Angela Barron
Nurnberger, John I, Jr
Nurnberger, John Ignatius, Sr
Quaid, Kimberly Andrea
Rebec, George Vincent
Simmons, James Edwin
Small, Iver Francis
Small, Joyce G
Stark, Paul

IOWA
Andersen, Arnold E
Andreasen, Nancy Coover
Benton, Arthur Lester
Cadoret, Remi Jere
Clancy, John
Comly, Hunter Hall
Jenkins, Richard Leos
Nelson, Herbert Leroy
Norris, Albert Stanley
Noyes, Russell, Jr
Stewart, Mark Armstrong
Tranel, Daniel T
Winokur, George

KANSAS
Bradshaw, Samuel Lockwood, Jr
Laybourne, Paul C
Menninger, Karl Augustus
Menninger, William Walter
Modlin, Herbert Charles
Simpson, William Stewart
Voth, Harold Moser

KENTUCKY
Adams, Paul Lieber
Elkes, Joel
Humphries, Laurie Lee
Landis, Edward Everett
Levy, Robert Sigmund
Sandifer, Myron Guy, Jr
Schwab, John J
Storrow, Hugh Alan
Surwillo, Walter Wallace
Teller, David Norton
Weisskopf, Bernard
Zolman, James F

LOUISIANA
Bradley, Ronald James
Brauchi, John Tony
Chin, Hong Woo
Daniels, Robert (Sanford)
Easson, William McAlpine
Epstein, Arthur William
Gallant, Donald
Heath, Robert Galbraith
Miles, Henry Harcourt Waters
Norman, Edward Cobb
Osofsky, Howard J
Palotta, Joseph Luke
Samuels, Arthur Seymour
Schober, Charles Coleman
Scrignar, Chester Bruno
Seltzer, Benjamin
Shushan, Morris
Straumanis, John Janis, Jr
Strauss, Arthur Joseph Louis

MAINE
Spock, Benjamin McLane

MARYLAND
Ader, Deborah N
Akiskal, Hagop S
Anderson, David Everett
Ascher, Eduard
Baker, Frank

Berlin, Fred S
Bever, Christopher Theodore
Blum, Robert Allan
Brauth, Steven Earle
Bridge, Thomas Peter
Brody, Eugene B
Burnham, Donald Love
Cantor, David S
Cath, Stanley Howard
Cohen, Robert Martin
Costa, Paul T, Jr
Dingman, Charles Wesley, II
DuPont, Robert L
Eckardt, Michael Jon
Fenton, Wayne S
Gershon, Elliot Sheldon
Glaser, Kurt
Glowa, John Robert
Godenne, Ghislaine D
Gold, Philip William
Goodwin, Frederick King
Gray, David Bertsch
Greenspan, Stanley Ira
Gruenberg, Ernest Matsner
Hendler, Nelson Howard
Hoehn-Saric, Rudolf
Hoffman, Julius
Holloway, Harry Charles
Imboden, John Baskerville
Johnson, David Norseen
Jones, Franklin Del
Klee, Gerald D'Arcy
Koslow, Stephen Hugh
Krantz, David S
Kurland, Albert A
Lake, Charles Raymond
Lemkau, Paul V
Marwah, Joe
Miller, Nancy E
Mosher, Loren Richard
Pare, William Paul
Park, Lee Crandall
Pickar, David
Pollin, William
Post, Robert M
Rankin, Joseph Eugene
Rapoport, Judith Livant
Resnik, Harvey Lewis Paul
Reynolds, Thomas De Witt
Ryback, Ralph Simon
Saavedra, Juan M
Schuster, Charles Roberts, Jr
Silbergeld, Sam
Suomi, Stephen John
Tamminga, Carol Ann
Wehr, Thomas A
Work, Henry Harcus
Wurmser, Leon
Yang, Bingzhi
Zatz, Martin

MASSACHUSETTS
Ayers, Joseph Leonard, Jr
Baldessarini, Ross John
Berman, Marlene Oscar
Blom, Gaston Eugene
Bloomberg, Wilfred
Bojar, Samuel
Chivian, Eric Seth
Chu, James Alexander
Coles, Robert
Coyle, Joseph T
Craig, John Merrill
Deutsch, Curtis Kimball
Eisenberg, Carola
Eisenberg, Leon
Epstein, Nathan Bernic
Frankel, Fred Harold
Galler, Janina Regina
Gamzu, Elkan R
Gold, Richard Michael
Goldings, Herbert Jeremy
Greenblatt, David J
Grinspoon, Lester
Gutheil, Thomas Gordon
Hartmann, Ernest Louis
Havens, Leston Laycock
Hebben, Nancy
Hobson, John Allan
Holzman, Philip S
Kagan, Jerome
Kleinman, Arthur Michael
Knapp, Peter Hobart
Kulka, Johannes Peter
Levy, Deborah Louise
Lipsitt, Don Richard
Locke, Steven Elliot
McCarley, Robert William
Mann, James
Mello, Nancy K
Mendelson, Jack H
Pierce, Chester Middlebrook
Pillard, Richard Colestock
Poussaint, Alvin Francis
Reich, James Harry
Riemer-Rubenstein, Delilah
Russo, Dennis Charles
Saxe, Leonard
Schildkraut, Joseph Jacob
Shader, Richard Irwin
Shore, Miles Frederick
Stotsky, Bernard A
Torda, Clara
Tsuang, Ming Tso

Volicer, Ladislav

MICHIGAN
Berent, Stanley
Bernardez, Teresa
Bloom, Victor
Curtis, George Clifton
Enzer, Norbert Beverley
Freedman, Robert Russell
Ging, Rosalie J
Greden, John F
Hawthorne, Victor Morrison
Hendrickson, Willard James
Luby, Elliot Donald
Margolis, Philip Marcus
Newman, Sarah Winans
Paschke, Richard Eugene
Pomerleau, Ovide F
Rainey, John Marion, Jr
Rosenzweig, Norman
Roth, Thomas
Sarwer-Foner, Gerald
Schorer, Calvin E
Silverman, Albert Jack
Uhde, Thomas Whitley
Waggoner, Raymond Walter
Watson, Andrew Samuel
Werner, Arnold

MINNESOTA
Abuzzahab, Faruk S, Sr
Clayton, Paula Jean
Dysken, Maurice William
Hanson, Daniel Ralph
Hartman, Boyd Kent
Jepson, William W
Levitan, Alexander Allen
Lucas, Alexander Ralph
Lykken, David Thoreson
Meier, Manfred John
Milliner, Eric Killmon
Mortimer, James Arthur
Overmier, J Bruce
Popkin, Michael Kenneth
Roberts, Warren Wilcox
Schofield, William
Swanson, David Wendell
Tyce, Francis Anthony
Yellin, Absalom Moses

MISSISSIPPI
Alford, Geary Simmons
Anderson, Kenneth Verle
Batson, Margaret Bailly
Halaris, Angelos
Suess, James Francis
Tourney, Garfield

MISSOURI
Cloninger, Claude Robert
Favazza, Armando Riccardo
Graham, Charles
Guze, Samuel Barry
Hudgens, Richard Watts
Justesen, Don Robert
Lamberti, Joseph W
Mattison, Richard
Murphy, George Earl
Natani, Kirmach
Olney, John William
O'Neal, Patricia L
Othmer, Ekkehard
Robins, Eli
Solomon, Kenneth
Thale, Thomas Richard
Weiss, James Moses Aaron
Wilkinson, Charles Brock

MONTANA
Lynch, Wesley Clyde

NEBRASKA
Bartholow, George William
Eaton, Merrill Thomas, Jr
Wigton, Robert Spencer

NEVADA
Hausman, William
Rahe, Richard Henry
Smith, Aaron

NEW HAMPSHIRE
Hermann, Howard T
Ritzman, Thomas A
Vaillant, George Eman

NEW JERSEY
Aronson, Miriam Klausner
Carlton, Peter Lynn
Chassan, Jacob Bernard
Gaudino, Mario
Gelfand, Jack Jacob
Gorodetzky, Charles W
Hankoff, Leon Dudley
Kohn, Herbert Myron
Layman, William Arthur
Manowitz, Paul
Millner, Elaine Stone
Moyer, John Allen
Murphree, Henry Bernard Scott
Niederland, William G
Perr, Irwin Norman
Pollack, Irwin W
Schleifer, Steven Jay

Psychiatry (cont)

Schwartz, Arthur Harold
Sugerman, Abraham Arthur
Swigar, Mary Eva
Verebey, Karl G
Weiss, Robert Jerome

NEW MEXICO
Berlin, Irving Norman
Elliott, Charles Harold
Levy, Leo
Rhodes, John Marshall
Uhlenhuth, Eberhard Henry
Van Orden, Lucas Schuyler, III

NEW YORK
Adler, Kurt Alfred
Arkin, Arthur Malcolm
Arlow, Jacob
Atkins, Robert W
Babineau, G Raymond
Baekeland, Frederick
Ballentine, Rudolph Miller
Barrett, James E
Bartus, Raymond T
Bell, James Milton
Bernstein, Alvin Stanley
Berry, Gail W(ruble)
Blass, John P
Blume, Sheila Bierman
Bodnar, Richard Julius
Bovbjerg, Dana H
Brahen, Leonard Samuel
Brodie, Jonathan D
Butler, Robert Neil
Cancro, Robert
Carlson, Eric Theodore
Carr, Edward Gary
Christ, Adolph Ervin
Clark, Julian Joseph
Cramer, Joseph Benjamin
Cushman, Paul, Jr
Daly, Robert Ward
David, Oliver Joseph
Davidson, Michael
Davis, Kenneth Leon
Doering, Charles Henry
Domantay, Antonio G
Engel, George Libman
Engelhardt, David M
Erlenmeyer-Kimling, L
Esser, Aristide Henri
Ferris, Steven Howard
Fieve, Ronald Robert
Fink, Max
Fisher, Saul Harrison
Fisher, Seymour
Flach, Frederic Francis
Forquer, Sandra Lynne
Freedman, Alfred Mordecai
Friedhoff, Arnold Jerome
Frosch, William Arthur
Gardner, Eliot Lawrence
Gaylin, Willard
Gibbs, James Gendron, Jr
Glass, David Carter
Glusman, Murray
Goldfried, Marvin R
Golub, Sharon Bramson
Gorzynski, Janusz Gregory
Ham, Richard John
Hamburg, Beatrix A M
Hamburg, David Alan
Hawkins, Linda Louise
Henn, Fritz Albert
Hershey, Linda Ann
Herz, Marvin Ira
Horwath, Ewald
Itil, Turan M
Kaplan, Harold Irwin
Kaplan, Helen Singer
Kegeles, Lawrence Steven
Kernberg, Otto F
Kestenbaum, Richard Steven
Klein, Donald Franklin
Klerman, Gerald L
Kolb, Lawrence Coleman
Kraft, Alan M
Krauss, Herbert Harris
Kupfermann, Irving
Lang, Enid Asher
Levis, Donald J
Levy, Norman B
Lifshitz, Kenneth
Loullis, Costas Christou
Malitz, Sidney
Marks, Neville
Mattson, Marlin Roy Albin
Mayerhoff, David I
Mesnikoff, Alvin Murray
Michels, Robert
Millman, Robert Barnet
Milman, Doris H
Nathan, Ronald Gene
Neugebauer, Richard
Olczak, Paul Vincent
Pardes, Herbert
Pasamanick, Benjamin
Phillips, Richard Hart
Rabe, Ausma
Rainer, John David
Ravitz, Leonard J, Jr
Reifler, Clifford Bruce

Rieder, Ronald Olrich
Rifkin, Arthur
Ritter, Walter Paul
Robinson, David Bancroft
Rogler, Lloyd Henry
Romano, John
Rothenberg, Albert
Sackeim, Harold A
Sadock, Benjamin
Sadock, Virginia A
Sager, Clifford J
Sankar, D V Siva
Schachter, Michael B
Schmale, Arthur H, Jr
Schwartz, Allan James
Shamoian, Charles Anthony
Sharpe, Lawrence
Siever, Larry Joseph
Small, S(aul) Mouchly
Smith, Charles James
Squires, Richard Felt
Stein, Marvin
Stein, Ruth E K
Steinglass, Peter Joseph
Stern, Marvin
Stewart, J W
Stokes, Peter E
Szasz, Thomas Stephen
Taketomo, Yasuhiko
Thaler, Otto Felix
Thomas, Alexander
Tryon, Warren W
Van praag, Herman M
Volavka, Jan
Wineburg, Elliot N
Wollman, Leo
Wynne, Lyman Carroll
Yolles, Stanley Fausst
Yozawitz, Allan
Zitrin, Arthur
Zitrin, Charlotte Marker
Zukin, Stephen R

NORTH CAROLINA
Anderson, William B
Blazer, Dan German
Brodie, Harlow Keith Hammond
Busse, Ewald William
Carmen, Elaine (Hilberman)
Carroll, Bernard James
Conners, Carmen Keith
Curtis, Thomas Edwin
Eason, Robert Gaston
Ellinwood, Everett Hews, Jr
Ewing, John Alexander
Fowler, John Alvis
Glaser, Frederick Bernard
Green, Robert Lee, Jr
Hall, William Charles
Halleck, Seymour Leon
Harris, Harold Joseph
Hawkins, David Rollo, Sr
Hicks, Robert Eugene
Hine, Frederick Roy
Krishnan, K Ranga Rama
Laupus, William E
Lewis, Mark Henry
Llewellyn, Charles Elroy, Jr
Logan, Cheryl Ann
Mathis, James L
Narasimhachari, Nedathur
Nenno, Robert Peter
Parker, Joseph B, Jr
Prange, Arthur Jergen, Jr
Reinhart, John Belvin
Rhoads, John McFarlane
Schiffman, Susan S
Smith, Charles E
Stolk, Jon Martin
Toole, James Francis
Verwoerdt, Adrian
Wang, Hsioh-Shan
Weiner, Richard D
Williams, Redford Brown, Jr
Wilson, William Preston
Zung, William Wen-Kwai

OHIO
Bernstein, Barbara Elaine
Berntson, Gary Glen
Block, Stanley L
Brugger, Thomas C
Cacioppo, John T
Dizenhuz, Israel Michael
Ferguson, Shirley Martha
Giannini, A James
Gregory, Ian (Walter De Grave)
Hagenauer, Fedor
Heller, Abraham
Hiatt, Harold
Kaplan, Stanley Meisel
Kramer, Milton
Lenkoski, L Douglas
McConville, Brian John
Macklin, Martin
Meltzer, Herbert Yale
Nasrallah, Henry A
Rojahn, Johannes
Ross, William Donald
Rowland, Vernon
Schubert, Daniel Sven Paul
Steele, Jack Ellwood
Titchener, James Lampton
Whitman, Roy Milton

OKLAHOMA
Deckert, Gordon Harmon
Donahue, Hayden Hackney
Hochhaus, Larry
Kosbab, Frederic Paul Gustav
Lovallo, William Robert
Miner, Gary David
Perez-Cruet, Jorge
Sengel, Randal Alan
Toussieng, Povl Winning

OREGON
Ary, Dennis
Bufford, Rodger K
Casey, Daniel Edward
Crawshaw, Ralph Shelton
Denney, Donald Duane
Lewy, Alfred James
Morgenstern, Alan Lawrence
Severson, Herbert H
Taubman, Robert Edward

PENNSYLVANIA
Amsterdam, Jay D
Arce, A Anthony
Auerbach, Arthur Henry
Barry, Herbert, III
Beck, Aaron Temkin
Berrettini, Wade Hayhurst
Brady, John Paul
Bridger, Wagner H
Clark, Robert Alfred
Cohen, Richard Lawrence
Cohen, Sheldon A
Detre, Thomas Paul
Dinges, David Francis
Doghramji, Karl
Eichelman, Burr S, Jr
Epstein, Alan Neil
Fagin, Claire M
Field, Howard Lawrence
Gershon, Samuel
Gottheil, Edward
Heller, Melvin S
Hill, Shirley Yarde
Josiassen, Richard Carlton
Kendall, Philip C
Kossor, Steven Albert
Kupfer, David J
Lakoski, Joan Marie
Leavitt, Marc Laurence
Leopold, Robert L
Lieberman, Daniel
Lief, Harold Isaiah
Madow, Leo
Mendelson, Myer
Michelson, Larry
Miller, Barry
Morris, Harold Hollingsworth, Jr
Myers, Jacob Martin
Needleman, Herbert L
O'Brien, Charles P
Oken, Donald
Orne, Martin Theodore
Overton, Donald A
Palmer, Larry Alan
Pinski, Gabriel
Plutzer, Martin David
Ragins, Naomi
Ray, William J
Reis, Walter Joseph
Reynolds, Charles F
Rickels, Karl
Roemer, Richard Arthur
Rubin, Robert Terry
Rucinska, Ewa J
Ruff, George Elson
Saul, Leon Joseph
Schachter, Joseph
Schechter, Marshall David
Settle, Richard Gregg
Shagass, Charles
Sonis, Meyer
Stunkard, Albert J
Turk, Dennis Charles
Van Kammen, Daniel Paul
Wadden, Thomas Antony
Werner, Gerhard
Whybrow, Peter Charles
Wolford, Jack Arlington
Zwerling, Israel

RHODE ISLAND
Carter, George H
Heath, Dwight B(raley)
McMaster, Philip Robert Bache
Marotta, Charles Anthony
Oxenkrug, Gregory Faiva

SOUTH CAROLINA
Ballenger, James Caudell
Kaiser, Charles Frederick
Lipinski, Joseph Floyd
McCurdy, Layton
Maricq, Hildegard Rand
Mellette, Russell Ramsey, Jr
Riggs, Benjamin C
Wilson, Marlene Ann

TENNESSEE
Aivazian, Garabed Hagpop
Ban, Thomas Arthur
Berney, Stuart Alan
Castelnuovo-Tedesco, Pietro

Friedman, Daniel Lester
Haywood, H Carl
Hollender, Marc Hale
Petrie, William Marshall
Strupp, Hans H
Thakar, Jay H
Wells, Charles Edmon

TEXAS
Blackburn, Archie Barnard
Bruce, E Ivan, Jr
Bryant, Stephen G
Cantrell, William Allen
Davis, David
Dilsaver, Steven Charles
Fabre, Louis Fernand, Jr
Faillace, Louis A
Fisher, Seymour
Forbis, Orie Lester, Jr
Gaitz, Charles M
Gardner, Russell, Jr
German, Dwight Charles
Giffen, Martin Brener
Gilliland, Robert McMurtry
Guynn, Robert William
Hansen, Douglas Brayshaw
Hatch, John Phillip
Hussian, Richard A
Johnson, Patricia Ann J
Kane, Francis Joseph, Jr
Kay, David Cyril
Knight, James Allen
Kraft, Irvin Alan
Leon, Robert Leonard
Maas, James Weldon
Martin, Jack
Meyer, George G
Peniston, Eugene Gilbert
Pennebaker, James W
Pokorny, Alex Daniel
Rafferty, Frank Thomas
Roffwarg, Howard Philip
Rosenthal, Saul Haskell
Sauerland, Eberhardt Karl
Schoenfeld, Lawrence Steven
Schoolar, Joseph Clayton
Simmons, Charles Edward
Ticku, Maharaj K
White, Robert B
Williams, Robert L
Wilmer, Harry A

UTAH
Bliss, Eugene Lawrence
Cole, Nyla J
Grosser, Bernard Irving
Ruhmann Wennhold, Ann Gertrude
Schenkenberg, Thomas
Wender, Paul H

VERMONT
Barton, Walter E
Brooks, George Wilson
Friedman, Matthew Joel
Huessy, Hans Rosenstock
Hughes, John Russell
Payson, Henry Edwards
Swett, Chester Parker

VIRGINIA
Abse, David Wilfred
Aldrich, Clarence Knight
Balster, Robert L(ouis)
Brand, Eugene Dew
Dovenmuehle, Robert Henry
Garnett, Richard Wingfield, Jr
Gottesman, Irving Isadore
Gullotta, Frank Paul
Pribram, Karl Harry
Sgro, J A
Spradlin, Wilford W
Stevenson, Ian
Volkan, Vamik

WASHINGTON
Anderson, Robert Authur
Bakker, Cornelis Bernardus
Bowden, Douglas Mchose
Dunner, David Louis
Garcia, John
Heston, Leonard L
McCabe, Donald Lee
Mason, Beryl Troxell
Reichler, Robert Jay
Scher, Maryonda E
Tucker, Gary Jay

WEST VIRGINIA
Bateman, Mildred Mitchell
Hein, Peter Leo, Jr
Kelley, John Fredric

WISCONSIN
Cleeland, Charles Samuel
Glover, Benjamin Howell
Harsch, Harold H
Headlee, Raymond
Jefferson, James Walter
Kepecs, Joseph Goodman
Prosen, Harry
Roberts, Leigh M
Spiro, Herzl Robert
Treffert, Darold Allen
Westman, Jack Conrad

MEDICAL & HEALTH SCIENCES / 491

ALBERTA
Baker, Glen Bryan
Dewhurst, William George
Pearce, Keith Ian
Sainsbury, Robert Stephen
Weckowicz, Thaddeus Eugene

BRITISH COLUMBIA
Crockett, David James
Freeman, Roger Dante
Laye, Ronald Curtis
Marcus, Anthony Martin
Margetts, Edward Lambert
Slade, H Clyde
Yonge, Keith A
Zaleski, Witold Andrew

MANITOBA
Adamson, John Douglas
Adaskin, Eleanor Jean
LeBow, Michael David
Sisler, George C
Varsamis, Ioannis

NOVA SCOTIA
Connolly, John Francis
Doane, Benjamin Knowles
Flynn, Patrick
Gold, Judith Hammerling
Hirsch, Solomon
Leighton, Alexander Hamilton
Richman, Alex

ONTARIO
Anderson, James Edward
Atcheson, J D
Beiser, Morton
Beninger, Richard J
Boag, Thomas Johnson
Caplan, Paula Joan
Cleghorn, Robert Allen
Donald, Merlin Wilfred
Garfinkel, Paul Earl
Hrdina, Pavel Dusan
Kutcher, Stanley Paul
Lapierre, Yvon Denis
Lee, Tyrone Yiu-Huen
Levine, Saul
Lipowski, Zbigniew J
Merskey, Harold
Offord, David Robert
Persad, Emmanuel
Powles, William Earnest
Rae-Grant, Quentin A
Rakoff, Vivian Morris
Sattar, Syed Abdus
Seeman, Mary Violette
Shamsie, Jalal
Silberfeld, Michel
Smith, Selwyn Michael
Stancer, Harvey C
Steinhauer, Paul David
Szyrynski, Victor
Witelson, Sandra Freedman

QUEBEC
Boulanger, Jean Baptiste
Cormier, Bruno M
Davis, John F(rederick)
Dongier, Maurice Henri
Ervin, Frank (Raymond)
Kravitz, Henry
Lal, Samarthji
Lehmann, Heinz Edgar
Lundell, Frederick Waldemar
Minde, Karl Klaus
Nair, Vasavan N P
Palmour, Roberta Martin
Pihl, Robert O
Tan, Ah-Ti Chu
Tousignant, Michel
Villeneuve, Andre
Young, Simon N

SASKATCHEWAN
Coburn, Frank Emerson
McDonald, Ian MacLaren
Richardson, J(ohn) Steven
Smith, Colin McPherson

OTHER COUNTRIES
Bartlett, James Williams, Jr
Belmaker, Robert Henry
Bogoch, Samuel
Hartocollis, Peter
Holman, Richard Bruce
Rutter, Michael L
Savage, Charles
Wettstein, Joseph G

Public Health & Epidemiology

ALABAMA
Alvarez, Jose O
Bailey, William C
Bridgers, William Frank
Habtemariam, Tsegaye
Heath, James Eugene
Hoff, Charles Jay
Holston, James L
Jamison, Homer Claude
King, William David
Klerman, Lorraine Vogel

Myers, Ira Lee
Oberman, Albert
Palmisano, Paul Anthony
Pass, Robert Floyd
Peacock, Peter N B
Schnaper, Harold Warren
Wallace, Dennis D
Windsor, Richard Anthony

ALASKA
Alter, Amos Joseph
Segal, Bernard

ARIZONA
Aickin, Mikel G
Bennett, Peter Howard
Burton, Lloyd Edward
Dezelsky, Thomas Leroy
Fritz, Roy Fredolin
Goodwin, Melvin Harris, Jr
Houser, Harold Byron
Lebowitz, Michael David
Moon, Thomas Edward
Nicolls, Ken E
Reinhard, Karl Raymond
Rittenbaugh, Cheryl K
Rowe, Verald Keith
Schneller, Eugene S
Silverstein, Martin Elliot

ARKANSAS
Brewster, Marjorie Ann

CALIFORNIA
Allen, LeRoy Richard
Aly, Raza
Anspaugh, Lynn Richard
Arthur, Susan Peterson
Ascher, Michael S
Austin, Donald Franklin
Barr, Allan Ralph
Barrett-Connor, Elizabeth L
Basch, Paul Frederick
Beard, Rodney Rau
Beck, Albert J
Belsky, Theodore
Benenson, Abram Salmon
Bernstein, Leslie
Black, Robert L
Blumberg, Mark Stuart
Brantingham, Charles Ross
Breslow, Lester
Brilliant, Lawrence Brent
Brook, Robert H
Buehring, Gertrude Case
Buffler, Patricia Ann Happ
Carlson, James Reynold
Chatigny, Mark A
Cohen, Kenneth Samuel
Constantine, Denny G
Cooke, Kenneth Lloyd
Cosman, Bard Clifford
Cretin, Shan
Daniels, Jeffrey Irwin
Davenport, Calvin Armstrong
Deedwania, Prakash Chandra
Detels, Roger
Dewey, Kathryn G
Duffey, Paul Stephen
Duhl, Leonard J
Dunn, Frederick Lester
Emmons, Richard William
Enstrom, James Eugene
Fallon, Joseph Greenleaf
Forghani-Abkenari, Bagher
Franti, Charles Elmer
Friedman, Gary David
Garland, Cedric Frank
Gill, Ayesha Elenin
Gofman, John William
Gong, William C
Gray, Gregory Edward
Hall, Thomas Livingston
Hansen, Howard Edward
Harris, Mary Styles
Haughton, James Gray
Henderson, Brian Edmond
Heyneman, Donald
Hill, Annlia Paganini
Hird, David William
Holder, Harold Douglas
Holman, Halsted Reid
Hook, Ernest
Horwitz, Marcus Aaron
Hoth, Daniel F
Hubbard, Eric R
Hubert, Helen Betty
Hunt, Isabelle F
James, Kenneth Eugene
Janda, John Michael
Jarvis, William Tyler
Jewell, Nicholas Patrick
Jordan, Brigitte
Kilburn, Kaye Hatch
King, Mary-Claire
Kissinger, David George
Kocol, Henry
Kraus, Jess F
Kritz-Silverstein, Donna
Lennette, Edwin Herman
Liskey, Nathan Eugene
Low, Hans
Mack, Thomas McCulloch
Maddy, Keith Thomas

Madison, Roberta Solomon
Magie, Allan Rupert
Merchant, Roland Samuel, Sr
Miller, Sol
Montgomery, Theodore Ashton
Moxley, John H, III
Nilsson, William A
Osborne, Michael Andrew
Paffenbarger, Ralph Seal, Jr
Palmer, Alan
Peterson, Jack Edwin
Petrakis, Nicholas Louis
Plank, Stephen J
Portnoy, Bernard
Puntenney, Dee Gregory
Reed, Dwayne Milton
Reeves, William Carlisle
Reisen, William Kenneth
Rice, Dorothy Pechman
Robinson, Henry William
Roemer, Milton Irwin
Rundel, Robert Dean
Salmon, Eli J
Schenker, Marc B
Schienle, Jan Hoops
Schmidt, Robert Milton
Schwabe, Calvin Walter
Scudder, Harvey Israel
Sensabaugh, George Frank
Sheneman, Jack Marshall
Shonick, William
Spivey, Gary H
Swan, Shanna Helen
Tager, Ira Bruce
Thomas, Claudewell Sidney
Thompson, David A(lfred)
Torfs, Claudine Pierette
Volz, Michael George
Voulgaropoulos, Emmanuel
Wallace, Helen M
Wang, Virginia Li
Washino, Robert K
Welch, James Lee
Werdegar, David
Whittenberger, James Laverre
Whorton, M Donald
Wingard, Deborah Lee
Winkelstein, Warren, Jr
Wood, Corinne Shear
Work, Telford Hindley
Wright, Harry Tucker, Jr
Wycoff, Samuel John
Wyzga, Ronald Edward

COLORADO
Barnes, Ralph Craig
Cada, Ronald Lee
Chase, Gerald Roy
Eickhoff, Theodore C
Francy, David Bruce
Gubler, Duane J
Jacobs, Barbara B
Kemp, Graham Elmore
McLean, Robert George
Nora, Audrey Hart
Poland, Jack Dean
Savage, Eldon P
Stallones, Lorann
Vetterling, John Martin

CONNECTICUT
Aitken, Thomas Henry Gardiner
Bailit, Howard L
Baltimore, Robert Samuel
Bhatt, Pravin Nanabhai
Cohart, Edward Maurice
Conrad, Eugene Anthony
Curnen, Mary G McCrea
Darling, George Bapst
Evans, Alfred Spring
Fischer, Diana Bradbury
Hinkle, Lawrence Earl, Jr
Horstmann, Dorothy Millicent
Jekel, James Franklin
Katz, Ralph Verne
Klimek, Joseph John
Lacouture, Peter George
Levine, Harvey Robert
Lieberman, James
Messing, Simon D
Murray, Francis Joseph
Niederman, James Corson
Novick, Alvin
Ostfeld, Adrian Michael
Peduzzi, Peter N
Roos, Henry
Salovey, Peter
Shope, Robert Ellis
Silver, George Albert
Smith, Abigail LeVan
Stitt, John Thomas
Stolwijk, Jan Adrianus Jozef
Sulzer-Azaroff, Beth
Tesh, Robert Bradfield
Viseltear, Arthur Jack

DELAWARE
Pell, Sidney

DISTRICT OF COLUMBIA
Ahmed, Susan Wolofski
Banta, James E
Briscoe, John
Bryant, Thomas Edward

Burris, James F
Chiazze, Leonard, Jr
Chopra, Joginder Gurbux
Dublin, Thomas David
Gebbie, Kristine Moore
Hamarneh, Sami Khalaf
Hanft, Ruth S
Howell, Embry Martin
Jacobsen, Frederick Marius
Joly, Daniel Jose
Kimbrough, Renate Dora
Landau, Emanuel
Lieberman, Edwin James
Miles, Corbin I
Miller, William Robert
Mullan, Fitzhugh
Novello, Antonia Coello
Olson, James Gordon
Pellegrino, Edmund Daniel
Pennington, Jean A T
Rao, Mamidanna S
Scheele, Leonard Andrew
Sexton, Ken
Soldo, Beth Jean
Speidel, John Joseph
Sphar, Raymond Leslie
Thiermann, Alejandro Bories
Thorne, Melvyn Charles
Vick, James

FLORIDA
Bevis, Herbert A(nderson)
Block, Irving H
Cohen, Pinya
Dougherty, Ralph C
Fackler, Martin L
Frankel, Jack William
Freudenthal, Ralph Ira
Hogan, Daniel James
Honeyman, Merton Seymour
Le Duc, James Wayne
Lyman, Gary Herbert
Mantero-Atienza, Emilio
Mercer, Thomas T
Mindlin, Rowland L
Orthoefer, John George
Plato, Phillip Alexander
Pletsch, Donald James
Porter, James Armer, Jr
Prather, Elbert Charlton
Prineas, Ronald James
Reich, George Arthur
Remmel, Randall James
Rosenbloom, Arlan Lee
Sinkovics, Joseph G
Smith, Albert Carl
Solon, Leonard Raymond
Sorensen, Andrew Aaron
Telford, Sam Rountree, Jr
Young, Martin Dunaway
Zusman, Jack

GEORGIA
Amirtharajah, Appiah
Baer, George M
Baranowski, Tom
Barbaree, James Martin
Bernstein, Robert Steven
Brachman, Philip Sigmund
Broome, Claire V
Bryan, Frank Leon
Cavallaro, Joseph John
Curran, James W
Dean, Andrew Griswold
Deitchman, Scott David
Dever, G E Alan
Dobrovolny, Charles George
Ellison, Lois Taylor
Falk, Henry
Foege, William Herbert
Francis, Donald Pinkston
Gangarosa, Eugene J
Glass, Roger I M
Glasser, John Weakley
Good, Robert Campbell
Gunn, Walter Joseph
Harris, Elliott Stanley
Hatheway, Charles Louis
Hawkins, Theo M
Heath, Clark Wright, Jr
Herrmann, Kenneth L
Hogue, Carol Jane Rowland
Hollowell, Joseph Gurney, Jr
Hopkins, Donald R
Hughes, James Mitchell
Johnson, Donald Ross
Jones, Gilda Lynn
Kagan, Irving George
Kahn, Henry Slater
Kaiser, Robert L
Kappus, Karl Daniel
Kaufmann, Arnold Francis
Kilbourne, Edwin Michael
Kutner, Michael Henry
LaMotte, Louis Cossitt, Jr
Margolis, Harold Stephen
Martin, Linda Spencer
Martinek, Robert George
Mathews, Henry Mabbett
Millar, John Donald
Mills, James Norman
Mitchell, Frank L
Morse, Stephen Allen
Mullis, Rebecca M

Public Health & Epidemiology (cont)

Mumtaz, Mohammad Moizuddin
Nguyen-Dinh, Phuc
Oakley, Godfrey Porter, Jr
Odend'hal, Stewart
Orenstein, Walter Albert
Pratt, Harry Davis
Reinhardt, Donald Joseph
Schantz, Peter Mullineaux
Schultz, Myron Gilbert
Schwartz, David Alan
Sellers, Thomas F, Jr
Shotts, Emmett Booker, Jr
Sudia, William Daniel
Thacker, Stephen Brady
Thun, Michael John
Trowbridge, Frederick Lindsley
Warren, McWilson

HAWAII
Char, Donald F B
Gilbert, Fred Ivan, Jr
Goodman, Madeleine Joyce
Grove, John Sinclair
Hankin, Jean H
Hartung, G(eorge) Harley
Lee, Richard K C
Marchette, Nyven John
Person, Donald Ames
Simeon, George John
Waslien, Carol Irene
Worth, Robert McAlpine

ILLINOIS
Brody, Jacob A
Carnes, Bruce Alfred
Cassel, Christine Karen
Cho, Yongock
Conibear, Shirley Ann
Davis, Harvey Virgil
Deangelis, Catherine D
Doege, Theodore Charles
Druyan, Mary Ellen
Dunn, Dorothy Fay
Ellis, Effie O'Neal
Farber, Marilyn Diane
Flay, Brian Richard
Foreman, Ronald Louis
Gastwirth, Bart Wayne
Hall, Stephen Kenneth
Hallenbeck, William Hackett
Hand, Roger
Hirsch, Lawrence Leonard
Jacobson, Marvin
Kochman, Ronald Lawrence
Layman, Dale Pierre
Lichter, Edward A
McFee, Donald Ray
Martin, Russell James
Masi, Alfonse Thomas
Moy, Richard Henry
Oleckno, William Anton
Olshansky, S Jay
Rosenblatt, Karin Ann
Rotkin, Isadore David
Slutsky, Herbert L
Stamler, Jeremiah
Stebbings, James Henry
Synovitz, Robert J
Thisted, Ronald Aaron
Winsberg, Gwynne Roeseler
Woods, George Theodore
Yogore, Mariano G, Jr

INDIANA
Beaver, Paul Chester
Bland, Hester Beth
Brown, Edwin Wilson, Jr
Glickman, Lawrence Theodore
Henderson, Hala Zawawi
Henzlik, Raymond Eugene
Hopper, Samuel Hersey
Hui, Siu Lui
Ivaturi, Rao Venkata Krishna
Jones, Russell K
Reed, Terry Eugene
Steinmetz, Charles Henry
Weinrich, Alan Jeffrey

IOWA
Beran, George Wesley
Berry, Clyde Marvin
Breuer, George Michael
Collins, Richard Francis
Colloton, John W
Cressie, Noel A C
Hausler, William John, Jr
Krafsur, Elliot Scoville
Orme-Johnson, David Wear
Remington, Richard Delleraine

KANSAS
Chin, Tom Doon Yuen
Jerome, Norge Winifred
Metzler, Dwight F
Neuberger, John Stephen
Novak, Alfred
Pierce, John Thomas
Thompson, Max Clyde
Tosh, Fred Eugene

KENTUCKY
Baumann, Robert Jay
Garrity, Thomas F
Hamburg, Joseph
Hochstrasser, Donald Lee
Hutchison, George B
Issel, Charles John
Lanska, Douglas John
Reed, Kenneth Paul
Stevens, Alan Douglas
Straus, Robert
Tamburro, Carlo Horace
Vandiviere, H Mac
Villarejos, Victor Moises
Weber, Frederick, Jr

LOUISIANA
Bertrand, William Ellis
Bradford, Henry Bernard, Jr
Diem, John Edwin
Ganley, James Powell
Hamrick, Joseph Thomas
Hensley, John Coleman, II
Jung, Rodney Clifton
Kern, Clifford H, III
Oalmann, Margaret Claire
Parent, Richard Alfred
Scott, Harold George
Soike, Kenneth Fieroe
Thompson, James Joseph
Vaughn, John B
Webber, Larry Stanford

MAINE
Doran, Peter Cobb
Gezon, Horace Martin
Paigen, Beverly Joyce
Stoudt, Howard Webster

MARYLAND
Abramson, Fredric David
Adkinson, Newton Franklin, Jr
Ahl, Alwynelle S
Aisner, Joseph
Alavanja, Michael Charles Robert
Alexander, Duane Frederick
Baker, Frank
Baker, Susan Pardee
Baldwin, Wendy Harmer
Barker, Lewellys Franklin
Bawden, Monte Paul
Beebe, Gilbert Wheeler
Berendes, Heinz Werner
Blattner, William Albert
Boice, John D, Jr
Brandt, Carl David
Bricker, Jerome Gough
Brubaker, Merlin L
Buck, Alfred A
Burkhalter, Barton R
Burton, George Joseph
Calabi, Ornella
Cantor, Kenneth P
Carney, William Patrick
Chen, Tar Timothy
Chiacchierini, Richard Philip
Chisolm, James Julian, Jr
Chou, Nelson Shih-Toon
Coelho, Anthony Mendes, Jr
Cohen, Bernice Hirschhorn
Comstock, George Wills
Conley, Veronica Lucey
Cornely, Paul Bertau
Correa Villasenor, Adolfo
Costa, Paul T, Jr
Curlin, George Tams
Cutler, Jeffrey Alan
Daza, Carlos Hernan
DeLaCruz, Felix F
Depue, Robert Hemphill
Der Simonian, Rebecca
Diamond, Earl Louis
Dodd, Roger Yates
Dragunsky, Eugenia M
Dubey, Jitender Prakash
Ducker, Thomas Barbee
Duggar, Benjamin Charles
Eaton, Barbra L
Edelman, Robert
Elkin, William Futter
Ellett, William H
Ensor, Phyllis Gail
Feldman, Jacob J
Ferencz, Charlotte
Fernie, Bruce Frank
Filner, Barbara
Finley, Joanne Elizabeth
Fisher, Gail Feimster
Fisher, Pearl Davidowitz
Foster, Willis Roy
Fozard, James Leonard
Freedman, Laurence Stuart
Ganaway, James Rives
Garruto, Ralph Michael
Gelberg, Alan
Gibson, Sam Thompson
Gillum, Richard Frank
Gluckstein, Fritz Paul
Goldberg, Irving David
Goldsmith, Peter Graham
Goldwater, William Henry
Gordis, Leon
Gori, Gio Batta
Greenhouse, Samuel William
Greenwald, Peter
Gruenberg, Ernest Matsner
Guthrie, Eugene Harding
Hairstone, Marcus A
Hallfrisch, Judith
Hamill, Peter Van Vechten
Harlan, William R, Jr
Harper, Paul Alva
Harris, Maureen Isabelle
Haynes, Suzanne G
Henderson, Donald Ainslie
Heyse, Stephen P
Higgins, Millicent Williams Payne
Hollingsworth, Charles Glenn
Hubbard, Donald
Hume, John Chandler
Jessup, Gordon L, Jr
Joly, Olga G
Kahn, Harold A
Kahrs, Robert F
Kalica, Anthony R
Kapikian, Albert Zaven
Kaslow, Richard Alan
Kaufmann, Peter G
Kawata, Kazuyoshi
Kessler, Irving Isar
Knapp, David Allan
Kneller, Robert William
Kolbye, Albert Christian, Jr
Krakauer, Henry
Kramer, Morton
Kroll, Bernard Hilton
Lachin, John Marion, III
Langenberg, Patricia Warrington
Lao, Chang Sheng
Lawrence, Dale Nolan
Lee, Yuen San
Legters, Llewellyn J
Lemkau, Paul V
Levine, David Morris
Levine, Richard Joseph
Lewis, Irving James
Lloyd, Douglas Seward
Locke, Ben Zion
Lundin, Frank E, Jr
McCauley, H(enry) Berton
Malone, Winfred Francis
Matanoski, Genevieve M
Meredith, Orsell Montgomery
Michelson, Edward Harlan
Micozzi, Marc S
Millar, Jack William
Miller, Robert Warwick
Mills, James Louis
Milton, Roy Charles
Minners, Howard Alyn
Molenda, John R
Monath, Thomas P
Monjan, Andrew Arthur
Moriyama, Iwao Milton
Morris, Joseph Anthony
Moscicki, Eve Karin
Nightingale, Stuart L
O'Beirne, Andrew Jon
Obrams, Gunta Iris
Omran, Abdel Rahim
Ory, Marcia Gail
Osborne, J Scott, III
Osman, Jack Douglas
Page, Norbert Paul
Paige, David M
Parkman, Paul Douglas
Parrish, Dale Wayne
Payne, William Walker
Pearson, Jay Dee
Phillips, Grace Briggs
Picciolo, Grace Lee
Pickle, Linda Williams
Plato, Chris C
Poirier, Miriam Christine Mohrhoff
Price, Thomas R
Purcell, Robert Harry
Quinnan, Gerald Vincent, Jr
Ravenholt, Reimert Thorolf
Reichelderfer, Thomas Elmer
Reinke, William Andrew
Rhoads, George Grant
Roberts, Doris Emma
Rosenberg, Saul H
Rosenstein, Marvin
Ross, Donald Morris
Rovelstad, Gordon Henry
Russell, Philip King
Sansone, Eric Brandfon
Saxinger, W(illiam) Carl
Schluederberg, Ann Elizabeth Snider
Schonfeld, Hyman Kolman
Schwartz, John T
Scott, Thomas Wallace
Seagle, Edgar Franklin
Seltser, Raymond
Shapiro, Sam
Sharrett, A Richey
Shelokov, Alexis
Shore, Moris Lawrence
Silverman, Charlotte
Sirken, Monroe Gilbert
Sloan, Michael Allan
Smith, Douglas Lee
Sommer, Alfred
Srivastava, Sudhir
Starfield, Barbara Helen
Stein, Harvey Philip
Steinwachs, Donald Michael
Stolley, Paul David
Strickland, George Thomas
Tarone, Robert Ernest
Tayback, Matthew
Taylor, Carl Ernest
Taylor, David Neely
Vaught, Jimmie Barton
Vermund, Sten Halvor
Waalkes, T Phillip
Wachs, Melvin Walter
Wallace, Gordon Dean
Walter, Eugene LeRoy, Jr
Walter, William Arnold, Jr
Wong, Dennis Mun
Yanagihara, Richard
Yodaiken, Ralph Emile
Yoder, Robert E
Young, Alvin L
Zelnik, Melvin
Zierdt, Charles Henry

MASSACHUSETTS
Barry, Joan
Berlandi, Francis Joseph
Bistrian, Bruce Ryan
Bures, Milan F
Castelli, William Peter
Clapp, Richard W
Cohen, Joel Ralph
Craven, Donald Edward
Czeisler, Charles Andrew
Darity, William Alexander
Dietz, William H
Eisenberg, Leon
Federman, Daniel D
Fineberg, Harvey V
First, Melvin William
Fiumara, Nicholas J
Frechette, Alfred Leo
Gentry, John Tilmon
George, Harvey
Gerety, Robert John
Goldberg, R J
Gottlieb, Marise Suss
Graham, John D
Hammond, Sally Katharine
Hattis, Dale B
Hershberg, Philip I
Hiatt, Howard Haym
Kosowsky, David I
Lachica, R Victor
Langmuir, Alexander D
Lavin, Philip Todd
Lemeshow, Stanley Alan
Leviton, Alan
Li, Frederick P
McCusker, Jane
MacMahon, Brian
Madoff, Morton A
Mata, Leonardo J
Newhouse, Joseph Paul
Ozonoff, David Michael
Pauker, Stephen Gary
Pierce, Edward Ronald
Rand, William Medden
Ratney, Ronald Steven
Reich, James Harry
Roberts, Louis W
Robinton, Elizabeth Dorothy
Rothman, Kenneth J
Roubenoff, Ronenn
Rubenstein, Abraham Daniel
Sartwell, Philip Earl
Sawin, Clark Timothy
Silverman, Gerald
Skrable, Kenneth William
Snyder, John Crayton
Speizer, Frank Erwin
Strong, Robert Michael
Telford, Sam Rountree, III
Thomas, Clayton Lay
Trichopoulos, Dimitrios V
Tsuang, Ming Tso
Tucker, Katherine Louise
VanPraagh, Richard
Webster, Edward William
Wegman, David Howe
Weiss, Scott T
Wray, Joe D
Wyon, John Benjamin
Yankauer, Alfred
Yerby, Alonzo Smythe

MICHIGAN
Adrounie, V Harry
Becker, Marshall Hilford
Cohen, Michael Alan
Craig, Winston John
Donabedian, Avedis
Guskey, Louis Ernest
Hawthorne, Victor Morrison
Higgins, Ian T
House, James Stephen
Humphrey, Harold Edward Burton, Jr
James, Sherman Athonia
Jensen, James Burt
Kay, Bonnie Jean
Krebs, William H
Maassab, Hunein Fadlo
Macriss, Robert A
Meeks, Robert G
Monto, Arnold Simon
Musch, David C(harles)
Mutch, Patricia Black

Paneth, Nigel Sefton
Powitz, Robert W
Rose, Gordon Wilson
Schottenfeld, David
Shea, Fredericka Palmer
Smith, Ralph G
Striffler, David Frank
Tilley, Barbara Claire
Viano, David Charles
Weg, John Gerard
Wegman, Myron Ezra
Wilmot, Thomas Ray
Wolman, Eric
Yates, Jon Arthur
Zemach, Rita
Zucker, Robert Alpert

MINNESOTA
Ackerman, Eugene
Barber, Donald E
Blackburn, Henry Webster, Jr
Diesch, Stanley L
Fan, David P
Fetcher, E(dwin) S(tanton)
Foreman, Harry
Gatewood, Lael Cranmer
Himes, John Harter
Jacobs, David R, Jr
Kurland, Leonard T
Lawrenz, Frances Patricia
Leon, Arthur Sol
Levitan, Alexander Allen
Luepker, Russell Vincent
Martens, Leslie Vernon
Melton, Lee Joseph, III
Mortimer, James Arthur
Roessler, Charles Ervin
Roy, Robert Russell
Schuman, Leonard Michael
Thawley, David Gorden
Tonn, Robert James
Verby, John E

MISSISSIPPI
Marshall, Douglas Lee
Watson, Robert Lee

MISSOURI
Bagby, John R, Jr
Black, Wayne Edward
Blenden, Donald C
Chapman, Ramona Marie
Domke, Herbert Reuben
Donnell, Henry Denny, Jr
Perry, Horace Mitchell, Jr
Spurrier, Elmer R
Weeks, Robert Joe
Weiss, James Moses Aaron
Wenner, Herbert Allan

NEBRASKA
Pearson, Paul (Hammond)
Thorson, James A

NEVADA
Dehne, Edward James

NEW HAMPSHIRE
Densen, Paul M
Hatch, Frederick Tasker
McCollum, Robert Wayne
Scrimshaw, Nevin Stewart
Wennberg, John E
Zumbrunnen, Charles Edward

NEW JERSEY
Baron, Hazen Jay
Bendich, Adrianne
Bowns, Beverly Henry
Breckenridge, Mary Barber
Canzonier, Walter J
Chance, Kenneth Bernard
Cross, Richard James
Elinson, Jack
Goel, Mahesh C
Greenberg, Michael Richard
Grinberg-Funes, Ricardo A
Hearn, Ruby Puryear
Iglewicz, Raja
Johnson, Mary Frances
Lederman, Sally Ann
Leventhal, Howard
Light, Donald W
Mallison, George Franklin
Millner, Elaine Stone
Quinones, Mark A
Schreibeis, William J
Taylor, Bernard Franklin
Weiss, Robert Jerome
Weiss, Stanley H

NEW MEXICO
Bechtold, William Eric
Crawford, Michael Howard
Eller, Charles Howe
Ettinger, Harry Joseph
Key, Charles R
Levy, Leo
Schottstaedt, William Walter
Wong, Siu Gum
Wood, Gerry Odell

NEW YORK
Abraham, Jerrold L
Agnew-Marcelli, G(ladys) Marie
Alderman, Michael Harris
Altman, Lawrence Kimball
Alvir, Jose Ma J
Axelrod, David
Bellak, Leopold
Blane, Howard Thomas
Brightman, I Jay
Bromet, Evelyn J
Brophy, Mary O'Reilly
Bross, Irwin Dudley Jackson
Brown, Lawrence S, Jr
Bullough, Vern L
Carpenter, David O
Cherkasky, Martin
Cipriano, Ramon John
Cohen, Irving Allan
Cohen, Joel Ephraim
Cohen, Leonard A
Colby, Frank Gerhardt
Collins, James Joseph
Conway, John Bell
Cort, Susannah
Coutchie, Pamela Ann
Craig, John Philip
Davies, Jack Neville Phillips
Davis, Karen Padgett
Deibel, Rudolf
Dornbush, Rhea L
Ely, Thomas Sharpless
Engle, Ralph Landis, Jr
Felman, Yehudi M
Finkel, Madelon Lubin
Forquer, Sandra Lynne
Frankle, Reva Treelisky
Freitag, Julia Louise
Frenkel, Krystyna
Frishman, William Howard
Garfinkel, Lawrence
Geiger, H Jack
Gerber, Linda M
Gibofsky, Allan
Goldstein, Inge F
Goswami, Meeta J
Greenhall, Arthur Merwin
Greenwald, Edward S
Haas, Jere Douglas
Habicht, Jean-Pierre
Harley, Naomi Hallden
Hauser, Willard Allen
Hipp, Sally Sloan
Horwath, Ewald
Imperato, Pascal James
Jahiel, Rene
Johnson, Kenneth Gerald
Keefe, Deborah Lynn
Kelsey, Jennifer Louise
Kilbourne, Edwin Dennis
Kim, Charles Wesley
Kline, Jennie Katherine
Landrigan, Philip J
Latham, Michael Charles
Lawton, Richard Woodruff
Leverett, Dennis Hugh
Levin, Morton Loeb
Lipson, Steven Mark
McCarthy, Eugene Gregory
McCarthy, James Francis
McCord, Colin Wallace
McCormack, Patricia Marie
McHugh, William Dennis
Matuszek, John Michael, Jr
May, Paul S
Mike, Valerie
Miller, Albert
Millman, Robert Barnet
Mooney, Richard T
Moor-Jankowski, J
Moss, Leo D
Neugebauer, Richard
Oates, Richard Patrick
Ottman, Ruth
Pasamanick, Benjamin
Pasternack, Bernard Samuel
Pearson, Thomas Arthur
Pechacek, Terry Frank
Piore, Nora
Priore, Roger L
Rabino, Isaac
Rambaut, Paul Christopher
Rasmussen, Kathleen Maher
Reader, George Gordon
Reifler, Clifford Bruce
Richie, John Peter, Jr
Robinson, Harry
Rogatz, Peter
Roman, Stanford A, Jr
Rosenfield, Allan
Sachdev, Om Prakash
Schecter, Arnold Joel
Schupf, Nicole
Shapiro, Raymond E
Sheehe, Paul Robert
Shore, Roy E
Sidel, Victor William
Siegel, Morris
Smith, Thomas Harry Francis
Soave, Rosemary
Socolar, Sidney Joseph
Spencer, Frank
Spiegel, Allen David
Stein, Ruth E K
Stephenson, Lani Sue
Stevens, Roy White
Sturman, Lawrence Stuart
Susser, Mervyn W
Tanner, Martin Abba
Thompson, John C, Jr
Tichauer, Erwin Rudolph
Trowbridge, Richard Stuart
Vianna, Nicholas Joseph
Vivona, Stefano
Wassermann, Felix Emil
Weiner, Matei
Weisburger, John Hans
Weissman, Myrna Milgram (Klerman)
Whelan, Elizabeth M
Winikoff, Beverly
Wood, Paul Mix
Wynder, Ernst Ludwig
Yates, Jerome William
Yolles, Tamarath Knigin
Zavon, Mitchell Ralph
Zimmer, James Griffith

NORTH CAROLINA
Anderson, Richmond K
Bergsten, Jane Williams
Blackwell, Floyd Oris
Blazer, Dan German
Boatman, Ralph Henry, Jr
Bradley, William Robinson
Bunn, William Bernice, III
Chapman, John Franklin, Jr
Cole, Jerome Foster
Costa, Daniel Louis
Coulter, Elizabeth Jackson
Decker, Clifford Earl, Jr
Dinse, Gregg Ernest
Fortney, Judith A
Fraser, David Allison
Freymann, Moye Wicks
Frothingham, Thomas Eliot
Gilfillan, Robert Frederick
Glaser, Frederick Bernard
Guess, Harry Adelbert
Hartzema, Abraham Gijsbert
Huang, Eng-Shang (Clark)
Hulka, Barbara Sorenson
Ibrahim, Michel A
LaFon, Stephen Woodrow
McClellan, Roger Orville
McPherson, Charles William
Marcus, Allan H
Miller, C Arden
Moggio, Mary Virginia
Rappaport, Stephen Morris
Richardson, Stephen H
Rogan, Walter J
Savitz, David Alan
Sloan, Frank A
Stern, A(rthur) C(ecil)
Suk, William Alfred
Tice, Raymond Richard
Udry, Joe Richard
Vakilzadeh, Javad
Weil, Thomas P
Wiener, Russell Warren
Willhoit, Donald Gillmor

NORTH DAKOTA
Klevay, Leslie Michael

OHIO
Alter, Joseph Dinsmore
Badger, George Franklin
Buncher, Charles Ralph
Burg, William Robert
Byard, Pamela Joy
Caldito, Gloria C
Caplan, Paul E
Clark, C Scott
Cook, Warren Ayer
Cooper, Dale A
Croft, Charles Clayton
Daniel, Thomas Mallon
Farrier, Noel John
Frazier, Todd Mearl
Friedland, Joan Martha
Geldreich, Edwin E(mery)
Gerson, Lowell Walter
Guo, Shumei
Hopps, Howard Carl
Jackson, Benita Marie
Jalil, Mazhar
Johnson, Ronald Gene
Jones, Paul Kenneth
Keller, Martin David
Kowal, Norman Edward
McNamara, Michael Joseph
Moeschberger, Melvin Lee
Neff, Raymond Kenneth
Neuhauser, Duncan B
Neville, Janice Nelson
Powers, Jean D
Reiches, Nancy A
Ross, John Edward
Rushforth, Norman B
Siervogel, Roger M
Smith, Jerome Paul
Sodeman, William A
Sprafka, Robert J

OKLAHOMA
Asal, Nabih Rafia
Caldwell, Glyn Gordon
Canter, Larry Wayne
Coleman, Ronald Leon
Folk, Earl Donald
Grubb, Alan S
Lynn, Thomas Neil, Jr
Shapiro, Stewart
Silberg, Stanley Louis
Walker, Bailus
Wolgamott, Gary

OREGON
Anderson, Carl Leonard
Ary, Dennis
Buist, A(line) Sonia
Dunnette, David Arthur
Elliker, Paul R
Greenlick, Merwyn Ronald
Horton, Aaron Wesley
Johnson, Larry Reidar
Morton, William Edwards
Osterud, Harold T
Radovsky, Frank Jay
Rossignol, Philippe Albert
Schaeffer, Morris
Severson, Herbert H

PENNSYLVANIA
Allen, Michael Thomas
Ball, John R
Beilstein, Henry Richard
Benarde, Melvin Albert
Brubaker, Clifford E
Buskirk, Elsworth Robert
Cronk, Christine Elizabeth
Denenberg, Herbert S
Drew, Frances L
Ferrer, Jorge F
Fischhoff, Baruch
Fisher, Edward Allen
Fletcher, Robert Hillman
Fletcher, Suzanne W
Haas, Charles Nathan
Houts, Peter Stevens
Hsu, Chao Kuang
Jungkind, Donald Lee
Kade, Charles Frederick, Jr
Kamboh, Mohammad Ilyas
Katz, Solomon H
Kendall, Norman
Kossor, Steven Albert
LaPorte, Ronald E
Levin, Simon Eugene
Levison, Matthew Edmund
Logue, James Nicholas
London, William Thomas
McGhan, William Frederick
Martin, Samuel Preston, III
Mason, Thomas Joseph
Mattison, Donald Roger
Meadows, Anna T
Miller, G(erson) H(arry)
Morehouse, Chauncey Anderson
Morgan, James Frederick
Mulvihill, John Joseph
Osborne, William Wesley
Pinski, Gabriel
Prier, James Eddy
Rogers, Fred Baker
Rubin, Benjamin Arnold
Rycheck, Russell Rule
Salazar, Hernando
Sazama, Kathleen
Shank, Kenneth Eugene
Shear, Charles L
Shoub, Earle Phelps
Singh, Balwant
Six, Howard R
Strom, Brian Leslie
Tepper, Lloyd Barton
Tokuhata, George K
Vogt, Molly Thomas
Wald, Niel
Waldron, Ingrid Lore
Weiss, Kenneth Monrad
Weiss, William
Wells, Adoniram Judson
Williams, Sankey Vaughan
Wing, Edward Joseph
Yu, Lucy Cha

RHODE ISLAND
Heath, Dwight B(raley)
Jones, Kenneth Wayne
LeBrun, Roger Arthur
Rossi, Joseph Stephen

SOUTH CAROLINA
Brown, Arnold
Charles, Edgar Davidson, Jr
Gelfand, Henry Morris
Johnson, James Allen, Jr
Keil, Julian E
Lewis, Robert Frank
McCutcheon, Ernest P
Oswald, Edward Odell
Richards, Gary Paul
Robson, John Robert Keith
Schuman, Stanley Harold
Thompson, Shirley Jean

SOUTH DAKOTA
Hodges, Neal Howard, II

TENNESSEE
Blaser, Martin Jack
Cullen, Marion Permilla

DISCIPLINE INDEX

Public Health & Epidemiology (cont)

Cypess, Raymond Harold
Dupont, William Dudley
Dysinger, Paul William
Federspiel, Charles Foster
New, John Coy, Jr
Scholtens, Robert George
Somes, Grant William
Zeighami, Elaine Ann

TEXAS
Alexander, Charles Edward, Jr
Baine, William Brennan
Binnie, William Hugh
Blangero, John Charles
Bryant, Stephen G
Bushong, Stewart Carlyle
Chow, Rita K
Cooper, Sharon P
Crandell, Robert Allen
Dahl, Elmer Vernon
Eppright, Margaret
Ford, Charles Erwin
Fultz, R Paul
Goldberg, Leonard H
Green, Hubert Gordon
Hanis, Craig L
Heidelbaugh, Norman Dale
Heimbach, Richard Dean
Hillis, Argye Briggs
Holguin, Alfonso Hudson
Hollinger, F(rederick) Blaine
Jenkins, Vernon Kelly
Kerr, George R
Kusnetz, Howard L
Labarthe, Darwin Raymond
Lee, Eun Sul
LeMaistre, Charles Aubrey
McCloy, James Murl
Moore, Donald Vincent
Murnane, Thomas George
Oseasohn, Robert
Park, Myung Kun
Parker, Roy Denver, Jr
Philips, Billy Ulyses
Prabhakar, Bellur Subbanna
Shekelle, Richard Barten
Smith, Jerome H
Steele, James Harlan
Troxler, Raymond George
Valdivieso, Dario
Vallbona, Carlos
Vernon, Ralph Jackson
Ward, Jonathan Bishop, Jr
Wilcox, Roberta Arlene

UTAH
Archer, Victor Eugene
Ensign, Paul Roselle
Gortatowski, Melvin Jerome
Horn, Susan Dadakis
Hunt, Steven Charles
Lee, Jeffrey Stephen
Melton, Arthur Richard
Moser, Royce, Jr
Nelson, Kenneth William
Shelby, Nancy Jane

VERMONT
Babbott, Frank Lusk, Jr
Costanza, Michael Charles
Falk, Leslie Alan
Haugh, Larry Douglas
Slack, Nelson Hosking
Terris, Milton
Waller, Julian Arnold

VIRGINIA
Attinger, Ernst Otto
Bailey, Robert Clifton
Bempong, Maxwell Alexander
Berkow, Susan E
Brignoli Gable, Carol
Brown, Robert Don
Ehrlich, S Paul, Jr
Emlet, Harry Elsworth, Jr
Fagan, Raymond
Held, Joe R
Hilcken, John Allen
Jurinski, Neil B(ernard)
Kilpatrick, S James, Jr
Kurtzke, John F, Sr
Lee, James A
Mahboubi, Ezzat
Maturi, Vincent Francis
Norman, James Everett, Jr
O'Brien, William M
Pao, Eleanor M
Potts, Malcolm
Rodricks, Joseph Victor
Sahli, Brenda Payne
Scott, David Bytovetzski
Spencer, Frederick J
Wands, Ralph Clinton
Wasti, Khizar
Weiss, William
White, Kerr Lachlan

WASHINGTON
Chrisman, Noel Judson
Day, Robert Winsor
Deyo, Richard Alden
Droppo, James Garnet, Jr
Emanuel, Irvin
Foy, Hjordis M
Gale, James Lyman
Grayston, J Thomas
Haviland, James West
Henderson, Maureen McGrath
Hurlich, Marshall Gerald
Johnstone, Donald Lee
Kathren, Ronald Laurence
Kenny, George Edward
Krieger, John Newton
Kuo, Cho-Chou
Lane, Wallace
Lee, John Alexander Hugh
Lovely, Richard Herbert
Omenn, Gilbert S
Perrin, Edward Burton
Peterson, Donald Richard
Rosenblatt, Roger Alan
Siscovick, David Stuart
Thomas, David Bartlett
Valanis, Barbara Mayleas
Weiss, Noel Scott
Woods, James Sterrett
Yamanaka, William Kiyoshi

WEST VIRGINIA
Bouquot, Jerry Elmer
Sheikh, Kazim

WISCONSIN
Barboriak, Joseph Jan
Blockstein, William Leonard
Dick, Elliot C
Farley, Eugene Shedden, Jr
Golubjatnikov, Rjurik
Klein, Ronald
Ladinsky, Judith L
Laessig, Ronald Harold
Lundquist, Marjorie Ann
McIntosh, Elaine Nelson
Southworth, Warren Hilbourne
Walker, Duard Lee

WYOMING
Blank, Carl Herbert
Eddy, David Maxon
Holbrook, Frederick R
Tabachnick, Walter J

PUERTO RICO
Rodriguez, Jose Raul

ALBERTA
Albritton, William Leonard
Athar, Mohammed Aqueel
Dixon, John Michael Siddons
Larke, R(obert) P(eter) Bryce
Munan, Louis
Thompson, Gordon William

BRITISH COLUMBIA
Anderson, Donald Oliver
Baird, Patricia A
Bates, David Vincent
Brantingham, Patricia Louise
Brantingham, Paul Jeffrey
Davison, Allan John
Green, Lawrence Winter
Mackenzie, Cortlandt John Gordon
McLean, Donald Millis
Petkau, Albert John

MANITOBA
Choi, Nung Won
Maniar, Atish Chandra

NEWFOUNDLAND
Orr, Robin Denise Moore
Segovia, Jorge

NOVA SCOTIA
Rautaharju, Pentti M
Richman, Alex
Stewart, Chester Bryant

ONTARIO
Anderson, G Harvey
Beaton, George Hector
Bertell, Rosalie
Bruce, W Robert
Buck, Carol Whitlow
Chambers, Larry William
Currie, Violet Evadne
Gibson, Rosalind Susan
Hachinski, Vladimir
Hewitt, David
Howe, Geoffrey Richard
Kraus, Arthur Samuel
Lang, Gerhard Herbert
Last, John Murray
Lees, Ronald Edward
Miller, Anthony Bernard
Righy, Charlotte Edith
Robertson, James McDonald
Room, Robin Gerald Walden
Roy, Marie Lessard
Silberfeld, Michel
Steele, Robert
Stemshorn, Barry William
Till, James Edgar
Todd, Ewen Cameron David
Tugwell, Peter
Wu, Alan SeeMing

QUEBEC
Aranda, Jacob Velasco
Bailar, John Christian, III
Becklake, Margaret Rigsby
Contandriopoulos, Andre-Pierre
Dugre, Robert
Frappier, Armand
Lippman, Abby
Maag, Urs Richard
Meyer, Francois
Page, Michel
Siemiatycki, Jack
Spitzer, Walter O
Talbot, Pierre J
Theriault, Gilles P
Tousignant, Michel
Trepanier, Pierre
Vobecky, Josef

OTHER COUNTRIES
Aalund, Ole
Back, Kenneth Charles
Bond, Marvin Thomas
Bryant, John Harland
Buzina, Ratko
Fraser, David William
Goldsmith, John Rothchild
Greene, Velvl William
Haschke, Ferdinand
Herman, Bertram
Hoadley, Alfred Warner
Jacobson, Bertil
Kaplan, Martin Mark
Kessler, Alexander
Kourany, Miguel
Mahler, Halfdan
Melhave, Lars
Rosen, Leon
Shuval, Hillel Isaiah
Siu, Tsunpui Oswald
Smith, Meredith Ford
Tanphaichitr, Vichai
Thiessen, Jacob Willem
Torun, Benjamin
Webster, Isabella Margaret
Xintaras, Charles

Surgery

ALABAMA
Blakemore, William Stephen
Bunch, Wilton Herbert
Curreri, P William
El Dareer, Salah
Gleysteen, John Jacob
Hankes, Gerald H
Heath, James Eugene
Horne, Robert D
Kahn, Donald R
Kirklin, John W
Lamon, Eddie William
Lloyd, L Keith
McCallum, Charles Alexander, Jr
Rodning, Charles Bernard
Smith, Frederick Williams
Waite, Peter Daniel
White, Lowell Elmond, Jr

ALASKA
Mills, William J, Jr

ARIZONA
Du Val, Merlin Kearfott
Hale, Harry W, Jr
Holloway, G Allen, Jr
Kleitsch, William Philip
Longley, B Jack
Malone, James Michael
Peltier, Leonard Francis
Peric-Knowlton, Wlatka
Schiller, William R
Silverstein, Martin Elliot
Stein, John Michael
Zukoski, Charles Frederick

ARKANSAS
Bard, David S
Boop, Warren Clark, Jr
Caldwell, Fred T, Jr
Campbell, Gilbert Sadler
Fazekas-May, Mary Alice
McGilliard, A Dare
Read, Raymond Charles

CALIFORNIA
Adams, John Edwin
Akeson, Wayne Henry
Andrews, Neil Corbly
Ascher, Nancy L
Bailey, Leonard Lee
Barker, Wiley Franklin
Benfield, John R
Bernstein, Eugene F
Bernstein, Leslie
Black, Kirby Samuel
Boyne, Philip John
Bradford, David S
Branson, Bruce William
Brantingham, Charles Ross
Brody, Garry Sidney
Carter, Paul Richard
Carton, Charles Allan
Chase, Robert A
Chatterjee, Satya Narayan
Cherrick, Henry M
Connolly, John E
Convery, F Richard
Cooper, Howard K
Cosman, Bard Clifford
Crowley, Lawrence Grandjean
DeBas, Haile T
Dev, Parvati
Eilber, Frederick Richard
Foltz, Eldon Leroy
Fonkalsrud, Eric W
Frey, Charles Frederick
Furnas, David William
Gaspar, Max Raymond
Gittes, Ruben Foster
Gius, John Armes
Gortner, Susan Reichert
Griffin, David William
Grimes, Orville Frank
Hadley, Henry Lee
Halasz, Nicholas Alexis
Harrison, Michael R
Harvey, J Paul, Jr
Hays, Daniel Mauger
Heifetz, Milton David
Higgins, George A, Jr
Hinshaw, David B
Hubbard, Eric R
Huertas, Jorge
Hunt, Thomas Knight
Jamplis, Robert W
Kaufman, Joseph J
Kirsch, Wolff M
Kurze, Theodore
Leake, Donald L
Lee, Sun
Lele, Padmakar Pratap
Lipscomb, Paul Rogers
Long, David Michael
Makowka, Leonard
Maloney, James Vincent, Jr
Moore, Wesley Sanford
Morton, Donald Lee
Murray, William R
Nelsen, Thomas Sloan
Newman, Melvin Micklin
Nickel, Vernon L
Ning, Shoucheng
Oberhelman, Harry Alvin, Jr
Orloff, Marshall Jerome
Parker, Harold R
Perry, Jacquelin
Peters, Richard Morse
Porter, Robert Willis
Preston, Frederick Willard
Richards, Victor
Roe, Benson Bertheau
Sadler, Charles Robinson, Jr
Sasaki, Gordon Hiroshi
Serkes, Kenneth Dean
Shoemaker, William C
Shumway, Norman Edward
Siposs, George G
Smith, Louis Livingston
State, David
Stemmer, Edward Alan
Stern, Martin
Stern, W Eugene
Street, Dana Morris
Sugar, Oscar
Thomas, Arthur Norman
Thompson, Ralph J, Jr
Tompkins, Ronald K
Trento, Alfredo
Trummer, Max Joseph
Turley, Kevin
Von Leden, Hans Victor
Wareham, Ellsworth Edwin
Way, Lawrence Wellesley
Weinreb, Robert Neal
Winet, Howard
Wise, Burton Louis
Wolfman, Earl Frank, Jr
Yamini, Sohrab
Youmans, Julian Ray
Zarem, Harvey A
Zawacki, Bruce Edwin

COLORADO
Boswick, John A
Eiseman, Ben
Jafek, Bruce William
Kindt, Glenn W
Lilly, John Russell
Moore, George Eugene
Paton, Bruce Calder

CONNECTICUT
Baldwin, John Charles
Ballantyne, Garth H
Brown, Paul Woodrow
Cole, Jack Westley
Collins, William Francis, Jr
Dumont, Allan E
Foster, James Henry
Glenn, William Wallace Lumpkin
Hayes, Mark Allan
Hlavacek, Robert Allen
Lewis, Jonathan Joseph
Lindskog, Gustaf Elmer

McKhann, Charles Fremont
Modlin, Irvin M
Owens, Guy
Pickett, Lawrence Kimball
Rafferty, Terence Damien
Shichman, D(aniel)
Southwick, Wayne Orin
Sumpio, Bauer E
Taylor, Kenneth J W
Topazian, Richard G
Wright, Hastings Kemper

DELAWARE
Adams, John Pletch

DISTRICT OF COLUMBIA
Balkissoon, Basdeo
Calhoun, Noah Robert
Depalma, Ralph G
Feller, William
Foegh, Marie Ladefoged
Geelhoed, Glenn William
Harmon, John W
Kobrine, Arthur
Koop, Charles Everett
Leffall, LaSalle Doheny, Jr
Luessenhop, Alfred John
May, Sterling Randolph
Potter, John F
Schwartz, Marshall Zane

FLORIDA
Burke, George William, III
Carey, Larry Campbell
Chandler, J Ryan
Cohen, Marc Singman
Colahan, Patrick Timothy
Conger, Kyril Bailey
Daicoff, George Ronald
Dunbar, Howard Stanford
Evans, Arthur T
Fackler, Martin L
Finlayson, Birdwell
Finney, Roy Pelham
Garcia-Bengochea, Francisco
Habal, Mutaz
Hayward, James Rogers
Kaiser, Gerard Alan
Ketcham, Alfred Schutt
Klopp, Calvin Trexler
Krizek, Thomas Joseph
Kurzweg, Frank Turner
Large, Alfred McKee
Legaspi, Adrian
Lilien, Otto Michael
Malinin, Theodore I
Pace, William Greenville
Persky, Lester
Peters, Thomas G
Pfaff, William Wallace
Ransohoff, Joseph
ReMine, William Hervey
Rhoton, Albert Loren, Jr
Riser, Wayne H
Ryan, Robert F
Smyth, Nicholas Patrick Dillon
Steenburg, Richard Wesley
Stephenson, Samuel Edward, Jr
Sypert, George Walter
Talbert, James Lewis
Wangensteen, Stephen Lightner
Waterhouse, Keith R
Woodward, Edward Roy
Zeppa, Robert

GEORGIA
Allen, Marshall B, Jr
Bliven, Floyd E, Jr
Brown, Robert Lee
Cavanagh, Harrison Dwight
Ellison, Robert G
Engler, Harold S
Humphries, Arthur Lee, Jr
Jurkiewicz, Maurice J
Ku, David Nelson
McGarity, William Cecil
Manganiello, Louis O J
Mansberger, Arlie Roland, Jr
Moretz, William Henry
Parrish, Robert A, Jr
Rawlings, Clarence Alvin
Sherman, Roger Talbot
Skandalakis, John Elias
Symbas, Panagiotis N
Tindall, George Taylor
Van De Water, Joseph M
Watne, Alvin Lloyd
Yancey, Asa G, Sr
Young, Henry Edward

HAWAII
Hong, Pill Whoon
Kokame, Glenn Megumi
Lance, Eugene Mitchell
McNamara, Joseph Judson
Mamiya, Richard T

ILLINOIS
Ausman, Robert K
Birch, Alan Grant
Block, George E
Bombeck, Charles Thomas, III
Brizio-Molteni, Loredana
Choukas, Nicholas C

Ebert, Paul Allen
Ernest, J Terry
Ferguson, Donald John
Geisler, Fred Harden
Greenlee, Herbert Breckenridge
Griffith, B Herold
Hanlon, C Rollins
Hardaway, Ernest, II
Harper, Paul Vincent
Hejna, William Frank
Hekmatpanah, Javad
Hines, James R
Keeley, John L
Kittle, Charles Frederick
Langston, Hiram Thomas
Levitsky, Sidney
Lopez-Majano, Vincent
Lurain, John Robert, III
Lyon, Edward Spafford
McKinney, Peter
Mackler, Saul Allen
Manohar, Murli
Mason, George Robert
Merkel, Frederick Karl
Moss, Gerald S
Mrazek, Rudolph G
Mullan, John F
Nahrwold, David Lange
Nyhus, Lloyd Milton
Paloyan, Edward
Prinz, Richard Allen
Raimondi, Anthony John
Ray, Robert Durant
Replogle, Robert Lee
Requarth, William H
Saclarides, Theodore John
Schlenker, James D
Schumer, William
Seed, Randolph William
Shields, Thomas William
Spurrier, Wilma A
Stein, Irving F, Jr
Sumner, David Spurgeon
Tobin, Martin John
Wetzel, Nicholas
Whisler, Walter William
Wolf, James Stuart
Zintel, Harold Albert
Zsigmond, Elemer K

INDIANA
Borgens, Richard Ben
Campbell, Robert Louis
Fryer, Minot Packer
Fuson, Robert L
Grayson, Merrill
Kalsbeck, John Edward
King, Harold
Lempke, Robert Everett
Mandelbaum, Isidore
Manders, Karl Lee
Mealey, John, Jr
Shumacker, Harris B, Jr
Weirich, Walter Edward
Yune, Heun Yung

IOWA
Al-Jurf, Adel Saada
Bonfiglio, Michael
Cooper, Reginald Rudyard
Ehrenhaft, Johann L
Grier, Ronald Lee
Gurll, Nelson
Lawton, Richard L
McLeran, James Herbert
Mason, Edward Eaton
Merkley, David Frederick
Pearson, Phillip T
Rossi, Nicholas Peter
Scott, William Edwin
Wass, Wallace M

KANSAS
Clawson, David Kay
Friesen, Stanley Richard
Hardin, Creighton A
Jewell, William R
Masters, Frank Wynne
Nelson, Stanley Reid
Schloerb, Paul Richard
Valk, William Lowell

KENTUCKY
Belker, Arnold M
Brower, Thomas Dudley
Costich, Emmett Rand
Dillon, Marcus Lunsford, Jr
Fingar, Victor H
Gross, David Ross
Lansing, Allan M
McRoberts, J William
Moore, Condict
Ochsner, John Lockwood
Polk, Hiram Carey, Jr
Shih, Wei Jen
Spratt, John Stricklin
Tobin, Gordon Ross

LOUISIANA
Albert, Harold Marcus
Boyce, Frederick Fitzherbert
Carter, Rebecca Davilene
Clemetson, Charles Alan Blake
Cohn, Isidore, Jr

Cook, Stephen D
DeCamp, Paul Trumbull
Etheredge, Edward Ezekiel
Hewitt, Robert Lee
Kline, David G
Krementz, Edward Thomas
Litwin, Martin Stanley
McDonald, John C
McKay, Douglas William
McLaurin, James Walter
Moulder, Peter Vincent, Jr
Nance, Francis Carter
Parkins, Charles Warren
Petri, William Henry, III
Richardson, Donald Edward
Rosenberg, Dennis Melville Leo
Rosenthal, J William
Schramel, Robert Joseph
Waldrep, Alfred Carson, Jr
Walton, Thomas Peyton, III
Webb, Watts Rankin

MAINE
Bredenberg, Carl Eric
Fackelman, Gustave Edward
Hiatt, Robert Burritt
Schimert, George
Swenson, Orvar

MARYLAND
Absolon, Karel B
Badder, Elliott Michael
Baker, Ralph Robinson
Barbul, Adrian
Baust, John G
Bering, Edgar Andrew, Jr
Blair, Emil
Colombani, Paul Michael
Djurickovic, Draginja Branko
Dubois, Andre T
Fuchs, James Claiborne Allred
Gann, Donald Stuart
Gott, Vincent Lynn
Handelsman, Jacob C
Hoopes, John Eugene
Javadpour, Nasser
Johnson, David Norseen
Lee, Lyndon Edmund, Jr
Letterman, Gordon Sparks
Liang, Isabella Y S
Linehan, William Marston
Long, Donlin Martin
Matthews, Leslie Scott
Moran, Walter Harrison, Jr
Morrow, Andrew Glenn
Mostwin, Jacek Lech
Riley, Lee Hunter, Jr
Rob, Charles G
Robinson, David Weaver
Rosenberg, Steven A
Salcman, Michael
Schmeisser, Gerhard, Jr
Scovill, William Albert
Smith, Gardner Watkins
Sommer, Alfred
Udvarhelyi, George Bela
Walker, Michael Dirck
Woods, Alan Churchill, Jr
Young, Eric D
Zimmerman, Jack McKay

MASSACHUSETTS
Abbott, William M
Alston, Theodore A
Atala, Anthony
Auchincloss, Hugh, Jr
Austen, W(illiam) Gerald
Banks, Henry H
Bougas, James Andrew
Brooks, John Robinson
Brown, Henry
Burke, John Francis
Byrne, John Joseph
Cady, Blake
Castaneda, Aldo Ricardo
Cohen, Jonathan
Cohn, Lawrence H
Cope, Oliver
Demling, Robert Hugh
Deterling, Ralph Alden, Jr
Doku, Hristo Chris
Duhamel, Raymond C
Egdahl, Richard H
Ellis, Franklin Henry, Jr
Fisher, John Herbert
Folkman, Moses Judah
Friedmann, Paul
Gaensler, Edward Arnold
Glimcher, Melvin Jacob
Glowacki, Julie
Goldberg, Sheldon Sumner
Goldsmith, Harry Sawyer
Goldwyn, Robert Malcolm
Hall, John Emmett
Harris, Melvyn H
Herrmann, John Bellows
Hume, Michael
Jackson, Benjamin T
Kosasky, Harold Jack
Kupiec-Weglinski, Jerzy W
Lavine, Leroy S
Leach, Robert Ellis
Localio, S Arthur
Loughlin, Kevin Raymond

MEDICAL & HEALTH SCIENCES / 495

McDermott, William Vincent, Jr
Malt, Ronald A
Meyer, Irving
Micheli, Lyle Joseph
Monaco, Anthony Peter
Moore, Francis Daniels
Murnane, Thomas William
Murray, Joseph E
O'Keefe, Dennis D
Patterson, William Bradford
Rheinlander, Harold F
Rousou, J A
Russell, Paul Snowden
Sachs, David Howard
Salzman, Edwin William
Schwartz, Anthony
Silen, William
Spellman, Mitchell Wright
Sweet, William Herbert
Tilney, Nicholas Lechmere
Turnquist, Carl Richard
VanPraagh, Richard
Walter, Carl
Watkins, Elton, Jr
White, Augustus Aaron, III
Yablon, Isadore Gerald
Yu, Yong Ming

MICHIGAN
Arbulu, Agustin
Bivins, Brack Allen
Cerny, Joseph Charles
Choe, Byung-Kil
Coon, William Warner
Coran, Arnold Gerald
DeWeese, Marion Spencer
Diaz, Fernando G
Fitzgerald, Robert Hannon, Jr
Fromm, David
Fry, William James
Greenfield, Lazar John
Kantrowitz, Adrian
Keller, Waldo Frank
Lindenauer, S Martin
Mahajan, Sudesh K
Moosman, Darvan Albert
Rosenberg, Jerry C
Senagore, Anthony J
Silbergleit, Allen
Silverman, Norman A
Sloan, Herbert
Smith, William S
Taren, James A
Thomas, Llywellyn Murray
Toledo-Pereyra, Luis Horacio
Turcotte, Jeremiah G
Walt, Alexander Jeffrey
Wilson, Robert Francis
Zamorano, Lucia Jopehina
Zuidema, George Dale

MINNESOTA
Buchwald, Henry
Caldwell, Michael D
Chou, Shelley Nien-Chun
Coventry, Mark Bingham
Dai, Xue Zheng (Charlie)
Danielson, Gordon Kenneth
Delaney, John P
Dennis, Clarence
Doughman, Donald James
French, Lyle Albert
Gilbertsen, Victor Adolph
Grage, Theodor B
Haines, Stephen John
Hitchcock, Claude Raymond
Humphrey, Edward William
Johnson, Einer Wesley, Jr
Kane, William J
Kaye, Michael Peter
Kelly, Patrick Joseph
Kelly, William Daniel
Leonard, Arnold S
Lofgren, Karl Adolph
Lovestedt, Stanley Almer
McCaffrey, Thomas Vincent
McQuarrie, Donald G
Najarian, John Sarkis
Nicoloff, Demetre M
Noer, Rudolf Juul
Ray, Charles Dean
Rohlfing, Stephen Roy
Ryan, Allan James
Sako, Yoshio
Sullivan, W Albert, Jr
Wallace, Larry J
West, Michael Allan

MISSISSIPPI
Andy, Orlando Joseph
Barnett, William Oscar
Hardy, James D
Holder, Thomas M
Hunter, Samuel W
Jones, Eric Wynn
McCaa, Connie Smith
Nelson, Norman Crooks
Wiser, Winfred Lavern

MISSOURI
Allen, William Corwin
Andrus, Charles Hiram
Ballinger, Walter F, II
Barner, Hendrick Boyer

Surgery (cont)

Bryant, Lester Richard
Callow, Allan Dana
Catalona, William John
Codd, John Edward
Francel, Thomas Joseph
Goldring, Sidney
Hershey, Falls Bacon
Kaiser, George C
Kaminski, Donald Leon
Liu, Maw-Shung
Major, Schwab Samuel, Jr
Moore, David Lee
Radford, Diane Mary
Scherr, David DeLano
Shealy, Clyde Norman
Silver, Donald
Smith, Kenneth Rupert, Jr
Stephenson, Hugh Edward, Jr
Stoneman, William, III
Ternberg, Jessie L
Tritschler, Louis George
Weeks, Paul Martin
Wells, Samuel Alonzo, Jr
Willman, Vallee L

MONTANA
Butler, Hugh C

NEBRASKA
Edney, James Augustine
Feldhaus, Richard Joseph
Hamsa, William Rudolph
Hinder, Ronald Albert
Hodgson, Paul Edmund
Lynch, Benjamin Leo
McLaughlin, Charles William, Jr
Stratta, Robert Joseph
Yonkers, Anthony J

NEVADA
Little, Alex G

NEW HAMPSHIRE
Crandell, Walter Bain
Kaiser, C William
Karl, Richard C
Naitove, Arthur
Ritzman, Thomas A
Rous, Stephen N

NEW JERSEY
Bregman, David
Brolin, Robert Edward
Fisher, Robert George
Gardner, Bernard
Hill, George James, II
Kaplan, Harry Arthur
Lazaro, Eric Joseph
Leyson, Jose Florante Justininane
MacKenzie, James W
Martin, Boston Faust
Najjar, Talib A
Neville, William E
Parsonnet, Victor
Rush, Benjamin Franklin, Jr
Salerno, Alphonse
Serlin, Oscar
Siegel, John H
Singh, Iqbal
Stefko, Paul Lowell
Swan, Kenneth G

NEW MEXICO
Barbo, Dorothy M
Davila, Julio C
Doberneck, Raymond C
Edwards, W Sterling
Morris, Don Melvin
Omer, George Elbert, Jr

NEW YORK
Ackerman, Norman Bernard
Adelson, Mark David
Adler, Richard H
Badalamente, Marie Ann
Bakay, Louis
Ballantyne, Donald Lindsay
Barie, Philip S
Barker, Harold Grant
Bassett, Charles Andrew Loockerman
Battista, Arthur Francis
Bauer, Joel
Beattie, Edward J
Boley, Scott Jason
Borden, Edward B
Breed, Ernest Spencer
Brennan, Murray F
Burdick, Daniel
Burt, Michael E
Clauss, Roy H
Cochran, George Van Brunt
Cook, Albert William
Day, Stacey Biswas
Del Guercio, Louis Richard M
DeWeese, James A
Dutta, Purnendu
Eisenberg, M Michael
Enquist, Irving Fritiof
Everhard, Martin Edward
Fein, Jack M
Feinman, Max L
Feustel, Paul J

Fortner, Joseph Gerald
Frankel, Victor H
Frater, Robert William Mayo
Furman, Seymour
Gage, Andrew Arthur
Gerst, Paul Howard
Gliedman, Marvin L
Godec, Ciril J
Goldstein, Louis Arnold
Golomb, Frederick M
Goulian, Dicran
Greengard, Olga
Gump, Frank E
Gumport, Stephen Lawrence
Harris, Matthew N
Harvin, James Shand
Herter, Frederic P
Hinshaw, J Raymond
Holman, Cranston William
Hornblass, Albert
Housepian, Edgar M
Hugo, Norman Eliot
Imparato, Anthony Michael
Jacobs, Richard L
Jacobson, Myron J
King, Robert (Bainton)
King, Thomas Creighton
Kinney, John Martin
Kottmeier, Peter Klaus
Landor, John Henry
Lane, Joseph M
Lane, Lewis Behr
Laufman, Harold
Lawson, William
Levenson, Stanley Melvin
Linkow, Leonard I
Litwak, Robert Seymour
Lord, Jere Williams, Jr
Lore, John M, Jr
Lowe, John Edward
Lowry, Stephen Frederick
McCormack, Patricia Marie
McSherry, Charles K
Madden, Robert E
Malis, Leonard I
Marchetta, Frank Carmelo
Matarasso, Alan
Menguy, Rene
Metcalf, William
Mindell, Eugene R
Mittelman, Arnold
Montefusco, Cheryl Marie
Morton, John Henderson
Nealon, Thomas F, Jr
Nelson, Curtis Norman
Nemoto, Takuma
Nicholas, James A
Olsson, Carl Alfred
Pankovich, Arsen M
Patterson, Russel Hugo, Jr
Peirce, Edmund Converse, II
Rabinowitz, Ronald
Ragins, Herzl
Rapaport, Felix Theodosius
Redo, Saverio Frank
Reemtsma, Keith
Reich, Theobald
Riehle, Robert Arthur, Jr
Rogers, Lloyd Sloan
Rohman, Michael
Rosenfeld, Isadore
Rovit, Richard Lee
Rubin, Gustav
Sako, Kumao
Santulli, Thomas V
Sawyer, Philip Nicholas
Schenk, Worthington G, Jr
Schlesinger, Edward Bruce
Schwartz, Seymour I
Shaftan, Gerald Wittes
Shedd, Donald Pomroy
Siffert, Robert S
Skinner, David Bernt
Slattery, Louis R
Smith, Donald Frederick
Soren, Arnold
Soroff, Harry S
Spotnitz, Henry Michael
Stark, Richard B
Stein, Bennett M
Stein, Theodore Anthony
Stinchfield, Frank E
Tice, David Anthony
Torre, Douglas Paul
Veith, Frank James
Waltzer, Wayne C
Whitsell, John Crawford, II
Wilder, Joseph R
Young, Morris Nathan

NORTH CAROLINA
Anlyan, William George
Archie, Joseph Patrick, Jr
Baker, Lenox Dial
Bell, William Harrison
Bevin, A Griswold
Bollinger, Ralph Randal
Buckwalter, Joseph Addison
Bucy, Paul Clancy
Clippinger, Frank Warren, Jr
Eifrig, David Eric
Georgiade, Nicholas George
Glower, Donald D, Jr
Goldner, Joseph Leonard

Hamit, Harold F
Hayes, John Terrence
Hightower, Felda
Howard, Donald Robert
Hudson, William Rucker
Johnson, George, Jr
Kettelkamp, Donald B
King, Lowell Restell
Klitzman, Bruce
Laros, Gerald Snyder, II
Lowe, James Edward
Lyerly, Herbert Kim
McCollum, Donald E
Machemer, Robert
Marble, Howard Bennett, Jr
Meredith, Jesse Hedgepeth
Nashold, Blaine S
Odom, Guy Leary
Peacock, Erle Ewart
Peete, William P J
Pennell, Timothy Clinard
Pories, Walter J
Postlethwait, Raymond Woodrow
Raney, Richard Beverly
Sabiston, David Coston, Jr
Scott, Stewart Melvin
Seigler, Hilliard Foster
Sheldon, George Frank
Stickel, Delford LeFew
Taylor, B Gray
Thomas, Colin Gordon, Jr
Thomas, Francis T
Watts, Charles D
White, Raymond Petrie, Jr
Wilcox, Benson Reid
Wilson, Frank Crane
Young, William Glenn, Jr

OHIO
Alexander, James Wesley
Berggren, Ronald B
Brodkey, Jerald Steven
Clifford, Donald H
DeWall, Richard A
Dobyns, Brown M
Elliott, Dan Whitacre
Ferguson, Ronald M
Gabel, Albert A
Geha, Alexander Salim
Gonzalez, Luis L
Hanto, Douglas W
Hasselgren, Per-Olof J
Heimlich, Henry Jay
Helmsworth, James Alexander
Herndon, Charles Harbison
Holden, William Douglas
Howard, John Malone
Hubay, Charles Alfred
Hull, Bruce Lansing
Hummel, Robert P
Hunt, William Edward
Husni, Elias A
Izant, Robert James, Jr
Joffe, Stephen N
Kilman, James William
McLaurin, Robert L
McQuarrie, Irvine Gray
Martin, Lester W
Montague, Drogo K
Peterson, Larry James
Rayport, Mark
Redman, Donald Roger
Resnick, Martin I
Rogers, Richard C
Rudy, Richard L
Schinagl, Erich F
Shons, Alan R
Stevenson, Jean Moorhead
Szczepanski, Marek Michal
Tucker, Harvey Michael
Vasko, John Stephen
Vogel, Thomas Timothy
Wilson, George Porter, III
Yashon, David

OKLAHOMA
Elkins, Ronald C
Farris, Bradley Kent
Miner, Gary David
Pollay, Michael
Williams, George Rainey

OREGON
Beals, Rodney K
Campbell, John Richard
DeWeese, David D
Edwards, James Mark
Fletcher, William Sigourney
Gallo, Anthony Edward, Jr
Hill, John Donald
Jacob, Stanley W
Peterson, Clare Gray
Starr, Albert

PENNSYLVANIA
Barker, Clyde Frederick
Beck, William Carl
Bell, Robert Lloyd
Beller, Martin Leonard
Black, Perry
Brighton, Carl T
Cooper, Donald Russell
Davis, Richard A
Deitrick, John E

Donawick, William Joseph
Edmunds, Louis Henry, Jr
Eggleston, Forrest Cary
Faich, Gerald Alan
Ferguson, Albert Barnett
Fineberg, Charles
Fisher, Bernard
Freese, Andrew
Futrell, J William
Gee, William
Grenvik, Ake N A
Griffen, Ward O, Jr
Grotzinger, Paul John
Hamilton, Ralph West
Harrison, Timothy Stone
Iwatsuki, Shunzaburo
Jannetta, Peter Joseph
Laing, Patrick Gowans
Langfitt, Thomas William
Laufer, Igor
Lotze, Michael T
Magargal, Larry Elliot
Magovern, George J
Marks, Gerald J
Matsumoto, Teruo
Miller, Leonard David
Moody, Robert Adams
Mullen, James L
Murphy, John Joseph
Nemir, Paul, Jr
O'Neill, James A, Jr
Pierce, William Schuler
Pirone, Louis Anthony
Raker, Charles W
Ramasastry, Sai Sudarshan
Randall, Peter
Reichle, Frederick Adolph
Rhoads, Jonathan Evans
Ritchie, Wallace Parks, Jr
Rombeau, John Lee
Rosemond, George P
Rothman, Richard Harrison
Salmon, James Henry
Sapega, Alexander Andrew
Shenkin, Henry A
Sigel, Bernard
Simmons, Richard Lawrence
Spatz, Sidney S
Starzl, Thomas E
Steel, Howard Haldeman
Stevens, Lloyd Weakley
Templeton, John Y, III
Tyson, Ralph Robert
Waldhausen, John Anton
Wallace, Herbert William
Wilberger, James Eldridge
Wolfson, Sidney Kenneth, Jr

RHODE ISLAND
Bland, Kirby Isaac
Greenblatt, Samuel Harold
Hopkins, Robert West
Karlson, Karl Eugene
Randall, Henry Thomas
Simeone, Fiorindo Anthony
Vezeridis, Michael Panagiotis

SOUTH CAROLINA
Anderson, Marion C
Bradham, Gilbert Bowman
Fitts, Charles Thomas
Friedman, Richard Joel
Greene, Frederick Leslie
Hochman, Marcelo Luis
Horger, Edgar Olin, III
Kempe, Ludwig George
Kolb, Leonard H
LeVeen, Harry Henry
Ning, John Tse Tso
Othersen, Henry Biemann, Jr
Perot, Phanor L, Jr
Putney, Floyd Johnson
Rittenbury, Max Sanford
Von Recum, Andreas F
Weidner, Michael George, Jr

TENNESSEE
Benson, Ralph C, Jr
Benz, Edmund Woodward
Byrd, Benjamin Franklin, Jr
Cox, Clair Edward, II
Dale, William Andrew
Dooley, Wallace T
Feman, Stephen Sosin
Giaroli, John Nello
Hall, Hugh David
Hansen, Axel C
Hendrix, James Harvey, Jr
Ingram, Alvin John
Kao, Race Li-Chan
Lynch, John Brown
McCoy, Sue
Meacham, William Feland
Merrill, Walter Hilson
Pate, James Wynford
Perry, Frank Anthony
Proctor, Kenneth Gordon
Randolph, Judson Graves
Reynolds, Vernon H
Robertson, James Thomas
Rosensweig, Jacob
Sawyers, John Lazelle
Scott, Henry William, Jr

TEXAS
Aust, J Bradley
Balagura, Saul
Balch, Charles Mitchell
Baxter, Charles Rufus
Beall, Arthur Charles, Jr
Bobechko, Walter Peter
Burdette, Walter James
Burton, Alexis Lucien
Clark, William Kemp
Conti, Vincent R
Cooley, Denton Arthur
Corriere, Joseph N, Jr
Crawford, Duane Austin
Cruz, Anatolio Benedicto, Jr
DeBakey, Michael Ellis
Derrick, John Rafter
Dillard, David Hugh
Dudrick, Stanley John
El-Domeiri, Ali A H
Ellett, Edwin Willard
Feola, Mario
Flatt, Adrian Ede
Franklin, Thomas Doyal, Jr
Fujitani, Roy Masami
Gildenberg, Philip Leon
Goldberg, Leonard H
Grossman, Robert G
Haley, Harold Bernard
Hall, Charles William
Hardaway, Robert M, III
Hartman, James T
Hecht, Gerald
Herbert, Morley Allen
Herndon, David N
Hinds, Edward C
Jackson, Francis Charles
Jackson, Gilchrist Lewis
Jennings, Paul Bernard, Jr
Jordan, George Lyman, Jr
Jordan, Paul H, Jr
Kahan, Barry D
Lewis, Stephen Robert
Lotzová, Eva
McClelland, Robert Nelson
McFee, Arthur Storer
McPhail, Jasper Lewis
Miller, Thomas Allen
Moody, Frank Gordon
Morris, George Cooper, Jr
Mountain, Clifton Fletcher
Musgrave, F Story
Myers, Paul Walter
Paulson, Donald Lowell
Pestana, Carlos
Peters, Paul Conrad
Playter, Robert Franklin
Pruitt, Basil Arthur, Jr
Radwin, Howard Martin
Rogers, Waid
Romsdahl, Marvin Magnus
Root, Harlan D
Roth, Jack A
Sakakini, Joseph, Jr
Sawaya, Raymond
Shires, George Thomas
Sparkman, Robert Satterfield
Spira, Melvin
Sponsel, William Eric
Stehlin, John Sebastian, Jr
Story, Jim Lewis
Thompson, James Charles
Thompson, Jesse Eldon
Tyner, George S
Urschel, Harold C, Jr
Waite, Daniel Elmer
Wolma, Fred J
Yollick, Bernard Lawrence

UTAH
Albo, Dominic, Jr
Anderson, Richard Lee
Castleton, Kenneth Bitner
Dixon, John Aldous
Kwan-Gett, Clifford Stanley
Maxwell, John Gary
Nelson, Russell Marion
Olsen, Don B
Olson, Randall J
Parkin, James Lamar
Snyder, Clifford Charles
Windham, Carol Thompson
Wolcott, Mark Walton

VERMONT
Coffin, Laurence Haines
Donaghy, Raymond Madiford Peardon
Foster, Roger Sherman, Jr
Minsinger, William Elliot

VIRGINIA
Adamson, Jerome Eugene
Cohen, I Kelman
Dammann, John Francis
Devine, Charles Joseph, Jr
Edgerton, Milton Thomas, Jr
Elmore, Stanley Mcdowell
Hayes, George J
Haynes, Boyd W, Jr
Horsley, John Shelton, III
Hurt, Waverly Glenn
Jane, John Anthony
Laskin, Daniel M
Lawrence, Walter, Jr

Laws, Edward Raymond, Jr
Lee, Hyung Mo
Lefrak, Edward Arthur
Lynn, Hugh Bailey
Muller, William Henry, Jr
Nolan, Stanton Peelle
Rodgers, Bradley Moreland
Rudolf, Leslie E
Sjogren, Robert W, Jr
Wayne, Jennifer Susan

WASHINGTON
Ansell, Julian S
Brockenbrough, Edwin C
Dennis, Melvin B, Jr
Fry, Louis Rummel
Gehrig, John D
Heimbach, David M
Hutchinson, William Burke
Kalina, Robert E
Karagianes, Manuel Tom
Kelly, William Albert
Kenney, Gerald
Krieger, John Newton
Loeser, John David
Malette, William Graham
Marchioro, Thomas Louis
Mitchell, Michael Ernst
Roberts, Theodore S
Schilling, John Albert
Strandness, Donald Eugene, Jr
Ward, Arthur Allen, Jr
Westrum, Lesnick Edward
Winterscheid, Loren Covart
Wyler, Allen Raymer

WEST VIRGINIA
Boland, James P
Malin, Howard Gerald
Nugent, George Robert
Warden, Herbert Edgar

WISCONSIN
Belzer, Folkert O
Condon, Robert Edward
Crowe, Dennis Timothy, Jr
Donegan, William L
Gingrass, Ruedi Peter
Javid, Manucher J
MacCarty, Collin Stewart
Meyer, Glenn Arthur
Mohs, Frederic Edward
Schulte, William John, Jr
Southwick, Harry W
Whiffen, James Douglass
Wolberg, William Harvey
Wussow, George C
Yale, Charles E
Young, William Paul

PUERTO RICO
Cruz-Korchin, Norma
Korchin, Leo

ALBERTA
Halloran, Philip Francis
Lakey, William Hall
Salmon, Peter Alexander
Scott, Gerald William

BRITISH COLUMBIA
Cleator, Iain Morrison
Harrison, Robert Cameron

MANITOBA
Ferguson, Colin C
Klass, Alan Arnold

NOVA SCOTIA
Perey, Bernard Jean Francois

ONTARIO
Baird, Ronald James
Bayliss, Colin Edward
Carroll, Samuel Edwin
Deitel, Mervyn
Downie, Harry G
Evans, Geoffrey
Farkas, Leslie Gabriel
Ferguson, Gary Gilbert
Ghent, William Robert
Gurd, Fraser Newman
Gutelius, John Robert
Harwood-Nash, Derek Clive
Heimbecker, Raymond Oliver
Huang, Bing Shaun
Keon, Wilbert Joseph
Lynn, Ralph Beverley
Macbeth, Robert Alexander
McCorriston, James Roland
McCredie, John A
Morley, Thomas Paterson
Pang, Cho Yat
Rappaport, Aron M
Saleh, Wasfy Seleman
Salter, Robert Bruce
Sirek, Anna
Sistek, Vladimir
Skoryna, Stanley C
Tasker, Ronald Reginald
Tator, Charles Haskell
Zingg, Walter

QUEBEC
Bentley, Kenneth Chessar
Chiu, Chu Jeng (Ray)
Collin, Pierre-Paul
Cruess, Richard Leigh
De Margerie, Jean-Marie
Entin, Martin A
Fazekas, Arpad Gyula
Feindel, William Howard
Gagnon, Claude
Germain, Lucie
Guttman, Frank Myron
MacLean, Lloyd Douglas
Mortola, Jacopo Prospero
Olivier, Andre
Shizgal, Harry M
Stratford, Joseph
Yamamoto, Y Lucas

SASKATCHEWAN
Shokeir, Mohamed Hassan Kamel

OTHER COUNTRIES
Cooke, T Derek V
Fujimoto, Shigeyoshi
Grossi, Carlo E
Hayry, Pekka Juha
Kennedy, John Hines
Poswillo, David E
Shiraki, Keizo
Yovich, John V

Veterinary Medicine

ALABAMA
Anderson, Peter Glennie
Bishop, Sanford Parsons
Clark, Carl Heritage
Cloyd, Grover David
Dalvi, Ramesh R
El Dareer, Salah
Farnell, Daniel Reese
Gray, Bruce William
Griswold, Daniel Pratt, Jr
Habtemariam, Tsegaye
Hankes, Gerald H
Heath, James Eugene
Horne, Robert D
Klesius, Phillip Harry
Knecht, Charles Daniel
Laster, William Russell, Jr
Lauerman, Lloyd Herman, Jr
Powers, Robert D
Ravis, William Robert
Sharma, Udhishtra Deva
Siddique, Irtaza H
Smith, Paul Clay
Vaughan, John Thomas
Walker, Donald F
Williams, Raymond Crawford
Williams, Theodore Shields

ARIZONA
Bjotvedt, George
Bone, Jesse Franklin
Chalquest, Richard Ross
Fisher, William Gary
Herrick, John Berne
Huber, William George
Mare, Cornelius John
Mulder, John Bastian
Perper, Robert J
Reinhard, Karl Raymond
Shively, James Nelson
Shull, James Jay

ARKANSAS
Moore, Gary Thomas
Skeeles, John Kirkpatrick

CALIFORNIA
Adams, Thomas Edwards
Altera, Kenneth P
Bailey, Cleta Sue
Bankowski, Raymond Adam
Bernoco, Domenico
Bidlack, Donald Eugene
Bissell, Glenn Daniel
Braemer, Allen C
Bullock, Leslie Patricia
Campbell, Kirby I
Cappucci, Dario Ted, Jr
Constantine, Denny G
Corbeil, Lynette B
Fowler, Murray Elwood
Giri, Shri N
Griffin, David William
Hird, David William
Holmberg, Charles Arthur
Hughes, John P
Hyde, Dallas Melvin
Jasper, Donald Edward
Jones, James Henry
Maddy, Keith Thomas
Morgan, Joe Peter
Muller, George Heinz
Parker, Harold R
Plymale, Harry Hambleton
Pritchard, William Roy
Rankin, Alexander Donald
Rhode, Edward A, Jr
Ridgway, Sam H
Riemann, Hans

Schwabe, Calvin Walter
Siegel, Edward T
Siegrist, Jacob C
Sinha, Yagya Nand
Steffey, Eugene P
Theilen, Gordon H
Yamamoto, Richard
Young, Robert, Jr
Zinkl, Joseph Grandjean

COLORADO
Allen, Larry Milton
Barber, Thomas Lynwood
Benjamin, Stephen Alfred
Bennett, Dwight G, Jr
Cockerell, Gary Lee
De Martini, James Charles
Gillette, Edward LeRoy
Glock, Robert Dean
Heath, Robert Bruce
Hilmas, Duane Eugene
Hoerlein, Alvin Bernard
Keefe, Thomas J
Kemp, Graham Elmore
Kingman, Harry Ellis, Jr
Larson, Kenneth Allen
Lebel, Jack Lucien
Lee, Arthur Clair
Lee, Robert Edward
Liddle, Charles George
Lumb, William Valjean
Park, Richard Dee
Ranu, Rajinder S
Rollin, Bernard Elliot
Smith, Ralph E
Tucker, Alan
Vetterling, John Martin

CONNECTICUT
Della-Fera, Mary Anne
Holzworth, Jean
Kersting, Edwin Joseph
Lieberman, James
Post, John E
Schaaf, Thomas Ken
Yang, Tsu-Ju (Thomas)

DELAWARE
Dohms, John Edward
Reed, Theodore H(arold)

DISTRICT OF COLUMBIA
Carlson, William Dwight
Crawford, Lester M
Dua, Prem Nath
Klein, Terry Allen
Miller, William Robert
Thiermann, Alejandro Bories
Wolfle, Thomas Lee
Zook, Bernard Charles

FLORIDA
Burridge, Michael John
Buss, Daryl Dean
Campbell, Clarence L, Jr
Carter, James Harrison, II
Clemmons, Roger M
Colahan, Patrick Timothy
Cooperrider, Donald Elmer
Courtney, Charles Hill
Decker, Winston M
Donovan, Edward Francis
Drummond, Willa H
Grafton, Thurman Stanford
Merritt, Alfred M, II
Moreland, Alvin Franklin
Murchison, Thomas Edgar
Orthoefer, John George
Porter, James Armer, Jr
Povar, Morris Leon
Riser, Wayne H
Samuelson, Don Arthur

GEORGIA
Brackett, Benjamin Gaylord
Brugh, Max, Jr
Clark, James Derrell
Cornelius, Larry Max
Crowell-Davis, Sharon Lynn
Edwards, Gaylen Lee
Finco, Delmar R
Hatch, Roger Conant
Jain, Anant Vir
Jones, Jimmy Barthel
Kaplan, William
Kaufmann, Arnold Francis
McClure, Harold Monroe
McMurray, Birch Lee
Miller, Paul George
Odend'hal, Stewart
Oliver, John Eoff, Jr
Patterson, William Creigh, Jr
Ragland, William Lauman, III
Rawlings, Clarence Alvin
Roberson, Edward Lee
Schantz, Peter Mullineaux
Shotts, Emmett Booker, Jr
Trim, Cynthia Mary
Williams, David John, III

HAWAII
DeTray, Donald Ervin

Veterinary Medicine (cont)

IDAHO
Frank, Floyd William

ILLINOIS
Ames, Edward R
Bennett, Basil Taylor
Brodie, Bruce Orr
Cera, Lee M
Davis, Lloyd Edward
Ferguson, James L
Flynn, Robert James
Ford, Thomas Matthews
Fritz, Thomas Edward
Gordon, Donovan
Hannah, Harold Winford
Hanson, Lyle Eugene
Hillidge, Christopher James
Khan, Mohammed Nasrullah
Koritz, Gary Duane
Manohar, Murli
Nath, Nrapendra
Patterson, Douglas Reid
Port, Curtis DeWitt
Reynolds, Harry Aaron, Jr
Ristic, Miodrag
Schiller, Alfred George
Small, Erwin
Tekeli, Sait
Thurmon, John C
Tripathy, Deoki Nandan
Upham, Roy Walter
Wagner, William Charles
Whitmore, Howard Lloyd
Woods, George Theodore

INDIANA
Amstutz, Harold Emerson
Babcock, William Edward
Brush, F(ranklin) Robert
Eisenbrandt, David Lee
Freeman, Arthur
Gale, Charles
Jones, Russell K
Matsuoka, Tats
Morter, Raymond Lione
Neher, George Martin
Pollard, Morris
Stump, John Edward
Van Sickle, David C
Weirich, Walter Edward
Williams, Robert Dee

IOWA
Adams, Donald Robert
Baker, Durwood L
Beran, George Wesley
Boney, William Arthur, Jr
Christensen, George Curtis
Cutlip, Randall Curry
Dougherty, Robert Watson
Frey, Merwin Lester
Ghoshal, Nani Gopal
Grier, Ronald Lee
Hughes, David Edward
Jackson, Larry LaVern
Kluge, John Paul
Leman, Allen Duane
Lundvall, Richard
McClurkin, Arlan Wilbur
Merkley, David Frederick
Miller, Janice Margaret Lilly
O'Berry, Phillip Aaron
Packer, R(aymond) Allen
Pearson, Phillip T
Roth, James A
Sweat, Robert Lee
Van Houweling, Cornelius Donald
Wass, Wallace M
Wood, Richard Lee

KANSAS
Anderson, Neil Vincent
Anthony, Harry D
Bartkoski, Michael John, Jr
Burch, Clark Wayne
Erickson, Howard Hugh
Gray, Andrew P
Kelley, Donald Clifford
Kennedy, George Arlie
McVey, David Scott
Minocha, Harish C
Mosier, Jacob Eugene
Oehme, Frederick Wolfgang
Stevens, David Robert
Upson, Dan W
Vorhies, Mahlon Wesley

KENTUCKY
Gochenour, William Sylva
Gross, David Ross
Gupta, Ramesh C
Issel, Charles John
Olds, Durward
Stevens, Alan Douglas
Tobin, Thomas

LOUISIANA
Beadle, Ralph Eugene
Besch, Everett Dickman
Hensley, John Coleman, II
Huxsoll, David Leslie
Stewart, T Bonner

Venugopalan, Changaram

MAINE
Donahoe, John Philip
Fackelman, Gustave Edward
Haynes, N Bruce
Myers, David Daniel

MARYLAND
Ahl, Alwynelle S
Bingham, Gene Austin
Chou, Nelson Shih-Toon
Das, Naba Kishore
Denniston, Joseph Charles
Dubey, Jitender Prakash
Earl, Francis Lee
Evarts, Ritva Poukka
Fayer, Ronald
Ganaway, James Rives
Gluckstein, Fritz Paul
Groopman, John Davis
Guest, Gerald Bentley
Hildebrandt, Paul Knud
Hoberg, Eric Paul
Holman, John Ervin, Jr
Joshi, Sewa Ram
Kahrs, Robert F
Kawalek, Joseph Casimir, Jr
Ladson, Thomas Alvin
LaSalle, Bernard
Liebelt, Annabel Glockler
Marquardt, Warren William
Mohanty, Sashi B
Moore, Roscoe Michael, Jr
Moreira, Jorge Eduardo
Norcross, Marvin Augustus
Olson, Leonard Carl
Page, Norbert Paul
Parkhie, Mukund Raghunathrao
Peterson, Irvin Leslie
Potkay, Stephen
Robens, Jane Florence
Russell, Robert John
Sarma, Padman S
Schricker, Robert Lee
Teske, Richard H
Thapar, Nirwan T
Thompson, Clarence Henry, Jr
Torres-Anjel, Manuel Jose
Vadlamudi, Sri Krishna
Vickers, James Hudson
Walton, Thomas Edward
Ward, Fraser Prescott
Weisbroth, Steven H
Whitehair, Leo A

MASSACHUSETTS
Anwer, M Sawkat
Cook, William Robert
Cotter, Susan M
Fleischman, Robert Werder
Foster, Henry Louis
Fox, James Gahan
Hinrichs, Katrin
Jonas, Albert Moshe
King, Norval William, Jr
Loew, Franklin Martin
Nielsen, Svend Woge
Papaioannou, Virginia Eileen
Pratt, George Woodman, Jr
Ross, James Neil, Jr
Schwartz, Anthony
Sevoian, Martin
Siegmund, Otto Hanns
Snoeyenbos, Glenn Howard
Tashjian, Robert John

MICHIGAN
Baker, John Christopher
Branstetter, Daniel G
Brieland, Joan Kathryn
Brown, Roger E
Chang, Timothy Scott
Cohen, Bennett J
Derksen, Frederik Jan
Dutta, Saradindu
Fadly, Aly Mahmoud
Haas, Kenneth Brooks, Jr
Keller, Waldo Frank
Lund, John Edward
Mather, Edward Chantry
Page, Edwin Howard
Piper, Richard Carl
Reddy, Chilekampalli Adinarayana
Ringler, Daniel Howard
Robinson, Norman Edward
Sawyer, Donald C
Sikarskie, James Gerard
Stinson, Al Worth
Stocking, Gordon Gary
Subramanian, Marappa G
Tasker, John B
Trapp, Allan Laverne
Watts, Jeffrey Lynn
Webster, Harris Duane
Witter, Richard L
Yancey, Robert John, Jr

MINNESOTA
Atluru, Durgaprasadarao
Case, Marvin Theodore
Diesch, Stanley L
Fletcher, Thomas Francis
Hansen, Mike

Johnson, Donald W
Kumar, Mahesh C
Lange, Robert Echlin, Jr
Larson, Vaughn Leroy
Manning, Patrick James
Muscoplat, Charles Craig
Nelson, Robert A
Osborne, Carl Andrew
Polzin, David J
Pomeroy, Benjamin Sherwood
Schlotthauer, John Carl
Sorensen, Dale Kenwood
Spurrell, Francis Arthur
Stoner, John Clark
Stowe, Clarence M
Wallace, Larry J

MISSISSIPPI
Jennings, David Phipps
Jones, Eric Wynn
Purchase, Harvey Graham

MISSOURI
Blenden, Donald C
Hahn, Allen W
Kohlmeier, Ronald Harold
McCune, Emmett L
Niemeyer, Kenneth H
Olson, Gerald Allen
Rosenquist, Bruce David
St Omer, Vincent Victor
Schmidt, Donald Arthur
Tritschler, Louis George
Wagner, Joseph Edward

MONTANA
Speer, Clarence Arvon
Young, David Marshall

NEBRASKA
Clemens, Edgar Thomas
Conrad, Robert Dean
Kolar, Joseph Robert, Jr
Kwang, Jimmy
Littledike, Ernest Travis
Schmitz, John Albert
Schneider, Norman Richard
Swift, Brinton L
Torres-Medina, Alfonso
White, Raymond Gene

NEVADA
Simmonds, Richard Carroll

NEW HAMPSHIRE
Allen, Fred Ernest
Langweiler, Marc

NEW JERSEY
Abrutyn, Donald
Benson, Charles Everett
Fabry, Andras
Fraser, Clarence Malcolm
Hayes, Terence James
Hessler, Jack Ronald
Jochle, Wolfgang
Johnson, Allen Neill
McCoy, John Roger
Medway, William
Saini, Ravinder Kumar
Shor, Aaron Louis
Stark, Dennis Michael
Tudor, David Cyrus
Wang, Guang Tsan
Willson, John Ellis

NEW MEXICO
Cummins, Larry Bill
Hahn, Fletcher Frederick
Hobbs, Charles Henry
Muggenburg, Bruce Al
Saunders, George Cherdron

NEW YORK
Antczak, Douglass
Babish, John George
Bloom, Stephen Earl
Callis, Jerry Jackson
Capps, Susan Glass
Center, Sharon Anne
Chishti, Muhammad Azhar
Fox, Francis Henry
Georgi, Jay R
Gilbert, Robert Owen
Gillespie, James Howard
Habel, Robert Earl
Hall, Charles E
Hard, Gordon Charles
Hillman, Robert B
Kallfelz, Francis A
Kirk, Robert Warren
Lowe, John Edward
Maestrone, Gianpaolo
Marshak, Robert Reuben
Moise, Nancy Sydney
O'Donoghue, John Lipomi
Phemister, Robert David
Pogue, John Parker
Poppensiek, George Charles
Postle, Donald Sloan
Quimby, Fred William
Reimers, Thomas John
Sack, Wolfgang Otto
Schlafer, Donald H

Scott, Fredric Winthrop
Smith, Donald Frederick
Som, Prantika
Tennant, Bud C
Thompson, John C, Jr
Tonelli, George
Zwain, Ismail Hassan

NORTH CAROLINA
Aronson, A L
Boorman, Gary Alexis
Breitschwerdt, Edward B
Clarkson, Thomas Boston
Coffin, David L
Curtin, Terrence M
Gilfillan, Robert Frederick
Gonder, Eric Charles
Hamm, Thomas Edward, Jr
Hamner, Charles Edward, Jr
Howard, Donald Robert
Johnson, George Andrew
Kohn, Frank S
Kuo, Eric Yung-Huei
Lay, John Charles
McClellan, Roger Orville
McConnell, Ernest Eugene
McPherson, Charles William
Noga, Edward Joseph
Page, Rodney Lee
Peiffer, Rober Louis, Jr
Pick, James Raymond
Popp, James Alan
Riviere, Jim Edmond
Sar, Madhabananda
Smallwood, James Edgar
Stoskopf, Michael Kerry
Tucker, Walter Eugene, Jr
Webb, Alfreda Johnson
Wright, James Francis

OHIO
Alden, Carl L
Bertram, Timothy Allyn
Boyle, Michael Dermot
Burt, James Kay
Clifford, Donald H
Diesem, Charles D
Dorn, Charles Richard
Gabel, Albert A
Gibson, John Phillips
Hull, Bruce Lansing
Jones, James Edward
Kreier, Julius Peter
Mattingly, Steele F
Muir, William W, III
Murdick, Philip W
Myer, Carole Wendy
Nokes, Richard Francis
Oglesbee, Michael Jerl
Olson, Carl Thomas
Ray, Richard Schell
Redman, Donald Roger
Rudy, Richard L
Saif, Linda Jean
Saif, Yehia M(ohamed)
Stuhlman, Robert August
Szczepanski, Marek Michal
Tarr, Melinda Jean
Tharp, Vernon Lance
Wallin, Richard Franklin
Wilson, George Porter, III
Wyman, Milton

OKLAHOMA
Alexander, Joseph Walker
Buckner, Ralph Gupton
Campbell, Gregory Alan
Confer, Anthony Wayne
Ewing, Sidney Alton
Faulkner, Lloyd (Clarence)
Friend, Jonathon D
Graves, Donald C
Hooper, Billy Ernest
Rolf, Lester Leo, Jr
Still, Edwin Tanner

OREGON
Blythe, Linda L
Hill, John Donald

PENNSYLVANIA
AcKerman, Larry Joseph
Andrews, Edwin Joseph
Babu, Uma Mahesh
Bovee, Kenneth C
Deubler, Mary Josephine
Eberhart, Robert J
Eichberg, Jorg Wilhelm
Fogleman, Ralph William
Fowler, Edward Herbert
Garman, Robert Harvey
Haskins, Mark
Herlyn, Dorothee Maria
Hsu, Chao Kuang
Jezyk, Peter Franklin
Lang, C Max
Liebenberg, Stanley Phillip
Live, Israel
McFeely, Richard Aubrey
Melby, Edward C, Jr
Miselis, Richard Robert
Morrow, David Austin, III
Onik, Gary M
Patterson, Donald Floyd

Peardon, David Lee
Pirone, Louis Anthony
Prasad, Suresh
Raker, Charles W
Rattan, Satish
Reid, Charles Feder
Rhodes, William Harker
Rozmiarek, Harry
Scott, George Clifford
Silverman, Jerald
Simmons, Donald Glick
Whitlock, Robert Henry
Wolfe, John Hall
Zarkower, Arian

SOUTH CAROLINA
Von Recum, Andreas F
Yager, Robert H

SOUTH DAKOTA
Hurley, David John
Kirkbride, Clyde Arnold

TENNESSEE
Armistead, Willis William
Bradley, Richard E
Cypess, Raymond Harold
Eiler, Hugo
Harvey, Ralph Clayton
Kant, Kenneth James
New, John Coy, Jr
Oliver, Jack Wallace
Sachan, Dileep Singh
Walburg, H E

TEXAS
Bailey, Everett Murl, Jr
Banks, William Joseph, Jr
Barnes, Charles M
Beaver, Bonnie Veryle
Bridges, Charles Hubert
Coleman, John Dee
Crandell, Robert Allen
Dubose, Robert Trafton
Edds, George Tyson
Ellett, Edwin Willard
Fidler, Isaiah J
Galvin, Thomas Joseph
Green, Ronald W
Haensly, William Edward
Hartsfield, Sandee Morris
Harvey, Roger Bruce
Heavner, James E
Hightower, Dan
Jennings, Paul Bernard, Jr
Kiel, Johnathan Lloyd
Kier, Ann B
Lees, George Edward
Loan, Raymond Wallace
Messersmith, Robert E
Meuten, Donald John
Murnane, Thomas George
Murthy, Krishna Kesava
Pakes, Steven P
Pierce, Kenneth Ray
Playter, Robert Franklin
Price, Alvin Audis
Roenigk, William J
Rosborough, John Paul
Smith, Malcolm Crawford, Jr
Steele, James Harlan
Szabuniewicz, Michael
Tizard, Ian Rodney
Whitford, Howard Wayne

UTAH
Blake, Joseph Thomas
Healey, Mark Calvin
Hoopes, Keith Hale
Larsen, Austin Ellis
McGill, Lawrence David
Roeder, Beverly Louise
Shelby, Nancy Jane
Shupe, James LeGrande
Thomas, Don Wylie

VIRGINIA
Blood, Benjamin Donald
Brown, Elise Ann Brandenburger
Colmano, Germille
Cordes, Donald Ormond
Cruthers, Larry Randall
Held, Joe R
Lee, Robert Jerome
McVicar, John West
Misra, Hara Prasad
Todd, Frank Arnold
Webb, Willis Keith
Wolfe, Barbara Ann

WASHINGTON
Allen, Julia Natalia
Baksi, Samarendra Nath
Brobst, Duane Franklin
Bustad, Leo Kenneth
Cho, Byung-Ryul
Dennis, Melvin B, Jr
Evermann, James Frederick
Gillis, Murlin Fern
Gorham, John Richard
Graves, Scott Stoll
Hardesty, Linda Howell
Hostetler, Roy Ivan
McDonald, John Stoner

Noel, Jan Christine
Ott, Richard L
Prieur, David John
Ragan, Harvey Albert
Ratzlaff, Marc Henry
Sande, Ronald Dean
Smith, Dean Harley
Van Hoosier, Gerald L, Jr
Whitney, Robert Arthur, Jr
Zamora, Cesario Siasoco
Zander, Donald Victor

WEST VIRGINIA
Dozsa, Leslie
Golway, Peter L
Olson, Norman O

WISCONSIN
Anand, Amarjit Singh
Berman, David Theodore
Cook, Mark Eric
Crowe, Dennis Timothy, Jr
Czuprynski, Charles Joseph
Dueland, Rudolf (Tass)
Easterday, Bernard Carlyle
Fahning, Melvyn Luverne
Hall, Robert Everett
Howard, Thomas Hyland
Larson, Lester Leroy
Nichols, Roy Elwyn
Price, Susan Gail
Splitter, Gary Allen
Will, James Arthur

WYOMING
Pier, Allan Clark

ALBERTA
Coppock, Robert Walter

BRITISH COLUMBIA
Booth, Amanda Jane
Salinas, Fernando A

NEW BRUNSWICK
Smith, Harry John

ONTARIO
Archibald, James
Atwal, Onkar Singh
Barnum, Donald Alfred
Batra, Tilak Raj
Belbeck, L W
Bouillant, Alain Marcel
Derbyshire, John Brian
Fisher, Kenneth Robert Stanley
Frederick, George Leonard
Hare, William Currie Douglas
Hulland, Thomas John
Kaushik, Azad
Lang, Gerhard Herbert
Miniats, Olgerts Pauls
Morrison, Spencer Horton
Nielsen, Klaus H B
Nielsen, N Ole
Robertson, James McDonald
Ruckerbauer, Gerda Margareta
Samagh, Bakhshish Singh
Savan, Milton
Shewen, Patricia Ellen
Slocombe, Joseph Owen Douglas
Stemshorn, Barry William
Wilkie, Bruce Nicholson
Willoughby, Russell A

PRINCE EDWARD ISLAND
Amend, James Frederick
Singh, Amreek

QUEBEC
Archambault, Denis
Barrette, Daniel Claude
Evering, Winston E N D
Lussier, Gilles L
Robert, Suzanne

SASKATCHEWAN
Blakley, Barry Raymond
Clayton, Hilary Mary
Leighton, Frederick Archibald

OTHER COUNTRIES
Aalund, Ole
Adldinger, Hans Karl
Baggot, J Desmond
Davidson, David Edward, Jr
Ishida, Yukisato
Johnsen, Dennis O
Lutsky, Irving
Makita, Takashi
Mitruka, Brij Mohan
Pan, In-Chang
Schmeling, Sheila Kay
Shore, Laurence Stuart
Usenik, Edward A
Yovich, John V

Other Medical & Health Sciences

ALABAMA
Borton, Thomas Ernest
Briles, David Elwood

Clark, Charles Edward
Crispens, Charles Gangloff, Jr
Duque, Ricardo Ernesto
Funkhouser, Jane D
Gay, Renate Erika
Grubbs, Clinton Julian
Hardy, Robert W
Lloyd, L Keith
Mundy, Roy Lee
Pass, Robert Floyd
Pohost, Gerald M
Ramey, Craig T
Reeves, R C
Richards, J Scott
Wilson, G Dennis

ALASKA
Hardin, James W
Mills, William J, Jr
Segal, Bernard

ARIZONA
Allen, John Jeffery
Appelgren, Walter Phon
Birkby, Walter H
Black, Dean
Butler, Lillian Catherine
Cate, Rodney Michael
Colburn, Wayne Alan
Eagle, Donald Frohlichstein
Goldman, Steven
Jeter, Wayburn Stewart
Kahn, Marvin William
Kazal, Louis Anthony
LaPointe, Leonard Lyell
Lohman, Timothy George
McKinley, Michael P
McQueen, Charlene A
Mossman, Kenneth Leslie
Nardella, Francis Anthony
Rowe, Verald Keith
Scott, Ralph Asa, Jr
Soren, David
Stini, William Arthur
Tong, Theodore G
Villanueva, Antonio R

ARKANSAS
Andreoli, Thomas E
Berky, John James
Burleigh, Joseph Gaynor
Kadlubar, Fred F
Light, Kim Edward
Sedelow, Walter Alfred, Jr
Slikker, William, Jr
Wenger, Galen Rosenberger
Wolff, George Louis

CALIFORNIA
Abrams, Herbert L
Abramson, Edward E
Akin, Gwynn Collins
Andre, Michael Paul
Andrews, Russell J
Applegarth, Adrienne P
Araujo, Fausto G
Assali, Nicholas S
Astrahan, Melvin Alan
Baker, Joffre Bernard
Banks, William Preas
Barnhart, James Lee
Basham, Teresa
Batchelder, William Howard
Beutler, Larry Edward
Bigler, William Norman
Blatt, Joel Martin
Brantingham, Charles Ross
Brasel, Jo Anne
Brubacher, Elaine Schnitker
Brummer, Elmer
Burnett, Bryan Reeder
Butcher, Larry L
Cailleau, Relda
Carlisle, Harry J
Cassel, Russell N
Castro, Joseph Ronald
Chao, Tony M
Chen, Fu-Tai Albert
Chen, Peter C Y
Chisari, Francis Vincent
Chu, William Tongil
Clever, Linda Hawes
Cohen, Malcolm Martin
Cohn, Stanton Harry
Conner, Brenda Jean
Cooper, Allen D
Cooper, Howard K
Cornford, Eain M
Courchesne, Eric
Cromwell, Florence S
Curry, Cynthia Jane
Davis, William Thompson
Derenzo, Stephen Edward
Dillmann, Wolfgang H
Disaia, Philip John
Dorr, Lawrence D
Ebbe, Shirley Nadine
Eckhert, Curtis Dale
Edwards, Ward Dennis
Eger, Edmond I, II
Elias, Peter M
Ellestad, Myrvin H
Epstein, John Howard
Epstein, Lois Barth

Fallon, James Harry
Fehling, Michael R
Feifel, Herman
Fessenden, Peter
Finn, James Crampton, Jr
Finston, Roland A
Fisher, Delbert A
Flacke, Joan Wareham
Friedman, Alan E
Friess, Seymour Louis
Fuchs, Victor Robert
Furtado, Victor Cunha
Geiselman, Paula J
Georghiou, George Paul
Gerich, John Edward
Giannetti, Ronald A
Gibson, Thomas Alvin, Jr
Gill, James Edward
Gluck, Louis
Gold, Richard Horace
Golde, David William
Goode, John Wolford
Gooding, Gretchen Ann Wagner
Gould, Robert George
Greene, John Clifford
Greenhouse, N Anthony
Greenleaf, John Edward
Griffin, John Henry
Griffiss, John McLeod
Grumbach, Melvin Malcolm
Haas, Gustav Frederick
Hahn, Theodore John
Hainski, Martha Barrionuevo
Hammerschlag, Richard
Harris, John Wayne
Harwood, Ivan Richmond
Hawkins, David A
Hislop, Helen Jean
Hoffman, Arlene Faun
Hoffman, Donald D
Hollenbeck, Clarie Beall
Holmes, Donald Eugene
Holtzman, Samuel
Hopf, Harriet Dudey Williams
House, John W
Houston, L L
Hubbard, Eric R
Jacobs, S Lawrence
Johnson, Cage Saul
Jones, Joie Pierce
Jordan, Stanley Clark
Jurris, Eric R
Keene, Clifford H
Keith, Robert Allen
Kempler, Walter
Kempter, Charles Prentiss
Kessler, Seymour
Kohn, Henry Irving
Kortright, James McDougall
Kosow, David Phillip
Kraft, Lisbeth Martha
Kripke, Daniel Frederick
Krippner, Stanley Curtis
Kritz-Silverstein, Donna
Kubler-Ross, Elisabeth
Largman, Corey
Lau, Kin-Hing William
Lee, David Anson
Levy, Ronald
Lindsey, Dortha Ruth
Litt, Lawrence
Liu, John
London, Ray William
Lopo, Alina C
Luft, Harold Stephen
McBride, William H
Makowka, Leonard
Marmor, Michael F
Marshall, Sally J
Mason, Dean Towle
Mastropaolo, Joseph
Meerdink, Denis J
Mills, Gary Kenith
Mitsuhashi, Masato
Mondon, Carl Erwin
Montgomery, Leslie D
Morewitz, Harry Alan
Moxley, John H, III
Mulla, Mir Subhan
Murphree, A Linn
Nelson, Walter Ralph
Nicholson, Arnold Eugene
Nickerson, Bruce Greenwood
Niklowitz, Werner Johannes
Nowlis, David Peter
Nydegger, Corinne Nemetz
Overstreet, James Wilkins
Pallak, Michael
Palmer, Alan
Palmer, Beverly B
Pardridge, William M
Parer, Julian Thomas
Pearson, Robert Bernard
Pelletier, Charles A
Peper, Erik
Peterlin, Boris Matija
Peterson-Falzone, Sally Jean
Pigiet, Vincent P
Pitas, Robert E
Polan, Mary Lake
Poland, Russell E
Ramachandran, Vilayanur Subramanian
Rapaport, Elliot
Raven, Bertram H(erbert)

Other Medical & Health Sciences (cont)

Reitz, Richard Elmer
Rheins, Lawrence A
Richman, David Paul
Richters, Arnis
Rosemark, Peter Jay
Rosenthal, Allan Lawrence
Rousell, Ralph Henry
Ryan, Stephen Joseph
Salel, Antone Francis
Salmon, Eli J
Samuel, Charles Edward
Schmidt, Robert Milton
Schweisthal, Michael Robert
Scrimshaw, Susan
Seibert, J A
Shanks, Susan Jane
Shapero, Sanford Marvin
Shapiro, David
Silagi, Selma
Smiles, Kenneth Albert
Smith, Michael R
Snead, O Carter, III
Sohn, Yung Jai
Soule, Roger Gilbert
Spanis, Curt William
Sperling, Jacob L
Spiro, Melford Elliot
Stanczyk, Frank Zygmunt
Stark, Allen Ross
Steckel, Richard J
Steffey, Eugene P
Strauch, Ralph Eugene
Sullivan, Stuart F
Swanson, Anne Barrett
Sweet, Benjamin Hersh
Syme, S Leonard
Taeusch, H William
Teng, Evelyn Lee
Teresi, Joseph Dominic
Thomas, Ralph Harold
Thompson, John Frederick
Tin-Wa, Maung
Tobis, Jerome Sanford
Torfs, Claudine Pierette
Toy, Arthur John, Jr
Tuckson, Reed V
Ulmer, Raymond Arthur
Urist, Marshall Raymond
Urquhart, John, III
Wade, Michael James
Walsh, John H
Waskell, Lucy A
Weliky, Norman
Whalen, Carol Kupers
Wieland, Bruce Wendell
Withers, Hubert Rodney
Wright, James Edward
Yager, Janice L Winter
Yagiela, John Allen
Yu, Jen

COLORADO
Balke, Bruno
Barton, Jeffrey W
Berge, Trygve O
Conger, John Janeway
Cordain, Loren
Cosgriff, Thomas Michael
Ellinwood, William Edward
Ellis, Frank Russell
Hager, Jean Carol
Hall, Alan H
Heim, Werner George
Lee, Arthur Clair
Leung, Donald Yap Man
McCalden, Thomas A
Pearson, John Richard
Riches, David William Henry
Schonbeck, Niels Daniel
Schrier, Robert William
Suinn, Richard M
Tucker, Alan
Verdeal, Kathey Marie
Weston, William Lee
Wilber, Charles Grady
Zimet, Carl Norman

CONNECTICUT
Alexander, Jonathan
Alpert, Nelson Leigh
Aune, Thomas M
Ballantyne, Bryan
Bartoshuk, Linda May
Berdick, Murray
Bohannon, Richard Wallace
Booss, John
Carroll, J Gregory
Chatt, Allen Barrett
Crean, Geraldine L
Dolan, Thomas F, Jr
Doshan, Harold David
Friedlaender, Gary Elliot
Grattan, James Alex
Kitahata, Luke Masahiko
Lynch, John Edward
Miller, Neal Elgar
Myers, Maureen
Potluri, Venkateswara Rao
Salovey, Peter
Siegel, Norman Joseph
Smith, Abigail LeVan
Steinmetz, Philip R
Sternberg, Robert Jeffrey
Sulzer-Azaroff, Beth
Verdi, James L
Warshaw, Joseph B
Wilhelm, Scott M
Zaret, Barry L
Zigler, Edward

DELAWARE
Davis, Leonard George
Frey, William Adrian
Johnson, Donald Richard
Krahn, Robert Carl
Law, Amy Stauber
Read, Robert E
Reinhardt, Charles Francis
Ruben, Regina Lansing
Steinberg, Marshall

DISTRICT OF COLUMBIA
Armaly, Mansour F
Atkins, James Lawrence
Bowen, Donnell
Chung, Ho
Drew, Robert Taylor
Eisner, Gilbert Martin
Franchini, Genoveffa
Gelmann, Edward P
Goldson, Alfred Lloyd
Hamarneh, Sami Khalaf
Holland, Christie Anna
Jones, Stanley B
Joseph, Stephen C
Kornhauser, Andrija
Kromer, Lawrence Frederick
Ladisch, Stephan
Larsen, Lynn Alvin
Ledley, Robert Steven
Leto, Salvatore
Lippman, Marc Estes
Melnick, Vijaya L
Milbert, Alfred Nicholas
Nightingale, Elena Ottolenghi
O'Connell, Daniel Craig
Pearlman, Ronald C
Perry, David Carter
Perry, Seymour Monroe
Pinkerson, Alan Lee
Pittman, Margaret
Rosenquist, Glenn Carl
Salu, Yehuda
Schwartz, Marshall Zane
Silver, Sylvia
Slaughter, Lynnard J
Stark, Nathan Julius
Tyner, C Fred
Vick, James
Vladeck, Bruce C
Weinberg, Myron Simon
Yeager, Henry, Jr
Zawisza, Julie Anne A

FLORIDA
Abitbol, Carolyn Larkins
Agarwal, Ram P
Amaranath, L
Arenberg, David (Lee)
Berkley, Karen J
Brim, Orville G, Jr
Cohen, David M
Cowan, Frederick Pierce
Dewanjee, Mrinal K
Fackler, Martin L
Faleschini, Richard John
Feldman, Alan Sidney
Fitzgerald, Lawrence Terrell
Frank, H Lee
Freudenthal, Ralph Ira
Goldstein, Mark Kane
Gorniak, Gerard Charles
Gosselin, Arthur Joseph
Gottlieb, Charles F
Gould, Anne Bramlee
Habal, Mutaz
Halpern, Ephriam Philip
Haynes, Duncan Harold
Henson, Carl P
Johnson, James Harmon
Kennedy, Robert Samuel
Kunce, Henry Warren
Lewis, Brian Kreglow
Lo, Hilda K
Manhold, John Henry, Jr
Margo, Curtis Edward
Meltzer, Martin Isaac
Mercer, Thomas T
Miller, James Woodell
Pargman, David
Patterson, Richard Sheldon
Peters, Thomas G
Pollock, Michael L
Routh, Donald Kent
Schein, Jerome Daniel
Silverstein, Herbert
Simpkins, James W
Simring, Marvin
Thornthwaite, Jerry T
Whitman, Victor
Wingo, Charles S

GEORGIA
Adams, David B
Adkison, Linda Russell
Allen, Delmas James
Allison, Jerry David
Baranowski, Tom
Brooks, John Bill
Buckalew, Louis Walter
Caplan, Daniel Bennett
Cavanagh, Harrison Dwight
Colborn, Gene Louis
Copeland, Robert B
David, Robert Michael
Dingledine, Raymond J
Faraj, Bahjat Alfred
Feldman, Stuart
Fridovich-Keil, Judith Lisa
Gordon, David Stewart
Greenberg, Raymond Seth
Gupton, Guy Winfred, Jr
Hatch, Roger Conant
Hersh, Theodore
Hugus, Barbara Hendrick
Inge, Walter Herndon, Jr
King, Frederick Alexander
Kokko, Juha Pekka
Krauss, Jonathan Seth
Lawley, Thomas J
Loring, David William
McDade, Joseph Edward
Martin, David Edward
Parthasarathy, Sampath
Porterfield, Susan Payne
Potter, David Edward
Rigdon, Raymond Harrison
Smith, David Fletcher
Sridaran, Rajagopala
Stevenson, Dennis A
Stone, Connie J
Taylor, Andrew T, Jr
Van Assendelft, Onno Willem
Wolf, Steven L
Wood, William C
Zaia, John Anthony

HAWAII
Diamond, Milton
Howarth, Francis Gard
Lance, Eugene Mitchell
Manly, Philip James
Nelson, Marita Lee
Yates, James T

IDAHO
McCune, Ronald William
Vestal, Robert Elden

ILLINOIS
Alexander, Kenneth Ross
Andreoli, Kathleen Gainor
Bakkum, Barclay W
Barenberg, Sumner
Bevan, William
Biller, Jose
Bitran, Jacob David
Boulos, Badi Mansour
Brutten, Gene J
Buckman, Robert W
Campbell, Suzann Kay
Carnes, Bruce Alfred
Cember, Herman
Cera, Lee M
Chandler, John W
Cohen, Louis
Cohen, Melvin R
Constantinou, Andreas I
Cook, L Scott
Debus, Allen George
Doege, Theodore Charles
Dunn, Dorothy Fay
Eriksen, Charles Walter
Ernest, J Terry
Fang, Victor Shengkuen
Flay, Brian Richard
Gamelli, Richard L
Gastwirth, Bart Wayne
Giometti, Carol Smith
Glisson, Silas Nease
Goebel, Maristella
Goldman, Allen S
Grdina, David John
Gross, Thomas Lester
Heller, Michael
Helseth, Donald Lawrence, Jr
Hermann, Edward Robert
Hoffman, William E
Ivankovich, Anthony D
Johnston, Patricia V
Kinders, Robert James
Kleinman, Kenneth Martin
Klocke, Francis J
Kochman, Ronald Lawrence
Kornel, Ludwig
Kulkarni, Bidy
Lapin, Gregory D
Lauterbur, Paul Christian
Lee, Jang Y
Lehman, Dennis Dale
Levy, Jerre Marie
Luborsky, Judith Lee
Lurain, John Robert, III
Lyon, Edward Spafford
McWhinnie, Dolores J
Mehta, Rajendra G
Melhado, L(ouisa) Lee
Meyer, Paul Reims, Jr
Moray, Neville Peter
Neugarten, Bernice L
O'Leary, Dennis Sophian
Olshansky, S Jay
Oscai, Lawrence B
Ratajczak, Helen Vosskuhler
Robertson, Abel Alfred Lazzarini, Jr
Rosen, Steven Terry
Saclarides, Theodore John
Sharma, Ram Ratan
Sherman, Laurence A
Solt, Dennis Byron
Spencer, Herta
Stevenson, G W
Strain, James E
Taber, David
Taswell, Howard Filmore
Taylor, D Dax
Taylor, Henry L
Thomas, Joseph Erumappettical
Vyborny, Carl Joseph
Weichselbaum, Ralph R
Woo, Lecon
Wynveen, Robert Allen
Zsigmond, Elemer K

INDIANA
Baird, William McKenzie
Bowers, Larry Donald
Bridges, C David
Bumol, Thomas F
Carlson, James C
Eigen, Howard
Friedman, Julius Jay
George, Robert Eugene
Gibbs, Frederic Andrews
Guth, S(herman) Leon
Hathaway, David Roger
Hayes, J Scott
Hendrix, Jon Richard
Hinsman, Edward James
Hui, Siu Lui
Jayaram, Hiremagalur N
LaFuze, Joan Esterline
McIntyre, John A
McKinney, Gordon R
Nahrwold, Michael L
Quist, Raymond Willard
Schoknecht, Jean Donze
Shoup, Charles Samuel, Jr
Smith, Gerald Duane
Tyhach, Richard Joseph
Van Etten, Robert Lee
Van Sickle, David C
Waller, Bruce Frank
Wellman, Henry Nelson
Ziemer, Paul L
Zimmerman, Neil J

IOWA
Al-Jurf, Adel Saada
Anderson, Charles V
Bar, Robert S
Bell, Edward Frank
Bonney, William Wallace
Di Bona, Gerald F
Gergis, Samir D
Goff, Jesse Paul
Hunninghake, Gary W
Jordan, Diane Kathleen
Levy, Leonard Alvin
Moorcroft, William Herbert
Orcutt, James A
Rhead, William James
Ruth, Royal Francis
Schrott, Helmut Gunther
Scott, William Edwin
Stewart, Mary E
Tranel, Daniel T
Ziegler, Ekhard E

KANSAS
Cooke, Allan Roy
Dawson, James Thomas
Dykstra, Mark Allan
Ferraro, John Anthony
Hinshaw, Charles Theron, Jr
Krishnan, Engil Kolaj
Maguire, Helen M
Norris, Patricia Ann

KENTUCKY
Budde, Mary Laurence
Ehmann, William Donald
Espinosa, Enrique
Essig, Carl Fohl
Farman, Allan George
Garrity, Thomas F
Giammara, Beverly L Turner Sites
Glassock, Richard James
Henrickson, Charles Henry
Kennedy, John Elmo, Jr
Porter, William Hudson
Randall, David Clark
Shih, Wei Jen
Straus, Robert
Syed, Ibrahim Bijli
Terango, Larry
Tietz, Norbert W
Urano, Muneyasu
Younszai, M Kabir
Zimmerman, Thom J

LOUISIANA
Bazan, Nicolas Guillermo
Beckerman, Robert Cy
Beuerman, Roger Wilmer
Cherry, Flora Finch
Chin, Hong Woo
Christian, Frederick Ade
Cook, Stephen D
Fermin, Cesar Danilo
Frazier, Claude Clinton, III
Jain, Sushil Kumar
Lertora, Juan J L
Leung, Wai Yan
Levitzky, Michael Gordon
McDonald, Marguerite Bridget
Manno, Barbara Reynolds
Manno, Joseph Eugene
Olson, Gayle A
Palmgren, Muriel Signe
Pelias, Mary Zengel
Raisen, Elliott
Riopelle, Arthur J
Roberts, James Allen
Shiau, Yih-Fu
Thompson, James Joseph
Watt, Edward William
Wolf, Thomas Mark
Xie, Jianming
Yaeger, Robert George

MAINE
Lamdin, Ezra
Mead, Jere
Norton, James Michael
Roderick, Thomas Huston
Swazey, Judith P

MARYLAND
Abergel, Chantel
Ader, Deborah N
Anderson, Richard Allen
Augustine, Robertson J
Baquet, Claudia
Baum, Siegmund Jacob
Becker, Bruce Clare
Becker, Lewis Charles
Beebe, Gilbert Wheeler
Blattner, William Albert
Blum, Bruce I
Blum, Robert Allan
Boldry, Robert C
Bresler, Jack Barry
Broder, Samuel
Brown, Kenneth Stephen
Bunyan, Ellen Lackey Spotz
Burton, George Joseph
Carr, Charles Jelleff
Chabner, Bruce A
Chiueh, Chuang Chin
Colburn, Theodore R
Coligan, John E
Cone, Richard Allen
Conley, Barbara A
Contrera, Joseph Fabian
Counts, George W
Coursey, Bert Marcel
Craig, Paul N
Curt, Gregory
Cutler, Gordon Butler, Jr
Dasler, Adolph Richard
David, Henry P
Davis, Cindy D
Domanski, Thaddeus John
Donham, Robert T
Doppman, John L
Duggar, Benjamin Charles
Dunning, Herbert Neal
Dyer, Robert Francis, Jr
Ebert, Ray F
Eckardt, Michael Jon
Edinger, Stanley Evan
Efron, Herman Yale
Engel, Bernard Theodore
Fahy, Gregory Michael
Fauci, Anthony S
Feigal, David William, Jr
Feldman, Robert H L
Ferris, Frederick L
Fex, Jorgen
Fishman, Elliot Keith
Ford, Leslie
Fozard, James Leonard
Friedman, Michael A
Gleich, Carol S
Goldberg, Andrew Paul
Gorden, Phillip
Gordis, Enoch
Grunberg, Neil Everett
Gutierrez, Peter Luis
Hackett, Joseph Leo
Hadley, Evan C
Hager, Gordon L
Hall, Beverly Fenton
Hamill, Peter Van Vechten
Hanig, Joseph Peter
Hanjan, Satnam S
Harkins, Rosemary Knighton
Harris, Curtis C
Hawkins, Michael John
Hazzard, DeWitt George
Heilman, Carol A
Hoak, John C
Hoffman, Julius
Hoofnagle, Jay H
Horan, Michael
Hubbell, John Howard
Johnson, Carl Boone
Johnson, Kenneth Peter
Johnston, Gerald Samuel
Kaper, James Bennett
Kaplan, Ann Esther
Kaplan, Stanley A
Katz, Stephen I
Kaufman, Joyce J
Kaufmann, Peter G
Kelty, Miriam C Friedman
Kessler, David
Killen, John Young, Jr
Kimes, Brian William
Klarman, Herbert E
Klinman, Dennis M
Klippel, John Howard
Knepper, Mark A
Kraemer, Kenneth H
Krantz, David S
Lakatta, Edward G
Laterra, John J
Lathers, Claire M
Lee, Che-Hung R
Leong, Kam W
Levine, Richard Joseph
Lichtenstein, Lawrence M
Liebelt, Annabel Glockler
Linehan, William Marston
Linnoila, Markku
Lowy, Douglas R
Lubet, Ronald A
Ludlow, Christy L
Luecke, Donald H
McKenna, Thomas Michael
McLaughlin, Jack A
McPherson, Robert W
Magrath, Ian
Maumenee, Irene Hussels
Mirsky, Allan Franklin
Mockrin, Stephen Charles
Mohla, Suresh
Murphy, Dennis L
Myslobodsky, M S
Nelson, James Harold
Nino, Hipolito V
Nissley, S Peter
Nussenblatt, Robert Burton
Okunieff, Paul
Osborne, J Scott, III
Osterberg, Robert Edward
Paskins-Hurlburt, Andrea Jeanne
Petty, William Clayton
Pikus, Anita
Pizzo, Philip A
Plaut, Marshall
Provine, Robert Raymond
Quraishi, Mohammed Sayeed
Ranney, J Buckminster
Reddi, A Hari
Riley, Matilda White
Roberts, William C
Roche, Lidia Alicia
Rodgers, Imogene Sevin
Sachs, Murray B
Schifreen, Richard Steven
Severns, Matthew Lincoln
Shoemaker, Dale
Shoukas, Artin Andrew
Shulman, N Raphael
Simpson, Ian Alexander
Smith, Philip Lees
Sowers, Arthur Edward
Spector, Novera Herbert
Spring, Susan B
Stolz, Walter S
Streicher, Eugene
Taylor, Lauriston Sale
Tearney, Russell James
Thompson, Donald Leroy
Tower, Donald Bayley
Triantaphyllopoulos, Eugenie
Tucker, Robert Wilson
Ulsamer, Andrew George, Jr
Umbreit, Thomas Hayden
Watters, Robert Lisle
Wehr, Thomas A
Weisburger, Elizabeth Kreiser
Wheeler, David M
Wilder, Ronald Lynn
Willoughby, Anne D
Wingate, Catharine L
Wintercorn, Eleanor Stiegler
Woods, Wilna Ann
Yoder, Robert E
Young, Eric D
Young, Gerald A

MASSACHUSETTS
Allukian, Myron
Alper, Seth L
Alpert, Joel Jacobs
Aronow, Saul
Banderet, Louis Eugene
Behrman, Richard H
Berman, Marlene Oscar
Berson, Eliot Lawrence
Bing, Oscar H L
Blank, Irvin H
Brenner, John Francis
Brown, Robert
Caldini, Paolo
Campbell, Charlotte Catherine
Cathcart, Edgar S
Charney, Evan
Chen, Bo Lan
Chernosky, Edwin Jasper
Clinton, Steven K
Cohen, Stephen Mark
Cormack, Allan MacLeod
Dehal, Shangara Singh
Deutsch, Thomas F
Dexter, Lewis
Fairbanks, Grant
Fein, Rashi
Fencl, Vladimir
Foster, Charles Stephen
Frelinger, Andrew L
Frisch, Rose Epstein
Gaudette, Leo Eward
Geisterfer-Lowrance, Anja A T
Gelman, Simon
Glowacki, Julie
Goldman, Harvey
Goyal, Raj K
Gutheil, Thomas Gordon
Hammer, Ronald Page, Jr
Hebben, Nancy
Hichar, Joseph Kenneth
Hirschman, Lynette
Holmes, Helen Bequaert
Hurd, Richard Nelson
Junghans, Richard Paul
Kagan, Jerome
Kanarek, Robin Beth
Kasper, Dennis Lee
Kressel, Herbert Yehude
Kusmik, William F
Lackner, James Robert
Landowne, Milton
Lee, Roland Robert
Lemeshow, Stanley Alan
Lidgard, Graham Peter
Lucey, Edgar C
Lux, Samuel E, IV
Madias, Nicolaos E
Mann, Jonathan M
Martin, Thomas George, III
Masek, Bruce James
Mason, Joel B
Milbocker, Michael
Milder, Fredric Lloyd
Modest, Edward Julian
Moreland, Robert Byrl
Morris, Robert
Muni, Indu A
Murrain, Michelle
Murray, Joseph E
Myers, Richard Hepworth
Naeser, Margaret Ann
Naimi, Shapur
Nathan, David G
Neumeyer, David Alexander
Olsen, Robert Thorvald
Palek, Jiri Frant
Parrish, John Albert
Pfeffer, Marc Alan
Philbin, Daniel Michael
Pike, Marilyn Cecile
Prout, Curtis
Rosenthal, Robert
Russo, Dennis Charles
Schaller, Jane Green
Scheuchenzuber, H Joseph
Schorr, Lisbeth Bamberger
Selker, Harry Paul
Siegal, Donald
Skrinar, Gary Stephen
Sledge, Clement Blount
Snyder, Benjamin Willard
Swets, John Arthur
Vogel, James Alan
Volpe, Joseph J
Wegman, David Howe
Wineman, Robert Judson
Wise, Robert J
Wolfe, Harry Bernard
Wyler, David J
Zapol, Warren Myron
Zebrowitz, Leslie Ann
Zhang, Dong Er

MICHIGAN
Allen, Cheryl
Al-Sarraf, Muhyi
Anderson, Thomas Edward
Antonucci, Tammy
Bach, Shirley
Baker, Laurence Howard
Balazovich, Kenneth J
Baumiller, Robert C
Berent, Stanley
Betz, A Lorris
Biermann, Janet Sybil
Brieland, Joan Kathryn
Buchtel, Henry Augustus, IV
Callewaert, Denis Marc
Chong, Kong Teck
Dijkers, Marcellinus P
Dumke, Paul Rudolph
Ferraro, Douglas P
Fisher, Leslie John
Floyd, John Claiborne, Jr
Gilbertsen, Richard B
Goodwin, Jesse Francis
Greene, Douglas A
Guttendorf, Robert John
Hagen, John William
Hakim, Margaret Heath
Harris, Gale Ion
Hawthorne, Victor Morrison
Hoff, Julian Theodore
Howard, Charles Frank, Jr
Hsu, Julie Man-Ching
Ilgen, Daniel Richard
James, Sherman Athonia
Keiser, Joan A
Klivington, Kenneth Albert
Krause, Brian Robert
Kurtz, Margot
Leopold, Wilbur Richard, III
Levine, Simon P
Levine, Steven Richard
Lichter, Paul Richard
Louis-Ferdinand, Robert T
Lusher, Jeanne Marie
McClary, Andrew
MacDonald, John Robert
McEwen, Charles Milton, Jr
Motiff, James P
Musch, David C(harles)
Nisbelt, Richard Eugene
Pelzer, Charles Francis
Pieper, David Robert
Piercey, Montford F
Porter, Arthur T
Punch, Jerry L
Rupp, Ralph Russell
Ruwart, Mary Jean
Schottenfeld, David
Senagore, Anthony J
Shaw, M(elvin) P
Sheagren, John Newcomb
Shlafer, Marshal
Siddiqui, Waheed Hasan
Sobota, Walter Louis
Stevenson, Harold William
Timmis, Gerald C
Toledo-Pereyra, Luis Horacio
Whitehouse, Frank, Jr
Zacks, James Lee
Zucker, Robert Alpert

MINNESOTA
Allen, David West
Anderson, W Robert
Athelstan, Gary Thomas
Barber, Donald E
Brown, David Mitchell
Butt, Hugh Roland
Chevalier, Peter Andrew
Dai, Xue Zheng (Charlie)
Darley, Frederic Loudon
Dewald, Gordon Wayne
Douglas, William Hugh
Dykstra, Dennis Dale
Hancock, Peter Adrian
Hovde, Ruth Frances
Iverson, Laura Himes
Keane, William Francis
Kornblith, Carol Lee
McCullough, John Jeffrey
McDermott, Daniel J
Manning, Patrick James
Mauer, S Michael
Michenfelder, John D
Moore, Sean Breanndan
Mortimer, James Arthur
Overmier, J Bruce
Pittelkow, Mark Robert
Regal, Jean Frances
Schofield, William
Slavin, Joanne Louise
Swingle, Karl Frederick
Timm, Gerald Wayne
Ulstrom, Robert A
Vallera, Daniel A
Vernier, Robert L
Vetter, Richard J
Wade, Michael George
White, James George
Winters, Ken C
Yellin, Absalom Moses

MISSISSIPPI
Alford, Geary Simmons
Fowler, Stephen C
Hall, Margaret (Margot) Jean
Jones, Eric Wynn
Kallman, William Michael
Lawler, Adrian Russell
Whitworth, Neil S
Wiser, Winfred Lavern

MISSOURI
Adams, Max David
Chapman, Ramona Marie
Clare, Stewart
Cook, Mary Rozella
Cope, John Raymond
Dilley, William G
Ferrendelli, James Anthony
Grubb, Robert Lee, Jr
Hageman, Gregory Scott
Hedlund, James L
Hirsch, Ira J
Hurst, Robert R(owe)
Joist, Heinrich J
Kaplan, Henry J
Kurtz, Michael E
Lubin, Bernard

502 / DISCIPLINE INDEX

Other Medical & Health Sciences (cont)

Moe, Aaron Jay
Mosher, Donna Patricia
Murrell, Hugh Jerry
Natani, Kirmach
Podrebarac, Eugene George
Reynolds, Robert D
Robins, Eli
Sauer, Gordon Chenoweth
Siegel, Marilyn J
Simchowitz, Louis
Smith, Carl Hugh
Smith, Robert Francis
Stenson, William F
Strunk, Robert Charles
Taub, John Marcus
Weber, Karl T
Zehr, Martin Dale
Zepp, Edwin Andrew

MONTANA
Combie, Joan D
Gajdosik, Richard Lee
Horner, John Robert
Strobel, David Allen

NEBRASKA
Badeer, Henry Sarkis
Benschoter, Reba Ann
Bernthal, John E(rwin)
Carmines, Pamela K(ay)
Cavalieri, Ercole Luigi
Heaney, Robert Proulx
Johansson, Sonny L
Kimberling, William J
Klemcke, Harold G
Kutler, Benton
Leuschen, M Patricia
Macaluso, Sr Mary Christelle
Nipper, Henry Carmack
Prioreschi, Plinio
Records, Raymond Edwin
Saski, Witold
Schneider, Norman Richard
Sketch, Michael Hugh
Sonderegger, Theo Brown
Spanier, Graham Basil

NEVADA
Bijou, Sidney William
May, Jerry Russell
Smith, Aaron

NEW HAMPSHIRE
Brous, Don W
Easstman, Alan

NEW JERSEY
Adelman, Steven J
Albu, Evelyn D
Alfano, Michael Charles
Alson, Eli
Apuzzio, J J
Armstrong, Lydia
Aronson, Miriam Klausner
Baker, Thomas
Banga, Harjit Singh
Bernstein, Alan D
Boyle, Joseph, III
Breckenridge, Mary Barber
Brubaker, Paul Eugene, II
Burgett, Michael
Chen, Ling-Sing Kang
Choy, Wai Nang
Dugan, Gary Edwin
Eisinger, Robert Peter
Falk, John L
Godfrey, John Joseph
Grebenau, Mark David
Grieco, Michael H
Griffo, James Vincent, Jr
Haft, Jacob I
Hakimi, John
Halpern, Myron Herbert
Harris, Jane E
Haughey, Francis James
Heveran, John Edward
Hey, John Anthony
Hite, Mark
Hockel, Gregory Martin
Houston, Vern Lynn
Ingoglia, Nicholas Andrew
Jacobs, Barry Leonard
Jerussi, Thomas P
Kapp, Robert Wesley, Jr
Khandwala, Atul S
Khavkin, Theodor
Kongsamut, Sathapana
Krauthamer, George Michael
Kruh, Daniel
Lappe, Rodney Wilson
Lehrer, Paul Michael
Leonard, Claire Marie
Lewis, Steven Craig
Leyson, Jose Florante Justininane
Light, Donald W
Lowndes, Herbert Edward
Lukacsko, Alison B
Machac, Josef
Mackerer, Carl Robert
Mechanic, David

Miller, George Armitage
Miner, Karen Mills
Natelson, Benjamin Henry
Newton, Paul Edward
Norvitch, Mary Ellen
O'Keefe, Eugene H
Petillo, Phillip J
Reuben, Roberta C
Rosenberg, Alberto
Schroeder, Steven A
Sen, Tapas K
Sherr, Allan Ellis
Siegel, Jeffry A
Slomiany, Amalia
Solis-Gaffar, Maria Corazon
Spitznagle, Larry Allen
Stanton, Robert E
Stein, Donald Gerald
Stern, William
Stillman, John Edgar
Stults, Frederick Howard
Tio, Cesario O
Unowsky, Joel
Van Pelt, Wesley Richard
Vogel, Veronica Lee
Watson, Richard White, Jr
Whitmer, Jeffrey Thomas
Wiggins, Jay Ross
Wislocki, Peter G

NEW MEXICO
Boecker, Bruce Bernard
Brann, Benjamin Sandford, IV
Davila, Julio C
Guziec, Lynn Erin
Lawrence, James Neville Peed
Mills, David Edward
Snipes, Morris Burton
Vogel, Kathryn Giebler
Werkema, George Jan
Yudkowsky, Elias B

NEW YORK
Ader, Robert
Alexander, Justin
Altar, Charles Anthony
Anderson, Albert Edward
Archibald, Reginald MacGregor
Baines, Ruth Etta
Barac-Nieto, Mario
Barish, Robert John
Barletta, Michael Anthony
Barrett, James E
Barzel, Uriel S
Baum, Howard Jay
Baum, John W
Bendixen, Henrik H
Bernstein, Alvin Stanley
Bigler, Rodney Errol
Bito, Laszlo Z
Bizios, Rena
Blanck, Thomas Joseph John
Bloodstein, Oliver
Bofinger, Diane P
Bovbjerg, Dana H
Bradley, Francis J
Brown, Jason Walter
Brownstein, Martin H
Butler, Katharine Gorrell
Caldwell, Peter
Callahan, Daniel
Carr, Edward Gary
Carsten, Arland L
Catalfamo, James L
Celada, Franco
Centola, Grace Marie
Choi, Haing U
Christman, David R
Chryssanthou, Chryssanthos
Coderre, Jeffrey Albert
Cohen, Bernard
Cohen, Elias
Cohen, Michael I
Conley, John Joseph
Conway, Walter Donald
Cort, Susannah
Costa, Erminio
Cunha, Burke A
Dahl, John Robert
Daly, John
Daly, Robert Ward
Darlington, Richard Benjamin
Davey, Frederick Richard
Davis, Carolyne K
Dornbush, Rhea L
Dumoulin, Charles Lucian
Edelstein, William Alan
Edwards, John Anthony
Ehrke, Mary Jane
Elbadawi, Ahmad
Ellis, Albert
Ellner, Paul Daniel
Fabricant, Catherine G
Faden, Howard
Fahey, Charles J
Famighetti, Louise Orto
Farah, Fuad Salim
Fawwaz, Rashid
Ferris, Steven Howard
Finando, Steven J
Finlay, Jonathan Lester
Fish, Jefferson
Fleischer, Norman
Fliedner, Leonard John, Jr

Forquer, Sandra Lynne
Friedberg, Carl E
Furlanetto, Richard W
Gering, Robert Lee
Gerstman, Hubert Louis
Giampietro, Philip Francis
Gintautas, Jonas
Glass, David Carter
Goldberger, Robert Frank
Goldfried, Marvin R
Gollon, Peter J
Gordon, Arnold J
Gordon, Wayne Alan
Goswami, Meeta J
Gray, William David
Greenhall, Arthur Merwin
Gumbiner, Barry M
Haber, Sonja B
Haddad, Heskel Marshall
Hainline, Louise
Halevy, Simon
Ham, Richard John
Haywood, Anne Mowbray
Hohnadel, David Charles
Holtz, Aliza
Hoo, Joe Jie
Huang, Alice Shih-Hou
Huang, Cheng-Chun
Hull, Andrew P
Jacobson, Myron J
Jan, Kung-Ming
Johnson, Marie-Louise T
Jones, Kevin F
Jung, Chan Yong
Kelman, Charles D
Killian, Carl Stanley
Kissileff, Harry R
Klemperer, Friedrich W
Komorek, Michael Joesph, Jr
Kotler, Donald P
Kotval, Pesho Sohrab
Krauss, Herbert Harris
Kronzon, Itzhak
Kuftinec, Mladen M
Lamberg, Stanley Lawrence
Lambertsen, Eleanor C
Law, Paul Arthur
Lempert, Neil
Lennon-Thompson, Doris
Lepow, Martha Lipson
Levenson, Stanley Melvin
Levine, Frederic M
Levitsky, David A
Lippmann, Heinz Israel
Lubic, Ruth Watson
Lythcott, George I
McCann, Michael Francis
McCarthy, Joseph Gerald
Maillie, Hugh David
Martin, Constance R
Masur, Sandra Kazahn
Matsuzaki, Masaji
Mausner, Leonard Franklin
Mayerhoff, David I
Mendelsohn, John
Mercier, Gustavo Alberto, Jr
Michaelson, Solomon M
Midlarsky, Elizabeth
Mizma, Edward John
Moline, Sheldon Walter
Movshon, J Anthony
Munne, Santiago
Nathan, Ronald Gene
Nelkin, Dorothy
Neu, Harold Conrad
Nevid, Jeffrey Steven
Okaya, Yoshi Haru
Olczak, Paul Vincent
O'Reilly, Richard
Orsini, Frank R
Ottenbacher, Kenneth John
Overweg, Norbert I A
Penn, Arthur
Pizzarello, Donald Joseph
Poppers, Paul Jules
Rafla, Sameer
Ranck, James Byrne, Jr
Ranu, Harcharan Singh
Reit, Barry
Rich, Kenneth Eugene
Richardson, Robert Lloyd
Ritch, Robert
Rivlin, Richard S
Roberts, Carlyle Jones
Roberts, Richard B
Rogler, Lloyd Henry
Ross, John Brandon Alexander
Rothman, David J
Sackeim, Harold A
Safai, Bijan
Salkin, Ira Fred
Schenk, Eric A
Schoenfeld, Cy
Shamos, Morris Herbert
Shank, Brenda Mae Buckhold
Shanzer, Stefan
Shapiro, Raymond E
Sidel, Victor William
Slatkin, Daniel Nathan
Smith, Thomas Harry Francis
Soave, Rosemary
Steigbigel, Roy Theodore
Stein, Ruth E K
Stull, G(eorge) A

Tatarczuk, Joseph Richard
Tichauer, Erwin Rudolph
Tulchin, Natalie
Varma, BAidya Nath
Verma, Ram S
Victor, Jonathan David
Villanueva, German Baid
Wadell, Lyle H
Wang, Theodore Sheng-Tao
Wasserheit, Judith Nina
Watson, James Dewey
Weber, David Alexander
Weinstein, Arthur
Weissman, Michael Herbert
Welle, Stephen Leo
Whitaker, James E
Williams, Christine
Wollman, Leo
Wright, Herbert N
York, James Lester
Yozawitz, Allan
Zaleski, Marek Bohdan
Zolla-Pazner, Susan Beth

NORTH CAROLINA
Arnold, Roland R
Barry, David Walter
Bowman, Marjorie Ann
Case, Verna Miller
Caterson, Bruce
Childers, Robert Warren
Churchill, Larry R
Ciancolo, George J
Cobb, Frederick Ross
Cochran, Felicia R
Cronenberger, Jo Helen
Devereux, Theodora Reyling
Dykstra, Linda A
Ehmann, Carl William
Emmerling, David Alan
Floyd, Carey E, Jr
Friedman, Mitchell
Gallagher, Margie Lee
Gilmore, Stuart Irby
Hallock, James A
Heck, Jonathan Daniel
Howard, Donald Robert
James, Karen K(anke)
Jaszczak, Ronald Jack
Jones, Lyle Vincent
Jorizzo, Joseph L
Kaplan, Berton Harris
Kirkbride, L(ouis) D(ale)
Klitzman, Bruce
Klotman, Paul
Knobeloch, F X Calvin
Krasny, Harvey Charles
Krishnan, K Ranga Rama
LaFon, Stephen Woodrow
Lockhead, Gregory Roger
Long, Robert Allen
Lowe, James Edward
Meredith, Jesse Hedgepeth
Misek, Bernard
Murdock, Harold Russell
Murray, William J
O'Callaghan, James Patrick
Olsen, Elise Arline
Penry, James Kiffin
Prazma, Jiri
Putman, Charles E
Shackelford, Ernest Dabney
Sneade, Barbara Herbert
Staddon, John Eric Rayner
Strauss, Harold C
Thomas, Francis T
Thomas, William Grady
Webber, Richard Lyle
Wiley, Albert Lee, Jr
Wilson, Ruby Leila
Young, Larry Dale
Yount, William J

OHIO
Anderson, Douglas K
Anderson, James Morley
Antar, Mohamed Abdelchany
Ball, Edward James
Bannan, Elmer Alexander
Berger, Melvin
Berntson, Gary Glen
Boulant, Jack A
Breglia, Rudolph John
Broge, Robert Walter
Brunengraber, Henri
Brunner, Robert Lee
Cacioppo, John T
Carter, James R
Carter, R Owen, Jr
Castile, Robert G
Chang, Jae Chan
Chen, Ker-Sang
Christensen, Julien Martin
Christoforidis, A John
Colbert, Charles
Coots, Robert Herman
Cope, Frederick Oliver
Cruz, Julio C
Daroff, Robert Barry
Dauchot, Paul J
Decker, Clarence Ferdinand
Degen, Sandra Joanne
DeNelsky, Garland
Doershuk, Carl Frederick

Dougherty, John A
Fass, Robert J
Fouke, Janie M
Frank, Thomas Paul
Fucci, Donald James
Glaser, Roger Michael
Graham, Robert M
Greenstein, Julius S
Guion, Robert Morgan
Hawk, William Andrew
Hinkle, George Henry
Hinman, Channing L
Hollander, Philip B
Holzbach, R Thomas
Hunt, Carl E
Jamasbi, Roudabeh J
Kang, Mohinder Singh
Kaplan, Arthur Lewis
Lass, Jonathan H
Levin, Jerome Allen
Lima, John J
Lydy, David Lee
McFadden, Edward Regis
McKee, Michael Geoffrey
McNamara, John Regis
Mantil, Joseph Chacko
Meites, Samuel
Miraldi, Floro D
Muir, William W, III
Mulick, James Anton
Myers, William Graydon
Neiman, Gary Scott
Nigrovic, Vladimir
Osipow, Samuel Herman
Ottolenghi, Abramo Cesare
Patton, Nancy Jane
Pinchak, Alfred Cyril
Reece, Robert William
Resnick, Martin I
Roche, Alexander F
Roessmann, Uros
Ross, John Edward
Rubenstein, Jack Herbert
Schmidt, Steven Paul
Smith, Carl Clinton
Smith, D Edwards
Somerson, Norman L
Starchman, Dale Edward
Stockman, Charles H(enry)
Strauss, Richard Harry
Subbiah, Ravi M T
Suskind, Raymond Robert
Tallman, Richard Dale (Junior)
Thames, Marc
Thomas, Donald Charles
Wallin, Richard Franklin
Warden, Glenn Donald
Warner, Ann Marie
Weiss, Harold Samuel
Wilkin, Jonathan Keith
Wodarski, John Stanley
Zoglio, Michael Anthony

OKLAHOMA
Folk, Earl Donald
Harrison, Aix B
Hochhaus, Larry
Letcher, John Henry, III
Lovallo, William Robert
Martin, Loren Gene
Melton, Carlton E, Jr
Min, Kyung-Whan
Revzin, Alvin Morton
Schaefer, Carl Francis
Schaefer, Frederick Vail
Schroeder, David J Dean
Schultz, Russell Thomas
Steffey, Oran Dean
Switzer, Laura Mae
Thomas, Ellidee Dotson

OREGON
Ary, Dennis
Bufford, Rodger K
Buist, A(line) Sonia
Buist, Neil R M
Burry, Kenneth A
Dooley, Douglas Charles
Fay, Warren Henry
Garwood, John Ernest
Gillis, John Simon
Greenberg, Barry H
Hackman, Robert Mark
Hayden, Jess, Jr
Jastak, J Theodore
Johnson, Larry Reidar
Maksud, Michael George
Matarazzo, Joseph Dominic
Parrott, Marshall Ward
Smith, Alvin Winfred
Vernon, Jack Allen

PENNSYLVANIA
Abdelhak, Sherif Samy
Allen, Michael Thomas
Aronstam, Robert S
Asbury, Arthur K
Aston, Richard
Axel, Leon
Bagasra, Omar
Ball, Edward D
Beavers, Ellington McHenry
Black, Martin
Blumberg, Baruch Samuel
Bochenek, Wieslaw Janusz

Bove, Alfred Anthony
Brubaker, Clifford E
Brunson, Kenneth W
Campbell, Phil G
Cander, Leon
Cardinal, John Robert
Carr, Brian I
Chan, Ping-Kwong
Chapman, John Donald
Clifford, Maurice Cecil
Cohn, Ellen Rassas
Cole, James Edward
Couch, Jack Gary
Cutler, Neal Evan
Dekker, Andrew
DeKosky, Steven Trent
Deliveria-Papadopoulos, Maria
Denenberg, Herbert S
Denis, Kathleen A
Diorio, Alfred Frank
Dobson, Margaret Velma
Doghramji, Karl
Edelstone, Daniel I
Etherton, Terry D
Fallon, Michael David
Fish, James E
Fisher, Aron Baer
Fisher, Robert Stephen
Foxx, Richard Michael
Fraser, Nigel William
Friend, Judith H
Fromm, Gerhard Hermann
Fuscaldo, Kathryn Elizabeth
Gehring, David Austin
Glick, Mary Catherine
Gordon, Philip Ray
Gorin, Michael Bruce
Greenberger, Joel S
Greene, Terry R
Grenvik, Ake N A
Haddad, John George
Hale, Creighton J
Harris, Dorothy Virginia
Harty, Michael
Hoek, Joannes (Jan) Bernardus
Holsclaw, Douglas S, Jr
Homan, Elton Richard
Hook, Jerry B
Houts, Peter Stevens
Isett, Robert David
Iwatsuki, Shunzaburo
Jensen, Gordon L
Johnson, Candace Sue
Jungkind, Donald Lee
Kalia, Madhu P
Keith, Jennie
Kendall, Philip C
Keshgegian, Albert Arakel
Kidawa, Anthony Stanley
Klein, Herbert A
Knutson, David W
Kresh, J Yasha
Kuiваniemi, S Helena
Kunz, Heinz W
Lebenthal, Emanuel
Leddy, Susan
Leeper, Dennis Burton
Leighton, Charles Cutler
Levine, Geoffrey
Liebenberg, Stanley Phillip
Lin, Jiunn H
Linask, Kersti Katrin
Lynn, Robert K
Magargal, Larry Elliot
Mansmann, Herbert Charles, Jr
Marazita, Mary Louise
Miller, Barry
Miller, Kenneth L
Minsker, David Harry
Moster, Mark Leslie
Moyer, Robert (Findley)
Nambi, Ponnal
Needleman, Herbert L
Neufeld, Gordon R
Orenstein, David M
Pack, Allan I
Piperno, Elliot
Porter, Sydney W, Jr
Prout, James Harold
Pruett, Esther D R
Rao, Vijay Madan
Reilly, Christopher F
Rescorla, Robert A
Rinaldo, Charles R
Roberts, James M
Robertson, Robert James
Rosenberg, Henry
Rozmiarek, Harry
Ruggieri, Michael Raymond
Ryan, James Patrick
Schidlow, Daniel
Schreiber, Alan D
Shoub, Earle Phelps
Smith, M Susan
Somers, Anne R
Sperling, Mark Alexander
Steplewski, Zenon
Studer, Rebecca Kathryn
Suzuki, Jon Byron
Tabachnick, Joseph
Tice, Linwood Franklin
Turk, Dennis Charles
Van Ausdal, Ray Garrison
Villafana, Theodore

Vogt, Molly Thomas
Walsh, Arthur Campbell
Waterman, Frank Melvin
Wein, Alan Jerome
Williams, Gerald D
Yousem, David Mark
Yu, Lucy Cha

RHODE ISLAND
Amols, Howard Ira
Barker, Barbara Elizabeth
Barnes, Frederick Walter, Jr
Cowett, Richard Michael
Flanigan, Timothy Palen
Gilman, John Richard, Jr
Hutzler, Leroy, III
Johanson, Conrad Earl
Lipsitt, Lewis Paeff
Rotman, Boris

SOUTH CAROLINA
Alley, Thomas Robertson
Apple, David Joseph
Bryan, Charles Stone
Gee, Adrian Philip
Johnson, James Allen, Jr
Kaiser, Charles Frederick
Patterson, C(leo) Maurice
Powell, Donald Ashmore
Rathbun, Ted Allan
Reinig, William Charles
Sanders, Samuel Marshall, Jr

TENNESSEE
Burish, Thomas Gerard
Buster, John Edmond
Carmack, Lee J
Cloutier, Roger Joseph
Deivanayagan, Subramanian
Fleischer, Arthur C
Gammage, Richard Bertram
Gupta, Ramesh C
Hubbell, Harry Hopkins, Jr
Jackson, Carl Marvin
Kaye, Stephen Vincent
Lawler, James E
Lenhard, Joseph Andrew
Long, Charles Joseph
Lott, Sam Houston, Jr
Love, Russell Jacques
McConnell, Freeman Erton
Miller, Mark Steven
Morgan, Karl Ziegler
Parker, Richard E
Rivas, Marian Lucy
Roberts, Lee Knight
Sittig, Dean Forrest
South, Mary Ann
Stone, William John
Strupp, Hans H
Studebaker, Gerald A
Sun, Deming
Tai, Douglas L
Tanner, Raymond Lewis
Thompson, Travis
Thurman, Gary Boyd
Vnencak-Jones, Cindy Lenore
Watson, Evelyn E
Wiley, Ronald Gordon
Wondergem, Robert

TEXAS
Abboud, Hanna E
Ali, Yusuf
Alperin, Jack Bernard
Armstrong, George Glaucus, Jr
Ashby, Jon Kenneth
Azizi, Sayed Ausim
Barnes, Jeffrey Lee
Benison, Betty Bryant
Bielamowicz, Mary Claire Kinney
Biles, Robert Wayne
Bonner, Hugh Warren
Bonte, Frederick James
Bruhn, John G
Buffone, Gregory J
Cardus, David
Chang, Tse-Wen
Chilian, William M
Convertino, Victor Anthony
Cooper, John C, Jr
Cottone, James Anthony
David, Yadin B
Deter, Russell Lee, II
Ellis, Kenneth Joseph
Fernandes, Gabriel
Fordtran, John Satterfield
Geoghegan, William David
German, Dwight Charles
German, Victor Frederick
Glatstein, Eli J
Gould, K Lance
Greeley, George H, Jr
Green, Joseph Barnet
Griffin, Kathleen (Mary)
Gwirtz, Patricia Ann
Harkless, Lawrence Bernard
Hatch, John Phillip
Henrich, Williams Lloyd
Holmquest, Donald Lee
Holtzman, Wayne Harold
Hong, Bor-Shyue
Hong, Waun Ki
Horger, Lewis Milton

Horton, Jureta
Horton, William A
Hurd, Eric R
Hussian, Richard A
Ickes, William K
Jackson, Gilchrist Lewis
Jenkinson, Stephen G
Johnson, John E, Jr
Johnson, Patricia Ann J
Kadekaro, Massako
Kalkwarf, Kenneth Lee
Karacan, Ismet
Kasi, Leela Peshkar
Kehrer, James Paul
Krohmer, Jack Stewart
Kuo, Lih
Kuo, Peter Te
Kutzman, Raymond Stanley
Lal, Harbans
Lane, John D
Lehmkuhl, L Don
Levin, Harvey Steven
Loubser, Paul Gerhard
McCullough, Laurence Bernard
MacDonald, Paul Cloren
Malloy, Michael H
Mavligit, Giora M
Mayer, Tom G
Meltz, Martin Lowell
Mendel, Julius Louis
Mikos, Antonios G
Minna, John
Patel, Anil S
Peniston, Eugene Gilbert
Pennebaker, James W
Posey, Daniel Earl
Poston, John Ware
Prentice, Norman Macdonald
Rehm, Lynn P
Riser, Mary Elizabeth
Roffwarg, Howard Philip
Rolston, Kenneth Vijaykumar Issac
Rudenberg, Frank Hermann
Ryan, Charles F
Sanders, Barbara
Schneider, Nancy Reynolds
Schoenfeld, Lawrence Steven
Shulman, Robert Jay
Siler-Khodr, Theresa M
Smith, Michael Lew
Sontheimer, Richard Dennis
Spence, Dale William
Stedman, James Murphey
Steffy, John Richard
Takatani, Setsuo
Tate, Charlotte Anne
Tomasovic, Stephen Peter
Triano, John Joseph
Tsin, Andrew Tsang Cheung
Wauquier, Albert
Willerson, James Thornton
Wilson, Russell H
Wolfenberger, Virginia Ann
Yung, W K Alfred

UTAH
Ailion, David Charles
Archer, Victor Eugene
Berenson, Malcolm Mark
Kohler, R Ramon
Lee, Steve S
Molteni, Richard A(loysius)
Moser, Royce, Jr
Olsen, Don B
Peters, Jeffrey L
Schiager, Keith Jerome
Stanley, Theodore H
Striefel, Sebastian
Westenskow, Dwayne R

VERMONT
Evans, John N
Falk, Leslie Alan
Halpern, William
Kaplow, Leonard Samuel
Pope, Malcolm H
Rosen, James Carl
Wooding, William Minor

VIRGINIA
Arora, Narinder
Asterita, Mary Frances
Auerbach, Stephen Michael
Barrett, Eugene J
Bolton, W Kline
Bonvillian, John Doughty
Bryant, Patricia Sand
Burger, Carol J
Chaudhuri, Tapan K
Conrad, Daniel Harper
DeSimone, John A
Detmer, Don Eugene
Furr, Aaron Keith
Gottesman, Irving Isadore
Grunder, Hermann August
Hicks, John T
Hoegerman, Stanton Fred
Howard-Peebles, Patricia Nell
Kim, Yong Il
Kripke, Bernard Robert
Kurtzke, John F, Sr
Lenhardt, Martin Louis
McCarthy, Charles R
McDonald, Gerald O

Other Medical & Health Sciences (cont)

Nagarkatti, Mitzi
Nixon, John V
Owen, Duncan Shaw, Jr
Payton, Otto D
Reppucci, Nicholas Dickon
Richards, Paul Bland
Roth, Karl Sebastian
Ruhling, Robert Otto
Scarr, Sandra Wood
Scheld, William Michael
Shukla, Kamal Kant
Squibb, Robert E
Sugerman, Harvey J
Talbot, Richard Burritt
Thorner, Michael Oliver
Walsh, Scott Wesley
Wands, Ralph Clinton
Wells, John Morgan, Jr
Wilson, John William
Wolfe, Barbara Ann

WASHINGTON
Anderson, Robert Authur
Bair, William J
Brunzell, John D
De Tornyay, Rheba
Evermann, James Frederick
Frank, Andrew Julian
Grootes-Reuvecamp, Grada Alijda
Hutchison, Nancy Jean
Jackson, Crawford Gardner, Jr
Johnson, John Richard
Jonsen, Albert R
Kalina, Robert E
Kelly, Gregory M
Kenney, Gerald
Kroll, Keith
Lakshminarayan, S
Larsen, Lawrence Harold
Ledbetter, Jeffrey A
McSweeney, Frances Kaye
Miller, Douglas Lawrence
Nelson, Iral Clair
Nelson, James Alonzo
Richardson, Michael Lewellyn
Root, Richard Kay
Roth, Gerald J
Salk, Darrell John
Schulz, John C
Shapley, James Louis
Shriver, M Kathleen
Smith, Victor Herbert
Soldat, Joseph Kenneth
Turner, Judith Ann
Wedlick, Harold Lee
Wehner, Alfred Peter
Williams, Douglas Edward
Wilson, Christopher
Woods, Nancy Fugate
Wyler, Allen Raymer

WEST VIRGINIA
Edelstein, Barry Allen
Mufson, Maurice Albert
Olenchock, Stephen Anthony
Reasor, Mark Jae

WISCONSIN
Agre, James C
Amrani, David L
Barschall, Henry Herman
Boerboom, Lawrence E
Brazy, Peter C
Byrne, Gerald I
Cohen, Bernard Allan
Crowe, Dennis Timothy, Jr
Diaz, Luis A
Friend, Milton
Goldstein, Robert
Hamsher, Kerry de Sandoz
Haughton, Victor Mellet
Hekmat, Hamid Moayed
Holden, James Edward
Kaufman, Paul Leon
Kent, Raymond D
Komai, Hirochika
Montgomery, Robert Renwick
Nagle, Francis J
Nellis, Stephen H
Shah, Dharmishtha V
Sheffield, Lewis Glosson
Sheldahl, Lois Marie
Terry, Leon Cass
Tormey, Douglass Cole
Vailas, Arthur C
Valdivia, Enrique
Vanderheiden, Gregg
Warltier, David Charles

WYOMING
Holbrook, Frederick R

ALBERTA
Albritton, William Leonard
Famiglietti, Edward Virgil, Jr
Jimbow, Kowichi
Jones, Richard Lee
King, Malcolm
Larke, R(obert) P(eter) Bryce
Leung, Alexander Kwok-Chu

Lorscheider, Fritz Louis
Mash, Eric Jay
Urtasun, Raul C
Wallace, John Lawrence

BRITISH COLUMBIA
Clement, Douglas Bruce
Cottle, Merva Kathryn Warren
Craig, Kenneth Denton
Cynader, Max Sigmund
Laszlo, Charles Andrew
Laye, Ronald Curtis
Lee, Chi-Yu Gregory
Milsum, John H
Singh, Raj Kumari
Thurlbeck, William Michael

MANITOBA
Adaskin, Eleanor Jean
Cowdrey, Eric John
Gerrard, Jonathan M
Rigatto, Henrique
Sinha, Helen Luella
Zelinski, Teresa A

NOVA SCOTIA
Byers, David M
Ellenberger, Herman Albert
Jones, John Verrier
Murray, T J
Tonks, Robert Stanley
Walker, Joan Marion

ONTARIO
Andrew, George McCoubrey
Basu, Prasanta Kumar
Birnboim, Hyman Chaim
Buick, Ronald N
Cafarelli, Enzo Donald
Cutz, Ernest
Czuba, Margaret
Drake, Charles George
Ferguson, Alastair Victor
Girvin, John Patterson
Goldenberg, Gerald J
Hare, William Currie Douglas
Harwood-Nash, Derek Clive
Ing, Harry
Kennedy, James Cecil
Leeper, Herbert Andrew, Jr
Lemaire, Irma
Logan, James Edward
McLachlan, Richard Scott
Mustard, James Fraser
Nielsen, Klaus H B
Osborne, Richard Vincent
Papp, Kim Alexander
Pfalzner, Paul Michael
Quinsey, Vernon Lewis
Rabinovitch, Marlene
Rapoport, Anatol
Rock, Gail Ann
Roland, Charles Gordon
Sibbald, William John
Silverman, Melvin
Sobell, Linda Carter
Sobell, Mark Barry
Till, James Edgar
Tolnai, Susan
Tyson, John Edward Alfred
Vallerand, Andre L

QUEBEC
Alavi, Misbahuddin Zafar
Aubry, Muriel
Burgess, John Herbert
Cantin, Marc
Castonguay, Andre
Chang, Thomas Ming Swi
Drouin, Gilbert
Dykes, Robert William
Glorieux, Francis Henri
Grassino, Alejandro E
Karpati, George
Landry, Fernand
Levy, Samuel Wolfe
Lippman, Abby
Liuzzi, Michel
Lussier, Gilles L
Malo, Jacques
Marisi, Dan(iel) Quirinus
Melzack, Ronald
Milner, Brenda (Atkinson)
Palayew, Max Jacob
Paris, David Leonard
Roskies, Ethel
Salas-Prato, Milagros
Serbin, Lisa Alexandra
Stewart, Jane
Tousignant, Michel

SASKATCHEWAN
Shokeir, Mohamed Hassan Kamel

OTHER COUNTRIES
Allessie, Maurits
Bader, Hermann
Bar-Yishay, Ephraim
Doll, Richard
Dunn, Christopher David Reginald
Fleckenstein, Albrecht
Garrard, Christopher S
Gomez, Arturo

Hall, Laurance David
Kelly, Paul Alan
Laura, Patricio Adolfo Antonio
Lekeux, Pierre Marie
Miquel, Jaime
Mitruka, Brij Mohan
O'Neill, Patricia Lynn
Quastel, Michael Reuben
Shimaoka, Katsutaro
Sterpetti, Antonio Vittorio
Stingl, Georg
Tsolas, Orestes
Valyasevi, Aree
Weil, Marvin Lee

PHYSICS & ASTRONOMY
Acoustics

ALABAMA
Borton, Thomas Ernest
Jones, Jess Harold

ALASKA
Roederer, Juan Gualterio

ARIZONA
Bowen, Theodore
Fliegel, Frederick Martin
Gatley, William Stuart
Lamb, George Lawrence, Jr
Wallace, C(harles) E(dward)

ARKANSAS
Horton, Philip Bish
Smith, E(astman)
Wold, Donald C

CALIFORNIA
Ahumada, Albert Jil, Jr
Armstrong, Donald B
Atchley, Anthony A
Aurand, Henry Spiese, Jr
Ayers, Raymond Dean
Bachman, Richard Thomas
Barmatz, Martin Bruce
Behravesh, Mohamad Martin
Booth, Newell Ormond
Bourke, Robert Hathaway
Brown, Maurice Vertner
Bucker, Homer Park, Jr
Chase, David Marion
Cho, Alfred Chih-Fang
Cho, Young-chung
Chodorow, Marvin
Coppens, Alan Berchard
Cummings, William Charles
De, Gopa Sarkar
Dempsey, Martin E(wald)
Dym, Clive L
Eargle, John Morgan
Edmonds, Peter Derek
Fisher, Frederick Hendrick
Flatte, Stanley Martin
Gales, Robert Sydney
Galloway, William Joyce
Garrett, Steven Lurie
Grace, O Donn
Green, Charles E
Green, Philip S
Greenfield, Eugene W(illis)
Halley, Robert
Henning, Harley Barry
Holliday, Dale Vance
Johnson, Robert A
Jones, Joie Pierce
Kagiwada, Reynold Shigeru
Kamegai, Minao
Keller, Joseph Bishop
Khoshnevisan, Mohsen Monte
Kinnison, Gerald Lee
Kino, Gordon Stanley
Kirchner, Ernst Karl
Knollman, Gilbert Carl
Lax, Edward
Lubman, David
McIntyre, Marie Claire
Mackenzie, Kenneth Victor
Marcellino, George Raymond
Marcus, John Stanley
Martin, Gordon Eugene
Martin, Jim Frank
Medwin, Herman
Meecham, William Coryell
Metherell, Alexander Franz
Meyer, Norman Joseph
Mikes, Peter
Mitzner, Kenneth Martin
Moe, Chesney Rudolph
Monkewitz, Peter Alexis
Morrow, Charles Tabor
Munson, Albert G
Nelson, Rex Roland
Neubert, Jerome Arthur
Nichols, Rudolph Henry
Piersol, Allan Gerald
Pierucci, Mauro
Potter, David Samuel
Powers, John Patrick
Reynolds, Michael David
Rott, Nicholas
Rudnick, Isadore
Salmon, Vincent

Sanders, James Vincent
Schacher, Gordon Everett
Scharton, Terry Don
Schenck, Harry Allen
Segal, Alexander
Slaton, Jack H(amilton)
Sutherland, Louis Carr
Thomas, Graham Havens
Thompson, Paul O
Thorpe, Howard A(lan)
Twersky, Victor
Vent, Robert Joseph
Warshaw, Stephen I
Watson, Earl Eugene
White, Ray Henry
Wiedow, Carl Paul
Williams, Jack Rudolph
Wilson, Oscar Bryan, Jr
Young, Robert William

COLORADO
Amme, Robert Clyde
Boggs, George Johnson
Brown, Edmund Hosmer
Clifford, Steven Francis
DeSanto, John Anthony
Geers, Thomas L
Singer, Howard Joseph
Sommerfeld, Richard Arthur
Tatarskii, Valerian I
Taylor, Kenneth Doyle
Verschoor, J(ack) D(ahlstrom)
White, James Edward

CONNECTICUT
Apfel, Robert Edmund
Arnoldi, Robert Alfred
Bell, Thaddeus Gibson
Burkhard, Mahlon Daniel
Cable, Peter George
Carter, G Clifford
Eby, Edward Stuart, Sr
Erf, Robert K
Fehr, Robert O
Gay, Thomas John
Ghen, David C
Hanrahan, John J
Hasse, Raymond William, Jr
Hemond, Conrad J(oseph), Jr
Howkins, Stuart D
Lai, Jai-Lue
Mellen, Robert Harrison
Paterson, Robert W
Peracchio, Aldo Anthony
Plona, Thomas Joseph
Reed, Robert Willard
Richards, Roger Thomas
Rubega, Robert A
Sen, Pabitra Narayan
Shirer, Donald Leroy
Tobias, Jerry Vernon
Von Winkle, William A
Williams, David G(erald)

DELAWARE
Gulick, Walter Lawrence
Mehl, James Bernard
Sohl, Cary Hugh

DISTRICT OF COLUMBIA
Baer, Ralph Norman
Berkson, Jonathan Milton
Borgiotti, Giorgio V
Corliss, Edith Lou Rovner
Corsaro, Robert Dominic
Gaumond, Charles Frank
Harris, Wesley Leroy
Hershey, Robert Lewis
Knight, John C (Ian)
Mayer, Walter Georg
Photiadis, Douglas Marc
Rojas, Richard Raimond
Rosenblum, Lawrence Jay
Segnan, Romeo A
Strasberg, Murray
Tolstoy, Alexandra
Wolf, Stephen Noll
Yaniv, Simone Liliane
Yen, Nai-Chyuan

FLORIDA
Beguin, Fred P
Blue, Joseph Edward
Clarke, Thomas Lowe
DeFerrari, Harry Austin
Dubbelday, Pieter Steven
Flax, Lawrence
Green, Eugene L
Groves, Ivor Durham, Jr
Henriquez, Theodore Aurelio
Hollien, Harry
Kinzey, Bertram Y(ork), Jr
Leblanc, Roger M
Mahig, Joseph
Malocha, Donald C
Mooers, Christopher Northrup Kennard
Olmstead, Paul Smith
Piquette, Jean C
Proni, John Roberto
Purdy, Alan Harris
Ramer, Luther Grimm
Rhodes, Richard Ayer, II
Rona, Donna C
Rothman, Howard Barry

Shih, Hansen S T
Siebein, Gary Walter
Silverman, Sol Richard
Tam, Christopher K W
Tanner, Robert H
Thomas, Dan Anderson
Ting, Robert Yen-Ying
Van Buren, Arnie Lee
Veinott, Cyril G
Vidmar, Paul Joseph
Wiebusch, Charles Fred
Williamson, Donald Elwin
Zalesak, Joseph Francis

GEORGIA
Ginsberg, Jerry Hal
Lau, Jark Chong
Nave, Carl R
Patronis, Eugene Thayer, Jr
Rogers, Peter H
Zinn, Ben T

HAWAII
Au, Whitlow W L
Frazer, L Neil
Learned, John Gregory
Yates, James T

IDAHO
Johnson, John Alan

ILLINOIS
Achenbach, Jan Drewes
Butler, Robert Allan
Dunn, Floyd
Elliott, Lois Lawrence
Frizzell, Leon Albert
Guttman, Newman
Haugen, Robert Kenneth
Mintzer, David
O'Brien, William Daniel, Jr
Rey, Charles Albert
Sathoff, H John
Schultz, Martin C
Solie, Leland Peter
Wey, Albert Chin-Tang
Yost, William A

INDIANA
Bernhard, Robert J
Brach, Raymond M
Cruikshank, Donald Burgoyne, Jr
Eggleton, Reginald Charles
Farringer, Leland Dwight
Hansen, Uwe Jens
Kelly-Fry, Elizabeth
Kent, Earle Lewis
Meeks, Wilkison (Winfield)
Rickey, Martin Eugene
Scheiber, Donald Joseph
Stanley, Gerald R
Trachman, Edward G
Tree, David R
Tubis, Arnold
Watson, Charles S

IOWA
Buck, Otto
Hanson, Roger James
Korpel, Adrianus
Papadakis, Emmanuel Philippos
Schwartz, Ralph Jerome
Small, Arnold McCollum, Jr
Thompson, Robert Bruce
Titze, Ingo Roland
Voots, Richard Joseph
Wilson, Lennox Norwood

KANSAS
Farokhi, Saeed
Ferraro, John Anthony

KENTUCKY
Brill, Joseph Warren
Cohn, Robert Warren
Edwards, Richard Glenn
Naake, Hans Joachim

LOUISIANA
Beeson, Edward Lee, Jr
Bradley, Marshall Rice
Chin-Bing, Stanley Arthur
Collins, Mary Jane
Erath, Louis W
Grimsal, Edward George
Head, Martha E Moore
Lafleur, Louis Dwynn
Morris, Halcyon Ellen McNeil
Slaughter, Milton Dean
Wenzel, Alan Richard

MAINE
LaCasce, Elroy Osborne, Jr
Shonle, John Irwin
Walkling, Robert Adolph

MARYLAND
Achor, William Thomas
Alers, George A
Andrulis, Marilyn Ann
Bell, James F(rederick)
Bennett, Charles L
Berger, Harold
Blake, William King

Blatstein, Ira M
Brownell, William Edward
Causey, G(eorge) Donald
Chimenti, Dale Everett
Cook, Richard Kaufman
Cornett, Richard Orin
Cornyn, John Joseph
Crump, Stuart Faulkner
Dickey, Joseph W
Eitzen, Donald Gene
Etter, Paul Courtney
Everstine, Gordon Carl
Feit, David
Gammell, Paul M
Gammon, Robert Winston
Gaunaurd, Guillermo C
Geneaux, Nancy Lynne
Gereben, Istvan B
Goepel, Charles Albert
Greenspon, Joshua Earl
Gupta, Vaikunth N
Harrison, George H
Hartmann, Bruce
Heydemann, Peter Ludwig Martin
Ho, Louis T
Johnson, William Bowie
Linzer, Melvin
McNicholas, John Vincent
Madigosky, Walter Myron
Martin, John J(oseph)
Menton, Robert Thomas
Moses, Edward Joel
Moyer, William C, Jr
Munson, J(ohn) C(hristian)
Palmer, Charles Harvey
Peppin, Richard J
Pierce, Harry Frederick
Potocki, Kenneth Anthony
Raabe, Herbert P(aul)
Raff, Samuel J
Reader, Wayne Truman
Rigden, John Saxby
Scharnhorst, Kurt Peter
Schulkin, Morris
Sevik, Maurice
Skinner, Dale Dean
South, Hugh Miles
Urick, Robert Joseph
Vande Kieft, Lawrence John
Weinstein, Marvin Stanley
Williams, Robert Glenn
Zankel, Kenneth L

MASSACHUSETTS
Barger, James Edwin
Baxter, Lincoln, II
Beranek, Leo Leroy
Bers, Abraham
Bogert, Bruce Plympton
Bolt, Richard Henry
Bose, Amar G(opal)
Breton, J Raymond
Burke, Shawn Edmund
Butler, John Louis
Callerame, Joseph
Cole, John E(mery), III
Dyer, Ira
Figwer, J(ozef) Jacek
Free, John Ulric
Frisk, George Vladimir
Garrelick, Joel Marc
Griesinger, David Hadley
Harrison, John Michael
Hawkins, W(illiam) Bruce
Hurlburt, Douglas Herendeen
Junger, Miguel C
Keast, David N(orris)
Kerwin, Edward Michael, Jr
Labate, Samuel
McCue, John Joseph Gerald
McElroy, Paul Tucker
Malme, Charles I(rving)
Milder, Fredric Lloyd
Morse, Robert Warren
Nelson, Donald Frederick
Nelson, Keith A
Noiseux, Claude Francois
Oliner, Arthur A(aron)
Pope, Joseph
Ryan, Robert Pat
Scharf, Bertram
Stephen, Ralph A
Stern, Ernest
Stevens, Kenneth N(oble)
Stowe, David William
Sulak, Lawrence Richard
Thompson, Charles
Tugal, Halil
Vignos, James Henry
White, James Victor

MICHIGAN
Birdsall, Theodore G
Blaser, Dwight A
Brown, Steven Michael
Gaerttner, Martin R
Gessert, Walter Louis
Goldstein, Albert
Hemdal, John Frederick
Johnson, Wayne Douglas
Moody, David Burritt
Nefske, Donald Joseph
O'Donnell, Matthew
Porter, James Colegrove

Reiher, Harold Frederick
Sawatari, Takeo
Stebbins, William Cooper
Sung, Shung H
Tosi, Oscar I
Weinreich, Gabriel
Wolf, Joseph A(llen), Jr
Wright, Wayne Mitchell
Yoder, Levon Lee

MINNESOTA
Anderson, Ronald Keith
Ivey, Elizabeth Spencer
Lea, Wayne Adair
Mueller, Rolf Karl
Van Doeren, Richard Edgerly
Walker, Charles Thomas
Ward, Wallace Dixon
Wild, John Julian

MISSISSIPPI
Bolen, Lee Napier, Jr
Breazeale, Mack Alfred
Bruce, John Goodall, Jr
Farwell, Robert William
Fleischer, Peter
Love, Richard Harrison
Morris, Gerald Brooks
Ramsdale, Dan Jerry

MISSOURI
Agarwal, Ramesh K
Conradi, Mark Stephen
Ellison, John Vogelsanger
Kibens, Valdis
Miller, James Gegan
Perisho, Clarence H(oward)
Weissenburger, Jason T

NEBRASKA
Eldred, Kenneth M

NEW HAMPSHIRE
Baumann, Hans D
Frost, Albert D(enver)
Seifer, Arnold David
Von-Recklinghausen, Daniel R

NEW JERSEY
Ballato, Arthur
Berkley, David A
Bhattacharya, Sabyasachi
Carter, Ashley Hale
Francis, Samuel Hopkins
Fretwell, Lyman Jefferson, Jr
Gaer, Marvin Charles
Goodfriend, Lewis S(tone)
Guernsey, Richard Montgomery
Hall, Joseph L
Henrikson, Frank W
Joyce, William B(axter)
Kam, Gar Lai
Kennedy, Anthony John
Kessler, Frederick Melvyn
Labianca, Frank Michael
Lee, Keun Myung
Lee, Peter Chung-Yi
Levinson, Stephen
Lunsford, Ralph D
Mitchell, Olga Mary Mracek
Morgan, Dennis Raymond
Parkins, Bowen Edward
Polak, Joel Allan
Prasad, Marehalli Gopalan
Quinlan, Daniel A
Raichel, Daniel R(ichter)
Rothleder, Stephen David
Scales, William Webb
Stickler, David Collier
Teaney, Dale T
Temkin, Samuel
Thurston, Robert Norton
Vahaviolos, Sotirios J
Wilson, Charles Elmer
Yi-Yan, Alfredo
Zipfel, George G, Jr

NEW MEXICO
Bullard, Edwin Roscoe, Jr
Frost, Harold Maurice, III
Herman, George
Kelly, Robert Emmett
Martin, Richard Alan
Mercer-Smith, James A
Reed, Jack Wilson
Roberts, Peter Morse
Schwarz, Ricardo
Sinha, Dipen N
Swift, Gregory William
Swingle, Donald Morgan
Visscher, William M

NEW YORK
Angevine, Oliver Lawrence
Bienvenue, Gordon Raymond
Blazey, Richard N
Breed, Henry Eltinge
Brody, Burton Alan
Carleton, Herbert Ruck
Cheney, Margaret
Cheung, Lim H
Frankel, Julius
Ganley, W Paul
Gavin, Donald Arthur

George, A(lbert) R(ichard)
Gerstman, Hubert Louis
Haines, Daniel Webster
Harris, Cyril Manton
Hiller, Lejaren Arthur
Hollenberg, Joel Warren
Ingenito, Frank Leo
Jacobson, Melvin Joseph
Katocs, Andrew Stephen, Jr
Khanna, Shyam Mohan
Klepper, David Lloyd
Kornreich, Philipp G
Kritz, J(acob)
Lang, William Warner
Lean, Eric Gung-Hwa
McDonald, Frank Alan
Maling, George Croswell, Jr
Melcher, Robert Lee
Meyer, Robert Jay
Nelson, David Elmer
Ozimek, Edward Joseph
Reiss, Diana
Sachse, Wolfgang H
Sackman, George Lawrence
Savkar, Sudhir Dattatraya
Schmid, Charles Ernest
Shiren, Norman S
Siegmann, William Lewis
Smith, Lowell Scott
Sutton, George Harry
Talham, Robert J
Tehon, Stephen Whittier
White, Frederick Andrew
Zwislocki, Jozef John

NORTH CAROLINA
Bowie, James Dwight
Brown, Richard Dean
Dixon, N(orman) Rex
Hageseth, Gaylord Terrence
Hutchins, William R(eagh)
Keltie, Richard Francis
Kremkau, Frederick William
Lawson, Dewey Tull
Royster, L(arry) H(erbert)
Soderquist, David Richard

OHIO
Adler, Laszlo
Carome, Edward F
Chu, Mamerto Loarca
Eby, Ronald Kraft
Ensminger, Dale
Ghering, Walter L
Greene, David C
Hagelberg, M(yron) Paul
Lehiste, Ilse
Marshall, Kenneth D
Martin, Daniel William
Mawardi, Osman Kamel
Moore, Thomas Joseph
Neiman, Gary Scott
Nichols, William Herbert
Oyer, Herbert Joseph
Proctor, David George
Richards, Walter Bruce
Rodman, Charles William
Schuele, Donald Edward
Stumpf, Folden Burt
Von Gierke, Henning Edgar
Yun, Seung Soo
Zaman, Khairul B M Q

OKLAHOMA
Boade, Rodney Russett
Brown, Graydon L
Coffman, Moody Lee
Major, John Keene

OREGON
Chartier, Vernon Lee
Drummond, James Edgar
Semura, Jack Sadatoshi
White, Donald Harvey

PENNSYLVANIA
Akay, Adnan
Coltman, John Wesley
DeSanto, Daniel Frank
Goldman, Robert Barnett
Grindall, Emerson Leroy
Hansen, Robert Jack
Harbold, Mary Leah
Harrold, Ronald Thomas
Josephs, Jess J
Lauchle, Gerald Clyde
Maynard, Julian Decatur
Meyer, Paul A
Panuska, Joseph Allan
Pearman, G Timothy
Pierce, Allan Dale
Prout, James Harold
Reid, John Mitchell
Rhodes, Donald Frederick
Schumacher, Robert Thornton
Shung, K Kirk
Sibul, Leon Henry
Smith, Wesley R
Spalding, George Robert
Stern, Richard
Thompson, William, Jr
Tittmann, Bernhard R
Toulouse, Jean
Varadan, Vasundara Venkatraman

PHYSICS & ASTRONOMY / 505

Acoustics (cont)
Whitman, Alan M
Wilson, Geoffrey Leonard
Young, In Min

RHODE ISLAND
Bartram, James F(ranklin)
Becken, Bradford Albert
Beyer, Robert Thomas
Browning, David Gunter
Dietz, Frank Tobias
Dinapoli, Frederick Richard
Ehrlich, Stanley L(eonard)
Ellison, William Theodore
Koch, Robert Michael
Lall, Prithvi C
Mellberg, Leonard Evert
Ng, Kam Wing
Petrou, Panayiotis Polydorou
Sherman, Charles Henry
Stepanishen, Peter Richard
Sullivan, Edmund Joseph

SOUTH CAROLINA
Manson, Joseph Richard
Uldrick, John Paul

TENNESSEE
Bell, Zane W
Lewis, James W L
Nabelek, Anna K
Stewart, James Monroe
Swim, William B(axter)
Wang, Taylor Gunjin

TEXAS
Anderson, Aubrey Lee
Angona, Frank Anthony
Ashby, Jon Kenneth
Baker, Dudley Duggan, III
Barnard, Garland Ray
Barthel, Romard
Birchak, James Robert
Blackstock, David Theobald
Boyer, Lester Leroy
Carroll, Michael M
Chung, Jing Yan
Connor, William Keith
DiFoggio, Rocco
Few, Arthur Allen
Finch, Robert D
Fountain, Lewis Spencer
Frazer, Marshall Everett
Griffy, Thomas Alan
Gruber, George J
Hampton, Loyd Donald
Hartley, Craig Jay
Hayre, Harbhajan Singh
Horton, Claude Wendell, Sr
Ickes, William K
Klein, William Richard
Kostoff, Morris R
Lengel, Robert Charles
Lopez, Jorge Alberto
McKinney, Chester Meek
Mitchell, Stephen Keith
Muir, Thomas Gustave, Jr
Nimitz, Walter W von
Packard, Robert Gay
Peters, Randall Douglas
Plemons, Terry D
Pope, Larry Debbs
Powell, Alan
Pray, Donald George
Rosenbaum, Joseph Hans
Shooter, Jack Allen
Simpson, S(tephen) H(arbert), Jr
Smith, R Lowell
Sparks, Cecil Ray
Truchard, James Joseph
Zemanek, Joseph, Jr

UTAH
Chabries, Douglas M
Dixon, Dwight R
Dudley, J Duane
Strong, William J
Symko, Orest George

VERMONT
Gould, Robert K
Nyborg, Wesley LeMars
Sachs, Thomas Dudley
Spillman, William Bert, Jr
Wu, Jun Ru

VIRGINIA
Bothwell, Frank Edgar
Briscoe, Melbourne George
Brooks, Thomas Furman
Burley, Carlton Edwin
Cantrell, John H(arris)
Carterette, Edward Calvin Hayes
Claus, Richard Otto
Davis, Charles Mitchell, Jr
Desiderio, Anthony Michael
Doles, John Henry, III
Dozier, Lewis Bryant
Duke, John Christian, Jr
Fei, Ding-Yu
Fuller, Christopher Robert
Gatski, Thomas Bernard
Gerlach, A(lbert) A(ugust)
Gilbert, Arthur Charles
Gottwald, Jimmy Thorne
Graver, William Robert
Hardin, Jay Charles
Hargrove, Logan Ezral
Heyman, Joseph Saul
Houbolt, John C(ornelius)
Hubbard, Harvey Hart
Hurdle, Burton Garrison
Johnson, Glen Eric
Kenyon, Stephen C
Kooij, Theo
Lenhardt, Martin Louis
Maestrello, Lucio
Masterson, Kleber Sanlin, Jr
Morrow, Richard Joseph
Myers, Michael Kenneth
Neubauer, Werner George
Orr, Marshall H
Parvulescu, Antares
Poon, Ting-Chung
Puster, Richard Lee
Raney, William Perin
Rosenthal, F(elix)
Sauder, William Conrad
Sebastian, Richard Lee
Seiner, John Milton
Smith, Carey Daniel
Spivack, Harvey Marvin
Stevens, Donald Meade
Stusnick, Eric
Sykes, Alan O'Neil
Tatro, Peter Richard
Ugincius, Peter
Watson, Robert Barden
Wyeth, Newell Convers
Yu, James Chun-Ying

WASHINGTON
Atneosen, Richard Allen
Chace, Alden Buffington, Jr
Crum, Lawrence Arthur
Feldman, Henry Robert
Forster, Fred Kurt
Hildebrand, Bernard Percy
Ingalls, Paul D
Ishimaru, Akira
Jacobs, Loyd Donald
Klepper, John Richard
Lipscomb, David M
Malone, Stephen D
Marston, Philip Leslie
Merchant, Howard Carl
Oncley, Paul Bennett
Peterson, Arnold (Per Gustaf)
Pichal, Henri Thomas, II
Richardson, Richard Laurel
Sandstrom, Wayne Mark
Sengupta, Gautam
Thorne, Richard Eugene

WEST VIRGINIA
Lass, Norman J(ay)

WISCONSIN
Elfner, Lloyd F
Eriksson, Larry John
Flax, Stephen Wayne
Fystrom, Dell O
Hind, Joseph Edward
Kent, Raymond D
Levy, Moises
Pillai, Thankappan A K

WYOMING
Roark, Terry P

ALBERTA
Ulagaraj, Munivandy Seydunganallur

BRITISH COLUMBIA
Greenwood, Donald Dean
Hughes, Blyth Alvin
Seymour, Brian Richard
Thomson, David James

MANITOBA
Lyons, Edward Arthur
Svenne, Juris Peteris

NOVA SCOTIA
McMahon, Garfield Walter
Roger, William Alexander

ONTARIO
Brammer, Anthony John
Chu, Wing Tin
Clarke, Robert Lee
Daigle, Gilles A
Embleton, Tony Frederick Wallace
Keeler, John S(cott)
Krishnappa, Govindappa
Kunov, Hans
Lew, Hin
Lipshitz, Stanley Paul
Prasad, S E
Ribner, Herbert Spencer
Richarz, Werner Gunter
Schneider, Bruce Alton
Shaw, Edgar Albert George
Stinson, Michael Roy
Stouffer, James L
Toole, Floyd Edward
Wong, George Shoung-Koon

QUEBEC
Madore, Guy Louis Joseph

OTHER COUNTRIES
Bies, David Alan
Kagawa, Yukio
Miyano, Kenjiro
Moore, James Edward, Jr
Orphanoudakis, Stelios Constantine
Sessler, Gerhard Martin
Sugaya, Hiroshi
Tolstoy, Ivan
Voutsas, Alexander Matthew (Voutsadakis)
Wasa, Kiyotaka
Weil, Raoul Bloch

Astronomy

ALABAMA
Abbas, Mian Mohammad
Baksay, Laszlo Andras
Boardman, William Jarvis
Buta, Ronald James
Curott, David Richard
Davis, John Moulton
Decher, Rudolf
Hermann, R(udolf)
Hinata, Satoshi
Krause, Helmut G L
Moses, Ray Napoleon, Jr
Sulentic, Jack William
Tandberg-Hanssen, Einar Andreas
Twieg, Donald Baker
Weisskopf, Martin Charles

ARIZONA
Ables, Harold Dwayne
Abt, Helmut Arthur
Angel, James Roger Prior
Barr, Lawrence Dale
Belton, Michael J S
Black, John Harry
Bougher, Stephen Wesley
Burrows, Adam Seth
Burstein, David
Cameron, Winifred Sawtell
Chaffee, Frederic H, Jr
Chapman, Clark Russell
Christy, James Walter
Cowley, Anne Pyne
Coyne, George Vincent
Crawford, David Livingstone
Dahn, Conard Curtis
Davis, Donald Ray
Dunkelman, Lawrence
Edlin, Frank E
Fitch, Walter Stewart
Foltz, Craig Billig
Fountain, John William
Franz, Otto Gustav
Gatley, Ian
Gehrels, Tom
Giclas, Henry Lee
Gordon, Mark A
Grandi, Steven Aldridge
Greeley, Ronald
Greenberg, Richard Joseph
Gregory, Brooke
Guetter, Harry Hendrik
Hagelbarger, David William
Hall, Richard Chandler
Harris, Hugh Courtney
Hartmann, William K
Hayes, Donald Scott
Hege, E K
Hilliard, Ronnie Lewis
Hoag, Arthur Allen
Hoffmann, William Frederick
Howard, Robert Franklin
Hubbard, William Bogel
Hunten, Donald Mount
Irwin, John (Henry) Barrows
Jacobson, Michael Ray
Joyce, Richard Ross
Kieffer, Hugh Hartman
Kinman, Thomas David
Kreidl, Tobias Joachim
Larson, Harold Phillip
Lebofsky, Larry Allen
Liebert, James William
Livingston, William Charles
Lockwood, George Wesley
Lucchitta, Baerbel Koesters
Lynds, Clarence Roger
Mahalingam, Lalgudi Muthuswamy
Massey, Philip Louis
Masursky, Harold
Mayall, Nicholas Ulrich
Melosh, Henry Jay, IV
Millis, Robert Lowell
Odell, Andrew Paul
O'Leary, Brian Todd
Pilachowski, Catherine A
Randall, Lawrence Kessler, Jr
Rieke, Carol Anger
Rieke, George Henry
Rieke, Marcia Jean
Roemer, Elizabeth
Rountree, Janet
Schaber, Gerald Gene
Sinton, William Merz
Soderblom, Laurence Albert
Sprague, Ann Louise
Strittmatter, Peter Albert
Strom, Robert Gregson
Swann, Gordon Alfred
Tifft, William Grant
Ulich, Bobby Lee
Vrba, Frederick John
Wasserman, Lawrence Harvey
Wehinger, Peter Augustus
Weidenschilling, Stuart John
Whitaker, Ewen A
White, Nathaniel Miller
White, Raymond E
Wildey, Robert Leroy
Wyckoff, Susan

ARKANSAS
Engle, Paul Randal
Grauer, Albert D
Rolleston, Aimar Andre
Sears, Derek William George

CALIFORNIA
Ahrens, Thomas J
Arens, John Frederic
Armstrong, John William
Aumann, Hartmut Hans-Georg
Babcock, Horace W
Backer, Donald Charles
Barry, Don Cary
Benemann, John Rudiger
Bergstralh, Jay Thor
Berman, Louis
Blinn, James Frederick
Bonsack, Walter Karl
Borucki, William Joseph
Bowyer, C Stuart
Bracewell, Ronald Newbold
Brodie, Jean Pamela
Brooks, Walter Lyda
Bunch, Theodore Eugene
Buratti, Bonnie J
Burbidge, Eleanor Margaret
Burbidge, Geoffrey
Carpenter, Roland LeRoy
Castle, Karen G
Catura, Richard Clarence
Chambers, Robert J
Chapman, Gary Allen
Chow, Brian G(ee-Yin)
Cohen, Marshall Harris
Collard, Harold Rieth
Cominsky, Lynn Ruth
Cordell, Bruce Monteith
Corwin, Harold G, Jr
Cruikshank, Dale Paul
Cudaback, David Dill
Danielson, George Edward, Jr
Davies, Merton Edward
Dinger, Ann St Clair
Doolittle, Robert Frederick, II
Drake, Frank Donald
Dressler, Alan Michael
Duke, Douglas
Edelson, Sidney
Elliott, Denis Anthony
Elson, Lee Stephen
Epps, Harland Warren
Faber, Sandra Moore
Filippenko, Alexei Vladimir
Fisher, Philip Chapin
Fliegel, Henry Frederick
Fraknoi, Andrew Gabriel
Fymat, Alain L
Gaposchkin, Peter John Arthur
Gault, Donald E
Gibson, James (Benjamin)
Gilvarry, John James
Gregorich, David Tony
Gulkis, Samuel
Hagar, Charles Frederick
Haisch, Bernhard Michael
Hamill, Patrick James
Hanson, Robert Bruce
Harris, Alan William
Heiles, Carl
Helin, Eleanor Kay
Herzog, Emil Rudolph
Hollenbach, David John
Horn, Linda Joyce
Hughes, Thomas Rogers, Jr
Ingersoll, Andrew Perry
Jackson, Bernard Vernon
Janssen, Michael Allen
Johnson, Hugh Mitchell
Johnson, Torrence Vaino
Jones, Burton Fredrick
Kammerud, Ronald Claire
King, Ivan Robert
Kirk, John Gallatin
Klein, Michael John
Klemola, Arnold R
Kliore, Arvydas J(oseph)
Koch, David Gilbert
Koo, David Chih-yuen
Kraft, Robert Paul
Kristian, Jerome
Krupp, Edwin Charles
Kuiper, Thomas Bernardus Henricus
Lampton, Michael Logan
Landecker, Peter Bruce
Lark, Neil LaVern
Linford, Gary Joe
Lonsdale, Carol Jean

Lopes-Gautier, Rosaly
Luu, Jane
McCormick, Philip Thomas
McFadden, Lucy-Ann Adams
Macy, William Wray, Jr
Malin, Michael Charles
Markowitz, Allan Henry
Martin, Terry Zachry
Matson, Dennis Ludwig
Mauche, Christopher W
Meinel, Aden Baker
Meinel, Marjorie Pettit
Miner, Ellis Devere, Jr
Morris, Mark Root
Mottmann, John
Muhleman, Duane Owen
Murray, Bruce C
Nelson, Burt
Nelson, Robert M
Neugebauer, Gerry
Newburn, Ray Leon, Jr
Newman, William I
Oke, John Beverley
Oliver, Bernard M(ore)
Olsen, Edward Tait
Orton, Glenn Scott
Osterbrock, Donald E(dward)
Palluconi, Frank Don
Peale, Stanton Jerrold
Pennypacker, Carlton Reese
Persson, Sven Eric
Peters, Geraldine Joan
Pollack, James Barney
Popper, Daniel Magnes
Prather, Michael John
Preston, George W, III
Preston, Robert Arthur
Reaves, Gibson
Resch, George Michael
Reynolds, Ray Thomas
Rhodes, Edward Joseph, Jr
Rickard, James Joseph
Robinson, Lloyd Burdette
Romanowicz, Barbara Anna
Rosen, Harold A
Rubin, Robert Howard
Rudnyk, Marian E
Russell, John Albert
Salanave, Leon Edward
Sandage, Allan Rex
Sandmann, William Henry
Sargent, Anneila Isabel
Scargle, Jeffrey D
Scattergood, Thomas W
Schaefer, Albert Russell
Schmidt, Maarten
Schopp, John David
Schulte, Daniel Herman
Schuster, William John
Sekanina, Zdenek
Shimabukuro, Fred Ichiro
Silk, Joseph Ivor
Simon, Lee Will
Simpson, Richard Allan
Slade, Martin Alphonse, III
Smith, Harding Eugene
Sovers, Ojars Juris
Spence, Harlan Ernest
Spinrad, Hyron
Standish, E Myles, Jr
Stauffer, John Richard
Stern, Albert Victor
Tarbell, Theodore Dean
Tenn, Joseph S
Terrile, Richard John
Toon, Owen Brian
Treffers, Richard Rowe
Trimble, Virginia Louise
Tucker, Wallace Hampton
Tyler, George Leonard
Underwood, James Henry
Unwin, Stephen Charles
Vedder, James Forrest
Villere, Karen R
Vogt, Steven Scott
Walker, Arthur Bertram Cuthbert, Jr
Walker, Merle F
Walker, Russell Glenn
Wang, Charles P
Wannier, Peter Gregory
Weaver, Harold Francis
Weissman, Paul Robert
Welch, William John
Wells, Ronald Allen
West, Robert A
Weston, Gary Steven
Westphal, James Adolph
Weymann, Ray J
Whitcomb, Stanley Ernest
Wilkening, Laurel Lynn
Wilson, William John
Wood, Fergus James
Woosley, Stanford Earl
Wray, James David
Wright, Frederick Hamilton
Wright, Melvyn Charles Harman
Yanow, Gilbert
Yeomans, Donald Keith
Yoder, Charles Finney
Young, Andrew Tipton
Young, Arthur
Zirin, Harold
Zook, Alma Claire
Zuckerman, Benjamin Michael

COLORADO
Begelman, Mitchell Craig
Bruns, Donald Gene
Conti, Peter Selby
Culver, Roger Bruce
Dietz, Richard Darby
Dulk, George A
Esposito, Larry Wayne
Garmany, Catharine Doremus
Langer, George Edward
Linsky, Jeffrey L
Little, Stephen James
Lynds, Beverly T
McCray, Richard A
Mansfield, Roger Leo
Munro, Richard Harding
Snow, Theodore Peck
Stencel, Robert Edward
Wackernagel, Hans Beat
Warwick, James Walter
Weihaupt, John George

CONNECTICUT
Ford, Clinton Banker
Ftaclas, Christ
Garfinkel, Boris
Herbst, William
Hoffleit, Ellen Dorrit
Houston, Walter Scott
Larson, Richard Bondo
Lu, Phillip Kehwa
Menke, David Hugh
Oemler, Augustus, Jr
Peterson, Cynthia Wyeth
Upgren, Arthur Reinhold, Jr
Van Altena, William F
Zeitlin, Bruce Allen
Zinn, Robert James

DELAWARE
Evenson, Paul Arthur
Herr, Richard Baessler

DISTRICT OF COLUMBIA
Appleby, John Frederick
Barker, Edwin Stephen
Bautz, Laura Patricia
Berendzen, Richard
Berg, Richard Allen
Bowers, Phillip Frederick
Boyce, Peter Bradford
Butler, Dixon Matlock
Cook, John William
Corbin, Thomas Elbert
Cruddace, Raymond Gibson
DeVorkin, David Hyam
Fiala, Alan Dale
Forman, Miriam Ausman
Fritz, Gilbert Geiger
Gursky, Herbert
Harrington, Robert Sutton
Hart, Richard Cullen
Hawkins, Gerald Stanley
Hughes, James Arthur
Janiczek, Paul Michael
Johnston, Kenneth John
Kaplan, George Harry
Klepczynski, William J(ohn)
McCarthy, Dennis Dean
Mariska, John Thomas
Matsakis, Demetrios Nicholas
Matson, Michael
Meekins, John Fred
Ohring, George
Parker, Robert Allan Ridley
Pascu, Dan
Pilcher, Carl Bernard
Rickard, Lee J
Rubin, Vera Cooper
Saunders, Ronald Stephen
Schweizer, Francois
Seidelmann, Paul Kenneth
Shulman, Seth David
Smith, Clayton Albert, Jr
Strand, Kaj Aage
Summers, Michael Earl
Van Flandern, Thomas C
Wang, Yi-Ming
Weiler, Edward John
Weiler, Kurt Walter
Withbroe, George Lund
Zimbelman, James Ray

FLORIDA
Bahng, John Deuck Ryong
Buchler, Jean-Robert
Burns, Michael J
Carr, Thomas Deaderick
Chen, Kwan-Yu
Cohen, Howard Lionel
Cooke, William Joe
Dermott, Stanley Frederick
Eichhorn, Heinrich Karl
Franklin, Kenneth Linn
Gottesman, Stephen T
Gustafson, Bo Ake Sture
Hett, John Henry
Hunter, James Hardin, Jr
Klock, Benny LeRoy
Leacock, Robert Jay
Lebo, George Robert
Markowitz, William
Merrill, John Ellsworth
Misconi, Nebil Yousif

Mulholland, John Derral
Oliver, John Parker
Pittman, Melvin Amos
Porter, Lee Albert
Rester, Alfred Carl, Jr
Roberts, Leonidas Howard
Smith, Haywood Clark, Jr
Stefanakos, Elias Kyriakos
Webb, James Raymond
Weinberg, Jerry L
Williams, Carol Ann
Wilson, Robert E
Wood, Frank Bradshaw

GEORGIA
Gudehus, Donald Henry
Jarvis, John Frederick
McAlister, Harold Alister
Powell, Bobby Earl
Sadun, Alberto Carlo
Shaw, James Scott

HAWAII
Boesgaard, Ann Merchant
Chambers, Kenneth Carter
Cowie, Lennox Lauchlan
Crowe, Richard Alan
Glaspey, John Warren
Gradie, Jonathan Carey
Hall, Donald Norman Blake
Heacox, William Dale
Herbig, George Howard
Kormendy, John
Krisciunas, Kevin L
Kron, Gerald Edward
McCord, Thomas Bard
McLaren, Robert Alexander
Meech, Karen Jean
Owen, Tobias Chant
Purton, Christopher Roger
Smith, Bradford Adelbert
Tokunaga, Alan Takashi
Tully, Richard Brent

IDAHO
Parker, Barry Richard
Patsakos, George

ILLINOIS
Bartlett, J Frederick
Carlson, Eric Dungan
Chamberlain, Joseph Miles
Chandrasekhar, Subrahmanyan
Crutcher, Richard Metcalf
Dickel, Helene Ramseyer
Harper, Doyal Alexander, Jr
Kaler, James Bailey
Kent, Stephen Matthew
Knappenberger, Paul Henry, Jr
Langebartel, Ray Gartner
Leibhardt, Edward
Lo, Kwok-Yung
McCormack, Elizabeth Frances
Nelson, Harry Ernest
Nissim-Sabat, Charles
Palmer, Patrick Edward
Palmore, Julian Ivanhoe, III
Peterson, Melbert Eugene
Roberts, Arthur
Saari, Donald Gene
Spinka, Harold M
Sweitzer, James Stuart
Taam, Ronald Everett
Truran, James Wellington, Jr
Ulmer, Melville Paul
Vandervoort, Peter Oliver
Webb, Harold Donivan
Yoss, Kenneth M

INDIANA
Beery, Dwight Beecher
Durisen, Richard H
Edmondson, Frank Kelley
Hrivnak, Bruce John
Hults, Malcom E
Jacobs, William Wescott
Poirier, John Anthony
Read, Robert H
Shore, Steven Neil
Sprague, Newton G
Wilson, Olin C(haddock)

IOWA
Basart, John Philip
Beavers, Willet I
Bowen, George Hamilton, Jr
Cadmus, Robert Randall, Jr
Frank, Louis Albert
Gurnett, Donald Alfred
Hartung, Jack Burdair, Sr
Hoff, Darrel Barton
Intemann, Gerald William
Jacob, Richard L
Mutel, Robert Lucien
Neff, John S
Riggs, Philip Shaefer
Smith, Paul Aiken
Spangler, Steven Randall
Struck-Marcell, Curtis John
Willson, Lee Anne Mordy

KANSAS
Anthony-Twarog, Barbara Jean
Shawl, Stephen Jacobs

Twarog, Bruce Anthony
Walters, Charles Philip
Williams, Dudley

KENTUCKY
Gwinn, Joel Alderson
Kielkopf, John F
Krogdahl, Wasley Sven
Powell, Smith Thompson, III

LOUISIANA
Landolt, Arlo Udell
Lee, Paul D
Perry, Charles Lewis
Ryan, Donald Edwin
Tohline, Joel Edward

MARYLAND
A'Hearn, Michael Francis
Ake, Thomas Bellis, III
Alexander, Joseph Kunkle
Beal, Robert Carl
Behannon, Kenneth Wayne
Bennett, Charles L
Bond, Howard Emerson
Brown, Robert Alan
Bruhweiler, Frederick Carlton, Jr
Brunk, William Edward
Buhl, David
Carbary, James F
Chaisson, Eric Joseph
Chartrand, Mark Ray, III
Chen, Hsing-Hen
Clark, Thomas Arvid
Conrath, Barney Jay
Crane, Patrick Conrad
Davidsen, Arthur Falnes
Deprit, Andre A(lbert) M(aurice)
Di Rienzi, Joseph
Doggett, LeRoy Elsworth
Dolan, Joseph Francis
Donn, Bertram (David)
Downing, Mary Brigetta
Dunham, David Waring
Erickson, William Clarence
Evans, Larry Gerald
Fallon, Frederick Walter
Feibelman, Walter A
Feldman, Paul Donald
Felten, James E(dgar)
Ford, Holland Cole
Gilliland, Ronald Lynn
Gull, Theodore Raymond
Hall, R(oyal) Glenn, Jr
Hart, Michael H
Harvel, Christopher Alvin
Haupt, Ralph Freeman
Hauser, Michael George
Heckman, Timothy Martin
Henry, Richard Conn
Hershey, John Landis
Hill, Jesse King
Hintzen, Paul Michael
Hobbs, Robert Wesley
Hollis, Jan Michael
Holman, Gordon Dean
Howard, Sethanne
Hubbard, Donald
Kaiser, Michael Leroy
Kaluzienski, Louis Joseph
Kerr, Frank John
Killen, Rosemary Margaret
Kondo, Yoji
Kostiuk, Theodor
Kowal, Charles Thomas
Kroeger, Richard Alan
Kundu, Mukul Ranjan
Lovas, Francis John
Lowman, Paul Daniel, Jr
Lufkin, Daniel Harlow
Malitson, Harriet Hutzler
Mead, Jaylee Montague
Milkey, Robert William
O'Keefe, John Aloysius
Ozernoy, Leonid M
Pearl, John Christopher
Perry, Peter M
Robertson, Douglas Scott
Roman, Nancy Grace
Schreier, Ethan Joshua
Shara, Michael M
Shivanandan, Kandiah
Smith, Joseph Collins
Stecher, Theodore P
Stecker, Floyd William
Stockman, Hervey S, Jr
Sturch, Conrad Ray
Teays, Terry John
Toller, Gary Neil
Tschunko, Hubert F A
Turnrose, Barry Edmund
Walborn, Nolan Revere
Warren, Wayne Hutchinson, Jr
Weber, Richard Rand
Webster, William John, Jr
Weinreb, Sander
Wells, Eddie N
Wende, Charles David
Westerhout, Gart
White, Richard Allan
Whitmore, Bradley Charles
Wilson, Andrew Stephen
Wood, Howard John, III
Zellner, Benjamin Holmes

Astronomy (cont)

MASSACHUSETTS
Albers, Henry
Ash, Michael Edward
Baliunas, Sallie Louise
Ball, John Allen
Bauer, Wendy Hagen
Bertschinger, Edmund
Birney, Dion Scott, Jr
Brooks, Edward Morgan
Canizares, Claude Roger
Cappallo, Roger James
Counselman, Charles Claude, III
Davis, Robert James
Dennis, Tom Ross
Downs, George Samuel
Edwards, Suzan
Elliot, James Ludlow
Eoll, John Gordon
Ewen, Harold Irving
Franklin, Fred Aldrich
Geary, John Charles
Geller, Margaret Joan
Gingerich, Owen (Jay)
Gordon, Kurtiss Jay
Greenstein, George
Grindlay, Jonathan Ellis
Harris, Daniel Everett
Hazen, Martha L(ocke)
Huchra, John P
Huguenin, George Richard
Irvine, William Michael
Jacchia, Luigi Giuseppe
Jaffe, Miriam Walther
Johnson, Douglas William
Joss, Paul Christopher
Kellogg, Edwin M
Kirshner, Robert Paul
Kleinmann, Douglas Erwin
Kwitter, Karen Beth
Lada, Charles Joseph
Lada, Elizabeth A
Latham, David Winslow
Layzer, David
Lilley, Arthur Edward
McCrosky, Richard Eugene
Marsden, Brian Geoffrey
Mattei, Janet Akyüz
Mayall, Margaret Walton
Mendillo, Michael
Moody, Elizabeth Anne
Moran, James Michael, Jr
Mumford, George Saltonstall
Murray, Stephen S
Myers, Philip Cherdak
Narayan, Ramesh
Niell, Arthur Edwin
Pasachoff, Jay M(yron)
Pedersen, Peder Christian
Pettengill, Gordon Hemenway
Ratner, Michael Ira
Roberts, David Hall
Rogers, Alan Ernest Exel
Salah, Joseph E
Schild, Rudolph Ernest
Schwartz, Daniel Alan
Seward, Frederick Downing
Shapiro, Irwin Ira
Shapiro, Ralph
Shawcross, William Edgerton
Silk, John Kevin
Snell, Ronald Lee
Staelin, David Hudson
Stone, Peter Hunter
Strom, Stephen
Tonry, John Landis
Toomre, Alar
VanSpeybroeck, Leon Paul
Wardle, John Francis Carleton
White, Richard Edward
Willner, Steven P
Wisdom, Jack Leach
Wood, David Belden

MICHIGAN
Aller, Margo Friedel
Atreya, Sushil Kumar
Berger, Beverly Kobre
Brockmeier, Richard Taber
Burek, Anthony John
Capriotti, Eugene Raymond
Christian, Larry Omar
Cowley, Charles Ramsay
Dunifer, Gerald Leroy
England, Anthony W
Haddock, Frederick Theodore, Jr
Houk, Nancy (Mia)
Kauppila, Walter Eric
Leonard, Charles Grant
MacAlpine, Gordon Madeira
Miller, Freeman Devold
Osborn, Wayne Henry
Richstone, Douglas Orange
Simkin, Susan Marguerite
Teske, Richard Glenn
Van Till, Howard Jay
Williams, John Albert

MINNESOTA
Davidson, Kris
Eckroth, Charles Angelo
Hakkila, Jon Eric
Humphreys, Roberta Marie

Luyten, Willem Jacob
Mathews, Robert Thomas
Rudnick, Lawrence
Whitehead, Andrew Bruce

MISSISSIPPI
Croft, Walter Lawrence
Lestrade, John Patrick

MISSOURI
Ashworth, William Bruce, Jr
Cowsik, Ramanath
Edwards, Terry Winslow
Geilker, Charles Don
McKinnon, William Beall
Nelson, George Driver
Peterson, Charles John
Schmitt, John Leigh
Schwartz, Richard Dean
Smith, William Hayden
Wolf, George William

MONTANA
Jeppesen, Randolph H

NEBRASKA
Billesbach, David P
Clements, Gregory Leland
Leung, Kam-Ching
Schmidt, Edward George
Taylor, Donald James
Underhill, Glenn

NEVADA
Shaffer, David Bruce
Weistrop, Donna Etta

NEW HAMPSHIRE
Baker, James Gilbert
Beasley, Wayne M(achon)
Cole, Charles N
Mook, Delo Emerson, II
Ryan, James Michael
Wegner, Gary Alan

NEW JERSEY
Bahcall, Neta Assaf
Boeshaar, Patricia Chikotas
Damen, Theo C
Fenstermacher, Robert Lane
Gott, J Richard, III
Hulse, Russell Alan
Knapp, Gillian Revill
Lee, Keun Myung
Levine, Jerry David
Maclennan, Carol G
Peebles, Phillip J
Penzias, Arno A(llan)
Pfeiffer, Raymond John
Pregger, Fred Titus
Schwarzschild, Martin
Spitzer, Lyman, Jr
Taylor, Joseph Hooton, Jr
Turner, Edwin Lewis
Wilkinson, David Todd
Williams, Theodore Burton
Wilson, Robert Woodrow

NEW MEXICO
Argo, Harold Virgil
Bignell, Richard Carl
Bleiweiss, Max Phillip
Burns, Jack O'Neal
Clark, Barry Gillespie
Clark, Robert Edward Holmes
Claussen, Mark J
Cox, Arthur Nelson
Dunn, Richard B
Gisler, Galen Ross
Hankins, Timothy Hamilton
Heiken, Grant Harvey
Linsley, John
MacCallum, Crawford John
Neidig, Donald Foster
Olson, Gordon Lee
Owen, Frazer Nelson
Reeves, Geoffrey D
Sanders, Walter L
Simon, George Warren
Sramek, Richard Anthony
Strong, Ian B
Tombaugh, Clyde W(illiam)
Wade, Campbell Marion
Zeilik, Michael

NEW YORK
Aveni, Anthony
Baez, Silvio
Baker, Norman Hodgson
Baum, Parker Bryant
Branley, Franklyn M
Burns, Joseph A
Campbell, Donald Bruce
Cheung, Lim H
Craft, Harold Dumont, Jr
Cummins, John Frances
Davis, Raymond, Jr
Dubisch, Russell John
Elmegreen, Debra Meloy
Forrest, William John
Gaffey, Michael James
Gelman, Donald
Giovanelli, Riccardo
Goebel, Ronald William

Gold, Thomas
Goldsmith, Paul Felix
Gross, Stanley H
Haynes, Martha Patricia
Helfand, David John
Huggins, Patrick John
Lattimer, James Michael
Lissauer, Jack Jonathan
Matsushima, Satoshi
Meisel, David Dering
Meltzer, Alan Sidney
Millikan, Allan G
Noonan, Thomas Wyatt
O'Donoghue, Aileen Ann
Philip, A G Davis
Prendergast, Kevin Henry
Reddy, Reginald James
Sagan, Carl
Schreurs, Jan W H
Sharpless, Stewart Lane
Simon, Michal
Squyres, Steven Weldon
Stull, John Leete
Talley, Robert Lee
Tavel, Judith Fibkins
Teiger, Martin
Terzian, Yervant
Thorsen, Richard Stanley
Veverka, Joseph F
Yahil, Amos

NORTH CAROLINA
Carney, Bruce William
Christiansen, Wayne Arthur
Davis, Morris Schuyler
Jenzano, Anthony Francis
Oberhofer, Edward Samuel
Rhynsburger, Robert Whitman
Shapiro, Lee Tobey

NORTH DAKOTA
Rathmann, Franz Heinrich
Seielstad, George A

OHIO
Anderson, Lawrence Sven
Annear, Paul Richard
Bidelman, William Pendry
Bishop, Edwin Vandewater
Bopp, Bernard William
Collins, George W, II
Ferguson, Dale Curtis
Ferland, Gary Joseph
Frogel, Jay Albert
Goedicke, Victor Alfred
James, Philip Benjamin
Keenan, Philip Childs
Keller, Geoffrey
Luck, Richard Earle
Malcuit, Robert Joseph
Mickelson, Michael Eugene
Mitchell, Walter Edmund, Jr
Morrison, Nancy Dunlap
Nahar, Sultana Nurun
Newsom, Gerald Higley
Osmer, Patrick Stewart
Patton, Jon Michael
Pearson, Jerome
Pesch, Peter
Peterson, Bradley M(ichael)
Protheroe, William Mansel
Richardson, David Louis
Rodman, James Purcell
Sellgren, Kristen
Stanger, Philip Charles
Stephenson, Charles Bruce
Strickler, Thomas David
Wagoner, Glen
Wing, Robert Farquhar
Young, Warren Melvin

OKLAHOMA
Cowan, John James
Rutledge, Carl Thomas
Shull, Peter Otto, Jr

OREGON
McDaniels, David K
McKinney, William Mark

PENNSYLVANIA
Augensen, Harry John
Bauer, Carl August
Blair, Grant Clark
Blitzstein, William
Boughn, Stephen Paul
Bowersox, Todd Wilson
Carpenter, Lynn Allen
Chambliss, Carlson Rollin
Cordova, France Anne-Dominic
Couch, Jack Gary
Feigelson, Eric Dennis
Gainer, Michael Kizinski
Garmire, Gordon Paul
Gatewood, George David
Guinan, Edward F
Heintz, Wulff Dieter
Holzinger, Joseph Rose
Jenkins, Edward Felix
Kiewiet de Jonge, Joost H A
Knacke, Roger Fritz
Koch, Robert Harry
Kozlowski, Richard William Hugh
Levitt, Israel Monroe

Lippincott, Sarah Lee
McCook, George Patrick
Partridge, Robert Bruce
Poss, Howard Lionel
Ramsey, Lawrence William
Schneider, Donald P
Seeds, Michael August
Sion, Edward Michael
Sitterly, Charlotte Moore
Stauffer, George Franklin
Usher, Peter Denis
Weedman, Daniel Wilson
Winkler, Louis

RHODE ISLAND
Pieters, Carle M

SOUTH CAROLINA
Adelman, Saul Joseph
Rust, Philip Frederick
Safko, John Loren

TENNESSEE
Burke, Edward Walter, Jr
Cordell, Francis Merritt
Daunt, Stephen Joseph
Eaton, Joel A
Hall, Douglas Scott
Heiser, Arnold M
Lorenz, Philip Jack, Jr
Tolk, Norman Henry
Verschuur, Gerrit L

TEXAS
Bash, Frank Ness
Benedict, George Frederick
Blanford, George Emmanuel, Jr
Chamberlain, Joseph Wyan
De Vaucouleurs, Gerard Henri
Douglas, James Nathaniel
Dufour, Reginald James
Duncombe, Raynor Lockwood
Evans, David Stanley
Evans, Neal John, II
Gomer, Richard Hans
Gott, Preston Frazier
Harrison, Marjorie Hall
Hemenway, Mary-Kay Meacham
Henize, Karl Gordon
Heymann, Dieter
Hildebrandt, Alvin Frank
Hinterberger, Henry
Hoffman, Alexander A J
Hoffman, Jeffrey Alan
Jefferys, William H, III
Kattawar, George W
King, Elbert Aubrey, Jr
Lambert, David L
Lewis, Ira Wayne
Mangum, Jeffrey Gary
Middlehurst, Barbara Mary
Morgan, Thomas Harlow
Nacozy, Paul E
Nather, Roy Edward
Nicks, Oran Wesley
O'Dell, Charles Robert
Plass, Gilbert Norman
Potter, Andrew Elwin, Jr
Robinson, Edward Lewis
Shapiro, Paul Robert
Smith, Harlan J
Synek, Miroslav (Mike)
Szebehely, Victor
Talent, David Leroy
Trafton, Laurence Munro
Tull, Robert Gordon
Vishniac, Ethan Tecumseh
Winningham, John David
Zook, Herbert Allen

UTAH
Christensen, Clark G
McNamara, Delbert Harold
Morris, Elliot Cobia

VERMONT
MacArthur, John Wood
Rankin, Joanna Marie
Wolfson, Richard L T

VIRGINIA
Beardall, John Smith
Bridle, Alan Henry
Broderick, John
Bunner, Alan Newton
Carpenter, Martha Stahr
Carr, Roger Byington
Chevalier, Roger Alan
Condon, James Justin
Dennison, Brian Kenneth
Donivan, Frank Forbes, Jr
DuPuy, David Lorraine
Evans, John C
Finke, Reinald Guy
Ford, William Kent, Jr
Fredrick, Laurence William
Friedman, Herbert
Gergely, Tomas Esteban
Goldstein, Samuel Joseph, (Jr)
Greisen, Eric Winslow
Heckathorn, Harry Mervin, III
Heeschen, David Sutphin
Hogg, David Edward
Hooper, William John, Jr

Howard, William Eager, III
Johnson, Donald Rex
Kerr, Anthony Robert
Kissell, Kenneth Eugene
Knowles, Stephen H
Kumar, Shiv Sharan
Langston, Glen Irvin
Lindenblad, Irving Werner
Liszt, Harvey Steven
Margrave, Thomas Ewing, Jr
Mast, Joseph Willis
Mayer, Cornell Henry
Mihalas, Barbara R Weibel
O'Connell, Robert West
Oesterwinter, Claus
Radoski, Henry Robert
Rea, Donald George
Roberts, Morton Spitz
Roberts, William Woodruff, Jr
Rudmin, Joseph Webster
Sarazin, Craig L
Schwiderski, Ernst Walter
Smith, Elske van Panhuys
Staib, Jon Albert
Thompson, Anthony Richard
Tolbert, Charles Ray
Turner, Barry Earl
Vanden Bout, Paul Adrian
Webber, John Clinton
Weinstein, Leonard Murrey
Wootten, Henry Alwyn

WASHINGTON
Baum, William Alvin
Belserene, Emilia Pisani
Bookmyer, Beverly Brandon
Bracher, Katherine
Brownlee, Donald E
Ellis, Fred E
Gammon, Richard Harriss
Gierasch, Peter Jay
Hodge, Paul William
Hunziker, Rodney William
Jenner, David Charles
Kells, Lyman Francis
Krienke, Ora Karl, Jr
Lutz, Thomas Edward
Mitchell, Robert Curtis
Quigley, Robert James
Stokes, Gerald Madison
Sullivan, Woodruff Turner, III
Szkody, Paula
Wallerstein, George
White, William Charles
Windsor, Maurice William

WEST VIRGINIA
Ghigo, Frank Dunnington
Hewitt, Anthony Victor

WISCONSIN
Anderson, Christopher Marlowe
Bless, Robert Charles
Campbell, Warren Adams
Churchwell, Edward Bruce
Code, Arthur Dodd
Cudworth, Kyle McCabe
Hobbs, Lewis Mankin
Kojoian, Gabriel
Morgan, William Wilson
Reynolds, Ronald J
Savage, Blair DeWillis
Schroeder, Daniel John
Shepherd, John Patrick George
Vilkki, Erkki Uuno
Woods, Robert Claude

WYOMING
Canterna, Ronald William
Grasdalen, Gary
Howell, Robert Richard
Roark, Terry P
Thronson, Harley Andrew, Jr

PUERTO RICO
Davis, Michael Moore
Lewis, Brian Murray
Turner, Kenneth Clyde

ALBERTA
Clark, Thomas Alan
Hube, Douglas Peter
Kwok, Sun
Milone, Eugene Frank

BRITISH COLUMBIA
Aikman, George Christopher Lawrence
Batten, Alan Henry
Burke, J Anthony
Fahlman, Gregory Gaylord
Hartwick, Frederick David Alfred
Higgs, Lloyd Albert
Hutchings, John Barrie
McCutcheon, William Henry
Morbey, Christopher Leon
Palmer, Leigh Hunt
Richardson, Eric Harvey
Richer, Harvey Brian
Scarfe, Colin David
Shuter, William Leslie Hazlewood
Stetson, Peter Brailey
Van den Bergh, Sidney

NEW BRUNSWICK
Faig, Wolfgang

NEWFOUNDLAND
Rochester, Michael Grant

NOVA SCOTIA
Bishop, Roy Lovitt
Turner, David Gerald
Welch, Gary Alan

ONTARIO
Andrew, Bryan Haydn
Bakos, Gustav Alfons
Barker, Paul Kenneth
Bertram, Robert William
Blackwell, Alan Trevor
Bolton, Charles Thomas
Broten, Norman W
Caldwell, John James
Chow, Yung Leonard
Clarke, James Newton
Clarke, Thomas Roy
Clement, Christine Mary (Coutts)
Dyer, Charles Chester
Evans, Nancy Remage
Feldman, Paul Arnold
Fernie, John Donald
Garrison, Robert Frederick
Griffith, John Sidney
Hagberg, Erik Gordon
Harris, William Edgar
Hogarth, Jacke Edwin
Hogg, Helen (Battles) Sawyer
Hughes, Victor A
Kaiser, Nicholas
Legg, Thomas Harry
Lester, John Bernard
McCallion, William James
MacLeod, John Munroe
MacRae, Donald Alexander
Marlborough, John Michael
Moffat, John William
Percy, John Rees
Seaquist, Ernest Raymond
Sida, Derek William
Vallée, Jacques P
Wehlau, Amelia W
Wehlau, William Henry

QUEBEC
Borra, Ermanno Franco
Demers, Serge
Moffat, Anthony Frederick John
Pineault, Serge Rene
Racine, Rene

SASKATCHEWAN
Kos, Joseph Frank
Naqvi, Saiyid Ishrat Husain

OTHER COUNTRIES
Arp, Halton Christian
Beckwith, Steven Van Walter
Bell Burnell, S Jocelyn
Blanco, Betty M
Blanco, Victor Manuel
Burton, William Butler
Butcher, Harvey Raymond, III
Contopoulos, George
Hagfors, Tor
Hewish, Antony
Kesteven, Michael
Kopal, Zdenek
Kozai, Yoshihide
Kunkel, William Eckart
Liller, William
McCarthy, Martin
Miley, George Kildare
Munch, G
Reitmeyer, William L
Schnopper, Herbert William
Shurman, Michael Mendelsohn
Van der Hulst, Jan Mathijs
Vander Vorst, Andre
Von Hoerner, Sebastian
Wampler, E Joseph
White, Simon David Manton
Zabriskie, Franklin Robert

Astrophysics

ALABAMA
Abbas, Mian Mohammad
Bailey, Wayne Lewis
Buta, Ronald James
Curott, David Richard
Davis, John Moulton
Decher, Rudolf
Dunn, Anne Roberts
Fishman, Gerald Jay
Hathaway, David Henry
Hinata, Satoshi
Hoover, Richard Brice
Krause, Helmut G L
Lundquist, Charles Arthur
Meegan, Charles Anthony
Moore, Ronald Lee
O'Dell, Stephen Lynn
Parnell, Thomas Alfred
Pomerantz, Martin Arthur
Sulentic, Jack William
Tan, Arjun

Weisskopf, Martin Charles

ALASKA
Deehr, Charles Sterling
Kan, Joseph Ruce
Lee, Lou-Chuang
Martins, Donald Henry
Roederer, Juan Gualterio
Stenbaek-Nielsen, Hans C
Swift, Daniel W

ARIZONA
Abt, Helmut Arthur
Adel, Arthur
Arnett, William David
Black, John Harry
Blitzer, Leon
Boynton, William Vandegrift
Burrows, Adam Seth
Burstein, David
Chaffee, Frederic H, Jr
Christy, James Walter
Dessler, Alexander Jack
De Young, David Spencer
Fan, Chang-Yun
Foltz, Craig Billig
Franz, Otto Gustav
Gatley, Ian
Giampapa, Mark Steven
Greenberg, Richard Joseph
Harris, Hugh Courtney
Hartmann, William K
Harvey, John Warren
Herbert, Floyd Leigh
Hill, Henry Allen
Hoffmann, William Frederick
Hsieh, Ke Chiang
Huffman, Donald Ray
Jefferies, John Trevor
Jokipii, Jack Randolph
Kreidl, Tobias Joachim
Kurz, Richard J
Kyrala, Ali
Leibacher, John W
Levy, Eugene Howard
Liebert, James William
Low, Frank James
Lutz, Barry Lafean
Massey, Philip Louis
Mayall, Nicholas Ulrich
Melia, Fulvio
Pacholczyk, Andrzej Grzegorz
Pilachowski, Catherine A
Rountree, Janet
Sandel, Bill Roy
Sonett, Charles Philip
Starrfield, Sumner Grosby
Swihart, Thomas Lee
Thompson, Rodger Irwin
Tifft, William Grant
Weekes, Trevor Cecil
Wehinger, Peter Augustus
Wildey, Robert Leroy
Wolff, Sidney Carne
Woolf, Neville John
Yelle, Roger V

ARKANSAS
Grauer, Albert D
Harter, William George
Rollefson, Aimar Andre
Wold, Donald C

CALIFORNIA
Aitken, Donald W, Jr
Allamandola, Louis John
Aller, Lawrence Hugh
Anderson, Kinsey Amor
Anderson, Milo Vernette
Armstrong, John William
Arons, Jonathan
Ashburn, Edward V
Aumann, Hartmut Hans-Georg
Band, David Louis
Barnard, John J
Barnes, Aaron
Barnes, Charles A
Barry, James Dale
Barwick, Steven William
Bauer, Daniel A
Baum, Peter Joseph
Becker, Robert Adolph
Bell, Graydon Dee
Benford, Gregory A
Berdjis, Fazlollah
Bergstralh, Jay Thor
Blake, J Bernard
Blandford, Roger David
Bloom, Elliott D
Bloxham, Laurence Hastings
Blumenthal, George Ray
Bodenheimer, Peter Herman
Bonsack, Walter Karl
Bratenahl, Alexander
Bridges, Frank G
Brown, Wilbur K
Brown, William Arnold
Bruner, Marilyn E
Buffington, Andrew
Burbidge, Eleanor Margaret
Burbidge, Geoffrey
Caldwell, David Orville
Carpenter, Roland LeRoy

Cassen, Patrick Michael
Castor, John I
Chapline, George Frederick, Jr
Chapman, Gary Allen
Chase, Lloyd Fremont, Jr
Chen, Liu
Chow, Tai-Low
Christy, Robert Frederick
Cloutman, Lawrence Dean
Cohen, Judith Gamora
Cohen, Ronald Herbert
Collard, Harold Rieth
Cominsky, Lynn Ruth
Coroniti, Ferdinand Vincent
Corwin, Harold G, Jr
Daub, Clarence Theodore, Jr
Davis, Leverett, Jr
Davis, Marc
De Gaston, Alexis Neal
Dewitt, Hugh Edgar
Dinger, Ann St Clair
Doolittle, Robert Frederick, II
Duke, Douglas
Dyal, Palmer
Erickson, Edwin Francis
Erickson, Stanley Arvid
Esposito, Pasquale Bernard
Everitt, C W Francis
Farmer, Crofton Bernard
Faulkner, John
Feynman, Joan
Filippenko, Alexei Vladimir
Fisher, Philip Chapin
Frazier, Edward Nelson
Frerking, Margaret Ann
Friedman, Richard M
Garrett, Henry Berry
Gerola, Humberto Cayetano
Goldreich, Peter
Goorvitch, David
Gould, Robert Joseph
Greenstein, Jesse Leonard
Gregorich, David Tony
Grossman, Allen S
Gulkis, Samuel
Gwinn, William Dulaney
Hackwell, John Arthur
Haisch, Bernhard Michael
Hall, Donald Eugene
Harless, William H
Harris, Alan William
He, Yudong
Heiles, Carl
Helliwell, R(obert) A(rthur)
Hollenbach, David John
Holzer, Robert Edward
Horowitz, Gary T
Imhof, William Lowell
Intriligator, Devrie Shapiro
Irvine, Cynthia Emberson
Jackson, Bernard Vernon
Jacobson, Allan Stanley (Bud)
Janssen, Michael Allen
Johnson, Fred M
Johnson, Hugh Mitchell
Jura, Michael Alan
Kagiwada, Harriet Hatsune Natsuyama
Karp, Alan H
Kennel, Charles Frederick
Klumpar, David Michael
Knobloch, Edgar
Knox, William Jordan
Koch, David Gilbert
Koga, Rokutaro
Koo, David Chih-yuen
Landman, Donald Alan
Larmore, Lewis
Lawrence, John Keeler
Lea, Susan Maureen
Lee, Edward Prentiss
Leich, Douglas Albert
Leighton, Robert Benjamin
Lillie, Charles Frederick
Lin, Robert Peichung
Linde, Andrei
Lindl, John D
Lingenfelter, Richard Emery
Litvak, Marvin Mark
Lombardi, Gabriel Gustavo
Lonsdale, Carol Jean
Lugmair, Guenter Wilhelm
Luthey, Joe Lee
Luu, Jane
McCormac, Billy Murray
McDonough, Thomas Redmond
McIlwain, Carl Edwin
McKee, Christopher Fulton
Mahoney, William A
Maloy, John Owen
Marcus, Philip Stephen
Markowitz, Allan Henry
Marti, Kurt
Mathews, Grant James
Mauche, Christopher W
Max, Claire Ellen
Mayfield, Earle Byron
Mazurek, Thaddeus John
Mendis, Devamitta Asoka
Metzger, Albert E
Mewaldt, Richard A
Meyerott, Roland Edward
Mihalov, John Donald
Miner, Ellis Devere, Jr
Morris, Mark Root

Astrophysics (cont)

Mozer, Forrest S
Muller, Richard A
Nash, Douglas B
Nelson, Jerry Earl
Neu, John Ternay
Neugebauer, Gerry
Newman, William I
Norman, Eric B
Oder, Frederic Carl Emil
O'Keefe, John Dugan
Olsen, Edward Tait
Orth, Charles Douglas
Osterbrock, Donald E(dward)
Parker, Winifred Ellis
Paulikas, George A
Payne, David Glenn
Pazich, Philip Michael
Peale, Stanton Jerrold
Peccei, Roberto Daniele
Pehl, Richard Henry
Peters, Geraldine Joan
Peterson, Laurence E
Phillips, Thomas Gould
Plavec, Mirek Josef
Pneuman, Gerald W
Preskill, John P
Price, Paul Buford, Jr
Primack, Joel Robert
Pritchett, Philip Lentner
Rauch, Richard Travis
Richards, Paul Linford
Rosen, Leonard Craig
Ross, Ronald Rickard
Rothschild, Richard Eiseman
Rouse, Carl Albert
Rubin, Robert Howard
Russell, Christopher Thomas
Sackmann, I Juliana
Sadoulet, Bernard
Sargent, Anneila Isabel
Sargent, Wallace Leslie William
Schindler, Stephen Michael
Scholl, James Francis
Schulz, Michael
Schuster, William John
Scott, Douglas
Sekanina, Zdenek
Shemansky, Donald Eugene
Silk, Joseph Ivor
Siscoe, George L
Slade, Martin Alphonse, III
Smith, Harding Eugene
Smith, Sheldon Magill
Smoot, George Fitzgerald, III
Spreiter, John R(obert)
Stella, Paul M
Stevenson, David John
Straka, William Charles
Sturrock, Peter Andrew
Swanson, Paul N
Talbot, Raymond James, Jr
Tarbell, Theodore Dean
Tarter, Curtis Bruce
Thompson, Thomas William
Thompson, Timothy J
Thorne, Kip Stephen
Thorne, Richard Mansergh
Timothy, John Gethyn
Toor, Arthur
Townes, Charles Hard
Trimble, Virginia Louise
Tsiang, Eugene Y
Tucker, Wallace Hampton
Unwin, Stephen Charles
Vampola, Alfred Ludvik
Van Hoven, Gerard
Villere, Karen R
Vitello, Peter A
Vogt, Rochus E
Wagner, Raymond Lee
Wagoner, Robert Vernon
Walker, Arthur Bertram Cuthbert, Jr
Walker, Raymond John
Walt, Martin
Wasson, John Taylor
Wax, Robert LeRoy
Weaver, William Bruce
Webber, Donald Salyer
Weissman, Paul Robert
Werner, Michael Wolock
Wertz, James Richard
West, Harry Irwin, Jr
Weston, Gary Steven
Whitcomb, Stanley Ernest
White, Robert Stephen
Whitford, Albert Edward
Wiggins, John Shearon
Wilson, James R
Wilson, William John
Witteborn, Fred Carl
Woodward, James Franklin
Woosley, Stanford Earl
Yen, Chen-Wan Liu
Young, Louise Gray
Zumberge, James Frederick
Zuppero, Anthony Charles
Zych, Allen Dale

COLORADO
Baur, Thomas George
Begelman, Mitchell Craig
Boggess, Albert

Bornmann, Patricia L(ee)
Brault, James William
Conti, Peter Selby
Culp, Robert D(udley)
Curtis, George William
Dryer, Murray
Dulk, George A
Eddy, John Allen
Faller, James E
Firor, John William
Fisher, Richard R
Garstang, Roy Henry
Gilman, Peter A
Hansen, Carl John
Hansen, Richard Olaf
Herring, Jackson Rea
Hildner, Ernest Gotthold, III
Holzer, Thomas Edward
House, Lewis Lundberg
Hummer, David Graybill
Hundhausen, Arthur James
Joselyn, Jo Ann Cram
Langer, George Edward
Linsky, Jeffrey L
Little, Stephen James
Liu, Joseph Jeng-Fu
Low, Boon-Chye
Malville, John McKim
Mankin, William Gray
Mizushima, Masataka
Rottman, Gary James
Sawyer, Constance B
Skumanich, Andrew P
Smith, Dean Francis
Snow, Theodore Peck
Speiser, Theodore Wesley
Stanger, Andrew L
Stebbins, Robin Tucker
Stokes, Robert Allan
Thomas, Richard Nelson
Vande Noord, Edwin Lee
Wieman, Carl E
Zweibel, Ellen Gould

CONNECTICUT
Demarque, Pierre
Ftaclas, Christ
Jakacky, John M
Kraft, David Werner
Larson, Richard Bondo
Rich, John Charles
Similon, Philippe Louis
Sofia, Sabatino

DELAWARE
Crawford, Michael Karl
Evenson, Paul Arthur
L'Heureux, Jacques (Jean)
MacDonald, James
Mullan, Dermott Joseph
Ness, Norman Frederick
Shipman, Harry Longfellow

DISTRICT OF COLUMBIA
Appleby, John Frederick
Apruzese, John Patrick
Beall, James Howard
Berman, Barry L
Bohlin, John David
Boss, Alan Paul
Brueckner, Guenter Erich
Caroff, Lawrence John
Cook, John William
Corliss, Charles Howard
Epstein, Gerald Lewis
Feldman, Uri
Forman, Miriam Ausman
Fritz, Gilbert Geiger
Goldstein, Jeffrey Jay
Harwit, Martin Otto
Holt, Alan Craig
Hubbard, Richard Forest
Jones, William Vernon
Kaplan, George Harry
Kinzer, Robert Lee
Kumar, Cidambi Krishna
Kurfess, James Daniel
McNeal, Robert Joseph
Mange, Phillip Warren
Mariska, John Thomas
Matsakis, Demetrios Nicholas
Meekins, John Fred
Pascu, Dan
Patel, Vithalbhai L
Prinz, Dianne Kasnic
Reeves, Edmond Morden
Rickard, Lee J
Schmerling, Erwin Robert
Schweizer, Francois
Serene, Joseph William
Shapiro, Anatole Morris
Share, Gerald Harvey
Summers, Michael Earl
Tsaol, Chen H
Wagner, William John
Walker, David N(orton)
Wang, Yi-Ming
Weiler, Kurt Walter
Withbroe, George Lund

FLORIDA
Buchler, Jean-Robert
Burns, Jay, III
Cohen, Howard Lionel

Detweiler, Steven Lawrence
Fry, James N
Gottesman, Stephen T
Gustafson, Bo Ake Sture
Halpern, Leopold (Ernst)
Hammond, Gordon Leon
Ipser, James Reid
Marshak, Robert Eugene
Mulholland, John Derral
Oelfke, William C
Pearton, Stephen John
Plendl, Hans Siegfried
Ramond, Pierre Michel
Rester, Alfred Carl, Jr
Smith, Alexander Goudy
Smith, Haywood Clark, Jr
Wang, Ru-Tsang
Webb, James Raymond
Wilson, Robert E

GEORGIA
Bridges, Alan L
Dod, Bruce Douglas
Gudehus, Donald Henry
Marks, Dennis William
Sadun, Alberto Carlo
Wiita, Paul Joseph

HAWAII
Boesgaard, Ann Merchant
Chambers, Kenneth Carter
Cowie, Lennox Lauchlan
Crowe, Richard Alan
Geballe, Thomas Ronald
Henry, Joseph Patrick
Herbig, George Howard
Learned, John Gregory
McLaren, Robert Alexander
Meech, Karen Jean
Mickey, Donald Lee
Orrall, Frank Quimby
Owen, Tobias Chant
Tully, Richard Brent

IDAHO
Hoagland, Gordon Wood
Parker, Barry Richard

ILLINOIS
Anders, Edward
Barnothy, Jeno Michael
Buscombe, William
Chandrasekhar, Subrahmanyan
Crutcher, Richard Metcalf
Davids, Cary Nathan
Dykla, John J
Eichten, Estia Joseph
Garcia-Munoz, Moises
Grossman, Lawrence
Harvey, Jeffrey Alan
Hildebrand, Roger Henry
Hill, Christopher T
Iben, Icko, Jr
Karim, Khondkar Rezaul
Kent, Stephen Matthew
Kolb, Edward William
Lamb, Donald Quincy, Jr
Lebovitz, Norman Ronald
Lo, Kwok-Yung
McKibben, Robert Bruce
Meyer, Peter
Mihalas, Dimitri
Miller, Richard Henry
Mouschovias, Telemachos Charalambous
Muller, Dietrich
Nezrick, Frank Albert
Novak, Giles
Oka, Takeshi
Olson, Edward Cooper
Oreglia, Mark Joseph
Palmer, Patrick Edward
Pandharipande, Vijay Raghunath
Parker, Eugene Newman
Pines, David
Pyle, K Roger
Redmount, Ian H
Roberts, Arthur
Rosner, Robert
Schramm, David N
Skadron, George
Smarr, Larry Lee
Smither, Robert Karl
Snyder, Lewis Emil
Spinka, Harold M
Taam, Ronald Everett
Truran, James Wellington, Jr
Turner, Michael Stanley
Watson, William Douglas
Webbink, Ronald Frederick
York, Donald Gilbert

INDIANA
Aprahamian, Ani
Burkhead, Martin Samuel
Durisen, Richard H
Gaidos, James A
Heinz, Richard Meade
Heydegger, Helmut Roland
Hrivnak, Bruce John
Jacobs, William Wescott
Johnson, Hollis Ralph
Lipschutz, Michael Elazar
Loeffler, Frank Joseph
LoSecco, John M

Mufson, Stuart Lee
Poirier, John Anthony
Shore, Steven Neil
Willmann, Robert B

IOWA
Beavers, Willet I
Bowen, George Hamilton, Jr
Cadmus, Robert Randall, Jr
Fix, John Dekle
Goertz, Christoph Klaus
Gurnett, Donald Alfred
Hagelin, John Samuel
Intemann, Gerald William
Lamb, Richard C
Neff, John S
Scudder, Jack David
Spangler, Steven Randall
Struck-Marcell, Curtis John

KANSAS
Armstrong, Thomas Peyton
Chu, Shih I
Donoghue, Timothy R
Twarog, Bruce Anthony

KENTUCKY
Elitzur, Moshe
Hawkins, Charles Edward
Kielkopf, John F
Pearson, Earl F

LOUISIANA
Chanmugam, Ganesar
Grenchik, Raymond Thomas
Hamilton, William Oliver
Lee, Paul D
Rense, William A
Tipler, Frank Jennings, III
Tohline, Joel Edward
Wefel, John Paul

MAINE
Beard, David Breed
Comins, Neil Francis
Hughes, William Taylor
McClymer, James P
Pilot, Christopher H

MARYLAND
Acuna, Mario Humberto
A'Hearn, Michael Francis
Allen, John Edward, Jr
Baker, Daniel Neal
Balasubrahmanyan, Vriddhachalam K
Bass, Arnold Marvin
Behannon, Kenneth Wayne
Bell, Roger Alistair
Bettinger, Richard Thomas
Birmingham, Thomas Joseph
Boggess, Nancy Weber
Bohlin, Ralph Charles
Boldt, Elihu (Aaron)
Bond, Howard Emerson
Bostrom, Carl Otto
Brasunas, John Charles
Bruhweiler, Frederick Carlton, Jr
Calame, Gerald Paul
Carbary, James F
Chaisson, Eric Joseph
Chan, Kwing Lam
Chapman, Robert DeWitt
Cheng, Andrew Francis
Cheng, Chung-Chieh
Chi, L K
Chiu, Hong-Yee
Cline, Thomas L
Connerney, John E P
Crane, Patrick Conrad
Crannell, Carol Jo Argus
Davidsen, Arthur Falnes
Dolan, Joseph Francis
Doschek, George A
Downing, Robert Gregory
Dubin, Maurice
Earl, James Arthur
Endal, Andrew Samson
Fainberg, Joseph
Feibelman, Walter A
Felten, James E(dgar)
Fichtel, Carl Edwin
Fischel, David
Fitz, Harold Carlton, Jr
Flasar, F Michael
Ford, Holland Cole
Fortner, Brand I
Fuhr, Jeffrey Robert
Gebbie, Katharine Blodgett
Gehrels, Neil
Gilliland, Ronald Lynn
Goldberg, Richard Aran
Goldstein, Melvyn L
Gull, Theodore Raymond
Gutsche, Graham Denton
Harrington, James Patrick
Hartle, Richard Eastham
Hartman, Robert Charles
Hashmall, Joseph Alan
Hauser, Michael George
Heckman, Timothy Martin
Henry, Richard Conn
Hildebrand, Bernard
Hills, Howard Kent
Hoffman, Robert A

Holman, Gordon Dean
Holt, Stephen S
Howard, Sethanne
Hu, Bei-Lok Bernard
Hunter, Stanley Dean
Jones, Frank Culver
Jordan, Stuart Davis
Kaluzienski, Louis Joseph
Kastner, Sidney Oscar
Kondo, Yoji
Kostiuk, Theodor
Krimigis, Stamatios Mike
Kroeger, Richard Alan
Krolik, Julian H
Krueger, Arlin James
Kyle, Herbert Lee
Lasker, Barry Michael
Leckrone, David Stanley
Leventhal, Marvin
Liu, Chuan Sheng
Lucke, Robert Lancaster
McEntire, Richard Willian
McIlrath, Thomas James
McNutt, Ralph Leroy, Jr
Maier, Eugene Jacob Rudolph
Manley, Oscar Paul
Maran, Stephen Paul
Mather, John Cromwell
Meredith, Leslie Hugh
Misner, Charles William
Mitler, Henri Emmanuel
Moos, Henry Warren
Mumma, Michael Jon
Mushotzky, Richard Fred
Noerdlinger, Peter David
Ogilvie, Keith W
Omidvar, Kazem
Ormes, Jonathan Fairfield
Orwig, Larry Eugene
Ostaff, William A(llen)
Ozernoy, Leonid M
Paddack, Stephen J(oseph)
Papadopoulos, Konstantinos Dennis
Pellerin, Charles James, Jr
Perry, Peter M
Poland, Arthur I
Powell, Edward Gordon
Ramaty, Reuven
Reames, Donald Vernon
Rodney, William Stanley
Rose, William K
Rosenberg, Theodore Jay
Routly, Paul McRae
Rust, David Maurice
Sari, James William
Schreier, Ethan Joshua
Shara, Michael M
Shivanandan, Kandiah
Silberberg, Rein
Sittler, Edward Charles, Jr
Smith, David Edmund
Smith, Richard Lloyd
Stecker, Floyd William
Stief, Louis J
Strobel, Darrell Fred
Teays, Terry John
Teegarden, Bonnard John
Thomas, Roger Jerry
Thompson, David John
Toller, Gary Neil
Trombka, Jacob Israel
von Rosenvinge, Tycho Tor
Warren, Wayne Hutchinson, Jr
Webster, William John, Jr
Weller, Charles Stagg, Jr
Wende, Charles David
Wentzel, Donat Gotthard
West, Donald K
White, Richard Allan
Wilkerson, Thomas Delaney
Williams, Donald J
Woodgate, Bruce Edward
Wright, James P
Yoon, Peter Haesung
Zalubas, Romuald

MASSACHUSETTS
Ahlen, Steven Paul
Arny, Thomas Travis
Avrett, Eugene Hinton
Baliunas, Sallie Louise
Ball, John Allen
Barnes, Arnold Appleton, Jr
Bechis, Kenneth Paul
Bell, Barbara
Bertschinger, Edmund
Bradt, Hale Van Dorn
Brecher, Aviva
Brecher, Kenneth
Bridge, Herbert Sage
Brown, Benjamin Lathrop
Brynjolfsson, Ari
Cameron, Alastair Graham Walter
Canizares, Claude Roger
Carleton, Nathaniel Phillips
Chang, Edward Shi Tou
Chernosky, Edwin Jasper
Cook, Allan Fairchild, II
Corey, Brian E
Crooker, Nancy Uss
Dalgarno, Alexander
Dupree, Andrea K
Edwards, Suzan
Eoll, John Gordon

Farhi, Edward
Fazio, Giovanni Gene
Field, George Brooks
Ford, Lawrence Howard
Geller, Margaret Joan
Gorenstein, Marc Victor
Gorenstein, Paul
Gregory, Bruce Nicholas
Grindlay, Jonathan Ellis
Guth, Alan Harvey
Habbal, Shadia Rifai
Harrison, Edward Robert
Huchra, John P
Huguenin, George Richard
Joss, Paul Christopher
Kalkofen, Wolfgang
Kalman, Gabor J
Kellogg, Edwin M
Kinoshita, Kay
Kirby, Kate Page
Kistiakowsky, Vera
Kohl, John Leslie
Krieger, Allen Stephen
Kurucz, Robert Louis
Kwan, John Ying-Kuen
Kwitter, Karen Beth
Lecar, Myron
Leiby, Clare C, Jr
Levine, Randolph Herbert
Lewin, Walter H G
Lightman, Alan Paige
Little-Marenin, Irene Renate
Marsden, Brian Geoffrey
Mendillo, Michael
Michael, Irving
Milbocker, Michael
Moody, Elizabeth Anne
Murray, Stephen S
Narayan, Ramesh
Noyes, Robert Wilson
Papagiannis, Michael D
Parkinson, William Hambleton
Pasachoff, Jay M(yron)
Press, William Henry
Rappaport, Saul A
Ratner, Michael Ira
Raymond, John Charles
Reasenberg, Robert David
Roberts, David Hall
Rothwell, Paul L
Sagalyn, Rita C
Schattenburg, Mark Lee
Schwartz, Daniel Alan
Seward, Frederick Downing
Shapiro, Irwin Ira
Sheldon, Eric
Shuman, Bertram Marvin
Smith, Peter Lloyd
Snell, Ronald Lee
Stark, Antony Albert
Stevens, Herbert H(owe), Jr
Strom, Stephen
Sulak, Lawrence Richard
Sutton, Christopher Sean
Tananbaum, Harvey Dale
Thaddeus, Patrick
Tonry, John Landis
Traub, Wesley Arthur
Vessot, Robert F C
Victor, George A
Vinti, John Pascal
Weber, Edward Joseph
Weinberger, Doreen Anne
Whipple, Fred Lawrence
Whitney, Charles Allen
Willner, Steven P
Wisdom, Jack Leach

MICHIGAN
Akerlof, Carl W
Berger, Beverly Kobre
Donahue, Thomas Michael
Fontheim, Ernest Gunter
Freese, Katherine
Hegyi, Dennis
Hiltner, William Albert
Jones, Lawrence William
Kaiser, Christopher B
Kingman, Robert Earl
Krisch, Jean Peck
Longo, Michael Joseph
MacAlpine, Gordon Madeira
Sears, Richard Langley
Stein, Robert Foster
Tarle, Gregory
Teske, Richard Glenn
Zimring, Lois Jacobs

MINNESOTA
Cahill, Laurence James, Jr
Davidson, Kris
Erickson, Kenneth Neil
Flower, Terrence Frederick
Gehrz, Robert Douglas
Hakkila, Jon Eric
Jones, Thomas Walter
Kuhi, Leonard Vello
Lysak, Robert Louis
McLerran, Larry Dean
Marshak, Marvin Lloyd
Rudaz, Serge
Rudnick, Lawrence
Stein, Wayne Alfred
Waddington, Cecil Jacob

Woodward, Paul Ralph

MISSISSIPPI
Cooksy, Andrew Lloyd
Lestrade, John Patrick

MISSOURI
Bieniek, Ronald James
Binns, Walter Robert
Bragg, Susan Lynn
Cowsik, Ramanath
Edwards, Terry Winslow
Friedlander, Michael Wulf
Hohenberg, Charles Morris
Huang, Justin C
Israel, Martin Henry
Katz, Jonathan Isaac
Kovacs, Sandor J, Jr
Manuel, Oliver K
Mashhoon, Bahram
Schwartz, Richard Dean
Walker, Robert Mowbray
Will, Clifford Martin

MONTANA
Acton, Loren Wilber
Caughlan, Georgeanne Robertson
Hiscock, William Allen

NEBRASKA
Leung, Kam-Ching
Sartori, Leo
Schmidt, Edward George
Simon, Norman Robert

NEVADA
Phaneuf, Ronald Arthur
Weistrop, Donna Etta

NEW HAMPSHIRE
Arnoldy, Roger L
Boley, Forrest Irving
Chupp, Edward Lowell
Forbes, Terry Gene
Hollweg, Joseph Vincent
Hudson, Mary Katherine
Lee, Martin Alan
Lockwood, John Alexander
Morris, Daniel Joseph
Ryan, James Michael
Torbert, Roy Banks

NEW JERSEY
Arthur, Wallace
Bahcall, John Norris
Bahcall, Neta Assaf
Bechis, Dennis John
Becken, Eugene D
Boeshaar, Patricia Chikotas
Daly, Ruth Agnes
Dicke, Robert Henry
Draine, Bruce T
Feit, Julius
Flannery, Brian Paul
Gautreau, Ronald
Goldston, Robert J
Goode, Philip Ranson
Gott, J Richard, III
Grek, Boris
Groth, Edward John, III
Gunn, James Edward
Hodges, Hardy M
Holt, Rush D(ew), Jr
Hut, Piet
Jenkins, Edward Beynon
Langer, William David
Lanzerotti, Louis John
Levine, Jerry David
Matilsky, Terry Allen
Mikkelsen, David Robert
Molnar, Michael Robert
Motz, Robin Owen
Ostriker, Jeremiah P
Owens, David Kingston
Paczynski, Bohdan
Pfeiffer, Raymond John
Raghavan, Pramila
Raghavan, Ramaswamy Srinivasa
Rogerson, John Bernard, Jr
Schwarzschild, Martin
Shipley, Edward Nicholas
Spergel, David Nathaniel
Spitzer, Lyman, Jr
Taylor, Harold Evans
Taylor, Joseph Hooton, Jr
Thorsett, Stephen Erik
Tishby, Naftali Z
Turner, Edwin Lewis
Tyson, J Anthony
Zimmermann, R Erik

NEW MEXICO
Ahluwalia, Harjit Singh
Altrock, Richard Charles
Arion, Douglas
Bame, Samuel Jarvis, Jr
Blake, Richard L
Browne, John Charles
Brownlee, Robert Rex
Burman, Robert L
Cahn, Julius Hofeller
Clark, Robert Edward Holmes
Claussen, Mark J
Cleveland, Bruce Taylor

Conner, Jerry Power
Coon, James Huntington
Cox, Arthur Nelson
Davey, William Robert
Davis, Cecil Gilbert
Deupree, Robert G
Dumas, Herbert M, Jr
Evans, John Wainwright, Jr
Feldman, William Charles
Forslund, David Wallace
Freyer, Gustav John
Fu, Jerry Hui Ming
Gosling, John Thomas
Greene, Arthur Edward
Gregory, Stephen Albert
Hills, Jack Gilbert
Hoffman, Cyrus Miller
Ingraham, Richard Lee
Jarmie, Nelson
Jones, Eric Manning
Keil, Stephen Lesley
Kemic, Stephen Bruce
King, David Solomon
Kopp, Roger Alan
Lilley, John Richard
Meier, Michael McDaniel
Moore, Elliott Paul
Mutschlecner, Joseph Paul
Newman, Michael J(ohn)
Olson, Gordon Lee
Petschek, Albert George
Price, Richard Marcus
Reedy, Robert Challenger
Reeves, Geoffrey D
Rokop, Donald J
Sanders, Walter L
Sandford, Maxwell Tenbrook, II
Schultz, Rodney Brian
Simon, George Warren
Stellingwerf, Robert Francis
Stone, Sidney Norman
Sweeney, Mary Ann
Tapia, Santiago
Terrell, N(elson) James
Van Riper, Kenneth Alan
Wallace, Richard Kent
Wayland, James Robert, Jr
Weaver, Robert Paul
Webber, William R
Westpfahl, David John
White, Paul C
Wing, Janet E (Sweedyk) Bendt
Winske, Dan
Zeilik, Michael
Zirker, Jack Bernard
Zurek, Wojciech Hubert

NEW YORK
Alpher, Ralph Asher
Baker, Norman Hodgson
Baltz, Anthony John
Bocko, Mark Frederick
Boyd, Robert William
Brink, Gilbert Oscar
Brown, Stanley Gordon
Cheung, Lim H
Chung, Kuk Soo
Cohen, Leon
Craft, Harold Dumont, Jr
DeNoyer, Linda Kay
Douglass, David Holmes
Dubisch, Russell John
Edelstein, William Alan
Elmegreen, Debra Meloy
Fishbone, Leslie Gary
Forrest, William John
Gill, Ronald Lee
Glassgold, Alfred Emanuel
Goebel, Ronald William
Goldsmith, Paul Felix
Grannis, Paul Dutton
Greisen, Kenneth I
Gross, Stanley H
Hahn, Richard Leonard
Hartmann, Francis Xavier
Helfand, David John
Helfer, Herman Lawrence
Houck, James Richard
Huggins, Patrick John
Kegeles, Lawrence Steven
Lasher, Gordon (Jewett)
Lattimer, James Michael
Lazareth, Otto William, Jr
Lissauer, Jack Jonathan
Lovelace, Richard Van Evera
Madey, Robert W
Mansfield, Victor Neil
Matsushima, Satoshi
Meisel, David Dering
Mohlke, Byron Henry
Moniot, Robert Keith
Motz, Lloyd
Nardi, Vittorio
Novick, Robert
O'Donoghue, Aileen Ann
Peak, David
Ratcliff, Keith Frederick
Remo, John Lucien
Rosenzweig, Carl
Salpeter, Edwin Ernest
Sato, Makiko
Savedoff, Malcolm Paul
Schmalberger, Donald C
Schulz, Helmut Wilhelm

Astrophysics (cont)

Schwartz, Melvin J
Seiden, Philip Edward
Shapiro, Stuart Louis
Sheppard, Ronald John
Smolin, Lee
Solomon, Philip M
Spergel, Martin Samuel
Spiegel, Edward A
Teukolsky, Saul Arno
Thomas, John Howard
Toner, John Joseph
Tryon, Edward Polk
Ullman, Jack Donald
Wolf, Henry
Yahil, Amos

NORTH CAROLINA
Blondin, John Michael
Brown, J(ohn) David
Christiansen, Wayne Arthur
Danby, John Michael Anthony
Siewert, Charles Edward
Simon, Sheridan Alan
Winkler, Linda Irene
York, James Wesley, Jr

NORTH DAKOTA
Berkey, Gordon Bruce

OHIO
Bidelman, William Pendry
Boyd, Richard Nelson
Chai, An-Ti
Chiu, Victor
Czyzak, Stanley Joachim
Delsemme, Armand Hubert
Ferguson, Dale Curtis
Ferland, Gary Joseph
Frogel, Jay Albert
Herbst, Eric
Jenkins, Thomas Llewellyn
Krauss, Lawrence Maxwell
Luck, Richard Earle
Macklin, Philip Alan
Peterson, Bradley M(ichael)
Preszler, Alan Melvin
Ptak, Roger Leon
Raby, Stuart
Rao, Kandarpa Narahari
Rodman, James Purcell
Sellgren, Kristen
Slettebak, Arne
Snider, Joseph Lyons
Steigman, Gary
Stoner, Ronald Edward
Winters, Ronald Ross
Witt, Adolf Nicolaus
Witten, Louis

OKLAHOMA
Branch, David Reed
Buck, Richard F
Cowan, John James
Hill, Stephen James
Merts, Athel Lavelle
Schroeder, Leon William
Shull, Peter Otto, Jr

OREGON
Isenberg, James Allen

PENNSYLVANIA
Axel, Leon
Beier, Eugene William
Blair, Grant Clark
Boughn, Stephen Paul
Carpenter, Lynn Allen
Cashdollar, Kenneth Leroy
Chambliss, Carlson Rollin
Cohen, Jeffrey M
Cordova, France Anne-Dominic
Duggal, Shakti Prakash
Feigelson, Eric Dennis
Frati, William
Frenklach, Michael Y
Garmire, Gordon Paul
Gaustad, John Eldon
Green, Louis Craig
Gross, Peter George
Guinan, Edward F
Hapke, Bruce W
Knacke, Roger Fritz
Lande, Kenneth
McCluskey, George E, Jr
Nagy, Theresa Ann
Quinn, Robert George
Ramsey, Lawrence William
Shen, Benjamin Shih-Ping
Sion, Edward Michael
Sitterly, Charlotte Moore
Soberman, Robert K
Steinhardt, Paul Joseph
Tobias, Russell Lawrence
Vila, Samuel Campderros

RHODE ISLAND
Brandenberger, Robert H
Gilman, John Richard, Jr
Lanou, Robert Eugene, Jr
Rossner, Lawrence Franklin
Timbie, Peter T
Widgoff, Mildred

SOUTH CAROLINA
Adelman, Saul Joseph
Clayton, Donald Delbert
Flower, Phillip John
Jenkins, Alvin Wilkins, Jr
Safko, John Loren
Yates, George Kenneth
Zechiel, Leon Norris

TENNESSEE
Barnes, Ronnie C
Bartelt, John Eric
Blass, William Errol
Eaton, Joel A
Fox, Kenneth
MacQueen, Robert Moffat
Thonnard, Norbert
Tolk, Norman Henry
Yau, Cheuk Chung

TEXAS
Allum, Frank Raymond
Auchmuty, Giles
Badhwar, Gautam D
Bering, Edgar Andrew, III
Black, David Charles
Cloutier, Paul Andrew
Criswell, David Russell
Deeming, Terence James
Dicus, Duane A
Duke, Michael B
Duller, Nelson M, Jr
Evans, Neal John, II
Fenyves, Ervin J
Freeman, John Wright, Jr
Graham, William Richard Montgomery
Harrison, Marjorie Hall
Haymes, Robert C
Heikkila, Walter John
Henize, Karl Gordon
Hill, Thomas Westfall
Hodges, Ralph Richard, Jr
Huebner, Walter F
Johnson, Francis Severin
Konradi, Andrei
Koplyay, Janos Bernath
Lewis, Ira Wayne
Liang, Edison Park-Tak
Lopez, Jorge Alberto
Mahajan, Swadesh Mitter
Mangum, Jeffrey Gary
Michel, F Curtis
Modisette, Jerry L
Morgan, Thomas Harlow
Nacozy, Paul E
Nather, Roy Edward
Nyquist, Laurence Elwood
O'Dell, Charles Robert
Page, Thornton Leigh
Palmeira, Ricardo Antonio Ribeiro
Pedigo, Robert Daryl
Robbins, Ralph Robert
Roeder, Robert Charles
Roeder, Robert Charles
Shapiro, Paul Robert
Smoluchowski, Roman
Su, Shin-Yi
Tajima, Toshiki
Talent, David Leroy
Tinsley, Brian Alfred
Trafton, Laurence Munro
Vishniac, Ethan Tecumseh
Weisheit, Jon Carleton
Wheeler, John Craig
Wilson, Thomas Leon
Zook, Herbert Allen

UTAH
Bergeson, Haven Eldred
Elbert, Jerome William
Jones, Douglas Emron
Kieda, David Basil
Lind, Vance Gordon
McDonald, Keith Leon
McNamara, Delbert Harold
O'Halloran, Thomas A, Jr
Price, Richard Henry
Schunk, Robert Walter
Stonehocker, Garth Hill
Taylor, Benjamin Joseph

VERMONT
Rankin, Joanna Marie
Winkler, Paul Frank
Wolfson, Richard L T

VIRGINIA
Adam, John Anthony
Adams, James Hall, Jr
Atalay, Bulent Ismail
Boozer, Allen Hayne
Bunner, Alan Newton
Calio, Anthony John
Chevalier, Roger Alan
Chubb, Talbot Albert
Dennison, Brian Kenneth
Dickman, Robert Laurence
Evans, John C
Gowdy, Robert Henry
Harvey, Gale Allen
Heckathorn, Harry Mervin, III
Howard, Russell Alfred
Jacobs, Kenneth Charles
Jaffe, Leonard

Kafatos, Minas
Kelch, Walter L
Kellermann, Kenneth Irwin
Kissell, Kenneth Eugene
Kniffen, Donald Avery
Koomen, Martin J
Lanzano, Paolo
Ludwig, George H
Lynch, John Thomas
Mayer, Cornell Henry
Mihalas, Barbara R Weibel
Moore, Garry Edgar
Morgan, Thomas Edward
Mullen, Joseph Matthew
O'Connell, Robert West
Oesterwinter, Claus
Opp, Albert Geelmuyden
Oster, Ludwig Friedrich
Radoski, Henry Robert
Rood, Robert Thomas
Sarazin, Craig L
Shapiro, Maurice Mandel
Sime, David Gilbert
Spagna, George Fredric, Jr
Thuan, Trinh Xuan
Tidman, Derek Albert
Toton, Edward Thomas
Turner, Barry Earl
Van Horn, Hugh Moody
Wallace, Lance Arthur
Wootten, Henry Alwyn
Youtcheff, John Sheldon

WASHINGTON
Anderson, Hugh Riddell
Bardeen, James Maxwell
Bodansky, David
Bookmyer, Beverly Brandon
Brodzinski, Ronald Lee
Campbell, Malcolm John
Despain, Lewis Gail
Ellis, Stephen Dean
Ely, John Thomas Anderson
Greene, Thomas Frederick
Haxton, Wick Christopher
Hoch, Richmond Joel
Holzworth, Robert H, II
Jenner, David Charles
Kouzes, Richard Thomas
Lake, George Russell
Laul, Jagdish Chander
Linnell, Albert Paul
Lutz, Julie Haynes
Margon, Bruce Henry
Olsen, Kenneth Harold
Peters, Philip Carl
Speer, Fridtjof Alfred
Szkody, Paula

WEST VIRGINIA
Ghigo, Frank Dunnington
Littleton, John Edward
Weldon, Henry Arthur

WISCONSIN
Allen, Bruce
Balantekin, Akif Baha
Cassinelli, Joseph Patrick
Code, Arthur Dodd
Doherty, Lowell Ralph
Fry, William Frederick
Gallagher, John Sill
Hobbs, Lewis Mankin
Kron, Richard G
Lawler, James Edward
McCammon, Dan
March, Robert Herbert
Mathis, John Samuel
Parker, Leonard Emanuel
Reynolds, Ronald J
Savage, Blair DeWillis
Scherb, Frank
Woods, Robert Claude

WYOMING
Howell, Robert Richard
Roark, Terry P
Thronson, Harley Andrew, Jr

PUERTO RICO
Lewis, Brian Murray

ALBERTA
Cook, Frederick Ahrens
Leahy, Denis Alan
Milone, Eugene Frank
Pinnington, Eric Henry
Rostoker, Gordon
Sreenivasan, Sreenivasa Ranga
Venkatesan, Doraswamy
Weale, Gareth Pryce
Wilson, William James Fitzpatrick

BRITISH COLUMBIA
Auman, Jason Reid
Batten, Alan Henry
Burke, J Anthony
Climenhaga, John Leroy
Fahlman, Gregory Gaylord
Galt, John (Alexander)
Hesser, James Edward
Horita, Robert Eiji
Menon, Thuppalay K
Morbey, Christopher Leon
Palmer, Leigh Hunt
Singh, Manohar
Underhill, Anne Barbara
Walker, Gordon Arthur Hunter
Wright, Kenneth Osborne

NEW BRUNSWICK
Hawkes, Robert Lewis

NOVA SCOTIA
Welch, Gary Alan

ONTARIO
Barker, Paul Kenneth
Bernath, Peter Francis
Bolton, Charles Thomas
Chau, Wai-Yin
Clement, Maurice James Young
Ewan, George T
Feldman, Paul Arnold
FitzGerald, Maurice Pim Valter
Garrison, Robert Frederick
Halliday, Ian
Henriksen, Richard Norman
Hruska, Antonin
Innanen, Kimmo A
Kronberg, Philipp Paul
Landstreet, John Darlington
Lester, John Bernard
Lowe, Robert Peter
McDiarmid, Ian Bertrand
McIntosh, Bruce Andrew
McNamara, Allen Garnet
Martin, Peter Gordon
Mitalas, Romas
Moffat, John William
Morton, Donald Charles
Ollerhead, Robin Wemp
Sutherland, Peter Gordon
Tremaine, Scott Duncan
Vallée, Jacques P
Varshni, Yatendra Pal
Wesson, Paul Stephen

PRINCE EDWARD ISLAND
Lin, Wei-Ching

QUEBEC
Aktik, Cetin
Borra, Ermanno Franco
Chaubey, Mahendra
Davies, Roger
Demers, Serge
Fontaine, Gilles Joseph
Michaud, Georges Joseph
Moffat, Anthony Frederick John
Pearson, John Michael
Pineault, Serge Rene
Tassoul, Jean-Louis
Tassoul, Monique

SASKATCHEWAN
Papini, Giorgio Augusto

OTHER COUNTRIES
Beckers, Jacques Maurice
Beckwith, Steven Van Walter
Bekenstein, Jacob David
Bell Burnell, S Jocelyn
Butcher, Harvey Raymond, III
Cernuschi, Felix
Chincarini, Guido Ludovico
Contopoulos, George
Crane, Philippe
Eggen, Olin Jeuck
Feinstein, Alejandro
Gao, Yi-Tian
Giacconi, Riccardo
Gil, Salvador
Greenberg, Jerome Mayo
Ho, Yew Kam
Icke, Vincent
Kozai, Yoshihide
Mark, James Wai-Kee
Munch, G
Parsignault, Daniel Raymond
Pottasch, Stuart Robert
Rees, Martin J
Samir, Uri
Schnopper, Herbert William
Schutz, Bernard Frederick
Seaton, Michael John
Sugiura, Masahisa
Vrebalovich, Thomas
Weil, Raoul Bloch
White, Simon David Manton
Wildman, Peter James Lacey

Atomic & Molecular Physics

ALABAMA
Bartlett, James Holly
Bauman, Robert Poe
Edlin, George Robert
Helminger, Paul Andrew
Howgate, David W
Izatt, Jerald Ray
Jones, Robert William
Pindzola, Michael Stuart
Rosenberger, Albert Thomas
Smith, Lewis Taylor
Stephens, Timothy Lee
Stettler, John Dietrich

Tidwell, Eugene Delbert
Tipping, Richard H
Varghese, Sankoorikal Lonappan

ALASKA
Deehr, Charles Sterling
Degen, Vladimir
Wentink, Tunis, Jr

ARIZONA
Bashkin, Stanley
Black, John Harry
Daw, Glen Harold
Dillingham, Thomas R
Engleman, Rolf, Jr
Fan, Chang-Yun
Farr, William Morris
Garcia, Jose Dolores, Jr
Gibbs, Hyatt McDonald
Hight, Ralph Dale
Jones, Roger C(lyde)
Lamb, Willis Eugene, Jr
Leavitt, John Adams
Lutz, Barry Lafean
Rafelski, Johann
Rez, Peter
Steimle, Timothy C
Stoner, John Oliver, Jr
Yelle, Roger V

ARKANSAS
Bronco, Charles John
Gupta, Rajendra
Hughes, Raymond Hargett
Jones, Charles E
Lieber, Michael
Vyas, Reeta

CALIFORNIA
Adams, Arnold Lucian
Alonso, Jose Ramon
Alvi, Zahoor M
Anderson, Charles Hammond
Antolak, Arlyn Joe
Armstrong, Baxter Hardin
Armstrong, Lloyd, Jr
Auerbach, Daniel J
Bach, David Rudolph
Ball, William Paul
Bardsley, James Norman
Baur, James Francis
Benjamin, Arlin James
Benton, Eugene Vladimir
Berkner, Klaus Hans
Bethune, Donald Stimson
Bloom, Arnold Lapin
Boehm, Felix H
Brewer, LeRoy Earl, Jr
Brewer, Richard George
Brown, Linda Rose
Brown, William Arnold
Burnett, Lowell Jay
Burton, Donald Eugene
Caird, John Allyn
Camparo, James Charles
Carman, Robert Lincoln, Jr
Chackerian, Charles, Jr
Chapline, George Frederick, Jr
Cheng, Kwok-Tsang
Chodos, Steven Leslie
Chu, Steven
Chutjian, Ara
Clark, Arnold Franklin
Cohen, Lawrence Mark
Conway, John George, Jr
Cook, Charles J
Crosley, David Risdon
Cutler, Leonard Samuel
Davis, James Ivey
Devoe, Ralph Godwin
Dowell, Jerry Tray
Dowling, Jerome M
Dows, David Alan
Doyle, Walter M
Dulgeroff, Carl Richard
Earhart, Richard Wilmot
Edwards, David Franklin
Ehlers, Kenneth Warren
Ewig, Carl Stephen
Fearn, Heidi
Feinberg, Benedict
Feng, Joseph Shao-Ying
Finston, Roland A
Fishburne, Edward Stokes, III
Fisher, Leon Harold
Fisk, George Ayrs
Franks, Larry Allen
Futch, Archer Hamner
Ganas, Perry S
Garrard, Thomas L
George, Simon
Gillen, Keith Thomas
Gillespie, George H
Gilmore, Forrest Richard
Glass, Alexander Jacob
Gould, Harvey Allen
Gruber, John B
Gwinn, William Dulaney
Hackel, Lloyd Anthony
Hagstrom, Stig Bernt
Hahn, Erwin Louis
Hanna, Stanley Sweet
Harrach, Robert James
Harris, David Owen

Harris, Dennis George
Heestand, Glenn Martin
Hemminger, John Charles
Herman, Frank
Hershey, Allen Vincent
Heusinkveld, Myron Ellis
Hill, Robert Matteson
Hrubesh, Lawrence W(ayne)
Hu, Chi-Yu
Huestis, David Lee
Jacobs, Ralph R
Jadusziiwer, Bernardo
Jaklevic, Joseph Michael
Janda, Kenneth Carl
Janney, Gareth Maynard
Jusinski, Leonard Edward
Karo, Arnold Mitchell
Kasai, Paul Haruo
Keiser, George McCurrach
Kelly, Paul Sherwood
Kelly, Raymond Leroy
Kim, Jinchoon
Klein, August S
Klein, David Joseph
Knight, Walter David
Knipe, Richard Hubert
Knize, Randall James
Kocol, Henry
Kulander, Kenneth Charles
Kunc, Joseph Anthony
Kunkel, Wulf Bernard
Kwok, Munson Arthur
Lam, Leo Kongsui
Landman, Donald Alan
Lapp, M(arshall)
Lee, Long Chi
Lee, Paul L
Lee, Yim Tin
Lester, William Alexander, Jr
Levenson, Marc David
Lichter, James Joseph
Linford, Gary Joe
Ling, Rung Tai
Lipeles, Martin
Litvak, Marvin Mark
Loda, Richard Thomas
Loew, Gilda Harris
Lombardi, Gabriel Gustavo
Lorents, Donald C
Luther, Marvin L
McFarlane, Ross Alexander
McGowan, James William
McKay, Dale Robert
McKenzie, Robert Lawrence
McKoy, Basil Vincent
McMaster, William H
Magnuson, Gustav Donald
Mahadevan, Parameswar
Maker, Paul Donne
Marrs, Roscoe Earl
Marrus, Richard
Mathis, Ronald Floyd
Meyerhof, Walter Ernst
Meyerott, Roland Edward
Milanovich, Fred Paul
More, Richard Michael
Muntz, Eric Phillip
Myers, Benjamin Franklin, Jr
Nanes, Roger
Neugebauer, Gerry
Newman, David Edward
Neynaber, Roy H(arold)
Olness, Dolores Urquiza
O'Malley, Thomas F
O'Malley, Thomas Francis
Palatnick, Barton
Park, Chul
Perel, Julius
Peterson, James Ray
Phillips, Edward
Plain, Gilbert John
Plock, Richard James
Post, Douglass Edmund
Prag, Arthur Barry
Prior, Michael Herbert
Rast, Howard Eugene, Jr
Reisler, Hanna
Rescigno, Thomas Nicola
Roeder, Stephen Bernhard Walter
Rosen, Mordecai David
Ross, Marvin Franklin
Rotenberg, Manuel
Rundel, Robert Dean
Salisbury, Stanley R
Satten, Robert A
Saxon, Roberta Pollack
Schaefer, Albert Russell
Schlachter, Alfred Simon
Schmieder, Robert W
Scott, Bruce L
Scott, Paul Brunson
Senitzky, Israel Ralph
Shemansky, Donald Eugene
Shore, Bruce Walter
Shugart, Howard Alan
Simon, Barry Martin
Sinton, Steven Williams
Smith, Felix Teissiere
Snow, William Rosebrook
Stalder, Kenneth R
Stearns, John Warren
Stephenson, David Allen
Stuart, George Wallace
Suchannek, Rudolf Gerhard

Swanson, William Paul
Syage, Jack A
Tang, Stephen Shien-Pu
Taylor, Howard S
Terhune, Robert William
Thoe, Robert Steven
Trajmar, Sandor
Trujillo, Stephen Michael
Utterback, Nyle Gene
Varney, Robert Nathan
Vroom, David Archie
Walker, Keith Gerald
Wallis, Richard Fisher
Waters, Joe William
Weber, Marvin John
Weissbluth, Mitchel
Whaling, Ward
Wieder, Harold
Wieder, Irwin
Woerner, Robert Leo
Woldseth, Rolf
Wood, Calvin Dale
Woodruff, Truman Owen
Woodward, Ervin Chapman, Jr
Worden, Earl Freemont, Jr
Young, Louise Gray
Zewail, Ahmed H
Zimmerman, Ivan Harold
Zook, Alma Claire

COLORADO
Adams, Gail Dayton
Amme, Robert Clyde
Beckmann, Petr
Beers, Yardley
Bills, Daniel Granville
Bollinger, John J
Brault, James William
Chappell, Willard Ray
Cooper, John (Jinx)
Drullinger, Robert Eugene
Dunn, Gordon Harold
Evenson, Kenneth Melvin
Fox, Michael Henry
Furcinitti, Paul Stephen
Gallagher, Alan C
Garstang, Roy Henry
Gau, John N
Greene, Christopher Henry
House, Lewis Lundberg
Howe, David Allan
Itano, Wayne Masao
Jespersen, James
Kiehl, Jeffrey Theodore
Leone, Stephen Robert
Mizushima, Masataka
Morgan, William Lowell
Mosburg, Earl R, Jr
Neumann, Herschel
Norcross, David Warren
O'Callaghan, Michael James
Oneil, Stephen Vincent
Parson, Robert Paul
Phelps, Arthur Van Rensselaer
Rinker, George Albert, Jr
Robinson, Hugh Gettys
Schowengerdt, Franklin Dean
Smith, Ernest Ketcham
Smith, Stephen Judson
Stephens, Jeffrey Alan
Sullivan, Donald Barrett
Wessel, William Roy
Whitten, Barbara L

CONNECTICUT
Akkapeddi, Prasad Rao
Antar, Ali A
Bartram, Ralph Herbert
Cable, Peter George
Chupka, William Andrew
Greenberg, Jack Sam
Greenwood, Ivan Anderson
Hayden, Howard Corwin
Hinchen, John J(oseph)
Jakacky, John M
Keith, H(arvey) Douglas
Kessel, Quentin Cattell
Lubell, Michael S
Malsky, Stanley Joseph
Michels, H(orace) Harvey
Monce, Michael Nolen
Moreland, Parker Elbert, Jr
Morgan, Thomas Joseph
Nath, Ravinder Katyal
Peterson, Cynthia Wyeth
Petersson, George A
Pollack, Edward
Smith, Winthrop Ware
Staker, William Paul
Stwalley, William Calvin
Ultee, Casper Jan
Wong, Shek-Fu

DELAWARE
Crawford, Michael Karl
Fou, Cheng-Ming
Morgan, John Davis, III
Sharnoff, Mark
Szalewicz, Krzysztof
Woo, Shien-Biau

DISTRICT OF COLUMBIA
Beard, Charles Irvin
Berman, Barry L

Borras, Caridad
Brown, Charles Moseley
Burkhalter, Philip Gary
Campillo, Anthony Joseph
Chiu, Lue-Yung Chow
Corliss, Charles Howard
Crandall, David Hugh
Crisp, Michael Dennis
Feldman, Uri
Haftel, Michael Ivan
Halpern, Joshua Baruch
Hessel, Merrill
Jacobs, Verne Louis
Jansen, Robert Werner
Kumar, Cidambi Krishna
Lucatorto, Thomas B
McMahon, John Michael
Mahon, Rita
Markevich, Darlene Julia
Misra, Prabhakar
Oertel, Goetz Kuno Heinrich
Phillips, Gary Wilson
Ramaker, David Ellis
Reeves, Edmond Morden
Reiss, Howard R
Scheel, Nivard
Taylor, Ronald D
Toburen, Larry Howard
Wagner, William John
White, John Arnold
Wing, William Hinshaw

FLORIDA
Alexander, Steven Albert
Bailey, Thomas L, III
Bartlett, Rodney Joseph
Brucat, Philip John
Burns, Jay, III
Burns, Michael J
Callan, Edwin Joseph
Darling, Byron Thorwell
Edmonds, Dean Stockett, Jr
Fitzgerald, Lawrence Terrell
Foner, Samuel Newton
Gibson, Henry Clay, Jr
Hammond, Gordon Leon
Hanson, Harold Palmer
Hooper, Charles Frederick, Jr
Johnson, Joseph Andrew, III
Killinger, Dennis K
Linder, Ernest G
Micha, David Allan
Monkhorst, Hendrik J
Mrozowski, Stanislaw W
Oelfke, William C
Rhodes, Richard Ayer, II
Sabin, John Rogers
Sheldon, John William
Shelton, Wilford Neil
Sherwood, Jesse Eugene
Skofronick, James Gust
Solon, Leonard Raymond
Sullivan, Neil Samuel
Szczepaniak, Krystyna
Van Zee, Richard Jerry
Zerner, Michael Charles

GEORGIA
Cheng, Wu-Chieh
Cooper, Charles Dewey
Edwards, Alan Kent
Flannery, Martin Raymond
Gole, James Leslie
Hutcherson, Joseph William
Manson, Steven Trent
Menendez, Manuel Gaspar
Nave, Carl R
Palmer, Richard Carl
Payne, Marvin Gay
Rao, Pemmaraju Venugopala
Roy, Rajarshi
Steuer, Malcolm F
Thomas, Edward Wilfrid
Uzer, Ahmet Turgay
Whetten, Robert Lloyd
Wood, Robert Manning
Zangwill, Andrew

HAWAII
Geballe, Thomas Ronald
Holmes, John Richard
McLaren, Robert Alexander

IDAHO
Denison, Arthur B
Harper, Henry Amos, Jr
Knox, John MacMurray

ILLINOIS
Ali, Naushad
Atoji, Masao
Averill, Frank Wallace
Barrall, Raymond Charles
Bennett, Edgar F
Berkowitz, Joseph
Berry, Henry Gordon
Bertoncini, Peter Joseph
Bonham, Russell Aubrey
Buck, Warren Louis
Childs, William Jeffries
Chu, Keh-Cheng
Cloud, William Max
Court, Anita
Day, Michael Hardy

Atomic & Molecular Physics (cont)

Dehmer, Joseph L(eonard)
Dehmer, Patricia Moore
De Rijk, Waldemar G
Dolecek, Elwyn Haydn
Donnally, Bailey Lewis
Druger, Stephen David
Dunford, Robert Walter
Eden, James Gary
Ellis, Donald Edwin
Fano, Ugo
Garcia-Munoz, Moises
Gislason, Eric Arni
Greene, John Philip
Hessler, Jan Paul
Holzberlein, Thomas M
Huebner, Russell Henry, Sr
Ingram, Forrest Duane
Inokuti, Mitio
Kang, Ik-Ju
Karim, Khondkar Rezaul
Kimura, Mineo
Klemm, Richard Andrew
McCormack, Elizabeth Frances
Makri, Nancy
Malik, Fazley Bary
Mallow, Jeffry Victor
Melendres, Carlos Arciaga
Nayfeh, Munir Hasan
Nieman, George Carroll
Oka, Takeshi
Palmer, Patrick Edward
Pratt, Stephen Turnham
Rhodes, Charles Kirkham
Sanders, Frank Clarence, Jr
Schatz, George Chappell
Secrest, Donald H
Sibener, Steven Jay
Smither, Robert Karl
Snyder, Lewis Emil
Spence, David
Tomkins, Frank Sargent
Treadwell, Elliott Allen
Verdeyen, Joseph T
Woodruff, William Lee
Yuster, Philip Harold
Zelac, Ronald Edward

INDIANA
Bentley, John Joseph
Bunker, Bruce Alan
Clikeman, Franklyn Miles
Grant, Edward R
Hults, Malcom E
Langhoff, Peter Wolfgang
Mozumder, Asokendu
Sapirstein, Jonathan Robert
Shupe, Robert Eugene
Tripathi, Gorakh Nath Ram

IOWA
Houk, Robert Samuel
Lutz, Robert William
Robinson, David

KANSAS
Andrew, Kenneth L
Bhalla, Chander P
Broersma, Sybrand
Chu, Shih I
Cocke, C Lewis
Curnutte, Basil, Jr
Dreschhoff, Gisela Auguste-Marie
Gray, Tom J
Greene, Frank T
Hagmann, Siegbert Johann
Harmony, Marlin D
Legg, James C
Lin, Chii-Dong
Richard, Patrick
Stockli, Martin P
Tran, Loc Binh
Weller, Lawrence Allenby

KENTUCKY
Bradley, Eugene Bradford
Clouthier, Dennis James
Kielkopf, John F
MacAdam, Keith Bradford

LOUISIANA
Chin-Bing, Stanley Arthur
Gibbs, Richard Lynn
Head, Charles Everett
Head, Martha E Moore
Kaldor, Andrew
Klasinc, Leo
Kumar, Devendra
McGuire, James Horton
Meckstroth, George R
Meriwether, John R
Perdew, John Paul
Rau, A Ravi Prakash
Scott, John Delmoth
Stockbauer, Roger Lewis
Zander, Arlen Ray

MAINE
Kingsbury, Robert Freeman

MARYLAND
Allen, John Edward, Jr
Bates, Lloyd M
Bauer, Ernest
Bax, Ad
Benesch, William Milton
Benson, Richard C
Beri, Avinash Chandra
Bhatia, Anand K
Birnbaum, George
Bryden, Wayne A
Buslik, Arthur J
Cavanagh, Richard Roy
Cave, William Thompson
Celotta, Robert James
Chaisson, Eric Joseph
Clark, Charles Winthrop
Cohen, Steven Charles
Coplan, Michael Alan
Criss, John W
Crosswhite, Henry Milton, Jr
Danos, Michael
De Planque, Gail
Dick, Charles Edward
Doering, John P
Doschek, George A
Elder, Robert Lee
Epstein, Gabriel Leo
Feldman, Paul Donald
Fuhr, Jeffrey Robert
Gammell, Paul M
Giacchetti, Athos
Ginter, Marshall L
Goodman, Leon Judias
Grant, David Graham
Hellwig, Helmut Wilhelm
Holt, Helen Keil
Hougen, Jon T
Hubbell, John Howard
Hudson, David Frank
Hudson, Robert Douglas
Hurst, Wilbur Scott
Jacox, Marilyn Esther
Jette, Archelle Norman
Judd, Brian Raymond
Julienne, Paul Sebastian
Kaplan, Alexander E
Kastner, Sidney Oscar
Kaufman, Victor
Kessler, Ernest George, Jr
Kessler, Karl Gunther
Kim, Yong-Ki
Klose, Jules Zeiser
Korobkin, Irving
Kostiuk, Theodor
Krisher, Lawrence Charles
Kuyatt, Chris E(rnie Earl)
Lafferty, Walter J
Land, David J(ohn)
LaVilla, Robert E
Leventhal, Marvin
Lide, David Reynolds, Jr
Lovas, Francis John
McClelland, Jabez Jenkins
McCoubrey, Arthur Orlando
McIlrath, Thomas James
McLaughlin, William Lowndes
Major, Fouad George
Mandelberg, Hirsch I
Martin, William Clyde
Meyer, Eric Stefan
Miller, Frank L
Mohr, Peter J
Moore, John Hays
More, Kenneth Riddell
Mulligan, Joseph Francis
Mumma, Michael Jon
Omidvar, Kazem
Ott, William Roger
Oza, Dipak H
Parr, Albert Clarence
Pearl, John Christopher
Phillips, William Daniel
Pierce, Elliot Stearns
Placious, Robert Charles
Plotkin, Henry H
Rasberry, Stanley Dexter
Redmon, Michael James
Rhee, Moon-Jhong
Roszman, Larry Joe
Ruffa, Anthony Richard
Saloman, Edward Barry
Sattler, Joseph Peter
Shakur, Asif Mohammed
Silver, David Martin
Silverstone, Harris Julian
Sinnott, George
Stone, Philip M
Strombotne, Richard L(amar)
Sugar, Jack
Swenberg, Charles Edward
Taylor, Barry Norman
Taylor, Lauriston Sale
Temkin, Aaron
Thomsen, John Stearns
Tilford, Shelby G
Van Brunt, Richard Joseph
Varma, Matesh Narayan
Venkatesan, Thirumalai
Weber, Alfons
Weber, Mark
Weiner, John
Weiss, Andrew W
White, Kevin Joseph
Wiese, Wolfgang Lothar
Wingate, Catharine L

MASSACHUSETTS
Allis, Willam Phelps
Andersen, Roy Stuart
Berney, Charles V
Bradford, John Norman
Bradley, Lee Carrington, III
Branscomb, Lewis McAdory
Brown, Benjamin Lathrop
Champion, Kenneth Stanley Warner
Chang, Edward Shi Tou
Chen, Sow-Hsin
Clough, Shepard Anthony
Cohen, Howard David
Crampton, Stuart J B
Douglas-Hamilton, Diarmaid H
Ezekiel, Shaoul
Feld, Michael S
Franzen, Wolfgang
French, Anthony Philip
Garing, John Seymour
George, James Z
Godyak, Valery Anton
Handelman, Eileen T
Haugsjaa, Paul O
Hawkins, Bruce
Heller, Eric Johnson
Hilborn, Robert Clarence
Huffman, Robert Eugene
Hunter, Larry Russel
Jones, Kevin McDill
Katayama, Daniel Hideo
Kelley, Paul Leon
Kellogg, Edwin M
King, John Gordon
Kirby, Kate Page
Kleppner, Daniel
Kohin, Barbara Castle
Kohl, John Leslie
Krotkov, Robert Vladimir
Landman, Alfred
Lin, Alice Lee Lan
Mandl, Alexander Ernst
Martin, Frederick Wight
Meal, Janet Hawkins
Michael, Irving
Morris, Robert Alan
Moulton, Peter Franklin
Myers, John Martin
Oberteuffer, John Amiard
Oettinger, Peter Ernest
Parkinson, William Hambleton
Pichanick, Francis Martin
Pipkin, Francis Marion
Posner, Martin
Pritchard, David Edward
Ramsey, Norman Foster, Jr
Robinson, Howard Addison
Rothman, Laurence Sidney
Ruskai, Mary Beth
Schneider, Robert Julius
Sellers, Francis Bachman
Smith, Peter Lloyd
Sternheim, Morton Maynard
Sun, Yan
Terry, James Layton
Traub, Wesley Arthur
Tuchman, Avraham
Vessot, Robert F C
Victor, George A
Viggiano, Albert
Von Hippel, Arthur R
Wachman, Harold Yehuda
Wang, Chia Ping
Weinberger, Doreen Anne
Winick, Jeremy Ross
Witherell, Egilda DeAmicis
Wittkower, Andrew Benedict
Yoshino, Kouichi
Zajonc, Arthur Guy

MICHIGAN
Beck, Donald Richardson
Berman, Paul Ronald
Bernius, Mark Thomas
Bernstein, Eugene Merle
Chupp, Timothy E
Davis, Lloyd Craig
Derby, Stanley Kingdon
Donahue, Thomas Michael
Duchamp, David James
Fekety, F Robert, Jr
Ferguson, Stephen Mason
Hase, William Louis
Kauppila, Walter Eric
Kim, Yeong Wook
Krimm, Samuel
Kubis, Joseph J(ohn)
Leroi, George Edgar
Miller, Carl Elmer
Parker, Paul Michael
Phelps, Frederick Martin, III
Rajnak, Katheryn Edmonds
Rand, Stephen Colby
Ressler, Neil William
Rimai, Lajos
Rogers, Jerry Dale
Rol, Pieter Klaas
Rothe, Erhard William
Rothschild, Walter Gustav
Sanders, Barbara A
Sands, Richard Hamilton
Sharp, William Edward, III
Slocum, Robert Richard
Smith, George Wolfram
Stein, Talbert Sheldon
Vaishnava, Prem P
Van Baak, David Alan
Waber, James Thomas
Weidman, Robert Stuart
Zimring, Lois Jacobs
Zitzewitz, Paul William
Zorn, Jens Christian

MINNESOTA
Cederberg, James W
Giese, Clayton
Greenlee, Thomas Russell
Hakkila, Jon Eric
Johnson, Walter Heinrick, Jr
Khan, Faiz Mohammad
Knutson, Charles Dwaine
McClure, Benjamin Thompson
Sautter, Chester A
Stuart, Robley Vane
Thomas, Bruce Robert
Truhlar, Donald Gene
Valley, Leonard Maurice

MISSISSIPPI
Cooksy, Andrew Lloyd
Monts, David Lee
Rayborn, Grayson Hanks

MISSOURI
Bieniek, Ronald James
Brackmann, Richard Theodore
Freerks, Marshall Cornelius
Leventhal, Jacob J
McFarland, Robert Harold
Madison, Don Harvey
Menne, Thomas Joseph
Palmer, Kent Friedley
Park, John Thornton
Plummer, Patricia Lynne Moore
Schearer, Laird D
Schupp, Guy
Taub, Haskell Joseph
Thomas, Timothy Farragut

MONTANA
Carlsten, John Lennart

NEBRASKA
Blotcky, Alan Jay
Burns, Donal Joseph
Burrow, Paul David
Cipolla, Sam J
Dowben, Peter Arnold
Fabrikant, Ilya I
Fairchild, Robert Wayne
Gale, Douglas Shannon, II
Jaecks, Duane H
Jones, Ernest Olin
Leichner, Peter K
Pan, Cheng
Rudd, Millard Eugene
Samson, James Alexander Ross
Starace, Anthony Francis

NEVADA
Altick, Philip Lewis
Farley, John William
Haun, Robert Dee, Jr
Moore, Edwin Neal
Phaneuf, Ronald Arthur

NEW HAMPSHIRE
Bel Bruno, Joseph James
Laaspere, Thomas
Lambert, Robert Henry
Phelps, James Parkhurst
Rieser, Leonard M
Wright, John Jay

NEW JERSEY
Ashkin, Arthur
Bjorkholm, John Ernst
Cecchi, Joseph Leonard
Collett, Edward
Fisanick, Georgia Jeanne
Freeman, Richard Reiling
Freund, Robert Stanley
Gabriel, Oscar V
Gottscho, Richard Alan
Hagstrum, Homer Dupre
Hall, Gene Stephen
Happer, William, Jr
Hill, Kenneth Wayne
Hulse, Russell Alan
Kelsey, Edward Joseph
Kugel, Henry W
McAfee, Kenneth Bailey, Jr
Martin, John David
Miles, Richard Bryant
Millman, Sidney
Mills, Allen Paine, Jr
Mittleman, Marvin Harold
Moses, Herbert A
Murnick, Daniel E
Pfeiffer, Raymond John
Phillips, Julia M
Ramsey, Alan T
Reder, Friedrich H
Redi, Olav
Rothberg, Lewis Josiah

Rousseau, Denis Lawrence
Roy, M S
Salwen, Harold
Seidl, Milos
Silfvast, William Thomas
Suckewer, Szymon
Tomaselli, Vincent Paul
Torrey, Henry Cutler
Wittenberg, Albert M
Wu, Zhen
Yurke, Bernard
Zipf, Elizabeth M(argaret)

NEW MEXICO
Ambrose, William Patrick
Barnes, John Fayette
Bartlett, Roger James
Beauchamp, Edwin Knight
Beckel, Charles Leroy
Becker, Wilhelm
Bellum, John Curtis
Bieniewski, Thomas M
Bingham, Felton Wells
Blake, Richard L
Bryant, Howard Carnes
Burr, Alexander Fuller
Cano, Gilbert Lucero
Cartwright, David Chapman
Chamberlin, Edwin Phillip
Chow, Weng Wah
Clark, Robert Edward Holmes
Coats, Richard Lee
Cohen, James Samuel
Cowan, Robert Duane
Cox, Arthur Nelson
Cross, Jon Byron
Cuderman, Jerry Ferdinand
Czuchlewski, Stephen John
Damerow, Richard Aasen
Doolen, Gary Dean
Dyer, Peggy Lynn
Funsten, Herbert Oliver, III
Gerardo, James Bernard
Greene, Arthur Edward
Hadley, George Ronald
Hill, Ronald Ames
Holland, Redus Foy
Jason, Andrew John
Jones, Michael E
Judd, O'Dean P
Kane, Daniel James
Kunz, Walter Ernest
Kyrala, George Amine
Ladish, Joseph Stanley
Lockwood, Grant John
Lyman, John L
MacArthur, Duncan W
MacCallum, Crawford John
McGuire, Eugene J
McKenzie, James Montgomery
McLeod, John
Maier, William Bryan, II
Mann, Joseph Bird, (Jr)
Mazarakis, Michael Gerassimos
Mjolsness, Raymond C
Ortiz, Joseph Vincent
Pack, Russell T
Palmer, Byron Allen
Patterson, Christopher Warren
Peek, James Mack
Price, Robert Harold
Redondo, Antonio
Robertson, Merton M
Sattler, Allan R
Schneider, Jacob David
Shaner, John Wesley
Sheffield, Richard Lee
Snider, Donald Edward
Steinhaus, David Walter
Sze, Robert Chia-Ting
Thompson, David Charles
Yates, Mary Anne
Younger, Stephen Michael

NEW YORK
Acrivos, Andreas
Arroe, Hack
Avouris, Phaedon
Becker, Kurt Heinrich
Bederson, Benjamin
Bendler, John Thomas
Bergeman, Thomas H
Borowitz, Sidney
Boyd, Robert William
Boyer, Donald Wayne
Bramlet, Roland C
Brink, Gilbert Oscar
Brown, Howard Howland, Jr
Budick, Burton
Carroll, C(lark) E(dward)
Chapman, Sally
Chaturvedi, Ram Prakash
Chen, James Ralph
Christensen, Robert Lee
Costa, Lorenzo F
Creasy, William Russel
Das, Tara Prasad
Derman, Samuel
Deutsch, John Ludwig
De Zafra, Robert Lee
Dodd, Jack Gordon, (Jr)
Eidson, William Whelan
Engelke, Charles Edward
Ezra, Gregory Sion

Fischer, C Rutherford
Fitchen, Douglas Beach
Fleischmann, Hans Hermann
Franco, Victor
Garvey, James F
Gavin, Donald Arthur
Gavin, Gerard Brennan
Gentner, Robert F
Gersten, Joel Irwin
Grobe, Rainer
Hagen, Jon Boyd
Halpern, Alvin M
Hameed, Sultan
Helbig, Herbert Frederick
Hendrie, Joseph Mallam
Hioe, Foik Tjiong
Hirsh, Merle Norman
Holbrow, Charles H
Honig, Arnold
Hurst, Robert Philip
Innes, Kenneth Keith
Jain, Duli Chandra
Janak, James Francis
Johnson, Brant Montgomery
Johnson, Philip M
Jones, Keith Warlow
Kaplan, Martin Charles
Kass, Robert S
Kato, Walter Yoneo
Koch, Peter M
Komorek, Michael Joesph, Jr
Kostroun, Vaclav O
Krieger, Joseph Bernard
Lerner, Rita Guggenheim
Levine, Alfred Martin
Levine, Melvin Mordecai
Lipari, Nunzio Ottavio
Li-Scholz, Angela
Lounsbury, John Baldwin
Lowder, Wayne Morris
Mandel, Leonard
Marino, Robert Anthony
Messmer, Richard Paul
Metcalf, Harold
Moe, George Wylbur
Morris, William Guy
Muenter, John Stuart
Mukhopadhyay, Nimai Chand
Osgood, Richard Magee, Jr
Pomilla, Frank R
Ries, Richard Ralph
Robinson, Edward J
Roellig, Leonard Oscar
Rohrig, Norman
Rosenberg, Leonard
Rustgi, Om Prakash
Sahni, Viraht
Scarl, Donald B
Scholz, Wilfried
Schreurs, Jan W H
Schroeder, John
Schweitzer, Donald Gerald
Solomon, Philip M
Soltysik, Edward A
Sternheimer, Rudolph Max
Stroke, Hinko Henry
Sumberg, David A
Takahashi, Hiroshi
Thieberger, Peter
Titone, Luke Victor
Tourin, Richard Harold
Uzgiris, Egidijus E
Vacirca, Salvatore John
Varanasi, Prasad
Venugopalan, Srinivasa I
Walmsley, Ian Alexander
Walter, William Trump
White, Frederick Andrew
Yaakobi, Barukh
Yencha, Andrew Joseph
Zaider, Marco A
Zhang, John Zeng Hui
Zucker, Martin Samuel

NORTH CAROLINA
Bonin, Keith Donald
Christian, Wolfgang C
Clegg, Thomas Boykin
Dixon, Robert Leland
Dotson, Allen Clark
Hegstrom, Roger Allen
Hubbard, Paul Stancyl, Jr
Johnson, Charles Edward
Joyce, James Martin
Kelly, John Henry
Lapicki, Gregory
Miller, Roger Ervin
Mowat, J Richard
Parker, George W
Patty, Richard Roland
Rabinowitz, James Robert
Risley, John Stetler
Rogosa, George Leon
Sayetta, Thomas C
Shafroth, Stephen Morrison
Stephenson, Harold Patty
Swofford, Robert Lewis
Vermillion, Robert Everett
Willis, Robert D

NORTH DAKOTA
Kluk, Edward
Moore, Vaughn Clayton
Rao, B Seshagiri

OHIO
Agard, Eugene Theodore
Bahr, Gustave Karl
Bhattacharya, Rabi Sankar
Brilliant, Howard Michael
Chen, Charles Chin-Tse
Dakin, James Thomas
De Lucia, Frank Charles
Ferland, Gary Joseph
Fox, Thomas Allen
Gangemi, Francis A
Garscadden, Alan
Hagee, George Richard
Heer, Clifford V
Henry, Ronald James Whyte
Herbst, Eric
Hunter, William Winslow, Jr
Kepes, Joseph John
Mickelson, Michael Eugene
Nahar, Sultana Nurun
Newsom, Gerald Higley
Ross, Charles Burton
Ruegsegger, Donald Ray, Jr
Russell, James Edward
Schectman, Richard Milton
Schlosser, Herbert
Schreiber, Paul J
Singleton, Edgar Bryson
Snider, Joseph Lyons
Stoner, Ronald Edward
Strickler, Thomas David
Williamson, William, Jr
Wood, David Roy
Woodard, Ralph Emerson
Yaney, Perry Pappas

OKLAHOMA
Buchanan, Ronnie Joe
Feiock, Frank Donald
Fowler, Richard Gildart
Huffaker, James Neal
Merts, Athel Lavelle
Miller, Thomas Marshall
Ryan, Stewart Richard
St John, Robert Mahard
Schmelling, Stephen Gordon
Waldrop, Morgan A
Westhaus, Paul Anthony

OREGON
Behringer, Robert Ernest
Carmichael, Howard John
Casperson, Lee Wendel
Ch'en, Shang-Yi
Crasemann, Bernd
Drake, Charles Whitney
Engelking, Paul Craig
Gilbert, David Erwin
Girardeau, Marvin Denham, Jr
Grover, James Robb
Hardwick, John Lafayette
Kocher, Carl A
Moseley, John Travis
Mossberg, Thomas William

PENNSYLVANIA
Bajaj, Ram
Berkowitz, Harry Leo
Biondi, Manfred Anthony
Cohen, Leonard David
Cole, Milton Walter
Davies, D K
Ernst, Wolfgang E
Fisher, David George
Galey, John Apt
Garbuny, Max
Gerjuoy, Edward
Gur, David
Harney, Robert Charles
Herman, Roger Myers
Hirsch, Robert George
Huennekens, John Patrick
Intemann, Robert Louis
Kellman, Simon
Kim, Yong Wook
Liberman, Irving
Lowry, Jerald Frank
MacLennan, Donald Allan
Norris, Wilfred Glen
Obermyer, Richard T
Polo, Santiago Ramos
Pratt, Richard Houghton
Ranck, John Philip
Rhodes, Donald Frederick
Sampson, Douglas Howard
Sheers, William Sadler
Siegel, Melvin Walter
Smith, Wesley R
Smith, Winfield Scott
Snyder, Harold Lee
Suntharalingam, Nagalingam
Taylor, Lyle Herman
Tredicce, Jorge Raul
Weiner, Brian Lewis
Willey, Daniel Robert
Williams, Willie, Jr
Withstandley, Victor DeWyckoff, III
Yarosewick, Stanley J
Yuan, Jian-Min

RHODE ISLAND
Mason, Edward Allen
Sherman, Charles Henry

SOUTH CAROLINA
Adelman, Saul Joseph
Cathey, LeConte
Chaplin, Robert Lee, Jr
Driggers, Frank Edgar
Graves, William Ewing
Haile, James Mitchell
Honeck, Henry Charles
Kendall, David Nelson
Malstrom, Robert Arthur
Singleton, Chloe Joi
Steiner, Pinckney Alston, III
Topp, Stephen V

SOUTH DAKOTA
Duffey, George Henry

TENNESSEE
Albridge, Royal
Allison, Stephen William
Alton, Gerald Dodd
Appleton, B R
Beene, James Robert
Bemis, Curtis Elliot, Jr
Blass, William Errol
Brau, Charles Allen
Burns, John Francis
Carlson, Thomas Arthur
Carman, Howard Smith, Jr
Chen, Chung-Hsuan
Christophorou, Loucas Georgiou
Compton, Robert Norman
Crawford, Oakley H
Cummings, Peter Thomas
Datz, Sheldon
Daunt, Stephen Joseph
Dittner, Peter Fred
Driskill, William David
Elston, Stuart B
Ewbank, Wesley Bruce
Feigerle, Charles Stephen
Fischer, Charlotte Froese
Fox, Kenneth
Garrett, William Ray
Goans, Ronald Earl
Gregory, Donald Clifford
Haglund, Richard Forsberg, Jr
Hefferlin, Ray (Alden)
Hubbell, Harry Hopkins, Jr
Hurst, G Samuel
Ice, Gene Emery
Isler, Ralph Charles
Jacobson, Harry C
James, David Randolph
Jones, Charles Miller, Jr
Keefer, Dennis Ralph
Klots, Cornelius E
Krause, Herbert Francis
Krause, Manfred Otto
Langley, Robert Archie
Lewis, James W L
Livingston, Ralph
Longmire, Martin Shelling
Macek, Joseph
McGregor, Wheeler Kesey, Jr
Mason, Arthur Allen
Meyer, Fred Wolfgang
Miller, John Cameron
Moak, Charles Dexter
Nalley, Samuel Joseph
Nestor, C William, Jr
Painter, Linda Robinson
Patterson, Malcolm Robert
Pegg, David John
Pinnaduwage, Lal Ariyaratna
Ricci, Enzo
Schmitt, Harold William
Sellin, Ivan Armand
Sheffield, John
Snell, Arthur Hawley
Staats, Percy Anderson
Tanner, Raymond Lewis
Tellinghuisen, Joel Barton
Thonnard, Norbert
Turner, James Edward
Vander Sluis, Kenneth Leroy
Wilson, Robert John
Young, Jack Phillip

TEXAS
Aldridge, Jack Paxton, III
Baker, Samuel I
Bengtson, Roger D
Borst, Walter Ludwig
Breig, Edward Louis
Brooks, Philip Russell
Browne, James Clayton
Chao, Jing
Chopra, Dev Raj
Church, David Arthur
Colegrove, Forrest Donald
Coulson, Larry Vernon
Decker, John P
Diana, Leonard M
Dunning, Frank Barrymore
Fetzer, Homer D
Fink, Manfred
Ford, Albert Lewis, Jr
Frommhold, Lothar Werner
Fry, Edward Strauss
Golden, David E
Graham, William Richard Montgomery
Hance, Robert Lee
Harvey, Kenneth C

PHYSICS & ASTRONOMY / 515

Atomic & Molecular Physics (cont)

Hatfield, Lynn LaMar
Herczeg, John W
Hubisz, John Lawrence, Jr
Huebner, Walter F
Hulet, Randall Gardner
Johnson, Raleigh Francis, Jr
Kenefick, Robert Arthur
Kouri, Donald Jack
Krohmer, Jack Stewart
Kyle, Thomas Gail
LeBlanc, Adrian David
Loyd, David Heron
McDaniel, Floyd Delbert, Sr
Mangum, Jeffrey Gary
Menzel, Erhard Roland
Meshkov, Sydney
Miller, Philip Dixon
Moore, C Fred
Parker, Cleofus Varren, Jr
Paske, William Charles
Powers, Darden
Quade, Charles Richard
Quarles, Carroll Adair, Jr
Rambow, Frederick H K
Redding, Rogers Walker
Robbins, Ralph Robert
Robinson, George Wilse
Schuessler, Hans A
Sharma, Suresh C
Sherrill, William Manning
Slocum, Robert Earle
Stuart, Joe Don
Synek, Miroslav (Mike)
Walters, Geoffrey King
Watson, Rand Lewis
Wehring, Bernard William
Weinstein, Roy
Weisheit, Jon Carleton
Weisman, R(obert) Bruce
Wells, Michael Byron
Woessner, Donald Edward
Wright, Ann Elizabeth

UTAH
Breckenridge, William H
Knight, Larry V
Larson, Everett Gerald
Shirts, Randall Brent

VERMONT
Dunham, Jeffrey Solon

VIRGINIA
Anderson, Richard John
August, Leon Stanley
Bloomfield, Louis Aub
Brill, Arthur Sylvan
Brown, Ellen Ruth
Champion, Roy L
Church, Charles Henry
Copeland, Gary Earl
Dardis, John G
Delos, John Bernard
Dharamsi, Amin N
Doverspike, Lynn D
Eubank, Harold Porter
Exton, Reginald John
Fox, Russell Elwell
Gallagher, Thomas Francis
Herm, Ronald Richard
Hoppe, John Cameron
Howard, Russell Alfred
Jalufka, Nelson Wayne
Johnson, Robert Edward
Junker, Bobby Ray
Kabir, Prabahan Kemal
Kelley, Ralph Edward
King, George, III
Kramer, Steven David
Kuang, Yunan
Lane, Neal F
Larson, Daniel John
Lee, Ja H
Liszt, Harvey Steven
Long, Edward Richardson, Jr
Mullen, Joseph Matthew
Pilloff, Herschel Sydney
Rainis, Albert Edward
Roy, Donald H
Sauder, William Conrad
Schneider, Barry I
Schoen, Richard Isaac
Sepucha, Robert Charles
Vuskovic, Leposava
Way, Kermit R
Wexler, Bernard Lester
Whitehead, Walter Dexter, Jr
Wodarczyk, Francis John
Ziock, Klaus Otto H

WASHINGTON
Adelberger, Eric George
Baird, Quincey Lamar
Baughcum, Steven Lee
Bichsel, Hans
Bierman, Sidney Roy
Braby, Leslie Alan
Braunlich, Peter Fritz
Carter, Leland LaVelle
Clark, Kenneth Courtright
Currah, Walter E
Dehmelt, Hans Georg
Dubois, Robert Dean
Eggers, David Frank, Jr
Feldman, Henry Robert
Forsman, Earl N
Fortson, Edward Norval
Geballe, Ronald
Glass, William A
Gouterman, Martin (Paul)
Hernandez, Gonzalo J
Jefferts, Keith Bartlett
McClure, J Doyle
McDermott, Mark Nordman
McDowell, Robin Scott
McElroy, William Nordell
Maki, Arthur George, Jr
Miller, John Howard
Palmer, Harvey Earl
Person, James Carl
Radziemski, Leon Joseph
Reinhardt, William Parker
Sheridan, John Roger
Soldat, Joseph Kenneth
Veit, Jiri Joseph
West, Martin Luther
Wilson, Walter Ervin

WISCONSIN
Anderson, Louis Wilmer
Brandenberger, John Russell
Dobson, David A
Doherty, Lowell Ralph
England, Walter Bernard
Fonck, Raymond John
Fystrom, Dell O
Greenebaum, Ben
Greenler, Robert George
King, Frederick Warren
Kobiske, Ronald Albert
Kurey, Thomas John
Lawler, James Edward
Lin, Chun Chia
Mallmann, Alexander James
Paliwal, Bhudatt R
Roesler, Fredrick Lewis
Sudhakaran, Gubbi Ramarao
Tonner, Brian P
Woods, Robert Claude

ALBERTA
Ali, M Keramat
Ali, Shahida Parvin
Bland, Clifford J
Freeman, Gordon Russel
Kisman, Kenneth Edwin
Klobukowski, Mariusz Andrzej
Lepard, David William
McClung, Ronald Edwin Dawson
Newbound, Kenneth Bateman
Pinnington, Eric Henry
Thorson, Walter Rollier

BRITISH COLUMBIA
Balfour, Walter Joseph
Brion, Christopher Edward
Coope, John Arthur Robert
Galt, John (Alexander)
Hesser, James Edward
Malli, Gulzari Lal
Measday, David Frederick
Ozier, Irving
Poll, Jacobus Daniel
Rieckhoff, Klaus E
Snider, Robert Folinsbee
Watton, Arthur

MANITOBA
Duckworth, Henry Edmison
Ens, E(rich) Werner
Kerr, Donald Philip
Mathur, Maya Swarup
Standing, Kenneth Graham
Tabisz, George Conrad

NEW BRUNSWICK
Grein, Friedrich
Kaiser, Reinhold
Lees, Ronald Milne
Sichel, John Martin
Thakkar, Ajit Jamnadas

NEWFOUNDLAND
Cho, Chung Won
Foltz, Nevin D
Reddy, Satti Paddi

NOVA SCOTIA
Coxon, John Anthony
Kusalik, Peter Gerard
Latta, Bryan Michael

ONTARIO
Atkinson, John Brian
Aziz, Ronald A
Baylis, William Eric
Bernath, Peter Francis
Brumer, Paul William
Campeanu, Radu Ioan
Carrington, Tucker
Chapman, George David
Colpa, Johannes Pieter
Costain, Cecil Clifford
Davies, John Arthur
Drake, Gordon William Frederic
Erickson, Lynden Edwin
Fraser, Peter Arthur
Ganza, Kresimir Peter
Garside, Brian K
Geiger, James Stephen
Goodings, John Martin
Gordon, Robert Dixon
Hackam, Reuben
Hagberg, Erik Gordon
Hebert, Gerard Rosaire
Herzberg, Gerhard
Hitchcock, Adam Percival
Holt, Richard A(rnold)
Howard-Lock, Helen Elaine
Hunt, James L
Krause, Lucjan
LeRoy, Robert James
Lew, Hin
Lowe, Robert Peter
McConkey, John William
McConnell, John Charles
McCourt, Frederick Richard Wayne
McKellar, Robert A
McLay, David Boyd
Marmet, Paul
Meath, William John
Morris, Derek
Mungall, Allan George
Penner, Glenn H
Raju, Govinda
Ramsay, Donald Allan
Rosner, Sheldon David
Sears, Varley Fullerton
Shepherd, Gordon Greeley
Sinha, Bidhu Bhushan Prasad
Stauffer, Allan Daniel
Stoicheff, Boris Peter
Vanier, Jacques
Vosko, Seymour H

PRINCE EDWARD ISLAND
Madan, Mahendra Pratap

QUEBEC
Beique, Rene Alexandre
Bose, Tapan Kumar
Knystautas, Emile J
Lee, Jonathan K P
Nadeau, Daniel
Salahub, Dennis Russell

OTHER COUNTRIES
Band, Yehuda Benzion
Bondybey, Vladimir E
Buckingham, Amyand David
Bunge, Carlos Federico
Cardona, Manuel
Cue, Nelson
Heinz, Ulrich Walter
Ho, Yew Kam
Huang, Keh-Ning
Kim, Dae Mann
Krumbein, Aaron Davis
Kwok, Hoi S
Lee, Tong-Nyong
Pitchford, Leanne Carolyn
Shafi, Mohammad
Toennies, Jan Peter
Yellin, Joseph

Electromagnetism

ALABAMA
Ellenburg, Janus Yentsch
Hubbell, Wayne Charles
Jones, Robert William
Kinzer, Earl T, Jr
McDonald, Jack Raymond
Mookherji, Tripty Kumar
Olson, Willard Paul
Passino, Nicholas R
Poularikas, Alexander D

ALASKA
Hunsucker, Robert Dudley

ARIZONA
Bickel, William Samuel
Chamberlin, Ralph Vary
Laurin, Pushpamala
Levy, Eugene Howard
Macleod, Hugh Angus
Reagan, John Albert
Stoner, John Oliver, Jr
Ziolkowski, Richard Walter

ARKANSAS
Eichenberger, Rudolph John
Mackey, James E

CALIFORNIA
Alvarez, Raymond Angelo, Jr
Arnold, James S(loan)
Astrahan, Melvin Alan
Baggerly, Leo L
Bajorek, Christopher Henry
Bardin, Russell Keith
Bethune, Donald Stimson
Bevc, Vladislav
Breidenbach, Martin
Bucker, Homer Park, Jr
Bush, Gary Graham
Buskirk, Fred Ramon
Caspers, Hubert Henri
Cautis, C Victor
Chao, Alexander Wu
Chase, Jay Benton
Chen, Tu
Cheng, Tsen-Chung
Cho, Young-chung
Cladis, John Baros
Clark, Arnold Franklin
Clark, Leigh Bruce
Clover, Richmond Bennett
Deacon, David A G
Dowell, Jerry Tray
Driscoll, Charles F
Duneer, Arthur Gustav, Jr
Dyal, Palmer
Eikrem, Lynwood Olaf
Evans, Todd Edwin
Farone, William Anthony
Fearn, Heidi
Feldman, Nathaniel E
Fialer, Philip A
Fisher, Leon Harold
Fraser-Smith, Antony Charles
Fried, Burton David
Fried, Walter Rudolf
Gamo, Hideya
George, Simon
Glass, Nathaniel E
Good, Roland Hamilton, Jr
Greifinger, Carl
Hahn, Erwin Louis
Harker, Kenneth James
Hausman, Arthur Herbert
Hayes, Claude Q C
Heinz, Otto
Hempstead, Robert Douglas
Hobbs, Willard Earl
Honey, Richard Churchill
Huang, C Yuan
Jackson, Gary Leslie
Jackson, John David
Jory, Howard Roberts
Jungerman, John (Albert)
Kalensher, Bernard Earl
Kinnison, Gerald Lee
Kip, Arthur F
Kirchner, Ernst Karl
Klein, Melvin Phillip
Konrad, Gerhard T(hies)
Kuehl, Hans H(enry)
Kulke, Bernhard
Lacey, Richard Frederick
Lambertson, Glen Royal
Lemke, James Underwood
Linford, Gary Joe
Ling, Rung Tai
Lipson, Joseph Issac
Lockhart, James Marcus
Love, Allan Walter
Ludwig, Claus Berthold
McCurdy, Alan Hugh
MacGregor, Malcolm Herbert
Margolis, Jack Selig
Mitzner, Kenneth Martin
Mo, Charles Tse Chin
Morris, Richard Herbert
Musal, Henry M(ichael), Jr
Nee, Soe-Mie Foeng
Nesbit, Richard Allison
Nevins, William McCay
Newcomb, William A
Newman, David Edward
Newman, John Joseph
Olney, Ross David
Papas, Charles Herach
Parker, Don
Passenheim, Burr Charles
Pellegrini, Claudio
Perry, Richard Lee
Peskin, Michael Edward
Poggio, Andrew John
Poynter, Robert Louis
Rabinowitz, Mario
Rahmat-Samii, Yahya
Romagnoli, Robert Joseph
Saito, Theodore T
Sample, Steven Browning
Sarwinski, Raymond Edmund
Schermer, Robert Ira
Sentman, Davis Daniel
Sharp, Richard Dana
Skomal, Edward N
Smith, Otto J(oseph) M(itchell)
Smith, Sheldon Magill
Smith, Todd Iversen
Snowden, Donald Philip
Spencer, James Eugene
Sperling, Jacob L
Stafsudd, Oscar M, Jr
Steinmetz, Wayne Edward
Stenzel, Reiner Ludwig
Stewart, Richard William
Stinson, Donald Cline
Tilles, Abe
Torgow, Eugene N
Tricoles, Gus P
Twersky, Victor
Tyler, George Leonard
Vernazza, Jorge Enrique
Vickers, Roger Spencer
Villeneuve, A(lfred) T(homas)
Wagner, Richard John

PHYSICS & ASTRONOMY / 517

Waterman, Alan T(ower), Jr
Weinstein, Berthold Werner
Weiss, Jeffrey Martin
Wickersheim, Kenneth Alan
Wiedow, Carl Paul
Winick, Herman
Wolff, Milo Mitchell
Woodward, James Franklin
Wyatt, Philip Joseph
Yeh, Cavour W
Yeh, Edward H Y
Yeh, Paul Pao

COLORADO
Bartlett, David Farnham
Beers, Yardley
Bennett, W Scott
Bernstein, Elliot R
Bussey, Howard Emerson
Chew, Herman W
Estin, Arthur John
Fickett, Frederick Roland
Galeener, Frank Lee
Goldfarb, Ronald B
Granzow, Kenneth Donald
Haydon, George William
Hill, David Allan
Hjelme, Dag Roar
Hogg, David Clarence
Hufford, George (Allen)
Huisjen, Martin Albert
Kerns, David Marlow
Kiehl, Jeffrey Theodore
Lubell, Jerry Ira
Ma, Mark T
Marks, Roger Bradley
Randa, James P
Rauch, Gary Clark
Rinker, George Albert, Jr
Stencel, Robert Edward
Tatarskii, Valerian I
Walden, Jack M

CONNECTICUT
Alpert, Nelson Leigh
Brinen, Jacob Solomon
Cheng, David H S
Close, Richard Thomas
Colthup, Norman Bertram
Glenn, William Henry, Jr
Liu, Qing-Huo
Sachdeva, Baldev Krishan
Snyder, John William
Taylor, Geoff W
Zandy, Hassan F

DELAWARE
Hurley, William Joseph
Jarrett, Howard Starke, Jr

DISTRICT OF COLUMBIA
Beard, Charles Irvin
Butcher, Raymond John
Colombant, Denis Georges
English, William Joseph
Florig, Henry Keith
Forester, Donald Wayne
Fusina, Robert Alan
Grossman, John Mark
Hafizi, Bahman
Jansen, Robert Werner
Jordan, Arthur Kent
Kelly, Francis Joseph
Lin-Chung, Pay-June
Manka, Charles K
Reilly, Michael Hunt
Reiss, Howard R
Schriever, Richard L
Wasylkiwskyj, Wasyl

FLORIDA
Baird, Alfred Michael
Bass, Michael
Bittner, Burt James
Burdick, Glenn Arthur
Darling, Byron Thorwell
De Lorge, John Oldham
Gustafson, Bo Ake Sture
Hagedorn, Fred Bassett
Johnston, Milton Dwynell, Jr
Kelso, John Morris
Killinger, Dennis K
Krakow, Burton
Leitner, Alfred
Millar, Gordon Halstead
Neuringer, Joseph Louis
Nunn, Walter M(elrose), Jr
Rawls, Walter Cecil, Jr
Ungvichian, Vichate
Vala, Martin Thorvald, Jr
Wang, Ru-Tsang

GEORGIA
Agrawal, Pradeep Kumar
Anantha Narayanan, Venkataraman
Carreira, Lionel Andrade
Currie, Nicholas Charles
Gaylord, Thomas Keith
Joy, Edward Bennett
Lawrence, Kurt C
Long, Maurice W(ayne)
Rodrigue, George Pierre
Ryan, Charles Edward, Jr
Simmons, James Wood

Turner, Janice Butler
Wang, Johnson Jenn-Hwa

HAWAII
Andermann, George
Hall, Donald Norman Blake
Holmes, John Richard

IDAHO
Strommen, Dennis Patrick

ILLINOIS
Basile, Louis Joseph
Brussel, Morton Kremen
Camras, Marvin
Chew, Weng Cho
Cole, Francis Talmage
Cooper, William Edward
Crabtree, George William
Curry, Bill Perry
Dreska, Noel
Evans, Kenneth, Jr
Gallagher, David Alden
Henderson, Giles Lee
Henneberger, Walter Carl
Hoshiko, Michael S
Knop, Charles M(ilton)
Koster, David F
Kustom, Robert L
Lari, Robert Joseph
Mechtly, Eugene A
Naylor, David L
Ng, King-Yuen
Parker, Eugene Newman
Peters, Robert Edward
Shenoy, Gopal K
Shepard, Kenneth Wayne
Smith, Richard Paul
Taflove, Allen
Tollestrup, Alvin V
Tomkins, Frank Sargent
Treadwell, Elliott Allen
Wah, Yau Wai

INDIANA
Farringer, Leland Dwight
Poirier, John Anthony

IOWA
Clem, John R
Gurnett, Donald Alfred
Johnston, David Carl
Lunde, Barbara Kegerreis, (BK)

KANSAS
Carpenter, Kenneth Halsey
Folland, Nathan Orlando
Long, Larry L
Sapp, Richard Cassell
Unz, Hillel
Weller, Lawrence Allenby
Williams, Dudley

KENTUCKY
Bakanowski, Stephen Michael
Wilt, Paxton Marshall

LOUISIANA
Chin-Bing, Stanley Arthur
Hilburn, John L
Imhoff, Donald Wilbur
Oberding, Dennis George

MAINE
Kingsbury, Robert Freeman

MARYLAND
Baker, Francis Edward, Jr
Bennett, Charles L
Bostrom, Carl Otto
Burke, Edward Raymond
Chang, Alfred Tieh-Chun
Chen, Wenpeng
Connerney, John E P
Driscoll, Raymond L
Eccleshall, Donald
Flory, Thomas Reherd
Gaunaurd, Guillermo C
Gordon, Daniel Israel
Granatstein, Victor Lawrence
Gray, Ernest Paul
Hochheimer, Bernard Ford
Hyde, Geoffrey
Jablonski, Daniel Gary
Jackson, William David
Jennings, Donald Edward
Johnson, William Bowie
Joiner, R(eginald) Gracen
Kuttler, James Robert
Ledley, Brian G
McDonald, Jimmie Reed
Marx, Egon
Mather, John Cromwell
Merkel, George
Parsegian, Vozken Adrian
Peredo, Mauricio
Phillips, William Daniel
Pirraglia, Joseph A
Raines, Jeremy Keith
Roth, Bradley John
Sattler, Joseph Peter
Schmugge, Thomas Joseph
Schumacher, Clifford Rodney
Sharma, Jagadish

Simmons, Joe Denton
Soln, Josip Zvonimir
Strenzwilk, Denis Frank
Tang, Cha-Mei
Taylor, Leonard S
Taylor, Robert Joseph
Tompkins, Robert Charles
Weinschel, Bruno Oscar
Whicker, Lawrence R
Williams, Conrad Malcolm

MASSACHUSETTS
Allen, Ryne C
Bekefi, George
Bers, Abraham
Bugnolo, Dimitri Spartaco
Burke, William J
Cohen, David
Connors, Robert Edward
Dhar, Sachidulal
Dionne, Gerald Francis
Dorschner, Terry Anthony
Drane, Charles Joseph, Jr
Gelman, Harry
Gianino, Peter Dominic
Giger, Adolf J
Godyak, Valery Anton
Graneau, Peter
Green, Jerome Joseph
Hooper, Robert John
Kincaid, Thomas Gardiner
Kolm, Henry Herbert
Lees, Wayne Lowry
Loewenstein, Ernest Victor
McBee, W(arren) D(ale)
Mack, Richard Bruce
Mailloux, Robert Joseph
Maloney, William Thomas
Maxwell, Emanuel
Mellen, Walter Roy
Myers, John Martin
Neuringer, Leo J
Newburgh, Ronald Gerald
Oberteuffer, John Amiard
Oliner, Arthur A(aron)
Osepchuk, John M
Parrott, Stephen Kinsley
Rabin, Monroe Stephen Zane
Ram-Mohan, L Ramdas
Rana, Ram S
Reed, F(lood) Everett
Rheinstein, John
Richter, Stephen L(awrence)
Rork, Eugene Wallace
Rotman, Walter
Schattenburg, Mark Lee
Schneider, Robert Julius
Sethares, James C(ostas)
Shane, John Richard
Shen, Hao-Ming
Silevich, Michael B
Silevitch, Michael B
Temkin, Richard Joel
Tzeng, Wen-Shian Vincent
Vanasse, George Alfred
Vignos, James Henry
Vittoria, Carmine
Weggel, Robert John
Weiss, Jerald Aubrey
Whitney, Cynthia Kolb
Wu, Tai Tsun
Yaghjian, Arthur David
Yngvesson, K Sigfrid

MICHIGAN
Asmussen, Jes
Banks, Peter Morgan
Beaman, Donald Robert
Bhattacharya, Pallab Kumar
Brailsford, Alan David
Bryant, John H(arold)
Clauer, C Robert, Jr
Derby, Stanley Kingdon
Donohue, Robert J
Gordon, Morton Maurice
Gustafson, Herold Richard
Jones, Frederick Goodwin
King, Stanley Shih-Tung
Kingman, Robert Earl
LaHaie, Ivan Joseph
Nagy, Louis Leonard
Ottova, Angela
Pollack, Gerald Leslie
Rand, Stephen Colby
Roessler, David Martyn
Rolnick, William Barnett
Schubring, Norman W(illiam)
Segall, Stephen Barrett
Senior, Thomas Bryan Alexander
Slavin, Andrei Nickolay
Smith, Kenneth Edward
Van Baak, David Alan

MINNESOTA
Brom, Joseph March, Jr
Crawford, Bryce (Low), Jr
Follingstad, Henry George
Hakkila, Jon Eric
Heltemes, Eugene Casmir
Lutes, Olin S
Roska, Fred James
Walters, John Philip

MISSISSIPPI
Cook, Robert Lee
Crow, Terry Tom
Miller, David Burke
Smith, Charles Edward
Taylor, Clayborne D
Winton, Raymond Sheridan

MISSOURI
Adawi, Ibrahim (Hasan)
Alexander, Ralph William, Jr
Bard, James Richard
Bell, Robert John
Brown, Harry Allen
Burgess, James Harland
DeFacio, W Brian
DuBroff, Richard Edward
Huang, Justin C
Huddleston, Philip Lee
Indeck, Ronald S
Leader, John Carl
Lind, Arthur Charles
Menne, Thomas Joseph
Muller, Marcel Wettstein
Smith, Robert Francis
Spielman, Barry
White, Warren D

MONTANA
Graham, Raymond

NEVADA
Rawat, Banmali Singh
Scott, William Taussig

NEW HAMPSHIRE
Buffler, Charles Rogers
Corum, Kenneth Lawrence
Frost, Albert D(enver)
Humphrey, Floyd Bernard
Simmons, Alan J(ay)

NEW JERSEY
Alig, Roger Casanova
Amitay, Noach
Amundson, Karl Raymond
Chand, Naresh
Cioffi, Paul Peter
Daniels, James Maurice
Dillon, Joseph Francis, Jr
Dutta, Mitra
Fish, Gordon E
Fleury, Paul A
Fork, Richard Lynn
Freeman, Richard Reiling
Gammel, George Michael
Grek, Boris
Hall, Herbert Joseph
Hershenov, B(ernard)
Koch, Thomas L
Mueller, Dennis
Passner, Albert
Rastani, Kasra
Reder, Friedrich H
Rossol, Frederick Carl
Rulf, Benjamin
Schrenk, George L
Schwering, Felix
Smith, Carl Hofland
Spencer, Edward G
Stevens, James Everell
Tenzer, Rudolf Kurt
Tsaliovich, Anatoly
Turner, Edward Harrison
Wang, Chao Chen
Whitman, Gerald Martin

NEW MEXICO
Adler, Richard John
Anderson, Robert Alan
Argo, Paul Emmett
Babb, David Daniel
Baum, Carl E(dward)
Bradshaw, Martin Daniel
Brook, Marx
Brownell, John Howard
Butler, Harold S
Chamberlin, Edwin Phillip
Chylek, Petr
Clark, Wallace Thomas, III
Cooper, Richard Kent
Dressel, Ralph William
Farmer, William Michael
Flicker, Herbert
Freeman, Bruce L, Jr
Frost, Harold Maurice, III
Godfrey, Brendan Berry
Gurbaxani, Shyam Hassomal
Herrmannsfeldt, William Bernard
Hoeft, Lothar Otto
Humphries, Stanley, Jr
Ladish, Joseph Stanley
Lindman, Erick Leroy, Jr
Lyo, SungKwun Kenneth
Maier, William Bryan, II
Miller, Edmund K(enneth)
O'Donnell, Edward Earl
Potter, James Martin
Rach, Randolph Carl
Reeves, Geoffrey D
Schriber, Stanley Owen
Schwarz, Ricardo
Searls, Craig Allen
Simon, George Warren

Electromagnetism (cont)

Taylor, Raymond Dean
Toepfer, Alan James
VanDevender, John Pace
Vittitoe, Charles Norman
Wiggins, Carl M
Wright, Thomas Payne

NEW YORK
Adams, Arlon Taylor
Barreto, Ernesto
Bernstein, Burton
Blumberg, Leroy Norman
Borrego, Jose M
Bottomley, Paul Arthur
Boyd, Robert William
Bradshaw, John Alden
Chen, Inan
Cheney, Margaret
Chrenko, Richard Michael
Costa, Lorenzo F
Cottingham, James Garry
Craft, Harold Dumont, Jr
Davis, Abram
Demerdash, Nabeel A O
Derman, Samuel
Diament, Paul
Garretson, Craig Martin
Geer, James Francis
Genack, Azriel Z
Goodwin, Arthur VanKleek
Haas, Werner E L
Hall, Dennis Gene
Hammer, David Andrew
Hartwig, Curtis P
Haus, Joseph Wendel
Hurst, Robert Philip
Kane, William Theodore
Kermisch, Dorian
Kitchen, Sumner Wendell
Kolb, Frederick J(ohn), Jr
Kramer, Stephen Leonard
Lauber, Thornton Stuart
Likharev, Konstantin K
McGuire, Thomas R
Mandel, Leonard
Marshall, Thomas C
Mihran, Theodore Gregory
Nelson, John Keith
Nisteruk, Chester Joseph
Oh, Byungdu
Ozimek, Edward Joseph
Packard, Karle Sanborn, Jr
Pflug, Donald Ralph
Plane, Robert Allen
Rezanka, Ivan
Roalsvig, Jan Per
Rohrlich, Fritz
Schlesinger, S Perry
Senus, Walter Joseph
Shapiro, Paul Jonathon
Shiren, Norman S
Simpson, Murray
Talley, Robert Lee
Viertl, John Ruediger Mader
Vortuba, Jan
Wolf, Edward Lincoln

NORTH CAROLINA
Aktas, Yildirin
Ji, Chueng Ryong
McRee, Donald Ikerd
Mink, James Walter
Prater, John Thomas
Straley, Joseph Ward
Tischer, Frederick Joseph

NORTH DAKOTA
Dewar, Graeme Alexander

OHIO
Bales, Howard E
Barranger, John P
Becker, Roger Jackson
Boughton, Robert Ivan, Jr
Compton, Ralph Theodore, Jr
Falkenbach, George J(oseph)
Hill, James Stewart
Holt, James Franklin
McLennan, Donald Elmore
Manning, Robert M
Mathews, Collis Weldon
Mettee, Howard Dawson
Morin, Brian Gerald
Pathak, Prabhakar H
Redlich, Robert Walter
Restemeyer, William Edward
Spitzer, Jeffrey Chandler
Strnat, Karl J
Stroud, David Gordon
Wigen, Philip E
Williamson, William, Jr

OKLAHOMA
Armoudian, Garabed
Baldwin, Bernard Arthur
Beasley, William Harold
Bilger, Hans Rudolf
Cohn, Jack
Dixon, George Sumter, Jr
Taylor, William L

OREGON
Casperson, Lee Wendel
Chartier, Vernon Lee
Csonka, Paul L
Drummond, James Edgar
Gilbert, David Erwin
Griffiths, David Jeffery
Kocher, Carl A
Yamaguchi, Tadanori

PENNSYLVANIA
Belohoubek, Erwin F
Berry, Richard Emerson
Carpenter, Lynn Allen
Charap, Stanley H
Coren, Richard L
Eberhardt, Nikolai
Falconer, Thomas Hugh
Grimes, Dale M(ills)
Havas, Peter
Heald, Mark Aiken
Hyatt, Robert Monroe
Jaggard, Dwight Lincoln
Judd, Jane Harter
Ku, Robert Tien-Hung
Lambeth, David N
Long, Howard Charles
Long, Lyle Norman
Mathews, John David
Matocha, Charles K
Miskovsky, Nicholas Matthew
Molter, Lynne Ann
Morabito, Joseph Michael
Schenk, H(arold) L(ouis), Jr
Seltzer, James Edward
Suhr, Norman Henry
Tahir-Kheli, Raza Ali
Thompson, Ramie Herbert
Thornburg, Donald Richard
Van Roggen, Arend
Williams, James Earl, Jr
Young, Frederick J(ohn)

RHODE ISLAND
Grossi, Mario Dario
Polk, C(harles)
Sullivan, Edmund Joseph

SOUTH CAROLINA
McNulty, Peter J
Pearson, Lonnie Wilson
Sheppard, Emory Lamar
Steiner, Pinckney Alston, III

TENNESSEE
Child, Harry Ray
Deeds, William Edward
Fuson, Nelson
Miller, Francis Joseph
Moak, Charles Dexter
Mosko, Sigmund W
Murakami, Masanori
Schwenterly, Stanley William, III
Shaw, Robert Wayne
Tarpley, Anderson Ray, Jr
Trivelpiece, Alvin William
Wikswo, John Peter, Jr

TEXAS
Andrychuk, Dmetro
Beissner, Robert Edward
Bronaugh, Edwin Lee
Brown, Glenn Lamar
Bussian, Alfred Erich
Chamberlain, Nugent Francis
Decker, John P
Fields, Reuben Elbert
Fung, Adrian K
Gavenda, John David
Heelis, Roderick Antony
Hurd, James William
Johnson, Francis Severin
Johnson, Raleigh Francis, Jr
Liemohn, Harold Benjamin
Ling, Hao
Long, Stuart A
Loos, Karl Rudolf
McIntyre, Peter Mastin
Neikirk, Dean P
Nevels, Robert Dudley
Olenick, Richard Peter
Parker, Jerald Vawer
Ranganayaki, Rambabu Pothireddy
Reiff, Patricia Hofer
Rester, David Hampton
Roberts, Thomas M
Rollwitz, William Lloyd
Sablik, Martin J
Sams, Lewis Calhoun, Jr
Sharp, A C, Jr
Shepley, Lawrence Charles
Sherry, Clifford Joseph
Slocum, Robert Earle
Sobey, Arthur Edward, Jr
Srnka, Leonard James
Stephenson, Danny Lon

UTAH
Chabries, Douglas M
Earley, Charles Willard
Edwards, W Farrell
Jeffery, Rondo Nelden
Miller, Akeley

VERMONT
Golden, Kenneth Ivan

VIRGINIA
Benesch, Jay F
Bogan, Denis John
Brown, Gary S
Calle, Carlos Ignacio
Cardman, Lawrence Santo
Choi, Junho
Connolly, John Irving, Jr
Dubin, Henry Charles
Fisher, Arthur Douglas
Folen, Vincent James
Gale, Harold Walter
Ganguly, Suman
Harp, William R, Jr
Harris, William Charles
Johnson, Charles Nelson, Jr
Johnson, Donald Rex
Leemann, Christoph Willy
McAdoo, John Hart
McWright, Glen Martin
Nuckolls, Joe Allen
Ofelt, George Sterling
Patterson, James Douglas
Rogowski, Robert Stephen
Sparrow, D(avid) A
Trizna, Dennis Benedict
Vilcins, Gunars
Watt, William Stewart

WASHINGTON
Booker, John Ratcliffe
Carlson, Frederick Paul
Cochran, James Alan
Craig, Richard Anderson
Currah, Walter E
Darling, Robert Bruce
Frasco, David Lee
Fremouw, Edward Joseph
Heacock, Richard Ralph
Holzworth, Robert H, II
Ishimaru, Akira
Lam, John Ling-Yee
Mullen, Anthony J
Nalos, Ervin Joseph
Olsen, Robert Gerner
Rimbey, Peter Raymond
Walton, Vincent Michael

WEST VIRGINIA
Bleil, David Franklin
Jefimenko, Oleg D

WISCONSIN
Churchwell, Edward Bruce
Hummert, George Thomas
Kobiske, Ronald Albert
Seshadri, Sengadu Rangaswamy
Wilson, Volney Colvin

WYOMING
McDonald, Francis Raymond
Rinehart, Edgar A

ALBERTA
James, Christopher Robert
Leahy, Denis Alan

BRITISH COLUMBIA
Arrott, Anthony
Bittner, John William
Jull, Edward Vincent
Kerr, J(ames) S(anford) Stephenson
Russell, Richard Doncaster
Thomson, David James

NEW BRUNSWICK
Verma, Ram D

NOVA SCOTIA
Bhartia, Prakash
Calkin, Melvin Gilbert

ONTARIO
Chow, Yung Leonard
Couture, Josaphat Michel
Ehrman, Joachim Benedict
Hackam, Reuben
Hawton, Margaret H
Holden, Thomas More
Lee-Whiting, Graham Edward
Moule, David
Mungall, Allan George
Pintar, M(ilan) Mik
Ryan, Dave
Vanier, Jacques

QUEBEC
Boivin, Alberic
Bolton, Richard Andrew Ernest
Bose, Tapan Kumar
Cabana, Aldee
Gagne, Jean-Marie
Ghosh, Asoke Kumar
Lorrain, Paul
Maruvada, P Sarma
Schwelb, Otto
Shkarofsky, Issie Peter
Subramanian, Sesha
Waksberg, Armand L

SASKATCHEWAN
Kos, Joseph Frank
Rummens, F H A

OTHER COUNTRIES
Azbel, Mark
Box, Michael Allister
Braginski, Aleksander Ignace
El-Bayoumi, Mohamed Ashraf
Gauster, Wilhelm Friedrich
Gindsberg, Joseph
Hagfors, Tor
Kagawa, Yukio
Kirby, Richard C(yril)
Kurokawa, Kaneyuki
Lapostolle, Pierre Marcel
Lindell, Ismo Veikko
Okamura, Sogo
Rozzi, Tullio
Shtrikman, Shmuel
Struzak, Ryszard G
Sugaya, Hiroshi
Vander Vorst, Andre
Yanabu, Satoru

Elementary Particle Physics

ALABAMA
Baksay, Laszlo Andras
Biggs, Albert Wayne
Coulter, Philip W
Edlin, George Robert
Harms, Benjamin C
Parker, Donald Lester
Wu, Xizeng

ARIZONA
Bowen, Theodore
Fisher, William Gary
Jenkins, Edgar William
Just, Kurt W
Kaufmann, William B
Nigam, Bishan Perkash
Rafelski, Johann
Scadron, Michael David
Thews, Robert L(eroy)

ARKANSAS
Harter, William George
Lieber, Michael

CALIFORNIA
Abrams, Gerald Stanley
Alvarez, Raymond Angelo, Jr
Anderson, Leonard Mahlon
Antolak, Arlyn Joe
Ash, William Wesley
Ashford, Victor Aaron
Aurilia, Antonio
Banner, David Lee
Barnett, R(alph) Michael
Bars, Itzhak
Bauer, Daniel A
Bennett, Glenn Taylor
Berardo, Peter Antonio
Berdjis, Fazlollah
Bingham, Harry H, Jr
Bintinger, David L
Birge, Robert Walsh
Blankenbecler, Richard
Blinn, James Frederick
Bloom, Elliott D
Breakstone, Alan M
Brown, Karl Leslie
Byers, Nina
Cahn, Robert Nathan
Caldwell, David Orville
Cary, Arthur Simmons
Cautis, C Victor
Chadwick, George Brierley
Chamberlain, Owen
Chanowitz, Michael Stephen
Chao, Alexander Wu
Chapline, George Frederick, Jr
Chau, Ling-Lie
Chow, Brian G(ee-Yin)
Cladis, John Baros
Conzett, Homer Eugene
Cork, Bruce
Cornwall, John Michael
Cottrell, Roger Leslie Anderton
Coward, David Hand
Coyne, Donald Gerald
Dahl, Orin I
Day, Thomas Brennock
Desai, Bipin Ratilal
Dorfan, David Elliot
Ecklund, Stanley Duane
Eisner, Alan Mark
Elioff, Thomas
Ely, Robert P, Jr
Epstein, Gary Martin
Erdmann, John Hugo
Finkelstein, Jerome
Fischer, Gerhard Emil
Flatte, Stanley Martin
Foote, James Herbert
Frautschi, Steven Clark
Freedman, Stuart Jay
Fryberger, David
Gaillard, Mary Katharine
Garwin, Charles A

Gearhart, Roger A
Gidal, George
Gillespie, George H
Godfrey, Gary Lunt
Goldhaber, Gerson
Good, Roland Hamilton, Jr
Gorelick, Jerry Lee
Gould, Harvey Allen
Gregorich, David Tony
Groom, Donald Eugene
Gross, Mark Warren
Gunion, John Francis
Haber, Howard Eli
Hall, Lawrence John
Hansen, Luisa Fernandez
Hartsough, Walter Douglas
He, Yudong
Helm, Richard H
Helmholz, August Carl
Hendry, George Orr
Heusch, Clemens August
Hewett, Joanne Lea
Hill, Robert Dickson
Hitlin, David G
Horowitz, Gary T
Innes, Walter Rundle
Jackson, John David
Jacobs, Peter Martin
Jones, Lewis Hammond, IV
Kadyk, John Amos
Kendall, Burton Nathaniel
Kenney, Robert Warner
Kernan, Anne
Kiskis, Joseph Edward, Jr
Ko, Winston Tai-Kan
Kohler, Donald Alvin
Kropp, William Rudolph, Jr
Krueger, Roland Frederick
Lambertson, Glen Royal
Leith, David W G S
Linde, Andrei
Lofgren, Edward Joseph
Loken, Stewart Christian
Loos, Hendricus G
Lu, Adolph
Lys, Jeremy Eion Alleyne
MacGregor, Malcolm Herbert
Madaras, Ronald John
Maglich, Bogdan
Mandelkern, Mark Alan
Manohar, Aneesh
Martin, Jim Frank
Marx, Jay Neil
Masek, George Edward
Mast, Terry S
Mathews, Grant James
Matis, Howard S
Mayer, Meinhard Edwin
Meyer, Thomas
Moe, Michael K
Monard, Joyce Anne
Morrison, Rollin John
Mozley, Robert Fred
Muller, Richard A
Murad, Emil Moise
Nauenberg, Michael
Nefkens, Bernard Marie
Nelson, Jerry Earl
Neugebauer, Gerry
Newman, David Edward
Noyes, H Pierre
Oddone, Piermaria Jorge
Pan, Yu-Li
Pang, Chan Yueh
Parker, Sherwood
Parker, Winifred Ellis
Paterson, James McEwan
Peccei, Roberto Daniele
Peck, Charles William
Perez-Mendez, Victor
Perl, Martin Lewis
Peskin, Michael Edward
Peterson, Jack Milton
Phinney, Nanette
Piccioni, Oreste
Plain, Gilbert John
Prescott, Charles Young
Preskill, John P
Price, Paul Buford, Jr
Primack, Joel Robert
Pugh, Howel Griffith
Quinn, Helen Rhoda Arnold
Rabin, Jeffrey Mark
Rarita, William Roland
Ratcliff, Blair Norman
Reines, Frederick
Richter, Burton
Roehrig, Jimmy Richard
Ronan, Michael Thomas
Ross, Marc Christopher
Ross, Marvin Franklin
Ross, Ronald Rickard
Salter, Robert Munkhenk, Jr
Schalk, Terry LeRoy
Schechter, Martin
Schroeder, Lee S
Schumann, Thomas Gerald
Schwarz, John Henry
Shapiro, Bernard Lyon
Shapiro, Gilbert
Sherden, David J
Shin, Ernest Eun-Ho
Siemann, Robert Herman
Silverman, Dennis Joseph

Slater, William E
Sobel, Henry Wayne
Spencer, Cherrill Melanie
Spencer, James Eugene
Spieler, Helmuth
Steiner, Herbert M
Stevenson, Merlon Lynn
Stork, Donald Harvey
Stowe, Keith S
Strovink, Mark William
Suri, Ashok
Suzuki, Mahiko
Swanson, Robert Allan
Swanson, William Paul
Taylor, Frank Eugene
Taylor, Richard Edward
Thompson, John Robert
Ticho, Harold Klein
Trilling, George Henry
Trippe, Thomas Gordon
Tsai, Yung Su
Van Bibber, Karl Albert
Vega, Roberto
Vernon, C(arl) Wayne
Waltz, Ronald Edward
Watt, Robert Douglas
Weinstein, Alan Jay
Weinstein, Berthold Werner
Weiss, Jeffrey Martin
Wenzel, William Alfred
Whitehead, Marian Nedra
Whitten, Charles A, Jr
Wilson, Perry Baker
Winkelmann, Frederick Charles
Witherell, Michael Stewart
Wojcicki, Stanley G
Wolff, Milo Mitchell
Wong, Chun Wa
Wong, David Yue
Wright, Jon Alan
Yager, Philip Marvin
Yearian, Mason Russell
Yellin, Steven Joseph
Zachariasen, Fredrik

COLORADO
Barut, Asim Orhan
Basri, Saul Abraham
Blackmon, Maurice Lee
Chang, Bunwoo Bertram
Chew, Herman W
DeGrand, Thomas Alan
Downs, Bertram Wilson, Jr
Heidenr, R D
Henley, John Raymond
Kiehl, Jeffrey Theodore
McAllister, Robert Wallace
Mahanthappa, Kalyana T
Mohapatra, Pramoda Kumar
Nauenberg, Uriel
Pawlicki, Anthony Joseph
Randa, James P
Sauer, Jon Robert
Smith, Richard Cecil

CONNECTICUT
Adair, Robert Kemp
Baltay, Charles
Chodos, Alan A
Greenberg, Jack Sam
Gürsey, Feza
Kasha, Henry
Kraybill, Henry Lawrence
Krisst, Raymond John
Liberman, Arthur David
Lubell, Michael S
MacDowell, Samuel Wallace
Shankar, Ramamurti
Sommerfeld, Charles Michael
Storrs, Charles Lysander
Zeller, Michael Edward

DELAWARE
Barnhill, Maurice Victor, III
Gaisser, Thomas Korff
Leung, Chung Ngoc

DISTRICT OF COLUMBIA
Berley, David
Fainberg, Anthony
Fricken, Raymond Lee
Kelly, Francis Joseph
Kinzer, Robert Lee
Matsakis, Demetrios Nicholas
O'Fallon, John Robert
Parke, William C
Photiadis, Douglas Marc
Riemer, Robert Lee
Shapero, Donald Campbell
Shapiro, Anatole Morris
Wallenmeyer, William Anton
Willard, Daniel

FLORIDA
Albright, John Rupp
Baer, Howard A
Curtright, Thomas Lynn
Dagotto, Elbio Ruben
Darling, Byron Thorwell
Duke, Dennis Wayne
Field, Richard D
Goldman, Joel Harvey
Hagopian, Vasken
Johnson, Joseph Andrew, III

Kennedy, Robin John
Kimel, Jacob Daniel, Jr
Lannutti, Joseph Edward
Marshak, Robert Eugene
Mintz, Stephen Larry
Nepomechie, Rafael Israel
Owens, Joseph Francis, III
Perlmutter, Arnold
Plendl, Hans Siegfried
Ramond, Pierre Michel
Sakmar, Ismail Aydin
Thorn, Charles Behan, III
Trubey, David Keith
Walker, James King
Woodard, Richard P

GEORGIA
Belinfante, Johan G F
Chou, T(su) T(eh)

HAWAII
Learned, John Gregory
Olsen, Stephen Lars
Peters, Michael Wood
Peterson, Vincent Zetterberg
Tata, Xerxes Ramyar
Yount, David Eugene

IDAHO
Luke, Robert A
Parker, Barry Richard

ILLINOIS
Abrams, Robert Jay
Albright, Carl Howard
Anderson, Kelby John
Ankenbrandt, Charles Martin
Appel, Jeffrey Alan
Auvil, Paul R, Jr
Ayres, David Smith
Baker, Winslow Furber
Bardeen, William A
Bart, George Raymond
Bartlett, J Frederick
Berge, Jon Peter
Berger, Edmond Louis
Block, Martin M
Brown, Bruce Claire
Brown, Charles Nelson
Burnstein, Ray A
Carey, David Crockett
Carrigan, Richard Alfred, Jr
Carroll, John Terrance
Chang, Shau-Jin
Cooper, John Wesley
Cooper, William Edward
Cossairt, Jack Donald
Curtis, Cyril Dean
Derrick, Malcolm
Eartly, David Paul
Edwards, Helen Thom
Eichten, Estia Joseph
Eisenstein, Robert Alan
Elwyn, Alexander Joseph
Errede, Steven Michael
Fields, Thomas Henry
Frisch, Henry Jonathan
Gelfand, Norman Mathew
Gladding, Gary Earle
Goldberg, Howard S
Gormley, Michael Francis
Griffin, James Edward
Groves, Thomas Hoopes
Hagstrom, Ray Theodore
Harvey, Jeffrey Alan
Hedin, David Robert
Hill, Christopher T
Hoff, Gloria Thelma (Alburene)
Hojvat, Carlos F
Holloway, Leland Edgar
Hyman, Lloyd George
Jensen, Douglas Andrew
Johnson, David
Johnson, Marvin Elroy
Johnson, Porter W
Jones, Lorella Margaret
Joshi, Umeshwar Prasad
Jostlein, Hans
Jovanovic, Drasko D
Kaplan, Daniel Moshe
Kephart, Robert David
Kolb, Edward William
Kovacs, Eve Veronika
Lennox, Arlene Judith
MacLachlan, James Angell
Malamud, Ernest I(lya)
Mantsch, Paul Matthew
Margulies, Seymour
Marriner, John P
Miller, Robert Carl
Moore, Craig Damon
Muller, Dietrich
Murphy, C(harles) Thornton
Nash, Edward Thomas
Nezrick, Frank Albert
Nodulman, Lawrence Jay
Oehme, Reinhard
Oreglia, Mark Joseph
Papanicolas, Costas Nicolas
Peoples, John, Jr
Pewitt, Edward Gale
Phillips, Thomas James
Pilcher, James Eric
Preston, Richard Swain

Price, Lawrence Edward
Pruss, Stanley McQuaide
Quigg, Chris
Raja, Rajendran
Redmount, Ian H
Rey, Charles Albert
Ringo, George Roy
Roberts, Arthur
Rosner, Jonathan Lincoln
Rubinstein, Roy
Sachs, Robert Green
Sard, Robert Daniel
Schluter, Robert Arvel
Schreiner, Philip Allen
Schultz, Peter Frank
Segel, Ralph E
Shochet, Melvyn Jay
Smart, Wesley Mitchell
Smith, Richard Paul
Solomon, Julius
Spinka, Harold M
Stanfield, Kenneth Charles
Stutte, Linda Gail
Sugano, Katsuhito
Sukhatme, Uday Pandurang
Swallow, Earl Connor
Swamy, Padmanabha Narayana
Teng, Lee Chang-Li
Thaler, Jon Jacob
Theriot, Edward Dennis, Jr
Thron, Jonathan Louis
Tollestrup, Alvin V
Toppel, Bert Jack
Toy, William W
Treadwell, Elliott Allen
Turkot, Frank
Uretsky, Jack Leon
Wagner, Robert G
Wah, Yau Wai
Ward, Charles Eugene Willoughby
Wattenberg, Albert
Wehmann, Alan Ahlers
White, Herman Brenner, Jr
Wicklund, Arthur Barry
Willis, Suzanne Eileen
Winston, Roland
Yamada, Ryuji
Yoh, John K
Yokosawa, Akihiko
Young, Donald Edward
Yovanovitch, Drasko D
Zachos, Cosmas K

INDIANA
Barnes, Virgil Everett, II
Bishop, James Martin
Brabson, Bennet Bristol
Cason, Neal M
Fischbach, Ephraim
Gaidos, James A
Garfinkel, Arthur Frederick
Gottlieb, Steven Arthur
Hanson, Gail G
Heinz, Richard Meade
Hendry, Archibald Wagstaff
Koetke, Donald D
Kuo, Tzee-Ke
Loeffler, Frank Joseph
LoSecco, John M
McIlwain, Robert Leslie, Jr
Martin, Hugh Jack, Jr
Miller, David Harry
Ogren, Harold Olof
Poirier, John Anthony
Ruchti, Randal Charles
Sapirstein, Jonathan Robert
Shephard, William Danks
Shibata, Edward Isamu
Tubis, Arnold
Westgard, James Blake
Willmann, Robert B
Wills, John G
Yoder, Neil Richard

IOWA
Firestone, Alexander
Hagelin, John Samuel
Hammer, Charles Lawrence
Intemann, Gerald William
Lamb, Richard C
Lassila, Kenneth Eino
Leacock, Robert A
Pursey, Derek Lindsay
Rizzo, Thomas Gerard
Rosenberg, Eli Ira
Smith, Paul Aiken
Young, Bing-Lin

KANSAS
Ammar, Raymond George
Davis, Robin Eden Pierre
Kwak, Nowhan
Munczek, Herman J
Ralston, John Peter

KENTUCKY
Elitzur, Moshe
Filaseta, John Eugene
France, Peter William

LOUISIANA
Courtney, John Charles
Haymaker, Richard Webb
Head, Charles Everett

Elementary Particle Physics (cont)

Head, Martha E Moore
Imlay, Richard Larry
Slaughter, Milton Dean

MAINE
Pilot, Christopher H

MARYLAND
Chang, Chung Yun
Charlton, Gordon Randolph
Chien, Chih-Yung
Comiso, Josefino Cacas
Domokos, Gabor
Dragt, Alexander James
Field, Herbert Cyre
Gates, Sylvester J, Jr
Greenberg, Oscar Wallace
Groves, Eric Stedman
Hauser, Michael George
Hildebrand, Bernard
Hu, Bei-Lok Bernard
Kacser, Claude
Kovesi-Domokos, Susan
Marx, Egon
Messing, Fredric
Mohapatra, Rabindra Nath
Peaslee, David Chase
Richardson, Clarence Robert
Romanowski, Thomas Andrew
Schewe, Phillip Frank
Schumacher, Clifford Rodney
Silberberg, Rein
Snow, George Abraham
Soln, Josip Zvonimir
Stecker, Floyd William
Taragin, Morton Frank
Toll, John Sampson
Yaes, Robert Joel
Zorn, Gus Tom

MASSACHUSETTS
Abbott, Laurence Frederick
Bar-Yam, Zvi H
Bensinger, James Robert
Brehm, John Joseph
Brenner, John Francis
Busza, Wit
Button-Shafer, Janice
Carter, Ashton Baldwin
Celmaster, William Noah
Chivukula, R Sekhar
Cook, LeRoy Franklin, Jr
Decowski, Piotr
Deser, Stanley
Dhar, Sachidulal
Donoghue, John Francis
Dowd, John P
Everett, Allen Edward
Faissler, William L
Farhi, Edward
Fazio, Giovanni Gene
Feld, Bernard Taub
Feldman, Gary Jay
Ford, Lawrence Howard
Friedman, Jerome Isaac
Friedman, Marvin Harold
George, James Z
Ginsberg, Edward S
Glashow, Sheldon Lee
Glaubman, Michael Juda
Godine, John Elliott
Goldberg, Hyman
Goldstone, Jeffrey
Guertin, Ralph Francis
Guth, Alan Harvey
Hafen, Elizabeth Susan Scott
Horwitz, Paul
Hulsizer, Robert Inslee, Jr
Hunter, Larry Russel
Jacobs, Laurence Alan
Jaffe, Robert Loren
Jagannathan, Kannan
Johnson, Walter Hudson
Kafka, Tomas
Kern, Wolfhard
Kinoshita, Kay
Kreisler, Michael Norman
Law, Margaret Elizabeth
Lipshutz, Nelson Richard
Mann, William Anthony
Marino, Richard Matthew
Matthews, June Lorraine
Mellen, Walter Roy
Milbocker, Michael
Milburn, Richard Henry
Miller, James Paul
Nath, Pran
Nicholson, Howard White, Jr
Oliver, William Parker
Osborne, Louis Shreve
Pipkin, Francis Marion
Pless, Irwin Abraham
Rabin, Monroe Stephen Zane
Ram-Mohan, L Ramdas
Ramsey, Norman Foster, Jr
Rebbi, Claudio
Redwine, Robert Page
Reucroft, Stephen
Rizzoli-Malanotte, Paola
Roberts, Bradley Lee

Rosenson, Lawrence
Rothwell, Paul L
Schneps, Jack
Shambroom, Wiliam David
Simmons, Elizabeth H
Strauch, Karl
Strelzoff, Alan G
Sulak, Lawrence Richard
Sutton, Christopher Sean
Ting, Samuel C C
Tuchman, Avraham
Uritam, Rein Aarne
Vaughn, Michael Thayer
Von Goeler, Eberhard
Wang, Chia Ping
Wijangco, Antonio Robles
Wilk, Leonard Stephen
Wu, Tai Tsun
Yamamoto, Richard Kumeo
Yamartino, Robert J

MICHIGAN
Abbasabadi, Alireza
Abolins, Maris Arvids
Akerlof, Carl W
Chen, Hsuan
Einhorn, Martin B
Francis, William Porter
Freese, Katherine
Gupta, Suraj Narayan
Gustafson, Herold Richard
Jones, Lawrence William
Krisch, Alan David
Krisch, Jean Peck
Longo, Michael Joseph
Miller, Robert Joseph
Neal, Homer Alfred
Overseth, Oliver Enoch
Pope, Bernard G
Raban, Morton
Reitan, Daniel Kinseth
Repko, Wayne William
Roe, Byron Paul
Rolnick, William Barnett
Rossi, Giuseppe
Savit, Robert Steven
Tarle, Gregory
Thun, Rudolf Paul
Tung, Wu-Ki
Wu, Dolly Y

MINNESOTA
Austin, Donald Murray
Courant, Hans Wolfgang Julius
Erickson, Kenneth Neil
Gasiorowicz, Stephen G
Heller, Kenneth Jeffrey
Jones, Roger Stanley
Marshak, Marvin Lloyd
Peterson, Earl Andrew
Rudaz, Serge

MISSISSIPPI
Lott, Fred Wilbur, III
Reidy, James Joseph

MISSOURI
Huang, Justin C

MONTANA
Robinson, Clark Shove, Jr

NEBRASKA
Finkler, Paul
Joseph, David Winram

NEVADA
Hudson, James Gary

NEW HAMPSHIRE
Shepard, Harvey Kenneth

NEW JERSEY
Banks, Thomas
Bechis, Dennis John
Brucker, Edward Byerly
Budny, Robert Vierling
Conway, John Stephen
Dao, Fu Tak
Devlin, Thomas J
Edwards, Christopher Andrew
Gabbe, John Daniel
Goldin, Gerald Alan
Goodman, Gerald Joseph
Goodman, Mark William
Hodges, Hardy M
Hughes, James Sinclair
Isbrandt, Lester Reinhardt
Kalelkar, Mohan Satish
Klebanov, Igor Romanovich
Koller, Earl Leonard
Kwo, Jueinai
Lemonick, Aaron
Li, Yuan
Loman, James Mark
McDonald, Kirk Thomas
Michelson, Leslie Paul
Piroue, Pierre Adrien
Plano, Richard James
Poucher, John Scott
Purohit, Milind Vasant
Raghavan, Ramaswamy Srinivasa
Raychaudhuri, Kamal Kumar
Reynolds, George Thomas

Rockmore, Ronald Marshall
Sannes, Felix Rudolph
Schivell, John Francis
Shapiro, Joel Alan
Shoemaker, Frank Crawford
Smith, Arthur John Stewart
Socolow, Robert H(arry)
Spergel, David Nathaniel
Stickland, David Peter
Watts, Terence Leslie
Weinrich, Marcel

NEW MEXICO
Baltrusaitis, Rose Mary
Barnes, Peter David
Bowman, James David
Brolley, John Edward, Jr
Bryant, Howard Carnes
Burleson, George Robert
Burman, Robert L
Coon, Sidney Alan
Cooper, Frederick Michael
Cooper, Martin David
Ginsparg, Paul H
Glass, George
Goldman, Terrence Jack
Gray, Edward Ray
Hoffman, Cyrus Miller
Hutson, Richard Lee
Ingraham, Richard Lee
Jarmie, Nelson
Johnson, Mikkel Borlaug
Jones, James Jordan
King, Nicholas S P
Leeper, Ramon Joe
MacArthur, Duncan W
Macek, Robert James
McNaughton, Michael Walford
Mischke, Richard E
Peng, Jen-chieh
Reinfelds, Juris
Robertson, Robert Graham Hamish
Sanders, Gary Hilton
Shafer, Robert E
Shively, Frank Thomas
Simmons, James E
Simmons, Leonard Micajah, Jr
Tanaka, Nobuyuki
Thiessen, Henry Archer
Toepfer, Alan James
Turner, Leaf
White, David Hywel
Williams, Robert Allen
Wolfe, David M
Zemach, Charles

NEW YORK
Amato, Joseph Cono
Aronson, Samuel Harry
Auerbach, Elliot H
Baltz, Anthony John
Behrends, Ralph Eugene
Berkelman, Karl
Bermon, Stuart
Bharadwaj, Prem Datta
Blumberg, Leroy Norman
Bodek, Arie
Bowick, Mark John
Bozoki, George Edward
Brown, Stanley Gordon
Bunce, Gerry Michael
Cassel, David Giske
Chang, Ngee Pong
Chen, James Ralph
Chien, Robert Edward
Chung, Kuk Soo
Chung, Suh Urk
Chung, Victor
Cole, James A
Creutz, Michael John
Das, Ashok Kumar
Dawson, Sally
Dean, Nathan Wesley
Dell, George F, Jr
Devons, Samuel
Duek, Eduardo Enrique
Eckert, Richard Raymond
Etkin, Asher
Feigenbaum, Mitchell Jay
Feingold, Alex Jay
Feldman, Lawrence
Ferbel, Thomas
Fernow, Richard Clinton
Foelsche, Horst Wilhelm Julius
Foley, Kenneth John
Fox, Geoffrey Charles
Franco, Victor
Friedberg, Carl E
Genova, James John
Gibbard, Bruce
Gittelman, Bernard
Goldberg, Marvin
Gollon, Peter J
Gordon, Howard Allan
Goulianos, Konstantin
Grannis, Paul Dutton
Greene, Arthur Franklin
Haggerty, John S
Hahn, Richard Leonard
Hand, Louis Neff
Harrington, Steven Jay
Hartmann, Francis Xavier
Herbert, Marc L
Hough, Paul Van Campen

Jain, Pyare Lal
Johnson, Brant Montgomery
Kalogeropoulos, Theodore E
Koplik, Joel
Kramer, Martin A
Kramer, Stephen Leonard
Kycia, Thaddeus F
Lee, Wonyong
Lee, Yong Yung
Leipuner, Lawrence Bernard
Lindenbaum, S(eymour) J(oseph)
Lindquist, William Brent
Lissauer, David Arie
Littauer, Raphael Max
Littenberg, Laurence Stephen
Lobkowicz, Frederick
Longacre, Ronald Shelley
Love, William Alfred
Lowenstein, Derek Irving
McDaniel, Boyce Dawkins
McDonough, James Edward
Mandel, Leonard
Mane, Sateesh Ramchandra
Marciano, William Joseph
Marx, Michael David
Mathur, Vishnu Sahai
Melissinos, Adrian Constantin
Mincer, Allen I
Mistry, Nariman Burjor
Moneti, Giancarlo
Morse, William M
Mukhopadhyay, Nimai Chand
Nagamiya, Shoji
Nelson, Charles Arnold
Okubo, Susumu
Orear, Jay
Ozaki, Satoshi
Paige, Frank Eaton, Jr
Parsa, Zohreh
Peierls, Ronald F
Pile, Philip H
Polychronakos, Venetios Alexander
Protopopescu, Serban Dan
Rahm, David Charles
Ratner, Lazarus Gershon
Rau, R Ronald
Rehak, Pavel
Ring, James Walter
Rohrlich, Fritz
Rosenzweig, Carl
Roth, Benjamin
Ruggiero, Alessandro Gabriele
Sachs, Allan Maxwell
Sadoff, Ahren J
Sakitt, Mark
Schamberger, Robert Dean
Schechter, Joseph M
Schlitt, Dan Webb
Schwarz, Cindy Beth
Sciulli, Frank J
Silverman, Albert
Slattery, Paul Francis
Smolin, Lee
Souder, Paul A
Spergel, Martin Samuel
Stachel, Johanna
Stearns, Brenton Fisk
Stoler, David
Stoler, Paul
Stone, Sheldon Leslie
Story, Harold S
Strand, Richard Carl
Sumner, Richard Lawrence
Sun, Chih-Ree
Swartz, Clifford Edward
Tannenbaum, Michael J
Taylor, Snowden
Thern, Royal Edward
Thorndike, Alan Moulton
Thorndike, Edward Harmon
Tigner, Maury
Trueman, Thomas Laurence
Tryon, Edward Polk
Tye, Sze-Hoi Henry
Vortuba, Jan
Wanderer, Peter John, Jr
Weinberg, Erick James
Wellner, Marcel
Weng, Wu Tsung
Woody, Craig L
Yamin, Samuel Peter
Yan, Tung-Mow
Yeh, Noel Kuei-Eng
Yuan, Luke Chia Liu
Zahed, Ismail
Zwanziger, Daniel

NORTH CAROLINA
Cotanch, Stephen Robert
Dayton, Benjamin Bonney
Dolan, Louise Ann
Dotson, Allen Clark
Fortney, Lloyd Ray
Frampton, Paul Howard
Fulp, Ronald Owen
Good, Wilfred Manly
Goshaw, Alfred Thomas
Han, Moo-Young
Ji, Chueng Ryong
Meisner, Gerald Warren
Petersen, Karl Endel
Spielvogel, Bernard Franklin
Wolf, Albert Allen

OHIO
Aubrecht, Gordon James, II
Duval, Leonard A
Fickinger, William Joseph
Gan, K K
Giamati, Charles C, Jr
Huwe, Darrell O
Johnson, Randy Allan
Kowalski, Kenneth L
Krauss, Lawrence Maxwell
Ling, Ta-Yung
Nussbaum, Mirko
Raby, Stuart
Reibel, Kurt
Robinson, Donald Keith
Rodman, James Purcell
Suranyi, Peter
Tanaka, Katsumi
Taylor, Beverley Ann Price
Witten, Louis

OKLAHOMA
Burwell, James Robert
Carlstone, Darry Scott
Howard, Robert Adrian
Kalbfleisch, George Randolph
Milton, Kimball Alan
Skubic, Patrick Louis

OREGON
Alston-Garnjost, Margaret
Brau, James Edward
Csonka, Paul L
Deshpande, Nilendra Ganesh
Griffiths, David Jeffery
Hwa, Rudolph Chia-Chao
Khalil, M Aslam Khan
Moravcsik, Michael Julius
Richert, Anton Stuart
Siemens, Philip John
Sivers, Dennis Wayne
Soper, Davison Eugene
White, Donald Harvey

PENNSYLVANIA
Beier, Eugene William
Carlitz, Robert D
Cleland, Wilfred Earl
Collins, John Clements
Cutkosky, Richard Edwin
Edelstein, Richard Malvin
Fetkovich, John Gabriel
Frankel, Sherman
Franklin, Jerrold
Frati, William
Goldschmidt, Yadin Yehuda
Highland, Virgil Lee
Hones, Michael J
Kanofsky, Alvin Sheldon
Kraemer, Robert Walter
Lande, Kenneth
Langacker, Paul George
Mann, Alfred Kenneth
Potter, Douglas Marion
Ridener, Fred Louis, Jr
Roskies, Ralph Zvi
Rovelli, Carlo
Russ, James Stewart
Selove, Walter
Shepard, Paul Fenton
Smith, Gerald A
Steinhardt, Paul Joseph
Sutton, Roger Beatty
Thompson, Julia Ann
White, George Rowland
Williams, Hugh Harrison

RHODE ISLAND
Cutts, David
Lanou, Robert Eugene, Jr
McCorkle, Richard Anthony
Widgoff, Mildred

SOUTH CAROLINA
Darden, Colgate W, III
Gurr, Henry S
Louttit, Robert Irving
Rosenfeld, Carl

TENNESSEE
Bartelt, John Eric
Beene, James Robert
Bugg, William Maurice
Condo, George T
Crater, Horace William
Hart, Edward Leon
Loebbaka, David S
Panvini, Robert S
Roos, Charles Edwin
Sears, Robert F, Jr
Ward, Bennie Franklin Leon
Weiler, Thomas Joseph
Young, Glenn Reid

TEXAS
Baggett, Neil Vance
Ball, Millicent (Penny)
Barton, Henry Ruwe, Jr
Bryan, Ronald Arthur
Chiu, Charles B
Clark, Robert Beck
Corcoran, Marjorie
Coulson, Larry Vernon
Cutler, Roger T

Dahl, Per Fridtjof
Danburg, Jerome Samuel
Dicus, Duane A
Dugan, Gerald Francis
Espinosa, James Manuel
Fenyves, Ervin J
Gaedke, Rudolph Meggs
Gleeson, Austin M
Huang, Huey Wen
Huson, Frederick Russell
Izen, Joseph M
Kirk, Thomas Bernard Walter
Ko, Che Ming
Lichti, Roger L
Lopez, Jorge Alberto
McIntyre, Peter Mastin
Mahajan, Swadesh Mitter
Meshkov, Sydney
Miller, Philip Dixon
Mutchler, Gordon Sinclair
Northcliffe, Lee Conrad
Olenick, Richard Peter
Olness, Fredrick Iver
Pedigo, Robert Daryl
Quarles, Carroll Adair, Jr
Rees, John Robert
Riley, Peter Julian
Sadler, Michael Ervin
Sanford, James R
Stevenson, Paul Michael
Stiening, Rae Frank
Teplitz, Vigdor Louis
Toohig, Timothy E
Trost, Hans-Jochen
Waggoner, James Arthur
Wagoner, David Eugene
Watson, Jerry M
Webb, Robert Carroll
Weinstein, Roy
Wilson, Thomas Leon

UTAH
Ball, James Stutsman
Bergeson, Haven Eldred
Elbert, Jerome William
Jolley, David Kent
Kieda, David Basil
Loh, Eugene C
O'Halloran, Thomas A, Jr

VIRGINIA
Aitken, Alfred H
Bardon, Marcel
Barrois, Bertrand C
Blanpied, William Antoine
Blecher, Marvin
Brenner, Alfred Ephraim
Bridgewater, Albert Louis, Jr
Buck, Warren W, III
Chang, Lay Nam
Cox, Bradley Burton
Cox, Richard Harvey
Diebold, Robert Ernest
Eckhause, Morton
Ficenec, John Robert
Ijaz, Mujaddid A
Isgur, Nathan
Jenkins, David A
Johnson, Rolland Paul
Kabir, Prabahan Kemal
Kayser, Boris Jules
Kuang, Yunan
Lynch, John Thomas
McFarlane, Kenneth Walter
Minehart, Ralph Conrad
Peacock, Richard Wesley
Perdrisat, Charles F
Sager, Earl Vincent
Schremp, Edward Jay
Shapiro, Maurice Mandel
Siegel, Robert Ted
Sinclair, Charles Kent
Small, Timothy Michael
Spivack, Harvey Marvin
Sundelin, Ronald M
Thacker, Harry B
Trower, W(illiam) Peter
Weber, Hans Jurgen
Welsh, Robert Edward
Winter, Rolf Gerhard
Ziock, Klaus Otto H

WASHINGTON
Albers, James Ray
Barrett, William Louis
Bichsel, Hans
Brodzinski, Ronald Lee
Burnett, Thompson Humphrey
Clifford, Howard James
Ellis, Stephen Dean
Ely, John Thomas Anderson
Fortson, Edward Norval
Harris, Roger Mason
Harrison, W Craig
Haxton, Wick Christopher
Lam, John Ling-Yee
Lord, Jere Johns
Lowenthal, Dennis David
Lubatti, Henry Joseph
Miller, Gerald Alan
Mockett, Paul M
Ruby, Michael Gordon
Sossong, Norman D
Wilkes, Richard Jeffrey

Williams, Robert Walter
Yaffe, Laurence G
Young, Kenneth Kong

WEST VIRGINIA
Weldon, Henry Arthur

WISCONSIN
Balantekin, Akif Baha
Barger, Vernon Duane
Durand, Bernice Black
Durand, Loyal
Ebel, Marvin Emerson
Erwin, Albert R
March, Robert Herbert
Morse, Robert Malcolm
Mortara, Lorne B
Pondrom, Lee Girard
Prepost, Richard
Reeder, Don David
Snider, Dale Reynolds
Wu, Sau Lan Yu

WYOMING
Kunselman, A(rthur) Raymond

ALBERTA
Kamal, Abdul Naim
Leahy, Denis Alan
Torgerson, Ronald Thomas
Umezawa, Hiroomi
Walton, Mark Allan

BRITISH COLUMBIA
Affleck, Ian Keith
Auld, Edward George
Bryman, Douglas Andrew
Davis, Charles Alan
Hausser, Otto Friedrich
Jennings, Byron Kent
Mason, Grenville R
Measday, David Frederick
Petch, Howard Earle
Picciotto, Charles Edward
Robertson, Lyle Purmal
Sample, John Thomas
Viswanathan, Kadayam Sankaran

MANITOBA
Van Oers, Willem Theodorus Hendricus

NEW BRUNSWICK
Armstrong, Robin L
Weil, Francis Alphonse

NOVA SCOTIA
Moriarty, Kevin Joseph

ONTARIO
Barton, Richard Donald
Buchanan, Gerald Wallace
Davis, Ronald Stuart
Dixit, Madhu Sudan
Edwards, Kenneth Westbrook
Ewan, George T
Frisken, William Ross
Hargrove, Clifford Kingston
Huschilt, John
Key, Anthony W
Lee, Hoong-Chien
Lindsey, George Roy
McDonald, Arthur Bruce
Mann, Robert Bruce
Mes, Hans
Nogami, Yukihisa
Prugovecki, Eduard
Resnick, Lazer
Romo, William Joseph
Sherman, Norman K
Simpson, John Joseph
Sundaresan, Mosur Kalyanaraman

QUEBEC
Depommier, Pierre Henri Maurice
Fuchs, Vladimir
Ho-Kim, Quang
Kalman, Calvin Shea
Kroger, Helmut Karl
Maciejko, Roman
Palfrey, Roger Grenville Eric
Patel, Popat-Lal Mulji-bhai
Ryan, David George
Taras, Paul
Terreault, Bernard J E J

SASKATCHEWAN
Kim, Dong Yun
Rangacharyulu, Chilakamarri

OTHER COUNTRIES
Afnan, Iraj Ruhi
Blumenfeld, Henry A
Booth, Norman E
De Bruyne, Pieter
Dominguez, Cesareo Augusto
Heinz, Ulrich Walter
Huang, Keh-Ning
Lipkin, Harry Jeannot
Luckey, Paul David, Jr
Ng, Tai-Kai
Nussenzveig, Herch Moyses
Paschos, Emmanuel Anthony
Powers, Richard James
Salam, Abdus

Satz, Helmut T G
Scott, Mary Jean
Settles, Ronald Dean
Steinberger, Jack
Wu, Yong-Shi

Fluids

ALABAMA
Davis, Anthony Michael John
Fowlis, William Webster
Hathaway, David Henry
Hermann, R(udolf)
Martin, James Arthur
Moore, Ronald Lee
Ramachandran, Narayanan
Rosenberger, Franz
Schutzenhofer, Luke A
Shih, Cornelius Chung-Sheng
Suess, Steven Tyler
Thompson, Byrd Thomas, Jr
Turner, Charlie Daniel, Jr
Visscher, Pieter Bernard
Yeh, Pu-Sen

ARIZONA
Boyer, Don Lamar
Fink, Jonathan Harry
Heinle, Preston Joseph
Kessler, John Otto
Levy, Eugene Howard
Rouse, Hunter
Wygnanski, Israel Jerzy

ARKANSAS
Fletcher, Alan G(ordon)

CALIFORNIA
Adam, Randall Edward
Alder, Berni Julian
Andersen, Wilford Hoyt
Banks, Robert B(lackburn)
Barmatz, Martin Bruce
Berger, Stanley A(llan)
Berryman, James Garland
Bingham, Harry H, Jr
Blackwelder, Ron F
Boercker, David Bryan
Bohachevsky, Ihor O
Bott, Jerry Frederick
Bowyer, J(ames) M(arston), Jr
Buckingham, Alfred Carmichael
Burden, Harvey Worth
Busse, Friedrich Hermann
Caflisch, Russel Edward
Cannell, David Skipwith
Carrigan, Charles Roger
Catton, Ivan
Chahine, Moustafa Toufic
Chang, Howard
Chapman, Gary Theodore
Chase, David Marion
Chen, Yang-Jen
Cheng, H(sien) K(ei)
Cheng, Ralph T(a-Shun)
Chin, Jin H
Chu, Peter Cheng
Cloutman, Lawrence Dean
Coles, Donald (Earl)
Concus, Paul
Corcos, Gilles M
Cross, Ralph Herbert, III
Cunningham, Mary Elizabeth
Daily, James W(allace)
Denn, Morton M(ace)
Dixon, Paul King
Driscoll, Charles F
Duff, Russell Earl
Ellis, Albert Tromly
Evans, Todd Edwin
Farber, Joseph
Foster, Theodore Dean
Frenzen, Christopher Lee
Fu, Lee Lueng
Garrett, Steven Lurie
Garwood, Roland William, Jr
Gibson, Carl H
Gilmore, Forrest Richard
Goldstein, Selma
Hackett, Colin Edwin
Hahn, Kyoung-Dong
Hershey, Allen Vincent
Hesselink, Lambertus
Higgins, Brian Gavin
Ho, Hung-Ta
Hoffmann, Jon Arnold
Hoover, William Graham
Horn, Kenneth Porter
Hornig, Howard Chester
Hsia, Henry Tao-Sze
Ingersoll, Andrew Perry
Jakubowski, Gerald S
Jing, Liu
Johnston, James P(aul)
Kamegai, Minao
Kaplan, Richard E
Kasper, Gerhard
Kelly, Robert Edward
Khurana, Krishan Kumar
Kim, J John
Knobloch, Edgar
Koh, Robert Cy
Kornfield, Julia Ann

Fluids (cont)

Kraus, Samuel
Kroeker, Richard Mark
Kulkarny, Vijay Anand
Kwok, Munson Arthur
Lange, Rolf
Langlois, William Edwin
Lapp, M(arshall)
Lapple, Charles E
Larock, Bruce E
Laurmann, John Alfred
Leith, Cecil Eldon, Jr
Lind, Maurice David
Lindley, Charles A(lexander)
Ling, Rung Tai
Liu, David Shiao-Kung
Lomax, Harvard
Loos, Hendricus G
McKenzie, Robert Lawrence
McMaster, William H
Malmuth, Norman David
Marcus, Bruce David
Marcus, Philip Stephen
Margolis, Stephen Barry
Martin, Elmer Dale
Massier, Paul Ferdinand
Mazurek, Thaddeus John
Mihalov, John Donald
Miller, James Angus
Mitchner, Morton
Munson, Albert G
Muntz, Eric Phillip
Naghdi, P(aul) M(ansour)
Nevins, William McCay
Newcomb, William A
Ozawa, Kenneth Susumu
Parkin, Blaine R(aphael)
Pierucci, Mauro
Platzer, Maximilian Franz
Plotkin, Allen
Powell, Thomas Mabrey
Putterman, Seth Jay
Radbill, John R(ussell)
Rathmann, Carl Erich
Redekopp, Larry G
Roberts, Paul Harry
Rosen, Mordecai David
Rosenkilde, Carl Edward
Roshko, Anatol
Rossow, Vernon J
Rudnick, Joseph Alan
Sabersky, Rolf H(einrich)
Sadeh, Willy Zeev
Saffman, Philip Geoffrey
Sanders, James Vincent
Schubert, Gerald
Seiff, Alvin
Shepherd, Joseph Emmett
Sifferman, Thomas Raymond
Singh, Rameshwar
Skalak, Richard
Somerville, Richard Chapin James
Spencer, Donald Jay
Spera, Frank John
Spreiter, John R(obert)
Steinberg, Daniel J
Sternberg, Joseph
Street, Robert L(ynnwood)
Stuart, George Wallace
Stuhmiller, James Hamilton
Sturtevant, Bradford
Surko, Clifford Michael
Synolakis, Costas Emmanuel
Tellep, Daniel M
Trigger, Kenneth Roy
Tulin, Marshall P(eter)
Utterback, Nyle Gene
Van Atta, Charles W
Van Dyke, Milton D(enman)
Vanyo, James Patrick
Vinokur, Marcel
Wang, H E Frank
Warren, Walter R(aymond), Jr
Wazzan, A R Frank
Wehausen, John Vrooman
Weichman, Peter Bernard
Welge, Henry John
Williams, Forman A(rthur)
Williams, John A(rthur)
Woolard, Henry W(aldo)
Wright, Frederick Hamilton
Wu, Lei
Wu, Theodore Yao-Tsu
Yuan, Sidney Wei Kwun
Ziegenhagen, Allyn James
Zukoski, Edward Edom

COLORADO
Arp, Vincent D
Benton, Edward Rowell
Blumen, William
Bruno, Thomas J
Clifford, Steven Francis
Dryer, Murray
Faddick, Robert Raymond
Gessler, Johannes
Gilman, Peter A
Goldburg, Arnold
Gosink, Joan P
Hanley, Howard James Mason
Hector, David Lawrence
Herring, Jackson Rea
Kantha, Lakshmi

Kassoy, David R
Kerr, Robert McDougall
Krantz, William Bernard
Krueger, David Allen
Lenschow, Donald Henry
Loehrke, Richard Irwin
Massman, William Joseph
Meiss, James Donald
Olien, Neil Arnold
O'Sullivan, William John
Owen, Robert Barry
Snyder, Howard Arthur
Tuttle, Elizabeth R
Weinstock, Jerome

CONNECTICUT
Apfel, Robert Edmund
Barnett, Mark
Burridge, Robert
Chriss, Terry Michael
Foley, William M
Gutierrez-Miravete, Ernesto
Lin, Jia Ding
Mandelbrot, Benoit B
Nigam, Lakshmi Narayan
Paskausky, David Frank
Peracchio, Aldo Anthony
Ramakrishnan, Terizhandur S
Reyna, Luis Guillermo
Similon, Philippe Louis
Smooke, Mitchell D
Sreenivasan, Katepalli Raju
Verdon, Joseph Michael
Von Winkle, William A
Wegener, Peter Paul

DELAWARE
Advani, Suresh Gopaldas
Greenberg, Michael D(avid)
Lauzon, Rodrigue Vincent
Miller, Douglas Charles
Schwartz, Leonard William
Seidel, Barry S(tanley)
Wu, Jin

DISTRICT OF COLUMBIA
Boris, Jay Paul
Colombant, Denis Georges
Fan, Dah-Nien
Huber, Peter William
Kalnay, Eugenia
Kao, Timothy Wu
Klebanoff, P(hilip) S(amuel)
Klein, Max
Oran, Elaine Surick
Pao, Hsien Ping
Picone, J Michael
Rockett, John A
Shen, Colin Yunkang
Sununu, John Henry
Watt, Hame Mamadov
Whang, Yun Chow
Widnall, Sheila Evans

FLORIDA
Anderson, Roland Carl
Bober, William
Buchler, Jean-Robert
Debnath, Lokenath
Dowdell, Rodger B(irtwell)
Howard, Louis Norberg
Hunter, James Hardin, Jr
Jacobson, Ira David
Johnson, Joseph Andrew, III
Krishnamurti, Ruby Ebisuzaki
Kumar, Pradeep
Kurzweg, Ulrich H(ermann)
Le Mehaute, Bernard J
Lindgren, E(rik) Rune
Mooers, Christopher Northrup Kennard
Nersasian, Arthur
Neuringer, Joseph Louis
Olson, Donald B
Sheldon, John William
Sheng, Yea-Yi Peter
Smith, Jerome Allan
Stern, Melvin Ernest
Tam, Christopher K W
Tiederman, William Gregg, Jr
Von Ohain, Hans Joachim
Wang, Hsiang
Weatherly, Georges Lloyd

GEORGIA
Fields, Gerald S
Hackett, James E
Lau, Jark Chong
Martin, Charles Samuel
Neitzel, George Paul
Strahle, Warren C(harles)
Wallace, James Robert
Zinn, Ben T

HAWAII
Cheng, Ping
Chuan, Raymond Lu-Po
Grace, Robert Archibald
Loomis, Harold George
Teng, Michelle Hsiao Tsing

IDAHO
Moore, Kenneth Virgil
Ramshaw, John David

ILLINOIS
Adams, John Rodger
Aref, Hassan
Barcilon, Victor
Beard, Kenneth Van Kirke
Berdichevsky, Alexander
Bond, Charles Eugene
Buckmaster, John David
Chung, Paul M(yungha)
Crawford, Roy Kent
Davis, Stephen H(oward)
Dolan, Paul Joseph, Jr
Fultz, Dave
Garcia, Marcelo Horacio
Georgiadis, John G
Hull, John R
Jacobi, Anthony Mark
Kath, William Lawrence
Kentzer, Czeslaw P(awel)
Kistler, Alan L(ee)
Kuo, Hsiao-Lan
Lebovitz, Norman Ronald
Matalon, Moshe
Maxwell, William Hall Christie
Michelson, Irving
Mokadam, Raghunath G(anpatrao)
Morkovin, Mark V(ladimir)
Olmstead, William Edward
Ottino, Julio Mario
Parker, Eugene Newman
Pearlstein, Arne Jacob
Pericak-Spector, Kathleen Anne
Peters, James Empson
Pierrehumbert, Raymond T
Rosner, Robert
Ryskin, Gregory
Sami, Sedat
Schlossman, Mark Loren
Simmons, Ralph Oliver
Strehlow, Roger Albert
Taam, Ronald Everett
Tankin, Richard S
Yuen, Man-Chuen

INDIANA
Agee, Ernest M(ason)
Ascarelli, Gianni
Gad-el-Hak, Mohamed
Goldschmidt, Victor W
Hoffman, Joe Douglas
McComas, Stuart T
Mueller, Thomas J
Nee, Victor W
Ng, Bartholomew Sung-Hong
Shore, Steven Neil
Snow, John Thomas
Thieme, Melvin T
Thompson, Howard Doyle
Tree, David R
Widener, Edward Ladd, Sr

IOWA
Chen, Lea-Der
Hering, Robert Gustave
Kennedy, John Fisher
Nariboli, Gundo A
Patel, Virendra C
Struck-Marcell, Curtis John
Wilson, Lennox Norwood

KANSAS
Braaten, David A
Connor, Sidney G
Farokhi, Saeed
Greywall, Mahesh S
Jackson, Roscoe George, II
Yu, Yun-Sheng

KENTUCKY
Stewart, Robert Earl
Wood, Don James

LOUISIANA
Baw, Philemon S H
Callens, Earl Eugene, Jr
Courter, R(obert) W(ayne)
Foote, Joe Reeder
Grimsal, Edward George
Hussey, Robert Gregory
Latorre, Robert George
Peattie, Robert Addison
Scott, Richard Royce
Slaughter, Milton Dean
Tohline, Joel Edward
Wenzel, Alan Richard

MAINE
Ingard, Karl Uno
McClymer, James P
McKay, Susan Richards

MARYLAND
Apel, John Ralph
Baer, Ferdinand
Bauer, Peter
Baum, Howard Richard
Bennett, Frederick Dewey
Blake, William King
Bournia, Anthony
Bungay, Peter M
Burns, Timothy John
Chadwick, Richard Simeon
Chan, Kwing Lam
Cleveland, William Grover, Jr
Cronvich, Lester Louis
Crump, Stuart Faulkner
Dick, Richard Dean
Domokos, Gabor
Eaton, Alvin Ralph
Ethridge, Noel Harold
Evans, David L
Ferrell, Richard Allan
Gessow, Alfred
Giddens, Don P(eyton)
Gray, Ernest Paul
Haberman, William L(awrence)
Herczynski, Andrzej
Huang, Thomas Tsung-Tse
Jacobs, Sigmund James
Kaiser, Jack Allen Charles
Lee, David Allan
Lugt, Hans Josef
McCormick, Michael Edward
McFadden, Geoffrey Bey
Mied, Richard Paul
Morgan, William Bruce
Moulton, James Frank, Jr
Nash, Jonathon Michael
Obremski, Henry J(ohn)
O'Brien, Vivian
Olson, Peter Lee
Pai, S(hih) I
Pirraglia, Joseph A
Potzick, James Edward
Randall, John Douglas
Rehm, Ronald George
Reilly, James Patrick
Robbins, Mark Owen
Rogers, David Freeman
Sandusky, Harold William
Schemm, Charles Edward
Schmid, Lawrence Alfred
Semerjian, Hratch G
Sengers, Jan V
Tachmindji, Alexander John
Traugott, Stephen C(harles)
Waldo, George Van Pelt, Jr
Wallace, James M
Webb, Richard A
Yeh, Tsyh Tyan
Yoon, Peter Haesung
Zien, Tse-Fou

MASSACHUSETTS
Abbott, Douglas E(ugene)
Abernathy, Frederick Henry
Appleton, John P(atrick)
Brenner, Howard
Brocard, Dominique Nicolas
Butman, Cheryl Ann
Chen, Sow-Hsin
Cipolla, John William, Jr
Cole, John E(mery), III
Cordero, Julio
Dewan, Edmond M
Dewey, C(larence) Forbes, Jr
Durgin, William W
Emmons, Howard W(ilson)
Fohl, Timothy
Goela, Jitendra Singh
Hallock, Robert B
Hecker, George Ernst
Hoffman, Richard Laird
Hoge, Harold James
Humi, Mayer
Kautz, Frederick Alton, II
Kesner, Jay
Kingston, George C
Klein, Milton M
Lien, Hwachii
Little, Sarah Alden
Madsen, Ole Secher
Martin, Paul Cecil
Mei, Chiang C(hung)
Mikic, Bora
Mollo-Christensen, Erik Leonard
Monson, Peter A
Murman, Earll Morton
Murphy, Peter John
Muthukumar, Murugappan
Neue, Uwe Dieter
Noiseux, Claude Francois
Ogilvie, T(homas) Francis
Phillies, George David Joseph
Pian, Carlson Chao-Ping
Reeves, Barry L(ucas)
Shapiro, Ascher H(erman)
Sonin, Ain A(nts)
Stanley, H(arry) Eugene
Stickler, David Bruce
Tokita, Noboru
Trefethen, Lloyd MacGregor
Vignos, James Henry
Wallace, James
Yamartino, Robert J
Yener, Yaman

MICHIGAN
Beck, Robert Frederick
Boyd, John Philip
Chock, David Poileng
Cutler, Warren Gale
Gulari, Esin
Habib, Izzeddin Salim
Hartman, John L(ouis)
Jacobs, Stanley J
Kordyban, Eugene S
Miklavcic, Milan

Prasuhn, Alan Lee
Roe, Philip Lawrence
Sichel, Martin
Smith, George Wolfram
Stansel, John Charles
Yih, Chia-Shun

MINNESOTA
Abraham, John
Arndt, Roger Edward Anthony
Beavers, Gordon Stanley
Calehuff, Girard Lester
Lysak, Robert Louis
Olson, Reuben Magnus
Reilly, Richard J
Silberman, Edward
Stefan, Heinz G
Uban, Stephen A
Valls, Oriol Tomas
Yuen, David Alexander
Zimmermann, William, Jr

MISSISSIPPI
Bernard, Robert Scales
Coleman, Neil Lloyd
Folse, Raymond Francis, Jr
Gordon, Carroll K
Green, Albert Wise
Horton, Thomas Edward, Jr
Pandey, Ras Bihari

MISSOURI
Bogar, Thomas John
Bower, William Walter
Chen, Ta-Shen
Elgin, Robert Lawrence
Gardner, Richard A
Hakkinen, Raimo Jaakko
Johnson, Darell James
Kibens, Valdis
Kotansky, D(onald) R(ichard)
Liu, Henry
Preckshot, G(eorge) W(illiam)
Roos, Frederick William
Spaid, Frank William
Venezian, Giulio

MONTANA
Hiscock, William Allen

NEW HAMPSHIRE
Baumann, Hans D
Hollweg, Joseph Vincent
Kantrowitz, Arthur (Robert)
Montgomery, David Campbell
Morduchow, Morris
Runstadler, Peter William, Jr
Sonnerup, Bengt Ulf osten

NEW JERSEY
Bachman, Walter Crawford
Barnes, Derek A
Bhattacharya, Sabyasachi
Blumberg, Alan Fred
Brown, Garry Leslie
Caffarelli, Luis Angel
Carlson, Curtis Raymond
Chaikin, Paul Michael
Choi, Byung Chang
Christodoulou, Demetrios
Daly, Ruth Agnes
Doby, Raymond
Edwards, Christopher Andrew
Fornberg, Bengt
Hohenberg, Pierre Claude
Hrycak, Peter
Huang, John S
Hudson, Steven David
Johnston, Christian William
Kangovi, Sach
Kolodner, Paul R
Kovats, Andre
Lee, Wei-Kuo
Levy, Marilyn
Lewellen, William Stephen
Miles, Richard Bryant
Nguyen, Hien Vu
Orszag, Steven Alan
Princen, Henricus Mattheus
Raichel, Daniel R(ichter)
Salwen, Harold
Simpkins, Peter G
Temkin, Samuel
Vann, Joseph M
Williams, Gareth Pierce
Yurke, Bernard
Zabusky, Norman J
Zebib, Abdelfattah M G

NEW MEXICO
Anderson, Robert Alan
Baggett, Lester Marchant
Bankston, Charles A, Jr
Benjamin, Robert Fredric
Bertin, John Joseph
Brackbill, Jeremiah U
Brownell, John Howard
Butler, Thomas Daniel
Cline, Michael Castle
Cox, Arthur Nelson
Daly, Bartholomew Joseph
Deal, William E, Jr
Deupree, Robert G
Erpenbeck, Jerome John

Farnsworth, Archie Verdell, Jr
Fickett, Wildon
Fradkin, David Barry
Fukushima, Eiichi
Glasser, Alan Herbert
Goldman, S Robert
Harlow, Francis Harvey, Jr
Hoffman, Nelson Miles, III
Johnson, James Daniel
Jones, Eric Manning
Julien, Howard L
Kashiwa, Bryan Andrew
Kopp, Roger Alan
Lam, Kin Leung
Lindman, Erick Leroy, Jr
Martin, Richard Alan
Mead, William C
Mjolsness, Raymond C
Mutschlecner, Joseph Paul
Orgill, Montie M
O'Rourke, Peter John
Price, Robert Harold
Ragan, Charles Ellis, III
Rose, Harvey Arnold
Selden, Robert Wentworth
Sicilian, James Michael
Sierk, Arnold John
Travis, John Richard
Wallace, Richard Kent
Wood, William Wayne
Wright, Thomas Payne
Zemach, Charles

NEW YORK
Ackerberg, Robert C(yril)
Acrivos, Andreas
Alpher, Ralph Asher
Archie, Charles Neill
Baker, Norman Hodgson
Blank, Albert Abraham
Brower, William B, Jr
Busnaina, Ahmed A
Chevray, Rene
Childress, William Stephen
De Boer, P(ieter) C(ornelis) Tobias
Feigenbaum, Mitchell Jay
Fesjian, Sezar
Foreman, Kenneth M
Fox, William
Gans, Roger Frederick
Geer, James Francis
George, William Kenneth, Jr
Glimm, James Gilbert
Goodman, Theodore R(obert)
Gould, Robert Kinkade
Halitsky, James
Hall, J(ohn) Gordon
Hartley, Charles LeRoy
Hollenberg, Joel Warren
Holmes, Philip John
Hwang, Shy-Shung
Irvine, Thomas Francis
Johnson, Wallace E
Kaplan, Paul
Keller, D Steven
Kocher, Haribhajan S(ingh)
Koplik, Joel
Kroeger, Peter G
Lai, W(ei) Michael
Lehman, August F(erdinand)
Leibovich, Sidney
Liu, Ching Shi
Liu, Philip L-F
Lumley, John L(eask)
McCrory, Robert Lee, Jr
Marrone, Paul Vincent
Moore, Franklin K(ingston)
Moretti, G(ino)
Nagamatsu, Henry T
O'Brien, Edward E
Parpia, Jeevak Mahmud
Reynolds, John Terrence
Rezanka, Ivan
Rudinger, George
Ruschak, Kenneth John
Savkar, Sudhir Dattatraya
Scala, Sinclaire M(aximilian)
Schwarz, Klaus W
Seyler, Charles Eugene
Shapiro, Paul Jonathon
Siegmann, William Lewis
Siggia, Eric Dean
Stahl, Laddie L
Staub, Fred W
Thomas, John Howard
Thompson, Philip A
Turcotte, Donald Lawson
Valentine, Daniel T
Wang, Lin-Shu
Ward, Lawrence W(aterman)
Weinbaum, Sheldon
Zakkay, Victor

NORTH CAROLINA
Amein, Michael
Arya, Satya Pal
Chandra, Suresh
Christiansen, Wayne Arthur
Doster, Joseph Michael
Dotson, Allen Clark
Edwards, John Auert
Griffith, Wayland C(oleman)
Janowitz, Gerald S(aul)
McRee, Donald Ikerd

Melton, Bill F
Mosberg, Arnold T
Yamamoto, Toshiaki

OHIO
Bodonyi, Richard James
Brazee, Ross D
Bridge, John F(loyd)
Burggraf, Odus R
Chung, Benjamin T F
Deissler, Robert G(eorge)
Edwards, David Olaf
Fenichel, Henry
Fitzmaurice, Nessan
Foster, Michael Ralph
Franke, Milton Eugene
Ghia, Kirti N
Goldstein, Marvin E
Hamed, Awatef A
Herbert, Thorwald
Jaikrishnan, Kadambi Rajgopal
Jeng, Duen-Ren
Keith, Theo Gordon, Jr
Kennedy, Lawrence A
Lawler, Martin Timothy
Leonard, Brian Phillip
Lin, Kuang-Ming
Ostrach, Simon
Pinchak, Alfred Cyril
Polak, Arnold
Prahl, Joseph Markel
Sajben, Miklos
T'ien, James Shaw-Tzuu
Yaney, Perry Pappas
Zaman, Khairul B M Q

OKLAHOMA
Ackerson, Bruce J
Blais, Roger Nathaniel
Cerro, Ramon Luis
Davies-Jones, Robert Peter
Fitch, Ernest Chester, Jr
Howard, Robert Adrian
Shell, Francis Joseph

OREGON
Donnelly, Russell James
Perry, Robert W(illiam)
Photinos, Panos John
Shirazi, Mostafa Ayat
Warren, William Willard, Jr

PENNSYLVANIA
Bau, Haim Heinrich
Blythe, Philip Anthony
Bona, Jerry Lloyd
Carelli, Mario Domenico
Carmi, Shlomo
Chow, Shin-Kien
Cohen, Ira M
Cole, Milton Walter
Crawford, John David
Cutler, Paul H
Dash, Sanford Mark
Daywitt, James Edward
DeSanto, Daniel Frank
Emrich, Raymond Jay
Farn, Charles Luh-Sun
Farouk, Bakhtier
Gollub, Jerry Paul
Hagen, Oskar
Hansen, Robert Jack
Hogenboom, David L
Jhon, Myung S
Kazakia, Jacob Yakovos
Kim, Yong Wook
Lauchle, Gerald Clyde
Lazaridis, Anastas
Liang, Shoudan
Long, Lyle Norman
McDowell, Maurice James
Macpherson, Alistair Kenneth
Mahaffy, John Harlan
Maher, James Vincent
Maynard, Julian Decatur
Merz, Richard A
Miskovsky, Nicholas Matthew
Pierce, Alan Dale
Ramezan, Massood
Ramos, Juan Ignacio
Reed, Joseph Raymond
Reischman, Michael Mack
Rockwell, Donald O
Settles, Gary Stuart
Smith, Wesley R
Tsou, F(u) K(ang)
Whirlow, Donald Kent
Wyngaard, John C
Zemel, Jay N(orman)

RHODE ISLAND
Dobbins, Richard Andrew
Karlsson, Sture Karl Fredrik
Kowalski, Tadeusz
Liu, J(oseph) T(su) C(hieh)
McEligot, Donald M(arinus)
Parmentier, Edgar M(arc)
Sibulkin, Merwin
Stratt, Richard Mark
Su, Chau-Hsing

SOUTH CAROLINA
Castro, Walter Ernest
Haile, James Mitchell

Przirembel, Christian E G
Yang, Tah Teh

TENNESSEE
Allison, Stephen William
Collins, Frank Gibson
Hayter, John Bingley
Hosker, Rayford Peter, Jr
James, David Randolph
Lee, Donald William
Lewis, James W L
Reddy, Kapuluru Chandrasekhara
Remenyik, Carl John
Swim, William B(axter)
Thomas, David Glen
Wang, Taylor Gunjin
Witten, Alan Joel

TEXAS
Anderson, Edward Everett
Aucoin, Paschal Joseph, Jr
Ball, Kenneth Steven
Bergman, Theodore L
Chang, Ping
Chevalier, Howard L
Chung, Jing Yan
Collins, Michael Albert
Cupples, Barrett L(eMoyne)
Darby, Ronald
Dodge, Franklin Tiffany
Eichhorn, Roger
Goldman, Joseph L
Haberman, Richard
Harpavat, Ganesh Lal
Hrkel, Edward James
Hurlburt, H(arvey) Zeh
Hussain, A K M Fazle
Hwang, Neddy H C
Koschmieder, Ernst Lothar
Liang, Edison Park-Tak
Lu, Frank Kerping
Maerker, John Malcolm
Markowski, Gregory Ray
Miller, Gerald E
Mitchell, James Emmett
Modisette, Jerry L
Porter, John William
Prats, Michael
Rajagopalan, Raj
Sample, Thomas Earl, Jr
Shapiro, Paul Robert
Sharma, Suresh C
Sparks, Cecil Ray
Swinney, Harry Leonard
Swope, Richard Dale
Truxillo, Stanton George
Voigt, Gerd-Hannes
Weatherford, W(illiam) D(ewey), Jr
Young, Fred M(ichael)

UTAH
Clyde, Calvin G(eary)
Hanks, Richard W(ylie)
Schunk, Robert Walter

VIRGINIA
Carlson, Harry William
Chang, Chieh Chien
Chubb, Scott Robinson
Coffey, Timothy
Crowley, Patrick Arthur
Csanady, Gabriel Tibor
Evans, John C
Evans, John Stanton, Jr
Fang, Ching Seng
Gale, Harold Walter
Gatski, Thomas Bernard
Grabowski, Walter John
Grossmann, William
Hardin, Jay Charles
Haussling, Henry Jacob
Hidalgo, Henry
Hoppe, John Cameron
Jaisinghani, Rajan A
Johnson, David Harley
Kroll, John Ernest
Kuo, Albert Yi-Shuong
Lai, Chintu (Vincent C)
Lee, Ja H
Lewis, Clark Houston
Lilleleht, L(embit) U(no)
Mihalas, Barbara R Weibel
Monacella, Vincent Joseph
Morton, Jeffrey Bruce
Muraca, Ralph John
Myers, Michael Kenneth
Pereira, Nino Rodrigues
Pierce, Felix J(ohn)
Pruett, Charles David
Renardy, Michael
Renardy, Yuriko
Roberts, A(lbert) S(idney), Jr
Roberts, William Woodruff, Jr
Roco, Mihail Constantin
Schwiderski, Ernst Walter
Seiner, John Milton
Selwyn, Philip Alan
Sepucha, Robert Charles
Simpson, Roger Lyndon
Sutton, George W(alter)
Tieleman, Henry William
Weinstein, Leonard Murrey
Witting, James M
Wood, Houston Gilleylen, III

Fluids (cont)

Yates, Charlie Lee
Yu, James Chun-Ying

WASHINGTON
Ahlstrom, Harlow G(arth)
Band, William
Bernard, Eddie Nolan
Booker, John Ratcliffe
Brown, Robert Alan
Christiansen, Walter Henry
Cokelet, Edward Davis
Criminale, William Oliver, Jr
Delisi, Donald Paul
Dixon, Robert Jerome, Jr
Duvall, George Evered
Dvorak, Frank Arthur
Erdmann, Joachim Christian
Forster, Fred Kurt
Fowles, George Richard
Fremouw, Edward Joseph
Frenzen, Paul
Harrison, Don Edmunds
Marston, Philip Leslie
Moore, Richard Lee
Orsborn, John F
Rattray, Maurice, Jr
Reyhner, Theodore Alison
Russell, David A
Sanford, Thomas Bayes
Schick, Michael
Stegen, Gilbert Rolland
Tung, Ka-Kit

WEST VIRGINIA
Ness, Nathan
Squire, William

WISCONSIN
Bird, R(obert) Byron
Bonazza, Riccardo
Doshi, Mahendra R
Hoopes, John A
Karkheck, John Peter
Kurath, Sheldon Frank
Meyer, Heribert
Schultz, David Harold
Woods, Frank Robert

PUERTO RICO
Khan, Winston

ALBERTA
Collins, David Albert Charles
De Krasinski, Joseph S
Freeman, Gordon Russel
Fu, Cheng-Tze
Gee, Norman
Rajaratnam, N(allamuthu)
Williams, Michael C(harles)

BRITISH COLUMBIA
Bechhoefer, John
Crozier, Edgar Daryl
Dewey, John Marks
Gartshore, Ian Stanley
Grace, John Ross
Hughes, Blyth Alvin
Krauel, David Paul
Parkinson, G(eoffrey) Vernon
Seymour, Brian Richard
Singh, Manohar

NEWFOUNDLAND
Morris, Claude C
Rochester, Michael Grant

NOVA SCOTIA
Kusalik, Peter Gerard

ONTARIO
Ahmed, Nasir Uddin
Baird, Malcolm Henry Inglis
Barker, Paul Kenneth
Barron, Ronald Michael
Boadway, John Douglas
Bragg, Gordon McAlpine
Chan, Yat Yung
Cowan, John Arthur
Donelan, Mark Anthony
Dunn, D(onald) W(illiam)
Egelstaff, Peter A
Irwin, Peter Anthony
James, David F
Krishnappan, Bommanna Gounder
Langford, William Finlay
Moffatt, William Craig
Moltyaner, Grigory
North, Henry E(rick) T(uisku)
Orlik-Ruckemann, Kazimierz Jerzy
Perlman, Maier
Pope, Noel Kynaston
Ray, Ajit Kumar
Ribner, Herbert Spencer
St-Maurice, Jean-Pierre
Svensson, Eric Carl
Vijay, Mohan Madan
Williams, Norman S W
Wilson, Kenneth Charles

QUEBEC
Chan, Sek Kwan
Dickinson, Edwin John

Eu, Byung Chan
Grmela, Miroslav
Lin, Sui
Marchand, Richard
Newman, B(arry) G(eorge)
Paidoussis, Michael Pandeli
Savage, Stuart B
Tassoul, Jean-Louis

SASKATCHEWAN
Tsang, Gee

OTHER COUNTRIES
Breslin, John P
Chwang, Allen Tse-Yung
Eichenberger, Hans P
Elad, David
Garcia-Colin, Leopoldo Scherer
Hama, Francis R(yosuke)
Hsieh, Din-Yu
Icke, Vincent
Kitaigorodskii, Sergei Alexander
Liu, Ta-Jo
Moore, James Edward, Jr
Romero, Alejandro F
Safran, Samuel A
Segel, Lee Aaron
Spence, Thomas Wayne
Vrebalovich, Thomas

Health Physics

CALIFORNIA
Hobbs, Willard Earl
McCall, Richard C

ILLINOIS
Murdoch, Bruce Thomas

IOWA
Lunde, Barbara Kegerreis, (BK)

KENTUCKY
Christensen, Ralph C(hresten)

MARYLAND
Mills, William Andy
Schwartz, Robert Bernard
Shleien, Bernard
Yaniv, Shlomo Stefan

NEW MEXICO
Davis, L(loyd) Wayne
Merritt, Melvin Leroy

NEW YORK
Fleischer, Robert Louis

OREGON
Kay, Michael Aaron

PENNSYLVANIA
Cohen, Bernard Leonard

SOUTH CAROLINA
Parkinson, Thomas Franklin

TENNESSEE
Bryan, Robert H(owell)
Darden, Edgar Bascomb
Gregory, Donald Clifford

WASHINGTON
Kathren, Ronald Laurence
Toohey, Richard Edward

MANITOBA
Jovanovich, Jovan Vojislav

ONTARIO
Rogers, David William Oliver

Mechanics

ALABAMA
Bailey, J Earl
Bowden, Charles Malcolm
Perkins, James Francis
Schutzenhofer, Luke A
Turner, Charlie Daniel, Jr

ARIZONA
Blitzer, Leon
Fraser, Harvey R(eed)
Lamb, Willis Eugene, Jr
Richard, Ralph Michael
Thompson, Lee P(rice)

ARKANSAS
Harter, William George

CALIFORNIA
Abey, Albert Edward
Baldwin, Barrett S, Jr
Barnett, David M
Behrens, H Wilhelm
Birkimer, Donald Leo
Buckingham, Alfred Carmichael
Burton, Donald Eugene
Casey, James
Caughey, Thomas Kirk
Chang, I-Dee

Chapman, Gary Theodore
Chen, Albert Tsu-Fu
Cheng, H(sien) K(ei)
Christensen, Richard Monson
Coles, Donald (Earl)
Corngold, Noel Robert David
Crebbin, Kenneth Clive
Dailey, C(harles) L(ee)
Fourney, M(ichael) E(ugene)
Gerpheide, John H
Gogolewski, Raymond Paul
Hall, Charles Frederick
Herrmann, Leonard R(alph)
Heusinkveld, Myron Ellis
Hodge, Philip G(ibson), Jr
Hoover, William Graham
Jennings, Paul C(hristian)
Kane, Thomas R(eif)
Kaplan, Abner
Keller, Joseph Edward, Jr
Kelly, Paul Sherwood
Knollman, Gilbert Carl
Kroeker, Richard Mark
Kubota, Toshi
Lambertson, Glen Royal
Lee, Martin Joe
Lee, Vin-Jang
Leitmann, G(eorge)
Lichtenberg, Allan J
Liepmann, H(ans) Wolfgang
Lin, Tung Hua
Lubliner, J(acob)
Lynch, David Dexter
McNiven, Hugh D(onald)
Mar, James Wah
Marble, Frank E(arl)
Meadows, Mark Allan
Miklowitz, Julius
Miller, James Avery
Mingori, Diamond Lewis
Moe, Mildred Minasian
Moore, Edgar Tilden, Jr
Mow, Maurice
Muki, Rokuro
Naghdi, P(aul) M(ansour)
Nielsen, Helmer L(ouis)
Passenheim, Burr Charles
Peale, Stanton Jerrold
Pearson, John
Pellegrini, Claudio
Pister, Karl S(tark)
Reissner, Eric
Roberts, Paul Harry
Rockwell, Robert Lawrence
Root, L(eonard) Eugene
Rosenberg, Reinhardt M
Roshko, Anatol
Sedgwick, Robert T
Sherman, Frederick S
Shi, Yun Yuan
Steele, Charles Richard
Tatro, Clement A(ustin)
Tsai, Stephen W
Vernon, Frank Lee, Jr
Westmann, Russell A
Williams, James Gerard
Willis, D(onald) Roger
Yang, An Tzu
Yang, Charles Chin-Tze
Yoder, Charles Finney
Zimmerman, Ivan Harold

COLORADO
Bartlett, David Farnham
Essenburg, F(ranklin)
Hoots, Felix R
Liu, Joseph Jeng-Fu
Meiss, James Donald
Webster, Larry Dale

CONNECTICUT
Arnoldi, Robert Alfred
Howkins, Stuart D
Jenney, David S
Johnston, E(lwood) Russell, Jr
Kenig, Marvin Jerry
Krahula, Joseph L(ouis)
Loo, Francis T C
Ojalvo, Irving U
Onat, E(min) T(uran)
Stetson, Karl Andrew
Strawderman, Wayne Alan

DELAWARE
Elliott, John Habersham
Fish, F(loyd) H(amilton), Jr
Huntsberger, James Robert
Nowinski, Jerzy L

DISTRICT OF COLUMBIA
Eggenberger, Andrew Jon
French, Francis William
Kelly, George Eugene
Klein, Vladislav
Liebowitz, Harold
Rockett, John A
Seidelmann, Paul Kenneth
Toridis, Theodore George

FLORIDA
Bose, Subir Kumar
Case, Robert Oliver
Cherepanov, Genady Petrovich
Coldwell, Robert Lynn

Drucker, Daniel Charles
Edson, Charles Grant
Eisenberg, Martin A(llan)
Fearn, Richard L(ee)
Foster, Irving Gordon
Hemp, Gene Willard
Huerta, Manuel Andres
Jacobson, Ira David
Lindgren, E(rik) Rune
Orth, William Albert
Roberts, Sanford B(ernard)
Sidebottom, Omar M(arion)
Taylor, C(harles) E(dwin)
Thomas, Dan Anderson
Ting, Robert Yen-Ying
Von Ohain, Hans Joachim
Zimmerman, Norman H(erbert)

GEORGIA
Fulton, Robert E(arle)
Goodno, Barry John
Hentschel, Hilary George
Smith, Cloyd Virgil, Jr
Ueng, Charles E(n) S(hiuh)

HAWAII
Hamada, Harold Seichi

ILLINOIS
Andriacchi, Thomas Peter
Baier, William H
Bazant, Zdenek P(avel)
Berdichevsky, Alexander
Bleier, Frank P(hilipp)
Cole, Francis Talmage
D'Souza, Anthony Frank
Dundurs, J(ohn)
Dykla, John J
Edelstein, Warren Stanley
Eubanks, Robert Alonzo
Mann, James Edward, Jr
Moore, Craig Damon
Mura, Toshio
Pericak-Spector, Kathleen Anne
Ramanathan, Ganapathiagraharam V
Smith, Richard Paul
Tang, James Juh-Ling

INDIANA
Farringer, Leland Dwight
Langhaar, Henry Louis
Scheiber, Donald Joseph
Snare, Leroy Earl

IOWA
Chandran, Krishnan Bala
Hartung, Jack Burdair, Sr
Kawai, Masataka
Liu, Young King
Stephens, Ralph Ivan
Weiss, Harry Joseph

KENTUCKY
Boyce, Stephen Scott
Gillis, Peter Paul
Tauchert, Theodore R

LOUISIANA
Peattie, Robert Addison
Puri, Pratap
Yannitell, Daniel W

MAINE
Brownstein, Kenneth Robert

MARYLAND
Bainum, Peter Montgomery
Bell, James F(rederick)
Berger, Robert Elliott
Burns, John Joseph, Jr
Burns, Timothy John
Cawley, Robert
Celmins, Aivars Karlis Richards
Charyk, Joseph Vincent
Dailey, George
Dally, James William
De Rosset, William Steinle
Deutch, John Mark
Dick, Richard Dean
Dickey, Joseph W
Dorfman, Jay Robert
Dragt, Alexander James
Edwards, Alan M
Feit, David
Flynn, Paul D(avid)
Gartrell, Charles Frederick
Gaunaurd, Guillermo C
Gillich, William John
Grebogi, Celso
Greenspon, Joshua Earl
Harris, Leonard Andrew
Herczynski, Andrzej
Hunston, Donald Lee
Kearsley, Elliot Armstrong
Khan, Akhtar Salamat
Kligman, Ronald Lee
Koh, Severino Legarda
Lee, David Allan
Liu, Han-Shou
McCormick, Michael Edward
Mermagen, William Henry
Moulton, James Frank, Jr
Murphy, Thomas James
Ostrom, Thomas Ross

Perrone, Nicholas
Rich, Harry Louis
Schuman, William John, Jr
Sevik, Maurice
Sikora, Jerome Paul
Simmons, John Arthur
Skop, Richard Allen
Wright, Thomas Wilson
Wu, Chih

MASSACHUSETTS
Brehm, John Joseph
Cappallo, Roger James
Chang, Tien Sun (Tom)
Cleary, Michael
Emmons, Howard W(ilson)
Greif, Robert
Junger, Miguel C
Kerrebrock, Jack Leo
Klein, Michael W
Leehey, Patrick
Liepins, Atis Aivars
McMahon, Thomas Arthur
Nash, William A(rthur)
Pallone, Adrian Joseph
Pian, Theodore H(sueh) H(uang)
Shu, Larry Steven
Siegel, Armand
Trilling, Leon
White, James Victor
Whitehead, John Andrews
Wisdom, Jack Leach
Yuan, Fan

MICHIGAN
Akcasu, Ahmet Ziyaeddin
Altiero, Nicholas James, Jr
Beltz, Charles R(obert)
Chace, Milton A
Chon, Choon Taik
Clark, Samuel Kelly
Denman, Harry Harroun
Goldstein, Steven Alan
Gordon, Morton Maurice
Hanson, Norman Walter
Hodgson, Voigt R
Hubbard, Robert Phillip
Im, Jang Hi
Lee, Sung Mook
Nefske, Donald Joseph
Reitan, Daniel Kinseth
Schmidt, Robert
Sung, Shung H
Webb, Paul
Whitman, Alan B
Wineman, Alan Stuart
Wu, Kuang Ming
Yang, Wei-Hsuin

MINNESOTA
Bayman, Benjamin
Fosdick, Roger L(ee)
Goodman, L(awrence) E(ugene)
Plunkett, Robert
Sethna, Patarasp R(ustomji)
Wilson, Theodore A(lexander)

MISSISSIPPI
Abdulrahman, Mustafa Salih
Bernard, Robert Scales
Bryant, Glenn D(onald)
Horton, Thomas Edward, Jr
Warsi, Zahir U A

MISSOURI
Cronin, James Lawrence, Jr
Gould, Phillip L
Hohenemser, Kurt Heinrich
Kitchin, Robert Walter
Liu, Henry
Murden, W(illiam) P(aul), Jr
Parry, Myron Gene
Venezian, Giulio

NEBRASKA
Wolford, James C

NEW HAMPSHIRE
Cranch, Edmund Titus
Montgomery, David Campbell

NEW JERSEY
Berman, Irwin
Borg, Sidney Fred
Brown, Garry Leslie
Caffarelli, Luis Angel
Cakmak, Ahmet Sefik
Chan, Paul C
Chen, Yu
Cheng, Sin-I
Coleman, Bernard David
Doby, Raymond
Eringen, Ahmed Cemal
Gilmer, George Hudson
Hayes, Wallace D(ean)
Marscher, William Donnelly
Monack, A(lbert) J(ames)
Neidhardt, Walter Jim
Pai, David H(sien)-C(hung)
Sak, Joseph
Simpkins, Peter G
Thurston, Robert Norton
Wang, Tsuey Tang
Wiesenfeld, Joel

Wissbrun, Kurt Falke

NEW MEXICO
Brannon, Paul J
Davison, Lee Walker
Engelke, Raymond Pierce
Genin, Joseph
Hecker, Siegfried Stephen
Hsu, Y(oun) C(hang)
Johnson, James Norman
Ju, Frederick D
Kraichnan, Robert Harry
Kuiper, Logan Keith
Norwood, Frederick Reyes
Passman, Stephen Lee
Reuter, Robert Carl, Jr
Sutherland, Herbert James
Tonks, Davis Loel
Travis, John Richard
Weigle, Robert Edward
Williamson, Walton E, Jr

NEW YORK
Allentuch, Arnold
Arin, Kemal
Armen, Harry, Jr
Bardwell, Steven Jack
Baron, Melvin L(eon)
Bendler, John Thomas
Bieniek, Maciej P
Boley, Bruno Adrian
Brunelle, Eugene John, Jr
Burns, Joseph A
Burns, Stephen James
Chiang, Fu-Pen
Cohen, Leon
Conners, Gary Hamilton
Conway, H(arry) D(onald)
Cowin, Stephen Corteen
De Zeeuw, Carl Henri
Eickelberg, W Warren B
Engel, Peter Andras
Fox, Edward A(lexander)
Gans, Roger Frederick
Garnet, Hyman R
Goldberg, Martin A
Gulati, Suresh T
Harris, Stewart
Hetnarski, Richard Bozyslaw
Holmes, Philip John
Johnson, Wallace E
Kahn, Elliott H
Koretz, Jane Faith
Krempl, Erhard
Krinsky, Samuel
Lance, R(ichard) H
Lane, Frank
Lee, Daeyong
Levy, Alan Joseph
Lissauer, Jack Jonathan
Liu, Ching Shi
Liu, Hao-Wen
Month, Melvin
Moon, Francis C
Mow, Van C
Nazem, Faramarz Franz
Pan, Huo-Hsi
Pao, Yih-Hsing
Ratner, Lazarus Gershon
Reddy, Reginald James
Reismann, Herbert
Ringlee, Robert J
Roskes, Gerald J
Sanders, W(illiam) Thomas
Steele, Theodore Karl
Stoker, James Johnston
Subudhi, Manomohan
Tasi, James
Tiersten, Harry Frank
Valyi, Emery I
Ward, Lawrence W(aterman)
Weinbaum, Sheldon
Yang, Ching Huan
Zabinski, Michael Peter

NORTH CAROLINA
Batra, Subhash Kumar
D'Arruda, Jose Joaquim
Douglas, Robert Alden
Dunn, Joseph Charles
Ely, John Frederick
Havner, Kerry S(huford)
Jawa, Manjit S
Ji, Chueng Ryong
Keltie, Richard Francis
Klenin, Marjorie A
McElhaney, James Harry
Pao, Chia-Ven
Roberts, Verne Louis
Schmiedeshoff, Frederick William
Stephenson, Harold Patty
Winkler, Linda Irene
Wright, William Vale

OHIO
Bailey, Cecil Dewitt
Bhattacharya, Amit
Bloom, Joseph Morris
Bundy, Francis P
Crum, Ralph G
Ferguson, Harry
Gangemi, Francis A
Ghia, Kirti N
Griffing, David Francis

Hart, David Charles
Kazarian, Leon Edward
Klingbeil, Werner Walter
Koo, Benjamin
Lin, Kuang-Ming
Liska, John W
Lovejoy, Owen
Macklin, Philip Alan
Nakamura, Shoichiro
Palazotto, Anthony Nicholas
Patton, Jon Michael
Richardson, David Louis
Robinson, David Nelson
Rollins, Roger William
St John, Douglas Francis
Schlegel, Donald Louis
Tsuei, Yeong Ging
Van Fossen, Don B
Whillans, Ian Morley
Wright, Harold E(ugene)

OKLAHOMA
Fowler, Richard Gildart
Lee, Lloyd Lieh-Shen
Lou, Alex Yih-Chung
Lubinski, Arthur
Rasmussen, Maurice L
Sack, Ronald Leslie

OREGON
Carmichael, Howard John
Ericksen, Jerald LaVerne
Semura, Jack Sadatoshi

PENNSYLVANIA
Antes, Harry W
Ayyaswamy, Portonovo S
Bahar, Leon Y
Beer, Ferdinand P(ierre)
Cheng, Kuang Liu
Chou, Pei Chi (Peter)
Crawford, John David
Daga, Raman Lall
Davies, Kenneth Thomas Reed
Erdogan, Fazil
Gainer, Michael Kizinski
Gunton, James D
Gurtin, Morton Edward
Hagen, Oskar
Havas, Peter
Hettche, Leroy Raymond
Kalnins, Arturs
Kazakia, Jacob Yakovos
Kiusalaas, Jaan
Liu, John K(ungfu)
McLennan, James Alan, Jr
Mortimer, Richard W(alter)
Moskowitz, Gordon David
Newman, John B(ullen)
Noll, Walter
Palusamy, Sam Subbu
Phares, Alain Joseph
Pierce, Allan Dale
Rose, Joseph Lawrence
Rovelli, Carlo
Smith, Gerald Francis
Swedlow, Jerold Lindsay
Wang, Albert Show-Dwo
Wayne, Clarence Eugene
Wei, Robert Peh-Ying
Williams, William Orville
Woo, Tse-Chien

RHODE ISLAND
Findley, William N(ichols)
Freund, Lambert Ben
Kolsky, Herbert
Wilson, Mason P, Jr

SOUTH CAROLINA
Hoppmann, William Henry, II

SOUTH DAKOTA
Winker, James A(nthony)

TENNESSEE
Crater, Horace William
Deeds, William Edward
Greenstreet, William (B) LaVon
Strauss, Alvin Manosh
Walker, Grayson Howard

TEXAS
Bedford, Anthony
Bowen, Ray M
Curtis, Frederick Augustus, Jr
Darby, Ronald
Everard, Noel James
Glaser, Frederic M
Goland, Martin
Hyman, William A
Juhasz, Stephen
Lindholm, Ulric S
Mallory, Merrit Lee
Narcowich, Francis Joseph
Ohnuma, Shoroku
Rand, James Leland
Schieve, William
Schneider, William Charles
Shepley, Lawrence Charles
Smith, Frederick Adair, Jr
Solberg, Ruell Floyd, Jr
Tucker, Richard Lee
Wah, Thein

Wang, Chao-Cheng
Wheeler, Lewis Turner
Yeh, Hsiang-Yueh

UTAH
Fife, Paul Chase

VERMONT
Wu, Jun Ru

VIRGINIA
Arkilic, Galip Mehmet
Burns, Grover Preston
Claus, Richard Otto
Debney, George Charles, Jr
Delos, John Bernard
Finke, Reinald Guy
Henneke, Edmund George, II
Jennings, Richard Louis
Kaplan, Marshall Harvey
Lewis, David W(arren)
Milligan, John H
Monacella, Vincent Joseph
Muraca, Ralph John
Nayfeh, Ali Hasan
Pilkey, Walter David
Porzel, Francis Bernard
Raju, Ivatury Sanyasi
Reifsnider, Kenneth Leonard
Renardy, Yuriko
Rokni, Mohammad Ali
Rosenthal, F(elix)
Sparrow, D(avid) A
Telionis, Demetri Pyrros
Vaughan, Christopher Leonard
Wang, C(hiao) J(en)
Watson, Nathan Dee

WASHINGTON
Bollard, R(ichard) John H
Cole, Robert Stephen
Ehlers, Francis Edward
Evans, Roger James
Filler, Lewis
Fowles, George Richard
Fyfe, I(an) Millar
Gupta, Yogendra M(ohan)
Kennet, Haim
Leaf, Boris
Parmerter, R Reid
Pilet, Stanford Christian
Riedel, Eberhard Karl
Roetman, Ernest Levane
Taya, Minoru
Walton, Vincent Michael
Wilson, Thornton Arnold

WISCONSIN
Brackenridge, John Bruce
Johnson, Arthur Franklin
McQuistan, Richmond Beckett
Morin, Dornis Clinton
Rang, Edward Roy
Symon, Keith Randolph
Van den Akker, Johannes Archibald
Washa, George William
Widera, Georg Ernst Otto

WYOMING
Boresi, Arthur Peter

PUERTO RICO
Blum, Lesser

ALBERTA
Ali, M Keramat
Ellyin, Fernand
Glockner, Peter G
Kumar, Shrawan
Murray, David William

BRITISH COLUMBIA
Graham, George Alfred Cecil
Hoffer, J(oaquin) A(ndres)
Lardner, Robin Willmott
Lind, Niels Christian
Parkinson, G(eoffrey) Vernon
Singh, Manohar

MANITOBA
Cohen, Harley

NOVA SCOTIA
Calkin, Melvin Gilbert
Sastry, Vankamamidi VRN

ONTARIO
Baronet, Clifford Nelson
Chapman, George David
Chau, Wai-Yin
Lennox, William C(raig)
Mansinha, Lalatendu
Olson, Allan Theodore
Roorda, John
Todd, Leonard

QUEBEC
Axelrad, D(avid) R(obert)
Fazio, Paul (Palmerino)
Li, Zi-Cai
Massoud, Monir Fouad
Paidoussis, Michael Pandeli
Savage, Stuart B
Subramanian, Sesha

Mechanics (cont)

OTHER COUNTRIES
Contopoulos, George
Hashin, Zvi
Kydoniefs, Anastasios D
Markovitz, Hershel
Tipans, Igors O

Medical Physics

CALIFORNIA
Anderson, Milo Vernette
Haymond, Herman Ralph
Katz, Barrett
McCall, Richard C
Mireshghi, Ali
Nelson, Richard David
Shrivastava, Prakash Narayan
Stevens, Lewis Axtell
Tsiang, Eugene Y

COLORADO
Adams, Gail Dayton

CONNECTICUT
Brenholdt, Irving R

FLORIDA
Andrew, Edward Raymond
Mauderli, Walter

ILLINOIS
Martin, Ronald Lavern
Spokas, John J
Stinchcomb, Thomas Glenn

MARYLAND
Rodgers, James Earl
Roth, Bradley John
Silver, David Martin
Sliney, David H

MASSACHUSETTS
Collins, James Joseph
Horowitz, Joseph
Sisterson, Janet M

MICHIGAN
Blosser, Henry Gabriel

MINNESOTA
Hobbie, Russell Klyver

MISSISSIPPI
LeBlanc, Michael H

NEVADA
Barth, Delbert Sylvester

NEW JERSEY
Roy, Herbert C

NEW MEXICO
Yarger, Frederick Lynn

NEW YORK
Bunch, Phillip Carter
Goodwin, Paul Newcomb
Liu, Mao-Zu

NORTH CAROLINA
DeGuzman, Allan Francis
Ekstrand, Kenneth Eric
Overton, John H

OHIO
Endorf, Robert J
Hangartner, Thomas Niklaus

OREGON
Frey, Bruce Edward

PENNSYLVANIA
Waterman, Frank Melvin

TENNESSEE
Gronemeyer, Suzanne Alsop

TEXAS
Bencomo, Jose A, Jr
Cranberg, Lawrence
Murphy, Paul Henry
Perry, Robert Riley
Richardson, Jasper E

UTAH
Berggren, Michael J

VERMONT
Wu, Jun Ru

VIRGINIA
Aqualino, Alan

WASHINGTON
Miller, Douglas Lawrence

WEST VIRGINIA
Douglass, Kenneth Harmon

WISCONSIN
DeLuca, Paul Michael, Jr
DeWerd, Larry Albert
Hendee, Wiliam R

BRITISH COLUMBIA
Johnson, R R
Kubik, Philip Roman

ONTARIO
Bernstein, Alan
Chettle, David Robert
Pfalzner, Paul Michael
Rogers, David William Oliver

QUEBEC
Pla, Conrado
Podgorsak, Ervin B
Schreiner, Lawrence John

Nuclear Structure

ALABAMA
Alford, William Lumpkin
Bartlett, James Holly
Regner, John LaVerne
Robinson, Edward Lee
Thaxton, George Donald

ARIZONA
Comfort, Joseph Robert
Donahue, Douglas James
Fan, Chang-Yun
Hill, Henry Allen
McCullen, John Douglas
McIntyre, Laurence Cook, Jr
Nigam, Bishan Perkash
Roy, Radha Raman
Scolman, Theodore Thomas
Yergin, Paul Flohr

ARKANSAS
Rollefson, Aimar Andre
Vyas, Reeta

CALIFORNIA
Adams, Ralph Melvin
Adelson, Harold Ely
Albergotti, J(esse) Clifton
Alonso, Carol Travis
Alonso, Jose Ramon
Alvarez, Raymond Angelo, Jr
Ames, Lawrence Lowell
Anderson, John D
Bach, David Rudolph
Baglin, John Edward Eric
Baisden, Patricia Ann
Ball, William Paul
Bardin, Tsing Tchao
Barengoltz, Jack Bennett
Barnes, Charles A
Bauer, Rudolf Wilhelm
Becker, John Angus
Bennett, Charles Lougheed
Bertin, Michael C
Bloom, Stewart Dave
Bodfish, Ralph E
Brady, Franklin Paul
Britt, Harold Curran
Brown, James Roy
Brown, Virginia Ruth
Burginyon, Gary Alfred
Buskirk, Fred Ramon
Cahill, Thomas A
Camp, David Conrad
Cerny, Joseph, III
Chadwick, George Brierley
Chalmers, Robert Anton
Chase, Lloyd Fremont, Jr
Chen, Chia Hwa
Clark, Arnold Franklin
Clark, William Gilbert
Collard, Harold Rieth
Conzett, Homer Eugene
Corman, Emmett Gary
Crandall, Walter Ellis
Creutz, Edward (Chester)
Crittenden, Eugene Casson, Jr
Diamond, Richard Martin
Dickinson, Wade
Dines, Eugene L
Dolan, Kenneth William
Draper, James Edward
Dudziak, Walter Francis
Dunlop, William Henry
Easton, W James
Edwards, David Franklin
Eisberg, Robert Martin
Erdmann, John Hugo
Fearn, Heidi
Ferguson, James Malcolm
Fisher, Thornton Roberts
Fricke, Martin Paul
Ganas, Perry S
Gearhart, Roger A
Gibson, Edward F
Gilbert, William Spencer
Glick, Harold Alan
Goldberg, Eugene
Goldwire, Henry C, Jr
Goodwin, Lester Kepner
Goosman, David R
Goss, John Douglas
Gould, Harvey Allen
Greiner, Douglas Earl
Groseclose, Byron Clark
Haddock, Roy P
Hafemeister, David Walter
Hanna, Stanley Sweet
Hansen, Luisa Fernandez
He, Yudong
Heckman, Harry Hughes
Heikkinen, Dale William
Heindl, Clifford Joseph
Hoff, Richard William
Holzer, Alfred
Howerton, Robert James
Hu, Chi-Yu
Hudson, Alvin Maynard
Hulet, Ervin Kenneth
Jacobs, Peter Martin
James, Ralph Boyd
John, Joseph
Jordan, Willard Clayton
Jungerman, John (Albert)
Kavanagh, Ralph William
King, Harriss Thornton
Knox, William Jordan
Kovar, Frederick Richard
Kull, Lorenz Anthony
Lanier, Robert George
Lark, Neil LaVern
Lee, I-Yang
Lee, Paul L
Lingenfelter, Richard Emery
Ludin, Roger Louis
Lugmair, Guenter Wilhelm
Lutz, Harry Frank
McCarthy, Mary Anne
MacGregor, Malcolm Herbert
Mann, Lloyd Godfrey
Margaziotis, Demetrius John
Marrs, Roscoe Earl
Marrus, Richard
Martinelli, Ernest A
Martz, Harry Edward, Jr
Mathews, Grant James
Matis, Howard S
Melgard, Rodney
Melkanoff, Michel Allan
Mendelson, Robert Alexander, Jr
Milne, Edmund Alexander
Moody, Kenton J
Morgan, James Frederick
Morton, John Robert, III
Myers, William D
Nefkens, Bernard Marie
Nero, Anthony V, Jr
Newman, David Edward
Nitschke, J Michael
Norman, Eric B
O'Dell, Jean Marland
Oehlberg, Richard N
Olness, Robert James
Orphan, Victor John
Parker, Vincent Eveland
Parker, Winifred Ellis
Peek, Neal Frazier
Pehl, Richard Henry
Perry, Joseph Earl, Jr
Peterson, Jack Milton
Plain, Gilbert John
Poppe, Carl Hugo
Poskanzer, Arthur M
Proctor, Ivan D
Pugh, Howel Griffith
Putnam, Thomas Milton, Jr
Randrup, Jorgen
Read, William George
Redlich, Martin George
Reines, Frederick
Reynolds, Glenn Myron
Reynolds, Harry Lincoln
Richardson, John Reginald
Rosenblum, Stephen Saul
Rosenfeld, Arthur H
Salaita, George N
Salisbury, Stanley R
Schweizer, Felix
Serduke, Franklin James David
Singletary, John Boon
Singletary, Lillian Darlington
Smith, Ronald James
Smoot, George Fitzgerald, III
Sobel, Henry Wayne
Spieler, Helmuth
Stevenson, John David
Stokstad, Robert G
Struble, Gordon Lee
Sugihara, Thomas Tamotsu
Symons, Timothy J
Tipler, Paul A
Tracy, James Frueh
Tucker, Allen Brink
Urone, Paul Peter
Van Bibber, Karl Albert
Verbinski, Victor V
Vogel, Peter
Walt, Martin
Weber, Hans Josef
Wegner, Harvey E
West, Harry Irwin, Jr
Weston, Gary Steven
Whitehead, Marian Nedra
Whitten, Charles A, Jr
Wilson, Robert Gray
Woldseth, Rolf
Wong, Chun Wa
Wood, Calvin Dale
Worth, Donald Calhoun
Wright, Byron Terry
Yap, Fung Yen
Yearian, Mason Russell
Yu, David U L
Zender, Michael J

COLORADO
Dooley, John Raymond, Jr
Haas, Francis Xavier, Jr
Herring, Jackson Rea
Kraushaar, Jack Jourdan
Lind, David Arthur
McAllister, Robert Wallace
Mobley, Ralph Claude
Moore, Benjamin LaBree
Peterson, Roy Jerome
Poirier, Charles Philip
Rost, Ernest Stephan
Shelton, Frank Harvey
Smythe, William Rodman
Spenny, David Lorin
Zafiratos, Chris Dan

CONNECTICUT
Adair, Robert Kemp
Albats, Paul
Antkiw, Stephen
Bockelman, Charles Kincaid
Bromley, D(avid) Allan
Cobern, Martin E
Firk, Frank William Kenneth
Greenberg, Jack Sam
Hill, David Lawrence
Hosain, Fazle
Kasha, Henry
Kendziorski, Francis Richard
Krisst, Raymond John
Kumar, Balasubramanian Shiva
Lange, Robert Carl
Levenstein, Harold
Lichtenberger, Harold V
Lubell, Michael S
Morrison, Richard Charles
Nath, Ravinder Katyal
Parker, Peter Donald MacDougall
Pisano, Daniel Joseph, Jr
Schweitzer, Jeffrey Stewart
Simpson, Wilburn Dwain
Slack, Lewis
Soufer, Robert
Stephenson, Kenneth Edward
Tao, Shu-Jen
Wolf, Elizabeth Anne

DELAWARE
Fou, Cheng-Ming
L'Heureux, Jacques (Jean)
Pittel, Stuart
Swann, Charles Paul

DISTRICT OF COLUMBIA
Berman, Barry L
Bingham, Robert Lodewijk
Brown, Louis
Cooper, Benjamin Stubbs
Crannell, Hall L
Erskine, John Robert
Fagg, Lawrence Wellburn
Hendrie, David Lowery
Kelly, Francis Joseph
Knight, John C (Ian)
Lambert, James Morrison
Lehman, Donald Richard
Mosher, David
Phillips, Gary Wilson
Reiss, Howard R
Richardson, Allan Charles Barbour
Ritter, James Carroll
Seglie, Ernest Augustus
Shon, Frederick John
Sober, Daniel Isaac
Stolovy, Alexander
Whetstone, Stanley L, Jr

FLORIDA
Chapman, Kenneth Reginald
Clark, Mary Elizabeth
Davis, Robert Houser
Dubbelday, Pieter Steven
Dunnam, Francis Eugene, Jr
Fletcher, Neil Russel
Fox, John David
Frawley, Anthony Denis
Harmon, G Lamar
Jin, Rong-Sheng
Kemper, Kirby Wayne
Llewellyn, Ralph A
Mintz, Stephen Larry
Peterson, Lennart Rudolph
Petrovich, Fred
Philpott, Richard John
Plendl, Hans Siegfried
Randall, Charles Addison, Jr
Rester, Alfred Carl, Jr
Reynolds, Bruce G
Riley, Mark Anthony
Robson, Donald
Sheline, Raymond Kay
Shelton, Wilford Neil
Sherwood, Jesse Eugene

Stetson, Robert Franklin
Tabor, Samuel Lynn
Weber, Alfred Herman
Willard, Harvey Bradford

GEORGIA
Baker, Francis Todd
Bass, William Thomas
Bowsher, Harry Fred
Boyd, James Emory
Braden, Charles Hosea
Dangle, Richard L
Ezell, Ronnie Lee
Hunter, Raymond Eugene
Keller, George Earl
Love, William Gary
Nelson, William Henry
Owens, William Richard
Prior, Richard Marion
Rao, Pemmaraju Venugopala
Russell, John Lynn, Jr
Steuer, Malcolm F
Thompson, Lewis Chisholm
Wyly, Lemuel David, Jr

IDAHO
Greenwood, Reginald Charles
Heath, Russell La Verne
Helmer, Richard Guy
McFarlane, Harold Finley
Reich, Charles William
Smith, J Richard
Vegors, Stanley H, Jr
Willmes, Henry
Young, Thomas Edward

ILLINOIS
Ahmad, Irshad
Back, Birger Bo
Barnothy, Jeno Michael
Bollinger, Lowell Moyer
Bowe, Joseph Charles
Braid, Thomas Hamilton
Bretscher, Manuel Martin
Brussel, Morton Kremen
Chu, Keh-Cheng
Cossairt, Jack Donald
Cronin, James Watson
Davids, Cary Nathan
Debevec, Paul Timothy
Dickerman, Charles Edward
Doerner, Robert Carl
Eisenstein, Robert Alan
Elwyn, Alexander Joseph
Fink, Charles Lloyd
Flynn, Kevin Francis
Garcia-Munoz, Moises
Goldman, David Tobias
Greene, John Philip
Guenther, Peter T
Gupta, Nand K
Hanson, Alfred Olaf
Hasan, Mazhar
Hatch, Albert Jerold
Henning, Walter Friedrich
Holland, Robert Emmett
Holt, Roy James
Hubbard, Lincoln Beals
Ingram, Forrest Duane
Jackson, Harold E, Jr
Jankus, Vytautas Zachary
Jeong, Tung Hon
Khoo, Teng Lek
Kuchnir, Franca Tabliabue
Kurath, Dieter
Land, Robert H
Langsdorf, Alexander, Jr
Laszewski, Ronald M
Lee, Tsung-Shung Harry
Malik, Fazley Bary
Marmer, Gary James
Martin, Ronald Lavern
Mateja, John Frederick
Meadows, James Wallace, Jr
Moore, Harold Arthur
Nathan, Alan Marc
Nolen, Jerry A, Jr
Pandharipande, Vijay Raghunath
Papanicolas, Costas Nicolas
Pardo, Richard Claude
Reiffel, Leonard
Schiffer, John Paul
Schmidt, Arthur Gerard
Schramm, David N
Segel, Ralph E
Seth, Kamal Kishore
Siefken, Hugh Edward
Silverstein, Edward Allen
Smith, Alan B
Smith, Donald Larned
Smith, James Hammond
Smither, Robert Karl
Snelgrove, James Lewis
Spinka, Harold M
Sturm, William James
Tollestrup, Alvin V
Treadwell, Elliott Allen
Trifunac, Alexander Dimitrije
Wah, Yau Wai
Wharton, William Raymond
Zeidman, Benjamin

INDIANA
Alvager, Torsten Karl Erik

Aprahamian, Ani
Bacher, Andrew Dow
Baker, David Thomas
Berners, Edgar Davis
Blackstead, Howard Allan
Browne, Cornelius Payne
Chagnon, Paul Robert
Darden, Sperry Eugene
Day, Benjamin Downing
Elmore, David
Friesel, Dennis Lane
Funk, Emerson Gornflow, Jr
Garg, Umesh
Goodman, Charles David
Grabowski, Zbigniew Wojciech
Harris, Samuel M(elvin)
Jacobs, William Wescott
Johnson, Orland Eugene
Jones, William Philip
Kemple, Marvin David
Kim, Yeong Ell
Kolata, James John
Lee, Shyh-Yuan
Londergan, John Timothy
Macfarlane, Malcolm Harris
Marshalek, Eugene Richard
Meyer, Hans-Otto
Mihelich, John William
Miller, Daniel Weber
Nann, Hermann
Ober, David Ray
Olmer, Catherine
Pollock, Robert Elwood
Rickey, Frank Atkinson, Jr
Rickey, Martin Eugene
Scharenberg, Rolf Paul
Schwandt, Peter
Singh, Prithe Paul
Stephenson, Edward James
Tendam, Donald Jan
Tubis, Arnold
Vigdor, Steven Elliot
Wellman, Henry Nelson

IOWA
Carlson, Richard Raymond
Duke, Charles Lewis
Graber, Harlan Duane
Hanson, Roger James
Hill, John Christian
Kelly, William H
Norbeck, Edwin
Pursey, Derek Lindsay
Stassis, Constantine
Weiffenbach, Conrad Venable
Zaffarano, Daniel Joseph

KANSAS
Creager, Charles Bicknell
Davidson, J(ohn) P(irnie)
Donoghue, Timothy R
Knapp, David Edwin
Legg, James C
Loper, Gerald D
Prosser, Francis Ware, Jr
Roberts, James Herbert
Sapp, Richard Cassell

KENTUCKY
Chalmers, Joseph Stephen
Gabbard, Fletcher
Kern, Bernard Donald
Laird, Christopher Eli
McEllistrem, Marcus Thomas
Maddox, William Eugene
Ouseph, Pullom John
Robertson, John David
Vourvopoulos, George
Weil, Jesse Leo
Yates, Steven Winfield

LOUISIANA
Bernard, Davy Lee
Buccino, Salvatore George
Chen, Isaac I H
Cosper, Sammie Wayne
Draayer, Jerry Paul
Durham, Frank Edington
Goodman, Alan Leonard
Rosensteel, George T
Yang, Chia Hsiung
Zander, Arlen Ray
Zganjar, Edward F

MAINE
Beard, David Breed
Hess, Charles Thomas

MARYLAND
Achor, William Thomas
Baker, Charles P
Barrett, Richard John
Beach, Eugene Huff
Bennett, Gary L
Bondelid, Rollon Oscar
Boye, Robert James
Buck, John Henry
Carlson, Allan David
Carter, Robert Sague
Chen, Lily Ching-Chun
Cialella, Carmen Michael
Clark, Gary Edwin
Cline, James E
Costrell, Louis

Coursey, Bert Marcel
Di Rienzi, Joseph
Eisenhauer, Charles Martin
Evans, Albert Edwin, Jr
Faust, William R
Fivozinsky, Sherman P
Fuller, Everett G(ladding)
Goldstein, Norman Phillip
Gossett, Charles Robert
Grayson, William Curtis, Jr
Griffin, Jerry J
Hayward, Raymond (W)ebster
Hess, Wilmot Norton
Hill, Lemmuel Leroy
Holmgren, Harry D
Hoppes, Dale DuBois
Hornyak, William Frank
Huttlin, George Anthony
Jones, Frank Culver
Kolstad, George Andrew
Krause, Thomas Otto
Lamaze, George Paul
Larson, Reginald Einar
Lee, Yung-Keun
McElhinney, John
Madansky, Leon
Malmberg, Philip Ray
Marshak, Harvey
Merkel, George
Miller, Frank L
Mitler, Henri Emmanuel
O'Connell, James S
Rasera, Robert Louis
Richardson, Clarence Robert
Rodney, William Stanley
Roos, Philip G
Schima, Francis Joseph
Silberberg, Rein
Smith, Gary Leroy
Temperley, Judith Kantack
Vanhoy, Jeffrey R
Walker, J Calvin
Wasson, Oren A
Young, Frank Coleman

MASSACHUSETTS
Ahlen, Steven Paul
Bainbridge, Kenneth Tompkins
Baker, Adolph
Bernstein, Aron
Blair, John Sanborn
Blatt, S Leslie
Bradford, John Norman
Brenner, Daeg Scott
Button-Shafer, Janice
Cameron, Alastair Graham Walter
Connors, Philip Irving
Decowski, Piotr
Denig, William Francis
Donnelly, Thomas William
Egan, James Joseph
Enge, Harald Anton
Fann, Huoo-Long
George, James Z
Glaubman, Michael Juda
Goldstone, Jeffrey
Goloskie, Raymond
Gorenstein, Paul
Grodzins, Lee
Guertin, Ralph Francis
Kegel, Gunter Heinrich Reinhard
Kennedy, Edward Francis
Klema, Ernest Donald
Kowalski, Stanley Benedict
Martens, Edward John
Mathur, Suresh Chandra
Matthews, June Lorraine
Michael, Irving
Milder, Fredric Lloyd
Miller, James Paul
Moniz, Ernest Jeffrey
Peterson, Gerald Alvin
Posner, Martin
Pullen, David John
Quinton, Arthur Robert
Redwine, Robert Page
Schneider, Robert Julius
Schwensfeir, Robert James, Jr
Scott, David Knight
Sellers, Francis Bachman
Seward, Frederick Downing
Sheldon, Eric
Sternheim, Morton Maynard
Todreas, Neil Emmanuel
Walker, James Frederick, Jr
Wang, Chia Ping
Wiggins, John
Williamson, Claude F
Yang, Tien-Fang
Zielinska Pfabe, Malgorzata

MICHIGAN
Agin, Gary Paul
Austin, Sam M
Beard, George B
Benenson, Walter
Beres, William Philip
Bernstein, Eugene Merle
Blue, Richard Arthur
Cormier, Thomas Michael
Crawley, Gerard Marcus
Dorenbusch, William Edwin
Ferguson, Stephen Mason
Frissel, Harry Frederick

Galonsky, Aaron Irving
Gelbke, Claus-Konrad
Gray, Walter Steven
Griffioen, Roger Duane
Hardie, Gerald
Harris, Gale Ion
Jolivette, Peter Lauson
Kashy, Edwin
Kocher, Charles William
Ludington, Martin A
Morrissey, David Joseph
Nadasen, Aruna
Riggs, Roderick D
Robbins, John Alan
Ronningen, Reginald Martin
Schneider, Eric West
Silver, Robert
Tarle, Gregory
Tickle, Robert Simpson
Wiedenbeck, Marcellus Lee
Zeller, Albert Fontenot

MINNESOTA
Agarwal, Som Prakash
Bayman, Benjamin
Dehnhard, Dietrich
Ellis, Paul John
Hintz, Norton Mark
Inman, Fred Winston
Johnson, Walter Heinrick, Jr
Steck, Daniel John

MISSISSIPPI
Croft, Walter Lawrence
Koshel, Richard Donald
Reidy, James Joseph
Whitlock, Lapsley Craig

MISSOURI
Brown, Ian McLaren
Brugger, Robert Melvin
Glascock, Michael Dean
Leader, John Carl

MONTANA
Jakobson, Mark John
Lee, Floyd Denman
Manhart, Robert (Audley)

NEBRASKA
Chakkalakal, Dennis Abraham
Cipolla, Sam J
Finkler, Paul
Gale, Douglas Shannon, II

NEVADA
Eccles, Samuel Franklin
Jobst, Joel Edward

NEW HAMPSHIRE
Couchell, Gus Perry
Fricke, Edwin Francis
Krone, Ralph Werner
Price, Glenn Albert
Rieser, Leonard M

NEW JERSEY
Arthur, Wallace
Cizewski, Jolie Antonia
Curtis, Thomas Hasbrook
Dietz, George Robert
Girit, Ibrahim Cem
Glashausser, Charles Michael
Goode, Philip Ranson
Groeger, Theodore Oskar
Kaita, Robert
Karp, Bennett C
Kloet, Willem M
Knox, Wayne H
Koller, Noemie
Kowalski, Ludwik
Kugel, Henry W
Makofske, William Joseph
Mekjian, Aram Zareh
Moshey, Edward A
Mueller, Dennis
Murnick, Daniel E
Pond, Thomas Alexander
Raghavan, Pramila
Raghavan, Ramaswamy Srinivasa
Rockmore, Ronald Marshall
Schilling, Gerd
Sharon (Schwadron), Yitzhak Yaakov
Sherr, Rubby
Spruch, Grace Marmor
Temmer, Georges Maxime
Wong, King-Lap
Zamick, Larry
Zolnowski, Dennis Ronald

NEW MEXICO
Amann, James Francis
Apt, Kenneth Ellis
Auchampaugh, George Fredrick
Baer, Helmut W
Baggett, Lester Marchant
Barnes, Peter David
Barr, Donald Westwood
Bendt, Philip Joseph
Bennett, Elbert White
Bingham, Felton Wells
Bowman, Charles D
Bowman, James David
Browne, John Charles

Nuclear Structure (cont)

Bunker, Merle E
Burleson, George Robert
Burman, Robert L
Butler, Gilbert W
Caldwell, John Thomas
Chambers, William Hyland
Close, Donald Alan
Coats, Richard Lee
Cochran, Donald Roy Francis
Coon, Sidney Alan
Cooper, Martin David
Cramer, James D
Davey, William George
Dingus, Ronald Shane
Diven, Benjamin Clinton
Doolen, Gary Dean
Drake, Darrell Melvin
Dressel, Ralph William
Dyer, Peggy Lynn
Ensslin, Norbert
Farrell, John A
Foster, Duncan Graham, Jr
Fowler, Malcolm McFarland
Friar, James Lewis
Froman, Darol Kenneth
Garvey, Gerald Thomas
Gibson, Benjamin Franklin, V
Ginocchio, Joseph Natale
Gordon, James Wylie
Graves, Glen Atkins
Grover, George Maurice
Haight, Robert Cameron
Halbleib, John A
Hardekopf, Robert Allen
Heffner, Robert H
Hemmendinger, Arthur
Hopkins, John Chapman
Hull, McAllister Hobart, Jr
Hurd, Jon Rickey
Jain, Mahavir
Jarmie, Nelson
Jett, James Hubert
Johnson, Mikkel Borlaug
Keaton, Paul W, Jr
Keepin, George Robert, Jr
Keyworth, George A, II
King, Nicholas S P
Koehler, Dale Roland
Krick, Merlyn Stewart
Lee, David Mallin
Liu, Lon-Chang
MacArthur, Duncan W
Macek, Robert James
McNally, James Henry
Maggiore, Carl Jerome
Malanify, John Joseph
Manley, John Henry
Marlow, Keith Winton
Mazarakis, Michael Gerassimos
Mischke, Richard E
Moore, Michael Stanley
Moss, Calvin E
Motz, Henry Thomas
Muir, Douglas William
Neher, Leland K
Nelson, Gerald Clifford
Nicholson, Nicholas
Nickols, Norris Allan
Nix, James Rayford
Obst, Andrew Wesley
Orth, Charles Joseph
Parker, Jack Lindsay
Peng, Jen-chieh
Perkins, Roger Bruce
Peterson, Robert W
Peterson, Rolf Eugene
Plassmann, Elizabeth Hebb
Plassmann, Eugene Adolph
Poore, Emery Ray Vaughn
Quinn, Warren Eugene
Ragan, Charles Ellis, III
Rahal, Leo James
Renken, James Howard
Richter, John Lewis
Robertson, Robert Graham Hamish
Sandmeier, Henry Armin
Sattler, Allan R
Schery, Stephen Dale
Schriber, Stanley Owen
Scott, Hugh Logan, III
Seagrave, John Dorrington
Seamon, Robert Edward
Shera, E Brooks
Sierk, Arnold John
Silbar, Richard R(obert)
Smith, Hastings Alexander, Jr
Southward, Harold Dean
Steffen, Rolf Marcel
Stein, William Earl
Stelts, Marion Lee
Summers, Donald Lee
Talbert, Willard Lindley, Jr
Tanaka, Nobuyuki
Tapphorn, Ralph M
Tesmer, Joseph Ransdell
Thiessen, Henry Archer
Veeser, Lynn Raymond
Vieira, David John
Wagner, Robert Thomas
Warner, Laurance Bliss
Watt, Bob Everett
Yates, Mary Anne

Young, Lloyd Martin
Young, Phillip Gaffney

NEW YORK
Alburger, David Elmer
Auerbach, Elliot H
Baltz, Anthony John
Bennett, Gerald William
Bernabei, Austin M
Bhat, Mulki Radhakrishna
Block, Robert Charles
Blumberg, Leroy Norman
Bond, Albert Haskell, Jr
Bond, Peter Danford
Bozoki, George Edward
Braun-Munzinger, Peter
Brown, Hugh Needham
Brown, Stanley Gordon
Burrows, Thomas Wesley
Carew, John Francis
Casten, Richard Francis
Chasman, Chellis
Chou, Tzi Shan
Choudhury, Deo C
Chrien, Robert Edward
Clark, David Delano
Cline, Douglas
Cokinos, Dimitrios
Corson, Dale Raymond
Cumming, James B
Davis, Harold Larue
Deady, Matthew William
Der Mateosian, Edward
Devons, Samuel
Dicello, John Francis, Jr
Divadeenam, Mundrathi
Dreiss, Gerard Julius
Duek, Eduardo Enrique
Eddy, Jerry Kenneth
Eidson, William Whelan
Engelke, Charles Edward
Favale, Anthony John
Feingold, Arnold Moses
Feldman, Lawrence
Fielder, Douglas Stratton
Foley, Kenneth John
Foreman, Bruce Milburn, Jr
Franco, Victor
Fulbright, Harry Wilks
Ganley, W Paul
Garber, Donald I
Garg, Jagadish Behari
Gavin, Gerard Brennan
Gill, Ronald Lee
Gonzalez, Robert Anthony
Graves, Robert Gage
Harkavy, Allan Abraham
Havens, William Westerfield, Jr
Hofer, Gregory George
Holbrow, Charles H
Holden, Norman Edward
Horoshko, Roger N
Huizenga, John Robert
Hurwitz, Henry, Jr
Johnson, Brant Montgomery
Kane, Walter Reilly
Kistner, Ottmar Casper
Kitchen, Sumner Wendell
Koltun, Daniel S
Kostroun, Vaclav O
Kraner, Hobart Wilson
Lancman, Henry
Lattimer, James Michael
Lee, Linwood Lawrence, Jr
Liebenauer, Paul (Henry)
Lipson, Edward David
Li-Scholz, Angela
Lubitz, Cecil Robert
McGrath, Robert L
McGuire, Stephen Craig
McLane, Victoria
Medicus, Heinrich Adolf
Mendell, Rosalind B
Millener, David John
Min, Kongki
Motz, Lloyd
Muether, Herbert Robert
Mughabghab, Said F
Mukhopadhyay, Nimai Chand
Nagamiya, Shoji
Neeson, John Francis
Olness, John William
Padalino, Stephen John
Parsa, Zohreh
Parsegian, V(ozcan) Lawrence
Paul, Peter
Pearlstein, Sol
Piel, William Frederick
Pile, Philip H
Raboy, Sol
Rafla, Sameer
Reber, Jerry D
Reynolds, John Terrence
Rezanka, Ivan
Ries, Richard Ralph
Roalsvig, Jan Per
Rustgi, Moti Lal
Sailor, Vance Lewis
Sandorfi, Andrew M J
Schneid, Edward Joseph
Scholz, Wilfried
Schulte, Robert Lawrence
Schwarzschild, Arthur Zeiger
Shakin, Carl M

Shore, Ferdinand John
Souder, Paul A
Sprouse, Gene Denson
Stachel, Johanna
Stoler, Paul
Stroke, Hinko Henry
Sun, Chih-Ree
Thieberger, Peter
Titone, Luke Victor
Tuli, Jagdish Kumar
Vortuba, Jan
Warburton, Ernest Keeling
Ward, Thomas Edmund
Wilson, Fred Lee
Zahed, Ismail
Zaider, Marco A

NORTH CAROLINA
Adelberger, Rexford E
Clator, Irvin Garrett
Clegg, Thomas Boykin
Cleveland, Gregor George
Cobb, Grover Cleveland, Jr
Cotanch, Stephen Robert
Ely, Ralph Lawrence, Jr
Floyd, Carey E, Jr
Good, Wilfred Manly
Gould, Christopher Robert
Gross, Edward Emanuel
Hubbard, Paul Stancyl, Jr
Jaszczak, Ronald Jack
Kalbach, Constance
Lewis, Harold Walter
Lindsay, James Gordon, Jr
Ludwig, Edward James
Mitchell, Gary Earl
Oberhofer, Edward Samuel
Park, Jae Young
Roberson, Nathan Russell
Rogosa, George Leon
Rolland, William Woody
Tilley, David Ronald
Waldman, Bernard
Walker, William Delany
Walter, Richard L
Weller, Henry Richard

OHIO
Agard, Eugene Theodore
Anderson, Bryon Don
Bassel, Robert Harold
Becker, Lawrence Charles
Brient, Charles E
Dillman, Lowell Thomas
Dollhopf, William Edward
Finlay, Roger W
Giamati, Charles C, Jr
Grimes, Steven Munroe
Hagee, George Richard
Hathaway, Charles Edward
Hausman, Hershel J
Hemsky, Joseph William
Jalbert, Jeffrey Scott
Jastram, Philip Sheldon
Jha, Shacheenatha
Julian, Glenn Marcenia
Kepes, Joseph John
Klingensmith, Raymond W
Lane, Raymond Oscar
Lemming, John Frederick
MacIntyre, William James
Madey, Richard
Madia, William J
Mantil, Joseph Chacko
Mayers, Richard Ralph
Ploughe, William D
Poggenburg, John Kenneth, Jr
Poth, James Edward
Rapaport, Jacobo
Ruegsegger, Donald Ray, Jr
Schneider, Ronald E
Segelken, Warren George
Seyler, Richard G
Sugarbaker, Evan Roy
Sund, Raymond Earl
Warner, Robert Edson
Winters, Ronald Ross
Wood, Galen Theodore
Wright, Louis E

OKLAHOMA
Bell, Robert Edward
Buchanan, Ronnie Joe
Cook, Charles Falk
Huffaker, James Neal
Major, John Keene
Petry, Robert Franklin
Polson, William Jerry
Souder, Wallace William

OREGON
Easterday, Harry Tyson
Goldhammer, Paul
Kelley, Raymond H
Krane, Kenneth Saul
Lefevre, Harlan W
McDaniels, David K
Madsen, Victor Arviel
Mather, Keith Benson
Orloff, Jonathan H
Overley, Jack Castle
Richert, Anton Stuart
Schecter, Larry
Siemens, Philip John

Wack, Paul Edward
Wetzel, Karl Joseph
White, Donald Harvey

PENNSYLVANIA
Ajzenberg-Selove, Fay
Balamuth, David P
Bartko, John
Bilaniuk, Oleksa-Myron
Bock, Charles Walter
Cohen, Leonard David
Connors, Donald R
Couch, Jack Gary
Daehnick, Wilfried W
Davies, Kenneth Thomas Reed
Devins, Delbert Wayne
Doub, William Blake
Dressler, Edward Thomas
Emmerich, Werner Sigmund
Erikson, J Alden
Feldmeier, Joseph Robert
Folk, Robert Thomas
Fortune, H Terry
Freed, Norman
Georgopulos, Peter Demetrios
Graetzer, Reinhard
Green, Lawrence
Guttmann, Mark
Hardy, Judson, Jr
Haskins, Joseph Richard
Kahler, Albert Comstock, III
Lippincott, Ezra Parvin
Lochstet, William A
Luce, Robert James
Middleton, Roy
Mooney, Edward, Jr
Ostrander, Peter Erling
Pacer, John Charles
Pinkerton, John Edward
Poss, Howard Lionel
Pratt, William Winston
Rhodes, Jacob Lester
Saladin, Jurg X
Sarram, Mehdi
Shwe, Hla
Siems, Norman Edward
Smith, Winfield Scott
Snedegar, William H
Sorensen, Raymond Andrew
Stallwood, Robert Antony
Thwaites, Thomas Turville
Van Patter, Douglas Macpherson
Walkiewicz, Thomas Adam
Wicker, Everett E
Zurmuhle, Robert W

RHODE ISLAND
Lall, Prithvi C
Levin, Frank S
Mecca, Stephen Joseph
Riesenfeld, Peter William
Sullivan, Edmund Joseph

SOUTH CAROLINA
Aull, Luther Bachman, III
Avignone, Frank Titus, III
Baumann, Norman Paul
Benjamin, Richard Walter
Blanpied, Gary Stephen
Brantley, William Henry
Edge, Ronald (Dovaston)
Hahn, Walter I
Hendrick, Lynn Denson
Keller, Eldon Lewis
O'Neill, George Francis
Palms, John Michael
Parks, Paul Blair
Preedom, Barry Mason
Roggenkamp, Paul Leonard
Saunders, Edward A
Sawers, James Richard, Jr
Turner, James David

SOUTH DAKOTA
Hein, Warren Walter

TENNESSEE
Akovali, Yurdanur A
Albridge, Royal
Appleton, B R
Bair, Joe Keagy
Ball, James Bryan
Beene, James Robert
Bemis, Curtis Elliot, Jr
Bertrand, Fred Edmond
Bingham, Carrol R
Blankenship, James Lynn
Burton, John Williams
Carter, Hubert Kennon
Collins, Warren Eugene
De Saussure, Gerard
Dickens, Justin Kirk
Ernst, David John
Ferguson, Robert Lynn
Garrett, Jerry Dale
George, Ted Mason
Gove, Norwood Babcock
Greenbaum, Elias
Halbert, Melvyn Leonard
Hamilton, Joseph H, Jr
Harvey, John Arthur
Horen, Daniel J
Johnson, Cleland Howard
Johnson, Noah R

Jones, Charles Miller, Jr
Kim, Hee Joong
Kozub, Raymond Lee
Lenhard, Joseph Andrew
Lide, Robert Wilson
Ludemann, Carl Arnold
McGowan, Francis Keith
Macklin, Richard Lawrence
MacPherson, Herbert Grenfell
Maienschein, Fred (Conrad)
Moak, Charles Dexter
Murphy, Brian Donal
Neiler, John Henry
Nestor, C William, Jr
Newman, Eugene
Obenshain, Felix Edward
Peelle, Robert W
Penny, Keith, Sr
Perey, Francis George
Pinajian, John Joseph
Pinkston, William Thomas
Plasil, Franz
Raman, Subramanian
Ramayya, Akunuri V
Riedinger, Leo Louis
Robinson, Russell Lee
Sayer, Royce Orlando
Schmitt, Harold William
Shriner, John Franklin, Jr
Shugart, Cecil G
Snell, Arthur Hawley
Spejewski, Eugene Henry
Toth, Kenneth Stephen
Trammell, Rex Costo
Turner, James Edward
Ventrice, Carl Alfred
Walter, F John
Wells, John Calhoun, Jr
Young, Gale
Young, Glenn Reid
Zucker, Alexander

TEXAS
Aldridge, Jack Paxton, III
Allum, Frank Raymond
Baker, Stephen Denio
Bendel, Warren Lee
Biedenharn, Lawrence C, Jr
Blue, Michael Henry
Bonner, Billy Edward
Bramblett, Richard Lee
Bronson, Jeff Donaldson
Class, Calvin Miller
Cook, Clarence Sharp
Cranberg, Lawrence
Diana, Leonard M
Duck, Ian Morley
Fetzer, Homer D
Fields, Reuben Elbert
Foster, Bruce Parks
Givens, Wyatt Wendell
Gnade, Bruce E
Green, Phillip Joseph, II
Haenni, David Richard
Hiebert, John Covell
Hudspeth, Emmett Leroy
Hurd, James W
Jacobson, Larry A
Johnson, Raleigh Francis, Jr
Kenefick, Robert Arthur
Ko, Che Ming
Kostoff, Morris R
Little, Robert Narvaez, Jr
Lodhi, Mohammad Arfin Khan
Lopez, Jorge Alberto
Loyd, David Heron
McDaniel, Floyd Delbert, Sr
McIntyre, John Armin
Maute, Robert Edgar
Meshkov, Sydney
Miller, Maurice Max
Miller, Philip Dixon
Mills, William Raymond
Mitra, Sunanda
Mize, Jack Pitts
Moore, C Fred
Morgan, Ira Lon
Mutchler, Gordon Sinclair
Northcliffe, Lee Conrad
Ray, Robert Landon
Rester, David Hampton
Smith, William Robert
Suttle, Andrew Dillard, Jr
Taylor, Morris Chapman
Terrell, Glen Edward
Trail, Carroll C
Tribble, Robert Edmond
Trost, Hans-Jochen
Vincent, Lloyd Drexell
Wang, Ken Hsi
Wehring, Bernard William
Weinstein, Roy
Wildenthal, Bryan Hobson
Wright, David Patrick
Zaidi, Syed Amir Ali

UTAH
Dixon, Dwight R
Hatch, Eastman Nibley
Hill, Max W
Jensen, Gary Lee
Lund, Mark Wylie
Mangelson, Nolan Farrin
Otteson, Otto Harry

Rogers, Vern Child
Stokes, Gordon Ellis

VERMONT
Arns, Robert George
Detenbeck, Robert Warren
Dunham, Jeffrey Solon

VIRGINIA
Abashian, Alexander
Artna-Cohen, Agda
Atalay, Bulent Ismail
August, Leon Stanley
Beach, Louis Andrew
Bernthal, Frederick Michael
Brandt, Richard Gustave
Buck, Warren W, III
Calle, Carlos Ignacio
Cardman, Lawrence Santo
Cohen, Leslie
Cook, Charles William
Cooper, Larry Russell
Crowley, Patrick Arthur
Dardis, John G
Davidson, Charles Nelson
Eck, John Stark
Engelder, Theodore Carl
Erb, Karl Albert
Eubank, Harold Porter
Gibbons, John Howard
Gilfoyle, Gerard Paul
Gotow, Kazuo
Grunder, Hermann August
Gugelot, Piet C
Hafele, Joseph Carl
Ijaz, Mujaddid A
Jenkins, David A
Joyner, Weyland Thomas, Jr
Kane, John Robert
King, George, III
Kossler, William John
Kramish, Arnold
Kuang, Yunan
Lankford, William Fleet
Lenz, George H
Lindgren, Richard Arthur
Long, Dale Donald
MacKinney, Arland Lee
Minehart, Ralph Conrad
Moore, Garry Edgar
Muehlhause, Carl Oliver
Orr, Marshall H
Peacock, Richard Wesley
Perdrisat, Charles F
Peters, Charles William
Petersen, Edward Leland
Phillips, E Alan
Reierson, James (Dutton)
Ritter, Rogers C
Roy, Donald H
Shapiro, Philip
Singh, Jag Jeet
Sloope, Billy Warren
Sobottka, Stanley Earl
Sparrow, D(avid) A
Stevens, Donald Meade
Stronach, Carey E
Sundelin, Ronald M
Thornton, Stephen Thomas
Townsend, Lawrence Willard
Ugincius, Peter
Weber, Hans Jurgen
Whitehead, Walter Dexter, Jr
Winter, Rolf Gerhard
Wolicki, Eligius Anthony
Zimmerman, Peter David
Ziock, Klaus Otto H

WASHINGTON
Adams, Harry
Adelberger, Eric George
Bertsch, George Frederick
Bodansky, David
Cramer, John Gleason
Davidson, Melvin G
Farwell, George Wells
Glasgow, Dale William
Glass, James Clifford
Gold, Raymond
Gore, Bryan Frank
Guy, Reed Augustus
Haxton, Wick Christopher
Hayward, Thomas Doyle
Ingalls, Paul D
Kerlee, Donald D
Kouzes, Richard Thomas
Lindenmeier, Charles William
Lindsay, Richard H
Mann, Frederick Michael
Miller, Gerald Alan
Newman, Darrell Francis
Nichols, Davis Betz
Pengra, James G
Porter, Leonard Edgar
Reeder, Paul Lorenz
Rehfield, David Michael
Rupaal, Ajit S
Schenter, Robert Earl
Slee, Frederick Watford
Sperry, Willard Charles
Tripard, Gerald Edward
Veit, Jiri Joseph
Warner, Ray Allen
Weitkamp, William George

Wilets, Lawrence
Wilkes, Richard Jeffrey
Wood, Donald Eugene
Youtz, Byron Leroy

WEST VIRGINIA
Arya, Atam Parkash

WISCONSIN
Balantekin, Akif Baha
Barschall, Henry Herman
Besch, Gordon Otto Carl
Bevelacqua, Joseph John
Blanchard, Converse Herrick
Chander, Jagdish
Dobson, David A
Ebel, Marvin Emerson
Ferry, James A
Knutson, Lynn D
Kroger, Larry A
Kurey, Thomas John
Lee, Kiuck
Nickles, Robert Jerome
Patterson, James Reid
Richards, Hugh Taylor
Stanwick, Glenn

WYOMING
DeVries, Ralph Milton
Kunselman, A(rthur) Raymond

ALBERTA
Bland, Clifford J
Lam, Sheung Tsing
McDonald, W John
Olsen, William Charles
Stinson, Glen Monette

BRITISH COLUMBIA
Axen, David
Bartholomew, Gilbert Alfred
Beer, George Atherley
Craddock, Michael Kevin
Cujec, Bibijana Dobovisek
Davis, Charles Alan
Dawson, Wilfred Kenneth
Fraser, John Stiles
Griffiths, George Motley
Hausser, Otto Friedrich
Jennings, Byron Kent
Jones, Garth
Martin, Peter Wilson
Mason, Grenville R
Measday, David Frederick
Moss, Gerald Allen
Neilson, George Croydon
Pepper, Thomas Peter
Robertson, Lyle Purmal
Sample, John Thomas
Vogt, Erich Wolfgang

MANITOBA
Barnard, John Wesley
Falk, Willie Robert
Svenne, Juris Peteris
Van Oers, Willem Theodorus Hendricus

NEW BRUNSWICK
Matthews, James Horace

NEWFOUNDLAND
Irfan, Muhammad

NOVA SCOTIA
Latta, Bryan Michael

ONTARIO
Aitken, James Henry
Alexander, Thomas Kennedy
Alford, William Parker
Andrews, Hugh Robert
Ball, Gordon Charles
Barton, Richard Donald
Batay-Csorba, Peter Andrew
Beukens, Roelf Pieter
Brannen, Eric
Cameron, John Alexander
Clarke, Robert Lee
Craig, Donald Spence
Davis, Ronald Stuart
Earle, Eric Davis
Egelstaff, Peter A
Ewan, George T
Geiger, James Stephen
Geiger, Klaus Wilhelm
Hagberg, Erik Gordon
Hardy, John Christopher
Horn, Dag
Ing, Harry
Johns, Martin Wesley
Johnson, Frederick Allan
Kennett, Terence James
King, James Douglas
Knowles, John Warwick
Koslowsky, Vernon Theodore
Kuehner, John Alan
Lachaine, Andre Raymond Joseph
Lee, Hoong-Chien
Litherland, Albert Edward
Logan, Brian Anthony
Lone, M(uhammad) Aslam
McDonald, Arthur Bruce
McNeill, Kenneth Gordon
Miller, Richard Graham

PHYSICS & ASTRONOMY / 529

Milton, John Charles Douglas
Nogami, Yukihisa
Ogata, Hisashi
Ollerhead, Robin Wemp
Preston, Melvin Alexander
Radford, David Clarke
Romo, William Joseph
Rowe, David John
Sargent, Bernice Weldon
Sherman, Norman K
Simpson, John Joseph
Sinha, Bidhu Bhushan Prasad
Sprung, Donald Whitfield Loyal
Stirling, Andrew John
Summers-Gill, Robert George
Sundaresan, Mosur Kalyanaraman
Taylor, Harry William
Volkov, Anatole Boris
Ward, David
Wolfson, Joseph Laurence
Wong, Samuel Shaw Ming

QUEBEC
Barrette, Jean
Chaubey, Mahendra
Del Bianco, Walter
De Takacsy, Nicholas Benedict
Eddy, Nelson Wallace
Goulard, Bernard
Hetherington, Donald Wordsworth
Kroger, Helmut Karl
Lee, Jonathan K P
Levesque, Rene J A
Mark, Shew-Kuey
Martin, William Macphail
Moore, Robert B
Paskievici, Wladimir
Pearson, John Michael
Pla, Conrado
Rioux, Claude
Robson, John Michael
Sharma, Ramesh C
St-Pierre, Claude
Taras, Paul

SASKATCHEWAN
Naqvi, Saiyid Ishrat Husain
Papini, Giorgio Augusto
Rangacharyulu, Chilakamarri
Shin, Yong-Moo

OTHER COUNTRIES
Afnan, Iraj Ruhi
Bolotin, Herbert Howard
Cue, Nelson
Delsanto, Pier Paolo
Gil, Salvador
Hansen, Ole
Huang, Keh-Ning
Koehler, Helmut A
Kubik, Peter W
Kwok, Hoi S
Ling, Samuel Chen-Ying
Lipkin, Harry Jeannot
Montague, John H
Parsignault, Daniel Raymond
Powers, Richard James
Radkowsky, Alvin
Robinson, Berol (Lee)
Van Der Meer, Simon
Yellin, Joseph

Optics

ALABAMA
Allan, Barry David
Bailey, Wayne Lewis
Barr, E(rnest) Scott
Blakney, William G G
Bloemer, Mark Joseph
Bowden, Charles Malcolm
Caulfield, Henry John
Christensen, Charles Richard
Cushing, Kenneth Mayhew
Dunn, Anne Roberts
Essenwanger, Oskar M
Gilbert, John Andrew
Haak, Frederick Albertus
Hartman, Richard Leon
Hoover, Richard Brice
Izatt, Jerald Ray
Jones, Robert William
Korsch, Dietrich G
Lee, Ching Tsung
McKnight, William Baldwin
Randall, Joseph Lindsay
Roberts, Thomas George
Rosenberger, Albert Thomas
Shealy, David Lee
Stettler, John Dietrich
Tidwell, Eugene Delbert
Vikram, Chandra Shekhar
Vohra, Yogesh K
Watson, Raymond Coke, Jr
Wild, Bradford Williston
Williamson, Arthur Elridge, Jr
Williamson, Ashley Deas

ALASKA
Deehr, Charles Sterling
Stenbaek-Nielsen, Hans C

Optics (cont)

ARIZONA
Angel, James Roger Prior
Bao, Qingcheng
Barrett, Harrison Hooker
Bartels, Peter H
Burke, James Joseph, Jr
Christy, James Walter
Daw, Glen Harold
Denton, M Bonner
Dereniak, Eustace Leonard
Falco, Charles Maurice
Franken, Peter Alden
Frieden, Bernard Roy
Gaskill, Jack Donald
Giampapa, Mark Steven
Gibbs, Hyatt McDonald
Green, Barry A
Harris, Hugh Courtney
Haynes, Munro K
Hill, Henry Allen
Hilliard, Ronnie Lewis
Hirleman, Edwin Daniel, Jr
Jacobson, Michael Ray
Kessler, Bernard V
McKenney, Dean Brinton
Macleod, Hugh Angus
Marathay, Arvind Shankar
Mayall, Nicholas Ulrich
Palmer, James McLean
Peyghambarian, Nasser
Powell, Richard Conger
Rasmussen, William Otto
Sarid, Dror
Scholl, Marija Strojnik
Schowengerdt, Robert Alan
Shack, Roland Vincent
Shannon, Robert Rennie
Shoemaker, Richard Lee
Slater, Philip Nicholas
Smith, David John
Stamm, Robert Franz
Turner, Arthur Francis
Vogelmann, Andrew Mark
Wolfe, William Louis, Jr
Wyant, James Clair

ARKANSAS
Eichenberger, Rudolph John
Gupta, Rajendra
Horton, Philip Bish
Leming, Charles William
Richardson, Charles Bonner
Salamō, Gregory Joseph
Singh, Surendra Pal
Smith, E(astman)

CALIFORNIA
Adams, Arnold Lucian
Ahumada, Albert Jil, Jr
Amirkhanian, Varouj David
Ammann, E(ugene) O(tto)
Aronowitz, Frederick
Ashby, Val Jean
Asmus, John Fredrich
Austin, Roswell W(allace)
Avizonis, Petras V
Babrov, Harold J
Baez, Albert Vinicio
Bailey, Ian L
Bajaj, Jagmohan
Bardsley, James Norman
Barry, James Dale
Bartling, Judd Quentin
Baur, James Francis
Becker, Randolph Armin
Beer, Reinhard
Bennett, Glenn Taylor
Bennett, Harold Earl
Bennett, Jean McPherson
Bernard, Douglas Alan
Bernstein, Uri
Bethune, Donald Stimson
Billings, Bruce Hadley
Bixler, Otto C
Bjorklund, Gary Carl
Bliss, Erlan S
Bloom, Arnold Lapin
Bloxham, Laurence Hastings
Boynton, Robert M
Bracewell, Ronald Newbold
Breckinridge, James Bernard
Brewer, LeRoy Earl, Jr
Burge, Dennis Knight
Caird, John Allyn
Cannell, David Skipwith
Carman, Robert Lincoln, Jr
Carver, John Guill
Ceglio, Natale Mauro
Cho, Young-chung
Chu, Steven
Cooper, Alfred William Madison
Cooper, Donald Edward
Coufal, Hans-Jurgen
Cover, Ralph A
Crittenden, Eugene Casson, Jr
Danielson, George Edward, Jr
Davies, Merton Edward
Davis, James Ivey
Davis, Jeffrey Arthur
Davis, Thomas Pearse
Deacon, David A G
Deckert, Curtis Kenneth
Dember, Alexis Berthold
Deri, Robert Joseph
Dessel, Norman F
Donovan, Terence M
Dowling, Jerome M
Doyle, Walter M
Duffield, Jack Jay
Eckstrom, Donald James
Edwards, David Franklin
Ellion, M Edmund
Elliott, Stuart Bruce
Enoch, Jay Martin
Epps, Harland Warren
Erdmann, John Hugo
Errett, Daryl Dale
Evtuhov, Viktor
Fahlen, Theodore Stauffer
Fairand, Barry Philip
Farmer, Crofton Bernard
Feichtner, John David
Feit, Michael Dennis
Fenner, Wayne Robert
Fisher, Philip Chapin
Fisher, Robert Alan
Fleck, Joseph Amadeus, Jr
Fouquet, Julie
Frank, Alan M
Fried, David L
Gamo, Hideya
Garren, Alper A
Gelbwachs, Jerry A
Geller, Myer
Gilmartin, Thomas Joseph
Gimlett, James I
Glass, Alexander Jacob
Glass, Nathaniel E
Goldsborough, John Paul
Goodman, Joseph Wilfred
Goosman, David R
Graham, Robert (Klark)
Green, Philip S
Grossman, Jack Joseph
Gruber, John B
Gudmundsen, Richard Austin
Gundersen, Martin Adolph
Hadeishi, Tetsuo
Halden, Frank
Hall, James Timothy
Han, Ki Sup
Harris, Dennis George
Haugen, Gilbert R
Hellwarth, Robert Willis
Helstrom, Carl Wilhelm
Henderson, David Michael
Henning, Harley Barry
Heritage, Jonathan Paul
Hesselink, Lambertus
Hilbert, Robert S
Hodges, Dean T, Jr
Holmes, Dale Arthur
Holzrichter, John Frederick
Honey, Richard Churchill
Hopkins, George William, II
Hsia, Yukun
Huang, C Yuan
Hudson, Richard Delano, Jr
Hunt, Arlon Jason
Hutchin, Richard Ariel
Jacobs, Ralph R
Janney, Gareth Maynard
Jansen, Michael
Jing, Liu
Johnson, Chris Alan
Johnston, Alan Robert
Johnston, George Taylor
Judd, Floyd L
Jusinski, Leonard Edward
Jutamulia, Suganda
Kahn, Frederic Jay
Kalma, Arne Haerter
Kapany, Narinder Singh
Karp, Arthur
Karunasiri, Gamani
Kelly, Raymond Leroy
Khoshnevisan, Mohsen Monte
Kino, Gordon Stanley
Kirk, John Gallatin
Knize, Randall James
Koehler, Wilbert Frederick
Krishnan, Kamala Sivasubramaniam
Kulander, Kenneth Charles
Lakkaraju, H S
Lampert, Carl Matthew
Lapp, M(arshall)
Larmore, Lewis
Leavy, Paul Matthew
Lee, Wai-Hon
Lehovec, Kurt
Levenson, Marc David
Levine, Bruce Martin
Li, Chia-Chuan
Lieber, Richard L
Lieberman, Robert Arthur
Lin, Anthony T
Linford, Gary Joe
Liu, Hua-Kuang
Liu, Jia-ming
Lombardi, Gabriel Gustavo
Lurie, Norman A(lan)
Lynch, David Dexter
McFarlane, Ross Alexander
McFee, Raymond Herbert
McKenzie, Robert Lawrence
McQuillen, Howard Raymond
Maker, Paul Donne
Manes, Kennneth Rene
Martin, Terry Zachry
Martinez-Uriegas, Eugenio
Massey, Gail Austin
Massie, N A (Bert)
Mast, Terry S
Mathis, Ronald Floyd
Maydan, Dan
Meier, Rudolf H
Meinel, Aden Baker
Meinel, Marjorie Pettit
Menzies, Robert Thomas
Meyers, Robert Allen
Morris, James Russell
Morris, Richard Herbert
Moss, Steven C
Muller, Richard A
Muller, Rolf Hugo
Mundie, Lloyd George
Murray, John Roberts
Nanes, Roger
Nee, Soe-Mie Foeng
Nicodemus, Fred(erick) E(dwin)
Noble, Robert Hamilton
O'Keefe, John Dugan
O'Loane, James Kenneth
Palatnick, Barton
Pan, Yu-Li
Park, Edward C(ahill), Jr
Partanen, Jouni Pekka
Passenheim, Burr Charles
Pavlopoulos, Theodore G
Perry, Richard Lee
Perry, Robert Nathaniel, III
Pertica, Alexander Jose
Pignataro, Augustus
Pines, Alexander
Porteus, James Oliver
Powers, John Patrick
Rancourt, James Daniel
Randall, Charles McWilliams
Randle, Robert James
Rast, Howard Eugene, Jr
Rawson, Eric Gordon
Reinheimer, Julian
Rennilson, Justin J
Reynolds, Michael David
Rice, Dennis Keith
Richman, Isaac
Richter, Thomas A
Rickard, James Joseph
Rockower, Edward Brandt
Rockwell, David Alan
Romagnoli, Robert Joseph
Rothrock, Larry R
Rowntree, Robert Fredric
Russell, Philip Boyd
Saito, Theodore T
Salanave, Leon Edward
Saltzman, Max
Sandefur, Kermit Lorain
Satten, Robert A
Schaefer, Albert Russell
Schawlow, Arthur Leonard
Schmars, William Thomas
Schulte, Daniel Herman
Scifres, Donald Ray
Sclar, Nathan
Senitzky, Israel Ralph
Seppala, Lynn G
Shank, Charles Vernon
Shelby, Robert McKinnon
Sheridon, Nicholas Keith
Sherman, George Charles
Shykind, David
Siegman, A(nthony) E(dward)
Silverstein, Elliot Morton
Small, James Graydon
Smiley, Vern Newton
Smith, Sheldon Magill
Smith, Todd Iversen
Soffer, Bernard Harold
Speck, David Ralph
Speen, Gerald Bruce
Spencer, Donald Jay
Spiro, Irving J
Spitzer, William George
Sprague, Robert Arthur
Stapelbroek, Maryn G
Starkweather, Gary Keith
Stegelmann, Erich J
Stierwalt, Donald L
Stockwell, Norman D
Stokowski, Stanley E
Stramski, Dariusz
Strand, Timothy Carl
Streifer, William
Swalen, Jerome Douglas
Sziklai, George C(lifford)
Talley, Robert Morrell
Tam, Andrew Ching
Tangonan, Gregory Ligot
Tarbell, Theodore Dean
Terhune, Robert William
Tin, Hla Ngwe
Titterton, Paul James
Todd, Terry Ray
Tooley, Richard Douglas
Treves, David
Tricoles, Gus P
Trigger, Kenneth Roy
Trolinger, James Davis
Tsiang, Eugene Y
Tubbs, Eldred Frank
Twersky, Victor
Unti, Theodore Wayne Joseph
Urbach, John C
Vance, Dennis William
Wada, James Yasuo
Wagner, Richard John
Wallerstein, Edward Perry
Wang, Jon Y
Wang, Michael Renxun
Warren, Walter R(aymond), Jr
Watkins, Robert Arnold
Webber, Donald Salyer
Weber, Marvin John
Webster, Emilia
Weinstein, Berthold Werner
Weissbluth, Mitchel
Wells, Willard H
Welsh, David Edward
Whitcomb, Stanley Ernest
Whitefield, Rodney Joe
Whitney, William Merrill
Wieder, Harold
Wiedow, Carl Paul
Winter, Donald Charles
Wittry, David Beryle
Woerner, Robert Leo
Wolff, Milo Mitchell
Wunderman, Irwin
Wyatt, Philip Joseph
Yablonovitch, Eli
Yamakawa, Kazuo Alan
Yap, Fung Yen
Yariv, A(mnon)
Yeh, Cavour W
Yeh, Edward H Y
Yeh, Pochi Albert
Yeh, Yea-Chuan Milton
Young, Andrew Tipton
Zhou, Simon Zheng
Zook, Alma Claire

COLORADO
Anderson, Dana Zachary
Baur, Thomas George
Brault, James William
Bruns, Donald Gene
Cathey, Wade Thomas, Jr
Cerni, Todd Andrew
Chang, Bunwoo Bertram
Chen, Di
Chew, Herman W
Chisholm, James Joseph
Churnside, James H
Clifford, Steven Francis
Dana, Robert Watson
Dichtl, Rudolph John
Eberhard, Wynn Lowell
Emery, Keith Allen
Fairbank, William Martin, Jr
Galeener, Frank Lee
Griboval, Paul
Hadley, Lawrence Nathan
Henley, John Raymond
Hjelme, Dag Roar
Itano, Wayne Masao
Jewell, Jack Lee
Johnson, Eric G, Jr
Lazzarini, Albert John
Lightsey, Paul Alden
McMahon, Thomas Joseph
Mankin, William Gray
Moddel, Garret R
O'Callaghan, Michael James
O'Sullivan, William John
Owen, Robert Barry
Phelan, Robert J, Jr
Robinson, Hugh Gettys
Sanmann, Everett Eugene
Sauer, Jon Robert
Schwiesow, Ronald Lee
She, Chiao-Yao
Smith, Archibald William
Smith, Richard Cecil
Smith, Stephen Judson
Staebler, David Lloyd
Tatarskii, Valerian I
Taylor, Kenneth Doyle
Young, Matt

CONNECTICUT
Akkapeddi, Prasad Rao
Antar, Ali A
Astheimer, Robert W
Bartram, Ralph Herbert
Boggs, Steven Allan
Brienza, Michael Joseph
Cable, Peter George
Cheo, Peter K
Crawford, John Okerson
DeMaria, Anthony John
Dolan, James F
Dreyfus, Marc George
Erf, Robert K
Glenn, William Henry, Jr
Haacke, Gottfried
Harrison, Irene R
Hufnagel, Robert Ernest
Hyde, Walter Lewis
Javidi, Bahram
Keith, H(arvey) Douglas
Leonberger, Frederick John
Morrison, Richard Charles
Nath, Dilip K

O'Brien, Brian
Pinsley, Edward Allan
Poultney, Sherman King
Rich, John Charles
Roychoudhuri, Chandrasekhar
Siegmund, Walter Paul
Snyder, John William
Stetson, Karl Andrew
Yoder, Paul Rufus, Jr

DELAWARE
Bierlein, John David
Daniels, William Burton
Garland, Charles E
Glasser, Leo George
Gulick, Walter Lawrence
Huppe, Francis Frowin
Ih, Charles Chung-Sen
Jansson, Peter Allan
Klemas, Victor V
Pontrelli, Gene J
Ross, William D(aniel)
Sharnoff, Mark

DISTRICT OF COLUMBIA
Aggarwal, Ishwar D
Anderson, Gordon Wood
Apruzese, John Patrick
Beran, Mark Jay
Brown, Charles Moseley
Brueckner, Guenter Erich
Campillo, Anthony Joseph
Daehler, Mark
Faust, Walter Luck
Feldman, Barry Joel
Gaumond, Charles Frank
Giallorenzi, Thomas Gaetano
Goldstein, Jeffrey Jay
Griscom, David Lawrence
Hafizi, Bahman
Hessel, Merrill
Hornstein, John Stanley
Jordan, Arthur Kent
Koo, Kee P
Lee, John Norman
Lin-Chung, Pay-June
Mahon, Rita
Marquardt, Charles L(awrence)
Marrone, Michael Joseph
Meekins, John Fred
Mendlowitz, Harold
Michels, Donald Joseph
Prinz, Dianne Kasnic
Quinn, Jarus William
Reintjes, John Francis, Jr
Schneider, Irwin
Shirk, James Siler
Sica, Louis
Weiler, Kurt Walter
Weinberg, Donald Lewis
Wieting, Terence James
Wing, William Hinshaw

FLORIDA
Adhav, Ratnakar Shankar
Amon, Max
Anderson, Roland Carl
Aubel, Joseph Lee
Ballard, Stanley Sumner
Bass, Michael
Beck, Warren R(andall)
Blatt, Joel Herman
Bolte, John R
Browder, James Steve
Cox, John David
Ewald, William Philip
Godwin, Robert O(wen)
Gordon, Howard R
Green, Eugene L
Gustafson, Bo Ake Sture
Helava, U(uno) V(ilho)
Hett, John Henry
Holst, Gerald Carl
Killinger, Dennis K
Kohane, Theodore
Lannutti, Joseph Edward
Levy, Norman Stuart
Li Kam Wa, Patrick
McAlpine, Kenneth Donald
Makowski, Lee
Moreno, Wilfrido A
Mukherjee, Pritish
Newman, Roger
Owens, James Samuel
Pendleton, Winston Kent, III
Pittman, Melvin Amos
Rosendahl, Gottfried R
Schmidt, Klaus H
Schneider, Richard Theodore
Shih, Hansen S T
Smith, Richard Carper
Stegeman, George I
Stickley, C(arlisle) Martin
Tobey, Frank Lindley, Jr
Todd, Hollis N
Van Ligten, Raoul Fredrik
Van Stryland, Eric William
Wang, Ru-Tsang
Williamson, Donald Elwin
Young, C(harles) Gilbert
Ziernicki, Robert S
Zory, Peter S, Jr

GEORGIA
Anantha Narayanan, Venkataraman
Gaylord, Thomas Keith
Goldman, John Abner
Grams, Gerald William
Griffiths, James Edward
Gudehus, Donald Henry
Hartman, Nile Franklin
Jokerst, Nan Marie
Kenan, Richard P
Korda, Edward J(ohn)
Meltzer, Richard S
O'Shea, Donald Charles
Owens, William Richard
Payne, Marvin Gay
Roy, Rajarshi
Saunders, Morton Jefferson
Shackelford, Robert G
Stevenson, James Rufus
Verber, Carl Michael
Welch, Roy Allen
Wiltse, James Cornelius
Yen, William Maoshung

HAWAII
Crooker, Peter Peirce
Geballe, Thomas Ronald
Holmes, John Richard
Learned, John Gregory

IDAHO
Johnston, Lawrence Harding
Knox, John MacMurray

ILLINOIS
Abella, Isaac D
Adrian, Ronald John
Ahmed, Wase U
Benjamin, Roland John
Buck, Warren Louis
Cohn, Gerald Edward
Curry, Bill Perry
Dlott, Dana D
Dolan, Paul Joseph, Jr
Eden, James Gary
Farhadieh, Rouyentan
Faries, Dillard Wayne
Hull, John R
Joshi, Umeshwar Prasad
Karara, H(oussam) M
Kath, William Lawrence
Leibhardt, Edward
McCormack, Elizabeth Frances
Merkelo, Henri
Naylor, David L
Primak, William L
Reft, Chester Stanley
Rhodes, Charles Kirkham
Ring, James George
Sciammarella, Caesar August
Stanford, George Stailing
Stark, Henry
Staunton, John Joseph Jameson
Verdeyen, Joseph T
Wey, Albert Chin-Tang
Winston, Roland
Wolfe, Gene H
Wong, Kam Wu
Wozniak, Wayne Theodore

INDIANA
Allen, Merrill James
Bakken, George Stewart
Bunch, Robert Maxwell
Bunker, Bruce Alan
Conklin, Richard Louis
Harris, Samuel M(elvin)
Hofstetter, Henry W
Khorana, Brij Mohan
Ramdas, Anant Krishna
Western, Arthur Boyd
Wilson, Olin C(haddock)

IOWA
Behroozi, Feredoon
Green, Robert Wood
Hekker, Roeland M T
Korpel, Adrianus
McClelland, John Frederick
Nelson, David Torrison
Olson, Dale Wilson

KANSAS
Long, Larry L
Sobczynski, Radek
Sorensen, Christopher Michael

KENTUCKY
Brill, Joseph Warren
Cook, Ancel Eugene
Kielkopf, John F
McGinness, William George, III
Nathan, Richard Arnold

LOUISIANA
Allen, Susan Davis
Azzam, Rasheed M A
Chin-Bing, Stanley Arthur
Harris, Ronald Wilbert
Witriol, Norman Martin

MAINE
Chonacky, Norman J
Hughes, William Taylor

Krueger, George Corwin
McClymer, James P
Pribram, John Karl

MARYLAND
Allen, John Edward, Jr
Baba, Anthony John
Birnbaum, George
Bjerkaas, Allan Wayne
Boone, Bradley Gilbert
Boyd, Derek Ashley
Brasunas, John Charles
Brill, Dieter Rudolf
Brown, Douglas Edward
Budgor, Aaron Bernard
Calame, Gerald Paul
Carlon, Hugh Robert
Chang, Ren-Fang
Chen, Wenpeng
Clark, Charles Winthrop
Condell, William John, Jr
Del Grosso, Vincent Alfred
Drummeter, Louis Franklin, Jr
Ellingson, Robert George
Elton, Raymond Carter
Embury, Janon Frederick, Jr
Feldman, Albert
Feldman, Paul Donald
Forman, Richard Allan
Gammon, Robert Winston
Gleason, Thomas James
Goepel, Charles Albert
Goldin, Edwin
Gravatt, Claude Carrington, Jr
Hardesty, George K(irwan) C(ollison)
Harrington, Marshall Cathcart
Harvey, Albert Bigelow
Hochheimer, Bernard Ford
Holland, Russell Sedgwick
Holt, Helen Keil
Huang, Jacob Wen-Kuang
Jablonski, Frank Edward
Jamieson, J(ohn) A(nthony)
Kaplan, Alexander E
Kessler, Karl Gunther
Kuyatt, Chris E(rnie Earl)
Larrabee, R(obert) D(ean)
Lettieri, Thomas Robert
Levinson, John Z
Maccabee, Bruce Sargent
McCally, Russell Lee
McClelland, Jabez Jenkins
McIlrath, Thomas James
McLaughlin, William Lowndes
Majkrzak, Charles Francis
Major, Fouad George
Malarkey, Edward Cornelius
Mandelberg, Hirsch I
Mason, Robert Paige
Massof, Robert W
Mather, John Cromwell
Meyer, Richard Arthur
Mielenz, Klaus Dieter
Mohr, Peter J
Monaldo, Francis Michael
Mumma, Michael Jon
Murphy, John Cornelius
Ott, William Roger
Palmer, Charles Harvey
Paul, Dilip Kumar
Perry, James Ernest
Phillips, William Daniel
Plotkin, Henry H
Poehler, Theodore O
Price, Michael Glendon
Rabin, Herbert
Ravitsky, Charles
Rhee, Moon-Jhong
Rosenstock, Herbert Bernhard
Sattler, Joseph Peter
Schubert, David Crawford
Shettle, Eric Payson
Shumaker, John Benjamin, Jr
Sikora, Jerome Paul
Simonis, George Jerome
Sliney, David H
Smith, James Alan
Steiner, Bruce
Stockman, Hervey S, Jr
Stuebing, Edward Willis
Sztankay, Zoltan Geza
Tropf, William Jacob
Tschunko, Hubert F A
Vande Kieft, Lawrence John
Van De Merwe, Willem Pieter
Venkatesan, Thirumalai
Walker, James Harris
Wang, Ting-I
Waynant, Ronald William
Weaver, Christopher Scot
Weber, Alfons
Weidner, Victor Ray
Weiner, John
Willett, Colin Sidney
Wood, Howard John, III
Young, Russell Dawson
Yu, Anthony Woonchiu
Zankel, Kenneth L
Zirkind, Ralph

MASSACHUSETTS
Aggarwal, Roshan Lal
Aldrich, Ralph Edward
Ameer, George Albert
Aronson, James Ries
Bansil, Rama
Bechis, Kenneth Paul
Benton, Stephen Anthony
Bloembergen, Nicolaas
Bluemel, Van (Fonken Wilford)
Bradley, Lee Carrington, III
Brown, Walter Redvers John
Caley, Wendell J, Jr
Chandler, Charles H(orace)
Chang, Edward Shi Tou
Chase, Charles Elroy, Jr
Cohen, Fredric Sumner
Cohn, Daniel Ross
Cronin-Golomb, Mark
Dakss, Mark Ludmer
Davidson, Gilbert
Decker, C(harles) David
Dewey, C(larence) Forbes, Jr
Dionne, Gerald Francis
Doane, Marshall Gordon
Dorschner, Terry Anthony
Douglas-Hamilton, Diarmaid H
Driver, Richard D
Eby, John Edson
Elmer, William B
Feld, Michael S
Fohl, Timothy
Fujimoto, James G
Furumoto, Horace Wataru
Garodnick, Joseph
George, James Z
Gianino, Peter Dominic
Goela, Jitendra Singh
Gschwendtner, Alfred Benedict
Gurski, Thomas Richard
Harte, Kenneth J
Hawkins, W(illiam) Bruce
Hearn, David Russell
Hilborn, Robert Clarence
Hope, Lawrence Latimer
Horowitz, Joseph
Horwitz, Paul
Howard, John Nelson
Hunter, Larry Russel
Hurlbut, Cornelius Searle, Jr
Itzkan, Irving
Jacob, Jonah Hye
Jacobsen, Edward Hastings
Jones, Robert Allan
Jones, Robert Clark
Karakashian, Aram Simon
Karo, Douglas Paul
Keicher, William Eugene
Kelley, Paul Leon
Kellogg, Edwin M
Kleppner, Daniel
Kolodzy, Paul John
Kramer, Robert
Krieger, Allen Stephen
Kupferberg, Lenn C
Kuppenheimer, John D, Jr
Langer, R M
Lees, David Eric Berman
Lemons, Thomas M
Levin, Robert Edmond
Liau, Zong-Long
Lis, Steven Andrew
Litzenberger, Leonard Nelson
Liu, Xichun
Loewenstein, Ernest Victor
Lu, Grant
Luther, Holger Martin
Mack, Michael E
Mahoney, William C
Maloney, William Thomas
Marcus, Stephen
Marino, Richard Matthew
Masso, Jon Dickinson
Mazur, Eric
Milbocker, Michael
Miller, Robert Charles
Minden, Henry Thomas
Miniscalco, William J
Miranda, Henry A, Jr
Moulton, Peter Franklin
Nebolsine, Peter Eugene
Nelson, Donald Frederick
Ogilvie, Robert Edward
Oliner, Arthur A(aron)
Pearson, Edwin Forrest
Phillips, Richard Arlan
Picard, Richard Henry
Plummer, William Torsch
Pulling, Nathaniel H(osler)
Rabin, Monroe Stephen Zane
Rao, Devulapalli V G L N
Reasenberg, Robert David
Rediker, Robert Harmon
Robinson, Howard Addison
Rogers, Howard Gardner
Rork, Eugene Wallace
Sasiela, Richard
Sekuler, Robert W
Shaughnessy, Thomas Patrick
Shepp, Allan
Silk, John Kevin
Singh, Shobha
Spears, David Lewis
Stoner, William Weber
Stowe, David William
Taylor, Raymond L
Temkin, Richard Joel
Traub, Alan Cutler

Optics (cont)

Traub, Wesley Arthur
Tsang, Dean Zensh
Tuchman, Avraham
Vanasse, George Alfred
Volkmann, Frances Cooper
Wallace, James
Walther, Adriaan
Warde, Cardinal
Weinberger, Doreen Anne
Wetzstein, H(anns) J(uergen)
Whitehouse, David R(empfer)
Wilk, Leonard Stephen
Williamson, Richard Cardinal
Willner, Steven P
Wolfe, Stephen Mitchell
Woskov, Paul Peter
Zajonc, Arthur Guy

MICHIGAN
Aleksoff, Carl Chris
Beglau, David Alan
Berman, Paul Ronald
Braithwaite, John Geden North
Brumm, Douglas B(ruce)
Chen, Hsuan
Cloud, Gary Lee
Cox, Mary E
Donohue, Robert J
Eesley, Gary
Fienup, James R
Gara, Aaron Delano
Kozma, Adam
Kubis, Joseph J(ohn)
LaHaie, Ivan Joseph
LaRocca, Anthony Joseph
Leith, Emmett Norman
Lewis, Alan Laird
Liu, Yi
Miller, Herman Lunden
Montgomery, George Paul, Jr
Nazri, Gholam-Abbas
Peterson, Lauren Michael
Peterson, Richard Carl
Phelps, Frederick Martin, III
Prostak, Arnold S
Rand, Stephen Colby
Ressler, Neil William
Rimai, Lajos
Roessler, David Martyn
Sawatari, Takeo
Schrenk, Walter John
Segall, Stephen Barrett
Sharp, William Edward, III
Trentelman, George Frederick
Turner, Robert
Upatnieks, Juris
Vaishnava, Prem P
Van Baak, David Alan
Weil, Herschel
Yang, Wen Jei
Zissis, George John

MINNESOTA
Berg, James Irving
Heinisch, Roger Paul
Hewitt, Frederick George
Hill, Brian Kellogg
Hocker, George Benjamin
Johnson, Edgar Gustav
Kruse, Paul Walters, Jr
Lee, Pui Kum
Lo, David S(hih-Fang)
Mikkelson, Raymond Charles
Ready, John Fetsch
Robinson, Glen Moore, III
Schumer, Douglas Brian
Valasek, Joseph
Valley, Leonard Maurice
Walker, Charles Thomas
Zeyen, Richard John

MISSISSIPPI
Breazeale, Mack Alfred
Cibula, William Ganley
Ferguson, Joseph Luther, Jr

MISSOURI
Anderson, Richard Alan
Bryan, David A
Huddleston, Philip Lee
Leader, John Carl
Leopold, Daniel J
Linder, Solomon Leon
Liu, Yu
Look, Dwight Chester, Jr
Palmer, Kent Friedley
Rigler, A Kellam
Schmitt, John Leigh

MONTANA
Carlsten, John Lennart
Tynes, Arthur Richard
Winthrop, John T

NEBRASKA
Ducharme, Stephen
Ma, Wen
Woods, Joseph
Zepf, Thomas Herman

NEVADA
Dunipace, Donald William
Farley, John William

NEW HAMPSHIRE
Adjemian, Haroutioon
Baker, James Gilbert
Ebersole, John Franklin
Kidder, John Newell
King, Allen Lewis
Meyer, James Wagner
Perovich, Donald Kole
Smith, F(rederick) Dow(swell)
Torbert, Roy Banks
Weiler, Margaret Horton

NEW JERSEY
Abeles, Joseph Hy
Andrekson, Peter A(vo)
Anthony, Philip John
Barrett, Joseph John
Bartolini, Robert Alfred
Bechis, Dennis John
Bechtle, Daniel Wayne
Behrens, Herbert Ernest
Biswas, Dipak R
Bjorkholm, John Ernst
Boyd, Gary Delane
Broer, Matthijs Meno
Brucker, Edward Byerly
Capasso, Federico
Carlson, Curtis Raymond
Castor, William Stuart, Jr
Celler, George K
Channin, Donald Jones
Chraplyvy, Andrew R
Chung, Yun C
Church, Eugene Lent
Cody, George Dewey
Collett, Edward
Collier, Robert Jacob
Dutta, Mitra
Edwards, Christopher Andrew
Field, Norman J
Fischer, Russell Jon
Fleury, Paul A
Fork, Richard Lynn
Giordmaine, Joseph Anthony
Glass, Alastair Malcolm
Goldstein, Robert Lawrence
Gottscho, Richard Alan
Graf, Hans Peter
Gray, Russell Houston
Gualtieri, Devlin Michael
Hernqvist, Karl Gerhard
Hubbard, William Marshall
Hwang, Cherng-Jia
Johnson, Leo Francis
Joyce, William B(axter)
Kaminow, Ivan Paul
Kaprelian, Edward Karnig
Kapron, Felix Paul
Kluver, J(ohan) W(ilhelm)
Knausenberger, Wulf H
Knight, Douglas Maitland
Koch, Thomas L
Koester, Charles John
Kolodner, Paul R
Kornstein, Edward
Koss, Valery Alexander
Levine, Barry Franklin
Li, Tingye
Liao, Paul Foo-Hung
Linke, Richard Alan
McKenna, James
Medley, Sidney S
Meyerhofer, Dietrich
Miles, Richard Bryant
Miller, Arthur
Miller, David A B
Mollenauer, Linn F
Morrow, Scott
Moshey, Edward A
Murnick, Daniel E
Nahory, Robert Edward
O'Gorman, James
Orlando, Carl
Palladino, Richard Walter
Partovi, Afshin
Passner, Albert
Personick, Stewart David
Pinczuk, Aron
Plummer, Dirk Arnold
Ramsey, Alan T
Rastani, Kasra
Raybon, Gregory
Reed, William Alfred
Rothberg, Lewis Josiah
Rubinstein, Charles B(enjamin)
Ruderman, Irving Warren
Schuler, Mathias John
Schulte, Harry John, Jr
Shand, Michael Lee
Siebert, Donald Robert
Silfvast, William Thomas
Smith, Peter William E
Snitzer, Elias
Stiles, Lynn F, Jr
Stone, Julian
Taylor, Gary
Thurston, Robert Norton
Tomaselli, Vincent Paul
Tomlinson, Walter John, III
Treu, Jesse Isaiah

White, Alan David
Whitman, Gerald Martin
Wiesenfeld, Jay Martin
Witkowski, Mark Robert
Wittenberg, Albert M
Wittke, James Pleister
Wolff, Peter A
Wood, Thomas H
Woodward, Ted K
Wu, Zhen
Wullert, John R, II
Xie, Ya-Hong
Yi-Yan, Alfredo
Yurke, Bernard

NEW MEXICO
Ackerhalt, Jay Richard
Ambrose, William Patrick
Anderson, L(awrence) K(eith)
Balog, George
Bean, Brent Leroy
Becker, Wilhelm
Bellum, John Curtis
Bieniewski, Thomas M
Bolie, Victor Wayne
Brannon, Paul J
Brueck, Steven Roy Julien
Bryant, Howard Carnes
Cahill, Paul A
Caves, Carlton Morris
Chen, Tuan Wu
Chow, Weng Wah
Chylek, Petr
Clark, Wallace Thomas, III
Czuchlewski, Stephen John
Depatie, David A
Devaney, Joseph James
Dumas, Herbert M, Jr
Farmer, William Michael
Flicker, Herbert
Gerardo, James Bernard
Gerstl, Siegfried Adolf Wilhelm
Giles, Michael Kent
Gourley, Paul Lee
Greiner, Norman Roy
Hahn, Yu Hak
Hanson, Kenneth Merrill
Hargis, Philip Joseph, Jr
Hessel, Kenneth Ray
Hill, Ronald Ames
Hillsman, Matthew Jerome
Jahoda, Franz C
Jamshidi, Mohammad Mo
Johnston, Roger Glenn
Jones, Eric Daniel
Judd, O'Dean P
Kane, Daniel James
Kopp, Roger Alan
La Delfe, Peter Carl
Ladish, Joseph Stanley
Land, Cecil E(lvin)
Leland, Wallace Thompson
Loree, Thomas Robert
Lyo, SungKwun Kenneth
McLeod, John
Mansfield, Charles Robert
Mauro, Jack Anthony
Mueller, Marvin Martin
Nereson, Norris (George)
Osinski, Marek Andrzej
Overhage, Carl F J
Parker, Joseph R(ichard)
Peercy, Paul S
Peterson, Otis G
Piltch, Martin Stanley
Quigley, Gerard Paul
Rayson, Gary D
Reichert, John Douglas
Rice, James Kinsey
Rigrod, William W
Robertson, Merton M
Robinson, C Paul
Schappert, Gottfried T
Seagrave, John Dorrington
Sheffield, Richard Lee
Shively, Frank Thomas
Snider, Donald Edward
Solem, Johndale Christian
Sollid, Jon Erik
Sorem, Michael Scott
Stone, Sidney Norman
Sze, Robert Chia-Ting
Tanaka, Nobuyuki
Telle, John Martin
Thompson, David Charles
Tsao, Jeffrey Yeenien
Turner, Leaf
Valencia, Valorie Sharron
Veeser, Lynn Raymond
Vieira, David John
Watt, Bob Everett
Wenzel, Robert Gale
Wiggins, Carl M
Yarger, Frederick Lynn
York, George William

NEW YORK
Agrawal, Govind P(rasad)
Alfano, Robert R
Axelrod, Norman Nathan
Ballantyne, Joseph M(errill)
Barrekette, Euval S
Becker, Kurt Heinrich
Bernstein, Burton

Billmeyer, Fred Wallace, Jr
Blaker, J Warren
Blazey, Richard N
Blumenthal, Ralph Herbert
Borrelli, Nicholas Francis
Boyd, Robert William
Breed, Henry Eltinge
Brehm, Lawrence Paul
Breneman, Edwin Jay
Brink, Gilbert Oscar
Brock, Robert H, Jr
Brody, Burton Alan
Burgmaier, George John
Carleton, Herbert Ruck
Chen, B(enjamin) T(eh-Kung)
Chen, Ying-Chih
Cheung, Lim H
Chiang, Fu-Pen
Choy, Daniel S J
Cohen, Martin Gilbert
Costa, Lorenzo F
Craxton, Robert Stephen
Crossmon, Germain Charles
Cummins, Herman Z
Cunningham, Michael Paul
Cupery, Kenneth N
Cusano, Dominick A
Das, Pankaj K
Delano, Erwin
Denk, Ronald H
DePalma, James John
Diamond, Fred Irwin
Doyle-Feder, Donald Perry
Dreyfus, Russell Warren
Duclos, Steven J
Duggin, Michael J
Eberly, Joseph Henry
Egan, Walter George
Eidson, William Whelan
Engelke, Charles Edward
Eyer, James Arthur
Fitchen, Douglas Beach
Folan, Lorcan Michael
Forsyth, James M
Foster, Kenneth William
Friedman, Helen Lowenthal
Gajjar, Jagdish T(rikamji)
Ganley, W Paul
Gelman, Donald
George, Nicholas
Gilmour, Hugh Stewart Allen
Goble, Alfred Theodore
Good, William E
Greenebaum, Michael
Hall, Dennis Gene
Hamblen, David Philip
Hanau, Richard
Harris, Jack Kenyon
Haus, Joseph Wendel
Herman, Irving Philip
Ho, Ping-Pei
Hoffman, Robert
Howard, Richard John
Howe, Dennis George
Hui, Sek Wen
Jacobsen, Chris J
Jupnik, Helen
Kash, Jeffrey Alan
Keck, Donald Bruce
Kermisch, Dorian
Khanna, Shyam Mohan
Kiang, Ying Chao
King, Marvin
Kingslake, Rudolf
Kinzly, Robert Edward
Kirtley, John Robert
Kirz, Janos
Kriss, Michael Allen
Kuehler, Jack D
Kurtz, Clark N
Lamberts, Robert L
LaMuth, Henry Lewis
La Russa, Joseph Anthony
Lax, Melvin
Lean, Eric Gung-Hwa
Lee, Yung-Chang
LeMay, Charlotte Zihlman
Levine, Alfred Martin
Levinson, Steven R
Lipson, Edward David
Litynski, Daniel Mitchell
Liu, Yung Sheng
Lynk, Edgar Thomas
MacAdam, David Lewis
McCamy, Calvin S
McKinstrie, Colin J(ohn)
McMahon, Donald Howland
Mahler, David S
Makous, Walter L
Maldonado, Juan Ramon
Manassah, Jamal Tewfek
Mandel, Leonard
Marchand, Erich Watkinson
Marcus, Michael Alan
Matulic, Ljubomir Francisco
Melcher, Robert Lee
Mendez, Emilio Eugenio
Metcalf, Harold
Millikan, Allan G
Milne, Gordon Gladstone
Moe, George Wylbur
Moore, Duncan Thomas
Morey, Donald Roger
Morris, George Michael

Newman, Jay Edward
Nicolosi, Joseph Anthony
Nyyssonen, Diana
Ockman, Nathan
Owens, James Carl
Parks, Harold George
Pearson, George John
Pellegrino, Francesco
Penney, Carl Murray
Pernick, Benjamin J
Persans, Peter D
Piech, Kenneth Robert
Pike, John Nazarian
Prasad, Paras Nath
Reddy, Reginald James
Remo, John Lucien
Robinson, David Zav
Robinson, Edward J
Roetling, Paul G
Rosenberg, Paul
Rosenberg, Robert
Rustgi, Om Prakash
Sanger, Gregory Marshall
Saunders, Burt A
Sayre, David
Scarl, Donald B
Schroeder, John
Scidmore, Wright H
Setchell, John Stanford, Jr
Shaw, Rodney
Shenker, Martin
Shone, Robert Tilden
Siegel, Benjamin Morton
Sinclair, Douglas C
Smith, Gail Preston
Smoyer, Claude B
Snavely, Benjamin Breneman
Soures, John Michael
Spiller, Eberhard Adolf
Srinivasan, Rangaswamy
Stannard, Carl R, Jr
Starr, John Edward
Stern, Miklos
Stinson, Douglas G
Stoler, David
Strome, Forrest C, Jr
Strong, Herbert Maxwell
Stroud, Carlos Ray
Teich, Malvin Carl
Thiel, Frank L(ouis)
Thompson, Brian J
Thompson, David Allen
Titone, Luke Victor
Tourin, Richard Harold
Vahey, David William
Vance, Miles Elliott
Venugopalan, Srinivasa I
Wald, Samuel Stanley
Walmsley, Ian Alexander
Walter, William Trump
Weber, Julius
Wertheimer, Alan Lee
Ziegler, Robert C(harles)

NORTH CAROLINA
Abbey, Kathleen Mary Kyburz
Almeida, Silverio P
Bonin, Keith Donald
Christian, Wolfgang C
Dow, Thomas Alva
Graham, Louis Atkins
Guenther, Bobby Dean
Kelly, John Henry
Lontz, Robert Jan
Mayes, Terrill W
Mink, James Walter
Paesler, Michael Arthur
Rochow, Theodore George
Stavn, Robert Hans
Swofford, Robert Lewis
Tsu, Raphael
Tucker, Paul Arthur
Vander Lugt, Anthony

NORTH DAKOTA
Rao, B Seshagiri
Soonpaa, Henn H

OHIO
Andonian, Arsavir Takfor
Andrus, Paul Grier
Berthold, John William, III
Beverly, Robert Edward, III
Burnside, Phillips Brooks
Campbell, Bernerd Eugene
Carome, Edward F
Cochran, William Ronald
Cross, Lee Alan
Curtis, Lorenzo Jan
Deck, Robert Thomas
Decker, Arthur John
Fenichel, Henry
Fernelius, Nils Conard
Fritsch, Klaus
Gamble, Francis Trevor
Ghosh, Sanjib Kumar
Grieser, Daniel R
Gustafson, Steven Carl
Hagelberg, M(yron) Paul
Harmon, William Lewis
Harris, Richard Jacob
Heer, Clifford V
Hengehold, Robert Leo
Johnson, Ray O

Kash, Kathleen
Kosel, Peter Bohdan
Lee, Haynes A
Lemke, Ronald Dennis
Lock, James Albert
McMillan, Robert McKee
Mallory, Willam R
Mickelson, Michael Eugene
Nahar, Sultana Nurun
Nam, Sang Boo
Nash, Harry Charles
Nestor, Ontario Horia
Nichols, William Herbert
Nielsen, Philip Edward
Palffy-Muhoray, Peter
Rockstroh, Todd Jay
Rodman, James Purcell
Rosenblatt, Charles Steven
Rutledge, Wyman Coe
Simon, William John
Spry, Robert James
Stith, James Herman
Stroud, David Gordon
Walker, Jearl Dalton
Walters, Craig Thompson
Wood, Van E(arl)

OKLAHOMA
Batchman, Theodore E
Bilger, Hans Rudolf
Feiock, Frank Donald
Grischkowsky, Daniel Richard
McKeever, Stephen William Spencer
Palmer, Bruce Robert
Wurth, Michael John

OREGON
Behringer, Robert Ernest
Carmichael, Howard John
Casperson, Lee Wendel
Csonka, Paul L
Engelbrecht, Rudolf S
Fairchild, Clifford Eugene
Hard, Thomas Michael
McIntyre, David H
Meyer-Arendt, Jurgen Richard
Mossberg, Thomas William
Photinos, Panos John
Solanki, Raj

PENNSYLVANIA
Abraham, Neal Broadus
Abramson, Edward
Bergmann, Ernest Eisenhardt
Bhagavatula, VijayaKumar
Bohren, Craig Frederick
Brandt, Gerald Bennett
Bricks, Bernard Gerard
Casasent, David P(aul)
Cashdollar, Kenneth Leroy
Cohen, Leonard David
Collings, Peter John
Coltman, John Wesley
Emmons, Larrimore Browneller
Ernst, Wolfgang E
Ferrone, Frank Anthony
Garbuny, Max
Gioggia, Robert Stephen
Goell, James E(manuel)
Hager, Nathaniel Ellmaker, Jr
Hansen, J Richard
Harney, Robert Charles
Hartzell, Kenneth R, Jr
Henisch, Heinz Kurt
Ivey, Henry Franklin
Jaggard, Dwight Lincoln
Johnston-Feller, Ruth M
Kirschner, Ronald Allen
Knerr, Reinhard H
Ku, Robert Tien-Hung
Kurtz, Stewart K
Liberman, Irving
Lowry, Jerald Frank
Molter, Lynne Ann
Narducci, Lorenzo M
Norris, Wilfred Glen
Ramsey, Lawrence William
Riedel, Ernest Paul
Riethof, Thomas Robert
Salama, Guy
Sciamanda, Robert Joseph
Settles, Gary Stuart
Smith, Winfield Scott
Sturdevant, Eugene J
Sucov, E(ugene) W(illiam)
Thompson, Julia Ann
Toulouse, Jean
Trainer, Michael Norman
Tredicce, Jorge Raul
Waring, Robert Kerr, Jr
White, George Rowland
Whitman, Alan M
Wiggins, Thomas Arthur
Wolfe, Reuben Edward
Yarosewick, Stanley J
Zhang, Qiming

RHODE ISLAND
Rao, Vasan N

SOUTH CAROLINA
Andringa, Keimpe
Berlinghieri, Joel Carl
Bourkoff, Etan

Inhaber, Herbert
Jones, Edwin Rudolph, Jr
McNulty, Peter J
Pearson, Lonnie Wilson
Roldan, Luis Gonzalez
Sheppard, Emory Lamar
Simon, Frederick Tyler
Zechiel, Leon Norris

TENNESSEE
Allison, Stephen William
Arakawa, Edward Takashi
Arlinghaus, Heinrich Franz
Bentley, Harry Thomas, III
Brau, Charles Allen
Daunt, Stephen Joseph
Davis, Montie Grant
Garrett, William Ray
Haglund, Richard Forsberg, Jr
Hayter, John Bingley
Hutchinson, Donald Patrick
Ice, Gene Emery
Jellison, Gerald Earle, Jr
Mason, Arthur Allen
Painter, Linda Robinson
Pegg, David John
Staats, Percy Anderson
Tolk, Norman Henry
Vander Sluis, Kenneth Leroy
Watson, Robert Lowrie

TEXAS
Albright, John Grover
Aldridge, Jack Paxton, III
Anderson, Edward Everett
Anderson, Wallace L(ee)
Bencomo, Jose A, Jr
Buckman, Alvin Bruce
Burger, Christian P
Busch, Kenneth Walter
Campbell, Charles Edgar
Cantrell, Cyrus D, III
Chivian, Jay Simon
Coleman, Howard S
Collins, Carl Baxter, Jr
Cooper, Howard Gordon
De Wit, Michiel
Duncan, Walter Marvin, Jr
Dunning, Frank Barrymore
Englund, John Caldwell
Fredin, Leif G R
Fry, Edward Strauss
Hagler, Marion O(tho)
Harvey, Kenneth C
Henize, Karl Gordon
Hu, Chia-Ren
Hulet, Randall Gardner
Kyle, Thomas Gail
MacFarlane, Duncan Leo
Mays, Robert, Jr
Mimmack, William Edward
Mitra, Sunanda
Pedrotti, Leno Stephano
Peters, Randall Douglas
Pitts, David Eugene
Potter, Robert Joseph
Rester, David Hampton
Schruben, Johanna Stenzel
Slocum, Robert Earle
Smith, James Lynn
Taboada, John
Talent, David Leroy
Tittel, Frank K(laus)
Walkup, John Frank
Wang, Paul Weily
Wells, Michael Byron
Wood, Lowell Thomas
Young, James Forrest
Zook, Herbert Allen

UTAH
Allred, David D
Benner, Robert E
Fowles, Grant Robert
Harris, Joel Mark
Howell, Barton John
Knight, Larry V
Lawton, Robert Arthur
Liou, Kuo-Nan
Lund, Mark Wylie
Luty, Fritz
Stevens, Richard Edward
Taylor, Philip Craig

VERMONT
Detenbeck, Robert Warren
Dunham, Jeffrey Solon
Foote, Vernon Stuart, Jr
Oughstun, Kurt Edmund
Smith, David Young
Spillman, William Bert, Jr

VIRGINIA
Alcaraz, Ernest Charles
Beardall, John Smith
Bloomfield, Louis Aub
Brockman, Philip
Browell, Edward Vern
Brown, Gary S
Bunner, Alan Newton
Burner, Alpheus Wilson, Jr
Carter, William Harold
Chun, Myung K(i)
Church, Charles Henry

Claus, Richard Otto
Corcoran, Vincent John
Crawford, George William
Davis, Charles Mitchell, Jr
Deepak, Adarsh
Doyle, Frederick Joseph
Farrukh, Usamah Omar
Fisher, Arthur Douglas
Flack, Ronald Dumont, Jr
Fox, Kenneth Richard
Gallagher, Thomas Francis
Gee, Sherman
Goldberg, Lawrence Spencer
Goodwin, Francis E
Grant, William B
Graver, William Robert
Hammer, Jacob Meyer
Hargrove, Logan Ezral
Hass, Georg
Heckathorn, Harry Mervin, III
Hoppe, John Cameron
Hunter, William Ray
Jacobs, Theodore Alan
Johnson, Charles Nelson, Jr
Kenyon, Stephen C
Kershenstein, John Charles
Kissell, Kenneth Eugene
Knudsen, Dennis Ralph
Kooij, Theo
Koomen, Martin J
Kramer, Steven David
Krassner, Jerry
Larson, Daniel John
Lee, Ja H
Liebenberg, Donald Henry
McAdoo, John Hart
McWright, Glen Martin
Major, Robert Wayne
Michalowicz, Joseph Victor
Neubauer, Werner George
Ni, Chen-Chou
Nicolai, Van Olin
Patterson, James Douglas
Phillips, E Alan
Pilloff, Herschel Sydney
Pohlmann, Juergen Lothar Wolfgang
Poon, Ting-Chung
Porter, Terence Lee
Ratches, James Arthur
Schnatterly, Stephen Eugene
Sepucha, Robert Charles
Silver, Meyer
Sime, David Gilbert
Spinrad, Richard William
Srikanth, Sivasankaran
Sutton, George W(alter)
VanDerlaske, Dennis P
Weichel, Hugo
Weinstein, Leonard Murrey
Wexler, Bernard Lester
Zarobila, Clarence Joseph

WASHINGTON
Bruckner, Adam Peter
Carlson, Frederick Paul
Clark, Kenneth Courtright
Craig, Richard Anderson
Darling, Robert Bruce
Dunham, Glen Curtis
Erdmann, Joachim Christian
Ewing, J J
Fortson, Edward Norval
Goddard, Murray Cowdery
Gunderson, Leslie Charles
Haueisen, Donald Carl
Henderson, Billy Joe
Hernandez, Gonzalo J
Hildebrand, Bernard Percy
Hoch, Richmond Joel
Ishimaru, Akira
Leney, Lawrence
McClure, J Doyle
McDermott, Mark Nordman
Marston, Philip Leslie
Morton, Randall Eugene
Nichols, Davis Betz
Radziemski, Leon Joseph
Russell, James Torrance
Safai, Morteza
Sargent, Murray, III
Sarles, Lynn Redmon
Soreide, David Christien
Stitch, Malcolm Lane
Tsang, Leung
Veit, Jiri Joseph
Veress, Sandor A

WISCONSIN
Botez, Dan
Cerrina, Francesco
Fystrom, Dell O
Greenler, Robert George
Hyzer, William Gordon
Lawler, James Edward
Mallmann, Alexander James
Reynolds, Ronald J
Saleh, Bahaa E A
Schroeder, Daniel John
Seshadri, Sengadu Rangaswamy
Sudhakaran, Gubbi Ramarao
Truszkowska, Krystyna
Van den Akker, Johannes Archibald
Wolf, Paul R

Optics (cont)

WYOMING
Inguva, Ramarao
Rinehart, Edgar A

ALBERTA
Fedosejevs, Robert
Offenberger, Allan Anthony
Pinnington, Eric Henry
Weale, Gareth Pryce

BRITISH COLUMBIA
Colbow, Kevin Michael
Morbey, Christopher Leon
Richardson, Eric Harvey
Rogers, Douglas Herbert

MANITOBA
Soliman, Afifi Hassan
Tabisz, George Conrad

NEW BRUNSWICK
Masry, Salem El

NOVA SCOTIA
Bishop, Roy Lovitt

ONTARIO
Aitken, G(eorge) J(ohn) M(urray)
Alcock, Alfred John
Bahadur, Birendra
Bedford, Ronald Ernest
Bernath, Peter Francis
Bertram, Robert William
Blachut, T(heodore) J(oseph)
Boivin, Louis-Philippe
Brach, Eugene Jenö
Brimacombe, Robert Kenneth
Brousseau, Nicole
Carman, Philip Douglas
Carswell, Allan Ian
Chapman, George David
Dixon, Arthur Edward
Dobrowolski, Jerzy Adam
Fjarlie, Earl J
Garside, Brian K
Geissinger, Hans Dieter
Howard-Lock, Helen Elaine
Jones, Alister Vallance
King, Gerald Wilfrid
Lit, John Wai-Yu
Mandelis, Andreas
Mavroyannis, Constantine
Measures, Raymond Massey
Moskovits, Martin
Nilson, John Anthony
Racey, Thomas James
Regan, David M
Remole, Arnulf
Robertson, Alan Robert
Rochon, Paul Leo
Shepherd, Gordon Greeley
Stoicheff, Boris Peter
Szabo, Alexander
Tam, Wing-Gay
Vanderkooy, John
Van Driel, Henry Martin
Willemsen, Herman William
Yevick, David Owen

QUEBEC
Arsenault, Henri H
Boivin, Alberic
Borra, Ermanno Franco
Bose, Tapan Kumar
Chin, See Leang
Delisle, Claude
Gagne, Jean-Marie
Izquierdo, Ricardo
Jordan, Byron Dale
Kiss, Laszlo Istvan
Lessard, Roger Alain
Maciejko, Roman
Meunier, Michel
Pawluczky, Romuald
Richard, Claude
Schwelb, Otto
Subramanian, Sesha
Waksberg, Armand L

OTHER COUNTRIES
Bottka, Nicholas
Bryngdahl, Olof
Buckingham, Amyand David
Cardona, Manuel
Eng, Sverre T(horstein)
French, William George
Friesem, Albert Asher
Gordon, Jeffrey Miles
Hansch, Theodor Wolfgang
Kafri, Oded
Kildal, Helge
Lee, Elhang Howard
Lee, Peter H Y
Loy, Michael Ming-Tak
McInerney, John Gerard
Malacara, Daniel
Marom, Emanuel
Martin, William Eugene
Nussenzveig, Herch Moyses
Rozzi, Tullio
Sirohi, Rajpal Singh
Snyder, Allan Whitnack

Stavroudis, Orestes Nicholas
Steffen, Juerg
Unger, Hans-Georg
Ushioda, Sukekatsu
Van Raalte, John A

Optoelectronics

ALABAMA
Bloemer, Mark Joseph
Kowel, Stephen Thomas
Nikles, David Eugene

ARIZONA
Kim, Matthew Hidong
Rountree, Janet

CALIFORNIA
Amirkhanian, Varouj David
Billman, Kenneth William
Bowers, John E
Bube, Richard Howard
Dines, Eugene L
Drake, Richard Paul
Fouquet, Julie
Goble, James H, Jr
Gruber, John B
Jones, Lewis Hammond, IV
Kerr, Donald M, Jr
Lam, Leo Kongsui
Long, Stephen Ingalls
Mathis, Ronald Floyd
Moerner, W(illiam) E(sco)
Musket, Ronald George
Paoli, Thomas Lee
Perl, Martin Lewis
Richter, Thomas A
Sheridan, Nicholas Keith
Swalen, Jerome Douglas
Urbach, John C
Weissman, Robert Henry
Wieder, Harry H

COLORADO
Anderson, Dana Zachary
Gould, Gordon
Henley, John Raymond

DELAWARE
Bragagnolo, Julio Alfredo
Golden, Kelly Paul

DISTRICT OF COLUMBIA
Hornstein, John Stanley

FLORIDA
Hummel, Rolf Erich
Kimble, Allan Wayne
Ludwig, Matthias Heinz

ILLINOIS
Loos, James Stavert
Osten, Donald Edward
Parker, John Calvin

MARYLAND
Boone, Bradley Gilbert
Degnan, John James, III
Passman, Sidney
Simonis, George Jerome
Steiner, Bruce
Yu, Anthony Woonchiu

MASSACHUSETTS
Alexander, Michael Norman
Dakss, Mark Ludmer
Donnelly, Joseph P(eter)
Hearn, David Russell
Litzenberger, Leonard Nelson
Liu, Xichun
Moody, Elizabeth Anne
Soref, Richard Allan
Tripathy, Sukant K

MINNESOTA
Nelson, Kyler Fischer
Tufte, Obert Norman

MISSOURI
Rode, Daniel Leon

NEBRASKA
Ducharme, Stephen

NEW JERSEY
Bosacchi, Bruno
Buch, Robert Michael
Chiu, Tien-Heng
Dubey, Madan
Eckert, John Andrew
Miller, David A B
Panayotatos, Paul
Shen, Tek-Ming
Snitzer, Elias
Tomlinson, Walter John, III
Tompsett, Michael F
Watson, Hugh Alexander
Zoltan, Bart Joseph

NEW MEXICO
Kane, Daniel James
Kurtz, Steven Ross
Osinski, Marek Andrzej

Rigrod, William W
Robertson, Merton M
Valencia, Valorie Sharron

NEW YORK
LeMay, Charlotte Zihlman
Liu, Yung Sheng
Owens, James Carl
Pellegrino, Francesco
Wolga, George Jacob
Yang, Kei-Hsiung
Yip, Kwok Leung

NORTH CAROLINA
Aktas, Yildirin
Periasamy, Ammasi

OHIO
Carome, Edward F
Land, Peter L
Lott, James Anthony
Ringel, Steven Adam
Yaney, Perry Pappas

OKLAHOMA
Grischkowsky, Daniel Richard

OREGON
Engelmann, Reinhart Wolfgang H

PENNSYLVANIA
Abramson, Edward
Asher, Sanford Abraham
Reynolds, Claude Lewis, Jr
Shuch, H Paul
Sturdevant, Eugene J

TEXAS
Glosser, Robert
Li, Shu

UTAH
Allred, David D
Lund, Mark Wylie

VIRGINIA
Knudsen, Dennis Ralph
Lawrence, David Joseph
Weichel, Hugo

WASHINGTON
Safai, Morteza
Sarles, Lynn Redmon

WISCONSIN
Sudhakaran, Gubbi Ramarao

BRITISH COLUMBIA
Booth, Ian Jeremy
Tiedje, J Thomas

ONTARIO
Dawson, Peter Henry
Dyment, John Cameron
Erickson, Lynden Edwin
Rochon, Paul Leo
Ruda, Harry Eugen
Sherman, Norman K
Willemsen, Herman William

QUEBEC
Waksberg, Armand L

OTHER COUNTRIES
Friesem, Albert Asher
Lee, Elhang Howard
Lueder, Ernst H
Weil, Raoul Bloch

Physics, General

ALABAMA
Alexander, Chester, Jr
Boardman, William Jarvis
Budenstein, Paul Philip
Buta, Ronald James
Carr, Howard Earl
Cooper, John Raymond
Curott, David Richard
Dalins, Ilmars
Decher, Rudolf
Dimmock, John O
Fowler, Charles Sidney
Franz, Frank Andrew
French, John Donald
Fullerton, Larry Wayne
Govil, Narendra Kumar
Harvey, Stephen Craig
Jones, Robert William
Jones, Stanley Tanner
Kaylor, Hoyt McCoy
Korsch, Dietrich G
MacRae, Robert Alexander
Mishra, Satya Narayan
Morton, Perry Wilkes, Jr
Parnell, Thomas Alfred
Passino, Nicholas R
Pearson, Colin Arthur
Pfeiffer, Robert S
Reid, William James
Rodgers, Aubrey
Roe, James Maurice, Jr
Rosenblum, William M

Rowell, Neal Pope
Stewart, Dorothy Anne
Stuhlinger, Ernst
Sulentic, Jack William
Tan, Arjun
Tandberg-Hanssen, Einar Andreas
Tohver, Hanno Tiit
Twieg, Donald Baker
Walker, William Waldrum
Wolin, Samuel
Young, John H

ALASKA
Bates, Howard Francis
Martins, Donald Henry
Morack, John Ludwig

ARIZONA
Allen, Edward Franklin
Baldwin, Ann Linda
Bedwell, Thomas Howard
Bennett, Peter A
Blitzer, Leon
Boettner, Edward Alvin
Broadfoot, Albert Lyle
Call, Reginald Lessey
Christy, James Walter
Cowley, John Maxwell
Curtis, David William
Dahl, Randy Lynn
Daw, Glen Harold
Dixon, Peggy A
Dunkelman, Lawrence
Erickson, Richard Ames
Evans, Robley D(unglison)
Franz, Otto Gustav
Gates, Halbert Frederick
Gregory, Brooke
Gullikson, Charles William
Hartig, Elmer Otto
Heaps, Melvin George
Hendrickson, Lester Ellsworth
Hodges, Carl Norris
Hoffmann, William Frederick
Hood, Lon(nie) Lamar
Jacobs, Stephen Frank
Jacobson, Michael Ray
Johnson, C(harles) Bruce
Kebler, Richard William
Kessler, John Otto
Kevane, Clement Joseph
Lamb, George Lawrence, Jr
Layton, Richard Gary
Lazar, Norman Henry
Macleod, Hugh Angus
Marzke, Robert Franklin
Meieran, Eugene Stuart
O'Leary, Brian Todd
Pokorny, Gerold E(rwin)
Robson, John William
Sarid, Dror
Scholl, Marija Strojnik
Senitzky, Benjamin
Smith, Harvey Alvin
Spence, John Charles
Stark, Royal William
Stearns, Martin
Swindle, Timothy Dale
Swink, Laurence N
Tillery, Bill W
Treat, Jay Emery, Jr
Tsong, Ignatius Siu Tung
Turner, Arthur Francis
Vogelmann, Andrew Mark
Wangsness, Roald Klinkenberg
Weaver, Albert Bruce
Willis, William Russell
Yelle, Roger V
Young, Robert A

ARKANSAS
Bond, Robert Levi
Eichenberger, Rudolph John
Engle, Paul Randal
Horton, Philip Bish
Leming, Charles William
McCarty, Clark William
Mackey, James E
Mink, Lawrence Albright
Prince, Denver Lee
Richardson, Charles Bonner
Rollefson, Aimar Andre
Sandefur, William Marion
Sharrah, Paul Chester
Wild, Wayne Grant
Zinke, Otto Henry

CALIFORNIA
Abers, Ernest S
Abraham, Farid Fadlow
Abrams, Richard Lee
Adams, John Andrew
Adler, Ronald John
Agnew, Harold Melvin
Aitken, Donald W, Jr
Alfven, Hannes Olof Gosta
Allen, Lew, Jr
Allen, Matthew Arnold
Allen, S James, Jr
Alvarez, Orlando
Alvarez, Raymond Angelo, Jr
Ames, Lawrence Lowell
Anderson, Charles Hammond
Anderson, John Thomas

Anderson, Milo Vernette
Anderson, Victor Charles
Anderson, Weston Arthur
Anspaugh, Bruce Edward
Archer, Douglas Harley
Armstrong, Baxter Hardin
Armstrong, Donald B
Arnold, James S(loan)
Ashby, Val Jean
Ashour-Abdalla, Maha
Assmus, Alexi Josephine
Atchison, F(red) Stanley
Atchley, Anthony A
Attwood, David Thomas, Jr
Augenstein, Bruno (Wilhelm)
Ayres, Wesley P
Bacher, Robert Fox
Badash, Lawrence
Baez, Albert Vinicio
Baggerly, Leo L
Bajorek, Christopher Henry
Bakun, William Henry
Ballam, Joseph
Bangerter, Roger Odell
Banner, David Lee
Bar-Cohen, Yoseph
Barker, William Alfred
Barnett, Tim P
Barr, Frank T(homas)
Barrett, Paul Henry
Bars, Itzhak
Bartling, Judd Quentin
Bate, George Lee
Baum, Dennis Willard
Baum, Peter Joseph
Baumeister, Philip Werner
Bechtel, James Harvey
Behravesh, Mohamad Martin
Bell, Alan Edward
Bell, Graydon Dee
Berdahl, Paul Hilland
Bernhardt, Anthony F
Bershader, Daniel
Bhaumik, Mani Lal
Binder, Daniel
Bingham, Harry H, Jr
Birge, Ann Chamberlain
Bjorken, James D
Blachman, Nelson Merle
Blades, John Dieterle
Bleakney, Walker
Blink, James Allen
Bohmer, Heinrich Everhard
Bonsack, Walter Karl
Bork, Alfred Morton
Bott, Jerry Frederick
Braun, Robert Leore
Braunstein, Rubin
Breidenbach, Martin
Bridges, William Bruce
Britt, Edward Joseph
Brodie, Jean Pamela
Brooks, Walter Lyda
Brouillard, Robert Ernest
Brown, Douglas Richard
Brown, George Stephen
Brown, Robert James Sidford
Brown, Robert Theodore
Brown, Sheldon (Jack)
Bruck, George
Burnett, Lowell Jay
Burns, Fred Paul
Bush, George Edward
Cabrera, Blas
Campbell, Edward Michael
Caren, Robert Poston
Carey, Frank Charles
Carlson, Richard Frederick
Carr, Robert H
Carron, Neal Jay
Carver, John Guill
Case, Lloyd Allen
Cassen, Patrick Michael
Catura, Richard Clarence
Chandler, Donald Ernest
Chang, David Bing Jue
Channel, Lawrence Edwin
Chase, Lloyd Fremont, Jr
Chase, Lloyd Lee
Chen, Chia Hwa
Chen, Mau Hsiung
Chiao, Raymond Yu
Chinowsky, William
Cho, Young-chung
Chow, Richard H
Christman, Arthur Castner, Jr
Chu, William Tongil
Churchill, Dewey Ross, Jr
Clark, Arnold Franklin
Clarke, David Richard
Clauser, John Francis
Cleland, Laurence Lynn
Clemens, Jon K(aufmann)
Clinton, James R
Cochran, Stephen G
Codrington, Robert Smith
Cole, Robert Kleiv
Coleman, Lamar William
Coleman, Philip Lynn
Commins, Eugene David
Cook, Thomas Bratton, Jr
Cooper, Eugene Perry
Coppens, Alan Berchard
Corey, Victor Brewer

Corruccini, Linton Reid
Costantino, Marc Shaw
Coufal, Hans-Jurgen
Cowan, Eugene Woodville
Cox, Aaron J
Craford, M George
Craig, Paul Palmer
Cranston, Frederick Pitkin, Jr
Crawford, Frank Stevens, Jr
Crawford, John Clark
Creighton, John Rogers
Crews, Robert Wayne
Crowe, Kenneth Morse
Cummings, Alan C
Curran, Donald Robert
Cutler, Cassius Chapin
Cutler, Leonard Samuel
Dahl, Harvey A
Dahlberg, Richard Craig
Dairiki, Ned Tsuneo
Dart, Sidney Leonard
D'Attorre, Leonardo
Davidson, Arthur L
Davis, James Ivey
Davis, Jay C
Davis, Leverett, Jr
Davis, Sumner P
Davis, Thomas Pearse
Daybell, Melvin Drew
Debs, Robert Joseph
Deleray, Arthur Loyd
Dember, Alexis Berthold
Denney, Joseph M(yers)
De Pangher, John
De Shazer, Larry Grant
Dessel, Norman F
Dickinson, Dale Flint
Dimeff, John
Dirks, Leslie C
Donaldson, John Riley
Doolittle, Robert Frederick, II
Dowell, Jerry Tray
Drake, Richard Paul
Driscoll, Charles F
Duggan, Michael J
Duneer, Arthur Gustav, Jr
Dunning, John Ray, Jr
Dyal, Palmer
Earhart, Richard Wilmot
Eby, Frank Shilling
Eckert, Hans Ulrich
Eckhardt, Gisela (Marion)
Eckhardt, Wilfried Otto
Edelson, Sidney
Edwards, Byron N
Eggen, Donald T(ripp)
Eimerl, David
Einarsson, Alfred W
Eisen, Fred Henry
Elings, Virgil Bruce
Elliott, Shelden Dougless, Jr
Elms, James Cornelius
Ely, Robert P, Jr
Enderby, Charles Eldred
Enns, John Hermann
Erlich, David C
Eschenfelder, Andrew Herbert
Esposito, Pasquale Bernard
Evans, John Ellis
Everitt, C W Francis
Fa'arman, Alfred
Farmer, Donald Jackson
Faulkner, John Edward
Feher, Elsa
Feher, George
Feinstein, Joseph
Ferris, Horace Garfield
Filippenko, Alexei Vladimir
Finkelstein, Robert Jay
Fisher, George Phillip
Fisher, John Crocker
Fisher, Leon Harold
Fisher, Philip Chapin
Fitzpatrick, Gary Owen
Flinn, Paul Anthony
Fonck, Eugene J
Forster, Harriet Herta
Forster, Kurt
Forward, Robert L
Foster, John Stuart, Jr
Fowler, Charles Arman, Jr
Fowler, William Alfred
Frank, Wilson James
Frankel, Richard Barry
Fredkin, Donald Roy
Fredrickson, John E
Fretter, William Bache
Fritchle, Frank Paul
Fu, Lee Lueng
Fullmer, George Clinton
Futch, Archer Hamner
Gailar, Owen H
Gardner, Wilford Robert
Garmire, Elsa Meints
Garren, Alper A
Garrick, B(ernell) John
Garrison, John Dresser
Garwin, Edward Lee
Getting, Ivan Alexander
Ghiorso, Albert
Ghose, Rabindra Nath
Giannini, Gabriel Maria
Gibson, Atholl Allen Vear
Gibson, James (Benjamin)

Gibson, Thomas Alvin, Jr
Giles, John Crutchlow
Giles, Peter Cobb
Gill, Stephen Paschall
Gillespie, George H
Gilmartin, Thomas Joseph
Ginzton, Edward Leonard
Glaser, Donald Arthur
Godfrey, Charles S
Goerz, David Jonathan, Jr
Gold, Richard Robert
Goodkind, John M
Goodstein, David Louis
Goss, Wilbur Hummon
Goubau, Wolfgang M
Greenberg, Allan S
Greenfield, Eugene W(illis)
Grinberg, Jan
Grismore, Roger
Hadley, James Warren
Hageman, Donald Henry
Hahn, Erwin Louis
Haisch, Bernhard Michael
Halbach, Klaus
Halden, Frank
Hall, Donald Eugene
Haloulakos, Vassilios E
Hamermesh, Bernard
Hamill, Patrick James
Hammond, Donald L
Hammond, Robert Hugh
Hansma, Paul Kenneth
Hanson, Merle Edwin
Harker, Kenneth James
Harrach, Robert James
Harris, Frank Bower, Jr
Harris, Sigmund Paul
Harrison, Don Edward, Jr
Hawryluk, Andrew Michael
Hayn, Carl Hugo
Heard, Harry Gordon
Hearst, Joseph R
Heiles, Carl
Helmer, John
Henderson, Sheri Dawn
Hendricks, Charles D(urrell), Jr
Herlin, Melvin Arnold
Herman, Paul Theodore
Heynick, Louis Norman
Hibbs, Andrew D
Higa, Walter Hiroichi
Hilker, Harry Van Der Veer, Jr
Hill, Robert Dickson
Hill, Robert Matteson
Hirsch, Jorge E
Hoch, Orion L
Hochstim, Adolf R
Hoffman, Marvin Morrison
Hogan, Clarence Lester
Hogan, William John
Hollenbach, David John
Holmstrom, Fred Edward
Homer, Paul Bruce
Hood, John Mack, Jr
Hope, Hakon
Horn, Linda Joyce
Housley, Robert Melvin
Hsia, Henry Tao-Sze
Huang, C Yuan
Hubbard, Harmon William
Hudson, Cecil Ivan, Jr
Hudson, Donald Robert
Hudson, George Elbert
Huebner, Albert Louis
Hughes, Gordon Frierson
Hughes, Thomas Rogers, Jr
Hunt, Angus Lamar
Hunter, Hugh Wylie
Hurlbut, Franklin Charles
Hurley, James P
Hutchin, Richard Ariel
Iberall, Arthur Saul
Ibser, Homer Wesley
Igo, George (Jerome)
Ikenberry, Dennis L
Jackson, Bernard Vernon
Jaduszliwer, Bernardo
Jaeger, Klaus Bruno
Janos, William Augustus
Jarrett, Steven Michael
Johnson, C Scott
Johnson, Horace Richard
Johnson, Quintin C
Jones, Clarence S
Jordan, Willard Clayton
Jory, Farnham Stewart
Joseph, Alfred S
Judge, Darrell L
Jungerman, John (Albert)
Kalmbach, Sydney Hobart
Kamegai, Minao
Kapany, Narinder Singh
Karlow, Edwin Anthony
Katen, Paul C
Kazan, Benjamin
Kehoe, Brandt
Kellers, Charles Frederick
Kelley, Charles Thomas, Jr
Kenealy, Patrick Francis
Kerth, Leroy Thomas
Kevles, Daniel Jerome
Khakoo, Murtadha A
Khan, Mahbub R
Khoshnevisan, Mohsen Monte

Kidder, Ray Edward
Kilkenny, Joseph David
Kim, Young Bae
Kimble, Harry Jeffrey
King, James, Jr
King, William Stanely
Kisliuk, Paul
Klein, August S
Klein, Christopher Francis
Klems, Joseph Henry
Knechtli, Ronald (C)
Knights, John Christopher
Knobloch, Edgar
Knollman, Gilbert Carl
Knopoff, Leon
Koehler, Thomas Richard
Kofler, Richard Robert
Koga, Rokutaro
Kohn, Walter
Kolb, Alan Charles
Kolin, Alexander
Kolsky, Harwood George
Kopelman, Jay B
Krall, Nicholas Anthony
Kraut, Edgar A
Krick, Irving Parkhurst
Kristian, Jerome
Kroeker, Richard Mark
Krueger, Robert William
Krueger, Roland Frederick
Kuhns, Ellen Swomley
Kulgein, Norman Gerald
Kulkarny, Vijay Anand
Kutzscher, Edgar Walter
Kvaas, T(horvald) Arthur
LaBerge, Walter B
Lam, Kai Shue
Lambertson, Glen Royal
Landecker, Peter Bruce
Lander, Richard Leon
Lane, Stephen Mark
Langmuir, David Bulkeley
Langmuir, Robert Vose
Lark, Neil LaVern
Larsen, Ted LeRoy
Laslett, Lawrence Jackson
Laughlin, Robert B
Layton, Thomas William
Lee, Edward Prentiss
Lee, Martin Joe
Leemans, Wim Pieter
Leff, Harvey Sherwin
Lesko, Kevin Thomas
Levie, Harold Walter
Levinthal, Elliott Charles
Levit, Lawrence Bruce
Lewis, Francis Hotchkiss, Jr
Lewis, Vance De Spain
Li, Seung P(ing)
Liebermann, Leonard Norman
Liebes, Sidney, Jr
Lim, Teong Cheng
Lindley, Charles A(lexander)
Lippa, Erik Alexander
Little, William Arthur
Lockhart, James Marcus
Loewe, William Edward
Lofgren, Edward Joseph
Logan, Joseph Granville, Jr
Long, John Vincent
Lonsdale, Carol Jean
Lorents, Donald C
Lorenz, Max R(udolph)
Lotspeich, James Fulton
Lowenstein, Carl David
Lucas, Joe Nathan
Lukasik, Stephen Joseph
Luu, Jane
Luxenberg, Harold Richard
Lyon, Waldo (Kampmeier)
McBride, John Barton
McCollum, Donald Carruth, Jr
McColm, Douglas Woodruff
MacDonald, Alexander Daniel
McFarlane, Ross Alexander
McGill, Thomas Conley, Jr
Machacek, Milos
McHugh, Stuart Lawrence
McIlwain, Carl Edwin
McIntyre, Adelbert
Mackenzie, Kenneth Victor
Mackin, Robert James, Jr
McManigal, Paul Gabriel Moulin
Maenchen, George
Mahadevan, Parameswar
Mailhiot, Christian
Maiman, Theodore Harold
Manasse, Roger
Maninger, Ralph Carroll
Maple, M Brian
Marcus, John Stanley
Marshall, J Howard, III
Marshall, Samuel Wilson
Marx, Kenneth Donald
Mate, C Matthew
Mather, Robert Laurance
Mathews, W(arren) E(dward)
Matsuda, Kyoko
Matthews, Dennis L
Matthews, Stephen M
Mauche, Christopher W
Maudlin, Lloyd Z
Maurer, Hans Andreas
May, Michael Melville

Physics, General (cont)

Maydan, Dan
Mayper, V(ictor), Jr
Mead, S Warren
Mechlin, George Francis, Jr
Meinel, Aden Baker
Meldner, Heiner Walter
Melvin, Mael Avramy
Mercereau, James Edgar
Merritt, Jack
Miller, Leonard Robert
Miller, Park Hays, Jr
Miller, Roger Heering
Mitchner, Morton
Mitescu, Catalin Dan
Moe, Chesney Rudolph
Moe, Osborne Kenneth
Moffeit, Kenneth Charles
Molmud, Paul
Montgomery, Leslie D
Montgomery, Michael Davis
Mooney, Walter D
Morgan, Harry Clark
Morinigo, Fernando Bernardino
Moriwaki, Melvin M
Morris, Donald E
Morris, James Russell
Muller, Thomas
Munsee, Jack Howard
Murray, John Roberts
Najar, Rudolph Michael
Nakada, Yoshinao
Nazaroff, William W
Neal, Richard B
Neilsen, Ivan Robert
Nellis, William J
Niemeyer, Sidney
Nierenberg, William Aaron
Nodvik, John S
Nosanow, Lewis H
Novotny, Vlad Joseph
Nuckolls, John Hopkins
O'Brien, Benedict Butler, Jr
Oddone, Piermaria Jorge
O'Dell, Austin Almond, Jr
Ohkawa, Tihiro
Olness, Robert James
Olney, Ross David
Orth, Charles Douglas
Osheroff, Douglas Dean
Owyang, Gilbert Hsiaopin
Packard, Martin Everett
Page, Robert Alan, Jr
Pake, George Edward
Palatnick, Barton
Palluconi, Frank Don
Palmer, Richard Everett
Panofsky, Wolfgang K H
Papanastassiou, Dimitri A
Park, Edward C(ahill), Jr
Parker, William Henry
Parks, Donald E
Patel, Chandra Kumar Naranbhai
Paterson, James McEwan
Pauling, Linus Carl
Peithman, Roscoe Edward
Pelizzari, Michael A
Pelka, David Gerard
Pellett, David Earl
Pennypacker, Carlton Reese
Perry, Joseph Earl, Jr
Perry, Richard Lee
Pertica, Alexander Jose
Phelps, Pharo A
Phillips, David T
Phillips, Thomas Joseph
Pignataro, Augustus
Pine, Jerome
Piserchio, Robert J
Platt, Joseph Beaven
Plesset, Milton Spinoza
Politzer, Hugh David
Powell, Ronald Allan
Prater, Ronald
Pripstein, Morris
Profet, Margie
Prosnitz, Donald
Puntenney, Dee Gregory
Quate, Calvin F(orrest)
Rabinowitz, Mario
Rabl, Veronika Ariana
Ragent, Boris
Randhawa, Jagir Singh
Reagan, Daryl David
Redlich, Martin George
Redmond, Peter John
Reinheimer, Julian
Remley, Marlin Eugene
Renzetti, Nicholas A
Resch, George Michael
Rhim, Won-Kyu
Rhodes, Edward Joseph, Jr
Richards, Paul Linford
Rider, Ronald Edward
Riggs, James W, Jr
Ritson, David Mark
Robinson, Lawrence Baylor
Rockower, Edward Brandt
Rockstad, Howard Kent
Rodden, Robert Morris
Rodeback, George Wayne
Rodriguez, Andres F
Rogers, Emery Herman
Rollosson, George William
Romagnoli, Robert Joseph
Romanowicz, Barbara Anna
Rooney, James Arthur
Rosen, Arthur Zelig
Rosenblum, Bruce
Rosencwaig, Allan
Rosengren, Jack Whitehead
Ross, Marvin
Rostoker, Norman
Rothwell, William Stanley
Roukes, Michael L
Rozsnyai, Balazs
Rubin, Robert Howard
Rudnick, Isadore
Rugge, Hugo R
Rumsey, Victor Henry
Rundel, Robert Dean
Russell, Thomas Paul
Rutledge, David B
Sadoulet, Bernard
Salaita, George N
Salem, Semaan Ibrahim
Salter, Robert Munkhenk, Jr
Samson, Sten
Sanders, Timothy D
Sands, Matthew
Sawatzky, Erich
Scheibe, Murray
Schimitschek, Erhard Josef
Schipper, Lee (Leon Jay)
Schneider, Harold O
Schrieffer, John Robert
Schuller, Ivan Kohn
Schultz, Cramer William
Schultz, Jonas
Schwarcz, Ervin H
Schwettman, Harry Alan
Schwinger, Julian Seymour
Schwirzke, Fred
Scifres, Donald Ray
Scott, John Campbell
Scott, Peter Leslie
Seibert, J A
Seidel, Thomas Edward
Seiden, Abraham
Seki, Hajime
Seppi, Edward Joseph
Sewell, Duane Campbell
Shakeshaft, Robin
Shanteau, Robert Marshall
Shapiro, Charles Saul
Shaw, Herbert John
Shelby, Robert McKinnon
Sherwin, Chalmers W
Shnider, Ruth Wolkow
Shreve, Ronald Lee
Simanek, Eugen
Simon, Ralph Emanuel
Singer, Stanley
Skinner, Loren Courtland, II
Smith, Albert Ernest
Smith, Edward John
Smith, George C
Smith, George Foster
Smith, Joe Nelson, Jr
Smith, Marianne (Ruth) Freundlich
Smith, Sheldon Magill
Smith, Todd Iversen
Smith, Walter Lee
Snyder, Conway Wilson
Snyder, Thoma Mees
Soffer, Bernard Harold
Sollfrey, William
Sooy, Walter Richard
Sorrels, John David
Speck, David Ralph
Spencer, Donald Jay
Spencer, Herbert W, III
Springer, Donald Lee
Stahl, Frieda A
Stalder, Kenneth R
Stambaugh, Ronald D
Stapp, Henry P
Stauffer, John Richard
Stella, Paul M
Stevenson, Merlon Lynn
Stewart, Gordon Ervin
Stewart, James Lloyd
Stone, Edward Carroll, Jr
Stone, Sam, III
Stone, William Ross
Strangeway, Robert Joseph
Straus, Thomas Michael
Strena, Robert Victor
Strominger, Andrew Eben
Stupian, Gary Wendell
Suhl, Harry
Surko, Clifford Michael
Sutcliffe, William George
Sutton, Paul McCullough
Swanson, Eric Richmond
Swanson, Paul N
Swenson, Donald Adolph
Szymanski, Paul Stephen
Taimuty, Samuel Isaac
Talbot, Raymond James, Jr
Tang, Homer H(o)
Targ, Russell
Taylor, Charles Joel
Taylor, Howard S
Taylor, Thomas Tallott, Sr
Teasdale, John G
Teller, Edward
Thatcher, Everett Whiting
Thomas, James H
Thomas, Mitchell
Thompson, Timothy J
Thompson, W P(aul)
Thorngate, John Hill
Tiller, William Arthur
Tobey, Arthur Robert
Tombrello, Thomas Anthony, Jr
Toon, Owen Brian
Toth, Robert Allen
Trenholme, John Burgess
Trigger, Kenneth Roy
Tripp, Robert D
Trorey, A(lan) W(ilson)
Tsai, Ming-Jong
Tucker, Wallace Hampton
Tuerpe, Dieter Rolf
Tuman, Vladimir Shlimon
Turneaure, John Paul
Turner, Eugene Bonner
Turner, John E
Tuul, Johannes
Unwin, Stephen Charles
Vajk, J(oseph) Peter
Van Atta, Lester Clare
Van Lint, Victor Anton Jacobus
Van Thiel, Mathias
Vardiman, Larry
Vawter, Spencer Max
Veigele, William John
Vernon, C(arl) Wayne
Vernon, Frank Lee, Jr
Victor, Andrew C
Vogt, Rochus E
Volpe, Richard Alan
Voreades, Demetrios
Vorhaus, James Louis
Vreeland, John Allen
Waddell, Charles Noel
Wahlig, Michael Alexander
Wainwright, Thomas Everett
Waiter, Serge-Albert
Wall, Leonard Wong
Wang, Michael Renxun
Wang, Shyh
Waniek, Ralph Walter
Warshaw, Stephen I
Washburn, Harold W(illiams)
Waterhouse, Richard (Valentine)
Waters, Rodney Lewis
Watson, Kenneth Marshall
Watson, Velvin Richard
Weiss, Max Tibor
Weissler, Gerhard Ludwig
Wenk, Hans-Rudolf
Westberg, Karl Rogers
Weston, Gary Steven
Wheaton, Elmer Paul
White, Richard Manning
Whitefield, Rodney Joe
Whitmer, Robert Morehouse
Whitney, Robert C
Whitney, William Merrill
Whittemore, William Leslie
Wilcox, Howard Albert
Wilcox, Thomas Jefferson
Wilkening, Dean Arthur
Williams, E(dgar) P
Williamson, Hugh A
Wilson, Walter Davis
Winkelmann, Frederick Charles
Wintroub, Herbert Jack
Witteborn, Fred Carl
Woehler, Karlheinz Edgar
Woerner, Robert Leo
Wong, K(wee) C
Wood, Calvin Dale
Woodbury, Eric John
Woodruff, Truman Owen
Woodworth, John George
Wooldridge, Dean E
Worden, Paul Wellman, Jr
Worth, Donald Calhoun
Wright, Frederick Hamilton
Wuerker, Ralph Frederick
Yadavalli, Sriramamurti Venkata
Yanow, Gilbert
Yodh, Gaurang Bhaskar
York, Carl Monroe, Jr
York, Herbert Frank
Young, Richard D
Yu, Simon Shin-Lun
Zhou, Simon Zheng
Zimmerman, Elmer Leroy
Zimmerman, George Beeden
Zubeck, Robert Bruce

COLORADO
Aamodt, Richard E
Anderson, Dana Zachary
Archambeau, Charles Bruce
Baird, Ramon Condie
Barnes, James Allen
Barrett, Charles Sanborn
Barth, Charles Adolph
Bartlett, Albert Allen
Beam, Kurt George, Jr
Beaty, Earl Claude
Beehler, Roger Earl
Begelman, Mitchell Craig
Bender, Peter Leopold
Bills, Daniel Granville
Bordner, Charles Albert, Jr
Bradley, Richard Crane
Bruce, Charles Robert
Bruns, Donald Gene
Burnett, Clyde Ray
Butler, William Albert
Clark, Noel A
Collins, Royal Eugene
Colvis, John Paris
Davis, Milford Hall
Derbyshire, William Davis
Duray, John R
Duvall, Wilbur Irving
Ellis, Homer Godsey
Engen, Glenn Forrest
Faller, James E
Ferguson, Eldon E
Fickett, Frederick Roland
Fox, Michael Henry
Franklin, Allan David
Gall, Robert Lawrence
Gille, John Charles
Gilman, Peter A
Gould, Gordon
Graetzer, Hans Gunther
Granzow, Kenneth Donald
Gray, James Edward
Griffith, Gordon Lamar
Hagerman, Donald Charles
Hall, John L
Hamerly, Robert Glenn
Hansen, Carl John
Harris, Richard Elgin
Heller, Marvin W
Hilt, Richard Leighton
Hjelme, Dag Roar
Hord, Charles W
Horton, Clifford E(dward)
Hurt, James Edward
Hust, Jerome Gerhardt
Iona, Mario
Johnson, Richard Harlan
Jordan, Albert Raymond
Kamper, Robert Andrew
Kashnow, Richard Allen
Kearney, Philip Daniel
Kemper, William Alexander
Kennedy, Patrick James
Koch, Herman William
Koldewyn, William A
Koontz, Philip G
Langer, George Edward
Lawrence, George Melvin
Lazzarini, Albert John
Leinbach, F Harold
Leisure, Robert Glenn
Levenson, Leonard L
Levine, Judah
McNutt, Douglas P
Marks, Roger Bradley
Muscari, Joseph A
Nesbitt, David John
Netzel, Richard G
Olson, John Richard
Owen, Robert Barry
Pawlicki, Anthony Joseph
Reichardt, John William
Ristinen, Robert A
Robinson, Hugh Gettys
Rutz, Richard Frederick
Sauer, Herbert H
Schramm, Raymond Eugene
Simon, Nancy Jane
Sites, James Russell
Slaughter, Maynard
Smith, Richard Cecil
Snyder, Wilbert Frank
Spence, William J
Spenny, David Lorin
Stebbins, Robin Tucker
Stein, Samuel Richard
Stern, Raul A(ristide)
Sullivan, Donald Barrett
Swenson, Hugo Nathanael
Tanttila, Walter H
Violett, Theodore Dean
Wahr, John Cannon
Ware, Walter Elisha
Watkins, Sallie Ann
Wells, Joseph S
White, Franklin Estabrook
Wineland, David J
Wollan, John Jerome
Wooldridge, Gene Lysle
Wyss, Walter
Zaidins, Clyde

CONNECTICUT
Anderson, Wilmer Clayton
Azaroff, Leonid Vladimirovich
Bannister, Peter Robert
Barker, Richard Clark
Bennett, William Ralph, Jr
Beringer, Robert
Best, Philip Ernest
Budnick, Joseph Ignatius
Burkhard, Mahlon Daniel
Chih, Chung-Ying
Cole, Henderson
Connell, Richard Allen
Cummerow, Robert Leggett
Damon, Dwight Hills
Davenport, Lee Losee
DeMaria, Anthony John
Fraser, J(ulius) T(homas)

Frueh, Alfred Joseph, Jr
Gilliam, Otis Randolph
Glenn, William Henry, Jr
Goldman, Jacob E
Green, Milton
Grubin, Harold Lewis
Halpern, Howard S
Hamilton, Douglas Stuart
Howkins, Stuart D
Hughes, Vernon Willard
Iachello, Francesco
Johnson, Bruce Virgil
Kattamis, Theodoulos Zenon
Klemens, Paul Gustav
Kraft, David Werner
Krisst, Raymond John
Lichten, William Lewis
Liebson, Sidney Harold
Lindquist, Richard Wallace
Lipschultz, Frederick Phillip
Locke, Stanley
Lu, Phillip Kehwa
Ma, Tso-Ping
MacDowell, Samuel Wallace
Marcus, Jules Alexander
Matthews, Lee Drew
Melcher, Charles L
Moran, Thomas Irving
Moreland, Parker Elbert, Jr
Nash, Harold Earl
Nelligan, William Bryon
Nudelman, Sol
Paolini, Francis Rudolph
Patterson, Elizabeth Chambers
Pease, Douglas M
Plona, Thomas Joseph
Pollock, Herbert Chermside
Poultney, Sherman King
Prober, Daniel Ethan
Rau, Richard Raymond
Rich, Leonard G
Richards, Roger Thomas
Roychoudhuri, Chandrasekhar
Russek, Arnold
Rydz, John S
Sachdev, Subir
Sandweiss, Jack
Seely, Samuel
Sherman, Harold
Shirer, Donald Leroy
Siegel, Lester Aaron
Sivinski, John A
Spencer, Domina Eberle (Mrs Parry Moon)
Strough, Robert I(rving)
Sung, Chong Sook Paik
Tobin, Marvin Charles
Tracy, Joseph Charles
Trousdale, William Latimer
Tucker, Edmund Belford
Waine, Martin
Wegener, Peter Paul
Wingfield, Edward Christian
Wolf, Elizabeth Anne
Zandy, Hassan F

DELAWARE
Barton, Randolph, Jr
Burgess, John S(tanley)
Case, John William
Chromey, Fred Carl
Cooper, Charles Burleigh
Davies, Robert Dillwyn
Ewing, Richard Dwight
Griffiths, David
Groff, Ronald Parke
Halprin, Arthur
Hardy, Henry Benjamin, Jr
Kent, Donald Wetherald, Jr
Knop, Harry William, Jr
Lutz, Bruce Charles
Maher, John Philip
Marcus, Sanford M
Mehl, James Bernard
Mrozinski, Peter Matthew
Robertson, Charles William, Jr
Shipman, Harry Longfellow
Weiher, James F
Wilson, Frank Charles

DISTRICT OF COLUMBIA
Appleman, Daniel Everett
Beasley, Edward Evans
Beckner, Everet Hess
Blankfield, Alan
Boris, Jay Paul
Brown, Harold
Callaham, Michael Burks
Carroll, Kenneth Girard
Cassidy, Esther Christmas
Chappell, Samuel Estelle
Chung, David Yih
Corden, Pierce Stephen
Corliss, Edith Lou Rovner
Daehler, Mark
Davis, Jack
Eagleson, Halson Vashon
Edelsack, Edgar Allen
Goldberg, Stanley
Haas, George Arthur
Hafizi, Bahman
Hendrie, David Lowery
Howard, Robert E
Hubler, Graham Kelder, Jr

Jhirad, David John
Kalmus, Henry P(aul)
Karle, Jerome
Keeny, Spurgeon Milton, Jr
Kelber, Charles Norman
Kelly, Henry Charles
Kepple, Paul C
Koon, Norman C
Litovitz, Theodore Aaron
McKay, Jack Alexander
Mandula, Jeffrey Ellis
Marcus, Gail Halpern
Michels, Donald Joseph
Montrose, Charles Joseph
Moon, Deug Woon
Needels, Theodore S
Ondik, Helen Margaret
Oran, Elaine Surick
Oswald, Robert B(ernard), Jr
Palmadesso, Peter Joseph
Peters, Gerald Joseph
Photiadis, Douglas Marc
Pickett, Warren Earl
Pikus, Irwin Mark
Prinz, Dianne Kasnic
Riegel, Kurt Wetherhold
Riemer, Robert Lee
Ritter, Enloe Thomas
Rittner, Edmund Sidney
Rogers, Kenneth Cannicott
Scheel, Nivard
Schildcrout, Michael
Schimmerling, Walter
Schrack, Roald Amundsen
Schweizer, Francois
Shapiro, Anatole Morris
Shih, Arnold Shang-Teh
Siry, Joseph William
Spohr, Daniel Arthur
Sprangle, Phillip Allen
Taylor, Ronald D
Tsao, Chen-Hsiang
Vardiman, Ronald G
Varnum, William Sloan
Wagner, Andrew James
Weiler, Kurt Walter
Weinstock, Harold
Wittmann, Horst Richard
Zuchelli, Artley Joseph

FLORIDA
Adhav, Ratnakar Shankar
Alonso, Marcelo
Andregg, Charles Harold
Ard, William Bryant, Jr
Aubel, Joseph Lee
Ballard, Stanley Sumner
Becker, Gordon Edward
Beguin, Fred P
Biver, Carl John, Jr
Bloch, Sylvan C
Bolemon, Jay S
Bolte, John R
Brandt, Bruce Losure
Buckwalter, Gary Lee
Bueche, Frederick Joseph
Burns, Jay, III
Calhoun, Ralph Vernon, Jr
Carr, Thomas Deaderick
Clapp, Roger Williams, Jr
Coggeshall, Norman David
Cook, Barnett C
Cooper, John Niessink
Cooper, Raymond David
Cox, Joseph Robert
Dalehite, Thomas H
Dam, Cecil Frederick
Dube, Roger Raymond
Edwards, Palmer Lowell
Fausett, Laurene van Camp
Fenna, Roger Edward
Ference, Michael, Jr
Field, George S
Field, Richard D
Fleddermann, Richard G(rayson)
Flowers, John Wilson
Forman, Guy
Foss, Martyn (Henry)
Gager, William Ballantine
Gianola, Umberto Ferdinando
Gielisse, Peter Jacob Maria
Green, Alex Edward Samuel
Green, Eugene L
Hamtil, Charles Norbert
Harrington, John Vincent
Herriott, Donald R
Herzog, Richard (Franz Karl)
Hirschberg, Joseph Gustav
Hock, Donald Charles
Huber, Oren John
Humphries, Jack Thomas
Hunt, Robert Harry
Ihas, Gary Gene
Jamerson, Frank Edward
Jensen, Richard Eugene
Jin, Rong-Sheng
Kendall, Harry White
Kerr, Robert Lowell
Keuper, Jerome Penn
Kiewit, David Arnold
Kinnmark, Ingemar Per Erland
Knowles, Harold Loraine
Krausche, Dolores Smoleny
Krc, John, Jr

Krishnamurti, Ruby Ebisuzaki
Kruschwitz, Walter Hillis
Kudman, Irwin
Lambe, Edward Dixon
Lebo, George Robert
Llewellyn, Ralph A
Ludeke, Carl Arthur
McClelland, John Albert
McLeroy, Edward Glenn
Mavroides, John George
Miller, Hillard Craig
Milton, James E(dmund)
Minkin, Jean Albert
Mitchell, Richard Warren
Mitchell, Robert D
Moulton, William G
Munk, Miner Nelson
Nelson, Edward Bryant
Oelfke, William C
Ohrn, Nils Yngve
Oleson, Norman Lee
Omer, Guy Clifton, Jr
Owens, James Samuel
Painter, Ronald Dean
Pardo, William Bermudez
Pendleton, Winston Kent, III
Perlmutter, Arnold
Pfahnl, Arnold
Potter, James Gregor
Reynolds, George William, Jr
Rhodes, Richard Ayer, II
Riley, Mark Anthony
Robertson, Harry S(troud)
Rosenzweig, Walter
Rudgers, Anthony Joseph
Schmidt, Klaus H
Shih, Hansen S T
Skramstad, Harold Kenneth
Smith, Malcolm (Kinmonth)
Stewart, Gregory Randall
Sundaram, Swaminatha
Tanner, David Burnham
Thomas, Dan Anderson
Thornton, J Ronald
Timme, Robert William
Toms, M Elaine
Tterlikkis, Lambros
Vollmer, James
Weber, Alfred Herman
Weller, Richard Irwin
Wells, Daniel R
Zaukelies, David Aaron
Zimmer, Martin F
Zinn, Walter Henry

GEORGIA
Ahrens, Rudolf Martin (Tino)
Anantha Narayanan, Venkataraman
Anderson, Robert Lester
Bomar, Lucien Clay
Boyd, James Emory
Burkhard, Donald George
Buzzelli, Edward S
Chandra, Kailash
Clement, Joseph D(ale)
Clemmons, John B
Cooper, Charles Dewey
Cramer, John Allen
Crawford, Vernon
Croft, George Thomas
Dixon, Robert Malcolm
Eichholz, Geoffrey G(unther)
Family, Fereydoon
Gallaher, Lawrence Joseph
Garrison, Allen K
Gersch, Harold Arthur
Goslin, Roy Nelson
Hardin, Clyde D
Harrison, Gordon R
Hart, Raymond Kenneth
Hartman, Nile Franklin
Hunt, Gary W
Hurst, Vernon James
Jackson, Prince A, Jr
Jarzynski, Jacek
Kiang, Chia Szu
Kollig, Heinz Philipp
McDaniel, Earl Wadsworth
Martin, David Willis
Nave, Carl R
Patronis, Eugene Thayer, Jr
Petitt, Gus A
Price, Edward Warren
Purcell, James Eugene
Rohrer, Robert Harry
Roy, Rajarshi
Sadun, Alberto Carlo
Simmons, James Wood
Stevenson, James Rufus
Williams, Joel Quitman
Winer, Ward Otis
Woodward, LeRoy Albert
Young, Robert Alan

HAWAII
Cence, Robert J
Chambers, Kenneth Carter
Crowe, Richard Alan
Dorrance, William Henry
Hayes, Charles Franklin
Holm-Kennedy, James William
Jayaraman, Aiyasami
McAllister, Howard Conlee
McFee, Richard

Meech, Karen Jean
Pakvasa, Sandip
Sogo, Power Bunmei
Steiger, Walter Richard
Stenger, Victor John

IDAHO
Browne, Michael Edwin
Carpenter, Stuart Gordon
Dolan, Thomas James
East, Larry Verne
Ford, Gilbert Clayton
Froes, Francis Herbert (Sam)
Hall, J(ames) A(lexander)
Harding, Samuel William
Harker, Yale Deon
Harmon, J Frank
Knox, John MacMurray
Lott, Layman Austin
Luke, Robert A
Majumdar, Debaprasad (Debu)
Marks, Darrell L
Murphey, Byron Freeze
Olson, Willard Orvin
Otting, William Joseph
Patsakos, George
Poenitz, Wolfgang P
Price, Joseph Earl
Reich, Charles William
Telschow, Kenneth Louis
Weyland, Jack Arnold
Wood, Richard Ellet

ILLINOIS
Abels, Larry L
Adler, Robert
Allen, Frank B
Altpeter, Lawrence L, Jr
Anderson, Scott
Anderson, William Raymond
Avery, Robert
Bailyn, Martin H
Behof, Anthony F, Jr
Bellamy, David
Block, Martin M
Bodmer, Arnold R
Boedeker, Richard Roy
Born, Harold Joseph
Borso, Charles S
Bowe, Joseph Charles
Bowen, Samuel Philip
Boyd, John William
Braid, Thomas Hamilton
Braundmeier, Arthur John, Jr
Brussel, Morton Kremen
Bushnell, David L
Carney, Rose Agnes
Chang, R P H
Cloud, William Max
Coester, Fritz
Cohn, Gerald Edward
Cole, Francis Talmage
Cooper, Duane H(erbert)
Cornell, David Allan
Coster, Joseph Constant
Crew, John Edwin
Crewe, Albert Victor
Crosbie, Edwin Alexander
Curry, Bill Perry
Curtis, Cyril Dean
Davis, A Douglas
Debrunner, Peter Georg
Dixon, Roger L
Doerner, Robert Carl
Drickamer, Harry George
Dudley, Horace Chester
Dunford, Robert Walter
Ebrey, Thomas G
Edwards, Helen Thom
Erber, Thomas
Erck, Robert Alan
Evans, William Paul
Fink, Joanne Krupey
Fish, Ferol F, Jr
Galayda, John Nicolas
Gerding, Dale Nicholas
Giese, Robert Frederick
Ginsberg, Donald Maurice
Goldwasser, Edwin Leo
Green, Donald Wayne
Greene, Laura H
Griem, Melvin Luther
Gupta, Nand K
Halperin, William Paul
Hannon, Bruce Michael
Harrington, Joseph Anthony
Hasdal, John Allan
Heaton, LeRoy
Herzenberg, Caroline Stuart Littlejohn
Hess, David Clarence
Huft, Michael John
Hull, John R
Hurych, Zdenek
Ivory, John Edward
Jackson, Edwin Atlee
Kaminsky, Manfred Stephan
Kaufmann, Elton Neil
Keren, Joseph
Ketterson, John Boyd
Klein, Miles Vincent
Knox, Jack Rowles
Koehler, James Stark
Krauss, Alan Robert
Krein, Philip T

Physics, General (cont)

Kruse, Ulrich Ernst
Lach, Joseph T
Lanzi, Lawrence Herman
Leonard, Byron Peter
LeSage, Leo G
Levi-Setti, Riccardo
Licht, Arthur Lewis
McAneny, Laurence Raymond
McCart, Bruce Ronald
Meeker, Ralph Dennis
Meyer, Axel
Meyer, Stuart Lloyd
Mills, Frederick Eugene
Moore, Paul Brian
Muller, Dietrich
Ng, King-Yuen
Nickell, William Everett
Noble, Gordon Albert
Noble, John Dale
O'Gallagher, Joseph James
Orr, John R
Pagnamenta, Antonio
Persiani, Paul J
Prohammer, Frederick George
Propst, Franklin Moore
Redman, William Charles
Rehn, Lynn Eduard
Renneke, David Richard
Rey, Charles Albert
Ring, James George
Roberts, Arthur
Roberts, John England
Roll, Peter Guy
Roothaan, Clemens Carel Johannes
Rossing, Thomas D(ean)
Rubin, Howard Arnold
Saboungi, Marie-Louise Jean
Saporoschenko, Mykola
Schillinger, Edwin Joseph
Schmidt, Charles William
Schramm, David N
Schweizer, Kenneth Steven
Shaffer, John Clifford
Sharma, Ram Ratan
Shen, Sin-Yan
Shih, Hsio Chang
Siegel, Jonathan Howard
Siegel, Stanley
Simpson, John Alexander
Singwi, Kundan Singh
Skaggs, Lester S
Smith, P Scott
Soule, David Elliot
Staunton, John Joseph Jameson
Stefanski, Raymond Joseph
Stenberg, Charles Gustave
Stevens, Fred Jay
Stinchcomb, Thomas Glenn
Strahm, Norman Dale
Stutz, Conley I
Sukhatme, Uday Pandurang
Sutton, David C(hase)
Tao, Rongjia
Throw, Francis Edward
Tomaschke, Harry E
Tomkins, Marion Louise
Turkot, Frank
Tuzzolino, Anthony J
Van Ginneken, Andreas J
Waddell, Robert Clinton
Wah, Yau Wai
Watson, Richard Elvis
Weaver, Allen Dale
Weertman, Johannes
Whalin, Edwin Ansil, Jr
Winstein, Bruce Darrell
Wolfson, James
Wolsky, Alan Martin
Wright, Sydney Courtenay
Wyld, Henry William, Jr
Wylie, Douglas Wilson
Yamanouchi, Taiji
Yntema, Jan Lambertus
Yoh, John K
Youtsey, Karl John
Zaromb, Solomon
Zhang, Shouyu

INDIANA
Allen, Merrill James
Alyea, Ethan Davidson, Jr
Beery, Dwight Beecher
Bent, Robert Demo
Biswas, Nripendra Nath
Blackstead, Howard Allan
Brill, Wilfred G
Cohen, Herbert Daniel
Conklin, Richard Louis
Coomes, Edward Arthur
Cowan, Raymond
Cox, Martha
Crable, George Francis
Crittenden, Ray Ryland
Cruikshank, Donald Burgoyne, Jr
Di Lavore, Philip, III
Durisen, Richard H
Dzierba, Alex R
Farringer, Leland Dwight
Feuer, Paula Berger
Flick, Cathy
Gaidos, James A
Gerritsen, Alexander Nicolaas

Gutay, Laszlo J
Hake, Richard Robb
Hale, Robert E
Hamilton-Steinraut, Jean A
Hartzler, Harrod Harold
Henninger, Ernest Herman
Henry, Hugh Fort
Henry, Robert Ledyard
Howes, Ruth Hege
Hrivnak, Bruce John
Hults, Malcom E
Johnson, Walter Richard
Jumper, Eric J
Kenney, Vincent Paul
Kesmodel, Larry Lee
Khorana, Brij Mohan
Klontz, Everett Earl
Kolata, James John
Kramer, Robert Allen
Leiter, Howard Allen
Loeffler, Frank Joseph
Lurie, Fred Marcus
Mastrototaro, John Joseph
Moloney, Michael J
Nann, Hermann
Palfrey, Thomas Rossman, Jr
Poorman, Lawrence Eugene
Reilly, James Patrick
Robinson, John Murrell
Schlueter, Donald Jerome
Shaffer, Lawrence Bruce
Snyder, Donald DuWayne
Sprague, Newton G
Taylor, Raymond Ellory
Thieme, Melvin T
Vigdor, Steven Elliot
Vinson, James S
Vondrak, Edward Andrew
Wolber, William George

IOWA
Azbell, William
Bowen, George Hamilton, Jr
Clotfelter, Beryl Edward
Denny, Wayne Belding
Frank, Louis Albert
Goertz, Christoph Klaus
Hodges, Laurent
Kernan, William J, Jr
Kirkham, Don
Kopp, Jay Patrick
Leung, Wai Yan
Macomber, Hilliard Kent
Moen, Allen LeRoy
Ostenson, Jerome E
Peterson, Francis Carl
Poppy, Willard Joseph
Pursey, Derek Lindsay
Soukoulis, Costas M
Stebbins, Dean Waldo
Van Allen, James Alfred

KANSAS
Bearse, Robert Carleton
Cocke, C Lewis
Dale, Ernest Brock
Ellsworth, Louis Daniel
Feaster, Gene R(ichard)
Gumnick, James Louis
Law, Bruce Malcolm
Pruitt, Roger Arthur
Senecal, Gerard
Stockli, Martin P
Thomas, James E
Twarog, Bruce Anthony
Unruh, Henry, Jr
Valenzeno, Dennis Paul
Wiseman, Gordon G
Witten, Maurice Haden
Wong, Kai-Wai

KENTUCKY
Carpenter, Dwight William
Cochran, Lewis Wellington
Davis, Thomas Haydn
Dillon, John A, Jr
Duncan, Don Darryl
Enzweiler, Raymond N
Feola, Jose Maria
Gwinn, Joel Alderson
Kadaba, Prasad Krishna
Merker, Stephen Louis
Moulder, Jerry Wright
Naake, Hans Joachim
Powell, Smith Thompson, III
Purdom, Ray Caldwell
Rollo, Frank David
Russell, Marvin W
Sinai, John Joseph
Watkins, Nancy Chapman

LOUISIANA
Bedell, Louis Robert
Bergeron, Clyde J, Jr
Bernard, William Hickman
Dempsey, Colby Wilson
Ebert, Paul Joseph
Edmonds, James D, Jr
Feldman, Martin
Fischer, David John
Hoy, Robert C
Johnson, Larry Don
Keiffer, David Goforth
King, Creston Alexander, Jr

Matlock, Rex Leon
Mitcham, Donald
Naidu, Seetala V
Neuman, Charles Herbert
Perdew, John Paul
Reynolds, Joseph Melvin
Riess, Karlem
Robert, Kearny Quinn, Jr
Smith, Ronald E
Stephenson, Paul Bernard
Ward, Truman L
Wefel, John Paul

MAINE
Bancroft, Dennison
Carr, Edward Frank
Chonacky, Norman J
Dudley, John Minot
Emery, Guy Trask
Gordon, Geoffrey Arthur
Grimes, Charles Cullen
Hoyt, Rosalie Chase
Hughes, William Taylor
Mitchell, H Rees
Otto, Fred Bishop
Ruff, George Antony
Semon, Mark David
Smith, Charles William, Jr
Snow, Joseph William
Tarr, Charles Edwin
Viette, Michael Anthony

MARYLAND
Achor, William Thomas
Acuna, Mario Humberto
Alers, Perry Baldwin
Alley, Reuben Edward, Jr
Anderson, J Robert
Arking, Albert
Arndt, Richard Allen
Aronson, Casper Jacob
Avery, William Hinckley
Babrauskas, Vytenis
Balasubrahmanyan, Vriddhachalam K
Barbe, David Franklin
Bargeron, Cecil Brent
Barnett, Bruce A
Barrows, Austin Willard
Bass, Arnold Marvin
Bay, Zoltan Lajos
Bennett, Charles L
Bennett, Lawrence Herman
Berg, Robert Frank
Bernard, William
Berning, Warren Walt
Bhagat, Satindar M
Biberman, Lucien Morton
Bjerkaas, Allan Wayne
Blake, Lamont Vincent
Blevins, Gilbert Sanders
Blewett, John P
Block, Stanley
Bonavita, Nino Louis
Bostrom, Carl Otto
Bowers, Robert Clarence
Branson, Herman Russell
Brodsky, Marc H
Brown, Samuel Heffner
Budgor, Aaron Bernard
Burlaga, Leonard F
Calame, Gerald Paul
Carlon, Hugh Robert
Carp, Gerald
Carter, Robert Emerson
Casella, Russell Carl
Cavalieri, Donald Joseph
Cavicchi, Richard E
Chen, Hsing-Hen
Chern, Engmin James
Chertock, George
Chu, Tak-Kin
Clark, Trevor H
Clarke, John F
Coakley, Kevin Jerome
Coates, R(obert) J(ay)
Coffman, John W
Cohen, Steven Charles
Conklin, Glenn Ernest
Conrath, Barney Jay
Cook, Richard Kaufman
Cornett, Richard Orin
Cramer, William Smith
Crane, Patrick Conrad
Crump, Stuart Faulkner
Culver, William Howard
Daniel, Charles Dwelle, Jr
Daniel, John Harrison
Dayhoff, Edward Samuel
Degnan, John James, III
Dehn, James Theodore
De Rosset, William Steinle
Deslattes, Richard D, Jr
Dick, Richard Dean
DiMarzio, Edmund Armand
Di Rienzi, Joseph
Dixon, Jack Richard
Dixon, John Kent
Dodge, William R
Dubin, Maurice
Dunlap, Brett Irving
Earhart, J Ronald
Earl, James Arthur
Eckerle, Kenneth Lee
Egner, Donald Otto

Eichelberger, Robert John
Elder, Samuel Adams
Elsasser, Walter M
Emch, George Frederick
Ernest, Michael Vance, Sr
Ethridge, Noel Harold
Falk, Charles Eugene
Fansler, Kevin Spain
Farrell, Richard Alfred
Field, Herbert Cyre
Finkelstein, Robert
Fiorito, Ralph Bruno
Fitzgerald, Edwin Roger
Fitzpatrick, Hugh Michael
Flook, William Mowat, Jr
Flory, Thomas Reherd
Fowler, Howland Auchincloss
Fristrom, Robert Maurice
Garroway, Allen N
Gasparovic, Richard Francis
Geckle, William Jude
Gelles, Isadore Leo
Gettner, Marvin
Gheorghiu, Paul
Gilliland, Ronald Lynn
Gion, Edmund
Glasser, Robert Gene
Gloeckler, George
Gluckstern, Robert Leonard
Goldberg, Benjamin
Goldfinger, Andrew David
Gonano, John Roland
Gordon, Gary Donald
Greenstone, Reynold
Griem, Hans Rudolf
Griffin, James J
Gutsche, Graham Denton
Hafstad, Lawrence Randolph
Hall, Forrest G
Hallgren, Richard E
Halpin, Joseph John
Harman, George Gibson, Jr
Harris, Forest K
Hart, Lynn W
Hartmann, Gregory Kemenyi
Hayward, Evans Vaughan
Heckman, Timothy Martin
Heinberg, Milton
Henry, Richard Conn
Heydemann, Peter Ludwig Martin
Hirschmann, Erwin
Hobbs, Robert Wesley
Hollinger, James Pippert
Holman, Gordon Dean
Horn, William Everett
Howard, Sethanne
Hudson, Ralph P
Hybl, Albert
Hynes, John Edward
Inglehart, Lorretta Jeanette
Irwin, George Rankin
Jablonski, Felix Joseph
Jacobs, Sigmund James
Jefferson, Donald Earl
Jen, Chih Kung
Jette, Archelle Norman
Johnston, Robert Ward
Kaeser, R S
Kagarise, Ronald Eugene
Kahn, Arnold Herbert
Kahn, Jack Henry
Kaiser, Jack Allen Charles
Kalil, Ford
Katcher, David Abraham
Katsanis, D(avid) J(ohn)
Kay, Mortimer Isaia
Kazi, Abdul Halim
Keating, Patrick Norman
Kelly, William Clark
Killen, Rosemary Margaret
Kim, Boris Fincannon
Kineke, J H, Jr
Kinsey, James Humphreys
Koblinsky, Chester John
Korobkin, Irving
Kotter, F(red) Ralph
Krasner, Sol H
Krauss, Morris
Kravitz, Lawrence C
Krumbein, Simeon Joseph
Kulp, Bernard Andrew
Kuriyama, Masao
Lashof, Theodore William
Lauriente, Mike
Lee, Ronald Norman
Leffel, Claude Spencer, Jr
Li, Ming Chiang
Link, John Clarence
Little, John Llewellyn
Liu, Han-Shou
Lubkin, Gloria Becker
Luebke, Emmeth August
Lundholm, J(oseph) G(ideon), Jr
Lyman, Ona Rufus
McCally, Russell Lee
McCoubrey, Arthur Orlando
McDonald, Frank Bethune
McElhinney, John
McKinney, John Edward
McLeod, Norman Barrie
McNutt, Ralph Leroy, Jr
Madden, Robert Phyfe
Mahoney, Francis Joseph
Majkrzak, Charles Francis

Mangum, Billy Wilson
Mann, Wilfrid Basil
Marcinkowski, M(arion) J(ohn)
Maxwell, Louis R
Messina, Carla Gretchen
Michaelis, Michael
Mielenz, Klaus Dieter
Miller, Gerald R
Moldover, Michael Robert
Monaldo, Francis Michael
Moon, Milton Lewis
More, Kenneth Riddell
Morgan, Bruce Harry
Motz, Joseph William
Mozer, Bernard
Munro, Ronald Gordon
Myers, Peter Briggs
Nakada, Minoru Paul
Ney, Wilbert Roger
Nossal, Ralph J
Ogilvie, Keith W
Oldfield, Edward Hudson
Oroshnik, Jesse
Ostrofsky, Bernard
Paik, Ho Jung
Park, Robert L
Parr, Albert Clarence
Patten, Raymond Alex
Peiser, Herbert Steffen
Perry, Peter M
Pevsner, Aihud
Piermarini, Gasper J
Potocki, Kenneth Anthony
Potyraj, Paul Anthony
Potzick, James Edward
Pouring, Andrew A
Powell, Cedric John
Prange, Richard E
Radcliffe, Alec
Rathbun, Edwin Roy, Jr
Ravitsky, Charles
Rawlins, Stephen Last
Reader, Joseph
Reames, Donald Vernon
Rector, Charles Willson
Reisse, Robert Alan
Richardson, John Marshall
Riedl, H Raymond
Roberts, James Richard
Robinson, Richard Carleton, Jr
Rosenbaum, Ira Joel
Roth, Bradley John
Roth, Michael William
Rubenstein, Albert Marvin
Ruff, Arthur William
Scannell, Edward Perry
Schindler, Albert Isadore
Schleif, Robert Ferber
Schmugge, Thomas Joseph
Schroeder, Frank, Jr
Schwee, Leonard Joseph
Seigel, Arnold E(lliott)
Shah, Shirish
Shotland, Edwin
Sicotte, Raymond L
Sikora, Jerome Paul
Slawsky, Zaka I
Snow, George Abraham
Soln, Josip Zvonimir
Solow, Max
Sorrows, Howard Earle
Spangler, Glenn Edward
Spornick, Lynna
Stecker, Floyd William
Stevens, Howard Odell, (Jr)
Stillman, Rona Barbara
Stone, Albert Mordecai
Strittmater, Richard Carlton
Stroh, William Richard
Strombotne, Richard L(amar)
Sucher, Joseph
Sutter, David Franklin
Swartzendruber, Lydon James
Sztankay, Zoltan Geza
Taylor, Barry Norman
Teays, Terry John
Toller, Gary Neil
Tousey, Richard
Townsend, John William, Jr
Tropf, William Jacob
Turnrose, Barry Edmund
Uppal, Parvez N
Varma, Matesh Narayan
Venkatesan, Thirumalai
Vette, James Ira
Von Bun, Friedrich Otto
Walker, Robert Lee
Walter, John Fitler
Wang, Theodore Joseph
Wang, Ting-I
Ward, Alford L
Wasilik, John H(uber)
Watson, Robert Tanner
Weber, Joseph
Weller, Charles Stagg, Jr
Whicker, Lawrence R
Whitmore, Bradley Charles
Willett, Colin Sidney
Wimenitz, Francis Nathaniel
Winslow, George Harvey
Wood, Howard John, III
Wood, Lawrence Arnell
Wyckoff, Harold Orville
Wyckoff, James M

Yin, Lo I
Young, Frank Coleman
Young, Russell Dawson
Zimmermann, Mark Edward
Zirkind, Ralph
Zucker, Paul Alan

MASSACHUSETTS
Abbott, Norman John
Afsar, Mohammed Nurul
Aisenberg, Sol
Aldrich, Ralph Edward
Allemand, Charly D
Altshuler, Edward E
Baker, Adolph
Band, Hans Eduard
Barbour, William E, Jr
Barrington, A(lfred) E(ric)
Baxter, Lincoln, II
Bedo, Donald Elro
Benedek, George Bernard
Bennett, Stewart
Bertozzi, William
Bjorkholm, Paul J
Bliss, David Francis
Block, Steven M
Bolt, Richard Henry
Boness, Michael John
Bowness, Colin
Brown, Fielding
Burke, Bernard Flood
Butler, James Preston
Caley, Wendell J, Jr
Callerame, Joseph
Cardon, Bartley Lowell
Carr, George Leroy
Carroll, Thomas Joseph
Caswell, Randall Smith
Cavallaro, Mary Caroline
Celmaster, William Noah
Chakrabarti, Supriya
Champion, Kenneth Stanley Warner
Champion, Paul Morris
Chang, Tien Sun (Tom)
Chase, Charles Elroy, Jr
Chretien, Max
Clancy, Edward Philbrook
Clark, George Whipple
Clark, Melville, Jr
Clifton, Brian John
Collins, James Joseph
Cook, LeRoy Franklin, Jr
Coor, Thomas
Coppi, Bruno
Costello, Ernest F, Jr
Cromer, Alan H
Dale, Brian
Davies, John A
Davis, Charles Freeman, Jr
Decowski, Piotr
Demos, Peter Theodore
Deutsch, Martin
Dimock, Dirck L
Dion, Andre R
Donoghue, John Francis
Doshi, Anil G
Douglas-Hamilton, Diarmaid H
Dunlap, William Crawford
Eklund, Karl E
Emshwiller, Maclellan
Emslie, Alfred George
Faissler, William L
Fante, Ronald Louis
Fariss, Thomas Lee
Feshbach, Herman
Fisher, Daniel S
Fletcher, Robert Chipman
Foulis, David James
Freedman, George
French, Robert Dexter
Freyre, Raoul Manuel
Friedman, Jerome Isaac
Frisk, George Vladimir
Gamota, George
Garelick, David Arthur
Gelman, Harry
Gentile, Thomas Joseph
Gericke, Otto Reinhard
Geschwind, Stanley
Glascock, Homer Hopson, II
Glassbrenner, Charles J
Glauber, Roy Jay
Godyak, Valery Anton
Goldsmith, George Jason
Goldstein, Jack Stanley
Golovchenko, Jene Andrew
Golowich, Eugene
Gordon, Joel Ethan
Greenstein, George
Greytak, Thomas John
Grisaru, Marcus Theodore
Guernsey, Janet Brown
Haber-Schaim, Uri
Hager, Bradford Hoadley
Hallock, Robert B
Harrison, Edward Robert
Hawkins, Bruce
Hayes, Dallas T
Hearn, David Russell
Heller, Ralph
Hellman, William S
Henry, Allan Francis
Heroux, Leon J
Heyda, Donald William

Hill, Albert G
Hinteregger, Hans Erich
Hohenemser, Christoph
Holton, Gerald
Holway, Lowell Hoyt, Jr
Hoop, Bernard, Jr
Horwitz, Paul
Hovorka, John
Hull, Gordon Ferrie, Jr
Hull, Robert Joseph
Humphrey, Charles Harve
Ingham, Kenneth R
Jackson, Francis J
Jacobsen, Edward Hastings
Jagannathan, Kannan
Janes, George Sargent
Jarrell, Joseph Andy
Jasperse, John R
Javan, Ali
Jensen, Barbara Lynne
Johnson, John Clark
Jones, Phillips Russell
Jones, Robert Allan
Kanter, Irving
Karas, John Athan
Karo, Douglas Paul
Keck, James Collyer
Keene, Wayne Hartung
Kelland, David Ross
Kendall, Henry Way
King, Ronold (Wyeth Percival)
King, William Connor
Kingston, Robert Hildreth
Kirsch, Lawrence Edward
Kleiman, Herbert
Klein, Claude N
Kleinmann, Douglas Erwin
Kleis, John Dieffenbach
Kleppner, Daniel
Kofsky, Irving Louis
Kohin, Roger Patrick
Koster, George Fred
Kovaly, John J
Krieger, Allen Stephen
Lai, Shu Tim
Lanza, Giovanni
Lanza, Richard Charles
Lawler, Jack (John) W
Lee, Roland Robert
Lees, Wayne Lowry
Leiby, Clare C, Jr
Lemnios, William Zachary
Lempicki, Alexander
Lewis, Henry Rafalsky
Lewis, Margaret Nast
Liuima, Francis Aloysius
Lowder, J Elbert
Lowndes, Robert P
Lynton, Ernest Albert
Lyons, Donald Herbert
McCarthy, Kathryn Agnes
McCune, James E
McElroy, Michael Brendan
Mack, Charles Lawrence, Jr
Mack, Michael E
Mahon, Harold P
Mailloux, Robert Joseph
Malkus, Willem Van Rensselaer
Maloney, William Thomas
Marino, Richard Matthew
Markstein, George Henry
Marrero, Hector G
Marshall, Donald James
Masi, James Vincent
Masso, Jon Dickinson
Mathieson, Alfred Herman
Maxwell, Emanuel
May, John Elliott, Jr
Mazur, John
Mellen, Walter Roy
Michelson, Louis
Milburn, Richard Henry
Mittler, Arthur
Montgomery, Donald Bruce
Moody, Elizabeth Anne
Morrison, Philip
Murray, Stephen S
Nebolsine, Peter Eugene
Nelson, Donald Frederick
Nowak, Welville B(erenson)
Obermayer, Arthur S
O'Connor, John Joseph
O'Neill, Edward Leo
Osepchuk, John M
Padovani, Francois Antoine
Pandorf, Robert Clay
Patrick, Richard Montgomery
Payne, Richard Steven
Peterson, Gerald Alvin
Petrasso, Richard David
Petschek, Harry E
Pettengill, Gordon Hemenway
Phillips, Richard Arlan
Pirri, Anthony Nicholas
Pless, Irwin Abraham
Poggio, Tomaso Armando
Post, Benjamin
Postol, Theodore A
Pound, Robert Vivian
Purcell, Edward Mills
Pyle, Robert Wendell, Jr
Raemer, Harold H
Ramsey, Norman Foster, Jr
Rao, K V N

Rediker, Robert Harmon
Redner, Sidney
Renner, Gerard W
Rheinstein, John
Richardson, Robert Esplin
Ring, Paul Joseph
Riseberg, Leslie Allen
Roberts, Louis W
Robinson, Howard Addison
Rogers, Howard Gardner
Romer, Robert Horton
Rosato, Frank Joseph
Rose, Peter Henry
Rossi, Bruno B
Rudenberg, H(ermann) Gunther
Safford, Richard Whiley
Sampson, John Laurence
Sandler, Sheldon Samuel
Sauro, Joseph Pio
Schenck, Hilbert Van Nydeck, Jr
Schild, Rudolph Ernest
Schorr, Marvin Gerald
Schwall, Robert E
Schwartz, Jack
Seibel, Frederick Truman
Shapira, Yaacov
Shapiro, Irwin Ira
Shaughnessy, Thomas Patrick
Shepherd, Freeman Daniel, Jr
Shiffman, Carl Abraham
Shirn, George Aaron
Skrable, Kenneth William
Smith, Alan Bradford
Smith, Luther W
Sodickson, Lester A
Soltzberg, Leonard Jay
Stanbrough, Jesse Hedrick, Jr
Stephen, Ralph A
Stone, Richard Spillane
Stoner, William Weber
Storer, James E(dward)
Stratton, Julius Adams
Strauss, Bruce Paul
Strimling, Walter Eugene
Strong, John (Donovan)
Stump, Robert
Sullivan, James Douglas
Swift, Arthur Reynders
Tancrell, Roger Henry
Taylor, Edwin Floriman
Thaddeus, Patrick
Tokita, Noboru
Tonry, John Landis
Toscano, William Michael
Tsipis, Kosta M
Tuchman, Avraham
Tugal, Halil
Turchinetz, William Ernest
Tzeng, Wen-Shian Vincent
Van Meter, David
Vaughn, Michael Thayer
Vignos, James Henry
Wachman, Harold Yehuda
Wadey, Walter Geoffrey
Walker, Christopher Bland
Waymouth, John Francis
Weiner, Stephen Douglas
Weiss, Jerald Aubrey
Weiss, Rainer
Weiss, Richard Jerome
Weisskopf, Victor Frederick
Whitaker, John Scott
White, Lawrence W
Wiederhold, Pieter Rijk
Wild, John Frederick
Willis, Charles Richard
Wilson, Richard
Winer, Bette Marcia Tarmey
Winston, Arthur William
Wong, Chuen
Worrell, Francis Toussaint
Yannas, Ioannis Vassilios
Yukon, Stanford P
Zamenhof, Robert G A
Zeiger, Herbert J
Zemon, Stanley Alan
Zimmerman, George Ogurek
Zombeck, Martin Vincent

MICHIGAN
Axelrod, Daniel
Baldwin, Keith Malcolm
Bardwick, John, III
Barnes, James Milton
Baxter, William John
Becchetti, Frederick Daniel
Blatt, Frank Joachim
Bleil, Carl Edward
Bornemeier, Dwight D
Bretz, Michael
Brockmeier, Richard Taber
Brown, Boyd Alex
Bryant, Barbara Everitt
Bucksbaum, Philip Howard
Capriotti, Eugene Raymond
Chang, Jhy-Jiun
Christian, Larry Omar
Chupp, Timothy E
Coucouvanis, Dimitri N
Cowen, Jerry Arnold
Crane, Horace Richard
DeGraaf, Donald Earl
Edwards, Thomas Harvey
Felmlee, William John

Physics, General (cont)

Fisk, Lennard Ayres
Foiles, Carl Luther
Fry, David Lloyd George
Gessert, Walter Louis
Graves, Bruce Bannister
Hagenlocker, Edward Emerson
Haidler, William B(ernard)
Haynes, Sherwood Kimball
Hazen, Wayne Eskett
Heilman, Warren Emanuel
Hendel, Alfred Z
Holter, Marvin Rosenkrantz
Inghram, Mark Gordon
Jacko, Michael George
Janecke, Joachim Wilhelm
Johnson, Wayne Jon
Kaiser, Christopher B
Kauppila, Walter Eric
Keeling, Rolland Otis, Jr
Keyes, Paul Holt
Kim, Yeong Wook
King, John Swinton
Koistinen, Donald Peter
Kolopajlo, Lawrence Hugh
Kruglak, Haym
Kullen, Mark K
LaRocca, Anthony Joseph
Laukonis, Joseph Vainys
Levich, Calman
Li, Chi-Tang
Lindsay, Robert Kendall
Lippmann, Seymour A
Liu, Yi
Loeber, Adolph Paul
Loh, Edwin Din
Malhiot, Robert
Marburger, Richard Eugene
Mayer, Frederick Joseph
Meitzler, Allen Henry
Menard, Albert Robert, III
Meyer, Donald Irwin
Montgomery, Donald Joseph
Musinski, Donald Louis
Nadasen, Aruna
Nichols, Nathan Lankford
Nims, John Buchanan
Nine, Harmon D
Nitz, David F
O'Neal, Russell D
Ottova, Angela
Peters, James John
Peterson, Edward Charles
Pollack, Gerald Leslie
Polmanteer, Keith Earl
Potnis, Vasant Raghunath
Pronko, Peter Paul
Robbins, John Alan
Roe, Byron Paul
Roessler, David Martyn
Rohrer, Douglas C
Rolnick, William Barnett
Roth, Richard Francis
Rowland, Sattley Clark
Sanders, Barbara A
Sanders, Theodore Michael, Jr
Schroeder, Peter A
Segall, Stephen Barrett
Sell, Jeffrey Alan
Shure, Fred C(harles)
Silver, Robert
Spence, Robert Dean
Stewart, Melbourne George
Summerfield, George Clark
Tanis, John Allen
Tibbetts, Gary George
Tiffany, Otho Lyle
Trentelman, George Frederick
Turley, Sheldon Gamage
Vander Velde, John Christian
Van Putten, James D, Jr
Vaz, Nuno A
Vincent, Dietrich H(ermann)
Ward, John Frank
Wasielewski, Paul Francis
Weinreich, Gabriel
Williamson, Robert Marshall
Wong, Victor Kenneth
Wright, Wayne Mitchell
Wu, Dolly Y
Yaniv, Zvi
Zietlow, James Philip

MINNESOTA
Aagard, Roger L
Anderson, Frances Jean
Anderson, Willard Eugene
Aspen, Frank
Bailey, Carl Leonard
Banaszak, Leonard Jerome
Blair, John Morris
Burkstrand, James Michael
Campbell, Charles Edwin
Casserberg, Bo R
Cederberg, James W
Challener, William Albert
Cohen, Arnold A
Cohen, Philip Ira
Collins, Robert Joseph
Cornelissen Guillaume, Germaine G
Dahlberg, Duane Arlen
DiBona, Peter James
Dickson, Arthur Donald

Dierssen, Gunther Hans
Eckroth, Charles Angelo
Flinner, Jack L
Flower, Terrence Frederick
Freier, Phyllis S
Fritts, Robert Washburn
Gallo, Charles Francis
Glick, Forrest Irving
Goldman, Allen Marshall
Graves, Wayne H(aigh)
Grimsrud, David T
Hamermesh, Morton
Hanson, Howard Grant
Heinisch, Roger Paul
Heltemes, Eugene Casmir
Hopkins, Don Carlos
Huang, Cheng-Cher
Johnson, Robert Glenn
Jorgenson, Gordon Victor
Keffer, Charles Joseph
Kiani, Mohammad F
Kroening, John Leo
Lesikar, Arnold Vincent
Lindgren, Gordon Edward
McClure, Benjamin Thompson
McKinney, James T
Mahmoodi, Parviz
Marquit, Erwin
Marshak, Marvin Lloyd
Mikkelson, Raymond Charles
Mitchell, Michael A
Mordue, Dale Lewis
Mueller, Rolf Karl
Nathan, Marshall I
Ney, Edward Purdy
Nier, Alfred Otto Carl
Norberg, Arthur Lawrence
Pepin, Robert Osborne
Pontinen, Richard Ernest
Pou, Wendell Morse
Reilly, Richard J
Reitz, Robert Alan
Rubens, Sidney Michel
Ruddick, Keith
Schmit, Joseph Lawrence
Scriven, L E(dward), (II)
Sipson, Roger Fredrick
Tang, Yau-Chien
Theisen, Wilfred Robert
Thornton, Arnold William
Waddington, Cecil Jacob
Warner, Raymond M, Jr
Watkins, Ivan Warren
Weyhmann, Walter Victor
Whitehead, Andrew Bruce
Winckler, John Randolph
Yasko, Richard N
Youngner, Philip Genevus
Zimmermann, William, Jr
Zoltai, Tibor

MISSISSIPPI
Bruce, John Goodall, Jr
Butler, Dwain Kent
Colonias, John S
Croft, Walter Lawrence
Cullen, Abbey Boyd, Jr
Green, Albert Wise
Grissinger, Earl H
Howell, Everette Irl
Lestrade, John Patrick
Lewis, Arthur B
Miller, Donald Piguet
Reidy, James Joseph
Shields, Fletcher Douglas

MISSOURI
Anderson, Harold D
Bogar, Thomas John
Carstens, John C
Corey, Eugene R
Cowan, David J
Fedders, Peter Alan
Friedlander, Michael Wulf
Fuller, Harold Q
Gard, Janice Koles
Geilker, Charles Don
Gingrich, Newell Shiffer
Grayson, Michael A
Hale, Barbara Nelson
Henson, Bob Londes
Hill, John Joseph
Hilton, Wallace Atwood
Hohenberg, Charles Morris
Holroyd, Louis Vincent
Hultsch, Roland Arthur
Johnson, Darell James
Kenyon, Allen Stewart
Klarmann, Joseph
Lowe, Forrest Gilbert
McFarland, Charles Elwood
McFarland, Robert Harold
McIntosh, Harold Leroy
Major, Schwab Samuel, Jr
Manson, Donald Joseph
Mashhoon, Bahram
Moss, Frank Edward
Nichols, Robert Ted
Nicholson, Eugene Haines
Norberg, Richard Edwin
Nothdurft, Robert Ray
Olson, Ronald E
Peery, Larry Joe
Phillips, James M

Phillips, Russell Allan
Philpot, John Lee
Querry, Marvin Richard
Rawlings, Gary Don
Riley, James A
Russell, George A
Scandrett, John Harvey
Schmidt, Bruno (Francis)
Schmidt, Paul Woodward
Shull, Franklin Buckley
Snoble, Joseph Jerry
Stacey, Larry Milton
Takano, Masaharu
Thro, Mary Patricia
Thurman, Robert Ellis, II
Tompson, Clifford Ware
Townsend, Jonathan
Tsoulfanidis, Nicholas
Waring, Richard C
White, Henry W
Zinner, Ernst K

MONTANA
Caughlan, Georgeanne Robertson
Hayden, Richard John
Irvine, Merle M
Kirkpatrick, Larry Dale
Lapeyre, Gerald J
Robinson, Clark Shove, Jr
Rugheimer, Norman MacGregor
Smith, Richard James

NEBRASKA
Byerly, Paul Robertson, Jr
Davies, Kennard Michael
Ducharme, Stephen
French, Walter Russell, Jr
Guenther, Raymond A
Hansen, Bernard Lyle
Harbison, Gerard Stanislaus
Hardy, John R
Hardy, Robert J
Jaswal, Sitaram Singh
Katz, Robert
Kirby, Roger D
Leichner, Peter K
Sachtleben, Clyde Clinton
Surkan, Alvin John
Swartzendruber, Dale
Woollam, John Arthur
Zepf, Thomas Herman

NEVADA
Barnes, George
Case, Clinton Meredith
Dunipace, Donald William
Frazier, Thomas Vernon
Hudson, James Gary
Jobst, Joel Edward
Kliwer, James Karl
Levine, Paul Hersh
Marsh, David Paul
Messenger, George Clement
Monroe, Eugene Alan
Odencrantz, Frederick Kirk
Prater, C(harles) D(wight)
Savage, Martha Kane

NEW HAMPSHIRE
Beasley, Wayne M(achon)
Comerford, Matthias F(rancis)
Davis, William Potter, Jr
Dawson, John Frederick
Doyle, William Thomas
Engstrom, Ralph Warren
Houston, Robert Edgar, Jr
Jolly, Stuart Martin
Kasper, Joseph F, Jr
Kaufmann, Richard L
Kidder, John Newell
King, Allen Lewis
Lewis, H(arold) Ralph
Lockwood, John Alexander
MacGillivray, Jeffrey Charles
McGinness, James E
McMickle, Robert Hawley
Mellor, Malcolm
Meyer, James Wagner
Mower, Lyman
Mulhern, John E, Jr
Pipes, Paul Bruce
Rendina, Rodney A
Taffe, William John
Temple, Peter Lawrence
Walsh, John Edmond

NEW JERSEY
Allen, Jonathan
Al-Salameh, Daniel Yousef
Arunasalam, Vickramasingam
Behrens, Herbert Ernest
Belt, Roger Francis
Bernstein, Alan D
Berthold, Joseph Ernest
Bishop, David John
Bjorkholm, John Ernst
Blumberg, William Emil
Blumenstock, Brent J
Bobeck, Andrew H
Bomke, Hans Alexander
Bornstein, Lawrence A
Brawer, Steven Arnold
Bridges, Thomas James
Briggs, George Roland
Brown, Joe Ned, Jr

Buchsbaum, Solomon Jan
Butler, Herbert I
Carr, Herman Yaggi
Chance, Morrell S
Channin, Donald Jones
Cherry, William Henry
Cizewski, Jolie Antonia
Cladis, Patricia Elizabeth Ruth
Cohen, Samuel Alan
Courtney-Pratt, Jeofry Stuart
Curran, Robert Kyran
Curtis, Robert K
Curtis, Thomas Hasbrook
Eisner, Philip Nathan
Elkholy, Hussein A
Epstein, Seymour
Fajans, Jack
Feinblum, David Alan
Ferrari, Lawrence A
Fisch, Nathaniel J
Fischer, Traugott Erwin
Fitch, Val Logsdon
Freund, Roland Wilhelm
Gautreau, Ronald
Geshner, Robert Andrew
Girit, Ibrahim Cem
Goldstein, Philip
Goldstein, Sheldon
Goodman, Gerald Joseph
Gordon, Eugene Irving
Gordon, James Power
Gottlieb, Melvin Burt
Grauman, Joseph Uri
Gruner, Sol Michael
Gunther-Mohr, Gerard Robert
Halemane, Thirumala Raya
Hall, Herbert Joseph
Hamann, Donald Robert
Harrington, David Rogers
Hartung, John
Hemmendinger, Henry
Hillier, James
Hinnov, Einar
Horn, David Nicholas
Hu, Pung Nien
Huang, John S
Irgon, Joseph
Jassby, Daniel Lewis
Jobes, Forrest Crossett, Jr
Johnson, Larry Claud
Kahng, Dawon
Kaugerts, Juris E
Kingery, Bernard Troy
Kogelnik, H W
Kojima, Haruo
Kraus, Hubert Adolph
Kronenberg, Stanley
Kurshan, Jerome
La, Sung Yun
Lang, David (Vern)
Langer, William David
Lau, Ngar-Cheung
Lebowitz, Joel Louis
Leupold, Herbert August
Levi, Barbara Goss
Levine, Jerry David
Libchaber, Albert Joseph
Lin, Dong Liang
Ling, Hung Chi
Maa, Jer-Shen
McCutchen, Charles Walter
Maclennan, Carol G
McPherson, Ross
McRae, Eion Grant
Manickam, Janardhan
Marcuse, Dietrich
Matey, James Regis
Mayhew, Thomas R
Miller, Gabriel Lorimer
Miller, Robert Alan
Mims, William B
Misra, Raj Pratap
Mitchell, John Peter
Mittleman, Marvin Harold
Nahory, Robert Edward
Nash, Franklin Richard
Natapoff, Marshall
Neidhardt, Walter Jim
Nelson, Terence John
Newsome, Ross Whitted
Nicholls, Gerald P
O'Gorman, James
Ollom, John Frederick
Ostermayer, Frederick William, Jr
Owens, David Kingston
Papantonopoulou, Aigli Helen
Pargellis, Andrew Nason
Penzias, Arno A(llan)
Peterson, George Earl
Pinczuk, Aron
Piroue, Pierre Adrien
Poate, John Milo
Polak, Joel Allan
Polk, Conrad Joseph
Pollock, Franklin
Polye, William Ronald
Pregger, Fred Titus
Presby, Herman M
Racette, George William
Rastani, Kasra
Reder, Friedrich H
Reisman, Otto
Riddle, George Herbert Needham
Robbins, Allen Bishop

Rose, Carl Martin, Jr
Salzarulo, Leonard Michael
Sauer, John A
Schilling, Gerd
Schlink, F(rederick) J(ohn)
Schmidt, George
Schneider, Martin V
Schubert, Rudolf
Shah, Atul A
Shastry, B Sriram
Shelton, James Churchill
Shyamsunder, Erramilli
Silfvast, William Thomas
Skorinko, George
Slusher, Richart Elliott
Slusky, Susan E G
Smith, Stephen Roger
Southgate, Peter David
Spruch, Grace Marmor
Stamer, Peter Eric
Steben, John D
Stephen, Michael John
Strauss, Walter
Subramanyam, Dilip Kumar
Summerfield, Martin
Tell, Benjamin
Tenzer, Rudolf Kurt
Thornton, William Andrus, Jr
Thouret, Wolfgang E(mery)
Tomlinson, Walter John, III
Towner, Harry H
Trambarulo, Ralph
Treiman, Sam Bard
Trimmer, William S(tuart)
Tung, Raymond T
Tzanakou, M Evangelia
Ullrich, Felix Thomas
Van Der Veen, James Morris
Varma, Chandra M
Vossen, John Louis
Wachtell, George Peter
Walsh, Peter
Watson, Hugh Alexander
Weimer, Paul Kessler
Weinrich, Marcel
Weir, William Thomas
Weissmann, Sigmund
Wieder, Sol
Wilkinson, David Todd
Wong, King-Lap
Wong, Yiu-Huen
Wood, Darwin Lewis
Woodward, J Guy
Worley, Robert Dunkle
Yevick, George Johannus

NEW MEXICO
Ackerhalt, Jay Richard
Adler, Richard John
Agnew, Lewis Edgar, Jr
Allred, John C(aldwell)
Alpert, Seymour Samuel
Anderson, Robert Alan
Baldwin, George C
Balestrini, Silvio J
Barfield, Walter David
Barker, Lynn Marshall
Barnaby, Bruce E
Begay, Frederick
Best, George Harold
Bleyl, Robert Lingren
Booth, Lawrence A(shby)
Bostick, Winston Harper
Bowles, Thomas Joseph
Bradbury, Norris Edwin
Brannon, Paul J
Brenton, June Grimm
Brook, Marx
Brower, Keith Lamar
Browne, John Charles
Brownell, John Howard
Broyles, Carter D
Bruce, Charles Wikoff
Burr, Alexander Fuller
Byrd, Roger Carl
Caldwell, Stephen E
Campbell, Katherine Smith
Chabai, Albert John
Chan, Kwok-chi Dominic
Clements, William Earl
Cobb, Donald D
Coburn, Horace Hunter
Colgate, Stirling Auchincloss
Conner, Jerry Power
Cowan, Maynard, Jr
Cress, Daniel Hugg
Cuthrell, Robert Eugene
Damerow, Richard Aasen
Davey, William George
Davis, William Chester
Daw, Harold Albert
Deal, William E, Jr
Dick, Jerry Joel
Dingus, Ronald Shane
Dressel, Ralph William
Duncan, Richard H(enry)
Eastman, Daniel Robert Peden
Elliott, Paul M
Evans, Winifred Doyle
Ewing, Ronald Ira
Feldman, Donald William
Flynn, Edward Robert
Fowler, Clarence Maxwell
Freeman, Bruce L, Jr

George, Michael James
Gibbs, Terry Ralph
Gillespie, Claude Milton
Ginsparg, Paul H
Godfrey, Thomas Nigel King
Godwin, Robert Paul
Goldstein, Louis
Goldstone, Philip David
Good, William Breneman
Graeber, Edward John
Graham, Robert Albert
Green, John Root
Griffin, Patrick J
Guenther, Arthur Henry
Hall, Jerome William
Harrison, Charles Wagner, Jr
Hartley, Danny L
Hatcher, Charles Richard
Hauer, Allan
Heckman, Richard Cooper
Heller, John Philip
Henins, Ivars
Hill, Benny Joe
Hillsman, Matthew Jerome
Hoeft, Lothar Otto
Hopson, John Wilbur, Jr
Hoyt, Harry Charles
Huestis, Stephen Porter
Hurd, Jon Rickey
Jacobson, Abram Robert
Janney, Donald Herbert
Jason, Andrew John
Johannes, Robert
Johnson, Ralph T, Jr
Jones, Merrill C(alvin)
Kelly, Robert Emmett
Kennedy, Jerry Dean
King, James Claude
King, L D Percival
Knapp, Edward Alan
Knief, Ronald Allen
Koehler, Dale Roland
Kraus, Alfred Andrew, Jr
Krohn, Burton Jay
Krupka, Milton Clifford
Kunz, Kaiser Schoen
La Delfe, Peter Carl
Langner, Gerald Conrad
Lanter, Robert Jackson
Lawrence, James Neville Peed
Lawrence, Raymond Jeffery
Layne, J Clyde Browning
Leavitt, Christopher Pratt
Lee, David Mallin
Lincoln, Richard Criddle
Lindstrom, Ivar E, Jr
Linsley, John
Liu, Lon-Chang
Luehr, Charles Poling
Lysne, Peter C
McClure, Gordon Wallace
McKibben, Joseph L
McLeod, John
Madden, Hannibal Hamlin, Jr
Maley, Martin Paul
Mansfield, Charles Robert
Matthews, John AJ
Matzen, Maurice Keith
Mautz, Charles William
Mead, William C
Menlove, Howard Olsen
Metzger, Daniel Schaffer
Miyoshi, Dennis S
Mogford, James A
Moss, Joel M
Mueller, Marvin Martin
Nagle, Darragh (Edmund)
Newman, Michael J(ohn)
Nickel, George H(erman)
Nielson, Clair W
Olson, Gordon Lee
Overton, William Calvin, Jr
Paciotti, Michael Anthony
Panitz, Janda Kirk Griffith
Parker, Jack Lindsay
Paul, Robert Hugh
Payton, Daniel N, III
Peng, Jen-chieh
Petschek, Albert George
Pigott, Miles Thomas
Plimpton, James D
Poukey, James W
Price, Robert Harold
Rabie, Ronald Lee
Ragan, Charles Ellis, III
Ram, Budh
Raymond, David James
Regener, Victor H
Ribe, Fred Linden
Rinehart, John Sargent
Robertson, Robert Graham Hamish
Roof, Raymond Bradley, Jr
Rosen, Louis
Sanders, Walter L
Schappert, Gottfried T
Scharn, Herman Otto Friedrich
Schelberg, Arthur Daniel
Schery, Stephen Dale
Schiferl, David
Schillaci, Mario Edward
Schlosser, Jon A
Schmidt, Helmut
Schreiber, R E

Schriber, Stanley Owen
Schultz, Rodney Brian
Schwarz, Ricardo
Seay, Glenn Emmett
Simmons, James E
Smith, Carl Walter
Smith, William Conrad
Spillman, George Raymond
Stellingwerf, Robert Francis
Stout, Virgil L
Stratton, William R
Stromberg, Thorsten Frederick
Strong, Ian B
Summers, Donald Lee
Swingle, Donald Morgan
Swinson, Derek Bertram
Sydoriak, Stephen George
Sze, Robert Chia-Ting
Talley, Thurman Lamar
Taschek, Richard Ferdinand
Taylor, Thomas Newton
Terrell, N(elson) James
Thorn, Robert Nicol
Trainor, Robert James
Travis, James Roland
Trela, Walter Joseph
Turi, Raymond A
Van Domelen, Bruce Harold
Van Riper, Kenneth Alan
Venable, Douglas
Vook, Frederick Ludwig
Vortman, L(uke) J(erome)
Voss, Donald E
Wackerle, Jerry Donald
Wallace, Jon Marques
Wallace, Richard Kent
Walsh, John M
Wangler, Thomas P
Warnes, Richard Harry
Weaver, Robert Paul
Webber, William R
Wenzel, Robert Gale
White, Paul C
Wilkening, Marvin H
Wilkins, Ronald Wayne
Winkler, Max Albert
Wolfe, Peter Nord
Wright, Thomas Payne
Wu, Adam Yu
Wylie, Kyral Francis
Yarger, Frederick Lynn

NEW YORK
Abkowitz, Martin A(aron)
Adams, Peter D
Altemose, Vincent O
Altman, Joseph Henry
Ames, Irving
Andrews, John W
Arnold, Stephen
Arons, Michael Eugene
Bak, Per
Bakis, Raimo
Barton, Mark Q
Batterman, Boris William
Bazer, Jack
Bean, Charles Palmer
Berry, Herbert Weaver
Bhargava, Rameshwar Nath
Bigelow, John Edward
Birnbaum, David
Blachman, Arthur Gilbert
Blessing, Robert Harry
Bloch, Alan
Blume, Martin
Bolon, Roger B
Booth, Eugene Theodore
Borst, Lyle Benjamin
Bouyoucos, John Vinton
Box, Harold C
Boyer, Donald Wayne
Braslau, Norman
Breed, Henry Eltinge
Brehm, Lawrence Paul
Brinkman, William F
Brody, Burton Alan
Brown, David Chester
Brown, Gerald E
Brown, Howard Howland, Jr
Burnham, Dwight Comber
Burns, Gerald
Cambey, Leslie Alan
Carrier, Gerald Burton
Carroll, C(lark) E(dward)
Casabella, Philip A
Cassel, David Giske
Chaudhari, Praveen
Chessin, Henry
Chiang, Joseph Fei
Chow, Che Chung
Christ, Norman
Chung, Victor
Churchill, Melvyn Rowen
Cipriano, Ramon John
Clark, David Lee
Cleare, Henry Murray
Cohen, Ezechiel Godert David
Cohen, Martin O
Cool, Terrill A
Coppens, Philip
Corfield, Peter William Reginald
Cotter, Maurice Joseph
Craighead, Harold G
Csermely, Thomas J(ohn)

Cusano, Dominick A
Damouth, David Earl
Danby, Gordon Thompson
Danielson, Lee Robert
Das, Ashok Kumar
Davey, Paul Oliver
Davitian, Harry Edward
Debiak, Ted Walter
Dempsey, Daniel Francis
DeNoyer, Linda Kay
Desilets, Brian H
DeTitta, George Thomas
Diamond, Joshua Benamy
Di Paola, Mario
Doering, Charles Rogers
Donaldson, Robert Rymal
Donnelly, Denis Philip
Donovan, John Leo
Douglass, David Holmes
Dove, Derek Brian
Duax, William Leo
Durkovic, Russell George
Edelstein, William Alan
Eisenstadt, Maurice
Elder, Fred Kingsley, Jr
Engelke, Charles Edward
Erlbach, Erich
Eshbach, John Robert
Espersen, George A
Fabricand, Burton Paul
Falk, Theodore J(ohn)
Farber, Elliot
Finch, Thomas Lassfolk
Fleisher, Harold
Forgacs, Gabor
Forrest, William John
Fossan, David B
Foster, Margaret C
Fowler, Alan B
Fox, Herbert Leon
Fullwood, Ralph Roy
Furst, Milton
Garber, Meyer
Garretson, Craig Martin
Garrett, Charles Geoffrey Blythe
Gasparini, Francis Marino
Gavin, Donald Arthur
Gelernter, Herbert Leo
Gelman, Donald
Ghosh, Arup Kumar
Giaever, Ivar
Glaser, Herman
Glickman, Walter A
Goertzel, Gerald
Goldberg, Conrad Stewart
Goldberg, Joshua Norman
Golden, Robert K
Goldhaber, Gertrude Scharff
Goldhaber, Maurice
Goldman, Vladimir J
Goldstein, Herbert
Good, Myron Lindsay
Good, William E
Gove, Harry Edmund
Graf, Erlend Haakon
Greisen, Kenneth I
Grumet, Alex
Gundlach, Robert William
Gunn, John Battiscombe
Hagen, Carl Richard
Hanau, Richard
Harke, Douglas J
Harper, Richard Allan
Harris, Donald R, Jr
Hart, Hiram
Hart, Robert John
Hartill, Donald L
Hartmann, George Cole
Hartmann, Sven Richard
Harvey, Alexander Louis
Hauptman, Herbert Aaron
Headrick, Randall L
Heberle, Juergen
Hecht, Eugene
Helfand, David John
Hendrie, Joseph Mallam
Henkel, Elmer Thomas
Henshaw, Clement Long
Hinson, David Chandler
Ho, Ping-Pei
Hoard, James Lynn
Hodgson, Rodney
Hogan, William
Honig, Arnold
Hopkins, Robert Earl
Horwitz, Nahmin
Howard, Richard John
Howe, David Allen
Huang, Jack Shih Ta
Huie, Carmen Wah-Kit
Hunter, Lloyd Philip
Hurd, Jeffery L
Jack, Hulan E, Jr
Jaffe, Bernard Mordecai
Jaffe, Morry
Janak, James Francis
Jarvis, James Gordon
Johnson, Peter David
Johnson, Peter Dexter
Johnson, Woodrow Eldred
Jona, Franco Paul
Jones, Keith Warlow
Joynson, Reuben Edwin, Jr
Kao, Yi-Han

Physics, General (cont)

Karz, Robert Stephen
Keck, Donald Bruce
Kelly, Martin Joseph
Kent, Henry Johann
Keyes, Robert William
Khare, Bishun Narain
Kinsella, John J
Kinsey, Kenneth F
Kirchgessner, Joseph L
Kirk, Harold Glen
Kiszenick, Walter
Klahr, Carl Nathan
Klaiber, George Stanley
Kleinman, Chemia Jacob
Koehler, Richard Frederick, Jr
Kohler, Robert Henry
Kouts, Herbert John Cecil
Kramer, Paul Robert
Krewer, Semyon E
Kriss, Michael Allen
Krongelb, Sol
Kunstadter, John W
Kuper, J B Horner
Ladd, John Herbert
Laibowitz, Robert (Benjamin)
Lang, Harry George
La Tourrette, James Thomas
Lawson, Kent DeLance
Layzer, Arthur James
Lea, Robert Martin
Lee, Charles Alexander
Lee, David Morris
Leigh, Richard Woodward
Levin, Eugene (Manuel)
Levitas, Alfred Dave
Li, Kelvin K
Li, Zheng
Litke, John David
Litynski, Daniel Mitchell
Lloyd, James Newell
Love, William Alfred
Lovejoy, Derek R
Lowder, Wayne Morris
McClain, John William
McGuire, Stephen Craig
McKinley, William Albert
Malone, Dennis P(hilip)
Marple, Dudley Tyng Fisher
Marrone, Paul Vincent
Mason, Max Garrett
Maurer, Robert Distler
May, John Walter
Mayer, James W(alter)
Mehran, Farrokh
Melkonian, Edward
Mendelsohn, Lawrence Barry
Mendez, Emilio Eugenio
Menes, Meir
Mermin, N David
Metcalf, Harold
Michael, Paul Andrew
Middleton, David
Miller, Allen H
Miller, Melvin J
Mishkin, Eli Absalom
Mistry, Nariman Burjor
Mitchell, Joan LaVerne
Mogro-Campero, Antonio
Morris, James Eliot
Morris, Thomas Wendell
Mueller, Alfred H
Murray, Cherry Ann
Mutter, Walter Edward
Nardi, Vittorio
Nation, John
Nelson, Clarence Norman
Newman, Jay Edward
Nicholson, William Jamieson
Nisteruk, Chester Joseph
Noonan, Thomas Wyatt
Nordin, Paul
O'Connor, Cecilian Leonard
O'Donoghue, Aileen Ann
Ogden, Philip Myron
Oja, Tonis
Okaya, Yoshi Haru
Orenstein, Albert
Ortel, William C(harles) G(ormley)
Otter, Mark William
Pai, Damodar Mangalore
Palmer, Robert B
Parratt, Lyman George
Parthasarathy, Rengachary
Paskin, Arthur
Passoja, Dann E
Pe, Maung Hla
Pearson, George John
Peckham, Donald Charles
Penfield, Robert Harrison
Phillips, Melba (Newell)
Pierucci, Olga
Pilcher, Valter Ennis
Pimbley, Walter Thornton
Piore, Emanuel Ruben
Platner, Edward D
Pomian, Ronald J
Pribil, Stephen
Prince, Jack
Pritchard, Robert Leslie
Prodell, Albert Gerald
Pullman, Ira
Rahm, David Charles

Raka, Eugene Cd
Range, R Michael
Rau, R Ronald
Rautenberg, Theodore Herman
Read, Albert James
Reilly, Edwin David, Jr
Reisinger, Joseph G
Reppy, John David
Resnick, Robert
Rezanka, Ivan
Richards, James Austin, Jr
Richardson, Robert Coleman
Rini, Frank John
Robinson, William Kirley
Robusto, C Carl
Roetling, Paul G
Rohr, Robert Charles
Romer, Alfred
Rosar, Madeleine E
Rosen, Paul
Rosenberg, Paul
Rowe, Raymond Grant
Rubin, Kenneth
Ruddick, James John
Russell, Allan Melvin
Ryan, Donald F
Sachs, Allan Maxwell
Sadoff, Ahren J
Samios, Nicholas Peter
Sampson, William B
Sarjeant, Walter James
Sato, Makiko
Scheib, Richard, Jr
Schell-Sorokin, Anita J
Schick, Jerome David
Schubert, Walter L
Schulman, Lawrence S
Schultheis, James J
Schwab, James Jerome
Schwartz, Ernest
Schwartz, Melvin
Schwartz, Melvin J
Seitz, Frederick
Seka, Wolf
Sells, Robert Lee
Senus, Walter Joseph
Serber, Robert
Sharbaugh, Amandus Harry
Shelupsky, David I
Shen, Samuel Yi-Wen
Siemann, Dietmar W
Silsbee, Henry Briggs
Silsbee, Robert Herman
Simon, William
Simpson, James Henry
Skiff, Peter Duane
Smart, James Samuel
Smilowitz, Bernard
Smith, Bernard
Smith, Gail Preston
Smith, George David
Smith, Lyle W
Smoyer, Claude B
Sobell, Henry Martinique
Sperber, Daniel
Spoelhof, Charles Peter
Springett, Brian E
Stachel, Johanna
Stannard, Carl R, Jr
Stearns, Robert L
Stenzel, Wolfram G
Stephenson, Thomas E(dgar)
Sternglass, Ernest Joachim
Stier, Paul Max
Stolov, Harold L
Stoner, George Green
Stover, Raymond Webster
Strassenburg, Arnold Adolph
Strnisa, Fred V
Strome, Forrest C, Jr
Strong, Herbert Maxwell
Strozier, John Allen, Jr
Struzynski, Raymond Edward
Stull, John Leete
Sumberg, David A
Swartz, Charles Dana
Swartz, Clifford Edward
Szydlik, Paul Peter
Talman, Richard Michael
Tan, Yen T
Tang, Chung Liang
Tavel, Judith Fibkins
Taylor, Jack Eldon
Taylor, Theodore Brewster
Teegarden, Kenneth James
Teichmann, Theodor
Teiger, Delmar Martin
Thiel, Frank L(ouis)
Thomas, Richard Alan
Thorndike, Edward Moulton
Tiersten, Martin Stuart
Titone, Luke Victor
Treanor, Charles Edward
Trexler, Frederick David
Triebwasser, Sol
Trischka, John Wilson
Vance, Miles Elliott
Van Steenbergen, Arie
Venugopalan, Srinivasa I
Vizy, Kalman Nicholas
Vosburgh, Kirby Gannett
Vought, Robert Howard
Wachnik, Richard Andre
Wagner, Peter Ewing

Waldrop, Ann Lyneve
Wang, Charles T P
Wang, Wen I
Webster, Harold Frank
Wessel, Gunter Kurt
Whetten, Nathan Rey
White, Frederick Andrew
Wiesner, Leo
Williams, Ross Edward
Williamson, Robert Samuel
Willis, William J
Wilson, Jack Martin
Wilson, Robert Rathbun
Winhold, Edward John
Wolga, George Jacob
Worthington, Thomas Kimber
Wu, Chien-Shiung
Wu, Tsu Ming
Wynne, James J
Yalow, A(braham) Aaron
Yang, Chen Ning
Young, James Roger
Young, Ralph Howard
Yuan, Luke Chia Liu
Zadoff, Leon Nathan
Zajac, Alfred
Zeitz, Louis
Ziegler, Robert C(harles)
Zwick, Daan Marsh

NORTH CAROLINA
Ade, Harald Werner
Baker, David Kenneth
Baltzer, Philip Keene
Bilpuch, Edward George
Bowers, Wayne Alexander
Briscoe, Charles Victor
Byrd, James William
Callihan, Alfred Dixon
Christian, Wolfgang C
Clark, Chester William
Clement, John Reid, Jr
Collins, Clifford B
Connolly, Walter Curtis
Daggerhart, James Alvin, Jr
Davis, William Robert
D'Eustachio, Dominic
Dotson, Marilyn Knight
Edwards, Harold Henry
Evans, Lawrence Eugene
Evans, Ralph Aiken
Fairbank, Henry Alan
Gould, Christopher Robert
Gray, Walter C(larke)
Griffith, Wayland C(oleman)
Gross, Edward Emanuel
Haase, David Glen
Hageseth, Gaylord Terrence
Hall, Peter M
Hocken, Robert John
Horie, Yasuyuki
Hovis, Louis Samuel
Howard, Clarence Edward
Ingram, Peter
Jacobson, Kenneth Allan
Jenzano, Anthony Francis
Jones, Creighton Clinton
Kratz, Howard Russel
Lado, Fred
Lawless, Philip Austin
Lawson, Dewey Tull
Lontz, Robert Jan
McCumber, Dean Everett
McEnally, Terence Ernest, Jr
Mann, David M
Manring, Edward Raymond
Mayes, Terrill W
Memory, Jasper Durham
Menius, Arthur Clayton, Jr
Meyer, Horst
Mitchell, Earl Nelson
Murray, Raymond L(eRoy)
Oates, Jimmie C
O'Foghludha, Fearghus Tadhg
Palmatier, Everett Dyson
Pollak, Victor Louis
Prater, John Thomas
Ramanan, V R V
Read, Floyd M
Reardon, Annya Joyce
Russell, Lewis Keith
Saby, John Sanford
Sayetta, Thomas C
Schetzina, Jan Frederick
Schmeltekopf, A
Seagondollar, Lewis Worth
Seevers, Delmar Oswell
Sharkoff, Eugene Gibb
Smith, Fred R, Jr
Snodgrass, Rex Jackson
Stephenson, Harold Patty
Stroscio, Michael Anthony
Sumney, Larry W
Suttle, Jimmie Ray
Thomas, Henry Coffman
Thomas, Llewellyn Hilleth
Thompson, Nancy Lynn
Varlashkin, Paul
Vermillion, Robert Everett
Wallingford, John Stuart
Walters, Mark David
Way, Katharine
Werntz, James Herbert, Jr
Wills, James E, Jr

Worley, Richard Dixon

NORTH DAKOTA
Bale, Harold David
Berkey, Gordon Bruce
Kluk, Edward
Kraus, Olen
Lykken, Glenn Irven
Muraskin, Murray
Sinha, Mahendra Kumar

OHIO
Agard, Eugene Theodore
Anderson, David Leonard
Andrus, Paul Grier
Anno, James Nelson
Ayyangar, Komanduri M
Bass, Robert Eugene
Becker, Lawrence Charles
Beer, Albert Carl
Bhattacharya, Ashok Kumar
Bishop, Edwin Vandewater
Bohn, Randy G
Boughton, Robert Ivan, Jr
Breitenberger, Ernst
Brown, L Carlton
Brown, Robert William
Burnside, Phillips Brooks
Byers, Walter Hayden
Campbell, Bernerd Eugene
Carome, Edward F
Carrabine, John Anthony
Cassim, Joseph Yusuf Khan
Cobb, Thomas Berry
Cochran, Thomas Howard
Cochran, William Ronald
Cook, James Robert
Cooke, Charles C
Crandall, Arthur Jared
Crandell, Merrell Edward
Crouch, Marshall Fox
Dahm, Arnold J
Damian, Carol G
Dickey, Frederick Pius
Dillman, Lowell Thomas
Drauglis, Edmund
Duncan, George Comer
Durand, Dominique M
Eck, Thomas G
Edwards, David Olaf
Elwell, David Leslie
Erdman, Robert J
Essig, Gustave Alfred
Farrell, David E
Fawcett, Sherwood Luther
Fernelius, Nils Conard
Fineberg, Herbert
Forman, Ralph
Foster, Edward Staniford, Jr
Frye, Glenn McKinley, Jr
Gaines, Gordon Bradford
Galm, James Michael
Ganguly, Bishwa Nath
Garrow, Robert Joseph
Garscadden, Alan
Greenberg, William Michael
Greene, David C
Greenslade, Thomas Boardman, Jr
Griffin, Charles Frank
Grimes, Steven Munroe
Gruber, H Thomas
Guth, Sylvester Karl
Hancock, George Whitmore
Hanzely, Stephen
Haque, Azeez C
Harpster, Joseph
Hart, John Birdsall
Hathaway, Charles Edward
Haybron, Ronald M
Henderson, H(oman) Thurman
Hsu, Hsiung
Hunt, Earle Raymond
Hunter, George Truman
Hunter, Joseph Lawrence
Hunter, William Winslow, Jr
Jekeli, Christopher
Jones, Dale Robert
Jossem, Edmund Leonard
Kelly, Kevin Anthony
Kereiakes, James Gus
Kime, Joseph Martin
Klein, Robert Herbert
Knecht, Walter Ludwig
Koch, Ronald Joseph
Kollen, Wendell James
Krauss, Lawrence Maxwell
Larson, Lee Edward
Laskowski, Edward L
Lutz, Arthur Leroy
McConville, G Terrence
McCown, Robert Bruce
McGraw, Delford Armstrong
Machlup, Stefan
Mallozzi, Philip James
Manning, Robert M
Marshall, Kenneth D
Merchant, Mylon Eugene
Milford, Frederick John
Miller, Don Wilson
Miller, Franklin, Jr
Miller, Harry Galen
Miller, James Roland
Miller, Raymond Edwin
Mochel, Virgil Dale

Nam, Sang Boo
Nelson, Richard Carl
Nelson, William Frank
Nielsen, Carl Eby
Nikkel, Henry
Nolan, James Francis
Notz, William Irwin
Palffy-Muhoray, Peter
Palmer, William Franklin
Palmieri, Joseph Nicholas
Peterson, Leroy Eric
Piper, Ervin L
Proctor, David George
Rao, Kandarpa Narahari
Rauckhorst, William H
Reneker, Darrell Hyson
Reynolds, Donald C
Richards, Walter Bruce
Riley, William Robert
Rollins, Roger William
Rothstein, Jerome
Rudy, Yoram
Rutledge, Wyman Coe
Sanderson, Richard Blodgett
Schlegel, Donald Louis
Schneider, James Roy
Seldin, Emanuel Judah
Serafini, John S
Shaffer, Wave H
Shaw, John H
Shetterly, Donivan Max
Silvidi, Anthony Alfred
Snider, John William
Soules, Jack Arbuthnott
Spielberg, Nathan
Stephenson, Francis Creighton
Stewart, Charles Neil
Sticksel, Philip Rice
Stith, James Herman
Strickler, Thomas David
Stumpf, Folden Burt
Sugarbaker, Evan Roy
Sundaralingam, Muttaiya
Tabakoff, Widen
Tanaka, Katsumi
Taylor, Beverley Ann Price
Toman, Karel
Tough, James Thomas
Uhrich, David Lee
Vanderburg, Vance Dilks
Vanderlind, Merwyn Ray
Van Heuvelen, Alan
Van Horn, David Downing
Versic, Ronald James
Von Meerwall, Ernst Dieter
Wada, Walter W
Wagoner, Glen
Wallis, Robert L
Weimer, David
Weinberg, Irving
Wheeler, Samuel Crane, Jr
White, John Joseph, III
Wickersham, Charles Edward, Jr
Williamson, William, Jr
Wilson, Charles Woodson, III
Wilson, Kenneth Geddes
Wittebort, Jules I
Witten, Louis
Wolfe, Paul Jay
Wood, Van E(arl)
Yu, Greta
Yun, Seung Soo
Zdanis, Richard Albert

OKLAHOMA
Albright, James Curtice
Bilger, Hans Rudolf
Boade, Rodney Russett
Crowe, Christopher
Davies-Jones, Robert Peter
Evens, F Monte
Fischbeck, Helmut J
Grischkowsky, Daniel Richard
Hartsuck, Jean Ann
Hoover, Gary McClellan
Kantowski, Ronald
Kuenhold, Kenneth Alan
Lange, James Neil, Jr
Leivo, William John
McKeever, Stephen William Spencer
Marks, L Whit
Mintz, Esther Uress
Naymik, Daniel Allan
Renner, John Wilson
Rutledge, Carl Thomas
Swartz, Charles
Thomsen, Leon
Troelstra, Arne
Weems, Malcolm Lee Bruce
Williams, Thomas Henry Lee
Yoesting, Clarence C

OREGON
Anderson, Tera Lougenia
Bates, David James
Beall, Robert Allan
Behringer, Robert Ernest
Belinfante, Frederik J
Bell, Anthony E
Blickensderfer, Robert
Bodegom, Erik
Brady, James Joseph
Brodie, Laird Charles
Broide, Michael Lynn

Burch, David Stewart
Charbonnier, Francis Marcel
Chezem, Curtis Gordon
Decker, Fred William
Evett, Arthur A
Evett, Jay Fredrick
Garwood, John Ernest
Graham, Beardsley
Griffiths, David John
Gurevitch, Mark
Hawkinson, Stuart Winfield
Henke, Burton Lehman
Hinrichs, Clarence H
Holmes, James Frederick
Hutzenlaub, John F
Jones, Robert Edward
Jones, William Barclay
Kolar, Oscar Clinton
Kunkle, Donald Edward
Lupton, John Edward
Martin, Robert Leonard
Martz, Dowell Edward
Matthews, Brian Wesley
Montague, Daniel Grover
Neher, H(enry) Victor
Nicodemus, David Bowman
Orrok, George Timothy
Overley, Jack Castle
Powell, John Leonard
Purbrick, Robert Lamburn
Rayfield, George W
Rempfer, Gertrude Fleming
Reynolds, Robert Eugene
Schrank, Glen Edward
Starr, Walter L(eRoy)
Stephas, Paul
Stringer, Gene Arthur
Swenson, Leonard Wayne
Taylor, Carson William
Towe, George Coffin
Zimmerman, Robert Lyman

PENNSYLVANIA
Abramson, Stanley L
Acheson, Willard Phillips
Alexander, Frank Creighton, Jr
Alstadt, Don Martin
Anderle, Richard
Anderson, Carl Einar
Anderson, Owen Thomas
Angello, Stephen James
Auerbach, Leonard B
Augensen, Harry John
Banks, Grace Ann
Barmby, David Stanley
Bell, Raymond Martin
Bennett, Lee Cotton, Jr
Benson, Brent W
Berger, Martin
Blasie, J Kent
Bose, Shyamalendu M
Boughn, Stephen Paul
Boyer, Robert Allen
Brody, Howard
Brown, Richard Leland
Bruns, Charles Alan
Burnett, Roger Macdonald
Burton, Larry C(lark)
Butkiewicz, Edward Thomas
Camarda, Harry Salvatore
Carfagno, Salvatore P
Carriker, Roy C
Cashdollar, Kenneth Leroy
Cheng, Kuang Liu
Cheston, Warren Bruce
Cohen, Anna (Foner)
Cook, Donald Bowker
Crawford, John David
Creagan, Robert Joseph
Cross, Leslie Eric
Dahlke, Walter Emil
Davies, D K
Davis, Jon Preston
DeGraef, Marc Julie
Devaty, Robert Philip
Dick, George W
Diehl, Renee Denise
Dowling, John
Duffin, Richard James
Edelstein, Richard Malvin
Eisner, Robert Linden
Elmore, William C
Engheta, Nader
Engler, Arnold
Esterling, Robert J
Fabish, Thomas John
Fahey, Paul Farrell
Feuchtwang, Thomas Emanuel
Fisher, David George
Fite, Wade Lanford
Fleming, Lawrence Thomas
Foderaro, Anthony Harolde
Foland, William Douglas
Friedberg, Simeon Adlow
Fuget, Charles Robert
Fugger, Joseph
Garfunkel, Myron Paul
Geller, Kenneth N
Georgopulos, Peter Demetrios
Gilstein, Jacob Burrill
Giltinan, David Anthony
Glickstein, Stanley S
Glusker, Jenny Pickworth
Goldberg, Norman

Goldman, Robert Barnett
Gordon, Richard Lee
Gottlieb, Milton
Graham, William Rendall
Haghighat, Alireza
Harbold, Mary Leah
Haruta, Kyoichi
Heid, Roland Leo
Hendrickson, Thomas James
Henry, Richard Warfield
Herczfeld, Peter Robert
Herman, Warren Nevin
Hickman, Warren David
Highland, Virgil Lee
Hogenboom, David L
Holland, Monte W
Hood, Richard Fred
Hower, Meade M
Hussain, Riaz
Jeffers, William Allen, Jr
Jeffrey, George Alan
Johnsen, Rainer
Kahn, David
Kang, Joohee
Kaplon, Morton Fischel
Kasner, William Henry
Keck, Winfield
Kendall, Bruce Reginald Francis
Khanna, Sardari Lal
Kicska, Paul A
Klein, Nelson Harold
Kleinsteuber, Tilmann Christoph Werner
Kline, Donald Edgar
Kozlowski, Richard William Hugh
Kremser, Thurman Rodney
Krutter, Harry
Kuo, Lawrence C
Lai, Ralph Wei-Meen
Lange, Nicholas A
LeLacheur, Robert Murray
Lochstet, William A
Long, Howard Charles
Longini, Richard Leon
Lord, Arthur E, Jr
Lowry, Jerald Frank
Lowry, Lewis Roy, Jr
Luce, Robert James
Luetzelschwab, John William
McAvoy, Bruce Ronald
McCarthy, Raymond Lawrence
McCaughey, Joseph M
McCubbin, Thomas King, Jr
McGinnis, Eugene A
MacInnis, Martin Benedict
Mara, Richard Thomas
Matcovich, Thomas J(ames)
Meakin, Paul
Mentzer, John R(aymond)
Miller, Alfred Charles
Miller, William Robert, Jr
Misra, Dhirendra N
Mitchell, Donald J
Mogus, Mary Ann
Monroe, James Lewis
Mooney, Edward, Jr
Mullins, William Wilson
Murphy, Bernard T
Music, John Farris
Muss, Daniel R
Newnham, Robert Everest
Nilan, Thomas George
Offenbacher, Elmer Lazard
Oneal, Glen, Jr
Ord, John Allyn
Osborne, Robert Kidder
Pack, John Lee
Peiffer, Howard R
Perrotta, Anthony Joseph
Plonsky, Andrew Walter
Pollack, Sidney Solomon
Prout, James Harold
Pruett, John Robert
Quarles, Gilford Godfrey
Rall, Waldo
Ramberg, Edward G
Raub, Harry Lyman, III
Rayne, John A
Reif, Frederick
Rickert, Russell Kenneth
Ross, Sidney
Runco, Paul D
Sakurai, Toshio
Schenk, H(arold) L(ouis), Jr
Schwerer, Frederick Carl
Sekerka, Robert Floyd
Sherwood, Bruce Arne
Shiono, Ryonosuke
Sittel, Karl
Skoner, Peter Raymond
Smith, Richard G(rant)
Smith, Warren LaVerne
Smith, Winfield Scott
Sokolowski, Henry Alfred
Sommers, Henry Stern, Jr
Southwick, Russell Duty
Soven, Paul
Spalding, George Robert
Stallwood, Robert Antony
Strauser, Wilbur Alexander
Sutter, Philip Henry
Swartz, John Croucher
Swendsen, Robert Haakon
Takeuchi, Kenji
Taylor, A

Taylor, Lyle Herman
Taylor, Merlin Gene
Tessler, George
Thrower, Peter Albert
Townsend, John Robert
Tozier, John E
Trainer, Michael Norman
Turner, Edward Felix, Jr
VanderVen, Ned Stuart
Vedam, Kuppuswamy
Verleur, Hans Willem
Voet, Donald Herman
Wagner, David Loren
Walchli, Harold E(dward)
Wales, Walter D
Waring, Robert Kerr, Jr
Wefers, Karl
Weidner, Richard Tilghman
Wheeler, Donald Bingham, Jr
White, George Charles, Jr
Williams, Willie, Jr
Wilson, Geoffrey Leonard
Witzig, Warren Frank
Young, Donald Reeder
Zener, Clarence Melvin

RHODE ISLAND
Geman, Stuart Alan
Gerritsen, Hendrik Jurjen
Gilbert, Barry Jay
Glanz, Peter K
Grossi, Mario Dario
Hayward, John T(ucker)
Kolsky, Herbert
Lanou, Robert Eugene, Jr
Letcher, Stephen Vaughan
Maxson, Donald Robert
O'Keefe, J George
Peck, Russell Allen, Jr
Williams, Arthur Olney, Jr

SOUTH CAROLINA
Adams, Louis W, Jr
Aharonov, Yakir
Anderson, Wallace Ervin
Bagchi, Amitabha
Bauer, Richard M
Blevins, Maurice Everett
Childers, Richard Lee
Clark, Hugh Kidder
Cronemeyer, Donald Charles
Farach, Horacio A
Henderson, Robert Edward
Highsmith, Phillip E
Jones, Gordon Ervin
Larsen, Miguel Folkmar
Michener, John William
Myhrer, Fred
Nevitt, Michael Vogt
Olds, Daniel Wayne
Palms, John Michael
Poole, Charles Patton, Jr
Porter, Hayden Samuel, Jr
Ray, John Robert
Rodesiler, Paul Frederick
Schuette, Oswald Francis
Stevens, James Levon
Sturcken, Edward Francis
Ulbrich, Carlton Wilbur
Wilson, Jerry D(ick)
Wilson, John Neville

SOUTH DAKOTA
Helsdon, John H, Jr
Jones, Robert William
Miller, Bruce Linn
Quist, Oren Paul
Thompson, John Darrell

TENNESSEE
Abraham, Marvin Meyer
Akers, Lawrence Keith
Ball, Raiford Mill
Barach, John Paul
Barr, Dennis Brannon
Bayuzick, Robert J(ohn)
Beene, James Robert
Bell, Zane W
Bentley, Harry Thomas, III
Berry, Lee Allen
Bertini, Hugo W
Biggerstaff, John A
Burke, Edward Walter, Jr
Callcott, Thomas Anderson
Carbone, Robert James
Charles, George William
Chen, Yok
Clark, Grady Wayne
Cohn, Hans Otto
Cole, Thomas Earle
Cordell, Francis Merritt
Culp, Frederick Lynn
Deeds, William Edward
Douglass, Terry Dean
Duckett, Kermit Earl
Dunlap, Julian Lee
Ellis, Eric Hans
England, Alan Coulter
Fields, David Edward
Gailar, Norman Milton
Garber, David H
Garber, Floyd Wayne
Gwin, Reginald
Haaland, Carsten Meyer

Physics, General (cont)

Harwell, Kenneth Edwin
Hefferlin, Ray (Alden)
Henry, Jonathan Flake
Holdeman, Jonas Tillman, Jr
Horton, Linda Louise Schiestle
Johnson, Carroll Kenneth
Jones, Charles Miller, Jr
Jones, Dallas Wayne
Jones, Ernest Addison
Jordan, Lawrence M
Keefer, Dennis Ralph
King, David Thane
Livingston, Robert Simpson
Longmire, Martin Shelling
Lott, Sam Houston, Jr
Love, Norman Duane
McGowan, Francis Keith
Maerker, Richard Erwin
Martini, Mario
Mayfield, Melburn Ross
Mihalczo, John Thomas
Modine, Franklin Arthur
Mook, Herbert Arthur, Jr
Morgan, Relbue Marvin
Mosko, Sigmund W
Mullaney, Paul F
Murphy, Brian Donal
Nielsen, Alvin Herborg
Nowlin, Charles Henry
Painter, Linda Robinson
Pare, Victor Kenneth
Parks, James Edgar
Patterson, Malcolm Robert
Pegg, David John
Phung, Doan Lien
Pietenpol, Jerry L
Pinnaduwage, Lal Ariyaratna
Pomerance, Herbert S(olomon)
Postma, Herman
Ricci, Enzo
Ritchie, Rufus Haynes
Rousseau, Cecil Clyde
Schweinler, Harold Constantine
Schwenterly, Stanley William, III
Sears, Robert F, Jr
Sekula, Stanley Ted
Smith, James H
Stelson, Paul Hugh
Stockdale, John Alexander Douglas
Stone, Robert Sidney
Taylor, Jack Howard
Thomson, John Oliver
Todd, Peter Justin
Trivelpiece, Alvin William
Urban, Eugene Willard
Wei, Chin Hsuan
White, James Wilson
Whitfield, Jack D
Whittle, Charles Edward, Jr
Wong, Cheuk-Yin
Wood, Richard Frost
Woollett, Albert Haines
Yakel, Harry L
Zehner, David Murray

TEXAS

Adair, Thomas Weymon, III
Allen, Linus Scott
Anderson, Miles Edward
Baker, Stephen Denio
Baker, Wilfred E(dmund)
Ball, Millicent (Penny)
Baltzer, Otto John
Barnes, Thomas Grogard
Beall, Charles
Blackwell, Robert Jerry
Blanford, George Emmanuel, Jr
Bledsoe, Horace Willie Lee
Blount, Charles E
Boehme, Hollis
Bolen, Max Carlton
Bowling, David Ivan
Bradfield, James E
Bramblett, Richard Lee
Bruce, Rufus Elbridge, Jr
Cheney, Monroe G
Chivian, Jay Simon
Chopra, Dev Raj
Chu, Paul Ching-Wu
Church, David Arthur
Clayton, Glen Talmadge
Closmann, Philip Joseph
Cochran, Robert Glenn
Colegrove, Forrest Donald
Cook, Billy Dean
Cook, Clarence Sharp
Cooley, Adrian B, Jr
Couchman, James C
Cranberg, Lawrence
Cude, Willis Augustus, Jr
Damewood, Glenn
Das Gupta, Kamalaksha
Davies, Emlyn B
Dawson, Horace Ray
Dean, Eugene Alan
Deering, William Dougless
De Lozanne, Alex L
Diana, Leonard M
Dollard, John D
Dugan, Gerald Francis
Duggan, Jerome Lewis
Dunlap, Henry Francis

Dutcher, Clinton Harvey, Jr
Elliott, Jerry Chris
Ellis, Jason Arundel
Fietz, William Adolf
Finch, Robert D
Foote, Robert S
Frazer, Marshall Everett
Gourley, Lloyd Eugene, Jr
Ham, Joe Strother
Hamm, William Joseph
Hampton, Loyd Donald
Hanson, William Bert
Harbaugh, Allan Wilson
Harwood, William H
Hasty, Turner Elilah
Haymes, Robert C
Herbert, Morley Allen
Herrmann, Ulrich Otto
Hickey, Jimmy Ray
Hildebrandt, Alvin Frank
Hodges, Sidney Edward
Jahns, Monroe Frank
Jen, Shen
Jennings, Michael Leon
Johnson, Richard Clayton
Kiehn, Robert Mitchell
Kinsey, Bernard Bruno
Ko, Che Ming
Kruse, Olan Ernest
Kusch, Polykarp
Levine, Duane Gilbert
Lysiak, Richard John
McKinney, Chester Meek
Mandeville, Charles Earle
Mann, Ralph Willard
Mark, Hans Michael
Matzner, Richard Alfred
Mayes, Billy Woods, II
Mays, Robert, Jr
Mechler, Mark Vincent
Melvin, Cruse Douglas
Mikeska, Emory Eugene
Miller, Robert DuWayne
Millett, Walter Elmer
Mills, John James
Mires, Raymond William
Mitchell, John Charles
Monostori, Benedict Joseph
Moore, Rufus Adolphus
Moseley, Harrison Miller
Mueller, Dennis W
Nolle, Alfred Wilson
Ohashi, Yoshikazu
O'Keefe, Dennis Robert
Olness, Fredrick Iver
Ong, Poen Sing
Osoba, Joseph Schiller
Pandey, Raghvendra Kumar
Parker, Cleofus Varren, Jr
Parker, Jerald Vawer
Peebles, Hugh Oscar, Jr
Phillips, Gerald C
Pizzo, Joseph Francis
Prigogine, Ilya
Provencio, Jesus Roberto
Puthoff, Harold Edward
Randolph, Philip L
Ranganayaki, Rambabu Pothireddy
Raskin, Donald
Rayburn, Louis Alfred
Reaves, Harry Lee
Richardson, Arthur Jerold
Rickard, James Alexander
Rigney, Carl Jennings
Robbins, Donald Eugene
Robbins, Kay Ann
Robinson, Joseph Dewey
Rodriguez, Argelia Velez
Roeder, Robert Charles
Rogers, William Alan
Roman, Bernard John
Rorschach, Harold Emil, Jr
Rosen, Simon Peter
Rosenbaum, Joseph Hans
Rubin, Richard Mark
Rudmose, H Wayne
Sablik, Martin J
Sadler, Michael Ervin
Sahasrabuddhe, Chintaman Gopal
Sandlin, Billy Joe
Saslow, Wayne Mark
Sawyer, Ralph Stanley
Scheie, Paul Olaf
Scherr, Charles W
Schumaker, Robert Louis
Schwetman, Herbert Dewitt
Schwitters, Roy Frederick
Scully, Marlan Orvil
Sewell, Kenneth Glenn
Shapiro, Paul Robert
Sharber, James Randall
Shepherd, Jimmie George
Shieh, Paulinus Shee-Shan
Siegfried, Robert Wayne, II
Skolnick, Malcolm Harris
Smith, Daniel Montague
Smith, R Lowell
Sobey, Arthur Edward, Jr
Spellicy, Robert L
Squire, Charles F
Stark, J(ohn) P(aul), Jr
Stebbins, Ronald Frederick
Stewart, Frank Edwin
Stubbs, Norris

Stucker, Harry T
Swinney, Harry Leonard
Sybert, James Ray
Synek, Miroslav (Mike)
Taboada, John
Taylor, Herbert Lyndon
Theobald, J Karl
Thomas, Estes Centennial, III
Thomas, Richard Garland
Thompson, James Chilton
Tittel, Frank K(laus)
Tittle, Charles William
Tompkins, Donald Roy, Jr
Trafton, Laurence Munro
Truxillo, Stanton George
Unterberger, Robert Ruppe
Uralil, Francis Stephen
VantHull, LORIN L
Vant-Hull, Lorin Lee
Walker, Robert Hugh
Wallick, George Castor
Webb, Theodore Stratton, Jr
White, Gifford
Wiggins, James Wendell
Wilheit, Thomas Turner
Wilkerson, Robert C
Williams, Robert Leroy
Wisseman, William Rowland
Young, Bernard Theodore
Young, David Thad
Youngblood, Dave Harper
Zimmerman, Carol Jean

UTAH

Allan, David Wayne
Baker, Kay Dayne
Benson, Alvin K
Bryner, John C
Dibble, William E
Dick, Bertram Gale
Dudley, J Duane
Dulock, Victor A, Jr
Eastmond, Elbert John
Gibbs, Peter (Godbe)
Grow, Richard W
Hall, Howard Tracy
Harris, Franklin Stewart, Jr
Hill, Armin John
Jamison, Ronald D
Jones, Merrell Robert
Kadesch, Robert R
Lee, James Norman
Lind, Vance Gordon
Loh, Eugene C
Megill, Lawrence Rexford
Merrill, John Jay
Rasband, S Neil
Ward, Roger Wilson
Wood, John Karl

VERMONT

Bertelsen, Bruce I(rving)
Clark, Donald Lyndon
Detenbeck, Robert Warren
Golden, Kenneth Ivan
Hutchins, Gudrun A
Lane, George H
MacArthur, John Wood
McIntire, Sumner Harmon
Pennebaker, William B, Jr
Scarfone, Leonard Michael
Walters, Virginia F
Wu, Jun Ru

VIRGINIA

Adams, Clifford Lowell
Aitken, Alfred H
Anthony, Lee Saunders
Atkins, Marvin C(leveland)
Berlincourt, Ted Gibbs
Berman, Alan
Bierly, James N, Jr
Bloomfield, Louis Aub
Bomse, Frederick
Boring, John Wayne
Bowker, David Edwin
Braddock, Joseph V
Brandt, Richard Gustave
Brown, Arlin James
Brown, Ellen Ruth
Bruno, Ronald C
Bunner, Alan Newton
Burner, Alpheus Wilson, Jr
Calle, Carlos Ignacio
Cameron, Louis McDuffy
Carpenter, Delma Rae, Jr
Carr, Roger Byington
Carter, William Walton
Caton, Randall Hubert
Chubb, Scott Robinson
Clay, Forrest Pierce, Jr
Cochran, Thomas B
Coleman, Robert Vincent
Collins, George Briggs
Connolly, John Irving, Jr
Crawford, George William
Crowley, Patrick Arthur
Davey, John Edmund
Davidson, Robert Bellamy
Dennison, Brian Kenneth
Desiderio, Anthony Michael
Dockery, John T
Dolezalek, Hans
Donaghy, James Joseph

Dubin, Henry Charles
Duykers, Ludwig Richard Benjamin
Dylla, Henry Frederick
Edlund, Milton Carl
Ehrlich, Robert
English, Bruce Vaughan
Epstein, Samuel David
Erb, Karl Albert
Featherston, Frank Hunter
Findlay, John Wilson
Flippen-Anderson, Judith Lee
Friedman, Herbert
Frost, Robert T
Gamble, Thomas Dean
Gergely, Tomas Esteban
Gillies, George Thomas
Goodwin, Francis E
Grandy, Charles Creed
Grant, Frederick Cyril
Grosch, Chester Enright
Gugelot, Piet C
Hafele, Joseph Carl
Ham, William Taylor, Jr
Harries, Wynford Lewis
Harrison, Mark
Hartline, Beverly Karplus
Harvalik, Zaboj Vincent
Hereford, Frank Loucks, Jr
Herwig, Lloyd Otto
Hilger, James Eugene
Hodge, Donald Ray
Hooper, William John, Jr
Hoppe, John Cameron
Hove, John Edward
Hoy, Gilbert Richard
Hurdle, Burton Garrison
Ifft, Edward M
Isaacson, Richard Allen
Ivanetich, Richard John
Johnson, Charles Minor
Johnson, Charles Nelson, Jr
Johnson, Rolland Paul
Jones, Terry Lee
Kalab, Bruno Marie
Keefe, William Edward
Kelly, Hugh P
Kelly, Peter Michael
Kernell, Robert Lee
Kiess, Edward Marion
King, George, III
Koomen, Martin J
Kossler, William John
Krafft, Joseph Martin
Kress, Kenneth A
Krueger, Peter George
Kuhlmann-Wilsdorf, Doris
Landes, Hugh S(tevenson)
Lapsley, Alwyn Cowles
Leemann, Christoph Willy
Leies, Gerard M
Leiss, James Elroy
Lewis, Clark Houston
Lilly, Arnys Clifton, Jr
Lowance, Franklin Elta
Lowitz, David Aaron
Lowry, Ralph A(ddison)
McLucas, John L(uther)
Major, Robert Wayne
Margrave, Thomas Ewing, Jr
Marsh, Howard Stephen
Melfi, Leonard Theodore, Jr
Minnix, Richard Bryant
Misner, Robert David
Mitchell, John Wesley
Mitchell, Richard Scott
Mo, Luke Wei
Molloy, Charles Thomas
Morgan, Joseph
Moritz, Barry Kyler
Moyer, Leroy
Mulder, Robert Udo
Murday, James Stanley
Neher, Dean Royce
Neubauer, Werner George
Neuendorffer, Joseph Alfred
Newbolt, William Barlow
Ni, Chen-Chou
Nikolic, Nikola M
Outlaw, Ronald Allen
Padgett, Doran William
Peacock, Richard Wesley
Pettus, William Gower
Porzel, Francis Bernard
Raney, William Perin
Reisler, Donald Laurence
Reiss, Keith Westcott
Remler, Edward A
Rice, Roy Warren
Ritter, Rogers C
Ritz, Victor Henry
Robertson, Randal McGavock
Rogers, Thomas F
Rosi, Fred
Rothenberg, Herbert Carl
Rudmin, Joseph Webster
Ruffine, Richard S
Ruvalds, John
Salinger, Gerhard Ludwig
Satterthwaite, Cameron B
Sauder, William Conrad
Saunders, Peter Reginald
Schone, Harlan Eugene
Scoggin, James F, Jr
Seiler, Steven Wing

Sepucha, Robert Charles
Sessler, John Charles
Shapiro, Maurice Mandel
Sime, David Gilbert
Simpson, John Arol
Sinclair, Rolf Malcolm
Singh, Jag Jeet
Sloope, Billy Warren
Smith, Elton S
Sobottka, Stanley Earl
Spagna, George Fredric, Jr
Spivack, Harvey Marvin
Staib, Jon Albert
Stauss, George Henry
Stevens, Donald Meade
Stewart, John Westcott
Stronach, Carey E
Taggart, Keith Anthony
Taylor, Gerald Reed, Jr
Taylor, Jackson Johnson
Teitler, Sidney
Thaler, William John
Thompson, Anthony Richard
Thorp, Benjamin A
Tidman, Derek Albert
Tiller, Calvin Omah
Tipsword, Ray Fenton
Ulrich, Dale V
Veazey, Sidney Edwin
Wagner, Richard Lorraine, Jr
Wagner-Bartak, Claus Gunter Johann
Wallace, Lance Arthur
Warren, Holland Douglas
Watson, Robert Barden
Weik, Martin H, Jr
Welch, Jasper Arthur, Jr
Welsh, Robert Edward
Wilsdorf, Heinz G(erhard) F(riedrich)
Winsor, Niels K
Wolfhard, Hans Georg
Woods, Roy Alexander
Yeager, Paul Ray

WASHINGTON
Anderson, Roger Harris
Astley, Eugene Roy
Auth, David C
Baker, Marshall
Barnes, Ross Owen
Bennett, Robert Bowen
Bjornerud, Egil Kristoffer
Bodansky, David
Brown, Bert Elwood
Brown, Robert Reginald
Bunch, Wilbur Lyle
Buske, Norman L
Cannon, William Charles
Carter, John Lemuel, Jr
Center, Robert E
Clayton, Eugene Duane
Cook, Victor
Dash, Jay Gregory
DeRemer, Russell Jay
Donaldson, Edward Enslow
Dunning, Kenneth Laverne
Dye, David L
Ellis, Fred E
Ewart, Terry E
Faris, John Jay
Farwell, George Wells
Ghose, Subrata
Gillis, Murlin Fern
Grootes, Pieter Meiert
Halpern, Isaac
Harrison, Melvin Arnold
Haslund, R L
Hayward, Thomas Doyle
Hildebrandt, Jacob
Hinman, George Wheeler
Hoverson, Sigmund John
Hunting, Alfred Curtis
Ingalls, Robert L
Kells, Lyman Francis
Kissinger, Homer Everett
Knapp, Robert Hazard, Jr
Krienke, Ora Karl, Jr
Lagergren, Carl Robert
Lathrop, Arthur LaVern
Leonard, Bowen Raydo, Jr
Lessor, Delbert Leroy
Livdahl, Philip V
Lloyd, Raymond Clare
Long, Daniel R
Lord, Jere Johns
Lorenz, Richard Arnold
Lowell, Sherman Cabot
Lundy, Richard Alan
McDermott, Lillian Christie
Mitchell, Robert Curtis
Mobley, Curtis Dale
Murphy, Stanley Reed
Nethercot, Arthur Hobart, Jr
Nicholson, Richard Benjamin
Nornes, Sherman Berdeen
Park, James Lemuel
Parks, George Kung
Paul, Ronald Stanley
Peck, Edson Ruther
Penning, John Russell, Jr
Peters, Philip Carl
Reudink, Douglas O
Richmond, James Kenneth
Riehl, Jerry A
Rothberg, Joseph Eli

Sandstrom, Donald Richard
Schmid, Loren Clark
Siciliano, Edward Ronald
Siegel, Andrew Francis
Smith, David Clement, IV
Snover, Kurt Albert
Soreide, David Christien
Speer, Fridtjof Alfred
Sperry, Willard Charles
Spillman, Richard Jay
Spinrad, Bernard Israel
Stenkamp, Ronald Eugene
Streib, John Fredrick
Tang, Kwong-Tin
Thomas, Montcalm Tom
Van Heerden, Pieter Jacobus
Vilches, Oscar Edgardo
Walske, M Carl
White, William Charles
Young, James Arthur, Jr

WEST VIRGINIA
Anderson, Gary Don
Brimhall, James Elmore
Ferer, Martin V
Fischer, Robert Lee
Ghigo, Frank Dunnington
Henry, Donald Lee
Jefimenko, Oleg D
Leech, H(arry) William
Manakkil, Thomas Joseph
Meloy, Thomas Phillips
Oberly, Ralph Edwin
Parker, David Hiram
Schramm, Robert William
Sheppard, David W
Tesser, Herbert
Vehse, William E

WISCONSIN
Anway, Allen R
Auchter, Harry A
Bailey, Sturges Williams
Barmore, Frank E
Beck, Donald Edward
Besch, Gordon Otto Carl
Brown, Bruce Elliot
Caston, Ralph Henry
Chander, Jagdish
Cybriwsky, Alex
Dittman, Richard Henry
Ekern, Ronald James
Fossum, Steve P
Friedman, William Albert
Gade, Sandra Ann
Garrett, Robert Ogden
Greene, Jack Bruce
Herb, Raymond George
Hershkowitz, Noah
Hyzer, William Gordon
Johnson, Arthur Franklin
Johnson, DeWayne Carl
Karkheck, John Peter
Kobiske, Ronald Albert
Kraushaar, William Lester
Lichtman, David
Lodge, Arthur Scott
McQuistan, Richmond Beckett
Mallmann, Alexander James
Meyer, Robert Paul
Morin, Dornis Clinton
Myers, Vernon W
Nutter, Gene Douglas
Papastamatiou, Nicolas
Pillai, Thankappan A K
Reynolds, Ronald J
Rollefson, Ragnar
Roys, Paul Allen
Schmieg, Glenn Melwood
Schneiderwent, Myron Otto
Schultz, Frederick Herman Carl
Seshadri, Sengadu Rangaswamy
Shinners, Carl W
Sikdar, Dhirendra N
Smith, Roy E
Spangler, John David
Stekel, Frank D
Walters, William Le Roy
Webb, Maurice Barnett
Wilson, Volney Colvin
Woods, Frank Robert
Zimm, Carl B

WYOMING
Edwards, William Charles
Muller, Burton Harlow
Pepin, Theodore John
Rosen, James Martin
Tauer, Jane E

PUERTO RICO
Lewis, Brian Murray
Muir, James Alexander
Weisz, Zvi S

ALBERTA
Bazett-Jones, David Paul
Challice, Cyril Eugene
Coombes, Charles Allan
Huzinaga, Sigeru
Kitching, Peter
Krouse, Howard Roy
Paranjape, Bhalachandra Vishwanath
Pinnington, Eric Henry

Rood, Joseph Lloyd

BRITISH COLUMBIA
Astbury, Alan
Barclay, John Arthur
Barker, Alfred Stanley, Jr
Barss, Walter Malcomson
Bechhoefer, John
Blewett, Myrtle Hildred
Bloom, Myer
Booth, Andrew Donald
Chapman, Robert Pringle
Crozier, Edgar Daryl
Dalby, Frederick William
Dewey, John Marks
Duffus, Henry John
Fahlman, Gregory Gaylord
Fisher, Paul Douglas
Gargett, Ann E
Gold, Andrew Vick
Harrower, George Alexander
Hart, Kenneth Howell
Howard, Betty
Kiefl, Robert Frances
Krauel, David Paul
Lobb, Donald Edward
Mackenzie, George Henry
Nafe, John Elliott
Palmer, Leigh Hunt
Sandhu, Harbhajan Singh
Seth, Rajinder Singh
Stewart, Robert William
Sullivan, Harry Morton
Sykes, Paul Jay, Jr
Turrell, Brian George
Viswanathan, Kadayam Sankaran
Weinberg, Fred
White, Bruce Langton
Williams, David Llewelyn

MANITOBA
Barber, Robert Charles
Barnard, John Wesley
Cowdrey, Eric John
Ens, E(rich) Werner
Giles, Robin Arthur
Hawthorne, Frank Christopher
Jovanovich, Jovan Vojislav
McDonnell, Francis Nicholas
McKee, James Stanley Colton
Morrish, Allan Henry
Smith, Roger M
Standil, Sidney
Svenne, Juris Peteris
Taylor, Peter
Woo, Chung-Ho

NEW BRUNSWICK
Edwards, Merrill Arthur
Sharp, Allan Roy
Vo-Van, Truong

NEWFOUNDLAND
Reddy, Satti Paddi

NOVA SCOTIA
Archibald, William James
Betts, Donald Drysdale
Boyle, Willard Sterling
Elliott, James Arthur
Langstroth, George Forbes Otty
March, Robert Henry
Murty, Dangety Satyanarayana
Nugent, Sherwin Thomas
Tillotson, James Glen
Welch, Gary Alan

ONTARIO
Ahern, Francis Joseph
Allin, Elizabeth Josephine
Ancsin, John
Armstrong, Robert A
Baird, Kenneth MacClure
Barrington, Ronald Eric
Bedford, Ronald Ernest
Bennett, W Donald
Blackwell, Alan Trevor
Blevis, Bertram Charles
Boivin, Louis-Philippe
Brammer, Anthony John
Burke, Dennis Garth
Campeanu, Radu Ioan
Carmichael, Hugh
Chau, Wai-Yin
Clarke, James Newton
Conkie, William R
Corbett, James Murray
Cowper, George
Daams, Herman
Daigle, Gilles A
Dauphinee, Thomas McCaul
Dolling, Gerald
Drury, Malcolm John
Dunn, D(onald) W(illiam)
Fewer, Darrell R(aymond)
Fleming, William Herbert
Forsyth, Peter Allan
Gait, Robert Irwin
Gattinger, Richard Larry
Geiger, James Stephen
Gough, Sidney Roger
Graham, Robert Lockhart
Greenblatt, Jack Fred
Habib, Edwin Emile

Hagberg, Erik Gordon
Harrison, John Patrick
Heasell, E(dwin) L(ovell)
Hedley, Mark Allen
Helbing, Reinhard Karl Bodo
Hepburn, John Duncan
Hughes, Victor A
Hunt, James L
Hutchison, Thomas Sherret
Isenor, Neil R
Ivey, Donald Glenn
Jackson, Ray Weldon
Jezak, Edward V
Ji, Guangda Winston
Johns, Harold E
Kaye, Brian H
Kelly, Paul James
Le Blanc, Marcel A R
Lee, Hoong-Chien
Legg, Thomas Harry
Locke, Jack Lambourne
Logan, Robert Kalman
Love, Hugh Morrison
Lumsden, Charles John
Lynch, Gerard Francis
Lyon, Gordon Frederick
McConkey, John William
McDonald, James Frederick
MacFarlane, John T
MacKenzie, Innes Keith
MacNaughton, Earl Bruce
Martin, Douglas Leonard
Millman, Barry Mackenzie
Moorcroft, Donald Ross
Morrison, Hugh MacGregor
Murty, Rama Chandra
Nicholls, Ralph William
Paul, Derek (Alexander Lever)
Plint, Colin Arnold
Prentice, James Douglas
Prince, Robert Harry
Pullan, Harry
Punsack, Arnold L
Raaphorst, G Peter
Racey, Thomas James
Redhead, Paul Aveling
Richardson, Mary Frances
Roach, Margot Ruth
Rowlands, John Alan
Rubin, G A
Rucklidge, John Christopher
Sadowski, Chester M
Sawicka, Barbara Danuta
Sayer, Michael
Schofield, Derek
Shaw, Edgar Albert George
Smith, Howard John Treweek
Stewart, Alec Thompson
Stinson, Robert Henry
Sullivan, Philip Albert
Taylor, John Gardiner Veitch
Templeton, Ian M
Tilley, Donald E
Timusk, Thomas
Topp, G Clarke
Tunnicliffe, Philip Robert
Vanier, Jacques
Van Wijngaarden, Arie
Vogan, Eric Lloyd
Walker, Michael Barry
Westwood, William Dickson
Whitehead, James Rennie
Whittier, Angus Charles
Wiederick, Harvey Dale
Willemsen, Herman William
Wong, George Shoung-Koon
Wood, Gordon Harvey

PRINCE EDWARD ISLAND
Jammu, K S
Meincke, P P M

QUEBEC
Banville, Bertrand
Beaulieu, Jacques Alexander
Boivin, Alberic
Claisse, Fernand
Cloutier, Gilles Georges
Demers, Serge
Donnay, Joseph Desire Hubert
Duguay, Michel Albert
Ghosh, Asoke Kumar
Grmela, Miroslav
Hedgcock, Frederick Thomas
Kerwin, John Larkin
Lederman, Frank L
Leroy, Claude
Madore, Guy Louis Joseph
Marcoux, Jules E
Mathieu, Roger Maurice
Moody, Harry John
Patera, J
Pineault, Serge Rene
Rioux, Claude
Roy, Guy
Slobodrian, Rodolfo Jose
Tremblay, André-Marie
Zuckermann, Martin Julius

SASKATCHEWAN
Amundrud, Donald Lorne
Kennedy, John Edward
Montalbetti, Raymon
Rangacharyulu, Chilakamarri

Physics, General (cont)

Skarsgard, Harvey Milton

OTHER COUNTRIES
Andres, Klaus
Bailey, Paul Townsend
Barbenel, Joseph Cyril
Bauer, Ernst Georg
Baur, Werner Heinz
Berlman, Isadore B
Blumenfeld, Henry A
Bohr, Aage
Box, Michael Allister
Broers, Alec N
Bryngdahl, Olof
Burnett, Jerrold J
Cernuschi, Felix
Chan, Albert Sun Chi
Charpak, Georges
Crane, Philippe
Deutch, Bernhard Irwin
Economou, Eleftherios Nickolas
Eland, John Hugh David
Esaki, Leo
Friedel, Jacques
Garcia-Colin, Leopoldo Scherer
Gil, Salvador
Glaberson, William I
Glover, Francis Nicholas
Golay, Marcel Jules Edouard
Hansch, Theodor Wolfgang
Hopfer, Samuel
Josephson, Brian David
Kendall, James Tyldesley
Kildal, Helge
Lapostolle, Pierre Marcel
Leppelmeier, Gilbert Williston
Liu, Chao-Han
Margaritondo, Giorgio
Margolis, Bernard
Martin, William Eugene
Nestvold, Elwood Olaf
Paul, Wolfgang
Rubbia, Carlo
Salk, Sung-Ho Suck
Saupe, Alfred (Otto)
Schlein, Peter Eli
Schopper, Herwig Franz
Schroeder, Manfred Robert
Settles, Ronald Dean
Shtrikman, Shmuel
Siegbahn, Kai Manne Borje
Telegdi, Valentine Louis
Tsernoglou, Demetrius
Ushioda, Sukekatsu
Van Der Meer, Simon
Von Klitzing, Klaus
Webster, William Merle
Wehner, Gottfried Karl
Weiser, Kurt
Wick, Gian Carlo
Woods, Alfred David Braine
Yellin, Joseph
Yu, Chyang John

Plasma Physics

ALABAMA
Askew, Raymond Fike
Barr, Thomas Albert, Jr
Chappell, Charles Richard
Fukai, Junichiro
Hinata, Satoshi
Miller, George Paul
Olson, Willard Paul
Poularikas, Alexander D
Roberts, Thomas George
Swanson, Donald G
Willenberg, Harvey Jack

ALASKA
Brown, Neal B
Hallinan, Thomas James
Hunsucker, Robert Dudley
Kan, Joseph Ruce
Lee, Lou-Chuang
Swift, Daniel W
Wescott, Eugene Michael

ARIZONA
Denton, M Bonner
Fan, Chang-Yun
Farr, William Morris
Herbert, Floyd Leigh
Kyrala, Ali
Melia, Fulvio
Sonett, Charles Philip

ARKANSAS
Grauer, Albert D
Hughes, Raymond Hargett
Wold, Donald C

CALIFORNIA
Albritton, James R
Alfven, Hannes Olof Gosta
Alonso, Carol Travis
Anderson, David Vincent
Arons, Jonathan
Auerbach, Jerome Martin
Bach, David Rudolph
Bailey, Vernon Leslie, Jr
Baldwin, David Ellis
Bandel, Herman William
Banner, David Lee
Bardsley, James Norman
Barnes, Aaron
Barr, William Lee
Barry, James Dale
Baum, Peter Joseph
Baur, James Francis
Bellan, Paul Murray
Benford, Gregory A
Berger, Richard Leo
Bevc, Vladislav
Birdsall, Charles Kennedy
Boercker, David Bryan
Boley, Charles Daniel
Borucki, William Joseph
Bratenahl, Alexander
Brown, Keith H
Burrell, Keith H
Byrne, Nelson
Campbell, Edward Michael
Ceglio, Natale Mauro
Chambers, Frank Warman
Chan, Vincent Sikhung
Chang, David Bing Jue
Chao, Alexander Wu
Chase, Jay Benton
Chen, Francis F
Chen, Liu
Chin, Maw-Rong
Chiu, Yam-Tsi
Cladis, John Baros
Cohen, Bruce Ira
Cohen, Lawrence Mark
Cohen, Ronald Herbert
Coleman, Lamar William
Coleman, Paul Jerome, Jr
Conn, Robert William
Cooper, Alfred William Madison
Corngold, Noel Robert David
Cornwall, John Michael
Coroniti, Ferdinand Vincent
Cover, Ralph A
Davies, John Tudor
Davis, James Ivey
Davis, Leverett, Jr
Dawson, John Myrick
Decyk, Viktor Konstantyn
De Gaston, Alexis Neal
De Groot, John S
Demetriades, Sterge Theodore
Dethlefsen, Rolf
Devoto, Ralph Stephen
Drake, Richard Paul
Driscoll, Charles F
Dulgeroff, Carl Richard
Eckhardt, Wilfried Otto
Ehler, Arthur Wayne
Ehlers, Kenneth Warren
Eltgroth, Peter George
Estabrook, Kent Gordon
Evans, Todd Edwin
Fa'arman, Alfred
Feinberg, Benedict
Felber, Franklin Stanton
Flamm, Daniel Lawrence
Foote, James Herbert
Forsen, Harold K(ay)
Fried, Burton David
Frieman, Edward Allan
Futch, Archer Hamner
Gajewski, Ryszard
Goldstein, Raymond
Goodman, Ronald Keith
Gould, Roy W(alter)
Green, Michael Philip
Greene, John M
Greenwood, James Robert
Guest, Gareth E
Haan, Steven W
Hackel, Lloyd Anthony
Hahn, Kyoung-Dong
Hall, Laurence Stanford
Hamilton, Gordon Wayne
Harker, Kenneth James
Hartman, Charles William
Hawryluk, Andrew Michael
Helmer, John
Helton, F Joanne
Hewett, Dennis W
Hinton, Frederick Lee
Hobbs, Willard Earl
Hofmann, Gunter August George
Holdren, John Paul
Holzer, Robert Edward
Hooper, Edwin Bickford, Jr
Ikezi, Hiroyuki
Jackson, Gary Leslie
Jancarik, Jiri
Jensen, Torkil Hesselberg
Johnson, Roy Ragnar
Joshi, Chandrashekhar
Kaiser, Thomas Burton
Kaplan, Daniel Eliot
Kaufman, Allan N
Kennel, Charles Frederick
Khurana, Krishan Kumar
Kilb, Ralph Wolfgang
Killeen, John
Kim, Jinchoon
King, Harry J
Kivelson, Margaret Galland
Knechtli, Ronald (C)
Knize, Randall James
Kohler, Donald Alvin
Kovar, Frederick Richard
Kruer, William Leo
Kuehl, Hans H(enry)
Kulke, Bernhard
Kunc, Joseph Anthony
Kunkel, Wulf Bernard
Landman, Donald Alan
Langdon, Alan Bruce
Lasinski, Barbara Forman
Lauer, Eugene John
Le, Guan
Lee, Edward Prentiss
Lee, Ping
Leonard, Stanley Lee
Leung, Ka-Ngo
Lichtenberg, Allan J
Lin, Robert Peichung
Litvak, Marvin Mark
Lohr, John Michael
McCurdy, Alan Hugh
McIlwain, Carl Edwin
McPherron, Robert Lloyd
McWilliams, Roger Dean
Maglich, Bogdan
Malmberg, John Holmes
Manes, Kennneth Rene
Marrs, Roscoe Earl
Matossian, Jesse N
Matsuda, Kyoko
Matsuda, Yoshiyuki
Matthews, Stephen M
Max, Claire Ellen
Maxon, Marshall Stephen
Mazurek, Thaddeus John
Meeker, Donald J
Mendis, Devamitta Asoka
Mihalov, John Donald
Mitchner, Morton
Mo, Charles Tse Chin
Moir, Ralph Wayne
Molvik, Arthur Warren
Moore, Barry Newton
Morales, George John
More, Richard Michael
Nee, Soe-Mie Foeng
Nevins, William McCay
Newcomb, William A
Olness, Robert James
Olney, Ross David
O'Neil, Thomas Michael
Orth, Charles Douglas
Osher, John Edward
Overskei, David Orvin
Papas, Charles Herach
Payne, David James
Pellinen, Donald Gary
Pontau, Arthur E
Porter, Gary Dean
Post, Douglass Edmund
Post, Richard Freeman
Pritchett, Philip Lentner
Rasor, Ned S(haurer)
Rathmann, Carl Erich
Rawls, John M
Rensink, Marvin Edward
Rhodes, Edward Joseph, Jr
Rognlien, Thomas Dale
Romagnoli, Robert Joseph
Rosen, Mordecai David
Rugge, Henry F
Russell, Christopher Thomas
Rynn, Nathan
Schulz, Michael
Scott, Franklin Robert
Sentman, Davis Daniel
Shaffner, Richard Owen
Shelley, Edward George
Shestakov, Aleksei Ilyich
Shotts, Wayne J
Simonen, Thomas Charles
Smith, Edward John
Smith, Gary Richard
Speck, David Ralph
Spence, Harlan Ernest
Sperling, Jacob L
Stalder, Kenneth R
Stallings, Charles Henry
Stearns, John Warren
Stenzel, Reiner Ludwig
Stewart, Gordon Ervin
Stoeckly, Robert E
Strangeway, Robert Joseph
Stuart, George Wallace
Surko, Clifford Michael
Tamano, Teruo
Thomassen, Keith I
Thompson, William Bell
Thorne, Richard Mansergh
Toor, Arthur
Trigger, Kenneth Roy
Tsiang, Eugene Y
Tsurutani, Bruce Tadashi
Turchan, Otto Charles
Unti, Theodore Wayne Joseph
Urquidi-MacDonald, Mirna
Van Hoven, Gerard
Vernazza, Jorge Enrique
Wada, James Yasuo
Wagner, Carl E
Walker, Raymond John
Walt, Martin
Waltz, Ronald Edward
Wang, Charles P
Wang, Paul Keng Chieh
Weinstein, Berthold Werner
Wessel, Frank J
Williams, Edward Aston
Wilson, Andrew Robert
Wong, Alfred Yiu-Fai
Yeh, Yea-Chuan Milton

COLORADO
Aamodt, Richard E
Chappell, Willard Ray
Cooper, John (Jinx)
Henley, John Raymond
Kaufman, Harold Richard
Law, William Brough
Lee, Yung-Cheng
Low, Boon-Chye
Meiss, James Donald
Morrison, Charles Freeman, Jr
Robertson, Scott Harrison
Singer, Howard Joseph
Stern, Raul A(ristide)
Weinstock, Jerome
Zhang, Bing-Rong

CONNECTICUT
Bernstein, Ira Borah
Cheng, David H S
Haught, Alan F
Hirshfield, Jay Leonard
Meyerand, Russell! Gilbert, Jr
Similon, Philippe Louis
Stein, Richard Jay

DELAWARE
Evenson, Paul Arthur

DISTRICT OF COLUMBIA
Apruzese, John Patrick
Baker, Dennis John
Bingham, Robert Lodewijk
Bodner, Stephen E
Colombant, Denis Georges
Corliss, Charles Howard
Crandall, David Hugh
Decker, James Federick
Dove, William Francis
Feldman, Uri
Gold, Steven Harvey
Grossman, John Mark
Haber, Irving
Hafizi, Bahman
Hart, Richard Cullen
Huba, Joseph D
Hubbard, Richard Forest
Joyce, Glenn Russell
Killion, Lawrence Eugene
Lampe, Martin
Mahon, Rita
Manheimer, Wallace Milton
Manka, Charles K
Mosher, David
Nelson, David Brian
Obenschain, Stephen Philip
Oktay, Erol
Palmadesso, Peter Joseph
Patel, Vithalbhai L
Picone, J Michael
Randolph, James E
Reilly, Michael Hunt
Ripin, Barrett Howard
Sethian, John Dasho
Stephanakis, Stavros John
Taylor, Ronald D
Walker, David N(orton)
Wang, Yi-Ming
Wieting, Terence James

FLORIDA
Ancker-Johnson, Betsy
Baird, Alfred Michael
Dufty, James W
Flynn, Robert W
Hooper, Charles Frederick, Jr
Howard, Bernard Eufinger
Jacobson, Jerry I
Johnson, Joseph Andrew, III
Jones, William Denver
Kribel, Robert Edward
Lamborn, Bjorn N A
Linder, Ernest G
Loper, David Eric
Miller, Hillard Craig
Pearton, Stephen John
Penney, Gaylord W
Robertson, Harry S(troud)
Stetson, Robert Franklin
Vollmer, James
Weber, Alfred Herman

GEORGIA
Chen, Robert Long Wen
Little, Robert
Stacey, Weston Monroe, Jr

IDAHO
Shaw, Charles Bergman, Jr
Woodall, David Monroe

ILLINOIS
Chang, R P H
Evans, Kenneth, Jr
Fink, Charles Lloyd

Hatch, Albert Jerold
Henneberger, Walter Carl
Massel, Gary Alan
Mouschovias, Telemachos Charalambous
Parker, Eugene Newman
Pyle, K Roger
Raether, Manfred
Skadron, George

INDIANA
Helrich, Carl Sanfrid, Jr
Parish, Jeffrey Lee

IOWA
Carpenter, Raymon T
Case, William Bleicher
D'Angelo, Nicola
Gurnett, Donald Alfred
Houk, Robert Samuel
Knorr, George E
Lonngren, Karl E(rik)
McClelland, John Frederick
Nicholson, Dwight Roy
Scudder, Jack David

KANSAS
Carpenter, Kenneth Halsey
Enoch, Jacob
Unz, Hillel

KENTUCKY
Hawkins, Charles Edward
McGinness, William George, III
Ulrich, Aaron Jack

LOUISIANA
Gibbs, Richard Lynn
Jung, Hilda Ziifle
Kumar, Devendra

MAINE
Beard, David Breed
Pilot, Christopher H

MARYLAND
Alexander, Joseph Kunkle
Baker, Daniel Neal
Behannon, Kenneth Wayne
Benson, Robert Frederick
Birmingham, Thomas Joseph
Book, David Lincoln
Boyd, Derek Ashley
Brace, Larry Harold
Carbary, James F
Dean, Stephen Odell
DeSilva, Alan W
Dragt, Alexander James
Embury, Janon Frederick, Jr
Evans, Larry Gerald
Faust, William R
Fuhr, Jeffrey Robert
Goldenbaum, George Charles
Goldstein, Melvyn L
Granatstein, Victor Lawrence
Gray, Ernest Paul
Grebogi, Celso
Greenblatt, Marshal
Greig, J Robert
Hartle, Richard Eastham
Heppner, James P
Hitchcock, Daniel Augustus
Holman, Gordon Dean
Huddleston, Charles Martin
Hutcherson, R Kenneth
King, Joseph Herbert
Liu, Chuan Sheng
Lupton, William Hamilton
McFadden, Geoffrey Bey
McLean, Edgar Alexander
McNutt, Ralph Leroy, Jr
Martin, John J(oseph)
Merkel, George
Meyer, Eric Stefan
Noerdlinger, Peter David
Northrop, Theodore George
Ott, Edward
Oza, Dipak H
Papadopoulos, Konstantinos Dennis
Peredo, Mauricio
Potemra, Thomas Andrew
Reilly, James Patrick
Rhee, Moon-Jhong
Rosenberg, Theodore Jay
Roszman, Larry Joe
Sari, James William
Singh, Amarjit
Sittler, Edward Charles, Jr
Skiff, Frederick Norman
Stamper, John Andrew
Stern, David P
Stone, Albert Mordecai
Stone, Philip M
Tang, Cha-Mei
Turner, James Marshall
Van Brunt, Richard Joseph
Wiese, Wolfgang Lothar
Wilkerson, Thomas Delaney
Wu, Ching-Sheng
Yoon, Peter Haesung
Young, Frank Coleman

MASSACHUSETTS
Allen, Ryne C
Allis, Willam Phelps

Bekefi, George
Berman, Robert Hiram
Bers, Abraham
Bugnolo, Dimitri Spartaco
Burke, William J
Catto, Peter James
Cohn, Daniel Ross
Cooke, Richard A
Cordero, Julio
Denig, William Francis
Douglas-Hamilton, Diarmaid H
Fougere, Paul Francis
Gelman, Harry
Godyak, Valery Anton
Graneau, Peter
Guss, William C
Habbal, Shadia Rifai
Hicks, Ross S
Jacob, Jonah Hye
Johnston, George Lawrence
Jose, Jorge V
Kalman, Gabor J
Kautz, Frederick Alton, II
Kreisler, Michael Norman
Krieger, Allen Stephen
Lai, Shu Tim
Lax, Benjamin
Leiby, Clare C, Jr
Litzenberger, Leonard Nelson
Marmar, Earl Sheldon
Marshall, Theodore
Oettinger, Peter Ernest
Petrasso, Richard David
Pian, Carlson Chao-Ping
Plumb, John Laverne
Porkolab, Miklos
Post, Richard S
Ratner, Michael Ira
Rogoff, Gerald Lee
Sigmar, Dieter Joseph
Silevich, Michael B
Silk, John Kevin
Spiegel, Stanley Lawrence
Temkin, Richard Joel
Terry, James Layton
Whitehouse, David R(empfer)
Wolfe, Stephen Mitchell
Woskov, Paul Peter
Yang, Tien-Fang

MICHIGAN
Asmussen, Jes
Banks, Peter Morgan
Beglau, David Alan
Berger, Beverly Kobre
Bleil, Carl Edward
Clauer, C Robert, Jr
Fontheim, Ernest Gunter
Getty, Ward Douglas
Gilgenbach, Ronald Matthew
Mayer, Frederick Joseph
Scott, Joseph Hurlong
Turner, Robert

MINNESOTA
Chanin, Lorne Maxwell
Heberlein, Joachim Viktor Rudolf
Kellogg, Paul Jesse
Lysak, Robert Louis
Stuart, Robley Vane

MISSISSIPPI
Ferguson, Joseph Luther, Jr
Johnston, Russell Shayne

MISSOURI
Leader, John Carl
McFarland, Robert Harold
Menne, Thomas Joseph
Painter, James Howard
Prelas, Mark Antonio
Rode, Daniel Leon
Shrauner, Barbara Abraham
Willett, Joseph Erwin

MONTANA
Bird, Harvey Harold
Rosa, Richard John

NEVADA
Phaneuf, Ronald Arthur

NEW HAMPSHIRE
Boley, Forrest Irving
Corum, Kenneth Lawrence
Forbes, Terry Gene
Hollweg, Joseph Vincent
Hudson, Mary Katherine
Lee, Martin Alan
Montgomery, David Campbell
Ryan, James Michael
Sonnerup, Bengt Ulf osten
Torbert, Roy Banks
White, Willard Worster, III

NEW JERSEY
Anandan, Munisamy
Barnes, Derek A
Bernabei, Stefano
Bol, Kees
Bond, Robert Harold
Budny, Robert Vierling
Bush, Charles Edward
Cecchi, Joseph Leonard

Chang, Tao-Yuan
Channon, Stephen R
Cohen, Samuel Alan
Davidson, Ronald Crosby
Ferrari, Lawrence A
Furth, Harold Paul
Gammel, George Michael
Grek, Boris
Grisham, Larry Richard
Grove, Donald Jones
Hawryluk, Richard Janusz
Hendel, Hans William
Hosea, Joel Carlton
Hsuan, Hulbert C S
Hulse, Russell Alan
Jardin, Stephen Charles
Jassby, Daniel Lewis
Johnson, John Lowell
Kaita, Robert
Karney, Charles Fielding Finch
Kugel, Henry W
Kulsrud, Russell Marion
Kunhardt, Erich Enrique
LaMarche, Paul Henry, Jr
Langer, William David
Lanzerotti, Louis John
Manickam, Janardhan
Mazzucato, Ernesto
Meade, Dale M
Mikkelsen, David Robert
Mills, Robert Gail
Motz, Robin Owen
Mueller, Dennis
Murnick, Daniel E
Okabayashi, Michio
Owens, David Kingston
Palladino, Richard Walter
Pargellis, Andrew Nason
Passner, Albert
Perry, Erik David
Plummer, Dirk Arnold
Ramsey, Alan T
Redi, Martha Harper
Reiman, Allan H
Rewoldt, Gregory
Rosen, Bernard
Rothleder, Stephen David
Rutherford, Paul Harding
Sauthoff, Ned Robert
Schilling, Gerd
Schivell, John Francis
Schmidt, George
Schmidt, John Allen
Schneider, Sol
Seidl, Milos
Silfvast, William Thomas
Sinnis, James Constantine
Spitzer, Lyman, Jr
Stevens, James Everell
Stix, Thomas Howard
Strachan, James Douglas
Suckewer, Szymon
Takahashi, Hironori
Tang, William Ming-Wu
Taylor, Gary
Valeo, Ernest John
Vickroy, David Gill
Von Goeler, Schweickhard E
Wadlow, David
Weinrich, Marcel
White, Alan David
White, Roscoe Beryl
Wong, King-Lap
Yamada, Masaaki
Zarnstorff, Michael Charles

NEW MEXICO
Ahluwalia, Harjit Singh
Arion, Douglas
Barnes, Daniel C
Benjamin, Robert Fredric
Bieniewski, Thomas M
Brackbill, Jeremiah U
Brownell, John Howard
Burns, Erskine John Thomas
Cartwright, David Chapman
Chamberlin, Edwin Phillip
Czuchlewski, Stephen John
Dietz, David
Doolen, Gary Dean
Dreicer, Harry
Engelhardt, Albert George
Finn, John McMaster
Forslund, David Wallace
Fradkin, David Barry
Freeman, Bruce L, Jr
Freese, Kenneth Brooks
Fries, Ralph Jay
Fu, Jerry Hui Ming
Gary, Stephen Peter
Gerwin, Richard A
Giovanielli, Damon V
Gisler, Galen Ross
Glasser, Alan Herbert
Godfrey, Brendan Berry
Goldstone, Philip David
Gosling, John Thomas
Greene, Arthur Edward
Gula, William Peter
Haberstich, Albert
Hammel, Jay Edwin
Henderson, Dale Barlow
Henins, Ivars
Hill, Ronald Ames

Hoffman, Nelson Miles, III
Humphries, Stanley, Jr
Ingraham, John Charles
Jahoda, Franz C
Judd, O'Dean P
Kindel, Joseph Martin
King, Nicholas S P
Kopp, Roger Alan
Kyrala, George Amine
Leeper, Ramon Joe
Leonard, Ellen Marie
Lilley, John Richard
Lindman, Erick Leroy, Jr
Linford, Rulon Kesler
Lyons, Peter Bruce
McComas, David John
McNally, James Henry
Martin, Thierry Dominique
Mason, Rodney Jackson
Matzen, Maurice Keith
Mazarakis, Michael Gerassimos
Mead, William C
Mjolsness, Raymond C
Morse, Fred A
Mueller, Marvin Martin
Phillips, James A
Phillips, John Lynch
Pickrell, Mark M
Pongratz, Morris Bernard
Price, Robert Harold
Ragan, Charles Ellis, III
Rahal, Leo James
Reeves, Geoffrey D
Rose, Evan Andrew
Rose, Harvey Arnold
Sawyer, George Alanson
Schappert, Gottfried T
Shaner, John Wesley
Sherwood, Arthur Robert
Solem, Johndale Christian
Stellingwerf, Robert Francis
Stratton, Thomas Fairlamb
Suydam, B R
Sweeney, Mary Ann
Thode, Lester Elster
Thomsen, Michelle Fluckey
Toepfer, Alan James
Touryan, Kenell James
Turner, Leaf
Tuszewski, Michel G
VanDevender, John Pace
Veeser, Lynn Raymond
Vittitoe, Charles Norman
Winske, Dan
Wright, Thomas Payne
Yates, Mary Anne
Yonas, Gerold
Younger, Stephen Michael

NEW YORK
Anderson, John M(elvin)
Auer, Peter Louis
Bardwell, Steven Jack
Barreto, Ernesto
Becker, Kurt Heinrich
Benenson, David Maurice
Boyer, Donald Wayne
Bradshaw, John Alden
Brown, Howard Howland, Jr
Coleman, John Howard
Craxton, Robert Stephen
Ellis, William Rufus
Farley, Donald T, Jr
Fleischmann, Hans Hermann
Gralnick, Samuel Louis
Gross, Robert Alfred
Hammer, David Andrew
Hickok, Robert L, Jr
Jensen, Betty Klainminc
Johnson, Brant Montgomery
Kaup, David James
Kranzer, Herbert C
Kusse, Bruce Raymond
Laghari, Javaid Rasoolbux
Lee, Yung-Chang
Levi, Enrico
Levine, Alfred Martin
Liboff, Richard L
Lovelace, Richard Van Evera
McCrory, Robert Lee, Jr
McKinstrie, Colin J(ohn)
Marshall, Thomas C
Mathad, Gangadhara S
Melnyk, Andrew Rostyslaw
Mendez, Emilio Eugenio
Morrone, Terry
Nagamatsu, Henry T
Nardi, Vittorio
Navratil, Gerald Anton
Rehak, Pavel
Rich, Joseph Anthony
Savkar, Sudhir Dattatraya
Schlesinger, S Perry
Seka, Wolf
Sen, Ujjal
Seyler, Charles Eugene
Simon, Albert
Soures, John Michael
Sudan, Ravindra Nath
Takacs, Gerald Alan
Toner, John Joseph
Tzoar, N
Webster, Harold Frank
Weitzner, Harold

Plasma Physics (cont)

Wharton, Charles Benjamin
Witting, Harald Ludwig
Worden, David Gilbert
Yaakobi, Barukh
Zucker, Martin Samuel

NORTH CAROLINA
Aktas, Yildirin
Christiansen, Wayne Arthur
Clegg, Thomas Boykin
Doggett, Wesley Osborne
Goldman, Leonard Manuel
Kratz, Howard Russel
McCormack, Francis Joseph
Masnari, Nino A
Vermillion, Robert Everett

OHIO
Andrews, Merrill Leroy
Beverly, Robert Edward, III
Dakin, James Thomas
Ferguson, Dale Curtis
Garscadden, Alan
Holt, James Franklin
Lauver, Milton Renick
Lee, Anthony
Leonard, Brian Phillip
Mallozzi, Philip James
Mawardi, Osman Kamel
Nestor, Ontario Horia
Nielsen, Carl Eby
Nielsen, Philip Edward
Pytte, Agnar
Rockstroh, Todd Jay
Stewart, Charles Neil
Turchi, Peter John
Walters, Craig Thompson

OKLAHOMA
Armoudian, Garabed
Duncan, Lewis Mannan
Fowler, Richard Gildart
Shull, Peter Otto, Jr

OREGON
Drummond, James Edgar

PENNSYLVANIA
Cheng, Chiang-Shuei
Cohen, Ira M
Cookson, Alan Howard
Davies, D K
Emmerich, Werner Sigmund
Ernst, Wolfgang E
Gurer, Emir
Heald, Mark Aiken
Karras, Thomas William
Katz, Lewis E
Kim, Yong Wook
Kimblin, Clive William
Klevans, Edward Harris
Kritz, Arnold H
Ku, Robert Tien-Hung
Lempert, Joseph
Lowry, Jerald Frank
Ludwig, Howard C
Machalek, Milton David
Macpherson, Alistair Kenneth
Mathews, John David
Moscatelli, Frank A
Poss, Howard Lionel
Quinn, Robert George
Radin, Shelden Henry
Rothman, Milton A
Shoaf, Mary La Salle
Sommer, Holger Thomas
Sucov, E(ugene) W(illiam)
Williams, James Earl, Jr
Wolf, Neil Stephan
Zollweg, Robert John

RHODE ISLAND
Logan, Joseph Skinner
McCorkle, Richard Anthony
Strauss, Walter A
Su, Chau-Hsing

SOUTH CAROLINA
Goode, Scott Roy
Hurren, Weiler R
Jenkins, Alvin Wilkins, Jr
Lee, Craig Chun-Kuo
Yarborough, William Walter, Jr

TENNESSEE
Alexeff, Igor
Baker, Charles Clayton
Barach, John Paul
Beasley, Cloyd O, Jr
Bell, Persa Raymond
Berry, Lee Allen
Carreras, Benjamin A
Crume, Enyeart Charles, Jr
Dory, Robert Allan
Garrison, George Walker
Glowienka, John Clement
Hedrick, Clyde Lewis, Jr
Hogan, John Thomas
Hutchinson, Donald Patrick
Isler, Ralph Charles
Kelley, George G
Langley, Robert Archie

McNally, James Rand, Jr
Milora, Stanley L
Murakami, Masanori
Pinnaduwage, Lal Ariyaratna
Rome, James Alan
Roth, J(ohn) Reece
Sheffield, John
Staats, Percy Anderson
Swain, David Wood
Thonnard, Norbert
Trivelpiece, Alvin William
Vander Sluis, Kenneth Leroy
Van Rij, Willem Idaniel
Vaslow, Dale Franklin
Veach, Allen Marshall
Ventrice, Carl Alfred
Walter, F John
Whealton, John Hobson
Zuhr, Raymond Arthur

TEXAS
Bengtson, Roger D
Bering, Edgar Andrew, III
Berk, Herbert L
Briggs, Richard Julian
Burch, James Leo
Cloutier, Paul Andrew
Economou, Demetre J
Eisner, Melvin
Frommhold, Lothar Werner
Gentle, Kenneth W
Hagler, Marion O(tho)
Hazeltine, Richard Deimel
Hill, Thomas Westfall
Hirsch, Robert Louis
Isham, Elmer Rex
Koemtzopoulos, C Robert
Kristiansen, Magne
Liang, Edison Park-Tak
Liemohn, Harold Benjamin
McCoy, James Ernest
Mahajan, Swadesh Mitter
Mitra, Sunanda
Newberger, Barry Stephen
Oakes, Melvin Ervin Louis
Parker, Jerald Vawer
Phillips, Perry Edward
Powers, Edward Joseph, Jr
Puthoff, Harold Edward
Ransom, C J
Robbins, Roger Alan
Salingaros, Nikos Angelos
Spencer, John Edward
Tajima, Toshiki
Voigt, Gerd-Hannes
Ware, Alan Alfred
Wolf, Richard Alan
Wong, Henry Vernon
Young, David Thad

UTAH
Gardner, Andrew Leroy
Hilton, John L
Knight, Larry V
Mason, Grant William
Schunk, Robert Walter

VERMONT
Foote, Vernon Stuart, Jr
Golden, Kenneth Ivan
Quinn, Robert M(ichael)
Wolfson, Richard L T

VIRGINIA
Armstrong, Carter Michael
Basu, Sunanda
Benesch, Jay F
Boozer, Allen Hayne
Brockman, Philip
Buck, Warren W, III
Chang, Chieh Chien
Coffey, Timothy
Cox, James Lester, Jr
Eubank, Harold Porter
Evans, John C
Ganguly, Suman
Grossmann, William
Herbst, Mark Joseph
Jalufka, Nelson Wayne
Krafft, Geoffrey Arthur
Langston, Glen Irvin
Lee, Ja H
Liebenberg, Donald Henry
Lynch, John Thomas
McAdoo, John Hart
Neil, George Randall
Nuckolls, Joe Allen
Ossakow, Sidney Leonard
Pereira, Nino Rodrigues
Radoski, Henry Robert
Robson, Anthony Emerson
Roy, Gabriel D
Seiler, Steven Wing
Soper, Gordon Knowles
Sutton, George W(alter)
Szuszczewicz, Edward Paul
Tidman, Derek Albert
Tsang, Kang Too
Vahala, George Martin
Winsor, Niels K
Wyeth, Newell Convers

WASHINGTON
Ahlstrom, Harlow G(arth)

Fremouw, Edward Joseph
Holzworth, Robert H, II
Jarboe, Thomas Richard
Lowenthal, Dennis David
MacVicar-Whelan, Patrick James
Moore, Richard Lee
Nicholson, Richard Benjamin
Pyle, Robert V
Radziemski, Leon Joseph
Siciliano, Edward Ronald

WEST VIRGINIA
Koepke, Mark E

WISCONSIN
Aita, Carolyn Rubin
Callen, James Donald
Conrad, John Rudolph
Emmert, Gilbert A
Fonck, Raymond John
Kerst, Donald William
Lawler, James Edward
Prager, Stewart Charles
Shohet, Juda Leon
Sprott, Julien Clinton
Symon, Keith Randolph
Wilson, Volney Colvin

ALBERTA
Fedosejevs, Robert
James, Christopher Robert
Liu, Dennis Dong
Offenberger, Allan Anthony
Sreenivasan, Sreenivasa Ranga

BRITISH COLUMBIA
Ahlborn, Boye
Barnard, Adam Johannes
Clements, Reginald Montgomery
Watanabe, Tomiya

NOVA SCOTIA
MacLatchy, Cyrus Shantz

ONTARIO
Alcock, Alfred John
Dixit, Madhu Sudan
Ehrman, Joachim Benedict
Goodings, John Martin
Hruska, Antonin
Laframboise, James Gerald
MacFarlane, John T
McNamara, Allen Garnet
Moorcroft, Donald Ross
Muldrew, Donald Boyd
Panarella, Emilio
Raju, Govinda
Ristic, Velimir Mihailo
Slutsky, Arthur
Stangeby, Peter Christian
St-Maurice, Jean-Pierre
Sundaresan, Mosur Kalyanaraman

QUEBEC
Bachynski, Morrel Paul
Bolton, Richard Andrew Ernest
Gregory, Brian Charles
Johnston, Tudor Wyatt
Marchand, Richard
Richard, Claude
Shkarofsky, Issie Peter
Stansfield, Barry Lionel
Terreault, Bernard J E J

SASKATCHEWAN
Hirose, Akira

OTHER COUNTRIES
Boxman, Raymond Leon
Choi, Duk-In
Chu, Kwo Ray
Crawford, Frederick William
Daly, Patrick William
Dewar, Robert Leith
Hagfors, Tor
Kapetanakos, Christos Anastasios
Krumbein, Aaron Davis
Lee, Peter H Y
Lee, Tong-Nyong
Mark, James Wai-Kee
Pitchford, Leanne Carolyn
Romero, Alejandro F
Samir, Uri
Vlases, George Charpentier
Wasa, Kiyotaka
Yanabu, Satoru

Semiconductor Devices

ALABAMA
Bloemer, Mark Joseph

ARIZONA
Kim, Matthew Hidong

CALIFORNIA
Benjamin, Arlin James
Bharat, Ramasesha
Biblarz, Oscar
Dapkus, Paul Daniel
Dines, Eugene L
Dorfan, David Elliot
Early, James M

Edwall, Dennis Dean
Fok, Samuel S(hiu) M(ing)
Gee, Allen
Gossard, Arthur Charles
Harris, James Stewart, Jr
Herb, John A
Huang, Di-Hui (David)
Hurrell, John Patrick
Jones, Lewis Hammond, IV
Karlsson, Eric Allan
Kilmer, Louis Charles
Knechtli, Ronald (C)
Long, Stephen Ingalls
Maserjian, Joseph
Nelson, Richard David
Paoli, Thomas Lee
Parker, Sherwood
Tatic-Lucic, Svetlana
Tsai, Chuang Chuang
Van de Walle, Chris G
Wang, Michael Renxun
Wieder, Harry H

COLORADO
Kanda, Motohisa
Noufi, Rommel
Oleszek, Gerald Michael
Trefny, John Ulric

CONNECTICUT
Reyna, Luis Guillermo

DELAWARE
Bragagnolo, Julio Alfredo

DISTRICT OF COLUMBIA
Anderson, Gordon Wood
Jansen, Robert Werner
Wittmann, Horst Richard

FLORIDA
DeLoach, Bernard Collins, Jr
Holloway, Paul Howard
Li Kam Wa, Patrick
Moreno, Wilfrido A
Pearton, Stephen John
Wade, Thomas Edward

ILLINOIS
Jerome, Joseph Walter
Kitchen, W J
Wang, Hwa-Chi

INDIANA
Stein, Frank S

IOWA
Tringides, Michael C

KANSAS
Folland, Nathan Orlando

MARYLAND
Carver, Gary Paul
Drew, Howard Dennis
Hua, Susan Zonglu
Kopanski, Joseph J
McClelland, Jabez Jenkins
Potyraj, Paul Anthony
Yu, Anthony Woonchiu

MASSACHUSETTS
Clarke, Edward Nielsen
Donnelly, Joseph P(eter)
Lyons, William Gregory
Minden, Henry Thomas
Skocpol, William John
Zheng, Xiao Lu

MICHIGAN
Partin, Dale Lee
Pavlidis, Dimitris
Pryor, Roger Welton

MINNESOTA
Tufte, Obert Norman

MISSOURI
Prelas, Mark Antonio
Rode, Daniel Leon

NEW JERSEY
Audet, Sarah Anne
Burrus, Charles Andrew, Jr
Chang, Chuan Chung
Chiu, Tien-Heng
Citrin, Paul H
Comizzoli, Robert Benedict
Cullen, Glenn Wherry
Dodabalapur, Ananth
Dubey, Madan
Eckert, John Andrew
Feldman, Leonard Cecil
Howard, Webster Eugene
Imhoff, Eugene A
Kurtz, Anthony David
Marcus, Robert Boris
Ong, Nai-Phuan
Ourmazd, Abbas
Panayotatos, Paul
Schneider, Sol
Shallcross, Frank V(an Loon)
Shen, Tek-Ming
Tompsett, Michael F

Vandenberg, Joanna Maria
Watson, Hugh Alexander

NEW MEXICO
Hoffbauer, Mark Arles
Mason, Rodney Jackson

NEW YORK
Arnold, Emil
Bulusu, Dutt V
Chen, Yu-Pei
Coleman, John Howard
Goldman, Vladimir J
Kinsella, John J
Kraner, Hobart Wilson
Kuo, Yue
Lee, Young-Hoon
LeMay, Charlotte Zihlman
Li, Zheng
Melnyk, Andrew Rostyslaw
Pliskin, William Aaron
Ruckman, Mark Warren
Schiff, Eric Allan
Thomas, J Earl
Wachnik, Richard Andre
Wie, Chu Ryang
Woodard, David W

NORTH CAROLINA
Burger, Robert M
Enquist, Paul Michael

OHIO
Connolly, Denis Joseph
Ringel, Steven Adam

OREGON
Arthur, John Read, Jr
Balko, Barbara Ann
Chrzanowska-Jeske, Malgorzata Ewa
Engelmann, Reinhart Wolfgang H

PENNSYLVANIA
Goldey, James Mearns
Kun, Zoltan Kokai
Licini, Jerome Carl
Reynolds, Claude Lewis, Jr
Schnable, George Luther
Wilkins, Cletus Walter, Jr
Withstandley, Victor DeWyckoff, III

SOUTH DAKOTA
Ross, Keith Alan

TENNESSEE
Haynes, Tony Eugene
Jellison, Gerald Earle, Jr

TEXAS
Banerjee, Sanjay Kumar
Chapman, Richard Alexander
Chu, Shirley Shan-Chi
Esquivel, Agerico Liwag
Garza, Cesar Manuel
Koemtzopoulos, C Robert
Menon, Venugopal Balakrishna
Overmyer, Robert Franklin
Rabson, Thomas A(velyn)
Singh, Vijay Pal
Spencer, Gregory Fielder

UTAH
Huber, Robert John

VERMONT
Anderson, R(ichard) L(ouis)

VIRGINIA
Joshi, Ravindra Prabhakar
Lawrence, David Joseph
Mitchell, Dean Lewis

WASHINGTON
Domansky, Karel

WISCONSIN
Mangat, Pawitterjit Singh

ALBERTA
Liu, Dennis Dong

BRITISH COLUMBIA
Colbow, Kevin Michael
Colbow, Konrad
Deen, Mohamed Jamal
Thewalt, Michael L W
Wallace, Philip Russell
Young, Lawrence

MANITOBA
Kao, Kwan Chi

ONTARIO
Bardwell, Jennifer Ann
Dawson, Peter Henry
Dyment, John Cameron
Graham, Michael John
Hurd, Colin Michael
Ruda, Harry Eugen
Witzman, Sorin

OTHER COUNTRIES
Lee, Elhang Howard

Solid State Physics

ALABAMA
Anderson, Elmer Ebert
Bloemer, Mark Joseph
Chen, An-Ban
Christensen, Charles Richard
Davis, Jack H
Doyle, William David
Fromhold, Albert Thomas, Jr
Harrell, James W, Jr
Hartman, Richard Leon
Kinzer, Earl T, Jr
Kroes, Roger L
Kwon, Tai Hyung
Lal, Ravindra Behari
McDonald, Jack Raymond
Mookherji, Tripty Kumar
Naumann, Robert Jordan
Pontius, Duane Henry
Reynolds, Robert Ware
Sibley, William Arthur
Thaxton, George Donald
Vohra, Yogesh K
Worley, S D

ARIZONA
Albrecht, Edward Daniel
Bao, Qingcheng
Benin, David B
Beyen, Werner J
Carpenter, Ray Warren
Chamberlin, Ralph Vary
De La Moneda, Francisco Homero
Delinger, William Galen
Dereniak, Eustace Leonard
Dow, John Davis
Emrick, Roy M
Everett, Paul Marvin
Falco, Charles Maurice
Farnsworth, Harrison E
Fliegel, Frederick Martin
Gibbs, Hyatt McDonald
Green, Barry A
Gregory, Brooke
Handy, Robert M(axwell)
Hanson, Roland Clements
Hickernell, Fred Slocum
Hohl, Jakob Hans
Hooper, Henry Olcott
Huffman, Donald Ray
Jackson, Kenneth Arthur
Johnson, Earnest J
Joyce, Richard Ross
Kim, Matthew Hidong
Leavitt, John Adams
Mahalingam, Lalgudi Muthuswamy
Marzke, Robert Franklin
Page, John Boyd, Jr
Parmenter, Robert Haley
Peyghambarian, Nasser
Powell, Richard Conger
Ren, Shang Yuan
Rez, Peter
Rutledge, James Luther
Sarid, Dror
Scholl, Marija Strojnik
Schroder, Dieter K
Scott, Alwyn C
Seraphin, Bernhard Otto
Smith, David John
Stark, Royal William
Stearns, Mary Beth Gorman
Stuart, Derald Archie
Tomizuka, Carl Tatsuo
Venables, John Anthony
Vuillemin, Joseph J
Wang, Edward Yeong

ARKANSAS
Horton, Philip Bish
McCloud, Hal Emerson, Jr
Wrobel, Joseph Stephen

CALIFORNIA
Adams, Arnold Lucian
Ahlers, Guenter
Akella, Jagannadham
Albert, Paul A(ndre)
Allen, Frederick Graham
Almasan, Carmen Cristina
Alves, Ronald V
Amelio, Gilbert Frank
Ames, Lawrence Lowell
Appelbaum, Joel A
Archer, Robert James
Arias, Jose Manuel
Aston, Duane Ralph
Astrue, Robert William
Atwater, Harry Albert
Aukerman, Lee William
Avrin, William F
Ayers, Raymond Dean
Bachrach, Robert Zelman
Baglin, John Edward Eric
Bagus, Paul Saul
Bajaj, Jagmohan
Bak, Chan Soo
Bardin, Russell Keith
Barmatz, Martin Bruce
Barnes, Gene A
Barnoski, Michael K
Baron, Robert
Barrera, Joseph S
Bartelink, Dirk Jan
Baskes, Michael I
Bate, Geoffrey
Bates, Clayton Wilson, Jr
Batra, Inder Paul
Bauer, Robert Steven
Bauer, Walter
Beardsley, Irene Adelaide
Beasley, Malcolm Roy
Bebb, Herbert Barrington
Becker, Milton
Bennett, Alan Jerome
Bennett, Glenn Taylor
Bennett, Harold Earl
Berdahl, Paul Hilland
Bernard, Douglas Alan
Berryman, James Garland
Bethune, Donald Stimson
Bharat, Ramasesha
Biedenbender, Michael David
Biegelsen, David K
Bienenstock, Arthur Irwin
Birecki, Henryk
Bliss, Erlan S
Blum, Fred A
Boster, Thomas Arthur
Bower, Frank H(ugo)
Bowman, Robert Clark, Jr
Bragg, Robert H(enry)
Bratt, Peter Raymond
Bridges, Frank G
Bron, Walter Ernest
Brown, Ronald Franklin
Bryden, John Heilner
Bube, Richard Howard
Bullis, William Murray
Burland, Donald Maxwell
Burmeister, Robert Alfred
Butler, Jack F
Calhoun, Bertram Allen
Cape, John Anthony
Carlan, Alan J
Caspers, Hubert Henri
Chakravarty, Sudip
Chase, Jay Benton
Chau, Cheuk-Kin
Chester, Marvin
Cheung, Jeffrey Tai-Kin
Childs, William Henry
Chin, Maw-Rong
Chodos, Steven Leslie
Chow, Paul C
Claassen, Richard Strong
Clark, William Gilbert
Clarke, John
Clemens, Bruce Montgomery
Cogan, Adrian Ilie
Cohen, Marvin Lou
Cohen, Richard Lewis
Coleman, Charles Clyde
Coleman, Lawrence Bruce
Colwell, Joseph F
Cooper, Donald Edward
Coufal, Hans-Jurgen
Crittenden, Eugene Casson, Jr
Curtis, Orlie Lindsey, Jr
Dalven, Richard
Dapkus, Paul Daniel
Davies, John Tudor
Davis, Jeffrey Arthur
De, Gopa Sarkar
Dember, Alexis Berthold
Deri, Robert Joseph
Deshpande, Shivajirao M
Detwiler, Daniel Paul
DeWames, Roger
Dixon, Paul King
Dobrov, Wadim (Ivan)
Doherty, Paul Michael
Domb, Ellen Ruth (Colmer)
Donovan, Terence M
Duffy, William Thomas, Jr
Duneer, Arthur Gustav, Jr
Dynes, Robert Carr
Eagleton, Robert Don
Eck, Robert Edwin
Eden, Richard Carl
Edwards, David Franklin
Emmett, John L
Eng, Genghmun
Engel, Jan Marcin
Everett, Glen Exner
Evtuhov, Viktor
Ewald, Arno Wilfred
Falicov, Leopoldo Maximo
Feng, Joseph Shao-Ying
Fenner, Wayne Robert
Ferber, Robert R
Fetter, Alexander Lees
Fetterman, Harold Ralph
Feuerstein, Seymour
Fink, Herman Joseph
Fisher, Robert Amos, Jr
Fletcher, Paul Chipman
Flinn, Paul Anthony
Foiles, Stephen Martin
Fong, Ching-Yao
Fouquet, Julie
Franks, Larry Allen
Frescura, Bert Louis
Fultz, Brent T
Gabriel, Cedric John
Gaines, Edward Everett
Gamble, Fred Ridley, Jr
Gathers, George Roger
Geballe, Theodore Henry
Gehman, Bruce Lawrence
Gershenzon, M(urray)
Gibson, Atholl Allen Vear
Gill, William D(elahaye)
Gilman, John Joseph
Gilvarry, John James
Glass, Nathaniel E
Gleckman, Philip Landon
Goddard, William Andrew, III
Goldberg, Ira Barry
Goldstein, Selma
Good, Roland Hamilton, Jr
Gossard, Arthur Charles
Goubau, Wolfgang M
Gould, Christopher M
Grant, Paul Michael
Grant, Ronald W(arren)
Green, Michael Philip
Greene, Alan Campbell
Gruber, John B
Gudmundsen, Richard Austin
Gundersen, Martin Adolph
Hackwood, Susan
Haegel, Nancy M
Hafemeister, David Walter
Hagenlocher, Arno Kurt
Hagstrom, Stig Bernt
Hahn, Erwin Louis
Haitz, Roland Hermann
Hall, James Timothy
Haller, Eugene Ernest
Hammond, Robert Bruce
Hanna, Stanley Sweet
Harper, Francis Edward
Harris, Charles Bonner
Harris, James Stewart, Jr
Harrison, Walter Ashley
Hartwick, Thomas Stanley
Heeger, Alan J
Hellman, Frances
Hemminger, John Charles
Hempstead, Robert Douglas
Herman, Ray Michael
Herring, William Conyers
Hill, Robert Matteson
Hiskes, Ronald
Holmstrom, Fred Edward
Holzrichter, John Frederick
Hone, Daniel W
Honnold, Vincent Richard
Hopfield, John Joseph
Housley, Robert Melvin
Hsieh, Yu-Nian
Huang, C Yuan
Huang, Di-Hui (David)
Hubert, Jay Marvin
Hunt, Arlon Jason
Hurrell, John Patrick
Hygh, Earl Hampton
Ikezi, Hiroyuki
Jaccarino, Vincent
Jacobson, Albert H(erman), Jr
Jansen, Frank
Jansen, Michael
Jeanloz, Raymond
Jeffries, Carson Dunning
Johanson, William Richard
Johnson, David Leroy
Johnson, Oliver
Johnson, Roy Ragnar
Johnson, William Lewis
Junga, Frank Arthur
Kagiwada, Reynold Shigeru
Kahn, Frederic Jay
Kalma, Arne Haerter
Kaplan, Daniel Eliot
Karunasiri, Gamani
Kay, Eric
Kedzie, Robert Walter
Kendrick, Hugh
Khan, Mahbub R
Khoshnevisan, Mohsen Monte
Kirchner, Ernst Karl
Kittel, Charles
Kittel, Peter
Klein, August S
Kleitman, David
Knight, Walter David
Knollman, Gilbert Carl
Koliwad, Krishna M
Kosai, Kenneth
Krikorian, Esther
Kroeker, Richard Mark
Kroemer, Herbert
Krupke, William F
Kumar, S
Kupfer, John Carlton
Kyser, David Sheldon
Lamb, Walter Robert
Langdon, Terence George
Lao, Binneg Yanbing
Larsen, Ted LeRoy
Lau, S S
Law, Hsiang-Yi David
Lax, Edward
Learn, Arthur Jay
Lee, Kenneth
Lehovec, Kurt
Leifer, Herbert Norman
Lerner, Lawrence S
Leung, Charles Cheung-Wan
Leventhal, Edwin Alfred

Solid State Physics (cont)

Leverton, Walter Frederick
Lewis, Robert Taber
Lind, Maurice David
Lindau, Evert Ingolf
Litvak, Marvin Mark
Liu, Jia-ming
Lockhart, James Marcus
Loda, Richard Thomas
Logan, James Columbus
Longo, Joseph Thomas
Lonky, Martin Leonard
Louie, Steven Gwon Sheng
Luehrmann, Arthur Willett, Jr
McCaldin, J(ames) O(eland)
McColl, Malcolm
Macfarlane, Roger Morton
MacLaughlin, Douglas Earl
McNutt, Michael John
Magnuson, Gustav Donald
Maker, Paul Donne
Maki, Kazumi
Maloney, Timothy James
Marrello, Vincent
Masters, Burton Joseph
Matare, Herbert Franz
Maxfield, Bruce Wright
Meike, Annemarie
Merz, James L
Meyer, Stephen Frederick
Miller, Allan Stephen
Mills, Douglas Leon
Minkiewicz, Vincent Joseph
Moerner, W(illiam) E(sco)
More, Richard Michael
Moriarty, John Alan
Morin, Francis Joseph
Moriwaki, Melvin M
Moss, Steven C
Muir, Arthur H, Jr
Musket, Ronald George
Naqvi, Iqbal Mehdi
Narayanamurti, Venkatesh
Nedoluha, Alfred K
Newsam, John M
Novotny, Vlad Joseph
Olness, Dolores Urquiza
Olson, Roy E
Onton, Aare
Orbach, Raymond Lee
Ozguz, Volkan Husnu
Page, D(errick) J(ohn)
Palmer, John Parker
Paquette, David George
Parker, William Henry
Parrish, William
Partanen, Jouni Pekka
Pearson, James Joseph
Pehl, Richard Henry
Perkins, Kenneth L(ee)
Persky, George
Peselnick, Louis
Petroff, Pierre Marc
Phillips, Thomas Gould
Phillips, William Baars
Philo, John Sterner
Pianetta, Piero Antonio
Picus, Gerald Sherman
Pines, Alexander
Pollak, Michael
Pontau, Arthur E
Portis, Alan Mark
Powell, Ronald Allan
Rabolt, John Francis
Radosevich, Lee George
Rasor, Ned S(haurer)
Redfield, David
Rehn, Victor Leonard
Reinheimer, Julian
Reiss, Howard
Renda, Francis Joseph
Reynolds, Richard Alan
Rice, Dennis Keith
Richards, Paul Linford
Richman, Isaac
Rockstad, Howard Kent
Rogers, Robert N
Rohy, David Alan
Rosenblum, Stephen Saul
Ross, Marvin Franklin
Rothrock, Larry R
Roukes, Michael L
Rowell, John Martin
Rudge, William Edwin
Rudnick, Joseph Alan
Saffren, Melvin Michael
Sager, Ronald E
Sashital, Sanat Ramanath
Satten, Robert A
Scalapino, Douglas J
Schaefer, Albert Russell
Schein, Lawrence Brian
Schmiedeshoff, George M
Schultz, Sheldon
Schwartz, Robert Nelson
Sclar, Nathan
Scott, John Campbell
Seidel, Thomas Edward
Sham, Lu Jeu
Shank, Charles Vernon
Shaw, Charles Alden
Shea, Michael Joseph
Shelton, Robert Neal
Shen, Yuen-Ron
Shepanski, John Francis
Sher, Arden
Shin, Ernest Eun-Ho
Shin, Soo H
Shing, Yuh-Han
Shore, Herbert Barry
Shotts, Wayne J
Shuh, David Kelly
Siegel, Irving
Siegel, Sidney
Silver, Arnold Herbert
Sinclair, Robert
Sladek, Ronald John
Slivinsky, Sandra Harriet
Smit, Jan
Smith, Joe Nelson, Jr
Snow, Edward Hunter
Snowden, Donald Philip
Somoano, Robert Bonner
Spicer, William Edward
Spieler, Helmuth
Spindt, Charles A (Capp)
Spitzer, William George
Sridhar, Champa Guha
Srour, Joseph Ralph
Stahl, Frieda A
Stapelbroek, Maryn G
Stella, Paul M
Stettner, Roger
Stevenson, David John
Stierwalt, Donald L
Stirn, Richard J
Stohr, Joachim
Stokowski, Stanley E
Stolte, Charles
Stupian, Gary Wendell
Suits, James Carr
Surko, Clifford Michael
Tabak, Mark David
Talley, Robert Morrell
Taylor, William H, II
Tennant, William Emerson
Terhune, Robert William
Thompson, Richard Scott
Thompson, Thomas Eaton
Thorsen, Arthur C
Torrance, Jerry Badgley, Jr
Tsai, Chuang Chuang
Tsai, Ming-Jong
Tseng, Samuel Chin-Chong
Tu, Charles Wuching
Vagelatos, Nicholas
Van de Walle, Chris G
Van Thiel, Mathias
Violet, Charles Earl
Vorhaus, James Louis
Wagner, Richard John
Walden, Robert Henry
Walker, William Charles
Wallis, Richard Fisher
Wang, Kang-Lung
Warburton, William Kurtz
Weber, Eicke Richard
Weber, Hans Josef
Weber, Marvin John
Webster, Emilia
Weichman, Peter Bernard
Weinberg, Elliot Hillel
Whaley, Katharine Birgitta
White, Ray Henry
Whitney, William Merrill
Wickersheim, Kenneth Alan
Wieder, Harold
Wieder, Harry H
Wild, Robert Lee
Wilkenfeld, Jason Michael
Will, Theodore A
Williams, Richard Stanley
Wilson, Barbara Ann
Wilson, William Dennis
Wittry, David Beryle
Woerner, Robert Leo
Wolf, Robert Peter
Wong, Joe
Wood, Charles
Woodruff, Truman Owen
Wooten, Frederick (Oliver)
Wu, Lei
Yablonovitch, Eli
Yamakawa, Kazuo Alan
Yang, Chung Ching
Yeh, Nai-Chang
Yessik, Michael John
Yu, Karl Ka-Chung
Yu, Peter Yound
Zeidler, James Robert
Zernow, Louis
Zucca, Ricardo
Zucker, Joseph
Zuleeg, Rainer
Zwerdling, Solomon

COLORADO

Adams, Richard Owen
Ahrenkiel, Richard K
Baldwin, Thomas O
Beale, Paul Drew
Benz, Samuel P
Blakeslee, A Eugene
Brock, George William
Bruns, Donald Gene
Carpenter, Steve Haycock
Ciszek, Ted F
Coffey, Mark William
DeGrand, Thomas Alan
Ekin, Jack W
Emery, Keith Allen
Fickett, Frederick Roland
Froyen, Sverre
Furtak, Thomas Elton
Galeener, Frank Lee
Gardner, Edward Eugene
Geller, Seymour
Goldfarb, Ronald B
Harris, Richard Elgin
Hermann, Allen Max
Kanda, Motohisa
Kazmerski, Lawrence L
Kremer, Russell Eugene
Lebiedzik, Jozef
Lefever, Robert Allen
Love, William F
McMahon, Thomas Joseph
Moddel, Garret R
Nahman, Norris S(tanley)
Nozik, Arthur Jack
O'Callaghan, Michael James
Pankove, Jacques I
Parson, Robert Paul
Patton, Carl E
Pitts, John Roland
Powell, Robert Lee
Price, John C
Quick, Nathaniel Richard
Raich, John Carl
Read, David Thomas
Rutz, Richard Frederick
Scales, John Alan
Schowengerdt, Franklin Dean
Scott, James Floyd
Siegwarth, James David
Sites, James Russell
Sliker, Todd Richard
Smith, David Reeder
Staebler, David Lloyd
Sullivan, Donald Barrett
Trefny, John Ulric
Westfall, Richard Merrill
Winder, Dale Richard
Zhang, Bing-Rong

CONNECTICUT

Akkapeddi, Prasad Rao
Bartram, Ralph Herbert
Best, Philip Ernest
Bursuker, Isia
Chang, Richard Kounai
Clapp, Philip Charles
Cordery, Robert Arthur
Dolan, James F
Gaertner, Wolfgang Wilhelm
Gaidis, Michael Christopher
Galligan, James M
Gardner, Fred Marvin
Haacke, Gottfried
Haag, Robert Edwin
Henrich, Victor E
Kappers, Lawrence Allen
Keith, H(arvey) Douglas
Kessel, Quentin Cattell
Klemens, Paul Gustav
Klokholm, Erik
Kraft, David Werner
Leonberger, Frederick John
Ma, Tso-Ping
Mandelbrot, Benoit B
Marcus, Jules Alexander
Markowitz, David
Marvin, Philip Roger
Melcher, Charles L
Morehead, Frederick Ferguson, Jr
Otter, Fred August
Peterson, Cynthia Wyeth
Peterson, Gerald A
Prober, Daniel Ethan
Rau, Richard Raymond
Read, Nicholas
Reed, Mark Arthur
Reed, Robert Willard
Reinberg, Alan R
Rollefson, Robert John
Sachdev, Subir
Schor, Robert
Shalvoy, Richard Barry
Shuskus, Alexander J
Siegmund, Walter Paul
Solomon, Peter R
Stein, Richard Jay
Stone, A Douglas
Taylor, Geoff W
Tracy, Joseph Charles
Wolf, Werner Paul

DELAWARE

Barteau, Mark Alan
Bierlein, John David
Bindloss, William
Boer, Karl Wolfgang
Chase, David Bruce
Christie, Phillip
Cole, G(eorge) Rolland
Cooper, Charles Burleigh
Crawford, Michael Karl
Daniels, William Burton
Dasgupta, Sunil Priya
Flippen, Richard Bernard
Hall, Robert B
Hegedus, Steven Scott
Hunsperger, Robert G(eorge)
Huppe, Francis Frowin
Ih, Charles Chung-Sen
Jarrett, Howard Starke, Jr
Johnson, Robert Chandler
McNeely, James Braden
Murray, Richard Bennett
Nielsen, Paul Herron
Onn, David Goodwin
Pontrelli, Gene J
Rogers, Donald B
Sharnoff, Mark
Silzars, Aris
Stiles, A(lvin) B(arber)
Swann, Charles Paul
Zumsteg, Fredrick C, Jr

DISTRICT OF COLUMBIA

Abraham, George
Anderson, Gordon Wood
Andreadis, Tim Dimitri
Batra, Narendra K
Bermudez, Victor Manuel
Borsuk, Gerald M
Campbell, Arthur B
Carlos, William Edward
Carter, Gesina C
Chung, David Yih
Civiak, Robert L
Crouch, Roger Keith
Davis, John Litchfield
Davis, Lance A(lan)
Della Torre, Edward
Dragoo, Alan Lewis
Edelstein, Alan Shane
Ehrlich, Alexander Charles
Feldman, Charles
Feldman, Joseph Louis
Ferguson, George Alonzo
Finger, Larry W
Forester, Donald Wayne
Francavilla, Thomas L(ee)
Friebele, Edward Joseph
Fusina, Robert Alan
Gaumond, Charles Frank
Goff, James Franklin
Gottschall, Robert James
Griscom, David Lawrence
Gubser, Donald Urban
Haas, George Arthur
Henry, Warren Elliott
Hobbs, Herman Hedberg
Jansen, Robert Werner
Kabler, Milton Norris
Kaplan, Raphael
Kertesz, Miklos
Kitchens, Thomas Adren
Krebs, James J(ohn)
Leibowitz, Jack Richard
Lin-Chung, Pay-June
McKay, Jack Alexander
Mao, Ho-Kwang
Marquardt, Charles L(awrence)
Marrone, Michael Joseph
Meijer, Paul Herman Ernst
Meyer, Ralph O
Michel, David John
Milton, Albert Fenner
Moon, Deug Woon
Ngai, Kia Ling
Nisenoff, Martin
Oosterhuis, William Tenley
Pande, Chandra Shekhar
Papaconstantopoulos, Dimitrios A
Parker, Robert Louis
Prinz, Gary A
Ramaker, David Ellis
Reinecke, Thomas Leonard
Rittner, Edmund Sidney
Roitman, Peter
Schneider, Irwin
Segnan, Romeo A
Senftle, Frank Edward
Serene, Joseph William
Shapero, Donald Campbell
Skelton, Earl Franklin
Sokoloski, Martin Michael
Thompson, Phillip Eugene
Tsang, James Chen-Hsiang
Tsang, Tung
Vold, Carl Leroy
Webb, Alan Wendell
Weinberg, Donald Lewis
Weinstock, Harold
White, Robert Marshall
Wieting, Terence James
Willis, James Stewart, Jr
Wilsey, Neal David
Wittmann, Horst Richard
Wolf, Stuart Alan

FLORIDA

Adams, Earnest Dwight
Andrew, Edward Raymond
Aubel, Joseph Lee
Bass, Michael
Birkhoff, Robert D
Brooker, Hampton Ralph
Browder, James Steve
Burdick, Glenn Arthur
Burns, Jay, III
Burns, Michael J
Dacey, George Clement

Dagotto, Elbio Ruben
DeLoach, Bernard Collins, Jr
Edwards, Palmer Lowell
Einspruch, Norman G(erald)
Faulkner, John Samuel
Foner, Samuel Newton
Fossum, Jerry G
Halder, Narayan Chandra
Halpern, Leopold (Ernst)
Hascicek, Yusuf Suat
Huebner, Jay Stanley
Hummel, Rolf Erich
Kumar, Pradeep
Lade, Robert Walter
Leblanc, Roger M
Leverenz, Humboldt Walter
Lind, David Melvin
Linder, Ernest G
Lindholm, Fredrik Arthur
Linz, Arthur
Ludwig, Matthias Heinz
McAlpine, Kenneth Donald
Malocha, Donald C
Manousakis, Efstratios
Melich, Michael Edward
Monkhorst, Hendrik J
Moulton, Grace Charbonnet
Moulton, William G
Mrozowski, Stanislaw W
Mukherjee, Pritish
Murphy, Joseph
Newman, Roger
Novotny, Mark A
Painter, Ronald Dean
Patterson, James Deane
Pearton, Stephen John
Pepinsky, Raymond
Reynolds, George William, Jr
Rhodes, Richard Ayer, II
Rikvold, Per Arne
Riley, Mark Anthony
Schlottmann, Pedro U J
Schulte, Alfons F
Shih, Hansen S T
Simmons, Joseph Habib
Skofronick, James Gust
Sproull, Robert Lamb
Stegeman, George I
Sullivan, Neil Samuel
Swartz, George Allan
Testardi, Louis Richard
Trickey, Samuel Baldwin
Van Sciver, Steven W
Vassamillet, Lawrence Francois
Von Molnr, Stephan
Wade, Thomas Edward
Wang, Yung-Li
Weber, Alfred Herman
Wille, Luc Theo
Wu, Jin Zhong
Yon, E(ugene) T
Zalesak, Joseph Francis
Ziernicki, Robert S

GEORGIA
Anderson, Robert Lester
Bacon, Roger
Bishop, Thomas Parker
Blue, Marts Donald
Chou, Mei-Yin
Copeland, John Alexander
De Mayo, Benjamin
DuVarney, Raymond Charles
Family, Fereydoon
Gaylord, Thomas Keith
Griffiths, James Edward
Hartman, Nile Franklin
Hathcox, Kyle Lee
Hetzler, Morris Clifford, Jr
Higgins, Richard J
Hsu, Frank Hsiao-Hua
Jokerst, Nan Marie
Korda, Edward J(ohn)
Landau, David Paul
Landman, Uzi
Long, Thomas Ross
McMillan, Robert Walker
Meltzer, Richard S
Pandey, Surendra Nath
Perkowitz, Sidney
Peterson, Keith L
Pollard, William Blake
Powell, Bobby Earl
Puri, Om Parkash
Rives, John Edgar
Rodrigue, George Pierre
Shand, Julian Bonham, Jr
Stanford, Augustus Lamar, Jr
Stevenson, James Rufus
Stombler, Milton Philip
Vail, Charles R(owe)
Yen, William Maoshung
Young, Robert Alan
Zangwill, Andrew

HAWAII
Gaines, James R
Pong, William

IDAHO
Buescher, Brent J
Iyer, Ravi
Kearney, Robert James
Sieckmann, Everett Frederick

Spencer, Paul Roger
Switendick, Alfred Carl

ILLINOIS
Aldred, Anthony T
Ali, Naushad
Anderson, Ansel Cochran
Askill, John
Atoji, Masao
Auvil, Paul R, Jr
Averill, Frank Wallace
Bader, Samuel David
Barnett, Scott A
Bishop, Stephen Gray
Bowen, Samuel Philip
Breig, Marvin L
Brodsky, Merwyn Berkley
Brown, Bruce Stilwell
Buck, Warren Louis
Bukrey, Richard Robert
Burnham, Robert Danner
Campbell, David Kelly
Carpenter, John Marland
Chan, Sai-Kit
Chang, Yia-Chung
Chen, Haydn H D
Chiang, Tai-Chang
Chung, Yip-Wah
Connor, Donald W
Crabtree, George William
Day, Michael Hardy
Dlott, Dana D
Dolan, Paul Joseph, Jr
Druger, Stephen David
Dunlap, Bobby David
Dutta, Pulak
Eades, John Alwyn
Eden, James Gary
Ehrlich, Gert
Ellis, Donald Edwin
English, Floyd L
Felcher, Gian Piero
Flynn, Colin Peter
Fradin, Frank Yale
Frank, Robert Carl
Freeman, Arthur Jay
Fritzsche, Hellmut
Gallagher, David Alden
Garg, Anupam
Garland, James W
Gibson, J Murray
Gilbert, Robert L
Girolami, Gregory Scott
Goodman, Gordon Louis
Granato, Andrew Vincent
Greene, Laura H
Guttman, Lester
Hess, Karl
Hessler, Jan Paul
Hill, Christopher T
Hines, Roderick Ludlow
Hinks, David George
Holonyak, N(ick), Jr
Hurych, Zdenek
Jesse, Kenneth Edward
Jorgensen, James D
Kallend, John Scott
Kannewurf, Carl Raeside
Karim, Khondkar Rezaul
Kimball, Clyde William
Kitchen, W J
Klemm, Richard Andrew
Koelling, Dale Dean
Kouvel, James Spyros
Krauss, Alan Robert
Kuchnir, Moyses
Lam, Nghi Quoc
Lange, Yvonne
Lazarus, David
Levin, Kathryn J
Levi-Setti, Riccardo
Liu, Liu
Liu, Yung Yuan
Macrander, Albert Tiemen
Mapother, Dillon Edward
Marks, Laurence D
Martin, Richard McFadden
Merkelo, Henri
Miller, Robert Carl
Mochel, Jack McKinney
Moncton, David Eugene
Mueller, Melvin H(enry)
Nagel, Sidney Robert
Naylor, David L
Noble, Gordon Albert
Noonan, John Robert
Olvera-dela Cruz, Monica
Parker, John Calvin
Preston, Richard Swain
Price, David Cecil Long
Primak, William L
Reft, Chester Stanley
Ren, Shang-Fen
Ring, James George
Rothman, Steven J
Rowland, Theodore Justin
Salamon, Myron B
Schlossman, Mark Loren
Schreiber, David Seyfarth
Schroeer, Juergen Max
Schultz, Arthur Jay
Sharma, Ram Ratan
Shenoy, Gopal K
Shepard, Kenneth Wayne

Sibener, Steven Jay
Siegel, Richard W(hite)
Sill, Larry R
Simmons, Ralph Oliver
Sinha, Shome Nath
Skov, Charles E
Slichter, Charles Pence
Smither, Robert Karl
Spector, Harold Norman
Stapleton, Harvey James
Stillman, Gregory Eugene
Stout, John Willard
Sturm, William James
Tuzzolino, Anthony J
Udler, Dmitry
Van Ostenburg, Donald Ora
Veal, Boyd William, Jr
Weertman, Julia Randall
Weissman, Michael Benjamin
Welsh, Lawrence B
Wey, Albert Chin-Tang
Wiedersich, H(artmut)
Wilson, Robert Steven
Windhorn, Thomas H
Wolf, Dieter
Wolfe, James Phillip
Wolfram, Thomas
Zitter, Robert Nathan

INDIANA
Ascarelli, Gianni
Bessey, William Higgins
Bray, Ralph
Bunch, Robert Maxwell
Bunker, Bruce Alan
Buschert, Robert Cecil
Colella, Roberto
Compton, Walter Dale
Croat, John Joseph
De Young, Donald Bouwman
Fan, Hsu Yun
Furdyna, Jacek K
Giordano, Nicholas J
Girvin, Steven M
Harland, Glen Eugene, Jr
Helrich, Carl Sanfrid, Jr
Honig, Jurgen Michael
Kaplan, Jerome I
Kesmodel, Larry Lee
Kissinger, Paul Bertram
Lundstrom, Mark Steven
MacDonald, Allan Hugh
MacKay, John Warwick
Melloch, Michael Raymond
Mullen, James G
Pearlman, Norman
Pierret, Robert Francis
Prohofsky, Earl William
Ramdas, Anant Krishna
Robinson, John Murrell
Rodriguez, Sergio
Skadron, Peter
Tomasch, Walter J
Tripathi, Gorakh Nath Ram
Van Zandt, Lonnie L
Western, Arthur Boyd
White, Samuel Grandford, Jr
Zimmerman, Walter Bruce

IOWA
Anderson, R(obert) M(orris), Jr
Barnaal, Dennis E
Behroozi, Feredoon
Bevolo, Albert Joseph
Clem, John R
Dalal, Vikram
Finnemore, Douglas K
Fishman, Louis
Fuchs, Ronald
Goertz, Christoph Klaus
Gschneidner, Karl A(lbert), Jr
Harmon, Bruce Norman
Hsu, David Kuei-Yu
Johnston, David Carl
Luban, Marshall
Lunde, Barbara Kegerreis, (BK)
Lynch, David William
McClelland, John Frederick
Morris, Robert Gemmill
Papadakis, Emmanuel Philippos
Schaefer, Joseph Albert
Schweitzer, John William
Snow, Joel A
Swenson, Clayton Albert
Thompson, Donald Oscar
Tringides, Michael C

KANSAS
Brothers, Alfred Douglas
Culvahouse, Jack Wayne
Dragsdorf, Russell Dean
Friauf, Robert J
Grosskreutz, Joseph Charles
Ho, James Chien Ming
Long, Larry L
Rahman, Talat Shahnaz
Sapp, Richard Cassell
Thomas, James E

KENTUCKY
Bordoloi, Kiron
Brill, Joseph Warren
Buckman, William Gordon
Connolly, John William Domville

PHYSICS & ASTRONOMY / 551

Davis, Thomas Haydn
DeLong, Lance Eric
Fox, Mary Eleanor
France, Peter William
Huang, Wei-Feng
Reucroft, Philip J
Straley, Joseph Paul
Subbaswamy, Kumble R(amarao)

LOUISIANA
Azzam, Rasheed M A
Bogle, Tommy Earl
Dahlquist, Wilbur Lynn
Deck, Ronald Joseph
Goodrich, Roy Gordon
Grenier, Claude Georges
Hoy, Robert C
Kurtz, Richard Leigh
Lafleur, Louis Dwynn
Marshak, Alan Howard
Naidu, Seetala V
O'Connell, Robert F
Perdew, John Paul
Piller, Herbert
Stockbauer, Roger Lewis
Vashishta, Priya Darshan
Veith, Daniel A
Watkins, Steven F
Zebouni, Nadim H

MAINE
Barnes, G Richard
Camp, Paul R
Grimes, Charles Cullen
McKay, Susan Richards
Nicol, James
Pribram, John Karl

MARYLAND
Abbundi, Raymond Joseph
Ahearn, John Stephen
Allgaier, Robert Stephen
Alperin, Harvey Albert
Bailey, Glenn Charles
Bardasis, Angelo
Bargeron, Cecil Brent
Berg, Norman J
Beri, Avinash Chandra
Bis, Richard F
Blessing, Gerald Vincent
Blue, James Lawrence
Blum, Norman Allen
Bohandy, Joseph
Boone, Bradley Gilbert
Broadhurst, Martin Gilbert
Brodsky, Marc H
Brodsky, Marc Herbert
Bryden, Wayne A
Buchner, Stephen Peter
Burke, Edward Raymond
Byer, Norman Ellis
Cacciamani, Eugene Richard, Jr
Callen, Earl Robert
Catchings, Robert Merritt, III
Cavanagh, Richard Roy
Celotta, Robert James
Charles, Harry Krewson, Jr
Chen, Wenpeng
Chern, Engmin James
Chesser, Nancy Jean
Chiang, Chwan K
Chien, Chia-Ling
Chimenti, Dale Everett
Clark, Alan Fred
Clark, Arthur Edward
Clark, Bill Pat
Cohen, Julius
Cohen, Marvin Morris
Colwell, Jack Harold
Conrad, Edward Ezra
Corak, William Sydney
Crane, Langdon Teachout, Jr
Cullen, James Robert
Davis, Guy Donald
Davisson, Charlotte Meaker
De Planque, Gail
DeWit, Roland
DiPirro, Michael James
Drew, Howard Dennis
Egelhoff, William Frederick, Jr
Embury, Janon Frederick, Jr
Feldman, Albert
Fontanella, John Joseph
Forbes, Jerry Wayne
Forman, Richard Allan
Franz, Judy R
Frazer, Benjamin Chalmers
Frederick, William George DeMott
Frederikse, Hans Pieter Roetert
Frey, Jeffrey
Gammon, Robert Winston
Garstens, Martin Aaron
Garver, Robert Vernon
Ghoshtagore, Rathindra Nath
Gleason, Thomas James
Glick, Arnold J
Glover, Rolfe Eldridge, III
Goldin, Edwin
Green, Robert E(dward), Jr
Greene, Richard Lorentz
Houston, Bland Bryan, Jr
Huang, Jacob Wen-Kuang
Hudson, Ralph P
Jablonski, Daniel Gary

Solid State Physics (cont)

Jette, Archelle Norman
Johnson, William Bowie
Joseph, Richard Isaac
Kaiser, Quentin C
Karulkar, Pramod C
Kirkendall, Thomas Dodge
Kirkpatrick, Theodore Ross
Klick, Clifford C
Kopanski, Joseph J
Kroeger, Richard Alan
Kumar, K Sharvan
Kuriyama, Masao
Langenberg, Donald Newton
Larrabee, R(obert) D(ean)
Lessoff, Howard
Lindmayer, Joseph
Lowney, Jeremiah Ralph
Lynn, Jeffrey Whidden
McCally, Russell Lee
McClelland, Jabez Jenkins
MacDonald, Rosemary A
McLean, Flynn Barry
Magno, Richard
Majkrzak, Charles Francis
Malmberg, Philip Ray
Mangum, Billy Wilson
Manley, Oscar Paul
Manning, Irwin
Manning, John Randolph
Maycock, John Norman
Melngailis, John
Metze, George M
Meyer, Eric Stefan
Minkowski, Jan Michael
Montalvo, Ramiro A
Moorjani, Kishin
Munasinghe, Mohan P
Murphy, John Cornelius
Neumann, Dan Alan
Oettinger, Frank Frederic
Palik, Edward Daniel
Pande, Krishna P
Pierce, Daniel Thornton
Poehler, Theodore O
Potyraj, Paul Anthony
Powell, Cedric John
Prask, Henry Joseph
Prince, Edward
Rabin, Herbert
Randolph, Lynwood Parker
Rasera, Robert Louis
Rector, Charles Willson
Reno, Robert Charles
Restorff, James Brian
Revesz, Akos George
Robbins, Mark Owen
Rosenstock, Herbert Bernhard
Rowe, John Michael
Ruffa, Anthony Richard
Rush, John Joseph
Scharnhorst, Kurt Peter
Schneider, Carl Stanley
Schulman, James Herbert
Schumacher, Clifford Rodney
Schwartz, Brian B
Seiler, David George
Sharma, Jagadish
Shaw, Robert William, Jr
Silverton, James Vincent
Simonis, George Jerome
Skrabek, Emanuel Andrew
Steiner, Bruce
Strenzwilk, Denis Frank
Summers, Geoffrey P
Sweeting, Linda Marie
Swenberg, Charles Edward
Taylor, Barry Norman
Thomson, Robb M(ilton)
Thurber, Willis Robert
Trevino, Samuel Francisco
Van Antwerp, Walter Robert
Vande Kieft, Lawrence John
Van Vechten, Deborah
Venables, John Duxbury
Waldo, George Van Pelt, Jr
Walker, J Calvin
Wang, Frederick E
Warner, Brent A
Warren, Wayne Hutchinson, Jr
Wasilik, John H(uber)
Waterstrat, Richard Milton
Webb, Richard A
Williams, Conrad Malcolm
Williams, Ellen D
Wittels, Mark C
Yakovenko, Victor Mikhailovich
Yedinak, Peter Demerton

MASSACHUSETTS
Addiss, Richard Robert, Jr
Adlerstein, Michael Gene
Aggarwal, Roshan Lal
Aldrich, Ralph Edward
Alexander, Michael Norman
Allen, Lisa Parechanian
Antal, John Joseph
Antonoff, Marvin M
Baird, Donald Heston
Bansil, Arun
Bell, Richard Oman
Berera, Geetha Poonacha
Berker, Ahmet Nihat
Birgeneau, Robert Joseph
Bloom, Jerome H(ershel)
Blum, John Bennett
Buss, Dennis Darcy
Button-Shafer, Janice
Carr, Paul Henry
Caslavsky, Jaroslav Ladislav
Chase, Charles Elroy, Jr
Chen, Joseph H
Clarke, Edward Nielsen
Clifton, Brian John
Collins, Aliki Karipidou
Cronin-Golomb, Mark
Davis, Charles Freeman, Jr
Davis, Luther, Jr
Di Bartolo, Baldassare
Dionne, Gerald Francis
Dorschner, Terry Anthony
Dresselhaus, Gene Frederick
Dresselhaus, Mildred S
Eby, John Edson
Ekman, Carl Frederick W
Fan, John C C
Feist, Wolfgang Martin
Fleming, Phyllis Jane
Foner, Simon
Free, John Ulric
Friedman, Lionel Robert
Gangulee, Amitava
Garth, John Campbell
Gianino, Peter Dominic
Gold, Albert
Gordon, Roy Gerald
Gould, Harvey A
Gregory, Bob Lee
Greytak, Thomas John
Gross, David John
Groves, Steven H
Guentert, Otto Johann
Guertin, Robert Powell
Gustafson, John C
Hallock, Robert B
Halperin, Bertrand Israel
Handelman, Eileen T
Harman, T(heodore) C(arter)
Harte, Kenneth J
Hartke, Jerome L
Haugsjaa, Paul O
Heiman, Donald Eugene
Hempstead, Charles Francis
Holland, Andrew Brian
Hope, Lawrence Latimer
Hsu, Sheng Teng
Jacobs, Laurence Alan
Jasperson, Stephen Newell
Jennings, Laurence Duane
Jirmanus, Munir N
Joannopoulos, John Dimitris
Johannessen, Paul Romberg
Jones, Richard Victor
Kalman, Gabor J
Karakashian, Aram Simon
Kaseta, Francis William
Kastner, Marc Aaron
Keil, Thomas H
Kelley, Paul Leon
Kennedy, Edward Francis
Khattak, Chandra Prakash
Klein, Michael W
Kleis, John Dieffenbach
Kolodziejski, Leslie Ann
Kumar, Kaplesh
Kupferberg, Lenn C
Langley, Kenneth Hall
Lanyon, Hubert Peter David
Lauer, Robert B
Lax, Benjamin
Lazay, Paul Duane
Lee, Patrick A
Lee, Roland Robert
Liau, Zong-Long
Lin, Alice Lee Lan
Linden, Kurt Joseph
Lipson, Herbert George
Lis, Steven Andrew
Litster, James David
Liu, Xichun
Livingston, James Duane
Loeb, Arthur Lee
Lu, Grant
Lyons, William Gregory
McColl, James Renfrew
Machta, Jonathan Lee
McNiff, Edward J, Jr
McWhorter, Alan L
Mahon, Harold P
Maloney, William Thomas
Malozemoff, Alexis P
Mandl, Alexander Ernst
Markiewicz, Robert Stephen
Marrero, Hector G
Martin, Frederick Wight
Martin, Paul Cecil
Masi, James Vincent
Mavretic, Anton
Maxwell, Emanuel
Mayburg, Sumner
Melngailis, Ivars
Meservey, Robert H
Meyer, Robert Bruce
Miller, Robert Charles
Minden, Henry Thomas
Miniscalco, William J
Montazer, G Hosein
Moodera, Jagadeesh Subbaiah
Moore, Thomas Warner
Moulton, Peter Franklin
Moustakas, Theodore D
Mozzi, Robert Lewis
Muthukumar, Murugappan
Nelson, Donald Frederick
Nelson, Keith A
Neuringer, Leo J
Oberteuffer, John Amiard
O'Connell, Richard John
Orlando, Terry Philip
Paul, William
Penchina, Claude Michel
Perry, Clive Howe
Pershan, Peter Silas
Phillips, Philip Wirth
Pierce, Camden Ballard
Praddaude, Hernan Camilo
Ram-Mohan, L Ramdas
Rana, Ram S
Rao, Devulapalli V G L N
Read, Philip Lloyd
Rediker, Robert Harmon
Rork, Eugene Wallace
Roth, Harold
Sagalyn, Paul Leon
Sage, Jay Peter
Sample, Howard H
Schwensfeir, Robert James, Jr
Seavey, Marden Homer, Jr
Shapiro, Edward K Edik
Shaughnessy, Thomas Patrick
Shull, Clifford Glenwood
Sichel, Enid Keil
Silvera, Isaac F
Simmons, Elizabeth H
Singh, Shobha
Smith, Arthur Clarke
Sneddon, Leigh
Sokoloff, Jeffrey Bruce
Soref, Richard Allan
Spaepen, Frans
Spears, David Lewis
Speliotis, Dennis Elias
Stanley, H(arry) Eugene
Stern, Ernest
Stevenson, Donald Thomas
Strandberg, Malcom Woodrow Pershing
Strauss, Alan Jay
Sun, Shuwei
Tancrell, Roger Henry
Tannenwald, Peter Ernest
Thun, Rudolf Eduard
Tinkham, Michael
Touger, Jerold Steven
Tripathy, Sukant K
Tsang, Dean Zensh
Tsaur, Bor-Yeu
Tustison, Randal Wayne
Ukleja, Paul Leonard Matthew
Unlu, M Selim
Vetterling, William Thomas
Wang, Christine A
Warde, Cardinal
Watt, James Peter
Weinberger, Doreen Anne
Weiss, Jerald Aubrey
Williamson, Richard Cardinal
Wuensch, Bernhardt J(ohn)
Yngvesson, K Sigfrid
Zimmerman, George Ogurek

MICHIGAN
Albers, Walter Anthony, Jr
Allen, James Ward
Bass, Jack
Beglau, David Alan
Bernius, Mark Thomas
Bhattacharya, Pallab Kumar
Bleil, Carl Edward
Bohm, Henry Victor
Brailsford, Alan David
Carlson, Edward H
Chang, Jhy-Jiun
Clarke, Roy
Crary, Selden Bronson
Current, David Harlan
Davis, Lloyd Craig
Devlin, John F
Dunifer, Gerald Leroy
Eagen, Charles Frederick
Evans, B J
Fisher, Galen Bruce
Frank, Max
Fuerst, Carlton Dwight
Gaerttner, Martin R
Golding, Brage, Jr
Graham, George Wade
Harrison, Michael Jay
Hartmann, William Morris
Hassold, Gregory Nahmen
Herbst, Jan Francis
Heremans, Joseph P
Hill, John Campbell
Hohnke, Dieter Karl
Jaklevic, Robert C
Johnson, Wayne Douglas
Johnson, Wayne Jon
Kaplan, Thomas Abraham
Katz, Ernst
Keem, John Edward
Kim, Changhyun
Kim, Yeong Wook
Kulkarni, Anand K
Kunz, Albert Barry
Kushida, Toshimoto
Logothetis, Eleftherios Miltiadis
McCarthy, Shaun Leaf
Marshall, Samson A
Meitzler, Allen Henry
Merlin, Roberto Daniel
Michel, Richard Edwin
Morgan, Caroline G
Mozurkewich, George
Mukherjee, Kalinath
Nazri, Gholam-Abbas
O'Donnell, Matthew
Ovshinsky, Iris M(iroy)
Ovshinsky, Stanford R(obert)
Pettersen, Howard Eugene
Pollack, Gerald Leslie
Potts, John Earl
Pryor, Roger Welton
Rand, Stephen Colby
Rimai, Lajos
Roessler, David Martyn
Rose, Frank Edward
Sander, Leonard Michael
Savit, Robert Steven
Schroeder, Peter A
Scott, Joseph Hurlong
Shaw, M(elvin) P
Sickafus, Edward N
Slavin, Andrei Nickolay
Slocum, Robert Richard
Smith, George Wolfram
Smith, John Robert
Sokolowski, Danny Hale
Srolovitz, David Joseph
Stewart, Melbourne George
Suits, Bryan Halyburton
Swets, Don Eugene
Tepley, Norman
Thomas, Robert L
Thorpe, Michael Fielding
Toth, Robert S
Vaishnava, Prem P
Van Till, Howard Jay
Van Zytveld, John Bos
Venkateswaran, Uma D
Waber, James Thomas
Wallace, William Donald
Weber, Willes Henry
Weidman, Robert Stuart
Weinreich, Gabriel
Wenger, Lowell Edward
Yee, Albert Fan
Zanini-Fisher, Margherita

MINNESOTA
Agarwal, Vijendra Kumar
Anderson, Gerald S
Berg, James Irving
Broshar, Wayne Cecil
Burkstrand, James Michael
Campbell, Charles Edwin
Chelikowsky, James Robert
Dahlberg, Earl Dan
Debe, Mark K
DiBona, Peter James
Eckroth, Charles Angelo
Ensign, Thomas Charles
Fuller, Richard M
Gallo, Charles Francis
George, Peter Kurt
Goldman, Allen Marshall
Hamm, Franklin Albert
Hartmann, Judith Brown
Katz, William
Keffer, Charles Joseph
Keister, Jamieson Charles
Korth, Michael Steven
Kotz, Arthur Rudolph
Kruse, Paul Walters, Jr
Lutes, Olin S
McClure, Benjamin Thompson
Meyer, Frank Henry
Mikkelson, Raymond Charles
Mitchell, William Cobbey
Morgan, Donald R
Mularie, William Mack
Nelson, Kyler Fischer
Noer, Richard Juul
Nussbaum, Allen
Schmit, Joseph Lawrence
Schuster, Sanford Lee
Smith, David Philip
Sydor, Michael
Thorland, Rodney Harold
Titus, William James
Tran, Nang Tri
Tsai, Ching-Long
Tufte, Obert Norman
Valasek, Joseph
Valls, Oriol Tomas
Walker, Charles Thomas
Weaver, John Herbert
Werner, Thomas R
Weyhmann, Walter Victor

MISSISSIPPI
Breazeale, Mack Alfred
Hughes, William E
Jaumot, Frank Edward, Jr
Lewis, Arthur B
Miller, David Burke
Pandey, Ras Bihari

Trehan, Rajender

MISSOURI
Adawi, Ibrahim (Hasan)
Alexander, Ralph William, Jr
Bell, Marvin Drake
Bell, Robert John
Brungs, Robert Anthony
Bryan, David A
Bryant, Paul James
Burgess, James Harland
Carlsson, Anders Einar
Ching, Wai-Yim
Conradi, Mark Stephen
Cowan, David Lawrence
Farmer, John William
Fedders, Peter Alan
Feldman, Bernard Joseph
Fisch, Ronald
Gerson, Robert
Gibbons, Patrick C(handler)
Greenlief, Charles Michael
Hagen, Donald E
Hale, Edward Boyd
Hensley, Eugene Benjamin
Hight, Robert
Hill, Dale Eugene
Hill, John Joseph
Holroyd, Louis Vincent
Indeck, Ronald S
Jacobsmeyer, Vincent Paul
Lamelas, Francisco Javier
Leopold, Daniel J
Long, Gary John
Martin, Frank Elbert
Meese, Jon Michael
Menne, Thomas Joseph
Muller, Marcel Wettstein
Northrip, John Willard
Phillips, Russell Allan
Rhyne, James Jennings
Rode, Daniel Leon
Ross, Frederick Keith
Rutledge, Harley Dean
Schupp, Guy
Shaw, Roger Walz
Sparlin, Don Merle
Sundfors, Ronald Kent
Switzer, Jay Alan
Taub, Haskell Joseph
Wagner, Robert G
Werner, Samuel Alfred
Wu, Cheng-Hsiao
Yelon, William B

MONTANA
Drumheller, John Earl
Hermanson, John Carl
Hess, Lindsay LaRoy
Lapeyre, Gerald J
Schmidt, Victor Hugo
Smith, Richard James

NEBRASKA
Billesbach, David P
Dowben, Peter Arnold
Ducharme, Stephen
Eckhardt, Craig Jon
Fabrikant, Ilya I
Fuller, Robert Gohl
Guenther, Raymond A
Liou, Sy-Hwang
Pearlstein, Edgar Aaron
Schadt, Randall James
Sellmyer, David Julian
Ullman, Frank Gordon
Weymouth, John Walter
Woods, Joseph
Zepf, Thomas Herman

NEVADA
Cathey, William Newton
Dienes, G J
Eichbaum, Barlane Ronald
Hallett, John
Johnson, Bruce Paul
Snaper, Alvin Allyn

NEW HAMPSHIRE
Christy, Robert Wentworth
Ge, Weikun
Hynes, Thomas Vincent
Ludman, Jacques Ernest
Pfluger, Clarence Eugene
Simon, Edward
Snively, Leslie O
Sturge, Michael Dudley
Temple, Peter Lawrence
Varnerin, Lawrence J(ohn)
Vernon, Robert Carey
Weiler, Margaret Horton

NEW JERSEY
Abeles, Benjamin
Abeles, Joseph Hy
Akselrad, Aline
Amick, James Albert
Anthony, Philip John
Ashworth, Harry Arthur
Bahr, Charles Chester
Bailey, William Charles
Ballato, Arthur
Ban, Vladimir Sinisa
Banks, Thomas

Baraff, Gene Allen
Batlogg, Bertram
Baughman, Ray Henry
Bean, John Condon
Bentz, Bryan Lloyd
Bernasek, Steven Lynn
Bevk, Joze
Bhalla, Ranbir J R Singh
Bharj, Sarjit Singh
Bhatt, Ravindra Nautam
Bhattacharya, Sabyasachi
Blanc, Joseph
Blicher, Adolph
Boorse, Henry Abraham
Borkan, Harold
Bosacchi, Bruno
Broer, Matthijs Meno
Brown, Rodney Duvall, III
Brown, Walter Lyons
Buck, Thomas M
Capasso, Federico
Caplan, Philip Judah
Carlson, David Emil
Carnes, James Edward
Catanese, Carmen Anthony
Celler, George K
Chadi, James D
Chaikin, Paul Michael
Chand, Naresh
Chang, Chuan Chung
Chen, Cheng Hsuan
Chen, Ho Sou
Chiu, Tien-Heng
Cho, Alfred Y
Chraplyvy, Andrew R
Chu, Sung Nee George
Chung, Yun C
Chynoweth, Alan Gerald
Citrin, Paul H
Cladis, Patricia Elizabeth Ruth
Cody, George Dewey
Connolly, John Charles
Cowley, Ernest Roger
Daly, Daniel Francis, Jr
Damen, Theo C
Daniels, James Maurice
D'Asaro, L Arthur
DeLuca, Robert D(avid)
Dentai, Andrew G
Derkits, Gustav
De Vegvar, Paul Geza Neuman
DiDomenico, Mauro, Jr
Dillon, Joseph Francis, Jr
Dingle, Raymond
Dixon, Richard Wayne
Dodabalapur, Ananth
Dolny, Gary Mark
Downs, David S
Dresner, Joseph
Dubey, Madan
Dutta, Mitra
Eckert, John Andrew
Eisenberger, Peter Michael
Enstrom, Ronald Edward
Feldman, Leonard Cecil
Fenstermacher, Robert Lane
Feuer, Mark David
Finke, Guenther Bruno
Fischer, Russell Jon
Fish, Gordon E
Fleming, Robert McLemore
Fleury, Paul A
Fonger, William Hamilton
Fork, Richard Lynn
Freeman, Richard Reiling
Fu, Hui-Hsing
Ghosh, Amal Kumar
Giordmaine, Joseph Anthony
Gittleman, Jonathan I
Glass, Alastair Malcolm
Golin, Stuart
Gossmann, Hans Joachim
Gourary, Barry Sholom
Greenblatt, Martha
Grest, Gary Stephen
Greywall, Dennis Stanley
Grupen, William Brightman
Gualtieri, Devlin Michael
Guenzer, Charles S P
Gummel, Hermann K
Gustafsson, Torgny Daniel
Gyorgy, Ernst Michael
Haase, Oswald
Haddon, Robert Cort
Hafner, Erich
Hagstrum, Homer Dupre
Hakim, Edward Bernard
Haldane, Frederick Duncan Michael
Halpern, Teodoro
Harbison, James Prescott
Hasegawa, Ryusuke
Hebard, Arthur Foster
Heilmeier, George Harry
Henry, Charles H
Hensel, John Charles
Herber, Rolfe H
Hershenov, B(ernard)
Hohenberg, Pierre Claude
Husa, Donald L
Hwang, Cherng-Jia
Imhoff, Eugene A
Jeck, Richard Kahr
Johnson, Leo Francis
Johnston, Wilbur Dexter, Jr

Josenhans, James Gross
Joyce, William B(axter)
Kahng, Dawon
Kaminow, Ivan Paul
Kapron, Felix Paul
Kasper, Horst Manfred
Koch, Thomas L
Kohn, Jack Arnold
Kokotailo, George T
Koller, Noemie
Kramer, Bernard
Kunhardt, Erich Enrique
Kunzler, John Eugene
Lang, David (Vern)
Langreth, David Chapman
Leath, Paul Larry
Levi, Anthony Frederic John
Levin, Edwin Roy
Li, Hong
Li, Yuan
Liang, Keng-San
Libowitz, George Gotthart
Lieb, Elliott Hershel
Lindenfeld, Peter
Lines, Malcolm Ellis
Logan, Ralph Andre
Loman, James Mark
Lunsford, Ralph D
Luryi, Serge
Lyons, Kenneth Brent
McLane, George Francis
MacRae, Alfred Urquhart
Madey, Theodore Eugene
Marezio, Massimo
Martinelli, Ramon U
Matey, James Regis
Mattheiss, Leonard Francis
Meiboom, Saul
Mette, Herbert L
Meyerhofer, Dietrich
Miller, David A B
Misra, Raj Pratap
Morris, Robert Craig
Murthy, N(arasimhaiah) Sanjeeva
Nachtergaele, Bruno Leo Zulma
Nahory, Robert Edward
Nassau, Kurt
O'Gorman, James
O'Keeffe, David John
Olsen, Gregory Hammond
Ong, Nai-Phuan
Ourmazd, Abbas
Owens, Frank James
Panayotatos, Paul
Panish, Morton B
Partovi, Afshin
Passner, Albert
Pei, Shin-Shem
Pfeiffer, Loren Neil
Phillips, James Charles
Phillips, Julia M
Pinczuk, Aron
Platzman, Philip M
Racette, George William
Raghavan, Pramila
Ramirez, Arthur P
Ravindra, Nuggehalli Muthanna
Raynor, Susanne
Reboul, Theo Todd, III
Riddle, George Herbert Needham
Ropp, Richard C
Rothberg, Gerald Morris
Rothberg, Lewis Josiah
Rowe, John Edward
Royce, Barrie Saunders Hart
Rudman, Peter S
Sabisky, Edward Stephen
Salwen, Harold
Schlam, Elliott
Schlueter, Michael Andreas
Schluter, Michael
Schneemeyer, Lynn F
Schubert, Rudolf
Seidl, Milos
Shacklette, Lawrence Wayne
Shah, Jagdeep C
Shand, Michael Lee
Shay, Joseph Leo
Sheng, Ping
Shumate, Paul William, Jr
Silbernagel, Bernard George
Sinclair, William Robert
Sinha, Sunil K
Skalski, Stanislaus
Slusky, Susan E G
Smith, Carl Hofland
Smith, George Elwood
Smith, Neville Vincent
Smith, Robert Owens
Solin, Stuart Allan
Solla, Sara A
Song, Jung Hyun
Spencer, Edward G
Spruch, Grace Marmor
Stach, Joseph
Stiles, Lynn F, Jr
Stolen, Rogers Hall
Stormer, Horst Ludwig
Sulewski, Paul Eric
Tauber, Arthur
Taylor, George William
Teaney, Dale T
Tecotzky, Melvin
Thaler, Barry Jay

Thomas, Gordon Albert
Thornton, C G
Tischler, Oscar
Tomaselli, Vincent Paul
Torrey, Henry Cutler
Tsui, Daniel Chee
Tully, John Charles
Tumelty, Paul Francis
Upton, Thomas Hallworth
Vanderbilt, David Hamilton
Van De Vaart, Herman
Van Roosbroeck, Willy Werner
Vehse, Robert Chase
Vella-Coleiro, George
Wachtman, John Bryan, Jr
Wagner, Sigurd
Walsh, Walter Michael, Jr
Walstedt, Russell E
Wang, Chen-Show
Wang, Tsuey Tang
Wenzel, John Thompson
Wertheim, Gunther Klaus
White, Alice Elizabeth
Wiegand, Donald Arthur
Wiesenfeld, Jay Martin
Williams, Brown F
Wittenberg, Albert M
Wolfe, Raymond
Wolff, Peter A
Wong, King-Lap
Wood, Thomas H
Woodward, Ted K
Wu, Chung Pao
Wu, Zhen
Xie, Ya-Hong
Yafet, Yako
Zaininger, Karl Heinz
Zucker, Melvin Joseph

NEW MEXICO
Albers, Robert Charles
Alldredge, Gerald Palmer
Ambrose, William Patrick
Anderson, Richard Ernest
Anderson, Robert Alan
Andrew, James F
Arko, Aloysius John
Arnold, George W
Barsis, Edwin Howard
Bartlett, Roger James
Beattie, Alan Gilbert
Beauchamp, Edwin Knight
Beckel, Charles Leroy
Bedell, Kevin Shawn
Benjamin, Robert Fredric
Borders, James Alan
Brandow, Baird H
Brice, David Kenneth
Brueck, Steven Roy Julien
Burr, Alexander Fuller
Butler, Michael Alfred
Castle, John Granville, Jr
Claytor, Thomas Nelson
Clogston, Albert McCavour
Cole, Edward Issac, Jr
Cooke, David Wayne
Dumas, Herbert M, Jr
Eckert, Juergen
Edwards, Leon Roger
Emin, David
Erickson, Dennis John
Ewing, Rodney Charles
Farnum, Eugene Howard
Follstaedt, David Martin
Fritz, Joseph N
Frost, Harold Maurice, III
Fukushima, Eiichi
Funsten, Herbert Oliver, III
Galt, John Kirtland
Giovanielli, Damon V
Gorham, Elaine Deborah
Gourley, Paul Lee
Grannemann, W(ayne) W(illis)
Green, Thomas Allen
Gubernatis, James Edward
Gurbaxani, Shyam Hassomal
Hammel, P Chris
Heffner, Robert H
Jennison, Dwight Richard
Johnson, Ralph T, Jr
Jones, Eric Daniel
Jones, Orval Elmer
Keller, William Edward
Kenkre, Vasudev Mangesh
Kepler, Raymond Glen
King, James Claude
Kreidl, Norbert J(oachim)
Kropschot, Richard H
Kurtz, Steven Ross
Land, Cecil E(lvin)
Lawson, Andrew Cowper, II
Linford, Rulon Kesler
Llamas, Vicente Jose
Loree, Thomas Robert
Lyo, SungKwun Kenneth
Maggiore, Carl Jerome
Martin, Thierry Dominique
Maurer, Robert Joseph
Miranda, Gilbert A
Mitchell, Terence Edward
Morosin, Bruno
Morris, Charles Edward
Moss, Marvin
Mueller, Fred Michael

Solid State Physics (cont)

Mueller, Karl Hugo, Jr
Myers, David Richard
Myers, Samuel Maxwell, Jr
Narath, Albert
Norris, Carroll Boyd, Jr
O'Rourke, John Alvin
Osbourn, Gordon Cecil
Osinski, Marek Andrzej
Panitz, Janda Kirk Griffith
Parkin, Don Merrill
Peercy, Paul S
Peterson, Otis G
Pettit, Richard Bolton
Picraux, Samuel Thomas
Pike, Gordon E
Pynn, Roger
Rauber, Lauren A
Redondo, Antonio
Richards, Peter Michael
Romig, Alton Dale, Jr
Salzbrenner, Richard John
Samara, George Albert
Schiferl, David
Schirber, James E
Schwalbe, Larry Allen
Schwarz, Ricardo
Schwoebel, Richard Lynn
Shaner, John Wesley
Silver, Richard N
Smith, Darryl Lyle
Smith, James Lawrence
Smith, William Conrad
Southward, Harold Dean
Taylor, Raymond Dean
Thacher, Philip Duryea
Thompson, Joe David
Tonks, Davis Loel
Trela, Walter Joseph
Tsao, Jeffrey Yeenien
Valencia, Valorie Sharron
Visscher, William M
Von Dreele, Robert Bruce
Vook, Frederick Ludwig
Weaver, Harry Talmadge
Westpfahl, David John
Whitmore, Stephen Carr
Willis, Jeffrey Owen
Wu, Xin Di
Yarnell, John Leonard

NEW YORK

Abkowitz, Martin Arnold
Adams, Peter D
Adler, Michael Stuart
Agrawal, Govind P(rasad)
Alben, Richard Samuel
Alfano, Robert R
Allen, Philip B
Alley, Phillip Wayne
Amato, Joseph Cono
Ambegaokar, Vinay
Amer, Nabil Mahmoud
Anantha, Narasipur Gundappa
Anderson, Wayne Arthur
Anthony, Thomas Richard
Arajs, Sigurds
Archie, Charles Neill
Arnold, Emil
Aschner, Joseph Felix
Ashcroft, Neil William
Aven, Manuel
Avouris, Phaedon
Bak, Per
Ballantyne, Joseph M(errill)
Bari, Robert Allan
Basavaiah, Suryadevara
Batterman, Boris William
Bednowitz, Allan Lloyd
Benenson, Raymond Elliott
Benumof, Reuben
Bermon, Stuart
Beshers, Daniel N(ewson)
Bhargava, Rameshwar Nath
Bickford, Lawrence Richardson
Bilderback, Donald Heywood
Blaker, J Warren
Bloch, Aaron Nixon
Bocko, Mark Frederick
Borrego, Jose M
Borrelli, Nicholas Francis
Bray, James William
Bright, Arthur Aaron
Brillson, Leonard Jack
Brinkman, William F
Brouillette, Walter
Brown, Dale Marius
Brown, Edmond
Brown, Ronald Alan
Budzinski, Walter Valerian
Buhrman, Robert Alan
Burkey, Bruce Curtiss
Burnham, Dwight Comber
Butler, Walter John
Cargill, George Slade, III
Carleton, Herbert Ruck
Carlson, Richard Oscar
Castner, Theodore Grant, Jr
Chang, Ifay F
Chen, Inan
Chen, Ying-Chih
Chen, Yu-Pei
Chow, Tat-Sing Paul

Chrenko, Richard Michael
Cluxton, David H
Coburn, Theodore James
Collins, Franklyn
Comly, James B
Connelly, John Joseph, Jr
Constable, James Harris
Conwell, Esther Marly
Corbett, James William
Costa, Lorenzo F
Cotts, Robert Milo
Cox, David Ernest
Crowe, George Joseph
Cummins, Herman Z
Cusano, Dominick A
Daman, Harlan Richard
Damask, Arthur Constantine
Das, Pankaj K
Das, Tara Prasad
Davenport, James Whitman
DeBlois, Ralph Walter
Dell, George F, Jr
Demuth, Joseph E
Dermit, George
Diamond, Joshua Benamy
Dienes, George Julian
Di Maria, Donelli Joseph
DiSalvo, Francis Joseph, Jr
DiStefano, Thomas Herman
Domingos, Henry
Dowell, Flonnie
Dreyfus, Russell Warren
Dropkin, John Joseph
Duclos, Steven J
Dudley, Michael
Eberhard, Jeffrey Wayne
Eckert, Richard Raymond
Edelglass, Stephen Mark
Edelheit, Lewis S
Edwards, Mark A
Engeler, William E
Feder, Ralph
Fischer, C Rutherford
Fiske, Milan Derbyshire
Fitchen, Douglas Beach
Fowler, Alan B
Frankel, Julius
Fuller, Lynn Fenton
Gambino, Richard Joseph
Ganley, W Paul
Genin, Dennis Joseph
Gersten, Joel Irwin
Ghezzo, Mario
Ghosh, Arup Kumar
Ghosh, Vinita Johri
Gibbs, Doon Laurence
Glasser, M Lawrence
Goland, Allen Nathan
Goldman, Vladimir J
Gray, Peter Vance
Greenbaum, Steven Garry
Griffiths, Clifford H
Grosewald, Peter
Grunwald, Hubert Peter
Gumbs, Godfrey Anthony
Gunter, Thomas E, Jr
Gupta, Devendra
Hall, Dennis Gene
Hall, John Jay
Hall, Robert Noel
Ham, Frank Slagle
Hamblen, David Philip
Harris, Erik Preston
Hart, Howard Roscoe, Jr
Hartstein, Allan Mark
Hatcher, Robert Douglas
Haus, Joseph Wendel
Hawkins, Gilbert Allan
Hayes, Timothy Mitchell
Hays, Dan Andrew
Headrick, Randall L
Heinz, Tony F
Henkels, Walter Harvey
Hewes, Ralph Allan
Himpsel, Franz J
Ho, John Ting-Sum
Holcomb, Donald Frank
Holtzberg, Frederic
Homan, Clarke Gilbert
Honig, Arnold
Houghton, Robert W
House, Gary Lawrence
Hovel, Harold John
Hriljac, Joseph A
Hsiang, Thomas Y
Huang, Ming-Hui
Hudson, John B(alch)
Huntington, Hillard Bell
Hurst, Robert Philip
Ingham, Herbert Smith, Jr
Isaacs, Leslie Laszlo
Jacobs, Israel S(amson)
Janak, James Francis
Jepsen, Donald William
Jona, Franco Paul
Jorne, Jacob
Juretschke, Hellmut Joseph
Kadin, Alan Mitchell
Kaplan, Martin Charles
Kaplan, Raymond
Kasprzak, Lucian A
Kawatra, Mahendra P
Keller, Seymour Paul
Keonjian, Edward

Ketchen, Mark B
Khosla, Rajinder Paul
Kingsley, Jack Dean
Kirkpatrick, Edward Scott
Kirtley, John Robert
Knox, Robert Seiple
Komiak, James Joseph
Koplik, Joel
Kornreich, Philipp G
Kressel, Henry
Krieger, Joseph Bernard
Kroger, Harry
Kuan, Tung-Sheng
Lakatos, Andras Imre
Laks, David Bejnesh
Landauer, Rolf William
Lanford, William Armistead
Lang, Norton David
Lauffenburger, James C
Lavine, James Philip
Lax, Melvin
Lee, Teh-Hsuang
Lee, Yung-Chang
Lemos, Anthony M
Levinson, Lionel Monty
Levy, Paul W(arren)
Levy, Peter Michael
Liboff, Richard L
Likharev, Konstantin K
Lin, Duo-Liang
Lindberg, Vern Wilton
Lipari, Nunzio Ottavio
Losee, David Lawrence
Lu, Toh-Ming
Lubberts, Gerrit
Luborsky, Fred Everett
Ludeke, Rudolf
Ludwig, Gerald W
Luntz, Myron
Lynn, Kelvin G
McCombe, Bruce Douglas
MacCrone, Robert K
McDonald, Frank Alan
MacDonald, Noel C
McGroddy, James Cleary
McKay, Kenneth Gardiner
McWhan, Denis B
Maldonado, Juan Ramon
Maltz, Martin Sidney
Mane, Sateesh Ramchandra
Marino, Robert Anthony
Marsh, Bruce Burton
Marwick, Alan David
Mayadas, A Frank
Mayer, James W(alter)
Melcher, Robert Lee
Mendez, Emilio Eugenio
Messmer, Richard Paul
Meyer, Robert Jay
Meyerson, Bernard Steele
Miller, William
Millet, Peter J
Mogro-Campero, Antonio
Montano, Pedro Antonio
Mooney, Patricia May
Morey, Donald Roger
Mort, Joseph
Mortenson, Kenneth Ernest
Moy, Dan
Murphy, James Joseph
Mutter, Walter Edward
Nettel, Stephen J E
Neuberger, Jacob
Neumark, Gertrude Fanny
Ning, Tak Hung
Nowick, A(rthur) S(tanley)
Ockman, Nathan
Oh, Byungdu
Oliver, David W
Owens, James Carl
Ozimek, Edward Joseph
Pantelides, Sokrates Theodore
Parks, Harold George
Parpia, Jeevak Mahmud
Passell, Laurence
Peak, David
Pell, Erik Mauritz
Persans, Peter D
Petrie, Sarah Elaine
Phelps, Daniel James
Philipp, Herbert Reynold
Piper, William Weidman
Pohl, Robert O
Pollak, Fred Hugo
Pollock, D(aniel) D(avid)
Pomerantz, Melvin
Pomian, Ronald J
Pope, Martin
Possin, George Edward
Prasad, Paras Nath
Price, Peter J
Privman, Vladimir
Pugh, Emerson William
Pytte, Erling
Redington, Rowland Wells
Reichert, Jonathan F
Renzema, Theodore Samuel
Resing, Henry Anton
Rice, Michael John
Rimai, Donald Saul
Ritsko, John James
Roellig, Leonard Oscar
Rose, Kenneth
Rosolowski, Joseph Henry

Roth, Laura Maurer
Ruckman, Mark Warren
Rudman, Reuben
Ruoff, Arthur Louis
Rustgi, Om Prakash
Saha, Bijay S
Sahni, Viraht
Sai-Halasz, George Anthony
Salzberg, Bernard
Sarachik, Myriam Paula
Satya, Akella V S
Schachter, Rozalie
Schenck, John Frederic
Schick, Kenneth Leonard
Schiff, Eric Allan
Schmidlin, Frederick W
Schmitt, Roland Walter
Scholz, Wilfried
Schroder, Klaus
Schroeder, John
Schwarz, Klaus W
Seiden, Philip Edward
Setchell, John Stanford, Jr
Shapiro, Sidney
Shapiro, Stephen Michael
Shen, Wu-Mian
Shirane, Gen
Shiren, Norman S
Sidran, Miriam
Siegel, Benjamin Morton
Sievers, Albert John, III
Siggia, Eric Dean
Silcox, John
Silsbee, Robert Herman
Silverman, Benjamin David
Singer, Barry M
Slack, Glen Alfred
Sliva, Philip Oscar
Slonczewski, John Casimir
Slowik, John Henry
Smith, Frederick William
Smith, Lowell Scott
Snavely, Benjamin Breneman
Snead, Clarence Lewis, Jr
Snyder, Lawrence Clement
Spergel, Martin Samuel
Springett, Brian E
Srinivasan, G(urumakonda) R
Stancampiano, Charles Vincent
Stannard, Carl R, Jr
Stephens, Peter Wesley
Stephenson, Thomas E(dgar)
Stern, Frank
Stinson, Douglas G
Story, Harold S
Strnisa, Fred V
Strongin, Myron
Szydlik, Paul Peter
Tan, Yen T
Taylor, Dale Frederick
Taylor, Jack Eldon
Temple, Victor Albert Keith
Tersoff, Jerry David
Thomas, Richard Alan
Thomas, Robert
Thompson, William A
Thygesen, Kenneth Helmer
Toner, John Joseph
Trevoy, Donald James
Turner, William Joseph
Tweet, Arthur Glenn
Uemura, Yasutomo J
Valsamakis, Emmanuel
Van Heyningen, Roger
Vanier, Peter Eugene
Vaughan, Michael Thomas
Venugopalan, Srinivasa I
Vidali, Gian Franco
Viertl, John Ruediger Mader
Vook, Richard Werner
Voss, Richard Frederick
Wajda, Edward Stanley
Walker, Edward John
Walker, Michael Stephen
Wang, Gwo-Ching
Wang, Xingwu
Watson, Richard E
Webb, Watt Wetmore
Webster, Harold Frank
Wei, Ching-Yeu
Weinstein, Bernard Allen
Welch, David O(tis)
Williams, Arthur Robert
Williamson, Samuel Johns
Wilson, Ronald Harvey
Wittmer, Marc F
Wolf, Edward Lincoln
Worden, David Gilbert
Yasar, Tugrul
Yip, Kwok Leung
Yoffa, Ellen June
York, Raymond A
Young, Ralph Howard
Yu, Ming Lun
Zhang, John Zeng Hui

NORTH CAROLINA

Ade, Harald Werner
Aspnes, David E
Bailey, Robert Brian
Bernholc, Jerzy
Berry, B(rian) S(hepherd)
Bowers, Wayne Alexander
Cain, Laurence Sutherland

Casey, Horace Craig, Jr
Clark, Chester William
Clark, Clifton Bob
Clarke, John Ross
Corliss, Lester Myron
Dy, Kian Seng
Etzel, Howard Wesley
Fair, Richard Barton
Fox, Bradley Alan
Giess, Edward August
Hacker, Herbert, Jr
Hall, Carol Klein
Hall, George Lincoln
Hauser, John Reid
Hernandez, John Peter
Holton, William Coffeen
Howell, James Milton
Hren, John J(oseph)
Iafrate, Gerald Joseph
Irene, Eugene Arthur
Kerr, William Clayton
Khan, Sultana
Koch, Carl Conrad
Lamorte, Michael Francis
Leamy, Harry John
Lucovsky, Gerald
Macdonald, James Ross
McNeil, Laurie Elizabeth
Mamola, Karl Charles
Markham, Jordan Jeptha
Masnari, Nino A
Massoud, Hisham Z
Matthews, N(eely) F(orsyth) J(ones)
Narayan, Jagdish
Paesler, Michael Arthur
Ramanan, V R V
Reid, F Joseph
Roberts, Louis Douglas
Rokoske, Thomas Leo
Rubloff, Gary W
Sandin, Thomas Robert
Sayers, Dale Edward
Schroeer, Dietrich
Shields, Howard William
Silver, Marvin
Slifkin, Lawrence
Smith, Charles Sydney, Jr
Stevens, John Gehret
Stroscio, Michael Anthony
Swanson, Max Lynn
Turner, Thomas Jenkins
Walters, Mark David
Wechsler, Monroe S(tanley)
Werntz, James Herbert, Jr
Whangbo, Myung Hwan
Wilkes, William Roy
Williams, Richard Taylor
Wortman, Jimmie J(ack)

NORTH DAKOTA
Dewar, Graeme Alexander
Kurtze, Douglas Alan
Schwalm, Mizuho K
Schwalm, William A
Soonpaa, Henn H

OHIO
Bagley, Brian G
Barkley, John R
Becker, Roger Jackson
Berney, Rex Leroy
Bhattacharya, Rabi Sankar
Blessinger, Michael Anthony
Boolchand, Punit
Boughton, Robert Ivan, Jr
Brandhorst, Henry William, Jr
Carbonara, Robert Stephen
Ceasar, Gerald P
Chen, Charles Chin-Tse
Cheng, Stephen Zheng Di
Chiu, Victor
Christensen, Stanley Howard
Clark, Alton Harold
Clark, William Alan Thomas
Cochran, William Ronald
Collings, Edward William
Compaan, Alvin Dell
Dalbec, Paul Euclide
Day, Stephen Martin
Detrio, John A
De Vries, Adriaan
DiCarlo, James Anthony
Doane, J William
Eby, Ronald Kraft
Edwards, David Olaf
Epstein, Arthur Joseph
Erdman, Robert J
Fenichel, Henry
Fernelius, Nils Conard
Forestieri, Americo F
Gamble, Francis Trevor
Garland, James C
Gordon, William Livingston
Gottling, James Goe
Grant, Roderick M, Jr
Griffing, David Francis
Grimes, Hubert Henry
Hambourger, Paul David
Hathaway, Charles Edward
Henderson, H(oman) Thurman
Hengehold, Robert Leo
Hirt, Andrew Michael
Hoffman, Richard Wagner
Holt, James Franklin

Hubin, Wilbert N
Jaworowski, Andrzej Edward
Joiner, William Cornelius Henry
Kash, Kathleen
Katz, J Lawrence
Kaufman, Miron
Kingston, David Lyman
Kosel, Peter Bohdan
Land, Peter L
Lawless, William N
Listerman, Thomas Walter
Look, David C
McGervey, John Donald
McGrath, James W
Middleton, Arthur Everts
Mier, Millard George
Mills, Robert Laurence
Miner, George Kenneth
Morin, Brian Gerald
Moss, Thomas H
Nam, Sang Boo
Nash, Harry Charles
Nestor, Ontario Horia
Newrock, Richard Sandor
Nielsen, Philip Edward
Noel, Gerald Thomas
Patton, Bruce Riley
Pechan, Michael J
Peterson, Timothy Lee
Petschek, Rolfe George
Pillai, Padmanabha S
Pinnick, Harry Thomas
Poynter, James William
Puglielli, Vincent George
Rai, Amarendra Kumar
Reynolds, Donald C
Rigney, David Arthur
Rollins, Roger William
Rosenblatt, Charles Steven
Sanford, Edward Richard
Schepler, Kenneth Lee
Schlosser, Herbert
Scholten, Paul David
Schreiber, Paul J
Schuele, Donald Edward
Segall, Benjamin
Seltzer, Martin S
Silvidi, Anthony Alfred
Simon, Henry John
Smith, Robert Emery
Spielberg, Nathan
Spry, Robert James
Steckl, Andrew Jules
Stevenson, Frank Robert
Stoner, Ronald Edward
Stroud, David Gordon
Styer, Daniel F
Suranyi, Peter
Swinehart, Philip Ross
Taylor, Barney Edsel
Taylor, Charles Emery
Taylor, Philip Liddon
Tenhover, Michael Alan
Trivisonno, Joseph, Jr
Wagoner, Glen
Weeks, Stephen P
White, John Joseph, III
Wiff, Donald Ray
Wigen, Philip E
Williams, Wendell Sterling
Wolff, Gunther Arthur
Wong, Anthony Sai-Hung
Wood, Van E(arl)
Yeo, Yung Kee
Yun, Seung Soo

OKLAHOMA
Batchman, Theodore E
Brumage, William Harry
Collins, Thomas C
Dixon, George Sumter, Jr
Doezema, Ryan Edward
Dreiling, Mark Jerome
Halliburton, Larry Eugene
Hartman, Roger Duane
Hendrickson, John Robert
Huffaker, James Neal
King, John Paul
Lafon, Earl Edward
Lauffer, Donald Eugene
McKeever, Stephen William Spencer
Major, John Keene
Martin, Joel Jerome
Waldrop, Morgan A
Wilson, Timothy M

OREGON
Abrahams, Sidney Cyril
Boedtker, Olaf A
Cohen, John David
Cutler, Melvin
Demarest, Harold Hunt, Jr
Engelbrecht, Rudolf S
Engelmann, Reinhart Wolfgang H
Ford, Wayne Keith
Gardner, John Arvy, Jr
Griffith, William Thomas
Howard, Donald Grant
McClure, Joel William, Jr
Montague, Daniel Grover
Nussbaum, Rudi Hans
Owen, Sydney John Thomas
Photinos, Panos John
Siemens, Richard Ernest

Solanki, Raj
Van Vechten, James Alden
Warren, William Willard, Jr
Wasserman, Allen Lowell
Wolfe, Gordon A
Yamaguchi, Tadanori

PENNSYLVANIA
Adda, Lionel Paul
Angello, Stephen James
Arora, Vijay Kumar
Artman, Joseph Oscar
Banavar, Jayanth Ramarao
Baratta, Anthony J
Barsch, Gerhard Richard
Baxter, Ronald Dale
Berger, Luc
Bhalla, Amar S
Biondi, Manfred Anthony
Bitler, William Reynolds
Bloomfield, Philip Earl
Boer, Karl Wolfgang
Bolef, Dan Isadore
Boltax, Alvin
Booth, Bruce L
Bose, Shyamalendu M
Bowlden, Henry James
Burstein, Elias
Callen, Herbert Bernard
Carr, Walter James, Jr
Caspari, Max Edward
Chaplin, Norman John
Charap, Stanley H
Cherry, Leonard Victor
Choyke, Wolfgang Justus
Chung, Tze-Chiang
Cohen, Alvin Jerome
Cole, Milton Walter
Collings, Peter John
Cutler, Paul H
Daniel, Michael Roger
Dank, Milton
Das, Mukunda B
Davis, R(obert) E(lliot)
DeGraef, Marc Julie
Deis, Daniel Wayne
Diehl, Renee Denise
Donahoe, Frank J
Drum, Charles Monroe
Dubeck, Leroy W
Dunn, Charles Nord
Egami, Takeshi
Emtage, Peter Roesch
Fabish, Thomas John
Feingold, Earl
Ferrone, Frank Anthony
Feuchtwang, Thomas Emanuel
Finegold, Leonard X
Fischer, John Edward
Fisher, David George
Flattery, David Kevin
Fonash, Stephan J(oseph)
Fowler, Wyman Beall, Jr
Fox, John Nicholas
Foxhall, George Frederic
Frankl, Daniel Richard
Freud, Paul J
Friedberg, Simeon Adlow
Fuchs, Walter
Gainer, Michael Kizinski
Gao, Quanyin
Garito, Anthony Frank
Girifalco, Louis A(nthony)
Gittler, Franz Ludwig
Goldburg, Walter Isaac
Goldey, James Mearns
Goldschmidt, Yadin Yehuda
Gollub, Jerry Paul
Goodwin, Charles Arthur
Griffiths, Robert Budington
Gurer, Emir
Harmer, Martin Paul
Harris, Arthur Brooks
Henisch, Heinz Kurt
Hensler, Donald H
Hess, Dennis William
Houlihan, John Frank
Hulm, John Kenneth
Ivey, Henry Franklin
Jain, Himanshu
Johnson, William Lewis
Jones, Clifford Kenneth
Kahn, David
Kakar, Anand Swaroop
Kampas, Frank James
Kang, Joohee
Kasowski, Robert V
Keffer, Frederic
Klein, Michael Lawrence
Kozlowski, Richard William Hugh
Kraitchman, Jerome
Krishnaswamy, S V
Kryder, Mark Howard
Lang, Lawrence George
Langer, Dietrich Wilhelm
Lannin, Jeffrey S
Larson, Donald Clayton
Lee, Richard J
Liang, Shoudan
Licini, Jerome Carl
Lowe, Irving J
Lu, Chih Yuan
Lu, Guangquan
McCammon, Robert Desmond

McGowan, William Courtney, Jr
McKelvey, John Philip
McKinstry, Herbert Alden
Macpherson, Alistair Kenneth
Mahajan, Subhash
Maher, James Vincent
Malmberg, Paul Rovelstad
Matolyak, John
Messier, Russell
Mihalisin, Ted Warren
Miller, Paul
Miskovsky, Nicholas Matthew
Molter, Lynne Ann
Morris, Marlene Cook
Mulay, Laxman Nilakantha
Muldawer, Leonard
Nathanson, H(arvey) C(harles)
Nigh, Harold Eugene
Nilan, Thomas George
Noreika, Alexander Joseph
Obermyer, Richard T
Oder, Robin Roy
Orehotsky, John Lewis
Pfrogner, Ray Long
Pierce, Russell Dale
Polo, Santiago Ramos
Post, Irving Gilbert
Prince, Morton Bronenberg
Przybysz, John Xavier
Rao, V Udaya S
Ray, Siba Prasad
Repper, Charles John
Reynolds, Claude Lewis, Jr
Richardson, Ralph J
Roeder, Edward A
Roelofs, Lyle D
Sabnis, Anant Govind
Schumacher, Robert Thornton
Shibib, M Ayman
Smits, Friedolf M
Spitznagel, John A
Steele, Martin Carl
Stein, Barry Fred
Steinhardt, Paul Joseph
Stewart, Glenn Alexander
Stroud, Jackson Swavely
Sucov, E(ugene) W(illiam)
Szepesi, Zoltan Paul John
Thomas, H Ronald
Thomas, R Noel
Thompson, Eric Douglas
Tittmann, Bernhard R
Toulouse, Jean
Wagner, George Richard
Walkiewicz, Thomas Adam
Waring, Robert Kerr, Jr
Watkins, George Daniels
Weidner, Richard Tilghman
Weiner, Brian Lewis
Weiner, Robert Allen
Weiss, Paul Storch
Wemple, Stuart H(arry)
White, Marvin Hart
Whiteman, John David
Willis, Roy F
Zhang, Qiming

RHODE ISLAND
Bray, Philip James
Estrup, Peder Jan Z
Glicksman, Maurice
Heller, Gerald S
Loferski, Joseph J
McCorkle, Richard Anthony
Mardix, Shmuel
Maris, Humphrey John
Nunes, Anthony Charles
Pelcovits, Robert Alan
Pickart, Stanley Joseph
Seidel, George Merle
Sherman, Charles Henry
Stiles, Phillip John
Tauc, Jan
Timbie, Peter T
Ying, See Chen

SOUTH CAROLINA
Bagchi, Amitabha
Bostock, Judith Louise
Chaplin, Robert Lee, Jr
Cronemeyer, Donald Charles
Darnell, Frederick Jerome
Datta, Timir
Edge, Ronald (Dovaston)
Faust, John William, Jr
Hamilton, John Frederick
Huray, Paul Gordon
Hurren, Weiler R
Jones, Edwin Rudolph, Jr
Keller, Frederick Jacob
Lathrop, Jay Wallace
Manson, Joseph Richard
Mosley, Wilbur Clanton, Jr
Moyle, David Douglas
Payne, James Edward
Payne, Linda Lawson
Poole, Charles Patton, Jr
Radford, Loren E
Saunders, Edward A
Sawers, James Richard, Jr
Sherrill, Max Douglas
Skove, Malcolm John
Steiner, Pinckney Alston, III
Stevens, James Levon

Solid State Physics (cont)

Stewart, William Hogue, Jr
Stillwell, Ephraim Posey, Jr
Vogel, Henry Elliott
Wood, James C, Jr

SOUTH DAKOTA
Ashworth, T
Quist, Oren Paul
Redin, Robert Daniel
Ross, Keith Alan
Tunheim, Jerald Arden
Vander Lugt, Karel L

TENNESSEE
Abraham, Marvin Meyer
Achar, B N Narahari
Appleton, B R
Arakawa, Edward Takashi
Arlinghaus, Heinrich Franz
Ball, Raiford Mill
Blankenship, James Lynn
Boatner, Lynn Allen
Borie, Bernard Simon
Budai, John David
Butler, William H
Cable, Joe Wood
Callcott, Thomas Anderson
Cardwell, Alvin Boyd
Child, Harry Ray
Christen, David Kent
Close, David Matzen
Coltman, Ralph Read, Jr
Dudney, Nancy Johnston
Feigerle, Charles Stephen
Fernandez-Baca, Jaime Alberto
Franceschetti, Donald Ralph
Garland, Michael McKee
Gruzalski, Greg Robert
Haglund, Richard Forsberg, Jr
Haynes, Tony Eugene
Huebschman, Eugene Carl
Jellison, Gerald Earle, Jr
Jenkins, Leslie Hugh
Kennedy, Eldredge Johnson
Kerchner, Harold Richard
Kliewer, Kenneth L
Langley, Robert Archie
Larson, Bennett Charles
Lenhert, P Galen
Lowndes, Douglas H, Jr
Lundy, Ted Sadler
McKee, Rodney Allen
Miller, James Robert
Moak, Charles Dexter
Mook, Herbert Arthur, Jr
Moon, Ralph Marks, Jr
Mostoller, Mark Ellsworth
Nicklow, Robert Merle
Oen, Ordean Silas
O'Neal, Thomas Norman
Painter, Gayle Stanford
Pennycook, Stephen John
Plummer, E Ward
Powell, Harry Douglas
Raudorf, Thomas Walter
Ritchie, Rufus Haynes
Roberto, James Blair
Robinson, Mark Tabor
Sales, Brian Craig
Schwenterly, Stanley William, III
Silberman, Enrique
Smith, Harold Glenn
Sonder, Edward
Springer, John Mervin
Srygley, Fletcher Douglas
Stephenson, Charles V
Stiegler, James O
Stocks, George Malcolm
Thompson, James R
Thomson, John Oliver
Tolk, Norman Henry
Trammell, Rex Costo
Wachs, Alan Leonard
Wang, Jia-Chao
Wang, Taylor Gunjin
Weeks, Robert A
Wendelken, John Franklin
White, Clark Woody
Wilkinson, Michael Kennerly
Ying, Zhiqiang Charles
Young, Frederick Walter, Jr
Zuhr, Raymond Arthur

TEXAS
Adair, Thomas Weymon, III
Allen, Roland Emery
Anderson, Robert E
Aton, Thomas John
Balasinski, Artur
Bartels, Richard Alfred
Barton, James Brockman
Bate, Robert Thomas
Birchak, James Robert
Black, Truman D
Bogardus, Egbert Hal
Bottom, Virgil Eldon
Bowen, Donald Edgar
Brient, Samuel John, Jr
Bucy, J Fred
Caflisch, Robert Galen
Carbajal, Bernard Gonzales, III
Celii, Francis Gabriel

Chivian, Jay Simon
Chopra, Dev Raj
Chu, Ting Li
Chu, Wei-Kan
Claiborne, Lewis T, Jr
Collins, Dean Robert
Collins, Francis Allen
Crossley, Peter Anthony
Deaton, Bobby Charles
De Lozanne, Alex L
De Wette, Frederick Willem
Dodds, Stanley A
Duncan, Walter Marvin, Jr
Dunning, Frank Barrymore
Dupuis, Russell Dean
Erskine, James Lorenzo
Estle, Thomas Leo
Fair, Harry David, Jr
Fontana, Robert E
Frensley, William Robert
Gavenda, John David
Gleim, Paul Stanley
Glosser, Robert
Goodenough, John Bannister
Goodman, David Wayne
Graham, William Richard Montgomery
Grobman, Warren David
Gryting, Harold Julian
Hamill, Dennis W
Hasty, Turner Elilah
Heinze, William Daniel
Heller, Adam
Herczeg, John W
Ho, Paul Siu-Chung
Holverson, Edwin LeRoy
Howard, Iris Anne
Hu, Chia-Ren
Hudson, Hugh T
Ignatiev, Alex
Isham, Elmer Rex
Kirk, Wiley Price
Klein, David L
Kleinman, Leonard
Kolesar, Edward S
Lacy, Lewis L
Lawson, Juan (Otto)
LeMaster, Edwin William
Leuchtag, H Richard
Levy, Robert Aaron
Lichti, Roger L
McCormick, William Devlin
McDavid, James Michael
McDonald, Perry Frank
Mackey, Henry James
Mahendroo, Prem P
Mao, Shing
Matzkanin, George Andrew
Mays, Robert, Jr
Medlin, William Louis
Merrill, John Raymond
Moiz, Syed Abdul
Moss, Simon Charles
Mueller, Dennis W
Naugle, Donald
Neikirk, Dean P
Nordlander, Peter Jan Arne
Pandey, Raghvendra Kumar
Parker, Sidney G
Paterson, James Lenander
Penz, P Andrew
Peters, Randall Douglas
Portnoy, William M
Potter, Robert Joseph
Rollwitz, William Lloyd
Rorschach, Harold Emil, Jr
Rubins, Roy Selwyn
Sablik, Martin J
Saslow, Wayne Mark
Schroen, Walter
Sears, Raymond Eric John
Seitchik, Jerold Alan
Sharma, Suresh C
Smith, Daniel Montague
Smith, James Lynn
Smoluchowski, Roman
Spencer, Gregory Fielder
Spencer, William J
Stratton, Robert
Sullivan, Jerry Stephen
Sybert, James Ray
Tasch, Al Felix, Jr
Thompson, Arthur Howard
Thompson, Bonnie Cecil
Uralil, Francis Stephen
Van Der Ziel, Jan Peter
Walters, Geoffrey King
Wang, Paul Weily
Wood, Lowell Thomas
Zrudsky, Donald Richard

UTAH
Ailion, David Charles
Allred, David D
Barnett, John Dean
Benson, Alvin K
Brandt, Richard Charles
Brophy, James John
Cohen, Richard M
Decker, Daniel Lorenzo
DeFord, John W
Evenson, William Edwin
Galli, John Ronald
Hansen, Wilford Nels
Jeffery, Rondo Nelden

Johnson, Owen W
Jones, Merrell Robert
Larson, Everett Gerald
Luty, Fritz
Mattis, Daniel Charles
Nelson, Homer Mark
Ohlsen, William David
Smith, Kent Farrell
Stokes, Harold T
Stringfellow, Gerald B
Symko, Orest George
Taylor, Philip Craig
Vanfleet, Howard Bay
Ward, Roger Wilson
Williams, George Abiah
Worlock, John M

VERMONT
Adler, Eric
Aitken, John Malcolm
Anderson, R(ichard) L(ouis)
Ellis-Monaghan, John J
Ferris-Prabhu, Albert Victor Michael
Foley, Edward Leo
Furukawa, Toshiharu
Lambert, Lloyd Milton, Jr
Muir, Wilson Burnett
Pires, Renato Guedes
Quinn, Robert M(ichael)
Smith, David Young
Warfield, George
Williams, Ronald Wendell

VIRGINIA
Amith, Avraham
Amstutz, Larry Ihrig
Bahl, Inder Jit
Bensel, John Phillip
Berlincourt, Ted Gibbs
Bishop, Marilyn Frances
Bloomfield, Louis Aub
Brandt, Richard Gustave
Brockman, Philip
Butler, Charles Thomas
Byvik, Charles Edward
Cantrell, John H(arris)
Catlin, Avery
Caton, Randall Hubert
Celli, Vittorio
Chang, Lay Nam
Chubb, Scott Robinson
Chun, Myung K(i)
Cook, Desmond C
Cooper, Larry Russell
Crane, Sara W
Crump, John C, III
Deaver, Bascom Sine, Jr
DeFotis, Gary Constantine
De Graaf, Adriaan M
Dharamsi, Amin N
Duesbery, Michael Serge
Duke, John Christian, Jr
Eck, John Stark
Faraday, Bruce (John)
Gilmer, Thomas Edward, Jr
Goodman, A(lvin) M(alcolm)
Hass, Georg
Hass, Marvin
Haworth, W(illiam) Lancelot
Hendricks, Robert Wayne
Hess, George Burns
Hilger, James Eugene
Hoffman, Allan Richard
Holt, William Henry
Hoy, Gilbert Richard
Huber, John G
Hudgins, Aubrey C, Jr
Imbrie, John Z
Jena, Purusottam
Jesser, William Augustus
Johnson, Charles Minor
Johnson, Robert Alan
Johnson, Robert Edward
Johnston, William Cargill
Joshi, Ravindra Prabhakar
Kellett, Claud Marvin
Kennedy, Andrew John
Kirkendall, Ernest Oliver
Kossler, William John
Kramer, Steven David
Lankford, William Fleet
Liebenberg, Donald Henry
Long, Jerome R
McBride, Duncan Eldridge
McIrvine, Edward Charles
Major, Robert Wayne
Marsh, Howard Stephen
Maycock, Paul Dean
Mielczarek, Eugenie V
Mitchell, Dean Lewis
Morris, Robert Carter
Nicoll, Jeffrey Fancher
Noakes, David Ronald
Park, Yoon Soo
Peters, Philip Boardman
Phillips, Donald Herman
Pollard, John Henry
Pritchard, Wenton Maurice
Rado, George Tibor
Ratchford, Joseph Thomas
Reynolds, Peter James
Rosenthal, Michael David
Rothenberg, Herbert Carl
Rothwarf, Frederick

Rubinstein, Mark
Sauder, William Conrad
Schnatterly, Stephen Eugene
Schriempf, John Thomas
Serway, Raymond A
Shur, Michael
Sigler, Julius Alfred, Jr
Silberglitt, Richard Stephen
Silver, Meyer
Sloope, Billy Warren
Streever, Ralph L
Stronach, Carey E
Taggart, G(eorge) Bruce
Wickman, Herbert Hollis
Williams, Clayton Drews
Wyeth, Newell Convers
Zallen, Richard

WASHINGTON
Baer, Donald Ray
Blakemore, John Sydney
Braunlich, Peter Fritz
Brimhall, J(ohn) L
Brown, Frederick Calvin
Claudson, T(homas) T(ucker)
Collins, Gary Scott
Craig, Richard Anderson
Darling, Robert Bruce
Dexter, Richard Norman
Domansky, Karel
Doran, Donald George
Dresser, Miles Joel
Dunham, Glen Curtis
Einziger, Robert E
Erdmann, Joachim Christian
Exarhos, Gregory James
Fain, Samuel Clark, Jr
Fischbach, David Bibb
Gillingham, Robert J
Glass, James Clifford
Gunsul, Craig J W
Gupta, Yogendra M(ohan)
Hinman, Chester Arthur
Howland, Louis Philip
Huang, Fan-Hsiung Frank
Johnson, Gordon Oliver
Johnson, Harlan Paul
Klepper, John Richard
Knotek, Michael Louis
Lam, John Ling-Yee
Lambe, John Joseph
Laramore, George Ernest
Lytle, Farrel Wayne
Macksey, Harry Michael
Matlock, John Hudson
Miles, Maurice Howard
Miller, John Howard
Olmstead, Marjorie A
Olsen, Larry Carrol
Quigley, Robert James
Rehr, John Jacob
Rimbey, Peter Raymond
Schick, Michael
Smith, Willis Dean
Spanel, Leslie Edward
Stern, Edward Abraham
Stuve, Eric Michael
Thouless, David James
Tsang, Leung
Tutihasi, Simpei
Vilches, Oscar Edgardo
Warren, John Lucius
Weber, William J
Willardson, Robert Kent
Yoshikawa, Herbert Hiroshi

WEST VIRGINIA
Bleil, David Franklin
Cooper, Bernard Richard
De Barbadillo, John Joseph
Franklin, Alan Douglas
Pavlovic, Arthur Stephen
Seehra, Mohindar Singh
Shanholtzer, Wesley Lee
Tewksbury, Stuart K

WISCONSIN
Aita, Carolyn Rubin
Anderson, James Gerard
Baum, Gary Allen
Bondeson, Stephen Ray
Botez, Dan
Buchholz, Jeffrey Carl
Burch, Thaddeus Joseph
Cerrina, Francesco
Deshotels, Warren Julius
Draeger, Norman Arthur
Ebel, Marvin Emerson
England, Walter Bernard
Gelatt, Charles Daniel, Jr
Gueths, James E
Huber, David Lawrence
Hyde, James Stewart
Johnson, Arthur Franklin
Lagally, Max Gunter
Larbalestier, David C
Levy, Moises
Lin, Chun Chia
Lindeberg, George Kline
Mangat, Pawitterjit Singh
Mendelson, Kenneth Samuel
Myers, Vernon W
Olson, Clifford Gerald
Rhyner, Charles R

Smith, Thomas Stevenson
Sorbello, Richard Salvatore
Stanwick, Glenn
Tonner, Brian P
Truszkowska, Krystyna
Wiley, John Duncan
Zimm, Carl B

PUERTO RICO
Bailey, Carroll Edward
Gomez-Rodriguez, Manuel

ALBERTA
Adler, John G
Apps, Michael John
Chatterjee, Ramananda
Deegan, Ross Alfred
Egerton, Raymond Frank
Haslett, James William
Kuntz, Garland Parke Paul
Liu, Dennis Dong
Rogers, James Stewart
Tuszynski, Jack A
Umezawa, Hiroomi
Weichman, Frank Ludwig
Woods, Stuart B

BRITISH COLUMBIA
Ballentine, Leslie Edward
Barclay, John Arthur
Barker, Alfred Stanley, Jr
Boal, David Harold
Booth, Ian Jeremy
Buckmaster, Harvey Allen
Clayman, Bruce Philip
Cochran, John Francis
Colbow, Kevin Michael
Colbow, Konrad
Crozier, Edgar Daryl
Dahn, Jeffery Raymond
Frindt, Robert Frederick
Haering, Rudolph Roland
Hardy, Walter Newbold
Irwin, John Charles
Kirczenow, George
Kubik, Philip Roman
Kudynska, Jadwiga
Martin, Peter Wilson
Morrison, Stanley Roy
Opechowski, Wladyslaw
Rieckhoff, Klaus E
Rogers, Douglas Herbert
Schwerdtfeger, Charles Frederick
Seth, Rajinder Singh
Thewalt, Michael L W
Tiedje, J Thomas
Viswanathan, Kadayam Sankaran
Wallace, Philip Russell
Watton, Arthur
Wortis, Michael

MANITOBA
Dong, Ronald Y
Gaunt, Paul
Loly, Peter Douglas
Southern, Byron Wayne
Vail, John Moncrieff
Woo, Chung-Ho

NEW BRUNSWICK
Banerjee, R L
Girouard, Fernand E
Leblanc, Leonard Joseph
Vo-Van, Truong
Wells, David Ernest

NOVA SCOTIA
Betts, Donald Drysdale
El-Masry, Ezz Ismail
Hunter, Douglas Lyle
Latta, Bryan Michael
Moriarty, Kevin Joseph
Roger, William Alexander
Simpson, Antony Michael
Steinitz, Michael Otto

ONTARIO
Alfred, Louis Charles Roland
Anderson, Anthony
Bahadur, Birendra
Baird, D C
Berezin, Alexander A
Berry, Robert John
Bertram, Robert William
Boswell, Frank William Charles
Brandon, James Kenneth
Brockhouse, Bertram Neville
Brodie, Don E
Brown, Ian David
Buyers, William James Leslie
Celinski, Olgierd J(erzy) Z(dzislaw)
Chamberlain, S(avvas) G(eorgiou)
Cocivera, Michael
Collins, Malcolm Frank
Colpa, Johannes Pieter
Datars, William Ross
Davies, John Arthur
Davison, Sydney George
Dixon, Arthur Edward
Dunlop, David John
Dyment, John Cameron
Eastman, Philip Clifford
Egelstaff, Peter A
Erickson, Lynden Edwin

Fawcett, Eric
Fenton, Edward Warren
Ferrier, Jack Moreland
Fujimoto, Minoru
Goodings, David Ambery
Griffin, Allan
Grindlay, John
Grodski, Juliusz Jan
Hawton, Margaret H
Hedgecock, Nigel Edward
Hitchcock, Adam Percival
Hoffstein, Victor
Holden, Thomas More
Holuj, Frank
Hunt, James L
Hurd, Colin Michael
Jackman, Thomas Edward
Jeffrey, Kenneth Robert
Joos, Béla
Jorch, Harald Heinrich
Kim, Soo Myung
Koffyberg, Francois Pierre
Lachaine, Andre Raymond Joseph
Lamarche, J L Gilles
Lee, Martin J G
Lee-Whiting, Graham Edward
Leslie, James D
Lipsett, Frederick Roy
Liu, Wing-Ki
Lockwood, David John
McAlister, Sean Patrick
Mandelis, Andreas
Martin, Douglas Leonard
Mendis, Eustace Francis
Nagi, Anterdhyan Singh
Nerenberg, Morton Abraham
Ogata, Hisashi
Ottensmeyer, Frank Peter
Pascual, Roberto
Perz, John Mark
Piercy, George Robert
Pintar, M(ilan) Mik
Popovic, Zoran
Powell, Brian M
Reddoch, Allan Harvey
Rochon, Paul Leo
Rolfe, John
Roulston, David J
Ruda, Harry Eugen
Salvadori, Antonio
Sawicka, Barbara Danuta
Sayer, Michael
Schlesinger, Mordechay
Sears, Varley Fullerton
Segel, Stanley Lewis
Selvakumar, Chettypalayam Ramanathan
Shewchun, John
Slater, Gary W
Song, Kong-Sop Augustin
Sonnenberg, Hardy
Stadnik, Zbigniew M
Stager, Carl Vinton
Stinson, Michael Roy
Straus, Jozef
Sundaresan, Mosur Kalyanaraman
Svensson, Eric Carl
Szabo, Alexander
Tam, Wing-Gay
Taylor, David Ward
Templeton, Ian M
Thomas, Raye Edward
Thompson, David Allan
Tong, Bok Yin
Tun, Zin
Vanderkooy, John
Van Driel, Henry Martin
Vincett, Paul Stamford
Vosko, Seymour H
Walton, Derek
Wang, Shao-Fu
Wei, L(ing) Y(un)
Yevick, David Owen
Zukotynski, Stefan

PRINCE EDWARD ISLAND
Madan, Mahendra Pratap

QUEBEC
Aktik, Cetin
Banville, Marcel
Bartnikas, Ray
Caille, Alain Emeril
Champness, Clifford Harry
Chaubey, Mahendra
Cochrane, Robert W
Crine, Jean-Pierre C
Fuchs, Vladimir
Grant, Martin
Harris, Richard
Izquierdo, Ricardo
Jordan, Byron Dale
Kahrizi, Mojtaba
McIntyre, Robert John
Madore, Guy Louis Joseph
Marchand, Richard
Masut, Remo Antonio
Meunier, Michel
Misra, Sushil
Morris, Stanley P
Salahub, Dennis Russell
Savard, Jean Yves
Sharma, Ramesh C
Strom-Olsen, John Olaf
Subramanian, Sesha

Terreault, Bernard J E J
Tremblay, André-Marie
Van Vliet, Carolyne Marina
Vijh, Ashok Kumar
Wertheimer, Michael Robert
Yelon, Arthur Michael

SASKATCHEWAN
Kos, Joseph Frank
Papini, Giorgio Augusto

OTHER COUNTRIES
Anantharaman, Tanjore Ramachandra
Azbel, Mark
Baratoff, Alexis
Bauer, Ernst Georg
Bottka, Nicholas
Braginski, Aleksander Ignace
Cardona, Manuel
Chandrasekhar, B S
Chang, L(eroy) L(i-Gong)
Choi, Sang-il
Cue, Nelson
Economou, Eleftherios Nickolas
Edmonds, James W
Eng, Sverre T(horstein)
Fischer, Albert G
Foglio, Mario Eusebio
Gauster, Wilhelm Belrupt
Greenfield, Arthur Judah
Haymet, Anthony Douglas-John
Heiblum, Mordehai
Hornbogen, Erhard
Irvine, Stuart James Curzon
Katz, Gerald
Keller, Jaime
Koch, J Frederick
Koehler, Helmut A
Kwok, Hoi S
Lee, Elhang Howard
Ling, Samuel Chen-Ying
Low, William
Lowdin, Per-Olov
Maas, Peter
McInerney, John Gerard
Margaritondo, Giorgio
Merz, Walter John
Miyano, Kenjiro
Moser, Frank
Mota, Ana Celia
Neel, Louis Eugene Felix
Ng, Tai-Kai
Nyburg, Stanley Cecil
Otsuka, Kazuhiro
Purbo, Onno Widodo
Rivier, Nicolas Yves
Safran, Samuel A
Salaneck, William R
Sessler, Gerhard Martin
Shimizu, Ken'ichi
Tantraporn, Wirojana
Tsong, Tien Tzou
Ushioda, Sukekatsu
Van Overstraeten, Roger Joseph
Von Klitzing, Klaus
Wagner, Timothy Knight
Wasa, Kiyotaka
Weil, Raoul Bloch
Wiser, Nathan
Woods, Alfred David Braine
Wu, Yong-Shi
Yanabu, Satoru
Zabel, Hartmut

Spectroscopy & Spectrometry

ALABAMA
Bailey, Wayne Lewis
Leonard, Kathleen Mary
McKannan, Eugene Charles
Miller, George Paul
Nordlund, Thomas Michael

ARIZONA
Bao, Qingcheng
Breed, William Joseph
Daw, Glen Harold
Giampapa, Mark Steven
Marzke, Robert Franklin
Peyghambarian, Nasser
Rez, Peter
Sprague, Ann Louise
Venables, John Anthony
Yelle, Roger V

ARKANSAS
Cardwell, David Michael
Muly, Emil Christopher, Jr

CALIFORNIA
Almasan, Carmen Cristina
Amirkhanian, Varouj David
Bajaj, Jagmohan
Biedenbender, Michael David
Bowman, Robert Clark, Jr
Bridges, Frank G
Brundle, Christopher Richard
Caird, John Allyn
Camp, David Conrad
Chambers, Robert J
Cummings, Alan C
Curtis, Earl Clifton, Jr
Davis, Sumner P

Donoho, David Leigh
Drake, Richard Paul
Fouquet, Julie
Fowler, Charles Arman, Jr
Garrard, Thomas L
Gat, Nahum
Hovanec, B(ernard) Michael
Ivanov, Igor C
Jacobs, Ralph R
Jeffries, Jay B
Jing, Liu
Jusinski, Leonard Edward
Kim, Jinchoon
Lakkaraju, H S
Leite, Richard Joseph
Lorents, Donald C
Martz, Harry Edward, Jr
Moerner, W(illiam) E(sco)
Monard, Joyce Anne
Musket, Ronald George
Nee, Soe-Mie Foeng
Onton, Aare
Palatnick, Barton
Passell, Thomas Oliver
Perel, Julius
Perkins, Willis Drummond
Pertica, Alexander Jose
Pianetta, Piero Antonio
Salaita, George N
Satten, Robert A
Seki, Hajime
Short, Michael Arthur
Shuh, David Kelly
Shykind, David
Smith, Lloyd
Stohr, Joachim
Strauss, Herbert L
Swalen, Jerome Douglas
Todd, Terry Ray
Townes, Charles Hard
Weber, Marvin John
Wessel, John Emmit
Wilson, Barbara Ann
Wu, Lei
Wu, Robert Chung-Yung

COLORADO
Eaton, Sandra Shaw
Goetz, Alexander Franklin Hermann
Harlan, Ronald A
Kinsinger, James A
Murphy, Robert Carl
O'Callaghan, Michael James
Zhang, Bing-Rong

CONNECTICUT
Chang, Ted T
Dolan, James F
Melcher, Charles L

DELAWARE
Daniels, William Burton
Dwivedi, Anil Mohan
Szalewicz, Krzysztof

DISTRICT OF COLUMBIA
Brown, Charles Moseley
Campillo, Anthony Joseph
Cook, John William
Corliss, Charles Howard
Goldstein, Jeffrey Jay
Hornstein, John Stanley
Meekins, John Fred
Misra, Prabhakar
Oosterhuis, William Tenley
Segnan, Romeo A

FLORIDA
Andrew, Edward Raymond
Block, Ronald Edward
Hammond, Gordon Leon
Kimble, Allan Wayne
Lind, David Melvin
Ludwig, Matthias Heinz
Mrozowski, Stanislaw W
Mukherjee, Pritish
Pittman, Melvin Amos
Searle, Norma Zizmer
Shelton, Wilford Neil
Xu, Renliang

GEORGIA
Anantha Narayanan, Venkataraman
De Sa, Richard John
Hart, Raymond Kenneth
Herget, William F
McMillan, Robert Walker
Meltzer, Richard S
Yen, William Maoshung

HAWAII
Crowe, Richard Alan

IDAHO
Wheeler, Gilbert Vernon

ILLINOIS
Ali, Naushad
Crumrine, David Shafer
Curry, Bill Perry
Cutnell, John Daniel
Dehmer, Patricia Moore
Dunford, Robert Walter
Levi-Setti, Riccardo

Spectroscopy & Spectrometry (cont)

McCormack, Elizabeth Frances
Melendres, Carlos Arciaga
Neuhalfen, Andrew J
Parker, John Calvin
Scheeline, Alexander
Schulz, Charles Emil
Yule, Herbert Phillip

INDIANA
Ascarelli, Gianni
Hults, Malcom E

KANSAS
Andrew, Kenneth L
Cocke, C Lewis
Long, Larry L
Macke, Gerald Fred
Sobczynski, Radek

LOUISIANA
Kurtz, Richard Leigh
Meriwether, John R
Scott, John Delmoth

MARYLAND
Benesch, William Milton
Benson, Richard C
Berger, Robert Lewis
Boyd, Derek Ashley
Brasunas, John Charles
Castle, Peter Myer
Criss, John W
Dick, Kenneth Anderson
Drew, Howard Dennis
Duignan, Michael Thomas
Elton, Raymond Carter
Ginter, Marshall L
Kastner, Sidney Oscar
Kostkowski, Henry John
Leapman, Richard David
Liang, Shoudeng
Martin, William Clyde
Neumann, Dan Alan
Neupert, Werner Martin
Phillips, William Daniel
Plotkin, Henry H
Saloman, Edward Barry
Sattler, Joseph Peter
Seliger, Howard Harold
Skiff, Frederick Norman
Suenram, Richard Dee
Warren, Wayne Hutchinson, Jr
Weber, Alfons
Weber, Mark
Wiese, Wolfgang Lothar

MASSACHUSETTS
Allen, Lisa Parechanian
Bansil, Rama
Brierley, Steven K
Carleton, Nathaniel Phillips
Clough, Shepard Anthony
Collins, Aliki Karipidou
Gardner, James A
Handelman, Eileen T
Knecht, David Jordan
Liu, Xichun
Miniscalco, William J
Oettinger, Peter Ernest
Pearson, Edwin Forrest
Picard, Richard Henry
Pritchard, David Edward
Rao, Devulapalli V G L N
Rupich, Martin Walter
Schattenburg, Mark Lee
Schempp, Ellory
Tiernan, Robert Joseph

MICHIGAN
Kim, Yeong Wook
Nyquist, Richard Allen
Peterson, Lauren Michael
Vaishnava, Prem P

MINNESOTA
DiBona, Peter James
Dickson, Arthur Donald
Steck, Daniel John

MISSISSIPPI
Monts, David Lee

MISSOURI
Bragg, Susan Lynn
Bryan, David A
Mori, Erik Jun
Sonnino, Carlo Benvenuto

NEBRASKA
Billesbach, David P
Fabrikant, Ilya I
Gross, Michael Lawrence
Jorgensen, T, Jr
Schadt, Randall James

NEVADA
Eastwood, DeLyle

NEW HAMPSHIRE
Ge, Weikun

Winn, John Sterling

NEW JERSEY
Barrett, Joseph John
Bentz, Bryan Lloyd
Broer, Matthijs Meno
Chung, Yun C
Citrin, Paul H
Damen, Theo C
Girit, Ibrahim Cem
Harris, Alexander L
Harris, Leonce Everett
Hill, Kenneth Wayne
Li, Hong
Marcus, Robert Troy
Platzman, Philip M
Sulewski, Paul Eric
Wadlow, David
Woodward, Ted K

NEW MEXICO
Ambrose, William Patrick
Apel, Charles Turner
Bieniewski, Thomas M
Blake, Richard L
Cartwright, David Chapman
Close, Donald Alan
Cox, Lawrence Edward
Hoffbauer, Mark Arles
Kelley, Gregory M
Lee, Kenneth
Loree, Thomas Robert
Palmer, Byron Allen
Parker, Jack Lindsay
Robertson, Merton M
Rokop, Donald J
Taylor, Raymond Dean
Thompson, David Charles
Thompson, Jill Charlotte
Valencia, Valorie Sharron
Werbelow, Lawrence Glen

NEW YORK
Becker, Kurt Heinrich
Clark, David Delano
Dreyfus, Russell Warren
Duclos, Steven J
Elder, Fred A
Folan, Lorcan Michael
Garvey, James F
Greenbaum, Steven Garry
Koutcher, Jason Arthur
Matienzo, Luis J
Mercer, Kermit R
Mizes, Howard Albert
Moy, Dan
Newman, Jay Edward
Pellegrino, Francesco
Persans, Peter D
Philips, Laura Alma
Pliskin, William Aaron
Reddy, Reginald James
Varanasi, Prasad
Walmsley, Ian Alexander

NORTH CAROLINA
Ade, Harald Werner
Bonin, Keith Donald
Periasamy, Ammasi

OHIO
Berthold, John William, III
Cochran, William Ronald
Hirt, Andrew Michael
Hwang, Soon Muk
Johnson, Ray O
Kash, Kathleen
Miyoshi, Kazuhisa
Yaney, Perry Pappas
Yu, Thomas Huei-Chung

OREGON
Park, Kwangjai
Solanki, Raj

PENNSYLVANIA
Allara, David Lawrence
Ernst, Wolfgang E
Gao, Quanyin
Gurer, Emir
Holm, Reimer
Kozlowski, Richard William Hugh
Liberman, Irving
Ryan, Frederick Merk
Schneider, Donald P
Steinbruegge, Kenneth Brian
Tingle, William Herbert
Tredicce, Jorge Raul
Willey, Daniel Robert

SOUTH CAROLINA
Berlinghieri, Joel Carl

TENNESSEE
Arlinghaus, Heinrich Franz
Boatner, Lynn Allen
Fisher, Charles J
Garrett, William Ray
Haywood, Frederick F
Hoffman, Kenneth Wayne
Jellison, Gerald Earle, Jr
Kiech, Earl Lockett
Ogg, Robert James
Rice, Walter Wilburn

Wood, Bobby Eugene
Young, Jack Phillip

TEXAS
Busch, Kenneth Walter
Celii, Francis Gabriel
Church, David Arthur
Cocke, David Leath
DiFoggio, Rocco
Fry, Edward Strauss
Glosser, Robert
Golden, David E
Hulet, Randall Gardner
Mangum, Jeffrey Gary
Maute, Robert Edgar
Pitts, David Eugene
Potter, Andrew Elwin, Jr
Quade, Charles Richard
Rabalais, John Wayne
Rhodes, John Rathbone
Rubins, Roy Selwyn
Smith, R Lowell
Waggoner, James Arthur
Wang, Paul Weily
Wood, Lowell Thomas

UTAH
Grey, Gothard C
Powers, Linda Sue

VERMONT
Dunham, Jeffrey Solon

VIRGINIA
Andrews, William Lester Self
Brill, Arthur Sylvan
Cook, Desmond C
Crawford, George William
Delos, John Bernard
Dharamsi, Amin N
Drew, Russell Cooper
Exton, Reginald John
Johnson, Charles Nelson, Jr
Kramer, Steven David
Zallen, Richard

WASHINGTON
Baer, Donald Ray
Brodzinski, Ronald Lee
Collins, Gary Scott
McDowell, Robin Scott
Olsen, Kenneth Harold

WEST VIRGINIA
Sandridge, Robert Lee
Seehra, Mohindar Singh
Wallace, William Edward, Jr
Zhang, Xiao

WISCONSIN
Brandenberger, John Russell
Hobbs, Lewis Mankin
Roesler, Frederick Lewis
Sudhakaran, Gubbi Ramarao
Wright, John Curtis

ALBERTA
Ali, M Keramat
Egerton, Raymond Frank
Gong, Peng
Jones, R Norman
Weale, Gareth Pryce

BRITISH COLUMBIA
Affleck, Ian Keith
Buckmaster, Harvey Allen
Colbow, Kevin Michael
Irwin, John Charles
Ozier, Irving
Thewalt, Michael L W

MANITOBA
Attas, Ely Michael
Dong, Ronald Y
Ens, E(rich) Werner

NEWFOUNDLAND
Reddy, Satti Paddi

ONTARIO
Anderson, Anthony
Brand, John C D
Carpenter, Graham John Charles
Dagg, Ian Ralph
French, J(ohn) Barry
Ganza, Kresimir Peter
Lockwood, David John
McKellar, Robert A
Pintar, M(ilan) Mik
Ryan, Dave
Stadnik, Zbigniew M
Stoicheff, Boris Peter
Szabo, Alexander
Williams, Norman S W

QUEBEC
Kahrizi, Mojtaba
Pawluczky, Romuald
Waksberg, Armand L

SASKATCHEWAN
Durden, David Alan

OTHER COUNTRIES
Band, Yehuda Benzion
Butcher, Harvey Raymond, III
Kafri, Oded
Turrell, George Charles

Superconductors

CALIFORNIA
Berdahl, Paul Hilland
Clarke, John
Cohen, Marvin Lou
Field, R Clive
Hurrell, John Patrick
Kihlstrom, Kenneth Edward
Luongo, Cesar Augusto
Portis, Alan Mark
Rabinowitz, Mario
Schermer, Robert Ira
Smith, Andrew Donovan
Smith, Todd Iversen

COLORADO
Benz, Samuel P
Coffey, Mark William
Trefny, John Ulric

CONNECTICUT
Read, Nicholas

DISTRICT OF COLUMBIA
Fuller-Mora, Wendy Webb
Gubser, Donald Urban
Henry, Warren Elliott
Nisenoff, Martin
Skelton, Earl Franklin

FLORIDA
Hascicek, Yusuf Suat
Jacobson, Jerry I
Novotny, Mark A

ILLINOIS
Crabtree, George William
Dolan, Paul Joseph, Jr
Garg, Anupam
Geiser, Urs Walter
Greene, Laura H
Schultz, Arthur Jay
Shen, Sinyan

INDIANA
Chancey, Charles Clifton
Honig, Jurgen Michael

IOWA
Behroozi, Feredoon
Johnston, David Carl

MARYLAND
Boone, Bradley Gilbert
Clark, Alan Fred
Danchenko, Vitaly
Drew, Howard Dennis
Grabow, Barry Edward
Lynn, Jeffrey Whidden
Wang, Ching-Ping Shih
Yakovenko, Victor Mikhailovich

MASSACHUSETTS
Lyons, William Gregory
Skocpol, William John
Zimmerman, George Ogurek

MICHIGAN
Blosser, Henry Gabriel
Chen, Juei-Teng
Hunt, Thomas Kintzing

MINNESOTA
Gallo, Charles Francis

NEW JERSEY
Bhattacharya, Sabyasachi
Denlinger, Edgar J
Lindenfeld, Peter
Ong, Nai-Phuan
Rosen, Carol Zwick
Vandenberg, Joanna Maria

NEW MEXICO
Bedell, Kevin Shawn
Hammel, P Chris
Heffner, Robert H
Laquer, Henry L
Liepins, Raimond
Smith, James Lawrence
Thompson, Joe David
Wu, Xin Di

NEW YORK
Budzinski, Walter Valerian
Burns, Stephen James
Chan, Siu-Wai
Hart, Howard Roscoe, Jr
Kadin, Alan Mitchell
Likharev, Konstantin K
Wang, Jui Hsin

OHIO
Connolly, Denis Joseph
Erdman, Robert J
Newrock, Richard Sandor

Peterson, Timothy Lee

OREGON
Sleight, Arthur William

PENNSYLVANIA
Bose, Shyamalendu M
Carr, Walter James, Jr
Obermyer, Richard T
Talvacchio, John

SOUTH CAROLINA
Cronemeyer, Donald Charles

SOUTH DAKOTA
Quist, Oren Paul

TENNESSEE
Christen, David Kent
Lubell, Martin S

TEXAS
Chen, In-Gann
Chu, Wei-Kan
Li, Shu
McIntyre, Peter Mastin
Rubins, Roy Selwyn
Weinstein, Roy

UTAH
Kieda, David Basil
Symko, Orest George

VIRGINIA
Benesch, Jay F
Gunzburger, Max Donald
Noakes, David Ronald
Pan, Shing-Kuo
Satterthwaite, Cameron B

BRITISH COLUMBIA
Kubik, Philip Roman

ONTARIO
Griffin, Allan
Nagi, Anterdhyan Singh

OTHER COUNTRIES
Chandrasekhar, B S
Mota, Ana Celia

Theoretical Physics

ALABAMA
Baird, James Kern
Coulter, Philip W
Fowler, Bruce Wayne
Hinata, Satoshi
Howgate, David W
Kinzer, Earl T, Jr
Krause, Helmut G L
McKinley, John McKeen
Perez, Joseph Dominique
Pindzola, Michael Stuart
Rush, John Edwin, Jr
Smalley, Larry L
Smith, Charles Ray
Stettler, John Dietrich
Sung, Chi Ching
Visscher, Pieter Bernard
Wu, Xizeng

ARIZONA
Barrett, Bruce Richard
Benin, David R
Burrows, Adam Seth
Carruthers, Peter A
Garcia, Jose Dolores, Jr
Green, Barry A
Hestenes, David
Hetrick, David LeRoy
Iverson, A Evan
Jacob, Richard John
Jokipii, Jack Randolph
Kyrala, Ali
Lovelock, David
Mahmoud, Hormoz Massoud
Maier, Robert S
Melia, Fulvio
Patrascioiu, Adrian Nicolae
Rafelski, Johann
Ren, Shang Yuan
Scadron, Michael David
Thews, Robert L(eroy)

ARKANSAS
Harter, William George
Hobson, Arthur Stanley
Lieber, Michael
Webb, Jerry Glen

CALIFORNIA
Abarbanel, Henry D I
Acheson, Louis Kruzan, Jr
Adler, Ronald John
Alonso, Carol Travis
Amster, Harvey Jerome
Antolak, Arlyn Joe
Armstead, Robert Louis
Arnush, Donald
Aron, Walter Arthur
Asendorf, Robert Harry
Augenstein, Bruno (Wilhelm)

Aurilia, Antonio
Bander, Myron
Barnett, R(alph) Michael
Bars, Itzhak
Beardsley, Irene Adelaide
Beni, Gerardo
Bennett, Alan Jerome
Berdjis, Fazlollah
Berman, Sam Morris
Bing, George Franklin
Birnir, Björn
Blunt, Martin Julian
Boley, Charles Daniel
Brodsky, Stanley Jerome
Brueckner, Keith Allan
Burke, William L
Burton, Donald Eugene
Buskirk, Fred Ramon
Byers, Nina
Chang, Howard How Chung
Chapline, George Frederick, Jr
Chase, David Marion
Chen, Joseph Cheng Yih
Chew, Geoffrey Foucar
Chow, Brian G(ee-Yin)
Chow, Paul C
Christy, Robert Frederick
Cladis, John Baros
Cohen, Marvin Lou
Cohen, Ronald Herbert
Conzett, Homer Eugene
Cook, Robert Crossland
Cooper, Ralph Sherman
Corman, Emmett Gary
Cornwall, John Michael
Cummings, Frederick W
Dalton, Francis Norbert
Dashen, Roger Frederick
Davies, John Tudor
De, Gopa Sarkar
Dedrick, Kent Gentry
Depp, Joseph George
Doniach, Sebastian
Dorn, David W
Drell, Sidney David
Dresden, Max
Drummond, William Eckel
Dzyaloshinskii, Igor Ekhielievich
Erickson, Glen Walter
Estabrook, Frank Behle
Falicov, Leopoldo Maximo
Faulkner, John
Fearn, Heidi
Feit, Michael Dennis
Felber, Franklin Stanton
Fernbach, Sidney
Fetter, Alexander Lees
Finkelstein, Jerome
Fletcher, John George
Foiles, Stephen Martin
Fowler, Thomas Kenneth
Frahm, Charles Peter
Frazer, William Robert
Frederickson, Glenn Harold
Fried, Burton David
Fulco, Jose Roque
Gaillard, Mary Katharine
Ganas, Perry S
Gaposchkin, Peter John Arthur
Garren, Alper A
Garrison, John Carson
Garrod, Claude
Garwin, Charles A
Gaskell, Robert Weyand
Gillespie, George H
Glass, Nathaniel E
Glendenning, Norman Keith
Goldberger, Marvin Leonard
Goldman, Stanford
Good, Roland Hamilton, Jr
Goodjohn, Albert J
Gould, Robert Joseph
Green, Joseph Matthew
Gregorich, David Tony
Gross, Mark Warren
Guest, Gareth E
Gunion, John Francis
Haber, Howard Eli
Haener, Juan
Hall, Laurence Stanford
Halpern, Francis Robert
Halpern, Martin B
Hamill, Patrick James
Harker, Kenneth James
Harris, Charles Bonner
Hartle, James Burkett
Helliwell, Thomas McCaffree
Herman, Frank
Herman, Zelek Seymour
Herring, William Conyers
Hewett, Joanne Lea
Hiskes, John Robert
Hollenbach, David John
Holliday, Dennis
Hone, Daniel W
Horowitz, Gary T
Huberman, Bernardo Abel
Huddlestone, Richard H
Hundley, Richard O'Neil
Jackson, John David
Judd, David Lockhart
Jusinski, Leonard Edward
Kalensher, Bernard Earl
Karunasiri, Gamani

Kasameyer, Paul William
Kaus, Peter Edward
Kershaw, David Stanley
Kikuchi, Ryoichi
Kiskis, Joseph Edward, Jr
Koga, Toyoki
Koonin, Steven Elliot
Korringa, Jan
Krieger, Stephan Jacques
Kroll, Norman Myles
Kulander, Kenneth Charles
Lambropoulos, Peter Poulos
Langer, James Stephen
Latter, Albert L
Latter, Richard
Le Levier, Robert Ernest
Lepore, Joseph Vernon
Lewis, Harold Warren
Linde, Andrei
Lindenberg, Katja
Lingenfelter, Richard Emery
Loos, Hendricus G
Louie, Steven Gwon Sheng
Lurie, Joan B
Lynch, David Dexter
McCormick, Philip Thomas
MacDonald, Gordon James Fraser
Maki, Kazumi
Mandelstam, Stanley
Maradudin, Alexei
Massey, Walter Eugene
Mathews, Grant James
Matlow, Sheldon Leo
Mayer, Meinhard Edwin
Mazurek, Thaddeus John
Meads, Philip Francis, Jr
Miller, Jack Culbertson
Moe, Mildred Minasian
Morawitz, Hans
More, Richard Michael
Moriarty, John Alan
Morrison, Harry Lee
Moszkowski, Steven Alexander
Murad, Emil Moise
Nedelsky, Leo
Nedoluha, Alfred K
Nelson, Eldred (Carlyle)
Nesbet, Robert Kenyon
Nevins, William McCay
Newcomb, William A
Norton, Richard E
Nosanow, Lewis H
Noyes, H Pierre
Olness, Robert James
O'Malley, Thomas F
O'Malley, Thomas Francis
Pardee, William Joseph
Payne, David Glenn
Pearlstein, Leon Donald
Peccei, Roberto Daniele
Perkins, Kenneth L(ee)
Perkins, Sterrett Theodore
Peskin, Michael Edward
Pincus, Philip A
Preskill, John P
Primack, Joel Robert
Pritchett, Philip Lentner
Rauch, Richard Travis
Redlich, Martin George
Ree, Francis H
Reid, Roderick Vincent, Jr
Rensink, Marvin Edward
Richardson, John Mead
Riddell, Robert James, Jr
Roberts, Charles A, Jr
Rockower, Edward Brandt
Rosenbluth, Marshall N
Rosenkilde, Carl Edward
Ross, Marvin Franklin
Rouse, Carl Albert
Rudnick, Joseph Alan
Runge, Richard John
Sachs, Rainer Kurt
Sadoulet, Bernard
Sarfatti, Jack
Sawyer, Raymond Francis
Saxon, David Stephen
Scalettar, Richard
Schechter, Martin
Schlafly, Roger
Schwartz, Charles Leon
Schwarz, John Henry
Schweizer, Felix
Scofield, James Howard
Scott, Bruce L
Seki, Ryoichi
Senitzky, Israel Ralph
Sessler, Andrew M
Sham, Lu Jeu
Shaw, Gordon Lionel
Shenker, Scott Joseph
Shin, Ernest Eun-Ho
Shore, Herbert Barry
Silk, Joseph Ivor
Simon, Barry Martin
Smith, Gary Richard
Steinert, Leon ALbert
Stevenson, David John
Stuart, George Wallace
Sugar, Robert Louis
Suzuki, Mahiko
Tamor, Stephen
Tarter, Curtis Bruce
Taylor, Howard S

Tenn, Joseph S
Thorne, Kip Stephen
Tin, Hla Ngwe
Tracy, Craig Arnold
True, William Wadsworth
Tsai, Yung Su
Van de Walle, Chris G
Vega, Robeto
Waggoner, Jack Holmes, Jr
Wagoner, Robert Vernon
Waltz, Ronald Edward
Warnock, Robert Lee
Weichman, Peter Bernard
Wells, Willard H
Wheelon, Albert Dewell
Wichmann, Eyvind Hugo
Wilcox, Charles Hamilton
Williams, James Gerard
Winter, Nicholas Wilhelm
Wong, Chun Wa
Woodruff, Truman Owen
Woods, Roger David
Wright, Jon Alan
Wulfman, Carl E
Yano, Fleur Belle
Yura, Harold Thomas
Zachariasen, Fredrik
Zee, Anthony
Zeleny, William Bardwell
Zhou, Simon Zheng
Zimmerman, Ivan Harold
Zumino, Bruno

COLORADO
Ashby, Neil
Augusteijn, Marijke Francina
Barut, Asim Orhan
Beale, Paul Drew
Beeman, David Edmund, Jr
Bierman, Arthur
Brittin, Wesley E
Brown, Edmund Hosmer
Brown, James T, Jr
Chang, Bunwoo Bertram
Chappell, Willard Ray
Chew, Herman W
Cole, Lee Arthur
DeGrand, Thomas Alan
DeSanto, John Anthony
Downs, Bertram Wilson, Jr
Froyen, Sverre
Geltman, Sydney
Hansen, Richard Olaf
Herring, Jackson Rea
Hundhausen, Arthur James
Iddings, Carl Kenneth
Krueger, David Allen
McCray, Richard A
Meiss, James Donald
Miller, James Gilbert
Mohapatra, Pramoda Kumar
Peterson, Robert Lee
Rainwater, James Carlton
Randa, James P
Rinker, George Albert, Jr
Sakakura, Arthur Yoshikazu
Scales, John Alan
Smith, Earl W
Stephens, Jeffrey Alan
Swanson, Lawrence Ray
Taylor, John Robert
Tuttle, Elizabeth R
Yeatts, Frank Richard

CONNECTICUT
Appelquist, Thomas
Baierlein, Ralph Frederick
Bartram, Ralph Herbert
Cable, Peter George
Chun, Kee Won
Cordery, Robert Arthur
Fenton, David George
Ftaclas, Christ
Goldstein, Rubin
Gürsey, Feza
Hadjimichael, Evangelos
Hahn, Yukap
Haller, Kurt
Herzenberg, Arvid
Hill, David Lawrence
Islam, Muhammad Munirul
Johnson, David Linton
Klemens, Paul Gustav
MacDowell, Samuel Wallace
McIntosh, John Stanton
Michels, H(orace) Harvey
Newton, Victor Joseph
Peterson, Gerald A
Picker, Harvey Shalom
Porter, William Samuel
Rawitscher, George Heinrich
Read, Nicholas
Russu, Irina Maria
Sachdev, Subir
Sen, Pabitra Narayan
Shankar, Ramamurti
Sommerfield, Charles Michael
Stone, A Douglas
Wardell, James C

DELAWARE
Chui, Siu-Tat
Glyde, Henry Russell
Hill, Robert Nyden

Theoretical Physics (cont)

Kerner, Edward Haskell
Leung, Chung Ngoc
Pittel, Stuart

DISTRICT OF COLUMBIA
Bergmann, Otto
Boss, Alan Paul
Brennan, James Gerard
Clinton, William L
Ehrlich, Alexander Charles
Eisenstein, Julian (Calvert)
Feldman, Joseph Louis
Haftel, Michael Ivan
Huba, Joseph D
Jordan, Arthur Kent
Kirwan, Donald Frazier
Klein, Lewis S
Lampe, Martin
Langworthy, James Brian
Lehman, Donald Richard
Lin-Chung, Pay-June
Lustig, Harry
McClure, Joseph Andrew, Jr
Meijer, Paul Herman Ernst
Mendlowitz, Harold
Palmadesso, Peter Joseph
Photiadis, Douglas Marc
Picone, J Michael
Rajagopal, Attipat Krishnaswamy
Reilly, Michael Hunt
Reinecke, Thomas Leonard
Reiss, Howard R
Rubin, Robert Joshua
Sáenz, Albert William
Schultz, Theodore David
Serene, Joseph William
Shapero, Donald Campbell
Trubatch, Sheldon L
Uberall, Herbert Michael
Werntz, Carl W
White, John Arnold

FLORIDA
Albright, John Rupp
Baird, Alfred Michael
Bose, Subir Kumar
Broyles, Arthur Augustus
Buchler, Jean-Robert
Curtright, Thomas Lynn
Dagotto, Elbio Ruben
Darling, Byron Thorwell
Desloge, Edward Augustine
Detweiler, Steven Lawrence
Dufty, James W
Duke, Dennis Wayne
Edwards, Steve
Faulkner, John Samuel
Field, Richard D
Fry, James N
Halpern, Leopold (Ernst)
Hart, Robert Warren
Hooper, Charles Frederick, Jr
Kazaks, Peter Alexander
Kimel, Jacob Daniel, Jr
Klauder, John Rider
Kumar, Pradeep
Kursunoglu, Behram N
Manousakis, Efstratios
Marshak, Robert Eugene
Mintz, Stephen Larry
Monkhorst, Hendrik J
Muttalib, Khandker Abdul
Nepomechie, Rafael Israel
Novotny, Mark A
Painter, Ronald Dean
Perlmutter, Arnold
Sabin, John Rogers
Sakmar, Ismail Aydin
Schlottmann, Pedro U J
Scruggs, Frank Dell
Spector, Richard M
Speisman, Gerald
Stegeman, George I
Sullivan, Neil Samuel
Thomas, Billy Seay
Thorn, Charles Behan, III
Trickey, Samuel Baldwin
Wille, Luc Theo
Williams, Gareth
Woodard, Richard P

GEORGIA
Abbe, Winfield Jonathan
Adomian, George
Batten, George L, Jr
Biritz, Helmut
Bowling, Arthur Lee, Jr
Chou, Mei-Yin
Chou, T(su) T(eh)
Dixon, Robert Malcolm
Duncan, Marion M, Jr
Family, Fereydoon
Finkelstein, David
Flannery, Martin Raymond
Fong, Peter
Ford, Joseph
Freiser, Marvin Joseph
Gallaher, Lawrence Joseph
Gatland, Ian Robert
Harrell, Evans Malott, II
Henkel, John Harmon
Jenkins, Hughes Brantley, Jr
Kenan, Richard P
Lee, M(onhe) Howard
Marks, Dennis William
Pandres, Dave, Jr
Payne, Marvin Gay
Pollard, William Blake
Sinha, Om Prakash
Strobel, George L
Tanner, James Mervil
Uzer, Ahmet Turgay
Uzes, Charles Alphonse
Valk, Henry Snowden
Wiita, Paul Joseph
Zangwill, Andrew

HAWAII
Tata, Xerxes Ramyar
Tuan, San Fu
Yamauchi, Hiroshi

IDAHO
Ivey, Jerry Lee
Majumdar, Debaprasad (Debu)
Morley, Gayle L
Parker, Barry Richard
Patsakos, George
Sieckmann, Everett Frederick
Switendick, Alfred Carl

ILLINOIS
Aref, Hassan
Avery, Robert
Axford, Roy Arthur
Bart, George Raymond
Baym, Gordon A
Benedek, Roy
Bodmer, Arnold R
Bowen, Samuel Philip
Brown, Laurie Mark
Campbell, David Kelly
Carhart, Richard Alan
Ceperley, David Matthew
Chan, Sai-Kit
Chang, Shau-Jin
Cohen, Stanley
Davis, A Douglas
Day, Michael Hardy
Dykla, John J
Eichten, Estia Joseph
Eubanks, Robert Alonzo
Evans, Kenneth, Jr
Fano, Ugo
Freund, Peter George Oliver
Gabriel, John R
Garg, Anupam
Gaylord, Richard J
Geroch, Robert Paul
Gilbert, Thomas Lewis
Golestaneh, Ahmad Ali
Hart, Harold Bird
Heaston, Robert Joseph
Henderson, George Asa
Hill, Christopher T
Inokuti, Mitio
Johnson, Porter W
Kadanoff, Leo P
Kimura, Mineo
Klemm, Richard Andrew
Kogut, John Benjamin
Kolb, Edward William
Kurath, Dieter
Lamb, Donald Quincy, Jr
Lamb, Frederick Keithley
Lange, Yvonne
Lawson, Robert Davis
Lellouche, Gerald S
Maclay, G Jordan
Malik, Fazley Bary
Mazenko, Gene Francis
Monahan, James Emmett
Myron, Harold William
Nambu, Yoichiro
Ng, King-Yuen
Oakes, Robert James
Oehme, Reinhard
Oono, Yoshitsugu
Pandharipande, Vijay Raghunath
Pappademos, John Nicholas
Peshkin, Murray
Pieper, Steven Charles
Pines, David
Quigg, Chris
Ravenhall, David Geoffrey
Redmount, Ian H
Rosner, Jonathan Lincoln
Sachs, Robert Green
Schult, Roy Louis
Schweizer, Kenneth Steven
Sharma, Ram Ratan
Siegert, Arnold John Frederick
Smarr, Larry Lee
Stack, John D
Sukhatme, Uday Pandurang
Swamy, Padmanabha Narayana
Teng, Lee Chang-Li
Thomas, Gerald H
Truran, James Wellington, Jr
Uretsky, Jack Leon
Uslenghi, Piergiorgio L
Wald, Robert Manuel
Warf, C Cayce, Jr
Witten, Thomas Adams, Jr
Wolf, Dieter
Wolynes, Peter Guy

Zachos, Cosmas K

INDIANA
Baker, Howard Crittenden
Balazs, Louis A P
Bose, Samir K
Capps, Richard H
Chancey, Charles Clifton
Day, Benjamin Downing
Dixon, Henry Marshall
Fuchs, Norman H
Gartenhaus, Solomon
Girvin, Steven M
Gottlieb, Steven Arthur
Hanson, Andrew Jorgen
Harris, Samuel M(elvin)
Kim, Yeong Ell
Langhoff, Peter Wolfgang
Lichtenberg, Don Bernett
Londergan, John Timothy
LoSecco, John M
MacDonald, Allan Hugh
Macfarlane, Malcolm Harris
McGlinn, William David
Mast, Cecil B
Meiere, Forrest T
Newton, Roger Gerhard
Overhauser, Albert Warner
Parish, Jeffrey Lee
Robinson, John Murrell
Rodriguez, Sergio
Salter, Lewis Spencer
Schaich, William Lee
Shore, Steven Neil
Swihart, James Calvin
Tubis, Arnold
Vasavada, Kashyap V
Walker, George Edward
Weston, Vaughan Hatherley
Wills, John G

IOWA
Byers, Ronald Elner
Carlson, Bille Chandler
Clem, John R
Jacob, Richard L
Klink, William H
Lassila, Kenneth Eino
Leacock, Robert A
Luban, Marshall
Marker, David
Payne, Gerald Lew
Pursey, Derek Lindsay
Rizzo, Thomas Gerard
Ross, Dennis Kent
Ruedenberg, Klaus
Schweitzer, John William
Snow, Joel A
Vary, James P
Williams, Stanley A

KANSAS
Folland, Nathan Orlando
Mainster, Martin Aron
Ralston, John Peter
Strecker, Joseph Lawrence
Weaver, Oliver Laurence
Welling, Daniel J

KENTUCKY
Chalmers, Joseph Stephen
Connolly, John William Domville
Elitzur, Moshe
Lehman, Guy Walter
MacKellar, Alan Douglas
Perry, Peter Anton
Subbaswamy, Kumble R(amarao)
Van Winter, Clasine
Yost, Francis Lorraine

LOUISIANA
Brans, Carl Henry
Callaway, Joseph
Carter, James Clarence
Draayer, Jerry Paul
Edmonds, James D, Jr
Head, Charles Everett
Matese, John J
O'Connell, Robert F
Perdew, John Paul
Purrington, Robert Daniel
Rau, A Ravi Prakash
Rosensteel, George T
Slaughter, Milton Dean
Tipler, Frank Jennings, III
Vashishta, Priya Darshan
Witriol, Norman Martin

MAINE
Beard, David Breed
Csavinszky, Peter John
McKay, Susan Richards
Morrow, Richard Alexander
Pilot, Christopher H
Semon, Mark David

MARYLAND
Atanasoff, John Vincent
Aviles, Joseph B
Bird, Joseph Francis
Bonavita, Nino Louis
Boye, Robert James
Brill, Dieter Rudolf
Burke, Edward Raymond

Campolattaro, Alfonso
Cawley, Robert
Chan, Kwing Lam
Cook, Richard Kaufman
Criss, Thomas Benjamin
Danos, Michael
Davis, Frederic I
Dickey, Joseph W
DiMarzio, Edmund Armand
Dorfman, Jay Robert
Drachman, Richard Jonas
Dragt, Alexander James
Einstein, Theodore Lee
Engle, Irene May
Feldman, Gordon
Ferrell, Richard Allan
Fivel, Daniel I
Frankel, Michael Jay
Franz, Judy R
Fulton, Thomas
Gadzuk, John William
Gates, Sylvester J, Jr
Gaunaurd, Guillermo C
Gilinsky, Victor
Glick, Arnold J
Gray, Ernest Paul
Grayson, William Curtis, Jr
Greenberg, Oscar Wallace
Griffin, James J
Grube, Geraldine Joyce Terenzoni
Guier, William Howard
Haig, Frank Rawle
Hakkinen, Sirpa Marja Anneli
Hayward, Raymond (W)ebster
Herrmann, Robert Arthur
Hu, Bei-Lok Bernard
Jones, George R
Judd, Brian Raymond
Kacser, Claude
Kaplan, Alexander E
Keenan, Thomas Aquinas
Kim, Chung W
Kirkpatrick, Theodore Ross
Korenman, Victor
Krause, Thomas Otto
Land, David J(ohn)
Libelo, Louis Francis
MacDonald, Rosemary A
MacDonald, William
Manning, Irwin
Markley, Francis Landis
Marx, Egon
Meckler, Alvin
Miller, James Taggert, Jr
Misner, Charles William
Mountain, Raymond Dale
O'Connell, James S
Oneda, Sadao
Ozernoy, Leonid M
Pati, Jogesh Chandra
Prats, Francisco
Redish, Edward Frederick
Robbins, Mark Owen
Robertson, Baldwin
Rosenstock, Herbert Bernhard
Roszman, Larry Joe
Rubin, Morton Harold
Rynasiewicz, Robert Alan
Sachs, Lester Marvin
Schmid, Lawrence Alfred
Schumacher, Clifford Rodney
Schwartz, Brian B
Sidhu, Deepinder Pal
Simmons, John Arthur
Soln, Josip Zvonimir
Steven, Alasdair C
Thirumalai, Devarajan
Toll, John Sampson
Valenzuela, Gaspar Rodolfo
Van Vechten, Deborah
Wallace, Stephen Joseph
Wang, Ching-Ping Shih
Webb, Richard A
Yaes, Robert Joel
Yakovenko, Victor Mikhailovich
Yedinak, Peter Demerton
Yoon, Peter Haesung
Zerilli, Frank J
Zucker, Paul Alan

MASSACHUSETTS
Aaron, Ronald
Altman, Albert
Antonoff, Marvin M
Argyres, Petros
Baker, Adolph
Bakshi, Pradip M
Baranger, Michel
Baublitz, Millard, Jr
Berker, Ahmet Nihat
Bernstein, Herbert J
Bertschinger, Edmund
Blankschtein, Daniel
Brynjolfsson, Ari
Burke, Edward Aloysius
Calusdian, Richard Frank
Carovillano, Robert L
Carpenter, Jack William
Carter, Ashton Baldwin
Charpie, Robert Alan
Chivukula, R Sekhar
Coleman, James Andrew
Coleman, Sidney Richard
Collins, Aliki Karipidou

Cook, LeRoy Franklin, Jr
Corinaldesi, Ernesto
Crowell, Julian
Dalgarno, Alexander
Deser, Stanley
Dewan, Edmond M
Donnelly, Thomas William
Donoghue, John Francis
Doyle, Jon
Drane, Charles Joseph, Jr
Durso, John William
Ehrenreich, Henry
Eisenbud, Leonard
Eyges, Leonard James
Farhi, Edward
Ford, Lawrence Howard
Freedman, Daniel Z
Friedman, Marvin Harold
Frosch, Robert Alan
Georgi, Howard
Godine, John Elliott
Goldberg, Hyman
Goldstone, Jeffrey
Graneau, Peter
Guertin, Ralph Francis
Gunther, Leon
Guth, Alan Harvey
Halperin, Bertrand Israel
Harrison, Ralph Joseph
Heller, Eric Johnson
Hilsinger, Harold W
Holstein, Barry Ralph
Horrigan, Frank Anthony
Huang, Kerson
Jackiw, Roman Wladimir
Jacobs, Laurence Alan
Jaffe, Robert Loren
Jagannathan, Kannan
Joannopoulos, John Dimitris
Johnson, Kenneth Alan
Johnston, George Lawrence
Jones, Robert Clark
Jose, Jorge V
Joss, Paul Christopher
Kalman, Gabor J
Kannenberg, Lloyd C
Kaplan, Irving
Karakashian, Aram Simon
Kelley, Paul Leon
Kerman, Arthur Kent
Kincaid, Thomas Gardiner
Klein, William
Krass, Allan S(hale)
Krumhansl, James Arthur
Kuhn, Thomas S
Lai, Shu Tim
Larsen, David M
Leiby, Clare C, Jr
Lightman, Alan Paige
Lomon, Earle Leonard
Low, Francis Eugene
Machacek, Marie Esther
Machta, Jonathan Lee
Malenka, Bertram Julian
Martin, Arthur Wesley, III
Martin, Paul Cecil
Mellen, Walter Roy
Melrose, Richard B
Mollow, Benjamin R
Moniz, Ernest Jeffrey
Moomaw, William Renken
Moses, Harry Elecks
Mullin, William Jesse
Negele, John William
Nelson, David Robert
Park, David Allen
Pendleton, Hugh Nelson, III
Pratt, George Woodman, Jr
Press, William Henry
Rebbi, Claudio
Redner, Sidney
Rizzoli-Malanotte, Paola
Roberts, David Hall
Robertson, David C
Salzman, George
Savickas, David Francis
Schloemann, Ernst
Schnitzer, Howard J
Schwebel, Solomon Lawrence
Schweber, Silvan Samuel
Segal, Irving Ezra
Sheldon, Eric
Shimony, Abner
Simmons, Elizabeth H
Sneddon, Leigh
Stachel, John Jay
Stanley, H(arry) Eugene
Sternheim, Morton Maynard
Stevens, Herbert H(owe), Jr
Tambasco, Daniel Joseph
Tangherlini, Frank R
Tisza, Laszlo
Tomljanovich, Nicholas Matthew
Towne, Dudley Herbert
Vassell, Milton O
Vaughn, Michael Thayer
Victor, George A
Vilenkin, Alexander
Villars, Felix Marc Hermann
Walker, James Frederick, Jr
Weaver, David Leo
Whitney, Cynthia Kolb
Wisdom, Jack Leach
Wong, Po Kee

Wootters, William Kent
Wu, Fa Yueh
Wu, Tai Tsun
Yos, Jerrold Moore
Young, James Edward
Zielinska Pfabe, Malgorzata

MICHIGAN
Abbasabadi, Alireza
Agin, Gary Paul
Berger, Beverly Kobre
Blass, Gerhard Alois
Boojum, Basil John
Brailsford, Alan David
Chang, Jhy-Jiun
Chen, Min-Shih
Davis, Lloyd Craig
Denman, Harry Harroun
Devlin, John F
Einhorn, Martin B
Favro, Lawrence Dale
Federbush, Paul Gerard
Fishman, Frank J, Jr
Ford, George Willard
Fradkin, David Milton
Francis, William Porter
Freese, Katherine
Garfinkle, David
Gay, Jackson Gilbert
Gordon, Morton Maurice
Gupta, Suraj Narayan
Harrison, Michael Jay
Kaskas, James
Kemeny, Gabor
Kovacs, Julius Stephen
Kromminga, Albion Jerome
Kuo, Pao-Kuang
Lee, Sung Mook
Lewis, Robert Richards, Jr
McManus, Hugh
Mahanti, Subhendra Deb
Morgan, Caroline G
Reitz, John Richard
Richter, Roy
Rolnick, William Barnett
Ross, Marc Hansen
Rossi, Giuseppe
Sander, Leonard Michael
Saperstein, Alvin Martin
Savit, Robert Steven
Schnitker, Jurgen H
Signell, Peter Stuart
Slavin, Andrei Nickolay
Smith, John Robert
Tomozawa, Yukio
Tung, Wu-Ki
Wu, Alfred Chi-Tai
Yao, York-Peng Edward
Yun, Suk Koo

MINNESOTA
Austin, Donald Murray
Bayman, Benjamin
Campbell, Charles Edwin
Casper, Barry Michael
Cassola, Robert Louis
Halley, James Woods, (Jr)
Jordan, Thomas Fredrick
Kim, Sung Kyu
Korth, Michael Steven
Lysak, Robert Louis
McLerran, Larry Dean
Meyer, Frank Henry
Rudaz, Serge
Schuldt, Spencer Burt
Suura, Hiroshi
Tait, William Charles
Titus, William James
Valls, Oriol Tomas
Wesley, Walter Glen

MISSISSIPPI
Corben, Herbert Charles
Lestrade, John Patrick

MISSOURI
Adawi, Ibrahim (Hasan)
Cheng, Ta-Pei
Ching, Wai-Yim
Clark, John Walter
Cochran, Andrew Aaron
DeFacio, W Brian
Delaney, Robert Michael
Fisch, Ronald
Gammel, John Ledel
Hagen, Donald E
Huang, Justin C
Jaynes, Edwin Thompson
Johnson, Darell James
Kovacs, Sandor J, Jr
Mashhoon, Bahram
Parks, William Frank
Parry, Myron Gene
Retzloff, David George
Shrauner, James Ely
Taub, Haskell Joseph

MONTANA
Hiscock, William Allen
Kinnersley, William Morris
Nordtvedt, Kenneth L
Swenson, Robert J

NEBRASKA
Chakkalakal, Dennis Abraham
Davies, Marcia A
Fabrikant, Ilya I
Flocken, John W
Goss, David
Kennedy, Robert E
Longley, William Warren, Jr
Underhill, Glenn
Zimmerman, Edward John

NEVADA
Levine, Paul Hersh
Scott, William Taussig
Winterberg, Friedwardt

NEW HAMPSHIRE
Huggins, Elisha R
Lawrence, Walter Edward
Shepard, Harvey Kenneth

NEW JERSEY
Adler, Stephen L
Aizenman, Michael
Alig, Roger Casanova
Anderson, James Leroy
Anderson, Philip Warren
Anderson, Terry Lee
Banks, Thomas
Becken, Eugene D
Bernstein, Jeremy
Blount, Eugene Irving
Bronzan, John Brayton
Callan, Curtis
Chadi, James D
Chan, Chun Kin
Christodoulou, Demetrios
Cohen, Morrel Herman
Cowley, Ernest Roger
Derkits, Gustav
Dohnanyi, Julius S
Gale, William Arthur
Gautreau, Ronald
Goldin, Gerald Alan
Goode, Philip Ranson
Gora, Thaddeus F, Jr
Gott, J Richard, III
Grant, James J, Jr
Gross, David (Jonathan)
Haldane, Frederick Duncan Michael
Helfand, Eugene
Hodges, Hardy M
Hohenberg, Pierre Claude
Jackson, Shirley Ann
Joyce, William B(axter)
Katz, Sheldon Lane
Kelsey, Edward Joseph
Kijewski, Louis Joseph
Klebanov, Igor Romanovich
Langreth, David Chapman
Lieb, Elliott Hershel
Lines, Malcolm Ellis
Lohse, David John
Lovelace, Claud William Venton
Luryi, Serge
McAfee, Walter Samuel
Mattheiss, Leonard Francis
Nachtergaele, Bruno Leo Zulma
Periwal, Vipul
Phillips, James Charles
Platzman, Philip M
Raichel, Daniel R(ichter)
Reiman, Allan H
Rockmore, Ronald Marshall
Salwen, Harold
Schiller, Ralph
Schluter, Michael
Shapiro, Joel Alan
Shenker, Stephen Hart
Solla, Sara A
Spergel, David Nathaniel
Tishby, Naftali Z
Torrey, Henry Cutler
Vanderbilt, David Hamilton
Von Hippel, Frank
Weinstein, Marvin
Wheeler, John Archibald
Wigner, Eugene Paul
Wilczek, Frank Anthony
Wojtowicz, Peter Joseph
Wolff, Peter A
Wong, Tang-Fong Frank
Zapolsky, Harold Saul
Zatzkis, Henry

NEW MEXICO
Albers, Robert Charles
Alldredge, Gerald Palmer
Baker, George Allen, Jr
Barnes, Daniel C
Bartel, Lewis Clark
Beckel, Charles Leroy
Becker, Wilhelm
Bedell, Kevin Shawn
Bergeron, Kenneth Donald
Bolsterli, Mark
Brandow, Baird H
Campbell, Laurence Joseph
Caves, Carlton Morris
Chandler, Colston
Chen, Tuan Wu
Chow, Weng Wah
Cobb, Donald D
Cohen, James Samuel

Coon, Sidney Alan
Cooper, Richard Kent
Coulter, Claude Alton
Davis, Cecil Gilbert
Devaney, Joseph James
Dietz, David
Doolen, Gary Dean
Du Bois, Donald Frank
England, Talmadge Ray
Erpenbeck, Jerome John
Evans, Foster
Feibelman, Peter Julian
Finley, James Daniel, III
Frank, Robert Morris
Friar, James Lewis
Gell-Mann, Murray
Gerstl, Siegfried Adolf Wilhelm
Gibbs, William Royal
Gibson, Benjamin Franklin, V
Ginocchio, Joseph Natale
Ginsparg, Paul H
Glasser, Alan Herbert
Goldman, Terrence Jack
Goldstein, John Cecil
Goldstein, Louis
Gordon, James Wylie
Gubernatis, James Edward
Gursky, Martin Lewis
Hadley, George Ronald
Hammel, P Chris
Heller, Leon
Hirt, Cyril William, Jr
Hull, McAllister Hobart, Jr
Ingraham, Richard Lee
Jennison, Dwight Richard
Johnson, James Daniel
Johnson, Mikkel Borlaug
Jones, Eric Daniel
Kenkre, Vasudev Mangesh
Laney, Billie Eugene
Lazarus, Roger Ben
Leon, Melvin
Lilley, John Richard
Liu, Lon-Chang
Lomanitz, Ross
Longley, H(erbert) Jerry
Lyo, SungKwun Kenneth
Martin, Thierry Dominique
Mascheroni, P Leonardo
Mayer, Harris Louis
Metropolis, Nicholas Constantine
Newby, Neal Dow, Jr
Newman, Michael J(ohn)
Nieto, Michael Martin
Nix, James Rayford
O'Donnell, Edward Earl
Peaslee, Alfred Tredway, Jr
Rach, Randolph Carl
Rauber, Lauren A
Reichert, John Douglas
Richards, Peter Michael
Schappert, Gottfried T
Schriber, Stanley Owen
Schultz, Rodney Brian
Sharp, David Howland
Sierk, Arnold John
Silver, Richard N
Simmons, Leonard Micajah, Jr
Slansky, Richard Cyril
Solem, Johndale Christian
Stephenson, Gerard J, Jr
Strottman, Daniel
Thompson, Samuel Lee
Turner, Leaf
Visscher, William M
Walker, James Joseph
Wallace, Richard Kent
Wienke, Bruce Ray
Winske, Dan
Younger, Stephen Michael
Zardecki, Andrew
Zurek, Wojciech Hubert

NEW YORK
Abrahamson, Adolf Avraham
Agrawal, Govind P(rasad)
Allen, Philip B
Ambegaokar, Vinay
Arin, Kemal
Aronson, Raphael (Friedman)
Ashcroft, Neil William
Bak, Per
Balazs, Nandor Laszlo
Baltz, Anthony John
Berger, Jay Manton
Bergmann, Peter Gabriel
Beth, Eric Walter
Bethe, Hans Albrecht
Birman, Joseph Leon
Boardman, John
Borowitz, Sidney
Bowick, Mark John
Boyer, Timothy Howard
Bray, James William
Brinkman, William F
Brown, Edmond
Brown, Ronald Alan
Butkov, Eugene
Buttiker, Markus
Carroll, C(lark) E(dward)
Chang, Ngee Pong
Chung, Victor
Clark, Edward Aloysius
Courant, Ernest David

Theoretical Physics (cont)

Craxton, Robert Stephen
Creutz, Michael John
Das, Ashok Kumar
Diamond, Joshua Benamy
Doering, Charles Rogers
Dover, Carl Bellman
Dreiss, Gerard Julius
Dubisch, Russell John
Duke, Charles Bryan
Eberly, Joseph Henry
Eger, F Martin
Emery, Victor John
Falk, Harold
Feigenbaum, Mitchell Jay
Feinberg, Gerald
Fishbone, Leslie Gary
Forgacs, Gabor
Fox, David
Francis, Norman
Franco, Victor
Friedberg, Richard Michael
Frisch, Harry Lloyd
Fuda, Michael George
Fujita, Shigeji
Gelman, Donald
Geoghegan, Ross
Gersten, Joel Irwin
Ghosh, Vinita Johri
Gillespie, John
Glassgold, Alfred Emanuel
Goldhaber, Alfred Scharff
Gottfried, Kurt
Greenberg, Howard
Greenberg, Newton Isaac
Greenberger, Daniel Mordecai
Grobe, Rainer
Gutzwiller, Martin Charles
Gyulassy, Miklos
Harrington, Steven Jay
Hart, Edward Walter
Harvey, Alexander Louis
Haus, Joseph Wendel
Hioe, Foik Tjiong
Inomata, Akira
Jabbur, Ramzi Jibrail
Jackson, Andrew D, Jr
Jepsen, Donald William
Kahana, Sidney H
Kahn, Peter B
Kaku, Michio
Kalos, Malvin Howard
Kaplan, Harvey
Kaup, David James
Kent, Henry Johann
Khuri, Nicola Najib
Kinoshita, Toichiro
Kirtley, John Robert
Kjeldaas, Terje, Jr
Klarfeld, Joseph
Knight, Bruce Winton, (Jr)
Koltun, Daniel S
Komar, Arthur Baraway
Koplik, Joel
Krieger, Joseph Bernard
Krinsky, Samuel
Kuo, Thomas Tzu Szu
Lang, Norton David
Lavine, James Philip
Lax, Melvin
Lee, Tsung Dao
Lee, Yung-Chang
Lemos, Anthony M
Levinger, Joseph S
Liboff, Richard L
Lin, Duo-Liang
Lindquist, William Brent
Lowenstein, John Hood
Lubitz, Cecil Robert
Lucey, Carol Ann
Luttinger, Joaquin Mazdak
McCoy, Barry
McDonald, Frank Alan
Malin, Shimon
Manassah, Jamal Tewfek
Mane, Sateesh Ramchandra
Mansfield, Victor Neil
Margulies, Marcel
Marr, Robert B
Mathur, Vishnu Sahai
Meyer, Robert Jay
Millener, David John
Millet, Peter J
Mizes, Howard Albert
Mould, Richard A
Mueller, Alfred H
Mukhopadhyay, Nimai Chand
Nardi, Vittorio
Nelkin, Mark
Nelson, Charles Arnold
Newman, Charles Michael
Newstein, Maurice
Noz, Marilyn E
O'Dwyer, John J
Ohanian, Hans C
Pais, Abraham
Pantelides, Sokrates Theodore
Parzen, George
Peak, David
Pearle, Philip Mark
Pease, Robert Louis
Peierls, Ronald F
Percus, Jerome K

Privman, Vladimir
Rafanelli, Kenneth R
Ram, Michael
Raman, Varadaraja Venkata
Ratcliff, Keith Frederick
Remo, John Lucien
Rice, Michael John
Richardson, Robert William
Rohrlich, Fritz
Rosenberg, Leonard
Rosenzweig, Carl
Roskes, Gerald J
Ruderman, Malvin Avram
Sachs, Mendel
Sakita, Bunji
Schlitt, Dan Webb
Schwartz, Melvin J
Seiden, Philip Edward
Seyler, Charles Eugene
Shalloway, David Irwin
Shimamoto, Yoshio
Siggia, Eric Dean
Silverman, Benjamin David
Sirlin, Alberto
Smith, Gerrit Joseph
Smith, Jack Howard
Smith, John
Smolin, Lee
Sobel, Michael I
Spergel, Martin Samuel
Spruch, Larry
Srinivasan, G(urumakonda) R
Stern, Frank
Tavel, Morton
Toner, John Joseph
Trigg, George Lockwood
Troubetzkoy, Eugene Serge
Trueman, Thomas Laurence
Tryon, Edward Polk
Tye, Sze-Hoi Henry
Wali, Kameshwar C
Weingarten, Donald Henry
Weisberger, William I
Wellner, Marcel
Weneser, Joseph
Werthamer, N Richard
Wilson, Fred Lee
Yennie, Donald Robert
Yoffa, Ellen June
Zwanziger, Daniel

NORTH CAROLINA

Adler, Carl George
Aktas, Yildirim
Bernholc, Jerzy
Brehme, Robert W
Brown, J(ohn) David
Choudhury, Abdul Latif
Ciftan, Mikael
Cotanch, Stephen Robert
Coulter, Byron Leonard
D'Arruda, Jose Joaquim
Dolan, Louise Ann
Dotson, Allen Clark
Fulp, Ronald Owen
Griffing, George Warren
Hall, George Lincoln
Han, Moo-Young
Iafrate, Gerald Joseph
Ji, Chueng Ryong
Katzin, Gerald Howard
Kerr, William Clayton
Klenin, Marjorie A
Lado, Fred
Lapicki, Gregory
Merzbacher, Eugen
Muller, Berndt
Park, Jae Young
Robl, Hermann R
Torquato, Salvatore
Tsu, Raphael
Whitlock, Richard T
Yang, Weitao
York, James Wesley, Jr

NORTH DAKOTA

Hassoun, Ghazi Qasim
Kluk, Edward
Kurtze, Douglas Alan
Schwalm, Mizuho K
Schwalm, William A

OHIO

Allender, David William
Arnett, Jerry Butler
Breitenberger, Ernst
Dilley, James Paul
Ellis, David Greenhill
Esposito, F Paul
Foldy, Leslie Lawrance
Goodman, Bernard
Gunther, Marian W J
Hanson, Harvey Myron
Kelly, Donald C
Kowalski, Kenneth L
Lang, Joseph Edward
Lock, James Albert
Macklin, Philip Alan
McLennan, Donald Elmore
Mainland, Gordon Bruce
Mallory, Willam R
Manning, Robert M
Mills, Robert Laurence
Montgomery, Charles Gray

Moroi, David S
Mulligan, Bernard
Nahar, Sultana Nurun
Nam, Sang Boo
O'Hare, John Michael
Onley, David S
Palffy-Muhoray, Peter
Patton, Bruce Riley
Pytte, Agnar
Raby, Stuart
Schlosser, Herbert
Schneider, Ronald E
Schreiber, Paul J
Stroud, David Gordon
Styer, Daniel F
Suranyi, Peter
Tamburino, Louis A
Taylor, Beverley Ann Price
Tenhover, Michael Alan
Thaler, Raphael Morton
Tobocman, William
Tuan, Tai-Fu
Whitesell, William James
Wiff, Donald Ray
Williamson, William, Jr
Witten, Louis
Wood, Van E(arl)
Wright, Louis E
Wright, William Robert

OKLAHOMA

Coffman, Moody Lee
Cohn, Jack
Collins, Thomas C
Cowan, John James
Feiock, Frank Donald
Huffaker, James Neal
Letcher, John Henry, III
Milton, Kimball Alan
Samuel, Mark Aaron
Sigal, Richard Frederick
Thomsen, Leon

OREGON

Belinfante, Frederik J
Carmichael, Howard John
Csonka, Paul L
Fontana, Peter R
Girardeau, Marvin Denham, Jr
Goswami, Amit
Hwa, Rudolph Chia-Chao
Isenberg, James Allen
Khalil, M Aslam Khan
Liu-Ger, Tsu-Huei
McClure, Joel William, Jr
McCollum, Gin
Madsen, Victor Arviel
Semura, Jack Sadatoshi
Siemens, Philip John
Sivers, Dennis Wayne
Skinner, Richard Emery

PENNSYLVANIA

Albano, Alfonso M
Amado, Ralph
Anderson, James B
Armenti, Angelo, Jr
Ashtekar, Abhay Vasant
Austern, Norman
Banavar, Jayanth Ramarao
Baranger, Elizabeth Urey
Becker, Stephen Fraley
Bloomfield, Philip Earl
Bludman, Sidney Arnold
Boer, Karl Wolfgang
Borse, Garold Joseph
Carlitz, Robert D
Carr, Walter James, Jr
Cohen, Jeffrey M
Cohen, Michael
Collins, John Clements
Cutkosky, Richard Edwin
Davidon, William Cooper
Davies, Kenneth Thomas Reed
Diaz, Mario Claudio
Emtage, Peter Roesch
Engheta, Nader
Feng, DaHsuan
Fleming, Gordon N
Ford, Kenneth William
Ford, William Frank
Franklin, Jerrold
Freed, Norman
Gerjuoy, Edward
Gibson, Gordon
Goldschmidt, Yadin Yehuda
Griffiths, Robert Budington
Grotch, Howard
Gunton, James D
Havas, Peter
Hoffman, Richard Bruce
Janis, Allen I(ra)
Jasnow, David Michael
Jhon, Myung S
Kazes, Emil
Kisslinger, Leonard Sol
Klein, Abraham
Langacker, Paul George
Larsen, Sigurd Yves
Li, Ling-Fong
Liang, Shoudan
Lim, Teck-Kah
Lubensky, Tom C
Mezger, Fritz Walter William

Miller, Irvin Alexander
Monroe, James Lewis
Neville, Donald Edward
Newman, Ezra
Novaco, Anthony Dominic
Pratt, Richard Houghton
Rabii, Sohrab
Roman, Paul
Rosen, Gerald Harris
Rothwarf, Allen
Rovelli, Carlo
Satz, Ronald Wayne
Segre, Gino C
Shaffer, Russell Allen
Sion, Edward Michael
Steinhardt, Paul Joseph
Tabakin, Frank
Taylor, Lyle Herman
Verhanovitz, Richard Frank
Weiner, Brian Lewis
Winicour, Jeffrey
Wolfenstein, Lincoln
Yuan, Jian-Min

RHODE ISLAND

Bonner, Jill Christine
Brandenberger, Robert H
Cooper, Leon N
Feldman, David
Fried, Herbert Martin
Gora, Edwin Karl
Kang, Kyungsik
Kaufman, Charles
McCorkle, Richard Anthony
Mellberg, Leonard Evert
Westervelt, Peter Jocelyn

SOUTH CAROLINA

Au, Chi-Kwan
Bostock, Judith Louise
Burt, Philip Barnes
Graben, Henry Willingham
Knight, James Milton
Lerner, Edward Clarence
Manson, Joseph Richard
Safko, John Loren

SOUTH DAKOTA

Duffey, George Henry
Hall, Harold Hershey
Jones, Robert William

TENNESSEE

Alsmiller, Rufard G, Jr
Barrett, John Harold
Becker, Richard Logan
Bloch, Ingram
Childers, Robert Wayne
Cook, James Marion
Davis, Harold Lloyd
Deeds, William Edward
Groer, Peter Gerold
Halbert, Edith Conrad
Harris, Edward Grant
Kliewer, Kenneth L
Lane, Eric Trent
Larson, Nancy Marie
Liu, Samuel Hsi-Peh
Mahan, Gerald Dennis
Mansur, Louis Kenneth
Nestor, C William, Jr
Oen, Ordean Silas
Painter, Gayle Stanford
Pinkston, William Thomas
Quinn, John Joseph
Ritchie, Rufus Haynes
Satchler, George Raymond
Turner, James Edward
Wang, Jia-Chao
Welton, Theodore Allen

TEXAS

Allen, Roland Emery
Antoniewicz, Peter R
Arnowitt, Richard Lewis
Bailey, James Stephen
Bassichis, William
Biedenharn, Lawrence C, Jr
Bohm, Arno
Brachman, Malcolm K
Breig, Edward Louis
Bryan, Ronald Arthur
Coker, William Rory
De Lozanne, Alex L
DeWitt, Bryce Seligman
DeWitt-Morette, Cecile
Dicus, Duane A
Doran, Robert Stuart
Englund, John Caldwell
Evans, Wayne Errol
Ford, Albert Lewis, Jr
Fulling, Stephen Albert
Gilman, Frederick Joseph
Griffy, Thomas Alan
Hannon, James Patrick
Hardcastle, Donald Lee
Herman, Robert
Hewett, Lionel Donnell
Hoffman, Alexander A J
Horton, Claude Wendell, Jr
Hu, Bambi
Hu, Chia-Ren
Huang, Huey Wen
Hubisz, John Lawrence, Jr

Ivash, Eugene V
Kim, Young Nok
Klein, Douglas J
Ko, Che Ming
Kobe, Donald Holm
Kouri, Donald Jack
Liang, Edison Park-Tak
Lipps, Frederick Wiessner
McCauley, Joseph Lee
McIntyre, Robert Gerald
Mahajan, Swadesh Mitter
Marlow, William Henry
Meshkov, Sydney
Miller, Bruce Neil
Minerbo, Gerald N
Moorhead, William Dean
Myles, Charles Wesley
Nordlander, Peter Jan Arne
Pettitt, Bernard Montgomery
Pooch, Udo Walter
Puthoff, Harold Edward
Rindler, Wolfgang
Robinson, Ivor
Roeder, Robert Charles
Roeder, Robert Charles
Ross, David Ward
Sablik, Martin J
Salingaros, Nikos Angelos
Schieve, William
Shepley, Lawrence Charles
Smith, Frederick Adair, Jr
Smith, Roger Alan
Stevenson, Paul Michael
Stratton, Robert
Sudarshan, Ennackel Chandy George
Swift, Jack Bernard
Trammell, George Thomas
Uralil, Francis Stephen
Voigt, Gerd-Hannes
Weinberg, Steven
Wilson, Thomas Leon
Winbow, Graham Arthur
Zook, Herbert Allen

UTAH
Ball, James Stutsman
Benson, Alvin K
Chatelain, Jack Ellis
Evenson, William Edwin
Gardner, John Hale
Harrison, Bertrand Kent
Hatch, Dorian Maurice
McDonald, Keith Leon
Price, Richard Henry
Shirts, Randall Brent
Sutherland, Bill

VERMONT
Oughstun, Kurt Edmund
Smith, David Young

VIRGINIA
Adam, John Anthony
Aitken, Alfred H
Asterita, Mary Frances
Atalay, Bulent Ismail
Barrois, Bertrand C
Bishop, Marilyn Frances
Bressler, Barry Lee
Brown, Ellen Ruth
Buck, Warren W, III
Burns, Grover Preston
Calle, Carlos Ignacio
Chambers, Charles MacKay
Chang, Lay Nam
Chubb, Scott Robinson
Coffey, Timothy
Coopersmith, Michael Henry
Debney, George Charles, Jr
De Wolf, David Alter
Dworzecka, Maria
Fowler, Michael
Gowdy, Robert Henry
Gross, Franz Lucretius
Hobart, Robert H
Hurley, William Jordan
Imbrie, John Z
Jacobs, Kenneth Charles
Jena, Purusottam
Kabir, Prabahan Kemal
Kelley, Ralph Edward
Kiefer, Harold Milton
McIrvine, Edward Charles
McKnight, John Lacy
Madan, Rabinder Nath
Mansfield, John E
Mayo, Thomas Tabb, IV
Noakes, David Ronald
Noble, Julian Victor
Ossakow, Sidney Leonard
Peacock, Richard Wesley
Pereira, Nino Rodrigues
Pokrant, Marvin Arthur
Reynolds, Peter James
Roper, L(eon) David
Sarazin, Craig L
Schremp, Edward Jay
Seaborn, James Byrd
Shur, Michael
Sparrow, D(avid) A
Taggart, G(eorge) Bruce
Tidman, Derek Albert
Townsend, Lawrence Willard
Trefil, James S

Trizna, Dennis Benedict
Ugincius, Peter
Van Ness, Kenneth E
Von Baeyer, Hans Christian
Walecka, John Dirk
Weber, Hans Jurgen
Werbos, Paul John
Williams, Clayton Drews
Williams, Harry Thomas
Wilson, John William
Zia, Royce K P
Zweifel, Paul Frederick

WASHINGTON
Albers, James Ray
Band, William
Barnett, Claude C
Boas, Mary Layne
Boulware, David G
Brown, Lowell Severt
Clifford, Howard James
Craig, Richard Anderson
Ellis, Stephen Dean
Galas, David John
Haxton, Wick Christopher
Henley, Ernest Mark
Henyey, Frank S
Howland, Louis Philip
Kerlick, George David
Lake, George Russell
Leaf, Boris
Lincoln, Charles Albert
Lindenmeier, Charles William
Ling, Daniel Seth, Jr
Miller, Gerald Alan
Miller, John Howard
Moore, Richard Lee
Nicholson, Richard Benjamin
Peters, Philip Carl
Puff, Robert David
Rehr, John Jacob
Rimbey, Peter Raymond
Schick, Michael
Siciliano, Edward Ronald
Thouless, David James
Warren, John Lucius
Wilets, Lawrence
Yaffe, Laurence G

WEST VIRGINIA
Cooper, Bernard Richard
Corum, James Frederic
Douglass, Kenneth Harmon
Levine, Arnold David

WISCONSIN
Balantekin, Akif Baha
Barger, Vernon Duane
Bevelacqua, Joseph John
Bincer, Adam Marian
Bruch, Ludwig Walter
Chow, Yutze
Cook, David Marsden
Durand, Bernice Black
Durand, Loyal
Ebel, Marvin Emerson
Epstein, Saul Theodore
Goebel, Charles James
Halzen, Francis
Huber, David Lawrence
Karkheck, John Peter
Kobiske, Ronald Albert
Leplae, Luc A
Lin, Chun Chia
Lind, Robert Wayne
Lubkin, Elihu
McVoy, Kirk Warren
Mendelson, Kenneth Samuel
Menzel, Wolfgang Paul
Myers, Vernon W
Parker, Leonard Emanuel
Schmieg, Glenn Melwood
Snider, Dale Reynolds
Sorbello, Richard Salvatore
Tani, Smio
Wackman, Peter Husting
Wang, Pao-Kuan

WYOMING
Grandy, Walter Thomas, Jr
Inguva, Ramarao
Schick, Lee Henry

PUERTO RICO
Gomez-Rodriguez, Manuel

ALBERTA
Capri, Anton Zizi
Israel, Werner
Kounosu, Shigeru
Leahy, Denis Alan
Ludwig, Garry (Gerhard Adolf)
Schiff, Harry
Stephenson, John
Takahashi, Yasushi
Torrence, Robert James
Tuszynski, Jack A
Umezawa, Hiroomi
Walton, Mark Allan

BRITISH COLUMBIA
Affleck, Ian Keith
Ballentine, Leslie Edward
Barrie, Robert

Boal, David Harold
Booth, Ian Jeremy
Cooperstock, Fred Isaac
Enns, Richard Harvey
Howard, Roger
Jennings, Byron Kent
Johnston, Robert R
Kaempffer, Frederick Augustus
Kirczenow, George
McMillan, Malcolm
Opechowski, Wladyslaw
Poll, Jacobus Daniel
Rastall, Peter
Unruh, William George
Van Kranendonk, Jan
Viswanathan, Kadayam Sankaran
Vogt, Erich Wolfgang
Volkoff, George Michael
Wallace, Philip Russell
Wortis, Michael

MANITOBA
Bhakar, Balram Singh
Loly, Peter Douglas
Southern, Byron Wayne
Svenne, Juris Peteris
Vail, John Moncrieff

NEWFOUNDLAND
Rendell, David H

NOVA SCOTIA
Betts, Donald Drysdale
Calkin, Melvin Gilbert
Dunn, Kenneth Arthur
Goble, David Franklin
Hunter, Douglas Lyle
Kennedy, Frederick James
Kreuzer, Hans Jurgen
Latta, Bryan Michael
Moriarty, Kevin Joseph
Needler, George Treglohan
Pink, David Anthony Herbert
Ravindra, Ravi

ONTARIO
Abdelmalek, Nabin N
Allen, James Roy
Baylis, William Eric
Berezin, Alexander A
Bielajew, Alexander Frederick
Billigheimer, Claude Elias
Campeanu, Radu Ioan
Chau, Wai-Yin
Coleman, Albert John
Couture, Josaphat Michel
Desai, Rashmi C
Ehrman, Joachim Benedict
Giles, Robin
Glass, Edward Nathan
Goodings, David Ambery
Griffin, Allan
Grindlay, John
Hardy, James Edward
Harvey, Malcolm
Hodgson, Richard John Wesley
Hoffstein, Victor
Horbatsch, Marko M
Hruska, Antonin
Hume, James Nairn Patterson
Huschilt, John
Jacobs, Allan Edward
Joos, Béla
Karl, Gabriel
Law, Jimmy
Lee, Hoong-Chien
Lee-Whiting, Graham Edward
Liu, Wing-Ki
Mann, Robert Bruce
Mavroyannis, Constantine
Mitalas, Romas
Nagi, Anterdhyan Singh
Nerenberg, Morton Abraham
Nogami, Yukihisa
Noolandi, Jaan
Nuttall, John
Paldus, Josef
Pathria, Raj Kumar
Preston, Melvin Alexander
Prugovecki, Eduard
Pugh, Robert E
Resnick, Lazer
Romo, William Joseph
Rowe, David John
Schlesinger, Mordechay
Sears, Varley Fullerton
Simpson, John Hamilton
Smith, Robert Clinton
Sokoloff, Jack
Sprung, Donald Whitfield Loyal
Talman, James Davis
Tam, Wing-Gay
Taylor, David Ward
Todd, Leonard
Towner, Ian Stuart
Trainor, Lynne E H
Varshni, Yatendra Pal
Vosko, Seymour H
Wang, Shao-Fu
Wesson, Paul Stephen
Whittington, Stuart Gordon
Wong, Samuel Shaw Ming

QUEBEC
Banville, Marcel
Bunge, Mario Augusto
Chaubey, Mahendra
Contogouris, Andreas P
Frank, Barry
Fuchs, Vladimir
Gauvin, J N Laurie
Goulard, Bernard
Grant, Martin
Grmela, Miroslav
Ho-Kim, Quang
Kroger, Helmut Karl
Lam, Harry Chi-Sing
LeTourneux, Jean
Morris, Stanley P
Paquette, Guy
Pearson, John Michael
Pineault, Serge Rene
Roth, Charles
Shkarofsky, Issie Peter
Szamosi, Geza
Tremblay, André -Marie
Van Vliet, Carolyne Marina

SASKATCHEWAN
Kim, Dong Yun
Papini, Giorgio Augusto
Rystephanick, Raymond Gary
Tomusiak, Edward Lawrence

OTHER COUNTRIES
Afnan, Iraj Ruhi
Andrej, Ladislav
Azbel, Mark
Band, Yehuda Benzion
Baratoff, Alexis
Bekenstein, Jacob David
Bohr, Aage
Burschka, Martin A
Cardona, Manuel
Carmeli, Moshe
Contopoulos, George
Delsanto, Pier Paolo
Dobson, Peter N, Jr
Dominguez, Cesareo Augusto
Eisenberg, Judah Moshe
Foglio, Mario Eusebio
Gauster, Wilhelm Friedrich
Hawking, Stephen W
Heinz, Ulrich Walter
Hertz, John Atlee
Ho, Yew Kam
Horwitz, Lawrence Paul
Huang, Keh-Ning
Keller, Jaime
Lande, Alexander
Lipkin, Harry Jeannot
Lowdin, Per-Olov
McKellar, Bruce Harold John
Miller, Arthur I
Ng, Tai-Kai
Nussenzveig, Herch Moyses
Owen, David A
Paschos, Emmanuel Anthony
Pethick, Christopher John
Rice, Thomas Maurice
Rivier, Nicolas Yves
Rosen, Nathan
Rozzi, Tullio
Satz, Helmut T G
Schutz, Bernard Frederick
Wiser, Nathan
Woo, Chia-Wei
Wu, Yong-Shi

Thermal Physics

ALABAMA
Askew, Raymond Fike
Gant, Fred Allan
Olson, Willard Paul
Ramachandran, Narayanan
Rosenberger, Franz
Yeh, Pu-Sen

ARIZONA
Benin, David B
Burrows, Adam Seth
Scholl, Marija Strojnik

ARKANSAS
Babcock, Robert E

CALIFORNIA
Alonso, Carol Travis
Armstrong, Baxter Hardin
Arnas, Ozer Ali
Barr, Kevin Patrick
Bennett, Glenn Taylor
Birecki, Henryk
Brewer, LeRoy Earl, Jr
Cannell, David Skipwith
Carrigan, Charles Roger
Catton, Ivan
Chin, Jin H
Coufal, Hans-Jurgen
Cowperthwaite, Michael
Diaz, Luis Florentino
Edwards, Donald K
Eggleston, Glen E
El-Saden, Munir Ridha
Fisher, Robert Amos, Jr

Thermal Physics (cont)

French, Edward P(erry)
Garrett, Steven Lurie
Geballe, Theodore Henry
Goldwire, Henry C, Jr
Gould, Christopher M
Gupta, Madhu Sudan
Hallet, Raymon William, Jr
Hellman, Frances
Herman, Ray Michael
Hoover, Carol Griswold
Hovingh, Jack
Hussain, Nihad A
Johanson, William Richard
Keizer, Joel Edward
Kittel, Peter
Klems, Joseph Henry
Kraus, Samuel
Krikorian, Oscar Harold
Kroemer, Herbert
Kroger, F(erdinand) A(nne)
Lax, Edward
Leff, Harvey Sherwin
Liu, Frederick F
Luongo, Cesar Augusto
Massier, Paul Ferdinand
Miner, Ellis Devere, Jr
Morizumi, S James
Morrison, Harry Lee
Nellis, William J
Nielsen, Helmer L(ouis)
O'Malley, Thomas F
Peskin, Michael Edward
Pomraning, Gerald C
Poppendiek, Heinz Frank
Reese, William
Rosen, Mordecai David
Sabersky, Rolf H(einrich)
Salerno, Louis Joseph
Sarwinski, Raymond Edmund
Schmiedeshoff, George M
Shepherd, Joseph Emmett
Speen, Gerald Bruce
Spencer, Donald Jay
Spokoyny, Felix E
Steinberg, Daniel J
Stowe, Keith S
Tien, C(hang) L(in)
Toor, Arthur
Urtiew, Paul Andrew
Vernon, Gregory Allen
Viswanathan, R
Weichman, Peter Bernard
Wieder, Harold
Yoshizumi, Shozo
Yuan, Sidney Wei Kwun

COLORADO
Arp, Vincent D
Bartlett, David Farnham
Beale, Paul Drew
Bruno, Thomas J
Daley, Daniel H
Froyen, Sverre
Krueger, David Allen
Loehrke, Richard Irwin
Oetting, Franklin Lee
Olien, Neil Arnold
Powell, Robert Lee
Radebaugh, Ray
Rainwater, James Carlton
Regenbrecht, D(ouglas) E(dward)
Roder, Hans Martin
Trefny, John Ulric
Urban, Thomas Charles

CONNECTICUT
De Rocco, Andrew Gabriel
Klemens, Paul Gustav
Morowitz, Harold Joseph
Morrison, Richard Charles
O'Brien, Robert L

DELAWARE
Daniels, William Burton
Ness, Norman Frederick
Sandler, Stanley I

DISTRICT OF COLUMBIA
Chang, Eddie Li
Kelly, Francis Joseph
Kitchens, Thomas Adren
Klein, Max
Meijer, Paul Herman Ernst
Moon, Deug Woon
Reinecke, Thomas Leonard
Serene, Joseph William
Spohr, Daniel Arthur
Sununu, John Henry
Weinstock, Harold
White, John Arnold

FLORIDA
Adams, Earnest Dwight
Barron, Saul
Buzyna, George
Edwards, Thomas Claude
Hooper, Charles Frederick, Jr
Hsieh, Chung Kuo
Kumar, Pradeep
Kurzweg, Ulrich H(ermann)
Lindholm, Fredrik Arthur
Millar, Gordon Halstead
Novotny, Mark A
Robertson, Harry S(troud)
Sullivan, Neil Samuel
Veziroglu, T Nejat
Vollmer, James

GEORGIA
Anderson, Robert Lester
Batten, George L, Jr
Colwell, Gene Thomas
Family, Fereydoon
Fox, Ronald Forrest
Kezios, Stothe Peter
Vachon, Reginald Irenee

IDAHO
Ramshaw, John David
Richardson, Lee S(pencer)

ILLINOIS
Abraham, Bernard M
Brewster, Marcus Quinn
Chao, B(ei) T(se)
Cooper, William Edward
Farhadieh, Rouyentan
Garg, Anupam
Jacobi, Anthony Mark
Kuchnir, Moyses
Lee, Anthony L
Liu, Yung Yuan
Mintzer, David
Mokadam, Raghunath G(anpatrao)
Olvera-dela Cruz, Monica
Quay, Paul Michael
Schlossman, Mark Loren
Shepard, Kenneth Wayne
Simmons, Ralph Oliver
Snelson, Alan
Tang, James Juh-Ling
Yuen, Man-Chuen

INDIANA
DeWitt, David P
Giordano, Nicholas J
Ho, Cho-Yen
Honig, Jurgen Michael
Keesom, Pieter Hendrik
Liley, Peter Edward
McComas, Stuart T
Viskanta, Raymond
Wark, Kenneth, Jr
Western, Arthur Boyd

IOWA
Chen, Lea-Der
Hering, Robert Gustave
Intemann, Gerald William
Johnston, David Carl
Luban, Marshall
Scudder, Jack David
Swenson, Clayton Albert

KANSAS
Gosman, Albert Louis
Ho, James Chien Ming
Sapp, Richard Cassell
Snyder, Melvin H(enry), Jr
Sorensen, Christopher Michael

KENTUCKY
Bakanowski, Stephen Michael
DeLong, Lance Eric
Mills, Roger Edward

LOUISIANA
Baw, Philemon S H
Rotty, Ralph M(cGee)

MARYLAND
Baba, Anthony John
Battin, William James
Baum, Howard Richard
Bennett, Gary L
Caveny, Leonard Hugh
Cezairliyan, Ared
Chang, Ren-Fang
Colwell, Jack Harold
Dorfman, Jay Robert
Ferrell, Richard Allan
Furukawa, George Tadaharu
Gammon, Robert Winston
Griffin, James J
Haberman, William L(awrence)
Hsia, Jack Jinn-Goe
Hudson, Ralph P
Kirkpatrick, Theodore Ross
MacDonald, Rosemary A
Marqusee, Jeffrey Alan
Mitler, Henri Emmanuel
Murphy, John Cornelius
Nash, Jonathon Michael
Parker, William James
Radermacher, Reinhard
Randall, John Douglas
Robbins, Mark Owen
Roszman, Larry Joe
Schmid, Lawrence Alfred
Schooley, James Frederick
Sengers, Jan V
Sengers, Johanna M H Levelt
Thomsen, John Stearns
Tong, Long Sun
Traugott, Stephen C(harles)
Waldo, George Van Pelt, Jr
Webb, Richard A
Weckesser, Louis Benjamin
Williams, Ellen D
Wu, Chih
Zirkind, Ralph

MASSACHUSETTS
Appleton, John P(atrick)
Bolsaitis, Pedro
Bowman, H(arry) Frederick
Cipolla, John William, Jr
Day, Robert William
Delvaille, John Paul
Emmons, Howard W(ilson)
Goldman, Ralph Frederick
Hallock, Robert B
Hatsopoulos, George Nicholas
Hoercher, Henry E(rhardt)
Hoge, Harold James
Jacobs, Laurence Alan
Jirmanus, Munir N
Jose, Jorge V
Kaufman, Larry
Klein, William
Liau, Zong-Long
Machta, Jonathan Lee
Maxwell, Emanuel
Mikic, Bora
Picard, Richard Henry
Reeves, Barry L(ucas)
Silvers, J(ohn) P(hillip)
Sonin, Ain A(nts)
Tiernan, Robert Joseph
Wang, Chia Ping
Yener, Yaman

MICHIGAN
Boyer, Raymond Foster
Chock, David Poileng
Edgerton, Robert Howard
Hartman, John L(ouis)
Heremans, Joseph P
Koenig, Milton G
LaRocca, Anthony Joseph
Menard, Albert Robert, III
Piccirelli, Robert Anthony
Smith, George Wolfram
Weinreich, Gabriel
Westrum, Edgar Francis, Jr
Yang, Wen Jei

MINNESOTA
Berg, James Irving
Bonne, Ulrich
Flower, Terrence Frederick
Goldman, Allen Marshall
Heberlein, Joachim Viktor Rudolf
Lutes, Olin S
Olson, Reuben Magnus
Valls, Oriol Tomas
Zimmermann, William, Jr

MISSISSIPPI
Eastland, David Meade
Smith, Allie Maitland

MISSOURI
Chen, Ta-Shen
Elgin, Robert Lawrence
Fisch, Ronald
Gardner, Richard A
Look, Dwight Chester, Jr
Viswanath, Dabir S
Whitney, Arthur Edwin, Jr

NEBRASKA
Kennedy, Robert E

NEW HAMPSHIRE
King, Allen Lewis
Lopez, Guido Wilfred

NEW JERSEY
Anthony, Philip John
Bachman, Walter Crawford
Bechis, Dennis John
Blair, David W(illiam)
Buteau, L(eon) J
Cody, George Dewey
Cotter, Martha Ann
Couchman, Peter Robert
Debenedetti, Pablo Gaston
Farkass, Imre
Geskin, Ernest S
Gittleman, Jonathan I
Glass, John Richard
Haindl, Martin Wilhelm
Holt, Vernon Emerson
Hrycak, Peter
Kolodner, Paul R
Lines, Malcolm Ellis
Makofske, William Joseph
Matey, James Regis
Rollino, John
Stillinger, Frank Henry
Tsonopoulos, Constantine
Zebib, Abdelfattah M G

NEW MEXICO
Bankston, Charles A, Jr
Dingus, Ronald Shane
Erpenbeck, Jerome John
Farnsworth, Archie Verdell, Jr
Grilly, Edward Rogers
Houghton, Arthur Vincent, III
Johnson, James Daniel
Julien, Howard L
Kropschot, Richard H
Laquer, Henry L
Martin, Richard Alan
Moss, Marvin
Neeper, Donald Andrew
Richards, Peter Michael
Rose, Harvey Arnold
Shaner, John Wesley
Sinha, Dipen N
Swift, Gregory William
Trela, Walter Joseph

NEW YORK
Abbott, Michael McFall
Alben, Richard Samuel
Allen, Philip B
Amato, Joseph Cono
Archie, Charles Neill
Begell, William
Budzinski, Walter Valerian
Daman, Harlan Richard
De Boer, P(ieter) C(ornelis) Tobias
Hall, J(ohn) Gordon
Hecht, Charles Edward
Henkel, Elmer Thomas
Holbrow, Charles H
Hurst, Robert Philip
Irvine, Thomas Francis
Karz, Robert Stephen
Kroeger, Peter G
LaPietra, Joseph Richard
Li, Chung-Hsiung
Loring, Roger Frederic
Pribil, Stephen
Privman, Vladimir
Rimai, Donald Saul
Ring, James Walter
Socolar, Sidney Joseph
Staub, Fred W
Stell, George Roger
Stiel, Leonard Irwin
Talley, Robert Lee
Thompson, Philip A
Torrance, Kenneth E(ric)
Tourin, Richard Harold
Zahed, Ismail

NORTH CAROLINA
Brown, J(ohn) David
Edwards, John Auert
Kratz, Howard Russel
Lado, Fred

NORTH DAKOTA
Dewar, Graeme Alexander
Kurtze, Douglas Alan

OHIO
Boughton, Robert Ivan, Jr
Bovenkerk, Harold P(aul)
Deissler, Robert G(eorge)
Edwards, David Olaf
Fenichel, Henry
Heer, Clifford V
Jacobs, Donald Thomas
Jones, Charles E(dward)
Kaufman, Miron
Keith, Theo Gordon, Jr
Keshock, Edward G
La Mers, Thomas Herbert
Lawler, Martin Timothy
Lemke, Ronald Dennis
Mayers, Richard Ralph
Moon, Tag Young
Petschek, Rolfe George
Pool, Monte J
Rockstroh, Todd Jay
Rosenblatt, Charles Steven
Sanford, Edward Richard
Schmitt, George Frederick, Jr
Scholten, Paul David
Schuele, Donald Edward
Styer, Daniel F
Thekdi, Arvind C
Wagoner, Glen
Walters, Craig Thompson

OKLAHOMA
Harden, Darrel Grover
Rutledge, Delbert Leroy

OREGON
Brodie, Laird Charles
Broide, Michael Lynn
Girardeau, Marvin Denham, Jr
Gokcen, Nev A(ltan)
Griffith, William Thomas
Schaefer, Seth Clarence
Semura, Jack Sadatoshi

PENNSYLVANIA
Arnold, James Norman
Boughn, Stephen Paul
Callen, Herbert Bernard
Carmi, Shlomo
Chan, Moses Hung-Wai
Chun, Sun Woong
Collings, Peter John
Farouk, Bakhtier
Finegold, Leonard X
Friedberg, Simeon Adlow

Goldschmidt, Yadin Yehuda
Griffiths, Robert Budington
Hager, Nathaniel Ellmaker, Jr
Hertzberg, Martin
Hogenboom, David L
Lazaridis, Anastas
Maher, James Vincent
Nagle, John F
Ramezan, Massood
Sturdevant, Eugene J
Toulouse, Jean
Tsou, F(u) K(ang)
Williams, Willie, Jr
Yao, Shi Chune
Zemel, Jay N(orman)

RHODE ISLAND
Fallieros, Stavros
McEligot, Donald M(arinus)
Timbie, Peter T
Wilson, Mason P, Jr

SOUTH CAROLINA
Gaines, Albert L(owery)

SOUTH DAKOTA
Ashworth, T

TENNESSEE
Boatner, Lynn Allen
Busey, Richard Hoover
Fontana, Mario H
Keefer, Dennis Ralph
LeVert, Francis E
Lewis, James W L
McGregor, Wheeler Kesey, Jr
Scott, Herbert Andrew
Wood, Bobby Eugene

TEXAS
Arumi, Francisco Noe
Bergman, Theodore L
Brostow, Witold Konrad
Caflisch, Robert Galen
Chao, Jing
Eichhorn, Roger
Holste, James Clifton
Howell, John Reid
Hrkel, Edward James
Huang, Huey Wen
Kirk, Wiley Price
Mack, Russel Travis
Prats, Michael
Rajagopalan, Raj
Robbins, Roger Alan
Spencer, Gregory Fielder
Swope, Richard Dale
Young, Fred M(ichael)

UTAH
Batty, Joseph Clair
Brown, Billings
Henderson, Douglas J

VERMONT
Schiffman, Robert A

VIRGINIA
Bartis, James Thomas
Dutt, Gautam Shankar
Exton, Reginald John
Goglia, Gennaro Louis
Johnson, David Harley
Liebenberg, Donald Henry
Milligan, John H
Reynolds, Peter James
Roberts, A(lbert) S(idney), Jr
Simpson, Roger Lyndon
Sutton, George W(alter)
Watson, Nathan Dee

WASHINGTON
Band, William
Erdmann, Joachim Christian
Johnson, B(enjamin) M(artineau)
Lam, John Ling-Yee
Leaf, Boris
Marston, Philip Leslie
Russell, David A

WISCONSIN
Boom, Roger Wright
Chang, Y Austin
Rooks, H Corbyn
Van den Akker, Johannes Archibald
Wilson, Volney Colvin
Zimm, Carl B

ALBERTA
Jessop, Alan Michael
Tuszynski, Jack A

NOVA SCOTIA
Hunter, Douglas Lyle
Kreuzer, Hans Jurgen

ONTARIO
Bedford, Ronald Ernest
Edwards, Martin Hassall
Hooper, F(rank) C(lements)
Kozicki, William
Kumaran, Mavinkal Kizhakkeveettil
Lachaine, Andre Raymond Joseph
Mandelis, Andreas

Martin, Douglas Leonard
Olson, Allan Theodore
Pathria, Raj Kumar
Pei, David Chung-Tze
Tarasuk, John David
Tyler, R(onald) A(nthony)
Valleau, John Philip
Witzman, Sorin

PRINCE EDWARD ISLAND
Madan, Mahendra Pratap

QUEBEC
Caille, Alain Emeril
Chan, Sek Kwan
Cochrane, Robert W
Kiss, Laszlo Istvan
Leigh, Charles Henry
Lin, Sui

SASKATCHEWAN
Kos, Joseph Frank

OTHER COUNTRIES
Chandrasekhar, B S
Garcia-Colin, Leopoldo Scherer
Jokl, Miloslav Vladimir
Mota, Ana Celia
Romero, Alejandro F
Safran, Samuel A
Sarmiento, Gustavo Sanchez
Shah, Mirza Mohammed

Other Physics

ALABAMA
Barr, E(rnest) Scott
Fowlis, William Webster
Hathaway, David Henry
Robinson, Edward Lee
Tan, Arjun
Werkheiser, Arthur H, Jr
Wilhold, Gilbert A

ARIZONA
Bickel, William Samuel
Galloway, Kenneth Franklin
O'Brien, Keran
Staley, Dean Oden
Twomey, Sean Andrew

CALIFORNIA
Adam, Randall Edward
Adams, Ralph Melvin
Adelson, Harold Ely
Alonso, Jose Ramon
Anderson, Leonard Mahlon
Anderson, Lloyd James
Andre, Michael Paul
Arnold, James Tracy
Ashford, Victor Aaron
Astrahan, Melvin Alan
Baglin, John Edward Eric
Banner, David Lee
Barengoltz, Jack Bennett
Barish, Barry C
Barnes, Charles A
Basler, Roy Prentice
Berkner, Klaus Hans
Bogle, Robert Worthington
Bowers, John E
Boyce, James B
Bradner, Hugh
Brodsky, Stanley Jerome
Brown, Keith H
Bryan, James Bevan
Cantow, Manfred Josef Richard
Carlson, Gary Wayne
Case, Kenneth Myron
Chao, Alexander Wu
Chau, Cheuk-Kin
Chemla, Daniel Simon
Cheng, David
Chester, Marvin
Chu, William Tongil
Clendenin, James Edwin
Clendenning, Lester M
Cohn, Stanton Harry
Cover, Ralph A
Cox, Donald Clyde
Cross, Michael Clifford
Cummings, Alan C
Deshpande, Shivajirao M
Dewitt, Hugh Edgar
Dickinson, William Clarence
Dobbins, John Potter
Donovan, Paul F(rancis)
Dorfan, Jonathan Manne
DuBridge, Lee Alvin
Elliott, David Duncan
Feldman, Jack L
Fillius, Walker
Fineman, Morton A
Fisk, Zachary
Frost, Robert Hartwig
Gaines, Edward Everett
Gajewski, Ryszard
Garwin, Charles A
Gillespie, Daniel Thomas
Good, Robert Howard
Goubau, Wolfgang M
Groce, David Eiben
Gutsche, Steven Lyle

Harrison, Don Edward, Jr
Hebel, Louis Charles
Heflinger, Lee Opert
Hickman, Albert Peet
Horn, Linda Joyce
Hornung, Erwin William
Hubbard, Edward Leonard
James, Ralph Boyd
Jones, Joie Pierce
Judd, Floyd L
Judge, Roger John Richard
Kaelble, David Hardie
Kamegai, Minao
Kittel, Peter
Klems, Joseph Henry
Knowles, Harrold B
Kohler, Donald Alvin
Koons, Stephen J
Krivanek, Ondrej Ladislav
Lal, Devendra
Le, Guan
Lee, Ronald S
Lerche, Richard Allan
Lieberman, Michael A
Litt, Lawrence
Liu, Fook Fah
Long, H(ugh) M(ontgomery)
Lu, Adolph
Lumsden, Michael A
Luth, Vera G
Ma, Joseph T
McGinnis, Carl L
MacKenzie, Kenneth R
Manalis, Melvyn S
Merker, Milton
Merriam, Marshal F(redric)
Miles, Ralph Fraley, Jr
Morewitz, Harry Alan
Muller, Richard A
Munch, Jesper
Murphy, Frederick Vernon
Myer, Jon Harold
Nelson, Raymond Adolph
Nelson, Walter Ralph
Newkirk, Lester Leroy
Oakley, David Charles
Ogier, Walter Thomas
Paige, David A
Parkins, William Edward
Parks, Lewis Arthur
Passenheim, Burr Charles
Phinney, Nanette
Piccioni, Oreste
Prag, Arthur Barry
Rabl, Veronika Ariana
Radosevich, Lee George
Randrup, Jorgen
Reisman, Elias
Richards, Lorenzo Adolph
Ride, Sally Kristen
Ridgway, Stuart L
Rockwood, Stephen Dell
Rodriguez, Andres F
Rothrock, Larry R
Rouse, Carl Albert
Sadoulet, Bernard
Salaita, George N
Sandmann, William Henry
Schermer, Robert Ira
Seibert, J A
Shacklett, Robert Lee
Shapiro, Mark Howard
Shepard, Roger N
Sheridon, Nicholas Keith
Siri, William E
Slater, Donald Carlin
Speen, Gerald Bruce
Spencer, James Eugene
Stahl, Ralph Henry
Stearns, Ronald C
Strangeway, Robert Joseph
Subramanya, Shiva
Thompson, John Robert
Tsai, Chuang Chuang
Tuul, Johannes
Warren, Walter R(aymond), Jr
Weaver, Harry Edward, Jr
Wiegand, Clyde E
Woodward, James Franklin
Zumberge, James Frederick

COLORADO
Allen, Joe Haskell
Aly, Hadi H
Anderson, Dana Zachary
Barrett, Charles Sanborn
Beylkin, Gregory
Carpenter, Donald Gilbert
Cerni, Todd Andrew
Czanderna, Alvin Warren
Dahl, Adrian Hilman
Dunn, Gordon Harold
Goldman, Martin V
Jewell, Jack Lee
Kunz, Peter Dale
Noufi, Rommel
Nozik, Arthur Jack
Pitts, John Roland
Wait, David Francis
Wieman, Carl E
Zimmerer, Robert W
Zoller, Paul

CONNECTICUT
Gillmor, Charles Stewart
Hamblen, David Gordon
Heller, John Herbert
Keith, H(arvey) Douglas
Marcus, Jules A
Monce, Michael Nolen
Powers, Joseph
Reed, Mark Arthur
Rosen, Mark David A
Sen, Pabitra Narayan
Stephenson, Kenneth Edward
Stevens, Joseph Charles

DELAWARE
Avakian, Peter
Cessna, Lawrence C, Jr
Coleman, Marcia Lepri
Cooper, Charles Burleigh
Gibson, Joseph W(hitton), Jr
Longworth, Ruskin
Onn, David Goodwin
Turner, Howard E

DISTRICT OF COLUMBIA
Batra, Narendra K
Briscoe, William John
Carlos, William Edward
Daehler, Mark
Faust, Walter Luck
Grossling, Bernardo Freudenburg
Haas, George Arthur
Holloway, John Thomas
Knudson, Alvin Richard
Lombard, David Bishop
Salu, Yehuda
Shapero, Donald Campbell
Soulen, Robert J, Jr
Wallenmeyer, William Anton

FLORIDA
Billica, Harry Robert
Cain, Joseph Carter, III
Cohen, Martin Joseph
Crow, Jack Emerson
Fitzgerald, Lawrence Terrell
Froehlich, Fritz Edgar
Hammer, Clarence Frederick, Jr
Heydenburg, N P
Kromhout, Robert Andrew
Mrozowski, Stanislaw W
Stewart, Burch Byron
Telkes, Maria
Wolfson, Bernard T

GEORGIA
Kotliar, Abraham Morris
Quisenberry, Dan Ray
Stevenson, Dennis A

IDAHO
Denison, Arthur B
Gasidlo, Joseph Michael

ILLINOIS
Abella, Isaac D
Abraham, Bernard M
Ascoli, Giulio
Bersted, Bruce Howard
Boldridge, David William
Bollinger, Lowell Moyer
Chang, Yia-Chung
De Volpi, Alexander
Eisenstein, Bob I
Elwyn, Alexander Joseph
Fisk, H Eugene
Geil, Phillip H
Haugen, Robert Kenneth
Herzenberg, Caroline Stuart Littlejohn
Hojvat, Carlos F
Kaminsky, Manfred Stephan
Kephart, Robert David
Leggett, Anthony J
Lennox, Arlene Judith
McInturff, Alfred D
McLeod, Donald Wingrove
Ng, King-Yuen
Nissim-Sabat, Charles
Perlow, Gilbert Jerome
Reft, Chester Stanley
Schulz, Charles Emil
Shepard, Kenneth Wayne
Smedskjaer, Lars Christian
Spokas, John J
Stanford, George Stailing
Sullivan, Jeremiah D
Voyvodic, Louis
Vyborny, Carl Joseph
Winston, Roland

INDIANA
Bellina, Joseph James, Jr
Cameron, John M
Carlson, James C
Carmony, Donald Duane
Foster, Charles C
George, Robert Eugene
Madden, Keith Patrick
Miller, Daniel Weber
Witt, Robert Michael

IOWA
Behroozi, Feredoon
Cruse, Richard M

Other Physics (cont)

Green, Robert Wood
Kasper, Joseph Emil
Macomber, Hilliard Kent

KANSAS
Grosskreutz, Joseph Charles
Stanton, Noel Russell

KENTUCKY
Almond, Peter R
Isihara, Akira
Syed, Ibrahim Bijli
Ulrich, Aaron Jack

LOUISIANA
Bourdillon, Antony John
Chen, Isaac I H
Erath, Louis W
Huggett, Richard William, Jr
Ponter, Anthony Barrie
Scott, L Max

MAINE
Snyder, Arnold Lee, Jr
Unertl, William Nelson

MARYLAND
Ambler, Ernest
Banerjee, Manoj K
Barker, John L, Jr
Barnes, John David
Beal, Robert Carl
Bean, Vern Ellis
Benson, Robert Frederick
Blaisten-Barojas, Estela Olga
Brodsky, Allen
Brown, Robert Alan
Carbary, James F
Chen, Henry Lowe
Coble, Anna Jane
Congel, Frank Joseph
Crissman, John Matthews
Danchenko, Vitaly
Das Gupta, Aaron
Davisson, James W
Dennis, Brian R
De Planque, Gail
Eckerman, Jerome
Fanconi, Bruno Mario
Faust, William R
Fava, Ronald Anthony
Fulmer, Glenn Elton
Galloway, William Don
Gibian, Gary Lee
Gloersen, Per
Guilarte, Tomas R
Hartmann, Bruce
Holdeman, Louis Brian
Holland, Russell Sedgwick
Hudson, Ralph P
Jordy, George Y
Kacser, Claude
Long, Gabrielle Gibbs
Matthews, David LeSueur
Maxwell, Louis R
Mazur, Jacob
Meyer, Eric Stefan
Penner, Samuel
Peredo, Mauricio
Richard, Jean-Paul
Sagdeev, Roald Zinnurovi
Schamp, Homer Ward, Jr
Schima, Francis Joseph
Seeger, Raymond John
Shakur, Asif Mohammed
Skrabek, Emanuel Andrew
Smith, Jack Carlton
Stern, David P
Sugai, Iwao
Temperley, Judith Kantack
Tholen, Albert David
Zorn, Gus Tom

MASSACHUSETTS
Abbott, Norman John
Bansil, Rama
Barish, Leo
Bar-Yam, Zvi H
Bechis, Kenneth Paul
Becker, Ulrich J
Behrman, Richard H
Byron, Frederick William, Jr
Caspar, Donald Louis Dvorak
Cohen, Howard David
Corey, Brian E
Crew, Geoffrey B
Dasari, Ramachandra R
Gabrielse, Gerald
Garth, John Campbell
Ginsberg, Edward S
Goitein, Michael
Gottschalk, Bernard
Hardt, David Edgar
Hawkins, W(illiam) Bruce
Heroux, Leon J
Hersh, John Franklin
Herzfeld, Judith
Hillel, Daniel
Hsu, Shaw Ling
Huffman, Fred Norman
Inglis, D R
Janney, Clinton Dales

Jirmanus, Munir N
Kupferberg, Lenn C
Laible, Roy C
Mazur, Eric
Newbower, Ronald Stanley
Post, Richard S
Quynn, Richard Grayson
Rosenson, Lawrence
Rubin, Lawrence G
Sisterson, Janet M
Strauss, Bruce Paul
Tedrow, Paul M
Tokita, Noboru
Young, James Edward

MICHIGAN
Goodrich, Max
Harris, Gale Ion
Heybey, Otfried Willibald Georg
Hunt, Thomas Kintzing
Johnson, Wayne Douglas
Kane, Gordon Leon
Lee, Robert W
Leffert, Charles Benjamin
Lynch, William Gregory
Malaczynski, Gerard W
Marhsall, Hal G
Meier, Dale Joseph
Muench, Nils Lilienberg
Savit, Robert Steven
Schreiber, Thomas Paul
Schumacher, Berthold Walter
Strickland, James Shive
Watson, John H L
Yeh, Gregory Soh-Yu
Zeller, Albert Fontenot

MINNESOTA
Andria, George D
King, Wendell L
Mitchell, William Cobbey
Premanand, Visvanatha
Stuart, Robley Vane

MISSOURI
Bender, Carl Martin
Burke, James Joseph
Ching, Wai-Yim
Hirsh, Ira Jean
Hohenberg, Charles Morris
Hurst, Robert R(owe)
Katz, Jonathan Isaac
Larson, James D
Martin, Frank Elbert
Schmidt, Paul Woodward
Schmitt, John Leigh
Taub, Haskell Joseph

MONTANA
Figueira, Joseph Franklin
Szpilka, Anthony M

NEBRASKA
Borchert, Harold R

NEVADA
Jobst, Joel Edward
Potter, Gilbert David

NEW HAMPSHIRE
Meyer, James Wagner

NEW JERSEY
Ashkin, Arthur
Austin, Robert Hamilton
Barr, William J
Bartnoff, Shepard
Burnett, Bruce Burton
Canter, Nathan H
Carlson, Curtis Raymond
Cooper, Douglas W
Forster, Eric Otto
Girit, Ibrahim Cem
Gray, Russell Houston
Hall, Herbert Joseph
Heimann, Peter Aaron
Jaffe, Michael
Johnson, Leo Francis
Kosel, George Eugene
Lindberg, Craig Robert
Lohse, David John
Ma, Z(ee) Ming
Madey, Theodore Eugene
Marcus, Robert Troy
Meissner, Hans Walter
Mikes, John Andrew
O'Malley, James Joseph
Ong, Chung-Jian Jerry
Padden, Frank Joseph, Jr
Patel, Jamshed R(uttonshaw)
Peebles, Phillip J
Polyakov, Alexander
Prevorsek, Dusan Ciril
Ransome, Ronald Dean
Reisner, John H
Shattes, Walter John
Shyamsunder, Erramilli
Sinclair, Brett Jason
Sprague, Basil Sheldon
Stanton, Robert E
Stiles, John Callender
Tank, David W
Thelin, Lowell Charles
Tyson, J Anthony

Van Dover, Robert B
Wadlow, David
Weimer, Katherine E
Wilchinsky, Zigmond Walter
Witten, Edward

NEW MEXICO
Belian, Richard Duane
Bendt, Philip Joseph
Bradbury, James Norris
Curro, John Gillette
Fenstermacher, Charles Alvin
Gosling, John Thomas
Gray, Edward Ray
Heldman, Julius David
Holm, Dale M
Jackson, Jasper Andrew, Jr
Jameson, Robert A
Johnson, James Daniel
Jones, James Jordan
Keller, Donald V
Kupferman, Stuart L
Ladish, Joseph Stanley
Lindman, Erick Leroy, Jr
MacArthur, Duncan W
Malik, John S
Mercer-Smith, James A
Miller, Guthrie
Miranda, Gilbert A
Morris, Christopher Lee
Oyer, Alden Tremaine
Pickrell, Mark M
Pigott, Miles Thomas
Ramsay, John Barada
Ranken, William Allison
Reedy, Robert Challenger
Ryder, Richard Daniel
Scott, Bobby Randolph
Seitz, Wendell L
Shafer, Robert E
Sheffield, Richard Lee
Sheffield, Stephen Anderson
Sherman, Robert Howard
Smith, Horace Vernon, Jr
Sprinkle, James Kent, Jr
Stokes, Richard Hivling
Suydam, B R
Taddeucci, Terry N
Taylor, Warren Egbert
Thiessen, Henry Archer
Thompson, Jill Charlotte
Young, Lloyd Martin

NEW YORK
Aronson, Raphael (Friedman)
Baker, John M
Barish, Robert John
Baum, Paul M
Baumel, Philip
Bayer, Raymond George
Beck, Harold Lawrence
Bennett, Gerald William
Berman, Lonny E
Bond, Peter Danford
Bozoki-Gombosi, Eva S
Breneman, Edwin Jay
Bunch, Phillip Carter
Callender, Robert
Campbell, Larry Enoch
Chow, Tsu-Sen
Chung, Victor
Cluxton, David H
Cohen, Beverly Singer
Cooper, Barbara Hope
Deady, Matthew William
Dermit, George
Eisenthal, Kenneth B
Erhardt, Peter Franklin
Feinberg, Robert Jacob
Fields, Alfred E
Folan, Lorcan Michael
Forgacs, Gabor
Freeouf, John L
Gallagher, William J
Garwin, Richard Lawrence
Gavin, Donald Arthur
Gibson, Walter Maxwell
Habicht, Ernst Rollemann, Jr
Hahn, Harald
Halket, Thomas D
Hamblen, David Philip
Hansen, James E
Hanson, Albert L
Hartman, Paul Leon
Henry, Arnold William
Hickey, Roger
Hobbs, Stanley Young
Hodge, Ian Moir
Hudak, Michael J
Hughes, Harold K
Johnson, Karen Elise
Jona, Franco Paul
Kambour, Roger Peabody
Kaplan, Robert A
Kaufman, Raymond
Knowles, Richard James Robert
Kornfield, Jack I
Kramer, Martin A
Kraner, Hobart Wilson
Lawson, Kent DeLance
Leipuner, Lawrence Bernard
LeVine, Micheal Joseph
Maldonado, Juan Ramon
Marcus, Michael Alan

Massa, Dennis Jon
Nicolosi, Joseph Anthony
O'Reilly, James Michael
Parsa, Zohreh
Pochan, John Michael
Prest, William Marchant, Jr
Ratcliff, Keith Frederick
Ring, James Walter
Sarko, Anatole
Sciulli, Frank J
Segal, Sheldon J
Shaevitz, Michael H
Slade, Paul Graham
Smith, Kenneth Judson, Jr
Soures, John Michael
Stearns, Brenton Fisk
Sullivan, Peter Kevin
Thieberger, Peter
Tuli, Jagdish Kumar
Tzoar, N
Verrillo, Ronald Thomas
Webb, Watt Wetmore
Wedding, Brent (M)
Willen, Erich H
Wright, Herbert N
Zwislocki, Jozef John

NORTH CAROLINA
Buchanan, David Royal
Callihan, Dixon
Childers, Robert Warren
Crawford-Brown, Douglas John
Dayton, Benjamin Bonney
Fornes, Raymond Earl
Frey, William Francis
Good, Wilfred Manly
Heffelfinger, Carl John
Hocken, Robert John
Johnson, G Allan
Leder, Lewis Beebe
McPeters, Arnold Lawrence
Moran, Paul Richard
Olf, Heinz Gunther
Tonelli, Alan Edward

NORTH DAKOTA
Prunty, Lyle Delmar

OHIO
Agard, Eugene Theodore
Bohm, Georg G A
Bohn, Randy G
Bundy, Francis P
Carpenter, Robert Leland
Colbert, Charles
Conant, Floyd Sanford
Dembowski, Peter Vincent
Dickens, Elmer Douglas, Jr
Eley, Richard Robert
Farlow, Gary C
Forster, Michael Jay
Friedman, Emil Martin
Fujimura, Osamu
Gelerinter, Edward
Griffing, David Francis
Hsu, William Yang-Hsing
Jaworowski, Andrzej Edward
Jezek, Kenneth Charles
Kanakkanatt, Sebastian Varghese
Keck, Max Johann
Lehr, Marvin Harold
Meadows, Brian T
Pillai, Padmanabha S
Ruegsegger, Donald Ray, Jr
Shroff, Ramesh N
Starchman, Dale Edward
Suranyi, Peter
Yun, Seung Soo

OKLAHOMA
Buchanan, Ronnie Joe
Burwell, James Robert
Meyerhoff, Arthur Augustus
Ryan, Stewart Richard
Stacy, Carl J

OREGON
Gerlach, Robert Louis
Orloff, Jonathan H

PENNSYLVANIA
Abraham, Neal Broadus
Baldino, Frank, Jr
Banavar, Jayanth Ramarao
Behrens, Ernst Wilhelm
Bergmann, Ernest Eisenhardt
Dhalla, Amar S
Burkhardt, Theodore W
DeBenedetti, Sergio
Engels, Eugene, Jr
Fetkovich, John Gabriel
Gilmore, Robert
Goldstein, E Bruce
Green, Barry George
Kliman, Harvey Louis
Knight, Stephen
Kurland, Robert John
Laws, Kenneth Lee
Moscatelli, Frank A
Page, Lorne Albert
Parker, James Henry, Jr
Pfeiffer, Heinz Gerhard
Pfrogner, Ray Long
Phares, Alain Joseph

Raymund, Mahlon
Shwe, Hla
Sittel, Karl
Stefanou, Harry
Sterk, Andrew A
Van Ausdal, Ray Garrison
Villafana, Theodore
Zehr, Floyd Joseph

RHODE ISLAND
Amols, Howard Ira
Cutts, David
Loferski, Joseph J

SOUTH CAROLINA
Goswami, Bhuvenesh C
Hay, Ian Leslie
Kimmel, Robert Michael
McCrosson, F Joseph
McNulty, Peter J
Manson, Joseph Richard
Parks, Paul Blair

TENNESSEE
Alexeff, Igor
Allen, William Richard
Bell, Zane W
Boye, Charles Andrew, Jr
Carreras, Benjamin A
Cooke, John F
Gabriel, Tony A
Gant, Kathy Savage
Garber, Floyd Wayne
Johnson, Elizabeth Briggs
Johnston, Archibald Currie
Lyon, James F
Morgan, Karl Ziegler
Painter, Linda Robinson
Shugart, Cecil G
Silver, Ernest Gerard
Watson, Evelyn E
Wendelken, John Franklin
Wood, Bobby Eugene
Yalcintas, M Guven
Zuhr, Raymond Arthur

TEXAS
Badhwar, Gautam D
Bering, Edgar Andrew, III
Behm, Arno
Dahl, Per Fridtjof
Ellis, Kenneth Joseph
Hillery, Herbert Vincent
Hirsch, John Michele
Hurd, James William
McCormick, William Devlin
Marlow, William Henry
Mendelson, Robert Allen
Nyquist, Laurence Elwood
Otis, John Noel
Patriarca, Peter
Rau, Carl
Sanchez, Isaac Cornelius
Sharber, James Randall
Sill, Alan F
Sobey, Arthur Edward, Jr
Stehling, Ferdinand Christian
Thornton, Kathryn C
West, Bruce Joseph
Yau, Wallace Wen-Chuan

UTAH
Galli, John Ronald
Hansen, Wilford Nels
Loh, Eugene C

VERMONT
Hayes, John William

VIRGINIA
Auton, David Lee
Barker, Robert Edward, Jr
Basu, Sunanda
Bruno, Ronald C
Chang, Chieh Chien
Davidson, Charles Nelson
Domingo, John J
Farmer, Barry Louis
Gabriel, Barbra L
Graham, Kenneth Judson
Kabir, Prabahan Kemal
Krafft, Geoffrey Arthur
Kranbuehl, David Edwin
Kubu, Edward Thomas
Leemann, Christoph Willy
Long, Dale Donald
Maune, David Francis
Northrop, John A
Sobottka, Stanley Earl
Strain, Charles A
Sundelin, Ronald M
Vogel, Richard W
Wincklhofer, Robert Charles

WASHINGTON
Buske, Norman L
Carter, Robert S
DeMoney, Fred William
Ely, John Thomas Anderson
Fisher, Darrell R
Johnson, John Richard
Kushmerick, Martin Joseph
Peterson, James Macon
Roesch, William Carl

Safai, Morteza
Styris, David Lee
Warren, John Lucius

WEST VIRGINIA
Franklin, Alan Douglas
Knight, Alan Campbell

WISCONSIN
Aita, Carolyn Rubin
Attix, Frank Herbert
Booske, John H
Day, Anthony R
Fry, William Frederick
Huston, Norman Earl
Mackie, Thomas Rockwell
Smith, William Leo
Truszkowska, Krystyna
Zagzebski, James Anthony
Zei, Dino

WYOMING
Doerges, John E

ALBERTA
Anger, Clifford D
Apps, Michael John
Kline, Loren W
Kounosu, Shigeru

BRITISH COLUMBIA
Bechhoefer, John
Crooks, Michael John Chamberlain
Fisher, Paul Douglas
Ward, Lawrence McCue
Weaver, Ralph Sherman

MANITOBA
Barnard, John Wesley

NEWFOUNDLAND
Irfan, Muhammad

ONTARIO
Alcock, Alfred John
Basinski, Zbigniew S
Beck, Alan Edward
Bigham, Clifford Bruce
Brand, John C D
Dastur, Ardeshir Rustom
Dunn, Andrew Fletcher
Fox, Daniel
Goodings, David Ambery
Greenstock, Clive Lewis
Harvey, John Wilcox
Hunt, John Wilfred
Iribarne, Julio Victor
Ivey, Donald Glenn
Jackman, Thomas Edward
Jones, Alun Richard
McDiarmid, Donald Ralph
Mann, Robert Bruce
Morris, Derek
Ollerhead, Robin Wemp
Prasad, S E
Preston-Thomas, Hugh
Racey, Thomas James
Sawicka, Barbara Danuta
Sharma, Sadhana R
Sinha, Bidhu Bhushan Prasad
Sonnenberg, Hardy
Sundararajan, Pudupadi Ranganathan
Whippey, Patrick William
Wood, Gordon Harvey

QUEBEC
Bunge, Mario Augusto
Kipling, Arlin Lloyd
Madore, Guy Louis Joseph
Rioux, Claude
Schreiner, Lawrence John
Utracki, Lechoslaw Adam

SASKATCHEWAN
Robertson, Beverly Ellis

OTHER COUNTRIES
Bednorz, Johannes Georg
Binnig, Gerd
Brown, Peter
Burschka, Martin A
Lapostolle, Pierre Marcel
Lee-Franzini, Juliet
Mott, Nevill Francis
Mueller, Karl Alexander
Neher, Erwin
Peterlin, Anton
Rabl, Ari
Rohrer, Heinrich
Scheiter, B Joseph Paul

OTHER PROFESSIONAL FIELDS

Education Administration

ALABAMA
Bailey, William C
Chastain, Benjamin Burton
Goldman, Jay
Goldman, Renitta Lee
Hill, Walter Andrew
Irwin, John David

Johnson, Brian John
Knecht, Charles Daniel
Lowry, James Lee
Meezan, Elias
Menaker, Lewis
Mullins, Dail W, Jr
Nair, Madhavan G
Roozen, Kenneth James
Segner, Edmund Peter, Jr
Smith, John Stephen
Warfield, Carol Larson
Watson, Raymond Coke, Jr
Wilson, G Dennis

ALASKA
Behrend, Donald Fraser
Kessel, Brina
Tilsworth, Timothy

ARIZONA
Childs, Orlo E
Cunningham, Richard G(reenlaw)
Goldner, Adreas M
Iserson, Kenneth Victor
Johnson, Vern Ray
Kaufman, C(harles) W(esley)
Metz, Donald C(harles)
Metzger, Darryl E
Mossman, Kenneth Leslie
Weaver, Albert Bruce
Witherspoon, James Donald

ARKANSAS
Deaver, Franklin Kennedy
Reed, Hazell
Tollett, James Terrell

CALIFORNIA
Allen, William Merle
Beljan, John Richard
Carlan, Audrey M
Carson, Virginia Rosalie Gottschall
Castles, James Joseph, Jr
Clague, William Donald
Clark, David Ellsworth
Cockrum, Richard Henry
Colson, Elizabeth F
Coyle, Bernard Andrew
Croft, Paul Douglas
Cromwell, Florence S
Crum, James Davidson
DeJongh, Don C
Dewey, Donald O
Estrin, Thelma A
Fischer, Robert Blanchard
Fisher, Leon Harold
Frank, Sidney Raymond
Greenspan, John Simon
Grobstein, Clifford
Hardy, Rolland L(ee)
Haughton, James Gray
Heiner, Douglas C
Heldman, Morris J
Henkin, Leon (Albert)
Heyneman, Donald
Hisserich, John Charles
Holten, Virginia Zewe
Hubbard, Richard W
Jakubowski, Gerald S
James, Philip Nickerson
Jen, Joseph Jwu-Shan
Johnson, Mark Scott
Kaplan, Richard E
Klinedinst, Paul Edward, Jr
Knapp, Theodore Martin
Krippner, Stanley Curtis
Lai, Kai Sun
Lanham, Richard Henry, Jr
L'Annunziata, Michael Frank
Laub, Alan John
Lave, Roy E(llis), Jr
Law, George Robert John
Leffler, Esther Barbara
Lein, Allen
McMasters, Alan Wayne
Maffly, Roy Herrick
Martin, Joseph B
Meriam, James Lathrop
Miller, Gary Arthur
Miller, William Frederick
Morris, John David
Moye, Anthony Joseph
O'Neill, R(ussell) R(ichard)
Orloff, Neil
Osborne, Michael Andrew
Phillips, William Baars
Picus, Gerald Sherman
Pierce, James Otto, II
Rebman, Kenneth Ralph
Robinson, Hamilton Burrows Greaves
Rogers, Robert N
Sample, Steven Browning
Saunders, Robert M(allough)
Shields, Loran Donald
Shull, Harrison
Smith, Wayne Earl
Spicher, Robert G
Spooner, Charles Edward, Jr
Stahl, Frieda A
Sugihara, Thomas Tamotsu
Triche, Timothy J
Vanderhoef, Larry Neil
Wagner, William Gerard
Watson, John Alfred

Watters, Gary Z
Watts, Malcolm S M
Wendel, Otto Theodore, Jr
Winer, Arthur Melvyn
Woyski, Margaret Skillman
Yagiela, John Allen
Zumberge, James Herbert

COLORADO
Ansell, George S(tephen)
Bond, Richard Randolph
Boudreau, Robert Donald
Briggs, William Egbert
Fraikor, Frederick John
Golden, John O(rville)
Hermann, Allen Max
Holley, Frieda Koster
Janes, Donald Wallace
Johnson, Arnold I(van)
Kano, Adeline Kyoko
Kassoy, David R
Kotch, Alex
Lamm, Warren Dennis
Netzel, Richard G
Niehaus, Merle Hinson
Ogg, James Elvis
Peters, Roger Paul
Regenbrecht, D(ouglas) E(dward)
Romig, Phillip Richardson
Schowengerdt, Franklin Dean
Stein, Robert G
Strom, Oren Grant
Thomas, William Robb

CONNECTICUT
Barone, John A
Brody, Harold David
Carroll, J Gregory
Clarke, George A
Darling, George Bapst
Eigel, Edwin George, Jr
Foster, James Henry
Genel, Myron
Gerald, Michael Charles
Haas, Thomas J
Hall, Newman A(rnold)
Hyde, Walter Lewis
Katz, Lewis
Kollar, Edward James
Krummel, William Marion
Parthasarathi, Manavasi Narasimhan
Singha, Suman
Strand, Richard Alvin
Work, Clyde E(verette)
Wright, Hastings Kemper
Wynn, Charles Martin, Sr
Wystrach, Vernon Paul

DELAWARE
Murray, Richard Bennett
Terss, Robert H
Wetlaufer, Donald Burton

DISTRICT OF COLUMBIA
Baker, Louis Coombs Weller
Berendzen, Richard
Bulger, Roger James
Chater, Shirley S
Coates, Anthony George
Frair, Karen Lee
Hawkins, Gerald Stanley
Huband, Frank L
Ison-Franklin, Eleanor Lutia
Kettel, Louis John
Lehman, Donald Richard
Lustig, Harry
Phillips, Don Irwin
Plowman, Ronald Dean
Raizen, Senta Amon
Robinowitz, Carolyn Bauer
Schultze, Charles L
Smith, Thomas Elijah
Spurlock, Langley Augustine
Wallis, W Allen
Webster, Thomas G

FLORIDA
Anderson, Melvin W(illiam)
Baker, Carleton Harold
Baum, Werner A
Bolte, John R
Brown, John Lott
Burdick, Glenn Arthur
Buss, Daryl Dean
Davison, Beaumont
Dickson, David Ross
Dubravcic, Milan Frane
Dwornik, Julian Jonathan
Fogel, Bernard J
Fry, Jack L
Grobman, Arnold Brams
Harris, Lee Errol
Hartley, Craig Sheridan
Hinkley, Robert Edwin, Jr
Jacobsen, Neil Soren
Jamrich, John Xavier
Jewett, Robert Elwin
Johnsen, Russell Harold
Kennelley, James A
Kline, Jacob
Korwek, Alexander Donald
Kribel, Robert Edward
Mason, D(onald) R(omagne)
Maue-Dickson, Wilma

Education Administration (cont)

Mecklenburg, Roy Albert
Nelson, Eldon Lane, Jr
Nelson, Gordon Leigh
Noble, Nancy Lee
Noble, Robert Vernon
Norris, Dean Rayburn
Peters, Thomas G
Rabbit, Guy
Revay, Andrew W, Jr
Rice, Stephen Landon
Russell, Terrence R
Smith, Wayne H
Spector, Bertram
Thomas, Dan Anderson
Weller, Richard Irwin
Wiebush, Joseph Roy
Wiltbank, William Joseph
Yavorsky, John Michael
Yerger, Ralph William
Zachariah, Gerald L
Zook, Harry David

GEORGIA
Adams, David B
Antolovich, Stephen D
Christenberry, George Andrew
Comeau, Roger William
Dragoin, William Bailey
Evans, Thomas P
Greenberg, Jerrold
Hawkins, Isaac Kinney
Hayes, John Thompson
Heric, Eugene LeRoy
Hooper, John William
Jones, William B(enjamin), Jr
Little, Robert Colby
Marks, Dennis William
Moretz, William Henry
O'Neil, Daniel Joseph
Papageorge, Evangeline Thomas
Satcher, David
Tedesco, Francis J
Threadgill, Ernest Dale
Watts, Sherrill Glenn
Weatherred, Jackie G
Wepfer, William J
Wilson, Donella Joyce

HAWAII
Carlson, John Gregory
Goodman, Madeleine Joyce
Hodgson, Lynn Morrison
Malmstadt, Howard Vincent
Watanabe, Daniel Seishi

IDAHO
Mech, William Paul
Price, Joseph Earl
Renfrew, Malcolm MacKenzie

ILLINOIS
Al-Khafaji, Amir Wadi Nasif
Alpert, Daniel
Andreoli, Kathleen Gainor
Arastoopour, Hamid
Auvil, Paul R, Jr
Bentley, Orville George
Bevan, William
Cleall, John Frederick
Davidson, Donald Miner, Jr
Dommert, Arthur Roland
Dunn, Dorothy Fay
Egan, Richard L
Evans, James Forrest
Frankenberg, Julian Myron
Funk, Richard Cullen
George, William Leo, Jr
Gilbert, Thomas Lewis
Giles, Eugene
Goldman, Jack Leslie
Haber, Meryl H
Heichel, Gary Harold
Hoshiko, Michael S
Jaffe, Philip Monlane
Jonas, Jiri
Kernis, Marten Murray
Kinsinger, Jack Burl
Kleinman, Kenneth Martin
Kraft, Sumner Charles
Laible, Jon Morse
Mapother, Dillon Edward
Martinec, Emil Louis
Munyer, Edward Arnold
Nair, Velayudhan
Pai, Anantha M
Pai, Mangalore Anantha
Sechrist, Chalmers Franklin, Jr
Seitz, Janice Ann
Sill, Larry R
Swoyer, Vincent Harry
Tang, Wilson H
Waterhouse, John P
Wilson, Gerald Gene

INDIANA
Alspaugh, Dale W(illiam)
Beering, Steven Claus
Beery, Dwight Beecher
Eckert, Roger E(arl)
Fredrich, Augustine Joseph
Guthrie, George Drake
Haffley, Philip Gene
Hagen, Charles William, Jr
Hansen, Arthur G(ene)
Hawks, Keith Harold
Hickey, William August
Hicks, John W, III
Hullinger, Ronald Loral
Kessler, David Phillip
Lorentzen, Keith Eden
Nagle, Edward John
Rand, Leon
Schaap, Ward Beecher
Smith, Charles Edward, Jr
White, Samuel Grandford, Jr

IOWA
Anderson, R(obert) M(orris), Jr
Burrill, Claude Wesley
Christensen, George Curtis
Collins, Richard Francis
Garfield, Alan J
Kelly, William H
Kluge, John Paul
Koppelman, Ray
Morriss, Frank Howard, Jr
Odgaard, A Jacob
Robinson, Robert George
Schafer, John William, Jr
Stebbins, Dean Waldo
Wilder, David Randolph
Williams, Fred Devoe

KANSAS
Connor, Sidney G
Crumb, Glenn Howard
Frost-Mason, Sally Kay
Jennings, Paul Harry
Lambson, Roger O
Larson, Vernon C
Locke, Carl Edwin, Jr
Loper, Gerald D
Parizek, Eldon Joseph
Roberts, James Arnold
Waxman, David
Zehr, John E

KENTUCKY
Boehms, Charles Nelson
Fardo, Stephen W
Farman, Allan George
Fleischman, Marvin
Gerhard, Earl R(obert)
Hanley, Thomas Richard
Harris, Henson
Hawkins, Charles Edward
Herron, James Dudley
Humphries, Asa Alan, Jr
Kraman, Steve Seth
Mitchell, George Ernest, Jr
Nash, David Allen
Peterson, Roy Phillip
Podshadley, Arlon George
Saxe, Stanley Richard
Vittetoe, Marie Clare

LOUISIANA
Birtel, Frank T
Caffey, Horace Rouse
Frohlich, Edward David
Hankins, B(obby) E(ugene)
Lembeck, William Jacobs
Maxfield, John Edward
Morris, Everett Franklin
Ruby, John Robert
Smith, Grant Warren, II
Sterling, Arthur MacLean
Teate, James Lamar
Walker, John Robert
Walsh, John Joseph
Wetzel, Albert John
Wilson, Raphael
Zander, Arlen Ray

MAINE
Bernstein, Seldon Edwin
Rosenfeld, Melvin Arthur

MARYLAND
Bateman, Barry Lynn
Bollum, Frederick James
Caret, Robert Laurent
Chartrand, Mark Ray, III
Cotton, William Robert
Dieter, George E(llwood), Jr
Dupont, Jacqueline (Louise)
Dutta, Samarendra N
Fowler, Earle Cabell
Gluckstern, Robert Leonard
Heller, Charles O(ta)
Henderson, Donald Ainslie
Hindle, Brooke
Holt, James Allen
Howard, Joseph H
Johnston, Robert Ward
Jones, Franklin Del
Karlander, Edward P
Kelley, John Francis
Knapp, David Allan
Krauss, Robert Wallfar
Langenberg, Donald Newton
Levchuk, John W
Lymangrover, John R
Mathieu, Richard D(etwiler)
Mayor, John Roberts
Miller, Gerald Ray
Montgomery, David Carey
Moorjani, Kishin
More, Kenneth Riddell
Mumford, Willard R
Nanzetta, Philip Newcomb
Ranney, Richard Raymond
Richardson, William C
Shelton, Robert Duane
Suskind, Sigmund Richard
Wallace, James M
Wilson, Donald Edward
Yarmolinsky, Adam
Young, Harold Henry

MASSACHUSETTS
Baird, Ronald C
Baldwin, William Russell
Barlas, Julie S
Bhargava, Hridaya Nath
Blatt, S Leslie
Bolz, R(ay) E(mil)
Cavallaro, Mary Caroline
Cheek, Dennis William
Coghlan, Anne Eveline
Cohen, Joel Ralph
Cohn, Lawrence H
Deck, Joseph Charles
Delaney, Patrick Francis, Jr
Fano, Robert M(ario)
Federman, Daniel D
Field, Hermann Haviland
Fuqua, Mary Elizabeth
Gold, Albert
Graham, Robert Montrose
Haworth, D(onald) R(obert)
Henry, Joseph L
Hershenson, Benjamin R
Hilferty, Frank Joseph
Hooper, James R(ipley)
Hulbert, Thomas Eugene
Hyatt, Raymond R, Jr
Kazimi, Mujid S
Koserksy, Donald Saadia
Kranich, Wilmer Leroy
Loew, Franklin Martin
Lynton, Ernest Albert
Mark, Melvin
Maser, Morton D
Morse, Robert Warren
Murnane, Thomas William
Nelson, J(ohn) Byron
Papageorgiou, John Constantine
Reintjes, J Francis
Robinson, Howard Addison
Seamans, R(obert) C(hanning), Jr
Skolnikoff, Eugene B
Snyder, John Crayton
Stinchfield, Carleton Paul
Stratton, Julius Adams
Taylor, Edwin Floriman
Taylor, Isaac Montrose
Tenca, Joseph Ignatius
Volkmann, Frances Cooper
Wedlock, Bruce D(aniels)
Weller, Paul Franklin
West, Arthur James, II
Wick, Emily Lippincott

MICHIGAN
Adrounie, V Harry
Banks, Peter Morgan
Baumiller, Robert C
Beaufait, Frederick W
Becker, Marshall Hilford
Bohm, Henry Victor
Bollinger, Robert Otto
Cerny, Joseph Charles
Conway, Lynn Ann
Courtney, Gladys (Atkins)
Cutler, Warren Gale
Dasgupta, Rathindra
Dauphinais, Raymond Joseph
Deuben, Roger R
Eliezer, Isaac
Fritts-Williams, Mary Louise Monica
Gilroy, John
Gray, Robert Howard
Johnson, John Harris
Khasnabis, Snehamay
Kiechle, Frederick Leonard
King, Herman (Lee)
Kingman, Robert Earl
Lewis, Alan Laird
Lomax, Ronald J(ames)
Lorenz, John Douglas
Lumsdaine, Edward
Magen, Myron S
Mather, Edward Chantry
Miller, M(urray) H(enri)
Nash, Edmund Garrett
Osborn, Wayne Henry
Reinke, David Albert
Rusch, Wilbert H, Sr
Saul, John Edward
Schneider, Michael J
Smith, Ellen Oliver
Smith, P(aul) Dennis
Sommers, Lawrence M
Soper, Jon Allen
Stocker, Donald V(ernon)
Strachan, Donald Stewart
Stynes, Stanley K
Tasker, John B
Tompkins, Curtis Johnston
Weg, John Gerard
Williams, Fredrick David
Wojcik, Anthony Stephen

MINNESOTA
Anderson, Sabra Sullivan
Barden, Roland Eugene
Cavert, Henry Mead
Conlin, Bernard Joseph
Giese, David Lyle
Lukasewycz, Omelan Alexander
Martens, Leslie Vernon
Miller, Gerald R
Miller, Jack W
Riley, Donald Ray
Rogers, Roy Steele, III
Romero, Juan Carlos
Sherman, Patsy O'Connell
Wolfe, Barbara Blair

MISSISSIPPI
Doblin, Stephen Alan
Foil, Robert Rodney
McKee, J(ewel) Chester, Jr
Oswalt, Jesse Harrell
Roark, Bruce (Archibald)
Scott, Charley
Smith, Allie Maitland

MISSOURI
Axthelm, Deon D
Bader, Kenneth L
Banakar, Umesh Virupaksh
Baumann, Thiema Wolf
Becker, Barbara
Brodmann, John Milton
Ericson, Avis J
Ferris, Deam Hunter
Festa, Roger Reginald
Hines, Anthony Loring
Johnson, James W(inston)
Kupchella, Charles E
Landiss, Daniel Jay
Malzahn, Ray Andrew
Mattox, Douglas M
Meyer, William Ellis
Mitchell, Henry Andrew
Myers, Richard F
Nikolai, Robert Joseph
Oetting, Robert B(enfield)
Oviatt, Charles Dixon
Richards, Graydon Edward
Rouse, Robert Arthur
Santiago, Julio Victor
Sauer, Harry John, Jr
Stoneman, William, III
Toom, Paul Marvin
Watson, John Leslie
Weiss, James Moses Aaron
Weldon, Virginia V
Wilson, Edward Nathan

MONTANA
Figueira, Joseph Franklin
Hess, Lindsay LaRoy
McRae, Robert James

NEBRASKA
Bahar, Ezekiel
Benschoter, Reba Ann
Clark, Ronald David
Gardner, Paul Jay
O'Brien, Richard Lee
Roche, Edward Browning
Schmitz, John Albert
Spanier, Graham Basil
Thorson, James A
Wehrbein, William Mead

NEVADA
Bohmont, Dale W
Leedy, Clark D
May, Jerry Russell
Merdinger, Charles J(ohn)
Smith, Robert Bruce

NEW HAMPSHIRE
Bernstein, Abram Bernard
Brown, Guendoline
Davis, William Potter, Jr
Dixon, Wallace Clark, Jr
Kaiser, C William
Melvin, Donald Walter
Price, Glenn Albert
Wixson, Eldwin A, Jr

NEW JERSEY
Albu, Evelyn D
Anderson, Robert E(dwin)
Arnold, Frederic G
Bartnoff, Shepard
Bassett, Alton Herman
Benedict, Joseph T
Buhks, Ephraim
Chance, Kenneth Bernard
Chizinsky, Walter
Colaizzi, John Louis
Colicelli, Elena Jeanmarie
DeProspo, Nicholas Dominick
Derucher, Kenneth Noel
Edelman, Norman H
Fahrenholtz, Susan Roseno

Fenster, Saul K
Field, Norman J
Finch, Rogers B(urton)
Friedman, Edward Alan
Gewirtz, Allan
Griffin, William Dallas
Griffo, James Vincent, Jr
Griminger, Paul
Hager, Mary Hastings
Houle, Joseph E
Leventhal, Howard
Lewandowski, Gordon A
Lucas, Henry C, Jr
Lynde, Richard Arthur
Merritt, Richard Howard
Miller, Edward
Pignataro, Louis J(ames)
Poiani, Eileen Louise
Posamentier, Alfred S
Robin, Michael
Schwartz, Arthur Harold
Sutman, Frank X

NEW MEXICO
Barbo, Dorothy M
Bhada, Rohinton(Ron) K
Bradshaw, Martin Daniel
Cano, Gilbert Lucero
Cornell, Samuel Douglas
Maez, Albert R
Monagle, John Joseph, Jr
Omer, George Elbert, Jr
Wood, Joseph George
Yudkowsky, Elias B

NEW YORK
Ahmad, Jameel
Alexander, Justin
Allentuch, Arnold
Armstrong, Donald
Aschner, Joseph Felix
Aull, Felice
Auston, David H
Baines, Ruth Etta
Ballentine, Rudolph Miller
Bartilucci, Andrew J
Belford, Julius
Berg, Daniel
Blackmon, Bobby Glenn
Bloch, Aaron Nixon
Brenner, Egon
Brody, Harold
Bruns, Peter John
Burge, Robert Ernest, Jr
Burke, G Thompson
Butler, Katharine Gorrell
Chabot, Brian F
Clark, Allan H
Cooke, Lloyd Miller
Davis, Carolyne K
Day, Stacey Biswas
De Carlo, Charles R
De France, Joseph J
DeLong, Stephen Edwin
Dizer, John Thomas, Jr
Eppenstein, Walter
Feisel, Lyle Dean
Gabriel, Mordecai Lionel
Gary, Nancy E
George, A(lbert) R(ichard)
George, Nicholas
Gilmore, Arthur W
Glass, David Carter
Goldberg, Louis J
Goldman, James Allan
Greaves, Bettina Bien
Greisen, Kenneth I
Griffith, Daniel Alva
Grosewald, Peter
Gussin, Arnold E S
Hood, Donald C
Hsu, Donald K
Hunte, Beryl Eleanor
Huntington, David Hans
Irwin, Lynne Howard
Johnson, Charles Richard, Jr
Kampmeier, Jack A
Kiser, Kenneth M(aynard)
Kletsky, Earl J(ustin)
Kuperman, Albert Sanford
Lanford, William Armistead
Litynski, Daniel Mitchell
McAllister, Harmon Carlyle, Jr
McClain, John William
McDonald, Donald
Matsushima, Satoshi
Mehaffey, Leathem, III
Mennitt, Philip Gary
Metz, Donald J
Miller, Russell Loyd, Jr
Mueller, Edward E(ugene)
Muench, Donald Leo
Murphy, James Joseph
Muschio, Henry M, Jr
Nathanson, Melvyn Bernard
Oates, Richard Patrick
Paldy, Lester George
Peters, Earl
Petersen, Ingo Hans
Polowczyk, Carl John
Postman, Robert Derek
Prutkin, Lawrence
Ranu, Harcharan Singh
Reifler, Clifford Bruce
Richards, Paul Granston
Riehle, Robert Arthur, Jr
Robusto, C Carl
Roman, Stanford A, Jr
Rosenberg, Robert
Rosenfeld, Norman Samuel
Sank, Diane
Savic, Michael I
Schneider, Allan Stanford
Schultze, Lothar Walter
Shapiro, Sidney
Simons, Gene R
Solomon, Seymour
Springer, Dwight Sylvan
Stafford, Magdalen Marrow
Stephens, Lawrence James
Streett, William Bernard
Tavel, Judith Fibkins
Thomas, Everett Dake
Thorpe, John A(lden)
Vidosic, J(oseph) P(aul)
Vinciguerra, Michael Joseph
Wiesenfeld, John Richard
Wilder, Joseph R
Wilkins, Bruce Tabor
Wittig, Kenneth Paul
Worden, David Gilbert

NORTH CAROLINA
Aronson, A L
Bailey, Kincheon Hubert, Jr
Bell, Norman R(obert)
Brotak, Edward Allen
Burnett, John Nicholas
Butts, Jeffrey A
Chin, Robert Allen
Daw, John Charles
Emmerling, David Alan
Hallock, James A
Hayek, Dean Harrison
Henning, Emilie D
Herrell, Astor Y
Hersh, Solomon Philip
Howard, Donald Robert
Hugus, Z Zimmerman, Jr
King, Kendall Willard
Land, Ming Huey
Lytle, Charles Franklin
Miller, Robert L
Mosier, Stephen R
Oblinger, Diana Gelene
Peirce, James Jeffrey
Richardson, Frances Marian
Ronzio, Robert A
Smith, Norman Cutler
Turinsky, Paul Josef
Werntz, James Herbert, Jr
Wilson, Lauren R
Winston, Hubert
Yarbrough, Charles Gerald
Yarger, William E

NORTH DAKOTA
Bares, William Anthony
Kuruganty, Sastry P
Sugihara, James Masanobu
Vasey, Edfred H

OHIO
Anderson, Ruth Mapes
Baker, Saul Phillip
Barsky, Constance Kay
Cassidy, Suzanne Bletterman
Collins, William John
Corbato, Charles Edward
Cox, Milton D
Curtis, Charles R
Daroff, Robert Barry
D'Azzo, John Joachim
Dobbelstein, Thomas Norman
Essman, Joseph Edward
Faiman, Robert N(eil)
Friedman, Morton Harold
Fucci, Donald James
Gamble, Francis Trevor
Goel, Prem Kumar
Gorse, Joseph
Greenstein, Julius S
Hales, Raleigh Stanton, Jr
Hutchinson, Frederick Edward
Jache, Albert William
Lee, Kai-Fong
Leonard, Billie Charles
McLean, Edward Bruce
Millard, Ronald Wesley
Moon, Donald Lee
Nuenke, Richard Harold
Papadakis, Constantine N
Ray, John Robert
Redmond, Robert F(rancis)
Robe, Thurlow Richard
Rubin, Stanley G(erald)
Stansloski, Donald Wayne
Stith, James Herman
Sweeney, Thomas L(eonard)

OKLAHOMA
Alexander, Joseph Walker
Batchman, Theodore E
Beasley, William Harold
Bell, Paul Burton, Jr
Brandt, Edward Newman, Jr
Campbell, John R(oy)
Carlile, Robert Elliot
Cassens, Patrick
Deckert, Gordon Harmon
Edmison, Marvin Tipton
Greene, Timothy J
Gunter, Deborah Ann
Hamm, Donald Ivan
Hooper, Billy Ernest
Kimpel, James Froome
Klatt, Arthur Raymond
Kriegel, Monroe W(erner)
Langwig, John Edward
McCollom, Kenneth A(llen)
Miranda, Frank Joseph
Owens, G(lenda) Kay
Parke, David W, II
Robinson, Jack Landy
Smith, Philip Edward
Swaim, Robert Lee
Turner, Wayne Connelly
Urban, Timothy I

OREGON
Allen, John Eliot
Arscott, George Henry
Brookhart, John Mills
Brown, Cynthia Ann
Clark, William Melvin, Jr
Davis, John R(owland)
Decker, Fred William
Griffith, William Thomas
Huber, Wayne Charles
Neiland, Bonita J
Owen, Sydney John Thomas
Scott, Peter Carlton
Van Vliet, Antone Cornelis
Weber, Lavern J
Zaerr, Joe Benjamin

PENNSYLVANIA
Andrews, Edwin Joseph
Baum, Joseph Herman
Beilstein, Henry Richard
Berlin, Cheston Milton
Bilgutay, Nihat Mustafa
Bluemle, Lewis W, Jr
Brooten, Dorothy
Cahir, John Joseph
Caputo, Joseph Anthony
Connors, Donald R
Cunningham, Joel L
Deitrick, John E
Fromm, Eli
Fuget, Charles Robert
Gaither, William Samuel
Gardiner, Keith M
Gidley, James Scott
Gotolski, William H(enry)
Gwiazda, S(tanley) J(ohn)
Herzfeld, Valerius E
Hinish, Wilmer Wayne
Ho, Thomas Inn Min
Keith, Jennie
Keyt, Donald E
Kidawa, Anthony Stanley
Kolmen, Samuel Norman
Kossor, Steven Albert
Kraybill, Edward K(ready)
Lang, Norma M
Larson, Russell Edward
Lee, John Denis
McMurtry, George James
Matsumoto, Teruo
Melvin, John L
Miklausen, Anthony J
Mullen, James L
Murray, Donald Shipley
Paolino, Michael A
Pyeritz, Reed E
Rackoff, Jerome S
Reischman, Michael Mack
Rizza, Paul Frederick
Roman, Paul
Sieber, James Leo
Siegel, Herbert S
Sovie, Margaret D
Tomezsko, Edward Stephen John
Walker, Augustus Chapman
Walker, Ruth Angelina
Waxman, Herbert Sumner
Williams, Willie, Jr
Wilson, Marjorie Price
Wolgemuth, Carl Hess
Wolke, Robert Leslie

RHODE ISLAND
Barnes, Frederick Walter, Jr
Clagett, Robert P
Gould, Mark D
Greer, David Steven
Hazeltine, Barrett
Hedlund, Ronald David
Pickart, Stanley Joseph
Rothman, Frank George
Worthen, Leonard Robert

SOUTH CAROLINA
Bates, G William
Bennet, Archie Wayne
Bourkoff, Etan
Calhoon, Stephen Wallace, Jr
Dieter, Richard Karl
Hahs, Sharon K
Hogan, Joseph C(harles)
Huray, Paul Gordon
Kimbler, Delbert Lee
Kinard, Frank Efird
Lawson, Benjamin F
Nerbun, Robert Charles, Jr
Sandrapaty, Ramachandra Rao
Sears, Harold Frederick
Wise, William Curtis
Worthington, Ward Curtis, Jr

SOUTH DAKOTA
Cook, David Edgar
Dunn, Jacquelin
Hilderbrand, David Curtis
Koepsell, Paul L(oel)

TENNESSEE
Adams, Dennis Ray
Armistead, Willis William
Ayers, Jerry Bart
Bonner, William Paul
Brady, Donald R
Burish, Thomas Gerard
Christman, Luther P
Demetriou, Charles Arthur
Fullerton, Ralph O
Garber, David H
Gotterer, Gerald S
Hunt, James Calvin
Jumper, Sidney Roberts
Keedy, Hugh F(orrest)
Kirksey, Howard Graden, Jr
Kozub, Raymond Lee
LeQuire, Virgil Shields
Lerner, Joseph
Lowe, James Urban, II
Lynch, Denis Patrick
McNeely, Robert Lewis
Malasri, Siripong
Massengill, Raymond
Miller, Paul Thomas
Moore, George Edward
Obear, Frederick W
Perry, Margaret Nutt
Prados, John W(illiam)
Reed, Peter William
Snyder, William Thomas
Trubatch, Janett
Uhrig, Robert Eugene
Volpe, Angelo Anthony
Waller, John Wayne
Williamson, Handy, Jr

TEXAS
Al-Saadoon, Faleh T
Bielamowicz, Mary Claire Kinney
Blank, Leland T
Brooks, Dorothy Lynn
Bruhn, John G
Bryan, George Thomas
Calhoun, John C, Jr
Chenoweth, R(obert) D(ean)
Chow, Rita K
Coleman, John Dee
Dickey, John Sloan, Jr
Finegold, Milton J
Flawn, Peter Tyrrell
Fujitani, Roy Masami
Gonzalez, Mario J
Gunn, Albert Edward
Haragan, Donald Robert
Hewett-Emmett, David
Hewitt, Roger R
Kalkwarf, Kenneth Lee
Largent, Max Dale
Linder, John Scott
Low, Morton David
Lytle, Michael Allen
McCloy, James Murl
Merrill, John Raymond
Mikiten, Terry Michael
Moiz, Syed Abdul
Murphy, Ruth Ann
Nelson, John Franklin
Noonan, James Waring
Parr, Christopher Alan
Pickard, Myrna Rae
Poage, Scott T(abor)
Quarles, John Monroe
Race, George Justice
Ramsey, Jerry Dwain
Robbins, Ralph Robert
Russell, B Don
Rutford, Robert Hoxie
Sanders, Barbara
Schofield, James Roy
Stephenson, Danny Lon
Storey, Arthur Thomas
Sweeney, Edward Arthur
Swope, Richard Dale
Trogdon, William Oren
Tucker, Woodson Coleman, Jr
Vines, Darrell Lee
Wakeland, William Richard
Walker, James Richard
Watt, Joseph T(ee), Jr
Whorton, Elbert Benjamin
Wilson, Bobby L
Wrotenbery, Paul Taylor
Wyllys, Ronald Eugene
Yoder Wise, Patricia S

UTAH
Barton, Cliff S
Fisher, Richard Forrest

Education Administration (cont)

Moser, Royce, Jr
Olson, Randall J
Striefel, Sebastian
Strozier, James Kinard
Welch, Garth Larry
Winicur, Daniel Henry
Wood, O Lew

VERMONT
Butler, Donald J(oseph)
Forcier, Lawrence Kenneth
Francis, Gerald Peter
Haugh, Larry Douglas
Lawson, Robert Bernard
McIntire, Sumner Harmon
Moyer, Walter Allen, Jr
Provost, Ronald Harold
Spearing, Ann Marie

VIRGINIA
Albrecht, Herbert Richard
Bishara, Michael Nageeb
Blair, Barbara Ann
Blouet, Brian Walter
Bordogna, Joseph
Brown, Gregory Neil
Brownlee, Paula Pimlott
Burstyn, Harold Lewis
Chambers, Charles MacKay
Clark, Mary Jane
Clarke, Alan R
Conger, William Lawrence
Conn, William David
Cooper, Earl Dana
Cross, Ernest James, Jr
Detmer, Don Eugene
Dobyns, Samuel Witten
Hall, Harvey
Hall, Philip Layton
Hench, Miles Ellsworth
Hill, Carl McClellan
Jaron, Dov
Kahne, Stephen James
Kugel, Robert Benjamin
Levine, Jules Ivan
McFarlane, Kenneth Walter
McIrvine, Edward Charles
Maune, David Francis
Moseley, Sherrard Thomas
Munday, John Clingman, Jr
Owens, Charles Wesley
Sigler, Julius Alfred, Jr
Smith, Elske van Panhuys
Spresser, Diane Mar
Starnes, William Herbert, Jr
Stein, Richard Louis
Stump, Billy Lee
Summerville, Richard Marion
Swiger, Louis Andre
Trindle, Carl Otis
Watt, William Joseph
Weaver, Warren Eldred
Weiss, Daniel Leigh
Whitesides, John Lindsey, Jr
Wiseman, Lawrence Linden

WASHINGTON
Albers, James Ray
Anderson, Charles Dean
Bhagat, Surinder Kumar
Chase, John David
Cochran, James Alan
Daniels, Patricia D
Frank, Andrew Julian
Frasco, David Lee
Glass, James Clifford
Hogness, John Rusten
Jackson, Crawford Gardner, Jr
Jones, Jerry Lynn
Kalina, Robert E
Kraft, Gerald F
Noel, Jan Christine
Norman, Joe G, Jr
Odegaard, Charles E
Rosse, Cornelius
Schrader, Lawrence Edwin
Smith, Samuel H
Steckler, Bernard Michael
Strandjord, Paul Edphil
Van der Werff, Terry Jay
Young, Paul Ruel
Zimmerman, Gary Alan

WEST VIRGINIA
Digman, Robert V
Friedman, Morton Henry
Stansbury, Harry Adams, Jr
Thornton, Stafford E

WISCONSIN
Biester, John Louis
Blockstein, William Leonard
Chang, Tsong-How
Davis, Thomas William
Drucker, William D
Greenebaum, Ben
Halgren, Lee A
Karkheck, John Peter
Long, Sally Yates
Moses, Gregory Allen

Outcalt, David L
Schopler, Harry A
Weinswig, Melvin H

WYOMING
Porter, A Duane

PUERTO RICO
Bonnet, Juan A, Jr
Gonzalez, Gladys
Haddock, Lillian
Lluch, Jose Francisco
Montes, Maria Eugenia

ALBERTA
Cook, David Alastair
Gee, Norman
Krueger, Peter J
Norford, Brian Seeley
Russell, Anthony Patrick

BRITISH COLUMBIA
Baird, Patricia A
Beirne, Bryan Patrick
Copes, Parzival
Gillis, Hugh Andrew
Kennedy, James M
Krauel, David Paul
Mindess, Sidney
Sheehan, Bernard Stephen
Ziemba, W T

MANITOBA
Henderson, James Stuart
Hunter, Norman Robert
Johnson, R(ichard) A(llan)

NEWFOUNDLAND
Gogan, Niall Joseph

NOVA SCOTIA
Hatcher, James Donald
McAlduff, Edward J
McDonald, Terence Francis
Oliver, Leslie Howard
Tonks, Robert Stanley
Wilson, George Peter

ONTARIO
Bayliss, Colin Edward
Bienenstock, John
Billigheimer, Claude Elias
Bodnar, Louis Eugene
Celinski, Olgierd J(erzy) Z(dzislaw)
Coleman, Albert John
Currie, Violet Evadne
Denzel, George Eugene
Fox, Joseph S
Ghista, Dhanjoo Noshir
Goldberg, David Myer
Ham, James M(ilton)
Hansson, Carolyn M
Henry, Bryan Roger
Key, Anthony W
Kinlough-Rathbone, Roelene L
LeRoy, Robert James
Okashimo, Katsumi
Preston, Melvin Alexander
Reza, Fazlollah M
Rothenbury, Raymand Albert
Savan, Milton
Sedra, Adel S
Sinclair, D G
Switzer, Clayton Macfie
Tyson, John Edward Alfred
Uffen, Robert James
Usborne, William Ronald
Valberg, Leslie S
Wright, Douglas Tyndall

QUEBEC
Cloutier, Gilles Georges
Cook, Robert Douglas
Cruess, Richard Leigh
Galanis, Nicolas
Hardy, Yvan J
Hetherington, Donald Wordsworth
Johns, Kenneth Charles
L'Ecuyer, Jacques
Marisi, Dan(iel) Quirinus
Moore, Sean
Shaw, Robert Fletcher

OTHER COUNTRIES
Bachmann, Fedor W
Barrett, Jerry Wayne
Maqusi, Mohammad
Mori, Wataru
Okamura, Sogo
Revillard, Jean-Pierre Remy
Scheiter, B Joseph Paul
Suematsu, Yasuharu
Toshio, Makimoto
Van Overstraeten, Roger Joseph
Watanabe, Mamoru
Whitten, Eric Harold Timothy

History & Philosophy of Science

ALABAMA
Barr, E(rnest) Scott
Chastain, Benjamin Burton

Fitzpatrick, J(oseph) F(erris), Jr
Glover, Elsa Margaret
Lewis, W David
Schneider, Susan Marguerite

ARIZONA
Baker, Victor Richard
Barnes, Charles Winfred
Black, Dean
Hartmann, William K
Ince, Simon
Iserson, Kenneth Victor
Linehan, Urban Joseph
Maienschein, Jane Ann
Pacholczyk, Andrzej Grzegorz
Rouse, Hunter
Swan, Cathy Wood

ARKANSAS
Berghel, Hal L

CALIFORNIA
Abegg, Victor Paul
Agnew, Leslie Robert Corbet
Alvarino, Angeles
Amara, Roy C
Assmus, Alexi Josephine
Aurilia, Antonio
Badash, Lawrence
Bahn, Gilbert S(chuyler)
Batchelder, William Howard
Beidelman, Richard Gooch
Bennett, John Francis
Berger, Ralph Jacob
Bertoldi, Gilbert LeRoy
Bertsch, Hans
Bibel, Debra Jan
Birnbaum, Michael Henry
Blumberg, Mark Stuart
Borchardt, Glenn (Arnold)
Bork, Alfred Morton
Cate, Robert Bancroft
Chen, Joseph Cheng Yih
Correia, John Sidney
Corwin, Harold G, Jr
Feferman, Solomon
Feifel, Herman
Fischer, Robert Blanchard
Gorman, Melville
Gortner, Susan Reichert
Gosselin, Edward Alberic
Grabiner, Judith Victor
Graves, Joseph L
Gregory, Joseph Tracy
Guerrini, Anita
Haen, Peter John
Haimes, Florence Catherine
Haisch, Bernhard Michael
Holeman, Dennis Leigh
Holmes, Ivan Gregory
Howe, George Franklin
Kauffman, George Bernard
Kellogg, Ralph Henderson
Kevles, Daniel Jerome
Krippner, Stanley Curtis
Laporte, Leo Fredric
Leff, Harvey Sherwin
Lerner, Lawrence S
Lippa, Linda Susan Mottow
Little, Angela C
Lofgren, Edward Joseph
Lombardi, Gabriel Gustavo
Lopes-Gautier, Rosaly
McCarthy, John Lockhart
McLaughlin, William Irving
Matlow, Sheldon Leo
Matuszak, Alice Jean Boyer
Maurer, E(dward) Robert
Merzbacher, Claude F
Miller, Donald Gabriel
Muller, Cornelius Herman
Mumme, Judith E
Norgaard, Richard Bruce
Noyes, H Pierre
Pang, Kevin
Pearl, Judea
Risse, Guenter Bernhard
Roth, Ariel A
Russell, Stuart Jonathan
Schelar, Virginia Mae
Schneiderman, Jill Stephanie
Schwabe, Calvin Walter
Shubert, Bruno Otto
Sime, Ruth Lewin
Tenn, Joseph S
Thaw, Richard Franklin
Towers, Bernard
Tschirgi, Robert Donald
Tuul, Johannes
Vincenti, Walter G(uido)
Way, E Leong
Weeks, Dennis Alan
Wells, Patrick Harrington
Wells, Ronald Allen
Woodward, James Franklin

COLORADO
Franklin, Allan David
Reinhardt, William Nelson
Rollin, Bernard Elliot
Watkins, Sallie Ann

CONNECTICUT
Conover, Lloyd Hillyard

Egler, Frank Edwin
Fenton, David George
Fraser, J(ulius) T(homas)
Fruton, Joseph Stewart
Gillmor, Charles Stewart
Hanson, Trevor Russell
Klein, Martin J(esse)
Kuslan, Louis Isaac
Levine, Robert John
Martinez, Robert Manuel
Menke, David Hugh
Patterson, Elizabeth Chambers
Salamon, Richard Joseph
Stevenson, Harlan Quinn
Summers, William Cofield
Suprynowicz, Vincent A
Viseltear, Arthur Jack
Wescott, Roger Williams

DELAWARE
Beer, John Joseph
Durbin, Paul Thomas
Ferguson, Raymond Craig
Pratt, Herbert T
Singleton, Rivers, Jr
Van Dyk, John William
Williams, Mary Bearden

DISTRICT OF COLUMBIA
Abler, Ronald Francis
Bellmer, Elizabeth Henry
Berendzen, Richard
Braunfeld, Peter George
Buzzelli, Donald Edward
Coates, Joseph Francis
Curtin, Frank Michael
DeVorkin, David Hyam
Earley, Joseph Emmet
Feldman, Martin Robert
Forman, Paul
Goldberg, Stanley
Handel, Mark David
Harwit, Martin Otto
Hazen, Robert Miller
Jones, Daniel Patrick
Lieberman, Edwin James
McLean, John Dickson
O'Hern, Elizabeth Moot
Scheel, Nivard
Veatch, Robert Marlin

FLORIDA
Albright, John Rupp
Cash, Rowley Vincent
Cooper, Raymond David
Hartman, John Paul
Leitner, Alfred
Linder, Ernest G
Mecklenburg, Roy Albert
Nunez, German
Rawls, Walter Cecil, Jr
Roberts, Leonidas Howard
Rubin, Richard Lee
Stern, William Louis
Vergenz, Robert Allan

GEORGIA
Dragoin, William Bailey
Loomis, Earl Alfred, Jr
Peacock, Lelon James
Saladin, Kenneth S

HAWAII
Krisciunas, Kevin L
Weems, Charles William

ILLINOIS
Amarose, Anthony Philip
Brown, Laurie Mark
Carozzi, Albert Victor
Debus, Allen George
Dinsmore, Charles Earle
Goldman, Jack Leslie
Gray, John Walker
Halberstam, Heini
Herzenberg, Caroline Stuart Littlejohn
Hirsch, Jerry
Mrtek, Robert George
Orland, Frank J
Pappademos, John Nicholas
Quay, Paul Michael
Rosen, Sidney
Schrage, Samuel
Spiroff, Boris E N
Spradley, Joseph Leonard
Stigler, Stephen Mack
Swanson, Don R
Sweitzer, James Stuart
Tuttle, Russell Howard
Wakefield, Ernest Henry
Yos, David Albert
Zusy, Dennis

INDIANA
Bellina, Joseph James, Jr
Cassidy, Harold Gomes
Dudley, Underwood
Helrich, Carl Sanfrid, Jr
Martin, Charles Wellington
Miller, Richard William
Zinsmeister, William John

IOWA
Kracher, Alfred

OTHER PROFESSIONAL FIELDS / 571

Patterson, John W(illiam)
Rider, Paul Edward, Sr
Ruth, Royal Francis

KANSAS
Coates, Gary Joseph
Keller, Leland Edward
Underwood, James Ross, Jr

KENTUCKY
Davidson, John Edwin
Engelberg, Joseph
Hamon, J Hill
Hunter, Norman W
Knoefel, Peter Klerner

LOUISIANA
Andrus, Jan Frederick
Goodman, Alan Leonard
Kak, Subhash Chandra
Morris, Everett Franklin
Tipler, Frank Jennings, III

MAINE
Long, Diana E
McEwen, Currier
Marcotte, Brian Michael
Whitten, Maurice Mason

MARYLAND
Anderson, Larry Douglas
Bell, James F(rederick)
Brieger, Gert Henry
Brush, Stephen George
Calinger, Ronald Steve
Chen, Tar Timothy
Condliffe, Peter George
Darden, Lindley
Doggett, LeRoy Elsworth
Dowling, Marie Augustine
Downing, Mary Brigetta
Eisenhart, Churchill
Embury, Janon Frederick, Jr
Felten, James E(dgar)
Fischer, Irene Kaminka
Gutsche, Graham Denton
Hanle, Paul Arthur
Harden, Victoria Angela
Hart, Michael H
Heilprin, Laurence Bedford
Hindle, Brooke
Joy, Robert John Thomas
Katz, Victor Joseph
Kippenberger, Donald Justin
McCauley, H(enry) Berton
Miles, Wyndham Davies
Mindel, Joseph
Mulligan, Joseph Francis
O'Grady, Richard Terence
Parascandola, John Louis
Parkinson, Claire Lucille
Parkinson, Claire Lucille
Porter, Beverly Fearn
Reupke, William Albert
Rynasiewicz, Robert Alan
Stern, David P
Suppe, Frederick (Roy)
Temkin, Owsei
Torres-Anjel, Manuel Jose
Tower, Donald Bayley
Weart, Spencer Richard

MASSACHUSETTS
Aldrich, Michele L
Andersen, Roy Stuart
Arganbright, Donald G
Breton, J Raymond
Brown, Gilbert J
Cavallaro, Mary Caroline
Cheek, Dennis William
Cohen, I Bernard
Cohen, Robert Sonne
Cohn, Lawrence H
Costanza, Mary E
Ehrenfeld, John R(oos)
Frondel, Clifford
Futrelle, Robert Peel
Gingerich, Owen (Jay)
Graham, Loren R
Harrison, Edward Robert
Hawkins, Thomas William, Jr
Herda, Hans-Heinrich Wolfgang
Hiebert, Erwin Nick
Hill, Victor Ernst, IV
Holmes, Helen Bequaert
Holton, Gerald
Hovorka, John
Howard, John Nelson
Kaufman, Martin
Kuhn, Thomas S
Lockwood, Linda Gail
Maddin, Robert
Mayr, Ernst
Mendelsohn, Everett Irwin
Ozonoff, David Michael
Platt, John Rader
Ramakrishna, Kilaparti
Rosenkrantz, Barbara G
Sawin, Clark Timothy
Schweber, Silvan Samuel
Senechal, Marjorie Lee
Sheldon, Eric
Shimony, Abner
Silverman, Sam M

Smith, Cyril Stanley
Stachel, John Jay
Stonier, Tom Ted
Struik, Dirk Jan
Tyler, H Richard
Vankin, George Lawrence
Webster, Eleanor Rudd
Weiner, Charles
Weininger, Stephen Joel
Weinstein, Alexander
Wertz, Dorothy C

MICHIGAN
Atkinson, James William
Blinn, Walter Craig
Davenport, Horace Willard
Davidovits, Joseph
Hagen, John William
Jones, Phillip Sanford
Kaiser, Christopher B
Katz, Ernst
Kopperl, Sheldon Jerome
Lopushinsky, Theodore
McManus, Hugh
Ramsay, Ogden Bertrand
Rollins-Page, Earl Arthur
Root-Bernstein, Robert Scott
Ruffner, James Alan
Schmidt, Robert
Toledo-Pereyra, Luis Horacio
Trosko, James Edward
Warren, H(erbert) Dale
Weymouth, Patricia Perkins
Zimring, Lois Jacobs

MINNESOTA
Abeles, Tom P
Keys, Thomas Edward
Layton, Edwin Thomas, Jr
Marquit, Erwin
Meyer, Frank Henry
Norberg, Arthur Lawrence
O'Rourke, Richard Clair
Rudolf, Paul Otto
Sapakie, Sidney Freidin
Schwartz, A(lbert) Truman
Shapiro, Alan Elihu
Sperber, William H
Stuewer, Roger Harry
Theisen, Wilfred Robert
Wilson, Leonard Gilchrist

MISSISSIPPI
Seymour, Raymond B(enedict)

MISSOURI
Adawi, Ibrahim (Hasan)
Allen, Garland Edward, III
Ashworth, William Bruce, Jr
Bieniek, Ronald James
Drennan, Ollin Junior
Ewan, Joseph (Andorfer)
Finger, Stanley
Klotz, John William
McCollum, Clifford Glenn
Miller, Jane Alsobrook
Peterson, Charles John
Sussman, Robert Wald
Tang-Martinez, Zuleyma
Twardowski, Zbylut Jozef
Watson, Richard Allan

MONTANA
McAllister, Byron Leon
McRae, Robert James
Rice, Richard Eugene
Winthrop, John T

NEBRASKA
Friedlander, Walter Jay
Harris, Holly Ann
Kennedy, Robert E
Zimmerman, Edward John

NEVADA
Merdinger, Charles J(ohn)
Scott, William Taussig

NEW HAMPSHIRE
Beasley, Wayne M(achon)
Corum, Kenneth Lawrence
Dixon, Wallace Clark, Jr
Ermenc, Joseph John
King, Allen Lewis
Schneer, Cecil J

NEW JERSEY
Abeles, Francine
Berendsen, Peter Barney
Chaiken, MarthaLeah
Chen, Thomas Shih-Nien
Chinard, Francis Pierre
Church, John Armistead
Colaizzi, John Louis
Cromartie, William James, Jr
Gershenowitz, Harry
Grob, Gerald N
Haindl, Martin Wilhelm
Holzmann, Gerard J
House, Edward Holcombe
Mark, Robert
Mihram, George Arthur
Prasad, Marehalli Gopalan
Rowe, John Edward

Rushton, Alan R
Schofield, Robert Edwin
Spruch, Grace Marmor
Tanner, Daniel
Widmer, Kemble
Wise, Matthew Norton
Woolf, Harry

NEW MEXICO
Bertin, John Joseph
Divett, Robert Thomas
Harris, Reece Thomas
Schufle, Joseph Albert
Zeilik, Michael

NEW YORK
Atwater, Edward Congdon
Barker, Kenneth Ray
Blumenthal, Ralph Herbert
Boeck, William Louis
Brown, Patricia Stocking
Bullough, Vern L
Carlson, Elof Axel
Cummins, John Frances
Dauben, Joseph W
Devons, Samuel
Donnelly, Denis Philip
Donovan, Arthur L
Edelglass, Stephen Mark
Edelson, Paul J
Eisele, Carolyn
Eisinger, Josef
Erlichson, Herman
Fitzgerald, Patrick James
Geiss, Gunther R(ichard)
Gorenstein, Shirley Slotkin
Gortler, Leon Bernard
Greaves, Bettina Bien
Harris, C(harles) Leon
Henderson, David Wilson
Heppa, Douglas Van
Holden, Norman Edward
Jahiel, Rene
Kanzler, Walter Wilhelm
Kennedy, Kenneth Adrian Raine
Kingslake, Rudolf
Kyburg, Henry
Lucey, Carol Ann
Lurkis, Alexander
Lutzker, Edythe
McClain, John William
Macklin, Ruth
Mansfield, Victor Neil
Merton, Robert K
Morey, Donald Roger
Neugebauer, Richard
Newman, Stuart Alan
Olczak, Paul Vincent
Packard, Karle Sanborn, Jr
Pence, Harry Edmond
Phillips, Esther Rodlitz
Pugh, Emerson William
Raman, Varadaraja Venkata
Rapaport, William Joseph
Ravitz, Leonard J, Jr
Rohrlich, Fritz
Romer, Alfred
Sauer, Charles William
Schoenfeld, Robert Louis
Segal, Sanford Leonard
Sharkey, John Bernard
Sharpe, William D
Smith, Gerrit Joseph
Smolin, Lee
Sparberg, Esther Braun
Sturges, Stuart
Tesmer, Irving Howard
Trigg, George Lockwood
Voelcker, Herbert B(ernhard)
Waldbauer, Eugene Charles
Westbrook, J(ack) H(all)

NORTH CAROLINA
Adler, Carl George
Burnett, John Nicholas
Fishman, George Samuel
Garg, Devendra Prakash
Gifford, James Fergus
Macon, Nathaniel
Miller, Robert L
O'Hara, Robert James
Petroski, Henry J
Roland, Alex
Shapere, Dudley
Stern, A(rthur) C(ecil)
Troyer, James Richard

OHIO
Becker, Lawrence Charles
Berman, Alex
Bernstein, Stanley Carl
Burnside, Phillips Brooks
Carter, Carolyn Sue
Dexter, Ralph Warren
Dirckx, John H
Eckert, George Frank
Eicher, John Harold
Fullmer, June Z(immerman)
Greenslade, Thomas Boardman, Jr
Kritsky, Gene Ralph
Kullman, David Elmer
Loud, Oliver Schule
Mulroy, Juliana Catherine
Murray, Thomas Henry

Pribor, Donald B
Stuckey, Ronald Lewis
Weinstock, Robert
Williams, John Paul

OKLAHOMA
Choike, James Richard
Nye, Mary Jo
Ogilvie, Marilyn Bailey
Owens, G(lenda) Kay
Ragep, F Jamil
Roller, Duane Henry DuBose
Taylor, Kenneth Lapham

OREGON
Clark, Mary Eleanor
Dews, Edmund
Douglas, Sarah A
Johnson, Dale E
Sherwood, Martha Allen
Spencer, James Brookes
Starr, Patricia Rae
Wirtz, John Harold

PENNSYLVANIA
Benfey, Otto Theodor
Bilaniuk, Oleksa-Myron
De Witt, Hobson Dewey
Dunham, William Wade
Foley, Thomas Preston, Jr
Freeman, Joseph Theodore
Gardiner, Keith M
Garfield, Eugene
Havas, Peter
Hounshell, David A
Janis, Allen I(ra)
Kiger, Robert William
Kraft, R(alph) Wayne
Kuserk, Frank Thomas
Litchfield, Carter
Marks, Louis Sheppard
Mayer, Joerg Werner Peter
Mercer, Sherwood Rocke
Mortimer, Charles Edgar
Nemeth, Edward Joseph
Patterson, Elizabeth Knight
Poss, Howard Lionel
Poupard, James Arthur
Rescher, Nicholas
Rosenberg, Charles E
Rosenstein, George Morris, Jr
Rothman, Milton A
Rovelli, Carlo
Small, Henry Gilbert
Stetson, Harold W(ilbur)
Stevens, Rosemary Anne
Tarr, Joel Arthur
Thackray, Arnold
Waterhouse, William Charles

RHODE ISLAND
Gehrenbeck, Richard Keith
Greenblatt, Samuel Harold
Simons, Roger Alan

SOUTH CAROLINA
Gurr, Henry S
Rembert, David Hopkins, Jr

SOUTH DAKOTA
Scott, George Prescott

TENNESSEE
Cordell, Francis Merritt
Freeman, Bob A
Freemon, Frank Reed
Gehrs, Carl William
Hanusa, Timothy Paul
Lorenz, Philip Jack, Jr
Otis, Marshall Voigt
Schweitzer, George Keene
Williamson, Handy, Jr

TEXAS
Avant, Patricia Kay Coalson
Burns, Chester Ray
Dahl, Per Fridtjof
Fearing, Olin S
Hillar, Marian
Howard, Charles
Jackson, Eugene Bernard
Kunkel, Harriott Orren
Markowski, Gregory Ray
Melrose, James C
Poage, Scott T(abor)
Schofield, James Roy
Stahl, Glenn Allan
Stranges, Anthony Nicholas
Taber, Willard Allen
Ward-McLemore, Ethel
Wood, Robert Charles

UTAH
Allred, David D
Earley, Charles Willard

VERMONT
Arns, Robert George
Hawthorne, Robert Montgomery, Jr
Rankin, Joanna Marie
Smith, Thomas David
Wolfson, Richard L T

History & Philosophy of Science (cont)

VIRGINIA
Albrecht, Herbert Richard
Bauer, Henry Hermann
Blanpied, William Antoine
Burian, Richard M
Burstyn, Harold Lewis
Daniel, Kenneth Wayne
Dobyns, Samuel Witten
Featherston, Frank Hunter
Fisher, Gordon McCrea
Gillies, George Thomas
Huang, H(sing) T(sung)
Jacobs, Kenneth Charles
Lee, John A N
McKnight, John Lacy
Nelson, Clifford M(elvin)
Reisler, Donald Laurence
Schneider, Philip Allen David
Teichler-Zallen, Doris
Weaver, Warren Eldred

WASHINGTON
Albers, James Ray
Benson, Keith Rodney
Buske, Norman L
Colcord, J E
Cunnea, William M
Damkaer, David Martin
Gerhart, James Basil
Heston, Leonard L
Kallaher, Michael Joseph
Kathren, Ronald Laurence
Kells, Lyman Francis
Levinson, Mark
Mills, B(lake) D(avid), Jr
Odegaard, Charles E
Orr, Jack Edward
Shepherd, Linda Jean
Skov, Niels A
Sullivan, Woodruff Turner, III

WEST VIRGINIA
Becker, Stanley Leonard
Elmes, Gregory Arthur
Fezer, Karl Dietrich
Gould, Henry Wadsworth
Hilloowala, Rumy Ardeshir

WISCONSIN
Balmer, Robert Theodore
Bodine, Robert Y(oung)
Ihde, Aaron John
Jungck, John Richard
Kitzke, Eugene David
Lee, Kathryn Adele Bunding
Oyen, Ordean James
Siegfried, Robert
Van Ryzin, Martina
Zei, Dino

ALBERTA
Milone, Eugene Frank
Osler, Margaret J

BRITISH COLUMBIA
Batten, Alan Henry

MANITOBA
Connor, Robert Dickson
Hall, Donald Herbert
LeBow, Michael David

NEWFOUNDLAND
Aldrich, Frederick Allen
Anderson, Hugh John
Rayner-Canham, Geoffrey William
Thomeier, Siegfried

NOVA SCOTIA
Cochkanoff, O(rest)
Farley, John
Mills, Eric Leonard
Murray, T J
Ravindra, Ravi

ONTARIO
Atkinson, George Francis
Berger, Jacques
Brose, David Stephen
Colpa, Johannes Pieter
Derbyshire, John Brian
Dorrell, Douglas Gordon
Gray, David Robert
Kutcher, Stanley Paul
Levere, Trevor Harvey
Lister, Earl Edward
Logan, Robert Kalman
Michael, Thomas Hugh Glynn
Nevitt, H J Barrington
Nicholls, Peter
Roland, Charles Gordon
Roth, René Romain
Singh, Rama Shankar
Slater, Keith
Sobell, Mark Barry
Thomas, Morley Keith
Usselman, Melvyn Charles

QUEBEC
Bates, Donald George

Cohen, Montague
Contandriopoulos, Andre-Pierre
Peters, Robert Henry
Sattler, Rolf
Shea, William Rene
Szamosi, Geza
Wolsky, Maria de Issekutz

OTHER COUNTRIES
Bader, Hermann
Brown, Richard Harland
Bye, Robert Arthur
Cartwright, Nancy L Delaney
D'Ambrosio, Ubiratan
Guimaraes, Romeu Cardoso
Lindell, Ismo Veikko
Lowrance, William Wilson, Jr
Miller, Arthur I
Schopper, Herwig Franz
Stavroudis, Orestes Nicholas
Trumpy, D Rudolf
Yamamoto, Mitsuyoshi

Research Administration

ALABAMA
Alford, Charles Aaron, Jr
Askew, Raymond Fike
Baksay, Laszlo Andras
Cataldo, Charles Eugene
Cloyd, Grover David
Davis, Elizabeth Young
Ferguson, Joseph Gantt
Finklea, John F
Hall, David Michael
Hill, Walter Andrew
Johnson, Loyd
Jolley, Homer Richard
Kandhal, Prithvi Singh
Macmillan, William Hooper
Marchant, Guillaume Henri (Wim), Jr
Meezan, Elias
Menaker, Lewis
Moore, Claude Henry
Parks, Paul Franklin
Piwonka, Thomas Squire
Prescott, Paul Ithel
Rhoades, Richard G
Rivers, Douglas Bernard
Rouse, Roy Dennis
Schnaper, Harold Warren
Singh, Shiva Pujan
Walsh, David Allan
Watson, Raymond Coke, Jr

ALASKA
Brown, Edward James
Deehr, Charles Sterling
Hameedi, Mohammad Jawed
LaPerriere, Jacqueline Doyle
Larson, Frederic Roger
Olsen, James Calvin
Rutledge, Gene Preston
Sweet, Larry Ross
Werner, Richard Allen

ARIZONA
Adickes, H Wayne
Andreen, Brian H
Barr, Ronald Edward
Boyd, George Addison
Carr, Clide Isom
Cox, Fred Ward, Jr
Cunningham, Richard G(reenlaw)
Formeister, Richard B
Furman, Robert Howard
Greenberg, Milton
Hause, Norman Laurance
Hoffmann, Joseph John
Johnson, Jack Donald
Jungermann, Eric
Kassander, Arno Richard, Jr
Kelley, Maurice Joseph
Koeneman, James Bryant
Leibacher, John W
McKlveen, John William
Mayall, Nicholas Ulrich
Metzger, Darryl E
Moon, Thomas Edward
Preston, Kendall, Jr
Ratkowski, Donald J
Sauer, Barry W
Shull, James Jay
Sridharan, Natesa S
Sundahl, Robert Charles, Jr
Thews, Robert L(eroy)
Wise, Edward Nelson

ARKANSAS
Cockerham, Lorris G(ay)
Hietbrink, Earl Henry
Keplinger, Orin Clawson
Mack, Leslie Eugene
Martin, Lloyd W
Norris, Arthur Robert
Odell, Norman Raymond
Warren, Lloyd Oliver
Wright, Bill C
Young, John Falkner

CALIFORNIA
Adey, William Ross
Al-Aish, Matti

Allen, LeRoy Richard
Allen, William Merle
Anderson, Paul Maurice
Ashcroft, Frederick H
Augood, Derek Raymond
Balzhiser, Richard E(arl)
Bartl, Paul
Bechtel, Stephen D, Jr
Beebe, Robert Ray
Beljan, John Richard
Bennett, Alan Jerome
Bennett, Harold Earl
Bevc, Vladislav
Billman, Kenneth William
Boehlert, George Walter
Bondar, Richard Jay Laurent
Bonham, Lawrence Cook
Bower, Frank H(ugo)
Braemer, Allen C
Brand, Samson
Brotherton, Robert John
Brownell, Robert Leo, Jr
Buchwald, Nathaniel Avrom
Butz, Robert Frederick
Caldwell, David Orville
Callaham, Robert Zina
Canham, John Edward
Carr, Robert Charles
Carter, Kenneth
Castro, George
Chang, Chih-Pei
Chartock, Michael Andrew
Chase, Jay Benton
Chester, Arthur Noble
Chmura-Meyer, Carol A
Cohen, Daniel Morris
Cohn, Theodore E
Coile, Russell Cleven
Coleman, Paul Jerome, Jr
Cox, Cathleen Ruth
Croft, Paul Douglas
Cuadra, Carlos A(lbert)
Davis, Jay C
Davis, Ward Benjamin
Detz, Clifford M
Dewey, Donald O
Dismukes, Robert Key
Dixon, Wilfrid Joseph
Dorough, Gus Downs, Jr
Duff, Russell Earl
Dulmage, William James
Ellis, Elizabeth Carol
Englert, Robert D
Epperly, W Robert
Erdman, Timothy Robert
Evtuhov, Viktor
Farley, Thomas Albert
Farone, William Anthony
Feinberg, Benedict
Feller, Robert S, Sr
Fisk, George Ayrs
Freise, Earl J
Frenzel, Lydia Ann Melcher
Friedman, Alan E
Friedman, Milton Joe
Funk, Glenn Albert
Gajewski, Ryszard
Ganz, Aaron
Gaynor, Joseph
Gentile, Ralph G
Gerber, Bernard Robert
Gilbert, Jerome B
Gilmer, David Seeley
Goh, Kean S
Goldman, Marvin
Govier, William Charles
Gritton, Eugene Charles
Groseclose, Byron Clark
Guest, Gareth E
Gundersen, Larry Edward
Gustavson, Marvin Ronald
Hackett, Nora Reed
Hadley, Jeffery A
Hagan, William Leonard
Hainski, Martha Barrionuevo
Hamlin, Kenneth Eldred, Jr
Hammar, Allan H
Harrison, Don Edward, Jr
Harrison, George Conrad, Jr
Hartsough, Walter Douglas
Hausmann, Werner Karl
Havemeyer, Ruth Naomi
Hayes, Thomas L
Heinrich, Milton Rollin
Heitkamp, Norman Denis
Hendricks, Charles D(urrell), Jr
Hennig, Harvey
Henry, David Weston
Hess, Patrick Henry
Hill, Richard William
Hollander, Jack M
Holliday, Dale Vance
Holtz, David
Hopkins, Ronald Murray
Howard, Gilbert Thoreau
Hu, Steve Seng-Chiu
Hudgins, Arthur J
Hugunin, Alan Godfrey
Hyde, Earl K
Ide, Roger Henry
Iglesia, Enrique
Iyer, Hariharaiyer Mahadeva
James, Kenneth Eugene
Johannessen, George Andrew

Johnson, Kenneth
Jones, Thomas V
Jorgensen, Paul J
Joyner, Powell Austin
Kahn, Raymond Henry
Kasper, Gerhard
Keith, Robert Allen
Kelman, Bruce Jerry
King, Alfred Douglas, Jr
King, Ray J(ohn)
Knapp, Theodore Martin
Koonce, Andrea Lavender
Kosow, David Phillip
Kramer, Henry Herman
Kuhns, Ellen Swomley
Kukkonen, Carl Allan
Larmore, Lewis
Lave, Roy E(llis), Jr
Lechtman, Max D
Lederer, C Michael
Legge, Norman Reginald
Letey, John, Jr
Levine, Mark David
Levinthal, Elliott Charles
Lindquist, Robert Henry
Lisman, Perry Hall
Lissaman, Peter Barry Stuart
Little, John Clayton
Lofgren, Edward Joseph
London, Ray William
Lukasik, Stephen Joseph
McCall, Chester Hayden, Jr
McDonough, Thomas Redmond
MacKay, Kenneth Donald
Marrone, Pamela Gail
Marto, Paul James
Mattice, Jack Shafer
May, Michael Melville
Meechan, Charles James
Merilo, Mati
Merten, Ulrich
Meyers, Robert Allen
Mikkelsen, Duane Soren
Miller, William Frederick
Morris, Fred(erick) W(illiam)
Musser, John H
Mysels, Karol J(oseph)
Neil, Gary Lawrence
Nesbit, Richard Allison
Newell, Gordon Wilfred
Orcutt, Harold George
Pake, George Edward
Parkinson, Bradford W
Payne, Holland I
Pearson, John
Pedersen, Leo Damborg
Peeters, Randall Louis
Perry, Robert Hood, Jr
Peterson, Laurence E
Pierce, James Otto, II
Poppoff, Ilia George
Prim, Robert Clay, III
Pugh, Howel Griffith
Ram, Michael Jay
Ranftl, Robert M(atthew)
Rapp, Donald
Richter, George Neal
Rossin, A David
Rowell, John Martin
Rowntree, Robert Fredric
Rudy, Thomas Philip
Rutan, Elbert L
Saffren, Melvin Michael
Sager, Ronald E
Sakagawa, Gary Toshio
Salsig, William Winter, Jr
Sanchez, Robert A
Saute, Robert E
Saxon, Roberta Pollack
Scheinok, Perry Aaron
Schonberg, Russell George
Schweiker, George Christian
Seibel, Erwin
Seppi, Edward Joseph
Shelton, Robert Neal
Silverman, Herbert Philip
Silverman, Jacob
Smith, Charles G
Sommers, William P(aul)
Spinrad, Robert J(oseph)
Statler, Irving C(arl)
Statt, Terry G
Stelzer, Lorin Roy
Stephanou, Stephen Emmanuel
Stirn, Richard J
Strena, Robert Victor
Strosberg, Arthur Martin
Sutherland, William Robert
Sutton, Paul McCullough
Sweeney, Lawrence Earl, Jr
Syvertson, Clarence A
Szego, Peter A
Szymanski, Paul Stephen
Tarbell, Theodore Dean
Taylor, Robert William
Thatcher, Everett Whiting
Thueson, David Orel
Trevelyan, Benjamin John
Turner, George Cleveland
Van Schaik, Peter Hendrik
Walker, Jimmy Newton
Walker, Joseph
Warner, Willis L
Waters, William E

OTHER PROFESSIONAL FIELDS / 573

Weir, William Carl
Welch, Robin Ivor
Welmers, Everett Thomas
Werth, Glenn Conrad
Whitehead, Marvin Delbert
Whitney, William Merrill
Wilbarger, Edward Stanley, Jr
Winick, Herman
Winter, David Leon
Worker, George F, Jr
Young, Robert, Jr
Younoszai, Rafi
Zander, Andrew Thomas
Zarem, Abe Mordecai
Zeren, Richard William
Ziemer, Robert Ruhl

COLORADO

Beck, Steven R
Boudreau, Robert Donald
Bourquin, Al Willis J
Boyd, L(andis) L(ee)
Broome, Paul W(allace)
Bull, Stanley R(aymond)
Burns, Denver P
Corby, Donald G
Czanderna, Alvin Warren
Davis, Wilford Lavern
Dusenberry, William Earl
Evans, Norman A(llen)
Fall, Michael William
Firor, John William
Fraikor, Frederick John
Gentry, Willard Max, Jr
Girz, Cecilia Griffith
Glock, Robert Dean
Gould, Gordon
Harrison, Monty DeVerl
Heberlein, Douglas G(aravel)
Hilmas, Duane Eugene
Hinners, Noel W
Horak, Donald L
Hutto, Francis Baird, Jr
Johnson, Janice Kay
Kniebes, Duane Van
Kotch, Alex
LaBounty, James Francis, Sr
Lameiro, Gerard Francis
Laughlin, Charles William
McKown, Cora F
McLean, Francis Glen
Micheli, Roger Paul
Morrison, Charles Freeman, Jr
Nauenberg, Uriel
Olien, Neil Arnold
Olson, George Gilbert
Pitts, Malcolm John
Purdom, James Francis Whitehurst
Scherer, Ronald Callaway
Schleicher, David Lawrence
Seibert, Michael
Siuru, William D, Jr
Stencel, Robert Edward
Streib, W(illiam) C(harles)
Sunderman, Duane Neuman
West, Anita
Westfall, Richard Merrill
Wilson, David George
Wyman, Charles Ely

CONNECTICUT

Bentley, William Ross
Bentz, Alan P(aul)
Berdick, Murray
Berets, Donald Joseph
Bormann, Barbara-Jean Anne
Brody, Harold David
Bronsky, Albert J
Brooks, Douglas Lee
Buchanan, Ronald Leslie
Burke, Carroll N
Burwell, Wayne Gregory
Carter, G Clifford
Clark, Hugh
Cobern, Martin E
Cole, Stephen H(ervey)
Conover, Lloyd Hillyard
Cooke, Theodore Frederic
Darling, George Bapst
Davenport, Lee Losee
DeVita, Vincent T, Jr
Doshan, Harold David
Duvivier, Jean Fernand
Eby, Edward Stuart, Sr
Emond, George T
Erlandson, Paul M(cKillop)
Fleming, James Stuart
Gunther, Jay Kenneth
Haacke, Gottfried
Haas, Ward John
Hand, Arthur Ralph
Hermann, Robert J
Hobbs, Donald Clifford
John, Hugo Herman
Kemp, Gordon Arthur
Kotick, Michael Paul
Leonard, Robert F
Levin, S Benedict
Lichtenberger, Harold V
Linke, William Finan
McKeon, James Edward
Marchand, Nathan
Megrue, George Henry
Meierhoefer, Alan W

Menkart, John
Milliken, Frank R
Moreland, Walter Thomas, Jr
Nelson, Roger Peter
O'Brien, Robert L
Ortner, Mary Joanne
Page, Edgar J(oseph)
Parthasarathi, Manavasi Narasimhan
Pierce, James Benjamin
Pinson, Ellis Rex, Jr
Pinson, Rex, Jr
Proctor, Alan Ray
Rauhut, Michael McKay
Ray, Verne A
Rishell, William Arthur
Robinson, Donald W(allace), Jr
Ryan, Richard Patrick
Schrage, F Eugene
Schurig, John Eberhard
Shorr, Bernard
Smith, Douglas Stewart
Snellings, William Moran
Stott, Paul Edwin
Suen, T(zeng) J(iueq)
Tangel, O(scar) F(rank)
Wheeler, Edward Stubbs
Wiseman, Edward H
Wood, David Dudley

DELAWARE

Baer, Donald Robert
Bancroft, Lewis Clinton
Chiu, Jen
Franta, William Alfred
Ginnard, Charles Raymond
Greenblatt, Hellen Chaya
Hannell, John W(eldale), Jr
Hess, Richard William
Hurley, William Joseph
Jackson, David Archer
Jackson, Thomas Edwin
Jefferson, Edward G
Johnson, Donald Richard
Johnson, Robert Chandler
Knipmeyer, Hubert Elmer
Lin, Kuang-Farn
Litchfield, William John
McClelland, Alan Lindsey
Montague, Barbara Ann
Naylor, Marcus A, Jr
Norling, Parry McWhinnie
Phillips, Brian Ross
Repka, Benjamin C
Rogers, Donald B
Schiroky, Gerhard H
Schroeder, Herman Elbert
Schwartz, Jerome Lawrence
Shambelan, Charles
Shotzberger, Gregory Steven
Sloan, Martin Frank
Soboczenski, Edward John
Steinberg, Marshall
Stopkie, Roger John
Williams, Ebenezer David, Jr
Woods, Thomas Stephen

DISTRICT OF COLUMBIA

Abrahamson, George R(aymond)
Apple, Martin Allen
Attaway, David Henry
Bachkosky, John M (Jack)
Baker, Charles Wesley
Baker, D(onald) James
Barker, Winona Clinton
Beckner, Everet Hess
Berley, David
Bingham, Robert Lodewijk
Boron, David J
Borsuk, Gerald M
Burris, James F
Bushman, John Branson
Chen, Davidson Tah-Chuen
Ching, Chauncey T K
Chung, Ho
Cowdry, Rex William
Davis, Lance A(lan)
Decker, James Federick
Diemer, F(erdinand) P(eter)
Diggs, John W
Dowdy, Edward Joseph
Fechter, Alan Edward
Fiske, Richard Sewell
Flora, Lewis Franklin
Fricken, Raymond Lee
Gaffney, Paul G, II
Goldfield, Edwin David
Goldin, Daniel S
Goldstein, Gerald
Good, Mary Lowe
Gould, Phillip
Green, Richard James
Greene, Michael P
Gress, Mary Edith
Hassan, Aftab Syed
Hedlund, James Howard
Holt, Alan Craig
Howell, Embry Martin
Hussain, Syed Taseer
Ison-Franklin, Eleanor Lutia
Johnston, Laurance S
Jones, Alice J
Jones, William Vernon
Katona, Peter Geza
Kimmel, Carole Anne

Korkegi, Robert Hani
Lehman, Donald Richard
Lichens-Park, Ann Elizabeth
Lombard, David Bishop
Mange, Phillip Warren
Mellor, John Williams
Murphy, Robert Earl
Nadolney, Carlton H
Novello, Antonia Coello
O'Fallon, Nancy McCumber
Panar, Manuel
Plowman, Ronald Dean
Rabin, Robert
Raizen, Senta Amon
Ramsey, Jerry Warren
Raslear, Thomas G
Read, Merrill Stafford
Reinecke, Thomas Leonard
Reisa, James Joseph, Jr
Riegel, Kurt Wetherhold
Rosen, Howard Neal
Rubinoff, Roberta Wolff
Scarr, Harry Alan
Schmidt, Berlie Louis
Schriever, Richard L
Schroeder, Anita Gayle
Scott, Roland Boyd
Smith, David Allen
Smith, Leslie E
Smith, Thomas Elijah
Sobotka, Thomas Joseph
Straf, Miron L
Tompkins, Daniel Reuben
Topper, Leonard
Wallenmeyer, William Anton
Weiss, Leonard
White, Irvin Linwood
Whitmore, Frank Clifford, Jr
Wile, Howard P
Wilsey, Neal David
Wilson, George Donald
Wittmann, Horst Richard
Wolfle, Thomas Lee
Zeizel, A(rthur) J(ohn)

FLORIDA

Abraham, William Michael
Amey, William G(reenville)
Aronson, M(oses)
Baker, Carleton Harold
Baum, Werner A
Berry, Robert Eddy
Biasell, Laverne R(obert)
Brotherson, Donald E
Bruner, Harry Davis
Burridge, Michael John
Callen, Joseph Edward
Coulter, Wallace H
Crompton, Charles Edward
Dacey, George Clement
Dailey, Charles E(lmer), III
De Lorge, John Oldham
Diaz, Nils Juan
Dubach, Leland L
Duke, Dennis Wayne
Einspruch, Norman G(erald)
Everett, Woodrow Wilson, Jr
Francis, Stanley Arthur
Friedlander, Herbert Norman
Friedman, Raymond
Gerber, John Francis
Gianola, Umberto Ferdinando
Gould, Anne Bramlee
Grobman, Arnold Brams
Grodberg, Marcus Gordon
Groupe, Vincent
Hall, David Goodsell, IV
Hampp, Edward Gottlieb
Harper, Verne Lester
Hayes, Edward J(ames)
Heggie, Robert
Henkin, Hyman
Heying, Theodore Louis
Hirsch, Donald Earl
Howsmon, John Arthur
Joftes, David Lion
Jones, Albert Cleveland
Joyce, Edwin A, Jr
Kauder, Otto Samuel
Kender, Walter John
Kiplinger, Glenn Francis
Klacsmann, John Anthony
Lannutti, Joseph Edward
Lawrence, Franklin Isaac Latimer
Long, Calvin H
McFadden, James Douglas
Malthaner, W(illiam) A(mond)
Medin, A(aron) Louis
Merritt, Alfred M, II
Murchison, Thomas Edgar
Nimmo, Bruce Glen
Nunnally, Stephens Watson
Owens, James Samuel
Peacock, Hugh Anthony
Pickard, Porter Louis, Jr
Rabbit, Guy
Randazzo, Anthony Frank
Rathburn, Carlisle Baxter, Jr
Rawls, Walter Cecil, Jr
Rice, Stephen Landon
Rich, Jimmy Ray
Riordan, Michael Davitt
Roesner, Larry A
Saunders, James Henry

Seugling, Earl William, Jr
Sharkey, William Henry
Sheppard, Albert Parker
Silver, Frank Morris
Simeral, William Goodrich
Smith, Wayne H
Splitter, Earl John
Spurlock, Jack Marion
Stephens, John C(arnes)
Stewart, Harris Bates, Jr
Stroh, Robert Carl
Weinberg, Jerry L
Wiltbank, William Joseph
Wise, Raleigh Warren
Wissow, Lennard Jay
Wood, William Otto

GEORGIA

Alberts, James Joseph
Barbaree, James Martin
Barreras, Raymond Joseph
Bodnar, Donald George
Broerman, F S
Buckalew, Louis Walter
Cassanova, Robert Anthony
Circeo, Louis Joseph, Jr
Dees, Julian Worth
Donahoo, Pat
Dragoin, William Bailey
Evans, Thomas P
Frere, Maurice Herbert
Goldsmith, Edward
Gollob, Lawrence
Grace, Donald J
Gunn, Walter Joseph
Harris, Elliott Stanley
Herbst, John A
Herrmann, Kenneth L
Ike, Albert Francis
Innes, David Lyn
Jackson, Curtis Rukes
Keller, Oswald Lewin
Koerner, T J
Kraeling, Robert Russell
Loomis, Earl Alfred, Jr
McClure, Harold Monroe
Malcolm, Earl Walter
Martin, Linda Spencer
Oliver, James Henry, Jr
Peck, Harry Dowd, Jr
Pence, Ira Wilson, Jr
Pollard, William Blake
Prince, M(orris) David
Reiss, Errol
Rossini, Frederick Anthony
Schutt, Paul Frederick
Schwartz, Robert John
Shlevin, Harold H
Smith, David Fletcher
Spauschus, Hans O
Stalford, Harold Lenn
Stevens, Ann Rebecca
Thompson, William Oxley, II
Threadgill, Ernest Dale
Vander Velde, George
Walls, Nancy Williams
Wepfer, William J
Wilson, Donella Joyce

HAWAII

Daniel, Thomas Henry
Duckworth, Walter Donald
Helfrich, Philip
Khan, Mohammad Asad
Malmstadt, Howard Vincent
Massey, Douglas Gordon
Matsuda, Fujio
Moberly, Ralph M
Swindale, Leslie D
Williams, David Douglas F
Yates, James T

IDAHO

Gittins, Arthur Richard
Griffith, W(illiam) A(lexander)
Gutzman, Philip Charles
Kleinkopf, Gale Eugene
Laurence, Kenneth Allen
Majumdar, Debaprasad (Debu)
Sarett, Lewis Hastings
Slansky, Cyril M
Vestal, Robert Elden
Wood, Richard Ellet

ILLINOIS

Abrams, Israel Jacob
Abrams, Robert A
Afremow, Leonard Calvin
Akin, Cavit
Altpeter, Lawrence L, Jr
Asbury, Joseph G
Auvil, Paul R, Jr
Baker, Winslow Furber
Barenberg, Sumner
Bartimes, George F
Bentley, Orville George
Bietz, Jerold Allen
Biljetina, Richard
Bollinger, Lowell Moyer
Brezinski, Darlene Rita
Brutten, Gene J
Bullard, Clark W
Burhop, Kenneth Eugene
Cafasso, Fred A

Research Administration (cont)

Cannon, Howard S(uckling)
Chapman, John R
Chaszeyka, Michael A(ndrew)
Chu, Daniel Tim-Wo
Cole, Francis Talmage
Collier, Donald W(alter)
Conn, Arthur L(eonard)
Danzig, Morris Juda
Davis, Edward Nathan
DiRienzo, Douglas Bruce
Ehrlich, Richard
Fenters, James Dean
Fish, Ferol F, Jr
Flay, Brian Richard
Ford, Richard Earl
Frank, James Richard
French, Allen Lee
Gealy, William James
Gillette, Richard F
Glaser, Janet H
Halaby, George Anton
Harris, Ronald David
Hauptschein, Murray
Hawkins, Neil Middleton
Herting, Robert Leslie
Heybach, John Peter
Holt, Donald Alexander
Huebner, Russell Henry, Sr
Jezl, James Louis
Johnson, Dale Howard
Kabisch, William Thomas
Kittel, J Howard
Klein, Robert L
Klinger, Lawrence Edward
Krug, Samuel Edward
LaFevers, James Ronald
Lederman, Leon Max
Leighton, Morris Wellman
Levin, Alfred A
Levy, Nelson Louis
Linde, Ronald K(eith)
Lofgren, Philip Allen
McLaughlin, Robert Lawrence
Mapother, Dillon Edward
Marchaterre, John Frederick
Markuszewski, Richard
Menke, Andrew G
Nair, Velayudhan
Nash, Kenneth Laverne
Nezrick, Frank Albert
Nutting, Ehard Forrest
Olson, Gregory Bruce
Pai, Mangalore Anantha
Pardo, Richard Claude
Pariza, Richard James
Pietri, Charles Edward
Plautz, Donald Melvin
Prabhudesai, Mukund M
Prakasam, Tata B S
Rafelson, Max Emanuel, Jr
Reilly, Christopher Aloysius, Jr
Richmond, James M
Rothfus, John Arden
Rowe, William Bruce
Rudnick, Stanley J
Rue, Edward Evans
Rundo, John
Rusinko, Frank, Jr
Sachs, Robert Green
Schaefer, Wilbur Carls
Schriesheim, Alan
Sekar, Raj
Shapiro, Stanley
Sheldon, Victor Lawrence
Shen, Sinyan
Sibbach, William Robert
Singiser, Robert Eugene
Smith, Leslie Garrett
Staats, William R
Stromberg, LaWayne Roland
Surgi, Marion Rene
Theriot, Edward Dennis, Jr
Thomas, Joseph Erumappettical
Thompson, Charles William Nelson
Tourlentes, Thomas Theodore
Trimble, John Leonard
Unik, John Peter
Wargel, Robert Joseph
Weinstein, Hyman Gabriel
Wise, Evan Michael
Wolsky, Alan Martin
Yackel, Walter Carl
Yarger, James G
Zeffren, Eugene
Zeiss, Chester Raymond

INDIANA

Amundson, Merle E
Amy, Jonathan Weekes
Andrews, Frederick Newcomb
Baumgardt, Billy Ray
Boyd, Frederick Mervin
Brinkmeyer, Raymond Samuel
Castleberry, Ron M
Cerimele, Benito Joseph
Cook, David Allan
Cravens, William Windsor
DeSantis, John Louis
Doering, Otto Charles, III
Fredrich, Augustine Joseph
Free, Helen M
Gunter, Claude Ray

Guthrie, George Drake
Ho, Cho-Yen
Jacobs, Martin Irwin
Judd, William Robert
Kammeraad, Adrian
Katz, Frances R
Khorana, Brij Mohan
Leander, John David
Leap, Darrell Ivan
McIntyre, Thomas William
McKeehan, Charles Wayne
Mazac, Charles James
Merritt, Doris Honig
Millar, Wayne Norval
Odor, David Lee
Ortman, Eldon Emil
Probst, Gerald William
Pyle, James L
Rao, Suryanarayana K
Sampson, Charles Berlin
Sannella, Joseph L
Stark, Paul
Strycker, Stanley Julian
Tarwater, Oliver Reed
Thompson, Gerald Lee
Wright, William Leland

IOWA

Arndt, Stephan
Austin, Tom Al
Clark, Robert A
Gaertner, Richard F(rancis)
Haendel, Richard Stone
Kelly, William H
Kluge, John Paul
Mather, Roger Frederick
Mengeling, William Lloyd
Morris, Robert Gemmill
O'Berry, Phillip Aaron
Odgaard, A Jacob
Reckase, Mark Daniel
Robinson, Robert George

KANSAS

Gerhard, Lee C
Grant, Stanley Cameron
Ham, George Eldon
Hanley, Wayne Stewart
Jacobs, Louis John
Lee, Joe
Leland, Stanley Edward, Jr
Lyles, Leon
Schroeder, Robert Samuel
Stevens, David Robert
Sutherland, John B(ennett)
Swanson, Lynn Allen
Von Rumker, Rosmarie
Zerwekh, Robert Paul

KENTUCKY

Broida, Theodore Ray
Cohn, David Valor
Harris, Patrick Donald
Huffman, Gerald P
Jain, Ravinder Kumar
Merker, Stephen Louis
Mitchell, Maurice McClellan, Jr
Podshadley, Arlon George
Price, Martin Burton
Smith, Gilbert Edwin
Smith, Terry Edward
Wells, William Lochridge

LOUISIANA

Arthur, Jett Clinton, Jr
Avent, Robert M
Barbee, R Wayne
Baril, Albert, Jr
Barras, Stanley J
Berni, Ralph John
Bradley, Ronald James
Caffey, Horace Rouse
Conrad, Franklin
Davis, Bryan Terence
DeMonsabert, Winston Russel
Eaton, Harvill Carlton
Gautreaux, Marcelian Francis
Goerner, Joseph Kofahl
Hale, Paul Nolen, Jr
Harrison, Richard Miller
Hastings, Robert Clyde
Hensley, Sess D
Hough, Walter Andrew
Huggett, Richard William, Jr
Mills, Earl Ronald
Moulder, Peter Vincent, Jr
O'Barr, Richard Dale
Polack, Joseph A(lbert)
Ristroph, John Heard
Sammarco, Paul William
Starrett, Richmond Mullins
Threefoot, Sam Abraham
Vail, Sidney Lee
Van Lopik, Jack Richard
Vashishta, Priya Darshan
Walker, John Robert
Wiewiorowski, Tadeusz Karol
Wollensak, John Charles
Zietz, Joseph R, Jr

MAINE

Hanks, Robert William
Hill, Marquita K
Kirkpatrick, Francis Hubbard

McKerns, Kenneth (Wilshire)
Marcotte, Brian Michael
Shipman, C(harles) William
Spencer, Claude Franklin

MARYLAND

Aamodt, Roger Louis
Abbey, Robert Fred, Jr
Albrecht, G(eorge) H(enry)
Alderson, Norris Eugene
Alexander, Joseph Kunkle
Anderson, William Averette
Andres, Scott Fitzgerald
Ansari, Aftab A
Atwell, Constance Woodruff
Baker, Carl Gwin
Baker, Charles P
Baldwin, Wendy Harmer
Bareis, Donna Lynn
Barnes, James Alford
Bastress, E(rnest) Karl
Batson, David Banks
Bausell, R Barker, Jr
Bawden, Monte Paul
Beattie, Donald A
Beaven, Vida Helms
Beisler, John Albert
Bellino, Francis Leonard
Benbrook, Charles M
Benson, Walter Roderick
Berman, Howard Mitchell
Blaunstein, Robert P
Bluhm, Leslie
Bollum, Frederick James
Boyd, Michael R
Brady, Donald Frederick, Jr
Brodsky, Marc Herbert
Brown, James Harvey
Brown, Paul Joseph
Bryan, Edward H
Buck, John Henry
Calabi, Ornella
Cave, William Thompson
Charlton, Gordon Randolph
Coates, Arthur Donwell
Coelho, Anthony Mendes, Jr
Coleman, James Stafford
Colvin, Burton Houston
Condell, William John, Jr
Condliffe, Peter George
Costa, Paul T, Jr
Crane, Langdon Teachout, Jr
Creveling, Cyrus Robbins, Jr
Crout, J Richard
Crump, Stuart Faulkner
Daniels, William Fowler, Sr
Das, Naba Kishore
Dashiell, Thomas Ronald
Dean, Donna Joyce
Denniston, Joseph Charles
Diness, Arthur M(ichael)
Dodd, Roger Yates
Domanski, Thaddeus John
Doschek, Wardella Molford
Duggar, Benjamin Charles
Dupont, Jacqueline (Louise)
Eaton, Alvin Ralph
Eccleshall, Donald
Efron, Herman Yale
Elkins, Earleen Feldman
Ellenberg, Susan Smith
Ellingwood, Bruce Russell
Ellwein, Leon Burnell
Engle, Robert Rufus
Ephremides, Anthony
Erickson, Duane Gordon
Evans, John V
Farmer, William S(ilas), Jr
Finkelstein, Robert
Fischer, Eugene Charles
Fisher, Dale John
Fisher, E(arl) Eugene
Fisher, Gail Feimster
Fowler, Earle Cabell
Fowler, Howland Auchincloss
Fozard, James Leonard
Frazer, Benjamin Chalmers
Freden, Stanley Charles
Fried, Jerrold
Frommer, Peter Leslie
Garman, John Andrew
Geller, Ronald G
Gessow, Alfred
Glenn, John Frazier
Golding, Leonard S
Goldstein, Robert Arnold
Goldwater, William Henry
Gonano, John Roland
Gonda, Matthew Allen
Gonzales, Ciriaco Q
Gordon, Stephen L
Graham, Bettie Jean
Granatstein, Victor Lawrence
Gravatt, Claude Carrington, Jr
Gray, David Bertsch
Gray, Paulette S
Green, Jerome George
Gupta, Ramesh K
Guss, Maurice Louis
Guttman, Helene Augusta Nathan
Haberman, William L(awrence)
Hallett, Mark
Hanna, Edgar E(thelbert), Jr
Harlan, William R, Jr

Hartmann, Gregory Kemenyi
Haseltine, Florence Pat
Hasselmeyer, Eileen Grace
Hausman, Steven J
Hayes, Joseph Edward, Jr
Hazzard, DeWitt George
Hellwig, Helmut Wilhelm
Herman, Samuel Sidney
Hiatt, Caspar Wistar, III
Hildebrand, Bernard
Hilmoe, Russell J(ulian)
Hindle, Brooke
Holt, James Allen
Holt, Stephen S
Howie, Donald Lavern
Hubbard, Van Saxton
Huddleston, Charles Martin
Irwin, David
Jacobson, Martin
Joiner, R(eginald) Gracen
Jones, Franklin Del
Kalica, Anthony R
Kalt, Marvin Robert
Karlander, Edward P
Katsanis, D(avid) J(ohn)
Kearney, Philip C
Kendrick, Francis Joseph
Ketley, Jeanne Nelson
Khachaturian, Zaven Setrak
Khan, Mushtaq Ahmad
Kirby, Ralph C(loudsberry)
Kneller, Robert William
Kolstad, George Andrew
Koslow, Stephen Hugh
Krakauer, Henry
Kroll, Bernard Hilton
Krotosk, Danuta Maria
Kushner, Lawrence Maurice
Lansdell, Herbert Charles
Leaders, Floyd Edwin, Jr
LeRoy, André François
Levine, Arthur Samuel
Liverman, James Leslie
Lymn, Richard Wesley
McElhinney, John
Mahle, Christoph E
Maier, Eugene Jacob Rudolph
Malotky, Lyle Oscar
Marsden, Halsey M
Maslow, David E
Masnyk, Ihor Jarema
Meredith, Leslie Hugh
Mermagen, William Henry
Metcalfe, Dean Darrel
Metzger, Henry
Mielenz, Klaus Dieter
Milkey, Robert William
Miller, Nancy E
Miller, Raymond Jarvis
Mishler, John Milton
Mockrin, Stephen Charles
Mohla, Suresh
Mohler, Irvin C, Jr
Moloney, John Bromley
Moorjani, Kishin
Moran, David Dunstan
Mordfin, Leonard
Morgan, John D(avis)
Muller, Robert E
Murphy, Donald G
Murray, William Sparrow
Myers, Lawrence Stanley, Jr
Nagle, Dennis Charles
Nelson, James Harold
Norvell, John Charles
Obrams, Gunta Iris
O'Donnell, James Francis
Orlick, Charles Alex
Ory, Marcia Gail
Osborne, J Scott, III
Oxman, Michael Allan
Pare, William Paul
Parkhie, Mukund Raghunathrao
Paskins-Hurlburt, Andrea Jeanne
Peiser, Herbert Steffen
Phelps, Creighton Halstead
Phillips, Grace Briggs
Piez, Karl Anton
Pinkus, Lawrence Mark
Plotkin, Henry H
Porter, Beverly Fearn
Postow, Elliot
Powers, Marcelina Venus
Prabhakar, Arati
Quraishi, Mohammed Sayeed
Rabin, Herbert
Randall, John Douglas
Randolph, Lynwood Parker
Ranhand, Jon M
Redish, Janice Copen
Remondini, David Joseph
Reuther, Theodore Carl, Jr
Richardson, Billy
Richardson, Clarence Robert
Richardson, John Marshall
Rochlin, Robert Sumner
Rudney, William Stanley
Rosenstein, Robert William
Rothman, Sam
Rueger, Lauren J(ohn)
Saloman, Edward Barry
Sansone, Eric Brandfon
Saville, Thorndike, Jr
Schiaffino, Silvio Stephen

Schindler, Albert Isadore
Schmidt, Jack Russell
Schneider, Carl Stanley
Schoman, Charles M, Jr
Sexton, Thomas John
Shalowitz, Erwin Emmanuel
Sharma, Dinesh C
Sheffield, Harley George
Sheridan, Philip Henry
Siegel, Elliot Robert
Simon, Robert Michael
Singh, Amarjit
Smith, Joseph Collins
Snow, George Abraham
Solomon, Neil
Sorrows, Howard Earle
Stamper, Hugh Blair
Stein, Harvey Philip
Stevens, Donald Keith
Stoolmiller, Allen Charles
Stuebing, Edward Willis
Swann, Madeline Bruce
Szego, George C(harles)
Tabor, Edward
Tai, Tsze Cheng
Talbott, Edwin M
Taylor, Barry Norman
Theuer, Paul John
Thompson, Robert John, Jr
Tilford, Shelby G
Tillman, Michael Francis
Tolman, Robert Alexander
Torres-Anjel, Manuel Jose
Tower, Donald Bayley
Triantaphyllopoulos, Eugenie
Turbyfill, Charles Lewis
Ulbrecht, Jaromir Josef
Valdes, James John
Vanderryn, Jack
Ventre, Francis Thomas
Vermund, Sten Halvor
Villet, Ruxton Herrer
Vocci, Frank Joseph
Wachs, Melvin Walter
Walker, Michael Dirck
Wallace, Craig Kesting
Walter, Donald K
Walter, Eugene LeRoy, Jr
Watkins, Clyde Andrew
Way, George H
Weisfeldt, Myron Lee
Wells, James Robert
White, Harris Herman
Wolff, David A
Wolff, John B
Wong, Kin Fai
Wright, James Roscoe
Yarmolinsky, Adam
Young, George Anthony

MASSACHUSETTS
Ahearn, James Joseph, Jr
Alpert, Ronald L
Anderson, Edward Everett
Beaton, Albert E
Berger, Harvey J
Bhargava, Hridaya Nath
Blatt, S Leslie
Bonkovsky, Herbert Lloyd
Busby, William Fisher, Jr
Camougis, George
Carleton, Nathaniel Phillips
Carton, Edwin Beck
Chang, Joseph Yoon
Chase, Fred Leroy
Cohn, Lawrence H
Collins, Carolyn Jane
Cox, Lawrence Henry
Cukor, Peter
Cushing, David H
Das, Pradip Kumar
Davis, Michael Allan
Feakes, Frank
Flagg, John Ferard
Fohl, Timothy
Friedman, Orrie Max
Futrelle, Robert Peel
Ganley, Oswald Harold
Garing, John Seymour
Girard, Francis Henry
Gregory, Eric
Griffith, Robert W
Hallock, Gilbert Vinton
Heyerdahl, Eugene Gerhardt
Hodgins, George Raymond
Holmes, Francis W(illiam)
Hunt, Ronald Duncan
Hurd, Richard Nelson
Hurlburt, Douglas Herendeen
Ingle, George William
Jonas, Albert Moshe
Kelland, David Ross
King, Norval William, Jr
Koehler, Andreas Martin
Kovatch, George
Leamon, Tom B
McBee, W(arren) D(ale)
Mack, Richard Bruce
Martin, Frederick Wight
Maser, Morton D
Milburn, Richard Henry
Miller, James Albert, Jr
Morbey, Graham Kenneth
Morrow, Walter E, Jr

Nauss, Kathleen Minihan
Papageorgiou, John Constantine
Parker, Henry Seabury, III
Petricciani, John C
Pharo, Richard Levers
Picard, Dennis J
Pulling, Nathaniel H(osler)
Riggi, Stephen Joseph
Roberts, Francis Donald
Rogers, William Irvine
Rohde, Richard Allen
Rohovsky, Michael William
Rosenkrantz, Harris
Roth, Harold
Rowley, Durwood B
Scadden, David T
Schaffel, Gerson Samuel
Seamans, R(obert) C(hanning), Jr
Sethares, James C(ostas)
Smith, Raoul Normand
Smith, William Edward
Staelin, David Hudson
Taylor, Isaac Montrose
Thorp, James Wilson
Vartanian, Leo
Wigley, Roland L
Williams, David Lloyd
Wilson, Linda S (Whatley)
Zalosh, Robert Geoffrey
Zebrowitz, Leslie Ann

MICHIGAN
Albers, Walter Anthony, Jr
Anderson, Amos Robert
Arrington, Jack Phillip
Axen, Udo Friedrich
Becker, Marshall Hilford
Bennett, Carl Leroy
Bens, Frederick Peter
Blurton, Keith F
Bohm, Henry Victor
Brown, William M(ilton)
Bryant, Barbara Everitt
Cantlon, John Edward
Caplan, John D(avid)
Conway, Lynn Ann
Cooper, Theodore
Cossack, Zafrallah Taha
Cotter, Richard
Decker, R(aymond) F(rank)
Dijkers, Marcellinus P
Elslager, Edward Faith
Eltinge, Lamont
Erbisch, Frederic H
Fawcett, Timothy Goss
Fonken, Gunther Siegfried
Forist, Arlington Ardeane
Galan, Louis
Gara, Aaron Delano
Graham, John
Grochoski, Gregory T
Hagen, John William
Harris, Gale Ion
Harris, Kenneth
Harwood, Julius J
Heberlein, Gary T
Herk, Leonard Frank
Herzig, David Jacob
Heyd, William Ernst
Hinman, Jack Wiley
Hoerger, Fred Donald
Howard, Charles Frank, Jr
Huebner, George J
Humphrey, Harold Edward Burton, Jr
Hunstad, Norman A(llen)
Isleib, Donald Richard
Janik, Borek
Jones, Frank Norton
Joshi, Mukund Shankar
Klivington, Kenneth Albert
Le Beau, Stephen Edward
Lewis, Alan Laird
Lustgarten, Ronald Krisses
Maasberg, Albert Thomas
McCann, Daisy S
Macriss, Robert A
Maxey, Brian William
Mohrland, J Scott
Moll, Russell Addison
Petrick, Ernest N(icholas)
Powers, William Francis
Punch, Jerry L
Quinn, Frank Hugh
Rausch, Doyle W
Reynolds, Herbert McGaughey
Reynolds, John Z
Scott, Joseph Hurlong
Senagore, Anthony J
Shah, Shirish A
Shea, Fredericka Palmer
Shore, Joseph D
Sokol, Robert James
Solomon, David Eugene
Soper, Jon Allen
Spilman, Charles Hadley
Spreitzer, William Matthew
Stevenson, Robin
Stucki, Jacob Calvin
Thurber, William Samuels
Tuesday, Charles Sheffield
Van Rheenen, Verlan H
Walters, Stephen Milo
Weisbach, Jerry Arnold
Weissman, Eugene Y(ehuda)

Wilks, John William
Wolman, Eric
Young, David Caldwell

MINNESOTA
Anderson, Donald E
Atchison, Thomas Calvin, Jr
Beebe, George Warren
Boelter, Don Howard
Calehuff, Girard Lester
Case, Marvin Theodore
Chevalier, Peter Andrew
Daggs, Ray Gilbert
Davidson, Donald
Dickie, John Peter
Evans, Roger Lynwood
Gumnit, Robert J
Holman, Ralph Theodore
Holmen, Reynold Emanuel
Janes, Donald Lucian
Jensen, Timothy B(erg)
Jones, Lester Tyler
Kohlstedt, Sally Gregory
Lawrenz, Frances Patricia
Leung, Benjamin Shuet-Kin
McDermott, Daniel J
McDuff, Charles Robert
Norberg, Arthur Lawrence
Paulson, David J
Pitcher, Eric John
Ray, Charles Dean
Reever, Richard Eugene
Sentz, James Curtis
Spilker, Bert
Tautvydas, Kestutis Jonas
Thalacker, Victor Paul
Thangaraj, Sandy
Thompson, Phillip Gerhard
Thompson, Roy Lloyd
Tihon, Claude
Uban, Stephen A
Wetherall, Neal T
Wolfe, Barbara Blair

MISSISSIPPI
Ballard, James Alan
Berry, Charles Dennis
Collins, Johnnie B
ElSohly, Mahmoud Ahmed
Ferer, Kenneth Michael
Foil, Robert Rodney
Helms, Thomas Joseph
Jones, Eric Wynn
Lloyd, Edwin Phillips
McGilberry, Joe H
Morgan, Eddie Mack, Jr
Paulk, John Irvine
Ranney, Carleton David
Smith, Allie Maitland
Steinberg, Martin H
Thompson, Joe Floyd

MISSOURI
Alexander, Charles William
Baumann, Thiema Wolf
Brodsky, Philip Hyman
Burton, Robert Main
Cook, Mary Rozella
Cramer, Michael Brown
Frankoski, Stanley P
Giarrusso, Frederick Frank
Goldberg, Herbert Sam
Hakkinen, Raimo Jaakko
Hedlund, James L
Hill, Jack Filson
Jester, Guy Earlscort
Keller, Robert Ellis
Kramer, Richard Melvyn
Lannert, Kent Philip
Leifield, Robert Francis
Lubin, Bernard
Moore, James Frederick, Jr
Munger, Paul R
Oliver, Larry Ray
Peters, Elroy John
Pfander, William Harvey
Rawlings, Gary Don
Richards, Graydon Edward
Robertson, David G C
Sauer, Harry John, Jr
Schott, Jeffrey Howard
Sharp, Dexter Brian
Sidoti, Daniel Robert
Teng, James
Thomas, Lewis Jones, Jr
Thornton, Linda Wierman
Toom, Paul Marvin
Weiss, James Moses Aaron
Yanders, Armon Frederick

MONTANA
Bay, Roger R(udolph)
Figueira, Joseph Franklin
Jackson, Grant D
McRae, Robert James

NEBRASKA
Gardner, Paul Jay
O'Brien, Richard Lee
Provencher, Gerald Martin
Vanderholm, Dale Henry

NEVADA
Brandstetter, Albin

Klainer, Stanley M
Leipper, Dale F
Merdinger, Charles J(ohn)
Nauman, Charles Hartley
Poziomek, Edward John
Simmonds, Richard Carroll
Smith, Aaron
Snaper, Alvin Allyn
Taranik, James Vladimir

NEW HAMPSHIRE
Cotter, Robert James
Egan, John Frederick
Petrasek, Emil John
Wolff, Nikolaus Emanuel
Zager, Ronald

NEW JERSEY
Addor, Roger Williams
Amory, David William
Armbrecht, Frank Maurice, Jr
Auerbach, Victor
Bahr, James Theodore
Bardoliwalla, Dinshaw Framroze
Barker, Richard Gordon
Baron, Hazen Jay
Barpal, Isaac Ruben
Barrett, Walter Edward
Bauer, Frederick William
Bendure, Raymond Lee
Besso, Michael M
Borg, Donald Cecil
Bornstein, Alan Arnold
Bullock, Francis Jeremiah
Burley, David Richard
Cardinale, George Joseph
Carney, Richard William James
Caroselli, Robert Anthony
Caruso, Frank San Carlo
Caudill, Reggie Jackson
Chipkin, Richard E
Conway, Paul Gary
Cutler, Frank Allen, Jr
De Sesa, Michael Anthony
DiSalvo, Walter A
Dugan, Gary Edwin
Dvornik, Dushan Michael
Eash, John T(rimble)
Eastman, David Willard
Eastwood, Abraham Bagot
Eckler, Paul Eugene
Edelstein, Harold
Enell, John Warren
English, Alan Taylor
Evers, William L
Fahey, John Leonard
Farrington, Thomas Allan
Feigenbaum, Abraham Samuel
Field, Norman J
Finch, Rogers B(urton)
Fishbein, William
Flam, Eric
Fost, Dennis L
Fowles, Patrick Ernest
Froyd, James Donald
Gavini, Muralidhara B
Giordmaine, Joseph Anthony
Glass, Michael
Goldenberg, David Milton
Good, Ralph Edward
Gordziel, Steven A
Grauman, Joseph Uri
Gregory, Richard Alan, Jr
Grenda, Victor J
Hanna, Samir A
Harding, Maurice James Charles
Hartop, William Lionel, Jr
Hayworth, Curtis B
Henrikson, Frank W
Henry, Sydney Mark
Herold, E(dward) W(illiam)
Heveran, John Edward
Hillier, James
Hirschberg, Erich
Hlavacek, Robert John
Hockel, Gregory Martin
Holt, Rush D(ew), Jr
Howell, Mary Gertrude
Hwang, Cherng-Jia
Jaffe, Jonah
Kachikian, Rouben
Kapron, Felix Paul
Kaufman, Arnold
Kaufman, Harold Alexander
Kaufmann, Thomas G(erald)
Keramidas, Vassilis George
King, John A(lbert)
Kline, Berry James
Kohn, Jack Arnold
Kohout, Frederick Charles, III
Koster, William Henry
Kronenthal, Richard Leonard
Leaning, William Henry Dickens
Letizia, Gabriel Joseph
Leventhal, Howard
Levy, Martin J Linden
Lukacsko, Alison B
Lunn, Anthony Crowther
Luryi, Serge
McDermott, Kevin J
Malkin, Martin F
Mandel, Lewis Richard
Marcus, Phillip Ronald
Mark, David Fu-Chi

Research Administration (cont)

Markle, George Michael
Marlin, Robert Lewis
Massa, Tobias
Matthijssen, Charles
Meisel, Seymour Lionel
Michel, Gerd Wilhelm
Migdalof, Bruce Howard
Mihalisin, John Raymond
Miller, Albert Thomas
Miller, Irwin
Miller, Robert H(enry)
Millman, Sidney
Moe, Robert Anthony
Morck, Roland Anton
Munger, Stanley H(iram)
Neiswender, David Daniel
O'Neill, Patrick Joseph
Opie, William R(obert)
Otterburn, Michael Storey
Paterniti, James R, Jr
Patrick, James Edward
Peets, Edwin Arnold
Pheasant, Richard
Pignataro, Louis J(ames)
Pinsky, Carl Muni
Polinsky, Ronald John
Robertson, Nat Clifton
Rodda, Bruce Edward
Rolle, F Robert
Rollmann, Louis Deane
Salvador, Richard Anthony
San Soucie, Robert Louis
Sawin, Steven P
Saxon, Robert
Schunn, Robert Allen
Scott, Mary Celine
Slater, Eve Elizabeth
Smagorinsky, Joseph
Smith, Robert Ewing
Smith, Stephen Roger
Solis-Gaffar, Maria Corazon
Steinberg, Eliot
Stockton, John Richard
Tanner, Daniel
Templeman, Gareth J
Tolman, Richard Lee
Toothill, Richard B
Torrey, Henry Cutler
Trice, William Henry
Trivedi, Nayan B
Tucci, Edmond Raymond
Vincent, Gerald Glenn
Vukovich, Robert Anthony
Wang, Chao Chen
Wang, Guang Tsan
Warters, William Dennis
Westoff, Charles F
Wiggins, Jay Ross
Wilbur, Robert Daniel
Wildnauer, Richard Harry
Woodruff, Hugh Boyd
Woolf, Harry

NEW MEXICO
Bahr, Thomas Gordon
Barr, Donald Westwood
Bhada, Rohinton(Ron) K
Brabson, George Dana, Jr
Cano, Gilbert Lucero
Chambers, William Hyland
Chapin, Charles Edward
Clauser, Milton John
Colp, John Lewis
Cooch, Frederick Graham
Danen, Wayne C
Darnall, Dennis W
Davis, L(loyd) Wayne
Del Valle, Francisco Rafael
Dreicer, Harry
Gilbert, Alton Lee
Giovanielli, Damon V
Goldstone, Philip David
Harrill, Robert W
Heffner, Robert H
Hughes, Jay Melvin
Icerman, Larry
Jones, Orval Elmer
Kropschot, Richard H
Lohrding, Ronald Keith
Macek, Robert James
Mauderly, Joe L
Maurer, Robert Joseph
Mayeux, Jerry Vincent
Molecke, Martin A
Monagle, John Joseph, Jr
Morse, Fred A
Overhage, Carl F J
Peterson, Otis G
Price, Richard Marcus
Pynn, Roger
Seay, Glenn Emmett
Selden, Robert Wentworth
Simmons, Gustavus James
Stokes, Richard Hivling
Taber, Joseph John
Tombes, Averett S
Venable, Douglas
Vigil, John Carlos
Vook, Frederick Ludwig
Westpfahl, David John
Westwood, Albert Ronald Clifton
White, Paul C

Wing, Janet E (Sweedyk) Bendt
Yonas, Gerold

NEW YORK
Abetti, Pier Antonio
Akers, Charles Kenton
Allentuch, Arnold
Amidon, Thomas Edward
Antzelevitch, Charles
Ausubel, Jesse Huntley
Axelrod, David
Berg, Daniel
Blackmon, Bobby Glenn
Blane, Howard Thomas
Blank, Robert H
Bloch, Aaron Nixon
Bogard, Terry L
Bolton, Joseph Aloysius
Brody, Harold
Burchfiel, James Lee
Butler, Katharine Gorrell
Carr, Edward Gary
Castellion, Alan William
Chanana, Arjun Dev
Cohen, Edwin
Comly, James B
Coppersmith, Frederick Martin
Cote, Wilfred Arthur, Jr
Cowell, James Leo
Dean, Nathan Wesley
Dickerman, Herbert W
Dietz, Russell Noel
Dolak, Terence Martin
Drosdoff, Matthew
Dubnick, Bernard
Dumbaugh, William Henry, Jr
Eschbach, Charles Scott
Fiske, Milan Derbyshire
Florman, Monte
Forero, Enrique
Gilmore, Arthur W
Goland, Allen Nathan
Goldsmith, Eli David
Gomory, Ralph E
Gordon, Arnold J
Gordon, Harry William
Gordon, Wayne Alan
Greenberg, Jacob
Greizerstein, Hebe Beatriz
Guman, William J(ohn)
Gutcho, Sidney J
Haase, Jan Raymond
Hamilton, Byron Bruce
Hard, Gordon Charles
Harding, Homer Robert
Harrison, James Beckman
Henderson, Thomas E
Hileman, Robert E
Hodgson, William Gordon
Holtzman, Seymour
Hutchings, William Frank
Insalata, Nino F
Irwin, Lynne Howard
Jain, Anrudh Kumar
Joshi, Sharad Gopal
Juby, Peter Frederick
Kato, Walter Yoneo
Katz, Maurice Joseph
Kelly, Thomas Michael
Kinzly, Robert Edward
Klotzbach, Robert J(ames)
Kohrt, Carl Fredrick
Kolaian, Jack H
Koppelman, Lee Edward
Kornfield, Jack I
Langworthy, Harold Frederick
Latimer, Clinton Narath
Levine, Stephen Alan
Levy, Mortimer
Lhila, Ramesh Chand
Lichtig, Leo Kenneth
Lommel, J(ames) M(yles)
Luther, Herbert George
Lyons, Joseph F
McCamy, Calvin S
McElligott, Peter Edward
McHugh, William Dennis
Makous, Walter L
Martell, Michael Joseph, Jr
Matsushima, Satoshi
Maurer, Gernant E
Meyer, Harry Martin, Jr
Midlarsky, Elizabeth
Miller, Melvin J
Morey, Donald Roger
Morrison, John Agnew
Murphy, Kenneth Robert
Mustacchi, Henry
Nolan, John Thomas, Jr
Paldy, Lester George
Pall, David B
Palmedo, Philip F
Pasco, Robert William
Pentney, Roberta Pierson
Perez-Alburene, Evelio A
Pittman, Kenneth Arthur
Reich, Ismar M(eyer)
Robertson, William, IV
Rogler, Lloyd Henry
Rorer, David Cooke
Rothenberg, Albert
Rubin, Isaac D
Salzberg, Bernard
Savic, Michael I

Schallenberg, Elmer Edward
Schneider, Robert Fournier
Schwartzman, Leon
Siebert, Karl Joseph
Singer, Barry M
Smoyer, Claude B
Spiegel, Allen David
Steffen, Daniel G
Sterman, Samuel
Sternfeld, Leon
Sternlicht, B(eno)
Streett, William Bernard
Strnisa, Fred V
Sturman, Lawrence Stuart
Talley, Charles Peter
Taylor, James Earl
Tessieri, John Edward
Thomann, Gary C
Thomas, J Earl
Topp, William Carl
Trowbridge, Richard Stuart
Tuite, Robert Joseph
Turnblom, Ernest Wayne
Upeslacis, Janis
Vogel, Philip Christian
Von Strandtmann, Maximillian
Walker, Theresa Anne
Weingarten, Norman C
Weiss, Martin Joseph
White, Donald Robertson
Wolfe, Roger Thomas
Woodall, Jerry M
Woodmansee, Donald Ernest
Wulff, Wolfgang
Yeung, Kwok Kam

NORTH CAROLINA
Ahearne, John Francis
Aronson, A L
Bartholow, Lester C
Blake, Thomas Lewis
Blumenkopf, Todd Andrew
Bogaty, Herman
Bredeck, Henry E
Burnett, John Nicholas
Callihan, Dixon
Carpenter, Benjamin H(arrison)
Carr, Dodd S(tewart)
Cole, Jerome Foster
Cornoni-Huntley, Joan Claire
Cote, Philip Norman
Cruze, Alvin M
Cubberley, Adrian H
Cundiff, Robert Hall
Daniels, William Ward
Daumit, Gene Philip
Elder, Joe Allen
Foley, Gary J
Glaser, Frederick Bernard
Glenn, Thomas M
Gross, Harry Douglass
Hamm, Thomas Edward, Jr
Harrison, Stanley L
Haynes, John Lenneis
Heck, Walter Webb
Holcomb, George Ruhle
Horvitz, Daniel Goodman
Iafrate, Gerald Joseph
Johnson, William Hugh
Johnston, John
King, Kendall Willard
Klarman, William L
Kull, Frederick Charles, Sr
Lewin, Anita Hana
Linton, Richard William
Loomis, Carson Robert
Miller, Daniel Newton, Jr
Mink, James Walter
Minnemeyer, Harry Joseph
Morse, Roy E
Mosier, Stephen R
Nader, John S(haheen)
Pagano, Joseph Stephen
Putman, Charles E
Reeber, Robert Richard
Reel, Jerry Royce
Rees, John
Roblin, John M
Roth, Roy William
Sander, William August, III
Schmiedeshoff, Frederick William
Schultz, Frederick John
Stier, Howard Livingston
Sud, Ish
Swenberg, James Arthur
Tapp, William Jouette
Tucker, Walter Eugene, Jr
Udry, Joe Richard
Vandemark, Noland Leroy
Wiley, Albert Lee, Jr
Williams, Myra Nicol
Wyman, George Martin
Wynne, Johnny Calvin

NORTH DAKOTA
Freeman, Thomas Patrick
Mathsen, Don Verden

OHIO
Adamczak, Robert L
Altan, Taylan
Altenau, Alan Giles
Anderson, Ruth Mapes
Ault, George Mervin

Badertscher, Robert F(rederick)
Bahnfleth, Donald R
Baker, Saul Phillip
Bales, Howard E
Bertelson, Robert Calvin
Bidelman, William Pendry
Bisson, Edmond E(mile)
Blaser, Robert U
Bonner, David Calhoun
Bridgman, Charles James
Broge, Robert Walter
Brunner, Gordon Francis
Carman, Charles Jerry
Ciricillo, Samuel F
Coleman, John Franklin
Connolly, Denis Joseph
Courchene, William Leon
Crable, John Vincent
Croxton, Frank Cutshaw
Curtis, Charles R
Daschbach, James McCloskey
De Vries, Adriaan
Doershuk, Carl Frederick
Duga, Jules Joseph
Durand, Edward Allen
Evans, Thomas Edward
Fabris, Hubert
Faiman, Robert N(eil)
Fawcett, Sherwood Luther
Finn, John Martin
Fordyce, James Stuart
Forestieri, Americo F
Frank, Thomas Paul
Geho, Walter Blair
Gerace, Michael Joseph
Germann, Richard P(aul)
Graham, Robert William
Grisaffe, Salvatore J
Hall, Franklin Robert
Hiles, Maurice
Hirt, Andrew Michael
Hoekenga, Mark T
Hollis, William Frederick
Holtkamp, Dorsey Emil
Hughes, Kenneth E(ugene)
Huston, Keith Arthur
Hutchinson, Frederick Edward
Jache, Albert William
Jaworowski, Andrzej Edward
Jayne, Theodore D
Johnson, Ralph Sterling, Jr
Kaufman, John Gilbert, Jr
Kawasaki, Edwin Pope
Kelley, Frank Nicholas
Kelly, Kevin Anthony
Kinsman, Donald Vincent
Landstrom, D(onald) Karl
Laskowski, Edward L
Leonard, Billie Charles
Leyda, James Perkins
Lydy, David Lee
McCracken, John David
McCune, Homer Wallace
McMillin, Carl Richard
McVean, Duncan Edward
Madia, William J
Martino, Joseph Paul
Massey, L(ester) G(eorge)
Meek, Violet Imhof
Millard, Ronald Wesley
Miraldi, Floro D
Moore, Thomas Joseph
Moss, Thomas H
Myhre, David V
Nestor, Ontario Horia
Nixon, Charles William
Ockerman, Herbert W
Olesen, Douglas Eugene
Olson, Walter T
Prince, Terry Jamison
Redmond, Robert F(rancis)
Richley, E(dward) A(nthony)
Roha, Max Eugene
Salkind, Michael Jay
Schlegel, Donald Louis
Schoch, Daniel Anthony
Schultz, Thomas J
Scozzie, James Anthony
Semler, Charles Edward
Seufzer, Paul Richard
Smith, James Edward, Jr
Snyder, Donald Lee
Stephan, David George
Stevenson, James Francis
Stone, Kathleen Sexton
Sweeney, Thomas L(eonard)
Tiernan, Thomas Orville
Townley, Charles William
Ungar, Edward William
Versic, Ronald James
Versteegh, Larry Robert
Vlcek, Donald Henry
Weeks, Stephen P
Wiff, Donald Ray
Williams, Josephine Louise
Wodarski, John Stanley
Wolaver, Lynn E(llsworth)
Wolf, Warren Walter
Woodard, Ralph Emerson
Woodward, James Kenneth
Wright, George Joseph

OKLAHOMA
Carlile, Robert Elliot

Collins, William Edward
Crafton, Paul A(rthur)
Davis, Robert Elliott
Edwards, Lewis Hiram
Faulkner, Lloyd (Clarence)
Gray, Samuel Hutchison
Green, Ray Charles
Hale, John Dewey
Henderson, Frederick Bradley, III
Hopkins, Thomas R (Tim)
Jayaraman, H
Johnson, Gerald, III
Johnson, Ronald Roy
Logan, R(ichard) S(utton)
McDevitt, Daniel Bernard
Millheim, Keith K
Miranda, Frank Joseph
Powell, Jerrel B
Robinson, Jack Landy
Rotenberg, Don Harris
Schroeder, David J Dean
Short, James N
Weakley, Martin LeRoy

OREGON

Arscott, George Henry
Bullock, Richard Melvin
Cereghino, James Joseph
Cooper, Glenn Adair, Jr
Davis, John R(owland)
Ehrmantraut, Harry Charles
Ethington, Robert Loren
Fleming, Bruce Ingram
Frakes, Rodney Vance
Hayden, Jess, Jr
Henderson, Robert Wesley
Heninger, Ronald Lee
Krueger, William Clement
Murphy, Thomas A
Norris, Logan Allen
Poulton, Charles Edgar
Schroeder, Warren Lee
Seil, Fredrick John
Severson, Herbert H
Siemens, Richard Ernest
Smiley, Richard Wayne
Thielges, Bart A
Walsh, Don
Walstad, John Daniel
Wick, William Quentin
Zaerr, Joe Benjamin

PENNSYLVANIA

Abdelhak, Sherif Samy
Alexander, Stuart David
Andrews, Edwin Joseph
Angeloni, Francis M
Angrist, Stanley W
Anouchi, Abraham Y
Antes, Harry W
Archer, David Horace
Bailey, Denis Mahlon
Bannon, James Andrew
Bartuska, Doris G
Beavers, Ellington McHenry
Beedle, Lynn Simpson
Beier, Eugene William
Berkoff, Charles Edward
Biebuyck, Julien Francois
Bockosh, George R
Bonewitz, Robert Allen
Bosshart, Robert Perry
Bramfitt, Bruce Livingston
Brenner, Gerald Stanley
Browning, Daniel Dwight
Bucher, John Henry
Buskirk, Elsworth Robert
Carley, Harold Edwin
Champagne, Paul Ernest
Chen, Michael S K
Ciccone, Patrick Edwin
Colman, Robert W
Corbett, Robert B(arnhart)
Daehnick, Wilfried W
Deckert, Fred W
Deitrick, John E
Derby, James Victor
De Tommaso, Gabriel Louis
Drake, B(illy) Blandin
Dromgold, Luther D
Ehrig, Raymond John
Feero, William E
Fleming, Lawrence Thomas
Foley, Thomas Preston, Jr
Forbes, Martin
Foti, Margaret A
Frank, William Benson
Frazier, James Lewis
Fritz, Lawrence William
Fromm, Eli
Frumerman, Robert
Gilstein, Jacob Burrill
Glaser, Robert
Gollub, Jerry Paul
Golton, William Charles
Gottscho, Alfred M(orton)
Graham, Roger Kenneth
Haines, William Joseph
Henderson, Robert E
Herman, Frederick Louis
Hollibaugh, William Calvert
Iezzi, Robert Aldo
Isett, Robert David
Jacobsen, Terry Dale

Jarrett, Noel
Johnson, Wayne Orrin
Keairns, Dale Lee
Kehr, Clifton Leroy
Kennedy, Flynt
Kidawa, Anthony Stanley
Kolmen, Samuel Norman
Kopchik, Richard Michael
Korchynsky, M(ichael)
Kornfield, A(lfred) T(heodore)
Kottcamp, Edward H, Jr
Kowalski, Conrad John
Krueger, Charles Robert
Kulakowski, Elliott C
Kunesh, Charles Joseph
Kwan, King Chiu
Leighton, Charles Cutler
Libsch, Joseph F(rancis)
Lidman, William G
Lopez, R C Gerald
Lynch, Thomas John
McCarthy, Raymond Lawrence
McCormick, Robert H(enry)
McEvoy, James Edward
Maclay, William Nevin
Martucci, John A
Matsumoto, Teruo
Mayer, George
Mehrkam, Quentin D
Michelson, Eric L
Miklausen, Anthony J
Miller, Barry
Morehouse, Chauncey Anderson
Morgan, Arthur I, Jr
Mottur, George Preston
Mulvihill, John Joseph
Nash, David Henry George
Oder, Robin Roy
Osborn, Elburt Franklin
Owens, Frederick Hammann
Packman, Albert M
Paul, Anand Justin
Pierson, Ellery Merwin
Podgers, Alexander Robert
Pool, James C T
Poos, George Ireland
Rackoff, Jerome S
Rall, Waldo
Ray, Eva K
Recktenwald, Gerald William
Robinson, Charles J
Roman, Paul
Roosa, Robert Andrew
Roseman, Arnold S(aul)
Rosenberg, Richard Martin
Ross, Sidney
Rushton, Brian Mandel
Saggiomo, Andrew Joseph
Santamaria, Vito William
Schipper, Arthur Louis, Jr
Schneider, Bruce E
Schweighardt, Frank Kenneth
Seiner, Jerome Allan
Settles, Gary Stuart
Silverman, Jerald
Smith, W Novis, Jr
Snow, Jean Anthony
Solt, Paul E
Steg, L(eo)
Stern, Richard
Suter, Stuart Ross
Taylor, Robert Morgan
Trippodo, Nick Charles
Varma, Asha
Voltz, Sterling Ernest
Wagner, J Robert
Walker, Eric A(rthur)
Warren, Alan
Wasylyk, John Stanley
Watson, William Martin, Jr
Weedman, Daniel Wilson
Weinryb, Ira
Whyte, Thaddeus E, Jr
Wilkins, Raymond Leslie
Wohleber, David Alan
Wolff, Ivan A
Yohe, Thomas Lester

RHODE ISLAND

Gilman, John Richard, Jr
Hedlund, Ronald David
Jarrett, Jeffrey E
Josephson, Edward Samuel
Mayer, Garry Franklin
Pickart, Stanley Joseph
Sage, Nathaniel McLean, Jr
Shipp, William Stanley
Silva, Armand Joseph
Spero, Caesar A(nthony), Jr

SOUTH CAROLINA

Bauer, Richard M
Bennet, Archie Wayne
Brown, Arnold
Dickerson, Ottie J
Dixon, Kenneth Randall
Doherty, William Humphrey
Dowler, William Minor
Failla, Patricia McClement
Felling, William E(dward)
Franklin, Ralph E
Hopper, Michael James
Huray, Paul Gordon
Irwin, Lafayette K(ey)

Keller, Eldon Lewis
Kinney, Terry B, Jr
Kittrell, Benjamin Upchurch
Krumrei, W(illiam) C(larence)
Miley, John Wulbern
Morris, J(ames) William
Nevitt, Michael Vogt
Pawson, Beverly Ann
Smith, Theodore Isaac Jogues
Swank, Robert Roy, Jr

SOUTH DAKOTA

Hamilton, Steven J
Smith, Paul Letton, Jr
Swiden, LaDell Ray
Sword, Christopher Patrick

TENNESSEE

Auerbach, Stanley Irving
Ayers, Jerry Bart
Benton, Charles Herbert
Berven, Barry Allen
Brau, Charles Allen
Bronzini, Michael Stephen
Bryan, Robert H(owell)
Coover, Harry Wesley, Jr
Cope, David Franklin
Dabbs, John Wilson Thomas
Dilworth, Robert Hamilton, III
Elowe, Louis N
Garber, David H
Garrison, George Walker
Genung, Richard Keith
Gossett, Dorsey McPeake
Gosslee, David Gilbert
Granoff, Allan
Harms, William Otto
Hasek, Robert Hall
Horen, Daniel J
House, Robert W(illiam)
Hunt, James Calvin
Isley, James Don
Jurand, Jerry George
Kozub, Raymond Lee
LeQuire, Virgil Shields
Lim, Alexander Te
Livingston, Robert Simpson
Lowndes, Douglas H, Jr
Lundy, Ted Sadler
Martini, Mario
Partain, Clarence Leon
Patel, Tarun K
Poutsma, Marvin Lloyd
Prairie, Michael L
Reed, Peter William
Richmond, Chester Robert
Robertson, David
Row, Thomas Henry
Schmidt, Thomas William
Schreiber, Eric Christian
Sittel, Chester Nachand
Southards, Carroll J
Thonnard, Norbert
Trivelpiece, Alvin William
Van Hook, Robert Irving, Jr
Veigel, Jon Michael
Wagner, Conrad
Wakefield, Troy, Jr
Wilkinson, Michael Kennerly
Williamson, Handy, Jr
Witherspoon, John Pinkney, Jr
Wymer, Raymond George

TEXAS

Adamski, Robert J
Alexander, William Carter
Allison, Jean Batchelor
Anthony, Donald Barrett
Aspelin, Gary B
Benedict, George Frederick
Blankenship, Lytle Houston
Blytas, George Constantin
Burton, Russell Rohan
Cannon, Dickson Y
Cargill, Robert Lee, Jr
Carlton, Donald Morrill
Cartwright, Aubrey Lee, Jr
Cassard, Daniel W
Chevalier, Howard L
Chu, Wei-Kan
Clark, D L
Coon, Julian Barham
Cooper, Howard Gordon
Costanzi, John J
Coulson, Larry Vernon
Curtin, Richard B
Curtin, Thomas J
Davis, Selby Brinker
Dear, Robert E A
Dietlein, Lawrence Frederick
Dvoretzky, Isaac
Earlougher, Robert Charles, Jr
Elder, Mark Lee
Fisher, Gene Jordan
Foster, Terry Lynn
Fowler, Robert McSwain
Franklin, Thomas Doyal, Jr
Gatti, Anthony Roger
Gipson, Robert Malone
Goins, William C, Jr
Goodwin, John Thomas, Jr
Gray, John Malcolm
Grobman, Warren David
Grogan, Michael John

Gully, John Houston
Gum, Wilson Franklin, Jr
Haase, Donald J(ames)
Hadley, William Owen
Hall, James Edison
Heavner, James E
Heggers, John Paul
Herbert, Stephen Aven
Hilde, Thomas Wayne Clark
Hoffman, Herbert I(rving)
Hollinger, F(rederick) Blaine
Iampietro, P(atsy) F
Johnson, Elwin L Pete
Jordan, Wayne Robert
Katz, Marvin L(aVerne)
Kennedy, Ken
Kennedy, Robert Alan
Kirk, Ivan Wayne
Kochhar, Rajindar Kumar
Koppa, Rodger J
Krautz, Fred Gerhard
Kutzman, Raymond Stanley
Lehmkuhl, L Don
Lifschitz, Meyer David
Linder, John Scott
Loan, Raymond Wallace
Lytle, Michael Allen
McDonald, Lynn Dale
Mack, Mark Philip
Mahendroo, Prem P
Martinez, J Ricardo
Mastromarino, Anthony John
Mefferd, Roy B, Jr
Merrill, John Raymond
Milliken, Spencer Rankin
Mitchell, James Emmett
Monti, Stephen Arion
Morton, Robert Alex
Musa, Samuel A
Nash, Donald Robert
Nelson, John Franklin
Nicks, Oran Wesley
O'Benar, John DeMarion
Parker, Harry W(illiam)
Patel, Anil S
Patton, Alton DeWitt
Phillips, Gerald C
Poddar, Syamal K
Poe, Richard D
Potter, Andrew Elwin, Jr
Purser, Paul Emil
Randolph, Philip L
Reinert, James A
Robinson, Alfred Green
Russell, B Don
Rutford, Robert Hoxie
Savit, Carl Hertz
Schuh, Frank J
Secrest, Everett Leigh
Sheppard, Louis Clarke
Shotwell, Thomas Knight
Smith, Curtis William
Smith, Larry
Spencer, Alexander Burke
Stewart, Bobby Alton
Sward, Edward Lawrence, Jr
Teller, Cecil Martin, II
Thompson, Granville Berry
Torp, Bruce Alan
Ulrich, Roger Steffen
Varsel, Charles John
Vitkovits, John A(ndrew)
Walton, Charles Michael
Warren, Kenneth Wayne
Weldon, William Forrest
Whorton, Elbert Benjamin
Williams, Darryl Marlowe
Wilson, Bobby L
Wilson, Peggy Mayfield Dunlap
Wimpress, Gordon Duncan, Jr
Wolf, Harold William
Wurth, Thomas Joseph

UTAH

Anderson, Douglas I
Case, James B(oyce)
Cox, Benjamin Vincent
Cramer, Harrison Emery
Germane, Geoffrey James
Healey, Mark Calvin
Joklik, G Frank
Krieger, Carl Henry
Lasater, Eric Martin
Phillips, Lee Revell
Sharp, Joseph C(ecil)
Straight, Richard Coleman
Walker, Donald I

VERMONT

Forcier, Lawrence Kenneth
Hoagland, Mahlon Bush
Lawson, Robert Bernard
Linnell, Robert Hartley
Lipson, Richard L
Spillman, William Bert, Jr
Steele, Richard

VIRGINIA

Albrecht, Herbert Richard
Auton, David Lee
Bardon, Marcel
Beach, Harry Lee, Jr
Berg, John Richard
Birely, John H

Research Administration (cont)

Borum, Olin H
Bottoms, Albert Maitland
Brandt, Richard Gustave
Broadbent, Noel Daniel
Brown, Richard Dee
Buchanan, Thomas Joseph
Burka, Maria Karpati
Burns, Allan Fielding
Cardman, Lawrence Santo
Carter, William Walton
Casazza, John Andrew
Cetron, Marvin J
Chambers, Charles MacKay
Chang, George Chunyi
Charvonia, David Alan
Cherniavski, John Charles
Childress, Otis Steele, Jr
Chulick, Eugene Thomas
Corwin, Gilbert
Courtney, Mark William
Cross, Ernest James, Jr
Crowley, Patrick Arthur
Cruthers, Larry Randall
Dane, Charles Warren
Daugherty, David M
Davis, Joel L
Denning, Peter James
DePoy, Phil Eugene
Dolezalek, Hans
Dorman, Craig Emery
Dorsey, Clark L(awler), Jr
Doyle, Frederick Joseph
Dvorchik, Barry Howard
Fabbi, Brent Peter
Fitzgerald, James Allen
Gargus, James L
Gaugler, Robert Walter
Gibbons, John Howard
Gibson, John E(gan)
Gillies, George Thomas
Gourley, Desmond Robert Hugh
Grafton, Robert Bruce
Grunder, Hermann August
Gulrich, Leslie William, Jr
Hall, Harvey
Hart, Dabney Gardner
Haworth, W(illiam) Lancelot
Hazlett, Robert Neil
Herwig, Lloyd Otto
Hill, Jim T
Horsburgh, Robert Laurie
Hortick, Harvey J
Houghton, Kenneth Sinclair, Sr
Howard, William Eager, III
Huggett, Clayton (McKenna)
Ijaz, Lubna Razia
Johnson, George Patrick
Johnson, Leonard Evans
Keever, David Bruce
Kelley, Ralph Edward
Kelly, Henry Charles
Kirby, Andrew Fuller
Kreipke, Merrill Vincent
Levy, Richard Allen
Luckenbach, Mark Wayne
Lundegard, Robert James
Lynch, Maurice Patrick
Lyon, Robert Lyndon
McAfee, Donald A
McClung, Andrew Colin
McIrvine, Edward Charles
McLaughlin, Gerald Wayne
Madan, Rabinder Nath
Malcolm, John Lowrie
Mandelberg, Martin
Marcus, Stanley Raymond
Martensen, Todd Martin
Masterson, Kleber Sanlin, Jr
Maune, David Francis
Milligan, John H
Mock, John E(dwin)
Morrison, David Lee
Muehlhause, Carl Oliver
Murino, Vincent S
Murray, Jeanne Morris
Ossakow, Sidney Leonard
Oster, Ludwig Friedrich
Padovani, Elaine Reeves
Patterson, Earl E(dgar)
Peake, Harold J(ackson)
Pixley, John Sherman, Sr
Pulling, Ronald W
Raab, Harry Frederick, Jr
Reynik, Robert John
Rickett, Frederic Lawrence
Robertson, Randal McGavock
Robinson, Bruce B
Romney, Carl Fredrick
Rosenthal, Michael David
Rozzell, Thomas Clifton
Schneck, Paul Bennett
Shapiro, Maurice Mandel
Shockley, Gilbert R
Sink, David Scott
Sloope, Billy Warren
Small, Timothy Michael
Smidt, Fred August, Jr
Sobieszczanski-Sobieski, Jaroslaw
Sobol, Stanley Paul
Squibb, Robert E
Streeter, Robert Glen
Swiger, Louis Andre

Thorp, Benjamin A
Tipper, Ronald Charles
Tucker, Richard Frank
Van Reuth, Edward C
Van Tilborg, André Marcel
Wallace, Lance Arthur
Wallace, Raymond Howard, Jr
Warfield, J(ohn) N(elson)
Webber, John Clinton
Weedon, Gene Clyde
West, Bob
Whitesides, John Lindsey, Jr
Wilkniss, Peter Eberhard
Williams, Patricia Bell
Wolff, Roger Glen
Wood, George Marshall
Wood, Leonard E(ugene)
Zornetzer, Steven F

WASHINGTON
Albaugh, Fred William
Bennett, Clifton Francis
Brown, Donald Jerould
Cykler, John Freuler
Danko, Joseph Christopher
De Goes, Louis
Dietz, Sherl M
Elfving, Donald Carl
Evans, Thomas Walter
Fetter, William Allan
Gaines, Edward M(cCulloch)
Gillis, Murlin Fern
Glass, William A
Hannay, Norman Bruce
Hinman, George Wheeler
Hopkins, Horace H, Jr
Johnson, John Richard
Kantowitz, Barry H
Katz, Yale H
Krier, Carol Alnoth
Landsberg, Johanna D (Joan)
Leng, Earl Reece
McMurtrey, Lawrence J
Mahlum, Daniel Dennis
Maloney, Thomas M
Monan, Gerald E
Morrill, Ralph Alfred
Nelson, Randall Bruce
Noel, Jan Christine
Norman, Joe G, Jr
Roos, John Francis
Sarles, Lynn Redmon
Schiffrin, Milton Julius
Schmid, Loren Clark
Smith, Orville Auverne
Smith, Samuel H
Taylor, Dean Perron
Toohey, Richard Edward
Tukey, Harold Bradford, Jr
Untersteiner, Norbert
Warren, John Lucius
Wolthuis, Roger A
Young, Paul Ruel
Zuiches, James J

WEST VIRGINIA
Bowdler, Anthony John
Calzonetti, Frank J
Chaty, John Culver
O'Mara, Michael Martin
Sherman, Paul Dwight, Jr
Smith, Joseph James

WISCONSIN
Barton, Kenneth Allen
Bishop, Charles Joseph
Casper, Lawrence Allen
Davis, Thomas William
Dickas, Albert Binkley
Ferguson, John Allen
Fleischer, Herbert Oswald
Freas, Alan D('Yarmett)
Friend, Milton
Goodman, Robert Merwin
Hendee, Wiliam R
Knous, Ted R
Koval, Charles Francis
McClenahan, William St Clair
Manning, James Harvey
Moore, Leonard Oro
Moses, Gregory Allen
Oehmke, Richard Wallace
Owens, John Michael
Parnell, Donald Ray
Riegel, Ilse Leers
Schopler, Harry A
Sedor, Edward Andrew
Sheppard, Erwin
Stackman, Robert W

WYOMING
Dorrence, Samuel Michael
Gloss, Steven Paul
Marchant, Leland Condo
Pier, Allan Clark
Smith, James Lee

PUERTO RICO
Bangdiwala, Ishver Surchand
Lluch, Jose Francisco
Sasscer, Donald S(tuart)

ALBERTA
Bird, Gordon Winslow

Butler, R(oger) M(oore)
Coppock, Robert Walter
Cormack, George Douglas
Krahn, Thomas Richard
McAndrew, David Wayne
McElgunn, James Douglas
Macqueen, Roger Webb
Madiyalakan, Ragupathy
Muir, Donald Ridley
Nasseri, Touraj
Neish, Gordon Arthur
Rennie, Robert John
Sonntag, Bernard H
Spence, Matthew W

BRITISH COLUMBIA
Copes, Parzival
Cox, Lionel Audley
Dewey, John Marks
Drew, T John
Dueck, John
Duncan, Douglas Wallace
Eaves, Allen Charles Edward
Franz, Norman Charles
Godolphin, William
Hatton, John Victor
Healey, Michael Charles
Hesser, James Edward
Muggeridge, Derek Brian
Struble, Dean L
Tingle, Aubrey James
Trussell, Paul Chandos
Weaver, Ralph Sherman
Wynne-Edwards, Hugh Robert

MANITOBA
Adaskin, Eleanor Jean
Angel, Aubie
Atkinson, Thomas Grisedale
Ayles, George Burton
Rosinger, Herbert Eugene
Singh, Ajit
Swierstra, Ernest Emke

NEW BRUNSWICK
Frantsi, Christopher
Kohler, Carl

NOVA SCOTIA
Bowen, William Donald
Collins, William Beck
Hooper, Donald Lloyd
Jamieson, William David
Wilson, George Peter

ONTARIO
Andrew, Bryan Haydn
Anstey, Thomas Herbert
Asculai, Samuel Simon
Babcock, Elkanah Andrew
Barica, Jan M
Baronet, Clifford Nelson
Bennett, W Donald
Biro, George P
Blevis, Bertram Charles
Bowman, C(lement) W(illis)
Burger, Dionys
Burton, Ian
Busk, Grant Curtis, Jr
Cartier, Jean Jacques
Challis, John Richard George
Clark, Ferrers Robert Scougall
Costain, Cecil Clifford
Dawson, Peter Henry
Dyne, Peter John
Ettel, Victor Alexander
Evans, John R
Ghista, Dhanjoo Noshir
Gillham, Robert Winston
Gold, Lorne W
Gowe, Robb Shelton
Halstead, Ronald Lawrence
Hansson, Carolyn M
Heacock, Ronald A
Hepburn, John Duncan
Hickman, John Roy
Hopton, Frederick James
Ingratta, Frank Jerry
Kavanagh, Robert John
Keys, John David
Kundur, Prabha Shankar
Landolt, Jack Peter
Lawford, George Ross
Lawrence, John
Lindsey, George Roy
McAdie, Henry George
McGeer, James Peter
Milton, John Charles Douglas
Moo-Young, Murray
Morand, Peter
Morton, Donald Charles
Mustard, James Fraser
Mutton, Donald Barrett
Olson, Arthur Olaf
Pigden, Wallace James
Prasad, S E
Pullan, George Thomas
Reid, Lloyd Duff
Rubin, Leon Julius
Sherman, Norman K
Sobell, Mark Barry
Stasko, Aivars B
Tugwell, Peter
Tyson, John Edward Alfred

Vollenweider, Richard A
Wade, Robert Simson

QUEBEC
Bailar, John Christian, III
Bordeleau, Lucien Mario
Buckland, Roger Basil
Cantor, Ena D
Chang, Thomas Ming Swi
Cloutier, Gilles Georges
Coffin, Harold Garth
Contandriopoulos, Andre-Pierre
Croctogino, Reinhold Hermann
Giroux, Guy
Giroux, Yves M(arie)
Guyda, Harvey John
Hardy, Yvan J
L'Archeveque, Real Viateur
Lavallee, Andre
Maclachlan, Gordon Alistair
Marisi, Dan(iel) Quirinus
Maruvada, P Sarma
Mavriplis, F
Nash, Peter Howard
Poirier, Louis
Wrist, Peter Ellis

SASKATCHEWAN
Kybett, Brian David
Strathdee, Graeme Gilroy

OTHER COUNTRIES
Baggot, J Desmond
Balk, Pieter
Baratoff, Alexis
Burkhardt, Walter H
Fitzgerald, John Desmond
Gelzer, Justus
Gowans, James L
Gregory, Peter
Hatzakis, Michael
Heiniger, Hans-Jorg
Humphrey, Albert S
Ince, A Nejat
Irgolic, Kurt Johann
Johnsen, Dennis O
Kessler, Alexander
Knutson, Lloyd Vernon
Kuroyanagi, Noriyoshi
Lammers, Wim
Metzger, Gershon
Modabber, Farrokh Z
Onoe, Morio
Packer, Leo S
Ray, Prasanta K
Renfrew, Robert Morrison
Spence, Thomas Wayne
Stebbing, Nowell
Struzak, Ryszard G
Suematsu, Yasuharu
Taylor, Howard Lawrence
Thissen, Wil A
Van Overstraeten, Roger Joseph
Van Raalte, John A
Ward, Gerald T(empleton)
Wennerstrom, Arthur J(ohn)
Winn, Edward Barriere

Resource Management

ALABAMA
Guthrie, Richard Lafayette
McCarl, Henry Newton
Moore, Fletcher Brooks
Rosene, Walter, Jr
Smith, John Stephen
Warfield, Carol Larson

ALASKA
Forest, Kathryn Jo

ARIZONA
Beck, John R(oland)
Dworkin, Judith Marcia
Harris, DeVerle Porter
Kay, Fenton Ray
Swalin, Richard Arthur
Wasson, James Walter

ARKANSAS
Mack, Leslie Eugene

CALIFORNIA
Appleman, Gabriel
Borsting, Jack Raymond
Brooks, William Hamilton
Bush, George Edward
Chang, David Bing Jue
Chen, Kao
Craib, Kenneth Bryden
Dawson, Kerry J
Doede, John Henry
Dowdy, William Louis
Earhart, Richard Wilmot
Greenberg, Bernard
Greene, Elias Louis
Hammond, Martin L
Hesse, Christian August
Hewston, John G
Hoffman, Howard Torrens
Hygh, Earl Hampton
Jansen, Henricus Cornelis
Kilgore, Bruce Moody

Koehler, Wilbert Frederick
L'Annunziata, Michael Frank
Little, John Clayton
McMasters, Alan Wayne
Mathias, Mildred Esther
Mooz, William Ernst
Norgaard, Richard Bruce
Olshen, Abraham C
Orcutt, Harold George
Partain, Gerald Lavern
Rabl, Veronika Ariana
Rickard, James Joseph
Ripley, William Ellis
Sakagawa, Gary Toshio
Schwabe, Calvin Walter
Shevel, Wilbert Lee
Smedes, Harry Wynn
Squires, Dale Edward
Stewart, Brent Scott
Sweeney, James Lee
Szekely, Ivan J
Urbach, John C
Vilkitis, James Richard
Walker, Warren Elliott
Welch, Robin Ivor

COLORADO
Asherin, Duane Arthur
Barber, George Arthur
Carnathan, Gilbert William
Fly, Claude Lee
Fox, Douglas Gary
Gentry, Donald William
LaBounty, James Francis, Sr
McKown, Cora F
Martin, Stephen George
Osborn, Ronald George
Parker, H Dennison
Steele, Timothy Doak
White, Gilbert Fowler

CONNECTICUT
Barone, John A
Bentley, William Ross
Blackwood, Andrew W
DeDecker, Hendrik Kamiel Johannes
Golden, Gerald Seymour
John, Hugo Herman
Walton, Alan George

DELAWARE
Andersen, Donald Edward
Busche, Robert M(arion)
Collette, John Wilfred
DeDominicis, Alex John
Habibi, Kamran
Kundt, John Fred
Stopkie, Roger John
Suarez, Thomas H

DISTRICT OF COLUMBIA
Beer, Charles
Blanchard, Bruce
Brown, Eleanor D
Carey, William Daniel
Green, Rayna Diane
Nickum, John Gerald
Radcliffe, S Victor
Ramsay, William Charles
Selin, Ivan
Smith, Richard S, Jr
Spruill, Nancy Lyon
Waters, Robert Charles
White, Irvin Linwood

FLORIDA
Barile, Diane Dunmire
Berry, Leonard
Burrill, Robert Meredith
Detrick, Robert Sherman
Fluck, Richard Conard
Jones, Albert Cleveland
Joyce, Edwin A, Jr
Koblick, Ian
Kruczynski, William Leonard
Malina, Marshall Albert
Means, D(onald) Bruce
Mesterton-Gibbons, Michael Patrick
Nell, Terril A
Noonan, Norine Elizabeth
Parker, Howard Ashley, Jr
Prager, Michael Haskell
Prince, Eric D

GEORGIA
Bozeman, John Russell
Ike, Albert Francis
Krochmal, Jerome J(acob)
Schultz, Donald Paul
Sheppard, David Craig
Ware, Kenneth Dale

HAWAII
Berg, Carl John, Jr
Clark, Allen LeRoy
Duckworth, Walter Donald
Flachsbart, Peter George
Helfrich, Philip
Hufschmidt, Maynard Michael
Liu, Wei
Morin, Marie Patricia
Singh, Amarjit

IDAHO
Austin, Carl Fulton
MacFarland, Craig George
Ulliman, Joseph James

ILLINOIS
Casella, Alexander Joseph
Girard, G Tanner
Herendeen, Robert Albert
Herricks, Edwin E
Linde, Ronald K(eith)
Lloyd, Monte
Rabb, George Bernard
Sasman, Robert T
Southern, William Edward
Winsberg, Gwynne Roeseler
Wise, Evan Michael
Wittman, James Smythe, III

INDIANA
Davis, Edgar Glenn
DeMillo, Richard A
Dube, David Gregory
McGowan, Michael James
Mercer, Walter Ronald

IOWA
Karlen, Douglas Lawrence
Lane, Orris John, Jr

KANSAS
Choate, Jerry Ronald
Coates, Gary Joseph
Kromm, David Elwyn

KENTUCKY
Kohnhorst, Earl Eugene

LOUISIANA
Bahr, Leonard M, Jr
Cordes, Carroll Lloyd

MAINE
Mobraaten, Larry Edward
Sherburne, James Auril

MARYLAND
Alter, Harvey
Andres, Scott Fitzgerald
Bawden, Monte Paul
Berkson, Harold
Costanza, Robert
Glenn, John Frazier
Goldenberg, Martin Irwin
Griswold, Bernard Lee
Hazzard, DeWitt George
Hodgdon, Harry Edward
Johnson, Frederick Carroll
Jones, LeeRoy G(eorge)
Karadbil, Leon Nathan
Kreysa, Frank Joseph
McCawley, Frank X(avier)
Miller, Raymond Jarvis
Morgan, John D(avis)
Mountford, Kent
Pattee, Oliver Henry
Rivkin, Maxcy
Schmidt, Jack Russell
Shalowitz, Erwin Emmanuel
Shands, Henry Lee
Shaw, Robert William, Jr
Sholdt, Lester Lance
Soto, Gerardo H
Talbott, Edwin M
Wachs, Melvin Walter
Yakowitz, Harvey

MASSACHUSETTS
Alpert, Peter
Bond, Robert Sumner
Bowley, Donovan Robin
Broadus, James Matthew
Clark, William Cummin
Godfrey, Paul Jeffrey
Hovorka, John
Hulbert, Thomas Eugene
Lapkin, Milton
Leahy, Richard Gordon
Meal, Harlan C
Morris, Robert
Petrovic, Louis John
Smith, Tim Denis
White, Alan Whitcomb

MICHIGAN
Armstrong, John Morrison
Bulkley, Jonathan William
Chappelle, Daniel Eugene
Edgerton, Robert Howard
Haidler, William B(ernard)
Haynes, Dean L
Meteer, James William
Niese, Jeffrey Neal
Northup, Melvin Lee
Patterson, Richard L
Sommers, Lawrence M
Weisbach, Jerry Arnold

MINNESOTA
Broderius, Steven James
Foose, Thomas John
Gregersen, Hans Miller
Johnson, Howard Arthur, Sr
Kuehn, Jerome H

Lange, Robert Echlin, Jr
Robertsen, John Alan
Silberman, Edward

MISSISSIPPI
Hodges, Donald Glenn

MISSOURI
Axthelm, Deon D
Craddock, John Harvey
Long, Lawrence William
Wiggers, Ernie P

MONTANA
Deibert, Max Curtis
McClelland, Bernard Riley

NEBRASKA
Grew, Priscilla Croswell Perkins

NEVADA
Davis, Phillip Burton
Hudig, Dorothy

NEW HAMPSHIRE
Dingman, Stanley Lawrence

NEW JERSEY
Ben-Israel, Adi
Curcio, Lawrence Nicholas
Eager, George S, Jr
Graham, Margaret Helen
Hort, Eugene Victor
Jain, Sushil C
Kesselman, Warren Arthur
Liebig, William John
Maxim, Leslie Daniel
Migdalof, Bruce Howard
Miller, Robert H(enry)
Semenuk, Nick Sarden
Widmer, Kemble
Zipf, Elizabeth M(argaret)

NEW MEXICO
Huey, William S
Hughes, Jay Melvin
Jett, James Hubert
Lee, William Wai-Lim
Schaefer, Dale Wesley
Seay, Glenn Emmett
Sitney, Lawrence Raymond
Thode, E(dward) F(rederick)
Vigil, John Carlos
Wilson, Lee

NEW YORK
Brooks, Robert R
Cole, Roger M
Cusano, Dominick A
Dykstra, Thomas Karl
Fey, Curt F
Forquer, Sandra Lynne
Ginzberg, Eli
Hennigan, Robert Dwyer
Katz, Maurice Joseph
Keck, Donald Bruce
Kelisky, Richard Paul
Kiviat, Erik
Klotzbach, Robert J(ames)
Lommel, J(ames) M(yles)
Lougeay, Ray Leonard
Maynard, Charles Alvin
Ries, Richard Ralph
Rosen, Stephen
Streett, William Bernard
Wilkins, Bruce Tabor

NORTH CAROLINA
Ahearne, John Francis
Andrews, Richard Nigel Lyon
Brockhaus, John Albert
Godschalk, David Robinson
Hamm, Thomas Edward, Jr
Holland, Marjorie Miriam
Kolenbrander, Lawrence Gene
Miller, Daniel Newton, Jr
Roblin, John M
Rulifson, Roger Allen
Smith, William Adams, Jr
Stephenson, Richard Allen
Van Pelt, Arnold Francis, Jr

OHIO
Adamczak, Robert L
Bisson, Edmond E(mile)
Blaser, Robert U
Doershuk, Carl Frederick
Hauser, Edward J P
Hickman, Howard Minor
Hiles, Maurice
Kochanowski, Barbara Ann
Lamp, Benson J
Mathews, A L
Ray, John Robert
Roth, Robert Earl
Schultz, Thomas J
Snyder, Donald Lee
Spanier, Edward J
Stambaugh, Edgel Pryce
Vassell, Gregory S
Vertrees, Robert Layman
Weeks, Stephen P

OKLAHOMA
Green, Ray Charles
Still, Edwin Tanner
Sturgeon, Edward Earl
Talent, Larry Gene
Wickham, M Gary

OREGON
Boyd, Dean Weldon
Demaree, Thomas L
Dews, Edmund
Doyle, Jack David
Pease, James Robert
Poulton, Charles Edgar
Ripple, William John
Van Vliet, Antone Cornelis

PENNSYLVANIA
Adovasio, James Michael
Danzon, Patricia M
Franco, Nicholas Benjamin
Frick, Neil Huntington
Gregg, Roger Allen
Grindel, Joseph Michael
Jordan, Andrew G
Lipsky, Stephen E
Mann, Stanley Joseph
Martucci, John A
Mezger, Fritz Walter William
Phelps, Lee Barry
Ramani, Raja Venkat
Sharer, Cyrus J
Waxman, Herbert Sumner
Woodruff, Kenneth Lee
Wuest, Paul J

RHODE ISLAND
Burroughs, Richard

SOUTH CAROLINA
Abercrombie, Clarence Lewis
Felling, William E(dward)
Smith, Theodore Isaac Jogues
Wilson, John Neville

SOUTH DAKOTA
Draeger, William Charles

TENNESSEE
Fordham, J(ames) Lynn
LeBlanc, Larry Joseph
McNutt, Charles Harrison
Wilson, James Larry

TEXAS
Arnold, J Barto, III
Crouse, Philip Charles
Darnell, Rezneat Milton
Eckelmann, Walter R
Gittings, Stephen Reed
Howe, Richard Samuel
Jackson, Eugene Bernard
Krishna, J Hari
Lacewell, Ronald Dale
Lehmann, Elroy Paul
Murnane, Thomas George
Schmandt, Jurgen
Stanford, Geoffrey
Sullivan, Arthur Lyon

UTAH
Downing, Kenton Benson
Krejci, Robert Henry
Lilieholm, Robert John

VERMONT
Smith, Thomas David

VIRGINIA
Awad, Elias M
Bayliss, John Temple
Berry, James G(ilbert)
Berryman, Jack Holmes
Casazza, John Andrew
Cavanaugh, Margaret Anne
Cetron, Marvin J
Chappelle, Thomas W
Cox, William Edward
Davis, William Spencer
Deal, George Edgar
Dillaway, Robert Beacham
Dobyns, Samuel Witten
Dragonetti, John Joseph
Gilbert, Arthur Charles
Hall, Otis F
Hamilton, Thomas Charles
Harowitz, Charles Lichtenberg
Hinckley, Alden Dexter
Hobeika, Antoine George
Jahn, Laurence R
Kiessling, Oscar Edward
Kirby, Andrew Fuller
Knap, James E(li)
Lynch, Maurice Patrick
McCormick-Ray, M Geraldine
McEwen, Robert B
Nealy, David Lewis
Sink, David Scott
Stone, Harris B(obby)
Streeter, Robert Glen
Talbot, Lee Merriam

WASHINGTON
Booth, Pieter Nathaniel

Resource Management (cont)

Brewer, William Augustus
Fluharty, David Lincoln
Gaines, Edward M(cCulloch)
Gilbert, Frederick Franklin
Hinman, George Wheeler
Marion, Wayne Richard
Michelsen, Ari Montgomery
Miles, Edward Lancelot
Roos, John Francis
Royce, William Francis
Strandjord, Paul Edphil
Verson, Scott

WEST VIRGINIA
Calzonetti, Frank J

WISCONSIN
Beatty, Marvin Theodore
Cain, John Manford
Sheppard, Erwin
Temple, Stanley A

WYOMING
Wheasler, Robert

PUERTO RICO
Wilson, Marcia Hammerquist

ALBERTA
Freeman, Milton Malcolm Roland
Gong, Peng
Lane, Robert Kenneth
Macpherson, Andrew Hall

BRITISH COLUMBIA
Bunce, Hubert William
Copes, Parzival
Griffiths, George Motley
Healey, Michael Charles
Marshall, Peter Lawrence
Newroth, Peter Russell
Perry, Richard Ian
Peterman, Randall Martin
Weintraub, Marvin
Welch, David Warren

NEW BRUNSWICK
Kohler, Carl

NEWFOUNDLAND
Bajzak, Denes

NOVA SCOTIA
Bowen, William Donald
Tremblay, Michael John
Vandermeulen, John Henri

ONTARIO
Ankney, C(laude) Davison
Banfield, Alexander William Francis
Burk, Cornelius Franklin, Jr
Hoffman, Douglas Weir
Krantzberg, Gail
Lavigne, David M
Lee, Hulbert Austin
McNairn, Heather Elizabeth
Marks, Charles Frank
Olson, Arthur Olaf
Powis, Alfred
Siddall, Ernest
Slocombe, Donald Scott
Stebelsky, Ihor
Watson, Jeffrey

QUEBEC
Bourchier, Robert James
Hardy, Yvan J

OTHER COUNTRIES
Dahl, Arthur Lyon
Khatib, Hisham M
Ni, I-Hsun

Science Administration

ALABAMA
Doorenbos, Norman John
Glaze, Robert P
Jolley, Homer Richard
Naumann, Robert Jordan
Six, Norman Frank, Jr

ALASKA
Proenza, Luis Mariano
Thomas, Gary Lee
Williamson, Francis Sidney Lanier

ARIZONA
Barnes, Charles Winfred
Bartocha, Bodo
Counts, Jon Milton
Gibbs, Robert John
Leibacher, John W
Ruhe, Carl Henry William
Washton, Nathan Seymour

ARKANSAS
Berky, John James
Havener, Robert D
Norris, Arthur Robert
Pogue, William Reid

CALIFORNIA
Adams, John Andrew
Andersen, Ronald
Ast, David Bernard
Atkinson, Richard C
Bechtel, Stephen D, Jr
Bienenstock, Arthur Irwin
Carlan, Audrey M
Chapman, Loring Frederick
Creutz, Edward (Chester)
Crum, James Davidson
Demetriades, Sterge Theodore
Drake, Frank Donald
Gatrousis, Christopher
Glass, Alexander Jacob
Hardy, Rolland L(ee)
Havemeyer, Ruth Naomi
Haygood, Margo Genevieve
Heindl, Clifford Joseph
Hill, Robert Matteson
Hoffman, Arlene Faun
Howd, Robert A
Hughes, Walter William, III
Irvine, Cynthia Emberson
Jakubowski, Gerald S
King, Sheldon Selig
Kushner, Arthur S
Legge, Norman Reginald
Lorance, Elmer Donald
Miller, Gary Arthur
Mode, V(incent) Alan
Moorhouse, Douglas C
Neel, James William
Neville, James Ryan
Oakley, David Charles
O'Donnell, Joseph Allen, III
Overby, Lacy Rasco
Parker, John William
Pitts, Wanna Dene
Putnam, Thomas Milton, Jr
Rickard, Corwin Lloyd
Ripley, William Ellis
Rubin, Barney
Sargent, Anneila Isabel
Seibel, Erwin
Seubold, Frank Henry, Jr
Silverman, Jacob
Simonen, Thomas Charles
Smith, James Payne, Jr
Steffan, Wallace Allan
Stewart, Donald Charles
Swanson, Anne Barrett
Swanson, Paul N
Terry, Richard D
Tschirgi, Robert Donald
Vermeulen, Theodore (Cole)
Wagner, William Gerard
Walden, Clyde Harrison

COLORADO
Allen, Joe Haskell
Cowart, Vicki Jane
Downs, William Fredrick
Hildner, Ernest Gotthold, III
Hinners, Noel W
Holtz, Wesley G
Kotch, Alex
Nelson, Bernard W
Niehaus, Merle Hinson
Shedlock, Kaye M
Strauss, Eric L

CONNECTICUT
Aaslestad, Halvor Gunerius
Buss, Leo William
Comer, James Pierpont
Ewing, Martin Sipple
Hopper, Paul Frederick
Lee, Douglas Scott
Luria, S(aul) M(artin)
Menke, David Hugh
Monahan, Edward Charles
Ortner, Mary Joanne
Remington, Charles Lee
Weinstein, Curt David
Welch, John F, Jr

DELAWARE
Bercaw, James Robert
Delp, Charles Joseph
Earle, Ralph Hervey, Jr
Marrs, Barry Lee
Scott, Thomas Russell
Sherman, John Walter

DISTRICT OF COLUMBIA
Abler, Ronald Francis
Alexander, Benjamin H
Allen, Robert Erwin
Barnes, John Maurice
Bautz, Laura Patricia
Beer, Charles
Berley, David
Bierly, Eugene Wendell
Boosman, Jaap Wim
Boyce, Peter Bradford
Burley, Gordon
Butter, Stephen Allan
Buzzelli, Donald Edward
Carlson, William Dwight
Carrigan, Richard Alfred
Chung, Ho
Civiak, Robert L
Cowdry, Rex William
Crawford, Lester M
Crouch, Roger Keith
Crum, John Kistler
Dragoo, Alan Lewis
Egeberg, Roger O
Finn, Edward J
Frank, Richard Stephen
Fuchs, Roland John
Gilbert, Myron B
Gist, Lewis Alexander, Jr
Gunn, John William, Jr
Hancock, Kenneth George
Hellmann, Max
Hendrie, David Lowery
Hudis, Jerome
Johnston, Laurance S
Jones, Stanley B
Kirwan, Donald Frazier
Kitchens, Thomas Adren
Kornhauser, Andrija
Lovejoy, Thomas E
Lustig, Harry
MacCracken, Michael Calvin
McEwen, Gerald Noah, Jr
Madden, Joseph Michael
Malcom, Shirley Mahaley
Maxwell, Paul Charles
Mog, David Michael
Nelson, David Brian
O'Fallon, John Robert
O'Hern, Elizabeth Moot
Rabin, Robert
Randolph, James E
Reining, Priscilla Copeland
Rhoden, Richard Allan
Robinowitz, Carolyn Bauer
Rodman, James Eric
Setlow, Valerie Petit
Shank, Fred R
Shon, Frederick John
Singer, Maxine Frank
Spurlock, Langley Augustine
Tape, Gerald Frederick
Vernikos, Joan
Wagner, William Anton
Wallenmeyer, William Anton
West, Robert MacLellan
Whetstone, Stanley L, Jr
Wood, Robert Winfield

FLORIDA
Arnold, Henry Albert
Arpke, Charles Kenneth
Baum, Werner A
Beckhorn, Edward John
Ben, Max
Brown, Ernest Benton, Jr
Cintron, Guillermo B
Clegern, Robert Wayne
Clement, Duncan
Cort, Winifred Mitchell
Diamond, Jacob J(oseph)
Dolan, John E
Frankel, Jack William
Giacco, Alexander Fortunatus
Heller, William Mohn
Joyce, Edwin A, Jr
Karp, Abraham E
Korwek, Alexander Donald
Leon, Kenneth Allen
Levy, Norman Stuart
Lillien, Irving
Mahadevan, Selvakumaran
Means, D(onald) Bruce
Menefee, Robert William
Nell, Terril A
Nelson, Gordon Leigh
Posner, Gerald Seymour
Rabbit, Guy
Randall, J Malcom
Robitaille, Henry Arthur
Ross, Fred Michael
Rumbach, William Ervin
Shaak, Graig Dennis
Wissow, Lennard Jay
Wood, William Otto

GEORGIA
Allen, Ernest Mason
Brent, J Allen
Buccino, Alphonse
Greenbaum, Lowell Marvin
Hughes, James Mitchell
Hugus, Barbara Hendrick
Jackson, Curtis Rukes
Johnson, Donald Ross
Long, Earl Ellsworth
Nethercut, Philip Edwin
Rich, Fredrick James
Richmond, Jonathan Young
Rogers, Peter H
Skeen, James Norman
Stevenson, James Rufus
Stone, Connie J
Stoneburner, Daniel Lee

HAWAII
Carlson, John Gregory
Duckworth, Walter Donald
Eldredge, Lucius G
Henry, Joseph Patrick
Johnson, Carl Edward

IDAHO
Leonard, John W

ILLINOIS
Ausman, Robert K
Bartimes, George F
Feigen, Larry Philip
Fields, Thomas Henry
Fossum, Robert Merle
Goldman, David Tobias
Griffith, James H
Hockman, Deborah C
Holleman, William H
Knapperberger, Paul Henry, Jr
LaFevers, James Ronald
Lanyon, Scott Merril
Lederman, Leon Max
Mafee, Mahmood Forootan
Nevalainen, David Eric
Patterson, Douglas Reid
Paulson, Glenn
Plautz, Donald Melvin
Prager, Denis Jules
Puchalski, Randall Francis
Rabinowitch, Victor
Ren, Shang-Fen
Seefeldt, Waldemar Bernhard
Semonin, Richard Gerard
Shortell, Stephen M
Simon, Jack Aaron
Stafford, Fred E
Stinchcomb, Thomas Glenn
Willoughby, William Franklin
Wolter, Kirk Marcus

INDIANA
Armstrong, James G
Cohen, Marlene Lois
Harvey, Thomas G
Maickel, Roger Philip
Shultz, Clifford Glen
Steinmetz, Charles Henry
Timma, Donald Lee

IOWA
Frey, Nicholas Martin
Hausler, William John, Jr
Koch, Donald Leroy
Lardner, James F
Snow, Joel A

KANSAS
Jenkinson, Marion Anne
Mills, Russell Clarence
Parizek, Eldon Joseph

KENTUCKY
DeLong, Lance Eric
Mihelich, John L

LOUISIANA
Defenbaugh, Richard Eugene
Kern, Clifford H, III
O'Barr, Richard Dale
Rabideau, Peter W
Strauss, Arthur Joseph Louis
Turner, Robert Eugene
Vanselow, Neal A

MAINE
Hayes, Guy Scull
Watling, Les

MARYLAND
Abramson, I Jerome
Alberts, Walter Watson
Altman, Philip Lawrence
Ambler, Ernest
Andres, Scott Fitzgerald
Asper, Samuel Phillips
Baer, Ledolph
Balinsky, Doris
Beattie, Donald A
Beaven, Vida Helms
Benbrook, Charles M
Bennett, William Ernest
Benton, George Stock
Bhatnagar, Gopal Mohan
Biersner, Robert John
Blattner, William Albert
Blumenthal, Herbert
Bram, Ralph A
Brodie, Harry Joseph
Bromberg, Eleazer
Brown, James Harvey
Budgor, Aaron Bernard
Bynum, Barbara S
Cain, Dennis Francis
Calder, John Archer
Carney, William Patrick
Carrico, Christine Kathryn
Carroll, James Barr
Chapin, Douglas Scott
Chartrand, Mark Ray, III
Chen, Philip Stanley, Jr
Clark, Gary Edwin
Cleland, Charles Frederick
Crouch, Madge Louise
Cutler, Jeffrey Alan
Daly, John Anthony
Dantzler, Herman Lee, Jr
Dean, Donna Joyce
Delappe, Irving Pierce
Demmerle, Alan Michael

Diggs, Carter Lee
Donlon, Mildred A
Douglass, Carl Dean
Eaton, Barbra L
Eaves, George Newton
Ebert, James David
Eisler, Ronald
Ensor, Phyllis Gail
Erickson, Duane Gordon
Ewing, June Swift
Fakunding, John Leonard
Falci, Kenneth Joseph
Feinman, Susan (Birnbaum)
Ferguson, Earl Wilson
Fisher, Kenneth D
Fivozinsky, Sherman P
Fleig, Albert J, Jr
Freden, Stanley Charles
Friedman, Mischa Elliot
Gartland, William Joseph
Gary, Robert
Gay, William Ingalls
Geller, Ronald G
Gerring, Irving
Gibson, Colvin Lee
Gobel, Stephen
Goldin, Edwin
Goldman, Dexter Stanley
Goldstein, Murray
Gonzales, Ciriaco Q
Gottlieb, Melvin Harvey
Graebert, Eric W
Graham, Joseph H
Gravatt, Claude Carrington, Jr
Green, Jerome George
Greenfeld, Sidney H(oward)
Grisamore, Nelson Thomas
Griswold, Bernard Lee
Gruber, Jack
Gryder, Rosa Meyersburg
Gueriguian, John Leo
Gull, Theodore Raymond
Hampar, Berge
Harding, Fann
Harter, Donald Harry
Hasselmeyer, Eileen Grace
Heinrich, Kurt Francis Joseph
Hempel, Franklin Glenn
Henry, Timothy James
Herczynski, Andrzej
Hertz, Harry Steven
Heydemann, Peter Ludwig Martin
Heydrick, Fred Painter
Hillery, Paul Stuart
Hilmoe, Russell J(ulian)
Hinman, Edward John
Hodge, David Charles
Holt, Stephen S
Hopkins, Homer Thawley
Horenstein, Evelyn Anne
Housewright, Riley Dee
Hsieh, Richard Kuochi
Hyatt, Asher Angel
James, John Cary
Jefferson, Donald Earl
Jenerick, Howard Peter
Johnson, Carl Boone
Johnston, Gerald Samuel
Joly, Olga G
Jordan, Craig Alan
Kalt, Marvin Robert
Kaluzienski, Louis Joseph
Kazi, Abdul Halim
Kelsey, Morris Irwin
Khachaturian, Zaven Setrak
Kingsbury, David Wilson
Kissman, Henry Marcel
Kline, Jerry Robert
Klombers, Norman
Kovach, Eugene George
Krauss, Robert Wallfar
Langenberg, Donald Newton
Lepovetsky, Barney Charles
Leshner, Alan Irvin
Levine, Alan Stewart
Longfellow, David G(odwin)
McCoubrey, Arthur Orlando
Masnyk, Ihor Jarema
Miller, Gerald Ray
Miller, Raymond Jarvis
Mishler, John Milton
Mohler, Irvin C, Jr
Morris, Rosemary Shull
Mumma, Michael Jon
Munn, John Irvin
Munson, J(ohn) C(hristian)
Murray, William Sparrow
Nayak, Ramesh Kadbet
Nekola, Mary Virginia
Newrock, Kenneth Matthew
Norvell, John Charles
O'Donnell, James Francis
Okunieff, Paul
Omata, Robert Rokuro
Pahl, Herbert Bowen
Parkhie, Mukund Raghunathrao
Paull, Kenneth Dywain
Pitlick, Frances Ann
Price, David Edgar
Price, Samuel
Quackenbush, Robert Lee
Ranney, J Buckminster
Rawlings, Samuel Craig
Rheinstein, Peter Howard

Rivera, Americo, Jr
Rodgers, Charles H
Rodgers, Imogene Sevin
Ross, Philip
Rotariu, George Julian
Sansone, Eric Brandfon
Sass, Neil Leslie
Schindler, Albert Isadore
Schreier, Ethan Joshua
Schwartz, Brian B
Schweizer, Malvina
Shafer, W Sue
Shore, Moris Lawrence
Smith, Douglas Lee
Spooner, Peter Michael
Sprott, Richard Lawrence
Steinwachs, Donald Michael
Stroud, Robert Church
Teays, Terry John
Tepper, Morris
Thompson, Warren Elwin
Tilford, Shelby G
Toll, John Sampson
Tyeryar, Franklin Joseph
Valega, Thomas Michael
Vocci, Frank Joseph
Watzman, Nathan
Waxdal, M J
Weissberg, Alfred
White, Harris Herman
White, Howard Julian, Jr
Williams, Conrad Malcolm
Wilson, Maureen O
Wingate, Catharine L
Wolff, David A
Wortman, Bernard
Yellin, Herbert
Zimmerman, Eugene Munro
Zink, Sandra

MASSACHUSETTS
Aldrich, Michele L
Allen, Thomas John
Askew, Eldon Wayne
Assaykeen, Tatiana Anna
Barlas, Julie S
Barnes, Arnold Appleton, Jr
Blaustein, Ernest Herman
Blendon, Robert Jay
Bonkovsky, Herbert Lloyd
Busby, William Fisher, Jr
Carlson, Herbert Christian, Jr
Ciappenelli, Donald John
Davidson, Gilbert
Ebert, Robert H
Garing, John Seymour
Gold, Albert
Grice, George Daniel, Jr
Gund, Peter Herman Lourie
Hilferty, Frank Joseph
Hilgar, Arthur Gilbert
Kelley, William S
Kosersky, Donald Saadia
LeBaron, Francis Newton
MacDougall, Edward Bruce
Moniz, Ernest Jeffrey
Morse, Robert Warren
Orlen, Joel
Poduska, John W, Sr
Reis, Arthur Henry, Jr
Riemer-Rubenstein, Delilah
Rosner, Anthony Leopold
Roth, Harold
Yerby, Alonzo Smythe

MICHIGAN
Braughler, John Mark
Costley, Gary E
Eliezer, Isaac
Haff, Richard Francis
Hagen, John William
Heustis, Albert Edward
Mallinson, George Greisen
Nagler, Robert Carlton
Paschke, John Donald
Solomon, Allen M
Stowell, Ewell Addison
Strickland, James Shive
Volz, Paul Albert
Weiss, Mark Lawrence

MINNESOTA
Haaland, John Edward
Kralewski, John Edward
Miller, Willard, Jr
Mulhausen, Robert Oscar
Nelson, Kyler Fischer
Norris, William C

MISSISSIPPI
Doblin, Stephen Alan
McClurkin, Iola Taylor
Regier, Lloyd Wesley
Sharpe, Thomas R

MISSOURI
Alexander, Charles William
Crosby, Marshall Robert
Hunn, Joseph Bruce
Kupchella, Charles E
Lipkin, David
Rash, Jay Justen
Rosebery, Dean Arlo
Santiago, Julio Victor

Stephens, Robert Eric
Vaughan, William Mace
Whitten, Elmer Hammond

MONTANA
Fritts, Steven Hugh
Hosley, Robert James

NEBRASKA
McClurg, James Edward

NEVADA
McElroy, James Lee

NEW HAMPSHIRE
Meader, Ralph Gibson
Wixson, Eldwin A, Jr

NEW JERSEY
Chizinsky, Walter
Coughran, William Marvin, Jr
Easterday, Otho Dunreath
Gale, George Osborne
Geshner, Robert Andrew
Hirsch, Robert L
Jaffe, Jonah
Kalm, Max John
Kasabach, Haig F
Meagher, Thomas Robert
Perhach, James Lawrence
Pinsky, Carl Muni
Scholl, Philip Jon
Shapiro, Stanley Seymour
Sutman, Frank X
Weiss, Marvin
White, Addison Hughson
Witherell, Peter Charles
Woolf, Harry

NEW MEXICO
Barbo, Dorothy M
Barnes, Peter David
Campbell, Laurence Joseph
Cornell, Samuel Douglas
Dinegar, Robert Hudson
Garcia, Carlos E(rnesto)
Howard, William Jack
Hughes, Jay Melvin
Monagle, John Joseph, Jr
North-Root, Helen May
Pappas, Daniel Samuel
Shannon, Spencer Sweet, Jr
Simmons, Leonard Micajah, Jr
Sivinski, Jacek Stefan
Stevenson, Robert Edwin
Todsen, Thomas Kamp
Venable, Douglas
Vook, Frederick Ludwig
Whetten, John T

NEW YORK
Allen, William F, Jr
Asch, Harold Lawrence
Berg, Daniel
Blaker, J Warren
Blatt, Sylvia
Bugliarello, George
Carr, Edward Gary
Chu, Richard Chao-Fan
Dolak, Terence Martin
Fickies, Robert H
Fox, Gerard F
Frechet, Jean M J
Glass, David Carter
Goland, Allen Nathan
Goldberger, Robert Frank
Goodman, Jerome
Holloman, John L S, Jr
Horowitz, Marilyn Stephens
King, Marvin
Lichter, Robert (Louis)
Loev, Bernard
Lovejoy, Derek R
Lowder, Wayne Morris
Lubic, Ruth Watson
Lulla, Jack D
McTernan, Edmund J
Manassah, Jamal Tewfek
Moloney, Thomas W
Montefusco, Cheryl Marie
Naughton, John
Przybylowicz, Edwin P
Rau, R Ronald
Reifler, Clifford Bruce
Rosen, Harry Mark
Rothstein, Robert
Schearer, Sherwood Bruce
Sharkey, John Bernard
Sheldon, Eleanor Bernert
Stanley, Edward Alexander
Sternfeld, Leon
Sturman, Lawrence Stuart
Szumski, Stephen Aloysius
Trasher, Donald Watson
Walker, Theresa Anne
Yates, Jerome William

NORTH CAROLINA
Burrous, Stanley Emerson
Claxton, Larry Davis
Cochran, George Thomas
Collins, Jeffrey Jay
Dickens, Charles Henderson
Etzel, Howard Wesley

Fairchild, Homer Eaton
Fjellstedt, Thorsten A
Flagg, Raymond Osbourn
Hadeen, Kenneth Doyle
Jameson, Charles William
Kelsey, John Edward
Kohn, Frank S
McPherson, Charles William
Maroni, Donna F
Rogosa, George Leon
Sander, William August, III
Sharkoff, Eugene Gibb
Walters, Douglas Bruce

NORTH DAKOTA
La Chance, Leo Emery

OHIO
Adamczak, Robert L
Beyer, William Hyman
Campbell, Colin
Dobbelstein, Thomas Norman
Kelley, Frank Nicholas
Mathews, David
Moore, Thomas Joseph
Sinner, Donald H
Wenzel, Richard Louis
White, Willis S, Jr

OKLAHOMA
Beasley, William Harold
Kimpel, James Froome
Lambird, Perry Albert
Neely, Stanley Carrell
Still, Edwin Tanner

OREGON
Beaulieu, John David
McQuate, Robert Samuel
Radovsky, Frank Jay

PENNSYLVANIA
Andes, Charles Lovett
Arrington, Wendell S
Bartuska, Doris G
Beedle, Lynn Simpson
Braunstein, David Michael
Brendlinger, Darwin
Coleman, Robert E
Cross, Leslie Eric
Hess, Eugene Lyle
Hoberman, Alfred Elliott
Kerstein, Morris D
Kulakowski, Elliott C
Moss, Melvin Lane
Perchonock, Carl David
Roosa, Robert Andrew
Rothman, Milton A
Skutches, Charles L
Smith, Gerald A
Smith, Stewart Edward
Spear, Jo-Walter
Sun, James Dean
Thomas, William J
Tobin, Thomas Vincent
Turoczi, Lester J
Tuthill, Harlan Lloyd
Varma, Asha
Waxman, Herbert Sumner
Weisfeld, Lewis Bernard
Wertheimer, Albert I
Wilson, Marjorie Price
Wolff, Ivan A

RHODE ISLAND
Gilman, John Richard, Jr
Nadolink, Richard Hughes

SOUTH CAROLINA
Aull, Luther Bachman, III
Blood, Elizabeth Reid
Dowler, William Minor
Jaco, Charles M, Jr
Nerbun, Robert Charles, Jr
Owen, John Harding
Rembert, David Hopkins, Jr
Thomas, Claude Earle

TENNESSEE
Bressler, Marcus N
Callcott, Thomas Anderson
Campbell, William B(uford), Jr
Horen, Daniel J
House, Robert W(illiam)
Kliewer, Kenneth L
Lowe, James Urban, II
Martin, Richard Blazo
Ramsey, Lloyd Hamilton
Row, Thomas Henry
Wachtel, Stephen Shoel
Wu, Ying-Chu Lin (Susan)

TEXAS
Baggett, Neil Vance
Barnes, Charles M
Bogard, Donald Dale
Caillouet, Charles W, Jr
Chu, Wei-Kan
Crawford, Stanley Everett
Farhataziz, Mr
Hartsfield, Henry Warren, Jr
Huebner, Walter F
Merrill, Joseph Melton
Miller, William

Science Administration (cont)

Peterson, Lysle Henry
Riese, Walter Charles Rusty
Spallholz, Julian Ernest
Sward, Edward Lawrence, Jr

UTAH
Cramer, Harrison Emery
Gay, Charles Wilford
Hansen, Dale J
McKell, Cyrus Milo
Madsen, James Henry, Jr
Olson, Randall J
Taylor, Philip Craig

VERMONT
Lawson, Robert Bernard
Moyer, Walter Allen, Jr

VIRGINIA
Auerbach, Stephen Michael
Barak, Eve Ida
Bardon, Marcel
Basu, Sunanda
Bernthal, Frederick Michael
Birely, John H
Blood, Benjamin Donald
Brosseau, George Emile, Jr
Brown, Richard Dee
Bryant, Patricia Sand
Burka, Maria Karpati
Callahan, James Thomas
Carr, Roger Byington
Clutter, Mary Elizabeth
Cooper, Earl Dana
Corwin, Gilbert
Daugherty, David M
Diebold, Robert Ernest
Dragonetti, John Joseph
Easter, Donald Philips
Fabbi, Brent Peter
Gaines, Alan McCulloch
Gardenier, John Stark
Gergely, Tomas Esteban
Greenfield, Richard Sherman
Grunder, Fred Irwin
Heebner, David Richard
Hill, Carl McClellan
Hixson, Susan Harvill
Huang, H(sing) T(sung)
Johnson, Leonard Evans
Katz, Alan Charles
Lapporte, Seymour Jerome
Lemp, John Frederick, Jr
Levine, Jules Ivan
Louttit, Richard Talcott
McGee, Henry A(lexander), Jr
Marshall, Harold George
Masters, Charles Day
Maybury, Robert H
Maycock, Paul Dean
Mock, John E(dwin)
Molnia, Bruce Franklin
Morgan, John Walter
Myers, Charles Edwin
Pitts, Nathaniel Gilbert
Ratchford, Joseph Thomas
Roscoe, Henry George
Rosenbaum, David Mark
Rosenthal, Michael David
Schad, Theodore M(acNeeve)
Schoen, Richard Isaac
Shetler, Stanwyn Gerald
Simpson, John Arol
Spindel, William
Steinberg, Marcia Irene
Suiter, Marilyn J
Tiedemann, Albert William, Jr
Torio, Joyce Clarke
Troxell, Terry Charles
Umminger, Bruce Lynn
Wagner-Bartak, Claus Gunter Johann
Watson, Robert Francis
Weiss, Daniel Leigh
Willett, James Delos
Zaborsky, Oskar Rudolf
Zwolenik, James J

WASHINGTON
Anderson, Donald Hervin
Curl, Herbert (Charles), Jr
Harlin, Vivian Krause
Holbrook, Karen Ann
Hutchinson, William Burke
Miles, Edward Lancelot
Rieke, William Oliver
Robertson, William O
Strandjord, Paul Edphil
Tenforde, Thomas SeBastian
Zuiches, James J

WEST VIRGINIA
Larson, Gary Eugene
Victor, Leonard Baker

WISCONSIN
Cadmus, Robert R
Cramer, Jane Harris
Dibben, Martyn James
Eckert, Alfred Carl, Jr
Huber, David Lawrence
Inman, Ross
Lowenstein, Michael Zimmer

Petersen, John Robert
Will, James Arthur

WYOMING
Meyer, Edmond Gerald

PUERTO RICO
McDowell, Dawson Clayborn

ALBERTA
McLeod, Lionel Everett

BRITISH COLUMBIA
Howard, John
Lancaster, George Maurice
Looney, Norman E

MANITOBA
Rosinger, Eva L J
Trick, Gordon Staples

NOVA SCOTIA
Cooke, Robert Clark
Simpson, Frederick James

ONTARIO
Bennett, W Donald
Blackwell, Alan Trevor
Casselman, Warren Gottlieb Bruce
Effer, W R
Forsdyke, Donald Roy
Greenaway, Keith R(ogers)
Holtslander, William John
Lister, Earl Edward
Lucas, Douglas M
Manchee, Eric Best
Michael, Thomas Hugh Glynn
Mustard, James Fraser
Smith, Charles Haddon
Solandt, Omond McKillop
Somers, Emmanuel
Stephenson, Norman Robert
Sturgess, Jennifer Mary
Watson, Michael Douglas

QUEBEC
Bourchier, Robert James
Burgess, John Herbert
Paulin, Gaston (Ludger)

OTHER COUNTRIES
Assousa, George Elias
Bell Burnell, S Jocelyn
Davis, William Jackson
Gwilt, John Ruff
Hayashi, Masao
Humphrey, Albert S
Johnsen, Dennis O
Kreider, Eunice S
Minnick, Danny Richard
Mori, Wataru
Routti, Jorma Tapio
Sellin, Lawrence C
Vrebalovich, Thomas

Science Communications

ALABAMA
Small, Robert James

ALASKA
Gedney, Larry Daniel

ARIZONA
Giampapa, Mark Steven
Handy, Robert M(axwell)
Hutchinson, Charles S, Jr
Pacholczyk, Andrzej Grzegorz
Paylore, Patricia Paquita
Tillery, Bill W
Washton, Nathan Seymour

ARKANSAS
Denton, James H

CALIFORNIA
Arnold, Harry L(oren), Jr
Ash, William Wesley
Barnett, R(alph) Michael
Bengelsdorf, Irving Swem
Benson, Harriet
Berg, Henry Clay
Blakeslee, Dennis L(auren)
Bourne, Charles Percy
Callaham, Robert Zina
Callahan, Joan Rea
Cuadra, Carlos A(lbert)
Davis, Ward Benjamin
Hardy, Edgar Erwin
Hearn, Walter Russell
McDonough, Thomas Redmond
Meyer, Carl Beat
Meyer, Gregory Carl
Novack, Gary Dean
O'Donnell, Joseph Allen, III
Poppoff, Ilia George
Rasmussen, Robert A
Rogers, Bruce Joseph
Rossbacher, Lisa Ann
Schneider, Meier
Schneider, Stephen Henry
Smith, Bradley Richard
Spiller, Gene Alan

Subramanya, Shiva
Zarem, Abe Mordecai

COLORADO
Allen, Joe Haskell
Barrett, Charles Sanborn
Craig, Roy Phillip
Davis, Milford Hall
Kreith, Frank
McKown, Cora F
Metzger, H Peter
Parker, H Dennison
Thompson, Starley Lee

CONNECTICUT
Alpert, Nelson Leigh
Aylesworth, Thomas Gibbons
Corning, Mary Elizabeth
Forman, J(oseph) Charles
Fraser, J(ulius) T(homas)
Gingold, Kurt
Grayson, Martin
Levine, Leon
Poincelot, Raymond Paul, Jr
Sullivan, Walter Seager
Tittman, Jay
Weinstein, Curt David

DELAWARE
Bartkus, Edward Peter
Cubberley, Virginia
Frankenburg, Peter Edgar
Hayek, Mason
Knodel, Elinor Livingston
Phillips, Brian Ross
Slade, Arthur Laird
Yamamoto, Y Stephen

DISTRICT OF COLUMBIA
Bowen, David Hywel Michael
Brown, Eleanor D
Carey, William Daniel
Carter, Gesina C
Cocks, Gary Thomas
Fainberg, Anthony
Gilbert, Myron B
Green, Rayna Diane
Hart, Charles Willard, Jr
Hines, Pamela Jean
Jones, Alice J
Lang, Roger H
Maxwell, Paul Charles
Vetter, Betty M
Vietmeyer, Noel Duncan
White, Irvin Linwood
Zawisza, Julie Anne A

FLORIDA
Barile, Diane Dunmire
Copeland, Richard Franklin
Ewart, R Bradley
Heller, William Mohn
Kinnmark, Ingemar Per Erland
Lacoste, Rene John
McKently, Alexandra H
Nell, Terril A
Oehser, Paul Henry
Robitaille, Henry Arthur
Sanford, Malcolm Thomas
Stang, Louis George

GEORGIA
Fletcher, James Erving
Hawkins, Isaac Kinney
Inge, Walter Herndon, Jr
Nethercut, Philip Edwin
Perkowitz, Sidney
Price, Edward Warren
Riggsby, Ernest Duward

IDAHO
Gesell, Thomas Frederick
Gittins, Arthur Richard

ILLINOIS
Barr, Susan Hartline
Berg, Dana B
Bernstein, Elaine Katz
Crawford, Susan Young
Evans, James Forrest
Kittel, J Howard
Koehler, Henry Max
Koenig, Michael Edward Davison
Lerman, Zafra Margolin
Lofgren, Philip Allen
Mafee, Mahmood Forootan
Newton, Stephen Bruington
Ren, Shang-Fen
Schaefer, Wilbur Carls
Swanson, Don R
Sweitzer, James Stuart
Throw, Francis Edward
Wilke, Robert Nielsen
Yos, David Albert

INDIANA
Cassidy, Harold Gomes
Conway, Hertsell S
Free, Helen M
Guthrie, Catherine Shirley Nicholson
Kory, Mitchell
Markee, Katherine Madigan
Tacker, Martha McClelland

IOWA
Arndt, Stephan
Hansman, Robert H
Stuckey, Richard E

KANSAS
Coates, Gary Joseph
Shortridge, Robert William

LOUISIANA
Courtney, John Charles
Davis, George Diament

MAINE
Comins, Neil Francis
Senders, John W
Spencer, Claude Franklin

MARYLAND
Altman, Philip Lawrence
Behar, Marjam Gojchlerner
Bishop, Walton B
Carton, Robert John
Chartrand, Mark Ray, III
Cosmides, George James
Crampton, Janet Wert
Crosby, Edwin Andrew
Downing, Mary Brigetta
Eisler, Ronald
Estrin, Norman Frederick
Gibson, Colvin Lee
Goodwin, Frederick King
Hader, Rodney N(eal)
Hammond, Allen Lee
Heilprin, Laurence Bedford
Hudson, Ralph P
Josephs, Melvin Jay
Kissman, Henry Marcel
Lide, David Reynolds, Jr
Lockard, J David
Lufkin, Daniel Harlow
McCarn, Davis Barton
Masys, Daniel R
Morgan, Walter L(eroy)
Moulton, James Frank, Jr
O'Brien, Paul J
Passman, Sidney
Redish, Janice Copen
Schewe, Phillip Frank
Siegel, Elliot Robert
Solomon, Neil
Spilhaus, Athelstan Frederick, Jr
Torres-Anjel, Manuel Jose
Wyckoff, James M
Zwanzig, Frances Ryder

MASSACHUSETTS
Chomsky, Noam
Claff, Chester Eliot, Jr
Cromer, Alan H
Ellingboe, James
Erickson, Alan Eric
Harris, Miles Fitzgerald
Illinger, Joyce Lefever
Karas, John Athan
Lichtenstein, Alice Hinda
Lightman, Alan Paige
Lipinski, Boguslaw
Mack, Charles Lawrence, Jr
Parker, Henry Seabury, III
Pasachoff, Jay M(yron)
Rickter, Donald Oscar
Shawcross, William Edgerton
Simpson, Rae Goodell
Taylor, Edwin Floriman

MICHIGAN
Bloom, Miriam
Crowther, C Richard
Kamrin, Michael Arnold
Moore, Eugene Roger
Ruffner, James Alan
Rycheck, Mark Rule

MINNESOTA
Anderson, Frances Jean
Ansari, Azam U
Daggs, Ray Gilbert
Lea, Wayne Adair

MISSISSIPPI
Ferguson, Mary Hobson

MISSOURI
Ucko, David A
Wrather, James Allen

NEBRASKA
Benschoter, Reba Ann

NEW JERSEY
Baum, Gerald A(llan)
Bendich, Adrianne
Chaiken, MarthaLeah
Cladis, Patricia Elizabeth Ruth
Colicelli, Elena Jeanmarie
Cromartie, William James, Jr
Frantz, Beryl May
Frost, David
Garone, John Edward
Goodman, Gerald Joseph
Hall, Luther Axtell Richard
Hay, Peter Marsland
Kasper, Horst Manfred

Levi, Barbara Goss
Lorenz, Patricia Ann
Miner, Karen Mills
Roper, Robert
Schindler, Max J
Seifert, Laurence C
Semenuk, Nick Sarden
Thelin, Lowell Charles
Weil, Benjamin Henry
Weinstein, Stephen B
Zipf, Elizabeth M(argaret)

NEW MEXICO
Amacher, Peter
Forscher, Bernard Kronman
Hill, Mary Rae
Otway, Harry John
Reifsnyder, William Edward

NEW YORK
Asimov, Isaac
Boardman, John
Brandwein, Paul Franz
Brown, Stanley Gordon
Christman, Edward Arthur
Cummins, John Frances
Daniels, John Maynard
Dauben, Joseph W
Day, Stacey Biswas
Feinberg, Robert Jacob
Fessenden-MacDonald, June Marion
Foreman, Bruce Milburn, Jr
Geiger, H Jack
Gerolimatos, Barbara
Goodman, Jerome
Gravani, Robert Bernard
Holtz, Aliza
Hoover, Peter Redfield
Immergut, Edmund H(einz)
Ingalls, James Warren, Jr
Insalata, Nino F
Kornberg, Fred
Leerburger, Benedict Alan
Li-Scholz, Angela
Mackay, Lottie Elizabeth Bohm
Mendelsohn, Naomi
Milton, Kirby Mitchell
Murphy, Eugene F(rancis)
Oliver, Gene Leech
Paldy, Lester George
Piel, Gerard
Rosen, Stephen
Rosenfeld, Jack Lee
Rothman, Francoise
Sage, Martin Lee
Sharpe, William D
Spiegel, Robert
Trasher, Donald Watson
Trigg, George Lockwood
Vroman, Leo
Watanabe, Myrna Edelman

NORTH CAROLINA
Boyers, Albert Sage
Carr, Dodd S(tewart)
Carson, Bonnie L Bachert
Dickens, Charles Henderson
Dixon, N(orman) Rex
Good, Wilfred Manly
Hess, Daniel Nicholas
Stubblefield, Charles Bryan
Zahed, Hyder Ali

OHIO
Bates, Robert Latimer
Brooman, Eric William
Centner, Rosemary Louise
Dehne, George Clark
Dirckx, John H
Fortner, Rosanne White
Krauss, Lawrence Maxwell
Losekamp, Bernard Francis
Martin, Scott McClung
Olson, Walter T
Platau, Gerard Oscar
Raciszewski, Zbigniew
Smith, James Edward, Jr
Turkel, Rickey M(artin)
Zaye, David F

OKLAHOMA
Collins, William Edward
Gorin, George

OREGON
Allen, John Eliot
Johnson, Dale E
Keltner, Llew
Moldenke, Alison Feerick
Pursglove, Laurence Albert

PENNSYLVANIA
Baechler, Charles Albert
Benfey, Otto Theodor
Casey, Adria Catala
Creasey, William Alfred
Fales, Steven Lewis
Foti, Margaret A
Freeman, Joseph Theodore
Garfield, Eugene
Gottlieb, Irvin M
Huth, Edward J
Jim, Kam Fook
Judge, Joseph Malachi

Ladman, Aaron J(ulius)
Russey, William Edward
Small, Henry Gilbert
Umen, Michael Jay
Vladutz, George E
Wolfe, Reuben Edward
Wolke, Robert Leslie
Wuest, Paul J

RHODE ISLAND
Zuehlke, Richard William

SOUTH CAROLINA
Manley, Donald Gene
Rothrock, George Moore
Wilson, Jerry D(ick)

SOUTH DAKOTA
Hamilton, Steven J
Leslie, Jerome Russell

TENNESSEE
Braunstein, Helen Mentcher
Dickens, Justin Kirk
Grant, Peter Malcolm
Horton, Charles Abell
Keedy, Hugh F(orrest)
Lasslo, Andrew
Partridge, Lloyd Donald

TEXAS
Albach, Roger Fred
Ashby, Jon Kenneth
Cunningham, R Walter
Dahl, Per Fridtjof
DeBakey, Lois
DeBakey, Selma
Jauchem, James Robert
Landolt, Robert George
Olenick, Richard Peter
Sakakini, Joseph, Jr

UTAH
Case, James B(oyce)
Downing, Kenton Benson

VIRGINIA
Allan, Frank Duane
Baker, Jeffrey John Wheeler
Boots, Sharon G
Briscoe, Melbourne George
Chappelle, Thomas W
Chulick, Eugene Thomas
Feldmann, Edward George
Frye, Keith
Kelley, John Michael
Kelly, Henry Charles
Lynd, Langtry Emmett
Naeser, Nancy Dearien
Scattergood, Leslie Wayne
Schoen, Robert
Shaub, Walter M
Streeter, Robert Glen
Von Baeyer, Hans Christian
Welt, Isaac Davidson

WASHINGTON
Scheffer, Victor B

WISCONSIN
Brock, Katherine Middleton
Buchanan-Davidson, Dorothy Jean
Dunwoody, Sharon Lee
Moffet, Hugh L
Riegel, Ilse Leers
Steinhart, Carol Elder

MANITOBA
Grant, Cynthia Ann

ONTARIO
Banfield, Alexander William Francis
Campbell, James Fulton
Clarke, Thomas Roy
Cook, David Greenfield
Halfon, Efraim
Hill, Martha Adele
Mackay, Rosemary Joan
Murphy, William Frederick
Sinha, Bidhu Bhushan Prasad
Smith, Lorraine Catherine
Wiebe, John Peter
Winter, Peter
Wood, Gordon Harvey

QUEBEC
Allen, Harold Don
Bordan, Jack
Buckland, Roger Basil
Davidson, Colin Henry
DuFresne, Albert Herman

SASKATCHEWAN
Irvine, Donald Grant
Larson, D Wayne

OTHER COUNTRIES
Bell Burnell, S Jocelyn
Biederman-Thorson, Marguerite Ann
De Bruyne, Pieter
Greene, Lewis Joel
Lloyd, Christopher Raymond
Siegel, Herbert

Science Education

ALABAMA
Baker, June Marshall
Baksay, Laszlo Andras
Barrier, J Wayne
Dean, Susan Thorpe
Doorenbos, Norman John
Falls, William Randolph, Sr
Glover, Elsa Margaret
McCarl, Henry Newton
Macmillan, William Hooper
Mullins, Dail W, Jr
Norman, Billy Ray
Rosenberger, Albert Thomas
Summerlin, Lee R
Thomas, Joseph Calvin
Watson, Jack Ellsworth

ALASKA
Beebee, John Christopher
Fahl, Charles Byron
Gedney, Larry Daniel
Hawkins, Daniel Ballou
Morrison, John Albert
Ragle, Richard Harrison
Smoker, William Williams

ARIZONA
Barnes, Charles Winfred
Bitter, Gary G
Bougher, Stephen Wesley
Cortner, Hanna Joan
Cunningham, Richard G(reenlaw)
Gates, Halbert Frederick
Iserson, Kenneth Victor
Justice, Keith Evans
Lawson, Anton Eric
Layton, Richard Gary
Lisonbee, Lorenzo Kenneth
Palmer, James McLean
Roemer, Elizabeth
Smith, David John
Steinbrenner, Arthur H
Swan, Cathy Wood
Swihart, Thomas Lee
Tillery, Bill W
Venables, John Anthony
Washton, Nathan Seymour
Willoughby, Stephen Schuyler
Witherspoon, James Donald

ARKANSAS
Anderson, Robbin Colyer
Barton, Harvey Eugene
Cockerham, Lorris G(ay)
Dodson, B C
Edson, James Edward, Jr
Eichenberger, Rudolph John
Fribourgh, James H
Hobson, Arthur Stanley
Leming, Charles William
Mackey, James E
Mullins, Charles Warren
Singh, Surendra Pal
Texter, E Clinton, Jr

CALIFORNIA
Anderson, Milo Vernette
Appleman, M Michael
Atkin, J Myron
Atkinson, Russell H
Baez, Albert Vinicio
Baggerly, Leo L
Barnett, R(alph) Michael
Bate, George Lee
Beard, Jean
Bennett, John Francis
Bharadvaj, Bala Krishnan
Billig, Franklin A
Bird, Harold L(eslie), Jr
Birge, Ann Chamberlain
Birnbaum, Michael Henry
Bloom, Elliott D
Bourne, Charles Percy
Buchsbaum, Ralph
Caldwell, David Orville
Chivers, Hugh John
Chow, Paul C
Cichowski, Robert Stanley
Clague, William Donald
Cogan, Adrian Ilie
Cohen, Nathan Wolf
Collins, Robert C
Corwin, Harold G, Jr
Cox, John William
Creutz, Edward (Chester)
Crum, James Davidson
D'Antoni, Hector Luis
Davis, Ward Benjamin
Dean, Richard Albert
Demond, Joan
Dessel, Norman F
Donnelly, Timothy C
Dunning, John Ray, Jr
Eisberg, Robert Martin
Feher, Elsa
Feign, David
Finney, Ross Lee, III
Fischer, Robert Blanchard
Fisher, Kathleen Mary Flynn
Fraknoi, Andrew Gabriel
Franklin, Gene F(arthing)
Fredenburg, Robert Love
Funk, Glenn Albert
Gardner, Marjorie Hyer
Garratty, George
Glick, Harold Alan
Gold, Marvin B
Gosselin, Edward Alberic
Greenstadt, Melvin
Gupta, Madhu Sudan
Haller, Eugene Ernest
Hanafee, James Eugene
Hartsough, Walter Douglas
Hayes, Janan Mary
Helmholz, August Carl
Henkin, Leon (Albert)
Hiel, Clement
Hiemenz, Paul C
Hoffmann, Jon Arnold
Hurd, Paul DeHart
Iltis, Wilfred Gregor
Irvine, Cynthia Emberson
Jarvis, William Tyler
Jendresen, Malcolm Dan
Johanson, William Richard
Johnson, Carl Emil, Jr
Johnson, Hugh Mitchell
Jungerman, John (Albert)
Juster, Norman Joel
Keeports, David
Kenealy, Patrick Francis
Kihlstrom, Kenneth Edward
Kim, Chung Sul (Sue)
Klinedinst, Paul Edward, Jr
Kocol, Henry
Korst, William Lawrence
Krippner, Stanley Curtis
L'Annunziata, Michael Frank
Lechtman, Max D
Leff, Harvey Sherwin
Lerner, Lawrence S
Liepman, H(ans) P(eter)
Lippa, Linda Susan Mottow
Lipson, Joseph Issac
Lopo, Alina C
Lorance, Elmer Donald
Lorenzen, Coby
Luehrmann, Arthur Willett, Jr
McDonough, Thomas Redmond
McPherron, Robert Lloyd
Mandelin, Dorothy Jane Bearcroft
Matis, Howard S
Mayse, Mark A
Morgan, Alan Raymond
Mussen, Eric Carnes
Myers, Andrew Gordon
Neidlinger, Hermann H
Norman, Eric B
Nunn, Robert Harry
Padian, Kevin
Pake, George Edward
Paselk, Richard Alan
Perl, Martin Lewis
Peterson, Laurence E
Petrucci, Ralph Herbert
Pettijohn, Richard Robert
Powell, Ronald Allan
Putz, Gerard Joseph
Quinn, Helen Rhoda Arnold
Rabin, Jeffrey Mark
Ranftl, Robert M(atthew)
Rateaver, Bargyla
Reynolds, Michael David
Rice, Robert Arnot
Rodriguez, Eloy
Romanowski, Christopher Andrew
Salanave, Leon Edward
Schelar, Virginia Mae
Schneiderman, Jill Stephanie
Schoenfeld, Alan Henry
Siebert, Eleanor Dantzler
Sime, Ruth Lewin
Smith, Richard Avery
Stollberg, Robert
Strickmeier, Henry Bernard, Jr
Stringall, Robert William
Swanson, Anne Barrett
Tanenbaum, Basil Samuel
Tenn, Joseph S
Thiessen, Richard E
Thomas, Barry
Thompson, Patrick W
Torgow, Eugene N
Trogler, William C
Tubbs, Eldred Frank
Turner, George Cleveland
Tuul, Johannes
Van Bibber, Karl Albert
Van Klaveren, Nico
Walker, Dan B
Walters, Richard Francis
Waters, William E
Webb, Leland Frederick
Wendel, Otto Theodore, Jr
White, Alvin Murray
White, Richard Manning
Whitney, Robert C
Wiley, Michael David
Will, Theodore A
Wulfman, Carl E
Yanow, Gilbert
Yu, Grace Wei-Chi Hu

COLORADO
Allen, Ernest E
Barber, Thomas Lynwood

Science Education (cont)

Bard, Eugene Dwight
Begelman, Mitchell Craig
Benton, Edward Rowell
Boudreau, Robert Donald
Coates, Donald Allen
Dixon, Robert Clyde
Druelinger, Melvin L
Edwards, Kenneth Ward
Gottschall, W Carl
Heikkinen, Henry Wendell
Henkel, Richard Luther
Hess, Dexter Winfield
Iona, Mario
Johnson, Janice Kay
Koldewyn, William A
Krausz, Stephen
Lauer, B(yron) E(lmer)
Lindauer, Ivo Eugene
Nauenberg, Uriel
Neumann, Herschel
Nisbet, Jerry J
Noble, Richard Daniel
Oleszek, Gerald Michael
Robinson, Hugh Gettys
Schonbeck, Niels Daniel
Shelton, Robert Wayne
Siuru, William D, Jr
Spenny, David Lorin
Trowbridge, Leslie Walter
West, Anita

CONNECTICUT
Braun, Phyllis C
Carroll, J Gregory
Connor, Lawrence John
Covey, Irene Mabel
Dix, Douglas Edward
Dolan, James F
Goodstein, Madeline P
Hanson, Earl Dorchester
Hurley, James Frederick
Marcus, Jules Alexander
Menke, David Hugh
Moyer, Ralph Owen, Jr
Nelson, Roger Peter
Patterson, Elizabeth Chambers
Poincelot, Raymond Paul, Jr
Rachinsky, Michael Richard
Ross, Donald Joseph
Salamon, Richard Joseph
Schile, Richard Douglas
Seitelman, Leon Harold
Votaw, Robert Grimm
Wolf, Elizabeth Anne
Wystrach, Vernon Paul

DELAWARE
Cole, G(eorge) Rolland
Herron, Norman
Islam, Mir Nazrul
Lipscomb, Robert DeWald
Shipman, Harry Longfellow
Singleton, Rivers, Jr
Skolnik, Herman
Tolman, Chadwick Alma
White, Harold Bancroft, III
Yamamoto, Y Stephen

DISTRICT OF COLUMBIA
Berendzen, Richard
Braunfeld, Peter George
Callanan, Margaret Joan
Cohn, Victor Hugo
Dowdy, Edward Joseph
Eisner, Milton Philip
Fechter, Alan Edward
Finn, Edward J
Frair, Karen Lee
Goldstein, Jeffrey Jay
Hassan, Aftab Syed
Kirwan, Donald Frazier
Lustig, Harry
Malcom, Shirley Mahaley
Mog, David Michael
Raizen, Senta Amon
Robinowitz, Carolyn Bauer
Rose, Raymond Edward
Rosenberg, Edith E
Scheel, Nivard
Voytuk, James A
Weintraub, Herbert D

FLORIDA
Ballard, Stanley Sumner
Bolte, John R
Bueche, Frederick Joseph
Burkman, Ernest
Busse, Robert Franklyn
Chase, Christine Davis
Cintron, Guillermo B
Clark, John F
Cohen, Howard Lionel
Cooke, William Joe
Davis, Jefferson Clark, Jr
DeLap, James Harve
Dudley, Frank Mayo
Elliott, Paul Russell
Faber, Shepard Mazor
Fausett, Laurene van Camp
Frank, Stanley
Fuller, Harold Wayne
Gilmer, Penny Jane

Gottesman, Stephen T
Grobman, Arnold Brams
Harris, Lee Errol
Huebner, Jay Stanley
Humphries, Jack Thomas
Jungbauer, Mary Ann
Katz, Albert Barry
Kelly, Richard Delmer
Kromhout, Robert Andrew
Leitner, Alfred
Lewis, Brian Kreglow
Lillien, Irving
Mellon, Edward Knox, Jr
Muller-Karger, Frank Edgar
Peters, Thomas G
Piquette, Jean C
Posner, Gerald Seymour
Pratt, Melanie M
Reynolds, George William, Jr
Riban, David Michael
Rowe, Mary Budd
Rumbach, William Ervin
Sanberg, Paul Ronald
Simring, Marvin
Solon, Leonard Raymond
Spencer, John Lawrence
Stewart, Harris Bates, Jr
Stewart, Herbert
Tucker, Gail Susan
Vergenz, Robert Allan
Waife, Sholom Omi
Weatherly, Georges Lloyd
West, Felicia Emminger
Wiebush, Joseph Roy
Yost, Richard A
Young, Frank Glynn

GEORGIA
Aust, Catherine Cowan
Barreras, Raymond Joseph
Bass, William Thomas
Blanchet, Waldo W E
Brieske, Thomas John
Butts, David
Carter, Eloise B
Fay, Alice D Awtrey
Garmon, Lucille Burnett
Grace, Donald J
Greenberg, Jerrold
Hamrick, Anna Katherine Barr
Hayes, John Thompson
Hayes, Willis B
Hunt, Gary W
Jones, Robert Leroy
Kilpatrick, Jeremy
Koballa, Thomas Raymond, Jr
Loomis, Earl Alfred, Jr
Marks, Dennis William
Murray, Joan Baird
Nosek, Thomas Michael
Price, Edward Warren
Reese, Andy Clare
Riggsby, Ernest Duward
Shrum, John W
Shutze, John V
Skypek, Dora Helen
Stokes, Jimmy Cleveland
Wepfer, William J

HAWAII
Brantley, Lee Reed
Demanche, Edna Louise
Henry, Joseph Patrick
Malmstadt, Howard Vincent

IDAHO
Bitterwolf, Thomas Edwin
Dykstra, Dewey Irwin, Jr
Jobe, Lowell A(rthur)
Mech, William Paul
Ulliman, Joseph James

ILLINOIS
Alpert, Daniel
Aprison, Barry Steven
Averill, Frank Wallace
Bond, Charles Eugene
Bowe, Joseph Charles
Brezinski, Darlene Rita
Burger, Ambrose William
Burnstein, Ray A
Carlson, Eric Dungan
Casella, Alexander Joseph
Clemans, Kermit Grover
Curtis, Veronica Anne
Dietz, Mark Louis
Dornhoff, Larry Lee
Egan, Richard L
Fast, Dale Eugene
Fisher, Ben
Forman, G Lawrence
Frank, Forrest Jay
Gamelli, Richard L
Garoian, George
George, William Leo, Jr
Goebel, Maristella
Greenberg, Elliott
Greenberg, Stephen Robert
Greene, John Philip
Hahn, Donald Richard
Halberstam, Heini
Hutchcroft, Alan Charles
Jaffe, Philip Monlane
Josefson, Clarence Martin

Lederman, Leon Max
Lennox, Arlene Judith
Lerman, Zafra Margolin
Lewin, Seymour Z
Lietz, Gerard Paul
Linde, Ronald K(eith)
Malamud, Ernest I(lya)
Mateja, John Frederick
Michael, Joel Allen
Mimnaugh, Michael Neil
Montgomery, Carla Westlund
Munyer, Edward Arnold
O'Donnell, Terence J
Perkins, Alfred J
Peters, James Empson
Puchalski, Randall Francis
Rakoff, Henry
Rappaport, David
Ren, Shang-Fen
Rosen, Sidney
Sawinski, Vincent John
Scheeline, Alexander
Schillinger, Edwin Joseph
Schmidt, Arthur Gerard
Schroeer, Juergen Max
Simmons, William Howard
Smith, Leslie Garrett
Solomon, Lawrence Marvin
Spivey, Bruce Eldon
Sweitzer, James Stuart
Treadway, William Jack, Jr
Uretsky, Jack Leon
Verderber, Nadine Lucille
Voelker, Alan Morris
Wagner, Gerald C
Wagreich, Philip Donald
Wessler, Max Alden
Wirszup, Izaak
Wolff, Robert John
Wolter, Kirk Marcus

INDIANA
Abegg, Carl F(rank)
Alton, Elaine Vivian
Baker, David Thomas
Bauer, Dietrich Charles
Beckman, Jean Catherine
Beery, Dwight Beecher
Bodner, George Michael
Brooks, Austin Edward
Bumpus, John Arthur
Caskey, Jerry Allan
Cunningham, Ellen M
Foos, Kenneth Michael
Grabowski, Sandra Reynolds
Grimley, Robert Thomas
Hathaway, David Roger
Hendrix, Jon Richard
Hoefer, Raymond H
Huffman, John Curtis
Humberd, Jesse David
Hutton, Lucreda Ann
Lorentzen, Keith Eden
McBride, Ralph Book
Machtinger, Lawrence Arnold
Mertens, Thomas Robert
Neie, Van E(lRoy)
Olmer, Catherine
Olson, John Bennet
Ovens, William George
Pohl, Victoria Mary
Postlethwait, Samuel Noel
Robinson, William Robert
Rothe, Carl Frederick
Spafford, Eugene Howard
Speece, Susan Phillips
Tacker, Willis Arnold, Jr
Uhlhorn, Kenneth W
Weinrich, Alan Jeffrey
Werking, Robert Junior
Widener, Edward Ladd, Sr
Wyma, Richard J
Zimmerman, John F

IOWA
Bowen, George Hamilton, Jr
Cater, Earle David
Dickson, James Sparrow
Hutton, Wilbert, Jr
Kelly, William H
Knaphus, George
Malone, Diana
Meints, Clifford Leroy
Rila, Charles Clinton
Schafer, John William, Jr
Speckard, David Carl
Watkins, Stanley Read
Yager, Robert Eugene

KANSAS
Albrecht, Mary Lewnes
Coates, Gary Joseph
Coles, Anna Bailey
Coltharp, Forrest Lee
Daniel, Thomas Bruce
Hanley, Wayne Stewart
Hansen, Merle Fredrick
Lanman, Robert Charles
Lee, Joe
Miller, Glendon Richard
Mills, Russell Clarence
Parrish, John Wesley, Jr
Pierson, David W

Urban, James Edward
Witten, Gerald Lee
Zollman, Dean Alvin

KENTUCKY
Bailey, Maurice Eugene
Blackburn, John Robert
Boggess, Gary Wade
Cooke, Samuel Leonard
Crawford, Thomas H
Davidson, John Edwin
Early, Joseph E
Engelberg, Joseph
Herron, James Dudley
Hoffelder, Ann McIntosh
Holler, Floyd James
Hunter, Norman W
Hussung, Karl Frederick
James, Mary Frances
Pearson, Earl F
Saxe, Stanley Richard
Shupe, Dean Stanley
Taylor, Morris D
Teller, David Norton
Tobin, Gordon Ross
Vittetoe, Marie Clare
Zimmerman, Thom J

LOUISIANA
Adams, Sam
Datta, Ratna
Grenchik, Raymond Thomas
Huggett, Richard William, Jr
Ledford, Thomas Howard
Moore, Thomas Stephen, Jr
Morris, Everett Franklin
Moulder, Peter Vincent, Jr
Norman, Edward Cobb
Pesson, Lynn L
Schiebout, Judith Ann
Zander, Arlen Ray

MAINE
Chonacky, Norman J
Comins, Neil Francis
Johnson, Robert Wells
Labov, Jay Brian

MARYLAND
Achor, William Thomas
Anand, Davinder K
Arem, Joel Edward
Baker, Carl Gwin
Benevento, Louis Anthony
Bishop, Walton B
Bond, Howard Emerson
Boyd, Derek Ashley
Boyd, Virginia Ann Lewis
Bunyan, Ellen Lackey Spotz
Calinger, Ronald Steve
Carver, Gary Paul
Chaisson, Eric Joseph
Colvin, Burton Houston
Farrell, Robert Edmund, Jr
Gause, Evelyn Pauline
Gillespie, Walter Lee
Grim, Samuel Oram
Gross, James Harrison
Hoster, Donald Paul
Karlander, Edward P
Kaufman, Samuel
Kelley, Russell Victor
Koch, William Frederick
Lankford, J(ohn) L(lewellyn)
Levitan, Herbert
Lockard, J David
MacDonald, William
Majkrzak, Charles Francis
Margolis, Simeon
Matyas, Marsha Lakes
Mindel, Joseph
Misner, Charles William
Mogil, H Michael
Molenda, John R
Mulligan, Joseph Francis
Murray, Laura
Myslinski, Norbert Raymond
Naibert, Zane Elvin
Norvell, John Charles
Obnibene, Frederick Peter
Odell, Lois Dorothea
Paddack, Stephen J(oseph)
Pancella, John Raymond
Potocki, Kenneth Anthony
Redish, Edward Frederick
Schwartz, Brian B
Smith, Douglas Lee
Snow, George Abraham
Sweeting, Linda Marie
Tepper, Morris
Toll, John Sampson
Williams, Conrad Malcolm
Wortman, Bernard
Yarbrough, Arthur C, Jr
Young, Frank Coleman

MASSACHUSETTS
Adams, David Lawrence
Baker, Adolph
Baldwin, William Russell
Berka, Ladislav Henry
Bernfield, Merton Ronald
Bernstein, Herbert J
Blatt, S Leslie

OTHER PROFESSIONAL FIELDS / 585

Borgatti, Alfred Lawrence
Brauner, Phyllis Ambler
Broitman, Selwyn Arthur
Bruno, Merle Sanford
Busza, Wit
Camougis, George
Carr, George Leroy
Caslavska, Vera Barbara
Cavallaro, Mary Caroline
Cheek, Dennis William
Chew, Frances Sze-Ling
Chivian, Eric Seth
Chou, Kuo-Chih
Ciappenelli, Donald John
Clapp, Richard W
Collins, Allan Meakin
Cook, LeRoy Franklin, Jr
Curran, David James
Dechene, Lucy Irene
Dennison, Byron Lee
Eisenberg, Murray
Farrington, John William
Figueras, John
Furbish, Francis Scott
Goldhor, Susan
Gregory, Bruce Nicholas
Haber-Schaim, Uri
Handelman, Eileen T
Hausman, Robert Edward
Hobey, William David
Horwitz, Paul
Israel, Stanley C
Jankowski, Conrad M
Jarret, Ronald Marcel
Johnson, Corinne Lessig
Joni, Saj-nicole A
Kamien, Ethel N
King, John Gordon
King, Norval William, Jr
Kinoshita, Kay
Kosersky, Donald Saadia
Li, Yao Tzu
Lightman, Alan Paige
Lockwood, Linda Gail
Lomon, Earle Leonard
Luther, Holger Martin
Mehta, Mahesh J
Merken, Melvin
Milburn, Richard Henry
Miller, Judith Evelyn
Morris, Robert
Moser, William Ray
Mumford, George Saltonstall
Olivo, Richard Francis
Olsen, Robert Thorvald
Paquette, Gerard Arthur
Parker, Henry Seabury, III
Pasachoff, Jay M(yron)
Paul, Igor
Peura, Robert Allan
Romey, William Dowden
Rosner, Anthony Leopold
Sauro, Joseph Pio
Schwartz, Judah Leon
Servi, I(talo) S(olomon)
Shuchat, Alan Howard
Simpson, Rae Goodell
Steinberg, Melvin Sanford
Stinchfield, Carleton Paul
Touger, Jerold Steven
Tuler, Floyd Robert
Wallace, Robert William
Watson, Fletcher Guard
Weininger, Stephen Joel
Wertz, Dorothy C
White, Richard Edward
Wilson, Philo Calhoun
Winter, Stephen Samuel
Wolfson, Adele Judith
Wright, Mary Lou

MICHIGAN
Atkinson, James William
Barney, Christopher Carroll
Becher, William D(on)
Blinn, Lorena Virginia
Blinn, Walter Craig
Brault, Margaret A
Breedlove, Charles B
Callewaert, Denis Marc
Cox, Mary E
Crane, Horace Richard
Crowther, C Richard
Czarnik, Anthony William
Douglass, Claudia Beth
Doyle, Daryl Joseph
Edgerton, Robert Howard
Eliezer, Naomi
Evans, B J
Haidler, William B(ernard)
Han, Seong S
Horvath, Ralph S(teve)
Jacoby, Ronald Lee
Jones, Phillip Sanford
Kamrin, Michael Arnold
Keinath, Steven Ernest
Kirschner, Stanley
Kruglak, Haym
Lewis, Ralph William
Lopushinsky, Theodore
McCarville, Michael Edward
Mallinson, George Greisen
Medlin, Julie Anne Jones
Menard, Albert Robert, III

Miller, William Anton
Motiff, James P
Murchison, Pamela W
Parks, Terry Everett
Poel, Robert Herman
Ramsay, Ogden Bertrand
Ray, Jesse Paul
Saperstein, Alvin Martin
Schneider, Michael J
Shappirio, David Gordon
Signell, Peter Stuart
Sleight, Stuart Duane
Snitgen, Donald Albert
Soper, Jon Allen
Spink, Gordon Clayton
Stowell, Ewell Addison
Tavano, Donald C
Vance, Irvin Elmer
Volz, Paul Albert
Weinshank, Donald Jerome
Weymouth, Patricia Perkins
Wright, Wayne Mitchell
Zitzewitz, Paul William

MINNESOTA
Bahn, Robert Carlton
Baker, Robert Charles
Barks, Paul Allan
Caskey, Michael Conway
Johnson, Louise H
Kerr, Norman Story
Knodel, Raymond Willard
Lawrenz, Frances Patricia
Lea, Wayne Adair
Leopold, Doreen Geller
Magnuson, Vincent Richard
Meeks, Benjamin Spencer, Jr
Mooers, Howard T(heodore)
Murdock, Gordon Robert
Noer, Richard Juul
Samuelson, Charles R
Schwartz, A(lbert) Truman
Steen, Lynn Arthur
Stein, Paul John
Welch, Wayne Willard

MISSISSIPPI
Angel, Charles
Bedenbaugh, Angela Lea Owen
Dunn, Dean A(lan)
Irby, Bobby Newell
Rayborn, Grayson Hanks
Rockhold, Robin William

MISSOURI
Baumann, Thiema Wolf
Berkland, Terrill Raymond
Bieniek, Ronald James
Brodmann, John Milton
Crosswhite, F Joe
De Buhr, Larry Eugene
Einhellig, Frank Arnold
Festa, Roger Reginald
Geilker, Charles Don
Hart, Richard Allen
Hill, Shirley Ann
Holloway, Thomas Thornton
Holmes, Neal Jay
Holroyd, Louis Vincent
Johannsen, Frederick Richard
Johnson, James W(inston)
Katti, Kattesh V
Levy, Ram Leon
Linder, Solomon Leon
McCollum, Clifford Glenn
Magruder, Willis Jackson
Moffatt, David John
Northrip, John Willard
Oshima, Eugene Akio
Overman, Ralph Theodore
Oviatt, Charles Dixon
Readnour, Jerry Michael
Ryan, Carl Ray
Santiago, Julio Victor
Schaefer, Joseph Thomas
Smith, Robert Eugene
Snoble, Joseph Jerry
Tsoulfanidis, Nicholas
Twining, Linda Carol
Ucko, David A
Wehrmann, Ralph F(rederick)
Whitten, Elmer Hammond
Wrather, James Allen

MONTANA
Amend, John Robert
Behan, Mark Joseph
Hess, Adrien LeRoy
Lott, Johnny Warren
McRae, Robert James
Parker, Keith Krom
Rice, Richard Eugene

NEBRASKA
Bahar, Ezekiel
Carr, James David
Francis, Charles Andrew
Fuller, Robert Gohl
Ikenberry, Richard W
Kenkel, John V
Mason, Stephen Carl
Provencher, Gerald Martin
Roark, James L
Sachtleben, Clyde Clinton

Thorson, James A
Wittke, Dayton D
Zebolsky, Donald Michael

NEVADA
Clements, Linda L
Davis, Phillip Burton

NEW HAMPSHIRE
Bilello, Michael Anthony
Blanchard, Robert Osborn
Bohm, Howard A
Fagerberg, Wayne Robert
Kidder, John Newell
Lahr, Charles Dwight
Simon, Edward
Treat, Donald Fackler

NEW JERSEY
Albin, Susan Lee
Arnold, Frederic G
Bergen, Catharine Mary
Blanck, Andrew R(ichard)
Bonacci, John C
Browne, Robert Glenn
Carlin, Ronald D
Cho, Soung Moo
Citron, Irvin Meyer
Cody, George Dewey
Colicelli, Elena Jeanmarie
Eisinger, Robert Peter
Essien, Francine B
Friedrich, Benjamin C
Frost, David
Gardiner, Lion Frederick
Gautreau, Ronald
Goldin, Gerald Alan
Goldstein, Martin
Griffin, William Dallas
Halpern, Teodoro
Hennings, George
Hoffman, Dorothea Heyl
Holt, Rush D(ew), Jr
Huber, Richard V
Hundert, Irwin
Hutcheon, Duncan Elliot
Jabbar, Gina Marie
Kingery, Bernard Troy
Klapper, Jacob
Kowalski, Stephen Wesley
Krishnamurthy, Subrahmanya
Kristol, David Sol
Krueger, Roger Warren
Landwehr, James M
Levi, Barbara Goss
Lindenfeld, Peter
McKay, Donald Edward
Mackway-Girardi, Ann Marie
Makofske, William Joseph
Malkin, Robert Allen
Marshall, Keith
Most, Joseph Morris
Navarra, John Gabriel
Paul, Edward W
Pickering, Miles Gilbert
Polk, Conrad Joseph
Posamentier, Alfred S
Pregger, Fred Titus
Ravindra, Nuggehalli Muthanna
Redden, Patricia Ann
Reynolds, Richard Clyde
Robin, Michael
Rosan, Alan Mark
Rosenthal, Judith Wolder
Rosenthal, Lee
Rouse, Robert S
Rulf, Benjamin
Schofield, Robert Edwin
Sharon (Schwadron), Yitzhak Yaakov
Sinclair, J Cameron
Sivco, Deborah L
Sutman, Frank X
Tanner, Daniel
Taylor, Harold Evans
Wangemann, Robert Theodore
Woolf, Harry

NEW MEXICO
Bleyl, Robert Lingren
Bowsher, Arthur LeRoy
Brabson, George Dana, Jr
Bravo, Justo Baladjay
Cheavens, Thomas Henry
Coburn, Horace Hunter
Cochran, Paul Terry
Duncan, Irma W
Foster, Duncan Graham, Jr
Giovanielli, Damon V
Groblewski, Gerald Eugene
Harrill, Robert W
Kidd, David Eugene
Knief, Ronald Allen
Manley, Kim
Miller, Edmund K(enneth)
Newby, Neal Dow, Jr
Price, Richard Marcus
Rayson, Gary D
Reilly, James Patrick
Shannon, Spencer Sweet, Jr
Woodfin, Beulah Marie
Yarger, Frederick Lynn
Zeilik, Michael

NEW YORK
Amato, Joseph Cono
Anderson, O(rvil) Roger
Bartilucci, Andrew J
Beal, Myron Clarence
Blumenthal, Ralph Herbert
Boynton, John E
Brady, James Edward
Brandwein, Paul Franz
Braun, Ludwig
Brink, Frank, Jr
Bunce, Stanley Chalmers
Burge, Robert Ernest, Jr
Carroll, Harvey Franklin
Cluxton, David H
Cohn, Deirdre Arline
Contento, Isobel
Cote, Wilfred Arthur, Jr
Cummins, John Frances
Darrow, Frank William
Donaldson, Robert Rymal
Druger, Marvin
Eppenstein, Walter
Farb, Edith H
Farrell, Margaret Alice
Feldman, Lawrence
Ferguson, David Lawrence
Fleishman, Bernard Abraham
Freedman, Aaron David
Geer, Ira W
Gering, Robert Lee
Gilmore, Arthur W
Greisen, Kenneth I
Greller, Andrew M
Gross, Jonathan Light
Grover, Paul L, Jr
Grunwald, Hubert Peter
Harke, Douglas J
Henderson, David Wilson
Henkel, Elmer Thomas
Henshaw, Clement Long
Hilborn, David Alan
Holbrow, Charles H
Horowitz, Marilyn Stephens
Horvat, Robert Emil
Hudson, John B(alch)
Irwin, Lynne Howard
Jackson, Kenneth William
Jacob, Klaus H
Jacobson, Willard James
Jaffe, Marvin Richard
Kaplan, Eugene Herbert
Katcoff, Seymour
Kingslake, Rudolf
Klir, George Jiri
Knee, David Isaac
Knox, Robert Seiple
Kramer, Alfred William, Jr
Labianca, Dominick A
Lane, Lewis Behr
Law, David Martin
Lawson, Kent DeLance
Lehrer, Gerard Michael
Lenchner, Nathaniel Herbert
Lindberg, Vern Wilton
Lindstrom, Gary J
Linkow, Leonard I
Lipsey, Sally Irene
McClain, John William
McGrath, John F
Mackay, Lottie Elizabeth Bohm
Madden, Robert E
Marino, Robert Anthony
Mayers, George Louis
Milmore, John Edward
Month, Melvin
Morrison, John Agnew
Murphy, James Joseph
Novak, Joseph Donald
Olczak, Paul Vincent
Orear, Jay
Osborn, H(arland) James
Paldy, Lester George
Parsegian, V(ozcan) Lawrence
Perry, J Warren
Petersen, Ingo Hans
Pierce, Carol S
Postman, Robert Derek
Prener, Robert
Raffensperger, Edgar M
Rapaport, William Joseph
Reed, George Farrell
Resnick, Robert
Reuss, Ronald Merl
Rich, Marvin R
Robinson, David Zav
Rockcastle, Verne Norton
Romer, Alfred
Rosenberg, Robert
Rosenberg, Warren L
Rousseau, Viateur
Russell, Charlotte Sananes
Sauer, Charles William
Scaife, Charles Walter John
Seago, James Lynn, Jr
Segal, Sanford Leonard
Shannon, Jerry A, Jr
Sharkey, John Bernard
Shilepsky, Arnold Charles
Siebert, Karl Joseph
Siemankowski, Francis Theodore
Siew, Ernest L
Sipe, Harry Craig
Smith, Alden Ernest

Science Education (cont)

Spelke, Elizabeth Shilin
Spielman, Harold S
Springer, Dwight Sylvan
Stanionis, Victor Adam
Stannard, Carl R, Jr
Stenzel, Wolfram G
Stephens, Lawrence James
Strassenburg, Arnold Adolph
Talley, Charles Peter
Tannenbaum, Harold E
Tanzer, Charles
Thorpe, John A(lden)
Thygesen, Kenneth Helmer
Tooney, Nancy Marion
Trimble, Robert Bogue
Vacirca, Salvatore John
Wald, Francine Joy Weintraub
Waltzer, Wayne C
Werner, Thomas Clyde
Whang, Sung H
Whittingham, M(ichael) Stanley
Wieder, Grace Marilyn
Wilson, Jack Martin
Wulff, Wolfgang
Wynter, Carlton Elleston, Jr
Zingaro, Joseph S

NORTH CAROLINA
Adelberger, Rexford E
Ayers, Caroline LeRoy
Bailey, Donald Etheridge
Beeler, Joe R, Jr
Blumenkopf, Todd Andrew
Boyers, Albert Sage
Bright, George Walter
Brotak, Edward Allen
Burford, Hugh Jonathan
Cain, Laurence Sutherland
Carroll, Felix Alvin, Jr
Clemens, Donald Faull
Dickens, Charles Henderson
Dough, Robert Lyle, Sr
Farrier, Maurice Hugh
Fyfe, James Arthur
Gains, Lawrence Howard
Glenn, Thomas M
Groszos, Stephen Joseph
Gupta, Ajaya Kumar
Hadzija, Bozena Wesley
Hartford, Winslow H
Hawkins, David Rollo, Sr
Herrell, Astor Y
Hoelzel, Charles Bernard
Hugus, Z Zimmerman, Jr
Hulcher, Frank H
Iafrate, Gerald Joseph
Keepler, Manuel
Kuenzler, Edward Julian
Levine, Solomon Leon
Lytle, Charles Franklin
Mattheis, Floyd E
Mayer, Eugene Stephen
Meyer, John Richard
Oberhofer, Edward Samuel
Perfetti, Patricia F
Petersen, Karl Endel
Read, Floyd M
Rees, John
Risley, John Stetler
Robinson, Gertrude Edith
Robinson, Kent
Rowlett, Russell Johnston, III
Russell, Henry Franklin
Saeman, W(alter) C(arl)
Sams, Emmett Sprinkle
Sandor, George N(ason)
Sheppard, Moses Maurice
Simon, Sheridan Alan
Smith, Fred R, Jr
Tyndall, Jesse Parker
Werntz, James Herbert, Jr
Willett, Hilda Pope

NORTH DAKOTA
Saari, Jack Theodore
Scoby, Donald Ray

OHIO
Barnes, H Verdain
Bernstein, Stanley Carl
Bidelman, William Pendry
Burrows, Kerilyn Christine
Carothers, Donna June
Carter, Carolyn Sue
Caughey, John Lyon, Jr
Coffin, Frances Dunkle
Cohen, Irwin
Cook, Lois
Damian, Carol G
Deal, Don Robert
DeVillez, Edward Joseph
Disinger, John Franklin
Dollhopf, William Edward
Fenner, Peter
Fortner, Rosanne White
Gamble, Francis Trevor
Gibbins, Betty Jane
Gonzalez, Paula
Gorse, Joseph
Gwinn, John Frederick
Hagelberg, M(yron) Paul
Hamed, Awatef A
Hauser, Edward J P
Heinke, Clarence Henry
Hern, Thomas Albert
Hill, James Stewart
Hoagstrom, Carl William
Hollis, William Frederick
Jacobs, Donald Thomas
Jaworowski, Andrzej Edward
John, George
Kagen, Herbert Paul
Kreidler, Eric Russell
Kritsky, Gene Ralph
Lamont, Gary Byron
Laughlin, Ethelreda R
Leonard, Billie Charles
Lichstein, Herman Carlton
Lutz, Arthur Leroy
McKenzie, Garry Donald
Mayer, Victor James
Meleca, C Benjamin
Miller, Franklin, Jr
Mulroy, Juliana Catherine
Murray, Thomas Henry
Pappas, Leonard Gust
Ploughe, William D
Pribor, Donald B
Priest, Joseph Roger
Proctor, David George
Richards, Walter Bruce
Ritchie, Austin E
Rogers, Charles Edwin
Saffran, Murray
Saul, Frank Philip
Schaff, John Franklin
Schmidt, Mark Thomas
Shanklin, James Robert, Jr
Sommer, John G
Stickney, Alan Craig
Stith, James Herman
Taylor, Beverley Ann Price
Taylor, Charles Emery
Taylor, John Langdon, Jr
Utgard, Russell Oliver
Wallis, Robert L
Weed, Herman Roscoe
Westneat, David French
Williams, John Paul
Wismar, Beth Louise
Wolfe, Paul Jay
Wyman, Bostwick Frampton

OKLAHOMA
Cassens, Patrick
Choike, James Richard
Conway, Kenneth Edward
Crane, Charles Russell
Eddington, Carl Lee
Graves, Victoria
Kimpel, James Froome
Koh, Eunsook Tak
Marek, Edmund Anthony
Martin, Loren Gene
Menzie, Donald E
Nalley, Elizabeth Ann
Owens, G(lenda) Kay
Renner, John Wilson
Richardson, Verlin Homer
Rutledge, Carl Thomas
Shmaefsky, Brian Robert
Stratton, Charles Abner
Weakley, Martin LeRoy
Woolever, Patricia
Yoesting, Clarence C

OREGON
Allen, John Eliot
Bandick, Neal Raymond
Bufford, Rodger K
Clark, Mary Eleanor
Clark, William Melvin, Jr
Cummins, Ernie Lee
Decker, Fred William
Griffith, William Thomas
Halko, Barbara Tomlonovic
Lendaris, George G(regory)
Long, James William
Matthes, Steven Allen
Moursund, David G
Overley, Jack Castle
Pizzimenti, John Joseph
Pursglove, Laurence Albert
Towe, George Coffin
Wick, William Quentin

PENNSYLVANIA
Alteveer, Robert Jan George
Bates, Thomas Fulcher
Benfey, Otto Theodor
Benson, Brent W
Berty, Jozsef M
Bilaniuk, Oleksa-Myron
Bortz, Alfred Benjamin
Botelho, Stella Yates
Brendlinger, Darwin
Brennan, Thomas Michael
Brungraber, Robert J
Burness, James Hubert
Byler, David Michael
Cahir, John Joseph
Carbonell, Jaime Guillermo
Carfagno, Salvatore P
Couch, Jack Gary
Diaz, Mario Claudio
Diehl, Renee Denise
Dubeck, Leroy W
Felty, Wayne Lee
Fowler, H(oratio) Seymour
Fox, John Nicholas
Gainer, Michael Kizinski
Glaser, Robert
Godfrey, Susan Sturgis
Goff, Christopher Godfrey
Gold, Lewis Peter
Goldstein, Frederick J
Gollub, Jerry Paul
Gottlieb, Irvin M
Greenberg, Herman Samuel
Grobstein, Paul
Gruenwald, Geza
Helfferich, Friedrich G
Henderson, Robert E
Hershey, Nathan
Ho, Thomas Inn Min
Hostetler, Robert Paul
Johnston, Gordon Robert
Kang, Joohee
Kerstein, Morris D
Khan, Parwaiz Ashraf Ali
Kissick, William Lee
Knaster, Tatyana
Kuserk, Frank Thomas
Larson, Russell Edward
Laws, Priscilla Watson
Lerner, Leonard Joseph
Libby, Carol Baker
Lipsky, Stephen E
Lokay, Joseph Donald
Melvin, John L
Milakofsky, Louis
Monroe, James Lewis
Moon, Thomas Charles
Moser, Gene Wendell
Muldawer, Leonard
Mullins, Jeanette Somerville
Murphy, Clarence John
Owens, Frederick Hammann
Pauline, Ronald Frank
Pitkin, Ruthanne B
Rhodes, Jacob Lester
Rothman, Milton A
Sagik, Bernard Phillip
Schattschneider, Doris Jean
Schearer, William Richard
Schmuckler, Joseph S
Shaffer, Russell Allen
Sherwood, Bruce Arne
Shoemaker, Richard Nelson
Shrigley, Robert Leroy
Silver, Edward A
Snyder, Evan Samuel
Sommer, Holger Thomas
Stetson, Harold W(ilbur)
Stinson, Richard Floyd
Thompson, Julia Ann
Torop, William
Twiest, Gilbert Lee
Veening, Hans
Vick, Gerald Kieth
Ward, Laird Gordon Lindsay
Waring, Robert Kerr, Jr
Weedman, Daniel Wilson
Welliver, Paul Wesley
Williamson, Hugh A
Wilson, Geoffrey Leonard
Wismer, Robert Kingsley
Young, Hugh David

RHODE ISLAND
Barnes, Frederick Walter, Jr
Fasching, James Le Roy
Hazeltine, Barrett
Heywood, Peter
Pickart, Stanley Joseph

SOUTH CAROLINA
Allen, Joe Frank
Alley, Thomas Robertson
Asleson, Gary Lee
Blevins, Maurice Everett
Carpenter, John Richard
Cheng, Kenneth Tat-Chiu
Darnell, W(illiam) H(eaden)
Dille, John Emmanuel
Gurr, Henry S
Hahs, Sharon K
Helms, Carl Wilbert
Hendrick, Lynn Denson
Kosinski, Robert Joseph
Longtin, Bruce
Montgomery, Terry Gray
Nerbun, Robert Charles, Jr
Pike, LeRoy
Roldan, Luis Gonzalez
Scott, Jaunita Simons
Sweeney, John Robert

SOUTH DAKOTA
Draeger, William Charles
Hilderbrand, David Curtis
Landborg, Richard John
Riemenschneider, Albert Louis

TENNESSEE
Armistead, Willis William
Ayers, Jerry Bart
Ball, Mary Uhrich
Ball, Raiford Mill
Baxter, John Edwards
Davidson, Robert C
Davis, Phillip Howard
Doody, John Edward
Driskill, William David
Edwards, Thomas F
Elowe, Louis N
Gee, Charles William
Greenberg, Neil
Herring, Carey Reuben
Kirksey, Howard Graden, Jr
Kliewer, Kenneth L
Kopp, Otto Charles
Kozub, Raymond Lee
Lane, Eric Trent
Loebbaka, David S
Lorenz, Philip Jack, Jr
Lowndes, Douglas H, Jr
Lyman, Beverly Ann
McGhee, Charles Robert
McNeely, Robert Lewis
Malasri, Siripong
Mathis, Philip Monroe
Mayfield, Melburn Ross
Moore, George Edward
Pafford, William N
Pegg, David John
Prados, John W(illiam)
Scott, Dan Dryden
Terry, Fred Herbert
Vo-Dinh, Tuan
Wilson, James Larry

TEXAS
Archer, Cass L
Bachmann, John Henry, Jr
Baine, William Brennan
Ball, Millicent (Penny)
Banerjee, Sanjay Kumar
Benedict, George Frederick
Bennett, Lloyd M
Blystone, Robert Vernon
Bossen, Douglas C
Bryan, George Thomas
Bullard, Truman Robert
Cargill, Robert Lee, Jr
Cody, John T
Craik, Eva Lee
Crawford, Gladys P
DeBakey, Lois
DeBakey, Selma
Duncombe, Raynor Lockwood
Dvoretzky, Isaac
Elliott, Jerry Chris
Funkhouser, Edward Allen
Hademenos, James George
Hassell, Clinton Alton
Hemenway, Mary-Kay Meacham
Hernandez, Norma
Hewett, Lionel Donnell
Hoage, Terrell Rudolph
Howe, Richard Samuel
Huebner, Walter F
Kasi, Leela Peshkar
Kesler, Oren Byrl
Krohmer, Jack Stewart
Landolt, Robert George
Lee, Addison Earl
Levy, Robert Aaron
Lindstrom, Marilyn Martin
Little, Robert Narvaez, Jr
Longacre, Susan Ann Burton
McBride, Raymond Andrew
Marcotte, Ronald Edward
Maxum, Bernard J
Melton, Lynn Ayres
Mize, Charles Edward
Moy, Mamie Wong
Nelson, John Franklin
Paske, William Charles
Pedigo, Robert Daryl
Pomerantz, James Robert
Quarles, Carroll Adair, Jr
Retzer, Kenneth Albert
Richardson, Richard Harvey
Robbins, Ralph Robert
Roebuck, Isaac Field
Rudenberg, Frank Hermann
Schimelpfenig, Clarence William
Schroeder, Melvin Carroll
Shumate, Kenneth McClellan
Stranges, Anthony Nicholas
Stubbs, Donald William
Summers, Richard Lee
Swartzlander, Earl Eugene, Jr
Sybert, James Ray
Tsin, Andrew Tsang Cheung
Vest, Floyd Russell
Wendel, Carlton Tyrus
Westmeyer, Paul
Whittlesey, John R B
Wood, Lowell Thomas

UTAH
Ailion, David Charles
Bergeson, Haven Eldred
Bunch, Kyle
Earley, Charles Willard
Gates, Henry Stillman
Grey, Gothard L
Hess, Joseph W, Jr
Lombardi, Paul Schoenfeld
Wilson, Byron J

VERMONT
Moyer, Walter Allen, Jr

VIRGINIA
Auerbach, Stephen Michael
Balster, Robert L(ouis)
Bauer, Henry Hermann
Benton, Duane Marshall
Bishop, Joseph Michael
Blair, Barbara Ann
Blatt, Jeremiah Lion
Cavanaugh, Margaret Anne
Chamot, Dennis
Davis, William Arthur
Decker, Robert Dean
Faulkner, Gavin John
Gough, Stephen Bradford
Hartline, Beverly Karplus
Hill, Carl McClellan
Horsburgh, Robert Laurie
Huddle, Benjamin Paul, Jr
Hudgins, Aubrey C, Jr
Hufford, Terry Lee
Ijaz, Lubna Razia
Jacobs, Kenneth Charles
Jones, Franklin M
Jones, Tappey Hughes
Jones, William F
Jordan, Wade H(ampton), Jr
Kernell, Robert Lee
Lamb, James C(hristian), III
Lilleleht, L(embit) U(no)
McCombs, Freda Siler
McEwen, Robert B
McFarlane, Kenneth Walter
Mahoney, Bernard Launcelot, Jr
Mandell, Alan
Maybury, Robert H
Mazzoni, Omar S
Morrell, William Egbert
Moyer, Wayne A
Myers, William Howard
Rosinski, Joanne
Rozzell, Thomas Clifton
Sigler, Julius Alfred, Jr
Slayden, Suzanne Weems
Smith, Elske van Panhuys
Smith, John Melvin
Spagna, George Fredric, Jr
Spresser, Diane Mar
Starnes, William Herbert, Jr
Stronach, Carey E
Suiter, Marilyn J
Sukow, Wayne William
Verell, Ruth Ann
Watson, Robert Francis
Wilber, Joe Casley, Jr
Willis, Lloyd L, II
Wiseman, Lawrence Linden

WASHINGTON
Armstrong, Frank C(larkson Felix)
Arons, Arnold Boris
Artru, Alan Arthur
Axtell, Darrell Dean
Barnard, Kathryn E
Benson, Keith Rodney
Brown, Willard Andrew
Cellarius, Richard Andrew
Clement, Stephen LeRoy
Crosby, Glenn Arthur
Daniels, Patricia D
Eustis, William Henry
Frasco, David Lee
Garland, John Kenneth
Gerhart, James Basil
Haight, Gilbert Pierce, Jr
Hall, Wayne Hawkins
Hardy, John Thomas
Heilenday, Frank W (Tod)
Hoisington, David B(oysen)
McDermott, Lillian Christie
McMichael, Kirk Dugald
Moore, Emmett Burris, Jr
Narayanan, A Sampath
Olstad, Roger Gale
Rehfield, David Michael
Richardson, Richard Laurel
Schmid, Loren Clark
Sheridan, John Roger
Shrader, John Stanley
Slesnick, Irwin Leonard
Smith, Sam Corry
Steckler, Bernard Michael
Szkody, Paula
Tabbutt, Frederick Dean
Wade, Leroy Grover, Jr
Waltar, Alan Edward
Whitmer, John Charles
Wilson, Archie Spencer

WEST VIRGINIA
Becker, Stanley Leonard
Dils, Robert James
Fezer, Karl Dietrich
Keller, Edward Clarence, Jr
Knorr, Thomas George
Malin, Howard Gerald
Matthews, Virgil Edison
Mufson, Maurice Albert
Richardson, Rayman Paul
Sperati, Carleton Angelo

WISCONSIN
Andrews, Oliver Augustus
Balsano, Joseph Silvio
Bast, Rose Ann
Burgess, Ann Baker
Chander, Jagdish
Dibben, Martyn James
Fitch, Robert McLellan
Gopal, Raj
Jeanmaire, Robert L
Johnson, J(ames) R(obert)
Jungck, John Richard
Koch, Rudy G
Lagally, Max Gunter
Lentz, Mark Steven
Montgomery, Robert Renwick
Moore, John Ward
O'Hearn, George Thomas
Patrick, Michael Heath
Peanasky, Robert Joseph
Pella, Milton Orville
Scheusner, Dale Lee
Schneiderwent, Myron Otto
Schroeder, Daniel John
Seeburger, George Harold
Smith, Roy E
Sommers, Raymond A
Southworth, Warren Hilbourne
Sprott, Julien Clinton
Stekel, Frank D
Stoever, Edward Carl, Jr
Waechter-Brulla, Daryle A
Walker, Duard Lee
Wright, Steven Martin

WYOMING
Laycock, William Anthony
Thronson, Harley Andrew, Jr

PUERTO RICO
McDowell, Dawson Clayborn
Ramos, Lillian
Rodriguez, Jose Raul

ALBERTA
Burwash, Ronald Allan
Hunt, Robert Nelson
Klassen, Rudolph Waldemar
Krueger, Peter J
McLeod, Lionel Everett
Martin, John Scott
Mash, Eric Jay
Plambeck, James Alan
Wieser, Helmut

BRITISH COLUMBIA
Bevier, Mary Lou
Godolphin, William
Lower, Stephen K
Paszner, Laszlo

MANITOBA
Carbno, William Clifford
Kao, Kwan Chi
LeBow, Michael David
Loly, Peter Douglas

NEW BRUNSWICK
Frantsi, Christopher
Whitla, William Alexander

NEWFOUNDLAND
Aldrich, Frederick Allen
Rayner-Canham, Geoffrey William
Rusted, Ian Edwin L H

NOVA SCOTIA
Borgese, Elisabeth Mann
Dunn, Kenneth Arthur
Wentzell, Peter Dale

ONTARIO
Anderson, Christine Michaela
Atkinson, George Francis
Buchanan, George Dale
Burton, John Heslop
Clarke, Thomas Roy
Davis, Harry Floyd, II
Gray, David Robert
Horton, Roger Francis
Jarzen, David MacArthur
Kendrick, Bryce
Key, Anthony W
Levere, Trevor Harvey
Mackay, Rosemary Joan
Montemurro, Donald Gilbert
Muller, Eric Rene
Nesheim, Michael Ernest
Nirdosh, Inderjit
Osmond, Daniel Harcourt
Parr, James Gordon
Percy, John Rees
Rangachari, Patangi Srinivasa Kumar
Rothman, Arthur I
Sandler, Samuel
Shurvell, Herbert Francis
Smith, Donald Alan
Stoicheff, Boris Peter
Tolnai, Susan
Ursino, Donald Joseph
Usselman, Melvyn Charles
VandenHazel, Bessel J
Vervoort, Gerardus
Waugh, Douglas Oliver William

Whippey, Patrick William
Wightman, Robert Harlan
Yan, Maxwell Menuhin

QUEBEC
Adamkiewicz, Vincent Witold
Alward, Ron E
Chagnon, Jean Yves
Cochrane, Robert W
Cook, Robert Douglas
Croctogino, Reinhold Hermann
Escher, Emanuel
Gurudata, Neville
Johns, Kenneth Charles
Marisi, Dan(iel) Quirinus
Pekau, Oscar A
Prichard, Roger Kingsley
Tomas, Francisco

SASKATCHEWAN
Larson, D Wayne
Rangacharyulu, Chilakamarri

OTHER COUNTRIES
Ahn, Tae In
Anantharaman, Tanjore Ramachandra
Cartwright, Hugh Manning
Chadwick, Nanette Elizabeth
Craig, John Frank
D'Ambrosio, Ubiratan
Davis, William Jackson
Fernandez-Cruz, Eduardo P
Fettweis, Alfred Leo Maria
Foecke, Harold Anthony
Gomez, Arturo
Lee, Joseph Chuen Kwun
Ling, Samuel Chen-Ying
McInerney, John Gerard
Matheny, Ellis Leroy, Jr
Moore, James Edward, Jr
Revillard, Jean-Pierre Remy
Rinehart, Frank Palmer
Robinson, Berol (Lee)
Scheiter, B Joseph Paul
Schwartz, Peter Larry
Stephens, Kenneth S
Suematsu, Yasuharu
Toshio, Makimoto
Unger, Hans-Georg

Science Policy

ALABAMA
Bridgers, William Frank

ARIZONA
Bartocha, Bodo
Bradley, Michael Douglas
Dworkin, Judith Marcia
Leibacher, John W
Swan, Cathy Wood
Szilagyi, Miklos Nicholas

ARKANSAS
Berghel, Hal L
Hobson, Arthur Stanley
Norris, Arthur Robert

CALIFORNIA
Akin, Gwynn Collins
Atkin, J Myron
Black, Robert L
Capron, Alexander Morgan
Chartock, Michael Andrew
Chow, Brian G(ee-Yin)
Craig, Paul Palmer
Dismukes, Robert Key
Djerassi, Carl
Gortner, Susan Reichert
Hardy, Edgar Erwin
Holdren, John Paul
Hollander, Jack M
Holtz, David
Howd, Robert A
Isherwood, Dana Joan
Kevles, Daniel Jerome
London, Ray William
Long, Franklin A
McKone, Thomas Edward
Orloff, Neil
Primack, Joel Robert
Rossin, A David
Rowell, John Martin
Schipper, Lee (Leon Jay)
Schneider, Stephen Henry
Shin, Ernest Eun-Ho
Shull, Harrison
Springer, Bernard G
Sutcliffe, William George
Swearengen, Jack Clayton
Swedler, Alan R
Szego, Peter A
Vajk, J(oseph) Peter
Whipple, Christopher George
York, Herbert Frank
Zeren, Richard William

COLORADO
Burns, Denver P
Byerly, Radford, Jr
Chamberlain, A(drian) R(amond)
Cohen, Robert
Giller, E(dward) B(onfoy)

McCray, Richard A
Rollin, Bernard Elliot
Thompson, Starley Lee
West, Anita

CONNECTICUT
Bloom, Barry Malcolm
Brooks, Douglas Lee
Corning, Mary Elizabeth
Fraser, J(ulius) T(homas)
Gillmor, Charles Stewart
Jackson, C(harles) Ian
Lubell, Michael S
Novick, Alvin
Reid, James Dolan

DELAWARE
Quisenberry, Richard Keith

DISTRICT OF COLUMBIA
Alic, John A
Apple, Martin Allen
Argus, Mary Frances
Baker, Louis Coombs Weller
Barber, Mary Combs
Beall, James Howard
Bingham, Robert Lodewijk
Blockstein, David Edward
Boyce, Peter Bradford
Burley, Gordon
Butter, Stephen Allan
Buzzelli, Donald Edward
Callaham, Michael Burks
Carey, William Daniel
Cassidy, Esther Christmas
Civiak, Robert L
Cole, Margaret Elizabeth
Decker, James Federick
Epstein, Gerald Lewis
Fainberg, Anthony
Fechter, Alan Edward
Florig, Henry Keith
Gaffney, Paul G, II
Good, Mary Lowe
Gough, Michael
Green, Rayna Diane
Hart, Richard Cullen
Hawkins, Gerald Stanley
Hemily, Philip Wright
Hershey, Robert Lewis
Hoagland, K Elaine
Houck, Frank Scanland
Hudis, Jerome
Jacobson, Michael F
Jhirad, David John
Keitt, George Wannamaker, Jr
Kelly, Henry Charles
Korkegi, Robert Hani
Lai, David Ying-lun
Lee, Alfred M
Logsdon, John Mortimer, III
Madden, Joseph Michael
Marcus, Michael Jay
Maxwell, Paul Charles
Mellor, John Williams
Menzel, Joerg H
Meserve, Richard A
Moore, John Arthur
Nadolney, Carlton H
Nightingale, Elena Ottolenghi
Novello, Antonia Coello
O'Fallon, John Robert
Offutt, Susan Elizabeth
Orlans, F Barbara
Parham, Walter Edward
Phillips, Don Irwin
Pikus, Irwin Mark
Policansky, David J
Powell, Justin Christopher
Putzrath, Resha Mae
Reisa, James Joseph, Jr
Riemer, Robert Lee
Rogers, Kenneth Cannicott
Setlow, Valerie Petit
Shrier, Adam Louis
Taylor, Ronald D
Topper, Leonard
Tucker, John Richard
Uman, Myron F
Urban, Edward Robert, Jr
Vernikos, Joan
Voytuk, James A
Weiss, Leonard
Werdel, Judith Ann
White, Irvin Linwood
Wolfle, Thomas Lee
Zawisza, Julie Anne A

FLORIDA
Ancker-Johnson, Betsy
Arnold, Henry Albert
Beckemeyer, Elizabeth Frances
Chartrand, Robert Lee
Harwell, Mark Alan
Huckaba, Charles Edwin
Lannutti, Joseph Edward
Mahadevan, Selvakumaran
Noonan, Norine Elizabeth
Suman, Daniel Oscar

GEORGIA
Boone, Donald Joe
Clifton, David S, Jr
Greenbaum, Lowell Marvin

588 / DISCIPLINE INDEX

Science Policy (cont)

Jackson, Curtis Rukes
Porter, Alan Leslie
Rossini, Frederick Anthony
Schutt, Paul Frederick

HAWAII
Craven, John P

IDAHO
Mills, James Ignatius

ILLINOIS
Alpert, Daniel
Amarose, Anthony Philip
Ban, Stephen Dennis
Bentley, Orville George
Bevan, William
Bullard, Clark W
Changnon, Stanley Alcide, Jr
Collier, Donald W(alter)
Cranston, Joseph William, Jr
Egan, Richard L
Fossum, Robert Merle
Gross, David Lee
Hubbard, Lincoln Beals
Huebner, Russell Henry, Sr
Kanter, Manuel Allen
LaFevers, James Ronald
Lehman, Dennis Dale
Lushbough, Channing Harden
Olshansky, S Jay
Papatheofanis, Frank John
Rabinowitch, Victor
Shen, Sin-Yan
Stafford, Fred E

INDIANA
Baumgardt, Billy Ray
Katz, Frances R
Leonard, Jack E
Ortman, Eldon Emil
Spafford, Eugene Howard

IOWA
Morris, Robert Gemmill
Snow, Joel A

KENTUCKY
Jain, Ravinder Kumar

MAINE
Fitch, John Henry

MARYLAND
Allgaier, Robert Stephen
Alter, Harvey
Anderson, William Averette
Bawden, Monte Paul
Beaven, Vida Helms
Benton, George Stock
Berger, Robert Elliott
Berkson, Harold
Budger, Aaron Bernard
Carroll, James Barr
Carton, Robert John
Chapin, Douglas Scott
Chen, Philip Stanley, Jr
Chernoff, Amoz Immanuel
Clark, Joseph E(dward)
Clark, Trevor H
Condliffe, Peter George
Daly, John Anthony
Dunning, Herbert Neal
Ebert, James David
Ewing, June Swift
Falk, Charles Eugene
Filner, Barbara
Fisher, Gail Feimster
Friedman, Abraham S(olomon)
Fry, John Craig
Gary, Robert
Geller, Ronald G
Goldwater, William Henry
Gonzales, Ciriaco Q
Hammond, Allen Lee
Heinrich, Kurt Francis Joseph
Heller, Stephen Richard
Herman, Samuel Sidney
Herrett, Richard Allison
Hildebrand, Bernard
Hilmoe, Russell J(ulian)
Holloway, Caroline T
Holt, James Allen
Kalt, Marvin Robert
Katcher, David Abraham
Kneller, Robert William
Kovach, Eugene George
Kushner, Lawrence Maurice
Langenberg, Donald Newton
Leshner, Alan Irvin
McCarn, Davis Barton
McRae, Vincent Vernon
Matyas, Marsha Lakes
Michaelis, Michael
Mielenz, Klaus Dieter
Miller, Hugh Hunt
Miller, Nancy E
Mills, William Andy
Myers, Peter Briggs
Nightingale, Stuart L
Pahl, Herbert Bowen
Passman, Sidney

Phelps, Creighton Halstead
Pitlick, Frances Ann
Porter, Beverly Fearn
Promisel, Nathan E
Rheinstein, Peter Howard
Richardson, John Marshall
Robbins, Robert John
Rosen, Milton W(illiam)
Rosenblum, Annette Tannenholz
Scarbrough, Frank Edward
Schulman, James Herbert
Schultz, Warren Walter
Seifried, Adele Susan Corbin
Shelton, Robert Duane
Shykind, Edwin B
Siegel, Elliot Robert
Silbergeld, Ellen K
Simon, Robert Michael
Singh, Amarjit
Smedley, William Michael
Smith, John Henry
Sorrows, Howard Earle
Spilhaus, Athelstan Frederick, Jr
Stanley, Ronald Alwin
Starfield, Barbara Helen
Stonehill, Elliott H
Tropf, Cheryl Griffiths
Ulbrecht, Jaromir Josef
Ventre, Francis Thomas
Vogl, Thomas Paul
Weiss, Charles, Jr
Whidden, Stanley John
Wortman, Bernard
Yakowitz, Harvey
Yarmolinsky, Adam
Young, Alvin L
Zimbelman, Robert George
Zwilsky, Klaus M(ax)

MASSACHUSETTS
Alpert, Peter
Baliunas, Sallie Louise
Bernstein, Herbert J
Bertera, James H
Bird, Stephanie J(ean)
Branscomb, Lewis McAdory
Breuning, Siegfried M
Broadus, James Matthew
Brooks, Harvey
Burris, John Edward
Carnesale, Albert
Carter, Ashton Baldwin
Clark, William Cummin
Cox, Lawrence Henry
De Neufville, Richard Lawrence
Foster, Charles Henry Wheelwright
Frosch, Robert Alan
Ganley, Oswald Harold
Graham, Loren R
Holmes, Helen Bequaert
Hornig, Donald Frederick
Karo, Douglas Paul
Krass, Allan S(hale)
Laster, Leonard
Lee, Kai Nien
Loew, Franklin Martin
McKnight, Lee Warren
Moore-Ede, Martin C
Ramakrishna, Kilaparti
Simpson, Rae Goodell
Skolnikoff, Eugene B
Stonier, Tom Ted
Tuler, Floyd Robert
Weiner, Charles
Wertz, Dorothy C
Wilson, Linda S (Whatley)

MICHIGAN
Bohm, Henry Victor
Evans, David Arnold
Hoerger, Fred Donald
Humphrey, Harold Edward Burton, Jr
Kamrin, Michael Arnold
McLaughlin, Renate
Mayer, Frederick Joseph
Montgomery, Donald Joseph
Saperstein, Alvin Martin
Schwing, Richard C
Viano, David Charles
Weissman, Eugene Y(ehuda)

MINNESOTA
Flower, Terrence Frederick
Johnson, Clark E, Jr
Perry, James Alfred
Steen, Lynn Arthur
Thomborson, Clark D(avid)

MISSISSIPPI
Alexander, William Nebel
ElSohly, Mahmoud Ahmed

MISSOURI
Johannsen, Frederick Richard
Lawless, Edward William
Morgan, Robert P
Wilson, Edward Nathan
Wilson, James Dennis

MONTANA
Bay, Roger R(udolph)

NEW HAMPSHIRE
Bernstein, Abram Bernard

Kantrowitz, Arthur (Robert)

NEW JERSEY
Brown, William Stanley
Brubaker, Paul Eugene, II
Hall, Homer James
Hay, Peter Marsland
Hazelrigg, George Arthur, Jr
Holt, Rush D(ew), Jr
Klapper, Jacob
Lederman, Peter B
Light, Donald W
Miller, Robert Alan
Pinsky, Carl Muni
Smagorinsky, Joseph
Socolow, Robert H(arry)
Steinberg, Eliot
Stiles, Lynn F, Jr
Tanner, Daniel
Temmer, Georges Maxime
Turner, Edwin Lewis
Willson, John Ellis

NEW MEXICO
Freese, Kenneth Brooks
Guenther, Arthur Henry
Icerman, Larry
Leonard, Ellen Marie
Lohrding, Ronald Keith
Otway, Harry John
Shipley, James Parish, Jr

NEW YORK
Allentuch, Arnold
Althuis, Thomas Henry
Ausubel, Jesse Huntley
Bedford, Barbara Lynn
Beyea, Jan Edgar
Bulloff, Jack John
Colligan, John Joseph
Corson, Dale Raymond
Edelson, Paul J
Fox, William
Geiger, H Jack
Goldfarb, Theodore D
Hardy, Ralph Wilbur Frederick
Jasanoff, Sheila Sen
Katz, Maurice Joseph
Komorek, Michael Joesph, Jr
Lubic, Ruth Watson
Lynn, Walter R(oyal)
McNeil, Richard Jerome
Mahoney, Margaret E
Manassah, Jamal Tewfek
Mike, Valerie
Month, Melvin
Mowshowitz, Abbe
Nathanson, Melvyn Bernard
O'Connor, Joel Sturges
Pechacek, Terry Frank
Rabino, Isaac
Rambaut, Paul Christopher
Richards, Paul Granston
Ries, Richard Ralph
Robertson, William, IV
Robinson, David Zav
Rosenberg, Barbara Hatch
Shaw, David Elliot
Swanson, Robert Lawrence
Swartz, Harry
Thorpe, John A(lden)
Voelcker, Herbert B(ernhard)
Walker, Theresa Anne
Wilkins, Bruce Tabor
Wilson, Fred Lee

NORTH CAROLINA
Ahearne, John Francis
Andrews, Richard Nigel Lyon
Bishop, Jack Belmont
Burger, Robert M
Damstra, Terri
Dickens, Charles Henderson
Garner, Jasper Henry Barkdoll
Gemperline, Margaret Mary Cetera
Holland, Marjorie Miriam
Mosier, Stephen R
Rees, John
Schroeer, Dietrich

OHIO
Chan, Yupo
Christensen, James Henry
Duga, Jules Joseph
Galloway, Ethan Charles
Moss, Thomas H
Mulick, James Anton
Sheskin, Theodore Jerome

OKLAHOMA
Crafton, Paul A(rthur)
Lambird, Perry Albert

OREGON
Bullock, Richard Melvin
Dews, Edmund
Linstone, Harold A
Moravcsik, Michael Julius
Newell, Nanette

PENNSYLVANIA
Coleman, Robert E
Cordova, France Anne-Dominic
Daehnick, Wilfried W

Dulka, Joseph John
Fischhoff, Baruch
Fritz, Lawrence William
Fromm, Eli
Gerjuoy, Edward
Greif, Karen Faye
Hager, Nathaniel Ellmaker, Jr
Humphrey, Watts S
Lang, Norma M
Liu, Andrew Tze-chiu
Lu, Chih Yuan
Morgan, M(illett) Granger
Nair, K Manikantan
Pool, James C T
Ray, Eva K
Reischman, Michael Mack
Shen, Benjamin Shih-Ping
Small, Henry Gilbert
Steg, L(eo)

SOUTH CAROLINA
Huray, Paul Gordon

SOUTH DAKOTA
Sword, Christopher Patrick
Wegman, Steven M

TENNESSEE
Bibb, William Robert
Cope, David Franklin
Garber, David H
Miller, Paul Thomas
Phung, Doan Lien
Sharples, Frances Ellen
Van Hook, Robert Irving, Jr
Veigel, Jon Michael

TEXAS
Albach, Roger Fred
Bartley, William Call
Cranberg, Lawrence
Dilsaver, Steven Charles
Elliott, Jerry Chris
Heavner, James E
Lytle, Michael Allen
Neureiter, Norman Paul
Phillips, Gerald C
Schmandt, Jurgen
Shotwell, Thomas Knight
Thomas, Estes Centennial, III
Walker, Laurence Colton

VIRGINIA
Barker, Robert Henry
Bartis, James Thomas
Basu, Sunanda
Berger, Beverly Jane
Berlincourt, Ted Gibbs
Bernthal, Frederick Michael
Birely, John H
Blanpied, William Antoine
Burnette, Mahlon Admire, III
Burstyn, Harold Lewis
Carpenter, James E(ugene)
Carpenter, Richard A
Chamot, Dennis
Conn, William David
Cox, William Edward
Denning, Peter James
De Simone, Daniel V
Dragonetti, John Joseph
Featherston, Frank Hunter
Feldmann, Edward George
Gaines, Alan McCulloch
Gibbons, John Howard
Gough, Stephen Bradford
Greenfield, Richard Sherman
Greer, William Louis
Hart, Dabney Gardner
Hurley, William Jordan
Ifft, Edward M
Jennings, Allen Lee
Johnson, George Patrick
Johnston, Pauline Kay
Kelly, Henry Charles
Kolsrud, Gretchen Schabtach
Kramish, Arnold
Krassner, Jerry
Ling, James Gi-Ming
Lynch, Maurice Patrick
Miller, William Lawrence
Morrison, David Lee
Munday, John Clingman, Jr
Newton, Elisabeth G
Owczarski, William A(nthony)
Pages, Robert Alex
Peake, Harold J(ackson)
Potts, Malcolm
Pry, Robert Henry
Ratchford, Joseph Thomas
Rodricks, Joseph Victor
Sabadell, Alberto Jose
Schad, Theodore M(acNeeve)
Schremp, Edward Jay
Shaub, Walter M
Smith, Bertram Bryan, Jr
Smith, Marcia Sue
Spindel, William
Strauss, Michael S
Sudarshan, T S
Teichler-Zallen, Doris
Tipper, Ronald Charles
Tolin, Sue Ann
Troxell, Terry Charles

Van der Vink, Gregory E
Watson, Robert Francis
Wilkinson, Christopher Foster
Wollan, David Strand
Zimmerman, Peter David
Zornetzer, Steven F
Zwolenik, James J

WASHINGTON
Banks, Norman Guy
Miles, Edward Lancelot
Smith, Stewart W
Szkody, Paula
Van der Werff, Terry Jay
Wenk, Edward, Jr
Wolfle, Dael (Lee)

WISCONSIN
Bleicher, Michael Nathaniel
Blockstein, William Leonard
Greger, Janet L
Inman, Ross
Steinhart, John Shannon

WYOMING
Meyer, Edmond Gerald
Mingle, John O(rville)
Thronson, Harley Andrew, Jr

BRITISH COLUMBIA
Dewey, John Marks
LeBlond, Paul Henri

MANITOBA
Hall, Donald Herbert
Rosinger, Eva L J
Trick, Gordon Staples

NEW BRUNSWICK
Pohle, Gerhard Werner

ONTARIO
Andrew, Bryan Haydn
Blevis, Bertram Charles
Cartier, Jean Jacques
Clark, Ferrers Robert Scougall
Costain, Cecil Clifford
Cupp, Calvin R
Fawcett, Eric
French, J(ohn) Barry
Godson, Warren Lehman
Heacock, Ronald A
Jackson, Ray Weldon
Kates, Josef
Keys, John David
Kruus, Jaan
McTaggart-Cowan, Patrick Duncan
Munroe, Eugene Gordon
Pigden, Wallace James
Read, D E
Sly, Peter G
Solandt, Omond McKillop
Sturgess, Jennifer Mary
Uffen, Robert James
Walsh, John Heritage
Wasi, Safia
Winder, Charles Gordon
Wright, Douglas Tyndall

QUEBEC
Bailar, John Christian, III
L'Archeveque, Real Viateur
L'Ecuyer, Jacques
McQueen, Hugh J

SASKATCHEWAN
Katz, Leon

OTHER COUNTRIES
Assousa, George Elias
Burnett, Jerrold J
Campbell, Virginia Wiley
D'Ambrosio, Ubiratan
Gowans, James L
Inose, Hiroshi
Lowrance, William Wilson, Jr
Luo, Peilin
Metzger, Gershon
Okamura, Sogo
Packer, Leo S
Richards, Adrian F
Saba, Shoichi
Schmell, Eli David
Schopper, Herwig Franz
Stebbing, Nowell
Thissen, Wil A
Van Andel, Tjeerd Hendrik

Technical Management

ALABAMA
Arnold, David Walker
Ayers, Orval Edwin
Cloyd, Grover David
Cochran, John Euell, Jr
Costes, Nicholas Constantine
Cowsar, Donald Roy
Davis, John Moulton
Dean, David Lee
Fowler, Bruce Wayne
Haak, Frederik Albertus
Hart, David R
Howgate, David W

Kandhal, Prithvi Singh
Lovingood, Judson Allison
Mookherji, Tripty Kumar
Moore, Fletcher Brooks
Morgan, Bernard S(tanley), Jr
Palmer, Robert Gerald
Piwonka, Thomas Squire
Regner, John LaVerne
Rekoff, M(ichael) G(eorge), Jr
Renard, Jean Joseph
Smith, Robert Earl
Vaughan, William Walton

ALASKA
Bennett, F Lawrence
Hulsey, J Leroy

ARIZONA
Adamson, Albert S, Jr
Adickes, H Wayne
Armstrong, Dale Dean
Batson, Raymond Milner
Carr, Clide Isom
Dickinson, Robert Earl
Dybvig, Paul Henry
Garrett, Lane Sayre
Gatley, William Stuart
Greenberg, Milton
Hoehn, A(lfred) J(oseph)
Johnson, Vern Ray
Jungermann, Eric
Klein, Morton Joseph
Liaw, Hang Ming
McGuirk, William Joseph
Pokorny, Gerold E(rwin)
Ramler, W(arren) J(oseph)
Randall, Lawrence Kessler, Jr
Rountree, Janet
Sears, Donald Richard
Seifert, Deborah Roeckner
Swalin, Richard Arthur
Swan, Cathy Wood
Swan, Peter A
Swink, Laurence N
Ulich, Bobby Lee
Upchurch, Jonathan Everett
Wasson, James Walter
Whaley, Thomas Patrick
Willson, Donald Bruce
Winder, Robert Owen

ARKANSAS
Berky, John James
Burchard, John Kenneth
Commerford, John D
Ervin, Hollis Edward
Glendening, Norman Willard
Hatfield, Efton Everett
Havener, Robert D
Keplinger, Orin Clawson
Lewis, Sherry M
Odell, Norman Raymond
Townsend, James Willis

CALIFORNIA
Addy, Tralance Obuama
Adelman, Barnet Reuben
Adelson, Harold Ely
Adicoff, Arnold
Agostini, Romain Camille
Albert, Richard David
Alper, Marshall Edward
Anderson, George William
Anjard, Ronald P, Sr
Appleman, Gabriel
Ashcroft, Frederick H
Aster, Robert Wesley
Atkinson, Russell H
Augenstein, Bruno (Wilhelm)
Baba, Paul David
Baker, Andrew Newton, Jr
Basch, Paul Frederick
Baugh, Ann Lawrence
Baumann, Frederick
Beebe, Robert Ray
Berkner, Klaus Hans
Bhanu, Bir
Bharadvaj, Bala Krishnan
Billig, Franklin A
Bilow, Norman
Birecki, Henryk
Bixler, Otto C
Block, Richard B
Bode, Donald Edward
Bolt, Robert O'Connor
Bonsack, Walter Karl
Brand, Samson
Brewer, George R
Brewer, LeRoy Earl, Jr
Brown, Tony Ray
Burnett, James R
Camp, David Conrad
Campbell, Bonita Jean
Campbell, Thomas Cooper
Canning, T(homas) F
Caren, Robert Poston
Carlan, Alan J
Carman, Robert Lincoln, Jr
Carr, Robert Charles
Caswell, John N(orman)
Chan, David S
Chan, Kwoklong Roland
Chang, Chih-Pei
Channel, Lawrence Edwin

Chase, Lloyd Fremont, Jr
Chern, Ming-Fen Myra
Chester, Arthur Noble
Chivers, Hugh John
Chodos, Steven Leslie
Clazie, Ronald N(orris)
Clegg, Frederick Wingfield
Cluff, Lloyd Sterling
Cockrum, Richard Henry
Cohen, Jack
Coile, Russell Cleven
Cole, Richard
Coleman, Lamar William
Comer, William Timmey
Compton, Leslie Ellwyn
Conant, Curtis Terry
Cook, Glenn Melvin
Copper, John A(lan)
Cox, Donald Clyde
Crandall, Ira Carlton
Crossley, Frank Alphonso
Culler, F(loyd) L(eRoy), Jr
Current, Jerry Hall
Dahlberg, Richard Craig
Danielson, George Edward, Jr
Davis, Louis E(lkin)
Davison, John Blake
Deckert, Curtis Kenneth
Depp, Joseph George
Detz, Clifford M
Digby, James F(oster)
Docks, Edward Leon
Donovan, Paul F(rancis)
Dowdy, William Louis
Durbeck, Robert C(harles)
Eck, Robert Edwin
Edamura, Fred Y
Edelson, Robert Ellis
Edgerley, Dennis A
Ehrlich, Richard
Elverum, Gerard William, Jr
Englert, Robert D
Epperly, W Robert
Erickson, Stanley Arvid
Ernst, Roberta Dorothea
Farber, Joseph
Farley, Thomas Albert
Farmer, Crofton Bernard
Fasang, Patrick Pad
Feinberg, Benedict
Feller, Robert S, Sr
Ferber, Robert R
Feuerstein, Seymour
Finn, James Crampton, Jr
Fischer, George K
Fishman, Norman
Fitzgerald, Thomas Michael
Fitzpatrick, Gary Owen
Fleischer, Allan A
Flora, Edward B(enjamin)
Forbes, Judith L
Forrest, James Benjamin
Forsberg, Kevin
Frazier, Edward Nelson
Frenzel, Lydia Ann Melcher
Fricke, Martin Paul
Funkhouser, John Tower
Gaynor, Joseph
Gearhart, Roger A
Gehman, Bruce Lawrence
Geminder, Robert
Gerwick, Ben Clifford, Jr
Gesner, Bruce D
Gilkeson, M(urray) Mack
Gill, James Edward
Gill, William D(elahaye)
Gillette, Dean
Glaser, Harold
Glover, Leon Conrad, Jr
Goerz, David Jonathan, Jr
Gogolewski, Raymond Paul
Goode, John Wolford
Goodjohn, Albert J
Graham, Dee McDonald
Greene, Elias Louis
Greenleaf, John Edward
Grier, Herbert E
Gritton, Eugene Charles
Groner, Gabriel F(rederick)
Guest, Gareth E
Gur, Turgut M
Gyorey, Geza Leslie
Hadley, James Warren
Hadley, Jeffery A
Hahn, Harold Thomas
Hainski, Martha Barrionuevo
Ham, Lee Edward
Hamilton, Carole Lois
Hanafee, James Eugene
Hansen, Grant Lewis
Hansmann, Douglas R
Harrison, George Conrad, Jr
Hartsough, Walter Douglas
Hartwick, Thomas Stanley
Harvey, A(lexander)
Hausmann, Werner Karl
Hayes, Thomas Jay, III
Heil, John F, Jr
Heitkamp, Norman Denis
Hilbert, Robert S
Hill, Robert
Hill, Russell John
Hinkley, Everett David
Hoch, Paul Edwin

Hodges, Dean T, Jr
Hodson, William Myron
Hoff, Richard William
Hoffman, Howard Torrens
Hoffman, Marvin Morrison
Holliday, Dale Vance
Hong, Ki C(hoong)
Hopponen, Jerry Dale
Hornig, Howard Chester
Horvath, Kalman
Howerton, Robert James
Hubert, Jay Marvin
Hudson, Cecil Ivan, Jr
Hughes, Robert Alan
Hugunin, Alan Godfrey
Hwang, Li-San
Ide, Roger Henry
Jacobs, Michael Moises
Jacobs, Ralph R
Jacobs, S Lawrence
Jacobson, Albert H(erman), Jr
Jacobson, Allan Stanley (Bud)
Jacobson, Marcus
Jaffe, Leonard David
James, Philip Nickerson
Janney, Gareth Maynard
Johannessen, George Andrew
Johnson, Conor Deane
Johnson, Mark Scott
Jones, E(dward) M(cClung) T(hompson)
Jones, Richard Lee
Jorgensen, Paul J
Judd, Stanley H
Kalvinskas, John J(oseph)
Kaplan, Richard E
Karn, Richard Wendall
Kassakhian, Garabet Haroutioun
Kay, Eric
Keaton, Michael John
Keller, Joseph Edward, Jr
Kennedy, James Vern
Kennel, John Maurice
Kerr, Donald M, Jr
King, John Mathews
King, Ray J(ohn)
Kissel, Charles Louis
Klestadt, Bernard
Knight, Patricia Marie
Koestel, Mark Alfred
Kropp, John Leo
Kukkonen, Carl Allan
Kull, Lorenz Anthony
Lagarias, John S(amuel)
Lampert, Carl Matthew
Lampert, Seymour
Larson, Edward William, Jr
Lavernia, Enrique Jose
Lechtman, Max D
Legge, Norman Reginald
Leibowitz, Lewis Phillip
Lewis, Arthur Edward
Lewis, Nina Alissa
Lewis, Robert Allen
Liang, Yola Yueh-o
Lichter, James Joseph
Liepman, H(ans) P(eter)
Lillie, Charles Frederick
Lippitt, Louis
Lissaman, Peter Barry Stuart
Livingston, Douglas Alan
Logan, James Columbus
Long, H(ugh) M(ontgomery)
Love, Sydney Francis
Lovelace, Alan Mathieson
Ludwig, Frank Arno
Lukasik, Stephen Joseph
Luongo, Cesar Augusto
Lurie, Norman A(lan)
McCall, Chester Hayden, Jr
McCarthy, John Lockhart
McCreight, Louis R(alph)
McDonald, David William
McDonald, William True
McEwen, C(assius) Richard
MacKay, Kenneth Donald
Mackin, Robert James, Jr
McLaughlin, William Irving
Malouf, George M
Manes, Kennneth Rene
Mantey, Patrick E(dward)
Marcus, Bruce David
Marcus, Rudolph Julius
Massier, Paul Ferdinand
Mathews, W(arren) E(dward)
May, Bill B
Meechan, Charles James
Meghreblian, Robert V(artan)
Meiners, Henry C(ito)
Meiser, Michael David
Merilo, Mati
Michaelson, Jerry Dean
Miller, Arnold
Miller, Jon Philip
Miller, William Frederick
Molinari, Robert James
Moore, Edgar Tilden, Jr
Morey, Booker Williams
Morris, Fred(erick) W(illiam)
Mow, C(hao) C(how)
Mulroy, Thomas Wilkinson
Nacht, Sergio
Nadler, Gerald
Neidlinger, Hermann H
Neil, Gary Lawrence

Technical Management (cont)

Ning, Robert Y
Nishi, Yoshio
O'Donnell, Ashton Jay
O'Donohue, Cynthia H
Oehlberg, Richard N
Offen, George Richard
O'Keefe, John Dugan
Opfell, John Burton
Orth, Charles Douglas
Page, D(errick) J(ohn)
Parks, Lewis Arthur
Parsons, Michael L
Paszyc, Aleksy Jerzy
Perkins, Willis Drummond
Perloff, David Steven
Perry, Joseph Earl, Jr
Persky, George
Peterson, Victor Lowell
Philipson, Lloyd Lewis
Phillips, Russell C(ole)
Pietrzak, Lawrence Michael
Pitcher, Wayne Harold, Jr
Pitts, Robert Gary
Porter, Clark Alfred
Potter, David Samuel
Pressman, Ada Irene
Putz, Gerard Joseph
Rabl, Veronika Ariana
Rachmeler, Martin
Rahmat-Samii, Yahya
Ram, Michael Jay
Ramspott, Lawrence Dewey
Ranftl, Robert M(atthew)
Rapp, Donald
Rast, Howard Eugene, Jr
Rauch, Richard Travis
Reardon, Edward Joseph, Jr
Remer, Donald Sherwood
Remley, Marlin Eugene
Resch, George Michael
Reynolds, Richard Alan
Rice, Dennis Keith
Richardson, John L(loyd)
Ripley, William Ellis
Ritzman, Robert L
Roberts, Charles Sheldon
Robertson, David W
Robertson, David Wayne
Robin, Allen Maurice
Rohm, C E Tapie, Jr
Romanowski, Christopher Andrew
Roney, Robert K(enneth)
Rosen, Leonard Craig
Rossen, Joel N(orman)
Roukes, Michael L
Rowe, Lawrence A
Rowell, John Martin
Rowntree, Robert Fredric
Rudavsky, Alexander Bohdan
Ruskin, Arnold M(ilton)
Sachdev, Suresh
Sager, Ronald E
Saito, Theodore T
Salsig, William Winter, Jr
Saltman, William Mose
Saute, Robert E
Schjelderup, Hassel Charles
Schmidt, Raymond LeRoy
Schooley, Caroline Naus
Schreiner, Robert Nicholas, Jr
Schurmeier, Harris McIntosh
Schwab, Michael
Schwartz, Daniel M(ax)
Schweiker, George Christian
Scott, Franklin Robert
Seling, Theodore Victor
Selinger, Patricia Griffiths
Serbia, George William
Sharma, Bhavender Paul
Sharpe, Roland Leonard
Shaw, Charles Alden
Shlanta, Alexis
Shotts, Wayne J
Siegel, Brock Martin
Siegel, Seymour
Siegfried, William
Siegman, Fred Stephen
Silverman, Herbert Philip
Silverman, Jacob
Siposs, George G
Skavdahl, R(ichard) E(arl)
Smiley, Vern Newton
Snow, John Thomas
Snyder, Charles Thomas
Snyder, Nathan W(illiam)
Sovish, Richard Charles
Spitzer, Irwin Asher
Sporer, Alfred Herbert
Srour, Joseph Ralph
Staudhammer, Peter
Stein, Richard Ballin
Steinberg, Gunther
Stevenson, Robin
Stockel, Ivar H(oward)
Strahl, Erwin Otto
Sundberg, John Edwin
Suri, Ashok
Sutcliffe, William George
Swain, Robert James
Swanson, Paul N
Swanson, Robert Nels
Swearengen, Jack Clayton
Sweeney, Lawrence Earl, Jr
Syvertson, Clarence A
Taft, David Dakin
Taylor, Charles Joel
Taylor, Robert William
Thomas, Frank J(oseph)
Tin, Hla Ngwe
Toepfer, Richard E, Jr
Torgow, Eugene N
Trevelyan, Benjamin John
Trippe, Anthony Philip
Uhl, Arthur E(dward)
Urbach, John C
Vagelatos, Nicholas
Vane, Arthur B(ayard)
Van Klaveren, Nico
Varon, Myron Izak
Vassiliou, Marius Simon
Victor, Andrew C
Viswanathan, R
Vlay, George John
Vorhaus, James Louis
Vroom, David Archie
Walker, Kelsey, Jr
Wallerstein, Edward Perry
Warburton, William Kurtz
Weber, Hans Josef
Weers, Jeffry Greg
Werth, Glenn Conrad
Wertheim, Robert Halley
Wertz, James Richard
Westerman, Edwin J(ames)
Wilcox, Charles Hamilton
Williams, Donald Spencer
Winick, Herman
Wolfe, Bertram
Woodbury, Eric John
Wright, Edward Kenneth
Yeomans, Donald Keith
Zarem, Abe Mordecai
Zeren, Richard William
Zwerdling, Solomon

COLORADO

Adams, James Russell
Barber, George Arthur
Barber, Robert Edwin
Bauer, Charles Edward
Broome, Paul W(allace)
Burdge, David Newman
Cada, Ronald Lee
Chamberlain, A(drian) R(amond)
Chisholm, James Joseph
Cleary, Michael E
Cochran, Leighton Scott
Cole, Lee Arthur
Cowart, Vicki Jane
Crockett, Allen Bruce
Davidson, Darwin Ervin
Dichtl, Rudolph John
Dixon, Robert Clyde
Dunn, Richard Lee
Evans, David Lane
Ferguson, William Sidney
Fickett, Frederick Roland
Fraikor, Frederick John
Gentry, Donald William
Giller, E(dward) B(onfoy)
Harlan, Ronald A
Harrington, Robert D(ean)
Havlick, Spenser Woodworth
Henkel, Richard Luther
Hill, Walter Edward, Jr
Johnson, Arnold I(van)
Kazmerski, Lawrence L
Keller, Frederick Albert, Jr
Koldewyn, William A
Kotch, Alex
Ladd, Conrad Mervyn
Lameiro, Gerard Francis
Legal, Casimer Claudius, Jr
McLean, Francis Glen
Mandics, Peter Alexander
Miner, Frend John
Munro, Richard Harding
Plows, William Herbert
Reichardt, John William
Renne, David Smith
Rothfeld, Leonard B(enjamin)
Sandstrom, Donald James
Smith, William Bridges
Steele, Timothy Doak
Thompson, John Stewart
Tischendorf, John Allen
Vande Noord, Edwin Lee
Wonsiewicz, Bud Caesar
Wyman, Charles Ely

CONNECTICUT

Agarwal, Vipin K
Andrade, Manuel
Auerbach, Michael Howard
Barney, James Earl, II
Bayney, Richard Michael
Berets, Donald Joseph
Berry, Richard C(hisholm)
Bronsky, Albert J
Buchholz, Allan C
Burwell, Wayne Gregory
Calabrese, Carmelo
Carter, G Clifford
Cassidy, John Francis, Jr
Collins, John Barrett
Corwin, H(arold) E(arl)
Coulter, Paul David
DeDecker, Hendrik Kamiel Johannes
Dobbs, Gregory Melville
Dooley, Joseph Francis
Drozdowicz, Zbigniew Maria
Duvivier, Jean Fernand
Emond, George T
Feinstein, Myron Elliot
Gans, Eugene Howard
Ghen, David C
Glomb, Walter L
Grattan, James Alex
Grayson, Martin
Greenwood, Ivan Anderson
Grossman, Leonard N(athan)
Haas, Ward John
Hanson, Trevor Russell
Harding, R(onald) H(ugh)
Hersh, Charles K(arrer)
Howard, Phillenore Drummond
Johnson, Loering M
Johnstone, John William, Jr
Juran, Joseph M
Kontos, Emanuel G
Landsman, Douglas Anderson
Linke, William Finan
Moreland, Parker Elbert, Jr
Nath, Dilip K
Paterson, Robert W
Peracchio, Aldo Anthony
Pinson, Rex, Jr
Rauhut, Michael McKay
Robinson, Donald W(allace), Jr
Sarada, Thyagaraja
Scherr, Allan L
Schwartz, Michael H
Shapiro, Warren Barry
Shuey, Merlin Arthur
Sinha, Vinod T(arkeshwar)
Staker, William Paul
Storrs, Charles Lysander
Stott, Paul Edwin
Strife, James Richard
Suen, T(zeng) J(iueq)
Tangel, O(scar) F(rank)
Thomas, Alvin David, Jr
Tittman, Jay
Tracy, Joseph Charles
Troutman, Ronald R
Wernau, William Charles
Wingfield, Edward Christian
Ziegler, William Arthur

DELAWARE

Abrahamson, Earl Arthur
Andersen, Donald Edward
Bartkus, Edward Peter
Boettcher, F Peter
Borden, James B
Bragagnolo, Julio Alfredo
Brown, Stewart Cliff
Busche, Robert M(arion)
Cantwell, Edward N(orton), Jr
Carr, John B
Chait, Edward Martin
Cherkofsky, Saul Carl
Chorvat, Robert John
Cone, Michael McKay
Daly, John Joseph, Jr
DeDominicis, Alex John
Delp, Charles Joseph
Dubas, Lawrence Francis
Eissner, Robert M
Erdmann, Duane John
Fleming, Richard Allan
Ginnard, Charles Raymond
Glasgow, Louis Charles
Gossage, Thomas Layton
Greenblatt, Hellen Chaya
Harvey, John, Jr
Holob, Gary M
Huppe, Francis Frowin
Hurley, William Joseph
Jack, John James
Johnson, Alexander Lawrence
Kane, Edward R
Kaplan, Ralph Benjamin
Ketterer, Paul Anthony
Knipmeyer, Hubert Elmer
Kolber, Harry John
Kramer, Brian Dale
Krone, Lawrence James
Lauzon, Rodrigue Vincent
McGee, Charles E
McKay, Sandra J
Marquardt, Donald Wesley
Maynes, Gordon George
Moyer, Calvin Lyle
Norling, Parry McWhinnie
Read, Robert E
Riggle, J(ohn) W(ebster)
Sammak, Emil George
Schappell, Frederick George
Schiewetz, D(on) B(oyd)
Schiroky, Gerhard H
Schwartz, Jerome Lawrence
Shambelan, Charles
Shellenbarger, Robert Martin
Shotzberger, Gregory Steven
Silzars, Aris
Slade, Arthur Laird
Stelting, Kathleen Marie
Stewart, Edward William
Tanner, David
Taylor, Robert Burns, Jr
Thayer, Chester Arthur
Ulery, Dana Lynn
Urquhart, Andrew Willard
Webb, Charles Alan
Wermus, Gerald R

DISTRICT OF COLUMBIA

Adey, Walter Hamilton
Alexander, Benjamin H
April, Robert Wayne
Bloch, Erich
Borsuk, Gerald M
Breen, Joseph John
Chen, Davidson Tah-Chuen
Cotruvo, Joseph Alfred
Davis, Lance A(lan)
Decker, James Frederick
Dowdy, Edward Joseph
Flora, Lewis Franklin
Forester, Donald Wayne
Gilbreath, William Pollock
Goldfield, Edwin David
Green, Richard James
Gunn, John William, Jr
Holloway, John Thomas
Holt, Alan Craig
Justus, Philip Stanley
Kordoski, Edward William
Larsen, Lynn Alvin
Lombard, David Bishop
Madden, Joseph Michael
Miles, Corbin I
Peters, Gerald Joseph
Reeves, Edmond Morden
Reisa, James Joseph, Jr
Schafrik, Robert Edward
Schroeder, Anita Gayle
Shaw, Robert R
Shrier, Adam Louis
Spurlock, Langley Augustine
Sterrett, Kay Fife
Tokar, Michael
Trubatch, Sheldon L
Wade, Clarence W R
Waters, Robert Charles
Weinberg, Myron Simon
Wilsey, Neal David
Wilson, George Donald

FLORIDA

Amey, William G(reenville)
Aronson, M(oses)
Baumann, Arthur Nicholas
Berry, Robert Eddy
Billica, Harry Robert
Bingham, Richard S(tephen), Jr
Bishop, Charles Anthony
Black, William Bruce
Callan, Edwin Joseph
Carrier, Steven Theodore
Chao, Raul Edward
Chryssafopoulos, Nicholas
Clark, Mary Elizabeth
Coppoc, William Joseph
Cortright, Edgar Maurice
Dailey, Charles E(lmer), III
Dannenberg, E(li) M
Davis, Duane M
DeHart, Arnold O'Dell
Donahue, Roger P
Doyon, Leonard Roger
Dreves, Robert G(eorge)
Everett, Woodrow Wilson, Jr
Feinman, J(erome)
Feldmesser, Julius
Frazier, Stephen Earl
Gash, Virgil Walter
Givens, Paul Edward
Gold, Edward
Griffith, John E(dward)
Haase, Richard Henry
Haeger, Beverly Jean
Hagenmaier, Robert Doller
Halpern, Ephraim Philip
Hammond, Charles Eugene
Haug, Arthur John
Heggie, Robert
Heying, Theodore Louis
Hirsch, Donald Earl
Horton, Thomas Roscoe
Huber, William Richard, III
Huckaba, Charles Edwin
Jackel, Simon Samuel
Karp, Abraham E
Kauder, Otto Samuel
Kennelley, James A
Kiewit, David Arnold
Klock, Benny LeRoy
Kolhoff, M(arvin) J(oseph)
Lacoste, Rene John
Lawrence, Franklin Isaac Latimer
Leon, Kenneth Allen
Long, Calvin H
Lowe, Ronald Edsel
McKently, Alexandra H
McSwain, Richard Horace
Mason, Kenneth M
Mendell, Jay Stanley
Mickley, Harold S(omers)
Migliaro, Marco William
Milton, Robert Mitchell
Myers, Richard Lee
Nimmo, Bruce Glen

Normile, Hubert Clarence
Nunnally, Stephens Watson
Oline, Larry Ward
Owen, Gwilym Emyr, Jr
Parker, Howard Ashley, Jr
Peterson, James Robert
Poet, Raymond B
Rosenthal, Henry Bernard
Seugling, Earl William, Jr
Sherman, Edward
Silver, Frank Morris
Stern, Milton
Stokes, Charles Anderson
Swart, William W
Tebbe, Dennis Lee
Van Ligten, Raoul Fredrik
Wiseman, Robert S
Wolsky, Sumner Paul
Yon, E(ugene) T
Young, C(harles) Gilbert
Ziernicki, Robert S

GEORGIA
Blanton, Jackson Orin
Camp, Ronnie Wayne
Clark, Allen Varden
Coyne, Edward James, Jr
Derrick, Robert P
Donahoo, Pat
Evans, Thomas P
Fineman, Manuel Nathan
Gambrell, Carroll B(lake), Jr
Goldsmith, Edward
Grace, Donald J
Griffin, Clayton Houstoun
Johnson, Donald Ross
Krajca, Kenneth Edward
Krochmal, Jerome J(acob)
Leffingwell, John C
Love, Jimmy Dwane
Lyons, Anthony Vincent
Meyer, Leon Herbert
Myers, Dirck V
Nethercut, Philip Edwin
Orofino, Thomas Allan
Owens, William Richard
Porter, Alan Leslie
Rossini, Frederick Anthony
Russell, John Lynn, Jr
Schulz, David Arthur
Shakun, Wallace
Shlevin, Harold H
Simmons, George Allen
Sizemore, Douglas Reece
Sommers, Jay Richard
Stone, John Austin
Thompson, William Oxley, II
Vachon, Reginald Irenee
Vicory, William Anthony
Ware, Kenneth Dale
Wiltse, James Cornelius
Woodall, William Robert, Jr
Yeske, Ronald A

HAWAII
Cruickshank, Michael James
Curtis, George Darwin
Khan, Mohammad Asad
Wylly, Alexander

IDAHO
Bickel, John Henry
Bills, Charles Wayne
Carpenter, Stuart Gordon
Christian, Jerry Dale
East, Larry Verne
Engen, Ivar A
Gesell, Thomas Frederick
Griffith, W(illiam) A(lexander)
Gutzman, Philip Charles
Lambuth, Alan Letcher
Wood, Richard Ellet

ILLINOIS
Agarwal, Ashok Kumar
Anderson, Arnold Lynn
Ankenbrandt, Charles Martin
Assanis, Dennis N
Ban, Stephen Dennis
Bartimes, George F
Bartlett, J Frederick
Bell, Charles Eugene, Jr
Bollinger, Lowell Moyer
Booth, David Layton
Bowles, William Allen
Brezinski, Darlene Rita
Broach, Robert William
Burrows, Brian William
Chorghade, Mukund Shankar
Christianson, Clinton Curtis
Chu, Daniel Tim-Wo
Ciriacks, Kenneth W
Clark, John Peter, III
Clark, Stephen Darrough
Collier, Donald W(alter)
Conn, Arthur L(eonard)
Danzig, Morris Juda
Davis, Edward Nathan
Deitrich, L(awrence) Walter
Devgun, Jas S
Dolgoff, Abraham
Dybel, Michael Wayne
Ehrlich, Donald Ian
Emmel, Robert Shafer
Empen, Joseph A
Falkenstein, Gary Lee
Feldman, Fred
Flinn, James Edwin
Flowers, Russell Sherwood, Jr
Frost, Brian R T
Galvin, Robert W
Gemmer, Robert Valentine
Giacobbe, F W
Gibbons, Larry V
Gillette, Richard F
Glaser, Milton Arthur
Goldman, Arthur Joseph
Griffith, James H
Hamer, Martin
Harris, Ronald David
Hauptschein, Murray
Heaston, Robert Joseph
Henning, Lester Allan
Hersh, Herbert N
Hoffman, Gerald M
Hojvat, Carlos F
Irvin, Howard H
Ivins, Richard O(rville)
Jarke, Frank Henry
Johari, Om
Johnson, Calvin Keith
Johnson, Dale Howard
Kamm, Gilbert G(eorge)
Kennedy, Albert Joseph
Kester, Dennis Earl
King, S(anford) MacCallum
Kitchen, W J
Klein, Robert L
Klinger, Lawrence Edward
Krein, Philip T
Kyle, Martin Lawrence
Layton, Terry North
Lemper, Anthony Louis
Linde, Ronald K(eith)
Linder, Louis Jacob
Lustig, Stanley
Mantsch, Paul Matthew
Marchaterre, John Frederick
Marmer, Gary James
Martinec, Emil Louis
Massel, Gary Alan
Mateles, Richard I
Maynard, Theodore Roberts
Menke, Andrew G
Meter, Donald M(ervyn)
Mielke, Robert L
Moore, John Fitzallen
Moser, Kenneth Bruce
Muhlenbruch, Carl W(illiam)
Osten, Donald Edward
Parker, John Calvin
Perino, Janice Vinyard
Pewitt, Edward Gale
Pietri, Charles Edward
Plautz, Donald Melvin
Prais, Michael Gene
Princen, Lambertus Henricus
Richmond, James M
Richmond, Patricia Ann
Rusinko, Frank, Jr
Scheie, Carl Edward
Sill, Larry R
Smith, Richard Paul
Smittle, Richard Baird
Spector, Leo Francis
Staats, William R
Steindler, Martin Joseph
Stetter, Joseph Robert
Studtmann, George H
Stull, Elisabeth Ann
Sutton, Lewis McMechan
Swoyer, Vincent Harry
Theriot, Edward Dennis, Jr
Thompson, Charles William Nelson
Thompson, David Jerome
Tiemeier, David Charles
Tomkins, Marion Louise
Tortorello, Anthony Joseph
Travelli, Armando
Trevillyan, Alvin Earl
Unik, John Peter
Wargel, Robert Joseph
Weil, Thomas Andre
Wensch, Glen W(illiam)
Wetegrove, Robert Lloyd
Whisler, Kenneth Eugene
Wingender, Ronald John
Wise, Evan Michael
Wittman, James Smythe, III
Wolf, Richard Eugene
Wolsky, Alan Martin
Wurm, Jaroslav
Wynveen, Robert Allen
Yao, Neng-Ping
Yarger, James G
Young, Donald Edward
Zeffren, Eugene

INDIANA
Bement, A(rden) L(ee), Jr
Case, Vernon Wesley
Contario, John Joseph
DeSantis, John Louis
Dube, David Gregory
English, Robert E
Free, Helen M
Gantzer, Mary Lou
Gorman, Eugene Francis
Gresham, Robert Marion
Halpin, Daniel William
Hoefer, Raymond H
Huber, Don Morgan
Hunt, Michael O'Leary
Hurrell, John Gordon
Jackson, John Eric
Jacobs, Martin Irwin
Janis, F Timothy
Kendall, Perry E(ugene)
Khorana, Brij Mohan
Klein, Howard Joseph
Koch, Kay Frances
Lazaridis, Nassos A(thanasius)
McAleece, Donald John
McIntyre, Thomas William
Markee, Katherine Madigan
Mazac, Charles James
Mercer, Walter Ronald
Morris, David Alexander Nathaniel
Morris, John F
Nirschl, Joseph Peter
Pepper, Daniel Allen
Pielet, Howard M
Putnam, Thomas Milton
Ranade, Madhukar G
Sannella, Joseph L
Shoup, Charles Samuel, Jr
Sieloff, Ronald F
Somerville, Ronald Lamont
Stroube, William Bryan, Jr
Tennyson, Richard Harvey
Thomas, Marlin Uluess
Trachman, George M
Tyhach, Richard Joseph
White, Samuel Grandford, Jr

IOWA
Anderson, R(obert) M(orris), Jr
Arndt, Stephan
Breuer, George Michael
Burrill, Claude Wesley
Edelson, Martin Charles
Egger, Carl Thomas
Eldin, Hamed Kamal
Haendel, Richard Stone
Hauenstein, Jack David
Hekker, Roeland M T
Kiser, Donald Lee
Kniseley, Richard Newman
Papadakis, Emmanuel Philippos
Riley, William F(ranklin)
Summers, William Allen, Jr

KANSAS
Bartkoski, Michael John, Jr
Cohen, Jules Bernard
Connor, Sidney G
Crowther, Robert Hamblett
Frazier, A Joel
Grant, Stanley Cameron
Hanley, Wayne Stewart
Hoffman, Jerry C
Jacobs, Louis John
McBean, Robert Parker
Miller, Glendon Richard
Roberts, James Arnold
Sutherland, John B(ennett)
Zerwekh, Robert Paul

KENTUCKY
Crowe, Richard Godfrey, Jr
Datta, Vijay J
Field, Jay Ernest
Gold, Harold Eugene
Goldknopf, Ira Leonard
Hantman, Robert Gary
Hawkins, Charles Edward
Kohnhorst, Earl Eugene
Lauterbach, John Harvey
Merker, Stephen Louis
Mihelich, John L
Squires, Robert Wright
Tran, Long Trieu
Wells, William Lochridge

LOUISIANA
Bauer, Dennis Paul
Bayer, Arthur Craig
Callens, Earl Eugene, Jr
Conrad, Franklin
DeLeon, Ildefonso R
Fischer, David John
Gibson, David Michael
Keys, L Ken
Kingrea, C(harles) L(eo)
Mills, Earl Ronald
Morris, Halcyon Ellen McNeil
Overton, Edward Beardslee
Parish, Richard Lee
Polack, Joseph A(lbert)
Shubkin, Ronald Lee
Smith, Robert Leonard
Stahly, Glenn Patrick
Stuntz, Gordon Frederick
Tewell, Joseph Robert
Wiewiorowski, Tadeusz Karol
Womack, Edgar Allen, Jr
Wrobel, William Eugene
Zaweski, Edward F
Zietz, Joseph R, Jr

MAINE
Hill, Marquita K

Shipman, C(harles) William
Wilms, Hugo John, Jr

MARYLAND
Abramson, Fredric David
Adamantiades, Achilles G
Adler, Sanford Charles
Anand, Davinder K
Anderson, W(endell) L
Anderson, William Averette
Atia, Ali Ezz Eldin
Baker, Francis Edward, Jr
Beach, Eugene Huff
Bernard, Dane Thomas
Bersch, Charles Frank
Blevins, Gilbert Sanders
Bournia, Anthony
Brady, Robert Frederick, Jr
Brodsky, Marc H
Bryan, Edward H
Buck, John Henry
Bugenhagen, Thomas Gordon
Butler, Louis Peter
Carrico, Christine Kathryn
Carver, Gary Paul
Cave, William Thompson
Charlton, Gordon Randolph
Chimenti, Dale Everett
Chmurny, Alan Bruce
Cipriano, Leonard Francis
Clark, Joseph E(dward)
Conklin, James Byron, Jr
Corak, William Sydney
Cornyn, John Joseph
Cotton, Ira Walter
Daniels, William Fowler, Sr
Degnan, John James, III
Deutsch, Robert W(illiam)
Eades, James B(everly), Jr
Edinger, Stanley Evan
Egner, Donald Otto
Eng, Leslie
Estrin, Norman Frederick
Evans, Larry Gerald
Fabic, Stanislav
Fairweather, William Ross
Feldman, Alfred Philip
Fink, Daniel J
Fischel, David
Fleig, Albert J, Jr
Flory, Thomas Reherd
Forman, Richard Allan
Frank, Carolyn
Frederick, William George DeMott
Gann, Richard George
Gardner, Sherwin
Gartrell, Charles Frederick
Garver, Robert Vernon
Gelberg, Alan
Gevantman, Lewis Herman
Gilmore, Thomas Meyer
Gleissner, Gene Heiden
Glick, J Leslie
Goldberg, Michael Ian
Goldschmidt, Peter Graham
Gonano, John Roland
Graebert, Eric W
Gupta, Ramesh K
Hader, Rodney N(eal)
Haffner, Richard William
Hamilton, Bruce King
Hammersmith, John L(eo)
Handwerker, Thomas Samuel
Hartmann, Gregory Kemenyi
Heath, George A(ugustine)
Heller, Charles O(ta)
Heller, Stephen Richard
Hellwig, Helmut Wilhelm
Heppner, James P
Hertz, Harry Steven
Holt, Stephen S
Hopenfield, Joram
Hopkins, Homer Thawley
Hsia, Jack Jinn-Goe
Huneycutt, James Ernest, Jr
Jackson, Benjamin A
Johnson, David Lee
Johnson, David Simonds
Jones, Theodore Charles
Kayser, Richard Francis
Kazi, Abdul Halim
Kiebler, John W(illiam)
Kirby, Ralph C(loudsberry)
Kirkpatrick, Diana (Rorabaugh) M
Koch, William Frederick
Ku, Jentung
Kutik, Leon
Lankford, J(ohn) L(lewellyn)
Leaders, Floyd Edwin, Jr
Leight, Walter Gilbert
Lettieri, Thomas Robert
Lieberman, Richard Barry
Lundholm, J(oseph) G(ideon), Jr
McBride, Gordon Williams
McElhinney, John
MacFadden, Kenneth Orville
McIlvain, Jess Hall
Mandelberg, Hirsch I
Marple, Stanley Lawrence, Jr
Michaelis, Michael
Miller, A Eugene
Miller, Raymond Earl
Mohler, William C
Morgan, Walter L(eroy)

Technical Management (cont)

Morgan, William Bruce
Morton, Joseph James Pandozzi
Moshman, Jack
Muller, Robert E
Mumford, Willard R
Nagle, Dennis Charles
Nash, Jonathon Michael
Orlick, Charles Alex
Ostaff, William A(llen)
Ostrofsky, Bernard
Penhollow, John O
Peppin, Richard J
Pierce, Harry Frederick
Poltorak, Andrew Stephen
Pomerantz, Irwin Herman
Potocki, Kenneth Anthony
Prentice, Geoffrey Allan
Quinn, Thomas Patrick
Randolph, Lynwood Parker
Rankin, Sidney
Rao, Gopalakrishna M
Rasberry, Stanley Dexter
Reuther, Theodore Carl, Jr
Rochlin, Robert Sumner
Rosen, Milton W(illiam)
Roth, Michael William
Saltman, Roy G
Schemm, Charles Edward
Schindler, Albert Isadore
Schleiter, Thomas Gerard
Schmidt, Jack Russell
Schroeder, Frank, Jr
Schulman, James Herbert
Schultz, Warren Walter
Schuman, William John, Jr
Schwoerer, F(rank)
Sinnott, George
Smith, Joseph Collins
Steiner, Henry M
Strombotne, Richard L(amar)
Stuebing, Edward Willis
Stuelpnagel, John Clay
Swanson, David Henry
Sztankay, Zoltan Geza
Talbott, Edwin M
Theuer, Paul John
Thompson, Robert John, Jr
Tropf, Cheryl Griffiths
Troutman, James Scott
Turnrose, Barry Edmund
Van De Merwe, Willem Pieter
Vanderryn, Jack
Ventre, Francis Thomas
Wacker, George Adolf
Walker, William J, Jr
Wallace, Alton Smith
Walsh, James Paul
Walter, John Fitler
Wang, Shou-Ling
Wende, Charles David
Williams, Jimmy Calvin
Wing, James
Yoho, Clayton W
Zehner, Lee Randall
Zink, Sandra
Zwilsky, Klaus M(ax)

MASSACHUSETTS
Adler, Ralph Peter Isaac
Ahern, David George
Allen, Thomas John
Avila, Charles Francis
Bailey, Fred Coolidge
Baliunas, Sallie Louise
Bass, Arthur
Bhargava, Hridaya Nath
Bonis, Laszlo Joseph
Brandler, Philip
Breslau, Barry Richard
Brown, Walter Redvers John
Buckler, Sheldon A
Buono, John Arthur
Callerame, Joseph
Carton, Edwin Beck
Cathou, Renata Egone
Clopper, Herschel
Close, R(ichard) N(orcross)
Cohen, Fredric Sumner
Colby, George Vincent, Jr
Das, Pradip Kumar
Deckert, Cheryl A
Dow, Paul C(rowther), Jr
Dudgeon, Dan Ed
Duffy, Robert A
Eaves, Reuben Elco
Eby, Robert Newcomer
Ehntholt, Daniel James
Eklund, Karl E
Fennelly, Paul Francis
Fix, Richard Conrad
Fortmann, Thomas Edward
Freedman, George
Frosch, Robert Alan
Gardner, Donald Murray
Gelb, Arthur
Gersh, Michael Elliot
Giger, Adolf J
Gordon, Bernard M
Gray, Douglas Carmon
Gregory, Bob Lee
Griffin, Paul Joseph
Griffin, William G
Griffith, Robert W
Guidice, Donald Anthony
Gund, Peter Herman Lourie
Haber, Stephen B
Hadlock, Charles Robert
Hanselman, Raymond Bush
Hertweck, Gerald
Herz, Matthew Lawrence
Heyerdahl, Eugene Gerhardt
Hockney, Richard L
Hodgins, George Raymond
Holmes, Douglas Burnham
Howard, Wilmont Frederick, Jr
Jalan, Vinod Motilal
Johnson, Aldie E(dwin), Jr
Kachinsky, Robert Joseph
Kapor, Mitchell David
Karo, Douglas Paul
Keast, Craig Lewis
Keast, David N(orris)
Kelley, Albert J(oseph)
Kelley, William S
Kepper, Robert Edgar
Kitchin, John Francis
Klein, Richard Morris
Kleinmann, Douglas Erwin
Kocon, Richard William
Kolodzy, Paul John
Kovatch, George
Kramer, Charles Edwin
Krause, Irvin
Lanzkron, Rolf W
Lavin, Philip Todd
Lederman, David Mordechai
Lee, D(on) William
Lee, Eric Kin-Lam
Levine, Randolph Herbert
Lien, Hwachii
Liggero, Samuel Henry
Luft, Ludwig
McClatchey, Robert Alan
Mack, Richard Bruce
Manley, Harold J(ames)
Manning, Monis Joseph
Maser, Morton D
Matsuda, Seigo
Mehta, Mahendra
Mehta, Mahesh J
Meyer, Vincent D
Miffitt, Donald Charles
Milligan, Terry Wilson
Muni, Indu A
Nelson, J(ohn) Byron
Neuringer, Leo J
Nickerson, Richard G
Norton, Robert Leo
Oberteuffer, John Amiard
Olsen, Kenneth
Plummer, William Torsch
Poirier, Victor L
Rhoads, Donald Cave
Riggi, Stephen Joseph
Roberts, Francis Donald
Rogers, William Irvine
Rosenthal, Richard Alan
Roth, Harold
Rubin, Lawrence G
Sagalyn, Paul Leon
Sahatjian, Ronald Alexander
Salant, Abner Samuel
Schaffel, Gerson Samuel
Schempp, Ellory
Scherer, Harold Nicholas, Jr
Seamans, R(obert) C(hanning), Jr
Serra, Jerry M
Servi, I(talo) S(olomon)
Shaffer, Harry Leonard
Shu, Larry Steven
Silevich, Michael B
Singleton, John Byrne
Sioui, Richard Henry
Sonnichsen, Harold Marvin
Sosnowski, Thomas Patrick
Spergel, Philip
Stanbrough, Jesse Hedrick, Jr
Stevens, J(ames) I(rwin)
Straub, Wolf Deter
Strelzoff, Alan G
Subramanian, Krishnamoorthy
Thornton, Richard D(ouglas)
Topping, Richard Francis
Towns, Donald Lionel
Tsandoulas, Gerasimos Nicholas
Turnquist, Carl Richard
Van Meter, David
Wagner, David Kendall
Wald, Fritz Veit
Wasserman, Jerry
Wentworth, Stanley Earl
Wheeler, Ned Brent
Whitehouse, David R(empfer)
Winett, Joel M
Winkelman, James W
Wood, David Belden
Yannoni, Nicholas

MICHIGAN
Aaron, Charles Sidney
Anderson, Amos Robert
Atwell, William Henry
Axen, Udo Friedrich
Ayres, Robert Allen
Bailey, Donald Leroy
Becher, William D(on)
Belmont, Daniel Thomas
Bens, Frederick Peter
Blaser, Dwight A
Blurton, Keith F
Bolling, Gustaf Fredric
Borcherts, Robert H
Braughler, John Mark
Brown, Steven Michael
Castenson, Roger R
Chandra, Grish
Chen, Kan
Chon, Choon Taik
Crummett, Warren B
Cunningham, Fay Lavere
Cutler, Warren Gale
Dasgupta, Rathindra
Decker, R(aymond) F(rank)
Dukes, Gary Rinehart
Eaton, Robert James
Evaldson, Rune L
Ezzat, Hazem Ahmed
Felmlee, William John
Flynn, John M(athew)
Francis, Ray Llewellyn
Frank, Max
Galan, Louis
Garvin, Donald Frank
Hahn, Russell H
Halberstadt, Marcel Leon
Harris, Kenneth
Hartman, John L(ouis)
Harwood, Julius Y
Holland, Steven William
Howe, Norman Elton, Jr
Jensen, David James
Jones, Frederick Goodwin
Joshi, Mukund Shankar
Kimball, Frances Adrienne
Klimisch, Richard L
Knoll, Alan Howard
Langvardt, Patrick William
Larson, John Grant
Lentz, Charles Wesley
Lett, Philip W(ood), Jr
Lorch, Steven Kalman
Macriss, Robert A
Malchick, Sherwin Paul
Maley, Wayne A
Medick, Matthew A
Mould, Peter Roy
Petrick, Ernest N(icholas)
Pincus, Jack Howard
Reynolds, John Z
Riley, Bernard Jerome
Rohde, Steve Mark
Rollinger, Charles N(icholas)
Rosinski, Michael A
Roth, Norman Gilbert
Salinger, Rudolf Michael
Sanders, Barbara A
Sard, Richard
Schumacher, Berthold Walter
Shah, Shirish A
Shahabuddin, Syed
Spreitzer, William Matthew
Spurgeon, William Marion
Stark, Forrest Otto
Stempel, Robert C
Stevens, Violete L
Stowe, Robert Allen
Thompson, James Lowry
Thurber, William Samuels
Tompkins, Curtis Johnston
Turley, June Williams
Van Rheenen, Verlan H
Walters, Stephen Milo
Weyenberg, Donald Richard
Whitfield, Richard George
Williams, Sam B
Wims, Andrew Montgomery

MINNESOTA
Atchison, Thomas Calvin, Jr
Bohon, Robert Lynn
Bolles, Theodore Frederick
Bonne, Ulrich
Britz, Galen C
Calehuff, Girard Lester
Case, Marvin Theodore
Davidson, Donald
Dinneen, Gerald Paul
Durnick, Thomas Jackson
Fleming, Peter B
Glewwe, Carl W(illiam)
Greene, Christopher Storm
Hapke, Bern
Haun, J(ames) W(illiam)
Heberlein, Joachim Viktor Rudolf
Hoffmann, Thomas Russell
Holmen, Reynold Emanuel
Jansen, Bernard Joseph
Johnson, Bryce Vincent
Johnson, Howard Arthur, Sr
Johnson, Lennart Ingemar
Joseph, Earl Clark, II
Katz, William
Keitel, Glenn H(oward)
Meeder, Jeanne Elizabeth
Nelson, Kyler Fischer
Penney, William Harry
Plovnick, Ross Harris
Premanand, Visvanatha
Pryor, Gordon Roy
Reich, Charles
Reinhart, Richard D
Reynolds, James Harold
Rodgers, Nelson Earl
Sackett, W(illiam) T(ecumseh), Jr
Sadjadi, Firooz Ahmadi
Schumer, Douglas Brian
Sheppard, John Richard
Sherck, Charles Keith
Spilker, Bert
Stearns, Eugene Marion, Jr
Talbott, Richard Lloyd
Tamsky, Morgan Jerome
Thangaraj, Sandy
Wade, James Joseph
Weiss, Douglas Eugene
Yapel, Anthony Francis, Jr
Zeyen, Richard John

MISSISSIPPI
Brunton, George Delbert
Cotton, Frank Ethridge, Jr
Ferer, Kenneth Michael
George, Clifford Eugene
Gupta, Shyam Kirti
Ramsdale, Dan Jerry

MISSOURI
Babcock, Daniel Lawrence
Bachman, Gerald Lee
Baltz, Howard Burl
Beaver, Earl Richard
Bodine, Richard Shearon
Bosanquet, Louis Percival
Carter, Don E
Craver, Clara Diddle (Smith)
Craver, John Kenneth
Dahl, William Edward
Deam, James Richard
Frankoski, Stanley P
Glauz, William Donald
Griffith, Virgil Vernon
Harland, Ronald Scott
Holloway, Thomas Thornton
Keller, Robert Ellis
Kohlmeier, Ronald Harold
Kozlowski, Don Robert
Kramer, Richard Melvyn
Krueger, Paul A
Kurz, James Eckhardt
Lannert, Kent Philip
Linder, Solomon Leon
Lyle, Leon Richards
Malik, Joseph Martin
Marshall, Lucia Garcia-Iniguez
Mattox, Douglas M
Morgan, Wyman
Munger, Paul R
Oliver, Larry Ray
Poynton, Joseph Patrick
Privott, Wilbur Joseph, Jr
Proctor, Stanley Irving, Jr
Radke, Rodney Owen
Rash, Jay Justen
Rinehart, Walter Arley
Rueppel, Melvin Leslie
Ryan, Carl Ray
Sarchet, Bernard Reginald
Scallet, Barrett Lerner
Schott, Jeffrey Howard
Schwartz, Henry Gerard, Jr
Sidoti, Daniel Robert
Styron, Clarence Edward, Jr
Thornton, Linda Wierman
Volz, William K(urt)
Wagenknecht, John Henry
Wehrmann, Ralph F(rederick)
Wohl, Martin H
Worley, Jimmy Weldon

MONTANA
Manhart, Robert (Audley)

NEBRASKA
James, Merlin Lehn
Kolar, Joseph Robert, Jr
Wittke, Dayton D

NEVADA
Fox, Neil Stewart
Haun, Robert Dee, Jr
Klainer, Stanley M
Mahadeva, Madhu Narayan
Wegst, Walter F, Jr

NEW HAMPSHIRE
Bahr, Karl Edward
Block, James A
Carreker, R(oland) P(olk), Jr
Creagh-Dexter, Linda T
Dancy, Terence E(rnest)
Egan, John Frederick
Ge, Weikun
Hunt, Everett Clair
Light, Thomas Burwell
Marion, Robert Howard
Petrasek, Emil John
Wang, Victor S F
Wolff, Nikolaus Emanuel
Zager, Ronald

NEW JERSEY
Adams, James Mills
Aldrich, Haven Scott
Andose, Joseph D

Andrews, Ronald Allen
Auerbach, Victor
Bardoliwalla, Dinshaw Framroze
Bartnoff, Shepard
Battisti, Angelo James
Beck, John Louis
Belanger, David Gerald
Bendure, Raymond Lee
Bernstein, Lawrence
Bieber, Harold H
Bieber, Herman
Bodin, Jerome Irwin
Bonacci, John C
Bond, Robert Harold
Borkan, Harold
Bornstein, Alan Arnold
Boucher, Thomas Owen
Bowen, J Hartley, Jr
Boyle, Richard James
Brandman, Harold A
Brill, Yvonne Claeys
Brofazi, Frederick R
Buckley, R Russ
Buhks, Ephraim
Burley, David Richard
Carlson, Curtis Raymond
Carnes, James Edward
Carter, Ashley Hale
Castenschiold, Rene
Chance, Ronald Richard
Chester, Arthur Warren
Chiao, Wen Bin
Chynoweth, Alan Gerald
Cipriani, Cipriano
Coombs, Robert Victor
Cornell, W(arren) A(lvan)
Cuffari, Gilbert Luke
Curcio, Lawrence Nicholas
Curry, Michael Joseph
Daly, Daniel Francis, Jr
Das, Santosh Kumar
Derucher, Kenneth Noel
Diegnan, Glenn Alan
DiSalvo, Walter A
Domeshek, S(ol)
Donovan, Richard C
Dorros, Irwin
Edelstein, Harold
Eisenberger, Peter Michael
Eisner, Mark Joseph
Falconer, Warren Edgar
Feldman, Leonard Cecil
Finch, Rogers B(urton)
Fischell, David R
Flam, Eric
Founds, Henry William
Fowles, Patrick Ernest
Francis, Samuel Hopkins
Franks, Richard Lee
Freedman, Henry Hillel
Friedman, Edward Alan
Froyd, James Donald
Garber, H(irsh) Newton
Garfinkel, Harmon Mark
Gierer, Paul L
Glass, Michael
Goffman, Martin
Goldemberg, Robert Lewis
Goulary, Barry Sholom
Grauman, Joseph Uri
Gregory, Richard Alan, Jr
Grenda, Victor J
Groeger, Theodore Oskar
Grubman, Wallace Karl
Grumer, Eugene Lawrence
Guenzer, Charles S P
Gyan, Nanik D
Hagel, Robert B
Halik, Raymond R(ichard)
Harding, Maurice James Charles
Harris, James Ridout
Hartung, John
Hatch, Charles Eldridge, III
Hay, Peter Marsland
Heath, Carl E(rnest), Jr
Heck, James Virgil
Heirman, Donald N
Hendrickson, Tom A(llen)
Herdklotz, John Key
Hlavacek, Robert John
Hong, Allan Jixian
Hong, Won-Pyo
Hubbard, William Marshall
Hundert, Murray Bernard
Hymans, William E
Inglis, James
Jaffe, Jonah
Jayant, Nuggehally S
Jennings, Carl Anthony
Johnston, John Eric
Kagan, Harvey Alexander
Kahn, Donald Jay
Kaufmann, Thomas G(erald)
Keramidas, Vassilis George
Kesselman, Warren Arthur
King, John A(lbert)
Knapp, Malcolm Hammond
Kobrin, Robert Jay
Krevsky, Seymour
Kronenthal, Richard Leonard
Kruh, Daniel
Ladner, David William
LaMastro, Robert Anthony
Langseth, Rollin Edward

Lasky, Jack Samuel
Lawson, James Robert
Leaning, William Henry Dickens
Lederman, Peter B
Leitz, Victoria Mary
Letterman, Herbert
Lewis, Arnold D
Liebe, Donald Charles
Liebig, William John
London, Mark David
Lorenz, Patricia Ann
Lyding, Arthur R
McCallum, Charles John, Jr
McDermott, Kevin J
McGuire, David Kelty
Mackerer, Carl Robert
Mandel, Andrea Sue
Manz, August Frederick
Martin, John David
Melveger, Alvin Joseph
Michaelis, Paul Charles
Michel, Gerd Wilhelm
Mikolajczak, Alojzy Antoni
Montana, Anthony J
Montgomery, Richard Millar
Moros, Stephen Andrew
Moshey, Edward A
Munger, Stanley H(iram)
Murthy, Andiappan Kumaresa Sundara
Murthy, Srinivasa K R
Mustafa, Shams
Nelson, Roger Edwin
Noone, Thomas Mark
Omar, Mostafa M
Opie, William R(obert)
Ortiz-Martinez, Aury
Pebly, Harry E
Perry, Erik David
Peters, Dale Thompson
Philipp, Ronald E
Popper, Robert David
Putter, Irving
Rau, Eric
Reboul, Theo Todd, III
Rein, Alan James
Ren, Peter
Robin, Michael
Robinson, Jerome David
Robrock, Richard Barker, II
Rorabaugh, Donald T
Rubinstein, Charles B(enjamin)
Sachs, Harvey Maurice
Sagal, Matthew Warren
Sausville, Joseph Winston
Sawin, Steven P
Scharfstein, Lawrence Robert
Schmidle, Claude Joseph
Schnapf, Abraham
Schoenfeld, Theodore Mark
Schutz, Donald Frank
Schwarz, Maurice Jacob
Schwenker, Robert Frederick, Jr
Sharp, Hugh T
Shelton, James Churchill
Sherman, Ronald
Siebert, Donald Robert
Sipress, Jack M
Snowman, Alfred
Sorenson, Wayne Richard
Spohn, Ralph Joseph
Stinson, Mary Krystyna
Strausberg, Sanford I
Sweed, Norman Harris
Tauber, Arthur
Thelin, Lowell Charles
Tompsett, Michael F
Toothill, Richard B
Tucci, Edmond Raymond
Van Saun, William Arthur
Vogt, Herwart Curt
Wangemann, Robert Theodore
Warters, William Dennis
Watson, Richard White, Jr
Weinzimmer, Fred
Wesner, John William
West, John M(aurice)
Westerdahl, Raymond P
White, Alice Elizabeth
Wildman, George Thomas
Wildnauer, Richard Harry
Witherell, Peter Charles
Wittick, James John

NEW MEXICO
Altrock, Richard Charles
Apt, Kenneth Ellis
Arion, Douglas
Asbridge, John Robert
Barr, Donald Westwood
Bartlit, John R(ahn)
Berrie, David William
Browne, John Charles
Burick, Richard Joseph
Butler, Harold S
Caldwell, John Thomas
Catlett, Duane Stewart
Chambers, William Hyland
Charles, Robert Wilson
Clark, Robert Paul
Clough, Richard H(udson)
Colp, John Lewis
Conner, Jerry Power
Cramer, James D
Cunningham, Paul Thomas

Danen, Wayne C
Davison, Lee Walker
Diegle, Ronald Bruce
Dreicer, Harry
Dumas, Herbert M, Jr
Erikson, Mary Jane
Farnum, Eugene Howard
Finch, Thomas Wesley
Franzak, Edmund George
Freese, Kenneth Brooks
Gancarz, Alexander John
Garcia, Carlos E(rnesto)
Gibbs, Terry Ralph
Gibson, Benjamin Franklin, V
Gillespie, Claude Milton
Goldstone, Philip David
Gover, James E
Gwyn, Charles William
Hansen, Wayne Richard
Hartley, Danny L
Henderson, Dale Barlow
Hill, Ronald Ames
Howell, Gregory A
Icerman, Larry
Jameson, Robert A
Janney, Donald Herbert
Jeffries, Robert Alan
Johnson, Ralph T, Jr
Jones, Kirkland Lee
Jones, Merrill C(alvin)
Jones, Orval Elmer
Keenan, Thomas K
King, Nicholas S P
Kjeldgaard, Edwin Andreas
Knief, Ronald Allen
Krupka, Milton Clifford
Kuswa, Glenn Wesley
Leonard, Ellen Marie
Lohrding, Ronald Keith
Lyons, Peter Bruce
McCulla, William Harvey
Marcy, Willard
Mason, Allen Smith
Matthews, R(obert) B(ruce)
Merritt, Melvin Leroy
Molecke, Martin A
Neeper, Donald Andrew
Newman, Michael J(ohn)
North-Root, Helen May
O'Donnell, Edward Earl
Osbourn, Gordon Cecil
Otway, Harry John
Pappas, Daniel Samuel
Peercy, Paul S
Perkins, Roger Bruce
Phister, Montgomery, Jr
Pynn, Roger
Renken, James Howard
Robinson, Ross Utley
Rohde, Richard Whitney
Ryder, Richard Daniel
Savage, Charles Francis
Schneider, Jacob David
Schueler, Donald G(eorge)
Searls, Craig Allen
Seay, Glenn Emmett
Selden, Robert Wentworth
Shipley, James Parish, Jr
Shore, Fred L
Sivinski, Jacek Stefan
Solem, Johndale Christian
Strniste, Gary F
Sullivan, J Al
Tapia, Santiago
Tapp, Charles Millard
Tesmer, Joseph Ransdell
Thode, E(dward) F(rederick)
Toepfer, Alan James
Van Domelen, Bruce Harold
Venable, Douglas
Wallace, Terry Charles, Sr
Welber, Irwin
White, Paul C

NEW YORK
Abate, Kenneth
Abbas, Daniel Cornelius
Abetti, Pier Antonio
Andersen, Kenneth J
Armstrong, John A
Bachman, Henry L(ee)
Bahary, William S
Batter, John F(rederic), Jr
Baum, Martin David
Beer, Sylvan Zavi
Benzinger, James Robert
Benzinger, William Donald
Bigelow, Sanford Walker
Birnbaum, David
Birnstiel, Charles
Blackmon, Bobby Glenn
Blaker, J Warren
Bolon, Roger B
Bottger, Gary Lee
Bray, James William
Britton, Marvin Gale
Broadwin, Alan
Buchel, Johannes A
Buhsmer, Charles P
Burnham, Dwight Comber
Chiang, Yuen-Sheng
Chiulli, Angelo Joseph
Christensen, Robert Lee
Christiansen, Donald David

Cohen, Edwin
Collins, James Joseph
Couch, Troy Lee
Coutchie, Pamela Ann
Dahm, Douglas Barrett
Daniels, John Maynard
Dehm, Richard Lavern
Dieterich, David Allan
Duke, David Allen
Edelheit, Lewis S
Eschbach, Charles Scott
Feit, Irving N
Feldman, Larry Howard
Fenrich, Richard Karl
Fey, Curt F
Fiedler, Harold Joseph
Fiske, Milan Derbyshire
Flannery, John B, Jr
Foley, Dennis Joseph
Foreman, Kenneth M
Formanek, R(obert) J(oseph)
Foucher, W(alter) D(avid)
Fuerst, Adolph
Gale, Stephen Bruce
Gambs, Gerard Charles
Gaudioso, Stephen Lawrence
Genin, Dennis Joseph
George, Philip Donald
Gerace, Paul Louis
Goldman, James Allan
Goldstein, Lawrence Howard
Golibersuch, David Clarence
Gomory, Ralph E
Goodwin, Arthur VanKleek
Gray, D Anthony
Green, Robert I
Grossman, Norman
Haas, David Jean
Haas, Werner E L
Haase, Jan Raymond
Haid, D(avid) A(ugustus)
Halket, Thomas D
Hall, Kimball Parker
Hall, Richard Travis
Hammer, Richard Benjamin
Handman, Stanley E
Harris, Bernard
Harrison, James Beckman
Haynes, Martha Patricia
Hedman, Dale E
Henkel, Elmer Thomas
Himpsel, Franz J
Hollenberg, Joel Warren
Hull, Andrew F
Indusi, Joseph Paul
Insalata, Nino F
James, Daniel Shaw
Janik, Gerald S
Jefferies, Michael John
Jensen, Betty Klainminc
Kagetsu, T(adashi) J(ack)
Kahn, Elliott H
Kammer, Paul A
Karz, Robert Stephen
Kasprzak, Lucian A
Kato, Walter Yoneo
Katz, Maurice Joseph
Kelisky, Richard Paul
Kelly, Thomas Joseph
Kelly, Thomas Michael
Kenett, Ron
Ketchen, Mark B
Khan, Paul
King, Marvin
Kirchmayer, Leon K
Kitchen, Sumner Wendell
Klijanowicz, James Edward
Klotzbach, Robert J(ames)
Kolaian, Jack H
Kolb, Frederick J(ohn), Jr
Korwin-Pawlowski, Michael Lech
Krieger, Gary Lawrence
Kuchar, Norman Russell
Landzberg, Abraham H(arold)
Langworthy, Harold Frederick
Laubach, Gerald D
Law, Paul Arthur
Lerman, Steven I
Litz, Lawrence Marvin
Logue, J(oseph) C(arl)
Longobardo, Anna Kazanjian
Lorenzo, George Albert
Lowder, Wayne Morris
Lurkis, Alexander
Luther, Herbert George
MacDowell, John Fraser
McElligott, Peter Edward
McGee, James Patrick
McHugh, James Anthony, Jr
Madey, Robert W
Marr, David Henry
Martens, Alexander E(ugene)
Mason, John Hugh
Meenan, Peter Michael
Merten, Alan Gilbert
Milton, Kirby Mitchell
Minnear, William Paul
Mort, Joseph
Mortenson, Kenneth Ernest
Mullen, Patricia Ann
Nelson, Robert Andrew
Nicolosi, Joseph Anthony
Ning, Tak Hung
Orlando, Charles M

Technical Management (cont)

Palmedo, Philip F
Parham, Margaret Payne
Patmore, Edwin Lee
Pearson, Glen Hamilton
Pedroza, Gregorio Cruz
Penwell, Richard Carlton
Przybylowicz, Edwin P
Randall, John D(el)
Rau, R Ronald
Reber, Raymond Andrew
Redington, Rowland Wells
Reich, Ismar M(eyer)
Rich, Kenneth Eugene
Richards, Jack Lester
Riffenburgh, Robert Harry
Rosen, Stephen
Rubin, Isaac D
Sanford, Karl John
Scala, Sinclaire M(aximilian)
Schachter, Rozalie
Schanker, Jacob Z
Schindler, Joe Paul
Schottmiller, John Charles
Schucker, Gerald D
Schulz, Helmut Wilhelm
Schwan, Thomas James
Schwartz, Herbert Mark
Sen, Ujjal
Setchell, John Stanford, Jr
Seward, Thomas Philip, III
Shaw, David Elliot
Sheeran, Stanley Robert
Shevack, Hilda N
Shuey, R(ichard) L(yman)
Sieg, Albert Louis
Siegart, William Raymond
Simons, Gene R
Siuta, Gerald Joseph
Slusarek, Lidia
Solomon, Jack
Springett, Brian E
Stanley, Edward Alexander
Sterman, Samuel
Stover, Raymond Webster
Strasser, Alfred Anthony
Strnisa, Fred V
Strojny, Norman
Stupp, Edward Henry
Sullivan, Michael Francis
Taub, Aaron M
Thieberger, Peter
Thompson, David Allen
Thompson, David Fred
Tuite, Robert Joseph
Tuli, Jagdish Kumar
Turnblom, Ernest Wayne
Uretsky, Myron
Vogel, Philip Christian
Vosburgh, Kirby Gannett
Wachtell, Richard L(loyd)
Wallis, Thomas Gary
Wang, Shu Lung
Werthamer, N Richard
Wertheimer, Alan Lee
Wien, Richard W, Jr
Wiese, Warren M(elvin)
Wolfe, Roger Thomas
Wood, Paul Mix
Wyman, Donald Paul
Yasar, Tugrul
York, Raymond A
Zdan, William
Zhou, Zhen-Hong

NORTH CAROLINA

Ahrens, Rolland William
Bockstahler, Theodore Edwin
Bode, Arthur Palfrey
Breuer, Charles B
Burger, Robert M
Bursey, Joan Tesarek
Carlson, William Theodore
Chin, Robert Allen
Conn, Paul Kohler
Cote, Philip Norman
Cubberley, Adrian H
Dibner, Mark Douglas
Dickerson, James Perry
Etter, Robert Miller
Fair, Richard Barton
Fytelson, Milton
Golden, Carole Ann
Gonder, Eric Charles
Harrison, Stanley L
Hurwitz, Melvin David
Iverson, F Kenneth
James, Karen K(anke)
Jones, Rufus Sidney
Jordan, Edward Daniel
Kanopoulos, Nick
Kirk, Wilber Wolfe
Kohn, Frank S
Lontz, Robert Jan
McCumber, Dean Everett
Min, Tony C(harles)
Moody, Max Dale
Mosier, Stephen R
O'Connor, Lila Hunt
Owens, Boone Bailey
Price, Howard Charles
Priestley, G T Eric
Pruett, Roy L

Quisenberry, Karl Spangler, Jr
Reisman, Arnold
Reynolds, John Hughes, IV
Richardson, Stephen Giles
Rideout, Janet Litster
Roblin, John M
Roth, Roy William
Rubloff, Gary W
Sargeant, Peter Barry
Saunders, Donald Roy
Stenger, William J(ames), Sr
Stubblefield, Charles Bryan
Sud, Ish
Tartt, Thomas Edward
Trenholm, Andrew Rutledge
Wasson, John R
Wehner, Philip
Wilkes, William Roy
Yeowell, David Arthur
Young, William Anthony

NORTH DAKOTA

Bares, William Anthony
Watt, David Milne, Jr

OHIO

Adamczak, Robert L
Alden, Carl L
Amjad, Zahid
Antler, Morton
Ault, George Mervin
Badertscher, Robert F(rederick)
Bahnfleth, Donald R
Baid, Kushalkumar Moolchand
Barsky, Constance Kay
Battershell, Robert Dean
Becker, Carter Miles
Bierlein, James A(llison)
Boggs, Robert Wayne
Bovenkerk, Harold P(aul)
Brandhorst, Henry William, Jr
Broge, Robert Walter
Brunner, Gordon Francis
Burte, Harris M(erl)
Campbell, Robert Wayne
Carman, Charles Jerry
Chamis, Alice Yanosko
Chan, Yupo
Christian, John B
Colby, Edward Eugene
Coleman, Marilyn A
Copeland, James Clinton
Cordea, James Nicholas
Crable, John Vincent
Dewar, Norman Ellison
Easterday, Jack L(eroy)
Erbacher, John Kornel
Faiman, Robert N(eil)
Feldman, Julian
Fordyce, James Stuart
Forestieri, Americo F
Forsyth, Thomas Henry
Fraker, John Richard
Frank, Thomas Paul
Freeberg, Fred E
Gage, Frederick Worthington
Galloway, Ethan Charles
Garscadden, Alan
Gerace, Michael Joseph
Ginning, P(aul) R(oll)
Golden, John Terence
Gorland, Sol H
Graef, Walter L
Grasselli, Jeanette Gecsy
Griffith, Walter M, Jr
Hannah, Sidney Allison
Harrington, Roy Victor
Hartsough, Robert Ray
Hein, Richard William
Henderson, Courtland M
Henry, William Mellinger
Holubec, Zenowie Michael
Ingram, David Christopher
Innes, John Edwin
Jackson, Curtis M(aitland)
Jayne, Theodore D
Johnson, Ralph Sterling, Jr
Katzen, Raphael
Keily, Hubert Joseph
Kelley, Frank Nicholas
Kinsman, Donald Vincent
Kircher, John Frederick
Kochanowski, Barbara Ann
Laughon, Robert Bush
Lenhart, Jack G
Letton, James Carey
Leyda, James Perkins
Lydy, David Lee
McCune, Homer Wallace
Macklin, Martin
McMillin, Carl Richard
McVean, Duncan Edward
Magee, Thomas Alexander
Mao, Hsiang-Kuen Mark
Marks, Alfred Finlay
Marshall, Kenneth D
Martino, Joseph Paul
Mathews, A L
Metcalfe, Joseph Edward, III
Michael, William R
Moncrief, Eugene Charles
Morgan, William R(ichard)
Mosser, John Snavely
Olesen, Douglas Eugene

Paynter, John, Jr
Pearson, Jerome
Piper, Ervin L
Platau, Gerard Oscar
Poggenburg, John Kenneth, Jr
Prince, Terry Jamison
Pugh, John W(illiam)
Purdon, James Ralph, Jr
Purvis, John Thomas
Razgaitis, Richard A
Replogle, Clyde R
Richley, E(dward) A(nthony)
Rodman, Charles William
Roehrig, Frederick Karl
Ross, John Edward
Schoch, Daniel Anthony
Schultz, Edwin Robert
Schultz, Thomas J
Schumm, Brooke, Jr
Schutta, James Thomas
Scozzie, James Anthony
Sharma, Shri C
Siefken, Mark William
Sinner, Donald H
Skees, Hugh Benedict
Soska, Geary Victor
Stadler, Louis Benjamin
Staker, Donald David
Stambaugh, Edgel Pryce
Standish, Norman Weston
Stephan, David George
Stephens, Marvin Wayne
Stotz, Robert William
Temple, Robert Dwight
Thomas, Alexander Edward, III
Tipton, C(lyde) R(aymond), Jr
Toeniskoetter, Richard Henry
Turchi, Peter John
Turner, Andrew
Ungar, Edward William
Updegrove, Louis B
Vassell, Gregory S
Ventresca, Carol
Versteegh, Larry Robert
Voigt, Charles Frederick
Wadelin, Coe William
Wakefield, Shirley Lorraine
Warshay, Marvin
Weber, Lester George
Weeks, Stephen P
Weeks, Thomas Joseph, Jr
Westerman, Arthur B(aer)
Wharton, H(arry) Whitney
Wickham, William Terry, Jr
Wigington, Ronald L
Wong, Anthony Sai-Hung
Yellin, Wilbur

OKLAHOMA

Carlile, Robert Elliot
Cowley, Thomas Gladman
Crafton, Paul A(rthur)
Davis, Robert Elliott
Ellington, Rex T(ruesdale), Jr
Evens, F Monte
Feiock, Frank Donald
Green, Ray Charles
Greenwood, Gil Jay
Hale, John Dewey
Hale, William Henry, Jr
Hopkins, Thomas R (Tim)
Howard, Robert Eugene
Johnston, Harlin Dee
Koerner, E(rnest) L(ee)
Kokesh, Fritz Carl
Lambird, Perry Albert
McCollom, Kenneth A(llen)
Monn, Donald Edgar
Paxson, John Ralph
Ripley, Dennis L(eon)
Rotenberg, Don Harvey
Schaffer, Arnold Martin
Shell, Francis Joseph
Still, Edwin Tanner
Stover, Dennis Eugene
Terrell, Marvin Palmer
West, John B(ernard)
Whitfill, Donald Lee
Zuech, Ernest A

OREGON

Autrey, Robert Luis
Brinkley, John Michael
Cleland, Franklin Andrew
Collins, John A(ddison)
Demaree, Thomas L
Ehrmantraut, Harry Charles
Fields, R(ance) Wayne
Hackleman, David E
Hansen, Hugh J
Hoffman, Michael G
Humphrey, J Richard
Jones, Donlan F(rancis)
Kocaoglu, Dundar F
Lendaris, George G(regory)
Orloff, Jonathan H
Preston, Eric Miles
Schroeder, Warren Lee
Scott, Peter Carlton
Shilling, Wilbur Leo
Wagner, David Henry
Walsh, Don

PENNSYLVANIA

Angeloni, Francis M
Bartish, Charles Michael Christopher
Bauman, Bernard D
Baumann, Dwight Maylon Billy
Baxter, Ronald Dale
Beedle, Lynn Simpson
Beilstein, Henry Richard
Beitchman, Burton David
Bell, Stanley C
Belohoubek, Erwin F
Berkoff, Charles Edward
Bingham, Richard Charles
Bowers, Klaus Dieter
Box, Larry
Bradley, James Henry Stobart
Braunstein, David Michael
Brendlinger, Darwin
Bretz, Philip Eric
Browning, Daniel Dwight
Bucher, John Henry
Burk, David Lawrence
Camp, Frederick William
Capaldi, Eugene Carmen
Cardarelli, Joseph S
Chen, Michael S K
Christie, Michael Allen
Colella, Donald Francis
Connor, James Edward, Jr
Connors, Donald R
Cook, Charles S
Corbett, Robert B(arnhart)
Craig, James Clifford, Jr
D'Alisa, Rose M
Dalton, Augustine Ivanhoe, Jr
Davis, Brian Clifton
Davis, Burl Edward
Derby, James Victor
Detwiler, John Stephen
Dicciani, Nance Katherine
Drake, B(illy) Blandin
Druffel, Larry Edward
Dumbri, Austin C
Durkee, Jackson L
Dyott, Thomas Michael
Ehrig, Raymond John
Einhorn, Philip A
Elder, James Franklin, Jr
Ellis, Jeffrey Raymond
Falcone, James Salvatore, Jr
Feero, William E
Fiore, Nicholas F
Frederick, David Eugene
Frumerman, Robert
Garfinkle, Barry David
Garrett, Thomas Boyd
Gertz, Steven Michael
Giannovario, Joseph Anthony
Giffen, Robert H(enry)
Gitlitz, Melvin H
Glessner, Alfred Joseph
Gluntz, Martin L
Golton, William Charles
Goodman, Alan Lawrence
Graybill, Donald Lee
Green, Lawrence
Griffith, Michael Grey
Hahn, Peter Mathias
Haines, William Joseph
Harkins, Thomas Regis
Herman, Frederick Louis
Herzog, Karl A
Herzog, Leonard Frederick, II
Hinton, Raymond Price
Hockswender, Thomas Richard
Hopson, Kevin Matthew
Howe, Richard Hildreth
Humphrey, Watts S
Jaffe, Donald
Jarrett, Noel
Johnson, Littleton Wales
Jones, Roger Franklin
Jordan, Andrew G
Judge, Joseph Malachi
Katz, Lewis E
Kazahaya, Masahiro Matt
Kehr, Clifton Leroy
Khan, Parwaiz Ashraf Ali
Kimblin, Clive William
Kingrea, James I, Jr
Kleinschuster, Jacob John
Koch, Ronald N
Korchak, Ernest I(an)
Kottcamp, Edward H, Jr
Krishnaswamy, S V
Kun, Kenneth Allan
Kunesh, Charles Joseph
Laemmle, Joseph Thomas
Lantos, P(eter) R(ichard)
Larson, Kenneth Curtis
Lavin, J Gerard
Leas, J(ohn) W(esley)
LeBlanc, Norman Francis
Lenox, Ronald Sheaffer
Levy, Bernard, Jr
Liu, Andrew Tze-chiu
Loprest, Frank James
MacDonald, Hubert C, Jr
McDonnell, Leo F(rancis)
Machalek, Milton David
Manuel, Thomas Asbury
Mao, Chung-Ling
Marmer, William Nelson
Martucci, John A

Mayer, George Emil
Merner, Richard Raymond
Meyer, Paul A
Mezger, Fritz Walter William
Michaelis, Arthur Frederick
Miller, Dale L
Moore, Walter Calvin
Mount, Lloyd Gordon
Moyer, Robert (Findley)
Murrin, Thomas J
Nedwick, John Joseph
Novak, Darwin Albert, Jr
Orphanides, Gus George
Owens, Frederick Hammann
Papariello, Gerald Joseph
Parish, Roger Cook
Paul, Anand Justin
Penman, Paul D
Petrich, Robert Paul
Pierson, Ellery Merwin
Pittman, G(eorge) F(rank), Jr
Prasad, Suresh
Procknow, Donald Eugene
Purdy, David Lawrence
Ramani, Raja Venkat
Redmount, Melvin B(ernard)
Reischman, Michael Mack
Richardson, Ralph J
Robinson, Charles Albert
Roseman, Arnold S(aul)
Rubin, Benjamin Arnold
Rush, James E
Rushton, Brian Mandel
Saggiomo, Andrew Joseph
Saluja, Jagdish Kumar
Santamaria, Vito William
Sayre, Clifford M(orrill), Jr
Schaller, Edward James
Schiffman, Louis F
Schulman, Marvin
Schwartz, Albert B
Scigliano, J Michael
Shellenberger, Donald J(ames)
Sheridan, John Joseph, III
Sherman, Anthony Michael
Shibib, M Ayman
Shoaf, Mary La Salle
Smith, Warren LaVerne
Snider, Albert Monroe, Jr
Sokolowski, Henry Alfred
Soler, Alan I(srael)
Spielvogel, Lawrence George
Spitznagel, John A
Stingelin, Ronald Werner
Stinson, William Sickman, Jr
Stone, Herman
Stryker, Lynden J
Sucov, E(ugene) W(illiam)
Sun, James Dean
Sykes, James Aubrey, Jr
Thomas, William J
Thompson, Ansel Frederick, Jr
Thompson, Sheldon Lee
Tobias, Philip E
Tobias, Russell Lawrence
Tuba, I Stephen
Tuthill, Harlan Lloyd
Tzodikov, Nathan Robert
Uher, Richard Anthony
Umen, Michael Jay
Vander Voort, George Frederic
Varma, Asha
Verhanovitz, Richard Frank
Vernon, John Ashbridge
Vidt, Edward James
Viest, Ivan M
Vorchheimer, Norman
Walker, Augustus Chapman
Watson, William Martin, Jr
Weber, Frank L
Weddell, George G(ray)
Weisfeld, Lewis Bernard
Weldes, Helmut H
White, George Rowland
Whiteman, John David
Wiehe, William Albert
Williams, James Earl, Jr
Zeigler, A(lfred) G(eyer)
Zinser, Edward John
Zweben, Carl Henry

RHODE ISLAND
Casey, John Edward, Jr
Donnelly, Grace Marie
Jarrett, Jeffrey E
Johnson, Gordon Carlton
Kawaters, Woody H
Logan, Joseph Skinner
Meschino, Joseph Albert
Poon, Calvin P C
Smith, Alan Jerrard
Spero, Caesar A(nthony), Jr
Von Riesen, Daniel Dean

SOUTH CAROLINA
Avegeropoulos, G
Benjamin, Richard Walter
Berger, Richard S
Boyd, George Edward
Gaines, Albert L(owery)
Garzon, Ruben Dario
Godfrey, W Lynn
Goldstein, Herman Bernard
Graulty, Robert Thomas

Hopper, Michael James
Jaco, Charles M, Jr
Joseph, J Walter, Jr
Krumrei, W(illiam) C(larence)
McCrosson, F Joseph
McDonell, William Robert
Morris, J(ames) William
O'Neill, John H(enry), Jr
Owen, John Harding
Parker, Julian E, III
Pawson, Beverly Ann
Pike, LeRoy
Przirembel, Christian E G
Robinson, Robert Earl
Terry, Stuart Lee
Tolbert, Thomas Warren
Townes, George Anderson
Wagner, William Sherwood

SOUTH DAKOTA
Cannon, Patrick Joseph
Koepsell, Paul L(oel)
Swiden, LaDell Ray
Wegman, Steven M

TENNESSEE
Adams, Robert Edward
Appleton, B R
Armentrout, Daryl Ralph
Baillargeon, Victor Paul
Baker, Charles Clayton
Baker, Merl
Baxter, John Edwards
Besser, John Edwin
Betz, Norman Leo
Bontadelli, James Albert
Boston, Charles Ray
Boudreau, William F(rancis)
Brewington, Percy, Jr
Clark, Richard Bennett
Cope, David Franklin
Cordell, Francis Merritt
Cowser, Kenneth Emery
Crane, Laura Jane
Daniel, Donald Clifton
Davidson, Robert C
Eads, B G
Fordham, J(ames) Lynn
Frye, Alva L(eonard)
Gant, Kathy Savage
Garber, Floyd Wayne
Gat, Uri
Grant, Peter Malcolm
Gray, Allen G(ibbs)
Gray, Theodore Flint, Jr
Guerin, Michael Richard
Harmer, David Edward
Hasek, Robert Hall
Haywood, Frederick F
Horak, James Albert
Horton, Linda Louise Schiestle
House, Robert W(illiam)
Jones, Charles Miller, Jr
Jones, Glenn Clark
Jones, Larry Warner
Kammann, Karl Philip, Jr
Keller, Charles A(lbert)
Kidd, George Joseph, Jr
Kraft, Edward Michael
Krutak, James John, Sr
Larkin, William (Joseph)
Lim, Alexander Te
Lotts, Adolphus Lloyd
Lubell, Martin S
Mansur, Louis Kenneth
Martin, Richard Blazo
Morrow, Roy Wayne
Murphy, William R
Naney, Michael Terrance
Pandeya, Prakash N
Patel, Tarun R
Pfuderer, Helen A
Phillips, Bobby Mal
Potter, John Leith
Prairie, Michael L
Quist, Arvin Sigvard
Rash, Fred Howard
Salk, Martha Scheer
Scherpereel, Donald E
Schmitt, Harold William
Schreiber, Eric Christian
Scroggie, Lucy E
Sittel, Chester Nachand
Smith, Michael James
Sorrells, Frank Douglas
Stiegler, James O
Stone, Robert Sidney
Thiessen, William Ernest
Van Winkle, Webster, Jr
Vaughen, Victor C(ornelius) A(dolph)
Weir, James Robert, Jr
West, Colin Douglas
Wilcox, William Jenkins, Jr
Witherspoon, John Pinkney, Jr
Wohlfort, Sam Willis
Woodard, Kenneth Eugene, Jr
Yarbro, Claude Lee, Jr
Zerby, Clayton Donald

TEXAS
Abernathy, Bobby F
Ahlberg, Dan Leander
Anthony, Donald Barrett

Arganbright, Robert Philip
Aspelin, Gary B
Bachmann, John Henry, Jr
Badhwar, Gautam D
Baker, James Haskell
Bare, Charles L
Barrow, Thomas D
Basham, Jerald F
Bendapudi, Kasi Visweswararao
Bettinger, George E
Bhuva, Rohit L
Blackwell, Lawrence A
Bland, William M, Jr
Bronaugh, Edwin Lee
Brown, James Michael
Burton, Russell Rohan
Campbell, William M
Cannon, Dickson Y
Canon, Roy Frank
Carlton, Donald Morrill
Cassard, Daniel W
Caswell, Gregory K
Chen, Peter
Collins, Dean Robert
Collipp, Bruce Garfield
Cory, William Eugene
Coulson, Larry Vernon
Cruse, Carl Max
Cupples, Barrett L(eMoyne)
Danburg, Jerome Samuel
Darwin, James T, Jr
David, Yadin B
Dear, Robert E A
Dillard, Robert Garing, Jr
Dimitroff, Edward
Dugan, Gerald Francis
Eckles, Wesley Webster, Jr
Elder, Mark Lee
Elliott, Jerry Chris
Engle, Damon Lawson
Faubion, Billy Don
Fisher, Gene Jordan
Frank, Steven Neil
French, Robert Leonard
Friberg, Emil Edwards
Gallagher, John Joseph, Jr
Gatti, Anthony Roger
Gerding, Thomas G
Gipson, Robert Malone
Glaspie, Donald Lee
Gray, John Malcolm
Grogan, Michael John
Gum, Wilson Franklin, Jr
Haase, Donald J(ames)
Hall, James Edison
Hebert, Joel J
Hirsch, Robert Louis
Ho, Clara Lin
Hoffman, Herbert I(rving)
Holmes, Larry A
Hopkins, George H, Jr
Hurd, James William
Ivey, E(dwin) H(arry), Jr
Johnson, David Russell
Johnson, Elwin L Pete
Johnson, Malcolm Pratt
Justice, James Horace
Kingsley, Henry A(delbert)
Krautz, Fred Gerhard
Kusnetz, Howard L
Kutzman, Raymond Stanley
Lawrence, Joseph D, Jr
Lichtenstein, Harris Arnold
Liemohn, Harold Benjamin
Linder, John Scott
Lord, Samuel Smith, Jr
Ludwig, Allen Clarence, Sr
Lytle, Michael Allen
Mack, Mark Philip
Mai, Klaus L(udwig)
Mansfield, Clifton Tyler
Mao, Shing
Marshall, Robert P(aul)
Matzkanin, George Andrew
Maute, Robert Lewis
Menon, Venugopal Balakrishna
Merrill, John Raymond
Musa, Samuel A
Myers, Paul Walter
Naismith, James Pomeroy
Noonan, James Waring
O'Connor, Rod
Overmyer, Robert Franklin
Perez, Ricardo
Petersen, Donald H
Pierce, Robert Wesley
Poddar, Syamal K
Posey, Daniel Earl
Potts, Mark John
Powell, Alan
Purser, Paul Emil
Purvis, Merton Brown
Rennie, Thomas Howard
Rhodes, Allen Franklin
Rhodes, John Rathbone
Ripperger, Eugene Arman
Robinson, Alfred Green
Roeger, Anton, III
Rosenquist, Edward P
Rudin, Bernard D
Russell, B Don
Scinta, James
Shilstone, James Maxwell, Jr

Smith, R Lowell
Smith, Rolf C, Jr
Southern, Thomas Martin
Spencer, Alexander Burke
Strieter, Frederick John
Stubbeman, Robert Frank
Tannahill, Mary Margaret
Teague, Abner F
Teller, Cecil Martin, II
Torp, Bruce Alan
Urdy, Charles Eugene
VantHull, Lorin L
Waid, Margaret Cowsar
Wakeland, William Richard
Ward, Donald Thomas
Watson, Jerry M
Widerquist, V(ernon) R(oberts)
Wilde, Garner Lee
Witterholt, Edward John
Wulfers, Thomas Frederick
Wurth, Thomas Joseph
Young, H(enry) Ben, Jr

UTAH
Ball, James Stutsman
Bramlett, William
Burton, Frederick Glenn
Butterfield, Veloy Hansen, Jr
Dulock, Victor A, Jr
Germane, Geoffrey James
Gortatowski, Melvin Jerome
Haslem, William Joshua
Kelsey, Stephen Jorgensen
Krejci, Robert Henry
Lee, Jeffrey Stephen
Lewis, Lawrence Guy
Madsen, James Henry, Jr
Marcus, Mark
Michaelson, S(tanley) D(ay)
Nelson, Mark Adams
Prater, John D
Rothenberg, Mortimer Abraham
Wardell, Joe Russell, Jr
Wood, O Lew

VERMONT
Arns, Robert George
Bhatt, Girish M
Smith, Thomas David
Spillman, William Bert, Jr
Steele, Richard

VIRGINIA
Aitken, Alfred H
Alcaraz, Ernest Charles
Almeter, Frank M(urray)
Awad, Elias M
Barmby, John G(lennon)
Beale, Guy Otis
Beardall, John Smith
Berlincourt, Ted Gibbs
Bolender, Carroll H
Brenner, Alfred Ephraim
Brewer, Dana Alice
Briscoe, Melbourne George
Brown, Ellen Ruth
Brown, Richard Dee
Buchanan, Thomas Joseph
Burley, Carlton Edwin
Burns, William, Jr
Bush, Norman
Calm, James M
Carstea, Dumitru
Carter, William Walton
Casazza, John Andrew
Cetron, Marvin J
Charvonia, David Alan
Cheng, George Chiwo
Childress, Otis Steele, Jr
Church, Charles Henry
Cooper, Earl Dana
Cooper, Henry Franklyn, Jr
Cross, Ernest James, Jr
Curtin, Brian Thomas
Davidson, Charles Nelson
Davidson, Robert Bellamy
Davis, William Spencer
Debney, George Charles, Jr
Dillaway, Robert Beacham
Doles, John Henry, III
Dorman, Craig Emery
Dorsey, Clark L(awler), Jr
Entzminger, John N
Estes, Edward Richard
Evans, David Arthur
Everett, Warren S
Fabrycky, Wolter J
Fedrick, James Love
Foy, C Allan
Gee, Sherman
Gergely, Tomas Esteban
Gerlach, A(lbert) A(ugust)
Gibbons, John Howard
Gibson, John E(gan)
Gilbert, Arthur Charles
Godbole, Sadashiva Shankar
Gold, Sydell Perlmutter
Goodwin, Francis E
Greenberg, Arthur Bernard
Greinke, Everett D
Hager, Chester Bradley
Hallanger, Lawrence William
Hanneman, Rodney E
Harris, Paul Robert

Technical Management (cont)

Hart, Dabney Gardner
Hartline, Beverly Karplus
Hartzler, Alfred James
Hauxwell, Gerald Dean
Heebner, David Richard
Hortick, Harvey J
Howard, William Eager, III
Hunt, V Daniel
Ijaz, Lubna Razia
Jacobs, Theodore Alan
James, W(ilbur) Gerald
Jimeson, Robert M(acKay), Jr
Kahn, Robert Elliot
Kelch, Walter L
Kirby, Andrew Fuller
Knap, James E(li)
Knudsen, Dennis Ralph
Krassner, Jerry
Krueger, Peter George
Lapporte, Seymour Jerome
Lassiter, William Stone
Lerner, Norman Conrad
McCutchen, Samuel P(roctor)
McGean, Thomas J
McGregor, Dennis Nicholas
McHale, Edward Thomas
MacKinney, Arland Lee
Mansfield, John E
Marcus, Stanley Raymond
Marsh, Howard Stephen
Mehta, Gurmukh D
Mengenhauser, James Vernon
Milici, Robert Calvin
Mock, John E(dwin)
Morrison, David Lee
Mulder, Robert Udo
Mullen, Joseph Matthew
Myers, Donald Albin
Myers, Thomas DeWitt
Neil, George Randall
Ossakow, Sidney Leonard
Osuch, Mary Ann V
Owczarski, William A(nthony)
Pages, Robert Alex
Peake, Harold J(ackson)
Perillo, Benjamin Anthony
Perry, Dennis Gordon
Pokrant, Marvin Arthur
Price, Byron Frederick
Pry, Robert Henry
Rader, Louis T(elemacus)
Rainer, Norman Barry
Ratchford, Joseph Thomas
Raymond, Dale Rodney
Rosenberg, Murray David
Sabadell, Alberto Jose
Sedriks, Aristide John
Seward, William Davis
Shaub, Walter M
Shelkin, Barry David
Shockley, Gilbert R
Siebentritt, Carl R, Jr
Sink, David Scott
Small, Timothy Michael
Smith, Carey Daniel
Sobel, Robert Edward
Spokes, G(ilbert) Neil
Stroup, Cynthia Roxane
Summers, George Donald
Thorp, Benjamin A
Tiedemann, Albert William, Jr
VanDerlaske, Dennis P
Waesche, R(ichard) H(enley) Woodward
Wallace, Raymond Howard, Jr
Watkins, Allen Harrison
Waxman, Ronald
Weedon, Gene Clyde
Welch, Jasper Arthur, Jr
West, Bob

Wilkniss, Peter Eberhard
Wilson, Herbert Alexander, Jr
Woisard, Edwin Lewis
Wood, Albert D(ouglas)
Wood, Leonard E(ugene)
Zakrzewski, Thomas Michael
Zilczer, Janet Ann

WASHINGTON

Anderson, Harlan John
Anderson, Richard Gregory
Booth, Pieter Nathaniel
Brehm, William Frederick
Brown, Donald Jerould
Bunch, Wilbur Lyle
Buonamici, Rino
Burns, Robert Ward
Butts, William Cunningham
Claudson, T(homas) T(ucker)
Clement, Stephen LeRoy
Cooper, Martin Jacob
Craine, Lloyd Bernard
Einziger, Robert E
Ewing, J J
Fetter, William Allan
Frank, Andrew Julian
Gahler, Arnold Robert
Goheen, Steven Charles
Goodstein, Robert
Gravitz, Sidney I
Gray, Robert H
Greager, Oswald Herman
Gunderson, Leslie Charles
Hanna, Jeff
Hinthorne, James Roscoe
Hirschfelder, John Joseph
Hueter, Theodor Friedrich
Johnson, B(enjamin) M(artineau)
Katz, Yale H
Kent, Ronald Allan
Krier, Carol Alnoth
Kruger, Owen L
Lago, James
McClure, J Doyle
McMurtrey, Lawrence J
MacVicar-Whelan, Patrick James
Martin, George C(oleman)
Matlock, John Hudson
Merchant, Howard Carl
Morton, Randall Eugene
Nelson, Randall Bruce
Newman, Darrell Francis
Nickerson, Robert Fletcher
Padilla, Andrew, Jr
Peterson, Warren Stanley
Pilet, Stanford Christian
Reid, Donald House
Riley, Robert Gene
Roetman, Ernest Levane
Rohrmann, Charles A(lbert)
Sanford, Thomas Bayes
Sarles, Lynn Redmon
Schiffrin, Milton Julius
Schmid, Loren Clark
Smilen, Lowell I
Stokes, Gerald Madison
Sutherland, Earl C
Templeton, William Lees
Thorne, Charles M(orris)
Tichy, Robert J
Warner, Ray Allen
Welliver, Albertus
Widmaier, Robert George
Yoshikawa, Herbert Hiroshi
Zimmermann, Charles Edward

WEST VIRGINIA

Beeson, Justin Leo
Breneman, William C
Bryant, George Macon
Chaty, John Culver

Chiang, Han-Shing
De Barbadillo, John Joseph
Decker, Quintin William
Derderian, Edmond Joseph
Fischer, Robert Lee
Lantz, Thomas Lee
O'Mara, Michael Martin
Smith, Joseph James

WISCONSIN

Adrian, Alan Patrick
Baetke, Edward A
Becker, Edward Brooks
Bevelacqua, Joseph John
Bisgaard, Søren
Bishop, Charles Joseph
Brill, Winston J
Cheng, Shun
Daub, Edward E
Dwyer, Sean G
Eckert, Alfred Carl, Jr
Ekern, Ronald James
Eriksson, Larry John
Ferguson, John Allen
Gopal, Raj
Gottschalk, Robert Neal
Hinkes, Thomas Michael
Horntvedt, Earl W
Hummert, George Thomas
Huss, Ronald John
Hynek, Robert James
Jared, Alva Harden
Johnson, J(ames) R(obert)
Kinsinger, Richard Estyn
Lowenstein, Michael Zimmer
McMurray, David Claude
Makela, Lloyd Edward
Naik, Tarun Ratilal
Oehmke, Richard Wallace
Price, Walter Van Vranken
Rechtin, Michael David
Savereide, Thomas J
Schultz, Jay Ward
Sedor, Edward Andrew
Smith, Michael James
Spiegelberg, Harry Lester
Sprott, Julien Clinton
Stackman, Robert W
Steudel, Harold Jude
Subach, Daniel James
Trischan, Glenn M
Whitehead, Howard Allan
Whyte, Donald Edward

WYOMING

Harnsberger, Paul Michael
Lawroski, Harry
Marchant, Leland Condo
Odasz, F(rancis) B(ernard), Jr
Schabron, John F
Tigner, James Robert

PUERTO RICO

Lluch, Jose Francisco

ALBERTA

Butler, R(oger) M(oore)
Fisher, Harold M
Neuwirth, Maria
Peirce, John Wentworth
Sears, Timothy Stephen
Silver, Edward Allan

BRITISH COLUMBIA

Becker, Edward Samuel
Cox, Lionel Audley
Duncan, Douglas Wallace
Mackenzie, George Henry
Wynne-Edwards, Hugh Robert

MANITOBA

Rosinger, Herbert Eugene
Saluja, Preet Pal Singh
Trick, Gordon Staples

ONTARIO

Agar, G(ordon) E(dward)
Bertram, Robert William
Biggs, Ronald C(larke)
Cohn-Sfetcu, Sorin
Crocker, Iain Hay
Cupp, Calvin R
Dawes, David Haddon
Dunford, Raymond A
Effer, W R
Fewer, Darrell R(aymond)
Ford, Richard Westaway
Graham, Richard Charles Burwell
Hancox, William Thomas
Handa, V(irender) K(umar)
Hemmings, Raymond Thomas
Henry, Roger P
Jackson, David Phillip
Kruus, Jaan
Kundur, Prabha Shankar
Lawford, George Ross
McGeer, James Peter
Morton, Donald Charles
Pundsack, Arnold L
Rauch, Sol
Reid, Lloyd Duff
Robertson, J(ohn) A(rchibald) L(aw)
Rummery, Terrance Edward
Sonnenberg, Hardy
Street, Kenneth Norman
Tang, You-Zhi
Tunnicliffe, Philip Robert
Van Driel, Henry Martin
Vincett, Paul Stamford
Wright, Joseph D
Yan, Maxwell Menuhin
Zytner, Richard G

QUEBEC

Mavriplis, F
Sood, Vijay Kumar

SASKATCHEWAN

Mikle, Janos J

OTHER COUNTRIES

Assousa, George Elias
Baker, Frank Weir
Balk, Pieter
Brown, Charles Eric
Burkhardt, Walter H
Butcher, Harvey Raymond, III
Chi, Chao Shu
Dawson, Frank G(ates), Jr
Dreze, Jacques H
Edmonds, James W
Ginis, Asterios Michael
Hedglin, Walter L
Jolly, Janice Laurene Willard
Kurokawa, Kaneyuki
Lahiri, Syamal Kumar
Mou, Duen-Gang
Packer, Leo S
Renfrew, Robert Morrison
Schmell, Eli David
Shafi, Mohammad
Slagel, Robert Clayton
Steffen, Juerg
Steichen, Richard John
Stephens, Kenneth S
Strong, Jerry Glenn
Struzak, Ryszard G
Voutsas, Alexander Matthew
 (Voutsadakis)
Ward, John Edward
Winn, Edward Barriere
Zaczepinski, Sioma